advise   planning   furnishing   support service

We design libraries

Photo Marie Jeanne Smets

As specialist for library
interior design we are
responsible for the detailed
planning and exclusive
furnishing of all types of
libraries, regardless of size
and location.
Whether you are in need of complete
furnishings or just an addition, public or
scientific library we can be the structure
and base to achieve your plans.
With international business relationships
the SCHULZ SPEYER Bibliothekstechnik AG
is one of the European market leaders with
distributors in nine countries.

SCHULZ SPEYER

SCHULZ SPEYER Bibliothekstechnik AG
Friedrich-Ebert-Straße 2a · D-67346 Speyer
Postfach 1780 · D-67327 Speyer
Tel.: +49 (0) 62 32 - 31 81-0
Fax: +49 (0) 62 32 - 31 81-800
sales@schulzspeyer.de
www.schulzspeyer.de

WORLD GUIDE TO LIBRARIES
2012

WORLD GUIDE TO LIBRARIES 2012

26th Edition

Volume 2
Libraries S–Z
Index

De Gruyter Saur

Editors
Marlies Janson, Helmut Opitz

ISBN 978-3-11-023548-7
e-ISBN 978-3-11-023549-4
ISSN 0936-0085

Library of Congress Control Number: 91643589

Bibliographic information published by the Deutsche Nationalbibliothek
The Deutsche Nationalbibliothek lists this publication in the Deutsche
Nationalbibliografie; detailed bibliographic data are available in the Internet
at http://dnb.d-nb.de.

Typesetting: bsix information exchange GmbH, Braunschweig
Printing: Strauss GmbH, Mörlenbach
∞ Printed on acid-free paper

Printed in Germany

www.degruyter.com

Contents

Suggestions on Use viii
Abbreviations x
List of Library Associations xii

Volume 1
Afghanistan 3
Albania 3
Algeria 3
American Samoa 4
Andorra 5
Angola 5
Anguilla 5
Antigua and Barbuda 5
Argentina 5
Armenia 10
Aruba 11
Australia 11
Austria 36
Azerbaijan 52

Bahamas 53
Bahrain 53
Bangladesh 53
Barbados 54
Belarus 54
Belgium 64
Belize 82
Benin 82
Bermuda 82
Bhutan 83
Bolivia 83
Bosnia and Herzegovina 83
Botswana 84
Brazil 84
Brunei 97
Bulgaria 97
Burkina Faso 101
Burundi 101

Cambodia 101
Cameroon 101
Canada 102
Cape Verde 134
Central African Republic 134
Chad 134
Chile 134
China 137
Colombia 155
Comoros 158
Congo, Democratic Republic 158
Congo, Republic 159
Cook Islands 160
Costa Rica 160
Côte d'Ivoire 160
Croatia 161
Cuba 164
Curaçao 166
Cyprus 166
Cyprus Turk 166
Czech Republic 166

Denmark 175

Dominica 183
Dominican Republic 183

Ecuador 184
Egypt 184
El Salvador 188
Eritrea 188
Estonia 188
Ethiopia 190

Falkland Islands 190
Fiji 190
Finland 191
France 200
French Guiana 243
French Polynesia 243

Gabon 243
Gambia 243
Georgia 243
Germany 245
Ghana 367
Gibraltar 369
Greece 369
Grenada 372
Guadeloupe 372
Guam 372
Guatemala 372
Guinea 373
Guinea-Bissau 373
Guyana 373

Haiti 373
Honduras 374
Hungary 374

Iceland 384
India 385
Indonesia 400
Iran 404
Iraq 405
Ireland 406
Isle of Man 410
Israel 410
Italy 415

Jamaica 467
Japan 468
Jordan 487

Kazakhstan 488
Kenya 490
Kiribati 491
Korea, Democratic People's Republic 491
Korea, Republic 491
Kuwait 493
Kyrgyzstan 493

Laos 494
Latvia 494
Lebanon 495
Lesotho 496

Contents

Liberia 496
Libya 496
Liechtenstein 496
Lithuania 497
Luxembourg 500

Macedonia 501
Madagascar 502
Malawi 502
Malaysia 503
Maldives 506
Mali 506
Malta 507
Marshall Islands 507
Martinique 507
Mauritania 508
Mauritius 508
Mayotte 508
Mexico 508
Micronesia 513
Moldova 513
Monaco 514
Mongolia 514
Montenegro 514
Morocco 514
Mozambique 516
Myanmar 516

Namibia 516
Nepal 517
Netherlands 517
New Caledonia 538
New Zealand 539
Nicaragua 543
Niger 543
Nigeria 543
Niue 547
Norway 547

Oman 556

Pakistan 557
Palau 559
Panama 559
Papua New Guinea 559
Paraguay 561
Peru 561
Philippines 563
Poland 566
Portugal 587
Puerto Rico 593

Qatar 595

Réunion 595
Romania 595
Russia 601
Rwanda 654

Volume 2

Samoa 657
San Marino 657
Sao Tome and Principe 657
Saudi Arabia 657
Senegal 657
Serbia 658
Seychelles 661
Sierra Leone 661
Singapore 662
Sint Maarten 663
Slovakia 663
Slovenia 668
Solomon Islands 673
South Africa 673
Spain 680
Sri Lanka 708
St. Lucia 710
St. Vincent and the Grenadines 710
Sudan 710
Suriname 711
Swaziland 711
Sweden 711
Switzerland 726
Syria 743

Taiwan 743
Tajikistan 746
Tanzania 746
Thailand 747
Togo 750
Tonga 750
Trinidad and Tobago 750
Tunisia 751
Turkey 751
Turkmenistan 753
Tuvalu 753

Uganda 753
Ukraine 754
United Arab Emirates 776
United Kingdom 776
United States of America 814
Uruguay 1121
Uzbekistan 1122

Vanuatu 1124
Vatican City 1124
Venezuela 1125
Vietnam 1126
Virgin Islands (U.S.) 1127

Yemen 1127

Zambia 1127
Zimbabwe 1128

INDEX
Alphabetical Index of Libraries 1131

World Guide to Libraries

Vol. 2

Libraries
S – Z

Index

Samoa

University Libraries, College Libraries

Avele College, Library, P.O. Box 45, *Apia*
5 000 vols   24379

University of the South Pacific – School of Agriculture, Library, Alafua Campus, Private Bag, *Apia*
T: +685 21671 ext 273; Fax: +685 22933
1966; Chris Nelson
Agricultural economics and education, appropriate technology; Pacific agriculture
22 000 vols; 1 000 curr per; 50 diss/theses; 3 000 govt docs; 100 maps; 100 microforms; 450 av-mat; 4 digital data carriers
libr loan; IAALD   24380

Government Libraries

Legislative Assembly, Parliamentary Library, Potu Tusi o Sui Usufono, Maota Fono, P.O. Box 1866, *Apia*
T: +685 21811, 21813; Fax: +685 21817
1972; Leata P. Fesili
6 000 vols; 16 curr per   24381

Public Libraries

Nelson Memorial Public Library, P.O. Box 598, *Apia*
T: +685 21028; Fax: +685 22933
1959; Ms Mataina Tuatagaloa Te'o
R.L. Stevenson Coll, Samoa Coll, Pacific Coll; 1branch libr, 1 bookmobile
90 000 vols   24382

San Marino

University Libraries, College Libraries

Università degli Studi, Biblioteca Universitaria, Contrada delle Mura, 16, 47890 *San Marino*
T: +378 882500; Fax: +378 882303;
E-mail: biblio@unirsm.sm; URL: www.unirsm.sm
1988; Dr. Gabriella Lorenzi
Fondo Young (Memory's Libr)
40 000 vols; 161 curr per; 1 mss; 12 incunabula
libr loan   24383

Government Libraries

Biblioteca di Stato, Contrada Omerelli, Palazzo Valloni, 47890 *San Marino*
T: +378 882248; Fax: +378 882295;
E-mail: biblioteca@omniway.sm
1858; Dr. Elisabetta Righi Iwanejko
Ctr of doc
11 000 vols; 500 curr per; 20 incunabula; 60 diss/theses; 700 rare bks
AIB, AIDA   24384

Sao Tome and Principe

Government Libraries

Assembleia Nacional, Biblioteca, Cx P 181, *São Tomé*
T: +239 1221927; Fax: +239 122835
1994; Ms Dr. Ligia Maria Duaresma de Ceita
1 000 vols   24385

Special Libraries Maintained by Other Institutions

Arquivo Histórico de São Tomé e Principe, CP 87, *São Tomé*
Maria Nazaré de Ceita
2 000 vols   24386

Centro de Documentação Técnica e Científica, *São Tomé*
T: +239 122585
Maria Rosário Assunção
45 000 vols; 2 000 curr per   24387

Ministério de Agricultura e Pesca, Biblioteca, CP 47, *São Tomé*
T: +239 22126
1973; Tomé de Sousa da Costa   24388

Saudi Arabia

National Libraries

King Fahd National Library, King Fahd Highway, *Riyadh*; P.O. Box 7572, Riyadh 11472
T: +966 1 4624888; Fax: +966 1 4645341; URL: www.kfnl.gov.sa
1989; Ali Sulaiman Al-Sowaine
Arabian gulf, Arabian peninsula, Kingdom of Saudi Arabia, Saudi publs, Saudi copyright law, Islam, Islamic coins, photos about Saudi Arabia, Arabic mss; Mss and Rare Books Coll, Saudi Diss Coll, Photogr Arch, Doc Coll; Dept of Publications
462 287 vols; 986 curr per; 4 377 mss; 8 incunabula; 10 094 diss/theses; 111 023 govt docs; 3 565 maps; 41 105 microforms; 17 238 sound-rec; 3 623 digital data carriers; 11 148 rare bks, 15 293 pamphlets, 23 356 coins, 89 636 postage stamps, 13 656 photos & paintings, 3 168 slides, 1 088 drawings, 932 press clippings, 4 332 videos, 37 realia
libr loan; ALA, ASIS, SAA, LA, IFLA, FID, SLA/Arabian Gulf Chapter, ICA, MELA   24389

University Libraries, College Libraries

King Fahd University of Petroleum and Minerals (KFUPM), Main Library, *Dhahran* 31261
T: +966 3 8603000; Fax: +966 3 8603018; E-mail: library@kfupm.edu.sa; URL: www.kfupm.edu.sa
1964; Dr. Ibrahim M. Al-Jabri
334 800 vols; 1 800 curr per; 549 752 microforms; 1 593 av-mat; 14 digital data carriers; 40 800 periodicals
libr loan; ALA, COMLIS   24390

King Abdul Aziz University, Central Library, Jamiat Al-Malik Abdulaziz, P.O. Box 80213, *Jeddah* 21589
T: +966 2 6452000, 6952481; Fax: +966 2 6400169; E-mail: library@kaau.edu.sa; URL: www.kaau.edu.sa
1967; Mohammed Ahmed Basager
Red Sea Coll, Al-Haramain Coll
500 000 vols
libr loan; IFLA   24391

Umm Al-Qura University, Central Library, P.O. Box 1629, *Mecca*
T: +966 2 5565621; Fax: +966 2 5565621; E-mail: cenlib@uqu.edu.sa; URL: www.uqu.edu.sa
1979; Dr. Abdul Muhsin bin Abdullah al Al-Sheikh
400 000 vols; 600 curr per; 4 000 mss; 2 712 diss/theses; 6 000 govt docs; 257 maps; 12 000 microforms; 3 digital data carriers
libr loan; IFLA   24392

– Taif Campus Library, Al-Saddad Rd, Shihar, P.O. Box 407-715, Taif, Mecca
T: +966 2 5564770; Fax: +966 2 7494477
1981; Muhammad Adil Usmani
40 000 vols   24393

Islamic University of Medina, Library, Al-Jamiat Al Islamiah, P.O. Box 170, *Medina*
T: +966 4 8474080; Fax: +966 4 8474560; URL: www.iu.edu.sa
1961
334 000 vols; 26 000 mss; 897 diss/theses   24394

King Abdul Aziz Military Academy, Library, P.O. Box 5969, *Riyadh* 11432
T: +966 1 4654244
1955
20 000 vols   24395

King Saud University, Prince Salman Central Library, P.O. Box 22480, *Riyadh* 11495
T: +966 1 467-6148/-6149/-6161; Fax: +966 1 4676162; E-mail: sfalogia@ksu.edu.sa; URL: www.ksu.edu.sa
1957; Dr. Sulaiman Saleh Al-Ogla
Arabian Peninsula Coll, al-Zirikli, Al al-Shaykh, Umar Hassan Coll
1 840 000 vols; 2 071 curr per; 20 187 mss; 12 492 diss/theses; 150 000 govt docs; 879 maps; 28 211 microforms; 18 690 av-mat; 3 700 sound-rec; 29 digital data carriers; 8 441 Rare publications, 13 943 Press Files
IFLA   24396

Government Libraries

Consultative Council, Majlis Ash-Shura Library, P.O. Box 63393, *Riyadh* 11516
T: +966 1 4821666; Fax: +966 1 4420146
1993; Abdul-Rahman al-Sarra
Islamic references, Arabic lit
23 000 vols; 300 curr per; 2 000 govt docs; 6 196 microforms; 851 av-mat
libr loan; IFLA   24397

Corporate, Business Libraries

Islamic Development Bank, IDB Library, P.O. Box 5925, *Jeddah* 21432
T: +966 2 6361400; Fax: +966 2 6366871; E-mail: hfoudeh@isdb.org.sa; URL: www.isdb.org
1975; Dr. H.M. Foudeh
Old Docs of Palestine Coll, Mss of Oman Coll; Member countries section, international and regional organization reports section, PC software and CDs section
41 000 vols; 823 curr per; 28 mss; 10 incunabula; 516 diss/theses; 26 000 govt docs; 96 maps; 3 670 microforms; 3 360 av-mat; 63 digital data carriers; 9 870 special mat rpt/bulletins
libr loan; IFLA, FID   24398

Special Libraries Maintained by Other Institutions

Arab Bureau for Education in the Gulf States, Library, P.O. Box 3908, *Riyadh* 11481
T: +966 1 4774644; Fax: +966 1 4783165
1975
30 000 vols   24399

British Council, Library, Al-Fazari Sq, Diplomatic Quarter, C-14, Third floor, P.O. Box 58012, *Riyadh* 11594
T: +966 1 4831818;
Fax: +966 1 4831717; E-mail: enquiry.riyadh@sa.britishcouncil.org; URL: www.britishcouncil.org/saudiarabia
1970
9 200 vols   24400

Institute of Public Administration, Library, P.O. Box 205, *Riyadh* 11141
T: +966 1 4768888; Fax: +966 1 4792136; E-mail: library@ipa.edu.sa; URL: www.ipa.edu.sa
1962; Mostafa M. Sadhan
Rare bks on Saudi Arabia in particular and Islam and Arabian Gulf States in general; Official publ; Saudi Gov. doc. Dpt
236 844 vols; 1 080 curr per; 1 476 diss/theses; 32 045 microforms; 712 digital data carriers; 4 624 official publs, 55 000 Saudi govt docs
libr loan   24401

King Abdul Aziz Foundation for Research and Archives, Library, P.O. Box 2945, *Riyadh* 11461
T: +966 1 4013444; Fax: +966 1 4013597; E-mail: info@darah.org.sa; URL: www.darah.org.sa
1972; Nasir Faris Al-khodairy
Private libr of the King (3 000 vols); Documentation ctr, special coll on King Abdul Aziz, Films and photographs archives, Manuscript Dept, Dept of special docs, oral history center
54 000 vols; 200 curr per; 3 500 mss; 2 500 diss/theses; 3 000 000 govt docs; 150 maps   24402

King Faisal Centre for Research and Islamic Studies, Library, P.O. Box 51049, *Riyadh* 11543
T: +966 1 4652255; Fax: +966 1 4659993
1983
142 041 vols; 2 176 curr per; 23 000 mss; 1 300 diss/theses; 40 maps; 84 500 microforms; 11 200 sound-rec   24403

Library of the King Abdul Aziz City for Science and Technology (KAACST), P.O. Box 6086, *Riyadh* 11442
T: +966 1 4883444; Fax: +966 1 4883756
1979; Dr. Muhammed Fasah Uddin
Computer and Information Science
50 000 vols; 320 curr per; 700 diss/theses; 150 govt docs; 350 av-mat; 60 sound-rec; 411 000 technical rpts on microfiche
FID   24404

Saudi Arabian Standards Organization, Information Centre, Imam Saud Ibn Abdulaziz ibn Mohamed Rd, P.O. Box 3437, *Riyadh* 11471
T: +966 1 4520000; Fax: +966 1 4520086; E-mail: sasoinfo@saso.org.sa; URL: www.saso.org.sa
1972; Mohammad Almeshari
10 000 vols; 650 000 standards 24405

Saudi Consulting House, Industrial Information Dept, Library, P.O. Box 1267, *Riyadh* 11431
T: +966 1 4484533;
Fax: +966 1 4481234; E-mail: library@suadiconsulting.org; URL: www.saudiconsulting.org
1967; Al Humaidi Al-Baqawi
Feasibility studies
11 000 vols; 121 curr per; 22 diss/theses; 2 000 govt docs; 15 maps; 500 microforms; 15 digital data carriers; Standards, specifications, industry, industrial management, industrial relations, annual repts, studies, conference proceedings
libr loan   24406

Public Libraries

King Abdul Aziz Library, *Medina*
T: +966 4 8232134; Fax: +966 4 8232126
1983; D. Abdulrahman Bin Suliman Almuziny
120 000 vols   24407

King Abdul Aziz Public Library, P.O. Box 86486, *Riyadh* 11622
T: +966 1 4911300; Fax: +966 1 4911949; URL: www.kapl.org.sa
1985; Faisal A. Al-Muaammar
King Abdul Aziz hist coll
275 000 vols; 1 100 curr per; 2 500 mss; 3 500 diss/theses; 5 000 av-mat; 53 000 hist docs on microform   24408

Senegal

National Libraries

Archives National du Sénégal, Bibliothèque, Immeuble administratif, Av Léopold Sedar Senghar, *Dakar*
T: +221 8217021, 8231088 est 595; Fax: +221 8217021; E-mail: Bdas@telecomplus.sn; URL: www.archivesdusenegal.gouv.sn/biblio.html
1913; Saliou Mbaye
26 000 vols
libr loan   24409

University Libraries, College Libraries

Université Cheikh Anta Diop de Dakar, Bibliothèque de l'Université (Cheikh Anta Diop University), BP 2006, *Dakar* RP
T: +221 8246981, 8250279; Fax: +221 8242379; E-mail: hsene@ucad.sn; URL: www.bu.ucad.sn
1957; Henri Sène
306 274 vols; 734 curr per; 36 mss; 72 380 diss/theses; 100 maps; 1 980 microforms; 25 av-mat; 25 digital data carriers
libr loan; IFLA, SCAULWA   24410

– Centre d'Etudes des Sciences et Techniques de l'Information, Médiathèque, Dakar-Fann, Dakar
T: +221 824-6875/-9366; Fax: +221 8242417; E-mail: gueyou@cesti.refer.sn
1967; Youssoupha Guèye
4 646 vols; 22 curr per; 483 diss/theses; 719 av-mat; 730 sound-rec; 6 digital data carriers
libr loan   24411

– Département de Géographie, Faculté des Lettres et Sciences humaines, Bibliothèque, Dakar – Fann
T: +221 8246270 ext 137
1959; Ousmane Diouf
2 500 vols; 12 curr per; 300 diss/theses; 300 av-mat; 600 aerial photos   24412

– Ecole de Bibliothécaires, Archivistes et Documentalistes, Bibliothèque, BP 3252, Dakar – Fann
T: +221 8257660; Fax: +221 8252883
1963; Daouda Diop
11 000 vols; 129 curr per; 136 diss/theses; 2 503 microforms; 150 av-mat; 10 digital data carriers
IFLA   24413

– Ecole Normale Supérieure, Bibliothèque, av du Président Bourguiba, BP 5036, Dakar
1962; Oumou Khaiiy Ly
46 000 vols
IFLA   24414

– Faculté des Sciences juridiques et économiques, Bibliothèque, Dakar – Fann
T: +221 8210134
Aida Sow
9 000 vols; 80 curr per   24415

– Institut de Médecine tropicale appliquée (IMTA), Bibliothèque, Dakar
8 000 vols   24416

– Institut fondamental d'Afrique noire, Bibliothèque, BP 206, Dakar – Fann
T: +221 8250090; E-mail: bibifan@ifan.refer.sn
Gora Dia
70 000 vols; 2 560 maps; 1 600 microforms; 8 200 brochures, 32 000 photos, 2 100 slides, 12 200 files of docs, 4 036 coll of periodicals   24417

Ecole National Supérieur d'Agriculture, Bibliothèque, BP 296 RP, *Thiès*
T: +221 9511257; Fax: +221 9511551; E-mail: ensath@telecomplus.sn
6 000 vols   24418

Ecole Supérieure Polytechnique, Centre de Thiès, Bibliothèque, BP A10, *Thiès*
T: +221 9511384; Fax: +221 9511476
1973; Philomène Faye
20 000 vols; 150 curr per
libr loan   24419

School Libraries

Ecole Militaire de Santé, Bibliothèque, Camp Dial Diop, BP 4042, *Dakar*
T: +221 822506 ext 478
7 000 vols   24420

Lycée Van Vollenhoven, Bibliothèque, Rue du 18 Juin, *Dakar*
1936
3 500 vols   24421

Lycée Charles de Gaulle, Bibliothèque, *Saint-Louis*
1920
3 500 vols   24422

Lycée de Jeunes Filles Ameth Fall, Bibliothèque, BP 1, *Saint-Louis*
T: +221 971000
18 000 vols; 10 curr per   24423

Government Libraries

Assemblée Nationale du Sénégal, Direction de la Recherche et de la Documentation, Pl Soweto, BP 86, *Dakar*
T: +221 8236708; Fax: +221 8236708; E-mail: assuat@primature.su
1956; M. Khayrou Cisse
4 800 vols; 30 curr per; 300 govt docs; 5 maps; 50 microforms; 10 digital data carriers
libr loan; ASBAD   24424

Direction des Eaux et Forêts, Bibliothèque, BP 1831, *Dakar* – Fann
T: +221 8320565; Fax: +221 8320426
Marie Therese Ndiohe
4 500 vols; 12 curr per   24425

Ecclesiastical Libraries

Centre Lebret, Bibliothèque, av Cheikh Anta Diop, BP 5098, *Dakar* – Fann
T: +221 8212608
1955; Emmanuel Diatta
Théologie, sciences juridiques
13 000 vols; 100 curr per
libr loan   24426

Monastère de Keur Moussa, Bibliothèque, BP 721, *Dakar*
T: +221 8363309; Fax: +221 8361617; E-mail: kmoussa@telecomplus.sn
1963; Louis-Marie Tressol
Théology (DTC), Sup. Dictionnaire de la Bible, Sources chrétiennes, Dictionnaire de Spiritualité
11 500 vols; 15 curr per; 35 diss/theses; 500 sound-rec   24427

Special Libraries Maintained by Other Institutions

Centre National de Recherches Agronomiques (CNRA), Bibliothèque, BP 53, *Bambey*
T: +221 9736050; Fax: +221 9736052; E-mail: isracnra@telecomplus.sn
6 700 vols   24428

Institut Sénégalais de Recherches Agronomiques (ISRA), Centre National de la Recherche Agronomique (CNRA), BP 53, *Bambey*
T: +221 973-6050/-6051/-6054; Fax: +221 9736052
1921; Rosalie Diouf
15 000 vols; 61 diss/theses; 20 maps; 500 microforms; 100 av-mat; 2 sound-rec; 6 digital data carriers; 400 periodicals
libr loan   24429

African Regional Centre for Technology / Centre Régional Africain de Technologie, Library, BP 2435, *Dakar*
T: +221 8237712; Fax: +221 8237713
African technodevelopment
6 000 vols   24430

Alliance Française, Bibliothèque, 3, Rue Parchappe, BP 1777, *Dakar*
T: +221 8210822; Fax: +221 8221225
1948; Bertrand Calmy
17 000 vols; 8 curr per; 20 sound-rec; 10 digital data carriers   24431

American Cultural Center, Library, Immeuble BIAO, pl d l'Indépendance, BP 49, *Dakar*
1950; Louise Diaw
5 000 vols; 49 curr per; Records, tapes, AV material   24432

British Senegalese Institute, Institut Senegalo-Britannique, Library, Rue du 18 juin, BP 35, *Dakar*
T: +221 8224023
1969; Mam Maty Cisse
10 900 vols; 37 curr per; 900 av-mat; 288 sound-rec   24433

Bureau Régional de l'UNESCO pour l'Education en Afrique (BREDA) / UNESCO Regional Office for Education in Africa, Bibliothèque, 12, av Roume, BP 3311, *Dakar*
1970
13 000 vols; 270 curr per; 655 microforms   24434

Centre Culturel Français, Bibliothèque, 96, rue Blanchot, BP 4003, *Dakar*
T: +221 8230320; Fax: +221 8212619; E-mail: ccf@sentoo.sn
1959; Mr M. Traore
Coll of francophone lit, Arts coll
40 000 vols; 80 curr per; 60 digital data carriers; 550 documentary films   24435

Centre Régional Africain de Technologie, Bibliothèque, BP 2435, *Dakar*
T: +221 8237712; Fax: +221 8237713
1977
6 000 vols   24436

Chambre de Commerce d'Industrie de la Region du Cap Vert, Service de documentation, BP 118, *Dakar*
T: +221 8217189
1973; Danielle Dieng
6 000 vols   24437

Chambre de Commerce d'Industrie et d'Agriculture de Dakar, Bibliothèque, pl de l'Indépendance, BP 118, *Dakar*
T: +221 8237189; Fax: +221 8239363; E-mail: cciad@sentoo.sn
1929; Séga Baldé
7 000 vols; 30 curr per; 50 digital data carriers
libr loan; ASBAD, ADBS   24438

Conseil pour le Développement de la Recherche Economique et Social en Afrique (CODESRIA), Bibliothèque, BP 3304, *Dakar*
T: +221 8230211
1983
5 000 vols; 300 curr per   24439

Direction des Mines et de la Géologie (DMG), Ministère du Développement industriel et de l'environment, Route de Ouakam, BP 1238, *Dakar*
4 000 vols; 1 000 curr per; 1 000 maps   24440

Institut Africain de Développement Economique et de Planification (IDEP), Bibliothèque, BP 3186, *Dakar*
T: +221 8234831; Fax: +221 8222964; E-mail: idep@sonatel.senet.net
1963; S.I. Odutene
25 000 vols; 1 220 curr per
libr loan; IFLA   24441

Institut de Recherche pour le Développement (IRD), Centre de Dakar, Library, Rte des Pères Maristes, BP 1386, *Dakar*
T: +221 8323480; Fax: +221 8324307; E-mail: durand@orstom.sn; URL: www.ird.sn
J.R. Durand
Soil biology, medical entomology, geology, hydrology, economics, agronomy, marine fisheries, microbiology
10 000 vols   24442

Institut Pasteur de Dakar, Bibliothèque, 36, av Pasteur, BP 220, *Dakar*
T: +221 8235181; Fax: +221 8238772; E-mail: pasteur.dakar@pasteur.sn; URL: www.pasteur.sn
1913; M. Papa Dieme
5 500 vols; 102 curr per; 35 diss/theses   24443

Laboratoire National de l'Elevage et des Recherches Vétérinaires, Bibliothèque, BP 2057, *Dakar* – Fann
T: +221 832065, 832067
1935; Khary Ndiaye
12 000 vols; 125 curr per   24444

Office de la Recherche Scientifique Outre-Mer (ORSTOM), Centre régional de documentation (CRDO), Bibliothèque, Route des Pères Maristes, BP 1386, *Dakar* – Fann
T: +221 8323480; Fax: +221 8324307
1955; M. Ndong
10 700 vols; 263 curr per
libr loan; IAMSLIC   24445

Radiodiffusion du Sénégal, Bibliothèque, 58, bd de la République, BP 1765, *Dakar*
1968
3 000 vols   24446

Service Météorologique de Hann, Ministère des Travaux publics, de l'urbanisme et des transports, Bibliothèque, *Dakar* – Fann
1 000 vols; 50 curr per   24447

SICAP Liberté, Bibliothèque, Rond-point Liberté, BP 1094, *Dakar*
1972
6 500 vols   24448

Centre Culturel Français, Bibliothèque, BP 368, *Saint-Louis*
T: +221 9611578; Fax: +221 9612223
1965; Herve Lenormand
18 000 vols; 76 curr per; 250 av-mat; 300 sound-rec   24449

Centre de Formation Pédagogique, Bibliothèque, av du Général de Gaulle, *Saint-Louis*
1962
2 500 vols   24450

Centre de Recherches et de Documentation du Sénégal (CRDS), Bibliothèque, Rue Neuville, Pointe Sud, BP 382, *Saint-Louis*
T: +221 9611050
1948; Ibrahima Niang
17 500 vols; 50 curr per   24451

Organisation pour la Mise en Valeur du Fleuve Sénégal (OMVS), Centre Régional de Documentation, Rue Duret, BP 383, *Saint-Louis*
T: +221 971488; Fax: +221 9641110; E-mail: Mbackegueye@hotmail.com; URL: www.omvs-hc.org
1970; Mbacké Gueye
Water resources, hydrology, agriculture
13 curr per; 30 000 maps; 57 000 microforms; 15 000 traités, 45 000 non traités, 12 index imprimés
libr loan; ASBAD   24452

Serbia

National Libraries

Narodna biblioteka Srbije (National Library of Serbia), Skerlićeva 1, 11000 *Beograd*
T: +381 11 2451242; Fax: +381 11 2451 289; E-mail: nbs@nb.rs; URL: www.nbs.rs
1832; Sreten Ugričić (Director), Vesna Injac (Deputy Director)
Old, rare and miniature books, Cartographic material, Engraving and Art materials, Photos and postcards, Printed music and sound records, Literary and other Manuscripts and Archival, Posters and small documentary material, Memorial libraries and Legacies, Cyrillic manuscripts; Bibliographic dept, Manuscript dept, Reference dept, Serial dept, Monographs dept, Special Collection dept
2 000 000 vols; 25 000 curr per; 35 000 e-journals; 75 000 e-books; 75 350 mss; 150 incunabula; 20 000 diss/theses; 120 000 govt docs; 36 000 music scores; 35 000 maps; 600 000 microforms; 120 000 av-mat; 30 000 sound-rec; 25 000 digital data carriers; 300 Cyrillic mss, 50 000 modern mss
libr loan; Serbian Library Association, Community of Serbian Parent Libraries, IFLA, CDNL, LIBER, CERL, CENL, EBLIDA, ABDOS, ELAG, TEL   24453

Biblioteka Matice srpske (Matica Srpska Library), ul. Matice srpske 1, 21000 *Novi Sad*
T: +381 21 420-199/-198; Fax: +381 21 528574; E-mail: bms@bms.ns.ac.rs; URL: www.bms.ns.ac.rs
1826; Miro Vuksanović
1 200 000 vols; 6 000 curr per; 671 mss; 17 incunabula; 5 000 diss/theses; 7 000 music scores; 5 500 maps; 1 300 microforms; 20 000 sound-rec; 3 digital data carriers; 35 000 rare bks, 696 000 items of separate material (maps, leaflets, posters, cassettes etc)
libr loan; IFLA, FID   24454

General Research Libraries

Srpska Akademija Nauka i Umetnosti, Biblioteka (Serbian Academy of Sciences and Arts), ul. Knez Mihailova 35, 11000 ***Beograd***
T: +381 11 2639120, 2027260; Fax: +381 11 2639120; E-mail: andjelka@bib.sanu.ac.rs; URL: www.sanu.ac.rs
1842; Milka Zečević
Arch of Serbian hist
1 250 000 vols; 12 500 curr per; 12 000 e-journals; 1 incunabula; 60 000 diss/theses; 1 562 maps; 4 689 microforms; 46 000 av-mat; 18 digital data carriers
libr loan; BDS 24455

University Libraries, College Libraries

Fakultet muzicke umetnosti, Biblioteka (Academy of Music), Kralja Milana 50, 11000 ***Beograd***
T: +38111 3620796; Fax: +38111 2643598; E-mail: biblio@fmu.bg.ac.rs; URL: www.fmu.bg.ac.rs/fmu_en/biblioteka.html
1937
110 000 vols; 60 curr per; 170 mss; 20 diss/theses; 82 000 music scores; 11 290 sound-rec; 104 digital data carriers
libr loan 24456

Institut za Pedagoška istraživanja, Biblioteka (Institute for Educational Research), Dobrinjska II/III, 11001 ***Beograd***
T: +381 11 642925; Fax: +381 11 38111
1959; Milena Dimoski
11 100 vols; 20 curr per
libr loan 24457

Institut za Uporedno Pravo, Biblioteka (Institute of Comparative Law), ul. Terazije 41, 11000 ***Beograd***
T: +381 11 3238198; Fax: +381 11 3233213
1956; Djurdjica Mučibabić
22 700 vols; 180 curr per; 114 mss
libr loan; IALL 24458

Univerzitet u Beogradu, Univerzitetska Biblioteka 'Svetozar Marković' (University of Belgrade), bul. Revolucije 71, p.p. 349, 11000 ***Beograd***
T: +381 11 3370509; Fax: +381 11 3370354; E-mail: stela@unilib.bg.ac.rs; URL: www.unilib.bg.ac.rs
1844; Dr. Dejan Ajdačić
UNESCO publs
1 450 000 vols; 1 200 curr per; 5 200 mss; 17 incunabula; 250 000 diss/theses; 200 maps; 1 000 microforms; 50 digital data carriers
libr loan; IFLA, ASLIB 24459

– Arhitektonski Fakultet, Biblioteka (Faculty of Architecture), bul. Revolucije 73, p.p. 866, 11000 Beograd
T: +381 11 322936; Fax: +381 11 324122
1897; Smiljka Soretić
Old architectural bks, urbanism
18 000 vols; 70 curr per; 550 diss/theses; 3 000 av-mat; 50 000 pictures
libr loan; IFLA, FID, BDS 24460

– Ekonomski Fakultet, Biblioteka (School of Economics), ul. Kamenička 6, 11000 Beograd
T: +381 11 3021148; Fax: +381 11 639560; E-mail: pds@one.ekof.bg.ac.rs; URL: www.ekof.bg.ac.sr
1937; Miloš Kokotović
85 000 vols; 55 curr per
libr loan 24461

– Fakultet Sporta i Fizičkog Vaspitanja, Biblioteka (Faculty of Sport and Physical Education), ul. Blagoja Parovica 156, 11000 Beograd
T: +381 11 3531041; Fax: +381 11 555474; E-mail: biblioteka@dif.bg.ac.rs; URL: www.dif.bg.ac.rs/
1946; Sida Bogosavljevic
Sports, sport medicine, physical education, recreation
25 457 vols; 8 curr per; 145 diss/theses
libr loan 24462

– Farmaceutski Fakultet, Centralna Biblioteka (Faculty of Pharmaceutics), ul. Dr. Subotića 8, 11000 Beograd
T: +381 11 684373
1947; Marija Prokić
15 000 vols; 275 curr per; 200 mss; 142 diss/theses 24463

– Filološki fakultet, Katedra za germanistiku, Biblioteka (German Seminar), Studentski trg 3, 11000 Beograd
T: +381 11 630096
1905; Milorad Sofronijević
31 000 vols 24464

– Fiolski fakultet, Katedra za Orijentalistiku, Biblioteka (Department for Oriental Studies), Akadamski trg 3, 11000 Beograd
URL: www.fil.bg.ac.yu
1926; Snezana Gencic
Oriental lit, culture, languages, hist
24 000 vols; 500 curr per; 30 diss/theses 24465

– Geografski Fakultet, Biblioteka (Faculty of Geography), Studentski trg 3, 11000 Beograd
T: +381 11 637421
1893; Gordana Stojanović
8 300 vols 24466

– Gradjevinski Fakultet, Biblioteka (Faculty of Civil Engineering), bul. Revolucije 73, 11000 Beograd
T: +381 11 3226667; Fax: +381 11 3220237
1948; Biljana Vukadinović
Building and construction
47 000 vols; 200 curr per; 190 diss/theses; 120 govt docs
libr loan 24467

– Institut Društvenih Nauk, Biblioteka (Institute of Social Sciences), Narodnog fronta 45, 11000 Beograd
1958
65 000 vols 24468

– Institute for Regional Geology and Paleontology, Faculty of Mining and Geology, Library, ul. Kamenička 6, Box 227, 11001 Beograd
T: +381 11 632166; Fax: +381 11 631137
1880; Milica Surla
76 789 vols; 964 curr per; 80 diss/theses; 2 310 maps 24469

– Institute of Mineralogy, Crystallography, Petrology and Geochemistry, Faculty of Mining and Geology, Library, Dušina 7, P.O. Box 162, 11000 Beograd
T: +381 11 3282111; Fax: +381 11 635217
1853; Živa Ilić
22 600 vols; 3 084 curr per; 216 mss; 84 diss/theses; 3 govt docs; 243 maps; 21 microforms
libr loan 24470

– 'Jaroslav Černi' Institute for the Development of Water Resources, Library, Jaroslava Černog 80, 11223 Beograd
T: +381 11 649113; Fax: +381 11 649414
1950; Vesna Koprivica
13 650 vols; 90 diss/theses; 2 000 govt docs; 20 maps; 2 digital data carriers 24471

– Mašinski Fakultet, Biblioteka (Faculty of Mechanical Engineering), 27. marta 80, 11000 Beograd
T: +381 11 3370-761/-266; Fax: +381 11 3370364; E-mail: mf@mas.bg.ac.rs; URL: www.mas.bg.ac.rs
1949; Milorad Milovančević
Documents of High School and University of Belgrade from XIX century
60 000 vols; 23 curr per; 100 mss; 2 720 diss/theses
libr loan; ZBUS 24472

– Matematički Zavod, Biblioteka (Institute of Mathematics), Studentski trg 3, 11000 Beograd
1894
15 000 vols 24473

– Medicinski Fakultet, Centralna Biblioteka (Faculty of Medicine), ul. Deligradska 35, 11000 Beograd
1946
25 000 vols 24474

– Odeljenja za Klasične Nauke, Fakultet Filozofski, Biblioteka (Department of Classical Studies), Cika Ljubina 18-20, 11000 Beograd
T: +381 11 639628; Fax: +381 11 639356
1875; Lucija Carević
Greek, Latin, linguistics
33 000 vols 24475

– Odeljenje za Filozofiju, Biblioteka (Department of Philosophy), ul. Čika Ljubina 18-20, 11000 Beograd
T: +381 11 3206215; Fax: +381 11 639356; E-mail: ljvujose@f.bg.ac.rs; URL: www.f.bg.ac.rs/sr-lat/filozofija/biblioteke
1947; Ljubiša Vujošević
25 000 vols; 10 curr per; 1 082 mss; 45 diss/theses; 20 govt docs
ZBUS 24476

– Poljoprivredni Fakultet, Biblioteka (Faculty of Agriculture), ul. Nemanjina 6, 11080 Zemun
T: +381 11 615363; Fax: +381 11 193659
1945; Gordana Jović
46 400 vols; 132 curr per
libr loan 24477

– Pravni Fakultet, Biblioteka (Faculty of Law), bul. Revolucije 67, 11000 Beograd
1911
110 000 vols 24478

– Rudarsko-Geološki Fakultet, Biblioteka (Faculty of Mining and Geology), ul. Djušina 7, 11000 Beograd
T: +381 11 338832; Fax: +381 11 3355392
1837; Lidija Matić
39 000 vols 24479

– Saobracajni fakultet, Biblioteka (Faculty of Transport and Traffic Engineering), Vojvode Stepe 305, 11000 Beograd
T: +381 11 3091370, 3976017 ext 370; Fax: +381 11 2466294; URL: www.sf.bg.ac.rs
1950; Slobodan Domadžić
Engineering Science, Railway Transport and Traffic, Road Transport and Traffic, Water Transport and Traffic, Air Transport and Traffic, Logistics, Materials Handling, Intermodal Freight Transport, Postal and Telecommunications Traffic, Transport Management; 4 legacy (366 books)
24 559 vols; 19 curr per; 240 diss/theses; 666 journal titles (6 164 vols), 3 539 graduation theses
libr loan; ZBUS 24480

– Seminar za francuski jezik i književnost, Biblioteka (French Seminar), Studentski trg 3, 11000 Beograd
T: +381 11 638716; Fax: +381 11 630039
1894; Prof. Mihailo Popovic
22 000 vols; 6 curr per; 53 diss/theses; 4 sound-rec 24481

– Seminar za Istoriju Jugoslovenske Književnosti, Biblioteka (Seminar for the History of Yugoslav Literature), Studentski trg 3, 11000 Beograd
1901
65 000 vols 24482

– Seminar za Južnoslovenske Jeziko, Biblioteka (Seminar of Southern Slavic Languages), Studentski trg 3, 11000 Beograd
T: +381 11 630039
1901; Nevenka Duric
111 000 vols 24483

– Slovenski Seminar, Biblioteka (Slavic Seminar), Studentski trg 3, 11000 Beograd
1945
18 000 vols 24484

– Sumarski Fakultet, Biblioteka (Faculty of Forestry), Kneza Višeslava 1, 11030 Beograd
T: +381 11 553-122/-117
1949; Milorad Šijak
39 600 vols; 1 500 curr per; 380 diss/theses
libr loan 24485

– Tehnološki Fakultet, Biblioteka (Faculty of Technology), ul. Karnedžijeva 4, 11000 Beograd
1910
12 000 vols 24486

– Veterinarski Fakultet Univerziteta, Centralna biblioteka (School of Veterinary Medicine), bul. JNA 18, 11000 Beograd
T: +381 11 685-666/-316; Fax: +381 11 685936
1936; Dušica Čavdarević, Ružica Komljenovič
23 400 vols; 732 curr per; 5 988 diss/theses 24487

Univerzitet u Nišu, Univerzitetska biblioteka 'Nikola Tesla', Kej Mike Paligorića 2, 18000 ***Niš***; P.O. Box 48, 18000 Niš
T: +381 18 523421; Fax: +381 18 523119; E-mail: info-ubn@ni.ac.rs; URL: www.ubnt.ni.ac.rs
1967; Zoran B. Živković
Law Information Centre
100 000 vols; 1 250 curr per; 35 000 e-journals; 40 000 e-books; 5 100 diss/theses; 200 sound-rec; 550 digital data carriers
libr loan; ZBUS, BDS, LIBER 24488

– Ekonomski fakultet, Biblioteka (Faculty of Economics), Trg kralja Aleksandra Ujedinitelja 11, 18000 Niš
T: +381 18 528622; E-mail: biblioteka@eknfak.ni.ac.rs; URL: www.eknfak.ni.ac.rs/src/Biblioteka.php
1961; Ljubomir Ilić
31 574 vols; 7 098 curr per; 787 diss/theses 24489

– Filozofski fakultet, Biblioteka (Faculty of Philosophy), Ćirila i Metodija 2, 18000 Niš
T: +381 18 47540; Fax: +381 18 46460; E-mail: bibl@filfak.ni.ac.rs; URL: www.filfak.ni.ac.rs/d_bibl.htm
1948; Brankica Milosavljević
59 300 vols; 108 curr per; 190 diss/theses
libr loan 24490

– Gradjevinski fakultet, Biblioteka (Faculty of Civil Engineering), Beogradska 14, 18000 Niš
T: +381 18 42066; Fax: +381 18 42758
1961; Mirjana Krstić
12 776 vols; 25 curr per
libr loan 24491

– Medicinski Fakultet, Centralna biblioteka (Faculty of Medicine), Bd Dr Zoran Djindjic 81, 18000 Niš
T: +381 18 326712; Fax: +381 18 338770; E-mail: milena@medfak.ni.ac.rs; URL: www.medfak.ni.ac.yu
1961; Milena Djordjević
15 000 vols; 37 curr per; 8 000 e-journals; 2 792 diss/theses; 25 digital data carriers
libr loan; KoBSON 24492

Univerzitet u Novom Sadu (University of Novi Sad), ***Novi Sad***
– Filozofski Fakultet, Biblioteka (Faculty of Philosophy), Dr Zorana Djindjica 2, 21000 Novi Sad
T: +381 21 4853986; Fax: +381 21 450873; E-mail: biblioteka@ff.uns.ac.rs; URL: www.ff.uns.ac.rs/biblioteka/biblioteka.html
1954; Gordana Vilotic
550 000 vols; 2 000 curr per; 400 diss/theses; 45 digital data carriers
libr loan; BDS, ZBUS 24493

– Pravni Fakultet, Biblioteka (Faculty of Law), Trg Dositeja Obradovicá 1, 21000 Novi Sad
T: +381 21 350377; Fax: +381 21 612720
1959; Radmila Dabanović
57 000 vols 24494

Government Libraries

Republicki Zavod za Statistiku, Biblioteka (Statistical Office of the Republic of Serbia), ul. Milana Rakića 5, 11000 ***Beograd***
T: +381 11 2412 922 ext. 251; Fax: +381 11 3617 438; E-mail: biblioteka@stat.gov.rs; URL: webrzs.stat.gov.rs/WebSite/
1951; Ljiljana Bascarevic
10 000 vols 24495

Savezni Zavod za Statistiku, Biblioteka (Federal Office of Statistics), ul. Kneza Miloša 20, P.O. Box 203, 11000 ***Beograd***
T: +381 11 681999; Fax: +381 11 642368
1944; Ljubica Keserović
145 000 vols; 715 curr per; 270 microforms
libr loan 24496

Corporate, Business Libraries

Narodni Bank FNRJ, Biblioteka (National Bank of Yugoslavia), bul. Revolucije 15, 11000 ***Beograd***
T: +381 11 3248841
1928; Nada Tomić
83 000 vols 24497

Special Libraries Maintained by Other Institutions

Arheološki Institut, Srpska Akademija Nauka i Umetnosti, Biblioteka (Archaeological Institute of the Serbian Academy of Sciences and Arts), Knez Mihajlova 35, 11000 ***Beograd***
T: +381 11 637191; Fax: +381 11 180189; E-mail: O.Ilic@ai.sanu.ac.rs; URL: www.ai.sanu.ac.rs
1947
Hist of art, architecture
32 600 vols; 746 curr per 24498

Arhiv Jugoslavije, Biblioteka, Vase Pelagica 33, 11000 ***Beograd***
T: +381 11 3690252; Fax: +381 11 3066635; E-mail: arhivyu@arhivyu.rs; URL: www.arhivyu.gov.rs
1950.; Suzana Srndovic
History, Yugoslavia
13 980 vols; 1 060 curr per
libr loan 24499

Arhiv Srbije, Biblioteka (Archive of Serbia), Karnegžijeva 2, 11000 ***Beograd***
T: +381 11 3370781; Fax: +381 11 3370246; E-mail: t.jovanovic@archives.org.rs; URL: www.archives.org.rs
1900; Tatjana Jovanović
75 000 vols; 50 curr per
libr loan 24500

Astronomska Opservatorija u Beogradu, Biblioteka (Astronomical Observatory of Belgrade), ul. Volgina 7, 11000 ***Beograd***
T: +381 11 3089063; Fax: +381 11 2419553; URL: www.astronomija.co.rs
1887
13 000 vols 24501

Botanički Zavod i Bašte, Biblioteka (Botanical Institute and Garden of the University of Belgrade), ul. Takovska 43, 11000 ***Beograd***
T: +381 11 767988; Fax: +381 11 638500
1853
25 000 vols 24502

British Council, Britanska Biblioteka, Library, Terazije 8/II, 11000 ***Beograd***
T: +381 11 3023800; Fax: +381 11 3023898; E-mail: info@britishcouncil.rs; URL: www.britishcouncil.org/serbia
1946
Social sciences, political sciences, law, media, British studies, English language
12 000 vols
libr loan 24503

Etnografski institut SANU, Biblioteka (Institute of Ethnography SASA), Kneza Mihaila 36/4, 11000 ***Beograd***
T: +381 11 2636804; Fax: +381 11 2638011; E-mail: eisanu@ei.sanu.ac.rs; URL: www.etno-institut.co.rs
1947; Biljana Milenković Vuković
17 847 vols; 903 curr per; 20 e-journals; 30 e-books; 7 mss; 26 diss/theses; 500 maps; 3 450 av-mat; 54 sound-rec; 13 digital data carriers; 117 rare books
libr loan; Zajednica biblioteka Srbije, Serbian Academic Library Association 24504

Etnografskij muzej u Beogradu, Biblioteka (Ethnographic Museum in Belgrade), Studentski trg 13, 11000 ***Beograd***; p.p. 357, 11000 Beograd
T: +381 11 3281888; Fax: +381 11 621284; E-mail: etnomuzej@yubc.net; URL: www.etnomuzej.co.yu
1901; Jordana Jovanovic
100 000 vols; 389 curr per; No 24505

Goethe-Institut, Bibliothek, Knez Mihailova 50, p.p. 491, 11000 ***Beograd***
T: +381 11 2622823; Fax: +381 11 636746; E-mail: bibl@belgrad.goethe.org; URL: www.goethe.de/ins/cs/bel/wis/bib/deindex.htm
1970; Bettina Radner
11 000 vols; 65 curr per; 780 av-mat; 1 300 sound-rec; 100 digital data carriers 24506

Institut Français de Serbie, Médiathèque, Zmaj Jovina 11, 11000 ***Beograd***
T: +381 11 3023600; Fax: +381 11 3023610; E-mail: mediatheque@ccf.org.rs; URL: www.ccf.org.rs/fr/01a.htm
1945; Catherine Krasojević
Music, Cinema, Comics, e.c. Local Country Department, Youth Department
18 500 vols; 70 curr per; 5 diss/theses; 1 200 av-mat; 1 600 sound-rec; 120 digital data carriers; 220 DVDs 24507

Institut Mihailo Pupin, Biblioteka, ul. Volgina 15, P.O. Box 15, 11060 ***Beograd***
T: +381 11 755171; Fax: +381 11 775870; E-mail: biblio@pupin.rs; URL: www.pupin.rs/
1948; Andjelka Mešiček
20 000 vols; 450 curr per; 200 digital data carriers
libr loan 24508

Institut za ekonomiku poljoprivrede, Biblioteka (Institute of Agricultural Economics), Cara Uroša 54, 11000 ***Beograd***
T: +381 11 632-779/-701
Vaska Kecojević
40 800 vols 24509

Institut za Medjunarodnu Politiku i Privredu, Biblioteka (Institute of International Politics and Economics), ul. Makedonska 25, 11000 ***Beograd***
T: +381 11 3373633; Fax: +381 11 3373835; E-mail: iipe@diplomacy.bg.ac.rs; URL: www.diplomacy.bg.ac.rs/
1948; Gordana Terzić
UN publs, int'l organizations, world economy, int'l workers movement, European Community
265 000 vols; 200 curr per
libr loan 24510

Institut za nuklearne nauke 'Vinča', Biblioteka (Institute of Nuclear Science ù Vinčaò), P.O. Box 522, 11000 ***Beograd***
T: +381 11 4440871
1948; Radmila Lučić
Physics, chemistry, mathematics, electronics, nuclear sciences, energy resources and technology, biology
27 500 vols; 150 diss/theses; 203 200 microforms 24511

Institut za Zaštitu Bilja i Životnu Sredinu, Biblioteka (Institute for Plant Protection and Environment), ul. Teodora Drajzera 7, 11000 ***Beograd***
T: +381 11 660049; Fax: +381 11 669860
1945; Dr. Mirko Draganič
18 700 vols; 650 curr per 24512

Jugoslovenska Kinoteka, Biblioteka (Yugoslav Film Archives), ul. Knez Mihailova 19, P.O. Box 67, 11000 ***Beograd***
T: +381 11 622555; Fax: +381 11 622555
1946; Rajka Miščević
19 040 vols; 85 curr per; 9 050 mss; 38 803 microforms; 150 000 clippings
libr loan 24513

Jugoslovenski Bibliografsko-Informacijski Institut, Biblioteka (Yugoslav Institute for Bibliography and Information), Terazija 26, 11000 ***Beograd***
T: +381 11 687836; Fax: +381 11 687760
1948; Tanja Ostojić
Bibliography of Yugoslavia (1950-current); International Exchange Centre
22 000 vols; 25 curr per; 1 digital data carriers
IFLA, FID, ISSN 24514

Matematički Institut, Srpska Akademija Nauka i Umjetnosti, Biblioteka (Institute of Mathematics of the Serbian Academy of Sciences and Arts), ul. Knez Mihajlova 35, p.p. 367, 11001 ***Beograd***
T: +381 11 630170; Fax: +381 11 186105; E-mail: branka@mi.sanu.ac.rs; URL: www.mi.sanu.ac.rs
1946; Branka Bubonja
36 000 vols; 550 curr per; 14 diss/theses; 125 digital data carriers 24515

Muzej Grada Beograda, Biblioteka (Belgrad City Museum), ul. Zmaj Jovina 1, P.O. Box 87, 11000 ***Beograd***
T: +381 11 637954
1945
15 600 vols; 300 rare bks 24516

Muzej Pozorišne Umjetnosti Srbije, Biblioteka (Museum of Performing Arts of Serbia), Gospodar Jevremova 19, 11000 ***Beograd***
T: +381 11 626630; Fax: +381 11 628920; E-mail: office@mpus.org.rs
1953; Irina Kikić
10 000 vols; 5 curr per 24517

Muzej primenjene umetnosti, Biblioteka (Museum of Applied Arts), Vuka Karadzica 18, 11 000 ***Beograd***
T: +381 11 2626494, 2626841; Fax: +381 11 629121; E-mail: andrijana.ristic@mpu.rs; URL: www.mpu.rs/
1950; Andrijana Ristić
24 732 vols; 972 curr per
libr loan 24518

Prirodnjački Muzej, Biblioteka (Natural History Museum), ul. Njegoševa 51, P.O. Box 401, 11000 ***Beograd***
T: +381 11 3442149; Fax: +381 11 3442265
1903; Marina Mučalica Milanović
64 000 vols; 1 130 curr per; 15 diss/theses; 1 001 maps
libr loan 24519

Republički zavod za zaštitu spomenika kulture, Biblioteka (Institute for the Protection of Cultural Monuments in Serbia), ul. Radoslava Grujica br. 11 (bivša Božidara Adžije), 11000 ***Beograd***
T: +381 11 2454786; Fax: +381 11 3441430; URL: www.heritage.sr.gov.yu/sr
1947; Ana Tešić
22 158 vols; 200 curr per; 16 digital data carriers; clippings (approx. 950 yearly) 24520

Savez Pedagoškik Društava Srbje, Biblioteka (Federation of Serbian Education Societies), ul. Terazije 26, 11000 ***Beograd***
1923
9 000 vols 24521

Srpska književna zadruga, Biblioteka (Serbian Literary Association), Srpskih vladara 19/I, 11000 ***Beograd***
T: +381 11 330305
1892
12 000 vols 24522

Vizantoloski Institut, Srpska Akademija Nauka i Umetnosti, Biblioteka (Institute of Byzantinology of the Serbian Academy of Sciences and Arts), ul. Knez Mihajlova 35, 11000 ***Beograd***
T: +381 11 637095; Fax: +381 11 637095
1948; Ljubomir Maksimović
9 400 vols; 8 331 periodicals 24523

Vojni Muzej, Biblioteka (Military Museum), Kalemegdan, 11000 ***Beograd***
T: +381 11 3343-441; Fax: +381 11 3344-915
1936; Ljuba Majstorović
National hist, archaeology hist of arms, national hist of art, War painting, conservation
16 500 vols; 397 curr per 24524

Vukov i Dositejev Muzej, Biblioteka (Vuk and Dositej Museum), ul. Gospodar Jevremova 21, 11000 ***Beograd***
T: +381 11 625161
1950; Ljiljana Čubrić
Cultural hist
12 500 vols
libr loan 24525

Zavod za Geološke i Geografske Istrazivanja 'Jovan Zujovic', Biblioteka (Jovan Zujovic Institute for Geological and Geographical Research), ul. Karadjordjeva 48, 11000 ***Beograd***
1931
42 000 vols 24526

Zavod za Zdravstvenu Zaštitu Srbije (Institute of Public Health of Serbia), ul. Dr Subotica 5, 11000 ***Beograd***
T: +381 11 684-566/-148; Fax: +381 11 685735
1924; Prof. Slobodan Antonović
42 000 vols; 700 curr per; 23 diss/theses
libr loan 24527

Zeljeznički Muzej, Biblioteka (Railway Museum), ul. Nemanjina 6, 11000 ***Beograd***
T: +381 11 3610334; Fax: +381 11 3616831; E-mail: muzej@yurail.co.yu
1950; Vesna Minić
10 000 vols; 5 curr per; 120 maps; 23 av-mat; 25 sound-rec; 500 clippings
libr loan 24528

Arhiv Vojvodine, Biblioteka (Vojvodina Archives), Dunavska 35, 21000 ***Novi Sad***
T: +381 21 21244; Fax: +381 21 22332
1926; Vesna Bašić
Regional hist, ethnography, archeology
14 000 vols; 522 curr per
libr loan 24529

Muzej Vojvodine, Biblioteka (Museum of Vojvodina), Dunavska 35-37, 21000 ***Novi Sad***
T: +381 21 420566; Fax: +381 21 25059; E-mail: muzejvojvodine1@nscable.net; URL: www.muzejvojvodine.org.rs
1947; Ljubica Kostić
old and rare bks till 1868
81 450 vols 24530

Srpska akademija nauka i umetnosti, Ogranak u Novom Sadu, Biblioteka (Serbian Academy of Sciences and Arts – Branch in Novi Sad), Nikole Pašića 6, 21000 ***Novi Sad***; P.O. Box 330, 21001 Novi Sad
T: +381 21 420210; Fax: +381 21 611750; E-mail: biblisanu@uns.ac.rs; URL: www.ogranak.sanu.ac.rs
1979; Branka Milic
19 818 vols 24531

Srpsko Narodno Pozorište, Biblioteka (Serbian National Theatre), Pozorišni trg 1, 21000 ***Novi Sad***
T: +381 21 621411 ext 686
1861; Slavka Vidaković
22 600 vols; 10 curr per; 20 000 journals, cat etc 24532

Državni Arhiv Sreza Valjevo, Biblioteka (State Archive), ul. Karadjordjeva 46, ***Valjevo***
1952
8 000 vols 24533

Narodni Muzej, Biblioteka (National Museum), ul. Andje Ranković 19, ***Vršac***
T: +381 13 822569; Fax: +381 13 822569; E-mail: muzejvrsac@hemo.net
1882; Mirjana Radujković
8 500 vols; 400 curr per
libr loan 24534

Public Libraries

Biblioteka Grada Beograda (Belgrade City Library), Knez Mihailova 56, 11000 ***Beograd***
T: +381 11 2024000; Fax: +381 11 2024033; E-mail: info@bgb.rs; URL: www.bgb.rs
1929; Yasmina Ninkov
Rare hist books about Belgrade, Belgrade's coll, art coll (23 079 bound vols)
1 700 000 vols; 1 250 curr per; 500 mss; 6 520 sound-rec
libr loan 24535

Biblioteka Milutin Bojić, Ilije Garašanina 5 (matično odeljenje), 11000 ***Beograd***
T: +381 11 3242418, 3233708; Fax: +381 11 3233708; URL: www.milutinbojic.org.rs
1957; Miro Maksimović
173 000 vols; 1 263 curr per; 100 digital data carriers 24536

Narodna Biblioteka Vuk Karadžic, Cirila i Metodija 2, **Beograd** – Zvezdara
T: +381 11 423483, 403018; Fax: +381 11 403018; URL: www.biblioteke.org.yu/zvezdara/index.html
190 000 vols   24537

Narodna Biblioteka Dr CorDe **Natoševic**, Trg slobode 2, **Indjija**
T: +381 22 551378; Fax: +381 22 551847; E-mail: info.biblioteka@indjija.net; URL: www.bibliotekaindjija.rs/code/navigate.php?Id=17
1946; Vesna Stepanovic
88 000 vols   24538

Kulturni Center Kannik, Matična knjižnica Kannik, Kolodvorska ul 2, 61240 **Kannik**
T: +381 61 831142
Matjaž Varšek
7 branch libs
57 260 vols   24539

Narodna Biblioteka Jovan Popović, Trg srpskih dobrovoljaca 57, 23300 **Kikinda**
T: +381 230 21458; Fax: +381 230 22553; URL: www.kikinda.com/biblioteka.html
1945; Milos Latinovic
140 000 vols; 20 curr per   24540

Narodna Biblioteka Vuk Karadžić, ul. Moše Pijade 10, **Kragujevac**
T: +381 34 32398; Fax: +381 34 64814
1846; Mila Stefanovich
270 000 vols; 40 curr per
libr loan   24541

Narodna Biblioteka, Trg oslobodenja 7, **Kula**
T: +381 25 722231; Fax: +381 25 722231; E-mail: bibliotekakula@yahoo.com
Snežana Tatalović
85 175 vols   24542

Biblioteka Dimitrije Tucovič, Hilandarska 2, 11550 **Lazarevac**
T: +381 11 8122997;
Fax: +381 11 8122997; URL: www.bibliotekalazarevac.org
1949; Radmila Bulatovic
670 other items
95 331 vols; 65 curr per; 250 sound-rec; 15 digital data carriers; Children's dept
libr loan   24543

Knjiž Lendava, Könyvtàr Lendva, Partizanska 10, 9220 **Lendava**
T: +381 69 75353; Fax: +381 69 75353
Žužana Žoldoš
10 branch libs
80 000 vols; 90 curr per; 1 800 av-mat; 140 sound-rec   24544

Narodna Biblioteka "Stevan Sremac", Borivoja Gojkovića br. 9, 18000 **Niš**
T: +381 18 511-403/-410; Fax: +381 18 511410; E-mail: n.vasic09@yahoo.com; URL: www.nbss.org.rs
1879; Nebojša Vasić
2404 old and rare books, 135 Serbian graphics; Childrens, Fine arts, Research, Local issue, Periodicals, Acquistion, Processing and cataloguing, Cultural activities, Central supervising, Binding, Accounting dept
208 441 vols; 261 curr per; 337 mss; 161 govt docs; 26 music scores; 10 maps; 781 microforms; 1 739 av-mat; 43 sound-rec; 167 pamphlets, 19 VR, 1171 invitations, 398 postcards, 9 albums, 218 exhibition catalogs, 26 touristic guides, 271 manifestation programmes
libr loan; BDS   24545

Narodna Biblioteka, ul. Branka Radulovića 1, **Pirot**
50 000 vols   24546

Narodna Biblioteka Ilja M. Petrović, Drinska 2, 12000 **Požarevac**
T: +381 12 221957; Fax: +381 12 221029; E-mail: bibliotekapo@ptt.rs; URL: www.bibliotekapozarevac.org
1847
112 000 vols; 775 curr per   24547

Narodna Biblioteka (Public Library Požega), Kralja Alexsandra 8, 31210 **Požega**
T: +381 31 811-525/-921; Fax: +381 31 811921; E-mail: pozega@ptt.yu; URL: www.biblioteke.org.yu
1869; Prof. Gordana Stevič
95 000 vols
libr loan   24548

Matična Biblioteka Vuk kKradžić (Central Library Vuk Karadžić), ul. Sandžac 01637kih brigada 2, **Prijepolje**
T: +381 33 712960; Fax: +381 33 711146; E-mail: info@biblioteka-prijepolje.com; URL: www.biblioteka-prijepolje.com
Hadija EDžigal
55 000 vols   24549

Narodna Biblioteka Žika Popović, Masarikova 18, 15000 **Šabac**
T: +381 15 325095, 322755
1847; Sonja Bokun-Ainič
200 000 vols   24550

Narodna Biblioteka (National Library of Smederevo), Karadordeva 5, 11300 **Smederevo**
T: +381 26 222270, 223740; Fax: +381 26 222270 ext 14; E-mail: knjigohraniteljica@nadlanu.com; URL: www.biblioteka-smederevo.org.rs
1846; Dragan Mrdaković
Coll of old and rare bks, the homeland coll; Foreign book department
202 575 vols; 56 curr per; 62 mss; 17 diss/theses; 72 maps; 10 av-mat; 52 digital data carriers
libr loan   24551

Gradska Biblioteka Karlo Bijelicki (Town Library Karlo Bijelicki), Trg cara Lazara 3, **Sombor**
T: +381 25 482827; Fax: +381 25 420266; E-mail: biblioso@eunet.rs; URL: www.biblioso.org.rs
1859; Miljana Zrnic
Coll of old and rare bks (739 bks), Periodicals coll
318 120 vols   24552

Biblioteka Gligorije Vozarević, Trg Milekica 3, **Sremska Mitrovica**
T: +381 22 223607, 221130
112 567 vols   24553

Narodna Biblioteka Dositej Obradović, Trg Dr Zorona Dindića b.b., 22300 **Stara Pazova**
T: +381 22 310440; Fax: +381 22 310543; E-mail: biblido@ptt.rs; URL: www.biblioteke.org.yu
1878; Ratko Colakovic
56 048 vols; 2 585 curr per; 25 digital data carriers
libr loan   24554

Gradska Biblioteka, Cara Dušana 2, 24000 **Subotica**
T: +381 24 553115; Fax: +381 24 521383; E-mail: subiblioteka@gmail.com; URL: www.subiblioteka.rs
1890
105 000 vols   24555

Narodna Biblioteka Bora Stankovic, Lole Ribara b.b., 17500 **Vranje**
T: +381 17 22963; Fax: +381 17 22963
119 120 vols   24556

Gradska Biblioteka, Svetosavski trg br. 2, 26300 **Vršac**
T: +381 13 822882; Fax: +381 13 822882; E-mail: biblioteka@hemo.net; URL: www.vrsac.com/kultura/gradska%20biblioteka.asp
1882; Vesna Zlaticanin
30 000 old german books, 1 633 french books
200 000 vols; 23 000 curr per; 576 mss; 2 000 music scores; 2 000 maps; Banatica dept   24557

Matična Biblioteka Svetozar Marković, Kumanovska 2, 19000 **Zaječar**
T: +381 19 22045; E-mail: zabiblio@verat.net; URL: www.biblioteke.org.yu
1944; Carmen Gnjech Mijovic
110 200 vols; 45 curr per; 27 maps; 135 av-mat; 205 sound-rec; 5 800 clippings, 650 posters, 150 picture postcards
libr loan   24558

Biblioteka Sveti Sava, Petra Zrinskog 8, **Zemun**
T: +381 11 2618146, 2106784; Fax: +381 11 2106784; E-mail: biblioteka.zemun@gmail.com
1825
240 000 vols   24559

Gradska Narodna Biblioteka Žarko Zrenjanin, Trg Slobode 2, 23000 **Zrenjanin**
T: +381 23 566210; Fax: +381 23 530744; E-mail: biblioteka@zrbiblio.org; URL: zrbiblio.org
1946
60 000 vols   24560

Seychelles

National Libraries

Seychelles National Library, Francis Rachel St, Mahe, **Victoria**; P.O. Box 45, Victoria
T: +248 321333 ext 8014; Fax: +248 323183; E-mail: natlib@seychelles.net; URL: www.national-library.edu.sc
1910; Ann-Mary Robert
Indian Ocean Collection (IOC), colls of: IMO (Int. Maritime Org.), ILO (Int. Labour Org.), FAO (Food and Agriculture Org.), UNESCO, drugs and alcohol; 4 branch libs, 1 regional reading center, 1 mobile libr
65 000 vols; 25 curr per; 3 diss/theses; 8 govt docs; 6 maps; 63 av-mat; 179 sound-rec; 27 digital data carriers; 350 other items
IFLA   24561

University Libraries, College Libraries

Seychelles Polytechnic, Library, P.O. Box 77, **Victoria**
T: +248 371188; Fax: +248 371545; E-mail: libpoly@sexchelles.net
1983; Marie-France Loze
Teachers reference, Seychelles Coll, Short Loan Coll, Career, Family Life
12 155 vols; 21 curr per; 1 govt docs; 10 digital data carriers
libr loan   24562

Sierra Leone

National Libraries

Sierra Leone Library Board, Rokel St, **Freetown**; P.O. Box 326, Freetown
T: +232 22 226993, 224297; E-mail: sielib2002@yahoo.com
1959; Victor B. Coker
Africana coll, text books coll; Children's dept
335 000 vols; 22 curr per; 10 diss/theses; 1 000 govt docs; 50 maps; 12 av-mat; 19 digital data carriers; pamphlets, clippings
libr loan; IFLA, COMLA, SLAALIS
Legal deposit libr, partly fulfilling functions of a national libr   24563

University Libraries, College Libraries

Bo Teachers' Training College, Library, P.O. Box 162, **Bo**
1967; Angela Koker
10 000 vols; 30 curr per   24564

Union College, Library, **Bunumbu**, via Segbwema
1933
9 000 vols; 70 curr per; 179 film strips, 36 films   24565

University of Sierra Leone, **Freetown**
– College of Medicine and Allied Health Sciences, Library, PMB, Freetown
T: +232 22 220758; Fax: +232 22 240583; E-mail: comahs@srl.healthnet.org
1988; Mrs L.N. M'jamtu-sie
WHO, Short loan, Audio cassettes, Sierra Leone
9 870 vols; 43 curr per; 23 diss/theses; 110 av-mat; 392 sound-rec; 92 digital data carriers; microforms (11 boxes)
libr loan; SLAALIS   24566

– Fourah Bay College, Library, Mount Aureol, P.O. Box 87, Freetown
T: +232 22 27260
1827; D.E. Thomas
UN doc coll, Sierra Leone coll
200 000 vols; 330 curr per; 1 500 microfilms
libr loan; IFLA, SCAULWA   24567

– Njala University College (NUC), Library, PMB, Freetown
E-mail: NUClib@sierratel.dl; URL: www.nuc-online.com
Mr A.N.T. Deen
35 000 vols; 50 curr per; 198 films
                                                  24568

– Technical Institute, Library, Congo Cross, Freetown
6 000 vols; 10 curr per   24569

Teachers' College, Library, **Port Loko**
1966; J.L. Metzger
6 000 vols; 80 curr per   24570

Government Libraries

Fisheries Division, Ministry of Agriculture and Natural Resources, Library, P.O. Box, **Freetown**
1 000 vols; 116 curr per   24571

House of Parliament, Library, Tower Hill, **Freetown**
T: +232 22 223915
1961; Jusu Sesay
3 000 vols; 12 curr per   24572

Judicial Library, High Court, **Freetown**
T: +232 22 24344
1935; Mrs L.N.M. M'Jamtu-sie
Criminal and civil law   24573

Ministry of Education, Library, New England, **Freetown**
2 000 vols; 10 curr per   24574

National Agricultural Documentation Centre (NADOC), Ministry of Agriculture, Natural Resources and Forestry, Library, Youyi Bldg, Brookfields, **Freetown**
T: +232 22 41500 ext 289
Unisa Sesay
5 branch libs
5 000 vols; 100 curr per   24575

State Law Office, Library, Guma Bldg, Laminah Sankoh St, **Freetown**
T: +232 22 26733
1960; O.L. Campbell
4 000 vols; 15 curr per
libr loan   24576

Department of Agriculture, Headquarters Library, **Njala**, via Mano
1930
3 000 vols   24577

Corporate, Business Libraries

Bank of Sierra Leone, Research Library, Siaka Stevens St, P.O. Box 30, **Freetown**
T: +232 22 226501; Fax: +232 22 224764; E-mail: copesafe@sierratel.sl
1964; Bridget S. Tucker, Samuel J. Sepha
4 500 vols; 25 000 journals, rpts, pamphlets, staff papers etc   24578

Special Libraries Maintained by Other Institutions

American Cultural Center, International Communication Agency, USA, Library, 8 Walpole St, **Freetown**
1959
4 600 vols; 65 curr per   24579

British Council, Library, Tower Hill, P.O. Box 124, **Freetown**
T: +232 22 222-223/-227; Fax: +232 22 224123; E-mail: enquiry@sl.britishcouncil.org; URL: www.britishcouncil.org/sierraleone
1967
Education, women studies, Sierra Leone
7 000 vols   24580

Geological Survey Division, Ministry of Mines, Library, Youyi Bldg, Brookfields, **Freetown**
T: +232 22 40740
16 000 vols; 600 curr per   24581

Sierra Leone Medical and Dental Association, Library, P.O. Box 850, **Freetown**
1961
3 000 vols
Shared with main hospital   24582

Singapore

National Libraries

National Library Board (NLB), Lee Kong Chian Reference Library, 100 Victoria St, **Singapore** 188064
T: +65 63323255; Fax: +65 66323248;
E-mail: ref@nlb.gov.sg; URL: www.nlb.gov.sg
1958; Christopher Chia
Southeast Asia Coll, Children's Asia Coll; Singapore Resource Libr
5 300 000 vols
libr loan; ALA, ASLIB, CLA, LA, IAOL, CONSAL, IFLA, LAS   24583

University Libraries, College Libraries

Nanyang Technological University, Library, Nanyang Av, **Singapore** 639798
T: +65 67911744; Fax: +65 67914637;
E-mail: wwwlib@ntu.edu.sg; URL: www.ntu.edu.sg/library
1991; Foo Kok Pheow
390 000 vols
libr loan; LAS   24584

National University of Singapore, Central Library, 12 Kent Ridge Crescent, **Singapore** 119275
T: +65 68742069; Fax: +65 67773571;
E-mail: clbsec@nus.edu.sg; URL: www.lib.nus.edu.sg
1905; Ms Jill Quah
Singapore/Malaysia Coll, Rare Book Coll
2 197 282 vols; 46 744 curr per; 60 742 diss/theses; 28 425 microforms; 19 423 av-mat; 1 207 digital data carriers
libr loan; IFLA, COMLA, CONSAL   24585

– Chinese Library, 10 Kent Ridge Crescent, Singapore 119275
T: +65 68742039; Fax: +65 67773571;
E-mail: clbgenl@nus.edu.sg; URL: www.lib.nus.edu.sg
1953; Mr Lee Ching Seng
405 010 vols; 970 curr per; 952 diss/theses; 735 microforms; 520 av-mat; 4 digital data carriers
libr loan   24586

– Hon Sui Sen Memorial Library, 1 Hon Sui Sen Dr, Singapore 117588
T: +65 68743131; Fax: +65 67786301;
E-mail: hlbfrm01@nus.edu.sg; URL: www.lib.nus.edu.sg
1987; Lim Bee Lum
Management, business, finance
84 000 vols; 2 414 curr per; 700 diss/theses; 2 000 microforms; 1 467 av-mat; 25 digital data carriers
libr loan   24587

– C. J. Koh Law Library, 10 Kent Ridge Crescent, Singapore 119260
T: +65 68742042; Fax: +65 67785521;
E-mail: llbfrm01@nus.edu.sg
1957; Mrs Thavamani Prem Kumar
Asean Coll
53 700 vols
libr loan; LAS, AALL, IALL   24588

– Medical Library, 14 Medical Dr, Singapore 117599
T: +65 68742500; Fax: +65 67782046;
E-mail: mlbfrm01@nus.edu.sg
1905; Mrs Ng Kim Leong
Singapore/Malaysia coll
63 244 vols; 4 340 curr per; 1 517 av-mat
libr loan   24589

– Science Library, 10 Science Dr 2, Singapore 117548
T: +65 68742454; Fax: +65 67786118;
E-mail: sclibrm01@nus.edu.sg
1986; Mr Ng Kok Koon
108 000 vols; 5 040 curr per; 1 426 av-mat; 136 digital data carriers
libr loan   24590

Ngee Ann Polytechnic, Library, 535 Clementi Rd, **Singapore** 599489
T: +65 64606290; Fax: +65 64668274;
E-mail: ltx@np.edu.sg; URL: www.np.edu.sg/library/library.html
1963; Lock Thi Xuan
International Business Coll, Multimedia Design and Development coll, Award-Winning Coll, Semiconductor Databooks Coll, Early Childhood Coll, Lifestyle Library Coll, National Education Coll; Quality Resource Coll
227 000 vols; 1 700 curr per; 73 733 av-mat; 8 027 sound-rec; 9 326 digital data carriers; 7 851 items of computer software
libr loan; Library Association (UK), ALA, LAS   24591

Singapore Polytechnic, Library, 500 Dover Rd, **Singapore** 139651
T: +65 67721516; Fax: +65 67721956;
E-mail: margie@sp.ac.sg; URL: www.sp.edu.sg/lib/libhome.htm
1958; Margie Teo Koon Neo
Business, Engineering, Technology
196 100 vols; 486 curr per; 5 762 av-mat; 596 sound-rec; 3 795 digital data carriers; 517 DVDs
libr loan; ASLIB, CILIP, LAS   24592

Temasek Polytechnic, Library and Information Resources Departement, 21 Tampines Av 1, **Singapore** 529757
T: +65 67805211; Fax: +65 67898196;
E-mail: asklib@tp.edu.sg; URL: spark.tp.edu.sg
1990; Esther Ong Wooi-Cheen
165 100 vols; 894 curr per; 46 diss/theses; 7 maps; 8 000 av-mat; 2 191 sound-rec; 1 193 digital data carriers
libr loan; LAS, ASLIB, CILIP, ALA, ACRL, LITA, RUSA   24593

School Libraries

Hwa Chong Junior College, Jing Xian Library, Bukit Batok St 34, **Singapore** 2365
T: +65 65623077; Fax: +65 65653114
1974; Loi Sai Bay
35 940 vols; 115 curr per   24594

Government Libraries

Attorney General's Chambers, Library and Resource Centre, 1 Coleman St 10-00, **Singapore** 179803
T: +65 63325952; Fax: +65 63325258;
E-mail: agc@agc.gov.sg; URL: www.agc.gov.sg/library/index.html
1948; Mary KP Ho
International Law-Treaties, Statutes
32 000 vols; 200 curr per; 1 000 microforms; 25 digital data carriers; 5 video cassette tapes
libr loan; ALA, SLA   24595

Defence Technology Tower Library, 2nd Storey, Tower A, Depot Rd, **Singapore** 109679
T: +65 63734992; Fax: +65 62755938
1975; N.N.
1 000 vols; 200 curr per; 300 diss/theses; 3 000 microforms
libr loan   24596

Ministry of Community Development and Sports, Resource Centre, Library, 512 Thomson Rd, No 13-00, MCD Bldg, **Singapore** 298136
T: +65 63548792; Fax: +65 62594535;
E-mail: Johan_ABDUL_RAHMAN@MCDS.gov.sg; URL: www.lib.gov.sg
1974; Johan A. Rahman
Singapore Coll
15 500 vols; 60 curr per; 50 diss/theses; 200 govt docs; 50 av-mat; 5 digital data carriers
libr loan   24597

Monetary Authority of Singapore MAS, Library, 10, Shenton Way, MAS Bldg, **Singapore** 0207
T: +65 62299320; Fax: +65 62299328;
E-mail: fcheo@mas.gov.sg
1971; Frances Cheo
115 000 vols; 1 016 curr per; 8 digital data carriers
libr loan   24598

Parliament Library, Parliament House, 1 High St, **Singapore** 179429
T: +65 63325506; Fax: +65 63325536
1955; Ms Khoo Sait Poh
20 000 vols; 200 curr per; 1 000 sound-rec
libr loan   24599

Port of Singapore Authority (PSA), Library, 02-01A PSA Vista, 20 harbour Dr, **Singapore** 099255
T: +65 67717304; Fax: +65 67740676;
E-mail: florabay@hq.psa.com.sg
1971; Flora Bay
Sea transport and traffic, port lit
16 000 vols; 500 curr per; 200 av-mat   24600

SAFTI Military Institut, Library, 500 Upper Jurong Rd, 01-03 **Singapore** 638364
T: +65 67997420; Fax: +65 67922780;
E-mail: kbeechin@starnet.gov.sg
1994; Koh Bee Chin
70 000 vols; 300 curr per; 16 000 av-mat; 300 sound-rec; 160 digital data carriers
libr loan   24601

Spring Singapore, Resource Library, PSB Bldg, 2 Bukit Merah Central, **Singapore** 159835
T: +65 62786666; Fax: +65 62786665;
E-mail: enterpriseone@spring.gov.sg; URL: www.spring.gov.sg
1972; Lee Suan Hiang
12 000 vols; 2 000 av-mat; 200 serial titles, 1 000 newsletters and annual rpts   24602

Subordinate Courts Library, Upper Cross St, **Singapore** 0105
T: +65 64355883
1975; Nooraeni Ahmad
8 600 vols; 65 curr per
libr loan   24603

Supreme Court Library, City Hall Bldg, St. Andrew's Rd, **Singapore** 178882
T: +65 63324091; Fax: +65 63324092
Seah Poh Geok
29 000 vols; 152 curr per; 5 microforms; 2 digital data carriers
libr loan   24604

Ecclesiastical Libraries

Singapore Bible College, Library, 9-15 Adam Rd, **Singapore** 1128
T: +65 64664677 ext 217; Fax: +65 64664980
1952; Dr. Ng Peh Cheng
Music coll
17 500 vols; 137 curr per; 30 diss/theses; 4 500 music scores; 400 microforms; 583 av-mat; 1 300 sound-rec; 4 digital data carriers
libr loan; LAS   24605

Corporate, Business Libraries

Development Bank of Singapore (DBS Bank), Resource Centre, DBS Bldg, 6 Shenton Way, **Singapore** 0106
T: +65 63215792; Fax: +65 65342886
1968; Shirley See-Toh
10 000 vols; 1 000 curr per
libr loan   24606

Donaldson & Burkinshaw's Library, 24 Raffles Pl, Clifford Centre, **Singapore** 0104
+65 65339422; Fax: +65 65333590
1874; Abu Bakar Jusoff
10 000 vols; 22 curr per
libr loan; LAS   24607

Special Libraries Maintained by Other Institutions

American Library Resource Centre, 30 Hill St, **Singapore** 0617
T: +65 62227033; Fax: +65 62243239
1950; Han Mui Lan
8 000 vols; 170 curr per; 7 034 incunabula; 3 090 pamphlets
libr loan   24608

Chung Hwa Free Clinic, Chinese Medical Library, 640 Lorong 4, Toa Payoh, **Singapore** 1231
T: +65 62563481
1957; Wong Peng
16 000 vols; 300 curr per   24609

Housing and Development Board, Library, HDB Centre, 3451, Jl Bukit Merah, **Singapore** 0315
T: +65 62796190; Fax: +65 62796097
1960; Lena Soh
Urban and Regional Planning, Housing, Architecture
14 800 vols; 110 curr per; 20 govt docs; 15 microforms; 400 av-mat
libr loan   24610

Institute of East Asian Philosophies, John Tung Memorial Library, National University of Singapore, Kent Ridge, **Singapore** 0511
T: +65 67792954; Fax: +65 67793409
1983; Lee Ching Seng
Hist, Philosophy, Religion, Sociology, Economy, Politics on East Asia
43 000 vols; 658 curr per   24611

Institute of Southeast Asian Studies, Library, 30 Heng Mui Keng Terrace, Pasir Panjang, **Singapore** 119614
T: +65 68702401; Fax: +65 67756184;
E-mail: iseaslib@iseas.edu.sg; URL: www.iseas.edu.sg/library.html
1968; Ch'ng Kim See
140 000 vols; 2 038 curr per; 822 maps; 159 252 microforms; 82 295 av-mat; 78 digital data carriers; 8 269 pamphlets, press cuttings of selective English language newspapers from Southeast Asia
libr loan; LAS, LA, ASLIB   24612

Institute of Technical Education, Library, 10 Dover Dr, **Singapore** 138683
T: +65 67720680; Fax: +65 67762172;
E-mail: siewgk@ite.edu.sg
1973; Siew Gek Kheng
20 000 vols; 350 curr per; 5 000 govt docs; 215 microforms; 4 500 av-mat; 350 sound-rec; 40 digital data carriers
libr loan   24613

Marine Fisheries Research Department, Library, 2 Perahu Rd, **Singapore** 498989
T: +65 65428455 ext 20; Fax: +65 65451483; E-mail: mfrdlibr@pacific.net.sg
1969; Ronnie Tan
31 200 vols; 334 curr per; 2 465 mss; 65 av-mat; 172 annual reports/annual statistics, 48 directories, 5 041 reprint, 5 694 technical serials   24614

National Heritage Board, National Museum Library, 93, Stamford Rd, **Singapore** 178897
T: +65 63323570; Fax: +65 63327998;
E-mail: tan_chor_koon@nhb.gov.sg
1960; Tan Chor Koon
21 000 vols; 85 curr per; 200 av-mat; 50 digital data carriers
libr loan   24615

PSB Information Centre, Singapore Productivity and Standards Board, Library, 2 Bukit Merah Central, **Singapore** 159835
T: +65 67729643; Fax: +65 67782774;
E-mail: ifd@psb.gov.sg
1963
Patents, productivity
20 000 vols; 500 curr per; 30 digital data carriers
libr loan; LAS   24616

Singapore Botanic Gardens, National Parks Board, Library, Cluny Rd, **Singapore** 259569
T: +65 64719921; Fax: +65 64754295; E-mail: christina_SOH@NPARKS.gov.sg; URL: www.nparks.gov.sg
1875; Christina Soh Jeng Har
231 000 vols; 1 000 curr per; 100 govt docs; 1 000 maps; 226 av-mat; 30 digital data carriers; 2 000 botanical artworks
LAS   24617

Singapore Police Force, Library, Thomson Rd, **Singapore** 1129
T: +65 62540000; Fax: +65 62555600
1969; Chang Phee Tsong
18 000 vols; 140 curr per; 200 course reports
libr loan   24618

Singapore Science Centre, Library, Science Centre Rd, **Singapore** 609081
T: +65 65603316; Fax: +65 65669533
1977; Teo Choon Jin
12 000 vols; 160 curr per
libr loan   24619

Singapore Sports Council, Library, 15 Stadium Rd, National Stadium, **Singapore** 397718
T: +65 63409518; Fax: +65 63409537; E-mail: library@ssc.gov.sg
1973; Lily Poh
15 000 vols; 350 curr per; 1 000 av-mat
libr loan; IASI, LAS   24620

Singapore Telecommunication Academy, Library, Tower 1, 9th storey, 1 Hillcrest Rd, **Singapore** 1128
T: +65 64600263; Fax: +65 64669855
1973; Lee Kwee Eng
20 000 vols; 350 curr per; 1 000 av-mat; 10 digital data carriers
libr loan; LAS   24621

Southeast Asian Ministers of Education Organization (SEAMEO), Regional Language Centre, Library, 30 Orange Grove Rd, **Singapore** 258352
T: +65 68857888; Fax: +65 67342753; E-mail: library@relc.org.sg; URL: www.relc.org.sg/libinfo.htm
1968; Yolanda Beh
Language education, linguistics, Southeast Asian languages, teaching English as a second or foreign language; RELC course projects and publs
55 014 vols; 308 curr per; 3 174 microforms; 5 500 av-mat; 3 258 sound-rec
libr loan; LAS   24622

Television Corporation of Singapore TCS, Central Library, Andrew Rd, **Singapore** 299939
T: +65 63503557; Fax: +65 63551201; E-mail: davidb@tcs.com.sg
1967; Sarminah Tamsir
RTS, SBC & TCS Television Archive
5 000 vols; 150 curr per; 1 500 microforms; 70 000 av-mat; 30 000 sound-rec; 50 digital data carriers; press clippings (chinese and english; Singapore heritage, personalities, current affairs)
LAS   24623

Public Libraries

Bukit Merah Community Library, National Library Board, Bukit Merah Central, **Singapore** 159462
T: +65 63755111; Fax: +65 63755128
1982; Janice Tan
250 000 vols; 380 curr per
libr loan   24624

Jurong Regional Library, 21 Jurong East Central 1, **Singapore** 609732
T: +65 6332 3255; Fax: +65 6665 0312; URL: www.lib.gov.sg/
1988
507 000 vols; 2 433 curr per; 22 000 microforms; 12 000 av-mat   24625

National Library – Ang Mo Kio Branch, 4300 Ang Mo Kio Av 6, **Singapore**, 569842
T: +65 65535000; Fax: +65 65535519; URL: www.lib.gov.sg
1985; Noor Aini Mohamed
286 000 vols; 282 curr per; 1 520 microforms; 8 000 av-mat; 630 sound-rec; 130 digital data carriers   24626

National Library – Bedok Branch, 21 Bedok North St 1, **Singapore** 469659
T: +65 62444902, 62444901; Fax: +65 62444917
1985; Chow Wun Han
253 000 vols; 300 curr per
libr loan   24627

National Library – Geylang East Branch, Geylang East Av 1, **Singapore** 1438
T: +65 67460966
Wong Heng
131 000 vols; 210 curr per   24628

National Library – Jurong East Branch, Jurong Town Hall Rd, **Singapore** 2260
T: +65 65676111; Fax: +65 65606506
1988; Olive Lee
123 000 vols; 200 curr per   24629

National Library – Marine Parade Branch, Marine Parade Rd, **Singapore** 1544
T: +65 64402111; Fax: +65 64400031
1978; Siti Hanifah Mustapha
305 000 vols; 347 curr per; 4 800 av-mat; 400 sound-rec   24630

National Library – Queenstown Branch, Margaret Drive, **Singapore** 0314
T: +65 64730066
1970; Lee-Chin Kuei Lian
240 200 vols; 212 curr per
libr loan   24631

Toa Payoh Community Library, Toa Payoh Central, **Singapore** 319191
T: +65 6354-5050/-5080; Fax: +65 63545067; E-mail: hajbee@nlb.gov.sg
1974; Mam Hajbee Abu Bakar
Braille Bk, Talking Bk
230 000 vols; 590 curr per; 4 999 av-mat; 2 523 sound-rec; 415 digital data carriers; 153 braille bks, 339 talking bks
libr loan   24632

Sint Maarten

Public Libraries

Philipsburg Jubileum Bibliotheek, Ch. E. W. Vogesstraat 12, P.O. Box 2, **Sint Maarten**
T: +599 542 2970; Fax: +599 542 5805; E-mail: info@stmaartenlibrary.org; URL: www.stmaartenlibrary.org/
Caribbean Coll
50 000 vols   24633

Slovakia

National Libraries

Slovenská Národná Knižnica (Slovak National Library), J.C. Hronského 1, 036 01 **Martin**
T: +421 43 4301802; Fax: +421 43 4301802; E-mail: snk@snk.sk; URL: www.snk.sk
1863; Dr. Dušan Katuščák
Exlibris, postcards, photos (portraits), phonographical rolls, daguerreotypes
4 404 375 vols; 2 286 curr per; 1 366 000 mss; 721 incunabula; 5 500 maps; 7 014 microforms; 24 693 av-mat; 41 000 sound-rec; 409 digital data carriers
libr loan; IFLA, CENL, ICAU, IAML   24634

General Research Libraries

Štátna vědecká knižnica Banská Bystrica (State Research Library Banská Bystrica), Lazovná No.9, 975 58 **Banská Bystrica**
T: +421 48 4155067, 4155111; Fax: +421 48 4124096; E-mail: svkbb@svkbb.sk; URL: www.svk-bb.eu
1926; Dr. Olga Lauková
Patents, standards, hist docs, European information; Musical dept
1 600 000 vols; 1 100 curr per; 6 e-journals; 474 e-books; 11 500 music scores; 1 200 maps; 1 700 microforms; 200 av-mat; 14 400 sound-rec; 300 digital data carriers
libr loan; SSK, SAK, IAML   24635

Slovenská Akadémia Vied (SAV), Ustredná knižnica (Slovak Academy of Sciences), Klemensova 19, 814 67 **Bratislava**
T: +421 2 52921733; Fax: +421 2 52921733; E-mail: podobova@up.upsav.sk; URL: www.savba.sk
1953; Marcela Horváthová
Hist bks coll, Publ of the World Bank; Dept of hist bks, Austrian libr
497 658 vols; 1 148 curr per; 30 000 mss; 45 incunabula; 415 microforms; 467 digital data carriers
SLA, Pro Scientia   24636

Univerzita Komenského v Bratislave, Akademická knižnica UK (Comenius University in Bratislava), Safarikovo nam. 6, 818 06 **Bratislava** 16
T: +421 2 59244447; E-mail: library@vili.uniba.sk; URL: vili.uniba.sk
1919; Dr. Daniela Gondová
UNO-Deposit coll, UNESCO-Deposit coll, NATO-Deposit coll; ISSN national agency
2 376 461 vols; 2 743 curr per; 2 098 mss; 450 incunabula; 8 500 diss/theses; 43 000 music scores; 4 000 maps; 4 500 microforms; 800 av-mat; 18 000 sound-rec; 100 digital data carriers
libr loan; IFLA, IAML, CERL   24637

Štátna vědecká knižnica (State Scientific Library), Hlavná 10, 042 30 **Košice**
T: +421 55 6226724; Fax: +421 55 6222331; E-mail: svkk@svkk.sk
1657/1946; Dr. Jan Gaspar
Dept of special founds – unit for the technical docs – unit of the audiovisual and another info sources; Dept of special founds, The hist found dept, Goethe Institute – german reading room
3 390 000 vols; 1 494 curr per; 45 incunabula; 21 300 music scores; 4 200 microforms; 182 av-mat; 8 906 sound-rec; 694 digital data carriers
libr loan; SLA   24638

Štátna vědecká knižnica (State Research Library), Hlavna 99, 081 89 **Prešov**
T: +421 91 7724175; Fax: +421 91 7724960; E-mail: kniznica@svkpo.sk; URL: www.svkpo.sk
1952; Anna Hudakova
451 150 vols; 1 037 curr per; 2 047 maps; 528 microforms; 91 av-mat; 432 sound-rec; 88 digital data carriers
libr loan; SAK   24639

University Libraries, College Libraries

Univerzita Mateja Bela, Knižnica (Matej Bel University), Tajovského 40, P.O. Box 285, 974 01 **Banská Bystrica**
T: +421 48 4465205; Fax: +421 48 4465207; E-mail: kniznica@umb.sk; URL: www.library.umb.sk
1954; Mária Krobotová
195 838 vols; 218 curr per; 250 music scores; 64 maps; 51 av-mat; 544 sound-rec; 79 digital data carriers
libr loan; SSK, SAK, Pro Scientia   24640

Ekonomická univerzita, Slovenská ekonomická knižnica (Slovak Economic Library), Dolnozemská cesta 1, 852 35 **Bratislava**
T: +421 2 62412293; Fax: +421 2 62412302; E-mail: ma@sekba.euba.sk; URL: www.euba.sk

1948; Dr. Renata Mártonová
OECD Coll, INF Coll, The World Bank Coll
350 723 vols; 281 curr per; 34 150 e-journals; 10 782 diss/theses; 932 digital data carriers
SSK, SAK, EBSLG, Pro Scientia   24641

Škola knihovníckych a informačných štúdí, Knižnica (School of Librarianship and Information Studies), Kadnárova 7, 834 14 **Bratislava**
T: +421 2 44871568; Fax: +421 2 44871042
1962; Dr. Tibor Trgiňa
10 200 vols; 12 curr per; 200 govt docs
SSK, SAK   24642

Slovak Postgraduate Academy of Medicine, Knižnica, Limbova 12, 833 03 **Bratislava**
T: +421 2 54774560; Fax: +421 2 373739
1962; Zuzana Onderov
16 400 vols; 149 curr per; 1 digital data carriers
libr loan   24643

Slovenská Akadémia Vied – Ustav experimentálnej farmakológie, Knižnica (Institute of Experimental Pharmacology, SAS), Dúbravská cesta 9, 842 16 **Bratislava**
T: +421 2 375928; fax: +421 2 375928; E-mail: exfasekr@savba.sk
1969; Veronika Bruderová
8 400 vols; 1 curr per
libr loan   24644

Slovenská Technická Univerzita v Bratislave (Slovak University of Technology in Bratislava), **Bratislava**
– Chemickotechnologická fakulta, Slovenská chemická knižnica (Faculty of Chemical and Food Technology), Radlinského 9, 812 37 Bratislava
T: +421 2 52495268; Fax: +421 2 42495265; E-mail: jozef.dzivak@stuba.sk; URL: www.chtf.stuba.sk
1955; Jozef Dzivák
43 700 vols; 230 curr per; 480 diss/theses
libr loan; SSK   24645

– Fakulta Architektúry, Študijné a informačné stredisko (Faculty of Architecture), Vazovova 5, 812 43 Bratislava
T: +421 2 57294323; Fax: +421 2 57294326; E-mail: prorektor.zahr@stuba.sk
1990; Darina Pawlusová
20 000 vols; 70 curr per
libr loan   24646

– Fakulta elektrotechniky a informatiky, Knižnica (Faculty of Electrical Engineering and Information Technology), Ilkovicova 3, 812 19 Bratislava
T: +421 2 60291301; Fax: +421 2 65420415; E-mail: kniznica@elf.stuba.sk; URL: www.elf.stuba.sk
1962; Maria Handzova
Electrical Engineering, Electronics, Computer Science, Information Technology, Telecommunications, Control
113 970 vols; 368 curr per; 4 317 diss/theses; 163 av-mat; 551 sound-rec; 388 digital data carriers; e-journals
libr loan; SAK, SSK   24647

– Kniznica a informacne stredisko Strojníckej fakulty, Nám Slobody 17, 812 31 Bratislava
T: +421 2 57296218; Fax: +421 2 52926625; E-mail: polcikova@sjf.stuba.sk; URL: www.stuba.sk
1963; Dr. Viera Polčíková
Mechanical engineering, computer science
120 000 vols; 105 curr per; 520 diss/theses; 35 av-mat; 192 digital data carriers
libr loan; LIBER, SAK, SSK   24648

– Materiálovo-technická fakulta, Ústredná knižnica (Faculty of Materials Science and Technology), Pavlinska 16, 917 24 Trnava
T: +421 33 32340; Fax: +421 33 32013
1986; Kvetoslava Rešetová
43 600 vols; 179 curr per
libr loan; SSK   24649

– Staverbna fakulta, Študijno-informačné stredisko (Faculty of Civil Engineering), Radlinského 11, 813 68 Bratislava
T: +421 2 364867; Fax: +421 2 367027
1960; Silvia Stasselová
105 000 vols; 203 curr per
libr loan; SAK 24650

Slovenská technická univerzita v Bratislave, Knižnica (Slovak University of Technology in Bratislava), Vazovova 5, 812 43 ***Bratislava*** 1
T: +421 257294111; Fax: +421 257294537; URL: www.stuba.sk
1938; Dr. Tibor Trgiňa
500 000 vols; 2 000 curr per; 2 500 diss/theses; 1 000 microforms; 2 000 pamphlets
libr loan 24651

Univerzita Komenského v Bratislave, Akademická knižnica UK (Comenius University in Bratislava), Safarikovo nam. 6, 818 06 ***Bratislava*** 16
T: +421 2 59244447; E-mail: library@vili.uniba.sk; URL: vili.uniba.sk
1919; Dr. Daniela Gondová
UNO-Deposit coll, UNESCO-Deposit coll, NATO-Deposit coll; ISSN national agency
2 376 461 vols; 2 743 curr per; 2 098 mss; 450 incunabula; 8 500 diss/theses; 43 000 music scores; 4 000 maps; 4 500 microforms; 800 av-mat; 18 000 sound-rec; 100 digital data carriers
libr loan; IFLA, IAML, CERL 24652

– Fakulta matematiky, fyziky a informatiky, Knižničné a edičné centrum (Faculty of Mathematics, Physics and Informatics), Mlynská dolina, pav. I, 842 48 Bratislava 4
T: +421 2 60295111; Fax: +421 2 65412305; E-mail: sd@fmph.uniba.sk; URL: www.fmph.uniba.sk/index.php?id=562
1989; Ester Prešnajderová
99 668 vols; 222 curr per; 900 diss/theses; 539 av-mat; 87 digital data carriers
libr loan 24653

– Fakulta telesnej vychovy a sportu, Knižnica (Faculty of Physical Education and Sport), nabr. arm. gen. L. Svobodu 9, 814 69 Bratislava
T: +421 2 0669921;
Fax: +421 2 5313327; E-mail: kniznica_mail@fsport.uniba.sk; URL: www.fsport.uniba.sk/index.php?id=1673
1960; Lubica Grebečlová
Physical education and sports, sports medicine
41 088 vols; 95 curr per; 8 283 diss/theses; 2 digital data carriers
libr loan 24654

– Farmaceutická fakulta, Ústredná knižnica (Faculty of Pharmacy), Kalinčiakova 8, 832 32 Bratislava; mail address: Ul. Odbojárov 10, 832 32 Bratislava
T: +421 2 50117-155/-156/-157;
Fax: +421 2 501170100; E-mail: hupkovam@fpharm.uniba.sk; URL: www.fpharm.uniba.sk
1952; Dr. Mária Kadnárová
85 277 vols; 114 curr per; 12 e-journals; 8 512 e-books; 10 064 diss/theses; 68 av-mat; 1 digital data carriers
libr loan; SLA 24655

– Filozofická fakulta, Ústredná knižnica (Faculty of Philosophy), Gondova 2, 818 01 Bratislava
T: +421 2 59244571; Fax: +421 2 52966016; E-mail: uk@fphil.uniba.sk; URL: vili.uniba.sk
1991; Dr. Daniela Gondová
360 000 vols; 480 curr per; 4 200 diss/theses; 6 digital data carriers
libr loan; SAK, SSK 24656

– Knižnica a študijné informačné stredisko, Jesseniová lekárska fakulta, Novomeského 7, 036 45 Martin
T: +421 43 4239549; Fax: +421 43 4133395; E-mail: svrkova@jfmed.uniba.sk; URL: www.jfmed.uniba.sk
1969; Dr. L'ubica Horáková
Medicine, Nursing
65 079 vols; 93 curr per; 2 e-journals; 166 diss/theses; 153 av-mat; 484 sound-rec
libr loan; SSK 24657

– Knižnica Právnickej fakulty (Faculty of Law Library), Šafárikovo nám. 6, 818 05 Bratislava
T: +421 2 59244213; E-mail: kniznica@flaw.uniba.sk; URL: www.flaw.uniba.sk/index.php?id=47
1991; Anna Budayová
Documents of the European Union; European Documentation Centre
100 000 vols; 207 curr per; 15 e-journals; 9 200 microforms; 2 digital data carriers
libr loan; SAK, IALL 24658

– Lekárska Fakulta, Akademická knižnica (Faculty of Medicine), Odborárske nám 14, 813 72 Bratislava
T: +421 2 59357437;
Fax: +421 2 59357630; E-mail: kniznica.info@fmed.uniba.sk; URL: www.fmed.uniba.sk/?kniznica
1954; Dr. Miriam Pekníková
WHO
185 690 vols; 438 curr per; 2 027 diss/theses; 39 av-mat; 51 digital data carriers
libr loan; EAHIL 24659

– Pedagogická fakulta, Akademická knižnica (Faculty of Education), Moskovská 2, 813 34 Bratislava
T: +421 2 50222402; E-mail: akademicka.kniznica@fedu.uniba.sk; URL: www.fedu.uniba.sk/index.php?id=128
1951; Ester Jaurová
100 000 vols; 116 curr per; 283 diss/theses; 69 av-mat
libr loan; SSK 24660

– Rímskokatolicka cyrilometodska bohoslovecka fakulta, Knižnica (Roman Catholic Faculty of Theology of Cyrill and Methodius), Kapitulská 26, 814 58 Bratislava
T: +421 2 54432396; Fax: +421 2 54430266; E-mail: kniznica@frcth.uniba.sk; URL: www.frcth.uniba.sk/index.php?id=1673
1992; Viliam Špánik
27 600 vols; 65 curr per; 2 mss; 75 diss/theses; 50 music scores; 1 digital data carriers 24661

Vysoká škola múzických umění, Ustredná knižnica a stredisko automatizácie (Academy of Music and Dramatic Arts), Zochova 1, 813 01 ***Bratislava***
T: +421 2 5312039; Fax: +421 2 5317081
1950; Dr. Elena Ružeková
75 000 vols; 210 curr per; 1 450 diss/theses; 2 800 music scores; 1 000 av-mat; 5 000 sound-rec; 800 clippings, 1 600 photos
libr loan; IAML, SSK 24662

Vysoká škola výtvarných umení Bratislava, Library (Academy of Fine Arts and Design), Hviezdoslavovo nám 18, 814 37 ***Bratislava***
T: +421 2 54431132; Fax: +421 2 54432340; E-mail: kniznica@afad.sk; URL: www.afad.sk
1949; Tatiana Vančová
32 500 vols; 63 curr per; 365 av-mat; 23 digital data carriers
libr loan 24663

Technická univerzita v Košiciach, Univerzitná knižnica (Technical University of Košice), Letná 9, 043 84 ***Košice***
T: +421 55 6022889; Fax: +421 55 6230438; E-mail: kniznica@lib.tuke.sk; URL: www.lib.tuke.sk
1952; Dr. Valéria Krokavcová
Engineering, mining, metallurgy, electrotechnics, civil engineering, economics
223 000 vols; 388 curr per; 57 688 mss; 430 diss/theses; 146 maps; 90 microforms; 4 digital data carriers
libr loan; SSK, SAK 24664

Univerzita Pavla Jozefa Šafárika v Košiciach, Univerzitná knižnica (Jozef Šafárik University in Košice), Moyzesova 9, 040 01 ***Košice***
T: +421 55 2341608; E-mail: garlib@upjs.sk; URL: www.upjs.sk
1963; Dr. Daniela Džuganová
European Documentation Centre
228 996 vols; 330 curr per; 12 e-journals; 3 469 diss/theses
SLA 24665

– Lekárská fakulta, Ústredná knižnica, Tr SNP 1, 040 66 Košice
T: +421 55 6428151; Fax: +421 55 6428151
1952
122 000 vols 24666

Univerzita veterinárskeho lekárstva, Knižnica (University of Veterinary Medicine), Komenského č. 73, 041 81 ***Košice***
T: +421 55 6332111; Fax: +421 55 6323666; E-mail: sekretariat@uvm.sk; URL: www.uvm.sk
1949; Sona Lemakova
99 604 vols; 226 curr per; 6 500 diss/theses; 90 av-mat; 33 sound-rec; 150 digital data carriers
libr loan; SSK, SAK 24667

Univerzita Konštantína Filozofa, Univerzitná Knižnica (Constantine the Philosopher University), Drazovska 4, 949 74 ***Nitra***
T: +421 37 6408096; Fax: +421 37 6408096; E-mail: lib@ukf.sk; URL: www.uk.ukf.sk
1960; Dr. Anezka Stribova
223 489 vols; 814 curr per; 8 e-journals; 26 289 diss/theses; 1 776 music scores; 552 maps; 212 av-mat; 925 sound-rec
libr loan; SAK 24668

Pedagogická a sociálna akadémia, Knižnica, Kmet'ovo stromoradie 5, 081 75 ***Prešov***
T: +421 51 7711294; Fax: +421 51 7711293; E-mail: skola@pasapo.sk; URL: www.pasapo.sk
28 000 vols 24669

Konzervatórium, Knihovna (Conservatoire), ul. Marxa-Engelka 12, 010 00 ***Žilina***
1952
17 500 vols 24670

Žilinská Univerzita v Žiline, Univerzitná knižnica (University of Žilina, University Library), ul.Vysokoskolakov 24, 011 84 ***Žilina***
T: +421 41 5131453; E-mail: Marta.Sakalova@ukzu.uniza.sk; URL: ukzu.uniza.sk
1961; Dr. Marta Sakalová
196 777 vols; 281 curr per; 864 diss/theses; 3 maps; 333 av-mat; 561 digital data carriers; 47 258 standards
libr loan; SSK, SAK 24671

Slovak Library of Forestry and Wood Sciences, Technical University in Zvolen, T.G.Masaryka 20, 96102 ***Zvolen***
T: +421 45 5206642; Fax: +421 45 5479942; E-mail: sldk@sldk.tuzvo.sk; URL: sldk.tuzvo.sk
1952; Alena Polácíková
Forestry, wood science and technology, ecology and environmental sciences, economics; Information Point of the Council of Europe
354 177 vols; 232 curr per; 12 000 e-journals; 9 mss; 1 incunabula; 12 388 diss/theses; 567 maps; 1 microforms; 651 av-mat; 533 sound-rec; 215 digital data carriers; 4 239 firm publs, 276 research reports, 387 travel reports, 15 570 standards, 7 989 reprints
libr loan; Slovak Libraries Association 24672

Government Libraries

Parlamentná knižnica a národnej rady slovenskej republiky (Parliamentary Library of the National Council of the Slovak Republic), Mudroňova 1, 831 01 ***Bratislava***
T: +421 2 59341277; Fax: +421 2 54415541; URL: www.nrsr.sk
1992; Georgina Gadusová
Standing orders and constitutions, parliamentary official docs, UNO docs
110 000 vols; 1 100 curr per; 50 digital data carriers; Parliamentary documents
SSK, IFLA 24673

Corporate, Business Libraries

Dopravoprojekt, Knižnica, Kominárská 2-4, 832 03 ***Bratislava*** 3
T: +421 2 50234111; Fax: +421 2 50234555; E-mail: director@dopravoprojekt.sk; URL: www.dopravoprojekt.sk
1949; Jana Hanúsková
Traffic engineering, roads, bridges, motorways, environmental engineering
8 700 vols; 68 curr per
libr loan 24674

SLOVNAFT A.S., Stredisko Informácií, Vlčie hrdlo 1, 824 12 ***Bratislava***
T: +421 2 40551111; Fax: +421 2 58599759; E-mail: info@slovnaft.sk; URL: www.slovnaft.sk/sk/
1983; Mária Dologová
Petrochemistry, plastics
62 000 vols; 228 curr per
libr loan; SSK 24675

Special Libraries Maintained by Other Institutions

Lekárska knižnica nemocnice F.D. Roosevelta (F.D. Roosevelt Hospital Medical Library), nám L. Svobodu 1, 975 17 ***Banská Bystrica***
T: +421 48 44123-41/-78; Fax: +421 48 4137240; E-mail: vkikova@mail.com
1961; Dr. Viera Kiková
Clinical medicine
17 204 vols; 131 curr per; 41 diss/theses; 31 av-mat; 43 digital data carriers
libr loan 24676

Literárne a hudobné múzeum Banská Bystrica, Štátna vědecká knižnica Banská Bystrica (Museum of Literature and Music Banská Bystrica), Lazovná 9, 975 58 ***Banská Bystrica***
T: +421 48 4155023; Fax: +421 48 4155023; E-mail: lhmbb@svkbb.sk; URL: www.svkbb.sk
1969; Dr. Mariana Bárdiová
Slovak lit, Slovac music, Musicology-Slovakia; Libr of Samuel Kollár
12 238 vols; 4 curr per; 1 incunabula; 4 diss/theses; 1 maps; 15 microforms; 386 av-mat; 1 418 sound-rec 24677

Múzeum Slovenského národného povstania, Knihovna (Museum of the Slovak National Uprising), Kapitulská č 23, 974 00 ***Banská Bystrica***
T: +421 48 4152070; Fax: +421 48 4123716; E-mail: muzeumsnp@isternet.sk; URL: www.muzeumsnp.sk
1955; Ján Stanislav
Slovak people during World War II
19 020 vols; 13 curr per 24678

Stredoslovenské múzeum (Museum of Central Slovakia), nám Slovenského národného postánia 4, 975 90 ***Banská Bystrica***
T: +421 48 4125896; Fax: +421 48 4155077; E-mail: smbb@stonline.sk; URL: www.stredoslovenske.muzeum.sk
1889; Dr. Zuzana Drugová
12 388 vols; 4 curr per
libr loan 24679

Slovenské banské múzeum (Slovak Mining Museum), 969 00 ***Banská Štiavnica***
T: +421 859 21541; Fax: +421 859 22764
1900; Ludmila Anqušová
300 old prints till 19th c
20 600 vols; 200 curr per; 30 mss; 33 incunabula; 50 maps; 83 000 microforms; 1 400 av-mat; 15 sound-rec
libr loan; SSK 24680

Múzeum Bojnice, Knihovna, 972 01 ***Bojnice***
T: +421 862 32624
1951; Márta Hudecová
Art hist, fine arts, crafts
8 000 vols; 32 curr per; 6 maps 24681

Nemocnica s Poliklinikou Prievidza, Knižnica (Prievidza Hospital and Polyclinic), 972 02 ***Bojnice***
T: +421 862 33619; Fax: +421 862 26619
Vladimír Gažík
13 400 vols 24682

Bibiana medzinárodný dom umenia pre deti, Centrum informatiky, dokumentácie a knižnica (Bibiana, International House of Art for Children), Panská 41, 815 39 *Bratislava*
T: +421 2 204671-01/-02; E-mail: kniznica@bibiana.sk; URL: www.bibiana.sk
1987; Hana Ondrejičková
Children's literature, Special literature on children's books and art for children, Libr's fond of Biennale of ill Bratislava 1967-, Honour List IBBY, The most beautiful books of Slovakia
18 500 vols; 45 curr per; 70 digital data carriers; reference materials about illustrators, authors and publishers of child lit (clippings, bibliographies, catalogs etc), bibliophile editions
SAK 24683

Botanický ústav – Slovenská Akadémia Vied (Institute of Botany, Slovak Academy of Sciences), Dúbravská cesta 14, 842 23 *Bratislava*
T: +421 2 54773507;
Fax: +421 2 54771948; E-mail: tatiana.mihalikova@savba.sk; URL: ibot.sav.sk/page/okniha.htm
1953; Dr. Eva Záletová
Dept of Geobotany
20 600 vols; 110 curr per
libr loan 24684

Dérerova Nemocnica s Poliklinikou, Knižnica (Dérer Hospital and Polyclinic), Limbová 5, 833 05 *Bratislava*
Eva Gubrická
10 300 vols 24685

Divadelný ústav, Knižnica (Theatre Institute), Jakubovo nám 12, 813 57 *Bratislava*
T: +421 2 59304710; Fax: +421 2 52931571; E-mail: kniznica@theatre.sk; URL: www.theatre.sk
1961; Viera Sadlonova
Slovak theatre publication, plays, reviews
30 000 vols; 57 curr per; 1 e-journals; 30 diss/theses; 20 music scores; 18 000 av-mat; 344 sound-rec; 283 digital data carriers
libr loan; SIBMAS 24686

Ekonomiky ústav SAV, Knižnica (Institute of Economic Research SAS), Šancova 56, 811 05 *Bratislava*
T: +421 2 52498214; Fax: +421 2 52495106; E-mail: ekonkniz@savba.sk; URL: www.ekonom.sav.sk
1959; František Brunner
Works and publications of the own institute – 2 150 bks
30 500 vols; 224 curr per; 620 diss/theses; 85 digital data carriers
24687

Fakultná nemocnica, Knižnica (University Hospital), Mickiewiczova 13, 813 49 *Bratislava*
T: +421 2 490611
1959; Dr. Mária Lietavová
25 500 vols; 185 curr per
libr loan 24688

Galéria mesta Bratislavy, Knižnica (Municipal Gallery), Mirbachov palác, Františkánske nám 11, 815 35 *Bratislava*
T: +421 2 54435102, 544351-8; Fax: +421 2 54432611; E-mail: gmb@gmb.sk; URL: www.gmb.bratislava.sk
1959; Ryšavá
13 055 vols; 5 curr per; 179 sound-rec
24689

Geodetický a Kartografický Ustav Bratislava, Odborové Informačné Stredisko (Bratislava Institute of Geodesy and Cartography), Chlumeckého 4, 827 45 *Bratislava*
T: +421 2 234822; Fax: +421 2 292563
Patrícia Sokáčová
Cadastral surveys
21 000 vols 24690

Geologický ústav Slovenskej akadémia vied -, Knižnica (Geological Institute of the Slovak Academy of Sciences), Dúbravská cesta 9, P.O. Box 106, 840 05 *Bratislava*
T: +421 2 54777076; Fax: +421 2 54777096; E-mail: geolinst@savba.sk
1956; Beata Čanakyová
Mineralogy

18 600 vols; 195 curr per; 759 diss/theses; 2 070 maps; 50 microforms
libr loan 24691

Goethe-Institut, Information Centre, Panenská 33, 814 82 *Bratislava*
T: +421 2 59204320;
Fax: +421 2 54413134; E-mail: bibl@bratislava.goethe.org; URL: www.goethe.de/bratislava
1992; Dr. Sylvia Kováčová
11 500 vols; 55 curr per; 820 av-mat; 1 450 sound-rec; 77 digital data carriers
libr loan; ISKO 24692

Historický ústav, Knižnica, Slovenská akadémia vied (Institute of Historical Studies, Slovak Academy of Sciences), Klemensova 19, 813 64 *Bratislava*
T: +421 2 52926321; URL: www.dejiny.sk
1953; Dr. Helena Třísková
Hist of Science and Technology
75 000 vols; 150 curr per; 465 diss/theses; 2 700 maps; 1 760 000 microforms; 950 pamphlets
libr loan; SSK 24693

Institute of Materials and Machine Mechanics of SAS, Knižnica (Slovak Academy of Sciences – Institute of Materials and Machine Mechanics), Račianska 75, 831 02 *Bratislava*
T: +421 2 44254751; Fax: +421 2 44253301; URL: www.umms.sav.sk
1980; Magdalena Sládková
Metal matrix composite mat, thin solid films, solidification of metals, fiber reinforced composites, dynamics of machines & structures, vibroisolation & noise reduction, vibration diagnostics & monitoring, dynamics of aggregates, materials and structures
13 000 vols; 35 curr per 24694

Institute of Preventive and Clinical Medicine, Library, Limbová 14, 833 01 *Bratislava*
T: +421 2 373965; Fax: +421 2 373906
Pharmacology, toxicology
23 000 vols 24695

Knižnica Geografického ústavu SAV (Institute of Geography – Slovak Academy of Sciences), Šefánikova 49, 81473 *Bratislava*
T: +421 2 57510184; Fax: +421 2 52491340; E-mail: geogkniz@savba.sk; URL: www.geography.sav.sk
1953; Dr. Eva Ďurišová
17 163 vols; 135 curr per; 48 diss/theses; 300 microforms; online databases available through the webpage of the central Library of SAS
libr loan 24696

Mestské muzeum, Knižnica (Municipal Museum), Primaciálne nám. 3, 815 18 *Bratislava*
T: +421 2 5334742; Fax: +421 2 5334631
1868; Monika Šurdová
20 700 vols 24697

Národné centrum zdravotníckych informácií – Slovenská lekárska knižnica (National Health Information Center – Slovak Medical Library), Lazaretská 26, 811 09 *Bratislava*
T: +421 2 57269217; Fax: +421 2 52968438; E-mail: sllk@sllk.gov.sk; URL: www.sllk.gov.sk
1951; Dr. Marta Žilová
WHO dept
140 860 vols; 255 curr per; 5 e-journals; 32 328 diss/theses; 261 digital data carriers
libr loan; EAHIL, SAK, SLA, Pro Scientia 24698

Národný onkologický ústav, Knižnica (National Cancer Institute), Klenova 1, 833 10 *Bratislava*
T: +421 2 3708622; Fax: +421 2 372601
Ľubica Šipošová
8 700 vols; 86 curr per
libr loan 24699

Národný ústav tuberkulózy a respiračných chorôb, Lekarska Knižnica (National TB and Respiratory Diseases Institute), Krajinska 91, 825 56 *Bratislava*
T: +421 2 2401555; Fax: +421 2 243622; E-mail: simonico@healthnet.sk; URL: www.healthnet.sk/resources/slovak/nutarch/nutarch.html

1952; Dr. Zuzana Simovicova
Tuberculosis, Respiratory and pulmonary diseases, Thoracic surgery
14 700 vols; 79 curr per; 210 diss/theses; 30 digital data carriers; 70 other items
libr loan 24700

Pamiatkovej ústav, Knižnica (Institute of Historic Monuments), Cesta na Červený most č. 6, 814 06 *Bratislava*
T: +421 2 54771901; Fax: +421 2 54775844; E-mail: ivo.stassel@gmail.com; URL: www.pamiatky.sk
1960; Halina Mojžišová
Restoration, architecture, archaeology, ethnography, hist, cultural heritage, photogrammetry, documentation
27 500 vols; 7 500 curr per 24701

Slovenská Akadémia Vied – Chemický ústav, Knižnica (Slovak Academy of Sciences – Institute of Chemistry), Dúbravská cesta 9, 842 38 *Bratislava*
T: +421 2 375000
Dr. Anna Vojtková
24 000 vols 24702

Slovenská Akadémia Vied – Elektrotechnický ústav, Knižnica (Slovak Academy of Sciences – Institute of Electrical Engineering), Dubravska cesta 9, 841 04 *Bratislava*
T: +421 2 54775806; Fax: +421 2 54775816; E-mail: elekanna@savba.sk; URL: nic.savba.sk/sav/inst/elu
1960; Anna Gömöryová
8 200 vols; 9 curr per; 536 diss/theses; 80 digital data carriers
libr loan 24703

Slovenská Akadémia Vied – Fyzikálny ústav SAV (Slovak Academy of Sciences – Institute of Physics), Dúbravská cesta 9, 842 28 *Bratislava*
T: +421 2 3782357; Fax: +421 2 376085; E-mail: fyziwich@savba.sk
1955; Iva Karasová
13 600 vols; 47 curr per; 115 diss/theses; 60 microforms; 3 digital data carriers
libr loan 24704

Slovenská Akadémia Vied – Geofyzikálny ústav (Slovak Academy of Sciences – Institute of Geophysics), Dúbravská cesta 9, 842 28 *Bratislava*
T: +421 2 3782126
Dr. Jana Kvapilíková
20 000 vols
libr loan 24705

Slovenská Akadémia Vied – Geofyzikálny ústav – Odbor fyziky atmosféry, Knižnica (Geophysical Institute of the Slovak Academy of Sciences), Dúbravská cesta 9, 842 28 *Bratislava*
T: +421 2 59410600;
Fax: +421 2 59410626; E-mail: geofbeta@savba.sk; URL: gpi.savba.sk/modules.php?name=BD&op=Lib
Alžbeta Radimáková
Geophysics, Geomagnetism, Seismology, Meteorology, Climatology, Atmospheric Environment
18 662 vols 24706

Slovenská akadémia vied – Jazykovedný ústav Ľ. Štúra, Knižnica (Slovak Academy of Sciences – Ľ. Štúra Institute of Linguistics), Panská 26, 813 64 *Bratislava*
T: +421 2 5443-1761/-1763; Fax: +421 2 54431756; E-mail: jazyuls@savba.sk; URL: www.juls.savba.sk
1953; Elena Flassiková
24 888 vols; 96 curr per; 87 diss/theses
24707

Slovenská Akadémia Vied – Matematický ústav (Slovak Academy of Sciences – Institute of Mathematics), Štefanikova 49, 814 73 *Bratislava*
T: +421 2 52497316; Fax: +421 2 52497316; E-mail: library@mat.savba.sk; URL: www.mat.savba.sk
1959; Silvia Zabadalová
23 799 vols; 11 348 curr per; 70 diss/theses; 212 digital data carriers
24708

Slovenská Akadémia Vied – Sociologický ústav, Knižnica (Slovak Academy of Sciences – Institute of Sociology), Klemensova 19, 813 64 *Bratislava*
T: +421 2 52926321; Fax: +421 2 52961312; E-mail: socukniz@savba.sk; URL: www.sociologia.sav.sk
1991; Adriana Cenká
10 600 vols; 45 curr per 24709

Slovenská Akadémia Vied – Umenovedný ústav, Knižnica (Slovak Academy of Sciences – Institute of Arts), Fajnorovo Nábr 7, 813 64 *Bratislava*
T: +421 2 330114
1953
29 800 vols; 26 curr per
IAML 24710

Slovenská Akadémia Vied – Ustav anorganickej chémie, Knižnica (Slovak Academy of Sciences – Institute of Inorganic Chemistry), Dúbravská cesta 9, 842 36 *Bratislava*
T: +421 2 375170; Fax: +421 2 373541
1953; Vladimira Peschlova
Physical chemistry, superconductivity, mineralogy, ceramics
14 500 vols; 50 curr per; 127 diss/theses
libr loan 24711

Slovenská Akadémia Vied – Ustav etnológie (Slovak Academy of Sciences – Institute of Ethnology), Jakubovo nám. 12, 813 64 *Bratislava*
T: +421 2 334925
1953; Zuzana Beňušková
14 300 vols; 118 curr per; 26 diss/theses
libr loan 24712

Slovenská Akadémia Vied – Ustav experimentálnej psychológie, Knižnica (Slovak Academy of Sciences – Institute of Experimental Psychology), Dúbravská cesta 9, 813 64 *Bratislava*
T: +421 2 375625; Fax: +421 2 375584; E-mail: expspro@savba.savba.sk
1953; K. Hornáčková
10 700 vols; 20 curr per; 3 diss/theses
libr loan 24713

Slovenská Akadémia Vied – Ustav hudobnej vedy, Knižnica (Slovak Academy of Sciences – Institute of Musicology), Dúbravská cesta 9, 841 04 *Bratislava* 4
T: +421 2 54773589;
Fax: +421 2 54773589; E-mail: musicology.library@savba.sk; URL: www.uhv.sav.sk
1953; Katarína Bánska
50 555 songs
14 120 vols; 26 curr per; 50 diss/theses; 500 microforms; 1 330 av-mat; 5 215 sound-rec; 20 000 photos
libr loan; IAML 24714

Slovenská Akadémia Vied – Ustav Krajinnej Ekológie, Knižnica (Slovak Academy of Sciences – Institute of Landscape Ecology), Štefánikova 3, P.O. Box 254, 814 99 *Bratislava*
T: +421 2 52493920; Fax: +421 2 52494508; E-mail: kniha-uke@savba.sk
1965; Katalín Kis-Csáji
Ecology, botany, zoology, landscape planning
20 630 vols; 96 curr per 24715

Slovenská Akadémia Vied – Ustav normálnej a patologickej fyziológie, Knižnica (Slovak Academy of Sciences – Institute of Normal and Pathological Physiology), Sienkiewiczova 1, 813 71 *Bratislava*
T: +421 2 52926271;
Fax: +421 2 52968516; E-mail: katarina.soltesova@savba.sk
Katarína Šoltésová
Dept of cardiovascular physiology, dept of neurophysiology
9 609 vols; 8 curr per; 22 diss/theses
libr loan 24716

Slovenská akadémia vied – Ustav počítačových systémov (Institute of Informatics), Dúbravská cesta 9, 845 07 *Bratislava*
T: +421 2 45911291; Fax: +421 2 54771004; E-mail: upsysekr@savba.sk; URL: www.ui.savba.sk
1991; Mária Postulková
14 500 vols; 35 curr per; 27 diss/theses; 30 microforms; 520 tech rpts
libr loan 24717

Slovenská Akadémia Vied – Ustav štátu a práva (Slovak Academy of Sciences – Institute of State and Law), Klemensova 19, 813 64 ***Bratislava***
T: +421 2 361833; Fax: +421 2 362325
1954; Darina Čelechovská
27 000 vols; 86 curr per
libr loan 24718

Slovenská Akadémia Vied – Virologický ústav, Mlynská dolina 1, 817 03 ***Bratislava***
T: +421 2 374000
Zdena Bartáková
27 400 vols 24719

Slovenská národná galéria, Knižnica SNG (Slovak National Gallery), Riečna 1, 815 96 ***Bratislava***
T: +421 2 20476322; Fax: +421 2 54433971; E-mail: kniznica@sng.sk; URL: www.sng.sk
1948; Katarína Czwitkovicsová
Fine Arts, architecture, art history; Library of Castle Strážky / Nehre / Zips
100 709 vols; 78 curr per; 45 mss; 1 incunabula; 242 maps; 13 082 microforms; 346 av-mat; 45 digital data carriers; 825 art reproductions, 242 hist maps
libr loan 24720

Slovenská pedagogická knižnica (Slovak Pedagogic Library), Hálova 16, 851 01 ***Bratislava*** 5
T: +421 2 62410992; Fax: +421 2 62410992; E-mail: postmast@spgk.sk; URL: www.spgk.sk
1956; Helena Pangrácová
Historical coll
292 336 vols; 811 curr per; 57 digital data carriers
libr loan; SSK, SAK 24721

Slovenská technická knižnica – Centrum vedecko-technických informácií SR (Slovak Centre of Scientific and Technical Information), Námestie slobody 19, 812 23 ***Bratislava***
T: +421 2 362419; Fax: +421 2 323527
1938; Ján Kurák
506 050 vols; 1 390 curr per; 86 782 standards, 238 798 patents, 25 850 trade publs
libr loan; FID, IATUL, LIBER, UNAL, Slovak Libraries Association, Union of Slovak Librarians 24722

Slovenské národné muzeum, Knižnica (Slovak National Museum), Vajanského nábrežie 2, 810 06 ***Bratislava***
T: +421 2 59349111; Fax: +421 2 52924344; E-mail: riaditel@snm.sk; URL: www.snm.sk
1964; Dr. Klaudia Bokesová
Archeology, hist, natural sciences
112 000 vols; 868 curr per; 31 mss; 1 incunabula
libr loan; SSK, IAML 24723

Štátny Zdravotný ústav, Knižnica (State Institute of Health), Trnavská 52, 826 45 ***Bratislava*** 29
T: +421 2 52961121 ext 208; Fax: +421 2 52961146
Anna Pagácová
Epidemiology, Hygiene, Preventiv Medicine, Public Health
14 393 vols; 36 digital data carriers 24724

SÚTN – Slovenský ústav technickej normalizácie, Knižnica (Slovak Institute for Standardization), Karloveská 63, 840 00 ***Bratislava*** 4
T: +421 2 60294314; Fax: +421 2 65411888; E-mail: studovna@sutn.gov.sk; URL: www.sutn.gov.sk
1993; Alica Domonkosová
National and international technical standards, catalogs of national and international technical standards
550 000 vols; 21 curr per; 2 e-journals; 6 digital data carriers 24725

Ústav merania SAV-ZIS (Institute of Measurement Science, Slovak Academy of Sciences), Dúbravská cesta 9, 841 04 ***Bratislava***
T: +421 2 59104570; Fax: +421 2 54775943; E-mail: umerzis@savba.sk; URL: www.um.savba.sk
Katarína Kozáková
Dynamic measurement of physical quantities in research experiments, development of contactless optical methods for measuring of shape parameters of optical and technical surfaces, neural networks and nonlinear dynamic systems for measurement of large objects space stability parameters, selected methods of NMR tomography, measuring methods and sensors of geometrical quantities, imaging systems based on NMR, modelling of heart electric field sources, structure of statistical models and creation of nonlinear statistical procedures, high sensitive system based on SQID sensors
10 473 vols; 119 curr per; 1 e-journals; 125 diss/theses
libr loan 24726

Ustav slovenskej literatúry / Ustav svetovej literatúry, Slovenská Akadémia Vied (Institute of Slovak Literature / Institute of World Literature), Konventná 13, 813 64 ***Bratislava***
T: +421 2 5441-3391/-2701; Fax: +421 2 54416025
1964; Katarina Veneniová
49 000 vols; 54 curr per; 180 diss/theses
libr loan 24727

Ustav stavebníctva a architektúry – Slovenská Akadémia Vied (Institute of Construction and Architecture – Slovak Academy of Sciences), Dúbravská cesta 9, 845 03 ***Bratislava***
T: +421 2 54773548; Fax: +421 2 372494; E-mail: usarvlad@savba.sk
1956; Monika Puťošova
45 800 vols; 240 curr per; 365 diss/theses; 116 microforms; 50 sound-rec
libr loan 24728

Výskumný ústav chemickej technológgie, Knižnica (Research Institute of Chemical Technology), Dimitrova 34, 836 03 ***Bratislava***
T: +421 2 68774
1947; Klačanská Zuzana
Additives for polymers, pesticides, rubber chemicals, hygiene and toxicology res
27 072 vols; 332 curr per; 74 diss/theses; 34 govt docs; 60 microforms; 63 sound-rec
libr loan 24729

Výskumný ústav káblov a izolantov – VUKI, a.s., Knižnica (Research Institute of Cables and Insulating Materials), Rybničná 38, 831 07 ***Bratislava***
T: +421 2 49213132; E-mail: vuki@vuki.sk; URL: www.vuki.sk
1924
132 000 vols 24730

Výskumný ústav vodného hospodárstva, Knižnica (Water Research Institute), nábr L. Svobodu 5, 812 49 ***Bratislava***
T: +421 2 59343340; E-mail: kniznica@vuvh.sk; URL: www.vuvh.sk
1952; Dr. Elvira Kováčiková
Archives
34 600 vols; 76 curr per; 35 av-mat; 64 digital data carriers; 5 000 research rpts, 2 350 standards 24731

Galéria Júliusa Jakobyho (Júliusa Jakobyho Gallery), Hlavná 27, 040 01 ***Košice***
T: +421 55 21187; Fax: +421 55 20340
Dr. Ladislav Zozuľák
10 000 vols 24732

Knižnica pre mládež mesta Košíc (Library for Children and Young of the City of Košice), Tajovského 9, 040 64 ***Košice***
T: +421 55 6222390; Fax: +421 55 6222476; E-mail: kmk@nextra.sk; URL: kmk.zvykni.sk
1955; Peter Halász
Nezabudka – work with disabled children
262 756 vols; 54 curr per; 327 music scores; 174 maps; 351 av-mat; 1 165 sound-rec; 44 digital data carriers
SSK, SAK 24733

Slovenské technické múzeum (Technical Museum), Leninóva 88, 043 82 ***Košice***
T: +421 55 24035
Greta Kminiaková
15 800 vols 24734

Slovenský kontrolný a skúšobný ústav polnohospodársky, Knihovna (Institute of Agriculture), Švermova 1, 040 00 ***Košice***
1900
14 700 vols 24735

Východoslovenské múzeum, Knižnica (Museum of Eastern Slovakia), Hviezdoslavova ul 3, 041 36 ***Košice***
T: +421 55 6220309; Fax: +421 55 6228696; E-mail: magdalena.zavatzka@gmail.com; URL: www.vsmuzeum.sk
1872; Magdalena Zavatzka
60 000 vols 24736

Tekovské múzeum, 934 69 ***Levice*** – hrad
T: +421 813 6312112; Fax: +421 813 6312866
1927; Danica Valachov
14 300 vols; 10 curr per
libr loan 24737

Slovenské múzeum ochrany prírody a jaskyniarstva (Slowak Museum of Nature Protection and Speleology), Školská 4, 031 80 ***Liptovský Mikuláš***
T: +421 44 5477212; Fax: +421 44 5514381; E-mail: copkova@smopaj.sk; URL: www.smopaj.sk
1930; Magda Čopková
21 730 vols; 81 curr per; 9 digital data carriers
libr loan 24738

Novohradské múzeum, Gottwaldovo nám. 3, 984 01 ***Lučenec***
T: +421 47 21980
Ján Rusznák
12 000 vols 24739

Martinská fakultná nemocicna, Lekárska knižnica (Martin Faculty Hospital), Kollárova 2, 036 59 ***Martin***
1935; Darina Točeková
28 000 vols
libr loan; SSK 24740

Slovenské národné literárne múzeum, Slovenská národná knižnica (Slovak National Literary Museum), Osloboditel'ov 11, 036 01 ***Martin***
T: +421 43 4134036; Fax: +421 43 4302374; E-mail: snlm@snk.sk; URL: www.snk.sk
1966; Jozef Beňovský
Art-Painting, Sculpture, Graphic Coll (10 300), Stage, costume, puppet design (736), Stage provision (1 433), Museum objects relating to hist of Slovak lit (5 440)
43 797 vols; 31 curr per; 7 084 av-mat; 3 291 sound-rec; 3 161 posters, 4 477 ephemeria, 651 literary text facsimile 24741

Slovenské národné múzeum – Etnografické múzeum, Knižnica (Ethnography Museum), Malá Hora 2, 036 80 ***Martin***
T: +421 43 4131011, 4131012; Fax: +421 43 4220290; E-mail: snm-em@stonline.sk; URL: www.snm-em.sk
1893; Dr. Mária Halmová
Herritage: Martin Benka Libr, Karol Plicka Libr, Jiří a Anna Horák Libr
54 776 vols; 65 curr per; 473 microforms; 28 digital data carriers
libr loan 24742

Slovenská Akadémia Vied – Archeologický ústav, Knižnica (Slovak Academy of Sciences – Archaeology Institute), Akademická ul č. 2, 949 21 ***Nitra***
T: +421 37 3573941; E-mail: nrauhalm@savba.sk
1953; Helena Halmešová
69 118 vols; 366 curr per; 95 diss/theses; 5 sound-rec; 50 digital data carriers 24743

Slovenská poľnohospodárska knižnica (Slovak Agricultural Library), Štúrova 51, P.O. Box 20/b, 949 59 ***Nitra***
T: +421 37 6517743; Fax: +421 37 6517743; E-mail: Beata.Bellerova@uniag.sk; URL: www.slpk.sk
1946; Beáta Bellérová
Slovak Technical Standards
539 000 vols; 513 curr per; 572 e-journals; 8 250 diss/theses; 170 microforms; 721 digital data carriers
IAALD 24744

Slovenské poľnohospodárske múzeum Nitra, Knižnica (Slovak Agricultural Museum Nitra), Dlhá 92, 950 50 ***Nitra***
T: +421 37 6572553, 6572554; Fax: +42 37 7333479; E-mail: muzeum@agrokomplex.sk; URL: www.muzeum.sk/muzeum/spm.htm
1966; Jana Décsiová
Hist of agriculture
42 635 vols; 477 curr per; 23 mss; 20 diss/theses; 25 microforms
libr loan 24745

Výskumný ústav živočišnej výroby – Oddelenie VTEI, Knižnica (Research Institute for Animal Production – Department of Information Systems and Publishing Activity), Hlohovská 2, 949 92 ***Nitra***
T: +421 37 6546357; Fax: +421 37 6546361; E-mail: ovti@scpv.sk
1950; Dr. Margarita Sirotkina
Special library of Prof. L. Landau, Deposit of award winning films from the international film festival 'Agrofilm'
25 424 vols; 110 curr per; 2 e-journals; 783 diss/theses; 70 av-mat
libr loan; SSK, IAALD 24746

Výskumný ústav rastlinnej výroby, Knižnica (Plant Production Research Center – Research Institute of Plant Production), Bratislavská cesta 122, 921 68 ***Piešťany***
T: +421 33 772-2311/-2312; Fax: +421 33 7726306; E-mail: vti@vurv.sk; URL: www.vurv.sk
1951; Dr. L'ubica Sedlárová
Plant breeding, agronomy, plant production, genetics,plant genetic resources, organic agriculture, plant protection
12 133 vols; 50 curr per; 5 e-journals; 20 digital data carriers; 1 155 research reports, 433 standards, 872 trip reports, 692 methodics, 620 lists of cultivars, 19 author bibliographies, 3 422 reprints
libr loan 24747

Podtatranské múzeum, Knižnica, Vajanského 72/4, 058 01 ***Poprad***
T: +421 52 7721924; Fax: +421 52 7721868; E-mail: sekretariat@muzeumpp.sk; URL: www.muzeum.sk/?obj=muzeum&ix=ptmp
1886; Kusnirákovа
Postcards, regional lit, posters
17 600 vols; 13 curr per; 101 mss; 409 maps; 43 av-mat; 150 sound-rec 24748

Výskumný ústav pre petrochémiu, Technická knižnica (Research Institute for Petrochemistry), Bojnická cesta 86, 971 04 ***Prievidza***
T: +421 46 31841; Fax: +421 46 32261
1951; Eva Hojčová
60 000 vols; 63 curr per; 105 diss/theses; 95 microforms
libr loan 24749

Gemersko-Malohontské Múzeum, Knižnica (Gemer-Malohont Museum), nám. M. Tompu, č. 24, 979 80 ***Rimavská Sobota***
T: +421 47 5632741; Fax: +421 47 5632730; E-mail: gmmuzeum@bb.telecom.sk; URL: www.gmmuzeum.sk
1882; Iveta Krnačová
Biology, hist, archeology, ethnology, art, regional hist, romani culture
43 163 vols; 44 curr per; 50 mss; 16 diss/theses; 12 maps; 33 microforms; 139 av-mat; 119 sound-rec; 97 digital data carriers; 1 321 pamphlets
libr loan; SAK 24750

Liptovské múzeum, Knižnica, Nám S. N. Hýroša 10, 034 50 ***Ružomberok***
T: +421 848 4322468; Fax: +421 848 4303532; E-mail: lipt.muzeum@post.sk; URL: www.liptovskemuzeum.sk
1912; Dr. Zuskinová Iveta
29 875 vols; 20 curr per; 63 mss; 1 incunabula; 906 maps; 129 av-mat; 113 sound-rec; 5 000 slides
libr loan 24751

Záhorské múzeum, Nám. Slobody 11, 909 01 ***Skalica***
T: +421 801 6644230; Fax: +421 801 6644230
Dr. Viera Drahošová
16 000 vols
libr loan 24752

Štátne múzeum ukrajinsko-rusínskej kultúry, Knižnica (State Museum of Ukrainian-Ruthenian Culture), Centralna 258, 089 01 ***Svidník***
T: +421 937 21365
1956; Dr. Miroslav Sopoliga
Ukrainian and Ruthenian lit, ethnography, hist
41 500 vols 24753

Slovenská Akadémia Vied – Astronomický ústav, Knižnica (Slovak Academy of Sciences – Astronomical Institute), 059 60 ***Tatranská Lomnica***
T: +421 527879159; Fax: +421 524467656; E-mail: library@astro.sk; URL: www.library.astro.sk/library.html
1943; Dr. Drahomir Chochol
9 800 vols; 33 curr per; 3 e-journals; 95 diss/theses; 30 maps; 500 microforms; 163 digital data carriers
libr loan 24754

Štátne lesy, Tatranského národného parku, Knižnica (State Forest of Tatra National Park), 059 60 ***Tatranská Lomnica***
T: +421 52 4467951-7; Fax: +421 52 4467124
1953; Jana Adamová
19 793 vols; 3 062 curr per; 1 e-journals; 2 av-mat; 3 digital data carriers 24755

Slovenské energetické strojárne, Technická knižnica (Slovak Power Engineering Works), Továrenská 210, 935 28 ***Tlmače***
T: +421 813 9262369; Fax: +421 813 921941
1951; Dr. Renáta Ráczová
13 000 vols; 98 curr per
libr loan; SSK 24756

Trenčianské muzeum, Knižnica (City Museum), Mierové nám 46, P.O. Box 120, 912 50 ***Trenčín***
T: +421 831 34431
1911; Dr. Leo Kužela
28 500 vols; 3 av-mat
libr loan 24757

Nemocnica s poliklinikou, Lekárska knižnica (Hospital), U. Spanyola 43, 012 07 ***Žilina***
T: +421 41 6870249; Fax: +421 41 34003
1957; Eva Kopalová
23 600 vols; 129 curr per 24758

Odvetvové informacné stredisko dopravy, Vyskumny ústav dopravny (Transport Library and Information Centre), Velky diel P.P.B-49, 011 39 ***Žilina***
T: +421 41 5652084; Fax: +421 41 5652084; E-mail: info@vud.sk; URL: www.vud.sk
1958; Božena Hlucháňová
40 000 vols; 104 curr per; 50 digital data carriers
libr loan; SAK 24759

Vutch-Chemitex Ltd.- Oddelenie VTEI, Knižnica (Department of STI), J. Milca 8, 011 68 ***Žilina***
T: +421 41 622418; Fax: +421 41 621704
1970; Valéria Čapeková
20 100 vols; 64 curr per
libr loan 24760

Národné Lesnícke Centrum, Knižnica (National Forest Centre), T. G. Masaryka 22, 960 92 ***Zvolen***
T: +421 45 5320316; Fax: +421 45 5314192; E-mail: informacie@nlcsk.sk; URL: www.nlcsk.sk
1898; Katarína Sládeková
Forest protection, forest environment, silviculture, forest management, game protection, monitoring of forest health condition, forest technology and engineering
77 153 vols; 122 curr per; 1 e-journals; 504 diss/theses; 28 digital data carriers; 884 proceedings from conferences, 1 203 rpts of journeys abroad, 2 721 reprints
libr loan; AGRIS 24761

Public Libraries

Verejná kniznica Mikuláša Kováča (Mikulas Kovac Public Library), ul. Lazovná 28, 974 01 ***Banská Bystrica***
T: +421 48 4154757; Fax: +421 48 4154757; E-mail: peter.klinec@vkmk.sk; URL: www.vkmk.sk
1973; Peter Klinec
140 000 vols; 100 curr per; 600 music scores; 170 av-mat; 4 700 sound-rec; 100 digital data carriers
libr loan; SSK, SAK 24762

Okresná knižnica Dávida Gutgesela, Radničné nám 1, 085 01 ***Bardejov***
T: +421 54 4722105; Fax: +421 54 4722714; E-mail: kniznica@gutgesel.sk; URL: www.gutgesel.sk
1925; Dr. Vasil Gula
73 498 vols; 96 curr per; 605 music scores; 730 av-mat; 3 421 sound-rec; 54 digital data carriers
SSK 24763

Archív hlavného mesta SR Bratislavy (Archives of the Capital of the Slovak Republic Bratislava), Gorhéko 5, 815 20 ***Bratislava***
T: +421 2 54433248; Fax: +421 2 54430848; E-mail: archiv.samb@vs.sk; URL: www.civil.gov.sk
1923; Anna Buzinkayová
98 000 vols; 240 curr per; 51 incunabula
libr loan; ICA 24764

Knižnica Bratislava Nové Mesto (Public Library Bratislava Nové Mesto), Pionierska 12, 831 02 ***Bratislava***
T: +421 2 44253985; Fax: +421 2 44254115; E-mail: kbnm@kbnm.sk; URL: old.kbnm.sk
1974; Dr. Irena Galátová
181 694 vols; 193 curr per; 1 108 music scores; 108 av-mat; 6 716 sound-rec
SAK 24765

Knižnica Ružinov (Public Library Ružinov), Tomášikova 25, 821 01 ***Bratislava***
T: +421 2 294751
1972; Dr. Darina Horáková
245 000 vols; 383 curr per
SSK 24766

Městská knižnica (Municipal Library), Klariská 16, 814 79 ***Bratislava***
T: +421 2 54435148;
Fax: +421 2 54435148; E-mail: bratislava@mestskakniznica.sk; URL: www.mestskakniznica.sk
1900; Juraj Sebesta
audio documents collection; Dpt for blind and sight-handiccaped people
291 885 vols; 175 curr per; 248 439 mss; 15 897 music scores; 1 087 maps; 2 199 av-mat; 23 327 sound-rec; 31 digital data carriers
libr loan; IFLA, INTAMEL, SAK, SSK 24767

Miestna knižnica Dúbravka (District Public Library Dúbravka), Sekurisova 12, 844 06 ***Bratislava***
T: +421 2 763105; Fax: +421 2 763105
1991; Mária Gryczová
100 000 vols; 42 curr per; 60 av-mat; 5 300 sound-rec 24768

Miestna knižnica Petržalka (Local Public Library Petržalka), Kutlikova 17, 851 02 ***Bratislava***
T: +421 2 62250087;
Fax: +421 2 62250259; E-mail: huska@kniznicapetrzalka.sk; URL: www.kniznicapetrzalka.sk
1985; Ernest Huska
227 280 vols; 95 curr per; 606 music scores; 6 579 sound-rec; 7 digital data carriers
libr loan 24769

Mestská knižnica (City Public Library), Ražusova 2, 977 01 ***Brezno***
T: +421 48 2991
1953; Jaroslav Šurina
68 000 vols; 37 curr per 24770

Okresná knižnica (Regional Public Library), 17. novembra 1258, 022 51 ***Čadca***
T: +421 824 21893
1920; Miroslav Golis
115 000 vols; 200 curr per; 1 187 govt docs; 5 122 music scores; 65 maps; 200 microforms; 3 189 av-mat; 11 780 sound-rec
libr loan; SSK 24771

Oravská Knižnica (Regional Public Library), S. Nováka 1763/2, 026 80 ***Dolný Kubín***
T: +421 435 863368, 862277; Fax: +421 435 863368; E-mail: orakult@stonline.sk; URL: web.stonline.sk/orakult
1924/1951; Dr. Milan Gonda
85 724 vols; 97 curr per; 4 e-journals; 53 music scores; 24 maps; 106 av-mat; 4 351 sound-rec; 86 digital data carriers
libr loan; SSK, SAK 24772

Mestská knižnica (City Public Library), Pionierska 396, 018 41 ***Dubnica nad Váhom***
T: +421 42 4425726; E-mail: kniznica@dubnica.sk; URL: dubnica.sk/kultura/mestska-kniznica
1951; Marta Zacikova
61 000 vols; 49 curr per; 5 digital data carriers
libr loan 24773

Okresná knižnica (Regional Public Library), Obchodná 788/1, 929 01 ***Dunajská Streda***
T: +421 709 522582
1953; Dezider Hadossy
104 000 vols; 214 curr per
SSK 24774

Okresná knižnica, Nám. V.I. Lenina, 924 00 ***Galanta***
T: +421 707 2920
68 000 vols 24775

Vihorlatská knižnica (Regional Public Library), Nám slobody 50, 066 80 ***Humenné***
T: +421 57 63517
1924; Michal Reby
142 000 vols; 120 curr per; 640 music scores; 215 av-mat; 3 300 sound-rec
libr loan; SSK, SLA 24776

Okresná knižnica (Regional Public Library), Eotvosova 35, 945 33 ***Komárno***
T: +421 36 60710; Fax: +421 36 64480
1953; Vlasta Amanová
224 000 vols; 176 curr per; 52 music scores; 857 sound-rec
libr loan 24777

Knižnica pre mládež mesta Košíc (Library for Children and Young of the City of Košice), Tajovského 9, 040 64 ***Košice***
T: +421 55 6222390; Fax: +421 55 6222476; E-mail: kmk@nextra.sk; URL: kmk.zvykni.sk
1955; Peter Halász
Nezabudka – work with disabled children
262 756 vols; 54 curr per; 327 music scores; 174 maps; 351 av-mat; 1 165 sound-rec; 44 digital data carriers
SSK, SAK 24778

Verejná knižnica Jána Bocatia (Public Library of Jan Bocatius), Hlavná 48, 042 61 ***Košice***
T: +421 55 6223291; Fax: +421 55 6223292; E-mail: vkjb@vkjb.sk; URL: www.vkjb.sk
1924; Klára Kernerová
Region's coll; Music dept of foreign languages
454 600 vols; 193 curr per; 12 mss; 2 incunabula; 3 200 music scores; 152 maps; 31 av-mat; 11 200 sound-rec; 11 digital data carriers; 652 clippings, 198 video-recordings, 830 pamphlets
libr loan; SLA 24779

Tekovská Knižnica (Tekovian Library), A. Sládkoviča č. 2, 934 68 ***Levice***
T: +421 813 6312802; Fax: +421 813 6312802; E-mail: tk@slovanet.sk
1951; Jana Holubcová
Music dept, bibliographical and informational dept
141 103 vols; 87 curr per; 1 178 music scores; 570 maps; 2 microforms; 80 av-mat; 2 986 sound-rec; 17 digital data carriers; 3 172 items of other mat
SSK, SAK 24780

Knižnica Gašpara Fejérpataky – Belopotockého, Štúrova 56, 031 80 ***Liptovský Mikuláš***
T: +421 44 5526216;
Fax: +421 44 5523084; E-mail: kniznicagfb@kniznicagfb.sk
1953; Marcela Feriančeková
93 000 vols; 72 curr per; 3 870 sound-rec; 97 digital data carriers
libr loan; SSK, SAK 24781

Okresná knižnica (Regional Public Library), ul. Kármaná 2, 984 01 ***Lučenec***
T: +421 47 22991; Fax: +421 47 22991
1851; Dr. Pavel Urbančok
130 000 vols; 195 curr per; 100 govt docs; 132 music scores; 300 av-mat; 92 sound-rec
libr loan; SSK 24782

Turčianska knižnica (Turiec Library in Martin), Divadelná 3, 036 01 ***Martin***
T: +421 43 4133590; Fax: +421 43 4133590; E-mail: kniznica@tkmartin.sk
1951; Katarína Vandlíková
Programmes, almanacs, plasters, annual rpts, all from the Region of Martin; 4 depts
144 991 vols; 12 curr per; 1 572 music scores; 1 700 av-mat; 3 212 sound-rec; 87 digital data carriers
SSK, SAK 24783

Zemplínska knižnica Gorazda Zvonického – Okresna knižnica (Regional Public Library), Štefánikova 20, 071 01 ***Michalovce***
T: +421 56 22341
1925; Monika Šleisová
127 700 vols; 137 curr per; 946 music scores; 360 maps; 360 av-mat; 2 549 sound-rec
libr loan; SSK, SAK 24784

Okresná knižnica (Regional Public Library), Samova 1, 950 51 ***Nitra***
1923
255 000 vols; 162 curr per
libr loan; SSK 24785

Knižnica Antona Bernolaka v Novych Zámkoch, Turecká 36, 940 01 ***Nové Zámky***
T: +421 35 6400385;
Fax: +421 35 6400386; E-mail: abkniznica@ba.telecom.sk
1951; Marta Szilagyi
5 branch libs
140 000 vols; 6 500 sound-rec; 230 digital data carriers 24786

Okresná knižnica (Regional Public Library), Holubyho 7, 902 01 ***Pezinok***
T: +421 33 412315
1953; Tatjana Kročková
95 500 vols; 141 curr per; 90 maps; 200 av-mat; 726 sound-rec
libr loan; SSK 24787

Podtatranská Knižnica (Regional Public Library), Podtatranská ul 1548/1, 058 01 ***Poprad***
T: +421 52 7729568; Fax: +421 52 7729568; E-mail: riaditel@kniznicapp.sk; URL: www.kniznicapp.sk
1952; Anna Balejová
Art dept
129 803 vols; 197 curr per; 4 000 govt docs; 641 music scores; 320 maps; 50 microforms; 750 av-mat; 3 562 sound-rec; 19 digital data carriers; 150 pamphlets, 10 000 clippings
libr loan; SAK, SSK 24788

Okresná knižnica (Regional Public Library), Štúrová 41/14, 017 45 ***Považská Bystrica***
T: +421 822 323320
1961; Eva Baculiková
176 000 vols; 255 curr per; 3 223 music scores
libr loan; SSK 24789

Knižnica P.O. Hviezdoslava v Prešove (Regional Public Library), Levočská č. 9, 080 99 ***Prešov***
T: +421 91 725393; Fax: +421 91 725393; E-mail: okpresov@vadium.sk; URL: www.sk/kniznica.presov
1925; Mária Koložváryová
257 377 vols; 136 curr per; 891 music scores; 143 av-mat; 5 268 sound-rec; 15 digital data carriers
libr loan; SSK, SAK 24790

Mestská knižnica (City Public Library), Štefániková ul, 020 01 ***Púchov***
T: +421 825 2488
51 500 vols; 29 curr per
libr loan 24791

Okresná knižnica (Regional Public Library), ul. SNP 27, 979 01 **Rimavská Sobota**
T: +421 866 21937; Fax: +421 866 24343
1952; Dr. Marta Čomajová
144 000 vols; 215 curr per; 231 av-mat; 2 778 sound-rec   24792

Okresná knižnica (Regional Public Library), Šafárikova 10, 048 01 **Rožňava**
T: +421 942 25597; Fax: +421 942 24591
Véra Ščuková
92 000 vols; 147 curr per
libr loan   24793

Okresná knižnica (Regional Public Library), Vajanského 28, 905 01 **Senica**
T: +421 802 2604
1936; Katarína Soukupová
91 300 vols; 142 curr per; 2 maps; 2 microforms; 597 av-mat; 1 377 sound-rec; 3 276 pamphlets, 9 691 clippings
libr loan; SSK   24794

Okresná knižnica (Regional Public Library), Zimná ul č. 58, 052 01 **Spišská Nová Ves**
T: +421 965 24757
1952; Dr. Stanislav Prochotský
Lit of region Spišská Nová Ves; Children and junior dept, Music dept, Non-fiction dept, Belles lettres dept
130 000 vols; 164 curr per; 1 232 music scores; 255 maps; 5 700 av-mat; 6 546 sound-rec; 177 slide coll, 5 video games for children
libr loan   24795

Okresná knižnica (Regional Public Library), Letná, 064 01 **Stará Ľubovňa**
T: +421 963 22210
1968; Mária Hnatková
Musical dept
50 000 vols; 100 curr per; 300 music scores; 128 av-mat; 3 876 sound-rec
SSK   24796

Okresná knižnica (Regional Public Library), ul. Čsl. armády 28, 955 35 **Topoľčany**
T: +421 815 24745
1920; Jozef Košík
96 000 vols; 100 music scores; 405 av-mat; 1 853 sound-rec
libr loan   24797

Okresná knižnica, M.R.Štefánika 53, 075 01 **Trebišov**
T: +421 56 668 9010; Fax: +421 56 672 2786; E-mail: sluzby@kniznicatv.sk; URL: www.kniznicatv.sk
1952; Mgr. Mária Brindzáková
51 000 vols; 23 curr per; 1 000 digital data carriers
libr loan; SSK   24798

Verejná knižnica Michala Rešetku v Trenčíne, Namestie SNP 2, 911 82 **Trenčín**
T: +421 32 7434267; Fax: +421 32 7435185; E-mail: vkmr@vkmr.sk; URL: www.vkmr.sk
1925; Lydia Brezova
230 000 vols; 398 curr per; 5 300 music scores; 588 maps; 440 av-mat; 3 531 sound-rec; 479 digital data carriers; 2 289 other items
libr loan; SSK, SAK   24799

Knižnica Juraja Fandlyho, Rázusova 1, 918 20 **Trnava**
T: +421 033 5511782;
Fax: +421 033 5511782; E-mail: kniznica@kniznicatrnava.sk; URL: www.kniznicatrnava.sk
1952; Dr. Emilia Diteova
Music Dept, Sound Dept for the Blind
243 649 vols; 443 curr per; 1 941 music scores; 18 microforms; 1 972 av-mat; 13 106 sound-rec; 24 digital data carriers
libr loan   24800

Okresná knižnica (Regional Public Library), Nemocničná 4, 990 01 **Veľký Krtíš**
1968; Dr. Štefan Turay
61 000 vols; 101 curr per
libr loan; SSK   24801

Okresná knižnica (Regional Public Library), M.R. Štefánika 875/200, 093 01 **Vranov na Topľou**
T: +421 57 22365
1954; Taťjana Koščová
74 900 vols; 78 curr per; 285 av-mat; 2 290 sound-rec
libr loan; SSK   24802

Mestská knižnica (City Public Library), ul. SNP 2, 937 01 **Želiezovce**
T: +421 36 7711000; Fax: +421 36 7711000; E-mail: libraryzeliezovce@kredit.sk
1907; Jana Beníková
Colls of Franz Schubert and Timea Major
55 800 vols; 9 curr per; 460 sound-rec; 11 digital data carriers   24803

Žilinská knižnica (Žilina Library), Bernolákova 47, 011 77 **Žilina**
T: +421 42 7232745; Fax: +421 41 7232765; E-mail: kniznica@zilinska-kniznica.sk; URL: www.zilinska-kniznica.sk
1955; Katarina Šušollaková
296 663 vols; 384 curr per; 8 341 sound-rec; 150 digital data carriers
libr loan; SAK   24804

Krajská knižnica Ľudovíta Štúra vo Zvolene (Regional Public Library of Ludovit Stur in Zvolen), Štúra 5, 960 82 **Zvolen**
T: +421 45 5331071; Fax: +421 45 5335335; E-mail: sluzby@kskls.sk; URL: www.kskls.sk
1922; Milota Torňošová
Regional docs of Zvolen; Dept of Regional Bibliography
181 500 vols; 331 curr per
libr loan; SSK, SAK   24805

Slovenia

National Libraries

Narodna in Univerzitetna Knjižnica, National and University Library, Turjaška 1, p.p. 259, 1000 **Ljubljana**
T: +386 1 2001-188/-209; Fax: +386 1 4257293; E-mail: info@nuk.uni-lj.si; URL: www.nuk.uni-lj.si
1774; Lenart Setinc
Mss coll, Music coll, Map and picture coll, AV coll, govt and official publ, EU publ; Information Centre for Librarians
2 385 621 vols; 4 500 curr per; 12 000 e-journals; 7 200 mss; 508 incunabula; 79 000 music scores; 241 700 maps; 3 900 microforms; 9 200 av-mat; 22 000 sound-rec; 1 600 digital data carriers; 550 000 miscellanea
libr loan; IFLA, Fid, CERL, ALA, IAML   24806

General Research Libraries

Slovenska akademija znanosti in umetnosti, Biblioteka (Slovenian Academy of Sciences and Arts), Novi trg 3-5, 1000 **Ljubljana**; p.p. 323, 1001 Ljubljana
T: +386 1 4706246; Fax: +386 1 4253462; E-mail: petra.vide@zrc-sazu.si; URL: www.sazu.si/biblioteka.html
1938; Petra Vide Ogrin
Dela akademikov (printed material of members (full, associate and corresponding ones) of Slovenian Academy of Sciences and Arts)
497 149 vols; 2 535 curr per; 21 e-journals; 136 mss; 2 incunabula; 3 854 maps; 853 microforms; 7 av-mat; 963 sound-rec; 364 digital data carriers; 9 630 photos
libr loan; IAML, IFLA, AALIB, ABDOS   24807

University Libraries, College Libraries

Visoka komercialna šola Celje – Fakulteta za komercialne in poslovne vede, Knjižnica, Lava 7, 3000 **Celje**
T: +386 3 4285544; Fax: +386 3 4285541; URL: www.fkpv.si/
Jožica Škorja
8 000 vols   24808

Akademija za Glasbo, Knjižnica, Biblioteka (Academy of Music), Stari trg 34, 1000 **Ljubljana**
T: +386 1 2427310; E-mail: ksenija.pozar@ag.uni-lj.si; URL: www.ag.uni-lj.si
1939; Ksenija Požar
11 370 vols; 4 curr per   24809

Faculty of Theology in Ljubljana, Poljanska 4, 1000 **Ljubljana**
1952 (1919)
60 000 vols   24810

Institut za metalne konstrukcije, Knjižnica (Institute of Metal Constructions), Mencingerjeva 7, 1000 **Ljubljana**
1956; Andra Žnidar
33 500 vols; 45 curr per; 820 microforms; 3 508 standards, 502 cat
libr loan   24811

Narodna in Univerzitetna Knjižnica, National and University Library, Turjaška 1, p.p. 259, 1000 **Ljubljana**
T: +386 1 2001-188/-209; Fax: +386 1 4257293; E-mail: info@nuk.uni-lj.si; URL: www.nuk.uni-lj.si
1774; Lenart Setinc
Mss coll, Music coll, Map and picture coll, AV coll, govt and official publ, EU publ; Information Centre for Librarians
2 385 621 vols; 4 500 curr per; 12 000 e-journals; 7 200 mss; 508 incunabula; 79 000 music scores; 241 700 maps; 3 900 microforms; 9 200 av-mat; 22 000 sound-rec; 1 600 digital data carriers; 550 000 miscellanea
libr loan; IFLA, Fid, CERL, ALA, IAML   24812

Pedagoški inšstitut pri Univerzi 'Eduarda Kardelja' v Ljubljana, Knjižnica (Educational Research Institute), Gerbičeva 62, 1000 **Ljubljana**
T: +386 1 4201246; Fax: +386 1 4201266; E-mail: pi.knjiznica@pei.si; URL: www2.arnes.si/~uljpeins
1975; Polona Ramsak
9 465 vols; 36 curr per; 3 e-journals; 594 diss/theses; 22 music scores; 15 av-mat; 31 sound-rec; 125 digital data carriers
libr loan   24813

Univerza v Ljubljani, **Ljubljana**
– Akademija za likovno umetnost, Knjižnica (University of Ljubljana Academy of Fine Arts), Erjavčeva 23, 1000 Ljubljana
T: +386 1 2512726; Fax: +386 1 2519071; E-mail: dekanat@aluo.uni-lj.si
1946; Dr. Nadja Zgonik
15 982 vols; 93 curr per; 810 diss/theses; 1 500 photo negatives of students' works
libr loan   24814

– Biološka Knjiznica (National Institute of Biology), Večma pot 111, 1000 Ljubljana
T: +386 1 4233388; Fax: +386 1 2573390, 2412980; E-mail: bioloska.knjiznica@bf.uni-lj.si; URL: www.nib.si/knjiznica
1962; Mira Horvat
Libr of Marine Biological Station
39 675 vols; 195 curr per; 1 761 diss/theses; 820 maps
libr loan; COBISS   24815

– Biološka knjižnica (The Biology Library), Večna pot 111, 1000 Ljubljana
T: +386 1 3203306; Fax: +386 1 2573390, 2412980; E-mail: bioloska.knjiznica@bf.uni-lj.si; URL: www.nib.si/eng/index.php/knjinica.html
1919; Barbara Černač   24816

– Biotehniska fakulteta, Centralna biotehniska knjiznica, BF-CBK (Biotechnical Faculty, Central Biotechnical Library), Jamnikarjeva 101, 1000 Ljubljana; p.p. 2995, 1001 Ljubljana
T: +386 1 3203043; Fax: +386 1 2565782; E-mail: cbk@bf.uni-lj.si; URL: www.bf.uni-lj.si/en/libraries/about/organisation/central-biotechnical-library.html
1947; Simona Juvan
branches: Agronomy department library, Biology department library, Food Science and Technology department library, Forestry and Renewable Forest Resources department library, Wood Science and Technology department library, Zootechnical department library
300 000 vols; 1 205 curr per; 2 082 diss/theses; 300 microforms; 5 av-mat; 40 digital data carriers
libr loan; COBISS, AGLINET, IAALD   24817

– Centralna ekonomska knjiznica (CEK), Ekonomska fakulteta (Central Economic Library), Kardeljeva ploščad 17, 1000 Ljubljana
T: +386 1 5892591; Fax: +386 1 5892698; E-mail: cek@ef.uni-lj.si; URL: www.ef.uni-lj.si/cek
1946; Ivan Kanič
EDC – European Documentation Centre
244 600 vols; 315 curr per; 127 e-journals; 250 e-books; 1 230 diss/theses; 175 av-mat; 9 digital data carriers
libr loan; EBSLG, ABDOS   24818

– Centralna tehniška knjižnica (Central Technological Library), Trg republike 3, 61000 Ljubljana
T: +386 1 2003400; Fax: +386 1 4256667; E-mail: post@ctk.uni-lj.si; URL: www.ctk.uni-lj.si
1949; Matjaž Žaucer
Natural hist, energetics, environment, civil engineering
150 000 vols; 600 curr per; 3 000 e-journals; 4 300 diss/theses; 3 952 govt docs; 20 microforms; 150 av-mat; 15 digital data carriers; 70 000 standards
libr loan; IATUL, IFLA, ASLIB, FID   24819

– Elektroinštitut Milan Vidmar, Knjižnica (Electrical Institute), Hajdrihova 2, 1000 Ljubljana
T: +386 1 4743601; Fax: +386 1 4743326; E-mail: info@eimv.si; URL: www.eimv.si
1948; Danila Božić
15 500 vols; 60 curr per   24820

– Fakulteta za arhitekturo, Knjižnica (Faculty of Architecture), Zoisova 12, 1000 Ljubljana
T: +386 1 1250448; Fax: +386 1 1264319
1975; Renata Stella-Čop
36 300 vols; 235 curr per   24821

– Fakulteta za elektrotehniko, in Fakultete za računalništvo in informatiko, Knjižnica (Faculty Library of Electrical and Computer Engineering), Tržaška cesta 25, 1000 Ljubljana
T: +386 1 4768416; Fax: +386 1 4264630; E-mail: fe-dekanat@fe.uni-lj.si; URL: www.fe.uni-lj.si
1920; Zdenka Oven
48 600 vols; 300 curr per; 231 diss/theses
libr loan   24822

– Fakulteta za farmacijo, Knjižnica, Aškerčeva 7, 1000 Ljubljana
T: +386 1 4769548; Fax: +386 1 4258031; E-mail: Borut.Toth@ffa.uni-lj.si; URL: www.ffa.uni-lj.si
1996; Borut Toth
10 500 vols; 63 curr per; 30 e-journals; 2 500 diss/theses
libr loan   24823

– Fakulteta za kemijo in kemijsko tehnologijo, Knjižnica (Faculty of Chemistry and Chemical Technology), Askerceva cesta 5, 1000 Ljubljana
T: +386 1 2419290; Fax: +386 1 2419291; E-mail: branko.skrinjar@fkkt.uni-lj.si; URL: www.fkkt.uni-lj.si/si/?693
1919; Branko Škrinjar
Chemistry, Chemical Engineering
26 620 vols; 80 curr per; 30 e-journals; 730 diss/theses   24824

– Fakulteta za matematiko in fiziko – Knjižnica za matematiko (Mechanical Library), Lepi pot 11, 1111 Ljubljana; p.p. 64, 1111 Ljubljana
T: +386 1 4250072;
Fax: +386 1 4250072; E-mail: knjiznica.mehanika@fmf.uni-lj.si; URL:

www.fmf.uni-lj.si
Zvezda Pecar
Soil mechanics, rock mechanics
11 700 vols; 13 curr per
libr loan 24825

– Fakulteta za pomorstvo in promet, Knjižnica (Faculty of Maritime Studies and Transport), Pot pomorščakov 4, 6320 Portorož
T: +386 5 6767140; Fax: +386 5 6747193; E-mail: knjiznica@fpp.uni-lj.si; URL: knjiznica.fpp.edu
1962; Igor Presl
20 000 vols; 150 curr per; 3 800 diss/theses; 70 sound-rec; 40 digital data carriers
libr loan; ZBDS, IFLA 24826

– Fakulteta za socialno delo, Knjižnica (Faculty of Social Work), Topniška 31, 1000 Ljubljana
T: +386 1 2809261; Fax: +386 1 2809270; E-mail: info@fsd.uni-lj.si; URL: www.fsd.si/knjiznica/
1957; Lidija Kunic
Social work, social policy, public welfare, psychology, psychiolry, research
32 000 vols; 95 curr per; 8 307 diss/theses; 30 av-mat; 55 digital data carriers
libr loan 24827

– Fakulteta za strojništvo, Knjižnica (Faculty of Mechanical Engineering), Aškerčeva 6, 1101 Ljubljana
T: +386 1 4771113; Fax: +386 1 2518567; E-mail: knjiznica@fs.uni-lj.si; URL: www.fs.uni-lj.si
1945; Tonija Zadnikar
86 557 vols; 231 curr per; 128 e-journals; 7 mss; 1 875 diss/theses; 67 microforms; 103 av-mat; 204 sound-rec; 8 086 standards
libr loan 24828

– Filozofska fakulteta, Knjiznica oddelka za umetnostno zgodovino, Knjižnica (Department of Art History), Askerceva 2, 1000 Ljubljana
T: +386 1 2411220; Fax: +386 1 2411211; E-mail: ida.kranjc@ff.uni-lj.si; URL: www.ff.uni-lj.si/umzgod/default.html
1921; Ida Macek Kranjc
14 500 vols; 54 curr per; 1 e-journals; 168 diss/theses; 40 000 av-mat; 40 digital data carriers; 12 000 photos
IFLA 24829

– Filozofska fakulteta – Knjižnica oddelkov za anglistiko in germanistiko (Department of English and Germanic Languages and Literatures), Aškerčeva 2, p.p. 580, 1000 Ljubljana
T: +386 1 2411352; Fax: +386 1 1259337; E-mail: ohk.ger@ff.uni-lj.si; URL: www.ff.uni-lj.si/OHK/default.htm
1922; Angelika Hribar
61 912 vols; 172 curr per; 999 diss/theses; 560 sound-rec
libr loan 24830

– Filozofska fakulteta – Oddelek za klasično filologijo, Knjižnica (Faculty of Arts, Department for Classical Philology), Aškerčeva 2, 1000 Ljubljana
T: +386 1 2411420; Fax: +386 1 2411421; E-mail: zala.rott@ff.uni-lj.si
1919
15 937 vols; 38 curr per; 16 diss/theses; 5 maps; 2 av-mat; 6 sound-rec; 6 digital data carriers
libr loan 24831

– Filozofska fakulteta – Oddelek za muzikologijo, Knjižnica (Department of Musicology), Aškerčeva 2, 1000 Ljubljana
T: +386 1 2411440; Fax: +386 1 4259337; E-mail: muzikologija@ff.uni-lj.si; URL: www.musicology.over.net/knjiznica.php
1962; Lidija Podlesnik
16 151 vols; 31 curr per; 29 diss/theses; 4 408 music scores; 34 av-mat; 4 078 sound-rec 24832

– Filozofska fakulteta – Oddelek za slovenistiko in Oddelek za slavistiko, Knjižnica (Faculty of Arts – Department of Slovene Studies and Department of Slavonic Languages and Literatures), Aškerceva 2, 1000 Ljubljana; P. p. 580, 1001 Ljubljana
T: +386 1 241 1302; Fax: +386 1 425 9337; E-mail: slav.knjiz@ff.uni-lj.si; URL: www.ff.uni-lj.si
1919; Anka Sollner Perdih
120 657 vols; 312 curr per; 3 e-journals; 2 074 diss/theses; 191 maps; 5 microforms; 86 av-mat; 880 sound-rec; 410 digital data carriers; 1 manuscript
libr loan; Slovenian library association 24833

– FNT – VTOZD Montanistika, Knjižnica (Department of Mining and Metallurgy), Aškerčeva 20, 1000 Ljubljana
T: +386 1 212 121
1919; Marja Stepisnik-Bogovčič
45 550 vols; 812 curr per; 1 393 diss/theses
libr loan 24834

– INDOK sluzba in knjiznica, Biotehniška fakulteta (Department of Wood Science and Technology), Rožna dolina, C. VIII/34, 1000 Ljubljana
T: +386 1 4231161; Fax: +386 1 2572297; E-mail: maja.cimerman@bf.uni-lj.si; URL: www.bf.uni-lj.si/les/knjiznica.html
1978; Maja Cimerman, Darja Vranjek
16 000 vols; 140 curr per; 205 diss/theses; 4 digital data carriers
libr loan 24835

– Insšitut za kriminologijo, Pravna fakulteta, Knjižnica (Institute of Criminology), Poljanski Nasip 2, 1000 Ljubljana
T: +386 1 4203242; Fax: +386 1 4203245; E-mail: inst.crim@pf.uni-lj.si; URL: www.pf.uni-lj.si.ik
1954; Ivanka Sket
23 400 vols; 80 curr per; 60 diss/theses; 15 digital data carriers
libr loan; WCJLN 24836

– Inštitut Jožef Stefan, Knjižnica (Jozef Stefan Institute), ul. Jamova 39, p.p. 3000, 1001 Ljubljana
T: +386 1 1773304; Fax: +386 1 1231569; E-mail: luka.sustersic@ijs.si; URL: libra.ijs.si
1952; Luka Šušteršič
Nuclear sciences, physics, chemistry, biochemistry, biophysics, ADD, artificial intelligence
88 500 vols; 450 curr per; 3 500 microforms; 20 digital data carriers
libr loan 24837

– Inštitut za ekonomska raziskovanja, Knjižnica (Institute for Economic Research), Kardeljeva ploščad 17, 1109 Ljubljana
T: +386 1 5303800; Fax: +386 1 5303810; E-mail: knjiznica@ier.si; URL: www.ier.si
Katja Mlinar-Gerbec
Economics
48 000 vols 24838

– Inštitut za mednarodno pravo in mednarodne odnose, Pravni fakulteta, Knjižnica (Institute of International Law and Relations), Poljanski nasip 2, 1000 Ljubljana
T: +386 1 4203131; Fax: +386 1 4203130; E-mail: dl-27.ozn@pf.uni-lj.si; URL: www.pf.uni-lj.si/povezave/institut-za-mednarodno-pravo/
1946; Marko Kos
Un docs and publs
18 200 vols
libr loan 24839

– Kemijski inštitut, Knjižnica (National Institute of Chemistry), ul. Hajdrihova 19, 1000 Ljubljana
T: +386 1 4760200; Fax: +386 1 4760300
1948; Lucija Kramberger
19 305 vols; 165 curr per; 2 e-journals; 1 888 govt docs; 550 microforms; 5 digital data carriers
libr loan 24840

– Matematična knjižnica (Mathematical Library), 3. nadstropje, Jadranska 19, p.p. 2964, 1000 Ljubljana
T: +386 1 4766558; Fax: +386 1 2517281; E-mail: matknjiz@fmf.uni-lj.si
1919; Maja Klavžar
50 670 vols; 360 curr per; 10 mss; 172 diss/theses; 780 microforms; 4 av-mat; 311 digital data carriers 24841

– Medicinska fakulteta, Centralna medicinska knjižnica (Faculty of Medicine – Central Medical Library), Vrazov trg 2, 1000 Ljubljana
T: +386 1 5437700; Fax: +386 1 5437701; E-mail: infocmk@mf.uni-lj.si; URL: www.mf.uni-lj.si/cmk
1945; Anamarija Rožić-Hristovski
198 368 vols; 950 curr per; 684 diss/theses; 25 digital data carriers
DBS, AMLA, EAHIL 24842

– Naravoslovnotehniška fakulteta, Oddelek za tekstilstvo, Knjižnica (Department of Textiles, Faculty of Natural Sciences and Engineering), Šžniška 5, p.p. 312, 1000 Ljubljana
T: +386 1 2003223; Fax: +386 1 4253175; URL: www.ntf.uni-lj.si/ot/index.php?page=static&item=488
1961; Mojca Kotar
Textile design, textile technology, graphic arts
17 747 vols; 106 curr per; 36 diss/theses; 14 av-mat; 16 digital data carriers
libr loan; Slovenian Library Association 24843

– Oddelek za agronomijo, Biotehniska fakulteta, Knjižnica (Agronomy Department, Biotechnical Faculty), Jamnikarjena 101, 1000 Ljubljana
T: +386 1 1231161; Fax: +386 1 1231088; E-mail: izposojabfagr@bf.uni-lj.si; URL: www.bf.uni-lj.si
1947; Prof. Franc Batič
42 000 vols; 281 curr per; 125 maps; 2 digital data carriers
libr loan; IAALD 24844

– Oddelek za arheologijo, Knjižnica (Department of Archeology), Aškerčeva 2, 1000 Ljubljana
T: +386 1 1234496; Fax: +386 1 1237220
1919; Dr. Milan Lovenjak
15 000 vols; 200 curr per; 42 diss/theses 24845

– Oddelek za etnologijo in kulturno antropologijo, Knjižnica (Department of Ethnology and Cultural Anthropology), Zavetiška 5, 1111 Ljubljana
T: +386 1 1234495; Fax: +386 1 1231220
1946; Mojca Račič-Simončič
14 400 vols; 180 curr per; 1 300 diss/theses; 130 maps; 10 av-mat; 2 sound-rec
libr loan 24846

– Oddelek za filozofijo, Knjižnica (Department of Philosophy), Aškerčeva 2, 1000 Ljubljana
T: +386 1 2411117; Fax: +386 1 4259337; E-mail: marjetka.scelkov@ff.uni-lj.si; URL: www.ff.uni-lj.si/oddelki/filo/vsebina/knjiznica.htm
1919; Marjetka Ščelkov
Meinong Libr
23 000 vols; 53 curr per; 72 diss/theses; 2 digital data carriers
libr loan 24847

– Oddelek za geografijo, Filozofska Fakulteta, Oddelek za Geografijo, Knjižnica (Department of Geography, Faculty of Arts), Aškerčeva 2, 1000 Ljubljana
T: +386 1 2411250; Fax: +386 1 4259337; URL: www.ff.uni-lj.si/geo
1922; Janja Turk
Slovene Geographical Society coll; Map collection
44 160 vols; 273 curr per; 4 845 diss/theses; 37 105 maps; 3 450 av-mat
libr loan 24848

– Oddelek za geologijo, Naravoslovnotehniška fakulteta, Knjižnica (Department of Geology), Aškerčeva 12, 1000 Ljubljana
T: +386 1 4704630; Fax: +386 1 4704560; URL: www2.ntf.uni-lj.si/kn/index.php?page=static&item=610&get.treerot=599
1920; Barbara Bohar Bobnar
Geology, mineralogy, petrology, paleontology, sedimentology, tectonics; Coll of geological maps
40 293 vols; 55 curr per; 739 diss/theses; 9 383 maps; 28 microforms; 700 av-mat; 10 digital data carriers
libr loan; IFLA 24849

– Oddelek za gozdarstvo, Biotehniška fakulteta, Gozdarski institut Slovenije (Forestry Library), Večna pot 2, 1000 Ljubljana
T: +386 1 2007843;
Fax: +386 1 2573589; E-mail: gozdarska.knjiznica@gozdis.si; URL: www.bf.uni-lj.si/gozdarstvo/knjiznica
1948; Maja Bozic
38 823 vols; 319 curr per; 168 e-journals; 2 748 diss/theses; 138 microforms
libr loan 24850

– Oddelek za lesarstvo, Biotehniška fakulteta, Knjižnica (Department of Wood Technology), Rožna dolina, Cesta VIII/34, 1000 Ljubljana
T: +386 1 4231161; Fax: +386 1 2572297; E-mail: darja.vranjek@bf.uni-lj.si; URL: les.bf.uni-lj.si/o-oddelku/organiziranost/knjiznica/osnovni-podatki/
Darja Vranjek
18 000 vols
DBS 24851

– Oddelek za primerjalno književnost in literarno teorijo, Knjižnica (Department of Comparative Literature and Literary Theory), Aškerčeva 2, 1000 Ljubljana
T: +386 1 2411382; Fax: +386 1 4259337; E-mail: Seta.Knop@ff.uni-lj.si; URL: www.ff.uni-lj.si/primknjz
1922; Seta Knop
15 700 vols; 37 curr per; 37 diss/theses; 43 sound-rec
libr loan 24852

– Oddelek za psihologijo, Filozofska fakulteta, Knjižnica (Department of Psychology), Aškerčeva 2, 1000 Ljubljana
T: +386 1 2411186;
Fax: +386 1 4259301; E-mail: psiholoska.knjiznica@ff.uni-lj.si; URL: 193.2.70.110
1951; Marjeta Longyka
16 072 vols; 43 curr per; 334 e-journals; 1 807 diss/theses; 319 digital data carriers
libr loan; DBS 24853

– Oddelek za romanske jezike in književnosti, Osrednja Humanistrina Knjiznica Filozofska Fakulteta (Romance Language and Literature Library), Aškerčeva 2, 1000 Ljubljana
T: +386 1 2411400; Fax: +386 1 1259337; E-mail: marjeta.prelesnik@ff.uni-lj.si; URL: www.ff.uni-lj.si
1921; Jožica Pirc
25 000 vols; 54 curr per 24854

– Oddelek za sociologijo, Filozofska fakulteta, Knjižnica (Department of Sociology), Aškerčeva 2, 1000 Ljubljana
T: +386 1 1769283; Fax: +386 1 1259337
1960; Alojz Cindrič
17 000 vols; 30 curr per; 20 diss/theses
libr loan 24855

– Oddelek za živilstvo, Biotehniška Fakulteta, Knjižnica in INDOK (Nutrition Department), Jamnikarjeva 101, 1000 Ljubljana
T: +386 1 1231161; Fax: +386 1 2660964; E-mail: ivica.hocevar@bf.uni-lj.si
Ivica Hočevar
10 000 vols 24856

– Oddelek za živinorejo, Biotehniška Fakulteta, Knjižnica (Department of Animal Husbandry), Groblje 3, 1230 Domžale
T: +386 61 713611; Fax: +386 61 721005
1978; Elizabeta Kmecl
Theses coll (5 700 vols)
21 300 vols; 321 curr per 24857

– Oddelek za zootehniko, Knjižnica (Zootechnical Department), Groblj 3, 1230 Domžale
T: +381 1 7217873; Fax: +381 1 7217873; E-mail: natasa.siard@bfro.uni-lj.si
1974; Nataša Siard
9 000 vols; 315 curr per; 774 diss/theses; 3 databases on CD-ROM
libr loan 24858

– Osrednja družboslovna knjižnica Jožeta Goričarja, Fakulteta za družbene vede (Central Social Sciences Library Jože Goričar), Kardeljeva ploščad 5, 1000 Ljubljana
T: +386 1 5805150; Fax: +386 1 5805107; E-mail: odkjg@fdv.uni-lj.si; URL: www.odk.fdv.uni-lj.si
1985; Mirjam Kotar
Publ of Council of Europe
166 248 vols; 408 curr per; 6 531 diss/theses
libr loan; ZBDS 24859

– Osrednja humanistična knjižnica, Filozofska Fakulteta (Central Humanities Library), Aškerčeva 2, 1000 Ljubljana
T: +386 1 2411000; Fax: +386 1 4259337; URL: www.ff.uni-lj.si/OHK
1996; Alenka Logar-Pleško
photo coll, coll of tourists guide bks of students organization
584 515 vols; 1 265 curr per; 1 mss; 24 797 diss/theses; 4 523 music scores; 37 749 maps; 8 microforms; 173 av-mat; 7 047 sound-rec; 465 digital data carriers; 9 463 sound recordings
libr loan; ZBDS 24860

– Pedagoška fakulteta, Knjižnica (Faculty of Education), Kardeljeva ploščad 16, 1000 Ljubljana
T: +386 1 5892-332/-334/-333; E-mail: knjiznica@pef.uni-lj.si
1949; Alja Smole Gašparovič
100 000 vols; 220 curr per; 120 diss/theses; 7 digital data carriers
libr loan 24861

– Pravna fakulteta, Knjižnica (Faculty of Law), Poljanski nasip 2, 1000 Ljubljana
T: +386 1 42032-31/-32/-82; E-mail: knjiznica@pf.uni-lj.si; URL: www.pf.uni-lj.si
1920; Anita Longo
115 000 vols; 656 curr per; 7 400 diss/theses; 49 digital data carriers 24862

– Teološka fakulteta v Ljubljani Enota v Mariboru, Teološka knjiznica Maribor (Faculty of Theology in Ljubljana, Dept Maribor), Slomškov trg 20, 2000 Maribor
T: +386 5908145; Fax: +386 5908147; E-mail: referat@teof.uni-lj.si; URL: www.teof.uni-lj.si
1859; Fanika Krajnc-Vrecko
Archeological coll, biblical museum
87 200 vols; 280 curr per; 21 e-journals; 48 mss; 94 incunabula; 91 diss/theses; 310 music scores; 660 av-mat; 370 sound-rec; 23 digital data carriers
libr loan; COBISS 24863

– Veterinarska fakulteta, Center za informatiko in knjiznica (Veterinary Medicine), Cesta v Mestni log 47, 1000 Ljubljana
T: +386 1 4779255; Fax: +386 1 1779225; URL: www.vf.uni-lj.si/vf/index.php?option=com_wrapper&Itemid=67
1947; M. Sc. Gita Grecs-Smole
Bibliography of doctors of veterinary medicne (DVM) in Slovenia
55 000 vols; 200 curr per; 1 e-journals; 142 diss/theses; 28 av-mat; 250 digital data carriers
libr loan; OCLC 24864

– Visoka šola za zdravstvo (School for Health Sciences), Poljanska cesta 26a, 1000 Ljubljana
T: +386 1 30011-11/-56/-57/-95; Fax: +386 1 3001119; E-mail: knjiznica@vsz.uni-lj.si; URL: www.vsz.uni-lj.si
1965; Vesna Denona
Nursing, sanitary engineering, occupational therapy, radiotherapy
13 400 vols; 78 curr per; 3 049 diss/theses
libr loan 24865

Pedagoška Akademija, Knjižnica (Academy of Education), Mladinska 9, 2000 ***Maribor***
Sonja Marić
24 681 vols; 152 curr per 24866

Univerza v Mariboru, Univerzitetna knjižnica (University of Maribor Library), Gospejna ul 10, p.p. 223, 2000 ***Maribor***
T: +386 2 2507400; Fax: +386 2 2526087; E-mail: ukm@uni-mb.si; URL: www.ukm.uni-mb.si
1903; Irena Sapac
841 099 vols; 5 011 curr per; 4 970 mss; 1 000 diss/theses; 11 500 music scores; 11 500 maps; 5 000 microforms; 4 300 av-mat; 3 500 sound-rec; 400 digital data carriers
libr loan; IFLA, LIBER 24867

– Ekonomsko-poslovna fakulteta, Knjižnica (Faculty of Economics and Business), Razlagova 14, 2000 Maribor
T: +386 2 2290237; Fax: +386 2 2290319; E-mail: library.epf@uni-mb.si; URL: www.epf.uni-mb.si
1960; Ines Gusel
70 911 vols; 895 curr per; 13 e-journals; 12 690 mss; 14 255 diss/theses; 4 maps; 1 542 sound-rec; 38 digital data carriers; e-degrees, e-specialised thesis, e-master's degrees, e-dissertations
libr loan; ZBDS 24868

– Fakulteta za kmetijstvo in biosistemske vede, Knjižnica (Faculty of Agriculture and Life Sciences), Rivola 10, 2311 Hoče
T: +386 2 3209000; Fax: +386 2 6161158; E-mail: fkbv@uni-mb.si
1960; Ksenija Škorjanc
25 000 vols; 170 curr per; 15 048 mss; 13 diss/theses; 4 digital data carriers 24869

– Fakulteta za organizacijske vede, Knjižnica in INDOK (Faculty of Business Management), Kidriceva 55/a, 4000 Kranj
T: +386 4 2374211; Fax: +386 4 2374399; E-mail: dekanat@fov.uni-mb.si; URL: oraport.fov.uni-mb.si
1973; Tone Perčič
41 000 vols; 282 curr per; 135 diss/theses 24870

– Pedagoška fakulteta, Knjižnica (Faculty of Education), Koroška c. 160, 2000 Maribor
T: +386 2 2293736; Fax: +386 2 2518180; URL: www.pfmb.uni-mb.si/
Mojca Garantini
93 077 vols; 385 curr per; 1 519 maps; 5 microforms
libr loan 24871

– Pravna fakulteta, Knjižnica (Faculty of Law), Mladinska ulica 9, 2000 Maribor
T: +368 2 2504249; E-mail: prfmb.knjiz@uni-mb.si; URL: www.pf.uni-mb.si/knjiznica
1961; Darja Čokl
29 000 vols; 131 curr per; 31 digital data carriers
libr loan 24872

– Tehniška fakulteta, Knjižnica (Faculty of Technical Sciences), Smetanova 17, 2000 Maribor
T: +386 2 2207562; Fax: +386 2 2207033; E-mail: ktfmb@uni-mb.si; URL: www.ktfmb.uni-mb.si
1962; Mojca Markovič
72 000 vols; 396 curr per; 1 014 diss/theses; 34 maps; 152 microforms; 14 av-mat; 8 sound-rec; 22 digital data carriers
libr loan; SAK 24873

Univerza v Novi Gorici, Knjižnica, Vipavska 13, 5000 ***Nova Gorica***
T: +386 5 3315220; Fax: +386 5 3315375; E-mail: vanesa.valentincic@p-ng.si; URL: www.p-ng.si
10 000 vols; 70 curr per
libr loan 24874

School Libraries

Osnovna šole Brezovica pri Ljubljani, Knjižnica (Primary School), Brezovica 183, 61351 ***Brezovica***
T: +381 1 3653002; Fax: +381 1 3655539; E-mail: info@osbrezovica.si; URL: www.osbrezovica.si
Polonca Raušl
92 000 vols 24875

Srednja tehniška šola, Šolska knjižnica, Pot na Lavo 22, 3000 ***Celje***
Štefan Volk
17 800 vols 24876

Srednja tekstilna, obutvena in gumarska šola Kranj, Knjižnica (School of Textiles, Footwear and Gum (Rubber)), Cesta Staneta Žagarja 33, 4000 ***Kranj***
T: +386 4 2041574
1930; Metka Sever
16 000 vols; 61 curr per; 50 maps; 267 av-mat; 214 sound-rec; 57 digital data carriers
libr loan; ZBDS 24877

Gimnazija Jožeta Plečnika, Knjižnica (Grammar School Jože Plečnik), Šubičeva 1, 1001 ***Ljubljana***
T: +386 1 2511182; E-mail: gjp@guest.arnes.si; URL: www.mtaj.si/gjp
1990; Marisa Zlobec-Skaza
10 200 vols; 46 curr per; 70 av-mat; 100 sound-rec 24878

Gimnazija Ledina, Knjižnica (Grammar School), Resljeva cesta 12, 1000 ***Ljubljana***
T: +386 1 4342220; E-mail: knjiznica@ledina.org; URL: www.s-gimled.lj.edus.si
1867; Zvezdana Tinta
40 000 vols; 45 curr per
libr loan 24879

Gimnazija Poljane, Knjižnica (High School Poljane), Strossmayerjeva 1, 1000 ***Ljubljana***
T: +386 1 2316073;
Fax: +386 1 2319650; E-mail: milena.bon@guest.arnes.si; URL: www.gimnazija-poljane.com
1900; Milena Bon
32 226 vols; 73 curr per; 23 150 e-journals; 2 diss/theses; 4 music scores; 77 maps; 443 av-mat; 598 sound-rec; 115 digital data carriers; 157 multimedia mat
ZBDS 24880

Srednja gradbena, geodetska in ekonomska sòla, Knjižnica (School of Economics, civil engineering), Dunajska 102, 1000 ***Ljubljana***
T: +386 1 5341071; Fax: +386 1 341071
1960; Tomaž Žitnik
17 000 vols; 30 curr per; 70 av-mat; 15 digital data carriers
libr loan 24881

Srednja gradbena šola Ivana Kavčiča, Knjižnica (Highschool of Civil Engineering 'Ivan Kavčič'), Kardeljeva ploščad 2, 1000 ***Ljubljana***
Ana Uršič-Pfeifer
29 183 vols 24882

Srednja trgovska šola, Knjižnica (Highschool of Trade), Poljanska 28 a, 1000 ***Ljubljana***
T: +386 1 3006858; E-mail: Majda.Papez@guest.arnes.si
Majda Papez
12 414 vols; 28 curr per; 60 maps; 91 av-mat; 46 digital data carriers 24883

II. Gimnazija Maribor, Knjižnica (II. Grammar School Maribor), Trg Miloša Zidanška 1, 2000 ***Maribor***
T: +386 2 3304443; Fax: +386 2 3304440; E-mail: knjiznica@druga.org
1952; Romana Fekonja
32 500 vols; 150 curr per; 30 av-mat; 20 sound-rec; 20 digital data carriers
libr loan; ZBDS 24884

PRVA Gimnazija Maribor, Knjižnica (Highschool of Social Sciences), Trg Generala Maistra 1, 2000 ***Maribor***
T: +386 2 2285300; Fax: +386 2 2285311; E-mail: knjiznica@prva-gimnazija.org; URL: www.prva-gimnazija.org
Metka Kostanjevec
22 500 vols; 50 maps; 100 av-mat; 250 sound-rec; 30 digital data carriers
COBISS 24885

Srednja kovinarska, strojna in metalurška šola Maribor, Knjižnica (Highschool of Metal, Mechanical and Metallurgical Engineering), Smetanova 18, 2000 ***Maribor***
Tatjana Bohak
23 000 vols 24886

Srednja šola elektrotehnične in računalniške usmeritve, knjižnica (Highschool of Electrical Engineering and Information Science), Smetanova 67, 2000 ***Maribor***
Branka Grašič
17 000 vols 24887

Srednja šola pedagoške in kulturne usmeritve, Knjižnica (Highschool of Education and Culture), Gospovsvetska 4, 2000 ***Maribor***
Jerica Rukavina
22 000 vols 24888

Srednja tekstilna in frizerska šola, Knjižnica (Highschool of Textiles and Hairdressing), Park mladih 8, 2000 ***Maribor***
T: +386 2 3323550; Fax: +386 2 3323550; URL: www2.arnes.si/~ssplbele/
Mojca Strnad
10 500 vols 24889

Srednja Šola, Knjižnica (Scientific Highschool Center), Delpinova 6, 5000 ***Nova Gorica***
1947; Genovefa Kuštrin
59 900 vols; 81 curr per; 5 diss/theses; 15 music scores; 2 digital data carriers
libr loan 24890

Government Libraries

Državni Zbor Republike Slovenije, Dokumentacijsko-knjižnični oddelek (Slovene Parliament), Šubičeva ul. 4, 1000 ***Ljubljana***
T: +386 1 4789730; Fax: +386 1 4789844
1963; Nataš Glavnik
10 000 vols; 320 curr per; 7 e-journals
libr loan; IFLA 24891

Ministrstvo za notranje zadeve RS, Specialna Knjiznica (Ministry of the Interior of the Republic of Slovenia), Štefanova 2, 1501 ***Ljubljana***
T: +386 1 4284234; Fax: +386 1 4285157; E-mail: library.mnz@gov.si; URL: www.mnz.gov.si/si/o_ministrstvu/specialna_knjiznica/
1955; Tatjana Kovse
Criminal Investigation, Criminology, Legislation, State Administration
29 000 vols; 287 curr per; 21 e-journals; 55 digital data carriers
libr loan; ZBDS 24892

Statistični Urad Republike Slovenije, Knjižnica (Statistical Office of the Republic of Slovenia), Vožarski pot 12, 1000 ***Ljubljana***
T: +386 1 2415104; E-mail: info.stat@gov.si; URL: www.stat.si
1945
26 400 vols; 128 curr per; 4 av-mat; 44 sound-rec; 28 digital data carriers
libr loan 24893

Vrhovno sodišče Republike Slovenije, Centralna pravosodna knjižnica (Supreme Court of the Slovenian Republic, Central Judical Library), Tavčarjeva 9/II, 1000 ***Ljubljana***
T: +386 1 3664306; Fax: +386 1 36607; E-mail: cpk.info@sodisce.si
1918; Alenka Jelenc Puklavec
50 000 vols; 130 curr per
libr loan 24894

Ecclesiastical Libraries

Samostan Stična-Knjižnica (Monastic Library), Stična 17, 1295 ***Ivančna Gorica***
1898; Franca Baraga
Cisterciensia
30 000 vols; 80 curr per 24895

Frančiškanski samostan in cerkev sv. Jakoba, Knjižnica (Franciscan Abbey), mail address: 1240 ***Kamnik***
URL: www.burger.si/Kamnik/SvJakob/Knjiznica/SLOKnjiznica.htm
1627
10 000 vols; 25 incunabula 24896

Frančiškanska knjižnica (Franciscan Library), Prešernov trg 4, 1000 ***Ljubljana***
T: +386 1 2429300;
Fax: +386 1 2429313; E-mail: miran.spelic@guest.arnes.si
1233; Miran Špelič
15th c psalters
68 400 vols; 54 curr per; 500 mss; 121 incunabula; 6 diss/theses; 200 music scores; 36 maps 24897

Semeniška knjižnica (Seminary Library), Dolničarjeva 4, 1000 ***Ljubljana***
1701; Marijan Smolik
Baroque opera libretti
59 000 vols; 120 curr per; 250 mss; 30 incunabula; 25 maps; 25 000 small religious graphics 24898

Skofijska knjižnica (Diocesan Library), Ciril-Metodov trg 4, 1000 *Ljubljana*
T: +386 1 2312593
1875; Dr. Marijan Smolik
40 500 vols; 120 curr per   24899

Škofijski arhiv, Knjižnica (Diocesan Archive), Koroška cesta 1/II, 2000 *Maribor*
T: +386 2 2517690; E-mail: skofijski.arhiv@slomsek.net
1509; Ilaria Montanar
22 000 vols   24900

Kapucinski samostan, Knjižnica (Capuchin Monastery), Kapucinski trg 1, 4220 *Škofja Loka*
T: +386 4 5120970; Fax: +386 4 5063004; E-mail: kapucini.loka@rkc.si; URL: www.kapucini.si
1707; Dr. Metod Benedik
Incunabula, bks 16-18 c (5 100)
25 000 vols; 28 curr per; 50 mss; 21 incunabula; 6 diss/theses; 5 maps; 5 av-mat; 5 sound-rec   24901

Corporate, Business Libraries

Železarna Jesenice, Strokovna Knjižnica (Iron and Steel Works Jesenice), Cesta Železarjev 8, 4270 *Jesenice*
1939; Jana Jamar
25 100 vols; 357 curr per
libr loan   24902

Slovenske železnice, d. o. o., Knjižnica (Slovenian Railways ltd., Library), Kolodvorska ul 11, 1506 *Ljubljana*
T: +386 1 2914206; Fax: +386 1 2914824; E-mail: knjiznica@slo-zeleznice.si; URL: www.slo-zeleznice.si
1965; Janez Drnovšek
50 000 vols; 250 curr per; 434 maps; 43 microforms; 2 av-mat; 400 digital data carriers
libr loan; Zveza bibliotekarskih društev Slovenije, Association of Slovene librarian societies   24903

Telekom Slovenije, Strokovna knjižnica, Pražakova 5, 1000 *Ljubljana*
T: +386 1 2346829;
Fax: +386 1 4131326; E-mail: teletrgovina.ljubljana@telekom.si; URL: www.telekom.si
1945; Milena Lovišček
Telecommunications Computers, Marketing Economy, Law
15 000 vols; 160 curr per; 3 800 vols
libr loan   24904

KRKA, d.d., Novo Mesto, Center za strokovno informatiko (KRKA Library and Information Services), Šmarješka cesta 6, 8501 *Novo Mesto*
T: +386 7 3312111; Fax: +386 7 3323987; E-mail: library@krka.si
1962; Angela Čuk
Chemical abstracts since 1919
17 000 vols; 460 curr per; 150 diss/theses; 27 digital data carriers
FID, EAHIL, ZBDS   24905

Slovenske Železarne – Metal Ravne d.o.o., Specialna knjižnica Ravne, Koroška cesta 14, 2390 *Ravne na Koroškem*
T: +386 2 8221131-5589; Fax: +386 2 8223013
1948; Danica Kobal
19 500 vols; 50 curr per
libr loan   24906

Special Libraries Maintained by Other Institutions

Posavski muzej Brežice, Knjižnica (Posavje Museum), Cesta prvih borcev 1, 8250 *Brežice*
T: +386 7 4961271;
Fax: +386 7 4660516; E-mail: ivan.kastelic@guest.arnes.si; URL: www.posavski-muzej.si
1949; Tomaž Teropšič
Local and regional hist, archeology
8 255 vols; 38 curr per; 30 mss; 9 incunabula; 1 diss/theses; 7 govt docs; 64 av-mat; 14 digital data carriers
libr loan   24907

Pokrajinski muzej, Biblioteka (Regional Museum), Muzejski trg 1, 3000 *Celje*
T: +386 3 4280950; Fax: +386 3 4280966; E-mail: knjiznica@pok-muzej-ptuj.si; URL: www.pok-muzej-ptuj.si/ang/knjiznica_ang.html
1882; Martin Steiner
9 792 vols   24908

Splošna bolnišnica, Medicinska knjižnica (Medical Library), Oblakova 5, 3000 *Celje*
T: +386 3 4233000; Fax: +386 3 4233666; URL: www.sb-celje.si
1962; Janja Korošec
8 996 vols; 180 curr per; 76 diss/theses; 32 digital data carriers
libr loan   24909

Univerzitetna klinika za pljučne bolezni in alergijo Golnik, Knjižnica (University Clinic of Respiratory and Allergic Diseases Golnik), Golnik 36, 4204 *Golnik*
T: +386 4 2569380; Fax: +386 4 2569117; E-mail: knjiznica@klinika-golnik.si; URL: www.klinika-golnik.si
1953; Anja Blazun
2 064 vols; 63 curr per; 48 e-journals; 20 e-books; 47 diss/theses; 10 govt docs; 2 av-mat; 61 digital data carriers
libr loan; DBS   24910

Gorenjski muzej, Knjižnica, Tavčarjeva 43, 4000 *Kranj*
Valerija Povšnar
11 000 vols   24911

Arhiv Republike Slovenije, Knjižnica (Archives of the Republic of Slovenia), Zvezdarska 1, 1127 *Ljubljana*
T: +386 1 2414218; Fax: +386 1 2414269; E-mail: alenka.hren@gov.si; URL: www.arhiv.gov.si
1945; Alenka Hren
Arch science
27 500 vols; 90 curr per; 1 av-mat; 71 digital data carriers
libr loan   24912

Center za mednarodno sodelovanje in razvoj knjižnica (Center for International Cooperation and Developing Special Library), Titova 104, 1000 *Ljubljana*
1972; Tanja Kovše
11 500 vols; 276 curr per   24913

Center za teatrologijo in filmologijo pri AGRFT (Center of Theater Studies and Cinematography at the Academy of Theater, Radio, Film and Television), Nazorjeva 3, 1000 *Ljubljana*
T: +386 1 2510412; Fax: +386 1 2510450; E-mail: knjiznica@agrft.uni-lj.si; URL: www.agrft.uni-lj.si
1945; Bojana Bajec
28 219 vols; 107 curr per; 6 438 diss/theses; 13 756 av-mat; 2 403 sound-rec; 529 243 archive coll, iconogr. of theater performances
libr loan; SIBMAS   24914

Delo d.d., Strokovna knjižnica ('Delo' Magazine Concern), Dunajska 5, 1000 *Ljubljana*
T: +386 1 4737375; Fax: +386 1 4737383; E-mail: infornavti@delo.si
1945; Jerica Mrak
Politics, law, business admin, 2 mil. of newspaper articles (in clippings)
24 202 vols; 499 curr per; 520 CDs
libr loan   24915

GIAM ZRC SAZU, Knjižnica (Anton Melik Geographical Institute of the Scientific Research Centre of the Slovenian Academy of Sciences and Arts (GI SRC SASA)), Gosposka ulica 13, 1000 *Ljubljana*; P.O. Box 306, 1000 Ljubljana
T: +386 1 4706355; Fax: +386 1 4257793; E-mail: maja.topole@zrc-sazu.si; URL: giam.zrc-sazu.si/?q=en
1946; Maja Topole
Geography, physical geography, social geography, earth sciences, statistics, cartography
32 000 vols; 700 curr per
libr loan   24916

Institut Français Charles Nodier, Francoski Inštitut Charles Nodier, Mediateka, Breg 12, 1000 *Ljubljana*
T: +386 1 2000517; Fax: +386 1 2000512; E-mail: biblio@institutfrance.si; URL: www.institutfrance.si/si/mediateka.php
Polona Končar
12 000 vols; 800 av-mat; 500 digital data carriers   24917

Inštitut za ekonomska raziskovanja, Knjižnica (Institute for Economic Research), Kardeljeva ploščad 17, 1000 *Ljubljana*
T: +386 1 5303876; Fax: +386 1 5303874; E-mail: knjiznica@ier.si; URL: www.ier.si
1955; Andreja Vide Hladnik
Commerce, management; World Bank Publications
49 000 vols; 120 curr per; 60 digital data carriers; Commerce, management, coll of World Bank pubs
libr loan   24918

Inštitut za geologijo, geotehniko, in geofiziko, Knjižnica (Institut of Geology, Geotechnics and Geophysics), Dimičeva 14, 1000 *Ljubljana*
T: +386 1 1682461; Fax: +386 1 1682557
1947; Andreja Šarabon-Bone
14 500 vols; 300 curr per   24919

Inštitut za narodnostna vprašanja, Knjižnica (Institute for Ethnic Studies), Erjavčeva 26, p.p. 1723, 1000 *Ljubljana*
T: +386 1 2001880; Fax: +386 1 2510964; E-mail: meta.cerar@inv.si; URL: www.inv.si
1944; Sonja Kurinčič Mikuž
Ethnicity, nationalism, minorities, human rights, bilingualism, assimilation, regionalism, international law, ethnic conflicts
40 000 vols; 270 curr per; 740 mss; 100 diss/theses; 1 086 maps; 152 000 clippings
libr loan; IFLA   24920

Institut za varilstvo, Knjižnica (Institute of Welding), Ptujska 19, 1000 *Ljubljana*
T: +386 1 2809400; Fax: +386 1 2809422; E-mail: ljwelding@guest.arnes.si; URL: www.i-var.si
1957; Agnes Brezovnik
Engineering, metallurgy, welding and allied processes
10 000 vols; 43 curr per; 30 diss/theses; 20 av-mat; 10 digital data carriers   24921

Kmetijski inštitut Slovenije, Knjižnica (Agricultural Institute of Slovenia), Hacquetova ulica 17, 1000 *Ljubljana*; p.p. 2553, 1001 Ljubljana
T: +386 1 2805262; Fax: +386 1 2805255; E-mail: knjiznica@kis.si; URL: www.kis.si/si/kis.web
1964; Lili Marincek
32 000 vols; 100 curr per; 5 e-journals; 53 diss/theses; 30 digital data carriers
libr loan; DBS   24922

Mestni muzej, Knjižnica (Municipal Museum), Gosposka 15, 1000 *Ljubljana*
T: +386 1 2412500; Fax: +386 1 2412540; E-mail: info@mgml.si; URL: www.mgml.si
1935; Hanna Trojanowska
10 200 vols; 60 curr per   24923

Moderna galerija, Knjižnica (Museum of Modern Art), Tomšičeva 14, 1000 *Ljubljana*
T: +386 1 2416828; Fax: +386 1 2514120; E-mail: k.apih@eunet.si; URL: www.mg-lj.si
1947; Katja Kranjc
62 700 vols; 210 curr per
libr loan; IFLA, ARLIS/NA, ARLIS/UK & Ireland   24924

Muzej novejše zgodovine Slovenije, Knjižnica (National Museum of Contemporary History), Celovška 23, 1000 *Ljubljana*
T: +386 1 3009610 ext 21/22; Fax: +386 1 4338244; E-mail: uprava@muzej-nz.si; URL: www.muzej-nz.si
1944; Marija Pečan
Modern hist studies, museological lit, original mat (unique partisan bks printed illegal, leaflets and posters from World War II), World War I; Newspaper Room
20 000 vols; 63 curr per; 110 mss; 880 maps; 135 digital data carriers
libr loan; IFLA, ZBDS   24925

Narodna galerija, Knjižnica (National Gallery), ul. Puharjeva 9, 1001 *Ljubljana*
T: +386 1 2415420; Fax: +386 1 2415403; E-mail: mateja_krapez@ng-slo.si; URL: www.ng-slo.si
1946; Mateja Krapez
95
33 000 vols; 73 curr per; 9 diss/theses; 114 av-mat; 26 sound-rec; 46 digital data carriers; 16 631 exhibition cat   24926

Narodni Muzej Slovenije, Knjižnica (National Museum of Slovenia), ul. Prešernova 20, 1000 *Ljubljana*
T: +386 1 2414468; Fax: +386 1 2414422; E-mail: anja.dular@nms.si; URL: www.nms.si
1821; Anja Dulár
Early Slovene prints (16-19th c), coll from 19th c, pamphlets from 19th c archival doc; Print room
145 000 vols; 320 curr per; 200 mss; 6 incunabula; 231 diss/theses; 260 maps; 10 microforms; Coll of pamphlets
libr loan; COBISS   24927

Onkološki inštitut Ljubljana, Specialna knjižnica in informacijski center za onkologijo (Institute of Oncology), Zaloška cesta 2, 1000 *Ljubljana*; P.O. Box 2217, 1001 Ljubljana
T: +386 1 5879373; Fax: +386 1 5879406; E-mail: knjiznica@onko-i.si; URL: www.onko-i.si/sl/raziskovanje_in_izobrazevanje/knjiznica/
1950; Matjaz Musek
Medical oncology, radiotherapy, cell biology, cancer research
20 804 vols; 229 curr per; 780 e-journals; 244 diss/theses; 30 av-mat; 35 digital data carriers
libr loan; ZBDS, EAHIL, EADI, SLAIS   24928

Pedagoška knjižnica zavoda za Šolstvo (Pedagogical library), Poljanska 28, 1000 *Ljubljana*
1955; Zdenka Jovanoska
24 000 vols
libr loan   24929

Planinska zveza Slovenije, Centralna planinska, Knjižnica (Slovene Alpine Library), Dvoržakova 9, 1000 *Ljubljana*
T: +386 1 4345686; Fax: +386 1 4345691; E-mail: alpinizem@pzs.si; URL: www.pzs.si
1956; Marija Kurnik
8 100 vols   24930

Prirodoslovni muzej Slovenije, Knjižnica (Slovenian Museum of Natural History), Prešernova 20, p.p. 290, 1001 *Ljubljana*
T: +386 1 2410944; Fax: +386 1 2410953; E-mail: biblioteka@pms-lj.si; URL: www2.pms-lj.si
1821; Alenka Jamnik
Ornithology, entomology, natural history, mammals
20 500 vols; 450 curr per; 2 e-journals; 30 diss/theses; 449 maps; 35 av-mat; 20 sound-rec; 27 digital data carriers
libr loan; ZBDS, DBL   24931

RTV Slovenija, Knjižnica (Radio and Television Slovenia), Tavčarjeva 17, 1000 *Ljubljana*
T: +386 1 1311333; Fax: +386 1 1334007
1947; Katjuša Strnad-Urek
24 980 vols; 17 curr per
libr loan   24932

Slovanska knjižnica (Slavic Library), Einspielerjeva 1, 1000 *Ljubljana*
T: +386 1 2363850;
Fax: +386 1 2363851; E-mail: slovanska.knjiznica@guest.arnes.si; URL: www.slovanskaknjiznica.si
1901/1946; Mateja Komel Snoj
graphic arts
142 000 vols; 250 curr per; 400 microforms; 70 digital data carriers
libr loan; ZBDS   24933

Slovenska filharmonija, Knjižnica (Slovenian Philharmonic), Kongresni trg 10, 1000 ***Ljubljana***
T: +386 1 2410800;
Fax: +386 1 2410900; E-mail: mateja.kralj@filharmonija.si; URL: www.filharmonija.si
1945; Anton Cimperman
70 000 vols; 1 500 music scores; 600 sound-rec
libr loan 24934

Slovenska Matica, Knjižnica (Slovenian Society), Kongresni trg 8, 1000 ***Ljubljana***
T: +386 1 4224342;
Fax: +386 1 4224343; E-mail: slovenskamatica@siol.net; URL: www.slovenska-matica.si
1864
10 000 vols 24935

Slovenski etnografski muzej, Knjižnica (Slovene Ethnographic Museum), Metelkova 2, 1000 ***Ljubljana***
T: +386 1 3008766; Fax: +386 1 3008736; E-mail: knjiznica@etno-muzej.si; URL: www.etno-muzej.si
1924; Mojca Račič Simončič
Clippings of Slovene newspapers on ethnology, museology and heritage
31 378 vols; 120 curr per
libr loan 24936

Slovenski gledališki muzej, Knjižnica (National Theatre Museum), Mestni trg 17, 1000 ***Ljubljana***
T: +386 1 2415800; Fax: +386 1 210142, 2415816; E-mail: sgm@guest.arnes.si
1952; Katarina Kocijančič
Slovene theatre hist; Arch, Phototèque, Phonotèque, AV
17 000 vols; 43 curr per; 16 500 posters, 330 stage and costume designs, 5 500 photos
libr loan 24937

Slovenski šolski muzej, Knjižnica (Slovenian School Museum), Plečnikov trg 1, 1000 ***Ljubljana***
T: +386 1 2513024;
Fax: +386 1 2513024; E-mail: solski.muzej@guest.arnes.si; URL: www.ssolski-muzej.si
1898; Tatjana Hojan
65 000 vols; 50 curr per; 17 000 docs 24938

Tehniški muzej Slovenije, Knjižnica (Technical Museum of Slovenia), Parmova 33, 1000 ***Ljubljana***
T: +386 1 4361606; Fax: +386 1 4361606; E-mail: info@tms.si; URL: www.tms.si
1951
Dept of Museum Documentation
12 475 vols; 35 curr per; 482 maps; 19 547 microforms; 355 av-mat; 25 sound-rec; 36 digital data carriers; 248 graphics, 80 playcards 24939

Uprava Republike Slovenije za kulturno dediščino (Cultural Heritage Office of the Republic of Slovenia), Plečnikov trg 2, 1000 ***Ljubljana***
T: +386 1 1259467; Fax: +386 1 213012
1913
13 000 vols 24940

Urbanistični inštitut, Indok/knjižnica (Urban Planning Institute), Trnovski pristan 2, 1127 ***Ljubljana***
T: +386 1 4201331; Fax: +386 1 4201330; E-mail: knjiznica@uirs.si; URL: www.urbinstitut.si
1960; Nevenka Kocijančič
34 720 vols; 271 curr per; 30 diss/theses; 2 govt docs; 4 000 maps; 3 digital data carriers; 13 000 plans
libr loan 24941

Zavod R. Slovenije za makroekonomske analize in razvoj, Knjižnica (Slovenian Macroeconomic Analysis and Development Institute), Gregorčičeva 27, 1000 ***Ljubljana***
T: +386 1 1782114; Fax: +386 1 1782070
1955; Jereb Janez
Public admin
25 000 vols; 125 curr per; 100 diss/theses; 1 500 govt docs
libr loan; COBISS 24942

Zavod za gradbeništvo Slovenije, Knjižnica (Slovenian National Building and Civil Engineering Institute), Dimičeva 12, 1001 ***Ljubljana***
T: +386 1 2804285; Fax: +386 1 2804236; E-mail: knjiznica@zag.si; URL: www.zag.si
1952; Metka Ljubešek
20 000 vols; 130 curr per; 30 diss/theses 24943

Zavod zu družbeno planiranje, Knjižnica (Institute for Social Planning), Gregorčičeva 25-27, 1000 ***Ljubljana***
1955; Janez Jereb
Pub Admin, law
20 500 vols 24944

Zgodovinski arhiv Ljubljana, Knjižnica (Historical Archive Ljubljana), Mestni trg 27, 1001 ***Ljubljana*** pp 1614
T: +386 1 3061301; Fax: +386 1 4264303; E-mail: nina.frakelj@zal-lj.si; URL: www.zal-lj.si
1898; Ziga Zeleznik
31 812 vols; 3 282 curr per; 36 av-mat; 2 sound-rec; 1 digital data carriers
libr loan 24945

Ekonomski institut Maribor, INDOK službe in strokovna knjižnica (Economic Institute of Maribor, Documentation Service and Special Library), Razlagova 22, 2000 ***Maribor***
T: +386 2 2226331; Fax: +386 2 2212597
1965; Kazimierz Wozniak
9 500 vols; 120 curr per
libr loan 24946

Medicinska knjižnica Univerzitetnega klinièenga Maribor (Medical Library of the General Hospital Maribor), Ljubljanska ulica 5, 2000 ***Maribor***
T: +386 2 3202066;
Fax: +386 2 3312393; E-mail: medknj.mb@guest.arnes.si
1950; Alenka Helbl
29 700 vols; 228 curr per; 74 e-journals; 55 diss/theses; 52 digital data carriers; Floppy disks, compact disks
libr loan 24947

Pokrajinski arhiv Maribor, Knjižnica (Maribor Regional Archive), Glavni trg 7, 2000 ***Maribor***
T: +386 2 2285021;
Fax: +386 2 2522564; E-mail: polde.mikec@pamb.pokarh-mb.si; URL: www.pokarh-mb.si
1929; Dr. Slavica Tovšak
Slovenian hist
15 000 vols; 40 curr per 24948

Pokrajinski muzej, Knjižnica (Regional Museum), ul. Grajska 2, 2000 ***Maribor***
T: +386 2 2211851; Fax: +386 2 2227777
1903; Peter Može
12 300 vols 24949

Goriški muzej, Knjižnica (Museum of Gorica Region), Grajska 1, 5000 ***Nova Gorica***
T: +386 5 3359811;
Fax: +386 5 3359820; E-mail: borut.koloini@guest.arnes.si
1965; Borut Koloini
27 000 vols; 900 curr per; 4 diss/theses; 265 av-mat; 3 digital data carriers
libr loan 24950

Pomorski muzej 'Sergej Mašera', Knjižnica (Maritime Museum 'Sergej Mašera'), Cankarjevo nabr. 3, 6330 ***Piran***
T: +386 5 6730620; Fax: +386 5 6732756; E-mail: muzej@pommuz-pi.si; URL: www2.arnes.si/~kppomm
1956; Igor Presl
Biblioteca Civica di Pirano, naval hist; Istrica
15 000 vols; 65 curr per; 850 maps
libr loan 24951

Inštitut za raziskovanje krasa ZRC SAZU, Knjižnica (Central Library of the Slovenian Academy of Sciences and Arts, Karst Research Institute), Titov trg 2, 6230 ***Postojna***
T: +386 5 7001900; Fax: +386 5 7001999; E-mail: izrk@zrc-sazu.si; URL: www.zrc-sazu.si/www/izrk/izrk-s.htm
1947; Maja Kranjc
36 000 vols; 610 curr per; 4 mss; 44 diss/theses; 46 av-mat; 6 digital data carriers
libr loan 24952

Loški muzej, Strokovna knjižnica (Museum of Skofja Loka), Grajska pot 13, 4220 ***Škofja Loka***
T: +386 4 5170400;
Fax: +386 4 5170412; E-mail: loski.muzej@guest.arnes.si; URL: www.loski-muzej.si
Mira Kalan
Personal libr of the Slovene writer Ivan Tavčar, libr of the noble family Wolkensperg, old prayer bks
18 200 vols; 143 curr per; 20 diss/theses; 20 av-mat 24953

Public Libraries

Knjižnica Brežice (Brežice Public Library), Trg izgnancev 12 b, 8250 ***Brežice***
T: +386 7 4962649; Fax: +386 749 62539; E-mail: knjiznicabre@bre.sik.si; URL: www.bre.sik.si
1946; Tea Bemkoč
88 565 vols; 4 020 curr per; 5 e-journals; 14 diss/theses; 489 govt docs; 85 music scores; 241 maps; 4 971 av-mat; 3 133 sound-rec; 321 digital data carriers; 186 multimedia
libr loan; ZBDS 24954

Osrednja knjižnica Celje (Celje Public Library), Muzejski trg 1a, 3000 ***Celje***
T: +386 3 4261710; E-mail: info@knjiznica-celje.si; URL: www.ce.sik.si
1973; Janko Germadnik
Vladimir Levstik Room
340 000 vols; 246 mss; 324 diss/theses; 2 000 music scores; 574 maps; 450 microforms; 26 sound-rec
libr loan 24955

Knjižnica Jožeta Udoviča Cerknica (Jože Udovič Library), Partizanska 22, 1380 ***Cerknica***
T: +386 1 7091078; Fax: +386 1 7097121; E-mail: info@cer.sik.si; URL: www.cer.sik.si
1972; Marija Hribar
Jože Udovič's private libr – regional studies
65 535 vols; 183 curr per; 55 mss; 229 maps; 28 microforms; 4 290 av-mat; 651 sound-rec; 115 digital data carriers; 5 215 clippings
libr loan; ZBDS 24956

Knjižnica Domžale (Public Library Domžale), Cesta Talcev 4, 1230 ***Domžale***
T: +386 1 7234119; Fax: +386 1 7225070; E-mail: info@dom.sik.si; URL: www.knjiznica-domzale.si
1904; Barbara Zupank Oberwalder
186 940 vols; 3 487 curr per; 4 e-journals; 488 music scores; 840 maps; 16 159 av-mat; 5 765 sound-rec; printed music, maps, AV mat, sound recordings, CD-ROM (9850) 24957

Knjižnica Grosuplje (Grosuplje Public Library), Adamičeva cesta 15, 1290 ***Grosuplje***
T: +386 1 7860020; Fax: +386 1 7860025; URL: www.gro.sik.si
1962; Roža Kek
Children's libr
90 000 vols; 159 curr per; 19 e-journals; 6 govt docs; 155 music scores; 247 maps; 724 av-mat; 707 sound-rec; 266 digital data carriers
libr loan 24958

Mestna knjižnica in Čitalnica Idrija (Public Library and Reading Room), ul. Sv Barbare 4-5, 5280 ***Idrija***
T: +386 5 3734060; Fax: +386 5 3734065; E-mail: knjiznicaidr@idr.sik.si; URL: www.idr.sik.si
1962; Milanka Trušnovec
99 500 vols; 197 curr per; 73 music scores; 86 maps; 2 911 av-mat; 649 sound-rec; 108 digital data carriers
libr loan; ZBDS 24959

Mestna knjižnica Izola (Public Library of Izola), Osvobodilne fronte 15, 6310 ***Izola***
T: +386 5 6631284;
Fax: +386 5 6631281; E-mail: knjiznica.izola@guest.arnes.si; URL: www.izo.sik.si
1958; Marina Hrs
56 000 vols; 147 curr per; 73 music scores; 64 maps; 425 av-mat; 616 sound-rec; 105 digital data carriers; 23 toys, 10 posters
libr loan; Union of Associations of Slovene Librarians 24960

Občinska knjižnica Jesenice (Public Library Jesenice), Trg Toneta Čufarja 4, 4270 ***Jesenice***
T: +386 4 5834200; Fax: +386 4 5834210; E-mail: info@knjiznica-jesenice.si; URL: www.knjiznica-jesenice.si
1954; Cvetka Tropenauer Martinčič
117 010 vols; 2 581 curr per; 1 e-journals; 62 music scores; 338 maps; 48 av-mat; 1 448 sound-rec; 961 digital data carriers
libr loan 24961

Knjižnica Kočevje (Public Library Kocevje), Trg Zbora Odposlancev 26, 1330 ***Kočevje***
T: +386 1 8931321; Fax: +386 1 8951936; E-mail: cirila.pekica@koc.sik.si; URL: www.knjiznica-kocevje.si/
1951; Cirila Pekica
55 000 vols; 105 curr per; 3 171 av-mat; 400 CD
libr loan; DBS 24962

Osrednja knjižnica Srečka Vilharja (Srečko Vilhar Central Library), Trg Brolo 1, 6000 ***Koper***
T: +386 5 6632600; Fax: +386 5 6632615; E-mail: amalia@kp.sik.si; URL: www.kp.sik.si
1951; Ivan Marković
Rare bks
285 000 vols; 554 curr per; 85 mss; 4 incunabula; 250 diss/theses; 1 358 music scores; 709 maps; 500 av-mat; 1 349 sound-rec; 22 digital data carriers
libr loan; ZBDS 24963

Osrednja knjižnica občine Kranj (Central Library Kranj), Tavčarjeva ul 41, 4000 ***Kranj***
T: +386 4 2013550; Fax: +386 4 2013560; E-mail: okk@kr.sik.si; URL: www.kr.sik.si
1960; Anatol Štern
338 000 vols; 943 curr per; 870 av-mat; 437 sound-rec
libr loan; ZBDS 24964

Knjižnica Lendava – Könyvtár Lendva, Glavna ulica 12, 9220 ***Lendava***
T: +386 2 57425-80/-82/-83; Fax: +386 2 5742584; E-mail: info@knjiznica-lendava.si; URL: www.knjiznica-lendava.si
Zsuzsana Žoldoš
126 000 vols; 179 curr per 24965

Knjižnica Bežigrad (Bežigrad Public Library), Einspielerjeva 1, 1000 ***Ljubljana***
T: +386 1 2363800; Fax: +386 1 2363813; E-mail: info@lj.sik.si; URL: www.mklj.si
1956; Igor Andrin
4 branch libs
175 000 vols; 333 curr per; 3 e-journals; 5 diss/theses; 87 maps; 8 051 av-mat; 14 425 sound-rec; 1 255 digital data carriers; 1 643 toys
libr loan; INTAMEL 24966

Knjižnica Jožeta Mazovca (Public Library), Zaloška 61, 1000 ***Ljubljana***
T: +386 1 5484510; Fax: +386 1 5484524; E-mail: moste@mklj.si; URL: www.mklj.si/index.php/enote-moste
1956; Janez Lah
130 000 vols; 244 curr per; 280 music scores; 200 maps; 2 800 av-mat; 7 800 sound-rec
libr loan 24967

Knjižnica Otona Župančiča (Ljubljana Metropolitan Library), Kersnikova 2, 1000 ***Ljubljana***
T: +386 1 6001300; Fax: +386 1 6001332; E-mail: info@koz.si; URL: www.koz.si
1981; Damijana Hainz
Aarch of children's bks in Slovene language, debate, life adulteducation, comic books, drug addict; Children's libr, Employment information centre, Youth info centre, Open learning centre
321 500 vols; 772 curr per; 197 music scores; 857 maps; 93 microforms; 8 107 av-mat; 9 840 sound-rec; 994 digital data carriers; artotheque 231
libr loan; IFLA, IBBY, ZBDS, UNAL 24968

Knjižnica Prežihov Voranc (Prezihov Voranc Library), Tržaška 47a, 1000 ***Ljubljana***
T: +386 1 2443574; Fax: +386 1 2443581; URL: www.vic.sik.si
1956; Milena Pinter
250 000 media items
libr loan 24969

Knjižnica Šiška (Šiška Library), Trg komandanta Staneta 8, 1000 ***Ljubljana***
T: +386 1 5193842; E-mail: knjiznicasis@lj-siska.sik.si; URL: www.lj-siska.sik.si
1956; Andreja Mervar
104 200 vols 24970

Splošna knjižnica Ljutomer, Glavni trg 1, 9240 ***Ljutomer***
T: +386 2 5841236; Fax: +386 2 5811758; E-mail: knjiznicaljt@lju.sik.si; URL: www.knjiznica-ljutomer.si
Silva Kosi
70 000 vols 24971

Mariborska knjižnica (Maribor Public Library), Rotovški Trg 2, 2000 ***Maribor***
T: +386 2 2352100; Fax: +386 2 2352127; E-mail: info@mb.sik.si; URL: www.mb.sik.si
1949; Dragica Turjak
A special study collection PKS, containing books of Slovene juvenile fiction, quality foreign picture books and specialized literature treating juvenile literature
582 988 vols; 7 099 curr per; 18 e-journals; 1 538 music scores; 2 524 maps; 22 173 av-mat; 10 196 sound-rec; 2 201 digital data carriers; 4 380 toys
libr loan; IFLA, ZBDS 24972

Pokrajinkska in študijska knjižnica (Regional Library), Zvezna ulica 10, 9000 ***Murska Sobota***
T: +386 2 5308110; Fax: +386 2 5308130; E-mail: joze.vugrinec@ms.sik.si; URL: www.ms.sik.si
1956; Jože Vugrinec
145 000 vols; 824 curr per; 413 mss; 2 diss/theses; 357 maps; 108 av-mat 24973

Goriška knjižnica Franceta Bevka (France Bevk Public Library), Trg Edvarda Kardelja 4, 5000 ***Nova Gorica***
T: +386 5 3309100; Fax: +386 5 3021688; E-mail: goriska.knjiznica@ng.sik.si; URL: www.ng.sik.si
1949; Rajko Slokar
F. Bevk's mss coll
282 511 vols; 564 curr per; 236 mss; 2 307 music scores; 815 maps; 97 microforms; 716 av-mat; 6 900 sound-rec; 6 digital data carriers; 13 multimedia
libr loan; ZBDS, UNAL 24974

Knjižnica Mirana Jarca (Miran Jarc Public Library), Rozmanova ul 28, 8000 ***Novo Mesto***
T: +386 7 3934600; Fax: +386 7 3934601; E-mail: knjiznicanm@nm.sik.si; URL: www.nm.sik.si
1946; Claudia Jerina Mestnik
Special Bogo Komelj Coll
412 567 vols; 200 curr per; 11 508 mss; 5 incunabula; 66 diss/theses; 3 030 music scores; 2 189 maps; 12 microforms; 2 271 av-mat; 2 508 sound-rec; 393 digital data carriers
libr loan 24975

Knjižnica Franca Ksavra Meška Ormož, Kolodvorska 9, 2270 ***Ormož***
T: +386 2 7415580; Fax: +386 2 7414489; E-mail: knjiznicaorm@orm.sik.si; URL: www.orm.sik.si
1963; Cirila Gabron-Vuk
Home documentation dept
70 000 vols; 250 curr per; 15 e-journals; 6 mss; 154 music scores; 340 maps; 2 microforms; 1 000 av-mat; 3 000 sound-rec; 2 000 digital data carriers; 50 old postcards
libr loan; ZBDS 24976

Knjižnica Bena Zupančiča (Beno Župančič Public Library), Trg padlih borcev 5, 6230 ***Postojna***
T: +386 5 7265073; Fax: +386 5 7203446; E-mail: sikpos@po.sik.si; URL: www.po.sik.si
1947; Tatjana Gorwik-Baraga
100 000 vols; 250 curr per; 20 music scores; 10 maps; 2 000 av-mat; 300 sound-rec; 20 digital data carriers
libr loan; IFLA 24977

Knjižnica Potrča Ptuj, Prešernova 33-35, 2250 ***Ptuj***
T: +386 2 27714800; Fax: +386 2 27714848; E-mail: kip@ptu.sik.si; URL: www.knjiznica-ptuj.si
1958; Matjaž Neudauer
400 000 media items; 762 curr per 24978

Knjižnica Radlje ob Dravi (Radlje ob Dravi Library), Koroška cesta 61a, 2360 ***Radlje ob Dravi***
T: +386 2 8880404; Fax: +386 2 8880403; E-mail: sikrdl@r-dr.sik.si; URL: www.knjiznica-radlje.si
1950; Slavica Potnik
58 683 vols; 1 511 curr per; 157 music scores; 177 maps; 806 av-mat; 579 sound-rec; 129 digital data carriers 24979

Knjižnica A. T. Linharta (Linhart Library), Gorenjska c 27, 4240 ***Radovljica***
T: +386 4 5373900; Fax: +386 4 5315840; E-mail: info@rad.sik.si; URL: www.rad.sik.si
1960; Božena Kolman Finžgar
108 000 vols; 161 curr per; 1 560 av-mat; 1 540 sound-rec; 50 digital data carriers
Public Library Association of Slovenia 24980

Koroška osrednja knjižnica dr. Franc Sušnik (Dr. Franc Sušnik Central Carinthian Library), Na gradu 1, 2390 ***Ravne na Koroškem***
T: +386 2 8705421; Fax: +386 2 8705430; E-mail: knjiznica@rav.sik.si; URL: www.rav.sik.si
1949; Majda Kotnik-Verčko
Local coll, Children dep, Mobile libr, 9 branch libs
238 712 vols; 1 115 curr per; 1 326 mss; 239 diss/theses; 1 399 music scores; 965 maps; 11 microforms; 720 av-mat; 1 719 sound-rec; 230 digital data carriers; 4 592 photos, 154 art collection, 788 pamphlets
libr loan; SLA 24981

Knjižnica Rogaška Slatina, Celjska cesta 13, 3250 ***Rogaška Slatina***
T: +386 3 8185780; Fax: +386 3 8185784; URL: www.r-sl.sik.si
61 000 vols 24982

Kosovelova Knjižnica, ul. Mirka Pirca 1, 6210 ***Sežana***
T: +386 5 7310031; Fax: +386 5 7310032; E-mail: kosovelova.knjiznica@guest.arnes.si; URL: www.sez.sik.si
1948; Nadja Mislej-Božič
130 000 vols; 105 curr per; 250 av-mat; 100 sound-rec
libr loan 24983

'Ivan Tavcar' Public Library, Solska 6, 4220 ***Škofja Loka***
T: +386 4 5112500; Fax: +386 4 5112510; E-mail: sik.it@guest.arnes.si; URL: www.knjiznicaskofjaloka.org
1945; Marija Lebar
161 591 vols; 175 curr per; 10 music scores; 205 maps; 1 171 av-mat; 2 689 sound-rec; 244 digital data carriers
libr loan 24984

Knjižnica Ivana Tavčarja Škofja Loka ('Ivan Tavčar' Public Library), Šolska 6, 4220 ***Škofja Loka***
T: +386 4 5112500; Fax: +386 4 5112510; E-mail: sik.it@guest.arnes.si
1945; Marija Lebar
5 branch libs
143 400 vols; 170 curr per; 845 av-mat; 1 660 sound-rec; 33 digital data carriers
libr loan 24985

Knjižnica Cirila Kosmača (Ciril Kosmač Public Library Tolmin), Tumov drevored 6, 5220 ***Tolmin***
T: +386 5 3811538; E-mail: tolmin@tol.sik.si; URL: www.tol.sik.si
1955; Viljem Leban
84 000 vols; 146 000 curr per; 2 mss; 1 000 av-mat; 1 361 sound-rec; 297 digital data carriers
libr loan; Slovenian Library Association 24986

Knjižnica Pavla Golie Trebnje (Library of Pavel Golja Trebnje), Kidričeva ul 2, 8210 ***Trebnje***
T: +386 7 3482111; Fax: +386 7 3482112; E-mail: knjiznica.trebnje@tre.sik.si; URL: www.tre.sik.si
1962; Milena Bon
local studies coll (photo albums, digital coll, paintings, illustrations)
75 825 vols; 1 063 curr per; 38 e-journals; 140 music scores; 283 maps; 2 100 av-mat; 996 sound-rec; 310 digital data carriers
libr loan 24987

Knjižnica dr. Toneta Pretnarja, Balos 4, 4290 ***Tržič***
T: +386 4 5923883; Fax: +386 4 5923881; E-mail: siktrz@trz.sik.si; URL: www.dbl-drustvo.si/knjiznica-trzic
1961; Marija Maršič
Mobile Libr, Department Bistrica
57 092 vols; 126 curr per; 3 e-journals; 1 music scores; 73 maps; 464 av-mat; 1 372 sound-rec; 226 digital data carriers
libr loan; ZBDS 24988

Knjižnica Velenje (Public Library Velenje), Šaleška 21, 3320 ***Velenje***
T: +386 3 8982550; Fax: +386 3 8982569; E-mail: postabralcev@vel.sik.si; URL: www.knjiznica-velenje.si
1934; Vlado Vrbič
162 000 vols; 340 curr per; 8 e-journals; 10 mss; 4 govt docs; 455 music scores; 970 maps; 5 500 av-mat; 4 270 sound-rec; 900 digital data carriers
libr loan; ZBDS 24989

Medobčinska matična knjižnica, Aškerčebva 9a, 3310 ***Žalec***
T: +386 3 7121252
Irena Štusej
Local hist (700 vols), homeland (Žalec docs)
58 700 vols; 161 curr per; 133 maps; 757 av-mat
libr loan 24990

Solomon Islands

National Libraries

National Library of Solomon Islands, Watts St, ***Honiara***; P.O. Box G28, Honiara
T: +677 27412; Fax: +677 22042
1974; Margaret Talasasa
Solomon Islands coll, Central ref coll; School, special, provincial libr advisory sections
100 000 vols; 50 curr per; 100 diss/theses; 200 av-mat; 100 digital data carriers; 2 000 pamphlets
IFLA, COMLA, LA, LAA, FLA, PNGLA 24991

University Libraries, College Libraries

Solomon-Islands College of Higher Education, Library Service, P.O. Box R113, ***Honiara***
T: +677 30111 ext 208; Fax: +677 30390
2 500 vols 24992

University of the South Pacific – Solomon Islands Centre Library, P.O. Box 460, ***Honiara***
T: +677 21307; Fax: +677 21287; E-mail: treadaway_j@usp.solomon.usp.ac.fj
1970; Julian Treadaway
9 000 vols; 850 curr per; 100 govt docs; 30 maps; 150 av-mat; 40 sound-rec
libr loan 24993

Government Libraries

Parliamentary Library, Parliament House, P.O. Box G 19, ***Honiara***
T: +677 21751, 21752; Fax: +677 23866
1976; John Laugolo
800 vols; 9 curr per 24994

South Africa

National Libraries

National Library of South Africa (NLSA) – Cape Town Division, 5 Queen Victoria St, 8001 ***Cape Town***; P.O. Box 496, 8000 Cape Town
T: +27 21 424-6320; Fax: +27 21 424-4848; E-mail: info@nlsa.ac.za; URL: aleph.nlsa.ac.za/
1818; Mr P. E. Westra
Grey Coll (mediaeval mss, early printed bks, first eds), Dessinian Coll (16th to 18th c bks), Africana Coll, Fairbridge Coll (rare bks), Muir Mathematical Coll (periodicals); Iconographs dept, Mss dept
750 000 vols; 8 300 curr per; 1 134 mss; 50 incunabula; 5 000 diss/theses; 50 000 govt docs; 20 000 maps; 12 900 microforms; 20 digital data carriers; 96 000 photos and slides
libr loan; IFLA 24995

National Library of South Africa (NLSA) – Pretoria Division, 228 Proes St, Private Bag X990, 0001 ***Pretoria***
T: +27 12 4019700; Fax: +27 12 3255984; E-mail: info@nlsa.ac.za; URL: www.nlsa.ac.za
1887; Dr. P.J. Lor
South and Southern Africa, legal deposit, United States official publs, league of Nations publs, UN publs, GATT-General Agreement of Trade and Tariffs, World Bank publs, International Labour Organisations publs, foreign official publs; South African Book Exchange Centre (SABEC), Study Services Centre
787 000 vols
libr loan; IFLA, SALIS 24996

General Research Libraries

Cape Provincial Library Service, Hospital and Chiappini Sts, P.O. Box 2108, ***Cape Town*** 8000
T: +27 21 41091111; Fax: +27 21 4197541
1955; N.F. van der Merwe
7 000 000 vols; 172 curr per; 48 000 microforms; 193 000 sound-rec; 62 000 art prints
libr loan 24997

University Libraries, College Libraries

University of Fort Hare, Library, PB X1314, ***Alice*** 5700
T: +27 40 6022011; Fax: +27 40 6531634; E-mail: dmc@ufh.ac.za; URL: www.ufh.ac.za/library
1916; Yoli Soul
Howard Pim Libr of Africana, Nursing Sciences branch libr
165 000 vols; 900 curr per; 250 av-mat; 600 sound-rec; 13 digital data carriers
libr loan 24998

Peninsula Technikon, Library, P.O. Box 1906, ***Bellville*** 7535, Cape Province
T: +27 21 9596911
1967
88 000 vols
libr loan 24999

University of the Western Cape, Library, Modderdam Rd, ***Bellville*** 7535
T: +27 21 959-3901; E-mail: servicedesk@uwc.ac.za; URL: www.uwc.ac.za
1989; Ellen Tise
Africana Coll, Afrikaans LitColl
263 000 vols; 1 335 curr per; 173 000 microforms; 56 823 bd journals
libr loan 25000

Central University of Technology, Free State, Library and Information Centre, 1 Park Road, 5800 ***Bloemfontein*** 9300; PB X20539, Bloemfontein 9300
T: +27 51 507 3141; Fax: +27 51 507 3468; URL: www.cut.ac.za/
1981; Trudie Venter
64 894 vols; 98 curr per; 184 diss/theses; 18 437 govt docs; 16 337 music scores;

16 maps; 348 av-mat; 103 sound-rec; 203 digital data carriers; 2157 other items
libr loan   25001

University of the Free State, Universiteit van die Vrijstaat – Yunivesithi Ya Freistata, Library und Information Services, 205 Nelson Mandela Drive, Park West, *Bloemfontein* 9301; P.O. Box 339, Bloemfontein 9300
T: +27 51 4012988; Fax: +27 51 4482879; E-mail: cm.bib@ufs.ac.za; URL: www.uovs.ac.za/faculties/index.php?FCode=12&DCode=431
1904; Mr C.R. Namponya
Dreyer-Africana Coll
576000 vols; 3462 curr per; 10 mss; 10000 diss/theses; 4000 govt docs; 5500 music scores; 300 maps; 3000 microforms; 3000 av-mat; 7000 sound-rec; 35 digital data carriers
libr loan   25002

– Frik Scott Medical Library, Faculty of Medicine, P.O. Box 2318, Bloemfontein 9300
T: +27 51 4053006; Fax: +27 51 4473222
1977; Mrs Radilene le Grange
35000 vols; 800 curr per; 40000 mss; 250 av-mat; 20 digital data carriers
libr loan; MLA   25003

– Qwaqwa Campus Library, PB X13, Phuthaditjaba 9866
T: +27 58 7130148; Fax: +27 58 7130148; E-mail: kokcj@it.uovs.ac.za
1982; Mr C.J. Kok
Africana dept (19 miles)
47000 vols; 425 curr per; 121 diss/theses; 51 av-mat; 31 sound-rec; 7 digital data carriers
libr loan   25004

Cape Technikon, Library Services, Tennant St, P.O. Box 652, *Cape Town* 8000
T: +27 21 4603227; Fax: +27 21 4603699
1923; J.A. Coetzee
Africana
52100 vols; 1261 curr per; 161 diss/theses; 300 maps; 200 microforms; 2875 av-mat; 19 sound-rec; 16 digital data carriers
IATUL   25005

University of Cape Town, Main Library, c/o Chancellor Oppenheimer Library, Lover's Walk, Rondebosch, 7700 *Cape Town*; Private Bag X3, 7701 Cape Town
T: +27 21 6503134; Fax: +27 21 6502965; E-mail: libraries@uct.ac.za; URL: www.lib.uct.ac.za
1905; Gwenda Thomas
Rare Books (some with fore-edge paintings), Kipling Collection, African Studies collections (including pre-1925 Africana, local imprints, and grey literature), Bolus Collection (antiquarian botanical works), Medical History Collection, Van Zyl Collection (antiquarian legal works), Manuscripts & Archives (including original material relating to South African history with a strong focus on the Western Cape, the Bleek and Lloyd Collection, the papers of Jack and Ray Simons, C. Louis Leipoldt, Pauline Smith and Olive Schreiner, African language collections, collections of Victorian and late 20th century architects, archives of the Black Sash (Cape Western Region), records of the Jewish community in the Western Cape and manuscript music scores of South African composers); African Studies Library, Manuscripts and Archives Dept, Government Publications Dept, Bolus Herbarium Library, Brand van Zyl Law Library, Built Environment Library, Health Sciences Library, Hiddingh Hall Library (fine arts & drama), Institute of Child Health Library, Jewish Studies Library, WH Bell Music Library.
1217800 vols; 28306 curr per; 63166 e-journals; 70 e-books; 1510 mss; 14000 diss/theses; 31600 govt docs; 22600 music scores; 3000 maps; 1100 microforms; 8400 av-mat; 8100 sound-rec
libr loan; IFLA, ALA   25006

– African Studies Library, Rondebosch 7700; Private Bag X3, Rondebosch 7701
T: +27 21 6503107; Fax: +27 21 6897568; E-mail: asl@uctlib.uct.ac.za; URL: www.lib.uct.ac.za/asl/
Sandy Shell
McMillan local hist press clippings coll
65000 vols; 500 curr per; 5000 diss/theses; 5000 maps; 10000 microforms; 1000 av-mat; 4 digital data carriers   25007

– W. H. Bell Music Library, College of Music, Lower Campus, Rosebank 7700
T: +27 21 6502624; Fax: +27 21 6502627; URL: www.lib.uct.ac.za/music
1943; Julie Strauss
Music Africana Coll
10000 vols; 67 curr per; 1000 mss; 40000 music scores; 300 microforms; 11100 sound-rec
libr loan   25008

– Brand van Zyl Law Library, Kramer Law School Bldg, Rondebosch 7700; PB X3, Rondebosch 7701
T: +27 21 6502708/-9; Fax: +27 21 6897568; URL: www.lib.uct.ac.za/law/
1948; Latifar Omar
Brand van Zyl Coll of Roman and Roman-Dutch antiquarian legal works, Gilfillan coll
30000 vols; 450 curr per
libr loan   25009

– Built Environment Library (Architecture Library), Centlivers Bldg, PB, Rondebosch 7700
T: +27 21 6502370; Fax: +27 21 6503127; E-mail: arl-bel@uct.ac.za
1951; Blythe Edwins
41100 vols; 109 curr per; 1200 diss/theses; 270 maps; 50 microforms; 5200 av-mat; 9 sound-rec
libr loan; SALIS   25010

– Health Sciences Library, Medical School Campus, Anzio Rd, Observatory 7925; PB, Rondebosch 7701
T: +27 21 4066138; Fax: +27 21 4482579; E-mail: lib-health@uct.ac.za; URL: www.lib.uct.ac.za/medical/
1928; Eugénie Söhnge
Med Hist Coll, Med Africana Coll
101000 vols; 710 curr per; 50 mss; 800 diss/theses; 15 microforms; 40 av-mat; 61 digital data carriers
libr loan; MLA   25011

– Hiddingh Hall Library, Hiddingh Campus, Orange St, Cape Town 8001
T: +27 21 480-7135; Fax: +27 21 4242889; URL: www.lib.uct.ac.za/hiddingh
1931; Jennie Underwood
Fine art, history of art, drama, new media
25000 vols; 40 curr per; 120 diss/theses; 7 microforms; 550 av-mat; 500 sound-rec; 18 digital data carriers
LIASA   25012

– Science and Engineering Library, Jagger Level, Immelman Bld, Rondebosch 7700
T: +27 21 6503115; Fax: +27 21 6897568; E-mail: Fiona.Jones@uct.ac.za
Fiona Jones
66871 vols; 2100 curr per   25013

Durban University of Technology, *Durban*
– M.L. Sultan Campus, Faculty of Commerce, Photograhy, Language & Translation Practice, Library, 41-43 Centenary Rd, P.O. Box 1334, Durban 4000
T: +27 31 3085246
1946; B.M. Patel
40000 vols   25014

Technikon Natal, Library, Mansfield Rd, P.O. Box 953, *Durban* 4000
T: +27 31 2042522;
Fax: +27 31 2042367; E-mail: muller@umfolozi.intech.ac.za
1907; N. Muller
Technologies, Health sciences incl. alternative medicine
86538 vols; 1351 curr per; 320 diss/theses; 10 microforms; 21000 av-mat; 200 sound-rec; 150 digital data carriers; 1132 pamphlets
libr loan; ALA, LA, LIASA, IATUL   25015

University of KwaZulu-Natal – Howard College Campus, EG Malherbe Library / Main Library, King George V Ave, *Durban* 4041
T: +27 31 2602317; Fax: +27 31 2602051; E-mail: trotter@admin.und.ac.za; URL: library.ukzn.ac.za
1922
Powell Coll of early science, Coll of works by Bantu and Negroes
170000 vols; 4261 curr per; 2422 diss/theses; 9710 govt docs
libr loan   25016

– Barrie Biermann Architectural Library, Howard College, 7th level, Denis Shepstone Bldg, King George V Ave, Durban 4041
T: +27 31 2602716; Fax: +27 31 2602051; URL: library.ukzn.ac.za
1969; Ms T.D. Shah
35000 vols; 150 curr per; 610 diss/theses; 4000 drawings
libr loan   25017

– Eleanor Bonnar Music Library, Francis Stock Bldg near Main Entrance Gate, King Georg V Av, Durban 4041
T: +27 31 2603044; Fax: +27 31 2602051; URL: library.ukzn.ac.za
1971; H.E. Gale
Malcolm Hunter Jazz Coll, Eleanor Bonnar Coll, Wilf Lowe Coll
35000 vols; 106 curr per; 20 mss; 230 diss/theses; 16247 music scores; 100 microforms; 120 av-mat; 8890 sound-rec; 4 digital data carriers
IAML   25018

– Medical Library, Medical School, 719 Umbilo Rd, Durban 4001
T: +27 31 260426-0/-1; Fax: +27 31 2604426; URL: library.ukzn.ac.za
1951; N.A. Russell
52000 vols; 618 curr per   25019

– G.M.J. Sweeney Law Library, Howard College Bldg, King George V Ave, Durban 4041
T: +27 31 2602541; Fax: +27 31 2602051; URL: library.ukzn.ac.at
1972
37000 vols   25020

University of KwaZulu-Natal – Westville Campus, Main Library, PB X54004, *Durban* 4000
T: +27 31 2044111; Fax: +27 31 2044383; E-mail: autharj@ukzn.ac.za; URL: www.library.ukzn.ac.za
1961; Mr M.M. Moodley
Fine arts, art hist, Africana, Indiana; Music Libr, Dental Libr
160000 vols; 1600 curr per; 7400 govt docs; 1100 microforms; 380 av-mat; 2700 sound-rec
libr loan; ALA   25021

Rhodes University, Library, *Grahamstown*, P.O. Box 184, 6140 Grahamstown 6140
T: +27 46 6038436; Fax: +27 46 6037310; E-mail: library@ru.ac.za; URL: www.ru.ac.za/library
1904; Gwenda Thomas
400000 vols; 650 curr per; 30000 e-journals; 4000 diss/theses; 3000 govt docs; 10000 music scores; 1000 av-mat; 800 digital data carriers
libr loan; ALA, LIASA   25022

Johannesburg College of Education, Library, 27 St. Andrews Rd, Parktown, *Johannesburg* 2193
T: +27 11 6421417; Fax: +27 11 6436312
ca 1946; J.B. Crow
Children's bks
150000 vols; 300 curr per; 2000 av-mat
libr loan; SALIS   25023

University of Johannesburg – Auckland Park Kingsway Campus Library, P.O. Box 524, Auckland Park, *Johannesburg* 2006
T: +27 11 4892171; Fax: +27 11 4892191; E-mail: info@rau.ac.za; URL: general.uj.ac.za/library/lidi/ujlic/home.htm
2005
N.P. Van Wyk Louw-Coll (lit and literary criticism), Boyazoglu-Coll (French Africana), Hist of Afrikaner on Witwatersrand
494000 vols; 44 digital data carriers; 700 internat databases
libr loan; ALA, UKSG, SCONUL (IFLA)   25024

University of the Witwatersrand, Wartenweiler Library, 1 Jan Smuts Av, *Johannesburg*; PB X1, Wits 2050
T: +27 11 71719-13/-14;
Fax: +27 11 4031421; E-mail: kalushi.kalushi@wits.ac.za; URL: web.wits.ac.za/Library
1934; Felix Ubogu
Gubbins, Humphreys and Jeffreys Coll of Africana, Landau Coll of Hebraica and Judaica, Sir Ernest Oppenheimer Institute of Portuguese Coll
1069000 vols; 8786 curr per; 100 incunabula; 4000 diss/theses; 30800 maps
libr loan   25025

– Architecture Library, 1 Jan Smuts Av, John Moffat Bld, Johannesburg; PB X1, Wits 2050
T: +27 11 717-1978/-1977; Fax: +27 11 4031421; E-mail: janie.johnson@wits.ac.za; URL: web.wits.ac.za/Library
1957; Janie Johnson
21000 vols; 100 curr per; 1000 diss/theses; 10 digital data carriers
libr loan   25026

– Biological and Physical Sciences Library (Biophy), 1 Jan Smuts Av, Oppenheimer Life Sciences Bldg, Johannesburg; PB X1, Wits 2050
T: +27 11 717-1960/-1962; Fax: +27 11 717-1969; E-mail: biophy@library.wits.ac.za; URL: web.wits.ac.za/Library
1962; Stephen R. Mitchell
Beilstein coll (Organic chem), Herbarium, Paine coll (hist of physics at UK)
Education science
25000 vols; 660 curr per; 60000 journal vols
libr loan   25027

– Education Library, Harold Holmes Library, 27 St Andrews Rd, Education Campus, Parktown; PB X1, Wits 2050
T: +27 11 7173239; E-mail: mark.sandham@wits.ac.za
Coll of South African law; international law and foreign law coll; law of Commonwealth jurisdictions and the United States
libr loan   25028

– Law Library, Oliver Schreiner Law Bldg, West Campus, Yale Rd, Braamfontein, Johannesburg; PB X3, Wits 2050
T: +27 11 7178504;
Fax: +27 11 7178511; E-mail: Cornelia.Bothma@wits.ac.za; URL: web.wits.ac.za/Library
Coll of South African law; international law and foreign law coll; law of Commonwealth jurisdictions and the United States
500 curr per
libr loan   25029

– Wits Health Sciences Library (WHSL), Wits Medical School, 7 York Rd, Parktown, Johannesburg; Box 351, Wits 2050
T: +27 11 7172348; Fax: +27 11 6438617; E-mail: webmaster@wits.ac.za; URL: web.wits.ac.za/Library
1926; Glenda Myers
Baragwanath lib, Strijdom libr, Dental libr
150000 vols; 750 curr per; 1000 diss/theses; 1 digital data carriers
libr loan; MLA, EAHIL   25030

University of Zululand, Library, PB X1001, *Kwa-Dlangezwa* 3886
T: +27 35 902-6462/-6463;
Fax: +27 35 902-6451; E-mail: lvahad@pan.uzulu.ac.za; URL: www.uzulu.ac.za
1960; Ms L. Vahed
Uzulu Coll, Development Studies, English Lit in Africa
303000 vols; 1400 curr per
libr loan   25031

Ramaano Mbulaheni Media Centre, P.O. Box 42, *Levubu* 0929
T: +27 15 5830560; Fax: +27 15 583-0562/-0561
1989; T.D. Maraga
Magazines on economics and agriculture; Mathematics, science, biology, english, education, engeneering, technical drawing depts
8200 vols; 6 diss/theses; 4 govt docs;

15 maps; 2 microforms; 6 digital data carriers; sound recordings, AV mat, videos, audio tapes
LIASA 25032

University of Limpopo, Medunsa Campus Library, Medical University of Southern Africa, P.O. Box 156, ***Medunsa*** 0204
T: +27 12 524323; Fax: +27 12 5600098; E-mail: hjones@medunsa.ac.za; URL: www.medunsa.ac.za/other/library
1977
Arch of Medunsa mat, Archaic bks, Africana medical mat, Cooper Coll of Medical Hist
60 000 vols; 581 curr per; 456 diss/theses; 6 maps; 34 microforms; 1 400 av-mat; 109 sound-rec; 102 digital data carriers; 175 slide sets
libr loan; LIASA 25033

North-West University, Library, Corner of Albert Luthuli and University Dr, PB X2046, ***Mmabatho*** 2735
T: +27 140 389-2017/-2118; E-mail: Jonathan.Nyebeleza@nwu.ac.za; URL: www.uniwest.ac.za/library/index.html
1980; Dudu Nkosi
Bophuthatswana coll
90 000 vols; 766 curr per; 52 microforms; 280 av-mat; 22 sound-rec
libr loan 25034

Oudtshoorn Teachers' College, Library, Park Rd, ***Oudtshoorn*** 6620
T: +27 443 3784
1964; J.J.G. Burger
25 000 vols; 120 curr per
libr loan 25035

University of KwaZulu-Natal – Edgewood Campus, Edminson Library, 1 Marianhill Rd, ***Pinetown***; PB X03, Ashwood 3605
T: +27 31 2603467; Fax: +27 31 2603658; URL: library.ukzn.ac.za
1966; Frances Roberts
Children's lit
85 000 vols; 200 curr per; 50 diss/theses; 400 maps; 22 000 av-mat; 700 sound-rec; 46 digital data carriers; 10 000 illustrations, 500 teaching packs, 350 workcards
libr loan; LIASA 25036

Nelson Mandela Metropolitan University, Library & Information Services, PB X6058, ***Port Elizabeth*** 6000
T: +27 41 5042281;
Fax: +27 41 5042280; E-mail: helena.fourie@nmmu.ac.za; URL: www.nmmu.ac.za/library
1965; Robert Pearce
Roman-Dutch Law
332 866 vols; 1 144 curr per; 11 437 diss/theses; 39 072 govt docs; 9 184 music scores; 87 maps; 2 537 microforms; 2 311 av-mat; 3 003 sound-rec; 25 digital data carriers
libr loan; LIASA, SCECSAL 25037

Port Elizabeth Technikon, Library Services, PB X6011, ***Port Elizabeth*** 6000
T: +27 41 5043239; Fax: +27 41 532545; URL: www.petech.ac.za/library
1881; M. Eales
Chemistry; 4 branch libs
66 856 vols; 693 curr per; 494 diss/theses; 3 641 av-mat; 204 digital data carriers
libr loan; LIASA 25038

Potchefstroom College of Education, Library, 37 Borcherd St, ***Potchefstroom*** 2520
T: +27 148 25116
1946; H. Venter
Model school coll
97 000 vols; 274 curr per; 41 microforms; 3 527 av-mat; 2 767 sound-rec; 6 462 pictures and illustrations, 10 021 reviews
libr loan 25039

Gold Fields Technobib, Library Services Technicon Pretoria, PB X680, ***Pretoria*** 0001
T: +27 12 3185240;
Fax: +27 12 3185485; E-mail: marinus@libmain.techpta.ac.sa
1934; M. Swanepoel
Arts dept, Natural Sciences dept, Horticulture dept, Polymer dept
59 000 vols; 1 200 curr per; 600 diss/theses; 1 000 music scores; 1 500 av-mat; 1 000 sound-rec; 10 digital data carriers
libr loan; IATUL 25040

University of Pretoria, Department of Library Services, Lynnwood Rd, Hatfield, P.O. Box 12411, Hatfield, ***Pretoria*** 0028
T: +27 12 420223-5/-6; Fax: +27 12 3625100; URL: www.up.ac.za
1908; Prof. E.D. Gerryts
Africana Coll
1 049 100 vols; 5 413 curr per; 26 000 music scores; 10 250 sound-rec
libr loan; IATUL 25041

– Music Library, Department of Library Sciences, Lynnwood Rd, Hatfield, P.O. Box 12411, Pretoria 0028
T: +27 12 4202317;
Fax: +27 12 3625175; E-mail: isobel.vandenwalt@up.ac.za
12 500 music scores; 6 500 sound-rec
libr loan 25042

– Veterinary Science Library, Academic Information Centre, Sir Arnold Theiler Bldg, PB X04, Onderstepoort, Pretoria 0110
T: +27 12 529800-7/-8/-9;
Fax: +27 12 5298302; E-mail: erica.vanderwesthuizen@up.ac.za
1974; Erica van der Westhuizen
12 700 vols; 320 curr per; 150 diss/theses; 800 av-mat
libr loan 25043

University of South Africa, Unisa Library, Muckleneuk Ridge, P.O. Box 392, ***Pretoria*** 0003
T: +27 12 4293111; Fax: +27 12 4293221; E-mail: Mbambtb@unisa.ac.za; URL: www.unisa.ac.za
1946; Dr. Buhle Mbambo-Thata
J.L. van Schaik Africana Coll, Cathedral Libs, German Africana Coll, Joubert Coll on Roman Dutch Law, Wagener Africana Coll, United Party Arch, Unisa Arch, Unisa Doc Ctr for African Studies, Verloren van Themaat Ctr for Internat Law
1 500 000 vols; 7 000 curr per; 750 mss; 8 841 diss/theses; 17 087 music scores; 4 169 maps; 241 448 microforms; 52 000 av-mat; 37 119 sound-rec; 1 520 digital data carriers; 381 diskettes, 49 realia
libr loan; IFLA, ALA, ALIA 25044

Baptist Theological College of Southern Africa, C.M. Doke Library, 260 Oak Av, ***Randburg***; P.O. Box 50710, Randburg 2125
T: +27 11 8860421; Fax: +27 11 8860453; URL: www.btc.co.za
1951; Anne Parker
C.M. Doke Coll, Baptist Union Arch
17 600 vols; 60 curr per; 190 diss/theses; 20 av-mat; 800 sound-rec
CILIP 25045

University of KwaZulu-Natal – Pietermaritzburg Campus, Cecil Renaud (Main) Library, King Edward Rd, PB X01, ***Scottsville*** 3209
T: +27 33 2605258; Fax: +27 33 2605260; URL: www.ukzn.ac.za
1912; Pravesh Sukram
St. Lawrence Coll Nataliana, Alan Paton Coll; Life Sciences Libr, Law Libr, Alan Paton Ctr, Univ Arch
440 000 vols; 2 597 curr per; 13 000 diss/theses; 734 maps; 6 110 microforms; 2 600 av-mat; 2 700 sound-rec; pamphlets
libr loan 25046

Helderberg College, Pieter Wessels Library, P.O. Box 22, ***Somerset West*** 7129
T: +27 21 8507558; Fax: +27 21 8507558; E-mail: wyoung@hbc.ac.za; URL: www.hbc.ac.za
1892; W.E. Young
Theology coll, Bible coll
70 000 vols; 220 curr per; 85 diss/theses; 529 maps; 12 microforms; 400 av-mat; 1 984 sound-rec; 99 digital data carriers; Eucational materials centre
libr loan; LIASA, ASDAL 25047

University of Limpopo – Turfloop Campus, Library, PB X1106, ***Sovenga*** LImpopo, 0727
T: +27 15 2689111; Fax: +27 15 2670152; E-mail: dorism@ul.ac.za; URL: www.unorth.ac.za/Library/
1959; Mrs M.M. Chuene
Africana, Govt docs; Law Dept, Management Sciences Dept, Theology Dept
170 000 vols; 2 700 curr per; 8 670 govt docs; 1 405 microforms; 1 404 av-mat; 50 CD-ROM titles
libr loan; IFLA 25048

Stellenbosch University, Library and Information Services, JS Gericke Library, JS Marais Square, Cnr Victoria and Ryneveld Sts, ***Stellenbosch***; PB X5036, Stellenbosch 7599
T: +27 21 808-4880; Fax: +27 21 808-4336; E-mail: etise@sun.ac.za; URL: library.sun.ac.za/
1895; Ellen Tise
Roman Dutch Law, Africana, D.F. Malan Coll; Mss dept, Rare Books dept
916 422 vols; 5 572 curr per; 6 800 govt docs; 39 044 music scores; 1 931 maps; 11 134 microforms; 3 194 av-mat; 14 751 sound-rec; 332 mss colls, 243 manuscript coll
libr loan; ALA, IFLA, LA, IATUL, ASLIB 25049

– Bellville Park Campus Information Centre, Carl Crontjé Drive, P.O. Box 610, Bellville 7535
T: +27 21 918427-0/-1/-2/-3; Fax: +27 21 9184113; E-mail: usbi@usb.sun.ac.za; URL: library.sun.ac.za/usbi/default.htm
Henriette Swart 25050

– Engineering and Forestry Library, Faculty of Engineering, PB X5036, Stellenbosch 7599
T: +27 21 8084978; Fax: +27 21 8082211; E-mail: ing_bib@sun.ac.za; URL: www.sun.ac.za//eng/engineering_library/default.html
1976; Corinna Truter
South African Bureau of Standards Publs, Forestry Pamphlet Coll (80 000)
75 000 vols; 580 curr per; 42 000 mss; 460 microforms; 350 av-mat; 13 sound-rec; 100 digital data carriers; 95 000 pamphlets, 4 000 standards 25051

– Health Sciences Library, Francie van Zyl Drive, Parow Valley, clinical bldg, 3rd fl, P.O. Box 19091, Tygerberg 7505
T: +27 21 9389368; Fax: +27 21 9337693; E-mail: healthinfo@sun.ac.za; URL: library.sun.ac.z//eng/tygerberg_library/default_tyg.html
1958; Linda Coetzee
32 000 vols; 700 curr per; 18 604 mss; 400 diss/theses; 50 microforms; 1 000 av-mat; 731 sound-rec; 31 digital data carriers
libr loan 25052

– Theology Library, Faculty of Theology, 171 Dorp St, PB X5036, Stellenbosch 7600
T: +27 21 8083508; Fax: +27 21 8084336; E-mail: bg1@sun.ac.za
1962; Beulah Gericke
Church Historical coll
51 924 vols; 263 curr per; 613 microforms; 250 av-mat; 33 sound-rec; 3 digital data carriers 25053

University of Venda, Library Services, PB X5050, ***Thohoyandou***, Limpopo Province 0950
T: +27 15 9628000; Fax: +27 15 9624749; E-mail: tshif@univen.ac.za; URL: www.univen.ac.za
Ms Mushoni Mulandzi 25054

University of Transkei (UNITRA), Library, PB X1, ***Umtata*** 5117
T: +27 471 3022501; Fax: +27 471 3022309; E-mail: nhlapo@getafix.utr.ac.za
1976; Mrs P.E. Ofori
Africana & Staff Research coll
45 000 vols; 992 curr per; 300 diss/theses; 24 microforms; 5 av-mat
libr loan; IFLA 25055

Vaal University of Technology, Library, PB X021, ***Vanderbijlpark*** 1900
T: +27 16 9509000; Fax: +27 16 9501203; URL: www.vut.ac.za
1966; Robert Pierce
45 000 vols; 454 curr per; 127 diss/theses; 133 maps; 501 av-mat; 42 digital data carriers; clippings, pamphlets etc (1 104 files), in-house index of articles (7 000 articles)
libr loan; LIASA 25056

Cape Peninsula University of Technology, Wellington Campus, M.J.L. Olivier Library, College St, ***Wellington*** 7655; PB X9, Wellington 7654
T: +27 21 8645200; E-mail: mollm@cput.ac.za; URL: www.cput.ac.za/
1963; Michiel Moll
Education, Agriculture, Business Informatics; Africana coll (3 500 vols)
75 000 vols; 60 curr per; 2 000 music scores; 160 maps; 2 000 av-mat; 1 500 sound-rec; 30 digital data carriers; pamphlets and clippings (200 files)
libr loan; LIASA 25057

Government Libraries

Supreme Court, Bloemfontein-Judges' Library, Fontein St, PB X20612, ***Bloemfontein*** 9300
T: +27 51 4478837; Fax: +27 51 4307041
Ms C.P. Oberholster
Old Authority coll
18 870 vols; 38 curr per; 5 diss/theses; 60 govt docs; 4 digital data carriers 25058

Supreme Court of Appeal, Library, Cnr Elizabeth St and President Brand St, P.O. Box 258, ***Bloemfontein*** 9300
T: +27 51 4472631; Fax: +27 51 4478098
1929; A.M. Street
Kotze Coll
44 898 vols; 44 curr per; 258 diss/theses; 159 govt docs; 2 digital data carriers 25059

High Court, Cape Town, Library, 35 Keerom St, ***Cape Town*** 8001; PB X9020, Cape Town 8000
T: +27 21 4802426; Fax: +27 21 230412
1840; Tanya Blake
Roman law
30 000 vols; 30 curr per; 2 digital data carriers
OSALL 25060

Library of Parliament, Parliament Bldg, Parliament St, P.O. Box 18, ***Cape Town*** 8000
T: +27 21 4032140; Fax: +27 21 4614331
1857; G. Swanepoel
Mendelssohn Coll of Africana, press cutting service, parliamentary and official docs, United Nations docs
350 000 vols; 2 700 curr per; 12 digital data carriers
libr loan; IFLA 25061

Department of Agriculture and Environmental Affairs – Free State Province, Library, Agricultural College, PB X01, ***Glen*** 9360
T: +27 5214 2051 ext 223;
Fax: +27 5214 2207; E-mail: hayesmm@glen1.agric.za
1919
8 000 vols; pamphlets
libr loan 25062

Supreme Court, Grahamstown, Library, 104 High St, PB X1011, ***Grahamstown*** 6140
1864; Harris
12 000 vols 25063

High Court, Witwatersrand Local Division, Library, Pritchard St, PB X7, ***Johannesburg*** 2000
T: +27 11 3328196; Fax: +27 11 3375162
1903; S. van Staden
41 000 vols; 70 curr per; 81 diss/theses; 12 govt docs 25064

Magistrate's Offices, Krause Library, Corner Fox and West Sts, PB X1, ***Johannesburg*** 2000
T: +27 11 8388346 ext 212
1939; M. Döman
9 000 vols; 18 curr per 25065

Metro Health Resource Centre, Library, Metropolitan Centre, P.O. Box 1477, ***Johannesburg*** 2000
T: +27 11 4077153; Fax: +27 11 4031069; E-mail: bmhlongo@mj.org.za
1971; B.Y. Mhlongo
16 000 vols; 70 curr per; 160 govt docs; 3 maps; 350 av-mat
libr loan 25066

Supreme Court, Kimberley, Library, Civic Ctr, Jan Smuts Bd, PB X5043, **Kimberley** 8300
T: +27 531 31441
1871
10 000 vols 25067

Natal Education Department, Library, 230 Prince Alfred St, PB X9055, **Pietermaritzburg** 3200
T: +27 33 51375
1902; J.L. Barker
25 000 vols; 196 curr per; 55 govt docs; 120 maps; 600 av-mat; 120 sound-rec
libr loan 25068

Supreme Court, Natal Provincial Division, Judges' Library, Commercial Rd, PB X9014, **Pietermaritzburg** 3200
T: +27 33 51835
1875; M.S. Verwey
11 000 vols; 175 curr per; 7 diss/theses; 645 govt docs 25069

North-West Agricultural Development Institute, Library, PB X804, **Potchefstroom** 2520
T: +27 18 299-6693/-6695; Fax: +27 18 2977135
1909; H.M.M. Harman
Grain Crops Research Inst, Agricultural Training Dept
11 600 vols; 660 curr per; 300 diss/theses; 1 000 govt docs; 30 maps; 2 microforms
libr loan 25070

Atomic Energy Corporation of South Africa Ltd, Library, Prelindaba, P.O. Box 582, **Pretoria** 0001
T: +27 12 3165211; Fax: +27 12 3165709
1959; C.N. van der Merwe
63 100 vols; 1 071 curr per; 60 000 microforms
libr loan 25071

Department of Agriculture, Forestry and Fisheries, Agriculture Place, 20 Beatrix Street, Arcadia, Pretoria, **Pretoria** 0002; Private Bag X388, Pretoria 0001
T: +27 12 3196830; Fax: +27 12 3197245; E-mail: daleenk@daff.gov.za; URL: www.daff.gov.za
1910; Daleen Koen
Africana, South African agricultural theses
100 000 vols; 100 curr per; 2 000 diss/theses; 165 000 govt docs; 800 microforms; 200 av-mat; 500 000 other items
libr loan 25072

Department of Arts, Culture, Science and Technology, Library, Schoeman St, Room 7133, Oranje-Nassau Bldg, PB X894, **Pretoria** 0001
T: +27 12 3378033; Fax: +27 12 3232720
1910; D. Mohlakwana
71 departmental publications
53 000 vols; 375 curr per; 42 000 mss; 21 diss/theses; 6 000 govt docs; 30 sound-rec
libr loan; IFLA, SLIS 25073

Department of Environmental Affairs and Tourism, Library, Federated Forum, 315 Pretorius St, PB X447, **Pretoria** 0001
T: +27 12 3103403; Fax: +27 12 3222682
1932; K.C. Prinsloo
Environmental matters, tourism
352 vols; 158 curr per; 4 034 mss; 44 diss/theses; 2 maps; 3 digital data carriers; 615 pamphlets 25074

Department of Finance, A. F. Corbett Library, Cnr Vermeulen and Andries Sts, African Eagle Bldg, P.O. Box 402, **Pretoria** 0001
T: +27 12 3155298; Fax: +27 12 3256006
ca 1918; L. van Rooyen
8 616 vols; 155 curr per; 16 500 mss; 35 diss/theses; 4 350 govt docs; 3 digital data carriers; law rept
libr loan 25075

Department of Foreign Affairs, Library, PB X152, **Pretoria** 0001
T: +27 12 351-1617/-1618/-1635; Fax: +27 12 3511651
Magazines in Black languages
9 000 vols; 182 curr per
libr loan 25076

Department of Health, Health Information Centre, Corner of Struben and Andries Sts, PB X828, **Pretoria** 0001
T: +27 12 3120000; Fax: +27 12 3264395
1919; Siva Ghetty
WHO; GELNET, Health Learning Mat
17 000 vols; 275 curr per; 89 diss/theses; 4 963 govt docs; 5 digital data carriers
libr loan 25077

Department of Justice, Library, Cnr Paul Kruger and Pretorius Sts, Presidia Bldg, PB X81, **Pretoria** 0001
T: +27 12 3151137; Fax: +27 12 3253034
S.C.W. Marais
Roberts Coll of Old Authorities
42 000 vols; 100 curr per
libr loan 25078

Department of Water Affairs and Forestry, Library, 173 Schoeman St, PB X313, **Pretoria** 0001
T: +27 12 299-2551/-2552; Fax: +27 12 3261780
1964; Mrs N. Bredekamp
14 220 vols; 218 curr per; 1 000 govt docs
libr loan 25079

South African Defence Force, Central Library, Potgieter St, PB X289, **Pretoria** 0001
T: +27 12 3551213; Fax: +27 12 3235613
1946; M.L. Fouché
24 000 vols; 50 curr per; 10 000 pamphlets
libr loan 25080

South African Police, Head Office, Library, Wachthis Ext., 231 Pretorius St, **Pretoria** 0002; PB X94, Pretoria 0001
T: +27 12 3101404; Fax: +27 12 3220266
1913
146 248 vols; 1 975 sound-rec; 315 audio books 25081

Supreme Court, Transvaal Provincial Division, Library, Vermeulen St, PB X67, **Pretoria** 0001
T: +27 12 3157564; Fax: +27 12 3157569
1902; H. Schutte
Wessels Coll
24 500 vols; 35 curr per; 2 500 mss; 1 320 govt docs; 3 digital data carriers
libr loan 25082

Ecclesiastical Libraries

St. Joseph's Theological Institute, Denis E. Hurley Library, Cedara, PB X6004, **Hilton** 3245
T: +27 33 3433293; Fax: +27 33 3431232; E-mail: stjoseph@nu.ac.za
1943; Hilary Russell
30 000 vols; 110 curr per
libr loan 25083

Lutheran Theological Seminary, Library, 5 km west of junction to Mapumulo on hwy R74 – 1 km south, PB X9206, **Mapumulo** 4470
1963; Marie K. Nelson
Missiological Institute, Evangelical Theology, African Theology, African independent churches; Pastoral Care and Counseling dept
15 500 vols; 46 curr per; 400 diss/theses; 933 microforms; 5 av-mat; 350 sound-rec; 200 pamphlets, 150 clippings
libr loan 25084

Dutch Reformed Church Archives – Pretoria, Transvaal and Central Africa, Library, 234 Visagie St, **Pretoria**, 0002; P.O. Box 433, Pretoria 0001
T: +27 12 3228900
Dr. C.S. Kotzé
10 000 vols 25085

St. John Vianney Seminary, Library, 179 Main St, **Waterkloof** 0181; P.O. Box 17128, Groenkloof 0027
T: +27 12 4602039; Fax: +27 12 4603596
1948; Bonaventure Hinwood
130 000 vols; 110 curr per; 1 incunabula; 50 sound-rec 25086

Dutch Reformed Theological School Stofberg, Library, PB 812, **Witsieshoek** 9870
T: +27 14382 139
1960; W. de W. van Velden
20 000 vols; 70 curr per; 50 pamphlets 25087

Corporate, Business Libraries

Ninham Shand Library, 81 Church St, **Cape Town** 8000; P.O. Box 1347, Cape Town 8001
T: +27 21 4245544; Fax: +27 21 4245588; E-mail: library@shands.co.za
ca 1968; Irma Liberty
Civil engineering consultants; Urban, regional and rural planning; Environmental consultants
10 000 vols; 60 curr per; 100 govt docs; 70 digital data carriers; 3 600 in-house technical rpts
libr loan; LIASA, SLIG 25088

South African Mutual Life Assurance Society, Information Resource Centre and Library, P.O. Box 5180, **Cape Town** 8000
T: +27 21 5093480; Fax: +27 21 5094676; URL: www.oldmutual.com
1845; Janet Lloyd
10 000 vols; 80 curr per 25089

Funda Centre, Library, 8642 Ramolongoane St, Zone 6, **Diepkloof**; P.O. Box 2056, Southdale 2135
T: +27 11 9381485; Fax: +27 11 9387439
1984; T.J.E. Twala
14 000 vols; 25 curr per; 10 maps; 30 av-mat
libr loan 25090

Kentron Library, Nellmapius Av, **Irene**, 1675; P.O. Box 7412, Hennopsmeer 0046
T: +27 12 6711143; Fax: +27 12 6711407
J.S. Kritzinger
Electronics
15 000 vols; 180 curr per; 300 diss/theses; 3 500 govt docs; 3 500 microforms; 1 000 av-mat; 1 000 trade pubs, 4 000 specifications 25091

Standard Bank of South Africa, Economic Research Division, Library, 9th Fl, 5 Simmonds St, P.O. Box 7725, **Johannesburg** 2001
T: +27 11 6369111; Fax: +27 11 6365617; URL: www.standardbank.co.za
1960; Pat Nichols
11 000 vols; 400 curr per
libr loan; SALA 25092

Simera, Technical Information Services Library, Central Library, Atlas Rd, P.O. Box 117, **Kempton Park** 1620
T: +27 11 9272270; Fax: +27 11 3951103
1967; M. Burger
International aviation legislative docs, Jane's All the World Aircraft
8 000 vols; 230 curr per; 100 diss/theses; 30 000 microforms; 150 digital data carriers
libr loan 25093

Anglo American Corporation of South Africa Ltd, Library, 45 Main St, P.O. Box 61587, **Marshalltown** 2107
T: +27 11 6385240; Fax: +27 11 632362
1948; Norma Roberts
Mining, earth sciences, law, business
35 515 vols; 1 545 curr per; 80 maps; 1 320 pamphlets
libr loan 25094

Steffen Robertson and Kirsten, Consulting Engineers, Library, P.O. Box 55291, **Northlands** 2116
T: +27 11 4411248; Fax: +27 11 8808086; E-mail: jhblibrary@srk.co.za
1977; S. King
10 000 vols; 70 curr per; 15 000 mss; 100 diss/theses; 1 000 govt docs; 3 000 pamphlets, slides
libr loan 25095

Hulett Aluminium Limited, Library, Edendale Rd, **Pietermaritzburg**, 3201; P.O. Box 74, Pietermaritzburg 3200
T: +27 33 3956127; Fax: +27 33 3956491; E-mail: Library@hulamin.co.za
1948; Elise Leibnitz
13 500 vols; 90 curr per 25096

CSIR, Information Services, Library, P.O. Box 395, **Pretoria** 0001
T: +27 12 8413278; Fax: +27 12 8414405
1947; E.E.S. Rooks
Chemical engineering, explosives technology, plastics technology
13 000 vols; 700 curr per; 1 000 maps; 30 digital data carriers; 25 000 pamphlets on microfiche, 600 periodical titles on microfilm
libr loan 25097

South African Iron and Steel Industrial Corporation Ltd, ISCOR Information Service, Library, Roger Dyason Rd, P.O. Box 450, **Pretoria** 0001
T: +27 12 2981111
1934; H.S. le Roux
British Standards, Patent Specifications
140 000 vols; 12 000 curr per; 15 000 standards, 65 000 patents
libr loan; ASLIB 25098

South African Reserve Bank, Library, 370 Church St, **Pretoria** 0002; P.O. Box 427, Pretoria 0001
T: +27 12 313-3911; Fax: +27 12 313-3197; E-mail: elna.franklin@gwise.resbank.co.za; URL: www.reservebank.co.za
1958; Elna Franklin
Annual Rpts of Central and commercial banks, annual rpts of govt depts, IMF working papers, IMF occasional papers, economic development
26 000 vols; 427 curr per; 180 diss/theses; 4 500 govt docs; 6 digital data carriers
libr loan 25099

Knight Piésold, Library, T.C. Watermeyer Centre, Cnr Rivonia Blvd and 10th Av, P.O. Box 221, **Rivonia** 2128
T: +27 11 8067028; Fax: +27 11 8067100
1974; J. Epstein
Water resources, hydrology, dams
8 000 vols; 40 curr per; 5 training videotapes
libr loan; SLIS 25100

Industrial Development Corporation of South Africa Ltd, Library and Information Service, 19 Fredman Dr, **Sandton**, 2199; P.O. Box 784055, Sandton 2146
T: +27 11 2693217; Fax: +27 11 2693116; URL: www.idc.co.za
1940; Janice M. Shipway
12 000 vols; 500 curr per; 200 govt docs; 5 digital data carriers
libr loan 25101

Sasol Technology Library, Process Department, Klasie Havenga Av, P.O. Box 1, **Sasolburg** 9570
T: +27 16 7082338; Fax: +27 16 7082341
1952
Petrochemicals, engineering; Research dept
50 000 vols 25102

Sasol Library, Synfuels Rd, PB X1034, **Secunda** 2302
T: +27 17 6192275; Fax: +27 17 6192495; E-mail: marietjie.marais@sasol.com; URL: www.sasol.com
1983; Mrs M.J. Marais
Petrochemicals, Petroleum, Chemicals
30 000 vols; 731 curr per; 300 diss/theses; 250 govt docs; 60 maps; 3 000 microforms; 2 000 av-mat; 49 sound-rec; 65 digital data carriers
libr loan; IFLA, LIASA 25103

Special Libraries Maintained by Other Institutions

CSIR – Mining Technology, Information Centre, Library, Kew Rd and Landau

Terrace, Richmond, P.O. Box 91230, **Auckland Park** 2006
T: +27 11 35840204; Fax: +27 11 4821214; E-mail: mvandeve@csir.co.za
1951; M.J. van Deventer
10 000 vols; 189 curr per; 400 mss; 450 microforms; 150 av-mat; 2 800 pamphlets
libr loan; SLIS   25104

EDULIS – Western Cape Education Library and Information Services, 15 Kruskal Av, **Bellville** 7530; PB X9099, Cape Town 8000
T: +27 21 9487504;
Fax: +27 21 9480748; E-mail: lmetcalf@pawc.wcape.gov.za
1859; L. Metcalfe
School libr, study and teaching
100 000 vols; 150 curr per; 21 000 av-mat
libr loan; LIASA, IASL   25105

FS Education Library, Free State Education Department, 31 Deale Rd, Bayswater, **Bloemfontein** 9301
T: +27 51 4365395; Fax: +27 51 4365395
E. Mostert
Education, philosophy, Psychology, Communication
13 109 vols; 68 curr per; 180 diss/theses; 47 maps; 1 microforms; 400 av-mat; 34 sound-rec
libr loan; LIASA   25106

Nasionale Afrikaanse Letterkundige Museum en Navorsingsentrum, Biblioteek, President Brand St, Old Government Bldg, PB X20543, **Bloemfontein** 9300
T: +27 51 4054011; Fax: +27 51 4054259
1973; M.M. van der Walt
Nienaber Afrikaans Lit coll, Van Schaik Mss coll, Lategan Afrikaans Lit coll, F.CL. Bosman Drama coll, Koos Prinsloo coll
100 000 vols; 160 curr per; 250 000 mss; 500 diss/theses; 5 000 music scores; 60 microforms; 120 av-mat; 2 630 sound-rec; 50 digital data carriers   25107

National Drama Library, Bloemfontein City Libraries, President Brand St, City Hall Bldg Bloemfontein, **Bloemfontein** 9300; P.O. Box 1029, Bloemfontein 9300
T: +27 51 4058254; Fax: +27 51 405 8604
1933; Laurette van Huyssteen
65 000 vols; 20 av-mat; 60 sound-rec
                          25108

National Museum Library, 36 Aliwal St, P.O. Box 266, **Bloemfontein** 9301
T: +27 51 4479609; Fax: +27 51 4476273; E-mail: ina@nasmus.co.za; URL: www.nasmus.co.za/
1954; Ina Marais
Natural hist
8 700 vols; 2 200 curr per; 60 diss/theses; 490 maps; 40 sound-rec
libr loan   25109

National Centre for Occupational Health, A.J. Orenstein Library for Industrial Medicine, P.O. Box 32492, **Braamfontein** 2017
T: +27 11 7241844
1963; S.K. Jugdan
Reprints on asbestos
8 000 vols; 200 curr per; 3 000 microforms
libr loan   25110

Iziko: Museums of Cape Town, Social History Library, 11-17 Church Sq, Parliament St, **Cape Town** 8001; P.O. Box 645, Cape Town 8000
T: +27 21 4618280; Fax: +27 21 4619592; URL: www.museums.org.za
1966; A.M. Greyling
Maritime hist coll
12 000 vols; 123 curr per; 109 maps; 257 av-mat; 34 sound-rec; 1 000 pamphlets, 200 glass negatives, 2 000 photos, 2 000 archival docs   25111

Jacob Gitlin Library, 88 Hatfield Street, **Cape Town** 8001
T: +27 21 5625088; Fax: +27 21 4658670; E-mail: gitlib@netactive.co.za
1959; Devis Iosifzon
Coll of Yiddish books (1 395 vols)
25 500 vols; 30 curr per; 854 av-mat; 258 sound-rec; 54 digital data carriers; 319 pamphlet boxes filled with documented archive material   25112

Marine and Coastal Management, Department of Environmental Affairs and Tourism, Gilchrist Library, 420 Foretrust House, Martin Hammerslag Rd, **Cape Town**; PB X2, Rogge Bay 8012
T: +27 21 4023249, 4023250; Fax: +27 21 4258635; URL: www.mcm-deat.gov.za
1930; B. Wessels
Oceanographic expedition rpt, Rare monographs coll
12 000 vols; 950 curr per; 210 maps; 600 microforms; 25 digital data carriers; 10 000 pamphlets
Ocean Data and Information Network for Africam (ODINAFRICA), IAMSLIC   25113

National Archives, Cape Town Archives Repository, Library, 72 Roeland St, **Cape Town** 8001; PB X9025, Cape Town 8000
T: +27 21 4624050; Fax: +27 21 4652960
1876; L. du Plessis
Jeffreys coll of Africana, local and regional hist
22 482 vols; 31 curr per; 180 diss/theses   25114

National Monuments Council, J.J. Oberholster Library, 111 Harrington St, **Cape Town**, 8001; P.O. Box 4637, Cape Town 8000
T: +27 21 4624502; Fax: +27 21 4624509
1978; K.J. Ayres
Conservation in built environment, town planning, shipwrecks, marine archeology, South African hist
10 000 vols; 90 curr per   25115

South African Museum, Library, 26 Queen Victoria St, P.O. Box 61, **Cape Town** 8000
T: +27 21 4243330; Fax: +27 21 4246716
1855; R.J. Krynauw
Oceanographic expedition reports
17 740 vols; 850 curr per; 100 diss/theses; 3 070 maps; 40 sound-rec; 8 digital data carriers; 43 370 pamphlets
libr loan; LIASA   25116

South African National Gallery, Library, P.O. Box 61, **Cape Town** 8000
T: +27 21 4674677; Fax: +27 21 4674679; E-mail: sjcasoojee@iziko.org.za; URL: www.iziko.org.za/sang/index.html
1955; Shaheeda Dante
Press Cuttings – Fine arts, mss
12 000 vols; 180 curr per; 700 mss; 250 diss/theses; 35 microforms; 35 000 av-mat; 17 digital data carriers; 3 000 pamphlets, 60 000 clippings
IFLA   25117

National Botanical Institute, Harry Molteno Library, PB X7, **Claremont** 7735
T: +27 21 7998712; Fax: +27 21 7620646
1913; Yvonne Reynolds
Botany, horticulture, ecology, conservation, environmental education
8 000 vols; 350 curr per; 30 microforms; 5 000 pamphlets
libr loan   25118

Durban Museum and Art Gallery, Library, Smith St, City Hall, P.O. Box 4085, **Durban** 4000
T: +27 31 3112264/9; Fax: +27 31 3112242; E-mail: strettonj@durban.gov.za; URL: www.durban.gov.za
ca 1900; B.E.A. Eisenhauer
Ornithology Dept, Mammalogy Dept, Entomology Dept, Fine Art, African Art
15 000 vols; 240 curr per; 110 diss/theses; 300 maps; 500 av-mat; 100 sound-rec; 20 digital data carriers; Reprints, pamphlets, photos
libr loan   25119

M. K. Gandhi Library, Bai Jerbhai Trust Bldg, Queen St, **Durban** 4001
T: +27 31 3056748
1940; R.J. Rustomjee
10 000 vols   25120

Oceanographic Research Institute, Library, P.O. Box 10712, Marine Parade, **Durban** 4056
T: +27 31 3373536; Fax: +27 31 3372132; E-mail: seaworld@dbn.lia.net; URL: www.ori.org.za
1961; A.B.D. Kleu
P.H. Boshoff coll

10 000 vols; 354 curr per; 100 diss/theses; 100 govt docs; 200 maps; 2 microforms; 4 100 av-mat; 1 digital data carriers; 12 000 reprints
libr loan; IAMSLIC, LIASA   25121

Sugar Milling Research Institute, Library, University of KwaZulu-Natal, Howard College Campus, **Durban** 4041
T: +27 31 2731300; Fax: +27 31 2731302; E-mail: mkort@smri.org; URL: www.smri.org
1949; M.J. Kort
Archief voor de Java/Nederlandsch-Indie Suikerindustrie
10 500 vols; 52 curr per; 20 diss/theses
libr loan   25122

East London Museum, Library, 319 Oxford St, **East London** 5201; P.O. Box 11021, Southernwood 5213
T: +27 43 7430686; Fax: +27 43 7433127
ca 1950; N.N.
The Papers of Sir Charles Crewe
10 000 vols; 100 curr per; 6 diss/theses; 150 maps; 3 000 av-mat; 15 000 photos   25123

Albany Museum, Library, Somerset St, **Grahamstown** 6140
1855
R.F. Lawrence Arachnology Coll, T.B. Bowker Africana Coll, Hist Documents Coll, Mary Pocock Botanical Coll
13 000 vols; 1 400 periodicals   25124

National English Literary Museum, PB 1019, **Grahamstown** 6140
T: +27 46 6227042; Fax: +27 46 6222582; E-mail: M.Hacksley@ru.ac.za; URL: www.rhodes.ac.za/nelm
1974; D.S. Landman
South African Engl lit, bks, mss; Education and Outreach Research Dept
17 500 vols; 150 curr per; 27 928 mss; 550 diss/theses; 50 microforms; 83 av-mat; 550 sound-rec; mss (361 linear m), multiple files under 5 504 headings (131,6 linear m), press clippings (44 m)
LIASA, ALA   25125

South African Library for the Blind, P.O. Box 115, **Grahamstown** 6140
T: +27 46 6227226; Fax: +27 46 6224645; E-mail: niojs@blindlib.org.za; URL: www.blindlib.org.za   25126

Brenthurst Library, Federation Rd, Parktown, 2193, P.O. Box 87184, **Houghton** 2041
T: +27 11 6466024; Fax: +27 11 4861651
1900s; M.D. Weiner
11 100 vols; 500 mss; 500 colls of art works
SALIS, SLIS   25127

South African Jewish Board of Deputies, Library, 2 Elray Str, Raedene, **Houghton** 2193; P.O. Box 87557, Houghton 2041
T: +27 11 6452500; Fax: +27 11 6452559; E-mail: sajbod@iafrica.com; URL: os2.iafrica.com/sajbod
1950; Prof. R. Musiker
SA Jewish hist
25 000 vols; 342 curr per; 50 diss/theses; 1 digital data carriers   25128

Animal Improvement Institute, Library, PB X2, **Irene** 1675
T: +27 12 6729111; Fax: +27 12 6651609
9 000 vols   25129

ARC Irene Campus Information Centre, Animal Nutrition and Animal Products Institute, Animal Improvement Institute, Olifantsfontein Rd, **Irene** 0062; PB X2, Irene 0062
T: +27 12 6719111; Fax: +27 12 6651605; URL: www.arc.agric.za
A.H.M. Labuschagne
Basil Ryder Coll, Int Dairy Federation publs
8 700 vols; 140 curr per
libr loan; IAALD   25130

Adler Museum of the History of Medicine, Faculty of Health Sciences, University of the Witwatersrand, Library, 7 York Rd, Parktown, **Johannesburg** 2193
T: +27 11 7172081; Fax: +27 11 7172081; URL: web.wits.ac.za/Academic/

Health/Entities/AdlerMuseum/
1962; Dr. L. Immelman
Coll of objects relating to SA medical history
10 000 vols
libr loan   25131

Bowman Gilfillan Hayman Godfrey Inc, Library, 28 Harrison St, P.O. Box 2439, **Johannesburg** 2000
T: +27 11 8362811; Fax: +27 11 8366909
E.M. Bourne
Taxation law
9 000 vols; 150 curr per
libr loan   25132

Electricity Supply Commission, Library, Megawatt Park, Maxwell Drive Sunninghill Ext. 3 Sandton, P.O. Box 1091, **Johannesburg** 2000
T: +27 11 8002403
1959; Hogben
45 000 vols; 1 200 curr per; 500 microforms
libr loan   25133

Jan Smuts House, SA Institute of International Affairs, Library, Jan Smuts Av, **Johannesburg**, 2001; P.O. Box 31596, Braamfontein 2017
T: +27 11 3392021;
Fax: +27 11 3392154; E-mail: 056jacky@witsvmb.wits.ac.za
1934; J.A. Kalley
22 000 vols; 600 curr per; 1 digital data carriers
libr loan   25134

Rand Club, Library, 33 Loveday St, **Johannesburg**, 2001; P.O. Box 1032, Johannesburg 2000
T: +27 11 8348311; Fax: +27 11 8341247
J. Brand
14 000 vols   25135

Rand Water Board Library, 522 Impala Rd, Glenvista, **Johannesburg** 2091; P.O. Box 1127, Johannesburg 2000
T: +27 11 6820473; Fax: +27 11 6820444
1958; L. Parker
9 000 vols; 200 curr per
libr loan   25136

Society of Advocates of South Africa, Johannesburg Bar Library, 1st Fl, Innes Chambers, Pritchard St, **Johannesburg** 2001
T: +27 11 3339471; Fax: +27 11 3330063
L. Davis
30 000 vols; 200 curr per; 50 diss/theses; 10 000 microforms; 5 digital data carriers; 200 pamphlets
libr loan; OSALL   25137

South African Brain Research Institute, 6 Campbell St, Waverly, **Johannesburg** 2192
T: +27 11 7862912; Fax: +27 11 7861766; E-mail: mag@africa.com
1981
10 000 vols   25138

South African Broadcasting Corporation, Reference Library, Henley Rd, Auckland Park, **Johannesburg** 2092; PB X3, Auckland Park 2006
T: +27 11 7142817; Fax: +27 11 7143106
1961; M.J. Strauss
30 000 vols; 550 curr per
libr loan   25139

South African Institute for Medical Research (SAIMR), Spenser Lister Memorial Library, De Korte St Hospital Hill, P.O. Box 1038, **Johannesburg** 2000
T: +27 11 4899000; Fax: +27 11 4899001
1916; L.D. Battaglia
15 000 vols; 520 curr per
libr loan; SALIS   25140

South African Institute of Chartered Accountants, Library, 7 Zulberg Close, Bruma Lake, **Johannesburg**, P.O. Box 59875, Kengray 2100
T: +27 11 6226655; Fax: +27 11 6223321
ca 1920; B. Massyn
8 200 vols; 67 curr per; 4 digital data carriers
libr loan   25141

South African Institute of International Affairs, Library, Jan Smuts House, University of the Witwatersrand, Braamfontein, *Johannesburg* 2017; P.O. Box 31596, Braamfontein, Johannesburg 2017
T: +27 11 3392021; Fax: +27 11 3392154; E-mail: jlibrary@saiia.org.za; URL: www.saiia.org.za
1934; A. M. Ramohlola
International Organizations Collection (INTORG)
15 000 vols; 300 curr per; 500 digital data carriers
libr loan
25142

Amathole Museum, Library, Cnr Albert Rd and Aôlexandra Rd, P.O. Box 1434, *King William's Town* 5600
T: +27 433 6424506; Fax: +27 433 6421569; E-mail: lloyd@huberta.ru.ac.za
Xhosa ethnography, local hist, Eastern Cape missionary hist, Eastern Cape German settler hist, Southern Africa mammalogy; Kitton Coll of Africana
8 200 vols; 150 curr per
25143

Kaffrarian Museum, Library, Cnr Albert Rd and Alexandra Rd, P.O. Box 1434, *King William's Town* 5600
T: +27 43 6424506; Fax: +27 43 6421569
1940s; N.N.
Mammalogy, Hist of Eastern Cape, Social Anthropology (Xhosas), Missionary Hist of Eastern Cape, Museology, Africana (2 153 vols, called Kitton Coll)
10 600 vols; 125 curr per; 8 microforms; 50 av-mat; 3 800 pamphlets, 3 newspaper titles
libr loan
25144

Chamber of Mines of South Africa, Library, 5 Holland St, P.O. Box 61809, *Marshalltown* 2107
T: +27 11 4987100; Fax: +27 11 8341884; URL: www.bullion.org.za
1891; Manil Kanniappen
Africana Coll
20 000 vols; 500 curr per; 100 maps; South African govt docs, govt gazettes
SLIS
25145

Grootfontein Agricultural Development Institute, Library, PB X529, *Middelburg* 5900
T: +27 483 21113; Fax: +27 483 24352; E-mail: bib1@karoo1.agric.za
1911; L. King
Farming in arid/semi-arid regions
9 000 vols; 279 curr per; 5 095 mss; 250 diss/theses; 1 640 govt docs
libr loan
25146

South African Sugar Association Experiment Station, Library, PB X02, *Mount Edgecombe*, Kwa Zulu Natal 4300
T: +27 31 5393205; Fax: +27 31 5395406
1925; P. Moodley
Slides coll
11 500 vols; 115 curr per; 35 diss/theses; 1 govt docs; 1 digital data carriers
25147

South African Astronomical Observatory Library, Observatory Rd, P.O. Box 9, *Observatory* 7935
T: +27 21 4470025; Fax: +27 21 4473639; E-mail: library@saao.ac.za; URL: www.saao.ac.za
1826; S. Davis
Astrophysics
20 000 vols; 300 curr per; 700 microforms; 500 av-mat; 200 digital data carriers; 2 000 astronomical photogr plates, 1 000 transparencies
25148

Onderstepoort Veterinary Institute, Agricultural Research Council, Library, Arnold Theiler Av, PB X05, *Onderstepoort* 0110
T: +27 12 5299279; Fax: +27 12 5299282; E-mail: library@moon.ovi.ac.za
1908; David A. Swanepoel
OP-Cana
51 500 vols; 130 curr per; 250 mss; 1 500 diss/theses; 200 govt docs; 1 maps; 17 microforms; 5 av-mat; 8 digital data carriers
libr loan; IAALD
25149

Natal Museum, Library, 237 Loop St, PB 9070, *Pietermaritzburg* 3200
T: +27 33 3451404; Fax: +27 33 3450561; E-mail: library@nmsa.org.za; URL: www.nmsa.org.za
1904; S.J. Hearne Perrin
Boswell Coll, Ogilvie Coll, Zumpt Coll, R.F. Lawrence Coll, Murray Coll
12 000 vols; 980 curr per; 1 895 mss; 890 maps; 250 microforms; 15 814 av-mat; 60 500 pamphlets, 18 500 press clippings
libr loan
25150

National Archives of South Africa, Pietermaritzburg Archives Repository, Library, 231 Pietermaritz St, PB X9012, *Pietermaritzburg* 3200
T: +27 33 3424712; Fax: +27 33 3944353
1935; U.P. Narrine
Natal Parliament Libr
32 600 vols; 18 curr per; 600 mss; 143 diss/theses; 32 600 govt docs; 1 643 maps; 226 microforms; 10 100 photos
25151

Old Mutual, Management Library, Mutual Park, Jan Smuts Dr, *Pinelands* 7405; P.O. Box 5180, Cape Town 8000
T: +27 21 5093480; Fax: +27 21 5094444
1950; L. Smith
Arch
20 000 vols; 500 curr per
libr loan
25152

Africa Institute of South Africa, Library, Nipilar House, Cor Hamilton & Vermeulen St, Arcadia, *Pretoria* 0083; P.O. Box 630, Pretoria 0001
T: +27 12 3286970; Fax: +27 12 3238153; E-mail: ai@ai.org.za; URL: www.ai.org.za
1960; Amanda Wortmann
Greenfield Coll, Pierre Botha Coll; Dept of Monitor and Current Affairs (files), Cartography (maps)
66 000 vols; 460 curr per; 3 000 maps; videos, photos, newspaper, clippings (1970-1980, 1981-2000)
libr loan; LIASA
25153

ARC-Institute for Soil, Climate and Water, Agricultural Research Council, Library, 600 Belvedere St, PB X79, *Pretoria* 0001
T: +27 12 3102500; Fax: +27 12 3231157; URL: www.arc.agric.za
1917; R. van Dyk
Soil colour charts
12 000 vols; 35 curr per; 7 038 mss; 300 diss/theses; 383 av-mat; 10 digital data carriers
libr loan
25154

ARC-Plant Protection Research Institute, Library, Roodeplaat, *Pretoria* South; PB X293, Pretoria 0001
T: +27 12 8080952; Fax: +27 12 8080321
1977; J.C. Combrink
Africana, mainly entomology, microbiology, herbicides, pesticides, weed research; Mycology, apiculture
17 500 vols; 145 curr per
25155

ARC-Roodeplaat Vegetable and Ornamental Plant Institute Library, Agricultural Research Council, Moloto/KwaMhlanga Rd, *Pretoria* 0001; PB X293, Pretoria 0001
T: +27 12 8419611; Fax: +27 12 8080844; E-mail: EJoubert@arc.agric.za; URL: www.arc.agric.za/
1963; Estia Joubert
Horticulture, plant diseases, vegetables; Publ by company research; Rpts on overseas trips by company research
19 000 vols; 54 curr per; 165 diss/theses; 26 govt docs; 5 maps; 2 digital data carriers
libr loan
25156

Blindiana Library, South African National Council for the Blind, 514 White St. Bailey's Muckleneuk, *Pretoria* 0181
T: +27 12 452-3811; Fax: +27 12 346-1177; E-mail: admin@sancb.org.za; URL: www.sancb.org.za
1976; Lilla Fourie
11 000 vols; 40 diss/theses; 20 av-mat; 10 sound-rec
libr loan
25157

Central Library Services, Meiring Naudé Rd, Scientia, P.O. Box 395, *Pretoria* 0001
T: +27 12 8413358; Fax: +27 12 3491154
1945; Margaret Lodder
110 000 vols; 2 500 curr per; 30 digital data carriers
libr loan; LA, SLA, ASIS, ALA, Aslib, EUSIDIC
25158

Council for Geoscience, Geological Survey of South Africa, Library, 280 Pretoria Rd, Silverton, PB X112, *Pretoria* 0001
T: +27 12 8411001; Fax: +27 12 8411076; URL: www.geoscience.org.za
1895; L. Niebuhr
Mineralogy, geochemistry, engineering, palaeontology, stratigraphy, geopyhsics, geomorphology of South Africa; Map libr
103 000 vols; 1 400 curr per; 525 diss/theses; 150 000 govt docs; 20 000 maps; 294 microforms; 3 digital data carriers
libr loan
25159

Council for Scientific and Industrial Research (CSIR), Central Library Services, P.O. Box 395, *Pretoria* 0001
T: +27 12 8412911; Fax: +27 12 3491154; E-mail: mlodder@csir.co.za
110 000 vols; 5 000 curr per; 22 000 pamphlets
25160

Gauteng Department of Education, LIS Education Library, 328 Van der Walt St, PB X290, *Pretoria* 0001
T: +27 12 3227685; Fax: +27 12 3227699
before 1900; E. Keller
Van Heerden Coll (old school bks), Textbook coll
240 000 vols; 300 curr per; 1 000 music scores; 66 maps; 10 000 av-mat; 2 600 sound-rec; 55 digital data carriers; 8 000 posters
ALA, LIASA
25161

Human Sciences Research Council (HSRC), Centre for Library and Information Services, 134 Pretorius St, PB 41, *Pretoria* 0001
T: +27 12 3022999; Fax: +27 12 3022933; E-mail: library@ludwig.hsrc.ac.za; URL: www.hsrc.ac.za
1969; Wanda Wessels, J.J. Knoetze
Africana (3 111 vols)
80 000 vols; 900 curr per; 2 000 diss/theses; 3 000 govt docs; 20 digital data carriers
LIASA
25162

National Archives Repository, Library, 24 Hamilton St, PB X236, *Pretoria* 0001
T: +27 12 3235300; Fax: +27 12 3235287
1919; Ms J.A. Pretorius
Jeffrey Coll, Ploeger Coll
15 000 vols; 116 curr per; 362 diss/theses; 7 500 pamphlets
25163

National Botanical Institute, Mary Gunn Library, 2 Cussonia Av, Brumeria, *Pretoria* 0002; PB X 101, Pretoria 0001
T: +27 12 8043200; Fax: +27 12 8043211
1912; E. Potgieter
rare valuable botanica, africana bks
13 210 vols; 314 curr per; 5 991 pamphlets
libr loan
25164

National Cultural History Museum Library, Afrika Window Bldg, 149 Visagie St, 0001 *Pretoria*; P.O. Box 28088, 0132 Sunnyside, Gauteng
T: +27 12 3246082; Fax: +27 12 3285173; E-mail: library@nchm.co.za
1969; Carin Bisschoff
E.G. Jansen Africana Coll
17 500 vols; 100 curr per; 5 000 mss; 120 diss/theses; 50 govt docs; 200 music scores; 300 maps; 250 av-mat; 1 000 pamphlets, 1 500 newspaper clippings
libr loan; LIASA
25165

Performing Arts Council Transvaal, Library, 320 Pretorius St, P.O. Box 566, *Pretoria* 0001
T: +27 12 285812
Peter Terry
André Huguenot coll
12 000 vols; 3 curr per; 2 000 mss; 300 sound-rec; 1 000 photos
libr loan
25166

Plan Medewerkers, Biblioteek, 373 Pretorius St, *Pretoria*, 0002; P.O. Box 1889, Pretoria 0001
T: +27 12 3203320; Fax: +27 12 3203324; E-mail: planassoc@icon.co.za
1964; Ms J. Joubert
Town and regional planning, town planning schemes; surveying, IDP, urban and rural planning, architecture
25 500 vols
libr loan
25167

Plant Protection Research Institute, Library, PB X134, *Pretoria* 0001
T: +27 12 8080952; Fax: +27 12 8081489
9 000 vols
25168

Society of Advocates of RSA CTPA, Oscar Galgut Library, Momentum Centre East Tower, 343 Pretorius St, *Pretoria* 0001
T: +27 12 3221511; Fax: +27 12 3221535
Rosalie Cloete
10 000 vols
OSALL
25169

South African Bureau of Standards (SABS), Library, 1 Dr Lategan Rd, Groenkloof, PB X191, *Pretoria* 0001
T: +27 12 4286633; Fax: +27 12 14286066
1946; Theresa Vermaak
Official journals of standardization
20 000 vols; 350 curr per; 16 000 pamphlets
libr loan
25170

Transvaal Education Department, Education Media Services, 328 Van der Walt St, PB X290, *Pretoria* 0001
T: +27 12 3227685; Fax: +27 12 3227699
J.A. Bierman
385 000 vols; 600 curr per; 30 000 av-mat
25171

Transvaal Museum, Library, Paul Krüger St, P.O. Box 413, *Pretoria* 0001
T: +27 12 3227632; Fax: +27 12 3227939
1892; Dinah van Driel
Africana Coll pertaining to natural science
20 000 vols; 800 curr per; 80 000 pamphlets
libr loan
25172

MINTEK Information Centre, 200 Hans Strijdom Dr, PB X3015, *Randburg* 2125
T: +27 11 7094259; Fax: +27 11 7934122; URL: www.mintek.co.za
1937; Ms M. Kanniappen
Metallurgy, minerals processing
10 200 vols; 386 curr per; 957 diss/theses; 208 microforms; 50 av-mat; 42 digital data carriers; 16 000 pamphlets; 5 660 patents; 7 100 news clippings; 19 000 Mintek rpts
libr loan; ASLIB
25173

Electoral Institute of Southern Africa, EISA library, 14 Park Rd, *Richmond Johannesburg*; P.O. Box 740, Auckland Park 2006
T: +27 11 381 6000; Fax: +27 11 482 6163; E-mail: jackie@eisa.org.za; URL: www.eisa.org.za
1995; Jackie Kalley
Elections, Democracy
12 000 vols; 50 curr per
25174

Royal Society of South Africa, Library, University of Cape Town, *Rondebosch* 7700; P.O. Box 594, Cape Town 8000
T: +27 21 6502543; Fax: +27 21 6502710; E-mail: royalsociety@uct.ac.za; URL: www.uct.ac.za/org/rssa
1877; Prof. J.R.E. Lutjeharms
33 000 vols
25175

Eskom Information Centre, Library, Maxwell Dr, Sunninghill Park, *Sandton*; P.O. Box 1091, Johannesburg 2000
T: +27 11 8002503; Fax: +27 11 8003903; E-mail: Sue.Cook@eskom.co.za; URL: www.eskom.co.za
1923; M.H. Hogben
42 631 vols; 800 curr per; 3 500 microforms; 1 046 av-mat; 142 sound-rec; 35 digital data carriers; 26 language courses
libr loan
25176

South African National Museum of Military History, Library, Erlswold Way, P.O. Box 52090, *Saxonwold* 2132
T: +27 11 6465513; Fax: +27 11 6465256; E-mail: milmus@icon.co.za
1966; B.L. de Lange
15 120 vols; 148 curr per; 1 950 maps; 80 000 av-mat; 256 sound-rec
libr loan; LIASA   25177

Institute for Maritime Technology, IMT Library, Martello Rd, *Simon's Town* 7995; P.O. Box 181, Simon's Town 7995
T: +27 21 7861092; Fax: +27 21 7863634; E-mail: nvr@imt.co.za; URL: www.imt.co.za
1976; Naomi van Rooyen
In-house reports
500 vols; 120 curr per; 8 311 mss; 4 030 diss/theses; 20 govt docs; 10 000 microforms; 150 sound-rec; 150 digital data carriers; environmental clippings
libr loan; LIASA National Film Library, SLIG   25178

San Reference Library, South African Navy, PB X11, *Simon's Town* 7995
T: +27 21 7873879; Fax: +27 21 7873282
1975; J.M. Calitz
9 300 vols; 95 curr per; 1 digital data carriers
libr loan   25179

Stevenson-Hamilton Memorial Library, Kruger National Park, P.O. Box 50, *Skukuza* 1350
T: +27 13 7355611; Fax: +27 13 7355138
1961; I.E. Grobler
13 150 vols; 280 curr per; 200 diss/theses; 25 maps
libr loan   25180

ARC-Infruitec Centre for Fruit Technology, Agricultural Research Council, Library, PB X5013, *Stellenbosch* 7599
T: +27 21 8093311; Fax: +27 21 8093400
1937; Karlien Coertzen
8 500 vols; 80 curr per
libr loan   25181

Public Libraries

Alberton Public Library, Alwyn Taljaard Rd, P.O. Box 4, *Alberton* 1450
T: +27 11 9079751; Fax: +27 11 9079751
1943; K. Kitching
207 000 vols; 113 curr per; 3 532 music scores; 1 283 sound-rec; 2 digital data carriers
libr loan   25182

Amanzimtoti Public Library, P.O. Box 26, *Amanzimtoti* 4125
T: +27 31 9134557, 9134559; Fax: +27 31 9134574
1935; D. Scorgie-Robinson
52 000 vols; 50 curr per; 150 av-mat; 2 049 sound-rec; 151 art prints
libr loan; LIASA   25183

Beacon Bay Public Library, P.O. Box 2001, *Beacon Bay* 5205
T: +27 43 472400
140 000 vols
libr loan   25184

Bellville Public Library, Carel van Aswegen St, *Bellville* 7530
T: +27 21 9182300; Fax: +27 21 9489313
1964; Ilze Swart
Pamphlet clippings coll
167 650 vols; 120 curr per; 1 300 av-mat; 2 295 sound-rec; 512 art prints
libr loan   25185

Benoni Public Library, Corner Tom Jones St and Elston Av, *Benoni* 1501; PB X060, Benoni 1500
T: +27 11 7416000; Fax: +27 11 4219422
1923; Patience Maisela
Benoniana special coll
140 000 vols; 139 curr per; 33 maps; 17 740 microforms; 5 968 sound-rec; 280 art prints, 3 656 pamphlets, 5 656 clippings
libr loan   25186

Bloemfontein City Libraries, 43 West Burger St, P.O. Box 1029, *Bloemfontein* 9300
T: +27 51 4058248; Fax: +27 51 4058604
1875; P.J. van der Walt
405 000 vols; 1 800 curr per; 6 200 maps; 3 200 microforms; 25 000 pamphlets
libr loan   25187

Bloemfontein Regional Library / Bloemfonteinse Streekbiblioteek, PB X20606, *Bloemfontein* 9300
T: +27 51 71993
1950; L. Orffer
500 000 vols   25188

Library and Information Services Directorate, Free State Provincial Government, Elizabeth St, C.R. Swart Bldg, PB X20606, *Bloemfontein* 9300
T: +27 51 4054681; Fax: +27 51 4033567; E-mail: lerouxaw@sac.fs.gov.za
1948; J.J. Schimper
4 000 000 vols; 400 curr per; 500 av-mat; 500 sound-rec
libr loan; LIASA   25189

Mmulakgoro Public Library, Mophete St, Phahameng, *Bloemfontein*; P.O. Box 16082, Mangaung 9307
T: +27 51 324776
1976
70 108 vols   25190

Boksburg Public Library, Trichardts St, P.O. Box 210, *Boksburg* 1460
T: +27 11 8934190
1905
Africana
80 000 vols
libr loan   25191

Brakpan Transitional Local Council, Library, Corner Park St and Kingsway Av, P.O. Box 15, *Brakpan* 1540
T: +27 11 7412025; Fax: +27 11 7404084
1929; Amanda van Heerden
69 300 vols; 50 curr per; 2 govt docs; 10 maps; 2 microforms; 900 av-mat; 2 400 sound-rec; 1 digital data carriers; 5 240 clippings
libr loan   25192

Cape Town City Libraries, Old Drill Hall, Parade St, *Cape Town*; P.O. Box 4728, Cape Town 8000
T: +27 21 4624400; Fax: +27 21 4615981
1952; H.C.F. Heymann
Art libr, children's lit reference coll; Municipal reference libr
1 521 629 vols; 312 curr per; 28 267 sound-rec; 5 digital data carriers
libr loan; INTAMEL, IFLA, LIASA, LA, ALA   25193

Milnerton Public Library, Pienaar Rd, *Cape Town* 7435; P.O. Box 35, Milnerton 7435
T: +27 21 5501130;
Fax: +27 21 5501261; E-mail: Mariethae@Blaauwberg.co.za; URL: www.blaauwberg.com
1968; Mariétha Eyssen
105 000 vols; 80 curr per; 1 300 music scores; 800 av-mat; 3 500 sound-rec; 78 digital data carriers; 900 files of pamphlets
libr loan; LIASA   25194

Carletonville Public Library, Emerald St, P.O. Box 3, *Carletonville* 2500
T: +27 1491 72131; Fax: +27 1491 91105
1939; R. E. Myburgh
78 485 vols; 48 curr per; 100 av-mat; 1 799 sound-rec; 90 educational toys
libr loan   25195

Lebowa National Library, PB X03, *Chuenespoort* 0745
T: +27 15 37137; Fax: +27 15 37149
1980; E.M. Tladi, R.M. Phaahla
Coll of Lebowa govt publ
15 000 vols   25196

Thekwini Municipal Library, Smith St, City Hall, P.O. Box 917, *Durban* 4000, Kwazuli-Natal
T: +27 31 3094405, 3112202; Fax: +27 31 3096033, 3112203
1853; Ramesh Jayaram
Don Africana Coll, Hillier Shakespeare Coll

1 275 733 vols; 602 curr per; 475 govt docs; 12 000 music scores; 4 030 maps; 3 500 microforms; 6 000 av-mat; 22 158 sound-rec
libr loan; ALA   25197

Buffalo City Library, Gladstone and Buxton Sts, *East London*, 5201; P.O. Box 652, East London 5200
T: +27 43 7224991; Fax: +27 43 7431729
1876; M.M. Davidson
201 000 vols; 100 curr per; 1 000 govt docs; 200 music scores; 500 maps; 10 250 microforms; 3 500 sound-rec; 2 500 pamphlets, 1 600 photos
libr loan; LIASA   25198

Edenvale Community Library, Ricardo Mulder Library, Van Riebeeck Av, *Edenvale* 1610; P.O. Box 25, Edenvale 1610
T: +27 11 4560212; Fax: +27 11 4525162; URL: www.edenvale.za.net
1955; Stella Wilckens
Africana Coll, Educational Toy Libr
85 000 vols; 1 100 curr per; 13 000 av-mat; 3 000 sound-rec; 80 digital data carriers; 520 files of clippings
libr loan; LIASA   25199

George Public Library, Caledon St, P.O. Box 19, *George* 6530
T: +27 44 8019287; Fax: +27 44 8733776
1840; S.M. Marais
82 986 vols; 52 curr per; 500 av-mat; 654 sound-rec; 438 art prints
LIASA   25200

Germiston Community Library, Cor. Cross, Queen and Hospital Sts, P.O. Box 246, *Germiston*, 1401
T: +27 11 8717242; Fax: +27 11 8717867
1909; S.J.W. Worsley
Cuttings, Africana, Rebecca, Ostroviac; Community, info and study dept
364 810 vols; 75 curr per; 23 av-mat; 585 sound-rec; 7 digital data carriers; 837 music CD's
libr loan   25201

Goodwood Libraries, Voortrekker Rd, BOB 100, *Goodwood* 7460
T: +27 21 5901555; Fax: +27 21 5901565
1961; Rosa Boshoff
Africana; Edgemead/Monte Vista branch libr
85 000 vols; 70 curr per; 500 av-mat; 300 sound-rec; 30 digital data carriers; pamphlets
libr loan; LIASA   25202

Grahamstown Public Library, 45-47 Hill St, P.O. Box 180, *Grahamstown* 6140
T: +27 46 6036040; Fax: +27 46 6229488
1842; M.J. Hartzenberg
80 000 vols; 20 curr per; 300 av-mat; 5 260 sound-rec
LIASA   25203

South African Library for the Blind, P.O. Box 115, *Grahamstown* 6140
T: +27 46 6227226; Fax: +27 46 6224645; E-mail: niojs@blindlib.org.za; URL: www.blindlib.org.za   25204

Howick Public Library, Main St, P.O. Box 5, *Howick* 3290
T: +27 3321 2124; Fax: +27 3321 304183
1894; J.A. Wynn-Thomas
146 460 vols; 23 curr per
libr loan
Affilated to Kwa-Zulu National Provincial Libr. Book exchanges every three months   25205

Johannesburg Public Library, Main Library, Library Gardens, Corner Fraser & Market Sts, *Johannesburg*, Gauteng 2001
T: +27 11 8363787; Fax: +27 11 8366607; E-mail: library@mj.org.za
1890; Ms E.J. Bevan
African Studies, Art, Music, Local Government, Photography, Early Children's Bks Coll
1 500 000 vols; 2 462 curr per; 58 483 music scores; 14 448 maps; 73 digital data carriers
libr loan; LIASA   25206

Kempton Park Public Library, Cnr C.R. Swart Drive / Pretoria Rd, P.O. Box 13, *Kempton Park* 1620
T: +27 11 9212150; Fax: +27 11 9750921
1956
Hist of Kempton Park
212 000 vols; 62 curr per; 200 av-mat; 2 566 sound-rec; 160 digital data carriers; 592 pamphlets
libr loan   25207

Kimberley Public Library, Chapel St, P.O. Box 627, *Kimberley* 8300
T: +27 531 806241; Fax: +27 531 8331954
1882; Mr F.H. van Dyk
Africana Libr (20 000 vols), Judy Scott Libr (43 000 vols)
156 000 vols; 160 curr per; 590 mss; 25 diss/theses; 1 350 govt docs; 1 900 maps; 55 microforms; 350 av-mat; 2 500 sound-rec; 620 digital data carriers; 12 000 photos, 1 270 pamphlets, 20 100 clippings
libr loan; LIASA   25208

King William's Town Public Library, Ayliff St, P.O. Box 33, *King William's Town* 5600
T: +27 43 6423391; Fax: +27 43 6423677
1861; Mrs P.A.C. Patrick
Blindlib coll (44 tapes)
62 000 vols; 3 curr per; 10 music scores; 200 av-mat; 433 sound-rec; 1 govt gazette, 99 artprints, 100 clippings
libr loan; LIASA   25209

Klerksdorp Public Library, Paul Kruger St, P.O. Box 99, *Klerksdorp* 2570
T: +27 18 29997
1893
H.M. Guest Africana Coll, Tienie Meter Medical Coll
84 000 vols; 63 curr per
libr loan   25210

Kroonstad Public Library, Cor. Hilland Steyn Sts, P.O. Box 302, *Kroonstad* 9500
T: +27 562 69245
1905; Ela Joubert
89 290 vols; 40 curr per; 5 082 sound-rec
libr loan   25211

Krugersdorp Public Library, Von Brandis St, P.O. Box 28, *Krugersdorp* 1740
T: +27 11 9512153; Fax: +27 11 6651780
1904; J. du Plooy
Indian Libr, Africana Sections
188 703 vols; 107 curr per; 3 985 sound-rec; 53 digital data carriers; 292 art reproductions
libr loan   25212

Snow Cruywagen Public Library, Wanders Av, P.O. Box 14, *Midddelburg* 1050
T: +27 132 497314; Fax: +27 132 432550
1945; E. Horn
Mhluzi branch libr, Eastdene branch libr
64 969 vols; 49 curr per; 423 sound-rec; 132 digital data carriers; 100 clippings, 1 set of microforms
libr loan   25213

North West Provincial Library Services, PB X2044, *Mmabatho* 2735
T: +27 140 22061; Fax: +27 140 22063
1978; N.B. Nomnga
1 600 000 vols   25214

Nelspruit Public Library, 45 Louis Trichardt St, P.O. Box 45, *Nelspruit* 1200
T: +27 13 759-2076/-2077/-2082; Fax: +27 13 7523447
1933; C.J. Pretorius
40 curr per; 55 000 mss; 5 diss/theses; 10 govt docs; 20 maps; 1 500 sound-rec; 50 pamphlets
libr loan; LIASA   25215

Newcastle Public Library, Scott St, PB X6621, *Newcastle* 2940
T: +27 3431 27211; Fax: +27 3431 21570
1915; E.P. Niemand
54 900 vols; 49 curr per; 90 av-mat; 1 189 sound-rec; 50 art prints
libr loan   25216

Orkney Public Library, Patmore Rd, PB X8, *Orkney* 2620
T: +27 18 4731451; Fax: +27 18 4730310
1942; R. Knoetze
Pamphlets, Tourist Information
54 000 vols; 30 curr per; 800 sound-rec
25217

C. J. Langenhoven Memorial Library, Voortrekker Rd, P.O. Box 35, *Oudtshoorn* 6620
T: +27 44 2723141; E-mail: elbe@oudtmun.co.za
Ms E. I. Kruyt
60 000 vols; 61 curr per; 260 av-mat; 550 sound-rec
libr loan
25218

Parow Public Library, Voortrekker Rd, P.O. Box 111, *Parow* 7500
T: +27 21 9302926; Fax: +27 21 925742
1938; C.J. Molier
58 000 vols; 80 curr per; 200 av-mat; 1 000 sound-rec
libr loan
25219

Parys Public Library, Philipst, P.O. Box 359, *Parys* 9585
T: +27 568 2131 ext 150; Fax: +27 568 2131 ext 147
1890s; R. Kotzé
55 987 vols; 14 curr per; 1 243 sound-rec; 1 000 clippings, 150 pamphlets
25220

Phalaborwa Public Library, H.F. Verwoerd Av, P.O. Box 67, *Phalaborwa* 1390
T: +27 15 7810111; Fax: +27 15 7810726
1962; N.S. Mokgaboki
Govt publs coll; Career information desk
24 curr per; 5 diss/theses; 30 govt docs; 25 music scores; 6 maps; 24 192 microforms; 841 sound-rec
libr loan
25221

KwaZulu-Natal Provincial Library Service, Prince Alfred St, *Pietermaritzburg* 3201; PB X9016, Pietermaritzburg 3200
T: +27 33 940241; Fax: +27 33 942237
1952; Dr. Rookaya Bawa
2 900 000 vols; 4 500 curr per; 31 000 av-mat; 37 000 sound-rec
libr loan
25222

Pinetown Public Library, Old Main Rd, P.O. Box 49, *Pinetown* 3600
T: +27 31 7182731; Fax: +27 31 7182732, 7182731
1864; Mrs I. Moodley
111 124 vols; 100 curr per; 200 mss; 2 647 av-mat; 252 digital data carriers; 973 games, govnt gazettes
libr loan
25223

Westville Public Library, Inner West City Council, P.O. Box 49, *Pinetown* 3600
T: +27 31 2037050; Fax: +27 31 2037045
1952; Penny Thompson
Children's Libr, Reference Dept, Audio-Visual Dept
64 591 vols; 51 curr per; 30 maps; 574 av-mat; 1 637 sound-rec; 21 digital data carriers; 259 art prints, 331 puzzles
libr loan; LIASA
25224

Polokwane City Library, 71 Hans van Rensburg St, *Polokwane* 0699; P.O. Box 111, Polokwane 0700
T: +27 15 2902155;
Fax: +27 15 2902150; E-mail: libraryenq@polokwane.org.za
1904; Koot Jacobs
Local hist coll
121 688 vols; 52 curr per; 5 mss; 7 diss/theses; 1 760 sound-rec; 51 digital data carriers; 259 art prints
libr loan; LIASA
25225

Nelson Mandela Metropolitan Municipality : Libraries, Market Sq, P.O. Box 66, *Port Elizabeth* 6000
T: +27 41 5061376; Fax: +27 41 5061390; URL: www.mandelametro.gov.za
1848; K. de Klerk
South African coll, John Owen Smith, maritime coll; Africana Libr, Reference Libr
600 000 vols; 100 curr per; 47 incunabula; 12 000 govt docs; 1 262 maps; 500 microforms; 2 000 av-mat;

6 000 sound-rec; 100 digital data carriers; 1 000 pamphlets, 700 jigsaw puzzles
libr loan, LIASA, IFLA
25226

Potchefstroom Community Library and Information Services, Gouwsstreet, *Potchefstroom* 2531; P.O. Box 113, Potchefstroom 2520
T: +27 18 299 5051; Fax: +27 18 294 8203
1909; E.E. Jansen van Rensburg
184 000 vols; 140 curr per; 100 music scores; 180 av-mat; 1 622 sound-rec; 5 digital data carriers
libr loan; LIASA
25227

City Library, Sammy Marks Sq, Church St, P.O. Box 2673, *Pretoria* 0001
T: +27 12 3088841; Fax: +27 12 3088955
1964; E.M. Hansen
Local Coll, Community Information
1 070 000 vols; 264 curr per; 3 841 microforms; 1 720 sound-rec
libr loan; LIASA
25228

Queensburgh Public Library, Rycroft Av Escombe, P.O. Box 39016, *Queensburgh* 4070
T: +27 31 441223; Fax: +27 31 442811
1961; Mr J.A.S. Welch
77 000 vols; 43 curr per; 450 av-mat; 769 sound-rec
libr loan; SALA
25229

Randburg Public Library, Hendrik Verwoerd Dr, *Randburg* 2194; PB X 1, Randburg 2125
T: +27 11 6862131; Fax: +27 11 6862643
1962; S.G. Nienaber
191 122 vols; 34 curr per; 60 av-mat; 589 sound-rec; 65 digital data carriers
libr loan
25230

Roodepoort City Library, 3rd Av, Florida Church Sq, PB 602, *Roodepoort* 1710
T: +27 11 6720433; Fax: +27 11 6721470
1920; R. Harmse
296 000 vols; 227 curr per; 100 mss; 100 diss/theses; 3 000 govt docs; 500 maps; 10 800 sound-rec; 20 digital data carriers
libr loan
25231

Rustenburg Public Library, Cor. Heystek and Smit St, P.O. Box 16, *Rustenburg* 0300
T: +27 14 590-3294/-3295; Fax: +27 14 5927596
1949; P.J.W. Louw
Rustenburg coll, Teenage section (bks + CD's), Large print bks, Study coll, Depot for SA Libr for the Blind (Grahamstown), Technikon SA, Depot for Damelin College
122 875 vols; 34 curr per; 2 diss/theses; 30 av-mat; 1 559 sound-rec; 8 digital data carriers; 210 pamphlets, 195 art prints, 8 newspapers
libr loan; LIASA
25232

Eastern Metropolitan Libraries, Corner West & Rivonia Rd, P.O. Box 78001, *Sandton* 2146
T: +27 11 8816440; Fax: +27 11 8816418
1970; Margaret Houliston
300 000 vols; 40 curr per; 500 govt docs; 100 av-mat; 700 sound-rec; 100 digital data carriers
libr loan; LIASA
25233

Sasolburg Public Library, John Vorster Rd, P.O. Box 60, *Sasolburg* 9570
T: +27 16 9760029 ext 2246; Fax: +27 16 9732191
1955; A.H. Duminy
150 000 vols; 25 curr per; 3 500 sound-rec; 2 250 pamphlets, 20 000 clippings
25234

Secunda Public Library, Central Business Centre, PB X1017, *Secunda* 2302
T: +27 17 6341166; Fax: +27 17 6341631
1977; P.M. Czanik
146 248 vols; 1 975 sound-rec; 40 digital data carriers; 315 audio bks
libr loan; LIASA
25235

Somerset West Public Library, Victoria St, *Somerset West* 7130; PB X22, Somerset West 7129
T: +27 21 8522235; Fax: +27 21 8516027
1958; A.M. Cloete
Africana, pamphlet coll; Reference section
75 000 vols; 74 curr per; 8 mss; 1 govt docs; 418 av-mat; 1 705 sound-rec; 392 art prints
libr loan
25236

Stilfontein Public Library, Bloem St, P.O. Box 20, *Stilfontein* 2550
T: +27 18 4841471, 4841472 ext 253, 254; Fax: +27 18 4842833
1952; Ms H.M. Brown
50 741 vols; 70 curr per; 707 records, 124 CDs, 85 audiobooks, 11 cassettes, 112 art prints, 326 clipping files, 4 379 clippings, 319 puzzles, 308 educational toys
libr loan
25237

Vanderbijlpark Public Library, Klasie Havenga St, P.O. Box 3, *Vanderbijlpark* 1900
T: +27 16 9505460; Fax: +27 16 9505264
1948; A.P. du Plooy
Clippings information pamphlets, files, Area clippings dept
204 118 vols; 140 curr per; 7 000 govt docs; 100 music scores; 30 mss; 980 microforms; 821 sound-rec; AV materials and sound recordings (5 000)
libr loan; LIASA
25238

Vereeniging Library, PB X047, *Vereeniging* 1930
T: +27 16 4503187, 4503189; Fax: +27 16 4224000
1912; J.M. Hammann
120 000 vols; 148 curr per; 49 maps; 6 microforms; 50 av-mat; 8 000 sound-rec; 40 digital data carriers; 1 000's of clippings and pamphlets
libr loan
25239

Verwoerdburg Public Library, Cantonments Rd, P.O. Box 14013, *Verwoerdburg* 0140
T: +27 12 621151; Fax: +27 12 6640909
1953; S. Coetsee
194 000 vols; 195 curr per; 4 000 sound-rec; 39 CD, 1 000 pamphlets
25240

Virginia Public Library, Virginia Gardens, PB X7, *Virginia* 9430
T: +27 57 23111; Fax: +27 57 2122885
1960; F.F. Linoko
55 900 vols; 50 curr per; 1 000 sound-rec
libr loan
25241

Welkom Public Library, Cor. Graat and Tulbach Sts, P.O. Box 186, *Welkom*, 9459
T: +27 171 913131; Fax: +27 171 32482
1950; A.A. Truber
150 000 vols; 221 curr per; 2 500 sound-rec
libr loan
25242

Westonaria Public Library, P.O. Box 19, *Westonaria* 1780
T: +27 11 7531121; Fax: +27 11 7532529
1955; E. Kuipers
67 267 vols; 40 curr per; 442 sound-rec; 115 digital data carriers; all govnt gazettes
25243

Witbank Public Library, Cnr Elizabeth Av and Hofmeyr St, P.O. Box 3, *Witbank* 1035
T: +27 135 906229; Fax: +27 135 906207
1937; Ms J. Rozmiarek
101 535 vols; 32 curr per; 229 sound-rec; 10 art prints
libr loan; LIASA
25244

Spain

National Libraries

Biblioteca Nacional de España (National Library of Spain), Paseo de Recoletos, 20-22, 28071 *Madrid*
T: +34 915807800; Fax: +34 915775634;

URL: www.bne.es
1712; Luis Racionero Grau
Libr of 'Real Sociedad Geografica', Coll Mendoza, Libr Luis Usoz, Libr José M. Asensio, Libr Boehl de Faber, Libr Augustín Durán, Coll Gayangos, Coll Graiño, Coll Sedó, Coll Campo Alange, Coll of 'Infante D. Fransisco de Paula' etc
8 000 000 vols
libr loan; ANABAD, IFLA, LIBER, CENL, ABINIA
25245

General Research Libraries

Biblioteca de Extremadura, Plaza Ibn Marwan, s/n, 06001 *Badajoz*
T: +34 924014484; E-mail: justo.vila@juntaextremadura.net; URL: www.bibliotecadeextremadura.es
Justo Vila Izquierdo
73 900 vols; 689 curr per; 7 incunabula; 35 av-mat; 1 260 sound-rec; 1 177 rare bks (16-18th c), 1 500 vols on local subjects
libr loan
25246

Biblioteca de Catalunya (Library of Catalonia), Carrer de l'Hospital, 56, 08001 *Barcelona*
T: +34 932702300; Fax: +34 932702304; E-mail: bustia@bnc.cat; URL: www.bnc.cat
1907; Dolors Lamarca
Cervantina, Bonsoms, Aguiló, Torres Amat, Catalan Theater, Verdaguer
3 000 000 vols; 50 000 curr per; 12 600 e-journals; 15 000 e-books; 7 300 mss; 660 incunabula; 47 000 music scores; 17 500 maps; 8 800 microforms; 27 000 av-mat; 160 000 sound-rec; 5 000 digital data carriers
libr loan; IFLA, AIB, LIBER, EBLIDA, IIPC, IASA, CERL, MLA, ARSC, AEDOM, COBDC, AAC
25247

Biblioteca de Andalucía, C/ Prof Sáinz Cantero, 6, 18002 *Granada*
T: +34 958026900; Fax: +34 958026937; E-mail: informa-cion.ba.ccul@juntadeandalucia.es; URL: www.juntadeandalucia.es/cultura/ba
1987; Francisco Javier Álvarez García
Bibliotecas de Luis Rosales y José Luis Cano, lit española
175 968 vols; 14 452 curr per; 5 e-journals; 38 mss; 484 music scores; 5 030 maps; 3 051 microforms; 3 500 av-mat; 12 644 sound-rec
libr loan; IFLA, EBLIDA
25248

Biblioteca de La Rioja, C/ Merced, 1, 26001 *Logroño*
T: +34 941211382; Fax: +34 941210536; E-mail: biblioteca@blr.larioja.org; URL: www.blr.larioja.org
1903; Lola Ramírez Domínguez
130 000 vols; 1 416 curr per; 13 mss; 28 incunabula; 311 music scores; 125 maps; 4 879 microforms; 12 174 av-mat; 4 228 sound-rec; 867 digital data carriers; 8 000 rare bks (16-18th c), 10 800 slides
libr loan
25249

Biblioteca Histórica Municipal, Conde Duque, 11, 28015 *Madrid*
T: +34 915885737; Fax: +34 914800520
1876; Ascensión Aguerri Martínez
Madrid, theater, music, Cervantes, Lope de Vega, bulls, bibliografia, paremiología
173 000 vols; 40 curr per; 7 100 mss; 14 incunabula; 1 700 music scores; 110 maps; 5 978 music scores
libr loan
25250

Biblioteca Regional de Madrid (Regional Library – Madrid), C/ Ramírez de Prado, 3, 28045 *Madrid*
T: +34 917208850, 917208860; Fax: +34 917208923; E-mail: biblio.regional@madrid.org; URL: www.madrid.org/bpcm
1984; Luisa I. Fernández Miedes
Colección Madrileña (Coll on Madrid); Patrimonio (Heritage, antique items), Materiales Especiales (Cartography, Graphic and AV materials), Hemeroteca (Periodicals and Serials)
693 789 vols; 4 509 curr per; 476 e-journals; 86 mss; 41 171 music scores; 15 523 maps; 19 167 microforms; 42 102 av-mat; 106 219 sound-rec; 16 876 digital data carriers; 216 709 other items
libr loan; IFLA, ANABAD, REBIUN
25251

Patrimonio Nacional, Real Biblioteca
– Palacio Real, C/ Bailén, s/n, 28071
Madrid
T: +34 914548732, 914548733;
Fax: +34 914548867; E-mail:
realbiblioteca@patrimonionacional.es; URL:
www.patrimonionacional.es/realbiblioteca
ca 1700; M.L. López-Vidriero
Spec coll libs: Gondomar, Ayala, Bruna,
Mayans y Siscar, Mansilla, Lameyer
267 500 vols; 66 curr per; 3 798 mss;
263 incunabula; 4 128 music scores;
3 280 maps; 22 300 microforms; 260
av-mat; 39 sound-rec; 25 digital data
carriers; 1 026 photo albums, 4 677 prints,
511 drawings, 237 slides   25252

Biblioteca Regional de Murcia, Av
Juan Carlos I, 17, 30008 *Murcia*
T: +34 96836-6599/-2500; Fax: +34
968366600; E-mail: brmu@listas.carm.es;
URL: www.bibliotecaregional.carm.es
1956
76 700 vols; 635 curr per; 4 579
microforms; 1 000 av-mat; 1 000
sound-rec
libr loan   25253

**Biblioteca de Asturias Ramón Pérez
de Ayala**, Pl de Daoiz y Velarde, 11,
33009 *Oviedo*
T: +34 985211397; Fax: +34 985207351;
E-mail: bibliast@asturias.org
1987; Milagros Garcia Rodriguez
Col Casimiro Cienfuegos, Col Antonio
García Oliveros, Col Enrique Rendueles,
Col Pérez de Ayala, Col José Manuel
Castanon, Col Familia Serrano, Col
Carmen Moieno, Col Luis Maria Canteli,
Col Enrique Vera, Col Ateneo de Oviedo
307 189 vols; 3 878 curr per; 225
mss; 1 080 music scores; 1 412 maps;
2 023 microforms; 975 av-mat; 15 699
sound-rec; 1 653 digital data carriers;
2 344 rare books (16-19th c)
IFLA   25254

Biblioteca General de Navarra, Pl San
Francisco, s/n, 31001 *Pamplona*
T: +34 948427797; Fax: +34 948427788;
E-mail: biblioteca.general@navarra.es
1940; Juan F. Elizari-Huarte
Historia de Navarra, Carlismo, Toros;
Fondo General Prestable, Fondo de
Consulta, Fondo de Referencia
289 694 vols; 1 600 curr per; 10 mss;
77 incunabula; 250 music scores; 312
maps; 625 microforms; 5 600 av-mat;
2 500 sound-rec; 300 digital data carriers;
14 100 brochures
libr loan   25255

Biblioteca Central de Cantabria, C/
Ruiz de Alda, 19, 39009 *Santander*
T: +34 942241550; Fax: +34 942241551;
E-mail: bcc@gobcantabria.es; URL:
bcc.cantabria.es
1839; José María Gutiérrez
64 600 vols; 54 curr per; 52 mss; 8
incunabula; 3 017 rare bks (16-18th c)
          25256

Biblioteca de Castilla-La Mancha,
Cuesta de Carlos V, s/n, 45071 *Toledo*
T: +34 925256680; Fax: +34 925253642;
E-mail: biblioclm@jccm.es; URL:
www.jccm.es/biblioclm
1771; Joaquin Selgas
Borbon-Lorenzana Coll, Javier Malagón
Barceló Coll, Antonio Buero Vallejo Coll,
Gianna Prodan Coll
314 235 vols; 485 curr per; 758 mss;
414 incunabula; 455 maps; 8 452 av-mat;
6 695 sound-rec; 1 212 digital data
carriers
libr loan   25257

Biblioteca Valenciana, Avinguda de la
Constitució, 284 (Monestir Sant Miquel
dels Reis), 46019 *Valencia*
T: +34 963874000; Fax: +34 963874037;
E-mail: bv@gva.es; URL: www.gva.es
1979
Regionalismo
159 000 vols; 7 497 curr per; 2 562 mss;
26 incunabula; 398 000 microforms
libr loan   25258

Biblioteca de Castilla y León, Pl de la
Trinidad, 2, 47003 *Valladolid*
T: +34 983358599; Fax: +34 983359040;
E-mail: bcl@jcyl.es; URL: www.bcl.jcyl.es
1992; Alejandro Carrión Gutiez
Bibl Jorge Guillén, Bibl Rosa Chacel,
Bibl Martín Abril, Bibl Francisco Pino,

Bibl Claudio Rodríguez, Bibl Gabino
Alejandro Carriedo
107 846 vols; 5 877 curr per; 14 mss;
683 diss/theses; 309 music scores;
3 400 maps; 550 microforms; 1 725
av-mat; 1 603 sound-rec; 930 digital
data carriers; 6 438 posters and pictures,
11 257 pamphlets   25259

University Libraries,
College Libraries

**Universidad de Castilla-La Mancha
– Campus de Albacete José Prat**,
Bibiloteca General, Pl de la Universidad,
2, 02071 *Albacete*
T: +34 967599212; Fax: +34 967599201;
E-mail: aaguilar@bibl-ab.uclm.es
1993
Social sciences, agronomy, forestry,
information science, engineering
67 500 vols; 759 curr per
libr loan   25260

Universidad de Alcalá, Biblioteca
General, C/ Trinidad, 1, 28801 *Alcalá
de Henares*, Madrid
T: +34 918854257; Fax: +34 918854206;
E-mail: juana.frias@uah.es; URL:
www.uah.es/biblioteca
1978; María del Carmen Fernández-
Galiano Peyrolón
Electronics
176 000 vols; 1 485 curr per
libr loan; SEDIC, ANABAD, IFLA, FID
          25261

– Biblioteca Central de Ciencias
Experimentales, Facultad de Medicina,
Campus Universitario, km 33.600, 28871
Alcalá de Henares, Madrid
T: +34 918854500; Fax: +34
918854580, 918854864; E-mail:
ib.experimentales@uah.es; URL:
www.uah.es/biblioteca
1977; María Isabel Dominguez Aroca
Pharmacy, Biology, Medicine
63 800 vols; 729 curr per; 2 000
diss/theses; 4 500 microforms; 38 digital
data carriers
libr loan; IFLA, FID, SEDIC, ANABAD,
EAHIL, REBIUN, MADROÑO   25262

– Biblioteca de Ciencias Económicas y
Empresariales, Pl de la Victoria, 3,
28802 Alcalá de Henares, Madrid
T: +34 918855151; Fax: +34 918855187;
E-mail: rosa.gallego@uah.es; URL:
www.uah.es/biblioteca/biblioteca/
economicas.html
1979; Rosa Gallego López
Documentos de Trabajo, Estadísticas
48 000 vols; 413 curr per
libr loan   25263

– Biblioteca de Derecho, C/ Libreros 17,
28801 Alcalá de Henares, Madrid
T: +34 918854300; Fax: +34 918854315;
E-mail: julieta.garcia@uah.es; URL:
www.uah.es
1975; Julieta García Morilla
51 850 vols
libr loan   25264

– Biblioteca de Filosofía y Letras, C/
Colegios 2, 28801 Alcalá de Henares,
Madrid
T: +34 918854453; Fax: +34 918854440;
E-mail: mangeles.arteta@uah.es; URL:
www.uah.es
1983; Maria Angeles Arteta Velasco
200 titulos Fondo Antiguo
68 900 vols
libr loan; REBIUN   25265

– Biblioteca de Magisterio, Madrid, s/n,
19001 Guadalajara
T: +34 949209761; Fax: +34 949209767;
E-mail: carmen.gallo@uah.es
1964; María Carmen Gallo Rolania
Fondo Antiguo (2 000 items)
21 130 vols
libr loan   25266

– Biblioteca del IUEN (Antiguo Cenua),
Antiguo Colegio de Trinitarios, C/
Trinidad 1, 28801 Alcalá de Henares,
Madrid
T: +34 918855303, 918855305; Fax: +34
918855319; E-mail: flor.fernandez@uah.es;
URL: www.uah.es
Florentina Fernández López
21 000 vols; 500 av-mat   25267

Universidad de Alicante, Servicio de
Información Bibliográfica y Documental,
Ctra de San Vicente del Raspeig s/n,
03690 *Alicante*
T: +34 96590-3400/-2524; Fax: +34
965903464; E-mail: Juan.Mesa@ua.es;
URL: www.ua.es/es/bibliotecas/index.html
1969; M. Remedios Blanes Gran
205 000 vols
libr loan   25268

– Biblioteca de Ciencias, Facultad de
Ciencias, Apdo de Correos 99, 03080
Alicante
T: +34 965903559; Fax: +34 965909747;
E-mail: remedios.nomdedeu@ua.es; URL:
www.ua.es/es/bibliotecas/SIBID/centros/
neo_ciencias/index.htm
1978; Remedios Nomdedeu Andreu
20 150 vols; 1 600 curr per; 772
diss/theses; 1 025 microforms
libr loan   25269

– Biblioteca de Ciencias Económicas y
Empresariales, Facultad de Ciencias
Económicas, Apdo de Correos, 99,
03080 Alicante
T: +34 965903400 ext 3130; Fax: +34
965903624; E-mail: Bibli.Economiques@
ua.es; URL: www.ua.es/es/bibliotecas/
SIBID/centros/neo_economicas/index.htm
1980; Beatriz Alberdi García
Statistics, Sociology, Economics, Business
Admin
30 000 vols; 857 curr per; 50 diss/theses;
35 maps; 700 microforms; 3 digital data
carriers
libr loan   25270

– Biblioteca de Derecho, Facultad de
Derecho, Apdo de Correos 99, 03080
Alicante
T: +34 965909337; Fax: +34 965903630;
E-mail: Bibli.Dret@ua.es; URL:
www.ua.es/es/bibliotecas/SIBID/centros/
neo_derecho/index.htm
1981; María Dolores Marás Menéndez
79 500 vols; 2 300 curr per; 63 e-
journals; 460 diss/theses; 10 500
microforms; 165 av-mat; 120 digital
data carriers
libr loan; IALL   25271

– Biblioteca de Educación, Apdo de
Correos 99, 03080 Alicante
T: +34 965903738; Fax: +34 965909482;
E-mail: Bibli.Magisteri@ua.es; URL:
www.ua.es/es/bibliotecas/SIBID/centros/
neo_educacion/index.htm
María Dolores García Ivars
Col Antigua de la Escuela Normal de
Alicante
50 000 vols; 471 curr per
libr loan   25272

– Biblioteca de Filosofía y Letras / Trabajo
Social, Edif. Biblioteca Central, Campus
de San Vicente del Raspeig, Apdo de
Correos 99, 03080 Alicante
T: +34 965903400 ext 2814, 3016;
E-mail: Bibli.FLTS@ua.es; URL:
www.ua.es/es/bibliotecas/SIBID/centros/
neo_letras/index.htm
María Victoria Játiva Miralles
Philsophy, philology, social work
153 700 vols; 192 curr per
libr loan   25273

– Biblioteca Politécnica, Óptica y
Enfermeria, Edif. Biblioteca General,
planta baja, Campus de San Vicente del
Raspeig, Apdo 99, 03080 Alicante
T: +34 965903400 (ext 3398); E-mail:
Bibli.EPS@ua.es; URL: www.ua.es/es/
bibliotecas/SIBID/centros/neo_politecnica/
index.htm
1994; María José Llorca García
16 000 vols; 322 curr per; 100 e-journals;
300 diss/theses
libr loan; REBIUN, MECANO, DOCUMAT,
SEDIC   25274

Universidad de Almería, Biblioteca,
Planta 1, despacho 1.211, Ctra.
Sacrmento s/n, Cañada de San Urbano
s/n, 04120 *Almería*
T: +34 950214052; E-mail:
efuentes@ual.es; URL: www.ual.es
1993; Encarnación Fuentes Melero
Journals dept; reference works dept
68 400 vols; 753 curr per; 10 diss/theses;
108 maps; 2 340 microforms; 1 525
av-mat; 240 sound-rec; 13 digital data
carriers
libr loan; ANABAD   25275

– Escuela Universitaria de Ciencias de
la Salud, Biblioteca, Ctra. Sacramento
s/n, La Cañada de San Urbano, 04120
Almería
T: +34 950015721
1977
Physiology, microbiology, genetics,
biochemistry
8 500 vols; 100 curr per   25276

Universidad de Extremadura, Biblioteca
Central de Badajoz, Ed. Rectorado,
Campus Universitaria, Av de Elvas, s/n,
06071 *Badajoz*
T: +34 924289310; Fax: +34 924289685;
E-mail: aferrer@unex.es; URL:
biblioteca.unex.es
1973; Ángeles Ferrer Gutiérrez
240 000 vols; 4 475 curr per
libr loan   25277

– Biblioteca Central de Badajoz, Edif.
Biblioteca Central de Badajoz, Av de
Elvas s/n, 06071 Badajoz
T: +34 924289310; Fax: +34 924289685;
E-mail: bibusuba@unex.es; URL:
biblioteca.unex.es   25278

– Biblioteca Central de Cáceres, Edif.
Biblioteca Central, Av de la Universidad,
s/n, 10071 Cáceres
T: +34 927257090; Fax: +34 927257097;
E-mail: biccusu@unex.es; URL:
biblioteca.unex.es
1999; Maria Teresa Mateos Fernández
          25279

– Biblioteca Centro Universitario de Mérida,
C/ Santa Teresa de Jornet, 38, 06800
Mérida
T: +34 924387498; Fax: +34 924303782;
E-mail: bibceum@unex.es; URL:
biblioteca.unex.es   25280

– Biblioteca Centro Universitario de
Plasencia, Av Virgen del Puerto, 10600
Plasencia
T: +34 927252100; Fax: +34 927425209;
URL: biblioteca.unex.es   25281

– Escuela de Ingenierías Agrarias,
Biblioteca, Ctra de Cáceres s/n, 06071
Badajoz
T: +34 924286218; Fax: +34 924286201;
E-mail: bibagra@unex.es; URL:
biblioteca.unex.es   25282

– Escuela de Ingenierías Industriales,
Biblioteca, Av de Elvas s/n, 06071
Badajoz
T: +34 924289637; Fax: +34 924289601;
E-mail: bibindu@unex.es; URL:
biblioteca.unex.es   25283

– Escuela Universitaria de Enfermería y
Terapia Ocupacional, Biblioteca, Ctra de
Trujillo s/n, 10071 Cáceres
T: +34 927257459; Fax: +34 927257459;
E-mail: bibenfer@unex.es; URL:
biblioteca.unex.es
Public Health, Occupational Safety and
Health   25284

– Facultad de Biblioteconomía, Av de Elvas
s/n, 06071 Badajoz
T: +34 924286408; E-mail: bibdocu@
unex.es; URL: biblioteca.unex.es
Library Sciences   25285

– Facultad de Ciencias del Deporte, Edif.
Biblioteca Central, Av de la Universidad,
s/n, 10071 Cáceres
T: +34 927257270; Fax: +34 927181022;
E-mail: bibdepo@unex.es; URL:
biblioteca.unex.es
Sports   25286

– Facultad de Derecho, Biblioteca, Ctra de
Trujillo, s/n, Campus Universitario, 10003
Cáceres
T: +34 927257270; Fax: +34 927181022;
E-mail: bibdere@unex.es; URL:
biblioteca.unix.es
1974
36 000 vols; 443 curr per   25287

– Facultad de Educación, Biblioteca, Av de
Elvas, s/n, 06071 Badajoz
T: +34 924289584; Fax: +34 924270214;
E-mail: bibeduc@unex.es; URL:
biblioteca.unex.es
1976; Esperanza Marina Serrano
35 000 vols; 500 curr per
libr loan   25288

– Facultad de Estudios Empresariales y Turismo, Biblioteca, Ctra de Trujillo s/n, 10071 Cáceres
T: +34 927257488; Fax: +34 927257481; E-mail: bibempre@unex.es; URL: biblioteca.unex.es
1974
Business management
9 000 vols; 150 curr per   25289

– Facultad de Medicina, Biblioteca, Av de Elvas, s/n, 06071 Badajoz
T: +34 924289433; Fax: +34 924272956; E-mail: bibmedi@unex.es; URL: biblioteca.unex.es
1976
11 000 vols; 450 curr per
libr loan   25290

– Facultad de Veterinaria, Biblioteca, Ctra de Trujillo s/n, 10071 Cáceres
T: +34 927251388; Fax: +34 927257110; E-mail: bibvete@unex.es; URL: biblioteca.unex.es
1983
9 700 vols; 425 curr per
libr loan   25291

Conservatori Municipal de Música de Barcelona, Biblioteca, Bruc 110-112, 08009 *Barcelona*
T: +34 934584302; Fax: +34 934593104; E-mail: biblioteca@cmmb.cat; URL: www.bcn.es/conservatori/cat/serveis_biblioteca.htm
1933; Esther Vilar T.
13 600 vols; 15 curr per; 10 000 music scores; 7 300 sound-rec   25292

Fundacion ESADE, Universidad Ramon Llull, Biblioteca, Av Predralbes, 60-62, 08034 *Barcelona*
T: +34 932806162; Fax: +34 932048105; E-mail: biblioteca@esade.edu
1959; María-José Marinón
Publs of EC, DECD, ILO, World Bank; European Documentation Center
80 000 vols; 600 curr per; 4 000 e-journals; 1 250 diss/theses; 600 av-mat
libr loan; EBSLG   25293

Fundació Privada Universitaria EADA, Centro de Documentacion, C/ Aragó, 204, 08011 *Barcelona*
T: +34 934520844; Fax: +34 933237317; E-mail: info@eada.edu; URL: www.eada.edu
1957; Carolina Sanmartín
Digital theses coll: 980 registers
13 990 vols; 175 curr per; 2 735 diss/theses; 582 av-mat; 25 sound-rec; 139 digital data carriers
libr loan; CDB   25294

Instituto de Estudios Superiores de la Empresa (IESE), Biblioteca (International Graduate School of Management), Av Pearson, 21, 08034 *Barcelona*
T: +34 932534200; Fax: +34 932534343; E-mail: biblioteca@iese.edu; URL: www.iese.edu/library
1958; Carina Huguet
46 800 vols; 4 100 curr per; 4 000 e-journals; 198 diss/theses
libr loan; EBSLG   25295

Universitat Abat Oliba CEU, Biblioteca, C/ Bellesguard, 30, 08022 *Barcelona*
T: +34 932040907; Fax: +34 932040907; E-mail: bibli@uao.es; URL: biblioteca.uao.es
1975; Elena Villatoro Boan
Economics, Law, Humanities
23 000 vols; 116 curr per; 930 e-journals; 30 diss/theses; 393 sound-rec; 200 digital data carriers
libr loan; REBIUN   25296

Universitat de Barcelona, Centre de Recursos per a l'Aprenentatge i la Investigació (University of Barcelona), Edifici Biblioteca, Baldiri Reixac, 2, 08028 *Barcelona*
T: +34 934034576; Fax: +34 934034577; E-mail: aferrer@ub.edu; URL: www.bib.ub.edu/
Adelaida Ferrer Torrens
Printed matters of Barcelona
2 000 000 vols; 38 000 curr per; 2 040 mss; 920 incunabula; 2 000 diss/theses; 1 433 maps; 720 microforms; 9 770 av-mat; 250 000 sound-rec; 500 engravings

– Biblioteca de Belles Arts (Fine Arts Library), Baldin Reixach, 2, 08028 Barcelona
T: +34 9340345-95/-95; E-mail: mmanadep@ub.edu; URL: www.bib.ub.edu/biblioteques/belles-arts/
1950; María Manadé
Fine arts, hist of cinematography, music and theatre
20 000 vols; 140 curr per; 281 diss/theses; 15 000 av-mat; 285 sound-rec; 7 250 slides
libr loan; IFLA   25297

– Biblioteca de Biblioteconomia (Information Sciences Library), C. Melcior de Palau, 140, 08014 Barcelona
T: +34 934035769; Fax: +34 934035768; E-mail: lougonzalez@ub.edu; URL: www.bib.ub.edu/biblioteques/biblioteconomia/
1915; Lourdes González
22 000 vols; 440 curr per; 368 diss/theses; 74 av-mat; 200 digital data carriers
libr loan; LA, ASLIB, REBIUN   25298

– Biblioteca de Biologia (Biology Library), Av Diagonal, 645, 08028 Barcelona
T: +34 934021567; Fax: +34 934021568; E-mail: psdbio@ub.edu; URL: www.bib.ub.edu/biblioteques/biologia/
1970; Teresa Pañell
Fons Ramon Margalef Lopez
30 000 vols; 230 curr per; 2 700 diss/theses; 700 av-mat; 450 digital data carriers
libr loan   25299

– Biblioteca de Dret (Law Library), Av Diagonal, 684, 08034 Barcelona
T: +34 934024456; Fax: +34 934024459; URL: www.bib.ub.edu/biblioteques/dret/
1959; Dolors Gutiérrez
United Natlons depository library, Centre d'Estudis Canandencs
90 500 vols; 1 869 curr per; 1 521 rare books (16th c until 1820)
libr loan   25300

– Biblioteca de Farmàcia (Pharmacy Library), Av Juan XXIII, s/n, 08028 Barcelona
T: +34 934021884; Fax: +34 934035980; E-mail: qmaicas@ub.edu; URL: www.bib.ub.edu/biblioteques/farmacia/
1965; Joaquima Maicas
13 000 vols; 162 curr per; 1 520 diss/theses   25301

– Biblioteca de Filosofia, Geografia i Història (Philosophy, Geography and History Library), C. Montalegre, 8, 08001 Barcelona
T: +34 934034584; Fax: +34 934034599; E-mail: aclavell@ub.edu; URL: www.bib.ub.edu/biblioteques/filosofia-geografia-historia/
Anna Clavell
Philosophy, geography, history, anthropology, archaeology, art history; Fons Ceddiq, Fons Martorell-Solanic
246 000 vols; 5 000 curr per
libr loan   25302

– Biblioteca de Física i Química (Physics and Chemistry Library), Av Diagonal, 647, 08028 Barcelona
T: +34 934021321; Fax: +34 934021323; E-mail: isabelpares@ub.edu; URL: www.bib.ub.edu/biblioteques/fisica-quimica/
1937; Isabel Parés
Chemicals, organic chemistry, physics, astronomy, pharmaceutical sciences
31 000 vols; 195 curr per; 4 000 e-journals; 4 000 diss/theses; 450 av-mat; 520 sound-rec
libr loan   25303

– Biblioteca de Geologia (Geology Library), C. Martí Franquès, s/n, 08028 Barcelona
T: +34 934021420; Fax: +34 934021421; E-mail: jcasadella@ub.edu; URL: www.bib.ub.edu/biblioteques/geologia/
1983; Jordi Casadella
16 500 vols; 1 054 curr per; 1 623 diss/theses; 12 000 maps; 59 digital data carriers; 17 000 aerial photos referring to Catalunya and Aragón
libr loan   25304

– Biblioteca de Lletres (Philology Library), Gran Via de las Cortes Catalanes, 585, 08007 Barcelona
T: +34 934035317 (Lletres), 934035325 (Romàniques), 934035326 (Hispàniques), 934035335 (Latí); Fax: +34 934035324; E-mail: turielj@ub.edu; URL: www.bib.ub.edu/biblioteques/lletres/
Josep Turiel
Languages and Literatures (Classical, Romance, Anglo-American, Germanic, Slavic, Arabic); Biblioteca Cirac, Llegat Bonner
270 000 vols; 1 681 curr per   25305

– Biblioteca de Matemàtiques (Library of Mathematics), Gran Via de les Corts Catalanes, 585, 08007 Barcelona
T: +34 934035549; Fax: +34 934021662; E-mail: cnavajas@ub.edu; URL: www.bib.ub.edu/biblioteques/matematiques/
1934; Carme Navajas
Mathematics, information science
25 200 vols; 783 curr per; 1 154 e-journals; 3 000 e-books; 1 071 diss/theses; 148 sound-rec
libr loan   25306

– Biblioteca de Medicina (Library of Medicine), C. Casanova, 143, 7 Planta, 08036 Barcelona
T: +34 934035718; Fax: +34 934039099; E-mail: cbacchetta@ub.edu; URL: www.bib.ub.edu/biblioteques/medicina/
1906; Coral Bacchetta
Rare bks (16th-19th c), Col Mira i López de Psiquiatría
40 000 vols; 3 136 curr per; 3 950 e-journals; 188 mss; 1 609 diss/theses; 1 093 microforms; 135 digital data carriers; 8 951 rare books
libr loan; IFLA   25307

– Biblioteca de Reserva (Rare Book and Manuscript Library), Gran Via de les Corts Catalanes, 585, 08007 Barcelona
T: +34 934025313; Fax: +34 934025324; E-mail: bib.reserva@ub.edu; URL: www.bib.ub.edu/biblioteques/reserva/
Neus Verger
Manuscripts, incunabula and printed books from 16th to 18th c   25308

– Biblioteca d'Economia i Empresa (Economics and Business Library), Av Diagonal, 690, 08034 Barcelona
T: +34 934034734; Fax: +34 934021967; E-mail: mcarmeferrer@ub.edu; URL: www.bib.ub.edu/biblioteques/economia-empresa/
2008; Carme Ferrer
Economics, Business Administration, Statistics, Sociology; Fons Ernest Lluch, Fons Jesús M. de Miguel, Fons Manuel Sacristán, Secció de Publicacions Estadístiques: Catalunya, Espanya, Internacionals
67 000 vols; 4 278 curr per; 1 060 statistics
libr loan
The library extends over 3 locations: Av Diagonal 690 (formerly Bibl d'Economiques, ca 45 000 vols), Av Diagonal 696 (formerly Bibl d'Empresarials, ca 18 000 vols), Edifici Tinent Coronel Valenzuela (formerly Bibl Campus Diagonal Nord, ca 6 000 vols)   25309

– Biblioteca Pavelló de la República (The Pavelló de la República Library), Av Cardenal Vidal i Barraquer, 34-36, 08035 Barcelona
T: +34 934285457; Fax: +34 934279371; E-mail: bibrepublica@ub.edu; URL: www.bib.ub.edu/en/libraries/pavello-republica/
1996; M. Lourdes Prades
Second Spanish Republic, Spanish Civil War, Franco Regime, transition to democracy, with a special emphasis on Catalonia
150 000 vols; 200 av-mat
libr loan   25310

– Campus de Bellvitge, Biblioteca (Bellvitge Campus), Feixa Llarga, s/n (Pavelló de Govern), 08007 L'Hospitalet de Llobregat
T: +34 934024211; E-mail: rangelet@ub.edu; URL: www.bib.ub.edu/biblioteques/campus-bellvitge/
1915; Rosa Angelet
Medicine, Public Health, Mental Health
27 300 vols; 800 curr per
libr loan; LA, ASLIB, REBIUN   25311

– CRAI Biblioteca de Campus Mundet (Resource Center for Learning and Research. Mundet Campus), Edifici de Llevant, Passeig de la Vall d'Hebron, 171, 08035 Barcelona
T: +34 934034750; Fax: +34 934029077; E-mail: bib_mundet@ub.edu; URL: www.bib.ub.edu/biblioteques/campus-mundet/
1995; Conxa Alvarez
Education, psychology, social work; Fons de reserva de l'Escola Normal de la Generalitat (1931-1939, fons de reserva de llibre de text antic (1789-1970), col de l'Aula de Poesia de Barcelona (2 000 works of contemporary poetry)
117 000 vols; 1 694 curr per; 500 e-journals; 2 700 diss/theses; 232 sound-rec; 500 digital data carriers; 350 tests
libr loan   25312

Universitat Politècnica de Catalunya, *Barcelona*
URL: bibliotecnica.upc.edu

– Biblioteca de Matemàtiques i Estadística, Pau Gargallo, 5, 08028 Barcelona
T: +34 934017020; E-mail: biblioteca.fme@upc.edu; URL: bibliotecnica.upc.edu/FME
Gemma Flaquer Fabregat
libr loan   25313

– Biblioteca de Nàutica, Pl del Palau, 18, 08003 Barcelona
T: +34 934017934; Fax: +34 934017910; E-mail: biblioteca.fnb@upc.edu; URL: bibliotecnica.upc.edu/FNB
1932; Jordi Serrano Muñoz
10 000 vols; 72 curr per; 30 mss; 80 diss/theses; 100 govt docs; 500 maps   25314

– Biblioteca del Campus de Terrassa, Pl del Campus, 1, 08222 Terrassa
T: +34 937398062; Fax: +34 937398187; E-mail: lluisa.Perona@upc.edu; URL: bibliotecnica.upc.edu/bib170
1995; Lluïsa Perona Gutiérrez
Textile eng and industry, paper eng, optics, environmental eng
32 000 vols; 244 curr per; 3 000 e-journals; 435 mss; 315 diss/theses; 83 microforms; 312 av-mat; 522 sound-rec; 500 digital data carriers
libr loan   25315

– Biblioteca del Campus del Baix Llobregat, C/ Esteve Terradas, 10, 08860 Castelldefels
T: +34 935523553; Fax: +34 935523558; E-mail: biblioteca.cbl@upc.edu; URL: bibliotecnica.upc.edu/BCBL/
libr loan   25316

– Biblioteca Rector Gabriel Ferraté, C/ Jordi Girona 1-3, Edifici BIB, 08034 Barcelona
T: +34 934015830; Fax: +34 934015600; E-mail: biblioteca.ferrate@upc.edu; URL: bibliotecnica.upc.edu/bib160
1996; Miquel Codina
Catalan poetry, Science fiction, Jazz music, Sustinaible human development
90 852 vols; 154 curr per; 8 947 e-journals; 3 760 diss/theses; 1 011 maps; 256 microforms; 1 178 av-mat; 200 sound-rec; 700 digital data carriers
libr loan   25317

– Escola d'Enginyeria d'Igualada, Biblioteca, Plaça del Rei, 15, 08700 Igualada
T: +34 938035300; Fax: +34 938031589; E-mail: biblioteca.euetii@upc.edu; URL: bibliotecnica.upc.es/bib860/
libr loan   25318

– Escola Politècnica Superior d'Edificació de Barcelona, Biblioteca d'Edificació, Av Doctor Marañón 44-50, 08028 Barcelona
T: +34 934016265; Fax: +34 934017019; E-mail: biblioteca.EPSEB@upc.edu; URL: bibliotecnica.upc.edu/EPSEB/
1960; Remei Garcia Martínez
Cartography Dept, Architectural' Heritage of catalonia
26 228 vols; 376 curr per; 25 diss/theses; 7 000 maps; 480 av-mat; 780 digital data carriers; 2 000 digitized theses   25319

– Escola Politècnica Superior d'Enginyeria de Manresa, Biblioteca del Campus Universitari de Manresa, Av Bases de Manresa, 7/11, 08242 Manresa, Barcelona
T: +34 938777221; Fax: +34 938777239; E-mail: montserrat.sala@upc.es; URL: bibliotecnica.upc.es/bib330/
1973; Montserrat Sala Torras
Health Sciences, Sustainable Development, Economics and Business Organization, Food engineering, Civil Engineering, Engineering of Materials, Chemical Engineering
24 233 vols; 458 curr per; 11 232 e-journals; 10 157 e-books; 62 mss; 200 diss/theses; 710 maps; 30 microforms; 596 av-mat; 10 sound-rec; 58 digital data carriers
libr loan 25320

– Escola Politècnica Superior d'Enginyeria de Vilanova i la Geltrú, Biblioteca (Technical School of Vilnanova i la Geltrú), Rambla de l'Exposició, 37, 08800 Vilanova i La Geltru, Barcelona
T: +34 938967712; Fax: +34 938967811; E-mail: BIBLIOTECA.EPSEVG@UPC.EDU; URL: bibliotecnica.upc.edu/bib340/
1954; Francesc Carnerero Gámiz
Opera, Technology and society, Fantastic and Terror Cinema, Science Fiction, Career integration, The environment; Cooking, EEES (European Space of Higher Education)
27 000 vols; 41 curr per; 8 360 e-journals; 133 diss/theses; 149 av-mat; 205 sound-rec; 200 digital data carriers; 106 standards
libr loan; Rebiun, CBUC, COBDC 25321

– Escola Tècnica Superior d'Arquitectura de Barcelona, Biblioteca, Diagonal, 649, 08028 Barcelona
T: +34 934011937; E-mail: biblioteca.ETSAB@upc.edu; URL: bibliotecnica.upc.edu/ETSAB/
1874; Neus Vilaplana Moreno
Colllection of 16th, 17th, 18th and 19th, cenntury books. Collection on 12.000 Architectural drawings. (online catalog bibliotecnica.upc.edu/arxiugraficETSAB/
45 000 vols; 197 curr per; 6 772 e-journals; 766 diss/theses; 1 206 maps; 4 720 microforms; 1 422 av-mat; 95 digital data carriers
libr loan; ABBA 25322

– Escola Tècnica Superior d'Arquitectura del Vallès, Biblioteca, Pere Serra, 1-15, 08190 Sant Cugat del Vallès
T: +34 934017844; Fax: +34 934017901; E-mail: biblioteca.etsav@upc.edu; URL: bibliotecnica.upc.edu/ARQUITECTURA-VALLES/
1977; Gloria Ramoneda Molins
10 500 vols; 115 curr per; 35 diss/theses; 500 maps
libr loan 25323

– Escola Técnica Superior d'Enginyeria Industrial de Barcelona, Biblioteca, Avinguda Diagonal, 647, 08028 Barcelona
T: +34 934016637; E-mail: biblioteca.etseib@upc.edu; URL: bibliotecnica.upc.edu/ETSEIB/
1932; Laia Alonso Cortina
Industrial, chemical, textile, mechanical, electrical and electonical engineering
8 000 vols; 72 curr per 25324

– Escola Universitària d'Enginyeria Tècnica Industrial de Barcelona, Biblioteca EUETIB, C. Comte d'Urgell, 08036 Barcelona
T: +34 934137252; Fax: +34 934137401; E-mail: toni.bejar@upc.edu; URL: bibliotecnica.upc.edu/ETSEIB/
Toni Béjar Farré
Industrial engineering
12 000 vols 25325

Universitat Pompeu Fabra, Campus de la Ciutadella, Biblioteca, Ramón Trias Fargas, 25-27, 08005 ***Barcelona***
T: +34 935421709; Fax: +34 935421799; E-mail: mar.biblioteca@upf.edu; URL: www.upf.edu/biblioteca
1990; Montserrat Espinós
Coll on Asian studies, INPUT (International Public Television) Coll, Coll Haas on hist of religions, Coll Chamber of Commerce of Barcelona
550 000 vols; 3 300 curr per; 10 000 e-journals
libr loan; IFLA, ALA 25326

Universitat Ramon Llull, ***Barcelona***
– Facultat de Ciències de la Comunicació, Biblioteca, Edifici I, 08193 Bellaterra (Cerdanyola del Vallès)
T: +34 932533096; Fax: +34 932533099; URL: ccc-web.uab.es/index.php
19 000 vols; 350 curr per 25327

– Facultat de Psicologia, Ciències de l'Educació i de l'Esport Blanquerna, Biblioteca, Cister 34, 08022 Barcelona
T: +34 932533030; Fax: +34 932533031; E-mail: susannavs@blanquerna.url.edu; URL: biblioteca.blanquerna.url.edu
1963; Susanna Vintró
Psychology, Education, Speech Therapy and Sport
44 789 vols; 568 curr per; 32 e-journals; 23 e-books; 165 diss/theses; 1 709 av-mat; 592 sound-rec
libr loan 25328

– Institut Borja de Bioética, Biblioteca y Centro de Documentación, Santa Rosa, 39-57, 08950 Esplugues de Llobregat (Barcelona)
T: +34 936006106; Fax: +34 936006110; E-mail: biblioteca-bioetica@ibb.hsjdbcn.org; URL: www.ibbioetica.org
1975; Marta Perpiñán Arias
Bioethics
15 000 vols; 198 curr per; 98 e-journals; 837 diss/theses; 62 govt docs; 60 digital data carriers
libr loan 25329

– Institut Químic de Sarrià, Servei de Biblioteca i Documentació Ernest Solvay, Via Augusta, 390, 08017 Barcelona
T: +34 932672005; Fax: +34 932056266; E-mail: roser.escude@iqs.edu; URL: biblioteca.iqs.es
1916; Roser Escudé
Chemistry, chemical engineering, economics and managment, marketing, industrial engineering, mechanics
65 000 vols; 280 curr per; 920 e-journals; 50 diss/theses; 10 digital data carriers
libr loan 25330

Universitat Autònoma de Barcelona, Biblioteca Universitària, Edifici N Planta 1, 08193 ***Bellaterra*** (Cerdanyola del Vallès)
T: +34 935811015; Fax: +34 935813219; E-mail: s.biblioteques@uab.es; URL: www.uab.cat/bib
1968; Joan Ramon Gomez Escofet
Biblioteca Carandell (economics), MARCA (Catalan periodical on modernism), MONES (catalan periodical on modernism), history of science
1 043 080 vols; 59 298 curr per; 696 e-journals; 43 mss; 13 679 diss/theses; 43 789 maps; 157 379 microforms; 38 274 av-mat; 8 809 sound-rec; 20 digital data carriers
libr loan; LIBER, ALA, UKSG 25331

– Biblioteca de Ciència i Tecnologia, Edifici C, Campus de la UAB, 08193 Bellaterra (Cerdanyola del Vallès)
T: +34 935811906; Fax: +34 935812010; E-mail: bib.ciencia.tecnologia@uab.cat; URL: www.uab.cat
1977
Chemistry, physics, biology, mathematics, geology, computer science
63 843 vols; 1 263 curr per; 2 000 mss; 2 400 diss/theses; 2 752 maps; 5 600 microforms; 1 165 av-mat; 499 digital data carriers
libr loan 25332

– Biblioteca de Ciències Socials, Edifici B, Campus de la UAB, 08193 Bellaterra (Cerdanyola del Vallès)
T: +34 935811801; Fax: +34 935812009; E-mail: bib.socials.adquisicions@uab.cat; URL: www.uab.cat.bib
1988; Montserrat Catafal
Law, economics, political science, sociology, statistics; Biblioteca Carandell (economic history); Statistics section
206 000 vols; 8 000 curr per; 11 000 e-journals; 4 500 e-books; 770 microforms; 430 av-mat; 28 sound-rec; 171 digital data carriers
libr loan; CBUC, REBIUN 25333

– Biblioteca de Comunicació i Hemeroteca General, Edifici N, Campus de la UAB, 08193 Bellaterra (Cerdanyola del Vallès)
T: +34 935814004; Fax: +34 935811227; E-mail: Antonia.Galceran@uab.es
1972; Antònia Galcerán
96 500 vols; 1 734 curr per; 10 000 e-journals; 7 387 mss; 7 076 microforms; 30 200 av-mat; 1 759 sound-rec; 904 digital data carriers
libr loan 25334

– Biblioteca de Medicina, Campus Universitari, Edifici M, 08193 Bellaterra (Cerdanyola del Vallès)
T: +34 935811918; Fax: +34 935811243; E-mail: bib.medicina@uab.es; URL: www.uab.cat/bib
1973; Cristina Arola
50 000 vols; 5 000 curr per; 1 618 diss/theses
libr loan 25335

– Biblioteca de Veterinària, Edifici V – Campus Universitari UAB, 08193 Bellaterra (Cerdanyola del Vallès)
T: +34 935811549; Fax: +34 935812006; E-mail: bib.veterinaria@uab.cat; URL: www.uab.cat/bib
1985
Hist coll, pre 1970, Mycology coll
28 000 vols; 650 curr per; 200 diss/theses; 190 maps; 300 microforms; 360 av-mat; 260 digital data carriers
libr loan; IAALD, EAHIL 25336

– Biblioteca d'Humanitas, Edifici L (Biblioteca), Edifici B (Sala de Revistes), Campus de la UAB, 08193 Bellaterra (Cerdanyola del Vallès)
T: +34 935812992 (Biblioteca), 935811288 (Sala de Revistes); Fax: +34 935812900; E-mail: MariaDolors.Dilme@uab.cat; URL: www.uab.cat
1969; Maria Dolors Dilmé
Fons Calders, Fons Goytisolo, Fons Lesfargues, Fons Rosenthal, Collecció asiàtica
300 000 vols; 5 280 curr per; 3 391 microforms; 2 494 av-mat; 2 469 sound-rec; 412 digital data carriers
libr loan 25337

– Biblioteca Universitària de Sabadell UAB, C/ Emprius, 2, 08202 Sabadell
T: +34 937287701; Fax: +34 937287726; E-mail: bib.sabadell@uab.cat; URL: www.uab.cat/bib
1928; Rosa M. Garcia Burdó
Coll Sala Montllor i Pujal, Local hist coll, Hemeroteca
130 000 vols; 300 curr per; 3 900 e-journals; 84 microforms; 1 940 av-mat; 208 sound-rec; 698 digital data carriers
libr loan; Collegi Oficial de Bibliotecaris i Documentalistes de Catalunya 25338

Universidad de Deusto, Biblioteca Universitaria, Ramón Rubial, 1, 48009 ***Bilbao***
T: +34 944139419; Fax: +34 944139424; E-mail: nieves.taranco@deusto.es; URL: www.biblioteca.deusto.es
1886; Nieves Taranco
900 000 vols; 11 792 curr per; 105 e-journals; 700 e-books; 21 incunabula; 80 digital data carriers; 12 000 rare books (16-18th c)
libr loan; IFLA, LIBER, REBIUN 25339

– Biblioteca – Sede de San Sebastián, Mundaiz, 50, 20012 San Sebastián, Guipozcoa
T: +34 943273100; Fax: +34 943292635; E-mail: web@deusto.es
1956; Carmen Navarrete
Basque Libr
78 000 vols; 586 curr per; 650 diss/theses; 254 microforms; 147 av-mat
libr loan 25340

Facultad Teológica del Norte de España, Biblioteca, Martínez del Campo, 10, 09003 ***Burgos***
T: +34 947267000; Fax: +34 947252025; E-mail: teologiaburgos@planalfa.es; URL: www.teologiaburgos.org/facultad/biblioteca.aspx
1897; Cándido Rubio
Theology 17th c; archive
100 000 vols; 400 curr per; 150 microforms; 1 digital data carriers; ca 4000 vols of the 16th c
ABIE 25341

Universidad de Burgos, Biblioteca Universitaria, Pza Infanta Doña Elena s/n, 09001 ***Burgos***
T: +34 947258000; Fax: +34 947258043; URL: www.ubu.es/biblioteca/
1995; Fernando Martin Rodriguez
133 000 vols; 3 502 curr per; 2 822 diss/theses; 261 maps; 219 microforms; 145 sound-rec; 1 926 digital data carriers
libr loan; REBIUN 25342

Real Conservatorio Profesional de Música "Manuel de Falla" de Cádiz, Biblioteca, C/ Marqués del Real Tesoro, 10, 11001 ***Cádiz***
T: +34 9562431-06/-07; Fax: +34 956243109; E-mail: gps12893@gmail.com; URL: conservatoriomanueldefalla.es
1971; Maria del Carmen Mañas Diaz
9 100 vols; 5 curr per; 20 diss/theses; 1 300 sound-rec; 800 slides 25343

Universidad de Cádiz, Servicio Central de Bibliotecas, C/ Dr. Marañón, 3, Edif. de Servicios Centrales "Andrés Segovia", 11002 ***Cádiz***
T: +34 956015275; Fax: +34 956015791; E-mail: scbdir@uca.es; URL: biblioteca.uca.es
1990; Enrique Juan González Conejero
511 685 vols; 1 629 curr per; 24 707 e-journals; 217 561 e-books; 378 maps; 4 216 microforms; 2 039 av-mat; 3 611 sound-rec
libr loan; REBIUN, CEBUA, GEUIN, ISTEC, UKSG, IATUL 25344

– Biblioteca de Ciencias de la Salud, C/ Dr. Marañón, 3, 11002 Cádiz
T: +34 956015871; Fax: +34 956015270; E-mail: biblioteca.csalud@uca.es; URL: www.uca.es
1993; Ana Remón Rodríguez
18 000 vols; 885 curr per; 1 000 e-journals; 140 diss/theses; 200 digital data carriers; 3 000 rare books (17-18th c) 25345

– Biblioteca de Ciencias Sociales y Juridicas, Duque de Nájera 8, 11002 Cádiz
T: +34 956015376; Fax: +34 956015382; E-mail: biblioteca.csociales@uca.es; URL: biblioteca.uca.es/sbuca/bibcsoc.htm
1970
27 000 vols
libr loan; AAB 25346

– Biblioteca de E.S. Ingeniería, C/ Chile 1, 11002 Casíz
T: +34 956015107; Fax: +34 956015382; E-mail: biblioteca.ingenieria@uca.es; URL: biblioteca.uca.es 25347

– Biblioteca de Humanidades, C/ Dr Gómez Ulla s/n, 11003 Cádiz
T: +34 956015627; Fax: +34 956015503; E-mail: biblioteca.humanidades@uca.es; URL: biblioteca.uca.es
1981
70 000 vols; 600 curr per; 240 diss/theses; 170 maps; 1 000 microforms; 100 sound-rec; 10 digital data carriers
libr loan; AAB 25348

– Biblioteca del Campus de Algeciras, Escuela Politécnica Superior de Algeciras, C/ Ramón Puyol s/n, 11202 Algeciras
T: +34 956028082; Fax: +34 956028089; E-mail: biblioteca.algeciras@uca.es; URL: biblioteca.uca.es 25349

– Biblioteca del Campus de Jerez, Av de la Universidad s/n, 11405 Jerez de la Frontera
T: +34 956037015; Fax: +34 956037077; E-mail: biblioteca.campusjerez@uca.es; URL: biblioteca.uca.es
1980
12 100 vols 25350

– Biblioteca del Campus Río San Pedro, Av República Saharaui s/n, 11510 Puerto Real
T: +34 956016304; Fax: +34 956016288; E-mail: biblioteca.riosanpedro@uca.es; URL: biblioteca.uca.es
1974
29 000 vols 25351

Universitat Jaume I, Biblioteca – Centro de Documentación, Campus del Riu Sec., 12071 ***Castelló de la Plana***
T: +39 964-728761/-728762; Fax: +39 964728778; E-mail: biblioteca@uji.es; URL: www.uji.es/cd
Vicent Falomir del Campo
155 000 vols
libr loan 25352

Universidad de Castilla-La Mancha – Biblioteca Universitaria, Av Camilo José Cela, s/n, 13071 ***Ciudad Real***
T: +34 926295300; Fax: +34 926295340; E-mail: biblioteca@uclm.es; URL: www.biblioteca.uclm.es
1985; Antonio L. Galán Gall
Biblioteca General de Albacete, de Ciudad Real, de Cuenca, de Toledo
838 037 vols; 10 496 curr per; 120 e-journals; 10 944 mss; 350 diss/theses; 3 912 maps; 2 561 microforms; 22 085 av-mat; 77 592 sound-rec
libr loan 25353

Conservatorio Profesional de Música de Córdoba, Biblioteca, Av del Mediterraneo s/n, 14011 ***Córdoba***
T: +34 957464302; Fax: +34 957463990; E-mail: secretario@cpmcordoba.com; URL: www.cpmcordoba.com/cms/Biblioteca/
1862; Antonio Ramirez Gaete
8 700 vols; 2 curr per; 6 000 music scores; 1 270 sound-rec 25354

Universidad de Córdoba, Biblioteca General, Campus de Rabanales s/n, 14071 ***Córdoba***
T: +34 957211009; Fax: +34 957218136; E-mail: bg1limam@uco.es; URL: www.uco.es
1978; María del Carmen Liñán Maza
Libr sci, libr automation, indexation, bibliography
300 000 vols; 4 008 curr per; 1 500 av-mat; 1 000 sound-rec 25355

– Biblioteca de Ciencias del Trabajo, Av de las Ollerías, 14071 Córdoba
T: +34 957212519; E-mail: bg3vigac@uco.es
Cristina Ruíz de Villegas García-Pelayo
Labor, Manpower 25356

– Biblioteca del Campus de Rabanales, Campus d Rabanales s/n, 14071 Córdoba
T: +34 957211016; Fax: +34 957218136; E-mail: bg3gugaj@uco.es; URL: www.uco.es
1999; Juan Alfredo Guzmán García
150 000 vols; 3 800 curr per; 16 000 e-journals; 105 diss/theses 25357

– Escuela Universitaria Politecnica, Biblioteca, Av Menendez Pidal, s/n, 14004 Córdoba
T: +34 957218326; Fax: +34 957218316
1976; Encarnita Baena Parejo
Computer science, industrial engineering
11 700 vols; 229 curr per; 741 diss/theses; 184 microforms; 23 av-mat; 50 sound-rec
libr loan 25358

– ETEA Biblioteca, Escritor Castilla Aguayo, 4, 14004 Córdoba
T: +34 957222171; Fax: +34 957222107; E-mail: biblioteca@etea.com; URL: www.etea.com
1963; Josep M. Margenat Peralta
62 000 vols; 776 curr per; 1 500 microforms; 200 av-mat; 244 sound-rec; 1 000 digital data carriers 25359

– Facultad de Arquitectura, Urbanismo y Diseño, Biblioteca Mario Fernandez Ordoñez, Av. Velez Sarsfield 264, 5000 Córdoba
T: +54 351 4332091; Fax: +54 351 4332092; URL: www.faudi.unc.edu.ar
1956; Sofia Gordillo
18 000 vols; 120 curr per; 130 diss/theses; 60 digital data carriers; 900 files 25360

– Facultad de Ciencias, Biblioteca, Av San Alberto Magno, s/n, 14004 Córdoba
T: +34 957218584; Fax: +34 957218606
1977; Manuela Ramírez Ponferrada
Physics, chemistry, biology
8 900 vols; 418 curr per; 870 microforms 25361

– Facultad de Ciencias de la Educación, Biblioteca, San Alberto Magno, s/n, 14071 Córdoba
T: +34 957218946; E-mail: bg1rocid@uco.es
Daniel Rodríguez Cibrián 25362

– Facultad de Derecho y CC. Económicas y Empresariales, Biblioteca, Puerta Nueva, s/n, 14071 Córdoba
T: +34 957218844; Fax: +34 957218902; E-mail: bg1fejam@uco.es
1983; Carmen Fernández Jaén
Law, Economy, Business Administration
41 000 vols; 450 curr per; 190 diss/theses; 3 microforms
libr loan 25363

– Facultad de Filosofía y Letras, Biblioteca, Pl Cardenal Salazar, s/n, 14071 Córdoba
T: +34 957218307; Fax: +34 957218788; E-mail: bg1puagr@uco.es
1972; Rosario Puerto Agüera
85 000 vols; 854 curr per; 500 diss/theses; 2 000 maps; 150 microforms; 5 240 av-mat 25364

– Facultad de Medicina, Biblioteca, Av Menéndez Pidal, s/n, 14004 Córdoba
T: +34 957218237; Fax: +34 957218229; E-mail: bg1cosam@uco.es
1972; Luisa Córdoba Sánchez
23 400 vols; 877 curr per; 306 diss/theses; 347 av-mat; 58 digital data carriers 25365

– Facultad de Veterinaria, Biblioteca, Campus Rabanales, 14071 Córdoba
T: +34 957211016; Fax: +34 957218666; E-mail: bg3eshim@uco.es; URL: www.uco.es/veterinaria/principal/servicios/biblioteca.html
1848; M.M. Cámara Aroca
Bks from 17th to 19th c
34 200 vols; 329 curr per; 619 diss/theses; 580 maps; 57 av-mat; 13 digital data carriers
libr loan; EAHIL, ANABAD, AAB 25366

Universidade da Coruña, Biblioteca Universitaria, Edif. Xoana Capdevielle, Campus de Elviña, s/n, 15071 ***A Coruña***
T: +34 981167000; Fax: +34 981167015; E-mail: informacion.bib@udc.es; URL: www.udc.es
1973; Ángeles Campos Rodríguez
670 000 vols; 7 600 curr per; 70 e-journals; 5 diss/theses 25367

– Escola Técnica Superior de Arquitectura, Biblioteca, A Zapateira s/n, 15650 A Coruña
T: +34 981167000 ext 5020/5001; Fax: +34 981167055; URL: www.udc.es/etsa/biblioteca/index.htm
1975; Araceli Sanjuán Pedreira
Architecture, Art, Construction, Hist, Urbanism
33 000 vols; 330 curr per; 6 e-journals; 35 diss/theses; 551 maps; 45 microforms; 1 565 av-mat; 147 digital data carriers
libr loan; ABBA 25368

– Escola Técnica Superior de Camiños, Canais e Portos, Biblioteca, Campus de Elviña, 15071 A Coruña
T: +34 981167000 ext 1460, 1461; Fax: +34 981167170; E-mail: caminos.bib@udc.es; URL: caminos.udc.es/servicios/biblioteca/
1991; María Pilar Roel Vilas
Applied Mathematics, Structures (Civil Engineering), Hydraulic Engineering, Port and Coastal Engineering, Pavements, Traffic and Transport Engineering, Water Management, Water distribution systems
libr loan 25369

– Escuela Universitaria Politécnico de Ferrol, Biblioteca, Ctra. de Aneiros, s/n – Serantes, 15405 Ferrol (A Corunña)
T: +34 981337400 ext 3044; E-mail: eup.bib@udc.es; URL: bibliotecaeup.cdf.udc.es
María Teresa Bárceba Varela de Limia
Marine engineering, engineering
27 000 vols
libr loan 25370

– Facultade de Ciencias, Biblioteca, Campus da Zapateira, s/n, 15071 A Coruña
T: +34 981167000 ext 2009/2054/2059; E-mail: ciencias.bib@udc.es; URL: ciencias.udc.es/biblioteca/
1972; Maria José Lobeiras Fernandez
Biology, Chemistry, Environmental Sciences
27 991 vols; 375 curr per; 388 diss/theses; 174 maps; 34 030 microforms; 372 av-mat; 228 digital data carriers; 165 slides
libr loan 25371

– Facultade de Ciencias da Educación, Biblioteca, Campus de Elviña, 15071 A Coruña
T: +34 981167100 ext 4609
1955; María Gutiérrez Calvete
Education, psychology
45 000 vols; 324 curr per; 83 sound-rec; 3 800 slides 25372

– Facultade de Ciencias Económicas e Empresariais, Biblioteca, C/ Almirante Lángara, s/n, 15011 A Coruña
T: +34 981167000 ext 4409; Fax: +34 981167110
1850; María José Parga Massa
9 740 vols; 140 curr per; 100 sound-rec; 12 digital data carriers
libr loan 25373

Universidad de Castilla-La Mancha – Colegio Universitario Cardenal Gil de Albornoz, Biblioteca, Av de los Alfares, 42, 16071 ***Cuenca***
T: +34 969232030
1972
Law, hist, demography
8 700 vols; 18 curr per; 25 maps; 100 sound-rec; 1 000 slides 25374

Universidad de Castilla-La Mancha – Escuela Universitaria del Profesorado de E.G.B. Fray Luis de León, Biblioteca, Astrana Marin, 6, 16002 ***Cuenca***
T: +34 969222456
1974
Education, hist, psychology
20 000 vols; 55 curr per; 48 maps; 250 av-mat; 1 800 slides 25375

Universidad Laboral de Eibar, Biblioteca, Otaola Iribidea, 29, 20000 ***Eibar***
T: +34 943718444
1968
16 100 vols 25376

Universidad Laboral 'José Antonio Girón', Biblioteca, ***Girón***, Asturias
1955
20 000 vols 25377

Universitat de Girona, Biblioteca, Pl Sant Domènec, 3, 17071 ***Girona***
T: +34 972418037; Fax: +34 972418243; E-mail: biblioteca.projectes@udg.edu; URL: biblioteca.udg.es; www.udg.edu/biblioteca
1972; Antònia Boix
Ferrater Mora (Philosophie), Vicens Vives (Hist), Bertrana (Lit), Education Hist, Robert Brian Tate (hispon), Anteiessores (ancient Judicial), ICHN (Natural Hist Serials), Social Studies of Saltwater fishing
272 973 vols; 3 240 curr per; 4 640 e-journals; 1 757 mss; 1 813 diss/theses; 283 music scores; 20 207 maps; 1 902 microforms; 10 268 av-mat; 1 446 sound-rec; 2 547 digital data carriers
libr loan; ALA, REBIUN, CBUC 25378

Universidad de Granada, Biblioteca Universitaria, Cuesta del Hospicio, s/n, 18071 ***Granada***
T: +34 958243053; Fax: +34 958243066; URL: biblioteca.ugr.es
1532; María José Ariza Rubio
Montenegro Coll (rare bks from 16th-18th c)
1 100 000 vols; 8 900 curr per; 118 e-journals; 1 500 mss; 50 incunabula; 15 000 diss/theses; 3 500 music scores; 6 150 maps; 37 230 microforms; 11 200 av-mat; 11 800 sound-rec
libr loan 25379

– Biblioteca de Medicina y CC de la Salud, Avenida de Madrid, s/n, 18012 Granada
T: +34 958242992; Fax: +34 958248970; E-mail: mdoloresfernandez@ugr.es; URL: biblioteca.ugr.es/pages/biblioteca_ugr/bibliotecas_centros/biosanitaria
1952; María Dolores Fernández García
Paramedical Training, Nursing
47 200 vols; 1 832 curr per; 3 diss/theses; 310 av-mat; 192 digital data carriers
libr loan 25380

– Biblioteca del Edificio de San Jerónimo, C/ Rector López Argüeta, s/n, 18001 Granada
T: +34 958244206; Fax: +34 958244213; E-mail: bibgesjeronimo@ugr.es; URL: biblioteca.ugr.es/pages/biblioteca_ugr/bibliotecas_centros/san_jeronimo
José Miguel Romero Pérez
Social Work, Labor
16 000 vols; 391 curr per; 213 av-mat; 232 digital data carriers 25381

– Biblioteca Politécnico, Campus Universitario de Fuentenueva, 18071 Granada
T: +34 958244162; Fax: +34 958242317; E-mail: carmenzea@ugr.es
Carmen Rosa Zea Espinar
Civil Engineering, Highway and Traffic Engineering, Chemistry
23 200 vols; 253 curr per; 200 av-mat; 940 digital data carriers; 600 photos/slides 25382

– Campus Universitario de Ceuta, Biblioteca, C/ Greco, 10, 51002 Ceuta
T: +34 956526118; Fax: +34 956526117; E-mail: bibceuta@ugr.es
1963; Lourdes Navarro González
24 000 vols; 169 curr per; 52 maps; 600 av-mat; 550 sound-rec; 380 digital data carriers
libr loan 25383

– Campus Universitario de Melilla, Biblioteca, Calle Santander, 1, 52071 Melilla
T: +34 952698740; Fax: +34 952698778; E-mail: bibgesmelilla@ugr.es
1970; Teresa Serrano Darder
31 000 vols; 602 curr per; 177 av-mat; 700 digital data carriers
libr loan 25384

– Colegio Máximo, Facultad de Comunicación y Documentación, Biblioteca, Campus Universitario de Cartuja, 18071 Granada
T: +34 958243944; Fax: +34 958249412; E-mail: rolivares@ugr.es; URL: fcd.ugr.es/pages/biblioteca/index
1986; Rafael Jesús Olivares Castillo
Dentistry
11 500 vols; 423 curr per; 15 e-journals; 126 diss/theses; 75 av-mat; 235 digital data carriers
libr loan; IFLA, AAB, ANABAD, SEDIC 25385

– Escuela Técnica Superior de Arquitectura, Biblioteca, Av Andalucía, 38, 18001 Granada
T: +34 958246114; Fax: +34 958246115; E-mail: luiscarizquierdo@ugr.es; URL: etsag.ugr.es/
1976; Luis Carlos Izquierdo Rivas
13 500 vols; 131 curr per; 268 av-mat; 354 digital data carriers; 3 500 photos/slides 25386

– Escuela Técnica Superior de Ingeniería Informática y Telecomunicación, Biblioteca, C/ Periodista Daniel Saucedo Aranda s/n, 18071 Granada
T: +34 958242806; Fax: +34 958242806; E-mail: rocioraya@ugr.es; URL: etsiit.ugr.es/pages/biblioteca/index
Rocío Raya Prida
Information Science, Telecommunications
21 000 vols; 196 curr per; 255 av-mat; 830 digital data carriers 25387

– Facultad de Bellas Artes, Biblioteca-Mediateca, Av de Andalucia, Edif. Aynadamar, 18071 Granada
T: +34 958242968; Fax: +34 958242713; E-mail: fbbaa@ugr.es; URL: www.bellasartesgranada.org/index.php/Biblioteca-Mediateca/39/0/
1985; Juan Francisco Herranz Navarra
17 000 vols; 185 curr per; 610 av-mat; 620 digital data carriers; 1 600 photos/slides
libr loan 25388

– Facultad de Ciencias, Biblioteca, Av de Fuentenueva, s/n, 18071 Granada
T: +34 958244020; Fax: +34 958244020
Carmen Berzosa Valencia
91 600 vols; 1 292 curr per; 670 maps; 1 930 microforms; 440 av-mat 25389

– Facultad de Ciencias de la Educación, Biblioteca, Campus Universitario Cartuja, 18071 Granada
T: +34 958243995; Fax: +34 958240615; E-mail: biblioce@ugr.es
Margarita Ramírez Reyes
71 300 vols; 640 curr per; 500 av-mat; 363 sound-rec; 705 digital data carriers
25390

– Facultad de Ciencias Economicas y Empresariales, Biblioteca, Campus Universitario de Cartuja, 18071 Granada
T: +34 958243698; Fax: +34 958242338
1944; Ascensión Vázquez Quero
Economics, Business Administration, Statistics
33 700 vols; 1 119 curr per; 79 diss/theses; 250 maps; 150 microforms; 430 av-mat; 40 digital data carriers
libr loan
25391

– Facultad de Ciencias Políticas y Sociología, Biblioteca, Calle Rector López Argüeta, s/n, 18071 Granada
T: +34 958244179; Fax: +34 958240896; E-mail: Isabelfajardo@ugr.es; URL: www.ugr.es/~ccpolsoc/
Isabel Fajardo Gómez
42 100 vols; 667 curr per; 260 microforms; 500 av-mat; 253 digital data carriers
25392

– Facultad de Derecho, Biblioteca, Plaza de la Universidad, s/n, 18171 Granada
T: +34 958243448; Fax: +34 958248938; E-mail: decanatoderecho@ugr.es
1862; Edelmira Suárez del Toro Rivero
Economics
149 000 vols; 2 053 curr per; 2 mss; 340 diss/theses; 18 200 microforms; 23 digital data carriers
libr loan
25393

– Facultad de Farmacia, Biblioteca, Campus de Cartuja, s/n, 18071 Granada
T: +34 958243822; Fax: +34 958246241; E-mail: bibfar03@ucartuja.ugr.es
1960; Dolores Aguaza Salvador
29 000 vols; 869 curr per; 1 000 diss/theses; 170 microforms; 35 av-mat; 240 digital data carriers
libr loan
25394

– Facultad de Filosofía y Letras, Biblioteca (Faculty of Philosophy and Letters), Campus Universitario de Cartuja, 18071 Granada
T: +34 958243563, 958243564, 958242850; Fax: +34 958249021; E-mail: cperegrin@ugr.es; URL: www.ugr.es/~bibliofl/
1858; Cristina Peregrín Pardo
Rare bks from 16th – 18th c (1 135 vols), Rare bks printed in 19th c (4 216 vols)
285 000 vols; 4 900 curr per; 1 844 diss/theses; 3 383 music scores; 5 200 maps; 5 850 microforms; 1 387 av-mat; 1 107 sound-rec; 769 digital data carriers
libr loan; REBIUN, CONSORCIO DE BIBLIOTECAS UNIVERSITARIAS DE ANDALUCIA (CEBUA)
25395

– Facultad de la Actividad Física y el Deporte, Biblioteca, Carretera de Alfacar, s/n, 18071 Granada
T: +34 958249424; Fax: +34 958249425; E-mail: bibcirdeportes@ugr.es
1984; Ana María Peregrín González
Sports, Physical Education; Coll of 'Centro de Estudios Olímpicos'
14 500 vols; 458 curr per; 6 000 e-journals; 154 diss/theses; 16 microforms; 175 av-mat; 6 sound-rec; 121 digital data carriers; 2 600 photos/slides
libr loan
25396

– Facultad de Psicología, Biblioteca, Campus Universitario de Cartuja, 18071 Granada
T: +34 958243777; Fax: +34 958242976; E-mail: secrepsico@ugr.es
Cristóbal Pasadas Ureña
35 600 vols; 609 curr per; 380 microforms; 1 230 av-mat; 125 digital data carriers
25397

– Facultad de Traducción e Interpretación, Biblioteca, Puentezuelas, 55, 18071 Granada
T: +34 958243485; Fax: +34 958243468; E-mail: bibgestraduccion@ugr.es
1980
Translations
18 100 vols; 367 curr per; 15 diss/theses;

9 000 microforms; 300 av-mat; 440 sound-rec; 240 digital data carriers; 500 photos/slides
libr loan
25398

– Hospital Real, Biblioteca, Cuesta del Hospicio, s/n, 18071 Granada
T: +34 958244256; Fax: +34 958244058
Inés María del Álamo Fuentes
Biblioteca de fondo histórico mulitdisciplinar (15th-19th c)
68 000 vols; 129 curr per; 950 microforms; 330 av-mat; 300 digital data carriers
25399

Universidad de Huelva, Biblioteca Central, Campus Universitario del Carmen, 21071 *Huelva*
T: +34 959219294; E-mail: director@biblio.uhu.es; URL: www.uhu.es/biblioteca/
José Carlos Villadóniga Gómez
260 000 vols; 4 755 curr per; 22 500 e-journals; 3 114 maps; 4 227 av-mat; 97 digital data carriers
25400

– Escuela Politécnica Superior, Campus La Rábida, Biblioteca, Ctra de Palos de la Frontera, 21071 Palos de la Frontera, Huelva
T: +34 959530245; Fax: +34 959350311
1959
8 000 vols; 1 curr per; 5 av-mat
25401

– Facultad de Ciencias Empresariales y Jurídicas, Campus La Merced, Biblioteca, Pl de la Merced, s/n, 21002 Huelva
T: +34 959284625; Fax: +34 959284306; E-mail: biblio_e@biblio.uhu.es; URL: www.uhu.es/biblioteca
1974
Business management, law, geography, hist
28 500 vols; 801 curr per; 17 microforms; 5 av-mat
libr loan
25402

Universidad de Jaén, Biblioteca, Campus Las Lagunillas, Edif. B-2, 23071 *Jaén*
T: +34 953212198; Fax: +34 953212543; E-mail: infobib@ujaen.es; URL: www.ujaen.es/serv/biblio
1993; Sebastián Jarillo Calvarro
204 737 vols; 3 329 curr per; 2 607 e-journals
libr loan; REBIUN
25403

Universidad de La Laguna, Biblioteca Universitaria, Campus de Guajara, 38204 *La Laguna*, S.C. Tenerife
T: +34 922317830; Fax: +34 922317451; E-mail: mmorales@ull.es; URL: www.bbtk.ull.es
1789; Luis Gonzalo Rey Pinzón
Canary Islands
550 000 vols; 8 100 curr per; 150 mss; 20 incunabula; 1 700 diss/theses; 5 000 maps; 3 000 microforms; 10 000 sound-rec
libr loan
25404

Universidad del País Vasco / Euskal Herriko Unibertsitatea, Biblioteca Central (University of the Basque Country), Sarriena, s/n, 48940 *Leioa*, Bizkaia
T: +34 946015125; Fax: +34 946013311; E-mail: llzgublc@lg.ehu.es; URL: www.biblioteka.ehu.es
1968; Carmen Guerra Blasco
Sección Estudios Vascos, Sección Historica de Literatura Infantil
873 120 vols; 7 180 curr per; 2 382 maps; 70 257 microforms; 6 878 av-mat; 4 386 sound-rec; 2 972 digital data carriers; 63 044 photos and slides
libr loan; ANABAD, IATUL
25405

– Campus de Alava, Biblioteca, Nieves Cano, 33, 01006 Vitoria-Gasteiz
T: +34 945013171; E-mail: vlgbibli@vc.ehu.es
Esperanza Iñurrieta Ambrosio
libr loan
25406

– Campus de Bizkaia, Biblioteca, Av Lehendakari Aguirre, 83, 48015 Bilbao
T: +34 946013999; E-mail: alberto.garciaf@ehu.es
Alberto García Fernández
libr loan
25407

– Campus de Gipuzkoa, Biblioteca, Campus de Ibaeta, s/n, 20009 San Sebastian
T: +34 943015869; E-mail: sqzpifem@sq.ehu.es
Marian Piñeiro Fernández
libr loan
25408

– Escuela Técnica Superior de Ingeniería, Ingeniaritza Goi Eskola Teknikoa, Biblioteca, Alameda de Urqijo, s/n, 48013 Bilbao
T: +34 946014104; Fax: +34 946014295; E-mail: bitidire@bi.ehu.es; URL: www.ehu.es/biblioteca
1956; Amaia Rola Isasi
Suscription to UNE, UIT-T and UIT-R standards, DIN standards translated to Spanish, ANSI/ASME standards, IEEE Standards
43 035 vols; 904 curr per; 607 diss/theses; 65 microforms; 122 av-mat; 671 digital data carriers
libr loan
25409

– Escuela Universitaria de Magisterio, Biblioteca, Plaza Oñati, 3, 20018 Donostia-San Sebastián, Guipozcoa
T: +34 943018491; Fax: +34 943018457; E-mail: sgrbibli@sc.ehu.es; URL: www.ehu.es/biblioteca
1980; Lourdes Arrien Marina
Education
50 380 vols; 387 curr per; 172 maps; 17 microforms; 645 av-mat; 560 sound-rec; 384 digital data carriers
25410

– Facultad de Bellas Artes, Arte Ederren Fakultateko, Biblioteca (Faculty of Fine Arts), B° Sarriena, s/n, 48940 Leioa, Bizkaia
T: +34 946012999; Fax: +34 946013377; E-mail: lbzbibli@lg.ehu.es
1970; Maria José Lertxundi Etxebarria
50 491 vols; 156 curr per; 264 diss/theses; 7 maps; 1 microforms; 2 655 av-mat; 445 sound-rec; 549 digital data carriers
libr loan; IFLA
25411

Universidad de Léon, Biblioteca General "San Isidro", Campus de Vegazana s/n, 24071 *León*
T: +34 987291133; Fax: +34 987291690; E-mail: buusu@unileon.es; URL: www5.unileon.es/bibportal/
1979; Santiago Asenso Rodriguez
458 170 vols; 4 295 curr per; 11 114 e-journals; 1 mss; 7 658 diss/theses; 7 161 maps; 869 microforms; 5 559 av-mat; 5 684 sound-rec; 254 digital data carriers
libr loan
25412

Universitat de Léon, Servei de Biblioteca i Documentació, Jaume II, 67, 25001 *Lleida*
T: +34 973003521; Fax: +34 973003518; E-mail: loli@sbd.udl.cat; URL: www.sbd.udl.cat
Loli Manciñeiras

– Biblioteca Cappont, Jaume II, 67, 25001 Lleida
T: +34 973003511; Fax: +34 973003512; E-mail: marta.planas@udl.cat; URL: www.sbd.udl.cat/sbd/cappont.html
Marta Planas
libr loan
25413

– Biblioteca Ciències de la Salut, Jaume II, 67, 25001 Lleida
T: +34 973702421; E-mail: csalut@sbd.udl.cat; URL: www.sbd.udl.cat/sbd/cs.html
Lauta Jové
Public Health
libr loan
25414

– Biblioteca ETSEA, Rovira Roure, 177, 25198 Lleida
T: +34 973702516; Fax: +34 973238264; E-mail: etsea@sbd.udl.cat; URL: www.bib.udl.cat
1976; Elionor Vilalta
34 000 vols; 445 curr per; 1 500 e-journals; 9 700 microforms
libr loan; REBIUM, MECANO
25415

– Biblioteca Lletres, Plaça Victor Siurana, 1, 25003 Lleida
T: +34 973702132; Fax: +34 973702122; E-mail: rosamaria.roso@udl.cat; URL: www.sbd.udl.cat/sbd/lletres.html
1976; Rosa Maria Rosó
76 000 vols; 620 curr per; 7 200 e-journals; 3 300 diss/theses; 100 maps; 40 013 microforms; 820 av-mat; 454 sound-rec; 617 digital data carriers; 3 928 legacies (Márius Torres (190), Samuel Gili Gaya (3738))
libr loan; CBUC, REBIUN
25416

Universidad de la Rioja, Biblioteca, Campus Universitario, 26006 *Logroño* (La Rioja)
T: +34 941299194; Fax: +34 941299193; E-mail: biblioteca@unirioja.es; URL: biblioteca.unirioja.es
1992; Marta Magriná Contreras
140 000 vols; 2 258 curr per; 89 diss/theses; 105 maps; 587 microforms; 5 400 av-mat; 529 sound-rec; 253 digital data carriers
libr loan
25417

Biblioteca Universitaria y Técnica Antón Ramírez, Caja de Ahorros y Monte de Piedad de Madrid, Guzmán el Bueno, 66, 28015 *Madrid*
T: +34 915443730
1977
23 000 vols; 78 curr per
25418

Casa do Brasil, Colegio Mayor Universitario, Biblioteca, Av Arco de la Victoria, s/n, 28040 *Madrid*
T: +34 914551560; Fax: +34 915435188; E-mail: adm@casadobrasil.es; URL: www.casadobrasil.es
1977; Cassio Roberto De Almeida Romano
Brazilian matters
8 500 vols; 90 curr per; 10 diss/theses; 500 govt docs; 1 050 av-mat; 5 digital data carriers
25419

Colegio Universitario de Estudios Financieros – CUNEF, Biblioteca, Serrano Anguita, 9 y 13, 28004 *Madrid*
T: +34 914489105
1976; Paula Vicente Garcia
10 400 vols; 430 curr per; 260 diss/theses
25420

Escuela Oficial de Correos y Telecommunicaciones, Biblioteca, Conde d Peñalver, 19, 28006 *Madrid*
T: +34 913962805; Fax: +34 913962747
1978
16 400 vols; 39 curr per; 105 av-mat; 820 sound-rec
25421

Escuela Superior de las Fuerzas Armadas (ESFAS), Biblioteca, Paseo de la Castellana, 61, 28046 *Madrid*
T: +34 913482500; Fax: +34 9134825-53/-54; E-mail: esfas@ceseden.es
1941
65 000 vols; 111 curr per
25422

Fundación Universitaria Española, Biblioteca, Alcalá, 93, 28009 *Madrid*
T: +34 914311122; Fax: +34 915767352; URL: www.fuesp.com/web/biblioteca/index.htm
1953; Isabel Balsinde Rodríguez
Archivos de Campomanes, Sáenz de Tejada, López Toro y Francos-Monroy
200 000 vols; 2 740 curr per; 1 835 mss; 790 diss/theses; 6 000 microforms; 8 000 av-mat; 1 332 enciclopedias
25423

Instituto Superior de Ciencias Catequéticas San Pio X, Biblioteca, La Salle Campus, C/ La Salle, 10, 28023 *Madrid*
T: +34 917401980; Fax: +34 913571730; E-mail: sia@lasallecampus.es; URL: www.lasallecampus.es
1955
Catechetical sciences, education
41 000 vols; 220 curr per; 466 diss/theses; 80 microforms; 1 720 av-mat; 57 819 sound-rec
libr loan
25424

Instituto Superior de Pastoral, Biblioteca, Juan XXIII, 3, 28040 *Madrid*
T: +34 915335200; Fax: +34 915340983; URL: instpast.upsa.es
1965; Sandra Sanchez H.
Pastoral theology, patristic studies
17 350 vols; 165 curr per; 203 diss/theses
25425

Real Conservatorio Superior de Música de Madrid, Biblioteca (Royal Conservatory of Madrid), C/ Dr Mata, 2, 28012 *Madrid*
T: +34 915392901; Fax: +34 915275822; E-mail: biblio.rcsmm@telefonica.net; URL: www.educa.madrid.org/csm.realconservatorio.madrid/
1831; Carlos J. Cosálves Lara
Archivo Rey Amadeo, Archivo Sociedad de Conciertos, Colección Uclés, Colección Stevenson, Colección of

manuscripts of Scarlatti, Boccherini, Rossini, Haydn, Albéniz, Tartini, Albéro etc.
20 000 vols; 800 curr per; 10 000 mss; 62 incunabula; 80 000 music scores; 1 600 microforms; 8 650 av-mat; 20 000 sound-rec; 130 800 sheet music, 62 music prints of 16th c
libr loan; AEDOM, AIBN, SEDEM, GALPIN 25426

Universidad Autónoma de Madrid, Biblioteca y Archivo de la UAM, Ctra de Colmenar, km 15, Cantoblanco, 28049 ***Madrid***
T: +34 913974399; Fax: +34 913975058; E-mail: servicio.biblioteca@uam.es; URL: biblioteca.uam.es
1968; Santiago Fernández Conti
Maps coll
840 000 vols; 18 000 curr per; 76 000 e-journals; 45 373 e-books; 12 195 diss/theses; 42 music scores; 28 000 maps; 8 162 microforms; 5 877 av-mat; 679 sound-rec; 29 379 photos, 8 711 CD-ROMs
libr loan; IFLA, LIBER, SEDIC, MADROÑO, REBIUN
Reading Room is open 24 hours 25427

– Archivo Central, Calle Einstein, 3, Edificio de Rectorado, 1ª planta, Campus de Cantoblanco, 28049 Madrid
T: +34 914975021; E-mail: archivo.central@uam.es; URL: biblioteca.uam.es/paginas/Archivo/indice.html
1987; José Luis López Fernández
Centro de Documentación Estadística, Centro de Documentación Europea (CEE) 25428

– Centro de Documentación Europea, Fac. de Ciencias Económicas y Empresariales (Módulo-III), C/ Francisco Tomás y Valiente, 5, Campus de Cantoblanco, 28049 Madrid
T: +34 914974797; Fax: +34 914975564; E-mail: cde@uam.es; URL: biblioteca.uam.es/paginas/Cent_Europea/cent_europea.html
María Sintes Olivar
Centro de Documentación Estadística, Centro de Documentación Europea (CEE) 25429

– Escuela Politécnica Superior, Biblioteca (Higher Polytechnical School), Fransisco Tomás y Valiente 11, Ciudad Universitaria de Cantoblanco, 28049 Madrid
T: +34 914972314; Fax: +34 914972317; E-mail: biblioteca.eps@uam.es; URL: biblioteca.uam.es/politecnica/default.html
Marisol Orta González-Orduña
libr loan 25430

– Facultad de Ciencias, Biblioteca Fernando González Bernáldez, Calle Isaac Newton, 11, Campus de Cantoblanco, 28049 Madrid
T: +34 914978353; Fax: +34 914974187; E-mail: biblioteca.ciencias@uam.es
1983; Teresa Gomez Nebreda
libr loan 25431

– Facultad de Ciencias Económicas y Empresariales, Biblioteca, Ctra de Colmenar, km 15, Cantoblanco, 28049 Madrid
T: +34 913942608; Fax: +34 913942426; E-mail: buc_cee@buc.ucm.es; URL: www.ucm.es/info/cee/biblioteca.htm
1975; Pepa Rodríguez
Centro de Documentación Estadística, Centro de Documentación Europea (CEE); Centro de Documentación Estadística
libr loan 25432

– Facultad de Derecho, Biblioteca, Calle Kelsen, 1, Campus de Cantoblanco, 28049 Madrid
T: +34 914978221; Fax: +34 914972586; E-mail: biblioteca.derecho@uam.es
1969; Nieves Martínez Maire 25433

– Facultad de Filosofía y Letras, Biblioteca de Humanidades, Calle Freud, 3, Campus de Cantoblanco, 28049 Madrid
T: +34 914975554; Fax: +34 914975064; E-mail: biblioteca.humanidades@uam.es; URL: biblioteca.uam.es/paginas/Humanidades.humanidades.html
1984; María Ángeles Martínez Frías
Cartoteca (cartoteca@uam.es) 25434

– Facultad de Formación de Profesorado y Educación, Biblioteca, Calle Francisco Tomás y Valiente, 3, Campus de Cantoblanco, 28049 Madrid
T: +34 914973975; Fax: +34 914978638; E-mail: biblioteca.fprofesorado@uam.es; URL: biblioteca.uam.es/paginas/Educacion/default.html
María Jesús Martínez Huelves 25435

– Facultad de Medicina, Biblioteca, Arzobispo Morcillo, s/n, 28029 Madrid
T: +34 914975483; Fax: +34 914975370; E-mail: biblioteca.medicina@uam.es; URL: biblioteca.uam.es/medicina/default.html
1970; Maria Pilar Barredo Sobrino
Biochemistry, biology, psychology
libr loan; EAHIL 25436

– Facultad de Psicología, Biblioteca, Cantoblanco, 28049 Madrid
T: +34 914974066; Fax: +34 914975215; E-mail: biblioteca.psicologia@uam.es; URL: biblioteca.uam.es/psicologia/default.html
1976; Carmen Lario Muñoz
Legado Rodríguez Sanabra, Archivo histórico: legados Lavín, Mallart, Pertejo, Dergam-Rupérez, Monasterio
36 000 vols; 222 curr per; 504 diss/theses; 282 av-mat
libr loan 25437

Universidad Carlos III de Madrid, Biblioteca María Moliner, Campus de Getafe, Av de Madrid, 126-128, 28903 ***Madrid***
T: +34 916249723; Fax: +34 916249783; E-mail: taladriz@db.uc3m.es; URL: www.uc3m.es/
1989; Margarita Taladriz Mas
90 000 vols; 1 416 curr per; 8 500 microforms; 110 av-mat; 589 sound-rec; 2 000 vols from 16-18th c
libr loan 25438

Universidad Complutense de Madrid, Biblioteca (Complutense University of Madrid), Edif. Multiusos 1. C, Profesor Aranugruen, s/n, Ciudad Universitaria, 28040 ***Madrid***
T: +34 913947852, 913947985; Fax: +34 913947849; E-mail: magan@buc.ucm.es; URL: www.ucm.es/BUCM
1499; José Antonio Magán Wals
Ancient books from 16th to 18th c (82 868 vols); Biblioteca Histórica 'Marquès de Valdecilla', Biblioteca Europea, Electronic Books: E-Libro, Digital Dissertations, Digital Ancient Books, etc.
2 721 013 vols; 42 000 curr per; 36 400 e-journals; 130 000 e-books; 30 000 mss; 727 incunabula; 34 104 diss/theses; 30 859 govt docs; 4 890 music scores; 40 303 maps; 16 200 microforms; 38 100 av-mat; 9 600 sound-rec; 81 504 digital data carriers
libr loan; IFLA, LIBER, UNICA, CERL, GEUIN, REBIUN, SEDIC, ANABAD, Consorcio Madroño 25439

– Biblioteca de Bellas Artes, Faculdad de Bellas Artes, C/ El Greco, 2, Ciudad Universitaria, 28040 Madrid
T: +34 913943626; Fax: +34 913941144; E-mail: buc_bba@buc.ucm.es; URL: www.ucm.es/BUCM/bba
1923; Angeles Vian Herrero
Drawings (artistic anatomy); Historical archive
34 000 vols; 248 curr per; 313 diss/theses; 36 microforms; 600 av-mat; 300 digital data carriers
libr loan 25440

– Biblioteca de Biológicas, C/ Jose Antonio Novais, 2, 28040 Madrid
T: +34 913945038; Fax: +34 913945038; E-mail: buc_bio@buc.ucm.es; URL: www.ucm.es/BUCM/bio/9550.php
1977; Alicia Sánchez Hontana
40 000 vols; 2 300 curr per; 522 diss/theses 25441

– Biblioteca de Filología Clásica, Facultad de Filología, Edif. A, 3a planta izq, Ciudad Universitaria, 28040 Madrid
T: +34 913945276; Fax: +34 913945289; E-mail: pimar@buc.ucm.es
1971; Pilar Martínez González
30 600 vols
libr loan 25442

– Biblioteca de Filología Hispánica y Romànica, Facultad de Filología, Ciudad Universitaria, Fac. de Filosofía, Edif. B, 28040 Madrid
T: +34 913946047; Fax: +34 913945746; E-mail: buc_fll@buc.ucm.es
Pilar Martínez González
Latin American Literature
19 000 vols; 400 curr per 25443

– Escuela de Relaciones Laborales, Biblioteca, C/ San Bernardo, 49, 28015 Madrid
T: +34 913946628; Fax: +34 913946626; E-mail: buc_rla@buc.ucm.es; URL: www.ucm.es/BUCM/erl/9778.php
Elvira Lázaro Godino
9 700 vols; 350 curr per
libr loan 25444

– Escuela Universitaria de Enfermería, Fisioterapia y Podología, Biblioteca, Fac. de Medicina, Pabellón II, Planta Sótano, Ciudad Universitaria, 28040 Madrid
T: +34 913941904; Fax: +34 913947173; E-mail: buc_enf@buc.ucm.es; URL: www.ucm.es/BUCM/enf/9683.php
Fuencisla Sanz Luengo
Public Health, Physical Therapy, Nutrition, Podology, Toxology
14 900 vols; 164 curr per
libr loan 25445

– Escuela Universitaria de Estadística, Biblioteca, Av Puerta de Hierro, s/n, 28040 Madrid
T: +34 913944035; E-mail: buc_est@buc.ucm.es; URL: www.ucm.es/BUCM/est/9416.php
Carmen Antón Luaces
Statistics
13 500 vols; 170 curr per
libr loan 25446

– Escuela Universitaria de Optica, Biblioteca, C/ Arcos de Jalón, 118, 28037 Madrid
T: +34 913946870; Fax: +34 913946885; E-mail: buc_opt@buc.ucm.es; URL: www.ucm.es/BUCM/opt/9696.php
1987; María Jesús Santurtún de la Hoz
Optometry, Contact Lens, Optics, Visual Science
10 200 vols; 133 curr per; 25 diss/theses; 176 av-mat; 61 sound-rec; 39 digital data carriers; 8 805 slides
libr loan 25447

– Escuela Universitaria de Trabajo Social, Biblioteca (School of Social Work), Campus de Somossaguas, Edif. Interfacultades, 28223 Pozuelo de Alarcón (Madrid)
T: +34 913942726; Fax: +34 913942743; E-mail: lillanas@buc.ucm.es; URL: www.ucm.es/BUCM/trs
1975; María Dolores Illanas Duque
39 830 vols; 622 curr per; 2 e-journals; 2 e-books; 49 mss; 9 maps; 10 microforms; 1 785 av-mat; 47 sound-rec; 637 digital data carriers; 113 other items
libr loan 25448

– Facultad de Ciencias de la Documentación, Biblioteca, C/ Santísima Trinidad, 37, 28010 Madrid
T: +34 913946672; Fax: +34 913946669; E-mail: buc_byd@buc.ucm.es; URL: www.ucm.es/BUCM/byd/9780.php
Manuel Vilariño Pardo
Archive, library, book and printing sciences
9 000 vols; 75 curr per
libr loan 25449

– Facultad de Ciencias de la Información, Biblioteca, Ciudad Universitaria, s/n, 28040 Madrid
T: +34 913942207; Fax: +34 913942012; E-mail: buc_inf@buc.ucm.es; URL: www.ucm.es/BUCM/inf/9795.php
1974; Antonio Moreno Cañizares
78 500 vols; 1 819 curr per; 7 840 e-journals; 900 diss/theses; 450 microforms; 3 000 av-mat; 600 sound-rec; 7 500 digital data carriers; 775 DVDs
libr loan 25450

– Facultad de Ciencias Económicas y Empresariales, Biblioteca, Campus de Somosaguas s/n, Pozuelo de Alarcón, 28023 Madrid
T: +34 913942608; Fax: +34 913942426; E-mail: buc_cee@buc.ucm.es; URL: www.ucm.es/info/cee/biblioteca.htm
1970; M. Luisa García-Ochoa Roldán
142 000 vols; 3 000 curr per; 8 000 diss/theses; 800 av-mat; 500 digital data carriers
libr loan 25451

– Facultad de Ciencias Físicas, Biblioteca, Campus de Moncloa, 28040 Madrid
T: +34 913944470; Fax: +34 913944644; E-mail: buc_fis@buc.ucm.es; URL: www.ucm.es/BUCM/fis/9544.php
1975; Raquel Benito Alonso
18 300 vols; 1 174 curr per; 300 diss/theses; 112 pamphlets, 103 offprints
libr loan 25452

– Facultad de Ciencias Geológicas, Biblioteca, C/José Antonio Novais, 2, Ciudad Universitaria, 28040 Madrid
T: +34 913944900; Fax: +34 913944887; E-mail: buc_geo@buc.ucm.es; URL: www.ucm.es/BUCM/geo/9555.php
1974; Javier García García
39 527 vols; 4 823 curr per; 1 252 diss/theses; 10 704 maps; 1 423 microforms; 441 av-mat; 353 digital data carriers; 5 400 photos, 6 314 pamphlets
libr loan 25453

– Facultad de Ciencias Matemáticas, Biblioteca, Plaza de Ciencias, no 3 (Ciudad Universitaria), 28040 Madrid
T: +34 913944653; Fax: +34 913944675; E-mail: efernandez@buc.ucm.es; URL: www.ucm.es/BUCM/mat/7243.php
1975; Emilio Fernandez González
Hist of mathematics
67 000 vols; 800 curr per; 1 200 diss/theses; 500 microforms; 150 digital data carriers
libr loan; DOCUMAT 25454

– Facultad de Ciencias Políticas y Sociología, Biblioteca, Campus de Somosaguas, 28223 Madrid
T: +34 913942968; Fax: +34 913942668; E-mail: buc_cps@buc.ucm.es; URL: www.ucm.es/BUCM/cps/9805.php
1943; Avelina Fernández Cruz
141 000 vols; 833 curr per; 4 digital data carriers
libr loan 25455

– Facultad de Ciencias Químicas, Biblioteca (Faculty of Chemistry), Ciudad Universitaria, 28040 Madrid
T: +34 913944252, 913944254; Fax: +34 913944253; E-mail: adelat@buc.ucm.es; URL: www.ucm.es/BUCM/qui/index.php
Adela Tercero Jiménez
33 316 vols; 1 057 curr per; 36 434 e-journals; 68 395 e-books; 3 000 diss/theses; 70 microforms; 5 651 digital theses, 75 853 digital books (Google project)
libr loan 25456

– Facultad de Derecho, Biblioteca, Ciudad Universitaria, 28040 Madrid
T: +34 913945616; E-mail: buc_der@buc.ucm.es
1957; Isabel Costales Ortiz
390 000 vols; 3 884 curr per; 30 mss; 2 incunabula; 13 800 diss/theses; 55 000 rare books (16-18th c), 30 000 pamphlets 25457

– Facultad de Education, Biblioteca, C/ Rector Royo Villanova, s/n, 28040 Madrid
T: +34 913946275; Fax: +34 913946277; E-mail: buc_edu@buc.ucm.es
1840; Ana Líter
150 000 vols; 530 curr per; 100 microforms; 50 av-mat; 50 digital data carriers
libr loan 25458

– Facultad de Farmacia, Biblioteca, Plaza de Ramón y Cajal s/n, 28040 Madrid
T: +34 913941780; Fax: +34 913941781; E-mail: buc_far@buc.ucm.es; URL: www.ucm.es/BUCM/far/9713.php
1806; Pilar Gómez Bachmann
Pharmacy, pharmacology, bio chemistry, botany, clinical tests, soil science, pharmacognosy, microbiology, food science, parasitology, analytic chemistry, organic chemistry, mineral chemistry, pharmaceutical chemistry, pharmacy hist, food engineering
59 700 vols; 3 554 curr per; 647 diss/theses; 115 microforms; 8 av-mat; 37 digital data carriers
libr loan 25459

– Facultad de Filología, Biblioteca, Ciudad Universitaria, Fac de Filosofía, Edif. A, 28040 Madrid
T: +34 913945356; E-mail: buc_fll@buc.ucm.es; URL: www.ucm.es/BUCM/fll/index.php
1943; Pilar Martínez González
480 000 vols; 5 052 curr per; 112 mss; 201 incunabula; 148 music scores; 5 200 microforms; 5 400 av-mat; 37 digital data carriers; 37 210 rare books (16-18th c), 4 000 offprints, 150 slides
libr loan 25460

– Facultad de Filosofía, Biblioteca, Edif. A, Ciudad Universitaria, 28040 Madrid
T: +34 913945257; Fax: +34 913945527; E-mail: buc_fsl@buc.ucm.es; URL: www.ucm.es/BUCM/fsl
1971; Cristina Arbós Ayuso
107 000 vols; 599 curr per; 314 diss/theses; 164 microforms; 200 av-mat; 13 sound-rec; 139 digital data carriers
libr loan; ANABAD, MADROÑO, REBIUN 25461

– Facultad de Geografía e Historia, C/ Profesor L. Aranguren s/n, Ciudad Universitaria, 28040 Madrid
T: +34 91394-6072/-6064; Fax: +34 913945976; E-mail: buc_ghi@buc.ucm.es; URL: www.ucm.es/BUCM/ghi/index.htm
273 237 vols; 2 946 curr per; 2 447 music scores; 16 744 maps; 1 167 microforms; 11 800 av-mat; 4 960 sound-rec; 344 digital data carriers
libr loan 25462

– Facultad de Informática, Biblioteca, C/ Profesor José García Santesmases, s/n (Ciudad Universitaria), 28040 Madrid
T: +34 913947521; Fax: +34 913947524; E-mail: buc_eis@buc.ucm.es; URL: www.ucm.es/BUCM/eis/9595.php
Manuel Antonio Martín Mota
24 600 vols; 646 curr per; 400 e-journals; 3 000 e-books; 1 400 av-mat; 3 200 sound-rec
libr loan 25463

– Facultad de Medicina, Biblioteca, Av de la Complutense, s/n, 28040 Madrid
T: +34 913941234; Fax: +34 913941428; E-mail: buc_med@buc.ucm.es; URL: www.ucm.es/BUCM/med/9070.php
1786; Juan Carlos Domínguez Martínez
Memorias de Balnearios (19th-20th c), Juntas Literarias del Real Colegio de San Carlos (18th); Depts of Public Health and History of Medicine
150 000 vols; 5 859 curr per; 15 500 mss; 66 incunabula; 5 000 diss/theses; 450 microforms; 80 av-mat; 36 digital data carriers
libr loan 25464

– Facultad de Odontología, Biblioteca, Ciudad Universitaria, s/n, 28040 Madrid
T: +34 913941919; Fax: +34 913941868; E-mail: buc_odo@buc.ucm.es; URL: www.ucm.es/BUCM/odo/9724.php
1979; María Angeles Lacasa Otín
11 000 vols; 661 curr per; 25 e-journals; 1 000 diss/theses; 29 microforms; 290 offprints, theses, 85 video-recordings
libr loan 25465

– Facultad de Psicología, Biblioteca, Campus de Somosaguas s/n, Pozuelo de Alarcón, 28223 Madrid
T: +34 913943202; Fax: +34 913943106; E-mail: buc_psi@buc.ucm.es; URL: www.ucm.es/BUCM/psi/9737.php
1972; Javier Fernández Iglesias
Psychiatry, psychology
66 100 vols; 193 curr per; 1 500 diss/theses; 1 122 microforms; 216 av-mat; 1 digital data carriers; 2 100 tests
libr loan 25466

– Facultad de Veterinaria, Biblioteca, Av Puerta de Hierro, s/n, 28040 Madrid
T: +34 913943828; Fax: +34 913943877; E-mail: buc_vet@buc.ucm.es; URL: www.ucm.es/bucm/vet/00.htm
1967; Cristina Ortiz Diez de Tortosa
4 493 monographs of the 19th c; Veterinary medicine, animal production, agriculture
38 453 vols; 2 053 curr per; 25 000 e-journals; 35 000 e-books; 3 147 diss/theses; 821 maps; 246 microforms; 541 av-mat; 26 sound-rec; 71 digital data carriers
libr loan 25467

– Instituto de Criminología, Biblioteca, Ciudad Universitaria. Facultad de Derecho, 28040 Madrid
T: +34 913945657; Fax: +34 913945411; E-mail: buc_icr@buc.ucm.es; URL: www.ucm.es/BUCM/icr/9952.php
1964
9 600 vols; 56 curr per; 35 diss/theses 25468

Universidad Internacional Menéndez Pelayo, Biblioteca, Isaac Peral, 23, 28040 ***Madrid***
T: +34 915495000
1948
15 500 vols 25469

Universidad Nacional de Eduación a Distancia, Biblioteca Central, Paseo Senda del Rey, 5, 28040 ***Madrid***
T: +34 913986631, 913986632; Fax: +34 913986694; E-mail: jburgoa@pas.uned.es; URL: biblioteca.uned.es/lenya/bibliuned/live/index.html
1972; María Jesús López Manzanedo
342 212 vols; 3 201 curr per; 1 500 e-journals; 2 641 diss/theses; 124 maps; 6 296 microforms; 4 325 av-mat; 8 472 sound-rec; 415 digital data carriers; 150 rare books (16-18th c)
libr loan; REBIUN 25470

Universidad Politécnica de Madrid, Av Ramiro Maeztu, 7, Ciudad Universitaria, 28040 ***Madrid***
E-mail: biblioteca.informacion@upm.es; URL: www.upm.es/institucional/UPM/Biblioteca
775 000 vols; 11 800 curr per; 20 000 e-books; 12 000 av-mat 25471

– Biblioteca Aeronáutica, Escuela Técnica Superior de Ingenieros Aeronáuticos / Escuela Universitaria de Ingeniería Técnica Aeronáutica, Pl del Cardenal Cisneros, 3, 28040 Madrid
T: +34 913366288; Fax: +34 913366288; E-mail: biblioteca.aeronauticos@upm.es; URL: aerobib.aero.upm.es
1967
50 000 vols; 235 curr per; 7 000 e-journals 25472

– Biblioteca Campus Sur, ETSI Topografía / EUIT Telecomunicación / EU Informática, Carretera de Valencia, km 7 – Complejo Politécnica de Vallecas, 28031 Madrid
T: +34 913367781 25473

– Centro de Documenación Europea, Biblioteca, C/ Doctor Federico Rubio y Gali, n° 11, bajo izda, 28039 Madrid
T: +34 915351628; E-mail: biblioteca.ceyde@upm.es
1985
EUROSTAT, European Community
8 500 vols; 6 003 curr per; 25 000 microforms
libr loan 25474

– Escuela de Minas, Biblioteca, Ríos Rosas, 21, 28003 Madrid
T: +34 913367031; Fax: +34 913367040; E-mail: cguio@minas.upm.es; URL: ww2.minas.upm.es/bib/Biblioteca-Minas.htm
1893; Carmen Guío Moreno
52 800 vols; 252 curr per; 126 mss; 218 diss/theses; 930 maps; 218 av-mat
libr loan 25475

– Escuela Técnica Superior de Arquitectura, Biblioteca, Av Juan de Herrera, 4, 28040 Madrid
T: +34 913366523; Fax: +34 915443817; E-mail: biblioteca.arquitectura@upm.es; URL: www.aq.upm.es/biblioteca
1895; Blanca Ruilope Urioste
85 000 vols; 300 curr per; 18 144 e-journals; 320 diss/theses; 1 500 govt docs; 25 390 maps; 99 907 microforms; 2 203 av-mat; 24 sound-rec; 9 391 digital data carriers; 2 500 offprints, 550 patents, 591 graphics, 8 000 rare books
REBIUN 25476

– Escuela Técnica Superior de Ingenieros Agrónomos, Biblioteca, Ciudad Universitaria, 28040 Madrid
T: +34 913365609; Fax: +34 915434879; E-mail: secretaria.director.agronomos@upm.es; URL: www.etsia.upm.es/ETSIAgronomos/Escuela/Servicios/ServBiblioteca
1855; Jesús Vázquez Minguela
17 dep. Importants files
50 000 vols; 130 curr per; 575 diss/theses; 2 050 maps; 2 205 av-mat; 4 digital data carriers; 5 200 rare books
libr loan 25477

– Escuela Técnica Superior de Ingenieros de Caminos, Canales y Puertos, Biblioteca, Ciudad Universitaria, 28040 Madrid
T: +34 913366739; E-mail: conchita@caminos.upm.es; URL: www.caminos.upm.es/servicios/presentacion.asp
1834; Concepción García Viñuela
Public works, canal construction, port installations
65 000 vols; 118 curr per; 129 mss; 589 diss/theses; 3 510 maps; 13 800 microforms; 680 av-mat; 306 digital data carriers
libr loan 25478

– Escuela Técnica Superior de Ingenieros de Telecomunicación, Biblioteca, Ciudad Universitaria, 28040 Madrid
T: +34 913367258; E-mail: biblioteca.etsit@upm.es; URL: www.etsit.upm.es/biblioteca.html
1965; Reyes Albo Sánchez-Bedoya
Social Implementation of Telecommunication Coll, Old spanish telecommunication periodicals, Fundesco Donation, Alcatel Donation
40 000 vols; 650 curr per; 20 000 e-journals; 4 000 diss/theses; 500 maps; 142 digital data carriers
libr loan; Madroño, REBIUN, IATUL 25479

– Escuela Técnica Superior de Ingenieros Industriales, Biblioteca, José Guitérrez Abascal, 2, 28006 Madrid
T: +34 913363076; Fax: +34 915618618; E-mail: biblioteca.industriales@upm.es; URL: www.etsii.upm.es/biblioteca
1905
30 000 vols; 230 curr per; 1 000 diss/theses; 500 maps; 225 av-mat; 83 000 standards 25480

– Escuela Técnica Superior de Ingenieros Navales, Biblioteca, Ciudad Universitaria, 28040 Madrid
T: +34 913367162; Fax: +34 915442149; E-mail: biblioteca.navales@upm.es; URL: www.etsin.upm.es
1940; José Ramón Rodríguez Fernández
21 509 vols; 107 curr per; 121 diss/theses; 12 maps; 538 sound-rec; 35 digital data carriers
libr loan 25481

– Escuela Universitaria de Arquitectura Técnica, Biblioteca, Av Juan de Herrera, 6 Ciudad Universitaria, 28040 Madrid
T: +34 913367578; Fax: +34 913367578; E-mail: biblioteca.arquitecturatecnica@upm.es; URL: www.euatm.upm.es/biblioteca/index.htm
1973
22 700 vols; 70 curr per; 319 av-mat; 2 digital data carriers; 840 standards
libr loan 25482

– Escuela Universitaria de Ingeniería Técnica Agrícola, Biblioteca, Ciudad Universitaria, 28040 Madrid
T: +34 913365448; Fax: +34 915493002; E-mail: chiruca.casado@upm.es; URL: www.agricolas.upm.es
1963; Mercedes Casado Vázquez
24 726 vols; 584 curr per; 2 870 diss/theses; 1 420 maps; 154 av-mat; 76 digital data carriers; 3 500 pamphlets
libr loan 25483

– Escuela Universitaria de Ingeniería Técnica de Obras Públicas, Biblioteca, Alfonso XII, 3, 28014 Madrid
T: +34 913367733; Fax: +34 913367958
1961
17 300 vols; 66 curr per 25484

– Escuela Universitaria de Ingeniería Técnica Forestal, Biblioteca, Av Ramiro de Maeztu, s/n, Ciudad Universitaria, 28040 Madrid
T: +34 913367540; Fax: +34 915446025; URL: www.forestales.upm.es/contenido.aspx?id=54
1957; Dolores Guio Moreno
20 300 vols; 79 curr per; 480 diss/theses; 1 000 maps; 1 000 sound-rec 25485

– Escuela Universitaria de Ingeniería Técnica Industrial, Biblioteca, Ronda de Valencia, 3, 28012 Madrid
T: +34 913367698; Fax: +34 913367711; E-mail: biblioteca.industrial@upm.es; URL: www.euiti.upm.es
1913
21 033 vols; 158 curr per; 5 600 microforms; 170 av-mat; 743 rare scientific and tech books of 17-18th c
libr loan 25486

– Facultad de Ciencias de la Actividad Física y del Deporte, Biblioteca, C/ Martín Fierro s/n, Ciudad Universitaria, 28040 Madrid
T: +34 913364046; E-mail: web.inef@upm.es; URL: www.inef.upm.es/INEF/Facultad/BibliotecaINEF
1959; Pilar Irureta-Goyena Sánchez
Sports, Physical Education
58 000 vols; 400 curr per; 400 rare books (16th c until 1830) 25487

– Facultad de Informática, Biblioteca, Campus de Montegancedo, s/n, Boadilla del Monte, 28660 Madrid
T: +34 913366916; E-mail: biblioteca.finformatica@upm.es; URL: www.fi.upm.es/?pagina=24 25488

Universidad Pontificia 'Comillas', Biblioteca, C/ Alberto Aguilera, 23, 28015 ***Madrid***
T: +34 915422800; Fax: +34 915596569; E-mail: maria@bib.upcomillas.es; URL: www.upcomillas.es/servicios/biblioteca/default.aspx
1892
Rare bks on theology and philosophy from the 16-18th c (18 500 vols)
492 000 vols; 10 300 curr per; 50 mss; 20 incunabula; 2 856 microforms; 68 av-mat; 675 digital data carriers
libr loan; REBIUN, ABIE 25489

Universitas Nebrissensis, Biblioteca, Pirineos, 55. Campus de la Dehesa de la Villa, 28040 ***Madrid***
T: +34 913116602; Fax: +34 913116613
1985
Law, economics, information science, European Union
8 000 vols; 79 curr per; 100 av-mat 25490

Hospital Clinico Universitario Ntra. Sra. de la Victoria, Biblioteca, Colonia Sta. Inés s/n°, 29010 ***Málaga***
T: +34 952649405; Fax: +34 952282182
1980; María Teresa García Ballesteros
'Bibliografía médica' (700 vols publ in 18th and 19th c, 2 450 vols publ until 1964); Bibl para pacientes
8 200 vols; 150 curr per; 200 govt docs; 50 microforms 25491

Universidad de Málaga, Biblioteca, Campus de Teatinos, s/n, 29071 ***Málaga***
T: +34 952132306; Fax: +34 952132342; E-mail: bblgral@uma.es; URL: www.uma.es/servicios/biblioteca
1972; Gregorio García Reche
512 910 vols; 5 249 curr per; 117 maps; 114 409 microforms; 2 213 av-mat; 1 198 sound-rec; 1 289 digital data carriers
libr loan; REBIUN 25492

– Biblioteca de Ciencias, Campus de Teatinos, s/n, 29071 Málaga
T: +34 952131999; Fax: +34 952132000; E-mail: bblcie@uma.es; URL: www.uma.es/ficha.php?id=62391&bbl=4
1972; Ana García Ferrer
Mathematics, physics, chemistry
10 000 vols; 224 curr per; 213 diss/theses 25493

– Biblioteca de Ciencias de la Comunicación, Campus de Teatinos, s/n, 29071 Málaga
T: +34 952133275; Fax: +34 952132914; E-mail: bblcccom@uma.es; URL: www.uma.es
Federico Arrebola
19 200 vols; 438 curr per; 2 microforms; 2 300 av-mat; 35 sound-rec 25494

– Biblioteca de Ciencias de la Educación y Psicología, Campus de Teatinos, s/n, 29071 Málaga
T: +34 952131326; Fax: +34 952131117; E-mail: bbledupsico@uma.es; URL: www.uma.es
Immaculada Urda González

92 500 vols; 1 582 curr per; 2 microforms
25495

– Biblioteca de Ciencias de la Salud,
Paseo de Martiricos, s/n, 29009 Málaga
T: +34 952133208; Fax: +34 952133208;
E-mail: bblsalud@uma.es; URL:
www.uma.es
Joaquín Aurrecoechea Fernández
9 470 vols; 204 curr per; 2 microforms
25496

– Biblioteca de Ciencias Económicas y
Empresariales, C/ Ejido, 6, 29011 Málaga
T: +34 952131148; Fax: +34 952137136;
E-mail: bblccee@uma.es; URL:
www.uma.es/ficha.php?id=62391&bbl=9
1965; Estrella Ayala Moscoso
Sociology, social hist, law
78 347 vols; 3 323 curr per; 248
diss/theses; 182 microforms; 270 av-mat;
211 sound-rec; 325 digital data carriers
libr loan
25497

– Biblioteca de Derecho Alejandro
Rodríguez Carrión, Campus de Teatinos
s/n, 29071 Málaga
T: +34 952132113; Fax: +34 952132338;
E-mail: mjcasado@uma.es; URL:
www.uma.es
1981; Maria Jose Casado Cañero
Sala de Fondo Antiguo Don Antonio
Ortega Carrillo de Albornoz (rare books
from 16th-19th c)
80 400 vols; 1 854 curr per
25498

– Biblioteca de Estudios Sociales y de
Comercio, C/ Francisco Trujillo s/n,
Ampliación de Campos de Teatinos,
29017 Málaga
T: +34 951952217; E-mail: bbltrabajo@
uma.es; URL: www.uma.es
1991; Clara Isabel Pérez Zalama
11 726 vols; 265 curr per; 50 digital data
carriers
libr loan
25499

– Biblioteca de Humanidades José Mercado
Ureña, Campus de Teatinos, s/n, 29071
Málaga
T: +34 952131697; Fax: +34 952131696;
E-mail: biblhum@uma.es; URL:
www.uma.es/ficha.php?id=62391&bbl=14
1970; Matilde Candil Gutierrez
Hist, linguística, lit, geografia, filosofia,
religión, hist del arte
101 134 vols; 1 184 curr per; 1
incunabula
libr loan
25500

– Biblioteca de Industriales y Politécnica,
C/ Doctor Ortíz Ramos, Ampliación del
Campo de Teatinos, s/n, 29071 Málaga
T: +34 951952494; E-mail: bblindupoli@
uma.es; URL: www.uma.es
1927; Carmen Ávila Maldonado
16 000 vols; 10 curr per
25501

– Biblioteca de Informática y
Telecomunicación, Campus de Teatinos,
s/n, 29071 Málaga
T: +34 952133328; Fax: +34 952133329;
E-mail: bbletsiitel@uma.es; URL:
www.uma.es
María López Bujalance
29 350 vols; 245 curr per; 2 microforms
25502

– Biblioteca de Medicina, Boulevard Louis
Pasteur, 32, 29071 Málaga
T: +34 952131525; Fax: +34 952131524;
E-mail: mvictoria@uma.es; URL:
www.uma.es
1973; María Victoria González
Pharmacy
17 000 vols; 88 curr per
25503

– Biblioteca del Turismo, Edif. de
Hostelería y Turismo, 29071 Málaga
T: +34 9521332-76/-77; Fax: +34
952136612; E-mail: bblturismo@uma.es;
URL: www.uma.es
Genoveva Lara Rodríguez
11 950 vols; 239 curr per; 2 microforms
25504

**Universidad Nacional de Educación
a Distancia**, Biblioteca, Lope de Vega,
1 Centro Cultural Feerico García Lorca,
29802 *Melilla*
T: +34 952681080; Fax: +34 952681468;
E-mail: infouned@adm.uned.es
1978
11 300 vols; 15 curr per; 760 maps; 152
av-mat; 1 200 sound-rec; 3 500 slides
25505

Universidad de Educación a Distancia,
Biblioteca, C/ Moreno de Vargas, 10,
06800 *Mérida*
T: +34 924315050; Fax: +34 924302556;
E-mail: admin.web@merida.uned.es; URL:
www.uned.es/ca-merida/Biblioteca.htm
1977
19 000 vols; 190 curr per; 1 mss
25506

Universidad Cardenal Herrera, CEU
Biblioteca (University Cardenal Herrera),
cl Luis Vives, 2, 46113 *Moncada*,
Valencia
T: +34 961369031; Fax: +34 961369030;
E-mail: biblio@uch.ceu.es; URL:
www.uch.ceu.es
1973; Elena Sauri Rodrigo
Ancient Books; Newspaper Library, Video
Library
85 168 vols; 1 680 curr per; 53 e-
journals; 127 diss/theses; 136 maps; 95
microforms; 846 av-mat; 112 sound-rec;
5 795 digital data carriers
libr loan; REBIUN
25507

Mondragon Unibertsitatea, Mondragon
Goi Eskola Politeknikoa, Biblioteca
(Mondragon High Polytechnic School),
Loramendi 4, Apdo 23, 20500
Mondragon
T: +34 943794700; Fax: +34 943791536;
E-mail: ovelez@eps.muni.es; URL:
biblioteka.mondragon.edu
1988; Obdulia Vélez Pérez
32 000 vols; 425 curr per; 1 700
diss/theses; 550 av-mat; 8 sound-rec;
300 digital data carriers
libr loan; REBIUN
25508

Universidad Rey Juan Carlos,
Biblioteca Universitaria, C/ Tulipán, s/n,
28933 *Móstoles* (Madrid)
T: +34 916647406, 916647414; Fax: +34
914887141; E-mail: biblioteca@urjc.es;
URL: www.urjc.es
1998; Ricardo González Castrillo
146 000 vols; 2 394 curr per; 8 600 e-
journals; 50 diss/theses; 3 000 govt
docs; 849 microforms; 1 244 av-mat; 291
sound-rec; 2 623 digital data carriers
libr loan; Consorcio Madroño, REBIUN,
IFLA
25509

– Facultad de Ciencias y Jurídicas
y Sociales, Biblioteca, Campus de
Vicálvaro, Paseo de los Artilleros, s/n,
28032 Madrid
T: +34 913019854, 913019860; Fax: +34
917751413; E-mail: info@urjc.es; URL:
www.fcjs.urjc.es/
Ms Soledad Vicente Rosillo
150 000 vols; 672 curr per; 10
microforms; 63 av-mat; 100 sound-rec;
152 digital data carriers
libr loan; IFLA
25510

**Conservatorio Superior de Música
de Murcia "Manuel Massotti Littel"**,
Biblioteca, Paseo del Malecón, 9, 30004
Murcia
T: +34 968294758; Fax: +34 968294756;
E-mail: josesanchez@csmmurcia.com;
URL: www.csmmurcia.com
1916; Cristina Pina Caballero
Bibl de Compositores Murcianos (printed
music), Fondo Eduardo Souan (900
scores)
8 000 vols; 53 curr per; 5 000 music
scores; 3 000 sound-rec
AIBM
25511

Universidad de Murcia, Biblioteca
Universitaria (Murcia University), Biblioteca
General, Santo Cristo, 1, 30001 *Murcia*;
mail address: Campus de Espinardo,
30100 Murcia
T: +34 868883339; Fax: +34 868887879;
E-mail: more@um.es; URL: www.um.es/
biblioteca/
1935; Lourdes Cobacho Gómez
Social Sciences, Law, Life Sciences,
Mathematical & Physical Sciences,
Health Sciences, Chemistry, Computer
Science. Humanities, Sports Sciences;
printed matters from 16-18th c, Catálogo
"Floridablanca", Catálogo General
de "Colección Histórica"; Centro de
Documentación Europea
793 519 vols; 9 860 curr per; 15 790 e-
journals; 26 642 e-books; 47 mss; 18
incunabula; 4 357 diss/theses; 32 govt
docs; 226 music scores; 757 maps;
18 152 microforms; 9 541 av-mat; 44 241
sound-rec; 7 700 digital data carriers;
14 272 other items
25505

libr loan; REBIUN, EUROPEAN
DOCUMENTATION CENTER
25512

– Biblioteca de Economía y Empresa,
Facultad de Economía y Empresa,
Campus de Espinardo, 30100 Murcia
T: +34 868887872; E-mail: borgo@um.es;
URL: www.um.es/biblio
1982
34 000 vols; 632 journals
libr loan
25513

– Biblioteca de Educación, Facultad de
Educación, Campus de Espinardo, 30100
Murcia
T: +34 968367042; E-mail: mpm@um.es
1845; Paloma Ródenas
41 800 vols; 288 curr per; 50 maps; 170
microforms; 90 av-mat; 547 sound-rec
libr loan
25514

– Biblioteca de Medicina, Facultad de
Medicina, Campus de Espinardo, 30100
Murcia
T: +34 868884241 (monday, wednesday,
friday), +34 868888171 (tuesday,
thursday); E-mail: mjlh@um.es
1970; Pilar Sabater Lorenzo
10 000 vols; 78 curr per; 370 diss/theses;
426 microforms
libr loan
25515

– Biblioteca de Química, Facultad de
Química, Campus de Espinardo, 30100
Murcia
T: +34 968367348; E-mail: mjtg@um.es
1946; María José Toral Garcia
17 000 vols; 100 curr per
25516

– Biblioteca Luis Vives (Ciencias del
Trabajo, Filosofía y Trabajo Social),
Edif. Luis Vives, Campus de Espinardo,
30100 Murcia
T: +34 968363442; Fax: +34 968367042;
E-mail: vivesbib@um.es
1982; Paloma Ródenas
Labour, philosophy, social work
30 700 vols; 441 curr per; 114 av-mat
25517

Universidad de Oviedo, Biblioteca
Central, C/ San Francisco, 1, 33003
Oviedo
T: +34 985104053; Fax: +34 985104110;
E-mail: biblio@uniovi.es; URL:
buo.uniovi.es
1608; Ramón Rodríguez Álvarez
Sect Cervantina, Sect Gineta, Sect
Asturias, Sect Raros, Sect Pliegos de
Teatro Español Antiguo
899 028 vols; 13 943 curr per; 39 744
e-journals; 31 869 e-books; 849 mss;
23 incunabula; 6 084 maps; 13 337
microforms; 8 867 av-mat; 1 315 rare
bks
libr loan; REBIUN, G9, IFLA, ADLUG
25518

– Biblioteca de Ciencias de la Educación,
Aniceto Sela, s/n, 33005 Oviedo
T: +34 985103215; Fax: +34 985103214;
E-mail: buoccee@uniovi.es; URL:
buo.uniovi.es
1989; Rosa Alvarez Huerta
22 000 vols; 525 curr per; 237
microforms; 372 av-mat; 186 slides,
2 500 e-journals vía Biblioteca de la
Universidad de Oviedo
libr loan; Rebiun
25519

– Biblioteca de Jovellanos, C/ Luis Moya
Blanco, 261/La Laboral, Ciudad de la
Cultura, 33203 Gijón
T: +34 985186976; Fax: +34 985186975
35 000 vols
libr loan
25520

– Biblioteca de Psicología, Plaza Feijoo,
s/n, 33003 Oviedo
T: +34 985104146; Fax: +34 985104126;
E-mail: rahuerta@uniovi.es; URL:
buo.uniovi.es
1979; Rosa Aluarez Huerta
18 000 vols; 375 curr per; 340
microforms; 178 av-mat
libr loan; Rebuin
25521

– Biblioteca de Química, C/ Julián Clavería,
8, 33006 Oviedo
T: +34 985103493; Fax: +34 985103470;
E-mail: ivaquero@uniovi.es; URL:
www.uniovi.net/Biblioteca/centros/quimica
1959; Isidoro Vaquero Piñero
10 000 vols; 400 curr per
25522

– Biblioteca de Tecnología y Empresa,
Campus de Viesques, Aulario Norte,
Carretera de Villaviciosa, s/n, 33203
Gijón
T: +34 985182328; Fax: +34 985182340;
E-mail: fgg@uniovi.es
1995; Fausto González González
20 000 vols; 200 curr per; 57 microforms;
72 sound-rec
libr loan
25523

– Escuela Técnica Superior de Ingenieros
de Minas, Biblioteca (School of Mines),
C/ Independencia, 13, 33004 Oviedo
T: +34 985104274; E-mail: gespfc@
epsig.uniovi.es; URL: www.etsimo.es
1963; Evaristo Alvarez Muñoz
20 000 vols; 900 curr per; 18 000 e-
journals; 55 diss/theses; 2 000 maps; 20
digital data carriers
libr loan; REBIUN, MECANO
25524

Universitat de les Illes Balears, Servei
de Biblioteca i Documentació, Edifici
Ramon Llull, Campus UIB, Cra. de
Valldemossa, km 7,5, 07122 *Palma
de Mallorca*
T: +34 971172508; Fax: +34
971172735; E-mail: sbdser@uib.es; URL:
biblioteca.uib.es
1978; Miquel Pastor
Fons cervantí J. Casasayas, Fons Joan
Mascaró Fornés, Fons Camilo J. Cela
Conde, La Guerra del Francès 1808 –
1814
765 974 vols; 4 224 curr per; 28 662 e-
journals; 23 065 e-books; 764 diss/theses;
1 330 music scores; 669 maps; 3 681
microforms; 3 637 av-mat; 1 196
sound-rec; 2 479 art catalogues, 9 141
clippings
libr loan; OCLC, IFLA
25525

– Biblioteca Anselm Turmeda, Edifici
Anselm Turmeda, Campus UIB, Cra.
de Valldemossa km 7,5, 07122 Palma de
Mallorca
T: +34 971173010; Fax: +34 971173003;
E-mail: cfdbib@uib.es
1972; Francesc Sastre
Mathematics, information science
32 500 vols; 776 curr per; 19 600
microforms; 130 av-mat
libr loan
25526

– Biblioteca Arxiduc Lluis Salvador, Edifici
Arxiduc Lluís Salvador, Campus UIB,
Cra. de Valldemossa km 7,5, 07122
Palma de Mallorca
T: +34 971172653; E-mail: sbdcpr@uib.es
Carme Pons
Tourism, hotel management
32 500 vols; 776 curr per; 19 600
microforms; 130 av-mat
25527

– Biblioteca Guillem Cifre de Colonya,
Edifici Guillem Cifre de Colonya; Campus
UIB, Cra. de Valldemossa km 7,5, 07122
Palma de Mallorca
T: +34 971172570; E-mail: vsbdmrr@
uib.es
1942; Martina Rotger
Education, psychology, social work, public
health
23 000 vols; 57 curr per; 41 microforms;
100 sound-rec
libr loan
25528

– Biblioteca Mateu Orfila i Rotger, Edifici
Mateu Orfila i Rotger, Campus UIB, Cra.
de Valldemossa km 7,5, 07122 Palma de
Mallorca
T: +34 971173212; E-mail:
vsbdmnf@uib.es
1972; Miquela Nadal
Biology, physics, chemistry, geography
25 000 vols; 601 curr per; 300
microforms; 112 av-mat; 50 sound-rec
25529

– Biblioteca Ramon Llull, Edifici Ramon
Llull, Campus UIB, Cra. de Valldemossa
km 7,5, 07122 Palma de Mallorca
T: +34 971173362; E-mail:
mercedes.duran@uib.es
1973; Mercedes Durán
History, art history, linguistics, literature,
philosophy, geography
68 000 vols; 600 curr per
25530

**Universidad de Las Palmas de
Gran Canaria**, Biblioteca General,
Edif. Central, Campus Universitario de
Tafira s/n, 35017 *Las Palmas* de Gran
Canaria
T: +34 9284586-70/-71;

Fax: +34 928457248; E-mail: mmartin@pas.ulpgc.es; URL: biblioteca.ulpgc.es/?q=big
María del Carmen Martín Marichal
119 000 media items 25531

– Biblioteca de Arquitectura, Edif. de Arquitectura, Campus Universitario de Tafira s/n, 35017 Las Palmas de Gran Canaria
T: +34 928451318; Fax: +34 928451395; E-mail: bu_ea@ulpgc.es; URL: biblioteca.ulpgc.es/?q=arq
1970; Delia Lopez Martin
42 000 media items 25532

– Biblioteca de Ciencias Básicas, Campus Universitario de Tafira s/n, 35017 Las Palmas de Gran Canaria
T: +34 92845-4403/-2921; Fax: +34 928452922; E-mail: bu_ecb@ulpgc.es; URL: biblioteca.ulpgc.es/?q=bas
Avelina Fernández Manrique de Lara
Biology, physics, chemistry, mathematics
20 000 media items 25533

– Biblioteca de Ciencias de la Salud (Health Science Library), Paseo Blas Cabrera Felipe “Físico” s/n, 35017 Las Palmas de Gran Canaria
T: +34 9284514-21/-53; Fax: +34 928451421; E-mail: bu_ecs@ulpgc.es; URL: biblioteca.ulpgc.es/?q=med
1974; Francisco Fumagallo Díaz-Llanos
Medicine, nursing, rehabilitation, physical therapy, biology
20 000 vols; 160 curr per; 348 diss/theses; 89 av-mat; 420 digital data carriers
libr loan 25534

– Biblioteca de Ciencias Económicas y Empresariales, Edif. Central de la Biblioteca Universitaria, Campus Universitario de Tafira s/n, 35017 Las Palmas de Gran Canaria
T: +34 92845-8675/-8690; Fax: +34 928457248; E-mail: bib_fcee@ulpgc.es; URL: biblioteca.ulpgc.es/?q=eco
1977; Héctor López Campos
16 500 media items
libr loan 25535

– Biblioteca de Ciencias Jurídicas, Edif. Central de la Biblioteca Universitaria, Campus Universitario de Tafira s/n, 35017 Las Palmas de Gran Canaria
T: +34 92845-8690/-8676; Fax: +34 928457245; E-mail: biblioteca_fcj@ulpgc.es; URL: biblioteca.ulpgc.es/?q=der
Héctor López Campos
Law
56 500 vols; 300 curr per 25536

– Biblioteca de Educación Física, Edif. de Educación Física, Campus Universitario de Tafira s/n, 35017 Las Palmas de Gran Canaria
T: +34 92845-8869/-8897; Fax: +34 928457348; E-mail: bu_fis@ulpgc.es; URL: biblioteca.ulpgc.es/?q=fis
Julio A. Martínez Morilla
Sports, physical education, psychology, education, medicine
16 500 media items 25537

– Biblioteca de Electrónica y Telecomunicación, Edif. Electrónica y Telecomunicación, Campus Universitario de Tafira s/n, 35017 Las Palmas de Gran Canaria
T: +34 92845-4558/-4559; Fax: +34 928454559; E-mail: bib_eet@ulpgc.es; URL: biblioteca.ulpgc.es/?q=tel
Alfonso Canella Prieto
18 700 media items 25538

– Biblioteca de Formación del Profesorado, Edif. de Formación del Profesorado, C/ Juana de Arco, 1, 35003 Las Palmas de Gran Canaria
T: +34 92845-2783/-8852/-8856/-8858; Fax: +34 928458852; E-mail: bu_efp@ulpgc.es; URL: biblioteca.ulpgc.es/?q=egb
1960; Luz Marina Acosta Peñate
61 000 media items
libr loan 25539

– Biblioteca de Humanidades, Edif. de Humanidades “Millares Carlo”, C/ Pérez del Toro, 1, 35004 Las Palmas de Gran Canaria
T: +34 92845-2783/-8852/-8856/-8858; Fax: +34 928458852; E-mail: bu_eh@ulpgc.es; URL: biblioteca.ulpgc.es/?q=hum
Luz Marina Acosta Peñate
Archeology, arts, cartography, language education, philology, geography, history, linguistics, literature, translation
126 800 media items 25540

– Biblioteca de Informática y Matemáticas, Edif. de Informática y Matemáticas, Campus Universitario de Tafira s/n, 35017 Las Palmas de Gran Canaria
T: +34 92845-8781/-8999/-4551; Fax: +34 928454551; E-mail: biblio@dma.ulpgc.es; URL: biblioteca.ulpgc.es/?q=inf
María Eugenia Rúa-Figueroa Hernández
19 700 media items 25541

– Biblioteca de Ingeniería, Edif. de Ingeniería, Campus Universitario de Tafira, 35017 Las Palmas de Gran Canaria
T: +34 928451868; Fax: +34 928454456; E-mail: bu_ei@ulpgc.es; URL: biblioteca.ulpgc.es/?q=ing
1902; Víctor P. Ramos Martín
Engineering, telecommunications, navigation, topography
34 000 vols; 75 curr per; 40 diss/theses; 560 av-mat
libr loan 25542

– Biblioteca de Veterinaria, Edif. de Veterinaria, Campus Universitario de Montaña Cardones, Autovía Las Palmas-Bañaderos, km 8, 35416 Arucas (Las Palmas)
T: +34 92845-1138/-1097; Fax: +34 928457241; URL: biblioteca.ulpgc.es/?q=vet
María del Carmen Iglesias López
10 700 media items 25543

Universidad de Navarra, Servicio de Bibliotecas (University of Navarra), Campus Universitario, Edif. de Bibliotecas, Apdo 177, 31080 ***Pamplona***
T: +34 948425600 ext 2495; Fax: +34 948178269; E-mail: biblioteca@unav.es; URL: www.unav.es/biblioteca/indice.html
1961; Dr. Víctor Sanz Santacruz
1 105 000 vols; 19 983 curr per; 32 916 e-journals; 56 000 e-books; 122 754 microforms; 8 776 av-mat; 5 674 sound-rec; 7 338 digital data carriers
IFLA, ANABAD, REBIUN, UKSG 25544

– Biblioteca de Ciencias, Apdo 177, 31080 Pamplona
T: +34 948425600 ext 6229; Fax: +34 948425650; E-mail: bibcielect@unav.es; URL: www.unav.es/biblioteca/ciendefinicion.html
1965; José Félix Villanueva Baquedano
135 000 vols; 4 287 curr per; 200 maps; 2 500 microforms; 300 av-mat; 35 digital data carriers
libr loan; IFLA 25545

– Biblioteca de la Clínica Universitaria, Clínica Universitaria, Edif. de Consultas, 8a planta, 31080 Pamplona
T: +34 948255400 ext (82) 4860; E-mail: bibcun@unav.es; URL: www.unav.es/biblioteca/clindefinicion.html
21 500 vols; 4 000 e-journals 25546

– Biblioteca de las Facultades Eclesiásticas, Apdo 177, 31080 Pamplona
T: +34 948425600 ext 2519; E-mail: bibteo@unav.es; URL: www.unav.es/biblioteca/huteodefinicion.html
8 400 vols 25547

– Escuela Técnica Superior de Ingenieros, Biblioteca, Paseo de Manuel Lardizabal, 13, 20018 San Sebastián, Guipozcoa
T: +34 943219877; Fax: +34 943311442; E-mail: mjuana@ceit.es; URL: www.biblioteca.ceit.es
1961; Marta de Juana
Electrical engineering, electronics, electronic systems, materials science, mechanical engineering, management, telecommunications
41 000 vols; 4 curr per; 2 000 e-journals; 460 diss/theses; 209 microforms; 780 sound-rec; 4 digital data carriers 25548

– IESE Business School, Library, Av Pearson 21, 08034 Barcelona
T: +34 93 2534200; Fax: +34 93 2534343; E-mail: biblioteca@iese.edu; URL: www.iese.edu/library
1958; Carina Huguet
50 000 vols; 910 curr per; 41 e-journals
libr loan; EBSLG 25549

Universidad Pública de Navarra, Biblioteca Universitaria (Public University of Navarra), Campus de Arrosadia, 31006 ***Pamplona***
T: +34 948169060; Fax: +34 948169069; E-mail: bupna@unavarra.es; URL: brocar.unavarra.es/biblio2.htm
1991; Guillermo Sánchez Martinez
Navarra pre 1936 imprints
215 000 vols; 2 092 curr per
libr loan 25550

Universidad Nacional de Educación a Distancia de Pontevedra, Biblioteca Centro Asociado, Portugal, 1, Urbanización Monte Porreiro, 36004 ***Pontevedra***
T: +34 986851850 ext 301; E-mail: biblioteca@pontevedra.uned.es; URL: www.uned.es/ca-pontevedra/bibliotecaFon.htm
1973; Carmen Fernandez-Valdes Martinez-Estellez
20 000 vols; 46 curr per; 24 diss/theses; 7 maps; 9 286 av-mat 25551

Universitario Francisco de Vitoria, Biblioteca, Ctra de Pozuelo-Majadahonda, km 1,8, 28223 ***Pozuelo de Alarcón***, Madrid
T: +34 917091448; Fax: +34 917091554; E-mail: biblioteca@ufv.es; URL: www.ufv.es/vuniversitaria.aspx?sec=39
1993; Rosa Salord Bertrán
Law, journalism, economics, business, cinema
50 000 vols; 305 curr per; 11 e-journals; 196 music scores; 2 000 microforms; 725 av-mat; 378 sound-rec; 1 698 digital data carriers
libr loan; SEDIC 25552

Universidad de Castilla-La Mancha – Escuela Universitaria del Profesorado de E.G.B., Biblioteca, Ronda de Calatrava, s/n, 13003 ***Real***
T: +34 926221047
Hist, lit, education
16 100 vols; 68 curr per 25553

Facultad de Teología de San Esteban PP. Dominicos, Biblioteca, Pl del Concilio de Trento, 4, Apdo 17, 37001 ***Salamanca***
T: +34 923215000; Fax: +34 923265480
1500; Emilio Garcia
100 000 vols; 250 curr per 25554

Universidad de Salamanca, Servicio de Archivos y Bibliotecas, Campus Miguel de Unamuno, Fac. de Derecho, Planta Sótano, 37007 ***Salamanca***
T: +34 923294400 ext 3568, 923294500 ext 3568; Fax: +34 923294503; E-mail: sabus@usal.es; URL: sabus.usal.es
1218; Severiano Hernandez Vicente
1 047 000 vols; 22 706 curr per; 2 280 mss; 483 incunabula; 9 140 maps; 120 373 microforms; 2 650 av-mat; 3 720 sound-rec; 1 020 digital data carriers
libr loan 25555

– Biblioteca Claudio Rodríguez del Campo de Zamora, Campus Viriato, Av de Requejo s/n, 49012 Zamora
T: +34 980545000 ext 3709; Fax: +34 980545005; E-mail: cframos@usal.es
1976; Carmen Flores Ramos
12 000 vols 25556

– Biblioteca de “Campus Canalejas”, Paseo de Canalejas, 169, 37008 Salamanca
T: +34 923294400; Fax: +34 923294609; URL: www.campus-oei.org/repertorio/101.htm
1974; José Martin García
50 500 vols; 769 curr per; 261 diss/theses; 388 microforms; 103 av-mat; 224 sound-rec; 247 digital data carriers
libr loan 25557

– Biblioteca Francisco de Vitoria, Campus Miguel de Unamuno, s/n, 37007 Salamanca
T: +34 923294400 ext 3140; E-mail: vito@usal.es; URL: campus.usal.es/vito
1996; Francisco José Sáenz de Valluerca López
Law, Sociology, Economy, Business Administration, Social Work, Philosophy, Management, Public Administration
210 000 vols; 20 000 curr per; 347 microforms; 1 500 av-mat; 1 500 sound-rec; 500 digital data carriers
libr loan 25558

– Campus de Ávila – Escuela Universitaria de Educación de Ávila, Biblioteca, C/ Madrigal, 3, 05003 Ávila
T: +34 923294400; E-mail: sabus@usal.es; URL: campus.usal.es/~turismo/biblioteca.html
1923; Consuelo Martín García
S.C. of Ávila
24 000 vols; 121 curr per; 115 maps; 5 840 av-mat; 427 sound-rec; 6 digital data carriers
libr loan 25559

– Facultad de Bellas Artes / Psicología, Biblioteca, Carretera de Toro s/n, 37005 Salamanca
T: +34 923294400; Fax: +34 923294604; URL: psi.usal.es/biblioteca
1979; Jesús García Benéitez
30 878 vols; 578 curr per; 285 diss/theses; 259 microforms; 413 av-mat; 166 sound-rec; 660 tests
libr loan 25560

– Facultad de Filología, Biblioteca, C/ Palominos, s/n, 37008 Salamanca
T: +34 923294400 ext 1705; Fax: +34 923294509; E-mail: aafs@usal.es; URL: campus.usal.es/~bibfilo/
Ángel A. Fernández Sevillano
227 100 vols; 3 223 curr per; 1 210 av-mat; 1 414 digital data carriers
libr loan 25561

– Facultad de Geografía e Historia, Biblioteca, C/ Cervantes, 3, 37007 Salamanca
T: +34 923294400; Fax: +34 923294512; E-mail: antolin@usal.es; URL: fgh.usal.es/infogeneralservicios.php
1991; Dr. Ana María García Antolín
115 000 vols; 2 500 curr per; 2 mss; 4 190 maps; 688 microforms; 637 av-mat; 1 605 sound-rec; 179 digital data carriers; 1 125 printed music, 1 136 closed periodicals
libr loan 25562

– Facultad de Medicina, Biblioteca de Medicina y Odontología, C/ Alonso el Sabio s/n, Campus Miguel de Unamuno, 37007 Salamanca
T: +34 923294500; Fax: +34 923294519; E-mail: helena@usal.es; URL: campus.usal.es/~bibmed/
Helena Martín Rodero
Medicine, Dentistry
15 000 vols; 300 curr per 25563

Universidad Pontificia de Salamanca, Biblioteca General, C/ Compañia, 5, 37002 ***Salamanca***
T: +34 923277118; Fax: +34 923277118; E-mail: biblioteca@upsa.es; URL: www.biblioteca.upsa.es
1940; D. Antonio García Madrid
Synodal coll Lamberto de Echeverría, Boletines Diocesanos de España; Theological and ecumenical libr, philosophical, pedag and psych dept, libr of Fac de Trilingüe, canon law dept
211 000 vols; 850 curr per; 220 mss; 4 incunabula; 500 diss/theses; 2 100 microforms; 120 sound-rec; 46 digital data carriers
libr loan; REBIUN 25564

Real Centro Universitario Escorial – María Cristina, Biblioteca, Paseo de los Alamillos, 2, 28200 ***San Lorenzo del Escorial*** (Madrid)
T: +34 918904545; Fax: +34 918906609; E-mail: contacto@rcumariacristina.com; URL: www.rcumariacristina.com
16 000 vols; 150 curr per 25565

Universidad de Cantabria, Biblioteca Universitaria, Av de los Castros, s/n, 39005 ***Santander***
T: +34 942201180; Fax: +34 943201183; URL: www.buc.unican.es
1987; Maria Jesús Sáiz
2 archives
373 000 vols; 1 975 curr per; 9 965 e-journals; 12 500 diss/theses; 1 900 maps; 8 000 microforms; 3 400 av-mat; 1 300 sound-rec; 4 400 digital data carriers
libr loan; IFLA, IATUL 25566

Instituto Teológico Compostelano, Biblioteca, Pl de La Inmaculada, 5, 15704 ***Santiago de Compostela***
T: +34 981586277; Fax: +34 981589916; E-mail: itc@alfaexpress.net; URL: www.archicompostela.org

Leonardo Lemos Montanet
Theology, philosophy, fenomenology, church hist, law, hist
70 000 vols; 250 curr per; 5 incunabula; 5 900 music scores; 12 digital data carriers; 2 000 other items 25567

Universidade de Santiago de Compostela, Biblioteca Universitaria, Franco s/n, 15782 ***Santiago de Compostela***
T: +34 981 563 100; Fax: +34 981 547 055; E-mail: bxsec@usc.es; URL: busc.usc.es
1550; María Virtudes Pardo Gómez
Biblioteca América; Biblioteca Gallega
1 637 938 vols; 7 779 curr per; 7 708 e-journals; 670 mss; 143 incunabula; 10 628 diss/theses; 6 020 maps; 106 131 microforms; 10 735 av-mat; 1 138 sound-rec; 9 259 digital data carriers; 5 305 photos
libr loan; LIBER, REBIUN 25568

– Biblioteca Concepción Arenal (Concepcion Arenal Library), Campus Vida, 15782 Santiago de Compostela
T: +34 8818 15210; Fax: +34 8818 15213; E-mail: derprest@usc.es; URL: busc.usc.es/A_Biblioteca/Puntos_de_servizo/dereito.asp
1504; José Francisco Méndez García
Law, political science and sociology; Intellectual Property coll, Labour Relations, International Relations; Centre of European Documentation
157 223 vols; 4 827 curr per; 235 e-journals; 750 diss/theses; 791 microforms; 58 av-mat; 269 digital data carriers
libr loan 25569

– Biblioteca da E.U. de Formación do Profesorado – Lugo, Escola Universitario de Formación do Profesorado, Av de Ramón Ferreiro, s/n, 27002 Lugo
T: +34 982285886, 982223996 (ext 21007, 21058); Fax: +34 982285886; E-mail: maxprado@usc.es; URL: busc.usc.es/A_Biblioteca/Puntos_de_servizo/max.asp
1850
30 300 vols; 278 curr per; 755 av-mat 25570

– Biblioteca da Facultade de Bioloxía, R./ Lope Gómez de Marzoa, Campus Universitario Sur, 15782 Santiago de Compostela
T: +34 981563100 (ext 13246, 13247, 13248); Fax: +34 981547063; E-mail: bioprest@usc.es; URL: busc.usc.es/A%5FBiblioteca/Puntos%5Fde%5Fservizo/bioloxia.asp
1983; Carmen Bermejo Díaz de Rábago
21 398 vols; 294 curr per; 83 e-journals; 1 270 diss/theses; 662 maps; 2 084 microforms; 61 av-mat; 1 308 digital data carriers
REBIUN 25571

– Biblioteca de Ciencias da Comunicación, Facultade de Ciencias da Comunicación, Av de Castelao, s/n, Campus Universitario Norte, 15704 Santiago de Compostela
T: +34 981563100 (ext 16502); Fax: +34 981547130; E-mail: perjuana@usc.es; URL: busc.usc.es/A_Biblioteca/Puntos_de_servizo/xornalismo.asp
Communication science
16 100 vols; 384 curr per 25572

– Biblioteca de Ciencias Económicas y Empresariais, Facultad de Ciencias Económicas y Empresariais, Av Burgo das Nacións, 15704 Santiago de Compostela
T: +34 98156310 (ext 11506); Fax: +34 981547065; E-mail: bueco@usc.es; URL: busc.usc.es/A_Biblioteca/Puntos_de_servizo/economica.asp
1967
Economics, management; sociology, demography
49 000 vols; 1 987 curr per; 170 diss/theses; 40 maps 25573

– Biblioteca de Farmacia, Facultade de Farmacia, Campus Universitario Sur, 15706 Santiago de Compostela
T: +34 981563100 (ext 14988); Fax: +34 981567066; E-mail: farprest@usc.es; URL: busc.usc.es/A_Biblioteca/Puntos_de_servizo/farmacia.asp
1985; Jose Francisco Mendez Garcia
Chemistry, microbiology
15 200 vols; 1 507 curr per; 224 diss/theses; 139 microforms; 45 av-mat; 3 digital data carriers 25574

– Biblioteca de Filoloxía, Facultade de Filoloxía, Av Castelao, s/n, Campus Universitario Norte, 15704 Santiago de Compostela
T: +34 981563100 (ext 11753); Fax: +34 981547067; E-mail: bufix@usc.es; URL: busc.usc.es/A_Biblioteca/Puntos_de_servizo/filoloxia.asp
1955; Pilar del Oro Trigo
114 000 vols; 1 564 curr per; 37 diss/theses; 3 maps; 1 123 microforms; 166 av-mat; 148 digital data carriers
libr loan 25575

– Biblioteca de Filosofía, Facultade de Filosofía, Praza de Mazarelos, s/n, 15782 Santiago de Compostela
T: +34 981563100 (ext 12500); E-mail: filvaz@usc.es; URL: busc.usc.es/A_Biblioteca/Puntos_de_servizo/filosofia.asp
1975
30 000 vols; 450 curr per; 60 diss/theses
libr loan 25576

– Biblioteca de Matemáticas, Facultade de Matemáticas, R./ Lope Gómez de Marzoa, Campus Universitario Sur, 15706 Santiago de Compostela
T: +34 981563100 (ext 13127, 13128, 13352); Fax: +34 981547069; E-mail: bumat@usc.es; URL: busc.usc.es/A_Biblioteca/Puntos_de_servizo/matematica.asp
1983
23 400 vols; 416 curr per 25577

– Biblioteca de Medicina e Odontoloxía, Facultade de Medicina, San Francisco, s/n, 15705 Santiago de Compostela
T: +34 981563100 (ext 12253-54); E-mail: bumed@usc.es; URL: busc.usc.es/A_Biblioteca/Puntos_de_servizo/medicina.asp
1975; Xosé A. Regos
Theses, videos, research coll, newspaper arch
16 000 vols; 2 308 curr per; 205 e-journals; 605 diss/theses; 6 084 microforms; 79 digital data carriers
libr loan 25578

– Biblioteca de Química, Facultade de Química, Av das Ciencias, s/n, 15706 Santiago de Compostela
T: +34 981563100 (ext 14275, 14276, 14277); Fax: +34 981547071; E-mail: quiprest@usc.es; URL: busc.usc.es/A_Biblioteca/Puntos_de_servizo/quimica.asp
1977
Biochemistry, biotechnology, chemical engineering, environmental chemistry
14 300 vols; 675 curr per; 1 700 diss/theses; 900 microforms
libr loan 25579

– Biblioteca de Xeografía e Historia, Facultade de Xeografía e Historia, Praza de Universidade, s/n, 15703 Santiago de Compostela
T: +34 981563100 (ext 12628-12687); Fax: +34 981547072; E-mail: hisara@usc.es; URL: busc.usc.es/A_Biblioteca/Puntos_de_servizo/historia.asp
1922
80 000 vols; 2 345 curr per; 436 diss/theses; 3 000 maps; 85 av-mat; 350 sound-rec; 15 000 slides 25580

– Biblioteca Intercentros de Físcia e Óptica, Facultade de Física, Campus Universitario Sur, 15706 Santiago de Compostela
T: +34 981563100 (ext 14071); Fax: +34 981547068; E-mail: bufis@usc.es; URL: busc.usc.es/A_Biblioteca/Puntos_de_servizo/fisica.asp
1981
9 600 vols; 479 curr per; 26 diss/theses; 128 microforms
libr loan 25581

– Biblioteca Intercentros de Psicoloxía e CC. Educación, Facultade de Ciencias da Educación, Av Xoán XXIII, 15704 Santiago de Compostela
T: +34 981563100 (ext 12019); Fax: +34 981547073; E-mail: exbprest@usc.es; URL: busc.usc.es/A%5FBiblioteca/Puntos%5Fde%5Fservizo/interpsicoeduc.asp
1969
41 000 vols; 1 000 curr per 25582

– Biblioteca Intercentros do Campus de Lugo, Edif. Biblioteca Intercentros, Campus Universitario de Lugo, 27002 Lugo
T: +34 982223996 (ext 23503, 23500); Fax: +34 982285934; E-mail: intdire@usc.es; URL: busc.usc.es/A_Biblioteca/Puntos_de_servizo/inter.asp
1973
Mathematics, physics, chemistry, agriculture, forestry, geology, engineering, biology, veterinary science, social sciences, humanities
81 000 vols; 1 568 curr per
libr loan 25583

– Biblioteca Xeral, Colexio de Fonseca, R/ Franco, s/n, 15782 Santiago de Compostela
T: +34 981563100 (ext 11092, 11080); Fax: +34 9815470-55/-70; E-mail: bxsec@usc.es; URL: busc.usc.es/A_Biblioteca/Puntos_de_servizo/xeral.asp
Fondo histórico, Fondo galego, Fondo hispanoamericano; humanities, social sciences, bibliographies and reference works, coll of newspaper clippings
243 000 vols; 8 240 curr per; 3 maps
libr loan 25584

Conservatorio Superior de Música "Manuel Castillo" de Sevilla, Biblioteca, C/ Baños, 48, 41002 ***Sevilla***
T: +34 954915630; Fax: +34 954374373; URL: www.csmsev.es
1934; Rafael Jorreto Lloves
8 000 vols; 11 curr per; 20 000 music scores; 32 rare bks, 812 records, 504 films 25585

Universidad de Sevilla, Biblioteca Universitaria de Sevilla, San Fernando, 4, Apdo 343, 41004 ***Sevilla***
T: +34 954551128; Fax: +34 954551135; E-mail: jmensaque@us.es; URL: bib.us.es
1505; Sonsoles Celestino Angulo
Rare bks of 16th c (8 000 vols)
1 299 230 vols; 14 558 curr per; 7 197 e-journals; 800 mss; 300 incunabula; 1 324 maps; 7 532 microforms; 8 188 av-mat; 3 337 sound-rec; 7 235 digital data carriers
libr loan; LIBER, ANABAD 25586

– Biblioteca Area Politécnica, Escuela Universitaria Politécnica, Virgen de África, 7, 41011 Sevilla
T: +34 954552819, 954550916; E-mail: bibeup@us.es; URL: bib.us.es/politecnica
1959; Consuelo Arahal Junco
26 000 vols; 170 curr per; 24 172 e-journals; 10 000 e-books; 13 microforms; 480 av-mat; 300 digital data carriers
libr loan 25587

– Biblioteca de Arquitectura, Escuela Técnica Superior de Arquitectura, Escuela Universitaria de Arquitectura Técnica, Av de Reina Mercedes, 2, 41012 Sevilla
T: +34 9545565-25/-26; Fax: +34 954556534; E-mail: anaisa@us.es; URL: bib.us.es/arquitectura/index-ides-idweb.html
1963; Elvira Ordóñez Cocovi
Architecture, arts, urban planning, civil engineering
17 000 vols; 77 curr per; 35 diss/theses; 525 maps; 24 av-mat
libr loan 25588

– Biblioteca de Bellas Artes, Facultad de Bellas Artes, Gonzalo Bilbao, 7-9, 41009 Sevilla
T: +34 954486499; Fax: +34 954486499; E-mail: paumol@us.es; URL: bib.us.es/bellas_artes/index-ides-idweb.html
1941; Paulina Molíno García
Obias del Siglo XIX
23 000 vols; 330 curr per; 2 e-journals; 259 diss/theses; 69 microforms; 294 av-mat; 210 digital data carriers
libr loan 25589

– Biblioteca de Comunicación, Facultad de Communicación, Américo Vespuccio, s/n, 41092 Isla de la Cartuja, Sevilla
T: +34 954486040, 954486041; Fax: +34 954486085; E-mail: bibcomu@us.es; URL: bib.us.es/bibcomunicacion.asp
1989; Maria Dolores Rodriguez Brito
Video coll, photographic coll
25 632 vols; 129 curr per; 56 diss/theses; 11 maps; 47 microforms; 5 302 av-mat; 692 sound-rec; 401 digital data carriers
libr loan 25590

– Biblioteca de Derecho y Ciencias del Trabajo, Fac. de Derecho, Fac. de Ciencias del Trabajo, C/ Enramadilla, 18-20, 41018 Sevilla
T: +34 954551216; Fax: +34 954556967; E-mail: hildam@us.es; URL: bib.us.es/derechoytrabajo/index-ides-idweb.html
1850; María Hilda Monar González
Philosophy, Social sciences, Economics, Labour, Law (General, Roman, International, Constitutional, Administrave, Penal, Canon, Labor, Social); Biblioteca digital Pixelegis
171 680 vols; 2 687 curr per; 1 314 e-journals; 2 697 e-books; 377 diss/theses; 1 974 av-mat; 1 340 digital data carriers
libr loan 25591

– Biblioteca de Económicas y Empresariales, Facultad de Económicas y Empresariales, Ramón y Cajal, 1, 41018 Sevilla
T: +34 954557517; Fax: +34 954556095; E-mail: bibeco3@us.es; URL: bib.us.es/economicas/index-ides-idweb.html
1973; Pilar Redondo López
Sociology, law
31 000 vols; 556 curr per; 100 diss/theses; 114 microforms; 3 digital data carriers 25592

– Biblioteca de Educación, Facultad de Ciencias de la Educación, Ciudad Jardín, 20-22, 41005 Sevilla
T: +34 954556539; Fax: +34 954551743; E-mail: angela-a@us.es; URL: bib.us.es/educacion/index-ides-idweb.html
1972; Angela Arévalo
60 000 vols; 274 curr per; 1 mss; 21 diss/theses; 29 maps; 715 microforms; 440 av-mat; 650 sound-rec; 1 digital data carriers
libr loan 25593

– Biblioteca de Empresariales, Escuela Universitaria de Estudios Empresariales, Av San Francisco Javier, s/n, 41018 Sevilla
T: +34 954552805; Fax: +34 954552804; E-mail: jmvinagre@us.es; URL: bib.us.es/empresariales/index-ides-idweb.html
José Manuel Vinagre Lobo
Business administration, management 25594

– Biblioteca de Farmacia, Facultad de Farmacia, Prof. García González, s/n, 41012 Sevilla
T: +34 9545567-14/-15; Fax: +34 9545567-14/-15; E-mail: biblfar@us.es; URL: bib.us.es/farmacia/index-ides-idweb.html
María Eugenia Díaz Pérez 25595

– Biblioteca de Humanidades, Facultad de Filología y Geografía e Historia, Palos de la Frontera, s/n, 41004 Sevilla
T: +34 954551503; Fax: +34 954551355; E-mail: vtejada@us.es; URL: bib.us.es/humanidades/index-ides-idweb.html
Victoria Tejada Enríquez
130 000 vols; 552 curr per; 2 incunabula; 20 maps; 293 microforms; 28 av-mat; 28 digital data carriers
libr loan 25596

– Biblioteca de Informática, E.T.S. Ingeniería Informática, Av Reina Mercedes, s/n, 41012 Sevilla
T: +34 95455-1696/-2760; E-mail: ngperez@us.es; URL: bib.us.es/informatica/index-ides-idweb.html
Natividad Gómez Pérez 25597

– Biblioteca de Ingeniería Técnica Agrícola, E.U. de Ingeniería Técnica Agrícola, Ctra. Utrera, km 1 (ant. Univ. Laboral), 41013 Sevilla
T: +34 95448644-0/-2; Fax: +34 954486436; E-mail: ngperez@us.es; URL: bib.us.es/informatica/index-ides-idweb.html
Leonardo Ponce Jiménez 25598

– Biblioteca de Ingenieros, Escuela Técnica Superior de Ingenieros, Camino de Los Descubrimientos, s/n, 41092 Sevilla
T: +34 954486135; Fax: +34 954486138; E-mail: magomez@us.es; URL: bib.us.es/ingenieros/index-ides-idweb.html
1966; Mercedes Aguilar Gómez
Telecommunications, mathematics, physics, chemistry, computational engineering, management science, operation research
50 036 vols; 389 curr per; 9 625 e-journals; 244 diss/theses; 5 000

microforms; 103 av-mat; 255 digital data carriers
libr loan; SEDIC 25599

– Biblioteca de Matemáticas, Facultad de Matemáticas, Tarfia, s/n, 41012 Sevilla
T: +34 954557920; Fax: +34 954557914; E-mail: bibmate@us.es; URL: bib.us.es/matematicas/index-ides-idweb.html
1978; José Maldonado Díaz
Mathematics, applied mathematics, statistics, computation
27 000 vols; 180 curr per; 400 diss/theses
libr loan; REBIUN, DOCUMAT 25600

– Biblioteca de Psicología y Filosofía, Fac. de Psicología, Fac. de Filosofía, Fac. de CC. de Educación (Pedagogía), Camilo José Cela, s/n, 41018 Sevilla
T: +34 9544557661; Fax: +34 954557662; E-mail: mgallego@us.es; URL: bib.us.es/psicologia/index-ides-idweb.html
1976/1994; María Gallego Gonzalez
Education, psychology, logic, ethics, philosophy
54 675 vols; 864 curr per; 50 e-journals; 700 diss/theses; 550 microforms; 122 av-mat; 89 digital data carriers; 500 tests
libr loan 25601

– Biblioteca de Química, Facultad de Química, Prof. García González, s/n, 41012 Sevilla
T: +34 954557136; Fax: +34 954557134; E-mail: paurora@us.es; URL: bib.us.es/quimica/index-ides-idweb.html
1978; Pilar Aurora Ruiz Martínez
Chemistry, geology, biology, agriculture, engineering, crystallography, environment
21 648 vols; 292 curr per; 1 300 diss/theses; 778 microforms; 10 sound-rec; 49 digital data carriers
libr loan 25602

– Facultad de Física, Biblioteca de Área de Fisica, Av de Reina Mercedes, s/n, 41012 Sevilla
T: +34 954552884; Fax: +34 954612097; E-mail: mjosegomez@us.es; URL: bib.us.es/fisica/index.ides-idweb.html
1981; María José Gómez Fernández
14 370 vols; 191 curr per; 22 509 e-journals; 462 diss/theses; 224 microforms; 194 av-mat; 87 sound-rec
libr loan 25603

Universidad Pablo de Olavide, Biblioteca, Edif. Juan Bautista Muñoz, Ctra. de Utrera, Km. 1, 41013 ***Sevilla***
T: +34 954349255; Fax: +34 954349257; E-mail: infobib@upo.es; URL: www.upo.es/biblioteca/
Carmen Baena Díaz
222 000 vols; 1 094 curr per; 26 700 e-journals; 242 000 e-books; 13 600 av-mat; 400 000 electronical docs
libr loan 25604

Universidad Laboral 'Francisco Franco', Biblioteca, Ctra de Salou, Apdo 119, ***Tarragona***
1956
27 000 vols 25605

Universitat Rovira i Virgili, Biblioteca, C/ de l'Escorxador, s/n, 43003 ***Tarragona***
T: +34 977559524; Fax: +34 977244256; URL: www.urv.es/biblioteca/index.html
1972; Mariantònia Aloguín Pallach
103 000 vols; 2 260 curr per; 700 diss/theses; 670 av-mat 25606

– Biblioteca Campus Bellissens (Campus Bellissens Library), Av Universitat, 1, 43204 Reus
T: +34 977759807; Fax: +34 977759808; E-mail: bib.bellissens@urv.cat; URL: www.urv.cat/biblioteca/biblioteques/campus_bellissens/index.html
1988; Ariadna Casals Garcia
Economics, Business Administration, Statistics, Mathematics, Architecture
44 118 vols; 771 curr per; 2 000 e-journals; 156 e-books; 65 diss/theses; 150 govt docs; 10 maps; 29 034 microforms; 162 av-mat; 193 sound-rec; 541 digital data carriers
libr loan; REBUIN, CBUC 25607

– Biblioteca d'Infermeria, Av de Roma, 15, 43005 Tarragona
T: +34 977250125; Fax: +34 977251424; E-mail: bibinf@bib.urv.es; URL: www.urv.es/biblioteca
1981; Carme Montcusí Puig
10 790 vols; 249 curr per; 15 maps; 540 av-mat; 31 sound-rec; 71 digital data carriers
libr loan 25608

– Campus Sescelades, Biblioteca, Av Països Catalans, 26, 43007 Tarragona
T: +34 977559614; Fax: +34 977559698; E-mail: esperanza.manera@urv.cat; URL: www.urv.cat/biblioteca/biblioteques/campus_sescelades/
1990; Esperanza Manera
20 500 vols; 429 curr per; 1 738 diss/theses; 250 av-mat; 1 400 digital data carriers
libr loan 25609

– Campus Sescelades Àrea d'Educació i Psicologia, Biblioteca, Av Països Catalans, u 26, 43007 Tarragona
T: +34 977559614; Fax: +34 977559698; URL: www.etse.urv.es/
1843; Xavier Adolfo Sevil Andujar
Coll of books for children
33 000 vols; 299 curr per; 427 diss/theses; 25 maps; 430 microforms; 1 060 av-mat; 260 sound-rec; 304 digital data carriers; didactic materials
libr loan 25610

– Secció de Lletres i Quimica, Biblioteca, Pl Imperial Tarraco, 1, 43005 Tarragona
T: +34 977559524; Fax: +34 977559597; E-mail: biblq@bib.urv.es; URL: www.urv.es/sgenerals/biblioteca
1972; Marta Sedó i Ramon
Chemistry, Hist, Anthropology
90 000 vols; 800 curr per; 500 diss/theses; 3 200 maps; 400 microforms; 380 av-mat; 70 sound-rec; 50 digital data carriers
libr loan; REBIUN, CBUC 25611

Universidad Nacional de Educación a Distancia, Biblioteca del Centro Asociado de la UNED de Teruel, Miguel Servet, 2, 44002 ***Teruel***
T: +34 978607189; Fax: +34 978601009; E-mail: biblioteca@teruel.uned.es; URL: www.uned.es
1985; Ana M Ubé González
20 000 vols; 150 curr per; 650 av-mat; 6 450 sound-rec; 300 digital data carriers
libr loan 25612

Universidad de Castilla-La Mancha – Campus de Toledo, Biblioteca General, Edif. de San Pedro Mártir, Cobertizo de San Pedro Mártir, 45071 ***Toledo***
T: +34 902204100; Fax: +34 902204130; E-mail: biblioteca.to@uclm.es; URL: www.biblioteca.uclm.es/dir_to.html
1969; Marta Navascues Palacio
84 812 vols; 530 curr per; 105 maps; 4 215 microforms; 110 sound-rec; 150 digital data carriers
libr loan 25613

Universidad de Castilla-La Mancha – Escuela Universitaria del Profesorade de E.G.B., Biblioteca, Av Barber, s/n, 45004 ***Toledo***
T: +34 925213716
1959
Education, psychology
17 700 vols; 162 curr per; 3 600 microforms; 94 av-mat; 1 360 sound-rec; 1 400 slides 25614

Escuela Universitaria del Profesorado de E.G.B., Biblioteca, Av Cristo Rey, 25, 23400 ***Ubeda***
T: +34 953796102
1944; Diego Blanco Lopez
Education, modern hist
40 005 vols; 83 curr per; 150 av-mat; 3 500 sound-rec; 600 digital data carriers; 5 000 slides 25615

Conservatorio Superior de Música de Valencia, Biblioteca, Camino de Vera, 29, 46022 ***Valencia***
T: +34 963605316; Fax: +34 963605701; URL: www.csmvalencia.es
1879; Pilar Moret Tordera
11 000 vols; 160 curr per; 6 336 music scores; 1 114 sound-rec 25616

Universidad Politécnica de Valencia, Biblioteca Central, Camino de Vera, s/n, Edif. 4L, 46022 ***Valencia***
T: +34 963877084; Fax: +34 963877089; E-mail: biblio@bib.upv.es; URL: www.upv.es/bib
1972; José Llorens
Agronomy, architecture, physics, informatics, telecommunications, fine arts
400 000 vols; 890 curr per; 5 000 diss/theses; 1 800 maps; 550 microforms; 1 830 av-mat; 1 500 patents
libr loan; ANABAD, FID, Liber 25617

Universitat de València, Servei de Biblioteques i Documentació (University of Valencia), C/ Universitat, 2, 46003 ***Valencia***
T: +34 963983140; Fax: +34 963864117; E-mail: Victoria.Garcia@uv.es; URL: Biblioteca.uv.es
1785; Victoria García Esteve
1 371 196 vols; 18 453 curr per; 137 e-journals; 173 220 e-books; 3 003 mss; 411 incunabula; 24 202 diss/theses; 354 music scores; 19 660 maps; 4 276 microforms; 4 554 av-mat; 2 640 sound-rec; 1 650 offprints
libr loan 25618

– Biblioteca de Ciències de la Salut – Sala de Medicina, Av Vicente Blasco Ibañez, 15, 46010 Valencia
T: +34 963864153; Fax: +34 963864174; E-mail: pisalud@uv.es
1891; Regina Pinilla Pérez de Tudela
Medicine, odontology
52 000 vols; 381 curr per; 2 160 diss/theses; 3 digital data carriers 25619

– Biblioteca de Ciències "Eduard Boscà", Dr Moliner, 50, 46100 Burjassot, Valencia
T: +34 963544158; Fax: +34 963544798; E-mail: rosario.ferre@uv.es
1991; Rosario Ferre Sanchis
Monographs, serials
55 500 vols; 792 curr per; 1 100 diss/theses; 1 000 microforms
libr loan 25620

– Biblioteca de Ciències Socials 'Gregori Mains', Av Tarongers s/n°, 46010 Valencia
T: +34 963828914; Fax: +34 963828735; E-mail: infobibsoc@uv.es
1999; Blanca Llopis Carles
Law, Economy, Management, Business, Social Sciences; Bibl Depositoria ONU
300 000 vols; 1 023 curr per; 8 000 e-journals; 1 maps; 90 av-mat; 16 digital data carriers
libr loan 25621

– Biblioteca de Psicologia i Esport, Av Blasco Ibañez, 21, 46010 Valencia
T: +34 963864534; Fax: +34 963864956; E-mail: psicoesp@uv.es
1981
11 400 vols; 195 curr per; 500 diss/theses; 111 av-mat; 188 sound-rec
libr loan 25622

– Biblioteca d'Humanitats Joan Reglà, C/ Artes Gráficas, 13, 46010 Valencia
T: +34 963864557; Fax: +34 963864558; E-mail: Humainformacion@uv.es; URL: biblioteca.uv.es/castellano/bibliotecas/de_campus/b_humanitats/b_humanidades.php
2002; Isabel Soler
Philosophy, languages and literatures, history, geography
349 081 vols; 4 942 curr per; 38 733 e-journals; 74 536 e-books; 1 mss; 9 824 maps; 690 microforms; 1 900 av-mat
REBIUN 25623

– Biblioteca Històrica (Historical Library), C/ Universitat, 2, 46003 Valencia
T: +34 963864556; Fax: +34 963983454; E-mail: Bibhistorica@uv.es; URL: biblioteca.uv.es/valenciano/bibliotecas/de_campus/b_historica/historica.php
1785; María Cruz Cabeza Sánchez-Albornoz
Spanish Civil War posters, children's drawings of the Spanish Civil War, coins, engravings
190 105 vols; 2 196 curr per; 3 089 mss; 347 incunabula; 342 diss/theses; 725 maps; 8 087 microforms; 737 av-mat; 465 sound-rec
libr loan; Rebiun, Europeana 25624

Universidad de Valladolid, Biblioteca General Reina Sofía (University of Valladolid), C/ Chancilleria, 6, 47003 ***Valladolid***
T: +34 983423029; Fax: +34 983423030; E-mail: biblioteca.reina.sofia@uva.es; URL: almena.uva.es/
1346; Carmen de Miguel Murado
Antiquarian Libr
816 000 vols; 16 061 curr per; 18 876 e-journals; 400 e-books; 520 mss; 351 incunabula; 1 816 maps; 22 354 microforms; 4 299 av-mat; 3 371 sound-rec; 363 digital data carriers
libr loan; LIBER, ANABAD, REBIUN 25625

– Área de Ciencias de la Salud, Biblioteca, Av de Ramón y Cajal, 7, 47005 Valladolid
T: +34 983423000 ext 4049; Fax: +34 983423947; URL: www.uva.es
1967; Francisco Rojo García
29 100 vols; 101 curr per; 664 diss/theses; 204 microforms; 4 396 av-mat
libr loan 25626

– Biblioteca de Ciencias (Library of Sciences), Doctor Mergelina, s/n, 47011 Valladolid
T: +34 983423939; Fax: +34 983423013; E-mail: bibmalu@uva.es
1979; María Luisa Pérez Cuadrado
Mathematics, physics, chemistry
36 141 vols; 266 curr per; 2 824 diss/theses; 123 av-mat; 40 digital data carriers
libr loan; DOCUMAT 25627

– Biblioteca Histórica de Santa Cruz, Plaza Santa Cruz 8, 47002 Valladolid
T: +34 983184271, 983184001; Fax: +34 983423234; E-mail: mariapilar.rodriguez.marin@uva.es; URL: www.bhsc.uva.es:8080/BHSC/
1484; Pilar Rodriguez Marin
Religion, humanities; Sección de Manuscritos, Sección de Incunables y Raros, Sección de Impresos, Sección de Legajos
31 600 vols; 528 mss; 353 incunabula; 114 microforms; 4 digital data carriers; 790 slides, 4 000 docs (19th c), 23 025 vols from 16-18th c 25628

– Campus de Soria, Biblioteca, Campus Universitario, s/n, 42004 Soria
T: +34 975129210; Fax: +34 975129401; E-mail: biblioteca.soria@uva.es
María del Carmen Sánchez Martínez
9 400 vols; 109 curr per
libr loan 25629

– Campus La Yutera, Biblioteca, Av de Madrid, 44, 34004 Palencia
T: +34 979108309; Fax: +34 979108302; E-mail: Biblioteca.palencia@uva.es; URL: www.uva.es/biblioteca.palencia
2001; Belén Burgos Nieto
Agriculture, forestry, education, labour
65 572 vols; 794 curr per; 17 diss/theses; 1 426 maps; 21 microforms; 976 av-mat; 309 sound-rec; 470 digital data carriers; 3 714 final year projects
libr loan 25630

– Centro de Documentación Europea, Biblioteca, Plaza de Santa Cruz, n° 5, 3ª planta, 47002 Valladolid
T: +34 983423767; Fax: +34 983423009; E-mail: mariaesperanza.serrano@uva.es; URL: www.cdoce.uva.es
Esperanza Serrano Fernández
libr loan 25631

– Escuela Técnica Superior de Arquitectura, Biblioteca, C/ Avda Salamanca, s/n, 47014 Valladolid
T: +34 983184945; Fax: +34 983423425; E-mail: biblioteca.arq@uva.es
1975; Beatriz García Posadas
20 000 vols; 326 curr per 25632

– Escuela Técnica Superior de Telecomunicaciones / Informática, Biblioteca, Campus Miguel Delibes, Camino del Cementerio, s/n, 47011 Valladolid
T: +34 983423000 ext 5599; Fax: +34 983423631; E-mail: piedad.casado@uva.es; URL: www.uva.es
Piedad Casado Fernández
libr loan 25633

- Escuela Universitaria de Estudios Empresariales, Biblioteca, C/ P° Prado de la Magdalena s/n, 47005 Valladolid
T: +34 983425020; Fax: +34 983423056; E-mail: biblioteca.emp@uva.es
1980; María Covadonga Matos Eguiluz
18315 vols; 360 curr per; 45 maps; 25 microforms
libr loan 25634

- Escuela Universitaria de Magisterio, Plaza de Colmenares, 1, 40001 Segovia
T: +34 921112210; Fax: +34 921112201; E-mail: biblioteca.segovia@uva.es; URL: www.uva.es
1975
Education, physical education, music
18000 vols; 205 curr per; 10 diss/theses; 430 maps; 9 microforms; 35 av-mat
libr loan 25635

- Facultad de Ciencias Económicas y Empresariales, Biblioteca, Av Valle de Esgueva, 6, 47011 Valladolid
T: +34 983423737; Fax: +34 983423299; E-mail: biblioteca.economicas@uva.es; URL: www.uva.es/biblioteca.economicas
1979; Clara Isabel Rincón Muñoz
51200 vols; 1694 curr per; 18200 e-journals; 965 e-books; 242 diss/theses; 56 microforms
libr loan 25636

- Facultad de Ciencias Sociales, Jurídicas y de la Comunicación, C/ Trinidad, 3, 40001 Segovia
T: +34 921112300; Fax: +34 921112302; E-mail: biblioteca.segovia@uva.es; URL: www.uva.es
Isabel Lecanda Meschede
Social sciences, communication, law
libr loan 25637

- Facultad de Derecho, Biblioteca, Pl de la Universidad, 1, 47002 Valladolid
T: +34 983424027; Fax: +34 983423012; E-mail: biblioteca.der@uva.es
1980; Camino Vega Fuertes
50700 vols; 360 curr per; 61 microforms
libr loan 25638

- Facultad de Filosofía y Letras, Biblioteca (Faculty of Philosophy and Letters), Plaza del Campus, s/n, 47011 Valladolid
T: +34 983423926; Fax: +34 983423007; E-mail: biblioteca.fyl@uva.es; URL: www.uva.es/biblioteca/filosofiayletras
1919; Luz Marina Fernández Rodriguez
283647 vols; 3193 curr per; 8625 maps; 34 av-mat; 1730 sound-rec
libr loan 25639

- INEA, Escuela Universitaria de Ingeniería Técnica Agrícola, Biblioteca, Camino Viejo de Simancas, km 4,5, 47008 Valladolid
T: +34 983235506; Fax: +34 983224869; E-mail: inea@inea.uva.es; URL: www.inea.uva.es/web/biblioteca/index.asp
1969; Clara Riva Sanchez
15000 vols; 125 curr per; 950 diss/theses; 100 maps; 62 digital data carriers
libr loan 25640

Universitat de Vic, Biblioteca, C. Sagrada Famlia, 7, 08500 ***Vic***
T: +34 938816170; Fax: +34 938815520; E-mail: biblioteca@uvic.cat; URL: www.uvic.cat/biblioteca
1977; Anna Andreu Molina
88957 vols; 1784 curr per; 4782 e-journals; 188 diss/theses; 667 maps; 4 microforms; 2728 av-mat; 647 sound-rec; 1586 digital data carriers
libr loan; CBUC, Rebiun 25641

Universidade de Vigo, Biblioteca Universitaria, University of Vigo, Campus Lagoas/Marcosende s/n, 36310 ***Vigo***
T: +34 986813854; Fax: +34 986813847; E-mail: dirbuv@uvigo.es; URL: www.uvigo.es/uvigo_gl/Administracion/Biblioteca/
560000 vols; 10000 curr per; 16000 e-journals; 110000 e-books; 12000 av-mat 25642

Universidad Europea de Madrid, Biblioteca universitaria Dulce Chacón, C/ Tajo, s/n, 28670 ***Villaviciosa de Odon***, Madrid
T: +34 912115247; E-mail: biblioteca@uem.es; URL: www.uem.es/biblioteca/index.html
Isabel Rico Rodriguez
78000 vols; 1241 curr per 25643

Universidad de Zaragoza, Biblioteca Universitaria (University of Zaragoza), Plaza de Paraíso, 4, 50005 ***Zaragoza***
T: +34 976761854; Fax: +34 976761857; E-mail: buz@unizar.es; URL: biblioteca.unizar.es
1583; Ramon Abad Hiraldo
23 Faculty and School Libs
1019781 vols; 27831 curr per; 109 e-journals; 1048 e-books; 419 mss; 406 incunabula; 3000 diss/theses; 622 music scores; 3697 maps; 86035 microforms; 4894 av-mat; 34047 sound-rec; 9713 digital data carriers; 274 Grabados
ANABAD, LIBER, IFLA 25644

- Biblioteca Biomédica, Facultad de Medicina, Domingo Miral, s/n, 50009 Zaragoza
T: +34 976761-000/-758; Fax: +34 976761702; E-mail: biblmed@unizar.es; URL: biblioteca.unizar.es/
Ana Romero Huerta
18700 vols; 2510 curr per; 220 e-journals; 150 microforms; 450 digital data carriers
libr loan; ANABAD 25645

- Biblioteca de Humanidades 'Maria Moliner', Facultad de Filosofia e Letras, Pedro Cerbuina, 12, 50009 Zaragoza
T: +34 976762676; Fax: +34 976761506; E-mail: mcantin@unizar.es; URL: biblioteca.unizar.es/
Matilde Cantin Luna
179000 vols; 4000 curr per; 807 sound-rec
libr loan 25646

- Biblioteca del Campus de Teruel, Ciudad Escolar, s/n, 44003 Teruel
T: +34 978618109; Fax: +34 978618104; E-mail: biteruel@unizar.es; URL: biblioteca.unizar.es/
1982; Carmen Montón Barea
Engineering
40000 vols
libr loan 25647

- Biblioteca Hypatia de Alejandría, Edif. Betancourt, Maria de Luna, 5, 50018 Zaragoza
T: +34 976761000 ext 5256; Fax: +34 976762189; E-mail: hypatia@unizar.es; URL: biblioteca.unizar.es/
2001; Ms Natividad Herranz Alfaro
Electric engineering, mechanical engineering, electronic engineering, chemical engineering, computer science
53100 vols
libr loan; REBIUN 25648

- Escuela Universitaria de Estudios Empresariales, Biblioteca, Edif. Lorenzo Normante, Maria de Luna, s/n, 50018 Zaragoza
T: +34 976762699; Fax: +34 976762740; E-mail: bibemp@unizar.es; URL: biblioteca.unizar.es/
Fondo bibliográfico de la antigua Escuela de Comercio
30000 vols
libr loan 25649

- Escuela Universitaria de Estudios Empresariales de Huesca, Biblioteca, Ronda Misericordia, 1, 22001 Huesca
T: +34 974239374; Fax: +34 974239375; E-mail: dbieueeh@unizar.es; URL: biblioteca.unizar.es/
1985; José Manuel Ubé González
19500 vols; 300 curr per
libr loan 25650

- Escuela Universitaria de Estudios Sociales, Biblioteca, Violante de Hungria, 23, 50009 Zaragoza
T: +34 976761000 ext 4515; Fax: +34 976761029; E-mail: bibeuesz@unizar.es; URL: biblioteca.unizar.es/
Jesús Gracia Ostáriz
22000 vols; 486 curr per
libr loan 25651

- Facultad de Ciencias, Biblioteca, C/ Pedro Cerbuna, 12, 50009 Zaragoza
T: +34 976762028; Fax: +34 976761050; E-mail: robertos@unizar.es; URL: biblioteca.unizar.es/ciencias
Roberto Soriano García
Physics, Chemistry, Mathematics, Earth Sciences
47710 vols; 3403 curr per; 306 e-journals; 1822 maps; 2 av-mat; 154 digital data carriers
libr loan 25652

- Facultad de Ciencias de la Salud y del Deporte, Biblioteca, Plaza Universidad, 3, 22002 Huesca
T: +34 974239397; Fax: +34 974239392; E-mail: bibfsd@unizar.es; URL: biblioteca.unizar.es/
1985; Maria Jesús Broto Martínez
26000 vols; 116 curr per
libr loan 25653

- Facultad de Ciencias Económicas y Empresariales, Biblioteca (Faculty of Economics and Business Sciences), Gran Via, 2, 50005 Zaragoza
T: +34 976761000 ext 4605; Fax: +34 976762500; E-mail: circu560@unizar.es; URL: biblioteca.unizar.es
1974; Ana Maria Pons León
70000 vols
libr loan 25654

- Facultad de Ciencias Humanidas y de la Educación, Biblioteca, Valentin Carderera, 4, 22003 Huesca
T: +34 974239353; Fax: +34 974239344; E-mail: bibfche@unizar.es; URL: biblioteca.unizar.es/
Maria Engracia Martin Valdunciel
29560 vols; 200 curr per
libr loan 25655

- Facultad de Derecho, Biblioteca, Pedru Cerbuna, 12, 50009 Zaragoza
T: +34 976761000 ext 3633; Fax: +34 976761499; E-mail: bibdere@unizar.es; URL: biblioteca.unizar.es/
1933; Sergio Grafiada Fernandez
140000 vols; 671 curr per; 51 e-journals; 198 diss/theses; 101 microforms; 5 legal databases
libr loan; IALL 25656

- Facultad de Educación, Biblioteca, San Juan Bosco, 7, 50009 Zaragoza
T: +34 976761000 ext 3391; Fax: +34 976762071; E-mail: bibeducz@unizar.es; URL: educacion.unizar.es/biblioteca.html
1976; Maria Pilar Arbe Serrano
47500 vols; 6 e-journals; 1 diss/theses
libr loan 25657

- Facultad de Veterinaria, Biblioteca, Miguel Servet 177, 50013 Zaragoza
T: +34 976761606; Fax: +34 976761612; E-mail: mjyusta@unizar.es; URL: biblioteca.unizar.es/
1938; María José Yusta Bonilla
Veterinary medicine, food sciences, agricultural economics
35000 vols; 217 curr per; 183 e-journals; 2648 diss/theses; 93 maps; 151 microforms; 792 av-mat
libr loan 25658

- Instituto de Ciencias de la Educación, Biblioteca, C/ Pedro Cerbuna, 12, 50009 Zaragoza
T: +34 976761990; Fax: +34 976761345; E-mail: bibice@unizar.es; URL: biblioteca.unizar.es
1970; Paz Miranda
15500 vols; 103 curr per; 2633 e-journals; 172 diss/theses; 47 microforms; 137 digital data carriers
libr loan 25659

School Libraries

IES Antonio Machado, Biblioteca, Alarpardo, s/n, 28806 ***Alcalá de Henares***, Madrid
T: +34 918892450; Fax: +34 918883916; E-mail: ies.machado.alcala@educa.madrid.org
1966; María Teresa Mena Guerrero
35000 vols; 15 curr per; 1059 diss/theses 25660

Colegio Sagrado Corazón de Jesús, Biblioteca Arrupe, Caspe, 25, 08010 ***Barcelona***
T: +34 933183704
1967
30000 vols; 135 curr per 25661

Collegi dels Jesuites de Casp, Biblioteca, Casp, 25, 08010 ***Barcelona***
T: +34 933183704
1983; Arturo Ruiz Fernández
30000 vols 25662

Centro de Enseñanzas Integradas, Biblioteca, Av de Quijotes, s/n, 10004 ***Cáceres***
T: +34 927249200; Fax: +34 927213705
1967; Maria Carmen Lagarejos Santano
17200 vols; 18 curr per; 130 av-mat 25663

Centro de Enseñanzas Integradas, Biblioteca, Ctra de Madrid, 46380 ***Cheste***, Valencia
T: +34 962510011
1972; Maria Angeles Vall Ojeda
23000 vols; 20 curr per
libr loan 25664

Centro de Enseñanzas Integradas, Biblioteca, Otaola Hiribidea, 29, 20600 ***Eibar***
T: +34 943718444
1968; Eladio Gonzalez Marin
17200 vols; 48 curr per; 160 diss/theses; 15 maps; 4608 av-mat
libr loan 25665

Escuela Andaluza de Salud Pública, Biblioteca, Campus Universitario de Cartuja, Apdo 2070, 18080 ***Granada***
T: +34 958027486; Fax: +34 958027488; E-mail: biblioteca.easp@juntadeandalucia.es; URL: www.easp.es
1984; Andra Manuel-Keenoy
Health promotion resource centre, 18 000 items (leaflets, posters, kits, etc) + 2 000 videos
59000 vols; 1000 curr per; 800 e-journals; 20 diss/theses; 4000 av-mat; 8 sound-rec; 200 digital data carriers
libr loan 25666

Centro de Enseñanzas Integradas, Biblioteca General, Salvador Allende, s/n, 15071 ***Haciadama-Elburgo***
1964; Juan Ramón Guerrero Saez
17500 vols; 34 curr per; 6505 av-mat 25667

Centro de Enseñanzas Integradas, Biblioteca, Ctra de Cuarte, ***Huesca***
T: +34 973244141
1967; Manuel Lozano Martinez
13000 vols; 517 curr per; 300 diss/theses; 325 av-mat; 325 sound-rec 25668

Centro Asociado a la UNED de Correos, Biblioteca, Conde de Peñalver 19, 28006 ***Madrid***
T: +34 913962805; Fax: +34 913962841
1913; Margarita Nieto Navarro
Law, economics, business
17226 vols; 39 curr per; 12 diss/theses; 104 av-mat; 813 sound-rec; 5 digital data carriers 25669

EOI Escuela de Negocios, Biblioteca, Gregorio del Amo, 6, Ciudad Universitaria, 28040 ***Madrid***
T: +34 913495600; Fax: +34 915542394; E-mail: informacion@eoi.es; URL: www.eoi.es/nw/publica/biblioteca.asp
1955; Rocio Casanueva
12000 vols; 150 curr per; 2000 diss/theses; 100 av-mat
libr loan 25670

Escuela de Guerra Naval, Biblioteca, Av de Burgos, 6, 28036 ***Madrid***
T: +34 913795500; Fax: +34 913795518
1925; Ana Jaqueti
13990 vols; 60 curr per
libr loan 25671

Escuela Nacional de Sanidad, Biblioteca, Sinesio Delgado 10, 28019 ***Madrid***
T: +34 913147989; Fax: +34 913150200
1924
12000 vols; 374 curr per; 3500 diss/theses
libr loan 25672

Escuela Oficial de Idiomas, Biblioteca, Jésús Maestro, s/n, 28003 ***Madrid***
T: +34 915335805
1950; Maria Teresa del Rio Cabrerizo
23500 vols; 168 curr per; 13 av-mat
ANABAD 25673

Escuela Politécnica del Ejército, Ministerio de Defensa, Biblioteca, Joaquín Costa, 6, 28002 ***Madrid***
T: +34 915618800
1942; Julian Santos D.
16200 vols; 480 curr per; 19 diss/theses 25674

Centro de Enseñanzas Integradas, Biblioteca, Av de Europa 28, 45005 ***Toledo***
T: +34 925223400
1972; Angela Felpeto Enriquez
13500 vols; 33 curr per 25675

Escuela Oficial de Idiomas de Valencia, Biblioteca Mediateca, Llano de Zaidia, 19, 46009 ***Valencia***
T: +34 963405022; Fax: +34 963492547; E-mail: eoival.biblio@edu.gva.es; URL: www.eoivalencia.es/es/biblioteca.html
13000 vols 25676

I.E.S. 'Universidad Laboral', Biblioteca Don Bosco, Av Principe de Asturias, 53, 49012 ***Zamora***
T: +34 980520100; Fax: +34 980513604; E-mail: ies.ul.biblioteca@arrakis.es
1957; Juan José Martín Herrero
36000 vols; 50 curr per; 1825 govt docs; 60 maps; 3000 av-mat; 20 digital data carriers; 17000 slides
libr loan 25677

Government Libraries

Diputación Almeria, Diputación Provincial, Biblioteca (Local and provincial library), Navarro Rodrigo, 17, 04001 ***Almería***
T: +34 950211174; Fax: +34 950269785; E-mail: biblioteca@dipalme.org; URL: www.dipalme.org
1981; Josefa Balsells Fernandez
Fondo Documental Juan Goytisolo
8134 vols; 1500 curr per; 100 diss/theses; 454 maps; 1030 microforms; 15 sound-rec; 228 digital data carriers; photos, 3000 postcards and drawings
AAB 25678

Centro de Formación de la Policía, Dirección General de la Policía, Biblioteca, Crta de Villacastín, s/n, 05002 ***Ávila***
T: +34 920226200
18000 vols; 152 curr per; 420 av-mat 25679

Ajuntament de Barcelona, Biblioteca General, Pl St. Jaume s/n., planta baixa, 08002 ***Barcelona***
T: +34 934027468; Fax: +34 934027469; URL: www.bcn.cat/bibliotecageneral
1984; M. Ràfales Caelles
15000 vols; 415 curr per; 141 e-journals; 870 microforms; 50 av-mat; 90 sound-rec; 80 digital data carriers 25680

Ajuntament de Barcelona – Centro de Documentación Estadística, Biblioteca, Pl Carles Pi i Sunyer, 8, 08002 ***Barcelona***
T: +34 934023403; Fax: +34 934023472
1987
10262 vols; 411 curr per; 133 maps 25681

Biblioteca de Comerç i Turisme, Generalitat de Catalunya (Library of Trade, Tourism and Consumer Affairs), Passeig de Gràcia, 105, 2ª planta, 08008 ***Barcelona***
T: +34 9348498-16/-17/-18; Fax: +34 934849822; E-mail: biblioteca.emo@gencat.cat; URL: www.gencat.cat/diue/serveis/biblioteca/
1985; Conxa Gonzalez Tascon
7000 vols; 400 curr per; 19 diss/theses; 18 microforms; 400 av-mat; 36 digital data carriers; 81 posters
COBDC 25682

Generalitat de Catalunya – Consell Consultiu, Biblioteca, Baixada Sant Miquel, 8, 08002 ***Barcelona***
T: +34 933176268; Fax: +34 933172821
1982; Emma Cots Segú
Administrative law, constitutional law, public law, legislation, jurisprudence
20100 vols; 208 curr per; 123 diss/theses; 19800 microforms; 17 digital data carriers
libr loan 25683

Generalitat de Catalunya – Departament de Governació, Secretaria General, Biblioteca, Pl St. Jaume, s/n, 08002 ***Barcelona***
T: +34 933178647
1982
10000 vols; 2040 curr per; 320 microforms
libr loan 25684

Generalitat de Catalunya – Departament de Sanitat, Negociat de Documentació i Biblioteca, Travessera de Les Corts, 131-159, 08028 ***Barcelona***
T: +34 932272939; Fax: +34 932272990; E-mail: biblioteca.sanitat@gencat.cat; URL: www.gencat.cat/salut/
1978; Gemma Serra i Mina
Public health
20400 vols; 404 curr per; 4 microforms; 4 digital data carriers
libr loan; EAHIL, CDB 25685

Parlament de Catalunya, Direcàs d'Estudis Parlamentaris, Area de Biblioteca, Parc de la Ciutadella, 08003 ***Barcelona***
T: +34 933046500; Fax: +34 933046549; URL: www.parlament-cat.net
1983; Francesc Pau i Vall
Grey lit, Int Organisations publ, Catalonian hist coll; Law Libr, General Libr
36703 vols; 1042 curr per
IFLA 25686

Servicio Histórico Militar, Biblioteca, Paseo de Colón, 14, 08002 ***Barcelona***
T: +34 933177500
14700 vols; 40 curr per 25687

Diputación Foral de Vizcaya, Biblioteca Provincial, C Astarloa, 10, 48008 ***Bilbao***
T: +34 944207702
1929
Bibliografía vasca, hemeroteca local
195000 vols; 627 curr per; 73 mss 25688

Diputación Provincial de Cáceres, Biblioteca, Plaza de Santa María, s/n, 10071 ***Cáceres***
T: +34 927255500; Fax: +34 927212409; E-mail: informacion@dip-caceres.es; URL: www.dip-caceres.es
1974; Antonia Maria Fajardo Caldera
Indice, Hemeroteca Legislativa; Archivo
22000 vols; 120 curr per; 138 music scores; 306 maps; 167 microforms; 59 sound-rec
ANABAD 25689

Zona Marítima del Mediterráneo, Ministerio de Defensa, Biblioteca, Muralla del Mar, 13, 30290 ***Cartagena***
T: +34 968502000
1900; Francisco Lopez Conesa
Law, mathematics, political sciences, naval engineering, astronomy, astrophysics
9100 vols; 23 curr per; 1100 standards
libr loan 25690

Diputación Provincial de Córdoba, Biblioteca, Pl de Colón s/n, 14071 ***Córdoba***
T: +34 957211100; E-mail: infodipu@dipucordoba.es; URL: www.dipucordoba.es
1982
9000 vols; 450 curr per; 350 microforms; 30 av-mat; 30 sound-rec 25691

Real Consulado de A Coruña, Biblioteca, Pl Pintor Sotomayor, 58, 15001 ***A Coruña***
T: +34 981206274
1806; Maria Luisa Babio Bescansa
Art hist, lit, fine arts
28000 vols 25692

Generalitat de Catalunya – Direcció General de l'Esport, Biblioteca, Av Països Catalans, 12, 08950 ***Esplugues (Barcelona)***
T: +34 933719011
1983; Maria Lluïsa Berasategui
Coll of bks of 19th and early 20th c; Sports Medicine Libr
18000 vols; 132 curr per; 18 diss/theses 25693

Cortes de Castilla y León, Sección de Documentación y Biblioteca, Castillo de Fuensaldaña, 47194 ***Fuensaldaña***
T: +34 983421532; Fax: +34 983421533
1983; María Victoria Juliá Dallo
12200 vols; 233 curr per; 20 maps; 37 microforms; 7 digital data carriers 25694

Biblioteca de Investigadores de la Provincia de Guadalajara, Diputación Provincial de Guadalajara, C/ Atienza 4, Centro S. Jose, 19003 ***Guadalajara***
T: +34 949887576; E-mail: b.investigadores@dguadalajara.es; URL: www.bipgu.es
1984; José Ramón Lopez de los Mozos
10600 vols; 100 curr per 25695

Asamblea de Madrid, Biblioteca, Pza. Asamblea de Madrid, 1, 28018 ***Madrid***
T: +34 917799554; Fax: +34 917799594; E-mail: egonzalez@asambleamadrid.es
1985; Eduardo González-Santander
Autonomous communities, Parliaments
25000 vols; 180 curr per; 7 microforms; 15 digital data carriers
IFLA 25696

Ayuntamiento de Madrid, Biblioteca Técnica, Pl Mayor, 27, 28012 ***Madrid***
T: +34 915882254
Col Legislativa de España, reglamentos y ordenanzas municipales, repertorios legislativos, administración local, derecho administrativo
10700 vols; 83 curr per
libr loan 25697

Biblioteca Central Militar, Instituto de Historia y Cultura Militar, Paseo de Moret, 3, 28008 ***Madrid***
T: +34 917808739; Fax: +34 917808704; E-mail: bcm@et.mde.es; URL: www.ejercito.mde.es
1932; Inocencia Soria
Military history, military and civil engineering, arts and crafts, military science; Military Music
250000 vols; 2300 curr per; 100 e-journals; 2111 mss; 1 incunabula; 200 diss/theses; 1300 music scores; 12500 microforms; 250 av-mat; 450 sound-rec; 350 digital data carriers
ANABAD, AEDOM 25698

Centro de Documentación del Transporte, Ministerio de Fomento, Biblioteca, Paseo de la Castellana, 67, C-217, 28071 ***Madrid***
T: +34 915977987; Fax: +34 915978453; E-mail: centrodoc@fomento.es; URL: www.fomento.gob.es
1987; María Concepción Sanz Bombín
Transport, Infrastructure, Communications
42000 vols; 1200 curr per; 40 e-journals; 2500 digital data carriers 25699

Centro de Estudios y Experimentación de Obras Públicas – Centro de Estudios de Técnicas Aplicadas, Biblioteca Central, Paseo de la Castellana, 67, Nuevos Ministerios, 28046 ***Madrid***
T: +34 913357286; Fax: +34 913357350
1959
Pollution, nuclear science as applied to public works, mathematics and computation, measuring techniques and instrumentation of public works
11300 vols; 685 curr per; 52700 microforms; 1800 standards 25700

Centro de Recursos Educativos, Área de Gobierno de Familia y Servicios Sociales, C/ José Ortega y Gasset, 100, 28006 ***Madrid***
T: +34 914801-289/-197; E-mail: drecursos@munimadrid.es; URL: www.munimadrid.es/educacion
Blanca Caballero María
8025 vols; 164 curr per; 1188 av-mat; 965 sound-rec 25701

Centro Estatal de Documentación e Información de Servicios Sociales, Ministerio de Trabajo y Asuntos Sociales, María de Guzmán, 52, 28003 ***Madrid***
T: +34 913633766; Fax: +34 913633889
1986; Luisa Martínez-Lázaro
33810 vols; 620 curr per
FID 25702

Centro Superior de Estudios de la Defensa Nacional (CESEDEN), Biblioteca (Spanish Center for National Defense Studies), Paseo de la Castellana, 61, 28046 ***Madrid***
T: +34 913482500; Fax: +34 9134825-53/-54; E-mail: ceseden@ceseden.es; URL: www.ceseden.es
1965; María de la Luz López
28246 vols; 123 curr per; 84 e-journals; 36 e-books; 2255 diss/theses; 6500 digital data carriers 25703

Comisión de las Comunidades Europeas, Centro de Documentación, Po. de la Castellana 46, 28046 ***Madrid***
T: +34 914315711; Fax: +34 914321411
1981
Series estadísticas de EUROSTAT
11100 vols; 646 curr per; 84000 microforms 25704

Comunidad de Madrid, Consejería de Economia y Hacienda, Servicio de Documentación y Publicaciones, Carrera de San Jerónimo, 13- Entreplanta, 28014 ***Madrid***
T: +34 915803622; Fax: +34 914208396; E-mail: centrodocumentacion.sanjeronimo@madrid.org; URL: www.madrid.org
1985; Teresa Gonzalez
Public administration, law
14000 vols; 185 curr per; 76 e-journals; 60 diss/theses; 20 maps; 8000 microforms; 20 av-mat; 300 digital data carriers; 200 offprints 25705

Congreso de los Diputados, Biblioteca, Floridablanca, s/n, 28014 ***Madrid***
T: +34 913906220; Fax: +34 913906765; E-mail: rosa.grau@sgral.congreso.es; URL: www.congreso.es
1811
Doc & bibliogr of the hist of parliament in Spain
100000 vols; 1500 curr per; 1 mss; 8 incunabula; 29 microforms; 257 rare bks (16-17th c)
libr loan; IFLA 25706

Consejería de Obras Públicas, Urbanismo y Transportes, Comunidad de Madrid, Biblioteca, C/ Maudes, 17, 28003 ***Madrid***
T: +34 915804480
1972; Maria Luisa Martínez Santiago
Urban management; Cartography dept, Photogr dept
27000 vols; 189 curr per; 20000 maps; 20000 photos
libr loan 25707

Consejo de Estado, Biblioteca, Mayor, 79, 28013 ***Madrid***
T: +34 915166256; Fax: +34 915166260; E-mail: archivo.biblioteca@consejo-estado.es; URL: www.consejo-estado.es
1880; Jorge Tarlea Lopez-Cepero
45000 vols; 220 curr per; 15 mss; 500 digital data carriers
libr loan 25708

Cuartel General de la Armada, Ministerio de Defensa, Biblioteca, Montalbán, 2, 28014 ***Madrid***
T: +34 915313912
1856
Marina, hist, militares
50300 vols; 587 curr per; 44 mss 25709

Cuartel General del Aire, Ministerio de Defensa, Biblioteca, Romero Robledo, 8, 28008 ***Madrid***
T: +34 914490700
1929; María del Carmen Lazaro y Corthay
21000 vols 25710

Cuartel General del Ejército, Ministerio de Defensa, Biblioteca, C de Alcalá, 51, 28014 ***Madrid***
T: +34 915212960
1941
24200 vols 25711

Dirección de Formación y Perfeccionamiento, Dirección General de la Policía, Biblioteca, Av Pio XII, 50, ***Madrid***
T: +34 917670955 ext 8719
1989; Martin Turrado Vidal
Forensic medicine, legislation, social problems, psychiatry, political sciences
8500 vols; 150 diss/theses 25712

Dirección General de la Guardia Civil, Biblioteca, Guzmán el Bueno, 110, 28003 ***Madrid***
T: +34 915340200
1951
10900 vols; 21 curr per 25713

Dirección General de lo Contencioso del Estado, Secretaría de Estado de Justicia, Biblioteca, Alcalá, 5, 28014 ***Madrid***
T: +34 915217304
1881
10000 vols 25714

Escuela Diplomática, Biblioteca, Paseo de Juan XXIII, 5, 28040 ***Madrid***
T: +34 915535300; Fax: +34 915347670; URL: www.mae.es
1954; Victoria Rodríguez López
Foreign affairs, internat law, inter nat economy, dilomacy; Pers libr of F.M. Castiella; UN publs, OECD publs
49 583 vols; 178 curr per; 2 399 diss/theses
libr loan 25715

Fábrica Nacional de Moneda y Timbre, Real Casa de la Moneda, Biblioteca, Jorge Juan, 106, 28071 ***Madrid***
T: +34 915666666; Fax: +34 915042943; URL: www.fnmt.es
1961; Rafael J. Feria y Pérez
21 200 vols; 5 846 curr per 25716

Instituto de Mayores y Servicios Sociales, Biblioteca IMSERSO, Av de la Ilustración, s/n – Ginzo de Limia, 58, 28029 ***Madrid***
T: +34 917033000; Fax: +34 901109899; E-mail: buzon@imserso.es; URL: www.imserso.es
1972; José Antonio Galego
Social services, elderly, disability, migration
22 528 vols; 252 curr per; 1 171 govt docs; 382 av-mat 25717

Instituto Nacional de Administración Pública, Biblioteca, Atocha 106, 28012 ***Madrid***
T: +34 913493281; Fax: +34 913493292; E-mail: webmaster@inap.map.es; URL: www.inap.map.es
1940; Enrique Orduña Rebollo
Public law, economics, administration
132 100 vols; 1 829 curr per; 60 mss; 1 933 diss/theses; 634 maps; 55 microforms; 2 427 pamphlets
libr loan 25718

Ministerio de Agricultura, Pesca y Alimentación, Biblioteca, Paseo Infanta Isabel, 1, 28014 ***Madrid***
T: +34 913475567; Fax: +34 913475618
1856; Cristina Garcia
100 000 vols; 2 200 curr per; 1 000 microforms; 25 000 pamphlets
libr loan 25719

Ministerio de Asuntos Exteriores, Biblioteca, Pl de la Provincia, 1, 28012 ***Madrid***
T: +34 913664800
1900; Miguel Santiago y Rodriguez
43 000 vols; 300 curr per; 500 mss; 1 000 maps 25720

Ministerio de Defensa, Centro de Documentación, Biblioteca, Paseo de la Castellana, 109, 28071 ***Madrid***
T: +34 912132444; Fax: +34 913955159; E-mail: cdoc@oc.mde.es; URL: www.mde.es
1990; Ana I. Cervantes Muñoz
Defence policy, Armed Forces, International Security, Geopolitics, Peacekeeping, Terrorism
35 000 vols; 1 922 curr per; 77 e-journals; 2 mss; 848 diss/theses; 3 music scores; 6 602 maps; 137 microforms; 650 av-mat; 70 sound-rec; 496 graphic materials, 60 000 analytical records
libr loan; SEDIC 25721

Ministerio de Economía y Hacienda, Biblioteca (Ministry of Economy and Public Finance), Paseo de la Castellana, 162, 28046 ***Madrid***
T: +34 915837400; Fax: +34 913493502
1893; Teresa Mezquita
75 000 vols; 598 journals 25722

Ministerio de Educación, Secretaria General Técnica, Biblioteca, San Agustín 5, 28014 ***Madrid***
T: +34 917748000; Fax: +34 917748026; E-mail: biblioteca@educacion.es
1912; Ernesto Calbet Rosello
80 000 vols; 1 800 curr per; 2 000 diss/theses; 14 000 microforms; 14 digital data carriers
libr loan 25723

Ministerio de Educación, Cultura y Deporte, Servicio de Documentación y Biblioteca, Plaza del Rey, 1, 28004 ***Madrid***
T: +34 917017217; Fax: +34 917017003
1982; Pilar Blanco Muñoz
Cultural policy, cultural economics, fine arts, archaeology, library sciences, heritage
22 616 vols; 519 curr per; 9 microforms; 560 av-mat; 136 sound-rec; 87 digital data carriers
libr loan 25724

Ministerio de Fomento, Biblioteca General, Paseo de la Castellana, 67, 28071 ***Madrid***
T: +34 91597-8838; E-mail: rthode@fomento.es
1959; Rosa M. Thode
Civil engineering, Administrative Law, Public administration, Architecture and urbanisme
43 381 vols; 320 curr per; 1 185 maps; 773 digital data carriers; 585 pamphlets
SEDIC 25725

Ministerio de Justicia, Biblioteca, San Bernardo, 45, 28015 ***Madrid***
T: +34 913902153; Fax: +34 913902141; E-mail: biblioteca@mjusticia.es
1884; María Adela Paz García
63 301 vols; 430 curr per; 183 e-journals 25726

Ministerio de la Presidencia, Servicio de Biblioteca y Documentación, Edif. Servicios Complejo de la Monclou, 28071 ***Madrid***
T: +34 913353173; Fax: +34 913353160
1871; Carmen Lazaro Corthay
67 000 vols; 944 curr per; 15 microforms; 2 digital data carriers
libr loan 25727

Ministerío de Medio Ambiente, Centro de Documentation Ambiental y Biblioteca General, Plaza San Juan de La Cruz, s/n, 28071 ***Madrid***
T: +34 915976253; Fax: +34 91 5975903; E-mail: biblioteca@mma.es; URL: www.mma.es
1976
14 600 vols; 57 000 microforms; 50 av-mat; 275 offprints
libr loan 25728

Ministerio de Sanidad y Consumo, Biblioteca Central, Paseo del Prado, 18-20, 28071 ***Madrid***
T: +34 9159611-28/-29; Fax: +34 915961239; E-mail: docbib@msc.es; URL: www.msc.es/biblioPublic/biblioDocum/home.htm
1986; Ramón Barga Hernandez
Medical Sciences pharmacology, legislation, human biology
20 000 vols; 600 curr per
libr loan 25729

Ministerio para las Administraciones Públicas, Servicio de Documentación, Alcala Galiano, 10, 28010 ***Madrid***
T: +34 915861201; Fax: +34 915861397
1980; Carlos Ibañez Montoya
Law, political sciences, public administration, administration law
14 600 vols; 704 curr per
libr loan 25730

Secretaría de Estado de Comercio, Ministerio de Comercio y Turismo, Biblioteca, Paseo de la Castellana, 160, 28046 ***Madrid***
T: +34 913493592; Fax: +34 913493502
1950
80 000 vols; 500 curr per; 11 000 microforms 25731

Secretaría de Estado de Interior, Ministerio del Interior, Biblioteca Central, Amador de los Ríos, 7, 28010 ***Madrid***
T: +34 915371362; Fax: +34 915371115
1852
31 000 vols; 304 curr per 25732

Senado, Cortes Generales, Biblioteca, Pl Marina Española, 8, 28071 ***Madrid***
T: +34 915381396; Fax: +34 915381646; E-mail: rosario.herrero@senado.es; URL: www.senado.es
1837; Rosario Herrero Gutièrrez
270 000 vols; 1 056 curr per; 202 mss; 208 microforms; 1 253 av-mat; 1 260 sound-rec
libr loan 25733

Tribunal Constitucional, Servicio de Documentación y Biblioteca, Domenico Scarlatti, 6, 28003 ***Madrid***
T: +34 914490400; Fax: +34 915449268
Jaime Nicolas Muniz
12 000 vols; 276 curr per 25734

Tribunal de Cuentas, Biblioteca, Fuencarral, 81, 28004 ***Madrid***
T: +34 914478701
1984; Isabel Urzaiz Gonzalez
15 100 vols; 35 curr per 25735

Tribunal Supremo, Biblioteca, Pl de la Villa de Paris, 1, 28004 ***Madrid***
T: +34 913196850
1918; Concepción Contel V.
9 800 vols; 82 curr per; 2 mss 25736

Escuela Naval Militar, Biblioteca Central, 36920 ***Marín***, Pontevedra
T: +34 986804731; Fax: +34 986804733
1943; Fernando Nieva Sánchez
12 638 vols; 29 curr per 25737

Diputación Provincial de Orense, Biblioteca, Rua do Progreso, 32, 32003 ***Ourense***, Prov Orense
T: +34 988385100; URL: www.depourense.es
1982; Francisco Javier Espino Domarco
18 000 vols; 50 curr per; 30 maps; 350 av-mat; 350 slides, 500 offprints 25738

Junta General del Principado de Asturias, Servicio de Biblioteca, Archivo y Documentación, Fruela, 17, 33071 ***Oviedo***
T: +34 985107515; Fax: +34 985107569
1987; Josefina Velasco Rozado
Archive
13 000 vols; 200 curr per; 60 microforms; 300 sound-rec 25739

Departamento de Presidencia. Asesoria Jurídica, Fondo Bibliográfico Central, Carlos III, 2, 31002 ***Pamplona***
T: +34 948107076; Fax: +34 948227673
1983; Elena Gimora Urdiain
8 000 vols; 70 curr per 25740

Parlamento de Navarra, Biblioteca, Navas de Tolosa, 1, 31002 ***Pamplona***
T: +34 948209206; Fax: +34 948210515; E-mail: ljfortun@parlamento-navarra.es; URL: www.parlamento-navarra.es
1985; Luis Javier Fortún Pérez de Ciriza
Legislation
12 000 vols; 150 curr per; 10 microforms 25741

Academia General del Aire, Ministerio de Defensa, Biblioteca, 30730 ***San Javier***, Murcia
T: +34 968570100
1945
Militarismo, matemáticas, física
15 000 vols; 16 curr per 25742

Audiencia Provincial de Santa Cruz de Tenerife, Biblioteca, Pza San Francisco 1, 38071 ***Santa Cruz de Tenerife***
T: +34 922247997
Esteban Emilio Hernandez Henriquez
14 300 vols 25743

Parlamento de Canarias, Documentacion, Biblioteca y Archivo, Teobaldo Power, 5, 38002 ***Santa Cruz de Tenerife***
T: +34 922473367; Fax: +34 922473411; E-mail: biblioteca@parcan.es; URL: www.parcan.es
1990
libr loan 25744

Parlamento de Cantabria, Servicio Biblioteca, Documentacion & Archivo, C/ Alta, 31-33, 39008 ***Santander***
T: +34 942241087; Fax: +34 942241072; E-mail: sbad@parlamento-cantabria.es; URL: www.parlamento-cantabria.es
1982; María Felisa Gallo Ruiz
Law
15 000 vols; 220 curr per
IFLA 25745

Consejería de Gobernación, Servicio de Documentación e Informacion, Pl Nueva, 4, planta baja, 41071 ***Sevilla***
T: +34 955041335; Fax: +34 954041311; URL: www.juntadeandalucia.es
1984; Gloria Verga Rodriguez
12 260 vols; 150 curr per; 33 govt docs; 10 650 microforms; 42 digital data carriers; 632 disks
libr loan; SEDIC, ANABAD 25746

Consejería de Salud, Junta de Andalucía, Biblioteca, Av de la Innovación, s/n, Edif. Arena 1, 41020 ***Sevilla***
T: +34 955006300; Fax: +34 955006328; URL: www.juntadeandalucia.es/salud
1984
Public health, administration, epidemiology
12 000 vols; 364 curr per
libr loan 25747

Hospital Universitario Virgen del Rocio, Biblioteca Central, Av Manuel Siurot, s/n, 41013 ***Sevilla***
T: +34 954248190; Fax: +34 954248111
1969; Isabel Montes del Olmo
10 210 vols
libr loan 25748

Parlamento de Andalucía, Servicio de Biblioteca, C/ San Juan de Ribera s/n, 41009 ***Sevilla***
T: +34 954592100 ext 2373, 2376, 2389, 2556, 2557; Fax: +34 954592103; E-mail: biblioteca@parlamentodeandalucia.es; URL: www.parlamentodeandalucia.es
1986; Carmen Jz-Castellanos
Constitutional law, Administrative law, Parliament, Legislation, Law of the autonomous regions, Economy, Contemporary history of Spain, Andalusian subject, Politics, Women
72 472 vols; 84 diss/theses; 15 music scores; 25 maps; 4 microforms; 4 592 av-mat; 11 sound-rec; 106 digital data carriers; 55 standards
libr loan; IFLA 25749

Ministerio de Defensa, Biblioteca Central, Cuesta San Servando S/N, 45090 ***Toledo***
T: +34 925247800 ext 4314 / 925247814; Fax: +34 925247818
1860
32 000 vols; 20 curr per 25750

Cortes Valencianas, Biblioteca, Pl Sant Llorenç, 4, 46003 ***Valencia***
T: +34 963876201; Fax: +34 963876202; E-mail: biblioteca@corts.es; URL: www.cortsvalencianes.es/OPAC/tlpccvv.html
1984; Julia Sevilla
30 696 vols; 538 curr per; 15 e-journals; 358 sound-rec; 476 digital data carriers
libr loan; IFLA 25751

Diputació Provincial de València, Biblioteca, Alboraya, 5, 46009 ***Valencia***
T: +34 963882705; Fax: +34 963882704; E-mail: vicente.boquera@dva.gva.es; URL: www.dival.es
1926
Cervantina
35 000 vols; 250 curr per 25752

Departamento de Educación, Universidades e Investigación, Biblioteca, Donostia, 1, 01010 ***Vitoria-Gasteiz***
T: +34 945018287, 945018456; Fax: +34 945018198; E-mail: huisedoc@ej-gv.es
1985
17 000 vols; 250 curr per; 700 diss/theses; 5 500 microforms; 110 digital data carriers 25753

Eusko-Legebiltzarra, Parlamento Vasco, Liburutegia – Biblioteca, Becerro de Bengoa, s/n, 01005 ***Vitoria-Gasteiz***
T: +34 945004000; Fax: +34 945004204; E-mail: estudios@parlam.euskadi.net; URL: www.parlamento.euskadi.net
1981; Andoni Iturbe Mach
Coll on Simón Bolívar, Public law, Constitutional law, Basques Studies; Archive, documentation
55 000 vols; 500 curr per; 30 e-journals; 250 mss; 1 incunabula; 32 music scores; 75 maps; 50 microforms; 250 digital data carriers
libr loan; IFLA, ICA 25754

Gobierno Vasco. Departamento de Justicia y Administración Pública, Servicio de Archivo, Biblioteca, Documentación y Publicaciones, Donostia-San Sebastián 1, 01010 ***Vitoria-Gasteiz***
T: +34 945 018 561; Fax: +34 945 018 709; E-mail: ejgvbiblioteka@ej-gv.es; URL: www.euskadi.net/ejgvbiblioteka
1981; Begoña de Urigüen
63 963 vols; 3 509 curr per; 35 e-journals; 351 govt docs; 56 music scores; 238 maps; 86 microforms; 440 av-mat; 70 sound-rec; 1 636 digital data carriers
libr loan; ANABAD, ALDEE 25755

Academia General Militar, Biblioteca de Oficiales, Ministerio de Defensa, Huesca, s/n, 50090 *Zaragoza*
T: +34 976294860
1932
23 400 vols; 41 curr per   25756

Cortes de Aragón, Servicio de Biblioteca, Archivo y Fondo Antiguo, Palacio de la Aljafería, 50071 *Zaragoza*
T: +34 976289528; Fax: +34 976289631;
URL: www.cortesaragon.es
1984; Maria Teresa Pelegrin Colomo
Spanish Civil War Coll
30 029 vols; 390 curr per; 62 e-journals; 242 mss; 1 incunabula; 20 diss/theses; 4 music scores; 65 maps; 63 microforms; 23 av-mat; 36 sound-rec; 216 digital data carriers; 4 852 rpts
libr loan; IFLA, ANABAD   25757

Ecclesiastical Libraries

Comunidad de Jesuítas, Biblioteca, Av de Denia, 86, 03016 *Alicante*
T: +34 965261600
1868
20 000 vols   25758

Seminario Diocesano, Biblioteca, Escultor Bañuls, 3, 03009 *Alicante*
T: +34 965234740
1979
Theology, patristic studies, Canon law, liturgics, pastoral theology
10 000 vols; 255 curr per   25759

Santuario de la Virgen del Carmen, Biblioteca, Larrea, s/n, 48340 *Amorebieta-Echano*, Vizcaya
T: +34 946730088
1712
8 000 vols   25760

Seminario Mayor Diocesano, Biblioteca, Pl Obispo D. Marcelo, s/n, 24700 *Astorga*, León
T: +34 987615118; Fax: +34 987615118;
E-mail: recseminarioas@planalfa.es
1799; Francisco Centeno Cristóbal
30 700 vols; 60 curr per   25761

Seminario Diocesano, Biblioteca, Inmaculada, 1, 05005 *Ávila*
Theology, Canon law, coll on local subjects
27 600 vols; 1 curr per; 500 vols from 16th c   25762

Seminario Conciliar, Biblioteca, Av del Ejército Español, 79, 22300 *Barbastro*, Huesca
T: +34 973311400
Catholic theology
18 000 vols; 37 curr per   25763

Arxiu Provincial de l'Escola Pia de Catalunya, Biblioteca, Ronda Sant Pau, 80, 08001 *Barcelona*
T: +34 934410004; Fax: +34 933296970;
E-mail: escolapia@escolapia.net; URL: www.escolapia.cat
1968; Joan Florensa Parés
coins, stamps, postcards coll
16 837 vols; 936 curr per; 14 incunabula; 18 diss/theses; 8 314 music scores; 234 microforms; 166 av-mat; 121 sound-rec; 36 digital data carriers
libr loan   25764

Biblioteca Balmes, Fundación Balmesiana, Durán y Bas, 9, 08002 *Barcelona*
T: +34 933177284; Fax: +34 933170498;
E-mail: biblioteca@balmesiana.org
1923; Ramon Corts Blay
Church hist
42 000 vols; 400 curr per   25765

Biblioteca Pública Episcopal del Seminari de Barcelona, Diputació 231, 08007 *Barcelona*
T: +34 934541607; Fax: +34 934541608;
E-mail: info@bibliotecaepiscopalbcn.org;
URL: www.bibliotecaepiscopalbcn.org
1792; Dr. Josep María Martí Bonet
450 000 vols; 750 curr per; 505 mss; 100 incunabula
libr loan   25766

Convento de Capuchinos, Biblioteca, Cardenal Vives I Tuto, 12-16, 08034 *Barcelona*
T: +34 932043458
1900; Eric Castell Mas
70 000 vols; 200 curr per   25767

Provincia Franciscana de Catalunya, Biblioteca Provincial, Santaló 80, 08021 *Barcelona*
T: +34 932092388; Fax: +34 932092648; E-mail: bibliotecaprovincial@franciscans.cat; URL: www.franciscans.cat
1957; fra Agustí Boadas, Josep M. Massana
XV-XVIth Century Catalan Law, Bible, American Hist, Franciscanism; Music dept
100 000 vols; 74 curr per; 312 mss; 175 incunabula; 2 000 parchments
libr loan   25768

Convento de Carmelitas, Biblioteca, Camino del Carmelo, 10, 48004 *Bilbao*
T: +34 944124811
1887
50 000 vols; 98 curr per   25769

Seminario Mayor Diocesano, Biblioteca, Av de los Quijotes, s/n, 10004 *Cáceres*
T: +34 927245400
1955
Theology, local subjects
20 500 vols; 178 mss; 2 incunabula   25770

Seminario Conciliar de San Bartolomé, Biblioteca, Compañía, s/n, 11005 *Cádiz*
T: +34 956282311; Fax: +34 956282009
1592; Jose Maria Bravo Aragón
Theology, local hist
20 000 vols; 30 curr per; 100 mss   25771

Santuario Virgen de Regla, Biblioteca Provincial, Pl de Regla, s/n, 11550 *Chipiona*, Cádiz
T: +34 956370189; Fax: +34 956374621
1910; Jesus España Delgado
Church hist, theology
50 000 vols; 80 curr per; 35 mss; 7 incunabula
ABIE   25772

Seminario Claretiano, Biblioteca, Ctra de Miraflores, 1, 28770 *Colmenar Viejo*, Madrid
T: +34 918450617; Fax: +34 918462973
1965
Theology
65 000 vols; 170 curr per   25773

Centro Fonseca, Biblioteca Fonseca, Compañía de Jesus, Fonseca, 8, 15004 *A Coruña*
T: +34 981216096; Fax: +34 981216861;
E-mail: biblioteca@jesgalicia.org
1979; José Luis Laredo, S.J.
Education, psychology, theology, literature; Hist, art and geography of Galicia
30 000 vols; 160 curr per; 3 mss; 2 incunabula; 150 music scores; 12 maps; 2 500 av-mat; 100 sound-rec; 300 digital data carriers   25774

Monasterio de San Jerónimo, Biblioteca, Monasterio de Yuste, s/n, 10430 *Cuacos de Yuste*, Cáceres
T: +34 927460530
1958
Theology, spirituality
12 000 vols; 64 curr per   25775

Seminario Conciliar de San Julián, Biblioteca, Pl de la Merced, 2, 16001 *Cuenca*
T: +351 969 211000
1584; Vicente Malabia Martínez
40 818 vols; 37 curr per; 150 mss; 79 incunabula; 11 diss/theses; 15 digital data carriers   25776

Seminario Diocesano de Bilbao, Biblioteca, C/ Larrauri 1, 2A, 48016 *Derio*, Vizcaya
T: +34 944541012
1956; Ander Manterola Aldecoa
Bible, Dogmatic Theology, Church history, Spirituality, Hagiography, Christian Piety, Liturgy, Canon Law, Catechetics, Collections
100 000 vols; 562 curr per; 22 mss; 13 incunabula; 440 microforms; 12 000 rare books (15th-18th c)
libr loan   25777

Biblioteca Seminari de Girona, Institut Superior de Ciencies Religioses de Girona, Pujada Sant Martí, 12, 17004 *Girona*; Ando 117, 17080 Gerona
1600; Jordi Reixach Masachs
Secció Gironina

80 000 vols; 47 curr per; 315 mss; 13 incunabula; 100 microforms
ABIE   25778

Arzobispado de Granada, Biblioteca de la Curia, Pl de Alonso Cano, s/n, 18001 *Granada*
T: +34 958216323; Fax: +34 958229725;
E-mail: arzgranada@planalfa.es
30 000 vols   25779

Facultad de Teología, Compañía de Jesús, Biblioteca (Theology Faculty Library of Granada), Campus Universitario de Cartuja, Apdo 2002, 18080 *Granada*
T: +34 958185252; Fax: +34 958162559;
E-mail: info@teol-granada.com; URL: www.teol-granada.com
1894; Gabriel Verd Conradi
250 000 vols; 5 000 curr per; 186 mss; 18 incunabula; 30 000 rare bks (before 1800)
libr loan; ABIE   25780

Seminario Diocesano, Biblioteca, Av de Santa Marta, 32, 21003 *Huelva*
T: +34 959224058
1954
15 000 vols; 120 curr per   25781

Obispado de Jaca, Biblioteca Diocesana, Obispo, 5, 22700 *Jaca*, Aragón
T: +34 974361841; Fax: +34 974355280;
E-mail: Objaca@planalfa.es
Theology, philosophy, Canon law, hist
20 268 vols; 36 curr per   25782

Seminario Diocesano, Fernando Chica Arellano, Biblioteca, C/ Juan Montilla, 1, 23002 *Jaén*
T: +34 953230023; Fax: +34 953233302
1660; Fernando Chica Arellano
Fondo Antiguo
19 800 vols; 48 curr per; 8 mss; 1 incunabula; 4 diss/theses; 10 music scores; 10 maps; 3 digital data carriers; 100 slides
AAD, AAB, Asociación de Bibliotecarios de la Iglesia en España   25783

Seminario Mayor, Biblioteca, Pl de Regla, 4, 24003 *León*
T: +34 987229412
1850
Theology, Canon law, philosophy, church hist
25 000 vols; 130 curr per   25784

Seminario Mayor, Biblioteca, Av Angel López Pérez, s/n, 27002 *Lugo*
T: +34 982220100
1954
Theology, religion, Canon law, church hist philosophy
30 500 vols; 98 curr per   25785

Biblioteca Cisneros de la Conferencia de Franziscanos, Padres Franziscanos, Joaquín Costa, 36, 28002 *Madrid*
T: +34 915619900
1941; Rafael Mota M.
Filosofía, historia, religión
12 000 vols; 304 curr per   25786

Casa Central de los Paules, Biblioteca, García de Paredes, 45, 28010 *Madrid*
T: +34 914466150
1883
Canon law philosophy
15 000 vols; 30 curr per   25787

Casa de Estudios de los Mercedarios, Biblioteca, Madre de Dios, 39, 28016 *Madrid*
T: +34 913593347
Liturgics, philosophy, hist
11 100 vols   25788

Comunidad de Agustinos de San Manuel y San Benito, Biblioteca, Columela, 12, 28002 *Madrid*
T: +34 915756006
1911
Theology, church hist, canon law
8 000 vols; 7 curr per   25789

Conferencia Episcopal Española, Biblioteca y Centro de Documentación, Añastro, 1, 28033 *Madrid*
T: +34 913439600; E-mail: conferenciaepiscopal@planalfa.es; URL: www.conferenciaepiscopal.es
1966; Crescencio Palomo Iglesias
8 000 vols   25790

Convento de Carmelitas Descalzos, Biblioteca, Triana, 9, 28016 *Madrid*
T: +34 913591661
1953
30 000 vols; 20 curr per   25791

Convento de Santo Domingo del Real, Biblioteca, Claudio Coello, 141, 28006 *Madrid*
T: +34 915617779
1915
Bibl de Teólogos Dominicos
12 300 vols; 33 curr per   25792

Facultad de Teología San Dámaso, Biblioteca, San Buenaventura, 9, 28005 *Madrid*
T: +34 913644900; E-mail: biblioteca@fsandamaso.es; URL: www.fsandamaso.es/facultad/biblioteca.htm
1906; Juan José Pérez-Soba Diez del Corral
Special Colls: Biblia poliglota; Minqe, documentos y dyarios de los concilios ecumenicos, patrologia oriega y latina, sources chretiens, etc; Special depts: Biblia, Catequetica, Liturgia, Ciencias de la Religion
15 000 vols; 660 curr per; 30 mss; 8 incunabula; 500 diss/theses; 35 sound-rec   25793

Instituto de Teología de la Vida Religiosa, Padres Claretianos, Biblioteca, Alvarez Mendizabel, 65, Duplicado, 28008 *Madrid*
T: +34 915401220; Fax: +34 915401226;
E-mail: bibliotecaclaret@planalfa.es
1972; Julio Escohotado Yuguero
70 000 vols; 200 curr per
ABIE   25794

Instituto Español de Misiones Extranjeras, Biblioteca, Ferrer del Rio, 17, 28028 *Madrid*
T: +34 913552342; Fax: +34 913563435; E-mail: info@ieme.org; URL: www.ieme.org
1950; J.M. Madruga
11 640 vols; 195 curr per; 1 incunabula; 15 diss/theses
ABIE   25795

Institutos Dominicos de Filología y Teología, Biblioteca, Apdo 61150, 28080 *Madrid*
T: +34 912024246
1958; José Vicente Rodrigo
45 700 vols; 245 curr per
libr loan   25796

Misioneros Combonianos, Biblioteca Mundo Negro, Arturo Soria, 101, 28043 *Madrid*
T: +34 914152412; Fax: +34 915635448;
E-mail: mundonegro@combonianos.com;
URL: www.mundonegro.com
1983
Missions, Africa, Latin America, developing countries
12 000 vols; 200 curr per   25797

Seminario Santa Catalina, Biblioteca, Praza do Seminario, 1, 27740 *Mondoñedo*
T: +34 982521000; Fax: +34 982507080
1787; Alfonso Morado Paz
Theology, philosophy, law
25 000 vols; 1 mss; 26 incunabula   25798

Abadia de Montserrat, Biblioteca, 08199 *Montserrat*
T: +34 938777766; Fax: +34 938777762; E-mail: biblioteca@bibliotecademontserrat.net; URL: www.bibliotecademontserrat.net
1029; Damià Roure
16th c rare books
345 970 vols; 1 062 curr per; 1 487 mss; 392 incunabula; 612 maps; 203 microforms; 18 000 engravings   25799

Instituto Teológico de Murcia, Padres Franciscanos, Biblioteca, Pl Beato Andrés Hibernon, 4, 30003 *Murcia*
T: +34 968239431; Fax: +34 968242397
1979; Antonio Mora Bernabé
Franciscanism, Teologia
61 010 vols; 427 curr per; 12 mss; 44 incunabula   25800

Comunidad de Canónigos Regulares Lateranenses, Biblioteca, Mortara, 1, 20560 *Oñati*
T: +34 943780311
1886
10 000 vols; 6 curr per; 100 rare bks (17-18th c) 25801

Santuario Nuestra Señora de Aranzazu, Padres Franciscanos, Biblioteca, 20567 *Oñati*, Guipuzcoa
T: +34 943780797; Fax: +34 943783370
1512; Candido Zubizarreta
80 000 vols; 100 curr per; 50 incunabula; 1 500 music scores 25802

Seminario Mayor, Biblioteca, Carretera del Deminario, 20, 32002 *Ourense*, Prov Orense
T: +34 988222750
1818
Religious education
30 000 vols; 187 curr per; 20 mss; 10 av-mat; 10 digital data carriers 25803

Seminario Metropolitano de Oviedo, Diocesis de Oviedo, Biblioteca, Prado Picón, 16, 33008 *Oviedo*
T: +34 985220897; Fax: +34 985223542
1948; Fernando Llenin Iglesias
70 000 vols; 618 curr per; 20 mss 25804

Monestir de La Real, Biblioteca Balear, 07010 *Palma de Mallorca*
T: +34 971250495
1933; Dr. Prof. Josep Amengual i Batle
60 000 vols; 50 curr per; 1 000 mss; 10 incunabula; 50 maps; 2 437 microforms 25805

Centro Teológico, Biblioteca, Campus Universitario de Tafira, 35017 *Las Palmas* de Gran Canaria
T: +34 928452878; Fax: +34 928452947
Maria Pino Tejera López
41 500 vols; 203 curr per; 80 mss; 1 incunabula; 7 300 rare books (16-18th c) 25806

Catedral de Pamplona, Biblioteca Capitular, Dormitalería, 5, planta baja, 31001 *Pamplona*
T: +34 948227950; E-mail: arbicapa@iglesianavarra.org
Theology, Canon law
150 mss; 130 incunabula; 750 rare books from 16th c, 11 365 vols from 17-19th c 25807

Curia Provincial de Hermanos Capuchinos de la Provincia de Navarra-Cantabria-Aragón, Biblioteca Central (Capuchins of the Province Navarra, Cantabria, Aragon), Errotazar, 42, 31014 *Pamplona*
T: +34 948382819; Fax: +34 948382210; E-mail: biblioteca@capuchinosnavarra.org
1606; Vidal Pérez de Villareal
Magnum bullarium romanum
64 000 vols; 169 curr per
libr loan 25808

Seminario Diocesano de Santo Tomás de Villanueva y del Santo Maestro Juan de Ávila, Biblioteca, Ctra de Porzuna, s/n, 13002 *Real*
T: +34 926226100
1878
Theology, philosophy
22 000 vols; 33 curr per 25809

Colegio San Estanislao de Kostka, Biblioteca, Paseo de San Antonio 14, 37003 *Salamanca*
T: +34923125000; Fax: +34923125195
1927; David Pérez Delgado, S.J.
Theology, spirituality, Greek & Roman classical literature
15 000 vols; 93 curr per; 12 incunabula 25810

Convento de Carmelitas Descalzos, Biblioteca, Zamora, 59, 37002 *Salamanca*
T: +34 923261160; Fax: +34 923268126
1901
30 000 vols; 70 curr per; 112 diss/theses 25811

Noviciado y Seminario Mayor de los Reparadores, Biblioteca Escolasticado, Maestro Tárrega, 6-12, 37008 *Salamanca*
T: +34 923213534; Fax: +34 923217828
1952
Theology, philosophy
11 000 vols; 36 curr per; 211 vols from 16-18th c 25812

Monasterio de Samos, Biblioteca, 27620 *Samos*, Lugo
T: +34 982546046
35 000 vols 25813

Monasterio Cisterciense de Osera, Biblioteca, Aldea Osera, s/n, 32740 *San Cristovo de Cea*, Orense
T: +34 988282004
20 000 vols 25814

Monasterio de San Juan de Poio, Padres Mercedarios, Biblioteca, 36995 *San Juan de Poio*, Pontevedra
T: +34 986770000
Ricardo Sanles Martinez
70 000 vols; 50 curr per; 3 000 music scores; 100 maps; 12 microforms; 5 000 av-mat; 15 000 pamphlets 25815

Abadía de Santa Cruz, Biblioteca, Valle de los Caídos, 28200 *San Lorenzo del Escorial*
T: +34 918960200
1958
27 200 vols; 163 curr per; 22 microforms; 367 av-mat; 1 000 sound-rec
libr loan 25816

Biblioteca Comunidad, Real Monasterio. PP. Augustinos, 28200 *San Lorenzo del Escorial* (Madrid)
T: +34 918905011; Fax: +34 918905421; E-mail: joserodriguezosa@hotmail.com
1885; José Rodriguez Díez
Literary collection from the 16th-17th c, Spanish literature; Theology and Philosophy dept
105 100 vols; 490 curr per; 59 mss; 82 incunabula; 200 diss/theses; 650 music scores; 10 maps; Archaeology coll (3 500 vols) 25817

Residencia de la Compañía de Jesús, Biblioteca, Garibay, 19, Donostia, 20004 *San Sebastián*
T: +34 943429720
1937
25 000 vols; 21 curr per; 1 mss 25818

Seminario Diocesano, Diocesis de San Sebastián, Biblioteca, Paseo de Heriz, 82. Barrio del Antiguo, Donostia, 20008 *San Sebastián*, Guipozcoa
T: +34 943213800
1955; Xabier Basurko Ulazia
960 000 vols; 300 curr per
libr loan 25819

Biblioteca Borja, Provincia Tarraconense de la Companyia de Jesús, Llaceres, 30, 08172 *Sant Cugat del Vallès*
T: +34 936741150; Fax: +34 935908111; E-mail: bib@bibliotecaborja.org; URL: www.bibliotecaborja.org
1864; Miquel Carbonell
Theology, religion, philosophy, fine art, Catholic orders: Jesuits; Special colls: Spirituality, Fine arts, Cinema
410 000 vols; 760 curr per; 981 mss; 41 incunabula; 3 000 diss/theses; 2 250 music scores; 110 maps; 282 microforms; 50 000 av-mat; 342 parchments
CBUC, REBIUN, ABIE 25820

Monestir de Solius, Biblioteca, Solius, s/n, 17246 *Santa Cristina d'Aro*
T: +34 972837084
Josep Peñarroya i Artola
8 000 vols 25821

Casa Provincial de los Paules, Biblioteca, Ctra de Madrid, 14, 37008 *Santa Marta de Tormes*, Salamanca
T: +34 923201188; Fax: +34 923201173
1957
Patrología Migne y Brepols, Fuentes del Vaticano II
43 500 vols; 4 600 curr per; 2 772 rare books from 16-18th c 25822

Comunidad de Franciscanos, Biblioteca, Campillo de San Francisco, 3, 15705 *Santiago de Compostela*
T: +34 981581600; Fax: +34 981571916; E-mail: biblioteca_ofmsan@hotmail.com
1240; Santiago Cepeda Iglesias
91 100 vols; 125 curr per; 72 incunabula; 25 000 pamphlets 25823

Abadía de Santo Domingo de Silos, Biblioteca, 09610 *Santo Domingo de Silos*, Burgos
T: +34 947390068; Fax: +34 947390033; URL: www.bibliotecadesilos.es
1880; P. Norberto Núñez
Liturgics, hist
182 000 vols; 370 curr per; 300 mss; 20 incunabula; 3 100 microforms; 300 digital data carriers; 500 rare bks (16-18th c) 25824

Casa del Obispado de Urgell, Biblioteca, Obispo Bell-Lloch, s/n, 25700 *La Seu d'Urgell*, Lérida
T: +34 973350500
1974
Dogmatic theology, ethics, Bible, spirituality, hist, lit
15 700 vols; 100 curr per; 152 mss; 80 incunabula; 600 vols from 16th c, 11 000 vols from 17-19th c 25825

Biblioteca del Arzobispado de Sevilla, Institucion Colombina, c/ Alemanes s/n, 41004 *Sevilla*
T: +34 954560769; Fax: +34 954562721; E-mail: direccionic@institucioncolombina.org
18th c; Nuria Casquete de Prado Sagrera
17 000 vols; 20 curr per; 30 mss 25826

Institucion Colombina, Biblioteca Capitular y Colombina, C/ Alemanes s/n, 41004 *Sevilla*
T: +34 954-560769; Fax: +34 954-562721; E-mail: bibliotecas@icolombina.es; URL: www.icolombina.es
13th c; Nuria Casquete de Prado Libr of Hernando Colón, son of of the explorer Cristóbal Colón (3 500 vols)
70 000 vols; 500 curr per; 1 800 mss; 1 270 incunabula; 70 music scores; 100 maps 25827

Seminario de Tarazona, Biblioteca, Teresa Cajal, s/n, 50500 *Tarazona*, Zaragoza
T: +34 976641912
13 000 vols; 18 curr per; 1 incunabula; 200 vols from 16th c, 7 500 vols from 17-19th c, 700 vols on local subjects 25828

Seminario Pontificio, Biblioteca, San Pablo, 4, 43003 *Tarragona*
T: +34 977232611; Fax: +34 977 248348; E-mail: arquebisbat@arquebisbattarragona.cat; URL: www.arquebisbattarragona.org
Theology, Canon law
60 000 vols; 4 incunabula; 45 000 rare bks (16-19th c) 25829

Provincia Franciscana de Castilla, Biblioteca San Juan de los Reyes, Reyes Católicos, 17, 45002 *Toledo*
T: +34 925223802
1967
Catholic church, Canon law
19 000 vols; 326 curr per; 2 incunabula; 48 rare bks from 16th c 25830

Seminario Mayor San Ildefonso, Biblioteca, Pl de san Andrés, 3, 45002 *Toledo*
T: +34 925225250
1890; Francisco María Fernández Jiménez
Philosophy, catholic theology, church hist; Special colls: Corpus Christianorum, Sources Chretiennes, Patrologie (Migne), Biblioteca de Autores Cristianos (BAC), Biblioteca Clásica Gredos, Diccionario de Spiritualité
80 000 vols; 152 curr per; 30 mss; 1 incunabula; 45 diss/theses; 62 vols from 16th c, 3 887 vols from 17-19th c 25831

Facultad de Teología de San Vicente Ferrer – Sección Dominicos, Biblioteca, Maestro Chapi 50, Apdo 136, 46900 *Torrente*
T: +34 961551750; Fax: +34 961564185; E-mail: bibliotecaprov.ar@dominicos.org
1964; Argimiro Velasco Delgado
219 100 vols; 621 curr per; 1 incunabula; 382 diss/theses; 926 microforms; 94 av-mat; 70 sound-rec; 18 digital data carriers 25832

Seminario Menor San Pelayo, Biblioteca, Calvo Sotelo, 4, 36209 *Tui*, Pontevedra
T: +34 986600051
1850
Patristic studies, Canon law
17 000 vols; 64 curr per; 1 mss; 178 rare bks from 16th c, 720 vols from 17th c 25833

Colegio Seminario del Corpus Christi el Patriarca, Biblioteca, Nave 1, 46003 *Valencia*
T: +34 963214209
1609
Religion, church music
24 000 vols; 40 curr per; 100 mss; 5 microforms 25834

Facultad de Teología de San Vicente Ferrer – Sección Diocesis, Biblioteca, Trinitarios, 3, 46003 *Valencia*
T: +34 963318774
1790; Emilio Aliaga Girbés
70 000 vols; 396 curr per; 23 incunabula; 79 diss/theses; 2 580 rare books (16th c), 2 000 books from 17-19th c 25835

Residencia Sagrado Corazón, Biblioteca Borja, Avda. San José de la Montaña, 15, 46008 *Valencia*
T: +34 963823300; Fax: +34 963820223; E-mail: residsacor@telefonica.net
1560; Juan Costa Catalá
30 000 vols; 39 curr per; 10 mss; 2 incunabula 25836

Estudio Teológico Agustiniano, Padres Agustinos, Biblioteca (Augustinian Library), Paseo de Filipinos, 47007 *Valladolid*
T: +34 983306800; Fax: +34 983397896; E-mail: biblioteca@agustinosvalladolid.org; URL: agustinosvalladolid.org
1967; Constantino Mielgo Fernández
Philippines, hist of Augustinians
165 000 vols; 495 curr per; 126 incunabula; 125 digital data carriers 25837

Monasterio Cisterciense de San Isidro de Dueñas, Biblioteca, 34200 *Venta de Baños*, Palencia
T: +34 979770701; Fax: +34 979770999; E-mail: ocsomcsi@planalfa.es; URL: www.planalfa.es/confer/cistercienses/isidoro
1891; Tomás Gallego Fernández
Patrística y Monástica
25 305 vols; 104 curr per; 43 mss; 22 maps; 153 microforms; 5 069 av-mat; 1 300 sound-rec; 50 digital data carriers 25838

Biblioteca Episcopal de Vic, C. Santa Maria, 1, 08500 *Vic*
T: +34 938894432; Fax: +34 938894807; E-mail: b.vic.episcopal@diba.es; URL: www.abev.net
1806; Dr. Miquel S. Gros
Church hist, local hist
42 500 vols; 23 curr per; 300 mss; 180 incunabula; 75 diss/theses; 2 000 vols from 16th c 25839

Monasterio de Santa María de la Vid, Orden de San Augustin, Biblioteca, Ctra de Soria, s/n, 09491 *La Vid*, Burgos
T: +34 947530510; Fax: +34 947530429; E-mail: biblioteca@monasteriodelvid.org; URL: www.monasteriodelavid.org
1866; José Serafín de la Hoz Veros
Philosophy, theology, canon law, civil law, hist, lit; coins coll
135 000 vols; 64 curr per; 5 mss; 23 incunabula 25840

Seminario Mayor Diocesano de San José, Biblioteca, Av de Madrid, s/n, 36200 *Vigo*, Pontevedra
T: +34 986412666
1959
Dogmatic theology, church hist, philosophy, ethics, patristic studies, spirituality, Canon law, music coll, coll on Galicia
15 100 vols; 326 curr per; 78 vols from 16th c, 7 000 vols from 17-19th c 25841

Comunidad de Jesuítas, Biblioteca, San Ignacio, 2, 06220 *Villafranca de los Barros*, Badajoz
T: +34 924524001; Fax: +34 924525909
1895
15 000 vols; 50 av-mat 25842

Monestir de Poblet, Biblioteca, Corona d'Aragó, 11, 43430 *Vimbodí*
T: +34 977870083
1940; Xavier Guanter i Roig
43 200 vols   25843

Convento de Carmelitas de la Orden de María del Monte Carmelo, Biblioteca, Manuel Iradier, 2-B, 01005 *Vitoria-Gasteiz*
T: +34 945231238; Fax: +34 945141095
1900
8 000 vols; 15 curr per   25844

Facultad de Teología del Norte de España, Sede de Vitoria, Biblioteca, Beato Tomás de Zumarraga, 37, Apartado 86, 01008 *Vitoria-Gasteiz*, Alava
T: +34 945214448; Fax: +34 945247857;
E-mail: eset@teologiavitoria.org; URL: www.teologiavitoria.org
1930; Javier Querejazu
195 700 vols; 400 curr per; 28 incunabula; 524 diss/theses; 1 000 microforms
libr loan   25845

Monasterio de San Salvador de Leyre, Biblioteca, 31410 *Yesa*, Navarra
T: +34 948884011; Fax: +34 948884230
1954; Francisco Javier Suarez Alba
40 000 vols; 100 curr per; 25 mss; 5 incunabula
ABIE, BETH   25846

Obispado de Zamora, Biblioteca Diocesana, Puerta del Obispo, s/n, 49001 *Zamora*
T: +34 980531802; Fax: +34 980531915
1983
20 000 vols; 132 curr per   25847

Cartuja de Aula Dei, Biblioteca, mail address: 50192 *Zaragoza*
T: +34 976450000; Fax: +34 976450001; URL: www.auladei.es
1901; Fr. Rafael Ma San Emeterio
Patrologia de Migne y Diccionarios, Acta Sanctorum de Bolandistas, Acta Apostólica Sedis, Bularios de Papas, Summa Artis
31 000 vols; 20 curr per; 15 mss; 10 diss/theses   25848

Comunidad de Franciscanos, Biblioteca, San Francisco, 2, 20800 *Zarautz*
T: +34 943830100
1905
12 000 vols; 230 curr per   25849

Corporate, Business Libraries

SENER Ingenieria y Sistemas, S.A., Servicio de Documentación, Av de Zugazarte, 56, 48930 *Las Arenas* (Vizcaya)
T: +34 944817500; Fax: +34 944817501; URL: www.sener.es
Eugenio Fresco Moral
14 000 vols; 500 curr per; 40 000 standards, 5 000 pamphlets   25850

ACERALIA, Centro de Documentación, Ap 90, 33480 *Aviles*, Asturias
T: +34 985126168; Fax: +34 985126152; URL: www.aceralia.es
1966; Javier Gancedo Verdasco
13 200 vols; 200 curr per   25851

Banco Exterior de España, Centro de Documentación Europea (Spanish Foreign Trade Bank), Paseo de Graci, 25, 08007 *Barcelona*
T: +34 934822970; Fax: +34 934822974
1989; Laura Lopez Lolina
27 000 vols; 240 curr per   25852

Gas Natural SDG S.A., Biblioteca, Av Porta de l'Angel, 20-22, 08002 *Barcelona*
T: +34 933180000
1976; Ramón Roca
10 000 vols   25853

Salvat Editores, S.A., Biblioteca, C/ Mallorca, 45, 08029 *Barcelona*
T: +34 902117547; Fax: +34 934955779; URL: www.salvat.com
1940
50 000 vols; 90 curr per   25854

Iberduero S.A., Centro de Documentación, Gardoqui, 8, 48008 *Bilbao*
T: +34 944151411
1972; Jesús Urbina Santamaria
11 600 vols; 517 curr per; 96 maps; 156 microforms; 21 567 standards
libr loan   25855

Biblioteca de Temas Gaditanos, Unicaja, C/ San Francisco, 26, 11004 *Cádiz*
T: +34 956286711
1974; Esperanza Salas Gallego
15 000 vols   25856

Empresa Nacional de Siderurgia (ENSIDESA), Centro de Documentación, Factoría de Veriña, Apdo 503, 33080 *Gijón*
T: +34 985320700
1966
Metallurgy, electricity, economics
13 100 vols; 215 curr per   25857

Banco de España, Biblioteca, C/ Alcalá, 48, 28014 *Madrid*
T: +34 913385058, 913385327; Fax: +34 913386041; E-mail: bibliobe@bde.es; URL: www.bde.es/servicio/biblio/biblio.htm
1964; Valentin Perez P.
300 000 vols; 4 700 curr per; 57 000 microforms
libr loan; IFLA, SEDIC, ANABAD   25858

Banco Urquijo, Biblioteca, Gran Vía 4, 28013 *Madrid*
T: +34 913372000
1914
8 000 vols; 370 curr per   25859

BBVA – Banco Bilbao Vizcaya Argentaria, Centro de Documentación, Paseo de la Castellana81, Ppiso 7, 28046 *Madrid*
T: +34 915378043; Fax: +34 915377807
1977; Teresa Vela
28 500 vols; 278 curr per; 36 microforms; 5 digital data carriers   25860

Bolsa Oficial de Comercio de Madrid, Biblioteca, Pl de La Lealtad, 1, 28014 *Madrid*
T: +34 915891321; E-mail: biblioteca@bolsamadrid.es
1980; Blas Calzada Terrados
Economics, law, politics
11 040 vols; 97 curr per; 11 diss/theses; 5 digital data carriers   25861

Empresa Nacional de Residuos Radioactivos S.A. (ENRESA), Biblioteca, Emilio Vargas, 7, 28043 *Madrid*
T: +34 915668100; Fax: +34 915668169; E-mail: registro@enresa.es
Safety Series, Tech Reports of IAEA, SKB, NRPB
16 000 vols; 160 curr per; 41 standards colls   25862

Empresarios Agrupados AIE, Biblioteca, Magallanes, 3, 28015 *Madrid*
T: +34 913098000; Fax: +34 915912655; E-mail: biblioteca@empre.es; URL: www.empre.es
1974; Lupe Sanz-Bueno
Nuclear engineering, technology
7 600 vols; 80 curr per; 40 000 microforms; 18 500 standards
Association of Esperantist Librarians, Comité de Documentación del Foro Nuclear   25863

IBERIA Líneas Aereas de España S.A., Centro de Documentación General, Velazquez, 130. Bloque 6. Semisotano, 28006 *Madrid*
T: +34 915877279; Fax: +34 915877471; E-mail: cdocumentacion@iberia.es
1980; Paloma Balbas Garate
12 081 vols; 596 curr per; 800 av-mat; 4 800 slides, 23 400 fotogr   25864

IBM – International Business Machines S.A.E., IBM Biblioteca, Santa Hortensia, 26-28, 28002 *Madrid*
T: +34 913977601; Fax: +34 915193987
1975; Charo Garribo
Mathematics
22 000 vols; 79 curr per; 10 maps; 5 000 microforms; 50 digital data carriers; 300 disketes
libr loan   25865

Televisión Española S.A. – Servicio de Documentación Audiovisual, Torrespaña. O'Donnell, 77, 28007 *Madrid*
T: +34 914091844
1970; Manuel Corral Baciero
Law, economics, political sciences
10 000 maps; 700 000 av-mat; 100 000 slides   25866

Televisión Española S.A. – Servicio de Documentación Escrita, Torrespaña. O'Donnell, 77, 28007 *Madrid*
T: +34 914091844
1978; Rafael Diaz Arias
Political science, sociology, law, economics, anthropology
10 400 vols; 1 350 microforms; 999 directories   25867

Special Libraries Maintained by Other Institutions

Instituto de Estudios Albacetenses, Biblioteca, Calle Monjas, s/n, 02005 *Albacete*; Apdo 404, 02080 Albacete
T: +34 967523046; Fax: +34 967523048; E-mail: biblioteca@iealbacetenses.com; URL: www.iealbacetenses.com
1985; Ramón Carrilero Martinez
Fondo Antiguo (works and docs since 16th c); Archivo Fotográfico (+7 000 photos); Bibl 'José Prat'
22 500 vols; 5 curr per; 15 mss; 5 diss/theses; 250 maps; 300 microforms; 100 av-mat; 75 sound-rec   25868

Museo de Albacete, Biblioteca Joaquín Sánchez Jiménez, Arcangel San Gabriel, s/n, 02002 *Albacete*
T: +34 967228307; Fax: +34 967229515
1978; Laura Martinez Marqueta
9 100 vols   25869

Caja de Ahorros del Mediterráneo – Biblioteca Gabriel Miró, Av de Ramón y Cajal, 5, 03003 *Alicante*
T: +34 965905661
1952
Humanities, oceanology, local hist, Mironiana, American studies
77 000 vols; 400 curr per; 350 microforms
libr loan   25870

Institución Cultural Santa Ana, Biblioteca, Ortega y Muñoz, 28, 06200 *Almendralejo*, Badajoz (Extremadura)
1976; José Angel Calero Carretero
50 000 vols; 25 curr per; 1 000 diss/theses; 100 maps; 700 microforms; 1 000 000 av-mat; 20 000 directories
                               25871

Estacíon Experimental de Zonas Aridas, Consejo Superior de Investigacions Científicas, Biblioteca, General Segura, 1, 04001 *Almería*
T: +34 950276400; Fax: +34 950277100; E-mail: isabel@eeza.csic.es; URL: www.eeza.csic.es/eeza/biblioteca.aspx
1947; Isabel Jiménez Borrajo
8 000 vols; 100 curr per; 500 maps
libr loan   25872

Centro de Estudios Extremeños, Biblioteca, C/ Godofredo Ortega Muñoz, 2, 06011 *Badajoz*
T: +34 924245722, 924245091; Fax: +34 924235908; E-mail: nnunez.ceex@dip-badajoz.es; URL: www.dip-badajoz.es
1927; Lucía Castellano Barrios
Revista de Estudios Extremeños (1927-), Colección centro de Estudios Extemeños (nr 1, 2 y 3)
18 000 vols; 1 147 curr per; 8 mss; 3 diss/theses; 5 maps; 384 microforms; 5 av-mat; 20 digital data carriers; Ancient coll (285 vols)
Confederación Española de Centros de Estudios Locales   25873

Real Sociedad Económica Extremeña de Amigos del País de Badajoz, Biblioteca, C/ Hernán Cortes, 1, 06002 *Badajoz*
T: +34 924207205; URL: www.biblioteca-rseap.org/
1830
11 000 vols   25874

Museu de Badalona, Biblioteca, Plaça de l'Assemblea de Catalunya, 1, 08911 *Badalona*
T: +34 933841750; Fax: +34 933841662; E-mail: info@museudebadalona.cat; URL: www.museudebadalona.cat/
Local hist, archeology, textile industry
9 000 vols; 415 curr per   25875

Hospital de Cruces, Biblioteca, Pl de Cruces, s/n, 48903 *Baracaldo*; Apdo 69, 48080 Bilbao
T: +34 946006125; Fax: +34 946006049; E-mail: biblioteca.cruces@osakidetza.net; URL: www.hospitalcruces.com
1970; Maria Asunción García Martin
Pathology, surgery
13 267 vols; 554 curr per; 1 348 e-journals; 85 diss/theses; 3 725 microforms; 1 522 av-mat; 32 digital data carriers
libr loan; IFLA, MLA, EAHIL, ANABAD, SEDIC, ALDEE   25876

Acadèmia de Ciències Mèdiques de Catalunya i Balears, Biblioteca, Carrer Major de Gran Caralleu 1-7, 08017 *Barcelona*
T: +34 932031050; Fax: +34 934188729; E-mail: academia@academia.cat; URL: www.academia.cat
1878; Ramona Casas
80 700 vols; 662 curr per; 2 470 slides
libr loan   25877

Archivo de la Corona de Aragón, Biblioteca, Almogàvars, 77, 08018 *Barcelona*
T: +34 934854285; Fax: +34 933001252; E-mail: felixdelafuente@mcu.es
1817; Felix de la Fuente
23 794 vols; 739 curr per; 587 mss; 51 incunabula; 23 maps; 105 microforms; 12 av-mat; 265 digital data carriers
ICA   25878

Arxiu Històric de la Ciutat de Barcelona, Biblioteca (The City Historical Archive Barcelona), Santa Llucía, 1, 08002 *Barcelona*
T: +34 933181195; Fax: +34 913178327; URL: www.bcn.es/arxiu/arxiuhistoric/
1922; Xavier Tarraubella Mirabet
135 800 vols; 15 466 curr per; 796 mss; 4 incunabula; 65 diss/theses; 40 music scores; 1 050 maps; 338 microforms; 500 sound-rec   25879

Arxiu Històric de Protocols de Barcelona, Biblioteca, Notariat 4, 08001 *Barcelona*
T: +34 933174800; URL: www.colnotcat.es/frames.asp?dir=3&op=0&idioma=esp
1862
40 000 vols; 186 curr per; 1 incunabula; 15 000 protocols (13th c)
libr loan   25880

Asociación de Prensa Profesional, Biblioteca, Rambla de Cataluña, 10, 08007 *Barcelona*
T: +34 933042582; Fax: +34 934121436; E-mail: asociacion@app.es; URL: www.app.es
1941
Mass media
8 200 vols; 24 curr per; 203 diss/theses
libr loan   25881

Associació d'Enginyers Industrials de Catalunya, Biblioteca, Vía Laietana, 39, 08003 *Barcelona*
T: +34 933192300; Fax: +34 933100681; E-mail: biblio@eic.cat; URL: www.eic.cat
1863; Marta Torelló Caballero
22 000 vols; 625 curr per; 85 digital data carriers   25882

Ateneo Barcelonès, Biblioteca (Barcelona Athenaeum Library), Canuda, 6, 08002 *Barcelona*
T: +34 933436120; Fax: +34 93 3171525; E-mail: biblioteca@ateneubcn.org; URL: biblioteca.ateneubcn.org
1860; Lluís Vicente
19th C library (History, Culture, Soziology, Literature, Philosophy), European culture reception in Catalonia, Creative Writing; Manuscripts, Rare and special collections, Serials, Archive
280 000 vols; 1 910 curr per; 1 400 mss; 24 incunabula; 50 diss/theses; 5 000 govt docs; 500 music scores; 30 maps; 400 av-mat; 100 sound-rec; 800 digital data

carriers; 10 000 offprints
libr loan; CCUC 25883

Biblioteca Circulant Braille, Organización Nacional de Ciegos – O.N.C.E., Calàbria, 60-66, 08015 ***Barcelona***
T: +34 933259200; Fax: +34 934249144; E-mail: soi@once.es; URL: www.once.es
1939; Jordi Pons Sala
11 200 vols
libr loan 25884

Biblioteca Infantil Santa Creu, Carrer de l'Hospital, 56, 08001 ***Barcelona***
T: +34 933025348; Fax: +34 933179492
1940; José Daza
Historical coll (4 000 vols)
30 000 vols; 80 curr per
libr loan; IFLA 25885

Biblioteca Pública Arús (Arus Public Library), Passeig de Sant Joan, 26, 08010 ***Barcelona***
T: +34 932325404; Fax: +34 932318341; E-mail: arus@bpa.es; URL: www.bpa.es
1895; Isabel Giner
Special colls: masonry, social movements hist, contemporary hist (18th-19th c), anarchism
68 000 vols; 100 curr per; 400 mss; 9 incunabula; 25 diss/theses; 700 music scores; 150 maps; 508 microforms
IALHI 25886

Cambra Oficial de Comerç, Indústria i Navegació de Barcelona, Biblioteca, Av Diagonal, 452, 08006 ***Barcelona***
T: +34 902448448; Fax: +34 934169301; E-mail: cambra@cambrabcn.org; URL: www.cambrabcn.es
1915; Xavier Cortez
98 400 vols
libr loan 25887

Centre d'Estudis Jurídics i Formació Especialitzada, Centro de Estudios Jurídicos y Formación Especializada, Departament Justícia, Generalitat de Catalunya, Biblioteca, C/ Ausiàs March, 40, 08010 ***Barcelona***
T: +34 932073114; Fax: +34 932076747; E-mail: biblioteca_cejfe.dj@gencat.net; URL: www20.gencat.cat
1985; Montserrat Garcia
Criminology, criminal law, psychology, legal and forensic psychology, forensic psychiatry, juvenile justice, child welfare
8 599 vols; 198 curr per; 355 av-mat; 7 digital data carriers
libr loan 25888

Centre Excursionista de Catalunya, Biblioteca, Paradís, 10, 08002 ***Barcelona***
T: +34 933152311; Fax: +34 933151408; URL: www.cec.cat
1876; Francesc Olivé
Folklore; local geogr a hist, Mountaineering
32 000 vols; 582 curr per; 6 200 maps 25889

Centro de Documentación Médica de Cataluña, Academia de Ciencias Médicas, Paseo de la Bonanova, 47, 08017 ***Barcelona***
T: +34 932127894
1984; Roser Cruells Serra
79 280 vols 25890

Centro d'Investigació i Desenvolupament, Biblioteca, Jordi Girone, 18-26, 08034 ***Barcelona***
T: +34 934006100; Fax: +34 932045900; E-mail: biblioteca@cid.csic.es
1967; Rosa Maria Ortadó
Chemistry, textiles, biology, physics, plastics, pharmacy, environment
22 747 vols; 524 curr per; 330 diss/theses
libr loan 25891

CIDOB – Centro de Investigación de Relaciones Internacionales y Desarrollo, Biblioteca, Elisabets, 12, 08001 ***Barcelona***
T: +34 933026495; Fax: +34 933022118; E-mail: biblioteca@cidob.org; URL: www.cidob.org/es/documentacion/biblioteca
1970; Anna Borrull Munt
Politics, international economis; archives of press clippings (1974 ff)
13 500 vols; 500 curr per; 10 000 govt docs; 356 maps; 6 microforms; 52 av-mat; 60 digital data carriers
IASSIST, EADI, REDIAL 25892

Cinema 'Delmiro de Caralt' de la Filmoteca, Department de Cultura de la Generalitat de Catalunya, Biblioteques, Portal de Santa Madrona, 6-8, 08008 ***Barcelona***
T: +34 933162780; Fax: +34 933162781
1924; Mercè Rueda i Tebé
1 250 cinema and movie apparatus
30 000 vols; 197 curr per; 50 mss; 30 diss/theses; 50 govt docs; 427 microforms; 5 500 av-mat; 125 sound-rec; 25 digital data carriers; 45 000 photos, press-bks, 20 000 press clipping dossiers, 7 500 posters
libr loan; SIBMAS, FIAF 25893

Il lustre Col legi d'Advocats de Barcelona, Biblioteca (Barcelona Bar association library), Mallorca, 283, 08037 ***Barcelona***
T: +34 934961880 ext. 3289; Fax: +34 934871128; E-mail: direcciobiblioteca@icab.es; URL: www.icab.cat
1833; Isabel Juncosa Ginestà, Patrícia Sanpera Izoard
Historical catalan law, roman law and consuetudinary law, collection formed by books between 11th to 19th C (ca. 100 000 vols)
3 500 000 vols; 1 447 curr per; 227 e-journals; 692 e-books; 200 mss; 28 incunabula; 1 diss/theses; 90 maps; 1 500 microforms; 1 041 digital data carriers; 38 000 digital images, 1 211 parchment and 10 000 allegatio iuris
libr loan; Consorci de Biblioteques Universitàries de Catalunya (CBUC), IALL 25894

Col.legi d'Aparelladors, Arquitectes Tècnics i Enginyers d'Edificació de Barcelona, Centre de Documentació, Bon Pastor, 5, 08021 ***Barcelona***
T: +34 932402380; Fax: +34 932402381; E-mail: biblioteca@apabcn.cat; URL: www.apabcn.cat
1959; Francesca Moya
Building and architecture, Spanish law; Standard Information Service
23 000 vols; 597 curr per; 7 e-journals; 66 diss/theses; 3 194 govt docs; 46 maps; 794 av-mat; 2 089 digital data carriers; 5 000 standards
COBDC 25895

Collegi d'Arquitectes de Catalunya, Demarcació de Barcelona, Biblioteca, Arcs, 1-3, 3ª planta, 08002 ***Barcelona***
T: +34 933067805; Fax: +34 934123964; E-mail: bibl.bcn@coac.cat; URL: www.coac.net/Barcelona
1890
57 500 vols; 200 curr per; 40 mss; 68 diss/theses; 560 av-mat; 10 digital data carriers
libr loan 25896

EAE Business School, Centro de Documentación y Recursos, C/ Aragó, 55, 08015 ***Barcelona***
T: +34 932278090; Fax: +34 933194436; E-mail: cdr@eae.es; URL: www.eae.es
1958; Andreu Bellot
Economy, finance, marketing, human resources, management
15 400 vols; 278 curr per; 100 e-journals; 1 246 diss/theses; 266 av-mat; 400 digital data carriers 25897

Fomento del Trabajo Nacional, Biblioteca, Via Layetana, 32-34, 08003 ***Barcelona***
T: +34 933105592
1771; Núria Sardà i Vidal
Foreign trade statistics of Spain, Diario de Barcelona, No 1 ff
72 400 vols; 1 070 curr per 25898

Fundació Antoni Tápies, Biblioteca, Aragó, 255, 08007 ***Barcelona***
T: +34 934870315; Fax: +34 934870009; E-mail: gdomenech@ftapies.com; URL: www.fundaciotapies.org
1990; Glòria Domènech
Oriental, African, Oceanic Art; Antoni Tàpies Arch
43 500 vols; 480 curr per; 500 av-mat; 150 digital data carriers
ARLIS/NA, Col.legi Oficial de Bibliotecaris-Documentalistes de Catalunya, Associaciò d'Arxivers de Catalunya 25899

Fundació Institut Amatller d'Art Hispànic, Biblioteca, Paseo de Gracia, 41, 08007 ***Barcelona***
T: +34 932160175; Fax: +34 934875827; E-mail: amatller@amatller.org; URL: www.amatller.org
1941; Santiago Alcolea Blanch
27 000 vols; 70 curr per; 300 000 photos 25900

Fundació Joan Miró, Centre d'Estudis d'Art Contemporani, Biblioteca, Parc de Montjuïc, 08038 ***Barcelona***
T: +34 934439492; Fax: +34 933298609; E-mail: biblioteca@bcn.fjmiro.es; URL: www.bcn.fjmiro.es
1975; Teresa Martí
Miro's bibliography, catalogs of the exhibitions of the Foundation, Josep Lluis Sert Archive
24 660 vols; 130 curr per; 3 e-journals; 502 mss; 10 diss/theses; 1 321 av-mat; 152 sound-rec; 256 digital data carriers; 250 posters, 236 press clipping dossiers
libr loan 25901

Fundació Josep Laporte, Biblioteca, Av San Antoni Ma. Claret, 171, 08041 ***Barcelona***
T: +34 934335040; Fax: +34 934335046; E-mail: biblioteca.fbjl@uab.cat; URL: www.fbjoseplaporte.org
2000; Sara Aguilera Castro
10 500 vols; 349 curr per; 840 e-journals
libr loan; EAHIL 25902

Fundació Pere Tarrés, Biblioteca, Santaló, 37, 08021 ***Barcelona***
T: +34 934152551; Fax: +34 932186590; E-mail: biblioteca@peretarres.org; URL: www2.peretarres.org/biblioteca
1992
Social education, environmental protection
50 000 vols; 600 curr per 25903

Goethe-Institut, Institut Alemany, Bibliothek (German Cultural Institute of Barcelona), Manso 24-28, 08015 ***Barcelona***
T: +34 932926006; Fax: +34 932926005; E-mail: info@barcelona.goethe.org; URL: www.goethe.de/barcelona
1955; Ramo Bellmann
10 347 vols; 50 curr per; 1 162 av-mat; 940 sound-rec; 167 digital data carriers
libr loan 25904

Institución Milà i Fontanals, CSIC, Departamento de Musicología, Biblioteca (Institution Mila and Fontanals), C/ Egipciaques, 15, 08001 ***Barcelona***
T: +34 934426576; Fax: +34 934430071; E-mail: ezpeleta@imf.csic.es; URL: www.imf.csic.es
1944; Marta Ezpeleta García
'Anselmo González dez Valle' Library (9 000 printed scores, 19th c); 'Fondo Reserva':= E:Bim (medieval musical parchments, Renaissance musical imprints, theoretical treatises on music ss. XVIII-XIX, etc)
9 000 vols; 50 curr per; 12 e-journals; 40 mss; 1 incunabula; 10 diss/theses; 9 000 music scores; 320 microforms; 500 av-mat; 1 000 sound-rec; 10 digital data carriers
libr loan; IAML 25905

L'Institut Agrícola, Biblioteca, Plaça Sant Josep Oriol, 4, 08002 ***Barcelona***
T: +34 933011636; E-mail: info@institutagricola.org; URL: www.institutagricola.org
1851; Mar Perez-Esteban
Revistas antiguas
15 200 vols; 200 curr per
libr loan 25906

Institut Cartogràfic de Catalunya, Cartoteca de Catalunya, Parc de Montjuïc, 08038 ***Barcelona***
T: +34 935671500; Fax: +34 935671567; E-mail: jfarres@icc.es; URL: www.icc.es
1986; Montserrat, Galera Monegal
32 497 vols; 365 curr per; 254 899 maps; 17 600 microforms; 212 digital data carriers; 90 144 photos 25907

Institut de Ciències de la Terra 'Jaume Almera', Consejo Superior de Investigaciones Científicas, Biblioteca, C/ Lluis Sole Sabaris, s/n, 08028 ***Barcelona***
T: +34 934021420; Fax: +34 934021421; E-mail: bib_almera@ija.csic.es; URL: www.ija.csic.es/gt/biblio/almerahome.html
1940; Carmen Losada-Fernández
55 000 vols; 450 curr per; 400 diss/theses; 12 000 maps; 500 microforms; 70 av-mat; 30 digital data carriers; 20 000 aerial photos
libr loan 25908

Institut del Teatre, Centre de Documentació i Museu de les Arts Escèniques, Plaça Margarida Xirgu, s/n, 08004 ***Barcelona***
T: +34 932273900; Fax: +34 932273939; E-mail: i.teatre@institutdelteatre.cat; URL: www.institutdelteatre.cat
1913; Montserrat Alvarez Mass
theatre architecture, theatre companies and enterprises, iconographies of both actors and performances, stage designs' sketches, designs for costumes, costumes, programmes, autographs, letters, Col. leccions Artur Sedó, Salvador Bonavia, Lluís Millà, Fernandez Guerra, Cotarelo, Montaner, Gelambert, Alfons Puig, Adrià Gual, bills and scenery maquettes; Newspaper libr
125 000 vols; 60 curr per; 5 400 mss; 200 sound-rec
SIBMAS 25909

Institut d'Estadística de Catalunya, Biblioteca, Via Laietana, 58, 08003 ***Barcelona***
T: +34 934120088; Fax: +34 934123145, 933180075; E-mail: biblio@idescat.es; URL: www.idescat.es
1960; Llorenç Birba i Fonts
144 706 vols; 3 957 curr per; 3 901 digital data carriers 25910

Institut d'Estudis Catalans – Servei de Documentació i Arxiu, Biblioteca, Carme, 47, 08001 ***Barcelona***
T: +34 932701620; Fax: +34 932701180; E-mail: lmiret@iec.cat; URL: www.iec.cat
1907; Eulàlia Miret Raspall
Catalan Studies, Catalan Science, Arxiu de l'Institut d'Estudis Catalans, Arxiu Mercè Rodoreda; Catalan Language and Lit Dept, Catalan Hist Dept
28 236 vols; 52 curr per; 3 e-journals; 800 linear metres of mss, 1 000 clippings 25911

Institut d'Estudis Nord-Americans, Biblioteca, Via Augusta, 123, 08006 ***Barcelona***
T: +34 932402858; Fax: +34 932020690; E-mail: ien@ien.es; URL: www.ien.es
1954; Judit Moya
10 000 vols; 121 curr per; 50 maps; 3 microforms; 880 av-mat; 100 sound-rec 25912

Institut Français, Mediateca, Moià, 8, 08006 ***Barcelona***
T: +34 935677790; Fax: +34 932006661; URL: www.ifbcn.cat
1919
Fine Arts, philosophy
13 000 vols; 40 curr per; 1 500 music scores; 1 500 av-mat; 1 500 sound-rec; 20 digital data carriers 25913

Institut Municipal d'Educació, Biblioteca Artur Martorell, Pl d'Espanya n° 1, 08014 ***Barcelona***
T: +34 933256262
1922; Esther Lopez P.
27 000 vols; 255 curr per
libr loan; IFLA 25914

Institut Nacional d'Educació Física de Catalunya, Biblioteca, Av de l'Estadi, s/n, Anella Olímpica de Montjuïc, 08038 ***Barcelona***
T: +34 934255445 ext 257, 258; Fax: +34 934263617; URL: www.inefc.cat
1975; Goretti Pascual Curià
Sports, sports medicine, sociology and psychology of sports, leisure, physical education
19 000 vols; 500 curr per; 250 diss/theses; 1 150 av-mat; 25 digital data carriers
libr loan; IASI 25915

Institut Politecnic de Formació Professional, Escola del Treball, Biblioteca, Urgell, 187, 08036 ***Barcelona***
T: +34 934309200
1919; Cristina Lidón Albiach
19 800 vols; 74 curr per; 150 av-mat
libr loan 25916

Instituto Català de la Salut, Centre Documental, Gran Via de les Corts Catalanes, 587-589, 08007 ***Barcelona***
T: +34 934824100; E-mail: atencioalciutada.ics@gencat.cat; URL: www.gencat.cat/ics/
1953; Maria Paula Fernandez-Gironés
Public health, social security, primary health care, health management
15 000 vols; 280 curr per
libr loan 25917

Instituto Nacional de Seguridad e Higiene en el Trabajo, Biblioteca del Centro de Información y Documentación, C/ Dulcet 2-10, 08034 ***Barcelona***
T: +34 932800102; Fax: +34 932803642; URL: www.insht.es
1973; Eulalia Carreras
OIT-CIS abstracts cards (1960-1973)
20 000 vols; 200 curr per; 30 diss/theses; 1 000 govt docs; 100 digital data carriers; 10 000 off-prints 25918

Istituto Italiano di Cultura, Biblioteca, Passatge Méndez Vigo, 5, 08009 ***Barcelona***
T: +34 934875306; Fax: +34 934874590; E-mail: iicbarcellona@esteri.it; URL: www.iicbarcellona.esteri.it/IIC_Barcellona
1947; Silvano Ferrari
11 600 vols; 87 curr per 25919

Museu d'Arqueologia de Catalunya, Biblioteca (Archaeological Museum of Catalonia), Passeig de Santa Madrona, 39-41, 08038 ***Barcelona***
T: +34 934232149; Fax: +34 9330033; E-mail: bibliotecamac.cultura@gencat.net; URL: www.mac.es
1935; Imma Albó Vidal de Llobatera
Classic Antiquity, Prehistory, Orient, Archaeology; Ethnography coll, Anthropology coll; Prehistory dept
40 635 vols; 702 curr per; 391 mss; 100 diss/theses; 200 govt docs; 500 maps; 60 sound-rec; 50 digital data carriers; 58 912 photos and cliché, 3 000 pamphlets
IFLA 25920

Museu de Ciències Naturals, Centre de Documentació, Passeig Picasso, s/n, 08003 ***Barcelona***
T: +34 932562-183/-214; Fax: +34 933199312; E-mail: bibmuseuciencies@bcn.cat
1916; Carles Curto Milà
Zoology, geology
11 600 vols; 1 650 curr per; 3 000 maps; 800 sound-rec; 5 400 rpts
libr loan 25921

Museu d'Història de la Ciutat, Biblioteca, C. Baixada de la Llibreteria, 7, 3r pis., 08002 ***Barcelona***
T: +34 932562106; Fax: +34 933150957; E-mail: jirigoyen@bcn.cat; URL: www.museuhistoria.bcn.es
1960; Maria B. Barrachina D.
8 200 vols; 220 curr per
libr loan 25922

Museu Etnològic, Biblioteca, Passeig Sta. Madrona, s/n, 08038 ***Barcelona***
T: +34 934246807; Fax: +34 934237364; URL: www.museuetnologic.bcn.es
1953; Gloria Serra Fernandez
Anthropology, ethnology, social sciences; Biblioteca Americanista del Doctor Lluís Pericot, Biblioteca Americanista del Dr. José Alcina (10 000 vols), Fons sobre Guinea Equatorial, Fons del Nord d'Àfrica, Japó i Equador
40 000 vols; 1 046 curr per; 24 mss; 120 maps
libr loan 25923

Museu Geològic del Seminari de Barcelona, Biblioteca, Diputació, 231, 08007 ***Barcelona***
T: +34 934541600; E-mail: almeracomas@hotmail.com; URL: www.bcn.es/medciencies/mgsb
1874; S. Calzada
13 070 vols; 62 curr per; 100 mss; 150 diss/theses; 1 875 maps; 1 250 av-mat
libr loan 25924

Museu Marítim, Consorci de les Drassanes de Barcelona, Biblioteca (Maritime Museum Barcelona), Av. Drassanes, s/n, 08001 ***Barcelona***
T: +34 933429920; Fax: +34 933187876; E-mail: bibliotecammb@gmail.com; URL: faust.vtlseurope.com:8088/
1942; Rosa Busquets Ferrer
Marine technology, naval engineering, science hist, technological hist
25 536 vols; 145 curr per; 2 500 maps; Navigational charts, pamphlets (aprox. 1 500)
libr loan 25925

Museu Nacional d'Art de Catalunya, Biblioteca General d'História de l'Art, Palau Nacional, Parc de Montjuïc, 08038 ***Barcelona***
T: +34 936220360; Fax: +34 936220383; E-mail: biblioteca@mnac.es; URL: www.mnac.es
1906; Rosa Reixats
Exhibition and hand catalogs
105 000 vols; 2 303 curr per; 2 incunabula; 270 microforms; 15 digital data carriers; newspaper clippings, ephemerial hand exhibition catalogs
COBDC, ARLIS/UK 25926

Orfeó Català, Biblioteca, Sant Francesc de Paula, 2, 08003 ***Barcelona***
T: +34 932957200; Fax: +34 932957210
1908; Josefina Sastre Canal
22 500 vols; 43 curr per; scores, photos, programs 25927

Organización Nacional de Ciegos, Delegación Territorial, Biblioteca, Calabria 66, 08015 ***Barcelona***
T: +34 933259200; Fax: +34 934249144
1939
Braille bks
11 300 vols 25928

Real Academia de Ciencias y Artes de Barcelona, Biblioteca (Royal Academy of Sciences and Arts of Barcelona), Rambla de Estudios, 115, 08002 ***Barcelona***
T: +34 933170536; Fax: +34 933043128; E-mail: info@racab.es; URL: www.racab.es/
1764; Jaume Casabo i Gispert
Geology; Historical Archives of the Academy
150 000 vols; 300 curr per; 2 500 maps 25929

Reial Academia de Bones Lletres, Biblioteca, C/ Bisbe Caçador, 3 (Plaça Sant Just), 08002 ***Barcelona***
T: +34 933150010; Fax: +34 933102349; E-mail: bones-lletres@boneslletres.cat; URL: www.boneslletres.cat
Hist, lit
30 000 vols; 1 500 curr per
libr loan 25930

Euskaltzaindia – Real Academia de la Lengua Vasca, Azkue Biblioteka (Royal Academy oh the Basque Language – Azkue Library), Pl Barria, 15, 48005 ***Bilbao***
T: +34 944152745; Fax: +34 944158144; E-mail: azkuebib@euskaltzaindia.net; URL: www.euskaltzaindia.net
1918; Jose Antonio Arana Martixa
Lacombe coll
72 000 vols; 2 035 curr per; 5 000 mss; 4 incunabula; 331 diss/theses; 303 maps; 10 215 microforms; 296 av-mat; 3 339 sound-rec; 3 238 digital data carriers
libr loan 25931

Ilustre Colegio de Abogados del Senorío de Vizcaya, Biblioteca, Rampas de Uribitarte, 3, 48001 ***Bilbao***
T: +34 944356200; Fax: +34 944356201; E-mail: administracion@icasv-bilbao.com; URL: www.icasv-bilbao.com/biblioteca.aspx
1986
8 000 vols; 73 curr per; 1 760 microforms 25932

Museo de Bellas Artes de Bilbao, Biblioteca, Pl del Museo, 2, 48011 ***Bilbao***
T: +34 944396142; Fax: +34 944396145; E-mail: biblioteca@museobilbao.com; URL: www.museobilbao.com
1985; Begoña Gonzalez Perez
20 650 vols; 127 curr per; 1 100 microforms; 22 820 av-mat
libr loan; ANABAD 25933

Sociedad Bilbaina, Biblioteca, Navarra, 1, 48001 ***Bilbao***
T: +34 944231407; Fax: +34 944230100; E-mail: info@sociedadbilbaina.com; URL: www.sociedadbilbaina.com
1839; José Antonio Larrinaga Bernardez
35 000 vols; 479 curr per; 6 incunabula; 7 maps
ANABAD 25934

Centro de Estudios Borjanos, Biblioteca-Archivo, Institución Fernando el Católico, Cervantes, 13, 50540 ***Borja***, Zaragoza
T: +34 976867402; Fax: +34 976852482; E-mail: cesbor@mundivia.es; URL: www.cesbor.com
1976
13 450 vols; 542 curr per; 40 diss/theses; 663 sound-rec; 294 digital data carriers; 12 850 photos, 3 645 diapositives, 4 659 brochures and posters 25935

Archivo Municipal de Burgos, Biblioteca Auxiliar, Ayuntamiento de Burgos, C/ Fernán Gonzáles 56, 09071 ***Burgos***
T: +34 947288812; Fax: +34 947288850; E-mail: archivo@aytoburgos.es
1984; Milagros Moratinos Palmoero
10 100 vols; 12 curr per; 9 mss; 1 incunabula; 18 maps; 4 900 microforms; 45 970 av-mat; 132 sound-rec
ANABAD 25936

Biblioteca Castilla y León, Real Monasterio de San Agustin, C/ Madrid 24, 09001 ***Burgos***
T: +34 947255758; E-mail: biblioteca@diputaciondeburgos.es
1981; María del Mar González Simón
Fondo Escolar Antiguo Biblioteca 'Manuel Machado' (en depósito), local hist, regional geography
30 000 vols; 30 curr per; 5 mss; 10 diss/theses; 10 govt docs; 400 maps; 10 av-mat; 38 sound-rec; 50 digital data carriers 25937

Biblioteca-Hemeroteca de Castilla y León y las Autonomías Españolas, C/ Progreso, 24-26, 5° H, 09002 ***Burgos***
T: +34 947266754; E-mail: servicio010@aytoburgos.es
1976; Prof. Federico Pérez
15 550 vols; 591 curr per; 125 sound-rec; 25 digital data carriers; 550 000 press clippings, 1 300 books on local and regional subjects 25938

Fundación Laboral Sonsoles Ballvé, Biblioteca, C/ La Bureba s/n (Campofrio), 09007 ***Burgos***
T: +34 947283103; Fax: +34 947283136; E-mail: fundacion.sonsolesballve@campofrio.es
1980; Manuel Ortega Porras
7 205 vols; 15 curr per; 21 e-journals; 20 diss/theses; 24 govt docs; 221 music scores; 651 maps; 325 av-mat; 595 sound-rec; 635 digital data carriers 25939

Real Monasterio de San Agustin, Archivo, Biblioteca Castilla y León, C/ Madrid, 24, 09001 ***Burgos***
T: +34 947255758; Fax: +34 947200750; E-mail: archivo@diputaciondeburgos.es; URL: www.diputaciondeburgos.es
1973; Floriano Ballesteros Caballero
24 300 vols; 161 curr per; 350 maps; 23 980 microforms; 15 111 av-mat 25940

Real Academia Provincial de Bellas Artes, Instituto de Academias de Andalucia, Biblioteca, Pza de Mina, s/n, 11004 ***Cádiz***
T: +34 9562101-34/-35; Fax: +34 956253415; E-mail: rapbac@hotmail.com
Francisco Ponce Cordones
8 000 vols; 500 curr per 25941

Seminario de Estudios Cerámicos, Centro de Documentación, Sargadelos, 27888 ***Cervo***
T: +34 981557841
1972
8 600 vols; 55 curr per; 10 diss/theses; 500 slides 25942

Museo Arqueológico de Córdoba, Biblioteca, Pza Jerónimo Paez, 7, 14003 ***Córdoba***
T: +34 957355520; Fax: +34 957355534; E-mail: museoarqueologicocordoba.ccul@juntadeandalucia.es
1960; M. Dolores Baena Alcántara
13 500 vols; 5 658 periodicals 25943

Museo de Bellas Artes de Córdoba, Biblioteca (Fine Arts Museum Cordoba Library), Pl del Potro, 1, 14002 ***Córdoba***
T: +34 957355550, 957355543; Fax: +34 957355548; E-mail: museobellasartescordoba.ccul@juntadeandalucia.es; URL: www.juntadeandalucia.es/cultura/museos/MBACO/
1981; Fuensanta García de la Torre
12 152 vols; 16 curr per 25944

Real Academia de Ciencias, Bellas Letras y Nobles Artes, Biblioteca, C/ Ambrosio de Morales, 9, 14003 ***Córdoba***
T: +34 957413168; Fax: +34 957413168; URL: www.racordoba.es/index.php/biblioteca.html
1810; José Cosano Mayano
Fondo Antiguo (Arab mss and 600 vols from 16th-18th c)
30 000 vols; 700 curr per; 15 music scores; 100 maps; 75 digital data carriers
libr loan 25945

Alianza Francesa de A Coruña, Biblioteca, C/ Real, 26 – 1°, 15003 ***A Coruña***
T: +34 981222356; Fax: +34 981222356; E-mail: afacoruna@mundo-r.com; URL: www.alliancefrancaise.es/
1927; Alain Rougilot
40 000 vols 25946

Archivo do Reino de Galicia, Xunta De Galicia. Consellería de Cultura e Deporte, Biblioteca auxiliar, Jardn De San Carlos, 15001 ***A Coruña***
T: +34 981209251; Fax: +34 981227094; E-mail: arq.reino.galicia@xunta.es; URL: www.xunta.es/conselle/cultura/patrimonio/arquivos/arquivo%20reino/index.html
1775; Gabriel Quiroga Barro
ADD. Ricardo Palmás; Restauration lab and microfilm lab
20 381 vols; 23 curr per; 1 e-journals; 8 mss; 7 incunabula; 4 diss/theses; 6 microforms; 94 av-mat; 355 sound-rec; 91 digital data carriers; 1808 closed periodicals, 121.413 boxes with govt. docs, 3 fimstrop colls 25947

Biblioteca Municipal de Estudos Locais (Local Studies Library), Rúa Duran Loriga, 10, 15003 ***A Coruña***
T: +34 981184386; Fax: +34 981184385; E-mail: bestudiosl@coruna.es; URL: www.coruna.es/bibliotecas
1991; Victoria Villanueva Pousa
Regional history from Galicia and A Coruña
41 606 vols; 1 068 curr per; 1 e-journals; 4 mss; 6 govt docs; 90 music scores; 380 maps; 11 microforms; 450 av-mat; 1 337 sound-rec; 246 digital data carriers; 583 minor publs, 2 571 posters, 9 clipping colls
libr loan 25948

Cámara Oficial de Comercio, Industria y Navegación, Centro de Documentación (La Corunne Chamber of Cominerce), C/ Alameda, 30, 1°, 15003 ***A Coruña***
T: +34 981216072; Fax: +34 981225208; E-mail: ccincoruna@camaras.org; URL: www.camaracoruna.com
1969; Pablo Garcia Pita
10 000 vols; 135 curr per; 12 diss/theses; 100 maps; 100 microforms; 800 av-mat 25949

Museo Arqueolóxico e Histórico, Castillo de San Antón, 15001 ***A Coruña***
T: +34 981210504; Fax: +34 981205994; E-mail: museo@sananton.org; URL: www.ctv.es/USERS/sananton
1979; Begõna bas López
8 000 vols; 345 curr per 25950

Real Academia Gallega, Biblioteca, RuaTabernas, 11, 15001 *A Coruña*
T: +34 981207308; Fax: +34 981207308; E-mail: secretaria@realacademiagalega.org; URL: www.realacademiagalega.org
1905; Xose Luis Axeitos Agrelo
Personal libraries of a numer of galician writers: Manuel Murguía, Emilia Pardo Bazán, Galo Salinas, Antóm Villar Ponte, Emilio González López, Luis Seoane etc.
55 000 vols; 110 curr per; 726 music scores; 65 maps; 80 microforms; 150 digital data carriers; 20 000 pamphlets, 1 000 files of manuscripts   25951

Euskal Biblioteka Labayru, Diocesis de Bilbao, Larrauri 1A-5a pl, 48160 *Derio* – Bizkaia
T: +34 944060171; Fax: +34 944060173; E-mail: idazkaritza@labayru.org; URL: www.labayru.org/biblioteka.html
1970; Ander Manterola Aldecoa
43 000 vols; 2 525 curr per; 100 mss; 17 diss//theses; 1 003 maps; 15 microforms; 225 digital data carriers; 1 200 slides   25952

Aranzadi Zientzia Elkartea, Liburutegia (Aranzadi Society of Sciences), Zorroagagaina, 11, 20014 *Donostia*
T: +34 943466142; Fax: +34 943455811; E-mail: liburutegia@aranzadi-zientziak.org; URL: www.aranzadi-zientziak.org
1947; Lourdes Ancin
Antropology, Archaeology
28 340 vols; 219 curr per
libr loan   25953

Museo de Arte e Historia, Biblioteca, San Agustinalde, 16, 48200 *Durango*
T: +34 946810312; URL: www.durango-udala.net
1986
11 000 vols; 62 curr per; 30 microforms   25954

Museo Arqueológico de Ibiza y Formentera, Biblioteca, Via Romana, 31, 07800 *Eivissa*, Islas Baleares
T: +34 971301771; Fax: +34 971303263; E-mail: maef@telefonica.net
1907; Jorge Fernández Gómez
Punic and Phoenician archaeology
20 147 vols; 217 curr per
libr loan   25955

Archivo Histórico Municipal, Biblioteca Auxiliar, Placeta Institut Vell, 1, 17004 *Girona*
T: +34 972221545; Fax: +34 972202694; E-mail: arximunicipal@ajgirona.org
1986; Joan Boadas i Raset
E.E. Schalit Libr (4 000 vols on Jewish hist and culture), www.ajgirona.org/call;
Jornades sobre imatge i recerca, conferences 'Esslesia: Pades'
17 448 vols; 1 353 curr per; 159 e-journals; 28 mss; 2 incunabula; 210 diss//theses; 20 000 music scores; 18 881 maps; 306 391 microforms; 1 089 633 av-mat; 20 000 sound-rec; 250 digital data carriers; 3 182 linear m of govt docs
ICA, CIA, La Gazette des Archives   25956

Collegi d'Arquitectes de Catalunya, Biblioteca de la Demarcació de Girona, Pl Catedral, 8 – Pia Almoina, 17004 *Girona*
T: +34 972412895; Fax: +34 972214151; E-mail: bibl.gir@coac.cat; URL: www.coac.net/girona
1975; Maite Rubiales Silguero
Antiquary and personal libraries of architects
21 600 vols; 104 curr per; 12 diss//theses; 47 maps; 464 av-mat; 406 digital data carriers; Antiquary (19 vols)
libr loan   25957

Escuela de Estudios Arabes (C.S.I.C.), Consejo Superior de Investigaciones Científicas, Biblioteca, Cuesta del Chápiz, 22, 18010 *Granada*
T: +34 958222290; Fax: +34 958224754; URL: www.eea.csic.es
1932; Miriam Font Ugalde
Islamic Civilization, Arabian Studies
14 000 vols; 124 curr per; 70 mss; 16 maps; 16 microforms; 81 digital data carriers; 13 e-journal subscriptions
libr loan   25958

Fundación Rodríguez-Acosta, Biblioteca, Callejón Niños del Rollo, 8, 18009 *Granada*
T: +34 958227497; Fax: +34 958210285; E-mail: info@fundacionrodriguezacosta.com; URL: www.fundacionrodriguezacosta.com
1972
Bibl particular de Manuel Gómez-Moreno Martínez y Raúl Alberto Diez
8 000 vols; 14 curr per   25959

Ilustre Colegio Notarial de Granada, Biblioteca, San Jerónimo, 50, 18001 *Granada*
T: +34 958202711; Fax: +34 958281122; URL: antigua.granada.notariado.org
1935; Manuel Pons González
10 000 vols; 200 curr per   25960

Patronato de la Alhambra y Generalife, Biblioteca y Archivo, C/ Real de la Alhambra, s/n, 18009 *Granada*
T: +34 958027945, 958027900; Fax: +34 958210235; URL: www.alhambra-patronato.es
1909; Encarnación Vilchez Ruiz
Alhambra y Generalife
12 993 vols; 395 curr per; 2 e-journals; 1 mss; 81 diss//theses; 4 microforms; 102 av-mat; 95 digital data carriers; 995 maps, plans and drawings, 17 037 photos, 1 764 transparencies
libr loan   25961

Instituto de Estudios Altoaragoneses, Biblioteca Azlor, Parque, 10, 22002 *Huesca*
T: +34 974294120; Fax: +34 974294122; E-mail: iea@iea.es; URL: www.iea.es
1950
48 200 vols; 1 200 curr per; 185 maps; 23 microforms; 184 av-mat   25962

Biblioteca, Archivo y Colección Arqueológica Municipal, Pl General Primo de Rivera 7 y 8, *Jerez de la Frontera*
1873; Carla Puerto Castrillón
Coll on horses, bullfighting, flamenco, incunabula, coll from 17th and 18th curies
75 000 vols   25963

Instituto de Astrofísica de Canarias, Biblioteca, C/ Via Lactea s/n, 38200 *La Laguna*
T: +34 922605-250/-248; Fax: +34 922605210; E-mail: biblio@iac.es; URL: www.iac.es/biblio/wwwbib1.htm
1985; Monique Gomez
13 000 vols; 230 curr per; 230 diss//theses; 100 microforms; 1 054 av-mat; 300 digital data carriers
libr loan; SEDIC   25964

Real Sociedad Económica de Amigos del País de Tenerife, Biblioteca, Calle de San Agustín, 23, *La Laguna*, Santa Cruz de Tenerife (Islas Canarias)
T: +34 922250010; Fax: +34 922257735; E-mail: biblioteca@rseapt.com; URL: www.rseapt.com
1777; Lorenzo Hernández-Abad González
Mss of Isles
14 500 vols; 1 035 curr per; 21 675 mss; 225 music scores; 20 maps; 5 microforms; 15 av-mat; 20 sound-rec; 20 digital data carriers; 600 photos   25965

Monasterio de Padres Benedictinos, Biblioteca, 20210 *Lazkao*, Guipuzcoa
T: +34 943880170; Fax: +34 943160868
1972; Jua José Agirre
Linguistica de 'Temática Vasca', hist local y regional, hist del Pais Vasco y Universal
45 000 vols; 2 730 curr per; 7 incunabula; 50 000 pamphlets
libr loan   25966

Caja España de Inversiones, Caja de Ahorros y Monte de Piedad, Centro de Documentación, Ordoño, II, 10, 24001 *León*
T: +34 987292500
Finance, law, economics, social sciences
14 300 vols; 285 curr per; 161 av-mat   25967

Instituto de Estudios Ilerdenses, Biblioteca, Plaça de la Catedral, 25002 *Lleida*
T: +34 973271500; Fax: +34 973274538; E-mail: fpiei@diputaciolleida.cat
1942; Maria Angeles Esforzado Viladrich
21 300 vols; 178 curr per; 5 240 books & pamphlets on the province of Lérida   25968

Instituto de Estudios Riojanos, Biblioteca, Portales, 2, 26601 *Logroño*, La Rioja
T: +34 941291181; Fax: +34 941291910; E-mail: biblioteca.ier@larioja.org
1977; María José Silvan Sada
1 485 postcards, 2 760 aerea photos
18 850 vols; 490 curr per; 570 mss; 1 incunabula; 139 diss//theses; 315 maps; 520 microforms; 7 770 av-mat; 84 sound-rec; 50 digital data carriers
libr loan; ANABAD   25969

Museo de La Rioja, Biblioteca (Museum of La Rioja Library), Plaza de San Agustín s/n, 26001 *Logroño*
T: +34 941291259; Fax: +34 941206821; E-mail: mteresa.alvarez@larioja.org
1979; Teresa Álvarez González
Hist, Art, Archaeology, Ethnology
12 091 vols; 90 curr per; 605 av-mat
ANABAD   25970

Museo Provincial de Lugo, Biblioteca, Praza da Soídade s/n, 27001 *Lugo*
T: +34 982242112; Fax: +34 982240240; E-mail: info@museolugo.org; URL: www.museolugo.org
14 100 vols; 300 curr per   25971

Agencia Española de Cooperación Internacional – Biblioteca Hispánica, Av Reyes Católicos, 4, 28040 *Madrid*
T: +34 915838524; Fax: +34 915838525
1946; Carmen Diez Hoyo
Graiño, Velarde, Chacón y Calvo
520 000 vols; 2 100 curr per; 440 incunabula; 1 300 diss//theses; 2 000 microforms; 30 digital data carriers
FID, IFLA, LIBER, SALALM   25972

Agencia Española de Cooperación Internacional para el Desarrollo (AECID) – Biblioteca Islámica Félix María Pareja, Ministerio de Asuntos Exteriores y de Cooperación (Islamic Library 'Felix Maria Pareja'), Av de Reyes Católicos, 4, 28040 *Madrid*
T: +34 915838156; Fax: +34 915838525; E-mail: biblioteca.islamica@aecid.es; URL: www.aecid.es
1954; Victoria Alberola
60 000 vols; 450 curr per; 20 e-journals; 15 mss; 55 diss//theses; 328 microforms; 282 av-mat; 197 digital data carriers
libr loan; IFLA, FID, MELCOM, SEDIC   25973

Agencia Nacional de Meteorología, Biblioteca, Ciudad Universitaria, Apdo 285, 28071 *Madrid*
T: +34 902531111; Fax: +34 915810264; URL: www.aemet.es/es/servicios/biblioteca
1887; Miguel Angel García Couto
20 400 vols; 190 curr per; 50 diss//theses; 2 150 microforms; 12 av-mat; 182 digital data carriers   25974

Archivo Histórico Nacional, Biblioteca Auxiliar, Serrano 115, 28006 *Madrid*
T: +34 917688500; Fax: +34 915631199; E-mail: ahn@mcu.es; URL: www.mcu.es
1898; Carolina Santamarina
4000 Antik Bände (XVI-XVIII Jahrhundert)
30 000 vols; 300 curr per; 2 incunabula; 150 digital data carriers
CJA   25975

Archivo Iberoamericano, Libreria-Editorial Cisneros, Joaquín Costa, 36, 28002 *Madrid*
T: +34 915619900; Fax: +34 915613990
1914; P. Enrique Chacón
Hist, philosophy, theology
12 000 vols; 400 curr per   25976

Asociación Española de Normalización y Certificación (AENOR), Centro de Información y Documentación y Biblioteca (Spanish Standards Association (AENOR), Génova 6, 28004 *Madrid*
T: +34 902102201; Fax: +34 913104032; E-mail: info@aenor.es; URL: www.aenor.es
Ana López

350 curr per; 50 digital data carriers; 700 000 standards   25977

Asociación Gran Peña, Biblioteca, C/ Gran Vía, 2, 28013 *Madrid*
T: +34 915224613
1883
50 000 vols   25978

Ateneo Científico, Literario y Artístico, Biblioteca, C/ del Prado, 21, 28104 *Madrid*
T: +34 914297442 ext. 118; Fax: +34 914297901; E-mail: biblioteca@ateneodemadrid.es; URL: www.ateneodemadrid.com
1835; N.N.
485 000 vols; 2 700 curr per; 322 microforms; 3 600 crystal positives
libr loan   25979

Biblioteca Central Militar, Instituto de Historia y Cultura Militar, Paseo de Moret, 3, 28008 *Madrid*
T: +34 917808739; Fax: +34 917808704; E-mail: bcm@et.mde.es; URL: www.ejercito.mde.es
1932; Inocencia Soria
Military history, military and civil engineering, arts and crafts, military science; Military Music
250 000 vols; 3 000 curr per; 100 e-journals; 2 111 mss; 1 incunabula; 200 diss//theses; 1 300 music scores; 12 500 microforms; 250 av-mat; 450 sound-rec; 350 digital data carriers
ANABAD, AEDOM   25980

Biblioteca de Pedagogía, Consejo Superior de Investigaciones Científicas, Pinar, 25, 28006 *Madrid*
T: +34 912619800
1943; Soledad Roldán Carrillo
83 000 vols; 180 curr per; 41 mss; 100 diss//theses   25981

Biblioteca Dominicos San Pedro Mártir, Apdo 61150, 28080 *Madrid*
T: +34 912024246; Fax: +34 917665684
1958; Maximiliano Rebollo
110 000 vols; 430 curr per   25982

Biblioteca Francisco de Zabálburu, Marqués del Duero, 7, 28001 *Madrid*
T: +34 915769066
1877
Archivo de Altamira, Archivo de D. José Ignacio Miró
18 000 vols   25983

Biblioteca General Medinaceli, Consejo Superior de Investigaciones Científicas, Duque de Medinaceli, 6, 28014 *Madrid*
T: +34 915854883; Fax: +34 915854878
1910; Dolores Corrons Rodriguez
61 500 vols   25984

Biblioteca Histórica Municipal, Conde Duque, 11, 28015 *Madrid*
T: +34 915885737; Fax: +34 914800520
1876; Ascensión Aguerri Martínez
Madrid, theater, music, Cervantes, Lope de Vega, bulls, bibliografia, paremiología
173 000 vols; 40 curr per; 7 100 mss; 14 incunabula; 1 700 music scores; 110 maps; 5 978 music scores
libr loan   25985

Biblioteca Musical, Ayuntamiento de Madrid, C/ Conde Duque, 9-11 2 planta, 28015 *Madrid*
T: +34 91 5885753; Fax: +34 91 5885840; E-mail: bibliomusical@munimadrid.es
1919; Aurora Rodríguez Martín
Music works based in Don Quijote by Cervantes
52 curr per; 100 mss; 50 000 music scores; 100 microforms; 10 800 sound-rec; 270 video records, 486 museum pieces, 276 instruments
AIBM   25986

British Council, Instituto Británico, Biblioteca, Paseo del General Martinez Campos, 31, 28010 *Madrid*
T: +34 913373551; Fax: +34 913373573; E-mail: sociedad@britishcouncil.es; URL: www.britishcouncil.org/spain
1941
Spanish coll; British studies, teachers' resource centre & education inf services
19 000 vols
libr loan   25987

Cámara Oficial de Comercio e Industria de Madrid, Biblioteca (Madrid Chamber of Commerce and Industry), Pedro Salinas, 11, 28043 *Madrid*
T: +34 915383845; Fax: +34 915383803; E-mail: bib1@camaramadrid.es; URL: www.camaramadrid.es
1887; María José Méndez Cachot
70 000 vols; 1 380 curr per; 4 000 microforms; 5 000 av-mat; 550 digital data carriers
libr loan   25988

Casa de Velázquez, Biblioteca, C/ Paul Guinard, 3, 28040 *Madrid*
T: +34 914551580; Fax: +34 914551625; URL: www.casadevelazquez.org/es/biblioteca/
1928; Philippe Bérato
History of Spain, Portugal and Latin America, Spanish literature, art, archeology, social sciences, geography
106 000 vols; 1 100 curr per; 77 mss; 1 470 maps; aerial photos of Spain
libr loan   25989

Centro de Ciencias Medioambientales, Consejo Superior de Investigaciones Científicas, Servicio Documentación y Biblioteca (Centre for Environmental Science), Serrano, 115-dpdo., 28006 *Madrid*
T: +34 917452500/-286 (ext)/-268 (ext); Fax: +34 915640800; E-mail: edafol@ccma.csic.es; URL: www.ccma.csic.es
1942; Clara M. Blanco Temprano
30 700 vols; 1 500 curr per; 3 100 diss/theses; 4 000 maps
libr loan   25990

Centro de Estudios Históricos, Consejo Superior de Investigaciones Científicas, Biblioteca, Duque de Medinaceli, 6, 28014 *Madrid*
T: +34 914290626; Fax: +34 913690940
1939; Pilar Lizan
Hist, art, archeology
273 400 vols; 2 280 curr per; 1 incunabula; 350 maps; 220 microforms; 8 digital data carriers
libr loan; LIBER, FID, ANABAD   25991

Centro de Estudios Jurídicos, Biblioteca, C/ Juan del Rosal, 2, 28040 *Madrid*
T: +34 914551670; Fax: +34 915431870; E-mail: infogral.cej@mju.es; URL: www.cej.justicia.es/cej/servlet/CEJServlet
1951
8 700 vols; 136 curr per   25992

Centro de Estudios Políticos y Constitucionales, Biblioteca, Pl de la Marina Española, 9, 28071 *Madrid*
T: +34 9142289-05/-42; Fax: +34 915419574; E-mail: docjur@cepc.es; URL: www.cepc.es
1939; Gloria Gómez del Pulgar Rodríguez de Segovia
82 700 vols; 1 830 curr per; 210 e-journals; 400 diss/theses; 740 microforms; 6 digital data carriers
libr loan   25993

Centro de Física Miguel A. Catalán, Consejo Superior de Investigaciones Científicas, Biblioteca (Miguel A. Catalán Physics Center (CFMAC) – Spanish National Research Council (CSIC)), Serrano 121-123, 28006 *Madrid*
T: +34 915616800; Fax: +34 915645557; E-mail: flora@cfmac.csic.es; URL: www.cfmac.csic.es/biblio/index.htm
1995; Flora Granizo Barrena
Sewction of Mathematics and Fundamental Physics, Section of Optics and Structure of Materials
11 098 vols; 580 curr per; 21 e-journals; 191 986 e-books; 10 mss; 500 diss/theses; 151 govt docs; 90 digital data carriers; Fundamental Physics, Optics and Structure of Materials
libr loan   25994

Centro de Información Documental de Archivos, Ministerio de Cultura, Biblioteca, C/ Santiago Rusiñol 8. Planta -2, 28040 *Madrid*
T: +34 915351984; 915348090; Fax: +34 915351973; E-mail: biblioteca.cida@mcu.es
1979; Rosa Chumillas
9 140 vols; 477 curr per; 113 microforms; 107 digital data carriers; 5 334 pamphlets, 20 VHS   25995

Centro de Información y Documentación Científica (CINDOC), Biblioteca de Ciencia y Tecnología (ICYT), Joaquín Costa, 22, 28002 *Madrid*
T: +34 915635482; Fax: +34 915642644; URL: www.cindoc.csic.es
1948; Carmen Vidal Perucho
Chemistry, biology, physics, medicine, pharmacy, engineering, information science, life science, documentation, socialscience, humanities; Aquaculture dept, Grey lit dept, Documentation dept
25 500 vols; 6 700 curr per; 2 100 diss/theses; 1 050 microforms; 200 digital data carriers
libr loan; IFLA, ASLIB, IATUL, EUSIDIC
                                       25996

Centro de Investigaciones Biológicas, Consejo Superior de Investigaciones Científicas, Biblioteca, Ramiro de Maeztu, 9, 28040 *Madrid*
T: +34 918373112; Fax: +34 915360432; E-mail: biblioteca@cib.csic.es; URL: www.cib.csic.es/es/servicio.php?iddepartamento=4
1958; Dr. Rafael Giraldo, Olvido Partearroyo
Biomedicine, cell biology, molecular microbiology, immunopharmacology, virology, biotechnology, etc
69 000 vols; 1 300 curr per; 1 150 diss/theses; 5 microforms; 40 digital data carriers
libr loan; SEDIC   25997

Centro de Investigaciones Energéticas, Medioambientales, Tecnológicas (CIEMAT), Centro de Documentación, Av Complutense, 22, 28040 *Madrid*
T: +34 913466000; Fax: +34 913466005
1959; Lope Diez García
Nuclear technology, renewable energies
150 000 vols; 2 000 curr per; 147 diss/theses; 629 790 microforms; 5 700 standards   25998

Centro de Investigaciones Sociológicas (CIS), Montalbán 8, 28014 *Madrid*
T: +34 915807600; URL: www.biblioteca.cis.es/ABSYS/abwebp.exe/LL1/G0
1965; Pilar Pinuaga
15 100 vols; 315 curr per
libr loan   25999

Centro de Tecnologías Físicas "Leonardo Torres Quevedo", Consejo Superior de Investigaciones Científicas, Biblioteca, Serrano, 144, 28006 *Madrid*
T: +34 915618803; Fax: +34 914117651; E-mail: biblioteca@cetef.csic.es; URL: www.cetef.csic.es/biblioteca.html
1946
Electrónica, telecomunicaciones, acústica, edafología
8 000 vols; 160 curr per; 420 diss/theses; 160 av-mat   26000

Centro Nacional de Biotecnología (CSIC), Biblioteca y Documentación, Campus Universidad Autónoma, Cantoblanco, 28049 *Madrid*
T: +34 95854511; Fax: +34 95854506; E-mail: bibliopi@cnb.uam.es; URL: www.cnb.csic.es/~biblio/biblioteca.html
Elisa Ana Morin   26001

Centro Nacional de Investigaciones Metalúrgicas (CENIM), Consejo Superior de Investigaciones Científicas, Biblioteca, Av Gregorio del Amo, 8, 28040 *Madrid*
T: +34 915538900; Fax: +34 915347425; E-mail: biblio@cenim.csic.es; URL: www.cenim.csic.es
1947; Ricardo Martínez de Madariaga
13 100 vols; 300 curr per; 175 diss/theses; 11 500 microforms; 8 digital data carriers; 16 000 standards, 80 patents
libr loan   26002

Centro Química Orgánica "Manuel Lora Tamayo", Consejo Superior de Investigaciones Científicas (CSIC), Biblioteca, C/ Juan de la Cierva, 3, 28006 *Madrid*
T: +34 915622900; Fax: +34 915644853; E-mail: biblioteca@cenquior.csic.es; URL: www.cenquior.csic.es/servicios/biblioteca.html

1966; Concepción Marín Pérez
5 branch libs
8 631 vols; 420 curr per; 56 diss/theses
libr loan   26003

Centro Superior de Estudios de la Defensa Nacional (CESEDEN), Biblioteca (Spanish Center for National Defense Studies), Paseo de la Castellana, 61, 28046 *Madrid*
T: +34 913482500; Fax: +34 9134825-53/-54; E-mail: ceseden@ceseden.es; URL: www.ceseden.es
1965; María de la Luz López
28 246 vols; 123 curr per; 84 e-journals; 36 e-books; 2 255 diss/theses; 6 500 digital data carriers   26004

Círculo de Bellas Artes, Biblioteca, Alcalá, 42, 28038 *Madrid*
T: +34 915324679
1885; Javier Martínez
Classic Spanish theatre
25 400 vols; 182 curr per   26005

Clínica Nuestra Señora de la Concepción, Fundación Jiménez Díaz, Biblioteca, Av Reyes Católicos, 2, 28040 *Madrid*
T: +34 915441600; Fax: +34 915448785; E-mail: vcalvo@fjd.es
1957; Valentin Calvo Herrero
8 000 vols; 570 curr per; 176 diss/theses
                                       26006

Colección de Prehistoria, Consejo Superior de Investigaciones Científicas (CSIC), Biblioteca, Serrano, 13, 28001 *Madrid*
T: +34 91576-7144/-1586; Fax: +34 914317746
1954
12 683 vols; 700 curr per; 1 198 maps; 415 microforms; 22 digital data carriers; 19 350 slides   26007

Colegio Oficial de Arquitectos de Madrid, Biblioteca, Barqillo, 12, 28004 *Madrid*
T: +34 915218200 ext 219; Fax: +34 915219265; E-mail: biblioteca@coam.org; URL: www.coam.org
1955; Paloma Barreiro Pereira
Colección José María Marañón
17 000 vols; 1 000 curr per; 40 mss
                                       26008

Comisión Nacional Española de Cooperación con la UNESCO, Biblioteca, Av de los Reyes Católicos, 4, 5a planta, 28040 *Madrid*
T: +34 915827936; Fax: +34 915827934; URL: www.aeci.es/unesco
Amalia Calvo Chalud
8 000 vols; 182 curr per; 80 microforms; 8 000 pamphlets   26009

Cruz Roja Española, Biblioteca, Rafael Villa, s/n, El Plantio, 28023 *Madrid*
T: +34 913354444; Fax: +34 913354455; E-mail: informa@cruzroja.es
1986
12 000 vols; 3 022 curr per; 500 av-mat; 88 000 photos, 4 mio docs   26010

Deutsches Archäologisches Institut Madrid / Instituto Arqueológico Alemán, Biblioteca (German Archaeological Institute Madrid), Calle de Serrano, 159, 28002 *Madrid*
T: +34 915610904; Fax: +34 915640054; E-mail: bibliothek@madrid.dainst.org; URL: www.dainst.org/index.php?id=3502
1943/1954; Dr. Dirk Paul Mielke
Archaeology of the Iberian Peninsula (Pre- and Protohistory, Roman Iberia, Late Antiquity, Early Christian and Islamic Iberia) and surrounding areas
70 000 vols; 650 curr per; 100 microforms; 30 digital data carriers
                                       26011

Escuela de Sanidad Militar, Biblioteca, Camino de Los Ingenieros, 6, 28047 *Madrid*
T: +34 914652600; Fax: +34 914667776
1898; Lloveres Rua-Figueroa
9 900 vols; 43 curr per; 50 av-mat; 50 sound-rec   26012

Facultad de Ciencias de la Actividad Física y del Deporte (INEF), Universidad Politécnica de Madrid, Biblioteca, Martín Fierro, s/n, 28040 *Madrid*
T: +34 915894045; Fax: +34 913364000; E-mail: biblioteca.inef@upm.es; URL: www.inef.upm.es/INEF
1967; Pilar Iruretagoyena
Sports, physical education (historical books from XVI c)
68 000 vols; 1 100 curr per; 55 e-journals; 900 diss/theses; 105 maps; 833 microforms; 1 200 av-mat; 35 sound-rec
libr loan; SEDIC, IASI   26013

Facultad de Teología 'San Dámaso', Biblioteca Alemana Görres, San Buenaventura, 9, 28005 *Madrid*
T: +34 913668508; E-mail: bibliotecagoerres@fsandamaso.es
1929, refounded 1955; Juan José Pérez-Soba
German and Spanish history and lit (18th-19th c)
26 000 vols; 21 curr per; 2 incunabula; 87 microforms
libr loan   26014

Filmoteca Española, C/ Magdalena, 10, 28012 *Madrid*
T: +34 914672600 ext 260; Fax: +34 914672611; E-mail: javier.herrera@mcu.es; URL: www.mcu.es
1953; Javier Herrera
Luis Buñuel coll, Spanish Civil War coll; Graphic materials archive, laboratory
24 500 vols; 240 curr per; 600 mss; 370 microforms; 8 000 av-mat; 5 500 sound-rec; 80 digital data carriers; 16/35 mm films (70 000 copies, 66 000 newsreels), 55 000 files of clipping, 400 000 photos, 35 000 posters, 75 000 press bks
FIAF, ACE   26015

Fundación Antonio Maura, Biblioteca, C/ Antonio Maura, 18 bajo, 28014 *Madrid*
T: +34 915312499; Fax: +34 915312499; E-mail: secretaria@fantoniomaura.org; URL: www.fantoniomaura.org
1970
Law, politics, hist
10 000 vols; 150 curr per   26016

Fundación Casa de Alba, Palacio de Liria, Biblioteca, Princesa, 20, 28008 *Madrid*
T: +34 915475302
Hist, lit
18 800 vols; 300 mss; 50 incunabula
                                       26017

Fundación de Investigaciones Marxistas, Biblioteca, Alameda, 5, 28014 *Madrid*
T: +34 914201388; Fax: +34 914202004; E-mail: fim@nodo50.org
Patricia González-Posada
Econimics, philosophy, politics, hist
10 000 vols; 117 curr per   26018

Fundación de las Cajas de Ahorros, Biblioteca (Foundation for Economic and Social Research), C/ Caballero de Gracia, 28, 28013 *Madrid*
T: +34 915965718; Fax: +34 915965796; E-mail: biblio@funcas.ceca.es; URL: www.funcas.ceca.es
1979; Valentin Edo-Hernández
100 000 vols; 677 curr per
libr loan; SEDIC   26019

Fundación de los Ferrocarriles Españoles, Biblioteca y Hemeroteca, Santa Isabel, 44, 28012 *Madrid*
T: +34 911511084; E-mail: biblioteca_ferroviaria@ffe.es; URL: www.ffe.es/documentacion/biblioteca.htm
1942; Ana Cabanes Martín
Railroads, History of tranports
35 000 vols; 2 740 curr per; 35 e-journals; 200 e-books; 66 diss/theses; 350 maps; 8 000 microforms; 344 av-mat; 10 digital data carriers; 300 posters, 1 820 pamphlets
libr loan; SEDIC, FESABID   26020

Fundación Fondo para la Investigación Económica y Social, Confederación Española de Cajas de Ahorros, Biblioteca, Alcalá, 27, 28014 *Madrid*
T: +34 915965000; URL: www.ceca.es
1968; Valentin Edo-Hernandez

Economics, banking and finfance
88 700 vols; 927 curr per; 9 000 microforms; 43 av-mat; 37 sound-rec; 1 266 digital data carriers
SEDIC, FID 26021

Fundación Francisco Largo Caballero, Biblioteca, C/ Antonio Grilo, 10, 28015 ***Madrid***
T: +34 915477990; Fax: +34 915423140
1978; Nuria Franco Fernández
Spanish labour and social movements, industrial relations, social sciences, workers migrations
15 651 vols; 2 159 curr per; 25 diss/theses; 8 microforms; 500 av-mat; 120 sound-rec; 7 000 photos, 630 posters, 2 000 slides
libr loan 26022

Fundación José Ortega y Gasset, Biblioteca, Fortuny, 53, 28010 ***Madrid***
T: +34 917004100; Fax: +34 917003530; E-mail: comunicacion@fog.es; URL: www.ortegaygasset.edu/contenidos.asp?id_s=64
1984; Enriqueta Olmo
Philosophy, hist of lit, pol science; Personal archive of José Ortega y Gasset
40 000 vols; 150 curr per; 87 diss/theses; 8 400 microforms; 2 000 sound-rec; 300 offprints
libr loan 26023

Fundación Juan March, Biblioteca Española de Música y Teatro Contemporáneos, Castelló, 77, 28006 ***Madrid***
T: +34 914354240; Fax: +34 915774170; E-mail: pfernandez@march.es; URL: www.march.es
1976; Paz Fernández
Contemporary Spanish Music, Performing Arts, Magic
57 957 vols; 99 curr per; 4 124 diss/theses; 18 435 music scores; 15 567 sound-rec; 58 672 press cuttings, 17464 photos, 9 514 theatre and music programs 26024

Fundación Pablo Iglesias, Biblioteca, Monte Esquinza, 30, 28010 ***Madrid***
T: +34 913104313; Fax: +34 913194585
1978; Aurelio Martin Najera
40 000 vols; 6 000 curr per; 150 000 mss
ANABAD, CIA 26025

Fundación Pablo VI, Biblioteca, Paseo Juan XXIII, 3, 28040 ***Madrid***
T: +34 915335200; Fax: +34 915535249; E-mail: comunicacion@upsam.com
1950
Sociología, política, economía, derecho
45 600 vols; 261 curr per; 60 sound-rec; 200 slides 26026

Fundación Pastor de Estudios Clásicos, Biblioteca, Serrano, 107, 28006 ***Madrid***
T: +34 915617236; Fax: +34 915634530; E-mail: fundacionpastor@telefonica.net; URL: fundacionpastor.e.telefonica.net/biblioteca.htm
1955; Esther Ortuño
13 500 vols; 48 curr per; 347 Greek and Coptic papyri
libr loan 26027

Fundación Sur – Departamento de África, Biblioteca, C/ Gaztambide, 31, 28015 ***Madrid***
T: +34 915441818; Fax: +34 915497787; E-mail: cidaf@planalfa.es; URL: www3.Planalfa.es/cidaf
1979; Rafael Sánchez Sanz
Africa: society, politics, economy, geography, history, art, culture, religion, philosophy, linguistics, science, technology, etc
18 020 vols; 60 curr per; 10 e-journals; 50 mss; 500 maps; 750 av-mat; 2 100 sound-rec; 54 digital data carriers; 6 000 pamphlets 26028

Hemeroteca Municipal de Madrid, C/ Conde Duque 9-11, 28015 ***Madrid***
T: +34 915885771; E-mail: hemeroteca@munimadrid.es; URL: www.munimadrid.es/hemeroteca
1918; Carlos Dorado Fernández
254 500 vols; 26 200 curr per; 1 incunabula; 8 300 microforms 26029

Ilustre Colegio de Doctores y Licenciados en Filosofía y Letras y Ciencias de la Comunidad de Madrid, Biblioteca, C/ Fuencarral, 101, 28004 ***Madrid***
T: +34 914471400; Fax: +34 914479056; E-mail: info@cdlmadrid.es; URL: www.cdlmadrid.es
1979; Marisa Montero Caldera
Filosofía, historia, geografía, educación
8 000 vols; 100 curr per; 200 av-mat 26030

Ilustre Colegio Notarial de Madrid, Biblioteca, C/ Juan de Mena, 9, 28014 ***Madrid***
T: +34 912130000; Fax: +34 912130048; E-mail: info@madrid.notariado.org; URL: madrid.notariado.org/?do=home
1920; José A. Molleda
30 000 vols; 270 curr per 26031

Ilustre Colegio Oficial de Médicos de Madrid, Biblioteca, Santa Isabel, 51, 28012 ***Madrid***
T: +34 91538510-0/-1; E-mail: info@icomem.es; URL: www.icomem.es
Maria José Serrano Zanon
8 000 vols; 200 curr per 26032

Institut Français, Instituto Francés, Médiathèque, Marqués de la Ensenada, 12, 28004 ***Madrid***
T: +34 917004827, 917004828; Fax: +34 913196401; E-mail: mediateca@ifmadrid.com; URL: www.ifmadrid.com
1913; Nathalie Lelong
25 000 vols; 100 curr per; 1 300 av-mat; 1 400 sound-rec; 50 digital data carriers 26033

Instituto Calasanz de Ciencias de la Educación, Escuelas Pias, Biblioteca, José Picón, 7, 28028 ***Madrid***
T: +34 917257200; Fax: +34 913611052; E-mail: info@icceciberaula.es; URL: www.icceciberaula.es
1967; Francisco Cubells Salas
36 000 vols; 2 700 curr per
libr loan 26034

Instituto de Ciencias de la Construcción Eduardo Torroja, Consejo Superior de Investigaciones Científicas, Biblioteca, Serrano Galvache, s/n, Apdo 19002, 28033 ***Madrid***
T: +34 913020440; Fax: +34 913020700; E-mail: biblioteca@ietcc.csic.es; URL: www.ietcc.csic.es/index.php?id=biblioteca&no_cache=1
1934; Dolores Fernandez Caballero
Colección "Monografías del Instituto Eduardo Torroja", Materiales de construcción, informática
23 100 vols; 433 curr per; 50 diss/theses; 2 000 norms
libr loan; ANABAD 26035

Instituto de Economía y Geografía, Consejo Superior de Investigaciones Científicas, Biblioteca, C/ Albasanz, 26-28, 28037 ***Madrid***
T: +34 916022300; Fax: +34 916022971; URL: www.ieg.csic.es
1986; Matilde Vilarroig Aroca
56 000 vols; 2 008 curr per; 100 diss/theses; 13 000 maps
libr loan 26036

Instituto de Estudios Fiscales, Biblioteca, Av Cardenal Herrera Oria, 378, 28035 ***Madrid***
T: +34 913398894, 913398965; Fax: +34 913398949; E-mail: biblioteca@ief.meh.es
1961; Javier Ramón Jiménez
Economics, law; Central Library (incl general coll, UNED coll, Sainz de Bujanda coll) and Regional Centres' libraries of Barcelona, Valencia, A Coruña and Sevilla
116 669 vols; 1 746 curr per; 91 e-journals; 73 e-books; 116 diss/theses; 126 microforms; 1 281 digital data carriers; 24 700 other items 26037

Instituto de Estudios Laborales y de la Seguridad Social, Biblioteca, Pio Baroja, 6, 28009 ***Madrid***
T: +34 914090941
1964
27 100 vols; 1 261 curr per
FID 26038

Instituto de Estudios Pedagógicos Somosaguas (IEPS), Biblioteca Castroverde, Vizconde de Matamala, 3, 28028 ***Madrid***
T: +34 913564404; Fax: +34 917267905; E-mail: ieps@ieps.es; URL: www.ieps.es
1979; Esther Gonzalez Rodriguez
Education, teacher training
26 000 vols; 77 curr per; 57 diss/theses; 15 av-mat 26039

Instituto de Estudios Turísticos, Centro de Documentación Turística de España, Biblioteca, C/ José Lázaro Galdiano, 6 – Entreplanta, 28036 ***Madrid***
T: +34 913433437; Fax: +34 913433440; E-mail: cdte@iet.tourspain.es; URL: www.iet.tourspain.es
1962; Emilia Alcelay Peinado
Journal 'Estudios Turísticos', Boletín Adquisiciones (trimestral), Boletín de Legislación Administrativa Turística (trimestral), Catálogo de Abstracts de Revistas (anual)
32 976 vols; 596 curr per; 188 diss/theses; 6 891 govt docs; 102 maps; 225 digital data carriers 26040

Instituto de Filosofía, Consejo Superior de Investigaciones Científicas, Biblioteca, Pinar, 25, 28006 ***Madrid***
T: +34 914111098 ext 279; Fax: +34 915645252; URL: www.ifs.csic.es/biblioteca/
1943; Ana Jimenez-Royo
Archivo Arangurev
55 000 vols; 186 curr per; 36 microforms; 12 digital data carriers
libr loan 26041

Instituto de Información y Documentación en Ciencias Sociales y Humanidades, Biblioteca, Pinar, 25, 28006 ***Madrid***
T: +34 912627755; Fax: +34 915645069
1971; Adelaida Roman Adman
35 600 vols; 1 600 curr per
libr loan; ASLIB, FID 26042

Instituto de la Juventud (INJUVE), Centro Nacional de Información y Documentación, José Ortega y Gasset, 71, 28010 ***Madrid***
T: +34 914019501; Fax: +34 914022194
1975; Francisco Canovas S.
16 000 vols; 450 curr per; 5 550 av-mat; 350 sound-rec 26043

Instituto de la Mujer, Centro de Documentación, Génova, 11, 1o dcha., 28004 ***Madrid***
T: +34 900191010; URL: www.migualdad.es/mujer
1984; Lorenzo Villanúa
Women's Studies
8 000 vols; 400 curr per; 18 diss/theses; 297 av-mat 26044

Instituto de Lenguas y Literaturas del Mediterráneo y Oriente Próximo, Consejo Superior de Investigaciones Científicas, Biblioteca, C/ Albasanz, 26-28, 1a planta, 28037 ***Madrid***
T: +34 915854835; Fax: +34 913690940; E-mail: b.gil@filol.csic.es; URL: www.filol.csic.es/ilc/
1985
Spanish and Hispanic languages and lit, Classical languages and lit, Arabic language and lit, Hebrew language and lit, Biblical Studies
75 000 vols; 101 mss; 1 500 maps; 240 sound-rec; 3 digital data carriers
libr loan 26045

Instituto de Valencia de Don Juán, Biblioteca, Calle Fortuny, 43, 28010 ***Madrid***
T: +34 913081848; Fax: +34 913197848; URL: www.artehistoria.com/genios/museos/518.htm
1920; Mrs Angeles Santos
Decorative arts, archeology, hist; Archivs de Felipe II y sus dos secretarías
30 000 vols; 70 curr per; 248 mss; 40 incunabula 26046

Instituto del Frío, Consejo Superior de Investigaciones Científicas, Biblioteca, C/ Jose Antonio Novais, 10, 28040 ***Madrid***
T: +34 915445607; Fax: +34 915493627; E-mail: azcona@if.csic.es; URL: www.if.csic.es
1951; Olga Ferrera
Meet and Fish, Metabolism and Nutrition, Engineering, Milk and Vegetables, USTA and USIE
11 737 vols; 150 curr per; 115 e-journals; 57 diss/theses; 14 microforms; 17 sound-rec; 45 digital data carriers
libr loan 26047

Instituto Egipcio de Estudios Islàmicos, Biblioteca (Cultural Office of the Egyptian Embassy in Madrid), Francisco de Asís Méndez Casariego, 1, 28002 ***Madrid***
T: +34 915636782; Fax: +34 915638640; E-mail: iegipcio@mundivia.es; URL: empresas.mundivia.es/iegipcio/home.html
1950; Ana del Puerto
Al-Andalus hist and lit; Mediterranean studies, Arab and Span lit
23 000 vols; 365 curr per; 134 mss; 92 diss/theses; 50 maps; 1 460 av-mat; 46 sound-rec; 1 400 slides 26048

Instituto Español de Comercio Exterior (ICEX), Centro de Documentación de Comercio Exterior, Paseo de la Castellana, 14, 28046 ***Madrid***
T: +34 913496100; Fax: +34 914316128; URL: www.icex.es
1982; Maria Arana
Foreign trade; International economics, world economy
10 000 vols; 750 curr per; 1 000 govt docs; 10 microforms; 50 digital data carriers
SEDIC 26049

Instituto Español de Oceanografía, Ministerio de Ciencia y Technología, Biblioteca/Centro de Documentación, Avenida del Brasil 31, 28020 ***Madrid***
T: +34 914175411; Fax: +34 915974770; E-mail: biblioteca@md.ieo.es; URL: www.ieo.es/biblioteca.html
1919; Eduardo Prieto Fernández de Toro
16 500 vols; 750 curr per; 75 diss/theses; 586 govt docs; 351 maps; 1 870 microforms; 168 digital data carriers
libr loan; EURASLIC 26050

Instituto Geográfico Nacional, Biblioteca y Cartoteca, General Ibañez Ibero, 3, Apdo 3007, 28003 ***Madrid***
T: +34 915333800; Fax: +34 915333444; URL: www.ign.es/ign/es/IGN/home.jsp
1870; Rosario Bordóns Escobar
Cartography, geodesy, photogrammetry, geophysics, seismology, teledetection
13 500 vols; 1 500 curr per; 80 000 maps; 90 microforms
libr loan 26051

Instituto Geológico y Minero de España, Biblioteca (Geological Survey of Spain), Rios Rosas, 23, 28003 ***Madrid***
T: +34 913495756; Fax: +34 913495844; E-mail: m.gutierrez@igme.es; URL: www.igme.es
1849; Margarita Gutierrez Garate
40 000 vols; 450 curr per; 120 e-journals; 30 e-books; 59 mss; 1 800 diss/theses; 6 500 govt docs; 18 000 maps
libr loan; SEDIC 26052

Instituto Histórico de la Marina / Museo Naval, Consejo Superior de Investigaciones Científicas, Biblioteca, C/ Juan de Mena, 2, 28071 ***Madrid***
T: +34 915238378; Fax: +34 913795056; E-mail: museonavalmadrid@fn.mde.es; URL: www.museonavalmadrid.com
1843; José María Zumalacárregui Calvo
12 000 vols 26053

Instituto Internacional, Biblioteca, Miguel Ángel, 8, 28010 ***Madrid***
T: +34 913081674; Fax: +34 913100964; E-mail: biblioteca@iie.es; URL: www.iie.es
1892; Noria Segui Ganzalez
Literature, art, history
80 000 vols; 80 curr per; 3 800 microforms; 1 335 av-mat; 152 sound-rec; 1 707 digital data carriers
libr loan; ANABAD, SEDIC 26054

Instituto Internacional en España, Biblioteca (International Institute), Miguel Angel, 8, 28010 ***Madrid***
T: +34 913081674; Fax: +34 913100964; E-mail: biblioteca@iie.es; URL: www.iie.es
1913; Maria Paz Lafuente
75 000 vols; 83 curr per; 3 800 microforms 26055

Instituto Nacional Administración Pública, Biblioteca, A tocka 106, 28012 ***Madrid***
T: +34 913493281, 913493236; Fax: +34 913493287; E-mail: Biblioteca@inap.map.es
1941; Enrique Orduña Rebollo
150 000 vols; 3 000 curr per; 110 mss; 68 maps; 1 216 microforms; 350 digital data carriers 26056

Instituto Nacional de Empleo, Biblioteca, Condesa Venadito, 9, 28027 ***Madrid***
T: +34 915859582
1964; José Vela Bonilla
Statistics, law, business management
14 600 vols; 230 curr per; 5 050 av-mat 26057

Instituto Nacional de Estadística, Biblioteca (National Statistical Office), Paseo de la Castellana, 183, 28046 ***Madrid***
T: +34 915839411; Fax: +34 915834889; E-mail: biblioteca@ine.es; URL: www.ine.es
1945
33 500 vols; 3 860 curr per 26058

Instituto Nacional de Gestión Sanitaria (INGESA), Biblioteca, C/ Alcalá, 56, 28071 ***Madrid***
T: +34 913380021; Fax: +34 913380024; E-mail: biblioteca@ingesa.mspsi.es
1985
Public health, law
6 100 vols; 22 curr per 26059

Instituto Nacional de Investigación y Tecnología Agraria y Alimentaria, Biblioteca (National Institute for Agriculture and Food Research and Technology), Ctra de la Coruña, km 7, 28040 ***Madrid***
T: +34 913471496; Fax: +34 913473938; E-mail: biblinia@inia.es; URL: www.inia.es
1935; Javier Romero Cano
Zoology, agriculture, forestry, food science, environment; AGRIS Center-FAO
40 956 vols; 4 000 curr per; 350 e-journals; 250 diss/theses; 20 000 microforms; 200 digital data carriers
libr loan; RIDA, IFLA, AGRIS, AGLINET 26060

Instituto Superior de Ciencias Morales, Padres Redentoristas, Biblioteca, Félix Boix, 13, 28036 ***Madrid***
T: +34 913453600; Fax: +34 913458679; E-mail: secretaria@iscm.edu; URL: www.iscm.edu/biblioteca.html
1971; Vicente Miranda Aliste
70 000 vols; 300 curr per; 500 diss/theses 26061

Istituto Italiano di Cultura, Biblioteca, C Mayor, 86, 28013 ***Madrid***
T: +34 91547-8603/-5205; Fax: +34 915422213; E-mail: biblioteca.iicmadrid@esteri.it; URL: www.iicmadrid.esteri.it/IIC_Madrid
1940; Marco Miele
40 000 vols; 100 curr per; 600 films, 9 600 slides 26062

Laboratorio de Geotécnia, Centro de Estudios y Experimentacón de Obras Públicas, Biblioteca, Alfonso XII, 3, 28014 ***Madrid***
T: +34 913357367; Fax: +34 913357322
1957; Encina Polo Polo
15 300 vols; 392 curr per; 400 maps; 270 microforms
ANABAD 26063

M.A.P.A. – Dirección General de Planificación y Desarrollo Rural, Biblioteca, Paseo de la Castellana, 112, 28071 ***Madrid***
T: +34 913471500; Fax: +34 914113770
1932; José Luis Cerezo Pombo
Official state bulletins
33 000 vols; 391 curr per; 5 mss; 1 190 diss/theses; 3 414 microforms 26064

Museo Arqueológico Nacional, Biblioteca (National Archaeological Museum), Serrano, 13, 28001 ***Madrid***
T: +34 915777912 ext 225; Fax: +34 914312757; E-mail: informacion.bibliografica.man@mcu.es; URL: man.mcu.es/biblioteca/biblioteca.html
1895; Rosa Chumillas Zamora
140 000 vols; 3 000 curr per
IFLA 26065

Museo Cerralbo, Biblioteca, Ventura Rodríguez, 17, 28008 ***Madrid***
T: +34 915591171
1924; Manuel Jorge Aragoneses
11 600 vols; 3 548 curr per
libr loan 26066

Museo de América, Biblioteca, Av Reyes Católicos, 6, 28040 ***Madrid***
T: +34 915492641; Fax: +34 915446742; E-mail: biblioteca@mamerica.mcu.es
1941
20 000 vols; 80 curr per 26067

Museo de la Farmacia Hispana, Facultad de Farmacia, U.C.M., Biblioteca, Fac. de Farmacia, U.C.M., Plaza de Ramón y Cajal s/n, 28040 ***Madrid***
T: +34 913941797; E-mail: museofar@farm.ucm.es
1951
10 000 vols 26068

Museo de San Isidro, Biblioteca, Plaza de San Andrés, 2, 28005 ***Madrid***
T: +34 913667415; Fax: +34 913645149; URL: www.munimadrid.es/museosanisidro
1961; Eduardo Salas Vazquez
Archaeology, hist of Madrid
10 576 vols; 1 770 curr per; 1 500 maps 26069

Museo del Ejército, Biblioteca, Méndez Nuñez 1, 28071 ***Madrid***
T: +34 915228977
1803
8 500 vols 26070

Museo del Trate, Centro de Investigation del Patriuonio Etnológico, Biblioteca, Av Juan de Herera 2, 28040 ***Madrid***
T: +34 915497150 ext 232; Fax: +34 915446970; E-mail: biblioteca@mt.mcu.es; URL: museodeltraje.mcu.es
1925; María Prego de Lis
24 523 vols; 680 curr per; 245 microforms; 570 av-mat; 2 475 sound-rec 26071

Museo Lázaro Galdiano, Biblioteca, Serrano, 122, 28006 ***Madrid***
T: +34 915616084
1951; Isabel Ibañez Losada
Archit, lit, art hist
39 000 vols; 1 063 curr per; 877 mss; 56 incunabula 26072

Museo Nacional Centro de Arte Reina Sofía, Centro de Documentación, Santa Isabel, 52, 28012 ***Madrid***
T: +34 914675062; Fax: +34 915281381; E-mail: miguel.valleinclan@cars.mcu.es
1990; Miguel del Valle-Inclán
100 000 vols; 130 curr per; 1 000 av-mat; 4 000 sound-rec; 7 000 slides, 15 000 exhibition cat, 6 000 folk music titles
libr loan 26073

Museo Nacional de Antropologia, Biblioteca, C/ Alfonso XII, 68, 28014 ***Madrid***
T: +34 915306418; Fax: +34 914677098; E-mail: antropologico@mna.mcu.es; URL: www.mcu.es
1895; Dolores Adellac
Antropología, Etnologia, Museologia
16 300 vols; 110 curr per; 30 diss/theses; 150 maps; 305 microforms; 71 av-mat; 155 sound-rec; 5 000 photos, 7 000 slides
libr loan 26074

Museo Nacional de Artes Decorativas, Biblioteca, C/ Montalbán, 12, 28014 ***Madrid***
T: +34 915326499 ext 202; E-mail: biblioteca.mnad@mcu.es; URL: mnartesdecorativas.mcu.es/biblioteca.html
1912; Isabel Moiño Campos
20 000 vols; 500 curr per 26075

Museo Nacional de Ciencias Naturales, Consejo Superior de Investigaciones Científicas, Biblioteca, José Gutierrez Abascal, 2, 28006 ***Madrid***
T: +34 914111328; Fax: +34 915645078; E-mail: biblioteca@mncn.csic.es; URL: www.mncn.csic.es/bibiblioteca.htm
1771; María Purificación Arribas García
Iconography coll (16th-17th c), ecology, entomology, zoology, evolutionary biology
61 000 vols; 1 056 curr per; 68 mss; 1 incunabula; 61 microforms
libr loan 26076

Museo Nacional de Reproducciones Artísticas, Biblioteca, Av de Juan Herrera, 2, 28040 ***Madrid***
T: +34 915497150; URL: mnreproduccionesartisticas.mcu.es/index-2.html
1878; M. J. Almagro Gorbea
8 583 vols; 300 curr per 26077

Museo Nacional del Prado, Biblioteca (Prado Museum Library), Casón del Buen Retiro, C/ Alfonso XII, 28, 28014 ***Madrid***
T: + 34 913302800, 902107077; E-mail: biblioteca@museodelprado.es; URL: www.museodelprado.es/pagina-principal/investigacion/biblioteca
1819; Javier Docampo
Rare books from 16-18th c
68 000 vols; 964 curr per; 802 e-journals; 103 mss; 2 incunabula; 173 diss/theses; 4 microforms; 460 av-mat; 97 sound-rec; 80 digital data carriers
libr loan 26078

Museo Postal y Telegráfico, Dirección General de Correos y Telégrafos, Biblioteca, Palacio de Comunicaciones. Montalban, s/n, 28070 ***Madrid***
T: +34 913962320; Fax: +34 913962129
Yolanda Estefanía
30 400 vols; 30 curr per; 55 mss; 485 maps; 500 av-mat 26079

Museo Romántico, Biblioteca, San Mateo, 13, 28004 ***Madrid***
T: +34 914481045; Fax: +34 914456940; E-mail: informacion.mromantico@mcu.es; URL: museoromantico.mcu.es/biblioteca.html
1924; Loreto Mampaso Boj
Romanticism, 19th c (fine arts, literature, history, etc)
20 000 vols; 12 curr per 26080

Oficina Española de Patentes y Marcas, Biblioteca (Patents and Trademarks Office), C/ Panamá, 1, 28071 ***Madrid***
T: +34 913495317; Fax: +34 913495380; URL: www.oepm.es
1989; Isabel Bertrán de Lis
Patent colls (29 000 000 patents, 6 400 000 patents on microfiche); Industrial Property Bulletins
6 500 vols; 53 curr per; 76 digital data carriers
SEDIC 26081

Organismo Autonomo Parques Nacionales, Biblioteca, C/ José Abascal, 41, 28003 ***Madrid***
T: +34 915468270; Fax: +34 915468267; E-mail: bibliotecaoapn@oapn.es
Ramon Hernandez Garcia
Fondo antiguo (2 167 vols from the 18th-19th c)
11 273 vols; 594 curr per; 85 av-mat; 40 digital data carriers
libr loan 26082

Organización de Estados Iberoamericanos (OEI), Biblioteca Digital/Centro de Recursos Documentales, Bravo Murillo 38, 28015 ***Madrid***
T: +34 915944382; Fax: +34 915943286; E-mail: ccalleja@oei.es; URL: www.oei.es/bibliotecadigital.htm
1953; Cristina Calleja Corpas
10 450 vols; 500 curr per; 1 e-journals; 270 e-books
libr loan 26083

Organización Nacional de Ciegos, Centro Bibliografico y Cultural, Biblioteca Central, C/ La Coruña, 18, 28020 ***Madrid***
T: +34 915894214; Fax: +34 915894225; E-mail: soi@once.es; URL: www.once.es
1965; Alberto Daudén Tallaví
Documentation and Translation Service (3 400 vols), Research Libr on Blindness
56 969 vols; 2 curr per; 75 diss/theses; 100 govt docs; 3 640 music scores; 111 av-mat; 11 029 sound-rec; 55 digital data carriers; 39 559 are braille bks, 31 398 are inkprints and 5 173 are braille music sheets
libr loan; IFLA 26084

Real Academia de Bellas Artes de San Fernando, Biblioteca, Alcala, 13, 28014 ***Madrid***
T: +34 915240884; Fax: +34 915319053; E-mail: secretaria@archivobiblioteca-rabasf.com; URL: www.archivobiblioteca-rabasf.com
1794; Irene Pintado Casas
Colls of engravings, drawings, architectural plans and projects
50 134 vols; 1 328 curr per; 657 mss; 1 incunabula; 468 diss/theses; 4 000 music scores; 500 maps; 257 microforms; 157 sound-rec; 99 digital data carriers; 4 039 docs, 3 150 engravings, 6 882 plans, 8 000 photos 26085

Real Academia de Ciencias Exactas, Físicas y Naturales, Biblioteca, Valverde, 22, 28013 ***Madrid***
T: +34 912212529; Fax: +34 915325716
1941; Elisa de la Torre
Coll Fernando de Castro (Medicine), Coll León Maroto (Physics)
26 000 vols; 5 000 curr per
libr loan 26086

Real Academia de Ciencias Morales y Políticas, Biblioteca, Pl de la Villa, 2, 28005 ***Madrid***
T: +34 917581505; Fax: +34 915481975; E-mail: biblioteca@racmyp.es; URL: www.racmyp.es
1858; Pablo Ramirez
112 000 vols; 900 curr per; 260 mss; 46 incunabula; 120 maps; 4 digital data carriers; 6 000 pamphlets 26087

Real Academia de Farmacia, Biblioteca, Calle de la Farmacia, 11, 28004 ***Madrid***
T: +34 915310307, 915223147; Fax: +34 915310306; E-mail: secretaria@ranf.com; URL: ranf.com
1932; Antonio R. Martínez Fernández
Farmacopeas
22 000 vols; 1 300 curr per; 600 diss/theses; 9 000 pamphlets
libr loan 26088

Real Academia de Jurisprudencia y Legislación, Biblioteca, Marqués de Cubas, 13, 28014 ***Madrid***
T: +34 915237161; E-mail: secretaria.rajyl@insde.es
1730; Carmen Crespo Tobarra
40 000 vols; 80 curr per; 15 mss; 200 digital data carriers; 1 200 rare books (16-18th c), online databases: Aranzadi, Tirant 26089

Real Academia de la Historia, Biblioteca, León, 21, 28014 ***Madrid***
T: +34 914290611; Fax: +34 914290704; E-mail: biblioteca@rah.es; URL: www.rah.es/biblioteca.htm
1738; Quintin Aldea Vaquero
Maps, engravings; Cartography dept
380 000 vols; 1 000 curr per; 9 060 mss; 198 incunabula 26090

Real Academia Española de la Lengua, Biblioteca, Felipe IV, 4, 28071 ***Madrid***
T: +34 914201478; Fax: +34 914200079; E-mail: biblioteca@rae.es; URL: www.rae.es
1713; Rosa Arboli
200 000 vols; 522 curr per; 641 mss; 42 incunabula; 2 200 microforms; 126 digital data carriers 26091

Real Academia Nacional de Medicina, Biblioteca, Arrieta, 12, 28013 ***Madrid***
T: +34 915479190; Fax: +34 915470320; E-mail: biblioteca@ranm.es; URL: www.ranm.es/biblioteca.html
1914; Ignacio Diaz-Delgado y Peñas
100 000 vols; 125 curr per; 1 720 mss; 450 diss/theses; 300 microforms 26092

Real Jardín Botánico, Consejo Superior de Investigaciones Científicas, Biblioteca, Plaza de Murillo, 28014 ***Madrid***
T: +34 915854669; Fax: +34 914200157; E-mail: inforjb@rjb.csic.es; URL: www.rjb.csic.es/jardinbotanico/jardin/
1781; Piedad Rodriguez-Piñero Fernández
401 Pre-Linnean works, 62 works by Linné; Archive of 18th c scientific expeditions
31 821 vols; 2 082 curr per; 15 000 mss; 3 incunabula; 310 diss/theses; 1 200 maps; 3 967 microforms; 10 000 av-mat; 14 digital data carriers; 23 769 pamphlets
libr loan; ANABAD 26093

Real Sociedad Económica Matritense de Amigos del País, Biblioteca, Pl de la Villa, 2, Torre de los Lujanes, 28005 ***Madrid***
T: +34 915418139; Fax: +34 915483413; E-mail: matritense@matritense.com; URL: www.economicamatritense.com

1775; Fabiola Azanza Santa Victoria
Agricultural Economics, history (18th c)
8 090 vols; 38 curr per; 49 mss; 364
sound-rec; 26 digital data carriers   26094

Real Sociedad Española de Historia Natural, Universidad Complutense de Madrid, Facultades de Biología y Geología, Biblioteca, Ciudad Universitaria, 28040 *Madrid*
T: +34 913945000; Fax: +34 913945000
1871; Margarita Costa
22 000 vols; 400 curr per; 100
diss/theses; 1 000 maps   26095

Real Sociedad Geográfica, Biblioteca, C/ Monte Esquinza, 41, 28010 *Madrid*
T: +34 913082477; Fax: +34 913082478; E-mail: secretaria@realsociedadgeografica.com; URL: www.realsociedadgeografica.com
1876
10 400 vols; 8 200 maps; 12 550 booklets   26096

Residencia de Estudiantes, Consejo Superior de Investigaciones Científicas (CSIC), Biblioteca, Pinar, 23, 28006 *Madrid*
T: +34 915636411; Fax: +34 915643890; E-mail: biblioteca@residencia.csic.es; URL: www.residencia.csic.es
1939
Education
76 900 vols; 737 curr per; 34 mss; 145 diss/theses; 4 220 offprints   26097

Sociedad General de Autores y Editores (S.G.A.E.), Centro de Documentación y Archivo (C.E.D.O.A.), Fernando VI, 4, 28004 *Madrid*
T: +34 913499622; Fax: +34 913499582; 913499710; E-mail: mgonzalez@sgae.es; URL: www.sgae.es
1899; Luz González Peña
Music, Opera, Zarzuela, Lit
12 500 vols; 2 curr per; 10 000 mss; 20 000 music scores; 2 000 av-mat; 9 000 sound-rec; 3 500 sheet music
AEDOM   26098

Centro Nacional de Alimentación, Biblioteca, Ctra de Majadahonda a Pozuelo, km 2, 28220 *Majadahonda*, Madrid
T: +34 916388078; E-mail: CNA@msps.es
1974; Ramón Barga Hernández
national documentation (1880-1982); international documentation (ISO, UNE, CEE, FAO, OMS, Codex Alimentarius Mundi)
9 400 vols; 527 curr per; 5 300 offprints
libr loan; Red Nacional de Bibliotecas   26099

Fundación Mapfre Estudios, Centro de Documentación, Monte del Pilar, s/n, El Plantio, 28023 *Majadahonda*, Madrid
T: +34 915812338; Fax: +34 913076641
1989
12 000 vols; 1 000 curr per; 40 000 av-mat   26100

Instituto de Salud Carlos III (Sede Majadahonda), Biblioteca, Ctra de Majadahonda a Pozuelo km 2,200, 28220 *Majadahonda*, Madrid
T: +34 918223100; Fax: +34 91 5097077; E-mail: oficina.informacion.auxiliar@isciii.es
1969; Rafael Najera Morrondo
47 000 vols; 340 curr per; 5 maps; 268 av-mat   26101

Instituto Andaluz del Deporte, Consejeria de Turismo, Comercio y Deporte, Biblioteca, Av. Santa Rosa de Lima, 5, 29007 *Málaga*
T: +34 951041918; Fax: +34 951041939; E-mail: biblioteca.iad.ctcd@juntadeandalucia.es
1984; Yolanda Biedma
Translating, publishing, (electronic) printing, AV production
11 000 vols; 451 curr per; 8 mss; 197 diss/theses; 2 govt docs; 2 microforms; 27 sound-rec; 154 digital data carriers; 624 videotapes
libr loan; SEDIC, IASI   26102

Ateneo Científico, Literario y Artístico, Biblioteca, Conde de Cifuentes, 25, 07703 *Maó*, Illes Balears
T: +34 971360553
1905
20 500 vols; 70 curr per   26103

Biblioteca Francesca Serrado, C/ Dr. Santiago Tifón, 21, 08107 *Martorelles*, El Vallès Oriental
T: +34 935704437; Fax: +34 935705964
1981
10 500 vols; 43 curr per   26104

Museo Nacional de Arte Romano, Biblioteca, José Ramón Melida, s/n, 06800 *Mérida*
T: +34 924311-912/-690; Fax: +34 924302006; E-mail: mnar@mnar.es
1986; A. Velazquez
Repertorio Bibliografico de Arqueologia Emntense (RBAE I y II); Fondo Antiguo (F.A.)
28 508 vols; 552 curr per; 85 diss/theses; 78 maps; 49 microforms; 90 av-mat; 50 sound-rec; 32 digital data carriers   26105

Casa-Museo Azorín, Biblioteca, Salamanca, 6, 03640 *Monóvar*, Alicante
T: +34 965470715; Fax: +34 965473034; E-mail: cmazorin@obs.cam.es; URL: www.obrasocial.cam.es/casamuseoazorin
1969
Spanish literature
17 600 vols   26106

Archivo Histórico Municipal de Murcia, Biblioteca Auxiliar, Palacio Almundi. Plano de S. Francisco, 8, 30004 *Murcia*
T: +34 968358735; Fax: +34 968218593; E-mail: biblioteca.archivo@ayto-murcia.es
Maria Angeles Jover Carrión
25 000 vols; 30 curr per; 4 incunabula; 900 maps; 7 microforms; 1 500 bulletins   26107

Museo de Bellas Artes de Asturias, Biblioteca (Museum of Fine Arts of Asturias), C/ Santa Ana, 1, 33003 *Oviedo*
T: +34 985213061; Fax: +34 985204232; E-mail: museobbaa@museobbaa.com; URL: www.museobbaa.com
1980; Teresa Caballero Navas
Coll of Asturian painters dossiers
17 000 vols; 100 curr per; 10 diss/theses; 86 av-mat; 100 digital data carriers; 250 slides, pamphlets   26108

Real Instituto de Estudios Asturianos, Biblioteca, Pl de Porlier, 9 – 1 planta, 33003 *Oviedo*
T: +34 984182801; E-mail: ridea@asturias.org; URL: ridea.org/biblioteca-real-instituto
1946; Ramón Rodríguez Álvarez
Photogr arch (3 000), hist arch (4 colls)
25 000 vols; 394 curr per; 10 mss
CSIC   26109

Arxiu del Regne de Mallorca, Archivo del Reino de Mallorca, Biblioteca, C/ de Ramón Llull, 3, 07001 *Palma de Mallorca*
T: +34 971725999; Fax: +34 971718781; E-mail: arm@arxregne.caib.es; URL: arxregne.caib.es
1851; Ricard Urgell Hernández
Coll of seals, docs in Arab and Hebrev, parchments, codices
13 442 vols; 193 curr per; 15 diss/theses; 2 633 microforms; 34 sound-rec; 90 digital data carriers; 7 000 m of Government docs
ANABAD   26110

Biblioteca de Cultura Artesana, Edifici de la Misericòrdia, Via Roma, 1, 07012 *Palma de Mallorca*
T: +34 971219536; Fax: +34 971219637; E-mail: bca@conselldemallorca.net; URL: www.conselldemallorca.net/biblioteques/bcart14.htm
1928; Francisco Riera Vayreda
Rare books (16-18th c), engravings
45 085 vols; 846 curr per; 7 mss; 250 music scores; 15 492 microforms; 1 628 av-mat; 407 sound-rec; 3 000 brochures   26111

Fundación Bartolomé March, Biblioteca, Conquistador, 13, 07001 *Palma de Mallorca*
T: +34 971711122; Fax: +34 971715600; E-mail: biblioteca@fundbmarch.es; URL: www.fundbmarch.es/biblio_e.htm
1970; Fausto Roldán Sierra
70 000 vols; 200 curr per; 2 000 mss; 25 incunabula; 500 microforms; 500 av-mat
libr loan   26112

Fundación Pilar y Joan Miró a Mallorca, Biblioteca, C/ Joan de Saridakis, 29, 07015 *Palma de Mallorca*
T: +34 971701420; Fax: +34 971702102; E-mail: biblioteca@fpjmiro.org; URL: miro.palmademallorca.es
1993; Aranzazu Miró
Contemporary Art
10 000 vols; 57 curr per; 3 diss/theses; 20 music scores; 180 av-mat; 270 sound-rec; 45 digital data carriers; 2 300 photos, 4 000 press clippings, 1 500 letters
libr loan   26113

Museo de Mallorca, Biblioteca Auxiliar, Portella, 5, 07001 *Palma de Mallorca*
T: +34 971717540; Fax: +34 971710483; E-mail: museudemallorca@dgcultur.caib.es
1976
8 615 vols; 315 curr per   26114

Societat Arqueológica Lulliana, Biblioteca, Mont Sió, 9, 07001 *Palma de Mallorca*
T: +34 971713912
1880; Josep Amengual i Batle Raimundus Lullus
25 000 vols; 114 curr per; 60 mss
CSIC   26115

Societat d'Història Natural de les Balears, Biblioteca, C/Margarida Xirgú 16, baixos, 07011 *Palma de Mallorca*
T: +34 971719667; Fax: +34 971719667; E-mail: biblioteca@shnb.org; URL: www.shnb.org
1950; Martin Llobera
24 000 vols; 400 curr per; 40 diss/theses; 2 govt docs; 300 maps; 40 sound-rec; 10 digital data carriers   26116

Casa de Colón, Biblioteca, Pl del Pilar, s/n, 35011 *Las Palmas* de Gran Canaria
T: +34 928-312384; Fax: +34 928331156; E-mail: casacolon@grancanaria.com; URL: www.grancanariacultura.com
1998; Candelaria Fumero Arucas
Emigration, food exchange between Europe and America, sugar cultivation; Special colls: Hist of America, Hist of the Canary Islands
35 500 vols; 400 curr per; 60 digital data carriers; videos, conference rpts, offprints, audio tapes   26117

Casa Museo Pérez Galdos, Biblioteca, C/ Cano, 2 y 6, 35002 *Las Palmas* de Gran Canaria
T: +34 928366976; Fax: +34 928373734; E-mail: perezgaldos@grancanaria.com; URL: www.casamuseoperezgaldos.com
1964
Lit
30 000 vols; 320 microforms   26118

Jardín Botánico Canario Viera y Clavijo, Sección de Documentación Científica, Biblioteca / Hemeroteca, Carretera del Centro Km 7, Apdo Correos 14 de Tafira Alta, 35017 *Las Palmas*, de Gran Canaria
T: +34 928219580; Fax: +34 928219581; E-mail: jardincanario@grancanaria.com
1975; Alfonso Luezas Hernandez
8 000 vols; 260 curr per; 7 diss/theses; 80 maps; 2 av-mat; 500 slides   26119

Museo Canario, Biblioteca, Doctor Verneau, 2, 35001 *Las Palmas* de Gran Canaria
T: +34 928336800; Fax: +34 928336801; E-mail: info@elmuseocanario.com; URL: www.elmuseocanario.com
1930; Lothar Siemens H.
Themes and authors relating to Canary Islands, Galdosiana
60 000 vols; 1 500 curr per; 10 000 mss; 32 encyclopedias, 80 dictionaries, 81 directories   26120

Archivo Real y General de Navarra, Biblioteca Auxiliar, C/ Dos de Mayo, s/n, 31001 *Pamplona*
T: +34 848424606; Fax: +34 848424611; E-mail: cultura.archivo.general@navarra.es; URL: www.navarra.es
1898; Pilar Los Arcos Sevillano
Archivística, Historia de Navarra
10 000 vols; 20 curr per
libr loan   26121

Colegio Oficial de Arquitectos Vasco-Navarro – Delegación Navarra, Biblioteca, Av del Ejército 2, 7a planta, 31002 *Pamplona*
T: +34 948206080; Fax: +34 948206073; E-mail: info@coavna.com; URL: www.coavna.com/biblioteca.asp?idioma=1
1974; Blanca Sagüés Arraiza
13 300 vols; 50 curr per; 40 maps; 2 150 av-mat   26122

Museo de Navarra, Biblioteca, Santo Domingo, 47, 31001 *Pamplona*
T: +34 948227831; Fax: +34 948211540; URL: www.cfnavarra.es/CULTURA/museo/index.html
1956; Ana Elena Redin Armañanzas
Archeology, art
11 600 vols   26123

Palacio de Peralada, Biblioteca y Archivo, Pl del Carme, s/n, 17491 *Peralada*, Gerona
T: +34 972538125; Fax: +34 972538087; E-mail: inespadrosa@castilloperalada.com
1882; Inés Padrosa
Cervantina, Books printed in Girona, Ejecutorias de Nobleza
80 000 vols; 25 curr per; 1 200 mss; 200 incunabula; 800 music scores; 120 maps; 90 microforms; 25 av-mat; 50 sound-rec; 15 digital data carriers
AAC   26124

Museo de Pontevedra, Biblioteca, Pasantería, 10, Apdo 104, 36002 *Pontevedra*
T: +34 986851455; Fax: +34 986840693; E-mail: secretaria@museo.depontevedra.es; URL: www.museo.depontevedra.es
1928; José Carlos Valle Perez
Arts, local hist, numismatics
102 000 vols; 4 010 curr per; 989 mss; 827 sound-rec; 150 872 photos, 5 632 slides   26125

Observatori de L'Ebre, Universitat Ramon Llull, Biblioteca, 43520 *Roquetes*, Tarragona
T: +34 977500511; Fax: +34 977504660; E-mail: mgenesca@obsebre.es; URL: www.obsebre.es
1904; María Genesca Sitjes
Geophysics, earth and space science; Astronomy, astrophysics, seismology, meterology, geomagnetism
55 000 vols; 2 259 curr per; 5 046 e-books; 75 diss/theses; 47 music scores; 15 microforms; 40 digital data carriers
libr loan   26126

Arxiu Històric de Sabadell, Indústria, 32-34, 08202 *Sabadell*, Barcelona
T: +34 937268777; Fax: +34 937275703; URL: www.sabadell.cat/websajsab/arxiu/
1981; Joan Comasòlivas i Font
9 600 vols; 230 curr per; 23 diss/theses; 1 400 maps; 12 000 microforms; 337 av-mat; 299 sound-rec; 9 digital data carriers; 30 000 photos
ICA, ANABAD, AAC   26127

Fundació Bosch i Cardellach, Biblioteca i Arxiu, C. Indústria, 18, 08201 *Sabadell*, Barcelona
T: +34 937258564; Fax: +34 937258564; E-mail: biblioteca@fbc.cat; URL: www.fbc.cat
1960; Magdalena Costa Vallés
Local hist
157 000 vols; 1 200 curr per; 5 000 mss; 2 diss/theses; 51 music scores; 1 500 maps; 40 000 spanish chapbooks (18th-20th c)
libr loan   26128

Casa Museo Unamuno, Biblioteca, C/ Libreros, 25, 37008 *Salamanca*
T: +34 923294400; Fax: +34 923294723; E-mail: unamuno@usal.es; URL: unamuno.usal.es/biblioteca.html
1976
Lit, personal libr of Unamuno (6 000 vols)
10 000 vols; 1 000 curr per   26129

Centro Internacional del Libro Infantil y Juvenil, Biblioteca y Centro de Documentación, Fundación German Sánchez Rupérez, Peña Primera, 14 y 16, 37002 *Salamanca*
T: +34 923269662; Fax: +34 923216317; E-mail: fgsr.salamanca@fundaciongsr.es; URL: www.fundaciongsr.es
1985; Dolores Gonzalez

Children's Libr, Youth Libr, Documentation and Research Centre on children and youth literature
99 856 vols; 236 curr per; 35 diss/theses; 218 av-mat; 6 871 sound-rec; 3 436 digital data carriers; 3 137 posters, slides
libr loan; Red Latinoamericana de Centros de Documentación de Literatura Infantil 26130

Real Instituto y Observatorio de la Armada, Ministerio de Defensa, Biblioteca, 11100 ***San Fernando***, Cádiz
T: +34 956599000, 956545599; Fax: +34 956599366; E-mail: biblio@roa.es
1753; Francisco José Gonzalez
hist archive
29 500 vols; 300 curr per; 4 incunabula; 3 485 maps 26131

Centro de Estudios Sociales de la Abadía de Santa Cruz del Valle de los Caídos, Biblioteca, Hospedería, Valle de los Caídos, 28200 ***San Lorenzo del Escorial***
T: +34 918905411 ext 112; Fax: +34 91 8905594; E-mail: scanteram@yahoo.es
1958; Santiago Cantera Montenegro
25 000 vols; 110 curr per 26132

Real Biblioteca del Monasterio, 28200 ***San Lorenzo del Escorial***, Madrid
T: +34 918903889; Fax: +34 918905421; E-mail: real.biblioteca@ctv.es
1565; José Luis del Valle Merimo
Art prints coll, Numismatic coll, Sound Arch, Drawings coll
75 000 vols; 6 000 e-journals; 5 000 mss; 650 incunabula; 2 000 Arabic, 2 090 Latin, 72 Hebrew and 580 Greek mss
libr loan 26133

Asociación para la Prevención de Accidentes, Biblioteca y Centro de Documentación, Portuetxe, 14, 20009 ***San Sebastián***
T: +34 943316203; Fax: +34 943316200; E-mail: apa@apa.es; URL: www.apa.es
1975; José Angel Ugarte Azcarate
11 000 vols; 160 curr per; 1 350 av-mat; 2 500 norms, 2 500 slides 26134

HABE Liburutegia, Biblioteca, Vitoria-Gasteiz 3, 4, 20018 ***San Sebastián***
T: +34 943022604; Fax: +34 943022601; E-mail: liburutegi@habe.org; URL: www.ikasbil.net/jetspeed/portal/media-type/html/user/anon/page/liburutegia
1982; Imanol Irizar
Multilingualism, Second Language Teaching, Sociolinguistics, Psycholinguistics, Basque Language
40 000 vols; 635 curr per; 320 diss/theses; 500 microforms; 650 av-mat; 1 500 sound-rec; 60 digital data carriers
libr loan 26135

SIIS Centro de Documentación y Estudios, Servicio Internacional de Información sobre Subnormales, C/ General Etxagüe, 10, 20003 ***San Sebastián***; Apdo de Correos 667, 20080 Donostia – San Sebastián
T: +34 943423656, 943423657; Fax: +34 943293007; E-mail: documentacion@siis.net; URL: www.siis.net
1972; Arantxa Mendieta
30 000 vols; 660 curr per; 275 e-journals; 150 av-mat
ELISAD 26136

Biblioteca Insular Especializada en Discapacidad, Sociedad Insular para la Promoción de las Personas con Discapacidad, C/ San Francisco, 102, Edif. Majona, 38001 ***Santa Cruz de Tenerife***
T: +34 922249199; Fax: +34 922244658; E-mail: biblioteca@sinpromi.es; URL: www.sinpromi.es/biblioteca
Handicapped
3 000 media items 26137

Cámara Oficial de Comercio, Industria y Navegación, Documentación y Biblioteca, Plaza de Velarde, 5 – 4a planta, 39001 ***Santander***
T: +34 942318305; Fax: +34 942314310; E-mail: teledocu-mentacion@camaracantabria.com; URL: www.camaracantabria.com/documentacion_biblioteca/biblioteca.php
María del Mar Cervera Perez
Statistics, business management
13 400 vols; 809 curr per; 6 diss/theses; 28 maps; 13 av-mat; 420 directories, 12 patents, 120 offprints 26138

Fundación Marcelino Botín, Biblioteca, Pedrueca, 1, 2a planta, 39003 ***Santander***
T: +34 942226072; Fax: +34 942360494; E-mail: biblioteca@fundacionbotin.org; URL: www.fundacionmbotin.org
1994
Modern art, music, history of Cantabria; Spanish medieval collections: Archivo Marcelino Sanz de Sautuola, Archivo Personal Federico Sopeña
20 000 media items; 297 curr per 26139

Hospital Universitario Marqués de Valdecilla, Instituto Nacional de la Salud, Biblioteca Marquesa de Pelayo, Av Valdecilla, s/n, 39008 ***Santander***
T: +34 942202539; Fax: +34 942202655
1929; Maria Francisca Ribes Cot
50 300 vols; 610 curr per; 128 diss/theses; 40 av-mat; 3 digital data carriers
libr loan; EAHIL, ANABAD, SEDIC 26140

Museo de Prehistoria y Arquelogía de Cantabria, Biblioteca, Casimiro Sainz, s/n, 39003 ***Santander***
T: +34 942207108; Fax: +34 942207106
1965
10 000 vols; 900 curr per; 6 000 av-mat; 10 000 brochures 26141

Instituto de Estudios Galegos "Padre Sarmiento", Biblioteca, Rua San Rogve 2, 15704 ***Santiago de Compostela***
T: +34 981540229; E-mail: sarmiento@cesga.es; URL: iegps-csic.blogspot.com
1944; Isabel Romaní Fariña
Microfilmoteca de doc gallega
21 000 vols; 535 curr per; 125 maps; 1 200 pamphlets 26142

Museo do Pobo Galego, Biblioteca, Santo Domingo de Bonaval, 15704 ***Santiago de Compostela***
T: +34 981583620; Fax: +34 981544840; E-mail: biblioteca@museodopobo.com; URL: www.museodopobo.es/servizos-biblioteca.php
1977; Rosa-María Méndez
Anthropology, ethnology, regional hist and social sciences; Colls: Taboada Chivite, Lorenzo Fernández y Fraguas Fraguas, Rodríguez Fraiz, Vicente Risco, Manuel Beiras, Bouza-Brey
37 000 vols; 1 450 curr per; 54 mss; 6 diss/theses; 10 music scores; 30 maps; 138 av-mat; 150 sound-rec; 100 digital data carriers 26143

Archivo General de Indias, Biblioteca (The Archive General of Indias Library), Av de la Constitución, s/n, 41004 ***Sevilla***
T: +34 954225158; Fax: +34 954219485
1933; María Teresa Molino García
Historia, arte, paleografía, archivistics, técnicas de investigación
33 120 vols; 2 126 curr per; 62 mss; 6 500 maps; 6 microforms; 130 digital data carriers 26144

Archivo Municipal, Biblioteca Auxiliar, c/ Almirante Apodaca, 6A, 41003 ***Sevilla***
T: +34 954596906; Fax: +34 954596920; E-mail: arhems@sevilla.org
1874; Rafael Cid Rodríguez
Reprografia (Reprography)
18 023 vols
AAB 26145

British Institute Library, Frederico Rubio, 12, ***Sevilla***
1947
9 000 vols 26146

Centro de Documentación de las Artes Escénicas de Andalucía, Biblioteca y Hemeroteca, C/ San Luis, 37, 41003 ***Sevilla***
T: +34 955040440; Fax: +34 955037344; E-mail: cdaea@juntadeandalucia.es; URL: www.juntadeandalucia.es/cultura/cdaea/
1992
Theatre, dance
34 000 vols; 867 curr per; 5 e-journals; 25 mss; 4 400 av-mat; 120 000 press clippings, posters, set and costume designs, photographs
libr loan; SIBMAS, ENICPA 26147

Centro de Información y Documentación Agraria, Biblioteca, Av de la Borbolla, 1, 41004 ***Sevilla***
T: +34 954410311; Fax: +34 954411888
1986
84 200 vols; 495 curr per; 12 diss/theses; 300 maps; 943 microforms; 1 030 av-mat; 983 slides, 4 900 offprints 26148

Escuela de Estudios Hispanoamericanos, Biblioteca, Alfonso XII, 16, 41002 ***Sevilla***
T: +34 954500974; Fax: +34 954224331; E-mail: bibescu@cica.es; URL: www.csic.es/cbic/hispano/hispano.htm
1942; Isabel Real Díaz
260 721 vols; 2 016 curr per; 416 maps; 130 microforms; 62 digital data carriers 26149

Fundación Casa Ducal de Medinaceli, Archivo, Pl de Pilatos, 1, ***Sevilla***
T: +34 954225055; Fax: +34 954224677
1700; Antonio Sanchez Gonzalez
arch
8 000 vols 26150

Hemeroteca Municipal de Sevilla, C/ Almirante Apodaca, 6A, 41003 ***Sevilla***
T: +34 954596-900/-to 908; Fax: +34 954596920; E-mail: arhems@sevilla.org
1932; Alfonso Braojos Garrido
Docs of the official committe Span-Am Expo of 1929; Photogr archive
25 800 vols; 30 curr per; 10 000 govt docs; 67 microforms; 100 000 av-mat 26151

Instituto Británico, Biblioteca, Federico Rubio, 14, 41004 ***Sevilla***
T: +34 954220240; Fax: +34 954501081; E-mail: info@ibsevilla.es; URL: www.ibsevilla.es
1944; Miranda Buttimore
13 000 vols; 13 curr per; 75 av-mat
libr loan 26152

Instituto de la Grasa, Consejo Superior de Investigaciones Cientifices, Biblioteca, Av Padre García Tejero 4, 41012 ***Sevilla***
T: +34 954611550; Fax: +34 954616790; E-mail: bibgrasa@cica.es; URL: www.ig.csic.es
1949; Ildefonso Martinez Fernández
8 500 vols; 152 curr per; 12 e-journals; 200 microforms; 2 digital data carriers
libr loan; CSIC 26153

Museo Arqueológico de Sevilla, Biblioteca, Pl de América, s/n, 41013 ***Sevilla***
T: +34 954232401; Fax: +34 954629542; E-mail: museoarqueologicosevilla.ccul@juntadeandalucia.es
1951; Carmen Martin Gomez
Archaeology, ancient hist, arts, museology
15 000 vols; 169 curr per; 44 av-mat; 12 digital data carriers 26154

Museo de Bellas Artes de Sevilla, Biblioteca, Plaza del Museo, 9, 41001 ***Sevilla***
T: +34 954786482; Fax: +34 954786490
1840
Coll of Ejecutorias de Hidalguía (11)
12 000 vols; 100 curr per 26155

Real Academia Sevillana de Buenas Letras, Biblioteca, C Abades, 12-14, 41004 ***Sevilla***
T: +34 954225200; Fax: +34 954225200; E-mail: SECRETARI-ARASBL@infonegocio.com; URL: www.academiasevillanadebuenasletras.org
1752; D. Antonio Collantes de Terán
13 895 vols; 262 curr per; 13 mss; 43 sound-rec 26156

Real Sociedad Económica Sevillana de Amigos del País, Biblioteca, Jesús de la Vera Cruz, 17, 41002 ***Sevilla***
T: +34 954228255
1979; Alfonso Bradios Garrido
Regional hist, Canon law
40 000 vols 26157

Archivo General de Simancas, Biblioteca Auxiliar, C/ Miravete, 8, 47130 ***Simancas***, Valladolid
T: +34 983590003; Fax: +34 983590311; URL: www.mcu.es/archivos/MC/AGS/index.html
1864; Jaime Sainz Guerra
21 500 vols; 686 curr per; 8 mss; 142 diss/theses; 123 maps; 60 sound-rec 26158

Museo Numantino, Biblioteca, Paseo del Espolón, 8, 42001 ***Soria***
T: +34 975221397; Fax: +34 975229872; E-mail: museo.soria@jcyl.es
1919; José L. Argente O.
Arqueología y Arte
10 300 vols; 950 curr per 26159

Biblioteca Azcona, Pl T. Cortes, 13, 31300 ***Tafalla***, Navarra
T: +34 948700053
Hist, heráldica, biografías, prensa
157 000 vols; 512 curr per; 5 diss/theses
libr loan 26160

Museu Nacional Arqueològic de Tarragona, Biblioteca, Av Ramón y Cajal, 82, 43005 ***Tarragona***
T: +34 977251515; Fax: +34 977252286; E-mail: biblioteca@mnat.es; URL: www.mnat.es
Anna Beltran Vallverdú
11 000 vols; 500 curr per; 33 mss; 25 digital data carriers 26161

Reial Societat Arqueologica Tarraconense, Biblioteca, Carrer Mayor, 35 pral., 43003 ***Tarragona***; Apdo 573, 43080 Tarragona
T: +34 977233789; Fax: +34 977233789; E-mail: informacio@arqueologica.org; URL: www.arqueologica.org/biblioteca.asp
Iberian, Roman, early Christian archeology
9 600 vols 26162

Instituto de Estudios Turolenses, Biblioteca, Apartado de correos 77, 44080 ***Teruel***
T: +34 978617860; Fax: +34 978617861; E-mail: ieturolenses@dpteruel.es
1983; Visitación García Valero
39 539 vols; 600 curr per; 125 diss/theses; 134 maps; 6 591 microforms; 10 103 av-mat; 103 sound-rec; 29 digital data carriers
libr loan 26163

Museo Provincial de Teruel, Diputación Provincial de Teruel, Biblioteca, Pl Fray Anselmo Polanco, s/n, 44001 ***Teruel***
T: +34 978600150; Fax: +34 978602832; URL: www.dpteruel.es/museo.htm
1956; Ana Andrés Hernández
Sotoca coll, coll from the Chemist of Alcala de la Selva, museology, construction and restoration
15 719 vols; 158 curr per; 21 mss; 99 diss/theses; 1 175 maps; 2 microforms; 115 av-mat; 20 sound-rec; 47 digital data carriers; 11 000 slides, 3 900 photos 26164

Sinagoga del Tránsito, Biblioteca, San Juan de Dios, 13, 45002 ***Toledo***
T: +34 925223665; Fax: +34 925215831; E-mail: museo@msefardi.mcu.es; URL: www.museosefardi.mcu.es
1964
Judaism, Hebraism, Sephardism
15 000 vols; 35 curr per; 50 microforms; 1 000 av-mat; 200 sound-rec; 40 digital data carriers 26165

Instituto Nacional de Técnica Aeroespacial (INTA), Biblioteca (National Institute of Aerospace Technology), Ctra de Ajalvir, km 4, 28850 ***Torrejón de Ardoz***, Madrid
T: +34 91520-1831/-2169; Fax: +34 915202049; E-mail: biblioteca@inta.es; URL: biblioteca.inta.es
1944; Eliseo Eliseo de la Fuente
45 000 vols; 122 curr per; 4 500 microforms; 1 506 av-mat 26166

Archivo del Reino de Valencia, Biblioteca Auxiliar, Alameda, 22, 46010 ***Valencia***
T: +34 963184550; Fax: +34 963184527; E-mail: arv@gva.es; URL: dglab.cult.gva.es/ArxiuRegne/index.htm
1930; Maria Cruz Farfán Navarro
9 800 vols; 250 curr per 26167

Biblioteca Musical de Compositores Valencianos, Pl de Maguncia, 1, 46018 ***Valencia***
T: +34 963525478 ext 4580, 4588, 4534; Fax: +34 963501321; E-mail: bcentral@valencia.es; URL: www.valencia.es
1962; Belén Gisbert Aguilar
Scores coll
11 453 vols; 5 curr per; 27 microforms; 412 sound-rec; 80 digital data carriers 26168

Centro de Estudios Norteamericano, Biblioteca, c& Aparisi y Guijarro 5-1, 46003 ***Valencia***
T: +34 963911573; Fax: +34-963911986; E-mail: infoamericana@hotmail.com; URL: www.casaamericana.org
1956; Pilar Ribelles
English lit
15 000 vols; 2 curr per; 25 maps 26169

Colegio Territorial de Arquitectos de Valencia, Biblioteca, Hernan Cortés, 19, 46004 ***Valencia***
T: +34 963510430; E-mail: biblioteca@ctav.es; URL: www.ctav.es
1941; Nestor Mir Planells
26 000 vols; 150 curr per; 24 diss/theses; 800 av-mat; 60 digital data carriers
ABBA 26170

Hemeroteca Municipal de Valencia, Pl de Maguncia, 1, 46018 ***Valencia***
T: +34 963525478 ext 4582; Fax: +34 963501321; E-mail: hemerotecam@valencia.es
1902; Isabel Guardiola Sellés
5 507 curr per; 47 microforms; 16 digital data carriers; 1 845 digitized periodicals belonging to the Navarro Cabanes collection (+ 42 other titles) 26171

Ilustre Colegio de Abogados de Valencia, Biblioteca, Glorieta, s/n, 46003 ***Valencia***
T: +34 963941880; E-mail: icav@icav.es; URL: www.icav.es
1759
9 000 vols; 142 curr per 26172

Instituto Valencià d'Art Modern (IVAM), Centro Julio Gonzalez, Biblioteca, Guillem de Castro, 118, 46003 ***Valencia***
T: +34 963863000; Fax: +34 963921094; E-mail: biblioteca@ivam.es; URL: www.ivam.es
1989; Victoria Goberna
Artists' arch
28 000 vols; 108 curr per; 258 sound-rec; 72 digital data carriers
libr loan; IFLA 26173

Museo de Bellas Artes de Valencia, Biblioteca (Museum of Fine Arts of Valencia), C/ San Pio V, 9, 46010 ***Valencia***
T: +34 963870315, 963870316; Fax: +34 963870301; E-mail: alfaro_ana@gva.es; URL: www.pre.gva.es/mbbaav/tlpmba.html
1984; Ana Alfaro, Rosa Rodríguez
25 179 vols; 848 curr per; 697 digital data carriers; 7 892 photo, 2 476 posters, 44 003 pamphlets, 85 760 press clippings
libr loan; UKIC, IIC, IPC, ARLIS 26174

Museo Nacional de Cerámica y Artes Suntuarias "González Martí", Biblioteca, Poeta Querol, 2, 46002 ***Valencia***
T: +34 963516392; Fax: +34 963513512; E-mail: informacion.mceramica@mcu.es; URL: mnceramica.mcu.es/biblioteca.html
1976; Teresa Estrela Castillo
Colección Mrio Blasco (4 500 vols on lit and hist), Colección Arrojo Muñoz (500 vols of ex libris plates), decorative arts, ex libris, hist
20 000 vols; 1 000 curr per; 85 mss; 10 diss/theses; 50 maps; 3 020 microforms; 687 av-mat
libr loan 26175

Museu de Prehistòria de València, Biblioteca, C/ Corona, 36, 46003 ***Valencia***
T: +34 963883600; Fax: +34 963883536; E-mail: bibliotecasip@dival.es; URL: www.museuprehistoriavalencia.es/biblioteca.html
1927; Consuelo Martín
Prehistory, Early Hist, Anthropology, Ethnology
60 000 vols; 580 curr per; 87 diss/theses; 200 maps; 4 microforms; 30 digital data carriers
IFLA 26176

Servicio de Investigación Prehistórica, Diputación Provincial, Biblioteca, C de la Corona, 36, 46003 ***Valencia***
T: +34 963883600; Fax: +34 3883536; E-mail: bibliotecasip@dva.gva.es
1927; Consuelo Martin
53 569 vols; 500 curr per; 87 diss/theses; 200 maps; 4 microforms; 30 digital data carriers
IFLA 26177

Archivo Histórico Provincial de Valladolid, Biblioteca Auxiliar, Avenida Ramon y Cajal 1, 47011 ***Valladolid***
T: +34 983255385; Fax: +34 983255385; E-mail: lasbalan@jcyl.es
1932; Angel Laso Ballesteros
11 764 vols; 194 curr per; 1 191 diss/theses; 7 maps; 15 av-mat; 66 digital data carriers 26178

Instituto Superior de Filosofía, Padres Domínicos, Biblioteca, Pl de San Pable, 4, 47011 ***Valladolid***
T: +34 983356699; E-mail: amcasado@dominicos.org; URL: www.dominicos.org/op/provinciaesp/estudiosisfil.htm
1970; Ignacio Erviti Yaben
708 000 vols; 1 250 curr per; 15 mss; 200 diss/theses; 450 microforms; 2 300 rare books (16-18th c) 26179

Museo Nacional Colegio de San Gregorio, Antiguo Museo Nacional de Escultura, Biblioteca, C/ Cadenas de San Gregorio, 1-3, 47011 ***Valladolid***
T: +34 983250375; Fax: +34 983259300; E-mail: mne@mcu.es; URL: museoescultura.mcu.es
8 000 vols; 60 curr per; 3 750 slides 26180

Centro Nacional de Educación Ambiental (CENEAM), Centro de Documentación, Paseo José María Ruiz-Dana, s/n, 40109 ***Valsain*** (Segovia)
T: +34 921473869; Fax: +34 921471746
1994
24 000 media items; 297 curr per 26181

Fundación Penzol, Biblioteca y Archivo Penzol, Casa da Cultura. Pza da Constitución, 2, 36202 ***Vigo***
T: +34 986226459; Fax: +34 986226459
1963; Francisco Fernandez del Riego
35 800 vols; 200 curr per; 90 mss; 140 diss/theses; 270 maps 26182

Museo 'Balaguer', Biblioteca, Av Victor Balaguer, s/n, 08800 ***Villanueva y Geltru***, Barcelona
T: +34 938154202; Fax: +34 938153684
1884; Montserrat Comas i Güell
50 000 vols; 1 600 curr per; 1 014 mss; 18 incunabula; 200 music scores; 750 maps; 4 000 ex-libris, 10 000 engravings, 10 000 photos, 5 000 menus 26183

Archivo Municipal, Ayuntamiento de Vitoria-Gasteiz, Biblioteca, Paseo de la Universidad, 1, 01005 ***Vitoria-Gasteiz***, Alava
T: +34 945161492; Fax: +34 945161507; E-mail: archivo.0107@vitoria-gasteiz.org
1850; María Pilar Arostegui Santiago
55 848 vols; 345 curr per; 39 mss; 3 microforms; 3 579 av-mat; 53 sound-rec 26184

Artium, Centro Museo Vasco de Arte Contemporáneo, Biblioteca, C/ Francia, 24, 01002 ***Vitoria-Gasteiz***
T: +34 945209000; Fax: +34 945209049; E-mail: museum@artium.org; URL: www.artium.org/museo_biblioteca_i.html#
Modern art, architecture, performing art, cinematography, design, comics, photography, catalogues
37 000 media items 26185

Fundación Sancho el Sabio, Centro de Documentación de la Cultura Vasca, Portal de Betoño, 23, 01013 ***Vitoria-Gasteiz***
T: +34 945253932; Fax: +34 945250219; E-mail: fs-inv@fsancho-sabio.es; URL: www.fsancho-sabio.es
1958; Carmen Gómez Pérez
Basque Country
59 000 vols; 7 284 curr per; 1 510 mss; 780 diss/theses; 1 200 maps; 1 240 microforms; 120 av-mat; 1 180 digital data carriers; 700 newspapers, 20 000 pamphlets, 14 000 photos 26186

Instituto Vasco de Administración Pública (IVAP), Biblioteca, Donostia-San Sebastiàn, 1, 01010 ***Vitoria-Gasteiz***
T: +34 945077641; E-mail: liburuteqia@ivap.es; URL: www.ivap.euskadi.net/
1982
Law, economics, political sciences, translation, Basque language, human resources
24 823 vols; 462 curr per; 40 e-journals; 40 sound-rec 26187

Museo de Bellas Artes de Alava, Diputación Foral de Alava, Biblioteca, Fray Francisco, 8, 01007 ***Vitoria-Gasteiz***, Alava
T: +34 945181918; Fax: +34 945181919; E-mail: museobellasartes@alava.net
1987; Sara González de Aspuru Hidalgo
12 400 vols; 117 curr per; 197 av-mat; 30 digital data carriers 26188

Biblioteca Moncayo, Mayor 62, Jarque, ***Zaragoza***
1972; Luis Marquina Marín
Aragonese authors, Aragon
15 000 vols 26189

Centro de Investigación y Tecnología Agroalimentaria de Aragón, Gobierno de Aragón (CITA), Biblioteca (Agrifood Research and Technology Centre of Aragon), Av Montañana, 930, 50059 ***Zaragoza***
T: +34 976716342; Fax: +34 976716343; E-mail: biblioteca.cita@aragon.es; URL: www.cita-aragon.es
1965; Marta Carracedo Martínez
Economics, horticulture, agriculture; Special depts: Fruticultura; Economía Agraria; Recursos Forestales; Producción Vegetal; Sanidad Vegetal; Producción Animal; Sanidad Animal; Suelos y Riegos; Calidad; Seguridad Alimentaria
21 653 vols; 1 055 curr per; 500 diss/theses; 24 000 microforms; 184 digital data carriers; 3 800 separatas, 6 900 publicaciones seriadas
libr loan 26190

Centro Pignatelli, Biblioteca, Paseo de la Constitución, 6, 50008 ***Zaragoza***
T: +34 976217217; Fax: +34 976230113; E-mail: biblioteca@centro-pignatelli.org; URL: www.centro-pignatelli.org/biblioteca.htm
Theology, education
50 000 vols; 73 curr per 26191

Estación Experimental de Aula Dei (CSIC), Consejo Superior de Investigaciones Científicas (CSIC), Unidad de Apoyo de Biblioteca y Documentación, Montañana, 177, Apdo 202, 50080 ***Zaragoza***
T: +34 976716047; Fax: +34 976716145; E-mail: bib_aula@eead.csic.es; URL: www.eead.csic.es/es/biblioteca_documentacion.html
1944; J. Carlos Martinez Gimenez
Agriculture, biology, botany
47 000 vols; 800 curr per
libr loan; IAALD 26192

Institución 'Fernando el Católico' de la Diputación Provincial, Biblioteca, Pl de España, 2, 50071 ***Zaragoza***
T: +34 976288879; Fax: +34 976288869; E-mail: biblioteca@ifc.dpz.es; URL: ifc.dpz.es
1943; Elvira Solano
85 900 vols; 2 400 curr per 26193

Museo de Zaragoza, Biblioteca, Pza de los Sitios, 6, 50001 ***Zaragoza***
T: +34 976222181, 976225682; Fax: +34 976222378; E-mail: museoza@aragon.es
1911; Maria Jesús Dueñas Jiménez
15 989 vols; 897 curr per; 1 incunabula; 350 diss/theses 26194

Museo e Instituto de Humanidades Camón Aznar, Caja de Ahorros de Zaragoza, Aragón y Rioja, Biblioteca, Espoz y Mina, 23, 50003 ***Zaragoza***
T: +34 976215351
1980; Pilar Camon Alvarez
Decor arts, fine arts
15 000 vols; 905 curr per; 20 diss/theses; 2 000 separates, 3 000 cat, 1 100 reference books, 5 000 offprints, 10 encyclopedias, 69 dictionaries 26195

Museo Pablo Gargallo, Biblioteca, Pl de San Felipe, 3, 50003 ***Zaragoza***
T: +34 976724922; Fax: +34 976392076
1982
Gargallo, sculpture
15 000 vols 26196

Public Libraries

Biblioteca Pública del Estado, San José de Calasanz, 14, 02002 ***Albacete***
T: +34 967590342; Fax: +34 967238209; E-mail: jdem0015@olmo.pntic.mec.es; URL: olmo.pntic.mec.es/~bpab/
1882; Juan Manuez de la Cruz Munoz
146 314 vols; 1 245 curr per; 9 mss; 6 incunabula; 89 music scores; 43 maps; 13 725 microforms; 2 406 av-mat; 4 063 sound-rec; 603 digital data carriers; 1 500 bks on local subjects, 3 000 rare bks (16-19th c)
libr loan 26197

Biblioteca Pública del Estado en Alicante "José Martínez Ruíz, 'Azorín", Paseíto de Ramiro, 15, 03002 ***Alicante***
T: +34 965206600; Fax: +34 965143658; E-mail: bpea@gva.es
1855
80 400 vols; 1 846 curr per; 50 mss; 1 incunabula; 85 av-mat; 4 800 sound-rec; 283 rare bks (16-18th c) 26198

Biblioteca Pública Provincial de Almería "Francisco Villaespesa", C/ Hermanos Machado, s/n, 04004 ***Almería***
T: +34 950175568; Fax: +34 950175584; E-mail: informacion.bp.al.ccul@juntadeandalucia.es
1947; Eudaldo Furtet Cabana
Local Coll
75 400 vols; 1 185 curr per; 4 mss; 22 diss/theses; 59 music scores; 250 maps; 1 298 microforms; 7 963 av-mat; 781 sound-rec; 170 digital data carriers
libr loan 26199

Biblioteca Pública Municipal de Alzira, Escoles Pies, 6, 46600 ***Alzira***, Valencia
T: +34 962417407; Fax: +34 962415756
1987
175 000 vols; 9 curr per 26200

Biblioteca Popular 'P. Fidel Fita' (Public Library), C/ Bonaire, 2, 08350 ***Arenys de Mar*** (Barcelona)
T: +34 937923253; Fax: +34 937922856; E-mail: b.arenysma.pff@diba.cat; URL: www.arenysdemar.cat/biblioteca
1957; Mercè Cussó i Cervera
Felix Cucurull, Salvador Espriu, Lluís Ferran de Pol, Libres en Braille, Esyllt T. Lawrence and other local authors
73 939 vols; 300 curr per; 5 diss/theses; 14 music scores; 160 maps; 5 552 microforms; 1 384 av-mat; 5 198 sound-rec; 450 digital data carriers; 9 550 posters
libr loan 26201

Biblioteca Publica 'Francesca Bonnemaison', Sant Pere més baix, 7, 08003 ***Barcelona***
T: +34 932680107; Fax: +34 933106401; E-mail: b.barcelona.fb@diba.es; URL: www.diba.es/biblioteques
1909; Anna Cabó
Fashion, women's studies, gastronomy
57 000 vols; 170 curr per 26202

Servei de Biblioteques, Diputació de Barcelona, Urgell, 187, 08036 ***Barcelona***
T: +34 934022545; Fax: +34 934022488; E-mail: s.biblioteques@diba.es; URL: www.diba.es/biblioteques
1918; Assumpta Bailac Puigdellívol
libr loan; IFLA 26203

Red de Bibliotecas Municipales de Bilbao, Biblioteca Central de Bidebarrieta, Bidebarrieta, 4, 48005 ***Bilbao***
T: +34 944156930; Fax: +34 944156438; E-mail: bidebarrieta@ayto.bilbao.net; URL: www.bilbao.net/biblioteca
1877; Felisa Sanz
Coll J. C. de Arriaga
306 965 vols; 1 720 curr per; 34 diss/theses; 1 404 music scores; 488 maps; 125 microforms; 17 832 av-mat; 22 149 sound-rec; 272 digital data carriers; 1 310 other items
libr loan 26204

Biblioteca Pública de Burgos, C/ Valladolid, 3, 09002 ***Burgos***
T: +34 947256419; Fax: +34 947202877; E-mail: bpburgos@jcyl.es
1871; María del Carmen Monje Maté
119 700 vols; 1 036 curr per; 81 mss; 119 incunabula; 15 maps; 889 microforms; 29 digital data carriers
libr loan 26205

Biblioteca Pública del Estado 'A. Rodríguez Moñino/M. Brey', c. Alfonso IX, 26, 10004 ***Cáceres***
T: +34 927627114; Fax: +34 927627121
1983; Ma Jesús Santiago Fernández
Regional and local coll, several personal legates
94 132 vols; 1 575 curr per; 88 mss; 10 incunabula; 80 maps; 14 719 microforms; 21 673 av-mat; 4 386 sound-rec; 1 134 digital data carriers; 7 141 rare bks (16-18th c)
libr loan 26206

Biblioteca Centro Cultural 'Ramon Alonso Luzzy', C/ Jacinto Benavente, 7, 30203 ***Cartagena***
T: +34 968128858; Fax: +34 968128853; E-mail: biblio@ayto-cartagena.es; URL: www.ayto-cartagena.es
1943; Maria Jose Mercader Blanco
280 400 vols; 213 curr per; 1 240 sound-rec; 388 digital data carriers 26207

Biblioteca Pública, Rafalafena 29, 12003 ***Castellón***
T: +34 96224309; Fax: +34 96236557; E-mail: bpec@gva.es
1848; Rosa Maria Diaz Salvador
Biblioteca Hemeroteca Fundacio' 1on Huguet
100 800 vols; 112 mss; 63 incunabula; 117 music scores; 20 maps; 1 248 microforms; 1 725 av-mat; 2 700 sound-rec; 470 digital data carriers; 4 699 pamphlets, 2 754 posters
libr loan; ANABAD 26208

Biblioteca Pública de Ciudad Real, C/ Prado, 10, 13002 ***Ciudad Real***
T: +34 926223310; Fax: +34 926231877; E-mail: bpcr@jccm.es; URL: www.bibliotecaspublicas.es/ciudadreal/index.jsp
1961
Bibliography of authors of the provinces of Ciudad Real and Castilla-La Mancha
64 400 vols; 350 curr per; 1 mss; 14 incunabula; 459 music scores; 14 300 av-mat; 750 sound-rec
libr loan; ANABAD 26209

Biblioteca Provincial de Córdoba, C/ Amador de los Ríos, s/n, 14004 ***Córdoba***
T: +34 957355500; Fax: +34 957355499; E-mail: direccion.bp.co.ccul@juntadeandalucia.es
1845; Francisco Javier del Río del Río
Departemento de Servicios Bibliotecarios Provinciales
108 571 vols; 608 curr per; 174 mss; 79 incunabula; 10 music scores; 48 maps; 117 microforms; 935 av-mat; 4 804 sound-rec; 640 digital data carriers; 3 000 posters, 1 556 slides
libr loan 26210

Biblioteca Pública Miguel González Garcés, C/ Miguel González Garcés 1, 15008 ***A Coruña***
T: +34 981170218; Fax: +34 981170218; E-mail: bpc@olmo.pntic.mec.es
1902; Laura González-Garcés Santiso
Legal depository
143 959 vols; 881 curr per; 1 mss; 108 music scores; 51 maps; 1 microforms; 2 505 av-mat; 2 368 sound-rec; 402 digital data carriers; 7 500 rare bks (16-18th c) and vols on local subjects
libr loan 26211

Diputación Provincial de a Coruña, Biblioteca, C/ Riego de Aguas, 37, 15001 ***A Coruña***
T: +34 981183342; Fax: +34 981183344; E-mail: biblioteca.info@dicoruna.es; URL: www.bibliotecaspublicas.es/bp-diputacion-coruna/
1845; Dolores Liaño Pedreira
Printed music (Fin S.XIX-Pp. S.XX): Colección Canuto Berea; Fondo Antiguo de Galicia; Baby's Libr, Music Libr
120 823 vols; 696 curr per; 17 mss; 32 000 music scores; 100 maps; 38 738 microforms; 5 370 av-mat; 3 297 sound-rec; 136 digital data carriers
libr loan; AIBM 26212

Biblioteca Pública del Estado 'Fermín Caballero', Glorieta González Palencia, 1, 16071 ***Cuenca***
T: +34 969241524; Fax: +34 969231244; E-mail: bp.cuenca@jccm.es
1846; Begoña Marlasca Gutiérrez
Donación Angel Blanc Soler (2 753 vols), Donación García Santiago (399 vols), Donación Rogelio Sanchiz (4 000 monografías y 109 pub.periódicas)
116 951 vols; 492 curr per; 10 mss; 1 incunabula; 10 820 av-mat; 2 776 sound-rec; 755 digital data carriers
libr loan; ANABAD 26213

Biblioteca Pública Municipal, Jovellanos, 21, 33205 ***Gijón***
T: +34 985181062; E-mail: bibliotecagijonsur@gijon.es; URL: www.bibliotecaspublicas.es/redgijon/index.jsp
1897; Ana Rodriguez Navarro
123 000 vols; 258 curr per; 20 mss; 3 incunabula; 528 maps; 3 170 av-mat; 2 872 sound-rec; 3 digital data carriers
libr loan 26214

Biblioteca Pública de Girona, Pl Hospital, 6, 17002 ***Girona***
T: +34 972202252; Fax: +34 972227695; E-mail: bpgirona.cultura@gencat.cat
1848; Teresa García Panadès
Fondos Antiguos de Monasterios Suprimidos
133 000 vols; 334 curr per; 153 mss; 168 incunabula; 217 maps; 8 microforms; 1 636 av-mat; 48 sound-rec; 18 332 postcards and calendars
libr loan 26215

Biblioteca Provincial de Granada, C/ Profesor Sainz Cantero, 6, 18002 ***Granada***
T: +34 958575650; Fax: +34 958575661; E-mail: informacion.bp.gr.ccul@juntadeandalucia.es
1933; Rosario Corral Quintana
68 000 vols; 305 curr per; 2 900 av-mat; 4 200 sound-rec; 450 digital data carriers
libr loan; AAB, AAD, IFLA 26216

Biblioteca Can Pedrals, C/ Espi i Grau, 2, 08400 ***Granollers***
T: +34 938793091; Fax: +34 938793296; E-mail: b.granollers.cp@diba.es; URL: www.diba.es/biblioteques/Guia/municipis/granollers/granollers1.htm
1995; Mercè Gasch i Rindor
Rare bks, Local coll
63 969 vols; 358 curr per; 222 mss; 43 music scores; 11 maps; 3 674 sound-rec; 824 digital data carriers; 1 829 video recordings, 264 DVDs
libr loan 26217

Biblioteca Pública del Estado en Guadalajara, Plaza de Dávalos s/n, 19001 ***Guadalajara***
T: +34 949234710; Fax: +34 949234723; E-mail: bpgu@jccm.es; URL: www.bibliotecaspublicas.es/guadalajara/index.jsp
1837; Blanca Calvo
102 500 vols; 1 520 curr per; 54 mss; 21 incunabula; 490 maps; 17 500 microforms; 2 067 av-mat; 5 003 sound-rec
libr loan; IFLA 26218

Biblioteca Pública del Estado, Av Martín Alonso Pinzón 16, 21003 ***Huelva***
T: +34 959247262; Fax: +34 959540148; E-mail: informacion.bp.hu.ccul@juntadeandalucia.es; URL: www.juntadeandalucia.es/cultura/opencms/export/bibliotecas/bibhuelva/
1957; Fernando M. González Ramón
51 502 vols; 318 curr per; 18 music scores; 141 maps; 12 microforms; 4 532 av-mat; 1 649 sound-rec; 418 digital data carriers; 1 395 other items
libr loan 26219

Biblioteca Pública del Estado, Av de los Pirineos, 2, 22004 ***Huesca***
T: +34 974239020; Fax: +34 974239021
1845; Rosario Fraile Gayarre
24 200 rare bks (16th-18th c), 2 265 vols on local and regional subjects
80 000 vols; 110 curr per; 135 mss; 130 incunabula; 12 227 microforms; 800 av-mat; 2 200 sound-rec; 200 digital data carriers
libr loan 26220

Biblioteca Central d'Igualada, Plaça de Cal Font, 08700 ***Igualada***
T: +34 938049077; Fax: +34 938043797; E-mail: b.igualada.c@diba.es; URL: www.aj-igualada.net; www.diba.es/biblioteques
1999; M. Teresa Miret
Local Coll
65 950 vols; 150 curr per; 5 e-journals; 1 970 av-mat; 2 838 sound-rec; 565 digital data carriers
libr loan; Xarxa de Biblioteques de la Diputació de barcelona 26221

Biblioteca Pública Provincial de Jaén, C/ Santo Reino, 1, 23003 ***Jaén***
T: +34 953313419; Fax: +34 953313403; E-mail: informacion.bp.ja.ccul@juntadeandalucia.es
1896; Fernando González Ramón
90 000 vols; 48 curr per; 6 mss; 2 incunabula; 1 054 av-mat; 3 198 sound-rec
libr loan 26222

Biblioteca Pública Municipal Central, Pl del Banco, 7-8, ***Jerez de la Frontera***
T: +34 956323300; Fax: +34 956323402
1873; Ramón Clavijo Provencio
Horses, bullfighting and flamenco
75 000 vols; 30 curr per; 125 mss; 1 incunabula; 107 av-mat; 200 sound-rec 26223

Biblioteca Pública del Estado en León, C/ Santa Nonia, 5, 24003 ***León***
T: +34 987206710; Fax: +34 987203025; E-mail: dieescal@jcyl.es
1844; Alfredo Diez Escobar
Codices, incunables, mss (16th c)
162 000 vols; 563 curr per; 27 mss; 22 incunabula; 143 diss/theses; 260 maps; 4 470 sound-rec
libr loan 26224

Biblioteca Pública, Rambla d'Aragó, 10, 25002 ***Lleida***
T: +34 973279070; Fax: +34 973279083; E-mail: bplleida.cultura@gencat.net; URL: cultura.gencat.net/bpl
1842; Carmen Ariche Axpe
109 552 vols; 1 493 curr per; 6 mss; 25 incunabula; 305 music scores; 19 844 microforms; 5 280 av-mat; 4 342 sound-rec; 867 digital data carriers; 2 Dq, 8 511 rare bks
libr loan 26225

Biblioteca Pública del Estado en Lugo, Av de Ramón Ferreiro, s/n, 27002 ***Lugo***
T: +34 982228525; Fax: +34 982228085; E-mail: consuelo.meirino.sanchez@xunta.es
1840; Consuelo Meiriño Sánchez
Col Montenegro, Col Menacho, Sección Braille (216 vols)
78 000 vols; 193 curr per; 25 mss; 4 incunabula; 525 av-mat; 640 sound-rec; 3 000 slides, 26 cat, 17 262 rare bks (16-18th c), 3 000 vols on local subjects 26226

Biblioteca Pública Moratalaz, C/ Corregidor Alonso Tobar, 5, 28030 ***Madrid***
T: +34 914394688; Fax: +34 914370234; E-mail: bib.moratalaz@madrid.org
1968
55 600 vols; 36 curr per; 5 230 av-mat; 3 244 sound-rec; 1 099 digital data carriers
libr loan 26227

Biblioteca Pública Retiro, C/ Doctor Esquerdo, 189, 28007 ***Madrid***, Retiro
T: +34 915019146; Fax: +34 915018130; E-mail: bib.retiro@madrid.org
1990
84 000 vols; 38 curr per; 8 313 av-mat; 6 468 sound-rec; 1 991 digital data carriers 26228

Biblioteca Pública Usera José Hierro, Av Rafaela Ybarra, 43, 28041 ***Madrid***, Usera
T: +34 914229501; Fax: +34 914691901; E-mail: bib.usera@madrid.org
1974
89 500 vols; 40 curr per; 7 893 av-mat; 6 596 sound-rec; 1 987 digital data carriers 26229

Bibliotecas Públicas de Madrid, Felipe el Hermoso, 4, 28010 ***Madrid***
T: +34 914459845, 914459782; Fax: +34 915940469
1915; Elena Sáizar Andrés
Books from the 19th c
690 000 vols; 900 curr per
libr loan; IFLA, INTAMEL, ANABAD, IBBY 26230

Organización Nacional de Ciegos, Centro Bibliografico y Cultural, Biblioteca Central, C/ La Coruña, 18, 28020 ***Madrid***
T: +34 915894214; Fax: +34 915894225; E-mail: soi@once.es; URL: www.once.es
1965; Alberto Daudén Tallaví
Documentation and Translation Service (3 400 vols), Research Libr on Blindness
56 969 vols; 2 curr per; 75 diss/theses; 100 govt docs; 3 640 music scores; 111 av-mat; 11 029 sound-rec; 55 digital data carriers; 39 559 are braille bks, 31 398 are inkprints and 5 173 are braille music sheets
libr loan; IFLA 26231

Biblioteca Pública del Estado, Av Europa, 49, 29003 ***Málaga***
T: +34 952344944; Fax: +34 952344972; E-mail: bpma.pi@olmo.pntic.mec.es
1895; Dr. José Salinero Portero
86 029 vols; 409 curr per; 32 mss; 4 incunabula; 6 400 av-mat; 1 121 sound-rec; 126 digital data carriers; 341 encyclopedias, 333 dictionaries
libr loan 26232

Cánovas del Castillo, Biblioteca, C/ Ollerías, s/n, 29012 ***Málaga***
T: +34 952133936; Fax: +34 952133966; URL: www.malaga2016.es/frmenu25b.php?instit=144
1950; María Sánchez García-Camba
Historia local y regional (de Málaga y Andalucía); Legado Temboury, especializado en historia y arte malagueños y civilización árabe. Consta de un archivo de notas manuscritas y mecanografiadas; biblioteca (1 581 vols) y archivo fotográfico (11 864)
57 571 vols; 1 177 curr per; 79 mss; 5 govt docs; 27 maps; 6 029 microforms; 3 400 av-mat; 781 sound-rec; 887 digital data carriers; 3 215 pamphlets, 11 864 photos, 10 electronic publs
libr loan 26233

Biblioteca Pública de Maó, Plaça Conquista, 8, 07701 ***Maó***, Illes Balears
T: +34 971369190; Fax: +34 971353717
1861; Juan Francisco Sanchez Nistaz
100 000 vols; 416 curr per; 70 mss; 85 incunabula; 5 727 microforms; 962 av-mat; 690 sound-rec; 5 digital data carriers
libr loan 26234

Biblioteca Popular de la Caixa d'Estalvis Laietana, Santa Teresa, 63, 08302 ***Mataró***, Barcelona
T: +34 937901572
1929
81 100 vols; 149 curr per 26235

Biblioteca Municipal, Tomas Perez, 3, 38300 ***La Orotava***, S.C. Tenerife
T: +34 922335804; E-mail: biblioteca@villadelaorotava.org
1985; Francisco Negrin de Ponte
50 000 vols 26236

Biblioteca Pública del Estado en Orense, Concello, 11, 32003 ***Ourense***
T: +34 988210700; Fax: +34 988510250; E-mail: bpou@olmo.pntic.mec.es
1847; Elisa Isabel López Rodríguez
Bibl Amor (specialized in Esperanto)
70 200 vols; 506 curr per; 18 av-mat; 260 sound-rec; 1 264 rare bks (16-19th c)
libr loan 26237

Biblioteca Pública del Estado en Palencia, C Eduardo Dato, 4, 34005 ***Palencia***
T: +34 979751100; Fax: +34 979751121; E-mail: izqberjo@jcyl.es
1936; José María Izquierdo Bertiz
Local studies
122 700 vols; 258 curr per; 3 mss; 24 incunabula; 119 maps; 1 764 microforms; 19 432 av-mat; 4 969 sound-rec; 7 digital data carriers; 2 000 pamphlets
libr loan; ANABAD 26238

Biblioteca Pública de l'Estat a Palma – Can Sales, Plaça Porta Santa Catalina, 24, 07012 ***Palma de Mallorca***
T: +34 971715239; Fax: +34 971715239; E-mail: bpcansales@bibliotecapalma.com
1847; María Lluch Alemanyy Mir
137 000 vols; 255 curr per; 1 388 mss; 718 incunabula; 114 av-mat
libr loan; ANABAD 26239

Biblioteca Insular, Cabildo Insular de Gran Canaria, Pl Hurtado de Mendoza, 3, 35002 *Las Palmas* de Gran Canaria
T: +34 928382672; Fax: +34 928364528; E-mail: bibliotecainsulargc@ grancanaria.com
1968
Fondo particular de Antonio Ballesteros (Hist), Fondo Fernando González (Lit), Fondo Orleans de Música, Fondo Canarias
100 000 vols; 51 curr per; 40 mss; 29 music scores; 7 000 sound-rec
libr loan   26240

Biblioteca Pública del Estado en Las Palmas, C/ Muelle de Las Palmas s/n, 35003 *Las Palmas* de Gran Canaria
T: +34 928432343; Fax: +34 928431019; E-mail: bibliolp.ceucd@ gobiernodecanarias.org; URL: www.bibliotecaspublicas.es/laspalmas/
1967
Colección de Nodos
110 000 vols; 53 curr per; 920 microforms; 544 av-mat; 1 900 sound-rec
libr loan   26241

Biblioteca Municipal de Ponferrada, Gran Vía Reino de León, 22, 24400 *Ponferrada*, León
T: +34 987412370; E-mail: biblioteca4@ponferrada.org; URL: www.bibliotecaspublicas.es/ponferrada/index.jsp
1955
53 100 vols; 328 curr per; 1 500 sound-rec
ANABAD   26242

Biblioteca Publica del Estado, C/ Alfonso XIII, 3, 36002 *Pontevedra*
T: +34 986850838; Fax: +34 986862127; E-mail: angeles.vazquez@ sauce.pntic.mec.es
1848; Angeles Vázquez Vaamonde
Colección Muruais, Sampedro, Soc. Arqueológica de Pontevedra, María Mendoza, Carlos Villar; Biblioteca Infantil
149 273 vols; 784 curr per; 20 mss; 10 incunabula; 35 music scores; 125 maps; 1 microforms; 3 094 av-mat; 3 644 sound-rec; 1 490 digital data carriers
libr loan; IFLA   26243

Centre de Lectura de Reus, C/ Major, núm. 15, 43201 *Reus*
T: +34 977773112; Fax: +34 977345152; E-mail: biblioteca@centrelectura.cat; URL: centrelectura.org
1859; Ramon Oteo Sans
Local Coll (7 000 vols); Magazine Libr
150 000 vols; 378 curr per; 32 mss; 18 incunabula; 935 govt docs; 1 100 music scores; 7 000 maps; 700 av-mat; 1 400 sound-rec; 600 digital data carriers; 3 000 rare bks (16-18th c), 350 pamphlets
                                             26244

Biblioteca Lambert Mata, C/ Vinyes, 1, 17500 *Ripoll*, Girona
T: +34 972700711; Fax: +34 972702253; E-mail: bibrip@ddgi.es; URL: www.bibliotecaripoll.blogspot.com/
1986
Fons Mata (16 000 vols)
50 000 vols; 92 curr per; 11 incunabula; 1 400 rare bks (16-17th c)   26245

Biblioteca Caixa de Sabadell (Library Caixa of Sabadell), D'en Font, 25, 08201 *Sabadell*
T: +34 937286678; Fax: +34 937298047
1928; Rosa M. Garcia Burdó
Coll Sala Montllor i Pujal
180 000 vols; 390 curr per; 975 av-mat; 125 digital data carriers
Col.legi Oficial de Bibliotecaris-Documentalistes de Catalunya   26246

Biblioteca Pública del Estado en Salamanca, C/ Compañía, 2, 37002 *Salamanca*
T: +34 923269317; Fax: +34 923269758; E-mail: bpsalamanca@jcyl.es
1933; Ramona Domínguez Sanjurjo
60 000 vols; 200 curr per; 1 800 av-mat; 3 500 sound-rec
libr loan   26247

Biblioteca Municipal de Donostia San Sebastián, Pl de la Constitución, 1, Donostia, 20003 *San Sebastián*
T: +34 943481156; Fax: +34 943431895; E-mail: donostiakoliburutegiak@ donostia.org

1874
60 000 vols; 418 curr per; 12 mss; 8 microforms; 25 av-mat; 30 sound-rec
                                             26248

Koldo Mitxelena Kulturunea, Diputacion Foral de Gipuzkoa, Urdaneta Kalea 9, 20006 *San Sebastián*
T: +34 943 112763;
Fax: +34 943 112765; E-mail: kmliburutegia@gipuzkoa.net; URL: www.kmliburutegia.net
1939; Francisco Javier Lopez Landatxe
Biblioteca Julio de Urquijo; Biblioteca Gabriel Celaya; Biblioteca Jose Maria Busca Isusi; Fondo Jose Francisco de Aizkibel; Biblioteca de Eusko Ikaskuntza; Fondo Apalategi; Fondo Carlos Santamaria; Biblioteca Alfonso Sastre y Eva Forest, Fondo Antonio Zavala
323 473 vols; 2 800 curr per; 85 e-journals; 6 651 mss; 3 incunabula; 807 music scores; 676 maps; 45 000 microforms; 10 116 av-mat; 12 891 sound-rec; 5 563 digital data carriers; 2 222 offprints, 3 297 graphic items
libr loan; Member of Amicus/Dobis Libis Users Group   26249

Biblioteca Pública del Estado en Santa Cruz de Tenerife (State Public Library of Santa Cruz de Tenerife), C/ Comodoro Rolín, 1, 38007 *Santa Cruz de Tenerife*
T: +34 922202202; Fax: +34 922206190; E-mail: bpetener-ife@gobiernodecanarias.org; URL: www.bibliotecaspublicas.es/tenerife
1971; Santiago Fariña Samblás
Personal Libr of Domingo Pérez Minik
114 990 vols; 2 762 curr per; 497 mss; 43 music scores; 17 maps; 11 190 av-mat; 11 951 sound-rec; 224 digital data carriers
libr loan; National Public Libraries Network   26250

Biblioteca Pública Municipal de Santa Cruz de Tenerife, José Murphy, 12, Apdo 976, 38002 *Santa Cruz de Tenerife*
T: +34 922243808; Fax: +34 922248873
1888
Canarian Bibliogr, canarian, spanish and foreign per publs
114 000 vols; 1 413 curr per; 748 mss
libr loan; ANABAD   26251

Biblioteca Menéndez Pelayo, Rubio, 6, 39001 *Santander*
T: +34 942234534; Fax: +34 942373766; E-mail: biblioteca-mp@ayto-santander.es
1915; Manuel Revuelta Sañudo
50 000 vols; 1 300 curr per; 400 mss
libr loan   26252

Biblioteca Municipal, Gravina, 4, 39001 *Santander*
T: +34 942203123; E-mail: bibmunicipal@ayto-santander.es
1877; Maria Teresa Gonzales Gonzales
Colección Manuscritos, Colección Cantabria, Fondo Antiguo
68 508 vols; 177 curr per; 1 514 mss; 12 incunabula; 6 diss/theses; 141 maps; 1 039 microforms
libr loan; UNAL   26253

Biblioteca Pública del Estado, Juan Bravo, 11, 40001 *Segovia*
T: +34 921463533; Fax: +34 921463523; E-mail: Luis.Garcia@bcl.jcyl.es
1842; Luis Garcia Méndez
113 230 vols; 439 curr per; 1 mss; 6 microforms; 6 200 av-mat; 1 784 sound-rec; 246 digital data carriers; 1 300 pamphlets, 3 589 rare bks (16-18th c)
libr loan   26254

Biblioteca Publica Provincial 'Infanta Elena', Avenida Maria Luisa, 8, 41013 *Sevilla*
T: +34 954712305; Fax: +34 954712284; E-mail: informa-cion.bp.se.ccul@juntadeandalucia.es; URL: www.juntadeandalucia.es/cultura/opencms/export/bibliotecas/bibsevilla/
1954; Juana Muñoz Choclán
100 542 vols; 657 curr per; 20 microforms; 11 897 av-mat; 293 sound-rec; 419 digital data carriers; 24 DVDs, 110 diskettes
libr loan; ANABAD   26255

Biblioteca Pública del Estado en Soria, Nicolas Rabal, 25, 42003 *Soria*
T: +34 975221800; Fax: +34 975229170; E-mail: fueleote@jcyl.es
1935; Teresa de la Fuente León
Local coll
93 600 vols; 260 curr per; 34 mss; 29 incunabula; 100 maps; 1 305 microforms; 3 170 av-mat; 14 200 sound-rec; 6 000 rare bks (16-18th c)
libr loan   26256

Biblioteca Pública, Fortuny, 30, 43001 *Tarragona*
T: +34 977240331; Fax: +34 977245312; E-mail: bptarragona.cultura@gencat.net
1846; Roser Lozano Diaz
Tarragona dept, Bibl Popular Diputación de Tarragona, bequest Marques de Montoliu, old bks dept
184 110 vols; 2 595 curr per; 318 mss; 230 incunabula; 140 music scores; 24 031 microforms; 8 383 sound-rec; 205 digital data carriers; 35 000 non book materials
libr loan   26257

Ateneo Mercantil de Valencia, Biblioteca, Pl del País Valenciano, 18, 46001 *Valencia*
T: +34 963525984; E-mail: ateneo@ateneomercantilvalencia.org; URL: www.ateneomercantilvalencia.org
1879
Local history
52 000 vols; 776 curr per; 16 mss
                                             26258

Biblioteca Pública de Valencia, Hospital, 13, 46001 *Valencia*
T: +34 963519996; Fax: +34 963516661; E-mail: bpv@gva.es; URL: www.gva.es/bpv
1916; Rafael Coloma
140 000 vols; 499 curr per; 100 maps; 2 425 microforms; 1 390 av-mat; 2 523 sound-rec; 42 digital data carriers
libr loan   26259

Biblioteca Pública Municipal Central, Pl de Maguncia 1, 46018 *Valencia*
T: +34 963525478 ext 4588, 4534, 4718; Fax: +34 963501321; E-mail: bcentral@ valencia.es; URL: www.valencia.es
1905; Belen Gisbert Aguilar
14 donaciones de ilustres valencianos (36 990 vols), 1.600 dibujos, grabados, Esperanto dept (670 vols), Biblioteca Deportiva (2 319 vols)
94 292 vols; 99 curr per; 1 840 mss; 9 incunabula; 545 microforms; 1 744 av-mat; 84 digital data carriers; 8 600 rare books   26260

Biblioteca Pública, Pl de la Trinidad, 2, 47003 *Valladolid*
T: +34 983358599; Fax: +34 983359040; E-mail: bpval@bcl.jcyl.es; URL: bpval.bcl.jcyl.es
1931; Alejandro Carrión Gútiez
Local studies (14 270 vols)
214 516 vols; 1 201 curr per; 340 music scores; 108 maps; 17 325 av-mat; 9 747 sound-rec; 3 127 digital data carriers
libr loan   26261

Biblioteca Central de Vigo, C/ Joaquín Yañez, 6, 36202 *Vigo*
T: +34 986229074; Fax: +34 986227342; E-mail: Bibli.Central.Vigo@Xunta.es
1942
Fondo bibliografico gallego de la Asociación Galega de Editores
118 300 vols; 660 curr per; 3 826 av-mat; 800 sound-rec; 324 digital data carriers
libr loan; ANABAD   26262

Biblioteca Pública Municipal Miguel Hernández, Pl de Santiago, 2, 03400 Villena *Villena*, Alicante
T: +34 965801150 ext 150; Fax: +34 965806146; E-mail: biblioteca.ayt@cv.gva.es
1982; Pilar Díaz Martínez
Personal libr of Juan Bautista Vilar
50 000 vols; 138 curr per; 1 375 sound-rec; 679 digital data carriers; 80 diskettes, 257 VHS, 106 DVDs
libr loan   26263

Biblioteca Pública del Estado en Vitoria-Gasteiz, Paseo de la Florida, 9, 01005 *Vitoria-Gasteiz*
T: +34 945181944; Fax: +34 945181945; E-mail: tcastro@alava.net
1861; Teresa Castro Legorburu

123 000 vols; 1 350 curr per; 13 mss; 41 av-mat; 1 950 sound-rec; 509 rare bks (16-18th c)
libr loan   26264

Biblioteca Pública del Estado, Pl de Claudio Moyano, s/n, 49001 *Zamora*
T: +34 980531551, 980533450; Fax: +34 988516032; E-mail: gondiamr@jcyl.es; URL: www.jcyl.es/bibliotecas
1846; Concepción González Díaz de Garayo
Rare book coll (1600-1801, 2 868 items), local documentation coll (13 364 items), Claudio Rodríguez coll (Poet, born in Zamora, 1934-1999, 620 items); Centro Provincial Coordinador de Bibliotecas: Professional advice and technical assistance to 23 Municipal Public Libraries and 4 mobile libraries
196 815 vols; 1 453 curr per; 38 e-journals; 12 mss; 10 incunabula; 129 diss/theses; 204 music scores; 230 maps; 32 microforms; 16 974 av-mat; 15 043 sound-rec; 2 957 digital data carriers; 3 701 posters
libr loan; Red de Bibliotecas Públicas del Estado (RABEL), Red de Bibliotecas Públicas de Castilla y León (RABEL)
                                             26265

Bibliotecas Públicas Municipales, Centro Coordinador de Bibliotecas / Centro Cívico Salvador Allende, Av. Miguel Servet 57, 50013 *Zaragoza*
T: +34 976724035; Fax: +34 976597850; E-mail: bibliotecas@zaragoza.es; URL: www.zaragoza.es
140 000 vols; 800 curr per; 600 maps; 36 000 microforms; 5 800 av-mat; 8 500 sound-rec; 1 700 digital data carriers
libr loan; ANABAD   26266

Sri Lanka

National Libraries

National Library and Documentation Centre, 14, Independence Ave, P.O. Box 1764, *Colombo* 07
T: +94 11 2698847; Fax: +94 11 2685201; E-mail: natlib@slt.lk; URL: www.natlib.lk
1990; Ms N. Mallawa Arachchi
Martin Wickremashinghe Coll, Unesco, Folk Culture, Book Development Project Deposit Copies, ISBN Deposit Copies, Braille Coll, Ola Leaf Coll, Science Coll
300 000 vols; 500 curr per; 828 mss; 695 diss/theses; 7 873 govt docs; 1 049 maps; 3 062 microforms; 927 av-mat; 102 sound-rec; 1 971 digital data carriers; 300 newspapers, 240 ola leaf
libr loan; IFLA, SLSTINET, ENLINET
                                             26267

University Libraries, College Libraries

Aquinas College of Higher Studies, Library, *Colombo* 8
T: +94 11 2694014; Fax: +94 11 2694709; E-mail: aqrector@panlanka.net
1954; Ben Nissangaratchie
Sri Lanka Coll
43 500 vols; 50 curr per; 100 mss; 460 diss/theses; 10 govt docs
libr loan   26268

Institute of Aesthetic Studies, Library, 21, Albert Crescent, *Colombo* 7
T: +94 11 2686071; Fax: +94 11 2686071
1974
Music, art, sculpture
18 000 vols   26269

National Institute of Business Management (NIBM), Library, 120/5 Wijerma Mawatha, *Colombo* 7
T: +94 11 2693404; Fax: +94 11 2693403; E-mail: nibmdg@slt.lk
1968; D.N. Champika Mahanthege
27 000 vols; 60 curr per; 15 maps; 100 av-mat; 10 sound-rec; 500 statistical publs, 197 research rpts, 2 180 project rpts, 260 newspaper clippings
libr loan   26270

Sri Lanka Law College, Library, 244 Hulftsdorp St, ***Colombo*** 12; P.O. Box 1501, Colombo
T: +94 11 1323759; Fax: +94 11 1436040; E-mail: Locwal@slt.lk
1874; S.D. Silva
11700 vols
libr loan 26271

University of Colombo, Library, 94 Cumaratunga Munidasa Mawatha, ***Colombo*** 3; P.O. Box 1698, Colombo
T: +94 11 2586432; Fax: +94 11 2583043; E-mail: scj@cmb.ac.lk; URL: www.lib.cmb.ac.lk
1921; Mrs S.C. Jayasuriya
Sri Lanka Coll, Research Coll and Rare books Coll; Sida/SAREC Library support
480000 vols; 600 curr per; 2000 e-journals; 1700 diss/theses; 1500 govt docs; 4000 microforms; 1500 av-mat; 50 digital data carriers; pamphlets, clippings
libr loan 26272

– Institute of Workers' Education, Library, University of Colombo, Cumaranatunga Munidasa Mawatha, P.O. Box 1557, Colombo 3
T: +94 1 500938
1979; R.M.S.N. Ratnayake
Special English Coll
18640 vols; 21 curr per; 450 diss/theses; 20 govt docs; 5 maps; 45 sound-rec
libr loan 26273

– Medical Faculty, Library, Kynsey Rd, Colombo 8
E-mail: medlib@cmb.ac.lk
1955; Mrs Geetha Yapa
Sri Lanka Bio Medical Lit
17605 vols; 230 curr per; 65 diss/theses; 7 microforms; 66 av-mat; 65 sound-rec
libr loan; HeLLIS 26274

Buddhist and Pali University of Sri Lanka, Library, Pitipana North, ***Homagama***
T: +94 12892331; Fax: +94 12892333; E-mail: bplib@sltnet.lk
1982; J.A. Amaraweera
Tripitaka-Pali Coll (Thai), Tripitaka-Pali Coll (Burmese), Tripitaka-CD, PTS Coll; Colombo HQ Library
26100 vols; 40 curr per; 250 mss; 145 diss/theses; 260 govt docs; 6 maps; 200 av-mat; 50 sound-rec; 120 digital data carriers; ola leaf bks
SLLA, ULASL 26275

University of Jaffna, Library, Thirunelvely, ***Jaffna***
T: +94 21 2222970; URL: jaffna.tripod.com/
1974; Mrs R. Pararajasingam
207 Ola mss; Medical Libr, Science Libr, Agriculture Libr, Fine Arts Libr, Siddha Libr
143794 vols; 470 curr per; 126 mss; 338 diss/theses; 6500 govt docs; 52 microforms; 41 sound-rec; 50 digital data carriers
libr loan; SALIS 26276

University of Moratuwa, Library, ***Katubedda***, Moratuwa
T: +94 11 2650301; Fax: +94 11 2650622; E-mail: umlibrary@lib.mrt.ac.lk; URL: www.lib.mrt.ac.lk
1972; Ms R.C. Kodikara
Sri Lanka Sc. & Tech Coll, UOM Archival Coll, UOM Thesis & Diss Coll, Arthur C. Clerk Coll, IDRC Coll, UOM Research Reports Coll, UOM Undergraduates Project Reports Coll
80000 vols; 300 curr per; 600 diss/theses; 100 maps; 100 microforms; 300 av-mat; 5 sound-rec; 115 digital data carriers; 750 pamphlets
libr loan 26277

University of Kelaniya, Library, Dalugama, ***Kelaniya***
T: +94 112 911918; Fax: +94 112 911918; E-mail: librarian@kln.ac.lk; URL: www.kln.ac.lk/library/index.htm
1959; Prof. Jayasiri Lankage
Ancient Sri Lankan Coin Coll, Tibetan Tripitaka Coll
124500 vols; 255 curr per; 120 diss/theses; 75 maps; 150 microforms
libr loan 26278

Technical College, Library, ***Matara***
T: +94 41 22565; Fax: +94 41 22565
1974; V. Hemananda
English for Specific Purposes
10560 vols; 7 curr per; 2 maps; 10 sound-rec; 4 digital data carriers 26279

University of Ruhuna, Main Library, ***Matara***
T: +94 41 2222681; Fax: +94 41 2222683; E-mail: postmaster@cc.ruh.ac.lk; URL: www.ruh.ac.lk/index.shtml
1978; K. Ananda
Prof Labrooy Coll, Prof Alawattagoda Premadasa Coll, Palitha Weeraman Coll, Sri Lanka Coll; Multimedia Unit
2000 vols; 175 curr per; 100 diss/theses; 200 govt docs; 50 maps; 5 av-mat; 40 sound-rec; 50 digital data carriers
libr loan 26280

– Faculty of Medicine, Library, P.O. Box 70, Galle
T: +94 9 34801; Fax: +94 9 22314; E-mail: mlibrary@mail.ewisl.net
1980; T. Kuruppu Arachchi
World Health Organization publs
11550 vols; 67 curr per; 5 diss/theses; 120 av-mat; 10 sound-rec; 100 digital data carriers; 125 pamphlets
libr loan; HeLLIS 26281

University of Sri Jayewardenepura, Vidyodaya Campus, Library, Gangodawilla, ***Nugegoda***
T: +94 1 852695; Fax: +94 1 852604; E-mail: vidana@sjp.ac.lk; URL: www.sjp.ac.lk/
1959; P. Vidanapathirana
Sri Lanka coll
100000 vols; 250 curr per; 100 diss/theses; 5000 govt docs; 2000 pamphlets
libr loan 26282

University of Peradeniya, Main Library, University Park, ***Peradeniya***; P.O. Box 35, Peradeniya
T: +94 81 2388678; Fax: +94 81 2388102; E-mail: librarian@pdn.ac.lk; URL: www.pdn.ac.lk
1921; Harrison Perera
Ceylon Coll, Deposit Coll, Environmental Coll; Agriculture Libr, Engineering Libr, Medical Libr, Science Libr, Vet Libr
756086 vols; 681 curr per; 4000 mss; 1731 diss/theses; 5200 govt docs; 250 maps; 750 microforms; 250 sound-rec; 100 digital data carriers; 350 drawings and paintings
libr loan; LAGBI 26283

– Faculty of Medicine, Medical Library, Peradeniya
T: +94 8 386008; Fax: +94 8 388678; E-mail: Sriyani@med.pdn.ac.lk
1972; Sriyani Perera
Medical, dental and veterinary sciences
30000 vols; 130 curr per; 50 diss/theses; 15 digital data carriers
libr loan; HELLIS, SLSTINET 26284

– Post Graduate Institute of Agriculture, Library, Old Galaha Rd, P.O. Box 55, Peradeniya
T: +94 8 388956; Fax: +94 8 388318; E-mail: ira@aglib.pdn.ac.lk
1952; Irangani Mudannayake
Sri Lanka Coll, ISNAR/IFPRI/IBPGR, Earl E. de Silva Coll, Y.D.A. Senanayake Coll, T. Jogaratnam Coll
35872 vols; 136 curr per; 485 diss/theses; 53 maps; 236 microforms; 150 av-mat; 29 digital data carriers; 3139 reprints
libr loan 26285

School Libraries

Hardy Technical College Library, ***Ampara***
T: +94 63 23485; Fax: +94 63 23485
1956; B.M. Jayasekara
20800 vols; 15 curr per; 10 govt docs; 10 maps
libr loan 26286

Technical College, Library, Brown Rd, ***Jaffna***; P.O. Box 4, Jaffna
1968; S. Thayanathan
15600 vols; 18 curr per; 4 mss; 5 diss/theses; 5 sound-rec
libr loan 26287

Jaffna College, Daniel Poor Library, ***Vaddukoddai***
T: +94 70 212531; E-mail: jafcol@SLtnet.lk
1876; S. P. Guasingam
65500 vols; 50 curr per; 15 mss; 27 diss/theses; 2 govt docs; 7 maps; 113 sound-rec 26288

Government Libraries

Archaeological Survey Department of Sri Lanka, Library, Sir Marcus Fernando Rd, ***Colombo*** 7
T: +94 11 2694727; Fax: +94 11 2696250; E-mail: gamini@sri.lanka.net
10000 vols 26289

Attorney General's Department, Library, P.O. Box 501, ***Colombo*** 12
1886; M.D.A. Diyasena
Sri Lanka Law
52000 vols; 20 curr per 26290

Department of Census and Statistics, Library and Documentation, 5th Floor, Unity Plaza, Galle Rd, ***Colombo*** 4
T: +94 11 2508819; Fax: +94 11 2697594; E-mail: library@statistics.gov.lk; URL: www.statistics.gov.lk
1947; M.B.M. Fairooz
ILO, FAO, WHO Coll, Sri Lanka Coll; UN div, Govt Publication div
10200 vols; 130 curr per; 20 diss/theses; 2400 govt docs; 175 maps; 30 digital data carriers; 13104 pamphlets, 5800 press clippings
libr loan; SLSTINET 26291

Department of Labour, Library, Labour Secretariat, ***Colombo*** 5
T: +94 11 2581141 ext 218
1923; C.L.S. Suriarachchi
31000 vols; 93 curr per 26292

Ministry of Foreign Affairs, Library, Republic Bldg, ***Colombo*** 1
1948; W.A. Vernon Boteju
10000 vols; 18 curr per 26293

Office of the Registrar of Books and Newspapers, Library, 7 Reid Av, ***Colombo*** 7
127000 vols 26294

Parliament of Sri Lanka, Library of Parliament, Sri Jayewardenepura Kotte, ***Colombo*** 1
T: +94 11 2777483; Fax: +94 11 2777227; E-mail: webmaster@parliament.lk; URL: www.parliament.lk
1927; Mr N.M.C. Thilakarathne
35500 vols; 51 curr per
APLAP 26295

Sri Lanka Ports Authority, Library, 45 Leyden Bastian Rd, ***Colombo*** 7
1952; M.D.H.K. Jayasinghe
8200 vols; 40 curr per; 15 diss/theses; 25 maps 26296

Centre for Industrial Technology Information Services, Industrial Development Board, Library, 615, Galle Rd, ***Katubedda***, Moratuwa
T: +94 1 605372; Fax: +94 1 607002; E-mail: citis@slt.lk
1969; Mrs. Nalini De Silva
Project rpts, machinery catalogs
14000 vols; 40 curr per; 5 maps; 10 av-mat; 6 digital data carriers
libr loan 26297

Ecclesiastical Libraries

Young Men's Christian Association, Library, 39 Bristol St, ***Colombo*** 1
1882; Keith Noyahr
15000 vols; 35 curr per 26298

Corporate, Business Libraries

Bank of Ceylon, Library, 4, Bank of Ceylon Mawatha, ***Colombo*** 01
T: +94 11 2446790-811 (22 lines); E-mail: boc@boc.lk
1974; P.S. Perera
8300 vols; 70 curr per; 130 subject files of Newspaper clippings, 240 unpublished docs
libr loan 26299

Central Bank of Sri Lanka, Information Department, 6th Fl, 30 Central Bank of Sri Lanka, Janadhipathi Mawatha, ***Colombo*** 01
T: +94 11 2477165; Fax: +94 11 2477712; E-mail: cbslgen@sri.lanka.net; URL: www.centralbanklanka.org
1951; C.K. Paranavithana
Sri Lanka Coll
72000 vols; 550 curr per; 50 maps
SLLA, CILIP 26300

Special Libraries Maintained by Other Institutions

Rubber Research Institute of Sri Lanka, Library, Dartonfield, ***Agalawatta***
T: +94 34 47426; Fax: +94 34 47427; E-mail: library@arri.ac.lk
1936; Kamani Perera
Rubber chemistry & technology section
8000 vols; 236 curr per; text-books
libr loan; SLSTINET, SLLA, IFLA 26301

Alliance Française, Library, 11 Barnes Pl, ***Colombo*** 7
T: +94 11 2694162
1954
12000 vols; 1000 sound-rec 26302

American Center Library, 44 Galle Rd, ***Colombo*** 3
T: +94 11 2421270
1949; M. Karunanayake
10000 vols; 142 curr per 26303

Bandaranayake Centre for International Studies, Library, Bauddhaloka Mawatha, ***Colombo*** 7
T: +94 11 2691175; Fax: +94 1 691176
1975; W.G.W. Ariyaratne
United Nations (UN) Documents
11000 vols; 100 curr per; 200 diss/theses; 8 govt docs
libr loan 26304

British Council, Library, 49 Alfred House Gardens, P.O. Box 753, ***Colombo*** 3
T: +94 11 4521593; Fax: +94 11 2587079; E-mail: library.cmb@britishcouncil.org; URL: www.britishcouncil.org/srilanka
1950
Science and technology, English language teaching
46250 vols; 52 curr per
libr loan 26305

Centre for Development Information, Department of National Planning, P.O. Box 1547, Galle Face Secretariat, ***Colombo*** 1
T: +94 11 2449378; Fax: +94 11 2448063; E-mail: cdinpd@sltnet.lk
1979; Anne Perera
Sri Lanka rpts
20000 vols; 210 curr per; 75 diss/theses; 2000 govt docs; 10 maps; 15 digital data carriers
libr loan 26306

Ceylon Chamber of Commerce, Foreign Trade/Information Division, 50 Navam Mawatha, ***Colombo*** 2
T: +94 11 2452183; Fax: +94 11 2437477; E-mail: pasantha@chamber.lk; URL: www.chamber.lk
1974; Pasantha Dissanayake
11000 vols 26307

Colombo National Museum, Library, Sir Marcus Fernando Mawatha, P.O. Box 854, ***Colombo*** 7
T: +94 11 2693314
1877
Govnt Oriental Libr
681220 vols; 4500 curr per; 3500 mss; 50 maps; 3772 palm leaf mss
libr loan 26308

Colombo Observatory, Library, Dept of Meteorology, Bauddhaloka Mawatha, ***Colombo*** 7
T: +94 11 2684746; Fax: +94 11 2691443; E-mail: meteo@slt.lk
1907
18000 vols 26309

Department of National Archives, Library, 7 Reid Av, ***Colombo*** 7
T: +94 11 2694523, 2696917; Fax: +94 11 2694419; E-mail: narchive@slt.lk
1902; Dr. K.D.G. Wimalaratne
Special private colls
19563 vols; 30 curr per 26310

Industrial Technology Institute, Information Services Centre, 363 Bauddhaloka Mawatha, P.O. Box 787, *Colombo* 7
T: +94 11 2698624; Fax: +94 11 2698642; E-mail: dilmani@iti.lk
1955; D.S.T. Warnasuriya
36 000 vols; 500 curr per; 200 diss/theses; 2 100 govt docs; 2 digital data carriers; 50 000 pamphlets, reprints of articles, standards and specs
libr loan; ASLIB   26311

Institute of Chartered Accountants of Sri Lanka, Library, 30 A, Malalasekera Mawatha, *Colombo* 7
T: +94 11 2586256; E-mail: icaweb@lanka.net
1959; E.S. Rajasingham
Encyclopeadias Brittanica, Acts of Ceylon, Legislative Enactments, Conference Papers
9 000 vols; 150 curr per; 30 diss/theses
libr loan   26312

Institution of Engineers, Sri Lanka, Library, 120/15 Wijerama Mawatha, *Colombo* 7
T: +94 11 2698426; Fax: +94 11 2699202; E-mail: iesl@slt.lk; URL: www.iesl.lk/
1906; Ms Mallika Yapa
9 000 vols; 75 curr per; 15 diss/theses; 100 govt docs; 10 maps; 50 digital data carriers
libr loan; Sri Lanka Library Association   26313

Marga Institute, Library, 93/10, Dutugemuna St, P.O. Box 601, Kirulapone, *Colombo* 6
T: +94 11 2828544, 2829051; Fax: +94 11 2828597; E-mail: marga@sri.lanka.net; URL: www.margasrilanka.org
1972; R. Wijegoonaratne
Krester Coll, World Bank Depository Coll, Conflict Resolution Coll
20 000 vols; 100 curr per; 1 000 govt docs; 100 maps; 20 digital data carriers   26314

Population Information Centre, Ministry of Plan Implementation No. 407, Bauddhaloka Mawatha, *Colombo* 7
1979; Sumithra Thalakadalie Geethananda
Nursing, public health, population, family planning
8 700 vols; 122 curr per; 5 diss/theses; 100 govt docs; 20 maps; 50 microforms; 245 av-mat; 10 sound-rec
libr loan; HeLLIS   26315

Post-Graduate Institute of Medicine, Library, 160, Morris Canal Rd, *Colombo* 8
T: +94 11 2693720
1979; D. Gunasekara
WHO publs, Sri Lanka coll
10 500 vols; 126 curr per; 200 diss/theses
libr loan; HeLLIS   26316

Royal Asiatic Society of Sri Lanka, Library, Mahaweli Centre and R.A.S. Bldg 96, Ananda Coomarswamy Mawatha, *Colombo* 7
T: +94 11 2699249
1845; B.E. Wijesuriya
8 000 vols; 12 curr per; 450 pamphlets   26317

Sri Lanka Institute of Development Administration, Library, 28/10 Longdon Place, *Colombo* 7
T: +94 11 2582181, 2582185
1966; Nanda Fernando
20 000 vols; 150 curr per   26318

Sri Lanka Medical Library, 6 Wijerama Mawatha SLMA House, *Colombo* 7
1844
24 000 vols
libr loan; HeLLIS   26319

United Nations Information Centre, Library, 204 Bauddhaloka Mawatha, *Colombo* 7
T: +94 11 2580691; Fax: +94 11 2581116; E-mail: Kumari.Wickramasinghe@undp.org
1961; Kumari S. Wickramasinghe
10 000 vols; 100 curr per; 50 digital data carriers   26320

Wiros Lokh Institute, Library, 81-1A Isipatana Mawatha, *Colombo* 5
T: +94 11 2580817; Fax: +94 11 2580721; E-mail: wiros@diamond.lanka.net
1981; Chandrika
Emerging Stock Markets, Urban Development, Sport Wrestling
9 000 vols; 10 curr per; 30 mss; 30 maps; 50 av-mat; 10 sound-rec; 50 digital data carriers   26321

National Institute of Health Sciences, Library and Documentation Centre, P.O. Box 28, *Kalutara*
T: +94 34 22264; Fax: +94 34 22319; E-mail: nihs@lanka.ccom.lk
1973; Srimathie Samarasinghe
Nursing, allied health, public health, biomedical sciences, pharmacy, Management, Primary Health Care, Social Sciences, Educational Sciences
11 560 vols; 35 curr per; 98 diss/theses; 23 av-mat; 50 sound-rec; 6 digital data carriers
libr loan; HeLLIS   26322

British Council, Information and Language Centre, Library, 88/3 Kotugodella Veediya, *Kandy*
T: +94 81 4473543; Fax: +94 81 2234284; E-mail: enquiries.kdy@britishcouncil.org; URL: www.britishcouncil.org/srilanka
1960
Sri Lanka; English Language Teaching Dept
20 000 vols
libr loan   26323

Coconut Research Institute, Library, Bandirippuwa Estate, *Lunuwila* 61150
T: +94 31 2255300; Fax: +94 31 2257391; E-mail: library@cri.lk; URL: www.cri.lk
1929; Mr J.M.D.T. Everard
Coconut cultivation, production & processing technology, staff pubis
22 000 vols; 56 curr per; 2 200 mss; 150 incunabula; 158 diss/theses; 300 govt docs; 15 maps; 7 055 microforms; 3 av-mat; 2 digital data carriers
libr loan; IAALD, SLSTINET   26324

Centre for Industrial Technology Information Services, Industrial Development Board, Library, 615 Galle Rd, Katubedda, *Moratuwa*
T: +94 1 605372; Fax: +94 1 607002; E-mail: citis@slt.lk; URL: www.nsf.ac.lk/idb
1969; N.N.
685 machinery catalogs, 239 project files, 112 project rpts, 35 fiction
14 500 vols; 155 curr per; 3 govt docs; 25 maps; 1 000 microforms; 25 av-mat; 95 digital data carriers; 3 023 rpts, 1 115 reprints
libr loan; SLSTINET   26325

Industrial Development Board Library, Industrial Information Service, 615 Galle Rd, Katubedda, *Moratuwa*
T: +94 1 505326; Fax: +94 1 607002; E-mail: citis@slt.lk
1969; Sharmini M. Tennekoon
15 000 vols   26326

Lanka Jatika Sarvodaya Shramadana Sangamaya, Sarvodaya Central Library, Damsak Mandira No 98, Rawatawatte Rd, *Moratuwa*
T: +94 1 647159; Fax: +94 1 646512; E-mail: sarsed@slt.ik
1961; Ms Chithra Weerawardana
Sarvodaya coll, Mahatma Gandhi Coll, Buddhism coll, Sri Lanka coll, non formal and adult education, Aids Resources Center, women development coll, child development coll, environment coll, youth coll
16 000 vols; 50 curr per; 12 diss/theses; 50 govt docs; 25 maps; 1 microforms; E.T. Compact Writer (English)   26327

Horticultural Research and Development Institute, Library, Gannoruwa, *Peradeniya*
T: +94 81 2388011
1965
16 500 vols   26328

Institute of Indigenous Medicine, University of Colombo, Library, Bandaranaike Pura, *Rajagiriya*
T: +94 1 692385; Fax: +94 1 697175; E-mail: iimcmb@lanka.ccom.lk
1929; S.A. Mendis

Medicine (Ayurveda, Unani, Siddha), medicinal plants, allied health, public health, pharmacy, dentistry; Ola leaf manuscript coll, medicinal plants coll, theses coll
12 500 vols; 30 curr per; 472 mss; 62 diss/theses; 112 govt docs; 3 maps; 55 microforms; 55 av-mat; 3 digital data carriers; 230 documents, 472 ola leaf manuscripts
libr loan; HeLLIS   26329

Tea Research Institute of Sri Lanka, Library, St Coombs, *Talawakele*
T: +94 70 521023; Fax: +94 70 528311
1928; Wasantha Illangantilake
Advisory, technology, biochemistry, plant breeding and physiology, agronomy, entomology and nematology div
15 000 vols; 250 curr per
libr loan; SLSTINET, AGRIS   26330

Public Libraries

Public Library, Municipal Council, *Jaffna*
T: +94 21 2019
1934; S. Thanabalasingham
Sri Lanka Coll (6 827 vols)
88 000 vols; 145 curr per; 60 diss/theses; 12 govt docs; 50 av-mat
libr loan   26331

D. S. Senanayake Memorial Public Library, Kandy Municipal Council (D. S. Senanayake Memorial Public Library), *Kandy*
T: +94 81 2223716; Fax: +94 81 2225638; E-mail: dssplk@sltnet.lk
1920; Mrs. P.K. Muwandeniya
Sri Lanka, Law, Environmental, Young Adults, Ayurveda, Martin Wickremasinghe Coll, D.S. Senanayake Coll, Welivita Saranankara Sangharaja Coll
191 457 vols; 155 curr per; 25 govt docs; 240 maps; 150 av-mat; 50 sound-rec; 4 digital data carriers   26332

St. Lucia

School Libraries

Sir Arthur Lewis Community College, Hunter J. Francois Library, Morne Fortune, *Castries*
T: +1 758 4522710; E-mail: jauguste@salcc.edu.lc; URL: www.salcc.edu.lc
1969; Jacqlyn Auguste
Curriculum Laboratory – Primary methodology
30 000 vols
libr loan; ACURIL, IASL   26333

Government Libraries

Parliament of Saint Lucia, Library, Government Bldgs, Laborie St, *Castries*
T: +1 758 4523856; Fax: +1 758 4525451; URL: www.stlucia.gov.lc
1991; Ms Lyndell Danzie
1 350 vols   26334

Special Libraries Maintained by Other Institutions

Organisation of Eastern Caribbean States (OECS), Documentation and Information Centre, Central Secretariat, The Morne Fortune, P.O. Box 179, *Castries*
T: +1 758 4522537; Fax: +1 758 4531628; E-mail: oesec@oecs.org; URL: www.oecs.org
1987; Swinburne Lestrade
10 000 vols   26335

St. Lucia Archives, P.O. Box 3060, La Clery, *Castries*
Margot Thomas
3 000 vols   26336

Public Libraries

Central Library of St. Lucia, Bourbon St, P.O. Box 103, *Castries*, Windward Islands
T: +1 758 4522875; Fax: +1 758 4527053; E-mail: clibrary@isis.org.lc; URL: www.education.gov.lc/lib/lib.htm
1927
St. Lucia coll, West Indian coll; 15 branch libs
159 170 vols; 120 curr per; 52 maps; 636 microforms
libr loan; ACURIL, COMLA   26337

St. Vincent and the Grenadines

University Libraries, College Libraries

St. Vincent Teachers' College, Library, *Arnos Vale*; P.O. Box 242, Kingstown
T: +1 784 4584611; Fax: +1 784 4564448; E-mail: svgteachcol@vincysurf.com
1964; Janice L. May
Teaching, education, curriculum management, children's lit, West Indian lit, African lit, professional lit
11 571 vols; 45 curr per; 1 729 diss/theses; 30 govt docs; 20 av-mat; 25 sound-rec; 10 digital data carriers; 40 pamphlets   26338

Government Libraries

National Documentation Centre, Administrative Centre, Bay St, *Kingstown*; P.O. Box 608, Kingstown
T: +1 784 4561689 ext 411; Fax: +1 784 4572022
1982; Gail B. Nurse
5 000 vols; 6 curr per   26339

Public Libraries

St. Vincent Public Library, Lower Middle St, *Kingstown*
T: +1 784 4572022; Fax: +1 784 4572022; E-mail: publiclibrary@caribsurf.com
1893; Joan L. O'Garro
CXC coll, General Coll, African Black study, French coll, Agricultural (CTA), Career, How to Books & the Dellimore Music coll, St Vincent & the Grenadines, West Indian & Sir Rupert John
260 000 vols; 150 curr per; 100 diss/theses; 250 govt docs; 277 music scores; 150 maps; 50 av-mat; 148 sound-rec; 25 digital data carriers; Recreational & Educational Videos, Clippings and Pamphlets
libr loan; ACURIL, COMLA   26340

Sudan

University Libraries, College Libraries

Institute of Education, Central Library, *Bakht er Ruda*
1945; Abdel Rahim Mohd Elfaki
Rare Arabic bks
28 000 vols; 200 govt docs   26341

College of Agricultural Studies, Department of Agricultural Sciences and Food Technology, Library, P.O. Box 71, *Khartoum* North
T: +249 11 611179; Fax: +249 11 774559
1954; Kamal Abbas
Food sciences, agricultural engineering
6 000 vols; 17 diss/theses; 140 govt docs; 6 maps   26342

Khartoum Polytechnic, Library, P.O. Box 407, *Khartoum*
T: +249 11 78922
1950; Dr. Abbas El Shazali
30 000 vols   26343

Neelain University Libraries, *Khartoum*
T: +249 11 776433;
Fax: +249 11 776338; E-mail:
elageed_seed.ahmed@hotmail.com
1993; Seed Ahmed A. o. Elageed
Commerce and trade
12 000 vols; 100 curr per; 865
diss/theses   26344

University of Juba, Library, P.O. Box
321/1, *Khartoum*
T: +249 11 222136; Fax: +249 11
222142; E-mail: zbimam@hotmail.com
1977; Prof. Gasim Osman Nur
Sudan coll, UN coll, Arabic coll,
Computer Science coll, Libr coll, Peace
coll; Arabic, English, Distance Education,
Peace and Development, Management
Science depts
150 000 vols; 800 curr per; 75
diss/theses; 500 govt docs; 1 200 maps;
50 microforms; 20 av-mat; 25 sound-rec;
500 slides   26345

University of Khartoum, Main Library,
P.O. Box 321, *Khartoum* 80295
T: +249 11 772601; Fax: +249 11
780295; URL: www.sudan.net/uk
1945; Dr. Mohammed Nouri Al Amin
Sudan and African coll; depository libr
for UN, FAO, ILO, WHO and UNESCO
publs
350 000 vols   26346

– Faculty of Agriculture and Veterinary
Science, Shambat Library, P.O. Box 71,
Khartoum North
1954
FAO coll
60 000 vols
IAALD   26347

– Faculty of Engineering and Architecture,
Library, P.O. Box 487, Khartoum
15 000 vols   26348

– Faculty of Medicine and Faculty of
Pharmacy, Campus Library, P.O. Box
102, Khartoum
T: +249 11 772224; E-mail:
lhassan@sud.Healthnet.org
1924; Isam Eldin Hassan Osman
Dentistry, biology
40 000 vols; 200 curr per; 300
diss/theses; 50 govt docs; 8 digital data
carriers
libr loan   26349

Ahfad University for Women, Main
Library (Maktabat El Hafeed), P.O. Box
167, *Omdurman*
T: +249 15 579111; Fax: +249 15
553363; E-mail: ahfad@sudanmail.net;
URL: www.ahfad.org
1966/1988; Dr. Asia Macawai Ahmed
Sudad coll; Sudan archives dept
120 000 vols; 174 curr per; 100
diss/theses; 250 govt docs; 25
microforms; 20 digital data carriers
libr loan   26350

**College for Arabic and Islamic
Studies**, Library, P.O. Box 328,
Omdurman
23 000 vols   26351

Higher Teachers' Training Institute,
Library, P.O. Box 406, *Omdurman*
12 000 vols; 52 curr per   26352

Omdurman Islamic University, Library,
P.O. Box 1221, *Omdurman*
T: +2491 87 556791
1901; A. Seed Osman Ahmad
Coll of Personal Libraries (3 222)
210 000 vols; 307 curr per; 100 mss;
4 782 diss/theses; 1 200 govt docs; 993
av-mat; 120 sound-rec; 311 digital data
carriers
libr loan; SALI   26353

Government Libraries

Ministry of Foreign Affairs, Library,
P.O. Box 873, *Khartoum*
8 000 vols; 110 curr per   26354

Special Libraries Maintained by Other Institutions

British Council, Library, 14 Abu Sinn St,
P.O. Box 1253, *Khartoum*
T: +249 183 780817, 777310;
Fax: +249 11 774935; E-mail:
info@sd.britishcouncil.org; URL:
www.britishcouncil.org/sudan
1963
10 070 vols
libr loan   26355

Centre Culturel Français, Library, Blue
Nile Insurance Bldg, Sharia EP Kasr,
P.O. Box 1568, *Khartoum*
T: +249 11 772837
1956; Wigdan Tahir
10 000 vols; 10 curr per; 400 av-mat
                                      26356

Educational Documentation Centre,
Library, P.O. Box 2490, *Khartoum*
T: +249 11 71898
1967; Ibrahim M.S. Shatir
20 000 vols; 65 curr per; 2 diss/theses;
2 000 govt docs; 40 pamphlets   26357

National Records Office, Library, P.O.
Box 1914, *Khartoum*
13 000 vols; 20 mio docs covering
Sudanese hist since 1870   26358

Agricultural Research Corporation,
Library, Gezira Agricultural Research
Station, P.O. Box 126, *Wad Medani*
T: +249 51 42226; Fax: +249 51 43213;
E-mail: arcsudan@sudanet.net
1931; Ahlam Ismail Musa
15 000 vols; 250 curr per; 20 000
pamphlets
IAALD   26359

Suriname

University Libraries, College Libraries

**Anton de Kom Universiteit van
Suriname**, Centrale Bibliotheek,
Universiteitscomplex, geb. 1, Leysweg
26, Postbus 9212, *Paramaribo*
T: +597 465558; Fax: +597 462291;
E-mail: adekbib@sr.net; URL:
www.uvs.edu
1968; Ine I.S. Tsai Meu Chong
Suriname coll
51 000 vols; 305 curr per; 1 400
diss/theses; 20 maps; 1 391 microforms;
14 av-mat
libr loan; ACURIL   26360

– Medische Bibliotheek, Geb. MWI,
Kernkampweg 5, Postbus 9212,
Paramaribo
T: +597 442215; E-mail: medbib@sr.net
1972; A. Ramkhelawan, D. Chin A. Fat
25 122 vols; 32 curr per; 5 045
diss/theses; 4 digital data carriers
libr loan; ACURIL   26361

Special Libraries Maintained by Other Institutions

Geologisch Mijnbouwkundige Dienst,
Bibliotheek, Kleine Waterstraat 2-6,
Paramaribo
T: +597 476215; Fax: +597 421533
1943
20 000 vols   26362

**Ministry of Agriculture, Animal
Husbandry and Fisheries**, Central
Library, letitia Vriesde laan 8, Postbus
160, *Paramaribo*
T: +597 472442; Fax: +597 478986;
E-mail: odlb.ond@sr.net
1904; Rajendra Chotoe
30 000 vols; 400 curr per; 200 mss;
200 diss/theses; 20 000 govt docs; 100
pamphlets
libr loan   26363

Stichting Surinaams Museum,
Bibliotheek, Abraham Crijnssenweg 1,
Postbus 2306, *Paramaribo*
T: +597 425871; Fax: +597 425881;
E-mail: museum@cq-link.sr
1954; N.N.
Surinamica
35 000 vols; 311 maps; periodicals, diss,
printed music   26364

Public Libraries

Stichting Cultureel Centrum Suriname,
Bibliotheek, Henck Arronstraat 112-114,
Postbus 1241, *Paramaribo*
T: +597 425871; Fax: +597 476516;
E-mail: stgccs@yahoo.com
1947; Marcella Augustuszoon
Suriname Coll, Coll of the Caribbean;
American Corner, Dutch Corner, Effendi
Ketwaru Public Music School, Bookmobile
850 000 vols; 7 curr per; 3 incunabula;
550 microforms; 250 av-mat; 5 000
sound-rec
libr loan; IFLA   26365

Swaziland

National Libraries

Swaziland National Library Service,
Headquarters and Mbabane Library, El
Watini Bldg, Grn Warner & Market St,
P.O. Box 1461, *Mbabane*
T: +268 4042633, 4042755,
4042722; Fax: +268 4043863; E-mail:
dijkunene@snls.gov.sz; URL: www.gov.sz/
home.asp?pid=763
1970; Mrs D.J. Kunene
Swaziana (1 500 vols), UN Coll (9 500
vols); National Libr Div, Public Libr Div;
Branch libs
250 000 vols; 400 curr per; 50
diss/theses; 300 govt docs; 10 music
scores; 100 maps; 40 microforms; 50
av-mat
libr loan; IFLA, LIASA, SWALA   26366

University Libraries, College Libraries

University of Swaziland, Library, PB 4,
Kwaluseni
T: +268 5184011; Fax: +268 5185276;
E-mail: mmavuso@uniswacc.uniswa.sz;
URL: library.uniswa.sz
1971; Makana R. Mavuso
95 000 vols
libr loan; COMLA, SCAUL, SCECSAL   26367

Swaziland College of Technology,
Library, P.O. Box 69, H 100 *Mbabane*
T: +268 4042681; Fax: +268 4044521;
E-mail: scot@africaonline.co.sz
1974
African writers series
17 000 vols
libr loan; SWALA   26368

Government Libraries

Economics/Statistics Library, P.O. Box
456/602, *Mbabane*
T: +268 43765
1976; E. Nxumalo
12 500 vols   26369

Sweden

National Libraries

Kungl. Biblioteket, Sveriges
nationalbibliotek (National Library of
Sweden), Humlegården, Postbox 5039,
10241 *Stockholm*
T: +46 8 4634000; Fax: +46 8 4634004;
E-mail: kungl.biblioteket@kb.se; URL:
www.kb.se
In the 16th C; Dr. Gunnar Sahlin
Old Swedish and Icelandic mss, Swedish
daily press on microfilm, national deposit
library, papers of Strindberg and Lagerlöf

e.g., Codex Gigas; LIBRIS, responible
for national library system; BIBSAM,
responsible for national (and international)
co-operation and development
4 000 000 vols; 22 000 curr per; 220 e-
journals; 65 000 mss; 1 500 incunabula;
170 000 maps; 10 000 000 small
prints/ephemera
libr loan; IFLA, ASLIB, SB, SCONUL
                                      26370

General Research Libraries

Stifts- och landsbiblioteket,
Länsbibliotek för Skaraborgs län,
Prubbatorget 1, Box 194, 532 32 *Skara*
T: +46 511 32000; Fax: +46 511 32069;
E-mail: skara.kommun@skara.se; URL:
www.skara.se
1938; Kerstin Andesson
Local hist
484 200 vols; 525 curr per; 300 mss; 30
incunabula; 10 000 microforms
IFLA, SB   26371

Almedalsbiblioteket, Stads- och
högskolebibliotek på Gotland,
Cramérgatan 5, Box 1121, 621 22 *Visby*
T: +46 498 299000;
Fax: +46 498 299011; E-mail:
almedalsbiblioteket@hgo.se; URL:
bibliotek.gotland.se
1865; Kerstin Simberg
Gotland coll
libr loan; SB   26372

University Libraries, College Libraries

Högskolan i Borås, Bibliotek och
läranderesurser (University of Borås),
Allégatan 1, mail address: 501 90 *Borås*
T: +46 33 4354050; Fax: +46 33
4354004; E-mail: biblioteket@hb.se; URL:
www.hb.se/blr
1972; Cotta Torhell
Special coll in libr and information
science; Dept of clothing and textile
160 500 vols; 591 curr per; 80 e-journals;
47 000 e-books; 18 digital data carriers
libr loan; IFLA, FID, ASLIB, SFIS, ALA,
SB   26373

Högskolan i Gävle, Biblioteket (University
of Gävle), Kungsbäcksvägen 47, 80176
Gavle
T: +46 26 648548; Fax: +46 26
648870; E-mail: registrator@hig.se; URL:
www.hig.se
Maivor Hallén
87 000 vols; 360 curr per; 3 400 e-
journals; 200 music scores; 100
microforms; 700 av-mat; 15 digital data
carriers
libr loan; SB   26374

Chalmers tekniska högskola,
Chalmers Library (Chalmers University
of Technology), Chalmers Tvärgata 1,
412 96 *Göteborg*
T: +46 31 7723737; Fax: +46 31
183544; E-mail: request@lib.chalmers.se;
URL: www.lib.chalmers.se
1829; Annika Sverrung
Stig Ekelöfs Library of works of the
histroy of electricity; Architectural Library
and Lindholmen Library
524 000 vols; 632 curr per; 145 e-
journals; 5 800 diss/theses
libr loan; IFLA, IATUL, SB   26375

– School of Architecture, Architectural
Library, Sven Hultins gata 6, 412 96
Göteborg
T: +46 31 7722413;
Fax: +46 31 7722419; E-mail:
abibmail@lib.chalmers.se; URL:
www.lib.chalmers.se/abib
1969; Elisabeth Kihlén
24 200 vols; 120 curr per
ARLIS/Norden   26376

Göteborgs Universitet,
Universitetsbiblioteket, Renströmsgatan
4, Box 222, 405 30 *Göteborg*
T: +46 31 7731739; Fax: +46 31
163797; E-mail: library@ub.gu.se; URL:
www.ub.gu.se
1891; Jon Erik Nordstrand
libr loan; IFLA, LIBER, LA, ALA, SB,
IASA   26377

– Biomedicinska biblioteket (Biomedical Library), Medicinaregatan 4, 413 90 Göteborg; Box 416, 405 30 Göteborg
T: +46 31 7733002; Fax: +46 31 7733746; E-mail: Gmloan@ub.gu.se; URL: www.ub.gu.se/Gm
1949; Erna Roos
Biosciences, medicine, dentistry
255 000 vols; 1 500 curr per
libr loan; EAHIL 26378

– Botanik- och miljöbiblioteket (Botanical and Environmental Library), Carl Skottsbergs gata 22 B, 413 19 Göteborg; Box 461, 405 30 Göteborg
T: +46 31 7732541; Fax: +46 31 7732544; E-mail: Gb.best@ub.gu.se; URL: www.ub.gu.se/bibliotek/Gb/
1969; Christina Wising
Dahl collection: older books in pomology and churchyard architecture; Nathorst Windahl: Mycology collection, older books
50 000 vols; 319 curr per; 755 e-journals; reports in environmental sciences, around 50 metres
libr loan; EBHL 26379

– Centralbiblioteket (Central Library), Renströmsgatan 4, Box 222, 405 30 Göteborg
T: +46 31 7731745; Fax: +46 31 7734411; E-mail: library@ub.gu.se; URL: www.ub.gu.se/bibliotek/G/
1891; Marianne Ladenvall
Humanities, Social Sciences; Women's Hist Coll, Snoilsky Coll, Ludvig Holberg Coll, deposit libr for Swedish publ; Dept of Gender Studies
2 000 000 vols; 8 400 curr per; 118 incunabula; 30 000 av-mat; 25 000 sound-rec; 1 000 shelf m of mss 26380

– Ekonomiska biblioteket (Economics Library), Vasagatan, 411 24 Göteborg; Box 670, 405 30 Göteborg
T: +46 31 7731431; Fax: +46 31 7734926; E-mail: Ge.Info@ub.gu.se; URL: www.ub.gu.se/Ge/Ge
1923; Agneta Olsson
Economics, economic hist, information technology, law, human geography, statistics; UN depositary library, European Community Documentation Center
200 000 vols; 1 952 curr per; 15 microforms; 15 digital data carriers
libr loan; IFLA, LIBER 26381

– Geovetenskapelige biblioteket (Earth Sciences Library), Guldhedsgatan 5A, Box 460, 405 30 Göteborg
T: +46 31 7732830; Fax: +46 31 7731986; E-mail: geovet.bibl@ub.gu.se; URL: www.ub.gu.se/Gg
libr loan 26382

– Högskolan för Design och Konsthantverk, Biblioteket (School of Design and Crafts), Box 131, 405 30 Göteborg
T: +46 31 7860000; Fax: +46 31 7734888; E-mail: info@hdk.gu.se; URL: www.hdk.gu.se
1848; Ulla-Britt Ban Wieslander
15 000 vols; 160 curr per
libr loan; IFLA 26383

– Iberoamerikanska samlingen (Iberoamerican collection), Seminariegatan 1D, 413 13 Göteborg Göteborg; P.O. Box 710, 405 30 Göteborg
T: +46 31 7861803; Fax: +46 31 7861804; E-mail: campuslinne@ub.gu.se; URL: www.ub.gu.se/samlingar/ibero/
1939; Marianne Ladenvall
The Iberian Peninsula and Latin America
40 000 vols; 114 curr per
libr loan; REDIAL, SALALM 26384

– Institutionen för kulturvård, Biblioteket, Bastionsplatsen 2, 411 08 Göteborg
T: +46 31 7734700; Fax: +46 31 7734703
Charlotte Hanner
16 000 vols; 141 curr per 26385

– Kurs- och tidningsbiblioteket (Undergraduate and Newspaper Library), Vasagatan 2A, 411 24 Göteborg; Box 600, 405 30 Göteborg
T: +46 31 7732720; Fax: +46 31 7732700; E-mail: Gk.Exp@ub.gu.se; URL: www.ub.gu.se/Gk
1995; Christina Wising
22 000 vols 26386

– Pedagogiska biblioteket (Education Library), Frölundagatan 118, Box 300, 405 30 Göteborg
T: +46 31 7732340; Fax: +46 31 7732338; E-mail: gp.lan@ub.gu.se; URL: www.ub.gu.se/Gp
1975; Birgitta Brown
Text book coll for elementary schools
90 000 vols; 272 curr per; 170 diss/theses; 263 av-mat
libr loan; IFLA, SB 26387

– Samhällsvetenskapliga biblioteket (Social Sciences Library), Sprängkullsgatan 21, 411 23 Göteborg; Box 714, 405 30 Göteborg
T: +46 31 7731004; Fax: +46 31 7731598
Jelena Dackén
50 000 vols; 470 curr per 26388

Konstfack, Biblioteket (University College of Art, Crafts and Design), LM Ericssons väg 14, 126 37 ***Hägersten***; Box 3601, 126 27 Stockholm
T: +46 8 4504124; Fax: +46 8 4504128; E-mail: biblioteket@konstfack.se; URL: www.konstfack.se
1844; Olle Holmquist
Martin's and Zickerman's picture coll
100 000 vols; 242 curr per; 100 incunabula
libr loan; ARLIS/Norden, SFIS 26389

Högskolan i Halmstad, Högskolebiblioteket (Halmstad University), Kristian IVs väg 3, Box 823, 301 18 ***Halmstad***
T: +46 35 167114; Fax: +46 35 148528; E-mail: bib@bib.hh.se; URL: www.hh.se/bibliotek
1982; Göran Ericson
Lit about Halland
62 000 vols; 847 curr per; 3 800 diss/theses; 93 music scores; 64 maps; 193 av-mat
libr loan 26390

Mittuniversitetet, Biblioteket (Mid Sweden University), Trädgårdsgatan 17, 87188 ***Härnösand***
T: +46 611 86011; Fax: +46 611 86186; E-mail: gunilla.genberg@miun.se; URL: www.bibmiun.se
1978; Gunilla Gemberg
Education, humanities, technology
60 000 vols; 300 curr per 26391

Karolinska Institutet, Universitetsbiblioteket (Karolinska Institute University Library), Alfred Nobels Allé 8, ***Huddinge***; mail address: 171 77 Stockholm
T: +46 8 52484000; Fax: +46 8 52484310; E-mail: ub@ki.se; URL: www.ki.se
Medical Science, Nursing, Dental Science
56 800 vols; 770 curr per; 10 500 diss/theses; 465 govt docs; 15 av-mat
libr loan; LIBER, UKSG, EAHIL, IFLA, Biological and Medical Sciences 26392

Södertörns högskola, Biblioteket, Alfred Nobels allé 11, Flemingsberg, Box 4155, 141 04 ***Huddinge***
T: +46 8 6084000; Fax: +46 8 6084010; E-mail: bibliotek@sh.se; URL: www.sh.se/bibliotek
1996; Louise Brunes
70 000 vols; 900 curr per; 5 000 diss/theses; 19 000 govt docs; databases, e-journals
libr loan; SB, IFLA 26393

Högskolan i Jönköping, Högskolebiblioteket (Jönköping University), Gjuterigatan 5, P.O. Box 1001, 551 11 ***Jönköping***
T: +46 36 101040; Fax: +46 36 100359; E-mail: bibl@hj.se; URL: www.bibl.hj.se
Inger Melin
Trade Fair Libr; Information Centre for Entrepreneurship, Information Centre for Foreign Law, European Doc Centre
200 000 vols; 650 curr per; 13 000 e-journals
libr loan; SB, SFIS, IFLA, IATUL 26394

Högskolan i Kalmar, Nygatan 18A, 391 82 ***Kalmar***
T: +46 480 446100; Fax: +46 480 446115; E-mail: info.bibl@hik.se; URL: www.hik.se/bibliotek
1977; Anders Rydquist
Social science
75 000 vols; 150 curr per; 7 500 e-journals; 5 000 diss/theses; 2 000 govt docs; 36 000 av-mat
libr loan 26395

Karlstads Universitet, Biblioteket (Karlstad University), Universitetsgatan 2, 651 87 ***Karlstad***
T: +46 54 7001000; Fax: +46 54 835450; E-mail: bibdisk@kau.se; URL: www.kau.se
Eva Arndt Kling
140 000 vols; 1 000 curr per 26396

Högskolan i Kristianstad (Kristianstad College), Elmetorpsv. 15, Hus 7, mail address: 291 88 ***Kristianstad***
T: +46 44 203059; Fax: +46 44 203053; E-mail: biblioteket@bibl.hkr.se; URL: www.hkr.se
Christina Jönsson Adrial
105 000 vols; 400 curr per; 7 000 e-journals; 3 000 diss/theses
libr loan 26397

– Hässleholms tekniska skola, Biblioteket, Stobygatan 7, 281 39 Hässleholm
T: +46 451 68653; Fax: +46 451 12082
Ingegerd Nohrstedt
50 000 vols; 110 curr per; 6 digital data carriers 26398

Linköpings Universitet, Universitetsbiblioteket (Linköping University), mail address: 581 83 ***Linköping***
T: +46 13 281000; Fax: +46 13 284424; E-mail: liub@bibl.liu.se; URL: www.bibl.liu.se
1969; Marianne Hällgren
European Doc Centre, Faroe Islands Coll, DDR-Kulturcentrum Coll
717 000 vols; 2 500 curr per; 10 000 e-journals; 5 200 maps; 31 000 microforms; 600 av-mat; 12 000 offprints
ARLIS/Norden, EAHIL, IATUL, IFLA, LIBER, MLA, NVBF, SB, UKSG, NAG/UK, IGeLU, WFOT 26399

– Hälsouniversitetets bibliotek (Health Science Library), University hospital, 581 85 Linköping
T: +46 13 221428; Fax: +46 13 221426; E-mail: hub@bibl.liu.se; URL: www.bibl.liu.se/kvartersbibl/HUB/hubstart.htm
1970; Eva Sofia Svensson
10 000 vols; 650 curr per; 300 av-mat; 6 digital data carriers
libr loan; EAHIL, MLA 26400

– Kvartersbibliotek A(KA), Ekonomi, Teknik, Hus A, ing 17, Valla, 581 83 Linköping
T: +46 13 281095; Fax: +46 13 282941
Ursula Nielsen
20 000 vols 26401

– Mediateket för lärarutbildningsarna, Östgötagatan 12, 581 83 Linköping
T: +46 13 282038; Fax: +46 13 281991
Marianne Auby
Education, psychology
30 000 vols; 300 curr per
libr loan 26402

Luleå Tekniska Universitet, Biblioteket (Luleå University of Technology), Betahuset, University Campus, 971 87 ***Luleå***
T: +46 920 491520; Fax: +46 920 492040; E-mail: lulelibrary@ltu.se; URL: www.ltu.se/
1964; Jenny Samuelsson
210 000 vols; 300 curr per; 32 000 e-journals
libr loan; SB, SFIS, ASLIB, IFLA, EUSIDIC, IATUL 26403

Lunds Universitet, Universitetsbibliotek, Helgonabacken, Box 134, 221 00 ***Lund***
T: +46 46 2220000;
Fax: +46 46 2224243; E-mail: universitetsbibliotekarie@lub.lu.se; URL: www.lub.lu.se
1671; Lars Bjørnshauge
Bibliotheca Gripenhielmiana, Taussig Coll of Schubert mss, Broman coll of Elsevier prints, De La Gardie coll of mss and prints; Newspaper Libr
3 200 000 vols; 11 800 curr per; 132 000 mss; 275 incunabula; 550 000 diss/theses; 90 200 microforms; 49 500 av-mat
libr loan; SB, IFLA, EUSIDIC, LIBER, FID 26404

– Asienbiblioteket (Asia Library), Scheelevägen 15 B, 223 63 Lund; mail address: Scheelevägen 15, 223 70 Lund
T: +46 46 2223043; Fax: +46 46 2223041; E-mail: bibliotek@ace.lu.se; URL: www.ace.lu.se/o.o.i.s/5816
1999; Iréne Sandell
22 000 vols; 93 curr per; 27 e-journals
libr loan 26405

– Ekologiska biblioteket (Library of Ecology), Sölvegatan 37, 223 62 Lund
T: +46 46 2223812; Fax: +46 46 2223811; E-mail: biol.bibl@ekol.lu.se; URL: www.bibl.biol.lu.se/index.html
1994; Åse Paulsson
Ecology, limnology, zoology
80 000 vols; 500 curr per; 1 200 e-journals; 1 000 diss/theses; 2 000 maps
libr loan 26406

– Ekonomihögskolans bibliotek, Forskningspolitiska Institutet (Research Policy Institute, Library of Economics and Management), Sölvegatan 16, 221 00 Lund; Box 117, 221 00 Lund
T: +46 46 2224807; Fax: +46 46 146986; E-mail: bitte.holm@ehl.lu.se; URL: www.fpi.lu.se/
1967; Siw Kourtzman
Technology and Innovation; Research Policy; Stevan Dedijer Intelligence collection
10 000 vols; 30 curr per; 558 e-journals; 27 000 e-books
libr loan 26407

– Filosofiska biblioteket, Kungshuset, Lundagård, Box 117, 221 00 Lund
T: +46 46 2223675; E-mail: filosofen@htbibl.lu.se
1994; Fredrik Eriksson
Philosophy, Cognitive Science
4 000 vols; 30 curr per; 100 e-journals; 1 000 e-books
libr loan 26408

– Geobiblioteket (Geolibrary), Sölvegatan 12, Geocentrum II, 223 62 Lund
T: +46 46 2223960; Fax: +46 46 2224419; E-mail: geobib@geobib.lu.se; URL: www.geobib.lu.se
1988; Robin Gullstrand
Human geography, physical geography, geology, mineralogy; Map Division
52 500 vols; 650 curr per; 1 digital data carriers
libr loan 26409

– Juridiska fakultetens bibliotek (Law Faculty Library), Lilla Gråbrödersg 6, 222 22 Lund; Box 207, 221 00 Lund
T: +46 46 2221007; E-mail: jurbibl@jur.lu.se; URL: www.jur.lu.se
Catarina Carlsson
Law; European Documentation Center
300 000 vols; 800 curr per; 45 microforms; 15 digital data carriers
libr loan; IALL 26410

– Kemicentrums bibliotek (Library of Chemistry and Chemical Engineering), Getingevägen 60, Box 124, 221 00 Lund
T: +46 46 2228339; Fax: +46 46 2224002; E-mail: bibliotek@kc.lu.se; URL: www.bibliotek.kc.lu.se
Åsa Lindblad
10 000 vols; 800 curr per
libr loan 26411

– Musikhögskolan i Malmö, Ystadvägen 25, Box 8203, 200 41 Malmö
T: +46 40 325446; Fax: +46 40 325490; E-mail: Britt.Roslund@mhm.lu.se; URL: www.mhm.lu.se/o.o.i.s/14509
Britt Roslund
30 000 vols; 60 curr per 26412

Social- och beteendevetenskapliga biblioteket, Lunds universitet (Library of Social and Behavioural Science), Allhelgona kyrgog. 14 J, Box 42, 221 00 ***Lund***
T: +46 46 2220990;
Fax: +46 46 2220991; E-mail: socbetbib@socbetbib.lu.se; URL: www.socbetbib.lu.se
1999; Karin Jönsson
60 000 vols; 300 curr per
libr loan 26413

Malmö högskola, Biblioteket (Malmö University), mail address: 205 06 ***Malmö***
T: +46 40 6657300; Fax: +46 40 6657301; E-mail: biblioteket@mah.se; URL: www.mah.se/bit
Jette Guldborg Petersen
168 020 vols; 651 curr per; 5 006 e-journals
libr loan; IFLA 26414

– Oral Health Library, Odontologiska biblioteket, Smedjegatan 16, 21421 Malmö; mail address: 20506 Malmö
T: +46 40 6658463; Fax: +45 40 925359; E-mail: odbiblioteket@mah.se; URL: www.mah.se/bibliotek
1947; Pablo Tapia
15 000 vols; 300 curr per
libr loan 26415

Vårdhögskolan i Malmö, Spårvägsgatan 9, 214 27 ***Malmö***
T: +46 40 343235
Britta Lindström
Social work
13 000 vols; 94 curr per 26416

World Maritime University, Library, Citadellsvägen 29, Box 500, 201 24 ***Malmö***
T: +46 40 356344; E-mail: library@wmu.se; URL: www.wmu.se
1983; David S. Moulder
Maritime industry, shipping; IMO Documents
14 000 vols; 350 curr per; 1 400 incunabula; 1 500 diss/theses; 300 av-mat; 35 digital data carriers
ASLIB, EURASLIC, IAMSLIC 26417

Örebro universitet, Universitetsbiblioteket, Fakultetsgatan 1, 701 82 ***Örebro***
T: +46 19 303240, 303245; Fax: +46 19 331217; E-mail: biblioteket@oru.se; URL: www.oru.se
1967; Elisabeth Andersson
327 098 vols; 793 curr per; 80 e-journals; 128 mss; 117 724 microforms; 4 279 av-mat; 8 000 sound-rec
libr loan; SB 26418

Vårdhögskolan i Örnsköldsvik, Biblioteket, Vallgatan 1, 891 89 ***Örnsköldsvik***
T: +46 660 89647
Lena Ekeborg
19 000 vols; 250 curr per 26419

Skeriabiblioteket, Skeria 5, 931 77 ***Skellefteå***
T: +46 910 585345; E-mail: catrin.andersson@ub.umu.se
Greta Zetterström
Technology, economics
8 000 vols; 150 curr per 26420

Ersta Sköndal Högskola, Campus Sköndal, Biblioteket (Ersta Sköndal University College), Herbert Widmans väg 5A, 128 64 ***Sköndal***
T: +46 8 55505160; Fax: +46 8 55505165; E-mail: anette.krengel@esh.se; URL: www.esh.se
Ann-Kristin Forsberg
Social work, Church music, Church Welfare work, Diakonie
50 000 vols; 75 curr per; 250 diss/theses; 400 sound-rec; 10 digital data carriers
libr loan 26421

Högskolan i Skövde, Biblioteket, Högskolevägen 1, 541 28 ***Skövde***
T: +46 500 448060; Fax: +46 500 448059; E-mail: biblioteket@his.se; URL: www.his.se/biblioteket
Tomas Lundén
Computer science, science, social science, business, technology, nursing; Austria Coll
100 000 vols; 1 000 curr per; 7 000 e-journals; 350 digital data carriers
libr loan 26422

Polishögskolans bibliotek (Swedish National Police Academy Library), Sorentorp, 170 82 ***Solna***
T: +46 8 4016679; Fax: +46 8 4016669; E-mail: library@phs.police.se; URL: www.polishogskolan.polisen.se
1980; Heelena Nilsson
Forensic sciences, police science, law
40 000 vols; 200 curr per; 16 digital data carriers
libr loan; SBS, SFIS 26423

Anna Lindh-biblioteket, Högskolebiblioteket för försvar, utrikes- och säkerhetspolitik, Drottning Kristinas väg 37, Box 27035, 102 51 ***Stockholm***
T: +46 8 55342560; Fax: +46 8 55342568; E-mail: info@annalindhbiblioteket.se; URL: www.annalindhbiblioteket.se
Eva Hesselgren Mortensen
Security Policy, Int Relations
185 000 vols; 325 curr per
libr loan; SBS, SFIS
Library of the Swedish National Defence Office and the Swedish Institute of International Affairs 26424

Ersta Sköndal högskola, Ersta högskolebibliotek, Stigbergsgatan 30, Box 111 89, 100 61 ***Stockholm***
T: +46 8 55505008; Fax: +46 8 55505080; E-mail: biblioteket.ersta@esh.se; URL: www.esh.se/bibliotek/om-ersta-bibliotek/
Ann-Kristin Forsberg
Nursing, Theology, Diaconal work
25 000 vols; 220 curr per 26425

GIH Biblioteket, Gymnastik- och idrottshögskolan (The Swedish School of Sport and Health Sciences), Lidingövägen 1, 114 33 ***Stockholm***; Box 5626, 114 86 Stockholm
T: +46 8 4022232; Fax: +46 8 4022281; E-mail: biblioteket@gih.se; URL: www.gih.se/biblioteket
1813; Mats Hagström
Gymnastics, Ling-gymnastics
60 000 vols; 200 curr per; 15 e-journals; 35 digital data carriers
libr loan; SB/IASI, NORSIB 26426

Handelshögskolan i Stockholm, Biblioteket (Stockholm School of Economics), Sveavägen 65, Box 6501, 113 83 ***Stockholm***
T: +46 8 7369702; Fax: +46 8 318213; E-mail: library@hhs.se; URL: www.hhs.se/library
1909; Marie-Louise Fendin
310 000 vols; 1 120 curr per; 7 600 microforms
libr loan; IFLA, EBSLG, SFIS 26427

Karolinska Institutet, Universitetsbiblioteket, Berzelius väg 8, 171 77 ***Stockholm***; Box 200, 171 77 Stockholm
T: +46 8 52484000; Fax: +46 8 52484320; E-mail: info@kib.ki.se; URL: kib.ki.se
1810; Per Olsson
Medical Science
673 000 vols; 3 100 curr per; 875 mss; 10 incunabula; 243 000 diss/theses
libr loan; IFLA 26428

Kungl. konsthögskolan, Arkitekturskolans bibliotek (Royal University College), Holmamiralens väg 2, Skeppsholmen, 111 49 ***Stockholm***; P.O. Box 16315, 103 26 Stockholm
T: +46 8 6144000; Fax: +46 8 6112113; E-mail: lindvall.hans@kkh.se; URL: www.kkh.se
Hans Lindvall
19 000 vols; 90 curr per 26429

Kungl. Tekniska högskolan, KTH Biblioteket (Royal Institute of Technology), Osquars backe 31, 100 44 ***Stockholm***
T: +46 8 7906000; Fax: +46 8 7907122; URL: www.lib.kth.se
1826; Gunnar Lager
History of Science and Technology; 5 branch libs: Studsvik Libr, Architecture, Aerotech, Forum, Mathematics Libr
850 000 vols; 800 curr per; 7 000 e-journals; 53 000 e-books; 4 800 diss/theses; 200 av-mat; 606 000 digital data carriers
libr loan; IATUL, ICSTI, IFLA, SCOTUL, SFIS 26430

– Arkitekturbiblioteket (Royal Institute of Technology, Architectural Library), Östermalmsgatan 26, 100 44 Stockholm
T: +46 8 7908558; Fax: +46 8 87909181; E-mail: archlib@lib.kth.se; URL: www.lib.kth.se
Margitta Kylberg
30 000 vols; 150 curr per
libr loan 26431

– Forumbiblioteket (Forumlibrary), Isafjordsgatan 39, 164 40 Kista; Forum 110, 164 40 Kista
T: +46 8 790 42 01; Fax: +46 8 752 99 06; E-mail: forumbiblioteket@lib.kth.se; URL: www.lib.kth.se
Tommy Westergren
17 000 vols; 90 curr per; 4 500 e-journals; 400 e-books 26432

– Matematikbiblioteket (Mathematics Library), 100 44 Stockholm
T: +46 8 7907209; Fax: +46 8 7907125; E-mail: math@lib.kth.se; URL: www.lib.kth.se
1995; Chris Druid
17 000 vols; 146 curr per; 42 av-mat; 40 digital data carriers
libr loan 26433

Lärarhögskolan i Stockholm, Högskolebiblioteket (Stockholm Institute of Education), Konradsbergsgatan 5A, P.O. Box 34103, 10026 ***Stockholm***
T: +46 8 7379832; Fax: +46 8 7379838; E-mail: larum@lhs.se; URL: www.lhs.se
Lena Olsson
232 000 vols; 450 curr per
libr loan 26434

Sophiahemmet högskola, Biblioteket (Sophiahemmet University College), Lindstedtsvägen 8, 114 86 ***Stockholm***; Box 5605, 114 86 Stockholm
T: +46 8 4062885; Fax: +46 8 102909; E-mail: biblioteket@sophiahemmethogskola.se; URL: www.sophiahemmethogskola.se
Eva Unemo Wahlfridsson
25 000 vols; 140 curr per; 350 diss/theses
libr loan; IFLA, SB, EAHIL 26435

Stockholms universitet, Biblioteket (Stockholm University), Universitetsvägen 14 D, Frescati, 106 91 ***Stockholm***
T: +46 8 162800; Fax: +46 8 157776; E-mail: wilhelm.widmark@sub.su.se; URL: www.sub.su.se
1978; Wilhelm Widmark
Colls Bergius, Linné, Swedenborg
2 900 000 vols; 21 900 curr per; 6 000 e-journals; 224 400 e-books; 12 100 mss; 15 incunabula; 1 013 diss/theses; 118 932 maps; 13 750 microforms; 300 digital data carriers
libr loan; IFLA, LIBER, SB 26436

– Biologibiblioteket, Arrheniuslaboratorierna, Hus C, Svante Arrhenius v 10-12, 106 91 Stockholm
T: +46 8 164100; Fax: +46 8 159907; E-mail: arrheniusbiblioteket@sub.su.se
Erik Leissner
Biochemistry
27 000 vols; 300 curr per 26437

– Botaniska biblioteket, Botaniska institutionen (Botanical Library), Lilla Frescativägen 5, 106 91 Stockholm
T: +46 8 163759; Fax: +46 8 165525; E-mail: botaniskabiblioteket@sub.su.se; URL: www.sub.su.se
Carin Rasa
55 000 vols; 81 curr per
libr loan 26438

– Geobiblioteket, Stockholms universitet, 106 91 Stockholm
T: +46 8 164847; Fax: +46 8 164969; E-mail: geo@sub.su.se
150 000 vols 26439

– Institutionen for Journalistik, Medier och Kommunikation, JMK Biblioteket (Department of Media, Journalism, Communication), Karlavägen 104, Box 278 61, 115 93 Stockholm
T: +46 8 164408; Fax: +46 8 6610304; E-mail: jmkbiblioteket@sub.su.se
Mats Hagström
18 000 vols; 300 curr per
libr loan 26440

– Kemiska biblioteket (Chemical Library), Svante Arrhenius väg 10-12, Frescati, 106 91 Stockholm
T: +46 8 162354; E-mail: kemiskabiblioteket@sub.su.se
1974
54 200 vols; 142 curr per; 1 digital data carriers
libr loan 26441

– Latinamerika-institutet, Biblioteket (Institute of Latin American Studies), Hus B, plan 5, mail address: 10691 Stockholm
T: +46 8 162887; Fax: +46 8 156582; E-mail: lai@sub.su.se; URL: www.lai.su.se
1951; Margareta Björling
50 000 vols; 300 curr per; 90 e-journals
libr loan; REDIAL, Red Europea de Información y Documentación sobre América Latina 26442

– Matematiska biblioteket (Mathematics Library), Kräftriket, Hus 6, 106 91 Stockholm
T: +46 8 164512; Fax: +46 8 6126717; E-mail: matematiskabiblioteket@sub.su.se
1962; Barbro Fernström
42 900 vols; 435 curr per; 1 digital data carriers
libr loan; IFLA 26443

Umeå Universitet, Universitetsbiblioteket (Umea University), Samhällsvetarhuset, mail address: 901 74 ***Umeå***
T: +46 90 7865693; Fax: +46 90 7867474; E-mail: laneexp@ub.umu.se; URL: www.ub.umu.se/ombiblioteket/hb.htm
1964; Kjell Jonsson
Östergren's coll, Curman's coll, Ekström coll; EDC (Europ Doc Supply Centre)
1 000 000 vols; 9 900 curr per; 300 microforms; 100 digital data carriers
libr loan 26444

– Medicinska biblioteket (Medical Library), University Hospital, 901 87 Umeå
T: +46 90 7852660; Fax: +46 90 7866678; E-mail: medbibl@ub.umu.se; URL: www.ub.umu.se
Ingrid Harnemo
5 700 vols; 951 curr per 26445

Sveriges lantbruksuniversitet (SLU), Ultunabiblioteket (huvudbibliotek) (Swedish University of Agricultural Sciences), Undervisningsplan 10, Box 7071, 750 07 ***Uppsala***
T: +46 18 671000; Fax: +46 18 672853; E-mail: bib-ulbest@slu.se; URL: www.slu.se/sv/bibliotek/
Snorre Rufelt
Agriculture, biosciences
379 578 vols; 1 157 curr per; 4 653 e-journals; 13 000 maps; 115 av-mat
libr loan 26446

– Hernquistbiblioteket (Hernquist Library), Malmgatan 10, Hus J, Box 234, 532 23 Skara
T: +46 511 67240; Fax: +46 511 67243; E-mail: hernquistbiblioteket@bibsk.slu.se; URL: www.bib.slu.se
1977; Beata Akersten
Veterinary History, food sciences
20 000 vols; 240 curr per; 4 000 e-journals; 1 000 mss; 630 diss/theses
libr loan; EVLG 26447

– Skogsbiblioteket (Forestry Library), 901 83 Umeå
T: +46 90 7868118; Fax: +46 90 7868125; E-mail: skogsbiblioteket@bibum.slu.se; URL: www.bib.slu.se
1918; Monica Danielsson
Coll of photos
135 000 vols; 1 800 curr per; 50 av-mat; 30 digital data carriers
libr loan; IFLA 26448

– Swedish University of Agriculturel Sciences – Alnarpsbiblioteket, Sundsvägen 6 C, Box 51, 230 53 Alnarp
T: +46 40 415050; Fax: +46 40 415058; E-mail: webbredaktionen@bib.slu.se; URL: www.slu.se/?id=64
1936; Lennart Hultin
Horticulture; Wooden Library, Örtengren Coll, nursery catalogues
100 000 vols; 1 100 curr per 26449

Uppsala Universitet, Universitetsbiblioteket (Uppsala University), Dag Hammarskjölds väg 1, Box 510, 751 20 ***Uppsala***
T: +46 18 4713900; Fax: +46 18 4713913; E-mail: info@ub.uu.se; URL: www.ub.uu.se
1620; Ulf Göranson
Codex Argenteus (Silver Bible), Carta Marina, Map of Mexico City (ca 1550), Bibl Walleriana (history of science and medicine), Bodoni Coll, Medieval manuscripts, Düben Coll (Baroque music), Collection of Swedish drawings and paintings, UN depository library, Swedish

legal deposit library
5 428 761 vols; 4 897 curr per; 30 738
e-journals; 326 347 e-books; 63 354 mss;
2 580 incunabula; 9 002 music scores;
406 453 maps; 61 099 microforms; 900
av-mat
libr loan; IFLA, LIBER, SB, SFIS, CERL
26450

– Ångströmbiblioteket (Angström Library),
Lägerhyddsvägen 1 / Regementsvägen
1, 752 37 Uppsala; Box 526, 751 20
Uppsala
T: +46 18 4712920; Fax: +46 18
4712922; E-mail: angstrombibl@ub.uu.se;
URL: www.ub.uu.se/linne/ang/
1990; Monica Bäck
22 000 vols; 260 curr per
libr loan 26451

– Biologibiblioteket, EBC, Norbyvägen 14,
mail address: Villavägen 9, 752 36
Uppsala
T: +46 8 4716434; Fax: +46 8 4712929;
E-mail: Biologi.Bibl@ub.uu.se; URL:
www.ub.uu.se/linne/bio/
Lena Vretblad
30 000 vols; 400 curr per
libr loan 26452

– BMC-biblioteken (BMC-library), Husargatan
3, Box 570, 751 23 Uppsala
T: +46 18 4714087; Fax: +46 18
4714563; E-mail: bmc.bibl@ub.uu.se;
URL: www.ub.uu.se/linne/bmc/
1968; Per Syrén
Biosciences, pharmacology, chemistry,
physiology, microbiology, immunology
88 000 vols; 675 curr per
libr loan 26453

– Bostadsforskningsbiblioteket (Library of
Housing Research), Rådhuset, Box 785,
801 29 Gävle
T: +46 26 4206565; Fax: +46 26
4206566; E-mail: bf.bibl@ub.uu.se; URL:
www.ub.uu.se/sam/bf/
1994; Ingalill Halvarsson
Housing, habitation, architecture
segregation, social planning and
urbanization
30 000 vols; 150 curr per
libr loan 26454

– Carolinabiblioteket (Carolina Library),
Carolina Rediviva, Dag Hammarskjölds
väg 1, 752 37 Uppsala; Box 510,
751 20 Uppsala
T: +46 18 4713900; Fax: +46 18
4713941; E-mail: carolinabibl@ub.uu.se;
URL: www.ub.uu.se
Marie Mosshammar
Humanities, theology, sociology, print and
electronic collections from 1851 to the
present 26455

– Dag Hammarskjöldbiblioteket (Dag
Hammarskjöld Library), Klostergatan 3,
Box 6508, 751 38 Uppsala
T: +46 18 4713350; Fax: +46 18
4713999; E-mail: dh@ub.uu.se; URL:
www.ub.uu.se/sam/dh/
Margareta Ärlebrand
United Nations, International Relations,
Peace and Conflict Research
30 000 vols; 300 curr per; 1 500
microforms 26456

– Ekonomikums bibliotek (Library for
Economic Sciences), Kyrkogårdsgatan
10, plan 2, 753 13 Uppsala; Box 513,
751 20 Uppsala
T: +46 18 4711460;
Fax: +46 18 4717955; E-mail:
infolib.ekonomikum@ub.uu.se; URL:
www.ub.uu.se/linne/med/
Agneta Ljunggren 26457

– Geobiblioteket (Earth Sciences Library),
Geocentrum, Villavägen 16, 752 36
Uppsala
T: +46 18 4713360; Fax: +46 18
4713365; E-mail: geo.bibl@ub.uu.se;
URL: www.ub.uu.se/linne/bio/
1984; Susanne Ehlin
Natural enviroment, Geology, Meteorology,
Geophysics, Hydrology, Building
constructions and materials
73 612 vols; 174 curr per; 203 e-journals;
120 377 maps
libr loan 26458

– ILU-biblioteket (Education and Teaching
Library), Seminariegatan 1, Box 2136,
750 02 Uppsala
T: +46 18 4712410; Fax: +46 18
4712407; E-mail: ilu-bibl@ub.uu.se; URL:
www.ub.uu.se/am/ilu
1965; Åsa Adebäck Lowén
Montessori Coll
165 000 vols; 120 curr per; 196 e-
journals; 126 av-mat; 330 digital data
carriers
libr loan; SB 26459

– Juridiska biblioteket (Law Library),
Klostergatan 3, Box 6508, 751 38
Uppsala
T: +46 18 4717850; Fax: +46 18
4713999; E-mail: jur.bibl@ub.uu.se; URL:
www.ub.uu.se/sam/jur
1962; Margareta Ärlebrand
European Documentation Center
184 095 vols; 236 curr per; 2 096 e-
journals
libr loan; IFLA 26460

– Karin Boye-biblioteket (Karin Boye
Library), Thunbergsvägen 3H, 752 38
Uppsala; Box 637, 751 26 Uppsala
T: +46 18 4717010; Fax: +46 18
4713941; E-mail: boyebibl@ub.uu.se;
URL: www.ub.uu.se/hum/karinboye/
index.cfm
Kerstin Mittelsson
Humanities, theology, sociology, literature
published up to 1962 26461

– Medicinska biblioteket (Medical Library),
Akademiska sjukhuset, Kunskapscentrum
Ingång 61, 751 85 Uppsala
T: +46 18 6113531; Fax: +46 18
503614; E-mail: medicinskabibl@ub.uu.se;
URL: www.ub.uu.se/en
Gunilla Sandstroem
SLA 26462

– Pedagogiska biblioteket (Library for
Educational Research and Studies),
Sturegatan 4, Box 2109, 750 02 Uppsala
T: +46 18 4711649; Fax: +46 18
4711658; E-mail: ped-bibl@ub.uu.se;
URL: www.ub.uu.se/sam/ped
Åsa Adebäck Lowén
22 000 vols; 160 curr per
libr loan 26463

Mälardalens högskola,
Högskolebiblioteket (Mälardalen
University), Högskoleplan 1, Box 832,
721 22 *Västerås*
T: +46 21 101344; Fax: +46 21 101340;
E-mail: biblioteket.vasteras@mdh.se; URL:
www.mdh.se/bib
Annsofie Oscarsson
112 000 vols; 420 curr per; 9 470 e-
journals; 456 music scores; 506 av-mat
libr loan; Swedish Library Ass, SFIS,
IFLA 26464

Växjö Universitet, Universitätsbibliotek,
Universitetsplatsen 2, 351 95 *Växjö*
T: +46 470 70-8400/-8680; Fax: +46 470
84523; E-mail: biblioteket@vxu.se; URL:
www.bib.vxu.se
1967; Mats Herder
235 000 vols; 850 curr per; 11 500 e-
journals
libr loan; IFLA 26465

Almedalsbiblioteket, Stads- och
högskolebibliotek på Gotland,
Cramérgatan 5, Box 1121, 621 22 *Visby*
T: +46 498 299000;
Fax: +46 498 299011; E-mail:
almedalsbiblioteket@hgo.se; URL:
bibliotek.gotland.se
1865; Kerstin Simberg
Gotland coll
libr loan; SB 26466

School Libraries

Runö folkhögskola, Biblioteket, Näsvägen
100, 184 92 *Åkersberga*
T: +46 8 55538000; Fax: +46 8
55538090; E-mail: run@runo.se; URL:
www.runo.se
Lasse Karlsson
70 000 vols; 200 curr per 26467

Björkängsgymnasiet, Biblioteket,
Klinikvägen 48, 504 57 *Borås*; mail
address: 501 80 Borås
T: +46 33 355803; Fax: +46 33 355801;
URL: www.utb.boras.se/uk/bj
1980
18 300 vols; 75 curr per; 300 av-mat; 50
digital data carriers
SB 26468

Stockholm School of Theology, Library,
Åkeshovsvägen 29, 168 39 *Bromma*
T: +46 8 56435725;
Fax: +46 8 56435706; E-mail:
birgitta.normosse@ths.se
1908; Birgitta Norrmosse
50 000 vols; 235 curr per
libr loan 26469

Brönnkyrka gymnasium, Biblioteket,
Tellusborgsvägen 10, 126 32 *Hägersten*
T: +46 8 189888
Gudrun Wängsell
40 000 vols; 75 curr per 26470

Tornedalsskolan, Gymnasiebibliotek,
Köpmangatan 21, mail address: 953 85
Haparanda
T: +46 922 15088; Fax: +46 922 10260;
E-mail: tornedalsskolan@haparanda.se;
URL: www.tskola.haparanda.se
Rickard Molarin
50 000 vols; 126 curr per 26471

Jönköping skolbibliotekscentral,
Linnégatan 2, 561 24 *Huskvarna*; Box
1002, 561 24 Huskvarna
T: +46 36106742; Fax: +46 36
107098; E-mail: elisabeth.fridsall-
emilsson@jonkoping.se; URL:
www.biblc.edu.jonkoping.se
Elisabeth Fridsäll-Emilsson
300 000 vols; 130 digital data carriers
IASL 26472

Nordiska folkhögskolan bibliotek
(Nordic Folk High School Library), Olof
Palmes väg 1, 442 31 *Kungälv*
T: +46 303 206216; E-mail:
eric.hellman@nordiska.fhsk.se; URL:
www.nordiska.fhsk.se
1947; Eric Hellman
20 000 vols; 120 curr per; 10 maps; 100
av-mat; 600 sound-rec; 1 digital data
carriers
libr loan 26473

Berzeliusskolan, Tekniskt-
naturvetenskapligt utbildningscentrum,
Biblioteket, Gustav Adolfsgatan 25, Box
3129, 580 03 *Linköping*
T: +46 13 207480; Fax: +46 13 207481;
E-mail: Sonja.Landkvist@linkoping.se;
URL: www.edu.linkoping.se/berzelius/
bibliotek.htm
Sonja Landkvist
15 000 vols; 230 curr per 26474

Värnhemsskolan, Gymnasiebibliotek,
Kungsgatan 44, 212 13 *Malmö*; Box
3506, 200 22 Malmö
T: +46 40 343155; Fax: +46 40 302118
Cecilia Forssman
40 000 vols; 120 curr per; 100 av-mat
26475

Fässbergsgymnasiets bibliotek,
Fiskargatan 4, 431 43 *Mölndal*
T: +46 31 677362; Fax: +46 31 677338
Barbro Holmstrand
30 000 vols; 115 curr per 26476

Platengymnasiet, Mediateket, Box 955,
591 29 *Motala*
T: +46 141 225677;
Fax: +46 141 219624; E-mail:
Platengymnasiet@motala.se
51 000 vols; 150 curr per 26477

Risbergska skolan, Bibliotek, Hagagatan
53, Box 31150, 701 35 *Örebro*
T: +46 19 216547; Fax: +46 19 216600;
E-mail: risbergska@orebro.se; URL:
www.risbergska.orebro.se
Karin Fredriksson
35 000 vols; 220 curr per 26478

LO-skolan Hasseludden,
Hamndalsvägen, 132 01 *Saltsjö-Boo*
T: +46 8 7478820
Birgitta Dahlander
17 000 vols; 150 curr per 26479

Adolf Fredriks musikklasser,
Skolbibliotek, Tegnergatan 44-46, 113 29
Stockholm
T: +46 8 50844874; URL: afmu.nu
Inga-Lill Guve-Snell
11 000 vols; 10 curr per 26480

Röda Korsets Högskola, Biblioteket,
Teknikringen 1, Box 55676, 102 15
Stockholm
T: +46 8 58751678; E-mail: biblioteket@
rkh.se; URL: biblioteket.rkh.se
Sabine Anderberg
16 000 vols; 175 curr per; 80 diss/theses
libr loan; SB 26481

Hedbergska skolan, Biblioteket,
Skolhusallén 6, 852 37 *Sundsvall*
T: +46 60 192780
Britta Torell
70 000 vols; 190 curr per 26482

Dragonskolans bibliotek, Dragongatan 1,
903 22 *Umeå*
T: +46 90 162450; Fax: +46 90 127308;
E-mail: biblioteket.dragonskolan@umea.se
1971; Kerstin Tägtsten
30 000 vols; 310 curr per; 1 000
sound-rec; 11 digital data carriers; 3 000
slides
SB 26483

Östra gymnasiet, Biblioteket,
Fridhemsvägen 2, 903 37 *Umeå*
T: +46 90 162907; E-mail:
christer.edeholt@umea.se; URL:
www.umea.se/ostra
1957; Christer Edeholt
35 000 vols; 120 curr per 26484

Government Libraries

Vägverket, Biblioteket (Swedish National
Road Administration), mail address:
781 87 *Borlänge*
T: +46 243 75059; Fax: +46 243 75717;
E-mail: vagverket.biblioteket@vv.se; URL:
www.vv.se
Anna Maria Magnusson
Rpt of Swedish National Road
Administration
35 000 vols; 460 curr per; 40 digital data
carriers
SFIS 26485

Försvarets Radioanstalt, Rörbyvägen
(Lovön), 178 93 *Drottningholm*; Box
301, 161 26 Bromma
T: +46 8 4714600; Fax: +46 8 4714853;
E-mail: fra@fra.se; URL: www.fra.se/
bibliotek.shtml
1942; Göran Rydeberg
15 000 vols; 200 curr per
libr loan 26486

Göteborgs stad, Traktörens
förvaltningsbibliotek (City of Gothenburg,
The Government Library), Köpmansgatan
20, Box 2554, 403 17 *Göteborg*
T: +46 31 612442; Fax: +46 31 133297;
URL: www.fastighetskontoret.goteborg.se
1979; Viveca Nyström
Lit about Gothenburg. Own departments
publ
25 000 vols; 170 curr per; 2 500 govt
docs; 5 av-mat; 15 digital data carriers
SB, SBS, SFIS 26487

Hovrätten för Västra Sverige, Biblioteket
(Court of Appeal for Western Sweden),
Packhusplatsen 6, Box 40, 401 20
Göteborg
T: +46 31 7012234;
Fax: +46 31 7742943; E-mail:
hovratten.vastrasverige@dom.se; URL:
www.vastrahovratten.dom.se
1948; Stefan Jacobson
Law
50 000 vols; 50 curr per 26488

Kammarrätten i Göteborg, Biblioteket
(Fiscal Court of Appeal), Box 1531,
401 50 *Göteborg*
T: +46 31 105000; Fax: +46 31 105144;
URL: www.kammarratten.goteborg.se
1972; Britt Bergh
Social insurance, social services, taxation
law
40 000 vols; 59 curr per
libr loan; SBS, SFIS 26489

Landstinget Västernorrland,
Förvaltningsbibliotek, Storgatan 1,
871 85 *Härnösand*
T: +46 611 80098; Fax: +46 611 80200;
E-mail: landstinget.vasternorrland@lvn.se;
URL: www.lvn.se
Vivi-Anne Höggvist
10 000 vols; 200 curr per 26490

Domstolsverket, Kyrkogata 34, 551 81 *Jönköping*
T: +46 36 155317; Fax: +46 36 165721
Hans Rundberg
Law, administration
12 000 vols; 40 curr per　　　　26491

Kammarrätten i Jönköping, Biblioteket (Administrative Court of Appeal), Slottsgatan 5, Box 2203, 550 02 *Jönköping*
T: +46 36 156917; Fax: +46 36 161968;
E-mail: kammarrattenijonkoping@dom.se;
URL: www.dom.se/kammarrattenijonkoping
1977; Elisabeth Nordström
Parliamentary papers, gov rpts
15 750 vols; 55 curr per　　　　26492

Landstinget i Kalmar län,
Förvaltningsbiblioteket, Strömgatan 13, Box 601, 391 26 *Kalmar*
T: +46 480 84876; Fax: +46 480 84197;
URL: www.ltkalmar.se
Eva Nilsson
12 000 vols; 150 curr per　　　　26493

Boverket, Biblioteket (National Board of Housing, Building and Planning), Drottninggatan 18, Box 534, 371 23 *Karlskrona*
T: +46 455 353138;
Fax: +46 455 15078; E-mail:
bengt.fernstedt@boverket.se; URL:
www.boverket.se
Bengt Fernstedt
Architecture, housing, urban planning
34 000 vols; 223 curr per　　　　26494

Länsstyrelsen i Norrbottens län,
Biblioteket, Stationsgatan 5, 971 86 *Luleå*
T: +46 920 96000; Fax: +46
920 228411; E-mail:
Norrbotten@lansstyrelsen.se; URL:
www.lansstyrelsen.se/norrbotten/Sv/Pages/
default.aspx
Pub Admin
70 000 vols; 100 curr per　　　　26495

Förvaltningsbiblioteket, Region Skåne,
Baravägen 1, Box 1, 221 00 *Lund*
T: +46 46153351; Fax: +46 46153472;
URL: www.skane.se/bibliotek
1970; Ann-Christine Hansson
Development, organization, public health, hygiene
8 000 vols; 35 curr per; 10 e-journals;
4 400 govt docs
libr loan; SB, SBS, SFIS　　　26496

Kriminalvården, Huvudkontoret, Biblioteket (Swedish Prison an d Probation Service), 601 80 *Norrköping*
T: +46 77 2280800;
Fax: +46 11 4963640; E-mail:
f.norrkoping@kriminalvarden.se; URL:
www.kriminalvarden.se
Dan Lönnblom
20 000 vols; 100 curr per
libr loan　　　　26497

Migrationsverket (Swedish Migration Board), Tegelvägsgatan 19A, *Norrköping*; mail address: 601 70 Norrköping
T: +46 11 156245; Fax: +46 11 156341;
E-mail: biblioteket@migrationsverket.se;
URL: www.migrationsverket.se
1976; Tommy Jansson
13 000 vols; 100 curr per; 20 e-journals
libr loan　　　　26498

Sjöfartsverket, Slottsgatan 82, 601 78 *Norrköping*
T: +46 11 191003; Fax: +46 11 189184;
E-mail: hk@sjofartsverket.se; URL:
www.sjofartsverket.se
Inter-Governmental Maritime Consultative Organization (IMCO), International Hydrographic Review and Bulletin, ship lists, dept publs
14 000 vols; 400 curr per　　　26499

Länsstyrelsen i Örebro län, Stortorget 22, 701 86 *Örebro*
T: +46 19 193809; Fax: +46 19 193010;
URL: www.t.lst.se
Inger Bennermyrf
Pub Admin
8 400 vols; 80 curr per　　　　26500

Försvarets materielverk, FMV
– Teknik för Sveriges säkerhet,
Biblioteket (Swedish Defence Material Administration), Banérgatan 62, 115 88 *Stockholm* 80
T: +46 8 7826781; Fax: +46 8
7825070; E-mail: bibliotek@fmv.se; URL:
www.fmv.se
1971; Bengt Kleijn, Åsa Jenslin
Aviation, defence technology
150 000 vols; 100 curr per; 2 500 microforms; 15 digital data carriers
libr loan　　　　26501

Högskoleverket, Biblioteket (Swedish National Agency for Higher Education), Birger Jarlsgatan 43, Box 7851, 103 99 *Stockholm*
T: +46 8 56308500; Fax: +46 8
56308550; E-mail: hsv@hsv.se; URL:
www.hsv.se
1976; Margareta Schaub
23 000 vols; 210 curr per; 65 diss/theses; 2 890 govt docs; 30 microforms
libr loan; SBS, SFIS　　　26502

Högsta domstolen, Biblioteket (Supreme Court of Sweden), Riddarhustorget 8, Box 2066, 103 12 *Stockholm*
T: +46 8 6176400; Fax: +46 8 6176521;
E-mail: hogsta.domstolen@dom.se; URL:
www.hogstadomstolen.se
Ann Charlotte Edenman
110 000 vols; 80 curr per; 4 e-journals
libr loan　　　　26503

Högsta förvaltningsdomstolens bibliotek (The Supreme Administrative Court Library), Birger Jarls Torg 13, 111 28 *Stockholm*; Box 2293, 103 17 Stockholm
T: +46 8 56167600; Fax: +46
8 56167820; E-mail: HFDD-bibliotek@dom.se
1909; Helen Otterhall
Yearbook of the Supreme Administrative Court (Högsta förvaltningsdomstolens årsbok / HFD)
45 000 vols; 70 curr per; 20 e-journals; 23 000 govt docs
IALL, SB, SFIS　　　26504

Jordbruksdepartementet, Drottninggatan 21, 103 33 *Stockholm*
T: +46 8 7631115; Fax: +46 8 206496
1978; Karin Kristofferson
17 500 vols; 400 curr per
libr loan　　　　26505

Kammarrätten i Stockholm, Bibliotek (Administrative Court of Appeal in Stockholm), Birger Jarls torg 13, Box 2302, 103 17 *Stockholm*
T: +46 8 7003800; Fax: +46 8 4118928
1945; Ulla Renberg
49 500 vols; 50 curr per; 2 750 maps
libr loan　　　　26506

Konkurrensverket, Biblioteket (Swedish Competition Authority), Sveavägen 167, mail address: 103 85 *Stockholm*
T: +46 8 7001600; Fax: +46 8 245543;
E-mail: biblioteket@kkv.se; URL:
www.konkurrensverket.se
1993
Competition
5 000 vols; 150 curr per; 4 digital data carriers　　　　26507

Länsstyrelsen i Stockholms län,
Biblioteket (County Administrative Board), Hantverkargatan 29, Box 22067, 104 22 *Stockholm*
T: +46 8 7854262; Fax: +46 8 6515295;
E-mail: stockholm@lansstyrelsen.se; URL:
www.ab.lst.se
1972; Elisabeth Frisén
Public administration
10 000 vols; 65 curr per
libr loan　　　　26508

Regeringskansliet, Biblioteket (Swedish Government Offices, Library), Drottninggatan 21, 103 33 *Stockholm*
T: +46 8 4054237; Fax: +46 8 4054979;
E-mail: centralbiblioteket@adm.ministry.se;
URL: www.sweden.gov.se/
Inger Jepsson
250 000 vols; 550 curr per　　　26509

Regeringskansliets Bibliotek (Swedish Government Offices), Drottninggatan 21, 10333 *Stockholm*
T: +46 8 4051000; E-mail:
centralbiblioteket@adm.ministry.se; URL:
www.regeringen.se; www.sweden.gov.se
1900; Inger Jeppson
Int relations, int law
90 000 vols; 300 curr per; 40 e-journals
IFLA　　　　26510

Riksdagsbiblioteket (Swedish Parliament Library), Storkyrkobrinken 7, mail address: 100 12 *Stockholm*
T: +46 20 555000; Fax: +46 8 7865870;
E-mail: biblioteket@riksdagen.se;
URL: www.riksdagen.se/templates/
R_SubStartPage___448.aspx
1851; Margareta Brundin
Swedish govt publs, legal depot; United Nations publs, depot libr; Publs of the European Union, depot libr; Archives for Swedish parliamentary documents
700 000 vols; 2 000 curr per
libr loan; IFLA, SFIS, SB　　　26511

Riksförsäkringsverket, Biblioteket (National Social Insurance Board), Adolf Fredriks Kyrkogata 8, 103 51 *Stockholm*
T: +46 8 7869000; Fax: +46 8 7869064;
URL: www.rfv.se
Alena Smidova
15 000 vols; 150 curr per　　　26512

Rikspolisstyrelsen, Biblioteket (National Swedish Police Board), Box 12256, 102 26 *Stockholm*
T: +46 8 4019988; Fax: +46 8 4019990
1965
Criminal investigation, police legislation, criminal law
15 000 vols; 300 curr per
SFIS　　　　26513

Riksrevisionen, Biblioteket (Swedish National Audit Office), Nybrogatan 55, 114 90 *Stockholm*; mail address: 114 90 Stockholm
T: +46 8 51714040;
Fax: +46 8 51714100; E-mail:
biblioteket@riksrevisionen.se; URL:
www.riksrevisionen.se
2003; Marguerite Frank
Accounting, economy, political science, government documents
25 000 vols; 100 curr per; 50 e-journals; 200 diss/theses; 10 000 govt docs
libr loan; SB, SFIS　　　26514

Skatteverkets bibliotek, Fatburen, 106 61 *Stockholm*
T: +46 8 6941945; Fax: +46 8 6418083;
URL: www.skatteverket.se
1987; Irene Rurling
Coll of Swedish National Tax Board publs
17 100 vols; 215 curr per; 20 e-journals; 1 000 govt docs; 80 sound-rec; 30 digital data carriers
libr loan; SFIS, SB　　　26515

Stadshusets Bibliotek, Stadshuset, 105 35 *Stockholm*
T: +46 8 50829095;
Fax: +46 8 50829110; E-mail:
susann.ronnholm@stadshuset.stockholm.se
1940; Susann Rönnholm
Social, health and welfare services; city council docs
60 000 vols; 100 curr per
libr loan　　　　26516

Statens kulturråd (Swedish National Council for Cultural Affairs), Långa raden 4, Skeppsholmen, Box 7843, 103 98 *Stockholm*
T: +46 8 51926400;
Fax: +46 8 51926499; E-mail:
kulturradet@kulturradet.se; URL:
www.kulturradet.se
1974
9 000 vols; 400 curr per
SB, IFLA, EBLIDA　　　26517

Statens strålskyddsinstitut, Biblioteket (Swedish Radiation Protection Institute), Karolinska sjukhuset, 171 16 *Stockholm*
T: +46 8 7297100; Fax: +46 8 7297108;
URL: www.ssi.se
1969; Rosmarie Wahlbeck
SSI rpts, IAEA's publs
12 000 vols; 150 curr per
libr loan; SBS, NVBF, Nordic Group of Radiation Protection Libraries, SFIS　　　26518

Statskontoret, Biblioteket, Fleminggatan 20, Box 8110, 104 20 *Stockholm*
T: +46 8 4544644; Fax: +46 8 7918972;
E-mail: statskontoret@statskontoret.se;
URL: www.statskontoret.se
Veronica Strid Sandberg
Information science, public administration, economics
9 000 vols; 160 curr per　　　26519

Stockholms läns landsting – Förvaltningsutskottets bibliotek,
Landstings Kontorets Bibliotek, Box 22550, 104 22 *Stockholm*
T: +46 8 7374229; Fax: +46 8 7374443
1971; Maria Hjorth
Administration, politics, public health, nursing
23 100 vols; 187 curr per
libr loan　　　　26520

Stockholms tingsrätt, Biblioteket (Stockholm City Court Library), Scheelegatan 7, Box 8307, 104 20 *Stockholm*
T: +46 8 561 65000;
Fax: +46 8 561 65011; E-mail:
stockholms.tingsratt.bibliotek@dom.se;
URL: www.stockholmstingsratt.se
Eva Ljungstedt
Law
16 000 vols; 50 curr per; 4 000 govt docs; 35 digital data carriers
Svensk Biblioteksförening　　　26521

Svea hovrätt, Biblioteket (Svea Court of Appeal), Birger Jarls torg 16, Box 2290, 103 17 *Stockholm*
T: +46 8 7003400; Fax: +46 8 7003438;
E-mail: svea.hovratt@dom.se; URL:
www.svea.se
Dan Lundin
Law
36 000 vols; 100 curr per　　　26522

Tekniska Nämndhuset, Biblioteket (Technical Office Building), Fleminggatan 4, 112 26 *Stockholm*
T: +46 8 50827144; Fax: +46 8 6506502
1964; Eva Wetterlind
Transport and traffic, building and construction, civil engineering, public works (highway and traffic engineering), city planning, architecture
30 000 vols; 350 curr per; 100 maps
libr loan　　　　26523

Court of Appeal for Southern Norrland, Library, Södra Tjärngatan 2, Box 170, 851 03 *Sundsvall*
T: +46 60 186600; Fax: +46 60 186839;
E-mail: hovratten.nedrenorrland@dom.se;
URL: www.hovrattenfornedrenorrland.se/
Per Helttunen
30 500 vols; 55 curr per　　　26524

Västra Götaland Region
Litteraturtjänst, Litteraturtjänst, 462 80 *Vänersborg*
T: +46 521 275000; Fax: +46 521
275600; URL: www.vgregion.se
Mona Suneson
Public Health, county council policies and administration, EU and European regions
10 000 vols; 100 curr per
libr loan　　　　26525

Länsstyrelsen i Västmanlands län,
Västra Ringvägen 1, 721 86 *Västerås*
T: +46 21 190500; Fax: +46 21 195135;
E-mail: vastmanland@lansstyrelsen.se;
URL: www2.vastmanland.lst.se
Gudrun Lundqvist
Pub Admin
10 000 vols; 185 curr per　　　26526

Corporate, Business Libraries

Saint-Gobain Isover AB, Biblioteket, Box 501, 260 50 *Billesholm*
T: +46 42 84000; Fax: +46 42
72314; E-mail: isover@isover.se; URL:
www.isover.se
Mats Lindquist
11 000 vols; 133 curr per　　　26527

AB Bahco Ventilation, Biblioteket, Stockholmsvägen, 199 01 *Enköping* 1
T: +46 171 22000
Ulf Brangfelt
Aeronautics, aviation, space technology
17 000 vols; 132 curr per　　　26528

VME Industries Sweden AB, Biblioteket, Tekniskt Centrum, 631 85 *Eskilstuna*
T: +46 16 151000; Fax: +46 16 152938
Lilian Hedman
10 000 vols; 155 curr per   26529

Stora Enso Research, Library, Södra Mariegatan 18, 791 80 *Falun*
T: +46 1046 80000; Fax: +46 23 17803;
E-mail: library.falun@storaenso.com; URL: www.storaenso.com
1917; Nils-Erik Engström
Paper technology
20 000 vols; 300 curr per; 51 av-mat; 45 digital data carriers
SFIS   26530

Götaverken Energy Systems, Biblioteket, Box 8734, 402 75 *Göteborg*
Monica Selestam
Marine technology, naval engineering
10 000 vols; 120 curr per   26531

Mannheimer Swartling Advokatbyrå – Göteborg, Biblioteket, Lilla Torget 1, Box 2235, 403 14 *Göteborg*
T: +46 31 3551600;
Fax: +46 31 3551601; E-mail: florence.wedberg@msa.se; URL: www.mannheimerswartling.se
Florence Wedberg
10 000 vols; 100 curr per   26532

Volvo Technology Corporation, Information Retrieval & Library, Dept. 6830, M1.4, 405 08 *Göteborg*
T: +46 31 3234039; Fax: +46 31 541346; E-mail: Biblioteket@volvo.com
1948; Ann-Britt Sedig
48 000 vols; 800 curr per; 300 e-books
SFIS, EUSIDIC   26533

Kanthal AB, Hans von Kantzows väg, Box 502, 734 27 *Hallstahammar*
T: +46 220 21733; Fax: +45 220 21193
Ulf Franzén
22 000 vols; 400 curr per   26534

Kabi Pharmacia Therapeutics AB, Biblioteket, Ringstorpsvägen, Box 941, 251 09 *Helsingborg*
T: +46 42 104000; Fax: +46 42 104096
Erik Helmer
16 000 vols; 500 curr per   26535

Höganäs AB, Biblioteket, Bruksgt, 263 83 *Höganäs*
T: +46 42 338000; Fax: +46 42 338150; E-mail: info@hoganas.com; URL: www.hoganas.com
1947; Ulf Carlbark
Chemical Engineering, Powder Metallurgy
12 000 vols; 120 curr per; 15 e-journals; 10 microforms; 20 digital data carriers
libr loan; SFIS   26536

AarhusKarlshamn Sweden AB, Library, 374 82 *Karlshamn*
T: +46 454 82201; Fax: +46 454 752025; E-mail: bibliotek@aak.com; URL: www.aarhuskarlshamn.com
Inga George
Patents
10 000 vols; 55 curr per; 10 e-journals; Patents
SLA, SFIS   26537

Luossavaara-Kiirunavaara AB (LKAB), Biblioteket, Fack, 981 01 *Kiruna* 1
Harald Bildt
8 000 vols; 230 curr per   26538

Syngenta Seeds AB, Biblioteket, Box 302, 261 23 *Landskrona*
T: +46 418 437125; Fax: +46 418 437155
Gunilla Björklund
Genetics, biotechnol, biochem, plant breeding seed technology
9 000 vols; 450 curr per; 100 diss/theses; 10 microforms
IAALD, SFIS, DTL   26539

Astra Draco AB, Biblioteket, Scheelvägen 12, Box 34, 221 00 *Lund*
T: +46 46 336000; Fax: +46 46 336933;
URL: www.astra.com
1978; Anna Carin Larssen
Old chemical lit, Coll of J.J. Berzelius works
8 000 vols; 550 curr per
libr loan   26540

Awapatent AB, Biblioteket, Box 5117, 200 71 *Malmö*
T: +46 40 985100; Fax: +46 40 260516;
URL: www.awapatent.com
G. Ekegren
10 000 vols   26541

Ferring Läkemedel AB, Gerjersgatan 2, Box 4041, 203 11 *Malmö*
T: +46 40 6916900; Fax: +46 40 6916995; E-mail: info@ferring.se; URL: www.ferring.se
Catharina Lindgren-Isberg
Chem, med, peptide
8 800 vols; 130 curr per
Eahil, SFIS   26542

Astra Hässle AB, Biblioteket, Körragatan 3, 431 83 *Mölndal* 1
T: +46 31 7761000; Fax: +46 31 7763726
1967; Karin Malmcrona-Friberg
Cardiovascular medicine, analytical chemistry, gastrointestinal medicine, organic chemistry
9 000 vols; 550 curr per; 1 000 diss/theses; 500 govt docs
libr loan; EAHIL, MLA, SFIS   26543

M-real Corporation, Corporate R & D, Library, Hömeborgsvägen 12, 891 80 *Örnsköldsvik*
T: +46 660 75012; Fax: +46 660 75981; E-mail: riikka.joukio@m-real.com; URL: www.m-real.com
1961; Ewa Kristmansson, Bengt Karlsson
Paper, chemistry, physics
85 curr per
libr loan; SFIS   26544

SSAB Oxelösund, Biblioteket, 613 80 *Oxelösund*
T: +46 155 254000; Fax: +46 155 254073; E-mail: info@ssabox.com; URL: www.ssabox.com
1961; Christina Tengroth
10 000 vols; 330 curr per
SFIS   26545

Perstorp AB, Biblioteket, Forskarvägen, 284 80 *Perstorp*
T: +46 435 38555; Fax: +46 435 38920; E-mail: berit.ekstroem@perstorp.com; URL: www.perstorp.se
Berit Ekström
Industrial chem
30 000 vols; 488 curr per   26546

Sandvik Steel AB, R & D Center, Biblioteket, 811 81 *Sandviken*
T: +46 26 264475; Fax: +46 26 264330
Monica Olofsson
Industrial chemistry
22 000 vols; 700 curr per
libr loan   26547

AssiDomän Skärblacka AB, 617 10 *Skärblacka*
T: +46 11 245536
Solveig Svensson
Paper, environment
10 000 vols; 130 curr per   26548

Boliden Mineral AB, Biblioteket, 932 81 *Skelleftehamn*
T: +46 910 773237; Fax: +46 910 773909; URL: www.boliden.se
1935; Björn Krikortz
26 000 vols; 120 curr per
SFIS   26549

MoDo Paper AB, Packaging Biblioteket, 617 10 *Skörblacka*
T: +46 11 235536; Fax: +46 11 59410
1970; Solveig Svensson
Forestry, industrial chemistry, pulp and paper
24 000 vols; 340 curr per
libr loan   26550

Astra Zeneca RoD Södertälje, Information Science & Library, 151 85 *Södertälje*
T: +46 8 55329190; Fax: +46 8 55328879
Dr. Elisabeth Kjellander
Pharmacy, med, chem
26 000 vols; 930 curr per; 10 digital data carriers
Eahil, Eusidic, MLA   26551

Scania Partner AB, Bibliotek, Tekniskt centrum, 151 87 *Södertälje*
T: +46 8 55385333; Fax: +46 8 55385350
1948; Bengt Sterner
Science and technology
40 000 vols; 700 curr per; 75 diss/theses; 200 govt docs; 100 maps; 15 digital data carriers
SBS, SB, SLA, SFIS   26552

Aftonbladet, Biblioteket, Vattugatan 12, 105 18 *Stockholm*
T: +46 8 7252000; URL: www.aftonbladet.se
14 000 vols; 224 curr per   26553

K-Konsult, Biblioteket, Liljehomsvägen 18, Box 47044, 100 74 *Stockholm*
T: +46 8 7758800; Fax: +46 8 7441623;
URL: www.kkonsult.se
1970; Bengt Starck
17 000 vols; 400 curr per   26554

Mannheimer Swartling Advokatbyrå – Stockholm, Biblioteket, Norrmalmstorg 4, Box 1711, 111 87 *Stockholm*
T: +46 8 50576500;
Fax: +46 8 50576501; URL: www.mannheimerswartling.se
Vera Stenberg
12 000 vols   26555

Pharmacia & Opjohn AB, Biblioteket & Dokumentation, Lindhagensgatan 133, 112 87 *Stockholm*
T: +46 8 6958000; Fax: +46 8 6954013
1950; Anders Löwenborg
28 000 vols; 900 curr per
libr loan; SB, SFIS, EAHIL   26556

Svensk Musik (Swedish Music Information Centre), Sandhamnsgatan 79, Box 27327, 102 54 *Stockholm*
T: +46 8 7838858; Fax: +46 8 7839510; E-mail: gustaf.bergel@mic.se; URL: www.mic.se
1965; Gustaf Bergel
Swedish popular music; EAM Hire coll, Edition Suecia (Scores), Phono Suecia (CDs)
75 curr per; 15 000 mss; 10 000 music scores; 9 857 sound-rec
IAML   26557

Volvo Aero Corporation, Biblioteket, Fack, 461 81 *Trollhättan*
T: +46 520 94294
Eva-Britt Fridell
36 800 vols; 292 curr per   26558

Special Libraries Maintained by Other Institutions

Sjukhuset i Arvika, Rackstavägen, 671 80 *Arvika*
T: +46 570 712000; Fax: +46 570 16789
Ulrika Sortelius
11 000 vols; 136 curr per   26559

Immigrant-institutet, Biblioteket (Immigrant Institute), Katrinedalsgatan 43, 504 51 *Borås*
T: +46 33136070; E-mail: migrant@immi.se; URL: www.immi.se
1973; Miguel Benito
Immigrant authors, immigrant journals
12 000 vols; 200 curr per
libr loan   26560

Södra Älvsborgs Sjukhus Borås, Medicinska biblioteket (Southern Älvsborg Hospital Borås), Brömhultsvägen 43, 501 82 *Borås*
T: +46 33 161957; Fax: +46 33 161958; E-mail: sas@vgregion.se; URL: sas.vgregion.se
1969; Lena Persson
10 000 vols; 400 curr per; 1 320 sound-rec
SFIS   26561

SP Sveriges Provnings- och Forskningsinstitut, Biblioteket (SP Swedish National Testing and Research Institute), Box 857, 501 15 *Borås*
T: +46 33 165000; Fax: +46 33 135502; E-mail: elisabeth.sandstrom@sp.se; URL: www.sp.se
1920; Elisabeth Sandström
NBS pubs, standards
17 000 vols; 900 curr per
libr loan; SFIS   26562

Banverket Biblioteket (Swedish National Rail Administration), Jussi Björlings väg 2, 781 85 *Borlänge*
T: +46 243 445000; Fax: +46 243 445548; URL: www.banverket.se
Björn Svahn
Railroad; UIC/ERRI-docs
12 000 vols; 400 curr per
libr loan   26563

Institut Mittag-Leffler, Biblioteket, Auravägen 17, 182 60 *Djursholm*
T: +46 8 6220560; Fax: +46 8 6220589;
E-mail: ragstedt@mittag-leffler.se; URL: www.mittag-leffler.se
Mikael Rågstedt
Mathematics
50 000 vols; 300 curr per   26564

Lasarettet i Enköping, Infcenter/Bibliotek (Enköping Hospital), Kungsgatan 71, Box 908, 745 25 *Enköping*
T: +46 171 418234; Fax: +46 171 418235; E-mail: infocenter.le@lul.se; URL: lul.se/templates/page.3611.aspx
Gunwor Weinmann
9 404 vols; 91 curr per; 1 200 e-journals   26565

Talboks- och punktskriftsbiblioteket (Swedish Library of Talking Books and Braille (TPB)), Sandsborgsvägen 52, mail address: 122 88 *Enskede*
T: +46 8 399350; Fax: +46 8 6599467;
E-mail: info@tpb.se; URL: www.tpb.se
1980; Ingar Beckman Hirschfeldt
66 000 talking book titles, 9 700 braille titles
IFLA, SB   26566

Dalens sjukhus, Åstorpsringen 6, 121 87 *Enskededalen*
T: +46 8 6487701; Fax: +46 8 394794
Björn Westberg
8 400 vols; 63 curr per   26567

Fortifikationsförvaltningen, Forskningsbyrån (Royal Swedish Fortifications Administration), Kungsgatan 43, 631 89 *Eskilstuna*
T: +46 16 154000; Fax: +46 16 133702
Inger Jöderberg
10 000 vols; 50 curr per
libr loan   26568

Mäarsjukhuset, Biblioteket, 631 88 *Eskilstuna*
T: +46 16 103897; Fax: +46 16 519822
Ingrid Frennesson
10 000 vols; 300 curr per   26569

Sjukhusbiblioteken i Sörmland, Mälarsjukhuset (The Hospital Libraries of Sörmland County Council), 63188 *Eskilstuna*
T: +46 16 104747; Fax: +46 16 519822 245967; E-mail: sjukhusbiblioteken@dll.se; URL: www.landstinget.sormland.se/sjukhusbiblioteken
Helena Kettner Rudberg
40 000 vols; 300 curr per; 10 e-journals
libr loan; SB   26570

Gällivare sjukhus, Biblioteket, 982 82 *Gällivare*
T: +46 970 19159; Fax: +46 970 19298
Olle Flemström
16 000 vols; 270 curr per   26571

Länssjukhuset Gävle-Sandvihen, Biblioteket, 801 87 *Gävle*
T: +46 26 154980; Fax: +46 26 154978;
E-mail: Biblioteket@lg.se
Marianne Hultén
21 000 vols; 350 curr per   26572

Lantmäteriverket, Biblioteket (National Land Survey), Lantmäterigatan 2, 801 82 *Gävle*
T: +46 26 633615; Fax: +46 26 687594;
URL: www.lantmateriet.se
Gertrud Wiking
20 000 vols; 230 curr per   26573

LOs fackliga utbildningar, Örenäs Folkhögskola, Biblioteket, Örenäs slott, 261 63 *Glumslöv*
T: +46 418 70230; Fax: +46 418 70631
Kent Bjuvegård
14 000 vols; 20 curr per   26574

Fiskeriverket (National Board of Fisheries), Ekelundsgatan 1, Box 423, 401 26 *Göteborg*
T: +46 31 7430374;
Fax: +46 31 7430444; E-mail: bibliotek@fiskeriverket.se; URL: www.fiskeriverket.se
1974; Britta Carlström
Coll of FAO publs
10 000 vols; 150 curr per
libr loan; EURASLIC   26575

Göteborgs Konstmuseum, Biblioteket (Art Museum), Götaplatsen, 412 56 ***Göteborg***
T: +46 31 3683500;
Fax: +46 31 3683526; E-mail: info.konstmuseum@kultur.goteborg.se;
URL: www.konstmuseum.goteborg.se
1861; Barbro Ahlfort
depts for painting, sculpture, prints and drawings from 1400 till present day
20 000 vols; 20 curr per; 20 000 photos, reproductions, clippings 26576

Göteborgs Naturhistoriska Museum, Biblioteket (Natural History Museum), Slottskogen, Box 7283, 402 35 ***Göteborg***
T: +46 31 7752401;
Fax: +46 31 129807; E-mail: info.naturhistoriska@vgregion.se; URL: www.gnm.se
1833; Anna Lindemark
Zoology, palaeontology, geology and natural history (mainly from 1850-1950), early accounts of arctic and antarctic travels, anthropology and ethnology
70 000 vols; 125 curr per; 50 e-books
libr loan 26577

Göteborgs Stadsmuseum, Biblioteket (City Museum of Gothenburg), Norra Hamngatan 12, 411 14 ***Göteborg***
T: +46 31 612770; Fax: +46 31 7743058;
URL: www.stadsmuseum.goteborg.se
Åsa Engström
70 000 vols; 100 curr per 26578

Landsarkivet i Göteborg, Biblioteket (The Regional Archives in Gothenborg), Geijersgatan 1, Box 19035, 400 12 ***Göteborg***
T: +46 10 4767800;
Fax: +46 10 4767820; E-mail: landsarkivet@landsarkivet-goteborg.ra.se;
URL: www.ra.se/gla
1911; Per Forsberg
40 000 vols; 118 curr per; 2 e-journals 26579

Röhsska museet, Biblioteket (Röhss museum of design and applied art), Vasagatan 37-39, Box 53178, 400 15 ***Göteborg***
T: +46 31 3683170; E-mail: info.designmuseum@kultur.goteborg.se
1916; Ingjerd Borén
35 000 vols; 15 curr per
ARLIS/Norden 26580

SIK – Institutet för Livsmedel och Bioteknik AB, Library (SIK – The Swedish Institute for Food and Biotechnology), Frans Perssonsväg, Box 5401, 402 29 ***Göteborg***
T: +46 31 355620; Fax: +46 31 833782;
E-mail: info@sik.se; URL: www.sik.se
1949; Birgitta Berg
15 000 vols; 200 curr per
SFIS 26581

Sjöfartsmuseet, Biblioteket (Maritime Museum), Karl Johansgatan 1-3, 414 59 ***Göteborg***
T: +46 31 612900; Fax: +46 31 246182;
E-mail: karl hellervik@kultur.goteborg.se;
URL: www.sjofartsmuseum.goteborg.se
1933; Gunnar Löwenstein
22 000 vols 26582

Sjömansbiblioteket (Swedish Seamen's Library), Rosenhill, Köpenhamnsgatan, 418 77 ***Göteborg***
T: +46 31 647971; Fax: +46 31 644430; E-mail: sjomansbiblioteket@sjofartsverket.se; URL: www.sjofartsverket.se/SeaTime/templates/STPageColumns____10455.aspx
1930; Anna Selme
Swedish fict about the sea, seamen, shipping
119 000 vols; 56 curr per; 12 000 sound-rec 26583

Språk- och folkminnesinstitutet, Dialekt-, ortnamns- och folkminnesarkivet i Göteborg, Biblioteket (Department of Dialectology, Onomastics and Folklore Research in Gothenburg), Vallgatan 22, 411 16 ***Göteborg***
T: +46 31 107530; Fax: +46 31 107537;
E-mail: DAG@sofi.se; URL: www.sofi.se
Annika Nordström
8 680 vols; 830 curr per; 442 sound-rec 26584

Statens museer för världskultur, Library (National Museums of World Culture), Södra vägen 54, 402 27 ***Göteborg***; Box 5306, 402 27 Göteborg
T: +46 31 632700; Fax: +46 31 632710;
E-mail: info@smvk.se; URL: www.smvk.se
1891; Jan Slavik
Erland Nordenskiölds etnografiska boksamling
25 000 vols; 80 curr per; 140 diss/theses; 400 maps; 44 av-mat; 548 sound-rec 26585

Härnösands sjukhus, Biblioteket, Box 1020, 871 82 ***Härnösand***
T: +46 611 84350; Fax: +46 611 84380
Åsa Hedström
10 000 vols; 42 curr per 26586

Murberget, Länsmuseet Västernorrland, Biblioteket, Murberget, Box 34, 87121 ***Härnösand***
T: +46 611 886 33;
Fax: +46 611 18730; E-mail: vivianne.hogqvist@murberget.se; URL: www.murberget.se
Vivi-Anne Högqvist
10 000 vols; 200 curr per; 1 e-journals
libr loan 26587

Hässleholms lasarett, Biblioteket, 281 25 ***Hässleholm***
T: +46 451 86004; Fax: +46 451 86003
Anna Grenthe Theander
11 000 vols 26588

Sjukhuset i Hässleholm, Sjukhusbiblioteket, Esplanadgt, Box 351, 281 25 ***Hässleholm***
T: +46 451 86000
Anna Grenthe-Theander
11 000 vols; 33 curr per 26589

Helsingborgs Lasarett, Bibliotheket, S. Vallgatan 5, 251 87 ***Helsingborg***
T: +46 42 102590; Fax: +46 42 184797;
E-mail: biblioteket.hbg@skane.se
1975; Anders Johansson
19 000 vols; 566 curr per
SB, TLS, SFIS 26590

Landsarkivet i Härnösand (Provincial Archives at Härnosand), Jonas Bures Plats, Box 161, 871 24 ***Hörnösand***
T: +46 611 347600; Fax: +46 611 347650; E-mail: landsarkivet@landsarkivet-harnosand.ra.se; URL: www.sa.se/hla
1935; Carl-Edvard Edvardsson
200 000 vols; 100 curr per; 4 000 mss; 10 000 diss/theses; 40 000 govt docs; 18 000 maps; 150 000 microforms; 300 sound-rec; 5 digital data carriers; 12 000 electronic archival units
ICA 26591

Hälsinglands museum, Biblioteket, Storgatan 31, 824 30 ***Hudiksvall***
T: +46 650 19609; Fax: +46 650 38186; E-mail: halsinglands.museum@hudiksvall.se; URL: www.hudiksvall.se
1860; Karin Abenius Falkstedt
8 000 vols; 30 curr per 26592

Hudiksvalls sjukhus, Biblioteket, 82481 ***Hudiksvall***
T: +46 650 92582; Fax: +46 650 92581; E-mail: biblioteket.hs@lg.se; URL: www.lg.se/sjukhusbiblioteken
Elinor Bergström-Falk
13 000 vols; 135 curr per; 1 000 e-journals
EiRA 26593

Ajtte, Svenskt Fjäll- och Samemuseum, Biblioteket (Ajtte, Swedish Mountain- and Sámi Museum), Kyrkogatan 3, Box 116, 962 23 ***Jokkmokk***
T: +46 971 17070; Fax: +46 971 12057; E-mail: info@ajtte.com; URL: www.ajtte.com
1989; Birgitta Edeborg
Sámi ethnography and languages
17 400 vols; 73 curr per; 3 e-journals; 580 diss/theses; 897 govt docs; 15 music scores; 48 maps; 1 microforms; 38 av-mat; 26 sound-rec; 13 digital data carriers; pamphlets
libr loan; MABBAS, SB 26594

Göta hovrätt, Biblioteket (Göta Court of Appeal), Hovrättstorget, 550 02 ***Jönköping***; Box 2223, 550 02 Jönköping
T: +46 36 156500; Fax: +46 36 156518;
E-mail: hgobibliotek@dom.se
Henrik Levin
Law rpts, Riksdag rpts
12 500 vols; 25 curr per; 4 e-journals; 270 diss/theses; 10 300 govt docs 26595

Jönköpings läns museum, Dag Hammarskjölds plats 1, Box 2133, 550 02 ***Jönköping***
T: +46 36 301800; Fax: +46 36 301818; E-mail: info@jkpglm.se; URL: www.jkpglm.se
1901
30 000 vols; 60 curr per 26596

Blekingesjukhuset, Biblioteket, Länsmansvägen, 374 80 ***Karlshamn***
T: +46 454 731000; E-mail: blekingesjukhuset@ltblekinge.se
Berit Bergman
8 600 vols; 10 curr per 26597

Länslasarettet, Sjukhusbiblioteket, 374 80 ***Karlshamn***
T: +46 454 35000
Berit Bergman
8 000 vols; 90 curr per 26598

Blekingesjukhusets bibliotek (Blekinge hospital library), Blekingesjukhuset, Vårdskolevägen 5, 371 41 ***Karlskrona***; mail address: 371 81 Karlskrona
T: +46 455 737825;
Fax: +46 455 737835; E-mail: biblioteket.blskna@ltblekinge.se
Lil Carleheden Ottoson
Patient information department
81 092 vols; 33 curr per; 955 e-journals
SB, SFIS, EAHIL 26599

Kungl. Örlogsmannasällskapet, Biblioteket (Royal Swedish Society of Naval Sciences), Admiralitetstorget 7, 371 30 ***Karlskrona***
T: +46 455 25991, 25993; Fax: +46 455 25991; E-mail: library@koms.se; URL: www.koms.se
1771; Rolf Edwardson
13 000 vols; 2 000 mss; 2 000 maps 26600

Marinmusei bibliotek (Naval Museum Library), Stumholmen, 371 32 ***Karlskrona***; P.O. Box 48, 37121 Karlskrona
T: +46 455 53916; Fax: +46 455 53949;
E-mail: registrator@maritima.se; URL: www.marinmuseum.se
1761; Lena Björk
22 000 vols; 60 curr per 26601

Centralsjukhuset, Sjukhusbiblioteket (Central Hospital), 651 85 ***Karlstad***
T: +46 54 615144; Fax: +46 54 615017;
E-mail: sjukhusbiblioteket.karlstad@liv.se;
URL: www.liv.se/sjukhusbibliotek
Jan Schützer
Consumer health information coll
30 000 vols; 300 curr per; 900 e-journals 26602

Karolinen Library, Våxnösgatan 10, 651 80 ***Karlstad***
T: +46 54 103000; Fax: +46 54 103393
Karin Pettersson
Behavioral sciences, biology, psychology
20 000 vols; 340 curr per
libr loan 26603

Värmlandsarkiv, Biblioteket, Hööksgatan 2, Box 475, 651 11 ***Karlstad***
T: +46 54 617730; Fax: +46 54 617731;
E-mail: varmlandsarkiv@regionvarmland.se;
URL: www.ra.se/vla
1972; Alain Droguet
Mining
25 000 vols; 60 curr per; 20 000 mss; 100 diss/theses; 25 maps; 10 sound-rec; 50 digital data carriers
libr loan; ICA 26604

Kristianstad centralsjukhuset, Biblioteket, J.A. Hedlundsväg 5, 291 85 ***Kristianstad***
T: +46 44 131000; Fax: +46 44 132668;
E-mail: centralsjukhuset@skane.se; URL: www.centralsjukhuset.nu
Eva Nilsson
Medicine, Psychology, Psychiatry
17 000 vols; 450 curr per 26605

Landskrona lasarett, Biblioteket, Vattenverksallén, Box 514, 261 24 ***Landskrona***
T: +46 418 74053; Fax: +46 418 74054
Inger Olsson
17 000 vols; 135 curr per 26606

Flygvapenmuseum, Biblioteket, Carl Cederströms gata, 581 98 ***Linköping***
T: +46 13 283697; Fax: +46 13 283557;
E-mail: biblioteket@flygvapenmuseum.se;
URL: www.flygvapenmuseum.se
Anita Sundgren
15 000 vols; 70 curr per 26607

Statens Väg- och Transportforskningsinstitut (VTI), Bibliotek och Informationscenter (BIC) (Swedish National Road and Transport Research Institute), 581 95 ***Linköping***
T: +46 13 204331; Fax: +46 13 141436;
E-mail: library@vti.se; URL: www.vti.se
1971; Birgitta Sandstedt
Swedish and International statistics and standards, report series from TRB and TRL
160 000 vols; 300 curr per; 60 e-journals; 880 microforms; 30 av-mat; 600 digital data carriers
libr loan; SBS, SFIS 26608

Stiftelsen Östergötlands Länsmuseum, Biblioteket, Raoul Wallenbergs plats, Box 232, 581 02 ***Linköping***
T: +46 13 230300; Fax: +46 13 140562;
E-mail: info@ostergotlandslansmuseum.se;
URL: www.ostergotlandslansmuseum.se/
Gunnel Mörkfors
22 200 vols; 65 curr per 26609

Brunnsviks Bokstuga (Library for Folk High School and Trade Union Education), Box 840, 771 28 ***Ludvika***
T: +46 240 667320; Fax: +46 240 667373; E-mail: info@brunnsvik.se; URL: www.brunnsvik.se/index2.html
1906; Christina Garbergs-Gunn
45 000 vols; 250 curr per
libr loan 26610

Norrbottens museum, Biblioteket, Hostvagen 7, plan 1, Box 266, 971 08 ***Luleå***
T: +46 920 243551; Fax: +46 920 243562; E-mail: helena.carlstrom@nll.se;
URL: www.norrbottensmuseum.se
1886; Helena Carlstrom
Lapp culture
18 000 vols; 100 curr per; 800 diss/theses; 1 200 pamphlets 26611

BTJ Företagsbibliotek, BTJ Sverige AB, Scheelevägen 18, 22 182 ***Lund***
T: +46 46 180000; Fax: +46 46 180125;
E-mail: btj@btj.se; URL: www.btj.com
Jan-Erik Lindström
Libraries
17 000 vols; 200 curr per 26612

Kulturen, Bibliotek (Culture History Museum), Karlins plats, Box 1095, 22104 ***Lund***
T: +46 46 350409; Fax: +46 46 350470;
E-mail: benjamin.vogel@kulturen.com;
URL: www.kulturen.com
Carl Gösta Bryve
44 500 vols; 59 curr per 26613

Landsarkivet i Lund, Bibliotek (Regional Archives in Lund), Dalbyvägen 4, Box 2016, 220 02 ***Lund***
T: +46 46 197000; Fax: +46 46 197070;
E-mail: landsarkivet@landsarkivet-lund.ra.se; URL: www.ra.se/lla
1903; Elisabeth Reuterswärd
40 curr per 26614

Havsfiskelaboratoriet, Fiskeriverket, Biblioteket (Swedish Board of Fisheries), Lysekil, Box 4, 453 21 ***Lysekil***
T: +46 31 7430300; Fax: +46 523 139 77; URL: www.fiskeriverket.se
1929; Karin Frohlund
Historical coll: 400 titles on marine fisheries and biology (mainly before 1900)
17 500 vols; 30 curr per; 89 e-journals; 165 diss/theses 26615

Malmö stadsarkiv, Biblioteket (Malmö City Archives), Isbergs gata 13, 211 19 ***Malmö***
T: +46 40 105300; Fax: +46 40 105328;
E-mail: stadsarkiv@malmo.se; URL: www.malmo.se/stadsarkivet
Helena Sjölin
17 000 vols; 90 curr per 26616

Malmö University Hospital, Medical
Library, 20 502 *Malmö*
T: +46 40 331187; Fax: +46 40
336214; E-mail: umas@skane.se; URL:
www.umas.se
1951; Eddie Fremer
Hist of Medicine (6 000 vol)
33 000 vols; 309 curr per; 6 100
diss/theses; 11 digital data carriers
libr loan; MLA, SB, SFIS   26617

Swerea IVF, Biblioteket (Swedish Institute
for Fibre and Polymer Research),
Argongatan 30, Box 104, 431 22
Mölndal
T: +46 31 7066300; Fax: +46 31
7066363; E-mail: ivf@swerea.se; URL:
www.ivf.se
1945; Håkan Damberg
Textiles, plastics, fibers, polymers;
Fiberspinning dept working with nano-
and bicomponent fibers and Tissue
Engineering
8 000 vols; 40 curr per; 200 diss/theses
SFIS   26618

Mora lasarett, Biblioteket, mail address:
792 85 *Mora*
T: +46 250 493070;
Fax: +46 250 15726; E-mail:
mora.biblioteket@ltdalarna.se
Lotta Svensk
15 000 vols; 115 curr per; 50 mss; 25
diss/theses; 250 av-mat; 170 sound-rec;
15 digital data carriers; 50 paintings
libr loan; SB   26619

Lasarettet, Biblioteket (Hospital at Motala),
Lasarettet, 591 85 *Motala*
T: +46 141 77506, 77206; Fax: +46 141
77890; E-mail: landstinget@lio.se
1972; Lena Savoca
Psychiatry dept
20 000 vols; 40 curr per; 100 diss/theses;
60 music scores; 25 maps; 300
av-mat; 659 sound-rec; 46 digital data
carriers; talking bks, pamphlets, patient's
information
libr loan; SB   26620

Arbetets museum, Biblioteket (Museum
of Work), Laxholmen, 602 21
Norrköping
T: +46 11 189800, 231734;
Fax: +46 11 182290; E-mail:
anncharlotte.persson@arbetetsmuseum.se;
URL: www.arbetetsmuseum.se
1992; Ann-Charlotte Persson
Museum of Work arch include worker's
memories and other documentations
of various occupations; From March
09: Archive of political cartonist Evert
Karlsson – EWK
12 000 vols; 160 curr per; 20 digital
data carriers; Slides of the museum of
work's exhibitions, Documentation of the
Museum of work's exhibitions on CD
          26621

Luftfartsverket, Vikboplan 11, 601 79
Norrköping
T: +46 11 192000; Fax: +46 11 192365;
URL: www.lfv.se
Terttu Hansen
EAC docs, ICAO docs
8 500 vols; 136 curr per   26622

Norrköpings Konstmuseum,
Biblioteket (Art Museum of Norrköping),
Kristinaplatsen, 602 34 *Norrköping*
T: +46 11 152600; Fax: +46 11 135897;
E-mail: konstmuseet@norrkoping.se; URL:
www.norrkoping.se/konstmuseet
Kerstin Holmer
Cat and per on 20th c art subjects
9 000 vols; 45 curr per   26623

**Sveriges meteorologiska och
hydrologiska institut (SMHI)**,
Biblioteket, Folkborgsvägen 1, 601 76
Norrköping
T: +46 11 4958000; Fax: +46 11
4958001; E-mail: library@smhi.se; URL:
www.smhi.se
Doris Pürkner-Landqvist
20 000 vols; 170 curr per
libr loan   26624

Örebro University Hospital, Medical
Library, Regionsjukhuset, 701 85 *Örebro*
T: +46 19 6021592;
Fax: +46 19 6117637; E-mail:
medicinska.biblioteket@orebroll.se; URL:
www.orebroll.se
1967; Margareta Landin

Medical sciences, nursing
16 000 vols; 600 curr per
libr loan; SFIS   26625

Örnsköldsviks sjukhus, Fackbiblioteket,
Box 700, 891 89 *Örnsköldsvik*
T: +46 660 89000; Fax: +46 660 89509
Ellinor Johansson
8 700 vols; 217 curr per   26626

Landsarkivet i Östersund, Biblioteket,
Arkivvägen 1, 831 31 *Östersund*
T: +46 63 108485; Fax: +46 63 121824;
E-mail: landsarkivet@landsarkivet-
ostersund.ra.se; URL: www.ra.se/ola
30 000 vols; 100 curr per   26627

Östersunds lasarett, Kyrkgt, 831 83
Östersund
T: +46 63 153192; Fax: +46 63 516459
Mona Lundström
25 200 vols; 321 curr per   26628

SVAR – Svensk arkivinformation,
Biblioteket, SVAR Forskarcentrum, Box
160, 880 40 *Ramsele*
T: +46 623 72500; Fax: +46 623
72505; E-mail: info@svar.ra.se; URL:
www.svar.ra.se
Bengt Erik Näsholm
115 000 vols   26629

Sandvikens sjukhus, Biblioteket,
Lasarettsleden, 811 89 *Sandviken*
T: +46 26 278186; Fax: +46 26 270106
Margareta Knutsson
12 600 vols; 150 curr per   26630

Sigtunastiftelsens Bibliotek, Manfred
Björkquists allé 2-4, Box 57, 193 22
Sigtuna
T: +46 8 59258906;
Fax: +46 8 59258917; E-mail:
bibliotek@sigtunastiftelsen.se; URL:
www.sigtunastiftelsen.se/bibliotek
1917; Svante Ögren
Christian religion, mystics
150 000 vols; 357 curr per; 4,5 mio
clippings
libr loan   26631

Skellefteå lasarett, Sjukhusbiblioteket
(Skelleftei Hospital Library), 931 86
Skellefteå
T: +46 910 771775;
Fax: +46 910 771778; E-mail:
sjukhusbiblioteket.skelleftea@vll.se; URL:
www.regionvasterbotten.se/det-haer-
goer-vi/laensbiblioteket/sjukhusbibliotek/
skelleftea.html
1989; Christer Felth
12 300 vols; 92 curr per   26632

Skoklosters Bibliotek, Skoklosters slott
(Library of Skokloster Castle), Skokloster
slott, 74696 *Skokloster*
T: +46 46 18386077; Fax: +46 46
18386446; E-mail: skokloster@lsh.se;
URL: www.skoklostersslott.se
1967; Elisabeth Westin Berg
Art, hist, 17th-18th c bks
22 000 vols; 55 incunabula
ARLIS/Norden   26633

Sjukhusbiblioteket, Rosenborgsgatan 6-
10, 152 86 *Södertälje*
T: +46 8 55024000
Helena Ljungqvist
17 800 vols; 16 curr per   26634

Södertälje sjukhus, Rosenborgsgatan
6-10, 152 86 *Södertälje*
T: +46 8 55024713; Fax: +46 8
55024054
Britt-Inger Nilsson
15 000 vols; 16 curr per   26635

Sollefteå sjukhus, Bibliotek, Skärvsta,
881 04 *Sollefteå*
T: +46 620 19190; Fax: +46 620 19638;
E-mail: biblioteket.solleftea@lvn.se; URL:
www.lvn.se/templates/Page___3040.aspx
10 000 vols; 150 curr per   26636

**Stockholm International Peace
Research Institute (SIPRI)**, Library and
Documentation Department, Signalistgatan
9, 169 70 *Solna*
T: +46 8 6559779; Fax: +46 8
6559733; E-mail: library@sipri.org; URL:
www.sipri.org/library
1966; Christine-Charlotte (Nenne) Bodell
International Security and Arms Control,
Conflict prevention and conflict resolution,
Military Technology and Arms Control,
Chemical and Biological Warfare, Military

Expenditure, Arms Transfers and Arms
Production
52 000 vols; 200 curr per; 10 e-journals;
4 000 digital data carriers
libr loan; SBS   26637

Swedish Centre for Terminology,
Library, Västra vägen 7B, Terrassen,
169 61 *Solna*
T: +46 8 4466600; Fax: +46 8 4466629;
E-mail: tnc@tnc.se; URL: www.tnc.se
1941; Urban Hurtig
Engineering, technology, terminology,
dictionaries
10 000 vols; 50 curr per
SFIS   26638

Arbetarrörelsens arkiv och bibliothek
(Labour Movement Archives and Library),
Upplandsgatan 4, Box 1124, 111 81
Stockholm
T: +46 8 4123900; Fax: +46 8
4123990; E-mail: info@arbark.se; URL:
www.arbark.se
1902; Karin Englund
Hist and ideological development of the
labour movement
125 000 vols; 370 curr per; 80 000 mss;
7 500 diss/theses; 6 000 microforms; 1 500
sound-rec; 50 digital data carriers; flags,
posters, photos
SB, SFIS, IALHI   26639

Arkitekturmuseet, Biblioteket (Swedish
Museum of Architecture), Skeppsholmen,
111 49 *Stockholm*
T: +46 8 58727000;
Fax: +46 8 58727070; E-mail:
bibliotek@arkitekturmuseet.se; URL:
www.arkitekturmuseet.se
1998; Lena Wranne
27 400 vols; 66 curr per; 410 diss/theses
libr loan; ARLIS/Norden   26640

Armémuseum, Biblioteket (Royal Army
Museum), Riddargatan 13, Box 14095,
104 41 *Stockholm*
T: +46 8 51956304;
Fax: +46 8 6626831; E-mail:
biblioteket@armemuseum.se; URL:
www.armemuseum.se
1879; Margaret Tainsh Beskow
Uniforms and weapons coll
50 000 vols; 120 curr per
libr loan; IFLA   26641

Bernadottebiblioteket (The Bernadotte
Library), Kungl. slottet, 111 30
Stockholm
T: +46 8 4026162; Fax: +46 8 4026168;
URL: www.royalcourt.se
1908; Göran Alm
Photo-arch from 1850, 16 mm films from
1897 onwards, Maps (early 19th c),
'Bernadotteana'; Art book coll from the
Royal collections
100 000 vols; 30 curr per; 10 mss; 1
incunabula; 5 000 music scores; 150
maps; 100 av-mat; 100 sound-rec 26642

CAN Bibliotek (Swedish Council for
Information on Alcohol and other
Drugs), Klara Norra Kyrkogata 34,
111 22 *Stockholm*; Box 70412, 107 25
Stockholm
T: +46 8 4124615; Fax: +46 8 104641;
E-mail: biblioteket@can.se; URL:
www.can.se
1906; Claes Olsson
Swedish reports on alcohol and narcotics,
Collection on Nordic temperance
literature, Swedish newspaper clippings
154 000 vols; 250 curr per; 25 e-journals;
800 e-books; 450 diss/theses; 10 digital
data carriers
libr loan; SB, Elisad, Salis, Substance
abuse librarians & information specialists
          26643

Cement och Betong Institutet,
Biblioteket (Swedish Cement and
Concrete Research Institute), Drottning
Kristinas vräg 26, 100 44 *Stockholm*
T: +46 8 6961100; Fax: +46 8 243137;
E-mail: bibl@cbi.se; URL: www.cbi.se
1950; Tuula Ojala
10 000 vols; 120 curr per   26644

Danderyds sjukhus Fackbibliotek
(Danderyd Hospital Professional Library),
Hus 26, plan 3, 182 88 *Stockholm*
T: +46 8 6555772; Fax: +46 8 7534065;
E-mail: fackbiblioteket@ds.se; URL:
www.ds.se/Web/NormalPage___136.aspx
1965; Bitte Lofgren
20 000 vols; 200 curr per; 900 e-journals
libr loan; SB   26645

Finlandsinstitutets Bibliotek
(The Finnish Institute in Sweden),
Snickarbacken 4, Box 1355, 111 83
Stockholm
T: +46 8 54521209;
Fax: +46 8 54521210; E-mail:
bibliotek@finlandsinstitutet.se; URL:
www.finlandsinstitutet.se
1894; Eivor Kotkamaa
12 000 vols; 130 curr per; 6 300 av-mat;
700 sound-rec; 40 digital data carriers
          26646

Folksam, Biblioteket (Folksam Group),
Bohusg 14, 116 67 *Stockholm*; mail
address: 106 60 Stockholm
T: +46 8 7726824; E-mail:
kerstin.wallstenius@folksam.se; URL:
www.folksam.se
1920; Kerstin Wallstenius
Insurance
1 500 vols; 100 curr per   26647

Goethe-Institut, Informationszentrum,
Bryggargatan 12A, 111 21 *Stockholm*
T: +46 8 4591205; Fax: +46 8 4591225;
E-mail: info@stockholm.goethe.org; URL:
www.goethe.de/stockholm
Eva Hackenberg
Information on Germany, Material on
german-related matters
9 578 vols; 38 curr per; 1 033 av-mat;
774 sound-rec; 107 digital data carriers
SB   26648

Hortus Botanicus Bergianus, Bibliotek,
Box 50017, 104 05 *Stockholm*
T: +46 8 162853; Fax: +46 8 6129005
Dag Ericson
Botany
30 000 vols; 106 curr per
libr loan   26649

Information Resource Center (IRC),
Embassy of the United States of
America, Dag Hammarskjölds Väg 31,
115 89 *Stockholm*
T: +46 8 7835300; Fax: +46 8 665
3303; E-mail: stockholmirc@state.gov;
URL: www.usemb.se/ircenter/index.html
Ian Bald
8 000 vols; 70 curr per   26650

Innventia AB, Information Centre,
Drottning Kristinas väg 61, Box 5604,
114 86 *Stockholm*
T: +46 8 6767000; Fax: +46 8 4115518;
E-mail: library@innventia.com; URL:
www.innventia.com
1945; Camilla Burman
Pulp, paper, printing, packaging, graphic
media, biofuel; Dard Hunter Coll
12 000 vols; 200 curr per; 4 e-journals;
1 500 diss/theses; 2 digital data carriers
libr loan; SFIS   26651

**Italienska Kulturinstitutet i Stockholm
'C.M. Lerici'**, Gördesgatan 14, 115 27
Stockholm
T: +46 8 54585760; E-mail:
bibliotek.iicstockholm@esteri.it; URL:
www.iicstockholm.esteri.it/IIC_Stoccolma
Laura Maurizio
21 000 vols   26652

Jernkontoret, Biblioteket (Swedish
Ironmasters' Association),
Kungsträdgårdsgatan 10, Box 1721,
111 87 *Stockholm*
T: +46 8 6791700; Fax: +46 8 6112089;
E-mail: office@jernkontoret.se; URL:
www.jernkontoret.se
1927; Yngve Axelsson
20 000 vols; 120 curr per; 50 mss
          26653

Judiska Biblioteket (The Jewish Library
in Stockholm), Wahrendorffsgatan 3,
111 47 *Stockholm*; Box 7427, 103 91
Stockholm
T: +46 8 58785834;
Fax: +46 8 58785858; E-mail:
judiska.biblioteket@jfst.se;
URL: www.jfst.se/verksamhet/
judiskabiblioteket.php?title=biblioteket
1867; Lars Raij
Books in Yiddish, Marcus Ehrenpreis
Coll
26 000 vols; 30 curr per; 100 diss/theses;
230 av-mat; 380 sound-rec
libr loan; Svensk Biblioteksförening
          26654

Juridiska biblioteket / Sveriges advokatsamfund, Laboratoriegatan 4, Box 27321, 102 54 ***Stockholm***
T: +46 8 4590320; Fax: +46 8 6600779; E-mail: biblioteket@advokatsamfundet.se; URL: www.juridiskabiblioteket.se
1935; Bitte Wölkert
60 000 vols; 150 curr per; 1 000 diss/theses; 10 000 govt docs; 20 digital data carriers 26655

Karolinska University Hospital, Biblioteket, 171 76 ***Stockholm***
T: +46 8 51774132; Fax: +46 8 332547; E-mail: biblioteket@karolinska.se; URL: www.karolinska.se/biblioteket
1986; Marie Källberg
11 300 vols; 25 curr per; 1 500 e-journals; 1 900 diss/theses
libr loan; EAHIL 26656

KF Library, Stadsgården 10, Box 15200, 104 65 ***Stockholm***
T: +46 8 7432621; Fax: +46 8 6443026; E-mail: michael.hagstrom@kf.se; URL: www.kf.se
1927; Michael Hagstroem
Robert Owen, Charles Fourier, Utopia, Consumer co-operation spec coll
90 000 vols; 300 curr per
SFIS 26657

Konstbiblioteket, Nationalmuseum / Moderna Museet (Art Library), Holmamiralens väg 2, på Skeppsholmen, Box 16176, 103 24 ***Stockholm***
T: +46 8 51954300;
Fax: +46 8 51954352; E-mail: konstbiblioteket@nationalmuseum.se; URL: www.nationalmuseum.se; www.modernamuseet.se
1867
320 000 vols; 175 curr per; more than 2 million clippings
libr loan; IFLA, ARLIS/Norden, SB 26658

Krigsarkivet (Swedish Military Archives), Banérgatan 64, 115 88 ***Stockholm***
T: +46 8 7824100; Fax: +46 8 7826976; E-mail: krigsarkivet@krigsarkivet.ra.se
1805; Kerstin Rundberg
250 000 vols; 250 curr per
libr loan; IFLA 26659

Kungl. Akademien för de fria konsterna, Bibliotek (Royal Swedish Academy of Fine Arts), Fredsgatan 12, Box 16317, 103 26 ***Stockholm*** 16
T: +46 8 232948; Fax: +46 8 7905924; E-mail: biblioteket@konstakademien.se; URL: www.konstakademien.se
1775; Anneli Ekström
17 600 engravings
45 000 vols; 65 curr per 26660

Kungl. Dramatiska Teatern, Arkiv och Bibliotek (Royal Dramatic Theatre), Nybrogatan 2, Box 5037, 102 41 ***Stockholm***
T: +46 8 6656194; Fax: +46 8 6656289; E-mail: dag.kronlund@dramaten.se; URL: www.dramaten.se
1970; Dag Kronlund
Theatre coll
10 000 vols
TLA 26661

Kungl. Skogs- och Lantbruksakademien, Biblioteket (Royal Swedish Academy of Agriculture and Forestry), Drottninggatan 95 B, Box 6806, 113 86 ***Stockholm***
T: +46 8 54547720; Fax: +46 8 54547730; E-mail: kslab@ksla.se; URL: www.kslab.ksla.se
1811; Lars Ljunggren
Agricultural hist, forest hist, gardening hist, garden hist, fishing and hunting hist of Europe and North America; 'Skogs- och lantbrukshistoriska meddelanden' (Publ on agricultural and forest hist) 1992 ff
98 000 vols; 300 curr per; 200 maps; 10 av-mat; 10 digital data carriers
libr loan; SBS, SFIS 26662

Landsorganisationen i Sverige, LO, Biblioteket (Swedish Confederation of Trade Unions), Barnhusgatan 18, 105 53 ***Stockholm***
T: +46 8 7962500; Fax: +46 8 245228; E-mail: mailbox@lo.se; URL: www.lo.se
Anneli Alriksson
31 000 vols; 170 curr per
libr loan 26663

Lantbrukarnas Riksförbund, Biblioteket (Federation of Swedish Farmers), Klara Ö. Kyrkogata 12, 105 33 ***Stockholm***
T: +46 8 573573; URL: www.lrf.se
1962; Margareta Norman
Nutrition
25 000 vols; 135 curr per
libr loan 26664

Livrustkammaren, Biblioteket (The Royal Armoury), Slottsbacken 3, 111 30 ***Stockholm***
T: +46 8 4023072; Fax: +46 8 207305; E-mail: charlotta.andersson@lsh.se; URL: www.livrustkammaren.se
1895; Elisabeth Westin Berg
Swedish Hist, Arms and Armoury, Equitation, Fencing, Hawking, Ceremony, Costume, Military Hist; The Emil Fick's Libr – Fencing and Equestrian Sports; the Thorwald Lindqvist's coll – Falconry, Hawking; The Max Dreger's coll – Artillery, Fortifications
43 000 vols; 14 curr per; 100 mss; 1 incunabula; 200 diss/theses; 150 maps; 20 000 photos of arms and armor, costume etc
libr loan; IFLA 26665

Medicinhistoriska museet Eugenia T-3, Biblioteket, Karolinska sjukhuset, 171 76 ***Stockholm***
T: +46 8 348620; Fax: +46 8 342415
1952; Eva Wahlberg Sandberg
5 000 vols; 10 curr per; 500 diss/theses; 500 govt docs; 10 maps; diapositives
libr loan 26666

Moderna Museet, Fotografibiblioteket, Holmamiralens väg 2, Skeppsholmen, Box 16382, 103 27 ***Stockholm***
T: +46 8 51955215; E-mail: fotografibiblioteket@modernamuseet.se; URL: www.modernamuseet.se
1971; Peter Schultz
Gernsheim Duplicate Coll; Photographic archives
25 000 vols; 130 curr per; video libr, clippings
libr loan; ARLIS/Norden 26667

Museum of Ethnography, Library, Djurgårdsbrunnsvägen 34, Box 27140, 102 52 ***Stockholm***
T: +46 8 51955000;
Fax: +46 8 51955070; E-mail: zsuzsanna.muller@etnografiska.se; URL: www.etnografiska.se
ca 1900; Zsuzsanna Müller
Sven Hedins ethnographic coll, Björn Strand N.A. Indian coll, Prof. Sigvald Linnés coll of American archeology and ethnographys
55 000 vols; 450 curr per 26668

Museum of Far Eastern Antiquities, Library (Far Eastern Library), Skeppsholmen, Box 16381, 103 27 ***Stockholm***
T: +46 8 51955-778/-777/-774;
Fax: +46 8 51955-775/-755; E-mail: biblioteket@ostasiatiska.se; URL: www.mfea.se
1986; Lars Fredriksson
Chinese, Japanese and Korean Lit; Dept of Art and Archaeology
140 000 vols; 1 500 curr per; 13 mss; 300 music scores; 100 maps
ARLIS/Norden 26669

Naturhistoriska riksmuseet, Biblioteket (Swedish Museum of Natural History), Frescativägen 40, Box 50007, 104 05 ***Stockholm***
T: +46 8 51954090;
Fax: +46 8 51954099; E-mail: marie.svensson@nrm.se; URL: www.nrm.se
1841; Marie Svensson
Entomology (45 000 reprints)
16 000 vols; 200 curr per 26670

Naturvårdsverket, Biblioteket (Swedish Environmental Protection Agency), Valhallavägen 195, 106 48 ***Stockholm***
T: +46 8 6981260; Fax: +46 8 6981400; E-mail: library@naturvardsverket.se; URL: www.naturvardsverket.se
1967; Susanne Borenberg
Environment, environmental pollution, environmental economics, environmental policy, sustainable development, climate change, toxicology, wildlife, noise, waste, natural resources, water
35 000 vols; 200 curr per; 2 300 e-journals
libr loan; SB 26671

Nordiska Museet, Bibliotek, Djurgårdsvägen 6-16, Box 27820, 115 93 ***Stockholm***
T: +46 8 51954633;
Fax: +46 8 51954765; E-mail: biblioteket@nordiskamuseet.se; URL: nordmlib.nordiskamuseet.se
1873; Anita Larson
Folklife studies, cultural hist, ethnology, handicraft
206 000 vols; 1 200 curr per
libr loan 26672

Patent- och registreringsverket, Biblioteket (Swedish Patent and Registration Office), Valhallavägen 136, Box 5055, 102 42 ***Stockholm***
T: +46 8 7822872; Fax: +46 8 6627215; E-mail: biblioteket@prv.se; URL: www.prv.se
1885; Susanne Berg
73 500 vols; 685 curr per; 12 000 e-journals; 60 mio patent specifications 26673

Postmusei Bibliotek (Postmuseum Library), Lilla Nygatan 6, Box 2002, 103 11 ***Stockholm***
T: +46 8 7811753; Fax: +46 8 205426; E-mail: biblioteket@posten.se; URL: www.postmuseum.posten.de
1943; Erik Hamberg
Philately and postal hist coll
56 000 vols; 195 curr per; 85 microforms
libr loan 26674

Riksarkivets bibliotek Marieberg-Arninge (The Library of the Swedish National Archives, Marieberg-Arninge), Fyrverkarbacken 13, Box 12541, 10229 ***Stockholm***
T: +46 10 4767259;
Fax: +46 10 4767120; E-mail: elisabet.sandstrom@riksarkivet.ra.se; URL: www.riksarkivet.se
1661; Elisabet Sandström
Heraldry, Sweden's Press Arch libr
140 000 vols; 350 curr per; 11 e-journals
libr loan 26675

Sida, Information Centre, Sveavägen 20, 105 25 ***Stockholm***
T: +46 8 6985585; Fax: +46 8 6985615; E-mail: info@sida.se; URL: www.sida.se
Eva Ehlin
Developing countries, East Europa
10 000 vols; 200 curr per; information materials on Sida's work 26676

SIFU – Statens Institut för Företagsutveckling, I&D-sektionen, Biblioteket (Swedish State Institute for Industrial Development), Olaf Palmes gata 31, Box 823, 101 36 ***Stockholm*** 4
URL: www.sifu.se
Poul H. Rasmussen
30 000 vols; 500 curr per 26677

Sjöhistoriska museet, Biblioteket (National Maritime Museum), Djurgårdsbrunnsvägen 24, Box 27131, 102 52 ***Stockholm***
T: +46 8 5195490; Fax: +46 8 51954949; E-mail: biblioteket@maritima.se; URL: www.sjohistoriska.se
1938; Gunvor Vretblad
50 000 vols; 154 curr per
libr loan; Bibsam 26678

Statens musikbibliotek (Music Library of Sweden), Torsgatan 19, 11321 ***Stockholm***; Box 16326, 10326 Stockholm
T: +46 8 51955412; Fax: +46 8 51955405; E-mail: exp@muslib.se; URL: www.muslib.se
1771; Anders Lönn
German Church coll (16-17th c), music libr of the Royal Opera (18-19th c), autographs of Swedish composers, Johan Helmich Roman coll, Boije Coll (guitar music)
52 933 vols; 1 631 curr per; 3 e-journals; 30 000 mss; 1 997 diss/theses; 393 530 music scores; 184 microforms; 164 av-mat; 460 sound-rec; 394 digital data carriers; Archives, Special Colls and Documentation
libr loan; IAML, IASA 26679

Statistiska centralbyråns bibliotek (Statistics Sweden Library), Karlavägen 100, 104 51 ***Stockholm***; Box 24300, 104 51 Stockholm
T: +46 8 6615261; Fax: +46 8 50694045; E-mail: library@scb.se; URL: www.scb.se
1858; Gunilla Lundholm
Official statistics of Sweden, Literature in statistical theory and methods, International statistics 1858-2008
230 000 vols; 2 041 e-journals; 750 digital data carriers; government docs
libr loan; IFLA 26680

Stockholms Spårvägsmuseum, Biblioteket, Tegelviksgatan 22, 116 41 ***Stockholm***
T: +46 8 4625531; Fax: +46 8 4625530; E-mail: sparvagsmuseet@sl.se; URL: www.sparvagsmuseet.sl.se
Carl Axel Petersson
10 000 vols 26681

Stockholms stadsarkiv, Biblioteket (Stockholm City Archives), Kungsklippan 6, Box 220 63, 104 22 ***Stockholm***
T: +46 8 50828333;
Fax: +46 8 50828301; E-mail: biblioteket.stadsarkivet@stockholm.se; URL: www.ssa.stockholm.se/sv/Om-Stadsarkivet/Stadsarkivets-bibliotek/
1900; Marie-Louise Andersson
Stockholmiana
130 000 vols; 170 curr per; 300 maps
libr loan; SB 26682

Stockholms Stadsmuseum, Biblioteket (City Museum of Stockholm), Ryssgården, Slussen, Box 15025, 104 65 ***Stockholm***
T: +46 8 50831600;
Fax: +46 8 50831699; URL: www.stadsmuseum.stockholm.se
1931; Sören Lindström
30 000 vols; 65 curr per; 1 mio photos
libr loan; SFIS 26683

Svenska Akademiens Nobelbibliotek (Nobel Library of the Swedish Academy), Källargränd 4, Box 2118, 103 13 ***Stockholm***
T: +46 8 55512552;
Fax: +46 8 55512599; E-mail: info@nobelbiblioteket.se; URL: www.nobelbiblioteket.se
1901; Lars Rydquist
Modern lit (fiction, poetry, drama), lit hist & criticism
220 000 vols; 160 curr per; 3 e-journals; 10 digital data carriers
libr loan; IFLA 26684

Svenska barnboksinstitutet (Swedish Institute for Children's Books), Odengatan 61, 113 22 ***Stockholm***
T: +46 8 54542050; Fax: +46 8 54542054; E-mail: biblioteket@sbi.kb.se; URL: www.sbi.kb.se
1965; Cecilia Östlund
Old and rare children's bks
88 073 vols; 170 curr per; 22 mss; 50 000 clippings, online-databases
libr loan; SB 26685

Svenska Filminstitutet, Biblioteket (Swedish Film Institute), Filmhuset, Borgvägen 1-5, Box 271 26, 102 52 ***Stockholm*** 27
T: +46 8 6651100; Fax: +46 8 6651450; E-mail: biblioteksexpeditionen@sfi.se; URL: www.sfi.se
Mats Skärstrand
Account-bks and inventories from early Swedish film companies, documentation dept of the clippings and the stills arch
45 000 vols; 250 curr per; 25 e-journals
FIAF, SB, SFIS 26686

Svenska kommunalarbetareförbundet (Swedish Municipal Workers' Union), Hagagatan 2, 104 32 ***Stockholm***; Box 19039, 104 32 Stockholm
T: +46 8 7283018; Fax: +46 8 318745; E-mail: carin.schultz@kommunal.se; URL: www.kommunal.se
Carin Schultz
Social sciences
10 000 vols; 130 curr per
SFIS 26687

Svenska Läkaresöllskapet, Biblioteket (Swedish Society of Medicine), Klara Östra Kyrkogata 10, Box 738, 101 35 ***Stockholm***
T: +46 8 4408898; Fax: +46 8 4408899
1807; Gunilla Sondén

Old medical bks, hist of medicine, medical ethics
35 000 vols; 20 curr per
libr loan   26688

Svenska Metallindustriarbetareförbundet, Biblioteket (Swedish Association of Metalworkers), Olof Palmes gata 11, 105 52 *Stockholm*
T: +46 8 7868000; Fax: +46 8 248674; E-mail: post.fk@metall.se; URL: www.metall.se
Ingela Beiming
9 000 vols; 116 curr per   26689

Svenskt visarkiv, Biblioteket (The Centre for Swedish Folk Music and Jazz Research), Torsgatan 19, Box 16326, 103 26 *Stockholm*
T: +46 8 51955488; Fax: +46 8 51955449; URL: www.visarkiv.se
1951; Karin Strand
32 000 vols; 110 curr per; 3 200 mss; 150 diss/theses; 8 000 music scores; 800 microforms; 17 400 av-mat; 48 300 sound-rec
IAML   26690

Sveriges Kommuner och Landsting, Biblioteket (Swedish Association of Local Authorities – Swedish Federation of County Councils), Hornsgatan 20, 118 82 *Stockholm*
T: +46 8 4527766; Fax: +46 8 6417798; E-mail: info@skl.se; URL: www.skl.se/
1968; Kenneth Åhlvik
60 000 vols; 180 curr per; 400 diss/theses
libr loan   26691

Sveriges Ornitologiska Förening, Biblioteket (Swedish Ornithological Society), Ekhagsvägen 3, 104 05 *Stockholm*
T: +46 8 6122530; Fax: +46 8 6122536; URL: www.sofnet.org
1945; Ante Strand
9 050 vols; 284 curr per; 45 sound-rec
                                           26692

Sveriges Radio Förvaltnings AB, Swedish Broadcasting Corporation, Biblioteket, Oxenstiernsgatan 20, 105 10 *Stockholm*
T: +46 8 7844130; Fax: +46 8 6633352; E-mail: biblioteket@srf.se; URL: www.srf.se
1942; Rosita Busch
Mass communications coll
70 000 vols; 175 curr per
libr loan; SB, SFIS   26693

Sveriges Teatermuseum, Biblioteket (Theatre Museum of Sweden), Kvarnholmsvägen 56, Nacka, Box 15417, 104 65 *Stockholm*
T: +46 8 55693111; Fax: +46 8 55693101; URL: www.teatermuseet.se
1922; Magnus Blomkvist
Arch of the Royal Swedish Theatres (1773-1985/86); Current repertory registration dept, DVD-documentation dept
72 000 vols; 51 curr per; 50 000 mss; 3 600 music score; 450 sound-rec; 95 digital data carriers; 1,5 mio photos, 1 mio clippings
SIBMAS   26694

Tekniska museets bibliotek (National Museum of Science and Technology, The Library), Museivägen 7, 115 93 *Stockholm*; Box 27842, 115 93 Stockholm
T: +46 8 4505640; Fax: +46 8 4505641; E-mail: biblioteket@tekniskamuseet.se; URL: www.tekniskamuseet.se
1924; Inger Björklund
Hist of Science and Technology, Industrial hist
50 000 vols; 80 curr per
libr loan; SB, SFIS   26695

Timbro/SFN, Library, Kungsgatan 60, Box 3037, 10361 *Stockholm*
T: +46 8 58789800;
Fax: +46 8 58789855; E-mail: kristina.von.unge@timbro.se; URL: www.timbro.se
1978; Kristina von Unge
Social Sciences, Political Science, Economics
10 000 vols; 100 curr per   26696

Vitterhetsakademiens bibliotek, Riksantikvarieämbetet (Library of the Royal Academy of Letters, History and Antiquities), Storgatan 41, Box 5405, 114 84 *Stockholm*
T: +46 8 51918000; Fax: +46 8 6633528; E-mail: bibl@raa.se; URL: www.raa.se/bibliotek
1786; Birgitta Enaeus-Andersson
Archeology, art hist, numismatics, preservation of cultural heritage
400 000 vols; 900 curr per; 400 e-journals; 200 e-books
libr loan; ARLIS/Norden, IFLA   26697

Kammarrätten i Sundsvall, Biblioteket, Trädgårdsgatan 6, Box 714, 85121 *Sundsvall*
T: +46 60 186600; Fax: +46 60 186652
Sture Nilsson
32 000 vols; 405 curr per   26698

Sjukhuset i Torsby, Biblioteket, Lasarettsgatan 4, Box 502, 685 00 *Torsby*
T: +46 560 12000
Staffan Fenander
8 000 vols; 151 curr per   26699

Trelleborg Industri AB, Biblioteket, Nygatan 102, 231 81 *Trelleborg*
T: +46 410 516 22; Fax: +46 410 16981
1903; Anne Wängberg
Rubber
15 000 vols; 40 curr per
libr loan; SFIS   26700

Norra Älvsborgs länssjukhus, Biblioteket (County Hospital), 461 85 *Trollhättan*
T: +46 520 91040; Fax: +46 520 91935
Kerstin Kalén
30 000 vols; 400 curr per
libr loan; SB   26701

Försvarets Forskningsanstalt, FOA, ABC-skydd, Biblioteket (National Defense Research Institute, Department of NBC defence), Cementvägen 20, 901 82 *Umeå*
T: +46 90 106600; Fax: +46 90 106800
1945; Carin Stenlund
18 000 vols; 253 curr per; 5 000 reports
libr loan   26702

Språk- och folkminnesinstitutet, Dialekt-, och ortnamns- och folkminnesarkivet i Umeå (DAUM), Biblioteket, Jägarvägen 18, Box 4056, 904 03 *Umeå*
T: +46 90 135815; Fax: +46 90 135820
Jan Nilsson
8 500 vols; 46 curr per   26703

Västerbottens museum, Biblioteket, Box 3183, 90304 *Umeå*
T: +46 90 171825; Fax: +46 90 779000; E-mail: info@vasterbottensmuseum.se; URL: www.vasterbottensmuseum.se
1943; Lars Holstein
30 000 vols; 520 curr per   26704

Akademiska sjukhuset, Allmänna biblioteket (Uppsala Academic Hospital), 751 85 *Uppsala*
T: +46 18 663536; Fax: +46 18 504315; E-mail: registrator.1@akademiska.se; URL: www.akademiska.se
1925; Maud Ekbom
Consumer Health information for patients and theirs families
25 000 vols; 58 curr per; 820 talking bks
SB   26705

Institutet för språk- och folkminnen, Biblioteket (Institute for Language and Folklore Research), Dag Hammarskjölds väg 19, Box 135, 751 04 *Uppsala*
T: +46 18 652160; Fax: +46 18 652165; E-mail: katharina.leibring@sofi.se; URL: www.sofi.se
1939; Katharina Leibring
Coll of recordings of Swedish dialects dating from 1930's onwards
80 000 vols; 280 curr per; 100 mss; 400 diss/theses; 1 000 maps; 12 000 sound-rec; 50 digital data carriers; 5 000 clippings, 10 000 offprints and pamphlets
                                           26706

Kungl. Vetenskaps-Societeten i Uppsala (Royal Society of Sciences at Uppsala), S. Larsgatan 1, 753 10 *Uppsala*
T: +46 18 131270; Fax: +46 18 131270; E-mail: kansli@vetenskapssocieteten uppsala.se; URL: www.vetenskapssocietetenuppsala.se
1710; N.N.
32 000 vols; 300 curr per   26707

Landsarkivet i Uppsala, Dag Hammarskjölds väg 19, Box 135, 751 04 *Uppsala*
T: +46 18 652100; Fax: +46 18 652103; E-mail: landsarkivet@landsarkivet-uppsala.ra.se; URL: www.ra.se/ula
1903; Björn Asker
11 700 vols; 80 curr per   26708

Nordiska Afrikainstitutet, Library (Nordic Africa Institute), Kungsgatan 38, 753 21 *Uppsala*; P O Box 1703, 751 47 Uppsala
T: +46 18 562270; Fax: +46 18 562290; E-mail: library@nai.uu.se; URL: www.nai.uu.se
1962; Åsa Lund Moberg
African government publications; Pamphlet collection
69 000 vols; 300 curr per; 127 e-journals; 230 e-books; 1 000 diss/theses; 10 000 govt docs; 1 000 maps; 600 microforms; 214 av-mat; 124 sound-rec; 35 digital data carriers; 400 files in Pamphlet collection
libr loan; SB, ELIAS, EADI, NorDoc
                                           26709

Sveriges geologiska undersökning (SGU), Biblioteket (Geological Survey of Sweden), Villavägen 18, Box 670, 751 28 *Uppsala*
T: +46 18 179396; Fax: +46 18 179306; E-mail: library@sgu.se; URL: www.sgu.se
1858; Martin Östling
geological maps from all countries
75 000 vols; 250 curr per; 50 000 maps
libr loan   26710

Landsarkivet i Vadstena, Biblioteket (Regional Archives in Vadstena), Slottet, Box 126, 592 23 *Vadstena*
T: +46 143 75300; Fax: +46 143 75337; E-mail: landsarkivet@landsarkivet-vadstena.ra.se; URL: www.ra.se/vala
Sven Malmberg
30 000 vols; 30 curr per   26711

Hjälpmedelsinstitutet, Biblioteket (Swedish Institute of Assistive Technology, Library), Sorteragatan 23, Box 510, 162 15 *Vällingby*
T: +46 8 6201700; Fax: +46 8 7392152; E-mail: biblioteket@hi.se; URL: www.hi.se/bibliotek
1968; Anne-Marie Tarnstrom
13 000 vols; 230 curr per; 544 diss/theses; 1 000 govt docs; 230 av-mat; 53 digital data carriers; Clippings, Disability from 1911 onwards, current video-recordings, disability, daily life, rehabilitation (240)
libr loan; EAHIL   26712

Vänersborgs museum, Biblioteket, Niklasborgs vägen 17, 46232 *Vänersborg*; Box 38, 46221 Vänersborg
T: +46 521 572676/-572670;
Fax: +46 521 19782; E-mail: vanersborgs.museum@vgregion.se; URL: www.kulturlagret.se; www.vanersborgsmuseum.se
1885; Clary Winberg, Peter Johansson
40 000 vols; 70 curr per   26713

Länsmuseet Varberg, Biblioteket, Fästningen, 432 44 *Varberg*
T: +46 340 82830; E-mail: biblioteket@lansmuseet.varberg.se; URL: www.lansmuseet.varberg.se
1916
20 000 vols; 150 maps; 4 000 av-mat; 60 sound-rec   26714

Sjukhuset i Varberg, Medicinska biblioteket (Hospital of Varberg, Medical Library), 432 81 *Varberg*
T: +46 340 481460;
Fax: +46 340 481833; E-mail: biblioteket.siv@lthalland.se; URL: www.lthalland.se/sbib
1972; Inga-Lill Jern Larsson
Medical books and papers
17 000 vols; 325 curr per; 17 e-journals; 1 800 av-mat
libr loan; SAB   26715

Centrallasarettet, Library (Central Hospital Library), 721 89 *Västerås*
T: +46 21 173440; Fax: +46 21 175223; URL: www.ltv.se
1970; Turla Söderholm Svensson
Medicine, nursing
29 000 vols; 400 curr per
SB   26716

Västerviks sjukhus, Biblioteket (Vastervik Hospital), 593 81 *Västervik*
T: +46 490 86132; Fax: +46 490 14210; E-mail: biblioteket.vastervik@ltkalmar.se; URL: www.pion.net
1969; Gunwor Weinmann
53 517 vols; 403 curr per; 331 diss/theses; 2 905 av-mat; 3 408 sound-rec; 48 digital data carriers
SB   26717

Psykiatriska Klinikerna, Med. Biblioteket (Psychiatric Clinic), Box 1223, 351 12 *Växjö*
T: +46 470 586470; Fax: +46 470 586475; URL: www.itkronoberg.se/psyk/bibl/bibl.htm
Gunilla Granat
Psychiatry
12 000 vols; 35 curr per   26718

Svenska Emigrantinstitutet, Biblioteket, Vilhelm Mobergs gata 4, Box 201, 351 04 *Växjö*
T: +46 470 20120; Fax: +46 470 39416; URL: www.swemi.se
Yngve Turesson
Swedish emigration
12 000 vols; 80 curr per   26719

Landsarkivet i Visby, Biblioteket (Regional Archives in Visby), Visborgsgatan 1, 621 57 *Visby*
T: +46 498 210514; Fax: +46 498 212955; E-mail: landsarkivet@landsarkivet-visby.ra.se; URL: www.ra.se/vila
1905; Tryggve Siltberg
10 000 vols; 15 curr per; 7 000 maps
                                           26720

Länsmuseet på Gotland, Biblioteket (The County Museum of Gotland), Mellangatan 19, 621 56 *Visby*
T: +46 498 292700;
Fax: +46 498 292729; E-mail: info@lansmuseetgotland.se; URL: www.lansmuseetgotland.se
1918; Birgitta Radhe
20 000 vols; 200 curr per
SB   26721

Riksutställningar, Biblioteket (The Library of the Swedish Travelling Exhibitions), Artillerigatan 33A, 62121 *Visby*; Box 1033, 62121 Visby
T: +46 498 790000;
Fax: +46 498 212720; E-mail: biblioteket@riksutstallningar.se; URL: www.riksutstallningar.se
Isabelle Wårfors
4 500 vols; 40 curr per   26722

Public Libraries

Åkersberga bibliotek, Stationsgränd 5, Box 504, 184 25 *Åkersberga*
T: +46 8 54081400;
Fax: +46 8 54069924; E-mail: akersberga.bibliotek@osteraker.se; URL: www.osteraker.se
Eva Bergström
107 000 vols; 214 curr per
libr loan   26723

Alingsås bibliotek, Södra Ringgatan 3, 441 81 *Alingsås*
T: +46 322 616000;
Fax: +46 322 636439; E-mail: biblioteket.alingsas@alingsas.se; URL: www.alingsas.se
1913; Ulla Forsén
220 880 vols; 770 curr per; 46 microforms; 285 av-mat; 6 450 sound-rec
libr loan; SB   26724

Älmhults bibliotek, Skolgatan 1, 343 31 *Älmhult*; Box 502, 343 23 Älmhult
T: +46 476 55220; Fax: +46 476 14817; E-mail: bibliotek@almhult.se; URL: www.almhult.se
Anette Ekström
85 000 vols; 200 curr per
SB   26725

Alvesta bibliotek, Folkets Hus, Albogatan 17, 342 80 ***Alvesta***
T: +46 472 15269, 15232; Fax: +46 472 15270; E-mail: biblioteket@alvesta.se; URL: www.alvesta.se
Lars-Göran Liliegren
104 000 vols; 220 curr per; 2 700 microforms; 200 av-mat; 700 sound-rec
libr loan; SB 26726

Älvsbyns kommunbibliotek, Skolgatan 23B, Box 124, 942 85 ***Älvsbyn***
T: +46 929 17000; Fax: +46 929 12633; URL: www.alvsbyn.se
Kjell Tegnelund
65 500 vols; 75 curr per 26727

Åmåls bibliotek, Kungsgatan 20, 662 21 ***Åmål***; Box 18, 662 21 Åmål
T: +46 532 17102; Fax: +46 532 16248; E-mail: biblioteket@amal.se; URL: www.amal.se
Kristin Bandgren
120 700 vols; 200 curr per 26728

Ånge folkbibliotek, Järnvägsgatan 3, mail address: 841 81 ***Ånge***
T: +46 690 15112; Fax: +46 690 61611; E-mail: ange@ange.se; URL: www.ange.se
Kjellåke Deldén
85 000 vols; 177 curr per 26729

Ängelholms stadsbibliotek, Stortorget, 262 80 ***Ängelholm***; Box 1070, 262 21 Ängelholm
T: +46 431 87040; Fax: +46 431 87630; E-mail: biblinfo@engelholm.se; URL: www.kommun.engelholm.se
1885; Mia Brinck
110 000 vols; 235 curr per; 350 microforms; 3 000 av-mat
libr loan; SB 26730

Arbogas stadsbibliotek, Kapellgatan 19B, 732 45 ***Arboga***
T: +46 589 87300; Fax: +46 589 18114
1938; Lennart Karlström
95 000 vols; 230 curr per; 275 microforms; 670 av-mat; 3 800 sound-rec; 150 digital data carriers 26731

Årjängs folkbibliotek, Torget 12, Box 903, 672 29 ***Årjäng***
T: +46 573 14285; Fax: +46 573 711176; E-mail: folkbiblioteket@arjang.se; URL: www.arjang.se
Owe Lindström
75 000 vols; 78 curr per 26732

Arvika bibliotek, Kyrkogata 39A, Box 956, 671 29 ***Arvika***
T: +46 570 81600; Fax: +46 570 80673; E-mail: biblioteket@arvika.se; URL: www.arvika.se
1919; Majken Magnusson
219 200 vols; 311 curr per; 200 music scores; 150 microforms; 150 av-mat; 2 500 sound-rec
SB 26733

Uppvidinge kommunbibliotek, Kyrkogata 7, 360 70 ***Åseda***
T: +46 474 47050; Fax: +46 474 47052; E-mail: aseda.bibliotek@uppvidinge.se; URL: www.uppvidinge.se
Sabine helmerding
60 000 vols; 63 curr per 26734

Åsele bibliotek, Lillgatan 2, 910 60 ***Åsele***
T: +46 941 14080; Fax: +46 941 14081; E-mail: biblioteket@asele.se; URL: www.asele.se
Elisabeth Öhman
50 500 vols; 70 curr per; 1 200 microforms; 13 av-mat; 1 154 sound-rec; 80 files of clippings
SB 26735

Askersunds kommunbibliotek, Lilla Bergsgatan 12a, 696 30 ***Askersund***; Box 39, 696 21 Askersund
T: +46 583 81095; Fax: +46 583 81076; E-mail: bibliotek@askersund.se; URL: www.askersund.se
Annika Restadh
61 000 vols; 130 curr per 26736

Åstorps bibliotek, Storgatan 28, Box 14, 265 21 ***Åstorp***
T: +46 42 64220; Fax: +46 42 64228; E-mail: biblioteket@astorp.se; URL: www.astorp.se/kulturfritid/biblioteken
1953; Jan Alfredsson
89 700 vols; 233 curr per; 495 music scores; 65 maps; 4 microforms; 77 av-mat; 1 754 sound-rec
libr loan; SB 26737

Åtvidabergs kommunbibliotek, Tilasplan 3C, Box 197, 597 24 ***Åtvidaberg***
T: +46 120 83197; Fax: +46 120 83195; E-mail: kommunbiblioteket@atvidaberg.se
Elisabeth Axelsson
50 000 vols; 80 curr per 26738

Avesta bibliotek, Kunsgatan 32, Box 505, 774 27 ***Avesta***
T: +46 226 645186; Fax: +46 226 645087; E-mail: bibliotek@avesta.se; URL: www.avesta.se/dokument/4/bibliotek/index.htm
196 200 vols; 207 curr per 26739

Håbo folkbibliotek, Bildningscentrum Jan Fridegård, Box 94, 74 622 ***Bålsta***
T: +46 171 52600; Fax: +46 171 56507; URL: www.habo.se
Malin Ögland
50 000 vols; 120 curr per 26740

Huvudbiblioteket Båstad, Lyckan 7, Box 1124, 269 22 ***Båstad***
T: +46 431 77088; Fax: +46 431 77093; E-mail: bastad.bibliotek@bastad.se; URL: www.bastad.se
Inga-Lill Ahlström
82 500 vols; 118 curr per 26741

Bengtsfors bibliotek, Tingshustorget 5, 666 31 ***Bengtsfors***; Box 24, 666 21 Bengtsfors
T: +46 531 526343; Fax: +46 531 10855; E-mail: bibliotek@bengtsfors.se
Lena grönlund
60 000 vols; 140 curr per 26742

Nordanstigs kommunbibliotek, Hundskinsvägen 1, Box 3, 820 70 ***Bergsjö***
T: +46 652 36200; Fax: +46 652 10066; E-mail: bergsjo@bibliotek.nordanstig.se; URL: www.nordanstig.se
Lena Gräntz
66 000 vols; 80 curr per
libr loan 26743

Bjuvs kommunbibliotek, Köpcentrum, Box 47, 267 21 ***Bjuv***
T: +46 42 85225; Fax: +46 42 85093; E-mail: biblioteket.bjuv@bjuv.se; URL: www.futurumbjuv.nu/bibliotek
Eva Andersen
77 300 vols; 202 curr per 26744

Bodens stadsbibliotek, Folkets Hus, Kunsgatan 51, 961 35 ***Boden***
T: +46 921 62480; Fax: +46 921 14223; E-mail: biblioteket@boden.se; URL: www.boden.se
1908; Jan-Olaf Frank
162 100 vols; 157 curr per 26745

Bollnäs bibliotek, Folkets Hus, Odengatan 17, 821 43 ***Bollnäs***
T: +46 278 25131; Fax: +46 278 15724; URL: www.dialogen.bollnas.se
1931; Anna-Maria Björklin
Carl von Linné-coll
125 000 vols; 210 curr per; 880 music scores; 40 microforms; 132 av-mat; 1 790 sound-rec
libr loan; SB 26746

Borås stadsbibliotek, Sturegatan 35, Box 856, 501 15 ***Borås***
T: +46 33 357621; Fax: +46 33 357675; E-mail: boras.stadsbibliotek@boras.se; URL: www.boras.se/kultur
1860; Tommy Olsson
720 800 vols; 643 curr per; 27 000 microforms; 12 100 av-mat; 11 000 sound-rec; 20 350 talking bks
SB, SFIS 26747

Huvudbiblioteket i Borgholm, Trädgårdsgatan 20, Box 116, 387 22 ***Borgholm***
T: +46 485 88070; Fax: +46 485 10481; E-mail: biblioteket@borgholm.se
Gunilla Lydmark
104 300 vols; 285 curr per 26748

Borlänge bibliotek, Sveatorget 5, 784 33 ***Borlänge***
T: +46 243 74389; Fax: +46 243 74685; E-mail: biblioteket@borlange.se; URL: www.borlange.se/kommun/bibliotek
1944; Kerstin Hed
Dalecarlia-Hällsjö coll, Strindberg coll, J.O. Wallin coll
210 000 vols; 400 curr per; 1 300 music scores; 1 000 microforms; 600 av-mat; 4 200 sound-rec; 1 600 bks on cassettes
libr loan; SB 26749

Östra Göinge bibliotek, Köpmannagatan 1, 280 60 ***Broby***
T: +46 44 7756140;
Fax: +46 44 7756296; E-mail: biblioteket.broby@ostragoinge.se; URL: www.ostragoinge.se/Kultur-Fritid/Bibliotek/
Maria Ehrenberg
81 000 vols; 81 curr per; 200 av-mat; 150 digital data carriers 26750

Bromölla folkbibliotek, Hermansens gatan 22, Box 6, 295 21 ***Bromölla***
T: +46 456 22222; Fax: +46 456 22263; E-mail: biblioteket@bromolla.se; URL: www.bromolla.se
Peter Alsbjer
75 300 vols; 172 curr per; 1 050 av-mat; 600 sound-rec; 20 digital data carriers
SLA 26751

Eda kommunbibliotek, Torget 1, Box 56, 673 22 ***Charlottenberg*** 1
T: +46 571 28230; Fax: +46 571 28240; E-mail: biblioteket@eda.se
1946; Ingalill Walander-Olsson
Coll of old photos and local newspaper clippings from 1930 until today
57 750 vols; 121 curr per; 10 govt docs; 500 music scores; 50 maps; 1 233 microforms; 300 av-mat; 860 sound-rec
libr loan; SB 26752

Danderyds bibliotek, Mörby Centrum, 182 31 ***Danderyd***; Box 114, 182 12 Danderyd
T: +46 8 7535164; Fax: +46 8 7535161
1901; Karin Garmer
198 000 vols; 297 curr per; 400 music scores; 1 900 av-mat; 30 digital data carriers
libr loan; SB, IFLA 26753

Gagnefs folkbibliotek, Södra Industrivägen 8, 785 61 ***Djurås***; mail address: 785 80 Gagnef
T: +46 241 15205; Fax: +46 241 15206; E-mail: bibliotek.djuras@gagnef.se; URL: www.gagnef.se
Liselotte Stöby-Ingvarsson
60 000 vols; 62 curr per 26754

Ekerö bibliotek, Roddargränd 5, Box 206, 178 00 ***Ekerö***
T: +46 8 56039301; Fax: +46 8 56035765; E-mail: bibliotek@ekero.se; URL: www.ekero.se
Barbro Persson
79 700 vols; 165 curr per 26755

Eksjö stadsbibliotek, Huvudbiblioteket, Kaserngatan 20B, Box 1001, 575 80 ***Eksjö***
T: +46 381 36100; Fax: +46 381 36147; E-mail: stadsbiblioteket@eksjo.se; URL: www.eksjo.se/biblioteket
Fredrik Hammar
96 800 vols; 200 curr per 26756

Emmaboda bibliotek, Götgatan 6, Box 67, 361 21 ***Emmaboda***
T: +46 471 18135; Fax: +46 471 18113; E-mail: viveca.alfelt@emmaboda.se; URL: www.emmaboda.se
1980; Viveca Alfelt
93 500 vols; 298 curr per 26757

Enköpings kommunbibliotek, Ågatan 29, Box 909, 745 25 ***Enköping***
T: +46 171 25510; Fax: +46 171 25311; E-mail: biblioteket@kommun.enkoping.se; URL: www.enkoping.se
1868; Anders Lohman
165 000 vols; 178 curr per; 484 microforms; 1 500 sound-rec 26758

Talboks- och punktskriftsbiblioteket (Swedish Library of Talking Books and Braille (TPB)), Sandsborgsvägen 52, mail address: 122 88 ***Enskede***
T: +46 8 399350; Fax: +46 8 6599467; E-mail: info@tpb.se; URL: www.tpb.se
1980; Ingar Beckman Hirschfeldt
66 000 talking book titles, 9 700 braille titles
IFLA, SB 26759

Stadsbiblioteket (City Library), Kriebsensgatan 4, 632 20 ***Eskilstuna***
T: +46 16 7105110; Fax: +46 16 132949; URL: www.eskilstuna.se/default2___152775.aspx
1925; Birgitta Widholm
Swedish Libr of Crime Fiction; Infopoint Europa
565 000 vols; 1 117 curr per; 17 400 govt docs; 4 300 music scores; 23 300 av-mat; 10 000 sound-rec; 200 digital data carriers 26760

Eslövs stadsbibliotek, Norregatan 9, Box 255, 241 24 ***Eslöv***
T: +46 413 62600; Fax: +46 413 61599; E-mail: stadsbiblioteket@eslov.se; URL: www.eslov.se/medborgarservice/bibliotek.3277.html
Torbjörn Lindell
73 700 vols; 202 curr per 26761

Fagersta bibliotek, Västmannavägen 12, 737 40 ***Fagersta***
T: +46 223 44240; Fax: +46 223 15404; E-mail: Fagersta.bibliotek@fagersta.se; URL: www.fagersta.se/bibliotek
Ulla Björklund
82 300 vols; 212 curr per; 200 microforms; 395 av-mat; 3 200 sound-rec
libr loan 26762

Falkenbergs bibliotek, Biblioteksgatan 2, 311 35 ***Falkenberg***
T: +46 346 886370; Fax: +46 346 886387; E-mail: biblioteket@falkenberg.se; URL: www.falkenberg.se/bibliotek
Magnus Nylén
152 150 vols; 161 curr per; 150 microforms; 260 av-mat; 5 000 sound-rec 26763

Falköpings bibliotek, St Sigfridsgatan 11, 521 81 ***Falköping***
T: +46 515 85040; Fax: +46 515 80158; URL: www.falkoping.se
1901; Ann-Christin Hermansson
154 000 vols; 253 curr per 26764

Falu stadsbibliotek, Länsbibliotek för Dalana, Kristinegatan 15, Stadshuset, 791 83 ***Falun***
T: +46 23 83335; Fax: +46 23 83300; E-mail: stadsbiblioteket@adm.falun.se
1907; Pia Witt
Coll of dalekarlian lit
400 000 vols; 582 curr per; 10 230 microforms; 10 700 sound-rec
libr loan; IFLA 26765

Länsbibliotek Dalarna, Stadshuset, 791 83 ***Falun***
T: +46 23 836 70; Fax: +46 23 833 93; E-mail: lansbiblioteket@falun.se; URL: www.ltdalarna.se
Lisa de Souza 26766

Filipstads bergslags bibliotek (Filipstad's Public Library), Stora torget 3D, 682 30 ***Filipstad***; Box 308, 682 27 Filipstad
T: +46 590 61352; Fax: +46 590 61596; E-mail: biblioteket@filipstad.se; URL: www.filipstad.se
1833; Jan Fröding
80 000 vols; 160 curr per; 100 maps; 50 microforms; 660 av-mat; 160 sound-rec; 60 digital data carriers
libr loan; SB 26767

Finspångs bibliotek, Kalkugnsvägen 11, 612 80 ***Finspång***
T: +46 122 85065; Fax: +46 122 85063; URL: www.finspong.se
1937; Eva Elmberg
90 000 vols; 120 curr per
libr loan 26768

Flens bibliotek, Drottninggatan 3, 642 37 ***Flen***
T: +46 157 19295; Fax: +46 157 19293
Per Hjertzell
120 000 vols; 250 curr per 26769

Forshaga folkbibliotek, Storgatan 43, Box 23, 667 21 ***Forshaga***
T: +46 54 172000; Fax: +46 54 172229; E-mail: bibliotek@forshaga.se
Kerstin Krona
70 000 vols; 80 curr per 26770

Gävle stadsbibliotek, Slottstorget 1, 801 30 *Gävle*; Box 801, 801 30 Gävle T: +46 26 178000; Fax: +46 26 688562; E-mail: info@gavlebibliotek.nu; URL: www.gavle.se/bibliotek 1907; Conny Persson 380 000 vols; 1 200 curr per; 9 300 music scores; 10 200 microforms; 27 600 av-mat libr loan; SB   26771

Gislaveds bibliotek, Gröna vägen 28, 332 33 *Gislaved* T: +46 371 81938; Fax: +46 371 81963; URL: www.gislaved.se Palmqvist Morgan 204 000 vols; 250 curr per   26772

Gnesta bibliotek, Huvudbiblioteket (Gnesta Public Library), Marieströmsgatan 3, 646 80 *Gnesta* T: +46 158 70294; Fax: +46 158 70295; E-mail: biblioteket@gnesta.se; URL: www.gnesta.se Ingrid Feldt 69 000 vols; 45 curr per; 4 000 sound-rec libr loan; Biblioteksföreningen   26773

Gnosjö folkbibliotek, Storgatan 8, 335 80 *Gnosjö* T: +46 370 331138; Fax: +46 370 331130; URL: www.gnosjo.se Mats Johansson 50 000 vols; 80 curr per libr loan   26774

Göteborgs stadsbibliotek (City Library of Gothenburg), Götaplatsen, Box 5404, 402 29 *Göteborg* T: +46 31 3683300; Fax: +46 31 3683598; URL: www.goteborg.se/bibliotek 1861; Christina Persson 292 400 vols; 23 600 curr per; 23 e-journals; 8 600 music scores; 491 maps; 39 500 sound-rec libr loan; INTAMEL, IFLA   26775

Götene kommunbibliotek, Centrumhuset, 533 80 *Götene* T: +46 511 346080; Fax: +46 511 59307; URL: www.gotene.se Annicka Berggren 64 600 vols; 111 curr per libr loan   26776

Grums kommun, 2. Biblioteket, Sveagatan 77-85, 664 80 *Grums* T: +46 555 42050; Fax: +46 555 42065; E-mail: bibliotek.kultur@grums.se; URL: www.grums.se Marie Valldor-Frykman 104 000 vols; 250 curr per libr loan   26777

Värmdö kommunbibliotek, 134 81 *Gustavsberg* T: +46 8 57038227; Fax: +46 8 57032048; E-mail: katarina.karlsson@varmdo.se; URL: www2.varmdo.se/Resource.phx/plaza/publica/invanare/bibliotek/index.htx 67 400 vols; 79 curr per   26778

Hagfors Bibliotek, Geijersholmsvägen 5, 683 80 *Hagfors* T: +46 563 18770; Fax: +46 563 18790; E-mail: biblioteket@hagfors.se 1974; Christina Pettersson 107 000 vols; 203 curr per; 200 music scores; 50 maps; 451 microforms; 80 av-mat; 2 695 sound-rec; 40 digital data carriers libr loan; SB   26779

Hallsbergs bibliotek, V. Storg 10, 694 30 *Hallsberg* T: +46 582 685363; Fax: +46 582 685372; E-mail: bibliotek.hallsberg@hallsberg.se; URL: www.hallsberg.se/biblioteket 1970; Ingrid Svedlund 110 000 vols; 200 curr per; 200 music scores; 50 digital data carriers; 600 CDs SB   26780

Hallstahammars bibliotek, Parkgatan 7, Box 3, 734 21 *Hallstahammar* T: +46 220 24110; Fax: +46 220 14731; URL: www.hallstahammar.se/bibliotek Mats Hansson 75 000 vols; 210 curr per   26781

Regionbibliotek Halland, Länsbibliotek för Hallands län, Kristian IV:s väg 1, 302 50 *Halmstad*; Box 538, 301 80 Halmstad T: +46 35 179861; Fax: +46 35 179898; E-mail: regionbibliotek.halland@halmstad.se; URL: www.regionhalland.se/regionbibliotek 1922; Anette Eliasson 718 av-mat libr loan   26782

Haninge bibliotek, Huvudbibliotek, Haninge kulturhus, Poseidons torg 8, 136 81 *Haninge* T: +46 8 6067000; Fax: +46 8 6068925; E-mail: bibliotekethanden@haninge.se; URL: www.haninge.se Karl-Heinz Mueller 100 000 vols   26783

Haparanda stadsbibliotek, Packhusgatan 4, 953 31 *Haparanda* T: +46 922 15067; E-mail: biblioteket@haparanda.se; URL: www.skola.haparanda.se Kristina Nyström Anttila 75 000 vols; 102 curr per   26784

Härnösands kommun bibliotek, mail address: 871 80 *Härnösand* T: +46 611 86530; Fax: +46 611 86186; E-mail: jan.wolf-watz@sambiblioteket.se; URL: www.sambiblioteket.se 1906; Jan Wolf-Watz 240 500 vols; 476 curr per; 130 mss; 1 600 maps; 3 000 microforms; 400 av-mat; 5 000 sound-rec; 4 400 talking bks, 265 shelf metres of govt docs libr loan; SB, SFIS   26785

Länsbiblioteket Västernorrland, Universitetsbacken 3, Box 1045, 87129 *Härnösand* T: +46 611 86551; Fax: +46 611 86550; E-mail: lansbib@ylb.se; URL: www.ylb.se   26786

Hässleholms stadsbibliotek, Vattugatan 18/Järnvägsgatan 23, Box 223, 281 23 *Hässleholm* T: +46 451 267100; URL: www.hassleholm.se/bibliotek 1937; Kristian Hedberg Göingesamling (Lokal coll) 227 400 vols; 548 curr per; 566 microforms; 7 924 av-mat; 4 030 sound-rec libr loan; SB   26787

Heby bibliotek, Centralgatan 1, Folkets Hus, 744 88 *Heby* T: +46 224 36215; Fax: +46 224 31662; URL: www.heby.se Kristina Alldahl 95 000 vols; 128 curr per; 2 000 microforms; 2 300 av-mat; 2 300 sound-rec; 10 digital data carriers libr loan; SB   26788

Hedemora stadsbibliotek, Hörkargatan 10, Box 201, 776 28 *Hedemora* T: +46 225 34184; Fax: +46 225 15333; E-mail: biblioteket@hedemora.se; URL: www.hedemorabibliotek.se/web/arena Lisa Isacson 83 000 vols; 285 curr per   26789

Helsingborgs stadsbibliotek, Bollbrogatan 1, 252 25 *Helsingborg* T: +46 42 106900; Fax: +46 42 240426; URL: biblioteket.helsingborg.se 1860; Eva Hedunger Ecological lit; 10 branch libs 500 000 vols; 600 curr per; 4 000 music scores; 900 av-mat; 16 400 sound-rec; 20 digital data carriers; 4 200 CDs, 12 200 talking bks, 2 700 videos libr loan; SB, IFLA   26790

Orusts kommunbibliotek, Hamntorget, 473 34 *Henån*; Box 84, 473 23 Henån T: +46 304 34250; Fax: +46 304 30487; E-mail: henan.bibliotek@orust.se; URL: www.orust.se Lars Jacobsson 79 900 vols; 189 curr per; 45 maps; 5 052 microforms; 720 av-mat; 2 653 sound-rec; 48 digital data carriers libr loan; SB   26791

Hofors Bibliotek, Storgatan 11, 813 30 *Hofors*; mail address: 813 81 Hofors T: +46 290 29379; Fax: +46 290 29028; E-mail: biblioteket@hofors.se; URL: www.hofors.se/kulturfritid/bibliotek 1898; Marika Näslund 50 000 vols; 44 curr per; 425 music scores; 750 av-mat; 2 800 sound-rec; 70 digital data carriers libr loan; SB   26792

Höganäs stadsbibliotek, Köpmansgatan 10, 263 38 *Höganäs* T: +46 42 337370; Fax: +46 42 337369; E-mail: biblioteket@hoganas.se; URL: www.hoganas.se 1860; Anna Wingårdh 113 000 vols; 225 curr per; 58 microforms; 828 av-mat; 825 sound-rec   26793

Högsby kommunbibliotek, Albert Engströms väg, 57980 *Högsby* T: +46 491 29159; Fax: +46 491 29295; E-mail: biblioteket@hogsby.se; URL: www.hogsby.se/Invaanare/Fritid-Kultur/Bibliotek Gunnar Westling 51 200 vols; 94 curr per libr loan   26794

Höörs bibliotek, Friluftsvägen 13, 234 30 *Höör* T: +46 413 28103; Fax: +46 413 28112; E-mail: biblioteket@hoor.se; URL: www.hoor.se Marianne Törnblad-Anderberg 55 000 vols; 50 curr per   26795

Hörby bibliotek, Vallgatan 7, 272 31 *Hörby* T: +46 415 18450; Fax: +46 415 18049; E-mail: biblioteket@horby.se; URL: www.horby.se 1865; Inga Brallberg Lennart Kjellgren Coll, Cederstrom books, Gunnar Persson Coll 67 430 vols; 132 curr per; 88 diss/theses; 356 music scores; 402 av-mat; 1 332 sound-rec libr loan   26796

Huddinge kummunbibliotek, Kommunalvägen 28, Box 1321, 141 27 *Huddinge* T: +46 8 53530500; Fax: +46 8 53530501; E-mail: huvudbibliotek@adm.huddinge.se; URL: www.huvudbiblioteket.huddinge.se 1966; Karin Fahller 130 000 vols; 250 curr per; 418 av-mat   26797

Hudiksvalls bibliotek, Rådhusparken, mail address: Långgatan 2, 824 52 *Hudiksvall* T: +46 650 19650; Fax: +46 650 38193; URL: www.hudiksvall.se/templates/Page___439.aspx Birgitta Fornstedt 90 000 vols; 250 curr per; 3 000 sound-rec; 100 language courses, 1000 talking bks   26798

Hultsfreds bibliotek, Stora torget, Box 503, 577 26 *Hultsfred* T: +46 495 16015; Fax: +46 495 16022; URL: www.hultsfred.se Anna-Maj Sundström 143 000 vols; 253 curr per   26799

Huskvarna bibliotek, Drottninggatan 19, Box 1002, 561 24 *Huskvarna* T: +46 36 106644; Fax: +46 36 107646 1904 93 000 vols; 250 curr per   26800

Järfälla folkbibliotek, Vöpnarstråket, 175 30 *Järfälla* T: +46 8 58029160; Fax: +46 8 58029174; E-mail: jakobsberg@folkbibliotek.jarfalla.se; URL: www.jarfalla.se 1914; Kaisa Nordberg 175 000 vols; 480 curr per; 260 microforms; 130 av-mat; 3 500 sound-rec; 16 digital data carriers libr loan   26801

Årebiblioteken, Norra vägen 21, Box 79, 830 05 *Järpen* T: +46 647 16960; Fax: +46 647 16965 Mikael Cederberg 80 000 vols; 122 curr per   26802

Jönköpings stadsbibliotek, Dag Hammarskjölds plats 1, Box 1029, 551 11 *Jönköping* T: +46 36 105583; Fax: +46 36 107744; E-mail: ulf.moberg@jonkoping.se; URL: lingonline.jonkoping.se 1916; Ulf Moberg Thesophic lit 580 000 vols; 700 curr per; 3 850 microforms; 17 600 av-mat; 6 809 sound-rec libr loan; SB   26803

Länsbibliotek Jönköping, Dag Hammarskjölds plats 1, 553 22 *Jönköping*; Box 1029, 551 22 Jönköping T: +46 36 105000; Fax: +46 36 105067; E-mail: lansbiblioteket@lj.se; URL: www.f.lanbib.se   26804

Kalix kommunbibliotek, Köpmannagatan 25, Box 10070, 952 27 *Kalix* T: +46 923 65162; Fax: +46 923 16714 Ewy Nilsson 107 500 vols; 200 curr per   26805

Kalmar stadsbibliotek, Tullslätten 4, Box 610, 391 26 *Kalmar* T: +46 480 450635; Fax: +46 480 450605; E-mail: pia.axeheim@kalmar.se; URL: www.kalmar.regionforbund.se 1922 370 000 vols; 650 curr per   26806

Karlsborgs kommunbibliotek, Storgatan 36, 546 32 *Karlsborg* T: +46 505 17030; Fax: +46 505 17031; E-mail: biblioteket@karlsborg.se; URL: www.karlsborg.se/kbg_Templates/Page___265.aspx Margaretha Rhenberg 58 600 vols; 167 curr per; 381 microforms; 629 av-mat; 518 sound-rec libr loan; SB   26807

Bibliotek & Kultur, Ågatan 48, 37438 *Karlshamn*; Box 40, 37421 Karlshamn T: +46 454 81230, 81240; Fax: +46 454 81246; E-mail: biblioteket@utb.karlshamn.se; URL: www.karlshamn.se/Bibliotek/ 1863; Suzanne Hammargren 215 000 vols; 407 curr per; 500 microforms; 4 040 av-mat; 1 400 sound-rec; 7 digital data carriers libr loan; SB   26808

Karlskogas stadsbibliotek, Kyrkbacken 9, 691 83 *Karlskoga* T: +46 586 61880; Fax: +46 586 56391; E-mail: biblioteket@karlskoga.se; URL: www.karlskoga.se 1850; Helesine Taylor 205 265 vols; 385 curr per   26809

Karlskrona stadsbibliotek, Länsbibliotek för Blekinge län, Stortorget 15-17, Box 320, 371 25 *Karlskrona* T: +46 455 83430; Fax: +46 455 25511 1906; Ulla-Marie Norrby Carlskrona Läse-Söllskaps Bibl 1793-1863 483 500 vols; 629 curr per; 9 570 music scores; 8 965 microforms; 705 av-mat; 8 105 sound-rec libr loan; SB   26810

Karlstads stadsbibliotek, Länsbibliotek för Värmlands län, V. Torggatan 26, Bibliotekshuset, 651 84 *Karlstad* T: +46 54 295873; Fax: +46 54 295879; E-mail: karlstadskommun@karlstad.se; URL: www.karlstad.se/ku/bibliotek/stadsbibl.shtml 1914; Dr. Kjell Fredriksson 250 000 vols; 750 curr per; 1 000 maps; 8 000 microforms; 18 500 sound-rec libr loan; SB, IFLA   26811

Länsbiblioteket i Värmland, a Torggatan 26, mail address: 651 84 *Karlstad* T: +46 54 7011000; Fax: +46 54 7011092 Anna Christina Rutquist   26812

Katrineholms bibliotek, Kulturhuset Ängeln, Djulögatan 27, 641 80 *Katrineholm* T: +46 150 576-30/-20; E-mail: biblioteket@katrineholm.se; URL: www.katrineholm.se 1905; Jan-Olof Larsson 130 000 vols; 400 curr per; 200 microforms; 200 av-mat; 2 000 sound-rec libr loan; AIBM   26813

Kävlinge bibliotek, Kvarngatan 17, 244 31 *Kävlinge*
T: +46 46 739473; Fax: +46 46 739484; E-mail: kavlinge.biblioteke@kavlinge.se
Anne-Marie Jönsson
50 000 vols; 100 curr per   26814

Kils bibliotek, Storgatan 41, Box 168, 655 23 *Kil*
T: +46 554 19166; Fax: +46 554 14629; E-mail: biblioteket@kil.se
Lisbeth Råman
58 300 vols; 153 curr per; 538 av-mat; 242 sound-rec
libr loan; SB   26815

Marks bibliotek, Huvudbiblioteket, Boråsvägen 42, 511 80 *Kinna*
T: +46 320 217090; Fax: +46 320 14886; E-mail: bibliotek@mark.se; URL: www.mark.se/
Lars-Gunnar Kristiansson, Margareta Hanning
228 000 vols; 175 curr per; 2 515 microforms; 100 av-mat; 6 500 sound-rec
SB   26816

Kiruna stadsbibliotek, Biblioteksgatan 2, 981 85 *Kiruna*
T: +46 980 70750; Fax: +46 980 70751; E-mail: biblio@kommun.kiruna.se; URL: www.kommun.kiruna.se/Kultur-och-bibliotek/Kiruna-stadsbibliotek/
Marie Engman-Suup
175 000 vols; 225 curr per   26817

Kinda bibliotek, Huvudbiblioteket, Torggatan 2, 590 40 *Kisa*
T: +46 494 19100; Fax: +46 494 19119; E-mail: biblioteket.kisa@kinda.se; URL: biblioteket.kinda.se
1987; Thomas C. Ericsson
Kindasamlingen
51 020 vols; 58 curr per; 10 diss/theses; 40 maps; 10 microforms; 469 av-mat; 243 sound-rec; 10 digital data carriers
libr loan; SB   26818

Klippans Bibliotek, Storgatan 9, 264 80 *Klippan*
T: +46 435 28177; Fax: +46 435 28181; E-mail: bibliotek@klippan.se; URL: www.klippan.se
Margareta Werke
97 700 vols; 260 curr per
libr loan   26819

Köpings stadsbibliotek, Folkets Hus, 731 41 *Köping*
T: +46 221 25182; Fax: +46 221 25198; E-mail: stadsbiblioteket@koping.se; URL: www.koping.se/bibliotek/HB/bibl.htm
1876; Henry Segerström
170 000 vols; 253 curr per; 1 076 sound-rec
libr loan; SB   26820

Biblioteket i Kopparberg, Konstmästaregatan 5, 714 30 *Kopparberg*
T: +46 580 80535; Fax: +46 580 10393; E-mail: bibliotek@ljusnarsberg.se; URL: www.ljusnarsberg.se/servicetjanster/kultur_bibliotek.asp?detta=kulturfritid
59 400 vols; 165 curr per   26821

Kramfors kommunbibliotek, Biblioteksgatan 8, Box 144, 872 23 *Kramfors*
T: +46 612 80255; Fax: +46 612 10054; E-mail: biblioteket.kommunen@kramfors.se; URL: www.biblioteken.kramfors.se
Ingrid Ohlsson
160 000 vols; 244 curr per   26822

Kristianstads stadsbibliotek, Föreningsgatan 4, 291 33 *Kristianstad*; mail address: Kultur och fritidsförvaltningen, 291 80 Kristianstad
T: +46 44 136710; E-mail: biblioteket@kristianstad.se; URL: www.kristianstad.se
1861
143 260 vols; 450 curr per; 4 200 sound-rec; 1 400 DVDs and VHS, 7 600 talking bks
libr loan   26823

Kristinehamns bibliotek, Tullportsgatan 13, 681 84 *Kristinehamn*
T: +46 550 88080; Fax: +46 550 13901; E-mail: biblioteket@kristinehamn.se; URL: www.kristinehamn.se
1860; Reidun Westman
165 000 vols; 230 curr per; 220 music scores; 352 microforms; 1 760 av-mat; 1 560 sound-rec
libr loan; SB   26824

Krokoms kommunbibliotek, Offerdalsvägen 8, 835 80 *Krokom*
T: +46 640 16150; Fax: +46 640 16155; E-mail: biblioteket.krokom@krokom.se; URL: www.z.lanbib.se/lanet_krokom.htm
114 000 vols; 60 curr per   26825

Kumla bibliotek, Ingång från Parkgatan, Fack 17, 692 80 *Kumla* 1
T: +46 19 588190; Fax: +46 19 588188; E-mail: biblioteket@kumla.se; URL: www.kumla.se
Carita Ehrstedt
142 000 vols; 300 curr per   26826

Kungälvs stadsbibliotek, Gymnasiegatan 3, 442 34 *Kungälv*
T: +46 303 99011; Fax: +46 303 19005; URL: www.kungalv.se/bibliotek
Elisabeth Lundgren
189 000 vols; 312 curr per   26827

Kungsbacka bibliotek, Kulturhuset Fyren, Borgmästaregatan 6, 434 32 *Kungsbacka*
T: +46 300 34185; Fax: +46 300 70469; E-mail: bibliotek@kungsbacka.se; URL: www.kungsbacka.se
Britta Särnmark
120 000 vols; 312 curr per   26828

Kungsörs stadsbibliotek, Drottninggatan 34, 736 31 *Kungsör*
T: +46 227 600210; URL: www.kungsor.se
51 000 vols; 130 curr per
libr loan   26829

Laholms bibliotek, Stadshuset, Humlegången 6, 312 80 *Laholm*
T: +46 430 15150; Fax: +46 430 15160; E-mail: biblioteket@laholm.se; URL: www.laholm.se
Börje Wall
161 000 vols; 300 curr per; 720 av-mat; 2 700 sound-rec; 10 digital data carriers
libr loan; SB   26830

Landskrona stadsbibliotek, Regeringsgatan 13, 26136 *Landskrona*
T: +46 418 473500; URL: www.landskronakultur.se/
1895; Margareta Rydhagen, Calle Sundewall
200 000 vols; 387 curr per; 11 000 av-mat; 9 000 sound-rec; 100 digital data carriers
libr loan; SB   26831

Laxå bibliotek, Huvudbiblioteket, Postgatan 13, 695 80 *Laxå*
T: +46 584 85196; Fax: +46 584 12312; E-mail: biblioteket@laxa.se; URL: www.laxa.se
Anne-Marie Alzén
60 000 vols; 116 curr per
SB   26832

Leksands bibliotek, Kyrkallén 3, Box 351, 793 27 *Leksand*
T: +46 247 80250; Fax: +46 247 12940
1867; Kersti Gullstrand Bergmark
70 000 vols; 200 curr per
libr loan; SB   26833

Lerums bibliotek, Alingsåsvägen 9, 443 80 *Lerum*
T: +46 302 521475; Fax: +46 302 521980; E-mail: biblioteket@lerum.se; URL: www.lerum.se/bibliotek
1918; Margareta Ryme
62 000 vols; 180 curr per; 550 av-mat; 2 000 sound-rec   26834

Lidingö stadsbibliotek, Stockholmsvägen 54F, Box 1244, 181 34 *Lidingö*
T: +46 8 7313771; Fax: +46 8 7313768; URL: Stadsbiblioteket@Lidingo.se; URL: www.lidingo.se/bibliotek
1859; Gunnel De Geer-Tolstoy
123 000 vols; 350 curr per; 2 000 govt docs; 4 459 av-mat; 5 667 sound-rec
libr loan; SB   26835

Lidköpings stadsbibliotek, Nya Stadens torg 5, Box 2403, 531 02 *Lidköping*
T: +46 510 770051; E-mail: stadsbiblioteket@lidkoping.se; URL: www.lidkoping.se/biblioteket
1862; Eva Mowitz
183 000 vols; 350 curr per   26836

Lindesbergs bibliotek, Norrtullstorget, 711 80 *Lindesberg*
T: +46 581 81140; Fax: +46 581 16571
Karin Blomqvist
170 200 vols; 170 curr per   26837

Linköpings stadsbibliotek, Östgötagatan 5, 581 19 *Linköping*; Box 1984, 581 19 Linköping
T: +46 13 206601; Fax: +46 13 206650; E-mail: stadsbiblioteket@linkoping.se; URL: www.linkoping.se/bibliotek
1926; Helena Selbing
The libr reseach dept contains the Diocesan libr
667 000 vols; 850 curr per; 2 300 mss; 130 incunabula; 4 240 microforms; 16 740 av-mat; 14 273 sound-rec; 2 700 digital data carriers
libr loan; SB, ALA, AIBM   26838

Ljungby bibliotek, Södra Järnvägsgatan 8, 341 83 *Ljungby*
T: +46 372 789200; Fax: +46 372 81949; E-mail: biblioteket@kommun.ljungby.se; URL: www.ljungby.se
1982; Anne-Lise Fäldt
200 000 vols; 218 curr per; 525 sound-rec; 2 224 digital data carriers
libr loan   26839

Ljusdals kommunbibliotek, Riotorget, Box 700, 827 80 *Ljusdal*
T: +46 651 18001; Fax: +46 651 12332; E-mail: bibliotek@ljusdal.se; URL: www.ljusdal.se
Kerstin Hassner
50 000 vols; 178 curr per
libr loan; IFLA   26840

Lomma folkbibliotek (Public Library), Strandvägen 96, 234 21 *Lomma*; Box 29, 234 21 Lomma
T: +46 40 6411373; Fax: +46 40 6411388; E-mail: lomma.bibliotek@lomma.se; URL: www.lib.lomma.se
Gun Andersson
100 000 vols; 110 curr per
SB   26841

Ludvika bibliotek, Carlavägen 24, 771 82 *Ludvika*
T: +46 240 86297; Fax: +46 240 86693; E-mail: ludvika.bibliotek@ludvika.se; URL: www.ludvika.se/bibliotek
1903; Bengt-Åke Kjell
200 000 vols; 381 curr per   26842

Luleå stadsbibliotek, Länsbibliotek för Norrbottens län, Kyrkogata, 971 79 *Luleå*
T: +46 920 294840; Fax: +46 920 435050; URL: www.lulea.se/bibliotek
1903; Gun-Britt Lindskog
460 000 vols; 700 curr per; 23 595 govt docs; 110 music scores; 1 243 maps; 12 780 microforms; 5 940 av-mat; 6 600 sound-rec
libr loan; SB, IFLA   26843

Norbottens länsbibliotek, Skeppsbrogatan 17, 971 79 *Luleå*
T: +46 920 453000; URL: www.norrbottenslansbibliotek.nu   26844

Lunds stadsbibliotek, Konst & Musik, St. Petri kyrkogata 6, Box 111, 221 00 *Lund*
T: +46 46 355990; Fax: +46 46 138515; E-mail: folkbiblioteken@lund.se
1864; Gunilla Herdenberg
494 000 vols; 1 200 curr per; 9 200 music scores; 280 maps; 2 740 av-mat; 12 461 sound-rec; 377 digital data carriers
libr loan; SB, IFLA   26845

Lycksele kommunbibliotek, Nora Torggatan 12, 921 31 *Lycksele*
T: +46 950 16821; Fax: +46 950 13401; E-mail: kommun@lycksele.se; URL: www.lycksele.se
Mona Lindgren
124 000 vols; 238 curr per   26846

Lysekils stadsbibliotek (Lysekil's City Library), Kungsgatan 18, 453 33 *Lysekil*
T: +46 523 19000; Fax: +46 523 19622; E-mail: stadsbiblioteket@lysekil.se
1959; Inger Sterneskog
Local coll
72 000 vols; 222 curr per; 400 music scores; 295 microforms; 800 av-mat; 1 410 sound-rec; 106 digital data carriers
libr loan; SB   26847

Gällivare folkbibliotek, Köpmangatan 6, Box 6, 983 21 *Malmberget*
T: +46 970 18163; Fax: +46 970 21109
Liselotte Lagnestig
130 800 vols; 171 curr per   26848

Malmö stadsbibliotek, Lånecentral, 205 81 *Malmö*
T: +46 40 6608500; Fax: +46 40 6608575; E-mail: info.stadsbiblioteket@malmo.se; URL: www.malmo.stadsbibliotek.org
1905; Gunilla Konradsson Mortin
District of Scania Coll, Hjalmar Gullberg Coll; County Libr for the Malmöhus County, Lending Center for Southern and Western Sweden
900 000 vols; 1 800 curr per; 15 500 music scores; 1 035 maps; 8 242 microforms; 15 986 av-mat; 31 150 sound-rec; 2 114 digital data carriers; 37 689 books on tape
libr loan; SB, IFLA   26849

Malungs kommunbibliotek, Lisagatan 32, Box 94, 782 22 *Malung*
T: +46 280 18300; Fax: +46 280 12163; E-mail: biblioteket@malung.se; URL: www.malung.se
ca 1860; Margaretha Hedblom
82 686 vols; 107 curr per; 110 av-mat; 1 010 sound-rec; 575 videorecordings, vhs, movies
libr loan   26850

Mariestads Stadsbibliotek, Drottninggatan 12, 54234 *Mariestad*
T: +46 501 63200; Fax: +46 501 63209; E-mail: info@mariestad.se; URL: www.mariestad.se
Gunilla Broberg
141 000 vols; 138 curr per; 25 digital data carriers
libr loan   26851

Markaryds bibliotek, Kulturhuset, Kungsgatan 1, Box 133, 285 23 *Markaryd*
T: +46 433 72105; Fax: +46 433 72213; URL: www.markaryd.se
Torsten Ekelund
69 100 vols; 200 curr per
SB   26852

Sigtuna kommuns bibliotek, Nymärsta kulle 1, 195 85 *Märsta*
T: +46 8 59126270; Fax: +46 8 59126271; E-mail: marstabibliotek@sigtuna.se; URL: www.sigtuna.se
Ingela Irlöv Edfast
105 000 vols; 385 curr per   26853

Melleruds kommunbibliotek, Södra Kungsgatan 7, Box 84, 464 22 *Mellerud*
T: +46 530 18200; Fax: +46 530 18206; E-mail: biblioteket@mellerud.se; URL: www.mellerud.se
Lennart Åberg
57 000 vols; 121 curr per   26854

Mjölby stadsbibliotek, Burensköldsvägen 13, Box 304, 595 23 *Mjölby*
T: +46 142 85105; Fax: +46 142 14520; E-mail: burensko@hotmail.com; URL: www.mjolby.se/bibliotek
Urban Bäckström
90 000 vols; 200 curr per   26855

Mölndals stadsbibliotek, Göteborgsvägen 19, 431 30 *Mölndal*
T: +46 31 3151600; Fax: +46 31 872713; E-mail: molndals.stadsbiblioteket@molndal.se; URL: bibliotek.molndal.se/molndal
1863; Margareta Ljunge
204 000 vols; 329 curr per   26856

Mölnlycke bibliotek, Biblioteksgatan 2, 435 30 *Mölnlycke*
T: +46 31 7246402; Fax: +46 31 7246442; E-mail: molnlycke.bibliotek@harryda.se; URL: www.harryda.se
Britta Särnmark
124 000 vols; 225 curr per   26857

Mönsterås kommunbibliotek, S. Tingsgatan 3, 383 21 *Mönsterås*
T: +46 499 17205; Fax: +46 499 17924; E-mail: bibliotek@kommun.monsteras.se; URL: www.monsteras.se
Carina Eskelin
104 000 vols; 224 curr per
libr loan   26858

Mora folkbibliotek, Köpmannagatan 4, Box 319, 792 25 *Mora*
T: +46 250 26779; Fax: +46 250 15482; E-mail: biblioteket@mora.se; URL: www.mora.se
Karl-Arne Sandvik
84 000 vols; 200 curr per   26859

Mörbylånga kommunbibliotek, Storgatan 2, 380 62 *Mörbylånga*
T: +46 485 47159; Fax: +46 485 47151; URL: www.mc.hok.se/~bib95olk/hemsida.html
Dag Jonasson
109 300 vols; 165 curr per   26860

Motala stadsbibliotek (Motala Public Library), Repslagaregatan 1, 591 29 *Motala*; Box 253, 591 23 Motala
T: +46 141 225000; Fax: +46 141 52103; E-mail: bibl.info@motala.se
1890; Margareta Alexanderson
139 000 vols; 286 curr per; 268 microforms; 662 av-mat; 1962 sound-rec; 17 digital data carriers   26861

Munkedals kommunbibliotek, Kommunhuset Forum, 455 80 *Munkedal*
T: +46 524 18100; Fax: +46 524 18107; E-mail: eva.hogmark@munkedal.se; URL: www.munkedal.se
65 000 vols; 190 curr per   26862

Nacka stadsbibliotek, Forum Nacka, Box 4150, 131 04 *Nacka*
T: +46 8 7189625; Fax: +46 8 7189617; URL: www.nacka.se/bibl/home.htm
1927; Börje Sjölund
228 000 vols; 625 curr per; 3 347 av-mat; 8 802 sound-rec
libr loan; SB   26863

Nora kommunbibliotek, Storgatan 16, Box 123, 713 23 *Nora*
T: +46 587 81290; Fax: +46 587 14176
1863; Peter Bernövall
71 500 vols; 198 curr per; 616 av-mat
SB   26864

Norbergs kommunbibliotek, Malmgatan 7, Box 25, 778 01 *Norberg*
T: +46 223 29140; Fax: +46 223 23700; URL: www.norberg.se
Bengt Raattamaa
55 000 vols; 110 curr per   26865

Nordmalings bibliotek, Kungsvägen 31, 914 32 *Nordmaling*; Box 47, 014 81 Nordmaling
T: +46 930 14175; Fax: +46 930 141 76; E-mail: biblioteket@nordmaling.se; URL: www.nordmaling.se
Agneta Forsberg
50 000 vols; 95 curr per; 500 microforms; 600 av-mat; 50 digital data carriers
Svensk Biblioteksförening   26866

Mediacentralen, Stadsbiblioteket, Södra Promenaden 105, Våning, Box 2113, 600 02 *Norrköping*
T: +46 11 153468; Fax: +46 11 128182; E-mail: mediacentralen@norrkoping.se; URL: www.nsb.norrkoping.se/mediacentralen/
Göran Nilsson
70 000 vols   26867

Norrköpings stadsbibliotek, Södra Promenaden 105, Box 2113, 600 02 *Norrköping*
T: +46 11 150000; Fax: +46 11 107461; E-mail: stadsbiblioteket@norrkoping.se; URL: www.nsb.norrkoping.se
1913; Conny Äng
500 000 vols; 500 curr per
libr loan   26868

Norrtälje stadsbiblioteket, Billborgsgatan 3, Box 805, 761 28 *Norrtälje*
T: +46 176 71444; Fax: +46 176 57438; E-mail: nbi@norrtalje.se; URL: www.norrtalje.se/templates/Page__2358.aspx
1905; Kerstin Ericsson
60 000 vols; 100 curr per; 527 microforms; 1 458 av-mat
libr loan; IFLA   26869

Norsjö bibliotek, Skolgatan 26, 935 00 *Norsjö*
T: +46 918 14255; Fax: +46 918 10486; URL: www.norsjo.net
Linnéa Dahlberg
50 000 vols; 125 curr per; 400 microforms; 20 digital data carriers   26870

Nössjö Stadsbibliotek, Mariagatan 2, 571 80 *Nössjö*
T: +46 380 78233; Fax: +46 380 75455
1906; Jan Holmquist
105 650 vols   26871

Nybro bibliotek, Storgatan 34, Box 113, 382 22 *Nybro*
T: +46 481 45460; Fax: +46 481 45462; E-mail: bibliotek@nybro.se; URL: www.nybro.se
Mats Johannson
120 000 vols; 350 curr per   26872

Nyköpings stadsbibliotek, Hospitalsgatan 4, 611 83 *Nyköping*
T: +46 155 248686; Fax: +46 155 248678; E-mail: info.biblioteket@nykoping.se; URL: www.nykoping.se
1927
275 600 vols; 250 curr per   26873

Nynäshamns bibliotek, Stadshusplatsen 3, 149 81 *Nynäshamn*
T: +46 8 52068310; Fax: +46 8 52016242; E-mail: biblioteket@nynashamn.se; URL: www.nynashamn.se
1912; Kerstin Frederiksson
66 400 vols; 110 curr per; 200 microforms; 1 000 av-mat; 4 500 sound-rec
libr loan; SB   26874

Olofströms bibliotek, Ö. Storgatan 24, Box 301, 293 24 *Olofström*
T: +46 454 93290; Fax: +46 454 92424; E-mail: biblioteket@olofstrom.se; URL: www.olofstrom.se
Elisabeth Jonsson
67 800 vols
libr loan   26875

Länsbiblioteket i Örebro län, Regionförbundet Örebro (Örebro County Library), Forsuarvägen 1, 70183 *Örebro*; mail address: 70183 Örebro
T: +46 19 6026364; Fax: +46 19 189829; E-mail: regionorebro@regionorebro.se; URL: www.regionorebro.se/blameny/kultur/lansbiblioteket
1936; Peter Alsbjer
400 000 vols; 40 curr per; 10 e-journals
SB, IFLA   26876

Örnsköldsviks stadsbibliotek, Lasarettsgatan 5, 891 33 *Örnsköldsvik*; Box 840, 891 18 Örnsköldsvik
T: +46 660 88020; Fax: +46 660 17561; URL: www.ornskoldsvik.se
Birgitta Bergfeldt
100 000 vols; 464 curr per   26877

Osby bibliotek, Ö. Järnvägsgatan 16, Box 13, 283 00 *Osby*
T: +46 479 18310; Fax: +46 479 18315; E-mail: bibliotek@osby.se; URL: www.osby.se/kommunal_info/utbildning/bibliotek/index.html
Torsten Ehelund
100 000 vols; 175 curr per
libr loan   26878

Oskarshamns stadsbiblioteket, Handverksgatan 18-20, Box 705, 572 28 *Oskarshamn*
T: +46 491 88011; Fax: +46 491 83245; E-mail: stadsbiblioteket@oskarshamn.se; URL: www.oskarshamn.se
1905; Ing-Britt Runermark
190 000 vols; 388 curr per; 220 av-mat; 3 300 sound-rec
libr loan   26879

Jämtlands Läns Bibliotek (Jämtland County Library), Rådhusgatan 25-27, 831 80 *Östersund*
T: +46 63 143050; Fax: +46 63 109840; E-mail: lansbibliotek@ostersund.se; URL: www.jlb.ostersund.se
1833; Bodil Köpsen
Canada collection, Zetterström library
500 000 vols; 700 curr per; 20 e-journals; 1 600 e-books; 20 incunabula; 4 000 music scores; 500 maps; 7 500 av-mat; 12 000 sound-rec; 200 digital data carriers
libr loan; SB   26880

Tornedalens bibliotek, Skolvägen 17, Box 18, 957 21 *Övertorneå*
T: +46 927 72182; Fax: +46 927 12054; E-mail: biblioteket@edu.overtornea.se; URL: www.overtornea.se/Snabbval/Bibliotek/
1928; Marita Mattsson Barsk
70 000 vols; 100 curr per   26881

Oxelösunds bibliotek, Höjdgatan 26, 613 81 *Oxelösund*
T: +46 155 38000; Fax: +46 155 38103
Malin Göthberg
70 000 vols; 165 curr per; 1 200 av-mat; 2 000 sound-rec   26882

Pajala kommunbibliotek, Biblioteksvägen 1, Box 13, 984 21 *Pajala*
T: +46 978 12150; Fax: +46 978 12133; E-mail: bibliotekschef@kommun.pajala.se; URL: www.pajala.se
Tunla Abraham
97 000 vols; 50 curr per
libr loan   26883

Partille bibliotek, Huvudbibliotek, Gamla Kronvägen 56, Box 602, 433 28 *Partille*
T: +46 31 7921011; Fax: +46 31 444632; URL: www.partille.se
Kristina Frizell
138 500 vols; 353 curr per   26884

Piteå bibliotek, Kultur Fritid, Olof Palmes gata 2, 941 33 *Piteå*; Box 721, 941 28 Piteå
T: +46 199 96800; Fax: +46 199 13893; E-mail: stadsbiblioteket@kf.pitea.se; URL: www.pitea.se/
1903; Margareta Pohjanen
250 000 vols; 300 curr per; 200 av-mat; 2 000 sound-rec
libr loan   26885

Robertsfors folkbibliotek, Skolgatan 6, 915 31 *Robertsfors*
T: +46 934 14024; Fax: +46 934 14026; E-mail: robbib@robertsfors.se
Lars Andersson
75 000 vols; 212 curr per   26886

Ronneby bibliotek, Kungsgatan 35, 372 37 *Ronneby*
T: +46 457 617477; E-mail: informationen@bibliotek.ronneby.se; URL: www.bibliotek.ronneby.se
Lena Brorsdotter
192 400 vols; 280 curr per   26887

Säffle bibliotek, Kanaltorget 3, 661 80 *Säffle*
T: +46 533 681610; Fax: +46 533 681598; E-mail: bibliotek@saffle.se; URL: www.saffle.se/biblweb/bibliotek/index.htm
Eva Björk
100 000 vols; 200 curr per
libr loan   26888

Sala stadsbibliotek, N. Esplanaden 5, 733 38 *Sala*
T: +46 224 55501; Fax: +46 224 17590; E-mail: stadsbibliotek@sala.se; URL: www.sala.se
100 000 vols; 155 curr per   26889

Sandvikens folkbibliothek, Folkets Hus, Köpmansgatan 5, Box 503, 811 30 *Sandviken*
T: +46 26 241611; Fax: +46 26 258361; E-mail: folkbiblioteket@sandviken.se; URL: www.sandviken.se
1938; Åsa Wirén Jonsson
149 000 vols; 517 curr per   26890

Säters bibliotek, Skolgränd 6, 783 30 *Säter*
T: +46 225 55186; Fax: +46 225 55137; E-mail: bibliotek@sater.se; URL: www.sater.se
Barbro Nordén Harbeck
670 000 vols; 156 curr per   26891

Sävsjö stadsbibliotek, Storgatan 4, 576 31 *Sävsjö*; Box 201, 576 24 Sävsjö
T: +46 382 15400; Fax: +46 382 15407; E-mail: biblioteket@savsjo.se; URL: www.savsjo.se
Haldis Oredsson
66 000 vols; 143 curr per   26892

Simrishamns bibliotek, Jörnvägsgatan 2, 272 80 *Simrishamn*
T: +46 414 819300; Fax: +46 414 10519; E-mail: biblioteket@simrishamn.se; URL: www.simrishamn.se/sv/kultur_fritid/bibliotek/
1896; Gunilla Janlert
160 000 vols; 240 curr per; 2 000 music scores; 1 100 microforms; 1 500 av-mat; 2 750 sound-rec; 50 digital data carriers; 2 300 talking bks
libr loan; SB   26893

Sjöbo bibliotek, Gamla Torg, mail address: Sjöbo kommun, 275 80 *Sjöbo*
T: +46 416 27160; Fax: +46 416 11792; E-mail: biblioteket@sjobo.se; URL: www.sjobo.se/templates/Page.aspx?id=983
Birgitta Miegel Sandborg
104 940 vols; 220 curr per   26894

Skellefteå bibliotek, Kanalgatan 73, Box 703, 931 27 *Skellefteå*
T: +46 910 736100; Fax: +46 910 736114; URL: www.skelleftea.se
1880; Robert Knudsson
Lit about Skellefteå
117 000 vols; 363 curr per; 1 800 microforms; 4 500 av-mat; 4 000 sound-rec
SB   26895

Skinnskattebergs bibliotek, Klockarbergsvägen 6, Box 103, 739 22 *Skinnskatteberg*
T: +46 222 45138; Fax: +46 222 45087; E-mail: biblioteket@skinnskatteberg.se
Carina Eriksson
58 746 vols; 108 curr per; 793 microforms
libr loan   26896

Hammarö kommunbibliotek (The Hammarö library), Folkets Hus, Mörmovägen 8, Box 46, 663 21 *Skoghall*
T: +46 54 515200; Fax: +46 54 515342; E-mail: biblioteket@hammaro.se; URL: www.hammaro.se
Karin Andreasson
The 'Värmland' coll
60 000 vols; 140 curr per
libr loan; SAB   26897

Skövde stadsbibliotek, Trädgårdsgatan 9, Box 404, 541 28 *Skövde*
T: +46 500 468570; Fax: +46 500 468084; URL: www.skovde.se
1931; Inger Davidsson
235 000 vols; 365 curr per; 100 maps; 1 000 av-mat
libr loan   26898

Skurups kommunbibliotek, Kyrkogata 31, mail address: Skurups kommun, 274 80 *Skurup*
T: +46 411 536071; Fax: +46 411 536080; E-mail: biblioteket@skurup.se; URL: www.skurup.se
59 600 vols; 99 curr per   26899

Smedjebackens bibliotek, Folkets Hus, Vasagatan 11, 777 30 *Smedjebacken*
T: +46 240 660280; Fax: +46 240 76763; E-mail: kultur@smedjebacken.se; URL: www.smedjebacken.se
Monica Borg
75 500 vols; 150 curr per   26900

Söderhamns Stadsbibliotek, Köpmangatan 11, 826 30 *Söderhamn*
T: +46 270 75510; Fax: +46 270 15814; E-mail: biblioteket@soderhamn.se; URL: www.soderhamn.se
Nils Rundkvist
182 170 vols; 250 curr per
libr loan   26901

Södertälje stadsbibliotek, St Ragnhildsgatan 2, 151 72 *Södertälje*; mail address: 15183 Södertälje
T: +46 755 21000; Fax: +46 755 21425; E-mail: stadsbiblioteket@sodertalje.se; URL: www.stadsbibl.sodertalje.se
1912; Wera Sundin
198 000 vols; 456 curr per   26902

Sollefteå kommunbibliotek, Huvudbiblioteket, Storgatan 59, 881 83 *Sollefteå*
T: +46 620 82501; Fax: +46 620 17615; URL: www.solleftea.se
Christina Johansson
60 000 vols; 230 curr per   26903

Sollentuna bibliotek, Huvudbibliotek, Aniaraplatsen, Box 63, 191 21 *Sollentuna*
T: +46 8 57921931; Fax: +46 8 57921935; E-mail: biblioteket@bibl.sollentuna.se; URL: www.sollentunabibliotek.nu
1924
168 000 vols; 347 curr per; 350 av-mat; 5 940 sound-rec
libr loan; SB, SFIS   26904

Solna stadsbibliotek, Solna centrum, Box 1049, 171 21 *Solna*
T: +46 8 7342000; Fax: +46 8 833278; E-mail: malin.norrby@solna.se; URL: www.solna.se
1914; Ms Malin Norrby
260 000 vols; 425 curr per; 2 000 microforms; 6 500 av-mat; 11 000 sound-rec   26905

Sölvesborgs bibliotek, Rundgatan 4, 294 31 *Sölvesborg*; mail address: Kommunen, 294 80 Sölvesborg
T: +46 456 816000; Fax: +46 456 10007; E-mail: biblioteket@solvesborg.se; URL: www.solvesborg.se
1908; Folke Frömmert
108 000 vols; 180 curr per; 50 diss/theses; 200 music scores; 50 maps; 100 microforms; 3 900 av-mat; 1 950 sound-rec; 100 digital data carriers
libr loan; SAB   26906

Staffanstorps bibliotek, Blekingevägen, Box 33, 245 80 *Staffanstorp*
T: +46 46 251265; Fax: +46 46 251259; E-mail: biblioteket@staffanstorp.se; URL: www.staffanstorp.se
1952; Marie-Louise Stensson
78 000 vols; 238 curr per; 176 av-mat; 1 870 sound-rec
libr loan; SB   26907

Stenungsunds bibliotek, Kulturhuset Fregatten, Box 187, 444 23 *Stenungsund*
T: +46 303 81880; Fax: +46 303 68012; E-mail: biblioteket@stenungsund.se; URL: www.stenungsund.se
Inger Thorsen
114 000 vols; 300 curr per   26908

Stockholms stadsbibliotek, Länsbibliotek för Stockholms län (Stockholm City and County Library), Asplundhuset Sveavägen 73, 113 80 *Stockholm*
T: +46 8 50831100;
Fax: +46 8 50831230; E-mail: inga.lunden@kultur.stockholm.se; URL: www.ssb.stockholm.se
1927; Inga Lundén
2 000 000 vols; 1 500 curr per; 160 000 av-mat
libr loan; IFLA, INTAMEL, SB   26909

Strängnäs kommunbibliotek, Eskilstunavägen 2, 645 80 *Strängnäs*
T: +46 152 29671; Fax: +46 152 29665; E-mail: fritid-ochkulturnamnden@strangnas.se; URL: www.strangnas.se
Monica Holm
107 400 vols; 162 curr per
libr loan   26910

Strömstads stadsbibliotek, Karlsgatan 17, 452 82 *Strömstad*
T: +46 526 19316; Fax: +46 526 19318; E-mail: stadsbibliotek@stromstad.se; URL: www.stromstad.se/bibliotek
Per Henrik Askeröd
73 300 vols; 115 curr per   26911

Strömsunds kommunbibliotek, Myrgatan 4, 833 35 *Strömsund*
T: +46 670 16275; Fax: +46 670 16175; E-mail: centralbiblioteket@stromsund.se; URL: www.stromsund.se
Gunnel Backlund
120 000 vols; 300 curr per   26912

Sundbybergs stadsbibliotek, Järnvägsgatan 24-26, 172 35 *Sundbyberg* 1
T: +46 8 7068480; Fax: +46 8 989028; E-mail: Gruppkonto.Huvudbiblioteketinfo@sundbyberg.se; URL: www.sundbyberg.se
1922; Ulla-Stina Jönsson
149 700 vols; 302 curr per   26913

Sundvalls stadsbibliotek, Packkusgatan 4, mail address: Kulturmagasinet, 851 96 *Sundsvall*
T: +46 60 191827; Fax: +46 60 125649; URL: www.sundsvall.se
1894; Kerstin Sjöström
408 000 vols; 513 curr per; 40 digital data carriers
libr loan   26914

Sunne bibliotek, Storgatan 22, mail address: Sunne kommun, 36, 686 80 *Sunne*
T: +46 565 16172; Fax: +46 565 16171; E-mail: biblioteket@sunne.se; URL: www.sunne.se
Berit Wester
97 000 vols; 220 curr per   26915

Surahammars folkbibliotek, Folketshus, Köpmangatan 26, Box 203, 735 31 *Surahammar*
T: +46 220 39090; Fax: +46 220 39087; E-mail: mariel.andersson@surahammar.se; URL: www.surahammar.se/kulturfritidturism/
Marie Andersson
89 000 vols; 180 curr per; 50 av-mat; 100 sound-rec; 10 digital data carriers
libr loan   26916

Svalövs folkbibliotek, Svalegatan 23, Box 73, 268 21 *Svalöv*
T: +46 418 475130; Fax: +46 418 62220; E-mail: biblioteket@svalov.se; URL: www.svalov.se
65 000 vols; 200 curr per   26917

Svedala bibliotek, Storgatan 27, 223 80 *Svedala*
T: +46 40 408203; Fax: +46 40 401710; E-mail: biblioteket.svedala@svedala.se; URL: www.svedala.se
Britt-Marie Sjöberg
51 700 vols; 198 curr per   26918

Härjedalens bibliotek, Medborgarhuset, 842 80 *Sveg*
T: +46 680 16122; Fax: +46 680 16105; E-mail: biblioteket.sveg@herjedalen.se; URL: www.herjedalen.se
Gunilla Feldt
142 700 vols; 158 curr per   26919

Bergs Bibliotek, Skolvägen 6, Box 73, 840 40 *Svenstavik*
T: +46 687 16060; Fax: +46 687 16206; E-mail: svenstarik.bibliotek@berg.se; URL: www.bergsbibliotek.se
65 000 vols; 90 curr per   26920

Täby Huvudbibliotek, Biblioteksgången 13, Box 152, 183 22 *Täby*
T: +46 8 7689715; Fax: +46 8 7681604; E-mail: hbibl@taby.se; URL: www.bibl.taby.se
1860
209 000 vols; 436 curr per   26921

Tibro bibliotek, Gymnasiegatan 29, 543 32 *Tibro*
T: +46 504 18250; Fax: +46 504 18263; E-mail: anna.ohlsson@tibro.se; URL: www.tibro.se/Kultur-fritid-och-idrott/Bibliotek/
Anna Ohlsson
50 000 vols; 200 curr per
libr loan   26922

Tierps folkbibliotek, Grevegatan 19, 815 40 *Tierp*
T: +46 293 10870; Fax: +46 293 13685; E-mail: biblioteket.tierp@tierp.se; URL: www1.tierp.se/kulturfritid/bibliotek/
Isabel Borrás
110 000 vols; 120 curr per   26923

Timrå kommunbibliotek, Skogsvägen 26, Box 44, 861 22 *Timrå*
T: +46 60 163215; Fax: +46 60 576137; E-mail: biblo@timra.se; URL: www.timra.se/templates/Page.aspx?id=2422
Anders Lohman
73 000 vols; 135 curr per; 500 maps; 1 500 microforms; 1 300 av-mat; 2 500 sound-rec; 10 digital data carriers
SB   26924

Tingsryds bibliotek, Skyttegatan 2, 362 22 *Tingsryd*
T: +46 477 44270; Fax: +46 477 31137; URL: www.bibliotek.tingsryd.se
Lennart Warsäter
105 700 vols; 98 curr per   26925

Täreboda kommunbibliotek, Kunsgatan 24, 545 30 *Töreboda*; Box 23, 545 21 Töreboda
T: +46 506 18264; Fax: +46 506 18267; URL: www.toreboda.se
Hans Thorsell
60 000 vols; 100 curr per
libr loan   26926

Torsås folkbibliotek, Applerumsgatan 2, Box 140, 385 25 *Torsås*
T: +46 486 4813910520
Lolita Persson
62 000 vols; 142 curr per   26927

Torsby Kommun, 25.
Kulturförvaltning/Bibliotek, Biografgatan 5, Box 504, 685 29 *Torsby*
T: +46 560 16108; Fax: +46 560 16105; E-mail: torsby.bibliotek@torsby.se; URL: www.torsby.se
Anita Vannevik
Archives and libr about the immigration from Finland to Sweden in the 1600th
110 500 vols; 96 curr per
libr loan   26928

Tranås stadsbibliotek, Storgatan 22, Box 1002, 573 82 *Tranås*
T: +46 140 68200; Fax: +46 140 19690
1906; Gunvor Aspviken
90 000 vols; 200 curr per; 620 microforms; 52 av-mat; 1 867 sound-rec
SB   26929

Tranemo bibliotek, Akerivägen 2, 514 33 *Tranemo*
T: +46 325 79290; Fax: +46 325 77258
Marianne Juhlin
63 200 vols; 54 curr per   26930

Trelleborgs bibliotek, C.B. Friisgatan 17-19, Box 93, 231 22 *Trelleborg*
T: +46 410 53180; Fax: +46 410 12885
1934; Christian Gisselquist
151 800 vols; 297 curr per   26931

Trollhättans stadsbibliotek (Trollhättan Public Library), Kungsgatan 25, 461 30 *Trollhättan*; Box 184, 461 24 Trollhättan
T: +46 520 497650;
Fax: +46 520 497755; E-mail: greg.church@trollhattan.se; URL: www.trollhattan.se
1904; Greg Church
197 000 vols; 350 curr per; 6 000 microforms; 14 000 av-mat; 14 000 sound-rec; 600 digital data carriers
SB   26932

Tumba bibliotek, Box 52, 147 21 *Tumba*
T: +46 8 5306-29/-30/-31; Fax: +46 8 53036527
1960; Inger Ekstrand
228 930 vols; 379 curr per; 2 600 av-mat; 1 127 sound-rec; 10 digital data carriers
libr loan; SB, IFLA   26933

Tyresö kommunbibliotek, Östangrönd 7, 135 21 *Tyresö*
T: +46 8 7429400; Fax: +46 8 7429490; URL: www.tyreso.se/bibliotek
1938; Lars Andersson
123 200 vols; 382 curr per; 253 av-mat; 110 sound-rec
libr loan; SB   26934

Ulricehamns bibliotek, Nämndhuset, Badhusgatan, mail address: Strandgatan 22, 523 86 *Ulricehamn*
T: +46 321 595153; E-mail: annalena.johansson@ulricehamn.se; URL: www.ulricehamn.se
Anna Lena Johansson
65 000 vols; 195 curr per   26935

Länsbiblioteket i Västerbotten (County Library of Västerbotten), Rådhusesplanaden 8, 901 78 *Umeå*
T: +46 90 7854581/-88; Fax: +46 90 770887; URL: www.vll.se/ac/lansbibliotek
Anette Sundbom   26936

Sveriges depåbibliotek, Umeå stadsbibliotek (City Library of Umeå), Nygatan 26 A, 901 78 *Umeå*; mail address: 901 78 Umeå
T: +46 90 163339; Fax: +46 90 770887; E-mail: sverigesdepabibliotek@umea.se; URL: www.sverigesdepabibliotek.se
1992; Inger Edebro Sikström
190 000 vols
libr loan; SB, IFLA   26937

Umeå stadsbibliotek, Lånecentralen (City Library of Umeå, ILL (interlibrary loan) Department), Rådhusesplanaden 6-8, 901 78 *Umeå*
T: +46 90 163372; Fax: +46 90 163383; E-mail: lc@umea.se; URL: www.bibliotek.umea.se
1903; Inger Edebro Sikström
664 000 vols; 1 000 curr per; 2 662 av-mat; 1 547 sound-rec; 200 digital data carriers; 30 000 films, AV materials, sound recordings and CD-ROM
libr loan; SB, IFLA   26938

Upplands Väsby bibliotek, Drabantvägen 7C, Väsby centrum, Box 68, 194 21 *Upplands Väsby*
T: +46 8 59097370;
Fax: +46 8 59073333; E-mail: biblioteket@upplandsvasby.se; URL: www.upplandsvasby.se
Anna-Clara Edin
86 500 vols; 253 curr per   26939

Länsbibliotek Uppsala (Uppsala County Library), Svartbäcksgatan 17, Box 643, 751 27 *Uppsala*
T: +46 18 611 66 21; Fax: +46 18 13 25 26; URL: /www.lul.se   26940

Uppsala stadsbibliotek (Uppsala City Library), Svartbäcksgatan 17, Box 643, 751 27 *Uppsala*
T: +46 18 7271700; Fax: +46 18 7270640; URL: www.uppsala.se/stadsbiblioteket
1906; Marie-Louise Riton
Uppsala and Uppland, regional hist
826 168 vols; 1 820 curr per
IFLA   26941

Vadstena bibliotek, Slottsgatan 10, 592 80 *Vadstena*
T: +46 143 15130; Fax: +46 143 15129; E-mail: biblioteket@vadstena.se
Jan Fribeg
59 000 vols; 225 curr per   26942

Vaggeryds bibliotek, Verner Malmstens väg, Box 183, 567 24 *Vaggeryd*
T: +46 393 78790; Fax: +46 393 78794; E-mail: vaggeryds.bibliotek@vaggeryd.se
Henrik Arvidsson
72 100 vols; 100 curr per   26943

Vallentuna bibliotek, Tuna torg 11, Box 104, 186 22 *Vallentuna*
T: +46 8 58785350; Fax: +46 8 58785338; E-mail: kultur@vallentuna.se; URL: www.vallentuna.se
Ingrid Ullman
94 000 vols; 255 curr per; 302 music scores; 143 av-mat; 2 251 sound-rec; 68 digital data carriers
libr loan   26944

Vänersborgs bibliotek, Köpmangatan 1, Box 77, 462 21 *Vänersborg*
T: +46 521 271411;
Fax: +46 521 271498; E-mail: biblioteket@vanersborg.se; URL: www.vanersborg.se
1910; Eva Boberg
130 300 vols; 237 curr per   26945

Vännäs kommunbibliotek, Ö. Jörnvägsgatan 10, 911 81 *Vännäs*
T: +46 935 14186; Fax: +46 935 14011; E-mail: biblioteket@vannas.se; URL: www.vannas.se
Ingalill Stenmark
76 200 vols; 174 curr per   26946

Vansbro kommunbibliotek, Medborgarhuset, 780 50 *Vansbro*
T: +46 281 75050; Fax: +46 281 75053; URL: www.vansbro.se
Ingvar Björk
57 600 vols; 100 curr per   26947

Vara folkbibliotek, Stora torget 5, 534 81 *Vara*
T: +46 512 31220; Fax: +46 512 31231; E-mail: biblioteken@vara.se; URL: www.vara.se
Gunlög Thorstensson
60 000 vols; 240 curr per   26948

Varbergs Kommun, Biblioteket, mail address: 432 80 *Varberg*
T: +46 340 88600; Fax: +46 340 697050; E-mail: biblioteket@kommunen.varberg.se; URL: www.varberg.se
1890/1980; Kristina Peterson
250 000 vols; 335 curr per; 1 500 music scores; 280 maps; 300 microforms; 840 av-mat; 5 800 sound-rec; 300 digital data carriers
libr loan; SB   26949

Värnamo kommunbibliotek, Järnvägsgatan 3, Box 414, 331 24 ***Värnamo***
T: +46 370 377149; E-mail: biblioteket@varnamo.se; URL: www.varnamo.se
Monica Lembke
100 000 vols; 200 curr per; 900 music scores; 600 av-mat; 7 500 sound-rec; 130 digital data carriers
libr loan; SB 26950

Västerås stad, proAros Kultur, Stadsbiblioteket (City Library), Biskopsgatan 2, 721 87 ***Västerås***; Box 717, 721 20 Västerås
T: +46 21 394600; Fax: +46 21 394680; E-mail: stadsbibliotek@vasteras.se; URL: www.bibliotek.vasteras.se
1952; Staffan Rune
Public libr with coll of bks about Västmanland and coll about the cathedral and diocese
530 000 vols; 650 curr per; 138 incunabula; 17 000 diss/theses; 9 000 govt docs; 5 600 music scores; 150 maps; 9 000 microforms; 3 000 av-mat; 50 000 sound-rec; 400 digital data carriers; 70 meters of mss
libr loan; SB 26951

Västerviks stadsbibliotek (Västervik City Library), Spötorget, Box 342, 593 24 ***Västervik***
T: +46 490 88777; Fax: +46 490 88761; E-mail: stadsbiblioteket@vastervik.se; URL: www.vastervik.se
1908; Anne Brandel
196 330 vols; 406 curr per; 200 music scores; 225 microforms; 1 930 av-mat; 6 520 sound-rec
libr loan 26952

Växjö bibliotek, Stadsbiblioteket, Västra Esplanaden 7, Box 1202, 351 12 ***Växjö***
T: +46 470 41420; Fax: +46 470 766992; E-mail: stadsbiblioteket@kommun.vaxjo.se; URL: www.vaxjo.se/bibliotek
1954
645 000 vols; 473 curr per 26953

Vellinge kommunbibliotek, Malmövägen 1, 235 36 ***Vellinge***
T: +46 40 425060; Fax: +46 40 425069; E-mail: vellinge.kommun@vellinge.se; URL: www.vellinge.se
116 000 vols; 212 curr per
libr loan 26954

Vetlanda bibliotek, Biblioteksgatan 5, mail address: 57480 ***Vetlanda***
T: +46 383 97450; Fax: +46 383 97006; E-mail: biblioteket@vetlanda.se; URL: www.vetlanda.se
Ann-Mari Forsberg
135 000 vols; 250 curr per 26955

Vilhelmina folkbibliotek, Folkets Hus, Postgatan 15, mail address: Torget 6, 912 81 ***Vilhelmina***
T: +46 940 14160; Fax: +46 940 12482; E-mail: kommunbiblioteket@vilhelmina.se; URL: www.kommun.vilhelmina.com/biblioteket
Lars Lagerström
50 000 vols; 130 curr per 26956

Vimmerby Bibliotek, Stadshuset, 598 81 ***Vimmerby***
T: +46 492 769095; Fax: +46 492 769096; E-mail: biblioteket@vimmerby.se; URL: www.vimmerby.se/templates/www_page___1047.aspx
Margret Halvardson
Astrid Lindgren-Coll under construction
124 500 vols; 140 curr per 26957

Vingåkers bibliotek, Bondegatan 6, mail address: 643 80 ***Vingåker***
T: +46 151 19190; Fax: +46 151 10734; E-mail: biblioteket@vingaker.se; URL: www.vingaker.se
50 000 vols; 110 curr per 26958

Gotlands länsbibliotek, Länsbibliothek för Gotland (Gotland County Library), Kung Magnus Väg 16, 621 81 ***Visby***
T: +46 498 269651; Fax: +46 498 269659; E-mail: karin.blomquist@hgo.se; URL: bibliotek.gotland.se
1865; Elisabeth Stenberg-Lundin
Gotland coll
565 076 vols; 760 curr per; 1 750 microforms; 2 360 sound-rec; 8 009 talking bks, 535 courses in foreign languages, art (103)
libr loan; SB 26959

Ystads bibliotek, Surbrunnsvägen 12, 271 80 ***Ystad***
T: +46 411 77290; Fax: +46 411 78115; E-mail: biblioteket@ystad.se; URL: www.ystad.se
1866; Ylva Amman
John Andrén Foundation (for research on Oscar Wilde, Aubrey Beardsley, T.E. Lawrence and Vaslav Nijinsky)
130 000 vols; 336 curr per
libr loan 26960

Switzerland

National Libraries

Schweizerische Nationalbibliothek / La Bibliothèque nationale suisse / Biblioteca nazionale svizzera (Swiss National Library), Hallwylstr. 15, 3003 ***Bern***
T: +41 313228935; Fax: +41 313228408; E-mail: info@nb.admin.ch; URL: www.nb.admin.ch
1895; Marie-Christine Doffey
Bibelslg Lüthi, Indicaslg Desai/Wyss, Schweizerisches Literaturarchiv LSA, Grafische Slg (Kleinmeisterslg; Gugelmann, Fotoslg, Eidgenössisches Archiv für Denkmalpflege (EAD), Archiv Daniel Spoerri), Archiv Neue Helvetische Gesellschaft; Libr of Swiss Gutenberg Museum, Libr of Swiss Soc for Family Research
2 825 012 vols; 15 406 curr per; 1 035 e-journals; 30 mss; 168 000 diss/theses; 69 685 music scores; 50 540 maps; 15 147 microforms; 19 012 av-mat
libr loan; IFLA, BIS, AIBM, LIBER, Memoriav, CENL, EDUG, The European Library, CDNL, OCLC, SNV, ISSN, DOC Deutsch, SWD Clearingstelle Schweiz, Consortium, GRD, ICOM Suisse, IIPC, Malvine, SIBMAS, SIGEGS, SKR/SCR 26961

General Research Libraries

Aargauer Kantonsbibliothek, Aargauerplatz, Postfach, 5001 ***Aarau***
T: +41 628352360; Fax: +41 628352369; E-mail: kantonsbibliothek@ag.ch; URL: www.ag.ch/kantonsbibliothek
1803; Dr. Ruth Wüst
Mykologie, Stenografie, Jagd, Militärbibl, Catholica, Predigerbibl, Zurlaubiana, Frank Wedekind, Sauerländer, Hans-Jakob Seiler, Ernst Widmer
679 500 vols; 271 curr per; 163 e-journals; 1 250 mss; 830 incunabula; 13 500 diss/theses; 1 200 govt docs; 728 music scores; 2 083 maps; 1 012 microforms; 2 906 av-mat; 85 digital data carriers; 25 000 brochures
libr loan; BIS, IFLA 26962

Kantonsbibliothek Uri, Bahnhofstr. 13, 6460 ***Altdorf***
T: +41 418752221; Fax: +41 418752226; E-mail: kantonsbibliothek@ur.ch; URL: www.kbu.ch
1952; Eliane Latzel
Uraniensia (lokaler Sammelauftrag)
130 000 vols; 350 curr per; 550 diss/theses; 270 music scores; 1 900 maps; 3 microforms; 600 av-mat; 5 700 sound-rec; 2 500 digital data carriers
libr loan; BIS 26963

Innerrhodische Kantonsbibliothek, Marktgasse 2, 9050 ***Appenzell***
T: +41 717889333; Fax: +41 717889339; E-mail: doris.ueberschlag@rk.ai.ch; URL: www.ai.ch/de/bildung/bibliotheken
1928; Doris Ueberschlag
Appencellensia (6 000 vol)
35 000 media items
libr loan; BIS, SAB 26964

Universitätsbibliothek Bern, Zentralbibliothek (University of Bern), Münstergasse 61, 3000 ***Bern*** 8; mail address: 3000 Bern 8
T: +41 316319211; Fax: +41 316319299; E-mail: info@ub.unibe.ch; URL: www.ub.unibe.ch
1528-1535; Marianne Rubli Supersaxo
Bernensia, Helvetica vor 1848, romangerman MA, MA, Archäologie, ornith Bibliothek Holzer, Richard-Wagner-Slg, Hermann-Rorschach-Archiv, Karten-Slg Ryhiner (alte Landkarten, v.a. 18. Jh), Dok zu Buchpflege u Papierrestauration; Swiss Eastern Europe Libr
4 234 000 vols; 53 100 e-journals; 292 000 e-books; 220 mss; 430 incunabula; 47 700 maps; 120 500 microforms; 33 400 av-mat; 1 200 digital data carriers; 430 000 other items
libr loan; BIS, LIBER 26965

Kantonsbibliothek Graubünden, Karlihofplatz, 7001 ***Chur***
T: +41 812572828; Fax: +41 812572153; E-mail: info@kbg.gr.ch; URL: www.kantonsbibliothek.gr.ch
1883; Dr. Christine Holliger
Raetica; Pastorale Libr of the Protestant National Church of Graubünden
360 000 vols; 1 500 curr per; 24 incunabula; 2 500 diss/theses; 17 000 govt docs; 3 000 music scores; 3 000 maps; 30 000 microforms; 17 000 av-mat; 10 000 sound-rec; 1 000 digital data carriers; 15 000 graphic documents (posters, graphic collections, picture postcards)
libr loan; BIS 26966

Kantonsbibliothek Thurgau, Promenadenstr. 12, 8501 ***Frauenfeld***
T: +41 527241888; Fax: +41 527241899; E-mail: kantonsbibliothek@tg.ch; URL: www.kantonsbibliothek.tg.ch
1805; Monika Mosberger
Thurgoviana, Cartusiana, Napoleonica
285 000 vols; 500 curr per; 5 200 mss; 641 incunabula; 450 music scores; 1 500 maps; 1 300 microforms; 500 av-mat; 1 300 sound-rec; 500 digital data carriers
libr loan; BIS 26967

Bibliothèque cantonale et universitaire Fribourg (BCU) / Kantons- und Universitätsbibliothek Freiburg (KUB), Rue Joseph-Piller 2, 1700 ***Fribourg***; CP 160, 1701 Fribourg
T: +41 263051333; Fax: +41 263051378; E-mail: bcu@fr.ch; URL: www.fr.ch/bcuf (french); www.fr.ch/kubf (deutsch)
1848; Dr. Martin Good
Friburgensia (2 700 vols), imprimés avant 1901 (190 000 vols), imprimés anciens, dont: Bibl de la Société Economique (26 000 vols), Bibl des Capucins de Fribourg et Bulle (28 000 vols), Bibl de Castella de Delley (3 100 vols), coll des photos (936 000); Depositories of the Fondation Archivum Helveto-Polonicum of the Société Dante Alighieri, of the Freiburger Naturwissenschaftliche Gesellschaft, and of the Schweizer Alpinclub (Freiburg)
2 413 000 vols; 5 000 curr per; 19 100 e-journals; 2 140 mss; 572 incunabula; 6 000 maps; 31 300 microforms; 7 700 av-mat; 9 500 sound-rec; 200 digital data carriers
libr loan; BIS, SAB/CLP, LIBER 26968

Bibliothèque de Genève, Promenade des Bastions, CP, 1211 ***Genève*** 4
T: +41 224182800; Fax: +41 224182801; E-mail: jean-charles.giroud@ville-ge.ch; URL: www.ville-ge.ch/bge
1562; Jean-Charles Giroud
Illuminierte Handschriften 14.-15. Jh, Reformationsgeschichte (Calvin), franz Lit d 18. Jh (Voltaire u Rousseau), Genf, Genfer Ablieferungspflicht, Geschichte
2 000 000 vols; 3 500 curr per; 100 000 mss; 470 incunabula; 25 000 maps; 10 000 microforms; 1 000 av-mat; 1 000 sound-rec; 500 digital data carriers; 100 000 posters, 100 000 iconographic representations
libr loan; BIS, AIBM 26969

Landesbibliothek des Kantons Glarus, Hauptstr. 60, Postfach 535, 8750 ***Glarus***
T: +41 556466321; Fax: +41 556466329; E-mail: landesbibliothek@gl.ch; URL: www.gl.ch
1761; Hanspeter Jörg
Glaronensia, Zwicky-Stiftung (Astrophysik u Morphologie), Blumer'sche Kartenslg, Bibliothek Prof. Arthur Dürst
140 000 vols; 500 curr per; 325 mss; 5 incunabula; 500 diss/theses; 2 500 govt docs; 5 000 music scores; 2 000 maps; 50 microforms; 2 800 av-mat; 4 300 sound-rec; 700 digital data carriers; 11 bequests
libr loan; BIS, SAB 26970

Université de Lausanne, Bibliothèque Cantonale et Universitaire (University of Lausanne), Bâtiment Unithèque, 1015 ***Lausanne*** – Dorigny; mail address: Place de la Riponne 6, 1014 Lausanne
T: +41 216924802; Fax: +41 216924845; E-mail: info-dorigny@bcu.unil.ch; URL: www.bcu-lausanne.ch
1537; Jeannette Frey
Orientalisme, cinéma, histoire de France, documentation vaudoise, musicologie, réserve précieuse, manuscrits, archives musicales
6 770 935 vols; 71 958 curr per; 114 271 e-books; 720 mss; 150 incunabula; 30 000 music scores; 3 841 maps; 70 400 microforms; 54 996 av-mat; 17 000 sound-rec; 285 digital data carriers
libr loan; AIBM, IFLA, LIBER, BIS, RERO, BiblioVaud 26971

Kantonsbibliothek Baselland, Emma Herwegh-Platz 4, 4410 ***Liestal***
T: +41 615526273; Fax: +41 615526968; E-mail: kantonsbibliothek@bl.ch; URL: www.kbl.ch
1838; Dr. Gerhard Matter
Baselbieter Schrifttum; Libr of 'Naturforschende Gesellschaft beider Basel'; Libr 'Burgenfreunde beider Basel'
181 000 vols; 950 curr per; 9 300 e-books; 16 mss; 330 govt docs; 3 400 music scores; 1 100 maps; 17 microforms; 32 940 av-mat; 18 600 sound-rec; 360 digital data carriers; 2 150 comics, 250 language courses
libr loan; SAB/GTB, BIS 26972

Biblioteca cantonale di Locarno, Republica e Cantone del Ticino: Dipartemento dell'educazione, della cultura e dello sport, Via Cappuccini 12, 6600 ***Locarno***; CP 1242, 6601 Locarno
T: +41 917597580; Fax: +41 917597599; E-mail: bclo-segr.sbt@ti.ch; URL: www.sbt.ti.ch/bclo/
1988; Andrea Ghiringhelli
Fondo Ruggero Leoncavallo, Fondo Virgilio Gilardoni, Fondo Waldes Keller, Fondo Pericle Patocchi, Fondo Felice Lattuada; Philosophy Dept, Music Dept, Cinema Dept
113 500 vols; 438 curr per; 216 music scores; 27 196 av-mat; 14 208 sound-rec; 103 digital data carriers
BBS, IFLA 26973

Biblioteca cantonale di Lugano, Viale Carlo Cattaneo 6, 6900 ***Lugano***
T: +41 918154611; Fax: +41 918154619; E-mail: bclu-infoprestito.sbt@ti.ch; URL: www.sbt.ti.ch/bclugano
1852; Dr. Gerardo Rigozzi
350 000 vols; 520 curr per; 200 incunabula; 750 digital data carriers
libr loan; BBS, IFLA 26974

Zentral- und Hochschulbibliothek Luzern, Sempacherstr. 10, 6003 ***Luzern***; Postfach, 6002 Luzern
T: +41 412285344; Fax: +41 412106255; E-mail: info@zhbluzern.ch; URL: www.zhbluzern.ch
1951; Dr. Ulrich Niederer
Katholische Theologie, Schweizer Geschichte, Rechtswissenschaft, Sozialwissenschaft, Lucernensia; Libr of Gen Music Society, Libr of Art Society, Citizen Libr, Canton Libr, University area, Special Coll (mss + prints before 1900, graphic docs / graphic coll, special music coll
989 272 vols; 2 610 curr per; 47 769 e-journals; 8 811 e-books; 2 736 mss; 850 incunabula; 144 diss/theses; 12 500 music scores; 4 817 maps; 36 968 microforms; 5 629 sound-rec; 1 996 digital data carriers; 130 449 prints and graphic docs, 289 online databases (license)
libr loan; AIBM, BIS, LIBER 26975

Biblioteca cantonale e del Liceo di Mendriso, Via Agostino Maspoli, CP 1162, 6850 ***Mendrisio***
T: +41 918159478; Fax: +41 918159479; E-mail: bcme-segr.sbt@ti.ch; URL: www.sbt.ti.ch/bcme
1979; Cristiano Lafferma
Scienze esatte e scienze naturali
59 030 vols; 365 curr per; 489 maps; 366 av-mat; 303 sound-rec; 159 digital data carriers
libr loan; BBS, SAB/CLP 26976

Bibliothèque publique et universitaire, Pl Numa-Droz 3, 2000 ***Neuchâtel***
T: +41 327177300; Fax: +41 327177309; E-mail: secretariat.bpu@unine.ch; URL: bpun.unine.ch/services/SetPret.htm
1788; Michel Schlup
Handschriften v J.J. Rousseau, Mme de Charriere, Geschichte d Buchs, Sozial- u Geistesgeschichte d 18. Jh, Neocomensia; Libr of the 'Schweizer Gesellschaft für Zeitmessung', Libr of the 'Naturwissenschaftliche Gesellschaft', Libr of the 'Geographische Gesellschaft'
551 322 vols; 1 980 curr per; 13 082 mss; 35 incunabula; 70 494 diss/theses; 3 498 music scores; 5 153 maps; 20 714 microforms; 691 av-mat; 505 sound-rec; 308 digital data carriers; 1 570 prints, 5 384 posters, 45 728 pamphlets
libr loan; BIS 26977

Bibliothèque cantonale jurassienne, Office de la culture, Rue Pierre-Péquignat 9, CP 64, 2900 ***Porrentruy*** 2
T: +41 324208410, 324208400; Fax: +41 324208499; E-mail: bibliotheque.cantonale@jura.ch; URL: www.jura.ch/occ/bicj
1982; Géraldine Rérat-Oeuvray
Fonds ancien (19 000 vols)
115 585 vols; 2 046 curr per; 573 mss; 216 incunabula; 279 microforms; 175 av-mat; 454 sound-rec
libr loan; BIS 26978

Kantonsbibliothek Schwyz, Rickenbachstr. 24, Postfach 264, 6431 ***Schwyz***
T: +41 418191908; Fax: +41 418191909; E-mail: kantonsbibliothek@sz.ch; URL: www.sz.ch/kantonsbibliothek
1870; Werner Büeler
100 000 vols; 165 curr per; 2 100 diss/theses; 700 maps; 1 825 av-mat; 2 575 sound-rec; 1 000 photos, 40 linear metres of mss
libr loan 26979

Médiathèque Valais (Bibliothèque cantonale) / Mediathek Wallis (Kantonsbibliothek), Avenue de Pratifori 18, CP 182, 1951 ***Sion***
T: +41 276064550; Fax: +41 276064554; E-mail: mv.sion@mediatheque.ch; URL: www.mediatheque.ch
1853; Damian Elsig
Vallesiana; Alpinismus
530 000 vols; 2 130 curr per; 90 mss; 140 incunabula; 13 000 music scores; 2 500 maps; 3 000 microforms; 80 000 av-mat; 35 000 sound-rec; 10 000 digital data carriers; 80 000 brochures, 75 000 postal cards, 20 000 posters, 120 000 press clippings
libr loan; BIS 26980

Zentralbibliothek Solothurn, Kantons-, Stadt- und Regionalbibliothek, Bielstr. 39, 4500 ***Solothurn***; mail address: 4502 Solothurn
T: +41 326276262; Fax: +41 326276200; E-mail: sekretariat@zbsolothurn.ch; URL: www.zbsolothurn.ch
1763/1882/1930; Peter Probst (Admin.), Verena Bider (Wissensch.)
Solodorensia, Grafische Slg, Handschriftensammlung, Privatnachlass-Slg, Inkunabelsammlung, Slg Altes Buch 1500-1900, Historische Musiksammlung, Exlibris, Sealsfieldiana; Music dept (sound archives), youth libr, hist music coll
970 000 vols; 600 curr per; 1 900 mss; 1 200 incunabula; 15 000 diss/theses; 17 000 music scores; 3 000 maps; 22 000 microforms; 799 av-mat; 64 000 sound-rec; 10 000 digital data carriers; 8 437 graphic docs (incl. 2 900 art prints, 4 900 posters)
libr loan; AIBM, BIS, SAB
The 'Zentralbibliothek Solothurn' is a regional library with a regional deposit agenda; it was merged from a town library from the 18th c (with holdings from family libraries) and a canton library from the 19th c (with holdings from secularized monastry libraries). Therefore the library is a mixture of a public and a scientific library. 26981

Kantonsbibliothek Vadiana, Notkerstr. 22, 9000 ***St.Gallen***
T: +41 712292321; Fax: +41 712292345; E-mail: kb.vadiana@sg.ch; URL: www.kb.sg.ch
1551; Dr. Dora Cornel
Sangallensien, Buchwissenschaften, Freimaurerei, Humanismus; St.Galler Zentrum für das Buch, Vadianische Sammlung
607 024 vols; 1 000 curr per; 67 e-journals; 6 847 e-books; 1 003 mss; 234 incunabula; 3 500 maps; 1 420 microforms; 82 251 av-mat; 900 sound-rec; 1 200 digital data carriers; 28 000 prints, 20 000 book plates, 25 300 autographs, 160 bequests
libr loan; BIS, LIBER 26982

Kantonsbibliothek Nidwalden, Engelbergstr. 34, 6371 ***Stans***
T: +41 416187302; Fax: +41 416187306; E-mail: kantonsbibliothek@nw.ch; URL: www.kantonsbibliothek.nw.ch
1970; Brigitte Durrer
Kanton Nidwalden (Nidwaldensia)
60 000 vols; 60 curr per; 1 e-journals; 150 mss; 60 incunabula; 180 diss/theses; 120 maps; 25 microforms; 600 av-mat; 300 sound-rec; 30 digital data carriers; 8 000 pamphlets and press clippings
libr loan; BIS 26983

Kantonsbibliothek Appenzell Ausserrhoden, Landsgemeindeplatz 1/7, 9043 ***Trogen***
T: +41 713436421; Fax: +41 713436429; E-mail: kantonsbibliothek@ar.ch; URL: www.ar.ch/kantonsbibliothek
1896; Dr. Heidi Eisenhut
Slg Carl Meyer, Nachlässe d Textilhandelsfamilie Zellweger, Schauwerk Sammlungsprojekt, Bibliothek und Archiv der Aeschbach-Stiftung
40 000 vols; 200 curr per; 540 mss; 82 incunabula; 65 diss/theses; 100 govt docs; 335 music scores; 135 maps; 340 microforms; 300 av-mat; 400 sound-rec; 150 digital data carriers; 14 000 picture docs, 10 000 pamphlets, 30 bequests
libr loan; BIS 26984

Stadt- und Kantonsbibliothek Zug, St.-Oswalds-Gasse 21, 6301 ***Zug***; Postfach 52, 6301 Zug
T: +41 41 7282313; Fax: +41 41 7282380; E-mail: bibliothek@stadtzug.ch; URL: www.bibliothekzug.ch
1836; Heinz Morf
Zuger Sammlung (regionale Dokumentation)
154 424 vols; 882 curr per; 1 incunabula; 1 663 microforms; 8 405 av-mat; 14 296 sound-rec; 2 755 digital data carriers; 2 028 dossiers with lowercases, 9 225 docs "Digitale Bibliothek", 23 161 photos of documentary value (inkl. maps)
libr loan; BIS, SAB, Memoriav 26985

Zentralbibliothek Zürich, Kantons-, Stadt- und Universitätsbibliothek (University of Zurich), Zähringerplatz 6, 8001 ***Zürich***
T: +41 442683100; Fax: +41 442683290; E-mail: zb@zb.uzh.ch; URL: www.zb.uzh.ch
1914; Prof. Susanna Bliggenstorfer
Bibl der Allgemeinen Musik-Gesellschaft Zürich, Fennica (Schweizer Vereinigung der Freunde Finnlands), Schweizer Alpenclub, hist Kinder- u Jugendbücher des Pestalozzianums Zürich, Bibl der Ornithologischen Gesellschaft Zürich, Nordamerika-Bibl; Graph coll, Mss dept, Map coll, Music dept, Old books coll, Libr Oskar R. Schlag
3 950 000 vols; 8 700 curr per; 40 000 e-journals; 10 000 e-books; 125 000 mss; 1 500 incunabula; 600 000 diss/theses; 100 000 music scores; 245 000 maps; 560 000 microforms; 40 000 sound-rec; 500 digital data carriers; 220 000 art prints, 5 000 photos, 160 000 picture postcards
libr loan; BIS, LIBER, AIBM, IFLA 26986

University Libraries, College Libraries

Fachhochschule Nordwestschweiz – Hochschule für Gestaltung und Kunst, Bibliothek, Bahnhofstr. 102, 5000 ***Aarau***
T: +41 62 8326680; Fax: +41 62 8326665; E-mail: bibliothek.hgk@fhnw.ch; URL: www.fhnw.ch/hgk/bibliothek/aarau
1999; Cornelia Zelger
Design, Medienkunst, Medientheorie Gebrauchsgrafik, Innenarchitektur, Architektur, Fotografie, Film, Moderne Kunst
8 209 vols; 94 curr per; 1 438 av-mat; 26 sound-rec; 397 digital data carriers
libr loan; AKMB 26987

Fachhochschule Nordwestschweiz – Hochschule für Soziale Arbeit, Mediothek, Thiersteinerallee 57, 4053 ***Basel***
T: +41 613372707; Fax: +41 613372720; E-mail: brigitte.forster@fhnw.ch; URL: www.fhnw.ch/sozialearbeit
1971; Brigitte Forster
16 500 vols; 65 curr per; 300 av-mat 26988

Musikak-Ademie der Stadt Basel, Bibliothek, Leonhardsstr. 6, 4051 ***Basel***; Postfach, 4003 Basel
T: +41 612645755; Fax: +41 612645756; E-mail: bibliothek@mab-bs.ch; URL: www.musik-akademie.ch/bibliothek
1867; Markus Erni
Interpretation alter Musik auf Originalinstrumenten
10 000 vols; 100 curr per; 5 e-journals; 130 000 music scores; 3 000 microforms; 18 000 sound-rec
AIBM 26989

Pädagogisches Zentrum Basel-Stadt (PZB), Erziehungsdepartement des Kantons Basel-Stadt, Bibliothek, Binningerstr. 6, 4051 ***Basel***
T: +41 612676837; Fax: +41 612676835; E-mail: bibliothek.pz@bs.ch; URL: www.pz.bs.ch/bibliothek
1973; Elisabeth Tschudi-Moser
Lernmittel, Bilderbücher, prämierte Kinder- und Jugendbücher, Archiv Lehrmittel ab Ende 18. Jh sowie Bilderbücher und prämierte Kinder- und Jugendbücher dem 20. Jh
79 018 vols; 385 curr per; 57 e-journals; 100 e-books; 96 diss/theses; 681 govt docs; 390 music scores; 336 maps; 16 326 av-mat; 1 187 sound-rec; 830 digital data carriers; picture material (128 units), 48 multimedia kits
BIS, SAB 26990

Universitäre Psychiatrische Kliniken Basel UPK, UPK-Bibliothek mit Pflegebibliothek, Wilhelm-Klein-Strasse 27, 4025 ***Basel***
T: +41 613255216; E-mail: info@upkbs.ch; URL: www.upkbs.ch
1886; Elke Liebel; Rita Machata
Sammlung Heinrich Meng ; Historische Bibliothek
25 000 vols; 58 curr per; 30 e-journals; 1 300 diss/theses; 3 digital data carriers
libr loan; BIS 26991

Universität Basel, Öffentliche Bibliothek (University of Basel), Schönbeinstr. 18-20, 4056 ***Basel***
T: +41 612673111; Fax: +41 612673103; E-mail: sekretariat-ub@unibas.ch; URL: www.ub.unibas.ch
nach 1460; Hannes Hug
Libs of 'Frey-Grynäisches Institut', 'Historische und Antiquarische Gesellschaft Basel', 'Basler Lesegesellschaft', 'Schweizer Akademie der Medizinischen Wissenschaften', 'Militärbibliothek Basel', and 'Schweizerische Musikforschende Gesellschaft'; archives of 'Allgemeine Musikgesellschaft'; medical libr; libr of 'Wirtschaftswissenschaftliches Zentrum' (WWZ); Swiss Ecnomy Archives (SWA)
4 132 000 vols; 9 351 curr per; 9 820 e-journals; 77 600 mss; 3 000 incunabula; 500 000 diss/theses; 60 000 music scores; 35 000 maps; 1 000 av-mat; 100 000 portraits, 70 000 ex libris
libr loan; BIS, LIBER, AIBM 26992

– Medizinbibliothek im Universitätsspital, Hebelstr. 20, 4031 Basel
T: +41 61 2652590; Fax: +41 61 2673191; E-mail: info-medb@unibas.ch; URL: www.ub.unibas.ch/medizinbibliothek/
1978; Christoph Wehrmüller
Pflegewissenschaft
30 000 vols; 450 curr per; 6 500 e-journals; 150 digital data carriers
libr loan; AGMB, Biomed 26993

– Bibliothek Altertumswissenschaft, Petersgraben 51 (1. Stock / Raum 104), 4051 Basel
T: +41 612671351; Fax: +41 612671330; E-mail: biblio-altertum@unibas.ch; URL: aw-o.philhist.unibas.ch/bibliothek-altertumswissenschaften/
Daniel Keller
Ägyptologie, Alte Geschichte, Gräzistik, Historisch-vergleichende Sprachwissenschaft, Klassische Archäologie, Latinistik, Ur- und Frühgeschichte 26994

– Biozentrum, Bibliothek, Klingelbergstr. 50/70, 4056 Basel
T: +41 612672041; Fax: +41 612672039; E-mail: bibliothek-biozentrum@unibas.ch; URL: www.biozentrum.unibas.ch/library/
1971; Simone Gloor
Biologie, Biophysik, Molukularbiologie
21 500 vols; 53 curr per; 23 e-journals 26995

– Botanisches Institut, Abt. Ökologie, Bibliothek, Schönbeinstr. 6, 4056 Basel
T: +41 612673500; Fax: +41 612672980; URL: www.ub.unibas.ch
20 000 vols; 20 curr per; 120 diss/theses 26996

– Departementsbibliothek Chemie, c/o Institut für organische Chemie, St. Johanns-Ring 19, 4056 Basel
T: +41 612671148; Fax: +41 612671103; E-mail: Bibliothek-Chemie@unibas.ch
1946; Claudia Wirthlin Diongue
25 000 vols; 120 curr per; 450 microforms
libr loan; BIS 26997

– Deutsches Seminar, Bibliothek, Nadelberg 4, Engelhof, 4051 Basel
T: +41 612673441; Fax: +41 612673440; E-mail: Judith.Sandhaas@unibas.ch; URL: www.germa.unibas.ch/bibliothek
1835; Judith Sandhaas
Nordic studies dept (12 000 vols)
40 000 vols; 70 curr per; 1 300 av-mat; 80 sound-rec; 30 digital data carriers 26998

– Englisches Seminar, Bibliothek, Nadelberg 6, 4051 Basel
T: +41 612672781; Fax: +41 612672780; E-mail: eva.sternkueker@unibas.ch
1927; Eva Sternküker
35 000 vols; 43 curr per; 100 av-mat; 200 sound-rec; 150 digital data carriers
BIS 26999

– Geographisches Institut, Bibliothek, Klingelbergstr. 27, 4056 Basel
T: +41 612673660; Fax: +41 612673651; E-mail: Heike.Schmidt@unibas.ch; URL: www.ub.unibas.ch/lib/ba/a198.htm
1912; Heike Schmidt
Libr of the Geographic Ethnological Society in Basel
30 000 vols; 300 curr per; 20 000 maps; 1 000 av-mat 27000

– Historisches Seminar, Bibliothek, Hirschgässlein 21, 4051 Basel
T: +41 612959668; Fax: +41 612959640; E-mail: Andrea.Leslie@unibas.ch; URL: www.histsem.unibas.ch
1887; Andrea Leslie
Eastern Europe Dept
50 000 vols; 75 curr per; 3 300 diss/theses; 8 digital data carriers 27001

– Institut für Iberoromanistik / Institut für Italianistik / Institut für Französische Sprach- und Literaturwissenschaft, Bibliothek, Maiengasse 51-53, 4056 Basel
T: +41 612671276; Fax: +41 612671286; E-mail: Patrizia.Gamarra@unibas.ch; URL: bibliomaiengasse.unibas.ch
1967; Patrizia Gamarra
50 000 vols; 100 curr per
libr loan; BIS 27002

– Institut für Physik, Bibliothek, Klingelbergstr. 82, 4056 Basel
T: +41 612673687; Fax: +41 612671349; E-mail: Francois.Erkadoo@unibas.ch
Prof. P. Oelhafen
10000 vols; 35 curr per 27003

– Institut für Psychologie, Bibliothek, Missionsstr. 62a, 4055 Basel
T: +41 61267 0572; E-mail: testothek-psycho@unibas.ch
1978; Prof. G. Steiner
Testothek
8000 vols; 85 curr per
libr loan
non-lending library 27004

– Institut für Soziologie, Bibliothek, Petersgraben 27, 4051 Basel
T: +41 612672819; Fax: +41 612672820; E-mail: Bibliothek-Soziologie@unibas.ch; URL: www.unibas.ch/soziologie
1968; Gabriela Degen
16604 vols; 75 curr per; 33 e-journals; 995 diss/theses; 52 av-mat; 24 digital data carriers
libr loan 27005

– Institut für Sport und Sportwissenschaften, Bibliothek, Birsstr. 320B, 4052 Basel
T: +41 613778760; Fax: +41 613778788; E-mail: bibliothek-sport@unibas.ch; URL: issw.unibas.ch
1861; Renate Leubin
9000 vols; 28 curr per
libr loan; AgSB 27006

– Institut für Umweltgeowissenschaften, Bibliothek, Bernoullistr. 30, 4056 Basel
T: +41 612673612; Fax: +41 612673613; E-mail: verena.scheuring@unibas.ch
1918; Verena Scheuring
libr loan 27007

– Juristische Fakultät, Bibliothek, Peter-Merian-Weg 8, mail address: 4002 Basel
T: +41 612672500; Fax: +41 612672509; E-mail: biblio-ius@unibas.ch; URL: www.ius.unibas.ch
1938; Giovanna Delbrück
110000 vols; 277 curr per; 282 e-journals; 18470 diss/theses
libr loan; IALL, BIS 27008

– Mathematisches Institut, Bibliothek, Rheinsprung 21, 4051 Basel
T: +41 612673990; Fax: +41 612673995; E-mail: Bibliothek-Math@unibas.ch; URL: www.math.unibas.ch
Prof. H. Kraft
18000 vols; 80 curr per 27009

– Musikwissenschaftliches Institut, Bibliothek (Institute of Musciology), Petersgraben 27, 4051 Basel
T: +41 612672800; Fax: +41 612672801; E-mail: sekretariat-mwi@unibas.ch; URL: mwi.unibas.ch/
1912; Dr. Simon Obert
Ethnomusikologie, Mittelalterliche Musikgeschichte, Neue Musik; Archive (microfilms)
41000 vols; 70 curr per; 7000 music scores; 10600 microforms; 35 av-mat; 7000 sound-rec; 6 digital data carriers; 41 slides 27010

– Orientalisches Seminar, Bibliothek, Maiengasse 51, 4055 Basel
T: +41 612672860; Fax: +41 612672864; E-mail: bibliothek-orientsem@unibas.ch; URL: www.orientsem.unibas.ch
1919; Dr. Monika Winet
Islamische Mystik, klassische persische und arabische Literatur, Hadith
13200 vols; 24 curr per; 20 maps; 10 av-mat; 12 sound-rec; 15 digital data carriers
libr loan 27011

– Philosophisches Seminar, Bibliothek, Nadelberg 6-8, 4051 Basel
T: +41 612672766; Fax: +41 612672769; E-mail: Susanne.Kress@unibas.ch; URL: philsem.unibas.ch/
Susanne Kress
20000 vols; 20 curr per; 60 sound-rec 27012

– Seminar für Klassische Philologie, Bibliothek, Nadelberg 6, 4051 Basel
T: +41 612672772; Fax: +41 612672771; E-mail: renate.leubin@unibas.ch; URL: pages.unibas.ch/klaphil.home.html
Griech u latein Sprache u Lit; Indo-Germanic Library
30000 vols; 25 curr per
non-lending library 27013

– Slavisches Seminar, Bibliothek, Nadelberg 4, 4051 Basel
T: +41 612673411; Fax: +41 612673411; E-mail: slavsem@unibas.ch; URL: www.slavistik.unibas.ch
1964; Prof. Andreas Guski
40000 vols; 20 curr per; 300 microforms; 500 av-mat 27014

– Theologische Fakultät, Bibliothek, Nadelberg 10, 4051 Basel
T: +41 612672901; Fax: +41 612672902; E-mail: bibliothek-theol@unibas.ch; URL: theolrel.unibas.ch/bibliothek/
1966; Susanne Schaub
Praktische Theologie, Hebräerbrief, Religionssoziologie, Religionsökonomie
52000 vols; 160 curr per; 15 e-journals; 80 av-mat; 10 digital data carriers
libr loan; BIS, VEBTIS
non-lending library 27015

Berner Fachhochschule – Hochschule der Künste Bern – Musikbibliothek (Bern University of Applied Sciences – Bern University of the Arts – Music Library), Papiermühlestr. 13j, 3014 ***Bern***; Postfach, 3000 Bern 22
T: +41 318483960; Fax: +41 318483998; E-mail: musikbibliothek@hkb.bfh.ch; URL: www.hkb.bfh.ch/de/campus/bibliotheken/musikbibliothek/
1917; Dr. Andrea Grandjean-Gremminger
Depositum des Frauenmusikforums Schweiz; Kirchenmusikalische Bibliothek der Reformierten Kirchen Bern-Jura-Solothurn
6000 vols; 60 curr per; 70 e-journals; 5 e-books; 700 mss; 45000 music scores; 100 av-mat; 5000 sound-rec
libr loan; IAML 27016

Berner Fachhochschule – Soziale Arbeit, Bibliothek, Hallerstr. 8, 3012 ***Bern***; mail address: Hallerstr. 10, 3012 Bern
T: +41 318483636; E-mail: bibliothek.soziale-arbeit@bfh.ch; URL: www.soziale-arbeit.bfh.ch
1976; Monika Schefer
Sozialarbeit, Sozialpädagogik
20140 vols; 118 curr per; 10 e-journals; 452 av-mat; 5 sound-rec; 21 digital data carriers
IDS Basel/Bern 27017

Berner Fachhochschule – Wirtschaft und Verwaltung, Bibliothek, Morgartenstr. 2c, 3014 ***Bern***; Postfach 305, 3000 Bern 22
T: +41 31 8483459; Fax: +41 31 8483401; E-mail: bibliothek.wirtschaft@bfh.ch; URL: www.wirtschaft.bfh.ch/de/campus/bibliothek.html
2007; Georg Graf
Betriebswirtschaft, Management, Informationsmanagement, Projektmanagement, Finance, Accounting, Wirtschaftsinformatik und Public Management (Verwaltung)
4500 vols; 80 curr per; 80 e-journals
BIS 27018

Hochschule der Künste Bern (HKB), Mediothek GKK, Fellerstr. 11, 3027 ***Bern***
T: +41 31 8483831; E-mail: susan.mengis@hkb.bfh.ch; URL: www.hkb.bfh.ch/mediothekengkk.html
2002; Susan Mengis Loretan
Design, Film, Fotografie, Grafik, Kultur und Kunst
16800 vols; 190 curr per; 1483 av-mat 27019

Universität Bern, Universitätsbibliothek Bern, Zentralbibliothek (University of Bern), Münstergasse 61, 3000 ***Bern*** 8; mail address: 3000 Bern 8
T: +41 316319211; Fax: +41 316319299; E-mail: info@ub.unibe.ch; URL: www.ub.unibe.ch
1528-1535; Marianne Rubli Supersaxo
Bernensia, Helvetica vor 1848, roman-german MA, MA, Archäologie, ornith Bibliothek Holzer, Richard-Wagner-Slg, Hermann-Rorschach-Archiv, Karten-Slg Ryhiner (alte Landkarten, v.a. 18. Jh), Dok zu Buchpflege u Papierrestauration; Swiss Eastern Europe Libr
4234000 vols; 53100 e-journals; 292000 e-books; 220 mss; 430 incunabula; 47700 maps; 120500 microforms; 33400 av-mat; 1200 digital data carriers; 430000 other items
libr loan; BIS, LIBER 27020

– Basisbibliothek Unitobler, Universitätsbibliothek Bern, Länggassstr. 49a, 3012 Bern; Postfach, 3000 Bern 9
T: +41 316314701; Fax: +41 316314705; E-mail: bto@ub.unibe.ch; URL: www.ub.unibe.ch/bto
1992; Katharina Steiner
Geisteswissenschaften, Sozialwissenschaften, Theologie
45000 vols; 7 curr per; 2000 av-mat; 1200 sound-rec; 480 digital data carriers
libr loan; BIS 27021

– Bibliothek Anglistik, Länggassstr. 49, 3000 Bern 9
T: +41 316318372; Fax: +41 316313636; E-mail: anglistik@ub.unibe.ch; URL: www.ens.unibe.ch
1898; Verena Breidenbach
Medieval English language and literature, Post-Colonial English literature, Modern English language and literature, American literature, Linguistics
43000 vols; 31 curr per; 72 e-journals; 212 diss/theses; 50 microforms; 140 av-mat; 50 sound-rec; 210 digital data carriers 27022

– Bibliothek Betriebswirtschaft, Engehaldenstr. 4, 3012 Bern
T: +41 316314204; Fax: +41 316314205; E-mail: info@ub.unibe.ch; URL: www.ub.unibe.ch/bbw/content/index_ger.html
1996; B. Scherrer
Die Bibliothek umfasst die Bestände des Inst. für Finanzmanagement, des Inst. für internationales Innovattionsmanagement mit der Abt. Quantitative Methoden, des Inst. für Marketing und Unternehmensführung mit den Abt. Marketing, Unternehmensführung und Consumer Behavior, des Inst. für Organisation und Personal sowie des Inst. für Unternehmensrechnung und Controlling mit den Abt. Controlling und Financial Accounting
30000 vols; 200 curr per; 10 e-journals 27023

– Bibliothek Erziehungswissenschaft, Muesmattstr. 27, 3012 Bern; mail address: 3000 Bern 9
T: +41 316313811; Fax: +41 316313773; E-mail: bew@ub.unibe.ch; URL: www.bew.unibe.ch
1988; Urs Trostel
75000 vols; 195 curr per; 45 e-journals; 1 mss; 2800 diss/theses; 350 av-mat; 300 sound-rec; 150 digital data carriers
libr loan 27024

– Bibliothek Exakte Wissenschaften, Sidlerstr. 5, 3012 Bern
T: +41 316318638; E-mail: bewi@ub.unibe.ch; URL: www.bewi.unibe.ch
1993; Jan Dirk Brinksma
Astronomie, Mathematik, Physik, Statistik, Versicherungsmathematik u Wissenschaftstheorie u -Geschichte
70000 vols; 300 curr per; 1000 diss/theses
libr loan 27025

– Bibliothek Germanistik, Unitobler, Länggassstr. 49, 3000 Bern 9
T: +41 316318301; Fax: +41 316318663; E-mail: michael.schlaefli@ub.unibe.ch; URL: www.ub.unibe.ch/germlib/
1885; Michael Schläfli
80000 vols; 150 curr per; 2000 diss/theses; 400 microforms; 500 av-mat; 300 sound-rec; 100 digital data carriers; 5000 brochures 27026

– Bibliothek Medizingeschichte, Bühlstr. 26, 3012 Bern; Postfach 753, 3000 Bern 9
T: +41 316318473; Fax: +41 316318491; E-mail: img@ub.unibe.ch; URL: www.mhi.unibe.ch
1963; Pia Burkhalter
Biografische Schriften, Anatomie, Pathologie, Ophtalmologie, Mikrobiologie, Zahnmedizin, Pädiatrie, Pharmakologie / Albrecht von Haller, Leben und Werk
56000 vols; 100 curr per; 3500 diss/theses; 100 microforms; 100 av-mat; 50 sound-rec
libr loan; BIS 27027

– Bibliothek Pflanzenwissenschaften, Institut für Pflanzenwissenschaften (Library Plant Sciences), Altenbergrain 21, 3013 Bern
T: +41 316314911; Fax: +41 316314942; E-mail: ips@ub.unibe.ch; URL: www.ub.unibe.ch/libips/content/index_ger.html
1860; Christine Dolder
9000 vols; 90 curr per; 12 e-journals; 16 microforms; 18 sound-rec
libr loan 27028

– Bibliothek Philosophie, Länggassstr. 49a, 3012 Bern; mail address: 3000 Bern 9
T: +41 316318055; Fax: +41 316313779; E-mail: philosophie@ub.unibe.ch; URL: www.philosophie.unibe.ch
Simone Hess
20000 vols; 60 curr per; 60 e-journals; 80 diss/theses 27029

– Bibliothek Romanistik, Länggassstr. 49, 3012 Bern; Postfach, 3000 Bern 9
T: +41 316318034; Fax: +41 316313818; E-mail: bibliorom@ub.unibe.ch; URL: www.bibliorom.unibe.ch
1981; Herr Andrea Müller
91000 vols; 222 curr per; 1400 av-mat; 450 sound-rec; 400 digital data carriers; 7000 photos
libr loan 27030

– Bibliothek Slavistik, Länggassstr. 49, 3000 Bern 9
T: +41 316318060; Fax: +41 316313990; URL: www.ub.unibe.ch/slavlib
1968; Prisca Kovač-Zurron
Kirchenslavisches Schrifttum
50700 vols; 120 curr per; 2 e-journals; 300 microforms; 40 sound-rec; 30 digital data carriers
libr loan 27031

– Bibliothek Sozialwissenschaften, Lerchenweg 36, 3000 Bern 9
T: +41 316314823; Fax: +41 316314834; E-mail: bsw@ub.unibe.ch; URL: www.bsw.unibe.ch
1992; Christine Wright
Politik, Soziologie, Medienwissenschaft und -geschichte, Wirtschaftssoziologie, Wissenschaftstheorie
60000 vols; 500 curr per; 80 e-journals; 20 digital data carriers; 100000 clippings (p.a.)
libr loan; BIS 27032

– Bibliothek Sprachwissenschaft, Länggass-Str. 49, 3000 Bern 9
T: +41 316318804; Fax: +41 316313603; E-mail: linguistik@ub.unibe.ch; URL: www.ub.unibe.ch/linglib
1961; Christine Zimmermann
Allgemeine und historisch-vergleichende Sprachwissenschaft, Psycholinguistik (speziell Spracherwerb), Indo-Iranistik, Zentralasien, Kaukasus (Sprachen und Kultur), Tibeto-birmanische Sprachwissenschaft, angewandte Linguistik, Sprachlehrmittel
50176 vols; 91 curr per; 9 e-journals; 2692 diss/theses; 2 music scores; 108 maps; 82 microforms; 86 av-mat; 514 sound-rec; 152 digital data carriers
libr loan 27033

– Bibliothek Vetsuisse Bern, Länggassstr. 120, 3012 Bern; Postfach, 3001 Bern
T: +41 31631-2216; Fax: +41 31631-2216; URL: www.vetbibl.unibe.ch
1805; Margrit Maeder
10000 vols; 50 curr per; 400 e-journals; 15000 diss/theses
non-lending library 27034

– Departement für Chemie und Biochemie (DCB), Bibliothek, Freiestr. 3, 3012 Bern
T: +41 316314262; Fax: +41 316318853; E-mail: biblio@dcb.unibe.ch; URL: dcbwww.unibe.ch/dcb-biblio
1892; Anna Owsianko, Ewa Szydlowski
Crystallography dept, Biochemistry dept
10000 vols; 110 curr per; 235 diss/theses; 3 microforms; 10 digital data carriers
libr loan 27035

– Departement für Christkatholische Theologie, Bibliothek, Länggassstr. 51, 3012 Bern
T: +41 316318240; Fax: +41 316313298; E-mail: christkath@ub.unibe.ch; URL: www.theol.unibe.ch/christkath/bibliothek.html
1905; Michael Schläfli
Alt-/Christkatholizismus, Katholizismus des 19. Jhdts, Anglikaner, Orthodoxie,

Ökumene, Liturgik, Ekklesiologie
12 500 vols; 60 curr per 27036

– Department Volkswirtschaftslehre, Bibliothek, Schanzeneckstr. 1, Postfach, 8573 Bern
T: +41 316318082; Fax: +41 316313992; E-mail: bibliothek@vwi.unibe.ch; URL: www.vwlib.unibe.ch/lib
1963; Eva Werner
30 000 vols; 180 curr per 27037

– Evangelisch-theologisches Departement, Bibliothek, Länggassstr. 51 / Unitobler, 3012 Bern
T: +41 316318068; Fax: +41 316318072; E-mail: evtheol@ub.unibe.ch; URL: www.theol.unibe.ch/evang/biblioinfo.html
1904; Martin Fischer, Eduard Wälchli
100 000 vols; 300 curr per; 8 000 diss/theses; 30 av-mat; 50 sound-rec; 82 digital data carriers
libr loan 27038

– Fachbereichsbibliothek Bühlplatz, Baltzerstr. 4, 3012 Bern; Postfach 874, 3000 Bern 9
T: +41 316314607; Fax: +41 316314121; E-mail: fbb@ub.unibe.ch; URL: www.ub.unibe.ch/fbb
1981; Ida-Maria Mäder
Biologie (spez. Zellbiologie und Zoologie), Medizin, Geologie; Map coll
110 000 vols; 280 curr per; 500 e-books; 2 470 diss/theses; 13 500 maps; 10 microforms; 330 av-mat; 7 sound-rec; 270 digital data carriers; 23 800 e-journals consortium
libr loan; BIS, AGMD 27039

– Geographisches Institut, Bibliothek Geographie, Hallerstr. 12, 3012 Bern
T: +41 316318861; Fax: +41 316318511; E-mail: biblio@giub.unibe.ch; URL: www.ub.unibe.ch/geolib/content/index_ger.html
1974; M. Lindt, R. Schnegg
10 000 vols; 150 curr per; 2 000 maps 27040

– Historisches Institut, Unitobler, Länggassstr. 49, 3000 Bern 9
T: +41 3163180918844, 316314801; Fax: +41 316314410; E-mail: geschichte@ub.unibe.ch; URL: www.ub.unibe.ch/histlib/content/
1870
80 000 vols; 200 curr per; 60 maps; 1 200 microforms
libr loan 27041

– Historisches Institut, Abt. Alte Geschichte und Epigraphik, Unitobler, Länggassstr. 49, 3000 Bern 9
T: +41 316318340; Fax: +41 316318342; E-mail: yvonne.zandolini@ub.unibe.ch; URL: www.hist.unibe.ch/content/bibliothek/index_ger.html
1835; Herr P. Nielson
Separata-Slg Prof G. Walser/römische Geschichte, Epigrafik
14 800 vols; 54 curr per; 500 maps; 8 digital data carriers; 600 photos, 200 slides, 1 000 transparencies 27042

– Institut für Archäologische Wissenschaften, Abt. Archäologie des Mittelmeerraumes, Bibliothek, Länggass-Str. 10, 3012 Bern
T: +41 316318992; Fax: +41 316314905; E-mail: christine.voegeli@iaw.unibe.ch; URL: www.iaw.unibe.ch/content/
1957; Dr. Martina Seifert
45 000 vols; 300 curr per; 1 300 microforms; 20 digital data carriers
libr loan 27043

– Institut für Archäologische Wissenschaften – Abt. Ur- und Frühgeschichte / Abt. Archäologie der Römischen Provinzen, Bibliothek, Bernastrasse 15a, 3005 Bern
T: +41 3135010-11/-10; Fax: +41 313501019; E-mail: agnese.spadini@ub.unibe.ch; URL: www.sfu.unibe.ch
1950; Agnese Spadini
Ur- und Frühgeschichte, Archäologie der Römischen Provinzen
50 000 vols; 365 curr per; 4 e-journals; 200 microforms; 180 av-mat
libr loan 27044

– Institut für Islamwissenschaft und Neuere Orientalische Philologie, Bibliothek, Falkenplatz 11, 3012 Bern
T: +41 316318892; Fax: +41 316318048; E-mail: maktaba@ub.unibe.ch; URL: www.islam.unibe.ch
1970; Prof. R. Schulze
12 000 vols; 81 curr per; 79 maps; 20 microforms; 9 av-mat; 259 sound-rec
libr loan 27045

– Institut für Klassische Philologie, Bibliothek, Länggass-Str. 49, 3000 Bern 9
T: +41 316318099; Fax: +41 316314486; E-mail: klassphilo@ub.unibe.ch; URL: www.kps.unibe.ch
1893; Regula Merz
42 000 vols; 103 curr per; 2 digital data carriers
libr loan; BIS 27046

– Institut für Musikwissenschaft, Bibliothek, Hallerstr. 5, 3012 Bern
T: +41 316318392; Fax: +41 316313459; E-mail: bibliothek@musik.unibe.ch; URL: www.musik.unibe.ch
1917; Renate Gygax
Block (bes slavische u nordische Musik), Gauss (Noten)
30 000 vols; 160 curr per; 13 e-journals; 1 500 diss/theses; 9 000 music scores; 150 microforms; 7 500 sound-rec; 80 digital data carriers
libr loan; BIS 27047

– Institut für Psychologie, Bibliothek, Muesmattstr. 45, 3000 Bern 9
T: +41 316314007; Fax: +41 316318212; E-mail: psychologie@ub.unibe.ch; URL: www.ub.unibe.ch/psychlib/content/
1911; George Sangiovanni
Tests
40 000 vols; 150 curr per; 1 700 diss/theses
BIS 27048

– Institut für Sozialanthropologie, Bibliothek, Länggassstr. 49a, 3000 Bern 9
T: +41 316318951; Fax: +41 316314212; E-mail: information@anthro.unibe.ch; URL: www.anthro.unibe.ch/content/bibliothek/index_ger.html
1968
19 000 vols; 140 curr per; 28 e-journals; 365 diss/theses; 6 maps; 4 700 microforms; 530 av-mat; 27 sound-rec; 27 digital data carriers
BIS 27049

– Institut für Sportwissenschaft (ISPW), Bibliothek, Bremgartenstr. 145, 3012 Bern
T: +41 316318321; Fax: +41 316314631; E-mail: info@ispw.unibe.ch; URL: www.ispw.unibe.ch
1969; Franziska Hofer
8 000 vols; 45 curr per
libr loan; BIS 27050

– Juristische Bibliothek Bern (JBB), Hochschulstr. 4, 3012 Bern
T: +41 316318267; Fax: +41 316318588; E-mail: jbb@ub.unibe.ch; URL: www.ub.unibe.ch/jbb/
1920; Bernhard Dengg
Recht, Welthandelsrecht; Eugen Huber Bibliothek; Juristische Forschungsbibliothek Bern (JFB), World Trade Institute
100 000 vols; 10 000 curr per; 500 e-journals; 9 500 diss/theses
VjBS 27051

– Universitäre Psychiatrische Dienste Bern (UPD), Universitätsklinik für Psychiatrie, Fachbibliothek WA, mail address: Bolligenstr. 111, 3000 Bern 60
T: +41 319309111; Fax: +41 319309404; E-mail: dkp@puk.unibe.ch; URL: www.puk.unibe.ch
1880; Barbara Kühne
23 000 vols; 60 curr per
libr loan 27052

– Universitätsspital-Bibliothek USB, Anna-Seiler-Haus, Inselspital 34, 3010 Bern; Postfach 63, 3010 Bern
T: +41 316322870; Fax: +41 316324991; E-mail: backoffice@usb.unibe.ch; URL: www.usb.unibe.ch
1966; Dr. Jürg Schenker, Barbara Winkelmann
100 000 vols; 500 curr per; 2 500 e-journals
libr loan 27053

Berner Fachhochschule – Technik und Informatik, Bibliothek Biel, Quellgasse 21, 2502 ***Biel***; Postfach, 2501 Biel
T: +41 32 3216460; Fax: +41 32 3216555; E-mail: biblio.ti-bi@bfh.ch; URL: www.ti.bfh.ch
Marguerite Egger
Departments: Automotive Engineering; Electrical Engineering and Communication Technology; Informatics; Mechanical Engineering; Micro Technology and Medical Technbology; Mathematics; Physics
15 000 vols; 120 curr per; 200 digital data carriers
libr loan 27054

Fachhochschule Nordwestschweiz – Pädagogische Hochschule, ***Brugg***
– Bibliothek Brugg, Baslerstr. 45, 5200 Brugg
T: +41 564600622; E-mail: mediothek.brugg.ph@fhnw.ch; URL: www.fhnw.ch/ph/bibliothek/brugg
Johannes M. Zaugg
Pädagogik, Psychologie, Erziehungswissenschaften, Deutschdidaktik, Sport/Bewegungswissenschaften
22 000 vols; 60 curr per; 150 e-journals; 50 maps; 1 000 av-mat; 400 sound-rec; 300 digital data carriers
libr loan 27055

– Institut Spezielle Pädagogik und Psychologie (ISP), Bibliothek, Elisabethenstr. 53, 4002 Basel
T: +41 612069069; Fax: +41 612711939; E-mail: mediothek.basel.isp.ph@fhnw.ch; URL: www.fhnw.ch/ph/isp
1971; Mirjam Oberholzer
Heilpädagogische Früherziehung/Heilpädagogik im Vorschulbereich (HFE/HPV), Unterrichtsmaterial (K-Reihe)
12 000 vols; 47 curr per; 100 av-mat; 20 sound-rec; 30 digital data carriers; 310 tests, 650 diploma theses 27056

– Mediothek, Kasernenstr. 31, 4410 Liestal
T: +41 619279178; E-mail: mediothek.liestal.ph@fhnw.ch; URL: www.fhnw.ch/ph/bibliothek/liestal
Thomas Schai
22 981 vols; 59 curr per; 11 music scores; 93 maps; 694 av-mat; 293 sound-rec; 105 digital data carriers 27057

– Mediothek für Schule und Bildung, Küttigerstr. 42, 5000 Aarau
T: +41 628389010; Fax: +41 628389019; E-mail: Mediothek.aarau.ph@fhnw.ch; URL: www.fhnw.ch/ph
1978/1985/1998; Isabel Dahinden
Pädagogik, Didaktik, Lehrmittel
26 000 vols; 164 curr per; 410 av-mat; 520 sound-rec; 420 digital data carriers; 128 multimedia kits 27058

– Mediothek Solothurn, Obere Sternengasse 7, 4500 Solothurn
T: +41 326279225; E-mail: mediothek.solothurn.ph@fhnw.ch; URL: www.fhnw.ch/ph/bibliothek/solothurn
1979
34 338 media items 27059

NTB Interstaatliche Hochschule für Technik Buchs, Bibliothek/Dokumentation, Werdenbergstr. 4, 9471 ***Buchs***
T: +41 817553351; Fax: +41 817565434; E-mail: bibliothek@ntb.ch; URL: www.ntb.ch/ueber-uns/dienste/bibliothek.html
1971; Susanne Brunschwiler, Hedwig Herres
Systemtechnik
30 000 vols; 300 curr per; 5 000 e-journals; 700 maps; 100 sound-rec; 300 digital data carriers
libr loan; DGI, BIS, BiB 27060

Berner Fachhochschule – Architektur, Holz und Bau/Technik und Informatik, Bibliothek, Pestalozzistr. 20, 3400 ***Burgdorf***; Postfach 1058, 3401 Burgdorf
T: +41 31 344264333; E-mail: biblio.ti-bu@bfh.ch; URL: www.ahb.bfh.ch/ahb/de/Schule/bibliotheken; www.ti.bfh.ch/index.php?id=975&L=0
Marguerite Egger
Architektur, Bauprozess- und Bauingenieurwesen, Elektro- und Maschinentechnik, Telekommunikation, Energietechnik
12 500 vols; 150 curr per; 85 digital data carriers 27061

Scuola universitaria professionale della Svizzera italiana (SUPSI), Biblioteca (SUPSI University of Applied Sciences and Arts of Southern Switzerland), Campus Trevano, 6952 ***Canobbio***
T: +41 586666381; Fax: +41 586666309; E-mail: biblioteca.dacd@supsi.ch; URL: www.biblioteca.supsi.ch
1972; Francesco Marvin
Architecture, Interior design, Visual communication, Conservation and restauration, Civil engineering; Dipartimento delle costruzioni e del territorio (DCT), Dipartimento di lavore sociale (DLS)
40 000 vols; 250 curr per; 300 maps; 1 400 av-mat; 40 sound-rec; 200 digital data carriers; 5 000 slides
libr loan; BIS 27062

Haute école de gestion de Genève, Infothèque, Route de Drize 7, Campus de Battelle, Bâtiment F, 4e étage, 1227 ***Carouge***
T: +41 223881825; Fax: +41 223881760; E-mail: infotheque@hesge.ch; URL: www.hesge.ch/heg/infotheque
2000
17 200 vols; 200 curr per; 15 000 e-journals; 140 digital data carriers 27063

Institut de Hautes Etudes en Administration Publique (IDHEAP), Bibliothèque, Route de la Maladière 21, 1022 ***Chavannes-près-Renens***
T: +41 215574002; Fax: +41 215574009; E-mail: Danielle.Javet@idheap.unil.ch; URL: www.idheap.ch
1981
Verwaltungswiss
10 000 vols; 100 curr per; 150 diss/theses
libr loan; BIS 27064

Theologische Hochschule Chur, Bibliothek, Alte Schanfiggerstr. 7, 7000 ***Chur***
T: +41 812549944; Fax: +41 812549998; E-mail: bibliothek@priesterseminar-thc.ch; URL: www.thchur.ch
1968; Prof. Michael Durst
62 000 vols; 142 curr per; 2 mss; 2 incunabula; 5 sound-rec; 10 digital data carriers 27065

Pädagogische Hochschule Freiburg, Didaktisches Zentrum & Fachbibliothek, Murtengasse 34, 1700 ***Freiburg***
T: +41 263057231; Fax: +41 263057239; E-mail: dz@edufr.ch; URL: www.hepfr.ch/dz
1970; Klaus Vonlanthen
Religion, Bibelkunde, Multimedia Atelier
17 000 vols; 70 curr per; 300 maps; 4 000 av-mat; 1 300 sound-rec; 750 digital data carriers; 175 games, 505 transparencies, 200 theme collls
BIS 27066

Université de Fribourg, Bibliothèque cantonale et universitaire Fribourg (BCU) / Kantons- und Universitätsbibliothek Freiburg (KUB), Rue Joseph-Piller 2, 1700 ***Fribourg***; CP 160, 1701 Fribourg
T: +41 263051333; Fax: +41 263051378; E-mail: bcu@fr.ch; URL: www.fr.ch/bcuf (french); www.fr.ch/kubf (deutsch)
1848; Dr. Martin Good
Friburgensia (2 700 vols), imprimés avant 1901 (190 000 vols), imprimés anciens, dont: Bibl de la Société Economique (26 000 vols), Bibl des Capucins de Fribourg et Bulle (28 000 vols), Bibl de Castella de Delley (3 100 vols), coll des photos (936 000); Depositories of the Fondation Archivum Helveto-Polonicum of the Société Dante Alighieri, of the Freiburger Naturwissenschaftliche Gesellschaft, and of the Schweizer Alpinclub (Freiburg)
2 413 000 vols; 5 000 curr per; 19 100 e-journals; 2 140 mss; 572 incunabula; 6 000 maps; 31 300 microforms; 7 700 av-mat; 9 500 sound-rec; 200 digital data carriers
libr loan; BIS, SAB/CLP, LIBER 27067

– Bibliothèque de la Faculté de droit (BFD), Miséricorde, salles 4010-4052, 1700 Fribourg; mail address: Av de l'Europe 20, 1700 Fribourg
T: +41 263008009, 263008010; Fax: +41 263009777; E-mail: bibl-BFD@unifr.ch; URL: www.unifr.ch/droit/bfd/fr/index.php
1941; Dr. Tudor Pop
Canon law dept
130 000 vols; 504 curr per; 224 e-journals; 54 digital data carriers
libr loan; VJBS 27068

– Bibliothèque de la Faculté des sciences (DOKPE), Rue du Musée 4, 1700 Fribourg
T: +41 263009261; Fax: +41 263009730; E-mail: dokpe@unifr.ch; URL: www.unifr.ch/dokpe/
2000; François Rappaz
20 000 vols; 450 curr per
libr loan 27069

– Bibliothèque de langues et littératures médiévales et modernes (BLL), Av de l'Europe 20 (MIS 2228), 1700 Fribourg
T: +41 263007856; E-mail: Bll@unifr.ch; URL: www.fr.ch/bcuf
1942; Sophie Mégevand
50 000 vols; 173 curr per; 30 microforms; 12 av-mat; 131 digital data carriers 27070

– Bibliothèque de l'Institut de droit européen, Bibliothek des Instituts für Europarecht, Av de Beauregard 11, 1700 Fribourg
T: +41 263008360; Fax: +41 263009776; E-mail: madeleine.dietrich@unifr.ch; URL: www.unifr.ch/euroinstitut/de/organisation/bibliothek
1995; Madeleine Bieri
9 000 vols; 90 curr per; 40 e-journals; 10 govt docs
libr loan 27071

– Bibliothèque de l'Institut de musicologie (MUS), Miséricorde, bureaux 2031/2033, 1700 Fribourg
T: +41 263007951; Fax: +41 263009700; E-mail: Luca.Zoppelli@unifr.ch; URL: www.fr.ch/bcuf/Dynamic.aspx?c=147
Delphine Vincent
30 000 vols 27072

– Bibliothèque de l'Institut du fédéralisme (IFF), Route d'Englisberg 7, 1763 Granges-Paccot
T: +41 263008125, 263008136; Fax: +41 263009724; URL: www.fr.ch/bcuf/Dynamic.aspx?c=640
1983; Christine Verdon
80 000 vols; 50 curr per 27073

– Bibliothèque de l'Institut interfacultaire de l'Europe orientale et central (IEO), Rte d'Englisberg 7, 1723 Granges-Paccot
T: +41 263007977/263007913; Fax: +41 363009697; URL: www.fr.ch/bcuf/Dynamic.aspx?c=763
2003; Jens Herlth
35 000 vols; 30 curr per 27074

– Bibliothèque de pédagogie curative (IPC-HPI), Rue Saint-Pierre-Canisius 21, 1700 Fribourg
T: +41 263007735; Fax: +41 263009749; E-mail: bibl-IPC@unifr.ch; URL: www.unifr.ch/spedu/
1938; Elisabeth Longchamp Schneider
24 000 vols; 124 curr per; 3 e-journals; 160 av-mat; 15 digital data carriers 27075

– Bibliothèque de pédagogie et psychologie (PSPE), Site Regina Mundi, Rue Faucigny 2, 1700 Fribourg
T: +41 263007680; Fax: +41 263009669; URL: www.fr.ch/bcuf/Dynamic.aspx?c=142
1979; Anne Devenoges
42 145 vols; 230 curr per; 150 digital data carriers
libr loan 27076

– Bibliothèque de sociologie, politiques sociales et travail social (STS), Rte des Bonnesfontaines 11, 1700 Fribourg
T: +41 263007798; Fax: +41 263009657; URL: www.fr.ch/bcuf/Dynamic.aspx?c=140
1993; Iris Thaler
16 900 vols; 108 curr per 27077

– Bibliothèque des langues étrangères (BLE) et Centre d'auto-apprentissage, Centre de langues et Domaine d'études Plurilinguisme et didactique des langues étrangères, Criblet 13, 1700 Fribourg
T: +41 263007958; Fax: +41 263009717; E-mail: veronica.gremaud@unifr.ch
1999; Veronica Gremaud-Rütsche (Bibliothèque), Peter Sauter (Centre d'auto-apprentissage)
Fremdsprachen, Mehrsprachigkeit; Allemand langue étrangère (DAF), Français langue étrangère (FLE), Anglais langue étrangère (EFL), Italien langue étrangère (ILS), Selbstlernzentrum Fremdsprachen (ca 30 langues)
22 800 vols; 130 curr per; 14 e-journals; 1 400 av-mat; 7 600 digital data carriers 27078

– Bibliothèque des Sciences de l'Antiquité (SCANT), Rue Pierre-Aeby 16, 1700 Fribourg
T: +41 26300-7834/-7838; Fax: +41 263009714; E-mail: Bibl-SCANT@unifr.ch; URL: www.unifr.ch/scant/bibliotheque/accueil.html
1985; Claire-Lyse Curty-Delley
Archéologie classique, hist ancienne, lit gréco-latine, Archéologie paléochrétienne et byzantine
31 200 vols; 5 curr per 27079

– Bibliothèque d'Histoire de l'art et de Philosophie (BHAP), Miséricorde, Av de l'Europe 20, 1700 Fribourg
T: +41 263007526; E-mail: Bibl-bha@unifr.ch; URL: www.fr.ch/bcuf/Dynamic.aspx?c=144
2001; Claire-Lyse Curty-Delley
25 454 vols; 73 curr per 27080

– Bibliothèque du Département de mathématiques (MATH), Rue du Musée 23, 1700 Fribourg
T: +41 263009180; Fax: +41 263009744; E-mail: claudia.kolly@unifr.ch; URL: www.unifr.ch/math
Jean-Pierre Gabriel
17 000 vols; 120 curr per 27081

– Bibliothèque interfacultaire d'histoire et théologie (BHT), Miséricorde, salles 4211-4215, 1700 Fribourg
T: +41 2630073-73(-74; Fax: +41 263009745; URL: www.fr.ch/bcuf/Dynamic.aspx?c=643
1977; Flavio G. Nuvolone
86 000 vols; 300 curr per
libr loan 27082

– Bibliothèque Pérolles2, Bibliothèque Économie, Société, Informatique et Sport, Boulevard de Pérolles 90, 1700 Fribourg
T: +41 263008211; Fax: +41 263009788; E-mail: bibl-bp2@unifr.ch; URL: www.unifr.ch/bp2
1950; Benoît Renevey
Sciences économiques, Sciences sociales; Informatique, Sciences du sport
85 000 vols; 420 curr per; 1 100 diss/theses; 250 digital data carriers 27083

Bibliothèque de Genève, Promenade des Bastions, CP, 1211 ***Genève*** 4
T: +41 224182800; Fax: +41 224182801; E-mail: jean-charles.giroud@ville-ge.ch; URL: www.ville-ge.ch/bge
1562; Jean-Charles Giroud
Illuminierte Handschriften 14.-15. Jh, Reformationsgeschichte (Calvin), franz Lit d 18. Jh (Voltaire u Rousseau), Genf, Genfer Ablieferungspflicht, Geschichte
2 000 000 vols; 3 500 curr per; 100 000 mss; 470 incunabula; 25 000 maps; 10 000 microforms; 1 000 av-mat; 1 000 sound-rec; 500 digital data carriers; 100 000 posters, 100 000 iconographic representations
libr loan; BIS, AIBM 27084

Conservatoire de musique, Bibliothèque, Place Neuve, 1204 ***Genève***; CP 5155, 1211 Genève 11
T: +41 227814517; Fax: +41 227814523; E-mail: biblio@cmusge.ch; URL: www.cmusge.ch/cmg/pages/bibliotheque/bibliotheque_catalogue.html
1835; Jacques Tchamkerten
Französische und russische Musik, Genfer und Schweizer Komponisten
20 000 vols; 250 curr per; 1 000 mss; 70 000 music scores; 3 000 microforms; 10 digital data carriers
AIBM 27085

Haute école d'art et de design Genève, Bibliothèque (Geneva University of Art and Design), Bd Helvétique 9, 1205 ***Genève***
T: +41 223885834; Fax: +41 223885801; E-mail: brigitte.oggier@hesge.ch; URL: head.hesge.ch/spip.php?rubrique32
1976; Brigitte Oggier
Arts plastiques, sculpture, peinture, dessin, audiovisuel, cinéma
13 512 vols; 44 curr per; 2 132 av-mat; 185 sound-rec 27086

Hôpital cantonal – Clinique d'ophtalmologie, Bibliothèque, 22, rue Alcide-Jentzer, 1205 ***Genève*** 14
T: +41 223828375; Fax: +41 223828382; E-mail: Bibliotheque.Ophtalmologie@hcuge.ch; URL: www.unige.ch
12 450 vols; 35 curr per
libr loan 27087

Hôpital cantonal – Département de chirurgie, Bibliothèque, Rue Micheli-du-Crest 24, 1211 ***Genève*** 4
T: +41 223727826; Fax: +41 223727956; URL: www.unige.ch/biblio
1900
8 000 vols; 130 curr per 27088

Hôpitaux universitaires de Genève – Departement de chirurgie, Bibliothèque, Rue Gabrielle-Perret-Gentil 4, 1211 ***Genève*** 14
T: +41 223727826; Fax: +41 223727666; E-mail: Mariette.Lapalud@unige.ch; URL: www.bibliochir-ge.ch
1900; Mariette Lapalud
8 000 vols; 130 curr per 27089

Hôpitaux universitaires de Genève -Service d'ophtalmologie, Bibliothèque, Rue Alcide-Jentzer, 1205 ***Genève***
T: +41 222828375; Fax: +41 223828382; E-mail: Bibliotheque.Ophtalmologie@hcuge.ch
12 500 vols; 35 curr per 27090

Institut de hautes études internationales et du développement (IHEID), Bibliothèque (Graduate Institute of International and Development Studies), P.O. Box 76, 1211 ***Genève*** 21
T: +41 229085888; Fax: +41 229086270; E-mail: Library@graduateinstitute.ch; URL: graduateinstitute.ch/corporate/contacts.html
1927; Yves Corpataux
Völkerbund, Vereinte Nationen, OCDE, 'New York Times_ auf Mikrofilm; OSGE Dept
150 000 vols; 901 curr per; 1 895 diss/theses; 20 digital data carriers
libr loan; BIS, GRD 27091

Université de Genève, DIS – Division de l'Information Scientifique (DIS), 24, Rue Général-Dufour, 1211 ***Genève*** 4
T: +41 223791000; Fax: +41 223797191; E-mail: dis@unige.ch; URL: www.unige.ch/biblio
1982; Véronique Hadengue-Dezael
1 200 000 vols; 3 500 curr per; 49 221 e-journals; 230 diss/theses; 769 microforms; 11 309 av-mat; 32 835 digital data carriers; 168 808 iconographic documents
AILIS, BIS, IFLA 27092

Conservatoire de Fribourg (COF), Bibliothèque, Rte Louis Braille 8, 1763 ***Granges-Paccot***
T: +41 263059958; Fax: +41 264666517
1904; Christiane Antoniazza Torche
1 500 vols; 9 000 music scores 27093

Hochschule Luzern – Technik und Architektur, Bibliothek (Lucerne University of Applied Sciences and Arts, School of Engineering and Architecture), Technikumstrasse 21, 6048 ***Horw***
T: +41 413493562; Fax: +41 413493960; E-mail: bibliothek.technik-architektur@hslu.ch; URL: www.hslu.ch/t-u_bibliothek
1958; André Sersa
Architektur, Bautechnik, Wirtschaftsingenieur, Elektrotechnik, Gebäudetechnik, Informatik, Maschinentechnik, Innenarchitektur
32 633 vols; 340 curr per; 7 000 e-journals; 500 e-books; 800 maps; 498 av-mat; 43 sound-rec; 178 digital data carriers; 2 500 Lucerne University theses and project works
libr loan 27094

Conservatoire de Lausanne HEM, Bibliothèque, Rue de la Grotte 2, 1003 ***Lausanne***; CP 5700, 1002 Lausanne
T: +41 213213524; Fax: +41 213213536; E-mail: bibliotheque@cdlhem.ch; URL: www.cdlhem.ch
1941; Olivier Gloor
Fonds Eugen Huber, Fonds du Musique chorale, Fonds Jacques Ehrhart
11 000 vols; 18 curr per; 250 diss/theses; 23 000 music scores; 600 sound-rec; 10 digital data carriers
BIS 27095

Ecole hôtelière de Lausanne, Bibliothèque, Le Chalet-à-Gobet, 1000 ***Lausanne*** 25
T: +41 21785-1243/-1239; Fax: +41 217851250; E-mail: bibliotheque@ehl.ch; URL: www.ehl.ch/ehl_library_ENG_HTML.htm
1998; Luisa Rossi
500 cartes de menus de restaurant
19 000 vols; 260 curr per; 28 000 e-journals; 680 av-mat; 500 digital data carriers
libr loan 27096

Ecole Polytechnique Fédérale de Lausanne, Bibliothèque, EPFL – Rolex Learning Center, 1015 ***Lausanne***; mail address: Station 20, 1015 Lausanne
T: +41 216932156; Fax: +41 216935100; E-mail: questions.bib@epfl.ch; URL: library.epfl.ch
1853/1969/1978; N.N.
Wissenschaft und Technologie
390 000 vols; 711 curr per; 10 400 e-journals; 33 000 e-books; 15 mss; 1 incunabula; 6 719 diss/theses; 664 maps; 300 microforms; 444 av-mat; 4 sound-rec; 2 600 digital data carriers; 6 265 courses, 3 456 diplomas, 431 ancient bks, 100 videos
libr loan; IFLA, BIS, LIBER, ASLIB, AILIS, ADBS 27097

Haute Ecole de Travail social et de la Santé – EESP – Lausanne, Centre de documentation, Chemin des Abeilles 14, 1010 ***Lausanne***
T: +41 216516265; Fax: +41 216516288; E-mail: biblio@eesp.ch; URL: www.eesp.ch
1974
Ergotherapie, Travail social, Animation, Education spécialisée
25 000 vols; 300 curr per; 3 300 diss/theses; 1 200 videos
BIS 27098

Haute école pédagogique, Bibliothèque-médiathèque, Av de Cour 33, 1014 ***Lausanne***
T: +41 213160916; Fax: +41 213160917; E-mail: bm@hepl.ch; URL: www.hepl.ch/index.php?id=bm
Véronique Avellan
Fonds historique en sciences de l'éducation
19 011 vols; 212 curr per; 1 514 av-mat; 993 sound-rec; 609 digital data carriers; 614 sets of diapositives, 755 professional biographies
libr loan; BIS 27099

Université de Lausanne, Bibliothèque Cantonale et Universitaire (University of Lausanne), Bâtiment Unithèque, 1015 ***Lausanne*** – Dorigny; mail address: Place de la Riponne 6, 1014 Lausanne
T: +41 216924802; Fax: +41 216924845; E-mail: info-dorigny@bcu.unil.ch; URL: www.bcu-lausanne.ch
1537; Jeannette Frey
Orientalisme, cinéma, histoire de France, documentation vaudoise, musicologie, réserve précieuse, manuscrits, archives musicales
6 770 935 vols; 71 958 curr per; 114 271 e-books; 720 mss; 150 incunabula; 30 000 music scores; 3 841 maps; 70 400 microforms; 54 996 av-mat; 17 000 sound-rec; 285 digital data carriers
libr loan; AIBM, IFLA, LIBER, BIS, RERO, BiblioVaud 27100

– Bibliothèque de Biologie, Bâtiment Biophore, 1015 Lausanne
T: +41 216924104; Fax: +41 216924105; E-mail: josiane.bonetti@unil.ch; URL: www.unil.ch/bibliobiol
1995; Josiane Bonetti
10 000 vols; 250 curr per; 50 maps
BIS 27101

– Bibliothèque de Droit et Sciences économiques (BDSE), 1015 Lausanne – Dorigny
T: +41 216924883; Fax: +41 216924885; E-mail: LaC3@bcu.unil.ch; URL: www.unil.ch/BCU
1977; Gérald Gavillet
70000 vols 27102

– Bibliothèque de la Riponne (BCU/R), Place de la Riponne 6, 1014 Lausanne
T: +41 21 3167880; Fax: +41 21 3167870; E-mail: info-riponne@bcu.unil.ch; URL: www.unil.ch/bcu
1537; Jeannette Frey
Documentation vaudoise, Phontohèque, Musicologie, Musique inprimée, Archives musicales, Dépôt légal
430000 vols; 2500 curr per; 36000 music scores; 7000 av-mat; 28000 sound-rec
libr loan; AIBM, IFLA, LIBER, BBS, RERO 27103

– Bibliothèque de l'Hôpital ophtalmique Jules Gonin, Av de France 15, CP 133, 1000 Lausanne 7
T: +41 216268560; Fax: +41 216268325; E-mail: madeleine.badan@fa2.ch; URL: www.asile-aveugles.ch
1844; Madeleine Badan
8500 vols; 57 curr per
libr loan; BIS, ABS 27104

– Bibliothèque des Cèdres, 7, chemin des Cèdres, 1004 Lausanne
T: +41 216480164; Fax: +41 216480166; E-mail: Marie-Pierre.Constant@bcu.unil.ch; URL: www.unil.ch/central/page2398_fr.html
1847; Marie-Pierre Constant
Theologie u Humanwiss 17.-21. Jh (wichtiger Altbestand: 17.-19. Jh); Kirchengeschichte; Kulturgesch
120000 vols; 150 curr per; 6700 diss/theses
libr loan; BIS 27105

– Bibliothèque des Sciences de la Terre (Earth Sciences Library), Batiment Anthropole, 1015 Lausanne
T: +41 216924301; Fax: +41 216924335; E-mail: Bibliost@unil.ch; URL: www.unil.ch/igp/page10580.html
1906; Catherine Schlegel Rey
30000 vols; 1127 curr per; 87 e-journals; 900 diss/theses; 12000 maps; 7800 microforms; 100 digital data carriers; 20 videos
libr loan 27106

– Bibliothèque scientifique commune UNIL-EPFL, Ecole Polytechnique Fédérale de Lausanne, EPFL-SB-SCGC-BCH, 1015 Lausanne
T: +41 216939800; Fax: +41 216939805; E-mail: biscom@epfl.ch; URL: biscom.epfl.ch
2005; Dr. Alain Borel
Chemie, Biotechnologie, Kriminologie, Kriminalität
52000 vols; 175 curr per; 2280 diss/theses; 420 microforms; 30 av-mat
libr loan; NEBIS 27107

– Bibliothèque Universitaire de Médecine (BiUM), Centre Hospitalier Universitaire Vaudois, Bugnon, 46, 1011 Lausanne
T: +41 213145082; Fax: +41 13145070; E-mail: bdfm@chuv.ch; URL: www.chuv.ch/bdfm
1980; Isabelle de Kaenel
100000 vols; 500 curr per; 15000 e-journals; 2000 diss/theses; 200 microforms; 2000 av-mat; 200 sound-rec; 50 digital data carriers; 25000 slides, 1000 videos
libr loan; BIS, EAHIL, IFLA 27108

– Département de Biologie Cellulaire et de Morphologie, Bibliothèque, Rue du Bugnon 9, 1005 Lausanne
T: +41 216925100; Fax: +41 216925105; URL: www.unil.ch/dbcm
9000 vols; 130 curr per 27109

– Siège de Dorigny, Unithèque, Unithèque, 1015 Lausanne – Dorigny
T: +41 216924802; Fax: +41 216924845; E-mail: info-Dorigny@bcu.unil.ch; URL: www.unil.ch/bcu
1982; Jeanette Frey
Sciences humaines
800000 vols; 5800 curr per; 2000 maps; 69000 microforms; 35 digital data carriers; 65000 vols of periodicals, 2000 videos
libr loan 27110

Pädagogische Hochschule FHNW, Standort Liestal, Mediothek, Kasernenstr. 31, 4410 ***Liestal***
T: +41 619279178; Fax: +41 619279166; E-mail: mediothek.liestal.ph@fhnw.ch; URL: aleph.unibas.ch
1984; Thomas Schai
24492 vols; 59 curr per; 183 maps; 280 av-mat; 293 sound-rec; 519 digital data carriers 27111

Franklin College, Library, Via Ponte Tresa 29, 6924 ***Lugano*** – Sorengo
T: +41 919852273; Fax: +41 919944117; E-mail: library@fc.edu; URL: www.fc.edu/
1970; Susan Gilfert
Art, hist, politics, economy, lit, philosophy, international studies
32000 vols; 171 curr per; 15 e-journals; 12 microforms; Fortune; Forbes; New York Times (1982 – present) in microfiches, Videos/DVDs, online access to full text resources
ALA, BIS, LIBER 27112

Università della Svizzera italiana, Biblioteca universitaria di Lugano, Via G. Buffi 13, 6900 ***Lugano***
T: +41 586664500; Fax: +41 586664509; E-mail: biblioteca@lu.unisi.ch; URL: www.bul.unisi.ch
1996; Giuseppe Origgi
61500 vols; 578 curr per; 7000 e-journals; 343 diss/theses; 362 digital data carriers
BIS 27113

Zentral- und Hochschulbibliothek Luzern, Sempacherstr. 10, 6003 ***Luzern***; Postfach, 6002 Luzern
T: +41 412285344; Fax: +41 412106255; E-mail: info@zhbluzern.ch; URL: www.zhbluzern.ch
1951; Dr. Ulrich Niederer
Katholische Theologie, Schweizer Geschichte, Rechtswissenschaft, Sozialwissenschaft, Lucernensia; Libr of Gen Music Society, Libr of Art Society, Citizen Libr, Canton Libr, University area, Special Coll (mss + prints before 1900, graphic docs / graphic coll, special music coll
989272 vols; 2610 curr per; 47769 e-journals; 8811 e-books; 2736 mss; 850 incunabula; 144 diss/theses; 12500 music scores; 4817 maps; 36968 microforms; 5629 sound-rec; 1996 digital data carriers; 130449 prints and graphic docs, 289 online databases (license)
libr loan; AIBM, BIS, LIBER 27114

Bundesamt für Sport, Sportmediathek, 2532 ***Magglingen***
T: +41 323276309; Fax: +41 323276408; E-mail: markus.kueffer@baspo.admin.ch; URL: www.sportmediathek.ch
1944; Markus Küffer
41000 vols; 250 curr per; 10 e-journals; 3000 diss/theses; 1200 microforms; 1500 av-mat
libr loan; BIS, IASI, AGSB 27115

Fachhochschule Nordwestschweiz – Hochschule für Architektur, Bau und Geomatik, Hochschulbibliothek Muttenz, Gründenstr. 40, 4132 ***Muttenz***
T: +41 614674444; Fax: +41 614674308; E-mail: bibliothek.muttenz@fhnw.ch; URL: www.fhnw.ch/habg/bibliothek
1970; Marianne Ingold
Architektur, Bauwesen, Geomatik
20000 vols; 150 curr per 27116

Bibliothèque publique et universitaire, Pl Numa-Droz 3, 2000 ***Neuchâtel***
T: +41 327177300; Fax: +41 327177309; E-mail: secretariat.bpu@unine.ch; URL: bpun.unine.ch/services/SetPret.htm
1788; Michel Schlup
Handschriften v J.J. Rousseau, Mme de Charriere, Geschichte d Buchs, Sozial- u Geistesgeschichte d 18. Jh, Neocomensia; Libr of the 'Schweizer Gesellschaft für Zeitmessung', Libr of the 'Naturwissenschaftliche Gesellschaft', Libr of the 'Geographische Gesellschaft'
551322 vols; 1980 curr per; 13082 mss; 35 incunabula; 70494 diss/theses; 3498 music scores; 5153 maps; 20714 microforms; 691 av-mat; 505 sound-rec; 308 digital data carriers; 1570 prints, 5384 posters, 45728 pamphlets
libr loan; BIS 27117

Université de Neuchâtel, Bibliothèque centrale, Av du 1er Mars 26, 2000 ***Neuchâtel***
T: +41 327181050; Fax: +41 327181001; E-mail: liliane.regamey@unine.ch; URL: www.unine.ch/biblio
1909; Liliane Regamey
400000 vols; 3200 curr per; 500 digital data carriers
libr loan; BIS 27118

– Bibliothèque de biologie, Rue Emile-Argand 11, 2007 Neuchâtel; CP 158, 2009 Neuchâtel
T: +41 327182090; Fax: +41 327183001; E-mail: pret.ne7@unine.ch; URL: www.unine.ch/bibliotheque/page23128.html
1909; Elisabeth Kuster, Josiane Pont (Biologie)
30000 vols; 400 curr per 27119

– Bibliothèque de chimie, Av de Bellevaux 51, 2000 Neuchâtel
T: +41 327182441; Fax: +41 327182511; E-mail: jocelyne.tissot@unine.ch
1909; Baise Dardel
12000 vols; 40 curr per
libr loan; BIS 27120

– Bibliothèque de droit, Av du 1er-Mars 26, 2000 Neuchâtel
T: +41 327181240; Fax: +41 327181241; E-mail: biblio.droit@unine.ch; URL: www.unine.ch/biblio/droit
1938; Slim Ben Younés
110000 vols; 400 curr per; 13 e-journals; 30 mss; 32000 diss/theses; 2000 govt docs; 120 microforms; 314 sound-rec; 100 digital data carriers
libr loan; IALL, VjBS 27121

– Bibliothèque de géologie et d'hydrogéologie, Rue Emile-Argand 11, CP 158, 2009 Neuchâtel
T: +41 327182091; Fax: +41 327182601; E-mail: pascale.pretre@unine.ch; URL: www2.unine.ch/biblio
1918 (Geology), 1968 (Hydrogeology); Pascale Prêtre
22000 vols; 50 curr per; 7 e-journals; 500 diss/theses; 5000 maps; 100 microforms; 7000 slides, 26000 offprints
libr loan 27122

– Bibliothèque de physique, Rue A.-L.-Breguet 1, 2000 Neuchâtel
T: +41 327182911; Fax: +41 327182901
1909; Nicole Derendinger
12000 vols; 44 curr per 27123

– Bibliothèque de théologie, Faubourg de l'Hôpital 41, 2000 Neuchâtel
T: +41 327181904; Fax: +41 327181901; E-mail: delphine.luder@unine.ch; URL: ww2.unine.ch/bibliotheque
1972; Delphine Luder
22845 vols; 94 curr per
libr loan 27124

– Bibliothèque des pasteurs, Fbg de l'Hôpital 41, 2000 Neuchâtel
T: +41 327181996; URL: www2.unine.ch/biblio/page4295.html
Cecilia Griener Hurley
90000 vols 27125

– Bibliothèque des sciences économiques, Rue de Pierre-à-Mazel 7, 2000 Neuchâtel
T: +41 327181340; Fax: +41 327181341; E-mail: biblio.se@unine.ch; URL: www.unine.ch/bibliotheque
1910; Natalie Brunner-Patthey
Depository Libr of European Community
80000 vols; 315 curr per; 50 e-journals; 1500 microforms; 1100 sound-rec
libr loan; BIS 27126

– Bibliothèque d'ethno, Rue St Nicolas 4, 2000 Neuchâtel
T: +41 327181713; Fax: +41 327181711; E-mail: raymonde.wicky@unine.ch; URL: www2.unine.ch/biblio/page3841.html
1945; Raymonde Wicky
20000 vols; 500 curr per; 400 sound-rec; 2000 slides, 120 videos
libr loan 27127

– Bibliothèques des lettres, Espace Louis-Agassiz 1, 2000 Neuchâtel
T: +41 327181759; Fax: +41 327181701; E-mail: biblio.lettres@unine.ch; URL: www2.unine.ch/biblio
1986; Patricia Jeanneret
255000 vols; 689 curr per; 190 digital data carriers
libr loan; BIS 27128

Fachhochschule Nordwestschweiz – Hochschule für Wirtschaft, University of Applied Sciences Northwestern Switzerland, Bibliothek Olten (University of Applied Sciences Northwestern Switzerland), Riggenbachstr. 16, 4600 ***Olten***
T: +41 622860135; Fax: +41 622860090; E-mail: bibliothek.olten@fhnw.ch; URL: www.fhnw.ch
1972; Marianne Hertner
Wirtschaft, Soziale Arbeit, Angewandte Psychologie
30000 vols; 250 curr per; 17 e-journals; 580 digital data carriers; Newspapers, videos, DVDs
libr loan 27129

HES-SO Valais, Médiathèque Sièrre, Rte de la Plaine 2, 3960 ***Sierre***; CP 112, 3960 Sierre
T: +41 276068927; Fax: +41 276068935; E-mail: biblio.sierre@hevs.ch; URL: www.hevs.ch/mediathequesierre
Tamara Soom
Wirtschaft, Tourismus, Informatik, Soziale Arbeit
20000 vols; 161 curr per; 210 maps; 140 av-mat; 30 digital data carriers
libr loan; BIS 27130

HES-SO Valais/Wallis, ***Sion***
– Bereich Ingenieurwissenschaften, Mediathek, 47, Rte du Rawyl, CP 2134, 1950 Sion 2
T: +41 276068580; Fax: +41 276068575; E-mail: biblio@hevs.ch; URL: mediatheques.hevs.ch/default.aspx?nolangue=1
1989; Karine Villettaz
13319 vols; 56 curr per; 8554 e-books; 1082 diss/theses; 383 av-mat; 6 sound-rec; 67 digital data carriers
libr loan; BIS, GVB 27131

– Médiathèque Santé-Social, Rue de Gravelone 5, 1950 Sion
T: +41 276064244; E-mail: biblio.sante-social@hevs.ch; URL: mediatheques.hevs.ch/
15000 vols; 100 curr per; 500 av-mat 27132

FHS St.Gallen, Hochschule für Angewandte Wissenschaften, ***St.Gallen***
– Fachbereich Gesundheit, Bibliothek, Tellstrasse 2, Postfach 664, 9001 St.Gallen
T: +41 712261504; Fax: +41 712261205; E-mail: bibliothek@fhsg.ch; URL: www.fhsg.ch/fhs.nsf/de/bibliothek-gesundheit
2006; Andrea Brenner
1000 vols; 5 curr per; 20 e-journals 27133

– Fachbereich Soziale Arbeit, Bibliothek, Industriestr. 35, 9400 Rorschach; mail address: 9401 Rorschach
T: +41 718444885; Fax: +41 718444850; E-mail: bibliothek@fhsg.ch; URL: www.fhsg.ch/fhs.nsf/de/bibliothek-soziale-arbeit
1996; Gregor Helg
15000 vols; 120 curr per; 1000 av-mat; 92 sound-rec
BIS 27134

Pädagogische Hochschule des Kantons St.Gallen, medienverbund.phsg, Notkerstr. 27, 9000 ***St.Gallen***
T: +41 712439471; Fax: +41 712439490; E-mail: mediathek@phsg.ch; URL: www.phsg.ch
2007; August Scherer-Hug
60000 vols; 170 curr per; 8 e-journals; 500 music scores; 4000 av-mat; 2500 sound-rec; 500 digital data carriers
BIS, SAB, SIKJM, MIPHD 27135

Universität St.Gallen – Hochschule für Wirtschafts-, Rechts- und Sozialwissenschaften, Bibliothek (University of St.Gallen – Graduate School of Business, Economics, Law and Social Sciences), Dufourstr. 50, 9000 ***St.Gallen***
T: +41 712242270; Fax: +41 712242294; E-mail: Bibliothek@unisg.ch; URL: www.biblio.unisg.ch
1898; Dr. Xaver Baumgartner
Utopia
460000 vols; 1400 curr per; 18000 e-journals; 70000 e-books; 49000

diss/theses; 100 maps; 500 microforms; 500 av-mat; 400 sound-rec; 1 000 digital data carriers
libr loan; BIS, EBSLG, IFLA, DGI 27136

– Institut für Finanzwissenschaft und Finanzrecht, Bibliothek, Varnbüelstr. 19, 9000 St.Gallen
T: +41 712242520; Fax: +41 712242670; E-mail: ladislava.staedler@unisg.ch; URL: www.iff.unisg.ch
1966
Steuerrecht, oeffentliche Verwaltung
10 000 vols; 25 curr per 27137

– Institut für Versicherungswirtschaft, Bibliothek, Kirchlistr. 2, 9010 St.Gallen
T: +41 712434050; Fax: +41 712434040; E-mail: andreina.zink@unisg.ch; URL: www.ivwhsg.ch
1949
Risiko-Management, Soziale Sicherheit, Versicherungswirtschaft
12 000 vols; 150 curr per 27138

– KMU-HSG Bibliothek, Bibliothek, Dufourstr. 40a, 9000 St.Gallen
T: +41 712247100; Fax: +41 712247101; E-mail: kmu-hsg@unisg.ch; URL: www.kmu.unisg.ch
1946; Heidi Fueglistaller
Klein- und Mittelunternehmen, Entrepreneurship, Familienunternehmen
15 000 vols; 100 curr per 27139

Fachhochschule Nordwestschweiz – Hochschule für Technik, Hochschulbibliothek für Technik und Wirtschaft, Klosterzelgstr. 2, 5210 ***Windisch***
T: +41 564624666 (Dieter Lang), +41 564624259 (Eva Karrer); E-mail: tw-bibliothek.windisch@fhnw.ch
Dieter Lang (Technik), Eva Karrer (Wirtschaft)
Automation, BWL, Elektrotechnik, Informatik/Wissenschaftsinformatik, Maschinenbau, Mathematik, Physik, Recht, Volkswirtschaft, Wirtschaftsethik
22 500 vols 27140

ZHAW Zürcher Hochschule für Angewandte Wissenschaften, Technikumstr. 9, Postfach, 8401 ***Winterthur***
T: +41 589347559; Fax: +41 589357559; E-mail: bibliothek@zhaw.ch; URL: www.zhaw.ch/biblio

– Hochschulbibliothek ZHAW, Bibliothek Angewandte Psychologie, Minervastr. 30, Postfach, 8032 Zürich
T: +41 589348410; Fax: +41 589348439; E-mail: psychologie.bibliothek@zhaw.ch; URL: www.zhaw.ch/de/zhaw/die-zhaw/bibliotheken.html
H. Quenzer, Kerstin Jaeger
Arbeits- und Organisationspsychologie, Entwicklungs- und Schulpsychologie, Klinische Psychologie, Rehabilitationspsychologie
15 000 vols; 200 curr per 27141

– Hochschulbibliothek ZHAW, Bibliothek Architektur und Bau, Tössfeldstr. 11, Halle 180, Postfach, 8401 Winterthur
T: +41 589347622; Fax: +41 589357622; E-mail: a.bibliothek@zhaw.ch; URL: www.zhaw.ch/de/zhaw/die-zhaw/bibliotheken.html
Wolfgang Giella
Architektur, Konstruktion, Städtebau
5 000 vols; 25 curr per 27142

– Hochschulbibliothek ZHAW, Bibliothek Gesundheit, Technikumstr. 71 / Eulachpassage, Postfach, 8401 Winterthur
T: +41 589346360; Fax: +41 589356360; E-mail: g.bibliothek@zhaw.ch; URL: www.zhaw.ch/de/zhaw/die-zhaw/bibliotheken.html
Wolfgang Giella
Gesundheitswesen, Gesundheitssoziologie, Pflege, Ergotherapie, Physiotherapie
4 000 vols; 94 curr per 27143

– Hochschulbibliothek ZHAW, Bibliothek Life Sciences und Facility Management, Grüntal 14, Postfach, 8820 Wädenswil
T: +41 589345999; Fax: +41 589345001; E-mail: lsfm.bibliothek@zhaw.ch; URL: www.zhaw.ch/de/zhaw/die-zhaw/bibliotheken.html
Hanspeter Quenzer, Rosmarie Schwager
Chemie, Biologie, pharmazeutische Technologie, Umweltwiss, urbaner Gartenbau, Landschafts- u Regionalentwicklung, Lebensmittel- u Getränketechnologie, Ernährungswiss, Facility Management u Betriebswirtschaft
14 000 vols; 300 curr per 27144

– Hochschulbibliothek ZHAW, Bibliothek Linguistik, Theaterstr. 15c / Mäander SM, Postfach, 8401 Winterthur
T: +41 589346260; Fax: +41 589356260; E-mail: l.bibliothek@zhaw.ch; URL: www.zhaw.ch/de/zhaw/die-zhaw/bibliotheken.html
Wolfgang Giella
Übersetzen/Dolmetschen, angewandte Medienwissenschaft und Sprache in Beruf und Bildung
12 000 vols; 150 curr per 27145

– Hochschulbibliothek ZHAW, Bibliothek Soziale Arbeit, Auenstr. 10, Postfach, 8600 Dübendorf 1
T: +41 589348686; Fax: +41 589348800; E-mail: s.bibliothek@zhaw.ch; URL: www.zhaw.ch/de/zhaw/die-zhaw/bibliotheken.html
Hanspeter Quenzer, Sylvia Wanke
Soziale Arbeit, Soziologie, Pschologie, Pschychologie, Recht und Philosophie
10 000 vols; 65 curr per 27146

– Hochschulbibliothek ZHAW, Bibliothek Technik, Technikumstr. 9, Geb. B, Postfach, 8401 Winterthur
T: +41 589347284; Fax: +41 589357284; E-mail: t.bibliothek@zhaw.ch; URL: www.zhaw.ch/de/zhaw/hochschulbibliothek/teilbibliotheken/bibliothek-technik.html
Wolfgang Giella, Hanspeter Quenzer
Aviatik, Verkehrssysteme, Chemie, Elektrotechnik, Informatik, Maschinenbau, Verfahrenstechnik, Mathematik, Physik
40 000 vols; 200 curr per 27147

– Hochschulbibliothek ZHAW, Bibliothek Wirtschaft und Recht, St. Georgenplatz 2 / Geb. SW, Postfach, 8401 Winterthur
T: +41 589347923; Fax: +41 589357923; E-mail: w.bibliothek@zhaw.ch; URL: www.zhaw.ch/de/zhaw/die-zhaw/bibliotheken.html
Wolfgang Giella
Betriebswirtschaft, Volkswirtschaft, Öffentliche Verwaltung und Recht
13 000 vols; 200 curr per 27148

Haute Ecole d'ingénierie et de gestion du Canton de Vaud (HEIG-VD), Bibliothèque, Route Cheseaux 1, 1400 ***Yverdon-les-Bains***; CP 587, 1401 Yverdon-les-Bains
T: +41 2445576391; Fax: +41 2445576404; E-mail: biblio@heig-vd.ch; URL: www.heig-vd.ch/biblio
1986; Tania Dutoit
Ingenieurwesen, Betriebswirtschaft
22 600 vols; 300 curr per; 3 000 e-journals; 50 maps; 10 av-mat; 10 sound-rec; 200 digital data carriers
libr loan; BIS 27149

Eidgenössisches Hochschulinstitut für Berufsbildung (EHB), Bibliothek, Kirchlindachstr. 79, 3052 ***Zollikofen***
T: +41 319103706; Fax: +41 319103701; E-mail: bibliothek@ehb-schweiz.ch; URL: www.ehb-schweiz.ch
1972; Beatrice Frick
7 700 vols; 130 curr per; gray Lit
BIS, BIB 27150

Eidgenössische Technische Hochschule Zürich, ETH-Bibliothek (Swiss Federal Institute of Technology Zurich), Rämistr. 101, 8092 ***Zürich***
T: +41 446322135; Fax: +41 446321087; E-mail: info@library.ethz.ch; URL: www.ethbib.ethz.ch
1855; Dr. Wolfram Neubauer
Alte Drucke, Archive u Nachlässe, Bilder u Karten, Reportsammlung
2 718 000 vols; 5 200 curr per; 11 200 e-journals; 48 000 e-books; 260 000 mss; 13 incunabula; 29 000 diss/theses; 400 000 maps; 2 213 000 microforms; 2 210 000 av-mat; 5 700 digital data carriers; 1 800 000 graphic docs, 142 databases, 113 000 abstracts a indizes
libr loan; IATUL, BIS FID, LIBER 27151

– Archiv für Zeitgeschichte, Bibliothek (Archives of Contemporary History), Hirschengraben 62, 8001 Zürich; mail address: 8092 Zürich
T: +41 446324003; Fax: +41 446321392; E-mail: afz@history.gess.ethz.ch; URL: www.afz.ethz.ch
1974; Dr. Gregor Spuhler
Außen- u Innenpolitik, Wirtschafts- u Außenhandelspolitik, Flüchtlingspolitik u Emigration, Rechtsextremismus u Antisemitismus, Mediengeschichte, politische Parteien und Bewegungen, Landesverteidigung u Sicherheitspolitik; Doc centre for jewish contemporary history, Doc centre for political history, Doc centre for economy and contemporary history
28 000 vols; 5 curr per; 40 maps; 1 100 microforms; 430 av-mat; 750 sound-rec; 510 private and institution stocks (3.000 linear m), 40 000 photos, 300 posters, 700 linear m of press clippings
VSA 27152

– Baubibliothek, HIL E2, 8093 Zürich – Hönggerberg
T: +41 446332906; Fax: +41 446331075; E-mail: baubib@library.ethz.ch; URL: www.ethbib.ethz.ch/dez/bau.html
1976; Markus Joachim
Architektur, Bauingenieurwesen, Kulturtechnik, Geodäsie, Kartografie, Raumplanung, Umweltwiss
100 000 vols; 580 curr per; 1 000 maps; 20 digital data carriers 27153

– Bibliothek Erdwissenschaften, Departement Erdwissenschaften, Sonneggstr. 5 / NO D 25.1, 8092 Zürich
T: +41 446323767; Fax: +41 446321288; E-mail: geobib@library.ethz.ch; URL: www.ethbib.ethz.ch/dez/erd.html
1856; H. Boedecker
Kartensammlung (Geologische Karten)
50 000 vols; 183 curr per; 800 diss/theses; 35 000 maps; 49 av-mat; 149 digital data carriers; rare books
libr loan; BIS 27154

– Bibliothek Informationstechnologie und Elektrotechnik, Gloriastr. 35, ETZ D 51, 8092 Zürich
T: +41 446325006; Fax: +41 446321188; E-mail: biblio@ee.ethz.ch; URL: www.ee.ethz.ch
1980; Hansruedi Missland
Elektrotechnik, Elektronik, Regeltechnik, Computer Hard- und Software, Kommunikations- und Informationstechnik
16 950 vols; 285 curr per 27155

– ETH-Bibliothek HDB, mail address: 8093 Zürich
T: +41 446332410; Fax: +41 446331037; E-mail: hdb@library.ethz.ch; URL: www.ethbib.ethz.ch/service/hdb.html
Elio Veraldi
Architektur, Bau- und Verkehrsingenieurwesen, Militaria
1 200 000 vols 27156

– Graphische Sammlung, Bibliothek, Rämistr. 101, HG E 52, 8092 Zürich
T: +41 446324046; Fax: +41 446321168; E-mail: info@gs.ethz.ch; URL: www.graphischesammlung.ch
1870; Dr. Michael Matile
Publikationen zu Handzeichnungen und Druckgrafik
16 500 vols; 10 curr per; 1 e-journals; 160 000 prints and drawings 27157

– Grüne Bibliothek, Universitätsstr. 16, CHN H 43, 8092 Zürich
T: +41 446328578; Fax: +41 446321590; E-mail: gruene_bibliothek@library.ethz.ch; URL: www.library.ethz.ch/de/Kontakt/Adressen-Oeffnungszeiten/Gruene-Bibliothek
2005; Manuela Schneider
Umweltwissenschaften, Botanik, Landwirtschaft, Forstwirtschaft, Tierzucht, Jagd, Fischerei, Nahrungsmittel, Ernährungswissenschaft, Lebensmitteltechnologie, Ökologie
50 000 vols; 650 curr per; 400 maps; 50 av-mat; 250 digital data carriers
libr loan 27158

– Informatikbibliothek, Departement Informatik, Computer Science Library, Universitätstr. 6, CAB G 41, 8092 Zürich
T: +41 446327207; E-mail: library@inf.ethz.ch; URL: www.library.inf.ethz.ch
Claudius Lüthi
Computerwiss, Informationswiss, angewandte Mathematik, Statistik, Computersicherheit
17 000 vols; 20 curr per; 250 diss/theses; 50 av-mat; 300 digital data carriers 27159

– Informationszentrum Chemie Biologie Pharmazie, Wolfgang-Pauli-Str. 10, mail address: HCI G5, 8093 Zürich
T: +41 44 6323066; Fax: +41 44 6321287; E-mail: infodesk@chem.ethz.ch; URL: www.infochembio.ethz.ch
1889; Dr. E. Zass
Electronic Libr, Museum
51 000 vols; 20 curr per; 2 400 e-journals; 12 700 e-books; 5 300 diss/theses; 35 000 microforms; 15 av-mat; 600 digital data carriers; 5 300 theses 27160

– Institut für Geschichte und Theorie der Architektur, Archivbibliothek, ETH Hönggerberg, HIL C 64, 8093 Zürich
T: +41 446332887; Fax: +41 446331253; E-mail: bibliothek@gta.arch.ethz.ch; URL: www.ethbib.ethz.ch/dez/bibleth/gta.html
Afra Häni
15 700 vols; 12 curr per; 670 vols of periodicals 27161

– Mathematik-Bibliothek, Departement Mathematik, Bibliothek (Mathematics Library), HG G7, Rämistr. 101, 8092 Zürich
T: +41 446323401; Fax: +41 446321085; E-mail: mathbib@math.ethz.ch; URL: www.math.ethz.ch/library
Monika Krichel
9 000 vols; 80 curr per; 14 e-journals; 200 diss/theses; 17 000 monographies and series
BIS 27162

– Physik-Bibliothek, Hönggerberg HPZ F10, 8093 Zürich
T: +41 446332560, 446332561; Fax: +41 446331354; E-mail: stampanoni@phys.ethz.ch; URL: www.phys.ethz.ch
1885; Anna Radi
Physik, Mathematik für die Physik, Geophysik
36 250 vols; 26 curr per; 200 vols of rare books
libr loan 27163

– Versuchsanstalt für Wasserbau, Hydrologie und Glaziologie, Bibliothek, Gloriastr. 37-39, ETH-Zentrum, VAW G 50-52, 8092 Zürich
T: +41 446324101; Fax: +41 446321192; E-mail: addor@vaw.baug.ethz.ch; URL: www.ethbib.ethz.ch/dez/bibleth/VAW.html
1930; Daniela Addor
17 894 vols; 25 curr per; 600 maps; 20 vols of rare books
libr loan 27164

Pädagogische Hochschule Zürich, Informationszentrum PHZH, Rämistr. 59, 8090 ***Zürich***
T: +41 433055555; Fax: +41 433055111; E-mail: info@phzh.ch; URL: www.phzh.ch

– Bibliothek Gymnasial-, Berufspädagogik und Weiterbildung (GBW), Beckenhofstr. 31, 8090 Zürich
T: +41 433056110; E-mail: bibliothek.gbw@phzh.ch; URL: iz.phzh.ch/content.35.r._Dz.html
2005; Barbara Aebi
Mittelschul- und Gymnasialpädagogik, Berufsbildung, Weiterbildung, Erwachsenenbildung, Allgemeinbildender Unterricht Berufsfachschulen, Geschichte der Weiterbildung
21 000 vols; 170 curr per
libr loan; BIS 27165

– Mediothek Zentrum (UPRAA), Rämistr. 59, 8001 Zürich; mail address: 8090 Zürich
T: +41 433056060; E-mail: mediothek.zentrum@phzh.ch; URL: iz.phzh.ch/
2002; Gisela Wieland
Pädagogik, Psychologie, Soziologie, Philosophie; Berufsberatung; Methematik, Informatik; Kunst und Gestaltung; Musik, Theater, Sport, Suchtprävention; Medienpädagogik, Wirtschaft, Recht
30 100 vols; 150 curr per; 5 000 e-journals; 12 000 av-mat; 4 400 sound-rec
libr loan 27166

Universität Zürich, Zentralbibliothek Zürich, Kantons-, Stadt- und Universitätsbibliothek (University of Zurich), Zähringerplatz 6, 8001 *Zürich*
T: +41 442683100; Fax: +41 442683290; E-mail: zb@zb.uzh.ch; URL: www.zb.uzh.ch
1914; Prof. Susanna Bliggenstorfer
Bibl der Allgemeinen Musik-Gesellschaft Zürich, Fennica (Schweizer Vereinigung der Freunde Finnlands), Schweizer Alpenclub, hist Kinder- u Jugendbücher des Pestalozzianums Zürich, Bibl der Ornithologischen Gesellschaft Zürich, Nordamerika-Bibl; Graph coll; Mss dept, Map coll, Music dept, Old books coll, Libr Oskar R. Schlag
3 950 000 vols; 8 700 curr per; 40 000 e-journals; 10 000 e-books; 125 000 mss; 1 500 incunabula; 600 000 diss/theses; 100 000 music scores; 245 000 maps; 560 000 microforms; 40 000 sound-rec; 500 digital data carriers; 220 000 art prints, 5 000 photos, 160 000 picture postcards
libr loan; BIS, LIBER, AIBM, IFLA
27167

– Anthropologisches Institut und Museum, Bibliothek, Winterthurerstr. 190, 8057 Zürich
T: +41 446355415; Fax: +41 446356804; E-mail: biblio@aim.uzh.ch; URL: www.aim.uzh.ch
1899; Ruth Haegi
8 100 vols; 30 curr per; 100 diss/theses; 6 200 av-mat; 120 digital data carriers; 6 100 slides 27168

– Archäologisches Institut, Bibliothek, Rämistr. 73, 8006 Zürich
T: +41 446342813, 446342815; Fax: +41 446344902; E-mail: palaczyk@archinst.uzh.ch; URL: www.archinst.uzh.ch
1868; Prof. Christoph Reusser
45 000 vols; 900 curr per; 3 000 microforms; 80 000 av-mat; 5 000 offprints
27169

– Bibliothek der Botanischen Institute, Zollikerstr. 107, 8008 Zürich
T: +41 446348414; Fax: +41 446348403; E-mail: biblbot@systbot.uzh.ch; URL: www.botlib.uzh.ch
1892; Elena Benetti
Systematische Botanik: Florenwerke u systemat Monografien weltweit; Struktur, Diversität, Phylogenie, Evolution d Pflanzen; Pflanzenbiologie: Entwicklungsbiologie u Molekularbiologie d Pflanzen, Pflanzenpathologie, Pflanzenphysiologie u Phytochemie, Mikrobielle Ökologie
48 615 vols; 103 curr per; 3 e-journals; 8 000 offprints
libr loan; BIS 27170

– Bibliothek für Betriebswirtschaft (BfB), Zentrale für Wirtschaftsdokumentation (ZWD), Plattenstr. 14, 8032 Zürich
T: +41 446342978; Fax: +41 446344995; E-mail: bfb@irc.uzh.ch; URL: www.irc.uzh.ch/bfb
1903; Frau K. Hertzberg-Schilling
Wirtschaftswissenschaften
34 000 vols; 215 curr per; 6 720 diss/theses
libr loan; BIS 27171

– Bibliothek für Volkswirtschaft (Department of Economics Library), Rämistr. 71, 8006 Zürich
T: +41 446342137; Fax: +41 446344982; E-mail: library@econ.uzh.ch; URL: www.soi.uzh.ch/library
1908; Barbara Jarrett
Allgemeine Mikroökonomie, allgemeine Makroökonomie, Wettbewerbstheorie und -politik, Industrial Organisation, Innovationsökonomie, Umweltökonomie, Verkehrsökonomie und -politik, Finanzwissenschaft, Monetäre Ökonomie, Statistik und Oekonometrie, Statistische Werke, Versicherungs- und Energieökonomie, Außenhandelstheorie und -politik; Holdings of the statistical seminar and holdings of the Economics seminar, Holdings of the empirical economic research institute
88 155 vols; 190 curr per; 8 000 e-journals; 2 000 e-books; 500 diss/theses
libr loan; BIS 27172

– Biochemisches Institut, Bibliothek, Winterthurerstr. 190, 8057 Zürich
T: +41 446355511; Fax: +41 446356805; E-mail: institut@bioc.uzh.ch; URL: www.bioc.uzh.ch
1931
9 400 vols; 135 curr per
libr loan 27173

– Deutsches Seminar, Bibliothek, Schönberggasse 9, 8001 Zürich
T: +41 446342555; E-mail: bibliods@ds.uzh.ch; URL: www.ds.uzh.ch/Bibliothek
1900; Silvia Meyer-Denzler
Deutschsprachige Literatur, Literaturwiss, germanische Philologie, Dialektologie, neuere Linguistik, Niederlandistik, Skandinavistik
120 000 vols; 200 curr per; 520 av-mat; 250 digital data carriers
BIS
non-lending library 27174

– Deutsches Seminar, Abteilung für Nordische Philologie, Bibliothek, Schönberggasse 9, 8001 Zürich
T: +41 446342515; Fax: +41 446344905; URL: www.opac.uzh.ch
1968; Thomas Seiler
35 500 vols; 100 curr per; 113 maps; 2 136 av-mat; 514 sound-rec 27175

– Englisches Seminar, Bibliothek, Plattenstr. 47, 8032 Zürich
T: +41 446343677; Fax: +41 446344908; E-mail: library@es.uzh.ch; URL: www.es.uzh.ch
1890; Katrin Eschenmoser
Commonwealth, Science Fiction
53 500 vols; 80 curr per; 9 e-journals; 100 microforms; 1 700 av-mat; 300 sound-rec; ca.1 000 separata = press clippings
BIS 27176

– Ethik-Zentrum, Institut für Sozialethik, Bibliothek (The Zurich University Centre for Ethics (ZUCE), Institute for Social Ethics), Zollikerstr. 117, 8008 Zürich
T: +41 446348521; Fax: +41 446348507; E-mail: skrieger@sozethik.uzh.ch; URL: www.ethik.uzh.ch/biblio.html
1964; Prof. J. Fischer
Nachlass Emil Brunner, Adolf Keller, Geschichte d religiösen Sozialismus
40 000 vols; 58 curr per; 315 offprints
libr loan 27177

– Ethnologisches Seminar, Bibliothek, Andreasstr. 15, 8050 Zürich
T: +41 446352236; Fax: +41 446352239; E-mail: library@ethno.uzh.ch; URL: www.ethno.uzh.ch
1971
28 000 vols; 130 curr per
libr loan 27178

– Forschungsbibliothek Jakob Jud, Romanisches Seminar, Zürichbergstr. 8, 8032 Zürich
T: +41 446343625; E-mail: filippon@rom.uzh.ch; URL: www.rose.uzh.ch/bibliothek/jakobjud.html
1952; Prof. Dr. Em. Jakob Th. Wüest
Romanische Sprachwissenschaft, insb. französische, italienische und rätoromanische Sprachgeschichte und Dialektologie; Sprachgeografie, Ortsnamenkunde, mittelalterliche Texte; Archiv von Jakob Jud (Aufnahmehefte und -bilder von den Mitarbeitern des AIS, Korrespondenz, Etymologische Karteikarten)
25 000 vols; 80 curr per; 500 sound-rec
27179

– Geographisches Institut, Bibliothek, Winterthurerstr. 190, 8057 Zürich
T: +41 446355115; E-mail: geobib@geo.uzh.ch; URL: www.geo.uzh.ch/en/library
um 1900; G. Seitz
Luftbilder, Satellitenbilder
450 000 vols; 250 curr per; 25 e-journals; 250 diss/theses; 16 000 maps; 20 000 clippings 27180

– Historisches Seminar, Bibliothek, Karl-Schmid-Str. 4, 8006 Zürich
T: +41 446342041; Fax: +41 446344913; E-mail: hsbib@hist.uzh.ch; URL: www.hist.uzh.ch/bibliotheken.html
1919; Dr. Corinne Pernet
130 000 vols; 210 curr per; 110 mss; 550 maps; 500 microforms; 10 000

av-mat
BIS 27181

– Institut für Empirische Wirtschaftsforschung, Bibliothek, Winterthurerstr. 30, 8006 Zürich; mail address: Blümlisalpstr. 10, 8006 Zürich
T: +41 446343705; Fax: +41 446344907; E-mail: bibliothek@iew.uzh.ch; URL: www.iew.uzh.ch
1970; Hanna Britt
35 000 vols; 40 curr per 27182

– Institut für Erziehungswissenschaft, Bibliothek, Freiestr. 36, 8032 Zürich
T: +41 446342758; Fax: +41 446344606; E-mail: bibl@ife.uzh.ch; URL: www.ife.uzh.ch/
1968; Esther Nellen
58 131 vols; 130 curr per; 4 e-journals; 600 av-mat
libr loan; BIS 27183

– Institut für Hermeneutik und Religionsphilosophie, Bibliothek, Kirchgasse 9, 8001 Zürich
T: +41 446344751; Fax: +41 446344991; E-mail: ihr-bibliothek@theol.uzh.ch; URL: www.uzh.ch
1962; Sibylle Bucher
Philosophische, theologische, fachübergreifende Hermeneutik, Nachschriften von Vorlesungen Schleiermachers
18 000 vols; 10 curr per; 1 digital data carriers 27184

– Institut für Mathematik, Bibliothek, Winterthurerstr. 190, 8057 Zürich
T: +41 446355841; Fax: +41 446355705; E-mail: bibliothek@math.uzh.ch; URL: www.math.uzh.ch/
1835; Christof Reiber
27 000 vols; 380 curr per; 350 e-journals; 384 diss/theses; 6 digital data carriers
libr loan; BIS 27185

– Institut für Politikwissenschaft, Bibliothek (Institute of Political Science), Affolternstrasse 56, 8050 Zürich
T: +41 446344140; Fax: +41 446344360; E-mail: bibliothek@ipz.uzh.ch; URL: www.ipz.uzh.ch/institut/bibliothek.html
1997; Christine Lercher
27 000 vols; 60 curr per 27186

– Institut für Populäre Kulturen, Bibliothek (Institute of Popular Culture Studies), Affolternstr. 56, 8050 Zürich
T: +41 446342433; Fax: +41 446344994; E-mail: bibliothek@ipk.uzh.ch; URL: www.ipk.uzh.ch
1951; Charlotte Bühlmann
Volkskunde, Alltagskulturen; Slg Populäre Literaturen und Medien; Dept Popular literatures and media
30 000 vols; 80 curr per; 5 e-journals; 200 av-mat; 100 sound-rec; 5 000 slides, 35 linear meters of pamphlets 27187

– Institut für Sonderpädagogik, Bibliothek, Hirschengraben 48, 8001 Zürich
T: +41 446343124; Fax: +41 446344941; URL: www.isp.uzh.ch/isp/dienstleistungen/bibliothek.htm
1974; Alice Fritschi
Archive
12 000 vols; 90 curr per; 600 diss/theses 27188

– IPMZ – Institut für Publizistikwissenschaft und Medienforschung, Bibliothek, Andreasstr. 15, 8050 Zürich
T: +41 446344663; Fax: +41 446344934; E-mail: bibliothek@ipmz.uzh.ch; URL: www.ipmz.uzh.ch
1923; Barbara Sommerauer Sidiali
Publizistik, Kommunikationswissenschaft, Massenmedien und Journalismus
24 415 vols; 110 curr per; 3 e-journals; 4 e-books; 110 av-mat
libr loan; BIS 27189

– Klassisch-Philologisches Seminar, Bibliothek, Rämistr. 68, 8001 Zürich
T: +41 446342041; Fax: +41 446344955; E-mail: bibliothek@klphs.uzh.ch; URL: www.uzh.ch/klphs
1861; Prof. Ulrich Eigler
38 000 vols; 60 curr per; 23 microforms; 15 av-mat; 5 sound-rec; 15 digital data carriers 27190

– Kunsthistorisches Institut, Bibliothek (Institute of Art History), Rämistr. 73, 8006 Zürich; mail address: Rämistr. 73, 8006 Zürich
T: +41 4463428-35/-34; Fax: +41 446344914; E-mail: bibliothek@khist.uzh.ch; URL: www.uzh.ch/khist/Bibliothek.html
1946; Susanne Blaser-Meier
Photo libr, Islam libr, East-Asian libr
100 000 vols; 188 curr per; 41 e-journals; 3 070 diss/theses; 29 microforms; 81 av-mat; 26 sound-rec; 135 digital data carriers; art catalogs
IFLA, BIS, AKMB 27191

– Medizinhistorisches Institut und Museum, Bibliothek, Hirschengraben 82, 8001 Zürich
T: +41 446342075; E-mail: mhbib@mhiz.uzh.ch; URL: www.mhiz.uzh.ch
1951; Monika Huber
Geschichte der Medizin und Pharmazie
77 000 vols; 20 curr per; 2 incunabula; 5 500 diss/theses; 10 microforms; 15 digital data carriers
libr loan; EAHIL, BIS 27192

– Mittellateinisches Seminar, Bibliothek, Karl-Schmid-Str. 4, 8006 Zürich
T: +41 446343846; Fax: +41 446344975; E-mail: turicense@gmail.com; URL: www.mls.uzh.ch/bibliotheca.html
1963
8 500 vols; 11 curr per; 110 microforms; 50 digital data carriers 27193

– Musikwissenschaftliches Institut, Bibliothek, Florhofgasse 11, 8001 Zürich
T: +41 446343822; Fax: +41 446344964; E-mail: oscar.gillieron@mwi.uzh.ch; URL: www.musik.uzh.ch
1938; Oscar Gilliéron
30 000 vols; 100 curr per; 2 e-journals; 6 000 music scores; 2 000 microforms; 30 av-mat; 3 200 sound-rec
AIBM, BIS 27194

– Nordamerika-Bibliothek, c/o Zentralbibliothek, Zähringerplatz 6, 8001 Zürich
T: +41 442683100; Fax: +41 442683290; E-mail: zb@zb.uzh.ch; URL: www.zb.uzh.ch
1970; B. Storchenegger
Amerikanische/Anglo-Kanadische Lit
93 000 vols; 157 curr per; 29 500 microforms; 6 digital data carriers
libr loan 27195

– Orientalisches Seminar, Bibliothek, Wiesenstr. 9, 8008 Zürich
T: +41 446340731; Fax: +41 446343692; E-mail: sekretariat@oriental.uzh.ch; URL: www.ori.uzh.ch
1961; Johannes Thomann
Arabisch, Persisch, Türkisch, Ägyptologie
25 000 vols; 81 curr per; 50 mss; 2 incunabula; 50 maps; 11 microforms; 40 av-mat; 3 digital data carriers
BIS 27196

– Ostasiatisches Seminar – Abendländische Bibliothek (Institute of East Asian Studies – Western Library), Zürichbergstr. 4, 8032 Zürich
T: +41 446343565; Fax: +41 446344921; E-mail: westlib@oas.uzh.ch; URL: www.ostasien.uzh.ch
1948; Doris Diener
Ostasien; Chinesische und Japanische Sprachen (alle Materialien in westlichen Sprachen)
35 850 vols; 315 curr per; 14 e-journals; 45 diss/theses; 4 maps; 700 microforms; 400 sound-rec; 30 digital data carriers; broschures 27197

– Ostasiatisches Seminar – Chinesische Bibliothek (Institute of East Asian Studies), Zürichbergstr. 4, 8032 Zürich
T: +41 446343567; Fax: +41 446344921; E-mail: siblio@oas.uzh.ch; URL: www.uzh.ch/ostasien
1948; Katharina Thölen, Wu Chia-hsün, Marc Winter
Slg Prof. R. P. Kramers, Slg Zhao Shuxia
80 000 vols; 203 curr per; 69 microforms; 900 av-mat; 68 sound-rec
EASL 27198

– Ostasiatisches Seminar – Japanologie, Japanische Bibliothek, Zürichbergstr. 4, 8032 Zürich
T: +41 446343566; Fax: +41 446344921; E-mail: jbibl@oas.uzh.ch; URL: www.uzh.ch/ostasien/?id=53
1968; Mariko Adachi
27 000 vols; 112 curr per; 70 maps; 2 400 av-mat; 160 sound-rec; 60 digital data carriers
libr loan; EAJRS, Arbeitskreis Japan-Bibliotheken 27199

– Paläontologisches Institut und Museum, Bibliothek, Karl Schmid-Str. 4, 8006 Zürich
T: +41 446342386; Fax: +41 446344923; E-mail: bibliothek@pim.uzh.ch; URL: www.pim.uzh.ch
1956; Jérôme Gapany
Paläontologie, Geologie, Zoologie, Botanik; Archiv der Schriften aus dem Institut
19 600 vols; 135 curr per; 10 e-journals; 30 diss/theses; 922 maps; 145 microforms; 15 digital data carriers; 5 000 slides, 30 600 offprints 27200

– Philosophisches Seminar, Bibliothek, Rämistr. 71, 8006 Zürich
T: +41 446344536; Fax: +41 446344539; E-mail: dacosta@philos.uzh.ch; URL: www.uzh.ch/philosophie
1931; Dr. Rainer Lambrecht
Hermann-Cohen-Archiv
44 800 vols; 100 curr per; 15 digital data carriers
libr loan 27201

– Physiologisches Institut, Bibliothek, Winterthurerstr. 190, 8057 Zürich
T: +41 446355015; Fax: +41 446356814; E-mail: pgysin@access.uzh.ch
1924; Ruzena Biehal
11 000 vols; 52 curr per
libr loan 27202

– Psychologisches Institut, Bibliothek, Binzmühlestrasse 14/2, 8050 Zürich
T: +41 446357171; E-mail: bibliothek@psychologie.uzh.ch; URL: www.psychologie.uzh.ch/bibliothek.html
1969; Regina Meier
20 000 vols; 195 curr per
libr loan; BIS 27203

– Rechtswissenschaftliches Institut, Bibliothek, Rämistr. 74/27, 8001 Zürich
T: +41 446343099; Fax: +41 446344382; E-mail: bibliothek@rwi.uzh.ch; URL: www.rwi.uzh.ch/bibliothek
1887; Franziska Gasser
Kriminologie, dt. u schweiz. Rechtsgeschichte, Völkerrecht
200 000 vols; 600 curr per; 10 e-journals; 8 000 diss/theses; 100 govt docs
libr loan; Vereinigung der juristischen Bibliotheken der Schweiz 27204

– Romanisches Seminar, Bibliothek, Zürichbergstr. 8, 8032 Zürich
T: +41 446343636; Fax: +41 446344940; E-mail: rose@rom.uzh.ch; URL: www.rose.uzh.ch/index.html
1900; Monique C. Furegati
Bibl Paul Claudel, Bibl Marcel Proust; Research Libr Prof Jakob Jud
160 937 vols; 369 curr per
BIS 27205

– Seminar für Allgemeine Sprachwissenschaft, Bibliothek, Plattenstr. 54, 8032 Zürich
T: +41 446342185; Fax: +41 446344357; E-mail: babel@spw.uzh.ch; URL: www.spw.uzh.ch
1975; Prof. Karen H. Ebert
Bedrohte Sprachen/Endangered Languages, Languages of Africa and South-Asia
14 900 vols; 90 curr per; 25 maps; 45 microforms; 510 sound-rec; 600 pamphlets 27206

– Seminar für Allgemeine und Vergleichende Literaturwissenschaft (AVL), Bibliothek, Plattenstr. 43, 8032 Zürich
T: +41 446343535; Fax: +41 446344971; E-mail: mnebel@komp.uzh.ch; URL: www.avl.uzh.ch
1969; Marianne Nebel
Slg Giedion (Surrealismus)
27 000 vols; 50 curr per; 812 diss/theses; 5 microforms; 10 sound-rec; 45 digital data carriers 27207

– Slavisches Seminar, Bibliothek, Plattenstr. 43, 8032 Zürich
T: +41 446343518; Fax: +41 446344971; E-mail: bibliothek@slav.uzh.ch; URL: www.slav.uzh.ch
1961; Prof. German Ritz
100 000 vols; 150 curr per; 450 microforms; 500 av-mat; 320 sound-rec
libr loan 27208

– Soziologisches Institut, Bibliothek (Institute of Sociology), Andreasstr. 15, 8050 Zürich
T: +41 446352388; Fax: +41 446352399; E-mail: bibsoz@soziologie.uzh.ch; URL: www.suz.uzh.ch/bibliothek
1968; Britta Biedermann
29 095 vols; 100 curr per; 50 digital data carriers
libr loan 27209

– Theologisches Seminar und Institut für Schweizerische Reformationsgeschichte und Institut für Hermeneutik und Religionsphilosophie, Bibliothek, Kirchgasse 9, 8001 Zürich
T: +41 446344715; Fax: +41 446344991; E-mail: bibliothek@theol.uzh.ch; URL: www.theol.uzh.ch
1933; Prof. Alois Rust
60 000 vols; 150 curr per; 100 microforms; 25 digital data carriers
libr loan; BIS 27210

– Universitätsklinik Balgrist und Schweizerisches Paraplegikerzentrum, Bibliothek, Forchstr. 340, 8008 Zürich
T: +41 443863716; Fax: +41 443861609; E-mail: ekuntschen@research.balgrist.ch
1912; Emmanuelle Kuntschen
500 vols; 7 curr per
libr loan; EAHIL 27211

– Völkerkundemuseum, Bibliothek, Pelikanstr. 40, 8001 Zürich
T: +41 446349031; Fax: +41 446349050; E-mail: kitab@vmz.unizh.ch; URL: www.musethno.uzh.ch/de/bibliothek/bibliothek.html
1889
36 000 vols; 180 curr per; 30 digital data carriers
libr loan 27212

– Zentrale für Wirtschaftsdokumentation, Bibliothek, Plattenstr. 14, 8032 Zürich
T: +41 446343911; Fax: +41 446344995; E-mail: bfb@irc.uzh.ch; URL: www.irc.uzh.ch/bfb
1910; Frau K. Hertzberg-Schilling
OECD, ADB
33 643 vols; 137 curr per; 3 800 govt docs; 4 000 business rpts, 120 000 newspaper clippings, 25 000 brochures
libr loan; BIS 27213

– Zentrum für Zahn-, Mund- und Kieferheilkunde, Bibliothek, Plattenstr. 11, 8032 Zürich
T: +41 44 6343302; Fax: +41 44 6344327; E-mail: illzu97@zzmk.uzh.ch; URL: www.dent.uzh.ch/bibliothek/
1912; Hildegard Eschle
Alle zahnmedizinische Dissertationen der Universität Zürich
10 000 vols; 126 curr per; 78 e-journals; 40 av-mat
libr loan; BIS, EAHIL 27214

Universität Zürich – Hauptbibliothek (University of Zurich – Main Library), Strickhofstr. 35, 8057 ***Zürich***
T: +41 446354731; Fax: +41 446356808; E-mail: hbz@hbz.uzh.ch; URL: www.hbz.uzh.ch
1980; Dr. Heinz Dickenmann
Naturwiss, Medizin; Forschungsbibliothek Irchel, Studienbibliothek Irchel, Medizinbibliothek Careum
330 000 vols; 1 500 curr per; 15 000 e-journals; 8 000 e-books; 6 000 diss/theses; 600 maps; 5 000 microforms; 1 000 av-mat; 12 000 digital data carriers
BIS, LIBER, EAHIL, AGMB, MLA 27215

Zürcher Hochschule der Künste, Medien- und Informationszentrum (Zurich University of the Arts), Ausstellungsstraße 60, 8005 ***Zürich***
T: +41 43 4464470; E-mail: miz.info@zhdk.ch; URL: miz.zhdk.ch
1875; Jan Melissen
Design, Kunst, Film, Fotografie, Musik, Theater, Elektroakustische Musik, Kulturvermittlung
130 000 vols; 220 curr per; 8 e-journals; 50 000 music scores; 14 000 av-mat; 15 000 sound-rec
NEBIS 27216

– Medien- und Informationszentrum MIZ, Bibliothek Florhofgasse (Media and Information Centre, Music Library), Florhofgasse 6, 8001 Zürich
T: +41 4465152; Fax: +41 4465130; E-mail: miz.musik@zhdk.ch; URL: miz.zhdk.ch
1876; Felix Falkner
Musica practica, Helvetica, elektro-akustische Musik
8 500 vols; 30 curr per; 2 e-journals; 47 000 music scores; 800 av-mat; 25 000 sound-rec; materials for orchestra and chorus
BIS, IAML 27217

School Libraries

WE'G Weiterbildungszentrum für Gesundheitsberufe, Bibliothek, Mühlemattstr. 42, 5000 ***Aarau***; Postfach, 5001 Aarau
T: +41 628375899; Fax: +41 628375860; E-mail: biblio@weg-edu.ch; URL: www.weg-edu.ch
1950; Silvia Rosser
Pflege, Spezialgebiete der Pflege, Pflegewissenschaft, Pflegemanagement, Pflegepädagogik, Spitex, Mütter- und Väterberatung, Geschichte der Pflege, Forschung, Gesundheitswissenschaft, Gesundheitswesen, Management, Soziologie, Psychologie, Pädagogik
10 500 vols; 90 curr per; 200 diss/theses; 420 av-mat; 100 sound-rec; 135 digital data carriers; 1 350 theses
BIS, EAHIL, AGMB 27218

Bibliothek für Gestaltung Basel, Spalenvorstadt 2, Postfach 612, 4003 ***Basel***
T: +41 612674500; Fax: +41 612674503; E-mail: andrea.schweiger@sfgbasel.ch; URL: www.sfgbasel.ch
1878; Andrea Schweiger
Architektur, Städtebau und Landschaftsarchitektur, Produkte- und Industriedesign, Fotografie, Kunsterziehung, Kunstgewerbe, Mode und Textildesign, Schmuck, Schrift- und Buchkunst, Video, Visuelle Kommunikation, Werbegrafik, zeitgenössische Kunst Stiftung Gartenbaubibliothek, Architekturtraktate 16.-20. Jh, Vorlagebücher 16.-19. Jh, Typografische Sammlung von Jan Tschichold, Handeinbände von Ignatz Wiemeler aus der Sammlung Richard Doetsch-Benziger, Nachlass Paul Hulliger zur Entwicklung der Schweizerischen Schulschrift, Schriftmeister-Bände und -Vorlagen, Entwicklungsgeschichte von Schrift und Buch
120 000 vols; 211 curr per; 51 e-journals; 2 500 govt docs; 10 000 av-mat; 1 000 digital data carriers; 25 000 photos, 15 000 art prints
libr loan; AKMB, BIS, SIGEGS, VMS 27219

Konservatorium für Musik Biel, Bibliothek, Bahnhofstr. 11, 2502 ***Biel***
T: +41 323228474; Fax: +41 323228471
1932; Dr. Roman Brotbeck
5 000 vols; 20 curr per; 50 000 music scores; 5 000 sound-rec
libr loan 27220

Bibliothèque des Arts et Métiers / Gewerbebibliothek, Derrière-les-Remparts 5, 1700 ***Fribourg***
T: +41 263052509; Fax: +41 263052600; E-mail: bam@edufr.ch; URL: admin.fr.ch/sfp/fr/pub/service/sfp_portrait_org/bam.cfm
1888; Sophie Rossel Imanze
20 000 vols; 200 curr per
libr loan; BIS 27221

Haute Ecole de la Santé la Source, Centre de documentation, Av Vinet 30, 1004 ***Lausanne***
T: +41 216413820; Fax: +41 216413838; E-mail: cedoc@ecolelasource.ch; URL: www.ecolelasource.ch/
1973
6 000 vols; 105 curr per; 166 e-journals; 1 000 av-mat; 1 200 theses, 300 pamphlets, 60 doc colls, 6 500 newspaper articles
libr loan; BIS 27222

Scuola Magistrale, Biblioteca, Piazza San Francesco 19, 6600 ***Locarno***
T: +41 917515114; Fax: +41 917521736
1890; Raoul Marconi
40 000 vols; 73 curr per; 800 microforms
libr loan; BIS 27223

Hochschule Luzern – Musik, Bibliotheken, Zentralstr. 18, 6003 ***Luzern***
T: +41 412260360; Fax: +41 412260371; E-mail: bibliothek.musik@hslu.ch; URL: www.hslu.ch/musik
1969; Bernadette Rellstab
14 950 vols; 30 curr per; 25 000 music scores; 1 700 av-mat; 12 800 sound-rec
ASCM, AIBM 27224

Biblioteca cantonale e del Liceo di Mendriso, Via Agostino Maspoli, CP 1162, 6850 ***Mendrisio***
T: +41 918159478; Fax: +41 918159479; E-mail: bcme-segr.sbt@ti.ch; URL: www.sbt.ti.ch/bcme
1979; Cristiano Lafferma
Scienze esatte e scienze naturali
59 030 vols; 365 curr per; 489 maps; 366 av-mat; 303 sound-rec; 159 digital data carriers
libr loan; BBS, SAB/CLP 27225

CFP – Technique, Centre de documentation, Av Louis-Bertrand 38, 1213 ***Petit-Lancy***
T: +41 223888805; Fax: +41 223888808; E-mail: olivier.rod@edu.ge.ch; URL: ge.ch/edufloraopac
1983; Olivier Rod
8 000 vols; 100 curr per; 300 av-mat; 100 sound-rec; 100 digital data carriers; Technical standards, technical documentation 27226

HSR Hochschule für Technik Rapperswil, Bibliothek, Oberseestr. 10, 8640 ***Rapperswil***
T: +41 552224531; Fax: +41 552224539; E-mail: bibliothek@hsr.ch; URL: www.hsr.ch
1972; Elisabeth Müller
Bauingenieurwesen, Elektrotechnik, Informatik, Landschaftsarchitektur, Maschinenbau, Raumplanung; Archive of "Stiftung für Landschaftsarchitektur"
30 000 vols; 160 curr per; 4 000 e-journals; 300 maps; 150 av-mat; 30 digital data carriers
libr loan; BIS 27227

Bildungszentrum Zofingen, Mediothek, Strengelbacherstr. 27, 4800 ***Zofingen***
T: +41 627455522; E-mail: mediothek.bzz@ag.ch
1977; Rosemarie Hess-Sarasin
32 527 vols; 87 curr per; 5 music scores; 438 maps; 2 198 av-mat; 1 230 sound-rec; 275 digital data carriers 27228

Government Libraries

Obergerichtsbibliothek, Kanton Aargau, Obere Vorstadt 38, 5000 ***Aarau***
T: +41 628353849; Fax: +41 628353949; URL: www.ag.ch/justizmanagement/de/pub/justizverwaltung/bibliothek.php
vor 1850; Eva Prieto
20 000 vols; 54 curr per 27229

Bundesamt für Sozialversicherungen, Dokumentation, Effingerstr. 20, 3003 ***Bern***
T: +41 313229208; Fax: +41 313227880; URL: www.bsv.admin.ch
1936; N.N.
Sozialrecht, Soziale Sicherheit, Sozialversicherung, Versicherungsrecht, öff Gesundheitswesen, Arbeitsrecht, Statistik; Family, generations, society questions
13 000 vols; 250 curr per; 10 e-journals; 115 000 govt docs
libr loan; BIS 27230

Bundesamt für Strassen (ASTRA), Bibliothek, Mühlestrasse 2, Ittigen, 3003 ***Bern***; mail address: 3003 Bern
Fax: +41 313232303; E-mail: info@astra.admin.ch; URL: www.astra.admin.ch
1974; Daniel Wüthrich
Straßenverkehrsrecht
80 000 vols; 250 curr per; 350 diss/theses; 20 000 govt docs 27231

Dokumentationsdienst der Bundesversammlung, Parlamentsbibliothek (Federal Assembly Documentation Service, Parliament Library), Bundeshaus West, 3003 *Bern*
T: +41 31 3223789; Fax: +41 31 3228297; E-mail: biblio@parl.admin.ch; URL: www.parlament.ch
2009; Jean-Claude Hayoz
Parlamentarismus, Recht, Internationales Recht, Internationale Politik, Staatspolitik, Europapolitik
111 320 vols; 500 curr per; 205 e-journals; 30 000 govt docs; 200 maps; electronic press archive (350 000 press clippings)
libr loan; IFLA 27232

Dokumentationsstelle für Wissenschaftspolitik, Bibliothek, Inselgasse 1, 3003 *Bern*
T: +41 313226856; Fax: +41 313239547; E-mail: hans-peter.jaun@swtr.admin.ch; URL: www.swtr.ch
1979; Hans-Peter Jaun
15 360 vols; 87 curr per; 50 e-journals; 51 digital data carriers 27233

Eidgenössisches Departement für Auswärtige Angelegenheiten, Generalsekretariat, Bibliothek, Bundeshaus West, 3003 *Bern*
T: +41 313223271; Fax: +41 313223237; URL: www.eda.admin.ch
1948; Martin Wyss
Public international law
80 000 vols; 270 curr per
libr loan; BIS 27234

Service des archives et de la documentation de la République et Canton du Jura, Rue du 24 septembre 2, 2800 *Delémont*
T: +41 324205042; Fax: +41 324205041; E-mail: jean-remy.chalverat@jura.ch; URL: www.jura.ch
1979; J.-R. Chalverat
12 000 vols; 230 curr per; 3 000 press files
AAS, BIS 27235

Tribunal fédéral / Schweizerisches Bundesgericht, Bibliothèque, Av du Tribunal-fédéral 29, 1000 *Lausanne* 14
T: +41 21 3189438;
Fax: +41 21 3233700; E-mail: bibliothek@bger.admin.ch; URL: www.bger.ch
1875; Michèle Lewis
70 000 vols; 600 curr per; 20 e-journals; 33 digital data carriers
BIS, ABJS 27236

Bundesamt für Gesundheit, Bibliothek, Schwarzenburgstr. 165, 3097 *Liebefeld*; mail address: 3003 Bern
T: +41 313248183; Fax: +41 313229507; E-mail: bibliothek@bag.admin.ch; URL: www.bag.admin.ch/dokumentation/zahlen/index.html?lang=de
1893; Robin Dutt
17 000 vols; 400 curr per; 60 e-journals; 43 digital data carriers
libr loan; BIS 27237

Kanton Basel-Landschaft, Archäologie und Museum, Bibliothek, Amtshausgasse 7, 4410 *Liestal*
T: +41 615526221; Fax: +41 615526960; E-mail: bettina.hunn@bl.ch; URL: www.kbl.ch
1837; Bettina Hunn
Mittelalterarchäologie, Museologie, Volkskunde
10 000 vols; 120 curr per
libr loan 27238

Bundesgericht, Bibliothek, Schweizerhofquai 6, 6004 *Luzern*
T: +41 414193648; Fax: +41 414193684; E-mail: kerstin.reiher@bger.admin.ch; URL: www.bger.ch
1918; Kerstin Reiher
Sozialversicherungen, Sozialrecht
15 500 vols; 120 curr per; 2 500 diss/theses; 1 000 govt docs 27239

Obergericht des Kantons Luzern, Bibliothek, Hirschengraben 16, 6002 *Luzern*
T: +41 412286262; Fax: +41 412286264; E-mail: Franz.Felder@lu.ch
1848; Ruth Aregger
10 000 vols 27240

Archives de l'Etat, 12, rue de la Collégiale, 2001 *Neuchâtel*
T: +41 328896040; Fax: +41 328896088; E-mail: OAEN@ne.ch
1898; Alexandre Dafflon
Hist, law, administration
20 000 vols
AAS, BIS 27241

Obergericht des Kantons Solothurn, Bibliothek, Amthaus 1, 4502 *Solothurn*
T: +41 313217325; E-mail: Erich.Wyss@bd.so.ch
Erich Wyss
12 000 vols
libr loan 27242

Obergericht des Kantons Zürich, Bibliothek, Hirschengraben 13/15, 8001 *Zürich*; Postfach 2401, 8021 Zürich
T: +41 442579191; Fax: +41 442579311; E-mail: andreas.keller@gerichte-zh.ch; URL: www.obergericht-zh.ch/
1867; Andi Keller
60 000 vols; 110 curr per; 8 000 diss/theses
BIS 27243

Ecclesiastical Libraries

Bibliothek im Basler Missionshaus, Missionsstr. 21, 4003 *Basel*
T: +41 612602241; Fax: +41 612602122; E-mail: marcus.buess@mission-21.org; URL: www.mission-21.org
1815; Marcus Buess
Kontextuelle Theologie, Missionswiss, Ethnologie und Kolonialgeschichte; Sondersammelgebiete: Pietismus u hist Bibeln; Archive
50 000 vols; 300 curr per; 200 diss/theses; 30 microforms; 200 av-mat; 1 100 hist journals, 1 500 hist maps, 40 000 photos, 200 videos 27244

Israelitische Gemeinde Basel, Karger Bibliothek, Leimenstr. 24, 4051 *Basel*
T: +41 612729850; Fax: +41 612799851; E-mail: edouard.selig@unibas.ch; URL: www.igb.ch
1906; Edouard Selig
30 000 vols; 10 curr per; 380 av-mat; 540 videos + DVDs 27245

Katholische Universitätsgemeinde, Bibliothek, Herbergsgasse 7, 4051 *Basel*
T: +41 612646363; Fax: +41 612646364; E-mail: kug@unibas.ch; URL: www.studentenhaus.ch
1931; Bruno Brantschen SI
15 000 vols; 54 curr per 27246

Stiftsbibliothek Beromünster, Stift 1, 6215 *Beromünster*
T: +41 9303277; E-mail: bibliothek@stiftberomuenster.ch
1820; Jakob Bernet
Beronensia, Lucernensia; Journals from the 19. centry
5 000 vols; 3 curr per; 30 mss; 110 incunabula 27247

Pilgermission St. Chrischona, Bibliothek, Chrischonarain 200, 4126 *Bettingen*
T: +41 616464435; Fax: +41 616464277; E-mail: gottfried.burger@chrischona.ch; URL: bibliothek.chrischona.ch
1840; Gottfried Burger
Protestantismus, Pietismus, Evangelikale Theologie, Freikirchen; Catechetics (ca 150 AV mat)
75 000 vols; 430 curr per; 60 music scores; 5 microforms; 160 av-mat; 30 sound-rec; 120 digital data carriers
BIS, VkwB, NEDS 27248

Convento Cappucini, Biblioteca, 6951 *Bigorio*
T: +41 91 94312-22/-23; Fax: +41 91 9434665; E-mail: bigorio@cappuccini.ch; URL: www.bigorio.ch
1535
8 000 vols 27249

CEDOFOR – Centre de documentation et de formation religieuses, Bibliothèque, 18, rue Jacques-Dalphin, 1227 *Carouge* /GE
T: +41 228274678, 8274686; Fax: +41 228274670; E-mail: bibliotheque@choisir.ch; URL: www.cedofor.ch
1975; Stjepan Kusar
Spiritualität, Theologie, Patrologie,

Religion, Literatur (Suisse Romande), Ältere und neuere theol. Encyclopädien und Lexika; Spezialsammlungen: Literatur über Gesellschaft Jesu (MHSI etc.), Migne (PG, PL), Mansi, Acta Syn. Conc Vaticani II, Bibliotheca Theubneriana, Realencyklopädie der classischen Altertumswissenschaft
50 000 vols; 141 curr per; 20 diss/theses
libr loan 27250

Centre orthodoxe du patriarcat œcuménique, Bibliothèque, Chemin de Chambésy 37, 1292 *Chambésy*
T: +41 227589860; Fax: +41 227589861; URL: www.centreorthodoxe.org/index.php
1970
15 000 vols; 500 curr per 27251

Bischöfliche Bibliothek Chur, Hof 19, 7000 *Chur*
T: +41 812586040; Fax: +41 812586001; E-mail: fischer@bistum-chur.ch; URL: www.bistum-chur.ch
Dr. Albert Fischer
Bistum Chur/Raetica; Romance libr (with Rhaeto-Romanic lit) 27252

Klosterbibliothek Disentis, Kloster, 7180 *Disentis*; Postfach 74, 7180 Disentis
T: +41 819296927; Fax: +41 819296901; E-mail: p.theo@kloster-disentis.ch
750/1880; P. Theo Theiler
Rätoromanica; Bibliotheca Romontscha, music libr
100 000 vols; 100 curr per; 5 000 music scores; 2 500 sound-rec
BIS 27253

Stiftsbibliothek Benediktinerabtei Einsiedeln, Kloster, 8840 *Einsiedeln*; Postfach, 8840 Einsiedeln
T: +41 554186314; Fax: +41 554186112; E-mail: kloster@kloster-einsiedeln.ch; URL: www.kloster-einsiedeln.ch
934; Prof. P. Odo Lang
Grafiken; Mss dept; music libr
230 000 vols; 132 curr per; 1 200 mss; 500 incunabula; 50 000 music scores; 500 microforms; 530 sound-rec
libr loan 27254

Stiftsbibliothek Benediktinerkloster Engelberg, Benediktinerkloster, 6390 *Engelberg*
T: +41 416396161; Fax: +41 416396113; E-mail: stiftsbibliothek@kloster-engelberg.ch; URL: www.kloster-engelberg.ch/kultur/stiftsbibliothek.htm
1120; P. Guido Muff
115 000 vols; 200 curr per; 1 000 mss; 360 incunabula 27255

Zisterzienserinnenabtei Eschenbach, Bibliothek, Freiherrenweg 11, 6274 *Eschenbach*
T: +41 41 4494000; Fax: +41 41 4494001; E-mail: zist.esch@bluewin.ch
Theologie, Cisterciensia
11 000 vols 27256

Convento dei Cappuccini, Biblioteca, Via Cantonale, 6760 *Faido*; CP 1261, 6760 Faido
T: +41 918661282; Fax: +41 918663113; E-mail: faido@cappuccini.ch; URL: www.cappuccini.ch/presenza/faido.htm
10 000 vols; 15 curr per; 20 incunabula 27257

Benediktinergemeinschaft Fischingen, Bibliothek, Kloster, 8376 *Fischingen*
T: +41 71 9787261, 9787273;
Fax: +41 71 9787280; E-mail: benediktiner@klosterfischingen.ch; URL: www.klosterfischingen.ch
1977; P. Benno Schildknecht
Theologie, Spiritualität, Kirchengeschichte und allgemeine Geschichte
29 000 vols 27258

Couvent des Capucins, Bibliothèque, 28, rue de Morat, Case postale 374, 1701 *Fribourg*
T: +41 263472323; Fax: +41 263472358; E-mail: fribourg@capucins.ch
1609; Norbert Sapin
P. Falck-Bibl
30 000 vols; 100 incunabula
libr loan 27259

Couvent des Cordeliers / Franziskanerkloster, Bibliothèque, Rue de Morat 6, CP 16, 1702 *Fribourg* 2
T: +41 263471160; Fax: +41 263471164; E-mail: couvcord-fr.ch@bluewin.ch
between 1300-1500 (1850); P. Supérieur
25 330 titres (sans les ouvrages de la bibliothèque moderne): dont 220 incunables, 660 titres du 16e s, 1 240 titres du 17e s et 2 600 titres du 18e s
25 600 vols; 300 mss; 220 incunabula
non-lending library 27260

Monastère Saint-Benoît de Port-Valais, Bibliothèque, Route de l'Eglise 38, 1897 *Le Bouveret*
T: +41 244812812; Fax: +41 244814398; E-mail: portier@abbaye-saint-benoit.ch; URL: www.abbaye-saint-benoit.ch
1931; François Huot osb
45 000 vols; 15 curr per
libr loan 27261

Biblioteca Salita dei frati, Salita dei Frati 4, 6900 *Lugano*
T: +41 91 9239188; Fax: +41 91 9238987; E-mail: bsf-segr.sbt@ti.ch; URL: www.sbt.ti.ch
1653; Luciana Pedroia
Ticinensia, alte Gebetbücher, ital Sprache u Lit
100 000 vols; 400 curr per; 27 incunabula; 400 microforms
libr loan; LIBER, AIB, BIS 27262

Schweizerische Kapuzinerprovinz, Bibliothek, Wesemlinstr. 42, 6006 *Luzern*
T: +41 414296755; Fax: +41 414296750; E-mail: Luzern@Kapuziner.org
1583; Oktavian Schmucki, Karl Flury
38 000 vols; 60 curr per; 439 incunabula; 200 maps; 87 vols of the early library from 1584, diss
libr loan 27263

Seminar St. Beat, Priesterseminar der Diözese Basel, Bibliothek, Adligenswilerstr. 15, 6006 *Luzern*
T: +41 414199191; Fax: +41 414199171; E-mail: priesterseminar@stbeat.ch
1878; Josef-Anton Willa
Exegese, Philosophie, Systematische Theologie, Kirchengeschichte, Liturgik, Pastoraltheologie, Katechetik, Spiritualität
20 000 vols; 34 curr per
BibliothekarInnen theologischer Institutionen in der Schweiz 27264

Benediktinerkloster Mariastein, Archiv und Bibliothek, Klosterplatz 1, 4115 *Mariastein*
T: +41 617351111; Fax: +41 617351103; E-mail: bibliothek@kloster-mariastein.ch; URL: www.kloster-mariastein.ch
ca 1100; Kilian Karrer
100 000 vols; 50 curr per; 40 mss; 60 incunabula
VSA 27265

Bibliothèque des Pasteurs, Fbg de l'Hôpital 41, 2000 *Neuchâtel*
T: +41 327181990; Fax: +41 327181901; E-mail: cecilia.griener@unine.ch; URL: www2.unine.ch/biblio/page3850.html
1538; Cecilia Griener Hurley
80 000 vols; 73 curr per; 4 000 mss; 10 incunabula; 300 diss/theses
RERO 27266

Kapuzinerkloster, Bibliothek, Postfach 1438, 8640 *Rapperswil*
T: +41 552205310; Fax: +41 552205303; E-mail: rapperswil@kapuziner.org; URL: www.klosterrapperswil.ch
1602; Br. Paul Meier
20 000 vols 27267

Séminaire international Saint Pie X, Bibliothèque, Ecône, 1908 *Riddes*
T: +41 273051080; Fax: +41 277443319; E-mail: info@seminaire-econe.ch; URL: www.seminaire-econe.ch
1970
Théologie catholique
40 000 vols; 60 curr per 27268

Couvent des Capucins, Bibliothèque, Rue Saint-François 10, 1890 *Saint-Maurice*
E-mail: st-maurice@capucins.ch; URL: www.capucins.ch
Théologie catholique, spécialement recueils de sermons et ascétisme
8 000 vols 27269

Benediktiner-Kollegium, Professorenheimbibliothek, Brünigstr. 177, 6060 *Sarnen*
T: +41 416665462; Fax: +41 416665469; E-mail: bibliothek@muri-gries.ch
1841; Beda Szukics
60 000 vols; 30 curr per; 120 mss; 45 incunabula
Arbeitsgemeinschaft Schweizerischer Stiftsbibliothekare   27270

Stiftsarchiv St.Gallen, Regierungsgebäude, 9001 *St.Gallen*
T: +41 582293823; Fax: +41 582293815; E-mail: info.stiftsarchiv@sg.ch; URL: www.stiftsarchiv.sg.ch
770
Handschriftenbibl der ehem Fürstabtei Pfäfers, Archiv des ehem adligen Damenstiftes Schänis, Nachlass Abt Pankraz Vorster
10 000 vols; 30 curr per; 3 000 mss   27271

Stiftsbibliothek St.Gallen, Klosterhof 6d, 9000 *St.Gallen*; Postfach, 9004 St.Gallen
T: +41 712273417; Fax: +41 712273418; E-mail: stibi@stibi.ch; URL: www.stiftsbibliothek.ch
719; Prof. Ernst Tremp
Mediävistik
170 000 vols; 225 curr per; 2 107 mss; 1 000 incunabula; 8 000 microforms; 2 000 av-mat; 100 sound-rec; 200 digital data carriers
libr loan; LIBER, BIS, AIBM   27272

Abbaye de Saint-Maurice, Bibliothèque, Avenue d'Agaune 15, CP 34, 1890 *St.-Maurice*
T: +41 244860404; Fax: +41 244860405; E-mail: biblio@stmaurice.ch; URL: www.amatus.ch
1694; Olivier Roduit
Théologie catholique, histoire, littérature, droit, sciences, philosophie
100 000 vols; 160 curr per; 15 mss; 31 incunabula   27273

Benediktinerinnenkloster Fahr (AG), Bibliothek, 8103 *Unterengstringen*
T: +41 17500753; Fax: +41 17400792
Mystik, Aszetik, hagiografische Literatur; Altbestand an Druckschrifften: 7 aus dem 16., 123 aus dem 17., 527 aus dem 18. und 846 aus dem 19. Jh
6 000 vols
non-lending library   27274

Benediktinerabtei St. Otmarsberg, Bibliothek, Postfach 135, 8730 *Uznach*
T: +41 55 2858111; Fax: +41 55 2858100; E-mail: abt.ivo@otmarsberg.ch; URL: www.abtei-uznach.ch
1919/1963; Dr. Ivo Auf der Maur
Missionaria, Ascetica, Exegetica, Monastica; Slg alter Drucke (456 Bde)
74 850 vols; 284 curr per; 1 649 e-journals; 300 diss/theses; 3 govt docs; 620 maps; 9 microforms; 6 sound-rec; 7 digital data carriers   27275

Bibliothèque du Centre diocésain (BCD) / Bibliothek des Diözesan-Zentrums (BDZ), Chemin Cardinal-Journet 3, 1752 *Villars-sur-Glâne*
T: +41 264263390; E-mail: bibliotheque.bcd@bluewin.ch
1795; Marie-Louise Zurkinden
Libr of Abbé Maurice Zundel
70 000 vols; 21 curr per
BIS   27276

Dominikanerinnenkloster Maria Zuflucht, Bibliothek, Im Städtli 29, 8872 *Weesen*
T: +41 55 6161625
Aszetik, Mystik und Liturgie
8 400 vols
non-lending Library   27277

Dominikanerinnenkloster St. Katharina, Bibliothek, Klosterweg 7, 9500 *Wil*
T: +41 71 9114647; Fax: +41 71 9110585; URL: www.klosterwil.ch
Aszetik, Geschichte, Pädagogik
18 000 vols   27278

Evangelisch-reformierte Landeskirche des Kantons Zürich, Bibliothek h50, Hirschengraben 50, 8001 *Zürich*
T: +41 442589273; Fax: +41 442589141; E-mail: bibliothek@zh.ref.ch; URL: www.bibliothekenderkirchen.ch

1998; Susanne Fretz
Religionspädagogik, Evangelische Kirchenmusik, Hymnologie, Liturgik; Hymn-book coll, picture book coll
12 000 vols; 70 curr per; 5 600 music scores; 250 av-mat; 220 sound-rec; 50 digital data carriers; games, objects   27279

Israelitische Cultusgemeinde Zürich (ICZ), Bibliothek, Lavaterstr. 29, 8002 *Zürich*; Postfach 282, 8027 Zürich
T: +41 442832250; Fax: +41 442832223; E-mail: bibliothek@icz.org; URL: www.icz.org
1939; Dr. Yvonne Domhardt
Hebräisch, Jiddisch, Judaistik, Judaica, Hebraica; Texte der Bibliothek des Breslauer Rabbinerseminars, Sonderkatalog (ab 2012 digital auf www.icz.org) zum "Schweizer Judentum (= Auswertung der Schweizer jüdischen Presse ab 1901ff)"
50 000 vols; 60 curr per; 450 diss/theses; 360 music scores; 600 av-mat; 130 sound-rec; 50 digital data carriers; Musical and personal bequest of Max Ettinger (1874-1951)
libr loan; BIS, IG WBS, VEBTIS   27280

Katholische Hochschulgemeinde aki, Bibliothek, Hirschengraben 86, 8001 *Zürich*
T: +41 442545463; Fax: +41 442545465; E-mail: bibliothek@aki-zh.ch
1920
Theologie, Kirchengeschichte
20 000 vols; 50 curr per
libr loan   27281

Corporate, Business Libraries

Basler Zeitung (BaZ), Information und Dokumentation, Aeschenplatz 7, Postfach, 4002 *Basel*
T: +41 616391422; Fax: +41 616310061; E-mail: doku@baz.ch; URL: www.baz.ch
1983; Roger Berger
Personen, Firmen, Länder, Sport, Wirtschaft, Basiliensa
8 821 vols; 10 curr per; 146 maps; 500 microforms; 60 digital data carriers; 2 mio press clippings, 500 000 electronic archived articles
VSA, BIS   27282

Lonza AG, Bibliothek, Münchensteinerstr. 38, 4002 *Basel*
T: +41 613168609; Fax: +41 613168733; E-mail: media@lonza.com; URL: www.lonza.com
1942; A. Hufschmid
11 000 vols; 500 curr per
libr loan   27283

CSL Behring AG, Bibliothek, Wankdorfstr. 10, 3000 *Bern* 22
T: +41 313445205; Fax: +41 313445555; E-mail: nadja.plavec@cslbehring.com; URL: www.cslbehring.com
1948; N. Plavec
Immunologie, Hämatologie, Transfusionsmedizin
10 000 vols; 97 curr per; 2 000 e-journals; 150 diss/theses
libr loan; BIS   27284

Ems-Chemie AG, Information und Dokumentation, Via Innovativa 1, 7013 *Domat* (Ems)
T: +41 816326485; Fax: +41 816327418; E-mail: welcome@emschem.com
1942; M. Maschek
18 000 vols; 275 curr per; 140 diss/theses; 4 400 microforms; 170 000 patent specifications
BIS   27285

Givaudan Schweiz AG, Bibliothek, Überlandstr. 138, 8600 *Dübendorf*
T: +41 18242241; Fax: +41 18214478
1970; Andreas Muheim
Chemie, Lebensmittelwiss
15 000 vols; 50 curr per; 1 500 e-journals; 200 diss/theses; 20 govt docs
BIS   27286

Leica Geosystems AG, Fachbibliothek, Heinrich-Wild-Strasse, 9435 *Heerbrugg*
T: +41 717273646; Fax: +41 717274719; URL: www.leica-geosystems.com
1924; Kristin Köppel
Luftbilder, Physik, Chemie
8 000 vols; 270 curr per; 2 500 slides, 18 000 patents
libr loan   27287

Nestlé Research Center, Bibliothèque Technique CRN, Vers-chez-les-Blanc, 1000 *Lausanne*; BP 44, 1000 Lausanne 26
T: +41 217858448; Fax: +41 217858552; E-mail: olivier.jordan@rdls.nestle.com
1967; Olivier Jordan
Food, Nutrition, Food Technology, Life Sciences
30 000 vols; 600 curr per   27288

Suva, Bibliothek Rösslimatt, Fluhmattstr. 1, Postfach, 6002 *Luzern*
T: +41 414195085; Fax: +41 414195958; URL: www.suva.ch
1918; Marianne Gallauer
Arbeitssicherheit, Gesundheitsschutz
12 000 vols; 650 curr per; 36 e-journals; 630 av-mat; 390 digital data carriers
BIS   27289

CILAG AG, Bibliothek, Hochstr. 201, 8205 *Schaffhausen*
T: +41 526309791; Fax: +41 526308090; E-mail: cilag@its.jnj.com; URL: www.cilag.ch
Chemie, Galenik, Medizin
8 000 vols; 170 curr per   27290

Oekopack AG, Lattigen, 3700 *Spiez*
T: +41 336546606; Fax: +41 336542889; E-mail: info@oekopack.ch; URL: www.oekopack.ch   27291

Syngenta Crop Protection Munchwilen AG, The Agribusiness Library, WST-810, 4332 *Stein*
T: +41 613232818; Fax: +41 613237559; URL: www.syngenta.com
M. Brodmann
Chemie (allg), Organ Chemie, Agro, Biologie
10 000 vols; 80 curr per
ASpB
Use only for members of staff   27292

Lonza AG, Bibliothek, Rottenstr. 6, 3930 *Visp*
T: +41 279485302; Fax: +41 279475302; URL: www.lonza.com
1946; Denise Rohr
Chemie
17 000 vols; 400 curr per; 130 e-journals; 30 e-books; 150 diss/theses; 25 microforms; 50 digital data carriers
BIS   27293

Rentenanstalt / Swiss Life, Bibliothek, General Guisan-Quai 40, 8002 *Zürich*
T: +41 442843378; Fax: +41 447116990; URL: www.swisslife.ch
1939; S. Ghelfi
26 000 vols; 460 curr per
libr loan; BIS   27294

Schweizerische Nationalbank, Bibliothek, Fraumünsterstr. 8, 8001 *Zürich*; mail address: 8022 Zürich
T: +41 446313284; Fax: +41 446318114; E-mail: library@snb.ch; URL: www.snb.ch
1907; Agnes Simonin
30 000 vols; 1 200 curr per   27295

Schweizerische Rückversicherungs-Gesellschaft, Knowledge & Information Management, Bibliothek, Mythenquai 50-60, 8022 *Zürich*; Postfach, 8022 Zürich
T: +41 432853025; Fax: +41 432854177; URL: www.swissre.com
1863
Versicherung und Rückversicherung
46 000 vols; 3 000 curr per; 100 maps; 500 av-mat; 50 digital data carriers
BIS   27296

Special Libraries Maintained by Other Institutions

Schweizerische Koordinationsstelle für Bildungsforschung (SKBF) / Centre Suisse de Coordination pour la Recherche en Education (CSRE), Bibliothek (Swiss Coordination Centre

for Educational Research), Francke-Gut, Entfelderstr. 61, 5000 *Aarau*
T: +41 628352390; Fax: +41 628352399; URL: www.skbf-csre.ch
1971; Peter Meyer
Erziehungswissenschaft, Psychologie, Soziologie, Bildungsforschung
10 000 vols; 100 curr per; 1 e-journals; 300 diss/theses; 10 av-mat; 10 digital data carriers; 5 000 non-trade publications   27297

Site et musée romain d'Avenches, Bibliothèque (Library of the Avenches Roman Museum), Av. Jomini 16, 1580 *Avenches*; CP 237, 1580 Avenches
+41 26 5573311; Fax: +41 26 5573313; E-mail: musee.romain@vd.ch; URL: www.aventicum.org
1964; Acacio Calisto
20 000 vols; 500 curr per; 100 mss; 100 diss/theses; 100 000 av-mat; 500 sound-rec
libr loan; BIS   27298

Antikenmuseum Basel und Sammlung Ludwig, Bibliothek, St. Albangraben 5, Postfach, 4010 *Basel*
T: +41 612011212; Fax: +41 612011210; E-mail: office@antikenmuseumbasel.ch; URL: www.antikenmuseumbasel.ch
1961; Ella van der Meijden
Ägyptische, griechische, etruskische, römische, byzantinische Kunst, Archäologie, Geschichte
30 200 vols   27299

Archäologie Schweiz, Bibliothek, Petersgraben 51, 4051 *Basel*; Postfach 116, 4003 Basel
T: +41 612613078; Fax: +41 612613076; E-mail: biblio@archaeologie-schweiz.ch; URL: www.archaeologie-schweiz.ch
1907; Dr. Urs Niffeler
16 600 vols; 300 curr per; 300 maps; 35 av-mat; 6 000 brochures, 5 000 photos   27300

Basler Afrika Bibliographien, Namibia Resource Centre – Southern Africa Library, Klosterberg 23, 4051 *Basel*; Postfach 2037, 4001 Basel
T: +41 612289333; Fax: +41 612289330; E-mail: bab@baslerafrika.ch; URL: www.baslerafrika.ch
1970; Luccio Schlettwein
Sammelschwerpunkt Namibia/südliches Afrika, Africana 16.-19. Jh, Bibliografien Afrika, Schweizer in Afrika; Hans. W. Debrunner Coll
34 000 vols; 200 curr per; 200 mss; 600 diss/theses; 1 000 govt docs; 1 500 maps; 300 av-mat; 200 sound-rec; 130 journals, brochures, press clippings, 100 000 hist photos, 3 000 posters   27301

Basler Papiermühle, Schweizerisches Museum für Papier, Schrift und Druck, Bibliothek, St. Alban-Tal 37, 4052 *Basel*
T: +41 612259090; Fax: +41 612720993; E-mail: info@papiermuseum.ch; URL: www.papiermuseum.ch
1980; Martin Kluge
Geschichte d Naturwissenschaften, Technikgeschichte, Wissenschaftsgeschichte, Drucktechnik, Papier, Papiertechnik, Schriftgeschichte, Typografie
8 000 vols; 80 maps; 30 av-mat; 500 photos   27302

Bibliothek für Gestaltung Basel, Spalenvorstadt 2, Postfach 612, 4003 *Basel*
T: +41 612674500; Fax: +41 612674503; E-mail: andrea.schweiger@sfgbasel.ch; URL: www.sfgbasel.ch
1878; Andrea Schweiger
Architektur, Städtebau und Landschaftsarchitektur, Produkte- und Industriedesign, Fotografie, Kunsterziehung, Kunstgewerbe, Mode und Textildesign, Schmuck, Schrift- und Buchkunst, Video, Visuelle Kommunikation, Werbegrafik, zeitgenössische Kunst Stiftung Gartenbaubibliothek, Architekturtraktate 16.-20. Jh, Vorlagebücher 16.-19. Jh, Typografische Sammlung von Jan Tschichold, Handeinbände von Ignatz Wiemeler aus der Sammlung Richard Doetsch-Benziger, Nachlass Paul Hulliger zur Entwicklung der Schweizerischen Schulschrift, Schriftmeister-Bände und -Vorlagen, Entwicklungsgeschichte von

Schrift und Buch
120 000 vols; 211 curr per; 51 e-journals;
2 500 govt docs; 10 000 av-mat; 1 000
digital data carriers; 25 000 photos,
15 000 art prints
libr loan; AKMB, BIS, SIGEGS, VMS
27303

Historisches Museum Basel, Bibliothek,
Steinenberg 4, 4051 *Basel*
T: +41 61 2058608; Fax: +41 61
2058601; E-mail: bibliothek@hmb.ch;
URL: www.hmb.ch
1854; Daniel Suter
Instrumentenkunde, Musikikonografie,
Kutschen u. Schlitten, Numismatik
24 000 vols; 40 curr per; auction
catalogues
BIS
27304

**Interkulturelle Bibliothek für Kinder
und Jugendliche (JUKIBU)**,
Elsässerstr. 7, 4056 *Basel*
T: +41 613226319; Fax: +41 613225320;
URL: www.jukibu.ch
1991; Maureen Senn-Carroll
Kinder- und Jugendbücher in über 50
Sprachen
21 000 vols; 200 av-mat; 300 sound-rec;
250 digital data carriers
SAB
27305

Kunsthalle Basel, Basler Kunstverein,
Bibliothek, Klostergasse 5, 4051 *Basel*
T: +41 612069906; Fax: +41 612069919;
E-mail: bibliothek@kunsthallebasel.ch;
URL: www.Kunsthallebasel.ch
1839; Heidrun Ziems
30 000 vols; 34 curr per
AKMB
27306

Kunstmuseum Basel, Bibliothek, St.
Alban-Graben 10, 4010 *Basel*
T: +41 612066270; Fax: +41
612066378; E-mail: Kunstmuseum-
Bibliothek@unibas.ch; URL:
www.kunstmuseumbasel.ch
1849; Rainer Baum
Geschichte der Malerei, Grafik und
Plastik seit Karl dem Großen bis zur
Gegenwart, besonders zur deutschen und
oberrheinischen Kunst des 15.-16. Jh und
zur Klassischen Moderne
163 000 vols; 240 curr per; 50 000
photos
libr loan; BIS
27307

**Museum der Kulturen und
Ethnologisches Seminar**, Bibliothek,
Münsterplatz 20 (Eingang: Schlüsselberg
17/Münsterplatz 19), 4001 *Basel*;
Postfach, 4001 Basel
T: +41 612665630; Fax: +41 612665605;
E-mail: mkb.biblio@bs.ch; URL:
www.mkb.ch/bib
1893; Elilsabeth Idris
Melanesien, Indonesien, Südamerika,
Textilien
80 000 vols; 400 curr per; 7 e-
journals; 80 microforms; 78 av-mat;
550 sound-rec; 291 digital data carriers;
6 500 offprints, 6 000 exhibition cat
libr loan; BIS
27308

Naturhistorisches Museum, Rütimeyer
Bibliothek, Augustinergasse 2, 4001
Basel
T: +41 612665500; Fax: +41 612665546;
E-mail: nmb@bs.ch; URL: www.nmb.bs.ch
1896; Christof Scherler
Acarologie, Anthropologie, Entomologie,
Geologie, Mineralogie, Osteologie,
Paläontologie, Zoologie
140 000 vols; 7 000 curr per; 5 000 maps
libr loan; BIS
27309

Paul Sacher Stiftung, Archiv und
Bibliothek, Münsterplatz 4, 4051 *Basel*
T: +41 612696644; Fax: +41 612619183;
E-mail: office-pss@unibas.ch; URL:
www.paul-sacher-stiftung.ch
1973
100 Nachlässe von Komponisten des 20.
Jh
35 000 vols; 100 curr per; 50 000 mss;
500 diss/theses; 12 000 music scores;
500 microforms; 3 000 av-mat; 5 000
sound-rec; 5 000 photos, 5 000 press
clippings, 1 000 catalogs
AIBM, BIS
27310

**Schweizerisches Institut für
Volkskunde**, Bibliothek, Spalenvorstadt
2, 4051 *Basel*; Postfach, 4001 Basel
T: +41 612671163; Fax: +41 612671163;
E-mail: sgv-sstp@volkskunde.ch; URL:
www.volkskunde.ch; katalog über
www.ub.unibas.ch
1937; Ernst J. Huber
Volkskunde, Volksmusik, Volkslied,
Bauernhaus, Schweiz; Volkslieder
Slg; Swiss folk song archives, center
for farmhouse research (c/o Kant,
Denkmalpflege, Sumpfstr 23, 6300 Zug,
www.bauernhausforschung.ch)
55 500 vols; 280 curr per; 650
diss/theses; 140 av-mat; 500 sound-rec
libr loan
Libr cat retrievable via:
aleph.unibas.ch/ALEPH
27311

**Schweizerisches Tropen- und
Public Health-Institut**, Bibliothek und
Information (Swiss Tropical and Pubic
Health Institute), Socinstr. 57, 4051
Basel; Socinstr. 57, Postfach, 4002
Basel
T: +41 612848223; Fax: +41 612848150;
E-mail: library-sti@unibas.ch; URL:
www.swisstph.ch
1946; Giovanni Casagrande
Nationale und Internationale Gesundheit,
Tropen- und Reisemedizin, Parasitologie,
Entwicklungszusammenarbeit,
Epidemiologie chronischer
Erkrankungen, Umweltepidemiologie und
Risikoabschätzung, Umweltexposition,
Gender und Gesundheit
8 494 vols; 88 curr per; 475 diss/theses;
300 maps; 14 500 av-mat; 228 digital
data carriers; 413 diploma works, and
Master theses, 1 600 photos, 12 000
slides
libr loan; BIS, EAHIL
27312

**Schweizerisches Wirtschaftsarchiv und
WWZ-Bibliothek**, Peter Merian-Weg 6,
Postfach, 4002 *Basel*
T: +41 612673220; Fax: +41 612673208;
E-mail: info-wwzb@unibas.ch; URL:
www.ub.unibas.ch/wwz-bibliothek-swa
1910; Irene Amstutz
Dokumentation zur schweizerischen
Wirtschaft, Wirtschaftspolitik und
Wirtschaftsgeschichte; Firmen- und
Verbandsarchive
260 000 vols; 4 700 curr per; 4 000
e-journals; 450 mss; 110 digital data
carriers; 2,3 mio press clippings, 3 800
current company rpts, 560 000 pamphlets
libr loan; ICA, BIS, VSA
27313

Staatsarchiv des Kantons Basel-Stadt,
Bibliothek, Martinsgasse 2, 4001 *Basel*
T: +41 612678601; Fax: +41
612676571; E-mail: stabs@bs.ch; URL:
www.staatsarchiv.bs.ch
1899; Dr. Hermann Wichers
Leichenreden (Nekrologe) 6 000
28 000 vols; 50 curr per; 300 diss/theses;
25 govt docs; 50 digital data carriers; ca
100 000 biographical and topographical
press clippings
BIS
27314

Archivio di Stato di Bellinzona,
Biblioteca, Viale S Franscini 30a, 6500
Bellinzona; CP 1070, 6501 Bellinzona
T: +41 918141321; Fax: +41 918141329;
E-mail: decs-asti@ti.ch; URL: www.ti.ch/
archivio
1881
30 000 vols; 354 curr per
27315

Centro didattico cantonale, Biblioteca e
documentazione, Viale Stefano Franscini
32, 6500 *Bellinzona*
T: +41 918146311; Fax: +41 918146319;
E-mail: decs-cdc@ti.ch
1973; Michele Tamagni
20 000 vols; 104 curr per; 2 000
sound-rec; 45 000 slides, 2 000 videos
27316

Alliance Sud, Dokumentation, Monbijoustr.
31, 3011 *Bern*; Postfach 6735, 3001
Bern
T: +41 313909337; E-mail:
dokumentation@alliancesud.ch; URL:
www.alliancesud.ch/de/dokumentation
1972
3 500 vols; 120 curr per; 30 lin mtr of
pressclippings
BIS, NEDS, IVEP
27317

**Bibliothek am Guisanplatz /
Bibliothèque Am Guisanplatz**,
Papiermühlestr. 21A, 3003 *Bern*
T: +41 313224243; Fax: +41 612671163;
E-mail: sue.stammbach@gs-vbs.admin.ch; URL:
www.guisanplatz.ch
1848; Sue Stammbach
Autografenslg Dr Georg Heberlein,
von Tschamer, Schafroth, Keller,
Militärpostkartensammlung;Militärwesen,
Zivilschutz, Katastrophenschutz und -
hilfe, Politik, Geopolitik, Sicherheitspolitik,
Umwelt, Verkehr, Energie, Finanzen und
Wirtschaft, Recht, Länderinformationen,
Kommunikation, Architektur, Bauten,
Logistik, Führungspsychologie
350 000 vols; 1 000 curr per; 15 e-
journals; 1 500 mss; 52 000 diss/theses;
21 000 govt docs; 25 000 maps; 1 800
microforms; 1 000 av-mat; 200 sound-rec;
100 digital data carriers; 7 bequests,
10 000 photos
libr loan; BIS
27318

**Burgerbibliothek Bern / Bibliothèque
de la Bourgeoisie de Berne**,
Münstergasse 63, 3011 *Bern*; Postfach,
3000 Bern 8
T: +41 313203333; Fax: +41 313203370;
E-mail: bbb@burgerbib.ch; URL:
www.burgerbib.ch
1951; Dr. Claudia Engler
Handschriftensammlungen: mittelalterliche
u neuzeitliche Bernensia u Helvetica,
Bongarsiana Codices, Grafik- u Fotoslg,
Fotodokumentation bernischer Porträts u
topografischer Ansichten
30 000 vols; 100 curr per; 2 100 linear
metres of mss
VSA
27319

**Eidgenössisches Büro für die
Gleichstellung von Frau und Mann
EBG**, Dokumentationsstelle (Federal
Office for Gender Equality FOGE),
Schwarztorstr. 51, 3003 *Bern*
T: +41 313229279; Fax: +41 313229281;
E-mail: dokumentation@ebg.admin.ch;
URL: www.gleichstellung-schweiz.ch
1988; Claudia Weilenmann
Genderforschung
8 000 vols; 30 curr per; 150 diss/theses;
1 500 govt docs
27320

Historisches Museum Bern, Abteilung
für Völkerkunde, Bibliothek, Helvetiaplatz
5, 3005 *Bern*
T: +41 313507711; Fax: +41 313507799;
E-mail: dir@bhm.ch; URL: www.bhm.ch
1894; Dr. Thomas Psota
Ältere Reisebeschreibungen, Ethnografie
(Türkei, Malediven, Islam, Indonesien),
Ethnologie
13 000 vols; 75 curr per; 10 000
brochures, 3 000 photos, 3 000 slides
libr loan
27321

Institut für Infektionskrankheiten,
Universität Bern, Bibliothek, Friedbühlstr.
51, Postfach 61, 3010 *Bern*
T: +41 316323256; Fax: +41 316324966;
URL: www.ifik.unibe.ch
1958
10 000 vols; 140 curr per
libr loan
27322

**Kunsthistorische Bibliothek des
Kunstmuseums Bern und des
Instituts für Kunstgeschichte der
Universität Bern**, Hodlerstr. 8, 3011
Bern
T: +41 316314746, 316314748;
E-mail: agatha.rihs@ikg.unibe.ch; URL:
www.e
1879, 1936; Agatha Rihs
130 000 vols; 350 curr per; 50 000 slides
27323

Naturhistorisches Museum, Bibliothek,
Bernastr. 15, 3005 *Bern*
T: +41 313507111; Fax: +41 313507499;
E-mail: youna.zahn@nmbe.ch; URL:
www.nmbe.ch
1832; Youna Zahn
Kynologie
22 000 vols; 500 curr per; 6 000 slides
libr loan; BIS
27324

PHBern, Institut für Bildungsmedien,
Helvetiaplatz 2, 3005 *Bern*
T: +41 31 3092811; Fax: +41 31
3092899; E-mail: info-ibm@phbern.ch;
URL: www.phbern.ch/bildungsmedien
1879; Gerhard Pfander
Unterrichtsmaterialien für die Volksschule,

Schulbücher, Pädagogik; Medien- und
Materialkoffer, Wandbilder
23 500 vols; 40 curr per; 3 400 av-mat;
740 sound-rec; 2 464 digital data carriers
BIS
27325

**Schweizerische Osteuropabibliothek
SOB**, Hallerstrasse 6, 3012 *Bern* 9
T: +41 316314170; Fax: +41 316314170;
E-mail: sob@ub.unibe.ch; URL:
www.ub.unibe.ch/sob
1959; Dr. Christophe v. Werdt
"Rossica Europeana": Russland im
Spiegel westlicher Druckerzeugnisse,
Karten und Grafiken aus dem 16.-19. Jh;
"Davoser Bibliothek": Russische Literatur
aus der Zeit um 1900; Zeitungsarchiv:
wichtigste osteuropäische Zeitungstitel aus
dem Zeitraum zwischen den 1940er und
1990er Jahren
176 000 vols; 230 curr per; 574 e-
journals; 162 mss; 560 maps; 7 655
microforms; 72 av-mat; 61 digital data
carriers
libr loan; ABDOS
27326

Schweizerische Theatersammlung,
Bibliothek, Schanzenstr. 15, 3008 *Bern*;
Postfach, 3001 Bern
T: +41 313015252; Fax: +41 313028525;
E-mail: info@theatersammlung.ch; URL:
www.theatersammlung.ch
1927; Dr. Heidy Greco-Kaufmann
Nachlass Appia
75 000 vols; 140 curr per; 500
diss/theses; 11 000 av-mat; 2 100
sound-rec; 300 digital data carriers;
750 000 press clippings, 20 000
photos/slides, 6 000 stage set sketches
and fancy dress sketches, 750 objects
(models, masks, dolls, paper theatre)
libr loan; SIBMAS, BIS
27327

**Schweizerischer Gewerkschaftsbund
SGB**, Bibliothek und Archiv (Swiss
federation of trade unions), Monbijoustr.
61, 3007 *Bern*; Postfach, 3000 Bern 23
T: +41 313770126; Fax: +41 313770102;
E-mail: biblio@sgb.ch; URL: www.sgb.ch/
1880; Dominique Moser-Brossy
Dokumentation, Archiv
30 000 vols; 290 curr per; 50 diss/theses;
10 sound-rec
BIS
27328

Schweizerisches Alpines Museum,
Bibliothek, Helvetiaplatz 4, 3005 *Bern*
T: +41 31 350 04 40;
Fax: +41 313510751; E-mail:
info@alpinesmuseum.ch; URL:
www.alpinesmuseum.ch
1903
Alpen, Berggebiete,
Alpinismus/Bergsteigen
8 000 vols; 10 curr per; 50 mss; 4 000
maps; 150 000 av-mat; 50 sound-rec
27329

Schweizerisches Bundesarchiv,
Bibliothek, Archivstr. 24, 3003 *Bern*
T: +41 313228989; Fax: +41 313227823;
E-mail: bundesarchiv@bar.admin.ch; URL:
www.admin.ch/bar
1798; Christine Moser
Schweizerischer Bundesstaat, Schweizer
Geschichte, Archivwissenschaft
25 000 vols; 300 curr per; 500
diss/theses; 200 govt docs; 80 000 maps;
20 000 microforms; 10 000 av-mat; 1 000
sound-rec; 200 digital data carriers; 10
mio pages of microforms, 800 hours of
16/35mm films, 10 000 hours of video,
15 000 hours of sound recordings, 40 000
m of archives of the Swiss Federal
Authority
ICA
27330

**Schweizerisches Literaturarchiv
(SLA)**, Bibliothek, c/o Schweizerische
Landesbibliothek, Hallwylstr. 15, 3003
Bern
T: +41 313229258; Fax: +41 313228463;
E-mail: arch.lit@nb.admin.ch; URL:
www.nb.admin.ch/sla/
1990; Dr. Thomas Feitknecht
Schweizer Lit (20.Jh.)
15 000 vols; 100 bequests, 1 000 m of
manuscripts
27331

**Staatsarchiv des Kantons Bern
/ Archives de l'Etat de Berne**,
Bibliothek, Falkenplatz 4, 3012 *Bern*;
Postfach 8424, 3001 Bern
T: +41 316335101; Fax: +41 316335102;
E-mail: info.stab@sta.be.ch; URL:

www.sta.be.ch/site/staatsarchiv
1843; Silvia Bühler
52 000 vols; 170 curr per; 180 govt docs; 30 000 maps; 22 000 m of archive material, 33 400 docs, 300 000 photos
VSA 27332

strasseschweiz, Verband des Strassenverkehrs FRS, Infothek, Mittelstr. 32, 3012 **Bern**
T: +41 313298080; Fax: +41 313298083; E-mail: info@strasseschweiz.ch; URL: www.strasseschweiz.ch
1945; H.J. Bardola
20 000 vols; 200 curr per 27333

Union Postale Universelle, Internationales Bureau des Weltpostvereins, Bibliothek, Weltpoststr. 4, 3000 **Bern** 15
T: +41 313503111; Fax: +41 313503110; E-mail: info@upu.int; URL: www.upu.int
1874; Andreas Kerll
300 000 vols; 180 curr per; 521 av-mat; 500 photos, 120 slides 27334

Institut œcuménique, Château de Bossey, Bibliothèque (Ecumenical Institute), Ch. Chenevière 2, 1279 **Bogis-Bossey**, P.O. Box 1000, 1299 Crans-près-Céligny
T: +41 229607396; Fax: +41 229607379; E-mail: library@wcc-coe.org; URL: library.oikoumene.org
1952; Waldvogel Andreas
35 200 vols; 100 curr per
libr loan; BIS, BETH, ATLA 27335

Fondation "Archivum Helveto-Polonicum", Bibliothèque, Beau chemin 27, 1722 **Bourguillon**
T: +41 263223354; Fax: +41 263251377; E-mail: sygnarskij@fr.ch; URL: www.fr.ch/bcu/bf/bourguillon/0.htm
1981; Jacek Sygnarski, Laurent Emery
Editions clandestines en Pologne, activités des émigrés polonais, relations polono-suisses, littérature polonaise
60 000 vols; 3 250 curr per; 50 mss; 300 diss/theses; 20 music scores; 20 maps; 5 microforms; 1 200 sound-rec; 10 000 photos, 10 000 press clippings, 5 000 different docs
libr loan; ABF 27336

Schweizerischer Bauernverband, Bibliothek, Laurstr. 10, 5201 **Brugg**
T: +41 564625111; Fax: +41 564415348; E-mail: info@sbv-usp.ch; URL: www.sbv-usp.ch
1897; Dr. Michel Gakuba
Agrarpolitik, landwirtschaftl Genossenschaftswesen, Bodenrecht, Raumplanung, Agrarsoziologie, Agrargeschichte, bäuerl Zivilrecht, EU-Integration, Umweltschutz
90 000 vols; 500 curr per; 50 mss; 2 300 diss/theses; 400 govt docs; 10 maps
BIS 27337

Afghanistan Institut und Afghanistan-Museum, Brühlstr. 2, 4416 **Bubendorf**
T: +41 61 9339877; Fax: +41 61 9339878; E-mail: info@afghanistan-institut.ch; URL: www.afghanistan-institut.ch
1975; Paul A. Bucherer-Dietschi
Phototheca Afghanica (Bilddokumentation ab ca 1880 mit ca 25 000 Einheiten), Lit d afghanischen Widerstandes
12 000 vols; 90 curr per; 50 mss; 200 diss/theses; 500 maps; 1 200 microforms; 300 av-mat; 400 autographs, 15 000 press clippings, government publ
libr loan 27338

Musée gruérien et Bibliothèque de Bulle, Rue de la Condémine 25, 1630 **Bulle**, CP 204, 1630 Bulle 1
T: +41 269161010; Fax: +41 269161011; E-mail: info@musee-gruerien.ch; URL: www.musee-gruerien.ch
1917; Isabelle Raboud-Schüle
Tissot-Slg, Bovet-Slg, Reichlen-Slg
60 000 vols; 130 curr per; 600 mss; 10 incunabula; 300 maps
libr loan; BIS 27339

Conservatoire et Jardin botaniques de la Ville de Genève, Bibliothèque, Chemin de l'Impératrice 1, CP 60, 1292 **Chambésy-Genève**
T: +41 224185200; Fax: +41 224185101; E-mail: biblio.cjb@ville-ge.ch; URL: www.ville-ge.ch/cjb/bibliotheque.php
1824; Patrick Perret

Werke aus d Zeit vor Linné (16. u 17. Jh), botanische Folios (18. Jh), Pflanzentaxonomie, Phyto-Ikonothek, Bio-Ikonothek
100 000 vols; 4 144 curr per; 227 e-journals; 60 000 mss; 1 incunabula; 2 200 maps; 140 microforms; 70 digital data carriers; 5 000 portraits 27340

Archives cantonales vaudoises, Rue de la Mouline 32, 1022 **Chavannes-près-Renens**
T: +41 213163711; Fax: +41 213163755; E-mail: info.acv@vd.ch; URL: www.musees-vd.ch/fr/acv
1798; Jean-Luc Wermeille
Nombreuses bibliothèques particulières classés comme fonds d'archives; Genealogy dept, Heraldry dept, Local history dept
28 500 vols; 700 curr per; 2 incunabula; 500 diss/theses; 4 000 govt docs; 300 maps; 5 microforms; 10 av-mat; 20 sound-rec; 7 400 other items
BIS 27341

HUG Hôpitaux universitaires de Genève – Belle-Idée, Bibliothèque centrale, Chemin du Petit-Bel-Air 2, 1225 **Chêne-Bourg**
T: +41 223054022; Fax: +41 223054162; E-mail: biblio-belle-idee@hcuge.ch; URL: biblio-belleidee.hug-ge.ch
1900; Geneviève Nicoud
25 000 vols; 300 curr per; 500 diss/theses; 4 digital data carriers
libr loan; BIS, AEIBS 27342

Institut dal Dicziunari Rumantsch Grischun (DRG), Ringstr. 34, 7000 **Chur**
T: +41 812846642; Fax: +41 812840204; E-mail: info@drg.ch; URL: www.drg.ch
1904
13 000 vols; 104 curr per; 1 635 slides, 24 250 photos 100 000 questionnaires 27343

Staatsarchiv Graubünden, Karlihofplatz, 7001 **Chur**
T: +41 812572803; Fax: +41 812572001; E-mail: info@sag.gr.ch; URL: www.staatsarchiv.gr.ch
1803; Dr. Silvio Margadant
Raetica (Lit. über Graubünden); Photo colls
12 000 vols; 20 curr per; 250 diss/theses; 30 govt docs; 5 000 linear metres of mss
VSA 27344

Fondation Martin Bodmer, 19-21 Route du Guignard, CP 7, 1223 **Cologny**
T: +41 227074433; Fax: +41 227074430; E-mail: info@fondationbodmer.ch; URL: www.fondationbodmer.org
1971; Prof. Charles Méla
Handschriften, seltene Bücher, Papyri
160 000 vols; 50 curr per; 1 640 mss; 280 incunabula; 1 000 diss/theses; 514 microforms
BIS 27345

Dokumentationsbibliothek der Landschaft Davos, Promenade 88, 7270 **Davos Platz**
T: +41 814130826; Fax: +41 814134074; E-mail: timothy.nelson@davos.gr.ch
1977; Dr. Timothy Nelson
Photo card file
10 500 vols; 8 curr per; 200 mss; 30 diss/theses; 300 govt docs; 15 music scores; 100 maps; 95 microforms; 100 av-mat; 20 sound-rec; 500 autographs, 8 000 photos, 45 vols of press clippings 27346

Kirchner Museum Davos, Fachbibliothek zur Kunstgeschichte des Expressionismus, Ernst Ludwig Kirchner Platz, 7270 **Davos Platz**
T: +41 81 4106300; Fax: +41 81 4106301; E-mail: info@kirchnermuseum.ch; URL: www.kirchnermuseum.ch
1992
Expressionismus, Ernst Ludwig Kirchner, Bildende Kunst der Klassischen Moderne
12 000 vols; 10 curr per; 60 diss/theses; 10 av-mat; 1 000 photos, 5 videos, 1 500 letters 27347

Dokumentation am Goetheanum, Bibliothek, Postfach, 4143 **Dornach**
T: +41 617064260; Fax: +41 617064401; E-mail: dokumentation@goetheanum.ch; URL: www.goetheanum.ch
1914; Uwe Werner
Anthroposophie
110 000 vols; 90 curr per; 1 000 music scores; 1 000 maps; 1 600 slides
libr loan; AaBuV 27348

Eidgenössische Materialprüfungs- und Forschungsanstalt (EMPA), Bibliothek, Überlandstr. 129, 8600 **Dübendorf**
T: +41 18234302; Fax: +41 18234579; E-mail: portal@empa.ch; URL: www.empa.ch
1952; J. Glas
13 000 vols; 340 curr per 27349

Lib4RI – Library for the Research Institutes within the ETH Domain: Eawag, Empa, PSI & WSL, Überlandstrasse 133, 8600 **Dübendorf**; mail address: Überlandstrasse 133, PO Box 611, 8600 Dübendorf
T: +41 587655700; Fax: +41 587655821; E-mail: info@lib4ri.ch; URL: www.lib4ri.ch
2011; Dr. Lothar Nunnenmacher
102 000 vols; 400 curr per; 8 000 e-journals; 16 000 e-books; 159 diss/theses; 1 200 maps; 100 000 microforms; 49 av-mat; 354 digital data carriers
libr loan 27350

Stiftung Bibliothek Werner Oechslin (Werner Oechslin Library Foundation), Luegeten 11, 8840 **Einsiedeln**
T: +41 554189040; Fax: +41 554189048; E-mail: info@bibliothek-oechslin.ch; URL: www.bibliothek-oechslin.ch
1998; Christoph Lanthemann
Architekturtheorie, Kulturgeschichte, Philosophie
60 000 vols; 25 curr per
non-lending library 27351

Forschungsanstalt Agroscope Reckenholz-Tänikon, Bibliothek, Tänikon 1, 8356 **Ettenhausen**
T: +41 523683131; Fax: +41 523651190; E-mail: doku@art.admin.ch; URL: www.agroscope.ch, www.maschinenkosten.ch, www.traktorentest.ch
1969; Alma Modes
Agrarwirtschaft, Landtechnik, Tierschutz
6 000 vols; 120 curr per
BIS, GBDL, IAALD 27352

Haute Ecole pédagogique, Centre Fribourgeois de Documentation Pédagogique (CFDP), Rue de Morat 36, 1700 **Fribourg**
T: +41 263057221; Fax: +41 263057229; E-mail: office.cfdp@edufr.ch; URL: www.hepfr.ch/cfdp
1972; Elisabeth Mauron-Hemmer
Pédagogie, psychologie, sciences de l'éducation, didactique, enseignement; Colls: jeux d'apprentissage, lectures suivies, appareils audiovisuels, matériel d'expérimentation, posters
60 000 vols; 150 curr per; 1 e-journals; 200 maps; 5 100 av-mat; 900 sound-rec; 350 digital data carriers
BIS 27353

Archives de la Ville de Genève, Bibliothèque, Palais Eynard 4, Rue de la Croix-Rouge, 1211 **Genève** 3
T: +41 224182990; Fax: +41 224182901; E-mail: archives@ville-ge.ch; URL: www.ville-ge.ch/archives
1986
3 000 vols; 55 curr per 27354

Archives d'Etat, Bibliothèque consultative, Rue de l'Hôtel-de-Ville 1, 1204 **Genève**; CP 3964, 1211 Genève 3
T: +41 223273395; Fax: +41 223273365; E-mail: archives@etat.ge.ch; URL: www.ge.ch/archives
1371; Pierre Flückiger
Verwaltungsarchive
25 000 vols; 250 curr per; 45 000 pamphlets
libr loan; RERO 27355

Bibliothèque Braille Romande et livre parlé, Association pour le Bien des Aveugles et malvoyants (Swiss French-Speaking Braille and Talking Books Library), Bourg-de-Four 34, 1204 **Genève**
T: +41 223177900; Fax: +41 223177911; E-mail: bbr@abage.ch; URL: www.abage.ch/bbr/bbr.aspx
1902; Anne Pillet
4 559 sound-rec; 1 612 digital data carriers; 3 206 braille titles
libr loan; IFLA, BIS, AIFBD 27356

Bibliothèque d'art et d'archéologie, Promenade du Pin 35, 1204 **Genève**
T: +41 224182700; Fax: +41 224182701; E-mail: info.baa@ville-ge.ch; URL: www.ville-ge.ch/baa
1910; Véronique Goncerut Estèbe
400 000 vols; 6 100 curr per; 30 000 e-journals; 20 mss; 2 incunabula; 534 microforms; 150 000 av-mat; 200 digital data carriers; 15 000 brochures, 12 000 press clippings, 70 000 exhibition cat, 70 000 auction cat, 400 videos
IFLA, BIS 27357

Bibliothèque Juive de Genève "Gérard Nordmann", Communauté Israelite de Genève, Rue Saint-Léger 10, 1205 **Genève**
T: +41 223178970; E-mail: markishj@comisra.ch; URL: biblio.comisra.ch/biblio
1945; Judith Markish
Bibl des jüdisch-theolog Seminars in Breslau
30 000 vols; 18 curr per
ALJHE, AJL 27358

Bibliothèque Musicale de la Ville de Genève, Maison des Arts du Grütli, 16, rue Général-Dufour, 1204 **Genève**
T: +41 224183580; Fax: +41 224183581; E-mail: bmus@ville-ge.ch; URL: www.ville-ge.ch/bmus
1962
Collection de programmes de concerts et d'ffiches du 19e siècle à nos
3 000 vols; 58 curr per; 750 mss; 35 000 music scores; 1 100 sound-rec; 50 digital data carriers
IAML, BIS 27359

Centre de ressources et de documentation pédagogiques (CRDP), Médiathèque, Rue des Gazomètres 5-7, CP 241, 1211 **Genève** 8
T: +41 223277700; Fax: +41 223277710; E-mail: patrick.johner@edu.ge.ch; URL: www.geneve.ch/crdp
1959; Emile-François Jeanneret
Education, Mass media
38 100 vols; 50 curr per; 880 diss/theses; 550 av-mat; 7 227 sound-rec; 768 digital data carriers; 13 150 videos, 238 DVDs, 4 800 slide-sets, 3 342 games, press clipping 27360

CERN – European Organization for Nuclear Research, Library, 1211 **Genève** 23
T: +41 227672444; Fax: +41 227672860; E-mail: corrado.pettenati@cern.ch; URL: library.cern.ch
1954; Corrado Pettenati
High-energy physics and its instrumentation, Particle Physics, Accelerators
50 000 vols; 400 curr per; 1 000 e-journals; 6 digital data carriers; 400 000 preprints and rpts
IFLA, IATUL, EUSIDIC 27361

Comité international de la Croix-Rouge / International Committee of the Red Cross / Internationales Kommitee vom Roten Kreuz, Bibliothèque, Av de la Paix 19, 1202 **Genève**
T: +41 227302030; Fax: +41 227302082; E-mail: lpa@icrc.org; URL: www.icrc.org
1863; Ms Callier
International Law
25 000 vols; 120 curr per; 30 e-journals; 50 mss; 300 diss/theses; 100 700 av-mat; 10 digital data carriers 27362

Conseil oecuménique des Églises (COE) (World Council of Churches, Library & Archives), Route de Ferney 150, 1211 **Genève**; P.O. Box 2100, 1211 Genève 2
T: +41 227916279; E-mail: library@wcc-coe.org; URL: library.oikoumene.org
1946; Andreas Waldvogel
Publications du COE; Histoire du mouvement oecuménique; Archives historiques du COE (133 linear m)
102 000 vols; 230 curr per; 30 e-journals; 500 000 microforms; 1 000 av-mat; 3 000

sound-rec
libr loan; BIS, ATLA, BETH, ICA, VSA
27363

Fédération des entreprises romandes Genève, Service de documentation, Rue de Saint-Jean 98, BP 5278, 1211 ***Genève*** 11
T: +41 227153258; Fax: +41 227153291; E-mail: doc@fer-ge.ch; URL: secure1.fer-ge.ch/wps/portal/fer-ge/Services/Documentation
1948; Chiara Crivelli
10 000 vols; 400 curr per; 10 e-journals; 500 000 clippings 27364

Institut et Musée Voltaire, Bibliothèque, Rue des Délices 25, 1203 ***Genève***
T: +41 223447133; Fax: +41 223451984; E-mail: institut.voltaire@ville-ge.ch; URL: www.ville-ge.ch/imv
1952; François Jacob
Voltaire u das 18. Jh
28 000 vols; 62 curr per; 1 e-journals; 10 000 mss; 10 maps; 350 microforms; 30 av-mat; 60 sound-rec; 91 digital data carriers; 5 linear metres of mss
BIS 27365

Institut international d'études sociales, Bibliothèque, 4 Route des Morillons, Case postale 6, 1211 ***Genève*** 22
T: +41 227996128; Fax: +41 227998542; E-mail: inst@ilo.org; URL: www.ilo.org/public/french/bureau/inst/
1960
10 000 vols; 490 curr per
libr loan 27366

International Labour Office (ILO), Bureau of Library and Information Services, Route des Morillons 4, 1211 ***Genève*** 22; mail address: 1211 Genève 22
T: +41 22 7998682; Fax: +41 22 7996516; E-mail: inform@ilo.org; URL: www.ilo.org/inform
1920; Ms Laurel Dryden
The ILO Library maintains one of the world's largest collections of official government legislation. The collection contains consolidated laws, statutes, decrees and gazettes for over 180 countries and territories in their original languages. The database Labordoc (http://labordoc.ilo.org) contains over 400,000 references to the world's most significant literature on labour and employment.
1 000 000 vols; 3 000 curr per; 268 e-journals; 3 000 diss/theses; 5 000 govt docs; 20 000 microforms; 1 500 digital data carriers
libr loan; ALA, ASLIB, EUSIDIC, IALL, IFLA, CILIP 27367

The Library in English, Rue de Monthoux 3, 1201 ***Genève***
T: +41 227328097; E-mail: library@thelibrary.ch; URL: www.thelibrary.ch
1930
14 000 vols; 8 curr per; 250 sound-rec
27368

Musée d'ethnographie, Bibliothèque, Bd Carl Vogt 65-67, 1205 ***Genève***; CP 191, 1211 Genève
T: +41 224184560; Fax: +41 224184551; E-mail: biblio.eth@ville-ge.ch; URL: www.ville-ge.ch/musinfo/ethg/index.php
1901; Annabelle Chanteraud
Bibl Georges Amoudruz, Bibl de la Société Suisse des Américanistes
45 000 vols; 220 curr per
BIS 27369

Musée d'histoire des sciences, Bibliothèque, Rue de Lausanne 128, 1202 ***Genève***
T: +41 224185068, 224186325; Fax: +41 224186301; E-mail: biblio.mhs@ville-ge.ch; URL: www.ville-ge.ch/mhs/bibliotheque.php
1964; Christelle Mougin
10 500 vols; 30 curr per; 1 000 mss
27370

Musée historique de la Réformation, c/o Université de Genève, 5 Rue de Candolle, 1211 ***Genève*** 4
T: +41 223797128; Fax: +41 223797133; E-mail: marianne.tsioli-bodenmann@ville-ge.ch; URL: www.unige.ch/ihr
1897; Marianne Tsioli
Histoire ecclésiastique, protestantisme, bible
8 500 vols; 150 mss; 1 incunabula; 190 microforms 27371

Muséum d'histoire naturelle, Bibliothèque, Route de Malagnou 1, 1208 ***Genève***; CP 6434, 1211 Genève 6
T: +41 224186325; Fax: +41 224186301; E-mail: biblio.mhn@ville-ge.ch; URL: www.ville-ge.ch/mhng/bibliotheque.php
1820; Christelle Mougin
80 000 vols; 1 000 curr per; 4 000 maps; 30 microforms; 120 sound-rec; 20 digital data carriers
BIS 27372

Service de la Recherche en Education (SRED), Bibliothèque, Quai du Rhône 12, 1205 ***Genève***
T: +41 225467166; Fax: +41 5467101; E-mail: christiane.pouly@edu.ge.ch; URL: www.ge.ch/sred/biblio/
1938
18 600 vols; 432 curr per; 33 e-journals; 50 av-mat; 150 sound-rec; 350 digital data carriers 27373

Société d'Histoire et d'Archéologie de Genève, Bibliothèque, Promenade des Bastions, 1211 ***Genève*** 4
T: +41 223208266; Fax: +41 223292855; E-mail: info@shag-geneve.ch; URL: www.shag-geneve.ch
1838; M. Droin
10 000 vols; 50 curr per; 1 000 mss
libr loan 27374

Télévision Suisse Romande, Service documentation et archives, Quai Ernest-Ansermet 20, CP 234, 1211 ***Genève*** 8
T: +41 227082020; Fax: +41 227089822; E-mail: albin.salamin@tsr.ch; URL: www.tsr.ch
1954; Albin Salamin
28 000 vols; 175 curr per; 255 000 av-mat; 20 digital data carriers; 230 000 videos, 1,1 mio press clippings, 80 000 slides, 303 000 photos 27375

Touring-Club Suisse, Section genevoise, Bibliothèque, Quai Gustave-Andor 2, 1207 ***Genève***
T: +41 227354653; Fax: +41 227354815; E-mail: reception@tcsge.ch; URL: www.tcs.ch/geneve/fr/home.html
1955
10 000 vols; 250 curr per; 2 144 av-mat
libr loan 27376

Union Internationale des Télécommunications / International Telecommunication Union, Service de la Bibliothèque et des Archives, Bâtiment Montbrillant, 6e étage, 2, rue de Varembé, 1202 ***Genève***; mail address: Place des Nations, 1211 Genève 20
T: +41 227306900; Fax: +41 227305326; E-mail: library@itu.int; URL: www.itu.int/library
1865; Kristine Clara
deposit library for ITU's documentary history
30 000 vols; 700 curr per; 9 000 e-journals; 5 000 govt docs; 300 maps; 100 digital data carriers
libr loan; ALLIS 27377

United Nations Conference on Trade and Development (UNCTAD), Reference Unit, Palais des Nations, 1211 ***Genève*** 10
T: +41 229076320; Fax: +41 229070195; E-mail: bob.cook@unctad.org; URL: www.unctad.org
1964; A. von Wartensleben
20 000 vols 27378

United Nations Office at Geneva, Library, Palais des Nations, 8-14 ave de la paix, 1211 ***Genève*** 10
T: +41 229173053; Fax: +41 229170157; E-mail: library@unog.ch; URL: www.unog.ch/library/start.htm
1919; Pierre Le Loarer
UN documents collection; League of Nations archive and documents; International peace movements collection; League of Nations archive
1 300 000 vols; 4 000 curr per; 6 000 e-journals; 7 000 diss/theses; 5 000 govt docs; 500 maps; 200 000 microforms; 100 av-mat; 50 sound-rec; 500 digital data carriers; 4 500 000 UN documents
libr loan; IFLA, AILIS, AIL, IALL 27379

World Health Organization (WHO) / Organisation Mondiale de la Santé, Library and Information networks for Knowledge, Av Appia 20, 1211 ***Genève*** 27
T: +41 227912062; Fax: +41 227914150; E-mail: library@who.int; URL: www.who.int/library
1946; Yvonne Grandbois
250 000 Dok (WHO u UN)
85 000 vols; 2 800 curr per; 200 000 docs
libr loan; IFLA, MLA, AILIS 27380

World Intellectual Property Organization (WIPO) / Organisation mondiale de la propriété intellectuelle (OMPI), Library, Chemin des Colombettes 34, 1211 ***Genève*** 20
Fax: +41 223388850; E-mail: library.mail@wipo.int; URL: www.wipo.int
1895; Laura Wray
Gewerblicher Rechtsschutz, Urheberrecht, Pflanzenschutz
30 000 vols; 300 curr per; 100 diss/theses; 10 digital data carriers; 15 000 press clippings
AILIS 27381

World Meteorological Organization (WMO) / Organisation météorologique mondiale (OMM), Technical Library, Av de la Paix 7 bis, CP 2300, 1211 ***Genève*** 2
T: +41 227308111; Fax: +41 227308181; E-mail: wmo@wmo.int; URL: www.wmo.int
1951; Zenebu Kefelew
40 500 vols; 200 curr per 27382

World Trade Organization / Organisation mondiale du commerce, Bibliothèque, Centre William Rappard, 154 rue de Lausanne, 1211 ***Genève***
T: +41 227395124; Fax: +41 227395795; E-mail: enquiries@wto.org; URL: www.wto.org
1952; Mrs Jany Barthel-Rosa
International trade law, economics, trade in goods, trade in services, trade and environment, statistics; GATT/WTO Archives; trade related intellectual property rights; trade and finance; Statistical section, Archive
30 000 vols; 1 500 curr per; 60 000 microforms; 50 digital data carriers
BIS, AILIS 27383

IUCN – The World Conservation Union, Library, Rue Mauverney 28, 1196 ***Gland***
T: +41 229990136; Fax: +41 229990002; E-mail: library@iucn.org; URL: www.iucn.org
1948; Frau Katherine Rewinkel El-Darwish
20 000 vols; 300 curr per; 250 av-mat; 20 digital data carriers
IFLA, AILIS, IAMSLIC 27384

Pädagogische Hochschule Thurgau, Medien- und Didaktikzentrum, Bibliothek, Unterer Schulweg, 8280 ***Kreuzlingen***
T: +41 716785696; Fax: +41 716785697; E-mail: mdz.bibliothek@phtg.ch; URL: www.phtg.ch
2008; Anita Thurnheer, Anja Strassburger
65 000 vols; 330 curr per; 12 e-journals; 200 maps; 12 000 av-mat; 1 500 sound-rec; 1 200 digital data carriers; 400 games, 1 400 media kits
BIS 27385

C. G. Jung-Institut Zürich, Bibliothek, Hornweg 28, 8700 ***Küsnacht***
T: +41 19141051; Fax: +41 19141050; E-mail: budai@junginstitut.ch; URL: www.junginstitut.ch
1948; Julia Budai
Bildarchiv: Zeichnungen von Patienten von C.G.Jung und Jolande Jacobi
16 000 vols; 18 curr per; 200 av-mat; 1 200 other items 27386

Bibliothèque des Jeunes, Rue de la Ronde 9, 2300 ***La Chaux-de-Fonds***
T: +41 329676852; Fax: +41 329676858; E-mail: Service.Bibliothequedesjeunes@ne.ch; URL: www.chaux-de-fonds.ch/bibliotheques
1953; Franziska Eggel Turtschy, Caroline Ruffieux
Reservé de livres d'images (1 500); Reservé de livres pour l'enfance et la jeunesse (600 livres du 19e et première moitié du 20e siècle),; Fonds de références sur la littérature pour l'enfance et la jeunesse (1 200)
44 183 vols; 60 curr per; 853 digital data carriers; 3 011 dossiers 27387

Musée d'histoire, Bibliothèque, Rue des Musées 31, 2300 ***La Chaux-de-Fonds***
T: +41 329676088; Fax: +41 327220762; E-mail: museehistoire.vch@ne.ch; URL: www.chaux-de-fonds.ch/museehistoire
1876; Sylviane Musy-Ramseyer
6 000 Postkarten und Fotos
4 000 vols; 16 curr per; 500 mss; 50 maps; 1 538 av-mat; 15 sound-rec; 2 500 pamphlets, 80 documentations
Due to renovation of the museum the library will be closed until 2013 27388

Musée d'histoire naturelle, Centre de documentation, Av Léopold-Robert 63, 2300 ***La Chaux-de-Fonds***
T: +41 329676071; Fax: +41 327220746; URL: www.mhnc.ch/default.asp/4-0-5285-8016-113-207-1/
Bibliothèque de référence (ouvrages de détermination) de plusieurs milliers de titres; Fonds Dr Albert Monard, Fonds Dr Georges Dubois (travaux personnels en parasitologie), Fonds Prof. Franz Allemann (géologie), Bibliothèque d'arachnologie du Prof. Legendre
18 curr per 27389

Archives de la Ville de Lausanne, Bibliothèque, Maupas 47, 1004 ***Lausanne*** 9
T: +41 213152121; Fax: +41 213152120; E-mail: archives@lausanne.ch; URL: www.lausanne.ch/archives
1892; Fabienne Chuat
Lausannensia, biographie, bibliographie, généalogie, histoire, géographie; 6 associated libs
30 000 vols; 800 curr per; 4 000 000 mss; 2 incunabula; 300 diss/theses; 30 000 govt docs; 180 000 maps; 2 600 microforms; 1 000 av-mat; 600 sound-rec
BIS 27390

Bibliothèque des Cèdres, Chemin des Cèdres 7, 1004 ***Lausanne***
T: +41 21 6480164; Fax: +41 21 6480166; E-mail: Marie-Pierre.Constant@bcu.unil.ch; URL: www.unil.ch/bcu/page17082_fr.html
Théologie, histoire de l'Eglise, Bibles, exégèse, théologie pratique (sermons, liturgies, psautiers), histoire, philosophie, langues et littératures, sciences et arts, Illuminés, théosophes, mystiques des 17e-18e
150 000 vols 27391

Bibliothèque Sonore Romande (BSR), Rue Genève 17, 1003 ***Lausanne***
T: +41 213211010; Fax: +41 213211019; E-mail: info@bsr-lausanne.ch
1976; Isabelle Albanese
Talking bks
10 curr per; 6 000 DAISY MP3-CDs
27392

Centre International de Recherches Sur l'Anarchisme (CIRA), Bibliothèque, Avenue de Beaumont 24, 1012 ***Lausanne***
T: +41 21 6524819; Fax: +41 21 6524819; E-mail: cira@plusloin.org; URL: www.anarca-bolo.ch/cira/
Anarchisme
20 000 vols 27393

Cinémathèque Suisse (SCL) / Schweizer Filmarchiv, Bibliothèque / Archives papeir (Swiss Film Archive), CP 5556, 1002 ***Lausanne***
T: +41 213152171; Fax: +41 2131152189; E-mail: nadia.roch@cinematheque.ch
1948; Nadia Roch
20 000 vols; 400 curr per; 100 mss; 70 000 av-mat; 2 mio photos, 800 000 press clippings, 35 bequests, 8 000 screenplays 27394

Institut suisse de droit comparé, Bibliothèque (Swiss Institute of Comparative Law), Dorigny, 1015 ***Lausanne***
T: +41 216924911; Fax: +41 216924949; E-mail: reference@isdc-dfjp.unil.ch; URL: isdc.ch
1981; Sadri Saieb
Ausländisches Recht, Internationales Privatrecht, Islamisches Recht, Chinesisches Recht

360 000 vols; 2 000 curr per; 7 000 e-journals; 40 digital data carriers
libr loan; IALL 27395

Institut suisse de prévention de l'alcoolisme et autres toxicomanies / Schweizerische Fachstelle für Alkohol- und andere Drogenprobleme, Bibliothèque, Av Ruchonnet 14, CP 870, 1001 ***Lausanne***
T: +41 213212957; Fax: +41 213212940; E-mail: biblio@sfa-ispa.ch; URL: www.sfa-ispa.ch
1902; Annick Lebreau
Alkoholismus, Drogenmissbrauch, Abhängigkeiten
8 500 vols; 60 curr per; 30 diss/theses; 5 series of government publ
BIS, SALIS, ELISAD 27396

International Olympic Committee, Library, Quai d'Ouchy 1, 1001 ***Lausanne***
T: +41 216216511; Fax: +41 216216512; E-mail: library@olympic.org; URL: www.olympic.org
1982; Ms Yoo-Mi Steffen
Olympische Spiele, Olympische Bewegung, Olympische Sportarten, Themen in Verbindung mit den Olympischen Spielen; Pierre de Coubertin (mss, works), Carl Diem Libr (hist of sports, German sports movement, Olympic games 1936), Usher Libr (archery), Bartrolli Libr (tennis)
24 000 vols; 412 curr per; 100 microforms; 130 digital data carriers
libr loan; BIS, IASI, IFLA, AILIS 27397

Musée de Zoologie, Bibliothèque, Place de la Riponne 6, 1005 ***Lausanne***; Place de la Riponne 6 – CP, 1014 Lausanne
T: +41 213163460; Fax: +41 213163479; E-mail: info.zoologie@vd.ch; URL: www.zoologie.vd.ch
Anne Freitag
Cryptozoology
8 400 vols; 170 curr per; 150 maps; 30 sound-rec 27398

Musée et Jardins Botaniques Cantonaux, Bibliothèque, Av de Cour 14 bis, 1007 ***Lausanne***
T: +41 21 3169982; Fax: +41 21 6164665; E-mail: info.botanique@vd.ch; URL: www.botanique.vd.ch/
1837; Dr. G. Müller
botanique systématique, géobotanique, horticulture, palynologie, archéobotanique, floristique, biologie végétale
15 000 vols; 115 curr per; 100 mss; 250 maps; 10 microforms; 500 photos, 5 000 slides, 3 000 autographs, 15 000 brochures
libr loan 27399

Musée monétaire cantonal, Bibliothèque, Pl de la Riponne 6, Palais de Rumine, 1014 ***Lausanne***
T: +41 213163990; Fax: +41 213163999; E-mail: musee.monetaire@vd.ch; URL: www.musees.vd.ch/musee-monetaire
1848; Anne Geiser, Cosette Lagnel
20 000 vols; 70 curr per; 200 mss; 100 microforms
libr loan; BIS 27400

Union interparlementaire, Bibliothèque, Chemin du Pommier 5, CP 330, 1218 ***Le Grand Saconnex*** / Genève
T: +41 229194150; Fax: +41 227333141; URL: www.ipu.org
1965; A.B. Johnsson
25 000 vols; 164 curr per
BIS, IFLA 27401

UNESCO – International Bureau of Education / Bureau International d'Education, Documentation Centre, Route des Morillons 15, 1218 ***Le Grand-Saconnex***; CP 199, 1211 Genève 20
T: +41 229177860; Fax: +41 229177801; E-mail: doc.centre@ibe.unesco.org; URL: www.ibe.unesco.org
1925
The IBE Documentation Centre houses the Archive of the IBE for the period 1925-1968 (computerized inventory); special coll of curricular materials
125 000 vols; 600 curr per; 15 000 govt docs; 450 000 microforms; archival periodicals coll (600 titles)
AILIS 27402

Museum Aargau, Schloss Lenzburg, Bibliothek, 5600 ***Lenzburg***
T: +41 628884840; Fax: +41 628884841; E-mail: schlosslenzburg@ag.ch; URL: www.ag.ch/lenzburg/de/pub/
1958; Clemens Moser
Kulturgeschichte
8 500 vols; 100 curr per; 50 diss/theses; 40 maps; 600 000 photos, 5 000 slides 27403

Staatsarchiv des Kantons Basel-Landschaft, Bibliothek, Wiedenhubstr. 35, mail address: 4410 ***Liestal***
T: +41 619267676; E-mail: staatsarchiv@bl.ch; URL: staatsarchiv.bl.ch
1832; Regula Nebiker
Akten des Kantons Basel-Landschaft und der früheren Landschaft Basel, Privatarchive von Personen und Institutionen aus dem Gebiet des heutigen Kantons BL
10 000 vols; 50 curr per; 500 maps; 9 000 microforms; 200 av-mat; 8 000 linear metres of government docs, historical fotos
libr loan; VSA, BBS 27404

Natur-Museum, Bibliothek, Kasernenplatz 5, 6003 ***Luzern***
T: +41 412285411; Fax: +41 412285406; E-mail: bibliothek.nml@lu.ch; URL: www.naturmuseum.ch
1978; Madeleine Herger
Geowissenschaften, Botanik, Zoologie, Ökologie, Umweltwissenschaften, Entomologie
8 000 vols; 200 curr per; 10 mss; 110 diss/theses; 200 maps; 1 000 av-mat; 30 sound-rec; Natural scientific offprints
BIS 27405

RomeroHaus, Forschungsbibliothek, Kreuzbuchstr. 44, 6006 ***Luzern***
T: +41 413757250; Fax: +41 413757275; E-mail: bibliothek@romerohaus.ch; URL: www.romerohaus.ch/bibliothek
1922; Dr. Ernstpeter Heiniger
Dritte-Welt-Theologie, Entwicklungswiss, kontextuelle Theologie, feministische Theologie, Missionswiss, katholische Theologie, Philosophie, Völkerkunde, Religionswiss
40 000 vols; 220 curr per; 1 mss
libr loan 27406

Staatsarchiv des Kantons Luzern, Schützenstr. 9, 6000 ***Luzern*** 7; Postfach 7853, 6000 Luzern 7
T: +41 412285365; Fax: +41 412286663; E-mail: staatsarchiv@lu.ch; URL: www.staatsarchiv.lu.ch
1803; Heidy Knüsel Zeller
30 000 vols; 200 curr per; 100 diss/theses; 300 govt docs; 150 maps; 5 sound-rec; 15 digital data carriers; 5 000 offprints
Informationsverbund Deutschschweiz (IDS) Luzern
non-lending library 27407

Verkehrshaus der Schweiz, Dokuzentrum, Sammlung/Archiv und Bibliothek (Swiss Museum of Transport), Lidostrasse 5, 6006 ***Luzern*** 9
T: +41 413704444; Fax: +41 413706168; E-mail: archive@verkehrshaus.ch; URL: www.verkehrshaus.ch
1959; Claudia Hermann
Transport und Verkehr; Dokumentationen zur Mobilitätsgeschichte der Schweiz, Archiv mit historischen Objekten wie Fotos, Plänen, Plakaten, Ansichtskarten etc
25 000 vols; 70 curr per
BBS 27408

Fondazione Gaudenzio e Palmira Giovanoli, Biblioteca, CP 107, 7516 ***Maloggia***
T: +41 818243239; Fax: +41 818243575; E-mail: info@salecina.ch; URL: www.salecina.ch
1974; Ursina Negrini Ganzoni
12 900 vols; 11 curr per; 100 mss; 80 sound-rec; 30 VHS, 1 000 photos 27409

Centro didattico cantonale, Via Madonna della Salute 24, 6900 ***Massagno***; CP 193, 6908 Massagno
T: +41 919665628; Fax: +41 919669755; E-mail: decs-cdc.massagno@ti.ch; URL: www.scuoladecs.ti.ch/cdc
1973
13 200 vols; 61 curr per; 600 av-mat; 750 sound-rec; 30 000 slides, 1 000 videos 27410

Organizzazione sociopsichiatrica cantonale, Biblioteca OSC, via Ag. Maspoli 6, 6850 ***Mendrisio***
T: +41 918165602; Fax: +41 918165609; E-mail: dss-osc.biblioteca@ti.ch; URL: www.ti.ch/osc
1986; Giuliana Schmid
Hist of psychiatry
10 500 vols; 84 curr per; 14 e-journals; 250 av-mat; 240 videos, 4 779 newspaper articles, 163 periodici morti 27411

Bibliothèque Municipale de Morges, Place du Casino 1, 1110 ***Morges***
T: +41 21 8049720; E-mail: bibliotheque@morges.ch; URL: www.morges.ch/biblio
Fonds ancien encyclopédique
45 000 vols 27412

Sportmuseum Schweiz, Bibliothek, Reinacherstr. 1-3, 4142 ***Münchenstein***
T: +41 612611221; Fax: +41 612611247; E-mail: info@sportmuseum.ch; URL: www.sportmuseum.ch
1945; Mike Gosteli
15 000 vols; 80 curr per; 5 000 av-mat; 25 000 prints, 500 000 photos, 1 000 posters 27413

Psychiatrische Klinik, Wissenschaftliche Bibliothek, Postfach 154, 8596 ***Münsterlingen***
T: +41 716864141; Fax: +41 716864035; URL: www.stgag.ch
1861
8 000 vols; 40 curr per; 18 e-journals
libr loan 27414

Bibliothek und Informationszentrum für Statistik / Bibliothèque et centre d'information pour la statistique, Espace de l'Europe 10, 2010 ***Neuchâtel***
T: +41 327136054; Fax: +41 327136903; E-mail: library@bfs.admin.ch; URL: www.statistik.admin.ch
1860; Norbert Lüber
Statistische Reihen europäischer Länder seit Mitte 19. Jahrhundert
140 000 vols; 1 200 curr per; 400 diss/theses; 550 maps; 250 microforms; 50 av-mat; 1 500 digital data carriers
libr loan; BIS 27415

Bibliothèque-Ludothèque Pestalozzi, Faubourg du Lac 1, 2000 ***Neuchâtel***
T: +41 327251000; E-mail: bib-pestalozzi@bluewin.ch; URL: neuchatel.ne.ch
1946; Antoinette Buerki
12 000 vols; 20 curr per; 1 200 av-mat; 500 sound-rec 27416

Glossaire des patois de la Suisse romande, Bibliothèque, Av DuPeyrou 4, 2000 ***Neuchâtel***
T: +41 327243680; Fax: +41 327243692; E-mail: gpsr@unine.ch; URL: www.glossaire-romand.ch/default.asp/2-0-4-6-6-1/
1898; Hervé Chevalley
Galloromanische Dialekte; Dialects of the French-speaking Switzerland dept
14 000 vols; 50 curr per; 1 000 mss; 100 diss/theses; 100 govt docs; 500 maps; 100 sound-rec; 3 000 press clippings, 5 000 brochures 27417

Institut de recherche et de documentation pédagogique, Bibliothèque, Faubourg de l'Hôpital 43, CP 556, 2002 ***Neuchâtel***
T: +41 328898600; Fax: +41 328896971; E-mail: documentation@irdp.ch; URL: www.irdp.ch
1970; Isabelle Deschenaux-Steullet
17 000 vols; 160 curr per 27418

Swiss Forum for Migration and Population Studies, Documentation Centre "Migrations", Fbg de l'Hôpital 106, 2000 ***Neuchâtel***
T: +41 327183936; Fax: +41 327183921; E-mail: aronne.watkins@unine.ch; URL: www2.unine.ch/sfm/page27816_en.html
1995
16 000 vols; 40 curr per; 15 e-journals; 100 av-mat 27419

Station de recherche Agroscope Changins-Wadenswil ACW, Bibliothèque, Route de Duillier, CP 1012, 1260 ***Nyon*** 1
T: +41 223634150; Fax: +41 223634155; URL: www.acw.admin.ch
1952; Mihaela Romoscanu
30 000 vols; 750 curr per
IAALD 27420

Association pour la Conservation du château d'Oron, Bibliothèque, Château d'Oron, CP 6, 1608 ***Oron-le-Châtel***
T: +41 219079051; Fax: +41 219079065; E-mail: chateau.oron@bluewin.ch; URL: www.swisscastles.ch/Vaud/Oron
Romans en langue française de 1750 à 1900
15 000 vols 27421

Château d'Oron, Bibliothéque, CP 6, 1608 ***Oron-le-Châtel***
T: +41 21 9079051; Fax: +41 21 9079065; E-mail: chateau.oron@bluewin.ch; URL: www.swisscastles.ch/Vaud/Oron/
Coll Encyclopédique, avec accent sur l'histoire et surtout les belles-lettres, roman français de la période révolutionnaire et de l'Empire
13 000 vols 27422

Forschungsanstalt Agroscope Liebefeld-Posieux ALP, Bibliothek, 1725 ***Posieux***
T: +41 264077111; Fax: +41 264077300; E-mail: gerhard.mangold@alp.admin.ch; URL: www.alp.admin.ch
1974; Gerhard Mangold
Milch- und Fleischproduktion, Tierernährung, Futtermittel, Veterinärmedizin, Fleischqualität, ökologische Landwirtschaft
30 000 vols; 320 curr per; 60 e-journals; 600 diss/theses; 300 maps; 150 av-mat; 80 digital data carriers; 2 000 slides
libr loan; GBDL, IAALD, BIS 27423

Institut Agricole de l'Etat de Fribourg, Service des bibliothèques et de la documentation, Route de Grangeneuve 31, 1725 ***Posieux***
T: +41 263055533; Fax: +41 263055504; E-mail: iag@fr.ch; URL: www.fr.ch/iag
1888
13 000 vols 27424

Polenmuseum Rapperswil, Bibliothek, Postfach 1251, 8640 ***Rapperswil***
T: +41 55 2101861; Fax: +41 55 2100662; E-mail: biblioteka@muzeum-polskie.org; URL: www.muzeum-polskie.org/biblioteka/
Anna Piotrowski
Polnische Geschichte und Kultur, Polen im Exil
25 000 vols 27425

Abegg-Stiftung, Bibliothek, Werner-Abegg-Str. 67, 3132 ***Riggisberg***
T: +41 318081223; Fax: +41 318081200; E-mail: bibliothek@abegg-stiftung.ch; URL: www.abegg-stiftung.ch
1967; Andrea Kälin, Beate Stucki-Kunze, Andrea Meuer
Kunstgewerbe, Textilkunst, Restaurierung
75 000 vols; 271 curr per; 21 000 microforms; 2 000 av-mat; 80 digital data carriers; 34 000 photos
AKMB, BIS 27426

Tibet-Institut, Bibliothek, Wildbergstrasse 10., 8486 ***Rikon***
T: +41 523831729; Fax: +41 523832095; E-mail: bibliothek@tibet-institut.ch; URL: www.tibet-institut.ch
1968; Renata Koller
Tibet, Buddhismus, Verhältnis Tibet-China, Tibeter im Exil; Press clippings since 1960
12 500 vols; 95 curr per; 12 diss/theses; 100 maps; 3 630 av-mat; 200 sound-rec; 35 digital data carriers; 2 500 photos
libr loan; BBS 27427

Bibliothek Schloss Arenenberg, Napoleonmuseum Thurgau, Schloss und Park Arenenberg, 8268 ***Salenstein***
T: +41 71 6633263; Fax: +41 71 6633261; E-mail: napoleonmuseum@tg.ch; URL: www.napoleonmuseum.tg.ch
1906; Christina Egli
Napoleonica; Beauharnais and Louis Napoleon Bonaparte Hortense library (ca. 1 000 vols)
25 000 vols; 37 curr per; 15 av-mat; 1 sound-rec 27428

Napoleonmuseum Thurgau, Bibliothek, Schloss und Park Arenenberg, Postfach 79, 8268 ***Salenstein***
T: +41 716633263; Fax: +41 716633261; E-mail: napoleonmuseum@tg.ch; URL: www.napoleonmuseum.tg.ch
1906; Christina Egli
Hist Libr of Queen Hortense and Prince Louis Napoléon; Napoleonica
25 000 vols; 30 curr per; 580 music scores; 180 maps; 25 av-mat; 3 bequests, 1 500 autographs, 5 000 photos, 12 000 art prints
SWB 27429

Observatoire de Genève, Bibliothèque, Chemin des Maillettes 51, 1290 ***Sauverny***
T: +41 22379221; Fax: +41 3792205; URL: www.unige.ch/sciences/astro/
1773; Claude Guidi
Hist Stiftung
20 100 vols; 342 curr per; 50 e-journals; 150 mss; 300 diss/theses; 509 microforms; 40 digital data carriers; 3 900 slides
EGAL 27430

Eisenbibliothek, Stiftung der Georg Fischer AG (Iron Library), Klostergut Paradies, 8252 ***Schlatt*** TG
T: +41 526312743; Fax: +41 526312755; E-mail: eisenbibliothek@georgfischer.com; URL: www.eisenbibliothek.ch
1948; Dr. Britta Leise
Monografien über Eisengewinnung und -verarbeitung, Gießereiwesen, Bergbau, Hüttenwesen, Geowissenschaften, Metallurgie, Werkstoffprüfung, Militaria, Archäologie, Kunst,- Technik- und Verkehrsgeschichte, Wirtschafts- und Sozialgeschichte, Architektur, Physik, Chemie; Enzyklopädien zur Technik; Grafische Slg; Landkartenslg
40 000 vols; 70 curr per; 57 mss; 5 incunabula; 700 diss/theses; 80 maps; 20 av-mat; 20 digital data carriers; 2 500 publs of companies of iron industry
BIS, VSA
non-lending library 27431

Staatsarchiv des Kantons Schwyz (STASZ), Bundesbriefmuseum, Bahnhofstr. 20, 6430 ***Schwyz***; Postfach 2201, 6431 Schwyz
T: +41 418192065; Fax: +41 418192069; E-mail: bundesbriefmuseum@sz.ch; URL: www.bundesbriefmuseum.ch
R. Rey
Kanton Schwyz
10 000 vols; 75 curr per; 1 700 govt docs; 1 000 maps; 1 700 microforms; 30 bequests, 14 000 photos, 2 400 slides, 4 000 art prints 27432

Schweizerische Vogelwarte, Bibliothek, 6204 ***Sempach***
T: +41 414629700; Fax: +41 414629710; E-mail: bibliothek@vogelwarte.ch; URL: www.vogelwarte.ch
1924; Dr. Christian Marti
Ornithologie
15 000 vols; 250 curr per; 250 diss/theses; 50 av-mat; 150 sound-rec; 30 digital data carriers
BIS 27433

Biblioteca Engiadinaisa, Postfach 113, 7514 ***Sils Maria***
T: +41 818265377; Fax: +41 818266377
1962; Kathrine Gerber
Raetica (bes Engadin)
15 000 vols; 20 curr per; 70 maps; 1 500 sound-rec; 200 digital data carriers
libr loan; BIS 27434

Musées cantonaux / Walliser Kantonsmuseen, Bibliothèque / Bibliothek, rue des châteaux 14, 1950 ***Sion***; CP 2244, 1950 Sion 2-Nord
T: +41 276064670; Fax: +41 276064674; URL: www.musees-valais.ch
1984
Kunst, Geschichte, Archäologie, Ethnografie, Numismatik, Naturgeschichte (Wallis betreffend)
15 000 vols; 300 curr per 27435

Frauenbibliothek und Fonothek Wyborada, Davidstr. 42, 9000 ***St.Gallen***; Postfach 2330, 9001 St.Gallen
T: +41 712226515; E-mail: info@wyborada.ch; URL: www.wyborada.ch
1987; Irene von Hartz
Feministische Theorie und Praxis, Lit von Frauen, Frauenkrimis; Sachbücher von und zu Frauen zu den Themen: Geschichte, Arbeit, Bildung, Politik, Gesellschaft, Recht, Körper u Psyche, Religionen, Kunst u Kultur, andere Kulturen, Frauenbewegung und Wissenschaften; nicht-sexistische Kleinkinderbücher
9 000 vols; 5 curr per; 5 av-mat; 509 sound-rec; 5 digital data carriers 27436

Kantonsspital St.Gallen, Bibliothek, Rorschacher Str. 95, 9007 ***St.Gallen***
T: +41 71262470; Fax: +41 71262871; URL: www.kssg.ch/
1975; Daniel Kauffmann
20 000 vols; 400 curr per
libr loan; BIS 27437

Staatsarchiv St.Gallen, Forschungsbibliothek, Regierungsgebäude, 9001 ***St.Gallen***
T: +41 71 2293205; Fax: +41 71 2293445; E-mail: info.staatsarchiv@sg.ch; URL: www.staatsarchiv.sg.ch
2006; Benno Hägeli
Geschichte, Kulturgeschichte, Quelleneditionen, Historische Hilfswissenschaften, Archivwissenschaften
49 500 vols; 306 curr per; 2 143 diss/theses; 50 govt docs; 10 av-mat; 20 digital data carriers
Pro Vadiana 27438

Textilbibliothek, Vadianstr. 2, 9000 ***St.Gallen***
T: +41 712221747; Fax: +41 712234239; E-mail: info@textilbibliothek.ch; URL: www.textilbibliothek.ch
1886; Regula Lüscher
Textilmusterbücher
22 500 vols; 80 curr per; 2 mio patterns, 60 000 loose-leaf patterns
BIS 27439

Schweizerisches Gastronomie-Museum, Bibliothek, Schloss Schadau, Seestr. 45, 3600 ***Thun***
T: +41 332231432; Fax: +41 332235432; E-mail: gastronomiemuseum@bluewin.ch; URL: www.thun.ch/d/gastronomiemuseum
1988
Hotellerie, Gastronomie, Ess- und Trinkkultur allgemein
7 500 vols; 10 curr per; 100 mss; 3 diss/theses; 20 000 menus
BIS 27440

Centre interrégional de perfectionnement (CIP) / Interregionales Fortbildungszentrum, Médiathèque, Les Lovières 13, 2720 ***Tramelan***
T: +41 324860670; Fax: +41 324860607; URL: www.cip-tramelan.ch
1991; Loïc Diacon
Pédagogie, formation des adultes, matérial didactique
25 000 vols; 180 curr per; 5 000 av-mat; 800 sound-rec; 60 digital data carriers; 300 sets of slides, 50 posters, 1 800 videos, 2 000 press clippings
BIS, CLP 27441

Paul Kläui-Bibliothek Uster, Zürichstr. 11, Postfach, 8610 ***Uster***
T: +41 44 9447224; Fax: +41 44 9447224; E-mail: stadtarchiv@stadt-uster.ch; URL: www.uster.ch
1967; Lucas Nicolussi
Geschichte der Zürcher Landschaft, insbesondere Zürcher Oberland; Historical documentation relating to Uster
20 000 vols; 70 curr per; 500 mss; 150 maps; 2 000 photos, 100 slides, 2 444 historical books 27442

Biblioteca Jaura, Museum Chasa Jaura, 7535 ***Valchava***
T: +41 818585155
Literatur über das Münstertal / Val Müstair. Raetoromanica. Raetica und Tyrolensia; hist Buchbestand mit rund 500 Titeln: 4 aus dem 16., 27 aus dem 17., 68 aus dem 18.und rund 400 Werke aus dem 19. Jh
4 500 vols 27443

Archiv zur Geschichte der schweizerischen Frauenbewegung, Gosteli Stiftung, Bibliothek, Altikofenstrasse 186, 3048 ***Worblaufen***
T: +41 319210222; Fax: +41 319217941; E-mail: info@gosteli-foundation.ch; URL: www.gosteli-foundation.ch
1982
Schweizer Frauenbewegung; Archivalien von Frauenorganisationen, Frauenverbänden und einzelnen Frauen, die in Politik, Wirtschaft, Bildung, Kultur, Gesellschaft und Familie eine wichtige Rolle gespielt haben
9 700 vols; 70 curr per; 8 diss/theses; 1 000 av-mat; 110 sound-rec; more than 300 association and personal archives, biographical dossiers, photos
VSA 27444

Maison d'Ailleurs, Musée de la science-fiction, de l'utopie et des voyages extraordinaires, Bibliothèque, Pl Pestalozzi 14, 1400 ***Yverdon-les-Bains***; CP, 1401 Yverdon-les-Bains
T: +41 244256438; Fax: +41 244256575; E-mail: maison@ailleurs.ch; URL: www.ailleurs.ch
1976; Patrick Gyger
35 000 vols; 30 curr per; 1 000 av-mat; 1 000 sound-rec; 10 000 photos 27445

Biblioteca Fundaziun Tscharner Zernez, Runatsch 146, 7530 ***Zernez***
T: +41 818561950; Fax: +41 818561950; E-mail: giontscharner@bluewin.ch
Rätoromanische Schriften der gesamten Romania. Literatur über das Engadin und Graubünden in verschiedenen Sprachen; Engadiner Autoren
8 000 vols 27446

Didaktisches Zentrum des Kantons Zug (DZ), Bibliothek, Hofstr. 15, 6300 ***Zug***
T: +41 417282930; Fax: +41 417282931; E-mail: dz-zug@datazug.ch; URL: www.dz-zug.ch
1980; Arlene Wyttenbach
22 000 vols; 50 curr per; 6 000 av-mat; 850 sound-rec; 500 digital data carriers; 155 models, 600 games
BIS 27447

Staatsarchiv des Kantons Zug, Bibliothek, Verwaltungszentrum an der Aa, Aabachstr. 5, Postfach 857, 6301 ***Zug***
T: +41 417285680; Fax: +41 417285689; E-mail: staatsarchivzug@allg.zg.ch; URL: www.zug.ch/staatsarchiv
Dr. Peter Hoppe
Zuger- u Schweizergeschichte
12 000 vols; 140 curr per; 6 000 govt docs; 100 personal archives and bequests
VSA 27448

Bibliothek MeteoSchweiz Zürich, Bundesamt für Meteorologie und Klimatologie (MeteoSchweiz) (Federal Office of Meteorology and Climatology), Krähbühlstr. 58, 8044 ***Zürich***
T: +41 442569235; Fax: +41 442569278; E-mail: bibliothek@meteoschweiz.ch; URL: www.meteoschweiz.ch
1881; Gregor Stork
Atmosphärenwiss (Meteorologie u Klimatologie)
22 000 vols; 74 curr per; 30 e-journals; 141 diss/theses; 5 000 govt docs; 2 300 av-mat; 40 sound-rec; coll of climatic and weather data, publs of World Meteorological Organization (WMO)
BIS, DGI 27449

Cinémathèque suisse / Dokumentationsstelle Zürich (Swiss Film Archive), Neugasse 10, 8005 ***Zürich***; Postfach 1215, 8031 Zürich
T: +41 438182465; Fax: +41 438182466; E-mail: cszh@cinematheque.ch; URL: cinematheque.ch
30-er Jahre; Bernadette Meier
Filmdatenbank mit 65 000 Filmtiteln
8 000 vols; 180 curr per; 8 000 digital data carriers; 80 000 dossiers, 200 000 photos, 200 000 files, 3 000 catalogs, 2 350 000 pressclippings
BIS 27450

Genealogie-Forschungsstelle, Kirche Jesu Christi HLT, c/o Kirche Jesu Christi der Heiligen der Letzten Tage, Herbstweg 120, 8050 ***Zürich***
T: +41 443201097; URL: www.familysearch.org
1894; Heiner Sannemann
2,2 mio geneological microfilm rolls (parish registers, registry office docs etc), 742 000 microfiches, 300 000 books and 4 500 periodicals (both only in Salt Lake City obtainable), 35,6 mio digital name authority files in ANCESTRAL FILE (AF), 725 mio digital individual name authority files in INTERNATIONAL GENEALOGICAL INDEX (IGI), 36 mio name autority files in PEDIGREE RESOURCE FILE
The name authority files are mainly persons who lived before 1920. Obtainable are docs from 110 countries (USA, CANADA, Europe, South-America, Asia and Africa). Worldwide 3 700 research centres in 88 countries. 27451

infoDoc RADIX, Stampfenbachstr. 161, 8006 ***Zürich***
T: +41 443604105; Fax: +41 443604114; E-mail: infodoc@radix.ch; URL: www.infodoc-radix.ch
1986; Diego Morosoli
WHO-Publikationen, Datenbank Integration von Migrant/-innen
7 000 vols; 40 curr per; 300 diss/theses; 500 govt docs; 2 microforms; 500 av-mat; 200 sound-rec; 250 digital data carriers; press clippings coll, online docs
libr loan; BIS 27452

Israelitische Cultusgemeinde Zürich (ICZ), Bibliothek, Lavaterstr. 29, 8002 ***Zürich***; Postfach 282, 8027 Zürich
T: +41 442832250; Fax: +41 442832223; E-mail: bibliothek@icz.org; URL: www.icz.org
1939; Dr. Yvonne Domhardt
Hebräisch, Jiddisch, Judaistik, Judaica, Hebraica; Texte der Bibliothek des Breslauer Rabbinerseminars, Sonderkatalog (ab 2012 digital auf www.icz.org) zum "Schweizer Judentum (= Auswertung der Schweizer jüdischen Presse ab 1901ff)"
50 000 vols; 60 curr per; 450 diss/theses; 360 music scores; 600 av-mat; 130 sound-rec; 50 digital data carriers; Musical and personal bequest of Max Ettinger (1874-1951)
libr loan; BIS, IG WBS, VEBTIS 27453

Kunsthaus Zürich, Bibliothek, Rämistrasse 45, 8001 ***Zürich***; Postfach, 8024 Zürich
T: +41 442538531; Fax: +41 442538651; E-mail: bibliothek@kunsthaus.ch; URL: opac.kunsthaus.ch
1787; Thomas Rosemann
Slg Dadaismus (542), Archiv der Zürcher Kunstgesellschaft und des Kunsthauses Zürich
245 171 vols; 333 curr per; 985 av-mat; 163 sound-rec; 260 digital data carriers; 4 bequests/archives
libr loan 27454

Museum Rietberg, Bibliothek, Villa Schönberg, Gablerstr. 14, 8002 ***Zürich***; mail address: Gablerstr. 15, 8002 Zürich
T: +41 442063131; Fax: +41 442063132; URL: www.rietberg.ch
1953; Katharina Thölen
Eastasian Art
15 500 vols; auction cat, offprints 27455

Museumsgesellschaft Zürich, Bibliothek, Limmatquai 62, 8001 ***Zürich***
T: +41 442545005; Fax: +41 442524409; E-mail: bibliothek@mug.ch; URL: www.museumsgesellschaft.ch
1834; Dr. Thomas Ehrsam
Belletristik (dt, engl, franz, ital), Sachliteratur
130 000 vols; 465 curr per; 20 mss; 400 sound-rec; 50 digital data carriers
BIS 27456

pro juventute, Bibliothek, Seehofstr. 15, Postfach, 8032 ***Zürich***
T: +41 442567777; Fax: +41 442567778; E-mail: info@projuventute.ch; URL: www.projuventute.ch
1920; Dr. R. Zwolanek
Jugend, Familie, Erziehung, Weiterbildung, Sozialarbeit, Gesundheit, Soziologie, Medienpädagogik
50 000 vols; 160 curr per; 20 av-mat
libr loan; BIS 27457

Pro Senectute Schweiz, Bibliothek und Dokumentation, Bederstr. 33, 8002 ***Zürich***
T: +41 442838981; Fax: +41 442838984; E-mail: bibliothek@pro-senectute.ch; URL: pro-senectute.ch/bibliothek
1975; Lisa Wyss Ribi

Gerontologie
22 500 vols; 250 curr per; 700 av-mat; 200 sound-rec; 100 digital data carriers; 100 offprints
BIS 27458

Schweizer Alpenclub (SAC), c/o Zentralbibliothek Zürich, Zähringerplatz 6, 8001 ***Zürich***
T: +41 442683100; Fax: +41 442683290; E-mail: zb@zb.uzh.ch; URL: www.zb.uzh.ch
1890; Thomas Germann
Depository in Central Libr Zurich
25 000 vols; 500 curr per; 3 000 panoramas 27459

Schweizer Fernsehen, Textdokumentation, Fernsehstr. 1-4, 8052 ***Zürich***
T: +41 443055288; Fax: +41 443055280
1960; Herbert Staub
Fernsehproduktionen Schweizer Fernsehen
20 000 vols; 70 curr per; 52 000 microforms; 220 000 av-mat; 2 mio press clippings (1965-1996)
BIS 27460

Schweizerdeutsches Wörterbuch, Bibliothek, Auf der Mauer 5, 8001 ***Zürich***
T: +41 442513676; Fax: +41 442513672; E-mail: info@idiotikon.ch; URL: idiotikon.ch
1862; Dr. Hans-Peter Schifferle
Geschichte der Mundartforschung
10 000 vols; 16 curr per; 120 mss; 250 diss/theses; 630 maps; 100 sound-rec; 500 pamphlets and calendars 27461

Schweizerische Bibliothek für Blinde und Sehbehinderte (SBS), Grubenstr. 12, 8045 ***Zürich***
T: +41 443333232; Fax: +41 443333233; E-mail: bibliothek@sbszh.ch; URL: www.sbs-online.ch
1904; Heinz Zysset
10 460 braille titles, 2 280 braille notes, 13 500 digital audio books (DAISY), 4 000 large-print editions, 30 digital audio magazines (DAYSY)
IFLA, BIS, MediBuS 27462

Schweizerischer Arbeitgeberverband, Bibliothek, Hegibachstr. 47, Postfach, 8032 ***Zürich***
T: +41 444211717; Fax: +41 444211718; E-mail: verband@arbeitgeber.ch; URL: www.arbeitgeber.ch
1908; Dr. Hans Reis
Sozialpolitik, Arbeitsfragen, Arbeitsrecht, Wirtschaftspublizistik
10 000 vols; 250 curr per 27463

Schweizerisches Institut für Kinder- und Jugendmedien (SIKJM), Bibliothek, Zeltweg 11, 8032 ***Zürich***
T: +41 432683904; Fax: +41 432683909; E-mail: bibliothek@sikjm.ch; URL: www.sikjm.ch
1968; Roger Meyer
Bettina Hürlimann-Slg, Elisabeth Waldmann-Slg, Johanna Spyri-Archiv
45 000 vols; 30 curr per; 2 700 mss; 1 300 photos, slides, posters
libr loan; BIS 27464

Schweizerisches Nationalmuseum, Landesmuseum Zürich, Bibliothek, Museumstr. 2, Postfach, 8021 ***Zürich***
T: +41 442186531; Fax: +41 442112949; E-mail: bibliothek@snm.admin.ch; URL: www.nationalmuseum.ch/bibliothek
1898; Doris Haben
Ur- und Frühgeschichte, Kunstgeschichte/Kunstgewerbe, Kunsthandwerk, Volkskunde, Militaria, Museumskunde, Konservierung, Restaurierung, Schweizer Geschichte
125 000 vols; 781 curr per; 65 av-mat; 15 sound-rec; 200 digital data carriers; auction cat, historical almanacs
libr loan; BIS, AKMB 27465

Schweizerisches Sozialarchiv, Stadelhoferstr. 12, 8001 ***Zürich***
T: +41 432688740; Fax: +41 432688759; E-mail: sozarch@sozarch.uzh.ch; URL: www.sozialarchiv.ch
1906; Dr. Anita Ulrich
Sozialwissenschaften, Gesellschaft, Politik, Sozialismus, Kommunismus, soziale Zustände und Bewegungen, Arbeiterbewegung, Sozialpolitik, Fürsorge, Arbeit, Kultur; archive dept, photo archive dept, documentation dept
228 323 vols; 1 476 curr per; 151 e-journals; 29 085 mss; 2 729 microforms; 1 685 av-mat; 1 725 digital data carriers; 8 528 boxes of press clippings, 6 199 boxes of brochures and pamphlets, 100 131 graphic docs
libr loan; BIS, VSA, IALHI, ICA, Memoriav 27466

SIK-ISEA Schweizerisches Institut für Kunstwissenschaft, Bibliothek (SIK-ISEA Swiss Institute for Art Research), Zollikerstr. 32, 8008 ***Zürich***; Postfach 1124, 8032 Zürich
T: +41 443885160; Fax: +41 443815250; E-mail: bibliothek@sik-isea.ch; URL: www.sik-isea.ch/bibliothek
1951; Regula Fischer
Schweizer Kunst, Kunst in der Schweiz; Library "Prof. Eduard Hüttinger / Schenkung Annette Bühler" (European art history, ca. 14 000 vols); photo archive; 20th and 21st c Swiss art
155 845 vols; 300 curr per; 10 e-journals; 804 av-mat; 15 778 auction catalogs, Dept "Documentation SIK-ISEA": artists' dossiers with millions of newspaper clippings, invitation cards, artists' bequests
libr loan; BIS, AKMB, IFLA 27467

Staatsarchiv des Kantons Zürich, Bibliothek, Winterthurerstr. 170, 8057 ***Zürich***
T: +41 446356911; Fax: +41 446356905; E-mail: staatsarchivzh@ji.zh.ch; URL: www.staatsarchiv.zh.ch
1837; Felix Stamm
Turicensia, Archivlit
30 000 vols; 200 curr per; 100 000 govt docs; 200 digital data carriers; 20 000 brochures
VSA, BIS 27468

Stadtarchiv Zürich, Bibliothek, Neumarkt 4, 8001 ***Zürich***
T: +41 442668646; Fax: +41 442668649; E-mail: stadtarchiv@zuerich.ch; URL: www.stadt-zuerich.ch/stadtarchiv
1798; Karin Beck
Turicensia
42 000 vols; 56 curr per; 100 av-mat; 100 digital data carriers; 350 000 press clippings
BIS, VSA, ICA 27469

Thomas-Mann-Archiv der Eidgenössischen Technischen Hochschule Zürich, Schönberggasse 15, 8001 ***Zürich***
T: +41 446324045; Fax: +41 446321254; E-mail: tma@tma.gess.ethz.ch; URL: www.tma.ethz.ch
1956; Dr. Thomas Sprecher
Lit Nachlass u Bibl Thomas Manns, Werkmss, Briefe, Werkausgaben, Übersetzungen, Sekundärlit
24 796 vols; 1 216 mss; 65 microforms; 578 av-mat; 406 sound-rec; 37 digital data carriers; 21 993 letters, 79 881 press clippings, 4 501 photos, 387 prints
BIS 27470

Public Libraries

Stadtbibliothek Baden, Mellinger Str. 19, 5400 ***Baden***
T: +41 562008320; Fax: +41 562008344; E-mail: stadtbibliothek@baden.ag.ch; URL: www.stadtbibliothek.baden.ch
1837; Pia-Maria Rutishauser
Balneologie, Regionalschriftum Baden/Aargau (Badensia), Kleinbadensia (ca. 600)
57 361 vols; 267 curr per; 10 e-books; 15 mss; 96 incunabula; 765 maps; 1 843 av-mat; 3 259 sound-rec; 382 digital data carriers
libr loan; BIS, SAB, SBD 27471

Allgemeine Lesegesellschaft Basel, Münsterplatz 8, 4051 ***Basel***; Postfach 1092, 4001 Basel
T: +41 612614349; E-mail: lesegesellschaft@balcab.ch; URL: www.lesegesellschaft-basel.ch
1787; Ruth Marzo
116 000 vols; 168 curr per; 10 govt docs; 252 sound-rec; 60 digital data carriers
BIS 27472

Stadtbibliothek Basel, Bibliothek Zentrum, Im Schmiedenhof 10, 4051 ***Basel***; Postfach, 4001 Basel
T: +41 612641111; Fax: +41 612641190; E-mail: info@stadtbibliothekbasel.ch; URL: www.stadtbibliothekbasel.ch
1807; Kurt Waldner
6 branch libs
199 530 vols; 610 curr per; 1 465 music scores; 4 293 maps; 7 840 av-mat; 40 765 sound-rec; 5 771 digital data carriers; 1 770 games and slides
BIS, SAB/GTB 27473

Biblioteca cantonale, Viale Stefano Franscini 30A, 6500 ***Bellinzona***
T: +41 918141500; Fax: +41 918141509; E-mail: bcb-segr.sbt@ti.ch; URL: www.sbt.ti.ch/BCB
1987; Andrea Ghiringhelli
Documentazione regionale ticinese (DRT), Centro documentazione sociale (CDS), Servizio audiovisivi (SAV)
160 000 vols; 1 000 curr per; 5 e-journals; 300 maps; 2 000 av-mat; 300 sound-rec; 500 digital data carriers
libr loan; BIS 27474

Bibliothek am Guisanplatz / Bibliothèque Am Guisanplatz, Papiermühlestr. 21A, 3003 ***Bern***
T: +41 313224243; E-mail: sue.stammbach@gs-vbs.admin.ch; URL: www.guisanplatz.ch
1848; Sue Stammbach
Autografenslg Dr Georg Heberlein, von Tschamer, Schafroth, Keller, Militärpostkartensammlung;Militärwesen, Zivilschutz, Katastrophenschutz und -hilfe, Politik, Geopolitik, Sicherheitspolitik, Umwelt, Verkehr, Energie, Finanzen und Wirtschaft, Recht, Länderinformationen, Kommunikation, Architektur, Bauten, Logistik, Führungspsychologie
350 000 vols; 1 000 curr per; 15 e-journals; 1 500 mss; 52 000 diss/theses; 21 000 govt docs; 25 000 maps; 1 800 microforms; 1 000 av-mat; 200 sound-rec; 100 digital data carriers; 7 bequests, 10 000 photos
libr loan; BIS 27475

Kornhausbibliothek, Kornhausplatz 18, 3011 ***Bern***; Postfach, 3000 Bern 7
T: +41 313271020; Fax: +41 313271021; E-mail: sekretariat@kornhausbibliotheken.ch; URL: www.kornhausbibliotheken.ch
1947; Margrit Dutt
Special libr for Design
291 600 vols; 300 curr per; 15 000 av-mat; 38 000 sound-rec; 4 400 digital data carriers
BIS, SAB 27476

SBD.bibliotheksservice ag / SSB.service aux bibliothèques sa / SSB.servizio per biblioteche sa / SSB.servetsch per bibliotecas sa, Zähringerstr. 21, 3012 ***Bern***; Postfach 8176, 3001 Bern
T: +41 313018266; Fax: +41 313026597; E-mail: info@sbd.ch; URL: www.sbd.ch
1969; Anita Büttiker
70 000 vols 27477

Stadtbibliothek / Bibliothèque de la Ville, Dufourstr. 26, 2502 ***Biel***; Postfach 3116, 2500 Biel 3
T: +41 323291100; Fax: +41 323291199; E-mail: biblio@bibliobiel.ch; URL: www.bibliobiel.ch
1765; Clemens Moser
220 000 vols; 266 curr per; 4 mss; 383 maps; 6 000 av-mat; 15 000 sound-rec; 2 100 digital data carriers
libr loan; BIS 27478

Mediathek Wallis, Schlossstr. 30, 3900 ***Brig-Glis***
T: +41 279230551; Fax: +41 279243613; E-mail: mw.brig@mediathek.ch; URL: www.brig-glis.ch/gemeinde/mediathek.php
85 000 vols; 210 curr per; 15 000 av-mat; 15 000 didactical media items
libr loan 27479

Stadtbibliothek der Burgergemeinde Burgdorf, Bernstr. 5, 3400 ***Burgdorf***
T: +41 344200070; Fax: +41 344200071; E-mail: info@bus-biblio.ch; URL: www.bus-biblio.ch
1729; Z. Kump
Hist Arch d Burgergemeinde Burgdorf
50 000 vols; 52 curr per; 3 000 av-mat; 2 000 sound-rec; 500 digital data carriers
libr loan; BIS, SAB 27480

Bibliothèque municipale de Carouge, Bd des Promenades 2 bis, 1227 ***Carouge***
T: +41 223078400; Fax: +41 223078406; E-mail: bibliotheque@carouge.ch; URL: www.carouge.ch/
1857
60 000 vols; 65 curr per; 100 DVDs 27481

Bündner Volksbibliothek, Arcas 1, 7000 ***Chur***
T: +41 812526140; E-mail: volksbibliothek.chur@spin.ch; URL: www.volksbibliothek.ch
1973; Anita Devonas
Children's and Young Adult Library
90 000 vols; 100 curr per; 5 000 sound-rec; 1 400 digital data carriers
Special service for school classes without school libraries in communities of Kanton Graubünden 27482

Bibliobus de l'Université populaire jurassienne, Rue de chêtre 36, CP 2038, 2800 ***Delémont*** 2
T: +41 324214010; Fax: +41 324214019; E-mail: bibliobus@bibliobus.ch; URL: www.bibliobus.ch
1977; Jean-Claude Guerdat
63 000 vols 27483

Bibliothèque Braille Romande et livre parlé, Association pour le Bien des Aveugles et malvoyants (Swiss French-Speaking Braille and Talking Books Library), Bourg-de-Four 34, 1204 ***Genève***
T: +41 223177900; Fax: +41 223177911; E-mail: bbr@abage.ch; URL: www.abage.ch/bbr/bbr.aspx
1902; Anne Pillet
4 559 sound-rec; 1 612 digital data carriers; 3 206 braille titles
libr loan; IFLA, BIS, AIFBD 27484

Bibliothèque de la Cité, Bibliothèques municipales de la Ville de Genève, Pl des Trois-Perdrix 5, 1204 ***Genève***; CP 3930, 1211 Genève 3
T: +41 224183250; Fax: +41 224183251; URL: www.ville-ge.ch/bmu/
1931; Isabelle Ruepp
613 200 vols; 250 curr per; 116 400 av-mat
libr loan; BIS 27485

Bibliothèque Musicale de la Ville de Genève, Maison des Arts du Grütli, 16, rue Général-Dufour, 1204 ***Genève***
T: +41 224183580; Fax: +41 224183581; E-mail: bmus@ville-ge.ch; URL: www.ville-ge.ch/bmus
1962
Collection de programmes de concerts et d'ffiches du 19e siècle à nos
3 000 vols; 58 curr per; 750 mss; 35 000 music scores; 1 100 sound-rec; 50 digital data carriers
IAML, BIS 27486

Société de lecture (SDL), Bibliothèque, Grand-Rue 11, 1204 ***Genève***
T: +41 223114590; E-mail: info@societe-de-lecture.ch; URL: www.societe-de-lecture.ch/index.php?id=20
1818
350 000 vols 27487

Regionalbibliothek Hochdorf, Brauiplatz 5, Postfach 564, 6281 ***Hochdorf***
T: +41 419103110; Fax: +41 419103115; E-mail: info@bibliothek-hochdorf.ch; URL: www.bibliothek-hochdorf.ch
20 000 vols; 10 000 further media items 27488

Bibliobus neuchâtelois, Rue Collège 90, 2300 ***La Chaux-de-Fonds***
T: +41 329680018; E-mail: Bibliobus.neuchatelois@rpn.ch
1974; Philippe Schindler
140 000 vols; 30 curr per; 1 500 av-mat; 5 500 sound-rec; 6 500 digital data carriers
BIS 27489

Bibliomedia, Bibliocentre de la Suisse Romande, Rue César-Roux 34, 1000 *Lausanne*
T: +41 213407030; Fax: +41 213407031; E-mail: lausanne@bibliomedia.ch; URL: www.bibliomedia.ch
1920; Laurent Voisard
100 000 vols
BIS   27490

Bibliothèque municipale de Lausanne, Pl Chauderon 11, 1003 *Lausanne*
T: +41 213156911; Fax: +41 213156007; E-mail: bml@lausanne.ch; URL: www.lausanne.ch/bibliotheque
1934; Frédéric Sardet
Livres d'art, de cuisine, d'ésotérisme, guides de voyage, romans policiers, bandes dessinées, théâtre
572 787 vols; 230 curr per; 20 mss; 472 sound-rec; 932 sound-rec; 40 000 press clippings
BIS, CLP   27491

Cercle litteraire, Bibliothèque, Pl Saint-François 7, 1000 *Lausanne*; mail address: 1002 Lausanne
1883
60 700 vols; 142 curr per; 105 maps
libr loan   27492

Bibliothèque de la Ville du Locle, Rue Daniel-Jean Richard 38, 2400 *Le Locle*; CP 816, 2400 Le Locle
T: +41 329388989; Fax: +41 329338987; E-mail: bvll@ne.ch
1830; Pierre-Yves Tissot
55 000 vols; 100 curr per; 50 mss; 1 incunabula; 150 maps; 520 microfilms; 320 av-mat; 100 sound-rec; 3 500 photos and postcards
BIS   27493

Stadtbibliothek Luzern (Public Library Lucerne), Löwenplatz 11, 6004 *Luzern*; mail address: Löwenplatz 10, 6004 Luzern
T: +41 414170707; Fax: +41 414170708; E-mail: info@bvl.ch; URL: www.bvl.ch
1969; Josef Birrer
63 000 vols; 150 curr per; 500 maps; 8 000 av-mat; 12 000 sound-rec; 2 000 digital data carriers
BIS, SAB   27494

Stadtbibliothek Olten, Hauptgasse 12, 4600 *Olten*
T: +41 622128955; Fax: +41 622128253; E-mail: stadtbibliothek@olten.ch; URL: www.bibliothekolten.ch
1898; Christoph Rast
Biografien; Libr 'Stiftung für Kunst des 19. Jahrhunderts'
150 000 vols; 160 curr per; 50 diss/theses; 1 000 govt docs; 700 maps; 2 000 av-mat; 1 600 sound-rec; 1 000 digital data carriers; press clipping coll from 1780
libr loan; BIS   27495

Kantonsbibliothek Obwalden, Grundacherhaus, 6060 *Sarnen*
T: +41 416601365; Fax: +41 416601437; E-mail: kantonsbibliothek@ow.ch; URL: www.kantonsbibliothek.ow.ch
1892; Regula Hirter
Obwalden, Niklaus von Flüe
60 000 vols; 100 curr per; 1 000 av-mat; 1 200 sound-rec
libr loan; BIS, SAB/GTB   27496

Stadtbibliothek Schaffhausen, Münsterplatz 1, 8200 *Schaffhausen*; Postfach 3368, 8201 Schaffhausen
T: +41 526248262; Fax: +41 526248205; E-mail: stadtbibliothek@stsh.ch; URL: www.bibliotheken-schaffhausen.ch
1636; Dr. René Specht
Johannes von Müller-Nachlass; Government Department Libr (theology), Libr 'Agnesenschütte'
253 700 vols; 650 curr per; 500 mss; 270 incunabula; 1 300 maps; 400 microforms; 3 300 av-mat; 5 800 sound-rec; 60 digital data carriers
libr loan; BIS   27497

Bibliomedia Schweiz, Rosenweg 2, 4500 *Solothurn*
T: +41 326249020; Fax: +41 326249028; E-mail: solothurn@bibliomedia.ch; URL: www.bibliomedia.ch
1920; Dr. Peter Wille
Bibliocenter Deutschschweiz in Solothurn, Bibliocentre de la Suisse romande in

Lausanne, Bibliocentro della Svizzera italiana in Biasca
563 500 vols; 1 800 av-mat; 5 000 sound-rec; 300 digital data carriers
BIS, SAB   27498

Zentralbibliothek Solothurn, Kantons-, Stadt- und Regionalbibliothek, Bielstr. 39, 4500 *Solothurn*; mail address: 4502 Solothurn
T: +41 326276262; Fax: +41 326276200; E-mail: sekretariat@zbsolothurn.ch; URL: www.zbsolothurn.ch
1763/1882/1930; Peter Probst (Admin.), Verena Bider (Wissensch.)
Solodorensia, Grafische Slg, Handschriftensammlung, Privatnachlass-Slg, Inkunabelsammlung, Slg Altes Buch 1500-1900, Historische Musiksammlung, Exlibris, Sealsfieldiana; Music dept (sound archives), youth libr, hist music coll
970 000 vols; 600 curr per; 1 900 mss; 1 200 incunabula; 15 000 diss/theses; 17 000 music scores; 3 000 maps; 22 000 microforms; 799 av-mat; 64 000 sound-rec; 10 000 digital data carriers; 8 437 graphic docs (incl. 2 900 art prints, 4 900 posters)
libr loan; AIBM, BIS, SAB
The 'Zentralbibliothek Solothurn' is a regional library with a regional deposit agenda; it was merged from a town library from the 18th c (with holdings from family libraries) and a canton library from the 19th c (with holdings from secularized monastry libraries). Therefore the library is a mixture of a public and a scientific library.   27499

St. Galler Freihandbibliothek, Katharinengasse 11, 9000 *St.Gallen*
T: +41 112236023; E-mail: info@freihandbibliothek.ch; URL: www.freihandbibliothek.ch
1969; Christa Oberholzer
40 125 vols; 119 curr per; 216 maps; 66 av-mat; 10 776 sound-rec; 5 208 digital data carriers
SAB-CLP   27500

Stadtbibliothek Thun, Bahnhofstrasse 6, 3600 *Thun*
T: +41 332258501/07; Fax: +41 332258508; E-mail: stadtbibliothek@thun.ch; URL: www.thun.ch/stadtbibliothek
1785; Jeanne Froidevaux Müller
Thunensia
50 000 vols; 80 curr per; 1 600 govt docs; 3 500 av-mat; 6 500 sound-rec; 500 digital data carriers; 17 000 brochures
libr loan; BIS, SAB
Attached: Talacker Libr   27501

Bibliothèque Médiathèque Municipale, Quai Perdonnet 33, 1800 *Vevey*
T: +41 219213349; Fax: +41 219214882; E-mail: biblio@vevey.ch; URL: www.vevey.ch/bibliotheque
1805; Christian Graf
Fonds Clarisse Francillon (5 000 vols), Fonds ancien (12 000 vols)
60 000 vols; 38 curr per; 3 000 sound-rec; 400 digital data carriers   27502

Gemeindebibliothek Wettingen, Mattenstr. 26, 5430 *Wettingen*
T: +41 564271110; Fax: +41 564300626; E-mail: bibliothek@wettingen.ag.ch; URL: www.wettingen.ch/de/verwaltung/aemter/?amt_id=2563
Marie-Madeleine Stöckli Wetzel
26 130 vols; 38 curr per; 270 maps; 4 300 sound-rec; 2 070 digital data carriers   27503

Stadtbibliothek Wil, Marktgasse 61, 9500 *Wil*
T: +41 719135333; URL: www.stadtwil.ch/desktopdefault.aspx/tabid-644/
30 000 media items
libr loan   27504

Winterthurer Bibliotheken, Obere Kirchgasse 6, 8402 *Winterthur*; Postfach 132, 8402 Winterthur
T: +41 522675145; Fax: +41 522675140; E-mail: stadtbibliothek@win.ch; URL: www.winbib.ch
1660; Dr. Hermann Romer
Africana, Vitodurania
1 117 733 media items; 1 000 curr per; 5 e-journals; 56 507 mss; 150 incunabula; 2 050 diss/theses; 16 827 music scores;

15 082 maps; 3 605 microforms; 17 998 av-mat; 38 867 sound-rec; 191 046 pictures, 66 261 heraldic figures
BIS, SAB   27505

Bibliothèque Publique d'Yverdon-les-Bains, Ancienne Poste 4, CP 217, 1401 *Yverdon-les-Bains*
T: +41 244251313; E-mail: biblio@yverdon-les-bains.ch; URL: biblio.ylb.ch/
1761; Cécile Vilas
Drucke aus Yverdon 17. u 18. Jh; Bibl von Elie Bertrand; wissenschaftl Werk von Prof R. Kasser (koptische Sprache u Lit); medizinhist Bestand; Bibl und Archiv der Oekonomischen Gesellschaft von Yverdon (gegründet 1761)
65 000 vols; 35 curr per; 60 mss; 2 incunabula
CLP, SIGEGS   27506

Stadtbibliothek Zofingen, Hintere Hauptgasse 20, 4800 *Zofingen*; Postfach, 4800 Zofingen
T: +41 627521653; E-mail: stadtbibliothek@zofingen.ch; URL: www.stadtbibliothek-zofingen.ch
1693; Cécile Vilas
Humanistenbriefe (400 handschriftliche Briefe, 16. Jh.), Handschriften, rund 27 000 gedruckte Werke des 15.-19. Jh
114 217 vols; 82 curr per; 400 mss; 54 incunabula; 592 maps; 3 588 av-mat; 3 791 sound-rec
BIS, SAB   27507

Stadt- und Kantonsbibliothek Zug, St.-Oswalds-Gasse 21, 6301 *Zug*; Postfach 52, 6301 Zug
T: +41 41 7282313; Fax: +41 41 7282380; E-mail: bibliothek@stadtzug.ch; URL: www.bibliothekzug.ch
1836; Heinz Morf
Zuger Sammlung (regionale Dokumentation)
154 424 vols; 882 curr per; 1 incunabula; 1 663 microforms; 8 405 av-mat; 14 296 sound-rec; 2 755 digital data carriers; 2 028 dossiers with lowercases, 9 225 docs "Digitale Bibliothek", 23 161 photos of documentary value (inkl. maps)
libr loan; BIS, SAB, Memoriav   27508

Pestalozzi-Bibliothek Zürich (PBZ) (Pestalozzi Library Zurich), Zähringerstrasse 17, 8001 *Zürich*; Postfach, 8022 Zurich
T: +41 442049696; Fax: +41 442049600; E-mail: info@pbz.ch; URL: www.pbz.ch
1896; Dr. Josephine Siegrist
284 000 vols; 8 600 curr per; 3 150 maps; 43 600 av-mat; 61 300 sound-rec; 23 700 digital data carriers; 2 700 games
libr loan; IFLA, BIX, IBL, SAB   27509

Syria

National Libraries

AL-Assad National Library, Malki St, P.O. Box 3639, *Damascus*
T: +963 11 3320803; Fax: +963 11 3320804; E-mail: contact@alassad-library.com; URL: www.alassad-library.com
1984; Ghassan Lahham
262 000 vols; 19 000 Arabic mss
IFLA   27510

University Libraries, College Libraries

University of Aleppo, Library, *Aleppo*
T: +963 21 236130; Fax: +963 21 229184
1960; Mustafa Jassoumeh
36 000 vols   27511

Arab Language Academy of Damascus, Library, P.O. Box 327, *Damascus*
T: +963 11 3713103; Fax: +963 11 3733363
1919
Arabic Islamic legacy, Arabic linguistic studies
15 000 vols; 500 mss   27512

Institut Français d'Etudes Arabes, Bibliothèque, P.O. Box 344, *Damascus*
T: +963 11 3330214; Fax: +963 11 3327887; E-mail: ifead@net.sy
1922; Genevieve Joly
Arabic and Islamic studies (hist, art and architecture, lit, religion)
70 000 vols; 1 260 curr per; 4 000 maps; 580 microforms   27513

University of Damascus, Library, P.O. Box 3003, *Damascus*
T: +963 11 2119840; URL: www.damasuniv.shern.net/english/index_en.htm
1919; Dr. Nizar Oyoun El-Soud
169 000 vols; 3 830 curr per; 25 mss; 1 750 diss/theses
libr loan   27514

Al-Baath University, Library, P.O. Box 77, *Homs*
T: +963 31 431440; Fax: +963 31 426716
1979; Lina Maasrani
63 000 vols   27515

Government Libraries

Directorate-General of Antiquities and Museums, Library, *Damascus*
1947
14 000 vols   27516

Ministry of Education, Library, *Damascus*
13 000 vols   27517

Special Libraries Maintained by Other Institutions

International Center for Agricultural Research in the Dry Areas (ICARDA), Library, P.O. Box 5466, *Aleppo*
T: +963 21 2213433; Fax: +963 21 2213490; E-mail: m.maliha@cgiar.org
1977; Nick Maliha
Arid areas, sustainable development, farming systems, breeding, biotechnology
9 000 vols; 500 curr per; 420 diss/theses; 28 digital data carriers
IAALD, AGLINET   27518

Centro Cultural Hispánico, Biblioteca, P.O. Box 224, *Damascus*
T: +963 11 714003
1957; Emilio Pérez Acosta
11 000 vols   27519

Public Libraries

Al Zahiriah, Bab el Barid, *Damascus*
T: +963 11 2112813
1880; Ms Sama El Mahassini
Coll of rare pre-1900 Arabic bks
100 000 vols; 13 000 mss; 50 000 periodicals   27520

Taiwan

National Libraries

National Central Library, 20 Chung Shan South Rd, *Taipei* 10040
T: +886 2 23619132; Fax: +886 2 23821489; URL: www.ncl.edu.tw
1933/1954; Dr. Fang Rung Juang
Chinese rare bks, Tunhang mss, Han insribed wooden slats, Stone rubbings, ancient maps; 1 branch libr, Resource and Information Center for Chinese Studies
2 000 000 vols; 58 840 curr per
libr loan; IFLA, FID, LAC, ALA, NLA, ASIS, AAS   27521

General Research Libraries

Academia Sinica, Library (Chinese Academy od Sciences), Nankang, *Taipei* 11529
T: +886 2 7822120; Fax: +886 2 7853847; URL: www.sinica.edu.tw/~libserv
1928
780 300 vols   27522

University Libraries, College Libraries

National Chiayi Teachers College, Library, 85 WenLong Mingsuin, *Chia-yi*, Hsien, 621
T: +886 5 2263411; Fax: +886 5 2264643
1960; Chang
185 140 vols; 889 curr per; 900 diss/theses; 50 maps; 297 815 microforms; 7 226 av-mat
libr loan   27523

National Chung Cheng University, Library, 168, University Rd, Min-Hsiung, 62102 *Chia-yi*
T: +886 5 2720411 ext. 15001; Fax: +886 5 2720527; E-mail: library@ccu.edu.tw; URL: www.lib.ccu.edu.tw
1989; Kuoyu Wang
887 052 vols; 6 418 curr per; 44 668 e-journals; 104 467 e-books; 14 475 diss/theses; 60 000 govt docs; 2 360 music scores; 1 348 maps; 551 859 microforms; 22 620 av-mat; 6 332 digital data carriers
libr loan; ILCA   27524

Taiwan Agricultural Research Institute, Library, 2 Min Chuan Rd, *Chia-yi* 60014
T: +886 5 2771341; Fax: +886 5 2773630
1919; Prof. Horng-Mau Chen
30 000 vols; 70 curr per; 55 diss/theses; 4 000 govt docs; 3 000 sound-rec
CLA   27525

Chung Yuan Christian University, Library, 22. Pu-jen. 320 Pu-Chungli, *Chungli*
T: +886 3 4563171; Fax: +886 3 4563160; URL: www.lib.cycu.edu.tw
1955
359 377 vols; 61 966 curr per; 9 483 e-journals; 10 000 periodicals
libr loan; TLA   27526

National Chiao-Tung University, Library, 1001 Ta Hsueh Rd, *Hsinchu* 30050
T: +886 35 712121; Fax: +886 35 714031; URL: www.lib.nctu.edu.tw
1960; Dr. Ruei-Chuan Chang
482 602 vols; 480 493 curr per; 6 500 e-journals; 31 002 av-mat   27527

National Tsing-Hua University, Library, 101, Sec 2, Kuang Fu Rd, *Hsinchu* 30043
T: +886 35742995; Fax: +886 3 5724034; URL: www.lib.nthu.edu.tw
1967; James T. Lin
585 928 vols; 2 302 curr per; 10 373 e-journals; 737 microforms; 48 645 av-mat; 199 digital data carriers   27528

National Dong Hwa University, Library, Da-Hsueh Rd 1, Sec 2, Shou Feng, *Hualien* 974, Taiwan
T: +886 3 8632838; Fax: +886 3 8632800; E-mail: library@mail.ndhu.edu.tw; URL: www.ndhu.edu.tw
1995
180 000 vols; 7 567 curr per; 8 500 av-mat   27529

Kaohsiung Medical College, Library, 100 Shin-chuan 1st Rd, *Kaohsiung* 80708
T: +886 7 3121101; Fax: +886 7 3210564; E-mail: lib@kmu.edu.tw; URL: www.kmu.edu.tw/~lib
1954
134 000 vols; 2 408 curr per
libr loan   27530

National Kaohsiung First University of Science and Technology, Library, 415 Chien-Kung Rd, *Kaohsiung* 807
T: +886 7 3814526; Fax: +886 7 3838435; URL: www.nkfust.edu.tw/english/1.html
1963
127 000 vols   27531

National Kaohsiung Normal University, Library, 116 Ho-ping 1st Rd, *Kaohsiung* 80284
T: +886 7 7172930; Fax: +886 7 7117002; E-mail: lib@kmu.edu.tw; URL: www.kmu.edu.tw
1970; Cecilia Shu-huei Sun
310 000 vols; 2 000 curr per
libr loan; CLA   27532

National Sun Yat-Sen University, Main Library, 70 Lien-hai Rd, *Kaohsiung* 804
T: +886 7 5252000 ext2426; Fax: +886 7 5252400; URL: www.lib.nsysu.edu.tw
573 326 vols; 5 976 curr per   27533

National Taiwan Ocean University, Library, 2 Peining Rd, *Keelung* 202
T: +886 2 24622192; Fax: +886 2 24620724; URL: www.lib.ntou.edu.tw
1964; Wen-ping Yang
Marine food science, aquaculture, electronic engineering, fisheries, oceanography, marine engineering, navigation, shipping management, naval architecture, river and harbor engineering, nautical technology
218 991 vols; 2 131 curr per
libr loan; LAC   27534

National Pingtung University of Science and Technology, Library, 1 Hseuh-Fu Rd, Neipu, *Ping-Tung*, Pingtung County 912
T: +886 8 7703202; Fax: +886 8 7740338; E-mail: king@mail.npust.edu.tw; URL: www.npust.edu.tw
1955
Animal and plant production, forest products technology, aquaculture, food science, agrobusiness management, civil engineering, environmental protectiopn, resources conserv ation technology, rural planning and landscaping
96 470 vols; 577 curr per   27535

China Medical College, Library, 91 Hsuehshih Rd, *Taichung* 404
T: +886 4 2053366; Fax: +886 4 2030969
1958; Chih-Chan Tsai
168 156 vols; 3 475 curr per; 9 573 e-journals; 30 901 av-mat
libr loan; Science & Technology Library Network   27536

Feng Chia University, Library, 100 Wenhwa Rd, Sea-T'un, *Taichung* 40724
T: +886 4 4517250; Fax: +886 4 4514907; E-mail: library@fcu.edu.tw; URL: www.lib.fcu.edu.tw
1961; Shu-ling Lin
469 000 vols   27537

National Chung-Hsing University, Chung-Cheng Memorial Library, 250 Kuokuang Rd, *Taichung* 40227
T: +886 4 22840290; Fax: +886 4 22873454; E-mail: wanjenchang@dragon.nchu.edu.tw; URL: www.lib.nchu.edu.tw
1946; Woei Lin
586 025 vols; 3 440 curr per; 21 e-journals; 10 483 diss/theses; 200 govt docs; 2 music scores; 1 301 maps; 271 208 microforms; 43 254 av-mat; 5 348 sound-rec; 247 digital data carriers
libr loan; CLA   27538

National Taichung Teachers College, Library, 140 Min-Sheng Rd, *Taichung* 40302
T: +886 4 2283181
1923; Lü ChuiQing
202 728 vols; 701 curr per; 26 e-journals; 2 077 maps; 258 971 microforms   27539

Tunghai University, Library, 181 Taichung Harbour Rd, Sec 3, *Taichung* 40704
T: +886 4 3590121; Fax: +886 4 3590361; URL: www.lib.thu.edu.tw
1958; Cheng-tung Lin
600 000 vols; 6 200 curr per   27540

National Cheng Kung University, University Library, 1 Ta- Hsueh Rd, *Tainan* 70101
T: +886 6 2757575 ext 65760; Fax: +886 6 2378232; E-mail: em81001@email.ncku.edu.tw; URL: www.lib.ncku.edu.tw
1927; Dr. Ming-Tzong Yan
Medical College Libr, Management Science Libr, Liberal Arts Libr
820 000 vols; 1 161 curr per
USBE, IATUL   27541

Chinese Culture University, Chung-Cheng Library, 55 Hwa Kang, Yangmingshan, *Taipei* 111
T: +886 2 28610511; Fax: +886 2 28615144; URL: www.pccu.edu.tw/intl/page/english/english04.htm¡tle01
1962
854 000 vols; 5 600 curr per   27542

Fu Hsing Kang College, Library, 70 Chung Yang N. Rd, Sec 2, *Taipei*
T: +886 2 891-2135/-2136; Fax: +886 2 8913140
1959; Lo Mou-Pin
Politics
157 000 vols   27543

Fu Jen Catholic University, Library, 510 Chung-cheng Rd, Hsinchuang shin, *Taipei* 24205
T: +886 2 29031111; Fax: +886 2 29017391; URL: lib.fju.edu.tw
1963; Dr. Chwen-chwen-Chang
699 000 vols
libr loan; ALA, LAC   27544

– Social Sciences Library, 510 Chung Cheng Rd, Hsinchuang, Taipei Hsien 24205
T: +886 2 2903-1111/-2670; Fax: +886 2 29089243; URL: lib.fju.edu.tw/sslib
1963; Tina M. Hsieh
257 000 vols; 1 208 curr per
libr loan   27545

National Cheng-Chi University, Library, 64, Chih-nan Rd, Section 2, *Taipei* 116
T: +886 2 29387090; Fax: +886 2 29390455; E-mail: clchen@nccu.edu.tw; URL: www.lib.nccu.edu.tw
1954; Prof. Jyi-Shane Liu
Social Sciences Info Ctr, The Institute of International Relations branch libr, Public & Business Administration Education Center's Libr
1 931 742 vols; 5 374 curr per; 320 e-journals; 30 000 diss/theses; 507 063 govt docs; 847 811 microforms; 18 547 av-mat; 7 603 sound-rec; 1 753 press clippings
libr loan; IFLA, CLA, CALISE, ILCA   27546

– Center for Public and Business Administration, Education Library, 187, Ching Hua St, Taipei 106
T: +886 2 3419151
1963; Cheng Shun-chan
54 000 vols   27547

– Institute of International Relations, Library, 64, Wan-Shou Rd, Taipei 11625
T: +886 2 939492; Fax: +886 2 2344920; E-mail: flsuen@nccu.edu.tw; URL: www.lib.nccu.edu.tw
1961; Kai-jen Hung
Newspaper clips
107 000 vols; 1 000 curr per; 36 296 microforms; 8 000 vols of bound newspapers
libr loan; LAC   27548

National Taipei College of Nursing, Library, 365 Ming-Te Rd, Peitou, *Taipei* 112
T: +886 2 28227101; Fax: +886 2 28205680; E-mail: jiun@ntcn.edu.tw; URL: www.ntcn.edu.tw/LIBRARY/index.htm
1953; Shwu-Jiun Wang
70 992 vols; 863 curr per; 1 135 microforms; 12 982 av-mat
libr loan; LAC   27549

National Taipei Teachers College, Library, 134 Ho-hing E Rd, Sec 2, P.O. Box 4444, *Taipei* 10659
T: +886 2 7321104
1962; Yin-huei Hsu
337 695 vols; 2 431 curr per   27550

National Taiwan College of Arts, Library, 59, Section 1, Ta Kuan Rd, Pan-chiao City, *Taipei* 220
T: +886 2 22722181; Fax: +886 2 29687563; E-mail: b0029@fall.ntca.edu.tw; URL: 140.131.21.7/eng/index.html
1957
104 000 vols; 503 curr per   27551

National Taiwan Institute of Technology, Library, 43 Keelung Rd, Sec 4, *Taipei* 10772
T: +886 2 27333141; Fax: +886 2 27376107; E-mail: lib@mail.ntust.edu.tw
1974
279 326 vols   27552

National Taiwan Normal University, Library, 162, Sec. 1, Ho-ping E. Rd, *Taipei* 10610
T: +886 2 23939740; Fax: +886 2 23937135; URL: www.lib.ntnu.edu.tw
1946; Heng-Cheng Liang
1 139 449 vols
libr loan; ALA, IFLA, CLA   27553

National Taiwan University, Library, 1 Roosevelt Rd, Sec 1, *Taipei* 106
T: +886 2 23632274; Fax: +886 2 23620886; E-mail: tul@ntu.edu.tw; URL: www.lib.ntu.edu.tw
1928; Ming-der Wu
11 coll on Chinese drama, Western Asia mat; Archives of Dutch East Indian Co, Tam-Hsin Archive
2 000 000 vols; 22 000 curr per
libr loan; LAC, ALA, ASIS   27554

– College of Law and Social Sciences and Public Health, Library, 21 Hsu-chou Rd, Taipei
T: +886 2 23519641
1949   27555

– Medical Library, 1 Jen Ai Rd, Sec 1, Taipei 10013, Taiwan
T: +886 2 2356-2207/-2208; Fax: +886 2 23938354, 23560831; E-mail: library@ha.mc.ntu.edu.tw; URL: ntuml.mc.ntu.edu.tw
1900; Huei-Chu Chang
202 415 vols; 1 598 curr per; 366 diss/theses; 2 622 microforms; 1 434 av-mat; 173 sound-rec; 205 digital data carriers; 542 pamphlets
libr loan; LAC   27556

Shih Hsin University, Main Library, 1 Lane 17, Sec 1, Mu-cha Rd, *Taipei* 116
T: +886 2 22368225 ext 256, 264; Fax: +886 2 22360429; E-mail: ysc@cc.shu.edu.tw; URL: www.shu.edu.tw/shu-e/htm/03administration/03-06library.htm
1956; Shien-chiang Yu
Comics dept
352 110 vols; 3 746 curr per; 286 maps; 28 776 microforms; 5 498 av-mat; 10 298 sound-rec; 40 digital data carriers
libr loan; LAI, STICA   27557

Soochow University, Library, 70 Lin-Hsi Rd, Wai-Shuang Hsi, Shihlin, *Taipei* 11102
T: +886 2 8819471 ext 5112; Fax: +886 2 8819471; URL: www.library.scu.edu.tw
1954; Joseph C.I. Wang
506 110 vols; 2 944 curr per; 4 715 govt docs; 483 maps; 5 958 microforms; 7 000 av-mat; 4 674 sound-rec; 40 digital data carriers; 206 pamphlets
LAC   27558

Taipei Institute of Technology, Library, 3, Sec 1, Shingsheng South Rd, *Taipei* 106
T: +886 2 27722171 ext 3100; Fax: +886 2 27762383; E-mail: library@ntut.edu.tw; URL: wwwlib.ntut.edu.tw
1948
112 000 vols   27559

Taipei Medical College, Library, 250 Wu Hsing St, *Taipei*
T: +886 2 27361661; Fax: +886 2 27386527; URL: library.tmu.edu.tw
1960; Chao-Chin Wei
83 000 vols   27560

Taipei Municipal Teacher's College, Library, 1 Aikuo West Rd, *Taipei* 10001
T: +886 2 3113040; Fax: +886 2 3831070; URL: lib.tmtc.edu.tw
1920; Liu Chen-Tzu
Children's lit, Mandarin lit
236 985 vols; 1 757 curr per; 300 maps; 73 709 microforms; 2 500 av-mat; 294 sound-rec; 10 digital data carriers
LAC   27561

Taipei National University of the Arts, Library, 1 Hsueh Yuan Rd, Peitou, *Taipei* 112
T: +886 2 28938798; Fax: +886 2 28938793; E-mail: master@library.tnua.edu.tw; URL: lib.tnua.edu.tw
1982; Tzu-An Yan
Music, musicology, theater, dance, performing arts, fine arts
425 209 vols; 610 curr per; 7 969 e-journals; 63 214 e-books; 25 149 microforms; 14 240 av-mat; 19 174 sound-rec; 651 digital data carriers; 2 903 posters, 539 art reproductions
libr loan; Library Association of the Republic of China (Taiwan), MLA   27562

Tamkang University, Chueh Sheng Memorial Library, 151 Yingchuan Rd, Tamsui, *Taipei* 25137
T: +886 2 26227821; Fax: +886 2 26226149; E-mail: library@mail.tku.edu.tw; URL: www.tku.edu.tw
1950; Prof. Hong-chu Huang
Spec Taiwan coll; Maritime Museum
880 221 vols; 2 672 curr per; 100 000 non-book mat
libr loan; IFLA, USBE, ALA, ASIS   27563

Tamsui Oxford University College, Library, 32 Yeh-li St, Tamsui County, *Taipei*
1965; Hwe-chin Hwang
Taiwanese and Japanese Literature
200 000 vols; 728 curr per; 30 diss/theses; 39 maps; 71 digital data carriers
libr loan; LAC   27564

Tatung Institute of Technology, Library, 40 Chungshan N Rd, Sec 3, *Taipei* 10451
T: +886 2 25925252; Fax: +886 2 25921813; URL: www.library.ttu.edu.tw
1957; Ru-Jou Hsieh
600 000 vols; 700 curr per
libr loan   27565

National Taitung Teachers College, Library, 684 Chunghua Rd, Sec 1, *Taitung* 95004
T: +886 2 28920498; Fax: +886 2 28944044
1947; Charlie Sun-pun Aos
66 800 vols
libr loan; LAC   27566

School Libraries

Chungcheng Military Academy, Library, Fengshan Post Office, 8894 / 9, *Kaohsiung* 830
30 000 vols   27567

Wen Tzao Ursuline Junior College of Modern Languages, Library, 900 Mintsu 1st Rd, *Kaohsiung* 807
T: +886 7 3426031 ext 207; Fax: +886 7 3508865
1966; Li-hsien Wang
101 041 vols; 510 curr per; 53 maps; 20 392 av-mat; 12 193 sound-rec; 921 digital data carriers
libr loan   27568

Private Cheng-Li College of Commerce, Library, 313 Wenhua Rd, Sec 1, *Panch'iao* 220
1965
14 800 vols   27569

National Taichung Institute of Commerce, Library, 129 Samming Rd, Sec III, *Taichung* 404
T: +886 4 2211181; Fax: +886 4 2211181 ext 2516; E-mail: web@www3.ntcic.edu.tw
1963; Liao Shih Kuan
156 337 vols; 595 curr per; 5 maps; 1 002 microforms; 2 290 sound-rec; 562 digital data carriers
libr loan   27570

Private Ching-i Girls' Institute for Literary Composition, Kai-Hsia Library, 170 Fuhsing Rd, Sec III, *Taichung* 400
1949
255 000 vols   27571

Private Sun Yat-Sen Medical School, Library, 113 Tach'ing Rd, Sec 2, *Taichung* 400
1960
14 000 vols   27572

Finance and Quartermaster School, Library, Chich'i Ts'uan, Panchiao Cheng, *Taipei*
1912
42 300 vols   27573

Private China College of Marine Technology, Library, 212 Yenp'ing North Rd, Sec 9, *Taipei* 111
T: +886 2 8122291
1966; Ann Hsiao
45 200 vols
CLA   27574

Private College of Practical Housekeeping, Library, 71 Peian Rd, Lane 501, *Taipei* 104
1958
54 200 vols   27575

Private Mingchih College of Technology, Library, 84 Kungchuan Rd, *Taipei* 243
1964
20 300 vols   27576

Chung Cheng College of Technology, Library, P.O. Box 7902-14, *T'aoyüan* 330
T: +886 3 3801126; URL: www.lib.ccit.edu.tw
1960
130 000 vols; 475 diss/theses
libr loan; Science & Technology Library Network   27577

Private Chien Hsing Junior College of Technology, Library, 29 Lungkang Rd, Sec 2, *T'aoyüan* 320
T: +886 34 583678
1966
58 350 vols   27578

Government Libraries

Parlamentary Secretariate of Taiwan Province, Chao-Chin Library, 734 Chungcheng Rd, *Taichung* 431
1962
38 300 vols   27579

Council for Economic Planning and Development, Executive Yuan, Library, 9th Fl 87, Nanking E.R. Sec 2, *Taipei* 10408
T: +886 2 5634944
1959; Shu-san Chang
Taiwan Economic development planning
32 000 vols
VLA, Library Consortium on Humanities and Social Science   27580

National Parliamentary Library, Legislative Yuan of Republic of China, 1 Chung-Shan South Rd, *Taipei* 100-40
T: +886 2 23585670; Fax: +886 2 23585672; E-mail: ly910@lyfw.ly.gov.tw; URL: npl.ly.gov.tw
1947; Karl Min Ku
Coll of newspaper clipping files; The Newspaper Clipping Dept
6 800 vols; 688 curr per; 960 diss/theses; 140 maps; 300 microforms; 38 digital data carriers; 30 kinds of government gazettes, 50 kinds of administrative rpts, 500 slides
libr loan; IFLA, ASIS, APLAP, IALL, ALA, CLA   27581

Ecclesiastical Libraries

China Evangelical Seminary, Library, 101 Ting Chou Rd Sec. 3, *Taipei* 100, Taiwan
T: +886 2 23659151; Fax: +886 2 23650225; E-mail: library@ces.org.tw; URL: www.ces.org.tw
1970; Mei-Chih Chen
Chinese Church Res
42 000 vols; 256 curr per; 140 maps; 86 microforms; 5 083 sound-rec; 6 digital data carriers
libr loan; ATLA   27582

Taiwan Nazarene Theological College, Library, 100 Sheng Ching Rd, Kuan Tu, *Taipei*, Taiwan 11272
T: +886 2 2858-3602/-5609; Fax: +886 2 28582611; E-mail: nazarene@ms18.hinet.net; URL: www.nazarene.org
1958; Ms Ching-Hwa Sun
9 000 vols; 30 curr per; 8 diss/theses; 30 govt docs   27583

Corporate, Business Libraries

Chinese Petroleum Corporation, Refining & Manufacturing, Research Center, Technical Library, 239 Min Sheng South Rd, *Chia-yi* 60036
T: +886 5 2224171; E-mail: techlib@rmrc.gov.tw
1977; Yu-chung Chou
SRI reports, NPRA reports, Chemical Abstracts (since 1907)
65 500 vols; 448 curr per; 358 mss; 800 diss/theses; 70 maps; 9 700 microforms; 98 av-mat; 44 sound-rec; 30 digital data carriers
libr loan; Science and Technical Library Network   27584

Taiwan Machinery Manufacturing Corp. Production Engineering Dept., Library and Technical Data Archives, 3 Taichi Rd, Hsiaokang, *Kaohsiung* 81235
T: +886 7 8020111; Fax: +886 7 8022129
1951; Y.N. Lin
Coll of Journal of Iron and Steel Institute (1911-1973), Metal Progress (1922-1970, 1972-1976, 1978-1985)
11 363 vols; 45 curr per; 2 maps; 151 av-mat; 22 sound-rec; 12 digital data carriers
libr loan   27585

Taiwan Sugar Research Institute, Library, 54 Sheng Chan Rd, *Tainan* 100
T: +886 62 2671911; Fax: +886 62 2685425
1932
47 000 vols   27586

Special Libraries Maintained by Other Institutions

Industrial Technology Research Institute, Library, 195 Chung-Hsing Rd, Sec 4, Chutung, *Hsinchu*
T: +886 35 820100; Fax: +886 35 820045
1981; C.H. Chung
130 000 vols
libr loan   27587

China Academy, Library, *Hwa Kang*, Yang Ming Shan
Sinology
450 000 vols   27588

Metal Industries Research and Development Centre, Library, 1001 Kaonan Highway, *Kaohsiung* 81103
T: +886 7 3513120; Fax: +886 7 3521528; E-mail: webmaster@mail.mirdc.org.tw; URL: www.mirdc.org.tw
1964; Mei R. Tsai
whole set of AWS Standards, ASTM Standards, CNS standards
16 214 vols; 454 curr per; 149 diss/theses; 36 maps; 158 av-mat; 96 sound-rec; 10 digital data carriers
libr loan   27589

Taiwan Fisheries Research Institute, Library, 199 Hou-Ih Rd, *Keelung* 220
T: +886 32 24622101; Fax: +886 32 24629388; E-mail: webmast@mail.tfrin.gov.tw; URL: www.tfrin.gov.tw
1933
16 000 vols
libr loan; IAMSLIC   27590

Academia Sinica – Institute of History and Philology, Fu Ssu-Nien Library, 130 Yenchiu Yüan Rd, Sec 2, *Nankang*, Taipei 11521
T: +886 2 26523136; Fax: +886 2 26523175; URL: lib.ihp.sinica.edu.tw/c/
1928; Juei-Hsiu Wu
400 000 vols; 1 200 curr per; 33 961 incunabula; 7 786 maps; 18 358 microforms; 216 sound-rec; 44 digital data carriers; 33 889 stone and bronze rubbings, 13 100 folk plays, 44 000 rare bks
libr loan; LAC   27591

Asian Vegetable Research and Development Center (AVRDC), Information and Documentation Services, 60 I-Min-Liao, P.O. Box 42, *Shanhua*, 741
T: +886 6 5837801; Fax: +886 6 5830009; URL: www.avrdc.org.tw
1972; Teng-Hui Hwang
Cabbage, Chinese cabbage, Garlic, Mungbean, Onion, Peppers, Shallot, Tomato, Soybean
28 760 vols; 642 curr per; 278 diss/theses; 84 maps; 11 microforms; 620 av-mat; 45 digital data carriers
libr loan; IAALD   27592

Taiwan Museum of Art, Library, 2 Wu-Chun West Rd, *Taichung* 403321
T: +886 4 3723552; Fax: +886 4 3721195; E-mail: artnet@art.tmoa.gov.tw; URL: sql.tmoa.gov.tw
1986
32 000 vols   27593

Academia Historica, Library, 406 Sec., 2 Pei Yi Rd, Hsintien, *Taipei*
T: +886 2 22171535; Fax: +886 2 22171640; E-mail: nha@academia.drnh.gov.tw; URL: www.drnh.gov.tw
1947
5 294 000 vols   27594

Academia Sinica – Institute of Chemistry, Library, 128 Yenchiu Yüan Rd, Sec 2, *Taipei* 115
T: +886 2 27898589; Fax: +886 2 26513416; E-mail: channel@gate.sinica.edu.tw; URL: www.chem.sinica.edu.tw/library/index_en.html
1958; Yeh
27 000 vols; 110 current print and online journal subscriptions   27595

Academia Sinica – Institute of Economics, Library, 128 Academia Rd Nangang, *Taipei* 11529
T: +886 2 2782-2791 ext 500; Fax: +886 2 2782-2791 ext 100; E-mail: library@econ.sinica.edu.tw; URL: www.econ.sinica.edu.tw/lib/English/mail_e_index.htm
1970; Chi-bin Liaw
35 758 vols; 743 curr per; 20 000 govt docs; 230 microforms; 3 digital data carriers
libr loan   27596

Academia Sinica – Institute of Ethnology, Library, 128 Yenchiu Yüan Rd, Sec 2, *Taipei* 115
T: +886 2 26523376; Fax: +886 2 26523378; E-mail: etchhy@gate.sinica.edu.tw; URL: www.sinica.edu.tw/ioe/tool/library/
1955; Chiang Hui-ying
Taiwan Contracts Materials in 1644-1945, Taiwan Genealogical Materials
141 208 vols; 459 curr per; 4 650 mss; 6 852 diss/theses; 4 568 govt docs; 6 842 maps; 1 036 microforms; 23 digital data carriers; 7 806 items
libr loan; LAC   27597

Academia Sinica – Institute of Mathematics, Library, 128 Yenchiu Yüan Rd, Sec 2, *Taipei* 115
T: +886 2 9851211 ext 440; Fax: +886 2 9832232; E-mail: masfkuo@ccvax.sinica.edu.tw; URL: www.math.sinica.edu.tw/library/
1947; Shu-Fen Kuo
30 422 vols; 1 200 curr per; 38 000 mss; 65 av-mat; 4 digital data carriers
libr loan   27598

Academia Sinica – Institute of Modern History, Library, 130 Yen-Chiu-Yüan Rd, Sec 2, P.O. Box 3337, *Taipei* 115
T: +886 2 27898290; Fax: +886 2 27898289; URL: www.sinica.edu.tw:8080/cgi-bin/jsu/brwhtm.cgi?o=dbrwhtm
1955; Jiu-Jung Lo
Mainland China Source Materials & Publ
188 633 vols; 724 curr per; 18 157 maps; 17 976 microforms; 71 sound-rec; 2 digital data carriers
libr loan   27599

Academia Sinica – Institute of Physics, Library, 130 Yenchiu Yüan Rd, Sec 2, *Taipei* 115
T: +886 2 7616075
1962; Su-ching Tsai
11 000 vols; 125 curr per
libr loan   27600

Academia Sinica – Life Science Library, 128 Yenchiu Yüan Rd, Sec 2, *Taipei* 115
T: +886 2 7822120; URL: www.botany.sinica.edu.tw
1996
23 000 vols; 1 500 microforms; 1 800 journal titles, 52 000 bound journals, 10 databases
libr loan   27601

Agricultural Science Information Center, 14 Wenchou St, 3rd Fl, P.O. Box 7-636, *Taipei*, Taiwan 10616
T: +886 2 23626222; Fax: +886 2 23632459
1978; Becky Chen
Coll of Agricultural Docs in Taiwan
8 000 vols; 638 curr per
libr loan; IFLA   27602

American Resource Center, Library, 54 Nanhai Rd, *Taipei* 100
T: +886 2 23327981; Fax: +886 2 23056757; E-mail: aitarc@mail.ait.org.tw
1979; Patricia Wang
9 000 vols; 200 curr per; 4 000 microforms; 700 digital data carriers
27603

Central Geological Survey, MOEA, Library, 2. Lane 109. Hua-Hsin Street. Chung-Ho, *Taipei* 235; P.O. Box 968, Taipei
T: +886 2 29462793; Fax: +886 2 29429291; E-mail: cgs@moeacgs.gov.tw; URL: www.moeacgs.gov.tw/english/index.jsp
1944
12 000 vols
27604

Chang Gung Memorial Hospital, Library, Nr.199. Tunghwa Rd, *Taipei*
T: +886 3 3281200; URL: www.cgmh.org.tw
1977; Meiling Hsieh
2 branch libs
20 000 vols; 630 curr per; 1 120 av-mat; 104 digital data carriers
libr loan; LAC, Science and Techn Netw
27605

Chinese Postal Museum, Library, 45 Chungkung South Rd, Sec 2, *Taipei* 100
T: +886 2 23945185
1966
Philately
45 000 vols
libr loan
27606

Council of Agiculture (COA), Library, 37 Nanhai Rd, *Taipei* 100
T: +886 2 3812991; Fax: +886 2 3310341; E-mail: coa@mail.coa.gov.tw; URL: www.coa.gov.tw
1984
18 000 vols; 88 curr per
27607

Examination Yüan Library, 1 Shihyüan Rd, *Taipei* 116
1931
28 500 vols
27608

Historical Research Commission of Taiwan Province, Library, 111 Yen Ping S Rd, *Taipei*
1949
20 200 vols
27609

Hong's Foundation for Education and Culture, 12 F-1, No 9, Sec 2, Roosevelt Rd, *Taipei*
T: +886 2 3965505; Fax: +886 2 3955009; URL: www.how.org.tw
1975; Celia C. Hong
Performing Arts (Chinese Opera), Folklore, Children lit; Tai Chung Sub-Libr (for children), Kao Hsiong Sub-Libr
12 800 vols
LAC
27610

Hsin Yi Foundation, Institute of Child Education, Resource Center, 3F, 75 Chung-Ching S Rd, Sec 2, *Taipei*
T: +886 2 3965303; Fax: +886 2 3965015; URL: www.hsin-yi.org.tw/hsin-yi/hsin-yi_pag4.asp
1988; Sing-Ju Ho
Childhood Education, Children's Lit, Caldecott Award Picturebooks
12 000 vols; 15 curr per; 1 182 pamphlets
LAC
27611

National Bureau of Standards, Ministry of Economic Affairs, Library, 185 Hsin-hai Rd, Sec 2, 3rd Fl, *Taipei* 106
T: +886 2 27380007; Fax: +886 2 27352656; URL: www.bsmi.gov.tw
1979; Ching-I Peng
Standards, metrology, patents and trademarks
20 000 vols; 500 curr per
libr loan; LAC
27612

National Institute for Compilation and Translation, Library, 247 Choushan Rd, *Taipei* 107
T: +886 2 33225558; Fax: +886 2 33225559; URL: www.nict.gov.tw
1932
60 000 vols
27613

National Museum of History, Library, 49 Nan Hai Rd, *Taipei* 10728
T: +886 2 23610270; Fax: +886 2 23610171; URL: www.nmh.gov.tw
1955
12 000 vols
27614

National Palace Museum Library, Wai-shuang-hsi, Shih-lin, *Taipei* 111
T: +886 2 28812021; Fax: +886 2 28821440; E-mail: npmlib@npm.gov.tw; URL: www.npm.gov.tw
1968; Chao-ling Sung
217 348 rare bks, 386 600 Ch'ing documents
112 700 vols; 1 135 curr per; 24 754 microforms; 3 156 genealogies on microfilm and assorted family histories, David Gredzens coll of 1 917 titles on the Himalayas, Central Asia, Tibet and the Orient
libr loan; ALA, CLA
27615

National Science Council, Library, 106 Ho-ping East Rd, Sec 2, *Taipei* 106
T: +886 2 27377501; Fax: +886 2 27377668; E-mail: nsc@nsc.gov.tw; URL: web.nsc.gov.tw
1959
24 000 vols
27616

Private Ancestral Temple Library of Wang Ch'ih, 2 Alley 14, Lane 64, Nanking W Rd, *Taipei* 102
12 000 vols
27617

Research Institute for Wines, Taiwan Tobacco and Wine Monopoly Bureau, Library, 13 Roosevelt Rd, Sec 6 Lane 142, *Taipei* 100
T: +886 2 9300375; Fax: +886 2 9300250
1972
9 311 vols; 174 curr per
libr loan
27618

Dr. Sun Yat-sen Library, 2F, 505 Jen Ai Rd, Sec 4, *Taipei* 116
T: +886 2 7297030; Fax: +886 2 7582460; URL: www.lib.nsysu.edu.tw
1927; Shaw Ming-Huang
Modern Chinese hist, Dr Sun Yat-sen, San Min Chu I
299 000 vols; 1 300 curr per; 53 570 incunabula
libr loan
27619

Tai Hsu Buddhist Library, 23 Chunghsiao E Rd, Sec I, *Taipei*
1960
11 100 vols
27620

Taiwan Forestry Research Institute, Library, 53 Nanhai Rd, *Taipei* 107
T: +886 2 23817107; Fax: +886 2 23142234
1945; Chen An-Chi
33 000 vols
27621

Tien Educational Center, Society of Jesus, Library, 22 Hsin Hai Rd, Sec I, *Taipei* 107
T: +886 2 3415865
1964; Fernando Mateos
Sinology, English language and lit, Sociology, Fine Arts, Hist, Bibliography, Religion, Theology
40 000 vols
LAC
27622

Veterans General Hospital, Library, 201 Shihp'ai Rd, Sec 2, *Taipei* 112
T: +886 2 8712121; Fax: +886 2 8757240; E-mail: library@vghtc.gov.tw; URL: www.vghtpe.gov.tw/doce/
1959; Josephine Y.T. Chuo
78 400 vols
libr loan; LAC, MLA
27623

Yun-Wu Library Foundation, 3-1 Lane 19, Hsing-Sheng South Rd, *Taipei* 10617
T: +886 2 7081574
1974
Social sciences, Humanities
59 000 vols
LAC
27624

Atomic Energy Council, Library, 80, Section 1, Chengkung Rd, *Yonghe*, Taipei County 23452
T: +886 2 82317919; Fax: +886 2 82317804; E-mail: public@aec.gov.tw; URL: www.aec.gov.tw
1955
Deposit libr at the Nat. Tsing Hua Univ of 36 000 vols ans 424 000 microcards
11 000 vols
27625

Public Libraries

Changhua County Library, 98 Kuangfu Rd, *Changhua* 500
T: +886 4 7510709; Fax: +886 4 7510753; URL: www4.cca.gov.tw
1936
59 000 vols
27626

Kaohsiung Municipal Library, 80 Minsheng 2nd Rd, *Kaohsiung* 800
T: +886 7 2112181; Fax: +886 7 28243418; URL: www.ksml.edu.tw
1945
81 500 vols
27627

Keelung City Library, 6 Lane 2, Yierh Rd, *Keelung* 200
1932
160 000 vols; 580 curr per
27628

Provincial Taichung Library, 291/3 Chingwu Rd, *Taichung* 400
1947
110 000 vols
27629

T'ainan City Library, 3 Kungyüan North Rd, *Tainan* 700
T: +886 6 2255146; Fax: +886 6 2210826; URL: www.tnml.tn.edu.tw
1919
62 800 vols
27630

Taipei Public Library, 125 Chien-Kuo S Rd, Sec 2, *Taipei* 107
T: +886 2 27552823; URL: www.tpml.edu.tw
1952
3 784 870 vols; 6 258 curr per
27631

T'aoyüan Hsien Cultural Center, Library, 21 Hsienfu Rd, *T'aoyüan* 330
URL: www.tyccc.gov.tw
1923
180 000 vols
27632

Tajikistan

National Libraries

Firdousi Tajik National Library, Rudaki, 36, 734025 *Dushanbe*
T: +992 372 274726; URL: www.nlc.gov.cn/old/nav/nlibs/tj/index.htm
1933; S. Mukhiddinov
2 990 000 vols; 443 036 curr per; 2 119 mss; 449 770 diss/theses; 62 370 music scores; 4 635 maps; 747 microforms
libr loan
27633

General Research Libraries

Tajik Academy of Sciences, Central Scientific Library, pr. Rudaki, 33, 734025 *Dushanbe*
T: +992 372 224224
1939; Dr. A.A. Aslitdinova
1 500 000 vols
27634

University Libraries, College Libraries

Pedagogical Institute, Library, Lenina, 222, *Chodshent*
T: +992 37765 62378
1932
173 000 vols
27635

Pedagogical Institute, Librarya, Rudaki, 105, *Dushanbe*
T: +992 372 41730
1931
329 000 vols; 200 mss
27636

Tajik Abu-Ali Ibn-Cina (Avicenna) State Medical Institute, Library, pr. Rudaki, 139, 734003 *Dushanbe*
T: +992 372 241253
1939
128 000 vols
27637

Tajik Agricultural University, Library, pr. Rudaki, 146, 734017 *Dushanbe*
T: +992 372 247207; Fax: +992 372 245808
1931; Helena Ivanovna Chaika
Agronomy, agrochemistry, agricultural economics, agricultural mechanization, hydraulic engineering, plant protection, veterinary science, zootechnics
400 000 vols
libr loan
27638

Tajik State University, University Library, pr. Rudaki, 17, 734025 *Dushanbe*
T: +992 372 233981
1948; Rustam Yarbabaev
1 039 000 vols
27639

Kuljab State University, Library, Sangak Safarov, 16, *Kuljab*
T: +992 37713 560
1948; Sattor Murodovich Murodov
200 000 vols; Theses, govt docs, printed music, maps
libr loan
27640

Special Libraries Maintained by Other Institutions

Central State Archives of the Tajik Republic, Library, Negmata Karabaeva, 38/1, *Dushanbe*
T: +992 372 39571
1934
8 815 vols
27641

Historical Institute, Tajik Academy of Sciences, Library, Istravshon, 18, *Dushanbe*
T: +992 372 21686
1954
15 000 vols; 25 curr per; 200 mss
27642

Institute for Oriental Studies, Tajik Academy of Sciences, Library, Parvin, 8, *Dushanbe*
T: +992 372 25921
1958
20 000 vols; 36 curr per; 4 000 mss
27643

Institute of Language and Literature, Tajik Academy of Sciences, Library, Rudaki, 21, 734025 *Dushanbe*
T: +992 372 21342
1951
13 000 vols; 30 curr per; 500 mss
27644

Kitobkhonai davlatii patentio tekhnikii Markazi millii patentu ittilosti Chumkhurii Tochikiston (State Patent and Technical Library of the National Center for Patent and Information (SPTL NCPI)), 14A Aini St, 734042 *Dushanbe*; P.O. Box 734042, 734042 Dushanbe
T: +992 372 2276517, 2216184; Fax: +992 372 217154; E-mail: librarypatenttj@mail.ru; URL: www.gptb.tj
1965; Boymurod Boev
Department of the Patent Infomation and Documentation
182 342 vols; 1 862 e-journals; 18 e-books; 1 900 theses; 6 542 govt docs; 8 music scores; 246 maps; 506 402 microforms; 32 av-mat; 18 sound-rec; 6 600 digital data carriers; 14 million patents, Patent electronic documents (DVD, CD, microfiche and microfilm), 219 023 periodicals
libr loan
27645

Tajik State Historical Museum, Library, ul. Aini, 31, 734012 *Dushanbe*
T: +992 372 231544
Hist, art and culture of the peoples of Tajikistan
14 000 vols
27646

Public Libraries

District Library, Lenina, 144, *Chorog*
1935
137 000 vols
27647

Tanzania

National Libraries

National Central Library, Tanzania Library Services Board, Bibi Titi Mohamed St, *Dar es Salaam*; P.O. Box 9283, Dar es Salaam
T: +255 22 2150923; Fax: +255 22 2151100; E-mail: tlsb@africaonline.co.tz
1963; Eliezer A. Mwinyimvua
UN publ and its agences coll; Library for the blind
650 000 vols; 30 curr per; 30 maps; 5

av-mat; 70 sound-rec; 60 digital data carriers
libr loan; IFLA, COMLA, SCECSAL, TLA
27648

University Libraries, College Libraries

Eastern and Southern African Management Institute, Library and Documentation Centre, P.O. Box 3030, *Arusha*
T: +255 27 2508385; Fax: +255 27 2508285; E-mail: library@esamihg.ac.tz
1975; Grace Lema
Gender and development, African management, Environmental management; World Bank Depository Libr
15 000 vols; 30 curr per; 102 diss/theses; 60 av-mat; 30 digital data carriers; 16 000 pamphlets
libr loan; TLA, SCECSAL, UNAL   27649

Ardhi Institute, Library, P.O. Box 35176, *Dar es Salaam*
T: +255 22 71263
J. Jengo
Archit, land management, urban a rural planning
15 000 vols; 140 curr per   27650

Dar es Salaam Institute of Technology, Library, Morogoro Rd, PB 2958, *Dar es Salaam*
T: +255 22 2150174; Fax: +255 22 2152504; E-mail: principaldit@intafrica.com
1957
21 000 vols   27651

Dar es Salaam School of Accountancy, Library, P.O. Box 9522, *Dar es Salaam*
T: +255 22 2851035
1973; Mathew B.B. Magunga
Accounting, materials management
24 000 vols; 5 curr per; 3 govt docs
27652

Institute of Finance Management, Library, Shaaban Robert St, P.O. Box 3918, *Dar es Salaam*
T: +255 22 23666; E-mail: ifm@costech.ctn.apc.org
1972; M. Mhina
World banks (3 000)
32 000 vols; 100 curr per; 30 diss/theses; 1 000 govt docs; 20 av-mat
libr loan; TLA   27653

University of Dar es Salaam, Library, P.O. Box 35092, *Dar es Salaam*
T: +255 22 410241; Fax: +255 22 410241; E-mail: libdirec@udsm.ac.tz; URL: www.udsm.ac.tz
1969; Prof. J. Nawe
East Africana Coll, UN Coll, Law Coll
750 000 vols; 850 curr per; 950 diss/theses; 800 microforms; 40 digital data carriers
libr loan; IATUL   27654

– **Muhimbili University College of Health Sciences**, Library, P.O. Box 65012, Dar es Salaam
T: +255 22 275004; Fax: +255 22 75448; E-mail: alfeo@uclas.ud.co.tz
1968; Salum Saidy
Tropical medicine research
80 000 vols; 350 curr per; 250 mss; 450 diss/theses; 675 govt docs; 500 microforms; 275 av-mat; 95 sound-rec; 670 pamphlets; 500 clippings
IFLA, TLA, AHILA, BMA   27655

Sokoine University of Agriculture, Sokoine National Agricultural Library, P.O. Box 3022, *Morogoro*
T: +255 23 2604639; Fax: +255 23 2604639; E-mail: snal@suanet.ac.tz; URL: www.suanet.ac.tz/
1984; F.W. Dulle
75 000 vols; 707 diss/theses; 500 govt docs; 52 digital data carriers
libr loan   27656

College of African Wildlife Management, Library, P.O. Box 3031, *Moshi*
1963; Nellie J. Mwandoloma
Tourism and ecotourism, protected areas management, environmental studies, wildlife conservation
10 000 vols; 39 curr per; 90 diss/theses; 14 govt docs; 400 maps; 2 600 av-mat; 5 digital data carriers
TLA   27657

Co-operative College, Library, Sokoine Rd, P.O. Box 474, *Moshi*
T: +255 27 275-54401/-54402; Fax: +255 27 2750806; E-mail: coopcolib@yahoo.com
1963; Rose Maro
Women in co-operatives; Co-operative Information Centre; Co-operative Information Centre, Gender Documentation Centre, Health Information Centre
32 000 vols; 120 curr per; 5 000 mss; 30 diss/theses; 110 govt docs; 5 maps; 400 pamphlets, 150 clippings
libr loan; Tanzania Academic and Research Libraries Association, TLA
27658

Institute of Development Management (IDM), Library, P.O. Box 1, *Mzumbe*, Morogoro
T: +255 23 602-4380/-4383/-4384; Fax: +255 23 4011; E-mail: idm@raha.com
1972; Matilda S. Kuzilwa
30 000 vols; 545 curr per; 500 diss/theses; 20 govt docs; 6 maps; 1 000 microforms; 12 av-mat; 20 sound-rec; 15 digital data carriers
libr loan; IFLA, TLA   27659

School Libraries

College of Business Education, Library, P.O. Box 1968, *Dar es Salaam*
T: +255 22 31056
1966; M.A. Kimambo
24 000 vols; 8 curr per; 5 mss; 16 maps
libr loan   27660

Kivukoni Academy of Social Sciences, Library, P.O. Box 9193, *Dar es Salaam*
T: +255 22 2820047
1961; Mr M.H. El-Naim
27 000 vols; 90 curr per; 20 diss/theses; 400 govt docs; 1 av-mat; 1 sound-rec
libr loan   27661

Korogwe Teachers Training College, Library, P.O. Box 533, *Korogwe*
1960; Ushalwande N. Mphere
38 000 vols; 6 curr per   27662

Marangu College of National Education, Library, P.O. Box 3080, *Moshi*
Hassan M. Liyoka
27 000 vols   27663

Tarime College of National Education, P.O. Box 199, *Tarime*
1975
40 100 vols   27664

Government Libraries

East African Community Secretariat, Library, AICC Kilimanjaro Wing, 5th Fl, P.O. Box 1096, *Arusha*
T: +255 27 2504253; Fax: +255 27 2504255; E-mail: eac@eachg.org
1999; Sarah Kagoda-Batuwa
2 000 vols; 20 curr per; 5 diss/theses; 700 govt docs; 5 maps; 10 av-mat; 10 sound-rec; 5 digital data carriers; 50 pamphlets,100 clippings   27665

Ministry of Planning and Economic Affairs, Library, P.O. Box 9242, *Dar es Salaam*
T: +255 22 29418
1969; Vincensio A. Kazimzuri
15 000 vols; 50 curr per; 1 000 govt docs; 100 maps
libr loan; TLA   27666

National Assembly of Tanzania, Library and Research, P.O. Box 941, *Dodoma*
T: +255 61 22-2761/-2765; Fax: +255 61 324218; E-mail: bunge@intafrica.com; URL: www.bungetz.org
1971; Kileo I. Nyambele
6 500 vols; 60 govt docs; 10 maps; newspapers (15 titles)
libr loan; TLA, APLESA   27667

Ecclesiastical Libraries

St. Charles Lwanga Senior Seminary, Library, P.O. Box 3522, *Dar es Salaam*
1978
15 000 vols   27668

St. Paul's Senior Seminary Kipalapala, P.O. Box 325, *Tabora*
T: +255 62 4073; Fax: +255 62 4641
1925; Marcel Boiuin
16 400 vols; 37 curr per   27669

Makumira University College, Library, P.O. Box 55, *Usa River*
T: +255 57 553634; Fax: +255 811 512070; E-mail: Library@Makumira.ac.tz; URL: www.Makumira.ac.tz
1947; Ndelilio Mbise
Religion, East Africa, newspaper clippings on Makumira publis, religions in Tanzania; Research Institute
36 000 vols; 70 curr per; 1 000 diss/theses; 80 maps; 15 microforms; 55 av-mat; 32 sound-rec; 35 digital data carriers
TLA   27670

Special Libraries Maintained by Other Institutions

Amani Medical Research Centre, Library, P.O. Box 4, *Amani*
1949; Michael K. Frank
11 000 vols; 46 curr per
libr loan   27671

Arusha International Conference Centre (AICC), Library, P.O. Box 3081, *Arusha*
URL: www.aicc.co.tz
1978; L.A. Nchimbi
Conference proceedings, Tanzania laws amended regularly
10 000 vols; 100 curr per
libr loan; TLA   27672

British Council, Library, Samora Machel Ave / Ohio St, P.O. Box 9100, *Dar es Salaam*
T: +255 22 211-6574/-6577; Fax: +255 22 211-2669; E-mail: info@britishcouncil.or.tz; URL: www.britishcouncil.org/tanzania
1952; Al-Amin Yusuph
Information about the UK resources coll, Education enquiry coll, Good governance, gender, health and English language teaching resources; Education Services Dept
13 000 vols; 56 curr per; 1 700 av-mat; 80 sound-rec; 60 digital data carriers
libr loan   27673

Civil Service Training Centre, Library, P.O. Box 2574, *Dar es Salaam*
T: +255 22 23547
Hilda Maganga
25 000 vols; 10 curr per   27674

Institute of Adult Education, Library, P.O. Box 20679, *Dar es Salaam*
T: +255 22 252115
Susan S. Kayetta
11 050 vols; 7 400 curr per   27675

Science and Technology Library, Tanzania Commission for Science and Technology, Kijitonyama Ali Hassan Mwinyi Rd, P.O. Box 4302, *Dar es Salaam*
T: +255 22 2700745 to 2700750; Fax: +255 22 275313, 275314; E-mail: maktabasayansi@hotmail.com
1972; Matthias M. M. Masawe
12 000 vols; 75 curr per; 500 mss; 100 diss/theses; 150 govt docs; 50 maps; 600 microforms; 300 av-mat; 400 sound-rec; 20 digital data carriers; 1 300 research rpts and reprints, 500 research proposals
libr loan; IFLA, COMLA, SCECSAL
Depository libr for all research conducted in or about Tanzania   27676

Tanzania Bureau of Standards, Library, P.O. Box 9524, *Dar es Salaam*
T: +255 22 490418
1976; Ally I.A. Msovu
20 000 vols   27677

Coffee Research and Experimental Station, Library, P.O. Box 3004, *Moshi*
T: +255 27 2755004; Fax: +255 27 2752877
1920
8 500 vols   27678

Kilimanjaro Christian Medical Centre (KCMC), Library, PB, *Moshi*
1971
10 000 vols   27679

Public Libraries

Kibaha Public Library, P.O. Box 30063, *Kibaha*
T: +255 52 402143; Fax: +255 52 402324
1970; A Thuman Mombokaleo
Local coll
25 000 vols; 400 curr per; 150 mss; 200 diss/theses; 300 govt docs; 100 maps
libr loan   27680

Mwanza Public Library, P.O. Box 1363, *Mwanza*
T: +255 68 2314
1968; Alli Abdallah Batta
50 000 vols; 110 curr per
libr loan; TLA, IFLA   27681

Thailand

National Libraries

National Library of Thailand, Samsen Rd, Dusit, *Bangkok* 10300
T: +66 2 2815212, 2815313; Fax: +66 2 2810263; E-mail: suwaksir@emisc.moe.go.th; URL: www.nlt.go.th/
1905; Mrs Suwakhon Siriwongworawat
Gilt bookcase, Thai traditional bks coll
2 454 746 vols; 2 567 curr per; 324 477 mss; 10 000 incunabula; 90 000 diss/theses; 250 000 govt docs; 1 200 music scores; 821 maps; 700 318 microforms; 5 960 av-mat; 6 400 sound-rec; 442 digital data carriers
libr loan; ISSN-SEA, CONSAL, CDNL
27682

University Libraries, College Libraries

American University Alumni, Language Center, Library, 179 Rajadamri Rd, *Bangkok* 10330
T: +66 2 2516988; Fax: +66 2 2554632; E-mail: kanda@mozart.inet.co.th
1952; Mrs. Kanda Lekhakul
AUA Chiangmai Libr
14 000 vols; 50 curr per; 8 govt docs; 14 maps; 53 microforms; 410 av-mat; 1 029 sound-rec; 5 digital data carriers
libr loan   27683

Assumption University, St. Gabriel's Library, Huamark Campus, *Bangkok* 10240
T: +66 2 3004543 ext 3402; Fax: +66 2 7191544; E-mail: library@au.edu; URL: library.au.ac.th
Suprata Sinchaisuk
Br Martin's Coll, Prof Dr Srisakdi Charmonman's Coll
92 000 vols   27684

Bangkok University, Central Library, Rama 4 Rd, Klong-Toey, *Bangkok* 10110
T: +66 2 3503500; Fax: +66 2 2496274; E-mail: savanee.c@bu.ac.th; URL: library.bu.ac.th
1962; Mrs Dr. Savanee Charoensri
200 000 vols; 15 000 av-mat   27685

Chulalongkorn University, Center of Academic Resources, Library, Soi Chulalongkorn 12, Phyathai Rd, *Bangkok* 10330
T: +66 2 2184084; Fax: +66 2 2161312; E-mail: library@sasin.chula.ac.th; URL: www.car.chula.ac.th
1910; Aree Phongchaisophon
Prince Chandaburi Coll, County files-Thailand coll
264 753 vols; 1 066 curr per; 19 011 diss/theses; 23 827 govt docs; 785 maps; 1 900 microforms; 12 805 av-mat; 50 sound-rec; 72 digital data carriers; 1 630 clippings, 13 320 pamphlets
libr loan; JAALD   27686

– Faculty of Arts, Library, Phya Thai Rd, Bangkok 10330
T: +66 2 218-4851/-4852/-4853; E-mail: lsupariy@chula.ac.th; URL: www.chula.ac.th
1954; Supariya Lulitananda
3530 rare bks
118000 vols; 209 curr per; 2140 diss/theses; 2515 govt docs; 16 maps; 1100 av-mat; 264 sound-rec; 37 digital data carriers; 213 pamphlets, 388 clippings
libr loan; TLA 27687

– Faculty of Commerce and Accountancy, Library, Phya Thai Rd, Bangkok 10500
T: +66 2 254-1811/-1824
Kulinee Vacharobol
20000 vols; 50 curr per; 100 pamphlets
libr loan 27688

– Faculty of Communication Arts, Library, Phya Thai Rd, Bangkok 10330
T: +66 2 2150871
Nitaya Lekkeaw
14000 vols; 85 curr per; 12 maps; 95 clippings, 612 pamphlets
libr loan 27689

– Faculty of Education, Library, Chula Soi 3, Bangkok 10500
T: +66 2 25108713
Uraiwan Tamanee
58000 vols; 210 curr per; 3800 diss/theses; 64 maps; 16 microforms; 340 sound-rec; 3000 pamphlets 27690

– Faculty of Engineering, Library, Phyathai Rd, Bangkok 10500
T: +66 2 2525001
1935; Patpuree Chongfeungprinya
IEEE coll
41000 vols; 193 curr per; 1106 diss/theses; 750 maps; 10 microforms; 29 av-mat
libr loan 27691

– Faculty of Medicine, Library, Phya Thai Rd, P.O. Box 2328, Bangkok 10500
T: +66 2 2527852
1944; Chira Intakosum
Hist of Medicine; Learning Resources Ctr
45000 vols; 590 curr per; 204 diss/theses; 1200 govt docs; 100 av-mat; 1525 sound-rec; 1000 pamphlets, 50 video rec
TLA 27692

– Faculty of Political Science, Library, Phya Thai Rd, Bangkok 10500
T: +66 2 2513131
1948; Narumon Kijpaisalrattana
19000 vols; 100 curr per; 321 diss/theses; 305 maps; 32 microforms
libr loan 27693

– Faculty of Science, Library, Phya Thai Rd, Bangkok 10330
T: +66 2 2527984
1948; Pritchajean Nakornthap
18500 vols; 512 curr per; 40 pamphlets 27694

– Faculty of Veterinary Science, Library, Yothi Rd, Bangkok 10500
T: +66 2 2520980
Pringsri Inkaninant
9000 vols; 381 curr per
libr loan 27695

– Thailand Information Center, Center of Academic Resources, Phya Thai Rd, Bangkok 10330
T: +66 2 218-2957/-2958/-2959; Fax: +66 2 2153617; E-mail: kultida@mail.car.chula.ac.th
1968; Kultida Boon-Itt
57000 vols; 35000 mss
libr loan 27696

Dhurakijpundit University, Library and Information Center, 110/1-4 Prachacheun Rd, Donmuang, ***Bangkok*** 10210
T: +66 2 954730029 ext 400; Fax: +66 2 5913155; E-mail: lib@sudhi.dpu.ac.th; URL: www.dpu.ac.th
1968; Prof. Kamala Roonguthai
150000 vols
libr loan; ALA 27697

Kasetsart University, Main Library, Bangkhen Campus, 50 Phaholyothin Rd, ***Bangkok*** 10903
T: +66 2 5792539; Fax: +66 2 5611369; URL: www.ku.ac.th
1943; Mrs Piboonsin Wanatapongse
Since 1980, the Kasetsart Univ Main Libr has been serving as the national center of Thailand for the AGRIS and CARIS, the internat information systems for agricultural science and technology of the Food and Agriculture Organization (FAO) of the United Nations, and for the IBIC (International Buffalo Information Center) of the Internat Development Research Center (IDRC) of Canada
250650 vols; 2025 curr per
libr loan; JAALD 27698

King Monghut's Institute of Technology Ladkrabang, Central Library, Chalongkrung Rd, Ladkrabang District, ***Bangkok*** 10520
T: +66 2 7392220, 7392221; Fax: +66 2 7392220, 7392221; E-mail: kbsuree@kmitl.ac.th; URL: www.lib.kmitl.ac.th
1960; Dr. Wattanachai Pongnak
King Mongkut's Coll
250000 vols; 1000 curr per; 3000 diss/theses; 5000 av-mat
libr loan 27699

– Faculty of Architecture, Ladkrabang Campus, Library, Hua Ta-Ké District, Lad Krabang, Bangkok
T: +66 2 3267320
Srirat Songchatchai
9000 vols; 147 curr per; 93 pamphlets
libr loan 27700

– Faculty of Engineering, Library, Hua Ta-ké District, Lad Krabang, Bangkok 10520
T: +66 2 3269080
1960; Puachanee Komoldit
18000 vols; 287 curr per; 140 microforms
libr loan 27701

King Mongkut's Institute of Technology Thonburi, KMIT Library and Documentation Center, 91 Sukswasdu 48, Bangmod, Rasdraburana, ***Bangkok*** 10140
T: +66 2 4276321; Fax: +66 2 4283533; E-mail: info@mod.lib.kmitt.ac.th; URL: www.kmutt.ac.th
1960; Mrs Aim-Orn Srinilta
142000 vols; 4000 curr per
libr loan; TLA 27702

Krirk University, Library, 43/1111 Ram-Indra Rd, Bangkhem, ***Bangkok*** 10220
T: +66 2 5523500; Fax: +66 2 5523511; E-mail: phakaphan@krirk.ac.th; URL: www.krirk.ac.th
68000 vols; 419 curr per 27703

National Institute of Development Administration, Library and Information Center, Klong Chan Bangkapi, ***Bangkok*** 10240
T: +66 2 3775070; Fax: +66 2 3759026; E-mail: sutannee@nida.nida.ac.th; URL: www.nida.ac.th
1966; Ms Sutannee Keesiri
200000 vols; 800 curr per; 10000 diss/theses; 410 maps; 4513 microforms; 82 av-mat; 751 sound-rec; 2 digital data carriers; 34350 clippings and pamphlets, 10300 docs, NIDA researches and term papers
libr loan; ALA, IFLA 27704

Rajamangala Institute of Technology, Bangkok Technical Campus, Library, 2 Nang Linchee Rd, ***Bangkok*** 10120
T: +66 2 2863991
1952
Engineering, agriculture
61100 vols 27705

Ramkamhaeng University, Central Library, Ramkhamhaeng Rd, Huamark, ***Bangkok*** 10240
T: +66 2 3108118; Fax: +66 2 3108022; E-mail: admin@ram1.ru.ac.th; URL: www.ru.ac.th
1971
495000 vols
libr loan 27706

– Bangna Campus, Library 2, Bangna-Trad Rd, Prakanong, Bangkok 10260
T: +66 2 3166473
M.L. Benjalaksana Navaratna
27800 vols
libr loan 27707

Siam University, University Library, 235 Petchkasem Rd, Phasi Charoen, ***Bangkok*** 10163
T: +66 2 4570068; Fax: +66 2 4573982; URL: www.siamu.ac.th
1973; Kannika Sirikhet
60000 vols; 150 curr per 27708

Silpakorn University, University Library – Thapra Palace Campus, Napralan Rd, ***Bangkok*** 10200
T: +66 2 2225885; Fax: +66 2 2225885; E-mail: chira@su.ac.th; URL: www.thapra.lib.su.ac.th
1964; Ms Chirayoo Dasri
Arts, architecture, design, archaeology; Coll of art exhibition cat; Faculty of Architecture Libr
127000 vols; 267 curr per; 800 diss/theses; 200 govt docs; 100 music scores; 4000 maps; 250 microforms; 1000 av-mat; 2500 sound-rec; 150 digital data carriers; 1600 posters, 1000 videos, 1000 pictures, 2000 clippings
libr loan 27709

Srinakharinwirot University, Central Library, Sukhumwit Rd 23, ***Bangkok*** 10110
T: +66 2 2584002, 2584003; Fax: +66 2 2604514; E-mail: library@psm.swu.ac.th; URL: www.swu.ac.th/lib
1954; Dr. Vira Supakit
Buddhism coll, Hist coll, Children coll
381390 vols; 950 curr per; 32529 diss/theses; 16532 govt docs; 300 maps; 673 microforms; 7000 av-mat; 6436 sound-rec; 3 digital data carriers; 5000 pictures, 335 laserdisks
libr loan; IFLA 27710

– Mahasarakham Library, Academic Resource Center, Mahasarakham
T: +66 43 711162; Fax: +66 2 721769; E-mail: arec@isan1.msu.ac.th
1968; Dr. Chaweelak Boonyakanchana
Isan (Northeastern of Thailand) coll; Isan Info Ctr
117166 vols; 1183 curr per; 90 mss; 11364 diss/theses; 1147 govt docs; 100 maps; 1112 microforms; 5279 av-mat; 2628 sound-rec; 44 digital data carriers; 5200 pamphlets, 22019 clippings
libr loan; TLA, IFLA 27711

– Songkhla Library, Songkhla
Jumrern Saengdongkhae
94000 vols; 45 maps; 1500 pamphlets 27712

Sripatum University, University Library, Bangkhen Campus, 61 Phaholyotin Rd, Jatujak, ***Bangkok*** 10900
T: +66 2 5799120; Fax: +66 2 5611721; E-mail: nimnuan@buntharik.spu.ac.th; URL: www.spu.ac.th
Prof. Lamoon Ruttakorn
62000 vols 27713

Thammasat University, University Libraries, 2 Prachand Rd, ***Bangkok*** 10200
T: +66 2 6133518, 6133501; Fax: +66 2 6235173; E-mail: tulib@alpha.tu.ac.th; URL: www.library.tu.ac.th
1934; Nualchawee Suthamwong
Awarded Books, Banned Books, Rare Books, TU Coll
1089375 vols; 4748 curr per; 56528 diss/theses; 1403 maps; 17200 microforms; 27034 av-mat; 7403 digital data carriers; 31 online databases
libr loan; TLA, IFLA, ALA, ASIS&T 27714

– Faculty of Commerce, Library, Prachan Rd, Bangkok
T: +66 2 2241339; Fax: +66 2 2264515; E-mail: keerati@alpha.tu.ac.th; URL: www.library.tu.ac.th
1967; Somsri Kirtivudhikul
AICPA Coll, annual report
32000 vols; 212 curr per; 1093 diss/theses; 30 microforms; 650 av-mat; 97 digital data carriers
libr loan 27715

– Faculty of Economics, Puey Ungphakorn Library, Prachan Rd, Bangkok 10200
T: +66 2 6133550; Fax: +66 2 6235147; E-mail: chantani@tu.ac.th; URL: www.library.tu.ac.th
1965; Chantani Phanishpon
Puey's Coll
80000 vols; 640 curr per; 8000 diss/theses; 10000 govt docs; 4 digital data carriers; 10000 docs, 5000 clippings and verticles files
libr loan 27716

– Faculty of Law, Library, Prachan Rd, Bangkok 10200
T: +66 2 2241381
1969; Suchitra Utamawatin
ASEAN coll
43100 vols; 140 curr per; 123 diss/theses; 232 pamphlets
libr loan 27717

– Faculty of Political Science, Library, Prachan Rd, Bangkok
T: +66 2 2216171 ext 158
1961; Aree Somboondamrongkul
19500 vols; 200 curr per; 1800 diss/theses
libr loan 27718

– Faculty of Social Administration, Library, Prachan Rd, Bangkok
T: +66 2 2216111
Boonthiwa Kulpaiboon
25000 vols; 210 curr per; 500 diss/theses; 75 subjects of clippings, 1509 pamphlets
libr loan 27719

– School of Journalism and Mass Communications, Library, Prachan Rd, Bangkok 10200
T: +66 2 2216111, 2216120
1954; Wanna Topibulpong
10000 vols; 245 curr per; 1200 diss/theses; 6000 rpts
libr loan 27720

– Thai Studies Research Information Center, Library, Prachan Rd, Bangkok 10200
T: +66 2 6235074; Fax: +66 2 2262112; E-mail: rtnd@alpha.tu.ac.th; URL: www.tu.ac.th
1971; Ratana Techamahachai
Research work and publ of Thai Khadi Research Institute, Tape and videos sponsored by the Institute
8000 vols; 21 curr per; 115 diss/theses; 295 microforms; 10 av-mat; 540 sound-rec; 4774 clippings 27721

University of the Thai Chamber of Commerce, Library, 126/1 Vibhavadi rangsit Rd, ***Bangkok*** 10320
T: +66 2 2761040; Fax: +66 2 2762126; E-mail: nitima@morakot.nectec.or.th; URL: www.utcc.ac.th
100000 vols; 6000 curr per 27722

Chiang Mai Teacher College, Library, Amphur Muang, Changwad, ***Chiang Mai*** 80000 vols 27723

Chiang Mai University, Central Library, 239 Huay Kaew Rd, Muang, ***Chiang Mai*** 50200
T: +66 53 221154; Fax: +66 53 222766; E-mail: libdir@chiangmai.ac.th; URL: www.lib.cmu.ac.th
1964; Prasit Malumpong
Lanna Museum and Art Gallery Project
655000 vols; 4858 curr per; 9600 av-mat
libr loan; ALA 27724

– Faculty of Agriculture Library, 130 Huay Kaew Rd, Chiang Mai 50002
T: +66 53 221699
1990; Prapapun Plaichjan
13000 vols; 500 curr per 27725

– Faculty of Dentistry, Library, Chiang Mai 50002
T: +66 53 221699
Wilawan Suvansamrit
8000 vols
libr loan 27726

– Faculty of Engineering Library, Chiang Mai 50200
T: +66 53 221699 ext 4116; Fax: +66 2 217287
1970; Kunawut Theamthong
Standards docs
21000 vols; 175 curr per; 200 diss/theses; 1500 govt docs; 70

microforms; 30 av-mat; 20 digital data
carriers; software disks (200 titles)
libr loan; TLA 27727

– Faculty of Humanities, Library, Chiang
Mai 50002
T: +66 53 221699
Choocheep Sutthima
17 000 vols; 51 curr per; 163 clippings;
21 pamphlets 27728

– Faculty of Medicine, Library, 110
Intravaroros, Chiang Mai
T: +66 53 945200 to 945208;
Fax: +66 2 945202; E-mail:
schotika@mail.med.cmu.ac.th; URL:
www.medicine.cmu.ac.th/library
1960; Suchada Chotikanont
Public health
37 324 vols; 421 curr per; 1 756
diss/theses; 9 325 govt docs; 10 500
microforms; 50 av-mat; 2 017 sound-rec;
170 digital data carriers; 6 774 slides
libr loan; ILA, MLA 27729

– Faculty of Social Sciences, Library, 130
Huaykaew Rd, Chiang Mai 50002
T: +66 53 221699
1971; Banchuen Tongpunchang
9 000 vols
libr loan 27730

– Tribal Research Library, 239 Huay Kaew
Rd, Chiang Mai 50200
T: +66 53 221933; Fax: +66 2 222494;
E-mail: tribal@loxinfo.co.th; URL:
www.chmai.com/tribal
1965; Thaworn Foofuang
8 200 vols; 20 curr per; 100 diss/theses;
1 100 govt docs; 100 maps; 100 av-mat;
50 sound-rec; 500 research rpts, 1 000
clippings, 1 000 artifacts on hilltribes 27731

Maejo University, Library, Sansai,
Chiang Mai 50290
T: +66 53 878038; Fax: +66 53
498861; E-mail: maria@mju.ac.th; URL:
www.mju.ac.th
1934; Dr. Boonrawd Supa-Udomlerk
76 500 vols; 858 curr per
libr loan; IAALD 27732

Northern Technical Institute, Library,
Huay Kaew Rd, *Chiang Mai*
1957
50 800 vols 27733

Payap University, Library, Amphur Miang,
Chiang Mai 50000
T: +66 53 304805; Fax: +66 53 241985;
E-mail: intexch@payab.ac.th; URL:
lib.payap.ac.th
1974; Mrs Suntree Ratayanant
125 000 vols 27734

Bang Phra Agricultural College,
Library, Bang Phra, *Chonburi*
1957
8 000 vols 27735

Burapha University, Library, Saen Sook,
Muang, *Chonburi* 20131
T: +66 38 390047; E-mail: pichan@
adm.buu.ac.th; URL: www.adm.buu.ac.th
Dr. Kwanchadil Pisalaphong
200 000 vols 27736

Khon Kaen University, Central Library,
123 Mittraphab Rd, *Khon Kaen* 40002
T: +66 43 20254-1/-2;
Fax: +66 43 202543; E-mail:
webmaster@library.kku.ac.th; URL:
www.library.kku.ac.th
1954; Assoc Prof. Dr. Sman Loipha
77 209 vols; 4 989 curr per; 23 e-
journals; 23 762 diss/theses; 5 govt docs;
16 music scores; 9 maps; 1 813 av-mat;
723 sound-rec
libr loan; TLA, PULINET, THAILIS 27737

– Faculty of Agriculture Library, Khon Kaen
40002
T: +66 43 238755
Supeeya Srikallaya
16 000 vols 27738

– Faculty of Engineering Library, Khon
Kaen 40002
T: +66 43 241331 ext 2169; Fax: +66 2
237604
1965; Mrs Walailak Saengwannakool
24 000 vols; 163 curr per; 117
diss/theses; 42 sound-rec; 1 950 rpts,
427 pamphlets, 1 215 standards
libr loan 27739

Tepsatri Teachers College, *Lopburi*
T: +66 36 411029
60 800 vols; 187 curr per; 203 clippings,
522 pamphlets 27740

Mahasarakham University, Library,
Muang District, *Mahasarakham* 44000
T: +66 43 722443; Fax: +66 43
721769; E-mail: sujin.b@msu.ac.th; URL:
www.library.msu.ac.th
1994
163 000 vols; 22 000 periodicals 27741

Mahidol University, Library and
Information Center, 25/25 Phutthamonthon
4, Salaya, *Nakornpathom* 73170
T: +66 2 4419741; Fax: +66 2 4419580;
E-mail: directli@mahidol.ac.th; URL:
www.li.mahidol.ac.th
Mrs Kannigar Chollampe 27742

– Faculty of Environment and Resource,
Library, 25/25 Phutthamonthon 4, Salaya,
Nakorn Chaisri, Nakornpathom 73170
T: +66 2 4419020
Aem-On P. Tuntawiroon
28 000 vols 27743

– Faculty of Nursing Library, Prannok Rd,
Bangkoknoi, Prannok Rd, Bangkok 10700
T: +66 2 41974680 ext 14002; Fax: +66
2 4128415; E-mail: lissj@mahidol.ac.th;
URL: www.li.mahidol.ac.th/LIBRARY/
nse.shtml
1981; Surang Sirorojsakul
Problem-Based Learning Coll; Hist of
Nursing Museum
28 000 vols; 145 curr per; 1 250
diss/theses; 580 govt docs; 6 digital
data carriers
libr loan 27744

– Faculty of Pharmacy Library, 447 Sri-
ayudhya Rd, Bangkok 10400
T: +66 2 6444495; Fax: +66 2 2474696;
E-mail: lisnc@mahidol.ac.th; URL:
www.202.28.141.1
1969; Sunee Nuichan
Researchs of 5th year pharmacy students
20 828 vols; 26 curr per; 454 diss/theses;
54 av-mat; 37 sound-rec; 38 digital data
carriers
libr loan; TLA 27745

– Faculty of Science, Library, Rama 6 Rd,
Bangkok 10400
T: +66 2 2455415
Puckvadee Songkroh
23 000 vols; 769 curr per; 13 diss/theses;
210 maps; 730 microforms; 543
pamphlets, 3 120 reprints
libr loan 27746

– Siriraj Medical Library, Siriraj Hospital,
Prannok Rd, Bangkok 10700
T: +66 2 4197466-80 ext 1401; Fax: +66
2 4128415; E-mail: lissj@mahidol.ac.th;
URL: www.li.mahidol.ac.th
1981; Prof. Teerachai Chantarojanasiri
Hist of Nursing Museum
23 000 vols; 199 curr per; 230 mss;
1 001 diss/theses; 850 govt docs; 8
maps; 43 sound-rec; 9 digital data
carriers
libr loan; TLA 27747

**Sukhothai Thammathirat Open
University (STOU)**, Office of
Documentation and Information, 9/9 Moo
9, Bang Phood, Pakkred, *Nonthaburi*
11120
T: +66 2 5032121; Fax: +66 2 5033607;
E-mail: dilawboo@stou.ac.th; URL:
www.odi.stou.ac.th
1978
651 000 vols; 1 930 curr per 27748

Rangsit University, University Library,
52/347 Paholyothin Rd, Tambon-Lakhok,
Pathum Thani 12000
T: +66 2 9972200; Fax: +66 2 5339470;
E-mail: info@rangsit.rsu.ac.th; URL:
www.rsu.ac.th
Uthai Dhutiyabhodhi
71 000 vols 27749

Naresuan University, Library,
Phitsanulok 65000
T: +66 55 261000; Fax: +66 55 261005;
URL: www.nu.ac.th
120 000 vols 27750

Huachiew Chalermprakiet University,
Library, 18/18 Bangna-Trad Rd,
Bangplee-, *Samutprakarn* 10540
T: +66 2 3126300; Fax: +66 2 3126237;
E-mail: webmaster@hcu.ac.th; URL:
www.hcu.ac.th
90 000 vols; 467 curr per 27751

Prince of Songkhla University, Library,
Pattani, *Songkhla* 94000
T: +66 74 348624; Fax: +66 74 331300
1964; Prates Kramut
John F. Kennedy Coll, Islamic Coll,
Buddhadasa Coll, Research publs
libr loan; TLA 27752

– Faculty of Medicine, Library, Hat-yai, P.O.
Box 84, Songkhla 90110
T: +66 74 212905; Fax: +66 2 212912;
E-mail: tiworanu@ratree.psu.ac.th
1976; Woranuch Tiautrakul
35 000 vols; 500 curr per; 60 diss/theses;
1 500 govt docs; 700 av-mat; 56 digital
data carriers
TLA 27753

School Libraries

Bangkok Patana School Library,
Resource Centre, 2/38 Moo 5, Soi
La Salle, Sukhumvit 105, Prakanong,
Bangkok
T: +66 2 3980200; Fax: +66 2 3993179
Michelle Mann
Children's bks, Teachers' reference bks
50 000 vols; 45 curr per; 230 maps;
1 350 av-mat; 270 sound-rec
SLA 27754

Government Libraries

**Department of Alternative Energy
Development and Efficiency (DEDE)**,
Library, 17 Ramá, 1 Rd Patumwan,
Bangkok 10330
T: +66 22230021-9; Fax: +66
22261416; URL: www.dede.go.th/dede/
index.php?id=118
1953
10 000 vols 27755

Department of Science Service,
Ministry of Science, Technology and
Environment, Scientific and Technological
Information Division, Library, Rama VI
Rd, Ratchathewi District, *Bangkok*
10400
T: +66 2 2455271; Fax: +66 2 2479468;
E-mail: info@dss.moste.go.th
1918; Mayuree Pongpudpunth
450 000 vols; 2 000 curr per
libr loan 27756

**FAO Regional Office for Asia and
the Pacific**, Pacific Library, Maliwan
Mansion, Phra Atit Rd, *Bangkok* 10200
T: +66 2 6974000; Fax: +66 2 6974445;
E-mail: FAO-RAP@fao.org; URL:
www.fao.org
Nid Swetarak
Agriculture, food sciences
20 000 vols; 300 curr per; 50 maps; 140
microforms 27757

Ministry of Education, Library and
International Documentation Centre,
External Relation Division, Rajdamnern
Av, *Bangkok* 10300
T: +66 2 2826796
9 500 vols 27758

Ministry of Interior, Library, Asdang Rd,
Bangkok
13 000 vols 27759

Ministry of Public Health, Library,
Samsen Rd Devavesm Palace, *Bangkok*
48 000 vols 27760

**National Economic and Social
Development Board**, Library, Thanon
Krungkasem, *Bangkok* 10 100
T: +66 2 2804085 ext 125
Porntip Pothong
12 000 vols; 150 curr per; 70 pamphlets,
150 clippings 27761

National Housing Authority Library,
Sukhapibal 1 Rd, Klongchan Bangkapi,
Bangkok 10240
T: +66 2 3775461
Noppawan Kaewkampol
Housing and human settlements
13 000 vols 27762

National Statistical Office, Library, Lan
Luang Rd, *Bangkok* 10100
T: +66 2 2820551; E-mail: yaiying@
nso.go.th; URL: www.nso.go.th
Chirajit Vichienchit
20 000 vols 27763

Office of Atomic Energy for Peace,
Library, Vibhavadi Rangsit Rd,
Chatuchuk, *Bangkok* 10900
T: +66 2 5795230; Fax: +66 2 5613021
1961
11 000 vols; 265 curr per 27764

**Office of the National Culture
Commission**, Library, Ratchadapisek
Rd, Huay Khwang, *Bangkok* 10320
T: +66 2 2470013; Fax: +66 2 2485841;
E-mail: culthai@hotmail.com; URL:
www.culture.go.th
1979
27 650 vols 27765

Payathai Curriculum Center, Library,
Sriayuthaya Rd, *Bangkok*
16 000 vols 27766

Ratha-Sapha, Hongsamud Rathasapha
(National Assembly), Thanon U-Thong
Nai, Dusit, *Bangkok* 10300
T: +66 2 2441078; Fax: +66 2 2441083
1974; Ms Wijitra Watcharaporn
35 000 vols; 200 curr per; 6 300 rpts 27767

The Royal Institute,
Ratchabandittayasathan, Library, Na Phra
Lan Rd, in The Grand Palace Grounds,
Bangkok 10200
T: +66 2 2214822; Fax: +66 2
2249910; E-mail: ripub@royal.go.th; URL:
www.royin.go.th
1933
Works of Fellows and Associate
Fellows of the Royal Institute
(Ratchabandittayasathan) 247 Books,
461 Articles
21 487 vols; 75 curr per; 920 govt docs;
280 maps
libr loan 27768

Corporate, Business Libraries

Baker & McKenzie Library, 25th Fl, 990
Rama IV Rd, *Bangkok* 10500
T: +66 2 6362000; Fax: +66 2 6362111
1978; Angsana Phuthanakul
Business and banking, taxation
10 000 vols; 52 curr per 27769

Bangkok Metropolitan Bank, Library,
8th Fl Suan Mali, Chalerm Khat 2,
Bangkok
T: +66 2 2230561 ext 296
13 500 vols; 57 curr per; 10 maps; 50
clippings, 100 pamphlets 27770

Bank of Thailand, Library and Archives,
P.O. Box 154, Hua Lampong Bulk
Posting Centre, *Bangkok* 10331
T: +66 2 2828909; Fax: +66 2 2835656;
E-mail: sirimas@bot.or.th
Siriam Srisomwong
40 000 vols; 450 curr per; 19 200
clippings, 19 600 pamphlets 27771

**Siam Cement Public Company
Limited**, Library, 1 Siam Cement Rd,
Bangsue, *Bangkok* 10800
T: +66 2 5863644; Fax: +66 2 5872199;
E-mail: info@cementhai.co.th; URL:
www.cementhai.co.th
Kannika Leangphibool
15 500 vols; 97 curr per; 70 maps; 37
microforms; 120 av-mat
libr loan 27772

Special Libraries Maintained by Other Institutions

Alliance Française, Médiathèque,
Bibliothèque, 29 Thanon Sathorn Tai,
Bangkok 10120
T: +66 2 2132122; Fax: +66 2 2132064;
E-mail: alliance-francaise@bkk.a-net.net.th
Michel Laval
15 000 vols; 30 curr per; 10 maps; 350
sound-rec; 40 digital data carriers 27773

Aua Language Center, Library, 179
Rajdamri Rd, *Bangkok* 10330
T: +66 2 2528170; Fax: +66 2 2516988
1969; M.L. Joy Nandhivajrin
17 500 vols; 300 curr per; 12 000
microforms 27774

Bangkok Metropolis Special Library,
4th Fl Bangkok Metropolis Hall, Dinsaw
Rd, *Bangkok*
T: +66 2 2244682
Kanchana Putwatana
18 500 vols; 80 curr per; 30 maps; 800
clippings, 200 pamphlets   27775

Book Development Centre, Educational
Materials Centre, Library, Ministry of
Education, Sukhumvit Rd, *Bangkok*
10110
Phan-Ngam Yamboonruang
rare book coll, Children's book coll,
Textbks and Supplementary readers,
Ministry of Education publ
97 000 vols; 342 maps; Pamphlets 30 000
libr loan   27776

**Economic and Social Commission
for Asia and the Pacific (ESCAP)**,
Library, United Nations Bldg,
Rajadamnern Av, *Bangkok* 10200
T: +66 2 2881799; Fax: +66 2 2883036;
E-mail: domingo-barker.unescape@un.org
1950; Evelyn Domingo-Barker
UN and ESCAP documentations
150 000 vols; 4 000 curr per; 200 maps;
4 000 microforms
libr loan; IFLA, CONSAL   27777

**Forest Products Research and
Development Division**, Royal Forest
Dept, Ministry of Agriculture and Co-
operatives, Library, 61 Paholyotin Rd,
Chatuchak, *Bangkok* 10900
T: +66 2 5794301
1935; Donnaya Klaipook
20 000 vols   27778

Goethe-Institut, German Cultural Institute,
Informationszentrum (Bibliothek), 18/1 Soi
Goethe, Sathorn 1, *Bangkok* 10120;
GPO Box 3027-3028, Bangkok 10501
T: +66 2 2870942, 2870944; Fax: +66 2
2871829; E-mail: goethevl@loxinfo.co.th;
URL: www.goethe.de/so/ban
1960; Angelika Chotikapanich, Ratchanok
Suwatthanabunpot
11 375 vols   27779

**Institute for the Promotion of
Teaching Science and Technology**,
Library, 924 Behind the Bangkok
Planetarium, Sukhumvit Rd, *Bangkok*
10110
T: +66 2 3924021
Noppakoon Saran
15 000 vols   27780

International Labour Organisation,
Regional Office for Asia, Library, United
Nations Bldg, Rajdamnern Av, P.O. Box
1759, *Bangkok* 10200
T: +66 2 2829161; Fax: +66 2 2801735
1969; Jirawan Aksornsuwan
Social welfare
25 700 vols; 345 curr per; 368
diss/theses; 18 maps; 12 av-mat
libr loan; CONSAL, TLA   27781

Marine Fisheries Division, Library, 89/1
Soi Sapanpla Yannawa Dist, *Bangkok*
12
T: +66 2 2114982
1967; Chutima Boolpipat
17 000 vols; 21 curr per
libr loan   27782

National Archives of Thailand, Fine
Arts Department, Library, Samsen Rd
Dusit, *Bangkok* 10300
T: +66 2 4248355, 2811599; Fax: +66 2
2815341
1952; Ms Korapin Taweta
1 450 vols; 616 867 curr per; 1 219
diss/theses; 316 130 govt docs; 24 069
maps; 1 434 036 av-mat; 1 738 sound-rec;
6 000 pamphlets, 10 356 clippings, 50 887
posters, 313 035 photos, 14 422 moving
pictures, 7 195 videos, 50 000 tv news,
3 000 documentary films   27783

National Education Commission,
Educational Information Division, Library,
Sukhothai Rd, *Bangkok* 10300
T: +66 2 2415152; Fax: +66 2 2431198
1960; Kalya Thamratnopkoon
20 000 vols; 245 curr per; 276
diss/theses; 8 000 govt docs; 25 maps;
290 microforms; 85 clippings, 170
pamphlets
libr loan   27784

National Energy Information Center,
17 Rama Rd, *Bangkok* 10500
T: +66 2 2210139
1979; Nonglak Boonjawatn
10 000 vols   27785

Rajavithi Hospital, Library, Rajavithi Rd,
Bangkok 10400
T: +66 2 2460052; URL:
www.rajavithi.go.th
Sodsri Pariyakul
22 000 vols   27786

Siam Society, Library, 131 Soi Asoke,
Sukhumvit 21, P.O. Box 65, *Bangkok*
10110
T: +66 2 6616470; Fax: +66 2 2583491
1904; Pinya Suwonachai
Rare bks on Thai hist, Palm leaf mss
27 000 vols
libr loan   27787

Teachers' Council of Thailand,
Kurusapha Library, Thanon Rachasimad,
Bangkok 3
T: +66 2 2817955
49 300 vols; 84 curr per; 10 maps
libr loan   27788

**Thailand Institute of Scientific and
Technological Research**, Thai National
Documentation Centre, 196 Phahonyotin
Rd, Bang Khen, *Bangkok* 10900
T: +66 2 5791121; Fax: +66 2 5794771;
URL: www.tistr.or.th
8 000 vols
libr loan   27789

**UNESCO Regional Office for
Education in Asia and the Pacific**,
Information Programmes and Services
(PIPS), P.O. Box 967, Prakhanong Post
Office, *Bangkok* 10110
T: +66 2 3910577; Fax: +66 2 3910866;
E-mail: eapinfo@unescobk.org
1961; Clive Wing
Development plans, education
45 000 vols; 400 curr per; 210 maps;
30 000 microforms; 900 av-mat; 112
sound-rec; 80 digital data carriers
libr loan   27790

**United Nations Asian Institute
for Economic Development and
Planning**, Library, P.O. Box 2-136,
Bangkok
22 000 vols   27791

**United Nations Children's Fund
(UNICEF)**, East Asia & Pacific Regional
Office, Library, 19 Phra Atit Rd,
Bangkok 10200
T: +66 2 2805931 ext 211; Fax: +66 2
2803563; E-mail: eapro@unicef.org
Ms Piyanart Supaphon
Social Sciences, Health, Food and
Nutrition, Child Survival & Development,
Women, Primary Health Care,
Urbanization, Water & Sanitation,
Population & Family Planning
30 000 vols; 200 curr per; 50 maps;
1 500 microforms; 500 pamphlets &
clippings
IFLA, CONSAL, TLA   27792

Tribal Research Institute, Library,
Amphur Muang, Chiang Mai University,
Chiang Mai 50002
T: +66 53 221933; Fax: +66 53 222494
1965; Thaworn Foofuang
Artifacts of the hill tribes on show in
Tribal Museum
8 600 vols; 25 curr per; 55 diss/theses;
300 govt docs; 50 maps; 500 av-mat;
20 sound-rec; 3 000 pamphlets, 5 000
clippings
LAT   27793

Asian Institute of Technology,
Center for Library and Information
Resources (CLAIR), Km 42 Paholyotin
Hwy, P.O. Box 4, *Klong Luang*,
Pathumthani 12120
T: +66 2 5245853; Fax: +66 2
5245870; E-mail: clair@ait.ac.th; URL:
www.clair.ait.ac.th
1959; Dr. Francis J. Devadason
Milton E. Bender, Jr coll, Wen-Jei Yang
Coll; Asian Info Ctr for Geotechnical
Engineering, International Ferrocement-,
Renewable Energy Resources Info Center
231 000 vols; 500 curr per; 8 400
diss/theses; 5 100 govt docs; 1 632 maps;
7 067 microforms; 1 231 av-mat; 402
sound-rec; 127 digital data carriers; 8 129
vertical files
libr loan   27794

Togo

National Libraries

Bibliothèque Nationale du Togo, Av de
la Victoire, BP 1002, *Lomé*
T: +228 216367; Fax: +228 221967
1969; Maboulah Wemmi-Agore Coulibaley
Arch de la colonisation allemande
18 050 vols; 74 curr per; 1 030
diss/theses; 15 digital data carriers
27795

University Libraries,
College Libraries

Ecole Normale Supérieure, Bibliothèque,
BP 7, *Atakpamé*
T: +228 400057
1972; Kodjo Fofogan Agbokou
11 000 vols; 15 curr per   27796

Université du Bénin, Bibliothèque
universitaire, BP 1515, *Lomé*
T: +228 254843; Fax: +228 250183;
URL: www.ub.tg
1970; E.E. Amah
70 000 vols; 230 curr per; 5 digital data
carriers
libr loan; SCAULWA, AIDBA   27797

Government Libraries

**Institut National d'Hygiène Ernst
Rodenwaldt**, Bibliothèque, BP 1396,
Lomé
T: +228 210633
1970; Miyoutouè Boukaya
10 000 vols; 20 curr per
libr loan   27798

Special Libraries
Maintained by Other
Institutions

Centre Culturel Français, Bibliothèque,
av du 24 Janvier, BP 2090, *Lomé*
T: +228 210232; Fax: +228 213442;
E-mail: cid@syfed.rg.refer.org
1962; M. Dominique Seurin
35 000 vols; 111 curr per; 1 900 av-mat;
4 digital data carriers
libr loan; ATBAD   27799

Goethe-Institut, Bibliothek, 25, Rue
Kokéti, Angle Rue d'Eglise, BP 914,
Lomé
T: +228 2210894; Fax: +228 2220777;
E-mail: information@lome.goethe.org;
URL: www.goethe.de/ins/tg/lom/inz/bib/
deindex.htm
1961; Eberhard Weller
9 395 vols   27800

Tonga

University Libraries,
College Libraries

Teacher Training College, Library,
Nuku'alofa
1944
5 000 vols   27801

**University of the South Pacific –
Tonga Centre Library**, P.O. Box 278,
Nuku'alofa
T: +676 29240 ext 55; Fax: +676 29249;
E-mail: judy@tonga.usp.ac.fj
1971; Judy Mailei
Agriculture, animal husbandry,
rural technology, small business
management, food and nutrition, women
in development, population and health,
adult education; Pacific Coll
9 000 vols; 100 curr per; 45 diss/theses;
500 govt docs; 50 maps; 40 microforms;
70 sound-rec
libr loan; TLA, COMLA   27802

Government Libraries

Ministry of Education, Library, Teachers
Training College, P.O. Box 123,
Nuku'alofa
T: +676 21588; Fax: +676 23829
1976; Mana Tuita
12 500 vols   27803

Trinidad and Tobago

National Libraries

**National Library and Information
System Authority of Trinidad and
Tobago**, 109 Abercromby St, *Port-of-
Spain*
T: +1 868 623-6962/-7278/-9673; Fax: +1
868 625-6096; E-mail: nalis@nalis.gov.tt
URL: library2.nalis.gov.tt
1945, reorganised 1998; Pamella Benson
West Indiana Coll; Heritage Library
Division, Educational Library Services
Division, Public Libraries Division,
Information Networks Division
442 000 vols; 529 curr per; 2 sound-rec;
51 digital data carriers; 198 audio-tapes,
110 video tapes, bound newspapers (600
vols), pamphlets (3 000), illustrations,
photos and calendars (300), 2 000
vertical files
ACURIL, IFLA   27804

University Libraries,
College Libraries

University of the West Indies, Main
Library, *Saint Augustine*
T: +1 868 6622002; Fax: +1 868
6629238; E-mail: mainlib@library.uwi.tt
URL: www.mainlib.uwi.tt
1926; Dr. Margaret Rouse-Jones
West Indian Coll, mss colls
395 000 vols; 41 189 curr per; 3 175
mss; 4 737 diss/theses; 1 220 maps;
49 193 microforms; 10 497 av-mat; 2 547
sound-rec; 94 digital data carriers; 10 060
items in vertical file (postcards, clippings
etc), 678 603 unbound serial parts
libr loan; ACURIL, AIBDA, ALA, ASLIB,
COMLA, FID, IFLA, IAALD, CILIP, LATT,
SALALM, ASIS   27805

– Faculty of Humanities and Education,
School of Education Library, Saint
Augustine
T: +1 868 6622002 ext 3338; Fax: +1
868 6626615
1973; Janet Fullerton Rawlins
West Indian materials in education
30 000 vols; 130 curr per; 1 300
diss/theses; 380 govt docs; 75 maps;
76 av-mat; 15 digital data carriers; 151
newspaper clipping files
libr loan; ACURIL   27806

– Institute of International Relations, Library,
Saint Augustine
T: +1 868 6453232, 6622002; Fax: +1
868 6639685; E-mail: iirt@fss.uwi.tt; URL:
www.uwi.tt/fss/iirt
1966; Cherill Farrell
Internat Relations, Internat Law,
Diplomacy, Internat Politics, Internat
Political Economy, Internat Trade,
Development and Third World Studies,
Latin American Studies, Caribbean
Studies; The Roy Prieswerk Coll
25 000 vols; 230 curr per; 65 diss/theses;
188 600 microforms; 20 digital data
carriers; 19 600 newspaper clippings
libr loan; LATT, CARALL   27807

Government Libraries

**Attorney General and Ministry of
Legal Affairs**, Library, 25-27 St Vincent
Street, *Port-of-Spain*; P.O. Box 595,
Port-of-Spain
T: +1 868 6247710; Fax: +1 868
6257339; E-mail: aglibrary@trinidad.net
1956; Sheryl Washington
Law, Civil Law, Constitutional Law,
Commonwealth Legislation, Criminal
Law; Director of Public Prosecutions,
Chief State Solicitor, Attorney General's
Chambers
12 700 vols; 50 curr per; 12 650 govt

docs; 75 digital data carriers; 5 200 pamphlets, vertical files on 275 subjects, 7 titles on diskettes (Total of 14 diskettes)
libr loan; CARALL   27808

Ministry of Energy, Library, Riverside Plaza, Level 7, P.O. Box 96, *Port-of-Spain*
T: +1 868 623-6708/-1006
1972; Cheryl Medford
Petroleum industry, geophysics, drilling, crude oil, natural gas, natural gas, economics, law
22 000 vols; 175 curr per; 300 microforms; 250 av-mat
libr loan; LATT   27809

Parliament Library, Red House, Abercromby St, *Port-of-Spain*
T: +1 868 6237055; Fax: +1 868 6254672; E-mail: parllib@trinidad.net; URL: www.ttparliament.org
1894; Caroline Kangalee
Federation of the West Indies documents, Royal Gazette 1865-1962, Council Papers + Minutes
8 150 vols; 49 curr per; 10 mss; 500 incunabula; 3 diss/theses; 15 200 govt docs; 1 500 maps; 70 av-mat; 56 sound-rec; 137 digital data carriers
LATT   27810

Special Libraries Maintained by Other Institutions

Institute of Marine Affairs, Library, Hilltop Lane, P.O. Box 3160, Carenage P.O., *Chaguaramas*
T: +1 868 634-4291/-4294; Fax: +1 868 6344433; E-mail: ottley@ima.gov.tt; URL: www.ima.gov.tt
1984; Tessa Ottley
Fisheries, aquaculture, environmental sci, coastal zone management, oceanography, marine law, coral reefs, beach formation
8 000 vols; 20 curr per; 12 diss/theses; 2 900 av-mat; 10 digital data carriers
IAMSLIC   27811

Trinidad and Tobago Bureau of Standards, Information Centre, Lot 1, Century Dr, Trincity Industrial Estate, Macoya, Tunapuna, P.O. Box 467, *Port-of-Spain*
T: +1 868 6628827; Fax: +1 868 6634335; E-mail: ttbs@carib-link.net; URL: www.ttbs.org.tt
1975; Frank Soodeen
Standards (nat, regional, internat), quality assurance, metrology, environment, industrial safety; World Trade Organisation Enquiry Point, Technical Help Exporters Unit
200 000 vols; 78 curr per; 20 diss/theses; 720 govt docs; 20 maps; 3 890 microforms; 68 av-mat; 19 sound-rec; 60 digital data carriers; 1 600 pamphlets, 105 video cassettes
libr loan; LATT   27812

Tunisia

National Libraries

Bibliothèque Nationale de Tunisie, Al-Maktaba Al-Watania (National Library of Tunisia), Bd 9 Avril 1938, BP 42, 1000 *Tunis*
T: +216 71572706, 71569477; Fax: +216 71572887; E-mail: Bibliotheque.Nationale@email.ati.tn; URL: www.bibliotheque.nat.tn
1885; Samia Kamarti
Doc and Info dept
1 500 000 vols; 40 000 mss; 1 000 maps; 400 microforms; 100 digital data carriers; 16 000 colls of periodicals
libr loan; IFLA   27813

University Libraries, College Libraries

Ecole Nationale des Ingenieurs de Gabes, Bibliothèque, Route de Medenine, BP 6029, *Gabes*
8 000 vols
libr loan   27814

Centre d'Etudes de Carthage, Bibliothèque, 33 rue Garibaldi, 1001 *Tunis*
T: +216 71343846
1957
22 000 vols; 40 curr per   27815

Ecole Nationale d'Administration (ENA), Bibliothèque, 24 av du Docteur Calmette, Mutuelle Ville, 1060 *Tunis*
T: +216 71848300; Fax: +216 71794188
1964; Makhlouf Chahrazed
53 000 vols; 500 curr per; 350 diss/theses; 50 maps   27816

Ecole Supérieure de Science et Technique, Bibliothèque, 5 rue Taha Hussein, *Tunis*
T: +216 71392559; Fax: +216 71391166
Sciences sociales, économie, sciences exactes, sciences de l'ingenieur, électricité, électronique
12 000 vols
libr loan   27817

Institut National de la Recherche Agronomique de Tunisie, Bibliothèque Centrale de l'INAT, 43 Av Charles Nicolle, 1082 Cité Mahrajène *Tunis*
T: +216 71287431; Fax: +216 71799391; E-mail: ae@cybertunis.zzn.com
1881; Abiadh Alouini Essia
Plant genetics, Animal husbandry, Cereal industry, Soil Physics and Soil Chemistry, Animal breeding, Eutomology
9 000 vols; 120 curr per; 1 024 diss/theses; 10 digital data carriers
libr loan   27818

Université des Lettres, des Arts et des Sciences humaines – Tunis I, *Tunis*
– Faculté des Lettres Manouba, Bibliothèque, Manouba, 2010 Tunis
T: +216 71601189; Fax: +216 71600910; E-mail: Jamel.Hajri@flm.RNU.tn
1986; Jameleddine Hajri
Lettres françaises, anglaises, italiennes, espagnols, russes et arabes; lettres allemandes, turques, coll de Geographie et de l'Histoire, de Linguistique, de Beaux-arts et de reference; coll 'La Pleiade' et 'Guillaume Budet'
68 000 vols; 66 curr per; 1 628 diss/theses; 13 056 microforms; 30 av-mat; 25 digital data carriers; cartographic service
libr loan   27819

School Libraries

Centre de Hautes Etudes d'Art Dramatique, Bibliothèque, *Hammamet*
25 000 vols   27820

Ecole des Beaux-Arts de Tunis, Bibliothèque, Route de l'Armée nationale, *Tunis*
90 000 vols   27821

Government Libraries

Chambre des Députés, Bibliothèque (House of Representatives), Le Bardo, 2000 *Tunis*
T: +216 71510200; Fax: +216 71514608
1956; Yasmina Hammami
10 000 vols   27822

Ministère du Développement économique, Direction des archives et de la documentation, Pl Ali Zouaoui, *Tunis* 1069
T: +216 71185155; E-mail: Elbey.ali@email.ati.tn
1996; El Bey Ali
Economic development, planning, privatization, public works, regional development, investment
20 000 vols; 91 curr per; 20 diss/theses; 520 govt docs; 170 maps; 15 sound-rec; 81 digital data carriers
libr loan   27823

Corporate, Business Libraries

Banque Centrale de Tunisie, Centre de documentation, Av Mohamed V, *Tunis*
11 000 vols; 650 curr per   27824

Special Libraries Maintained by Other Institutions

Institut de Formation Pedagogique, Bibliothèque, 43 rue de la Liberté, *Le Bardo*, Tunis
1982; Mohamed Abdeljaoued
10 000 vols; 190 curr per
libr loan   27825

Institut National des Sciences et Technologies de la Mer (INSTM), Bibliothèque, 2025 *Salammbô*
T: +216 71730420; Fax: +216 71732622
1924; Ben Mahmond Chiraz
30 000 vols; 1 500 curr per; 500 mss; 2 incunabula; 200 diss/theses; 300 maps
                                            27826

Arab League Documentation and Information Centre (ALDOC), Library, 37 av Khereddine Pacha, *Tunis*
T: +216 71890100; Fax: +216 71781801
1982
League of Arab States Documents, United Nations Documents
66 000 vols   27827

Arab League Educational, Cultural and Scientific Organization (ALECSO), Library, Av Mohamed V, BP 1120, *Tunis*
T: +216 71784466; Fax: +216 71784965; E-mail: Alesco-Library@Email.ati.tn
1970; Ben Aissa Slaheddine
Publs of ALECSO (1 400)
7 000 vols; 200 curr per; 300 govt docs; 120 digital data carriers
IFLA, AFLI   27828

British Council, Library, 87 Av Mohamed V, 1002 *Tunis* Belvédère
T: +216 71848588; Fax: +216 71893066; E-mail: info@tn.britishcouncil.org; URL: www.britishcouncil.org/tunisia
1962
10 000 vols   27829

Centre de Diffusion Culturelle et de Documentation Pédagogique de Tunis, Bibliothèque, 87, av de la Liberté, *Tunis*
T: +216 71285111
Yves Saget
9 000 vols   27830

Centre de Documentation Nationale, Bibliothèque, 4 rue Ibn Nadim, Cité Montplaisir, Belvédère, 1002 *Tunis*
T: +216 71780357; Fax: +216 71792241
1966; Abdelbaki Daly
Documentation de Presse sur la Tunisie (tous sujets); Le monde arabe et les grandes questions internationales
8 000 vols; 2 000 curr per; 40 diss/theses; 10 000 photos, 500 reference works, 12 000 press clippings
IFLA, FID   27831

Centre d'Etudes et de Recherches Economiques et Sociales (CERES), Bibliothèque, 23 rue d'Espagne, *Tunis*
T: +216 71330338; Fax: +216 71338103; E-mail: Moncef.hergli@ceres.rnrt.tn
1962; Hèdia Chinkaoui
Journal officiel de la République Tunisienne (1960-1999)
12 000 vols; 313 curr per; 29 maps
                                            27832

Institut de Presse et des Sciences de l'Information, Bibliothèque, 7, Impasse Med. Bachrouch, Montfleury, 1008 *Tunis*
T: +216 71335216, 71335228; Fax: +216 71348596
1973; Mr A. Benkhedder, Mr C. Frigut
Documentation Research Dept, Periodical Room
11 000 vols; 120 curr per; 726 diss/theses; 50 digital data carriers; Les mémories de fin d'etudes (1 128), Les mémoires DESS (13)
libr loan; IFLA, FID   27833

Institut d'Economie Quantitative, Bibliothèque, 27 rue de Liban, 1002 *Tunis*
T: +216 71784336; Fax: +216 71787034; E-mail: aicardes@avicenne.rnrt.tn
1964; Moncef ABBES
11 800 vols; 300 curr per; 58 diss/theses; 1 digital data carriers
libr loan   27834

Institut des Belles Lettres Arabes (IBLA), Bibliothèque, 12 rue Jemaa El-Haoua, 1008 *Tunis*
T: +216 71560133; Fax: +216 71572683; E-mail: ibla@gnet.tn
1960; André Ferre
32 000 vols; 236 curr per   27835

Institut National d'Archéologie et d'Arts (INAA), Bibliothèque, 4 pl du Château, 1008 *Tunis*
T: +216 71261622
1957; Thabet Hayet
12 000 vols; 700 curr per; 38 diss/theses; 300 maps
libr loan   27836

Institut National des Sciences de l'Education, Bibliothèque, 17 rue d'Irak, Belvédère, 1002 *Tunis*
T: +216 71287722; Fax: +216 71795423
30 000 vols; 50 curr per   27837

Institut National du Patrimoine, Library, 4 pl du Château, 1008 *Tunis*
T: +216 71261622
1957
25 000 vols   27838

Institut Supérieur de Gestion (ISG), Bibliothèque, 2 rue Ibn Khaldoun, *Tunis*
8 000 vols; 250 curr per   27839

Office National des Mines, Direction du Service Géologique, Documentation & Bases de Données (National Office of Mines – Tunisia), 24, rue 8601, La Charguia, 2035 *Tunis*; BP 215, 1080 Tunis
T: +216 71787366, 71798699; Fax: +216 71794016; E-mail: chefservice.onm@email.ati.tn
Moncef Amri
Earth sciences
9 500 vols; 500 curr per; 750 diss/theses; 600 govt docs; 1 000 maps; 1 000 microforms; 10 000 rpts
libr loan   27840

Société Tunisienne de Banque, Bibliothèque, 1 rue Habib Thamour, *Tunis*
8 000 vols; 450 curr per   27841

Public Libraries

Bibliothèques publiques, 39 rue Asdrubal, Lafayette, 1002 *Tunis*
T: +216 71782552; Fax: +216 71797752
1965; Ali Fettahi
Public libs, children's libs, local and community libs, mobile libs
2 746 700 vols   27842

Turkey

National Libraries

Millî Kütüphane (National Library of Turkey), Baskanligi Bahçelievler Son Dura, 06490 *Ankara*
T: +90 312 2126200; Fax: +90 312 2230451; E-mail: bilgi@mkutup.gov.tr; URL: www.mkutup.gov.tr
1946; Mr Celal Tok
959 024 vols; 6 844 curr per; 21 833 mss; 6 736 music scores; 6 742 maps; 105 083 microforms; 1 485 av-mat; 12 136 sound-rec; 49 252 pamphlets, 1 012 paintingtables, 419 atlasses, 3 799 stamps, 517 600 periodicals
libr loan; IFLA, FID, LIBER   27843

General Research Libraries

Beyazit Devlet Kütüphanesi (Beyazit State Library), Imaret Sok 18, *Istanbul* – Beyazit
T: +90 212 5222488; Fax: +90 212 5261133
1882; Yusuf Tavaci
Mustafa Kemal Atatürk and his reformations, libr and information science; Dept of talking bks for the blind
500 000 vols; 19 800 curr per; 11 119 mss; 5 000 incunabula; 153 music scores; 100 maps; 95 microforms; 32 961 photos and postcards, 10 000 posters   27844

National Library of Izmir, Mill
Kütüphane Caddesi, no 39, **Konak**,
Izmir
1912; Ali Riza Atay
350 000 vols; 1 310 curr per; 4 200 mss
27845

University Libraries, College Libraries

Ankara Üniversitesi, Kütüphanesi
(University of Ankara), Tandogan
Meydani, 06100 **Ankara** – Diskapi
T: +90 312 2236070; URL:
www.ankara.edu.tr/kutuphane/English/
1933; Dr. H. Sekine Karakas
750 000 vols; 250 curr per
libr loan
27846

Bilkent Üniversitesi Kütüphanesi,
Kütüphanesi (Bilkent University), Bilkent,
06800 **Ankara**
T: +90 312 2664472;
Fax: +90 312 2664391; E-mail:
librdirector@bilkent.edu.tr; URL:
library.bilkent.edu.tr
1986; Dr. Phyllis L. Erdogan
Halil Inalcik hist coll (Ottoman hist),
Turkish Plastic Arts Arch; US Docs
Full Depository since 1992, World Bank
Regional Depository since 1999, Europ
Doc Centre since 2001; branch libr (15
653 vols)
427 200 vols; 3 157 curr per; 21 060
e-journals; 561 mss; 3 106 diss/theses;
110 286 govt docs; 11 649 music scores;
1 251 maps; 132 328 microforms; 3 010
av-mat; 351 sound-rec; 4 782 digital data
carriers; 12 678 clippings, 571 pamphlets,
1 634 posters, 6 585 art exhibition
invitations, 184 art reproductions,
1 813 diskettes, 472 postcards, 1 813
offprint/reprints
libr loan; IFLA, LIBER, ALA, MELA,
IAML, IATUL, CILIP
27847

Gazi Üniversitesi, Merkez Kütüphanesi
(Gazi University, Central Library),
Tekniokullar, 06500 **Ankara**
T: +90 312 2022900; Fax: +90 312
2022938; E-mail: canatali@gazi.edu.tr;
URL: www.lib.gazi.edu.tr
1982; Tünsel Canatali
146 000 vols; 1 300 curr per; 5 600
diss/theses; 3 500 microforms; 147
av-mat; 77 sound-rec; 215 digital data
carriers; 34 064 slides
libr loan; IFLA
27848

Orta Doğu Teknik Üniversitesi,
Kütüphanesi (Middle East Technical
University), Ismet Inönü Bulvari, 06531
Ankara
T: +90 312 210-2780/-2782;
Fax: +90 312 2101119; E-mail:
lib-hot-line@metu.edu.tr; URL:
www.lib.metu.edu.tr
1956; Prof. Bulent Karasozen
Special Coll for the Blind
145 000 vols; 1 872 curr per; 10 125
diss/theses; 167 maps; 500 microforms;
469 av-mat; 50 sound-rec; 985 digital
data carriers; 2 500 pamphlets
libr loan; IATUL
27849

Akdeniz Üniversitesi, Kütüphanesi,
Kampüs Dumlupinar Bulvari, 07058
Antalya
T: +90 242 2274430;
Fax: +90 242 2275540; E-mail:
webmaster@akdeniz.edu.tr; URL:
www.akdeniz.edu.tr
1985; Fehamet Gülcehre
Medicine
25 000 vols; 372 curr per; 738
diss/theses; 150 govt docs; 5 digital
data carriers
libr loan; IFLA
27850

Ege Üniversitesi Rektörlüğü, Kütüphane
ve Dokümantasyon, Kütüphanesi (Aegean
University Central Library), Daire
Başkanlığı, 35040 **Bornova**, Izmir
T: +90 232 3881100; Fax: +90 232
3881100
1961; Nurcan Baykal
120 000 vols; 500 curr per; 6 030
diss/theses; 8 digital data carriers
libr loan
27851

Uludağ Üniversitesi, Merkez
kütüphanesi (Uludağ University), Görükle
Kampüsü Merkez Kütüphane, 16059
Bursa
T: +90 224 4428536; Fax: +90 224
4428012; E-mail: narat@uludag.edu.tr;
URL: kutuphane.uludag.edu.tr
1975; Neşe Arat
123 351 vols; 492 curr per; 16 000 e-
journals; 4 053 diss/theses; 3 digital data
carriers
libr loan
27852

Hacettepe Üniversitesi Kütüphaneleri
(Hacettepe University Libraries), University
Medical Library, 06100 **Hacettepe**
/ Ankara
T: +90 312 3051061; Fax: +90 312
311799; E-mail: library@hacettepe.edu.tr;
URL: www.library.hacettepe.edu.tr
1967; Ayşen Küyük
Hacettepe University Medical Library,
06100 Hacettepe, Ankara / Beytepe
Library, 06532 Beytepe, Ankara /
Hacettepe University Conservatory Library,
06532 Beşevler, Ankara
86 926 vols; 326 curr per; 29 273
e-journals; 64 825 e-books; 9 250
diss/theses; 1 627 music scores; 1 366
av-mat; 1 491 sound-rec; 1 200 000
electronic theses
libr loan; MLA, IFLA, TKD
27853

Boğaziçi Üniversitesi, Kütüphanesi
(University of Bosphorus), Bebek, 80815
Istanbul
T: +90 212 3581540 ext 1739; Fax: +90
212 2575016; E-mail: bulib@boun.edu.tr;
URL: www.library.boun.edu.tr
1863; Prof. Gün Kut
Near East coll, Braille coll
375 000 vols; 1 400 curr per; 77 mss;
5 839 diss/theses; 5 761 govt docs; 231
maps; 19 410 microforms; 756 av-mat;
3 042 sound-rec; 45 digital data carriers
libr loan; ALA, LA, IFLA, ASLIB
27854

**Istanbul Devlet Mühlendislik ve
Mimarlik Akademisi**, Kütüphanesi
(Istanbul State Academy of Engineering
and Architecture), Vildiz, **Istanbul**
T: +90 212 610220
1911; Hulya Albayrak
45 000 vols
27855

Istanbul Teknik Üniversitesi, Mustafa
Inan Kütüphanesi (Istanbul Technical
University), Ayazağa, Maslak, 80626
Istanbul
T: +90 212 2853596; Fax: +90 212
2853302; E-mail: kutuphane@itu.edu.tr;
URL: www.library.itu.edu.tr
1773; Ayhan Kaygusuz
500 000 vols; 749 curr per; 103 e-
journals; 51 817 e-books; 19 190
diss/theses; 800 govt docs; 3 342 music
scores; 304 maps; 3 022 av-mat; 5 756
sound-rec
libr loan; ANKOS
27856

İstanbul Üniversitesi, Kütüphane ve
Dokümantasyon Daire Başkanligi (Istanbul
University Library and Documentation
Centre), Beyazıt, **Istanbul**
T: +90 212 5140380, 5281248;
Fax: +90 212 5111219; E-mail:
webmaster@istanbul.edu.tr; URL:
www.istanbul.edu.tr
1924; Prof. Meral Alpay
30 000 vols printed in Arabic letters, 800
photo albums (1880-1910), Bibliography
of Istanbul Univ Publs; Rare book div,
Museum div, Faculty libs div
1 500 000 vols; 13 100 curr per;
18 606 mss; 16 000 incunabula; 38 000
diss/theses; 212 music scores; 1 390
maps; 822 microforms; 32 600 av-mat; 50
digital data carriers; 34 000 photos, 596
paintings of old scripts
libr loan; IFLA
27857

– Deontoloji ve Tıp Tarihi Anabilim
Dalı, I.Ü. Cerrahpaşa Tıp Fakültesi,
Kütüphanesi (Medical Ethics and Medical
History Department Library), Tıp Tarihi
Enstitüsü, 34303 Istanbul
T: +90 212 5861509; Fax: +90 212
6320050
1969; Gülten Dinç
16 000 vols; 10 curr per; 741 mss; 50
diss/theses; 10 maps; 2 000 av-mat; 220
sound-rec; 2 000 slides
libr loan
27858

– Fen Fakültesi Matematik Bölümü Orhan
Şerafettin İçen Kütüphanesi (Faculty
of Science Department of Mathematics
Orhan Serafettin İçen Library), Vezneciler,
34118 Istanbul
T: +90 212 4555700, 15316;
Fax: +90 212 5190834; E-mail:
sozobek@istanbul.edu.tr; URL:
www.istanbul.edu.tr/fen/kutupmat.php
1933; Mrs Semra Özöbek
9 831 vols; 441 curr per; 216 diss/theses
27859

– Hukuk Fakultesi, Kütüphanesi (Faculty of
Law), Beyazıt, 34115 Istanbul
T: +90 212 4400115;
Fax: +90 212 4400116; E-mail:
hukuk_dekan@istanbul.edu.tr
1950; Ms-Ilknur Durmaz
Economy, hist, philosophy
104 211 vols; 173 curr per; 8 mss; 907
diss/theses; 5 govt docs; 1 digital data
carriers; 22 062 periodicals
libr loan
27860

– Iktisat Fakultesi, Kütüphanesi (Faculty of
Economics), 34452 Istanbul
T: +90 212 5140300
1936; Aysegül Ardic
Econometrics, statistics, business, finance,
economic hist, politics, public sector,
industrial relations, economics
52 000 vols; 150 curr per; 1 500
diss/theses
libr loan
27861

Koç University, Suna Kýraç Library,
Rumeli Feneri Kampusu, Rumeli Feneri
Yolu, Sariyer, 80860 **Istanbul**
T: +90 212 3381318; Fax: +90 212
3381321; E-mail: libinformation@ku.edu.tr;
URL: www.ku.edu.tr/main.php?lang=en
1993; Jane Ann Lindley
107 403 vols; 82 curr per; 18 e-journals;
122 diss/theses; 23 860 av-mat; 11 629
sound-rec; 2 296 digital data carriers
libr loan; IFLA, LIBER
27862

**Marmara Universitesi Merkez
Kutuphanesi** (Marmara University
Library), Goztepe Kampusu Kadikoy,
34722 **Istanbul**
T: +90 216 3484379;
Fax: +90 216 3484379; E-mail:
kutuphane@marmara.edu.tr; URL:
library.marmara.edu.tr
1982; Sevinc Kazaz
98 000 vols; 294 curr per; 11 402
diss/theses; 8 music scores
27863

– İlâhiyat Fakültesi, Kütüphane (Faculty of
Theology), Bağlarbaşı-Üsküdar, 81150
Istanbul
T: +90 216 3410297; Fax: +90 216
3410298
1959; Dr. Ziya Demir
50 000 vols; 50 curr per; 1 000 mss; 100
diss/theses; 100 maps; 500 microforms;
30 sound-rec
27864

Yıldız Teknik Üniversitesi, Kütüphanesi,
Yıldız Kampüsü, 80750 **Istanbul**
T: +90 212 2597070; Fax: +90 212
2608094; URL: www.yildiz.edu.tr
1937; Ulya Akbayrak
Engineering, life sciences, economics,
architecture, Turkish hist
60 000 vols; 33 curr per; 2 500
27865

Dokuz Eylül Üniversitesi, Kütüphanesi
(Ninth September University), Cumhuriyet
Bul 144, 35160 Buca **Izmir**
T: +90 232 4535072; Fax: +90 232
4535599; E-mail: hale.baltepe@deu.edu.tr;
URL: www.deu.edu.tr
1991; Hale Baltepe
235 400 vols
TKD
27866

Selçuk Üniversitesi, Kütüphanesi, **Konya**
T: +90 332 3521000; Fax: +90 332
3520998; URL: www.selcuk.edu.tr
1977; Dr. Ali Özgökmen
199 431 vols; 13 575 curr per; 30 mss;
2 875 diss/theses; 25 digital data carriers
libr loan
27867

Cumhuriyet Üniversitesi, Kütüphanesi,
Kampüs, 58140 **Sivas**
T: +90 346 2261566; Fax: +90 346
2261513
1976; Filiz Dener
Paul Valery coll
24 000 vols; 385 curr per; 726
diss/theses
libr loan
27868

Karadeniz Teknik Üniversitesi,
Kütüphanesi, 61080 **Trabzon**
T: +90 462 3772201; Fax: +90 462
3253245; E-mail: library@ktu.edu.tr; URL:
www.ktu.edu.tr
1963
Ottoman Turkish language
100 000 vols
27869

Government Libraries

TBMM Kütüphane, Dokümantasyon
ve Tercüme Müdürlüğü (Turkish
Grand National Assembly Library and
Documentation Centre), Bakanliklar,
06543 **Ankara**
T: +90 312 4206835; Fax: +90 312
4207540; E-mail: library@tbmm.gov.tr;
URL: www.tbmm.gov.tr
1920; Ismet Baydur
Documentation of parliamentary debates,
Turkish law div, newspaper, minutes
and periodical documentation, Research
service
324 586 vols; 1 002 curr per; 30 e-
journals; 500 mss; 782 diss/theses;
24 049 govt docs; 221 maps; 6 366
microforms; 198 av-mat; 157 digital data
carriers
libr loan; IFLA, TKD, LIBER, APLAP,
TKD
27870

Yükseköğretim Kurulu Baskanligi,
Dokumantasyon Merkezi (Higher
Education Council), 06539 **Ankara** –
Bilkent
T: +90 312 2987451; Fax: +90 312
2664745; E-mail: dokuman@yok.gov.tr;
URL: www.yok.gov.tr
1985/1996; Mrs Zesrin Esensoy
National deposit libr for theses and diss
110 000 diss/theses
27871

Special Libraries Maintained by Other Institutions

British Council, Library, Karum Merkezi,
No 437, 06700 **Ankara**
T: +90 312 4553600;
Fax: +90 312 4553636; E-mail:
customer.services@britishcouncil.org.tr;
URL: www.britishcouncil.org.tr
1942; Arhan Isin
13 660 vols; 85 curr per; 1 microforms;
1 250 av-mat; 1 640 sound-rec; 18 digital
data carriers
libr loan; ASLIB
27872

**British Institute of Archaeology
at Ankara**, Library, Tahran Cad 24,
Kavaklıdere, 06700 **Ankara**
T: +90 312 4280330; Fax: +90 312
4280331; E-mail: yeran@biaatr.org; URL:
www.biaatr.org
1948; Dr. Yaprak Eran
Epigraphic arch, Photogr arch; Bone,
seed and pottery collections
43 757 vols; 350 curr per; 67 diss/theses;
336 govt docs; 1 575 maps; 389
microforms; 26 digital data carriers
27873

Goethe-Institut, Infozentrum/Bibliothek,
Atatürk Bulvari 131, 06640 **Ankara**
T: +90 312 4251436;
Fax: +90 312 4180847; E-mail:
kutuphane.ankara.goethe.org; URL:
www.goethe.de/ankara
Nico Sandfuchs
11 076 vols; 1 e-journals; 544 av-mat;
1 005 sound-rec; 181 digital data carriers
27874

**Maden Tetkik ve Arama Genel
Müdürlüğü**, Kütüphanesi (General
Directorate of Mineral Research and
Exploration), Ismet Inönü Bulvari, **Ankara**
T: +90 312 2873430 ext 2277; Fax: +90
312 2879188; E-mail: library@mta.gov.tr;
URL: www.mta.gov.tr
1935; Gönül Kocer
Mineralogy, petrology, earth sciences
180 000 vols; 518 curr per; 540
diss/theses
27875

Tmmob Jeoloji Mühendisleri Odası,
Geology Library (Chamber of Geological Engineers of Turkey), PK 464, Yenişehir, 06444 *Ankara*
T: +90 312 4343601; Fax: +90 312 4342388; E-mail: jmo@jmo.org.tr
1974
Geology, sedimentology, mining, petroleum, hydrogeology, mineralogy, petrology, stratigrafy
14 000 vols   27876

Türk Tarih Kurumu, Kütüphanesi (Turkish Historical Society), Kızılay Sokak, 1, *Ankara*
T: +90 312 3102368; Fax: +90 312 3101698; E-mail: library@ttk.org.tr; URL: www.ttk.org.tr
1931; Neşecan Uysal
226 015 vols; 692 curr per; 9 e-journals; 1 216 mss; 854 maps; 285 microforms; 8 av-mat; 99 digital data carriers; 166 translations, 431 photocopies   27877

ULAKBIM Cahit Arf Bilgi Merkezi,
TUBITAK (The Scientific and Technical research Council of Turkey) (ULAKBIM – National Academic Network and Cahit Information Center), YOK Binasi, 06539 *Bilkent, Ankara*
T: +90 312 2989302; Fax: +90 312 2989393; E-mail: ulakbim@ulakbim.gov.tr; URL: www.ulakbim.gov.tr
1983; Assistant Prof. Tugrul Yilmaz
4 500 Research Project Rpts (supported by The Scientific and Technical Research Council of Turkey) and National Mediacal Database (includes only national medical titles); Nat and Internat Doc Supply Dept, ULAKNET (Nat Academic Network)
2 300 curr per; 2 700 e-journals; 10 000 periodical titles (cancelled, ceased and current)
IFLA   27878

Orient-Institut der Deutschen Morgenländischen Gesellschaft Istanbul (OII), Bibliothek (Orient Institute of the German Oriental Society, Istanbul Branch), Susam Sokak 16/18, 80060 *Cihangir,* Istanbul
T: +90 212 2521983, 2936067; Fax: +90 212 2496359; E-mail: oiist@oidmg.org; URL: www.oidmg.org
1987; Dr. Wolf-Dieter Lemke
Oriental research (Turgological), Ottoman, Modern Turkey
25 000 vols; 112 curr per   27879

Selimiye Kütüphanesi, *Edirne*
1575; Mrs Özlem Ağirgan
33 000 vols; 3 172 mss   27880

Marmara Scientific and Industrial Research Centre, Library, P.O. Box 21, 41470 *Gebze* – Kocaeli
T: +90 262 6412300; Fax: +90 262 6412309
1972; Mrs Nilgün Avcioğlu
90 000 vols; 746 curr per   27881

Halil Hamit Paşa Kütüphani, *Isparta*
1783; Mahmut Kayici
20 200 vols; 850 mss   27882

Ayasofya (Saint Sophia) Museum,
Kütüphanesi, Sultan Ahmet, *Istanbul*
1934; M. Hâdi Altay
8 500 vols   27883

British Council, Library, Maya Akar Centre, P.O. Box 16, Besiktas, 81690 *Istanbul*
T: +90 212 3555657;
Fax: +90 212 3555658; E-mail: customer.services@britishcouncil.org.tr; URL: www.britishcouncil.org/turkey
1942; Meral Kirkali
13 935 vols; 41 curr per; 4 963 av-mat; 749 sound-rec; 402 digital data carriers
ASLIB   27884

Deniz Bilimleri ve İşletmeci,
Kütüphanesi (Institute of Marine Sciences and Management), Müsküle Sok. Vefa 1, 34470 *Istanbul*
T: +90 216 5282539
1933
11 300 vols   27885

Deutsches Archäologisches Institut,
Abteilung Istanbul (German Archeological Institute), Ayazpaşa Camii Sk 48, 80090 *Istanbul* – Gümüşsuyu
T: +90 212 252-3490;
Fax: +90 212 252-3491; E-mail: bibliothek@istanbul.dainst.org; URL: www.dainst.de
1929; M. Wagner
54 000 vols; 290 curr per; 450 maps   27886

Goethe-Institut, Bibliothek/Informationszentrum, Yeni Çarsi Cad 52, Beyoglu, 34431 *Istanbul*
T: +90 212 2492009; Fax: +90 212 2525214; E-mail: info@istanbul.goethe.org; URL: www.goethe.de/istanbul
1961; Ursula Wester
15 000 vols; 40 curr per; 1 000 av-mat; 1 000 sound-rec; 320 digital data carriers
libr loan   27887

Goğrafya Enstitüsü, Kütüphanesi (Geographical Institute), Müsküle sok. Vefa, *Istanbul*
1933; Prof. Ajun Kurter
11 000 vols   27888

Hollanda Tarihve Arkeoloji Enstitüsü,
Kütüphanesi (Netherlands Historical and Archaeological Institute), Istiklâl Cad 393, Beyoğlu, *Istanbul;* PK 132, Beyoğlu, 80072 Istanbul
T: +90 212 2939283; Fax: +90 212 2513846
1958; Yasemin Erdemli
12 500 vols; 147 curr per; 80 maps   27889

Institut Français, Médiathèque – Centre d'information (French Cultural Center of Istanbul), Istiklal Caddesi, 8, 34435 Taksim *Istanbul*
T: +90 212 3938-111/-122/-123;
Fax: +90 212 2444495; E-mail: mediatheque.istanbul@ifturquie.org; URL: www.infist.org/mediatheque/index.html
1945; Nikoleta Bouilloux-Lafont
Books on the Turkey in French, School-books for learning French, French books translated into Turkish; Children's dept
23 000 vols; 60 curr per; 1 600 av-mat; 3 300 sound-rec; 107 digital data carriers
libr loan   27890

Institut Français d'Etudes Anatoliennes d'Instanbul, Bibliothèque, Palais de France, Beyoğlu, PK 54, 80072 *Istanbul*
T: +90 212 2441717; Fax: +90 212 2528091
1930
16 000 vols   27891

Istanbul Arkeoloji Müzeleri, Kütüphanesi (Istanbul Archaeological Museums), Gülhane, 34400 *Istanbul*
T: +90 212 5207740; Fax: +90 212 5274300
1902; Havva Koç, Ferda Albek
80 000 vols; 600 curr per; 2 000 mss; 52 incunabula   27892

Istanbul Deniz Müzesi, Kütüphanesi (Istanbul Naval Museum), Beşiktaş, *Istanbul*
T: +90 212 2610040; Fax: +90 212 2606038
1897; Yildiz Tuğ
The Ottoman Archive
20 000 vols; 220 mss; 5 incunabula; 626 maps; 220 microforms; 22 436 archive files (19th-20th c Ottoman Empire period), Also 25 million documents   27893

Istituto Italiano di Cultura, Italyan Kültür Merkezi, Kütüphanesi, Mesrutiyet Cad 161, 80050 *Istanbul*
T: +90 212 2939848; Fax: +90 212 2510748
Adelia Rispoli
11 000 vols   27894

Süleymaniye Kütüphanesi Müdürlüğü,
Ayse Kadin Hamami Sok 35, Beyazit, *Istanbul*
T: +90 212 5206460; Fax: +90 212 5206462
1557; Muammer Ülker
Waqf and special coll; Pathological Service
113 100 vols; 66 117 mss; 458 valuable calligraph samples in the shaps of framed inscriptions
libr loan   27895

Topkapi Sarayi Muzesi, Kütüphanesi (Topkapi Palace Museum), Sultanahmet, 34400 *Istanbul*
T: +90 212 5120480; Fax: +90 212 5224422
1928; Dr. Filiz Çağman
Archaeology, hist
21 300 vols; 19 800 mss   27896

Türk Tıp Tarihi Kurumu, Kütüphanesi (Turkish Medical History Society), I.Ü. Cerrapaş Tıp Fakültesi, Deontoloji ve Tıp Tarihi Anabilim Dah, Cerrahpaş, *Istanbul*
T: +90 212 5861509; Fax: +90 212 6320050
1938
70 000 vols   27897

Türkiye Askeri Müzesi, Kütüphanesi (Military Museum of Turkey), Harbiye, *Istanbul*
T: +90 212 2332720; Fax: +90 212 2404889
1726
12 000 vols   27898

Türkiye Diyanet Vakfi – İslâm Araştırmaları Merkezi, Kütüphanesi ve Dokümantasyon Müdürlüğü (Turkish Religious Foundation Centre for Islamic Studies), Bağlarbaşı, Gümüşyolu Cad No. 40, 81200 *Istanbul*
T: +90 216 4740850; Fax: +90 216 4740874; E-mail: kutuphane@isam.org.tr; URL: www.isam.org.tr
1984; Fatıh Çardakli
Islam, social sciences, religion, theology; complete arch of the İstanbul Şerye Sicilleri on microfilm
150 000 vols; 2 166 curr per; 1 400 diss/theses; 200 maps; 1 082 microforms
libr loan   27899

Türk Dil Kurumu, Kütüphanesi (Institution of Turkish Language), Atatürk Bulvaŕy 217, 06680 *Kavaklŕdere/Ankara*
T: +90 312 4286100; Fax: +90 312 4285288; E-mail: kitaplik@tdk.org.tr; URL: tdkkitaplik.org.tr/; www.tdk.gov.tr
1932; Nilberk Yemisci
43 916 vols; 82 curr per; 666 mss; 586 diss/theses; 34 sound-rec   27900

Public Libraries

Atatürk Kitaphgı, Mete Caddesi 45, Taksim, *Istanbul*
T: +90 212 2495683; Fax: +90 212 2517972
1929
184 000 vols   27901

Istanbul II Halk Kütüphanesi (Istanbul Public Library), Bayezit Meydanı, *Istanbul*
Fax: +90 212 5173811
1939
Turkish newspaper colls since 19th c (21 200 vols)
156 000 vols; 6 543 mss   27902

Turkmenistan

National Libraries

National Library of Turkmenistan, pl Karla Marksa, 744000 *Ashkhabad*
T: +993 12 253254; Fax: +993 12 257311
1895; N.A. Kurbanov
5 500 000 vols; 2 891 curr per; 366 mss; 16 200 maps   27903

General Research Libraries

Academy of Sciences of Turkmenistan, Central Scientific Library, Gogolya 17, 744000 *Ashkhabad*
T: +993 12 356571
1941; Almaz B. Jazberdiev
2 130 000 vols; 1 621 curr per; 1 015 microforms   27904

Republican Scientific and Technical Library of Turkmenistan, prosp. Svobody 106, *Ashkhabad*
Z.I. Cherepanova
900 000 vols   27905

University Libraries, College Libraries

Turkmen Agricultural Institute, Library, Pervomayskaya, 62, 744000 *Ashkhabad*
T: +993 12 42522
1930
136 000 vols   27906

Turkmen State Medical Institute, Library, ul. Shaumyana 58, 744000 *Ashkhabad,* GSP-19
T: +993 12 254096
1932
191 000 vols   27907

Turkmen State University, Library, Prosp. Lenina 31, 744005 *Ashkhabad*
T: +993 12 351159; E-mail: math3@online.tm; URL: www.tacistm.org/tempus/tgu.htm
1950; A.T. Vorobeva
542 000 vols; 468 curr per   27908

Turkmen Pedagogical Institute, Library, ul. Karla Marksa, 62, *Chardzhou*
1949
335 000 vols; 464 curr per   27909

Special Libraries Maintained by Other Institutions

National Museum of Turkmenistan, Library, Archibald av 30, 744000 *Ashkhabad*
T: +993 12 519020; Fax: +993 12 519022; E-mail: museum@online.fm
1899; Mammentnurow Owezmuhammet
History, ethnology, art, reference books, natural science dept
9 300 vols; 12 curr per   27910

Tuvalu

Government Libraries

Parliamentary Library, Private Mail Bag, Vaiaku, *Funafuti*
T: +688 20250; Fax: +688 20253
1984
600 vols; 12 curr per   27911

Uganda

University Libraries, College Libraries

East African School of Library and Information Science, Makerere University, Makerere University Library, *Kampala;* P.O. Box 7062, Kampala
T: +256 41 531530; Fax: +256 41 235002; E-mail: direct@easlis.mak.ac.ug.
1972; Edith Z. Lutwama
11 050 vols; 30 curr per; 500 diss/theses; 2 maps; 50 microforms; 38 digital data carriers; News letters, Magazines
ULA, SCECSAL, COMLA, IFLA   27912

Kyambogo University, Library, P.O. Box 1, Kyambogo, *Kampala*
T: +256 41 220273; Fax: +256 41 220464; E-mail: iteklib@afsat.com
1953; Justin N. Kiyimba
Higher teacher education; Dept of Special Education
60 000 vols; 65 curr per; 20 000 diss/theses
libr loan; ULA, IFLA   27913

Makerere University, Main Library, P.O. Box 7062, *Kampala*
T: +256 41 531041, 531042; Fax: +256 41 540374; E-mail: universitylibrarian@mulib.mak.ac.ug; URL: mulib.mak.ac.ug
1940; James Mugasha
East Africa colls, Uganda legal deposit; Faculty libs of Medicine, Technology, Education, Social Sciences, Farm Management
700 000 vols
libr loan   27914

– Makere Medical School, Albert Cook Library, P.O. Box 7072, Kampala
T: +256 41 534149; Fax: +256 41 530024
1946; Eunice N.N. Sendikadiwa
Albert Cook Coll
55 000 vols; 210 curr per   27915

Uganda Martyrs University, Archbishop Kiwanuka Library, Nkozi Hill, Off Masaka Rd, 83 km West of Kampala, Nkozi Village, P.O. Box 5498, *Kampala*
T: +256 38 410611, 481 21894, 21895, 21896; Fax: +256 38 410100; E-mail: umu@umu.ac.ug; URL: www.fiuc.org/umu
1993; Gertrude Kayaga Mulindwa
25 000 vols; 600 curr per; 330 diss/theses; 150 govt docs; 2 music scores; 15 maps; 60 digital data carriers; 350 pamphlets
libr loan; ULA   27916

Arapai Agricultural College, Library, Soroti, P.O. Box 203, *Soroti*
10 000 vols; 600 curr per   27917

Government Libraries

Cabinet Office Library, P.O. Box 7168, *Kampala*
T: +256 41 254881; Fax: +256 41 235459
1920; Herbert R. Kiboole
10 000 vols   27918

High Court, Library, P.O. Box 7085, *Kampala*
T: +256 41 233422; Fax: +256 41 243947
1930; Julius Peter Torach
18 000 vols; 25 curr per
libr loan; ULA   27919

Uganda Management Institute, Deposit Library and Documentation Center, Plot 44/52 Jinja Rd, Lugogo, 256-041 *Kampala*; P.O. Box 20131, 256-041 Kampala
T: +256 41 259722; Fax: +256 41 259581; E-mail: umi@starcom.co.ug
1969; Mary Basaasa Muhenda
Parliamentary debates (Hansard), gazettes, statutes, bills etc
12 500 vols; 80 curr per; 1 000 diss/theses; 7 000 govt docs; 25 maps
ULA   27920

Ecclesiastical Libraries

Aquinas Library, Katigondo National Seminary, P.O. Box 232, *Masaka*
T: +256 481 20094; Fax: +256 481 20514; E-mail: katigond@africaonline.co.ug
1911; Willem Kragting
Lugudde Peter, mss section
23 463 vols; 67 curr per; 235 mss; 1 175 diss/theses; 170 maps; 3 microforms; 50 av-mat; 40 sound-rec
CLA   27921

Special Libraries Maintained by Other Institutions

Animal Health Research Centre, Veterinary Department, Library, P.O. Box 24, *Entebbe*
T: +256 42 20192
1926; H.R. Kiboole
14 000 vols; 150 curr per; 100 diss/theses; 21 000 govt docs; 200 maps; 120 annual reports, 21 000 reprints
libr loan; IAALD   27922

Geological Survey and Mines Department, Documentation Center, P.O. Box 9, *Entebbe*
T: +256 41 320-559/-656; Fax: +256 41 320634; E-mail: gsurvey@starcom.co.ug
1919; John Odida
UN published rpts on mineral commodities; Petroleum exploration project dept
23 000 vols; 30 curr per; 55 diss/theses; 60 govt docs; 390 maps   27923

Geological Survey Museum and Library, P.O. Box 9, *Entebbe*
T: +256 42 20656; Fax: +256 42 20364
1919; John Odida
Mineral and rock samples
35 000 vols; 30 curr per; 3 000 mss; 32 diss/theses; 400 maps; 3 digital data carriers; 3 900 periodicals
libr loan   27924

Forest Department Library, P.O. Box 7124, *Kampala*
T: +256 41 347085; Fax: +256 41 347086
1904; W.M. Bwiruka
20 000 vols
libr loan; IPA   27925

Makerere Institute of Social Research, Library, P.O. Box 16022, *Kampala*
T: +256 41 554582; Fax: +256 41 532821
1958; B.M. Kawesa
10 000 vols; 300 curr per; 3 000 mss   27926

Namulonge Agricultural and Animal Production Research Institute, Agricultural Research Information System (ARIS), Library, P.O. Box 7084, *Kampala*
T: +256 41 341554; Fax: +256 42 21070; E-mail: naari@naro.bushnet.net
1950; Innocent Rugambwa
6 000 vols; 1 500 curr per; 10 diss/theses; 600 govt docs; 10 maps; 8 digital data carriers
IAALD   27927

Uganda Coffee Development Authority, Library, Coffe House, Plot 35, Jinja Rd, *Kampala*
T: +256 41 256940; Fax: +256 41 236994; E-mail: ucdajc@swiftuganda.com
James E. Chiria
10 000 vols   27928

Public Libraries

National Library of Uganda (NLU), Plot 50 Buganda Road, P.O. Box 4262, *Kampala*
T: +256 41 4254661, 4233633; Fax: +256 41 4348625; E-mail: admin@nlu.go.ug; URL: www.nlu.go.ug
1964/2003; P. Birungi
20 district libs; 5 rural community libs
IFLA, COMLA, ULA   27929

Ukraine

National Libraries

Nationalna biblioteka Ukrainy im. V. I. Vernadskoho (V. I. Vernadsky National Library of the Ukraine), pr. 40-richchya Zhovtnia, 3, 03039 *Kyiv*
T: +38 44 5248136; Fax: +38 44 5243398, 5241770, 5255602; E-mail: library@nbuv.org.ua; URL: www.nbuv.gov.ua
1919; Aleksei Semyonovich Onishchenko
Presidential Fund
1 500 000 media items
libr loan; IFLA   27930

Nationalna biblioteka Ukrainy im. V. I. Vernadskoho (filiala 2) (V. I. Vernadsky National Library of the Ukraine (Branch No 2)), Vernadskoho Akademika bul 79, 03142 *Kyiv*
T: +380 44 4242321   27931

General Research Libraries

Derzhavna oblasna universalna biblioteka (State Regional Universal Library), Baidi Vishnevetskoho, 8, 18000 *Cherkasy*
T: +380 472 473242
1954; Liliya Pavlivna Kapkaeva
1 900 000 vols; 352 curr per   27932

Derzhavna oblasna universalna naukova biblioteka (State Regional Universal Scientific Library), Kooperativna, 13, 61000 *Kharkiv*
1951; Marina Vasilivna Timonina
196 000 vols; 243 curr per   27933

Kharkivska derzhavna naukova biblioteka im. V.G. Korolenka (Kharkiv State Scientific V. Korolenko Library), prov. Korolenka, 18, 61003 *Kharkiv*
T: +380 57 7311101; E-mail: kharkivlibrary@yandex.ru; URL: korolenko.kharkov.ua
1886; Liliya Pavlivna Neznamova
Austrian Libr, German Reading-Hall, Council of Europe's Human Rights Libr, Ukrainian-Canadian Libr
6 700 000 vols; 1 950 curr per; 2 000 mss; 9 incunabula; 600 133 diss/theses; 38 400 govt docs; 105 996 music scores; 3 693 maps; 30 348 microforms; 343 av-mat; 5 600 sound-rec; 50 000 rare bks (16-18th c), 1 508 200 patents, 34 683 standards
libr loan; IFLA   27934

Kirovogradska oblasna universalna naukova biblioteka im. D.I. Chizhevski (D.I. Chizhevsky Regional Universal Research Library), vul. K. Marksa, 24, 25050 *Kirovograd*
T: +380 522 226579; Fax: +380 522 244619
1898; Lidiya Ivanivna Demeschenko
820 000 vols; 621 curr per; 20 500 music scores; 30 maps; 2 900 av-mat; 6 600 sound-rec   27935

Derzhavna naukovo-pedahohichna biblioteka Ukrainy im. V. O. Sukhomlynskoho (State Scientific and Pedagogical Library of Ukraine), vul. M.Berlinskoho 9, 04060 *Kyiv*
T: +380 44 4672214; Fax: +380 44 4403548; E-mail: dnpb@i.ua; URL: www.library.edu-ua.net
1999; Paula Rohova
V. O. Sukhomlynski Documentary Fund, Fund of Library Science Documents, Fund of Textbooks for the Different Levels of Educational Establishments of the 19-20th c, Periodical Collection of the 20th c
543 750 vols; 400 curr per; 2 300 diss/theses; 400 music scores; 162 sound-rec; 7 500 other items, 250 vols of Eyeryman's Libr (Britannica, Americana, World Book) 10 430 offprints, 124 951 periodicals
libr loan   27936

Derzhavna naukovo-tekhnichna biblioteka Ukrainy (Scientific and Technical Library of the Ukraine), Horkoho vul 180, 03680 *Kyiv*
T: +380 44 5282338
1935
4 000 000 vols; 15,5 mio periodicals and patents   27937

Luganska oblasna universalna naukova biblioteka im. O.M. Gorkoho (Lugansk Regional Universal Scientific Library named after M. Gorky), Radinska, 78, 91000 *Lugansk*
T: +380 642 532570; Fax: +380 642 532570; URL: www.library.lg.ua/ukr/
1897; Inna Pavlivna Rybiantseva
Dept of rare bks, Art Lit dept, Foreign Language dept
1 030 000 vols; 1 500 curr per; 15 000 music scores; 640 av-mat; 7 600 sound-rec   27938

Derzhavna oblasna universalna naukova biblioteka im. O. Pchilki (Pchilka State Regional Universal Scientific Library), Shopena, 11, 43000 *Lutsk*
T: +380 3300 24237
1940; Galina Sergiivna Kukharska
625 000 vols; 784 curr per   27939

Derzhavna oblasna universalna naukova biblioteka (State Regional Universal Scientific Library), pl Galitska, 10, 79061 *Lviv*
T: +380 32 2724609
1940; Vasil Dmitrovich Siman
800 000 vols; 736 curr per   27940

Lvivska naukova biblioteka im V. Stefanyka, Natsionalna Akademiya Nauk Ukrainy (Lviv National V. Stefanyk Scientific Library of Ukraine), vul. Stefanika, 2, 79000 *Lviv*
T: +380 32 2725720; Fax: +380 32 2729147; E-mail: acquis@lsl.lviv.ua
1940; Dr. Myroslav M. Romaniuk
Old and Rare Books Coll, Ukrainica Coll,

Art Coll, Manuscript Coll
6 514 212 vols; 35 e-journals; 131 401 mss; 49 incunabula; 63 200 diss/theses; 134 996 music scores; 15 425 maps; 29 240 microforms; 106 av-mat; 820 sound-rec; 306 digital data carriers; 2,13 mio periodicals, 32 641 other items
libr loan   27941

Oblasna universalna naukova biblioteka im. Gmyrov (Gmyrov Regional Universal Scientific Library), Moskovska, 9, *Mykolaiv* 54038
T: +380 512 352578; Fax: +380 512 471272
1881; Nadiya Fedorivna Bogza
Literature in foreign languages (278 362 vols)
501 292 vols; 654 curr per; 32 835 music scores; 149 maps; 741 av-mat; 9 140 sound-rec; 366 digital data carriers; 3 966 brochures
libr loan; Ukrainian Library Association   27942

Derzhavna oblasna universalna naukova biblioteka (State Regional Universal Scientific Library), Troitska, 49/51, 65045 *Odesa*
T: +380 482 225069
1920; Yuliana S. Omelchenko
1 200 000 vols; 986 curr per
libr loan   27943

Nauchnaya biblioteka im. Gorkoho (Gorky Scientific Library), Pastera, 13, 65000 *Odesa*
T: +380 482 230252
1830; Olga f. Botushanska
4 400 000 vols; 1 500 curr per; 1 000 mss; 4 600 music scores; 1 300 maps; 1 200 microforms; 100 000 rare books   27944

Oblasna universalna naukova biblioteka im. Kotlarevskoho (Kotlarevsky Regional Universal Scientific Library), Lenina, 17, 36600 *Poltava*
T: +380 532 271861
1894; Valeri D. Zagorulko
637 000 vols; 680 curr per   27945

Krymska respublikanska universalna naukova biblioteka im. I.Ya. Franko (Crimean Republican Universal Scientific Library), vul. Gorkoho, 10, 95000 *Simferopol*
T: +380 652 276319
1890; Lyudmila M. Drozdova
841 000 vols; 527 curr per
libr loan   27946

Oblasna universalna naukova biblioteka im. N.K. Krupskoi (Sumy State Scientific Regional Library), vul. Geroiv Stalingrada, 10, 40000 *Sumy*
T: +380 542 222473; Fax: +380 542 225397; E-mail: sumy_lib@mail.ru; URL: sumylib.iatp.org.ua
1939; Tetyana Frolova
Valuable and rare bks
596 141 vols; 466 curr per; 2 e-journals; 17 894 music scores; 264 maps; 6 883 microforms; 6 793 av-mat; 7 203 sound-rec; 39 digital data carriers   27947

Derzhavna oblasna universalna naukova biblioteka (State Regional Universal Scientific Library), pr. 40-richchya Zhovtnia, 16, 88000 *Uzhgorod*
T: +380 3100 23998
1945; Lyudmila Z. Grigash
593 000 vols; 455 curr per   27948

Vinnytski oblasni universalny naukovi bibliotetsi im. K.A. Timiryazeva (Timiryazev Vinnitsa Regional Scientific Library), Soborna, 73, 21100 *Vinnytsya*
T: +380 432 322034; Fax: +380 432 352394; URL: www.library.vinnitsa.com
1907; Andrei I. Luchko
872 700 vols; 912 curr per; 26 661 music scores; 1 817 maps; 205 av-mat; 5 668 sound-rec
libr loan   27949

Derzhavna oblasna universalna naukova biblioteka im. Gorkoho (Gorky State Regional Universal Scientific Library), pr. Lenina, 142, 69095 *Zaporizhzhya*
T: +380 612 624591
1905; Inna Pavlivna Stepanenko
Art dept, technical lit dept, dept of foreign language lit
1 350 000 vols; 1 054 curr per; 597 av-mat; 4 084 sound-rec   27950

University Libraries, College Libraries

Donbas Institute of Mining and Metallurgy, Library, pr. Lenina, 16, 349104 *Alchevsk*
T: +380 6442 23123; Fax: +380 6442 26887
1957; Tetyana Ivanivna Lavrukhina
690 000 vols; 5 diss/theses; 500 govt docs; 20 sound-rec; 14 digital data carriers; 11 000 periodicals
libr loan   27951

Derzhavny pedahohichny instytut, Biblioteka (State Teacher Training Institute), vul. Shmidta, 4, 71116 *Berdyansk*
T: +380 6153 32298
1932; Lidiya Yosipovna Voina
224 000 vols; 187 curr per   27952

Bilotserkivsky derzhavny agrarny universytet, Biblioteka (Bila Tserkva Agricultural University), Soborka Vili 8/1, 09117 *Bila Tserkva*, Kyivska obl
T: +380 4463 51288; URL: www.btsau.kiev.ua
1920; Zoya Mikolaevna Denisenko
433 000 vols; 281 curr per; 8 000 mss
27953

Inzhenerno-tekhnologichny instytut, Biblioteka (Polytechnical Institute), Shevchenka, 460, 18006 *Cherkasy*
T: +380 472 430513
1960; Larisa Filimonivna Tovkach
236 000 vols; 398 curr per   27954

B. Khmelnytsky Cherkasy State University, Library, Shevchenka, 81, 18031 *Cherkasy*
T: +380 472 472142; Fax: +380 472 472233
1930; Nataliya Y. Silka
442 000 vols; 318 curr per   27955

Chernigivsky derzhavny tekhnologichny universytet (Chernigiv State Technological University), Shevchenka, 95, 14000 *Chernigiv*
T: +380 4622 952502; Fax: +380 4622 34244; E-mail: cancel@stu.cn.ua; URL: www.stu.cn.ua
1960; Lina Semenivna Tatarenko
242 000 vols; 87 curr per; 1 350 microforms; 25 digital data carriers
libr loan; Ukrainian Library Association
27956

Derzhavny pedahohichny universytet, Biblioteka (State Teacher Training University), vul. Sverdlova, 53, 14013 *Chernigiv*
1916
612 000 vols; 264 curr per   27957

Bukovinski derzhavni medichny universitet, Biblioteka (Bucovinian State Medical University), pl Teatralna, 2, 58000 *Chernivtsi*
T: +380 372 553754; Fax: +380 372 555811; E-mail: inter@bsmu.edu.ua; URL: www.bsmu.edu.ua/
1944; Valentina G. Tsimbal
341 000 vols; 451 curr per; 600 diss/theses   27958

Derzhavny universytet, Biblioteka (Chernivtsi State University), Lesi Ukrainy, 23, 58012 *Chernivtsi*
T: +380 372 229391
1875; Andri Y. Voloshchuk
1 722 000 vols; 300 music scores   27959

Derzhavny tekhnichny universytet, Biblioteka (State Technical University), Dniprobudivska, 2, 51918 *Dniprodzerzhinsk*
T: +380 5692 51576
1920; Alla M. Orlova
Machine building, automation, industrial electronics, metallurgy, heating, power stations environmental protection
450 000 vols; 374 curr per   27960

Derzhavna medychna akademiya, Biblioteka (State Medical Academy), Dzerzhinskoho, 9, *Dnipropetrovsk* 44
1946; Lyudmila K. Loktionova
458 000 vols; 240 curr per   27961

Derzhavna metallurgina akademiya Ukrainy, Biblioteka (National Metallurgical Academy of Ukraine), pr. Gagarina, 4, 49600 *Dnipropetrovsk*
T: +380 562 410244; Fax: +380 562 474461
1930; S.N. Fakhrutdinova
453 296 vols; 170 curr per; 1 921 diss/theses; 10 digital data carriers
27962

Derzhavny agrarny universytet, Biblioteka (State Agricultural University), vul. Voroshilova, 25, *Dnipropetrovsk*
T: +380 562 463126
1922; Zinaida Y. Dudchenko
328 000 vols; 251 curr per   27963

Dnipropetrovsky natsionalny universytet, Naukova biblioteka (Dnipropetrovsk State University), vul. Kozakova, 8, 49050 *Dnipropetrovsk*
T: +380 562 469213
1918; Svetlana V. Kubishkina
118 000 vols; 946 curr per; 1,8 mio periodicals   27964

Dnipropetrovsky natsionalny universytet zaliznichnoho transportu (Dnipropetrovsk State Technical University of Railway Transport), vul. Akademika Lazaryana, 2, 49010 *Dnipropetrovsk*
T: +380 562 471892; Fax: +380 562 471866; E-mail: lib@b.diit.edu.ua; URL: www.diit.edu.ua/
1930; Konstyantin Y. Kornev
808 000 vols; 780 curr per   27965

Dnipropetrovsky universytet ekonomiky ta prava, Biblioteka (Dnipropetrovsk University of Economics and Law), vul. Naberezhna Lenina, 18, 49600 *Dnipropetrovsk*
T: +380 56 3703621, 7911921; URL: www.duep.edu
Ms S.V. Cherginets   27966

National Mining University of Ukraine, Biblioteka, pr. Marksa, 19, 49027 *Dnipropetrovsk*
T: +380 562 470766; Fax: +380 562 470766; E-mail: dfr@nmuu.dp.ua
1899; Olga N. Nefedova
910 000 vols; 550 curr per   27967

Prydniprovska State Academy of Civil Engineering and Architecture, Library, vul. Chernyshevskoho, 24a, 49600 *Dnipropetrovsk*
T: +380 562 452372; Fax: +380 562 470788; E-mail: dik@pgasa.dp.ua
1930
600 000 vols   27968

Ukrainian State University of Chemical Technologies, Library, pr. Gagarina, 8, 49600 *Dnipropetrovsk*
T: +380 562 469990; E-mail: ugxtu@dicht.dp.ua
1930; Nelly F. Maksimova
730 000 vols; 946 curr per; 365 000 other materials   27969

Derzhavna konservatoriya im. Prokofeva, Biblioteka (Prokofiev State Conservatoire), vul. Artema, 44, 83086 *Donetsk*
T: +380 622 911459
1964
108 000 vols; 209 curr per; 17 200 music scores   27970

Donetsk National University of Economy and Trade named after M. Tugan-Baranovsky, Library, vul. Shchorsa, 31, 83050 *Donetsk*
T: +380 622 3351029; Fax: +380 622 3351029
1920; Tatyna Petrovna Tkachenko
Economics, commerce, trade, food technology and equipment
700 000 vols; 9 500 e-books; 350 diss/theses; 70 000 periodical items
UBA   27971

Donetsk State Medical University, Library, pr. Ilicha, 16, 83003 *Donetsk*
T: +380 622 955372
1930; Galina K. Kabardina
381 000 vols; 535 curr per; 810 mss; 25 500 microforms   27972

Donetsky natzionalny universytet, Naukova biblioteka (Donetsk national University), universytetska, 24, 83055 *Donetsk*
T: +380 622 992378; Fax: +380 622 927112; E-mail: bdongu@dongu.donetsk.ua; URL: library.dongu.donetsk.ua
1937; N.A.Karyagina
1 078 750 vols; 1 361 curr per; 2 900 mss; 38 616 diss/theses; 127 digital data carriers
libr loan; ULA   27973

Instytut pidvishchennya kvalifikatsiyi pratsivnikiv profesino-tekhnichnoyi osviti Ukrainy, Biblioteka (Ukrainian Professional Schools Personnel Advanced Training Institute), Kuibysheva, 31a, 83101 *Donetsk*
T: +380 622 662133
1979
95 000 vols; 90 curr per   27974

Pedahohichny instytut im. I. Franko, Biblioteka (Franko Teacher Training Institute), L. Ukrainki, 2, *Drogobych*
T: +380 3244 23775
1940; Mariya M. Dmitriv
558 000 vols; 270 curr per; 8 000 sheet music   27975

Lvivsky silskogospodarsky instytut, Biblioteka (Lviv Agricultural Institut), 292040 *Dublyany*, Zhovkivski rn, Lvivska obl.
1946; Lyubov A. Pinda
521 000 vols; 251 curr per; 220 000 other items   27976

Derzhavny pedahohichny instytut, Biblioteka (State Teacher Training Institute), Radyanska, 24, *Glukhiv*
1874; Galina A. Stolbisheva
330 000 vols; 320 curr per   27977

Gorlovka State Pedagogical Institute of Foreign Languages, Library, ul. Rudakova, 25, 338001 *Gorlovka*, Donetskaya obl.
T: +380 6242 46501
1956
213 000 vols   27978

Ivano-Frankivsky derzhavny medychny universytet, Biblioteka (Ivano-Frankivsk State Medical University), Halytska, 2, 76000 *Ivano-Frankivsk*
1944; Miroslava O. Pshik
109 000 vols; 197 curr per   27979

Ivano-Frankivsky natsionalny tekhnichny universytet nafti / gazu, Biblioteka (Ivano-Frankivsk National Technical of Oil and Gas), Karpatska, 15, 76019 *Ivano-Frankivsk*
T: +380 342 243149; Fax: +380 342 242139; E-mail: library@ifdtung.if.ua; URL: www.ifdtung.if.ua
1967
668 000 vols; 572 curr per   27980

Prikarpatsky natsionalny universytet im. Vasilya Stefanika, Naukova biblioteka (Stefanik Prikarpaty State University, Scientific Library), vul. Shevchenka, 57, 76018 *Ivano-Frankivsk*
T: +380 342 596043, 596110; Fax: +380 342 231574; E-mail: inst@pu.if.ua; URL: www.pu.if.ua
1944
554 000 vols; 395 curr per; 5 000 sheet music   27981

Pedahohichny instytut, Biblioteka (Teacher Training Institute), Ogienka, 61, *Kamyanets-Podilsk*
T: +380 3849 26637
1921
758 000 vols; 255 curr per; 19 500 sheet music   27982

Silskogospodarsky instytut, Biblioteka (Agricultural Institute), Shevchenka, 13, 68392 *Kamyanets-Podilsk*
T: +380 3849 25218
1955; Liliya I. Malina
492 000 vols; 467 curr per   27983

Derzhavna akademiya miskoho gospodarstva, Biblioteka (Kharkiv State Academy of Municipal Services), vul. Revolyutsiyi, 12, 61002 *Kharkiv*
1930; Petro M. Kuznetsov
890 000 vols; 132 curr per   27984

Derzhavny ekonomichny universytet, Biblioteka (State University of Economics), pr. Lenina, 9a, 61059 *Kharkiv*
1930; Valentina Grigorivna Morozova
Management, machine building, metallurgical industry, chemical industry
680 000 vols; 203 curr per   27985

Gosudarstvenny tekhnichesky universytet radioelektroniki, Biblioteka (State National University of Radioelectronics), pr. Lenina, 14, 61166 *Kharkiv*
T: +380 57 7021488; Fax: +380 57 7021013; E-mail: library@kture.kharkov.ua; URL: www.lib.kture.kharkov.ua
1930; Tamara B. Krishenko
Radio engineering, communications, computing science, automatic control systems, multichannel communication, medical instrumentation, economics
615 000 vols; 1 000 curr per; 867 diss/theses; 4 250 microforms; 25 sound-rec; 115 digital data carriers
libr loan; Ukrainian Library Association
27986

Kharkiv State Academy of Culture, Library, Bursatsi Uzviz, 4, 61003 *Kharkiv*
T: +380 57 7127440; Fax: +380 57 7128105; E-mail: SLP@ic.ac.kharkov.ua
1947; Larisa P. Semenenko
works for the teaching staff of the Kharkiv state academy of culture – 2 300
316 659 vols; 120 curr per; 805 incunabula; 40 diss/theses; 1 064 govt docs; 17 876 music scores; 8 av-mat; 6 sound-rec; 66 digital data carriers
libr loan; IFLA, Association of Modern Library Technologies   27987

Kharkiv State Academy of Design, LIbrary, Chervonopraporna, 8, 61002 *Kharkiv*
E-mail: academy@design.kharkiv.com; URL: www.design.kharkiv.com
1921; Nina Viktorivna Soboleva
118 000 vols; 490 curr per; 600 govt docs
libr loan   27988

Kharkivsky derzhavny medichny universytet, Biblioteka (Kharkiv National Medical University), pr. Lenina, 4, 61022 *Kharkiv*
T: +380 57 7050721; Fax: +380 57 7050721; E-mail: info@ksmu.kharkov.ua; URL: ksmu.kharkov.ua
1920; Izabella Sh. Ivanova
Hist of medicine
817 070 vols; 318 curr per; 40 mss; 47 000 diss/theses; 25 513 govt docs; 19 045 microforms; 6 digital data carriers
27989

Kharkivsky derzhavny tekhnichny universytet budivnytstva i arkhitektury, Biblioteka (Kharkiv State Technical University of Construction and Architecture), Sumska, 40, 61002 *Kharkiv*
T: +380 57 7000112; Fax: +380 57 7000250; E-mail: office@kstuca.kharkov.ua; URL: www.kstuca.kharkov.ua
1944; Tamara M. Raspopova
492 000 vols; 300 curr per; 4 500 maps
27990

Kharkivsky natsionalny avtomobylno-dorozhny universytet, Biblioteka (Kharkiv National Automobile and Highway University), vul. Petrovskoho, 25, 61002 *Kharkiv*
T: +380 57 7003865; Fax: +380 57 7003866; E-mail: admin@khadi.kharkov.ua; URL: www.khadi.kharkov.ua
1930; Fedir Demidyuk
402 000 vols; 300 curr per; 36 000 periodicals   27991

Kharkivsky natsionalny pedagogichny universytet, Biblioteka (Kharkiv National Pedagogical University), Artema, 29, 61002 *Kharkiv*
T: +380 57 7171034; E-mail: root@pu.ac.kharkov.ua
1811; Ivan Fyodorovich Turko
610 000 vols   27992

Kharkivsky natsionalny universytet im. V. N. Karazina, Tsentralna naukova biblioteka (V. N. Karazin Kharkiv National University), pl Svobody, 4, 61077 ***Kharkiv***
T: +380 57 7075420; Fax: +380 57 7051255; E-mail: cnb@univer.kharkov.ua; URL: www.library.univer.kharkov.ua
1804; Irina Zhuravlyova
Art bks, rare bks
3 500 000 vols; 1 310 curr per; 5 e-journals; 1 000 mss; 19 incunabula; 49 000 diss/theses; 2 600 maps; 50 000 microforms; 3 000 av-mat
libr loan 27993

Natsionalna jurydychna akademiya im. Yaroslava Mydroho, Biblioteka (National Law Academy), vul. Petrovskogo, 77, 61000 ***Kharkiv***
T: +380 57 7041158
Nadiya Petrivna Pasmor
950 000 vols; 216 curr per; 40 000 incunabula; 760 diss/theses; 32 250 govt docs; 20 maps; 7 digital data carriers
libr loan 27994

Natsionalny aerokosmicheski universitet im. N.E. Zhukovskogo / "Zhukovski aviatsionny institut", Scientific and Technical Library (Zhukovsky Institute of Aviation Engineering), vul. Chkalova, 17, 61070 ***Kharkiv***
T: +380 57 7074009; Fax: +380 57 3151131; E-mail: khai@khai.edu; URL: www.khai.edu
1930; Olena M. Varvarova
911 166 vols; 1 299 curr per; 546 diss/theses
libr loan 27995

Natsionalny farmatsevtichny universytet, Biblioteka (National University of Pharmacy), vul. Pushkinska, 53, 61002 ***Kharkiv***
T: +380 57 7063071; Fax: +380 57 7063071; E-mail: mail@ukrfa.kharkov.ua; URL: www.ukrfa.kharkov.ua
1921; Viktoriya N. Shchepinova
300 000 vols; 400 diss/theses; 2 393 microforms; 90 000 pamphlets, 40 000 periodicals
libr loan 27996

Natsionalny tekhnichesky universytet / "Kharkovski politekhnichesky institut", Nauchno-tekhnicheskaya biblioteka (National Technical University "Kharkovy Politechnical Institute"), Krasnoznamennaya, 16, 61001 ***Kharkiv***
T: +380 57 7076361, 7076888; Fax: +380 57 7076601; E-mail: bibl@kpi.kharkov.ua; URL: library.kpi.kharkov.ua
1885; Larisa Semenenko
1 561 198 vols; 721 curr per; 3 651 diss/theses; 4 207 microforms; 147 sound-rec; 49 digital data carriers; 286 603 periodicals, 515 photocopies
libr loan 27997

Tekhnichny universytet silskoho gospodarstva, Biblioteka (Kharkiv State Agricultural Technical University), 44, Artema St, 61002 ***Kharkiv***
T: +380 57 7329963, 7164173; Fax: +380 57 7003914; E-mail: khstua@lin.com.ua; URL: www-library.univer.kharkov.ua/himesh.htm
1931; Tetyana V. Novikova
341 000 vols; 112 curr per; 20 digital data carriers
libr loan 27998

Ukrainska inzhenerno-pedahohichna akademiya, Biblioteka (Ukrainian Engineering and Teacher Training Academy), Universytetska, 16, 61003 ***Kharkiv***
T: +380 57 7337897; E-mail: director@bibl.uipa.kharkov.ua; URL: library.uipa.kharkov.ua
1958; Natalia Nikolaevna Nikolaenko
960 000 vols; 590 curr per; 6 e-journals; 1 327 diss/theses
libr loan 27999

Khersonsky derzhavny universytet, Biblioteka (Kherson State University), 40 rokiv Zhovtnya, 27, 73000 ***Kherson***
T: +380 552 226262, 326705; Fax: +380 552 226705; E-mail: office@kspu.kherson.ua; URL: www.kspu.kherson.ua
1917; Yuliya Vasilivna Kuznetsova
Rare bks (19th-20th c)
400 000 vols; 372 curr per 28000

Silskokhospodarsky instytut, Biblioteka (Agricultural Institute), pl Oleksandrivska, 2, 73006 ***Kherson***
1924; Nataliya Y. Barsukova
300 000 vols; 282 curr per 28001

Sudomekhanichny tekhnikum, Biblioteka (Shipbuilding College), Ushkakova, 20, 73000 ***Kherson***
1946
74 000 vols 28002

Tekhnichesky universytet, Biblioteka (University of Technology), Berislavske shosse, 24, 73008 ***Kherson***
1960; Lyudmila Anatolivna Kashchenko
353 000 vols; 170 curr per; 159 diss/theses 28003

Khmelnitsky National University, Library, Kamyanetska 110/1, 29016 ***Khmelnitski***
T: +380 3822 22266; Fax: +380 3822 23265; E-mail: library@mailhub.tup.km.ua; URL: library.tup.km.ua
1962; Valentina Petritska
Fund of foreign lit
650 000 vols; 35 344 curr per; 35 e-journals; 1 340 diss/theses; 276 sound-rec; 276 digital data carriers
libr loan 28004

Derzhavny pedahohichny instytut im. Pushkina, Biblioteka (Pushkin State Teacher Training Institut), vul. Shevchenka, 1, 25050 ***Kirovograd***
T: +380 522 229564
1930; Inna Vasilivna Svoren
636 000 vols; 200 curr per 28005

Derzhavny pedahohichny instytut, Biblioteka (State Teacher Training Institute), pr. Gagarina, 54, 324086 ***Krivi Rig***
T: +380 564 715821
1930; Nelya Grigorivna Smirnova
501 000 vols; 217 curr per 28006

Krivorozhsky gornodobyvayushchi instytut, Biblioteka (Krivoi Ore Mining Institut), Pushkinska vul, 37, ***Krivi Rig***
T: +380 564 294811
1922; Nina Mikhailivna Kyudryashova
1 040 000 vols; 1 500 curr per; 1 500 mss; 958 diss/theses; 813 microforms 28007

Derzhavna konservatoriya im. P.I. Chaikovskoho, Biblioteka (P.I. Chaikovsky State Conservatoire), vul. Marksa, 1-3, ***Kyiv***
T: +380 44 2292856
1913; Olena Ivanivna Boglei
198 403 vols; 85 curr per; 5 700 mss; 80 000 music scores; 8 000 sound-rec 28008

Derzhavne khoreografichne uchilishche, Biblioteka (State College of Dance of Ballet), Shamrila, 4, 252112 ***Kyiv***
1940
Music hist, musicology, pedagogy
53 000 vols 28009

Derzhavne muzichne uchilishche, Biblioteka (State College of Music), L. Tolstoho, 31, 03032 ***Kyiv***
T: +380 44 2206212
1890
85 000 vols; 45 curr per
libr loan 28010

Derzhavny instytut udoskonalennya likariv, Biblioteka (State Institute of Advanced Medical Training), Dorogozhitska, 9, ***Kyiv***
T: +380 44 4414038
1918; Olena Vitaliivna Smakhtina
192 000 vols; 147 curr per 28011

Kiev Polytechnic Institute, Library, pr. Peremogi, 37, 03056 ***Kyiv***
T: +380 44 2746913; Fax: +380 44 2740954
1898
2 500 000 vols 28012

Kyiv Economic Institute of Management, Library, vul. Nemirovicha-Danchenko, 2, 01011 ***Kyiv***
T: +380 44 2948266; Fax: +380 44 2904193
1930; Alla Vasilivna Golidbina
Food sciences, textiles, timber
1 015 300 vols; 814 curr per 28013

Kyivsky natsionalny linguistichny universytet, Biblioteka (Kiev National Linguistic University), vul. Chervonoarmiska, 73, 01650 ***Kyiv***
T: +380 44 2271419; Fax: +380 44 2276788
1948; Ludmila I. Gupalova
624 457 vols; 345 curr per; 690 diss/theses
libr loan 28014

Kyivsky natsionalny torgovelno-ekonomichny universytet, Biblioteka (Kyiv National University of Trade and Economics), Kioto, 19, 02156 ***Kyiv***
T: +380 44 5137215; URL: www.knteu.kiev.ua
1966; Liliya Sergiivna Shestopalova
1 500 000 vols; 500 curr per 28015

Natsionalny agrarny universytet, Biblioteka (National Agricultural University), Heroyiv Oborony, 15, 03041 ***Kyiv***
T: +380 44 527-8233; E-mail: rectorat@nauu.kiev.ua; URL: www.nauu.kiev.ua
1954; Viktor V. Dolgopolov
486 000 vols; 557 curr per; 3 740 mss 28016

Natsionalny aviatsiny universytet, Biblioteka (National Aviation University), pr. Kosmonavta Komarova, 1, 03058 ***Kyiv***
T: +380 44 4067901; Fax: +380 44 4973141; E-mail: post@nau.edu.ua; URL: www.nau.edu.ua
1947; Lyudmila Vasilivna Potapova
2 000 000 vols; 627 curr per 28017

Natsionalny medychny universytet im. O.O. Bogomoltsya, Biblioteka (National Medical University), vul. Zoologichna, 1, 01601 ***Kyiv***
T: +380 44 4544928; URL: www.nmu.edu.ua/biblio.php
Larisa Mikhailivna Drozdova
400 000 vols 28018

Natsionalny universytet fizychnoho vykhovannya i sportu Ukrainy, Biblioteka (National University of Physical Education and Sport of Ukraine), Fizkulturi, 1, 03650 ***Kyiv***
T: +380 44 2875141, 2897651; Fax: +380 44 2876191; E-mail: sportlib@ukr.net
1930; Tatyana Mikolaivna Poltavets
268 000 vols; 16 631 curr per; 382 e-books; 615 diss/theses; 56 sound-rec; 21 629 other materials 28019

Naukova biblioteka im. M. Maksymovycha, Kyivskoho natsionalnoho universytetu im. Taras Shevchenka (Maksymovych Scientific Library of the Taras Shevchenko Kyiv National University), vul. Volodymyrska, 58, 01033 ***Kyiv***
T: +380 44 2393244; Fax: +380 44 2393440; E-mail: director@library.univ.kiev.ua; URL: www.library.univ.kiev.ua
1834; Oleksandr Hryhorovych Kyrylenko
Coll of rare books (7 000), Coll of Univ authors'scientific works (8 000); Educational lit dept, Rare books dept, Specialized holdings according to faculties specialization (18), Fiction holdings for students, Reading room of Ukrainian lang and lit
3 500 000 vols; 15 000 curr per; 9 incunabula; 10 982 diss/theses; 45 digital data carriers
libr loan 28020

Pedahohichny universytet im. M. Dragomanova, Biblioteka (M. Dragomanov National Pedagogical University), Pirogova, 9, 01030 ***Kyiv***
T: +380 44 2163860
Elga Volodimirivna Tatarchuk
1 430 000 vols; 470 curr per; 1,3 mio other materials 28021

Tekhnichny universytet budivnitstva i arkhitekturi, Biblioteka (Technical University of Architecture and Construction), pr. Povitroflotski, 31, ***Kyiv***
T: +380 44 2729588
1930
585 000 vols; 352 curr per 28022

Tsentralny instytut podvishchennya kvalifikatsiyi kerivnykh kadriv osviti, Biblioteka (Advanced Teacher Training Institute), Artema, 52a, 03053 ***Kyiv***
T: +380 44 2113915
1953
100 000 vols; 51 curr per 28023

Ukrainian Academy of Foreign Trade, Library, Chigorina, 57a, 01042 ***Kyiv***
T: +380 44 2694446; Fax: +380 44 2692555
1946; Larisa Ivanivna Krinina
80 000 vols; 102 curr per; 550 mss 28024

Ukrainsky derzhavny universytet kharchovykh tekhnologi, Biblioteka (Ukrainian State University of Food Industry), Volodimirska, 68, ***Kyiv***
T: +380 44 2206500; Fax: +380 44 2207266
1930
739 000 vols; 384 curr per 28025

Ukrainsky transportny universytet, Navchalno-bibliotechny korpus (Ukrainian Transport University), Kikvidze, 42, 01101 ***Kyiv***
T: +380 44 2952427
1944; Lyudmila Volodimirivna Kovalenko
424 000 vols; 270 curr per 28026

Universytet ekonomiky ta prava "KROK", Biblioteka (University of Economics and Law), vul. Lagerna, 30-32, 03113 ***Kyiv***
T: +380 44 4556984; Fax: +380 44 4556981; E-mail: library@krok.edu.ua; URL: www.krok.edu.ua
Galina Ivanivna Demidenko
120 000 vols; 160 curr per 28027

Universytet 'Kyivo-Mohilyanska Akademiya', Biblioteka (University 'Kiev-Mohyla Academy'), Skovorody, 2, 24070 ***Kyiv***
T: +380 44 4166055; Fax: +380 44 4636783; E-mail: library@ukma.kiev.ua; URL: www.ukma.kiev.ua
1992; Tetyana Yaroshenko
14 private colls by Briokhovetsky, Starytsky, Paslavsky, Holovach, Tereshenko, Rinberg, Pavlychko, Zuevsky, etc
250 000 vols; 500 diss/theses; 1 000 av-mat; 1 000 sound-rec; 150 digital data carriers; 15 000 periodical
libr loan; IFLA, Ukrainian Library Association 28028

T. Shevchenko Lugansk State Pedagogical University, Library, Oboronna, 2, 91011 ***Lugansk***
T: +380 642 530243
1924; Tamara O. Bezverkhnya
580 000 vols; 361 curr per; 8 900 music scores 28029

Silskogospodarsky instytut, Biblioteka (Agricultural Institute), 91008 ***Lugansk***
T: +380 642 952151
1921; Valentina M. Fesechko
246 000 vols; 413 curr per
libr loan 28030

Skhidnoukrainsky derzhavny universytet, Biblioteka (East Ukrainian State University), kvartal Molodizhni, 20a, 91034 ***Lugansk***
T: +380 642 461230; Fax: +380 642 461364
1921; Vira Dmitrivna Kholod
700 000 vols; 892 curr per
libr loan 28031

Academy of Veterinary Medicine, Library, Pekarska, 50, 79010 ***Lviv***
T: +380 32 2756784; Fax: +380 32 2756785
1881
333 000 vols; 311 curr per; 300 mss 28032

Derzhavny instytut fizychnoyi kulturi, Biblioteka (State Institute of Physical Education), Kostyushka, 11, 79000 ***Lviv***
T: +380 32 2725769
1946; Tamara Ivanivna Karpenko
Recreation, sports, anatomy, pedagogy, psychology, biochemistry
314 000 vols; 132 curr per 28033

Derzhavny universytet Lvivska politekhnika, Biblioteka (Lviv Polytechnic University), vul. St. Bandery, 12, 79646 *Lviv*
T: +380 32 2724733
1844; Oleksandr V. Shishka
3 000 000 vols; 1 800 curr per   28034

Lvivska Akademiya Mystetstv, Naukova Biblioteka (Lviv Academy of Arts), vul. Goncharova, 38, 79011 *Lviv*
T: +380 32 2761412
1946; Svitlana Cherepanova
Ukrainian and World Art, modern training appliances of academy's structure, philosophic-cultural lit, foreign Ukrainica
100 000 vols; 49 curr per; 9 diss/theses; 10 sound-rec
libr loan   28035

Lvivsky natsionalny universytet im. Ivana Franka, Biblioteka (Ivan Franko National University of Lviv), vul. Dragomanova, 5, 79601 *Lviv*
T: +380 32 2756001; E-mail: libf@libr.franko.lviv.ua; URL: www.lnu.edu.ua/indexu.html
1608; V.K. Potaichuk
2 500 000 vols; 2 000 curr per; 1 914 mss; 50 incunabula   28036

M. Lysenko Higher State Music Institute, Library, vul. O. Nyzhankivskoho, 5, 79005 *Lviv*
T: +380 32 2743106; Fax: +380 32 22723613
1940
196 000 vols; 189 curr per; 11 000 mss; 70 800 music scores   28037

Oblasna naukovo-pedahohichna biblioteka (Regional Pedagogical Scientific Library), Zelenaya, 24, 79005 *Lviv*
T: +380 32 2754121
1926
409 500 vols; 168 curr per
libr loan   28038

Torgovo-ekonomichny instytut, Biblioteka (Institute of Commerce and Economics), Tugan-Baranovskoho, 10, 79008 *Lviv*
T: +380 32 2797640
1939; Olena Sergiivna Marchuk
500 000 vols; 511 curr per
libr loan   28039

Ukrainska akademiya drukarstva, Biblioteka (Ukrainian Academy of Printing), vul. Podholsko, 19, 79020 *Lviv*
T: +380 32 2422340; Fax: +380 32 2527168; E-mail: uad@uad.lviv.ua; URL: www.uad.lviv.ua/modules/xfsection/article.php?category=80
1945; Dina G. Dubova
384 000 vols; 583 curr per   28040

Zooveterinarny instytut, Biblioteka (Veterinary Institute), 62341 *P.O. Mala Danylivka*, Derhachivsky Rayon, Kharkiv obl
T: +380 5763 57448; Fax: +380 5763 32176
1960; Galina V. Sviridenko
428 000 vols; 275 curr per   28041

Pryazovsky derzhavny tekhnichny universytet, Biblioteka (Pryazovsky State Technical University), prov Respubliki, 7, 87500 *Mariupol*, Donetskaya obl.
T: +380 629 233416; Fax: +380 629 529924; E-mail: admin@pstu.edu; URL: www.pstu.edu
1929
595 000 vols; 405 curr per
libr loan   28042

Derzhavny pedahohichny instytut, Biblioteka (State Teacher Training Institute), vul. Lenina, 10, *Melitopol*
T: +380 6142 42301
1921; Valentina Fyodorivna Kolesnichenko
330 000 vols; 182 curr per
libr loan   28043

Tavriska derzhavna agrotekhnichna akademiya, Biblioteka (Tavrisk State Agricultural Academy), pr. Bogdana Khmelnitskoho, 18, *Melitopol*, Zaporozhskaya obl.
T: +380 6142 21345
1932; Lyudmila O. Petrova
Agricultural equipment, civil engineering, automation
330 000 vols; 166 curr per; 220 mss   28044

Derzhavny pedahohichny instytut, Biblioteka (State Teacher Training Institute), vul. Rozy Lyuksemburg, 24, *Mykolaiv*
T: +380 512 77356
1913; Tamara V. Mazurenko
436 000 vols; 155 curr per; 4 000 sheet music   28045

Mykolaivsky derzhavny agrarny universitet, Biblioteka (Mykolaiv State Agrarian University), Karpenka, 73, 54029 *Mykolaiv*
T: +380 512 341140; E-mail: BIBL_mdau@mail.ru; URL: www.mdau.mk.ua
1984; Olga Pustova
210 957 vols; 179 curr per; 197 895 e-books; 495 diss/theses; 509 govt docs; 10 maps; 700 digital data carriers; 377 offprints
libr loan   28046

Ukrainsky derzhavny morskoyi tekhnichi universytet, Biblioteka (Ukrainian Shipbuilding Technical University), Geroiv Stalingrada, 9, *Mykolaiv*
T: +380 512 323754
1920; Lyudmila Ivanivna Grabovenko
Transactions and rpts of the institute, automation, welding, motor ships, ocean industry, sea technology, mechanics, machinery
740 000 vols; 320 curr per; 500 diss/theses   28047

Pedahohichny instytut, Biblioteka (Teacher-Training Institute), Kropivnyanskoho, 2, *Nizhin*
T: +380 4631 22457
1820; Nataliya O. Lenchenko
753 000 vols; 360 curr per; 3 000 sheet music
libr loan   28048

Derzhavna akademiya budivnitstva ta arkhitekturi, Biblioteka (State Academy of Architecture and Construction), vul. Didrikhsona, 4, 65029 *Odesa*
T: +380 482 233342; Fax: +380 482 323229
1930; Galina Y. Sirotkina
600 000 vols; 280 curr per   28049

Derzhavna akademiya kholoda, Biblioteka (Odessa State Academy of Refrigeration), vul. Dvoryanska, 1/3, 65026 *Odesa*
T: +380 482 209166
1924
487 143 vols; 175 curr per   28050

Derzhavna konservatoriya im. Nezhdanovoi, Biblioteka (Nezhdanov State Conservatory of Music), Ostrovidova, 63, *Odesa*
T: +380 482 268372
1950
100 000 vols; 1 500 curr per; 68 500 music scores   28051

Derzhavny gidrometeorologichny instytut, Biblioteka (Odessa Hydrometeorological Institute), Lvivska, 15, 65016 *Odesa*
T: +380 482 636209; Fax: +380 482 636308
1932; Lidiya Vasilivna Puchkova
250 000 vols; 223 curr per
libr loan   28052

Derzhavny medychny universytet im. N.I. Pirogova, Biblioteka (Odessa State Medical University), 2, Valikhovskyi Lane, 65026 *Odesa*
T: +380 482 232005; E-mail: exp_c_osmu@paco.net
1903; Lidiya Ivanivna Epishkina
433 000 vols; 2 300 curr per   28053

Derzhavny universytet im. I.I. Mechnikova, Biblioteka (Mechnikov State University), Preobrazhenska, 24, *Odesa*
T: +380 482 260401
1817; Svitlana M. Staritska
3 510 000 vols; 1 177 curr per
libr loan   28054

Odeska derzhavna akademiya kharchovykh tekhnologi, Biblioteka (Odessa National Academy of Food Technologies), vul. Kanatna, 112, 65039 *Odesa*
T: +380 482 291188; Fax: +380 482 252925; E-mail: kaprelyants@paco.net; URL: www.onaft.edu.ua
1922; Lyudmila F. Sinyakova
600 000 vols; 583 curr per   28055

Odeska natsionalna morska akademiya, Biblioteka (Odessa National Maritime Academy), vul. Didrikhsona, 8, korp. 2, 65029 *Odesa*
T: +380 48 7775774; Fax: +380 48 2345287; E-mail: info@ma.odessa.ua; URL: www.ma.odessa.ua
1947; Galina M. Vasileva
450 000 vols; 700 curr per   28056

Odeski gosudarstvenny ekonomicheski universitet, Biblioteka (Odessa State Economic University), Preobrazhenska, 8, *Odesa*
T: +380 482 231103; E-mail: rector@oseu.edu.ua; URL: www.oseu.edu.ua
1921
359 000 vols; 240 curr per   28057

Odesky natsionalny morsky universytet, Scientific Library (Odessa State Marine University), Mechnikova, 34, 65029 *Odesa*
T: +380 482 233528; Fax: +380 482 236033; E-mail: office@onmu.odessa.ua; URL: www.osmu.odessa.ua
1930; Galina Skoptsova
Naval engineering, marine technology, sea transport, managment
700 000 vols; 500 curr per; 1 420 diss/theses; 12 govt docs
libr loan   28058

Odesky natsionalny politekhnychny universytet, Biblioteka (Odessa National Polytechnic University), Shevchenka, 1, Bldg Nr. 15, 65044 *Odesa*
T: +380 48 7348307; E-mail: library@lib.opu.ua; URL: www.library.opu.ua
1918; Svetlana G. Banokina
1 323 735 vols; 1 500 curr per; 1 286 diss/theses; 31 sound-rec; 251 462 periodical items
libr loan   28059

Silskogospodarsky instytut, Biblioteka (Agricultural Institute), Sverdlova, 99, 65039 *Odesa*
T: +380 482 296503
1921; Svitlana S. Dzhagunova
280 000 vols; 680 curr per   28060

Ukrainska derzhavna akademiya zvyazku im. Popova, Biblioteka (Popov Ukrainian State Academy of Telecommunications), Kuznechna, 1, *Odesa*
T: +380 482 237332
1930
919 000 vols; 350 curr per   28061

K. Ushynsky South Ukrainian Pedagogical University, Library, Komsomolska, 26, 65091 *Odesa*
T: +380 482 234098; Fax: +380 482 7325103
1920; Evgeniya S. Kukhta
395 000 vols; 270 curr per; 6 000 sheet music   28062

Kooperativny instytut, Biblioteka (Institute of Commerce), vul. Kovalya, 3, 36601 *Poltava*
T: +380 532 220472; Fax: +380 532 279160
1974; Svitlana V. Sadova
446 000 vols; 235 curr per
libr loan   28063

Pedahohichny instytut im. V.G. Korolenko, Biblioteka (Korolenko Teacher-Training), vul. Ostrogradskoho, 2, 36000 *Poltava*
T: +380 532 225650
1914; Lidiya I. Kostenko
517 000 vols; 351 curr per   28064

Poltavsky natsionalny tekhnichi universytet, Biblioteka (Poltava National Technical University), Pershotravnivi pr, 24, 36601 *Poltava*
T: +380 532 2500621; Fax: +380 532 222850
1930; Valentine O. Sidorenko
414 000 vols; 680 curr per; 110 diss/theses
libr loan   28065

Ukrainian Medical Stomatological University, Library, Shevchenka, 23, 36024 *Poltava*
T: +380 532 222566; E-mail: inostr@umsa.pl.ua
1934; Nina Y. Filipenko
323 000 vols; 230 curr per
libr loan   28066

Pedahohichne uchilishche, Biblioteka (Teacher Training College), Peremogi, 170, 251350 *Priluki*
1915
612 000 vols; 315 curr per   28067

Derzhavny pedahohichny instytut, Biblioteka (State Teacher Training Institute), vul. Ostafova, 37, *Rivne*
T: +380 362 26044
1945; Lidiya P. Pabat
410 600 vols; 49 curr per
libr loan   28068

Oblastnoho instytuta povyishenniya kvalifikatsiyi pedahohicheskykh kadrov, Biblioteka (Regional Advanced Teacher Training Institute), Topolyeva, 74, 33000 *Rivne*
T: +380 362 237130
1964; Angelika Lyevosyuk
66 000 vols; 40 curr per   28069

Ukrainsky instytut inzheneriv vodnoho gospodarstva, Biblioteka (Ukrainian Institute for Water Management Engineers), vul. Soborna, 11, 33000 *Rivne*
T: +380 362 222197; Fax: +380 362 222197
1922; Anatoli P. Feshchuk
600 000 vols; 400 curr per; 600 mss   28070

Sevastopol State Technical University, Library, Striletska bukhta, Studentske mistechko, 99053 *Sevastopol*
T: +380 692 244120; E-mail: root@sevgtu.sebastopol.ua
1960; Vira O. Popova
755 000 vols; 628 curr per
libr loan   28071

Crimean State University of Agriculture, Library, 95492 *Simferopol*
T: +380 652 263377, 263339; Fax: +380 652 227451
1931; Ganna P. Sushenko
475 000 vols; 400 curr per; 2 900 mss
libr loan   28072

Derzhavny universytet im. Frunze, Biblioteka (Frunze State University), Yaltinska, 4, *Simferopol* 36
T: +380 652 233907
1918; Viktoriya I. Dryagina
776 000 vols; 1 500 curr per   28073

Krymsky medychny instytut, Biblioteka (Crimean Medical Institute), bul. Lenina 5/7, 95670 *Simferopol*
T: +380 652 291696
1931; Galina K. Prokopchuk
545 000 vols; 1 300 curr per   28074

Derzhavny pedahohichny instytut, Biblioteka (State Teacher Training Institute), Batyuka, 19, 84200 *Slovyansk*
1939; Nina M. Zemlyanska
476 000 vols; 200 curr per
libr loan   28075

Derzhavny pedahohichny instytut im. Makarenka, Biblioteka (Makarenko State Teacher Training Institute), Romenska, 87, *Sumy*
T: +380 542 299384
1930; Zoya M. Gorova
449 000 vols; 441 curr per
libr loan   28076

Sumy State University of Agriculture, Library, Kirova, 160, 40021 *Sumy*
T: +380 542 223715; Fax: +380 542 223715; E-mail: admin@sau.sumy.ua
1977; Lyubov G. Sushchenko
206 000 vols; 220 curr per
libr loan   28077

Medychny instytut, Biblioteka (Medical Institute), Ruska, 12, *Ternopil*
T: +380 3522 54577
1957; Nina Y. Grigoreva
330 000 vols; 210 curr per   28078

Ternopilskiy nicionalniy ekonomichniy universytet, Biblioteka (Ternopil State Economic University), Lvivska 11, 46004 *Ternopil*
T: +380 352 533976; Fax: +380 352 331102; E-mail: kasio@ukr.net; URL: www.library.tane.edu.ua
1971; Kazimir Voznyy
325 000 vols; 323 curr per; 1 102 e-journals; 1 439 e-books; 28 069 incunabula; 371 diss/theses; 1 439 microforms; 1 439 sound-rec   28079

Derzhavny pedahohichny universytet, Biblioteka (State Teacher Training University), vul. K. Marksa, 2, *Uman*
T: +380 4744 2529
1930
284 000 vols; 228 curr per   28080

Uman State Agrarian Academy, Library, PO Sofiyivka-5, Institutska, 1, 20305 *Uman*, Cherkasy obl.
T: +380 4744 52205; Fax: +380 4744 53170
1844; Ganna P. Stolyarenko
200 000 vols; 225 curr per
libr loan   28081

Uzhgorodsky natsionalny universytet, Biblioteka (Uzhgorod National University), Pidhirna, 46, 88000 *Uzhgorod*
T: +380 3122 33341; Fax: +380 3122 33341; E-mail: admin@univ.uzhgorod.ua
1945; Olena I. Pochekushova
1 160 000 vols; 1 000 curr per   28082

Pedahohichny universytet, Biblioteka, Ostrozhskoho, 32, 21100 *Vinnytsya*
T: +380 432 265198; Fax: +380 432 263302; E-mail: vspu-lib@mail.ru
1913; Valentina S. Bilous
528 312 vols; 337 curr per; 53 e-books; 77 diss/theses; 74 av-mat
libr loan   28083

Vinnytsky natsionalny medichny universytet im. N.I. Pirogova, Biblioteka (Vinnitsa National Memorial Medical University), vul. Pirogova, 56, 21018 *Vinnytsya*
T: +380 432 323507;
Fax: +380 432 322773; E-mail: admission@vsmu.vinnica.ua; URL: vnmu.vn.ua
1934; Lina V. Maevska
65 000 vols; 110 curr per; 33 925 diss/theses
libr loan   28084

Vinnytsky natsionalny tekhnichny universytet (Vinnitsa National Technical University), Khmelnitske shose, 95, 21100 *Vinnytsya*
T: +380 432 440387; Fax: +380 432 465772; E-mail: vstu@vstu.vinnica.ua; URL: www.vstu.edu.ua/ua/
1960; Galina Chalovskaya
764 000 vols; 245 curr per; 3 470 mss; 328 diss/theses; 25 186 govt docs; 135 000 periodicals   28085

Derzhavny instytut vdoskonalennya likariv im. Gorkoho, Biblioteka (State Institute of Advanced Medical Training), bul. Vintera, 20, 69096 *Zaporizhzhya*
T: +380 612 570524
1927
541 000 vols; 486 curr per
libr loan   28086

Oblastnoy instytut vdoskonalennya vchiteliv, Biblioteka (Regional Institute of Advanced Teacher Training), 40-richchya Radyanskoyi Ukrainy, 57a, 69600 *Zaporizhzhya*
T: +380 612 346711
1939
65 000 vols; 68 curr per
libr loan   28087

Zaporizka derzhavna inzhenerna akademiya, Biblioteka (Zaporizhzhya State Engineering Academy), Lenina, 226, 69006 *Zaporizhzhya*
T: +380 61 2238207; E-mail: library@zgia.zp.ua; URL: www.library.zgia.zp.ua/ukr/index.php
1960; Oleksandra Vasilivna Pazyuk
455 027 vols; 176 curr per; 500 govt docs; 40 digital data carriers
libr loan   28088

Zaporizky derzhavny tekhnichny universytet, Biblioteka (Zaporizhzhya State Technical University), vul. Zhukovskoho, 64, 69600 *Zaporizhzhya*
T: +380 612 642506;
Fax: +380 612 642141; E-mail: interdep@zstu.zaporizhzhe.ua
1920; Lyudmila I. Kozuryatskaya
Machine building
1 000 000 vols; 626 curr per   28089

Zaporizky derzhavny universytet, Biblioteka (Zaporizhzhya State University), vul. Zhukovskoho, 66, 69063 *Zaporizhzhya*
T: +380 612 642932; URL: www.zsu.zaporizhzhe.ua
1930; Valentina O. Gerasimova
692 000 vols; 673 curr per; 200 maps
libr loan   28090

State Academy of Agriculture and Ecology, Library, Stary bul, 7, 10001 *Zhytomyr*
T: +380 412 374931; Fax: +380 412 221402
1922
370 000 vols   28091

Zhytomyr Engineering and Technological Institute, Library, Chernyakhivskoho, 103, 10005 *Zhytomyr*
T: +380 412 241425
1960
127 000 vols; 97 curr per   28092

School Libraries

Medichne uchilishche, Biblioteka (Nursing College), Karastoianovoi, 17/9, 261400 *Berdichiv*
1937
45 000 vols   28093

Medichne uchilishche, Biblioteka (Nursing College), Pushkinska, 27, 270029 *Bilgorod Dnistrovski*
T: +380 4849 22709; Fax: +380 4849 22709
1945; Luidmila Petrovna Slobodyanyuk
41 000 vols; 500 curr per   28094

Chernigivsky basovy medychny koledzh, Biblioteka (Chernihiv Medical College), Piatnitskaia, 42, 14000 *Chernigiv*
T: +380 462 775046; Fax: +380 462 775046; E-mail: chbmc@ukr.net; URL: chbmc.com.ua
1881; Lyudmyla Mykolaivna Kulyck
56 000 vols   28095

Medichne uchilishche no 1, Biblioteka (Nursing School no 1), Geroiv Stalingrada, 23, 49104 *Dnipropetrovsk*
1944
65 000 vols   28096

Politekhnikum, Biblioteka (Polytechnical Professional School), Chelyuskintsiv, 159, 83055 *Donetsk*
T: +380 622 930417
1927
71 000 vols; 95 curr per   28097

Srednya spetsialna shkola militsiyi, Biblioteka (Donetsk Institute of Internal Affairs), Zasyadka, 13, 83054 *Donetsk*
T: +380 622 553016; Fax: +380 622 574758
1962; Lubov Andreevna Olenich
150 000 vols; 6 750 curr per; 500 mss; 150 diss/theses
libr loan   28098

Tekhnikum promislovoyi avtomatiki, Biblioteka (Industrial Automation Professional School), Gorkoho, 163, 83055 *Donetsk*
T: +380 622 910669
1930
85 000 vols; 83 curr per   28099

Muzychne uchilishche, Biblioteka (Music College), Sichovykh striltsiv, 44/b, 76000 *Ivano-Frankivsk*
T: +380 340 22365
1940
51 000 vols; 304 curr per
libr loan   28100

Medichne uchilishche, Biblioteka (Nursing College), 30 rokiv Peremogi, 9, 281900 *Kamyanets-Podilsk*
T: +380 3849 21049
1891
53 000 vols   28101

Kooperativny tekhnikum, Biblioteka (College of Commerce), pr. Ushakova, 60, 73026 *Kherson*
1928
67 000 vols; 54 curr per   28102

Medichne uchilishche, Biblioteka (Nursing College), Perekopska, 164a, 73008 *Kherson*
1872
148 000 vols   28103

Morekhidne uchilishche im. Shmidta, Biblioteka (Schmidt Marine Transport College), 40-richchya Zhovtnya, 25, 73013 *Kherson*
Sea transport, navigation, economics
113 000 vols; 45 curr per   28104

Tekhnikum mekhanizatsiyi silskoho gospodarstva, Biblioteka (College of Agricultural Engineering), pr. Pravdy, 70a, 25028 *Kirovograd*
T: +380 522 553384
1946; Nera L. Geychenko
75 000 vols; 43 curr per   28105

Medichne uchilishche, Biblioteka (Nursing College), Pavlika, 1, 285200 *Kolomiya*
1947
60 000 vols
libr loan   28106

Medichne uchilishche, Biblioteka (Nursing School), Semashka, 14, 324051 *Krivi Rig*
T: +380 564 231838
1930
68 000 vols   28107

Elektromekhanichny tekhnikum zaliznichnoho transportu im. M. Ostrovskoho, Biblioteka (Ostrovsky Electrical and Mechanical Railway Transport Professional School), pr. Povitroflotski, 35, 03037 *Kyiv*
1953
85 000 vols; 45 curr per
libr loan   28108

Medichne uchilishche no 1, Biblioteka (Medical College no 1), Blyukhera, 6, 03128 *Kyiv*
T: +380 44 4494558
1842; Olena Anatolivna Leshchenko
103 000 vols
libr loan   28109

Medichne uchilishche no 4, Biblioteka (Medical College no 4), Bratislavska, 5, 02166 *Kyiv*
T: +380 44 5187668
1965; Antonina O. Goroshchenko
87 000 vols; 47 curr per   28110

Mekhaniko-metalurginy tekhnikum, Biblioteka (Mechanic-Metallurgical Professional School), Kharkivske shosse, 15, 03160 *Kyiv*
T: +380 44 5593977
1972
93 000 vols   28111

Pedahohichne uchilishche no 2, Biblioteka (Primary School Teachers Training College no 2), bul. Davidova, 18/2, 03154 *Kyiv*
1979
71 000 vols; 64 curr per   28112

Profesiyne-tekhnichne uchilishche metalistiv no 2, Biblioteka (Metallurgy Professional School no 22), Polovetska, 49, 03107 *Kyiv*
1962
54 000 vols; 27 curr per   28113

Profesiyne-tekhnichne uchilishche zaliznichnoho transportu no 17, Biblioteka (Railway Transport Professional School no 17), Furmanova, 1/5, 03049 *Kyiv*
T: +380 44 2233510
1946
79 000 vols; 50 curr per   28114

Respublikansky zaochny avtotransportny tekhnikum, Biblioteka (Automobile Transport Professional School), Vasilkivska, 20, 03040 *Kyiv*
T: +380 44 2632395
1944
99 000 vols; 23 curr per   28115

Spetsialna muzichna shkola-internat im. M.V. Lisenka, Biblioteka (Lisenko Special Music School), Shamrila, 4, 03112 *Kyiv*
T: +380 44 4469051
1935
103 000 vols; 30 curr per   28116

Tekhnikum elektronnykh priladiv, Biblioteka (Electronic Devices Professional School), P. Lumumbi, 17, 03042 *Kyiv*
T: +380 44 2692509
1920
Electronics, telecommunications, microelectronics
85 000 vols; 120 curr per   28117

Tekhnikum gromadskoho kharchuvannya, Biblioteka (Public Catering Professional School), Chigorina, 57, 03042 *Kyiv*
T: +380 44 2698955
1964; Valentina Zakharivna Shelest
Food industry
90 000 vols; 62 curr per   28118

Tekhnikum legkoyi promislovosti, Biblioteka (Light Industry Professional School), I. Kudri, 29, 03042 *Kyiv*
1956; Valentina Zakharivna Shelest
Sewing technology, hosiery
84 000 vols; 88 curr per   28119

Tekhnikum radioelektroniki, Biblioteka (Radioelectronics Professional School), Lvivska pl, 14, 03053 *Kyiv*
T: +380 44 2217486
1960; Lidiya Matviivna Umanska
80 000 vols; 41 curr per   28120

Tekhnikum transportnoho budivnitstva, Biblioteka (Railway Construction Professional School), Vinnitska, 10, 03151 *Kyiv*
1920
Construction engineering, transport engineering
74 000 vols; 80 curr per   28121

Tekhnikum zaliznichnoho transportu, Biblioteka (Railway Transport Professional School), prov Ostrovskoho, 16, 03035 *Kyiv*
1948
65 000 vols; 90 curr per   28122

Torgovli koledzh, Biblioteka (Trade College), Lvivska, 2/4, 03115 *Kyiv*
T: +380 44 4441132
1964
59 000 vols; 33 curr per   28123

Medichne Uchilishche, Biblioteka (Nursing School), Antipova, 5, 322911 *Nikopol*
T: +380 5662 10536
1960; Lyudmila N. Serikova
36 000 vols; 32 curr per   28124

Tekhnikum gidromelioratsiyi, mekhanizatsiyi ta elektrifikatsiyi silskoho gospodarstva, Biblioteka (Agricultural Engineering and Land Reclamation College), Gorkoho, 1, 326840 *Nova Kakhovka*
T: +380 5549 44812
1957; Valentina A. Melnik
119 000 vols; 112 curr per
libr loan   28125

Medichne uchilishche no 3, Biblioteka (Nursing College no 3), Barznova, 10, 65029 *Odesa*
1957
88 000 vols   28126

Medichne uchilishche, Biblioteka (Nursing College), Komsomolska, 51a, 36011 *Poltava*
T: +380 532 270614
1925; Irina A. Kovtun
57 000 vols   28127

Medichne uchilishche, Biblioteka (Nursing College), Miryushchenka, 53, 33017 *Rivne*
1957
42 000 vols   28128

Medichne uchilishche, Biblioteka (College of Medicine), Shtormova, 27, 99000 *Sevastopol*
Medicine and related subjects
33 000 vols   28129

Medichne uchilishche, Biblioteka (Nursing School), Pirogova, 57, 21100 *Vinnytsya*
T: +380 432 350659
1921
60 000 vols   28130

Farmatsevtichne uchilishche, Biblioteka (Pharmaceutical College), Chernyakhivskoho, 99, 10005 *Zhytomyr*
T: +380 412 202380
1938
60 000 vols; 75 curr per   28131

Medichne uchilishche, Biblioteka (Nursing College), Marksa, 52, 10013 *Zhytomyr*
T: +380 412 373015
1945
81 000 vols; 87 curr per   28132

Government Libraries

Derzhkharchoprom Ukrainy, Biblioteka (Ukrainian Ministry of the Food Industry), B. Grinchenka, 1, 03001 *Kyiv*
T: +380 44 2288229
1948; Lyudmila V. Uvarova
495 000 vols; 32 curr per; 77 000 patents
28133

Natsionalna parlamentska biblioteka Ukrainy (National Parliamentary Library of Ukraine), Hrushevsky Str, 1, 01001 *Kyiv*
T: +380 44 2788512; Fax: +380 44 2788512; E-mail: office@nplu.org; URL: www.nplu.kiev.ua
1866; Tamara I. Vylegzhanina
Cyrill print bks (16th-18th c), Civil print bks (1725-1860), Hist of Ukrainian Kossaks, Ukrainian scientist A.I. Biletski's book coll, Ukrainian biologist S. Siropolk coll; Rare and Valuable Bks Dept, Foreign Lit Dept, Mechanisation and Automatisation of Libr Processes Dept, Information Center for Culture, Ukrainian-Canadian Parliament Information Centre
3 900 000 vols
libr loan; Ukrainian Library Association
28134

Corporate, Business Libraries

AO Alacheevsky metalurgiyny zavod, Biblioteka (Alacheevsk Metallurgical Plant Joint-Stock Company), pr. K. Marksa, 1, 349123 *Alchevsk*
T: +380 6642 93315
Tamara Petrivna Karpenko
357 000 vols; 271 curr per   28135

Khimvolokno company, Biblioteka (Public Joint-Stock company Chernigov enterprise Khimvolokno), Shchorsa, 72a, 14011 *Chernigiv*
T: +380 4622 160323; Fax: +380 4622 101451
1960; Oliksandra Stipanivna Gusak
40 000 vols; 80 curr per   28136

Vibronichno obednannya Azot, Biblioteka (Azot Industrial Corporation), Gorobtsya, 1, 51909 *Dniprodzerzhinsk*
T: +380 5692 38152
1936; Olga Mikhailivna Nikulina
Chemical engineering, fertilizers, nitrogen products
37 000 vols; 139 curr per   28137

Nizhnodniprovsky truboprokatny zavod, Biblioteka (Nizhnodniprovsk Pipe Rolling Plant, Library), Stoletova, 21, *Dnipropetrovsk*
T: +380 562 207207
1945
219 000 vols; 200 curr per   28138

Orendno obedannya Dniprovazhmash, Biblioteka (Dnipr Rollingstock Construction Industrial Corportion), Sukhi ostriv, 3, *Dnipropetrovsk*
T: +380 562 592434
1946
114 000 vols; 70 curr per   28139

Orendno pidpriemstvo Dniproshina, Biblioteka (Dniproshina Tyre Industrial Corporation), Krotova, 24, *Dnipropetrovsk*
T: +380 562 983381
1961/1991
Rubber technology
98 000 vols   28140

Metalurginy zavod, Biblioteka (Metallurgical Plant), Tkachenka, 122, 83062 *Donetsk*
T: +380 622 612814
1944
132 000 vols   28141

Vibronichno obednannya Donetskgirmash, Biblioteka (Donetsk Mining Engineering Industrial Corporation), Tkachenka, 189, 83005 *Donetsk*
T: +380 622 614659
1947
65 000 vols; 146 curr per   28142

Vibronichno obednannya Tochmash, Biblioteka (Tochmash Precision Machine Building Industrial Corportion), Tochmash, 83007 *Donetsk*
T: +380 622 514186
Galina Khristoforovna Bezborodova
41 400 vols; 24 curr per
libr loan   28143

Korporatsiya Ukrlisprom, Biblioteka (Ukrainian Corporation of Timber Technology), Panfilovtsiv, 48, 76015 *Ivano-Frankivsk*
T: +380 342 42033
1961
126 000 vols   28144

Kamish-Burunsky zalizorudny kombinat, Biblioteka (Kamish-Burunsk Iron More Corporation), Ordzhonikidze, 11, 98313 *Kerch*
T: +380 6561 36066
1950
Metallurgy, ore deposits, mining, metal technology, electrotechnology, mathematics
108 000 vols   28145

Traktorny zavod, Biblioteka (Tractor Plant), pr. Moskovski, 275, 61007 *Kharkiv*
1931; Valentina M. Frundina
patent department
540 000 vols; 100 curr per; 500 govt docs
libr loan   28146

Bavovnyany kombinat, Biblioteka (Cotton Industrial Corporation), 50-richchya SRSR, 73008 *Kherson*
1956; Nadezhda E. Grishko
37 400 vols; 91 curr per   28147

Vibronichno obednannya kombainovi zavod (Combine Building Industrial Corporation), Teraspilska, 1, 73026 *Kherson*
1945
Machine building, tractors, economics
86 000 vols; 153 curr per   28148

Vibronichno obednannya Chervona zirka, Biblioteka (Chervona zirka Industrial Corporation), Medvedeva, 1, 25050 *Kirovograd*
T: +380 522 278323; Fax: +380 522 274511
1909; Anna Lukinichna Lozovaya
Agriculture, civil engineering, technology, hist, politics, mathematics
87 080 vols; 39 curr per; 85 076 av-mat
28149

Zavod Avtosklo, Biblioteka (Autoglass Plant), Shmita, 20, 85000 *Kostyantynivka*
1950; Valeri V. Dudnik
Glass industry
62 000 vols; 77 curr per   28150

Vibronichno obednannya NDI proektuvannya vazhkoho mashinobuduvannya, Biblioteka (Welding and Machine Building Industrial Corporation), Ordzhonikidze, 4, 343913 *Kramatorsk*
T: +380 6264 39133
1957; Valentina M. Popova
51 700 vols; 30 curr per
libr loan   28151

Vibronichno obednannya Novokramatorsky mashinobudivny zavod, Biblioteka (Novokramatorsk Machine Building Plant Industrial Corporation, Library), Ordzhonikidze, 21, 343905 *Kramatorsk*
T: +380 6264 48279
1934
202 000 vols
libr loan   28152

Kombinat Krivorizhstal, Biblioteka (Krivorizhstal Industrial Corporation), Kosiora, 32, 324200 *Krivi Rig*
T: +380 564 717727
1934
Metallurgy and related subjects
362 000 vols; 240 curr per   28153

Tsentralny girnichno-zbagachuvalny kombinat, Biblioteka (Central Mining and Ore Rectification Industrial Corporation), 324018 *Krivi Rig*
T: +380 564 518920
1960
73 000 vols
libr loan   28154

Aktsionerno tovarishchestvo 'Naukovo-doslidny instytut po teplu ta gazopostachannyu, kompleksnomu blagoustriyu mist i sil Ukrainy', Biblioteka (Ukrainian Cities and Villages Complex Modernization, Heat and Gas Supply Research Institute Ltd), Turgenevska, 38, *Kyiv*
T: +380 44 2164492
1959; Nina M. Tsirkun
75 000 vols; 70 curr per   28155

Budivelny tekhnikum, Biblioteka (Construction Professional School), Stadionna, 2/10, 03049 *Kyiv*
1946; Oksana Anatolivna Nadolenko
Building materials
58 000 vols; 35 curr per   28156

Derzhavny komunalny proektny kompleks Kyivproekt, Biblioteka (Kievproekt State Municipal Construction Corporation), B. Khmelnitskoho, 16/22, 03001 *Kyiv*
T: +380 44 2212004
1936; Nataliya M. Tereshchenko
Architecture, electric technology, hydrotechnology, sanitary equipment, geology, geodesy, automation
106 000 vols; 160 curr per   28157

Mototsikletny zavod, Biblioteka (Motorcycle Plant), simi Khokhlovikh, 8, 03119 *Kyiv*
T: +380 44 2119924
1950; Irina O. Glinka
57 000 vols; 26 curr per   28158

Naukovo-virobnichno obednannya kompleksnoyi avtomatizatsiyi ta mekhanizatsiyi tekhnologiyi Kamet, Biblioteka (Kamet Automated Technology Scientific and Industrial Corporation), per. Peremogi, 65, 03062 *Kyiv*
T: +380 44 4429211
1941
Robotics, computer technology, welding, metal processing, mechanical engineering
113 000 vols; 107 curr per   28159

Ukrainsky naukovo-doslidny instytut spirtu i biotekhnologiyi prodovolchykh produktiv 'Biospirtprod', Nauchno-tekhnicheskaya biblioteka (Ukrainian Research and Production Corporation of Alcohol and Food Biotechnology), prov Babushkina, 3, 03190 *Kyiv*
T: +380 44 4490446
1958; Ms Belous
44 000 vols; 160 curr per   28160

Virobnichno obednannya im. Artema, Biblioteka (Artem Industrial Corporation), 04050 *Kyiv*
T: +380 44 2118648
1943; Lyudmila O. Domaeva
Machine building, metal cutting and pressing
113 000 vols; 234 curr per   28161

Vibronichno obednannya Lugansksteplovoz, Biblioteka (Lugansk Diesel Locomotive Industrial Corporation), pl R. Lyuksemburg, 91002 *Lugansk*
T: +380 642 528441
1924
177 000 vols; 75 curr per   28162

Avtobusny zavod, Biblioteka (Automobile Factory), Striyska, 45, 79618 *Lviv*
T: +380 32 2653098
1952
99 000 vols
libr loan   28163

Metalurgiyny kombinat im. Kirova, Biblioteka (Kirov Metallurgical Plant), Zhdanova, 21, 339001 *Makiyivka*
T: +380 632 92789
1920; Valeri V. Dudnik
469 000 vols; 1 200 curr per   28164

Girnichno-zbagachuvalny kombinat, Biblioteka (Dressing Integrated works Research Library), pl Lenina, 1, 53407 *Marganets*
T: +380 5665 22346
1970; Galina L. Kalashnik
92 000 vols; 67 curr per
libr loan   28165

Pivdennotrubny zavod, Biblioteka (Pipe Rolling Plant), pr. Lenina, 36, 322901 *Nikopol*
T: +380 5662 284759
1935
19 000 vols
28166

Metalurgiyny zavod, Biblioteka (Metallurgical Plant), Suchkova, 115, 322010 *Novomoskovsk*
1933
Pipe technology
79 000 vols   28167

Vibronicho obedannya Sevastopolsky morsky zavod, Biblioteka (Sevastopol Marine Plant Industrial Corporation), 99000 *Sevastopol*
T: +380 692 363024
Machine building and related subjects
177 000 vols; 732 curr per
libr loan   28168

Vibronichno obednannya Stekloplastik, Biblioteka (Plastic and Glass Industrial Corporation), Stekloplastik, 349940 *Severodonetsk*, Luganska obl.
T: +380 6452 95071
1924
Glass fibers, polymers
105 000 vols; 245 curr per
libr loan   28169

Mashinobudivno vibronichno obednannya, Biblioteka (Machine Building Industrial Corporation), Gorkoho, 58, 40004 *Sumy*
T: +380 542 297225
1930
112 000 vols; 112 curr per
libr loan   28170

Zavod chistykh mataliv, Biblioteka (Pure Metals Plant), Zavodskaya, 3, 27500 *Svitlovodsk*
T: +380 5236 23525; Fax: +380 5236 23710
1962
Semiconductors, christallography, physics, machine building, solid state physics
85 000 vols; 98 curr per
libr loan   28171

Aktsionernoe tovarishchestvo instytut reabilitatsiyi Sechenova, Biblioteka (Sechenov Rehabilitation Institute Ltd), Polikurovska, 25, 98203 *Yalta*
T: +380 654 327591
1972; Tetyana M. Endeko
67 000 vols   28172

J/V 'AvtoZAZ-Daewoo', Scientific-technical library, Lenin avenue-8, 69600 *Zaporizhzhya*, MSP-630
T: +380 612 138092; Fax: +380 612 138092
1948; Valentina I. Skazova
63 368 vols; 33 729 curr per
libr loan   28173

Korporatsiya Zaporizhtransformator, Biblioteka (Transformator Industrial Corporation), Zaporizhtransformator, 69600 *Zaporizhzhya*
T: +380 612 593474
1948
Metal technology, metal cutting, transformers, industrial management, economics
109 000 vols; 167 curr per
libr loan   28174

Mashinobudivno konstruktorsko byuro Progres (Progres Machine Building Design Office), Progres, 69064 *Zaporizhzhya*
T: +380 612 614585
1953
108 000 vols; 140 curr per
libr loan   28175

Vibronichno obednannya Motor-Sich, Biblioteka (Motor Industrial Corporation), Motor-Sich, 69064 **Zaporizhzhya**
T: +380 612 604013
1950
Machine building, electronics, technology, natural science
277 000 vols; 428 curr per
libr loan   28176

Zavod Dniprospetsstal, Biblioteka (Steel Works Dniprospetsstal), Prydvenne shose 81, 69008 **Zaporizhzhya**
T: +380 612 134296; Fax: +380 612 131780
1949
Metallurgy, ferrous metals, casting, machine building
82 000 vols; 106 curr per
libr loan   28177

Special Libraries Maintained by Other Institutions

Alupka State Palace and Park Preserve, Library, Dvortsove shosse, 10, **Alupka**
T: +380 654 722951
1921
10 000 vols   28178

Nuchno-issledovatelsky instytut solyanoyi promyshlennosti, Nauchno-tekhnicheskaya biblioteka (Research Institute of Salt Industry), vul. Artema, 35, **Artemivsk**, Donetskaya obl.
T: +380 6274 3764
1947
49 000 vols; 207 curr per; 1 000 mss   28179

Uchbovo-vibronichno obednannya Ukrainskoho tovaristva slipikh, Biblioteka (Industrial Corporation and Professional School of the Ukrainian Association of the Blind), Profinterna, 186, 343400 **Artemivsk**
T: +380 6274 23843
Talking books, records
32 000 vols   28180

M. F. Ivanov Institute for Animal Husbandry in Steppe Regions 'Askania Nova', Biblioteka, 75230 **Askaniya-Nova**, Khersonskoi obl.
T: +380 5538 61396; Fax: +380 5538 22675
1921; O.G. Kovalyk
'Ascania Nova' Reserve
132 000 vols; 55 curr per; 6 173 mss; 174 diss/theses
libr loan   28181

Bakhchisarai Historical and Cultural State Preserve, Library, Rechnaya str 133, 94805 **Bakhchisarai**, Krymskaya obl
T: +380 6554 42881; Fax: +380 6554 42881
1917; Larisa Stogova
12 300 vols; 20 curr per; 15 diss/theses; 60 govt docs; 5 maps
libr loan   28182

Zakarpatsky Agroindustrial Production, Library, 295520 **Bakhta**, Zakarpatskaya obl., Beregovo rayon
1989
Agricultural economics
40 000 vols   28183

Berdyanskaya gorodskaya tekhnicheskaya biblioteka (Berdyansk City Technical Library), vul. Dzerzhinskoho, 36/27, **Berdyansk**
T: +380 6153 322607
1961; T.V. Shcheblanova
400 000 vols; 1 351 curr per   28184

Gosudarstvenny arkhiv Zakarpatskoyi oblasti, Nauchno-spravochnaya biblioteka (State Archives of the Transkarpatian District), Lukachevskaya, 5, **Beregovo**
1946
25 500 vols   28185

Instytut ptitsevodstva, Ukrainskaya Akademiya Nauk, Biblioteka (Poultry Research Institute of the Ukrainian Academy of Science), 313410 **Borki**, Zmiiv District, Kharkiv
T: +380 5747 34439; Fax: +380 5747 34958
1932; Alla P. Ponomaryova
Encyclopedias; Patents division, standards division
51 500 vols; 50 curr per; 300 mss; 575 diss/theses; 55 govt docs; 50 microforms
libr loan   28186

Ukrainsky nauchno-issledovatelsky instytut zemledeliya, Yuzhnoe otdelenie, Biblioteka (Research Institute of Agriculture), 252205 **Chabany**, Kyivska obl.
T: +380 4463 2662277; Fax: +380 4463 2662025
1934
Crop husbandry, agricultural chemicals
66 000 vols; 279 curr per; 2 100 mss   28187

Oblasna medychna biblioteka (Regional Medical Library), Zhovtneva, 148, 18000 **Cherkasy**
T: +380 472 479065
1954; Valentina V. Nedorizova
152 000 vols; 252 curr per   28188

Gosudarstvenny arkhiv Chernigovskoi oblasti, Nauchno-spravochnaya biblioteka (State Archives of the Chernigov District), vul. Frunze, 2, **Chernigiv**
1923
13 000 vols   28189

Nauchno-issledovatelsky instytut mashin dlya proizvodstva sinteticheskykh volokon, Nauchno-tekhnicheskaya biblioteka (Institute for Textile Machines), Kotsyubinskoe shosse, 1, 14011 **Chernigiv**
T: +380 4622 920306
1960
57 000 vols; 140 curr per   28190

NDI silsko-gospodarskoyi mikrobiologiyi, Ukrainska akademiya agrarnykh nauk (Research Institute of Agricultural Microbiology, Ukrainian Academy of Agricultural Sciences), vul. Shevchenko, 97, 14027 **Chernigiv**
T: +380 4622 31749; Fax: +380 4622 37015
1962; Nataliya P. Bogdan
15 874 vols; 23 curr per; 960 diss/theses   28191

Oblasnaya medychna biblioteka (Regional Medical Library), vul. Lyubechskaya, 7-b, 14000 **Chernigiv**
T: +380 4622 699596; Fax: +380 4622 40137; E-mail: irinaivanovna7@rambler.ru
1933; Nadesda Dmitrievna Yaremenko
192 000 vols; 100 curr per; 116 300 incunabula; 27 250 diss/theses; 821 microforms
libr loan   28192

Oblasny istorychny muzei im. V. Tarnovskogo, Biblioteka (V.Tarnovsky Historical Museum), vul. Gorkogo, 4, 14006 **Chernigiv**
T: +380 4622 72650, 72793
1925
15 000 vols   28193

Chernivetsky kraeznavchy muzei, Biblioteka (Chernivtsi Local Museum), vul. Kobylyanskoi, 28, 58000 **Chernivtsi**
T: +380 372 224489
1944
10 000 vols; 23 curr per   28194

Derzhavny arkhiv Chernivetskoi oblasti, Biblioteka (State Archives of Chernivtsi Oblast), 20 Stasiuka St, 58001 **Chernivtsi**
T: +380 372 2578654; Fax: +380 372 2578654
1944; Yuri Pavlovich Lyapunov
51 700 vols; 30 curr per; 3 400 booklets   28195

Oblasna medychna biblioteka (Regional Medical Library), K. Marksa, 14, 58000 **Chernivtsi**
1945; Khorni P. Gusak
169 000 vols; 330 curr per   28196

Dniprodzerzhinsk Museum of Town History, Library, 51900 **Dniprodzerzhinsk**
T: +380 5692 31110
1931; Klaudia Demchenko
10 000 vols   28197

Derzhavny instytut fizychnoyi kulturi ta sportu, Biblioteka (State Institute of Physical Training and Sports), nab Peremogi, 10, 49094 **Dnipropetrovsk**
T: +380 562 472506
1938; Larisa Fedorivna Krivoshnik
87 000 vols; 178 curr per   28198

Gosudarstvenny arkhiv Dnepropetrovskoyi oblasti, Nauchno-spravochnaya biblioteka (Dnepropetrovsk State Archives), vul. K. Libknekht, 89, **Dnipropetrovsk**
T: +380 562 938090; Fax: +380 562 938090
1922; Tatyana N. Chentsova
11 500 vols; 2 curr per   28199

Grain Farming Institute UAAS, Ukrainian Academy of Agricultural Sciences, Scientific Library, vul. Dzerzhinskoho, 14, 49600 **Dnipropetrovsk**
T: +380 562 471433, 567 454449; Fax: +380 562 362618
1930; Ludmila M. Belokon
43 223 vols; 4 e-books; 1 188 mss; 5 045 diss/theses; 33 010 periodical items
libr loan   28200

Instytut geotekhnicheskoyi mekhaniki, Natsionalna Akademiya Nauk Ukrainy, Biblioteka (Institute of Geotechnical Mechanics), Simferopolskaya, 2a, 49095 **Dnipropetrovsk**
T: +380 562 471356
1962; Ganna P. Ivanova
107 000 vols; 180 curr per
libr loan   28201

Instytut po proektirovaniyu metalurginykh zavodov, Biblioteka (Research and Planning Institute for Metallurgical Industry), nab Lenina, 17, **Dnipropetrovsk**
T: +380 562 412291
1946
99 000 vols; 291 curr per   28202

Instytut prirodokoristuvannya ta ekologiyi, Natsionalna Akademiya Nauk Ukrainy, Biblioteka (Institute of Problems on Nature Management and Ecology), Moskovska, 6, **Dnipropetrovsk**
T: +380 562 453043; Fax: +380 562 785778
1982; Aleksandr Romanovski
15 000 vols; 15 curr per   28203

Instytut tekhnichnoyi mekhaniki, Natsionalna Akademiya Nauk Ukrainy (Mechanical Engineering Institute), Leshko-Popelya, 15, 49000 **Dnipropetrovsk**
T: +380 562 472474; Fax: +380 562 473413
1968; Lyubov Ivanivna Shmalko
113 184 vols; 102 curr per; 52 digital data carriers
libr loan   28204

Naukovo-doslidny instytut gastroenterologiyi, Nauchnaya biblioteka (Gastroenterology Research Institute), pr. Pravdy, 96, 49037 **Dnipropetrovsk**
T: +380 562 270557
1965; Olga V. Litvin
47 500 vols; 176 curr per   28205

Oblasna medychna biblioteka (Regional Medical Library), Dzerzhinskoho, 8, **Dnipropetrovsk**
T: +380 562 448188
1930; Nelli Mikhalivna Mochala
233 000 vols; 287 curr per; 125 000 other materials   28206

Pridniprovska dorozhna naukovo-tekhnichnia biblioteka (Dnipr Railways Scientific and Technical Library), pr. K. Marksa, 108, **Dnipropetrovsk**
T: +380 562 504681
1929; Raisa G. Brilyova
247 300 vols; 74 400 periodicals   28207

State Tube Research Institute, Library, vul. Pisarzhevskoho, 1A, 320600 **Dnipropetrovsk**
T: +380 562 461192; Fax: +380 562 461192
1937
84 000 vols   28208

Tsentr normativnoyi dokumentatsiyi z metalurgiyi, Tsentralna biblioteka (Metallurgical Technical Documentation Centre), vul. Dzerzhinskoho, 23, **Dnipropetrovsk**
T: +380 562 442122
1954; Tamara N. Ermilova
1 500 000 vols; 506 curr per; 256 500 patents, 1 700 periodicals   28209

Ukrainsky gosudarstvenny proektny instytut Tyazhpromelektroproekt, Biblioteka (Ukrainian State Institute of Heavy Electrical Engineering), vul. Lenina, 41, **Dnipropetrovsk**
T: +380 562 445508
1943; K.N. Nesterenko
500 000 vols; 143 curr per   28210

Botanichny sad, Natsionalna Akademiya Nauk Ukrainy, Biblioteka (Botanical Garden), Illicha, 110, 83059 **Donetsk**
T: +380 622 941280; Fax: +380 622 941280, 946157
1965; Ludmila Evgenievna Strunina
46 467 vols; 30 curr per; 16 diss/theses; 50 govt docs; 4 maps; 4 av-mat
libr loan   28211

Derzhavny komitet Ukrainy po vugilnoyi promislovosti, Biblioteka (Ukrainian State Coal Industry Board), pr. Illicha, 89, 83003 **Donetsk**
T: +380 622 903532
1966/1992
80 000 vols; 400 curr per   28212

Derzhavny proektno-konstruktorsky ta eksperementalny instytut kompleksnoyi mekhanizatsiyi shakht, Biblioteka (State Complex Mining Mechanisation Research and Planning Institute), vul. Artema, 157, **Donetsk** 48
T: +380 622 550424
1944
85 000 vols; 183 curr per   28213

Derzhavny proektny instytut Dongiproshakht, Biblioteka (Meaning Research and Planning Institute), vul. Artema, 125, 83055 **Donetsk**
T: +380 622 24725
1930
86 000 vols; 88 curr per   28214

Donetska Zaliznitsa, Dorozhna nauchno-tekhnichna biblioteka (Donetsk Railway, Railway Scientific and Technical Library), Gorna, 4, 83018 **Donetsk**
T: +380 622 312620
1937
382 000 vols; 58 curr per   28215

Donetsky komertsiny instytut, Biblioteka (Donetsk Institute of Commerce), Shevchenka, 30, 83017 **Donetsk**
T: +380 622 953426
1920; Lyudmila Vasilivna Oleinikova
565 000 vols; 258 curr per; 240 000 other materials   28216

Fizyko-tekhnichny instytut im. O. Galkina, Natsionalna Akademiya Nauk Ukrainy, Biblioteka (Galkin Physical and Technological Institute), Lyuksemburg, 72, 83114 **Donetsk**
T: +380 622 537326
1965; Valentina I. Barashko
Physics, mathematics, mechanics
153 000 vols; 490 curr per; Microfilms   28217

Gigienichny tsentr profilaktiky travmatizmu, Biblioteka (Centre of Traumatology and Hygiene), Chelyuskintsev, 163, 83015 **Donetsk**
T: +380 622 929623
1955; Nataliya Y. Grigoreva
56 000 vols; 192 curr per   28218

Instytut ekonomiky promislovosti, Natsionalna Akademiya Nauk Ukrainy, Biblioteka (Institute of Industrial Economics), universytetska, 77, 83048 **Donetsk**
T: +380 622 550468; Fax: +380 622 555312, 559400
1963; Nadezhda I. Vishnevskaya
Social and economic development of

towns, labour safety, economic law
57 000 vols; 200 curr per; 500 diss/theses; 475 microforms
libr loan 28219

Instytut fizyko-organichnoyi khimiyi i vuglekhimiyi im. Litvinenka, Natsionalna Akademiya Ukrainy, Biblioteka (Litvinenko Physical and Organic Chemistry and Coal Chemistry Institute), Lyuksemburg, 70, 83114 ***Donetsk***
T: +380 622 558004; Fax: +380 622 553542
1967; Nina Egorivna Dosta
135 000 vols; 341 curr per 28220

Instytut Pivdendiprogaz, Dovidkovo-informatsiny fond (Institute of Natural Gas, Information Centre), Artema, 169g, 83121 ***Donetsk***
T: +380 622 582090
1948
Metallurgy, machine building, welding, economics, computers
341 000 vols; 322 curr per 28221

Instytut prikladnoyi matematyky i mekhaniki, Natsionalna Akademiya Ukrainy, Biblioteka (Applied Mathematics and Mechanics Institute), Lyuksemburg, 74, 83114 ***Donetsk***
T: +380 622 510148
1965; Valentina Ivanivna Minenko
73 000 vols 28222

Instytut tekhnichnoyi ekologiyi, Natsionalna Akademiya Ukrainy, Biblioteka (Industrial Ecology Institute), bul. Shevchenka, 25, 83017 ***Donetsk***
T: +380 622 951793
1991
Ecology, biology, economics
70 000 vols; 200 curr per 28223

JSC 'Institute Yuzhniyigiprogaz', Library, Artyom str 169g, 83121 ***Donetsk***
T: +380 622 560075; Fax: +380 622 582067; E-mail: root@ungg.donetsk.ua
1933; Galina Lugova
38 000 vols; 40 curr per; 88 000 incunabula; 331 diss/theses; 30 maps 28224

Nauchno-issledovatelski, proektno-konstruktorsky i tekhnologichesky instytut vzryvozashchishchennoho i rudnichnoho elektrooborudovaniya, Nauchno-tekhnicheskaya biblioteka (Institute of Electrical Explosion Protected Equipment for the Mining, Gas and Oil Industries), vul. 50 Gvardiskoyi Diviziyi, 17, 83052 ***Donetsk***
T: +380 622 40171
1957; R.Y. Kubrak
Depository of lit on explosion protection (7 000 vols)
58 000 vols; 293 curr per; 4 000 bibliogr 28225

Naukovo-doslidny insitut girnichoryatuvalnoyi spravi, Biblioteka (Mine Rescue Equipment Research Institute), Artema, 157, 83048 ***Donetsk***
T: +380 622 555405
1945
Mining, electrical engineering, medicine, economics, physics
97 000 vols; 286 curr per 28226

Naukovo-doslidny instytut chornoyi metalurgiyi, Biblioteka (Ferrous Metallurgy Research Institute), Shevchenka, 26, 83017 ***Donetsk***
T: +380 622 949337
1960
97 000 vols; 462 curr per 28227

Naukovo-doslidny instytut gorichnoyi mekhaniky im. M.M. Fyodorova, Biblioteka (Fyodorov Research Institute of Mining), Teatralny per, 7, 83055 ***Donetsk***
T: +380 622 930859
1957
57 000 vols; 467 curr per; Microfilms 28228

Naukovo-doslidny instytut reaktivov ta khimichno chistykh materialov dlya elektronnoyi tekhniki, Biblioteka (Chemically Pure Reagents for Electronics Industry Research Institute), Bakinskykh Komisariv, 17a, 83096 ***Donetsk***
T: +380 622 599574
1959
58 000 vols; 230 curr per
libr loan 28229

Naukovo-doslidny instytut vibukhozakhishchenoho elektrobladnannya, Biblioteka (Electrical Equipment for the Mining Industry Research Institute), 50 Gvardiskoyi diviziyi, 17, 83054 ***Donetsk***
T: +380 622 949076
1957
159 000 vols; 150 curr per 28230

Naukovo-doslidny vugilny instytut, Biblioteka (Coal Research Institute), vul. Artema, 114, 83048 ***Donetsk***
T: +380 622 551531
1946
118 000 vols; 204 curr per; 105 100 mss 28231

Oblasnaya naukova medychna biblioteka (Regional Scientific Medical Library), bul. Pushkina, 26, 83055 ***Donetsk***
T: +380 622 926190
1931; Katerina G. Panechko
312 000 vols; 543 curr per; 38 243 diss/theses
libr loan 28232

Oblastnoyi kraevedchesky muzei, Nauchnaya biblioteka (Regional Local Lore Museum), Chelyuskintsev 189a, 83048 ***Donetsk*** 48
T: +380 622 553474
1944; Vitali I. Ovinnitsov
33 500 vols; 50 curr per; 430 av-mat
libr loan 28233

Proektny i naukovo-doslidny instytut prombud NDI proekt, Biblioteka (Research Institute of Industrial Engineering), universytetskaya, 112, 83004 ***Donetsk***
T: +380 622 581359
1958
105 000 vols; 80 curr per; 1 000 mss; 500 music scores 28234

Tsentralno byuro naukovo-teknichnoyi informatsiyi Vugilnoyi Promyslovosti Ukrainy, Biblioteka (Ukrainian Ministry of Coal Industry Scientific Information Centre), Serova, 56, 83017 ***Donetsk***
T: +380 622 951109
Raisa Leonidivna Boldireva
714 000 vols; 698 curr per; 132 000 patents 28235

Golovna mezhspilkova biblioteka (Main Trade Union Library), Lenina, 5, 343820 ***Enakieve***
T: +380 6252 26595
1983
122 000 vols; 85 curr per 28236

Karadag Natural Reserve, Ukrainyan National Academy of Sciences, Biblioteka (Kovalevsky Southern Seas Biology Institute – Karadag Branch), p/v Kurortne, 334876 ***Feodosiya***
T: +380 6562 38337; Fax: +380 6562 38331
1914; Valentina Yuriedna Lapchenko
Marine biology, oceanography, ichthyology, botany, biochemistry, biophysics, paleontology, T. Vyasemski coll
70 000 vols; 55 curr per; 372 maps; 141 microforms; 1 digital data carriers; 258 pamphlets 28237

Kraeznavchi muzei, Biblioteka (Regional Studies Museum), Lenina, 11, 334800 ***Feodosiya***
T: +380 6562 30277
1811
17 000 vols; 20 curr per 28238

NDI lubyanykh kultur, Biblioteka (Institute of Bast Crops), vul. Lenina, 45, 245130 ***Glukhiv***
T: +380 5444 22135; Fax: +380 5444 22643
1931; Irina Volodymirivna Tregubenko
58 000 vols; 150 curr per; 120 diss/theses; 440 pamphlets, 305 translations
libr loan 28239

Derzhavny arkhiv Ivano-Frankivskoyi oblasti, Biblioteka (Ivano-Frankivsk State Archives), Gryunvaldskaya, 3, 76000 ***Ivano-Frankivsk***
T: +380 340 22770
1944; Ms M. Rymaryk
16 600 vols 28240

Oblasna naukova medychna biblioteka (Regional Scientific Medical Library), Getmana Mazepi, 101, 76000 ***Ivano-Frankivsk***
T: +380 340 22110
1945; Mariya Arkhipivna Kovrizhina
217 000 vols; 480 curr per 28241

Tsentr narodnoyi tvorchosti, Biblioteka (Ukrainian Folklore Centre), Nezalezhnosti, 12, 76000 ***Ivano-Frankivsk***
T: +380 340 22543
1946; Mikola Pavlovich Znishchenko
48 000 vols 28242

Ukrainsky naukovo-doslidny instytut girskoho lisivnichtva, Biblioteka (Ukrainian Research Institute for Mountain Forestry), Grushevskoho vul, 31, ***Ivano-Frankivsk***
T: +380 342 25631
1965; Lidiya Lushchak
45 700 vols; 157 curr per 28243

Gosudarstvenny arkhiv Khmelnitskoyi oblasti, Nauchno-spravochnaya biblioteka (State Archives of the Khmelnitski District), vul. Rozy Lyuksemburg, 15a, ***Kamyanets-Podilsk***
58 100 vols 28244

Shevchenkovsky natsionalny zapovednik v Kaniva, Biblioteka (Shevchenko Memorial National Park in Kaniv), 19002 ***Kaniv***, Cherkasska Region
T: +380 4736 32368; Fax: +380 4736 32086; E-mail: tarasovagora@ukr.net
1939; Natasha Gayer
Sarkisov-Seresini collection, national and world classical literature
23 306 vols; 2 697 curr per; 2 828 periodicals
libr loan 28245

Morskoyi gidrofizychesky instytut, Biblioteka (Marine Hydrophysical Institute), Simeiz, ***Katsiveli***, Krymskaya obl.
1937
37 000 vols; 354 curr per 28246

Derzhavny istoriko-kulturny zapovednik, Biblioteka (State Historical Reserve), Sverdlova, 7, 98300 ***Kerch***
T: +380 6561 20475; Fax: +380 6561 20475
1826/1985; Anzhela V. Tereshchenko
20 000 vols; 94 curr per; 173 diss/theses; 850 reprints 28247

Scientific Technical Library of YugNIRO, Inquire Information Fund, Sverdlova, 2, 98300 ***Kerch***
T: +380 6561 21065; Fax: +380 6561 61627; E-mail: yugniro@kerch.com.ua; URL: yugniro.crimea.com
1921; Olga Sokolova
Oceanography, marine sciences, fish and fisheries; Confidential dept (inner service only)
24 621 vols; 207 curr per; 1 450 diss/theses; 60 govt docs; 684 maps; 914 microforms; 7 375 av-mat; 29 720 periodical items
libr loan; EURASLIC 28248

Derzhavna akademiya tekhnologiyi ta organizatsiyi kharchuvaniya, Biblioteka (State Public Catering Academy), Klochkovskaya, 333, 61051 ***Kharkiv***
1967; Lyudmila D. Levshina
360 000 vols; 156 curr per 28249

Derzhavna naukova medychna biblioteka (State Medical Scientific Library), pl Poeziyi, 5, 61676 ***Kharkiv***
1861; Sergei Ivanovich Gribov
1 140 000 vols; 660 curr per
libr loan 28250

Fizyko-tekhnichny instytut nizkykh temperatur im. B. Verkina, Biblioteka (B. Verkin Institute for Low Temperature Physics and Engioneering), pr. Lenina, 47, 61103 ***Kharkiv***
T: +380 57 3402223; Fax: +380 57 3403370; E-mail: ilt@ilt.kharkov.ua; URL: www.ilt.kharkov.ua
1960; Larisa Zimenko
211 486 vols; 516 curr per 28251

Fizyko-tekhnichny instytut Ukrainy, Biblioteka (Physical Engineering Institute), vul. Akademicheska, 1, 61108 ***Kharkiv*** 24
1934
91 000 vols; 423 curr per; Microfilms 28252

Institute of Animal Science of the Ukrainyan Academy of Agrarian Sciences, Scientific Library (Institute of Livestock Breeding in Forest-Steppe and Wood-Lands), p/o Kulinichi, 62404 ***Kharkiv***
T: +380 57 74039-94; E-mail: it_uaan@bk.ru
1929; Victoria Kunetz
96 500 vols; 800 curr per; 12 000 e-journals; 40 e-books; 677 diss/theses; 4 av-mat; 40 sound-rec; 500 rare books, 38 947 periodical items
libr loan 28253

Instytut eksperimentalnoyi i klinicheskoyi veterinarnoyi meditsiny, Ukrainskaya akademiya agrarnykh nauk, Nauchnaya biblioteka (Institute for Experimental and Clinical Veterinary Medicine), Pushkinskaya, 83, 61023 ***Kharkiv***
1923; Anna Vasilevna Shemaeva
Veterinary medicine, biology; Branches in Rovne and Odessa
164 000 vols; 224 curr per; 168 diss/theses; 150 govt docs; 45 microforms
libr loan 28254

Instytut gruntoznavstva i agrokhimiyi im. A.N. Sokolovskoho, Biblioteka (A.N. Sokolovski Institute of Agricultural Chemistry and Soil Cultivation), Chaykovskoho, 4, 61024 ***Kharkiv***
1956; Tamary I. Kisil
55 000 vols; 115 curr per
libr loan 28255

Instytut medichnoi radiologii im. S.P. Grigoreva, Biblioteka (Grigoriev Institute for Radiology), vul. Pushkinskaya, 82, 61024 ***Kharkiv***
T: +380 57 7041065; Fax: +380 57 7000500; E-mail: imr@ukr.net; URL: www.imr.kharkov.ua
1920; Lyudmila Stepanivna Onoprienko
70 000 vols 28256

Instytut monokrystaliv, Natsionalna Akaemiya Nauk Ukrainy, Biblioteka (Institute for Single Crystals), pr. Lenina, 60, 61001 ***Kharkiv***
T: +380 57 3410166; Fax: +380 57 3409343; E-mail: info@isc.kharkov.com; URL: www.isc.kharkov.com
1962/2000; Larisa Eliseeva
Crystallography, physica status solidi, chemical physics, functional mat, lumenescence, scintillators, radiation instruments
125 541 vols; 670 curr per; 19 873 mss; 1 130 diss/theses; 977 govt docs; 7 105 microforms
libr loan 28257

Instytut problem kriobiologiyi ta kriomeditsini, Natsionalna Akademiya Nauk Ukrainy, Biblioteka (Institute for Problems of Cryobiology and Cryomedicine), Pereyaslivska, 23, 61015 ***Kharkiv***
T: +380 57 3734143; Fax: +380 57 3733084; E-mail: cryo@online.kharkov.ua; URL: www.cryo.org.ua/ipk_eng/history.html
1973; Tetyana Nikolaivna Timchenko
Biology, biophysics, biochemistry, immunology, pathology
53 000 vols; 80 curr per 28258

Instytut problem mashinobuduvannya, Natsionalna Akademiya Nauk Ukrainy, Biblioteka (Institute of Machine Building), Pozharskoho, 2/10, 61046 ***Kharkiv***
1944; Lyubov Anatoliivna Kulik
154 000 vols 28259

Instytut radiofizyky i elektroniky, Natsionalna Akademiya Nauk Ukrainy, Biblioteka (Radiophysics and Elektronics Research Institute), Proskuri, 12, 61085 ***Kharkiv***
1955; Larysa Ivanovna Chalova
126 000 vols; 510 curr per; 110 diss/theses; 9 010 microforms
libr loan 28260

Instytut tvarinnitstva, Ukrainska akademiya agrarnykh nauk, Biblioteka (Institute of Animal Husbandry), p/o Kulynychi, 61120 *Kharkiv*
1930
64 800 vols; 90 curr per   28261

Instytut yuzhgiprotsement, Nauchno-tekhnicheskaya biblioteka (Institute of Civil Engineering), Sumska vul, 40, 61002 *Kharkiv*
Lidiya K. Tolstaya
218 000 vols; 91 curr per   28262

Kharkiv Institute of Agricultural Mechanization and Electrification, Library, vul. Artema, 44, 61078 *Kharkiv*
1929
150 000 vols   28263

Kharkiv State Academy of Railway Transport, Library, pl Feierbakha, 7, 61050 *Kharkiv*
URL: www.kart.edu.ua
1930
700 000 vols   28264

Kharkiv State Institute of Arts, Library, pl Sovetskoy Ukrainy, 11/13, 61003 *Kharkiv*
100 000 vols   28265

Kharkivsky khudozhni muzei, Biblioteka (Kharkiv Art Museum), vul. Radnarkomivska, 11, 61002 *Kharkiv*
1934
18 000 vols   28266

Muzichno-teatralna biblioteka im. Stanislavskoho (Stanislavsky Music and Theatre Library), Inzhenerny per, 1, 61882 *Kharkiv*
1956
122 000 vols; 102 curr per; 9 000 music scores; 750 sound-rec   28267

NDI elektromashinobuduvannya, Biblioteka (Machine Building and Electrical Engineering Research Institute), Pavlova, 82, 61831 *Kharkiv*
1956
92 000 vols; 295 curr per   28268

NDI gigieny pratsi i profesiynykh zakhvoryuvan, Biblioteka (Industrial Hygiene and Occupational Diseases Research Institute), vul. Trinklera, 6, 61022 *Kharkiv*
1923; Nadiya V. Shevtsova
36 547 vols; 96 curr per   28269

NDI metaliv, Biblioteka (Research Institute of Metal Technology), Darvina, 20, 61002 *Kharkiv*
1928; Lyubov Anatoliivna Kulik
Ferrous metals, metal technology
130 000 vols; 410 curr per   28270

NDI mikrobiologiyi ta immunologiyi im. I.I. Mechnikova, Biblioteka (Mechnikov Microbiology and Immunology Institute), Pushkinska, 14, 61057 *Kharkiv*
1886; Tetyana Oleksiivna Rezak
virosology, genetics, Biochemistry, communicable diseases, epidemiology
53 700 vols; 242 diss/theses   28271

NDI organizatsiyi i mekhanizatsiyi shakhtnoho budivnitstva, Biblioteka (State Scientific and Research Institute for Organization and Mechanization of Mine Building), Otakara Yarosha, 18, 61045 *Kharkiv*
Lyudmila M. Alekseeva
115 000 vols; 20 000 curr per   28272

NDI ortopediyi i travmatologiyi im. M.I. Sitenka, Biblioteka (Orthopaedics and Traumatology Research Institute), Pushkinska, 80, 61024 *Kharkiv*
1907; Ganna Semyonivna Veligora
40 500 vols; 74 curr per   28273

NDI roslinnitstva, selektsiyi ta genetiki, Ukrainska akademiya agrarnykh nauk, Biblioteka (Plant Breeding, Selection and Genetics Research Institute), pr. Moskovski, 142, 61060 *Kharkiv*
1910; Valentina M. Ozereleva
51 000 vols; 850 curr per; 330 mss   28274

NDI zagalnoyi i nevidkladnoyi khirurgiyi, Biblioteka (General and Emergency Surgery Research Institute), Balakireva, 1, 61018 *Kharkiv*
1930; Svitlana V. Fastova
40 000 vols; 64 curr per   28275

Proektno-izyskatelny i nauchno-issledovatelsky institut gidroproekt im. S.Ya. Zhuka, Nauchno-tekhnicheskaya biblioteka (Research and Construction Institute of Hydroengineering S.Ya. Zhuk), pr. Lenina, 9, *Kharkiv* 59
1943
781 000 vols; 98 curr per   28276

Radioastronomichny instytut, Natsionalna Akademiya Nauk Ukrainy, Biblioteka (Institute of Radio Astronomy), Chervonopraporna, 4, 61002 *Kharkiv*
T: +380 57 7061415; E-mail: rai@ri.kharkov.ua; URL: www.ri.kharkov.ua
1980; Liudmyla Mykolaivna Zakharenko
72 950 vols; 588 curr per; 1 155 diss/theses; 13 158 microforms; 115 av-mat   28277

Research Institute of Endocrinology and Hormone Chemistry, Library, vul. Artema, 10, 61002 *Kharkiv*
1919
44 800 vols   28278

SSIA Metrology, Library, vul. Mironositskaya, 40, 61078 *Kharkiv*
V.A. Saraeva
18 700 vols; 80 curr per   28279

State Research and Design Institute of Basic Chemistry, Library, vul. Mironositska, 25, 61002 *Kharkiv*
T: +380 57 7000123;
Fax: +380 57 7004825; E-mail: office@niochim.kharkov.ua
1923
185 000 vols   28280

State Scientific Centre of Drugs, Library, vul. Astronomicheska, 33, 61085 *Kharkiv*
1920; Elena L. Kopaneva
30 000 vols; 2 000 microforms; 703 276 periodicals
libr loan   28281

Ukrainsky gosudarstvenny nauchno-issledovatelski uglekhimichesky instytut, Nauchno-tekhnicheskaya biblioteka (Ukrainian State Research Institute for Carbochemistry), 7 Vesnina Street, 61023 *Kharkiv*
T: +380 57 7041318; Fax: +380 57 7006906, 7041323; E-mail: post@ukhin.org.ua; URL: www.ukhin.org.ua
1931; Ludmila Minova
Cokemaking, carbochemistry, coal preparation, waste water treatment, pollution control; Depositary of carbochemistry
153 000 vols; 16 curr per; 8 100 mss; 90 diss/theses; 5 000 microforms
libr loan   28282

Ukrainsky nauchno-issledovatelsky instytut prirodnykh gazov (Ukrniigas), Nauchno-tekhnicheskaya biblioteka (Ukrainian Natural Gas Research Institute), nab Krasnoshkilna, 20, 61125 *Kharkiv*
T: +380 57 7300362
1959; Natalia A. Stephanovich
Patent dept
64 800 vols; 35 curr per; 31 500 av-mat   28283

Ukrainsky naukovo-doslidny instytut masel i zhirov, Nauchno-tekhnicheskaya biblioteka (Ukrainian Research Institute of Oils and Fats), pr. Dzyuby, 2a, 61019 *Kharkiv*
1949; O. Gurtovai
Food sciences, nutrition research, margarine production, soap manufacture, hydrogenation
35 000 vols; 165 curr per   28284

Ukrainyan Research Institute of Dermatology and Venereology, Library, vul. Chernyshevska, 7/9, 61057 *Kharkiv*
1924
40 000 vols   28285

Ukrainyan Research Institute of Experimental and Clinical Psychoneurology, Scientific Medical Library, Akad. Pavlova, 46, 61068 *Kharkiv*
1926; L.M. Markova
37 900 vols; 102 curr per; 3 600 diss/theses
libr loan   28286

Kraeznavchi muzei mista, Biblioteka (Kherson Local Museum), Lenina, 9, 73000 *Kherson*
1890; Tamara Koralchuk
39 000 vols; 100 curr per   28287

NDI zroshuvanoho zemlerobstva, Ukrainska akademiya agrarnykh nauk, Biblioteka (Agricultural Irrigation Research Institute), Beseslavske shosse, sel Naddnirpryanske, 73029 *Kherson*
1889; Nataliya Petrivna Matsko
Arch unique lit ((1 650 samplers), irrigated farming, crop and seed breeding
78 000 vols; 156 curr per; 203 mss; 196 diss/theses
libr loan   28288

Oblasna naukova medychna biblioteka (Regional Scientific Medical Library), Suvorova, 27, 73000 *Kherson*
T: +380 552 264226; E-mail: medlib@public.kherson.ua
1945; Dizhur Tetyana
Hist of medicine
17 900 vols; 160 curr per; 31 930 diss/theses; 15 digital data carriers
libr loan   28289

Oblasna medychna biblioteka (Regional Medical Library), Shevchenka, 46, 29000 *Khmelnitski*
T: +380 382 69392
1945; Larisa M. Yaremchuk
132 000 vols; 252 curr per   28290

Instytut silskogospodarskoho maschinobuduvannya, Biblioteka (Agricultural Engineering Institute), Pravdi, 70a, 25050 *Kirovograd*
T: +380 522 593541
1960
420 000 vols; 220 curr per   28291

Oblasna naukova medychna biblioteka (Regional Scientific Medical Library), Vasilini, 7, 25050 *Kirovograd*
T: +380 522 249609
1945
135 000 vols; 254 curr per   28292

Krymskaya gosudarstvannaya silskokhozyaistvennaya opytnaya stantsiya, Biblioteka Krasnogvardeiskoho raiona (Crimea State Agricultural Experimental Station), 334061 *Klepinino*, Krymskaya obl.
T: +380 6556 95274
1924; Lubov T. Tumareva
65 800 vols   28293

Muzei narodnoho mystetstva gutsulshchini ta pokuttya im. J. Kobrynskoho, Biblioteka (Kolomiya State Museum of Folk Art), vul. Teatralna, 25, 78200 *Kolomiya*, Ivano-Frankivska obl.
T: +380 3433 24404, 27891; Fax: +380 3433 23912; E-mail: kmh@yes.net.ua
1926; Kalyna Pyatnichuk
100 bks of religious lit of 17-18 c
9 000 vols; 15 curr per; 20 mss; 200 govt docs; 10 maps; 200 000 microforms; 50 av-mat; 65 sound-rec   28294

V. Dokuchayev Kharkiv State University of Agriculture, Library, *P.O. Komunist-1*, Kharkiv oblast
T: +380 572 997123; Fax: +380 572 997777
1816/1991; Lyudmila M. Polyakh
602 000 vols; 363 curr per   28295

Dombaska derzhavna mashino-budivna akademiya, Biblioteka (Dombas State Machine Building Academy), Shkadinova, 76, 343916 *Kramatorsk*
T: +380 6264 52215
1953
475 000 vols; 428 curr per
libr loan   28296

Instytut Mekhanobchermet, Biblioteka (Metallurgy and Mechanics Institute), Televiziyna, 3, 324087 *Krivi Rig*
1958
87 000 vols; 203 curr per   28297

Naukovo-doslidny girnichorudny instytut, Biblioteka (Mining Research Institute), pr. Gagarina, 57, *Krivi Rig*, Dnepropetrovsk obl.
T: +380 564 780612; Fax: +380 564 742848
1933
100 000 vols; 241 curr per   28298

Naukovo-doslidny instytut gigieny pratsy i profzakhviryuvan, Biblioteka (Industrial Hygiene and Occupational Diseases Research Institute), 20 partzizdu, 29, 324096 *Krivi Rig*
T: +380 564 532185
1955
50 000 vols; 143 curr per   28299

Naukovo-doslidny instytut girnichorud-noho mashinobuduvannya, Biblioteka (Institute for the Construction of Ore Mining Machines), Kharitonova, 1a, 299271 *Krivi Rig*
T: +380 564 299271
1955
87 000 vols; 200 curr per   28300

Research and Design Institute for the Enrichment and Agglomeration of Ferrous Metal Ores, Library, vul. Televiziyna, 3, 324039 *Krivi Rig*
T: +380 564 716136; Fax: +380 564 714842
1956
65 000 vols   28301

Biblioteka natsionalnoi spilky pysmennykhiv Ukrainy (Library of the National Centre of Writers), Bankova vul 2, 01024 *Kyiv*
T: +380 44 2538412
1945
Ukrainian and foreign lit, lit studies
125 000 vols; 90 curr per   28302

Bogomoletz Institute of Physiology, National Academy of Sciences of the Ukraine, Library, vul. Bogomoltsa, 4, 01024 *Kyiv*
T: +380 44 2562418; Fax: +380 44 2562000; URL: biph.kiev.ua/
1931; Dr. Alla Shevko
Special Bogomoletz Coll
90 000 vols; 130 curr per; 3 045 diss/theses; 354 microforms; microfilms
libr loan   28303

Derzhavna istorichna biblioteka Ukrainy (DIBU) (Ukrainian National Historical Library), Sichnevoho povstannia vul 21, korp 24, 01015 *Kyiv*
T: +380 44 2802874
1939; Leri Leonidovich Makarenko
750 000 vols; 292 curr per   28304

Derzhavna naukova arkhitekturno-budivelna biblioteka (Zabolotnov State Architecture and Construction Scientific Library), Kontraktova pl., 4, 04070 *Kyiv*
T: +380 44 4176387; Fax: +380 44 4250310; URL: www.dnabb.org
1944; Galina Anatoliivna Voitsekhivska
Building and construction, civil engineering, architecture; Rare bks dept, standards dept
410 000 vols; 220 curr per; 250 diss/theses
libr loan   28305

Derzhavna naukova silskohospodarska biblioteka UAAN, Ukrainska Akademiya Agrarnekh Nauk (State Scientific Agricultural Library), Heroiv Oborony vul 10, 03127 *Kyiv*
T: +380 44 5278075
1921; Romuald Iosifovich Tselinski
Forestry, veterinary science
960 992 vols; 3 485 curr per; 26 463 diss/theses; 3 477 microforms; 12 digital data carriers
libr loan; IAALD   28306

Derzhavna naukovo-pedahohichna biblioteka Ukrainy im. V. O. Sukhomlynskoho (State Scientific and Pedagogical Library of Ukraine), vul. M.Berlinskoho 9, 04060 *Kyiv*
T: +380 44 4672214; Fax: +380 44 4403548; E-mail: dnpb@i.ua; URL: www.library.edu-ua.net
1999; Paula Rohova
V. O. Sukhomlynski Documentary Fund, Fund of Library Science Documents, Fund of Textbooks for the Different Levels of Educational Establishments of the 19-20th c, Periodical Collection of the 20th c
543 750 vols; 400 curr per; 2 300 diss/theses; 940 music scores; 162 sound-rec; 7 500 other items, 250 vols of Eyeryman's Libr (Britannica, Americana, World Book) 10 430 offprints, 124 951 periodicals
libr loan   28307

Derzhavny instytut proektuvannya zavodov silskogospodarskogo mashinobuduvannya, Biblioteka (State Agricultural Engineering Planning Institute), bul. L. Ukrainki, 25, 252133 ***Kyiv***
T: +380 44 2950103
1945; Galina M. Muromtseva
85 000 vols 28308

Derzhavny proektny instytut po tsivilnomu i promislovomu budivnitstvu, Biblioteka (State Civil Engineering and Industrial Construction Planning Institute), Gogilivska, 22/24, 03054 ***Kyiv***
T: +380 44 2166043
1946
88 000 vols; 86 curr per 28309

Filial Ukrainskoyi Derzhavny akademiyi zvyazku im. O. Popova, Biblioteka (Popov Ukrainian National Telecommunications Academy, Branch Library), Solomyanska, 7, 03052 ***Kyiv***
T: +380 44 2714093
1970; Nataliya Dmitrivna Ishchenko
188 000 vols; 76 curr per 28310

Gigienichi tsentr, Ministerstva okhorony zdorovya Ukrainy, Biblioteka (Ukrainian Ministry of Health – Centre of Hygiene), Popudrenka, 50, 252660 ***Kyiv***
T: +380 44 5595018
1989; Zoya Viktorivna Rozsadina
102 213 vols; 18 curr per
libr loan 28311

O. O. Chuyko Institute of Surface Chemistry, Library, 17 Generala Naumova St, 03164 ***Kyiv***
T: +380 44 4241135; Fax: +380 44 4243567; E-mail: info@isc.gov.ua; URL: www.isc.gov.ua
1986; Rulova Nadežda
30 000 vols; 59 diss/theses 28312

Instytut Agrokhimfarmproekt, Biblioteka (Agrochemical Research Institute), Prazka, 5, 01660 ***Kyiv***
T: +380 44 5595518
1959
64 000 vols; 25 curr per 28313

Instytut arkheologiyi, Natsionalna Akademiya Nauk Ukrainy, Biblioteka (Institute of Archeology), Gerojev Stalingrada, 12, 04210 ***Kyiv***
T: +380 44 4189191; Fax: +380 44 4183306; E-mail: sekretar@iananu.kiev.ua; URL: www.iananu.kiev.ua
1921; Viktoriya A. Kolenikova
116 446 vols; 33 160 curr per 28314

Instytut bioorganichnoyi khimiyi ta naftokhimiyi, Biblioteka (Institute of Biochemistry and Petrochemistry), vul. Leontovicha, 9, 03094 ***Kyiv***
T: +380 44 5435356
1927
53 000 vols; 144 curr per 28315

Instytut botaniky im. M.G. Kholodnoho, Natsionalna Akademiya Nauk Ukrainy, Biblioteka (Kholodny Botany Institute), Velika Zhitomirska, 28, 01601 ***Kyiv***
T: +380 44 2123221
1921; Nataliya Grigorivna Nikityuk
118 000 vols; 157 curr per 28316

Instytut derzhavi i prava im. V.M. Koretskoho, Natsionalna Akademiya Nauk Ukrainy, Biblioteka (Koretsky Institut of State and Law), Triokhsvyatitelska, 4, 01601 ***Kyiv***
T: +380 44 2285155; Fax: +380 44 2285474
1949; L.V. Galzhevska
Coll of the academian V.M. Koretsko
80 600 vols; 65 curr per; 5 925 diss/theses; 18 microforms 28317

Instytut ekonomiki, Nauchnaya biblioteka (Institute of Economics), vul. Panasa Mirnoho 26, 262011 ***Kyiv***
T: +380 44 2901589; Fax: +380 44 2908663
1936; Iryna O. Sinchylo
62 000 vols; 232 curr per
libr loan 28318

Instytut eksperimentalnoyi patologiyi, onkologiyi i radiobiologiyi im. R.E. Kavetskoho, Natsionalna Akademiya Nauk Ukrainy, Biblioteka (R.E. Kavetsky Institute of Experimental Pathology, Oncology and Radiobiology), Vasylkivska, 45, 03022 ***Kyiv***
T: +380 44 2590271; Fax: +380 44 2581656; E-mail: nauka@onconet.kiev.ua; URL: www.onconet.kiev.ua
1960; Polina M. Shkatula
Biochemistry, physiology, oncology, cell biology, cancer research, radiology, nuclear medicine, pathology
75 000 vols; 140 curr per; 1 000 diss/theses 28319

Instytut elektrodinamiki, Natsionalna Akademiya Nauk Ukrainy, Biblioteka (Institute of Electrodynamics), pr. Peremogi, 56, 03080 ***Kyiv***
T: +380 44 460361
1944; Larisa F, Perekhrest
230 000 vols; 1 000 curr per 28320

Instytut elektrozvaryuvannya im. E. Patona, Biblioteka (Paton Electric Welding Institute), vul. Bozhenko, 11, 03680 ***Kyiv***
T: +380 44 2270777; Fax: +380 44 2680486; E-mail: office@paton.kiev.ua
1934; Diamara S. Pshenichnikova
260 167 vols; 510 curr per; 7 000 mss; 909 diss/theses
ULA 28321

Instytut filosofiyi im. H.S. Skovorody, Natsionalna Akademiya Nauk Ukrainy, Biblioteka (H.S. Skovoroda Institute of Philosophy), vul. Triokhasviatytelska, 4, 01001 ***Kyiv***
T: +380 44 2284178; Fax: +380 44 2286366; E-mail: melishkevychl@ukr.net
1946; Olga Lukashchuk
Philosophy, logics, ethics, aesthetics, A. Fokht's coll
41 836 vols; 70 curr per; 44 000 mss; 1 037 diss/theses; 259 microforms; 1 600 pamphlets, 3 500 synopsis of theses, 17 954 periodical items
libr loan; UBA 28322

Instytut fizyky, Natsionalna Akademiya Nauk Ukrainy, Biblioteka (Institute of Physics), pr. Nauki, 144, 01650 ***Kyiv***
T: +380 44 2650595
1929
286 000 vols; 1 694 curr per 28323

Instytut fizyky napivprovidnikiv, Natsionalna Akademiya Nauk Ukrainy, Biblioteka (Semiconductor Physics Institute), pr. Nauki, 45, 03028 ***Kyiv***
T: +380 44 2656276; Fax: +380 44 2658342
1960; E.V. Tarasenko
Physics of semiconductors, physics of semiconductor surface, optoelectronics, photoelectric phenomena, microelectronics
54 360 vols; 93 curr per; 636 diss/theses
libr loan 28324

Instytut fizyologiyi raslin i genetiki, Biblioteka (Plant Physiology and Genetics Institute), Vasilkivska, 31/17, 03022 ***Kyiv***
T: +380 44 2649922
1946
76 000 vols; 132 curr per 28325

Instytut gaza, Natsionalna Akademiya Nauk Ukrainy, Biblioteka (Institute of Natural Gas), Degtyarivska, 39, 03113 ***Kyiv***
T: +380 44 4568142; Fax: +380 44 4568830
1950; N.G. Mishina
110 000 vols; 250 curr per; 85 diss/theses; 1 869 microforms; Microfilms
libr loan; ULA 28326

Instytut gerontologiyi, Ministerstvo Zdravookhraneniya Ukrainy, Nauchnaya biblioteka (Academy of Medical Sciences of Ukraine Inst. of Gerontology), Vyshgorodskaya, 67, 04114 ***Kyiv***
T: +380 44 4310562; Fax: +380 44 4329956
1958; Valentina Kostyuchenko
18 200 vols; 50 curr per; 232 diss/theses; 2 221 microforms; 7 360 reprints
libr loan 28327

Instytut gidromekhaniki, Natsionalna Akademiya Nauk Ukrainy, Biblioteka (Hydromechanics Institute), vul. Zhelyabova, 8/5, 03057 ***Kyiv***
T: +380 44 4410934
1926; Zoya Vasilivna Taranovska
84 000 vols; 407 curr per 28328

Instytut gidrotekhniky i melioratsiyi, Ukrainska Akademiya Agramykh Nauk, Biblioteka (Hydrotechnology and Land Reclamation Research Institute), Vasilkivska, 37, 03022 ***Kyiv***
1933; Lyudmila G. Zaika
77 000 vols; 132 curr per 28329

Instytut istoriyi Ukrainy, Natsionalna Akademiya Nauk Ukrainy (Institute of History of Ukraine), Grushevskoho, 4, 01001 ***Kyiv***
T: +380 44 2298432; Fax: +380 44 2296362
1936; Lyudmila Yukivna Mukha
125 000 vols; 400 curr per; 750 diss/theses; 165 maps
libr loan 28330

Instytut klitinnoyi biologiyi ta genetichnoyi inzheneriyi, Natsionalna Akademiya Nauk Ukrainy, Biblioteka (Cell Biology and Genetic Engineering Institute), Zabolotnoho, 148, 03143 ***Kyiv***
T: +380 44 2661567
1991; Svetlana S. Povalova
15 000 vols 28331

Instytut literatury im. T.G. Shevchenka, Natsionalna Akademiya Nauk Ukrainy, Biblioteka (Shevchenko Institute of Literature), Hrushevskoho, 4, 04210 ***Kyiv***
T: +380 44 2290503
1926
145 000 vols; 83 curr per; 200 mss 28332

Instytut matematiki, Natsionalna Akademiya Nauk Ukrainy, Biblioteka (Mathematical Institute), Tereshchenkivska, 3, 01601 ***Kyiv*** – 4
T: +380 44 2345150; Fax: +380 44 2352010; E-mail: institute@imath.kiev.ua
1934; Ganna Ivanivna Vradi
150 000 vols; 374 curr per 28333

Instytut mekhaniki, Natsionalna Akademiya Nauk Ukrainy, Biblioteka (Institute of Mechanics), Nesterova, 3, 03057 ***Kyiv***
T: +380 44 4417754
1963; Lyudmila Vlasivna Perova
97 000 vols 28334

Instytut metallofizyky, Natsionalna Akademiya Nauk Ukrainy, Biblioteka (Metal Physics Institute), pr. Vernadskoho, 01680 ***Kyiv***
T: +380 44 441021
1950; Tamara Evgeniivna Charupa
100 000 vols; 160 curr per 28335

Instytut mikrobiologiyi i virusologiyi im. D.K. Zabolotnoho, Natsionalna Akademiya Nauk Ukrainy, Biblioteka (Zabolotny Institute of Microbiology and Virology), Zabolotnoho, 154, 03143 ***Kyiv***
T: +380 44 2669924; Fax: +380 44 2662379; E-mail: smirnov@imv.kiev.ua; URL: www.imv.kiev.ua
1930; Svitlana Mikhailivna Serdyuk
Coll from the private libr of the founder of the Institute: academician D.K. Zabolotny
115 576 vols; 148 curr per; 371 diss/theses; 21 maps; 4 001 pamphlets, 289 clippings
libr loan 28336

Instytut mistetstvoznavstva, folkloru ta etnologiyi im. M.T. Rilskoho, Natsionalna Akademiya Nauk Ukrainy, Biblioteka (Rilsky Institute of Arts, Folklore and Ethnography), Grushevskoho, 4, 01001 ***Kyiv***
T: +380 44 2295029
1936; Olga Oleksandrivna Roenko
musicologists M.O. Grinchenko, M.M. Gordijchuk
96 630 vols; 455 curr per; 35 diss/theses; 4 788 music scores
libr loan 28337

Instytut molekulyarnoyi biologiyi ta genetiki, Natsionalna Akademiya Nauk Ukrainy, Biblioteka (Molecular Biology and Genetics Institute), Zabolotnoho, 150, 03143 ***Kyiv***
T: +380 44 2660739
1969; Tetyana Igorevna Biatova
96 100 vols; 550 curr per; 267 diss/theses; 3 maps; 2 765 microforms; 2 500 synopsis of theses, 250 pamphlets
libr loan 28338

Instytut movoznavstva im. O.O. Potebni, Natsionalna Akademiya Nauk Ukrainy, Biblioteka (Potebnya Institute of Linguistics), Grushevskoho, 4, 01601 ***Kyiv***
T: +380 44 2291793
1930; Ganna S. Rizhenko
106 000 vols; 670 curr per 28339

Instytut natsionalnykh vidnosin i politologiyi, Natsionalna Akademiya Nauk Ukrainy, Biblioteka (Institute of National Politics), Kutuzova, 8, 03011 ***Kyiv***
T: +380 44 2956998
1928; Nataliya Pavlivna Biryukova
Political underground publs (pre 1917)
120 000 vols 28340

Instytut natverdykh materialiv im. V.M. Bakulya, Natsionalna Akademiya Nauk Ukrainy, Biblioteka (Bakul Institute for Superhard Materials), Avtozavodska, 2, 04074 ***Kyiv***
T: +380 44 4688639
1956; Nadiya Ivanivna Kolodnitska
Solid state physics, physical chemistry, powder metallurgy, crystallography, mechanics
112 000 vols; 14 curr per; 250 diss/theses
libr loan 28341

Instytut neyrokhirurgiyi im. akademika A. P. Romodanova AMN Ukrainy (Institute of Neurosurgery named after A.P. Romodanov), vul. Manuyilskoho, 32, 04050 ***Kyiv***
T: +380 44 4839198; Fax: +380 44 4839573; E-mail: neuro.kiev@gmail.com; URL: www.neuro.kiev.ua
1950
70 000 vols; 370 diss/theses 28342

Instytut organicheskoyi khimiyi, Natsionalna Akademiya Nauk Ukrainy, Biblioteka (Institute of Organic Chemistry), Murmanskaya, 5, 02660 ***Kyiv*** – 94
T: +380 44 5510637; Fax: +380 44 5436843
1928; Mrs O.M. Grebelnik
49 000 vols; 242 curr per; 1 600 mss; 570 diss/theses; 163 microforms
libr loan 28343

Instytut problem energoberezhennya, Natsionalna Akademiya Nauk Ukrainy, Biblioteka (Energy Conservation Institute), Andriyivska, 19, 03070 ***Kyiv***
T: +380 44 4163169
1977; Nataliya Grigoriivna Litoshenko
Thermal and mass exchange, energetics, plasma physics
42 000 vols 28344

Instytut problem litya, Natsionalna Akademiya Nauk Ukrainy, Biblioteka (Institute for the Problems of Iron-Casting), vul. Vernadskoho, 34/1, 01680 ***Kyiv***
T: +380 44 443150
1958
156 000 vols; 492 curr per; 700 mss 28345

Instytut problem modeluvannya v ener getitsi, Biblioteka (Institute of the Simulation Problems in Power engineering), Gen. Naumova, 15, 03164 ***Kyiv***
T: +380 44 4441063; Fax: +380 44 4440586; E-mail: svetlana@ipme.Kiev.ua; URL: www.ipme.kiev.ua
1981; Nelly Vasilivna Chernyshova
Libr od acad G.E. Puxov
35 000 vols; 24 000 curr per; 83 mss; 913 diss/theses; 620 govt docs; 3 816 microforms
libr loan 28346

Instytut problem reestratsii informatsii, Natsionalna Akademiya Nauk Ukrainy, Biblioteka (Institute for Information Recording), Shpaka, 2, 03113 *Kyiv*
T: +380 44 4412167; Fax: +380 44 2417233; URL: www.ipri.kiev.ua
1989; Valentina Yakivna Biba
4 000 vols; 22 curr per; 7 diss/theses; 3 govt docs; 8 digital data carriers   28347

Instytut svitovoyi ekonomiky ta mizhnarodnykh vidnosin, Natsionalna Akademiya Nauk Ukrainy, Biblioteka (Institute of World Economy and Politics), Leontovicha, 5, 03030 *Kyiv*
T: +380 44 2243901
1978; Lyudmila Grigorivna Kvitkovska
38 000 vols   28348

Instytut tekhnichnoyi teplofizyky, Natsionalna Akademiya Nauk Ukrainy, Biblioteka (Thermophysical Engineering Institute), vul. Zhelyabova, 2a, 03057 *Kyiv*
T: +380 44 4417134
1947; Lyudmila D. Oliynikova
136 000 vols; 242 curr per   28349

Instytut teoretichnoi fizyky im. M.M. Bogolyubova, Natsionalna Akademiya Nauk Ukrainy, Biblioteka (Bogolyubov Instiute for Theoretical Physics), Metrolohichna, 14/b, 03680 *Kyiv*
T: +380 44 4921423; Fax: +380 44 5265998; E-mail: itp@bitp.kiev.ua; URL: www.bitp.kiev.ua
1966
156 000 vols; 320 curr per; 243 diss/theses   28350

Instytut tsukrovykh buryakiv, Ukrainska Akademiya Agrarnykh Nauk, Biblioteka (Sugar Beet Institute), Klinichna, 25, 03110 *Kyiv*
T: +380 44 2775355; Fax: +380 44 2775366
1922; Alexandra Minert
Sugar beet, wheat, agronomy
28 494 vols; 26 783 curr per; 545 diss/theses; 169 microforms; 2 602 pamphlets
libr loan   28351

Instytut yadernykh doslidzhen, Natsionalna Akademiya Nauk Ukrainy, Biblioteka (Nuclear Research Institute), pr. Nauki, 47, 01650 *Kyiv*
T: +380 44 2654380; Fax: +380 44 2654463; E-mail: interdep@kinr.kiev.ua
1970; Olena M. Storozhenko
81 000 vols; 768 curr per; 265 diss/theses
libr loan   28352

Kiev I.K. Karpenko-Kary State Institute of Theatrical Art, Library, Yaroslavov Val, 40, 01034 *Kyiv*
T: +380 44 2120200
1918
72 000 vols   28353

Kiev Research Institute of Otolaryngology, Library, vul. Zoologichna, 3, 03057 *Kyiv*
T: +380 44 2132202; Fax: +380 44 2137368
1960
33 000 vols   28354

V. P. Komissarenko Institue of Endocrinology and Metabolism, Academy of Medical Sciences of Ukraine, Library, vul. Vyshgorodska, 69, *Kyiv*
T: +380 44 4310254; Fax: +380 44 4303718
1965; Aleksander A. Statsenko
Radiobiology, biochemistry, physiology, immunology, pharmacology
18 510 vols; 792 curr per; 78 mss; 233 diss/theses; 343 microforms; 4 003 brochures
libr loan   28355

Kyivsky muzei russkoho iskusstva, Biblioteka (Museum of Russian Art in Kyiv), vul. Tereshchenkivska, 9, 01004 *Kyiv*
T: +380 44 2248288; Fax: +380 44 2246107
1922; Irena Paysova
19 500 vols; 75 curr per   28356

Legal Library (City Trade Union Committee, Main Library), Kostolna, 3, 01001 *Kyiv*
T: +380 44 2287940; Fax: +380 44 2284740
1994; Halyna Polozova
80 000 vols; 143 curr per; 32 000 govt docs   28357

Movno-informatsiyny fond, Natsionalna Akademiya Nauk Ukrainy, Biblioteka (Linguistics Information Centre), 40-richchya Zovtnya, 3, 03039 *Kyiv*
T: +380 44 2674859
1991; Larisa O. Tereshchenko
15 000 vols   28358

Muzychno-teatralna oblasna biblioteka (Regional Music and Theatrical Library), Mykhailivska, 9, 01001 *Kyiv*
T: +380 44 2783641
1959
50 000 vols; 30 curr per; 24 000 music scores; 2 200 sound-rec   28359

National Academy of Fine Arts and Architecture, Natsionalna Akademia obrazotvorchogo mistetstva i arkhitekturi, Library, vul. Smirnova-Lastochkina, 20, 04053 *Kyiv*
T: +380 44 2721428; Fax: +380 44 2721540
1917; Olena F.Sokolova
155 000 vols; 100 diss/theses; 100 sound-rec; 14 000 periodical items
UBA   28360

Natsionalna naukova medychna biblioteka Ukrainy (State Scientific Medical Library of Ukraine), Tolstoho, 7, 01033 *Kyiv*
T: +380 44 2345197; Fax: +380 44 2351135; E-mail: medlib@library.gov.ua; URL: www.library.gov.ua
1930; Raisa I. Pavlenko
Biology, Physiology, Radiology, Psychology, Veterinary Medicine; Dept of Patent Documentation
40 000 vols; 600 curr per; 3 e-journals; 3 incunabula; 1 300 diss/theses; 170 govt docs; 150 music scores; 21 675 microforms; 15 600 av-mat; 530 sound-rec; 320 digital data carriers
libr loan; Ukrainian Library Association, Association of the Libraries of Ukraine   28361

Natsionalny botanichny sad im. M.M. Hryshka, Biblioteka (Grishko National Botanical Garden), Timiryazevska, 1, 01014 *Kyiv*
T: +380 44 2854105
1944; Liudmila Isakova
38 905 vols; 87 curr per; 38 973 mss; 215 diss/theses
Ukrainian Library Association   28362

Nauchno-doslidny instytut budivelnykh konstruktsi, Biblioteka (Building Structures Research Institute), Klimenka, 5/2, 01680 *Kyiv*
T: +380 44 2762365; Fax: +380 44 2766269
1964; Zhanna M. Vorontsova
30 000 vols; 45 curr per; 72 000 patents   28363

Nauchno-issledovatelsky instytut gematologyi i perelivaniya krovi, Biblioteka (Hematology and Blood Transfusion Research Institute), Maksima Berlinskoho, 12, 04060 *Kyiv*
T: +380 44 4403166; Fax: +380 44 4407222
1946; E.N. Karasyova
20 000 vols   28364

Naukovo-doslidny insitut urologiyi ta nefrologiyi, Akademiya Medychnykh Nauk Ukrainy, Biblioteka (Urology and Nephrology Research Institute), Yu. Kotsyubinskoho, 9 A, 03053 *Kyiv*
T: +380 44 2163905; Fax: +380 44 2446862
1965; Nadiya Ivanivna Kozlyuk
47 820 vols; 112 curr per; 316 diss/theses   28365

Naukovo-doslidny instytut ekonomiki, Ministerstvo ekonomiky Ukrainy, Biblioteka (Ukrainian Ministry of Economics – Research Institute of Economics), bul. Druzhbi Norodiv, 28, 01601 *Kyiv*
T: +380 44 2969742
1962; Neonila Ivanivna Kashka
Environmental protection, machine building, industrial technology
95 000 vols; 248 curr per   28366

Naukovo-doslidny instytut elektromekhanichnykh priladiv, Biblioteka (Electro-mechanical Devices Research Institute), Kramskoho, 27, 03142 *Kyiv*
T: +380 44 4448939
1961
Metal technology, physics, mathematics, economics
122 000 vols   28367

Naukovo-doslidny instytut epidemiologi ta infektsinykh zakhvoryuvan im. L.V. Gromashevskoho, Akademiya Medychnykh Nauk Ukrainy, Biblioteka (Gromashevsky Epidemiology and Infectious Deseases Research Institute), Protasiv uzviz, 4, 03038 *Kyiv*
T: +380 44 2692391
1896; Olga Andriivna Dvoina
81 000 vols; 193 curr per   28368

Naukovo-doslidny instytut farmakologiyi ta toksikologiyi, Biblioteka (Institute of Pharmacology and Toxicology), vul. Ezhena Pote, 14, 03057 *Kyiv*
T: +380 44 4468042
1945; Nataliya M. Bortnik
80 000 vols; 100 curr per; 600 mss   28369

Naukovo-doslidny instytut kardiologiyi im. M.D. Strazheska, Akademiya Medychnykh Nauk Ukrainy, Biblioteka (M.D. Strazhesky Cardiology Research Institute), vul. Narodnoho Opolchenia, 5, 03151 *Kyiv*
T: +380 44 2776577; Fax: +380 44 2287272
1936; Olga Ivanivna Melnik
65 000 vols   28370

Naukovo-doslidny instytut klinichnoyi ta eksperimentalnoyi khirurgiyi, Akademiya Medychnykh Nauk Ukrainy, Biblioteka (Clinical and Experimental Surgery Research Institute), Geroiv Sevastopolya, 30, 03165 *Kyiv*
T: +380 44 4839620
1896
Cardiosurgery
25 000 vols   28371

Naukovo-doslidny instytut meditsiny pratsi, Akademiya Medychnykh Nauk Ukrainy, Biblioteka (Health and Safety at Work Research Institute), vul. Saksakanskoho, 75, 03033 *Kyiv*
T: +380 44 2201719; Fax: +380 44 2206677
1988
42 000 vols; 87 curr per   28372

Naukovo-doslidny instytut onkologiyi ta radiologiyi, Biblioteka (Radiology and Oncology Research Institute), vul. Lomonosova, 33/43, 03022 *Kyiv*
T: +380 44 26675467
1920; Zinaida V. Nazarenko
27 000 vols; 67 curr per   28373

Naukovo-doslidny instytut pedahogiky Ukrainy, Biblioteka (Ukrainian Research Institute of Pedagogy), Tryokhsvyatitelskaya, 8, 01601 *Kyiv*
T: +380 44 2252320
1926
500 000 vols; 218 curr per   28374

Naukovo-doslidny instytut ribnigo gospodarstva, Biblioteka (Institute for Fisheries), Obukhivska, 135, 252164 *Kyiv*
T: +380 44 4525086; Fax: +380 44 646685
1964; Olga Mikhalivna Dubnik
58 000 vols; 33 curr per; 1 321 mss; 1 698 diss/theses; 4 govt docs   28375

Naukovo-doslidny instytut zakhistu roslin, Ukrainska Akademiya Agrarnykh Nauk, Biblioteka (Plant Protection Research Institute), Vasilkivska, 33, 03022 *Kyiv*
T: +380 44 2631370
1947; Lyudmila Petrivna Shelikhova
40 000 vols; 105 curr per; 600 mss   28376

Naukovo-doslidny ta proektno-konstruktorsky instytut Energoproekt, Biblioteka (Power Engineering Design Experimental and Research Institute), pr. Peremogi, 4, 03135 *Kyiv*
T: +380 44 2747943
1947; Katerina Oleksiivna Krasnokutska
68 000 vols; 80 curr per   28377

Naukovo-proektny instytut po transportu prirodnoho gaza, Biblioteka (Natural Gas Transportation Research and Planning Institute), Kuibisheva, 20, 03023 *Kyiv*
T: +380 44 2278759
1944
53 000 vols; 388 curr per   28378

Naukovo-virobnichno obednannya 'Instytut avtomatiki', Biblioteka (Automation Institute Scientific and Industrial Corporation), P. Karkotsa, 22, 03107 *Kyiv*
T: +380 44 2119245
1957; Larisa Oleksiivna Kvashina
190 000 vols; 452 curr per   28379

Physico-Technological Institute of Metals and Alloys, Library, pr. Vernadskoho, 34/1, 01680 *Kyiv*
T: +380 44 4443515; Fax: +380 44 4441210
1958
200 000 vols   28380

Pivdenno-Zakhidna Zaliznitsa, Biblioteka (South-Western Railways), Franka, 19, 03030 *Kyiv*
T: +380 44 2234893
1898; Viktoriya Viktorivna Chuiko
50 000 vols; 193 curr per   28381

Rada po vivchennyu produktivnykh sil Ukrainy, Natsionalna Akademiya Nauk Ukrainy, Biblioteka (Council for Labor Efficiency Improvement), bul. Shevchenka, 60, 03032 *Kyiv*
T: +380 44 2169182
1961; Tetyana G. Alyanova
55 000 vols; 144 curr per   28382

Research and Development Institute for Municipal Facilities and Services, Library, vul. Uritskoho, 35, 03035 *Kyiv*
T: +380 44 2763189; Fax: +380 44 2768121
1963
75 000 vols   28383

Research Institute of the Sewn Goods Industry, Library, U. P. Lyubchenko 15, 252022 *Kyiv*
T: +380 44 268-5541; Fax: +380 44 268-6457
1961
22 800 vols; 6 180 curr per; 200 diss/theses; 155 microforms; 3 200 pamphlets   28384

Spetsializovana evreiska biblioteka im. O. Shvartsmana (Shvartsman Specialized Jewish Library), Stritenska, 4, 04025 *Kyiv*
T: +380 44 2123918
1976
Jewish hist and traditions, Hebrew
27 000 vols; 113 curr per   28385

Tekhnologichny instytut moloka ta myasa, Ukrainska Akademiya Agrarnykh Nauk, Biblioteka (Technological Institute of Milk and Meat), Raskovoi, 4a, 02002 *Kyiv*
T: +380 44 5173825; Fax: +380 44 5170228; E-mail: timm@fm.com.ua; URL: www.timm.kiev.ua
1960; Elena Bosko
40 000 vols; 30 curr per
libr loan   28386

Tsentralna spetsializovana biblioteka dlia slipykh im. M. Ostrovskoho (Central Library for the Blind), Pechersky uzviz, 5, 01601 *Kyiv*
T: +380 44 2350063; Fax: +380 44 2350063; E-mail: csbs@ukrpost.net
1936; Yuri M. Vishnyakov
Education of the Blind, Braille books
214 510 vols; 72 curr per; 18 100 e-books; 480 music scores; 83 230 sound-rec; 30 digital data carriers; 1 860 offprints
libr loan; National Ukrainian Library Association   28387

Ukrainsky derzhavny instytut proektuvannya mist, Biblioteka (Ukrainian Urban Planning Research Institute), bul. L. Ukrainki, 26, 03133 *Kyiv*
T: +380 44 2962515
1930
94 000 vols; 109 curr per   28388

Ukrainsky gosudarstvenny muzei teatralnoho i muzykalnoho iskusstva i kinematografiyi, Biblioteka (Ukrainian State Museum of Theatrical and Musical Arts and Cinematography), vul. Sichnevoho Povostanya, 21/24, 03015 *Kyiv*
T: +380 44 2905131
1923; Elena Galushkevich
29 000 vols; 30 curr per; 279 music scores; 895 sound-rec
libr loan   28389

Ukrainsky nauchno-issledovatelsky instytut po spetsialnym vidam pechati, Nauchno-tekhnicheskaya biblioteka (Ukrainian Research Institute for Special Printing Processes), Kioto, 25, 253660 *Kyiv*
T: +380 44 5130472; Fax: +380 44 5138281
1961; Ludmila Shevchuk
Graphic arts
22 300 vols; 114 curr per   28390

Ukrainsky naukovi tsentr radiatseinoyi meditsini, Biblioteka (Ukrainian Research Centre of Radiation Medicine), Melnikova, 53, 03050 *Kyiv*
T: +380 44 2130637; Fax: +380 44 2137202
1986; Lyudmila Anatolivna Dovgan
200 000 vols   28391

Ukrainsky naukovo-doslidny instytut shtuchnykh volokon s doslidnim virobnitstvom, Biblioteka (Ukrainian Man-made Fibres Research and Experimental Institute), Magnitogorska, 1/b, 253094 *Kyiv*
T: +380 44 5515705
1959; Tamara Ivanivna Molot
85 000 vols; 255 curr per   28392

Ukrainian Research Institute of Traumatology and Othopaedics, Library, vul. Vorovskoho, 27, 03054 *Kyiv*
T: +380 44 2164249; Fax: +380 44 2164462
1919
57 000 vols   28393

Ukrainyan Research Institute of Water Management and Ecological Problems, Library, Inzhenerny prov, 4B, 03010 *Kyiv*
T: +380 44 2900302; Fax: +380 44 2900302
1974
10 000 vols   28394

UkrNIIB, Library (Ukrainian Paper Research Institute), vul. Renezov, 18/7, 01133 *Kyiv*
T: +380 44 2952166; Fax: +380 44 2952166
1931; Sumyna
21 200 vols; 5 625 mss; 750 diss/theses; 4 235 govt docs; 30 100 periodicals
libr loan   28395

UKRNIIPLASTMASH, Ukrainsky nauchno-issledovatelsky konstruktorny instytut po razrabotke mashin i oborudovaniya dlya pererabotky plastmass, reziny i iskusstvennoyi kozhi, Biblioteka (Ukrainian Research and Development Institute of Plastics, Rubber and Artificial Leather Engineering), vul. I. Shevtsova, 1, 252113 *Kyiv*
T: +380 44 4464196; Fax: +380 44 4464429
1961; Sofia Andreevna Kulinich
Machine-building; Industry of high-molecular materials; Rubber industry; Industry of plasties
54 600 vols; 17 curr per; 2 800 govt docs
libr loan   28396

Vniikhimproekt instytut, Library, vul. M. Raskovoi, 11, 253002 *Kyiv*
T: +380 44 5170581; Fax: +380 44 5171518
1970
48 200 vols   28397

Zonalny naukovo-doslidny instytut esperimentalnoho proedtuvannya zhitovykh i gromadskykh sporud, Scientific and Technical Library (Ukrainian Zonal Scientific and Research Design Institute of Civil Engineering), bul. L. Ukrainki, 26, 01133 *Kyiv*
T: +380 44 2964668; Fax: +380 44 2957481
1963; Elizaveta A. Nechaeva
125 000 vols; 66 curr per
libr loan   28398

Lugansk State Medical University, Library, 50 kvartal oborony Luganska, 7a, 91045 *Lugansk*
T: +380 642 548403; Fax: +380 642 532036; URL: www.lsmu.com
1956; Lyudmila Vasilivna Rogova
329 000 vols; 366 curr per   28399

Oblasna naukova medychna biblioteka (Regional Scientific Medical Library), 11 liniya, 3, 91055 *Lugansk*
T: +380 642 527410
1945; Valentina Kalyuzhnaya, Elena Bloshenko
228 000 vols; 120 curr per   28400

Derzhavny arkhiv Volynskoyi oblasti (Volynsk Regional State Archives), Veteraniv, 21, 43024 *Lutsk*
T: +380 3300 57995
1940
42 000 vols; 242 curr per; 440 rare books (16-18th c)   28401

Industrialny instytut, Biblioteka (Industrial Institute), Lvivska, 75, 43018 *Lutsk*
T: +380 3322 68070; Fax: +380 3322 64840
1966; Elizaveta N. Kost
Sociological and economical depts
180 000 vols; 100 curr per; 10 diss/theses
libr loan   28402

Volynskaya oblastnaya nauchna ta meditsinska biblioteka (Medical Research Library), Suvorova, 5, *Lutsk*
T: +380 3322 43593
1945; Vitali Sergiyovich Sokolyuk
Hist of medicine
115 000 vols; 183 curr per
libr loan   28403

Volynsky oblastnoyi kraevedchesky muzei, Biblioteka (District Museum of Regional Studies), vul. Shopena, 20, 43025 *Lutsk*, Volynskaya obl.
T: +380 3322 45619; Fax: +380 3322 43412
1929; Valentina Petrenko
24 000 vols; 75 curr per   28404

Derzhavny medichny instytut, Biblioteka (State Medical Institute), Sichovykh Striltsiv, 6, *Lviv*
T: +380 32 2727015
1939; Nataliya M. Kurnat
587 000 vols; 188 curr per   28405

Derzhavny prirodoznavchi muzei, Natsionalna Akademiya Nauk Ukrainy, Biblioteka (State Museum of the Natural History), Teatralna, 18, 79008 *Lviv*
T: +380 32 2728917; Fax: +380 32 2742307; E-mail: office@museum.lviv.net
18th c; Irina Igorivna Panskikh
68 108 vols; 17 curr per; 124 mss; 106 diss/theses   28406

Fizyko-mekhanichny instytut im. Karpenka, Natsionalna Akademiya Nauk Ukrainy, Biblioteka (Karpenko Physics and Mechanics Institute), Naukova, 5, 79601 *Lviv*
T: +380 32 2654281
1951; Irina S. Klapkin
89 000 vols; 208 curr per   28407

Institut Ukrainoznavstva imeni Ivana Kripyakevicha NAN Ukraini, Biblioteka (Institute of Ukrainian Studies im. I. Kripyakevich), vul. Kozelnytska, 4, 79026 *Lviv*
T: +380 32 2707022; Fax: +380 32 2707021; E-mail: inukr@inst-ukr.lviv.ua; URL: www.inst-ukr.lviv.ua
1951; Natalia Nakonechna
Card files for Hist Dictionary of Ukrainian Language, Oral Hist Tapes on Dissident Movement in Soviet Ukraine
26 200 vols; 67 curr per; 80 diss/theses; 100 sound-rec   28408

Instytut fizyologiyi i biokhimiyi tvaryn, Ukrainska Akademiya Agrarnykh Nauk, Biblioteka (Institute for Research in Physiology and Biochemistry of Animals), Stusa, 38, 79034 *Lviv*
T: +380 32 22425192
1960; Maria V. Oliynyk
25 000 vols; 80 curr per; 4 530 mss; 243 diss/theses
libr loan   28409

Instytut geologiyi i geokhimiyi goryuchykh kopalin, Natsionalna Akademiya Nauk Ukrainy, Biblioteka (Institute of Geology and Geochemistry of Raw Fuel Materials), Naukova, 3a, 79053 *Lviv*
T: +380 32 2637144
1951; Irina G. Kolesnik
73 000 vols; 261 curr per   28410

Instytut narodoznavstva, Natsionalna Akademiya Nauk Ukrainy, Biblioteka (Ethnography Institute), pr. Svobodi, 15, 79000 *Lviv*
T: +380 32 2727012
1874; Vira Mikhalivna Yablonska
Applied arts, folklore
60 000 vols; 70 curr per
libr loan   28411

Instytut prikladnykh problem mekhaniky i matematiky im. Pidstrigacha, Natsionalna Akademiya Nauk Ukrainy, Biblioteka (Pidstrigach Applied Mathematics and Mechanics Institute), Naukova, 3/b, 79601 *Lviv*
T: +380 32 2395424
1961; Lyubov G. Zaprutska
66 000 vols; 310 curr per
libr loan   28412

Instytut regional'nych dostidjen, Natsionalna Akademiya Nauk Ukrainy, Biblioteka (Institut of regional research), Kozelnyts'ka str, 4, 79008 *Lviv*
T: +380 32 2427068; Fax: +380 32 2427168
1964; Galina V. Volf
54 300 vols; 25 curr per; 245 diss/theses   28413

Instytut tuberkulozu, Biblioteka (Tuberculosis Institute), Sychiv, 79000 *Lviv*
T: +380 32 2423449
1945; I. Stasiuk
Respiratory diseases
42 000 vols; 33 curr per
libr loan   28414

Istorichny muzei, Biblioteka (Historical Museum), pl Rynok, 4/6, 79008 *Lviv*
T: +380 32 2720671 ext 237; Fax: +380 32 2743314
1893; Zita Stolyar
13 500 vols; 18 curr per   28415

Kyivsky NDI gematologiyi i perelivannya krovi, Lvivsky filial, Biblioteka (Kiev Haematology and Blood Transfusion Research Institute – Lviv Branch), Pushkina, 45, 79044 *Lviv*
T: +380 32 252259
1946; Adel A. Akhmedova
12 000 vols; 143 curr per; 21 000 periodicals
libr loan   28416

Lviv Art Gallery, Biblioteka, vul. Stefanika, 3, 79000 *Lviv*
T: +380 32 2614448; Fax: +380 32 2614448; E-mail: lv.galery@mail.ru
1907; Oksana Maksymenko
30 915 vols; 4 340 periodical items   28417

Lviv Scientific-Research Institute Epidemiology and Hygiene, Ministry of Public Health of Ukraine, Research Libraries, Zelena, 12, 79005 *Lviv*
T: +380 32 22762832, 22601200; Fax: +380 32 22762831; E-mail: valentinas2002@ukr.net
1945; Valentina Smonytskal
21 289 vols; 7 009 curr per; 92 diss/theses
libr loan; Ukrainian Library Association   28418

Lviv State Institute of Applied and Decorative Art, Library, vul. Kubiyovycha, 38, 79011 *Lviv*
T: +380 32 2761477
1946
90 000 vols   28419

Lvivsky mezhgaluzevoyi tsentr naukovo-tekhnichnoyi informatsiyi i propagandy, Biblioteka (Lviv Scientific and Technical Information Popularisation Center), 700-richchya Lviva, 57, 79601 *Lviv*
T: +380 32 2523280
1962; Irina Petrivna Galitska
Machine building, coal,oil and gas industry, electrical engineering, light industry, food industry, construction materials
643 000 vols; 1 000 curr per
libr loan   28420

Naukova medychna biblioteka (Medical Scientific Library), Ruska, 20, 79008 *Lviv*
T: +380 32 2725800
1944; Svitlana S. Lozinska
209 000 vols; 170 curr per; 900 mss   28421

Naukovo-doslidny instytut spadkovoyi patologiyi, Biblioteka (Research Institute of Hereditary Pathology), Lysenka, 31a, 79000 *Lviv*
T: +380 32 2765499; Fax: +380 32 2753844
1940; Svitlana Mamchuz
28 000 vols   28422

Naukovo-vibronichne obedenanne Avtovazhmash, Biblioteka (Automobile and Railway Transport Industrial Corporation), Shevchenka, 323, 79069 *Lviv*
T: +380 32 2334291
1965
Metal technology, mechanics, machine buiding, cranes, mathematics, physics
58 000 vols; 107 curr per
libr loan   28423

Ukrainsky nauchno issledovatelsky instytut po vidam pechati, Biblioteka (Ukrainian Research Institute for the Printing Industry), Volodimiya Velikoho, 4, 79026 *Lviv*
T: +380 32 2342111; Fax: +380 32 2634255
1932; Marina Pashulya
52 000 vols   28424

Ukrainyan State Geological Research Institute, Scientific-Technical Library, pl Mitskevicha, 8, 79000 *Lviv*
T: +380 32 2726522; Fax: +380 32 2725614
1956; Stepanova
3 500 reports on scientific-research works; geological funds
32 500 vols; 80 curr per; 1 110 diss/theses; 2 govt docs; 82 maps
libr loan   28425

MakNII – State Makeyevka Safety in Mines Research Institute, Scientific and Technical Library, Likhachova, 60, 86108 *Makeyevka*
T: +380 622 903654; E-mail: maknii@tr.dn.ua
1927; Velena Aleksandrovna Chertushkina
95 100 vols; 110 curr per; 160 diss/theses
libr loan   28426

Instytut sadivnitstva, Tekhnikum Krimskoyi doslidnoyi stantsiyi, Biblioteka (Institute of Horticulture – Crimean Research Station and Professional School), Shkilna, 334105 *Malenke*
1828; Lyudmila L. Bereza
85 000 vols; 60 curr per   28427

Instytut zroshuvanoho sadivnytstva, Naukovo-tekhnichna biblioteka (Institute of Irrigation, Fruit Growing), Vakulenchuka, 99, 72311 *Melitopol*
T: +380 6192 31320; Fax: +380 6192 31378
1943; L. A. Zahurska
biology, agriculture, engineering dept
22 852 vols; 4 diss/theses; 21 080 periodical items, 1 720 offprints   28428

Instytut impulsnykh protsesiv i tekhnologi, Natsionalna Akademiya Nauk Ukrainy, Biblioteka (Institute of Pulse Research and Engineering), pr. Zhovtnevi, 43, 54018 *Mykolaiv*
T: +380 512 58-7138; Fax: +380 512 226140; E-mail: iipt@iipt.com.ua; URL: www.iipt.com.ua
1963; Tamara M. Lugova
Mechanics, metallurgy, metal technology, machine building

133 000 vols; 430 diss/theses; 1 000 govt docs; 4 411 microforms
libr loan 28429

Kraeznavchi muzei, Biblioteka (Museum of Regional Studies), Dekabristov, 32, 54000 ***Mykolaiv***
T: +380 512 371698
1913
Regional hist, archaeology, museology
14 000 vols; 37 curr per
libr loan 28430

Oblasna naukova medychna biblioteka (Regional Scientific Medical Library), Chervonykh mayovshchikiv, 12, 54058 ***Mykolaiv***
T: +380 512 313592
1939; Olena O. Parkhomova
130 000 vols; 143 curr per
libr loan 28431

Oblasnoyi derzhavny arkhiv, Biblioteka (State Regional Archives), Vaslyaeva, 43, 54044 ***Mykolaiv***
T: +380 512 214039
1925
16 000 vols 28432

Myronivka Institute of Wheat, Research Library, Tsentralna St, 08853 ***Myronivka***, Kyivska obl.
T: +380 4474 74135; Fax: +380 4474 74446; E-mail: mwheats@ukr.net
1911; Lyudmyla Ivantsova
Crop husbandry
25 159 vols; 1 diss/theses; 14 345 periodical items
libr loan 28433

Krimska astrofizychna observatoriya, Natsionalna Akademiya Nauk Ukrainy, Biblioteka (Crimean Astrophysical Observatory), 98409 ***Nauchne***, AR Krym
T: +380 6554 71177; Fax: +380 6554 40704; E-mail: Library@crao.crimea.ua; URL: www.crao.crimea.ua
1908; Elena Kostylova
172 572 vols; 1 065 curr per; 171 diss/theses; 2 705 microforms; 2 499 other items
libr loan 28434

NDI Zemlerobstva i tvarinnitstva zakhidnoho regionu Ukrainy, Biblioteka (Western Ukraine Research Institute of Crop Growing and Animal Husbandry), 292084 ***Obroshine***, Pustomitivski
1956; Lyubov A. Pinda
115 000 vols; 282 curr per; 1 060 mss 28435

Fizyko-khimichny instytut im. Bogatskoho, Natsionalna Akademiya Nauk Ukrainy, Biblioteka (Bogasky Institute of Physical Chemistry), Chernomorska doroga, 86, 65080 ***Odesa***
T: +380 482 618131
1911
133 000 vols
libr loan 28436

Institute of Stomatology of AMS of Ukraine, Biblioteka (Stomatology Research Institute), 11, Richelierivska st, 65026 ***Odesa***
T: +380 482 224823; Fax: +380 482 348168; E-mail: stomat@paco.net
1928; Alla Ischakova
Orthopedics, physiology, cancer research
68 000 vols; 59 curr per; 3 275 diss/theses 28437

Instytut problem rinku i ekonomiko-ekologichnykh doslidzhen, Natsionalna Akademiya Nauk Ukrainy, Biblioteka (Marketing, Economics and Ecology Research Institute), Frantsuzky bl, 29, 65044 ***Odesa***
T: +380 482 254125
1971; Aida Yakivna Borodina
Sociology, cybernetics
68 000 vols
libr loan 28438

Naukovo-doslidny instytut ochnykh zakhvoryuvan i tkanevoyi terapiyi im. Filatova, Biblioteka (Filatov Research Institute of Ophthalmology), Frantsuzky bul, 49-51, 65061 ***Odesa***
T: +380 482 603772; Fax: +380 482 684851
1936; Tamila A. Marisheva
Medicine, biology
80 000 vols; 10 curr per 28439

Oblasna naukova medychna biblioteka (Regional Scientific Medical Library), Vorobova, 5, 65077 ***Odesa***
T: +380 482 201527
1930; Valentina E. Malyarenko
127 000 vols; 180 curr per 28440

Odesky arkheologichny muzei, Biblioteka (Odessa Archaeological Museum), Lanzheronovskaya, 4, 65026 ***Odesa***
T: +380 487 220171; E-mail: dvgn@eurocom.od.ua
1825; Galina P. Ukrainska
30 939 vols; 20 curr per; 1 incunabula; 10 microforms
libr loan 28441

Odesky khudozhni muzei, Biblioteka (Fine Arts Museum of Odessa), vul. Sofiska, 5A, 65026 ***Odesa***
T: +380 482 238272; Fax: +380 482 238393; URL: museum.odessa.net/fineartsmuseum
1899; Lidiya Kalmanovskaya
15 000 vols 28442

Odessa Museum of Western and Eastern Art, Fine Arts Library, vul. Pushkinska, 9, 65026 ***Odesa***
T: +380 482 224815; Fax: +380 482 246747
1920; Lena Sadykova
14 908 vols; 500 curr per; 12 incunabula; 6 av-mat 28443

Odessa Research Institute of Virology and Epidemiology, Library, vul. Yubileina, 6, 65031 ***Odesa***
T: +380 482 330338; Fax: +380 482 330338
1886
20 000 vols 28444

Ukrainsky naukovo-doslidny institut medychnoi reabilitatsii ta kurortologii, Biblioteka (Ukranian Research Institute for Medical Rehabilitation and Resort Therapy), No 6, Lermontovsky Pereulok, 65014 ***Odesa***
T: +380 482 223568; Fax: +380 482 223568; E-mail: mvik@kurort.odessa.net
1928; Nadiya Tkhoryevskaya
Medical rehabilitation, balneology, electrotherapy
28 617 vols; 6 315 e-books; 70 mss; 355 diss/theses; 12 800 periodical items
libr loan; UBA 28445

Ukrainsky selektsiyno-genetichny instytut, Biblioteka (Ukrainian Institute of Plant Breeding and Genetics), Ovidiopolska doroga, 3, 65036 ***Odesa***
T: +380 482 694707
1934; Elena F. Svirskaya
107 000 vols; 462 curr per 28446

Gosudarstvenny arkhiv Poltavskoyi oblasti, Nauchno-spravochnaya biblioteka (State Archives of the Poltava District), vul. Pushkina, 18/24, ***Poltava***
T: +380 532 220566
1944; V.V. Kozotenko
Life and work of the Ukrainian writers P. Mirny, J. Kotljarevski, V.G. Korolenko, N.V. Gogol
17 800 vols 28447

Gravimetrichna observatoriya, Natsionalna Akademiya Nauk Ukrainy, Biblioteka (Gravimetric Observatory), vul. Myasoedova, 27/29, 36029 ***Poltava***
T: +380 532 272039
1926; Efrosiniya Ivanovna Bobovoz
25 400 vols; 40 curr per; 108 incunabula 28448

Kraeznavchi muzei, Biblioteka (Museum of Regional Studies), pl Lenina, 2, 36020 ***Poltava***
T: +380 532 225738
1944; Ganna O. Medvedeva
Life and work of the Ukrainian writers P. Mirny, J. Kotlyarevski, V.G. Korolenko, N.V. Gogol
80 000 vols; 10 curr per 28449

Nauchno-issledovatelsky instytut svinovodstva, Biblioteka (Pig Breeding Research Institute), Shvedskaya mogila, 36006 ***Poltava***
T: +380 532 220604; Fax: +380 532 222753
1930; Lyudmila N. Sherednik
54 000 vols; 42 curr per; 330 mss 28450

NDI Emalkhimmash, Biblioteka (Emalkhimmash Chemical Engineering Research Institute), Frunze, 153, 36002 ***Poltava***
T: +380 532 2103256
1964; Zoya A. Velichko
Polymers, enamels, industrial ceramics
60 700 vols; 35 curr per 28451

Oblasna naukova medychna biblioteka (Regional Scientific Medical Library), Pushkina, 133, 36001 ***Poltava***
T: +380 532 275021
1938; Valentina V. Slipchenko
341 000 vols; 232 curr per 28452

Poltava Consumers' Co-operative Institute, Library, vul. Kovalya, 3, 36601 ***Poltava***
T: +380 532 220929; Fax: +380 532 274542
1974
390 000 vols 28453

State Scientific Agrarian Library of UAAS, Shvedska Mogila 1, 36013 ***Poltava***
T: +380 532 527419; Fax: +380 532 522753; E-mail: pigbreeding@ukr.net
1946; Lyudmila N. Cherednyk
Pig Breeding
54 800 vols; 141 diss/theses; 40 microforms; 20 digital data carriers; 15 300 periodical items
libr loan 28454

NDI tekhnologiyi mashinobuduvannya, Biblioteka (Machine Building Technology Research Institute), Novorosiyska, 16, 266018 ***Rivne***
T: +380 362 36396
1973
97 000 vols; 72 curr per
libr loan 28455

Oblasna naukova medychna biblioteka (Regional Scientific Medical Library), vul. Kotlyarevskogo, 2, 33028 ***Rivne***
T: +380 362 266811; Fax: +380 362 635243; E-mail: ronmb@ukr.net
1945; Petro R. Gorun
Medicine, medical history
191 899 vols; 218 curr per; 389 diss/theses; 66 642 periodical items
libr loan 28456

NDI organichnykh napivprovodnkiv i barvnikiv, Biblioteka (Organic Semiconductors Research Institute), Lenina, 57, 349870 ***Rubizhne***
T: +380 253 57733
1927; M.M. Dugina
294 000 vols; 200 curr per
libr loan 28457

Institute of Vegetable and Melon Growing, Ukrainian Academy of Agrarian Sciences, Library, ***Selektsionny***, Kharkovskaya obl.
T: +380 572 429191; Fax: +380 572 429191
1947; L. Romanenko
9 167 vols; 35 curr per; 2 000 mss; 163 diss/theses; 250 govt docs
libr loan 28458

Instytut Biologiyi Yuzhnykh morei, National Academy of Sciences of Ukraine, Nauchnaya Biblioteka (Institute of Biology of the Southern Seas), 2, Nakhimov Ave, 99011 ***Sevastopol***
T: +380 692 544110; Fax: +380 692 557813; E-mail: akimova@ibss.iuf.net; URL: www.ibss.org.ua/Default.aspx?tabid=169
1871; Olga Akimova
A.L. Behning, Max Hartmann, V.M. Rilov, S.A. Zernov, T.S. Petipa, V.A. Vodyanitskij colls
155 000 vols; 80 curr per; 3 250 mss; 4 300 microforms; 67 digital data carriers; 6 100 reprints
libr loan; EURASLIC 28459

Khersones Museum of History and Archaeology, Library, 99045 ***Sevastopol***
1860
20 000 vols 28460

Miska naukova medychna biblioteka (City Scientific Medical Library), Velika Morska, 31, 99000 ***Sevastopol***
T: +380 692 525304
1981; Valentina O. Olkhova
61 000 vols
libr loan 28461

Morskaya biblioteka im. admirala M. P. Lazareva, Sevastopolski Dom ofitserov ChF RF, pr. Nakhimova, 7, 99011 ***Sevastopol***
T: +380 692 543315; E-mail: mb_sev@mail.ru
1822; Nikolai Ivanovich Krasnolitski
History of the Black Sea Fleet, history of the Crimean War, Redki Fond (18th-20th c, 28 000 vols), Tarvika (history of the Crimea), Sevastopoliana (history of Sevastopol)
251 363 vols; 266 curr per; 251 363 e-books; 350 mss; 18 diss/theses; 300 govt docs; 70 music scores; 150 maps; 40 microforms; 110 av-mat; 250 sound-rec; 200 digital data carriers; 743 offprints, autographs 28462

Morskoyi gidrofizychesky instytut, National Academy of Sciences of Ukraine, Nauchnaya Biblioteka (Scientific Library of the Marine Hydrophysical Institute), 2, Kapitanskaya Street, 99011 ***Sevastopol***
T: +380 692 544201; E-mail: ocean@alpha.mhi.iuf.net; URL: www.mhi.iuf.net
1944; Alevtina Kalinina
102 000 vols; 87 curr per
libr loan 28463

Sevastopol Machine Engineering Institute, Library, Streletsky Bay, Studgorodok, 99053 ***Sevastopol***
T: +380 690 243590; Fax: +380 690 244530
1963
850 000 vols 28464

Naukovo-vibronichno obednannya Impuls, Biblioteka (Automatic Control Systems Scientific and Industrial Corporation), pl Pobedy, 2, ***Severodonetsk***, Luganska obl.
T: +380 6452 27715; Fax: +380 6452 41323
1954
Automation, computers
120 000 vols; 320 curr per
libr loan 28465

State Research and Design Institute of Chemical Engineering, Library, vul. Dzerzhinskoho, 1, 349940 ***Severodonetsk***, Luganska obl.
T: +380 6452 23388; Fax: +380 6452 25042
1950
97 000 vols 28466

State Research & Design Institute of Chemical Engineering, Library (Methanol Technology Planning and Research Institute), vul. Vilesova, 1, 349940 ***Severodonetsk***, Luganska obl.
T: +380 6452 99748
1951
Chemical engineering
122 000 vols; 516 curr per
libr loan 28467

Vibronichno obednannya Azot, Biblioteka (Nitrogen Industrial Corporation), Khimikiv, 5, ***Severodonetsk***, Luganska obl.
T: +380 6452 94469
1951
143 000 vols; 330 curr per
libr loan 28468

Instytut efiromaslichnykh i lekarstvennykh rasteni, Biblioteka (Institute of Essential Oil and Medicinal Plants), Kievskaya, 150, 330620 ***Simferopol***
T: +380 652 223405; Fax: +380 652 223419
1965; Irina G. Skachkova
Essential oil crops, their production and sale, agricultural machinery
172 000 vols; 398 curr per; 1 200 mss; 36 400 microforms; 300 pamphlets
libr loan 28469

Instytut mineralnykh resursov, Biblioteka (Ukrainian State Institute of Mineral Resourses), Kirova, 47/2, ***Simferopol***
T: +380 652 297148; Fax: +380 652 275264
1948; Luchanskaya
Mineralogy, ecology, geology, geophysics, deposits, dressing of useful minerals
120 000 vols; 271 curr per; 1 300 mss; 300 diss/theses; 250 maps; 50

microforms
libr loan 28470

Krymsky kraevedchesky muzei,
Biblioteka 'Tavrika' (Scientific 'Taurica'
Library), Gogolya, 14, 95000 *Simferopol*
T: +380 652 252511; Fax: +380 652
252511
1873; Nina Kolesnikova
A.L. Berte-Delogard coll, books on
emperor Nikolai II, archeology, nature
and culture of Crimea
40 500 vols; 44 curr per; 18 mss; 340
govt docs; 6 music scores; 128 maps; 5
microforms; 2 digital data carriers; view
of crimea (200 copies), people of crimea
(70 copies) 28471

**Krymsky tsentr naukovo-tekhnichnoyi
informatsiyi**, Tsentralna biblioteka
(Crimean Centre of Scientific and
Technical Information), Yaltinska, 20,
95640 *Simferopol*
T: +380 652 233857
1961; Akim M. Gasymov
77 000 vols; 100 curr per 28472

**Viddil okhorony zdorovya radi
ministriv Respubliky Krym**, Medychna
biblioteka (Health Ministry of Crimean
Republic – Medical Library), Gorkoho, 3,
95000 *Simferopol*
T: +380 652 277777
1946; Larisa O. Ermolenko
304 000 vols; 460 curr per 28473

Oblasna naukova medychna biblioteka
(Regional Scientific Medical Library),
Petropavlivska, 105, 40030 *Sumy*
T: +380 542 22331
1945; Alina I. Gromova
336 000 vols; 390 curr per 28474

**Scientific Research Stock Company
'NIEMAS'**, Library (Pressing Equipment
Research Institute), Kursky av. 6, 40020
Sumy
T: +380 542 266462; Fax: +380 542
225306
1966; Olga Ivanovna Syntik
137 000 vols; 820 curr per; 148
diss/theses; 1 282 microforms
libr loan 28475

**Research Institute of Agriculture
of the Central Chernozem Zone
named after V.V. Dokuchaev**, Russian
Academy of Agricultural Sciences, Library,
397463 *Talovaya*, Voronezhskaya obl.
T: +7 47352 45433; Fax: +7 47352
45537; E-mail: niishtc@mail.ru
1939; Larisa Bondareva
Melioration in agroforestry, selection of
agricultural crops, agricultural economics,
plant and fruit growing, cattle breeding
81 000 vols; 98 curr per 28476

**Gosudarstvenny arkhiv Ternopilskoyi
oblasti**, Biblioteka (Ternopil State
Archives), Getman Sagaidachny, 14,
46001 *Ternopil*
T: +380 3522 52767; Fax: +380 3522
28618
1939; Galina R. Kurys
33 000 vols; 6 200 curr per 28477

Oblasna naukova medychna biblioteka
(Regional Scientific Medical Library),
Ruska, 29, 46011 *Ternopil*
T: +380 3522 50769
1945; Oleksandra R. Zlivko
228 000 vols; 163 curr per 28478

Oblasna naukova medychna biblioteka
(Regional Scientific Medical Library),
Peremogi, 27, 88000 *Uzhgorod*
T: +380 3100 38060
1946
177 000 vols; 220 curr per 28479

Oblasna naukova medychna biblioteka
(Regional Medical Research Library), Lva
Tolstoho, 11, 21100 *Vinnytsya*
T: +380 432 326613
1933
150 000 vols; 263 curr per 28480

**Vinnytsky tsentr naukovo-tekhnichnoyi
informatsiyi**, Biblioteka (Vinnitsa
Centre of Scientific and Technical
Information), Khmelnitske shosse, 25,
21016 *Vinnytsya*
T: +380 432 446157; Fax: +380 432
438036
1957; Bella A. Sokol
495 000 vols; 514 curr per 28481

**Ukrainsky naukovo-doslidny instytut
mekhanizatsiyi ta elektrifikatsiyi
silskoho gospodarstva**, Biblioteka
(Agricultural Electrification and
Mechanization Research Institute),
40-richchya Peremogi, 8, 332260
Yakimivka
T: +380 6131 94433
1912; Lyubov Vasilivna Novoseltseva
54 000 vols; 90 curr per 28482

Derzhavny Nikitsky botanichny sad,
Biblioteka (Nikita Botanical Garden),
98267 *Yalta*, Krymska republika
T: +380 654 335535; Fax: +380 654
335386
1812; Ms N.G. Lobova
Biology of fruit, ornamental, essential oil
plants, phytopathology, plant protection,
agroecology, nature conservation
213 000 vols; 2 400 curr per; 152
diss/theses; 442 microforms
libr loan 28483

Institute for Vine and Wine, Research
Library (Magarach Viticulture and
Oenology Research Institute), Kirova,
31, 98600 *Yalta*
T: +380 654 325591; Fax: +380 654
230608
1928; Larisa Klimova-Donchuk
Viticulture, economy of agriculture
and food production, mechanization
of agriculture, biology, chemistry, field
management; depository storage fund (37
600), newspaper abstracts (30 volumes)
114 120 vols; 260 curr per; 820 mss;
328 diss/theses; 780 govt docs; 35
maps; 880 microforms 28484

Instytut mekhanizatsiyi tvarinnitstva,
Akademiya Agrarnykh Nauk Ukrainy,
Biblioteka (Institute of Mechanization
in Animal Husbandry), Khortitsya,
Zaporizhzhya
T: +380 612 605344
1945; Lyudmila O. Anokhina
Civil engineering
74 700 vols; 370 curr per
libr loan 28485

Oblasna naukova medychna biblioteka
(Regional Scientific Medical Library),
Chervonogvardiska, 38, 69600
Zaporizhzhya
T: +380 612 646614
1945; Lyudmila O. Dikalova
248 000 vols 28486

**Zaporizky tsentr naukovo-tekhnichnoyi
informatsiyi**, Biblioteka (Zaporizhzhya
Centre of Scientific and Technical
Information), pr. Lenina, 77,
Zaporizhzhya
T: +380 612 641463
1961; Oleksandrea I. Osipenko
Metallurgy, machine building, chemistry,
timber industry, light industry
606 000 vols; 305 curr per; 517 500
microforms
libr loan 28487

Oblasna naukova medychna biblioteka
(Regional Scientific Medical Library),
Chervonoho Khresta, 3, 10013
Zhytomyr
T: +380 412 373083
1945; Klara F. Kominarets
167 000 vols; 255 curr per 28488

Public Libraries

**Alushtinska miska TsBS, Tsentralna
bibioteka im. S.M. Sergeeva-
Tsenkoho** (Alushta City Centralized
Library System, Sergeev-Tsenky Main
Library), Lenina, 20, 98270 *Alushta*
T: +380 6560 33486
79 000 vols; 138 curr per 28489

**Apostolivska raionna TsBS,
Tsentralna bibiloteka** (Apostolove
Regional Centralized Library System,
Main Library), 322450 *Apostolove*
T: +380 5656 91659
1947
60 000 vols; 70 curr per 28490

**Balakliska raionna TsB S, Tsentralna
biblioteka** (Balakliya Regional
Centralized Library System, Main Library),
Shevchenka, 1/2, 313810 *Balakliya*
T: +380 5749 52117
1921
65 000 vols; 125 curr per 28491

**Barska raionna TsBS, Tsentralna
biblioteka** (Bar Regional Centralized
Library System, Main Library), 50-richchya
Zhovtnya, 30, 288600 *Bar*
T: +380 4366 1536
1946
84 600 vols; 186 curr per; 4 govt docs;
297 music scores; 15 maps; 28 av-mat;
32 sound-rec
libr loan 28492

**Barvinkovska raionna TsBS,
Tsentralna biblioteka** (Barvinkove
Regional Centralized Library System,
Main Library), Lenina, 9, 313650
Barvinkove
T: +380 5757 2914
1946
52 000 vols; 102 curr per 28493

**Belogirska raionna TsBS, Tsentralna
biblioteka** (Belogirsk Regional
Centralized Library System, Main Library),
Chaban-Zae, 26, 334140 *Belogirsk*
T: +380 6559 92864
1975; Tetyana O. Shanaeva
400 000 vols; 70 curr per
libr loan 28494

**Berdichivska miska TsBS, Tsentralna
biblioteka** (Berdichiv City Centralized
Library System, Main Library), Sverdlova,
10, 261400 *Berdichiv*
T: +380 4143 21423
1919
80 000 vols; 184 curr per 28495

**Berdyanska miska TsBS, Tsentralna
biblioteka** (Berdyansk City Centralized
Library System, Main Library), Pushkina,
14, 71140 *Berdyansk*
T: +380 6153 34044
1892
Regional studies
165 000 vols 28496

**Beregovska raionna TsBS, Tsentralna
biblioteka** (Beregove Regional
Centralized Library System, Main Library),
pl Geroiv, 7a, 295510 *Beregove*
T: +380 3141 24351
1948
112 000 vols; 111 curr per
libr loan 28497

**Berezhanska raionna TsBS, Tsentralna
biblioteka** (Berezhani Regional
Centralized Library System, Main
Library), 40-richchya Zhovtnya, 3, 283150
Berezhani
T: +380 3548 21452
1939
58 000 vols; 107 curr per 28498

**Bershadska raionna TsBS, Tsentralna
biblioteka** (Bershad Regional
Centralized Library System, Main Library),
Chervonorarmiska, 6, 288540 *Bershad*
T: +380 4365 22596
1919
94 000 vols; 122 curr per 28499

**Bilotserkivska miska TsBS, Tsentralna
biblioteka** (Bila Tserkva City Centralized
Library System, Main Library), Lenina pl,
4/27, 56400 *Bila Tserkva*
T: +380 4463 51434
1930; Petro Ivanivich Krasnozhon
130 000 vols; 151 curr per 28500

**Bilozerkska raionna TsBS, Tsentralna
biblioteka** (Bilozerka Regional
Centralized Library System, Main Library),
K. Marksa, 89, 326300 *Bilozerka*
T: +380 5547 22176
1922
53 000 vols; 136 curr per 28501

**Bobrinetska raionna TsBS, Tsentralna
biblioteka** (Bobrinets Regional
Centralized Library System, Main Library),
Lenina, 78, 317220 *Bobrinets*
T: +380 5257 21792
1946
60 000 vols; 117 curr per 28502

**Bogodukhivska raionna TsBS,
Tsentralna biblioteka** (Bogodukhiv
Regional Centralized Library System,
Main Library), pl Lenina, 1, 312320
Bogodukhiv
T: +380 5758 23236
1928
76 000 vols; 84 curr per 28503

**Boguslavska raionna TsBS, Tsentralna
biblioteka** (Boguslav Regional
Centralized Library System, Main Library),
Ostriv, 1, 09700 *Boguslav*
T: +380 4461 51165
1920; Tatyana V. Kosyushko
56 354 vols; 386 curr per; 545
incunabula; 120 govt docs; 209 music
scores; 34 maps; 1 189 sound-rec 28504

**Borislavska raionna TsBS, Tsentralna
biblioteka** (Borislav Regional Centralized
Library System, Main Library),
Shevchenka, 20, 293760 *Borislav*
T: +380 3248 4549
1940
78 000 vols; 153 curr per 28505

**Borodyanska raionna TsBS,
Tsentralna biblioteka** (Borodyanka
Regional Centralized Library System,
Main Library), *Borodyanka*
T: +380 4477 53307
1936; Galina Grigorivna Nekrutenko
64 000 vols; 171 curr per
libr loan 28506

**Borivska raionna TsBS, Tsentralna
biblioteka** (Borova Regional Centralized
Library System, Main Library), Miru, 12,
312670 *Borova*
T: +380 5759 61976
1979
55 000 vols; 168 curr per 28507

**Brodivska raionna TsBS, Tsentralna
biblioteka** (Brodi Regional Centralized
Library System, Main Library), pl
Svobodi, 3, 80600 *Brody*
T: +380 3266 43114
1947
62 000 vols; 99 curr per 28508

**Brovarska raionna TsBS, Tsentralna
biblioteka** (Brovari Regional Centralized
Library System, Main Library), Kirova, 36,
255020 *Brovari*
T: +380 4494 40595
1947; Mariya F. Dichakivska
78 000 vols; 129 curr per 28509

**Buchatska raionna TsBS, Tsentralna
biblioteka** (Buchach Regional
Centralized Library System, Main Library),
Halytska, 21, 48400 *Buchach*
T: +380 3544 21215; Fax: +380 3544
21393; E-mail: bibl@buc.tr.ukrtel.net
1946
65 000 vols; 104 curr per 28510

**Buska raionna TsBS, Tsentralna
biblioteka** (Busk Regional Centralized
Library System, Main Library),
Grushevskoho, 5, 292050 *Busk*
T: +380 3264 21520
1944
64 000 vols; 92 curr per 28511

Vognetrivkivi kombinat, Biblioteka
profkomu (Railway Wagons Industrial
Corporation, Trade Union Library),
Komsomolska, 1, 383440 *Chasiv Yar*
1950/1975
108 000 vols; 107 curr per 28512

**Chechelniksa raionna TsBS,
Tsentralna biblioteka** (Chechelnik
Regional Centralized Library System,
Main Library), Lenina, 29, 288600
Chechelnik
T: +380 4363 42441
1947
56 000 vols; 160 curr per 28513

**Chemerovetska raionna TsBS,
Tsentralna biblioteka** (Chemerivtsi
Regional Centralized Library System,
Main Library), Lenina, 38, 281670
Chemerivtsi
T: +380 3859 91379
1930
63 000 vols; 136 curr per 28514

**Cherkaska miska TsBS, Tsentralna
biblioteka** (Cherkasi City Centralized
Library System, Main Library), Uritskoho,
200, 18002 *Cherkasy*
T: +380 472 459191
1963
100 000 vols; 130 curr per 28515

Oblasna biblioteka dlya yunatstva (Regional Youth Library), Ilina, 285, 18002 *Cherkasy*
T: +380 472 476500
1977; Olena Antoniona Fedorenko
Art dept
106 794 vols; 410 govt docs; 2 738 music scores; 26 maps; 300 av-mat; 206 sound-rec; 36 639 periodicals
libr loan   28516

Oblasnaya biblioteka dlya ditei (Regional Children's Library), Kirova, 26, 18000 *Cherkasy*
T: +380 472 457075
1954; Olga Petrivna Dubova
249 000 vols; 135 curr per   28517

Chernigivska miska TsBS, Tsentralna biblioteka (Chernigiv City Centralized Library, Main Library), Kirponosa, 22, 14000 *Chernigiv*
T: +380 4622 75156
1949
128 000 vols; 260 curr per   28518

Derzhavna oblasna universalna biblioteka (State Regional universal Library), Lenina, 41, 14000 *Chernigiv*
T: +380 4622 74563
1877
755 000 vols; 580 curr per; 16 000 music scores; 270 maps; 6 600 av-mat   28519

Oblasna biblioteka dlya ditei (Regional Children's Library), Rokosovskoho, 22a, 14032 *Chernigiv*
T: +380 4622 32761
1938; Nataliya Ivanivna Lisenko
133 000 vols; 109 curr per   28520

Oblasna biblioteka dlya yunatstva (Regional Youth Library), Shevchenka, 63, 14027 *Chernigiv*
T: +380 4622 35051
1978
128 000 vols; 180 curr per   28521

Chernivetska miska TsBS, Tsentralna biblioteka (Chernivtsi City Centralized Library System, Main Library), Golovna, 162, 58018 *Chernivtsi*
1950
65 000 vols; 56 curr per   28522

Derzhavna oblasna universalna biblioteka (State Regional Universal Library), vul. Kobylyanskoi, 47, 58000 *Chernivtsi*
1940
675 000 vols; 620 curr per   28523

Oblasna biblioteka dlya ditei (Regional Children's Library), Shevchenka, 29, 58000 *Chernivtsi*
1944; Olena Seminivna Zaichkova
126 000 vols; 67 curr per   28524

Chernyakhivska raionna TsBS, Tsentralna biblioteka (Chernyakhiv Regional Centralized Library System, Main Library), Zhitomirska, 2, 261030 *Chernyakhiv*
T: +380 4134 51603
1944
83 000 vols; 130 curr per   28525

Chervonogradska raionna TsBS, Tsentralna biblioteka (Chervonograd Regional Centralized Library System, Main Library), Bandery, 11, 292210 *Chervonograd*
T: +380 249 24357
1957; G.D. Sachovska
73 200 vols; 63 curr per   28526

Chornukhinska raionna TsBS, Tsentralna biblioteka (Chornukhi Regional Centralized Library System, Main Library), Lenina, 47, 37100 *Chornukhi*
T: +380 5340 51374; E-mail: Lib_ch@ukrpost.ua
1918; Nikola Ivanovich Bulda
195 564 vols; 75 curr per; 569 av-mat; 122 sound-rec   28527

Debaltsevska raionna TsBS, Tsentralna biblioteka (Debaltseve Regional Centralized Library System, Main Library), Radyanska, 65, 343810 *Debaltseve*
T: +380 6249 22560
1947
85 800 vols; 93 curr per   28528

Dikanska raionna TsBS, Tsentralna biblioteka (Dikanka Regional Centralized Library System, Main Library), Lenina, 119, 315100 *Dikanka*
T: +380 5351 91373
1936; Lyubov V. Gulanova
66 000 vols; 110 curr per   28529

Dniprodzerzhinska mezhspilkova TsBS, Tsentralna biblioteka (Dniprodzerzhinsk Trade Union Centralized Library System, Main Library), pl Gagarina, 5, 51925 *Dniprodzerzhinsk*
T: +380 5692 399866
1933
320 000 vols; 243 curr per   28530

Dniprodzerzhinska raionna TsBS, Tsentralna biblioteka (Dniprodzerzhinsk Regional Centralized Library System, Main Library), Sirovtsya, 14, 51908 *Dniprodzerzhinsk*
T: +380 5692 38619
1933
203 000 vols; 259 curr per   28531

OAO DniproAzot, Profspilkova biblioteka (DniproAzot Joint-Stock Company, Trade Union Library), Stovbi, 3, 51909 *Dniprodzerzhinsk*
T: +380 5692 71135
1930; Elena E. Zaritskaya
73 000 vols; 97 curr per   28532

Derzhavna oblasna biblioteka dlya ditei (State Regional Children's Library), Voroshilova, 9, 49060 *Dnipropetrovsk*
T: +380 562 461296
1927; Taisiya Gavrilivna Kravchenko
162 000 vols; 124 curr per   28533

Derzhavna oblasna biblioteka dlya yunatstva im. M. Svetlova (M. Svetlov State Regional Youth Library), Komsomolska, 60, 320060 *Dnipropetrovsk*
T: +380 562 445380
1977; Alla Andriivna Rudnik
Art dept
179 770 vols; 274 curr per; 5 180 music scores; 1 000 av-mat; 8 000 sound-rec
libr loan   28534

Dnipropetrovska oblasna universalna naukova biblioteka (Dnipropetrovsk Regional Scientific Library), vul. Yu. Savchenko, 10, 49006 *Dnipropetrovsk*
T: +380 562 423119; Fax: +380 562 423119; E-mail: library@libr.dp.ua; URL: www.libr.dp.ua
1834; Nadiya M. Titova
Patent stock; Patent, arts, lit, foreign languages lit, regional studies depts
2 700 000 vols; 1 800 curr per; 30 000 govt docs; 18 000 music scores; 1 350 microforms; 900 av-mat; 3 500 sound-rec
libr loan; Dnipropetrovs'k Library Association   28535

Dnipropetrovska raionna TsBS, Tsentralna biblioteka (Dnipropetrovsk Regional Centralized Library System, Main Library), Moskovska, 1, 49000 *Dnipropetrovsk*
T: +380 562 450557
1947
598 000 vols; 345 curr per   28536

Mashinobudivny zavod, Biblioteka profkomu (Machine Building Plant, Trade Union Library), Budivelnikiv, 36, 49059 *Dnipropetrovsk*
T: +380 562 992295
1952; Lyudmila Ivanivna Dubovska
78 000 vols; 75 curr per   28537

Nizhnodniprovsky truboprokatny zavod, Biblioteka-filial no 1 (Nizhnodniprovsk Pipe Rolling Plant, Library Branch no 1), Stoletova, 3, *Dnipropetrovsk*
T: +380 562 207956
1933
82 000 vols; 114 curr per   28538

Pridneprovska raionna elektrostantsia, Biblioteka profkomu (Dnipr Regional Electric Power Station, Trade Union Library), Kosmonavta Volkova, 6, 49112 *Dnipropetrovsk*
T: +380 562 959528
1952
77 000 vols   28539

Truboprokatny zavod, Biblioteka profkomu (Pipe Rolling Plant, Trade Union Library), pr. Svobodi 98a korp 1, 49019 *Dnipropetrovsk*
T: +380 562 596159
1988; Vera Viktorivna Shapoval
53 500 vols; 81 curr per   28540

Zavod im. Petrovskoho, Profspilkova biblioteka (Petrovsky Plant, Trade Union Library), Kalinina, 47, 49009 *Dnipropetrovsk*
T: +380 562 423251
1933
185 000 vols; 52 curr per   28541

Dobropilska raionna TsBS, Tsentralna biblioteka (Dobropillya Regional Centralized Library System, Main Library), Pershotravneva, 121, Pervomaiskaya, 343120 *Dobropillya*
T: +380 6277 24606
1978
51 000 vols; 82 curr per   28542

Derzhavna oblasna biblioteka dlya ditei (State Regional Children's Library), Artema, 84, 83055 *Donetsk*
T: +380 622 938276
1931; Valentina I. Vyazova
213 000 vols; 131 curr per   28543

Derzhavna oblasna biblioteka dlya yunatstva (State Regional Youth Library), Shevchenka, 3, 83055 *Donetsk*
T: +380 662 933549
1979; Valentina D. Nikulina
113 260 vols; 189 curr per   28544

Derzhavna oblasna universalna naukova biblioteka im. N. K. Krupskoi (N.K. Krupskaia Donetsk Regional General Scientific Library), vul. Artema, 84, 83055 *Donetsk*
T: +380 622 3372930, 3353079; Fax: +380 622 3350179; E-mail: postmaster@library.donetsk.ua; URL: www.library.donetsk.ua
1926; Lyudmila A. Novakova
Rare bks from the 17th-19th c (8 893 vols)
1 686 323 vols; 681 293 diss/theses; 18 122 govt docs; 193 many music scores; 291 maps; 153 av-mat; 6 781 sound-rec; 108 digital data carriers; 5 211 pamphlets; 493 236 periodicals
libr loan; Ukrainian Library Association, Library Assembly of Eurasia   28545

Donetska miska TsBS dlya ditei, Tsentralna biblioteka (Donetsk City Centralized Children's Library System, Main Library), Kuibysheva, 221, 83122 *Donetsk*
T: +380 622 935354
1979
76 000 vols; 74 curr per   28546

Donetska miska TsBS dlya doroslikh, Tsentralna biblioteka (Donetsk City Centralized Adult Library System, Main Library), Chelyuskintsiv, 123, 83000 *Donetsk*
T: +380 622 921592
1974
51 000 vols; 110 curr per   28547

Palats kuturi im. I. Franka, Tsentralna biblioteka (Franko Recreation Centre), Kirova, 145, 83037 *Donetsk*
T: +380 622 772656
1927; Lyudmila Andriivna Pisnyak
70 000 vols; 22 curr per   28548

Profkom metalurgiinoho zavodu, Oblasna basova biblioteka profspilok (Metallurgical Plant Trade Union Committee, Regional Trade Union Library), Kuibysheva, 67, 83045 *Donetsk*
T: +380 622 660384
1927
197 000 vols; 256 curr per   28549

Drogobitska miska TsBS, Tsentralna biblioteka (Drogobich City Central Library), Shevchenka, 27, 82120 *Drogobych*
T: +380 3244 21628
Lyudmila Litvin
105 300 vols; 49 curr per; 7 000 govt docs; 32 200 music scores; 18 maps; 179 av-mat; 3 sound-rec   28550

Drogobitska raionna TsBS, Tsentralna biblioteka (Drogobich Regional Centralized Library System, Main Library), 22 Sichnya, 28, 82160 *Drogobych*
T: +380 3244 20692
1970
62 000 vols; 165 curr per   28551

Raionna TsBS, Tsentralna biblioteka (Dubno District Centralized Library System, Main Library), Dubenska, 35600 *Dubno*
T: +380 3656 43152
1949; Ludmila Ivanivna Vitrenko
52 247 vols; 52 curr per; 132 govt docs; 157 music scores; 4 maps; 6 av-mat; 165 sound-rec; 129 press cuttings   28552

Dunaivetska raionna TsBS, Tsentralna biblioteka (Dunaivtsi Regional Centralized Library System, Main Library), Lenina, 9, 281780 *Dunaivtsi*
T: +380 3858 21959
1939
82 000 vols; 250 curr per   28553

Dvorichanska raionna TsBS, Tsentralna biblioteka (Dvorichna Regional Centralized Library System, Main Library), Gvardisky Diviziyi, 35, 312720 *Dvorichna*
T: +380 5750 72162
1921
52 000 vols; 105 curr per   28554

Dzerzhinska raionna biblioteka (Dzerzhinsk Regional Library), 50-richchya Zhovtnya, 3, 261630 *Dzerzhinsk*
T: +380 6247 91487
54 000 vols; 132 curr per   28555

Dzerzhinska raionna TsBS, Tsentralna biblioteka (Dzerzhinsk Regional Centralized Library System, Main Library), 50 rokiv Zhovtnya, 34, 343550 *Dzerzhinsk*
T: +380 6247 32244
1947
82 000 vols; 82 curr per   28556

Shakhta im. Dzerzhinskoho, Biblioteka profkomu (Dzerzhinsky Mine, Trade Union Library), 50-richchya Zhovtnya, 20, 340045 *Dzerzhinsk*
T: +380 6247 32174
1929
77 000 vols; 79 curr per   28557

Dzhankoiska raionna TsBS, Tsentralna biblioteka (Dzhankoi Regional Centralized Library System, Main Library), Krimska, 55, 334010 *Dzhankoi*
T: +380 6564 32038
1945
78 000 vols; 20 curr per
libr loan   28558

Tsentralna biblioteka dlya ditei (Main Children's Library), Tolsoho, 11, 334010 *Dzhankoi*
T: +380 6564 31078
1950
52 000 vols; 49 curr per   28559

Energodarska miska TsBS, Tsentralna biblioteka (Energodar City Centralized Library System, Main Library), pr. Budivelnikiv, 21, 332608 *Energodar*
T: +380 6139 33189
1987
82 000 vols; 154 curr per   28560

Evpatoriska raionna TsBS, Tsentralna biblioteka im. Pushkina (Evpatoriya Regional Centralized Library System, Pushkin Main Library), Buslaevikh, 23, 334320 *Evpatoriya*
T: +380 6569 32462
1916; Galina Proskuzina
98 000 vols; 359 curr per
libr loan   28561

Fastivska raionna TsBS, Tsentralna biblioteka (Fastiv Regional Centralized Library System, Main Library), Tolsoho, 8, 255530 *Fastiv*
T: +380 4465 51511
1947; Katerina Oleksiivna Kruglyak
88 000 vols; 184 curr per   28562

Feodosiska raionna TsBS, Tsentralna biblioteka im. A. Grina (Feodosiya Regional Centralized Library System, Grin Main Library), Kirova, 2, 334800 *Feodosiya*
T: +380 6562 30958
1897; Tamara Rudomazina
101 000 vols; 64 curr per; 140 av-mat; 95 sound-rec   28563

Gadyatska raionna TsBS, Tsentralna biblioteka (Gadyach Regional Centralized Library System, Main Library), 50-richchya Zhovtnya, 19, 315870 *Gadyach*
T: +380 53542 22862
1861
66 000 vols; 162 curr per   28564

Gaisinska raionna TsBS, Tsentralna biblioteka (Gaisin Regional Centralized Library System, Main Library), Pl Miru, 1, 287500 *Gaisin*
T: +380 4394 42284
1905; Galina Grigorivna Gavrish
74 690 vols; 102 curr per   28565

Gaivoronska raionna TsBS, Tsentralna biblioteka (Gaivoron Regional Centralized Library System, Main Library), Kirova, 17, 317600 *Gaivoron*
T: +380 5254 22784
1934; Galina Mikhalivna Droben
77 000 vols; 33 curr per   28566

Galitska raionna TsBS, Tsentralna biblioteka (Galich Regional Centralized Library System, Main Library), Osmomisla, 1, 285100 *Galich*
1946
64 000 vols; 88 curr per   28567

Glibotska raionna TsBS, Tsentralna biblioteka (Gliboka Regional Centralized Library System, Main Library), Lenina, 91, 275500 *Gliboka*
T: +380 3734 22134
1948
70 000 vols; 60 curr per   28568

Globinska raionna TsBS, Tsentralna biblioteka (Globine Regional Centralized Library System, Main Library), Lenina, 178, 315960 *Globine*
T: +380 5365 21291
1946
60 000 vols; 118 curr per   28569

Gorlivska raionna TsBS, Tsentralna biblioteka (Gorlivka Regional Centralized Library System, Main Library), pr. Peremogi, 132a, 338038 *Gorlivka*
T: +380 6242 28170
1947
108 000 vols; 185 curr per   28570

Mashinobudivny zavod, Biblioteka profkomu (Machine Building Plant, Trade Union Library), Katerinicha, 12, 348003 *Gorlivka*
T: +380 6242 74911
1922
69 000 vols; 92 curr per   28571

Gorodenkivska raionna TsBS, Tsentralna biblioteka (L. Martovych Govodenka Central District Library), Boguna, 7, 285800 *Gorodenka*
T: +380 3430 22764
1946; Oleksandra Ivanevna Gudzik
64 000 vols; 88 curr per   28572

Gorodnyanska raionna TsBS, Tsentralna biblioteka (Gorodnya Regional Centralized Library System, Main Library), Lenina, 8a, 251510 *Gorodnya*
T: +380 4645 21651
1947
53 000 vols; 160 curr per   28573

Gorodokska raionna TsBS, Tsentralna biblioteka (Gorodok Regional Centralized Library System, Main Library), Zhovtneva, 53, 281630 *Gorodok*
T: +380 3231 91152
1933
92 000 vols; 178 curr per   28574

Gorokhivska raionna TsBS, Tsentralna biblioteka (Gorokhiv Regional Centralized Library System, Main Library), Shevchenka, 14, 264020 *Gorokhiv*
T: +380 3379 21750; E-mail: gor-libr@ukr.net; URL: gorlibr.hmarka.net
1940; Lyubov Y. Lishchishchina
480 000 vols; 90 curr per   28575

Grebinkivska raionna TsBS, Tsentralna biblioteka (Grebinka Regional Centralized Library System, Main Library), 50-richchya Zhovtnya, 4, 315470 *Grebinka*
T: +380 5359 91604
1950; Galina Fedorivna Chernish
50 000 vols; 120 curr per   28576

Gulyaipilska raionna TsBS, Tsentralna biblioteka (Gulyaipole Regional Centralized Library System, Main Library), pl Petrovskoho, 2, 332830 *Gulyaipole*
T: +380 6145 2777
1895
76 000 vols; 163 curr per   28577

Gusyatinska raionna TsBS, Tsentralna biblioteka (Gusyatin Regional Centralized Library System, Main Library), Nalivaika, 1, 283260 *Gusyatin*
T: +380 3557 22102
1940
62 000 vols; 106 curr per   28578

Simferopolska raionna TsBS, Tsentralna biblioteka (Simferopol Regional Centralized Library System, Main Library), K. Marksa, 77, 334080 *Gvardiyske*, Simferopolski rn
T: +380 652 323033
1927
55 000 vols; 190 curr per   28579

Ichnyanska raionna TsBS, Tsentralna biblioteka (Ichnya Regional Centralized Library System, Main Library), Lenina, 22, 251320 *Ichnya*
T: +380 4633 21779
1946
57 000 vols; 77 curr per   28580

Illinetska raionna TsBS, Tsentralna biblioteka (Illinitsi Regional Centralized Library System, Main Library), Lenina, 19, 287400 *Illinitsi*
T: +380 4377 23688
1965
50 000 vols; 92 curr per   28581

Irpinska raionna TsBS, Tsentralna biblioteka (Irpin Regional Centralized Library System, Main Library), 3a, 255710 *Irpin*
T: +380 4497 56521
1947; Olena Grigorivna Tsiganenko
82 000 vols; 145 curr per   28582

Derzhavna oblasnaya biblioteka dlya yunatstva (State Regional Youth Library), Nezalezhnosti, 12, 76000 *Ivano-Frankivsk*
T: +380 340 22615
1978; Lyudmila Vasilivna Aseeva
93 000 vols; 10 curr per   28583

Ivano-Frankivsk City Centralized Library System, Main Library (Ivano-Frankivsk City Centralized Library System, Main Library), Korolya Danili, 16, 76010 *Ivano-Frankivsk*
T: +380 340 45273
1954; Ganna Medvid
475 853 vols; 141 curr per; 115 govt docs; 4 sound-rec
libr loan   28584

Oblasna universalna naukova biblioteka im. I. Franka (Scientific District Library I. Franko), Chornovola, 22, P.O. Box 400, 76014 *Ivano-Frankivsk*
T: +380 342 750132; Fax: +380 342 32189; E-mail: libifua@gmail.com; URL: www.lib.if.ua
1939; Lyudmyla Babi
477 165 vols; 800 curr per; 1 e-journals; 52 maps; 367 sound-rec; 141 068 periodical items
UBA   28585

Izyaslavska raionna TsBS, Tsentralna biblioteka (Izyaslav Regional Centralized Library System, Main Library), Lenina, 13, 281200 *Izyaslav*
T: +380 3852 52045
1930
84 000 vols; 101 curr per   28586

Izyumska raionna TsBS, Tsentralna biblioteka (Izyum Regional Centralized Library System, Main Library), pr. Lenina, 49, 313850 *Izyum*
T: +380 5743 24067
1977
51 000 vols; 132 curr per   28587

Kakhovska raionna TsBS, Tsentralna biblioteka (Kakhovka Regional Centralized Library System, Main Library), K. Marksa, 79, 326800 *Kakhovka*
T: +380 5536 36123
1901
82 000 vols; 88 curr per   28588

Kalinivkska raionna TsBS, Tsentralna biblioteka (Kalinivka Regional Centralized Library System, Main Library), Dzerzhinskoho, 22, 287060 *Kalinovka*
T: +380 4396 21992
1920
90 000 vols; 202 curr per   28589

Kalushka raionna TsBS, Tsentralna biblioteka (Kalush Regional Centralized Library System, Main Library), Pidvalna, 6, 77300 *Kalush*
T: +380 3472 25140
1944; Svitlana Sokulska
57 800 vols; 21 curr per; 2500 govt docs; 500 music scores
libr loan   28590

Kamyanets-Podilska miska TsBS, Tsentralna Biblioteka (Kamyanets-Podilsk City Centralized Library System, Main Library), Chkalova, 3, 281900 *Kamyanets-Podilsk*
T: +380 3849 24452
163 000 vols   28591

Kamyanets-Podilska raionna TsBS, Tsentralna biblioteka (Kamyanets-Podilsk Regional Centralized Library System, Main Library), Volodarskoho, 2, 281900 *Kamyanets-Podilsk*
T: +380 3849 91044
1948
71 000 vols; 118 curr per   28592

Kamyanka-Buzka raionna TsBS, Tsentralna biblioteka (Kamyanka-Buzka Regional Centralized Library System, Main Library), Khomina, 4, 292100 *Kamyanka-Buzka*
T: +380 4732 51462
1949
67 000 vols; 76 curr per   28593

Kamyanka-Dniprovska raionna TsBS, Tsentralna biblioteka (Kamyanka-Dniprovska Regional Centralized Library System, Main Library), Naberezhna, 78, 332600 *Kamyanka-Dniprovska*
T: +380 6138 25308
1938
50 000 vols; 186 curr per   28594

Karlivska raionna TsBS, Tsentralna biblioteka (Karlivka Regional Centralized Library System, Main Library), Lenina, 95, 315720 *Karlivka*
T: +380 5346 22451
1977; Raisa Satanovska
56 000 vols; 73 curr per   28595

Kegichivska raionna TsBS, Tsentralna biblioteka (Kegichivka Regional Centralized Library System, Main Library), Voloshina, 76, 313310 *Kegichivka*
T: +380 5755 21094
1946
55 000 vols; 130 curr per   28596

Kerchenska raionna TsBS, Tsentralna biblioteka im. Belinskoho (Kerch Regional Centralized Library System, Belinsky Main Library), Dubinina, 9/19, 98300 *Kerch*
T: +380 6561 23493
1856; Lyudmila Y. Popova
1 600 000 vols; 110 curr per   28597

Chervonozavodska raionna TsBS mista Kharkiva, Tsentralna biblioteka (Chervonozavod Regional Centralized Library System of Kharkiv City, Main Library), Kirova, 20, 61140 *Kharkiv*
1927
81 000 vols   28598

Derzhavna oblasna biblioteka dlya yunatstva (State Regional Youth Library), 50-richchya VLKSM, 49/8, 61120 *Kharkiv*
1937; Elizaveta K. Fateeva
112 000 vols; 159 curr per   28599

Dzerzhinska raionna TsBS mista Kharkiva, Tsentralna biblioteka (Dzerzhinsk Regional Centralized Library System of Kharkiv City, Main Library), Danilevskoho, 34, 61058 *Kharkiv*
1933
170 000 vols; 198 curr per   28600

Leninska raionna TsBS mista Kharkova, Tsentralna biblioteka (Lenin Regional Centralized Library System of Kharkiv City, Main Library), Unnativ, 6, 61177 *Kharkiv*
1946
87 000 vols; 243 curr per   28601

Oblasna biblioteka dlya ditei (Regional Children's Library), Artema, 43, 61078 *Kharkiv*
Oleksandr Petrovich Trokhimenko
163 000 vols; 110 curr per   28602

Khartsizka raionna TsBS, Tsentralna biblioteka (Khartsizk Regional Centralized Library System, Main Library), prov Nikolenka, 1, 343700 *Khartsizk*
T: +380 6257 49520
1950
117 000 vols   28603

Bavovnyany kombinat – Biblioteka profkomu (Cotton Industrial Corporation – Trade Union Library), Tekstilnikov, 4, 73002 *Kherson*
1956
108 000 vols; 46 curr per   28604

Gidrometeorologichny tekhnikum, Biblioteka profkomu (Professional School of Hydrometeorology, Trade Union Library), Dzerzhinskoho, 11, 73000 *Kherson*
1944
51 000 vols; 38 curr per   28605

Khersonska miska TsBS, Tsentralna biblioteka (Kherson City Centralized Library System, Main Library), K. Marksa, 97, 73000 *Kherson*
T: +380 552 240211
1968; Olga Ustimenko
106 000 vols; 250 curr per; 10 000 music scores; 7 555 sound-rec   28606

O. Honchar Scientific Regional Public Library, Dnipropetrovska, 2, 73024 *Kherson*
T: +380 552 264029; Fax: +380 552 226448; E-mail: library@tlc.ks.ua; URL: www.lib.kherson.ua
1872; Nadezhda Korotun
Miniature bks, letters of writers, Ukrainian authors edited abroad, coll of rare prerevolutionary bks, bks of the war period (1941-1945)
966 560 vols; 243 022 curr per; 521 mss; 37 940 music scores; 3 100 maps; 206 microforms; 8 372 av-mat; 15 622 sound-rec; 1 119 digital data carriers; 176 213 patents
libr loan; Ukrainian Library Association   28607

Oblasna biblioteka dlya ditei (Regional Children's Library), Chervonostudentska, 21, 73000 *Kherson*
T: +380 552 494171; Fax: +380 552 491200; E-mail: vbpi@library.kherson.ua; URL: www.library.kherson.ua
1924; Anna Bardashevska
Internet Centre
99 724 vols; 233 curr per; 417 music scores; 18 maps; 218 av-mat; 944 sound-rec; 77 digital data carriers; 2 503 other items
libr loan; ULA   28608

Oblasna biblioteka dlya yunatstva (Regional Youth Library), Dimitrova, 14a, 73020 *Kherson*
1980; Nadiya Ivanivna Kovtun
64 000 vols; 305 curr per   28609

Derzhavna oblasna universalna naukova biblioteka (State Regional Universal Library), Teatralna, 28, *Khmelnitski*
T: +380 382 64631
1901; Stanislava L. Karvan
682 000 vols; 760 curr per   28610

Oblasna biblioteka dlya ditei (Regional Children's Library), Svobodi, 51, 29000 *Khmelnitski*
T: +380 3822 67277
1937; Olena Oleksiivna Kirichuk
137 000 vols; 136 curr per   28611

Oblasna biblioteka dlya yunatstva (Regional Youth Library), Soborna, 33, 29013 *Khmelnitski*
T: +380 3822 69451; Fax: +380 3822 68456
1977; Tamara M. Tanchyk
170 000 vols; 169 curr per
libr loan   28612

Khmilnitska raionna TsBS, Tsentralna biblioteka (Khmilnik Regional Centralized Library System, Main Library), Stolyarchuka, 4, 288600 *Khmilnik*
T: +380 4395 2658
1946
61 000 vols; 110 curr per   28613

Khorolska raionna TsBS, Tsentralna biblioteka (Khorol Regional Centralized Library System, Main Library), K. Marksa, 61, 315910 *Khorol*
T: +380 5362 91584
1924
67 000 vols; 83 curr per   28614

Khotinska raionna TsBS, Tsentralna biblioteka (Khotin Regional Centralized Library System, Main Library), Franka, 2, 275360 *Khotin*
1946
65 000 vols; 57 curr per   28615

Kirovogradska miska TsBS, Tsentralna biblioteka (Kirovograd City Centralized Library System, Main Library), pl Druzhbi narodiv, 6, 25050 *Kirovograd*
T: +380 522 248917; Fax: +380 522 236020
1977; Adelina Posypaiko
80 000 vols; 85 curr per
Ukrainian Library Association   28616

Oblasna biblioteka dlya ditei im. Gaidara (Gaidar Regional Children's Library), Shevchenko, 5/22, 25050 *Kirovograd*
T: +380 522 25524
1921; Dr. Svitlana Petrivna Mikhlova
193 000 vols; 120 curr per   28617

Kobelytska raionna TsBS, Tsentralna biblioteka im. Zalki (Kobelyaki Regional Centralized Library System, Zalko Main Library), Lenina, 33, 315250 *Kobelyaki*
T: +380 5343 91041
1971; Valentina S. Kolodichka
56 000 vols; 75 curr per   28618

Kolomiyska raionna TsBS, Tsentralna biblioteka (Kolomiya Regional Centralized Library System, Main Library), Tsentralna, 21a, 285200 *Kolomiya*
T: +380 3433 23514
1980
67 000 vols; 87 curr per
libr loan   28619

Komsomolska raionna TsBS, Tsentralna biblioteka (Komsomolsk Regional Centralized Library System, Main Library), Mira, 22, 39800 *Komsomolsk*
T: +380 5348 21768; Fax: +380 5348 21098
1967; Zoya G. Treshova
201 800 vols; 102 curr per; 6 173 govt docs; 930 music scores; 19 maps; 130 av-mat; 502 sound-rec; 38 200 pamphlets
libr loan   28620

Konotopska raionna TsBS, Tsentralna biblioteka (Konotop Regional Centralized Library System, Main Library), 19, 245130 *Konotop*
T: +380 5447 42149
1944
50 000 vols; 178 curr per   28621

Korostenska raionna TsBS, Tsentralna biblioteka (Korosten Regional Centralized Library System, Main Library), Franka, 5, 260100 *Korosten*
T: +380 4142 43036
1946; Galina O. Taranyuk
516 000 vols; 42 curr per   28622

Korostishivska raionna TsBS, Tsentralna biblioteka (Korostishiv Regional Centralized Library System, Main Library), Chervona pl, 20, 261220 *Korostishiv*
T: +380 4130 35315
1969 000 vols; 65 curr per
libr loan   28623

Metalurgiyny zavod, Biblioteka profkomu (Metallurgical Plant, Trade Union Library), Proletarska, 144, 85001 *Kostyantynivka*
T: +380 6272 34293
1935; Liliya M. Bubnova
70 000 vols; 30 curr per   28624

Tsentralna miska publichna biblioteka im. M. Gorkogo, Misky viddin kultury, bulvar Kosmonavtiv, 11, 85113 *Kostyantynivka*
T: +380 6272 27006; Fax: +380 6272 27006; E-mail: konstlib@ukr.net; URL: konstlib.net
1943; Tetyana Oratovska
186 328 vols; 290 curr per; 8 e-journals; 156 music scores; 26 maps; 57 av-mat; 104 sound-rec; 11 822 brochures, 185 199 periodical items
libr loan; UBA   28625

Kotelevska raionna TsBS, Tsentralna biblioteka (Kotelva Regional Centralized Library System, Main Library), Ostrovskoho, 1, 315160 *Kotelva*
T: +380 5350 91260
1908; Taisiya Ivanivna Ruchitsya
55 000 vols; 106 curr per   28626

Kovelska raionna TsBS, Tsentralna biblioteka (Kovel Regional Centralized Library System, Main Library), Chkalova, 2, 264410 *Kovel*
T: +380 252 23396
Galina M. Bozhik
710 000 vols; 44 curr per
libr loan   28627

Kramatorska miska TsBS, Tsentralna biblioteka (Kramatorsk City Centralized Library System, Main Library), Kantemirivtsev, 16, 343929 *Kramatorsk*
T: +380 6264 71404
1937; Elena Grigorivna Pomoz
850 000 vols; 540 curr per; 5 616 music scores; 327 maps; 119 microforms; 1 766 av-mat; 2 087 sound-rec; 320 prints (18-19th c)
libr loan   28628

Starokramatorsky zavod mashinobudivannya, Biblioteka profkomu (Starokramatorsk Machine Building Plant, Trade Union Library), Shkilna, 9, 343900 *Kramatorsk*
T: +380 6264 21383
1935
86 000 vols; 70 curr per   28629

Vibronichno obednannya Mashinobudivny zavod, Biblioteka profkoma (Machine Building Plant Industrial Corporation, Trade Union Library), Marata, 5, 343901 *Kramatorsk*
T: +380 6264 30588
1934; Lyudmila M. Boiko
317 000 vols; 135 curr per   28630

Kremenetska raionna TsBS, Tsentralna biblioteka (Kremenets Regional Centralized Library System, Main Library), Dubenska, 107, 283280 *Kremenets*
T: +380 3546 22340
69 000 vols; 90 curr per   28631

Kreminska raionna TsBS, Tsentralna biblioteka (Kreminna Regional Centralized Library System, Main Library), pl Krasnaya, 6, 349850 *Kreminna*
T: +380 54 31333
1940; Zoya Bborisovna Oleinik
64 800 vols; 65 curr per   28632

Biblioteka-filial no 1 (Library Branch no 1), pl Artema, Palats kulturi, 324036 *Krivi Rig*
T: +380 564 444314
1945
52 000 vols; 83 curr per   28633

Biblioteka-filial no 14 (Library Branch no 14), Panasa Mirnoho, 28, 324026 *Krivi Rig*
T: +380 564 211180
1960
53 000 vols; 65 curr per   28634

Biblioteka-filial no 16 (Library Branch no 16), Cherkasova, 324079 *Krivi Rig*
1964
64 000 vols; 148 curr per   28635

Kombinat Krivorizhstal – Biblioteka profkomu (Krivorizhstal Industrial Corporation, Trade Union Library), Revolyutsiyna, 21, 324000 *Krivi Rig*
T: +380 564 715807
1935
63 000 vols; 118 curr per   28636

Krivorizhka miska TsBS, Tsentralna biblioteka (Krivi Rig Regional Centralized Library System, Main Library), Rokosovskoho, 36, 324027 *Krivi Rig*
T: +380 564 746640
1934; Nadiya Illivna Gvozdyova
Art dept, patent dept
668 000 vols; 280 curr per; 2 300 music scores; 18 000 av-mat; 1 500 sound-rec; 1,7 mio patents, 254 400 pamphlets
libr loan   28637

Krivorizhka profspilka metalurgiv, Tsentralna biblioteka (Krivi Rig Trade Union of Metallurgy Workers, Main Library), pr. Gagarina, 27a, 324027 *Krivi Rig*
T: +380 564 260519
1946
87 000 vols; 127 curr per   28638

Krizhopilska raionna TsBS, Tsentralna biblioteka (Krizhopil Regional Centralized Library System, Main Library), Lenina, 1, 288400 *Krizhopil*
T: +380 4374 21240
1946
64 000 vols; 145 curr per   28639

Krustska raionna TsBS, Tsentralna biblioteka (Krust Regional Centralized Library System, Main Library), Repina, 3, 295600 *Krust*
T: +380 3142 22046
1945
64 000 vols; 137 curr per   28640

Kuibishevska raionna TsBS, Tsentralna biblioteka (Kuibisheve Regional Centralized Library System, Main Library), Lenina, 57, 332910 *Kuibisheve*
T: +380 6147 91572
1945
57 000 vols; 103 curr per   28641

Kupyanska raionna TsBS, Tsentralna biblioteka (Kupyask Regional Centralized Library System, Main Library), Lenina, 4, 312640 *Kupyansk*
T: +380 5742 53342
1951
66 000 vols; 116 curr per
libr loan   28642

Biblioteka dlya ditei im. A. Gaidara (A. Gaidar Children's Library), Geroiv Stalingrada, 51/b, 252200 *Kyiv*
T: +380 44 4135127
1947; Lidiya A. Volnova
54 000 vols; 56 curr per   28643

Biblioteka dlya ditei im. G. Kotovskoho (G. Kotovsky Children's Library), Mate Zalke, 3, 252211 *Kyiv*
T: +380 44 4186709
1920; Nataliya Andriivna Koval
50 000 vols; 66 curr per   28644

Biblioteka dlya ditei im. K. Chukovskoho (K. Chukovsky Children's Library), pr. Pravdi, 88/b, 254208 *Kyiv*
T: +380 44 4343011
Lyudmila Ivanivna Vladiko
71 000 vols; 73 curr per   28645

Biblioteka dlya ditei im. M. Kotsyubinskoho (M. Kotsyubinsky Children's Library), pr. 40-richchya Zhovtnya, 97a, 03022 *Kyiv*
T: +380 44 2617861
1947; Nataliya Markivna Lakhovska
51 000 vols; 18 curr per   28646

Biblioteka dlya ditei im. N. Zabili (N. Zabil Children's Library), pr. Nauki, 4, 03039 *Kyiv*
T: +380 44 2656277
1947; Olena S. Andreeva
56 000 vols   28647

Biblioteka dlya ditei im. O. Pirogovskoho (O. Pirogovsky Children's Library), Krivonosa, 19, 03037 *Kyiv*
T: +380 44 2765894
1949; Larisa P. Shevchenko
50 000 vols; 57 curr per   28648

Biblioteka dlya ditei im. P. Usenka (P. Usenko Children's Library), Serafimovicha, 7, 252152 *Kyiv*
T: +380 44 2506083
1972; Tetyana Fyodosivna Fedorenko
57 000 vols; 16 curr per   28649

Biblioteka dlya ditei im. P. Verdhigori (P. Verdhigori Children's Library), bul. Lepse, 34, 252126 *Kyiv*
T: +380 44 4837045
1951; Tetyana Ivanivna Kushpi
57 000 vols; 126 curr per   28650

Biblioteka dlya ditei im. Yu. Gagarina (Gagarin Children's Library), Kurchatova, 9/21, 253166 *Kyiv*
T: +380 44 5187363
1971; Lyudmila O. Prishchepa
70 000 vols; 67 curr per   28651

Biblioteka dlya ditei no 115 (Children's Library no 115), Balzaka, 28, 253225 *Kyiv*
T: +380 44 5156170
1977; Nadiya Olifirivna Prikhodko
53 000 vols; 100 curr per   28652

Biblioteka dlya yunatstva no 11 (Youth Library no 11), Marshala Grechka, 20a, 252136 *Kyiv*
T: +380 44 4436281
1963
55 000 vols; 113 curr per   28653

Biblioteka I. Kudri (I. Kudrya Library), Kutuzova, 14, 252133 *Kyiv*
T: +380 44 2952692
1956; Olena K. Tilkovets
57 000 vols; 108 curr per   28654

Biblioteka im. B. Kuchera (B. Kucher Library), 9/21 Kurchatov St, 253166 *Kyiv*
T: +380 44 5198980
1967; Vira Pavlivna Vergun
51 000 vols; 139 curr per   28655

Biblioteka im. Gertsena (Gertsen Library), Krasnova, 12, 252165 *Kyiv*
T: +380 44 4443301
before 1917
55 000 vols; 131 curr per   28656

Biblioteka im. K. Simonova (K. Simonov Library), Kurnatovskoho, 9, 253125 *Kyiv*
T: +380 44 5146930
1977
62 000 vols; 137 curr per   28657

Biblioteka im. Makhtumkuli (Makhtumkuli Library), Zodchikh, 6, 252194 *Kyiv*
T: +380 44 4750437
1969
58 000 vols; 128 curr per   28658

Biblioteka im. N. Gogolya (N. Gogol Library), Vasilkivska, 136, 252150 *Kyiv*
T: +380 44 2691473
1904; Eleonora Sverdlova
N. Gogol coll
64 000 vols; 155 curr per; 700 pamphlets, 18 clippingfiles   28659

Biblioteka im. O. Bloka (O. Blok Library), Gnata Yuri, 5, 252148 *Kyiv*
T: +380 44 4780104
1950
101 000 vols; 211 curr per   28660

Biblioteka im. O. Novikova-Priboya (O. Novikov-Priboy Library), Novgorodska, 5, 03020 *Kyiv*
T: +380 44 2720002
1950
50 000 vols; 152 curr per   28661

Biblioteka im. P. Mirnoho (P. Mirny Library), Kondratyuka, 4/b, 252200 *Kyiv*
T: +380 44 4321471
1970
70 000 vols; 109 curr per   28662

Biblioteka im. V. Yana (V. Yan Library), pr. Miru, 13, 253660 *Kyiv*
T: +380 44 5593361
1953
63 000 vols; 222 curr per   28663

Biblioteka M. Rilskoho (M. Rilsky Library), Kitaivska, 83, 03028 *Kyiv*
T: +380 44 2655230
1953; Olena Kostyantinivna Tikovets
64 000 vols; 109 curr per   28664

Biblioteka mizhnarodnoho tsentru kultury ta mystetstv profspilnok Ukrainy (Ukrainian Trade Union Federation's International Centre of Culture and Art Library), Instytutska, 1, 01001 *Kyiv*
T: +380 44 2787952
1957; Margarita Semyonivna Rudneva
130 000 vols; 210 curr per   28665

Biblioteka no 5, Gagarina, 19/30, 253094 *Kyiv*
T: +380 44 5527035
1955
58 000 vols; 164 curr per   28666

Biblioteka no 13, pr. Vidradni, 14/45, 03061 *Kyiv*
T: +380 44 4837304
1968
60 000 vols; 126 curr per   28667

Biblioteka-Knizhkova svitlitsya dlya ditei, Zodchikh, 30/6, 252194 *Kyiv*
T: +380 44 4746261
1970; Lidiya Grigorivna Kukharska
58 000 vols; 66 curr per   28668

Darnitska raionna TsBS m. Kyiva, Tsentralna biblioteka im. V. Mayakovskoho (Darnitski Region Central Library System of the City of Kiev – Mayakovsky Main Library), Entuziastiv, 11, 253147 *Kyiv*
T: +380 44 5173281
1950; Galina Prokhirivna Mazur
61 000 vols; 149 curr per   28669

Derzhavna biblioteka Ukrainy dlya ditei (State Library of Ukraine for Children), Baumana, 60, 03190 *Kyiv*
T: +380 44 4426587; Fax: +380 44 4426587; E-mail: chl@chl.kiev.ua; URL: www.chl.kiev.ua
1966; Anastasiya S. Kobzarenko
Coll of rare children's bks; Rare children's editions dept, Art dept
443 097 vols; 300 curr per; 4 808 music scores; 386 maps; 6 505 av-mat; 13 791 sound-rec; 166 digital data carriers; 248 pamphletes
libr loan; Ukrainian Library Association, Ukrainian Children's Librarians Association   28670

Derzhavna biblioteka Ukrainy dlya yunatstva (State Library for Youth of Ukraine), pr. Golosijivsky, 122, 03127 *Kyiv*
T: +380 44 2635334; Fax: +380 44 2635334
1975; Georgi Saprykin
330 409 vols; 800 curr per; 4 146 music scores; 18 820 av-mat; 8 294 sound-rec; 76 digital data carriers
libr loan; Ukrainian Library Association   28671

Derzhavna oblasna biblioteka dlya ditei im. D. U. Dobroi (Dobra State Regional Children's Library), Rusanivska nab. 12, 253147 *Kyiv*
T: +380 44 2292856
1909; Mikola Pawlovich Znitshchenko
D.U. Dobra coll
266 500 vols; 25 curr per; 3 939 music scores; 2 133 av-mat; 18 sound-rec
libr loan   28672

Derzhavny instytut kulturi, Biblioteka (State Institute of Culture), Shchorsa, 36, 252195 *Kyiv*
T: +380 44 2696138
1956; Katerina Dmitrivna Golovkina
Art, folklore, music, bibliography, libr studies
450 000 vols; 248 curr per   28673

Dniprovska raionna TsBS m. Kyiva, Tsentralna biblioteka im. P. Tichiny (Dniprovski Region Central Library System of the City of Kiev – P. Tychina Main Library), Lunacharskoho, 24, 253097 *Kyiv*
T: +380 44 5174170
1966
97 000 vols; 138 curr per   28674

Kharkivska raionna TsBS m. Kyiva, Trsentralna biblioteka im. V. Stusa (Kharkiv Region Central Library System of the City of Kiev – V. Stus Main Library), Sumskoho, 4a, 253098 *Kyiv*
T: +380 44 5509001
1953
111 000 vols; 262 curr per
libr loan   28675

Leningradska raionna TsBS m. Kyiva, Tsentralna biblioteka Svichado (Leningrad Region Central Library System of Kiev – Svichado Main Library), bul. P. Polana, 13v, 03170 *Kyiv*
T: +380 44 4751010
1938; Valentina M. Timashova
88 000 vols; 193 curr per
libr loan   28676

Moskovska raionna TsBS m. Kyiva, Tsentralna biblioteka im. M. Nekrasova (Moscow Region Central Library System of the City of Kiev – M. Nekrasov Main Library), Bubnova, 9, 03022 *Kyiv*
T: +380 44 2633300
1929; Galina Pitrivna Stanislavchuk
80 000 vols; 254 curr per   28677

Naukovo-virobnicho obednannya Bolshovik, Tsentralna biblioteka (Bolshevik Scientific and Industrial Corporation, Main Library), Industrialna, 2, 01680 *Kyiv*
T: +380 44 4461724
1934; Viktoriya Ivanivna Vashchuk
155 000 vols; 436 curr per   28678

Obednannya Gospkomunobslugov-uvannya miskoyi Radi narodnykh deputativ,, Biblioteka (Municipal Services Administration), Khreschatik, 36, 03044 *Kyiv*
T: +380 44 2212335
1949; Vera Tikhonivna Polosina
90 000 vols   28679

Obolonski raion TsBS m. Kyiva, Tsentralna biblioteka im. A. Pushkina (Minsk Region Central Library System of the City of Kiev – A. Pushkin Main Library), pr. Obolonski, 16, 04205 *Kyiv*
T: +380 44 4102471; Fax: +380 44 4102473; E-mail: ludvik.07@mail.ru; URL: ocls.kyivlibs.org.ua
1947; Ludvik Natali
177 000 vols
libr loan   28680

Pecherska raionna TsBS m. Kyiva, Tsentralna biblioteka im. M. Saltikova-Shchedrina (Pechersk Region Central Library System of the City of Kiev – Saltikov-Shchedrin Main Library), bul. L. Ukrainki, 7, 03133 *Kyiv*
T: +380 44 2244321
1954; Katerina Fyodorivna Kalinovska
121 000 vols; 288 curr per   28681

Pivdenno-Zakhidna Zaliznitsa – Profkom Darnitskoho vagono-remontnoho zavodu, Biblioteka (Pivdenno-Zakhidna Railway – Railway-carriage Works), Alma-Atinska, 109, 03092 *Kyiv*
T: +380 44 5529900
1954
Fiction
63 000 vols; 21 curr per
libr loan   28682

Podilska raionna TsBS m. Kyiva, Tsentralna biblioteka im. I. Franka (Podil Region Central Library System of the City of Kiev – I. Franko Main Library), Frunze, 117, 04073 *Kyiv*
T: +380 44 4350110
1943; Antonina Mikhailivna Oshkalo
113 000 vols; 273 curr per   28683

Profkom orendnoho pidpriemstva Rostok, Biblioteka (Rostok Ltd, Trade Union Library), Garmatna, 26/2, Palats kulturi, 03067 *Kyiv*
T: +380 44 4410831
1960; Valentina Grigorivna Belavina
50 000 vols   28684

Profkom virobnichnoho obedinennya im. Artema, Biblioteka (Artem Industrial Corporation Trade Union Committee), Tatarska, 1, 03107 *Kyiv*
T: +380 44 2119645
1945; Nataliya S. Goncharuk
112 000 vols; 149 curr per   28685

Publichna biblioteka imeny Lesi Ukrainki (Lesya Ukrainka Public Library), vul. Turgenivska, 83-85, 04050 *Kyiv*
T: +380 44 4860146; Fax: +380 44 4821334; E-mail: reference@lucl.kiev.ua; URL: lucl.lucl.kiev.ua
1943; Lyudmila Ivanivna Kovalchuk
483 000 vols; 251 curr per; 2 244 av-mat
libr loan; Ukrainian Library Association   28686

Radyanska raionna TsBS m. Kyiva, Tsentralna biblioteka im. M. Kostomarova (Radyansk Region Central Library System of the City of Kiev, M. Kostomarov Main Library), Shcherbakova, 51v, 03111 *Kyiv*
T: +380 44 4425317

1957/1992; Sergei Bondar
108 500 vols; 157 curr per
libr loan   28687

Starokievska raionna TsBS m. Kyiva, Tsentralna biblioteka im. E. Pluznika (Starokiev Region Central Library System of the City of Kiev – E. Pluzhnik Main Library), Prorizna, 15, 03003 *Kyiv*
T: +380 44 2280787
1948
69 000 vols; 308 curr per
libr loan   28688

Tsentralna miska biblioteka im. T. Shevchenko dlya ditei Kyeva (Shevchenko Central City Children's Library), pr. Peremogi, 25, 03055 *Kyiv*
T: +380 44 2362119; Fax: +380 44 2362119, 2380879; E-mail: shevlib@kyivlibs.org.ua
1918; Nadiya Ivanivna Bezruchko
90 000 vols; 163 curr per; 300 av-mat; 3 000 sound-rec; Depts of: junior services, 5-9 forms, aesthetic education, competition of children's reading skills, newest information technologies
libr loan; Association of Children Libraries in Ukraine   28689

Tsentralna spetsializovana biblioteka dlia slipykh im. M. Ostrovskoho (Central Library for the Blind), Pechersky uzviz, 5, 01601 *Kyiv*
T: +380 44 2350063; Fax: +380 44 2350063; E-mail: csbs@ukrpost.net
1936; Yuri M. Vishnyakov
Education of the Blind, Braille books
214 510 vols; 72 curr per; 18 100 e-books; 480 music scores; 83 230 sound-rec; 30 digital data carriers; 1 860 offprints
libr loan; National Ukrainian Library Association   28690

Vatutinska raionna TsBS m. Kyiva, Tsentralna biblioteka (Vatutinski Region Central Library System of the City of Kiev, Main Library), Draizetra, 6, 253217 *Kyiv*
T: +380 44 5469044
Tetyana Ivanivna Kurilo
57 000 vols; 402 curr per   28691

Zaliznichna raionna TsBS m. Kyiva, Tsentralna biblioteka im. F. Dostoevskoho (Zaliznichni Region Central Library System of the City of Kiev – Dostoevsky Main Library), Osviti, 14, 03037 *Kyiv*
T: +380 44 2439152
1946; Orisya Ivanivna Chukhri
112 000 vols; 241 curr per   28692

Zavod Leninska Kuzhya, Profspilkova biblioteka (Lenin Kuznya Plant, Trade Union Library), Dlektrikiv, 2, 04176 *Kyiv*
T: +380 44 4169351
1918; Tatyana O. Polyakova
110 000 vols   28693

Zhovtnevska raionna TsBS m. Kyiva, Tsentralna biblioteka im. M. Lermontova (Zhovtnevski Region Central Library System of the City of Kiev – Lermontov Main Library), Geroiv Sevastopolya, 22, 03065 *Kyiv*
T: +380 44 4832045
1923
87 000 vols; 243 curr per
libr loan   28694

Lanovetska raionna TsBS, Tsentralna biblioteka (Lanivtsi Regional Centralized Library System, Main Library), Mishchenka, 8, 283000 *Lanivtsi*
T: +380 3549 21573
1943
68 000 vols; 116 curr per   28695

Lebedinska raionna TsBS, Tsentralna biblioteka (Lebedin Regional Centralized Library System, Main Library), pl Voli, 27, 245440 *Lebedin*
T: +380 5445 64433
1938; Sofya Vasilivna Kostyuk
549 000 vols; 70 curr per   28696

Lipovetskaya tsentralnaya biblioteka (Lipovets Regional Centralized Library System, Main Library), Shevchenka, 2, 22500 *Lipovets*, Vinnitskoi oblasti
T: +380 4358 21745; E-mail: liblipov@rambler.ru
1897; Varvara Stanislavovna Lypko
48 000 vols; 120 curr per
libr loan   28697

Lisichanska raionna TsBS, Tsentralna biblioteka (Lisichansk Regional Centralized Library System, Main Library), Lenina, 94, 349920 *Lisichansk*
T: +380 6451 23231
1980
67 000 vols; 157 curr per
libr loan   28698

Litinska raionna TsBS, Tsentralna biblioteka (Litin Regional Centralized Library System, Main Library), Lenina, 7, 287300 *Litin*
T: +380 4370 21750
1928
51 000 vols   28699

Lokachinska raionna TsBS, Tsentralna biblioteka (Lokachi Regional Centralized Library System, Main Library), Radyanska, 1, 264120 *Lokachi*
T: +380 3374 21211
1944
61 000 vols; 90 curr per   28700

Lokhvitska raionna TsBS, Tsentralna biblioteka (Lokhvitsya Regional Centralized Library System, Main Library), Peremogi, 10, 315810 *Lokhvitsya*
T: +380 5356 31806
1872; Nataliya M. Bilik
85 000 vols; 166 curr per
libr loan   28701

Luganska miska TsBS, Tsentralna biblioteka (Lugansk City Centralized Library System, Main Library), kvartal Volkova, 1m, 91057 *Lugansk*
T: +380 642 477165
1978
54 000 vols; 125 curr per
libr loan   28702

Oblasna biblioteka dlya ditei (Regional Children's Library), Radyanska, 78, 91016 *Lugansk*
T: +380 642 538246
1935; Galina A. Omelchenko
151 000 vols; 88 curr per   28703

Oblasna biblioteka dlya yunatstva (Regional State Youth Library), Tarassa Shevchenka, 4, 91055 *Lugansk*
T: +380 642 551712; E-mail: biblio2001@ukr.net
1980; Svitlana Oleksiivna Aladzhalyan art dept, analitical dept, sector of automation and marketing, sector of ukrainian culture, youth creation and leisure dept
126 424 vols; 238 curr per; 1 379 music scores; 5 000 av-mat; 4 159 sound-rec; 36 digital data carriers   28704

Derzhavna oblasna biblioteka dlya ditei (State Regional Children's Library), pr. Voli, 37, 43010 *Lutsk*
T: +380 3300 42497
1947; Klavdiya S. Blashchuk
159 000 vols; 95 curr per   28705

Derzhavna oblasna biblioteka dlya yunatstva (State Regional Youth Library), pr. Voli, 2, 43000 *Lutsk*
T: +380 3300 23995
1977; Mariya M. Makh
125 000 vols; 191 curr per   28706

Lutska raionna TsBS, Tsentralna biblioteka (Lutsk Regional Centralized Library System, Main Library), L. Ukrainki, 56, 43001 *Lutsk*
T: +380 3300 23275
1956
55 000 vols; 91 curr per   28707

Oblasna mezhspilkova biblioteka (Regional Trade Union Library), pr. Molodi, 4, 43024 *Lutsk*
T: +380 3300 53623
1961
321 000 vols; 208 curr per
libr loan   28708

Derzhavna oblasna biblioteka dlya ditei (State Regional Children's Library), Vinnichenka, 1, 79004 *Lviv*
T: +380 32 2727856; URL: www.lodb.org.ua
1939; V. Kosonogova
147 000 vols; 5 100 curr per; 900 music scores; 9 000 microforms; 11 000 sound-rec; 10 digital data carriers
Ukrainian Library Association, Association of Polish Libraries   28709

Derzhavna oblasna biblioteka dlya yunatstva (State Regional Youth Library), pl Runok, 9, 79008 *Lviv*
T: +380 32 2720684
1976; Yaroslav S. Kotsur
170 000 vols; 134 curr per   28710

Lvivska miska TsBS dlya ditei, Tsentralna biblioteka (Lviv City Centralized Children's Library System, Main Library), Okunevskoho, 3, 79000 *Lviv*
T: +380 32 2523108
1946; Valentina F. Galushko
145 000 vols; 99 curr per
libr loan   28711

Lvivska miska TsBS, Tsentralna biblioteka im. L. Ukrainki (Lviv City Centralized Library System, L. Ukrainki Main Library), Mulyarska, 2a, 79009 *Lviv*
T: +380 32 2720581
1947
238 000 vols; 197 curr per
libr loan   28712

Poltavska raionna TsBS, Tsentralna biblioteka (Poltava Regional Centralized Library System, Main Library), 315024 *Machukhi*, Poltavski rn
1946; Olga Ivanivna Lashko
63 000 vols; 72 curr per
libr loan   28713

Makarivska raionna TsBS, Tsentralna biblioteka (Makariv Regional Centralized Library System, Main Library), Frunze, 35, 255100 *Makariv*
T: +380 4478 51104
Valentina M. Yovenko
76 000 vols; 170 curr per
libr loan   28714

Makiyivska miska TsBS, Tsentralna biblioteka im. Gorkoho (Makiivka City Centralized Library System, Gorky Main Library), Lenina, 73/19, 339000 *Makiyivka*
T: +380 632 63241
1929
156 000 vols; 263 curr per
libr loan   28715

Metalurgiiny kombinat, Biblioteka profkomu (Metallurgical Industrial Corporation, Trade Union Library), Palats kulturi, 339001 *Makiyivka*
T: +380 6232 92220
1923/1980; Oliksandra V. Popova
154 000 vols; 98 curr per; 1 400 sound-rec
libr loan   28716

Maloviskivska raionna TsBS, Tsentralna biblioteka (Mala Viska Regional Centralized Library System, Main Library), Zhovtneva, 80, 317430 *Mala Viska*
T: +380 5258 21237
1933
53 000 vols; 97 curr per
libr loan   28717

Manevytska raionna TsBS, Tsentralna biblioteka (Manevichi Regional Centralized Library System, Main Library), Nezaleshnosti, 18, 264810 *Manevichi*
T: +380 3376 21131
1945
51 000 vols; 38 curr per   28718

TsBS profkomu girnichno-zbagachuvalnoho kombinatu, Tsentralna biblioteka (Mining and Ore Rectification Corporation Trade Union Committee Centralized Library System, Main Library), pl Lenina, 1, 53407 *Marganets*
T: +380 5665 21082
1954; Nina Petrovna Kovalenko
67 000 vols; 25 curr per   28719

Marinska mizhvisomcha TsBS, Tsentralna biblioteka (Marinka Inter Union Centralized Library System, Main Library), pr. Voroshilova, 29, 342500 *Marinka*
T: +380 6212 51762
1947
68 000 vols; 95 curr per
libr loan   28720

Koksokhimichny zavod, Biblioteka profkomu (Coke Chemical Technology Plant, Trade Union Library), Metalugriv, 52, 87500 *Mariupol*
T: +380 6292 332106
1945
66 000 vols; 59 curr per   28721

Mariupolska miska TsBS, Tsentralna biblioteka im. Korolenka (Mariupol City Centralized Library System, Korolenko Main Library), pr. Lenina, 93/b, 87500 *Mariupol*
T: +380 6292 344323
1904
154 000 vols; 180 curr per
libr loan   28722

Metalurgiiny kombinat, Biblioteka profkomu (Metallurgical Industrial Corporation, Trade Union Library), Semashka, 16, 87504 *Mariupol*
T: +380 6292 63706
1918
187 000 vols; 123 curr per
libr loan   28723

Melitopolska miska TsBS, Tsentralna biblioteka (Melitopol City Centralized Library System, Main Library), pr. Peremogi, 1, 72315 *Melitopol*
T: +380 6142 22953
1904
141 000 vols; 276 curr per
libr loan   28724

Mezhivska raionna TsBS, Tsentralna biblioteka (Mezhova Regional Centralized Library System, Main Library), pr. Karla Marksa, 22, 323300 *Mezhova*
T: +380 5670 91983
1935
54 000 vols; 138 curr per
libr loan   28725

Mikhailivska raionna TsBS, Tsentralna biblioteka (Mikhailivka Regional Centralized Library System, Main Library), 50 rokiv Radyanskoyi vladi, 10, 332240 *Mikhailivka*
T: +380 6132 91234
1944
50 000 vols; 24 curr per
libr loan   28726

Slovyanska DRES, Biblioteka profkomu (Slovyansk DRES, Trade Union Library), pl Radyanska, 343245 *Mikolaivka*
T: +380 6262 43285
1954
76 000 vols   28727

Milovska mezhvidomcha TsBS, Tsentralna bibliotka (Milove Regional Centralized Library System, Main Library), vul. Lenina, 6, 92500 *Milove*
T: +380 6465 91176
1913
189 386 vols; 20 curr per; 20 govt docs; 13 sound-rec
libr loan   28728

Melitopolska raionna TsBS, Tsentralna biblioteka (Melitopol Regional Centralized Library System, Main Library), Yuzhnaya, 1, 332386 *Mirne*
56 000 vols; 70 curr per   28729

Mlinivska raionna TsBS, Tsentralna biblioteka (Mlinivsk Regional Centralized Library System, Main Library), Kirova, 17, 265110 *Mlinivsk*
T: +380 3659 22134
1944
60 000 vols; 170 curr per
libr loan   28730

Mogiliv-Podolska raionna TsBS, Tsentralna biblioteka (Mogiliv-Podolski Regional Centralized Library System, Main Library), Lenina, 28, 288700 *Mogiliv-Podolski*
T: +380 4369 23211
1940
94 000 vols
libr loan   28731

Mostiska raionna TsBS, Tsentralna biblioteka (Mostiska Regional Centralized Library System, Main Library), Grushevskoho, 21, 292570 *Mostiska*
T: +380 3234 41195
1977
51 000 vols; 88 curr per
libr loan   28732

Mukachevska miska TsBS, Tsentralna biblioteka (Mukachevo City Centralized Library System, Main Library), Dukhnovicha, 1, 295400 *Mukachevo*
T: +380 3131 22046
1929
85 000 vols; 131 curr per
libr loan   28733

Murovanokurilovetska raionna TsBS, Tsentralna biblioteka (Murovani Kurilivtsi Regional Centralized Library System, Main Library), Lenina, 41, 288650 *Murovany Kurilivtsi*
1937
77 000 vols; 122 curr per
libr loan   28734

Kyivsky derzhavny instytut kulturi, Kulturno-ovsitny fakultet, Biblioteka (Kiev State Institute of Culture, Mikolaiv Branch), Dekabristiv, 17, 54021 *Mykolaiv*
Pedagogy, music, folklore, psychology
114 000 vols; 155 curr per
libr loan   28735

Mikolaivska miska TsBS dlya doroslikh, Tsentralna biblioteka (Mikolaiv City Centralized Library System, Main Library), Potomkinska, 143a, 54055 *Mykolaiv*
T: +380 512 240110
1905
200 000 vols; 160 curr per
libr loan   28736

Mikolaivska raionna TsBS, Tsentralna biblioteka (Mikolaiv Regional Centralized Library System, Main Library), Grushevskoho, 6, 54040 *Mykolaiv*
T: +380 512 31134
1965
55 000 vols; 78 curr per
libr loan   28737

Mykolaivska miska TsBS dlya ditei (Mikolaiv City Centralized Children's Library System, Main Libarary), pr. Lenina, 173, 54003 *Mykolaiv*
T: +380 512 552929; Fax: +380 512 550604; E-mail: library@mksat.net
1922; Nataliya Semilet
507 244 vols; 720 curr per; 8 547 av-mat; 4 497 sound-rec; 62 digital data carriers; 125 058 other items
libr loan   28738

Oblasna biblioteka dlya ditei im. O. Lyagina (Lyagin Regional Library for Children), Spaska, 66, 54001 *Mykolaiv*
T: +380 512 356045
1921; Olena Kuzmivna Karpenko
159 774 vols; 250 curr per; 30 govt docs; 722 music scores; 12 maps; 8 642 av-mat; 4 542 sound-rec; 486 pamphlets   28739

Oblasna biblioteka dlya yunatstva (Regional Youth Library), Velika Morska, 92, 54001 *Mykolaiv*
T: +380 512 355443
1977; Larisa I. Rudnitskaya
arts dept, performance programs dept, youth libr
100 000 vols; 199 curr per; 1 800 music scores; 2 427 sound-rec   28740

Vibronichno obednannya Chernomorsky sudobudivny zavod, Biblioteka profkomu (Chernomorski Ship Building Industrial Corporation, Trade Union Library), Industrialna, 1, 54011 *Mykolaiv*
T: +380 512 372011
1922
190 000 vols; 92 curr per
libr loan   28741

Myrgorodska raionna TsBS, Tsentralna biblioteka (Mirgorod Regional Centralized Library System, Main Library), Nezalezhnosti, 20/17, 37600 *Myrgorod*
T: +380 5355 54773
1978; Katerina V. Odai
61 653 vols; 160 curr per; 159 av-mat; 9 291 pamphlets   28742

Nadvirnyanska raionna TsBS, Tsentralna biblioteka (Nadvirna Regional Centralized Library System, Main Library), Mazepi, 20, 285700 *Nadvirna*
T: +380 3475 23477
1945
61 000 vols; 120 curr per
libr loan   28743

Nedrigailivska raionna TsBS, Tsentralna biblioteka (Nedrigailiv Regional Centralized Library System, Main Library), Shevchenka, 3, 245980 *Nedrigailiv*
T: +380 5455 51034
1900
56 000 vols; 130 curr per
libr loan   28744

Nemirivska raionna TsBS, Tsentralna biblioteka (Nemiriv Regional Centralized Library System, Main Library), Nekrasova, 15, 287200 *Nemiriv*
T: +380 4931 22752
1947; Elena Koral
70 000 vols; 167 curr per
libr loan   28745

Nikopolska raionna TsBS, Tsentralna biblioteka (Nikopol Regional Centralized Library System, Main Library), Sverdlova, 27, 322911 *Nikopol*
T: +380 5662 11330
1906
143 000 vols; 144 curr per   28746

Nizhnogirska raionna TsBS, Tsentralna biblioteka (Nizhnogirsk Regional Centralized Library System, Main Library), Lenina, 7, 334750 *Nizhnogirsk*
T: +380 6550 21007
1936
51 000 vols; 123 curr per
libr loan   28747

Nosivska raionna TsBS, Tsentralna biblioteka (Nosivka Regional Centralized Library System, Main Library), Dzerzhinskoho, 1, 251120 *Nosivka*
T: +380 4642 21372
1945
65 000 vols; 80 curr per
libr loan   28748

Novokakhovska raionna TsBS, Tsentralna biblioteka (Nova Kakhovka Regional Centralized Library System, Main Library), Pionerska, 26, 326840 *Nova Kakhovka*
T: +380 5549 45359
1953
430 250 vols; 348 curr per; 479 music scores; 1 076 av-mat
libr loan   28749

Novoushitska raionna TsBS, Tsentralna biblioteka (Nova Ushitsya Regional Centralized Library System, Main Library), Lenina, 18, 281720 *Nova Ushitsya*
T: +380 3847 21307
1928
59 000 vols; 90 curr per   28750

Novovodolazka raionna TsBS, Tsentralna biblioteka (Nova Vodolaga Regional Centralized Library System, Main Library), Lenina, 1, 313020 *Nova Vodolaga*
T: +380 5740 22409
1921
52 000 vols; 53 curr per
libr loan   28751

Novgorod-Siverska raionna TsBS, Tsentralna biblioteka (Novgorod-Siverski Regional Centralized Library System, Main Library), Maistrenka, 4, 251780 *Novgorod Siverski*
T: +380 4658 22836
1979
65 000 vols; 82 curr per
libr loan   28752

Starobeshivska DRES, Biblioteka profkomu (Starobeshivsk DRES, Trade Union Library), Shkilna, Palats kulturi, 342422 *Novi Svit*
1954
68 000 vols; 100 curr per
libr loan   28753

Novoarkhangelska raionna TsBS, Tsentralna biblioteka (Novoarkhangelsk Regional Centralized Library System, Main Library), Pushkina, 42, 317530 *Novoarkhangelsk*
T: +380 5255 21091
1980
68 000 vols; 178 curr per
libr loan   28754

Novograd-Volinska raionna TsBS, Tsentralna biblioteka im. L. Ukrainki (Novograd-Volinski Regional Centralized Library System, L. Ukrainki Library), Radyanska, 16, 260500 *Novograd-Volinski*
T: +380 4141 52546
1946
91 000 vols; 90 curr per
libr loan   28755

Novomoskovska raionna TsBS, Tsentralna biblioteka (Central library of Novomoskovsk), Radyanska, 37, 322010 *Novomoskovsk*
T: +380 5612 24278
1947; Vira Micolayevna Kovnir
Children libr, dept of art, 4 branch libs
130 290 vols; 68 curr per
libr loan   28756

Novopskovska raionna TsBS, Tsentralna biblioteka (Novopskov Regional Centralized Library System, Main Library), Lenina, 33, 349670 *Novopskov*
T: +380 243 91140
1929; Raisa G. Kadatskaya
351 000 vols; 36 curr per
libr loan   28757

Novoselitska raionna TsBS, Tsentralna biblioteka (Novoselitsya Regional Centralized Library System, Main Library), Lenina, 55, 275210 *Novoselitsya*
T: +380 3733 2584
1948
66 000 vols; 55 curr per
libr loan   28758

Novoukrainska raionna TsBS, Tsentralna biblioteka (Novoukrainka Regional Centralized Library System, Main Library), Lenina, 67, 317320 *Novoukrainka*
T: +380 5251 22353
1930
69 000 vols; 190 curr per
libr loan   28759

Obukhivska raionna TsBS, Tsentralna biblioteka (Obukhiv Regional Centralized Library System, Main Library), Lenina, 14, 255400 *Obukhiv*
T: +380 4472 53762
1929; Marina M. Shkuratyana
54 000 vols; 120 curr per
libr loan   28760

Ochakivska Tsentralna Biblioteka (Ochakiv Central Library), Lenina 20, 57500 *Ochakiv*
T: +380 5154 24137, 22121; E-mail: ochakovbiblioteka1@yandex.ua; URL: libr.ochakiv.info
1930; Galina Skripnichenko
Coll of rare books issued between 1874 and 1947 (129 vols)
57 600 vols; 61 curr per; 11 500 e-journals; 5 govt docs; 20 maps; 17 microforms; 250 av-mat; 430 sound-rec; 10 digital data carriers; 35 sets of thematic clippings on the history of the Ochakiv Area
libr loan   28761

Derzhavna oblasna biblioteka dlya ditei (State Regional Children's Library), Radyanskoyi Armiyi, 64, 65045 *Odesa*
T: +380 482 256882
1976
160 000 vols
libr loan   28762

Derzhavna oblasna bilbioteka dlya yunatstva (State Regional Youth Library), Koroleva, 46, 65114 *Odesa*
T: +380 482 476627
1976; Lyudmila F. Ponomarenko
110 000 vols; 230 curr per; 1 100 govt docs; 2 500 music scores; 210 av-mat; 1 200 sound-rec   28763

Odeska miska TsBS, Tsentralna biblioteka im. I. Franka (Odesa City Centralized Library System, Franko Main Library), prov Knizhkovi, 1a, 65007 *Odesa*
T: +380 482 258400
1891; L.P. Denisenko
236 000 vols; 3 110 curr per
libr loan   28764

Odeska zaliznitsa, Biblioteka (Odessa Railways), Mizikevicha, 44, 65005 *Odesa*
T: +380 482 274037
1936
Railway transport, fiction, popular science, Ukrainian lit and hist
319 000 vols; 106 curr per
libr loan   28765

Oleksandrivska raionna TsBS, Tsentralna biblioteka (Oleksandrivka Regional Centralized Library System, Main Library), Lenina, 58, 317800 *Oleksandrivka*
1934
70 000 vols; 103 curr per
libr loan   28766

Oleksandriyska miska TsBS, Tsentralna biblioteka (Oleksandriya City Centralized Library System, Main Library), Chervonoarmiyska, 41, 317800 *Oleksandrivka*
1892
101 000 vols; 240 curr per   28767

TsBS profkomu girnichno-zbagachuvalnoho kombinatu, Tsentralna biblioteka (Mining and Ore Rectification Corporation Trade Union Committee Centralized Library System, Main Library), Kalinina, 39, 322960 *Ordzhonikidze*
T: +380 5667 3032
1952
99 000 vols; 149 curr per
libr loan   28768

Orekhivska raionna TsBS, Tsentralna biblioteka (Orekhiv Regional Centralized Library System, Main Library), Leninskykh Kursantiv, 41, 332700 *Orekhiv*
T: +380 6141 33374
1936
60 000 vols; 182 curr per
libr loan   28769

Ostrozka raionna TsBS, Tsentralna biblioteka (Rivne Regional Centralized Library System, Ostrog District Main Library), Manuilskoho, 31, 265620 *Ostrog*
T: +380 3654 23456
1939; Taisia Ivanivna Karpova
54 000 vols; 50 curr per; 320 govt docs; 180 music scores; 20 maps
libr loan   28770

Ovrutska raionna TsBS, Tsentralna biblioteka im. A.S. Malishka (Ovruch Regional Centralized Library System, A.S. Malishka Main Library), Lenina, 6, 260000 *Ovruch*
T: +380 4183 32292
1920
119 000 vols; 145 curr per
libr loan   28771

Pavlogradska miska TsBS, Tsentralna biblioteka (Pavlograd City Centralized Library System, Main Library), K. Marksa, 67, 323000 *Pavlograd*
T: +380 5672 60456
1861
146 000 vols; 140 curr per   28772

Pereyaslav-Khmelnitska raionna TsBS, Tsentralna biblioteka (Pereyaslav-Khmelnitski Regional Centralized Library System, Main Library), Skovorody, 60, 256110 *Pereyaslav-Khmelnitski*
T: +380 3822 52786
1907; Svitlana V. Kharina
53 000 vols; 100 curr per
libr loan   28773

Pervomaiska miska TsBS, Tsentralna biblioteka (Pervomaisk City Centralized Library System, Main Library), pl Lenina, 1, 329810 *Pervomaisk*, Mikolaivska oblast
T: +380 5161 42789
1893
53 000 vols; 95 curr per
libr loan   28774

Pervomaiska raionna TsBS, Tsentralna biblioteka (Pervomaiske Regional Centralized Library System, Main Library), Zhovtneva, 69, 334920 *Pervomaiske*
1923
53 000 vols; 95 curr per
libr loan   28775

Pervomaiska raionna TsBS, Tsentralna biblioteka (Pervomaiski Regional Centralized Library System, Main Library), Lenina, 20, 313450 *Pervomaiski*
T: +380 5748 23036
1948
68 000 vols; 219 curr per
libr loan   28776

Petropavlivska raionna TsBS, Tsentralna biblioteka (Petropavlivka Regional Centralized Library System, Main Library), Radyanska, 70, 323400 *Petropavlivka*
T: +380 5671 21136
1936
60 000 vols; 117 curr per
libr loan   28777

Petrivska raionna TsBS, Tsentralna biblioteka (Petrove Regional Centralized Library System, Main Library), Ilicha, 45, 317220 *Petrove*
T: +380 5237 91552
1939; Oleksandra Ro. Zhovtyuk
85 000 vols; 87 curr per
libr loan   28778

Pidvolochiska raionna TsBS, Tsentralna biblioteka (Pidvolochisk Regional Centralized Library System, Main Library), Shevchenka, 7, 283000 *Pidvolochisk*
T: +380 3543 21038
1948
60 000 vols; 58 curr per   28779

Pishchanska raionna TsBS, Tsentralna biblioteka (Pishchanka Regional Centralized Library System, Main Library), Mayakovskaya, 1, 288370 *Pishchanka*
T: +380 4368 21240
1978
59 000 vols; 125 curr per   28780

Pogrebishchenska raionna TsBS, Tsentralna biblioteka (Pogrebishche Regional Centralized Library System, Main Library), Kooperativna, 32, 287600 *Pogrebishche*
T: +380 4378 22450
1927
82 000 vols; 157 curr per   28781

Pologivska raionna TsBS, Tsentralna biblioteka (Pologi Regional Centralized Library System, Main Library), prov Stantsionni, 3, 332800 *Pologi*
T: +380 6165 21081
1948
66 000 vols; 59 curr per   28782

Polonska raionna TsBS, Tsentralna biblioteka (Polonne Regional Centralized Library System, Main Library), Lenina, 95, 281000 *Polonne*
T: +380 3843 22558
1946
76 000 vols; 135 curr per   28783

Oblasna biblioteka dlya ditei im. Panasa Mirnoho (Panas Myrny Poltava Regional Public Library for Children), Gagarina, 5, 36011 *Poltava*
T: +380 532 272041
1938; Leonid G. Chobitko
155 000 vols; 100 curr per; 3 900 music scores; 1 500 av-mat; 2 400 sound-rec
libr loan   28784

Oblasna biblioteka dlya yunatstva im. Gonchara (Regional Youth Library), Engelsa, 25a, 36038 *Poltava*
T: +380 532 6764-31/-02; Fax: +380 532 6764-03/-31; E-mail: pobugonchara@ukr.net
1976; Nina M. Dudenko
148 589 vols; 400 curr per; 1 146 music scores; 92 microforms; 303 av-mat; 1 002 sound-rec; 3 398 records
libr loan   28785

Poltavska miska TsBS, Tsentralna biblioteka (Poltava City Centralized Library System, Main Library), Almazna, 6/11, 36021 *Poltava*
T: +380 532 231560
1978; Tamila A. Duzenko
106 000 vols; 140 curr per
libr loan   28786

Popasnyanska raionna TsBS, Tsentralna biblioteka (Popasna Regional Centralized Library System, Main Library), Pervomaiska, 5a, 349980 *Popasna*
T: +380 6474 20408
1947; Lidiya A. Sorokolat
52 000 vols; 35 curr per
libr loan   28787

Popilnyanska raionna TsBS, Tsentralna biblioteka (Popilnya Regional Centralized Library System, Main Library), Lenina, 12, 261710 *Popilnya*
T: +380 235 5760
1948; Svetlana O. Adamchuk
605 000 vols; 39 curr per
libr loan   28788

Priazovska raionna TsBS, Tsentralna biblioteka (Priazovsk Regional Centralized Library System, Main Library), Lenina, 30, 332340 *Priazovsk*
T: +380 6133 32137
1945
53 000 vols
libr loan   28789

Prilutska raionna TsBS, Tsentralna biblioteka (Priluki Regional Centralized Library System, Main Library), Radyanska, 35, 251350 *Priluki*
T: +380 4637 31384
1944
56 000 vols; 50 curr per
libr loan   28790

Primorska raionna TsBS, Tsentralna biblioteka (Primorsk Regional Centralized Library System, Main Library), Kirova, 95, 72100 *Primorsk*
T: +380 6137 72307
1977; Ms O. V. Parkhomenko
60 000 vols
libr loan   28791

Radekhivska raionna TsBS, Tsentralna biblioteka (Radekhiv Regional Centralized Library System, Main Library), Lvivska, 14, 292130 *Radekhiv*
T: +380 255 21338
1945; Galina M. Kovalchuk
51 600 vols; 128 curr per
libr loan   28792

Radomishlska raionna TsBS, Tsentralna biblioteka (Radomishl Regional Centralized Library System, Main Library), prov Shkilni, 10, 263140 *Radomishl*
T: +380 4132 44181
1944
94 000 vols; 154 curr per
libr loan   28793

Rakhivska raionna TsBS, Tsentralna biblioteka (Rakhiv Regional Centralized Library System, Main Library), Gorkoho, 1, 295800 *Rakhiv*
T: +380 3132 22650
1976
71 000 vols; 109 curr per
libr loan   28794

Reshetilivska raionna TsBS, Tsentralna biblioteka (Reshetilivka Regional Centralized Library System, Main Library), Lenina, 19, 315690 *Reshetilivka*
T: +380 5363 92802
1978; Tetyana A. Gorbach
55 000 vols
                                       28795

Derzhavna oblasna biblioteka (State Regional Library), Pl Korolenka, 6, 33000 *Rivne*
T: +380 362 21174
1940; Valentina Yaroshchuk
570 000 vols; 520 curr per; 1 200 diss/theses; 7 790 music scores; 222 av-mat; 27 043 pamphlets
libr loan   28796

Derzhavny instytut kulturi, Biblioteka (State Institute of Culture), Tolstoho, 3, 33028 *Rivne*
T: +380 362 64772
1979
Pedagogy, music, theatre, art, psychology, folklore
105 000 vols; 110 curr per
libr loan   28797

Oblasna biblioteka dlya ditei (Regional Children's Library), Petlyuri, 27, 33028 *Rivne*; P.O. Box 121, 33028 Rivne
T: +380 362 221404; Fax: +380 362 221404; E-mail: childlib@cis.rv.ua
1940; Larysa Lisova
148 477 vols; 127 curr per; 7 568 music scores; 98 maps; 7 079 av-mat   28798

Regional Youth Library, Kyivska, 18, 33002 *Rivne*
T: +380 3622 230298; Fax: +380 3622 230298; E-mail: molody@ukr.net; URL: www.molody.ukrwest.net
1976; Maria Tarasivna Verbets
171 501 vols; 120 curr per; 50 e-books; 12 govt docs; 3 040 music scores; 178 av-mat; 2 971 sound-rec
libr loan; UBA   28799

Rivnenska miska TsBS, Tsentralna biblioteka (Rivne City Centralized Library System, Main Library), 266027 *Rivne*
T: +380 362 33538
1947; Anatoli P. Fishchuk
165 000 vols; 94 curr per
libr loan   28800

Rogatinska raionna TsBS, Tsentralna biblioteka (Rogatin Regional Centralized Library System, Main Library), Franka, 8, 285140 *Rogatin*
T: +380 3435 21993
1977
63 000 vols
libr loan   28801

Rokitnyanska raionna TsBS, Tsentralna biblioteka (Rokitne Regional Centralized Library System, Main Library), Pershotravneva, 7, 256710 *Rokitne*
T: +380 4462 52153
1947; Alla P. Krichuk
55 000 vols; 87 curr per
libr loan   28802

Rovenkivska raionna TsBS, Tsentralna biblioteka (Rovenki Regional Centralized Library System, Main Library), Lenina, 112, 349230 *Rovenki*
T: +380 6433 20070
1882; Valentina Hlopkina
88 000 vols; 35 curr per   28803

Rozdolnenska raionna TsBS, Tsentralna biblioteka (Rozdolne Regional Centralized Library System, Main Library), Lenina, 5a, 334370 *Rozdolne*
T: +380 253 91351
1948; Lidiya I. Stavisyuk
50 000 vols; 57 curr per
libr loan   28804

Rozhnyativska raionna TsBS, Tsentralna biblioteka (Rozhnyativ Regional Centralized Library System, Main Library), pl Yednosti, 6, 285500 *Rozhnyativ*
T: +380 3474 20735
1944; Mariya Vasylivna Rybchak
53 000 vols; 54 curr per; 36 govt docs; 25 maps; 601 av-mat; 52 sound-rec; 35 files of newspaper clippings
libr loan; Ukrainian Library Association
28805

Rubizhanska raionna TsBS, Tsentralna biblioteka (Rubizhne Regional Centralized Library System, Main Library), Lenina, 51, 349870 *Rubizhne*
T: +380 6453 57733
1975
91 000 vols; 178 curr per
libr loan   28806

Sakhnovshchinska raionna TsBS, Tsentralna biblioteka (Sakhnovshchina Regional Centralized Library System, Main Library), Pivdennovokzalna, 4, 313210 *Sakhnovshchina*
T: +380 5762 21169
1927
54 000 vols; 97 curr per
libr loan   28807

Sambirska raionna TsBS, Tsentralna biblioteka (Sambir Regional Centralized Library System, Main Library), Bachinskikh, 5, 292610 *Sambir*
T: +380 3236 53249
1947
58 000 vols; 130 curr per
libr loan   28808

Sarnenska raionna TsBS, Tsentralna biblioteka (Sarni Regional Centralized Library System, Main Library), Radyanska, 34, 265450 *Sarni*
T: +380 3655 23221
1981
80 000 vols; 90 curr per
libr loan   28809

Sevastopolskaya tsentralnaya biblioteka im. L.M. Tolstogo (Tolstoi Sevastopol Central City Library), Lenina, 51, 99011 *Sevastopol*
T: +380 692 5444733; Fax: +380 692 5444733; E-mail: sevtolib@library.iuf.net; URL: www.sevtolib.iuf.net
1901; Tamara O. Essin
Megaproject "Pushkin Library", coll from "Alexander Nevky's fund", Association of the Moscow Publishers, coll "Universal Scientific Russian Book", book coll on St. Petersburg, coll from the Forum of young Ukraine leaders, The Millenium Library, Crimean War (from British Council), coll of Sabre-Svitlo
278 173 vols; 820 curr per; 9 e-journals; 39 e-books; 11 596 govt docs; 300 music scores; 87 maps; 23 av-mat; 2 095 sound-rec
libr loan; UBA, BAE, ABU, LIBCOM, EBNIT (International Association of Users and Digital Libraries and New Information Technologies)   28810

Shakhtarska miska TsBS, Tsentralna biblioteka (Shakhtarsk City Centralized Library System, Main Library), Zhuravlivka, 2, 343720 *Shakhtarsk*
T: +380 6255 46043
1953
50 000 vols; 90 curr per
libr loan   28811

Shchorska raionna TsBS, Tsentralna biblioteka (Shchors Regional Centralized Library System, Main Library), Striletskoyi diviziyi, 8, 251530 *Shchors*
T: +380 4654 21491
1936
52 000 vols; 197 curr per
libr loan   28812

Shepetivska raionna TsBS, Tsentralna biblioteka (Shepetivka Regional Centralized Library System, Main Library), K. Marksa, 54, 281040 *Shepetivka*
T: +380 3840 55118
1919
90 000 vols; 135 curr per   28813

Shirokivska raionna TsBS, Tsentralna biblioteka (Shiroke Regional Centralized Library System, Main Library), Lenina, 113, 322830 *Shiroke*
T: +380 5657 21552
1920
68 000 vols; 69 curr per
libr loan   28814

Shostlinska miska TsBS, Tsentralna biblioteka (Shostka City Centralized Library System, Main Library), Miru, 11, 245110 *Shostka*
T: +380 5449 62173
1945; Olga I. Evtushenko
65 000 vols; 128 curr per   28815

Krymska respublikanska biblioteka dlya yunatstva (Crimean Republican Youth Library), Kechkemetska, 94a, 95000 *Simferopol*
T: +380 652 228695
1977; Lyubov O. Gerasimova
98 000 vols; 180 curr per
libr loan   28816

Simferopolska raionna TsBS dlya doroslikh, Tsentralna biblioteka im. Trenova (Simferopol Regional Centralized Library System, Trenev Main Library), Lyuksemburg, 2/1, 95000 *Simferopol*
T: +380 652 274235
1948; Maya M. Grekhova
628 000 vols; 192 curr per
libr loan   28817

Sinelnikovska raionna TsBS, Tsentralna biblioteka (Sinelnikove Regional Centralized Library System, Main Library), Vikonkomivska, 30, 323110 *Sinelnikove*
T: +380 5615 22013
1945
77 000 vols; 85 curr per
libr loan   28818

Skolivska raionna TsBS, Tsentralna biblioteka (Skole Regional Centralized Library System, Main Library), Galitskoho, 54, 293600 *Skole*
T: +380 3251 4549
1949
52 000 vols; 54 curr per
libr loan   28819

Slavutska raionna TsBS, Tsentralna biblioteka (Slavuta Regional Centralized Library System, Main Library), Dzerzhinskoho, 40, 281070 *Slavuta*
T: +380 3842 22790
1935
90 000 vols; 103 curr per   28820

Joint-stock company 'Armaturno-izolyatorny zavod-Energia', Trade Union Library (Steel Reinforcement and Insulation Plant, Trade Union Library), Marksa, 45, 84112 *Slovyansk*
T: +380 6262 33374
1946; Maya Romanovna Kryviakina
80 000 vols; 45 curr per   28821

Slovyanska miska TsBS, Tsentralna biblioteka (Slovyansk City Centralized Library System, Main Library), pl Zhovtnevoyi revolyutsiyi, 2a, 84222 *Slovyansk*
T: +380 6262 37155
1918
85 000 vols; 140 curr per
libr loan   28822

Smilyanska raionna TsBS, Tsentralna biblioteka (Smila Regional Centralized Library System, Main Library), Lenina, 26, 258410 *Smila*
T: +380 4733 22322
1924
130 000 vols; 185 curr per
libr loan   28823

Snyatinska raionna TsBS, Tsentralna biblioteka (Snyatin Regional Centralized Library System, Main Library), Voevodi Kosnyatina, 41, 285300 *Snyatin*
T: +380 3476 21152
1946
67 000 vols; 87 curr per
libr loan   28824

Sokiryanska raionna TsBS, Tsentralna biblioteka (Sokiryani Regional Centralized Library System, Main Library), Lenina, 31, 275000 *Sokiryani*
T: +380 3739 21151
1947
57 000 vols; 52 curr per
libr loan   28825

Sosnitska raionna TsBS, Tsentralna biblioteka (Sosnitsya Regional Centralized Library System, Main Library), Khmelnitskoho, 28, 251630 *Sosnitsya*
T: +380 4655 21247
1872
54 000 vols; 114 curr per
libr loan   28826

Sovetska raionna TsBS, Tsentralna biblioteka (Sovetske Regional Centralized Library System, Main Library), Pervomaiska, 33, 334730 *Sovetske*
T: +380 251 91190
1920; Svitlana G. Traynina
362 000 vols; 200 curr per; 1 040 av-mat
28827

Stakhanovska raionna TsBS, Tsentralna biblioteka (Stakhanov Regional Centralized Library System, Main Library), Lenina, 17, 349700 *Stakhanov*
T: +380 6444 32773
1933; Valentina M. Kochetkova
338 000 vols; 87 curr per
libr loan   28828

Starovizhivska raionna TsBS, Tsentralna biblioteka (Stara Vizhivka Regional Centralized Library System, Main Library), Pl Lenina, 4, 264720 *Stara Vizhivka*
T: +380 3346 21156
1944
53 000 vols; 130 curr per
libr loan   28829

Starobilska raionna TsBS, Tsentralna biblioteka (Starobilsk Regional Centralized Library System, Main Library), Komunariv, 36, 349600 *Starobilsk*
T: +380 6461 21792
1896; Vera P. Dyuyzina
50 000 vols; 77 curr per
libr loan   28830

Starokostyantynivska raionna TsBS, Tsentralna biblioteka (Starokostyantyniv Regional Centralized Library System, Main Library), Lenina, 14, 281100 *Starokostyantyniv*
T: +380 3854 22108
1975
88 000 vols; 295 curr per   28831

Stavishchenska raionna TsBS, Tsentralna biblioteka (Stavishche Regional Centralized Library System, Main Library), Radyanska, 38, 256500 *Stavishche*
T: +380 4464 52357
1933
66 000 vols; 111 curr per
libr loan   28832

Striyska miska TsBS, Tsentralna biblioteka (Striy City Centralized Library System, Main Library), Shevchenka, 63, 293500 *Striy*
T: +380 45 52025
1940; Lyudmila T. Yaremko
148 000 vols; 110 curr per
libr loan   28833

Glukhivska raionna TsBS, Profspilkova biblioteka (Glukhiv Regional Centralized Library System, Trade Union Library), Gorkoho, 23/1, 40044 *Sumy*
T: +380 542 76048
1936
195 000 vols; 130 curr per
libr loan   28834

Oblasna biblioteka dlya ditei (Regional Children's Library), Petropavlivska, 51, 40030 *Sumy*
T: +380 542 20082
1921; Maya Petrivna Ferberova
183 000 vols; 91 curr per
libr loan   28835

Sumska miska TsBS, Tsentralna biblioteka (Sumy City Library System, Main Library), Kooperativna, 6, 40000 *Sumy*
T: +380 542 222128; Fax: +380 542 212610
1950; Ludmila Stadnichenko
182 000 vols; 214 curr per; 1 318 music scores; 1 263 av-mat; 1 242 sound-rec
libr loan; ULA   28836

Svitlovodska miska TsBS, Tsentralna biblioteka (Svitlovodsk Regional Centralized Library System, Main Library), Primorska, 54, 27504 *Svitlovodsk*
T: +380 5236 25221
1957; Elizaveta M. Yalovaya
86 532 vols; 63 curr per; 67 govt docs; 895 music scores; 42 maps
libr loan   28837

Teofipolska raionna TsBS, Tsentralna biblioteka (Teofipol Regional Centralized Library System, Main Library), Lenina, 14, 281550 *Teofipol*
T: +380 3844 91536
1923
57 000 vols; 180 curr per   28838

Teplitska raionna TsBS, Tsentralna biblioteka (Teplik Regional Centralized Library System, Main Library), Lenina, 39, 287590 *Teplik*
T: +380 4367 21586
1949
53 000 vols; 113 curr per
libr loan   28839

Terebovlyanska raionna TsBS, Tsentralna biblioteka (Terebovlya Regional Centralized Library System, Main Library), Vasilka, 105, 283400 *Terebovlya*
T: +380 3551 22244
1945; Ludmila Evgenivna Kruchinina
58 000 vols; 114 curr per
libr loan 28840

Derzhavna oblasna biblioteka (Regional Library), bul. Shevchenka, 15, 46000 *Ternopil*
T: +380 352 225264; Fax: +380 352 225264
1939; Vasil I. Vitenko
557 296 vols; 441 curr per; 34 music scores; 1 502 sound-rec
libr loan 28841

Mezhspilkova biblioteka (Trade Union Library), Medova, 2, 46000 *Ternopil*
T: +380 3522 52562
1978
Fiction, Ukrainian lit and hist, regional studies, trade unions, law
227 000 vols; 55 curr per
libr loan 28842

Oblasna biblioteka dlya ditei (Regional Children's Library), Kopernika, 17a, 46001 *Ternopil*
T: +380 3522 228718, 224665
1940; Nadiya Noritska
157 000 vols; 120 curr per 28843

Oblasna biblioteka dlya yunatstva (Regional Youth Library), Nechaya, 35, 46003 *Ternopil*
T: +380 3522 59705
1980; Miroslava V. Khmurich
94 000 vols 28844

Ternopilska miska TsBS, Tsentralna biblioteka (Ternopil City Centralized Library System, Main Library), Franka, 21, 46001 *Ternopil*
T: +380 3522 26459
1946
70 000 vols; 100 curr per 28845

Tetiyivska raionna TsBS, Tsentralna biblioteka (Tetiiv Regional Centralized Library System, Main Library), Lenina, 54, 256560 *Tetiyiv*
T: +380 4460 51303
1976; Nataliya I. Troyanska
73 000 vols; 108 curr per
libr loan 28846

Tivrivska raionna TsBS, Tsentralna biblioteka (Tivriv Regional Centralized Library System, Main Library), Lenina, 49, 287140 *Tivriv*
T: +380 4373 21783
1987
55 000 vols; 113 curr per
libr loan 28847

Tlumachanska raionna TsBS, Tsentralna biblioteka (Tlumach Regional Centralized Library System, Main Library), Vinnichenka, 2, 285070 *Tlumach*
T: +380 3479 22132
1946
50 000 vols
libr loan 28848

Tokmatska raionna TsBS, Tsentralna biblioteka (Tokmak Regional Centralized Library System, Main Library), Revolyutsiyna, 10, 332530 *Tokmak*
T: +380 6178 21172
1945
75 000 vols
libr loan 28849

Tomashpilska raionna TsBS, Tsentralna biblioteka (Tomashpil Regional Centralized Library System, Main Library), Leninskoho komsomolu, 1, 288220 *Tomashpil*
T: +380 4392 21565
1946; Valentina Nagirnyak
68 500 vols; 27 curr per 28850

Torezka miska TsBS, Tsentralna biblioteka (Torez City Centralized Library System, Main Library), 4 mikroraion, 20, 347340 *Torez*
T: +380 6254 32021
1932
80 000 vols; 154 curr per
libr loan 28851

Trostyanetska raionna TsBS, Tsentralna biblioteka (Trostyanets Regional Centralized Library System, Main Library), 20 partsizdu, 1, 288330 *Trostyanets*
1977
77 000 vols; 208 curr per
libr loan 28852

Truskavetska raionna TsBS, Tsentralna biblioteka (Truskavets Regional Centralized Library System, Main Library), Drogobitska, 12, 293780 *Truskavets*
T: +380 3247 51755
1952
83 000 vols; 183 curr per
libr loan 28853

Tsarichanska raionna TsBS, Tsentralna biblioteka (Tsarichanka Regional Centralized Library System, Main Library), Teatralna, 16, 322340 *Tsarichanka*
T: +380 5610 91664
67 000 vols; 115 curr per 28854

Tulchinska raionna TsBS, Tsentralna biblioteka (Tulchin Regional Centralized Library System, Main Library), 50-richchya SPSP, 8, 288300 *Tulchin*
T: +380 4360 21447
1917
76 000 vols; 250 curr per 28855

Tyachivska raionna TsBS, Tsentralna biblioteka (Tyachiv Regional Centralized Library System, Main Library), Lenina, 18, 295710 *Tyachiv*
T: +380 3132 1080
1976
61 000 vols
libr loan 28856

Umanska raionna TsBS, Tsentralna biblioteka (Uman Regional Centralized Library System, Main Library), Zhovtneva, 1, 258900 *Uman*
T: +380 4744 58526
1897
107 000 vols; 141 curr per
libr loan 28857

Derzhavna oblasna biblioteka dlya ditei im. Vakarova (Vakarov State Regional Children's Library), Zhovtneva, 20, 88018 *Uzhgorod*
T: +380 3100 3515
1945; Svitlana G. Timoshko
112 000 vols; 89 curr per 28858

Derzhavna oblasna biblioteka dlya yunatstva im. Vaidi (Vaida State Regional Youth Library), Kapitulna, 25, 88000 *Uzhgorod*
T: +380 3100 35057
1981; Vira I. Korlyar-Gubina
51 000 vols; 218 curr per 28859

Uzhgorodska miska TsBS, Tsentralna biblioteka (Uzhgorod City Centralized Library System, Main Library), Teatralna, 20, 88000 *Uzhgorod*
T: +380 3100 37245
1945
64 000 vols; 119 curr per
libr loan 28860

Vasilkivska raionna TsBS, Tsentralna biblioteka (Vasilkiv Regional Centralized Library System, Main Library), Zhovtneva, 2, 255130 *Vasilkiv*
T: +380 4471 51632
1931; Galina A. Prokopenko
79 000 vols; 111 curr per
libr loan 28861

Velikobagachanska raionna TsBS, Tsentralna biblioteka (Velika Bagachka Regional Centralized Library System, Main Library), Lenina, 1, 315640 *Velika Bagachka*
T: +380 5345 91205
1948; Grigori D. Kolinko
55 000 vols; 123 curr per 28862

Velikooleksandrivska raionna TsBS, Tsentralna biblioteka (Velika Oleksandrivka Regional Centralized Library System, Main Library), Lenina, 145, 326130 *Velika Oleksandrivka*
T: +380 5574 21098
1898
62 000 vols; 151 curr per
libr loan 28863

Verkhnodniprovska raionna TsBS, Tsentralna biblioteka (Verkhnodniprovsk Regional Centralized Library System, Main Library), pr. Lenina, 31, 322570 *Verkhnodniprovsk*
T: +380 5618 31359
1875
146 000 vols; 50 curr per 28864

Veselivska raionna TsBS, Tsentralna biblioteka (Vesele Regional Centralized Library System, Main Library), Lenina, 140, 332560 *Vesele*
T: +380 6136 21159
1946
60 000 vols
libr loan 28865

Vilnyanska raionna TsBS, Tsentralna biblioteka (Vilnyansk Regional Centralized Library System, Main Library), Radyanska, 1, 332000 *Vilnyansk*
T: +380 6143 22775
1940; Vera I. Erginyan
51 000 vols; 215 curr per
libr loan 28866

Vilshanska raionna TsBS, Tsentralna biblioteka (Vilshanska Regional Centralized Library System, Main Library), Komsomolska, 5, 317680 *Vilshanka*
57 000 vols; 86 curr per
libr loan 28867

Derzhavna oblasna biblioteka dlya ditei im. Franka (Franko State Regional Children's Library), Lenina, 24, 21100 *Vinnytsya*
T: +380 432 322549
1928; Sergei V. Gorbulinski
146 000 vols; 98 curr per
libr loan 28868

Derzhavna oblasna biblioteka dlya yunatstva (State Regional Youth Library), Tolstoho, 22, 21100 *Vinnytsya*
T: +380 432 358357
1980; Vasil M. Dubovi
100 000 vols; 10 curr per
libr loan 28869

Vinnytska miska TsBS, Tsentralna biblioteka im. Bevza (Vinnitsa City Centralized Library System, Bevz Main Library), Khmelnitske shosse, 21100 *Vinnytsya*
T: +380 432 325020
1972
269 000 vols; 283 curr per 28870

Kyivo-Svyatoshinska raionna TsBS, Tsentralna biblioteka (Kievo-Svyatoshinsk Regional Centralized Library System, Main Library), Lesya Ukrainka St, 33, 255500 *Vishneve*
T: +380 4498 52151
1950; Irina O. Pivovarova
690 000 vols; 200 curr per; 50 music scores; 125 maps; 261 av-mat; 3 digital data carriers; 170 video-recordings
libr loan 28871

Vizhnitska raionna TsBS, Tsentralna biblioteka (Vizhnitsya Regional Centralized Library System, Main Library), Lenina, 36, 275640 *Vizhnitsya*
T: +380 3730 91706
1947
62 000 vols; 57 curr per
libr loan 28872

Volnovaska raionna TsBS, Tsentralna biblioteka (Volnovakha Regional Centralized Library System, Main Library), 1 Travnya, 5, 342300 *Volnovakha*
T: +380 6214 41452
1944
109 000 vols; 158 curr per
libr loan 28873

Volochiska raionna TsBS, Tsentralna biblioteka (Volochisk Regional Centralized Library System, Main Library), Kirova, 80, 281370 *Volochisk*
T: +380 3845 21532
1944
99 000 vols; 90 curr per 28874

Volodarska raionna TsBS, Tsentralna biblioteka (Volodarka Regional Centralized Library System, Main Library), Voroshilova, 20, 256530 *Volodarka*
T: +380 4469 50274
1947; Tamara V. Kuksa
70 000 vols; 72 curr per
libr loan 28875

Volodarsko-Volinska raionna TsBS, Tsentralna biblioteka (Volodarsk-Volinsky Regional Centralized Library System, Main Library), Marksa, 21, 261010 *Volodarsk-Volinski*
T: +380 6216 21334
1924
50 000 vols; 62 curr per
libr loan 28876

Volodimir-Volinska raionna TsBS, Tsentralna biblioteka (Volodimir-Volinsk Regional Centralized Library System, Main Library), Kopernika, 3, 264940 *Volodimir-Volinsk*
T: +380 3342 24614
1940
55 000 vols; 92 curr per 28877

Vovchanska raionna TsBS, Tsentralna biblioteka (Vovchansk Regional Centralized Library System, Main Library), Gagarina, 2, 329600 *Vovchansk*
T: +380 5741 22286
1892
60 000 vols
libr loan 28878

Voznesenska mezhivdomcha TsBS, Tsentralna biblioteka (Voznesensk Centralized Library System, Main Library), Shevchenka, 8, 329600 *Voznesensk*
T: +380 5134 44908
1935
85 000 vols; 189 curr per
libr loan 28879

Yakimivska raionna TsBS, Tsentralna biblioteka (Yakimivka Regional Centralized Library System, Main Library), Teatralna, 9, 332260 *Yakimivka*
T: +380 6131 91989
1920
63 000 vols; 91 curr per
libr loan 28880

Yaltinska raionna TsBS, Tsentralna biblioteka im. Chekhova (Yalta Regional Centralized Library System, Chekhov Main Library), Morka, 8, 98200 *Yalta*
T: +380 654 322200
1897; T.V. Fyodorova
141 000 vols; 266 curr per; 1 000 govt docs
IFLA 28881

Yampilska raionna TsBS, Tsentralna biblioteka (Yampil Regional Centralized Library System, Main Library), Lenina, 108, 288240 *Yampil*
1944
76 000 vols; 154 curr per
libr loan 28882

Zaleshchitska raionna TsBS, Tsentralna biblioteka (Zaleshchiki Regional Centralized Library System, Main Library), Ukrainska, 66, 283540 *Zaleshchiki*
T: +380 3554 22007
1947
71 000 vols; 55 curr per 28883

Alyuminevi kombinat, Biblioteka profkomu (Aluminium Plant, Trade Union Library), Metalurgiv, 1a, 69032 *Zaporizhzhya*
T: +380 612 25236
1952
236 000 vols 28884

Derzhavna oblasna biblioteka dlya ditei (State Regional Children's Library), Sportivna, 14, 69006 *Zaporizhzhya*
T: +380 612 26094
1936; Raisa N. Kirpa
122 000 vols; 167 curr per; 8 455 music scores; 2 984 av-mat; 2 481 sound-rec; 43 pamphlets
libr loan; IFLA 28885

Derzhavna oblasna biblioteka dlya yunatstva (State Regional Youth Library), pr. Lenina, 210, 69037 *Zaporizhzhya*
T: +380 612 325951
1976; Tetyana V. Petrenko
103 000 vols; 127 curr per 28886

Korporatsiya Zaporizhtransformator – Biblioteka profkomu (Transformator Industrial Corporation, Trade Union Library), Krimlivska, 14, 69015 *Zaporizhzhya*
T: +380 612 593046
1949
55 000 vols; 70 curr per 28887

Vibronichno obednannya Motor-Sich, Biblioteka profkomu (Motor Industrial Corporation, Trade Union Library), Krasna, 23, 69000 *Zaporizhzhya*
T: +380 612 614294
1947
194 000 vols   28888

Zaporizhstal metalurgiyny kombinat, Biblioteka profkomu (Zaporizhstal Metallurgical Corporation, Trade Union Library), 40-richchya Radyanskoyi Ukrainy, 17, 69006 *Zaporizhzhya*
T: +380 612 23772
1947
104 000 vols; 101 curr per   28889

Zaporizka miska TsBS, Tsentralna biblioteka (Zaporizhzhya City Centralized Library System, Main Library), Gogolya, 66, 69002 *Zaporizhzhya*
T: +380 617 642148; Fax: +380 617 641618
1952; Larisa Dolyna
148 000 vols; 190 curr per   28890

Zaporizka raionna TsBS, Tsentralna biblioteka (Zaporizhzhya Regional Centralized Library System, Main Library), Istomina, 4, 69086 *Zaporizhzhya*
T: +380 612 578064
1945
50 000 vols
libr loan   28891

Zastavnivska raionna TsBS, Tsentralna biblioteka (Zastavna Regional Centralized Library System, Main Library), Nezalezhnosti, 88, 275330 *Zastava*
T: +380 3737 22207
1947
61 000 vols; 57 curr per
libr loan   28892

Zdolbunivska raionna TsBS, Tsentralna biblioteka (Zdolbuniv Regional Centralized Library System, Main Library), 40-richchya Peremogi, 39, 265640 *Zdolbuniv*
T: +380 3652 4712
1945
64 000 vols; 67 curr per
libr loan   28893

Zhidachivska raionna TsBS, Tsentralna biblioteka (Zhidachiv Regional Centralized Library System, Main Library), Shevchenka, 3, 293300 *Zhidachiv*
T: +380 3239 32296
1948
58 000 vols; 65 curr per
libr loan   28894

Zhmerinska raionna TsBS, Tsentralna biblioteka (Zhmerinka Regional Centralized Library System, Main Library), Lenina, 3, 288020 *Zhmerinka*
T: +380 4396 21505
1935
78 000 vols; 136 curr per
libr loan   28895

Zhovtovodska miska TsBS, Tsentralna biblioteka (Zhovti Vodi City Centralized Library System, Main Library), 50 Rokiv Komsomolu, 23, 52200 *Zhovti Vody*
T: +380 5652 332633
1971
58 000 vols; 151 curr per
libr loan   28896

Derzhavna oblasna biblioteka dlya ditei (State Regional Children's Library), Pushkinska, 36, 10000 *Zhytomyr*
T: +380 412 378441
1938; Olga G. Lepilkina
135 000 vols   28897

Derzhavna oblasna biblioteka dlya yunatstva (State Regional Youth Library), Kotovskoho, 9, 10000 *Zhytomyr*
T: +380 412 374021
1979; Olga G. Lepilkina
80 000 vols; 124 curr per   28898

Derzhavna oblasna universalna naukova biblioteka (State Regional Universal Scientific Library), Novi bul, 4, 10001 *Zhytomyr*
T: +380 412 378432
1866; Valeriya F. Trokhimenko
611 000 vols; 749 curr per   28899

Zhytomyrska miska TsBS, Tsentralna biblioteka im. Zemlyaka (Zhitomir City Centralized Library System, Zemlyak Central Library), Partizanska, 3, 10001 *Zhytomyr*
T: +380 412 362463, 360500
1947; Igor Saknenko-Bereznyak
402 497 vols; 63 curr per; 12 074 govt docs; 39 109 music scores; 2 690 sound-rec
libr loan   28900

Znamyanska raionna TsBS, Tsentralna biblioteka (Znamyanka Regional Centralized Library System, Main Library), Mayakovskoho, 40, 317060 *Znamyanka*
T: +380 5233 51413
1932
62 000 vols; 120 curr per
libr loan   28901

Zolochivska raionna TsBS, Tsentralna biblioteka (Zolochiv Regional Centralized Library System, Main Library), Pravdi, 2, 312220 *Zolochiv*, Lvivska obl.
T: +380 5764 91254
1916
62 000 vols; 20 curr per
libr loan   28902

Zolochivska raionna TsBS, Tsentralna biblioteka (Zolochiv Regional Centralized Library System, Main Library), Ternopilska, 1, 293100 *Zolochiv*, Lvivska obl.
T: +380 5764 33543
1946
74 000 vols; 130 curr per
libr loan   28903

United Arab Emirates

National Libraries

National Library, Zayed 1st street, P.O. Box 2380, *Abu Dhabi*
T: +971 2 215300; Fax: +971 2 6217472
1981; Jumaa Alqubaisi
Sheikh Faleh Libr; UN Publs (Deposit Ctr)
800 000 vols
libr loan; IFLA   28904

University Libraries, College Libraries

Higher Colleges of Technology, Learning Resource Services, 32nd St, Khalidiya, 00000 *Abu Dhabi*; P.O. Box 32092, 00000 Abu Dhabi
T: +971 2 6922563; Fax: +971 2 6812258; E-mail: luinda.lilley@hct.ac.ae; URL: www.hct.ac.ae/library/aspx/index.aspx
1988; Paul Mace
Engineering, business, English language training, health sciences, computer technology, communications technology
340 294 vols; 514 curr per; 29 e-journals; 15 000 av-mat; 1 800 sound-rec; 4 000 digital data carriers; online database subscriptions
IUG   28905

United Arab Emirates University, University Libraries, P.O. Box 15551, Al-Ain, *Abu Dhabi*
T: +971 3 5043200; Fax: +971 3 666975; E-mail: h.alulama@uaeu.ac.ae; URL: www.uaeu.ac.ae
1977; Dr. Hesam Mohammed Sultan Al-Ulama
Environment coll; 6 branch libs
300 000 vols; 2 000 curr per; 900 mss; 500 diss/theses; 50 000 microforms; 100 av-mat; 30 digital data carriers; ERIC Documents
libr loan   28906

Zayad University, Library and Learning Resources, P.O. Box 19282, *Dubai*
T: +971 4 4021143; Fax: +971 4 4021007; E-mail: pat.wand@zu.ac.ae; URL: www.zu.ac.ae/library
1998; Patricia A. Wand
Coll of Arabic and English Newspapers, Arabic and English Special Collection Materials, Emirates Collection; Learning Enhancement Center, Curriculum Resource Center

95 000 vols; 517 curr per; 12 760 e-journals; 2 679 av-mat; 325 digital data carriers
libr loan; IFLA, SLA – Arabian Gulf Chapter   28907

Special Libraries Maintained by Other Institutions

Presidential Court, Centre for Documentation and Research, Airport Rd, Near International Exhibition Center, P.O. Box 5884, *Abu Dhabi*
T: +971 2 4449500; Fax: +971 2 4444306; E-mail: DG@cdr.gov.ae; URL: www.cdr.gov.ae
1968
Arabian Gulf Coll
12 000 vols; 80 curr per; 30 diss/theses; 500 maps; 6 045 microforms; 320 sound-rec
AFLI   28908

British Council, Library, Tariq bin Zaid St (near Rashid Hospital), P.O. Box 1636, *Dubai*
T: +971 4 3370109; Fax: +971 4 3370703; E-mail: information@au.britishcouncil.org; URL: www.britishcouncil.org/uae
1970
11 000 vols   28909

Public Libraries

Dubai Municipality Public Libraries, P.O. Box 67, *Dubai*
T: +971 4 262788; Fax: +971 4 266226
1962; Mohammed Jassim Al-Oraidi
Coll of Islamic art bks
140 000 vols; 537 curr per   28910

United Kingdom

National Libraries

National Library of Wales / Llyfrgell Genedlaethol Cymru, *Aberystwyth* SY23 3BU
T: +44 1970 632800; Fax: +44 1970 615709; URL: www.llgc.org.uk
1907; Andrew Green
Sir John Williams, Univ College Coll, St Asaph Cathedral Libr Coll, Castell Gwyn, Borg, Peniarth MSS, Church in Wales Records and arch of nonconformist denominations, Records of Great Sessions in Wales and special coll on Euclid, folklore and egyptology
4 500 000 vols; 20 000 curr per; 32 e-journals; 75 000 mss; 220 incunabula; 37 000 diss/theses; 60 000 govt docs; 265 000 music scores; 1 300 000 maps; 40 000 microforms; 25 200 av-mat; 22 750 sound-rec; 45 digital data carriers; 4 mio archives, 750 000 photogr images, 300 wallcharts
SCONUL, IFLA, ASLIB, CILIP, CURL, RLG, ICA, BRA, IASA   28911

National Library of Scotland, George IV Bridge, *Edinburgh* EH1 1EW
T: +44 131 6233700; Fax: +44 131 6233701; E-mail: enquiries@nls.uk; URL: www.nls.uk
1925; Martyn Wade
Early Scottish bks Coll, Scottish Gaelic Coll, theology, mountaineering, rare bks and mss; Scottish Screen Archive, John Murray Archive
8 000 000 vols; 25 000 curr per; 120 000 mss; 600 incunabula; 1 000 000 govt docs; 250 000 music scores; 1 600 000 maps; 3 600 sound-rec; 32 000 films
libr loan; SCONUL, IFLA, ASLIB, LIBER   28912

The British Library, The Board Headquarters, 96 Euston Rd, *London* NW1 2DB
T: +44 870 444 1500 or 20 74127332; Fax: +44 20 74127340; E-mail: Customer-Services@bl.uk; URL: www.bl.uk
1753/1973; Lynne Brindley
13 310 000 vols; 99 500 curr per; 5 500 e-journals; 314 100 mss; 155 400 diss/theses; 1 592 000 music scores; 4 311 000 maps; 10 153 000 microforms; 1 558 000 sound-rec; 110 000 charters, 19 800 seals, 55 833 000 patents
IFLA, LIBER, CURL, SCONUL, RLG, IAML, CDNL, CENL, CERL   28913

The British Library – Asia, Pacific and Africa Collections, 96 Euston Rd, *London* NWI 2DB
T: +44 20 7412-7873; Fax: +44 20 74127641; E-mail: apac-prints@bl.uk; URL: www.bl.uk
1753; Graham Shaw
Archives: East India Company (1600-1858), Board of Control or Board of Commissioners for the Affairs of India (1784-1858), India Office (1858-1947), Burma Office (1937-1948)
900 000 vols; 8 e-journals; 40 000 maps; 65 000 oriental mss, 120 000 vols of oriental periodicals and newspapers
libr loan   28914

The British Library – Humanities Reading Rooms, 96 Euston Rd, *London* NW1 2DB
T: +44 20 7412 7000; Fax: +44 20 7412 7609; URL: www.bl.uk
John Tuck
9 000 000 vols   28915

The British Library – Librarianship & Information Sciences Service (LIS), 96 Euston Rd, *London* NW1 2DB
T: +44 20 74127676; Fax: +44 20 74127691; E-mail: lis@bl.uk; URL: www.bl.uk
1902; Michael Stringer
Libr Science, Information Science
1 500 vols; 100 curr per; 350 e-journals; 3 500 diss/theses; 100 av-mat
ALA, ASLIB, IAML, CILIP   28916

The British Library – Manuscripts Collections, 96 Euston Rd, *London* NW1 2DB
T: +44 20 74127513; Fax: +44 20 74127745; URL: www.bl.uk
1753; Brett Dolman
Illuminated mss incl 7th c Lindisfame Gospels, 14th c Luttrell Psalter; early gospels incl 4th c Codex Sinaiticus; early mss of Beowulf, Chaucer etc; 'Magna Carta'; other hist, literary, antiquarian, theatrical colls
100 000 vols; 5 000 maps; 100 000 bound mms, 100 000 charters and rolls, 3 100 papyri, 4 000 ostraca, 18 000 detached seals and casts of seals, incl. 300 000 mss incl 5 000 maps
SCONUL   28917

The British Library – Map Collections, 96 Euston Rd, *London* NW1 2DB
T: +44 20 7412 7702; Fax: +44 20 7412 7780; URL: www.bl.uk/collections/maps.html
1867; Peter Barber
Ordnance Survey mapping since 1800, Antiquarian maps and atlases to 15th cent. incl King George 111 Topographical and Maritime Collections; Reference material: Atlases, Gazetteers, History of Cartography
50 000 vols; 200 curr per; 20 000 mss; 40 diss/theses; 6 250 000 maps; 16 000 microforms; 500 av-mat; Collection of pamphlets and offprints of articles on the History of Cartography, 3 digitized map collections   28918

The British Library – Music Collections, 96 Euston Rd, *London* NWI 2DB
T: +44 20 7412 7772; Fax: +44 20 7412 7751; URL: www.bl.uk/collections/music/music.html
1753; R.J. Chesser
Royal Music Libr, Paul Hirsch Coll, Stefan Zweig Coll, Royal Philharmonic Society Arch
150 000 vols; 100 000 mss; 1 600 000 music scores
IAML, MLA   28919

The British Library – Newspaper Library, Colindale Ave, ***London*** NW9 5HE
T: +44 20 74127353; Fax: +44 20 74127379; E-mail: newspaper@bl.uk; URL: www.bl.uk/collections/newspapers.html
1932; Ed King
Microfilm set of the Burney coll (1603-1818), Chatham House Press Cuttings coll, Microfiche colls: Belgian Underground Press in WWII 1939-1945, Dutch Underground Press 1940-1945, Early American Newspapers, Northern Ireland Political Lit, 1966-89
660 000 vols; 3 000 curr per; 370 000 microforms; 45 digital data carriers
IFLA 28920

The British Library – Science Technology & Business Collections Development Policy, 96 Euston Rd, ***London*** NW1 2DB
T: +44 20 7412-7494/-7496; Fax: +44 20 7412-7217/-7495; URL: www.bl.uk/collections/science.html
1855 28921

The British Library – Sound Archive Information Service, 96 Euston Rd, ***London*** NW1 2DB
T: +44 20 74127676; Fax: +44 20 74127441; URL: www.bl.uk/soundarchive
1955; Mr A.C. Jewitt
Oral hist coll, International Music (world/ethnic music), Drama and literature recordings, Wildlife Sounds, Music recordings
8 850 vols; 120 curr per; 4 050 microforms; 23 000 av-mat; 1 313 000 sound-rec; 4 digital data carriers
IASA 28922

University Libraries, College Libraries

Aberdeen College, Gallowgate Centre, Library, Gallowgate, ***Aberdeen*** AB25 1BN
T: +44 1224 642138; Fax: +44 1224 612001; E-mail: k.hilton@abcol.ac.uk; URL: www.abcol.ac.uk/facilities/library/
1959; Kelly hilton
Financial studies, communication, media studies, art & design, building, social studies, service industries, office technology & computing, engineering
30 000 vols; 219 curr per; 2 591 av-mat; 808 sound-rec; 709 digital data carriers
libr loan; CILIP, Grampian Information 28923

Robert Gordon University, Library Service, Garthdee Rd, ***Aberdeen*** AB10 7QE
T: +44 1224 263450; Fax: +44 1224 263460; E-mail: library@rgu.ac.uk; URL: www.rgu.ac.uk
1938; E.M. Dunphy
Early modern works on architecture, landscape gardening, interior decoration
247 826 vols; 1 523 curr per; 2 755 e-journals; 12 digital data carriers
libr loan; CILIP, ASLIB, ALA, ARLIS/UK & Ireland, BBSLG, SLIC, UKSG, BIC, NAG, SHINE 28924

University of Aberdeen, Library and Historic Collections, Queen Mother Library, Meston Walk, ***Aberdeen*** AB24 3UE
T: +44 1224 273600; Fax: +44 1224 273956; E-mail: library@abdn.ac.uk; URL: www.abdn.ac.uk/diss/library
1495; Ms Chris Banks
Biesenthal Coll (Judaica), MacBean Coll (Jacobite), Gregory Coll (hist of science and medicine), George Washington Wilson Photogr arch; Law, Medicine, Education, European Documentation Centre, Official Publications
1 200 000 vols; 8 700 curr per; 15 000 e-journals; 3 000 mss; 250 incunabula; 12 000 diss/theses; 100 000 govt docs; 18 000 music scores; 20 000 maps; 9 000 microforms; 2 500 av-mat; 2 000 sound-rec; 3 000 digital data carriers; 40 000 George Washington Wilson photogr glass plates
libr loan; SCONUL, LIBER, CILIP, SLA, SLIC, SCURL, RLG 28925

– Medical Library, Foresterhill, Aberdeen AB25 2ZD
T: +44 1224 552488; Fax: +44 1224 685157; E-mail: medlib@abdn.ac.uk; URL: www.abdn.ac.uk/diss/library/geninfo/sites/medical
1971; Keith Nockels
Coll of pre-1900 medical bks
55 000 vols; 700 curr per; 3 000 e-journals; 1 950 diss/theses; 1 000 microforms; 500 av-mat; 100 digital data carriers
libr loan 28926

Aberystwyth University / Prifysgol Aberystwyth, Hugh Owen Library, Penglais Campus, ***Aberystwyth*** SY23 3DZ
T: +44 1970 622400; Fax: +44 1970 622404; E-mail: is@aber.ac.uk; URL: www.aber.ac.uk/en/is/index.html
1872; Ms Rebecca Davies
Humanities, Social sciences, Law, Biological and earth sciences; Rudler Coll (Geology), Duff Coll (Classics), Scott Blair Coll (deposited by the British Society of Rheology), Diplomatic Documents Coll, European Documentation Centre, Horton Collection, (children's material), Appleton Collection (Victorian printing)
800 000 vols; 6 000 curr per; 10 000 e-journals; 10 000 mss; 9 000 diss/theses; 60 000 govt docs; 3 000 music scores; 56 000 microforms; 1 620 av-mat; 60 digital data carriers; 18 000 pamphlets
libr loan; CILIP, SCONUL, IFLA 28927

– Institute of Biological, Environmental and Rural Sciences (IBERS), Stapledon Library, Gogerddan, Aberystwyth SY23 3EB
T: +44 1970 823000; Fax: +44 1970 828357; E-mail: gogstaff@aber.ac.uk; URL: www.aber.ac.uk/en/ibers
1919; Steve Smith
Ecology, grassland agronomy, cereal & grass breeding, plant and cell genetics and biotechnology, sustainable agriculture, Animal Nutrition; Staff publications
40 000 vols; 800 curr per; 2 000 e-journals; 350 diss/theses; 100 govt docs; 800 maps; 10 microforms; 20 digital data carriers; Report Coll
ASLIB, WLA, CILIP, UKSG, NAG, AHIS 28928

– Old College Library, Old College, King St, Aberystwyth SY23 2AX
T: +44 1970 622130; Fax: +44 1970 622122; E-mail: epd@aber.ac.uk; URL: www.inf.aber.ac.uk/locations/libraries.asp#oc
E.P. Davies
Spec Subj: Celtic studies, education, drama, Welsh language and lit; Colls: Thomas Jones Coll, AU Music Coll
40 000 vols 28929

– Thomas Parry Library, Llanbadarn Campus, Aberystwyth SY23 3AS
T: +44 1970 622412; Fax: +44 01970 621868; E-mail: parrylib@aber.ac.uk; URL: www.inf.aber.ac.uk/
1964; Dr. M.J. Davies
Libr sciences, Clearinghouse for IFLA council papers, Welsh colls, Children's lit colls, agriculture, countryside management
150 000 vols; 1 300 curr per; 2 000 e-journals; 2 000 mss; 1 000 diss/theses; 1 000 maps; 20 000 microforms; 20 000 av-mat; 2 000 sound-rec; 50 digital data carriers; 6 000 libr building plans, 6 000 press clippings
libr loan; ALA, ASLIB, FID, IFLA, CILIP, CLA, LAA, COMLA 28930

Accrington and Rossendale College, Sandy Lane centre, Library, Sandy Lane, ***Accrington*** BB5 2AW
T: +44 1254 354041; Fax: +44 1254 354151; URL: www.accross.ac.uk
60 000 vols
libr loan 28931

University College of St Martin, St Martin's Services Ltd, Ambleside Library, Rydal Rd, ***Ambleside*** LA22 9BB
T: +44 15394 30274; Fax: +44 15394 30371
1950; Janet Henderson
Education, hist, geography, mathematics, natural hist, botany, zoology, outdoor education, psychology, sociology, art, children's lit; Charlotte Mason Archive, Archive of P.N.E.U.
88 000 vols; 180 curr per; 7 613 incunabula; 189 diss/theses; 150 govt docs; 230 maps; 170 microforms; 977 av-mat; 295 sound-rec; 12 digital data carriers
libr loan 28932

Scottish Agricultural College, W.J. Thomson Library, Auchincruive, ***Ayr*** KA6 5HW
T: +44 1292 525209; Fax: +44 1292 525211; E-mail: libraryau@sac.ac.uk; URL: www.sac.ac.uk
1899; E.P. Muir
Conservation, environmental management, leisure, food technology, agriculture, horticulture
20 000 vols; 170 curr per; 2 283 e-journals; 568 e-books; 400 diss/theses; 300 govt docs; 250 maps; 350 av-mat; 20 digital data carriers
libr loan; SALG, SALCTG, AHIS 28933

Bangor University, Library & Archives Service, College Rd, ***Bangor*** LL57 2DG
T: +44 1248 382981; Fax: +44 1248 382979; E-mail: library@bangor.ac.uk; URL: www.bangor.ac.uk/library
1911; David Learmont
Sir Frank Brangwyn Art Collection, Talfourd-Jones Coll (botany, zoology), Cathedral libr Coll, Welsh Libr Coll
820 000 vols; 10 000 curr per; 22 000 e-journals; 4 000 e-books; 500 000 mss; 5 incunabula; 12 000 diss/theses; 17 800 music scores; 1 400 maps; 3 500 microforms; 3 130 av-mat; 6 800 sound-rec
libr loan; SCONUL, WHELF, CILIP 28934

Barnsley College, Library, Old Mill Lane, ***Barnsley*** S70 2AX
T: +44 1226 216334; Fax: +44 1226 298514; URL: www.barnsley.ac.uk
1932; Lesley Dickinson
85 000 vols; 200 curr per; 4 e-journals; 100 diss/theses; 50 music scores; 80 digital data carriers
libr loan; CILIP 28935

Bath Spa University, Sion Hill Library, 8 Somerset Place, ***Bath*** BA1 5HB
T: +44 1225 875763; Fax: +44 1225 427080; E-mail: sh-library-enquiries@bathspa.ac.uk; URL: www.bathspa.ac.uk
1946; Helen Rayner
50 000 vols; 116 curr per; 153 av-mat; 50 000 slides
libr loan; ARLIS/UK & Ireland, CILIP 28936

University of Bath, Library, Claverton Down, ***Bath*** BA2 7AY
T: +44 1225 385000; Fax: +44 1225 386229; E-mail: library@bath.ac.uk; URL: www.bath.ac.uk/library
1966; Howard D. Nicholson
Pitman Coll (hist of shorthand), Bath A. West Coll (hist of agriculture), Watkins Coll (Steam engineering)
488 000 vols; 1 800 curr per; 8 000 e-journals; 10 000 diss/theses; 200 music scores; 225 maps; 44 000 microforms; 200 av-mat; 55 digital data carriers
libr loan; CILIP, SCONUL, IATUL 28937

Belfast Institute of Further and Higher Education, Learning and Teaching Resources, College Sq East, Room A16, ***Belfast*** BT1 6DJ
T: +44 28 90265016; Fax: +44 28 90265001; URL: www.belfastinstitute.ac.uk
Alan Dummigan
6 site libs
45 000 vols; 385 curr per; 200 av-mat; 200 sound-rec; 100 digital data carriers; 100 multimedia packs 28938

Queen's University Belfast, University Library, Main Library, University Rd, ***Belfast*** BT7 1LS
T: +44 28 90975020; Fax: +44 28 90973072; E-mail: library@qub.ac.uk; URL: www.qub.ac.uk/lib
1849; John Gormley
Hibernica, Thomas Percy, Sir Hamilton Harty
1 274 135 vols; 8 005 curr per; 11 645 e-journals; 7 000 mss; 17 incunabula; 12 947 diss/theses; 9 854 music scores; 23 maps; 1 328 microforms; 2 102 av-mat; 198 sound-rec; 4 610 digital data carriers
libr loan; SCONUL, LIBER 28939

– AFBI Library, Newforge Lane, Belfast BT9 5PX
T: +44 28 90255227; Fax: +44 28 90255400; E-mail: afbi.library@afbini.gov.uk; URL: www.afbilib.qub.ac.uk
2007; K. Latimer
Research libr, Horticultural and plant breeding libr
32 000 vols; 350 curr per; 250 e-journals; 100 av-mat; 60 digital data carriers
libr loan; SCONUL, IAALD, IFLA 28940

– Northern Ireland Health and Social Services, Institute of Clinical Science, Medical Library, Mulhouse building, Mulhouse road, Belfast BT12 6DP
T: +44 28 90632500; Fax: +44 28 9035038; E-mail: g.creighton@qub.ac.uk
1954; Gaynor Creighton
Medical hist, works by Northern Ireland doctors
100 000 vols; 1 350 curr per
CILIP 28941

St Mary's University College, 191 Falls Rd, ***Belfast*** BT12 6FE
T: +44 28 90268317; E-mail: f.jones@smucb.ac.uk; URL: www.stmarys-belfast.ac.uk
1900; Felicity Jones
Magee collection
100 000 vols; 404 curr per; 34 e-journals; 317 diss/theses; 587 av-mat; 2 100 sound-rec; 4 digital data carriers; 4 000 multimedia packs
libr loan; SCONUL 28942

Stranmillis University College, Library, ***Belfast*** BT9 5DY
T: +44 28 90384310; Fax: +44 28 90663682; E-mail: l.chambers@stran.ac.uk; URL: www.stran-ni.ac.uk
1950
80 000 vols; 400 curr per; 5 000 microforms; 400 av-mat; 1 000 sound-rec
libr loan; CILIP, SLA 28943

University of Ulster, Belfast Campus, Library, York St, ***Belfast*** BT15 1ED
T: +44 28 90267268; Fax: +44 28 90267278; E-mail: m.khorshidian@ulster.ac.uk; URL: library.ulster.ac.uk/
1969; Marion Khorshidian
Arts, design, architecture, modern languages, history, business
53 927 vols; 184 curr per; 19 144 e-journals; 3 553 e-books; 111 microforms; 95 000 av-mat; 50 digital data carriers
libr loan; ARLIS/UK & Ireland 28944

Bishop Burton College, Library, ***Beverley*** HU17 8QG
T: +44 1964 553041; Fax: +44 1964 553101
J. Godwin
30 000 vols; 350 curr per; 12 e-journals; 2 200 diss/theses; 60 maps; 500 av-mat; 52 sound-rec; 100 digital data carriers; 8 000 pamphlets
CILIP, COLRIC 28945

Wolverhampton College, Wellington Road Campus, Learning Centre, Wellington Rd, ***Bilston*** WV14 6BT
T: +44 1902 821054; Fax: +44 1902 821101; E-mail: mail@wolvcoll.ac.uk; URL: www.wolverhamptoncollege.ac.uk
1962; Valerie Bigford
Post-16 Education
53 693 vols; 200 curr per; 200 maps; 8 000 av-mat; 60 digital data carriers
libr loan 28946

Aston University, Library & Information Services, Aston Triangle, ***Birmingham*** B4 7ET
T: +44 121 2044525; Fax: +44 121 2044530; E-mail: library@aston.ac.uk; URL: www.aston.ac.uk/lis/
1898; Dr. Nick R. Smith
209 000 vols; 500 curr per; 7 100 e-journals; 40 700 e-books; 3 000 diss/theses
libr loan; SCONUL, ASLIB, CILIP 28947

Birmingham City University, Library & Learning Recources, Perry Barr, ***Birmingham*** B42 2SU
T: +44 121 3815300; Fax: +44 121 3562875; URL: library.bcu.ac.uk
Judith Andrews
Teaching Practice Library (children's lit. 38 000 vols); Royal College of Organists

coll
733 000 vols; 2 600 curr per; 4 000 e-journals; 3 000 e-books; 78 000 govt docs; 71 000 music scores; 8 000 microforms; 123 000 av-mat; 9 200 sound-rec; 9 764 digital data carriers
libr loan; CILIP, ASLIB 28948

– Birmingham Conservatoire Library, Paradise Place, Birmingham B3 3HG
T: +44 121 3315914;
Fax: +44 121 3315906; E-mail: conservatoire.library@uce.ac.uk; URL: library.bcu.ac.uk
1854; F. J. Firth
Printed music of Granville Bantock, Birmingham Flute, Society Coll, Arch of Birmingham School of Music
8 000 vols; 70 curr per; 10 e-journals; 100 mss; 80 diss/theses; 109 000 music scores; 10 microforms; 152 av-mat; 10 000 sound-rec; 10 digital data carriers
libr loan; BLLD, IAML 28949

Newman College, Library, Genners Lane, Bartley Green, *Birmingham* B32 3NT
T: +44 121 4761181 ext 2208;
Fax: +44 121 4761196; E-mail: library@newman.ac.uk
1968; Caroline Rock
Special education, religious education, hist of the English West Midlands, local hist coll, J.H. Newman coll
80 000 vols; 250 curr per; 182 microforms; 14 075 av-mat; 1 020 sound-rec; 100 digital data carriers; 1 078 pamphlets, 304 doc folders, 78 comp software
BLDSC 28950

Orchard Learning Resources Centre, University of Birmingham, Weoley Park Rd, Selly Oak, *Birmingham* B29 6LL; mail address: Hamilton Dr, Weoley Park Rd, Selly Oak, Birmingham B29 6QW
T: +44 121 4158454; Fax: +44 121 4158476; E-mail: olrc@bham.ac.uk; URL: www.olrc.bham.ac.uk
1997; Dorothy Vuong
Mingana Coll (Arabic and Syriac mss), Rendel Harris Coll (Greek papyri), Harold Turner Coll on New Religious Movements, religions, esp Christianity, Islam, Judaism, area and social studies
150 000 vols; 800 curr per; 3 000 mss; 350 diss/theses; 200 maps; 6 000 microforms; 32 digital data carriers; 500 papyri
libr loan; ABTAPL, SCOLMA, MELCOM 28951

Queen's Foundation, Library, Somerset Rd, *Birmingham* B15 2QH
T: +44 121 4522621; Fax: +44 121 4548171; E-mail: library@queens.ac.uk; URL: www.queens.ac.uk
1929; Michael Gale
17th-18th c Anglican and Methodist hist coll
50 000 vols; 70 curr per; 300 e-journals; 35 diss/theses; 46 av-mat; 4 digital data carriers
ABTAPL 28952

South Birmingham College, Hall Green Campus Library, Colebank Rd, *Birmingham* B28 8ES
T: +44 121 6945000; Fax: +44 121 6945007; URL: www.sbirmc.ac.uk
1961; J. A. Duffus
16 000 vols; 250 curr per; 1 500 microforms; 300 av-mat; 30 digital data carriers
libr loan 28953

University of Birmingham, Information Services, Main Library, Edgbaston, *Birmingham* B15 2TT
T: +44 121 4145828; Fax: +44 121 4714691; E-mail: library@bham.ac.uk; URL: www.is.bham.ac.uk/mainlib
1880; Ms Michele Shoebrifge
Chamberlain Arch, CMS Arch, Avon Papers, Harriet Martineau Papers, Papers of Francis Brett Young, Sir Oswald Mosley, Noel Coward, Charles Masterman, Mingana Coll libs of St Mary's Church of Warwick, James Rendell Harris Coll
2 655 251 vols; 12 500 curr per; 9 000 e-journals; 3 000 000 mss; 62 incunabula; 20 400 diss/theses; 65 000 music scores; 12 000 maps; 90 000 microforms; 11 000 av-mat; 15 000 sound-rec; 150 digital data carriers

libr loan; SCONUL, ASLIB, CURL, RLG 28954

– Barber Fine Art Library, Barber Institute of Fine Arts, Edgbaston, Birmingham B15 2TS
T: +44 121 4147334; Fax: +44 121 4145853; E-mail: D.Pulford@bham.ac.uk; URL: www.barberart.bham.ac.uk/index.htm
1940; David Pulford
Coll of sale catalogs from 1 680 onwards
50 000 vols; 100 curr per; 40 000 music scores; 250 microforms; 12 000 av-mat; 12 digital data carriers
libr loan; IAML 28955

– Barnes Library, Medical School, Edgbaston, mail address: Birmingham B15 2TT
T: +44 121 4143567; Fax: +44 121 4145855; E-mail: ba-lib@bham.ac.uk; URL: www.is.bham.ac.uk/barnes
Jean Scott
140 000 vols; 800 curr per; 500 diss/theses; 1 000 govt docs 28956

– Education Library, Edgbaston, Birmingham B15 2TT
T: +44 121 4144869; Fax: +44 121 4714691; E-mail: edlib@bham.ac.uk
1949; Ms D. Vuong
Hist of special education; Hist coll of children's bks
100 000 vols; 300 curr per; 65 e-journals; 2 500 diss/theses; 1 000 govt docs; 60 av-mat; 5 digital data carriers; 10 000 pamphlets
libr loan; BLDSC, WMRLB, LISE 28957

– Harding Law Library, Edgbaston; P.O. Box 363, Birmingham B15 2TT
T: +44 121 4145865; E-mail: g.g.watkins@bham.ac.uk; URL: www.is.bham.ac.uk
1927; G.G. Watkins
55 000 vols; 250 curr per
BIALL, IALL, AALL 28958

Blackburn College, Library, Feilden St, *Blackburn* BB2 1LH
T: +44 1254 292190; Fax: +44 1254 695265; URL: www.blackburn.ac.uk
1955; Sheila Goodman
50 000 vols; 320 curr per; 6 000 av-mat; 28 digital data carriers
libr loan 28959

Blackpool and The Fylde College, Learning Resource Centre, Ashfield Rd, Bispham, *Blackpool* FY2 0HB
T: +44 1253 352352; Fax: +44 1253 356127; URL: www.blackpool.ac.uk
1958; C. McAllister
Maritime-Fleetwood Resource Centre
50 000 vols; 300 curr per; 500 av-mat
libr loan; COLRIC 28960

University of Bolton, Eagle Learning Support Centre, Learning Support Services, Chadwick St, *Bolton* BL2 1JW
T: +44 1204 903563; E-mail: k.senior@bolton.ac.uk; URL: www.lss.bolton.ac.uk
1982; Mrs K. Senior
110 000 vols 28961

University of Bradford, J.B. Priestley Library, *Bradford* BD7 1DP
T: +44 1274 233400; Fax: +44 1274 233398; E-mail: library@bradford.ac.uk; URL: www.brad.ac.uk/library/index.php
1966; Grace Hudson
600 000 vols; 2 450 curr per; 18 000 diss/theses; 600 maps; 30 000 microforms; 500 av-mat
SCONUL, IATUL 28962

University of Brighton, Aldrich Library, Cockcroft Building, Lewes Road, *Brighton* BN2 4GJ
T: +44 1273 642760;
Fax: +44 1273 642988; E-mail: AskAldrich@brighton.ac.uk; URL: www.brighton.ac.uk/is
1963; Steve Newman
200 000 vols; 1 500 curr per; 30 000 e-journals; 3 000 e-books; 550 microforms; 21 000 av-mat; 1 000 sound-rec
libr loan; CILIP, ASLIB, SCONUL 28963

– Aldrich Library, Cockcroft Bldg, Lewes Rd, Moulsecoomb, Brighton BN2 4GJ
T: +44 1273 642760; Fax: +44 1273 642988; URL: library.brighton.ac.uk
1970; Terry Hanson
197 600 vols; 1 775 curr per; 23 000 e-journals; 2 000 e-books; 500 diss/theses; 12 microforms
libr loan; CILIP, ASLIB 28964

– Falmer Library, Falmer, Brighton BN1 9PH
T: +44 1273 643569; Fax: +44 1273 643560; E-mail: askfalmer@brighton.ac.uk
Keith Baxter
156 720 vols; 646 curr per 28965

– St Peter's House Library, Information Services, 16-18 Richmond Place, Brighton BN2 9NA
T: +44 1273 643221; Fax: +44 1273 607532; E-mail: Asksph@brighton.ac.uk; URL: www.brighton.ac.uk
1877; Louise Tucker
Arts, design, humanities
100 436 vols; 250 curr per; 1 000 microforms; 150 000 av-mat; 2 200 sound-rec
ARLIS/UK & Ireland, CILIP 28966

– Welkin Library, Carlisle Rd, Eastbourne BN20 7SN
T: +44 1273 643822; Fax: +44 1273 643829
1979; Mike Ainscough
Teaching resources coll
45 000 vols; 300 curr per; 25 diss/theses; 25 maps; 3 026 av-mat; 4 000 sound-rec; 2 digital data carriers
CILIP 28967

University of Sussex, University Library, Falmer, *Brighton* BN1 9QL
T: +44 1273 678163; Fax: +44 1273 678441; E-mail: library@sussex.ac.uk; URL: www.susx.ac.uk/library
1961; Deborah Shorley
French Commune of 1871, Leonard and Virginia Woolf mss, Rudyard Kipling papers, Kingsley Martin papers, publs of British Pressure Groups, New Statesman Arch 1943-88; Mass-Observation Archive
750 000 vols; 3 000 curr per; 70 mss; 20 incunabula; 4 000 diss/theses; 20 000 music scores; 3 000 av-mat; 10 000 sound-rec
libr loan; SCONUL, ASLIB 28968

– British Library for Development Studies, Institute of Development Studies, Andrew Cohen Bldg, Falmer, Brighton BN1 9RE
T: +44 1273 915659; Fax: +44 1273 621202; E-mail: blds@ids.ac.uk; URL: www.ids.ac.uk
1966; Julie Brittain
Developing countries, Economic and social development; Contemporary (1966-) govt publs of most African, Asian and Latin American countries; UN & UNESCO depository libs
200 000 vols; 12 000 curr per; 1 000 e-journals; 150 e-books; 110 diss/theses; 50 000 govt docs
libr loan; SCOLMA, M25 Consortium, IFLA, CILIP 28969

– SPRU – Science and Technology Policy Research, Keith Pavitt Library, Falmer, Brighton BN1 9QE
T: +44 1273 678178; Fax: +44 1273 685865; E-mail: m.e.winder@sussex.ac.uk; URL: www.sussex.ac.uk/spru
1966; Barbara Merchant
40 000 vols; 300 curr per
libr loan; ASLIB 28970

University of Bristol, University Library, Tyndall Ave, *Bristol* BS8 1TJ
T: +44 117 928 8005; Fax: +44 117 925 5334; E-mail: library@bris.ac.uk; URL: www.bris.ac.uk/is
1909; Peter King
Brunel papers, Penguin coll, early science, early English novels, geology, Liberal party history, University of Bristol archive, Pinney Coll (West Indies)
1 400 000 vols; 9 000 curr per; 139 000 microforms; 1 700 av-mat; 100 digital data carriers
libr loan; SCONUL, CILIP, CURL 28971

– School of Medical Sciences, Library, University Walk, Bristol BS8 1TD
T: +44 117 9287945; Fax: +44 117 9290185; E-mail: library@bristol.ac.uk/is
1892; Jennifer Scherr
110 000 vols; 950 curr per; 100 e-journals 28972

University of the West of England, Library Services, Frenchay Campus, Caldharbour Lane, *Bristol* BS16 1QY
T: +44 117 3282404;
Fax: +44 117 3282407; E-mail: customer.services@uwe.ac.uk; URL: www.uwe.ac.uk/library
1969; Cathy Rex
National Institute of Social Work Libr, George Budden Music Coll
723 543 vols; 14 223 curr per; 16 629 e-journals; 154 e-books; 22 000 av-mat; 67 digital data carriers; videos, slides, bones, extensive coll of electronic sources of information
libr loan; CILIP, SCONUL, UK Libraries Plus, SWRLIN, Avon University Libraries in Cooperation (AULIC) 28973

– Faculty of Art, Media and Design, Bower Ashton Library, Kennel Lodge Rd, Bristol BS3 2JT
T: +44 117 3284750; Fax: +44 117 3284745; URL: www.uwe.ac.uk/library/info/bower_ashton
Geoff Cole
Artists Book coll
30 000 vols; 200 curr per; 50 diss/theses; 20 microforms; 90 000 av-mat; 1 000 sound-rec; 150 digital data carriers; 2 000 files of cuttings
CILIP, ARLIS/UK & Ireland 28974

UWE Bristol, Glenside Campus, Library, Blackberry Hill, *Bristol* BS16 1DD
T: +44 117 3288404; Fax: +44 117 3288402; URL: www.uwe.ac.uk/library
1991; Caroline Plaice
Nursing, midwifery, radiography, social work, occupational therapy, healt and social care, public health
60 000 vols; 1 000 curr per; 700 e-journals; 1 200 av-mat; 150 anatomical models
CILIP, SCONUL, BLDSC 28975

Wesley College Library, College Park Drive, Henbury Road, *Bristol* BS10 7QD
T: +44 117 9591200; Fax: +44 117 9501277; E-mail: librarian@wesley-college-bristol.ac.uk; URL: www.wesley-college-bristol.ac.uk
1945; Michael Brealey
Methodist Church Music Society Library, Shapland Patristic Collection, tract and pamphlet collection on Methodist and other church controversies, Mss collection relating to history of Methodism in 18th and 19th centuries
37 000 vols; 65 curr per; 2 000 mss; 68 diss/theses; 2 000 music scores
ABTAPL 28976

Bromley College of Further and Higher Education, Library, Rookery Lane, *Bromley* BR2 8HE
T: +44 20 82957024; Fax: +44 20 82957099
1959; J.A. Murdoch
30 000 vols; 200 curr per; 200 maps; 170 av-mat; 34 digital data carriers
libr loan; SEAL 28977

Hertford Regional College, Broxbourne Centre Library, Turnford, *Broxbourne* EN10 6AE
T: +44 1992 411400; Fax: +44 1992 411522; E-mail: library.brox@hertreg.ac.uk; URL: www.hertreg.ac.uk
1965; Ian Radcliffe
25 000 vols; 130 curr per; 100 maps; 3 650 av-mat; 210 sound-rec; 100 digital data carriers
libr loan; CILIP 28978

University of Buckingham, University Library, Hunter St, *Buckingham* MK18 1EG
T: +44 1280 814080; Fax: +44 1280 820270; E-mail: library@buckingham.ac.uk; URL: www.buckingham.ac.uk
1973; Louise Hammond
90 000 vols; 250 curr per; 3 000 e-journals; 30 e-books; 100 diss/theses; 500 sound-rec; 40 digital data carriers
libr loan; CILIP, BIALL 28979

– Denning Law Library, Hunter St, Buckingham MK18 1EG
T: +44 1280 828207; Fax: +44 1280 828288; E-mail: library@buckingham.ac.uk; URL: www.buckingham.ac.uk/library
1977; Louise Hammond
Law, business, humanities, science; Edward Legg local history, Beloff newspaper cuttings and books
93 000 vols; 257 curr per; 16 e-journals; 307 diss/theses; 30 maps; 259 av-mat
libr loan; BIALL, SCONUL, JISC   28980

– James Meade Library of Economics, Hunter St, Buckingham MK18 1EG
T: +44 1280 814080; Fax: +44 1280 820312
Swee Har Newell, Louise Hammond
95 000 vols; 400 curr per
libr loan   28981

Royal Military Academy Sandhurst, Central Library, *Camberley* GU15 4PQ
T: +44 1276 63344 ext 2367; Fax: +44 1276 412538; URL: www.sandhurst.mod.uk/tour/library.htm
A.A. Orgill
180 000 vols; 220 curr per   28982

Anglia Ruskin University, University Library, East Rd, *Cambridge* CB1 1PT
T: +44 1223 363271 ext 2301; Fax: +44 1223 352973; E-mail: nicky.kershaw@anglia.ac.uk; URL: libweb.anglia.ac.uk/
1989; Nicky Kershaw
libr loan   28983

Clare College, Forbes Mellon Library, Memorial Court, *Cambridge* CB3 9AJ
T: +44 1223 333202; Fax: +44 1223 765560; E-mail: library@clare.cam.ac.uk; URL: www.clare.cam.ac.uk/academic/libraries/index.html
Anne C. Hughes
Cecil Sharp Mss
28 000 vols; 13 curr per; 17 music scores; 6 sound-rec; 15 digital data carriers   28984

Gonville and Caius College, Library, Trinity St, *Cambridge* CB2 1TA
T: +44 1223 332419; E-mail: Library@cai.cam.ac.uk; URL: www.cai.cam.ac.uk/library/index.php
1349; D. Abulafia
Branthwaite Libr, Charles Wood mss
80 000 vols; 84 curr per; 900 mss; 100 incunabula; 600 music scores; 200 microforms; 300 sound-rec; 5 digital data carriers   28985

Needham Research Institute, East Asian History of Science Library, 8 Sylvester Rd, *Cambridge* CB3 9AF
T: +44 1223 311545; Fax: +44 1223 362703; E-mail: jm10019@cam.ac.uk; URL: www.nri.org.uk
1976; John P.C. Moffett
History of traditional East Asien science, technology and medicine
30 000 vols; 150 curr per; 100 e-journals; 500 diss/theses; 100 maps; 50 microforms; 20 000 offprints   28986

University of Cambridge, Cambridge University Library, West Rd, *Cambridge* CB3 9DR
T: +44 1223 333000; Fax: +44 1223 333160; E-mail: library@lib.cam.ac.uk; URL: www.lib.cam.ac.uk
1400; P.K. Fox
Cambridge Coll, Ely Diocesan and chapter records, Taylor-Schechter Genizah Coll, Chinese oracle bones, Royal Commonwealth Society Coll, Acton Libr, Bible Society Libr, Royal Greenwich Observatory Arch, University Arch
8 000 000 vols; 46 000 curr per; 10 000 e-journals; 157 200 mss; 4 800 incunabula; 31 000 diss/theses; 1 151 000 maps; 1 800 000 microforms; 25 655 sound-rec; 6 500 digital data carriers; 825 000 pamphlets
libr loan; SCONUL, ASLIB, CILIP, RLUK, RLG, LIBER   28987

– Cambridge Union Society, University of Cambridge, Keynes Library, 9a Bridge St, Cambridge CB2 1UB
T: +44 1223 741289; Fax: +44 1223 566444; E-mail: info@cus.org; URL: www.cus.org
1815; Pat Aske
Fiction, Biography, Cambridge history, English Literature, Politics, Travel ; Fairfax Rhodes Coll (Art and Lit)

30 000 vols; 7 curr per; 3 300 music scores; 550 maps; 424 sound-rec; 343 digital data carriers
CILIP   28988

– Central Science Library, Benet St, Cambridge CB2 3PY
T: +44 1223 334742; Fax: +44 1223 334748; E-mail: lib-csl-inquiries@lists.cam.ac.uk; URL: www.lib.cam.ac.uk/CSL
1819; Yvonne Nobis
Agriculture, animal husbandry, nineteenth century periodicals and textbooks
138 754 vols; 1 459 curr per
libr loan; IATUL   28989

– Centre of African Studies Library, Mond Building, Free School Lane, Cambridge CB2 3RF
T: +44 1223 334398; Fax: +44 1223 769325; E-mail: afrlib@cam.ac.uk; URL: www.african.cam.ac.uk/library/library.html
1965; Marilyn Glanfield
Interdiciplinary (Arts, Humanities, Social Sciences) – Sub saharan Africa ; Donated collections from academics, colonial officers etc.
35 000 vols; 70 curr per; 100 maps; 500 av-mat
libr loan; SCOLMA   28990

– Centre of Latin American Studies, Mill Lane Library, 8 Mill Lane, Cambridge CB2 1RX
T: +44 1223 335398; Fax: +44 1223 335397; E-mail: jac46@cam.ac.uk; URL: www.latin-american.cam.ac.uk/library/index.html
Wendy Thurley
12 000 vols; 25 curr per; 206 diss/theses
ACLAIIR   28991

– Centre of South Asian Studies, Library, Faculty Rooms, Laundress Lane, Cambridge CB2 1SD
T: +44 1223 338094; Fax: +44 1223 316913; E-mail: webmaster@s-asian.cam.ac.uk; URL: www.s-asian.cam.ac.uk
1964; Ms R.M. Rowe
Coll of personal papers of Europeans in India, Modern South Asian studies
26 000 vols; 180 curr per; 400 diss/theses; 1 000 govt docs; 300 maps; 9 000 microforms; 300 sound-rec; 600 colls of mss   28992

– Christ's College, Library, Cambridge CB2 3BU
T: +44 1223 334950; Fax: +44 1223 339557; E-mail: library@christs.cam.ac.uk; URL: www.christs.cam.ac.uk/library/
1505; Alexander Gavin
Lesingham Smith Coll (16th & 17th c math and sci bks), Robertson Smith Coll (oriental bks), John Milton Coll
80 000 vols; 36 curr per; 50 mss; 55 incunabula; 4 000 music scores
CILIP   28993

– Classical Faculty and Museum of Classical Archaeology, Library, Sidgwick Av, Cambridge CB3 9DA
T: +44 1223 335154; Fax: +44 1223 331794; E-mail: library@classics.cam.ac.uk; URL: www.classics.cam.ac.uk/library/library.html
1884; L.K. Bailey
Coll William Martin Leake, large coll of maps on mediterranean region, large coll of 35mm slides, Greek and Roman epigraphy
50 000 vols; 280 curr per   28994

– Cory Library, University Botanic Garden, Cory Lodge, Bateman St, Cambridge CB2 1JF
T: +44 1223 336265; Fax: +44 1223 336278; E-mail: enquiries@botanic.cam.ac.uk; URL: www.botanic.cam.ac.uk
Prof. J.S. Parker
Horticulture, taxonomy; coll of nurserymen's catalogs
9 000 vols; 60 curr per; 5 000 av-mat; 2 900 pamphlets   28995

– Department of Biochemistry, Colman Library, Tennis Court Rd, Cambridge CB2 1QW
T: +44 1223 333613; Fax: +44 1223 333345; E-mail: librarian@bioc.cam.ac.uk; URL: www.bioc.cam.ac.uk/library/index.html
1924; Dr. Robin Hesketh
8 714 vols; 127 curr per
libr loan; BLL   28996

– Department of Clinical Veterinary Medicine, Library, Madingley Rd, Cambridge CB3 0ES
T: +44 1223 337633; Fax: +44 1223 337610; E-mail: lel1000@cam.ac.uk; URL: www.vet.cam.ac.uk/library.html
1955; Mrs L.E. Leonard
Sir John Hammond Coll (nutrition)
12 000 vols; 241 curr per; 134 diss/theses; 33 digital data carriers; 9 000 mss
libr loan; BLL   28997

– Department of Experimental Psychology, Library, Downing St, Cambridge CB2 3EB
T: +44 1223 333554; Fax: +44 1223 333564; E-mail: library@psychol.cam.ac.uk
Dr. A. Dickinson
Psychopathology; Maccurdy Psychopathology Libr
14 300 vols; 114 curr per; 99 diss/theses
BLDSC   28998

– Department of History and Philosophy of Science, Whipple Library, Free School Lane, Cambridge CB2 3RH
T: +44 1223 334547; Fax: +44 1223 334554; E-mail: hpslib@hermes.cam.ac.uk; URL: www.hps.cam.ac.uk
1944; Tim Eggington
Works of Robert Boyle, phrenology coll, R.S. Whipple coll of pre-1900 scientific works, Landmarks of Science coll of microprints, Foster Pamphlet Coll (physiology), Wellcome Iconographic coll (videodisk), Gerd Buchdahl Kant Coll & archive
35 316 vols; 108 curr per; 3 mss; 6 incunabula; 369 diss/theses; 10 microforms; 11 av-mat; 10 sound-rec; 8 digital data carriers; 7 000 pamphlets and offprints, 3 linear metres of manuscripts   28999

– Department of Pathology, Kanthack and Nuttall Library, Tennis Court Rd, Cambridge CB2 1QP
T: +44 1223 333698; Fax: +44 1223 765781; E-mail: librarian@path.cam.ac.uk; URL: www.path.cam.ac.uk/~library
Dr. T.D.K. Brown
8 042 vols; 44 curr per
libr loan   29000

– Department of Physics, Rayleigh Library, Cavendish Laboratory, JJ Thomson Ave, Cambridge CB 3 0HE
T: +44 1223 337414; Fax: +44 1223 363263; E-mail: librarian@phy.cam.ac.uk; URL: www.phy.cam.ac.uk/cavendish/library
1922; Ms N. Huntic
Napier Shaw Libr (Meteorology), Maxwell Coll, History of Science Coll
20 250 vols; 69 curr per; 60 digital data carriers
libr loan   29001

– Department of Physiology Library, Downing St, Cambridge CB2 3EG
T: +44 1223 333821; Fax: +44 1223 333840; E-mail: cer34@cam.ac.uk; URL: www.physiol.cam.ac.uk/misc/library.htm
1880; N.N.
Hist coll of science bks, Sir Michael Foster Coll, Keith Lucas Coll, Sir Joseph Barcroft Coll, Dr. W.H. Gaskell Coll
14 670 vols; 55 curr per   29002

– Department of Zoology, Balfour and Newton Libraries, Downing St, Cambridge CB2 3EJ
T: +44 1223 336648; Fax: +44 1223 336676; E-mail: library@zoo.cam.ac.uk; URL: www.zoo.cam.ac.uk/library/index.html
1882; Ms C.M. Castle MCLIP
Ornithology
42 257 vols; 183 curr per; 3 incunabula; 473 diss/theses; 104 915 reprints
BLL   29003

– Divinity Faculty, Library, West Rd, Cambridge CB3 9BS
T: +44 1223 763040; E-mail: divlib@hermes.cam.ac.uk; URL: www.divinity.cam.ac.uk/library.html
1861; Dr. P. Dunstan
Feltoe bequest (liturgy bks), Lightfoot Libr (libr of bishop J.B. Lightfoot who died 1889), church hist, biblical studies, Judaism, Islam, Buddhism, Hinduism, patristics, philosophy of religion, science and religion, sociology of religion, liturgy, ethics
58 500 vols; 40 curr per; 25 e-journals;

2 000 pamphlets
ABTAPL   29004

– Earth Sciences Library, Downing St, Cambridge CB2 3EQ
T: +44 1223 333429; Fax: +44 1223 333450; E-mail: libraryhelp@esc.cam.ac.uk; URL: www.esc.cam.ac.uk/guide.html
E. Tilley
Black Coll, Bulman Coll, O. Fisher Coll, Harker Coll, N.F. Hughs Coll, H. Jeffreys Coll, Macfadyen Coll, Sedgwick Coll, Tilley Coll, Whittington Coll, Glaessner Coll
16 000 vols; 2 700 curr per; 150 e-journals; 500 diss/theses; 50 000 maps; 50 digital data carriers; 900 rare bks
libr loan   29005

– Emmanuel College, Library, St Andrew's St, Cambridge CB2 3AP
T: +44 01223 334233; E-mail: library@emma.cam.ac.uk
1584; Dr. H.C. Carron
Archbishop William Sancroft Libr, W.C. Bishop Libr (liturgy), Graham Watson
75 000 vols; 70 curr per; 400 mss; 100 incunabula   29006

– Engineering Department, Library, Trumpington St, Cambridge CB2 1PZ
T: +44 1223 332626; Fax: +44 1223 332662; E-mail: cued-library@eng.cam.ac.uk; URL: www.eng.cam.ac.uk
1956; H.M. McOwat
Engineering, electronics, information sciences, control, structures, soil mechanics, fluid mechanics, materials science, CAD
50 000 vols; 300 curr per; 2 500 diss/theses
ASLIB   29007

– English Faculty Library, 9 West Rd, Cambridge CB3 9DP
T: +44 1223 335077; E-mail: efllib@hermes.cam.ac.uk; URL: lib.english.cam.ac.uk
Elizabeth Tilley
80 000 vols; 92 curr per; 440 av-mat; 1 330 sound-rec; 3 000 transparencies
CILIP   29008

– Faculty of Architecture and History of Art, Library, 1 Scroope Terrace, Cambridge CB2 1PX
T: +44 1223 332953; Fax: +44 1223 332960; E-mail: library@aha.cam.ac.uk; URL: www.arct.cam.ac.uk/library.html
1910; Madeleine Brown
2 000 rare bks
34 000 vols; 106 curr per; 1 000 diss/theses; 7 av-mat; 4 sound-rec; 12 digital data carriers
libr loan; Arlis, Arclib   29009

– Faculty of Education, Library & Information Service, 184 Hills Rd, Cambridge CB2 2PQ
T: +44 1223 767700; Fax: +44 1223 767602; E-mail: library@educ.cam.ac.uk; URL: www.educ.cam.ac.uk/library
Angela Cutts
Coll of Children's Fiction
99 000 vols; 226 curr per; 450 diss/theses; 1 500 microforms; 1 000 av-mat; 210 sound-rec; 1 digital data carriers
LISE   29010

– Faculty of History, Seeley Historical Library, West Rd, Cambridge CB3 9EF
T: +44 1223 335335; Fax: +44 1223 335968; E-mail: seeley@hist.cam.ac.uk; URL: www.hist.cam.ac.uk/library/
1807; Dr. Linda Washington
Hadley coll of bks on the french rev and the Napoleonic era
75 000 vols; 132 curr per; 648 diss/theses; 4 602 microforms; 395 av-mat; 10 sound-rec; 10 digital data carriers; 5 238 pamphlets   29011

– Faculty of Oriental Studies, Library, Sidgwick Av, Cambridge CB3 9DA
T: +44 1223 335111, 335112; Fax: +44 1223 335110; E-mail: library@ames.cam.ac.uk; URL: www.ames.cam.ac.uk
1935; F. Simmons
Special coll of early printed bks in oriental subjects, Chinese, Japanese, Indian lit, hist, archeology, Arabic, Hebrew, egyptology, assyriology, Arch

of Oriental Scholars
60 000 vols; 250 curr per; 40 e-journals; 1 000 maps; 500 av-mat; 350 sound-rec; 5 digital data carriers; personal papers of Oriental scholars connected with Cambridge archive 29012

– Faculty of Philosophy, Library, Sidgwick Av, Cambridge CB3 9DA
T: +44 1223 762939; Fax: +44 1223 335091; URL: www.phil.cam.ac.uk/library.html
M. Pellegrino
14 000 vols; 31 curr per; 2 digital data carriers 29013

– Faculty of Social and Political Sciences, Library, Free School Lane, Cambridge CB2 3RQ
T: +44 1223 334522; Fax: +44 1223 334550; E-mail: sps-library@lists.cam.ac.uk; URL: www.sps.cam.ac.uk/library/index.html
Julie Nicholas
30 000 vols; 61 curr per; 43 e-journals; 1 000 pamphlets 29014

– Fitzwilliam College, Library, Storey's Way, Cambridge CB3 0DG
T: +44 1223 332042; Fax: +44 1223 477976; E-mail: library@fitz.cam.ac.uk; URL: www.fitz.cam.ac.uk/library/index.jsp
1967; Christine E. RobertsLewis
Park Bequest (International Law)
40 000 vols; 75 curr per; 300 music scores; 50 av-mat; 30 digital data carriers
BCA 29015

– Fitzwilliam Museum, Department of Manuscripts & Printed Books, Library, Trumpington St, Cambridge CB2 1RB
T: +44 1223 332900; Fax: +44 1223 332923; E-mail: fitzmuseum-enquiries@lists.cam.ac.uk; URL: www.fitzmuseum.cam.ac.uk
1816; Stella Panayotova
Illuminated mss, music mss, literary mss, historical mss, autograph letters, private press bks, fine bks; Founder's Libr
250 000 vols; 450 curr per; 40 000 mss; 750 incunabula; 600 music scores; 400 maps; 1 200 microforms; 30 digital data carriers 29016

– Genetics Library, Downing St, Cambridge CB2 3EH
T: +44 1223 333973; Fax: +44 1223 333992; E-mail: library@gen.cam.ac.uk; URL: www.gen.cam.ac.uk/library
Naomi Davies
Drosophila offprint coll
10 150 vols; 100 curr per; 3 e-journals; 115 600 pamphlets
libr loan 29017

– Geography Library, Downing Place, Cambridge CB2 3EN
T: +44 1223 333391; Fax: +44 1223 333392; E-mail: library@geog.cam.ac.uk; URL: www.geog.cam.ac.uk/library/
1903; Robert Carter
Clark Coll (exploration)
27 000 vols; 450 curr per; 50 diss/theses; 785 govt docs; 50 microforms; 30 av-mat; 20 sound-rec; 10 000 offprints, rpts, pamphlets
BLDSC 29018

– Girton College, Library, Huntingdon Rd, Cambridge CB3 0JG
T: +44 1223 338970; Fax: +44 1223 339890; E-mail: library@girton.cam.ac.uk; URL: www-lib.girton.cam.ac.uk
1869; Frances Gandy
Hebrew mss Coll, Womens Suffrage Coll, Hist of Women's higher education (Mss & published material)
95 000 vols; 125 curr per 29019

– Haddon Library of Archaeology & Anthropology, Haddon Library, Downing St, Cambridge CB2 3DZ
T: +44 1223 333505; Fax: +44 1223 333503; E-mail: haddon-library@lists.cam.ac.uk; URL: www.archanth.cam.ac.uk/library
1920; Aidan Baker
Book and offprint colls from the libs of Alfred Haddon, James Frazer, William Ridgeway, Charles McBurney, A.H. Pitt-Rivers, Miles Burkitt, J.M. De Navarro, G.H.S. Bushnell, Grahame Clark; Cambridge Antiquarian Soc Libr
67 159 vols; 299 curr per; 32 e-journals; 1 000 e-books; 1 000 diss/theses; 17 000 pamphlets
libr loan; CILIP 29020

– Institute of Astronomy, Library, The Observatories, Madingley Rd, Cambridge CB3 0HA
T: +44 1223 337537; Fax: +44 1223 337523; E-mail: ioalib@ast.cam.ac.uk; URL: www.ast.cam.ac.uk/library/
1823; Mr M. Hurn
John Couch Adams coll, pamphlet & offprint coll, Observatory publs
22 700 vols; 250 curr per; 812 diss/theses; 57 digital data carriers
libr loan; BLL 29021

– Institute of Continuing Education, Library, Madingley Hall, Madingley, Cambridge CB23 8AQ
T: +44 1954 280-280/-206; Fax: +44 1954 280200
1924; B. Pemberton
56 872 vols 29022

– Institute of Criminology, Radzinowicz Library, Sidgwick Avenue, Cambridge CB3 9DT
T: +44 1223 335375;
Fax: +44 1223 335356; E-mail: crimlib@hermes.cam.ac.uk; URL: crim.cam.ac.ut.uk/library
1959; Mary Gower
ESRC Qualidata Arch material
60 000 vols; 250 curr per; 50 e-journals; 10 000 govt docs; 18 digital data carriers; 20 000 pamphlets
BIALL, IALL, WCJLN 29023

– Jesus College, Quincentenary Library, Jesus Lane, Cambridge CB5 8BL
T: +44 1223 339451; Fax: +44 1223 324910; E-mail: Quincentenary-library@jesus.cam.ac.uk; URL: www.jesus.cam.ac.uk
Ms R.K. Watson
Quincentenary Libr, Old Libr, T.R. Malthus Libr
50 000 vols; 40 curr per; 90 mss 29024

– King's College, Library, King's Parade, Cambridge CB2 1ST
T: +44 1223 331232; Fax: +44 1223 331891; E-mail: library@kings.cam.ac.uk; URL: www.kings.cam.ac.ak/library
1441; P.M. Jones
Keynes Libr; Modern Archs, Rowe Music Libr
130 000 vols; 72 curr per; 554 mss; 200 incunabula; 200 diss/theses; 15 000 music scores; 200 microforms; 200 sound-rec
IAML 29025

– Lucy Cavendish College, Library, Lady Margaret Rd, Cambridge CB3 0BU
T: +44 1223 332183; Fax: +44 1223 332178; E-mail: lcc-admin@lists.cam.ac.uk; URL: www.lucy-cav.cam.ac.uk
C.A. Reid
20 000 vols; 10 curr per 29026

– Magdalene College, Old Library, Cambridge CB3 0AG
T: +44 1223 332100; Fax: +44 1223 332187
1492; N.G. Jones
The Pepys Libr (3 000 vols)
18 000 vols 29027

– Maitland Robinson Library of Downing College, Regent St, Cambridge CB2 1DQ
T: +44 1223 335352; URL: www.dow.cam.ac.uk
1800; Karen Lubarr
17th century historical pamphlets English civil war, Bowtell mss, Bowtell bks, Cuttle Coll, Richmond, Fisher, Graystone
50 000 vols; 35 curr per; 1 000 govt docs; 1 000 music scores; 50 maps; 100 microforms; 20 av-mat; 100 digital data carriers
CILIP 29028

– Marshall Library of Economics, Sidgwick Av, Cambridge CB3 9DB
T: +44 1223 335217; E-mail: marshlib@econ.cam.ac.uk; URL: www.econ.cam.ac.uk/marshlib/index.html
1925; Rowland Thomas
Alfred Marshall letters
65 000 vols; 350 curr per; 51 e-journals; 75 photos
libr loan 29029

– Materials Science and Metallurgy Library, Pembroke St, Cambridge CB2 3QZ
T: +44 1223 334318; Fax: +44 1223 334567; E-mail: library@msm.cam.ac.uk; URL: www.msm.cam.ac.uk/library
Dr. K.M. Knowles
14 200 vols; 71 curr per; 50 e-journals; 1 000 diss/theses; 10 microforms 29030

– Mill Lane Lecture Halls, Mill Lane Library, Mill Lane, Cambridge CB2 1RX
T: +44 1223 337110; Fax: +44 1223 337130; E-mail: wt10000@cam.ac.uk; URL: www.landecon.cam.ac.uk/library.htm
1962; W. Thurley
Agricultural economics, planning, real estate finance
22 000 vols; 240 curr per; 240 diss/theses; 1 000 govt docs
libr loan 29031

– Modern and Medieval Languages Library, Sidgwick Av, Cambridge CB3 9DA
T: +44 1223 335041;
Fax: +44 1223 335062; E-mail: mmllib@hermes.cam.ac.uk; URL: www.mml.cam.ac.uk/library
1900; Dr. A.E. Cobby
Beit Libr (German research coll)
110 200 vols; 43 curr per; 3 000 av-mat; 750 sound-rec; 200 digital data carriers
libr loan 29032

– Murray Edwards College – Rosemary Murray Library, Huntingdon Rd, Cambridge CB3 0DF
T: +44 1223 762202;
Fax: +44 1223 763110; E-mail: library@newhall.cam.ac.uk; URL: www.murrayedwards.cam.ac.uk/exploring/rosemarymurraylibrary/rosemarymurraylibrary/
1954; Alison M. Wilson
Women's studies, Eleonora Duse Coll
65 000 vols; 11 curr per; 609 e-books; 10 mss; 2 microforms; 300 av-mat; 10 sound-rec; 40 digital data carriers 29033

– Newnham College Library, Sidgwick Avenue, Cambridge CB3 9DF
T: +44 1223 335740; E-mail: librarian@newn.cam.ac.uk; URL: www.library.newn.cam.ac.uk
1882; D.K. Hodder
90 000 vols; 74 curr per; 9 mss; 16 incunabula 29034

– Pembroke College, Library, Cambridge CB2 1RF
T: +44 1223 338100; Fax: +44 1223 338163; E-mail: lib@pem.cam.ac.uk; URL: www.pem.cam.ac.uk
1347; T.R.S. Allan
Thomas Gray Coll, Christopher Smart Coll, Stores Papers
40 000 vols; 25 curr per; 317 mss; 90 incunabula; 250 music scores 29035

– Pendlebury Library of Music, West Rd, Cambridge CB3 9DP
T: +44 1223 335182; Fax: +44 1223 335183; URL: www.mus.cam.ac.uk
1880; Anna Pensaert
Picken Gift (early Bach editions), Walter Emery bequest (Bach sources on microfilm)
21 000 vols; 48 curr per; 140 diss/theses; 30 000 music scores; 1 500 microforms; 15 000 sound-rec; 1 500 pamphlets
MLA, IMS, RMA, IAML 29036

– Physiology, Development and Neuroscience Library, Downing St, Cambridge CB2 3EG
T: +44 1223 333821; Fax: +44 1223 333840; E-mail: library@pdn.cam.ac.uk
1920; Margaret Wilson
Anatomy, Physiology, Neuroscience, Developmental Biology, Embryology; Boyd Histological Coll, West Suffolk General Hospital Coll
10 860 vols; 118 curr per; 1 digital data carriers
libr loan 29037

– Plant Sciences Library, Downing St, Cambridge CB2 3EA
T: +44 1223 333930;
Fax: +44 1223 333953; E-mail: library@plantsci.cam.ac.uk; URL: www.plantsci.cam.ac.uk
Dr. D. Coomes
Simpson Coll of early local floras
20 100 vols; 274 curr per; 12 digital data carriers 29038

– Pure Mathematical and Mathematical Statistics Library, Top Fl, Mill Lane Lecture Rooms, Cambridge CB2 1RX
T: +44 1223 337998
1969; P.T. Johnstone
23 800 vols; 64 curr per; 80 diss/theses; 3 760 offprints 29039

– Queens' College, Library, Cambridge CB3 9ET
T: +44 1223 335549; Fax: +44 1223 335522; E-mail: keb36@cam.ac.uk; URL: www.quns.cam.ac.uk
1448; K.E. Begg
Thomas Smith, Erasmus, Isaac Milner, David Hughes, Cohen Coll of Latin American Lit
70 000 vols; 40 curr per; 50 mss; 20 incunabula; 20 digital data carriers; 500 vols 18th c pamphlets 29040

– Ridley Hall Library, Ridley Hall Rd, Cambridge CB3 9HG
T: +44 1223 741080
Rev. Dr Jeremy Begbie
18th c private libr of Robert Cecil
13 000 vols; 53 curr per 29041

– Scott Polar Research Institute, University of Cambridge, Library, Lensfield Road, Cambridge CB2 1ER
T: +44 1223 336552; Fax: +44 1223 336549; E-mail: library@spri.cam.ac.uk; URL: www.spri.cam.ac.uk
1920; Mrs H.E. Lane
Polar archives (mss and photos); Picture Library
150 000 vols; 740 curr per; 15 e-journals; 10 000 mss; 725 diss/theses; 20 music scores; 25 000 maps; 1 600 microforms; 630 av-mat; 30 sound-rec; 75 digital data carriers; 12 000 items
libr loan; Polar Libraries Colloquy 29042

– Selwyn College, Library, Grange Rd, Cambridge CB3 9DQ
T: +44 1223 335880; Fax: +44 1223 335837; E-mail: lib@sel.cam.ac.uk; URL: www.sel.cam.ac.uk
1882; Mrs Sarah Stamford
Bishop Selwyn arch; College arch
50 000 vols; 40 curr per; 3 000 19th C. Theological Pamphlets 29043

– Sidney Sussex College, Library, Sidney St, Cambridge CB2 3HU
T: +44 1223 338852; Fax: +44 1223 338884; E-mail: librarian@sid.cam.ac.uk; URL: www.sid.cam.ac.uk/life/lib/
1596; N.N.
Cromwell Coll; 18th-19th Century Mathematical Book Coll, Taylor Mathematical Libr
36 000 vols; 70 curr per; 106 mss; 75 music scores; 500 sound-rec; 60 digital data carriers
CILIP, Bliss Classification Association (BCA) 29044

– Squire Law Library, 10 West Rd, Cambridge CB3 9DZ
T: +44 1223 330077; Fax: +44 1223 330048; E-mail: dfw1003@cam.ac.uk; URL: www.squire.law.cam.ac.uk
1904; David Wills
130 000 vols; 2 000 curr per 29045

– St Catharine's College, Library, Cambridge CB2 1RL
T: +44 1223 338343; Fax: +44 1223 338340; E-mail: librarian@caths.cam.ac.uk; URL: www.caths.cam.ac.uk/library
1473; Dr. Richard S.K. Barnes
Thomas Sherlock Bequest, H.J. Chaytor Coll (medieval romance lit), Smith Coll (Darwinalia)
64 000 vols; 60 curr per; 1 000 e-books; 25 mss; 15 incunabula; 290 av-mat; 1 000 sound-rec 29046

– St John's College, Library, St John's St, Cambridge CB2 1TP
T: +44 1223 338662; Fax: +44 1223 337035; E-mail: library@joh.cam.ac.uk; URL: www.joh.cam.ac.uk/library/
1511; Dr. Mark Nicholls
John Couch Adams (papers), Cecil Beaton (letters, diaries), Rollo Brice Smith (modern handpress & limited ed bks), Samuel Butler Coll, Soulden Lawrence (law), Samuel Parr Coll, W.F. Smith (Rabelais lit), Sir Harold Jeffreys (papers), Sir Fred Hoyle (papers), James Wood Coll, William Wordsworth, G.U. Yule, Photogr & Prints of College, Biographical Arch of all members of

College
150 000 vols; 150 curr per; 10 000 mss; 270 incunabula; 500 music scores; 400 maps; 100 microforms; 1 000 av-mat; 1 000 sound-rec; 10 digital data carriers; 1 500 photogr glass negatives
FOL 29047

– Trinity College, Library, Cambridge CB2 1TQ
T: +44 1223 338488; Fax: +44 1223 338532; E-mail: trin-lib@lists.cam.ac.uk; URL: www.trin.cam.ac.uk/library
1546; D.J. McKitterick
William Whewell Coll, Tennyson notebks, Wittgenstein notebks, Hebrew mss, Capell Coll of Shakespeareana. Rothschild Libr (18th c bks)
300 000 vols; 250 curr per; 2 500 mss; 750 incunabula; 100 diss/theses; 250 music scores 29048

– Trinity Hall, Library, Trinity Lane, Cambridge CB2 1TJ
T: +44 1223 332546; Fax: +44 1223 332537; E-mail: amh55@cam.ac.uk
1350; Mrs A. Hunt, Dr A.C. Lacey
Canon law, civil law; Coll of European 16th c law bks, Larman bequest
20 000 vols; 32 curr per; 40 mss; 30 incunabula; 100 pamphlets 29049

– University Medical Library, Addenbrooke's Hospital, Hills Rd, P.O. Box 111, Cambridge CB2 2SP
T: +44 1223 336750; Fax: +44 1223 331918; E-mail: pbm2@cam.ac.uk; URL: www.his.path.cam.ac.ik/library/library.htm
P.B. Morgan
30 000 vols; 2 500 curr per; 3 e-journals; 1 300 diss/theses; 110 maps; 350 av-mat; 500 sound-rec; 3 700 pamphlets 29050

– Ward and Perne Libraries, Peterhouse, Trumpington St, Cambridge CB2 1RD
T: +44 1223 338218; Fax: +44 1223 37578; E-mail: lib@pet.cam.ac.uk; URL: www.pet.cam.ac.uk
1284; M.S. Golding
50 000 vols; 280 mss; 90 incunabula; 12 digital data carriers 29051

– Westminster College, Library, Madingley Rd, Cambridge CB3 0AA
T: +44 1223 741084;
Fax: +44 1223 300765; E-mail: admin@westminster.cam.ac.uk; URL: www.westminster.cam.ac.uk
1899; Dr. Peter McEnhill
Cheshunt college: correspondence of Selina Countess of Huntingdon, Elias Libr of Hymnology, Hebrew fragments from Cairo Genizeh
30 000 vols; 36 curr per; 50 mss 29052

– Wolfson College, Lee Library, Barton Rd, Cambridge CB3 9BB
T: +44 1223 335965; Fax: +44 1223 3353937; E-mail: library@wolfson.cam.ac.uk; URL: www.wolfson.cam.ac.uk
1994; Hilary Pattison
18 000 vols; 40 curr per; 2 microforms 29053

Canterbury Christ Church University College, Library Services, North Holmes Rd, ***Canterbury*** CT1 1QU
T: +44 1227 782352;
Fax: +44 1227 767530; E-mail: library.enquiries@canterbury.ac.uk; URL: www.canterbury.ac.uk/library/
1962; Pete Ryan
Historical children's literature, Elizabeth Gaskill
300 000 vols; 900 curr per; 3 000 music scores; 4 000 microforms; 3 000 av-mat; 2 000 sound-rec; 30 digital data carriers
libr loan 29054

University of Kent at Canterbury, Templeman Library, ***Canterbury*** CT2 7NU
T: +44 1227 823570; Fax: +44 1227 823984; E-mail: library-enquiry@kent.ac.uk; URL: www.kent.ac.uk/library/templeman/
1963; M.M. Coutts
Pettingell coll (mss and printed plays), John Crow coll (pre 1 800 bks), political cartoons, British Govt publs, 19th c drama coll; Europ Doc Centre, Cartoon Centre
850 000 vols; 3 700 curr per; 1 300 mss; 2 200 diss/theses; 95 000 govt docs; 1 000 music scores; 2 000 maps; 120 000 microforms; 86 000 av-mat; 500 sound-rec; 120 digital data carriers
libr loan; BLL, SCOLMA, SCONUL, NAG, ARLIS/UK & Ireland 29055

Cardif University, University Library, Colum Drive, P.O. Box 430, ***Cardiff*** CF10 3XT
T: +44 29 20874795; Fax: +44 29 20371921; E-mail: library@cardiff.ac.uk; URL: www.cf.ac.uk
1988; P.J. Martin
EC and UN deposit libr, Tennyson coll, Salisbury libr; Company Info Service, Welsh Music Information Service
750 000 vols
SCONUL 29056

Cardiff University, Trevithick Library, P.O. Box 430, ***Cardiff*** CF24 3AA
T: +44 2920 874286; Fax: +44 2920 874209; E-mail: trevliby@cardiff.ac.uk; URL: www.cf.ac.uk/insrv/libraries/trevithick/index.html
1989; Ruth Thornton
Engineering -civil, electrical, machanical; metallurgy, minteral exploitation; physics; computer science
100 000 vols; 490 curr per; 750 diss/theses; 250 govt docs; 100 maps; 50 microforms; 100 digital data carriers
libr loan 29057

Cardiff University, B2 Wales College of Medicine, Biology, Life and Health Sciences, Duthie Library, Heath Park, ***Cardiff*** CF4 4XN
T: +44 29 20742874;
Fax: +44 29 20743651; E-mail: duthieliby@Cardiff.ac.uk; URL: www.cf.ac.uk/insrv/libraries/duthie/index.html
1931; Mr SJ Pritchard
Hist Coll; Velindre Hospital Libr, Cochrane Libr, llandough Hospital, Cooke Dental Libr, School of Nursing Studies Libr, School of Healthcare Studies Libr
155 000 vols; 1 350 curr per; 50 mss; 2 200 diss/theses; 56 maps; 19 microforms; 500 av-mat; 450 digital data carriers
libr loan; SCONUL, EAHIL, RLG, AWHL, CILIP 29058

Royal Welsh College of Music and Drama, Library, Castle Grounds, Cathays Park, ***Cardiff*** CF10 3ER
T: +44 29 20342854; Fax: +44 29 20391304; E-mail: AgusJM@rwcmd.ac.uk
1975; Judith Agus
24 000 vols; 90 curr per; 22 000 music scores; 4 000 sound-rec; 10 digital data carriers; 330 video recordings, 60 DVDs
IAML 29059

University of Wales Institute, Cardiff, Library, Western Ave, Llandaff, ***Cardiff*** CF5 2YB
T: +44 29 20416240; Fax: +44 29 20416908; URL: www.uwic.ac.uk
1976; Paul Riley
580631 vols; 1 000 curr per; 9 500 e-journals; 1 000 diss/theses; 50 microforms; 90 000 av-mat; 25 sound-rec; 30 digital data carriers
libr loan; SCONUL, ARLIS/UK & Ireland, WHELF 29060

Carlisle College, Library, Victoria Place, ***Carlisle*** CA1 1HS
T: +44 1228 822760; E-mail: Library@carlisle.ac.uk; URL: www.carlisle.ac.uk
1950; Sally Frost
16 000 vols; 40 curr per; 250 govt docs; 50 maps; 350 av-mat; 40 sound-rec; 30 digital data carriers
libr loan; COFHE, CILIP, COLRIC 29061

University of Cumbria, Faculty of the Arts, Brampton Road Library, Brampton Rd, ***Carlisle*** CA3 9AY
T: +44 1228 400312; Fax: +44 1228 514491; URL: www.cumbria.ac.uk
C. Daniel
Cumberland & Westmorland Antiquarian & Archaeological Society Libr Coll
27 000 vols; 200 curr per; 763 e-journals; 350 av-mat; 30 000 slides
libr loan; Information North 29062

Anglia Ruskin University, University Library, Queens Building, Bishop Hall Lane, ***Chelmsford*** CM1 ISQ
T: +44 1245 683757; Fax: +44 1245 683149; E-mail: Diane.Hilton@anglia.ac.uk; URL: libweb.anglia.ac.uk/
Nicola Kershaw
330 000 vols; 1 200 curr per; 6 e-journals; 3 300 diss/theses; 16 500 microforms; 22 000 av-mat; 80 sound-rec; 550 press clippings 29063

Writtle College, Library, ***Chelmsford*** CM1 3RR
T: +44 1245 424200; Fax: +44 1245 420456; E-mail: thelibrary@writtle.ac.uk; URL: www.writtle.ac.uk
1969; Rachel Hewings
Agriculture, horticulture
48 000 vols; 450 curr per; 200 diss/theses; 250 maps; 480 av-mat; 6 digital data carriers
libr loan; ALLCU, SCONUL 29064

University of Gloucestershire, Pittville Campus Library, Albert Rd, ***Cheltenham*** GL52 3JG
T: +44 1242 532210; Fax: +44 1242 532207; URL: www.glos.ac.uk
1964; D. Thompson, A. Jeffery
Artists' books coll
50 000 vols; 150 curr per; 31 000 mss; 50 diss/theses; 2 000 av-mat; 20 digital data carriers; 96 000 Slides
BLL, SWRLB, ARLIS/UK & Ireland 29065

Chester College of Higher Education, Library, Parkgate Road, ***Chester*** CH1 4BJ
T: +44 1244 513304;
Fax: +44 1244 511325; E-mail: library.enquiries@chester.ac.uk; URL: www.chester.ac.uk/lr/
1956; Christine Stockton
200 000 vols
libr loan; IATUL 29066

West Cheshire College, Handbridge Library and Learning Centre, Eaton Rd, Handbridge, ***Chester*** CH4 7ER
T: +44 1244 670574, 670579; Fax: +44 1244 670584; E-mail: library@west-cheshire.ac.uk; URL: www.west-cheshire.ac.uk
C. Martin
10 000 slides, 5 000 pamphlets, 700 videos
40 000 vols; 250 curr per; 200 maps; 700 av-mat; 200 sound-rec; 100 digital data carriers; 10 000 slides, 5 000 pamphlets
libr loan; COFHE 29067

University of Chichester, Learning Resource Centre, Bishop Otter Campus, College Lane, ***Chichester*** PO19 6PE
T: +44 1243 816089; Fax: +44 1243 816096; URL: www.chi.ac.uk
1997; Scott Robertson
Ted Walker Coll, Gerard Young Coll, Bishop Kemp Coll, Chichester Theological Collection
271 000 vols; 652 curr per; 52 e-journals; 47 000 e-books; 500 diss/theses; 2 000 music scores; 100 microforms; 15 000 av-mat; 1 200 sound-rec
libr loan; SCONUL 29068

Royal Agricultural College, Hosier Library, Stroud Rd, ***Cirencester*** GL7 6JS
T: +44 1285 652531; Fax: +44 1285 889844; E-mail: library@rac.ac.uk; URL: rac.ac.uk
1847; Peter Brooks
Land management, Rural Development, Agribusiness; Coll of rare/antique bks on agriculture and land management (ca. 2 000)
30 000 vols; 600 curr per; 70 000 e-books; 615 diss/theses; 100 maps; 600 av-mat; 40 digital data carriers
libr loan; ALLCU, CILIP, SCONUL 29069

Stephenson College, Library, Bridge Rd, ***Coalville*** LE6 2QR
T: +44 1530 836136;
Fax: +44 1530 814253; E-mail: Deniser@stephensoncoll.ac.uk; URL: www.stephensoncoll.ac.uk
1920; Ms D. Rossell
Local History, Coalville, Leicestershire
20 000 vols; 118 curr per; 80 digital data carriers
libr loan 29070

University of Essex, Albert Sloman Library, Wivenhoe Park, ***Colchester*** CO4 3SQ
T: +44 1206 873333; Fax: +44 1206 872289; E-mail: robert@essex.ac.uk; URL: libwww.essex.ac.uk
1963; R. Butler
Latin American studies 82 000 vols; Russian and Soviet studies 80 000 vols; US studies 57 000 vols, rare books and named coll 26 000 vols
840 000 vols; 7 329 curr per; 120 000 govt docs; 2 000 maps; 143 000 microforms; 2 200 av-mat; 3 000 sound-rec; 600 digital data carriers; 446 linear m of mss
SCONUL, OCLC 29071

Royal Forest of Dean College, Learning Resource Centre, Five Acres Campus, Berry Hill, ***Coleford*** GL16 7JT
T: +44 1594 838522; Fax: +44 1594 837497
1964; Mrs J.A. Offord
local information to Forest of Dear, Gloucestershire
26 000 vols; 70 curr per; 200 maps; 10 digital data carriers
libr loan 29072

University of Ulster Coleraine Campus, Library, Cromore Rd, ***Coleraine*** BT52 1SA
T: +44 28 70324345; Fax: +44 28 70324928; URL: www.ulst.ac.uk/library
1968; Stephanie McLaughlin
Henry Davis Gift (rare bks), Stelfox Coll (nat hist), Headlam-Morley (World War I), Galbraith Coll (Mediaeval hist), Morris (Irish studies), John Hewitt (Anglo-Irish lit); Europ Doc Centre
260 000 vols; 2 100 curr per
CILIP, LAI, SCONUL, ASLIB 29073

Coventry University, Lanchester Library, Gosford St, ***Coventry*** CV1 5DD
T: +44 24 76887575; Fax: +44 24 76887525; E-mail: lbx022@coventry.ac.uk; URL: www.coventry.ac.uk
1970; Patrick Noon
Frederick Lanchester Coll; Europ Doc Centre, Patents info coll
350 000 vols; 2 000 curr per; 12 000 e-journals; 600 diss/theses; 20 000 govt docs; 3 500 music scores; 15 000 microforms; 20 000 av-mat; 66 000 ills, 110 000 slides
libr loan; SCONUL, ARLIS/UK & Ireland, BIALL 29074

University of Warwick, Library, Gibbet Hill Rd, ***Coventry*** CV4 7AL
T: +44 24 76523033; Fax: +44 24 76524211; E-mail: library@warwick.ac.uk; URL: library.warwick.ac.uk
1963; Anne Bell
Modern Records Centre, BP Archive
1 000 000 vols; 13 000 curr per; 9 000 e-journals; 6 000 diss/theses; 1 500 maps; 161 000 microforms; 40 000 av-mat; 900 sound-rec; 100 digital data carriers
libr loan; SCONUL, CURL, RLG, ASLIB 29075

Cranfield University, Kings Norton Library, ***Cranfield*** MK43 0AL
T: +44 1234 754444;
Fax: +44 1234 752391; E-mail: hazel.woodward@cranfield.ac.uk; URL: www.cranfieldlibrary.cranfield.ac.uk
1946; Hazel Woodward
98 000 vols; 4 489 curr per; 3 765 e-journals; 16 700 diss/theses; 2 300 microforms; 850 digital data carriers; 122 000 technical aerospace rpts
ASLIB, CILIP, AIAA 29076

Croydon College, Library, College Rd, ***Croydon*** CR9 1DX
T: +44 20 87605843; E-mail: library@croydon.ac.uk; URL: www.croydon.ac.uk/library
1957; Thomas Butler
Art and Design
45 673 vols; 80 curr per; 4 500 e-books; 20 maps; 3 750 av-mat; 82 digital data carriers
libr loan 29077

Darlington College, Learning Resource Centre, Central Park, Haughton Rd, ***Darlington*** DLI I1DR
T: +44 1325 503249; Fax: +44 1325 503000; E-mail: enquire@darlington.ac.uk; URL: www.darlington.ac.uk

1961; Ms M.L. Dack
15 000 vols; 120 curr per; 1 000 e-journals; 4 diss/theses; 2 000 govt docs; 50 music scores; 250 maps; 700 av-mat; 50 sound-rec; 130 digital data carriers
CILIP, BUFVC 29078

Britannia Royal Naval College, Library, ***Dartmouth*** TQ6 0HJ
T: +44 1803 677279; Fax: +44 1803 677015
1905; R.J. Kennell
Simmons Coll of World War 1 and World War 2 (2 500 vols)
52 000 vols; 180 curr per 29079

University of Derby, Kedleston Road Library, Kedleston Rd, ***Derby*** DE22 1GB
T: +44 1332 59120-5/-6; Fax: +44 1332 622767; URL: ulib.derby.ac.uk/library
1977; Richard Maccabee
19th c children's books, Derbyshire and East Midlands
260 000 vols; 2 000 curr per; 7 000 av-mat; 2 000 sound-rec; 60 digital data carriers; 90 000 slides
libr loan; SCONUL 29080

Doncaster College, Learning Resource Centre, The Hub, Doncaster College, Chappell Drive, ***Doncaster*** DN1 2RF
T: +44 1302 553713; Fax: +44 1302 553559; E-mail: enquiry.desk@don.ac.uk; URL: www.don.ac.uk
1976; Ann Hill
Business studies, art and design, management, social science, local industries, engineering, ICT, intermedia and performance arts, hair, beauty, floristry, fashion
200 000 vols; 791 curr per
libr loan; SINTO, BL, ASLIB, COLRIC 29081

Dudley College of Technology, Library, The Broadway, ***Dudley*** DY1 4AS
T: +44 1384 363353; Fax: +44 1384 363311
1955
64 000 vols; 170 curr per; 31 e-journals; 2 000 av-mat
BLL 29082

Dumfries and Galloway College, Library, Bankend Road, ***Dumfries*** DG1 4FD
T: +44 1387 734323; Fax: +44 1387 734040; E-mail: info@dumgal.ac.uk; URL: www.dumgal.ac.uk
A. Guthrie
25 000 vols; 70 curr per; 200 maps; 1 120 av-mat; 100 sound-rec; 30 digital data carriers; 50 files of clippings, 1 000 pamphlets
libr loan; SLIC 29083

University of Abertay Dundee, Bernard King Library, Bell St, ***Dundee*** DD1 1HG
T: +44 1382 308833; Fax: +44 1382 308877; E-mail: library@abertay.ac.uk; URL: www.abertay.ac.uk
1964; Michael Turpie
121 097 vols; 273 curr per; 5 621 e-journals; 672 e-books; 1 234 diss/theses; 17 music scores; 850 maps; 1 665 av-mat; 310 sound-rec; 2 436 digital data carriers; 8 327 pamphlets, 267 computer games; Archives (753), Local coll (1 321), Faculty of Procurators (2 126)
libr loan; SCONUL, SCURL 29084

University of Dundee, Library and Learning Centre, ***Dundee*** DD1 4HN
T: +44 1382 384087; Fax: +44 1382 386228; E-mail: library@dundee.ac.uk; URL: www.dundee.ac.uk/library
1967; R. Parsons
Nicoll coll (fine art), Leng coll (Scottish philos), William Lyon Mackenzie coll (Canadiana), Thoms coll (mineralogy), Kinnear local coll, Brechin Diocesan lib
500 000 vols; 3 000 curr per; 3 mss; 6 incunabula
libr loan; BLL, SCONUL, CILIP, SCURL 29085

Carnegie College, George Lauder Library, Halbeath, ***Dunfermline*** KY11 5DY
T: +44 1383 845000;
Fax: +44 1383 845001; E-mail: library@carnegiecollege.ac.uk; URL: www.carnegiecollege.ac.uk
1899; Tom MacMaster
18 000 vols; 50 curr per; 10 e-journals; 400 av-mat; 20 sound-rec; 2 digital data carriers; Some British standards
libr loan; SLIC, CILIPS, COLRIC, TAFLIN 29086

Durham University, University Library, Stockton Rd, ***Durham*** DH1 3LY
T: +44 191 3343042;
Fax: +44 191 3342971; E-mail: main.counter@durham.ac.uk; URL: www.dur.ac.uk/library
1833; Mr Jon Purcell
European Doc Centre, Middle East Doc Unit
820 000 vols; 3 000 curr per
libr loan; BLDSC 29087

– Education Library, Leazes Rd, Durham DH1 1TA
T: +44 191 3348137;
Fax: +44 191 3248311; E-mail: educ.library@durham.ac.uk; URL: www.dur.ac.uk/library/
1963; Mr J. Purcell
50 000 vols; 250 curr per; 2 e-journals
libr loan 29088

– Palace Green Library, Palace Green, Durham DH1 3RN
T: +44 191 3342932; Fax: +44 191 3342942; E-mail: pg.library@durham.ac.uk; URL: www.dur.ac.uk/library/
1834; Sheila Hingley
Renaissance medicine, cryptography, hymnology, modern literary mss, Sudan; Law dept, Music dept
177 000 vols; 352 curr per; 9 000 mss; 204 incunabula; 860 diss/theses; 5 000 music scores; 800 maps; 300 microforms; 5 000 photos, 1 000 topographical and portrait prints 29089

Ushaw College, Library, ***Durham*** DH7 9RH
T: +44 191 3738516; E-mail: alistair.macgregor@ushaw.ac.uk; URL: www.ushaw.ac.uk
1808; Alistair MacGregor
Papers and correspondence of John Lingard and Nicholas Wiseman, Horace Mann Coll (medieval church hist); Weld Bank parish libr
60 000 vols; 60 curr per; 50 mss; 150 incunabula; 23 diss/theses; 6 maps; 3 sound-rec; 6 000 pamphlets 29090

Edinburgh College of Art, Library, Lauriston Place, ***Edinburgh*** EH3 9DF
T: +44 131 221 6033; Fax: +44 131 221 6033; E-mail: w.smith@eca.ac.uk; URL: eca.lib.ed.ac.uk
1907; Wilson Smith
Architectural Reference, Fine Art, Architecture, Applied Art, Landscape Architecture, Town & Country Planning, Housing; Art Design Libr, Environmental Studies Libr
120 000 vols; 450 curr per; 6 000 diss/theses; 10 400 govt docs; 5 000 maps; 150 000 av-mat; 128 sound-rec; 100 digital data carriers
libr loan; CILIP, ARLIS/UK & Ireland, ARCLIB, SLIC, SALCTG 29091

Heriot-Watt University, Library, Riccarton, ***Edinburgh*** EH14 4AS
T: +44 131 4513570; Fax: +44 131 4513164; E-mail: m.l.breaks@hw.ac.uk; URL: www.hw.ac.uk/library
1821; Michael Breaks
Brewing, optoelectronics, petroleum and offshore engineering
170 000 vols; 2 600 curr per; 5 000 e-journals; 3 900 diss/theses
libr loan; SCONUL, IATUL 29092

– Martindale Library, Scottish Borders Campus, Netherdale, Galashiels TD1 3HF
T: +44 1896 892185; Fax: +44 1896 758965; URL: www.hw.ac.uk/sbc/library
1964; Peter Sandison
'Co-operative coll at Heriot-Watt University, Historic coll of books, journals, and records about the co-operative movement
16 000 vols; 170 curr per; 50 diss/theses; 30 microforms; 1 300 av-mat; 70 digital data carriers
libr loan; CILIP, Aslib, IATUL 29093

Moray House College, Holyrood Campus, Library, Holyrood Rd, ***Edinburgh*** EH8 8AQ
T: +44 131 5586193; Fax: +44 131 5573458
1907; Denny Colledge
102 000 vols; 684 curr per; 470 microforms; 24 000 av-mat
libr loan 29094

Napier University, Learning Information Services, Sighthill Court, ***Edinburgh*** EH11 4BN
T: +44 131 4553558;
Fax: +44 131 4553566; E-mail: international@napier.ac.uk; URL: nulis.napier.ac.uk
1966; C.J. Pinder
Edward Clark Coll, War Poets Coll; Napier Univ Libr consists of 8 campus libs in Edinburgh, Melrose and Livingston
200 000 vols; 3 000 curr per; 5 mss; 100 incunabula; 200 diss/theses; 1 000 music scores; 3 500 av-mat; 4 000 sound-rec; 100 digital data carriers
libr loan; CILIP, ASLIB 29095

Royal College of Surgeons of Edinburgh, Nicolson St, ***Edinburgh*** EH8 9DW
T: +44 131 5271600; Fax: +44 131 5576406; E-mail: library@rcsed.ac.uk; URL: www.library.rcsed.ac.uk
1505; Ms Marianne Smith
Sir James Young Simpson papers, Joseph Lister papers, Conan Doyle and Joseph Bell papers, many other medical historical individuals' materials
33 000 vols; 123 curr per; 20 000 mss; 50 incunabula; 200 diss/theses; 1 000 av-mat; 200 sound-rec; 500 digital data carriers
libr loan 29096

SAC – Scottish Agricultural College, Agriculture Library, Peter Wilson Building, Kings Buildings, West Mains Road, ***Edinburgh*** EH9 3JG
T: +44 131 5354116; Fax: +44 131 5354246; E-mail: LibraryEd@sac.ac.uk; URL: www.sac.ac.uk/library
1945; Dawn McQuillan
100 000 vols; 300 curr per; 50 e-journals; 500 diss/theses; 150 av-mat; 10 digital data carriers; 12 000 pamphlets
libr loan; CILIP, IAALD, SALG 29097

University of Edinburgh, University Library, George Sq, ***Edinburgh*** EH8 9LJ
T: +44 131 6503384; Fax: +44 131 6679780; E-mail: library@ed.ac.uk; URL: www.lib.ed.ac.uk
1580; Ian R.M. Mowat
Hist of Scotland and the UK, European hist and lit, coll of early printed bks, mediaeval and modern Western and Oriental mss
2 456 320 vols; 9 633 curr per; 516 e-journals; 27 405 diss/theses; 106 705 maps; 225 690 microforms; 4 000 av-mat
libr loan; SCOLMA, SCONUL, IFLA, LIBER, SCOLCAP 29098

– Law & Europa Library, Old College, South Bridge, Edinburgh EH8 9YL
T: +44 131 6502046; Fax: +44 131 6506343; URL: www.lib.ed.ac.uk
1959; Elizabeth Stevenson
114 000 vols; 612 curr per; 20 000 microforms; 10 av-mat; 30 digital data carriers
libr loan; BIALL, SLLG, EIA, EUDUG 29099

– New College Library, Mound Place, Edinburgh EH1 2LU
T: +44 131 6508957; Fax: +44 131 6507952; E-mail: New.College.Library@ed.ac.uk
1843; Christine Love-Rodgers
234 000 vols; 438 curr per; 3 200 mss; 100 incunabula
libr loan; ABTAPL, ATLA, Hebraica Libr Group 29100

– Psychiatry Library, Dept of Psychiatry, Royal Edinburgh Hospital, Morningside Park, Edinburgh EH10 5HF
T: +44 131 5376285
Ms W. Mill
Hist of medicine
10 000 vols; 106 curr per; 333 diss/theses; 1 200 pamphlets 29101

– Reid Music Library, Alison House, Nicolson Sq, Edinburgh EH8 9DF
T: +44 131 6502436;
Fax: +44 131 6502425; E-mail: reid.music.library@ed.ac.uk; URL: www.lib.ed.ac.uk/lib/sites/music.shtml
1847; Teresa Jones
Weiss coll of Beethoven material, Tovey coll, Kenneth Leighton mss, Dallapiccola complete printed music
25 000 vols; 75 curr per; 100 mss; 120 diss/theses; 65 000 music scores; 9 000 sound-rec; 15 digital data carriers
IAML 29102

– Robertson Engineering and Science Library, The King's Bldgs, West Mains Rd, Edinburgh EH9 3JF
T: +44 131 6505666; Fax: +44 131 6506702; E-mail: j.flemington@ed.ac.uk; URL: www.lib.ed.ac.uk/sites/robe.shtml
Jenny Flemington
562 800 vols; 1 565 curr per; 99 mss; 2 152 diss/theses; 747 maps
libr loan 29103

– Royal (Dick) School of Veterinary Studies, Library, Summerhall, Edinburgh EH9 1QH
T: +44 131 6506175; Fax: +44 131 6506594; E-mail: dick.vetlib@ed.ac.uk; URL: www.lib.ed.ac.uk/lib/sites/vetl.shtml
1951; A. Kennett
Hist of veterinary medicine; Ctr for Tropical Veterinary Medicine Libr and Easter Bush Vet Centre Lib (both located in Easter Bush, Roslin, Midlothian)
40 000 vols; 452 curr per; 157 mss; 577 diss/theses; 15 microforms; 249 av-mat
libr loan 29104

– Scottish Studies Library, 27-29 George Square, Edinburgh EH8 9LD
T: +44 131 650 3060; E-mail: scottish.studies@ed.ac.uk; URL: www.lib.ed.ac.uk/sites/scostud.shtml
1951
8 000 vols; 250 curr per; 10 000 prints, 9 000 seperate sheets
libr loan 29105

University of Exeter, Library, Stocker Rd, ***Exeter*** EX4 4PT
T: +44 1392 263869; Fax: +44 1392 263871; E-mail: library@exeter.ac.uk; URL: www.ex.ac.uk/library
1937; Michele Shoebridge
Literary figures of the South-West; Bill Douglas and Peter Jewell coll of cinema history
1 200 000 vols; 3 500 curr per; 7 000 e-journals; 10 000 mss; 6 incunabula; 3 000 diss/theses; 30 000 microforms; 120 000 av-mat; 8 000 sound-rec; 50 digital data carriers
libr loan; SCONUL, SCOLMA, LIBER, IATUL, CILIP 29106

– Devon and Exeter Institution Library, 7 Cathedral Close, Exeter EX1 1EZ
T: +44 1392 251017; URL: www.ex.ac.uk/library/devonex.html
1813; T. Gardner
South West (Cornwall, Devon, Dorset, Somerset)
36 000 vols; 100 curr per; 30 mss; 1 incunabula; 1 000 maps; 1 500 pamphlets, 250 files of clippings
Association of Independent Libraries (AIL) 29107

– St Luke's Library, Heavitree Rd, Exeter EX1 2LU
T: +44 1392 264785; Fax: +44 1392 264784; E-mail: library@exeter.ac.uk; URL: www.ex.ac.uk
Roy Davies
Education; Nat curriculum archive
130 000 vols; 380 curr per; 3 508 e-journals; 2 000 diss/theses; 12 digital data carriers
BL, LISE 29108

Falmouth College of Arts, Library and Information Services, Woodlane, ***Falmouth*** TR11 4RA
T: +44 1326 213815; Fax: +44 1326 211205; E-mail: library@falmouth.ac.uk
1964; Roger C. Towe
Cornwall artist's, artists' bks, contemporary artists, exhibition cat
20 000 vols; 170 curr per; 1 500 av-mat; 4 150 sound-rec; 70 000 slides, clipping files
libr loan; ARLIS/UK & Ireland 29109

Surrey Institute of Art and Design LLRC, Library, Falkner Road, ***Farnham*** GU9 7DS
T: +44 1252 892709; Fax: +44 1252 892725; E-mail: library@ucreative.ac.uk; URL: www.ucreative.ac.uk
19th c; Rosemary Lynch (Director of LLCs), Gwynneth Wilkey (College Librarian)
85 000 vols; 307 curr per

libr loan; ARLIS/UK & Ireland, SCONUL, UKSG, NAG 29110

Blackpool and The Fylde College, Fleetwood Nautical Campus Library, Broadwater, ***Fleetwood*** FY7 8JZ
T: +44 1253 352352 ext 4035
1965; C. MacDermott
Maritime studies
12 000 vols; 100 curr per; 1 000 govt docs; 2 000 av-mat
libr loan 29111

Gateshead College, Learning Centre, Library, Quarryfield Rd, Baltic Business Quarter, ***Gateshead*** NE8 3BE
T: +44 191 4902249; Fax: +44 191 4902313; URL: www.gateshead.ac.uk
1957; Louise Docherty
36 100 vols; 50 curr per; 1 556 av-mat; 18 digital data carriers
BL, CILIP, COFHE 29112

Central College of Commerce, Library, 300 Cathedral St, ***Glasgow*** G1 2TA
T: +44 141 5523941 ext 2263; Fax: +44 141 5525514; E-mail: information@central-glasgow.ac.uk; URL: www.centralcollege.ac.uk
1962; Kirsteen J. Dowie
27 500 vols; 160 curr per; 300 av-mat; 32 digital data carriers 29113

Glasgow Caledonian University, University Library, City campus, Cowcaddens Rd, ***Glasgow*** G4 0BA
T: +44 141 2731000; Fax: +44 141 3313005; E-mail: cbsliaison@gcal.ac.uk; URL: www.learningservices.gcal.ac.uk/library
1992; N.N.
Caledonian Coll, David Donald Coll, Gallacher Memorial Library, George Johannes Coll, James Martin Milligan Coll, John Lenihan Coll on the History and Philosophy of Science, Kevin Morrison Coll, Mike Scott Coll, Norman & Janey Buchan Coll, Norrie McIntosh Coll, Queen's College Special Coll, Samuel Stewart Coll, Sandy Hobbs Coll, Scottish Northern Book Distribution Co-operative Coll, Spanish Civil War Coll, William Kemp Coll, William Taylor Coll; Heatherbank Museum of Social Work, University Archives, Special Coll, Centre for Political Song
220 000 vols; 1 700 curr per; 10 000 e-journals; 370 diss/theses; 500 govt docs; 350 maps; 2 500 microforms; 350 av-mat; 1 000 sound-rec; 450 digital data carriers; 1 000 pamphlets
libr loan; ASLIB, SCONUL 29114

Glasgow College of Nautical Studies, Library, 21 Thistle St, ***Glasgow*** G5 9XB
T: +44 141 5652582; Fax: +44 141 5652599; E-mail: resources@gcns.ac.uk
1969; M. Scalpello
Navigation, marine engineering, telecommunications, electronics, instrumentation and control, dangerous cargoes, seamanship, yachting, maritime hist and law, merchant shipping
17 000 vols; 170 curr per; 1 000 govt docs; 80 maps; 9 digital data carriers; 300 British standards
libr loan; CILIP, SLIC 29115

Glasgow Dental Hospital and School, James Ireland Library, 378 Sauchiehall St, ***Glasgow*** G2 3JZ
T: +44 141 2119705; Fax: +44 141 3312798; E-mail: library@dental.gla.ac.uk; URL: www.gla.ac.uk/schools/dental/
1949; Ms B.W. Rankin
Hist of dentistry
12 000 vols; 160 curr per; 182 diss/theses; 200 govt docs; 85 av-mat; 10 digital data carriers; 542 pamphlets, 16 tape/slide programmes
libr loan 29116

Glasgow School of Art, Library, 167 Renfrew St, ***Glasgow*** G3 6RQ
T: +44 141 3534550; Fax: +44 141 3534670; E-mail: c.nicholson@gsa.ac.uk; URL: www2.gsa.ac.uk/library
1846; Catherine Nicholson
80 000 vols; 300 curr per; 500 diss/theses; 60 000 av-mat; 200 digital data carriers
libr loan; ARLIS/UK & Ireland, SLIC 29117

Royal College of Physicians and Surgeons, 232-242 St. Vincent St, ***Glasgow*** G2 5RJ
T: +44 141 2273234; Fax: +44 141 2211804; E-mail: library@rcpsg.ac.uk; URL: www.rcpsg.ac.uk
1698; Carol Parry
Mac Ewen Coll on surgery, Ross Coll on tropical medicine, coll of 19th c medical essays
28 000 vols; 70 curr per; 12 e-journals; 200 mss; 5 incunabula
EAHIL 29118

Royal Scottish Academy of Music and Drama, Library, 100 Renfrew St, ***Glasgow*** G2 3DB
T: +44 141 3324101; Fax: +44 141 2708353; E-mail: library@rsamd.ac.uk; URL: www.rsamd.ac.uk
1929; Gordon Hunt
16 700 vols; 76 curr per; 10 mss; 86 000 music scores; 46 microforms; 528 av-mat; 13 483 sound-rec; 15 digital data carriers; 1 103 orchestral sets, 784 choral sets
IAML, SIBMAS, BLDSC, SCONUL, GALT 29119

Stow College Learning Centre, 43 Shamrock Street, ***Glasgow*** G4 9LD
T: +44 1413321786; Fax: +44 141 3325207; E-mail: lvaughan@stow.ac.uk; URL: www.stow.ac.uk
1963; Linda Vaughan
Engineering, Music Technology, Antiques, COLEG learning packs
13 000 vols; 40 curr per; 14 000 e-journals; 3 600 e-books; 400 govt docs; 75 music scores; 140 av-mat; 50 sound-rec; 50 digital data carriers; Offprint collections on topics e.g. sound recording, mental health, bridge construction, disability and teaching
libr loan; SLIC 29120

University of Glasgow, University Library, Hillhead Street, ***Glasgow*** G12 8QE
T: +44 141 3306704; Fax: +44 141 3304952; E-mail: library@lib.gla.ac.uk; URL: www.lib.gla.ac.uk/AboutLibrary/
1963; Mrs Chris Bailey
Psychology, pedagogy, orthopedagogy, sociology, psychiatry, social work
12 000 vols; 150 curr per; 33 000 mss; 1 000 incunabula; 12 545 diss/theses; 250 000 govt docs; 1 606 music scores; 50 880 maps; 411 142 microforms; 1 122 av-mat; 4 000 sound-rec; 90 digital data carriers; 80 000 pamphlets, 5 000 photos, 1 600 e-journals, 160 other on-line subscriptions, 1 753 electronic media
libr loan; ASLIB, CILIP, CURL, BIALL, SCOLMA, SCOLUG, IAML 29121

– Adam Smith Library, Adam Smith Bldg, 40 Bute Gardens, Hillhead, Glasgow G12 8RT
T: +44 141 3305648; E-mail: adamsmith@lib.gla.ac.uk; URL: www.lib.gla.ac.uk/AboutLibrary/adam.shtml
1968; Kerr Ross
22 177 vols; 7 curr per; 1 257 diss/theses 29122

– Chemistry Branch, Library, Joseph Black Bldg, Glasgow G12 8QQ
T: +44 1413305502; E-mail: library@lib.gla.ac.uk 29123

– Russian and East European Studies Library, Hillhead St, Glasgow G12 8QE
T: +44 141 3306735; Fax: +44 141 3304952; E-mail: t.konn@lib.gla.ac.uk
1948; T. Konn-Roberts
CEEBIS (Central and East European Business Information Service)
80 000 vols; 600 curr per; 300 microforms; 5 digital data carriers
COSEELIS 29124

– Veterinary School, James Herriot Library, Garscube Estate, Bearsden, Glasgow G61 1QH
T: +44 141 3305708; Fax: +44 141 3304888; E-mail: vetlib@lib.gla.ac.uk; URL: www.lib.gla.ac.uk
M. Findlay
17 130 vols; 83 curr per; 1 600 e-journals; 160 other on-line subscriptions 29125

University of Strathclyde, Andersonian Library, Curran Bldg, 101 St. James' Rd, ***Glasgow*** G4 0NS
T: +44 141 5484620; Fax: +44 141 5523304; E-mail: library@strath.ac.uk; URL: www.lib.strath.ac.uk
1796; Michael Roberts
John Anderson Coll, Young Coll (alchemy, chemistry), Laing Coll (mathematics), Robertson Coll (economic and social history of West of Scotland), annual company reports; Archives & Special Colls
938 000 vols; 1 164 curr per; 26 986 e-journals; 126 294 e-books; 2 000 mss; 12 340 diss/theses; 75 000 govt docs; 600 music scores; 2 400 maps; 201 000 microforms; 1 000 av-mat; 50 sound-rec; 250 digital data carriers; 4 000 company reports, 300 market research reports
libr loan; CILIP, IATUL, OCLC, SCONUL, SCURL 29126

– Jordanhill Library, 76 Southbrae Drive, Glasgow G13 1PP
T: +44 141 9503300; Fax: +44 141 9503150; E-mail: jordanhill.library@strath.ac.uk; URL: www.strath.ac.uk/jhlibrary
1845
185 641 vols; 13 946 av-mat 29127

Oldham College Library, Learning Resources Unit, Rochdale Rd, Oldham, ***Greater Manchester*** OL9 6AA
T: +44 161 6245214; Fax: +44 161 7854234; E-mail: info@oldham.ac.uk; URL: www.oldham.ac.uk/library
1967; Margarete Wood
14 500 vols; 93 curr per
libr loan; CoFHE, CoLRIC, BLA 29128

Guildford College of Further and Higher Education, Library and Resources Centre, Stoke Park, ***Guildford*** GU1 1EZ
T: +44 1483 448500; Fax: +44 1483 448600; E-mail: library@guildford.ac.uk; URL: www.guildford.ac.uk/ServicesFacilities/LRC/Welcome.aspx
1943; Diana Marshall
Health care, education, business, tourism
50 000 vols; 400 curr per; 2 000 govt docs; 50 music scores; 1 200 av-mat; 60 digital data carriers
libr loan; CILIP 29129

Guildford Institute Library, Ward St, ***Guildford*** GU1 4LH
T: +44 1483 562142; Fax: +44 1483 451034; E-mail: library@guildford-institute.org.uk; URL: www.guildford-institute.org.uk
1834; Liz Markwell
Prints of Guildford and surrounding area (1750-1850)
14 000 vols; 2 diss/theses; 40 music scores; pamphlets, albums of ephemera
AIL 29130

University of Surrey, University Library, George Edwards Building, ***Guildford*** GU2 7XH
T: +44 1483 689287; Fax: +44 1483 689500; E-mail: library-enquiries@surrey.ac.uk; URL: www.surrey.ac.uk/Library
1968; Robert Hall
Shepard coll, European Doc Centre, National Resource Centre for Dance
400 000 vols; 3 000 curr per; 2 500 mss; 9 000 diss/theses; 3 500 music scores; 200 maps; 6 000 microforms; 500 sound-rec; 55 digital data carriers
libr loan; CILIP, SCONUL, IATUL, Sconul Research Extra, M25, UK Libraries Plus 29131

University of Hertfordshire, Learning and Information Services, College Lane, ***Hatfield*** AL10 9AB
T: +44 1707 284678; Fax: +44 1707 284666; E-mail: d.martin@herts.ac.uk; URL: www.herts.ac.uk/lis
1952; D.E. Martin
600 000 vols; 27 000 curr per
SCONUL, IFLA, IATUL, ASLIB 29132

Henley Management College, Library, Greenlands, ***Henley-on-Thames*** RG9 3AU
T: +44 1491 418823; Fax: +44 1491 418896; E-mail: library@henleymc.ac.uk; URL: www.henleymc.ac.uk
1948; Nightingale
Urwick papers, market research rpt
10 000 vols; 100 curr per; 13 000 e-journals; 2 500 diss/theses; 210 av-mat; 11 digital data carriers
libr loan; ASLIB, BBSLG, EBSLG 29133

Buckinghamshire New Unversity, Learning Resource Centre, Library, Queen Alexandra Road, ***High Wycombe*** HP11 2JZ
T: +44 1494 522141; Fax: +44 1494 450774; E-mail: hwlib@bucks.ac.uk; URL: www.bucks.ac.uk/library
1961
Regional Furniture Arch
195 934 vols; 2 693 curr per; 1 000 av-mat
libr loan; ASLIB 29134

Havering College of Further and Higher Education, Library, 42 Ardleigh Green Rd, ***Hornchurch*** RM11 2LL
T: +44 1708 462857; Fax: +44 1708 462788; E-mail: LRCAG@havering-college.ac.uk
1964; Audrey Stranders
54 045 vols; 210 curr per; 25 e-journals; 15 e-books; 476 diss/theses; 1 366 av-mat; 173 sound-rec; 907 digital data carriers; 474 pamphlets
CILIP, COLRIC 29135

University of Huddersfield, The Library and Computing Centre, Queensgate, ***Huddersfield*** HD1 3DH
T: +44 1484 473888; Fax: +44 1484 472385; URL: www.hud.ac.uk
1841; Sue White
Huddersfield Labour Party Arch, G.H. Wood Coll, Huddersfield Mechanic Institute Coll
430 000 vols; 2 445 curr per; 9 000 e-journals; 25 000 music scores; 712 maps; 24 560 microforms; 9 000 av-mat; 6 000 sound-rec
IATUL, SCONUL, CILIP 29136

University of Hull, Brynmor Jones Library, Cottingham Rd, ***Hull*** HU6 7RX
T: +44 1482 466581; Fax: +44 1482 466205; E-mail: libhelp@hull.ac.uk; URL: www.hull.ac.uk/lib
1929; Dr. Richard G. Heseltine
South-East Asia Coll, Labour Hist Coll, Philip Larkin Coll
999 000 vols; 5 200 curr per; 13 000 e-journals; 6 900 diss/theses; 6 000 music scores; 66 600 maps; 28 500 microforms; 4 800 av-mat; 4 000 sound-rec; 500 digital data carriers; 2 600 metres of mss
libr loan; SCONUL 29137

– Keith Donaldson Library, Scarborough Campus, Filey Rd, Scarborough YO11 3AZ
T: +44 1723 357277; Fax: +44 1723 357328; E-mail: libhelp-scar@hull.ac.uk; URL: www.hull.ac.uk/lib
1994; Juliet Crowther
libr loan; Sconul 29138

Suffolk College, Library, ***Ipswich*** IP4 1LT
T: +44 1423 296362; E-mail: deirdregriffin@suffolk.ac.uk; URL: www.suffolk.ac.uk/support4learners/index.htm
1961; Deirdre Griffin
80 000 vols; 500 curr per; 100 microforms; 5 000 av-mat; 100 sound-rec; 5 digital data carriers
libr loan; ARLIS/UK & Ireland 29139

West Thames College, Resources Centre, London Rd, ***Isleworth*** TW7 4HS
T: +44 020 83262308; E-mail: library.services@west-thames.ac.uk; URL: www.west-thames.ac.uk
1964; Karen Bewen-Chappell
Local Hist Coll
55 000 vols; 200 curr per; 300 microforms; 2 000 av-mat; 100 digital data carriers
ARLIS/UK & Ireland, BL, CILIP, ALA 29140

Keele University Library, ***Keele*** ST5 5BG
T: +44 1782 733535; Fax: +44 1782 734502; E-mail: libhelp@lib.keele.ac.uk; URL: www.keele.ac.uk
1949; Paul Reynolds
William Blake Coll, Izaak Walton Coll, Warrilow Coll, Le Play (Sociology) Coll; European Documentation Centre

580 000 vols; 2 829 curr per; 7 000 e-journals; 150 000 mss; 2 681 diss/theses; 8 000 music scores; 30 000 maps; 11 000 microforms; 1 500 av-mat; 3 108 sound-rec; 126 digital data carriers; 10 000 pamphlets
libr loan; SCONUL 29141

Kidderminster College, Library, Hoo Rd, ***Kidderminster*** DY10 1LX
T: +44 1562 820811; Fax: +44 1562 748504; URL: www.kidderminster.ac.uk
1963; J.A. Edwards
Careers Libr
23 004 vols; 94 curr per; 80 diss/theses; 276 maps; 1 170 av-mat; 462 sound-rec; 58 digital data carriers; 798 videocassettes
libr loan 29142

SPSA Scottish Police College, Learning Resources Centre, Tulliallan Castle, ***Kincardine***, Fife FK10 4BE
T: +44 1259 732073;
Fax: +44 1259 732285; E-mail: library@tulliallan.pnn.police.uk; URL: www.tulliallan.police.uk
1954; Polly St.-Aubyn
Police science (police management, leadership, structures, criminal investigation etc), Scots criminal law, criminology; Association of Chief Police Officers in Scotland (ACPOS) Professional Development Programme; Scottish Police Museum
14 600 vols; 56 curr per; 14 e-journals; 30 diss/theses; 80 av-mat; +60 digital language programs
libr loan; NLS, BLDSC 29143

College of West Anglia, Library, Tennyson Ave, ***King's Lynn*** PE30 2QW
T: +44 1553 761144; URL: www.col-westanglia.ac.uk
John Ross
50 000 vols; 100 curr per; 1 200 av-mat 29144

Kingston University, Library, Penrhyn Rd, ***Kingston upon Thames*** KT1 2EE
T: +44 20 85477101; Fax: +44 20 85477111; E-mail: library@kingston.ac.uk; URL: www.kingston.ac.uk/library
1970; Graham Bulpitt
Mineralogical Society Libr, Vane Ivanovic Libr, Iris Murdoch Coll
500 000 vols; 2 000 curr per; 16 000 e-journals; 3 000 diss/theses; 13 000 govt docs; 6 000 music scores; 1 500 maps; 33 000 microforms; 5 000 sound-rec; 1 000 digital data carriers; 190 000 slides
libr loan; CILIP, M25, UKLP, ARLIS/UK & Ireland, NAG 29145

– Knights Park Learning Resources Centre, Library Services for the Faculty of Design, Knights Park, Kingston upon Thames KT1 2QJ
T: +44 20 85477057; Fax: +44 20 85477011
1963; A. Kent
coll of original source mat, French art and architecture of the early 20th c
60 000 vols
libr loan; ARLIS/UK & Ireland, IFLA, CILIP, ASLIB 29146

– Roehampton Vale Library, Friars Av, London SW15 3DW
T: +44 20 85477903; Fax: +44 20 85477800; URL: www.kingston.ac.uk/library-media
Mechanical and aeronautical engineering
15 000 vols; 100 curr per
libr loan 29147

Adam Smith College, Library, St. Brycedale Ave, ***Kirkcaldy*** KY1 1EX
T: +44 1592 223040, 223411;
Fax: +44 1592 640225; E-mail: library@adamsmith.ac.uk; URL: www.adamsmithcollege.ac.uk
1963
28 000 vols; 450 av-mat; 18 sound-rec; 153 digital data carriers
libr loan 29148

University of Wales Lampeter, Library, ***Lampeter*** SA48 7ED
T: +44 1570 424798; Fax: +44 1570 423875; E-mail: library@lamp.ac.uk; URL: www.lamp.ac.uk/library
1822; Dr. A. Prescott
Tracts Coll, Welsh Coll
200 000 vols; 1 000 curr per; 120 mss; 67 incunabula
libr loan; SCONUL, ASLIB 29149

Lancaster and Morecambe College, Library, Morecambe Rd, ***Lancaster*** LA1 2TY
T: +44 1524 66215; Fax: +44 1524 843078; URL: www.lmc.ac.uk/home/index.php
1960; A.C. Wilson
Art and Design
22 000 vols; 110 curr per; 275 maps; 20 digital data carriers 29150

Lancaster University, Library, Bailrigg, ***Lancaster*** LA1 4YH
T: +44 1524 592535; Fax: +44 1524 65719; E-mail: library@lancaster.ac.uk; URL: libweb.lancs.ac.uk
1963; Clare Powne
Jack Hylton Archive, Quaker Coll, Business Hist Coll, Socialist Coll
615 000 vols; 850 curr per; 7 200 e-journals; 7 260 e-books; 145 mss; 10 700 diss/theses; 204 300 govt docs; 17 300 music scores; 8 100 maps; 1 800 microforms; 90 000 av-mat; 9 000 sound-rec; 1 100 digital data carriers
libr loan; SCONUL, ASLIB, CILIP, UKSG, NAG 29151

University of Cumbria, Harold Bridges Library, Bowerham Rd, ***Lancaster*** LA1 3JD
T: +44 1524 384243; Fax: +44 1524 384588; E-mail: liblan@cumbria.ac.uk; URL: www.cumbria.ac.uk/Services/lis/home.aspx
1963; Margaret Weaver
Teaching practice resources
370 679 vols; 900 curr per; 700 diss/theses; 1 500 av-mat; 1 500 sound-rec
SCONUL, UK Libraries Plus 29152

Leeds College of Building, Library, North St, ***Leeds*** LS2 7QT
T: +44 113 2226098; Fax: +44 113 2226001; E-mail: library@lcb.ac.uk; URL: library.lcb.ac.uk/Heritage/Default.htm
1960; Anne-Mary Ingleheam
Health & Safety, project files, careers, history of construction archive
20 000 vols; 150 curr per; 15 e-books; 200 av-mat; 54 digital data carriers 29153

Leeds College of Music, Library, 3 Quarry Hill, ***Leeds*** LS2 7PD
T: +44 113 2223458; E-mail: C.Marsh@lcm.ac.uk; URL: www.lcm.ac.uk/about-lcm/library
Jay Glasby
Jazz archive
40 000 vols 29154

Leeds Metropolitan University, The Civic Quarter Library, Woodhouse Lane, ***Leeds*** LS1 3HE
T: +44 113 2832600;
Fax: +44 113 2836779; E-mail: infodesk.lc@leedsmet.ac.uk; URL: www.leedsmet.ac.uk/lis/lss
1960; Philip Payne
Europ Doc Centre, Coll of West Yorkshire Society of Architects; West Yorkshire European Information Centre
368 000 vols; 2 121 curr per; 7 457 e-journals; 2 900 maps; 100 digital data carriers; diss, microforms, sound recordings
libr loan; SCONUL, UK Libraries Plus, CILIP 29155

Trinity and All Saints' College, Library, Brownberrie Lane, Horsforth, ***Leeds*** LS18 5HD
T: +44 113 2837244; Fax: +44 113 2837200; URL: www.leedstrinity.ac.uk/services/library
1966; Elizabeth Murphy
Yorkshire studies, education, humanities, sport, psychology, media, management, film and television studies
140 000 vols; 450 curr per; 400 e-journals; 60 microforms; 2 300 av-mat; 900 digital data carriers
libr loan; CILIP, SCONUL 29156

University of Leeds, Brotherton Library, ***Leeds*** LS2 9JT
T: +44 113 2335501;
Fax: +44 113 2335561; E-mail: libraryenquiries@leeds.ac.uk; URL: www.leeds.ac.uk/library
1875; Margaret Coutts
Brotherton Coll (rare bks and mss), Anglo-French Coll, Icelandic and Scandinavian Coll, Blanche Leigh and John F. Preston Coll (cookery bks), Harold Whitaker Coll (atlases and maps), Roth Coll (post-Biblical Judaica)
2 600 000 vols; 9 000 curr per; 4 054 mss; 36 721 microforms; 4 000 av-mat; 621 digital data carriers
libr loan; SCONUL, ASLIB, CILIP, CURL, RLG 29157

– Edward Boyle Library, Leeds LS2 9JT
T: +44 113 2335540;
Fax: +44 113 2335539; E-mail: eblcounter@library.leeds.ac.uk
Engineering, science
300 000 vols 29158

– Health Sciences Library, Leeds LS2 9JT
T: +44 113 3435549; Fax: +44 113 3434381
1977; A. Collins
Medical Libr, Dental Libr, Leeds General Infirmary Healthcare Libr, High Royds Healthcare Libr
100 000 vols; 1 000 curr per
libr loan; EAHIL 29159

De Montfort University, Kimberlin Library, The Gateway, ***Leicester*** LE1 9BH
T: +44 116 2577165; Fax: +44 116 2577046; E-mail: karnold@dmu.ac.uk; URL: www.library.dmu.ac.uk
1937; Kathryn Arnold
623 000 vols; 5 151 curr per; 500 diss/theses
libr loan; SCONUL 29160

– Polhill Campus Library, Polhill Av, Bedford MK41 9EA
T: +44 1234 351671; Fax: +44 1234 217738
Diana Saulsbury
Hockcliffe Coll (early children's bks)
140 000 vols; 720 curr per 29161

University of Leicester, University Library, University Rd, P.O. Box 248, ***Leicester*** LE1 9QD
T: +44 116 2522031; Fax: +44 116 2522066; E-mail: library@leicester.ac.uk; URL: www.le.ac.uk/library
1921; Christine Fyfe
English local hist, mathematical assn, Orton Papers
1 000 000 vols
SCONUL, ASLIB, CILIP 29162

– Clinical Sciences Library, Leicester Royal Infirmary, P.O. Box 65, Leicester LE2 7LX
T: +44 116 2523104; Fax: +44 116 2523107; E-mail: clinlib@le.ac.uk
1978; Louise Jones
Libr of the Leicester Medical Society
41 000 vols; 550 curr per; 1 000 govt docs; 100 av-mat
libr loan 29163

– School of Education, Library, 21 University Rd, Leicester LE1 7RF
T: +44 116 2523739; Fax: +44 116 2525798
1949; R.W. Kirk
Coll of 19th c children's fiction and school primers
90 000 vols; 200 curr per; 50 maps; 100 av-mat; 200 sound-rec; 300 ills and photos, 300 posters
libr loan; LISE 29164

Bishop Grosseteste University College, Sibthorp Library, Newport, ***Lincoln*** LN1 3DY
T: +44 1522 583790; Fax: +44 1522 530243; E-mail: library-enquiries@bishopg.ac.uk; URL: www.bishopg.ac.uk/library
1886; Emma Sansby
College Archive
140 000 vols; 250 curr per; 600 diss/theses; 2 500 govt docs; 698 av-mat; 1 073 sound-rec; 169 digital data carriers
libr loan; BL 29165

Liverpool Hope, Sheppard-Worlock Library, Hope Park, P.O. Box 95, ***Liverpool*** L16 9JD
T: +44 151 2912000; Fax: +44 151 2912037; URL: www.hope.ac.uk
Ms L. Taylor
250 000 vols; 900 curr per
libr loan; SCONUL 29166

Liverpool Institute of Higher Education, Library, Woolton Rd, P.O. Box 6, ***Liverpool*** L16 8ND
T: +44 151 7373549
1981; P.F. Capewell
230 000 vols; 600 curr per; 120 diss/theses; 2 000 govt docs; 6 000 music scores; 100 maps; 800 microforms; 60 200 av-mat; 1 000 sound-rec; 12 000 slides
libr loan 29167

Liverpool John Moores University, Avril Robarts Learning Resource Centre, 79 Tithebarn St, ***Liverpool*** L2 2ER
T: +44 151 2314022; URL: www.livjm.ac.uk
1997; J. Ainsworth
Computing and Mathematics, Health, Social Science
935 000 vols; 2 607 curr per; 2 000 diss/theses; 10 000 govt docs; 5 000 maps; 37 000 microforms; 76 000 av-mat; 56 digital data carriers
libr loan; UK Libraries Plus 29168

University of Liverpool, University Library, Chatham St, P.O. Box 123, ***Liverpool*** L69 3DA
T: +44 151 7942673; Fax: +44 151 7942681; E-mail: c.kay@liverpool.ac.uk; URL: www.liv.ac.uk/library/libhomep.html
1892; Phil Sykes
Gypsy Lore Society, Rathborne Papers, Science Fiction Coll, Knowsley Pamphlets, C19 Children's Books, Liverpool Learned Societies, Liverpool Poets, Private Press Books
1 900 000 vols; 5 000 curr per; 2 500 e-journals; 400 mss; 250 incunabula
libr loan; RLUK, SCONUL, NOWAL, CILIP 29169

Architectural Association (AA), Library, 34-36 Bedford Square, ***London*** WC1B 3ES
T: +44 20 7887 4035; Fax: +44 20 4414 0782; E-mail: hsklar@aaschool.ac.uk; URL: aaschool.ac.uk/library
1862; Hinda F. Sklar
Architecture, building, garden cities, exhibitions; Architectural Association Archives
46 467 vols; 139 curr per; 10 mss; 1 incunabula; 1 403 diss/theses; 1 000 maps; 6 digital data carriers; 3 000 articles, pamphlets, catalogs.
libr loan; BLL, ARLIS/UK & Ireland, ARLIS/NA, ARCLIB 29170

Camberwell College of Arts, University of the Arts London, Library, 43-45 Peckham Rd, ***London*** SE5 8UF
T: +44 20 75146349;
Fax: +44 20 75146324; E-mail: e.ocallaghan@camberwell.arts.ac.uk; URL: www.arts.ac.uk/library
1898; Ms Liz Kerr
Walter Crane, Poster art
50 000 vols; 160 curr per
BNBC 29171

Central Saint Martins College of Art & Design, University of the Arts London, Southampton Row, Library, 16 John Islip St., ***London*** SW1P 4JU
T: +44 20 75147773; Fax: +44 20 75147785; E-mail: mr-lib@linst.ac.uk; URL: www.arts.ac.uk/library/chelsea.htm
1964; E. Ward
Exhibition cat, artists' bks, American art of 20th c, multiples Afro-Caribbean & Asian, British Art; Slide libr
70 000 vols; 280 curr per; 200 diss/theses; 200 microforms; 400 av-mat; 200 sound-rec; 30 digital data carriers; 160 000 slides, 8 000 clippings, 27 000 pamphlets
ARLIS/UK & Ireland, ARLIS/NA 29172

Central School of Speech and Drama, Library, Embassy Theatre, 64 Eton Ave, ***London*** NW3 3HY
T: +44 20 75523942; Fax: +44 20 77224132; E-mail: library@cssd.ac.uk; URL: www.cssd.ac.uk/about_central/lis.htm
Peter Collett
30 000 vols; 81 curr per; 850 av-mat; 850 sound-rec; 500 digital data carriers; 7 500 slides
SIBMAS, SCONUL, UK Libraries Plus 29173

Central St Martins College of Art and Design, The London Institute, Library, Southampton Row, ***London*** WC1B 4AP
T: +44 20 75147037; Fax: +44 20 75147033; URL: www.linst.ac.uk/library
1896; Ms Pat Christie
Fashion files, materials and products coll; Slide Lib
80 000 vols; 200 curr per; 800 diss/theses; 3 000 av-mat; 40 digital data carriers
libr loan; ARLIS/UK & Ireland 29174

Chelsea College of Art & Design, University of the Arts London, Library, 16 John Islip St., ***London*** SW1P 4JU
T: +44 20 75147773; Fax: +44 20 75147785; E-mail: mr-lib@linst.ac.uk; URL: www.arts.ac.uk/library/chelsea.htm
1964; E. Ward
Exhibition cat, artists' bks, American art of 20th c, multiples Afro-Caribbean & Asian, British Art; Slide libr
70 000 vols; 280 curr per; 200 diss/theses; 200 microforms; 400 av-mat; 200 sound-rec; 30 digital data carriers; 160 000 slides, 8 000 clippings, 27 000 pamphlets
ARLIS/UK & Ireland, ARLIS/NA 29175

College of North East London, Centenary Learning Centre, High Rd, ***London*** N15 4RU
T: +44 20 84423014; Fax: +44 20 84423091; URL: www.conel.ac.uk
1948; Ms J. Dunster
54 000 vols; 110 curr per; 500 av-mat; 40 digital data carriers; 50 online learning mat
BLL, COLRIC 29176

College of North West London, Learning Resources Centre, Priory Park Rd, Willesden, ***London*** NW10 2XD
T: +44 20 82085329; Fax: +44 20 82085321
1960; Beatrice McAdam
42 000 vols; 204 curr per; 1 e-journals 29177

Guildhall School of Music and Drama, Library, Barbican, ***London*** EC2Y 8DT
T: +44 20 76282571; Fax: +44 20 72569438; E-mail: connect@gsmd.ac.uk; URL: www.gsmd.ac.uk
Kate Eaton
Appleby Guitar Music Coll, Goossens Oboe Music Coll, Harris Opera coll
35 000 vols; 70 curr per; 50 000 music scores; 800 av-mat; 6 000 sound-rec; 5 digital data carriers
libr loan; IAML, SIBMAS 29178

Imperial College London Library, Central Library, South Kensington Campus, ***London*** SW7 2AZ
T: +44 20 75948820;
Fax: +44 20 75948876; E-mail: a.brown@imperial.ac.uk; URL: www.imperial.ac.uk/library
Deborah Shorley 29179

– Aeronautics Department Library, South Kensington campus, London SW7 2AZ
T: +44 20 75945069; Fax: +44 20 75848120; E-mail: ae.office@imperial.ac.uk 29180

– Charing Cross Campus Library, St Dunstan's Rd, London W6 8RP
T: +44 20 75940755; Fax: +44 20 75940851 29181

– Chelsea & Westminster Campus Library, Fulham Rd, London SW10 9NH
T: +44 20 87468107;
Fax: +44 20 87468215; E-mail: librarycw@imperial.ac.uk; URL: www3.imperial.ac.uk/library/usethelibrary/cw
Annan coll (history of metal mining and metallurgy, 1495 to present), Charing Cross coll, Chelsea and Westminster Hospital early medical books, Early geology books (1565-1909), H.G. Wells coll, Early life sciences books (1733-1978), John M. Corin memorial coll (English literature, 1824-1961), Materials history of science coll, National Heart and Lung Institute historical colls, St Mary's Hospital colls, T.H. Huxley book coll (natural history and life sciences, 1681-1927) 29182

– Chemical Engineering and Chemical Technology Department Library, South Kensington campus, London SW7 2AZ
T: +44 20 75945598; Fax: +44 20 75945604 29183

– Department of Civil and Environmental Engineering, Civil Engineering Department Library, Room 402 (Level 4), Skempton Bldg, South Kensington Campus, London SW7 2AZ
T: +44 20 75946007; Fax: +44 20 72252716; E-mail: n.lau@imperial.ac.uk; URL: www3.imperial.ac.uk/civilengineering/library
23 500 vols; 2 000 diss/theses 29184

– Hammersmith Campus Library, Du Cane Rd, London W12 0NN
T: +44 20 83833246;
Fax: +44 20 83832195; E-mail: lib.hamm@imperial.ac.uk; URL: www3.imperial.ac.uk/library/usethelibrary/hammersmith 29185

– Mathematics Department Library, South Kensington Campus, London SW7 2AZ
T: +44 20 75948542; Fax: +44 20 75948517
19 000 vols 29186

– Mechanical Engineering Department Library, South Kensington campus, London SW7 2AZ
T: +44 20 75947166; Fax: +44 20 75948517 29187

– Royal Brompton Campus Library, Dovehouse St, London SW3 6LY
T: +44 20 73518150; Fax: +44 20 73518117 29188

– Silwood Park Campus Library, Silwood Park campus, Ascot, Berkshire SL 5 7TE
T: +44 1491 829112; Fax: +44 1491 829123 29189

– St Mary's Campus Library, Norfolk Place, London W2 1PG
T: +44 20 75943692; Fax: +44 20 74023971; E-mail: sm-lib@imperial.ac.uk; URL: www3.imperial.ac.uk/library/usethelibrary/sm
9 000 vols 29190

King's College London, University of London, ***London***
– Chancery Lane Maughan Library & Information Services Centre, Chancery Lane, London WC2A 1LR
T: +44 20 78482424; Fax: +44 20 78482277; URL: www.kcl.ac.uk/iss
1831; Margaret Harries, Vivien Robertson
Humanities, law, physical science, engineering; Special colls: Marsden Libr, Box Libr, Whealstone Libr, Liddell-Hart Arch, Adam Arch, Portugese, Modern Greek, War Studies, Medical Law, Medical Ethics
800 000 vols; 1 580 curr per; 4 500 e-journals; 10 incunabula; 50 digital data carriers; 3 300 linear m of mss
libr loan; IFLA, IATUL, LIBER, CURL, SCONUL, ASLIB, EAHIL, M25 29191

– Denmark Hill Weston Education Centre ISC, Cutcombe Rd, London SE5 9RJ
T: +44 20 7848554-1/-2; Fax: +44 20 78485550; URL: www.kcl.ac.uk/iss
1831; Margaret Haines, Rodney Amis
Medicine and dentistry; Hist coll
11 000 vols; 600 curr per; 1 incunabula; 50 digital data carriers
libr loan; IFLA, IATUL, LIBER, CURL, SCONUL, ASLIB, EAHIL 29192

– Guy's New Hunt's House Information Services Centre, New Hunts House, Guy's Campus, London SE1 1UL
T: +44 20 78486600; Fax: +44 20 78486743; URL: www.kcl.ac.uk/library
1831; E.A. Bell
Medecine, dentistry, biomedical science
81 250 vols; 640 curr per; 3 000 e-journals; 50 digital data carriers
IFLA, IATUL, LIBER, CURL, SCONUL, ASLIB, EAHIL 29193

– St Thomas' Medical Library, St Thomas' Hospital, Sherrington Bldg, Lambeth Palace Rd, London SE1 7EH
T: +44 20 71883740; Fax: +44 20 74013932; URL: www.kcl.ac.uk/library
1831; E.A. Bell, Angela Gunn
13 500 vols; 340 curr per; 50 digital data carriers
IFLA, IATUL, LIBER, CURL, SCONUL, ASLIB, EAHIL 29194

– Waterloo Campus Information Services Centre, Franklin-Wilkins Bldg, 150 Stamford St, London SE1 9NH
T: +44 20 78484378; Fax: +44 20 78484290; E-mail: issenquiry@kcl.ac.uk; URL: www.kcl.ac.uk/iss
1831; V. Robertson
Health and life sciences, education, nursing, management
180 200 vols; 1 190 curr per; 3 000 e-journals; 420 av-mat; 280 digital data carriers
IFLA, IATUL, LIBER, CURL, SCONUL, ASLIB, EAHIL 29195

London Business School, Library, 25 Taunton Pl, ***London*** NW 1; mail address: Regent's Park, London NW1 4SA
T: +44 20 70007620; Fax: +44 20 70007601; E-mail: library@london.edu; URL: www.london.edu/library
1966; Helen Edwards
Corporate Libr
19 000 vols; 650 curr per; 87 e-journals; 250 diss/theses; 130 govt docs; 12 000 microforms; 1 350 av-mat; 30 000 annual company rpts
libr loan; ASLIB, EBSLG, BBSLG 29196

London College of Communication, Library and Learning Resources, Elephant + Castle, ***London*** SE1 6SB
T: +44 20 75146527; Fax: +44 20 75176597; URL: www.arts.ac.uk/library
1894; Elizabeth Davison
Cinema, applied arts, design, printing, printing hist coll
100 000 vols; 600 curr per; 4 000 diss/theses; 41 000 microforms; 100 000 av-mat; 400 sound-rec; 5 digital data carriers
BLL, LASER, ARLIS/UK & Ireland 29197

London College of Fashion, University of the Arts London, Library, 20 John Princes St, ***London*** W1G 0BJ
T: +44 20 75147453;
Fax: +44 20 75147580; E-mail: library@fashion.arts.ac.uk; URL: www.arts.ac.uk/library/lcf.htm
1963; Diane Mansbridge
2 000 Fashion Designer Files & Illustration Files
43 000 vols; 150 curr per; 50 diss/theses; 26 microforms; 30 sound-rec; 14 digital data carriers; 1 000 videocassettes, 19 000 slides
libr loan; BLL, ARLIS/UK & Ireland 29198

London Metropolitan University, ***London***
– Commercial Road Library, 41-71 Commercial Rd, London E1 1LA
T: +44 20 73201867; Fax: +44 20 73201177; URL: www.londonmet.ac.uk/
Cathy Phillpotts
Art, media, design
39 000 vols; 400 curr per; 35 000 slides 29199

– Tower Hill Library, 100 Minories, London EC3N 1JY
T: +44 20 73201767; Fax: +44 20 73202766
Phil Jones
Business, management, civil aviation, computing
50 000 vols; 60 curr per 29200

– The Women's Library, 25 Old Castle St, London El 7NT
T: +44 20 73202222;
Fax: +44 20 73201177; E-mail: moreinfo@thewomenslibrary.ac.uk; URL: www.londonmet.ac.uk/thewomenslibrary/
1926
Josephine Butler Society Library, Papers relating to women's suffrage, women's liberation
264 000 vols; 1 425 curr per; 1 100 diss/theses; 300 music scores; 774 microforms; 77 500 av-mat; 900 sound-rec; 27 digital data carriers
CILIP; ASLIB, EAHIL, ARLIS Uk & Ireland 29201

London School of Economics and Political Science (LSE), British Library of Political and Economic Science, 10 Portugal St, ***London*** WC2A 2HD
T: +44 20 79557229;
Fax: +44 20 79557454; E-mail: library.information.desk@lse.ac.uk; URL: www.lse.ac.uk/library
1896; Ms Jean M. Sykes
Mss of modern radical movements, Coll for US Federal docs and UN, European Union, Historical Pamphlets and statistics
1 000 000 vols; 10 000 curr per; 1 400 e-journals; 14 019 diss/theses; 350 000 govt docs; 137 digital data carriers; 1 400 mss colls, 1 008 microform collas
libr loan 29202

London School of Hygiene and Tropical Medicine, University of London, Library, Keppel St, ***London*** WC1E 7HT
T: +44 20 79272276; Fax: +44 20 79272273; E-mail: library@lshtm.ac.uk; URL: www.lshtm.ac.uk/library
1899; Caroline Lloyd
Reece (smallpox & vaccination), Sir Ronald Ross Arch, Sir Patrick Manson Coll, Historical Coll on Public Health and Tropical Medicine
72 000 vols; 800 curr per; 18 e-journals; 2 000 mss; 1 430 diss/theses; 220 maps; 120 av-mat; 10 digital data carriers; 29 000 pamphlets
libr loan; EAHIL, SCONUL, M25, UKSG 29203

London School of Jewish Studies (LSJS), Library, Schaller House, 44A Albert Rd, ***London*** NW4 2SJ
T: +44 20 82036427; Fax: +44 20 82036420; E-mail: info@lsjs.ac.uk; URL: www.lsjs.ac.uk
1860; Erla Zimmels
New West End Synagoge Coll
100 000 vols; 100 curr per; 150 mss; 30 diss/theses; 500 music scores; 950 sound-rec
libr loan; Hebraica Libraries Group 29204

London South Bank University, Perry Library, 250, Southwark Bridge Rd, ***London*** SE1 6NJ
T: +44 20 78156625; Fax: +44 20 7815; E-mail: library@lsbu.ac.uk; URL: www.library.lsbu.ac.uk
1949; James Bethan
300 000 vols; 1 300 curr per; 11 600 av-mat; 50 digital data carriers; 36 700 transparencies 29205

– Perry Library, 250 Southwark Bridge Rd, London SE1 6NJ
T: +44 20 78156604; Fax: +44 20 78156629; E-mail: library@lsbu.ac.uk; URL: www.library.lsbu.ac.uk/
Michael Veitch
Early books on architecture
100 000 vols; 40 000 slides 29206

Middlesex University, Learning Resources, Sheppard Library, Nort London Business Park, Oakleigh Road South, ***London*** NW4 4BT
T: +44 20 84115234; Fax: +44 20 84115163; URL: www.lr.mdx.ac.uk/index.htm
1977; W.A.J. Marsterson
Bernie Grant Trust, Future Histories, Runnymede Trust, Hall Carpenter arch, John Lansdown
528 353 vols; 1 633 curr per; 30 000 e-journals; 3 000 diss/theses; 7 500 music scores; 385 000 av-mat; 11 000 sound-rec; 1 900 digital data carriers
libr loan; CILIP, SCONUL 29207

Morley College, Library, 61 Westminster Bridge Rd, ***London*** SE1 7HT
T: +44 20 74501828;
Fax: +44 20 79284074; E-mail: elaine.andrews@morleycollege.ac.uk; URL: www.morleycollege.ac.uk
1890; Elaine Andrews
16 150 vols; 42 curr per; 8 508 music scores; 1 037 av-mat; 1 660 sound-rec; 128 digital data carriers; Scrapbooks of Morley college history, prospectuses of Morley College, books relating to Morley College history and personalities, collected history of Morley Theatre School by former student, folders of ephemera on tutors, College Art Collections and buildings 29208

Ravensbourne College of Design and Communication, Study Zone, 6 Penrose Way, Greenwich Peninsular, ***London*** SE10 0EW
T: +44 20 30403769; E-mail: studyzone@rave.ac.uk

1976; April Yasamee
Decorative Arts, Design, Digital Design, Mass Media, Communication Studies
18000 vols; 120 curr per; 100 e-journals; 62 e-books; 55 diss/theses; 500 av-mat
SCONUL 29209

RCVS Charitable Trust Library, Belgravia House, 62-64 Horseferry Rd, ***London*** SW1P 2AF
T: +44 20 72020752; Fax: +44 20 72020751; E-mail: library@rcvstrust.org.uk; URL: www.rcvstrust.org.uk
1844
Hist of farriery, horsemanship and early veterinary med
30000 vols; 250 curr per; 900 e-journals; 500 diss/theses
AHIS 29210

Regent's College, Library, Inner Circle, Regent's Park, ***London*** NW1 4NS
T: +44 20 74877547; Fax: +44 20 74877545; E-mail: collinsm@regents.ac.uk; URL: www.regents.ac.uk
Mary Collins
Psychotherapy, counselling
40000 vols; 380 curr per; 500 diss/theses; 200 sound-rec; 12 databases
libr loan; CILIP, ASLIB 29211

Royal Academy of Dramatic Art, Library, 18 Chenies St, ***London*** WC1E 7PA
T: +44 20 76367076; E-mail: library@rada.ac.uk; URL: www.rada.ac.uk/about-rada/library
1904; James Thornton
Ivo Currall Shaw Coll (450 items), Nuryev U-matic tape coll
30000 vols; 14 curr per; 1500 av-mat; more than 10000 plays, many rare and out of print 29212

Royal Academy of Music, Library, Marylebone Rd, ***London*** NW1 5HT
T: +44 20 78737323; Fax: +44 20 78737322; E-mail: library@ram.ac.uk; URL: www.ram.ac.uk
1822; Kathryn Adamson
Henry Wood Orchestral Libr, Otto Klemperer Scores Lib, Barbirolli arch, Sullivan arch, R.J.S. Stevens Coll, David Munrow Coll, Mosco Carner Coll, Robert Spencer Coll, Foyle Menuhin Arch
100000 vols; 100 curr per; 2000 mss; 90000 music scores; 450 microforms; 120 av-mat; 6000 sound-rec; 3 digital data carriers; 7000 orchestral and choral sets
IAML, MLA 29213

Royal College of Art, Library, Kensington Gore, ***London*** SW7 2EU
T: +44 20 75904224; Fax: +44 20 75904217; E-mail: library@rca.ac.uk; URL: www.rca.ac.uk
1953; Darlene Maxwell
Colour reference libr, comprehensive coll on colour, art hist, fine arts, decorative arts, design, architecture, photography, computer related design
70000 vols; 120 curr per; 903 diss/theses; 3500 av-mat; 110000 slides
libr loan; ARLIS/UK & Ireland, SCONUL, UK Libraries Plus 29214

Royal College of Music, Library, Prince Consort Rd, ***London*** SW7 2BS
T: +44 20 75914325; Fax: +44 20 75914326; E-mail: library@rcm.ac.uk; URL: www.rcm.ac.uk
1883; Pamela Thompson
Sacred Harmonic Society, Concerts of Ancient Music, Heron-Allen Coll, Maurice Frost Coll, and many others (see www.rcm.ac.uk/library)
350000 vols; 400 curr per; 50 e-journals; 20000 mss; 160 diss/theses; 240000 music scores; 5430 microforms; 325 av-mat; 25000 sound-rec; 15 digital data carriers; 305
MLA, IAML 29215

Royal College of Nursing, Library, 20 Cavendish Sq, ***London*** W1G 0RN
T: +44 20 76473610; Fax: +44 20 76473420; E-mail: rcn.library@rcn.org.uk; URL: www.rcn.org.uk
1921; Ms J. Cheeseborough
Arch coll (Edinburgh) 100000 items; Steinberg Coll of Nursing Research
72000 vols; 350 curr per; 700 e-journals; 400 e-books; 1200 diss/theses; 50 microforms; 500 av-mat; 20 sound-rec 29216

Royal College of Organists, Library, P.O. Box 56357, ***London*** SE16 7XL
T: +44 5600 767208; E-mail: admin@rco.org.uk; URL: www.rco.org.uk
1864; Andrew McCrea
34000 vols 29217

Royal College of Physicians of London, Library, 11 St Andrew's Place, ***London*** NW1 4LE
T: +44 20 72241539;
Fax: +44 20 74863729; E-mail: infocentre@rcplondon.ac.uk; URL: www.rcplondon.ac.uk/library
1518; Julie Beckwith
Marquis of Dorchester libr, Evan Bedford libr of cardiology, hist of medicine; Heritage centre, information centre, medical education resource centre
50000 vols; 100 curr per; 8 e-journals; 104 incunabula; 115 microforms; 3000 av-mat; 172 sound-rec; 17000 portrait photos and prints, 430 linear metres of manuscripts
libr loan; BLDSC 29218

Royal College of Surgeons of England, Library, 35-43 Lincoln's Inn Fields, ***London*** WC2A 3PE
T: +44 20 7869-6555/-6556; Fax: +44 20 74054438; E-mail: library@rcseng.ac.uk; URL: www.rcseng.ac.uk/library
1823; Mrs T. Knight
Special collections on: www.rcseng.ac.uk/library/collections/
160000 vols; 300 curr per; 80 e-journals; 5000 mss; 50 incunabula; 200 diss/theses; 500 av-mat; 100 digital data carriers; 30000 pamphlets catalogued online, 3000 engravings, 2000 bookplates
CILIP: Health Libraries Group; UKEIG; UKSG 29219

Royal National Institute for the Deaf, University College London, Library, 330-332 Gray's Inn Rd, ***London*** WC1X 8EE
T: +44 20 79151553; Fax: +44 20 79151443; E-mail: midlib@ucl.ac.uk; URL: www.ucl.ac.uk/library/RNID
Alex Stagg
Audiology, sign language, deaf studies, communication orders, hearing science
20000 vols; 200 curr per; 200 diss/theses; 500 govt docs 29220

Thames Valley University, Learning Resource Centres, Library, St Myry's Rd, Ealing, ***London*** W5 5RF
T: +44 20 85792000; Fax: +44 20 85661353; URL: www.tvu.ac.uk
1928; John Wolstenholme
EC Reference Centre
230000 vols; 1084 curr per; 8964 music scores; 72 microforms; 10371 av-mat; 5178 sound-rec; 100 digital data carriers; 45000 slides
libr loan; ASLIB, BIALL, Learning Resources Development Group, CILIP, School Library Association, Scottish Libr, IFLA, SCONUL, M25, EIA 29221

Trinity College of Music, Jerwood Library of the Performing Arts, King Charles Court, Old Royal Naval College, Greenwich, ***London*** SE10 9JF
T: +44 20 83053951; E-mail: library@tcm.ac.uk; URL: www.tcm.ac.uk
Claire Kidwell
Music, Theatre Hist, Jazz, British composers, Hist live sound recordings; Special colls: Mander & Mitchenson Theatre Coll, Centre for Young Musicians Libr, Music Preserved, British Music Society Arch, Antonio de Almeida Coll, Sir Frederick Bridge Libr, etc
60000 vols; 60 curr per; 950 mss; 60000 music scores; 8000 sound-rec; 14 digital data carriers
libr loan; IAML, SCONUL 29222

UCL Eastman Dental Institute Library, 256 Gray's Inn Road, ***London*** WC1X 8LD
T: +44 20 34561045; E-mail: ic@eastman.ucl.ac.uk; URL: www.eastman.ucl.ac.uk/departments/others/library/index.html
1947; Medwenna Buckland
Hist of dentistry coll
8000 vols; 70 curr per; 100 e-journals; 50 av-mat; 100 digital data carriers; web based databases
libr loan; CILIP 29223

University College London, University of London, Library, Gower St, ***London*** WC1E 6BT
T: +44 20 73807700; Fax: +44 20 73807373; E-mail: library@ucl.ac.uk; URL: www.ucl.ac.uk/library
1828; Dr. Paul Ayris
Graves Early Sci Libr, Whitley Stokes Celtic Libr, George Orwell Arch, James Joyce Centre, C.K. Ogden Libr, Geologists' Assn Libr, London Math Soc Libr, Royal Statistical Inc Libr, Folklore Soc Libr, Royal Hist Soc Libr, Mocatta Libr, Filtration Society Libr
1500000 vols; 8000 curr per; 160 incunabula; 633 microforms; 1637 av-mat
libr loan; SCONUL, CILIP, RLG, CURL 29224

– Bartlett Built Environment Library, Wates House, 22 Gordon St, London WC1H 0QB
T: +44 20 76794900; Fax: +44 20 76797373; E-mail: library@ucl.ac.uk; URL: www.ucl.ac.uk/Library
Mrs S.M. Tonkin, Mrs C.E. Fletcher
38000 vols; 270 curr per; 50 av-mat; 20 digital data carriers
libr loan 29225

– Cruciform Library, Gower St, London WC1E 6AU
T: +44 20 76796079; Fax: +44 20 76796981; E-mail: clinscilib@ucl.ac.uk; URL: www.ucl.ac.uk/library
1907; K. Cheney
Sir Robert Carswell Coll (2000 pathological drawings)
30000 vols; 200 curr per; 50 av-mat 29226

– Institute of Child Health, Friends of the Children of Great Ormond Street Library, 30 Guilford St, London WC1N 1EH
T: +44 20 72429789 ext 2424;
Fax: +44 20 78310488; E-mail: library@ich.ucl.ac.uk; URL: www.ich.ucl.ac.uk/library
1946; J.M. Clarke
Pediatrics, child care, child development, pediatric nursing, international child health; Source collection
10000 vols; 260 curr per; 100 av-mat 29227

– Institute of Orthopaedics, Library, Royal National Orthopaedic Hospital Trust, Brockley Hill, Stanmore HA7 4LP
T: +44 20 89095351; Fax: +44 20 89095390; E-mail: orthlib@ucl.ac.uk; URL: www.ucl.ac.uk/library/iorthlib.shtml
1946; P.F. Smith
Orthopedics, Musculoskeletal Science; Coll of hist orthopaedic bks
15000 vols; 150 curr per; 150 diss/theses; 230 av-mat
libr loan 29228

– UCL School of Slavonic and East European Studies, Library, Senate House, Malet St, London WC1H 0BW
T: +44 20 76798702; Fax: +44 20 76798710; E-mail: ssees-library@ssees.ucl.ac.uk; URL: www.ssees.ucl.ac.uk/library/
1915; Lesley Pitman
Moses Gaster (Rumanian Lit.), Béla Iványi-Grünwald (Hungary), Russ.-Orthodox Church in London
360000 vols; 1100 curr per
libr loan 29229

University of Greenwich, Information and Library Services, Bexley Road, ***London*** SE9 2PQ
T: +44 20 83319656;
Fax: +44 20 83318756; E-mail: d.heathcote@greenwich.ac.uk; URL: www.gre.ac.uk/lib/index.html
650000 vols
libr loan; ASLIB, SCONUL 29230

– Avery Hill Library, Mansion Site, Bexley Rd, London SE9 2PQ
T: +44 20 83318484; Fax: +44 20 83319645; E-mail: r.m.moon@gre.ac.uk; URL: www.gre.ac.uk/offices/ils/ls/services/lib/ahlib
Rosemary Moon
100000 vols 29231

University of London, Senate House Library, Senate House, Malet St, ***London*** WC1E 7HU
T: +44 20 78628500; Fax: +44 20 78628480; E-mail: enquiries@shl.lon.ac.uk; URL: www.ull.ac.uk
1838; David Pearson
Goldsmiths' Libr of Economic Lit, Durning-Lawrence Libr, Sterling Libr, Porteus Libr, Eliot-Phelips Coll (+ 260000 items altogether); English libr, palaeography libr, romance lang libr, psychology libr, philosophy libr, area studies, music libr, history libr
2000000 vols; 5500 curr per; 971 mss; 130 incunabula; 84000 diss/theses; 70000 govt docs; 36000 music scores; 67000 maps; 9500 sound-rec; 50 digital data carriers; 265 DVDs, pamphlet/broadside etc coll
libr loan; ASLIB, CERL, CURL, SCONUL, IFLA, LIBER, RLG, FID 29232

– Birkbeck College, Library, Malet St, London WC1E 7HX
T: +44 20 76316239;
Fax: +44 20 76316066; E-mail: libhelp@library.bbk.ac.uk; URL: www.bbk.ac.uk/lib
1823; Philippa Dolphin
Trevelyan Memorial Libr, Canadian Studies Libr
libr loan; SCONUL, M25 29233

– Courtauld Institute of Art, Library, Somerset House, Strand, London WC2R 0RN
T: +44 20 78482701;
Fax: +44 20 78482887; E-mail: booklib@courtauld.ac.uk; URL: www.courtauld.ac.uk
1933; Antony Hopkins
90000 vols; 230 curr per; 2500 diss/theses; 200 microforms; 20 digital data carriers; 60000 pamphlets, off-prints and other small unbound items
ARLIS/UK & Ireland, SCONUL, RLG 29234

– Goldsmiths' College, Information Services, Rutherford Information Services Building, New Cross, London SE14 6NW
T: +44 20 79197150; Fax: +44 20 79197165; E-mail: library@gold.ac.uk; URL: www.goldsmiths.ac.uk
1906; N.N.
260000 vols; 1200 curr per; 1000 diss/theses; 10000 music scores; 1439 maps; 10000 microforms; 160000 av-mat; 11000 sound-rec; 375 digital data carriers
CILIP, ARLIS/UK & Ireland, IAML, SEAL 29235

– Heythrop College, Library, Kensington Sq, London W8 5HN
T: +44 20 77954250; Fax: +44 20 77954253; E-mail: library@heythrop.ac.uk; URL: www.heythrop.ac.uk
1614; Rev. Christopher Pedley, S.J.
Religion, Theology, Philosophy, Church Hist
200000 vols; 420 curr per; 40 incunabula; 50 diss/theses; 5000 microforms; 10 digital data carriers
libr loan; ABTAPL, SCONUL 29236

– Institute for the Study of the Americas, School of Advanced Study, Library, 35 Tavistock Sq, London WC1H 9HA
T: +44 20 78628501; Fax: +44 20 78628971; E-mail: americas@sas.ac.uk; URL: www.sas.ac.uk/americas
1966; Sarah Pink
Latin American political pamphlets, British Union Catalogue of Latin Americana (to 1988), News Sources
21435 vols; 323 curr per; 12 e-journals; 551 diss/theses; 50 maps; 500 microforms; 330 av-mat; 10 digital data carriers; 3000 working papers
ACLAIIR, SALALM, REDIAL 29237

– Institute of Advanced Legal Studies, School of Advanced Study, Library, 17 Russell Sq, London WC1B 5DR
T: +44 20 78625800; Fax: +44 20 78625770; E-mail: ials@sas.ac.uk; URL: ials.sas.ac.uk
1947; Jules Winterton
Commonwealth Law Libr from Foreign & Commonwealth Office, Records of Legal Education
260000 vols; 3000 curr per; 7500 e-journals; 900 diss/theses; 52000

microforms; 5 av-mat; 164 sound-rec; 88 digital data carriers
SCOLMA, BIALL, IALL, AALL, CALL, ASLIB, OSALL   29238

– Institute of Classical Studies, Library, Senate House, Malet St, London WC1E 7HU
T: +44 20 78628709; Fax: +44 20 78628735; E-mail: colin.annis@sas.ac.uk; URL: icls.sas.ac.uk/library/Home.htm
1880; Colin H. Annis
Homer coll, Greek and Roman philosophy, Religion, Art, Numismatics, Etruscan civilisation, Minoan-Mycenaean civilisation, Greek and Roman epigraphy and papyrology, Greek and Roman hist, Hist of the Early Church
130 000 vols; 675 curr per; 12 e-journals; 28 mss; 500 diss/theses; 1 040 maps; 3 162 microforms; 6 800 av-mat; 51 sound-rec; 220 digital data carriers; Press cuttings relating to Classics
SCONUL, CURL, M25
Also joint libr of the Hellenic and Roman Societies   29239

– Institute of Commonwealth Studies, School of Advanced Study, Library, 28 Russell Sq, London WC1B 5DS
T: +44 20 7862 8844; Fax: +44 20 7862 8820; E-mail: icommlib@sas.ac.uk; URL: commonwealth.sas.ac.uk/library.htm
1949; David Clover
In addition to our arch: Commonwealth bibliography, Commonwealth political party docs, Caribbean, Canadian and southern African mat, West India Committee Libr, Official publs, Census & Statistical Rpts, Grey lit
200 000 vols; 900 curr per; 250 mss; 80 000 govt docs; 450 maps; 1 000 microforms; 60 digital data carriers; numerous electronic sources
SCOLMA, SALG, CILIP, ANZLAG   29240

– Institute of Education, Newsam Library, 20 Bedford Way, London WC1H 0AL
T: +44 20 76126080; Fax: +44 20 76126093; E-mail: lib.enquiries@ioe.ac.uk; URL: www.ioe.ac.uk/library
1902; Stephen Pickles
Hans Coll. (Comparative educ), Assistant Masters Assn arch, World Education Fellowship arch, National Union of Women Teachers arch, Schools' Council arch, German educational reconstruction arch, schoolbks and classroom mat
350 000 vols; 1 000 curr per; 2 000 e-journals; 10 000 diss/theses; 1 000 microforms; 1 700 av-mat; 400 sound-rec; 90 digital data carriers
libr loan; IFLA, CILIP, SCONUL   29241

– Institute of Germanic Studies, Library, 29 Russell Sq, London WC1B 5DP
T: +44 20 7862 8967; Fax: +44 20 78628970; E-mail: igslib@sas.ac.uk; URL: www.sas.ac.uk/igs
1950; W. Abbey
German language and lit, German philosophy, German theatre and film, Exile studies (1933-1945); Priebsch-Closs Coll, Gundolf Coll, English Goethe Society Libr
90 000 vols; 330 curr per; 25 000 mss; 13 incunabula; 22 000 microfilms; pamphlets, newspaper cuttings
CURL, SCONUL, M25   29242

– Institute of Historical Research, School of Advanced Study, Library, Senate House, Malet St, London WC1E 7HU
T: +44 20 78628760; Fax: +44 20 78628762; E-mail: ihr.library@sas.ac.uk; URL: www.history.ac.uk/library/index.html
1921; Jennifer Higham
171 000 vols; 400 curr per; 50 e-journals; 2 000 diss/theses; 210 microforms; 30 digital data carriers   29243

– Institute of Psychiatry, Library, De Crespigny Park, Denmark Hill, London SE5 8AF
T: +44 20 79193204; Fax: +44 20 77034515
1948; M. Guha
Guttman-Maclay coll on Art and Psychiatry, Mayer Gross coll on the hist of psychiatry
32 000 vols; 350 curr per; 1 000 diss/theses; 27 000 reprints
libr loan; BLL   29244

– Joint Library of Ophthalmology, Moorfields Eye Hospital & UCL Institute of Ophthalmology, 11-43 Bath Street, London EC1V 9EL
T: +44 20 7608 6814; E-mail: ophthlib@ucl.ac.uk; URL: www.ucl.ac.uk/ioo/The_Joint_Library
1947; Deborah Heatlie
Historical book collection on ophthalmology and related subjects dating from 1585.; some drawings, ophthalmic instruments and papers from Edward Nettleship collection
23 000 vols; 119 curr per; 160 av-mat; 30 digital data carriers; thousands of e-journals and e-books
libr loan; AVSL   29245

– London Hospital Medical College, St Bartholomew's and the Royal London School of Medicine and Dentistry, White Chapel Site Library, Turner St, London E1 2AD
T: +44 20 72957115; Fax: +44 20 72957113
1887; P.S. Hockney
Alumnus coll, forensic medicine coll
38 000 vols; 320 curr per; 400 diss/theses; 300 av-mat
libr loan; BMA   29246

– Queen Mary and Westfield College, Library, Mile End Rd, London E1 4NS
T: +44 20 77753300; Fax: +44 20 89810028; E-mail: b.murphy@qmw.ac.uk
1887; Brian Murphy
European Union, arts, social sciences, law, engineering, mathematics, computer science, medicine
570 000 vols; 3 000 curr per
libr loan; SCONUL, M25, ASLIB   29247

– Royal Holloway Library, Royal Holloway and Bedford New College, University of London, Egham Hill, Egham TW20 0EX
T: +44 1784 443334; Fax: +44 1784 477670; E-mail: Library@rhul.ac.uk; URL: www.rhul.ac.uk/information-services/library/
1849; John Tuck
A.V.Coton Coll on dance, Robert Simpson Soc Arch, Dom Anselm Hughes Libr, Busk Hist of Science Coll, Dawson Coll of N.Z. Literature, Herringham Coll, Oliver Coll, Sargant Benson Botanical Coll, S.E. Asia Geology Libr, Sir Alfred Sherman papers, Tuke Italian Coll; Music libr
510 850 vols; 1 767 curr per; 7 000 e-journals; 4 incunabula; 3 290 diss/theses; 8 050 govt docs; 14 324 music scores; 1 735 microforms; 1 698 av-mat; 5 200 sound-rec; 1 160 digital data carriers; 17 409 other items
libr loan; CILIP, ASLIB, SCONUL, M25   29248

– Royal Veterinary College, Library, Royal College St, London NW1 0TU
T: +44 20 74685162; E-mail: sjackson@rvc.ac.uk; URL: www.rvc.ac.uk
1791; Simon Jackson
Hist coll of 2 000 vols on early vet lit, Veterinary Museum
30 000 vols; 259 curr per; 300 diss/theses; 8 microforms; 80 av-mat; 3 digital data carriers
ASLIB, EAHIL, AHIS (UK & Ireland)   29249

– School of Oriental and African Studies, Library, Thornhaugh St, Russell Sq, London WC1H 0XG
T: +44 20 78984163; Fax: +44 20 78984159; E-mail: libenquiry@soas.ac.uk; URL: www.soas.ac.uk/library
1917; Keith Webster
Int African Bibliographic Missionary Arch, Hardyman Madagascar coll
1 000 000 vols; 4 800 curr per; 2 800 mss; 50 000 maps; 57 000 microforms; 6 223 av-mat; 1 200 sound-rec; 30 digital data carriers; 50 000 pamphlets, 25 000 photos
SCONUL, IFLA, SCOLMA   29250

– St George's Hospital Medical School, Library, HUnter Wing, Cranmer Terrace, London SW17 0RE
T: +44 20 87255466; Fax: +44 20 87674696
1836; Susan Gove
Sir Benjamin Brodie Coll, St. George's Hist Coll
150 000 vols; 780 curr per; 800 diss/theses; 400 av-mat
libr loan; EAHIL   29251

– University Marine Biological Station Millport (UMBSM), Library, Millport KA28 0EG
T: +44 1475 530581; Fax: +44 1475 530601; E-mail: kstevens@udcf.gla.ac.uk; URL: www.gla.ac.uk/acad/marine/lib
1897; K.H. Stevenson
Marine technology, fisheries, reprints of Marine Biology, Marine taxonomy
5 000 vols; 55 curr per; 46 diss/theses; 200 maps; 10 000 reprints   29252

– Warburg Institute, School of Advanced Study, Library, Woburn Sq, London WC1H 0AB
T: +44 20 78628935; Fax: +44 20 78628939; E-mail: warburg.library@sas.ac.uk; URL: warburg.sas.ac.uk
late 1890s/incorporated in the Univ of London in 1944; Jill Kraye
352 993 vols; 944 curr per; 57 e-journals; 12 mss; 15 incunabula; 447 diss/theses; 4 020 microforms; 636 digital data carriers   29253

University of Roehampton, Roehampton Lane Learning Resources Centre, Erasmus House, Roehampton Lane, *London* SW15 5PU
T: +44 20 83923770; Fax: +44 20 83923026; E-mail: library@roehampton.ac.uk; URL: www.roehampton.ac.uk/
1975; Sue Clegg
Wesley Hist Soc. Libr, Sharpe Coll. (children's bks), Ruskin Coll. Froebel Archive (education)
490 000 vols; 1 500 curr per; 2 145 e-journals
libr loan; SCONUL   29254

University of Westminister, Cavendish Campus Library, 115 New Cavendish St, *London* W1W 6UW
T: +44 20 79115000 ext 3627 or 3613; Fax: +44 20 79115093; E-mail: e.salter@westminster.ac.uk; URL: www.wmin.ac.uk/library
1970; Elaine Salter
280 000 vols; 2 100 curr per; 1 000 diss/theses; 10 000 govt docs; 1 000 maps; 500 microforms; 5 000 av-mat; 500 sound-rec
ASLIB   29255

– Little Titchfield Street Library, Regent Campus, 4-12 Little Titchfield St, London W1P 7FH
T: +44 20 79115000 ext 2537; Fax: +44 20 79115846; E-mail: e.salter@wmin.ac.uk
Elaine Salter
Social sciences, humanities, business, alternative health sciences
80 000 vols; 40 digital data carriers; 24 000 periodicals   29256

– Marylebone Campus Library, 35 Marylebone Rd, London NW1 5 LS
T: +44 20 79115000 ext 3212; Fax: +44 20 79115058; URL: www.wmin.ac.uk/library
Jane Harrington
Built environment, management
70 000 vols; 400 curr per; 3 000 e-journals
ASLIB, BBSLG   29257

– Regent Campus Library, 4-12 Little Titchfield St, London W1W 7UW
T: +44 20 79115000 ext 2537; Fax: +44 20 79115894; URL: www.wmin.ac.uk/page-11674
Elaine Salter
Dictionaries – mono/bi/multi-lingual, Capital Punishment Studies coll
86 000 vols; 450 curr per; 4 000 e-journals; 50 diss/theses; 500 av-mat; 500 sound-rec; 50 digital data carriers   29258

Wimbledon School of Art, Library, Merton Hall Rd, *London* SW19 3QA
T: +44 20 84085027; Fax: +44 20 84085050; E-mail: info@wimbledon.ac.uk; URL: www.wimbledon.ac.uk/
1940
Costume
28 000 vols; 101 curr per; 10 000 microforms; 3 000 av-mat; 700 sound-rec; 7 CD-ROM subscriptions
ARLIS/UK & Ireland, SIBMAS, TIG   29259

Loughborough University, University Library, Ashby Rd, *Loughborough* LE11 3TU
T: +44 1509 222360; Fax: +44 1509 222361; E-mail: R.Jenkins@lboro.ac.uk; URL: www.lboro.ac.uk/library
1952; Ruth Jenkins
600 000 vols; 4 000 curr per; 19 000 e-journals; 1 000 maps; 5 500 microforms; 125 av-mat; 33 000 pamphlets, e journals
libr loan; IATUL, SCONUL, ASLIB, CILIP, UKSG   29260

University of Bedfordshire, Learning Resources, Park Sq, *Luton* LU1 3JU
T: +44 1582 743488; Fax: +44 1582 489325; URL: lrweb.beds.ac.uk
1957; T.P. Stone
Hockcliffe Coll of Early Children's Books (1685)
300 000 vols; 1 500 curr per; 20 000 e-journals; 450 diss/theses; 25 microforms; 5 000 av-mat; 500 sound-rec
libr loan; ASLIB, UKSG, NAG, M25 Consorting, SCONUL   29261

University for the Creative Arts, Library, Oakwood Park, *Maidstone* ME16 8AG
T: +44 1622 621120; Fax: +44 1622 621100; E-mail: librarymaid@ucreative.ac.uk; URL: www.ucreative.ac.uk/library
Vanessa Crane
Artists' books coll, Video art coll, Bandes Dessinées coll
236 000 vols; 650 curr per; 600 maps; 200 microforms; 5 500 av-mat; 50 digital data carriers; 70 000 slides
libr loan; ARLIS/UK & Ireland, ARCLIB, CILIP   29262

Manchester Metropolitan University, All Saints Library, Oxford Rd, *Manchester* M15 6BH
T: +44 161 2476646; Fax: +44 161 2476347; E-mail: g.r.barry@mmu.ac.uk; URL: www.mmu.ac.uk/services/library
1970; Gill Barry
Book Design, North West Film Arch; Site libs at Didsbury, Aytoun, Hollings, Elizabeth Gaskell, Crewe and Alsager
1 000 000 vols; 4 148 curr per
libr loan; ALA, ASLIB, IFLA, CILIP, SCONUL, NAG, UKSG   29263

– Aytoun Library, Aytoun St, Manchester M1 3GH
T: +44 161 2473093
Ms K. Morrison
Accounting, finance, business, econ, languages
50 000 vols   29264

– Elizabeth Gaskell Library, Hathersage Rd, Manchester M13 0JA
T: +44 612476132
Ms A. Mackenzie
Health care, psychology and speech pathology, physiotherapy
150 000 vols   29265

– Hollings Library, Old Hall Lane, Fallowfield, Manchester M14 6HR
T: +44 161 2476119; Fax: +44 161 2476670; E-mail: hollings-lib-enq@mmu.ac.uk
Ian Harter
Clothing techn, food techn, hotel a catering, home studies
25 000 vols
libr loan   29266

Nazarene Theological College, Library, Dene Rd, Didsbury, *Manchester* M20 2GU
T: +44 161 4381922; Fax: +44 161 4480275; E-mail: library@nazarene.ac.uk
1944; Donald Maciver
Theology, church hist, biblical studies, philosophy, Wesley studies
31 000 vols; 70 curr per; 64 diss/theses; 1 000 music scores; 125 sound-rec
ABTAPL   29267

Royal Northern College of Music, Library, 124 Oxford Road, *Manchester* M13 9RD
T: +44 161 9075243; Fax: +44 161 2737611; E-mail: library@rncm.ac.uk; URL: www.library.rncm.ac.uk
1973; Anna Wright
Rawsthorne Mss, Brodsky Arch, John Ogdon Mss, Philip Jones Brass Ensemble arch, Pitfield Arch, Walter & Ida Carroll Arch, Arthur Butterworth

Mss, Manchester European Wind Libr (MEWL), Rothwell Coll, Eva Turner coll, Sir Charles Groves scores, John Golland Mss, Philip Newman Arch, RMCM Arch, RNCM Coll of Historic Musical Instruments
100 000 vols; 85 curr per; 20 e-journals; 27 e-books; 1 000 mss; 100 diss/theses; 100 govt docs; 60 000 music scores; 200 microforms; 900 av-mat; 37 000 sound-rec; 10 digital data carriers
libr loan; IAML 29268

University of Manchester, John Rylands University Library, Oxford Rd, *Manchester* M13 9PP
T: +44 161 2753751; Fax: +44 161 2737488; URL: rylibweb.man.ac.uk
1851; William G. Simpson
Althorp Libr, Bullock Coll, Christie Libr, Med Soc Libr, Christian Brethren Coll, Dante Coll, Deag Eduaction Coll, Marmorstein Coll, Methodist Arch, Partington Coll, Unitarian College Coll
3 700 000 vols; 8 372 curr per; 1 187 483 mss; 5 000 incunabula; 333 641 microforms; 21 735 av-mat; 9 830 sound-rec; 225 digital data carriers
libr loan; ASLIB, SCONUL 29269

– Ashburne Hall Library, Old Hall Lane, Manchester M14 6HP
T: +44 161 2242835
1927; E.B. French
Morley
16 000 vols
libr loan 29270

– Institute of Science and Technology, Library, Sackville St, P.O. Box 88, Manchester M60 1QD
T: +44 161 2004924; Fax: +44 161 2004941; URL: www.umist.ac.uk/library
1825; M.P. Day
Joule Coll
320 000 vols; 6 900 curr per; 2 mss; 22 113 diss/theses; 2 530 microforms; 443 av-mat; 1 460 digital data carriers; 1 260 software packages
libr loan; SCONUL, IATUL, CILIP 29271

– Manchester Business School, Eddie Davies Library, MBS WEst, Booth St West, Manchester M15 6PB
T: +44 161 2756507; Fax: +44 161 2756505; E-mail: libdesk@mbs.ac.uk; URL: www.mbs.ac.uk/corporate/libraryservices/index.aspx
1966; Kathy Kirby
Company Rpt and Accounts
10 000 vols; 400 curr per; 3 000 diss/theses; 6 000 microforms; 10 digital data carriers; 11 000 working papers
libr loan; BBSLG, EBSLG, AAIM 29272

Cleveland College of Art and Design, Library, Green Lane, Linthorpe, *Middlesbrough* TS5 7RJ
T: +44 1642 821441; Fax: +44 1642 823467
1966; Ann Kanyon
25 000 vols; 150 curr per; 55 microforms; 210 000 av-mat; 300 sound-rec; 20 digital data carriers; 3 000 ills
libr loan; ARLIS/UK & Ireland 29273

Teesside University, Library and Information Services, Borough Rd, *Middlesbrough* TS1 3BA
T: +44 1642 342100; Fax: +44 1642 342190; E-mail: enquiries@tees.ac.uk; URL: lis.tees.ac.uk
Ian C. Butchart
250 000 vols; 1 890 curr per; 31 000 microforms; 430 av-mat; 45 600 transparencies
libr loan 29274

Open University, The Library and Learning Resources Centre, Walton Hall, *Milton Keynes* MK7 6AA
T: +44 1908 653138; Fax: +44 1908 653571; URL: www.open.ac.uk/library
1969; Nicky Whitsed
Jennie Lee papers, Betty Boothroyd papers, Walter Perry papers, Geoffrey Vickers papers, Maureen Oswin papers, distance education teaching materials, Kauvel Collection on the History of Mathematics
196 000 vols; 2 000 curr per; 12 000 e-journals; 4 000 diss/theses; 100 000 govt docs; 500 music scores; 500 maps; 100 000 microforms; 130 000 av-mat; 10 488 sound-rec; 200 digital

data carriers; 130 linear metres of mss, 9 800 video cassettes
libr loan; CILIP, ASLIB, SCONUL 29275

Queen Margaret University, Learning Resource Centre, Queen Margaret University Drive, *Musselburgh* EH21 6UU
T: +44 131 474000; Fax: +44 131 474001; E-mail: lrchelp@qmu.ac.uk; URL: www.qmu.ac.uk/lb
1875; Jo Rowley
University Archives
130 000 vols; 720 curr per; 70 e-journals; 45 000 e-books; 200 diss/theses; 400 av-mat
libr loan 29276

Newcastle University, Robinson Library, Back Jesmond Rd, *Newcastle upon Tyne* NE2 4HQ
T: +44 191 2227662; Fax: +44 191 2226235; E-mail: lib-readerservices@ncl.ac.uk; URL: www.ncl.ac.uk/library
1871; Dr. T.W. Graham
Gertrude Bell Coll (Middle East), Pybus Coll (hist of med), Robert White Coll, Trevelyan Papers, Runciman Papers; Medical and Dental Libr, Law Libr, Libr of Japanese Science and Technology
1 000 000 vols; 5 000 curr per; 700 e-journals; 3 500 mss; 6 800 diss/theses; 80 000 govt docs; 500 music scores; 5 000 maps; 6 200 microforms; 2 600 av-mat; 1 000 sound-rec; 4 000 transparencies, 100 other on-line subsciptions
libr loan; BLL, LISE, SCONUL 29277

– Law Library, 22-24 Windsor Terrace, Newcastle upon Tyne NE1 7RU
T: +44 191 2227944; Fax: +44 191 2226235; E-mail: lib-law@ncl.ac.uk; URL: www.ncl.ac.uk/library/law/
1977; Wayne Connolly
36 000 vols; 227 curr per; 3 000 e-journals; 167 e-books; 5 000 govt docs
BIALL 29278

– Medical School, Walton Library, Framlington Place, Newcastle upon Tyne NE2 4HH
T: +44 191 2227550; Fax: +44 191 2228102; E-mail: lib-walton-rs@ncl.ac.uk; URL: www.ncl.ac.uk/library/walton/
1984; Erika Gavillet
40 000 vols; 90 curr per; 40 e-books; 800 diss/theses; 240 av-mat; 17 sound-rec; 75 digital data carriers
libr loan 29279

– School of Architecture, Library, 21 Ellison Terr., Newcastle upon Tyne NE1 7RU
T: +44 191 2226025; Fax: +44 191 2226311
D.A. Nichol
10 000 vols; 60 curr per 29280

– School of Education, Brian Stanley Library, St Thomas St, Newcastle upon Tyne NE1 7RU
T: +44 191 2226575
25 000 vols; 200 curr per 29281

Isle of Wight College, Learning Resources Centre, Medina Way, *Newport, Isle of Wight* PO30 5TA
T: +44 1983 526631 ext 201; Fax: +44 1983 521707
1957; Lynne Christopher
31 000 vols; 300 curr per; 5 500 av-mat; 40 digital data carriers
libr loan; CILIP 29282

Harper Adams University College, The Bamford Library, *Newport, Shropshire* TF10 8NB
T: +44 1952 820280; Fax: +44 1952 815391; E-mail: libhelp@harper-adams.ac.uk; URL: www.harper-adams.ac.uk
1901; Kathryn Greaves
Hist poultry book coll
46 506 vols; 891 current periodicals (of which 316 are electronic)), AV mat
libr loan; ALLCU 29283

University of Ulster, The Library, Shore Rd, *Newtownabbey* BT37 0QB
T: +44 28 90366370; E-mail: online@ulster.ac.uk; URL: www.ulster.ac.uk
1984; Mrs E. Urquhart
Henry Davis Coll, John Hewitt Coll, Irish Room Coll, Henry Morris Coll, The Headlam-Morley Coll, Francis Stuart Coll,

Denis Johnston Coll, The George Shiels Coll, Stelfox and Carrothers Coll, Magee Community Coll
823 145 vols; 16 305 curr per; 14 000 e-journals; 8 255 diss/theses; 23 670 microforms; 120 000 Art and Design slides
libr loan; SCONUL, IAML 29284

Moulton College, Learning Resources Centre, West St, Moulton, *Northampton* NN3 1RR
T: +44 1604 491131 ext 222; Fax: +44 1604 491127; E-mail: lrcenquiries@moulton.ac.uk; URL: www.moulton.ac.uk
1921; Karen Acham
Agriculture, horticulture, arboriculture, country side management, construction, equestrian, sport science, furniture, interior design, floristry, animal care + welfare
25 000 vols; 200 curr per; 90 e-journals; 1 015 av-mat; 200 digital data carriers
libr loan; COLRIC, ALLCU 29285

University College Northampton, Rockingham Library, Park Campus, Boughton Green Rd, *Northampton* NN2 7AL
T: +44 1604 735500; Fax: +44 1604 718819; URL: library.northampton.ac.uk
1975; Ms H.J. Johnson
Leather coll, wastes management coll
233 000 vols; 900 curr per; 2 300 e-journals; 1 182 diss/theses; 1 330 music scores; 300 maps; 200 microforms; 90 digital data carriers
libr loan 29286

Norwich City College of Further and Higher Education, Library, Ipswich Rd, *Norwich* NR2 2LJ
T: +44 1603 660011
1951; B.R. Derbyshire
Media Resources Ctr
60 000 vols; 615 curr per; 30 diss/theses; 5 000 microforms; 6 000 av-mat
libr loan; ASLIB 29287

Norwich School of Art and Design, Library, Francis House, 3-7 Redwell St, *Norwich* NR2 4SN
T: +44 1603 610561; Fax: +44 1603 615728; E-mail: info@nsad.ac.uk; URL: www.nsad.ac.uk
Tim Giles
30 000 vols; 100 curr per; 145 000 av-mat; 100 sound-rec; 40 digital data carriers
libr loan; ARLIS/UK & Ireland, CILIP 29288

University of East Anglia, Library, *Norwich* NR4 7TJ
T: +44 1603 592421; Fax: +44 1603 591010; E-mail: library@uea.ac.uk; URL: www.lib.uea.ac.uk
1962; Jonathan colam
Holloway Coll, Zuckerman Arch, Pritchard Papers, Hill Arch, Everest Arch, Illustrated bks, Military Science (predominantly 1776-1815), Abbott Coll, Ketton-Crame Coll (local hist), Kimber Coll (local lit)
800 000 vols; 2 800 curr per; 8 000 e-journals; 35 000 mss; 5 000 diss/theses; 10 000 govt docs; 9 500 music scores; 35 000 maps; 54 000 microforms; 10 000 av-mat; 10 000 sound-rec; 600 digital data carriers
libr loan; SCONUL, SPARC, UKSG, NAG, BIALL 29289

Nottingham Trent University, Library and Information Services, Burton St, *Nottingham* NG1 4BU
T: +44 115 9418418; Fax: +44 115 9415380; E-mail: cor.web@ntu.ac.uk; URL: www.ntu.ac.uk
1948; Elizabeth Lines
380 000 vols; 2 500 curr per
libr loan; ASLIB 29290

Nottingham University Hospitals NHS Trust, City Campus, Medical Library and Information Service, Postgraduate Medical Education Centre, Hucknall Rd, *Nottingham* NG5 1PB
T: +44 115 9691169, ext: 55735; Fax: +44 115 962 7741; E-mail: library@nuh.nhs.uk; URL: www.nuh.nhs.uk/nch/library/
1972; Maria Nolan
8 000 vols; 160 curr per; 250 av-mat; 5

digital data carriers
libr loan 29291

The People's College, Resources Centres, Maid Marian Way, *Nottingham* NG1 6AB
T: +44 115 9128636; Fax: +44 115 9128600; URL: www.peoples.ac.uk
J.D. Sutcliffe
30 000 vols; 210 curr per; 2 microforms; 400 av-mat; 20 digital data carriers
libr loan 29292

University of Nottingham, Hallward Library, University Park, *Nottingham* NG7 2RD
T: +44 115 9514557; URL: www.nottingham.ac.uk/is
1928; R.E. Oldroyd
D.H. Lawrence Coll, Porter Coll (ornith), Briggs Coll (early educ mat), East Midlands hist, Jacob Coll (medical hist), Benedikz Coll (Icelandic); Hallward Libr, Science and Engineering Libr, Greenfield Medical Libr, Agricultural and Food Sciences Libr, Djanogly LRC, Derby Royal Infirmery (Nursing and Midwifery), Kings Mill Hospital (Nursing and Midwifery), Univ of Nottingham in Malaysia, Kuala Lumpur Libr
1 000 000 vols; 5 000 curr per; 3 000 000 mss; 100 000 govt docs; 20 000 music scores; 100 000 microforms; 170 000 pamphlets
libr loan; SCONUL, CILIP, ASLIB, BLDSC, BIALL, IAML 29293

– George Green Library of Science and Engineering, University Park, Nottingham NG7 2RD
T: +44 115 9514581; Fax: +44 115 9514578; URL: www.nottingham.ac.uk/library
1964; N.N.
150 000 vols; 1 000 curr per; 10 000 e-journals; 60 000 e-books; 4 500 diss/theses; 100 microforms; 100 av-mat; 200 digital data carriers
libr loan; SCONUL, CURL, RLG 29294

– Greenfield Medical Library, Nottingham NG7 2UH
T: +44 115 9709441; Fax: +44 115 9709449; E-mail: library-medical-enquiries@nottingham.ac.uk; URL: www.nottingham.ac.uk/is/location/library/greenfield.phtml
1968; W.J. Stanton
Nursing, Hist of Medicine, Reports and Series
105 000 vols; 7 500 curr per; 1 296 diss/theses; 92 av-mat; 65 digital data carriers
libr loan; CURL, RLG 29295

– James Cameron-Gifford Library of Agricultural and Food Sciences, Sutton Bonington Campus, Loughborough LE12 5RD
T: +44 115 9516390; Fax: +44 115 9516389; URL: www.nottingham.ac.uk/is
1908; S.M. Bennett
Early agricultural bks
40 000 vols; 450 curr per; 450 diss/theses; 30 av-mat 29296

Edge Hill University, Learning Resource Centre, St. Helens Rd, *Ormskirk* L39 4QP
T: +44 1695 575171; Fax: +44 1695 579997; URL: www.edgehill.ac.uk
1902; Sue Roberts
188 000 vols; 1 000 curr per; 11 000 e-journals
libr loan; CILIP 29297

Orpington College, Library, The Walnuts, *Orpington* BR6 0TE
T: +44 1689 899712 ext 212; Fax: +44 1689 877949; E-mail: enquiries@orpington.ac.uk; URL: www.orpington.ac.uk
1972; Sue Kitchener
30 000 vols; 70 curr per; 30 digital data carriers
libr loan 29298

All Souls College, University of Oxford, Codrington Library, High St, *Oxford* OX1 4AL
T: +44 1865 279318; Fax: +44 1865 279299; E-mail: codrington.library@all-souls.ox.ac.uk; URL: www.all-souls.ox.ac.uk
1716; Prof. I.W.F. Maclean
Mil hist Coll., Letters of Junius, Peerage

Cases
180 000 vols; 271 curr per; 413 mss;
338 incunabula
libr loan (but books cannot be lent)
29299

Brasenose College, University of Oxford,
Library, Radcliffe Square, **Oxford**
OX1 4AJ
T: +44 1865 277827; Fax: +44
1865 277-822/-831; E-mail:
library@bnc.ox.ac.uk; URL:
www.bnc.ox.ac.uk
1664; Dr. E H Bispham
60 000 vols; 91 curr per; 52 mss; 2 320
incunabula; 55 music scores
UKSG
29300

Magdalen College, Library, High St,
Oxford OX1 4AU
T: +44 1865 276045; Fax: +44 1865
276057; URL: www.magd.ox.ac.uk
1458; Dr. Christine Y. Ferdinand
Late medieval English hist coll
110 000 vols; 98 curr per
29301

Oxford and Cherwell Valley College,
Millstream Centre, Oxpens Rd, **Oxford**
OX1 1SA
T: +44 1865 551961; Fax: +44 1865
551386; E-mail: oxfordlibrary@ocvc.ac.uk;
URL: www.ocvc.ac.uk
1960; Rob Collier
44 117 vols; 200 curr per; 1 767 av-mat;
192 digital data carriers
29302

Oxford Brookes University, University
Library, Headington, **Oxford** OX3 0BP
T: +44 1865 483156; Fax: +44 1865
483998; E-mail: library@brookes.ac.uk;
URL: www.brookes.ac.uk/library/
1958; Dr. H.M. Workman
John Fuller Coll of 6500 bks and
pamphlets on catering, hospitality
and gastronomy; Tourism coll, 19th c
children's bks, National Brewing library,
Andre Deutsch coll, Oxfordshire Society
of Architects Coll, Harold Fullard Coll of
Atlases, Medical Sciences Video Archive,
Museum of Modern Art Oxford Coll,
Dorset House Archive, Welfare Coll, Sally
Croft Collection, Booker Prize Archive,
Jane Grigson Collection
462 984 vols; 2 149 curr per
BLDSC, SCONUL, CILIP
29303

Ruskin College, Library, Walton St,
Oxford OX1 2HE
T: +44 1865 554331; Fax: +44 1865
554372; E-mail: library@ruskin.ac.uk
1899; Valerie Moyses
Social and Labour Hist (20th c British);
MacColl Seeger coll of folk song and
protest music
45 000 vols; 175 curr per; 50 e-journals;
1 000 diss/theses; 600 govt docs; 500
av-mat; 400 sound-rec; 400 boxes of
historical pamphlets
libr loan; CILIP
29304

University of Oxford, Bodleian Library,
Broad St, **Oxford** OX1 3BG
T: +44 1865 277000;
Fax: +44 1865 277182; E-mail:
enquiries@bodley.ox.ac.uk; URL:
www.bodley.ox.ac.uk
1602; Sarah E. Thomas
Radcliffe Science Libr, Bodleian Law Libr,
Bodleian Japanese Libr, Indian Inst Libr,
Oriental Inst Libr, Commonwealth and
African Studies Libr, Vere Harmsworth
Libr
7 400 000 vols; 16 000 curr per; 180 000
mss; 6 500 incunabula; 500 000 music
scores; 1 240 000 maps; 1 000 000
microforms
libr loan; CILIP, SCONUL, IFLA, CURL,
OCLC
29305

– Balliol College, Library, Oxford OX1 3BJ
T: +44 1865 277709; Fax: +44 1865
277803; E-mail: library@balliol.ox.ac.uk;
URL: www.balliol.ox.ac.uk/welcome-to-
balliol
1263; Dr. Penelope Bulloch
Ancient History, Biochemistry, Biological
Sciences, Classics, Computing, Economic
and Social History, Economics,
Engineering, English Language and
Literature, Modern Languages and
Literatures, History, Inorganic Chemistry,
Organic Chemistry, Physical Chemistry,
International Relations, Law, Management,
Mathematics, Philosophy; Physics, Politics
120 000 vols; 100 curr per; 450 mss; 20
incunabula
29306

– Bodleian Japanese Library, 27 Winchester
Rd, Oxford OX2 6NA
T: +44 1865 284506;
Fax: +44 1865 284500; E-mail:
japanese@bodleian.ox.ac.uk; URL:
www.bodleian.ox.ac.uk/bjl
1993; I.K. Tytler
1 000 Japanese mss, early printed books
125 000 vols; 638 curr per; 100 mss;
100 diss/theses; 200 govt docs; 50
maps; 200 microforms; 400 av-mat; 200
sound-rec; 60 digital data carriers; 500
other items
EAJRS, JLG
29307

– Bodleian Law Library, St Cross Bldg,
Manor Rd, Oxford OX1 3UR
T: +44 1865 271462;
Fax: +44 1865 271475; E-mail:
law.library@bodley.ox.ac.uk; URL:
www.ouls.ox.ac.uk/law
1964; Ruth Bird
400 000 vols; 1 500 curr per; 500
diss/theses; 11 900 microforms
libr loan; BIALL, ALLA, NELLCO, AALL
29308

– Bodleian Library of Commonwealth and
African Studies at Rhodes House, South
Parks Rd, Oxford OX1 3RG
T: +44 1865 270908;
Fax: +44 1865 270912; E-mail:
rhodes.house.library@bodleian.ox.ac.uk;
URL: www.ouls.ox.ac.uk/rhodes
1928; Lucy McCann
International Relations, Organizations,
African Studies (Africana); Scicluna
collection on Malta, papers of the Anti-
Slavery Society, papers of the United
Society for the Propagation of the
Gospel, papers of the Anti-Apartheid
Movement
320 000 vols; 2 360 curr per; 4 000 mss;
1 200 diss/theses; 10 000 govt docs;
1 200 microforms
libr loan; SCOLMA
29309

– Campion Hall Library, Brewer St, Oxford
OX1 1QS
T: +44 1865 286106
1896; Revd Peter Edmonds
G.M. Hopkins mss
50 000 vols; 40 curr per
29310

– Christ Church Library, Oxford OX1 1DP
T: +44 1865 276169; E-mail: library@
chch.ox.ac.uk; URL: www.chch.ox.ac.uk/
library
1563; Janet McMullin
150 000 vols; 200 curr per; 500 mss;
100 incunabula; 500 music scores; 500
maps; 300 microforms; 37 000 pre-1800
imprints
29311

– Corpus Christi College, Library, Merton
St, Oxford OX1 4JF
T: +44 1865 276744; Fax: +44 1865
276767; E-mail: library.staff@ccc.ox.ac.uk;
URL: www.ccc.ox.ac.uk/p/Library-and-
Archives/
1517; Joanna Snelling
Classical Languages and Lit, Hist,
Philosophy
80 000 vols; 90 curr per; 565 mss; 278
incunabula; 500 microforms
CILIP
29312

– Department for Continuing Education,
Library, Rewley House, 1 Wellington Sq,
Oxford OX1 2JA
T: +44 1865 270454;
Fax: +44 1865 270309; E-mail:
enquiries@conted.ox.ac.uk; URL:
www.conted.ox.ac.uk
1926; Mrs A.E. Rees
80 000 vols; 110 curr per; 80 diss/theses;
2 500 music scores; 50 maps; 735
av-mat; 20 sound-rec; 20 digital data
carriers; 10 000 slides, Local Studies Coll
29313

– Department of Education, Library, 15
Norham Gardens, Oxford OX2 6PY
T: +44 1865 274028;
Fax: +44 1865 274027; E-mail:
library@education.ox.ac.uk; URL:
www2.ouls.ox.ac.uk/edstud
1921; Kate Williams
Pre-1918 govt publs
52 000 vols; 350 curr per; 100 e-journals;
1 700 diss/theses; 3 400 govt docs; 15
microforms; 250 av-mat; 30 sound-rec;
150 digital data carriers
libr loan; LISE, CILIP
29314

– Department of Engineering Science,
Library, Parks Rd, Oxford OX1 3PJ
T: +44 1865 273193; Fax: +44 1865
273010
Ms A. Greig
23 000 vols; 150 curr per
libr loan
29315

– Department of Zoology, Edward Grey
Institute of Field Ornithology, Alexander
Library, South Parks Rd, Oxford
OX1 3PS
T: +44 1865 271143; Fax: +44 1865
271142; URL: www.ouls.ox.ac.uk/isbes/
zoology
Mrs Sophie Wilcox
Ornithology
22 394 vols; 500 curr per; 430
diss/theses; 82 700 pamphlets, mss
29316

– English Faculty Library, St Cross Bldg,
Manor Rd, Oxford OX1 3UQ
T: +44 1865 271050; Fax: +44
1865 271054; E-mail: efl-
enquiries@bodleian.ox.ac.uk; URL:
www.bodleian.ox.ac.uk/english
1914; Sue Usher
Languages and Literatures, Anglo-
American Languages and Literatures;
Old Icelandic Coll, Wilfred Owen Coll
(Owen's bks & mss, plus photos and
letters)
105 000 vols; 80 curr per; 1 080 e-
journals; 350 e-books; 300 av-mat; 300
sound-rec; 200 mss (Wilfred Owen), 2
Old Icelandic mss, 400 audio cassettes
(Shakespeare, poetry)
29317

– Exeter College Library, Turl Street,
Oxford OX1 3DP
T: +44 1865 279657; Fax: +44 1865
279630; E-mail: library@exeter.ox.ac.uk;
URL: www.lib.ox.ac.uk/libraries/guides/
EXE.html
Juliet Chadwick
71 000 vols; 189 mss; 75 incunabula;
107 music scores
29318

– Faculty of Music, Library, St. Aldate's,
Oxford OX1 1DB
T: +44 1865 276148; Fax: +44 1865
286260; E-mail: library@music.ox.ac.uk;
URL: www.ouls.ox.ac.uk/music/
1620; John Wagstaff
Howes Folk Music Coll, Wellesz Coll on
Byzantine Music
23 000 vols; 60 curr per; 65 e-journals;
150 diss/theses; 19 500 music scores;
250 microforms; 150 av-mat; 18 000
sound-rec; 5 digital data carriers; 1 150
pamphlets, 30 DVDs
libr loan; IAML, CILIP, RMA, MLA
29319

– Green College, Library, Radcliffe
Observatory, Woodstock Rd, Oxford
OX2 6HG
T: +44 1865 274788; Fax: +44 1865
274796; URL: www.green.ox.ac.uk
1979; Gill Edwards
Medicine, Social Sciences, Business
Studies
8 000 vols; 24 curr per; 3 digital data
carriers
29320

– Green Templeton College, Management
and Business Studies Library, Woodstock
Rd, Oxford OX2 6HG
T: +44 1865 274791;
Fax: +44 1865 274796; E-mail:
Debra.farrell@gtc.ox.ac.uk; URL:
www.gtc.ox.ac.uk
1964; Gill Edwards
Pierre Wack Scenarios and Futures
Collection; Retail (Oxford Institute of
Retail Management)
4 200 vols; 15 curr per; 3 600 e-journals
libr loan; BBSLG
The access to the College Libr is
restricted
29321

– Harris Manchester College, Library,
Mansfield Rd, Oxford OX1 3TD
T: +44 1865 281472; Fax: +44 1865
271012; E-mail: librarian@hmc.ox.ac.uk;
URL: www.hmc.ox.ac.uk
1786; Sue Killoran
History of Protestant Dissent secifically
Unitarianism
70 000 vols; 25 curr per; 370 mss;
2 incunabula; 40 diss/theses; 10
microforms; 20 av-mat; 20 sound-rec;
30 digital data carriers
libr loan; ABTAPL
29322

– Hertford College, Library, Catte St,
Oxford OX1 3BW
T: +44 1865 279409; Fax: +44 1865
279466; E-mail: library@hertford.ox.ac.uk;
URL: www.hertford.ox.ac.uk/library
1909; S. Griffin
Magdalen Hall Libr (18th c coll)
51 000 vols; 40 curr per; 40 mss; 5
incunabula; 50 music scores; 25 digital
data carriers
29323

– Institute of Economics and Statistics,
Library, St. Cross Bldg, Manor Rd,
Oxford OX1 3UL
T: +44 1865 271072; Fax: +44 1865
271094
1935; M.G. Robb
120 000 vols; 2 000 curr per; 2 000
microforms; 250 digital data carriers
libr loan; ASLIB
29324

– Jesus College, Library, Turl St, Oxford
OX1 3DW
T: +44 1865 279704; E-mail: librarian@
jesus.ox.ac.uk; URL: www.jesus.ox.ac.uk/
library
1628; Owen McKnight
Celtic Coll, Material relating to T.E.
Lawrence, Antiquarian printed books
(Fellows' Library)
60 000 vols; 100 curr per; 200 mss; 44
incunabula
Historic Libraries Forum
29325

– Keble College, Library, Parks Rd, Oxford
OX1 3PG
T: +44 1865 272797; Fax: +44 1865
272705; E-mail: library@keble.ox.ac.uk;
URL: www.keble.ox.ac.uk/about/library
1876; Mrs M.A. Sarosi
57 000 vols; 90 curr per; 87 mss; 97
incunabula; 20 diss/theses
CILIP
29326

– Lady Margaret Hall Library, Norham
Gardens, Oxford OX2 6QA
T: +44 1865 274361;
Fax: +44 1865 270708; E-mail:
roberta.staples@lmh.ox.ac.uk
R. Staples
70 000 vols; 95 curr per
29327

– Lincoln College, Library, Turl St, Oxford
OX1 3DR
T: +44 1865 279831; E-mail: info@
lincoln.ox.ac.uk; URL: www.lincoln.ox.ac.uk
1427; Fiona Piddock
40 000 vols; 70 curr per
29328

– Mansfield College, Library, Mansfield Rd,
Oxford OX1 3TF
T: +44 1865 270975;
Fax: +44 1865 270970; E-mail:
alma.jenner@mansfield.ox.ac.uk; URL:
www.mansfield.ox.ac.uk
1886; Alma Jenner
Congregational Church hist, bibl studies,
patristics; Theology libr, Law libr,
Principal libr
32 000 vols; 25 curr per; 30 mss; 1
incunabula; 40 diss/theses
ABTAPL
29329

– Merton College, Library, Merton St,
Oxford OX1 4JD
T: +44 1865 276380;
Fax: +44 1865 276361; E-mail:
julia.walworth@merton.ox.ac.uk; URL:
www.merton.ox.ac.uk
1264; Dr. Julia Walworth
Book and papers of F.H. Bradley,
Brenchley coll of T.S. Eliot, Coll
of drawings, mss and bks by Max
Beerbohm, Blackwell Collection, extensive
archive relating to college administration
and estates
80 000 vols; 100 curr per; 328 mss; 100
incunabula
29330

– New College, Library, Holywell St, Oxford
OX1 3BN
T: +44 1865 279580;
Fax: +44 1865 279590; E-mail:
naomi.vanloo@new.ox.ac.uk; URL:
www.new.ox.ac.uk/The_Library/
Welcome.php
1379; Naomi van Loo
Seton Watson Coll on East European
history, World War 1+2
100 000 vols; 30 curr per; 388 mss;
332 incunabula; 500 music scores; 360
microforms; 200 av-mat; 100 digital data
carriers
CILIP
29331

– Nuclear and Astrophysics Laboratory, Library, Keble Rd, Oxford OX1 3RH
T: +44 1865 273421; Fax: +44 1865 273418
1890; P.J. Gledhill
28 800 vols; 121 curr per; 800 diss/theses; 2 200 maps; 1 406 microforms; 1 800 35mm slides   29332

– Nuffield College, Library, New Rd, Oxford OX1 1NF
T: +44 1865 278550; Fax: +44 1865 278621; E-mail: library-enquiries@nuffield.ox.ac.uk; URL: www.nuff.ox.ac.uk/library
1937; Elizabeth Martin
Archives of GDH and Margaret Cole, Lord Cherwell, William Cobbett special colls
90 000 vols; 500 curr per; 800 diss/theses; 125 000 govt docs; 25 000 archive items (personal papers), IALHI
Shared access to Oxford University e-resources   29333

– Oriel College, Library, Oriel Sq, Oxford OX1 4EW
T: +44 1865 276558; E-mail: library@oriel.ox.ac.uk; URL: www.oriel.ox.ac.uk
1326; Marjory Szurko
Lord Leigh's Libr
100 000 vols; 200 curr per; 100 mss; 34 incunabula; 100 music scores; 37 microforms   29334

– Oriental Institute, Library, Pusey Lane, Oxford OX1 2LE
T: +44 1865 278202; Fax: +44 1865 278204; E-mail: library@orinst.ox.ac.uk; URL: www.bodley.ox.ac.uk/dept/oriental/oil.htm
1963; Anthony D. Hyder
Chinese, Japanese, Korean, Islamic, Jewish and South Asian studies, Oriental art and architecture
100 000 vols
SCONUL   29335

– Oxford Centre for Hebrew and Jewish Studies, Library, Yarnton Manor, Oxford OX5 1PY
T: +44 1865 377946; Fax: +44 1865 375079; E-mail: muller.library@ochjs.ac.uk; URL: www.ochjs.ac.uk/library/index.html
1974; Dr. Piet van Boxel
90 000 vols
libr loan   29336

– Oxford Centre for Mission Studies, Library, St Philip's & St James', Woodstock Rd, P.O. Box 70, Oxford OX2 6HB
T: +44 1865 556071; Fax: +44 1865 510823; E-mail: ocms@ocms.ac.uk; URL: www.ocms.ac.uk/
13 000 vols   29337

– Pembroke College, McGowin Library, Pembroke Sq, Oxford OX1 1DW
T: +44 1865 276409; Fax: +44 1865 276418; E-mail: library@pmb.ox.ac.uk; URL: www.pmb.ox.ac.uk/Students/Library_Archives/index.php
Lucie Walker
Chandler (Aristotelia), Samuel Johnson Coll
40 000 vols; 80 curr per; 30 mss; 140 music scores; 5 microforms; 6 av-mat; 6 digital data carriers   29338

– Philosophy Faculty, Library, Philosophy Centre, 10 Merton St, Oxford OX1 4JJ
T: +44 1865 276927;
Fax: +44 1865 276932; E-mail: philosophy.library@bodleian.ox.ac.uk; URL: www.bodleian.ox.ac.uk/philosophy
Dr. Hilla Wait
28 000 vols; 10 curr per; 3 mss; 3 incunabula; 30 av-mat   29339

– Plant Sciences Library, Library and Information Service, South Parks Rd, Oxford OX1 3RB
T: +44 1865 275087; E-mail: enquiries.plant@ouls.ox.ac.uk; URL: www.ouls.ox.ac.uk/isbes/plants
1610; Roger A. Mills
Oxford Forest Information Service – world forestry lit; Sherard Coll – Pre-Linnaean taxonomic lit
200 000 vols; 2 000 curr per; 400 maps; 4 000 microforms
libr loan; ASLIB, BLDSC, IFLA, IAALD   29340

– Queen's College, Library, High St, Oxford OX1 4AW
T: +44 1865 279130; Fax: +44 1865 790819; E-mail: library@queens.ox.ac.uk; URL: www.queens.ox.ac.uk/library/
1341; Amanda Saville
Peet Memorial Libr (egyptology), c. 100 000 pre-1850 bks, c. 50 medieval mss
150 000 vols; 100 curr per; 1 000 e-journals; 800 mss; 400 incunabula; 100 diss/theses; 100 govt docs; 300 music scores; 30 maps; 400 microforms; 30 av-mat; 50 sound-rec; 50 digital data carriers
libr loan   29341

– Radcliffe Science Library, Parks Rd, Oxford OX1 3QP
T: +44 1865 272800;
Fax: +44 1865 272821; E-mail: enquiries.rsl@ouls.ox.ac.uk; URL: www.ouls.ox.ac.uk/rsl
1714; Dr. Judith Palmer
Sir Henry Acland pamphlet coll, Offprint colls of E.B. Tylor, G.H. Hardy and W. Le Gros Clark; Radcliffe Library records
900 000 vols; 5 000 curr per; 100 mss; 10 incunabula; 25 000 diss/theses; 1 000 maps
libr loan; ASLIB, IFLA, IATUL   29342

– Regent's Park College, Library, Pusey St, Oxford OX1 2LB
T: +44 1865 288120; Fax: +44 1865 288121; E-mail: library@regents.ox.ac.uk; URL: www.rpc.ox.ac.uk/index.php?pageid=126
1810; Andrew Hudson
Baptist Missionary Society Archives (on deposit); Baptist Union of Great Britain minute books; David Nicholls Collection; Angus Library (Baptist hist and arch)
50 000 vols; 150 curr per; 1 incunabula; 150 diss/theses; 1 500 music scores; 100 maps; 250 microforms; 32 600 av-mat; 10 sound-rec; 5 digital data carriers; 20 000 pamphlets,access to all electronic publs networked within Univ. of Oxford, 700 metres of boxes with manuscripts
CILIP, ABTAPL, SEMLAC   29343

– Sackler Library, 1 St John Street, Oxford OX1 2LG
T: +44 1865 288190; Fax: +44 1865 278098; E-mail: enquiries@saclib.ox.ac.uk; URL: www.saclib.ox.ac.uk
ca 1901; James Legg
Archaeology, art hist, classical studies, numismatics, egyptology; Special Colls: Grenfell & Hunt papyrological coll, Romano-British papers of Sir I.A. Richmond, Haverfield Arch, Wind Coll; Heberden Coin Room Libr
250 000 vols; 2 000 curr per; 1 000 mss; 300 diss/theses; 200 maps; 100 microforms; 100 digital data carriers   29344

– School of Geography and the Environment, Library, Mansfield Rd, Oxford OX1 3TB
T: +44 1865 285070; Fax: +44 1865 275885; E-mail: enquiries@geog.ox.ac.uk; URL: www.geog.ox.ac.uk
1887; L.S. Atkinson
Oxford Mountaineering Libr, Marjorie Sweeting Karst Coll
50 000 vols; 200 curr per; 170 diss/theses; 6 000 govt docs; 65 000 maps; 20 digital data carriers; 6 000 slides, 4 000 air photos
libr loan   29345

– Social Science Library, Centre for Advanced Studies In The Social Sciences, Manor Rd, Oxford OX3 CMS
T: +44 1865 271093; Fax: +44 1865 271072; E-mail: library@ssl.ox.ac.uk; URL: www.ssl.ox.ac.uk
2004; Margaret Robb
200 000 vols; 800 curr per
SCOLMA, IIALD, EADI, CILIP   29346

– Somerville College, Library, Woodstock Rd, Oxford OX2 6HD
T: +44 1865 270694; Fax: +44 1865 270620; E-mail: library@some.ox.ac.uk; URL: www.some.ox.ac.uk
1879; Anne Manuel
John Stuart Mill Libr, Amelia B Edwards mss coll, Percy Withers mss coll, Vernon Lee mss, Margaret Kennedy mss
120 000 vols; 74 curr per   29347

– St Anne's College, Library, Woodstock Rd, Oxford OX2 6HS
T: +44 1865 274811; Fax: +44 1865 274899; E-mail: library@st-annes.ox.ac.uk; URL: www.st-annes.ox.ac.uk/study/undergraduate/library.html
1895; Dr. Smith
110 000 vols; 79 curr per; 20 000 pamphlets, 200 photos   29348

– St Antony's College, Library, 62 Woodstock Rd, Oxford OX2 6JF
T: +44 1865 274480;
Fax: +44 1865 274526; E-mail: rosamund.campbell@sant.ox.ac.uk; URL: www.sant.ox.ac.uk/
1950; Rosamund Campbell
Russia, Latin American, Middle East coll
110 000 vols; 250 curr per   29349

– St Edmund Hall Library, St Edmund Hall, Queens Lane, Oxford OX1 4AR
T: +44 1865 279062; E-mail: blanca.martin@seh.ox.ac.uk; URL: www.lib.ox.ac.uk/libraries/guides/EDM.html
1970; Blanca Martin
Emden Coll (1st and 2nd World War Naval Hist), Aularian Coll
50 000 vols; 170 curr per; 14 mss; 3 incunabula; 4 diss/theses; 30 govt docs; 100 music scores; 12 av-mat; 200 sound-rec; 12 digital data carriers; 2 560 photos, 1 000 postcards of WW I
ALA, Oxford Bibliograph Soc, National Acquisitions Group, Historic Libraries Forum   29350

– St Hilda's College, Library, Cowley Place, Oxford OX4 1DY
T: +44 1865 276848; E-mail: maria.croghan@st-hildas.ox.ac.uk
1893; M. Croghan
50 000 vols; 100 curr per   29351

– St Hugh's College, Library, St. Margaret's Rd, Oxford OX2 6LE
T: +44 1865 274900; Fax: +44 1865 274912; E-mail: deborah.quare@st-hughs.ox.ac.uk; URL: www.st-hughs.ox.ac.uk
1886; Deborah C. Quare
Oxford Movement
92 000 vols; 100 curr per; 50 sound-rec   29352

– St John's College, Library, St. Giles', Oxford OX1 3JP
T: +44 1865 277300;
Fax: +44 1865 277435; E-mail: library@fyfield.sjc.ox.ac.uk; URL: www.sjc.ox.ac.uk
1555; Dr. P.M.S. Hacker
A.E. Housman Coll, Muniments, Robert Graves Trust Coll, mss, early printed books
100 000 vols; 96 curr per; 370 mss; 145 incunabula; 4 digital data carriers   29353

– St Peter's College, Library, New Hall Inn St, Oxford OX1 2DL
T: +44 1865 278900; E-mail: library@spc.ox.ac.uk; URL: www.spc.ox.ac.uk/
Dr. David Johnson
40 000 vols; 60 curr per   29354

– Taylor Institution Library, St Giles', Oxford OX1 3NA
T: +44 1865 278158, 278161;
Fax: +44 1865 278165; E-mail: tay-enquiries@bodleian.ox.ac.uk; URL: www.bodleian.ox.ac.uk/taylor/
mid 1840s; Taylor Librarian: Mr James Legg, Librarian in Charge: Ms Amanda Peters
Modern Languages (other than English) & literatures, primarily French, Italian, German, Spanish, Portuguese, Russian and East European languages, Modern Greek, Celtic, Albanian, Basque, Linguistics and Philology; Special colls: Voltaire and the French Enlightenment; Latin America, Yiddish, Francophone African, Canadian, North and Sub-Saharan African, literature of the former GDR; bks on Anglo-German relations and Goethezeit, Luther Flugschriften and the Fiedler coll; Dante and Futurist holdings; Golden Age lit, Strachan coll, Slavonic & Modern Greek colls: Morfill, Nevill Forbes (Russ.), Dawkins (Gk), Hasluck (Albanian) colls; Slavonic & Modern Greek: at 47 Wellington Square, OX1 2JF
600 000 vols; 1 300 curr per; 1 500 mss; 56 incunabula; 20 000 microforms; 1 660 av-mat; 125 sound-rec; 90 digital data carriers
libr loan
Includes former Modern Languages Faculty Libr   29355

– Trinity College, Library, The Library, Trinity College, Oxford OX1 3BH
T: +44 1865 279863;
Fax: +44 1865 279902; E-mail: sharon.cure@trinity.ox.ac.uk; URL: www.trinity.ox.ac.uk/college/library/
Sharon Cure
56 000 vols   29356

– University College, Library, Oxford OX1 4BH
T: +44 1865 276621; Fax: +44 1865 276987
1292; Dr. T.W. Child
50 000 vols; 110 curr per; 100 mss; 3 incunabula; 65 av-mat; 40 digital data carriers   29357

– University Laboratory of Physiology, Library, Parks Rd, Oxford OX1 3PT
T: +44 1865 272524; Fax: +44 1865 272469; URL: www.physiol.ox.ac.uk/Library
Sherington Libr Coll for the Hist Neuroscience
12 900 vols; 90 curr per; 196 diss/theses; 10 av-mat; 1 sound-rec; 38 digital data carriers; Newspaper clippings, Reprints & Letters   29358

– Wadham College, Library, Parks Rd, Oxford OX1 3PN
T: +44 1865 277900; Fax: +44 1865 277937; E-mail: library@wadh.ox.ac.uk
1613; Prof. R.W. Fiddian
60 000 vols; 90 curr per   29359

– Wolfson College, Floersheimer Library, Linton Rd, Oxford OX2 6UD
T: +44 1865 274100; Fax: +44 1865 274125; E-mail: library@wolfson.ox.ac.uk
Mrs Fiona Elizabeth Wilkes
Ancient hist, classics, oriental religions
30 000 vols; 3 curr per
libr loan   29360

– Worcester College, Library, Worcester St, Oxford OX1 2HB
T: +44 1865 278300; Fax: +44 1865 278387; URL: www.lib.ox.ac.uk/libraries/guides/WOR.html
1714; Dr. J. Parker
Jacobean Restoration Drama, archit drawings by Inigo Jones, English Civil War mat
80 000 vols; 82 curr per; 300 mss; 70 incunabula   29361

– Wycliffe Hall Library, 54 Banbury Rd, Oxford OX2 6PW
T: +44 1865 274204; Fax: +44 1865 274215; E-mail: library@wycliffe.ox.ac.uk; URL: www.wycliffe.ox.ac.uk
1877; Chris Leftley
Church History, Christianity, Evangelical Anglicanism, Bible Studies, Doctrine, Pastoral Theology, Spirituality, Science and Religion
21 000 vols; 100 curr per; 25 e-journals; 20 diss/theses; 200 av-mat; 400 sound-rec; 20 digital data carriers
libr loan; ABTAPL, CILIP   29362

South Devon College, Learning Resource Centre, Library, Long Rd, *Paignton* TQ4 7EJ
T: +44 1803 540551; Fax: +44 1803 400701; E-mail: lrc@southdevon.ac.uk; URL: www.southdevon.ac.uk/
1951; D.M. Harper
40 000 vols; 200 curr per; 286 av-mat; 48 sound-rec; 70 digital data carriers   29363

University of the West of Scotland, University Libraries, *Paisley* PA1 2BE
T: +44 141 8483758; Fax: +44 141 8483761; E-mail: info@uws.ac.uk; URL: www.paisley.ac.uk

– Ayr Campus Library and Learning Resource Center, Beech Grove, Ayr KA8 0SR
T: +44 1292 886345; Fax: +44 1292 886288; E-mail: libraryayr@uws.ac.uk; URL: www.uws.ac.uk/schoolsdepts/library/
1964; Gordon Hunt
Business; commercial music; education; media; nursing; poetry
60 000 vols; 250 curr per; 650 diss/theses; 200 govt docs; 200 maps; 400 microforms; 3 000 av-mat; 1 000

sound-rec; 300 digital data carriers; 2500 wallcharts
libr loan; CILIP   29364

– Hamilton Campus Library, Almada St, Hamilton ML3 OJB
T: +141 1698 894424; Fax: +141 1698 286856; URL: www.paisley.ac.uk/about/hamilton/facilities.asp
80 000 vols   29365

– Robertson Trust Library and Learning Resource Centre, Paisley Campus, Paisley PA1 2BE
T: +44 141 8483000; E-mail: library@uws.ac.uk; URL: www.paisley.ac.uk
1998; Teresa Gilbert
200 000 vols   29366

– Royal Alexandra Hospital Library, Corsebar Rd, Paisley PA2 9PN
T: +44 141 3147178; Fax: +44 141 8874962; E-mail: ruth.robinson4@nhs.net; URL: www.nhsggc.org.uk
1986; Ruth Robinson
Health management, medicine, nursing, sociology
21 000 vols; 200 curr per
SHINE   29367

Pershore Group of Colleges, Hundlip College, Holmie Lacy College, Library, Avonbank, Pershore WR10 3JP
T: +44 1386 552443 ext 2242; Fax: +44 1386 556528
1965; Jane Keightley
Botany, Horticulture, Garden Design, Arboriculture, Worcestershire Beekeeping Coll
20 000 vols; 200 curr per; 200 diss/theses; 400 av-mat; 250 digital data carriers
libr loan; ALLCU, CILIP, ALA   29368

College of Further Education Plymouth, Library and Information Service, Kings Rd, Devonport, Plymouth PL1 5QG
T: +44 1752 385366; Fax: +44 1752 385343
Hilary A. Rees
50 000 vols; 350 curr per; 1 200 av-mat   29369

Plymouth College of Art and Design, Library, Tavistock Place, Plymouth PL4 8AT
T: +44 1752 203412; Fax: +44 1752 203444; E-mail: lharding@pcad.ac.uk
1962; Linda Harding
Communications, crafts, fashion, graphic design, photography, printing media, art
21 000 vols; 117 curr per; 200 diss/theses; 100 maps; 1 microforms; 300 av-mat; 20 digital data carriers; 300 pamphlets, 2 000 clippings
ARLIS/UK & Ireland, BLDSC, NAG, SWRLS, BUFVC   29370

University College Plymouth of St Mark and St John, Library, Derriford Rd, Plymouth PL6 8BH
T: +44 1752 636100 ext 4206; Fax: +44 1752 636712; E-mail: wevans@marjon.ac.uk; URL: www.marjon.ac.uk
1840; F.A. Clements
134 000 vols; 1 120 curr per; 565 e-journals; 1 750 diss/theses; 600 maps; 1 750 microforms; 4 250 av-mat; 2 150 sound-rec; 250 digital data carriers
CILIP, ASLIB, SCONUL   29371

University of Plymouth, Information and Learning Services, Drake Circus, Plymouth PL4 8AA
T: +44 1752 232296; Fax: +44 1752 232340; URL: www.plym.ac.uk
1963; Bob Sharpe
616 000 vols; 3 234 curr per
libr loan; ASLIB, CILIP   29372

– Exeter Campus Library, Topsham Rd, Exeter EX2 6ES
T: +44 1752 587428; URL: www.plymouth.ac.uk/pages/view.asp?page=11957
1854; Judith Cartwright
Finre art, graphic design, media
25 000 vols; 350 curr per
libr loan; ARLIS/UK & Ireland   29373

– Seale-Hayne Campus Library, Ashburton Rd, Newton Abbot TQ12 6NQ
T: +44 1626 325828; Fax: +44 1626 325836; E-mail: ablackman@plymouth.ac.uk
1920; Ms A.J. Blackman
Devon Libr Services Special Reserve: Veterinary Science, Dartington Amenity Research Trust Libr
55 000 vols; 400 curr per; 100 diss/theses; 100 maps; 100 microforms; 500 av-mat; 20 sound-rec
libr loan   29374

University of Glamorgan, Learning Resources Centre, Llantwit Rd Treforest, Pontypridd CF37 1DL
T: +44 1443 482625; Fax: +44 1443 482629; E-mail: smorgan1@glam.ac.uk; URL: www.glam.ac.uk/lrc/
1963; Jeremy Atkinson
Centre for the Study of Welsh Writing in English
224 000 vols; 1 745 curr per; 6 000 e-journals; 250 diss/theses; 500 maps; 20 838 microforms; 17 279 av-mat; 1 000 sound-rec; 80 digital data carriers
libr loan; SCONUL, CILIP, ASLIB   29375

Bournemouth University, The Sir Michael Cobham Library, Talbot Campus, Fern Barrow, Poole BH12 5BB
T: +44 1202 965044; Fax: +44 1202 965475; E-mail: jascott@bournemouth.ac.uk; URL: www.bournemouth.ac.uk
1957; Jill Beard
Wedlake and Greening: Archaeology; Ernst and Young: Taxation and Revenue Law; BBC Radio 4 Analysis Programme; Broadcasting Audience Research; Independent Local Radio Programme Sharing Digitization Project Archive; Segrue Journalism Coll; TV Times Project; Printed and electronic sources covering history of broadcasting, public relations and advertising; product design, interior design and engineering
268 295 vols; 542 curr per; 65 000 e-journals; 90 000 e-books; 1 000 diss/theses; 1 968 microforms; 7 725 av-mat; 695 digital data carriers
libr loan; CILIP, ASLIB, SCONUL, HATRICS, IFLA, LIBER   29376

University of Portsmouth, University Library, Cambridge Rd, Portsmouth PO1 2ST
T: +44 23 92843228; Fax: +44 23 92843233; E-mail: library@port.ac.uk; URL: www.port.ac.uk/library
1953; Roisin Gwyer
400 000 vols; 3 000 curr per; 20 000 e-journals; 35 000 e-books
libr loan   29377

Myerscough College, Library, Myerscough Hall, Bilsborrow, Preston PR3 0RY
T: +44 1995 642122; Fax: +44 1995 642333; E-mail: mailbox@myerscough.ac.uk; URL: www.myerscough.ac.uk
1894; Jon Humfrey
John Shildrick Memorial coll, equine science, turf technology, agricultural mechanisation, veterinary science, equine science
45 000 vols; 110 curr per; 500 diss/theses; 300 govt docs; 412 maps; 2 microforms; 850 av-mat; 100 digital data carriers; 3 000 pamphlets
libr loan; CILIP, ALLCU   29378

University of Central Lancashire, UCLan Library, St Peter's Sq, Preston PR1 2HE
T: +44 1772 201201; E-mail: LWadsworth@uclan.ac.uk; URL: www.uclan.ac.uk/library/cireulate/contactpreston.htm
1940; Kevin Ellard
Joseph Livesey Coll, W.E. Moss Coll, Wainwright Coll
300 000 vols; 2 100 curr per; 100 mss; 1 160 maps; 56 000 microforms; 2 200 av-mat; 48 160 transparencies, e-journals
libr loan   29379

University of Reading, Library, Whiteknights, P.O. Box 223, Reading, Berks RG6 6AE
T: +44 118 3788770; Fax: +44 118 3786636; E-mail: library@reading.ac.uk; URL: www.reading.ac.uk/library

1893; Julia Munro
Europ Doc Centre
1 200 000 vols; 9 778 curr per; 7 881 e-journals; 2 800 mss; 4 388 microforms; 10 000 av-mat; 4 650 sound-rec; 30 digital data carriers; 81 000 pamphlets
libr loan; BLDSC, LISE, SCONUL, CILIP, UKSG, NAG   29380

– Museum of English Rural Live Library, Redlands Road, Reading RG1 5EX
T: +44 118 3788660; Fax: +44 118 9751264; E-mail: merl@reading.ac.uk; URL: www.reading.ac.uk/merl
1966; Peter McShane
G.E. Fussell Coll (agricultural hist), Edgar Thomas Coll (agricultural economics), Dairy Coll, Milling & Baking Coll; Archives and business records, photogr coll
100 000 vols; 80 curr per; 120 diss/theses; 10 000 govt docs; 70 maps; 100 microforms; 5 500 av-mat; 300 sound-rec; 1 digital data carriers; 7 000 pamphlets
ALLCU   29381

University of West London, Crescent Road library, Crescent Road, Reading RG1 5RQ
T: +44 118 943 5855; Fax: +44 118 9675561; E-mail: barbara.moye@uwl.ac.uk; URL: library.uwl.ac.uk/
1965; Barbara Moye
70 000 vols; 250 curr per; 250 diss/theses; 4 000 av-mat; 500 sound-rec; 100 digital data carriers   29382

North East Worcestershire College, Redditch Campus, Library, Redditch B98 8DW
T: +44 1527 572519; Fax: +44 1527 572557
M. Gain
29 500 vols; 200 curr per; 80 av-mat; 1 digital data carriers
libr loan   29383

East Surrey College, Learning Resource Centre, Library, Gatton Point North, Claremont Road, Redhill RH1 2JX
T: +44 1737 772611; Fax: +44 1737 768641; URL: www.esc.ac.uk
Emily Hildred
Art
16 951 vols; 150 curr per; 900 av-mat; 78 digital data carriers   29384

Richmond, The American International University in London, Cyril Taylor Library, Queens Rd, Richmond, Surrey TW10 6JP
T: +44 20 83328210; Fax: +44 20 83323050; E-mail: library@richmond.ac.uk; URL: www.richmond.ac.uk/resources/library/index.asp
1972; Frank Trew
Asa Briggs Coll (810 vols), Haward Care Coll
75 000 vols; 250 curr per; 77 microforms; 2 000 av-mat; 50 digital data carriers
ASLIB, CILIP, ALA   29385

Rotherham College of Arts and Technology, Library Learning Centre, Eastwood Lane, Rotherham S65 1EG
T: +44 1709 722741; Fax: +44 1709 360765; E-mail: ctaylor@Rotherham.ac.uk; URL: www.rotherham.ac.uk
1954; Colin Tyalor
25 000 vols; 100 curr per; 6 000 e-journals; 40 maps; 2 000 av-mat; 200 sound-rec; 100 digital data carriers
libr loan; SINTO   29386

University of Salford, Information Services Division, Clifford Whitworth Bldg, Salford M5 4WT
T: +44 161 2955846; Fax: +44 161 2955888; E-mail: ils-servicedesk@salford.ac.uk; URL: www.isd.salford.ac.uk/library
1957; Mr A.M. Lewis
Walter Greenwood Coll, Bridgewater Est arch, Stanley Houghton Mss, Canal Duke Coll
662 066 vols; 4 379 curr per; 12 200 e-journals
libr loan; CILIP, ASLIB   29387

Salisbury College, The Learning Resources Centre, Library, Southampton Rd, Salisbury SP1 2LW
T: +44 1722 344325; Fax: +44 1722 344345; E-mail: Library@salisbury.ac.uk; URL: www.salisbury.ac.uk
1957; Janet Beauchamp
Local History Coll
21 000 vols; 100 curr per; 8 000 e-journals; 500 av-mat; 100 digital data carriers
CILIP, BL   29388

Rother Valley College of Further Education Library, Doe Quarry Lane, Dinnington, Sheffield S31 7NH
T: +44 114 2550550 ext 144; Fax: +44 114 2550504
N.M. Roome
30 000 vols; 200 curr per   29389

Sheffield Hallam University, Adsetts Centre, Eric Mensforth Library, City Campus, Howard St, Sheffield S1 1WB
T: +44 114 2252330; Fax: +44 114 2533859; E-mail: learning.centre@shu.ac.uk; URL: www.shu.ac.uk/services/lc
Graham Bulpitt
498 000 vols; 2 600 curr per
libr loan   29390

– Psalter Lane Campus Library, Psalter Lane, Sheffield S11 8UZ
T: +44 114 2252721; Fax: +44 114 2252717; E-mail: learning.centre@shu.ac.uk; URL: www.shu.ac.uk/services/lc
1847; C. Abson
Fine arts, art hist, photography, film, video art, design
30 000 vols; 200 curr per; 6 000 microforms; 13 500 av-mat; 100 sound-rec   29391

University of Sheffield, Main Library, Sheffield S10 2TN
T: +44 114 2227290; E-mail: library@sheffield.ac.uk; URL: www.shef.ac.uk/library
1897; Martin J. Lewis
Architecture Colls, East Asian Studies Colls; Geography Planning and Landscape Libr, St. George's Libr (serves the Faculty of Engineering and the School of Management)
1 200 000 vols; 4 500 curr per; 80 000 pamphlets
libr loan; SCONUL, IFLA, CILIP, RLG, CURL   29392

– Health Sciences Library, Royal Hallamshire Site, Glossop Rd, Sheffield S10 2JF
T: +44 114 2712030; Fax: +44 114 2261167; E-mail: lib-hsl@sheffield.ac.uk; URL: www.shef.ac.uk/~lib
1978; Martin Lewis
Medicine, nursing, dentistry
50 000 vols; 450 curr per; 3 995 e-journals; 65 e-books; 150 av-mat
libr loan   29393

Rose Bruford College of Theatre & Performance, Learning Resources Centre, Burnt Oak Lane, Sidcup, Kent DA15 9DF
T: +44 20 83082626; Fax: +44 20 83080524; E-mail: library@bruford.ac.uk; URL: www.bruford.ac.uk
Catherine Beach
Clive Barker collection, Noel Greig Archive, David Bolland Kathakali collection; Stanislavski Centre
50 224 vols; 67 curr per; 7 e-journals; 14 770 av-mat; 1 658 sound-rec; 25 digital data carriers
SCONUL, Theatre Information Group (TIG)   29394

MOD RAF College Cranwell, Royal Air Force College, Library, Sleaford NG34 8HB
T: +44 1400 266219; Fax: +44 1400 262532; E-mail: college011@btconnect.com; URL: www.cranwell.raf.mod.uk
1920; Miss M Guy
Aeronautics, aviation, space technology, military science, RAF history, Sir Frank Whittle coll, T E Lawrence coll; RAF College archive
170 000 vols; 300 curr per; 30 000 microforms; 4 800 av-mat; 20 digital data carriers; 20 000 reports
libr loan   29395

South Tyneside College, Library, St. George's Ave, ***South Shields*** NE34 6ET
T: +44 0191 4273605; E-mail: mharam@stc.ac.uk; URL: www.stc.ac.uk
1956; Margaret Haram
Marine Engineering, Nautical Studies
40535 vols; 158 curr per; 10 e-journals; 3000 e-books; 583 av-mat; 100 sound-rec; 200 digital data carriers
libr loan; BLA 29396

Southampton Solent University, Mountbatten Library, East Park Terrace, ***Southampton*** SO14 0YN
T: +44 23 80319342; Fax: +44 23 80319248; E-mail: library.enquiries@solent.ac.uk; URL: www.solent.ac.uk/library
1964; R. Burrell
Yacht and Boat design coll, Godden coll, Rome coll, Ken Russell coll, Consumer Culture coll
230000 vols; 2000 curr per; 3782 e-journals; 230 diss/theses; 90 maps; 67000 av-mat; 1800 sound-rec; 370 digital data carriers
libr loan; HATRICS, SCONUL, SLIC 29397

University of Southampton, Hartley Library, Highfield, ***Southampton*** SO17 1BJ
T: +44 23 80592180; Fax: +44 23 80593007; E-mail: libenqs@soton.ac.uk; URL: www.soton.ac.uk/library
1862; Dr. Mark Brown
Parkes Libr (Jewish-Christian relations), Perkins Agricultural Libr, Cope Coll (Local hist), Archs of Anglo-Jewish material and of 19th and 20th c political and military hist
1750000 vols; 3750 curr per; 13000 e-journals; 4500 e-books; 6000000 mss; 5000 sound-rec; 50000 photos
libr loan; SCONUL, Research Libraries UK, LIBER 29398

– Hartley Library, University of Southampton, Highfield, Southampton SO17 1BJ
T: +44 23 22180; Fax: +44 23 23007; E-mail: M.L.Brown@soton.ac.uk; URL: www.soton.ac.uk/library/about/hl/index.html
Mark Brown
Biomedical Sciences
60000 vols
libr loan 29399

– Health Services Library, Mailpoint 883, Southampton General Hospital, Tremona Road, Southampton SO16 6YD
T: +44 23 80796547; Fax: +44 23 80798939; E-mail: E.M.Robertson@soton.ac.uk; URL: www.soton.ac.uk/library/about/hsl/index.html
1971; Elizabeth Robertson
86000 vols
libr loan 29400

– National Oceanographic Library, Southampton Oceanography Centre, Waterfront Campus, European way, Southampton SO14 3ZH
T: +44 23 80596116; Fax: +44 23 80596115; E-mail: nol@noc.soton.ac.uk; URL: www.library.soton.ac.uk/nol
1949; J. Stephenson
Discovery Investigations, Challenger Society coll
500100 vols; 1000 curr per; 6000 maps; 300 digital data carriers; 110000 reports, 105000 reprints
libr loan; IAMSLIC, EURASLIC, CILIP 29401

Oaklands College, St Albans City Campus, St Peter's Rd, ***St Albans***, Herts AL1 3RX
T: +44 1727 737122; Fax: +44 1727 737126
1957; Linda Julian
36000 vols; 350 curr per; 150 diss/theses; 200 maps; 2683 av-mat; 239 sound-rec; 50 digital data carriers
libr loan 29402

St Andrew's College, Library, North Street, ***St Andrews*** KY16 9TR
T: +44 1334 462281; Fax: +44 1334 2282; URL: www.library.st-andrews.ac.uk/
Mr Jon Purcell
Finzi Collection; Valentine Photogr Coll
54000 vols; 1342 curr per; 3843 e-journals; 5800 av-mat; 12 digital data carriers
libr loan; CILIP, SALCTG, SLA 29403

University of St Andrews, Library, North St, ***St Andrews*** KY16 9TR
T: +44 1334 462281; Fax: +44 1334 462282; E-mail: lis.library@st-and.ac.uk; URL: www-library.st-and.ac.uk
1411
Von Hügel (philosophy), J.D. Forbes (sci), G.H. Forbes (theol), Bishop Low (theol), Donaldson (classics), Valentine (photos)
920000 vols; 2800 curr per; 100000 mss; 160 incunabula; 4200 diss/theses; 60000 govt docs; 5500 music scores; 3100 maps; 105000 microforms; 100 av-mat; 85 digital data carriers; 300000 photos
libr loan; SCONUL, SLIC 29404

Hastings College of Arts and Technology, Main Library, Archery Rd, ***St Leonards-on-Sea*** TN38 0HX
T: +44 1424 442222; Fax: +44 1424 721763; URL: www.hastings.ac.uk/intranet/resources/library/library.asp
1960; Sarah Eatwell
16-19 Library
23000 vols; 50 curr per; 4000 av-mat; 600 sound-rec; e-journals via Infotrac
libr loan; IsNTO 29405

North Hertfordshire College, Learning Resources, Monkswood Way, ***Stevenage*** SG1 1LA
T: +44 1438 443079; Fax: +44 1438 443021; E-mail: learningresources@nhc.ac.uk
1958; Miss R. Daniell
60000 vols; 300 curr per; 6500 av-mat
libr loan; LASER 29406

University of Stirling, Library, ***Stirling*** FK9 4LA
T: +44 1786 467235; Fax: +44 1786 466866; E-mail: library@stir.ac.uk; URL: www.is.stir.ac.uk
1967; Mark Toole
Walter Scott, James Hogg, Watson and Tait Labour Hist, John Grierson Arch, Bibliography Centre, Lindsay Anderson Archive, David Daiches; some special collection materials are inaccessible until August 2010 due to a major refurbishment project (for more up-to-date details see the website)
477000 vols; 3000 curr per; 311000 e-journals; 1200 e-books; 98 mss; 5000 diss/theses; 3000 govt docs; 50 maps; 2000 microforms; 700 av-mat; 100 sound-rec; 750 digital data carriers
libr loan; CILIP, SCONUL 29407

Stockport College Library+, Wellington Rd South, ***Stockport*** SK1 3UQ
T: +44 161 9583471; Fax: +44 161 9583469; E-mail: library+@stockport.ac.uk
1964; Louse Clark
56000 vols; 286 curr per; 500 av-mat; 100 sound-rec; 1880 digital data carriers; picture files (5000 items), 23 online resource subscriptions
libr loan; BLDSC 29408

Staffordshire University, Information Services Thompson Library, College Rd, ***Stoke-on-Trent*** ST14 7BX
T: +44 1782 294369; Fax: +44 1782 295799; E-mail: libraryhelpdesk@staffs.ac.uk; URL: www.staffs.ac.uk/uniservices/infoservices
1971; Mrs E.A. Hart
10 special colls
300000 vols; 2000 curr per; 2000 e-journals; 1200 diss/theses; 150000 govt docs; 100 maps; 462 microforms; 6500 av-mat; 490 sound-rec; 70 digital data carriers; 2000 pamphlets, 5000 art catalogs
libr loan; ASLIB, CILIP 29409

– Nelson Library, Beaconside, P.O. Box 368, Stafford ST18 0YU
T: +44 1785 353236; Fax: +44 1785 251058; URL: www.staffs.ac.uk
S. Taylor
Computier science, engineering, health
71500 vols; 500 curr per; 40 digital data carriers 29410

Stoke-on-Trent College – Burslem Campus Library, Moorland Rd, Burslem, ***Stoke-on-Trent*** ST6 1JJ
T: +44 1782 603142
G. Plant
30000 vols; 120 curr per 29411

Stoke-on-Trent College – Learning Resources Centre, New Library Building, Cauldon Campus, Stoke Rd, Shelton, ***Stoke-on-Trent*** ST4 2DG
T: +44 1782 208208; Fax: +44 1782 603669; URL: www.stokecoll.ac.uk
1967; Lesley Harvey
Construction and Property Information Centre, Business Workshop, Care Workshop, Languages Workshop
60000 vols; 420 curr per; 310 maps; 150 microforms; 2000 av-mat; 1340 sound-rec; 70 digital data carriers
libr loan 29412

University of East London, West Ham Precint Library, University House, Romford Rd, ***Stratford*** E15 4LZ
T: +44 20 82234224; Fax: +44 20 82234273; E-mail: chopra@uel.ac.uk; URL: www.uel.ac.uk
1936; Paul Chopra
Mathematics, physics, chemistry, psychology, medicine; Industrial Psychology Dept
75000 vols; 220 curr per; 300 diss/theses 29413

– Barking Campus Library, Longbridge Rd, Dagenham RM8 2AS
T: +44 208 2232619; Fax: +44 208 2232804; E-mail: library@uel.ac.uk; URL: www.uel.ac.uk/lss/index.htm
Paul Chopra
Arts, social sciences, business, engineering
248000 vols; 1400 curr per
libr loan; CILIP 29414

– Duncan House Library, High St, London E15 2JB
T: +44 20 82233346; Fax: +44 20 88493659; E-mail: maureen1@uel.ac.uk
Maureen Azubike
Business, management, finance; East London local authority annual rpts, local authority development plans, firms annual rpts
32500 vols; 500 curr per; 80 diss/theses; 100 sound-rec
libr loan 29415

Strode College Learning Centre, Church Rd, ***Street*** BA16 0AB
T: +44 1458 844410; Fax: +44 1458 844415; URL: www.strode-college.ac.uk
1973; Chris Bull
30000 vols; 200 curr per; 1126 av-mat; 26 sound-rec; 254 digital data carriers 29416

University of Sunderland, Information Services Library, Chester Rd, ***Sunderland*** SR1 3SD
T: +44 191 5152900; Fax: +44 191 5152904; E-mail: library@sunderland.ac.uk; URL: www.library.sunderland.ac.uk
Prof. Andrew McDonald
Humanities, Fine Arts, Law
300000 vols; 1600 curr per; 3000 govt docs
libr loan; NRLB 29417

– Ashburne Library, Ashburne House, Backhouse Park, Ryhope Rd, Sunderland SR2 7EF
T: +44 191 5152119; Fax: +44 191 5153166; E-mail: jan.dodshon@sunderland.ac.uk; URL: www.library.sunderland.ac.uk
1961; Janice Dodshon
Art and design
11500 vols; 150 curr per; 66000 av-mat; 70000 35mm slides
libr loan; ARLIS/UK & Ireland 29418

Merton College Library, Morden Park, London Rd, ***Surrey*** SM4 5QX
T: +44 20 84086406; Fax: +44 20 84086666; E-mail: library@merton.ac.uk; URL: www.merton.ac.uk
1961; Dr. Kwasi Darko-Ampem
Musical instrument technology, motor cycle technology, children and families
30000 vols; 90 curr per; 17 e-journals; 1400 av-mat; 50 sound-rec; 20 digital data carriers
CILIP, COLRIC, ALISS 29419

Oscott College, Glancey Library, Chester Rd, ***Sutton Coldfield*** B73 5AA
T: +44 121 3215069; Fax: +44 121 3215002; E-mail: g.boylan@oscott.org; URL: www.oscott.net/
1794; Gerard Boylan
Catholic theology, cathechesis, Biblical studies, religion, and philosophy; Recusant Libr, a coll of 16000 books and pamphlets of Catholic interest dating from ca 1470-1840. Access by appointment only, please contact the curator on +44 121 3215069
60000 vols; 82 curr per; 30 e-journals; 226 mss; 30 incunabula; 2 diss/theses; 10 govt docs; 30 av-mat; 30 sound-rec; 20 digital data carriers; 600 pamphlets (16-19th C)
ABTAPL 29420

Swansea Institute of Higher Education, Library & Learning Support Service, Mount Pleasant, ***Swansea*** SA1 6ED
T: +44 1792 481000; Fax: +44 1792 481085; E-mail: enquiry@sihe.ac.uk; URL: www.sihe.ac.uk
1976; J A Lamb
175000 vols; 900 curr per; 300 diss/theses; 10000 av-mat; 100 digital data carriers
libr loan; CILIP 29421

Swansea University, Library & Information Services, Singleton Park, ***Swansea*** SA2 8PP
T: +44 1792 295697; Fax: +44 1792 295851; E-mail: library@swansea.ac.uk
1920; C.M. West
South Wales Miners' Arch, Salmon Pamphlet Coll on Educ a Welsh Affairs, Llewellyn Bequest of bks and pamphlets on Wales and the Border, Rush Rhees coll of Wittgenstein-related bks
800000 vols; 3100 curr per; 9 e-journals; 600 mss; 9300 diss/theses; 5000 govt docs; 510 music scores; 610 maps; 1000 microforms; 2500 av-mat; 350 sound-rec; 1000 digital data carriers
CILIP, SCONUL, Endeavor Informations Systems 29422

Defence College of Management and Technology, Cranfield University, Library, Shrivenham, ***Swindon*** SN6 8LA
T: +44 1793 785743; Fax: +44 1793 785555; E-mail: libraryenquiry.cu@defenceacademy.mod.uk; URL: diglib.shrivenham.cranfield.ac.uk
1864; Dr. D. Rossiter
Defence Rpts Section
90000 vols; 440 curr per; 250 maps; 900 av-mat; 40000 rpts
SWRLS 29423

Somerset College, Integrated Learning Centres, Wellington Road, ***Taunton*** TA1 5AX
T: +44 1823 366469; Fax: +44 1823 366751; E-mail: ilc@somerset.ac.uk; URL: www.somerset.webhoster.co.uk/
1960; Jolanta Peters
60000 vols; 400 curr per; 750 diss/theses; 100 maps; 40000 av-mat; 400 sound-rec; 25 digital data carriers
libr loan; BL, SWRLS 29424

Stockton Riverside College, Library, Harvard Avenue, ***Thornaby-on-Tees*** TS17 6FB
T: +44 1642 865472; Fax: +44 1642 865470; URL: www.stockton.ac.uk
1956; Mr J. E. Casey
20000 vols; 150 curr per; 6 e-journals; 30 music scores; 200 maps; 800 av-mat; 50 sound-rec; 100 digital data carriers
MLA North East 29425

West Kent College, Library, Brook St, ***Tonbridge*** TN9 2PW
T: +44 1732 358101; Fax: +44 1732 771415; E-mail: enquiries@wkc.ac.uk; URL: www.wkc.ac.uk
1965; Jean Heyes
4 Satellite Learning Centres
30650 vols; 220 curr per; 500 govt docs; 200 maps; 1600 av-mat; 650 sound-rec; 40 digital data carriers; 40 newspaper cuttings files, 1500 Illustrations
libr loan; COFHE, LASER 29426

Dartington College of Arts, Library, ***Totnes*** TQ9 6EJ
T: +44 1803 861651; Fax: +44 1803 863569; E-mail: library@dartington.ac.uk; URL: www.dartington.ac.uk
1976; Chris Pressler
36358 vols; 166 curr per; 25 e-journals; 936 diss/theses; 100 govt docs; 40195 music scores; 10 maps; 8900 av-mat;

4 000 sound-rec; 37 digital data carriers
libr loan; IAML, SIBMAS, ARLIS/UK &
Ireland, CILIP   29427

St Mary's University College, Learning
Resources Centre, Waldegrave Rd,
Strawberry Hill, *Twickenham* TW1 4SX
T: +44 20 82404097; Fax: +44 20
82404270; E-mail: enquiry@smuc.ac.uk;
URL: www.smuc.ac.uk
1850; Mary Lanigan
Anthony West archive
143 000 vols; 500 curr per; 400
diss/theses; 578 microforms; 430 av-mat;
140 sound-rec; 15 digital data carriers
libr loan; CILIP, ASLIB, NBL   29428

Brunel University, University Library,
Kingston Lane, *Uxbridge* UB8 3PH
T: +44 1895 274000; Fax: +44 1895
203263; E-mail: library@brunel.ac.uk;
URL: www.brunel.ac.uk/life/study/library
1966; Beryl-Anne Thompson
Transport Hist Coll (including
Clinker/Garnett Colls), working class
autobiographies, railway hist, HMSO
publs, ordnance survey
347 000 vols; 3 000 curr per
libr loan; SCONUL   29429

North Tyneside College, Library,
Embleton Ave, *Wallsend* NE28 9NJ
T: +44 191 2295000 ext 5243,
5445; Fax: +44 191 2295301; URL:
www.ntyneside.ac.uk
1964; G.S. Rutherford
Library online learning resources
18 500 vols; 85 curr per; 125 govt docs;
650 av-mat; 100 sound-rec; 15 digital
data carriers; 500 information files, e-
journals
libr loan; BLDSC, ASLIB, CILIP,
NEMLAC, NRLB   29430

Walsall College, Learning Resource
Centre, St. Paul's St, *Walsall* WS1 1XN
T: +44 1922 657091; E-mail:
info@walcat.ac.uk; URL: www.walcat.ac.uk
1962; Emma Green
Business, Care, Construction, Creative
Arts, Engneering, Hair & Beauty,
Hospitality, Tourism & Leisure, Science &
ICT, Supported Learning
70 000 vols; 200 curr per; 1 e-journals
libr loan; COFHE/CILIP/JISC   29431

Warwickshire College, Library, Moreton
Morrell Centre, Moreton Morrell,
Warwick CV35 9BL
T: +44 1926 318000; Fax: +44 1926
318111; E-mail: enquiries@warkscol.ac.uk;
URL: www.warkscol.ac.u
1982; O.E. Unwin
Equine studies, horticulture, greenkeeping,
woodland game rearing, small annal care
13 000 vols; 170 curr per; 124
diss/theses; 630 av-mat; 2 digital data
carriers; 15 000 slides
CADIG, ALLCU   29432

**Sandwell College of Further and
Higher Education**, Library and Learning
Resources Centre, Woden Rd South,
Wednesbury WS10 0PE
T: +44 121 556-6000/- 6603; Fax: +44
121 2536661
1940; Anne Hughes
Metallurgy, business administration,
management, automotive engineering,
automobiles, casting coll, further
education curriculum, deposit ctr,
industrial technologies, resource ctr for
West Midlands and Liaisson
90 000 vols; 600 curr per; 3 300
diss/theses; 6 000 govt docs; 400
maps; 58 microforms; 1 474 av-mat;
222 sound-rec; 200 digital data carriers
libr loan; WMRLB, BLL, MISLIC   29433

University of Winchester, Martial Rose
Library, Sparkford Road, *Winchester*
SO22 4NR
T: +44 1962 827306;
Fax: +44 1962 827443; E-mail:
libenquiries@winchester.ac.uk; URL:
www.winchester.ac.uk/library
1840; David Farley
250 000 vols; 600 curr per; 2 000
e-journals; 200 diss/theses; 4 000
microforms; 2 000 sound-rec
libr loan; ASLIB, SCONUL   29434

Winchester School of Art, University
of Southampton, Library, Park Av,
Winchester SO23 8DL
T: +44 2380 596982; E-mail:
askwsa@soton.ac.uk; URL:
www.soton.ac.uk/library/
1965; Linda Newington
Design Hist study coll, Artists' Book coll;
Montse Stanley coll; Richard Rutt coll;
Janer Arnold coll
40 000 vols; 200 curr per; 15 e-journals;
1 000 diss/theses; 120 000 av-mat; 69
digital data carriers
libr loan; ARLIS/UK & Ireland   29435

Isle College, Library, Ramnoth Rd,
Wisbech PE13 0HY
T: +44 1945 582561 ext 214; Fax: +44
1945 582706
Ruth Kenyon
Covers Lib
16 000 vols; 80 curr per; 110 av-mat; 60
digital data carriers
libr loan   29436

University of Wolverhampton,
Department of Learning Resources,
ML Block, Stafford St, *Wolverhampton*
WV1 1NJ
T: +44 1902 322302;
Fax: +44 1902 322668; E-mail:
learningcentredirect@wlv.ac.uk; URL:
www.wlv.ac.uk/lib
1969; Mary E. Heaney
Europ Doc Centre
582 329 vols; 14 487 curr per; 11 876
e-journals
libr loan; ASLIB, CILIP, SCONUL, LIBER
                                  29437

– Compton Learning Centre, Compton
Park, Compton Rd West, Wolverhampton
WV3 9DX
T: +44 1902 323642; Fax: +44 1902
323702; URL: www.wlv.ac.uk
Oliver Pritchard
Business, management
30 000 vols; 200 curr per
libr loan   29438

– Harrison Learning Centre, Wolverhampton
Campus, St Peter's Sq, Wolverhampton
WV1 1RH
T: +44 1902 323560; Fax: +44 1902
322194
Irene Orridge
485 000 vols; 3 503 curr per   29439

University of Worcester, Library,
Henwick Grove, *Worcester* WR2 6AJ
T: +44 1905 855414; Fax: +44 1905
855132; E-mail: a.hannaford@worc.ac.uk;
URL: www.worcester.ac.uk
1946; Ms A. Hannaford
Historical children's literature collections,
Kays Archive
180 000 vols; 1 204 curr per; 360 e-
journals; 8 000 e-books; 200 microforms;
300 av-mat; 300 sound-rec
libr loan; CILIP   29440

Glyndŵr University, Library, Mold Rd,
Wrexham LL11 2AW
T: +44 1978 293250;
Fax: +44 1978 293435; E-mail:
enquirydesk@glyndwr.ac.uk; URL:
www.glyndwr.ac.uk
1993; Paul Jedrrett
117 000 vols; 350 curr per; 750 e-books;
400 maps; 400 av-mat; 92 digital data
carriers
libr loan; BLDSC   29441

Askham Bryan College, Learning
Resource Centre, Askham Bryan, *York*
YO23 3FR
T: +44 1904 772234; Fax: +44 1904
772287; URL: www.askham-bryan.ac.uk
1947; Julie Amery
Garden design, agriculture, horticulture,
land management, animal management,
arboriculture, equine, floristry,
greenkeeping, agricultural engineering,
business, countryside management,
veterinary nursing, sports surface, food
33 500 vols; 300 curr per; 20 e-journals;
450 diss/theses; 400 maps; 1 500 av-mat;
30 language packages, 400 learning
packages
ALLCU, COLRIC   29442

University of York, J.B. Morrell &
Raymond Burton Libraries, Heslington,
York YO10 5DD
T: +44 1904 433865; Fax: +44 1904
433866; URL: www.york.ac.uk/services/
library
1962; Stephen Town
Mirfield Coll, Halifax Parish Librr,
Slaithwaite Parish Librr, Wormald Coll,
Dyson Coll, (17th & 18th c Engl lit),
Poetry Society Librr, Newbold, Eliot, Elton
Milner White, Milnes-Walker, Newton,
Vickers Coll
800 000 vols; 3 000 curr per; 7 000 e-
journals; 6 000 diss/theses; 500 maps;
131 000 microforms; 2 500 av-mat; 1 200
sound-rec; 40 digital data carriers
SCONUL, CILIP, LIBER   29443

– King's Manor Library, Exhibition Sq, York
YO1O 2EP
T: +44 1904 433969; Fax: +44 1904
433949; E-mail: lib-km@lists.york.ac.uk
1950; P. Haywood
Town and country planning, garden and
landscape, medieval studies, 18 c studies
35 000 vols; 130 curr per
ARLIS/UK & Ireland   29444

York St John University, Fountains
Learnig Centre, Lord Mayor's Walk, *York*
YO31 7EX
T: +44 1904 886700; Fax: +44 1904
876324; E-mail: library@yorksj.ac.uk;
URL: www.yorksj.ac.uk/library/learningcent
1845; A. Chalcraft
Victorians childrens books, York Religious
Education Centre, Yorkshire Film Arch,
Comenius Centre
190 000 vols; 800 curr per; 500 e-
journals; 400 govt docs; 1 000 music
scores; 250 microforms; 1 500 av-mat;
2 000 sound-rec; 100 digital data carriers;
Videotape Computer Software
libr loan; ASLIB, CILIP, SCONUL   29445

School Libraries

Acton College, Library, Mill Hill Rd,
Acton W3 8OX
T: +44 1993 2344
1952; A. Ray
25 000 vols; 140 curr per; 1 000 av-mat
libr loan; BLL   29446

South Trafford College, Library,
Manchester Rd, West Timperley,
Altrincham WA14 5PQ
T: +44 161 9524600; Fax: +44 161
9524672
Rachel Smith
32 000 vols; 300 curr per; 50 maps; 750
av-mat; 100 sound-rec; 15 digital data
carriers
libr loan   29447

Civil Service College, Library,
Sunningdale Park, *Ascot*, Berkshire
SL5 0QE
T: +44 1344 634286
A.J. Drewett
35 000 vols; 250 curr per   29448

Northumberland College, Learning
Resource Centre, College Rd,
Ashington NE63 9RG
T: +44 1903 841200;
Fax: +44 1903 841201; E-mail:
Fiona.Middlemist@northland.ac.uk; URL:
www.northland.ac.uk/facilities.htm
1957; Fiona E. Middlemist
Social welfare, education
50 000 vols; 100 curr per; 2 500 av-mat;
20 digital data carriers
NRLB   29449

Basingstoke College of Technology,
Library, Worting Rd, *Basingstoke*
RG21 1TN
T: +44 1256 354141; Fax: +44 1256
306444; URL: www.bcot.ac.uk
1953; M. Maloney
25 000 vols; 102 curr per; 100 av-mat
libr loan; SWRLS, BLLD, HATRICS   29450

Wirral Metropolitan College, Library,
Conway Park Campus, Europa Boulevard,
Birkenhead CH41 4NT
T: +44 151 5517701; URL: wmc.ac.uk/
1982
Building and construction, catering and
hotel management, management
60 000 vols; 300 curr per; 200
microforms; 1 000 av-mat
libr loan; BLL   29451

Brooklyn Technical College, Library,
Aldridge Rd, *Birmingham* B44 8NE
1957; S. Harper
10 000 vols; 108 curr per   29452

B.I.C.C. – College Library, Grove Bldg,
Great Horton Rd, *Bradford* BD7 1 AY
T: +44 1274 753331; Fax: +44 1274
394810
1964; Margaret Chapman
88 500 vols; 844 curr per
BLL, YRLB   29453

Bridgwater College Library, Bath Rd,
Bridgwater TA6 4PZ
T: +44 1278 455464; Fax: +44 1278
444363
1959; Margaret Harwood
30 000 vols; 300 curr per; 1 000 av-mat;
21 digital data carriers; 648 learning
packages
libr loan; SWRLB   29454

Canterbury College of Technology,
Library, New Dover Rd, *Canterbury*
CT1 3AJ
T: +44 1227 766081
1964; S.J. Bennett
Civil engineering, mining
30 000 vols; 152 curr per; 60 maps; 70
av-mat
libr loan   29455

**Mid-Kent College of Higher and
Further Education**, Library, Horsted,
Maidstone Rd, *Chatham* ME5 9UQ
T: +44 1634 830633 ext 2142; Fax: +44
1634 830224
J. Loose
99 000 vols; 566 curr per; 30 digital
data carriers; Access to Internet, e-mail,
Software
libr loan; BLL, LASER, SEAL   29456

**Gloucestershire College of Arts and
Technology**, Library, 73 The Park,
Cheltenham GL50 2RR
T: +44 1242 28021
1958; Sheila Appleton
Tourism, Health & Safety
30 000 vols; 250 curr per; 500 maps; 25
microforms; 150 av-mat; 45 sound-rec
libr loan; BLL   29457

Chesterfield College, Libraries –
Learning Centres, Infirmary Rd,
Chesterfield S41 7NG
T: +44 1246 500550; Fax: +44 1246
550587; URL: www.chesterfield.ac.uk
1958; Mary Dawson
30 000 vols; 133 curr per; 10 e-journals;
100 maps; 2 000 av-mat; 300 sound-rec;
100 pamphlets
COLRIC, SINTO   29458

Wiltshire College Lackham, Library,
Lacock, *Chippenham* SN15 2NY
T: +44 1249 466814; Fax: +44 1249
444474
1948; Stella M. Vain
Agriculture, agronomy, horticulture
22 000 vols; 190 curr per; 2 e-journals;
800 av-mat; 43 digital data carriers
libr loan; ALLCU, BLDSC   29459

Coatbridge College, Library, Kildonan St,
Coatbridge ML5 3LS
T: +44 1236 422316
1960; Sam Brunton
Science and technology, engineering
19 000 vols; 150 curr per; 500 govt
docs; 200 maps; 1 500 av-mat; 400
sound-rec
libr loan; BLLD   29460

Coleg Llandrillo Cymru, Library
(Llandrillo Technical College), Llandudno
Rd, *Colwyn Bay* LL28 4HZ
T: +44 1492 54666 ext 284; Fax: +44
1492 544391
1964; Dr. A. Eynon
Catering and hotel management,
engineering
40 000 vols; 250 curr per; 4 000 av-mat
libr loan   29461

Dewsbury College, Library, Halifax Rd,
Dewsbury WF13 2AS
T: +44 1924 465916; Fax: +44 1924
457047
Abbas Bismillah
Art and design resources
30 000 vols; 180 curr per; 1 900 av-mat;
300 sound-rec; 60 digital data carriers
libr loan   29462

South Kent College, Library, Maison Dieu Rd, *Dover* CT16 1DH
T: +44 1304 204573; Fax: +44 1304 204573
1972; Paul Martyn North
55 000 vols; 200 curr per; 300 maps; 350 av-mat; 200 sound-rec; 100 digital data carriers; 1 000 pamphlets, 5 000 clippings
libr loan   29463

Dundee College of Further Education, Library, 30 Constitution Rd, *Dundee* DD3 8LF
T: +44 1382 834813
33 000 vols   29464

South Kent College, Library, Shorncliffe Rd, *Folkestone* CT20 2TZ
T: +44 1303 858340;
Fax: +44 1303 858400; E-mail: webmaster@southkent.ac.uk; URL: www.southkent.ac.uk
P.M. North
55 000 vols; 200 curr per; 200 maps; 300 av-mat; 200 sound-rec; 200 digital data carriers; 2 000 pamphlets   29465

Glasgow College of Building and Printing, Laird Library, 60 N Hanover St, *Glasgow* G1 2BP
T: +44 141 5664132; Fax: +44 141 3325170; URL: www.gcbp.ac.uk/library
Catherine Kearney
Construction, printing industries
21 000 vols; 175 curr per; 25 e-journals; 200 govt docs; 50 maps; 200 av-mat; 100 digital data carriers; 1 000 pamphlets
libr loan; CILIP, SLIC   29466

Calderdale College, Learning Resources Centre, Francis St, *Halifax* HX1 3UZ
T: +44 1422 399312; Fax: +44 1422 399320; E-mail: info@calderdale.ac.uk; URL: www.calderdale.ac.uk
1958; Kenneth Pole
42 000 vols; 70 curr per; 10 e-journals; 92 diss/theses; 3 772 av-mat; 250 sound-rec; 600 digital data carriers; 1 500 pamphlets   29467

Harlow College, Library, Velizy Ave, *Harlow* CM20 3LM
T: +44 1279 868136; Fax: +44 1279 868260; E-mail: reception@harlow-college.ac.uk; URL: www.harlow-college.ac.uk
1960; D. Monk
48 000 vols; 250 curr per; 150 govt docs; 300 av-mat; 400 sound-rec; 50 digital data carriers
libr loan   29468

Herefordshire College of Technology, Learning Resources Centre, Library, Folly Lane, *Hereford* HR1 1LS
T: +44 1432 352235; Fax: +44 1432 365348; E-mail: lrc@hct.ac.uk
1960; Mary Hilder
Building and construction, civil eng, mechanical eng, food and catering, business studies
31 000 vols; 117 curr per; 89 diss/theses; 525 av-mat; 92 sound-rec; 48 digital data carriers
libr loan; CILIP   29469

North Herts College, Library, Cambridge Rd, *Hitchin* SG4 0JD
T: +44 1462 2351
1975; Patrick Ryan
Catering and hotel management
35 000 vols; 250 curr per
libr loan; HERTIS   29470

Huddersfield Technical College Library, New North Rd, *Huddersfield* HD1 5NN
T: +44 1484 536521; Fax: +44 1484 511885
1973; L. Rich
Current educational developments
30 000 vols; 260 curr per; 200 maps; 1 000 av-mat; 120 items of computer software
libr loan   29471

South East Derbyshire College, Library, Field Rd, *Ilkeston* DE7 5RS
T: +44 115 8492049; Fax: +44 115 8492121; URL: www.sedc.ac.uk
1961; H.M. Gascoyne
32 000 vols; 130 curr per; 24 e-journals; 150 maps; 1 610 av-mat; 900 sound-rec; 340 digital data carriers
BL   29472

Liverpool Community College, Old Swan Centre, Library, Broadgreen Rd, *Liverpool* L13 5SQ
T: +44 151 2523008
1963; Patrick Cox
Textiles
13 000 vols; 108 curr per; 50 digital data carriers
libr loan; LADSIRLAC   29473

Liverpool School of Tropical Medicine, Donald Mason Library, Pembroke Place, *Liverpool* L3 5QA
T: +44 151 7053221; Fax: +44 151708 7088733; E-mail: dmlib@liv.ac.uk; URL: www.liv.ac.uk/lstm/support_services/library/index.htm
1898; C.M. Deering
Ross Coll of Mat on the Hist of Tropical Medicine
13 000 vols; 260 curr per
libr loan   29474

City and Islington College, Spring House, 6-38 Holloway Rd, *London* N7 8JL
T: +44 20 76973492; Fax: +44 20 76973470; E-mail: principal@candi.ac.uk; URL: www.candi.ac.uk
1972; Heinke Wild
Applied optics
80 000 vols; 100 curr per; 12 microforms; 1 500 av-mat; 800 sound-rec; 30 digital data carriers
libr loan; Library Association UK   29475

College of Arms, Library, Queen Victoria St, *London* EC4V 4BT
T: +44 20 72482762
30 000 vols   29476

Ealing Hammersmith and West London College, Learning Centre, Gliddon Rd, *London* W14 9BL
T: +44 20 87411688; Fax: +44 20 87412491; URL: www.wlc.ac.uk
1975; Sue Bull
115 000 vols; 250 curr per; 10 000 av-mat; 100 digital data carriers   29477

Hackney College, Library, Brooke House, Kenninghall Rd, *London* E5 8BP
T: +44 20 89858484; Fax: +44 20 89854845
1974; Anne Trevett
Engineering, building construction, horology, foundry and pattern making, housing management
110 000 vols; 400 curr per; 2 800 av-mat
libr loan; LASER   29478

Lewisham College, Breakspears Learning Centre, Lewisham Way, *London* SE4 1UT
T: +44 20 86920353 ext 3086; Fax: +44 20 86949163
1946; Ms. M. Beverton
Management, hotel management, civil engineering, engineering
75 000 vols; 250 curr per; 2 000 govt docs; 400 av-mat; 400 sound-rec; 20 digital data carriers; Computer Software (200)
libr loan   29479

South Thames College, Library, Wandsworth High St, *London* SW18 2PP
T: +44 20 89187161; Fax: +44 20 89187136
1956; Ms J.R. Forde
78 900 vols; 536 curr per; 2 240 av-mat; 131 digital data carriers   29480

St Paul's School, Walker Library, Lonsdale Rd, Barnes, *London* SW13 9JT
T: +44 20 87465413;
Fax: +44 20 87465353; E-mail: Librarian@stpaulsschool.org.uk
1509; Mrs A. Aslett
John Milton Coll, G.K. Chesterton Coll, Edward Thomas Coll
32 500 vols; 60 curr per; 120 av-mat; 60 sound-rec; 50 digital data carriers
CILIP, SLA   29481

Trafford College, Library, Talbot Rd, Stretford, *Manchester* M32 0XH
T: +44 161 8867000; E-mail: enquiries@trafford.ac.uk; URL: library.trafford.ac.uk
1958; J. Temple
Chemical Engineering, Electronics
18 000 vols; 200 curr per; 2 000 govt docs; 400 maps; 2 microforms; 100 av-mat; 40 digital data carriers
CILIP   29482

West Nottinghamshire College, Learning Resource Centre, Derby Rd, *Mansfield* NG18 5BH
T: +44 1623 413644; Fax: +44 1623 623063; URL: www.wnc.ac.uk/lr
1965; Sue Sproston
40 200 vols; 260 curr per; 12 e-journals; 68 maps; 1 200 av-mat; 60 sound-rec; 142 digital data carriers; visual reference file
libr loan   29483

Newcastle College, Parsons Library, Rye Hill Campus, Scotswood Rd, *Newcastle upon Tyne* NE4 7SA
T: +44 191 2004018; Fax: +44 191 2004100
1966; Fiona Forsythe
Local Studies, Music, Construction, Art, Food and Catering, Computing
130 000 vols; 300 curr per; 1 400 av-mat; 50 digital data carriers
libr loan   29484

Mid Cheshire College of Further Education, Library, Chester Rd, Hartford, *Northwich* CW8 1LJ
T: +44 1606 720646; Fax: +44 1606 75101; E-mail: library@midchesh.ac.uk; URL: www.midchesh.ac.uk
1962; Lindsay Wallace
33 000 vols; 65 curr per; 272 av-mat; 413 sound-rec; 256 digital data carriers   29485

Bedales Memorial Library, Bedales School, Steep, *Petersfield* GU32 2DG
T: +44 1730 300100; Fax: +44 1730 300500; E-mail: admin@bedales.org.uk
1921; Anne Archer
English language; hist; Arts + Crafts movement
36 000 vols; 5 curr per; 18 digital data carriers   29486

Bournemouth and Poole College, Learning Resources Service, North Rd, Parkstone, *Poole* BH14 0LS
T: +44 1202 205804; Fax: +44 1202 205477; E-mail: rpeden@bpc.ac.uk; URL: www.thecollege.co.uk/elibrary/
1950; Rosemary Peden
55 000 vols; 218 curr per; 33 e-journals; 3 000 av-mat; 73 digital data carriers
CILIP   29487

Highbury College, Learning Centre, Cosham, *Portsmouth* PO6 2SA
T: +44 23 92313213; Fax: +44 23 92371972; E-mail: library@highbury.ac.uk
1963; S. Stevenson
43 000 vols; 300 curr per; 3 000 microforms; 650 av-mat; 75 sound-rec; 15 digital data carriers
libr loan; SWRLS, HATRICS   29488

Hopwood Hall College, Rochdale Campus, Learning Resource Centre, St. Mary's Gate, *Rochdale* OLI2 6RY
T: +44 1706 345346; Fax: +44 1706 41426; E-mail: enquiries@hopwood.ac.uk; URL: www.hopwood.ac.uk
1959; Angela Appleby (Rochdale Campus), Marilyn Redwood (Middleton Campus)
Art, costume, business studies; Hopwood Hall College Libr (Rochdale Rd, Middleton M24 6XH, T: (0161) 6437560, Fax 6432114, E-mail: lindsay.wallace@hopwood.ac.uk
60 000 vols; 150 curr per; 30 digital data carriers
libr loan; BLL   29489

Sherborne School Library, Abbey Rd, *Sherborne* DT9 3AP
T: +44 1935 810559; Fax: +44 1935 810426; E-mail: vac@sherborne.org; URL: www.sherborne.org
Victoria A. Chayton
17th-18th c hist, theology and classics, naval hist coll, Old Shirburnian coll; Arch
36 000 vols; 80 curr per; 6 e-journals; 250 sound-rec; 270 digital data carriers
libr loan; CILIP, SLA, IASL   29490

Shrewsbury College of Arts and Technology, Learning Resource Centre, London Rd, *Shrewsbury* SY2 6PR
T: +44 1743 342350; Fax: +44 1743 342343; URL: www.shrewsbury.ac.uk
1958; Gillian Taylor
Complete set of British standards, technical indexes: construction / engineering; Art libr
42 000 vols; 260 curr per; 300 govt docs; 20 maps; 600 av-mat; 200 sound-rec; 180 digital data carriers
libr loan; Learning Resources Development Group, COLRIC, COFHE   29491

Shrewsbury School Library, The Schools, *Shrewsbury* SY3 7BA
T: +44 1743 280595; Fax: +44 1743 243107; E-mail: MMM@shrewsbury.org.uk; URL: www.shrewsbury.org.uk
1606; J.B. Lawson
Taylor Libr (7 500 vols), Charles Darwin, Samuel E. Butler
40 000 vols; 100 mss; 70 incunabula   29492

Warley College of Technology, Library, Crocketts Lane, *Smethwick* B66 3BU
T: +44 121 5584121
T. Hildred
140 000 vols; 800 curr per
libr loan; MISLIC   29493

Solihull College, Library, Blossomfield Rd, *Solihull* B91 1SB
T: +44 121 641702; Fax: +44 121 6787200; E-mail: central.library@derby.gov.uk; URL: www.derby.gov.uk/libraries
1961; C.J. Strudwick
42 000 vols; 405 curr per; 300 microforms; 1 200 av-mat; pamphlets, clippings, 2 000 art slides
CILIP   29494

St Helens College, Library, Water St, *St Helens* WA10 1PP
T: +44 1744 623225; Fax: +44 1744 623007; E-mail: library@sthelens.ac.uk; URL: www.sthelens.ac.uk
1964; Barry Jones
Motor vehicle engineering, management
64 000 vols; 325 curr per; 17 e-journals; 150 av-mat; 5 digital data carriers
libr loan   29495

Wiltshire College Trowbridge, Library, College Rd, *Trowbridge* BA14 0ES
T: +44 1225 766241; Fax: +44 1225 777148
1959; Mrs G. Thomas
20 000 vols; 150 curr per; 75 digital data carriers   29496

Richmond upon Thames College, Library, Egerton Rd, *Twickenham* TW2 7SJ
T: +44 20 86078356; Fax: +44 20 87449738; URL: www.richmond-utcoll.ac.uk
1937; Helen Berry
55 000 vols; 250 curr per; 40 diss/theses; 200 music scores; 1 000 maps; 570 microforms; 27 000 av-mat; 1 560 sound-rec; 25 digital data carriers; 3 000 other items
libr loan; BLL   29497

Wakefield College Library, Margaret St, *Wakefield* WF1 2DH
T: +44 1924 789789;
Fax: +44 1924 789340; E-mail: h.sherwood@wakefield.ac.uk; URL: www.wakefield.ac.uk
1966; Helen Sherwood
45 000 vols; 150 curr per; 12 e-journals; 2 000 av-mat
libr loan; CoLRiC   29498

Hertford Regional College, Ware Centre, Library, Scotts Road, *Ware* SG12 9JF
T: +44 1992 411977;
Fax: +44 1992 411978; E-mail: library.ware@hertreg.ac.uk; URL: www.hrc.ac.uk
1947
30 000 vols; 140 curr per
HERTIS   29499

West Herts College, Learning Centre Service, Hempstead Road, *Watford* WD17 3EZ
T: +44 1923 812554; Fax: +44 1923 812557
1945; Anne Harris
Futher Education
54 000 vols; 140 curr per; 60 e-journals; 200 diss/theses; 500 av-mat; 50 sound-rec; 50 digital data carriers   29500

De Havilland College, The Campus, Library, ***Welwyn Garden City*** AL8 6AH
T: +44 17073 26318
1978; B.C. Willgoss
36 000 vols; 205 curr per; 3 700 slides, 150 filmstrips, 200 cassettes
libr loan; HERTIS 29501

Weymouth College, Library, Cranford Avenue, ***Weymouth*** DT4 7LQ
T: +44 1305 208820; Fax: +44 1305 208912; E-mail: library@weymouth.ac.uk; URL: www.weymouth.ac.uk
1958; Liz Hayman
40 000 vols; 250 curr per
libr loan; SWRLS 29502

Eton College, Library, Eton, ***Windsor*** SL4 6DB
T: +44 1753 671221;
Fax: +44 1753 671244; E-mail: collections@etoncollege.org.uk
1440; N.N.
Material relating to hist of Eton (Ms & printed & schoolbks), 18 th c art and antiquities, early bindings, Thomas Hardy, Robert and Elizabeth Barrett Browning, Edward Gordon Craig
70 000 vols; 15 curr per; 1 000 mss; 206 incunabula; 200 maps; 200 sound-rec; ca. 100 000 autograph letters 29503

Worcester College of Technology, Library, Deansway, ***Worcester*** WR1 2JF
T: +44 1905 725579; Fax: +44 1905 725600; E-mail: library@wortech.ac.uk; URL: www.wortech.ac.uk/StudyCentres/
1966; W. Party
Art Dept
41 430 vols; 75 curr per; 8 e-journals; 3 000 e-books; 3 890 av-mat; 86 sound-rec
libr loan; CILIP 29504

Worsley College of Further Education, Library, Walkden Rd, ***Worsley*** M28 4QD
T: +44 161790 2730
1958; H. Hunt
17 000 vols; 110 curr per; 4 000 archives, 400 AV materials
libr loan 29505

Northbrook College Sussex, Library, Littlehampton Rd, ***Worthing*** BN12 6NU
T: +44 1903 606060; E-mail: enquiries@nbcol.ac.uk; URL: www.northbrook.ac.uk
1960; Julian J.G. Millerchip
45 000 vols; 250 curr per; 10 000 av-mat; 25 digital data carriers; 1 500 clippings
libr loan; ARLIS/UK & Ireland 29506

York College of Arts and Technology, Library, Tadcaster Rd, Dringhouses, ***York*** YO2 1UA
T: +44 1904 704141
Barbara Hull
30 000 vols; 300 curr per 29507

Government Libraries

Department of Agriculture and Rural Development, Library and Information Service, Upper Newtownards Rd, Dundonald House, Room 615, ***Belfast*** BT4 3SB
T: +44 28 90524401; Fax: +44 28 90525546; E-mail: library@dardni.gov.uk; URL: www.dardni.gov.uk
Janice Ewing
50 000 vols; 450 curr per; 50 e-journals; 5 000 govt docs
IAALD 29508

Department of Enterprise Trade and Investment, Netherleigh, Massey Ave, ***Belfast*** BT4 2JP
T: +44 28 90529555; Fax: +44 28 90529286; E-mail: foi@detini.gov.uk
Ruth Menary
22 000 vols; 200 curr per 29509

Northern Ireland Assembly, Library, Parliament Bldgs, Stormont, ***Belfast*** BT4 3XX
T: +44 28 90521250;
Fax: +44 2890521922; E-mail: issuedesk.library@niassembly.gov.uk
1921; N.N.
Irish hist and politics, particularly Northern Ireland, Northern Ireland official publications and legislation
19 000 vols; 90 curr per; 68 e-journals; 8 000 govt docs; 100 maps; 15 microforms; 17 digital data carriers; Subscribe to 32 other online servers
libr loan; CILIP, IFLA, BIALL, ASLIB 29510

Northern Ireland Assembly Library, Room 141, Parliament Bldgs, Stormont, ***Belfast*** BT4 3XX
T: +44 28 90521250;
Fax: +44 28 92521922; E-mail: issuedesk.library@niassembly.gov.uk
1921; N.N. Contact: Mr G.D. Woodman
Govt and politics, Irish history, Northern Ireland
18 000 vols; 150 curr per; 180 e-journals; 4 e-books; 200 000 govt docs; 15 microforms; 20 digital data carriers
libr loan; CILIP, IFLA, BIALL 29511

Northern Ireland Housing Executive, Library Information Services, 2 Adelaide St, ***Belfast*** BT2 8PB
T: +44 28 90318021; Fax: +44 28 90318024; E-mail: library@nihe.gov.uk; URL: www.nihe.gov.uk
Margaret Gibson
Housing Executive Publications
12 000 vols; 200 curr per; 48 digital data carriers; 2 000 pamphlets
libr loan 29512

Scottish Executive Library and Information Service, Spur, Saughton House, Broomhouse Dr, ***Edinburgh*** EH11 3XD
T: +44 131 2444272;
Fax: +44 131 2444545; E-mail: jane.mackenzie@scotland.gsi.gov.uk
1939; Jane Mackenzie
130 000 vols; 1 400 curr per; 50 e-journals; 60 000 govt docs; 500 maps; 100 microforms; 25 digital data carriers; 10 000 pamphlets
libr loan; CILIP, ASLIB, Forum for Interlending, SHINE, SLLG, UKSG 29513

Forestry Commission, Library, Forest Research Station, Alice Holt Lodge" ***Farnham*** GU10 4LH
T: +44 1420 22255; Fax: +44 1420 23653; E-mail: library@forestry.gsi.gov.uk; URL: www.forestry.gov.uk or www.forestresearch.gov.uk
1919; Ms C A Oldham
12 000 vols; 300 curr per; 120 diss/theses; 15 microforms; 190 av-mat; 17000 pamphlets
libr loan; EBHL 29514

Civil Aviation Authority (CAA), Library and Information Centre, Aviation House, ***Gatwick Airport South*** RH6 OYR
T: +44 1293 573725; Fax: +44 1293 573181; URL: www.caa.co.uk
Vagn Pedersen
60 000 vols; 850 curr per; 220 av-mat; 150 digital data carriers; 10 000 rpts
CILIP 29515

Department for Communities and Local Government, Library and Information Service, Zone 2/H24, Ashdown House, 123 Victoria St, ***London*** SW1E 6DE
T: +44 20 79443039; Fax: +44 20 79446098; URL: www.communities.gov.uk
1970; Carol Gokce
Coll of development plans and environmental statements
CILIP 29516

Department for Environment, Food and Rural Affairs, Information Resource Centre, Ergon House, 17 Smith Square, ***London*** SW1P 3JR
T: +44 20 72386575;
Fax: +44 20 72386609; E-mail: defra.library@defra.gsi.gov.uk; URL: www.defra.gov.uk
1889; Katie Woolf
Department for Environment, Food and Rural Affairs publications
2 000 vols; 50 curr per; 30 e-journals; 100 digital data carriers
libr loan; IAALD 29517

Department for Work and Pensions, Library and Information Services, Room 114 The Adelphi, 1-11 John Adam Street, ***London*** WC2N 6HT
T: +44 20 77122500;
Fax: +44 20 79628491; E-mail: Library.services@dwp.gsi.gov.uk; URL: www.dwp.gov.uk
Graham Monk
Moth Deposit of Photographs, Fenton Collection of Leaflets
35 500 vols; 185 curr per; 35 e-journals; 60 e-books; 200 diss/theses; 15 000 govt docs; 250 maps; 238 av-mat; 6 sound-rec; 900 digital data carriers; 350 other items
libr loan 29518

Department of Health, Library, Skipton House, 80 London Rd, ***London*** SE1 6LH
T: +44 20 72104580; Fax: +44 20 79721609; URL: www.open.gov.uk/doh/dhome.htm
1834; Mrs P.L. Bower
200 000 vols; 2 000 curr per 29519

Department of Trade and Industry – Legal Library & Information Centre, 10A Victoria St, ***London*** SW1H ONN
T: +44 20 72153054;
Fax: +44 20 72153535; E-mail: dti.enquiries@dti.gsi.gov.uk; URL: www.dti.gov.uk
N.A. Hasker
Commercial law
30 000 vols; 150 curr per; 30 digital data carriers
BIALL
Open to departmental staff only 29520

Home Office, Library and Information Team, Seacole Building, 2 Marsham Street, ***London*** SW1P 4DF
T: +44 20 70354848;
Fax: +44 20 70354745; E-mail: public.enquiries@homeoffice.gsi.gov.uk; URL: www.homeoffice.gov.uk
1782; P.D. Griffiths
40 000 vols; 2 000 curr per; 100 000 microforms 29521

House of Commons, Department of the Library, ***London*** SW1A 0AA
T: +44 20 72194272; Fax: +44 20 72195839; URL: www.parliament.uk
1818; John Pullinger
Public Information Office
155 000 vols; 1 500 curr per; 800 e-journals; 1 000 000 govt docs; 2 000 maps; 500 microforms; 200 av-mat; 54 digital data carriers
CILIP, ASLIB, IFLA 29522

House of Lords, Library, Westminster, ***London*** SW1A 0PW
T: +44 20 72195242; Fax: +44 20 72196396; E-mail: hllibrary@parliament.uk
1826; Dr. Elizabeth Hallam Smith
120 000 vols; 500 curr per; 45 e-journals; 10 000 govt docs; 70 microforms; 45 digital data carriers; 15 000 pamphlets
IFLA 29523

Law Commission Library, Still House, 11 Tothill St, ***London*** SW1H 9LJ
T: +44 20 33340200;
Fax: +44 20 33340201; E-mail: library@lawcommission.gsi.gov.uk; URL: www.lawcom.gov.uk
1965; Keith Tree
55 000 vols 29524

Ministry of Defence, Whitehall Library, 3/5 Great Scotland Yard, Whitehall, ***London*** SW1A 2HW
T: +44 20 72184445; Fax: +44 20 72185413; URL: www.mod.uk
1989; R.H. Searle
Defence technology, computers and data processing, electronics, mathematics, physics, management sciences, hist coll on the hist of the 3 armed services
1 000 000 vols; 1 100 curr per; 50 000 govt docs; 1 500 microforms; 35 digital data carriers
libr loan 29525

National Statistics Information & Library Services, Office for National Statistics, 1 Drummand Gate, ***London*** SW1V 2QQ
T: +44 20 75336262; Fax: +44 20 75336261; E-mail: info@ons.gov.uk
1997; John Birch
30 000 vols; 200 curr per; 10 000 govt docs; 200 maps; 100 microforms; 12 digital data carriers
libr loan; ASLIB 29526

Supreme Court Library, Royal Courts of Justice, Department for Constitutional Affairs, Strand, ***London*** WC2A 2LL
T: +44 20 79476587; Fax: +44 20 79477935; URL: www.dca.gov.uk
1970; Julia Robertson
Court of Appeal (Civil Division) Transcripts 1951-
300 000 vols; 200 curr per
BIALL 29527

H. M. Treasury and Cabinet Office, Library and Information Service, 1 Horse Guards Rd, ***London*** SW1A 2HQ
T: +44 20 72705290; Fax: +44 20 72705290; E-mail: library@hm-treasury.gov.uk; URL: www.hm-treasury.gov.uk
Mr R. Simpson
54 000 vols
libr loan 29528

Treasury Solicitor's Library, 1 Kemble Street, ***London*** WC2B 4TS
T: +44 20 72102937; Fax: +44 20 72103058; E-mail: bvinfo@tsol.gsi.gov.uk; URL: www.tsol.gov.uk/
Evelyn Stevens
British legal material
25 000 vols
libr loan 29529

United Nations Information Centre, Library, Millbank Tower, 21st Fl, 21/24 Millbank, ***London*** SW1P 4QH
T: +44 20 76301981; Fax: +44 20 79766478
1946; Alexandra McLeod
UN Publs, FAO Publs, WHO Publs
8 500 vols
libr loan 29530

The Centre for Environment, Fisheries and Aquaculture Science, Lowestoft Laboratory, Pakefield Rd, ***Lowestoft*** NR33 0HT
T: +44 1502 562244; Fax: +44 1502 513865; E-mail: lowlibrary@cefas.co.uk; URL: www.cefas.co.uk
1919; S.L. Carter
Fisheries research, hydrography, oceanography, marine biology, aquatic pollution
35 000 vols; 1 200 curr per; 100 e-journals
libr loan; EURASLIC, IAMSLIC 29531

Prison Service College Library, PSC Newbold Revel, ***Rugby*** CV23 0TH
T: +44 1788 804119; Fax: +44 1788 804114
1960; C. Fell
18th and 19th c official rpts on prisons, crime, transportation to Australia, etc
20 000 vols; 70 curr per
libr loan 29532

Ecclesiastical Libraries

Edgehill Theological College, Library, 9 Lennoxvale, ***Belfast*** BT9 5BY
T: +44 28 90686936; Fax: +44 28 90687204; E-mail: libr@edgehillcollege.org
1928; S. Edgar
18 000 vols; 12 curr per; 15 diss/theses; 5 av-mat
ABTAPL 29533

Union Theological College, Gamble Library, 108 Botanic Ave, ***Belfast*** BT7 1JT
T: +44 28 90205093; Fax: +44 28 90205099; E-mail: librarian@union.ac.uk
1873; Stephen Gregory
F.J. Paul coll (church hist), Magee college pamphlets coll, Assembly's college pamphlets coll
66 000 vols; 115 curr per; 120 mss; 200 diss/theses; 4 microforms; 105 av-mat; 290 sound-rec; 8 digital data carriers; 20 000 pamphlets
ABTAPL 29534

Woodbrooke Quaker Study Centre Library, 1046 Bristol Rd, ***Birmingham*** B29 6LJ
T: +44 121 4725171;
Fax: +44 121 4725173; E-mail: library@woodbrooke.org.uk; URL: www.woodbrooke.org.uk
1903; Ian Jackson
Quaker studies – 17th c religious tracts
40 000 vols; 150 curr per; 24 diss/theses; 3 050 microforms; 50 av-mat; 150 sound-rec
ABTAPL, CILIP, BTLG 29535

Trinity College, Library, Stoke Hill, ***Bristol*** BS9 1JP
T: +44 117 9682803; Fax: +44 117 9687470; URL: www.trinity-bris.ac.uk
1932; Susan L. Brown
41 447 vols; 82 curr per; 448 diss/theses; 13 maps; 118 microforms; 187 av-mat; 218 sound-rec; 28 digital data carriers; 4 920 pamphlets
libr loan; BLDSC, ABTAPL, CILIP 29536

United Reformed Church History Society, Westminster College, Madingley Rd, ***Cambridge*** CB3 0AA
T: +44 1223 741300; Fax: +44 1223 300765; E-mail: hw374@cam.ac.uk
1972
17th c religious pamphlets (England)
6 000 vols; 5 curr per
libr loan 29537

Canterbury Cathedral Library, The Precincts, ***Canterbury*** CT1 2EH
T: +44 1227 865287; Fax: +44 1227 865222; E-mail: library@canterbury-cathedral.org; URL: canterbury-cathedral.org/history/libraries.html
597; Mr Keith M.C. O'Sullivan
Howley-Harrison coll (ecclesiastical topics, Bibles, travel, natural hist, slavery and abolitionist movement), Preston-next-Wingham parish libr (Bray cabinet), Elham parish libr, Mendham coll of Catholic and anti-Catholic wrintings, vocal and instrumental scores of Canterbury Catch Club, antiquarian sequence of St. Augustine's Missionary College libr, Stephan Hunt coll of Bibles
52 000 vols; 6 curr per; 100 mss; 400 incunabula; 75 diss/theses; 1 000 music scores; 100 maps; 500 microforms
CLAA 29538

Carlisle Cathedral Library, 7 The Abbey, ***Carlisle*** CA3 8TZ
T: +44 1228 548151;
Fax: +44 1228 547049; E-mail: office@carlislecathedral.org.uk
1691; D. Jenkins
Religion, theology, history
8 000 vols; 1 incunabula 29539

Durham Cathedral Library, The College, ***Durham*** DH1 3EH
T: +44 191 3862489; E-mail: Library@durhamcathedral.co.uk; URL: www.durhamcathedral.co.uk/library
995
European univ theses 16-18 c, Anglo-Saxon and medieval mss, early mss and printed music, antiquarian mss; Archdeacon Sharp Libr of modern theology in English, Meissen Libr of German language Protestant theology
50 000 vols; 30 curr per; 360 mss; 60 incunabula; 200 music scores
Bibliograph Soc, CLAA 29540

Free Church College, Library, The Mound, ***Edinburgh*** EH1 2LS
T: +44 131 2265286;
Fax: +44 131 2200597; E-mail: dmacleod@freescotcoll.ac.uk
Prof. Donald Macleod
30 000 vols; 30 curr per; 20 000 pamphlets 29541

Gillis Centre Library, 100 Strathearn Rd, ***Edinburgh*** EH9 1BB
T: +44 131 6238939;
Fax: +44 131 6238944; E-mail: gilliscentre@staned.org.uk; URL: www.gilliscentre.org.uk
1953; Philip J. Kerr
20 000 vols; 15 curr per
ABTAPL 29542

Scottish Catholic Archives, Columba House, 16 Drummond Place, ***Edinburgh*** EH3 6PL
T: +44 131 5563661;
Fax: +44 131 5563661; E-mail: archivists@scottishcatholicarchives.org.uk; URL: www.scottishcatholicarchives.org
1958; Andrew Nicoll
Papers of families in NE Scotland (1366-1840), Archbishop James Beaton's personal papers (1540-1603), Diocesan archives: St Andrews + Edinburgh; Dunkeld; Argyll + Isles; Galloway; Motherwell; Schottenkloester-Scots Benedictine Abbeys in Germany 12th-19th c; records of Scots colleges in Paris, Douai, Rome and Spain 16th c onwards
10 000 diss/theses; 10 curr per; 2 incunabula; 1 000 music scores; 100 maps; 150 microforms; 100 av-mat; 15 sound-rec; 6 digital data carriers; 250 000 letters, 1 000 linear m of mss
libr loan; SUSCAG 29543

Exeter Cathedral Library, Dean & Chapter of Exeter Cathedral, West Wing, The Palace, Palace Gate, ***Exeter*** EX1 1HX
T: +44 1392 495954; Fax: +44 1392 285986 (Mark 'fao Library'); E-mail: library@exeter-cathedral.org.uk; URL: www.exeter-cathedral.org.uk/Admin/Library.html
11th century; P.W. Thomas
Early printed medicine and science books, tracts (English Civil War period), Cook Collection, Harington Collection
20 000 vols; 10 curr per; 100 mss; 18 incunabula; 5 diss/theses; 50 music scores; 100 maps; 30 microforms; 300 av-mat; 5 sound-rec; 60 000 cathedral archives, 3 000 pamphlets
CLAA 29544

Catholic Central Library, St Michael's Abbey, Farnborough Rd, ***Farnborough*** GU14 7NQ
T: +44 1252 543818; URL: www.catholic-library.org.uk
1907; Mrs Joan Bond
Thomas Merton coll
75 000 vols; 65 curr per 29545

Hereford Cathedral Library and Archives, Hereford Cathedral, ***Hereford*** HR1 2NG
T: +44 1432 374226, 374225;
Fax: +44 1432 374220; E-mail: office@herefordcathedral.org; URL: www.herefordcathedral.org
c. 11th c; Nick Baker
Historic collections include the Chained Library, All Saints' Chained Library, More Parochial Library and Lady Hawkins School Library
15 000 vols; 23 curr per; 1 e-journals; 227 mss; 56 incunabula; 20 diss/theses; 300 music scores; 100 maps; 300 microforms; 1 000 av-mat; 20 sound-rec; 10 digital data carriers
CLAA, ABTAPL, AMARC 29546

Mount Saint Bernard Abbey, Coalville, ***Leicester*** LE67 5UL
T: +44 1530 832-298;
Fax: +44 1530 814608; E-mail: collectanea@btopenworld.com
1835; Br. Erik Varden
600 bks from 17th and 18th c
35 000 vols; 30 curr per; 500 music scores; 250 sound-rec
libr loan 29547

Lichfield Cathedral Library, 19B The Close, ***Lichfield*** WS13 7LD
T: +44 1543 306100; Fax: +44 1543 306109; E-mail: enquiries@lichfield-cathedral.org; URL: www.lichfield-cathedral.org
1673; P. J. Wilcox
10 000 vols; 200 mss; 100 incunabula; 20 diss/theses; 500 music scores
CLA 29548

Lincoln Cathedral Library, Minster Yard, ***Lincoln*** LN2 1PX
T: +44 1522 561640; E-mail: librarian@lincolncathedral.com; URL: www.lincolncathedral.com
1092; Dr. Nicholas Bennett
Medieval manuscripts, 17th c printed pamphlets, 19th c religious tracts
20 000 vols; 12 curr per; 350 mss; 100 incunabula; 40 diss/theses; 60 music scores; 220 music manuscripts 29549

British Orthodox Church, Coptic Orthodox Patriarchate of Alexandria, Library, 10 Heathwood Gardens, ***London*** SE7 8EP
T: +44 20 88543090, 82447888;
E-mail: boc@gotadsl.co.uk; URL: www.britishorthodox.org
1944; Father Gregory Tillett
Orthodoxy (Eastern & Oriental Orthodox); Liturgy
10 000 vols; 25 curr per; 100 000 mss; 10 diss/theses; 10 microforms; 3 000 pamphlets, 3 500 photos
ABTAPL 29550

Congregational Library, 14 Gordon Square, ***London*** WC1H 0AG
T: +44 20 73873727
1831; Dr. D.L. Wykes
50 000 vols; 20 incunabula; 10 000 pamphlets 29551

London Oratory Library, The Oratory, Brompton Rd, ***London*** SW7 2RP
T: +44 20 78080900; Fax: +44 20 75841095; URL: www.brompton-oratory.org.uk
1854; Dr. U.M. Lang
English Catholic recusant history
40 000 vols 29552

Religious Society of Friends in Britain, Library, Friends House, Euston Rd, ***London*** NW1 2BJ
T: +44 20 76631135; Fax: +44 20 76631001; E-mail: library@quaker.org.uk; URL: www.quaker.org.uk/library
1673; Beverley Kemp
Coll on anti-slavery and peace; conscientious objection; relief help for famine and war victims
56 000 vols; 150 curr per; 6 000 mss; 150 diss/theses; 100 maps; 900 microforms; 15 000 prints, 20 000 pamphlets 29553

Spurgeon's College Library, South Norwood Hill, ***London*** SE25 6DJ
T: +44 20 86530850;
Fax: +44 20 87710959; E-mail: enquiries@spurgeons.ac.uk; URL: www.spurgeons.ac.uk
1856; Judith C. Powles
Theology, archive relating to the life and work of Charles Haddon Spurgeon, 1834-1892
60 000 vols; 90 curr per; 324 diss/theses; 180 av-mat; 145 sound-rec
ABTAPL 29554

St Paul's Cathedral, Library, Chapter House, ***London*** EC4M 8AD
T: +44 20 72468345;
Fax: +44 20 72483104; E-mail: library@stpaulscathedral.org.uk
J.J. Wisdom
16 000 vols; 12 000 pamphlets 29555

Westminster Abbey, Muniment Room and Library, East Cloister, Westminster Abbey, ***London*** SW1P 3PA
T: +44 20 76544830; Fax: +44 20 76544827; E-mail: library@westminster-abbey.org; URL: www.westminster-abbey.org
1623; T.A. Trowles
Church history, Church art and architecture; Camden pamphlets, Oldaker Coll of English Bookbindings, Coronation Coll
18 000 vols; 4 curr per; 60 mss; 60 incunabula; 10 diss/theses; 300 music scores; 40 microforms; 5 000 av-mat; 200 sound-rec; 5 000 transparencies and negatives 29556

Dr. Williams's Library, 14 Gordon Square, ***London*** WC1H 0AR
T: +44 20 73873727; E-mail: enquiries@dwlib.co.uk; URL: www.dwlib.co.uk
1729; Dr. D.L. Wykes
Religion, theology, philosophy, church hist, social hist; George Henry Lewes Coll, New College, Christopher Walton Coll; Norman H. Baynes Byzantine Coll
300 000 vols; 1 000 curr per; 2 000 mss; 28 incunabula; 30 diss/theses; 100 microforms; early nonconformist pamphlets
CILIP 29557

Luther King House Library, Luther King House, Brighton Grove, Rusholme, ***Manchester*** M14 5JP
T: +44 161 2492514; Fax: +44 161 2489201; E-mail: library@lkh.co.uk; URL: www.lutherkinghouse.org.uk
1985; Rachel Eichhorn
Free Church Hist, Community Studies, Missiology, Denominational Hist, Spirituality
30 000 vols; 200 curr per; 2 e-journals; 100 e-books; 30 diss/theses; 60 av-mat; 50 sound-rec; 10 digital data carriers
libr loan; ABTAPL, LCF, THUG 29558

London School of Theology, Library, Green Lane, ***Northwood*** HA6 2UW
T: +44 1923 456190; Fax: +44 1923 456001; E-mail: library@lst.ac.uk; URL: www.lst.ac.uk
1943; Alan Linfield
Centre for Islamic Studies and Muslim-Christian Relations (CIS)
50 000 vols; 200 curr per; 250 diss/theses; 400 music scores; 200 sound-rec; 8 digital data carriers
libr loan; ABTAPL 29559

Norwich Cathedral Library, 12 The Close, ***Norwich*** NR1 4DH
T: +44 1603 218443; E-mail: library@cathedral.org.uk
1096; The Revd. Canon Dr. Peter Doll
Theology; Swaffham Parish Libr (450 vols)
25 000 vols; 2 mss; 7 incunabula; 800 pamphlets 29560

Crowther Mission Education Centre Library, Watlington Rd, ***Oxford*** OX4 6BZ
T: +44 1865 787552; Fax: +44 1865 776375; E-mail: ken.osborne@cms-uk.org; URL: www.cms-uk.org
1800; Ken Osborne
Church Mission Society's Max Warren Coll, i.e. CMS's pre 1946 books and periodicals
30 000 vols; 370 curr per; 1 e-journals; 77 diss/theses; 500 maps; 2 microform collections
libr loan; ABTAPL 29561

Cathedral Library, Rochester Cathedral, The Precinct, ***Rochester*** ME1 1SX
T: +44 1634 843366; Fax: +44 1634 401410
604; P.H. D'A. Lock
Phyllis Ireland Coll (Norman Britain, esp ref to Battle Abbey), Photogr collection – Rochester Cathedral topics
9 000 vols; 12 curr per; 7 av-mat; 15 sound-rec 29562

Syon Abbey, Library, ***South Brent*** TQ10 9JX
T: +44 1364 72256
9 000 vols; 15 curr per 29563

Thorold & Lyttelton Library, Diocese of Winchester, 9 The Close, ***Winchester*** SO23 9LS
T: +44 1962 844644;
Fax: +44 1962 841815; E-mail: reception@winchester.anglican.org
1906
Lockton Liturgy Bequest
9 000 vols; 4 curr per
ABTAPL 29564

York Minster Library, Dean's Park, ***York*** YO1 7JQ
T: +44 1904 625308, 611118;
Fax: +44 1904 611119; URL: libcatalogue.york.ac.uk/F.york.ac.uk
1414; John Powell
Yorkshire local hist, Liturgical hist, English civil war tracts, pre-1801 humanities, Yorkshire topographical prints; York Minster Archives, Conservation Studio
120 000 vols; 50 curr per; 400 mss; 115 incunabula; 400 music scores; 200 maps; 50 microforms; 50 av-mat; 2 digital data carriers
CLAA 29565

Corporate, Business Libraries

Arjo Wiggins Ltd, Information Services, Butlers' Court, ***Beaconsfield*** HP9 1RT
T: +44 1494 652213; Fax: +44 1494 652290
1956; S.E. Taylor
9 000 vols; 120 curr per; 400 govt docs; 4 digital data carriers; 4 600 engl transl from scientific Journ on pulp and papermaking, monogrs, rpts on papermaking
libr loan 29566

Unilever Research Laboratory, Colworth House, Sharnbrook, ***Bedford*** MK44 1LQ
T: +44 1234 781781
1960; Sheila Dunne
Blount Coll of poultry lit
31 000 vols; 700 curr per; 200 diss/theses; 3 000 govt docs; 2 000 microforms
libr loan; ASLIB, CILIP 29567

GEC-Marconi Information Centre, Lyon Way, Frimley Rd, *Camberley* Surrey, GU16 5EX
T: +44 1276 696357; Fax: +44 1276 696470
1967; D. Picken
Information technology
20 000 vols; 350 curr per; 5 000 microforms; 50 digital data carriers; 10 000 other items
libr loan; Aslib   29568

BAE Systems Advanced Technology Centres, Library, West Hanningfield Rd, Great Baddow, *Chelmsford* CM2 8HN
T: +44 1245 242394;
Fax: +44 1245 242388; E-mail: baddow.library@baesystems.com
1910; A.C. Jones
Telecommunications, aeronautics, aviation, space technology, Marconi Hist Coll
10 000 vols; 500 curr per; 21 000 rpts
ASLIB, BL, Essex County Library   29569

Rhone-Poulenc Rorer Ltd, Library, Rainham Rd South, *Dagenham* RM10 7XS
T: +44 20 89193491; Fax: +44 20 89192637
1935; E. Elliott-Jay
Medicinal chemistry, organic chemistry, pharmacology, biochemistry
25 000 vols; 300 curr per; 150 diss/theses
libr loan; ASLIB   29570

Highlands & Islands Enterprise, Library, Cowan House, Inverness Retail & Business Park, *Inverness* IV2 7GF
T: +44 1463 244409; Fax: +44 1463 244351; E-mail: library@hient.co.uk; URL: www.hie.co.uk
1969; N.N.
15 000 vols; 150 curr per; 10 e-journals; 3 000 reports commissioned by Hie + Highlands + Islands development board (before 1991)
libr loan; ASLIB   29571

Arup Information and Library Services, 13 Fitzroy St, *London* W1T 4BQ
T: +44 20 7755-3271; Fax: +44 20 77552126; E-mail: arup.library@arup.com; URL: www.arup.com
1959; James Griffith
Civil engineering, structural engineering, construction, consulting engineering, architectural design; Photogr coll, map and aerial photogr coll
38 000 vols; 670 curr per; 30 digital data carriers; 40 000 pamphlets   29572

Bank of England Information Centre, Threadneedle St, *London* EC2R 8AH
T: +44 20 7601-4715;
Fax: +44 20 76014356; E-mail: informationcentre@bankofengland.co.uk; URL: www.bankofengland.co.uk
1931; Ms P.A. Hope
Central Bank Reports
75 000 vols; 1 500 curr per
CILIP, ASLIB, SLA   29573

Imperial Chemical Industries plc, Library, 20 Manchester sq, *London* W1U 3AN
T: +44 20 70095000; Fax: +44 20 77985830
S.A.T. Russell
30 000 vols; 1 000 curr per; 200 maps; 10 000 microforms; 4 000 pamphlets
                                               29574

Institute of Directors, Information Centre, 123 Pall Mall East, *London* SW1Y 5ED
T: +44 20 7766 8778; Fax: +44 20 7766 2642; E-mail: reception123@iod.com; URL: www.iod.com
Anna Burmajster
Directors and boardroom practice
10 000 vols; 50 curr per; 15 digital data carriers   29575

Mail Newspapers Ltd, Reference Library, Northcliffe House, Derry St, Kensington, *London* W8
T: +44 20 79386000
G.S. Johnson
10 000 vols   29576

Syngenta Library CTL, Alderley Park, *Macclesfield* SK10 4TJ
T: +44 1625 515441; Fax: +44 1625 517314
Mrs S. Cotton
Toxicology
70 000 vols; 100 curr per; 2 000 govt docs   29577

Wyeth Laboratories, Huntercombe Lane South, Taplow, *Maidenhead* SL6 0PH
T: +44 1628 414723; Fax: +44 1628 414813
N.N.
11 000 vols; 300 curr per; 1 200 govt docs; 2 000 pamphlets
libr loan; OCLC   29578

United Distillers p.l.c., International Technological Services, Library, Glenochil, *Menstrie* FK11 7ES
T: +44 131 5296803; Fax: +44 131 5296807
1945; Elizabeth Anne Lauchlan
company technical rpts since 1945
13 000 vols; 112 curr per; 50 diss/theses; 1 500 govt docs; 50 maps; 50 microforms; 100 av-mat; 20 sound-rec; 10 digital data carriers
libr loan; CILIP   29579

Corus UK Limited, Information and Library Services, Swinden Technology Centre, Moorgate, *Rotherham* S60 3AR
T: +44 1709 820166;
Fax: +44 1709 825464; E-mail: STC.library@corusgroup.com
1933; Christine Rawson
Metallurgy, Specifications Index
20 000 vols; 400 curr per; 27 000 pamphlets
SINTO   29580

Random House Group, Archive and Library, 1 Cole St, Crown Park, *Rushden* NN10 6RZ
T: +44 207 8408801;
Fax: +44 1933 419428; E-mail: jrose@randomhouse.co.uk; URL: archive.randomhouse.co.uk
1987; Jean Rose
Author Files for Random House inprints
950 000 vols   29581

Kvaerner Metals Davy Ltd, Library, Prince of Wales Rd, *Sheffield* S9 4EX
T: +44 114 2449971; Fax: +44 114 2914011
1830; Ms V.H. Hawksley
Mechanical engineering, iron and steel industry
17 400 vols; 45 curr per; 20 diss/theses; 150 govt docs; 50 maps; 100 digital data carriers; 5 000 manufacturer's cat, 4 000 british and foreign standards
libr loan; CILIP   29582

RWE npower, Information Centre, Windmill Hill Business Park, *Swindon* SN5 6PB
T: +44 1793 892565;
Fax: +44 1793 892994; E-mail: suzanne.botter@rwenpower.com
1990; Suzanne Botter
Hist of electricity generation in England and Wales
16 000 vols; 75 curr per   29583

BMT Group Limited, Library, Goodrich House, 1 Waldegrave Road, *Teddington* TW11 8LZ
T: +44 20 8943 5544; Fax: +44 20 8943 5347; E-mail: dgriffiths@bmtmail.com; URL: www.bmt.org
David Griffiths
1 000 vols; 50 curr per; 1 e-journals; 100 diss/theses; 500 govt docs; 50 maps; 50 microforms; 20 digital data carriers
libr loan; CILIP   29584

Smithkline Beecham Pharmaceuticals, Welwyn Library, The Frythe, *Welwyn Garden City* AL6 9AR
T: +44 1438 782000; Fax: +44 1438 782570
1952; J. Borutan
30 000 vols; 425 curr per
libr loan; ASLIB   29585

Yorkreco Information Centre, Nestec York Ltd, P.O. Box 201, *York* YO1 1XY
T: +44 1904 602421; Fax: +44 1904 604887
1917; Maxine Nicholson
Food sciences, nutrition
10 000 vols; 500 curr per
libr loan   29586

Special Libraries Maintained by Other Institutions

James McBey Art Reference Library, Aberdeen Art Gallery, Schoolhill, *Aberdeen* AB9 1FQ
T: +44 1224 523700; Fax: +44 1224 632133; E-mail: info@aagm.co.uk; URL: www.aagm.co.uk
1961; Jeff Evans
James McBey Ex Libris
12 000 vols; 65 curr per; 20 microforms; 20 400 av-mat; 150 sound-rec
ARLIS/UK & Ireland   29587

Macaulay Land Use Research Institute, Library, Craigiebuckler, *Aberdeen* AB15 8QH
T: +44 1224 318611;
Fax: +44 1224 311556; E-mail: L.Robertson@macaulay.ac.uk; URL: www.macaulay.ac.uk
1930; Lorraine Robertson
Forestry, geology, agriculture, chemistry
14 000 vols; 300 curr per
Scottish Agricultural Librarians Group, Grampian Information   29588

Marine Scotland Marine Laboratory, Library, 375 Victoria Rd, *Aberdeen* AB11 9DB
T: +44 1224 876544; Fax: +44 1224 295309; E-mail: ml_library@marlab.ac.uk
1899; Helen McGregor
Ogilvie Coll on Diatomaceae
90 000 vols; 200 curr per; 150 e-journals; 1 000 maps; 5 000 microforms; 30 av-mat; 10 000 patents, rpts, British standards
libr loan; EURASLIC   29589

Rowett Research Institute, Reid Library, Greenburn Rd, Bucksburn, *Aberdeen* AB21 9SB
T: +44 1224 712751; Fax: +44 1224 715349; E-mail: library@rri.sari.ac.uk
1922; Mary Mowat
Nutrition, animal husbandry
25 000 vols; 130 curr per; 700 diss/theses; 200 govt docs; 50 microforms; 28 000 pamphlets
libr loan; BLL, ASLIB   29590

CCFE Fusion Library, Culham Centre for Fusion Energy, *Abingdon* OXON OX14 3DB
T: +44 1235 466347; Fax: +44 1235 466507; E-mail: helen.bloxham@ccfe.ac.uk
1962; Helen Bloxham
Plasma physics, controlled nuclear fusion
21 000 vols; 30 curr per; 28 e-journals; 40 diss/theses; 2 000 microforms; 20 av-mat; 30 digital data carriers; pamphlets, culled articles
libr loan; CILIP   29591

Animal Health and Veterinary Laboratories Agency, Library, New Haw, *Addlestone*, Surrey KT15 3NB
T: +44 1932 357314;
Fax: +44 1932 357608; E-mail: enquiries@vla.defra.gsi.gov.uk; URL: www.defra.gov.uk/vla/vla/vla_library.htm
Mrs H. Hulse
Staff publications
10 000 vols; 400 curr per; 60 e-journals; 250 diss/theses; 300 maps; 3 000 av-mat; 7 digital data carriers
libr loan; AHIS   29592

Prince Consort's Library, Knollys Rd, *Aldershot* GU11 1PS
T: +44 1252 349381; Fax: +44 1252 349382
1860
Mil hist coll of Prince Albert
65 000 vols; 120 curr per; 100 maps; 20 digital data carriers
libr loan   29593

Alnwick Castle Library, Estates Office, Alnwick Castle, *Alnwick* NE66 1NQ
T: +44 1665 510777 ext 141
Colin Shrimpton
Family and estate papers
90 000 vols   29594

The Armitt Collection, Library, Rydal Rd, *Ambleside* LA22 9BL
T: +44 15394 31212; Fax: +44 15394 31313; E-mail: info@armitt.com; URL: www.armitt.com
1912; Tanja Flower

Charlotte Mason Arch, Section of Fell & Rock Club Libr, Early Guidebooks to the Lakes, Beatrix Potter, Kurt Schwitters, John Ruskin, Josefina de Vasconcellos, Harriet Martineau colls, Oral History Archive
11 500 vols; 20 curr per; 2 000 mss; 4 diss/theses; 300 maps
AIL   29595

Freshwater Biological Association, Library, The Ferry Landing, Far Sawrey, *Ambleside* LA22 0LP
T: +44 15394 42468; Fax: +44 15394 46914; E-mail: lis@fba.org.uk; URL: www.fba.org.uk
1929; Hardy Schwamm
Fritsch Collection of Algal Illustrations, Archive of Freshwater data & images; Freshwaterlife Ecological Information Service
9 000 vols; 250 curr per; 500 e-journals; 450 diss/theses; 300 maps; 250 microforms; 50 digital data carriers
BIASLIC, EURASLIC, IAMSLIC, WILG   29596

Armagh Observatory, Library, College Hill, *Armagh* BT61 9DG
T: +44 28 37522928; Fax: +44 28 37527174; E-mail: jmf@arm.ac.uk; URL: climate.arm.ac.uk; star.arm.ac.uk
1790; John McFarland
Astronomy
25 000 vols; 50 curr per; 16 e-journals; 100 mss; 1 incunabula; 120 diss/theses; 100 av-mat; 5 000 slides, photos   29597

Chartered Institute of Building, Library and Information Service, Englemere, King's Ride, *Ascot*, Berkshire SL5 7TB
T: +44 1344 630707; Fax: +44 1344 630764; E-mail: lis@ciob.org.uk; URL: www.ciob.org.uk
1970; Caroline Collier
Professional interview reports
12 000 vols; 160 curr per; 75 diss/theses; 1 500 govt docs; 160 av-mat; 50 sound-rec; 10 digital data carriers; 2 000 pamphlets
ASLIB   29598

Hannah Research Institute, Library, Hannah Research Park, *Ayr* KA6 5HL
T: +44 1292 674000; Fax: +44 1292 674003; URL: www.hri.sari.ac.u
1928; E.J. Barbour
Agriculture, food, nutrition
9 000 vols; 150 curr per; 100 diss/theses; 1 000 pamphlets, 500 photos
libr loan   29599

Countryside Council for Wales / Cyngor Cefn Gwlad Cymru, HQ Library, Hafod Elfyn, Penrhos Road, *Bangor* LL57 2BQ
T: +44 1248 385522; Fax: +44 1248 385510; E-mail: library@ccw.gov.uk; URL: www.ccw.gov.uk
1991; D. Lloyd
Wales coll
22 000 vols; 200 curr per; 50 e-journals; 80 sound-rec; 55 digital data carriers
                                              29600

Bowes Museum, Museum Reference Library, *Barnard Castle* Durham; mail address: Barnard Castle DL12 8NP
T: +44 1833 690606;
Fax: +44 1833 637163; E-mail: info@bowesmuseum.org.uk; URL: www.bowesmuseum.org.uk
1869
John Bowes ms, arch, local hist
10 000 vols; 25 curr per; 5 000 mss; 50 incunabula; 5 music scores   29601

Barnsley District General Hospital, Staff Library, Education Centre, Gawber Rd, *Barnsley* S75 2EP
T: +44 1226 777973; Fax: +44 1226 770365
1976; R. C. Merrill
14 000 vols; 140 curr per
libr loan   29602

North Devon Athenaeum, Library, The Square, *Barnstaple* EX32 8LN
1888; G.A. Morris
40 000 vols
libr loan; SWRLS, BLL   29603

Royal United Hospital NWS Trust Library, Postgraduate Centre, Royal United Hospital, ***Bath*** BA1 3NG
T: +44 1225 824897; Fax: +44 1225 316575
1968; D.S. Rumsey
Medical hist, rheumatology
13 000 vols; 300 curr per
libr loan; HATRICS 29604

B. P. Library of Motoring, National Motor Museum, Reference Library, ***Beaulieu*** SO42 7ZN
T: +44 1590 614652;
Fax: +44 1590 612655; E-mail: motoring.library@beaulieu.co.uk; URL: www.beaulieu.co.uk
1952; Malcolm Thorne
Automotive engineering, automobiles, transport and traffic, motor cycling, motor sport, St. John Nixon coll; Film libr, photographic libr
50 000 vols; 300 curr per; 250 maps
CILIP 29605

Wellcome Foundation, Central Library, Langley Court, ***Beckenham*** BR3 3BS
T: +44 20 86582211; Fax: +44 20 86397103
S. Williams
50 000 vols; 1 000 curr per; 100 diss/theses; 5 000 govt docs; 12 digital data carriers
libr loan 29606

The Bar Library, 91 Chichester Street, mail address: ***Belfast*** BT1 3JQ
T: +44 28 90241523;
Fax: +44 28 90231850; E-mail: niamh.burns@barcouncil-ni.org.uk; URL: www.barlibrary.com
Niamh Burns
85 000 vols; 100 curr per; 20 digital data carriers 29607

Linen Hall Library, 17 Donegall Sq North, ***Belfast*** BT1 5GD
T: +44 28 90321707; Fax: +44 28 90438586; E-mail: info@linenhall.com; URL: www.linenhall.com
1788; John Gray
Irish coll, Northern Ireland Political coll, Blackwood coll (genealogy), early Ulster and Belfast print, Theatre in Ulster; Theatre and Performing Arts Dept
200 000 vols; 1 000 curr per; 600 mss; 2 000 maps; 2 500 microforms
BLL 29608

Ulster Museum, Botanic Gardens, ***Belfast*** BT9 5AB
T: +44 28 90383000; Fax: +44 28 90383003; URL: www.nmni.com/um
1831; Margaret Quine
30 000 vols; 180 curr per; 100 mss; 200 maps; 15 000 photogr negatives 29609

Birmingham and Midland Institute, The Birmingham Library, 9 Margaret St, ***Birmingham*** B3 3BS
T: +44 121 2363591; Fax: +44 121 2124577; E-mail: admin@bmi.org.uk; URL: www.bmi.org.uk
1779; Mrs Sheila Utley
100 000 vols; 13 100 sound-rec
AIL 29610

Birmingham Heartlands Hospital, Heartlands Education Centre, Library, Bordesley Green East, ***Birmingham*** B9 5SS
T: +44 121 7666611 ext 4790; Fax: +44 121 7736897
1974; Lesley Allen
18 000 vols; 130 curr per; 5 digital data carriers
libr loan 29611

Birmingham Law Society, Library, 8 Temple St, ***Birmingham*** B2 5BT
T: +44 121 6336902;
Fax: +44 121 6333507; URL: www.birminghamlawsociety.co.uk/library_about.asp
1818; Hilary C. Boucher
40 000 vols; 45 curr per; 10 av-mat
British and Irish Association of Law Librarians, CILIP 29612

Nuffield House, Queen Elizabeth Medical Centre, Trust Library, Edgbaston, ***Birmingham*** B15 2TH
T: +44 121 6978266; Fax: +44 121 6978300; E-mail: qelibrary@uhb.nhs.uk
1999; Ursula Ison
10 000 vols; 120 curr per; 50 e-journals
libr loan; CILIP 29613

Royal Society for the Prevention of Accidents (RoSPA), Information Centre, RoSPA House, 28 Calthorpe Rd, Edgbaston, ***Birmingham*** B15 1RP
T: +44 121 2482000; Fax: +44 121 2482001; E-mail: infocentre@rospa.com; URL: www.rospa.com
1917; Dr. Ibidapo Oketunji
Road Safety, Health and Safety Management, Risk Management, Home and Product Safety, Occupational Safety, Water and Leisure Safety
24 730 vols; 59 curr per; 22 e-journals; 12 maps; 400 av-mat; 59 sound-rec; 15 digital data carriers
libr loan; CILIP 29614

Blackburn Royal Infirmary, Education Centre Library, Bolton Rd, ***Blackburn*** BB2 3LR
T: +44 1254 294308; Fax: +44 1254 294065
1971; Mrs C. L. Riley
8 000 vols; 120 curr per; 100 av-mat; 8 digital data carriers 29615

WRc plc, Library, Frankland Rd, ***Blagrove*** SN5 8YF
T: +44 1793 865154; Fax: +44 1793 865001
S. Gardner
Water and waste water treatment, process engineering
8 000 vols; 100 curr per
libr loan 29616

Pilgrim Hospital, Library, Sibsey Rd, ***Boston*** PE21 9QS
T: +44 1205 364801 ext 2272
A.L. Willis
Health care
11 150 vols; 240 curr per 29617

Information Centre, BSRIA Ltd, Information Centre, Old Bracknell Lane West, ***Bracknell*** RG12 7AH
T: +44 1344 465571; Fax: +44 1344 465605; E-mail: information@bsria.co.uk; URL: www.bsria.co.uk
1955; Clare Sinclair
Mechanical and electrical services for buildings (heating, ventilating, air conditioning, plumbing and sanitation, lighting and power, communications and transport)
6 000 vols; 100 curr per; 30 000 pamphlets 29618

Booth Museum of Natural History, Library, 194 Dyke Rd, ***Brighton*** BN1 5AA
T: +44 1273 292777; E-mail: boothmuseum@brighton-hove.gov.uk; URL: www.brighton-hove-rpml.org.uk/Museums/boothmuseum/Pages/home.aspx
1890; Dr. Gerald Legg
Natural History
18 000 vols; 10 curr per; 200 mss; 4 diss/theses; 200 maps; 1 microforms; 40 000 photographs (35 mm) 29619

Audit Commission, Information Service, Nicholson House, Lime Kiln Close, Stoke Gifford, ***Bristol*** BS12 6SU
T: +44 117 9757809; Fax: +44 117 9790552; URL: www.audit-commission.gov.uk
Dawn Witherden
Auditing, local govt, health, law, management, accounting
19 000 vols; 200 curr per; 15 000 govt docs
libr loan; ASLIB 29620

Suffolk Record Office, Local Studies Library, Bury St Edmunds Branch, 77 Raingate St, ***Bury St Edmunds*** IP33 2AR
T: +44 1284 352352; E-mail: bury.ro@libher.suffolkcc.gov.uk; URL: www.suffolkcc.gov.uk/leisureandculture/
Cullum coll – a typical 'family library' of the 19th c gentleman (4 000 vols on travel, topography, social and political sciences, natural history, philosophy, geneology etc)
20 000 vols; 54 curr per; 6 diss/theses; 12 sound-rec; 10 digital data carriers 29621

Concrete Information Ltd, Riverside House, Library, Riverside House, 4 Meadows Business Park, Station Approach Blackwater, ***Camberley*** Surrey, GU17 9AB
T: +44 1276 607140;
Fax: +44 1276 607141; E-mail: enquiries@concreteinfo.org; URL: www.concreteinfo.org
1935; E.A.R. Trout
Advanced Concrete Technology Reports
10 500 vols; 400 curr per; 50 diss/theses; 1 000 microforms; 20 digital data carriers; 51 000 rpts
libr loan; ASLIB 29622

The Babraham Institute, Library, Babraham Research Campus, ***Cambridge*** CB22 3AT
T: +44 1223 496235;
Fax: +44 1223 496027; E-mail: jennifer.maddock@babraham.ac.uk
1948; J.R. Maddock
21 000 vols; 2 curr per; 4 774 e-journals; 398 diss/theses
libr loan; CILIP, RESCOLINC, UKSG 29623

British Antarctic Survey, Natural Environment Research Council, Library, High Cross, Madingley Rd, ***Cambridge*** CB3 0ET
T: +44 1223 2214000; Fax: +44 1223 362616; E-mail: baslib@bas.ac.uk; URL: www.antarctica.ac.uk
1976; J. Milton
Geology, geophysics, glaciology, meteorology, climatology, upper atmosphere physics, botany, zoology, marine biology of Antarctica and the Southern ocean
11 500 vols; 300 curr per; 400 diss/theses; 200 microforms; 16 000 reprints
libr loan; Polar Libraries Colloquy 29624

Cambridge Refrigeration Technology, Library, 140 Newmarket Road, ***Cambridge*** CB5 8HE
T: +44 1223 461352;
Fax: +44 1223 461522; E-mail: dgoddard@crtech.demon.co.uk; URL: www.crtech.co.uk
1945; Mrs G.D. Goddard
CA Hanvals, CA Seminar Papers, IIR Proceedings, Annexes
10 000 vols; 70 curr per; 27 000 mss; 30 diss/theses; 5 000 govt docs; 50 maps; 100 microforms; 100 av-mat; 200 digital data carriers; 5 000 pamphlets, 1 000 photos, 1 000 slides
libr loan; BL 29625

Corpus Christi College, Parker Library, Trumpington St, ***Cambridge*** CB2 1RH
T: +44 1223 338025; Fax: +44 1223 339041; E-mail: parker-library@corpus.cam.ac.uk; URL: www.corpus.cam.ac.uk/parker-library
1579; Dr. Christopher de Hamel
Manuscript studies; medieval history, theology and literature, particularly of England pre-1066; Matthew Parker Coll (incl 40 Anglo-Saxon mss), Stokes Coll (Jews in England), Lewis Coll (coins) – now housed in the Fitzwilliam Museum, Cambridge, Perowne Coll (orders of the church)
10 000 vols; 30 curr per; 637 mss; 120 incunabula; 500 microforms; 4 000 av-mat; 20 digital data carriers; Journal offprints and other pamphlets relating to the manuscript coll
libr loan; AMARC 29626

Tyndale House, Biblical Research Library, 36 Selwyn Gardens, ***Cambridge*** CB3 9BA
T: +44 1223 566604;
Fax: +44 1223 566608; E-mail: librarian@tyndale.cam.ac.uk; URL: www.tyndale.cam.ac.uk/
1944; Dr. Elizabeth Magba
Biblical studies, Judaism, Ancient Near East
42 000 vols; 200 curr per; 97 e-journals; 260 diss/theses; 200 maps; 150 av-mat; 40 digital data carriers
ABTAPL 29627

Institute of Heraldic and Genealogical Studies, 79-82 Northgate, ***Canterbury*** CT1 1BA
T: +44 1227 768664; Fax: +44 1227 765617; E-mail: ihgs@ihgs.ac.uk; URL: www.ihgs.ac.uk
1961; S. Bulson
Heraldic Manuscripts
20 000 vols; 150 curr per; 300 maps; 20 000 microforms; 400 digital data carriers 29628

Cardiff Naturalists' Society, Library, c/o National Museum of Wales, Cathays Park, ***Cardiff*** CF10 3NP
T: +44 29 20573202; Fax: +44 29 20573216
John Robert Kenyon
Natural hist, archeology, local hist
12 000 vols; 70 curr per 29629

National Museum of Wales, Library, Cathays Park, ***Cardiff*** CF10 3NP
T: +44 29 20573202; E-mail: library@museumwales.ac.uk
1912; John Robert Kenyon
Science and Technology Tomlin coll (conchology), Willoughby Gardner coll (early natural hist), Vaynor coll (astronomy and physical sciences early bks)
200 000 vols; 1 700 curr per; 2 incunabula; 60 diss/theses; 1 000 maps; 200 microforms; 70 digital data carriers
ARLIS/UK & Ireland, IFLA 29630

Sain Ffagan: Amguedda Werin Cymru – Llyfrgell / St Fagans: National History Museum Library, St Fagans, ***Cardiff*** CF5 6XB
T: +44 29 20573446; Fax: +44 29 20573490; URL: www.museumwales.ac.uk/en/196/
1969; Niclas L. Walker
30 000 vols; 250 curr per; 107 diss/theses; 100 maps; 34 microforms; 238 av-mat; 9 000 sound-rec; 12 000 slides, 130 000 photos, mss coll
BL 29631

Education Centre, Cumberland Infirmary, Library, Cumberland Infirmary, ***Carlisle*** CA2 7HY
T: +44 1228 814878, 814879; Fax: +44 1228 814843
1986; Pauline Goundry
Medicine, nursing, NHS
10 000 vols; 201 curr per; 12 av-mat; 8 sound-rec; 28 digital data carriers; 140 video recordings
libr loan 29632

Royal Engineers Library, Brompton Barracks, ***Chatham*** ME4 4UG
T: +44 1634 822221; E-mail: mail@re-library.co.uk; URL: www.remuseum.org.uk
1813; Miss C.E. Hughes
Hist of military engineering; hist of corps of Royal Engineers
35 000 vols; 175 curr per; 1 000 mss; 1 000 diss/theses; 2 000 maps; 600 000 photos (dating from 1850 to the present), "The Sapper", magazine on microfiche (1895 to present)
BLL 29633

Countryside Agency, Library, John Dower House, Crescent Place, ***Cheltenham*** GL50 3RA
T: +44 1242 521381; Fax: +44 1242 584270
Jean V. Bacon
Countryside conservation, informal countryside recreation
20 000 vols; 400 curr per; 30 000 av-mat
ASLIB 29634

Weald and Downland Open Air Museum, The Armstrong Library, Singleton, ***Chichester*** PO18 0EU
T: +44 1243 811363; Fax: +44 1243 811475; E-mail: office@wealddown.co.uk; URL: www.wealddown.co.uk
1981; R. Harris
Vernacular architecture, rural life and crafts, building materials and techniques, windmills, watermills, plumbing and leadwork
12 000 vols; 30 curr per; 40 diss/theses; 500 maps; 3 000 photos, 100 000 slides 29635

National Radiological Protection Board, Library, ***Chilton*** OX11 0RQ
T: +44 1235 822649; Fax: +44 1235 833891; URL: www.hpa.org.uk
1970; David Perry
ICRP, NCRP, ICRU publs
70 000 vols; 100 curr per; 60 000 reports 29636

The Chartered Institute of Logistics and Transport (UK), Logistics and Transport Centre, Library, Earlstrees Court, Earlstrees Rd, ***Corby*** NN17 4AX
T: +44 1536 740112;
Fax: +44 1536 740102; E-mail: peter.huggins@ciltuk.org.uk; URL:

www.ciltuk.org.uk
1999; Peter Huggins
On-line library catalogue, Business Source Corporate – electronic journals database, UK/EIRE Reference Centre – electronic journals database, Mint UK database of 3300000 UK companies, Pressdisplay – database of global newspapers, Croner-i – database of legislature and compliance information
12000 vols; 150 curr per; 4 e-journals; 200 diss/theses; 1500 govt docs; 200 av-mat; 20 digital data carriers
ASLIB 29637

Chartered Management Institute, Management House, Cottingham Rd, ***Corby*** NN17 1TT
T: +44 1536 207400;
Fax: +44 1536 401013; E-mail: mic.enquiries@managers.org.uk; URL: www.managers.org.uk
1947; Nick Parker
30000 vols; 23 curr per; 35 e-journals; 64 e-books 29638

Walsgrave Hospital, Medical Library, Clifford Bridge Rd, ***Coventry*** CV2 2DX
T: +44 24 76602020 ext 8455
Eileen Edward
10000 vols; 150 curr per 29639

Cowes Library and Maritime Museum, Beckford Rd, ***Cowes*** PO31 7SG, Isle of Wight
T: +44 1983 823433; Fax: +44 1983 823841
1941; Joyce Blizzard
Maritime coll, (yachting, boat building)
26000 vols; 50 curr per; 465 incunabula; 800 av-mat; 10000 photos, 200 slides
libr loan 29640

TRL Library & Information Centre, TRL Ltd, Nine Mile Ride, ***Crowthorne*** RG40 3GA
T: +44 1344 770203; Fax: +44 1344 770193; E-mail: enquiries@trl.co.uk; URL: www.trl.co.uk
1933; Stuart Benjamin
Special colls: own TRL reports and staff papers, publications from US Transportation Research Board OECD Transport Research Programme
30000 vols; 250 curr per; 20 e-journals; 50 diss/theses; 500 govt docs; 250 maps; 70000 microforms; 50 av-mat; 80000 pamphlets
libr loan; ASLIB, CILIP
Lead library in the OECD International Transportation Research Documentation database (ITRD) 29641

Wiltshire Archaeological and Natural History Society, Wiltshire Heritage Library, 41 Long St, ***Devizes*** SN10 1NS
T: +44 1380 727369;
Fax: +44 1380 722150; E-mail: l.haycock@wiltshireheritage.org.uk
1853; Lorna Haycock
Colt Hoare, Benett, Cunnington, Everett, Hungerford, Britton, Story-Maskelyne
11600 vols; 85 curr per; 4000 mss; 25 diss/theses; 2 music scores; 1500 maps; 500 microforms; 20000 prints, 35 cuttings books 29642

Medical Research Council, Library, Harwell, ***Didcot*** OX11 0RD
T: +44 1235 841000; Fax: +44 1235 841200; E-mail: m.bulman@har.mrc.ac.uk
M.J. Bulman
10000 vols; 80 curr per; 50 e-journals; 1 digital data carriers
RESCOLINC 29643

Science and Technology Facilities Council, Rutherford Appleton Laboratory, Library, Harwell Science and Innovation Campus, ***Didcot*** OX11 0QX
T: +44 1235 445384; Fax: +44 1235 446403; E-mail: council@stfc.ac.uk
1961; Mrs D. Franks
ESA reports
36000 vols; 412 curr per; 830 e-journals; 260 diss/theses; 42000 reports + preprints 29644

Doncaster Royal Infirmary, Medical and Professional Library, Doncaster Royal Infirmary, Armthorpe Road, ***Doncaster*** DN2 5LT
T: +44 1302 553118;
Fax: +44 1302 553250; E-mail: doncaster.medicallibrary@dbh.nhs.uk
1968; M.E. Evans
12500 vols; 170 curr per; 1500 e-journals; 1 digital data carriers; 10 electronic databases
libr loan 29645

QinetiQ Winfrith Information Centre, Library, Rm 46, Bldg A22; Winfrith Technology Centre, ***Dorcherster*** DT2 8XJ
T: +44 1305 212218; URL: www.qinetiq.com
W. Gubbels
18000 vols; 70000 rpts
CILIP 29646

Dorset County Museum, High West St, ***Dorchester*** DT1 1XA
T: +44 1305 262735;
Fax: +44 1305 257180; E-mail: enquiries@dorsetcountymuseum.org; URL: www.dorsetcountymuseum.org/library
1846; Jennifer Martindale
William Barnes, archeology, photos, Thomas Hardy Memorial Coll, Sylvia Townsend Warner & Valentine Ackland Coll, Powys Coll
30000 vols; 120 curr per; 10000 mss; 30 diss/theses; 50 music scores; 1000 maps; 20 av-mat; 50 sound-rec; 5000 letters 29647

Scottish Crop Research Institute, Library, Invergowrie, ***Dundee*** DD2 5DA
T: +44 1382 562731; Fax: +44 1382 562426; E-mail: library@scri.ac.uk; URL: www.scri.ac.uk
1920; Sarah Collier
Agriculture, agronomy, plant diseases, plant protection, crop husbandry, plant breeding
20000 vols; 50 curr per; 5000 e-journals; 200 diss/theses; 450 maps; 220 microforms; 11 digital data carriers; 31000 slides, 5000 pamphlets
libr loan; UKSG, SALG 29648

SOC – Scottish Ornithologists' Club, Waterston Library, Aberlady, ***East Lothian*** EH32 OPY
T: +44 1875 871330; Fax: +44 1875 871035; E-mail: library@the-soc.org.uk; URL: www.the-soc.org.uk
1959; Mrs K. Bidgood
5000 vols; 200 curr per 29649

East Malling Research Library, New Rd, ***East Malling*** ME19 6BJ
T: +44 1732 843833; Fax: +44 1732 849067
1920; Jean Hodges
Plant science, plant breeding and biotechnology, specialising in fruit, woodland and crops
40000 vols; 200 curr per; 115 diss/theses; historic publs on horticultural practices from 17th c
ASLIB, UKSG 29650

Advocates Library, Faculty of Advocates, Parliament House, ***Edinburgh*** EH1 1RF
T: +44 131 2605683;
Fax: +44 131 2605663; E-mail: inqdesk@advocates.org.uk; URL: www.advocates.org.uk
1689; Andrea Longson
Roman-Dutch law Coll, Coll of French customary laws, Dieterichs Coll, 18th c session papers coll, Abbotsford library
200000 vols; 1500 curr per; 16 incunabula; 6 av-mat; 34 digital data carriers
BIALL, SLLG 29651

British Geological Survey, Scottish Regional Office, Library, Murchison House, West Mains Rd, ***Edinburgh*** EH9 3LA
T: +44 131 6671000; Fax: +44 131 6682683; E-mail: mhlib@bgs.ac.uk; URL: www.bgs.ac.uk
1872; R.P. McIntosh
150000 vols; 500 curr per; 400 mss; 15000 maps; 200 microforms; 5000 photos 29652

Common Services Agency, Information and Statistics Division, Trinity Park House, South Trinity Rd, ***Edinburgh*** EH5 3SE
T: +44 131 5518775; Fax: +44 131 5511392
1976; A.H. Jamieson
13500 vols; 270 curr per
libr loan 29653

Health Scotland Library, NHS Health Scotland, The Priory, Canaan Lane, Edinburgh, ***Edinburgh*** EH10 4SG
T: +44 131 5365581; Fax: +44 131 5365502; E-mail: nhs.healthscotland-library@nhs.net; URL: www.healthscotland.com/library
1960; Sharon Jamieson
coll of older health education materials from the Scottish Health Education Group and the Health Education Board for Scotland
8000 vols; 100 curr per; 4500 e-journals; 50 diss/theses; 1000 govt docs; 100 digital data carriers
libr loan; Shine 29654

Institut Français, Bibliothèque (French Institute), 13 Randolph Crescent, ***Edinburgh*** EH3 7TT
T: +44 131 2255366; Fax: +44 131 2200648; E-mail: library@ifecosse.org.uk; URL: www.ifecosse.org.uk
1946; Ms A.-M. Usher
13546 vols; 50 curr per; 1321 av-mat; 2207 sound-rec; 630 digital data carriers
libr loan 29655

National Gallery of Scotland, Library, The Mound, ***Edinburgh*** EH2 2EL
T: +44 131 624 6501; Fax: +44 131 220 0917; E-mail: nginfo@nationalgalleries.org; URL: www.nationalgalleries.org
1950; Penelope Carter
35000 vols; 50 curr per; 30000 photos
ARLIS/UK & Ireland 29656

National Monuments Record of Scotland, RCAHMS, Library, John Sinclair House, 16 Bernard Terrace, ***Edinburgh*** EH8 9NX
T: +44 131 6621456; Fax: +44 131 6621477; URL: www.rcahms.gov.uk
1908; Diana Murray
Architectural drawings, archaeological survey and excavation drawings and manuscripts, ground and aerial photos
40000 vols; 90 curr per; 38000 mss; 250 diss/theses; 70000 maps; 1500000 photos, 100000 slides, 400000 drawings 29657

National Museums Scotland, Library, Chambers St, ***Edinburgh*** EH1 1JF
T: +44 131 2474137; Fax: +44 131 2474311; E-mail: library@nms.ac.uk; URL: www.nms.ac.uk
1781; Mark Glancy
J.A. Harvie-Brown, W.S. Bruce, Society of Antiquaries of Scotland papers and mss; Includes the libs of the Royal Museum of Scotland (1854), former Museum of Antiquities of Scotland (1780), and National War Museum of Scotland which was formerly the Scottish United Services Museum (1930)
300000 vols; 760 curr per; 80 e-journals; 5000 mss; 3 incunabula; 400 maps; 500 microforms; 100 sound-rec; 30 digital data carriers
libr loan; ASLIB, ARLIS/UK & Ireland, SCURL 29658

National War Museum of Scotland, Library, The Castle, ***Edinburgh*** EH1 2NG
T: +44 131 2474409; Fax: +44 131 2253848; E-mail: library@nms.ac.uk
1930; Sarah Dallman
12000 vols; 100 curr per; 6000 mss; 200 maps; 5 microforms; 100 sound-rec; 10000 prints 29659

Royal Botanic Garden Library, 20 A Inverleith Row, ***Edinburgh*** EH3 5LR
T: +44 131 5527171; Fax: +44 131 2482901; E-mail: library@rbge.org.uk; URL: www.rbge.org.uk
1670; Jane Hutcheon
Pre-linneau coll, herbals and early medical, agricultural and horticultural lit; Libr of the Botanical Soc Edinburgh, Cleghorn Memorial Libr, Plinian Society Libr, Wernerion Nat Hist Soc Libr
175000 vols; 1490 curr per; 40 e-journals; 90000 mss; 1 incunabula; 150 diss/theses; 5000 govt docs; 2000 maps; 50000 microforms; 15000 av-mat; 25 digital data carriers; 120000 cuttings, 5000 nurseryman's cat
CBHL, EBHL 29660

Royal Observatory, UK ATC (Part of STFC), Library, Blackford Hill, ***Edinburgh*** EH9 3HJ
T: +44 131 6688395; Fax: +44 131 6688264; E-mail: library@roe.ac.uk; URL: www.roe.ac.uk/roe/library/index.html
1896; Karen Moran
Crawford Collection; Historical Archives
90000 vols; 100 curr per; 20 e-journals; 150 mss; 100 incunabula; 100 diss/theses; 300 maps; 120 digital data carriers
RESCOLINC 29661

Scottish Law Commission, Library, 140 Causewayside, ***Edinburgh*** EH9 1PR
T: +44 131 6682131; Fax: +44 131 6624900; URL: www.scotlawcom.gov.uk
1965; N.G.T. Brotchie
20000 vols; 100 curr per 29662

Scottish National Gallery of Modern Art, Library and Archive, Dean Gallery, 73 Belford Rd, ***Edinburgh*** EH4 3DS
T: +44 131 6246253;
Fax: +44 131 6237126; E-mail: gmalibrary@nationalgalleries.org; URL: www.nationalgalleries.org
1960; Jane Furness
The Roland Penrose Arch & Libr, the Gabrielle Keiller Arch & Libr, (both Dada and Surrealism); archives of Scottish artists; special and artists' books coll
58000 vols; 27 curr per; 3000 mss
ARLIS/UK & Ireland, Society of Archivists 29663

Scottish Poetry Library, 5 Crichton's Close, Canongate, ***Edinburgh*** EH8 8DT
T: +44 131 5572876; Fax: +44 131 5578393; E-mail: reception@spl.org.uk; URL: www.spl.org.uk
1984; Julie Johnstone
Scottish poetry in Scots, Gaelic, English, international poetry; Edwin Morgan Archive
25000 vols; 70 curr per; 150 music scores; 100 av-mat; 3000 sound-rec; 10 digital data carriers; 4000 press clippings, 1500 pamphlets, 100 Braille and large print items
libr loan; SLIC 29664

Signet Library, Parliament Sq, ***Edinburgh*** EH1 1RF
T: +44 131 2254923; Fax: +44 131 2204016; E-mail: library@wssociety.co.uk; URL: www.signetlibrary.co.uk
1722; A.R. Walker
Roughead coll on trials
70000 vols; 50 curr per; 100 mss; 12 digital data carriers; Session papers
BIALL 29665

Society of Solicitors in the Supreme Courts of Scotland, The S.S.C. Library, 11 Parliament Sq, ***Edinburgh*** EH1 1RF
T: +44 131 2256268;
Fax: +44 131 2252270; E-mail: enquiries@ssclibrary.co.uk; URL: www.ssclibrary.co.uk
1784; Christine Wilcox
Scots law
15000 vols; 50 curr per; 500 govt docs; 15 digital data carriers 29666

CABI Europe – UK, Library, Bakeham Lane, ***Egham*** TW20 9TY
T: +44 1491 829080; Fax: +44 1491 829100; E-mail: l.ragab@cabi.org; URL: www.cabi.org
1920; Lesley Ragab
Plant pathology, fungal taxonomy, general and applied mycology and plant bacteriology, nematode taxonomy, plant nematodes, ecology
20000 vols; 450 curr per; 15000 microforms; 150000 reprints 29667

Exeter Health Library, First Fl, Peninsula Medical School Bldg, Barrack Rd, ***Exeter*** EX2 5DW
T: +44 1392 406800; Fax: +44 1392 406728; E-mail: Medlib@ex.ac.uk; URL: www.ex.ac.uk/eml
1813; V.B. Newton
17th and 18th c medical bks in Exeter cathedral libr
16728 vols; 126 curr per
libr loan 29668

National Meteorological Library and Archive, Met Office, Fitzroy Rd, ***Exeter*** EX1 3PB
T: +44 1392 884841; Fax: +44 1392 885681; E-mail: metlib@metoffice.gov.uk; URL: www.metoffice.gov.uk/corporate/library/index.html
1870; Sara Osman
200 000 vols; 350 curr per; 5 000 microforms; 8 000 av-mat; 50 sound-rec; 100 digital data carriers; 50 000 pamphlets, 2 000 000 weather charts
ASLI 29669

Medway NHS Foundation Trust, Trust Library, Windmill Rd, ***Gillingham*** ME7 5NY
T: +44 1634 407820; Fax: +44 1634 845640
Carla Wearing
Health and Social care, Medicine, Nursing
17 000 vols; 120 curr per; 150 av-mat; 9 digital data carriers
libr loan; Kent Surrey & Sussex Library & Knowledge Services team 29670

Alliance Française, Library, 2-3 Park Circus, ***Glasgow*** G3 6AX
T: +44 141 3314080; Fax: +44 141 3394224; E-mail: biblio@afglasgow.org.uk; URL: www.afglasgow.org.uk
1946; Carole Jacquet
11 000 vols; 37 curr per; 50 maps; 970 av-mat; 1 000 sound-rec; 151 digital data carriers
libr loan 29671

Baillie's Library, The Mitchell Library, North St, ***Glasgow*** G3 7DN
T: +44 141 2872999; Fax: +44 141 2872815
1887; F. MacPherson
Foulis Press bks, MacLean Soc Libr, Glasgow coll, Bennett coll (art), Caithness coll
20 000 vols; 60 mss; 1 incunabula; 200 maps 29672

Glasgow Royal Infirmary Library and E-Learning Centre, North Glasgow Division, NHS Greater Glasgow and Clyde, 10 Alexandra Parade, ***Glasgow*** G31 2ER
T: +44 141 2115975; Fax: +44 141 2114802; E-mail: gri-library@ggc.scot.nhs.uk; URL: www.nhsggc.org.uk/content/default.asp?page=s287_1
1964
Patient Centred Coll
12 000 vols; 236 curr per; 4 000 e-journals; 150 govt docs; 280 av-mat; 180 digital data carriers
CILIP, ASHSL 29673

IDOX Information Service, Tontine House, 8 Gordon St, ***Glasgow*** G1 3PL
T: +44 141 5741920; Fax: +44 141 248 9433; E-mail: iu@idoxplc.com; URL: www.idoxplc.com
1973; Christine Johnston
Scottish local plans, housing plans, environmental policy, local economic development
50 000 vols; 520 curr per; 10 000 govt docs; 6 digital data carriers; 40 statistical series
ASLIB 29674

James Bridie Library, South Glasgow University Hospitals, Langside Rd, ***Glasgow*** G42 9TY
T: +44 141 2015760; Fax: +44 141 2015759
Shona MacNeilage
8 000 vols; 150 curr per
libr loan; SHINE 29675

Royal Faculty of Procurators in Glasgow, Library, 12 Nelson Mandela Place, ***Glasgow*** G2 1BT
T: +44 141 332 3593; Fax: +44 141 332 4714; E-mail: library@rfpg.org; URL: www.rfpg.org
1857; John McKenzie
Hill Collection (local history); Sheriff Court Library
32 000 vols; 155 curr per; 10 e-journals; 50 maps
libr loan; BIALL SLLG 29676

Scottish Enterprise, Knowledge Exchange, Library, 150 Broomielaw, Atlantic Quay, ***Glasgow*** G2 8LU
T: +44 141 2282997;
Fax: +44 141 2282589; E-mail: gail.rogers@scotent.co.uk; URL: www.scottish-enterprise.com
Ms G. Rogers
10 000 vols; 1 000 curr per; company annual rpts 29677

Royal Horticultural Society, The Wisley Library, RHS Garden Wisley, ***Guildford*** GU2 4ES
T: +44 1483 212428; E-mail: library@rhs.org.uk; URL: www.rhs.org.uk
1867; Barbara Collecott, Brent Elliott
21 000 vols; 400 curr per; Nursery catalogs
EBHL 29678

Surrey Archaeological Society, Library, Castle Arch" ***Guildford*** GU1 3SX
T: +44 1483 532454;
Fax: +44 1483 532454; E-mail: librarian@surreyarchaeology.org.uk; URL: www.surreyarchaeology.org.uk
1854; R. Hughesdon
10 000 vols; 170 curr per; 500 mss; 24 diss/theses; 1 500 maps; 100 microforms; 4 digital data carriers; pamphlets, paintings, postcards
libr loan; BLLD 29679

Cleveland Scientific Institution, c/o Mr Rodger, 6 Kirkdale, ***Guisborough***, Cleveland TS14 8EX
T: +44 1287 638173; URL: www.the-csi.org.uk
1921
Mechanical engineering, metallurgy
20 000 vols; 30 curr per 29680

Rothamsted Research Library, BBSRC, ***Harpenen*** AL5 2JQ
T: +44 1582 763133; Fax: +44 1582 760981; E-mail: res.library@bbsrc.ac.uk; URL: www.rothamsted.bbsrc.ac.uk
1913; Liz Allsopp
Soils and plant nutrition; agriculture; agronomy; entomology; crop protection; biomathematics; early works on agriculture 1471-1840; livestock prints and paintings 1780-1910
80 000 vols; 35 curr per; 2 mss; 13 incunabula; 450 diss/theses; 400 maps
libr loan 29681

John Squire Library, Northwick Park and St Marks Hospitals, North West London Hospitals NHS Trust, Watford Rd, ***Harrow*** HA1 3UJ
T: +44 20 88693322;
Fax: +44 20 88693326; E-mail: library@johnsquirelibrary.org.uk; URL: www.johnsquirelibrary.org.uk/
1971; M.J. Kendall
20 000 vols; 280 curr per; 14 000 mss; 50 digital data carriers
libr loan 29682

Gladstone's Library, ***Hawarden*** CH5 3DF
T: +44 1244 532350; Fax: +44 1244 520643; E-mail: enquiries@gladlib.org; URL: www.gladstoneslibrary.org
1896; Peter Francis
Gladstone's personal libr, theol, philos, hist (19th c), pre-1800 coll, Bishop Moorman Franciscan libr, Benson Judaica
150 000 vols; 200 curr per; 250 000 mss; 350 diss/theses; 700 microforms; 50 000 pamphlets
ABTAPL, AIL 29683

Tun Abdul Razak Research Centre (Tarrc), Library, Brickendonbury, ***Hertford*** SG13 8NL
T: +44 1992 584966; Fax: +44 1992 554837; E-mail: general@tarrc.co.uk; URL: www.tarrc.co.uk
1938; Kristina Lawson
30 000 vols; 82 curr per; 40 000 mss; 50 diss/theses; 50 maps; 200 microforms; 20 digital data carriers; 90 000 docs 29684

Ulster Folk and Transport Museum, Library, Cultra, ***Holywood*** BT18 0EU
T: +44 28 90428428; Fax: +44 28 90428728
1959; D. Roger Dixon
Connell Coll, Irish social and economic hist
25 000 vols; 450 curr per; 3 500 mss; 35 diss/theses; 700 govt docs; 70 music scores; 1 000 maps; 850 microforms; 210 000 av-mat; 13 500 sound-rec; 2 000 Pamphlets, 22 000 Ships plans and charts Complete set of Lloyds's Registers 29685

National Police Library, NPIA, Bramshill, ***Hook***, Hampshire RG27 0JW
T: +44 1256 602650; Fax: +44 1256 602285; E-mail: library@npia.pnn.police.uk; URL: www.npia.police.uk
1948; Patricia Hughes
Annual rpts of H.M. Inspectors of Constabulary from 1858, Police Review from 1893, Police & Constabulary Almanac 1861
42 000 vols; 250 curr per; 250 e-journals; 800 diss/theses; 240 maps; 460 av-mat; 330 digital data carriers; 11 200 pamphlets (including multiple copies, 10600 unique titles), Times newspaper on microfilm (1900-1951)
libr loan; WCJLN 29686

Hull Medical Library, Postgraduate Education Centre, Hull Royal Infirmary, Anlaby Rd, ***Hull*** HU3 2JZ
T: +44 1482 28541; Fax: +44 1482 586587
D.I. Thompson
15 000 vols; 215 curr per 29687

Highland Health Sciences Library, University of Stirling, Old Perth Rd, ***Inverness*** IV2 3FG
T: +44 1463 705269; Fax: +44 1463 713471
1971; Rebecca B. Higgins
Highland Health Board archival mat from 1790 onwards; 6 branch libs
9 000 vols; 300 curr per; 20 diss/theses; 900 av-mat; 4 digital data carriers
libr loan; ASHSL, PLCS, Grampian Information 29688

Scottish Natural Heritage, Library Services, Great Glen House, Leachkin Rd, ***Inverness*** IV3 8NW
T: +44 1463 725290; Fax: +44 1463 725067; E-mail: library@snh.gov.uk
1992; Paul Longborrom
Nature and landscape conservation, land use and recreation especially in relation to Scotlands natural heritage
69 000 vols; 639 curr per; 67 e-journals; 226 e-books
libr loan 29689

Suffolk Record Office, Local Studies Library, Ipswich Branch, Gatacre Rd, ***Ipswich*** IP1 2LQ
T: +44 1473 584541;
Fax: +44 1473 584533; E-mail: ed.button@libher.suffolkcc.gov.uk; URL: www.suffolkcc.gov.uk/libraries_and_heritage
Ed Button
16 000 vols; 70 curr per 29690

Scottish Natural History Library, Foremount House, ***Kilbarchan*** PA10 2EZ
T: +44 1505 702419
1970; Dr. J.A. Gibson
Everything ever publ on Scottish natural hist
150 000 vols; 500 curr per; 100 mss 29691

E. A. Hornel Library, Broughton House, National Trust For Scotland, High St, ***Kirkcudbright*** DG6 4JX
T: +44 1557 330437;
Fax: +44 1557 330437; E-mail: broughtonhouse@nts.org.uk
1919; James Allan
South West Scotland, South West Scotland authors, Robert Burns, Scottish ballads, Thomas Carlyle, J.M. Barrie, William Macmath
20 000 vols; 2 000 mss; 10 diss/theses; 200 music scores; 300 maps; 200 photos, 1 400 photogr plates, 1 000 prints 29692

Leatherhead Food International, Library, Randalls Rd, ***Leatherhead*** KT22 7RY
T: +44 1372 822280;
Fax: +44 1372 822268; E-mail: library@leatherheadfood.com; URL: www.leatherheadfood.com
1946; G.R. Ford
500 curr per 29693

Leeds Library, 18 Commercial St, ***Leeds*** LS1 6AL
T: +44 113 2453071
1768; Geoffrey Forster
Victorian publs, 18th and 19th century travel
135 000 vols; 50 curr per; 50 maps
AIL 29694

Royal Armouries Library, Armourie's Drive, ***Leeds*** LS10 1LT
T: +44 113 2201832; E-mail: library@armouries.org.uk; URL: www.armouries.org.uk
1965; P.A.B. Abbott
Antiquarian books of fencing, Military manuals, Arms & Armour Auction Sales Catalogs, Royal Small Arms Factory arch (Enfield); Photogr Libr, Museum Arch
25 000 vols; 160 curr per; 3 mss; 20 diss/theses; 10 000 govt docs; 50 microforms; 1 500 000 av-mat; 10 digital data carriers; 5 000 pamphlets, 10 000 unbound periodicals and cat
libr loan 29695

Yorkshire Archaeological Society, Library, Claremont 23 Clarendon Rd, ***Leeds*** LS2 9NZ
T: +44 113 2457910;
Fax: +44 113 2441979; E-mail: yas.secretary@googlemail.com; URL: www.yas.org.uk
1863; Robert Frost
Wakefield Court Rolls 1274-1925, coll of 18th & 19th c religious tracts, civil war tracts, heraldry, antiquarian mss
40 000 vols; 300 curr per; 2 500 mss; 500 maps; 10 000 microforms 29696

National Youth Agency, Library, Eastgate House, 19-23 Humberstone Rd, ***Leicester*** LE5 3GJ
T: +44 116 2427350; Fax: +44 116 2427444; E-mail: nya@nya.org.uk; URL: www.nya.org.uk
1973; Jo Poultney
Youth service, Young people, Youth affairs
25 000 vols; 250 curr per; 150 av-mat; 200 training kits
libr loan 29697

Sussex Archaeological Society, Barbican House 169 High St, ***Lewes*** BN7 1YE
T: +44 1273 405138; E-mail: library@sussexpast.co.uk; URL: www.sussexpast.co.uk
1853; Esme Evans
40 000 vols; 130 curr per; 40 incunabula; 250 diss/theses; 100 music scores; videos, tapes, archaeological architectural, ecclesiastical & domestic plans of sites & buildings, photos, brass rubbings, fich-reader
CILIP 29698

Lincoln County Hospital, Hospital Library, Greetwell Rd, ***Lincoln*** LN2 5QY
T: +44 1522 573940; Fax: +44 1522 573954
L. Church
10 000 vols; 150 curr per; 40 diss/theses; 20 digital data carriers
libr loan 29699

Lincolnshire Archives, Foster Library, St. Rumbold St, ***Lincoln*** LN2 5AB
T: +44 1522 526204, 782040;
Fax: +44 1522 530047; E-mail: lincolnshire_archive@lincolnshire.gov.uk; URL: www.lincolnshire.gov.uk/archives
20 000 vols; 55 curr per; 50 diss/theses; 13 000 pamphlets 29700

Athenaeum Liverpool Library, Church Alley, ***Liverpool*** L1 3DD
T: +44 151 7097770, 7020404;
Fax: +44 151 7090418; E-mail: library@athena.force9.net; URL: www.athena.force9.co.uk
1797; V. deP. Roper
William Roscoe coll, Eshelby coll of bks on Yorkshire, Gladstone coll of bks on local hist, Blanco White coll, Liverpool directories 1766-1970, large coll of local maps, Liverpool playbills, English 18th c plays
50 000 vols; 40 curr per; 100 mss; 20 incunabula; 200 maps
AIL 29701

Broadgreen Hospital, Education Centre, Library, Thomas Dr, ***Liverpool*** L14 3LB
T: +44 151 2826447; Fax: +44 151 2826988; URL: www.rlbuht.nhs.uk
1992; Ms J.E.McKie
10 000 vols; 147 curr per; 800 e-journals; 60 av-mat; pamphlets
libr loan 29702

Health and Safety Executive, Knowledge Centre, I.G. Redgrave Court, Merton Rd, Merseyside, ***Liverpool*** L20 7HS
T: +44 151 9514382;
Fax: +44 151 9513674; E-mail: knowledgecentre@hse.gsi.gov.uk; URL: www.hse.gov.uk
1977; Sue Valentine
Occupational health, industrial safety, responses to consultative docs, civil nuklear power incl transcripts from public inquiries
60 000 vols; 800 curr per; 200 e-journals; 25 mss; 20 000 microforms; 10 000 pamphlets, 10 000 rpts
libr loan; LADSIRLAC 29703

Liverpool Medical Institution, Library, 114 Mount Pleasant, ***Liverpool*** L3 5SR
T: +44 151 7099125; Fax: +44 151 7072810; E-mail: library@lmi.org.uk; URL: www.lmi.org.uk
1779; Adrienne Mayers
History of medicine
40 000 vols; 165 curr per
libr loan 29704

Merseyside Maritime Museum, Records Centre, Albert Dock, ***Liverpool*** L3 4AA
T: +44 151 4784418; Fax: +44 151 4784590
1986; H. Threlfall
Emigration, slavery
10 000 vols; 30 curr per; 30 000 maps; 294 microforms; 400 sound-rec; 3 000 slides, 250 000 photos, 4 000 pamphlets 29705

Proudman Oceanographic Laboratory, Library, Joseph Proudman Bldg, Brownlow St, ***Liverpool*** L3 5DA
T: +44 151 7954800; Fax: +44 151 7954801; E-mail: pollib@pol.ac.uk; URL: www.pol.ac.uk
Ms J. Martin
Physical oceanograpgy, tides, ocean dynamics, sedimentology, underwater acoustics, sea level
40 000 vols; 75 curr per; 6 e-journals; 400 diss/theses; 25 digital data carriers
libr loan 29706

Science Fiction Foundation Collection, University of Liverpool Library, c/o University of Liverpool Library, P.O. Box 123, ***Liverpool*** L69 3DA
T: +44 151 7942696; Fax: +44 151 7942681; E-mail: asawyer@liverpool.ac.uk; URL: www.sfhub.ac.uk
1970; Andy Sawyer
Ramsey Campbell Archive, John Brunner Archive, Steph Baxter Archive
30 000 vols; 130 science fiction magazines 29707

Prince Philip Hospital, Multidisciplinary Library, Dafen, ***Llanelli*** SA14 8QF
T: +44 1554 749301; Fax: +44 1554 749301
1911; Ann Leeuwerke
9 500 vols; 109 curr per; 500 govt docs; 150 av-mat; 20 digital data carriers
libr loan; AWHL, AWHILES 29708

Academy of Medical Sciences, Library, 41 Portland Place, ***London*** W1B 1QH
T: +44 20 31762150; E-mail: info@acmedsci.ac.uk; URL: www.acmedsci.ac.uk
1949; Dr. D.J. Chadwick
Biosciences
8 000 vols; 30 curr per; medical biographies (850 vols) 29709

Age Concern England (ACE), Library, Astral House, 1268 London Rd, ***London*** SW16 4ER
T: +44 20 87657200; Fax: +44 20 87657211; E-mail: ace@ace.org.uk; URL: www.ageconcern.org.uk
Cherry-Ann Dowling
Social policy, older people
10 000 vols; 150 curr per 29710

Alpine Club, Library, 55 Charlotte Rd, ***London*** EC2A 3QF
T: +44 20 76130745; Fax: +44 20 76130755; E-mail: library@alpine-club.org.uk; URL: www.alpine-club.org.uk/
1861; Margaret Ecclestone
25 000 vols; 100 curr per; 5 000 pamphlets, 20 000 photos, 5 000 slides 29711

Anti-Slavery International, Library, Thomas Clarkson House, The Stableyard, Broomgrove Rd, ***London*** SW9 9TL
T: +44 20 75018939;
Fax: +44 20 77384110; E-mail: j.howarth@antislavery.org; URL: www.antislavery.org
1839; Jeff Howarth
Belgian Congo, Colonial Africa, indigenous peoples, child labour, forced labour, debt bondage, trafficking,
10 000 vols 29712

Association of Commonwealth Universities (ACU), Reference Library, Woburn House, 20-24 Tavistock Sq, ***London*** WC1H 9HF
T: +44 20 73806700; Fax: +44 20 73872655; E-mail: info@acu.ac.uk; URL: www.acu.ac.uk
1913; Mr N.P. Mulhern
Higher education
18 000 vols
CILIP 29713

The Athenaeum, Library, 107 Pall Mall, ***London*** SW1Y 5ER
T: +44 20 79304843; Fax: +44 20 78394114; E-mail: library@hellenist.org.uk
1824; Ms K.O. Walters
Basil Hall Coll, Morton-Pitt Coll, Gibbon Coll, Dreyfus Coll; Boer War; Tractarian Controversy Coll
60 000 vols; 80 curr per 29714

Austrian Cultural Forum, Library, 28 Rutland Gate, ***London*** SW7 1PQ
T: +44 20 72257300; Fax: +44 20 72250470; URL: www.austria.org.uk/culture
1956; Melita Essenko
German and Germanic Languages and Literatures, Collection on Austria
9 000 vols 29715

Bishopsgate Library, 230 Bishopsgate, ***London*** EC2M 4QH
T: +44 207 3929270;
Fax: +44 207 3929275; E-mail: library@bishopsgate.org.uk; URL: www.bishopsgate.org.uk
1895; Stefan Dickers
London, labour, co-operation, freethought, humanism; London Collection, George Howell Collection (Labour History), London Co-operative Society archive, George Jacob Holyoake Collection, Charles Bradlaugh Papers, Freedom Press Library, Raphael Samuel Archive, British Humanist Association archive, Rationalist Association archive, Stop the War Coalition archive, Bernie Grant archive, Andrew Roth archive, Lesbian and Gay Newsmedia Archive
100 000 vols; 350 curr per; 2 500 maps; 5 000 microforms; 200 av-mat; 250 sound-rec; 50 digital data carriers; 25 000 pamphlets and press cuttings an East London hist, 100 manuscript colls
Association of Independent Libraries, International Association of Labour History Institutions, Society for the Study of Labour History, Socialist History Society 29716

Bloomsbury Healthcare Library, 52 Gower St, ***London*** WC1E 6EB
T: +44 20 73809097; Fax: +44 20 74365111
Michael Larkin
20 000 vols; 200 curr per 29717

British Architectural Library, Royal Institute of British Architects, 66 Portland Place, ***London*** W1B 1AD
T: +44 20 75805533; Fax: +44 20 76311802; E-mail: info@inst.riba.org; URL: www.architecture.com
1834; Ruth H. Kamen
Handley-Read (Victorian decorative arts) Modern movement, Early printed bks 1478-1840; Drawings Coll (610 000 drawings)
135 400 vols; 700 curr per; 700 000 mss; 500 diss/theses; 1 000 govt docs; 1 000 microforms; 200 av-mat; 300 sound-rec; 20 digital data carriers; 700 000 photos
libr loan; ARLIS/UK & Ireland, ASLIB, ARCLIB 29718

British Broadcasting Corporation, BBC Information and Archives, Library, G067, Broadcasting House, Portland Pl, ***London*** W1A 1AA
T: +44 20 15572425; Fax: +44 20 15572128; URL: www.bbc.co.uk
1927; Alison Heighton
150 000 vols; 1 500 curr per; 26 mio press cuttings
LASER, ASLIB, CILIP 29719

British Dental Association, Library, 64 Wimpole St, ***London*** W1G 8YS
T: +44 20 75634545; Fax: +44 20 79356492; E-mail: infocentre@bda.org; URL: www.bda.org
1920; Roger Farbey
11 000 vols; 250 curr per; 300 av-mat; 100 digital data carriers; 4 000 pamphlets, computer assisted learning 29720

British Film Institute, National Library, 21 Stephen St, ***London*** W1T 1LN
T: +44 20 72551444; Fax: +44 20 74362338; E-mail: library@bfi.org.uk; URL: www.bfi.org.uk
1934; David Sharp
Cinematography, motion pictures, television, mass media, communication study
71 000 vols; 300 curr per; 15 e-journals; 750 diss/theses; 2 000 govt docs; 500 000 microforms; 850 sound-rec; 35 digital data carriers; 600 'named' collections (excluding scripts and pressbooks),1 528 music cue sheets, 12 music scores
SCONUL, ARLIS, AUKML 29721

British Geological Survey, London Office, Natural History Museum, Cromwell Rd, South Kensington, ***London*** SW7 5BD
T: +44 20 75894090; Fax: +44 20 75848270; E-mail: bgslondon@bgs.ac.uk; URL: www.bgs.ac.uk
1986; Miss C Tombleson
Geology, palaeontology, earth sciences, environmental geology, hazards, economic geology; Coll of geological maps of the British Isles, Coll of BGS technical/research rpts
15 000 vols; 140 curr per; 4 e-journals; 20 000 maps; 3 digital data carriers; Sale Coll of Bos and Some, Now Bos publications 29722

The British Library – Asia, Pacific and Africa Collections, 96 Euston Rd, ***London*** NWI 2DB
T: +44 20 7412-7873; Fax: +44 20 74127641; E-mail: apac-prints@bl.uk; URL: www.bl.uk
1753; Graham Shaw
Archives: East India Company (1600-1858), Board of Control or Board of Commissioners for the Affairs of India (1784-1858), India Office (1858-1947), Burma Office (1937-1948)
900 000 vols; 8 e-journals; 40 000 maps; 65 000 oriental mss, 120 000 vols of oriental periodicals and newspapers
libr loan 29723

The British Library – Humanities Reading Rooms, 96 Euston Rd, ***London*** NW1 2DB
T: +44 20 7412 7000; Fax: +44 20 7412 7609; URL: www.bl.uk
John Tuck
9 000 000 vols 29724

The British Library – Librarianship & Information Sciences Service (LIS), 96 Euston Rd, ***London*** NW1 2DB
T: +44 20 74127676; Fax: +44 20 74127691; E-mail: lis@bl.uk; URL: www.bl.uk
1902; Michael Stringer
Libr Science, Information Science
1 500 vols; 100 curr per; 350 e-journals; 3 500 diss/theses; 100 av-mat
ALA, ASLIB, IAML, CILIP 29725

The British Library – Manuscripts Collections, 96 Euston Rd, ***London*** NW1 2DB
T: +44 20 74127513; Fax: +44 20 74127745; URL: www.bl.uk
1753; Brett Dolman
Illuminated mss incl 7th c Lindisfarne Gospels, 14th c Luttrell Psalter; early gospels incl 4th c Codex Sinaiticus; early mss of Beowulf, Chaucer etc; 'Magna Carta'; other hist, literary, antiquarian, theatrical colls
100 000 vols; 5 000 maps; 100 000 bound mms, 100 000 charters and rolls, 3 100 papyri, 4 000 ostraca, 18 000 detached seals and casts of seals, incl. 300 000 mss incl 5 000 maps
SCONUL 29726

The British Library – Map Collections, 96 Euston Rd, ***London*** NW1 2DB
T: +44 20 7412 7702; Fax: +44 20 7412 7780; URL: www.bl.uk/collections/maps.html
1867; Peter Barber
Ordnance Survey mapping since 1800, Antiquarian maps and atlases to 15th cent. incl King George 111 Topographical and Maritime Collections; Reference material: Atlases, Gazetteers, History of Cartography
50 000 vols; 200 curr per; 20 000 mss; 40 diss/theses; 6 250 000 maps; 16 000 microforms; 500 av-mat; Collection of pamphlets and offprints of articles on the History of Cartography, 3 digitized map collections 29727

The British Library – Music Collections, 96 Euston Rd, ***London*** NWI 2DB
T: +44 20 7412 7772; Fax: +44 20 7412 7751; URL: www.bl.uk/collections/music/music.html
1753; R.J. Chesser
Royal Music Libr, Paul Hirsch Coll, Stefan Zweig Coll, Royal Philharmonic Society Arch
150 000 vols; 100 000 mss; 1 600 000 music scores
IAML, MLA 29728

The British Library – Newspaper Library, Colindale Ave, ***London*** NW9 5HE
T: +44 20 74127353; Fax: +44 20 74127379; E-mail: newspaper@bl.uk; URL: www.bl.uk/collections/newspapers.html
1932; Ed King
Microfilm set of the Burney coll (1603-1818), Chatham House Press Cuttings coll, Microfiche colls: Belgian Underground Press in WWII 1939-1945, Dutch Underground Press 1940-1945, Early American Newspapers, Northern Ireland Political Lit, 1966-89
660 000 vols; 3 000 curr per; 370 000 microforms; 45 digital data carriers
IFLA 29729

The British Library – Science Technology & Business Collections Development Policy, 96 Euston Rd, ***London*** NW1 2DB
T: +44 20 7412-7494/-7496; Fax: +44 20 7412-7217/-7495; URL: www.bl.uk/collections/science.html
1855 29730

The British Library – Sound Archive Information Service, 96 Euston Rd, ***London*** NW1 2DB
T: +44 20 74127676; Fax: +44 20 74127441; URL: www.bl.uk/soundarchive
1955; Mr A.C. Jewitt
Oral hist coll, International Music (world/ethnic music), Drama and literature recordings, Wildlife Sounds, Music recordings
8 850 vols; 120 curr per; 4 050 microforms; 23 000 av-mat; 1 313 000 sound-rec; 4 digital data carriers
IASA 29731

British Medical Association, Library, BMA House, Tavistock Sq, ***London*** WC1H 9JP
T: +44 20 73836625; Fax: +44 20 73882544; E-mail: bma-library@bma.org.uk; URL: www.bma.org.uk
Jacky P. Berry
30 000 vols; 2 500 curr per; 60 e-journals; 1 500 e-books; 1 500 av-mat
libr loan 29732

The British Museum, Central Library, Great Russell St, ***London*** WC1B 3DG
T: +44 20 73238000;
Fax: +44 20 73238480; E-mail: librariesandarchives@britishmuseum.org; URL: www.thebritishmuseum.ac.uk
Joanna Bowring 29733

The British Museum – Centre of Anthropology, Anthropology Library, Great Russell Street, *London* WC1B 3DG
T: +44 20 73238031; Fax: +44 20 73238049; E-mail: AnthropologyLibrary@ thebritishmuseum.ac.uk
Jan Ayres
Former Royal Anthropological Institute Libr, Christy Libr, Sir Eric Thompson Libr
125 000 vols; 1 500 curr per
ASLIB, SCONUL, SCOLMA, CILIP
29734

BSI Knowledge Centre, Library, 389 Chiswick High Rd, *London* W4 4AL
T: +44 20 89967004;
Fax: +44 20 89967005; E-mail: knowledgecentre@bsigroup.com; URL: www.bsigroup.com
1942; Bethan Carter
Collection of current and withdrawn British Standards, European, International and Overseas Standards, BSI Publications
75 curr per; 10 e-journals; 100 govt docs
29735

Cancer Research UK, Library, 61 Lincoln's Inn Fields, *London* WC2A 3PX
T: +44 20 72693206; Fax: +44 20 72693084; E-mail: lib.info@cancer.org.uk; URL: info.cancerresearchuk.org
1902; Julia Chester
10 000 vols; 300 curr per
libr loan; ASLIB
29736

Centre for Policy on Ageing, Library, 25-31 Ironmonger Row, *London* EC1V 3QP
T: +44 20 75536500; Fax: +44 20 75536501; E-mail: gcrosby@cpa.org.uk; URL: www.cpa.org.uk
1972; Gillian Crosby
Averil Osborn arch, Michael Whitelaw coll
49 500 vols; 200 curr per; 20 diss/theses; 1 800 govt docs; 1 digital data carriers
29737

Chartered Institute of Management Accountants, Library and Information Service, 26 Chapter St, *London* SW1P 4 NP
T: +44 20 88492251; E-mail: cima.contact@cimaglobal.com; URL: www.cimaglobal.com
1919; C. de Vidas
Cost and management accounting financial management, internal audit and control
10 000 vols; 175 curr per
libr loan; ASLIB
29738

Chartered Institute of Personnel and Development, Library and Information Services, 151 The Broadway, Wimbledon, *London* SW19 1JQ
T: +44 20 82166210; E-mail: lis@cipd.co.uk; URL: www.cipd.co.uk
1966; Barbara Salmon
Human resource management, learning, training and development
8 250 vols; 120 curr per; 350 e-journals
CILIP
29739

Child Accident Prevention Trust (CAPT), Resource Centre, Clerk's Ct, 4th Fl, 18-20 Farringdon Lane, *London* EC1R 3HA
T: +44 20 76083828; Fax: +44 20 76083674; E-mail: safe@capt.org.uk; URL: www.capt.org.uk
Ms K. Pordage
10 000 vols
29740

CII Knowledge Services, The Chartered Insurance Institute, 20 Aldermanbury, *London* EC2V 7HY
T: +44 20 74174415, 74174416; Fax: +44 20 79720110; E-mail: knowledge@cii.co.uk; URL: www.cii.co.uk/ knowledge
1934; Hannah West, Adam Parkinson
Insurance, Actuarial science, Risk, Financial Services; Coll of Insurance policies; ephemera; mss
23 000 vols; 1 200 curr per; 320 diss/theses; 75 sound-rec; 100 digital data carriers
ASLIB
29741

Commonwealth Secretariat, Library and Archives, Marlborough House, Pall Mall, *London* SW1Y 5HX
T: +44 20 7747 6164; Fax: +44 20 7747 6168; E-mail: library@commonwealth.int; URL: www.thecommonwealth.org
1965; David Blake
Commonwealth Secretariat archives (30 year rule applies) Commonwealth Secretariat
25 000 vols; 1 000 curr per; 50 e-journals; 17 000 mss; 5 000 govt docs
libr loan; ASLIB, SCOLMA
29742

Competition Commission, Information Centre, Victoria House, Southampton Row, *London* WC1B 4AD
T: +44 20 72710100; Fax: +44 20 72710367; E-mail: info@cc.gsi.gov.uk; URL: www.competition-commission.org.uk
L.J. Fisher
29743

Council for the Protection of Rural England (CPRE), Library and Information Unit, 128 Southwark St, *London* SE1W 0SW
T: +44 20 79812800; Fax: +44 20 79812899; E-mail: info@cpre.org.uk; URL: www.cpre.org.uk
1996; Ms H. Morris
Archival holdings at the Rural History Centre, University of Reading, UK
18 000 vols; 120 curr per; 9 812 govt docs; 1 760 files/records folders of work done by the organistaion
29744

The Egypt Exploration Society, Library, 3 Doughty Mews, *London* WC1N 2PG
T: +44 20 72422266;
Fax: +44 20 74046118; E-mail: chris.naunton@ees.ac.uk; URL: www.ees.ac.uk
1882; Christopher Naunton
Egyptology, archaeology, ancient hist, ancient languages
22 000 vols; 50 curr per; Pamphlets, catalogues
29745

Energy Institute, Library and Information Service, 61 New Cavendish St, *London* W1G 7AR
T: +44 20 74677100; Fax: +44 20 72551472; E-mail: lis@energyinst.org; URL: www.energyinst.org
1914; Catherine Cosgrove
Petroleum industry, standards and codes of practice API, ASTM, BSI
15 000 vols; 200 curr per; 150 maps; 70 av-mat; 100 digital data carriers
libr loan; BLDSC, SLA, CILIP
29746

English Folk Dance and Song Society, Vaughan Williams Memorial Library, 2 Regent's Park Rd, *London* NW1 7AY
T: +44 20 74852206; Fax: +44 20 72840523; E-mail: library@efdss.org; URL: library.efdss.org; www.efdss.org/ library
1930; Malcolm Taylor
Folk, traditional, and popular songs, ballads, dancing, customs and festivals, tales and legends, dialects, Musical instruments, Instrumental music, Street literature, EFDSS, Folk revival, Music history, Biographies; manuscript collections and field recordings of Britisch folk culture and elements of British based cultures in other lands, particularly North America and Ireland.
12 000 vols; 150 curr per; 1 e-journals; 20 mss; 35 diss/theses; 200 microfilms; 150 av-mat; 6 000 sound-rec; 20 digital data carriers; 12 000 photos and slides
libr loan; IAML
29747

English Speaking Union, Page Memorial Library, Dartmouth House, 37 Charles St, *London* W1J 5ED
T: +44 20 75291550; Fax: +44 20 74956108; E-mail: library@esu.org; URL: www.esu.org
1927; Gill Hale
Archival coll, newspaper clippings coll
12 500 vols; 33 curr per; 5 av-mat; 60 sound-rec; 5 digital data carriers
CILIP
29748

The Geological Society of London, Library, Burlington House, Piccadilly, *London* W1J 0BG
T: +44 20 74320999; Fax: +44 20 74393470; E-mail: library@geolsoc.org.uk; URL: www.geolsoc.org.uk
1809; S. Meredith
Papers of Roderick Impey Murchison (1792-1871), Papers of George Bellas Greenough (1778-1855), Maps and other mss of William Smith (1769-1839)
300 000 vols; 700 curr per; 80 e-journals; 40 000 maps
libr loan
29749

German Historical Institute, Library, 17 Bloomsbury Sq, *London* WC1A 2NJ
T: +44 20 73092019, 73092022; Fax: +44 20 73092069, 73092072; E-mail: library@ghil.ac.uk; URL: www.ghil.ac.uk/library.html
1976; Dr. Michael Schaich
German hist – middle ages to contemporary hist, Anglo-German relations
76 000 vols; 200 curr per; 22 e-journals; 20 maps; 75 microforms; 100 digital data carriers; Archive of the Anglo-German Foundation
ASpB
29750

Goethe-Institut, Bibliothek, 50 Princes Gate Exhibition Rd, *London* SW7 2PH
T: +44 20 75964040;
Fax: +44 20 75940230; E-mail: library@london.goethe.org; URL: www.goethe.de/london
1958; Elisabeth Pyroth
German Studies; German as a foreign language dept
14 000 vols; 80 curr per; 10 e-books; 2 100 av-mat; 2 200 sound-rec; 200 digital data carriers
CILIP
29751

Gray's Inn Library, 5, South Sq, Gray's Inn, *London* WC1R 5ET
T: +44 20 74587822;
Fax: +44 20 74587850; E-mail: Library.Information@graysinn.org.uk; URL: www.graysinn.org.uk
1555; T.L. Thom
Francis Bacon
45 000 vols; 270 curr per; 64 mss
BIALL
29752

Greater London Authority, Information Services, City Hall, The Queen's Walk, *London* SE1 2AA
T: +44 20 79834672; Fax: +44 20 79834674; E-mail: isinfo@london.gov.uk; URL: www.london.gov.uk
2000; Annabel Davies
150 000 vols; 500 curr per; 85 e-journals
libr loan
29753

Health Promotion Information Centre (HPIC), Trevelyan House, 30 Great Peter St, *London* SW1 2BY
T: +44 20 74131995; Fax: +44 20 74131834
Ms C. Herman
16 000 vols; 400 curr per; 2 500 av-mat; 150 sound-rec; 1 000 pamphlets
ASLIB
29754

Health Protection Agency, Colindale Library, 61 Colindale Ave, *London* NW9 5HT
T: +44 20 82004400; Fax: +44 20 82007875
1946; Ms M.A. Clennett
Infectious diseases, medical microbiology
41 000 vols; 250 curr per
EAHIL, UKSG
29755

High Commission of India, India House Library, Aldwych, *London* WC2B 4NA
T: +44 20 76323166; Fax: +44 20 76323204; URL: hcilondon.net
1924; Mr Murari Lan, Ms Travis
Politics, Literature, Arts, Indian Fiction, Foreign Affairs (from 1950 –); Mahatma Gandhi Coll
19 000 vols; 7 curr per
29756

Highgate Literary and Scientific Institution, Library, 11 South Grove, *London* N6 6BS
T: +44 20 83403343; Fax: +44 20 83405632; E-mail: librarian@hlsi.net; URL: www.hlsi.net
1839; Margaret Mackay
London, local area, Samuel Taylor Coleridge Coll, John Betjeman Coll; Archives (varied types)
25 000 vols; 20 curr per; 50 sound-rec
Association of Independent Libraries
29757

Hispanic & Luso Brazilian Council, Canning House Library, 2 Belgrave Sq, *London* SW1X 8PJ
Fax: +44 20 78389258; E-mail: library@canninghouse.com; URL: www.canninghouse.com
1943; Alan Biggins
R.B. Cunninghame-Graham Coll, W.H. Hudson Coll, George Canning Coll, Rare bks of history, travel, literature etc. (1 000 vols, 17th – 20th c), Latin American and Iberian Studies
65 000 vols; 40 curr per; 50 maps; 350 av-mat; 100 sound-rec; 20 digital data carriers
CILIP, ACLAIIR, REDIAL, SALALM
29758

Honourable Society of the Middle Temple, Library, Middle Temple Lane, *London* EC4Y 9BT
T: +44 20 74274830;
Fax: +44 20 74274831; E-mail: library@middletemple.org.uk; URL: www.middletemplelibrary.org.uk
1641; Vanessa Hayward
European Communities; Capital Punishment; USA law; American Libr, European Communities Libr
140 000 vols; 200 curr per
BIALL
29759

The Horniman Museum and Gardens, The Horniman Library, Forest Hill, 100 London Rd, *London* SE23 3PQ
T: +44 20 82918681;
Fax: +44 20 82915506; E-mail: enquiry@horniman.ac.uk; URL: www.horniman.ac.uk
1901; A. Yasamee
30 000 vols; 120 curr per; 13 e-journals; 350 maps; 700 av-mat; 1 600 sound-rec
ARLIS/UK & Ireland, Historic Libraries Forum, M25 Consortium of Academic Libraries, Museums Libraries and Archives Group
29760

Huguenot Library, UCL, mail address: Gower St, *London* WC1E 6BT
T: +44 20 76795199; E-mail: library@huguenotsociety.org.uk; URL: www.huguenotsociety.org.uk/library
1885; Miss L Gwynn
Books and archives relating to the Huguenots, Walloons, Palatines, Waldensians
5 000 vols; 25 curr per; 10 e-books; 3 400 mss; 20 diss/theses; 50 microforms; 10 av-mat; some early pamphlets relating to the history of the Huguenots, as well as newscuttings, genealogical material and ephemera
29761

Hulton Getty Picture Collection, Library, Unique House, 21-31 Woodfield Rd, *London* W9 2BA
T: +44 20 72662662; Fax: +44 20 72663154; E-mail: info@getty-images.com
M. Butson
77 000 vols; 2 650 curr per; 10 000 maps; 1 mio slides, 14 mio pictures
29762

Imperial War Museum, Department of Printed Books, Lambeth Rd, *London* SE1 6HZ
T: +44 20 74165342; E-mail: collections@iwm.org.uk; URL: www.iwm.org.uk
1917; Richard Golland
World War I and II, women in World War I
150 000 vols; 400 curr per; 600 diss/theses; 15 000 maps; 4 500 microforms; 100 digital data carriers; 25 000 pamphlets, 15 000 vols of periodicals
CILIP, ASLIB
29763

Inner Temple Library, Inner Temple, *London* EC4Y 7DA
T: +44 20 77978217;
Fax: +44 20 75836030; E-mail: library@innertemple.org.uk; URL: www.innertemplelibrary.org.uk
1505; Margaret Clay
Petyt Mss
70 000 vols; 500 curr per; 10 000 mss; 20 incunabula; 7 digital data carriers; 2 000 pamphlets
BIALL, ASLIB, NAG, UKSG, CILIP
29764

Institut Français du Royaume-Uni, Mediatheque, 17 Queensberry Place, ***London*** SW7 2DT
T: +44 20 70731350;
Fax: +44 20 70731363; E-mail: library@ambafrance.org.uk; URL: www.institut-francais.org.uk
1910; Anne-Elisabeth Buxtorf
Free French coll, 'old' stock from 17th, 18th, 19th century France, periodicals (old stock and back copies) Mss Denis Saurat, Easy French coll; Child Libr (15 000) quick information service, studies in France desk
51 000 vols; 110 curr per; 2 e-journals; 7 500 mss; 3 800 av-mat; 5 200 sound-rec; 2 250 digital data carriers; 80 press cuttings
libr loan; CILIP, ABF 29765

Institute of Chartered Accountants in England & Wales, Chartered Accountants Trust for Education and Research, Library & Information Service, Moorgate Place, Chartered Accountants Hall, P.O. Box 433, ***London*** EC2P 2BJ
T: +44 20 79208620; Fax: +44 20 79208621; E-mail: library@icaew.com; URL: www.icaew.com/library
1880; S.P. Moore
Hist Accounting coll
42 000 vols; 250 curr per; 100 e-journals; 250 e-books; 50 diss/theses; 1 000 microforms; 500 av-mat; 10 digital data carriers
libr loan; SLA, CILIP, CIG 29766

Institute of Materials, Minerals and Mining, David West Library, 1 Carlton House Terrace, ***London*** SW1Y 5DB
T: +44 20 74517360; Fax: +44 207 4517406; E-mail: hilda.kaune@iom3.org; URL: www.iom3.org
1869; Hilda Kaune
Ceramics, including Glass (limited), Fiberglass Technology, Refractories, Tests and Measurements, Materials Science, Metallurgy, Plastics, Polymers, Geology, Mining, Composites, Packaging, Clay Technology; Wood Science: Hist Coll (biographies, portraits, photos, mss, memorabilia, etc); Percy's Papers; Sir Henry Bessemer Memorabilia; Technical drawings; Institute's records; Hume-Rothery Coll; J.S. Jeans Coll and Dunlop/BTR Research Rpts, Minerals Industry Historical Collection
65 000 vols; 1 030 curr per; 15 mss; 1 000 maps; 2 000 microforms 29767

Institute of Psychoanalysis, Library, 112A Shirland Rd, ***London*** W9 2EQ
T: +44 20 75635008; Fax: +44 20 75635001; E-mail: library@iopa.org.uk; URL: www.psychoanalysis.org.uk/library.htm
1924; Saven Morris
22 000 vols; 212 curr per; 37 e-journals 29768

Institution of Civil Engineers, Library, Great George St Westminster, ***London*** SW1P 3AA
T: +44 20 72227722; Fax: +44 20 79767610; E-mail: library@ice.org.uk; URL: www.ice.org.uk
1820; M.M. Chrimes
Arch of the Inst of Civil Eng, Vulliamy Horological Libr, W.A. Fairhurst Coll, J.G. James Coll, MacKenzie Coll, Telford mss, Rennie papers
120 000 vols; 970 curr per; 200 e-journals; 400 mss; 3 incunabula; 200 diss/theses; 10 000 govt docs; 2 500 maps; 5 000 microforms; 31 000 av-mat; 1 500 digital data carriers; 3 000 drawings
libr loan; ASLIB 29769

Institution of Engineering and Technology, Library, Savoy Place, ***London*** WC2R 0BL
T: +44 20 73445461; Fax: +44 20 73448467; E-mail: libdesk@theiet.org; URL: www.theiet.org/library
1880; John Coupland
Sir Francis Ronalds Coll, S.P. Thompson Coll
300 000 vols; 848 curr per; 8 500 e-journals; 1 000 e-books; 3 000 govt docs; 300 av-mat; 500 sound-rec; 2 000 digital data carriers
libr loan
Houses the Libr of the British Computer Society 29770

Institution of Mechanical Engineers, Information and Library Service, 1 Birdcage Walk, ***London*** SW1H 9JJ
T: +44 20 79731274; Fax: +44 20 72228762; E-mail: library@imeche.org; URL: www.imeche.org
1847; Sarah Rogers
George Stephenson Letters, James Nasmyth, Lord Hinton I.K. Brunel and David Joy
130 000 vols; 400 curr per; 15 e-journals; 600 e-books
libr loan; SLA, CILIP 29771

Institution of Structural Engineers, Library, 11 Upper Belgrave Street, ***London*** SW1X 8BH
T: +44 20 7201 9105; Fax: +44 20 7201 9118; E-mail: library@istructe.org; URL: www.istructe.org/library
1929; Rob Thomas
12 360 vols; 150 curr per; 100 mss; 70 diss/theses; 2 000 govt docs; 150 microforms; 200 av-mat; 12500 reports
libr loan; Aslib 29772

Instituto Cervantes, Library, 102 Eaton Square, ***London*** SW1W 9AN
T: +44 20 7201 0756/57; Fax: +44 20 7235 0329; E-mail: biblon@cervantes.es; URL: londres.cervantes.es
1946; David Carrión
Spanish and Latin American Studies
27 000 vols; 60 curr per; 3 000 sound-rec; 3 000 digital data carriers; 2 000 other items
libr loan; CILIP, ACLAIIR 29773

International Coffee Organization, Library, 22 Berners St, ***London*** W1P 4DD
T: +44 20 75808591; Fax: +44 20 75806129; E-mail: info@ico.org; URL: www.ico.org
1963; Celsius A. Lodder
20 000 vols; 250 curr per
ASLIB 29774

International Institute of Communications, Library, 2 Printers Yard, 90A The Broadway, ***London*** SW19 1RD
T: +44 20 84170600; Fax: +44 20 84170800; E-mail: enquiries@iicom.org; URL: www.iicom.org
15 000 vols; 200 curr per 29775

International Sugar Organisation, Library, One canada Sq, Canary Wharf, ***London*** E14 5AA
T: +44 20 75131144;
Fax: +44 20 75131146; E-mail: Publications@isosugar.org; URL: www.isosugar.org
1968
10 000 vols; 3 digital data carriers 29776

Italian Institute, Library, 39 Belgrave Sq, ***London*** SW1X 8NX
T: +44 20 73964425; Fax: +44 20 72354618; E-mail: library@italcultur.org.uk; URL: www.italcultur.org.uk
1950; Ms M. Reidy
Dante Alighieri
23 000 vols; 123 curr per; Coll of Italian videos
libr loan; BLL 29777

Keats House, Memorial Library, London Metropolitan Archives, 40 Northampton Road, ***London*** EC1R 0HB
T: +44 20 73323820;
Fax: +44 20 78339136; E-mail: ask.lma@corpoflondon.gov.uk; URL: www.cityoflondon.gov.uk/keats
1925; Dr. D. Jewkins
John Keats
9 000 vols; 20 curr per; 1 000 mss; 1 000 pamphlets, 800 photos, 100 slides
Visits only by appointment, reference libr 29778

Kings College London, St Thomas' House Information Service Centre, Lambeth Palace Rd, ***London*** SE1 7EH
T: +44 20 7188370; Fax: +44 20 71888358; E-mail: issenquiry@kcl.ac.uk; URL: www.kcl.ac.uk/iss
1951; Karen Stanton
Medicine, Nursing, Midwifery
15 000 vols; 100 curr per; 14 000 e-journals; 500 diss/theses; 300 govt docs; 10 av-mat
libr loan; CILIP 29779

King's Fund Information & Library Service, 11-13 Cavendish Sq, ***London*** W1G 0AN
T: +44 20 73072568;
Fax: +44 20 73072805; E-mail: library@kingsfund.org.uk; URL: www.kingsfund.org.uk/library
1960; Ray Phillips
30 000 vols; 300 curr per; 2 000 govt docs
EAHIL, CILIP, Health Management Information Consortium, CHILL 29780

Laban Library and Archive, Creekside, ***London*** SE8 3DZ
T: +44 20 84699533; Fax: +44 20 86918400; E-mail: library@laban.org; URL: www.laban.org
1982; Ralph Cox
Peter Williams Coll, Laban Coll, Shirley Wynne Coll, Peter Brinson Coll, News clippings collection about contemporary dance, 1960s onwards
17 000 vols; 65 curr per; 678 diss/theses; 30 music scores; 120 microforms; 3 190 av-mat; 800 sound-rec; 12 digital data carriers
libr loan; SIBMAS, TIG 29781

Lambeth Palace Library, Lambeth Palace, ***London*** SE1 7JU
T: +44 20 78981400; Fax: +44 20 79287932; E-mail: lpl.staff@c-of-e.org.uk; URL: www.lambethpalacelibrary.org
1610; Dr. R.J. Palmer
Hist of Church of England, Oecumenism, liturgy, English political hist
200 000 vols; 100 curr per; 5 000 mss; 200 incunabula; 11 800 maps; 2 300 microforms; 50 000 pamphlets, very large archival colls for the Church of England
CILIP 29782

Law Society of England & Wales, Library, 113 Chancery Lane, Law Society's Hall, ***London*** WC2A 1PL; mail address: 113 Chancery Lane, London WC2A 1PL
T: +44 20 73205946, 0870 6062511;
Fax: +44 20 78311687; E-mail: library@lawsociety.org.uk; URL: www.lawsociety.org.uk/productsandservices/libraryservices.law
1828; Chris Holland
English legal materials
65 000 vols; 400 curr per; 5 e-journals; 14 microforms 29783

Lincolns Inn Library, Holborn, ***London*** WC2A 3TN
T: +44 20 72424371;
Fax: +44 20 74041864; E-mail: library@lincolnsinn.org.uk; URL: www.lincolnsinn.org.uk
1474; Guy Holborn
170 000 vols; 390 curr per; 20 e-journals; 2 000 mss; 25 incunabula; 6 000 diss/theses; 47 000 microforms; 50 digital data carriers; 1 334 vols, pamphs and tracts 29784

Linnean Society of London, Learned Society, Library, Burlington House, Piccadilly, ***London*** W1J OBF
T: +44 20 74344479; Fax: +44 20 72879364; E-mail: library@linnean.org; URL: www.linnean.org
1788; Lynda Brooks
Portraits, medals, drawings, Herbaria, Carl von Linné
90 000 vols; 300 curr per; 4 e-journals; 1 000 mss
ASLIB 29785

London Fire and Emergency Planning Authority, Library and Information Resource Centre, 520 Hampton House, 20 Albert Embankment, ***London*** SE1 7SD
T: +44 20 75876340; Fax: +44 20 75876086; E-mail: libraryservices@london-fire.gov.uk; URL: www.london-fire.gov.uk
Gail Parlane
10 000 vols; 105 curr per; 5 e-journals; 30 diss/theses; 500 digital data carriers; 3 subscription databases
CILIP 29786

London Library, 14 St. James's Sq, ***London*** SW1Y 4LG
T: +44 20 79307705;
Fax: +44 20 77664766; E-mail: membership@londonlibrary.co.uk; URL: www.londonlibrary.co.uk
1841; Inez Lynn
1 000 000 vols; 700 curr per; 50 microforms; 15 digital data carriers; 35 000 pamphlets
libr loan; AIL 29787

London Metropolitan Archives, City of London, Library, 40 Northampton Rd, ***London*** EC1R 0HB
T: +44 20 73323820;
Fax: +44 20 78339136; E-mail: ask.lma@cityoflondon.gov.uk; URL: www.cityoflondon.gov.uk
1889; Mick Scott
John Burns Coll relating to London
110 000 vols; 200 curr per; 15 000 maps; 400 000 photos, 40 000 prints
CILIP, Society of Archivists, ASLIB 29788

London Transport Museum Library, 39 Wellington St, Covent Garden, ***London*** WC2E 7BB
T: +44 20 75657280; E-mail: enquiry@ltmuseum.co.uk; URL: www.ltmuseum.co.uk
Caroline Warhurst
Frank Pick collection, Reinohl collection
13 000 vols; 500 curr per; 200 maps; 4 600 pamphlets, 3 4-drawer filing cabinets
CILIP 29789

Marx Memorial Library, 37A Clerkenwell Green, ***London*** EC1R 0DU
T: +44 20 72531485; Fax: +44 20 72516039
1933; Tish Collins
J D Bernal Peace Coll, John Williamson US Labour Movement American Coll, Spanish Civil War and Int Brigade Coll, Klugmann early radicals and chartists
150 000 vols; 120 curr per; 100 mss; 30 diss/theses; 100 govt docs; 50 music scores; 10 microforms; 50 av-mat; 40 sound-rec; 3 digital data carriers; 15 000 photos 29790

Marylebone Cricket Club, Library, Lord's Ground, ***London*** NW8 8QN
T: +44 20 76168559;
Fax: +44 20 76168659; E-mail: neil.robinson@mcc.org.uk
1893; Stephen E.A. Green
15 000 vols; 20 curr per; 1 000 mss; 20 diss/theses; 10 govt docs; 50 music scores; 20 maps; 100 sound-rec 29791

Museum of London, 150 London Wall, ***London*** EC2Y 5HN
T: +44 8704443852;
Fax: +44 8704443853; E-mail: library@museumoflondon.org.uk; URL: www.museumoflondon.org.uk
1977; Sally Brooks
The W.G. Bell coll (the plague & the Great Fire of London), The Sir Richard Tangye coll (Cromwelliana), Warwick Wroth scrapbooks (pleasure gardens); Printed Ephemera department containing Whitefriars glass, Kiralfy and suffragette collections
40 000 vols; 70 curr per; 20 mss; 700 maps 29792

National Aerospace Library, The Hub, Farnborough Business Park, ***London*** W1J 7BQ
T: +44 252 701038, 701039; E-mail: hublibrary@aerosociety.com
2008
Aviation History Journals, ARC and NACA Technical Reports, Royal Aircraft Establishment (RAE) Reports, Technical Reports at the National Aerospace Library, SBAC Minutes
10 000 vols 29793

National Army Museum, Library, Royal Hospital Rd, ***London*** SW3 4HT
T: +44 20 77300717; Fax: +44 20 78236573; E-mail: info@nam.ac.uk; URL: www.national-army-museum.ac.uk
1960; Michael Ball
British military hist, Indian army to 1947, Colonial armies
51 000 vols; 260 curr per; 1 e-journals; 800 mss; 30 000 govt docs; 100 music scores; 300 maps; 400 microforms; 3 000 sound-rec; 25 000 prints, 650 000 photos 29794

National Art Library (NAL), Victoria and Albert Museum, Cromwell Rd, South Kensington, ***London*** SW7 2RL
T: +44 20 79422400;
Fax: +44 20 79422401; E-mail: nal.enquiries@vam.ac.uk; URL: www.vam.ac.uk/nal
1837; Joha Meriton
Special colls: Dyce and Forster (gen lit), Clements (armorial bindings), Eugène Piot (bks on pageants), Hutton (fencing, swordsmanship), Larionov, Linder bequest (Beatrix Potter material), Rakoff (comics); Archive of Art and Design
1 000 000 vols; 2 114 curr per; 140 incunabula; 150 diss/theses; 350 digital data carriers; 300 illuminated codices, 60 000 exhibition cat, 50 000 sales cat, 1 million archival docs
IFLA, ARLIS/UK & Ireland, CILIP, M25 29795

National Children's Bureau, Library, 8 Wakley St, ***London*** EC1V 7QE
T: +44 20 78436000; Fax: +44 20 78436007; E-mail: library@ncb.org.uk; URL: www.ncb.org.uk
1965; Nicola Hilliard
15 000 vols; 350 curr per; 1 e-journals; 1 500 govt docs; 5 digital data carriers
ASLIB, CILIP 29796

National Gallery, Libraries and Archives, Trafalgar Sq, ***London*** WC2N 5DN
T: +44 20 77472830; Fax: +44 20 77472892; E-mail: lad@ng-london.org.uk
1869; Elspeth J. Hector
80 000 vols; 150 curr per; 5 e-journals; 2 incunabula 29797

National Institute for Medical Research, Library & Information Service, The Ridgeway, Mill Hill, ***London*** NW7 1AA
T: +44 20 88162228; Fax: +44 20 88162230; E-mail: library@nimr.mrc.ac.uk; URL: www.nimr.mrc.ac.uk/library
1920; Frank Norman
Molecular Biology, Cell Biology, Neuroscience, Structural Biology, Developmental Biology; Medical Research Council publications, NIMR Scientific Archives
20 000 vols; 10 curr per; 1 500 e-journals; 10 e-books; 400 diss/theses; 500 govt docs
CHILL, RESCOLINC, EAHIL, UKSG 29798

National Institute for Social Work, 5 Tavistock Place, ***London*** WC1H 9SN
T: +44 20 73879681; Fax: +44 20 73877968
1961; J. McTernan
20 000 vols; 230 curr per
libr loan; ASLIB 29799

National Institute of Economic and Social Research, Library, 2 Dean Trench St, Smith Sq, ***London*** SW1P 3HE
T: +44 20 76541907; Fax: +44 20 76541900; E-mail: library@niesr.ac.uk; URL: www.niesr.ac.uk
1938; Patricia Oliver
5 000 vols; 300 curr per; 2 000 govt docs; 10 digital data carriers 29800

National Maritime Museum, Caird Library, Greenwich, ***London*** SE10 9NF
T: +44 20 83126516; Fax: +44 20 83126599; E-mail: library@nmm.ac.uk; URL: www.nmm.ac.uk
1934; Jill Davies
Maritime hist, art and technology, piracy, shipwrecks, crew lists, Lloyd's surveys, papers of naval personnel, admiralty and navy board recs, master certificates, navy list, Lloyd's List, Lloyd's Register
120 000 vols; 200 curr per; 55 000 mss; 18 incunabula; 2 000 microforms; 50 av-mat; 50 digital data carriers; 5 397 linear metres of mss, 4 miles mss
ASLIB, MA 29801

National Portrait Gallery, Heinz Archive and Library, 39-40 Orange St, ***London*** WC2H 0HE
T: +44 20 73216617;
Fax: +44 20 73060056; E-mail: archiveenquiry@npg.org.uk; URL: www.npg.org.uk/live/archive.asp
Robin Francis
Reference coll of portraits, George Scharf Lib incl sketchbooks and notebooks
35 000 vols; 65 curr per; 12 e-journals; 300 000 photos and illustrates, 50 mss coll, 900 autograph letters
ARLIS/UK & Ireland 29802

National Society for the Prevention of Cruelty to Children (NSPCC), Safeguarding Information and Library Services, Weston House, 42 Curtain Road, ***London*** EC2A 3NH
T: +44 20 78252706; E-mail: library@nspcc.org.uk; URL: www.nspcc.org.uk/inform
1974; Karen Childs Smith
Child protection, child abuse and neglect, child abuse prevention
10 000 vols; 100 curr per; 20 e-journals; 280 av-mat
ALISS, CHILL 29803

Natural History Museum, Library & Information Services, Cromwell Rd, ***London*** SW7 5BD
T: +44 20 79425460; Fax: +44 20 79425559; E-mail: library@nhm.ac.uk; URL: www.nhm.ac.uk
1881; Graham Higley
Linnaeus Coll, Owen Coll, Walsingham Coll, Murray Coll, Tweeddale Coll, Rothschild Library, Kohler Darwin Coll, Wallase Coll
1 000 000 vols; 9 230 curr per; 256 e-journals; 100 000 mss; 100 incunabula; 150 000 maps; 5 000 microforms; 5 000 av-mat; 50 sound-rec; 500 digital data carriers; 500 000 drawings/artworks
libr loan; RLG 29804

New Zealand High Commission Library, 80 Haymarket, ***London*** SW1Y 4TQ
T: +44 20 79308422;
Fax: +44 20 78394580; URL: www.newzealandhc.org.uk
Cate Maccreade
8 000 vols 29805

Polish Institute and Sikorski Museum, Reference Library, 11 Leopold Rd, ***London*** W5 3PB; mail address: 20 Princes Gate, London SW7 1PT
T: +44 20 89926057; Fax: +44 20 89926057
1965; Aleksander J. Szkuta
Poland, Hist of Polish Forces, Mainly 1st and 2nd World War
20 000 vols; 5 curr per; 40 000 microforms 29806

The Polish Library, Polish Social and Cultural Association, 238-246 King St, ***London*** W6 0RF
T: +44 20 87410474;
Fax: +44 20 87417724; E-mail: library@polishlibrary.co.uk
1942; Jadwiga Szmidt
Polish emigré publs, Conradiana, Polish underground, 'Solidarność', uncensored publs, bookplates; CCL (Central Circulating Library) – lending books to British libraries
156 000 vols; 4 622 curr per; 1 990 mss; 150 diss/theses; 2 086 music scores; 1 383 maps; 49 000 photos, 1 744 printed ephemera
COSEELIS, SKMABPZ 29807

Rockefeller Medical Library, University College London & University College London Hospitals Foundation Trust, UCL Institute of Neurology, National Hospital for Neurology & Neurosurgery, Queen Sq, ***London*** WC1N 3BG
T: +44 20 78298709; E-mail: library@ion.ucl.ac.uk; URL: www.iðn.ucl.ac.uk/library/
1950; Louise Shepherd
Historical Colls and Archives (approx 4 000 monographs, 8 000 reprints, 600 case notes, 1 500 films, artifacts and photos). These are related to neurology, neurosurgery, neuroscience and the history of the National Hospital, the Institute of Neurology and Queen Square
21 000 vols; 120 curr per; 200 diss/theses
UMHLG 29808

Royal Academy of Arts, Library, Burlington House, Piccadilly, ***London*** W1J 0BD
T: +44 20 7300 5740;
Fax: +44 20 7300 5765; E-mail: library@royalacademy.org.uk; URL: www.royalacademy.org.uk
1768; Adam Waterton
Historic Book Collection – ca. 12,000 Books which entered the Royal Academy of Arts Library since its Foundation in 1768 to 1920; Illustrated Book Collection – ca. 2000 books illustrated or designed by British artists
40 000 vols; 15 curr per; 2 e-journals; 10 000 mss
ARLIS/UK & Ireland 29809

Royal Aeronautical Society, Library, 4 Hamilton Place, ***London*** W1J 7BQ
T: +44 20 76704362;
Fax: +44 20 76704359; E-mail: brian.riddle@aerosociety.com; URL: www.aerosociety.com
1866; Brian L. Riddle
Coll of material relating to the development and recent technical advances in aeronautics, aviation and aerospace technology; Cuthbert-Hodgson, Poynton and Maitland colls of early ballooning, airships and other early aeronautical material; letters, papers and mss of Sir George Cayley, John Stringfellow, Wilbur Wright, Orville Wright, Katharine Wright, Lawrence Hargrave, Major B.F.S. Baden-Powell, C.G. Grey and F.S. Barnwell, files of papers relating to the hist of the society itself; photographic glass lantern slide, lithographic coll of aviation images (over 100 000) from the early days of ballooning, airships and before; civil and military aircraft, space, rocketry, missiles through to modern technology aircraft of today, incl a number of portrait photogr of aviation personalities
30 000 vols; 1 000 curr per; 2 000 mss; 10 diss/theses; 40 000 govt docs; 20 music scores; 10 microforms; 50 av-mat; 100 sound-rec; 100 000 photos, 120 000 slides, 6 000 pamphlets and clippings, 20 000 tech reports 29810

Royal Air Force Museum, Library, Hendon, Grahame Park Way, ***London*** NW9 5LL
T: +44 20 82052266;
Fax: +44 20 82001751; E-mail: research@rafmuseum.org; URL: www.rafmuseum.org
1963; P.J.V. Elliott
100 000 vols; 300 curr per; 3 000 maps; 5 000 microforms 29811

Royal Arsenal (West), James Clavell Library, Warren Lane, Woolwich, ***London*** SE18 6ST
T: +44 20 83127125; E-mail: paule@firepower.org.uk; URL: www.firepower.org.uk
1840; Mr M.P. Evans
First World War regimental war diaries, Second World War regimental war diaries, British artillery equipments handbooks, post war regimental records; Photograph coll
22 000 vols; 40 curr per; 3 450 mss; 200 maps; 4 000 microforms; 40 digital data carriers 29812

Royal Asiatic Society Library, Royal Asiatic Society of Great Britain and Ireland, 14 Stephenson Way, ***London*** NW1 2HD
T: +44 20 73884539;
Fax: +44 20 73919429; E-mail: library@royalasiaticsociety.org; URL: www.royalasiaticsociety.org
1823; Kathy Lazenbatt
Storey coll of Persian Literature, Sir Richard Burton coll
80 000 vols; 400 curr per; 1 500 mss; 450 maps; 2 000 paintings & drawings 29813

Royal Astronomical Society, Library, Burlington House, Piccadilly, ***London*** W1V 0NL
T: +44 20 77344582; Fax: +44 20 74940166; E-mail: info@ras.org.uk; URL: www.ras.org.uk
1820; P.D. Hingley
Geophysics
27 000 vols; 270 curr per
libr loan; BLLD 29814

Royal Automobile Club (RAC), Library, 89 Pall Mall, ***London*** SW1Y 5HS
T: +44 20 77473398;
Fax: +44 8704606285; E-mail: library@royalautomobileclub.co.uk; URL: www.royalautomobileclub.co.uk
1898; Trevor G. Dunmore
Motoring history; Royal Automobile Club – Badges Coll; Club Archive, Churchill College, Cambridge, UK
15 000 vols; 150 curr per; 125 maps; 250 microforms; 10 av-mat; 2 digital data carriers; In-house motoring catalog
SLA, CILIP, MLA London 29815

Royal College of Obstetricians and Gynaecologists, Information Services, 27 Sussex Place, ***London*** NW1 4RG
T: +44 20 77726309; Fax: +44 20 72628331; E-mail: library@rcog.org.uk; URL: www.rcog.org.uk/what-we-do/information-services
1933; Lucy Reid
Historical coll, college archive, deposited archives, museum collection
14 000 vols; 200 curr per; 170 e-journals; 10 e-books; 5 mss; 2 incunabula
libr loan 29816

Royal Geographical Society (with the Institute of British Geographers), Library, 1 Kensington Gore, ***London*** SW7 2AR
T: +44 20 75913040; Fax: +44 20 75913001; URL: www.rgs.org
1830; E. Rae
special colls: Fordham (roadbks), Brown (Morocco), Feilden (Polar), Hotz (Persia); Archives, picture library, Map room
150 000 vols; 800 curr per; 1 000 000 maps; 500 000 images in the picture libr
libr loan 29817

Royal Horticultural Society, Lindley Library, 80 Vincent Sq, ***London*** SW1P 2PE
T: +44 20 78213050;
Fax: +44 20 78283022; E-mail: library.london@rhs.org.uk; URL: www.rhs.org.uk/learning/library
1867; Barbara Collecott
Gardening, garden hist, horticulture, gardens; Nursery cats, Hist gardening journals
50 000 vols; 400 curr per; 2 e-journals; 150 mss; drawings
EBHL 29818

Royal Institute of International Affairs, Chatham House Library, Chatham House, 10 St. James's Sq, ***London*** SW1Y 4LE
T: +44 20 79575723;
Fax: +44 20 79575710; E-mail: libenquire@chathamhouse.org.uk; URL: www.chathamhouse.org.uk/library
1920; Mary Bone
International Relations, Organizations; International Law; RIIA Press Clippings coll 1924-1997, Chatham House archives
140 000 vols; 300 curr per; 20 diss/theses; 45 000 govt docs; 300 maps; 600 microforms; 35 digital data carriers; 5 mio clippings
EINIRAS, AUKML, COSEELIS 29819

Royal Institution of Chartered Surveyors, Library, 12 Great George St, Parliament Sq, ***London*** SW1P 3AD
T: +44 870 3331600; Fax: +44 207 3343784; E-mail: library@rics.org; URL: www.rics.org/library
1868; Cathy Linacra
Building and construction, surveying, photogrammetry, valuation, measurement, property, real estate, land management
30 000 vols; 400 curr per; 10 000 e-journals; 800 diss/theses; 4 000 govt docs; 50 maps; 300 av-mat; 200 sound-rec; 20 digital data carriers; 2 000 hist bks
libr loan; BLL 29820

Royal Institution of Great Britain, Library, 21 Albemarle St, ***London*** W1S 4BS
T: +44 20 74092992; Fax: +44 20 76293569; E-mail: kdodd@ri.ac.uk; URL: www.rigb.org
1799; N.N.
Science and technology, science hist, technological hist, mss of 19th and 20th c scientists connected with The Royal Institution C. Davy, Faraday, Tyndall, Dewar, W.H. and W.L. Bragg
90 000 vols; 100 curr per 29821

Royal Institution of Naval Architects, Denny Library, 10 Upper Belgrave St, *London* SW1X 8BQ
T: +44 20 72354622; Fax: +44 20 72595912; E-mail: hq@rina.org.uk; URL: www.rina.org.uk
1860; J. Adam
Marine technology and engineering, naval architecture, oceanography, Scott (naval archit 1600)
10 000 vols; 100 curr per; 100 govt docs; 20 digital data carriers   29822

Royal Pharmaceutical Society of Great Britain, 1 Lambeth High Street, *London* SE1 7JN
T: +44 20 75722300; Fax: +44 20 75722499; E-mail: library@rpsgb.org; URL: www.rpsgb.org
1841
Proprietary books (worldwide); Pharmacopoeias (worldwide); Early Printed Collection (including the Hanbury Library, and a collection of herbals)
65 000 vols; 250 curr per; 7 e-journals; 5 000 mss; 5 incunabula; 350 diss/theses; 500 govt docs; 200 av-mat; 40 digital data carriers; 15 000 pamphlets
libr loan; BLL, ASLIB, EAHIL   29823

Royal Society, Centre for History of Science, 6-9 Carlton House Terrace, *London* SW1Y 5AG
T: +44 20 74512606;
Fax: +44 20 79302170; E-mail: library@royalsociety.org; URL: royalsociety.org
1660; Keith Moore
Science hist, technological hist, biogr info on fellows of the Royal Society, science policy
145 000 vols; 300 curr per; 10 e-journals; 120 000 mss; 50 incunabula; 100 diss/theses; 12 000 govt docs; 200 maps; 600 microforms; 100 sound-rec; 40 digital data carriers; artefacts, portraits, engravings, busts, 2 000 ft of manuscripts
libr loan   29824

Royal Society for the Encouragement of Arts, Manufactures and Commerce (RSA), Library, 8 John Adam St, *London* WC2N 6EZ
T: +44 20 74516874; Fax: +44 20 78395805; E-mail: library@rsa.org.uk; URL: www.rsa.org.uk
1754; Julie Cranage
Early Libr (500 bks published before 1830); Arch (from 1754 to present)
10 000 vols; 30 curr per; 10 000 mss; 100 digital data carriers
libr loan; BLL   29825

Royal Society of Chemistry, Library and Information Centre, Burlington House, Piccadilly, *London* W1J OBA
T: +44 20 74378656; Fax: +44 20 72879798; E-mail: library@rsc.org; URL: www.rsc.org/library
1841; Nigel Lees
Nathan coll on explosives
70 000 vols; 650 curr per; 20 mss; 100 microforms; 22 digital data carriers; 8 000 images in historical chemistry
ASLIB   29826

Royal Society of Medicine, Library, 1 Wimpole St, *London* W1G OAE
T: +44 20 72902934;
Fax: +44 20 72902939; E-mail: collection.management@rsm.ac.uk; URL: www.rsm.ac.uk
1805; Wayne Sime
Hugh Diamond photos, Hist Coll, Portrait Coll (Doctors/Scientists)
50 000 vols; 1 300 curr per; 856 e-journals; 500 mss; 20 incunabula; 2 000 govt docs; 4 digital data carriers
libr loan; MLA, EAHIL   29827

Royal Statistical Society, University College London, Science Library, D.M.S. Watson Bldg, Gower St, *London* WC1E 6BT
T: +44 20 73877050 ext 2628; Fax: +44 20 73807373
1834; Dilip Chatarji
18 000 vols; 40 curr per
libr loan   29828

The Saison Poetry Library, Level 5, Royal Festival Hall, *London* SE1 8XX
T: +44 20 79210943;
Fax: +44 20 79210607; E-mail: info@poetrylibrary.org.uk; URL:

www.poetrylibrary.org.uk
1953; Chris McCabe, Miriam Valencia
Press cuttings on contemporary poetry, recordings of poets reading their work, Sylvia Townsend Warner Bequest, Poster Poems, Poem Cards; Childrens Poetry dept, Poetry in Education dept
100 000 vols; 2 000 curr per; 5 e-journals; 20 diss/theses; 20 250 av-mat; 18 000 sound-rec; 30 digital data carriers
libr loan; CILIP, ASLIB   29829

Science Museum Library, Imperial College Rd, South Kensington, *London* SW7 5NH
T: +44 20 79424242;
Fax: +44 20 79424243; E-mail: smlinfo@sciencemuseum.org.uk; URL: www.sciencemuseum.org.uk/library
1883; Rupert Williams
Science hist, technological hist, medical hist, Science Museum publications
120 000 vols; 80 curr per; 50 e-journals; 1 e-books; 1 000 mss; 50 diss/theses; 400 maps; 4 000 microforms; 80 av-mat; 50 sound-rec; 50 vido rec
libr loan; CILIP   29830

Society for Cooperation in Russian and Soviet Studies, 320 Brixton Rd, *London* SW9 6AB
T: +44 20 72742282; Fax: +44 20 72743230; E-mail: ruslibrary@scrss.org.uk; URL: www.scrss.org.uk
1924; J. Cunningham
Art Book & Photo Coll
35 000 vols; 200 music scores; 100 maps; 60 000 av-mat; 1 000 sound-rec; 1 000 pamphlets, newspaper clippings   29831

Society of Antiquaries of London, Library, Burlington House, Piccadilly, *London* W1J OBE
T: +44 20 74797084; Fax: +44 20 72876967; E-mail: library@sal.org.uk; URL: www.sal.org.uk
1707; Ms H.L. Rowland
Archaeology, especially British, architectural history and the decorative arts (especially medieval), heraldry and older works on British local history
150 000 vols; 850 curr per; 1 000 mss; 51 incunabula; 30 000 prints and drawings
libr loan   29832

Society of Genealogists, Library, 14 Charterhouse Bldgs, Goswell Rd, *London* EC1M 7BA
T: +44 20 72518799; Fax: +44 20 72501800; E-mail: librarian@sog.org.uk; URL: www.sog.org.uk
1911; Susan Gibbons
Family hist, Topography, Biography, Local hist, Heraldry
105 000 vols; 500 curr per; 6 e-journals; 20 000 mss; 2 500 maps; 23 000 microforms; 22 av-mat; 200 sound-rec; 2 000 digital data carriers   29833

Sound and Music, Library, Somerset House, The Strand, *London* WC2R 1LA
T: +44 20 77591800; E-mail: info@soundandmusic.org; URL: www.bmic.co.uk
1967; M. Greenall
Background information on composers, press clippings, programme notes, publishers' leaflets, biographies, lists of works, videos
32 000 vols; 6 diss/theses; 32 000 music scores; 30 av-mat; 16 000 sound-rec
IAML, IAMIC   29834

St Bride Library, Corporation of London Libraries, Bride Lane, Fleet St, *London* EC4Y 8EE
T: +44 20 73534660; Fax: +44 20 75837073; URL: www.stbride.org
1895; Nigel Roche
Eric Gill Coll, Coll of manuals on shorthand, incl 3 000 bks on shorthand writing, Type specimens coll, broadsides, printing trade documents
50 000 vols; 250 curr per   29835

Tate Library and Archive, Millbank, *London* SW1P 4RG
T: +44 20 78878838;
Fax: +44 20 78878901; E-mail: reading.rooms@tate.org.uk; URL: www.tate.org.uk
1950; Jane Bramwell
Artists' books, British art 16th c to

present, modern and contemporary art (British and international) See also Tate Archive's collection – archive of 20th c British art; Tate Gallery Records; Tate Archive and Gallery Records are related departments
250 000 vols; 250 curr per; 170 000 exhibition catalogs, plus auction catalogs, press cuttings, printed ephemera
libr loan; ARLIS/NA, ARLIS/UK & Ireland, SCONUL, OCLC, COPAC   29836

Tavistock and Portman NHS Trust, Tavistock Library, 120 Belsize Lane, *London* NW3 5BA
T: +44 20 74473776; Fax: +44 20 74473734; E-mail: library@tavi-port.org; URL: www.tavi-port.org
1946; Angela Douglas
24 000 vols; 300 curr per; 600 diss/theses; 20 microforms; 500 av-mat; 350 sound-rec
libr loan; BLDSC, ASLIB, CILIP   29837

Tower Hamlets Local History Library and Archives, 277 Bancroft Rd, *London* E1 4DQ
T: +44 20 73641290;
Fax: +44 20 73641292; E-mail: localhistory@towerhamlets.gov.uk; URL: www.ideastore.co.uk
1965; N.N.
10 000 vols; 40 curr per; 50 diss/theses; 2 500 maps; 2 000 microforms; 5 050 av-mat; 120 sound-rec; 50 digital data carriers; 9 000 pamphlets, 200 cubic m of archives   29838

Tower Hamlets Schools Library Service, English St, *London* E3 4TA
T: +44 20 73646428; Fax: +44 20 73646422; URL: www.towerhamlets-sls.org.uk
1991; Gillian Harris
170 000 vols; 30 curr per   29839

V&A Theatre & Performance Department Library (formerly Theatre Museum Library), V&A Museum, Blythe House, 23 Blythe St, *London* W14 0QX
T: +44 20 79422697;
Fax: +44 20 74719864; E-mail: tmenquiries@vam.ac.uk; URL: www.vam.ac.uk/collections/theatre_performance
1924; Ms C. Hudson
Performing arts, Theatre, Drama, Dance, Opera, Rock and Pop, Puppets, Circus, Pantomime, Musicals; Stage designs, photos, posters, costumes playbills, theatre programmes, newspaper cuttings; National Video Archive of Performance
150 000 vols; 30 curr per; 500 av-mat; 300+ named special colls (www.backstage.ac.uk)
SIBMAS   29840

Visit Britain Library, Thames Tower, Blacks Rd, *London* W6 9EL
T: +44 20 85633011; Fax: +44 20 85633391; URL: www.staruk.org.uk
1969; Gaynor Evans
17 000 vols; 100 curr per; 500 maps
ASLIB, CILIP   29841

Wellcome Library for the History and Understanding of Medicine, Wellcome Trust, 183 Euston Rd, *London* NW1 2BE
T: +44 20 76118722; Fax: +44 20 76118360; E-mail: library@wellcome.ac.uk; URL: library.wellcome.ac.uk
1895; Frances Norton
Oriental Coll, Iconographic Coll, Early Printed Bks, Arch and Manuscripts
600 000 vols; 837 curr per; 150 e-journals; 20 000 mss; 715 incunabula; 650 microforms; 1 179 av-mat; 888 sound-rec; 70 digital data carriers; 100 000 prints, drawings, paintings, photos, films, 50 000 pamphlets   29842

Westminster Music Library, Westminster City Libraries, 160 Buckingham Palace Road, *London* SW1W 9UD
T: +44 20 76414292;
Fax: +44 20 76414281; E-mail: musiclibrary@westminster.gov.uk; URL: www.westminster.gov.uk/libraries
1948; Ruth Walters
Pre 1800 coll of printed music
18 000 vols; 134 curr per; 800 mss; 28 000 music scores; 10 digital data carriers; Orchestral Sets
libr loan; IAML   29843

Wiener Library Institute of Contemporary History, 4 Devonshire St, *London* W1W 5BH
T: +44 20 76367247;
Fax: +44 20 74366428; E-mail: library@wienerlibrary.co.uk; URL: www.wienerlibrary.co.uk
1933; Katharina Hübschmann
Holocaust, Third Reich, antisemitism, facism; Eyewitness accounts and family docs from Holocaust survivors and refugees, illegal ant-Nazi pamphlets, press arch re. the Holocaust, its causes and legacies; Photo Arch
60 000 vols; 200 curr per; 2 e-journals; 1 200 mss; 150 diss/theses; 10 000 microforms; 3 000 av-mat; 50 digital data carriers; 12 000 photos
M25 Consortium of Academic Libraries   29844

Zoological Society of London, Library, Regent's Park, *London* NW1 4RY
T: +44 20 74496293; Fax: +44 20 75865743; E-mail: library@zsl.org; URL: www.zsl.org
1826; A. Sylph
Zoology, animal conservation; Arch of the Zoological Society of London; zoo rpts, magazines and newsletters from around the world
160 000 vols; 1 300 curr per; 50 e-journals; 50 diss/theses; Zoological Record
libr loan   29845

Institution of Gas Engineers and Managers (IGEM), Library, Charnwood Wing, Ashby Rd, *Loughborough* LE11 3 GH
T: +44 1509 282728; Fax: +44 1509 283110; E-mail: general@igem.org.uk; URL: www.igem.org.uk
1869; Anita Witten
10 000 vols; 100 curr per; 10 000 reports
libr loan   29846

Wardown Park Museum, Wardown Park, *Luton* LU2 7HA
T: +44 1582 546722;
Fax: +44 1582 546763; E-mail: Elizabeth.adey@Lutonculture.com
1931; Dr. Elizabeth Adey
Bagshawe
12 000 vols; 12 curr per; 50 mss; 10 diss/theses; 20 govt docs; 100 music scores; 100 maps; 200 av-mat; 200 sound-rec   29847

Centre for Kentish Studies, Sessions House, County Hall, *Maidstone* ME14 1XQ
T: +44 1622 694363; Fax: +44 1622 694379; E-mail: archives@kent.gov.uk
P. Rowsby
35 000 vols   29848

DERA Malvern, Library, St. Andrews Rd, *Malvern* WR14 3PS
T: +44 1684 894616; Fax: +44 1684 894148
David Little
Electronics, radar, electro-optics, signal processing, microwaves
40 000 vols; 700 curr per; 3 000 maps; 100 digital data carriers; 100 000 reports
libr loan; ASLIB   29849

Centre for Local Economic Strategies, Express Networks, Library, 1 George Leigh St, *Manchester* M4 5DL
T: +44 161 2367036; Fax: +44 161 2361891; E-mail: info@cles.org.uk; URL: www.cles.org.uk
1986; Victoria Bradford
13 000 vols; research and commitee rpts   29850

Chetham's Library, Long Millgate, *Manchester* M3 1SB
T: +44 161 8347961;
Fax: +44 161 8395797; E-mail: librarian@chethams.org.uk; URL: www.chethams.org.uk
1653; Michael R. Powell
Halliwell-Phillipps Broadside Coll (3 000 items), John Byrom Coll (3 000 vols)
100 000 vols; 51 curr per; 800 mss; 90 incunabula; 1 000 music scores; 150 maps; 60 microforms; 2 000 av-mat; 100 000 clippings, 7 000 pamphlets   29851

Christie Hospital NHS Trust, Kostoris Medical Library, Wilmslow Rd, Withington, **Manchester** M20 4BX
T: +44 161 4463452; Fax: +44 161 4463454; E-mail: enquiries@christie-tr.nwest.nhs.uk; URL: www.christie.nhs.uk
1932; Steve Glover
Oncology, palliative care
13 500 vols; 230 curr per; 12 digital data carriers   29852

National Co-operative Archive, Co-operative College, Library, Holyoake House, Hanover Street, **Manchester** M60 0AS
T: +44 161 2462926; Fax: +44 161 2462946; URL: www.co-op.ac.uk/our-heritage/national-co-operative-archive/
1911; Gillian Lonergan
Robert Owen, G.J. Holyoake a Edward Owen Greening correspond a doc(s), co-operative women's guild, Rochdale equitable pioneers society, hist of co-operative movement, credit unions, co-operative Party, co-operative Youth Groups, co-operative Societies-minutes & records, Christian Socialists, National co-operative Oral History Archive, National co-operative film Archive
12 000 vols; 200 curr per; 30 000 mss   29853

Portico Library, 57 Mosley St, 1st Floor, **Manchester** M2 3HY
T: +44 161 2366785;
Fax: +44 161 2366803; E-mail: librarian@theportico.org.uk; URL: www.theportico.org.uk
1806; Emma Marigliano
Travel and Tourism, North West Fiction, 19th century literature, architecture, travel, biography, history, topography, reference, 'Polite Literature', Local History and North West English fiction
25 000 vols; 40 curr per; 6 diss/theses; Manuscript and printed catalogues (with supplements) of holdings between 1810 and 1892, Minute Books from 1806, Issue Books 1851-1906, Visitors Books, Receipts and Invoices from 1806, Archive of some works and correspondence of Tinsley Pratt (former Librarian), assorted other archives relating to the Library Association of Independent Libraries, Historic Libraries Forum   29854

Corus UK Limited, Teesside Technology Centre, Library & Information Services, Grangetown, P.O. Box 11, **Middlesbrough** TS6 6UB
T: +44 1642 467144;
Fax: +44 1642 460321; E-mail: Carol.Patton@corusgroup.com
Mrs C. Patton
Engineering, metallurgy
10 000 vols; 300 curr per; 1000 pamphlets, iron and steelmaking topics conference proceedings   29855

Literary and Philosophical Society of Newcastle upon Tyne, Library, 23 Westgate Rd, **Newcastle upon Tyne** NE1 1SE
T: +44 191 2320192;
Fax: +44 191 2612885; E-mail: litphil.library@btinternet.com
1793; E.A. Pescod
140 000 vols; 156 curr per; 100 maps; 12 000 sound-rec; 5 500 pamphlets
libr loan; CILIP, AIL   29856

Natural History Society of Northumbria, c/o Great North Museum: Hancock, Barras Bridge, **Newcastle upon Tyne** NE2 4PT
T: +44 191 2223555;
Fax: +44 191 2322177; E-mail: gnmlibrary@twmuseums.org.uk; URL: www.nhsn.ncl.ac.uk/resources-library.php
1829; Dr. L. Jessop
Thomas Bewick watercolour and pencil drawing coll, Natural history archives relating to the society from 1829 – to date, Albany Hancock Nudibranch drawings
12 000 vols; 157 curr per; 300 mss; 17 diss/theses; 290 maps; 303 pamphlets, 1 631 offprints   29857

Society of Antiquaries of Newcastle upon Tyne, Library, Black Gate, Castle Garth, **Newcastle upon Tyne** NE1 1RQ
T: +44 191 2615390
1825; D. Peel

15 000 vols; 75 curr per
libr loan   29858

Glan Hafren NHS Trust, Library, The Friars, Friars Rd, **Newport, South Wales** NP9 4EZ
T: +44 1633 238134; Fax: +44 1633 238123
1951; Ms J.L. Grey-Lloyd
Arch of research reports papers, articles written by trust staff
13 000 vols; 161 curr per; 20 digital data carriers; 5 000 books + pamphlets, 200 videotapes
AWHL, BLDSC, CILIP, AWHILES   29859

Princess Marina Library,
Northamptonshire Teaching Primary Care NHS Trust, Princess Marina Hospital, 3 Alexandra Close, **Northampton** NN5 6UH
T: +44 1604 595267;
Fax: +44 1604 586056; E-mail: pmhlibrary@northants.nhs.uk
G. Meades
Mental health, learning disabilities, nursing, health promotion
10 000 vols; 117 curr per; 450 leaflets & posters, health promotion resources (1 620 items)
libr loan   29860

The Operations Centre Library, Norwich Bioscience Institutes, Colney, **Norwich** NR4 7UH
T: +44 1603 450670; Fax: +44 1603 450045; E-mail: toc.library@bbsrc.ac.uk; URL: www.nbi.bbsrc.ac.uk
1910; Kate West
Botanical bks, hist of genetics coll
20 000 vols; 1 000 curr per; 500 e-journals; 1 000 diss/theses; 40 000 reprints; 10 000 pamphlets
CILIP   29861

British Geological Survey, Research Knowledge Services, Kingsley Dunham Centre, **Nottingham** NG12 5GG
T: +44 115 9363205; Fax: +44 115 9363200; E-mail: libuser@bgs.ac.uk; URL: www.bgs.ac.uk
1840; Ken Hollywood
Geology, paleontology, nat repository of geological arch (25 000), photos of British regional geology and scenery (75 000), Murchison and Ramsay pamphlet coll
500 000 vols; 11 500 curr per; 15 000 mss; 100 diss/theses; 200 000 maps; 500 microforms; 6 digital data carriers; 40 000 pamphlets
ASLIB, CILIP   29862

e.on UK plc, Power Technology, Library and Information Centre, Ratcliffe on Soar, **Nottingham** NG11 0EE
T: +44 115 9362360; Fax: +44 115 9362711; E-mail: sue.seal@eon-uk.com; URL: www.eon-uk/powertechnology
1961; Susan Seal
Metallurgy, corrosion, chem eng, power station, world electricity, utility information
30 000 vols; 100 curr per; 10 e-journals; 200 diss/theses; 400 maps; 16 000 microforms; 800 av-mat; 60 digital data carriers; 40 000 pamphlets
Aslib   29863

Nottingham Subscription Library Ltd, Bromley House, Angel Row, **Nottingham** NG1 6HL
T: +44 115 9473134
1816; Ms J.V. Wilson
Hist, theology, poetry, local hist, travel
31 000 vols
Available to the public for reference only, by prior appointment   29864

Nottinghamshire Healthcare NHS Trust, Duncan Macmillan House, Staff Library, Porchester Rd, **Nottingham** NG3 6AA
T: +44 115 9691300 ext 11186;
Fax: +44 115 9691882; E-mail: dmhstafflibrary@nottshc.nhs.uk
1960; Victoria Boskett
Psychiatry, mental health, learning disability
11 000 vols; 63 curr per; pamphlets, AV materials
libr loan   29865

Oldham Local Studies and Archives, 84 Union St, **Oldham** OL1 1DN
T: +44 161 9114654;
Fax: +44 161 9114664; E-mail: local.studies@oldham.gov.uk; URL: www.oldham.gov.uk/community/local.studies
1885; T. Berry
13 600 vols; 13 curr per; 4 500 mss; 3 008 maps; 3 527 microforms; 136 sound-rec; 33 digital data carriers; 19 000 photos, 8 000 pamphlets   29866

Ulster American Folk Park, Centre for Migration Studies, Mellon Rd, Castletown, **Omagh** BT78 5QY
T: +44 28 82256315; Fax: +44 28 82242241; E-mail: centremigstudies@ni-libraries.net; URL: www.qub.ac.uk/cms
1981; Christine Johnston
Emigration, Ulster folk-life
12 000 vols; 30 curr per; 48 diss/theses; 1 349 maps; 120 microforms; 117 sound-rec; 46 digital data carriers; 40 video recordings
Library & Information Services Council, Northern Ireland   29867

Pilkington European Technical Centre Lathom, Information Resource Centre, Library, Hall Lane, Lathom, **Ormskirk** L40 5UF
T: +44 1695 54306; Fax: +44 1695 54366
1960; D Bennett
Glass technology
10 000 vols; 250 curr per; 200 diss/theses; 5 000 microforms; 30 000 reports   29868

Institute of Actuaries, Library, Napier House, 4 Worcester St, **Oxford** OX1 2AW
T: +44 1865 268208; Fax: +44 1865 268211; E-mail: libraries@actuaries.org.uk; URL: www.actuaries.org.uk
1848; Sally Grover
Historic collection on Actuarial Science
13 000 vols; 200 curr per; 20 mss; 1 000 govt docs
CILIP, ASLIB   29869

Museum of the History of Science, University of Oxford, Library, Old Ashmolean Bldg, Broad St, **Oxford** OX1 3AZ
T: +44 1865 277278; Fax: +44 1865 277288; E-mail: library@mhs.ox.ac.uk; URL: www.mhs.ox.ac.uk
1924; Gemma Wright
Hist of scientific instruments, Lewis Evans coll, R.T. Gunther coll, H.E. Stapleton coll, Univ Observatory coll, Radcliffe Tracts, photogr coll, Royal microscopical soc coll
13 000 vols; 20 curr per; 2 000 mss; 4 incunabula; 30 diss/theses; 70 microforms; 15 sound-rec; 11 digital data carriers; 5 000 pamphlets, 2 000 engravings, 2 000 photos   29870

Oxford University Museum of Natural History, The Hope and Arkell Libraries, Parks Rd, **Oxford** OX1 3PW
T: +44 1865 272982;
Fax: +44 1865 272970; E-mail: stella.brecknell@oum.ox.ac.uk; URL: www.oum.ox.ac.uk
1860; Stella Brecknell
Papers of eminent British 19th and 20th c entomologists and geologists, paleontology, paleobotany
11 500 vols; 114 curr per; 60 000 entomological offprints, 20 000 catalogued geological offprints   29871

Pitt Rivers Museum, University of Oxford, Balfour Library, South Parks Rd, **Oxford** OX1 3PP
T: +44 1865 270939; Fax: +44 1865 270943; E-mail: library@prm.ox.ac.uk; URL: www.prm.ox.ac.uk/balfour.html
1884; Mark Dickerson
Pitt Rivers Museum Photographic Archive, 125 000 images of ethnographical interest dating from 1850 onwards. Ms Coll: Papers & Correspondence of a number of leading figures in the hist of anthropology Inc. Prof. E.B. Tylor & B. Spencer
30 000 vols; 150 curr per; 100 diss/theses; 770 microforms; 10 sound-rec; 12 000 pamphlets
Museums Association   29872

Dr. Pusey Memorial Library, Pusey House, **Oxford** OX1 3LZ
T: +44 1865 278415; Fax: +44 1865 278416; URL: www.puseyhouse.org.uk
1884; W.E.P. Davage
Church history, patristics, liturgy; Oxford movement arch
80 000 vols; 20 curr per   29873

Centre for Ecology and Hydrology, Library, Edinburgh Research Station, Bush Estate, **Penicuik** EH 26 0QB
T: +44 131 4458512; Fax: +44 131 4453943; E-mail: sjpr@ceh.ac.uk; URL: www.ceh.ac.uk
2000; Stephen Prince
Hurst Coll, Rofe Coll
25 000 vols; 600 curr per; 1 e-journals; 150 diss/theses; 8 000 govt docs; 1 000 maps; 6 digital data carriers; 20 000 staff papers
libr loan; ASLIB, IAMSLIC, EURASLIC, RESCOLINC   29874

Morrab Library, Morrab House, Morrab Gardens, **Penzance** TR18 4DA
T: +44 1736 364474; Fax: +44 1736 364474; URL: www.morrablibrary.co.uk
1818; Mrs A. Read
Cornish book Coll, Dawson Arch of Napoleonic prints; Photographic Arch
40 000 vols; 10 000 photos, subject-cornish
AIL   29875

Royal Geological Society of Cornwall, Alverton St, **Penzance** TR18 2QR
1814; G.J. Shrimpton
10 000 vols   29876

Royal Scottish Geographical Society, Library, Lord John Murray's House, 15-17 North Port, Blackfriars, **Perth** PH1 5LU
E-mail: enquiries@rsgs.org; URL: www.rsgs.org
1884
Isobel Wylie Hutchison Coll, J.T. Coppock Coll, Scottish National Antarctic Expedition Coll
20 000 vols; 200 curr per; 60 000 maps; photogr coll   29877

English Nature (Nature Conservancy Council for England), Library, Northminster House, **Peterborough** PE1 1UA
T: +44 1733 455094;
Fax: +44 1733 568834; E-mail: enquiries@naturalengland.org.uk; URL: www.english-nature.org.uk
1949
40 000 vols; 1 500 curr per
libr loan   29878

Marine Biological Association, National Marine Biological Library, Citadel Hill, **Plymouth** PL1 2PB
T: +44 1752 633266; Fax: +44 1752 633102; E-mail: nmbl@mba.ac.uk; URL: www.mba.ac.uk/nmbl/
1888; Linda Noble
Marine Pollution, Arch of the Marine Biological Assn
50 000 vols; 1 100 curr per; 75 000 mss; 1 250 maps; 2 300 microforms; 75 000 pamphlets and reprints
IAMSLIC, EURASLIC, BIASLIC   29879

Plymouth Proprietary Library, Alton Terrace, 111 North Hill, **Plymouth** PL4 8JY
T: +44 1752 660515
1810; John R. Smith
17 000 vols; 1 diss/theses
AIL   29880

Royal Naval Museum, Library, HM Naval Base (PP66), **Portsmouth** PO1 3NM
T: +44 23 92723795;
Fax: +44 23 92723942; E-mail: library@royalnavalmuseum.org; URL: www.royalnavalmuseum.org
1990; Allison Wareham
Part of Admiralty Libr, Monographs Coll (60 000 vols) and Admiralty Libr mss coll
70 000 vols; 30 curr per; mss held by RNM (Contact Curator of mss on 023-9272-7577)
SLA, CILIP   29881

Lancashire Record Office, Bow Lane, **Preston** PR1 2RE
T: +44 1772 533039;
Fax: +44 1772 533050; E-mail:
record.office@lancashire.gov.uk; URL:
www.archives.lancashire.gov.uk
1940; B. Jackson
Local and regional history, genealogy
33 476 vols; 40 curr per; 20 e-journals;
7 000 000 mss; 7 000 maps; 2 000
microforms; 15 sound-rec   29882

Atomic Weapons Establishment (AWE), Library, Aldermaston, **Reading** RG7 4PR
1951; A.E.G. Willson
28 000 vols; 180 curr per; 450 000
microforms
ASLIB   29883

British Gas, Exploration and Production Information Centre, 100 Thames Valley Park Dr, **Reading** RG6 1PT
T: +44 118 9353222; Fax: +44 118
9353484; URL: www.bg-group.com
D. Fairnbairn
50 000 vols; 400 curr per   29884

Philips Research Laboratories, Library, Cross Oak Lane, **Redhill** RH1 5HA
T: +44 1293 815432; Fax: +44 1293
815500
S. Camp
20 000 vols; 300 curr per; 15 diss/theses;
400 maps; 4 000 microforms; 20 av-mat;
20 digital data carriers
libr loan; ASLIB   29885

The National Archives, Resource Centre & Library, Kew, **Richmond**, Surrey TW9 4DU
T: +44 20 88763444; URL:
www.nationalarchives.gov.uk
1838; Natalie Ceeney
History from 11th c to the present day;
The official archive for England, Wales
and the central UK government
65 000 vols; 300 curr per; 30 e-journals;
30 microforms; 100 digital data carriers;
100 miles of government docs
CILIP, LIBER, ICA   29886

Royal Botanic Gardens, Kew, Library, Art and Archives, **Richmond**, Surrey TW9 3AE
T: +44 20 83325414; Fax: +44 20
83325430; E-mail: library@kew.org; URL:
www.kew.org/library/index.html
1852; Christopher Mills
Linnean Coll, Adams Diatoms Coll, Arch,
Kewensia; prints and drawings
200 000 vols; 1 300 curr per; 292 e-
journals; 18 e-journals; 1 512 diss/theses;
11 000 maps; 10 000 microforms; 40
digital data carriers; 150 000 pamphlets,
500 portraits, 200 000 botanical
illustrations, 7m sheets of mss in 4 600
colls
libr loan; CBHL, EBHL   29887

Working Class Movement Library, 51, The Crescent, **Salford** M5 4WX
T: +44 161 7363601; Fax: +44 161
7374115; E-mail: enquiries@wcml.org.uk;
URL: www.wcml.org.uk
Lynette Cawthra
Labour history, Thomas Paine, Chartism,
the Peterloo Massacre, Irish politics
and history, women's emancipation,
conscientious objectors, 1926 General
Strike, Spanish Civil War, Christian
Socialism, Co-operation
35 000 vols; Pamphlets, archives, photos,
plays, poetry, songs, banners, posters,
badges, cartoons   29888

Chemical and Biological Defence Establishment, Library, Porton Down, **Salisbury** SP4 0JQ
T: +44 1980 613414; Fax: +44 1980
613970
Mrs P. Goddard
Chemistry, biology, chemical and
biological warfare
9 000 vols; 300 curr per; 5 digital data
carriers
libr loan; BLDSC   29889

Health Protection Agency Porton Down, Library, Porton Down, **Salisbury** SP4 0JG
T: +44 1980 612711; Fax: +44 1980
612818; E-mail: porton.library@hpa.org.uk;
URL: www.hpa.org.uk; www.camr.org.uk
Mrs S. Goddard
10 000 vols; 80 curr per   29890

Salisbury NHS Foundation Trust, Library and Information Service, Salisbury District Hospital, **Salisbury** SP2 8BJ
T: +44 1722 336262 ext 4432; Fax: +44
1722 339690; E-mail: library.office@
salisbury.nhs.uk
J. Lang
18 000 vols; 200 curr per; 125 av-mat   29891

Royal Society for the Protection of Birds, Library, The Lodge, **Sandy** SG19 2DL
T: +44 1767 680551; Fax: +44 1767
692365; E-mail: ian.dawson@rspb.org.uk;
URL: www.rspb.org.uk
I. Dawson
W.H. Hudson Coll
11 000 vols; 20 curr per; RSPB archive   29892

Shropshire Archives, Castle Gates, **Shrewsbury** SY1 2AQ
T: +44 1743 255350;
Fax: +44 1743 255355; URL:
www.shropshirearchives.org.uk
1885; M. McKenzie
Local hist, genealogy, geography;
Parochial libs coll
50 000 vols; 100 curr per; 3 e-journals;
700 mss; 125 diss/theses; 9 000 maps;
50 000 microforms; 160 sound-rec; 100
digital data carriers; 6 000 pamphlets,
25 000 photos, 1 500 slides, 8km of
shelving occupied   29893

National Institute for Biological Standards and Control, Library, Blanche Lane, **South Mimms** EN6 3QG
T: +44 1707 641000; E-mail:
enquiries@nibsc.ac.uk; URL:
www.nibsc.ac.uk
1978; S.P. Johnson
Medicine, biology, vaccines, hormones,
blood coagulation factors, antibiotics,
immunology
10 000 vols; 150 curr per; 20 diss/theses;
1 000 govt docs; 20 microforms; 10
av-mat; 2 digital data carriers
libr loan; CILIP, ASLIB   29894

Royal South Hants Hospital, Staff Library, Newtown, **Southampton** SO9 4PE
T: +44 23 80631743 ext 2714
R. Noyes
10 000 vols; 354 curr per   29895

Royal Marines Museum, Reference Library, **Southsea** PO4 9PX
T: +44 23 92819385;
Fax: +44 23 92838420; E-mail:
archive@royalmarinesmuseum.co.uk
1958; Matthew Grant Little
388 Personal Diaries 1775-1982;
Photographic Library, Landing Craft Plans
& Technical Data Coll
20 000 vols; 40 000 mss; 20 000 govt
docs; 100 music scores; 400 maps;
2 000 microforms; 70 av-mat; 150
sound-rec; 1 000 pamphlets, 400 000
photos/illustrations, 2 000 slides   29896

Royal Entomological Society, Library, The Mansion House, Chiswell Green Lane, **St Albans**, Herts AL2 3NS
T: +44 1727 899387; Fax: +44 1727
894797; E-mail: lib@royensoc.co.uk; URL:
www.royensoc.co.uk/about_library.shtml
1833; Val McAtear
Correspondence of (19th c)
entomologists, Rothschild coll of flea
reprints
11 000 vols; 750 curr per; 58 000 reprints
libr loan; BLL   29897

Société Jersiaise, Lord Coutanche Library, 7, Pier Rd, **St Helier**, Jersey, Channel Islands JE2 4XW
T: +44 1534 730538; Fax: +44 1534
888262; E-mail: library@societe-
jersiaise.org; URL: www.societe-
jersiaise.org
1873; Brenda Ross
Regional hist, genealogy
30 000 vols; 56 curr per; 30 000 mss;
60 diss/theses; 10 000 govt docs; 300
music scores; 500 maps; 50 microforms;
15 av-mat; 100 sound-rec; 10 digital data
carriers   29898

Priaulx Library, Candie Rd, **St Peter Port**, Guernsey, Channel Islands GY1 1UG
T: +44 1481 721998; Fax: +44 1481
713804; E-mail: priaulx.library@gov.gg
URL: www.priaulxlibrary.co.uk
1889; A. Le C. Bennett
Local and family hist, militaria, rare
books, newspapers, Carel Toms
photograph collection; Guernsey French
30 000 vols; 10 curr per; 1 000 mss;
10 incunabula; 50 diss/theses; 300
maps; 350 microforms; 1 000 av-mat;
20 sound-rec; Channel island newspapers
(1791-present), clippings, prints and
drawings, photographs, ephemera
libr loan; LASER   29899

William Salt Library, Staffordshire and Stoke on Trent Archive Service, 19 Eastgate Street, **Stafford** ST16 2LZ
T: +44 1785 278372;
Fax: +44 1785 278414; E-mail:
william.salt.library@staffordshire.gov.uk;
URL: www.staffordshire.gov.uk/salt
1872; Joanna Terry   29900

National Library for the Blind – NLB, Far Cromwell Rd, Bredbury, **Stockport** SK6 2SG
T: +44 161 3552000; Fax: +44 161
3552098
1882; Helen Brazier
Moon bks, large-type bks, Braille bks,
Braille music scores
36 000 vols; 12 curr per; 30 000 music
scores; 120 maps
libr loan; CILIP, IFLA, ASLIB   29901

University Hospital of North Tees, Library, Hardwick, **Stockton-on-Tees** TS19 8PE
T: +44 1642 624789; E-mail:
medical.library@nth.nhs.uk
1974; J. Blenkinsopp
16 000 vols; 200 curr per; 1 000 e-
journals
libr loan   29902

Mellor Memorial Library, CERAM Research Ltd, Queens Rd, Penkhull, **Stoke-on-Trent** ST4 7LQ
T: +44 1782 764444; Fax: +44 1782
412331; E-mail: enquiries@ceram.com;
URL: www.ceram.com
1959; Ann Pace
Ceramics Technology
12 000 vols; 300 curr per; 7 000 av-mat;
60 000 pamphlets   29903

The Shakespeare Centre Library, Shakespeare Birthplace Trust, Henley St, **Stratford-upon-Avon** CV37 6QW
T: +44 1789 201813; Fax: +44 1789
296083; URL: www.shakespeare.org.uk/
library
1862; Sylvia Morris
Stoker coll of Henry Irving, Royal
Shakespeare Company archive;
Localhistory coll in Shakespeare
Birthplace Trust Records Office
56 000 vols; 45 curr per; 900 mss; 1
incunabula; 10 diss/theses; 200 music
scores; 500 microforms; 180 av-mat;
1 000 sound-rec; 200 digital data carriers;
300 000 photos, 20 000 prints, 10 000
playbills, 1 500 designs, 250 vols press
clippings, 8 000 theatre programmes
CILIP, SIBMAS, TIG   29904

English Heritage Library, The English House, Fire Fly Avenue, **Swindon** SN2 2EH
T: +44 1793 414632; E-mail:
library@english-heritage.org.uk; URL:
www.english-heritage.org.uk
1983; Felicity Gilmour
Historic preservation, archaeology;
Mayson Beeton coll of old books and
prints (London)
100 000 vols; 1 000 curr per; 123
e-journals; 2 000 govt docs; 100
microforms; 50 digital data carriers
ARLIS/UK & Ireland, ARCLIB   29905

Science Museum Library, Hackpen Lane, Wroughton, **Swindon** SN4 9NS; mail address: Swindon SN4 9LT
T: +44 1793846222;
Fax: +44 1793798021; E-mail:
smlwroughton@sciencemuseum.org.uk;
URL: www.sciencemuseum.org.uk/library
1883; Rupert Williams
Rare books on Science, technology and
medicine, early bks on veterinary science

(the Comben Coll), arch
500 000 vols; 350 curr per; 50 e-journals;
1 e-books; 1 000 mss; 5 incunabula;
400 maps; 4 000 microforms; 80 av-mat;
80 000 pamphlets and catalogs
libr loan; CILIP   29906

Somerset Archaeological and Natural History Society, Somerset Studies Library, Paul St, **Taunton** TA1 3XZ
T: +44 1823 340300;
Fax: +44 1823 340301; E-mail:
somstud@somerset.gov.uk; URL:
www.sanhs.org
1849; A. Nix
40 000 vols; 90 curr per; 4 000 slides,
6 000 photos   29907

National Physical Laboratory, Library, Bldg 27, Queens Rd, **Teddington** TW11 0LW
T: +44 20 89436417; Fax: +44 20
86140424; E-mail: library@npl.co.uk;
URL: www.npl.co.uk
1900; Ms B.M. Sanger
Physics, eng
65 000 vols; 300 curr per   29908

Ironbridge Gorge Museum Trust, Library, Ironbridge, **Telford** TF8 7DQ
T: +44 1952 432141; Fax: +44 1952
432237; E-mail: library@ironbridge.org.uk;
URL: www.ironbridge.org.uk
J. Powell
Industrial archeology, technology
30 000 vols; 100 curr per; 5 000 photos   29909

Natural History Museum, Rothschild Library and Ornithology Library, Akeman St, **Tring** HP23 6AP
T: +44 207 9426156; Fax: +44 207
9426150; E-mail: omlib@nhm.ac.uk; URL:
www.nhm.ac.uk/library/index.html
1892; Mr Graham Higley
Hist coll of bird lit, general natural hist
and early travel works
70 000 vols; 420 curr per; 41 e-journals;
350 mss; 1 incunabula; 50 diss/theses;
750 maps; 100 microforms; 1 000 av-mat;
15 sound-rec; 20 digital data carriers
2 330 original drawings, 50 000 reprints
ASLIB, IFLA, FID, SCONUL, LIBER   29910

The Courtney Library, The Royal Institution of Cornwall, River St, **Truro** TR1 2SJ
T: +44 1872 272205;
Fax: +44 1872 240514; E-mail:
RIC@royalcornwallmuseum.org.uk; URL:
www.royal.cornwallmuseum.org.uk
1818; Angela Broome
Henderson coll (hist and antiquities of
Cornwall), Doble coll (hagiography),
Shaw coll (Cornish methodism); Dunn,
Davies (shipbuilding/wreck coll); Philbrick
(Cornish post office packet service);
Nance: (Cornish language, literature and
folklore)
45 000 vols; 55 curr per; 40 000 mss;
105 incunabula; 32 diss/theses; 50 music
scores; 440 maps; 800 microforms; 5
av-mat; 43 sound-rec; 35 digital data
carriers; 50 000 photos
SOA, CILIP   29911

Yorkshire Libraries & Information (YLI) Music and Drama Service to Groups, Balne Lane, **Wakefield** WF2 0DQ
T: +44 1924 302210; Fax: +44 1924
302245; E-mail: lib.yliill@wakefield.gov.uk
1974; N.N.
Music and drama, joint fiction reserve
authors N-S
15 curr per; 550 000 music scores;
90 000 playtexts   29912

Longleat House, Library, **Warminster Wiltshire** BA12 7NN
T: +44 1985 844400; Fax: +44 1985
844885; E-mail: enquiries@longleat.co.uk;
URL: www.longleat.co.uk
1541; Dr. Kate Harris
Mediaeval MSS, late 17C-early 18C
pamphlets, French Revolution pamphlets,
Balkan affairs, natural hist; Libr. of
Beriah Botfield
40 000 vols; 108 mss; 150 incunabula;
300 music scores; 430 maps; archives   29913

Science and Technology Facilities Council, Chadwick Library, Daresbury Laboratory, Keckwick Lane, ***Warrington*** WA4 4AD
T: +44 01925 603397; Fax: +44 01925 603779; E-mail: librarydl@stfc.ac.uk; URL: www.stfc.ac.uk/e-Science/services/22463.aspx
1966; Deborah Franks
170 curr per; 500 e-journals; 200 diss/theses
libr loan 29914

Horticulture Research International, Library, ***Wellesbourne*** CV 35 9EF
T: +44 1789 470382; Fax: +44 1789 470552
Clare Singleton
16 000 vols; 500 curr per
libr loan; ASLIB 29915

South Staffordshire Medical Centre, Library, New Cross Hospital, ***Wolverhampton*** WV10 0QP
T: +44 1902 643109
J.H. Paterson
10 893 vols; 230 curr per 29916

Plunkett Foundation, Library, The Quadrangle, ***Woodstock*** OX20 1LH
T: +44 1993 810730; Fax: +44 1993 810849; E-mail: info@plunkett.co.uk; URL: www.plunkett.co.uk
1914; Elodie Malhomme
Sir Horace Plunkett's letters and diaries, Irish Homestead
12 000 vols; 100 curr per; 30 diss/theses; 40 av-mat; 10 000 rpts and pamphlets 29917

Central Science Laboratory, Information Centre, Library, Sand Hutton, ***York*** YO4 1LZ
T: +44 1904 462272; Fax: +44 1904 462111; E-mail: science@csl.gov.uk
A.M. Cassels
Stored products
15 000 vols; 1 000 curr per; 12 digital data carriers; 22 500 pamphlets
libr loan; ASLIB 29918

National Railway Museum, Search Engine – the National Railway Museum's library and archive centre, Leeman Rd, ***York*** Y026 4 XJ
T: +44 1904 686235;
Fax: +44 1904 611112; E-mail: search.engine@nrm.org.uk; URL: www.nrm.org.uk
1975; Karen Baker
20 000 vols; 800 curr per; 40 mss; 50 diss/theses; 400 govt docs; 300 maps; 35 000 microforms; 120 av-mat; 650 sound-rec; 1 000 000 engineering drawings; 1 500 000 million images, 300 archive colls; 4 100 works of art 29919

Yorkshire Museum, Museum Street, ***York*** YO1 7FR
T: +44 1904 687687;
Fax: +44 1904 687662; URL: www.yorkshiremuseum.org.uk
1823; Camilla Nichol
geology, natural history, prehistoric, classical and mediaeval archaeology, numismatics and the decorative arts
35 000 vols; 37 curr per; 1 e-journals; 100 mss; 1 000 incunabula 29920

Public Libraries

Rhonoda-Cynon-Taf Libraries, Aberdare Library, Green St, ***Aberdare*** CF44 7AG
T: +44 1685 880050; Fax: +44 1685 881181
1904; Gill Evans
29 branch libs
200 000 vols; 146 curr per; 10 mss; 10 diss/theses; 620 music scores; 772 maps; 272 microforms; 358 av-mat; 6 125 sound-rec; 34 vol newspaper clippings, 16 cassettes local hist videos 29921

Aberdeen City Council, Library and Information Services, Central Library, Rosemount Viaduct, ***Aberdeen*** AB16 5LP
T: +44 1224 652500;
Fax: +44 1224 641985; E-mail: centlib@aberdeencity.gov.uk; URL: www.aberdeencity.gov.uk
1892; Fiona Clark
Special collections: Walker coll (463 items), J.M. Henderson (700 items), Cosmo Mitchell (221), Dance, photogr colls e.g. G.W. Wilson (15 000 images in total); 17 branch libs
504 626 vols; 600 curr per; 126 mss; 5 954 maps; 76 786 microforms; 1 280 av-mat; 56 157 sound-rec; 135 digital data carriers; 15 000 photos
libr loan; CILIP, SLIC, Grampian Information 29922

Ceredigion County Library, Corporation St, ***Aberystwyth*** SY23 2BU
T: +44 1970 633703;
Fax: +44 1970 625059; E-mail: reception@ceredigion.gov.uk; URL: www.ceredigion.gov.uk/libraries
1947; William H. Howells
Local History
280 000 vols
libr loan 29923

Alderney Library, Off Church Street, ***Alderney***, Channel Islands GY9 3TE
T: +44 1481 824178; E-mail: info@alderneylibrary.org; URL: www.alderneylibrary.org
Mrs J Birmingham
50 000 vols; 2 curr per
libr loan 29924

Alloa Library, Clackmannanshire Libraries, 26-28 Drysdale St, ***Alloa*** FK10 1JL
T: +44 1259 722262; Fax: +44 1259 219469; E-mail: libraries@clacks.gov.uk; URL: www.clacksweb.org.uk/culture/alloalibrary
1921; David A. Hynd
Walter Murray Local Studies Coll; Clackmannanshire Arch
55 000 vols; 80 curr per; 1 000 maps; 2 800 microforms; 5 000 av-mat; 5 000 sound-rec; 2 798 digital data carriers; 1 216 software items
libr loan; EARL 29925

North Ayrshire Council, Headquarters Library, 39-41 Princes St, ***Ardrossan*** KA22 8BT
T: +44 1294 469682; Fax: +44 1294 604236; E-mail: Ardrossanlibrary@north-ayrshire.gov.uk; URL: www.north-ayrshire.gov.uk
1996; Christine Campbell
Robert Burns, Alexander Wood Memorial coll, Ayrshire topography and families; 17 branch libs
265 000 vols; 16 curr per; 1 000 mss; 388 maps; 85 microforms; 1 300 av-mat; 22 000 sound-rec
libr loan 29926

Southern Education and Library Board, Library Headquarters, 1 Markethill Rd, ***Armagh*** BT60 1NR
T: +44 28 37525353; Fax: +44 28 37526879
1973; Andrew Morrow
Irish hist coll; Divisional headquarters in Dungannon, Portadown, Newry, 23 branch libs
1 360 000 vols; 525 curr per; 10 000 govt docs; 1 000 music scores; 4 000 maps; 5 045 microforms; 23 200 av-mat; 39 000 sound-rec; 1 500 pictures
libr loan; CILIP 29927

Tameside Metropolitan District Council, Libraries Service Unit, Sustainable Communities, Wellington Rd, ***Ashton-under-Lyne*** OL6 6DL
T: +44 161 3423673; URL: www.tameside.gov.uk/libraries
1882; C. Simensky
Area libs in Ashton-under-Lyne, Denton, Droylsden, Hyde, Stalybridge, 6 branch libs
400 000 vols; 869 curr per; 1 627 maps; 11 710 sound recordings
libr loan; CILIP 29928

Buckinghamshire County Council, County Library, County Hall, Walton St, ***Aylesbury*** HP20 1UU
T: +44 1296 383549; Fax: +44 1296 382259; E-mail: library@buckscc.gov.uk; URL: www.buckscc.gov.uk/libraries
1918; Bob Strong
Centre for Buckinghamshire Studies; Regional libs in Aylesbury, Wycombe
805 032 vols; 507 curr per; 13 361 music scores; 4 023 maps; 29 456 microforms; 43 940 sound-rec; 10 051 digital data carriers; 44 157 DVDs
libr loan; CILIP 29929

South Ayrshire Council, Carnegie Library, 12 Main St, ***Ayr*** KA8 8ED
T: +44 01292 286385; Fax: +44 01292 611593; URL: www.south-ayrshire.gov.uk
1893; Aileen J. Cowan
Robert Burns coll, South Ayrshire Local Coll; 12 libr, 1 art gallery, Scotish and Local Hist Libr, Cyber Centre
323 000 vols; 50 curr per; 2 000 maps; 1 000 microforms; 3 600 sound-rec
libr loan; SLIC 29930

North Eastern Education and Library Board, Neelb Library Service H.Q, 25-31 Demesne Ave, ***Ballymena*** BT43 7BG
T: +44 28 25664100; Fax: +44 28 25632038; URL: www.neelb.org
1922; Mrs P. Valentine
Local studies coll, area libr HQ, Ballymena, Ballymoney special coll; Divisional libs in Coleraine, Carrickfergus, Ballymena, 37 branch libs, 10 mobile libs
975 695 vols; 3 606 maps; 9 414 microforms; 4 080 av-mat; 4 174 sound-rec; 464 digital data carriers
libr loan; CILIP 29931

South Eastern Education and Library Board, Library Headquarters, Windmill Hill, ***Ballynahinch*** BT24 8DH
T: +44 28 97566400; Fax: +44 28 97565072
1940; Mrs B. Porter
Irish coll; Area libs in Belfast, Downpatrick, Holywood, 25 branch libs
1 200 000 vols
libr loan; ASLIB, CILIP 29932

Barking Central Library, Unit 53, Vicarage Field Shopping Centre, Ripple Rd, ***Barking*** IG11 8DQ
T: +44 20 87241313; Fax: +44 20 87241314; E-mail: reference@lbbd.gov.uk
1889; Ann Laskey
Local hist, laser: printing, publishing, journalism; 10 branch libs
560 000 vols
libr loan; LASER, EARL, CILIP 29933

Barnsley Central Library, Shambles St, ***Barnsley*** S70 2JF
T: +44 1226 773930; Fax: +44 1226 773955
1890; Mr S. Bashforth
Archives
166 500 vols; 145 curr per; 30 000 govt docs; 1 250 music scores; 2 534 maps; 20 700 microforms; 12 500 av-mat; 6 301 sound-rec; 42 digital data carriers; 292 metres shelf space of manuscripts
libr loan; ASLIB, CILIP 29934

East Renfrewshire Libraries and Infromation Service, District Libraries, Library Headquarters, Glen Street, ***Barrhead*** G78 1QA
T: +44 141 577 3500; Fax: +44 141 577 3501; URL: www.eastrenfrewshire.gov.uk/libraries
1996; Liz McGettigan
135 831 vols; 4 623 sound-rec; 226 digital data carriers; Adult Language Courses: 264; Spoken word: 5 132; Video and DVD: 3 186
libr loan; CILIP 29935

Bath and North East Somerset Council, Bath Central Library, 19 The Podium, Northgate St, ***Bath*** BA1 5AN
T: +44 1225 787400;
Fax: +44 1225 787426; E-mail: bathlibraries@bathnes.gov.uk; URL: www.bathnes.gov.uk
Julia Fieldhouse
Local studies coll on the hist of the city of Bath and surrounding district
181 894 vols; 468 curr per; 811 mss; 154 music scores; 4 567 maps; 35 090 microforms; 2 640 av-mat; 7 852 sound-rec; 147 digital data carriers
libr loan 29936

Bedfordshire Libraries, County Hall, Cauldwell St, ***Bedford*** MK42 9AP
T: +44 1234 363222; Fax: +44 1234 228921; URL: www.bedfordshire.gov.uk
1925; Barry S. George
John Bunyan Libr, Fowler Coll; 17 branch libs
690 000 vols; 1 000 curr per; 6 200 av-mat; 25 000 sound-rec; 336 digital data carriers
libr loan; CILIP 29937

Belfast Public Library, Central Library, Royal Ave, ***Belfast*** BT1 1EA
T: +44 28 90509150; Fax: +44 28 90312819, 90312886; E-mail: infobelb@ni-libraries.net; URL: www.belb.org.uk
1888; L. Houston
F.J. Bigger Coll (Irish hist); 20 branch libs
1 400 000 vols; 650 curr per; 35 000 music scores; 8 805 maps; 82 637 microforms; 26 400 av-mat; 57 000 sound-rec; 57 000 slides, 660 photos, 350 mss, 5 700 postcards
libr loan; CILIP, ASLIB, IFLA, INTAMEL 29938

Bexley Library Service, Central Library, Townley Rd, ***Bexleyheath*** DA6 7HJ
T: +44 20 83037777;
Fax: +44 20 83037058; E-mail: libraries.els@bexley.gov.uk; URL: www.bexley.gov.uk/library/lib-central.html
1899; Mr F.V. Johnson
Local colls; 13 branch libs
600 000 vols; 500 curr per; 30 000 govt docs; 2 000 maps; 60 000 microforms; 5 000 av-mat; 7 000 sound-rec; 150 digital data carriers
libr loan; CILIP 29939

Stockton-on-Tees Borough Council, Education, Leisure & Cultural Services Dept., Bibliographical Services, Ieast Precinct, ***Billingham*** TS23 2JZ
T: +44 1642 358592; Fax: +44 1642 358501; E-mail: sbl@stockton.gov.uk
1877; Ms A. Barker
Local/family hist, adult lending, reference, childrens
82 000 vols; 80 curr per; 200 music scores; 1 129 maps; 20 microforms; 472 av-mat; 5 500 sound-rec; 10 digital data carriers
CILIP 29940

Wirral Metropolitan Borough Council, Central Library, Borough Rd, ***Birkenhead*** CH41 2XB
T: +44 151 6538932; Fax: +44 151 6537320; E-mail: janesterling@wirral-libraries.net; URL: www.wirral-libraries.net
1856; Owen Roberts
Area libs in Bebington, Wallasey, 21 branch libs
663 487 vols; 106 curr per; 6 391 mss; 7 224 maps; 13 976 microforms; 3 494 av-mat; 30 848 sound-rec; 542 digital data carriers 29941

Birmingham Central Library, Chamberlain Sq, ***Birmingham*** B3 3HQ
T: +44 121 3034511;
Fax: +44 121 3034458; E-mail: central.library@birmingham.gov.uk; URL: www.birmingham.gov.uk/centrallibrary.bcc
1861; Mrs V.M. Griffiths
Shakespeare coll, Milton coll, Cervantes coll
3 200 000 vols; 10 200 curr per; 6 112 mss; 134 incunabula; 50 000 maps; 370 821 microforms; 379 000 av-mat; 123 612 sound-rec; 610 000 photos, 177 000 postcards, 5 000 company rpts, 6 000 sheets circuit diagrams, 8 000 000 patents, 140 000 standards/specifications
libr loan; ASLIB, CILIP, IFLA, INTAMEL 29942

Bolton Central Library, Central Library, Le Mans Crescent, ***Bolton*** BL1 1SE
T: +44 1204 333173; Fax: +44 1204 332225; URL: www.bolton.gov.uk/libraries
1853; Marguerite Gracey
Walt Whitman Coll, Thompson Coll (19th c maps), Albinson Coll (18 c maps), Bill Naughton Archive, Bolton Archives and Local Studies; Regional libs in Farnworth, Bolton, Westhoughton
455 740 vols; 350 curr per; 2 106 mss; 6 148 maps; 46 000 microforms; 1 700 av-mat; 13 340 sound-rec; 5 933 digital data carriers
libr loan; CILIP, ASLIB 29943

Sefton Metropolitan District Council, Libraries, Magdalen House, 30 Trinity Rd, ***Bootle*** LZ0 3NT
T: +44 151 9342376;
Fax: +44 151 9342370; E-mail: library.service@leisure.sefton.gov.uk; URL: www.sefton.gov.uk/libraries
J. Hilton
Area libs, 13 branch libs, 1 mobile libr
553 200 vols; 185 curr per; 10 028 music scores; 134 microforms; 103 av-mat;

4 540 sound-rec; 803 digital data carriers
29944

Bradford Metropolitan Council,
Central Library, Prince's Way, **Bradford**
BD1 1NN
T: +44 1274 433600;
Fax: +44 1274 395108; E-mail:
public.libraries@bradford.gov.uk; URL:
www.bradford.gov.uk/
1871; Ian Watson
Brontë Coll, Federer Coll, Philip Snowden
Coll; 3 area libs, 31 branch libs, 3
mobile libs
886 000 vols; 960 curr per; 50
diss/theses; 10 000 govt docs; 29 642
music scores; 15 900 maps; 18 000
microforms; 7 700 av-mat; 37 000
sound-rec; 300 digital data carriers
libr loan; CILIP, IAML   29945

**Bridgend Library and Information
Service**, Coed Parc, Park St, **Bridgend**
CF31 4BA
T: +44 1656 754810; Fax: +44 1656
645719; E-mail: blis@bridgend.gov.uk;
URL: www.bridgend.gov.uk
1996; J.C. Woods
Local Studies and Family Hist; 13 branch
libs, 2 mobile libs, 1 Reference and
Information Centre
292 244 vols; 150 curr per; 193 mss; 29
diss/theses; 3 041 music scores; 2 656
maps; 32 811 microforms; 4 000 av-mat;
8 072 sound-rec; 76 digital data carriers;
8 000 photos, ephemera coll (4 000 items)
libr loan; CILIP, EARL   29946

Somerset County Council, Library
Administration, Mount St, **Bridgwater**
TA6 3ES
T: +44 1278 451201; Fax: +44 1278
452787; E-mail: rnfroud@somerset.gov.uk;
URL: www.somerset.gov.uk/libraries
1918; R.N. Froud
Laurence Housman Coll, Aircraft
coll, Somerset Studies Lib, Historical
Children's Literature Coll
676 572 vols; 150 curr per; 137 000
music scores; 3 479 maps; 22 770
sound-rec; 436 digital data carriers;
13 622 spoken word, 15 943 video and
DVDs
libr loan   29947

Brighton Central Library, Vantage Point,
New England St, **Brighton** BN1 2GW
T: +44 1273 290800; Fax: +44 1273
296951; E-mail: libraries@brighton-
hove.gov.uk; URL: www.citylibraries.info
1872; Amanda Bagville
Bloomfield Coll (23 000 collectors
editons), Lewis Coll (2 000 fine art bks in
foreign languages), Mathews Coll (4 000
Hebrew and Oriental lit), Wolseley Coll
(4 000 letters and bks of Field Marshall
Wolseley), Halliwell-Phillips Coll (1 000
works on Shakespeare), Clericetti Coll
(570 early works in Italian), Cobden Coll
(3 100 bks and pamphlets on c. 19th
French and English agriculture), Elllliott
Coll (3 000 theological works)
200 000 vols; 80 curr per; 55 mss;
31 incunabula; 20 000 govt docs;
8 000 music scores; 1 500 maps; 2 500
microforms; 2 000 sound-rec; 45 digital
data carriers; 25 000 other items
libr loan; IFLA, ALA   29948

Bristol City Council, Central Library,
College Green, **Bristol** BS1 5TL
T: +44 117 9037200(switchboard),
9037202 (Ref. Dept); Fax: +44
117 9221081; E-mail:
Refandinfo@bristol.gov.uk; URL:
www.bristol.gov.uk/libraries
N.N:
Vincent Stuckey Lean Coll, Emanuel
Green Coll, Thomas Chatterton Coll,
Private Press Coll, Braikenridge Coll;
Music Library
184 249 vols; 266 curr per; 300 mss;
100 diss/theses; 3 500 maps; printed
music coll, map coll, CDs, videos and
DVDs, available for loan
libr loan; CILIP, ASLIB   29949

Bromley Central Library, High St,
Bromley BR1 1EX
T: +44 20 84609955;
Fax: +44 20 83139975; E-mail:
central.library@bromley.gov.uk
1965; David Brockurst
H.G. Wells Coll, Walter De la Mare Coll,
Crystal Palace Coll, general hist; Area

libs in Beckenham, Orpington, 11 branch
libs
889 000 vols; 444 curr per; 3 602 maps;
33 113 microforms; 5 883 av-mat; 52 975
sound-rec
libr loan; ASLIB, LASER, SEAL   29950

**Blaenau Gwent Country Borough
Council**, Central Depot, Library,
Barleyfields Industrial Estate, **Brynmar**
NP23 4YF
T: +44 1495 355311; Fax: +44 1495
312357
M Jones
130 000 vols; 120 maps; 2 000 av-mat;
5 000 sound-rec; 400 digital data carriers;
150 playstations
libr loan   29951

Bury Central Library, Manchester Rd,
Bury BL9 0DG
T: +44 161 2535873; Fax: +44 161
2535857; URL: www.bury.gov.uk/
index.aspx?articleid=5916
1901; Diana Sorrigan
6 branch libs, visual impairment unit,
hearing impairment unit
313 104 vols; 113 curr per; 354
mss; 30 diss/theses; 3 036 maps;
3 511 microforms; 4 000 av-mat; 9 200
sound-rec
libr loan; CILIP   29952

Central Library Cambridge, Grand
Arcade, Level 1, **Cambridge**; mail
address: 7 Lion Yard, Cambridge
CB2 3QD
T: +44 345 045 5225;
Fax: +44 1223 717088; E-mail:
your.library@cambridgeshire.gov.uk; URL:
www.cambridgeshire.gov.uk/library
1925; Christine May (Acting Head of
Service)
Cambridgeshire Collection (local studies),
Arthur Rackham Collection (illustrated
books)
1 350 000 vols; 2 159 curr per; 20 e-
journals; 8 500 music scores; 16 000
maps; 38 075 microforms; 7 750 av-mat;
65 000 sound-rec; 1 070 digital data
carriers
libr loan; CILIP, ASLIB   29953

Cardiff Council, Central Library, The
Hayes, **Cardiff** CF10 1FL
T: +44 2920382116; E-mail:
centrallibrary@cardiff.gov.uk; URL:
www.cardiff.gov.uk/libraries
1861; Elspeth Morris
Welsh, Local Studies; 17 branch libs
1 030 000 vols; 481 curr per; 4 000 mss;
140 incunabula; 2 006 maps; 6 301
microforms; 26 306 sound-rec; 3 800
digital data carriers
libr loan   29954

Cumbria County Council, Heritage
Services, Arroyo Block, The
Castle, **Carlisle** CA3 8XF
T: +44 1228 607295;
Fax: +44 1228 607299; E-mail:
Information@cumbriacc.gov.uk; URL:
www.cumbria.gov.uk
1974; J.D. Hendry
6 group libs
1 300 000 vols; 256 curr per
libr loan; CILIP, ASLIB   29955

Carmarthenshire County Libraries,
Carmarthen Public Library, St. Peter's St,
Carmarthen SA31 1LN
T: +44 1267224824; Fax: +44
1267 221839; E-mail:
libraries@carmarthenshire.gov.uk
1996; William Philipps
3 Area libs, 30 branch libs, 5 mobile libs
517 500 vols; 117 curr per; 2 e-journals;
3 312 mss; 56 diss/theses; 7 671 music
scores; 3 968 maps; 4 620 microforms;
9 100 av-mat; 21 800 sound-rec; 350
digital data carriers
libr loan   29956

Milton Keynes Council, Central Library,
555 SilburyBd, Saxon Gate East,
Central Milton Keynes MK9 3HL
T: +44 1908 254050; Fax: +44 1908
254089; E-mail: Central.Library@milton-
keynes.gov.uk; URL: www.mkweb.co.uk/
library_services
1981; Teresa Carroll
144 275 vols; 239 curr per; 846 music
scores; 3 328 maps; 40 605 microforms;
76 621 av-mat; 10 829 sound-rec; 621
digital data carriers
libr loan   29957

Essex County Council, Essex Libraries,
Goldlay Gardens, **Chelmsford**
CM2 0EW
T: +44 845 6037628; E-mail:
answers.direct@essex.gov.uk; URL:
www.essex.gov.uk/libraries
1926; Susan Carragher
Harsnett Coll, Castle Coll, National Jazz
Arch, Victorian Studies Coll, Performing
Arts Coll; 74 Public libs, Local Studies,
Music, Science & Technology Business,
Social Science
2 034 070 vols; 639 curr per; 24 161 e-
journals; 24 161 govt docs; 21 109 music
scores; 24 000 maps; 74 775 av-mat;
89 975 sound-rec; 2 297 digital data
carriers; current periodicals, microforms
libr loan; ASCEL, BIC, NAG, ENQUIRE
29958

Cheshire County Council, Libraries
and Archives Service, Goldsmith House,
Hamilton Pl, **Chester** CH1 1SE
T: +44 1244 602424; Fax: +44 1244
602805; URL: www.cheshire.gov.uk
Ian Dunn
1 800 000 vols; 625 curr per; 10 495
maps; 7 074 microforms; 19 695 av-mat;
43 445 sound-rec
CILIP, ASLIB   29959

West Sussex County Council, Library
Service Administration Centre, Tower St,
Chichester PO19 1QJ
T: +44 1243 777352;
Fax: +44 1243 531610; E-mail:
chichester.library@westsussex.gov.uk
1925; R.A. Kirk
Area libs in Worthing, Crawley, Chichester,
35 branch lib, Local studies; Area libs in
Worthing, Crawley, Chichester, 35 branch
libs
1 270 000 vols; 800 curr per
libr loan; ASLIB, CILIP   29960

West Dunbartonshire Libraries,
Clydebank Library, Dumbarton Rd,
Clydebank G81 1XH
T: +44 141 9521416; Fax: +44 141
9518275
1912; Susan Carragher
148 000 vols; 95 curr per; 300 av-mat;
10 000 sound-rec
libr loan   29961

North Lanarkshire Council, Coatbridge
Library, Academy St, **Coatbridge**
ML5 3AW
T: +44 1236 424150; Fax: +44 1236
437997
1905; John Fox
412 000 vols; 2 175 av-mat; 11 630
sound-rec   29962

Coventry City Council, Central Library,
Smithford Way, **Coventry** CV1 1FY
T: +44 24 76832314; Fax: +44 24
76832315; URL: www.coventry.gov.uk
1868; Richard Munro
Bartleet Coll (bicycles), Automobile Coll,
Tom Mann Centre for Trade Union and
Labour Studies, George Eliot, Angela
Brazil
480 000 vols; 250 curr per; 26 500
sound-rec; 386 digital data carriers
CADIG, CILIP   29963

Croydon Libraries, London Borough of
Croydon, Central Library, Katharine St,
Croydon CR9 1ET
T: +44 20 87266900;
Fax: +44 20 82531004; E-mail:
libraries@croydon.gov.uk; URL:
www.croydon.gov.uk
1888; Aileen Cahill
460 000 vols; 12 e-journals; 9 500 av-mat;
24 000 sound-rec   29964

**Cumbernauld & Kilsyth District
Council**, District Library, 8 Allander
Walk, **Cumbernauld** G67 1EE
T: +44 1236 725664; Fax: +44 1236
458350
1975; Jean Dawson
4 branch libs
234 000 vols
libr loan; CILIP   29965

**Cumnock and Doon Valley District
Library**, Lugar, **Cumnock** KA18 3JQ
T: +44 1290 22111; Fax: +44 1290
22461
1975; Stuart C. Brownlee
Ayrshire coll, Boswell coll, Burns coll; 13
branch libs
227 000 vols
libr loan   29966

**Monmouthshire Libraries and
Information Service**, County Hall,
Cwmbran NP44 2HX
T: +44 1633 644550;
Fax: +44 1633 644545; E-mail:
infocentre@monmouthshire.gov.uk; URL:
libraries.monmouthshire.gov.uk
1996; Ann Jones
Chepstow coll; 6 community libs, 1
mobile libr
161 185 vols; 49 curr per; 14 e-journals;
1 000 maps; 500 microforms; 3 500
av-mat; 7 000 sound-rec
libr loan   29967

Midlothian District Council, Dalkeith
Library, White Hart St, **Dalkeith**
EH22 1AE
T: +44 131 6632083
1975; Ms J. Fergus
8 branch libs
195 200 vols
libr loan; CILIP   29968

**Doncaster Metropolitan District
Council**, Central Library, Waterdale,
Doncaster DN1 3JE
T: +44 1302 734305;
Fax: +44 1302 369749; E-mail:
Reference.library@doncaster.gov.uk; URL:
www.doncaster.gov.uk/library
1869
Business Libr, Local Studies Libr; 25
branch libs
631 000 vols; 390 curr per; 8 700 maps;
42 500 sound-rec
libr loan; YHJLS   29969

Dorset County Council, County Library
HQ, Colliton Park, **Dorchester** DT1 1XJ
T: +44 1305 225000; Fax: +44 1305
224344
1920; P. Leivers
Thomas Hardy Coll, Powys Coll, Dorset
Coll, T.E. Lawrence Coll; Divisional HQS
at Dorchester Ferndown, Blandford +
Weymouth, 34 Libraries, 5 Mobiles, 5
Prisons
85 000 vols
libr loan; CILIP   29970

Dudley Metropolitan Borough Council,
Dudley Library, St. James's Rd, **Dudley**
DY1 1HR
T: +44 1384 815560, 815557;
Fax: +44 1384 815543; E-mail:
libraries@dudley.gov.uk; URL:
www.dudley.gov.uk/libraries
1884; Kate Millin
4 area libs
1 000 000 vols; 200 curr per; 5 e-
journals; 4 000 maps; 1 000 microforms;
7 400 av-mat; 118 100 sound-rec; 1 000
video cassetter
libr loan; WESLINK, CILIP   29971

West Dunbartonshire Council, Libraries
& Museums Headquarters, 19 Poplar Rd,
Dumbarton G82 2RJ
T: +44 1389 608042;
Fax: +44 1389 608044; E-mail:
library.headquarters@west-
dunbarton.gov.uk; URL: www.wdcweb.info/
1996; Gill Graham
9 branch libs, 1 mobile libr
335 845 vols; 92 curr per; 91 mss;
950 music scores; 1 108 maps; 9
297 microforms; 496 av-mat; 14 890
sound-rec; 30 digital data carriers
libr loan; CILIP   29972

Dumfries and Galloway Libraries,
Information and Archives, Catherine St,
Dumfries DG1 1JB
T: +44 1387 252070; Fax: +44 1387
260292; E-mail: libs&i@dumgal.gov.uk
1904; J.H.Goldie
Dumfries and Galloway Coll, Dumfries
Burns Club Coll, Frank Miller Coll
(ballads), R.C. Reid Genealogical Coll;
25 District Libs, Schools Libr Service
340 000 vols; 127 curr per; 1 350 maps;
109 826 microforms; 10 690 av-mat;
16 137 sound-rec
CILIP, SLA   29973

Dundee City Council, Central Library,
The Wellgate, **Dundee** DD1 1DB
T: +44 1382 431500; Fax: +44 1382
431558; E-mail: central.library@
leisureandculturedundee.com; URL:
www.leisureandculturedundee.com/library/
central
1869; Mrs M. Methven
Wighton coll of old Scottish music;

Art & Music Dept, Commerce & Technology Dept, Local Hist Divisions, 13 Neighbourhood Libs
791 500 vols; 240 curr per
libr loan; CILIP, ASLIB, IAML 29974

Dunfermline Carnegie Library, 1 Abbot Street, ***Dunfermline*** KY12 7NL
T: +44 1383 602365; Fax: +44 1383 602307; E-mail: Dunfermline.library@fife.gov.uk
1883; Dorothy Browse
Murison Burns Coll, Reid Mss Coll, Carnegie Coll; Main Lib, 15 branch libs, 1 mobile lib
290 000 vols; 40 curr per; 1 500 maps; 2 000 microforms; 6 315 sound-rec; 40 digital data carriers
libr loan 29975

Durham County Council, Durham Clayport Library, Millennium Place, ***Durham*** DH1 1WA
T: +44 191 3864003; Fax: +44 191 3860379; URL: www.durham.gov.uk
June Gowland
libr loan 29976

South Lanarkshire Council, Central Library, 40 The Olympia Centre, ***East Kilbride*** G74 1PG
T: +44 1355 220046; Fax: +44 1355 229365
1975; Ms D. Barr
3 divisional libs
269 100 vols; 60 curr per; 80 mss; 33 000 govt docs; 1 500 music scores; 1 383 maps; 11 174 microforms; 90 av-mat; 26 778 sound-rec; 4 000 photos
libr loan; CILIP, ASLIB 29977

The Moray Council, Department of Educational Services, High St, ***Elgin*** IV30 1BX
T: +44 1343 563398; Fax: +44 1343 563478; E-mail: campbea@moray.gov.uk; URL: www.moray.gov.uk
1975; G.A. Campbell
Wittet Coll, Falconer papers, coll on early census returns, whisky industry, Doig Coll
259 996 vols; 88 curr per; 13 873 microforms; 8 048 av-mat; 15 236 sound-rec; 22 604 photos, 842 art prints and paintings
libr loan; CILIP 29978

Enfield Town Library, Thomas Mardy House, 39 London Rd, ***Enfield*** EN2 605
T: +44 20 83798391;
Fax: +44 20 83798401; E-mail: central.library@enfield.gov.uk; URL: www.enfield.gov.uk
1894; Claire Lewis
Linguistics special coll; Area libs in Edmonton, Palmers Green, Enfield, 11 branch libs, Business libs, 2 mobile libs, 2 Open Learning Center
761 657 vols; 210 curr per; 113 mss; 600 music scores; 7 361 maps; 6 004 microforms; 14 332 av-mat; 32 556 sound-rec; 300 digital data carriers
CILIP, IAML, LASER, NBL 29979

Devon Libraries, Great Moor House, Bittern Rd, Sowton, ***Exeter*** EX2 7NL
T: +44 1392 384315; Fax: +44 1392 384316; E-mail: devlibs@devon.gov.uk; URL: www.devon.gov.uk/libraries
1924; Ciara Eastell
Coll of early children's bks, early printed bks, brass rubbings, bookplates, Pocknell shorthand collection, Napoleonic material; Westcountry Studies Libr, Railway Studies Libr
1 209 563 vols; 352 curr per; 1 750 mss; 1 incunabula; 15 727 music scores; 3 735 maps; 51 microforms; 30 363 av-mat; 65 601 sound-rec
libr loan 29980

Falkirk Council Library Services, Victoria Buildings, Queen St, ***Falkirk*** FK2 7AF
T: +44 1324 506800;
Fax: +44 1324 506801; E-mail: library.support@falkirk.gov.uk; URL: www.falkirk.gov.uk
1996
Aeneas Mackay Coll; 7 libs
327 394 vols; 785 curr per; 18 mss; 2 055 maps; 8 838 microforms; 3 642 av-mat; 8 765 sound-rec; 470 digital data carriers; 18 mtr mss
libr loan; CILIP, CILIPS 29981

Angus Council, Cultural Services, County Bldgs, Market St, ***Forfar*** DD8 3WF
T: +44 1307 461460; Fax: +44 1307 466220; E-mail: cultural@angus.gov.uk; URL: www.angus.gov.uk
J.C. Ewing Scottish Coll; 7 area libs
286 000 vols; 110 curr per; 5 460 maps; 44 121 microforms; 6 000 av-mat; 18 000 sound-rec; 692 digital data carriers; 3 000 glass negatives, 4 000 lantern slides, 15 000 photos, over 100 000 images
libr loan 29982

Gateshead Library, Prince Consort Rd, ***Gateshead*** NE8 4LN
T: +44 191 4338400;
Fax: +44 191 4777454; E-mail: enquiries@gateshead.gov.uk; URL: www.gateshead.gov.uk/libraries
1885; Ann Borthwick
Area libs in Birtley, Gateshead, Whickham, Blaydon, Leam Lane
258 233 vols; 350 curr per; 100 mss; 2 855 av-mat; 32 306 sound-rec; 100 digital data carriers
libr loan; CILIP 29983

East Dunbartonshire Libraries, 2 West High St, Kirkintilloch, ***Glasgow*** G66 1AD
T: +44 141 7754501;
Fax: +44 141 7760408; E-mail: libraries@eastdunbarton.gov.uk; URL: www.eastdunbarton.gov.uk
1996; David Kenvyn, Frances Macarthur
Archive collections held within Information & Archives; Information and Archives at Kirkintilloch and Bearsden, 2 mobile libraries, 8 branch libraries
210 000 vols; 50 curr per; 14 e-journals; 600 e-books; 20 000 govt docs; 3 000 maps; 800 microforms; 50 000 av-mat; 15 400 sound-rec; Ephemera, newspaper extracts, information handouts, local history indexes, published leaflets
libr loan; CILIP 29984

The Mitchell Library, North St, ***Glasgow*** G3 7DN
T: +44 141 2872999; Fax: +44 141 2872815; E-mail: archives@csglasgow.org; URL: www.mitchelllibrary.org/vm
1877; Bridget Mc Connell
Scottish poetry, 18th c English music, Robert Burns, Scottish Regimental Histories, North British Locomotive Coll; 32 branch libs, Children's Serv Dept
2 135 curr per; 46 incunabula; 44 599 music scores; 29 258 maps; 3 924 sound-rec; 418 digital data carriers; 381,5 linear m of mss, 8 mio patent specifications, UK Govt Papers 1715 since, Scottish Parliament 12th-18th C.
libr loan; IFLA, CILIP, IAML, INTAMEL 29985

Leicestershire County Council, Library and Information Services HQ, County Hall, ***Glenfield*** LE3 8SS
T: +44 116 2657374; Fax: +44 116 2657370
1974; P. Oldroyd
Stretton Coll (railway hist); Area libs in Leicester, Loughborough, Oakham, Hinckley, 28 branch libs
1 700 000 vols; 1 200 curr per; 105 000 sound-rec
libr loan; IFLA 29986

Gloucestershire County Council, County Library, Quayside House, Shire Hall, ***Gloucester*** GL1 2HY
T: +44 1452 425048;
Fax: +44 1452 425042; E-mail: Customerservices@gloucestershire.gov.uk; URL: www.gloscc.gov.uk
Colin Campbell
Local Studies Coll, Civil War Coll, Stow; Strategic libs in Cheltenham, Gloucester, Cinderford, Tewkesbury, Circencester, Stroud, 33 other local libs
847 219 vols; 187 curr per; 14 191 music scores; 12 295 av-mat; 44 557 sound-rec; 720 digital data carriers; 1 135 language learning packs
libr loan; WESLINK, CILIP, SWRLS, ARLIS/UK & Ireland, EARL 29987

Inverclyde Libraries, Inverclyde Council, Clyde Sq, ***Greenock*** PA15 1NA
T: +44 1475 712323;
Fax: +44 1475 712339; E-mail: Library.central@inverclyde.gov.uk; URL: www.inverclyde.gov.uk/Libraries
1996; Ms S. Macdougall
8 branch libs
300 000 vols; 20 curr per; 2 000 music scores; 250 maps; 500 microforms; 2 000 av-mat; 12 000 sound-rec; 100 digital data carriers
CILIP 29988

Grimsby Central Library, North East Lincolnshire Council, Town Hall Sq, ***Grimsby*** DN31 1HG
T: +44 1472 323600; Fax: +44 1472 323618; E-mail: Lib@nelincs.gov.uk; URL: www.nelincs.gov.uk/libraries
1968; A.S. Hipkins
Skelton Collection of 19th century local posters and handbills, Ruhleben collection – First World War; 10 branch libs
458 541 vols; 230 curr per; 12 e-journals; 11 000 govt docs; 3 851 music scores; 8 390 maps; 330 microforms; 32 297 sound-rec; 79 digital data carriers; Local History collection contains newspaper cuttings, pamphlets and ephemera
29989

East Lothian Library Services, Library & Museum HQ, Dunbar Rd, ***Haddington*** EH41 3PJ
T: +44 1620 828200;
Fax: +44 1620 828201; E-mail: libraries@eastlothian.gov.uk; URL: www.eastlothian.gov.uk/libraries
1800; Alison Hunter
Local coll
179 741 vols; 185 curr per; 1 394 music scores; 1 557 maps; 16 309 sound-rec
libr loan; CILIP 29990

Calderdale Metropolitan Borough Council, Central Library, Northgate House, Northgate, ***Halifax*** HX1 1UN
T: +44 1422 392630;
Fax: +44 1422 392615; E-mail: libraries@calderdale.gov.uk; URL: www.calderdale.gov.uk/leisure/libraries/branches/central-library.html
1883; M. Stone
District libs in Elland, Brighouse, Hebden Bridge, Sowerby Bridge, Todmorden, 27 branch libs
417 000 vols; 153 curr per; 2 210 maps; 3 896 microforms; 7 220 av-mat; 14 712 sound-rec; 186 digital data carriers
libr loan 29991

Hamilton District Council, Central Library, 98 Cadzow St, ***Hamilton*** ML3 6HQ
T: +44 1698 282323
1975; W. McCoubrey
Papers from the Hamilton estates
339 700 vols
libr loan; CILIP, ASLIB 29992

London Borough of Harrow, Civic Centre Library, Station Rd, P.O. Box 4, Civic Centre, ***Harrow***, Middlesex HA1 2UU
T: +44 20 84241055;
Fax: +44 20 84241971; E-mail: civiccentre.library@harrow.gov.uk; URL: www.harrow.gov.uk
1971; J.E. Pennells
2 central and 9 branch libs
424 000 vols; 287 curr per; 177 mss; 5 000 maps; 20 900 microforms; 26 400 sound-rec; 17 000 pamphlets
libr loan; CILIP 29993

Hartlepool Borough Council, Central Library, 124 York Rd, ***Hartlepool*** TS26 9DE
T: +44 1429 272905;
Fax: +44 1429 275685; E-mail: infodesk@hartlepool.gov.uk; URL: www.publiclibrary.org.uk/hartlepool
Mrs S. Atkinson
84 700 vols; 7 118 sound-rec; 28 digital data carriers
libr loan; CILIP 29994

Hertfordshire County Council, Central Resources Library, New Barnfield Centre, Travellers Lane, ***Hatfield*** AL10 8XG
T: +44 1438 737333;
Fax: +44 1707 281514; E-mail: centralresources.library@hertscc.gov.uk; URL: www.hertsdiret.org/libsleisure/libraries
1925; A. Roberston
Regional libs in Stevenage, Welwyn Garden City, Hemel Hempstead, Hoddesdon, St Albans, Watford, 45 Community libs
2 300 000 vols; 2 500 curr per; 100 diss/theses; 50 000 govt docs; 200 000 music scores; 47 000 maps; 32 000 microforms; 110 000 sound-rec; 450 digital data carriers
CILIP, ASLIB, EBLIDA 29995

Pembrokesire County Library, Dew St, ***Haverfordwest*** SA61 1SU
T: +44 1437 775244; Fax: +44 1437 767092; E-mail: haverfordwest-library@pembrokeshire.gov.uk; URL: www.pembrokeshire.gov.uk/libraries
1924; Neil Bennett
Francis Green Coll (local history transcripts and genealogical notes)
323 490 vols; 62 curr per; 1 441 maps; 5 647 microforms; 10 188 sound-rec; 44 digital data carriers; 244 vols local newscuttings
libr loan 29996

Hounslow Library Network (CIP), Libraries and Heritage, High St, ***Hounslow*** MIDDX TW3 1ES
T: +44 845 4562800; Fax: +44 845 4562880
Linda Simpson
388 066 vols; 300 curr per; 2 700 maps; 15 738 microforms; 12 925 av-mat; 35 035 sound-rec
libr loan; CILIP 29997

Kirklees Metropolitan District Council, Huddersfield Library, Princess Alexandra Walk, ***Huddersfield*** HD1 2SU
T: +44 1484 221959, 221960; Fax: +44 1484 221952; E-mail: hudlib.office@kirklees.gov.uk; URL: www.kirklees.gov.uk/
J. Drake
4 area libs
1 160 000 vols
libr loan; CILIP 29998

Knowsley Metropolitan Borough Council, Huyton Library, Civic Way, ***Huyton*** L36 9GD
T: +44 151 443 3734;
Fax: +44 151 443 3739; E-mail: huyton.library.dlcs@knowsley.gov.uk; URL: www.knowsley.gov.uk
1997; Pauline Taylor
8 branch libs
80 000 vols; 50 curr per; 260 mss; 3 000 maps; 40 microforms; 300 av-mat; 150 digital data carriers
libr loan; CILIP 29999

London Borough of Redbridge, Central Library, Clements Rd, ***Ilford*** IG1 1EA
T: +44 20 87082414;
Fax: +44 20 87082431; E-mail: central.library@redbridge.gov.uk; URL: www.redbridge.gov.uk
1909; M. Timms
Social services, photography; 9 branch libs
600 000 vols
CILIP, ASLIB, SLA 30000

The Highland Council, Highland Libraries, 31A Harbour Rd, ***Inverness*** IV1 1UA
T: +44 1463 235713; Fax: +44 1463 236986
1975; S.C. Brownlee
Fraser-MacKintosh Coll, Inverness Kirk Session, Inverness Gaelic Soc
420 000 vols; 100 curr per; 200 mss; 300 govt docs; 850 maps; 750 microforms; 6 000 sound-rec
libr loan; CILIP, EARL, SLIC, SLA 30001

East Ayrshire Council, Llibrary, Registration and Information Services, 14 Elmbank Ave, ***Kilmarnock*** KA1 3BU
T: +44 1563 554300; Fax: +44 1563 554311; E-mail: libraries@east-ayrshire.gov.uk; URL: www.east-ayrshire.gov.uk
1909; Gerard Cairns
Ayrshire Coll, Robert Burns Coll, John Galt Coll, Braidwood Coll, James Boswell Coll, James Keir Hardie Coll, Buchanan Bequest; 21 Community libs, 2 mobile libs, 1 district hist centre
365 554 vols; 70 curr per; 1 e-journals; 1 995 music scores; 732 maps; 2 235 microforms; 9 230 av-mat; 11 382 sound-rec; 6 514 digital data carriers
libr loan; CILIPS, NAG, SLIG 30002

Kingston upon Hull City Council, Hull Central Reference Library, Albion St, ***Kingston upon Hull*** HU1 3TF
T: +44 1482 210055, 210066, 223344; Fax: +44 1482 616858; E-mail: reference.library@hullcc.gov.uk; URL:

www.hullcc.gov.uk
1901; Michelle Alford
Slavery & William Wilberforce, Winifred Hotby, Andrew Marvell, whaling, Napoleon; Central libr and 13 branch libs
815 996 vols; 1 130 curr per; 13 920 music scores; 10 886 maps; 56 927 microforms; 3 489 av-mat; 26 417 sound-rec; 478 digital data carriers; 9 321 illustrations
ASLIB   30003

Kingston Libraries, Royal Borough of Kingston upon Thames, Fairfield Rd, *Kingston upon Thames* KT1 2PS
T: +44 20 85476400;
Fax: +44 20 85476401; E-mail: library.admin@rbk.kingston.gov.uk; URL: www.kingston.gov.uk/libraries
1903; Barbara Lee
Virtual Training Suite: a collection of ca. 1000 training courses available for free on LAN; Lifelong Learning Team (info: antonio.rizzo@rbk.kingston.gov.uk)
250 000 vols; 160 curr per; 17 e-journals
libr loan; CILIP   30004

Fife Council, Libraries Cultural Services, Town House, *Kirkcaldy* KY1 1XW
T: +44 1592 412878; Fax: +44 1592 412750
1975; I. Whitelaw
3 area libs, 8 group libs
libr loan; CILIP, SLIC   30005

Fife Council Libraries, 16 East Fergus Place, *Kirkcaldy* KY1 1XT
T: +44 1592 583204; Fax: +44 1334 412941; URL: www.fife.gov.uk
1994; Dorothy Browse
185 000 vols; 42 curr per; 3 000 maps; 6 000 microforms; 11 200 av-mat; 1 000 sound-rec; 5 000 photos
CILIP, SLIC   30006

East Dunbartonshire Council, William Patrick Library, 2 West High St, *Kirkintilloch* G66 1ADE
T: +44 141 7665666;
Fax: +44 141 7660408; E-mail: libraries@eastdunbarton.gov.uk
1975; Don Martin
7 branch libs
93 000 vols
libr loan; CILIP   30007

The Orkney Library and Archive, 44 Junction Rd, *Kirkwall*, Orkney KW15 1AG
T: +44 1856 873166;
Fax: +44 1856 875260; E-mail: enquiries@orkneylibrary.org.uk; URL: www.orkneylibrary.org.uk
1683
Dr Marwick Coll, Robert Rendall Coll, Tom Kent Photogr Coll
120 000 vols; 82 curr per; 25 diss/theses; 2 100 maps; 1 157 microforms; 2 500 av-mat; 2 700 sound-rec; 1 350 digital data carriers; 650 m of archives, 45 000 photogr archive
libr loan; CILIP, SLA   30008

Leeds Central Library, Municipal Bldg, Calverley St, *Leeds* LS1 3AB
T: +44 113 2476016; Fax: +44 113 2478421; URL: www.leeds.gov.uk
1884; Catherine Blanshard
Judaica, Early Gardening Bks, Leeds and Yorkshire, Military Hist; Depts: music, fine arts, patents; local hist, business information, reference
750 000 vols; 1 000 curr per; 25 incunabula; 1 450 music scores; 13 000 maps; 5 000 microforms; 48 000 sound-rec; 50 digital data carriers
libr loan; CILIP, YHJLS   30009

Shetland Islands Council, Shetland Library, Lower Hillhead, *Lerwick*, Shetland ZE1 0EL
T: +44 1595 743868;
Fax: +44 1595 694430; E-mail: shetlandlibrary@shetland.gov.uk; URL: www.shetland-library.gov.uk
1923; Silvia Crook
Local history maps, Icelandic Saga lit, Saga lit, local history
106 800 vols
libr loan   30010

East Sussex County Council, Libraries, Southdown House, 44 St. Anne's Crescent, *Lewes* BN7 1SQ
T: +44 345 6080190; Fax: +44 1273 481261; URL: www.eastsussex.gov.uk/libraries/
1974; Arison Merriman
Major reference libs at Hastings and Eastbourne
1 100 000 vols; 700 curr per; 27 000 music scores; 4 000 maps; 8 000 microforms; 40 000 av-mat; 7 500 sound-rec; 400 digital data carriers
libr loan; ASLIB, LIBRARY CAMPAIGN, NAG   30011

Lincoln Central Library, Free School Lane, *Lincoln* LN2 1EZ
T: +44 1522 782010;
Fax: +44 1522 535882; E-mail: lincoln_library@lincolnshire.gov.uk; URL: www.lincolnshire.gov.uk
1895; John Pateman
Sir Joseph Banks Letters, Sir Joseph Banks Ills, Tennyson Coll
170 000 vols; 130 curr per; 10 000 music scores; 2 200 maps; 36 500 microforms; 4 663 av-mat; 1 026 sound-rec; 400 digital data carriers
libr loan; ASLIB, CILIP   30012

Liverpool City Council, Central Library, William Brown St, *Liverpool* L3 8EW
T: +44 151 2335835;
Fax: +44 151 2335886; E-mail: refbt.central.library@liverpool.gov.uk; URL: www.liverpool.gov.uk
1850; Ms J. Little
2 000 000 vols
libr loan   30013

Llanelli Borough Council, Library, Vaughan St, *LLanelli* SA15 3AS
T: +44 1554 773538
1896; D.F. Griffiths
400 000 vols; 350 curr per; 30 000 govt docs; 15 000 music scores; 4 000 maps; 40 000 microforms; 10 000 av-mat; 8 000 sound-rec
libr loan; CILIP   30014

Isle of Anglesey County Council, Llangefni Library, Lôn-y-Felin, *Llangefni* LL77 7RT
T: +44 1248 752099; Fax: +44 1248 750197; URL: www.ynysmon.gov.uk
Mr D.H. Evans
10 branch libs, Mobile Libr Service, Housebound Service, Schools Libr Service
libr loan   30015

Barbican Centre, Barbican Library, Silk St, *London* EC2Y 8DS; P.O. Box, London EC2Y 8DS
T: +44 20 76380569;
Fax: +44 20 76382249; E-mail: barbicanlib@cityoflondon.gov.uk; URL: www.cityoflondon.gov.uk/barbicanlibrary
1982; John Lake
Arts, Finance, Management, Crime Fiction, London Collection; Music libr
119 203 vols; 179 curr per; 9 e-journals; 15 908 music scores; 2 225 maps; 10 067 av-mat; 16 078 sound-rec; 28 digital data carriers
libr loan; CILIP   30016

Barnet Libraries, London Borough of Barnet, Cultural Services, Bldg 4, North London Business Park, Oakleigh Road South, *London* N11 1NP
T: +44 20 83597770; Fax: +44 870 8896804; URL: www.barnet.gov.uk/
1965; P. Usher
Sociology Coll; 16 branch libs, 1 mobile libr, Home delivery service
579 356 vols; 328 curr per; 15 158 av-mat; 35 785 sound-rec; 7 011 digital data carriers   30017

City Business Library, Aldermanbury, *London* EC2V 7HH
T: +44 20 73321812; E-mail: cbl@cityoflondon.gov.uk; URL: www.cityoflondon.gov.uk/citybusinesslibrary
1873/1970; Ms G. Considine
Company info sources, product and industry research mat, company reports of UK companies, UK and International company info, product and industry research, import/export data
15 000 vols; 650 curr per; 20 e-journals; 10 digital data carriers   30018

Guildhall Library, City of London Libraries, Aldermanbury, *London* EC2V 7HH
T: +44 20 7332-1868/-1870;
Fax: +44 20 76003384; E-mail: guildhall.library@cityoflondon.gov.uk; URL: www.cityoflondon.gov.uk/guildhalllibrary
1824; David Pearson
Horology, gardening, food and wine, hist of London, Business history, St. Thomas More Coll, Charles Lamb Coll, Gresham College Music Coll, John Wilkes Coll, Samuel Pepys Coll, Lloyd's marine Coll
210 000 vols; 1 500 curr per; 95 000 mss; 71 incunabula; 400 diss/theses; 110 000 govt docs; 350 music scores; 15 000 microforms; 15 digital data carriers; 60 000 ephemera, 1 mio other items
CILIP, IFLA, INTAMEL   30019

Lambeth Reference and Information Services, Tate Library (Brixton), Brixton Oval, *London* SW2 1JQ
T: +44 20 79261067; Fax: +44 20 79261070
120 000 vols; 320 curr per; 300 maps; 20 microforms; 20 digital data carriers   30020

London Borough of Ealing, Central Library, 103 Ealing Broadway Centre, *London* W5 5JY
T: +44 20 85673670; Fax: +44 20 88402351; URL: www.ealing.gov.uk
1965; B.E. Cope
G.D.H. Cole, Austin Dobson and Peal Natural Hist Coll, Selborne Soc Libr, electrical and electronic engineering, Martinware Pottery Coll, Indic Languages, automatic control engineering, telecommunications; Area libr in Acton, Southall, Greenford, West Ealing, 7 branch libs
213 000 vols; 200 curr per; 2 000 mss; 10 000 music scores; 250 maps; 4 000 microforms; 700 av-mat; 6 000 sound-rec; 22 digital data carriers
LASER   30021

London Borough of Greenwich, Library Support Services, Plumstead Library, 232 Plumstead High St, *London* SE18 1JL
T: +44 20 83174466; Fax: +44 20 83174868
1905
Sport spec coll
484 500 vols
libr loan; CILIP, ASLIB, LASER   30022

London Borough of Hackney, Libraries, Hackney Town Hall, Mare St, *London* E8 1EA
T: +44 20 83562539;
Fax: +44 20 85333712; E-mail: libraries@hackney.gov.uk; URL: www.hackney.gov.uk/cl-libraries-branches.htm
1908; A. Whittle
7 branch libs
299 293 vols; 558 curr per; 1 449 maps; 2 251 microforms; 4 951 av-mat; 46 865 sound-rec; 429 digital data carriers; 40 000 rec, 11 000 casettes
libr loan; CILIP, LASER, ARL   30023

London Borough of Hammersmith and Fulham, Hammersmith Library, Shepherds Bush Rd, *London* W6 7AT
T: +44 20 8753823; Fax: +44 20 87533815; URL: www.lbhf.gov.uk
1887; Nigel Bouttell
5 other libs, 1 mobile lib
540 000 vols; 652 curr per; 2 050 mss; 100 000 govt docs; 10 000 music scores; 3 256 maps; 15 193 microforms; 4 440 av-mat; 47 046 sound-rec
libr loan; BIALL, ABTAPL   30024

London Borough of Islington, Central Library, 2 Fieldway Crescent, *London* N5 1PF
T: +44 20 76196900;
Fax: +44 20 76196906; E-mail: library.informationunit@islington.gov.uk; URL: www.islington.gov.uk
1904; Ms R. Dyle
9 Branch Libraries, Home Library Service, New Horizons Community Libraries
312 084 vols
libr loan; CILIP, ASLIB   30025

London Borough of Lambeth, Lambeth Libraries, Art and Archives, 1st Floor, Blue Star House, 234-244 Stockwell Road, *London* SW9 9SP
T: +44 20 79266060;
Fax: +44 20 79260751; E-mail: libraries@lambeth.gov.uk; URL: www.lambeth.gov.uk/libraries
D. Jones
535 000 vols   30026

London Borough of Lewisham, Education and Community Services, Town Hall, Catford, *London* SE6 4RU
T: +44 20 83148024;
Fax: +44 20 83143039; E-mail: libraries@lewisham.gov.uk; URL: www.lewisham.gov.uk/LeisureAndCulture/Libraries
1900; Ms J.M. Newton
514 100 vols; 465 curr per; 4 320 maps; 16 054 microforms; 1 072 av-mat; 41 428 sound-rec; 8 digital data carriers
libr loan; LASER, SEAL   30027

London Borough of Newham, East Ham Library, High St South, East Ham, *London* E6 4EL
T: +44 20 84721430; Fax: +44 20 85578845
1892; R. McMaster
510 600 vols
libr loan; CILIP, UK   30028

London Borough of Southwark, Southwark Education & Leisure, 15 Spa Rd, *London* SE16 3QW
T: +44 20 75251993; Fax: +44 20 75251505
19th c; A. Olsen
17 district libs
500 000 vols; 460 curr per; 1 579 maps; 3 237 microforms; 5 000 av-mat; 59 000 sound-rec; 90 digital data carriers
libr loan; LASER, SEAL   30029

London Borough of Tower Hamlets, Bancroft Library, 277 Bancroft Rd, *London* E1 4DQ
T: +44 20 89804366; Fax: +44 20 89834510
1965; Ms A. Cunningham
North American lit (12 000 vols), French, German and Portuguese lit (23 000 vols), Chinese and Vietnamese lit (4 500 vols), Indic lit (12 500 vols); 12 branch libs
500 000 vols; 450 curr per; 800 mss; 4 800 govt docs; 3 500 maps; 5 500 microforms; 650 av-mat; 33 000 sound-rec
libr loan; LASER   30030

London Borough of Waltham Forest, Central Library, High St, *London* E17 7JN
T: +44 20 85203017;
Fax: +44 20 85209645; URL: www.walthamforest.gov.uk/index/leisure/libraries.htm
1893; C. Richardson
European Public Information Relay (Level 2); 11 branch libs
802 475 vols; 205 curr per; 3 050 maps; 1 250 microforms; 3 445 av-mat; 28 809 sound-rec
libr loan; CILIP, European Public Information Relay, European Information Association   30031

London Borough of Wandsworth, Leisure and Amenity Services Department, Town Hall, Wandsworth High St, *London* SW18 2PU
T: +44 20 88716364; Fax: +44 20 88717630
Jane Allen
3 area libs, 11 branch libs
899 000 vols; 800 curr per; 9 000 av-mat; 45 000 sound-rec; 300 digital data carriers   30032

Royal Borough of Kensington and Chelsea, Central Library, Phillimore Walk, *London* W8 7RX
T: +44 20 79372542;
Fax: +44 20 73612976; E-mail: information.services@rbkc.gov.uk; URL: www.rbkc.gov.uk
1965; J. McEachen
Biography, genealogy and heraldry spec colls
618 000 vols; 600 curr per; 3 000 maps; 11 200 microforms; 3 000 av-mat; 50 000 sound-rec
libr loan; CILIP, ALA, COMLA, IAML, ASLIB   30033

Royal London Hospital Patients Library, Royal London Hospital, Whitechapel, **London** E1 1BB
T: +44 20 3777000 x 3495
1940; H.P. Raimes
60 000 vols   30034

Upper Norwood Joint Library, 39-41 Westow Hill, Upper Norwood, **London** SE19 1TJ
T: +44 20 86702551;
Fax: +44 20 86705468; E-mail:
info@uppernorwoodlibrary.org; URL:
www.uppernorwoodlibrary.org/
1900; Bradley Millington
Local history coll, especially Norwood and the Crystal Palace (Hyde Park and Norwood sites); J.B. Wileon coll about Norwood; Gerald Massey coll – Victorian poet, socialist, spiritualist and Egyptologist
50 000 vols; 36 curr per; 5 mss; 100 govt docs; 2 000 music scores; 250 maps; 2 microforms; 1 500 sound-rec; 20 digital data carriers; 300 pamphlets   30035

Wandsworth Borough Council, Leisure and Amenity Services Department, Libraries, Town Hall, Wandsworth High St, **London** SW18 2PU
T: +44 20 88716364;
Fax: +44 20 88717630; E-mail:
libraries@wandsworth.gov.uk; URL:
www.wandsworth.gov.uk/Home/
LeisureandTourism/Libraries
Ms J. Allen
European hist
787 000 vols; 334 curr per; 47 000 av-mat
libr loan   30036

Westminster City Council, Westminster Libraries and Archives, 13th Fl, Westminster City Hall, 64 Victoria St, **London** SW1E 6QP
T: +44 20 77982496; Fax: +44 20 77983404; URL: www.westminster.gov.uk/libraries/
1857; David Ruse
William Blake Gillow Arch, Sherlock Holmes Coll; 13 branch libs
998 419 vols; 508 curr per; 3 062 mss; 235 000 govt docs; 55 428 music scores; 7 840 maps; 40 339 sound-rec; 100 digital data carriers; 15 968 other items
libr loan; CILIP, ASLIB, IAML, INTAMEL   30037

Wood Green Central Library, Central Library, High Rd, Wood Green, **London** N22 6XD
T: +44 20 84892700
1890; Gill Harvey
Heath Robinson Coll, Bruce Castle Museum – Local Hist Coll; 3 area libs, 7 branch libs, 1 mobile libr
534 847 vols; 290 curr per
libr loan; LASER, CILIP   30038

Kent County Council, Kent Libraries, Springfield, **Maidstone** ME14 2LH
T: +44 1622 696511; Fax +44 1622 690897; URL: www.kent.gov.uk/libs
1921; Cath Anley
Railway coll; County Central Libr and 12 main town centre libs
2 886 000 vols
libr loan; CILIP, ASLIB, Museums Association   30039

Manchester City Council, Central Library, St. Peter's Sq, **Manchester** M2 5PD
T: +44 161 2341900; Fax: +44 161 2341963; URL: www.manchester.gov.uk
1852; Vicky Rosin
Newman Flowers coll of Handel mss; Commercial Libr, Europ Information Unit, Social Sciences Libr, Technical Libr, Manchester Archives and Local Studies, Arts Libr, Henry Watson Musik Libr, Chinese Libr, Language and Lit Libr
2 100 000 vols; 2 400 curr per; 300 000 music scores; 32 000 maps; 312 000 microforms; 35 000 av-mat; 101 000 sound-rec; 50 digital data carriers; 43 000 pictures, 2 900 m of mss   30040

Derbyshire County Council, Libraries, County Hall, **Matlock** DE4 3AG
T: +44 1629 580000; Fax: +44 1629 585363; URL: www.derbyshire.gov.uk/leisure/libraries/
1923; M.J. Molloy
Local studies, printed music, drama sets

for performance; Central libr and 12 other main libs
1 600 000 vols
libr loan   30041

Merthyr Tydfil Central Library, High St, **Merthyr Tydfil** CF47 8AF
T: +44 1685 723057;
Fax: +44 1685 370690; E-mail:
library.services@merthyr.gov.uk
1935; G. James
Aberfas Disaster, Iron/Steel and Coal Industries; 2 branch libs
180 200 vols; 30 curr per; 300 mss; 90 incunabula; 35 diss/theses; 655 govt docs; 700 music scores; 360 maps; 1 100 microforms; 1 500 av-mat; 1 075 sound-rec; 128 digital data carriers
libr loan   30042

Middlesbrough Borough Council, Central Library, Victoria Sq, **Middlesbrough** TS1 2AY
T: +44 1642 263397; Fax: +44 1642 263354
Chrys Mellar
Local hist coll
1 100 000 vols
libr loan   30043

Flintshire Library and Information Services, County Hall, **Mold** CH7 6NW
T: +44 1352 704400;
Fax: +44 1352 753662; E-mail:
mobileoffice@flintshire.gov.uk; URL:
www.flintshire.gov.uk
1926; Lawrence M. Rawsthorne
Arthurian Coll: 2 400 items relating to King Arthur, Index to births, marriages and deaths in England and Wales 1837-1983, overseas index to births, marriages and deaths
300 000 vols; 180 curr per; 800 music scores; 2 000 maps; 38 000 microforms; 1 600 av-mat; 9 000 sound-rec; 400 digital data carriers
CILIP, WLA   30044

London Borough of Merton, Libraries and Heritage Service, Civic Centre, London Rd, **Morden** SM4 5DX
T: +44 20 85453783;
Fax: +44 20 85453237; E-mail:
library.enquiries@merton.gov.uk; URL:
www.merton.gov.uk/libraries
1887; Ingrid Lackajis
Merton local studies (includes William Morris, Nelson)
407 686 vols; 171 curr per; 12 e-journals; 9 851 microforms; 11 220 av-mat; 21 787 sound-rec; 100 digital data carriers
libr loan; CILIP   30045

Northumberland County Council, County Central Library, The Willows, **Morpeth** NE61 1TA
T: +44 1670 534518;
Fax: +44 1670 534513; E-mail:
ask@northumberland.gov.uk
1924; D.E. Bonser
Northern Poetry Libr film coll
635 000 vols; 180 curr per; 30 000 govt docs; 11 000 music scores; 5 000 maps; 18 000 microforms; 1 000 av-mat; 21 000 sound-rec
libr loan; CILIP   30046

North Lanarkshire Council, Motherwell Library, 35 Hamilton Rd, **Motherwell** ML1 3BZ
T: +44 1698 332626; Fax: +44 1698 332624
1906; J. Fox
2 area libs
318 230 vols; 253 curr per; 4 400 mss; 90 incunabula; 275 diss/theses; 275 govt docs; 198 music scores; 770 maps; 26 662 microforms; 5 115 av-mat; 36 903 sound-rec; 508 pictures
CILIP   30047

City Library, Princess Sq, **Newcastle upon Tyne** NE99 1DX; P.O. Box 88, Newcastle upon Tyne NE991DX
T: +44 84500200336;
Fax: +44 191 2774137; E-mail:
tony.durcan@newcastle.gov.uk; URL:
www.newcastle.gov.uk/libraries
1880; Tony Durcan
UK and European Patents, Tyneside unidentified flying objects, society sighting records, Thomlinson coll (18th c bks), Joseph Cowan coll (19th c radicalism), Bewick coll (book engravings); 20 branch libs, central libs: lending libr, reference

and information libr, local studies libr
760 400 vols; 500 curr per; 100 000 govt docs; 18 000 music scores; 1 000 maps; 10 000 sound-rec
libr loan; INTAMEL   30048

Lord Louis Library, Orchard St, **Newport, Isle of Wight** PO30 1LL
T: +44 1983 527655; Fax: +44 1983 825972; URL: www.iwight.com/thelibrary
1904; R. Jones
Isle of Wight Local Studies, Isle of Wight Maritime
362 000 vols
libr loan; HATRICS   30049

Newport Central Library, John Frost Sq, **Newport, South Wales** NP20 1PA
T: +44 1633 656656;
Fax: +44 1633 222615; E-mail:
central.library@newport.gov.uk
Mrs Gill John
Special collections: Monmouthshire Collection, Haines Collection, Arthur Machen Collection, Chartist Collection
367 441 vols; 15 mss; 3 074 maps; 9 097 microforms; 1 810 av-mat; 11 467 sound-rec; 193 digital data carriers
libr loan; NAG, WRLS, CILIP   30050

North Tyneside Metropolitan District Council, Central Library, Northumberland Sq, **North Shields** NE30 1QU
T: +44 191 2005424; Fax: +44 191 2006118
1870; Mrs J. Stafford
Area libs in Wallsend, Whitley Bay, 13 branch libs
435 000 vols; 470 curr per; 24 117 sound-rec; 7 019 recordings
libr loan; CILIP, ASLIB, LIBRARY CAMPAIGN   30051

North Yorkshire County Council, Adult and Community Services, Library and Community Services, Library HQ, 21 Grammar School Lane, **Northallerton**, North Yorkshire DL6 1DF
T: +44 1609 533800;
Fax: +44 1609 780793; E-mail:
libraries@northyorks.gov.uk; URL:
www.northyorks.gov.uk/libraries
1974; Julie Blaisdale (Assistant Director Library and Community Services)
Unne photogr coll; Main reference, information, local hist and genealogical colls in Harrogate, Northallerton and Scarborough, 45 branch libs, 12 mobile libs
839 083 vols; 490 curr per; 7 909 maps; 63 858 microforms; 66 911 av-mat; 41 153 sound-rec
libr loan; CILIP, ASLIB   30052

Northamptonshire Libraries and Information Service, John Dryden House 8-10 The Lakes, **Northampton**, Northants NN4 7DD; P.O. Box 216, Northampton, Northants NN1 1Ba
T: +44 1604 236236;
Fax: +44 1604 237937; E-mail:
nlis@northamptonshire.gov.uk; URL:
www.northamptonshire.gov.uk
1876; E.W. Wright
Local studies – central library
1 105 000 vols; 2 108 curr per; 109 840 sound-rec; 496 digital data carriers
libr loan; CILIP   30053

Norfolk County Council Library and Information Service, County Hall, Martineau Lane, **Norwich** NR1 2UA
T: +44 1603 222049; Fax: +44 1603 222422; E-mail: libraries@norfolk.gov.uk; URL: www.library.norfolk.gov.uk
1850; Jennifer Holland
Illustrator Coll of Brock Brothers of Cambridge, Colman Libr, Taylor-Bell slide coll, American Memorial Arch, City Libr Coll, Shipdham Parish Libr Coll; 46 branch libs incl 2 shop libs, 12 mobile libs, Norfolk and Norwich Millennium Library, Thetford, Yarmouth and Kings Lynn
1 950 000 vols; 677 curr per; 33 mss; 30 incunabula; 5 000 maps; 64 300 microforms; 23 600 sound-rec; 40 digital data carriers
libr loan; CILIP   30054

Oldham Library and Lifelong Learning Centre, Greaves St, **Oldham** OL1 1AL
T: +44 161 7708000; E-mail:
oldham.library@oldham.gov.uk; URL:
www.oldham.gov.uk

1883; Andrea Ellison
Talking Books Service, Housebound Service
412 000 vols; 143 curr per
libr loan; CILIP, ASLIB, SLA   30055

Aberdeenshire Libraries, Meldrum Meg Way, **Oldmeldrum**, Aberdeenshire AB51 0GN
T: +44 1651 872707;
Fax: +44 1651 872142; E-mail:
alis@aberdeenshire.gov.uk; URL:
www.aberdeenshire.gov.uk/libraries
Anne Harrison
George MacDonald Coll, Middleton Coll, Lewis Gillies Local History Coll, Morrison Postcard Coll, Watson Slide Coll
623 000 vols; 263 curr per; 6 e-journals; 1 000 e-books; 16 000 microforms; 36 500 sound-rec
libr loan   30056

Western Education and Library Board, Library Headquarters, 1 Spillars Pl, **Omagh** BT78 1HL
T: +44 28 82244821; Fax: +44 28 82246716
1927; Russell T.A. Farrow
21 branch libs
1 400 000 vols; 450 curr per; 4 471 maps; 3 834 microforms; 35 530 sound-rec
CILIP   30057

Perth and Kinross Council, A. K. Bell Library, York Place, **Perth** PH2 8EP
T: +44 1738 444949; Fax: +44 1738 477010
1975; M.C.G. Moir
William Soutar's Libr, Athol coll of music, Brough art coll; 10 branch libs, 3 mobile libs
400 000 vols; 150 curr per; 20 mss; 5 diss/theses; 500 music scores; 400 maps; 1 100 microforms; 20 000 sound-rec
CILIP, SLA   30058

Plymouth Central Library, Drake Circus, **Plymouth** PL4 8AL
T: +44 1752 305923; Fax: +44 1752 305929; E-mail: library@plymouth.gov.uk; URL: www.plymouthlibraries.info
1876; Alasdair MacNaughtan
Holcenberg Coll of Jewish life; Plymouth/SWRLS Music Coll of performing sets; Naval Studies Coll
500 000 vols; 60 curr per; 60 mss; 50 diss/theses; 10 000 music scores; 3 000 maps; 5 000 microforms; 5 000 av-mat; 10 000 sound-rec
libr loan   30059

Portsmouth City Council, Norrish Central Library, Central Library Reference Department, Guildhall Sq, **Portsmouth** PO1 2DX
T: +44 23 92819311; Fax: +44 23 92839855; URL: www.portsmouth.gov.uk
Jackie Painting
Europ Relay Centre
200 000 vols; 250 curr per; 1 000 UK company rpts
libr loan   30060

Lancashire County Council, Bibliographical Services, Bowran St, **Preston** PR1 2UX; P.O. Box 61, County Hall, Preston PR1 8RJ
T: +44 1772 534008;
Fax: +44 1772 534880; E-mail:
enquiries@lancashire.gov.uk; URL:
www.lancashire.gov.uk/libraries
1924; Mr D.G. Lightfoot
Dr Shephard Libr, Fuller Maitland Coll, Spencer Coll; Area libs in Accrington, Burnley, Chorley, Lancaster, Preston, 84 static service points + 13 mobile services
2 386 475 vols; 5 e-journals; 360 mss; 87 000 music scores; 46 794 maps; 220 917 microforms; 28 000 av-mat; 72 000 sound-rec; 2 000 digital data carriers; 1 981 subscriptions to 781 periodical titles
libr loan; IFLA, CILIP, BL, ASLIB   30061

Reading Borough Libraries, Abbey Sq, **Reading** RG1 3BQ
T: +44 118 9015950;
Fax: +44 118 9015954; E-mail:
info@readinglibraries.org.uk; URL:
www.readinglibraries.org.uk
Mr A. Dane
Toy library; 7 branch libs
261 633 vols; 189 curr per; 23 772 music scores; 4 492 maps; 8 126 microforms;

4 626 av-mat; 13 604 sound-rec
libr loan; CILIP   30062

Redcar and CLeveland Borough Council, Redcar Central Library, Coatham Rd, *Redcar* TS10 1RP
T: +44 1642 472162; Fax: +44 1642 492253
I.L. Wilson
58 000 vols   30063

Richmond Lending Library, Little Green, *Richmond*, Surrey TW9 1QL
T: +44 20 89400981;
Fax: +44 20 89407516; E-mail: richmond.library@richmond.gov.uk; URL: www.richmond.gov.uk
1880; Ms S. Kirkpatrick
Sir Richard Burton Coll, Alexander Pope Coll, Douglas Sladen Coll, Capt. George Vancouver Coll; 4 branch libs
395 000 vols; 400 curr per; 1 000 mss; 5 000 music scores; 29 544 av-mat; 24 188 sound-rec
libr loan; CILIP, LASER   30064

Rochdale Metropolitan Borough Council, Wheatsheaf Library, Baillie St, *Rochdale* OL16 1JZ
T: +44 1706 924900;
Fax: +44 1706 924992; E-mail: library.service@rochdale.gov.uk; URL: www.rochdale.gov.uk/libraries
1974; Mrs S.M. Sfrijan
Internat Co-operation Coll
472 000 vols; 250 curr per; 5 035 maps; 1 036 microforms; 8 200 sound-rec
libr loan; CILIP   30065

London Borough of Havering, Central Library, St. Edward's Way, *Romford* RM1 3AR
T: +44 1708 432389; Fax: +44 1708 432391; E-mail: info@havering.gov.uk; URL: www.havering.gov.uk
1965; Robert Worcester
9 branch libs
400 000 vols; 62 curr per
libr loan   30066

Rotherham Metropolitan Borough Council, Central Library and Arts Centre, Walker Place, *Rotherham* S65 1JH
T: +44 1709 823611;
Fax: +44 1709 823699; E-mail: bernard.murphy@rotherham.gov.uk; URL: www.rotherham.gov.uk
1880; Bernard Murphy
15 community libs, 2 mobile libs
288 931 vols; 134 curr per; 3 519 music scores; 2 258 maps; 11 416 av-mat; 18 095 sound-rec; 210 digital data carriers
libr loan; CILIP   30067

Argyll & Bute Council Library Information Service, Library HQ, Highland Ave, *Sandbank* PA23 8PB
T: +44 1369 703214; Fax: +44 1369 705797; E-mail: andy.ewan@argyll-bute.gov.uk
1996; Andrew I. Ewan
13 branch libs, 5 mobile libs
212 000 vols; 54 curr per; 1 185 maps; 3 000 microforms; 13 000 sound-rec
libr loan; SLIC, CILIP   30068

Scottish Borders Council, HQ Library, St. Marys Mill, *Selkirk* TD7 5EW
T: +44 1750 20842;
Fax: +44 1750 22875; E-mail: libraries@scotborders.gov.uk
1975; M. Menzies
Area libs in Hawick, Peebles, Galashiels
309 891 vols; 4 000 items of filmstrips, maps, rec
libr loan; CILIP   30069

Sheffield Libraries, Archives and Information, Surrey St, *Sheffield* S1 1XZ
T: +44 114 2734712;
Fax: +44 114 2735009; E-mail: libraries@sheffield.gov.uk; URL: www.sheffield.gov.uk/libraries
1856; Janice Maskort
Fairbank Coll, Carpenter Coll, Arundel Castle Mss, Wentworth Woodhouse Muniments, British and foreign metal standards, Worldwide Coll of technical trade lit on metals, Patents Coll, World Metal Index, Alan Rouse mountaineering coll
893 582 vols; 750 curr per; 500 000 mss; 28 063 av-mat; 34 489 sound-rec; 30 digital data carriers

libr loan; SINTO, EARL, YLI, NAG, CILIP   30070

Shropshire Council, Shropshire Libraries, Shirehall, Abbey Foregate, *Shrewsbury* SY2 6NW
T: +44 1743 255000; Fax: +44 1743 255050; URL: www.shropshire.gov.uk/library.nsf
1925; James Anthony
Midlands Fiction, West Midlands Creative lit coll, Welsh coll at Oswestry libr, Bridgnorth hist coll, visnomy impaired coll; Area libs in Bridgnorth, Shrewsbury, Oswestry
483 761 vols; 556 curr per; 4 620 maps; 13 959 microforms; 11 056 av-mat; 21 297 sound-rec; 500 digital data carriers; 100 m shelf space of mss
libr loan   30071

Solihull Metropolitan Borough Council, Central Library, Homer Rd, *Solihull* B91 3RG
T: +44 121 7046965; Fax: +44 121 7046991
1947; Nigel Ward
Solihull joined coll photos, BSA Arch; Central libr Solihull, 11 branches, 1 mobile libr
609 700 vols; 152 curr per; 4 659 maps; 30 388 microforms; 4 079 av-mat; 14 067 sound-rec; 143 digital data carriers
libr loan   30072

South Tyneside Metropolitan Borough Council, Central Library, Prince Georg Sq, *South Shields* NE33 2PE
T: +44 191 4271818; Fax: +44 191 4558085; E-mail: reference.library@s-tyneside-mbc.gov.uk
1974; D. Abbott
7 branch libs
324 507 vols; 181 curr per; 1 320 maps; 17 636 sound-rec
libr loan; ASLIB, CILIP   30073

Metropolitan Borough of St Helens, Central Public Library, The Gamble Building, Victoria Sq, *St Helens* WA10 1DY
T: +44 1744 456951; Fax: +44 1744 20836; E-mail: criu@sthelens.gov.uk
1894; J. Roughley
Area libs in Billinge, Haydock, Newton-le-Willows, Rainford, Rainhill, Parr. Thatto Heath Sutton
235 000 vols; 183 curr per
libr loan; CILIP   30074

Jersey Library, Halkett Pl, *St Helier*, Jersey, Channel Islands JE2 4WH
T: +44 1534 448700; Fax: +44 1534 448730; E-mail: je.library@gov.je; URL: www.gov.je/library
1742; Mrs Pat Davis
Jersey local studies collection; 1 branch library, 1 mobile library, schools resources service, housebound and home service, 1 open learning centre
225 000 vols; 200 curr per; 20 mss; 13 incunabula; 12 diss/theses; 5 000 govt docs; 2 000 music scores; 500 maps; 1 500 microforms; 2 000 sound-rec; 200 digital data carriers
libr loan; IAML   30075

Guille-Allès Library, Market St, *St Peter Port*, Guernsey, Channel Islands GY1 1HB
T: +44 1481 720392; Fax: +44 1481 712425; E-mail: ga@library.gg; URL: www.library.gg
1882; Ms M.J. Falla
90 000 vols; 76 curr per; 3 000 music scores; 2 000 av-mat; 7 000 sound-rec
libr loan; SWRLS   30076

Staffordshire County Council, Library + Information Services, 16 Martin St, *Stafford* ST16 2LG
T: +44 1785 278312; Fax: +44 1785 278319
1916; O. Spencer
Area libs in Cannock, Burton upon Trent, Lichfield, Newcastle, Stafford, Tamworth, 43 libraries plus 11 mobile libraries
1 013 000 vols; 671 curr per; 13 000 maps; 51 000 microforms; 22 500 av-mat; 23 000 sound-rec; 3 000 digital data carriers
libr loan; Midlands On-Line User Group, CILIP, NAG, EIA   30077

North Lamarkshire Council, Libraries and Information Services, Cultural Services Division, Buchanan Business Park, Cumbermauld Rd, *Stepps* G33 6HR; P.O. Box 14, Motherwell ML1 3BZ
T: +44 141 3041843; Fax: +44 141 3041902; URL: www.northlan.gov.uk
1853; John Fox
Hurst Nelson, Hamilton of Dalzell, Orbiston Papers; 6 mobile libs, 24 libs
607 337 vols; 9 059 curr per; 6 diss/theses; 250 govt docs; 180 music scores; 934 maps; 18 102 av-mat; 31 264 sound-rec; 331 digital data carriers
libr loan; CILIP, SLIC   30078

Stirling Council, Library HQ, Borrowmeadow Rd, Springkerse Industrial Estate, *Stirling* FK7 7TN
T: +44 1786 432383; Fax: +44 1786 432395
Allan Gillies
15 branch libs, 2 bookmobiles, housebound service
340 000 vols; 95 curr per; 500 maps; 20 microforms; 5 932 av-mat; 2 153 sound-rec; 24 digital data carriers
libr loan; CILIP   30079

Stockport Metropolitan Borough Council, Central Library, Wellington Rd South, *Stockport* SK1 3RS
T: +44 845 6444307; E-mail: libraries@stockport.gov.uk; URL: www.stockport.gov.uk/libraries
1875; Martin Roberts
13 branch libs
528 000 vols; 430 curr per; 2 500 maps; 3 500 microforms
libr loan; CILIP, Aslib   30080

Stoke-on-Trent Libraries, Information and Archives, City Central Library, Bethesda St, Hanley, *Stoke-on-Trent* ST1 3RS
T: +44 1782 238455;
Fax: +44 1782 238434; E-mail: central.library@stoke.gov.uk; URL: www.stoke.gov.uk
Margret Green
Solon Coll on the history and development of pottery and ceramics
100 000 vols; 100 curr per; 500 maps; 2 000 sound-rec
libr loan   30081

Western Isles Libraries, Comhairle nan Eilean Siar, 19 Cromwell St, *Stornoway* HS1 2DA
T: +44 1851 708631; Fax: +44 1851 708676; URL: www.cne-siar.gov.uk
1975; Robert Eaves
Local hist, Scottish Gaelic; 3 mobile libs, 5 area libs
130 000 vols; 35 curr per; 130 mss; 30 diss/theses; 2 500 maps; 1 900 microforms; 3 500 av-mat; 10 000 sound-rec
libr loan; CILIP, CILIPS   30082

Trafford Metropolitan Borough Council, Education, Arts and Leisure Department, Libraries, Trafford Town Hall, Talbot Rd, *Stretford* M32 0YZ; P.O. Box 20, Stretford M33 1AH
T: +44 161 9121212; Fax: +44 161 9124639
1974; R.G. Luccock
Area Libs in Altrincham, Sale, Stretford, Urmston, 10 branch libs, Mobile Libr, School Libr Service, Housebound Readers Service
592 000 vols; 158 curr per; 22 000 av-mat; 13 000 sound-rec; 253 pictures
libr loan   30083

City of Sunderland Metropolitan District Council, City Library and Arts Centre, 28-30 Fawcett St, *Sunderland* SR1 1RE
T: +44 191 5141235;
Fax: +44 191 5148444; E-mail: user@edcom.sunderland.gov.uk; URL: www.sunderland.gov.uk
1858; Jane F. Hall
Area libr in Washington, 19 branch libs, 2 mobile libs
450 129 vols; 116 curr per; 52 mss; 5 859 maps; 277 microforms; 1 306 av-mat; 14 817 sound-rec; 173 digital data carriers
libr loan; CILIP   30084

Sutton Central Library, London Borough of Sutton Library Service, St. Nicholas Way, *Sutton* SM1 1EA
T: +44 20 87704700;
Fax: +44 20 87704777; E-mail: sutton.library@sutton.gov.uk; URL: www.sutton.gov.uk
1934; T. Knight
8 branch libs
407 983 vols; 318 curr per; 31 914 govt docs; 8 934 music scores; 2 490 maps; 4 986 microforms; 16 280 av-mat; 31 113 sound-rec; 721 digital data carriers
libr loan
Stock figures relate to the whole library service   30085

City and County of Swansea, Libraries, County Mall Oystermouth Rd, *Swansea* SA1 3SN
T: +44 1792 636430;
Fax: +44 1792 636235; E-mail: swansea.libraries@swansea.gov.uk; URL: www.swansea.gov.uk/libraries
1887; M. Allen
Dylan Thomas coll, Roland Williams coll, Rowland Williams coll, Deffett Francis coll; Area libr in Swansea, 19 branch libs
606 750 vols; 173 curr per; 50 incunabula; 4 306 maps; 3 669 microforms; 30 500 sound-rec   30086

Torbay Library Services, Torquay Library, Lymington Rd, *Torquay* TQ1 3DT
T: +44 1803 208300; Fax: +44 1803 208311; E-mail: tqreflib@torbay.gov.uk; URL: www.torbay.gov.uk/libraries
1938; Katie Lusty
3 branch libs
216 000 vols; 79 curr per; 5 e-journals; 1 500 maps; 10 microforms
libr loan   30087

Wiltshire County Council, Libraries and Heritage HQ, Bythesea Rd, *Trowbridge* BA14 8BS
T: +44 1225 713700;
Fax: +44 1225 713993; E-mail: libraryenquiries@wiltshire.gov.uk; URL: www.wiltshire.gov.uk
1919; P. Palmer
6 special colls; 30 branch and 4 mobile libs
752 283 vols; 214 curr per; 7 118 maps; 51 064 microforms; 24 000 av-mat; 34 269 sound-rec; 2 205 digital data carriers; 80 000 photos
libr loan; CILIP, SWRLS   30088

County Reference & Information Library, Cornwall Library Service, Truro Library, Union Place, *Truro* TR1 1EP
T: +44 1872 279205;
Fax: +44 1872 223772; E-mail: truro.library@cornwall.gov.uk; URL: www.cornwall.gov.uk/Library/
Mr R. Gould
Cornish Studies Libr, Redruth Performing Arts Libr, St. Austell and art book coll at Penzance and St. Ives Libr
772 551 vols
libr loan   30089

London Borough of Hillingdon, Uxbridge Central Library, 14-15 High St, *Uxbridge* UB8 1HD
T: +44 1895 250600; Fax: +44 1895 811164
1965; Mrs T. Grimshaw
3 area libs
563 850 vols
libr loan; CILIP, ASLIB, LASER   30090

Wakefield Metropolitan District Council, Library Headquarters, Balne Lane, *Wakefield* WF2 0DQ
T: +44 1924 302210;
Fax: +44 1924 302245; E-mail: lib.admin@wakefield.gov.uk; URL: www.wakefield.gov.uk/CultureAndLeisure/libraries/default.htm
C.J. MacDonald
George Gissing and J.S. Fletcher Coll, Henry Moore Coll; 7 major libs
487 806 vols; 170 curr per; 350 000 music scores; 2 235 maps; 2 969 microforms; 5 500 av-mat; 16 000 sound-rec; 380 digital data carriers
libr loan   30091

Walsall Metropolitan District Council, Central Library, Lichfield St, *Walsall* WS1 1TR
T: +44 1922 653121; Fax: +44 1922 722687
1859; I. Everall
19 branch libs, 1 touriste travel infopoint, 1 urban mobile, 1 special needs mobile
550 000 vols; 300 curr per; 2 900 maps; 1 000 microforms; 900 av-mat; 15 200 sound-rec; 50 digital data carriers
libr loan; ELA   30092

Warwickshire County Council, Information Service, Barrack St, *Warwick* C34 4TH
T: +44 1926 412166; Fax: +44 1926 412471; E-mail: librarieslearningand-culture@warwickshire.gov.uk; URL: www.warwickshire.gov.uk/libraries
1924
George Eliot Coll; Divisional libs at Warwick, Atherstone, Leamington Spa, Stratford on Avon, Nuneaton, Rugby and 28 other libraries
899 000 vols; 605 curr per; 16 000 music scores; 18 700 maps; 50 000 microforms; 5 300 av-mat; 37 000 sound-rec
libr loan; CILIP   30093

Brent Library Service, Administrative Headquarters, Chesterfield House, 4th Fl, *Wembley* HA9 7RW
T: +44 20 89373144;
Fax: +44 20 89373008; E-mail: karen.tyerman@brent.gov.uk
1894   30094

Environment and Culture Libraries, Arts and Heritage, Ground Floor East, Brent House, High Rd, *Wembley* Middlesex HA9 6BZ
T: +44 20 89373144;
Fax: +44 20 89373008; E-mail: sue.mckenzie@brent.gov.uk; URL: www2.brent.gov.uk/library.nsf
Sue McKenzie
13 area libs
628 000 vols
libr loan; CILIP   30095

Nottinghamshire County Council, Libraries, Archives & Information, Community Services, County Hall, 4th Fl, *West Bridgford* NG2 7QP
T: +44 115 9774201;
Fax: +44 115 9772428; E-mail: contactlibraries@nottscc.gov.uk; URL: www.nottinghamshire.gov.uk
1974; D. Lathrope
D.H. LawrenceColl, Byron Coll, Robin Hood Coll
1 142 000 vols; 621 curr per; 100 000 music scores; 21 800 maps; 66 250 microforms; 14 900 av-mat; 75 250 sound-rec
libr loan   30096

Sandwell Metropolitan Borough Council, Central Library, High St, *West Bromwich* B70 8DZ
T: +44 121 5694904; Fax: +44 121 5259465
1874; K.W. Heyes
18 branch libs
858 000 vols; 287 curr per
libr loan; CILIP, ASLIB   30097

North Somerset Council, Weston Library, The Boulevard, *Weston-super-Mare* BS23 1PL
T: +44 1934 636638; Fax: +44 1934 413046; E-mail: weston.library@n-somerset.gov.uk
1901; Nigel Kelly
North Somerset hist
113 000 vols; 13 curr per; 1 970 maps; 27 687 microforms; 4 642 sound-rec; 42 digital data carriers
libr loan; BLA   30098

Wigan Library, Redgate Rd, Bryn, *Wigan* WN4 8DT
T: +44 1942 827621; Fax: +44 1942 827640; E-mail: Wigan.Library@wlct.org; URL: www.wiganmbc.gov.uk
1878; Taryn Pearson
Area libs in Wigan, Leigh, Hindley, 13 branch libs
660 000 vols; 162 curr per
libr loan; CILIP, ASLIB, Society of Archivists   30099

Hampshire County Council, Winchester Reference Library, 81 North Walls, *Winchester* SO23 8BY
T: +44 1962 826666; Fax: +44 1962 856615; URL: www.hants.gov.uk/library
1925; Paul H. Turner
Dickens Coll, Pitt Coll, Naval Coll, Maritime Coll, Military Coll, Railway Coll, Aircraft Coll; 4 divisional libs
3 200 000 vols; 1 901 curr per; 100 000 govt docs; 43 535 maps; 244 569 microforms; 971 av-mat; 84 850 sound-rec
libr loan; CILIP, ASLIB, HATRICS, SWRLS   30100

Witham Library, 18 Newland St, *Witham* CM8 2AQ
T: +44 1376 519625; Fax: +44 1376 501913
1981; Ms. I. Probert
Drama coll (40 000 vols)
85 000 vols; 80 curr per; 300 music scores; 500 maps; 600 av-mat; 2 500 sound-rec
libr loan   30101

Wolverhampton City Council, Libraries and Information Services, Central Library, Snow Hill, *Wolverhampton* WV1 3AX
T: +44 1902 552025;
Fax: +44 1902 552024; E-mail: libraries@wolverhampton.gov.uk
1869; Mrs K. Lees
19 branch libs
555 925 vols; 220 curr per; 18 e-journals; 5 406 av-mat; 20 637 sound-rec; 21 341 talking books
libr loan; CILIP, IAML   30102

Worcestershire Library and Information Service, Libraries and Information Service, Cultural Services, County Hall, Spetchley Road, Worcester, *Worcester* WR5 2NP
1974; Cathy Evans
District libs in Redditch, Kidderminster, Worcester, Evesham, Malvern, Bromsgrove, 16 branch libs
900 000 vols
libr loan; Cilip   30103

City of York Libraries, York Library, Museum St, *York* YO1 7DS
T: +44 1904 655631; Fax: +44 1904 611025; E-mail: lending@york.gov.uk; URL: www.york.gov.uk
1893; Fiona Williams
Informations Technology Resource Centre, Business Resource Centre, Local Studies Library, General Reference and Lending Departments
165 000 vols; 243 curr per; 4 500 music scores; 5 300 maps; 36 microforms; 600 av-mat; 4 500 sound-rec; 21 digital data carriers; 9 500 illustrations, 1 500 pamphlets
libr loan   30104

United States of America

National Libraries

Library of Congress, 101 Independence Ave at First St SE, *Washington*, DC 20540
T: +1 202 7075444; Fax: +1 202 7071925; URL: www.loc.gov
1800; Dr. James H. Billington
Gutenberg Bible, Mss of Eminent Americans, Papers of the first 23 Presidents; 23 branch libs
31 000 000 vols; 4 350 000 maps; 3 800 000 microforms; 2 200 000 sound-rec
libr loan   30105

Library of Congress – African & Middle Eastern Division, Jefferson Bldg, Rm 220, 101 Independence Ave SE, *Washington*, DC 20540-4820
T: +1 202 7077937; Fax: +1 202 2523180; E-mail: amed@loc.gov; URL: www.loc.gov/rr/amed
Mary Jane Deeb
600 000 vols   30106

Library of Congress – Asian Division, Jefferson Bldg, Rm 149, *Washington*, DC 20540-4610
T: +1 202 7075420; Fax: +1 202 7071724; URL: www.loc.gov/rr/asian
Peter R. Young
3 000 000 vols; 14 900 curr per   30107

Library of Congress – Music, James Madison Memorial Bldg, Rm LM 113, *Washington*, DC 20540-4710
T: +1 202 7075503; Fax: +1 202 7070621; URL: www.loc.gov/rr/perform
Susan Vita
22 000 000 vols   30108

Library of Congress – National Library Service for the Blind & Physically Handicapped, 1291 Taylor St NW, *Washington*, DC 20542
T: +1 202 7075100; Fax: +1 202 7070712; E-mail: nls@loc.gov; URL: www.loc.gov/nls
Frank Kurt Cylke
30109

Library of Congress – Prints & Photographs Division, 101 Independence Ave SE, Rm LM 337, *Washington*, DC 20540-4730
T: +1 202 7076394; Fax: +1 202 7076647; URL: www.loc.gov/rr/print
Shirley Berry
14 000 000 vols   30110

Library of Congress – Rare Book & Special Collections Division, Thomas Jefferson Bldg, Deck A, *Washington*, DC 20540
T: +1 202 7075434; Fax: +1 202 7074142; URL: www.loc.gov/rr/rarebook
Mark G. Dimunation
800 000 vols   30111

Library of Congress – Science, Technology & Business Division, Sci Reading Rm, John Adams Bldg, Rm 508, *Washington*, DC 20540-4750
T: +1 202 7071205; Fax: +1 202 7071925; URL: www.loc.gov/rr/scitech
Ronald Bluestone
American Nat Standards; British, Chinese, French, German & Japanese Colls; OSRD Reports on World War II Res & Development (Dept of Energy, Dept of Defense, Nat Aeronautics & Space Administration & Nat Tech Info Service)
3 750 000 vols   30112

General Research Libraries

New York State Library, State Education Department, Cultural Education Center, Empire State Plaza, *Albany*, NY 12230
T: +1 518 4731189; Fax: +1 518 4866880; E-mail: nyslweb@mail.nysed.gov; URL: www.nysl.nysed.gov
1818; Bernard Margolis
Dutch Colonial Records, New York State Political & Social Hist, Shaker Coll; 1 branch libr
2 599 000 vols; 13 796 curr per; 202 000 maps; 1 200 000 microforms; 100 sound-rec; 100 digital data carriers
30113

New York State Library, Talking Book & Braille Library, Empire State Plaza, Cultural Education Ctr Basement, *Albany*, NY 12230
T: +1 518 4745935; Fax: +1 518 4861957
1896; Jane Somers
780 000 vols   30114

Maine State Library, LMA Bldg, 230 State St, *Augusta*, ME 04333
T: +1 207 2875441; Fax: +1 207 2875638; E-mail: reference.desk@maine.gov; URL: www.maine.gov/msl/
1839; N.N.
Genealogy, Hist; Maine, Indians, Conservation (Baxter Coll); 1 branch libr
282 000 vols; 644 curr per; 202 000 microforms; 42 147 av-mat; 430 sound-rec
30115

Texas State Library & Archives Commission, 1201 Brazos St, *Austin*, TX 78701; P.O. Box 12927, Austin, TX 78711-2927
T: +1 512 4635455; Fax: +1 512 4635436; E-mail: info@tsl.state.tx.us; URL: www.tsl.state.tx.us
1909; Peggy D. Rudd
Hist Coll, mss, maps; Professional Librarianship Coll; Broadside Coll; Texas & Federal Govt Docs Coll; 1 branch libr
2 000 000 vols; 352 curr per; 28 000 e-books
libr loan   30116

Texas State Library & Archives Commission – Talking Book Program, 1201 Brazos, *Austin*, TX 78711-1938; P.O. Box 12927, Austin, TX 78711-2927
T: +1 512 4635458; Fax: +1 512 9360685; E-mail: tbp.services@tsl.state.tx.us; URL: www.TexasTalkingBooks.org
1931; Ava M. Smith
Spanish Coll, cassettes; Texas Coll, cassettes
795 000 vols; 710 660 av-mat; 22 650 Large Print bks, 710 660 Talking Bks
30117

State Library of Louisiana, 701 N Fourth St, *Baton Rouge*, LA 70802-5232; P.O. Box 131, Baton Rouge, LA 70821-0131
T: +1 225 3424915; Fax: +1 225 2194725; E-mail: admin@state.lib.la.us; URL: www.state.lib.la.us
1925; Rebecca Hamilton
Louisiana; Louisiana hist, Huey Long photos
596 000 vols; 860 curr per; 17 810 large print bks
libr loan; ALA, PLA   30118

State Library of Louisiana – Services for the Blind & Physically Handicapped, 701 N Fourth St, *Baton Rouge*, LA 70802; P.O. Box 131, Baton Rouge, LA 70821-0131
T: +1 225 3424944; Fax: +1 225 3426817; E-mail: sbph@state.lib.la.us; URL: www.state.lib.la.us
1933; Margaret C. Harrison
Braille, Louisiana Cassettes, Descriptive Videos
100 000 vols; 10 000 large print bks
30119

North Dakota State Library, 604 East Blvd Ave, Dept 250, Liberty Memorial Bldg, *Bismarck*, ND 58505-0800
T: +1 701 3282492; Fax: +1 701 3282040; E-mail: statelib@nd.gov; URL: ndsl.lib.state.nd.us
1907; Doris Ott
North Dakota Coll
197 960 vols; 137 curr per; 9 000 govt docs; 14 649 sound-rec; 49 digital data carriers; 50 000 Talking Bks
libr loan   30120

State Library of Massachusetts, George Fingold Library, State House, Rm 341, 24 Beacon St, *Boston*, MA 02133
T: +1 617 7272590; Fax: +1 617 7275819; URL: www.state.ma.us/lib
1826; Elvernoy Johnson
Law, Legis hist, Political science, Public administration; Americana, early Massachusetts imprints; Massachusetts Hist & Biogr, atlases, city directories, mss, maps; Massachusetts State House Coll, photos & prints; New England Hist; Revolutionary War Broadsides
825 000 vols; 1 845 curr per; 7 196 maps; 700 000 microforms; 15 digital data carriers
libr loan   30121

Nevada State Library & Archives, 100 N Stewart St, *Carson City*, NV 89701-4285
T: +1 775 6843360; Fax: +1 775 6843330; URL: www.dmla.clan.lib.nv.us/docs/nsla
1859; Daphne DeLeon
Nevada Coll, Nevada Newspaper Coll (microform, retrospective), US Bureau of the Census Data Ctr; 1 branch libr
67 000 vols; 150 curr per
libr loan   30122

Nevada State Library & Archives – Regional Library for the Blind & Physically Handicapped, 100 N Stewart St, ***Carson City***, NV 89701-4285
T: +1 775 6843354; Fax: +1 775 6843355; URL: www.nevadaculture.org
1968; Keri Putnam
Nevada Authors
77 000 vols; 2 digital data carriers
libr loan 30123

Wyoming State Library, 2800 Central Ave, ***Cheyenne***, WY 82002
T: +1 307 7776339; Fax: +1 307 7776289; E-mail: refdesk@wyo.gov; URL: will.state.wy.us
1871; Lesley Boughton
Business & management, Libr science, Wyoming authors
500 000 vols; 348 curr per; 1 470 av-mat; 821 sound-rec; 88 Journals, 415 Large Print Bks
libr loan; ALA 30124

South Carolina State Library, 1430-1500 Senate St, ***Columbia***, SC 29201; P.O. Box 11469, Columbia, SC 29211
T: +1 803 7348026; Fax: +1 803 7348676; E-mail: reference@statelibrary.sc.gov; URL: www.statelibrary.sc.gov
1943; David Goble
Nonfiction, Southern hist; Eric Coll, Foundation Ctr Regional Coll, South Carolina Coll, South Carolina Govt Publs
275 000 vols; 2 123 curr per; 396 987 av-mat; 19 650 Large Print Bks, 400 000 Talking Bks
libr loan 30125

South Carolina State Library, Talking Book Services, 1430 Senate St, ***Columbia***, SC 29201-3710; P.O. Box 821, Columbia, SC 29202-0821
T: +1 803 7344611; Fax: +1 803 7344610; E-mail: tbsbooks@statelibrary.sc.gov; URL: www.state.sc.us/scsl/bph
1973; Guynell Williams
South Caroliniana, cassettes, descriptive videotapes
349 000 vols; 45 160 av-mat; 16 800 Large Print Bks, 45 160 Talking Bks 30126

State Library of Ohio, 274 E First Ave, Ste 100, ***Columbus***, OH 43201
T: +1 614 6446950; Fax: +1 614 6447004; E-mail: refhelp@sloma.state.oh.us; URL: www.library.ohio.gov
1817; Joanne Budler
Rare Bks
723 000 vols; 902 curr per; 7 000 e-books; 105 digital data carriers
libr loan 30127

New Hampshire State Library, 20 Park St, ***Concord***, NH 03301-6314
T: +1 603 2712144; Fax: +1 603 2712205; E-mail: nhslill@library.state.nh.us; URL: www.state.nh.us/nhsl
1716; Michael York
Hist Children's Bks, Lincoln Coll, New Hampshire Govt & Hist, New Hampshire Imprints, New Hampshire Maps; 1 branch libr
551 000 vols; 264 curr per; 1 131 av-mat; 35 digital data carriers; 8 141 mss
libr loan 30128

New Hampshire State Library – Talking Book Services, 117 Pleasant St, Gallen State Office Park, Dolloff Bldg, ***Concord***, NH 03301-3852
T: +1 603 2713429; Fax: +1 603 2718370; E-mail: marilyn.stevenson@dcr.nh.gov; URL: www.state.nh.us/nhsl/talkbks
1970; Marilyn Stevenson
80 000 vols; 24 curr per; 460 av-mat; 96 520 sound-rec; 6 180 Large Print Bks 30129

State Library of Iowa, 1112 E Grand Ave, ***Des Moines***, IA 50319
T: +1 515 2814105; Fax: +1 515 2816191; URL: www.statelibraryofiowa.org
1838; Mary Wegner
Attorney General Opinions, Bar Assn Proceedings, Iowa State Publs, Medical Coll
364 000 vols; 1 228 curr per
libr loan 30130

State Library of Pennsylvania, Forum Bldg, 607 South Dr, ***Harrisburg***, PA 17120-0600; mail address: Pennsylvania Dept of Education, 333 Market St, Harrisburg, PA 17126-1745
T: +1 717 7835950; Fax: +1 717 7832070; E-mail: alubrecht@pa.gov; URL: www.statelibrary.state.pa.us
1745; Caryn J. Carr
Jansen-Shirk Bookplate Coll; Central PA Genealogy; PA Newspapers – historic & current; PA imprints, 1689-1850; PA Colonial Assembly Coll; Jansen Mss Letters Coll
996 000 vols; 1 282 curr per; 530 780 govt docs; 7 908 maps; 3 554 000 microforms; 1 211 sound-rec; 5 903 digital data carriers; 215 Bks on Deafness & Sign Lang
ALA 30131

Connecticut State Library, 231 Capitol Ave, ***Hartford***, CT 06106
T: +1 860 7576520; Fax: +1 860 7576559; E-mail: isref@cslib.org; URL: www.cslib.org
1854; Kendall Wiggin
Govt, law, state hist, polital science; Cemetery inscriptions; Charter of 1662; colt firearms; Connecticut – aerial photogr survey maps; Arch & hist mss; census records; church, town & vital records; fraternal orders; legislative transcripts; newspapers; state statutes; governors' portraits; law rpts; legislative reference; map coll; medals, coins & Indian relics; military records & war posters; Connecticut Shelf Clock Coll; Old Houses of Connecticut; state & local hist; 3 branch libs
1 236 000 vols; 17 900 e-journals; 50 digital data carriers; 150 bks on deafness & Sign Lang
libr loan 30132

Montana State Library, 1515 East Sixth Ave, ***Helena***, MT 59620-1800
T: +1 406 4443115; Fax: +1 406 4440204; E-mail: dstaffeldt@mt.gov; URL: msl.mt.gov
1929; Darlene Staffeldt
Geographic Info System, MT Natural Resource Index, Natural Resource Info System, Water Info System, Talking Bk Libr
56 000 vols; 123 curr per; 50 e-journals; 324 e-books; 151 576 av-mat; 760 sound-rec; 3 090 digital data carriers; 180 Braille Vols
libr loan 30133

Indiana State Library, 140 N Senate Ave, ***Indianapolis***, IN 46204-2296
T: +1 317 2323675; Fax: +1 317 2323728; URL: www.in.gov/library
1825; Roberta L. Brooker
Hist, Indiana, Libr & info science; Americana (Holliday Coll), Genealogy (Darrach Coll of Indianapolis Pub Libr), Hymn Books (Levering Sunday School), Indiana Academy of Science, Libr, Shorthand & Typewriting (Strachan Coll)
2 191 000 vols; 16 908 curr per; 43 100 av-mat; 12 100 Large Print Bks, 43 100 Talking Bks
libr loan 30134

Missouri State Library, Wolfner Library for the Blind & Physically Handicapped, 600 W Main St, ***Jefferson City***, MO 65101-1532; P.O. Box 387, Jefferson City, MO 65102-0387
T: +1 573 7518720; Fax: +1 573 5262985; E-mail: wolfner@sos.mo.gov; URL: www.sos.mo.gov/wolfner
1924; Dr. Richard J. Smith
360 000 vols; 100 curr per; 76 140 av-mat; 403 000 sound-rec; 10 digital data carriers; 3 020 Large Print Bks, 41 830 Braille Vols 30135

Missouri State Library (MOSL), James C Kirkpatric State Information Ctr, Rm 200, 600 W Main St, ***Jefferson City***, MO 65101-1532; P.O. Box 387, Jefferson City, MO 65102-0387
T: +1 573 7510970; Fax: +1 573 5261142; E-mail: moslill@sos.mo.gov; URL: www.sos.mo.gov/library/
1907; Margaret M. Conroy
State govt
126 000 vols; 270 curr per; 3 840 e-journals; 4 000 e-books; 130 av-mat
libr loan 30136

Alaska State Library – Alaska Historical Collections, 333 Willoughby Ave, ***Juneau***, AK 99801; P.O. Box 110571, Juneau, AK 99811-0571
T: +1 907 4652925; Fax: +1 907 4652990; E-mail: asl.historical@alaska.gov; URL: www.eed.state.ak.us/lam
1900; Gladi Kulp
Alaskana (Wickersham Coll); Alaska-Arctic Research; Alaska Juneau Mining Company Records, Alaska Packers Association Records, Marine Hist (L H Bayers), doc; Russian American Coll; Russian Hist-General and Military, (Dolgopolov Coll); Salmon Canneries, (Alaska Packers Assn Records); Trans-Alaska Pipeline Impact; Winter & Pond Photogr Coll; Vinokouroff Coll
37 000 vols; 400 mss; 2 000 maps; 6 000 microforms 30137

Alaska State Library – Archives & Museums, 333 Willoughby Ave, State Office Bldg, 8th Flr, ***Juneau***, AK 99801; P.O. Box 110571, Juneau, AK 99811-0571
T: +1 907 4652988; Fax: +1 907 4652151; E-mail: aslanc@alaska.gov; URL: www.library.state.ak.us
1957; Linda Thibodeau
Education, Alaska Hist, Libr & info science, State government; Alaska Hist (Wickersham Coll of Alaskana, Alaska Marine Hist (L H Bayers Coll), Salmon Canneries, Trans-Alaska Pipeline Impact; 3 branch libs
110 000 vols; 350 curr per; 390 av-mat; 500 sound-rec 30138

Library of Michigan, 702 W Kalamazoo St, ***Lansing***, MI 48915; P.O. Box 30007, Lansing, MI 48909
T: +1 517 3731580; Fax: +1 517 3734480; URL: www.michigan.gov/hal
1828; Nancy Robertson
Genealogy, Law, Michigan, Public policy, Talking bks; Federal & Michigan Doc Coll, Michigan Resources Coll, Rare Bk Room
4 678 120 vols; 13 236 curr per 30139

Nebraska State Library, State Capitol Bldg, 3rd Flr South, Rm 325, P.O. Box 98931, ***Lincoln***, NE 68509-8931
T: +1 402 4713189; Fax: +1 402 4711011; E-mail: nsc.lawlibrary@nebraska.gov; URL: www.court.nol.org
1854; Marie Wiechman
Law
132 000 vols 30140

Arkansas State Library, One Capitol Mall, 5th Flr, ***Little Rock***, AR 72201-1085
T: +1 501 6822053; Fax: +1 501 6821531; URL: www.asl.lib.ar.us
1935; Carolyn Ashcraft
Reference, Business & management, Computer science, US industries; Arkansiana, Blind & Physically Handicapped, CIS microfiche, Library & Information Science (Prof Coll), Patent Depository; 1 branch libr
492 000 vols; 2 500 curr per
libr loan 30141

State of Vermont Department of Libraries, 109 State St, ***Montpelier***, VT 05609-0601
T: +1 802 8283261; Fax: +1 802 8281481; E-mail: dol_central@mail.dol.state.vt.us; URL: dol.state.vt.us
Martha Reid
Joseph L Wheeler Libr Sci Coll
675 000 vols; 1 000 curr per 30142

Tennessee State Library & Archives, 403 Seventh Ave N, ***Nashville***, TN 37243-0312
T: +1 615 7412764; Fax: +1 615 7416471; E-mail: reference.tsla@state.tn.us; URL: www.tennessee.gov/tsla
1854; Jeanne Sugg
Genealogy Coll, Popular Sheet Music (Rose Music), Southeastern US Maps, Tennessee Newspapers, Tennessee County & Public Records
822 000 vols; 329 curr per; 13 e-books; 221 557 av-mat 30143

Washington State Library, 6880 Capitol Blvd S, ***Olympia***, WA 98501-5513; P.O. Box 42460, Olympia, WA 98504-2460
T: +1 360 7045213; Fax: +1 360 5867575; URL: www.secstate.wa.gov/library
1853; Jan B. Walsh
Washington Newspapers, microflm; Washington Authors; Washington State Docs; 17 branch libs
1 608 000 vols; 265 curr per; 401 444 av-mat
libr loan 30144

Arizona State Library, Archives & Public Records, 1700 W Washington, Rm 200, ***Phoenix***, AZ 85007
T: +1 602 9264035; Fax: +1 602 2567983; URL: www.lib.az.us
1864; GladysAnn Wells
Arizona, Genealogy, Law; Federal & State Docs, State Law Libr
1 228 000 vols; 513 curr per; 131 000 microforms; 424 av-mat; 490 digital data carriers
libr loan 30145

South Dakota State Library, 800 Governors Dr, ***Pierre***, SD 57501-2294
T: +1 605 7735071; Fax: +1 605 7736962; E-mail: library@state.sd.us; URL: library.sd.gov
1913; Dan Siebersma
177 000 vols; 557 curr per; 692 000 microforms; 557 sound-rec; 336 digital data carriers
libr loan 30146

Rhode Island State Library, 82 Smith St, State House, Rm 208, ***Providence***, RI 02903
T: +1 401 2222473; Fax: +1 401 2223034; E-mail: tevans@sos.ri.gov; URL: www.sos.ri.gov/library/
1852; Thomas R. Evans
Law, Legislation, Local hist; Rhode Island Acts & Resolves, 1750-Present; Indexes to Rhode Island Legislation, 1758-Present; Rhode Island House & Senate Journals, 1908-Present; Legislative Reference
157 000 vols; 90 curr per; 1 mss; 140 000 govt docs; 100 maps; 30 000 microforms; 25 av-mat; 15 sound-rec; 45 digital data carriers; 500 other items
libr loan; ALA 30147

State Library of North Carolina, 109 E Jones St, ***Raleigh***, NC 27601; mail address: 4640-45 Mail Service Ctr, Raleigh, NC 27699-4640
T: +1 919 8077400; Fax: +1 919 7338748; URL: statelibrary.dcr.state.nc.us/
1812; Mary L. Boone
Demographics, North Carolina, Statistics
171 000 vols; 416 curr per
libr loan 30148

The Library of Virginia, 800 E Broad St, ***Richmond***, VA 23219-8000
T: +1 804 6923777; Fax: +1 804 6923556; URL: www.lva.virginia.gov
1823; Sandra G. Treadway
Virginia & Southern Hist, Virginia Newspapers, Virginia State Docs, Virginia Public Records, Virginia Maps, Genealogy Coll, Confederate Imprints, Virginia Broadsides, Virginia Picture Coll, Sheet Mucic
819 000 vols; 794 curr per
libr loan 30149

California State Library, Library & Courts Bldg 1, 914 Capitol Mall, Rm 220, ***Sacramento***, CA 95814; P.O. Box 942837, Sacramento, CA 94237-0001
T: +1 916 6540266; Fax: +1 916 6540241; E-mail: cslsirc@library.ca.gov; URL: www.library.ca.gov/
1850; N.N.
1 327 000 vols; 4 067 curr per; 10 000 Electronic Media & Resources
libr loan 30150

Oregon State Library, 250 Winter St NE, ***Salem***, OR 97301-3950
T: +1 503 3784243; Fax: +1 503 5858059; E-mail: library.help@state.or.us; URL: oregon.gov/osl
1905; James B. Scheppke
Family Hist (Genealogy Coll), bks, micro; Oregon Hist (Oregoniana Coll); State & Federal Govt, bks, doc
63 000 vols; 301 curr per; 193 mss; 10 470 maps; 518 000 microforms;

127 353 av-mat; 253 sound-rec; 1 683 digital data carriers; 127 353 Talking Bks
libr loan   30151

Oregon State Library Talking Book & Braille Services, 250 Winter St NE, **Salem**, OR 97301-3950
T: +1 503 378-5389; Fax: +1 503 5858059; E-mail: susan.b.westin@state.or.us; URL: www.oregon.gov/osl/tbabs
1932; Susan Westin
Descriptive Video Coll
158 000 vols
libr loan   30152

Utah State Library Division, 250 N 1950 W, Ste A, **Salt Lake City**, UT 84116-7901
T: +1 801 715-6777; Fax: +1 801 715-6767; E-mail: dmorris@utah.gov; URL: www.library.utah.gov
1957; Donna Jones Morris
Local Utah Hist Coll; 17 bookmobiles
44 941 vols; 26 curr per; 233 av-mat; 249 sound-rec
libr loan   30153

Utah State Library Division – Program for the Blind & Disabled, 250 N 1950 West, Ste A, **Salt Lake City**, UT 84116-7901
T: +1 801 7156789; Fax: +1 801 7156767; E-mail: blind@utah.gov; URL: blindlibrary.utah.gov
1957; Bessie Oakes
Mormon Lit Coll, Western Books Coll
470 000 vols; 140 curr per; 1 830 High Interest/Low Vocabulary Bks, 30 Bks on Deafness & Sign Lang   30154

California State Library – Sutro Library, 480 Winston Dr, **San Francisco**, CA 94132
T: +1 415 7314477; Fax: +1 415 5579325; E-mail: sutro@library.ca.gov; URL: www.library.ca.gov
1913; Haleh Motiey
Ancient Hebrew Mss and Scrolls Coll; English Hist and Lit Colll; Genealogy and Local Hist Coll; Hist of Printing and Book Illustrations Coll; Hist of the Pure and Applied Sciences Coll; Mexican Hist and Lit Coll; Papers of Sir Joseph Banks Coll; Voyages and Travel Coll
150 000 vols
libr loan   30155

New Mexico State Library, 1209 Camino Carlos Rey, **Santa Fe**, NM 87507
T: +1 505 4769700; Fax: +1 505 4769701; URL: www.stlib.state.nm.us
1929; Susan Oberlander
Govt, Public policy; Southwest Resources, bks, doc & per; 1 branch libr
200 000 vols; 2 000 curr per; 510 av-mat; 18 digital data carriers; 1,29 mio docs
libr loan   30156

Illinois State Library, Gwendolyn Brooks Bldg, 300 S Second St, **Springfield**, IL 62701-1976
T: +1 217 7827573; Fax: +1 217 7854326; E-mail: islinformationline@ilsos.net; URL: www.cyberdriveillinois.com/departments/library/home.html
1839; Anne Craig
Govt, Political science; Illinois Authors Coll; US Patent, Trademark & Maps; Talking Bk & Braille Service
1 845 000 vols; 1 025 curr per
libr loan   30157

Florida Department of State – Division of Library & Information Services, State Library of Florida, 500 S Bronough St, R A Gray Bldg, **Tallahassee**, FL 32399-0250
T: +1 850 2456600; Fax: +1 850 2456735; E-mail: info@dos.state.fl.us; URL: dlis.dos.state.fl.us/index.cfm
1845; Judith A. Ring
Florida Coll
684 090 vols; 1 391 curr per
libr loan   30158

State Library of Kansas, State Capitol Bldg, 300 SW 10th Ave, Rm 343-N, **Topeka**, KS 66612; mail address: 300 SW Tenth St, Rm 343-N, Topeka, KS 66612
T: +1 785 2963296; Fax: +1 785 2966650; E-mail: infodesk@kslib.info; URL: www.kslib.info

1855; Christie P. Brandau
Social sciences, Govt; Kansas Legislative Mat; Talking Book Service, Literacy Service
200 000 vols; 250 curr per; 50 000 av-mat; 25 Bks on Deafness & Sign Lang
libr loan   30159

New Jersey State Library, 185 W State St, **Trenton**, NJ 08618; P.O. Box 520, Trenton, NJ 08625-0520
T: +1 609 2782640; Fax: +1 609 2782647; E-mail: refdesk@njstatelib.org; URL: www.njstatelib.org
1796; Norma E. Blake
Jerseyana
1 452 000 vols; 2 781 curr per; 11 000 e-books
libr loan   30160

New Jersey State Library, New Jersey Library for the Blind & Physically Handicapped, 2300 Stuyvesant Ave, **Trenton**, NJ 08618; P.O. Box 501, Trenton, NJ 08625-0501
T: +1 609 5304000; Fax: +1 609 5306384; E-mail: njlbh@njstatelib.org; URL: www.njlbh.org
1968; Adam Szczepaniak Jr
763 000 vols; 90 curr per; 48 000 av-mat; 32 430 Braille vols, 12 Special Interest Per Sub, 150 Bks on Deafness & Sign Lang, 18 920 Large Print Bks, 45 400 Talking Bks   30161

University Libraries, College Libraries

Northern State University, Beulah Williams Library, 1200 S Jay St, **Aberdeen**, SD 57401-7198
T: +1 605 6262645; Fax: +1 605 6262473; E-mail: mulvaney@northern.edu; URL: www.northern.edu/library
1901; Dr. J. Philip Mulvaney
Business, Education; South Dakota Hist, Harriet Montgomery Water Resources Coll
201 630 vols; 150 curr per
libr loan   30162

Abilene Christian University, Margaret & Herman Brown Library, 221 Brown Library, ACU Box 29208, **Abilene**, TX 79699-9208
T: +1 325 6742344; Fax: +1 325 6742202; URL: www.acu.edu/library
1906; Dr. Mark Tucker
Donner Libr of Americanism, Sewell Bible Libr, Bibles, Herald of Truth Radio & Television Arch, Burleson Congressional Papers, Church Hist & Arch Mat, Robbins Railroad Coll, Austin Taylor Hymn Book Coll
499 000 vols; 2 417 curr per; 2 113 e-journals; 34 000 e-books; 9 250 maps; 3 500 av-mat; 5 310 sound-rec; 500 digital data carriers; 400 Bks on Deafness & Sign Lang, 33 Large Print Bks
libr loan   30163

Hardin-Simmons University, Richardson Library, 2341 Hickory St, **Abilene**, TX 79698; P.O. Box 16195, Abilene, TX 79698-6195
T: +1 325 6701578; Fax: +1 325 6778351; URL: rupert.alc.org/library
1891; Alice Specht
Local Hist Photogr Arch; Printing of Carl Hertzog; Southwest Hist, bks & micro; Texana; Betty Woods Rare Book Coll; Lee & Lunell Hemphill Business Reading Room
298 000 vols; 664 curr per; 35 000 e-journals; 27 000 e-books; 2 500 av-mat; 9 400 sound-rec; 200 digital data carriers; 30 000 Journals
libr loan   30164

McMurry University, Jay-Rollins Library, Sayles Blvd & S 14th, **Abilene**, TX 79697; mail address: McMurry Sta, Box 218, Abilene, TX 79697-0218
T: +1 325 7934692; Fax: +1 325 7934930; URL: www.mcm.edu/academic/depts/library/libraryhome.htm
1923; Joe W. Specht
20th c American popular culture, African-American studies, Religious studies, Spanish-American lit; Hunt Libr of Texana & Southwest; E L & A W Yeats Coll
137 000 vols; 580 curr per; 500 av-mat; 835 sound-rec; 6 digital data carriers
                                             30165

Pennsylvania State University, Abington College Library, 1600 Woodland Rd, **Abington**, PA 19001
T: +1 215 8817462; Fax: +1 215 8817423; URL: www.libraries.psu.edu/abington
1950; Samuel R. Stormont
69 000 vols; 163 curr per; 139 av-mat
                                             30166

East Central University, Linscheid Library, 1100 E 14th St, **Ada**, OK 74820-6999
T: +1 580 3105374; Fax: +1 580 4363242; URL: www.ecok.edu
1909; Dr. Adrianna Lancaster
176 000 vols; 702 curr per
libr loan   30167

Ohio Northern University, Heterick Memorial Library, 525 S Main St, **Ada**, OH 45810
T: +1 419 7722185; Fax: +1 419 7721927; URL: www.onu.edu/library
1915; Paul Logsdon
Ohio Northern Univ Authors
256 000 vols; 5 500 curr per
libr loan   30168

– Taggart Law Library, 525 S Main St, Ada, OH 45810
T: +1 419 7722239; Fax: +1 419 7721875; URL: www.law.onu.edu/library.html
Nancy A. Armstrong
211 000 vols; 3 023 curr per   30169

Adrian College, Shipman Library, 110 S Madison St, **Adrian**, MI 49221
T: +1 517 2643828; Fax: +1 517 2643748; URL: www.adrian.edu/library
1859
Lincolniana (Piotrowski-Lemke), Methodist (Detroit Conference Arch), Women's Studies (Microcard Edition)
132 000 vols; 571 curr per; 56 000 microforms; 1 600 av-mat; 750 sound-rec; 25 digital data carriers   30170

Siena Heights College, Library, 1247 E Siena Heights Dr, **Adrian**, MI 49221-1796
T: +1 517 2647150; Fax: +1 517 2647711; URL: www.sienaheights.edu/library.aspx
1919; Dr. Robert W. Gordon
Art & architecture
113 000 vols; 352 curr per   30171

International College, J.W. Cook Memorial Library, 99-860 Iwaena St, **Aiea**, HI 96701
T: +1 808 4844045; Fax: +1 808 4848887; URL: www.hits.edu
1971; Beverly Vallejo-Sanderson
Religion; Anthropology & Missiology Coll, Theology Coll
21 000 vols; 42 curr per; 453 microforms; 73 av-mat; 244 sound-rec   30172

University of South Carolina – Aiken, Gregg-Graniteville Library, 471 University Pkwy, **Aiken**, SC 29801
T: +1 803 6486851; Fax: +1 803 6413302; URL: library.usca.edu
1961; Jane H. Tuten
May Coll of Southern Hist, Dept of Energy Public Docs Coll, Gregg-Graniteville Hist Files
136 000 vols; 1 328 curr per; 4 000 e-books; 740 CDs   30173

University of Akron, University Libraries, 315 Buchtel Mall, **Akron**, OH 44325-1701
T: +1 330 9726275; Fax: +1 330 9725106; URL: www.uakron.edu/libraries/index.php
1872; Cheryl Kern-Simirenko
Propaganda Coll, B-26 Arch, The Arch of the Hist of of American Psychology & the American Res Ctr
1 260 000 vols; 15 266 curr per; 3 000 e-books; 270 Electronic Media & Resources   30174

– Science and Technology Library, 104 Auburn Science Center, Akron, OH 44325-3907
T: +1 330 9727195; Fax: +1 330 9727033; URL: www.uakron.edu:80/library
1967; Sherri Edwards
ASC Rubber Division Papers
95 000 vols; 2 300 curr per   30175

New Mexico State University at Alamogordo, Townsend Library, 2400 N Scenic Dr, **Alamogordo**, NM 88310
T: +1 505 4393650; Fax: +1 505 4393657; E-mail: library@nmsua.nmsu.edu; URL: alamo.nmsu.edu/library
1974; Dan Kammer
52 650 vols; 250 curr per
libr loan; ALA, ACRL   30176

Adams State College, Nielsen Library, 208 Edgemont Ave, **Alamosa**, CO 81102-2373
T: +1 719 5877781; Fax: +1 719 5877590; E-mail: asclib@adams.edu; URL: www.library.adams.edu
1925; Dianne L. Machado
Education, Hist, Law, Latin America; Colorado Room, San Luis Valley Hist
130 000 vols; 404 curr per; 1 213 maps; 639 000 microforms; 1 577 av-mat; 2 499 sound-rec; 15 digital data carriers; 191 art reproductions, 1 600 archives
libr loan   30177

Albany Law School, Schaffer Law Library, 80 New Scotland Ave, **Albany**, NY 12208
T: +1 518 4452340; Fax: +1 518 4725842; URL: www.albanylaw.edu
1851; Robert T. Begg
Anglo-American Law Coll
285 000 vols; 1 666 curr per; 950 sound-rec; 38 digital data carriers
libr loan   30178

Albany Medical College, Schaffer Library of Health Sciences, 47 New Scotland Ave, MC 63, **Albany**, NY 12208
T: +1 518 2625530; Fax: +1 518 2625820; E-mail: library@mail.amc.edu; URL: www.amc.edu/academic/library
1928; Enid Geyer
149 000 vols; 3 810 curr per   30179

Albany State University, James Pendergrast Memorial Library, 504 College Dr, **Albany**, GA 31705-2796
T: +1 229 4304799; Fax: +1 229 4304803; E-mail: library@asurams.edu; URL: asuweb.asurams.edu/asu/pendergrast
1903; LaVerne McLaughlin
Hist & Lit (Black Studies), Libr of American Civilization, US Govt Census Data
193 000 vols; 323 curr per; 27 000 e-books; 500 000 microforms
libr loan   30180

College of Saint Rose, Neil Hellman Library, 392-396 Western Ave, **Albany**, NY 12203
T: +1 518 4542155; Fax: +1 518 4542897; E-mail: refdesk@mail.strose.edu; URL: library.strose.edu
1920; Peter Koonz
Education, Special education; College Arch, Curriculum libr
202 000 vols; 925 curr per; 200 000 microforms; 630 sound-rec; 11 digital data carriers   30181

University at Albany, State University of New York, University Libraries, 1400 Washington Ave, **Albany**, NY 12222-0001
T: +1 518 4423613; Fax: +1 518 4423567; URL: library.albany.edu/
1844; Frank D'Andraia
German Intellectual Emigre Coll, Arch for Public Affairs & Policy, Children's Hist Lit Coll
2 094 000 vols; 5 187 curr per; 4 830 e-journals; 5 000 e-books; 5 633 av-mat; 2 720 sound-rec; 26 460 digital data carriers; 220 Bks on Deafness & Sign Lang
libr loan   30182

– Governor Thomas E. Dewey Graduate Library for Public Affairs & Policy, 135 Western Ave, Albany, NY 12222
T: +1 518 4423691; Fax: +1 518 4425780; E-mail: dewref@albany.edu; URL: library.albany.edu/dewey
1981; Barbara Jean Via
123 000 vols; 636 curr per   30183

– Science Library, 1400 Washington Ave, Albany, NY 12222
T: +1 518 4373945; Fax: +1 518 4373952; URL: library.albany.edu/science/
Gregg Baron
350 000 vols   30184

Albion College, Stockwell – Mudd Libraries, 602 E Cass St, ***Albion***, MI 49224-1879; mail address: 611 E. Porter St, Kellog Ctr 4692, Albion, MI 49224
T: +1 517 6290382; Fax: +1 517 6290504; URL: www.albion.edu/library
1835; Dr. John P. Kondelik
Liberal arts; Albion Americana, Albion College Arch, Bible Coll, M.F.K. Fisher, Modern Literary First Editions, Western Michigan Conference of United Methodist Church Arch
368 000 vols; 2 091 curr per; 1 360 e-journals; 6 000 e-books; 3 000 av-mat; 2 500 sound-rec; 2 000 digital data carriers; 40 Bks on Deafness & Sign Lang
libr loan 30185

University of New Mexico – Zimmerman Library, 1900 Roma NE, ***Albuquerque***, NM 87131; mail address: One University of New Mexico, MSC 05-3020, Albuquerque, NM 87131-0001
T: +1 505 2772003; Fax: +1 505 2774097; E-mail: libinfo@unm.edu; URL: elibrary.unm.edu
1892; Martha Bedard
Indian Affairs (Glenn Leonidas Emmons & Michael Steck Coll), papers; Indians (Doris Duke Foundation Coll AIM arch); Land Records (Maxwell Land Grant Company, United States Soil Conservation Service Reports); Center for Southwest Research, Popular Culture Coll
2 100 000 vols; 13 000 curr per
libr loan; ARL 30186

– Centennial Science & Engineering Library, MSC05 3020, 1 University of New Mexico, Albuquerque, NM 87131-0001
T: +1 505 2774858; Fax: +1 505 2770702; E-mail: cselref@unm.edu; URL: elibrary.unm.edu/csel/
1988
Map & Geographic Info Ctr (MAGIC); Patent Depot (1833 to present)
345 000 vols; 2 000 curr per; 180 000 maps; 1 400 000 microforms; 7 029 air photos 30187

– Fine Arts Library, MSC 05 3020, One University of New Mexico, Albuquerque, NM 87131
T: +1 505 2772355; Fax: +1 505 2777134; E-mail: falref@unm.edu; URL: elibrary.unm.edu/falref
Dena Kinney
230 000 vols; 250 curr per; 69 000 e-books 30188

– Health Sciences Library + Informatics Center, Albuquerque, NM 87131-5686
T: +1 505 2722311; Fax: +1 505 2725350; E-mail: RefLib@salud.unm.edu; URL: hsc.unm.edu/library
1963; Holly Shipp Buchanan
Indian Health Papers; Southwest & New Mexico Medicine, media; World Health; Health Historical Archives
178 992 vols; 1 640 curr per; 1 282 govt docs; 623 microforms; 2 075 av-mat; 560 sound-rec; 90 digital data carriers
libr loan; MLA, AAHSL, AMIA 30189

– Law Library, 1117 Stanford Dr NE, Albuquerque, NM 87131-1441
T: +1 505 2776236; Fax: +1 505 2770068; URL: lawschool.unm.edu/lawlib/index.php
1948; Carol Parker
American Indian Law Coll, Land Grant Law Coll, Mexican & Latin American Legal Mat
413 000 vols; 3 168 curr per; 741 000 microforms; 22 digital data carriers; 12 000 court records & briefs
libr loan 30190

– William J. Parish Memorial Business & Economics Library, One University of New Mexico, MSC05 3020, Albuquerque, NM 87131-1496
T: +1 505 2775912; Fax: +1 505 2779813; URL: elibrary.unm.edu/parish
Susan C. Awe
Economics, management
185 000 vols; 5 000 curr per; 8 000 e-books 30191

Alcorn State University, J. D. Boyd Library, 1000 ASU Dr, ***Alcorn State***, MS 39096-7500
T: +1 601 8776350; Fax: +1 601 8773885; URL: jdboyd.alcorn.edu
1871; Jessie B. Arnold
Agriculture, Education; Alcorn Arch
287 000 vols; 1 046 curr per; 19 000 e-books; 487 maps; 351 000 microforms; 639 av-mat; 814 sound-rec; 14 digital data carriers; 32 Slides, 170 Overhead Transparencies 30192

Louisiana State University at Alexandria, James C. Bolton Library, 8100 Hwy 71 S, ***Alexandria***, LA 71302
T: +1 318 7673975; Fax: +1 318 4736556; URL: library.lsua.edu
1960; Albert Jules Tate III
Local Hist; Louisiana Newspapers; United States Census Records, microfilm
166 000 vols; 1 541 curr per; 57 000 e-books
libr loan; ALA, SOLINET 30193

Alfred University, Herrick Memorial Library, One Saxon Dr, ***Alfred***, NY 14802
T: +1 607 8712184; Fax: +1 607 8712299; URL: herrick.alfred.edu
1857; Steve Crandall
American hist, Behav science, Social sciences; British Lit & Hist (Openhym Coll), William Dean Howells (Frechette Coll), Nazi Germany (Waid Coll)
150 000 vols; 550 curr per; 28 000 e-journals; 27 000 e-books
libr loan 30194

– Scholes Library of Ceramics, New York State College of Ceramics at Alfred University, Two Pine St, Alfred, NY 14802-1297
T: +1 607 8712492; Fax: +1 607 8712349; URL: scholes.alfred.edu
1947; Carla Conrad Johnson
Building Materials (McBurney Coll); Ceramics (Barringer Coll); Glass (Hostetter, Silverman, & Shand Coll); Metals & Materials (Spretnak Coll); Ceramic Art (NCECA Arch), (Charles F Binns Papers) & John & Mae McMahon memorabilia
104 000 vols; 699 curr per; 5 040 diss/theses; 60 095 govt docs; 202 maps; 60 000 microforms; 860 av-mat; 307 sound-rec; 40 digital data carriers; 629 microtexts, 158 890 slides, 1 422 pamphlets, 420 trade cat, 30 laser discs, 875 pictures
libr loan; ALA 30195

State University of New York – College of Technology, Walter C. Hinkle Memorial Library, ***Alfred***, NY 14802
T: +1 607 5874313; Fax: +1 607 5874351; URL: web.alfredstate.edu/library
1911; David G. Haggstrom
Agriculture, Allied health, Business, Engineering; Western New York Hist Coll
64 000 vols; 293 curr per; 72 000 microforms; 12 digital data carriers
libr loan 30196

Alliant International University, Los Angeles Campus Library, 1000 S Fremont Ave, Unit 5, ***Alhambra***, CA 91803
T: +1 626 2842777; Fax: +1 626 2841682; URL: library.alliant.edu
1969; Stephanie Ballard
Mental health, Psychology, Women's studies; Psychology Fine Editions (Abbott Kaplan Memorial Coll)
26 000 vols; 71 curr per; 1 813 e-journals; 5 000 e-books; 2 705 diss/theses; 2 000 microforms; 641 sound-rec; 84 digital data carriers; 214 psychological tests
libr loan 30197

Grand Valley State University, University Libraries, One Campus Dr, ***Allendale***, MI 49401-9403
T: +1 616 3312630; Fax: +1 616 3312895; E-mail: refdesk@gvsu.edu; URL: www.gvsu.edu/library
1962; Lee VanOrsdel
US Geological Survey Maps, Limited Edition Series, Lincoln & the Civil War Coll, Michigan, Michigan Novels
634 000 vols; 5 000 curr per
libr loan 30198

Cedar Crest College, Cressman Library, 100 College Dr, ***Allentown***, PA 18104-6196
T: +1 610 6063543; Fax: +1 610 7403769; URL: library.cedarcrest.edu
1867; MaryBeth Freeh
Women's studies; Women in the US, bks, journals; American Poetry, bks, journals; Social Work, bks, journals
141 000 vols; 2 647 curr per; 2 000 e-books; 12 000 microforms; 1 340 av-mat; 4 077 sound-rec; 34 digital data carriers 30199

Muhlenberg College, Harry C. Trexler Library, 2400 Chew St, ***Allentown***, PA 18104-5586
T: +1 484 6643500; Fax: +1 484 6643511; E-mail: refdesk@muhlenberg.edu; URL: www.muhlenberg.edu/library
1867; Joyce Hommel
Muhlenberg Family Coll, Pennsylvania German Coll, Samuels Sheet Music Coll, Ray R Brennan Map Coll, Sam Slowall Coll, Paul McHale Papers
227 000 vols; 14 760 curr per; 18 993 e-journals; 6 incunabula; 100 diss/theses; 281 330 govt docs; 63 000 music scores; 137 000 microforms; 4 406 av-mat; 4 557 sound-rec; 3 270 digital data carriers; 70 Bks on Deafness & Sign Lang
libr loan 30200

Mount Union College, Library, 1972 Clark Ave, ***Alliance***, OH 44601-3993
T: +1 330 8234140; Fax: +1 330 8233963; URL: www.muc.edu/article/archive/127/
1846; Robert Garland
Education, Sports medicine; Greek & Latin Classics (Charles Sutherin), Graphic Arts (Shilts Rare Bks)
243 000 vols; 941 curr per; 396 846 govt docs; 39 000 microforms; 6 132 sound-rec; 150 digital data carriers; 1 135 slides
libr loan 30201

Alma College, Library, 614 W Superior St, ***Alma***, MI 48801
T: +1 989 4637229; Fax: +1 989 4638694; URL: library.alma.edu
1889; Carol A. Zeile
College Arch
252 000 vols; 1 200 curr per
libr loan; OCLC 30202

Sul Ross State University, Bryan Wildenthal Memorial Library, PO Box C-109, ***Alpine***, TX 79832-0001
T: +1 432 8378123; Fax: +1 432 8378400; URL: www.libit.sulross.edu
1920; Don Dowdey
Arch of the Big Bend
221 000 vols; 1 245 curr per
libr loan 30203

Theosophical University Library, 2416 N Lake Ave, ***Altadena***, CA 91001; P.O. Box C, Pasadena, CA 91109-7107
T: +1 626 7988020; Fax: +1 626 7984749; E-mail: tslibrary@theosociety.org
1919; James T. Belderis
Christian origins, Egypt, Greek & Roman, India, Pre-Columbian America, Theosophy
45 000 vols; 30 curr per 30204

Southern Illinois University – School of Dental Medicine, Biomedical Library, SIUE Campus Box 1111, 2800 College Ave, ***Alton***, IL 62002
T: +1 618 4747277; Fax: +1 618 4747270; URL: www.siue.edu/lovejoylibrary/biomed/contacts.shtml
1970; Lois Ridenour
35 000 vols; 151 curr per 30205

Pennsylvania State University – Altoona College, Altoona Campus, Robert E. Eiche Library, 3000 Ivyside Park, ***Altoona***, PA 16601-3760
T: +1 814 9495256; Fax: +1 814 9495520; URL: www.aa.psu.edu
1939; Timothy Wherry
Drama & The Dance (Cutler Coll), Drama on Records (Buzzard Coll), Lincoln Coll (Klevan Coll)
90 000 vols 30206

Western Oklahoma State College, Learning Resources Center, 2801 N Main St, ***Altus***, OK 73521
T: +1 580 4777770; Fax: +1 580 4777777; E-mail: librarian@wosc.edu; URL: www.wosc.edu/index.php?page=library-lrc
1926; Tony Hardman
32 530 vols 30207

Northwestern Oklahoma State University, J.W. Martin Library, 709 Oklahoma Blvd, ***Alva***, OK 73717
T: +1 580 3278574; Fax: +1 580 3278501; URL: www.nwosu.edu/library/index.html
1897; Susan Jeffries
William J Mellor Coll of Indian artifacts, bks, paintings, sculpture, stereoptican slides, cylinder records & player; Children's Lit Coll
159 000 vols; 1 405 curr per; 125 000 Gov Pubs, 325 000 Items on Microfiche 30208

Georgia Southwestern State University, James Earl Carter Library, 800 Georgia Southwestern State University Dr, ***Americus***, GA 31709
T: +1 229 9312259; Fax: +1 229 9312265; E-mail: libcirc@canes.gsw.edu; URL: www.gsw.edu/~library/index.html
1928; Vera J. Weisskopf
Education; Third World Studies Coll
144 000 vols; 516 curr per; 27 000 e-books
libr loan 30209

Iowa State University, University Library, 302 Parks Library, ***Ames***, IA 50011-2140
T: +1 515 2948073; Fax: +1 515 2945525; URL: www.lib.iastate.edu
1869; Olivia M. A. Madison
Agriculture, Hist of Science; American Arch of the Factual Film, American Arch of Veterinary Medicine, Arch of Women in Science & Eng, Evolution/Creation Arch, Hist of Science & Technology, Regional Hist, Soil Conservation, Statistical Arch
2 462 000 vols; 32 775 curr per; 128 426 maps; 3 015 000 microforms; 13 953 sound-rec; 478 digital data carriers; graphic mat (721 324)
libr loan; ARL, OCLC, BCR 30210

Amherst College, Robert Frost Library, ***Amherst***, MA 01002
T: +1 413 5422373; Fax: +1 413 5422662; URL: www.amherst.edu/library
1821; Sherre L. Harrington
Theater; Amherst College Hist Coll, American Poetry & Theater Coll, English Lit & Theater Coll
1 078 000 vols; 5 798 curr per
libr loan 30211

– Keefe Science Library, Amherst, MA 01002
T: +1 413 5428112; Fax: +1 413 5422662; URL: www.amherst.edu/library/depts/science
1968; Susan Kimball
53 000 vols 30212

– Vincent Morgan Music Library, Amherst, MA 01002
T: +1 413 5422387; URL: www.amherst.edu/library/depts/music
Jane Beebe
31 000 vols 30213

Daemen College, Library, 4380 Main St, ***Amherst***, NY 14226-3592
T: +1 716 8398243; Fax: +1 716 8398475; E-mail: fcarey@daemen.edu; URL: my.daemen.edu/library
1948; Francis Carey
98 000 vols; 589 curr per; 38 000 e-books
libr loan 30214

Hampshire College, Harold F. Johnson Library Center, 893 West St, ***Amherst***, MA 01002-5001
T: +1 413 5596691; Fax: +1 413 5595419; E-mail: library@hampshire.edu; URL: www.hampshire.edu/library
1970; Susan Dayall
Environmental studies, Films & filmmaking, Gender studies, Public policy, Third World
135 000 vols; 28 558 curr per 30215

University of Massachusetts at Amherst, W.E.B. Du Bois Library, 154 Hicks Way, ***Amherst***, MA 01003-9275
T: +1 413 5451370; Fax: +1 413 5771634; URL: www.library.umass.edu
1865; Jay Schafer
Afro-American, Agriculture, Ethnology, Geography, Latin America, Massachusetts, Natural hist, New

England; Benjamin Smith Lyman Papers & Japanese Coll, Broadside Press, French Revolution Coll, Harvey Swados Papers, Mass Govt Publs, Robert Francis Coll, Slavery Pamphlets, W E B Du Bois Papers, W B Yeats
5 900 000 vols; 15 427 curr per; 114 082 maps; 2 487 000 microforms; 5 703 363 other items
libr loan 30216

– Biological Sciences, Morrill Science Ctr, Amherst, MA 01003
T: +1 413 5456739; Fax: +1 413 5771531; E-mail: mschmidt@library.umass.edu; URL: www.library.umass.edu
1963; Maxine Schmidt
Arthur Cleveland Bent Coll (ornithology), Guy Chester Crampton Coll (entomology and evolutionary biology)
100 000 vols; 1 460 curr per; 125 000 maps 30217

– Science and Engineering Library, Lederle Graduate Res Ctr, Amherst, MA 01003
T: +1 413 5451370; Fax: +1 413 5771534; E-mail: pborrego@library.umass.edu; URL: www.library.umass.edu
1971
300 000 vols; 900 000 tech rpts on microfiche, US patents (1950 to present) 30218

Alaska Resources Library & Information Services (ARLIS), Library Bldg, 3211 Providence Dr, Ste 111, ***Anchorage***, AK 99508-4614
T: +1 907 7867677; Fax: +1 907 7867652; E-mail: reference@arlis.org; URL: www.arlis.org
1997; Cathy Vitale
Cultural res, Natural resources; US Doc Depository
200 000 vols; 700 curr per; 185 e-journals; 43 117 mss; 450 incunabula; 3 343 diss/theses; 1 390 maps; 6 000 microforms; 13 000 sound-rec; 16 digital data carriers
libr loan; ALA, SAA 30219

University of Alaska Anchorage – Consortium Library, 3211 Providence Dr, ***Anchorage***, AK 99508-8176
T: +1 907 7861871; Fax: +1 907 7861834; URL: www.consortiumlibrary.org
1973; Stephen Rollins
Health sciences; Alaskana & Polar Regions Coll, Arch & Mss Coll, Music Coll
730 000 vols; 3 700 curr per; 5 000 e-books; 5 437 maps; 440 000 microforms; 200 digital data carriers; 2 991 cubic feet of archival items
libr loan 30220

University of Alaska, Anchorage – Environment & Natural Resources Institute, Arctic Environmental Information & Data Center Library, 707 A St, ***Anchorage***, AK 99501
T: +1 907 2572732; Fax: +1 907 2572707; E-mail: anjaa@uaa.alaska.edu
1972; Judy Alward
8 800 vols; 5 digital data carriers; 11 500 gov publ & microforms 30221

Anderson College, Johnston Memorial Library, 316 Boulevard, ***Anderson***, SC 29621
T: +1 864 2312050; Fax: +1 864 2312191; E-mail: library@andersonuniversity.edu; URL: www.andersonuniversity.edu/library/resources.html
1911; Kent Millwood
77 000 vols; 33 000 e-books; 57 000 mss; 5 000 microforms; 800 sound-rec; 80 digital data carriers
libr loan; ALA 30222

Anderson University, Robert A. Nicholson University Library, 1100 E Fifth St, ***Anderson***, IN 46012-3495
T: +1 765 6414286; Fax: +1 765 6413850; URL: library.anderson.edu
1917; Dr. Janet L. Brewer
Religion, Criminal law, Nursing; Archs
182 000 vols; 827 curr per; 1 000 e-books; 13 digital data carriers 30223

Massachusetts School of Law Library, 500 Federal St, ***Andover***, MA 01810
T: +1 978 6810800; Fax: +1 978 6816330; URL: www.mslaw.edu/directory_library.htm
Judith Wolfe
81 000 vols; 276 curr per
libr loan 30224

Trine University, Perry T. Ford Memorial Library, 300 S Darling St, mail address: One University Ave, ***Angola***, IN 46703-1764
T: +1 260 6654162; Fax: +1 260 6654283; E-mail: brewerk@trine.edu; URL: www.trine.edu/
1962; Kristina Brewer
Engineering, Business, Teacher education; NACA, NASA Publs, Hershey Coll, NATO Advisory Group for Aerospace Res & Development (AGARD) Publs, NATO Res & Technology Organization Publs
74 580 vols; 345 curr per; 147 e-journals; 6 100 e-books; 383 maps; 2 951 microforms; 4 360 av-mat; 1 263 sound-rec; 6 digital data carriers 30225

Trine University, Sponsel Library & Information Services, Center for Technology & Online Resources, One University, ***Angola***, IN 46703
T: +1 260 6654162; Fax: +1 260 6654272; URL: www.trine.edu/lis
Kristina Brewer
62 000 vols; 282 curr per; 6 000 e-books 30226

Pacific Union College, W.E. Nelson Memorial Library, One Angwin Ave, ***Angwin***, CA 94508-9705
T: +1 707 9656241; Fax: +1 707 9656504; URL: www.library.puc.edu
1882; Adu Worku
Natural science; Pitcairn Islands Study Ctr, Seventh-day Adventist Study Ctr, Ellen G. White Study Ctr
142 000 vols; 812 curr per; 2 300 e-books; 3 000 av-mat; 4 631 sound-rec
libr loan 30227

Ave Maria School of Law Library, 3475 Plymouth Rd, ***Ann Arbor***, MI 48105
T: +1 734 8278030; Fax: +1 734 3021475; URL: www.avemarialaw.edu/library
Janice Selberg
200 000 vols; 4 501 curr per 30228

Concordia University, Zimmerman Library, 4090 Geddes Rd, ***Ann Arbor***, MI 48105-2797
T: +1 734 9957353; Fax: +1 734 9957405; E-mail: ls@cuaa.edu; URL: cuaa.edu/library
1963; Keith Upton
Classics, French language and lit (Denkinger), Hist of science
113 000 vols; 146 curr per; 14 420 e-journals; 16 000 microforms; 325 av-mat; 5 245 sound-rec; 4 digital data carriers; 15 Overhead Transparencies 30229

University of Michigan, University Library, 818 Hatcher Graduate Library, ***Ann Arbor***, MI 48109-1205
T: +1 734 7649356; Fax: +1 734 7635080; URL: www.lib.umich.edu
1817; Paul N. Courant
American Society of Information Sciences, Food & Agriculture Organizations, OAS, WHO, Human Relations Area Files
7 291 000 vols; 63 329 curr per 30230

– Alfred Taubman Medical Library, 1135 E Catherine, Ann Arbor, MI 48109-2038
T: +1 734 9361403; Fax: +1 734 7631473; E-mail: medical.library@umich.edu; URL: www.lib.umich.edu/taubman
1920; Jane Blumenthal
Hist of Medicine (Crummer, Pilcher & Warthin Colls)
399 000 vols; 2 digital data carriers
libr loan 30231

– Art, Architecture & Engineering Library, Duderstadt Ctr, 2281 Bonnisteel Blvd, Ann Arbor, MI 48109-2094
T: +1 734 6475735; Fax: +1 734 7644487; URL: www.lib.umich.edu/aael/
1903; Catherine B. Soehner
Architecture, Art, Design, Engineering, Urban planning; Visual Resources Coll, Univ of Michigan Engineering College unpublished tech rpts
663 000 vols; 1 000 maps; 100 digital data carriers; 100 000 slides, 477 prints, 1 500 photos, 2 000 drawings 30232

– Asia Library, Hatcher Library, 920 N University St, Rm 418, Ann Arbor, MI 48109-1205
T: +1 734 7640406; Fax: +1 734 6472885; URL: www.lib.umich.edu/asia
1948; Dr. Jidong Yang
China, Far East, Japan, Korea; Ch'ing Arch, Gaimosho Arch, GB PRO Files on China, Hussey Papers, Japanese Diet Procs, URI Files
690 000 vols; 657 diss/theses; 80 000 microforms; 33 sound-rec; 472 digital data carriers
libr loan; CEAL 30233

– Dentistry Library, 1100 Dental Bldg, 1011 N University Ave, Ann Arbor, MI 48109-1078
T: +1 734 7641526; Fax: +1 734 7644477; E-mail: dentistry.library@umich.edu; URL: www.lib.umich.edu/dentlib
1875; P. F. Anderson
Rare bks
63 000 vols 30234

– Fine Arts Library, 260 Tappan Hall, 519 S State St, Ann Arbor, MI 48109-1357
T: +1 734 7645405; Fax: +1 734 7645408; E-mail: finearts@umich.edu; URL: www.lib.umich.edu/finearts/
1949; Deirdre Spencer
Hist of art
96 000 vols; 5 digital data carriers; 32 VF drawers 30235

– Harlan Hatcher Graduate Library, 209 Hatcher N, 920 N University, Ann Arbor, MI 48109-1205
T: +1 734 7640400; URL: www.lib.umich.edu/grad
Laurie Alexander
English lit, Hist, Military hist
3 606 000 vols 30236

– Information & Library Studies Library, 300 Hatcher N, Ann Arbor, MI 48109-1205
T: +1 734 9363038; E-mail: dhayward@umich.edu
Donna Hayward
Award Winning Children's & Young Adult Bks; Hist Juvenile Bks
63 260 vols; 589 curr per; 701 diss/theses; 7 473 microforms; 20 av-mat; 22 digital data carriers 30237

– Museums Library, 2500 Museums Bldg, 1108 Geddes Rd, Ann Arbor, MI 48109-1079
T: +1 734 7640467; Fax: +1 734 7643829; URL: www.lib.umich.edu/museums
1928; Scott Martin
Anthropology, Botany, Natural hist, Paleontology, Zoology
129 000 vols; 2 digital data carriers 30238

– Music Library, School of Music, 3239 Moore Bldg, Ann Arbor, MI 48109-2085
T: +1 734 7642512; Fax: +1 734 7645097; E-mail: music.library@umich.edu; URL: www.lib.umich.edu/music/
1941; Charles Reynolds
American Sheet Music, Music & Musicology Coll (17th-19th C), Women Composers Coll
146 000 vols; 3 000 microforms; 28 100 sound-rec; 7 digital data carriers
libr loan; MLA, IAML 30239

– Public Health Library & Information Division, 109 S Observatory, M2030 SPH II, Ann Arbor, MI 48109-2029
T: +1 734 7635109; Fax: +1 734 7639851; E-mail: phli@umich.edu; URL: www.sph.umich.edu/phisa/
1943; Helen Look
Education
82 000 vols; 10 digital data carriers 30240

– Shapiro Science Library, 3175 Shapiro Library, 919 S University Ave, Ann Arbor, MI 48109-1185
T: +1 734 7647490; Fax: +1 734 7639813; URL: www.lib.umich.edu/science/
1992; Catherine Soehner
Astronomical Maps, Rare Book Coll
484 000 vols; 19 digital data carriers 30241

– Shapiro Science Library – Undergraduate Library, 919 S University Ave, Ann Arbor, MI 48109-1185
T: +1 734 7634141; Fax: +1 734 7646849; E-mail: undergrad.library@umich.edu; URL: www.lib.umich.edu/ugl
Nadia Lalla
204 000 vols 30242

– Social Work Library, B700 School of Social Work Bldg, 1080 S University, Ann Arbor, MI 48109-1106
T: +1 734 7645169; E-mail: social.work.library@umich.edu; URL: www.lib.umich.edu/socwork
1958; Jennifer Nason Davis
47 000 vols; 2 digital data carriers 30243

– Special Collections Library, 711 Hatcher Graduate Library, South Bldg, 7th Flr, 920 N University, Ann Arbor, MI 48109-1205
T: +1 734 7649377; Fax: +1 734 7649368; E-mail: special.collections@umich.edu; URL: www.lib.umich.edu/spec-coll
Peggy Daub
220 000 vols 30244

University of Michigan, ***Ann Arbor***
– Bentley Historical Library, 1150 Beal Ave, Ann Arbor, MI 48109-2113
T: +1 734 7643482; Fax: +1 734 9361333; E-mail: bentley.ref@umich.edu; URL: www.umich.edu/~bhl/
1935; Francis X. Blouin Jr
Michigan; Philippine Islands, Sino-American Relations, Printed & Mss Holdings on Temperance & Prohibition, Architectural Coll
60 000 vols; 110 curr per; 28 900 mss; 10 370 maps; 4 000 microforms; 2 096 av-mat; 12 361 sound-rec; 38 000 linear feet of mss & arch, 32 VF drawers, 1 500 000 photos
libr loan; RLG, OCLC, SAA 30245

– Kresge Business Administration Library, Stephen M Ross School of Business, 701 Tappan St, K3330, Ann Arbor, MI 48109-1234
T: +1 734 7641375; Fax: +1 734 7643839; E-mail: kresge_library@umich.edu; URL: www.bus.umich.edu/kresgelibrary
1925; Corey Seeman
144 284 vols; 1 552 curr per; 64 000 e-journals; 12 000 e-books; 500 diss/theses; 147 000 microforms; 1 250 av-mat; 100 sound-rec; 3 digital data carriers
libr loan; Academic Business Library Directors, SLA 30246

– Law Library, 801 Monroe St, Ann Arbor, MI 48109-1210
T: +1 734 7644252; Fax: +1 734 9363884; URL: www.law.umich.edu/library/info/pages/default.aspx
1859; Margaret A. Leary
941 000 vols; 9 733 curr per; 4 digital data carriers 30247

– Transportation Research Institute Library, 2901 Baxter Rd, Ann Arbor, MI 48109-2150
T: +1 734 7642171; Fax: +1 734 9361081; E-mail: bsweet@umich.edu; URL: www.umtri.umich.edu
1966; Bob Sweet
Automotive engineering
110 000 vols; 300 curr per 30248

– William L. Clements Library, 909 S University Ave, Ann Arbor, MI 48109-1190
T: +1 734 7642347; Fax: +1 734 6470716; E-mail: clements.library@umich.edu; URL: www.clements.umich.edu/
1922; J. Kevin Graffagnino
American hist
77 000 vols; 34 curr per; 773 mss; 5 228 maps; 2 000 microforms 30249

Bard College, Stevenson Library, One Library Rd, ***Annandale-on-Hudson***, NY 12504; P.O. Box 5000, Annandale-on-Hudson, NY 12504-5000
T: +1 845 7587359; Fax: +1 845 7585701; URL: www.bard.edu/library
1860; Jeffrey Katz
Bardiana, publs by Bard faculty & alumnae; Hudson Valley Hist; Hannah Arendt & Heinrich Bluecher Coll, Hudson

Valley Hist
280 000 vols; 1 075 curr per 30250

– Center for Curatorial Studies, Library, PO Box 5000, Annandale-on-Hudson, NY 12504-5000
T: +1 845 7587567; Fax: +1 845 7582442; E-mail: ccslib@bard.edu; URL: www.bard.edu/ccs/library
Susan Leonard
19 000 vols; 40 curr per 30251

– Levy Economics Institute, Library, Blithewood Ave, Annandale-on-Hudson, NY 12504
T: +1 845 7587729; Fax: +1 845 7581149
Willis C. Walker
10 000 vols; 70 curr per 30252

St. John's College, Greenfield Library, 60 College Ave, ***Annapolis***, MD 21401; P.O. Box 2800, Annapolis, MD 21404-2800
T: +1 410 2956928
1696; Andrea Lamb
Philosophy, classical studies, hist of science; Annapolitan Libr (Bray Coll, Rev Thomas); Prettyman Coll of Signed and Inscribed Books
120 000 vols; 123 curr per; 5 incunabula; 5 000 microforms; 1 331 sound-rec; 9 digital data carriers; 2 164 slides
libr loan 30253

US Naval Academy, Nimitz Library, 589 McNair Rd, ***Annapolis***, MD 21402-5029
T: +1 410 2936900; Fax: +1 410 2936909; URL: www.usna.edu/library
1845; Richard Hume Werking
Naval science & hist, Military science; Electricity & Magnetism (Benjamin Coll); Naval Hist & Seapower; Physics (Albert A. Michelson Coll); Somers Submarine Coll; Steichen Photogr Coll, bks & photos; US Navy Mss Colls; Guggenheim Coll; Naval Academy Archive
611 000 vols; 2 805 curr per; 120 sound-rec; 89 digital data carriers; 68 674 govt docs 30254

Lebanon Valley College, Bishop Library, 101 N College Ave, ***Annville***, PA 17003-1400
T: +1 717 8676977; Fax: +1 717 8676979; URL: www.lvc.edu/library
1867; Frank Mols
Social sciences, Music; Pennsylvania German (Hiram Herr Shenk), Early Iron Industry (C B Montgomery)
192 000 vols; 700 curr per; 12 000 e-books; 2 005 sound-rec; 10 digital data carriers 30255

Lawrence University, Seeley G. Mudd Library, 113 S Lawe St, ***Appleton***, WI 54911-5683; P.O. Box 599, Appleton, WI 54912-0599
T: +1 920 8326758; Fax: +1 920 8326967; E-mail: reference@lawrence.edu; URL: www.lawrence.edu/library
1850; Peter J. Gilbert
Lincoln Coll
395 000 vols; 1 787 curr per; 21 digital data carriers
libr loan 30256

Humboldt State University, Library, One Harpst St, ***Arcata***, CA 95521-8299
T: +1 707 8264889; Fax: +1 707 8265590; URL: library.humboldt.edu
1913; Dr. Ray Wang
Childrens Coll, Hist (Humboldt County Coll, Humboldt State Univ Arch)
595 000 vols; 1 714 curr per; 23 000 e-books; 385 879 govt docs; 23 043 maps; 12 006 sound-rec; 21 digital data carriers
libr loan 30257

Henderson State University, Huie Library, 1100 Henderson, ***Arkadelphia***, AR 71999-0001
T: +1 870 2305292; Fax: +1 870 2305365; URL: library.hsu.edu
1890; Robert Yehl
Graphic Novel as Lit, Southwest Arkansas Indians, Arkansas State Docs Coll, Southwest Arkansas Hist Coll, Oceania Coll, Hist of comics, Blackmon Aviation
256 000 vols; 1 478 curr per; 240 e-journals; 1 032 diss/theses; 3 301 govt docs; 4 123 music scores; 91 maps; 215 000 microforms; 5 084 sound-rec; 738 digital data carriers; 21 590 full text journals through database subscriptions
libr loan; ALA, ACRL 30258

Ouachita Baptist University, Riley-Hickingbotham Library, 410 Ouachita, OBU-3742, ***Arkadelphia***, AR 71998-0001
T: +1 870 2455119; Fax: +1 870 2455245; URL: www.obu.edu
1886; Ray Granade
Music; Arkansas Hist
145 040 vols; 1 032 curr per; 307 083 microforms; 7 557 av-mat; 4 065 sound-rec; 259 660 docs
libr loan 30259

Marymount University, Emerson G. Reinsch Library, 2807 N Glebe Rd, ***Arlington***, VA 22207-4299
T: +1 703 2841649; Fax: +1 703 2841685; E-mail: library@marymount.edu; URL: www.marymount.edu/lls
1950; Dr. Zary Mostashari
Economics (Gertrude Hoyt Memorial Coll), Ireton Inspiration Reading
204 000 vols; 31 082 curr per; 1 100 e-journals; 295 e-books 30260

University of Texas at Arlington Library, 702 Planetarium Pl, ***Arlington***, TX 76019; P.O. Box 19497, Arlington, TX 76019-0497
T: +1 817 2723394; Fax: +1 817 2725797; URL: www.library.uta.edu
1895; Dr. Gerald D. Saxon
Jenkins Garrett Libr (Texana & Mexican War), Virginia Garrett Cartographic Hist Libr, Fort Worth Star-Telegram Arch, Fort Worth News-Tribune Arch, Basil Clemons Photogr Coll, Robertson Colony Coll, Yucatan Arch
1 185 000 vols; 4 210 curr per; 32 320 e-journals; 271 000 e-books
libr loan 30261

School of Islamic and Social Sciences, Ibn Khaldun Library, 45150 Russell Branch Pwy, Ste 303, ***Ashburn***, VA 20147-2902
T: + 1 571 2230500; Fax: +1 571 2230544
1996; Alaa El-Talmas
Hadith, Tafsir, Quran, Education; Isma'il al-Faruqi's Libr
35 000 vols; 110 curr per 30262

University of North Carolina at Asheville, D. Hiden Ramsey Library, One University Heights, CPO 1500, ***Asheville***, NC 28804-8504
T: +1 828 2516111; Fax: +1 828 2516012; URL: www.unca.edu/library
1928; James Robert Kuhlman
Liberal arts; Harrison Coll of Early American Hist, Mss & Photogr Coll, docs the hist of Western North Carolina, Peckham Coll of WWI Narratives, Univ Arch
390 000 vols; 4 810 curr per; 43 090 e-journals; 25 000 e-books; 4 460 av-mat; 2 440 sound-rec; 980 digital data carriers
libr loan 30263

Ashland University, University Library, 509 College Ave, ***Ashland***, OH 44805-3796
T: +1 419 2895400; Fax: +1 419 2895422; E-mail: library@ashland.edu; URL: www.ashland.edu/library
1878; William B. Weiss
American Studies (Libr of American Civilization); Bibles; 19th c English Lit (Andrews Special Bks Coll), 1st editions; 19th c Hist Children's Lit (Lulu Wood Coll), 1st editions
276 000 vols; 900 curr per
libr loan 30264

Northland College, Dexter Library, 1411 Ellis Ave, ***Ashland***, WI 54806-3999
T: +1 715 6821279; Fax: +1 715 6821693; URL: library.northland.edu
1892; Julia Trojanowski
74 000 vols; 250 curr per 30265

Randolph-Macon College, McGraw-Page Library, 204 Henry St, ***Ashland***, VA 23005; P.O. Box 5005, Ashland, VA 23005-5505
T: +1 804 7527200; Fax: +1 804 7527345; URL: www.rmc.edu/library
1830; Dr. Virginia E. Young
18th C European Culture (Casanova & Goudar), Intellectual Hist of the Colonial South, Virginia Methodism, Henry Miller, Southern Hist
182 000 vols; 801 curr per; 194 000 microforms
libr loan; OCLC 30266

Southern Oregon University, Lenn & Dixie Hannon Library, 1250 Siskiyou Blvd, ***Ashland***, OR 97520-5076
T: +1 541 5526823; Fax: +1 541 5526429; E-mail: library@sou.edu; URL: www.sou.edu/library
1926; Teresa Montgomery
Shakespeare-Renaissance (Bailey), Southern Oregon Hist, Adrienne Lee Ferte Memorial Coll
310 000 vols; 2 028 curr per; 280 000 govt docs; 30 094 maps; 785 000 microforms; 5 600 av-mat; 527 sound-rec; 44 digital data carriers
libr loan 30267

Kent State University, Ashtabula Campus Library, 3431 W 13th St, ***Ashtabula***, OH 44004-2298
T: +1 440 9644239; Fax: +1 440 9644271; URL: www.ashtabula.kent.edu/library/index.cfm
1961; Kevin Deemer
50 000 vols; 150 curr per 30268

Neumann College, Library, One Neumann Dr, ***Aston***, PA 19014-1298
T: +1 610 5585545; Fax: +1 610 4591370; E-mail: library@neumann.edu; URL: www.neumann.edu/academics/library.asp
1965; Tiffany McGregor
Curriculum, Franciscan Studies, Delaware County Tax Records, Betty Newman Coll
90 000 vols; 400 curr per; 2 000 e-books; 300 av-mat; 4 045 sound-rec; 5 digital data carriers
libr loan; ALA 30269

Benedictine College, Library, 1020 N Second St, ***Atchison***, KS 66002-1499
T: +1 913 3607608; Fax: +1 913 3607622; URL: www.benedictine.edu/library
1858; Steven Gromatzky
Religion, Hist, Education, Agriculture; Belloc, Chesterton, Church Fathers (Abbey Coll), Gerontology (Jay Gatson Coll), Monasticism, Philosophy (Ture Snowden Coll)
369 000 vols; 300 curr per
libr loan 30270

Athens State University, University Library, 407 E Pryor St, ***Athens***, AL 35611; mail address: 300 N Beaty St, Athens, AL 35611
T: +1 256 2166650; Fax: +1 256 2336547; E-mail: patsy.naves@athens.edu; URL: www.athens.edu/library
1842; Dr. Robert Burkhardt
Local Hist, Religion (Artifacts of Bible Land), Rare Books
109 000 vols; 317 curr per; 48 000 e-books 30271

Concord University, J. Frank Marsh Library, 1000 Vermillion St, ***Athens***, WV 24712; P.O. Box 1000, Athens, WV 24712-1000
T: +1 304 3845371; Fax: +1 304 3847955; E-mail: library@concord.edu; URL: library.concord.edu
1872; Stephen D. Rowe
F. Wells Goodykoontz Holograph Coll, Goodykoontz Autogr Coll, West Virginia Coll
155 000 vols; 227 curr per; 1 000 av-mat; 1 000 sound-rec; 1 200 digital data carriers
libr loan 30272

Ohio University, Vernon R. Alden Library, 30 Park Pl, ***Athens***, OH 45701-2978
T: +1 740 5930981; Fax: +1 740 5932708; URL: www.library.ohiou.edu/find/index.html
1804; Scott Seaman
Cornelius Ryan World War II Papers, Morgan Hist of Chemistry, Southeast Asia, Edmund Blunden Coll of Romantic & Georgian Lit, Gilbert and Ursula Farfel coll of 450 incunabula leaves from bks written between 1450-1500
2 468 000 vols; 25 557 curr per; 1 959 461 govt docs; 7 832 music scores; 170 949 maps; 2 580 000 microforms; 18 533 sound-rec; 486 digital data carriers; 12 711 Videotapes, 24 721 Slides, 46 DVDs
libr loan; IFLA, CRL, ASIS, LOEX, OCLC 30273

– Health Sciences Library, Alden Library, 3rd Flr, Athens, OH 45701
T: +1 740 5932680; Fax: +1 740 5934693; E-mail: HealthInfo@ohio.edu; URL: www.library.ohiou.edu/hsl/
1977; Connie Flores
Osteopathic Medicine Coll
103 175 vols; 941 curr per; 16 852 microforms; 152 digital data carriers
libr loan 30274

– Hwa-Wei Lee Center for International Collections, Southeast Asia Collection, Park Place, Alden Library, Athens, OH 45701-2978
T: +1 740 5932658; Fax: +1 740 5971879; E-mail: ferrier@ohio.edu; URL: www.library.ohiou.edu/cic/
1999; Jeff Ferrier
Malay World, especially Malaysia, Singapore, Brunei, Indonesia; Overseas Chinese
221 327 vols; 9 068 curr per; 46 455 microforms; 85 digital data carriers
libr loan; CORMOSEA 30275

– Mahn Center for Archives & Special Collections, Library, Vernon R Alden Library, 30 Park Pl, Fifth Flr, Athens, OH 45701-2978
T: +1 740 5932710; Fax: +1 740 5932708
William Kimok
54 000 vols; 115 curr per 30276

– Music-Dance Library, Robert Gidden Hall, Fifth Flr, Athens, OH 45701-2978
T: +1 740 5934255; Fax: +1 740 5939190
Holly Oberle
50 000 vols 30277

Tennessee Wesleyan College, Merner-Pfeiffer Library, 23 Coach Farmer Dr, ***Athens***, TN 37303; P.O. Box 40, Athens, TN 37371-0040
T: +1 423 7465250; Fax: +1 423 7465272; E-mail: library@twcnet.edu; URL: www.twcnet.edu/library
1857; Sandra Clariday
Methodist Church Hist (Cooke Memorial Coll)
110 000 vols; 809 curr per; 30 000 e-books; 3 000 microforms; 3 169 sound-rec; 6 digital data carriers
libr loan; ALA, ACRL 30278

University of Georgia, University Libraries, ***Athens***, GA 30602-1641
T: +1 706 5428460; Fax: +1 706 5424144; URL: www.libs.uga.edu
1800; William Gray Potter
Music, Georgiana, Rare Book Coll, Confederate Imprints Coll, Georgia Authors, 19th & 20th C Politics, Records Management Department & Univ Arch, Georgia Newspaper Coll
4 029 000 vols; 67 268 curr per
libr loan 30279

– Georgia Agricultural Experiment Stations Library, 305 Riverbend Rd, Athens, GA 30602
T: +1 706 5420817; Fax: +1 706 5424144; URL: www.uga.edu
100 000 vols 30280

– Science Library, Athens, GA 30602-7412
T: +1 706 5420698; Fax: +1 706 5426523; E-mail: sciref@arches.uga.edu; URL: www.libs.uga.edu/science/science.html
1968; Lucy M. Rowland
750 000 vols; 4 903 curr per 30281

University of Georgia – Alexander Campbell King Law Library, Law Library Annex A203, 225 Herty Dr, ***Athens***, GA 30602-6018
T: +1 706 5426591; Fax: +1 706 5426500; URL: www.law.uga.edu
1860; Ann E. Puckett
The Louis B Sohn Libr on Int Studies
375 000 vols; 6 984 curr per 30282

Menlo College, Bowman Library, 1000 El Camino Real, ***Atherton***, CA 94027-4301
T: +1 650 5433826; Fax: +1 650 5433833; URL: www.menlo.edu/library; www.rosie.menlo.edu
1962; C. Brigid Welch
Business & management, Mass communications, Liberal arts
65 000 vols; 175 curr per; 3 000 e-books 30283

Art Institute of Atlanta, Library, 6600 Peachtree-Dunwoody Rd, 100 Embassy Row, **Atlanta**, GA 30328-1635
T: +1 770 6894885; Fax: +1 770 3949800; E-mail: gmeier@aii.edu; URL: www.aii.edu
1949; Gayle Meier
Design, Fashion, Culinary arts, Media arts
46 000 vols; 173 curr per; 4 000 av-mat
30284

Atlanta College of Art, Woodruff Arts Center, Library, 1280 Peachtree St NE, **Atlanta**, GA 30309
T: +1 404 7335020; Fax: +1 404 7335312; URL: www.aca.edu
1931; Moira Steven
Artists' Book Coll; Rare Book Coll
25 000 vols; 185 curr per; 400 av-mat; 500 sound-rec; 1 500 artists bks, 3 000 vertical files
libr loan; ARLIS/NA, ALA
30285

Atlanta University Center Inc, Robert W. Woodruff Library, 111 James P Brawley Dr SW, **Atlanta**, GA 30314
T: +1 404 9782025; Fax: +1 404 5775158; URL: www.auctr.edu
1964; Loretta Parham
The Afro-American Experience, Southern US Hist, Mat by and about People of African Descent, Lincoln Coll
400 000 vols; 1 419 curr per
30286

Emory University, Robert W. Woodruff Library, 540 Asbury Circle, **Atlanta**, GA 30322-2870
T: +1 404 7276861; Fax: +1 404 7270805; URL: web.library.emory.edu/
1915; Richard Luce
Methodism, Confederate Hist, Southern Hist & Lit, African-American Hist & Culture, British & Irish Lit, Communism, Arch
3 185 000 vols; 37 779 curr per
libr loan
30287

– Health Sciences Center Library, 1462 Clifton Rd NE, Atlanta, GA 30322
T: +1 404 7278727; Fax: +1 404 7279821; URL: www.emory.edu/WHSCL
1923; Sandra Franklin
Hist of Medicine in Georgia
214 595 vols; 2 000 curr per; 1 534 av-mat
30288

– Pitts Theology Library, Candler School of Theology, 505 Kilgo Circle, Atlanta, GA 30322-2810
T: +1 404 7274166; Fax: +1 404 7271219; E-mail: libmpg@emory.edu; URL: www.pitts.emory.edu
1914; Dr. Matt Patrick Graham
Wesleyana Coll, Cardinal Henry Edward Manning Libr, Richard C. Kessler Reformation Coll, Arch & Mss Coll
547 000 vols; 1 462 curr per; 110 000 microforms; 3 digital data carriers; 6 700 rpts
30289

Emory University School of Law, Hugh F. MacMillan Law Library, 1301 Clifton Rd, **Atlanta**, GA 30322
T: +1 404 7276826; Fax: +1 404 7272202; URL: www.law.emory.edu/law/library
1916; Terry Gordon
European Union Depository Coll
401 000 vols; 4 053 curr per
30290

Georgia Institute of Technology, Library & Information Center, 704 Cherry St, **Atlanta**, GA 30332-0900
T: +1 404 8944511; Fax: +1 404 8948190; URL: www.library.gatech.edu
1885; Catherine Murray-Rust
2 525 000 vols; 44 487 curr per; 1 091 000 govt docs; 179 840 maps; 3 800 000 microforms; 1 044 digital data carriers; 54 000 pamphlets, 5 000 000 patents, 2 500 000 tech rpts, 72 214 slides, 29 000 photos
libr loan
30291

– College of Architecture, Library, 225 North Ave NW, Atlanta, GA 30332-0900
T: +1 404 8944877; Fax: +1 404 8940572; E-mail: kathy.brackney@library.gatech.edu; URL: www.library.gatech.edu/architect/
Kathryn S. Brackney
Neel Reid Coll, Rare Bks
37 500 vols; 136 curr per; 200 av-mat; 3 digital data carriers; 70 600 slides 30292

Georgia State University, **Atlanta**
– College of Law, Library, University Plaza, 140 Decatur St, Atlanta, GA 30303; P.O. Box 4008, Atlanta, GA 30302-4008
T: +1 404 4139100; E-mail: njohnson@gsu.edu; URL: law.gsu.edu/library
1982; Nancy P. Johnson
44 500 vols; 3 500 curr per; 5 digital data carriers
libr loan
30293

– William Russell Pullen Library, 100 Decatur St SE, Atlanta, GA 30303-3202
T: +1 404 4639943; Fax: +1 404 6512476; E-mail: libcrn@gsu.edu; URL: www.library.gsu.edu
1931; Charlene S. Hurt
Georgia Women's Coll, Georgia Govt Doc Project, Labor Hist, Johnny Mercer Coll
1 345 000 vols; 12 238 curr per 30294

Mercer University Atlanta, Monroe F. Swilley Jr Library, 3001 Mercer University Dr, **Atlanta**, GA 30341
T: +1 678 5476280; Fax: +1 678 5476270; E-mail: swilley_ref@mercer.edu; URL: swilley.mercer.edu
1968; Judith D. Brook
Education, Business, Theology, Pharm; British & American Lit Coll, 18th-19th c, 1st ed
147 000 vols; 900 curr per; 4 000 e-journals; 1 017 000 microforms; 50 digital data carriers
30295

Morehouse School of Medicine, Multi-Media Center, 720 Westview Dr SW, **Atlanta**, GA 30310-1495
T: +1 404 7521530; Fax: +1 404 7557318; URL: www.msm.edu/library/
1978; Cynthia L. Henderson
75 000 vols; 450 curr per; 450 e-books
libr loan
30296

Oglethorpe University, University Library, Philip Weltner Library, 4484 Peachtree Rd NE, **Atlanta**, GA 30319
T: +1 404 3648511; Fax: +1 404 3648517; E-mail: asalter@oglethorpe.edu; URL: library.oglethorpe.edu
1916; Anne Salter
James E Oglethorpe Coll; Sidney Lanier Coll
147 000 vols; 745 curr per; 3 650 microforms; 100 av-mat; 1 500 sound-rec; 6 digital data carriers 30297

Auburn University, **Auburn**
– The Library of Architecture, Design & Construction, Dudley Hall Commons, Auburn, AL 36849
T: +1 334 8441752; Fax: +1 334 8441756; E-mail: childgb@auburn.edu; URL: www.lib.auburn.edu/architecture
1952; G. Boyd Childress
44 000 vols; 85 curr per; 75 000 slides
30298

– Ralph Brown Draughon Library, 231 Mell St, Auburn, AL 36849
T: +1 334 8441737; Fax: +1 334 8444424; URL: www.lib.auburn.edu/
Bonnie MacEwan
3 017 000 vols; 36 395 curr per; 285 000 e-books
30299

– Veterinary Medical Library, 101 Greene Hall, Auburn, AL 36849-5606
T: +1 334 8441749; Fax: +1 334 8441758; URL: www.vetmed.auburn.edu/index.pl/library
1971; Cindy Mitchell
Veterinary Acupuncture Coll, Epidemiology & Hist Coll
30 000 vols; 550 curr per
30300

Augusta State University, Reese Library, 2500 Walton Way, **Augusta**, GA 30904-2200
T: +1 706 7371745; Fax: +1 706 6674415; E-mail: reference@aug.edu; URL: www.aug.edu/library/
1957; William N. Nelson
Cumming Family Papers, Edison Marshall Papers, Local Hist (Richmond County Hist Society)
480 000 vols; 6 283 curr per; 3 291 sound-rec; 969 digital data carriers
libr loan
30301

Medical College of Georgia, Robert B. Greenblatt MD Library, 1459 Laney-Walker Blvd, **Augusta**, GA 30912-4400
T: +1 706 7213441; Fax: +1 706 7212018; URL: www.mcg.edu/library
1834; David King
19th C Libr, Greenblatt Arch, Medical Artifacts, Landmarks in Modern Medicine Coll
51 000 vols; 3 988 curr per
30302

Paine College, Collins Callaway Library, 1235 15th St, **Augusta**, GA 30901-3105
T: +1 706 8218253; Fax: +1 706 8218698; URL: www.paine.edu
1991
Black Hist (Martin Luther King, Jr Coll)
64 000 vols; 207 curr per
30303

Aurora University, Charles B. Phillips Library, 347 S Gladstone, **Aurora**, IL 60506-4877
T: +1 630 8447583; Fax: +1 630 8443848; URL: www.aurora.edu/academics/library/index.html
1893; John Law
Education, Nursing, English lit, American Indians, Social service; Jenks Coll (Adventism), Ritzman Coll (Civil War), Prouty & Perry Memorial Coll (Elizabethan Hist & Theatre)
99 000 vols; 210 curr per; 16 digital data carriers
libr loan
30304

Wells College, Louis Jefferson Long Library, Main St, **Aurora**, NY 13026-0500
T: +1 315 3643351; Fax: +1 315 3643412; URL: www.wells.edu/library/li1.htm
1868; Jeri L. Vargo
Chemistry-Physics Coll, Economics Coll, Fine Arts Coll, Hist Coll, Philosophy Coll, Americana Coll, Wells Fargo Express Co
253 000 vols; 411 curr per; 11 000 microforms; 370 sound-rec
libr loan
30305

Houston-Tillotson College, Downs-Jones Library, 900 Chicon St, **Austin**, TX 78702-3430
T: +1 512 5053078; Fax: +1 512 5053190; URL: www.htc.edu
1891; Patricia Quarterman
Afro-American Hist (Schomburg Coll), Religion (Heinsohn Coll)
91 000 vols; 301 curr per
30306

Saint Edwards University, Scarborough-Phillips Library, 3001 S Congress Ave, **Austin**, TX 78704-6489
T: +1 512 4488474; Fax: +1 512 4488737; URL: libr.stewards.edu
1889; Thomas W. Leonhardt
155 000 vols; 1 185 curr per
30307

University of Texas at Austin, Perry-Castaneda Library, Perry-Castaneda Library, 101 E 21st St, **Austin**, TX 78713; P.O. Box P, Austin, TX 78713-8916
T: +1 512 4954350; Fax: +1 512 4954283; URL: www.lib.utexas.edu
1883; Dr. Fred M. Heath
Alexander Architectural Archive
8 482 000 vols; 48 096 curr per; 48 100 mss; 319 026 maps; 6 424 000 microforms; 173 373 sound-rec; 6 791 digital data carriers; 7 274 313 graphic items mss items
IFLA, ALA, ARL, CRL, CLIR, RLG, SALALM
30308

– Architecture and Planning Library, Mail Code S5430, BTL 200, Austin, TX 78713-8916; P.O. Box P, Austin, TX 78713-8916
T: +1 512 4954623; URL: www.lib.utexas.edu/apl/
Janine Henri
88 000 vols; 240 curr per
30309

– Center for American History Library, SRH 2-101, D1100, University of Texas at Austin, Austin, TX 78712
T: +1 512 4954518; Fax: +1 512 4954542; URL: www.cah.utexas.edu
Don E. Carleton
150 000 vols; 103 curr per
30310

– Chemistry Library, Welch Hall 2.132, Austin, TX 78713
T: +1 512 4954600; URL: www.lib.utexas.edu/chem
1883; David Flaxbart
Food science, Chemical engineering, Human nutrition
92 000 vols; 50 microforms; 4 digital data carriers
libr loan
30311

– Classics Library, Waggener Hall 1, Austin, TX 78713
T: +1 512 4954690
1967; Gina Giovannone
27 000 vols
30312

– East Asian Library Program, PO Box P, PCL4-114, S5431, Austin, TX 78713-8916
T: +1 512 4954325
Meng-fen Su
China, Japan, Korea
104 000 vols
30313

– Fine Arts Library, Doty Fine Arts Bldg 3-200, 23rd & Trinity, Austin, TX 78713; mail address: One University Sta S5437, Austin, TX 78712
T: +1 512 4954480; Fax: +1 512 4954490; URL: www.lib.utexas.edu/fal
1979
Hist Music Recordings Coll
300 000 vols; 600 curr per; 50 000 music scores; 5 000 microforms; 1 840 av-mat; 58 980 sound-rec; 31 slides, audio recordings
libr loan
30314

– Harry Ransom Humanities Research Center, Harry Ransom Ctr, 300 W 21st St, Austin, TX 78712; University of Texas, P.O. Box 7219, Austin, TX 78713-7219
T: +1 512 4718944; Fax: +1 512 4712899; E-mail: info@hrc.utexas.edu; URL: www.hrc.utexas.edu
1957; Dr. Thomas F. Staley
Wrenn Libr (English Lit), Aitken & Griffith Libs (17th-18th C Lit), Lutcher Stark Coll (Romantic & Victorian Writers), Rare Bk Coll, Parsons Coll (incunabula, printing hist, art, & early Louisiana imprints), Hanley Libr (20th c mss), Knopf Libr, Modern French Lit
750 000 vols; 276 curr per; 12 000 microforms; 37 000 linear feet of mss, 5 000 000 photogr images, 100 000 works of visual art
30315

– Jamail Center for Legal Research Library, University of Texas School of Law, 727 E Dean Keeton St, Austin, TX 78705-3224
T: +1 512 4717735; Fax: +1 512 4710243; URL: tarlton.law.utexas.edu
1886; Roy M. Mersky
Law in Popular Culture, Rare Bks
1 012 000 vols; 7 492 curr per; 476 e-journals; 3 incunabula; 1 128 000 microforms; 2 295 sound-rec; 14 digital data carriers
libr loan; ALA, AALL, SLA, IALL 30316

– Life Science (Biology, Pharmacy) Library, Main Bldg 220, Austin, TX 78713
T: +1 512 4954630; Fax: +1 512 4954638; URL: www.lib.utexas.edu/lsl/
Nancy Elder
Plant Taxonomic Lit Coll
220 000 vols; 1 800 curr per; 32 000 microforms; 1 032 opaques 30317

– McKinney Engineering Library, One University Sta S5435, ECJ 1.300, Austin, TX 78712
T: +1 512 4954511; Fax: +1 512 4954507; E-mail: englib@lib.utexas.edu; URL: www.lib.utexas.edu/engin
Susan Ardis
165 000 vols
30318

– Middle Eastern Library Program, 21st & Speedway, Austin, TX 78713; P.O. Box P, Austin, TX 78713-8916
T: +1 512 4954322; Fax: +1 512 4954296
Abazar Sepehri
Islamic world
319 000 vols
30319

– Nettie Lee Benson Latin American Collection, Sid Richardson Hall 1-108, Austin, TX 78713-8916
T: +1 512 4954568; E-mail: blac@lib.utexas.edu; URL: www.lib.utexas.edu/benson/

contacts.html
1921; Ann Hartness
Author Colls of Sor Juana Ines de la Cruz, Joaquin Fernandez de Lizardi, Alfonso Reyes, Jose Angel Gutierrez, Julian Samora Libr, Jose Toribio Medina & Many Literary Figures of Mexico, Argentina, Brazil, Chile & Peru, Afro-Jamiacan Folklore, Mexican Cultural Hist
889 000 vols; 19 442 maps; 37 000 microforms; 1 582 sound-rec; 32 711 photos, mss (2 500 linear feet)
libr loan; SALALM 30320

– Physics-Mathematics-Astronomy Library, Robert L Moore Hall 4-200, Austin, TX 78713
T: +1 512 4954610; Fax: +1 512 4954611; E-mail: pma@lib.utexas.edu; URL: www.lib.utexas.edu/pma
Molly White
112 000 vols; 317 microforms 30321

– Population Research Center Library, Main Bldg 1800, G1800, Austin, TX 78712
T: +1 512 4718332; Fax: +1 512 4714886; URL: www.prc.utexas.edu
1971; Mark Hayward
Int Census Coll, World Fertility Survey
35 000 vols; 30 file drawers 30322

– South Asian Library Program, PCL 3 313, Austin, TX 78713
T: +1 512 4954329; Fax: +1 512 4954397
1976; Merry Burlingham
Censuses & Gazeteers of British India, Univ of Pennsylvania Sanskrit Mss Coll
300 000 vols 30323

– Undergraduate Library, Flawn Academic Ctr 101AF, S5443, Austin, TX 78713
T: +1 512 4954467; Fax: +1 512 4954340; URL: www.lib.utexas.edu/ugl/
Damon Jaggars
151 000 vols 30324

– Walter Geology Library, Geology 4-202, Austin, TX 78713-8916
T: +1 512 4954680; Fax: +1 512 4954102; E-mail: georequests@lib.utexas.edu; URL: www.lib.utexas.edu/geo
Dennis Trombatore
Tobin Geologic Maps Coll
134 000 vols; 48 000 maps; 18 000 microforms; 18 digital data carriers 30325

– Wasserman Public Affairs Library, Sid Richardson Hall 3.243, S5442, Austin, TX 78712-1282
T: +1 512 4954400; Fax: +1 512 4714697; E-mail: lib-hr@lib.utexas.edu; URL: www.lib.utexas.edu/pal
Stephen Littrell
250 000 vols 30326

Ave Maria University, Canizaro Library, 5251 Donahue St., ***Ave Maria***, FL 34142
T: +1 239 2802557; Fax: +1 734 4824187; E-mail: library@avemaria.edu; URL: www.avemaria.edu/MajorsPrograms/Library.aspx
Sarah Beiting
30 000 vols; 50 curr per 30327

Babson College, Horn Library, 231 Forest St, ***Babson Park***, MA 02457-0310
T: +1 781 2394596; Fax: +1 781 2395226; E-mail: library@babson.edu; URL: library.babson.edu
1919; Hope N. Tillman
Business & management, Economics; Hinckley Coll (sailing & transportation), Sir Isaac Newton Coll
114 000 vols; 409 curr per; 134 e-journals; 58 000 e-books; 347 000 microforms; 1 400 av-mat; 1 024 sound-rec; 287 digital data carriers
libr loan 30328

Webber International University, Grace & Roger Babson Learning Center, 1201 State Rd 17, ***Babson Park***, FL 33827; P.O. Box 97, Babson Park, FL 33827-0097
T: +1 863 6382937; Fax: +1 863 6382778; E-mail: library@webber.edu; URL: www.webber.edu
1927; Sue Dunning
Business; Civil War Coll
36 000 vols; 70 curr per; 21 000 microforms; 271 av-mat; 1 097 sound-rec; 50 digital data carriers; 7 495 slides
ALA 30329

California State University, Walter W. Stiern Library, 9001 Stockdale Hwy, ***Bakersfield***, CA 93311-1022
T: +1 661 6543172; Fax: +1 661 6543238; E-mail: emontoya@csub.edu; URL: www.csub.edu/library/
1970; Rodney M. Hersberger
462 000 vols; 1 208 curr per
libr loan 30330

Baker University, Collins Library, 518 Eighth St, ***Baldwin City***, KS 66006-0065; P.O. Box 65, Baldwin City, KS 66006-0065
T: +1 785 5948585; Fax: +1 785 5946721; E-mail: reference@bakeru.edu; URL: www.bakeru.edu/library
1858; Kay Bradt
Bibles (Quayle); United Methodist Hist Arch; Baker Univ Arch
85 000 vols; 520 curr per; 8 000 e-books; 26 459 govt docs; 1 863 av-mat; 1 230 sound-rec; 90 digital data carriers
libr loan 30331

Coppin State College, Parlett Moore Library, 2500 W North Ave, ***Baltimore***, MD 21216-3698
T: +1 410 9513400; Fax: +1 410 9513430; URL: www.coppin.edu/library
1900; Mary Wanza
Biology, Nursing, Education, Social sciences; Black Studies, Maryland
78 000 vols; 705 curr per 30332

Goucher College, Julia Rogers Library, 1021 Dulaney Valley Rd, ***Baltimore***, MD 21204
T: +1 410 3376360; Fax: +1 410 3376419; E-mail: jrogers@goucher.edu; URL: www.goucher.edu/library
1885; Nancy Magnuson
Jane Austen (Alberta H. Burke Coll), Hist of Costume, J.W. Bright Coll, B.S. Corrin & C.I. Winslow Political Memorablia & Political Humor Coll, Southern Women During the Civil War (Passano Coll), Goucher College Arch, Sara Haardt Coll, H.L. Mencken Coll, Mark Twain (Eugene Oberdorfer Coll)
310 000 vols; 800 curr per; 20 000 e-journals; 1 000 e-books; 80 Electronic Media & Resources 30333

Johns Hopkins University – The Sheridan Libraries, 3400 N Charles St, ***Baltimore***, MD 21218
T: +1 410 5168325; Fax: +1 410 5165080; URL: www.library.jhu.edu
1876; Winston Tabb
Byron Coll, Kurrelmeyer German Lit Coll, Abram G. Hutzler Coll of Economic Classics, Levy Sheet Music Coll, Sidney Lanier Coll, Birney Anti-Slavery Coll, Tudor & Stuart Coll of English Lit, Loewenberg Coll of German Drama
2 619 000 vols; 17 757 curr per; 7 000 e-journals 30334

– Abraham M. Lilienfeld Memorial Library, Bloomberg School of Public Health, School of Hygiene & Public Health, 624 N Broadway, 9th Flr, Baltimore, MD 21205-1901
T: +1 410 9553028; Fax: +1 410 9550200; URL: www.welch.jhu.edu
1963; Susan Rohner
32 000 vols; 174 curr per 30335

– George Peabody Library, 17 E Mount Vernon Pl, Baltimore, MD 21202
T: +1 410 6598179; Fax: +1 410 6598137; URL: archives.mse.jhu.edu:8000
1857
Cervantes: Editions of Don Quixote Coll, Voyages and travels, Natural hist; Hist of science + technology; Religions
252 000 vols; 20 curr per; 2 mss; 50 incunabula 30336

– John Work Garrett Library, Evergreen House, 4545 N Charles St, Baltimore, MD 21210
T: +1 410 5160889; Fax: +1 410 5167202; URL: archives.mse.jhu.edu
1952
16th & 17th c English Lit, 15th c Bks, Early Travel Exploration, Natural Hist, Architecture, Bibles, Bks Printed before 1700 Relating to Maryland, Laurence H. Fowler Architectural Coll, Hofman Bible Coll, Signers of the Declaration of Independence
29 000 vols 30337

– William H. Welch Medical Library, 1900 E Monument St, Baltimore, MD 21205-2113
T: +1 410 9553411; Fax: +1 410 9550985; URL: www.welch.jhu.edu
1929; Nancy K. Roderer
Hist of the Diseases of the Chest & Vaccination (Henry Jacobs Coll), Hist of Nursing (Florence Nightingale Coll), Med Education, AV Coll
397 000 vols; 2 432 curr per
libr loan 30338

Johns Hopkins University-Peabody Conservatory of Music, Arthur Friedheim Library, One E Mount Vernon Pl, ***Baltimore***, MD 21202-2397
T: +1 410 6598100; Fax: +1 410 6850657; URL: www.peabody.jhu.edu/lib
1866; Robert Follet
J.. Thomas Coll, Enrico Caruso Coll, Jazz Colls
114 000 vols; 240 curr per; 250 diss/theses; 2 000 microforms; 22 000 sound-rec; 14 digital data carriers; 2 000 mss, 6 boxes of clippings, 8 VF drawers
libr loan 30339

Maryland Institute College of Art, Decker Library, 1401 Mount Royal Ave, ***Baltimore***, MD 21217; mail address: 1300 Mount Royal Ave, Baltimore, MD 21217
T: +1 410 2252311; Fax: +1 410 2252316; URL: www.mica.edu/library
1826; Sherri Faaborg
19th c Prints & Art Bks (Lucas Coll)
52 000 vols; 300 curr per; 9 VF drawers of plates, 18 VF drawers of pamphlets, 70 000 slides 30340

Morgan State University, Morris A. Soper Library & Information Technology Center, 1700 E Cold Spring Lane, ***Baltimore***, MD 21251
T: +1 443 8853488; URL: library.morgan.edu
1867; Karen A. Robertson
Afro-American Hist & Life (Beulah M. Davis Special Colls Rm); Quaker & Slavery (Forbush Coll), bk, mss; Papers of Emeritus President (D.O.W. Holmes Papers, Martin D. Jenkins Papers), letters; Negro Employment in WWII (Emmett J. Scott Coll), letters; Correspondence of Late Poet & Editor (W. S. Braithwaite Coll), letters, & papers
310 000 vols; 1 500 curr per; 295 000 microforms; 60 digital data carriers 30341

Ner Israel Rabbinical College, Library, 400 Mount Wilson Lane, ***Baltimore***, MD 21208
T: +1 410 4847200; Fax: +1 410 4843060
1933; Avrohom S. Shnidman
Biblical Commentaries, Responsa, Talmudic Laws, Hebrew Newspapers of European Communities 1820-1937
23 000 vols; 18 curr per 30342

Sojourner-Douglass College, Walter P. Carter Library, 500 N Caroline St, ***Baltimore***, MD 21205
T: +1 410 2760306; Fax: +1 410 6751810; E-mail: oali@host.sdc.edu; URL: www.sdc.edu
1976; Sadiq Omowali Ali
Africa & African American Hist (Old Rare Books Coll)
22 000 vols; 80 curr per 30343

University of Baltimore, Langsdale Library, 1420 Maryland Ave, ***Baltimore***, MD 21201
T: +1 410 8374260; Fax: +1 410 8374319; E-mail: langcirc@ubalt.edu; URL: langsdale.ubalt.edu
1926; Lucy Holman
Arch of Society of Colonial Wars, Steamship Hist Society of American Coll, WMAR-TV Film Arch
181 000 vols; 450 curr per; 207 diss/theses; 47 456 govt docs; 726 maps; 1 664 digital data carriers
libr loan 30344

– Law Library, 1415 Maryland Ave, Baltimore, MD 21201
T: +1 410 8374554; Fax: +1 410 8374570; URL: law.ubalt.edu/lawlib
1925; Will Tress
305 000 vols; 3 500 curr per; 674 000 microforms; 465 av-mat; 113 sound-rec; 4 digital data carriers 30345

University of Maryland, Baltimore, *Baltimore*
– Health Sciences & Human Services Library, 601 W Lombard St, Baltimore, MD 21201
T: +1 410 7067928; Fax: +1 410 7063101; URL: www.hshsl.umaryland.edu/
1813; M. J. Tooey
Crawford Medical Hist Coll; Cordell Medical Hist Coll; Grieves Dental Hist Coll; Hist Bk Colls in Pharmacy, Social Work, & Nursing
365 000 vols; 2 400 curr per; 10 digital data carriers
libr loan 30346

– Thurgood Marshall Law Library, 501 W Fayette St, Baltimore, MD 21201-1768
T: +1 410 7063240; Fax: +1 410 7068354; URL: www.law.umaryland.edu/marshall
1843; Barbara Gontrum
German & French Civil Law Coll
354 000 vols 30347

University of Maryland, Baltimore County, Albin O. Kuhn Library & Gallery, 1000 Hilltop Circle, ***Baltimore***, MD 21250
T: +1 410 4552354; Fax: +1 410 4551061; E-mail: wilt@umbc.edu; URL: www.umbc.edu/library
1966; Dr. Larry Wilt
Photogr Coll, American Society for Microbiology Arch, Society for Invitro Biology Arch, American Type Culture Coll, Rosenfeld Science Fiction Res Colls, Children's Science Coll
950 000 vols; 4 200 curr per; 3 000 e-books; 5 000 maps; 1 025 000 microforms; 29 134 sound-rec; 1 500 digital data carriers; 2 000 High Interest/Low Vocabulary Bks 30348

Husson College, Library, One College Circle, ***Bangor***, ME 04401-2999
T: +1 207 9417188; Fax: +1 207 9417989; URL: www.husson.edu
1947; Amy Averre
35 000 vols; 500 curr per 30349

University of Maine at Augusta, University College Library, Bangor Campus, 124 Eastport Hall, 128 Texas Ave, ***Bangor***, ME 04401
T: +1 207 2627902; Fax: +1 207 2627901
1968; Judith Nottage
24 000 vols; 195 curr per; 2 000 microforms
libr loan 30350

College of the Atlantic, Thorndike Library, 109 Eden St, ***Bar Harbor***, ME 04609-1198
T: +1 207 2885015; Fax: +1 207 2882328; URL: www.coa.edu/thorndikelibrary
1972; Jane Hultberg
Environmental studies, Evolution, Botany, Horticulture, Philosophy; Philip Darlington Coll on Evolution, Humanities (R. Amory Thorndike Coll), Science Hist (Thomas S & Mary Hall Coll)
40 000 vols; 250 curr per 30351

University of Wisconsin Baraboo-Sauk County, T.N. Savides Library, 1006 Connie Rd, ***Baraboo***, WI 53913
T: +1 608 3555251; Fax: +1 608 3555291; URL: www.baraboo.uwc.edu/library/libraryinfo.htm
1968; Mark L. Rozmarynowski
37 000 vols; 110 curr per; 6 229 govt docs; 132 music scores; 77 microforms; 3 369 sound-rec; 10 digital data carriers; 11 800 slides, 13 art reproductions
libr loan 30352

Union College, Weeks Townsend Memorial Library, 310 College St, Campus Box D-21, ***Barbourville***, KY 40906-1499
T: +1 606 5461243; Fax: +1 606 5461239; E-mail: library@unionky.edu; URL: www.unionky.edu/library
1879; Tara L. Cooper
Genealogy; Kentucky Hist
113 000 vols; 339 curr per; 103 000 e-books; 215 000 microforms; 488 av-mat; 531 sound-rec; 31 digital data carriers; 14 959 slides, 45 overhead transparencies 30353

Oklahoma Wesleyan University, Library, 2201 Silver Lake Rd, ***Bartlesville***, OK 74006-6299
T: +1 918 3356285; Fax: +1 918 3356220; E-mail: libraryadulted@hotmail.com; URL: www.okwu.edu/library
1958; Wendell Thompson
94000 vols; 182 curr per; 15000 e-books 30354

Lyon College, Mabee-Simpson Library, 2300 Highland Rd, ***Batesville***, AR 72501-3699
T: +1 870 3077205; Fax: +1 870 3077279; URL: library.lyon.edu
1872; Dean Covington
Arkansas (Regional Studies Center), John Quincy Wolf Folklore Coll
147000 vols; 406 curr per; 24100 e-journals; 22000 e-books; 2200 maps; 3600 av-mat; 1100 sound-rec
libr loan 30355

Louisiana State University, University Libraries, Louisiana State University, 295 Middleton Library, ***Baton Rouge***, LA 70803
T: +1 225 5782058; Fax: +1 225 5785723; URL: www.lib.lsu.edu
1860; Jennifer Cargill
Louisiana & Lower Mississippi Coll; McIlhenny Natural Hist Coll; Rare Bk Coll
3213000 vols; 18695 curr per; 17270 e-journals; 43000 e-books; 4190000 microforms
OCLC, SOLINET, ARL, ASERL, SPARC 30356

– Carter Music Resources Center, 202 Middleton Library, Baton Rouge, LA 70803-3300
T: +1 225 5784674; Fax: +1 225 5786825; URL: www.lib.lsu.edu/music
Lois Kuyper-Rushing
58000 vols; 90 curr per; 1873 digital data carriers 30357

– Chemistry Library, 301 Williams Hall, Baton Rouge, LA 70803
T: +1 225 5782530; Fax: +1 225 5782760
Beilstein Coll, Gmelin Coll, Sadtler Spectra Coll, Landolt-Bernstein Coll, Chemical Abstracts
64250 vols; 700 curr per; 18659 microforms; 18 digital data carriers 30358

– Design Resource Center Library, 104 Design Bldg, Baton Rouge, LA 70803-7010
T: +1 225 5780280; Fax: +1 225 5780280; E-mail: smooney@lsu.edu; URL: www.lib.lsu.edu/lib/design.html
1959; Sandra T. Mooney
13853 vols; 100 curr per; 50 maps; 115 av-mat; 8 digital data carriers; 6 VF drawers of clippings, 200 blueprints
libr loan 30359

– Education Resources, 227 Middleton Library, Baton Rouge, LA 70803-3300
T: +1 225 5782349; Fax: +1 225 5786992; URL: www.lib.lsu.edu/edu/er/
Peggy Chalaron
Instructional Mat
29000 vols; 200 av-mat; 60 digital data carriers 30360

– Paul M. Hebert Law Center, One E Campus Dr, Baton Rouge, LA 70803-1000
T: +1 225 5784041; Fax: +1 225 5785773; E-mail: info@law.lsu.edu; URL: www.law.lsu.edu/library
Dragomir Cosanici
827000 vols; 12229 curr per 30361

– School of Veterinary Medicine Library, Skip Bertman Dr, Baton Rouge, LA 70803-8414
T: +1 225 5789796; Fax: +1 225 5789798; E-mail: library@vetmail.lsu.edu; URL: www.vetmed.lsu.edu/library
1974; Christine Mitchell
Reprint Coll on Parasitology, 1865-1972
47000 vols; 600 curr per; 5 digital data carriers 30362

Southern University, John B. Cade Library, 167 Roosevelt Steptoe Ave, ***Baton Rouge***, LA 70813-0001
T: +1 225 7714990; Fax: +1 225 7714113; URL: www.lib.subr.edu
1880; Emma Bradford Perry
Arch & Shade Coll
407000 vols; 1675 curr per; 200 e-journals; 16000 e-books; 1065 maps; 429000 microforms; 12878 sound-rec; 422 digital data carriers; 600 Electronic Media & Resources
libr loan 30363

– Art & Architecture Library, School of Architecture, Engineering West Bldg, 2nd Flr, Baton Rouge, LA 70813
T: +1 225 7713290; Fax: +1 225 7714709
Lucille Bowie
8000 vols; 40 curr per 30364

– Oliver B. Spellman Law Library, 56 Roosevelt Steptoe, Baton Rouge, LA 70813; P.O. Box 9294, Baton Rouge, LA 70813-9294
T: +1 225 7712146; Fax: +1 225 7716254; URL: www.sulc.edu/library/library.htm
Ruth J. Hill
Civil rights
475000 vols; 989 curr per 30365

Duke University – Nicholas School of the Environment and Earth Sciences, Pearse Memorial Library, Duke Univ Marine Lab, 135 Duke Marine Lab Rd, ***Beaufort***, NC 28516-9721
T: +1 252 504-7510; Fax: +1 252 504-7648; E-mail: marlib@duke.edu; URL: library.duke.edu/marine/
1938; David Talbert
23000 vols; 60 curr per; 4 digital data carriers 30366

Lamar University, Mary & John Gray Library, 211 Redbird Lane, ***Beaumont***, TX 77705; P.O. Box 10021, Beaumont, TX 77710-0021
T: +1 409 8801898; Fax: +1 409 8802318; URL: library.lamar.edu
1923; Dr. Christina Baum
Cookery Coll; Peter Wells Texana Coll
432000 vols; 1800 curr per; 470 Bks on Deafness & Sign Lang
libr loan 30367

Geneva College, McCartney Library, 3200 College Ave, ***Beaver Falls***, PA 15010-3599
T: +1 724 8476563; Fax: +1 724 8476687; E-mail: gmoran@geneva.edu; URL: www.geneva.edu
1931; Gerald D. Moran
Reformed Presbyterian Church (Covenanter Coll); Personal Libr & Papers of Dr Clarence Macartney; (Macartney Coll), bks & unpublished papers; Early American Imprints, microcard; Libr of American Civilization, microfiche
173030 vols; 857 curr per; 25 e-journals; 16150 av-mat; 1810 sound-rec; 223050 Electronic Media & Resources
libr loan 30368

Oregon National Primate Research Center, McDonald Library, 505 NW 185th Ave, ***Beaverton***, OR 97006
T: +1 503 6905311; Fax: +1 503 6905243; URL: www.ohsu.edu/library/primate.shtml
Denise Urbanski
Primates
15000 vols; 81 curr per 30369

Mountain State University, University Library, 609 S Kanawha St, ***Beckley***, WV 25801; P.O. Box 9003, Beckley, WV 25802
T: +1 304 9291528; Fax: +1 304 9291665; URL: www.mountainstate.edu/current/library/
Judy Jean Altis
Child Lit, West Virginia
96000 vols; 157 curr per; 973 e-books 30370

Arkansas State University – Beebe, Abington Memorial Library, 1000 W Iowa St, P.O. Box 1000, ***Beebe***, AR 72012
T: +1 501 8828207; Fax: +1 501 8828233; E-mail: circ@asub.edu; URL: www.asub.edu/library/
1929; Carolyn Powers
Arkansas
76000 vols; 485 curr per; 35000 e-journals; 32000 e-books; 4700 av-mat 30371

University of Florida, Belle Glade – Everglades Research & Education Center, 3200 E Palm Beach Rd, ***Belle Glade***, FL 33430-8003
T: +1 561 9931500; Fax: +1 561 9931582; E-mail: klkr@ufl.edu; URL: erec.ifas.ufl.edu
1926; Kathleen Krawchuk
Agriculture, Turf grass
12000 vols; 100 curr per 30372

Bellevue University, Freeman-Lozier Library, 1000 Galvin Rd S, ***Bellevue***, NE 68005
T: +1 402 5577307; Fax: +1 402 5575427; E-mail: library@bellevue.edu; URL: www.bellevue.edu/resources/library.asp
1966; Robin R. Bernstein
Social & Behavioral Sci (Human Relations Area Files Coll), microfiche; Microfiche College Cats
113000 vols; 37677 curr per; 43000 e-books 30373

City University, Vi Tasler Library, 150-120th Ave NE, ***Bellevue***, WA 98005-3019; mail address: 11900 NE First St, Bellevue, WA 98005-3030
T: +1 425 7093440; Fax: +1 425 7093455; E-mail: library@cityu.edu; URL: www.cityu.edu/library
1973
Business management, Education
24000 vols; 1561 curr per 30374

Western Washington University, Mabel Zoe Wilson Library, 516 High St, ***Bellingham***, WA 98225
T: +1 360 6503076; Fax: +1 360 6503044; URL: www.library.wwu.edu
1899; Jerry Boles
Canadian, Mongolian Coll
800000 vols; 4900 curr per 30375

Belmont Abbey College, Abbot Vincent Taylor Library, 100 Belmont-Mt Holly Rd, ***Belmont***, NC 28012
T: +1 704 8256748; Fax: +1 704 8256743; E-mail: donaldbeagle@bac.edu; URL: www.belmontabbeycollege.edu
1876; Donald Beagle
Hist, Religious studies; Autographed Bks; Benedictine Coll, Napoleonic Coll, North & South Carolina Coll, Valuable Bks from 15-18th c brought by monks from Europe
130000 vols; 600 curr per; 116521 microforms; 1845 sound-rec; 40 digital data carriers; 104562 microforms of per
libr loan 30376

Notre Dame de Namur University, Gellert Library, 1500 Ralston Ave, ***Belmont***, CA 94002-1908
T: +1 650 5083747; Fax: +1 650 5083697; E-mail: library@ndnu.edu; URL: www.ndnu.edu/Gellertlibrary
1852; Dr. Klaus Musmann
91000 vols; 500 curr per; 455000 microforms; 7420 sound-rec
libr loan 30377

Beloit College, Colonel Robert H. Morse Library & Richard Black Information Center, 731 College St, ***Beloit***, WI 53511-5595
T: +1 608 3632567; Fax: +1 608 3632487; URL: www.beloit.edu/~libhome
1849; Charlotte Patriquin
Beloit Poetry Journal Coll, Pacifism & Nonviolence (M L King Coll), World Order (Cullister Coll)
257000 vols; 1859 curr per
libr loan 30378

University of Mary Hardin-Baylor, Townsend Memorial Library, 900 College St, UMHB Sta, Box 8016, ***Belton***, TX 76513-2599
T: +1 254 2954637; Fax: +1 254 2954642; E-mail: library@umhb.edu; URL: umhblib.umhb.edu
1845; Denise Karimkhani
Baptist hist, University arch
233000 vols; 928 curr per; 92 e-journals; 27000 e-books; 1308 music scores; 161 maps; 46000 microforms; 455 av-mat; 1755 sound-rec; 1032 digital data carriers; 27109 NetLibr e-book coll
libr loan; TLA 30379

Bemidji State University, A.C. Clark Library, 1500 Birchmont Dr NE, ***Bemidji***, MN 56601-2600
T: +1 218 7553342; Fax: +1 218 7552051; URL: www.bemidjistate.edu/library
1919; Ron Edwards
American Indian hist, Northern Minnesota hist; National Indian Education Assn Coll
459000 vols
libr loan 30380

Oak Hills Christian College, Cummings Library, 1600 Oakhills Rd SW, ***Bemidji***, MN 56601-8832
T: +1 218 7518670 Ext 246; Fax: +1 218 7518825; E-mail: oakhills@oakhills.edu; URL: www.oakhills.edu
1946
Religious studies, Philosophy
25000 vols; 160 curr per
libr loan; ACL 30381

Bennington College, Crossett Library, One College Dr, ***Bennington***, VT 05201-6001
T: +1 802 4404737; Fax: +1 802 4404580; E-mail: library@bennington.edu; URL: bennington.edu/library
1932; Oceana Wilson
Literary Reviews; Photography
100000 vols; 275 curr per; 50 maps; 7000 microforms; 899 sound-rec; 8500 Electronic Media & Resources
libr loan 30382

Lake Michigan College, William Hessel Library, 2755 E Napier Ave, ***Benton Harbor***, MI 49022
T: +1 269 9278605; Fax: +1 616 9276656; URL: www.lakemichigancollege.edu
1946
Lake Michigan College Arch
50540 vols; 7350 curr per 30383

Baldwin-Wallace College, Ritter Library, 57 E Bagley Rd, ***Berea***, OH 44017
T: +1 440 8262206; Fax: +1 440 8268558; URL: www.bw.edu/academics/libraries/ritter
1845; Dr. Patrick J. Scanlan
Education, Hist, Music; Folksongs (Harry E Ridenour Coll); Paul & Josephine Mayer Rare Bk Coll; Religion (Methodist Hist Coll) bks, artifacts
200000 vols; 800 curr per
libr loan 30384

– Riemenschneider Bach Institute Library, Merner-Pfeiffer Hall, 49 Seminary St, Berea, OH 44017-2088; mail address: 275 Eastland Rd, Berea, OH 44017-2088
T: +1 440 8262207; Fax: +1 440 8268138; E-mail: bachinst@bw.edu; URL: www.bw.edu/academics/libraries/bach
1969; Dr. Melvin P. Unger
Riemenschneider's Bach Coll; JS Bach & Contemporaries, bks, mss; Hans David's Books of Musik & books about music, mainly from the Baroque Era Coll; Emmy Martin's Rebound Scores; Tom Villella Record Coll
13000 vols; 27 curr per; 19 mss; 6963 music scores; 280 microforms; 9294 sound-rec; 1329 rare vault-held items
ALA 30385

Berea College, Hutchins Library, 100 Campus Dr, CPO Library, ***Berea***, KY 40404
T: +1 859 9853275; Fax: +1 859 9853912; E-mail: reference_desk@berea.edu; URL: www.berea.edu/library/library.html
1870; Anne Chase
Berea Arch, mss; Shedd-Lincoln Coll; Weatherford-Hammond Appalachian Coll; Rare Books; Appalachian Sound Arch
386000 vols; 850 curr per; 149 e-journals; 67000 e-books
libr loan 30386

Graduate Theological Union, Library, 2400 Ridge Rd, ***Berkeley***, CA 94709
T: +1 510 6492502; Fax: +1 510 6492508; E-mail: library@gtu.edu; URL: library.gtu.edu
1969; Robert Benedetto
Ecumenical and interreligious activity in the western United States & the Pacific Rim; new religious movements
468000 vols; 1582 curr per; 3375 diss/theses; 186000 microforms; 1180

av-mat; 6031 sound-rec; 74 digital data carriers; 62093 Journals
libr loan; ATLA, OCLC, ALA, ACRL
30387

University of California at Berkeley, 245 Doe Library, *Berkeley*, CA 94720-6000
T: +1 510 6427365; Fax: +1 510 6438476; URL: www.lib.berkeley.edu
1871; Thomas C. Leonard
Letters, Literary Mss & Scrapbks of Samuel Clemens; Music Hist; Radio Carbon Date Cards, Photogr Plates, Rubbings, Univ Arch photos, Aerial photos, VF mat
10 000 000 vols; 89 948 curr per; 404 655 maps; 5 380 000 microforms; 1 235 digital data carriers
libr loan
30388

– Anthropology Library, 230 Kroeber Hall, Berkeley, CA 94720-6000
T: +1 510 6422400; Fax: +1 510 6439293
1956; Suzanne Calpestri
69 000 vols; 2 000 microforms; 1 934 pamphlets
30389

– Bancroft Library, Berkeley, CA 94720-6000
T: +1 510 6423781; Fax: +1 510 6427589; E-mail: bancref@library.berkeley.edu; URL: bancroft.berkeley.edu
Charles Faulhaber
600 000 vols
30390

– Center for Chinese Studies, 2223 Fulton, Basement, Berkeley, CA 94720
T: +1 510 6426510; Fax: +1 510 6423817; E-mail: ccs@berkeley.edu; URL: ieas.berkeley.edu/ccs/
1959; Kevin O'Brien
KEIO Coll; Hatano Coll; Union Res Institute Classified File on Contemporary China, 1949-1975; Chen Cheng Coll; Wen Shih Tzu Liao Coll; Videotape Coll of TV Programs from the People's Republic of China; People's Univ Reprint Series of Selected Articles; Hsin Hsien Chih Gazetter Coll
54 000 vols; 40 drawers of microfilm, 4 250 items in VF drawers
30391

– Chemistry Library, 100 Hildebrand Hall, Berkeley, CA 94720-6000
T: +1 510 6423753; Fax: +1 510 6439041; E-mail: chem@library.berkeley.edu; URL: www.lib.berkeley.edu/chem
1948; Mary Ann Mahoney
Russian Monogr & Serials Obtained on Exchange; US Chemical Patents
70 000 vols; 37 000 microforms; 322 pamphlets
30392

– Earth Sciences & Maps Library, 50 McCone Hall, Berkeley, CA 94720-6000
T: +1 510 6422997; Fax: +1 510 6436576; URL: www.lib.berkeley.edu/eart
Fatemah Van Buren
118 000 vols; 2 000 curr per
30393

– Earthquake Engineering Research Center Library, 453 Richmond Field Sta, 1301 S 46th St, Richmond, CA 94804
T: +1 510 6653419; Fax: +1 510 6653456; E-mail: eerclibrary@berkeley.edu; URL: nisee.berkeley.edu/library
1972; Charles James
Earthquake Eng Abstracts Database, Steinbrugge Image Coll, Godden Int Structural Slide Libr
56 000 vols
30394

– East Asian Library, Durant Hall, Rm 208, Berkeley, CA 94720-6000
T: +1 510 6422556; Fax: +1 510 6423817; E-mail: eal@library.berkeley.edu; URL: www.lib.berkeley.edu/eal/
Peter Zhou
796 000 vols
30395

– Education Psychology Library, 2600 Tolman Hall, Berkeley, CA 94720-6000
T: +1 510 6428224; URL: lib.berkeley.edu/edp/
1924; Ron Heckart
Children's & Young Adult Lit Coll
185 000 vols; 802 curr per; 575 000 microforms
30396

– Environmental Design Library, 210 Wurster Hall, Berkeley, CA 94720-6000
T: +1 510 6424818; Fax: +1 510 6428266; URL: www.lib.berkeley.edu/ENVI
1903; Elizabeth D. Byrne
Beatrix Jones Farrand Coll (Early & Rare Landscape Architectural Hist)
211 000 vols; 750 curr per; 10 000 microforms
30397

– Ethnic Studies Library, 30 Stephens Hall, MC 2360, Berkeley, CA 94720-2360
T: +1 510 6431234; Fax: +1 510 6438433; E-mail: esl@library.berkeley.edu; URL: eslibrary.berkeley.edu
1969; Lillian Castillo-Speed
Chicano Retrospective Newspaper Coll; Chicano Art Color Transparencies; Chicano Posters; Annual Rpts of the Commissioner of Indian Affairs, 1849-1949; Survey of the Conditions of the Indians of the US, 1929-1944; Chinese American Res Coll
69 000 vols; 2 700 curr per; 1 500 microforms; 206 av-mat; 700 sound-rec; 5 000 slides, 150 linear feet of archives
30398

– Fong Optometry & Health Sciences Library, 490 Minor Hall, Berkeley, CA 94720-6000
T: +1 510 6421020; Fax: +1 510 6438600; URL: www.lib.berkeley.edu/opto/
1949; Bette Anton
13 000 vols
30399

– Giannini Foundation of Agricultural Economics Library, 248 Giannini Hall, Berkeley, CA 94720-3310
T: +1 510 6427121; Fax: +1 510 6438911; URL: are.berkeley.edu/library
1930; Susan Grabarino
21 000 vols; 700 curr per
30400

– Harmer E. Davis Transportation Library, 412 McLaughlin Hall, MC 1720, Berkeley, CA 94720-1720
T: +1 510 6423604; Fax: +1 510 6429180; E-mail: itslib@uclink.berkeley.edu; URL: www.lib.berkeley.edu/itsl
1948; Rita Evans
San Francisco Bay Area Transit Coll
149 000 vols; 1 500 maps; 135 000 microforms; 15 VF drawers of newspaper clippings
30401

– Institute of Governmental Studies, Library, 109 Moses Hall, Ground Flr, Berkeley, CA 94720-2370
T: +1 510 6421472; Fax: +1 510 6430866; E-mail: igsl@berkeley.edu; URL: www.igs.berkeley.edu/library
Nick Robinson
437 000 vols
30402

– Institute of Industrial Relations Library, 2521 Channing Way, MC 5555, Berkeley, CA 94720-5555
T: +1 510 6421705; Fax: +1 510 6426432; E-mail: iirl@library.berkeley.edu; URL: iir.berkeley.edu
1945; Terence Huwe
64 000 vols; 3 000 uncat items
30403

– Jean Gray Hargrove Music Library, Hargrove Music Library, Berkeley, CA 94720-6000
T: +1 510 6422623; Fax: +1 510 6428237; URL: www.lib.berkeley.edu/musi
1947; Manuel Erviti
Connick & Romberg Opera Colls, Music Mss of the 11th-20th c, Chambers Campanology Coll, Archival Colls, Cortot Opera, Italian Instrumental Music, Opera Libretti
165 000 vols; 13 000 microforms; 46 000 sound-rec
30404

– Kresge Engineering Library, 110 Bechtel Engineering Ctr, Berkeley, CA 94720-6000
T: +1 510 6423532; Fax: +1 510 6436771; URL: www.lib.berkeley.edu/engi
Camille Wanat
236 000 vols; 1 804 curr per
30405

– Law Library, 225 Boalt Hall, Berkeley, CA 94720
T: +1 510 6424044; Fax: +1 510 6435039; URL: www.law.berkeley.edu/library
1912; Robert Berring
Colby Coll (Mining Law), Robbins Coll (Civil & Medieval Law), Anglo-American, Foreign & Int Law
560 000 vols; 8 250 curr per; 662 000 microforms; 963 sound-rec; 175 mss, 400 000 court briefs & theses
30406

– Marian Koshland Bioscience & Natural Resources Library, 2101 VLSB, No 6500, Berkeley, CA 94720-6500
T: +1 510 6422531; URL: www.lib.berkeley.edu/bios/
1930; Beth Weil
Holl Coll of Cookbks, Rare Book Coll (17th- to 19th-c Natural Hist)
500 000 vols; 6 500 curr per; 26 000 microforms; 50 633 pamphlets & reprints
30407

– Mathematics-Statistics Library, 100 Evans Hall, Berkeley, CA 94720-6000
T: +1 510 6423381; Fax: +1 510 6428257; E-mail: math@library.berkeley.edu; URL: www.lib.berkeley.edu/math
1959; Brian Quigley
87 000 vols
30408

– Physics Library, 351 LeConte Hall, Berkeley, CA 94720-6000
T: +1 510 6423122; Fax: +1 510 6428350; URL: www.lib.berkeley.edu/phys/
Susan Koskinen
48 000 vols
30409

– Public Health Library, 42 Warren Hall, No 7360, Berkeley, CA 94720-7360
T: +1 510 6422511; Fax: +1 510 6427623; E-mail: publ@library.berkeley.edu; URL: www.lib.berkeley.edu/publ/
1955; Deborah Jan
93 000 vols; 1 000 microforms; 10 000 pamphlets
30410

– Social Welfare Library, 227 Haviland Hall, Berkeley, CA 94720-6000
T: +1 510 6424432; Fax: +1 510 6431476; URL: www.lib.berkeley.edu/socw
Ron Heckart
33 000 vols; 200 curr per
30411

– South-Southeast Asia Library Service, 120 Doe Library, Berkeley, CA 94720-6000
T: +1 510 6423095; Fax: +1 510 6438817; URL: www.lib.berkeley.edu/sseal
1970; Virginia Jing-yi Shih
Ghadar Party Coll; Nepal Coll; Thai Coll; Modern Hindi, Indonesian & Malay Lit
400 000 vols
30412

– Thomas J. Long Business & Economics Library, Haas School of Business, Rm S352, Berkeley, CA 94720-6000
T: +1 510 6420400; Fax: +1 510 6438476; URL: www.lib.berkeley.edu/BUSI/
1964; Milton G. Ternberg
159 000 vols; 1 090 000 microforms
30413

– Water Resources Center Archives, 410 O'Brien Hall, Berkeley, CA 94720
T: +1 510 6422666; Fax: +1 510 6421943; E-mail: waterarc@library.berkeley.edu; URL: www.lib.berkeley.edu/WRCA
Linda Vida
109 000 vols
30414

Andrews University, James White Library, 1400 Library Rd, *Berrien Springs*, MI 49104-1400
T: +1 269 4713506; Fax: +1 269 4716166; URL: www.andrews.edu/library
1962; Lawrence Onsager
Seventh Day Adventist Church Hist (Heritage Ctr), Environmental Design Res
750 000 vols; 2 800 curr per; 43 533 maps; 684 000 microforms; 67 422 sound-rec; 40 digital data carriers
libr loan
30415

– Architectural Resource Center, Architecture Bldg, Berrien Springs, MI 49104-0450
T: +1 616 4712417; E-mail: demskyk@andrews.edu; URL: www.andrews.edu/library/arc/
Kathleen M. Demsky
24 500 vols
30416

– Music Materials Center, Hamel Hall, Berrien Springs, MI 49104
T: +1 616 4713114; E-mail: mack@andrews.edu; URL: www2.Andrews.edu/~mack
Linda Mack
12 500 vols
30417

Bethany College, Mary Cutlip Center for Library and Information Technology Services, T. W. Phillips Memorial Library, 300 Main St, *Bethany*, WV 26032
T: +1 304 8297321; Fax: +1 304 8297333; E-mail: library@bethanywv.edu; URL: www.bethanywv.edu/library/
1841; Dr. Mary-Bess Halford
Alexander Campbell & Christian Church-Disciples of Christ (Alexander Campbell Arch), Ornithology (Brooks Bird Club Coll), Upper Ohio Valley Coll
125 970 vols; 161 curr per; 10 350 e-journals; 55 000 e-books; 4 060 av-mat
libr loan; PALINET
30418

Southern Nazarene University, R.T. Williams Learning Resources Center, 4115 N College, *Bethany*, OK 73008
T: +1 405 4916351; Fax: +1 405 4916355; URL: www.snu.edu
1920; Jan Reinbold
Hymnological Coll, Ross Hayslip Bible Coll, Signatures (John E. Moore Letter Coll)
103 000 vols; 200 curr per; 17 digital data carriers; 168 000 microforms of per
libr loan
30419

Uniformed Services University of the Health Sciences, Learning Resource Center, 4301 Jones Bridge Rd, *Bethesda*, MD 20814-4799
T: +1 301 2953350; Fax: +1 301 2953795; E-mail: ref@lrcm.usuhs.mil; URL: www.lrc.usuhs.mil
1976; Janice Powell Muller
Military medicine
42 000 vols
libr loan
30420

Lehigh University, Fairchild-Martindale Library, Fairchild-Martindale Library, 8A E Packer Ave, *Bethlehem*, PA 18015-3170
T: +1 610 7583030; Fax: +1 610 7586524; URL: www.lehigh.edu/library
1865; Susan Cady
Bayer Galleria of Rare Books, Lehigh Coll; Linderman Library
1 146 000 vols; 12 000 curr per; 617 000 govt docs; 2 000 digital data carriers
libr loan
30421

Montserrat College of Art, Paul M. Scott Library, 23 Essex St, *Beverly*, MA 01915; P.O. Box 26, Beverly, MA 01915-0026
T: +1 978 9214242; Fax: +1 978 9224268; URL: www.montserrat.edu/academics/library.php
Cheri Coe
12 000 vols; 75 curr per
30422

Academy of Motion Picture Arts & Sciences, Margaret Herrick Library, 333 S La Cienega Blvd, *Beverly Hills*, CA 90211
T: +1 310 2473020; Fax: +1 310 6575193; URL: www.oscars.org
1927; Linda Harris Mehr
42 000 vols; 200 curr per; 7 mio photos, 8 000 scripts, 1 505 VF drawers of clippings & production stills
30423

University of New England Libraries, Jack S. Ketchum Library, 11 Hills Beach Rd, *Biddeford*, ME 04005
T: +1 207 6022361; Fax: +1 207 6025922; E-mail: library@une.edu; URL: www.une.edu/library
1831; Stew MacLehose
Education, Life Sci, Medicine; New England Osteopathic Heritage Ctr
150 000 vols; 40 000 curr per; 2 200 e-journals; 6 000 e-books; 1 610 av-mat; 430 sound-rec; 480 digital data carriers
libr loan
30424

Ferris State University, Ferris Library for Information, Technology & Education, 1010 Campus Dr, ***Big Rapids***, MI 49307-2279
T: +1 231 5913602; Fax: +1 231 5913724; E-mail: reference@ferris.edu; URL: www.ferris.edu/library
1884; Leah Monger
Univ Arch, Woodbridge N Ferris Papers, Northwest Michigan
316 000 vols; 811 curr per; 15 540 e-journals; 59 000 e-books; 95 diss/theses; 73 202 govt docs; 7 000 maps; 3 347 000 microforms; 1 380 av-mat; 390 sound-rec; 340 digital data carriers
libr loan 30425

Montana State University – Billings Library, 1500 University Dr, ***Billings***, MT 59101-0298
T: +1 406 6571662; Fax: +1 406 6572037; URL: www.msubillings.edu/Library
1927; Jane L. Howell
Billings, Yellowstone County & Eastern Montana, mss; Montana & Western Hist (Dora C White Memorial Coll); USGS Montana Maps Coll
339 000 vols; 1 713 curr per
libr loan 30426

Rocky Mountain College, Paul M. Adams Memorial Library, 1511 Poly Dr, ***Billings***, MT 59102-1796
T: +1 406 6571087; Fax: +1 406 6571085; E-mail: ill@rocky.edu; URL: www.library.rocky.edu
1878; Bill Kehler
68 000 vols; 375 curr per; 1 000 e-books; 182 music scores; 200 sound-rec; 18 digital data carriers
libr loan 30427

State University of New York at Binghamton, University Libraries, Vestal Pkwy E, ***Binghamton***, NY 13902; P.O. Box 6012, Binghamton, NY 13902-6012
T: +1 607 7772345; Fax: +1 607 7774347; URL: library.lib.binghamton.edu
1946; John M. Meador Jr
French Colonial Hist (William J Haggerty Coll), Max Reinhardt Arch & Libr, Local Hist Coll, Music (Frances R Conole Arch), Arthur Schnitzler Arch, William Klenz Libr & Music Coll
2 380 000 vols; 81 959 curr per; 36 460 e-journals; 5 000 e-books; 5 050 sound-rec; 1 100 digital data carriers
libr loan 30428

– Science Library, Vestal Pkwy E, Binghamton, NY 13902
T: +1 607 7772218; Fax: +1 607 7772274; E-mail: skuster@binghamton.edu; URL: library.lib.binghamton.edu
1973; Charlotte Skuster
200 000 vols; 1 400 curr per 30429

Birmingham-Southern College, Charles Andrew Rush Learning Center & N. E. Miles Library, 900 Arkadelphia Rd, ***Birmingham***, AL 35254; P.O. Box 549020, Birmingham, AL 35254-0001
T: +1 205 2264744; Fax: +1 205 2264743; URL: www.bsc.edu
1856; Charlotte Ford
Americana; Alabama Authors, Alabama Hist, Alabama Methodism, Branscomb Coll (Women)
260 000 vols; 540 curr per; 42 000 e-books
libr loan 30430

Samford University, University Library, 800 Lakeshore Dr, ***Birmingham***, AL 35229
T: +1 205 7262748; Fax: +1 205 7264009; E-mail: library@samford.edu; URL: www.samford.edu/library
1841; Kimmetha Herndon
William H Brantley Coll, Albert E Casey Coll, Douglas C McMurtrie Coll, John Ruskin Coll, John Masefield Coll, Alfred Tennyson Coll, Lafcadio Hearn Coll
401 000 vols; 1 216 curr per
libr loan 30431

– Lucille Stewart Beeson Law Library, 800 Lakeshore Dr, Birmingham, AL 35229
T: +1 205 7262714; Fax: +1 205 7262644; URL: www.lawlib.samford.edu
1847; Gregory K. Laughlin
203 000 vols; 2 029 curr per 30432

University of Alabama at Birmingham, ***Birmingham***
– Mervyn H. Sterne Library, 917 13th St S, Birmingham, AL 35205; mail address: 1530 Third Ave S, SL 172, Birmingham, AL 35294-0014
T: +1 205 9346360; Fax: +1 205 9756230; URL: www.mhsl.uab.edu
Dr. Jerry W. Stephens
1 536 000 vols; 1 677 curr per; 53 000 e-books 30433

– University of Alabama at Birmingham, Lister Hill Library of the Health Sciences, 1700 University Blvd, Birmingham, AL 35294-0013; mail address: 1530 3rd Ave S LHL251, Birmingham, AL 35294-0013
T: +1 205 9342356; Fax: +1 205 9756498; URL: www.uab.edu/lister
1945; T. Scott Plutchak
History of Anesthesia
313 000 vols; 1 263 curr per; 27 242 e-journals; 2 188 mss; 29 microforms; 1 207 sound-rec; 909 digital data carriers
libr loan; MLA, AAHSL, CONBLS, NAAL 30434

University of Mary, Welder Library, 7500 University Dr, ***Bismarck***, ND 58504-9652
T: +1 701 3558070; Fax: +1 701 3558255; URL: www.umary.edu
1959; Cheryl M. Bailey
70 000 vols; 525 curr per; 3 000 microforms; 2 500 sound-rec 30435

Virginia Polytechnic Institute & State University Libraries, Newman Library, Drill Field Dr, ***Blacksburg***, VA 24062-9001; P.O. Box 90001, Blacksburg, VA 24062-9001
T: +1 540 2316170; Fax: +1 540 2313946; URL: www.lib.vt.edu
1872; Eileen Hitchingham
Arch of American Aerospace Exploration, Arch of Norfolk & Western Railway, Arch of Southern Railway Precedessors, Sherwood Anderson Book Coll, Heraldry, Science Fiction Magazines, Hist of Technology; Southwest Virginiana, Western Americana, American Civil War Coll
2 299 000 vols; 35 596 curr per; 399 525 govt docs; 132 769 maps; 6 090 000 microforms; 15 270 av-mat; 18 664 sound-rec; 7 250 digital data carriers; 3 700 linear feet of mss
libr loan 30436

– Art and Architecture Library, Cowgill Hall, 3rd Flr, Blacksburg, VA 24062
T: +1 540 2319271; URL: www.lib.vt.edu/services/branches/artarch
Heather Gendron
77 000 vols 30437

– Geosciences Library, 3040 Derring Hall, Blacksburg, VA 24061-0421
T: +1 540 2316101; Fax: +1 540 2319263; E-mail: Lener@vt.edu; URL: www.lib.vt.edu
1972; Edward Lener
32 000 vols; 275 curr per; 700 diss/theses; 16 500 maps; 23 500 microforms; 20 000 aerial photos 30438

– NVC Resource Center, 7054 Haycock Rd, Falls Church, VA 22043
T: +1 703 5388340; E-mail: dcash@vt.edu; URL: www.lib.vt.edu
Debbie Cash
12 483 vols 30439

– Veterinary Medicine Library, Phase III Duck Pond Dr, Blacksburg, VA 24061-0442
T: +1 540 2316610; Fax: +1 540 2317367; URL: www.lib.vt.edu/services/branches/vetmed
Victoria T. Kok
13 000 vols 30440

Dana College, C.A. Dana Life Library, 2848 College Dr, ***Blair***, NE 68008-1099
T: +1 402 4267912; Fax: +1 402 4267332; URL: www.dana.edu/library
1884; Thomas S. Nielsen
Social Work, Education; Danish Hist & Lit, Danish Immigrant Arch, Opera (Lauritz Melchior Memorial Coll)
132 000 vols; 207 curr per; 30 000 e-books; 7 000 Electronic Media & Resources
libr loan 30441

Dominican College, Library, 480 Western Hwy, ***Blauvelt***, NY 10913-2000
T: +1 845 3598188; Fax: +1 845 3592313; URL: www.dc.edu
1957; John Barrie
103 000 vols; 490 curr per 30442

Bloomfield College, Library, Liberty St & Oakland Ave, ***Bloomfield***, NJ 07003
T: +1 973 7489000 ext 337; Fax: +1 973 7433998; E-mail: danilo_figueredo@bloomfield.edu; URL: bloomfield.edu/academics/library/default.aspx
1869; Danilo H. Figueredo
Education, Multicultural diversity
64 000 vols; 375 curr per; 4 500 microforms; 100 av-mat; 910 sound-rec; 3 digital data carriers
libr loan 30443

Illinois Wesleyan University, The Ames Library, One Ames Plaza, ***Bloomington***, IL 61701-7188; P.O. Box 2899, Bloomington, IL 61702-2899
T: +1 309 5563350; Fax: +1 309 5563706; E-mail: ask_us@iwu.edu; URL: www.iwu.edu/library/
1850; Karen Schmidt
Music, Nursing; Political Science & Govt (Leslie Arends Coll), 20th C Lit (Gernon Coll)
335 000 vols; 1 046 curr per
libr loan 30444

Indiana University Bloomington, University Libraries, Herman B. Wells Library 234, Herman B Wells Library 234, 1320 E Tenth St, ***Bloomington***, IN 47405-1801
T: +1 812 8554673; Fax: +1 812 8552576; E-mail: libref@indiana.edu; URL: www.libraries.iub.edu
1824; Patricia A. Steele
English Hist, Austrian Hist, American Revolution, British Plays, Hist of Science & Medicine, English & American Lit, Western Americana
6 770 000 vols; 70 370 curr per; 3 623 000 microforms
libr loan 30445

– Business SPEA Library, SPEA 150, 1315 E Tenth St, Bloomington, IN 47405
T: +1 812 8551957; Fax: +1 812 8553398; E-mail: libbus@indiana.edu; URL: www.libraries.iub.edu/index.php?pageId=77
1926; Steven Sowell
100 000 vols; 118 000 microforms; 100 web-based databases 30446

– Chemistry Library, Chemistry C003, 800 E Kirkwood Ave, Bloomington, IN 47405-7102
T: +1 812 8559452; Fax: +1 812 8556611; E-mail: libchem@indiana.edu
1941; Roger Beckman
48 000 vols; 63 000 microforms 30447

– Education Library, Wright Education 1160, 201 N Rose St, Bloomington, IN 47405-1006
T: +1 812 8568590; Fax: +1 812 8568593; E-mail: libeduc@indiana.edu
1955; Gwendolyn Pershing
Curriculum Mat, Children's Lit Coll
88 000 vols; 535 000 microforms 30448

– Fine Arts Library & Slide & Digital Imgae Library, Fine Arts Museum, 1133 E Seventh St, Bloomington, IN 47405
T: +1 812 8554597; Fax: +1 812 8553443; E-mail: libart@indiana.edu
1939; Tony White
Artist's Bks
130 000 vols; 325 curr per; 30 000 microforms; 150 av-mat; 240 sound-rec; 50 000 digital data carriers; 58 000 photos 30449

– Geography & Map Library, Student Bldg 015, 701 E Kirkwood Ave, Bloomington, IN 47405
T: +1 812 8551108; Fax: +1 812 8554919; E-mail: libgm@indiana.edu
1946; Brian Winterman
Sanborn Fire Insurance Maps of Indiana Cities
20 000 vols; 300 000 maps; 5 000 microforms 30450

– Geology Library, Geology 603, 1001 E Tenth St, Bloomington, IN 47405
T: +1 812 8551494; Fax: +1 812 8556614; E-mail: libgeol@indiana.edu; URL: www.libraries.iub.edu/index.php?pageId=82
1871; Lou Malcomb
Geology, geophysics, paleontology, geochemistry
124 000 vols; 819 incunabula; 325 000 maps; 50 000 microforms; 625 digital data carriers
libr loan 30451

– Health, Physical Education and Recreation Library, HPER 029, 1025 E Seventh St, Bloomington, IN 47405
T: +1 812 8554420; Fax: +1 812 8556778; E-mail: libhper@indiana.edu
Jian Liu
23 000 vols; 12 000 microforms
libr loan 30452

– Journalism Library, Ernie Pyle Hall, 940 E Seventh St, Bloomington, IN 47405-7108
T: +1 812 8559247; Fax: +1 812 8550901; E-mail: libjourn@indiana.edu
1975; Grace Jackson-Brown
Ernie Pyle Coll, Roy Howard Arch
27 000 vols 30453

– Life Sciences Library, Jordan Hall A304, 1001 E Third St, Bloomington, IN 47405-7005
T: +1 812 8558947; E-mail: liblife@indiana.edu
Roger Beckman
126 000 vols
libr loan 30454

– Lilly Library Rare Books & Manuscripts, 1200 E Seventh St, Bloomington, IN 47405-5500
T: +1 812 8552452; Fax: +1 812 8553143; E-mail: liblilly@indiana.edu; URL: www.indiana.edu/~liblilly
1960; Breon Mitchell
Mendel Latin American Coll, Ellison Far West Coll, Elisabeth Ball Children's Lit Coll, Oakleaf Lincoln Coll, J.K. Lilly Coll, Wendell Willkie Papers
408 000 vols; 500 microforms; 15 500 linear feet of mss 30455

– Neal-Marshall Black Culture Center Library, Neal-Marshall Ctr, Rm A113, 275 N Jordan, Bloomington, IN 47405
T: +1 812 8553237; Fax: +1 812 8564558; E-mail: bcclib@indiana.edu
Deloice Holliday
African-American culture and history
8 000 vols 30456

– Optometry Library, Optometry 301, 800 E Atwater Ave, Bloomington, IN 47405
T: +1 812 8558629; Fax: +1 812 8556616; E-mail: libopt@indiana.edu; URL: www.libraries.iub.edu/index.php?pageId=91
1968; Douglas Freeman
22 000 vols; 4 000 microforms 30457

– Swain Hall Library, Swain Hall West 208, 727 E Third St, Bloomington, IN 47405-7105
T: +1 812 8552758; Fax: +1 812 8555533; E-mail: libswain@indiana.edu
1940; Robert Noel
Astronomy, Computer science, Mathematics, Physics; Cyclotron Facility
101 000 vols 30458

– William & Gayle Cook Music Library, Simon Music Library & Recital Ctr M160, 200 S Jordan Ave, Bloomington, IN 47405
T: +1 812 8552970; Fax: +1 812 8553843; E-mail: libmus@indiana.edu
1921; Philip Ponella
377 000 vols; 318 000 music scores; 18 000 microforms; 132 000 sound-rec 30459

Indiana University – Indiana Institute on Disability & Community, Library, 2853 E Tenth St, ***Bloomington***, IN 47408-2696
T: +1 812 8559396; Fax: +1 812 8559630; E-mail: cedir@indiana.edu; URL: www.iidc.indiana.edu/cedir
Sharon Soto
Disability awareness
10 000 vols 30460

Indiana University – Research Institute for Inner Asian Studies, Library, Indiana University, Goodbody Hall 344, 1011 E Third St, *Bloomington*, IN 47405-7005
T: +1 812 8551605; Fax: +1 812 8557500; E-mail: rifias@indiana.edu; URL: www.indiana.edu/~rifias
1963; Devin DeWeese
Central Asian Arch/Tibetan Coll
9 000 vols; 26 curr per; 2 000 microforms
30461

Indiana University – School of Law Library, Maurer School of Law, 211 S Indiana Ave, *Bloomington*, IN 47405
T: +1 812 8556404; Fax: +1 812 8557099; URL: www.law.indiana.edu/lawlibrary
1842; Colleen K. Pauwels
Indiana Court of Appeals Briefs, Indiana Supreme Court Briefs & Records, US Supreme Court Briefs & Records, Rare Bks
711 000 vols; 7 775 curr per; 1 102 000 microforms; 543 av-mat; 627 sound-rec; 308 digital data carriers
30462

Northwestern Health Science University, Greenawalt Library, 2501 W 84th St, *Bloomington*, MN 55431-1599
T: +1 952 8855419; Fax: +1 952 8843318; E-mail: library@nwhealth.edu; URL: www.youseemore.com/nhsu/
1966; Della Shupe
Chiropractic Journals, Arch of Chiropractic Lit
11 500 vols; 315 curr per; 16 500 microforms; 1 292 sound-rec; 20 digital data carriers; 16 VF drawers, 486 videotapes, 340 slide sets
libr loan
30463

Bloomsburg University of Pennsylvania, Harvey A. Andruss Library, 400 E Second St, *Bloomsburg*, PA 17815-1301
T: +1 570 3894204; Fax: +1 570 3893066; URL: www.library.bloomu.edu
1839; Wayne Mohr
Art Exhibit Catalogs; Bloomsburg Univ Arch; Covered Bridges Newberry and Caldecott Awards (Elinor R. Keefer Coll)
471 000 vols; 1 713 curr per; 11 686 govt docs; 430 maps; 1 970 000 microforms; 1 500 av-mat; 8 080 sound-rec; 1 197 digital data carriers
libr loan
30464

Blue Mountain College, Guyton Library, 201 W Main St, *Blue Mountain*, MS 38610; P.O. Box 160, Blue Mountain, MS 38610-0160
T: +1 662 6854771; Fax: +1 662 6859519; E-mail: library@bmc.edu; URL: www.bmc.edu/library_minisite.asp
1875; Sue Ann Owens
Liberal arts
46 000 vols; 180 curr per; 33 000 e-books
30465

Bluefield College, Easley Library, 3000 College Dr, *Bluefield*, VA 24605
T: +1 276 3264269; Fax: +1 276 3264288; E-mail: library@bluefield.edu; URL: www.bluefield.edu/library/
1922; Nora Lockett
McKenzie Memorial Religion Coll, Barbour Coll
56 000 vols; 250 curr per
30466

Bluefield State College, Wendell G. Hardway Library, 219 Rock St, *Bluefield*, WV 24701
T: +1 304 3274054; Fax: +1 304 3274203; URL: www.bluefieldstate.edu/library.htm
1895; Joanna M. Thompson
BSC Arch
72 000 vols; 89 curr per
30467

Bluffton University, Musselman Library, One University Dr, *Bluffton*, OH 45817-2104
T: +1 419 3583262; Fax: +1 419 3583384; E-mail: referencedesk@bluffton.edu; URL: www.bluffton.edu/library
1899; Mary Jean Johnson
Robert Frost Coll; Mennonite Hist, Culture, & Theology; Mennonite Hist Libr, College Arch
156 000 vols; 700 curr per; 4 500 maps; 116 000 microforms; 50 sound-rec; 300 digital data carriers; 100 charts, pamphlet file
libr loan; ALA
30468

University of South Carolina at Beaufort, Library, One University Blvd, *Bluffton*, SC 29909-6085
T: +1 843 2088022; Fax: +1 843 2088296; URL: www.sc.edu/beaufort/library/
1959; Ellen E. Chamberlain
South Carolina hist
71 000 vols; 141 curr per; 25 000 e-books
libr loan
30469

Snead State Community College, Virgil B. McCain Learning Resource Center, 220 N Walnut, *Boaz*, AL 35957
T: +1 256 8404173; E-mail: jmiller@snead.edu; URL: www.snead.edu/library
1935; John Miller
Alabama Authors (Borden Deal, Babs Deal, William B Huie, William Heath, Elise Sanguinetti, Thomas Wilkerson), Alabama Coll
38 275 vols; 222 curr per
30470

Florida Atlantic University, S.E. Wimberly Library, 777 Glades Rd, *Boca Raton*, FL 33431; P.O. Box 3092, Boca Raton, FL 33431-0992
T: +1 561 2976911; Fax: +1 561 2972282; E-mail: lyref@fau.edu; URL: www.fau.edu/library
1961; William Miller
Jaffe Collection of Books as Aesthetic Objects, Writings of Florida author Theodore Pratt, Fraiberg Judaica Colls, Judaica Sound Archs
1 203 000 vols; 1 466 curr per; 5 667 e-journals; 305 000 e-books; 976 755 govt docs; 37 022 maps; 1 620 000 microforms; 2 269 sound-rec; 2 457 digital data carriers
libr loan; CRL
30471

Lynn University, University Library, 3601 N Military Trail, *Boca Raton*, FL 33431-5598
T: +1 561 2377254; Fax: +1 561 2377074; URL: www.lynn.edu/library
1963; Charles L. Kuhn
Humanities, Business, Management, Fashion, Geriatrics, Retailing
100 000 vols; 350 curr per
30472

Gardner-Webb University, Dover Memorial Library, 110 S Main St, *Boiling Springs*, NC 28017; P.O. Box 836, Boiling Springs, NC 28017-0836
T: +1 704 4063050; Fax: +1 704 4064623; URL: www.library.gardner-webb.edu
1928; Valerie M. Parry
Religion; Thomas Dixon Coll, Church Baptist Curriculum Laboratory Coll
223 000 vols; 450 curr per; 24 000 e-books
libr loan
30473

Boise State University, Albertsons Library, 1865 Campus Lane, *Boise*, ID 83725-1430; mail address: 1910 University Dr, Boise, ID 83725-1430
T: +1 208 4263756; Fax: +1 208 4261394; URL: library.boisestate.edu
1932; Marilyn Moody
Business and Management, Education; Mss Coll, Rare Books, Idaho Coll, De Groot Coll (Hemingway), Artists Bks, Historica Scholastica, Idaho Writers Arch, Photo Coll, Univ Arch
659 000 vols; 3 065 curr per
libr loan
30474

Southwest Baptist University, Harriett K. Hutchens Library, 1600 University Ave, *Bolivar*, MO 65613
T: +1 417 3281621; Fax: +1 417 3281652; URL: www.sbuniv.edu/library/
1878; Edward W. Walton
Butler Baptist Heritage Coll, Libr of American Civilization – Microbk Coll, Christian Education Resource Lab
178 000 vols; 407 curr per; 9 830 e-journals; 16 000 e-books; 415 000 microforms; 2 520 av-mat; 3 990 sound-rec; 400 digital data carriers
libr loan
30475

1903; Mary Reichel
Education; Justice-Query Instructional Resources Center, William Eury Appalachian Coll
823 000 vols; 5 306 curr per; 759 703 govt docs; 7 966 maps; 1 436 000 microforms; 25 469 sound-rec; 1 230 digital data carriers; 81 220 Electronic Media & Resources
libr loan
30476

Frank Phillips College, James W. Dillard Library, P.O. Box 5118, *Borger*, TX 79008-5118
T: +1 806 4574200 ext 733; Fax: +1 806 4574230; E-mail: klane@fpctx.edu; URL: www.fpctx.edu
1948; Karen Lane
35 000 vols; 138 curr per
libr loan
30477

Berklee College of Music, Stan Getz Library, 1140 Boylston St, *Boston*, MA 02215
T: +1 617 7472258; Fax: +1 617 7472050; URL: www.library.berklee.edu
1964; Paul Engle
Jazz & Rock Music Coll
35 710 vols; 150 curr per; 15 754 music scores; 1 000 microforms; 2 423 av-mat; 18 745 sound-rec; 32 digital data carriers
30478

Boston Baptist College Library, 950 Metropolitan Ave, *Boston*, MA 02136
T: +1 617 3643510 ext 216; E-mail: ftatro@boston.edu; URL: www.boston.edu
1976; Fred Tatro
Religous Resources 1800's to present
57 000 vols; 4 000 e-journals
30479

The Boston Conservatory, Albert Alphin Music Library, Eight The Fenway, *Boston*, MA 02215-4099
T: +1 617 9129131;
Fax: +1 617 9129101; URL: www.bostonconservatory.edu/programs/library.html
1867; Jennifer Hunt
Jan Veen & Katrine Amory Hooper Memorial Dance Coll, Jan Veen & Katrine Amory Hooper Memorial Art Coll, James Pappoutsakis Memorial Coll, Katherine Rossi Memorial Coll
40 000 vols; 85 curr per
30480

Boston University, *Boston*, MA 02215

– African Studies Library, 771 Commonwealth Ave, Boston, MA 02215
T: +1 617 3533726; Fax: +1 617 3581729; URL: www.bu.edu/library/asl/index.html
David Westley
African Govt Docs
200 000 vols; 425 curr per; 50 000 govt docs; 500 maps; 6 000 microforms; 5 000 pamphlets & ephemera
30481

– Alumni Medical Library, 715 Albany St L-12, Boston, MA 02118-2394
T: +1 617 6384232; Fax: +1 617 6384233; URL: medlib.bu.edu
1848; Dr. David S. Ginn
160 000 vols; 4 000 curr per; 190 e-books; 5 digital data carriers
libr loan
30482

– Frederick S. Pardee Management Library, Boston University School of Management" Boston, MA 02215
T: +1 617 3534304; Fax: +1 617 3534307; URL: www.bu.edu/library/management
1997; Arlyne Ann Jackson
92 000 vols; 453 curr per; 20 000 e-journals; 436 sound-rec
30483

– Mugar Memorial Library, 771 Commonwealth Ave, Boston, MA 02215
T: +1 617 3533710; Fax: +1 617 3532084; URL: www.bu.edu/library
1839; Robert Hudson
African Studies, Americana to 1920 (Mark & Llora Bortman Coll), Art of the Printed Book, H G Wells Coll, Browning Coll, G B Shaw Coll, Hist of Nursing, Lincolniana, Liszt Coll, 19th C English Lit, Military Hist, Pascal Coll, Robert Frost Coll, Theodore Roosevelt Coll, Whitman Coll
2 396 000 vols; 34 214 curr per; 52 224 av-mat; 52 230 Talking Bks
libr loan
30484

– Music Library, 771 Commonwealth Ave, Boston, MA 02215
T: +1 617 3533705; Fax: +1 617 3532084; URL: web.bu.edu/library/music/index.htm
Holly E. Mockovak
Arthur Fiedler Coll, Byzantine Res Coll on Egon Wellesz, Boston Symphony Orchestra, Franz Liszt Coll, Roger Voisin Coll of Trumpet Repertoire
73 500 vols; 350 curr per; 5 000 music scores; 4 000 microforms; 52 000 sound-rec
libr loan
30485

– Pappas Law Library, School of Law, 765 Commonwealth Ave, Boston, MA 02215
T: +1 617 3533151; Fax: +1 617 3535995; URL: www.bu.edu/lawlibrary
1872; Marlene Aldeman
Anglo-American Law Coll, Health Law, Banking Law, Int Law
650 000 vols; 10 digital data carriers
30486

– Pickering Educational Resources Library, Two Sherborn St, Boston, MA 02215
T: +1 617 3533734; Fax: +1 617 3536105; URL: www.bu.edu/library/education
1949; Linda Plunket
Standardized Psychological & Educational Tests, Curriculum Guides, Children & Young Adult Lit
20 000 vols; 100 curr per; 188 diss/theses; 7 000 microforms; 90 av-mat; 10 digital data carriers
libr loan; ALA
30487

– School of Theology Library, 745 Commonwealth Ave, 2nd Flr, Boston, MA 02215
T: +1 617 3533034; Fax: +1 617 3580699; E-mail: sthlib@bu.edu; URL: www.bu.edu/sth/sthlibrary
1839; Jack W. Ammerman
American Guild of Organists Libr, Nuttler-Metcalf Hymnal Coll, Rogers Hymnal Coll, Kimball Bible Coll, Massachusetts Bible Society Coll, Woodward Coll of Oriental Art, New England Hist Society Coll, Jesuitica Coll, Carmelite Coll
146 000 vols; 545 curr per; 760 incunabula; 5 800 music scores; 4 maps; 27 000 microforms; 68 av-mat; 780 sound-rec; 9 digital data carriers; 960 linear feet of archives
libr loan; ATLA
30488

– Science & Engineering Library, 38 Cummington St, Boston, MA 02215
T: +1 617 3539474; Fax: +1 617 3533470; E-mail: selill@bu.edu; URL: www.bu.edu/library/sel
1983; Paula Carey
85 000 vols; 1 800 curr per; 80 000 serial vols
30489

Boston University – Stone Science Library, 675 Commonwealth Ave, Rm 440, *Boston*, MA 02215
T: +1 617 3535679; Fax: +1 617 3535358; E-mail: nparveen@bu.edu; URL: www.bu.edu/library/stone/stone.html
1988; Nasim Momen
Archaeology, geography, geology; geography (George K. Lewis Coll), space science, Apollo missions (NASA Photo Arch), ballon aerial photos
10 000 vols; 225 curr per
30490

Emerson College, Library, 120 Boylston St, *Boston*, MA 02116-4624
T: +1 617 8248339; Fax: +1 617 8247817; E-mail: circulation@emerson.edu; URL: www.emerson.edu/library
1880; Mickey Zemon
Mass communications, Theatre; Boston Herald Theatre Clipping Coll
151 000 vols; 17 612 curr per; 8 000 e-books; 6 000 microforms; 1 998 sound-rec; 16 digital data carriers; 240 Electronic Media & Resources
30491

Emmanuel College, Cardinal Cushing Library, 400 The Fenway, *Boston*, MA 02115
T: +1 617 7359927; Fax: +1 617 7359763; URL: www1.emmanuel.edu/library
1919; Dr. Susan E. von Daum Tholl
Theology, Lit, Art, Women's studies; 17th C Biblical Commentaries
96 000 vols; 800 curr per; 10 microforms
libr loan; ALA
30492

Appalachian State University, Carol Grotnes Belk Library, 218 College St, *Boone*, NC 28608; P.O. Box 32026, Boone, NC 28608-2026
T: +1 828 2622818; Fax: +1 828 2623001; URL: www.library.appstate.edu
libr loan

Fisher College, Library, 118 Beacon St, ***Boston***, MA 02116
T: +1 617 2368875; Fax: +1 617 6704426; E-mail: library@fisher.edu; URL: www.fisher.edu/library
1903; Joshua Van Kirk McKain
Fashion, Hospitality, Travel, Physical therapy
25 000 vols; 99 curr per
libr loan 30493

Massachusetts College of Art, Morton R. Godine Library, 621 Huntington Ave, ***Boston***, MA 02115-5882
T: +1 617 8797150; Fax: +1 617 8797110; E-mail: reference@massart.edu; URL: www.massart.edu/library
1873; Paul Dobbs
Art, Art hist, Films & filmmaking; Art Education, Design, College Arch
99 350 vols; 442 curr per; 710 microforms; 8 700 microforms of per, 117 960 slides 30494

Massachusetts College of Pharmacy & Health Sciences, Sheppard Library, 179 Longwood Ave, ***Boston***, MA 02115-5896
T: +1 617 7322803; Fax: +1 617 2781566; URL: www.mcphs.edu/libraries
1823; Richard Kaplan
Hist of Pharmacy, College Arch
20 000 vols; 700 curr per; 31 000 microforms; 2 digital data carriers 30495

Massachusetts School of Professional Psychology (MSPP) Library, 221 Rivermoor St, ***Boston***, MA 02132
T: +1 617 3276777; Fax: +1 617 3274447; E-mail: library@mspp.edu; URL: www.mspp.edu
Matthew Kramer
Psychopharmacology, psychotherapy
10 000 vols; 25 curr per 30496

New England College of Optometry, Library, 424 Beacon St, ***Boston***, MA 02115
T: +1 617 5875589; Fax: +1 617 5875573; E-mail: library@neco.edu; URL: www.neco.edu/library
1894; Cindy Hutchison
Hist of Optometry Coll
18 000 vols; 238 curr per; 118 av-mat; 152 sound-rec; 27 digital data carriers; 11 443 slides, 5 181 charts 30497

New England Conservatory of Music – Harriet M. Spaulding Library, 33 Gainsborough St, ***Boston***, MA 02115
T: +1 617 5851250;
Fax: +1 617 5851245; URL: www.newenglandconservatory.edu
1867; Jean Morrow
New England Composers; Preston Coll of Musicians' Letters; Firestone Hour Coll Music; Vaughn Monroe Coll of Camel Caravan, scores; Idabell Firestone Audio Library
142 000 vols; 295 curr per; 5 000 mss; 250 diss/theses; 500 microforms; 50 000 sound-rec; 1 000 other items 30498

New England Conservatory of Music – Idabelle Firestone Audio Library, 290 Huntington Ave, ***Boston***, MA 02115
T: +1 617 5851255;
Fax: +1 617 5851245; URL: www.newenglandconservatory.edu/libraries
Jean Morrow
American Composers' Manuscripts, Early Jazz Recordings
26 000 vols; 300 curr per 30499

New England School of Law Library, 154 Stuart St, ***Boston***, MA 02116-5687
T: +1 617 4227282; Fax: +1 617 4227303; URL: www.nesl.edu/library
1917; Anne M. Acton
Standard American Law Libr Coll, micro, & AV
340 000 vols; 3 100 curr per; 100 av-mat; 573 sound-rec; 50 digital data carriers 30500

Northeastern University, ***Boston***
– School of Law Library, 400 Huntington Ave, Boston, MA 02115
T: +1 617 3733332; Fax: +1 617 3738705; URL: www.slaw.neu.edu/library
1898
Abolition of Death & Capital Punishment (Sara Ehrmann Coll); Pappas Public Interest Law Coll
397 000 vols; 165 000 microforms 30501

– Snell Library, 320 Snell Library, 360 Huntington Ave, Boston, MA 02115
T: +1 617 3737088; Fax: +1 617 3738681; URL: www.library.neu.edu
1898; William M. Wakeling
Glenn Cray/Casa Loma Orchestra Swing, original scores & rec; Freedom House Coll
977 000 vols; 6 196 curr per; 113 000 e-books
libr loan 30502

Simmons College, Beatley Library, 300 The Fenway, ***Boston***, MA 02115-5898
T: +1 617 5212784; Fax: +1 617 5213093; E-mail: library@simmons.edu; URL: www.simmons.edu/libraries
1899; Daphne Harrington
Libr & info science, Social service, Women's studies; Career Resource Mat, Children's Lit (Knapp Coll), Simmons College Arch
214 000 vols; 1 748 curr per; 40 000 e-books
libr loan 30503

Suffolk University, Mildred F. Sawyer Library, 73 Tremont St, ***Boston***, MA 02108-2770
T: +1 617 5738535; Fax: +1 617 5738756; E-mail: sawlib@suffolk.edu; URL: www.suffolk.edu/sawlib/sawyer.htm
1906; Rebecca Fulweiler
Afro-American Lit Coll
139 862 vols; 227 curr per; 68 503 e-books; 115 diss/theses; 144 700 microforms; 560 av-mat
libr loan 30504

– Law Library, 120 Tremont St, Boston, MA 02108-4977
T: +1 617 3051614; URL: www.law.suffolk.edu/library
1906; Elizabeth McKenzie
211 000 vols; 6 620 curr per; 718 000 microforms; 58 av-mat; 141 sound-rec; 32 digital data carriers
libr loan 30505

Tufts University, Health Sciences Library, 145 Harrison Ave, ***Boston***, MA 02111-1843
T: +1 617 6366705; Fax: +1 617 6364039; E-mail: hhsl@tufts.edu; URL: www.library.tufts.edu/hsl/about
1906; Eric Albright
158 000 vols; 972 curr per; 11 000 microforms; 830 sound-rec; 1 digital data carriers; 292 slide titles
libr loan; AAHSLD, MLA, AMIA 30506

University of Massachusetts Boston, Joseph P. Healey Library, 100 Morrissey Blvd, ***Boston***, MA 02125-3300
T: +1 617 2875910; Fax: +1 617 2875955; E-mail: library.admin@umb.edu; URL: www.lib.umb.edu
1965; Daniel Ortiz
Massachusetts Society for the Prevention of Cruelty to Children Coll, Urban Planning Aid Coll, Boston Children's Aid Society Coll, Dorchester House Coll, New England Free Press Coll, Thompson's Island Coll, William Joiner Center Coll on the Vietnam War
587 000 vols; 773 000 microforms; 1 826 sound-rec; 59 digital data carriers
libr loan 30507

Wentworth Institute of Technology, Alumni Library, 550 Huntington Ave, ***Boston***, MA 02115-5998
T: +1 617 9894040; Fax: +1 617 9894091; URL: www.wit.edu/Library/
1904; Walter T. Punch
Hist of Technology Coll; Arch of the American Society for Engineering Education – Engineering Technology Divisions, Lufkin Mechanical Engineering Coll
75 000 vols; 500 curr per; 13 000 microforms; 7 digital data carriers
libr loan 30508

Wheelock College, Library, 132 The Riverway, ***Boston***, MA 02215-4815
T: +1 617 8792220; Fax: +1 617 8792408; E-mail: reference@wheelock.edu; URL: www.wheelock.edu/library/index.asp
1889; Brenda Ecsedy
Early Childhood Curriculum Resource Coll; Hist of Kindergarten in the US; Rare & hist children's lit (US & Great Britian)
92 000 vols; 536 curr per; 10 000 e-books; 17 digital data carriers
libr loan 30509

Naropa University, Allen Ginsberg Library, 2130 Arapahoe Ave, ***Boulder***, CO 80302
T: +1 303 2464668; Fax: +1 303 2454636; E-mail: library@naropa.edu; URL: library.naropa.edu
1974; N.N.
Buddhism, Psychology, Contemporary poetry; Tibetan Religious Texts, Modern Poetry Coll
28 000 vols; 131 curr per; 2 000 e-books; 1 000 av-mat; 5 000 sound-rec; 7 500 slides
libr loan 30510

University of Colorado at Boulder, University Libraries, 1720 Pleasant St, 184 UCB, ***Boulder***, CO 80309-0184
T: +1 303 4927521; Fax: +1 303 4921881; URL: ucblibraries.colorado.edu
1877; James F. Williams II
Juvenile Lit (Epstein & Block Colls); Hist of Silver (Leavens Coll); Human Area Relations File; Human Rights; Labor Arch; Miller Milton Coll; Mountaineering (John J. Jerome Hart Coll); Tippit Photobook Coll; Western Hist
3 843 000 vols; 30 350 curr per; 83 000 e-journals; 974 116 govt docs; 50 000 music scores; 208 605 maps; 6 508 000 microforms; 63 400 sound-rec; 11 000 digital data carriers; 42 300 linear ft of mss, 58 890 Music Scores
libr loan 30511

– Archives Department, Norlin Library, 184 UCB, Boulder, CO 80309-0184
T: +1 303 4927242; Fax: +1 303 4923960; E-mail: arv@colorado.edu; URL: ucblibraries.colorado.edu/archives/index.htm
1917; Bruce P. Montgomery
Colorado hist + politicians, labor, citizen activism, women's hist, environmentalism, peace, 19th & 20th c US West
21 000 vols; 900 mss; 12 300 diss/theses; 1 500 maps; 1 000 microforms; 3 880 av-mat; 250 sound-rec; 21 000 linear ft of mss, 10 000 linear ft of univ. archives, 2 000 linear ft of newspapers, 350 000 photos 30512

– Art & Architecture Library, Norlin Library, Campus Box 184, Boulder, CO 80309-0184
T: +1 303 492-3966; Fax: +1 303 492-0935; E-mail: Meredith.Kahn@Colorado.edu; URL: ucblibraries.colorado.edu/art/index.htm
1966; Meredith Kahn
Artist's books; art exhibition cat
100 000 vols; 500 curr per; 12 311 microforms
libr loan 30513

– East Asian Library, Norlin Library, 1720 Pleasant St, Boulder, CO 80309; University Libraries East Asian Library, 184 UCB, Boulder, CO 80309-0184
T: +1 303 4928822; Fax: +1 303 4921881; E-mail: asianlib@colorado.edu; URL: ucblibraries.colorado.edu/eastasian/index.htm
1989; N.N.
Chinese Language
75 000 vols; 180 curr per; 2 140 e-journals; 500 e-books; 250 sound-rec
libr loan 30514

– Equity Diversity & Education Library, School of Education, Campus Box 249, Rm 344, Boulder, CO 80309-0249
T: +1 303 4923359; Fax: +1 303 4927090; E-mail: educlib@colorado.edu; URL: education.colorado.edu/library
Suzanne Sawyer-Ratliff
25 000 vols; 15 curr per 30515

– Gemmill Engineering Library, Mathematics Bldg, Rm 135, 184 UCB, Boulder, CO 80309-0184
T: +1 303 4925396; Fax: +1 303 4926488; E-mail: engref@colorado.edu; URL: ucblibraries.colorado.edu/engineering/index.htm
1966; Jack Maness
ANSI & ASTM Standards coll
155 000 vols; 91 000 microforms
libr loan 30516

– Government Publications Library, Norlin Library, 1720 Pleasant St, Boulder, CO 80309; University Libraries Government Publications, 184 UCB, Boulder, CO 80309-0184
T: +1 303 4928834; Fax: +1 303 4921881; E-mail: govpubs@colorado.edu; URL: ucblibraries.colorado.edu/govpubs/index.htm
Jennifer Gerke
825 000 vols 30517

– Howard B. Waltz Music Library, University Libraries Music Library, N250, 184 UCB, Boulder, CO 80309-0184
T: +1 303 4928093; Fax: +1 303 7350100; E-mail: mus@colorado.edu; URL: ucblibraries.colorado.edu/music/index.htm
1959; Laurie Sampsel
American Music Research Center, Perry Como Coll, Lumpkin Folk Song, Cecil Effinger, Normen Lockwood, George Lynn, Eric Katz, Calif. Mission Music, Moravian Music
74 000 vols; 60 400 music scores; 810 av-mat; 45 000 sound-rec
libr loan 30518

– Jerry Crail Earth Sciences & Map Library, Benson Earth Science Bldg, Campus Box 184, Boulder, CO 80309-0184
T: +1 303 4926133; Fax: +1 303 7354879; E-mail: suzanne.larsen@colorado.edu; URL: www-libraries.colorado.edu
Suzanne Larsen
Map coll
44 031 vols; 325 curr per; 53 420 microforms
libr loan 30519

– Natural Hazard Center Library (Institute of Behavioral Science), 482 UCB, Boulder, CO 80309-0482
T: +1 303 4925787; Fax: +1 303 4922151; E-mail: hazlib@colorado.edu; URL: www.colorado.edu/hazards
Wanda Headley
25 000 vols; 21 curr per 30520

– Oliver C. Lester Library of Mathematics & Physics, Duane Physical Labs G-140, 184 UCB, Boulder, CO 80309-0184
T: +1 303 4928231; Fax: +1 303 7355355; URL: ucblibraries.colorado.edu/mathphysics/index.htm
1963; Suzanne T. Larsen
80 000 vols; 1 000 curr per; 1 000 govt docs
libr loan 30521

– Science Library, Norlin Library, 1720 Pleasant St, Boulder, CO 80309; University Libraries Science Library, 184 UCB, Boulder, CO 80309-0184
T: +1 303 4925136; Fax: +1 303 4921881; E-mail: sci@colorado.edu; URL: ucblibraries.colorado.edu/science/index.htm
1940; David Fagerstrom
121 000 vols
libr loan 30522

– Special Collections Department, Norlin Library, 1720 Pleasant St, 184 UCB, Boulder, CO 80309-0184
T: +1 303 4923910; Fax: +1 303 4921881; URL: ucblibraries.colorado.edu/specialcollections/index.htm
Deborah Hollis
Colorado, Anlo-American literature, natural history, fine printing, book arts
78 000 vols 30523

– William M. White Business Library, Koelbel Bldg, Leeds College of Business, Boulder, CO 80309-0184
T: +1 303 4923194; Fax: +1 303 7350333; URL: ucblibraries.colorado.edu/business/index.htm
Carol Krismann
Douglas H. Buck Financial Records Coll
33 000 vols; 202 000 microforms; 200 sound-rec; 25 digital data carriers
libr loan 30524

University of Colorado at Boulder – William A. Wise Law Library, 2450 Kittredge Loop Dr, 402 UCB, ***Boulder***, CO 80309-0402
T: +1 303 4924945; Fax: +1 303 4922707; E-mail: lawlib@colorado.edu; URL: lawpac.colorado.edu
1892; Barbara A. Bintliff
Commonwealth & Foreign Law

700 000 vols
libr loan; AALL 30525

Olivet Nazarene University, Benner Library & Resource Center, One University Ave, ***Bourbonnais***, IL 60914-2271
T: +1 815 9395354; Fax: +1 815 9395170; URL: library.olivet.edu
1909; Kathryn R. Boyens
Theology, Education, Nursing; John Wesley Coll, Arch of Olivet Univ, Geological Maps
166 000 vols; 601 curr per; 11 360 e-journals; 36 000 e-books; 11 002 maps; 246 000 microforms; 2 599 av-mat; 14 890 sound-rec; 147 digital data carriers 30526

Bowie State University, Thurgood Marshall Library, 14000 Jericho Park Rd, ***Bowie***, MD 20715
T: +1 301 8603850; Fax: +1 301 8603848; URL: www.bowiestate.edu
1937; Dr. Richard Bradberry
Art; Afro-American Experience, Slave Doc, Rare Bks, Maryland Subject, Univ Hist
244 000 vols; 732 curr per 30527

Bowling Green State University, University Libraries, 204 William Jerome Library, ***Bowling Green***, OH 43403-0170
T: +1 419 3722856; Fax: +1 419 3727996; E-mail: libhelp@bgsu.edu; URL: www.bgsu.edu/colleges/library
1913; N.N.
Popular Culture Coll, Popular Recordings (Music Libr)
2 416 045 vols; 4 520 curr per; 553 digital data carriers; 2 118 662 microforms of per
libr loan 30528

– Center for Archival Collections, Jerome Library, 5th Flr, Bowling Green, OH 43403
T: +1 419 3722411; Fax: +1 419 3720155; E-mail: archive@bgsu.edu; URL: www.bgsu.edu/colleges/library
1968; Ann Jenks
Genealogy, Hist preservation, State hist, Labor hist; Lud Ashley Papers, Delbert Latta Papers, Sam Pollock Labor Coll, Rare Bks (Ray Bradbury)
23 590 vols; 156 curr per; 20 000 microforms; 200 000 photos 30529

– Curriculum Resource Center Library, Bowling Green, OH 43403-0170
T: +1 419 3722956; Fax: +1 419 3727996; E-mail: sbushong@bgnet.bgsu.edu; URL: www.bgsu.edu/colleges/library
1967; Sara Bushong
Children's & Young Adults lit
73 950 vols; 9 curr per; 3 503 av-mat; 4 468 pamphlets, 2 012 publishers' cat, 107 tests, 9 037 pictures 30530

– Frank Ogg Science & Health Library, Mathematical Sciences Bldg, Rm 308, Bowling Green, OH 43403
T: +1 419 3722591; Fax: +1 419 3726817; URL: www.bgsu.edu/colleges/library
1962; Mary Keil
282 500 vols; 1 525 curr per 30531

– Government Documents, Jerome Library 1st Flr, Bowling Green, OH 43403
T: +1 419 3722142; URL: www.bgsu.edu/colleges/library/services/govdocs/index.html
Coleen Parmer
700 000 vols 30532

– Historical Collections of the Great Lakes, Jerome Library 6th Flr, Bowling Green, OH 43403
T: +1 419 3729612; Fax: +1 419 3720155; E-mail: archive@bgsu.edu; URL: www.bgsu.edu/colleges/library
1983; Robert Graham
Great Lakes Hist & Maritime Hist, Naval Architecture
8 000 vols; 70 curr per; 894 microforms; 250 000 marine eng drawings & tracings, 2 000 linear feet of mss, 538 linear feet of newspaper clippings, 4 500 pamphlets, 130 000 photos, 2 500 navigational charts, 3 500 slides 30533

– Music Library & Sound Recordings Archives, Jerome Library, 3rd Flr, Bowling Green, OH 43403
T: +1 419 3722307; Fax: +1 419 3727996; URL: www.bgsu.edu/colleges/library/music/music.html
1967; Susannah Cleveland
College of Musical Arts Coll
34 830 vols; 100 curr per 30534

– Ray & Pat Browne Popular Culture Library, Jerome Library, 4th Flr, Bowling Green, OH 43403
T: +1 419 3722450; Fax: +1 419 3727996; E-mail: ndown@bgnet.bgsu.edu; URL: www.bgsu.edu/colleges/library/pcl/pcl.html
1969; Nancy Down
Allan & John Saunders Coll, E T Ned Guymon Detective Fiction Coll, H James Horovitz Science Fiction Coll
130 000 vols; 122 curr per; 165 mss; 325 microforms; 200 av-mat; 55 000 comic bks, 1 500 linear feet of mss, 9 800 pulp mag, 8 700 scripts 30535

Western Kentucky University, University Libraries, Helm-Cravens Library Complex, 1906 College Heights Blvd, No 11067, ***Bowling Green***, KY 42101-1067
T: +1 270 7453951; Fax: +1 270 7456422; E-mail: library.web@wku.edu; URL: www.wku.edu/library
1907; Dr. Michael Binder
Rare Kentuckiana, Mammoth Cave, Kentucky Writers, Civil War, Shakers, Ohio Valley
784 000 vols; 3 912 curr per; 2 400 000 microforms; 356 sound-rec; 725 digital data carriers 30536

– Educational Resources Center, Tate Page Hall, Rm 366, 1906 College Heights Blvd, No 31031, Bowling Green, KY 42101
T: +1 270 7454659; Fax: +1 270 7454553
Roxanne Spencer
Education
12 000 vols 30537

– Kentucky Library & Museum, 1400 Kentucky St, Bowling Green, KY 42101-3479; mail address: 1906 College Heights Blvd, No 11092, Bowling Green, KY 42101-1092
T: +1 270 7452592; Fax: +1 270 7456264; URL: www.wku.edu/library/kylm
Timothy J. Mullin
70 000 vols; 10 curr per 30538

Montana State University – Bozeman, Roland R. Renne Library, Centennial Mall, ***Bozeman***, MT 59717; P.O. Box 173320, Bozeman, MT 59717-3320
T: +1 406 9943171; Fax: +1 406 9942851; URL: www.lib.montana.edu
1893; Tamara Miller
Montana Hist, Burton K Wheeler Coll, Yellowstone National Park, Haynes Coll, M. L. Wilson Agricultural Hist Coll, Leggat-Donahoe Northwest Coll, Abraham Lincoln Coll, Montana Architectural Drawings Coll
712 000 vols; 8 757 curr per; 5 190 e-journals
libr loan 30539

University of Pittsburgh at Bradford, T. Edward & Tullah Hanley Library, 300 Campus Dr, ***Bradford***, PA 16701
T: +1 814 3627610; Fax: +1 814 3627688; URL: www.library.pitt.edu/brad/hanley.html; www.upb.pitt.edu/offices_services/library/library.htm
1963; Trisha A. Morris
Montaigne, French Lit (Lowenthal Coll)
95 000 vols; 350 curr per
libr loan 30540

Central Lakes College, Jon Hassler Library, 501 W College Dr, ***Brainerd***, MN 56401
T: +1 218 8558000, 8180; Fax: +1 218 8558179; E-mail: lkellerm@clcmn.edu; URL: www.clcmn.edu/library
1938; Larry M. Kellerman
American Indian Coll, Local Govt, Scandinavian Coll
32 000 vols 30541

School for International Training, Donald B. Watt Library, Kipling Rd, ***Brattleboro***, VT 05302; P.O. Box 676, Brattleboro, VT 05302-0676
T: +1 802 2583354; Fax: +1 802 2583248; E-mail: library@sit.edu; URL: www.sit.edu/library
1967; Amy Beth
Area studies, Social sciences, Behav sciences, Bicultural studies, Bilingual studies, Education, Environmental studies; Int Organization Files, Masters Studies Essays, Language & Culture Coll
35 000 vols; 12 000 curr per; 15 000 e-books; 150 maps; 8 000 microforms; 10 digital data carriers
libr loan 30542

Long Island University, Brentwood Campus Library, 100 Second Ave, ***Brentwood***, NY 11717
T: +1 631 2735112; Fax: +1 631 2735198; URL: www.liu.edu
1972; Joong Suk Kim
56 000 vols; 273 curr per 30543

Brevard College, J.A. Jones Library, One Brevard College Dr, ***Brevard***, NC 28712-4283
T: +1 828 8848268; Fax: +1 828 8845424; E-mail: library@brevard.edu; URL: www.brevard.edu/library
1934; Michael M. McCabe
60 000 vols; 200 curr per; 19 000 e-journals; 65 000 e-books; 1 523 music scores; 47 maps; 3 000 microforms; 1 000 av-mat; 2 637 sound-rec; 9 digital data carriers
libr loan; SOLINET, ALA 30544

University of Bridgeport, Magnus Wahlstrom Library, 126 Park Ave, ***Bridgeport***, CT 06604-5620
T: +1 203 5764745; Fax: +1 203 5764791; E-mail: reference@bridgeport.edu; URL: www.bridgeport.edu/library/
1927; Diane Mirvis
Health Sciences Hist Coll, McKew Parr Coll (Exploration), Starr Coll (Lit & Art), Lincolniana, Schwartzkopf Coll (American Socialism & Communism), Volk-Pattberg Coll (Printing), Edwin Stanton Coll (Southeast Asia)
216 000 vols; 692 curr per; 16 690 e-journals; 8 000 e-books; 1 000 000 microforms; 254 sound-rec; 105 digital data carriers 30545

Bridgewater College, Alexander Mack Memorial Library, 402 E College St, ***Bridgewater***, VA 22812
T: +1 540 8285411; Fax: +1 540 8285482; URL: www.bridgewater.edu/departments/library
1880; Andrew Pearson
Genealogy, Local hist; Rare Bibles & Church of the Brethren Mat
181 000 vols; 650 curr per; 1 700 av-mat; 1 075 sound-rec; 480 Electronic Media & Resources
libr loan 30546

Bridgewater State College, Clement C. Maxwell Library, Ten Shaw Rd, ***Bridgewater***, MA 02325
T: +1 508 6971706; Fax: +1 508 5316103; E-mail: libraryweb@bridgew.edu; URL: www.bridgew.edu/library
1840; Michael Somers
Education, Liberal Arts; Albert G Boyden Coll of Early American Textbooks; Bridgewaterana; Children's Coll; Dicken's Coll; Educational Resources Info Center; Libr of American Civilization; Abraham Lincoln Coll; Theodore Roosevelt Coll; Standardized Tests; Tests in Microfiche
300 000 vols; 1 100 curr per; 500 diss/theses; 16 988 maps; 710 000 microforms; 21 digital data carriers
libr loan 30547

King College, The E.W. King Library, 1350 King College Rd, ***Bristol***, TN 37620
T: +1 423 6524716; Fax: +1 423 6524871; E-mail: library@king.edu; URL: www.king.edu/library
1867; Julie A. Roberson
Religious studies, Missiology, Classics, Lit; Early Church & Presbyterian Hist
94 000 vols; 463 curr per
libr loan 30548

Roger Williams University, University Library, One Old Ferry Rd, ***Bristol***, RI 02809
T: +1 401 2543112; Fax: +1 401 2540818; URL: www.rwu.edu/library
1946; Elizabeth Peck Learned
Rhode Island Hist; State Census Coll; Roger Williams Information/Hist
223 000 vols; 23 127 curr per; 4 000 e-books; 630 maps; 30 000 microforms; 4 180 sound-rec; 12 digital data carriers; 58 659 Slides
libr loan; NELINET, OCLC 30549

– Architecture Library, One Old Ferry Rd, Bristol, RI 02809-2921
T: +1 401 2543679
John Schlinke
20 000 vols; 200 curr per 30550

– School of Law Library, Ten Metacom Ave, Bristol, RI 02809-5171
T: +1 401 2544548; Fax: +1 401 2544543; URL: law.rwu.edu/library/
Gail Winson
Law
302 000 vols; 3 564 curr per 30551

Virginia Intermont College, J.F. Hicks Memorial Library, 1013 Moore St, ***Bristol***, VA 24201
T: +1 276 4667960; URL: www.vic.edu/library/
1884; Jonathan Tallman
64 000 vols; 70 curr per; 28 000 e-books 30552

State University of New York – College at Brockport, Drake Memorial Library, 350 New Campus Dr, ***Brockport***, NY 14420-2997
T: +1 585 3955667; Fax: +1 585 3955651; URL: www.brockport.edu/library/
1860; Frank M. Wojcik
Early American Imprints, 1639-1800-Readex; Early English Books, 1475-1700; Libr of American Civilization; Libr of English Lit
643 000 vols; 1 910 curr per; 14 822 e-journals; 4 000 e-books; 17 500 maps; 1 967 000 microforms; 7 680 av-mat; 5 700 sound-rec; 75 digital data carriers
libr loan 30553

Albert Einstein College of Medicine, D. Samuel Gottesman Library, 1300 Morris Park Ave, ***Bronx***, NY 10461
T: +1 718 4303122; Fax: +1 718 4308795; URL: library.aecom.yu.edu
1955; Judie Malamud
230 000 vols; 4 000 e-journals; 650 e-books 30554

City University of New York (CUNY), Lehman College Library, 250 Bedford Park Blvd W, ***Bronx***, NY 10468-1589
T: +1 718 9608576; Fax: +1 718 9608952; URL: www.lehman.edu/library/library2.htm
1931; Dr. Rona Lynn Ostrow
Liberal Arts Coll, Bronx Hist Coll, José Luis Ponce de León Coll on the Spanish Civil War and Exile
563 000 vols; 1 548 curr per; 621 000 microforms
libr loan; ALA 30555

College of Mount Saint Vincent, Elizabeth Seton Library, 6301 Riverdale Ave, ***Bronx***, NY 10471-1093
T: +1 718 4053395; Fax: +1 718 6012091; URL: www.mountsaintvincent.edu/library2/index.htm
1910; William Perrenod
125 000 vols; 800 curr per
libr loan 30556

Fordham University, Walsh Library, 441 E Fordham Rd, ***Bronx***, NY 10458-5151
T: +1 718 8173570; Fax: +1 718 8173582; URL: www.library.fordham.edu
1841; Dr. James P. McCabe
American Revolution & Early Federal Americana (Charles Allen Munn Coll), Arts & Architecture (Gambosville Coll), William Cobbett Coll, Crimes (McGarry Coll), French Revolution (Joseph Givernaud Coll), Gaelic (McGuire-McLees Coll), Jesuitica, Vatican (Barberini Coll), Arts & Architecture (Gambosville Coll)
2 200 000 vols; 2 300 curr per; 355 000 e-books
libr loan 30557

Mercy College – Bronx Campus Library, 1200 Waters Pl, ***Bronx***, NY 10461
T: +1 718 6788850; Fax: +1 718 6788668; E-mail: libbx@mercy.edu
Michele Lee
30 000 vols 30558

Monroe College, Thomas P. Schnitzler Library, 2468 Jerome Ave, ***Bronx***, NY 10468
T: +1 718 9336700; Fax: +1 718 5844242; URL: www.monroecoll.edu
Shawn Kaba
1 000 000 vols; 350 curr per 30559

State University of New York – SUNY Maritime College, Stephen B. Luce Library, Six Pennyfield Ave, Fort Schuyler, ***Bronx***, NY 10465
T: +1 718 4097231; Fax: +1 718 4097256; E-mail: library@sunymaritime.edu; URL: www.sunymaritime.edu/library
1946; Constantia Constantinou
Maritime hist, Marine Casualty Reports, Maritime Research (Technical Reports), Marine Society of New York – arch, Sailors' Snug Harbor – arch
85 000 vols; 363 curr per; 588 diss/theses; 13 754 govt docs; 964 maps; 13 000 microforms; 804 av-mat; 1 014 sound-rec; 135 digital data carriers; 8 824 pamphlets
libr loan 30560

Concordia College, Scheele Memorial Library, 171 White Plains Rd, ***Bronxville***, NY 10708
T: +1 914 3379300; Fax: +1 914 3954893; URL: www.concordia-ny.edu/pages/library_home.htm
1881; James E. Corbly
Religion, Teacher education; Libr of American Civilization, Libr of English Lit
74 000 vols; 350 curr per; 32 000 microforms; 877 av-mat; 4 283 sound-rec; 96 overhead transparencies
libr loan; ALA 30561

Sarah Lawrence College, Esther Raushenbush Library, One Mead Way, ***Bronxville***, NY 10708
T: +1 914 3952474; Fax: +1 914 3952473; E-mail: library@slc.edu; URL: www.slc.edu/library/
1926; Charling Chang Fagan
Sarah Lawrence College Faculty Coll, Bessie Schoenberg Dance Coll
283 000 vols; 938 curr per; 938 maps; 17 483 sound-rec; 18 digital data carriers; 115 815 other items
libr loan 30562

South Dakota State University, Hilton M. Briggs Library, N Campus Dr, Box 2115, ***Brookings***, SD 57007-1098
T: +1 605 6885106; Fax: +1 605 6886133; E-mail: blref@sdstate.edu; URL: lib.sdstate.edu
1884; Dr. David Gleim
South Dakota Hist
644 000 vols; 1 830 curr per; 25 500 e-journals; 14 000 e-books; 350 000 govt docs; 745 000 microforms; 1 905 av-mat; 5 053 sound-rec; 250 digital data carriers
libr loan 30563

Hellenic College & Holy Cross Greek Orthodox School of Theology, Archbishop Iakovos Library, 50 Goddard Ave, ***Brookline***, MA 02445-7496
T: +1 617 8501243; Fax: +1 617 8501470; E-mail: jcotsonis@hchc.edu; URL: www.hchc.edu
1937; Very Rev Dr. Joachim Cotsonis
Contemporary Greek Lit; Contemporary Orthodox Theologians; Modern Greek Theology, Rare Greek Books, Archival Material; Archbishop Iakovos Coll
75 000 vols; 723 curr per; 3 mss; 310 music scores; 863 microforms; 2 700 av-mat; 100 sound-rec; 12 digital data carriers
libr loan; ATLA, ACRL, ALA 30564

Newbury College, Library, 150 Fisher Ave, ***Brookline***, MA 02445-5747; mail address: 129 Fisher Ave, Brookline, MA 02445-5796
T: +1 617 7307070; Fax: +1 617 7307239; E-mail: library@newbury.edu; URL: www.newbury.edu/library
1961; Peter G. Obuchan
36 000 vols; 189 curr per; 4 000 e-journals; 7 000 e-books; 42 000 microforms; 670 av-mat; 284 slides
libr loan; ALA 30565

Boricua College, Special Collections Library, 186 N Sixth St, ***Brooklyn***, NY 11211
T: +1 718 7822200; Fax: +1 718 7822050; URL: www.boricuacollege.edu
1973; Liza Rivera
Puerto Rico Coll
15 000 vols; 50 curr per 30566

Brooklyn College, Library, 2900 Bedford Ave, ***Brooklyn***, NY 11210-2889
T: +1 718 9514414; Fax: +1 718 9514799; E-mail: refdesk@brooklyn.cuny.edu; URL: ait.brooklyn.cuny.edu; library.brooklyn.cuny.edu
1930; Dr. Barbra Buckner Higginbotham
Academic Freedom, Brooklyn College Arch (incl student publs), Brooklyniana, Colonial, Ethiopian & Somalian Hist
1 299 000 vols; 20 025 curr per; 15 000 e-books
libr loan 30567

Brooklyn Law School Library, 250 Joralemon St, ***Brooklyn***, NY 11201
T: +1 718 7807975; Fax: +1 718 7800369; URL: www.brooklaw.edu/library
1901; Sara Robbins
International law
544 000 vols; 2 237 curr per; 90 000 microforms; 63 digital data carriers
libr loan; ALA, AALL 30568

Charles Evans Inniss Memorial Library, 1650 Bedford Ave, ***Brooklyn***, NY 11225-2010
T: +1 718 2704880; Fax: +1 718 2705182; URL: www.mec.cuny.edu/library
1970; Madeline Ford
American Culture Series (PCMI Coll); American Fiction Series; Black Hist & Culture; Libr of American Civilization Coll
117 000 vols; 450 curr per; 62 000 microforms; 698 av-mat; 5 839 sound-rec; 2 200 slides, 398 Overhead Transparencies 30569

Long Island University, Brooklyn Library, One University Plaza, ***Brooklyn***, NY 11201-9926
T: +1 718 4881081; Fax: +1 718 7804057; URL: www.brooklyn.liu.edu/cwis/bklyn/library/home.htm
1927; Dr. Constance Woo
19th & 20th c Black Social & Economic Docs (Eato Aid Society Coll, William Hamilton Relief Society Coll & New York African Society for Mutual Relief Coll), Urban Architecture & City Planning (Robert Weinberg Coll)
274 000 vols; 1 412 curr per
libr loan 30570

Polytechnic University, Bern Dibner Library of Science & Technology, Five MetroTech Ctr, ***Brooklyn***, NY 11201-3840
T: +1 718 2603530; Fax: +1 718 2603756; E-mail: blibrary@poly.edu; URL: www.poly.edu/library
1854; Jana Stevens Richman
Hist of Science & Technology, Paint & Surface Coatings (Mathiello Memorial Coll)
185 000 vols; 2 643 diss/theses; 62 000 microforms; 52 digital data carriers; 58 000 microtexts
libr loan 30571

Pratt Institute Library, 200 Willoughby Ave, ***Brooklyn***, NY 11205-3897
T: +1 718 6363704; Fax: +1 718 3994401; URL: library.pratt.edu
1887; Pat Cutright
Architecture, Art, Libr science; Fine Printing
203 000 vols; 800 curr per; 61 000 microforms; 750 av-mat; 118 digital data carriers; 100 000 slides, 216 000 art reproductions 30572

Saint Francis College, McGarry Library, 180 Remsen St, ***Brooklyn***, NY 11201
T: +1 718 4895307; Fax: +1 718 5221274; E-mail: library@stfranciscollege.edu; URL: www.stfranciscollege.edu
1884; James P. Smith
Kennedy Coll; Curriculum Libr
126 000 vols; 573 curr per; 680 av-mat; 1 370 sound-rec; 7 digital data carriers
libr loan 30573

Saint Joseph's College, McEntegart Hall Library, 222 Clinton Ave, ***Brooklyn***, NY 11205-3697
T: +1 718 6366860; Fax: +1 718 6367250; E-mail: mcentegart@sjcny.edu; URL: libraries.sjcny.edu
1916; William Meng
Local New York Hist Coll
110 000 vols; 235 curr per; 266 av-mat; 3 109 sound-rec; 4 digital data carriers
libr loan 30574

State University of New York – Brooklyn Educational Opportunity Center, 111 Livingston St, Ste 306, ***Brooklyn***, NY 11201
T: +1 718 8023314; Fax: +1 718 8023332
1968; Denese Mars
African-American hist
10 000 vols; 25 curr per 30575

State University of New York Health Science Center at Brooklyn, Medical Research Library of Brooklyn, 395 Lenox Rd, ***Brooklyn***, NY 11203; mail address: 450 Clarkson Ave, P.O. Box 14, Brooklyn, NY 11203
T: +1 718 2707401; Fax: +1 718 2707468; URL: library.downstate.edu
1860; Richard M. Winant
53 000 vols; 2 452 curr per; 85 e-books
libr loan 30576

Long Island University – C.W. Post Campus, B. Davis Schwartz Memorial Library, 720 Northern Blvd, ***Brookville***, NY 11548
T: +1 516 2992307; Fax: +1 516 2994169; URL: www.liu.edu/cwis/cwp/library/libhome.htm
1955; Dr. Donald L. Ungarelli
Tax; Auction Cat, Henry James First Eds 1900-50 Children's Fiction, Eugene & Carlotta O'Neill Private Libr, Theodore Roosevelt Assn Coll, Lord Coll of Sporting Bks
1 200 000 vols; 4 000 curr per; 51 maps; 337 000 microforms; 1 184 av-mat; 4 924 sound-rec; 5 digital data carriers; 27 649 slides, 239 overhead transparencies, 565 charts, 983 art reproductions 30577

– Center for Business Research Library, 720 Northern Blvd, 25-A, Brookville, NY 11548
T: +1 516 2992832; Fax: +1 516 2994170; E-mail: cbr@cwpost.liu.edu
Martha Cooney
20 000 vols; 250 curr per 30578

University of Texas at Brownsville & Texas Southmost College, Oliveira Memorial Library, 80 Fort Brown, ***Brownsville***, TX 78521
T: +1 956 5448220; Fax: +1 956 5445495
1926; Douglas M. Ferrier
Brownsville, Cameron County, Lower Rio Grande Valley & Northeast Mexico Hist
175 000 vols; 7 000 curr per; 35 000 e-journals; 8 000 e-books
libr loan; ALA, TLA 30579

Howard Payne University, Walker Memorial Library, 1000 Fisk Ave, ***Brownwood***, TX 76801
T: +1 325 6462502; Fax: +1 325 6498904; E-mail: library@hputx.edu; URL: ww1.hputx.edu\library
1889; Nancy Anderson
Burress Genealogical Coll
121 000 vols; 598 curr per; 45 000 e-books; 472 av-mat; 1 880 sound-rec 30580

Bowdoin College, Library, 3000 College Sta, ***Brunswick***, ME 04011-8421
T: +1 207 7253227; Fax: +1 207 7253083; URL: library.bowdoin.edu
1794; Sherrie Bergman
Abbot Coll, Arctic Coll, Carlyle Coll, Hawthorne Coll, Huguenot Coll, Longfellow Coll, Senator George J. Mitchell Papers, Maine
1 009 000 vols; 9 121 curr per
libr loan 30581

Bryn Athya College, Swedenborg Library, 2925 College Dr, ***Bryn Athyn***, PA 19009; P.O. Box 740, Bryn Athyn, PA 19009-0740
T: +1 267 5022536; Fax: +1 267 5022637; E-mail: library@brynathyn.edu; URL: www.brynathyn.edu/library
1877; Carroll C. Odhner
Religion (Swedenborgiana); Scientific Bks (Published in 16th, 17th & 18th Centuries)
109 000 vols; 152 curr per; 2 incunabula; 3 000 microforms; 118 av-mat; 472 sound-rec; 14 digital data carriers
libr loan; PALINET, OCLC, ALA 30582

Bryn Mawr College, Mariam Coffin Canaday Library, 101 N Merion Ave, ***Bryn Mawr***, PA 19010-2899
T: +1 610 5265276; Fax: +1 610 5267480; URL: www.brynmaw.edu/library
1885; Elliott Shore
Goodhart Medieval Coll, Dillingham & Monegal Coll of Latin American Hist & Lit, Adelman Coll, Castle Coll (Botany & Ornithology), McBride Coll (Asian & African Studies), Mss Coll, Graphics Coll
891 445 vols; 1 712 curr per; 5 240 Electronic Media & Resources
libr loan 30583

– Lois & Reginald Collier Science Library, 101 N Merion Ave, Bryn Mawr, PA 19104-2899
T: +1 610 5265118; Fax: +1 610 5267464; E-mail: tfreedma@brynmawr.edu; URL: www.brynmawr.edu/library
1890; Terri Freedman
Pennsylvania Geologic & Topographic Publs, Guidebks of Geological Societies
94 000 vols; 455 curr per; 1 300 microforms; 15 000 surveys & rpts 30584

– Rhys Carpenter Library for Art, Archaeology, and Cities, 101 N Merion Ave, Bryn Mawr, PA 19010-2899
T: +1 610 5267912; Fax: +1 610 5267911; URL: www.brynmawr.edu/library
1930; Camilla MacKay
113 000 vols; 150 curr per; 200 microforms 30585

Harcum College, Library, 750 Montgomery Ave, ***Bryn Mawr***, PA 19010-3476
T: +1 610 5266062; Fax: +1 610 5266086; URL: www.harcum.edu/cs_lib_harcum_library.aspx
1915; Ann Ranieri
40 000 vols; 232 curr per; 920 sound-rec; 34 digital data carriers 30586

West Virginia Wesleyan College, Annie Merner Pfeiffer Library, 59 College Ave, ***Buckhannon***, WV 26201
T: +1 304 4738059; Fax: +1 304 4738888; E-mail: librarian@wvwc.edu; URL: www.wvwc.edu/lib
1890; Paula L. McGrew
Religios studies, Civil War hist; Jones Lincoln Coll, Pearl S Buck Mss, West Virginia Hist Arch
120 000 vols; 447 curr per; 13 650 e-journals; 65 000 e-books; 58 maps; 4 000 microforms; 140 digital data carriers; 172 art reproductions
libr loan 30587

Southern Virginia University, Von Canon Library, One University Hill Dr, ***Buena Vista***, VA 24416
T: +1 540 2614234; Fax: +1 540 2618496; URL: www.svu.edu
1900; Duane E. Wilson
Latter-day Saints (Mormon) Coll, Melville Coll, Local (city and county) Hist Coll
250 000 vols; 95 curr per; 28 diss/theses; 280 music scores; 200 maps; 4 000 microforms; 800 av-mat; 750 sound-rec; 15 digital data carriers
libr loan; SOLINET, VLA 30588

Canisius College, Andrew L. Bouwhuis Library, 2001 Main St, ***Buffalo***, NY 14208-1098
T: +1 716 8882901; Fax: +1 716 8882887; URL: library.canisius.edu
1870; Dr. Joel A. Cohen
Jesuitica Coll
280 000 vols; 445 curr per; 8 000 e-books
libr loan 30589

D'Youville College, Montante Family Library, 320 Porter Ave, ***Buffalo***, NY 14201-1084
T: +1 716 8297618; Fax: +1 716 8297770; E-mail: refdesk@dyc.edu; URL: www.dyc.edu/library
1908; Leon Shkolnik
Education Coll (Curriculum Libr)
100 000 vols; 726 curr per; 3 500 av-mat
libr loan 30590

Medaille College, Library, 18 Agassiz Circle, ***Buffalo***, NY 14214
T: +1 716 8802283; Fax: +1 716 8849638; URL: www.medaille.edu/library
1925; Ilona Middleton
Animal Health Technology, Elementary Education Coll, Rare Bks on Buffalo Hist Coll
57 000 vols; 320 curr per; 20 000 e-books 30591

State University of New York – College at Buffalo, E.H. Butler Library, 1300 Elmwood Ave, ***Buffalo***, NY 14222-1095
T: +1 716 8786303; Fax: +1 716 8783134; E-mail: library@buffalostate.edu; URL: www.buffalostate.edu/library
1910; Maryruth F. Glogowski
Children's Author (Lois Lenski Coll), Courier-Express Coll, Creative Studies Coll, Elementary & Secondary Curriculum, Historical Children's Books (Hertha Ganey Coll), Hist Textbooks (Kempke-Root Coll), Isaac Klein Papers, Jazz (William H. Talmadge Coll), Polish Community Coll, Selig Adler Jewish Arch Coll, Tom Fontana Coll
658 000 vols; 17 269 curr per; 5 000 e-books; 508 maps; 958 000 microforms; 6 910 av-mat; 7 921 sound-rec; 165 digital data carriers
libr loan 30592

University at Buffalo – State University of New York, University Libraries, 433 Capen Hall, ***Buffalo***, NY 14260-1625
T: +1 716 6452813; Fax: +1 716 6453844; URL: ublib.buffalo.edu/libraries
1922; Barbara Von Wahlde
American Popular Lit, Arch, Poetry Coll, Science & Engineering Coll, First Editions, R.L. Stevenson Libr, Dylan Thomas, Health Sciences, Poetry Coll
3 330 000 vols; 32 796 curr per; 193 225 maps; 3 800 000 microforms; 390 digital data carriers
libr loan 30593

– Architecture & Planning Library, Hayes Hall, Buffalo, NY 14214-3087
T: +1 716 8293505; Fax: +1 716 8292780; E-mail: askasl@buffalo.edu; URL: ublib.buffalo.edu/libraries/units/apl
Deborah Koshinsky
24 000 vols; 123 curr per 30594

– Charles B. Sears Law Library, John Lord O'Brian Hall, Buffalo, NY 14260-1110
T: +1 716 6452047; Fax: +1 716 6453860; E-mail: asklaw@buffalo.edu; URL: ublib.buffalo.edu/libraries/units/law
1887; James Milles
Law (John Lord O'Brian Coll), Morris L Cohen Rare Book Coll, United Nations Coll
288 000 vols; 7 207 curr per; 666 000 microforms; 68 digital data carriers 30595

– Health Sciences Library, Abbott Hall, 3435 Main St, Buffalo, NY 14214-3002
T: +1 716 8293900; Fax: +1 716 8292211; E-mail: askhsl@buffalo.edu; URL: ublib.buffalo.edu/libraries/units/hsl
1846; Dr. Gary D. Byrd
Hist of Medicine Coll; Media Resources Ctr
352 000 vols; 2 990 curr per; 24 000 microforms; 2 959 av-mat; 130 sound-rec; 143 digital data carriers; 840 slide programs
libr loan; AAHSL, MLA, AMIA 30596

– Lockwood Memorial Library, 235 Lockwood Library, Buffalo, NY 14260-2200
T: +1 716 645-2814; Fax: +1 716 645-2820; E-mail: library@buffalo.edu; URL: ublib.buffalo.edu/libraries/units/lml
Karen Smith
American Popular Lit, Polish Coll
1 500 000 vols; 400 digital data carriers 30597

– Music Library, 104 Baird Hall, Buffalo, NY 14260-4700
T: +1 716 6452924; Fax: +1 716 6453609; E-mail: musique@buffalo.edu; URL: ublib.buffalo.edu/libraries/units/music
1970; Nancy Nuzzo
Hist of Music Librarianship; Arnold Cornelissen and Ferdinand Praeger Mss Colls; Concert Programs; Music Antiquarian & Auction Cats; Archives of Yvar Mikhashoff, Jan Williams, Morton Feldman, Center of the Creative and Performing Arts, June in Buffalo, North American New Music Festival
55 000 vols; 1 400 curr per; 76 000 music scores; 8 000 microforms; 300 av-mat; 32 400 sound-rec; 51 digital data carriers; 2 100 slides
libr loan 30598

– Oscar A. Silverman Undergraduate Library, 107 Capen Hall, Buffalo, NY 14260-2200
T: +1 716 6452944; Fax: +1 716 6453067; E-mail: library@buffalo.edu; URL: ublib.buffalo.edu/libraries/units/ugl
Glendora Jackson-Cooper
109 744 vols; 686 curr per; 7 digital data carriers 30599

Campbell University, Carrie Rich Memorial Library, 191 Main St, ***Buies Creek***, NC 27506-0098; P.O. Box 98, Buies Creek, NC 27506-0098
T: +1 910 8931466; Fax: +1 910 8931470; E-mail: reference@campbell.edu; URL: www.lib.campbell.edu
1887; Borree Kwok
Pharmacy, Religion; American Hist (Libr of American Civilization/Sabin), Trust & Estate Coll
224 000 vols; 487 curr per; 16 800 e-journals; 23 000 e-books 30600

– School of Law Library, 113 Main St, Buies Creek, NC 27506; P.O. Box 458, Buies Creek, NC 27506-0158
T: +1 910 8931796; Fax: +1 910 8931829; URL: www.law.campbell.edu/infores
1976; Olivia L. Weeks
196 000 vols; 2 584 curr per; 27 digital data carriers; 71 148 vols in microform 30601

Woodbury University, University Library, 7500 Glenoaks Blvd, ***Burbank***, CA 91510-1099; P.O. Box 7846, Burbank, CA 91510-7846
T: +1 818 2525202; Fax: +1 818 7674534; URL: web3.woodbury.edu/library/index.html
1884; Nedra Peterson
Business & management, Int business, Art, Architecture, Interior design, Fashion marketing & design; Law (Dr John C. Hogan Coll), Masters Theses, Senior Papers
67 000 vols; 307 curr per; 3 000 e-books; 125 maps; 94 000 microforms; 678 av-mat; 703 sound-rec; 42 digital data carriers; 14 142 slides 30602

University of Vermont & State Agricultural College, Bailey Howe Library, 538 Main St, ***Burlington***, VT 05405-0036
T: +1 802 6562020; Fax: +1 802 6564038; URL: library.uvm.edu
1800; Mara Saule
Geography, Foreign Affairs, Linguistics & Ecology (George Perkins Marsh Coll), Civil War (Howard-Hawkins Coll); Lit & Personal Correspondence (Dorothy Canfield Fisher Coll), Vermontiana
1 392 000 vols; 3 500 curr per
libr loan 30603

– Dana Medical Library, University of Vermont, Medical Education Ctr, 81 Colchester Ave, Burlington, VT 05405
T: +1 802 6562200; Fax: +1 802 6560762; URL: library.uvm.edu/dana
1917; Marianne Burke
Hist of Medicine Coll, Consumer Health Coll
124 000 vols; 3 157 curr per; 5 digital data carriers
libr loan 30604

Kent State University, Geauga Campus Library, 11411 Claridon-Troy Rd, ***Burton***, OH 44021-9535
T: +1 440 8343722; Fax: +1 440 8340919; E-mail: library@geauga.kent.edu; URL: www.geauga.kent.edu
1976; Mary Hricko
18 000 vols 30605

Montana Tech of The University of Montana, Montana Tech Library, 1300 W Park St, ***Butte***, MT 59701-8997
T: +1 406 4964281; Fax: +1 406 4964133; URL: www.mtech.edu/library
1900; Ann St Clair
US Trademark & Patent Coll
80 000 vols; 414 curr per; 31 270 e-journals; 3 000 e-books; 80 385 maps; 247 000 microforms; 3 380 sound-rec; 95 digital data carriers
libr loan; ALA 30606

Albertson College of Idaho, N.L. Terteling Library, 2112 Cleveland Blvd, ***Caldwell***, ID 83605-4432
T: +1 208 4595506; Fax: +1 208 4595299; E-mail: library@collegeofidaho.edu; URL: www.collegeofidaho.edu/academics/library/default.asp?ID=academics
1891; Christine Schutz
Humanities, Curriculum; Children's Lit
190 000 vols; 340 curr per
libr loan 30607

Caldwell College, Jennings Library, 120 Bloomfield Ave, ***Caldwell***, NJ 07006-6195
T: +1 973 6183564; Fax: +1 973 6183360; URL: jenningslibrary.caldwell.edu
1939; Peter Panos
Religious studies, Women's studies, Curriculum mat; Grover Cleveland Coll of American Hist
123 000 vols; 841 curr per
libr loan 30608

San Diego State University, Imperial Valley Campus Library, 720 Heber Ave, ***Calexico***, CA 92231-0550
T: +1 760 7685633; Fax: +1 760 7685525; URL: www.ivcampus.sdsu.edu/library/
1959; William Payne
Law, Bilingual education, psychology, Social service; Border Coll
115 000 vols; 150 curr per 30609

California University of Pennsylvania, Louis L. Manderino Library, 250 University Ave, ***California***, PA 15419-1394
T: +1 724 9384049; Fax: +1 724 9385901; URL: www.library.cup.edu
1852; Douglas A. Hoover
390 000 vols; 9 314 curr per
libr loan 30610

Harvard University, University Library, 1341 Massachusetts Ave, Wadsworth House, ***Cambridge***, MA 02138
T: +1 617 4953650; Fax: +1 617 4950370; E-mail: administration@hulmail.harvard.edu; URL: www.hul.harvard.edu
1638; Robert Darnton
15 943 000 vols; 8 449 000 microforms
libr loan 30611

– Andover-Harvard Theological Library, 45 Francis Ave, Divinity School, Cambridge, MA 02138
T: +1 617 4955788; Fax: +1 617 4964111; E-mail: reference@hds.harvard.edu; URL: www.hds.harvard.edu/library
1816; Frances O'Donnell
Unitarian Universalist and other liberal churches
485 000 vols; 2 310 curr per; 60 000 microforms
libr loan; ATLA 30612

– Arnold Arboretum Horticulture Library, 125 Arborway, Jamaica Plain, MA 02130
T: +1 617 5241718; Fax: +1 617 5241418; E-mail: arbweb@arnarb.harvard.edu; URL: www.arboretum.harvard.edu
Sheila Connor
40 000 vols 30613

– Arthur & Elizabeth Schlesinger Library on the History of Women, Three James St, Cambridge, MA 02138-3766; mail address: 10 Garden St, Cambridge, MA 02138
T: +1 617 4958647; Fax: +1 617 4968340; E-mail: slref@radcliffe.edu; URL: www.radcliffe.edu/schles
1943; Nancy F. Cott
Women's studies, Culinary hist
80 000 vols; 230 curr per; 50 av-mat 30614

– Baker Library, Harvard Business School, Ten Soldiers Field Rd, Boston, MA 02163
T: +1 617 4956040; Fax: +1 617 4966909; URL: www.library.hbs.edu
1908; Mary Lee Kennedy
Corporate Rpts, Hist Records, Kress Libr of Business & Economics, Baker Classification Coll, HBS Arch, Vanderblue Adam Smith Coll
638 000 vols; 5 615 curr per; 1 076 000 microforms 30615

– Blue Hill Meteorogical Observatory Library, Pierce Hall, 29 Oxford St, Cambridge, MA 02138
T: +1 617 4952836; Fax: +1 617 4959837; E-mail: library@deas.harvard.edu; URL: library.deas.harvard.edu
1885; Martha F. Wooster
Center Publs
115 000 vols; 583 curr per 30616

– Center for Hellenic Studies Library, 3100 Whitehaven St NW, Washington, DC 20008
T: +1 202 7454414; Fax: +1 202 7971540; E-mail: chs@fas.harvard.edu; URL: chs.wrlc.org
1961; Thomas Temple Wright
Greek & Latin Authors in Original Languages
60 000 vols; 300 curr per; 11 000 microforms 30617

– Center for Population Studies Library, 665 Huntington Ave, Rm 1-1111, Boston, MA 02115
T: +1 617 4321234; Fax: +1 617 5660365; E-mail: scoit@hsph.harvard.edu
Sarah Coit
12 000 vols; 60 curr per; 109 diss/theses; 16 maps; 6 digital data carriers 30618

– Chemistry Library, Department of Chemistry & Chemical Biology, 12 Oxford St, Cambridge, MA 02138
T: +1 617 4954076; Fax: +1 617 4950788; URL: www.chem.harvard.edu/library
1875; Marcia L. Chapin
57 000 vols; 250 curr per 30619

– Child Memorial & English Tutorial Library, Widener, Rm Z, Cambridge, MA 02138
T: +1 617 4954681
Eric Idsvoog
17 000 vols 30620

– Davis Center for Russian & Eurasian Studies Fung Library, Knafel Bldg, Concourse Level, 1737 Cambridge St, Cambridge, MA 02138
T: +1 617 4960485; Fax: +1 617 4960091; URL: hcl.harvard.edu/libraries/#fung
1948; Hugh K. Truslow
Harvard Project on the Soviet Social System, Schedule A
20 000 vols; 115 curr per 30621

– Dumbarton Oaks Research Library & Collection, 1703 32nd St NW, Washington, DC 20007
T: +1 202 3396400; Fax: +1 202 6250279; URL: www.doaks.org
1936; Sheila Klos
Byzantine studies, Pre-Columbian studies, Landscape architecture
215 000 vols; 860 curr per
libr loan 30622

– Eda Kuhn Loeb Music Library, Harvard College Library, Music Bldg, Harvard University, Cambridge, MA 02138
T: +1 617 4952794; Fax: +1 617 4964636; E-mail: muslib@fas.harvard.edu; URL: hcl.harvard.edu/loebmusic
1956; Virginia Danielson
Arch of World Music, Isham Memorial Libr
238 000 vols; 350 curr per; 650 diss/theses; 25 000 microforms; 1 500 av-mat; 80 000 sound-rec; 50 digital data carriers
libr loan 30623

– Ernst Mayr Library, 26 Oxford St, Cambridge, MA 02138
T: +1 617 4952475; E-mail: crinaldo@oeb.harvard.edu
1861; Constance Rinaldo
Natural hist, zoology, paleontology, biodiversity, evolution and ecology; Rare Bks
274 000 vols 30624

– Fine Arts Library, Harvard College Library, Fogg Art Museum, 32 Quincy St, Cambridge, MA 02138
T: +1 617 4953374; Fax: +1 617 4964889; URL: hcl.harvard.edu/libraries/#fal
1962; Mary Clare Altenhofen
Classical Art & Archaeology, Islamic Architecture, Italian Renaissance Painting, German Expressionism, Wendell Portrait Coll
330 000 vols; 8 000 microforms; 282 boxes of pamphlets & cat, 132 linear feet of arch mat & mss, 1 800 000 slides & photos 30625

– Frances Loeb Library, Harvard Graduate School of Design, 48 Quincy St, Gund Hall, Cambridge, MA 02138
T: +1 617 4961304; Fax: +1 617 4965929; URL: www.gsd.harvard.edu/library/
1900; Hugh Wilburn
Architecture, Landscape architecture, Urban planning; Le Corbusier Res Coll, John Charles Olmsted Coll, Charles Eliot Coll, Cluny Coll, Hugh Stubbins Arch, The Work of Dan Kiley, Edward Larrabee Barnes Coll, Josep Luis Sert Coll, Ferrari-Hardoy Coll, Eleanor Raymond Coll, CIAM Coll
282 000 vols; 1 650 curr per; 10 000 maps; 4 digital data carriers; 20 000 architectural drawings, 184 020 slides 30626

– Francis A. Countway Library of Medicine, Boston Medical Library – Harvard Medical Library, Boston Med Libr-Harvard Med Libr, Ten Shattuck St, Boston, MA 02115
T: +1 617 4322136;
Fax: +1 617 4320693; URL: www.countway.med.harvard.edu
1964; Dr. Isaac Kohane
Hist of Medicine; European Bks, 16th to 19th c; English Bks, 1475-1800; American Bks, 1668-1870; 14th c Medical Hebraica & Judaica; Nat Arch of Medical Illust; Coll of Medical Medals & Portraits
695 000 vols; 2 428 curr per; 177 e-books 30627

– George David Birkhoff Mathematical Library, Science Ctr 337, One Oxford St, Cambridge, MA 02138
T: +1 617 4952147
Nancy Milller
13 000 vols; 40 curr per; 1 050 rpts 30628

– Godfrey Lowell Cabot Science Library, Harvard College Library, Science Center, One Oxford St, Cambridge, MA 02138
T: +1 617 4955353; Fax: +1 617 4955324; E-mail: cabref@fas.harvard.edu; URL: hcl.harvard.edu/cabot
1973; Lynne M. Schmelz
400 000 vols 30629

– Gordon McKay Library, Division of Engineering Applied Sciences, School of Engineering & Applied Sciences, Pierce Hall, 29 Oxford St, Cambridge, MA 02138
T: +1 617 4952836; Fax: +1 617 4959837; E-mail: library@seas.harvard.edu; URL: library.seas.harvard.edu
1919; Martha F. Wooster
Computer science, Electrical & mechanical engineering, Microbiology, Optics, Physics, Water resources; Division Publs
106 000 vols; 260 000 microforms; 42 av-mat; 100 500 monogr, 4 000 Division reprints 30630

– Gutman Library-Research Center, Graduate School of Educ, Sixth Appian Way, Cambridge, MA 02138
T: +1 617 4953423; Fax: +1 617 4950540; URL: www.gse.harvard.edu/library
1920; Joseph Gabriel
18th & 19th c Textbks, Reading Coll, Children & Television, US Public School Rpts & Private School Cat, Educational Software
202 000 vols; 1 300 curr per; 435 000 microforms 30631

– Harry Elkins Widener Memorial Library, Harvard Yard, Cambridge, MA 02138
T: +1 617 4952411; Fax: +1 617 4950403; URL: hcl.harvard.edu/widener
1915
Lit, hist, linguistics, classical and modern languages, folklore, bibliography, economics, philosophy, psychology, history of science and technology, hist of social sciences
3 000 000 vols
libr loan 30632

– Harvard College Library (Headquarters in Harry Elkins Widener Memorial Library), Widener Library, Rm 110, Cambridge, MA 02138
T: +1 617 4952425; Fax: +1 617 4964750; URL: www.hcl.harvard.edu; www.hcl.harvard.edu/widener
1915; Nancy M. Cline
Author Colls, Hofer Graphic Arts Coll, Printing & Graphic Arts, Theatre Coll, Trotsky Arch, Farnsworth Recreational Reading Room, Winsor Memorial Map Room, Woodberry Poetry Room
10 000 000 vols; 4 850 000 microforms 30633

– Harvard Forest Library, 324 N Main St, Petersham, MA 01366
T: +1 978 7243302; Fax: +1 978 7243595; E-mail: hflib@fas.harvard.edu; URL: harvardforest.fas.harvard.edu
1907
29 000 vols 30634

– Harvard Map Collection, Harvard College Library, Pusey Library, Cambridge, MA 02138
T: +1 617 4952417; Fax: +1 617 4960440; E-mail: maps@harvard.edu; URL: hcl.harvard.edu/libraries/mc
Harvard Geospatial Libr
400 000 maps, 6 000 atlases 30635

– Harvard-Yenching Library, Harvard College Library, Two Divinity Ave, Cambridge, MA 02138
T: +1 617 4952756; Fax: +1 617 4966008; URL: hcl.harvard.edu/harvard-yenching/
1928; James Cheng
Japanese language, Tibetan language, Vietnam; Humanities & Social Sciences relating to China, Japan, and Korea, Rare Chinese Bks & Mss, Chinese Rubbings, Tibetan & Mongolian Tripitaka, Manchu Publs, Nakhi Mss, Vietnamese Publs, Tiananmen Arch
952 000 vols; 5 000 curr per; 70 000 microforms 30636

– Herbert Weir Smyth Classical Library, Widener, Rm E, Cambridge, MA 01238
T: +1 617 4954027; Fax: +1 617 4966720
Kathleen Comaen
8 000 vols; 25 curr per 30637

– History Department Library, Robinson Hall, Cambridge, MA 02138
T: +1 617 4952556; Fax: +1 617 4963425; E-mail: history@fas.harvard.edu; URL: lib.harvard.edu/libraries/0045.html
1964; Michael McCormick
10 000 vols; 33 curr per 30638

– History of Science Library – Cabot Science Library, Science Ctr, One Oxford St, Cambridge, MA 02138
T: +1 617 4955355; Fax: +1 617 4955324; E-mail: cabref@fas.harvard.edu; URL: www.hcl.harvard.edu/cabot
1973; Lynne Schmelz
24 000 vols 30639

– Houghton Library-Rare Books & Manuscripts, Harvard College Library, Houghton Library, Cambridge, MA 02138
T: +1 617 4952441; Fax: +1 617 4951376; E-mail: houghref@fas.harvard.edu; URL: www.hcl.harvard.edu/houghton
1942; William P. Stoneman
600 000 vols; 200 curr per; 50 000 av-mat; 4,5 mio mss, 5 mio theatre ephemera 30640

– John F. Kennedy School of Government Library, 79 John F Kennedy St, Cambridge, MA 02138
T: +1 617 4951300; Fax: +1 617 4951972; URL: www.ksg.harvard.edu/library
1978; Ellen Isenstein
Government, Management, International affairs; Belfor Ctr for Science, Ctr for Science and International Affairs
60 000 vols; 1 500 curr per; 41 000 microforms; 1 digital data carriers; 4 000 working papers
libr loan; OCLC 30641

– John K. Fairbank Center for East Asian Research Library, 625 Massachusetts Ave, Cambridge, MA 02139; mail address: Coolidge Hall, 1737 Cambridge St, Cambridge, MA 02138
T: +1 617 4955753; Fax: +1 617 4959976; URL: hcl.harvard.edu/libraries/#fung
1963; Nancy Hearst
Chinese language, English language; China, 1949 to present
40 000 vols; 200 curr per 30642

– Lamont Library-Undergraduate, Harvard College Library, Main Harvard Yard, 11 Quincy St, Boston, MA 02138
T: +1 617 4952452; Fax: +1 617 4963692; E-mail: lamref@fas.harvard.edu; URL: hcl.harvard.edu/lamont
Heather E. Cole
200 000 vols 30643

– Law School Library, Langdell Hall, 1545 Massachusetts Ave, Cambridge, MA 02138
T: +1 617 4953170; Fax: +1 617 4954449; URL: www.law.harvard.edu/library
1817; Harry S. Martin III
Dunn English Common Law Coll, Violett Coll of French Legal Hist, Olivart Int Law Coll, de Becker Japanese Law Coll
1 689 000 vols; 15 596 curr per; 1 500 000 microforms; 37 digital data carriers; 1 500 000 mss 30644

– Littauer Library, Harvard College Library, Lamont Library, Level B, Cambridge, MA 02138
T: +1 617 4952106; Fax: +1 617 4965570; URL: hcl.harvard.edu/libraries/sp
1939; Diane Geraci
Economics, Statistics, Government, Manpower & industrial relations; Slichter Industrial Relations Coll
428 000 vols; 894 curr per 30645

– Minda de Gunzburg, Center for European Studies Library, 27 Kirkland St at Cabot Way, Cambridge, MA 02138
T: +1 617 4954303; Fax: +1 617 4958509; E-mail: ces@fas.harvard.edu
1969; Paul Dzus
13 000 vols; 120 curr per 30646

– Physics Research Library, 450 Jefferson Laboratory, 17 Oxford St, Cambridge, MA 02138
T: +1 617 4952878; Fax: +1 617 4950416; E-mail: library@physics.harvard.edu; URL: www.physics.harvard.edu/prl/
1931; Marina D. Werbeloff
29 000 vols 30647

– Robbins Library of Philosophy, Emerson Hall 211, Harvard University, Dept of Philosophy, 25 Quincy St, Cambridge, MA 02138
T: +1 617 4952194; Fax: +1 617 4952192; URL: www.fas.harvard.edu/~phildept/robbins.html
Jason A. Pannone
Kierkegaard Coll, Bechtel Coll
10 000 vols; 51 curr per 30648

– Social Relations-Sociology Library, William James Hall, Rm 101, 33 Kirkland St, Cambridge, MA 02138
T: +1 617 4953838; URL: lib.harvard.edu
Richard E. Kaufman
28 000 vols; 162 curr per 30649

– Tozzer Library, Harvard College Library, 21 Divinity Ave, Cambridge, MA 02138
T: +1 617 4951481; Fax: +1 617 4962741; URL: hcl.harvard.edu/tozzer
1866; Gregory A. Finnegan
Biological & cultural anthropology, Linguistics, Prehistoric archeology; Latin American Archeology, Ethnology & Linguistics
260 000 vols; 1 581 curr per 30650

Lesley University, Eleanor DeWolfe Ludcke Library, 30 Mellen St, ***Cambridge***, MA 02138-2790
T: +1 617 3498872; Fax: +1 617 3498849; E-mail: liboff@mail.lesley.edu; URL: www.lesley.edu/library
1909; Patricia Payne
Education, Special education, Therapy, Psychology, Feminism, Art; Children's Coll, Curriculum Mat, Educational Test
100 000 vols; 700 curr per; 785 000 microforms; 2 000 av-mat; 1 325 sound-rec; 42 000 slides 30651

Longy School of Music, Bakalar Music Library, One Follen St, ***Cambridge***, MA 02138
T: +1 617 8760956 ext 1540; Fax: +1 617 3548841; E-mail: roy.rudolph@longy.edu; URL: www.longy.edu
1915; Roy Rudolph
Baroque Dance (Margaret Daniels-Girard Coll); Nadia Baulanger, E. Power Biggs & other 20th C Longy Faculty (Longy Arch Coll)
20 500 vols; 25 curr per; 15 000 music scores; 65 av-mat; 8 000 sound-rec; 50 digital data carriers
libr loan 30652

Massachusetts Institute of Technology Libraries, Office of the Director, 160 Memorial Dr, ***Cambridge***, MA 02142; mail address: 77 Massachusetts Ave, Office of the Director, Rm 14S-216, Cambridge, MA 02139-4307
T: +1 617 2535655; Fax: +1 617 2531690; URL: libraries.mit.edu
1862; Ann Wolpert
Early Hist of Aeronautics, Architecture & Planning, Civil Engineering, Drawings by Charles Bulfinch & Benjamin Latrobe, 19th c Glass Manufacture in US, Maps, Linguistics, Early Works in Mathematics & Physics, Microscopy, 17th-19th Century, Spectroscopy
2 742 000 vols; 22 312 curr per; 117 914 maps; 2 338 000 microforms; 3 413 av-mat; 23 953 sound-rec; 9 972 digital data carriers; 16 734 linear feet of mss, 450 272 slides
libr loan
Non-lending libr (except to Museum staff, volunteers and local teachers). Public welcome to research on-site 30653

– Aeronautics & Astronautics Library, 77 Massachusetts Ave, Rm 33-111, Cambridge, MA 02139
T: +1 617 2535665; Fax: +1 617 2533256; E-mail: barbaraw@mit.edu; URL: libguides.mit.edu/aero
1941; Barbara Williams
Publs of Nat Advisory Committee for Aeronautics & Nat Aeronautics & Space Administration; Complete Set of Institute of the Aeronautical/Aerospace Sciences, American Rocket Society, American Institute of Aeronautics & Astronautics Tech Papers
98 500 vols; 476 curr per; 638 diss/theses; 1 043 maps; 351 934 microforms; 143 av-mat; 105 digital data carriers; 50 226 tech rpts 30654

– Barker-Engineering Library, 77 Massachusetts Ave, Rm 10-500, Cambridge, MA 02139
T: +1 617 2530968; Fax: +1 617 2585623; E-mail: tag@mit.edu; URL: libraries.mit.edu/barker
Tracy Gabridge
283 829 vols; 304 956 microforms; 998 av-mat; 1 381 computer files 30655

– Dewey-Social Sciences & Management Library, ES3-100 – 30 Wadsworth St, Cambridge, MA 02139
T: +1 617 2535676; Fax: +1 617 2530642; E-mail: dewey@mit.edu; URL: libraries.mit.edu/dewey
1938; Catherine A. Friedman
Industrial Relations Hist Docs, UN Hist Docs, Corp Financial Rpts, OECD Publs, World Bank Publs
560 380 vols; 7 508 curr per; 8 531 diss/theses; 1 645 maps; 460 382 microforms; 140 av-mat; 17 sound-rec; 27 digital data carriers; 25 593 tech rpts, 123 101 pamphlets, 3 421 computer files 30656

– Humanities Library, 77 Massachusetts Ave, Rm 14S-200, Cambridge, MA 02139 T: +1 617 2535683; Fax: +1 617 2533109; E-mail: tat@mit.edu; URL: libraries.mit.edu/humanities Theresa A. Tobin 255 864 vols; 1 871 curr per; 582 diss/theses; 739 maps; 53 639 microforms; 110 av-mat; 137 sound-rec; 569 digital data carriers; 2 390 tech rpts, 2 059 pamphlets, 183 slides   30657

– Institute of Archives & Special Collections, 77 Massachusetts Ave, Rm 14N-118, Cambridge, MA 02139 T: +1 617 2535136; Fax: +1 617 2587305; URL: libraries.mit.edu/archives Megan Sniffin-Marinoff 159 570 vols; 550 mss; 85 919 diss/theses; 160 395 microforms; 510 arch colls   30658

– Lewis Music Library, 77 Massachusetts Ave, Rm 14E-109, Cambridge, MA 02139-4307 T: +1 617 2535636; Fax: +1 617 2533109; E-mail: jsbanks@mit.edu; URL: libraries.mit.edu/music Jennifer S. Banks 42 160 vols; 308 curr per; 26 703 music scores; 1 058 microforms; 407 av-mat; 18 082 sound-rec; 19 digital data carriers   30659

– Science Library, 77 Massachusetts Ave, Rm 14S-100, Cambridge, MA 02139 T: +1 617 2535671; Fax: +1 617 2536365; E-mail: libraries.mit.edu/esl/science/index.html Jennifer S. Banks Chemical engineering, Mathematics, Nuclear engineering, Physics 300 540 vols; 2 848 curr per; 2 946 diss/theses; 77 790 maps; 309 327 microforms; 16 av-mat; 4 digital data carriers; 53 849 tech rpts   30660

MIT Science Fiction Society Library, 84 Massachusetts Ave, W20-473, *Cambridge*, MA 02139-4307 T: +1 617 2585126; E-mail: mitsfs@mit.edu; URL: www.mit.edu/ ~mitsfs/ 1949; T. C. Skinner Foreign Language materials, Non-English Science Fiction, amateur fanzines 60 000 vols; 14 curr per; 70 mss; 1 000 microforms; 20 sound-rec   30661

Radcliffe Institute for Advanced Study, Harvard University, Arthur & Elizabeth Schlesinger Library on the History of Women in America, Three James St, *Cambridge*, MA 02138-3766; mail address: 10 Garden St, Cambridge, MA 02138 T: +1 617 4958647; Fax: +1 617 4968340; E-mail: slref@radcliffe.edu; URL: www.radcliffe.edu/schles 1943; Nancy F. Cott Cookery, Women's Rights Coll 80 000 vols; 230 curr per; 6 000 microforms; 2 236 av-mat; 2 542 sound-rec; Mss 8 522 linear feet libr loan   30662

Rutgers, The State University of New Jersey, *Camden* – Camden Law Library, 217 N Fifth, Camden, NJ 08102-1203 T: +1 856 2256172; URL: lawlibrary.rutgers.edu 1926; Anne Dalesandro Coll of Soviet Legal Mat 440 000 vols; 4 624 curr per; 618 000 microforms; 40 000 gov docs   30663

– Paul Robeson Library, 300 N Fourth St, Camden, NJ 08102-1404 T: +1 856 2256033; Fax: +1 856 2256428; URL: www.libraries.rutgers.edu/ ru/rul/libs/robeson/lib/index.shtml 1951; Gary A. Golden US & State Doc Dept 256 000 vols; 1 464 curr per; 70 digital data carriers; 96 000 doc bd   30664

University of Medicine & Dentistry of New Jersey, Camden Campus Library, Cooper University Hospital Library, One Cooper Plaza, *Camden*, NJ 08103 T: +1 856 3422525; Fax: +1 856 3429588; URL: www4.umdnj.edu/ camlbweb Barbara Miller 23 000 vols; 200 curr per   30665

Campbellsville University, Montgomery Library, One University Dr, *Campbellsville*, KY 42718-2799 T: +1 270 7895024; Fax: +1 270 7895336; E-mail: library@campbellsville.edu; URL: www.campbellsville.edu 1906; John R. Burch Jr Religion, US Civil War; College Arch 86 000 vols; 900 curr per   30666

Culver-Stockton College, Carl Johann Memorial Library, One College Hill, *Canton*, MO 63435 T: +1 573 2886321; Fax: +1 573 2886615; URL: www.culver.edu/library 1853; Sharon K. Upchurch Business & management, Hist, Religious studies; American Freedom Studies, Hist of Christian Church (Disciples of Christ Coll), Hist & Lit of Missouri & Midwest (Mark Twain Coll), Midwest Americana (Johann Coll) 165 000 vols; 196 curr per; 6 000 e-books   30667

Malone University Library, 2600 Cleveland Ave NW, *Canton*, OH 44709 T: +1 330 4718314; Fax: +1 330 4546977; E-mail: reference@malone.edu; URL: www3.malone.edu 1892; Dr. Joseph A. McDonald Friends Libr (Quakers mat), Evangelical Friends Church-Eastern Region Arch, China & India Friend missionary mss, Malone University Archives 181 420 vols; 809 curr per; 43 401 e-journals; 43 782 e-books; 70 024 govt docs; 2 476 music scores; 3 224 maps; 692 124 microforms; 3 094 av-mat; 7 059 sound-rec; 286 digital data carriers libr loan   30668

St. Lawrence University, Owen D. Young Library, 23 Romoda Dr, *Canton*, NY 13617 T: +1 315 2295485; Fax: +1 315 2295729; URL: www.stlawu.edu/library 1856; Bart M. Harloe Irving Bacheller Coll; Nathaniel Hawthorne (Milburn Coll), Northern New York Hist, David Parish & Family, Poetry (Benet Coll), Robert Frost Coll 576 000 vols; 10 000 curr per libr loan   30669

State University of New York – College at Canton, Southworth Library, 34 Cornell Drive, *Canton*, NY 13617 T: +1 315 3867228; Fax: +1 315 3867931; E-mail: bucher@canton.edu; URL: www.canton.edu/library 1948; Molly Mott 60 750 vols; 330 curr per libr loan   30670

West Texas A&M University, Cornette Library, University Dr & 26th St, *Canyon*, TX 79016; WTAMU Box 60748, Canyon, TX 79016-0001 T: +1 806 6512229; Fax: +1 806 6512213; URL: www.wtamu.edu/library 1910; Paul Coleman Southwestern Americana, Western Americana, English Lit & Hist 1 096 000 vols; 19 022 curr per libr loan   30671

Southeast Missouri State University, Kent Library, One University Plaza, *Cape Girardeau*, MO 63701 T: +1 573 7942230; Fax: +1 573 6512666; URL: library.semo.edu 1873; Ed Buis Brodsky Coll of William Faulkner Mat, Charles Harrison Rare Bks Coll, Regional Hist Coll 388 000 vols; 2 457 curr per; 11 942 maps; 1 930 av-mat; 1 020 sound-rec   30672

Southern Illinois University, Carbondale, Delyte W. Morris Library, 605 Agriculture Dr, Mailcode 6632, *Carbondale*, IL 62901 T: +1 618 4533374; Fax: +1 618 4538100; URL: www.lib.siu.edu 1869; David H. Carlson Irish Literary Renaissance, Modern American Philosophy, James Joyce, Lawrence Durrell 2 800 000 vols; 36 000 curr per   30673

– Law Library, Lesar Law Bldg, 1150 Douglas Dr, Carbondale, IL 62901; Mailcode 6803, Carbondale, IL 62901 T: +1 618 4533798; Fax: +1 618 4538728; E-mail: lawlib@siu.edu; URL: www.law.siu.edu/lawlib 1973; Doug Lind Self-Help Legal Coll, Rare Law Bks, Lincoln As A Lawyer, Water Quality, Mining Law 215 000 vols; 1 052 curr per; 311 000 microforms; 376 av-mat; 1 163 sound-rec; 63 digital data carriers; 4 88 slides libr loan   30674

Blackburn College, Lumpkin Library, 700 College Ave, *Carlinville*, IL 62626 T: +1 217 8543231; Fax: +1 217 8543231; URL: www.blackburn.edu/ workprogram/library.asp 1862; Andrew Ott 80 000 vols; 100 curr per   30675

Dickinson College, Waidner-Spahr Library, 333 W High St, *Carlisle*, PA 17013-2896; mail address: College & High St, Carlisle, PA 17013 T: +1 717 2451397; Fax: +1 717 2451439; E-mail: library@dickinson.edu; URL: lis.dickinson.edu/library/ 1784; Eleanor Mitchell American hist, East European hist, Russian hist & lit, Int business; James Buchanan Coll, letters, mss; Moncure Conway Coll, letters; John Drinkwater Coll, bks, letters, mss; John F Kennedy Coll, artifacts, bks, per; Marianne Moore Coll, bks, letters; Isaac Norris Coll; Joseph Priestly Coll, bks, mss; Carl Sandburg Coll, bks, letters, mss; Eli Slifer Coll, letters; Bible Coll 479 000 vols; 12 500 curr per libr loan   30676

Pennsylvania State University – Dickinson School of Law, Montague Law Library, 1170 Harrisburg Pike, *Carlisle*, PA 17013-1617 T: +1 717 2405267; Fax: +1 717 2405127; URL: www.dsl.psu.edu/library 1834; Steven Hinckley 451 000 vols; 1 482 curr per; 11 000 mss; 75 000 govt docs; 29 000 microforms; 300 av-mat; 100 sound-rec; 120 digital data carriers libr loan; AALL   30677

US Army War College, Library, 122 Forbes Ave, *Carlisle*, PA 17013-5220 T: +1 717 2454298; Fax: +1 717 2453323; URL: www.carlisle.army.mil/ library/ Bohdan Kohutiak Official Military publs, UN docs 285 000 vols; 1 100 curr per libr loan; ALA   30678

University of West Georgia, Irvine Sullivan Ingram Library, University of West Georgia, 1601 Maple St, *Carrollton*, GA 30118 T: +1 678 8396354; Fax: +1 678 8396511; URL: www.westga.edu/~library 1933; E. Lorene Flanders Georgia Political Heritage 405 000 vols; 1 228 curr per; 28 000 e-books; 313 mss; 976 diss/theses; 218 810 govt docs; 19 847 maps; 1 037 000 microforms; 500 sound-rec; 18 digital data carriers   30679

California State University Dominguez Hills, Educational Resources Center, 1000 E Victoria St, *Carson*, CA 90747 T: +1 310 2433758; Fax: +1 310 5164219; URL: www.csudh.edu 1963; Sandra Parham American Best Sellers (Claudia Buckner Coll), Arch of California State Univ Syst 434 000 vols; 669 curr per libr loan   30680

John A. Logan College, Learning Resources Center, 700 Logan College Rd, *Carterville*, IL 62918 T: +1 618 9853741 ext 8338; E-mail: library@jalc.edu; URL: www.jalc.edu/lrc/ library/ 1968; Dr. Linda Barrette Illinois, Genealogy, Nursing; John A Logan Memorial Coll 69 870 vols; 450 curr per   30681

Panola College, M.P. Baker Library, 1109 W Panola St, *Carthage*, TX 75633 T: +1 903 6932052; Fax: +1 903 6931115; E-mail: library@panola.edu; URL: www.panola.edu/library.htm 1947; Zeny Jett East Texas Documents & Genealogies, oral hist; Health/Science Coll 33 071 vols; 306 curr per; 60 music scores; 100 maps; 37 microforms; 6 427 av-mat; 476 sound-rec; 85 digital data carriers; 86 pamphlets, 463 clippings libr loan; ALA, ACRL   30682

Casper College, Goodstein Foundation Library, 125 College Dr, *Casper*, WY 82601 T: +1 307 2682269; Fax: +1 307 2682682; E-mail: cspcref@caspercollege.edu; URL: www.caspercollege.edu/library/index.html 1945 Hist of Wyoming, Natrona & Casper County, bks, maps, per, VF 81 000 vols; 653 curr per   30683

Maine Maritime Academy, Nutting Memorial Library, Pleasant St, Box C-1, *Castine*, ME 04420 T: +1 207 3262263; Fax: +1 207 3262261; URL: library.mma.edu 1941; Brent Hall Maritime hist, Military hist 89 000 vols; 347 curr per; 2 504 maps; 4 059 charts (nautical) libr loan; OCLC, NELINET   30684

Castleton State College, Calvin Coolidge Library, 178 Alumni Dr, *Castleton*, VT 05735 T: +1 802 4681256; Fax: +1 802 4681475; URL: www.castleton.edu/library 1787; Sandy Duling Vermontiana; ERIC Coll 160 000 vols; 500 curr per libr loan   30685

Southern Utah University, Gerald R. Sherratt Library, 351 W University Blvd, *Cedar City*, UT 84720 T: +1 435 5867947; Fax: +1 435 8658531; URL: www.li.suu.edu 1897; Diana T. Graff Opera Scores & Bks of the 19th C (Victorian Room); Shakespeare Coll; Southern Paiute Indian Coll; Southern Utah Hist Coll 233 000 vols; 21 337 curr per; 6 000 e-books; 674 000 microforms; 467 sound-rec; 476 digital data carriers libr loan; ALA   30686

University of Northern Iowa, Rod Library, 1227 W 27th St, *Cedar Falls*, IA 50613-3675 T: +1 319 2732838; Fax: +1 319 2732913; E-mail: libill@uni.edu; URL: www.library.uni.edu 1876; Marilyn J. Mercado Art, Music, Education; Grassley Papers, American Fiction, Univ Arch 794 000 vols; 2 205 curr per; 1 350 e-journals; 6 000 e-books; 40 779 maps; 903 000 microforms; 5 000 av-mat; 22 220 sound-rec; 470 digital data carriers libr loan   30687

Northwood University, Cedar Hill Campus Library, 1114 W FM 1382, *Cedar Hill*, TX 75104 T: +1 972 2935436; Fax: +1 972 2937026; URL: www.northwood.edu 1967; Kaethryn Duncan Advertising, Business Management, Fashion Merchandising, Hotel & Restaurant, Management, Automotive Marketing 10 000 vols; 160 curr per; 50 av-mat; 7 digital data carriers   30688

Coe College, Stewart Memorial Library, 1220 First Ave NE, *Cedar Rapids*, IA 52402-5092 T: +1 319 3998586; Fax: +1 319 3998019; URL: www.library.coe.edu 1900; Richard Doyle Works by Paul Engle, Wm Shirer Mss 265 000 vols; 2 200 curr per; 21 000 e-books libr loan   30689

Mount Mercy College, Busse Library, 1330 Elmhurst Dr NE, ***Cedar Rapids***, IA 52402-4797
T: +1 319 3686465; Fax: +1 319 3639060; E-mail: library@mtmercy.edu; URL: www.mtmercy.edu/busselibrary.html
1958; Marilyn Murphy
Nat League for Nursing
139 000 vols; 600 curr per
libr loan 30690

Cedarville University, Centennial Library, 251 N Main St, ***Cedarville***, OH 45314-0601
T: +1 937 7667840; Fax: +1 937 7662337; E-mail: library@cedarville.edu; URL: www.cedarville.edu/dept/ls
1887; Lynn Alan Brock
Theological studies, Baptist hist; English Bible Coll, Limited Edition Bk Club Coll, Michael Dewine Congressional Papers
175 000 vols; 973 curr per; 6 900 e-journals; 14 000 e-books; 3 120 av-mat; 1 970 sound-rec; 540 digital data carriers; 190 Electronic Media & Resources
libr loan 30691

Desales University, Trexler Library, 2755 Station Ave, ***Center Valley***, PA 18034
T: +1 610 2821100; Fax: +1 610 2822342; URL: www.desales.edu/library/
1965; Debbie Malone
Roman Catholic Religion, Hist, Nursing; St Francis De Sales, American Theatre (John Y Kohl Coll), St Thomas More
141 000 vols; 541 curr per 30692

Southern Wesleyan University, Claude R. Rickman Library, 916 Wesleyan Dr, ***Central***, SC 29630-9748; P.O. Box 1020, 907 Wesleyan Dr, Central, SC 29630-1020
T: +1 864 6445060; Fax: +1 864 6445904; E-mail: library@swu.edu; URL: www.swu.edu/library
1906; Robert E. Sears
Genealogical Coll (upstate South Carolina families), Wesleyan Hist Coll
106 000 vols; 515 curr per 30693

New York Institute of Technology, Central Islip Library, PO Box 9029, ***Central Islip***, NY 11722-9029
T: +1 631 3483321; Fax: +1 631 3483094; URL: www.nyit.edu/library
1984; Rosemary S. Feeney
Architecture, Culinary
41 000 vols; 300 curr per; 6 000 e-books 30694

Touro College, Jacob D. Fuchsberg Law Center Library, 225 Eastview Dr, ***Central Islip***, NY 11722-4539
T: +1 631 4212320; Fax: +1 631 4215386; URL: www.tourolaw.edu
1980; April Schwartz
Foreign & Int Law, Jewish Law
415 000 vols; 1 300 curr per 30695

Chadron State College, Reta E. King Library, 300 E 12th St, ***Chadron***, NE 69337
T: +1 308 4326271; URL: www.csc.edu
1911; Milton Wolf
Farrar, Hulm, Madrid Photography Coll, Graves Photography Colls
215 000 vols; 550 curr per 30696

Wilson College, John Stewart Memorial Library, 1015 Philadelphia Ave, ***Chambersburg***, PA 17201-1285
T: +1 717 2622008; Fax: +1 717 2637194; URL: www.wilson.edu
1869; Kathleen Murphy
175 000 vols; 312 curr per
libr loan 30697

University of North Carolina at Chapel Hill, Walter Royal Davis Library, 208 Raleigh St, Campus Box 3900, ***Chapel Hill***, NC 27514-8890
T: +1 919 9621151; Fax: +1 919 9624451; E-mail: reference@unc.edu; URL: www.lib.unc.edu/davis
1789; Sarah Michalak
North Carolina Coll, Southern Hist Coll, Rare Bk Coll, Documenting the American South (http://docsouth.unc.edu); Depts: Health Sciences, Law, Math/Physics, Planning, Biology, Geological Sciences, Information and Library Science, Chemistry, Music, Art, Undergraduate
6 000 000 vols; 41 000 curr per; 19 079 015 mss; 330 373 maps; 4 648 000 microforms; 145 081 sound-rec
libr loan; ALA, ARL, IFLA 30698

– Brauer Library (Math-Physics), 365 Phillips Hall, CB No 3250, Chapel Hill, NC 27599-3250
T: +1 919 9622323; Fax: +1 919 9622568
1920; Zahra Kamarei
86 000 vols; 2 000 microforms; 178 sound-rec; 208 CDs 30699

– Chapin Library (City & Regional Planning), 211 New East, CB No 3140, Chapel Hill, NC 27599-3405
T: +1 919 9624770; URL: www.lib.unc.edu/planning
1949; Susan Martin
15 000 vols; 6 000 microforms; 63 av-mat; 8 880 docs, 3 856 slides 30700

– Couch Biology Library (Botany Section), 301 Coker Hall, CB No 3280, Chapel Hill, NC 27599
T: +1 919 9623783; Fax: +1 919 8438393; URL: www.lib.unc.edu/biology
1926; William Burk
36 000 vols; 2 088 maps; 16 000 microforms; 9 002 mycological reprints 30701

– Couch Biology Library (Zoology Section), 213 Wilson Hall, CB No 3280, Chapel Hill, NC 27599
T: +1 919 9622264; Fax: +1 919 8438393; URL: www.lib.unc.edu/biology
1940; David Romito
31 000 vols; 128 microforms 30702

– Geological Sciences Library, 121 Mitchell Hall, CB No 3315, Chapel Hill, NC 27599-3315
T: +1 919 9622386; Fax: +1 919 9664519; URL: www.lib.unc.edu/geolib
Miriam Kennard
Geology, invertebrate paleontology, physical oceanography, geophysics
48 000 vols; 850 curr per; 43 000 maps; 28 000 microforms; 300 digital data carriers 30703

– Health Sciences Library, 355 S Columbia St, Chapel Hill, NC 27599; Campus Box 7585, Chapel Hill, NC 27599-7585
T: +1 919 9662111; Fax: +1 919 9665592; URL: www.hsl.unc.edu
1952; Carol G. Jenkins
Hist of Health Sciences Coll, Rare Bks, Learning Resources
311 000 vols; 4 193 curr per; 43 000 microforms; 20 digital data carriers
libr loan 30704

– Highway Safety Research Center Library, 730 Airport Rd, CB No 3430, Chapel Hill, NC 27599
T: +1 919 9628701; Fax: +1 919 9628710; E-mail: metucker@email.unc.edu; URL: www.hsrc.unc.edu
1970; Mary Ellen Tucker
North Carolina Traffic Data Coll
40 000 vols; 125 curr per; 55 000 microforms; 5 digital data carriers 30705

– Institute of Government Library, Knapp-Sanders Bldg, CB No 3330, Chapel Hill, NC 27599-3330
T: +1 919 9664172; Fax: +1 919 9664762; E-mail: library@iogmail.iog.unc.edu; URL: www.sog.unc.edu/library
1930; Alex Hess III
16 000 vols; 20 651 pamphlets 30706

– Joseph Curtis Sloane Art Library, 102 Hanes Art Ctr, CB No 3405, Chapel Hill, NC 27599-3405
T: +1 919 9622397; Fax: +1 919 9620722; URL: www.lib.unc.edu/art
1958; Joshua Hockensmith
100 000 vols; 300 diss/theses; 450 microforms; 37 digital data carriers; artist vertical file, museum vertical file
libr loan 30707

– Kathrine R. Everett Law Library, UNC Law Library, 160 Ridge Rd, CB No 3385, Chapel Hill, NC 27599-3385
T: +1 919 9621321; Fax: +1 919 8437810; URL: www.library.law.unc.edu
1923; Anne Klinefelter
Native American Law Coll
315 000 vols; 6 277 curr per; 653 e-journals; 948 000 microforms; 2 000 sound-rec; 1 170 digital data carriers
libr loan 30708

– Kenan Library (Chemistry), Wilson Library, 2nd Level, CB No 3290, Chapel Hill, NC 27599
T: +1 919 9621188; Fax: +1 919 9622388; URL: www.lib.unc.edu/kenan
1885
Venable Rare Bk Coll
57 000 vols; 3 000 microforms 30709

– Music Library, 300 Wilson Library, Campus Box 3906, Chapel Hill, NC 27514-8890
T: +1 919 9661113; Fax: +1 919 8430418; URL: www.lib.unc.edu/music
1935; Philip Vandermeer
Opera Coll, Hist of Music Theory, The Sonata, 19th C American Sheet Music Coll, American Shape-Note Tunebook Coll, Italian Opera Libretti Coll
54 000 vols; 8 000 microforms; 81 000 Music Scores 30710

– Robert B. House Undergraduate Library, 203 South Rd, CB No 3942, Chapel Hill, NC 27514-8890
T: +1 919 9621355; Fax: +1 919 9622697; URL: www.lib.unc.edu/house
Leah Dunn
85 000 vols 30711

– School of Information and Library Science Library, 114 Manning Hall, CB No 3360, Chapel Hill, NC 27599
T: +1 919 9628361; Fax: +1 919 9628071; E-mail: library@ils.unc.edu; URL: ils.unc.edu/itrc/library
1931; Rebecca Vargha
Childrens fiction and nonfiction coll
86 000 vols; 50 diss/theses; 414 sound-rec; 118 digital data carriers
ALA, MLA 30712

Charleston Southern University, L. Mendel Rivers Library, 9200 University Blvd, ***Charleston***, SC 29406; P.O. Box 118087, Charleston, SC 29423-8087
T: +1 843 8637946; Fax: +1 843 8637947; URL: www.csuniv.edu/library/index.html
1966; David Mash
250 000 vols; 412 curr per; 24 268 e-journals; 23 000 e-books; 143 306 govt docs; 424 maps; 217 000 microforms; 2 845 sound-rec; 18 digital data carriers
libr loan; ALA, SCLA 30713

The Citadel, Daniel Library, 171 Moultrie St, ***Charleston***, SC 29409-6140
T: +1 843 9532569; Fax: +1 843 9535190; URL: www.citadel.edu/library
1842; Angie W. LeClercq
Academic, General, Military science, South Carolina; Hardin Coll of German Lit; Citadel Publs
244 000 vols; 514 curr per; 28 200 e-journals; 650 e-books; 18 000 govt docs; 1 111 000 microforms; 569 digital data carriers
libr loan 30714

College of Charleston, Addlestone Library, 205 Calhoun St, ***Charleston***, SC 29401-3519; mail address: 66 George St, Charleston, SC 29424
T: +1 843 9538001; Fax: +1 843 9536319; URL: www.cofc.edu/~library
1785; David Cohen
Colonial Reading Habits (Ralph Izard Coll), Pigeons (Wendell Levi Coll), South Carolina, US Senator Burnet Rhett Maybank Papers, Bk Arts; Marine Resources Dept
599 000 vols; 4 429 curr per; 24 440 e-journals; 68 000 e-books; 4 099 Journals
libr loan 30715

Eastern Illinois University, Booth Library, 600 Lincoln Ave, ***Charleston***, IL 61920
T: +1 217 5816072; Fax: +1 217 5817534; URL: www.eiu.edu/~booth
1896; Dr. Allen Lanham
Dorothy Hansen Theatre Organ Coll, Remo Belli Int Percussion Libr; Media Services
1 045 000 vols; 5 292 curr per; 9 400 e-journals; 223 422 govt docs; 14 951 music scores; 27 726 maps; 1 334 000 microforms; 23 754 sound-rec; 815 digital data carriers; 1 538 other items
libr loan 30716

Medical University of South Carolina, Library, 171 Ashley Ave, Ste 300, ***Charleston***, SC 29425-0001; P.O. Box 250403, Charleston, SC 29425-0403
T: +1 843 7922372; Fax: +1 843 7927947; URL: www.library.musc.edu
1824; Thomas G. Basler
Hist of Medicine (Waring Hist Libr Coll), Micro-Circulation (Melvin M Knisely Coll)
209 000 vols; 737 e-books; 22 000 microforms; 2 300 av-mat; 64 sound-rec; 2 digital data carriers; 642 Slides
libr loan 30717

University of Charleston, Schoenbaum Library, 2300 MacCorkle Ave SE, ***Charleston***, WV 25304-1099
T: +1 304 3574780; Fax: +1 304 3574715; E-mail: librarian@ucwv.edu; URL: www.ucwv.edu/library
1888; Lynn Sheehan
Civil War (Gorman Coll), Early American Hist (John Allen Kinnaman Coll), Presidential Biogr (James David Barber Coll), Kendall Vintroux Political Cartoons, Appalachian Culture Coll
119 000 vols; 303 curr per 30718

West Virginia University, Health Sciences Library, 3110 MacCorkle Ave SE, ***Charleston***, WV 25304
T: +1 304 3471285; Fax: +1 304 3471288; URL: www.hsc.wvu.edu/charleston/library
1974; Patricia Dawson
25 000 vols; 360 curr per; 500 av-mat; 400 sound-rec; 16 000 journals 30719

Johnson C. Smith University, James B. Duke Memorial Library, 100 Beatties Ford Rd, ***Charlotte***, NC 28216
T: +1 704 3716740; Fax: +1 704 3783524; E-mail: refdesk@jcsu.edu; URL: www.jcsu.edu
1867; Inja Hong
Black Life & Lit (Schomburg Coll), Economics Governmental Hist (Calvin Hoover Coll), Hist & Governmental Biography (Neimeyer Coll); Earl A. Johnson Coll (Black Studies & Judaica & Art)
74 000 vols; 328 curr per 30720

Queens College, Everett Library, 1900 Selwyn Ave, ***Charlotte***, NC 28274-0001
T: +1 704 3377127; Fax: +1 704 3372517; URL: www.queens.edu/library
1857; Dr. Carol Walker Jordan
North Carolina Coll, Hicks Asia Coll, Queens College Memorabilia & Arch
131 000 vols; 632 curr per
libr loan 30721

University of North Carolina at Charlotte, J. Murrey Atkins Library, 9201 University City Blvd, ***Charlotte***, NC 28223-0001
T: +1 704 6872416; Fax: +1 704 6872232; URL: library.uncc.edu
1946; Carole Runnion
Business, Management; 17th & 18th c English Drama, Contemporary, Social & Political Hist of Charlotte & Mecklenburg County, NC state & county docs
970 000 vols; 34 486 curr per; 24 000 e-books
libr loan 30722

University of Virginia, Alderman Library, PO Box 400114, ***Charlottesville***, VA 22904-4114
T: +1 434 9243021; Fax: +1 434 9241431; E-mail: library@virginia.edu; URL: www.lib.virginia.edu/
1819; Karin Wittenborg
American Authors Colls, Evolution Coll, Finance, Folklore, Kafka, French Renaissance Lit, Virginia Hist, Int Law, Medieval Mss, Middle East Coll, Modern Art, Natural Hist, Thomas Jefferson Papers
2 538 000 vols 30723

– Albert & Shirley Small Special Collections Library, PO Box 400113, Charlottesville, VA 22904-4110
T: +1 434 2431776; Fax: +1 434 9244968; URL: www.lib.virginia.edu/small/
N.N.
292 000 vols 30724

– Arthur J. Morris Law Library, 580 Massie Rd, Charlottesville, VA 22903-1789
T: +1 434 9243519; Fax: +1 434 9822232; E-mail: lawlibref@virginia.edu; URL: www.law.virginia.edu/library
1826; Taylor Fitchett
Newlin Coll on Oceans Law & Policy, John Bassett Moore Coll of Int Law
890 000 vols; 80 152 govt docs; 214 000 microforms 30725

– Astronomy Library, Charles L Brown Sci & Eng Library, 264 Astronomy Bldg, 530 McCormick Rd, Charlottesville, VA 22904; P.O. Box 400330, Charlottesville, VA 22904-4330
T: +1 434 9243921; Fax: +1 434 9244337; URL: www.lib.virginia.edu/science/scilibs/astr-lib.html
Carla H. Lee
14 000 vols 30726

– Biology-Psychology Library, 290-A Gilmer Hall, 485 McCormick Rd, Charlottesville, VA 22903; P.O. Box 400400, Charlottesville, VA 22904-4400
T: +1 434 9825260; Fax: +1 434 9825626; URL: www.lib.virginia.edu/science/scilibs/bio-psych-lib.html
Sandi Dulaney
32 000 vols 30727

– Charles L. Brown Science & Engineering Library, Clark Hall, Charlottesville, VA 22903-3188; P.O. Box 400124, Charlottesville, VA 22904-4124
T: +1 434 9243628; Fax: +1 434 9244338; E-mail: sciref@virginia.edu; URL: www2.lib.virginia.edu/brown/
Carla Lee
224 000 vols 30728

– Chemistry Library, 259 Chemistry Bldg, Charlottesville, VA 22903-2454; P.O. Box 400315, Charlottesville, VA 22904-4315
T: +1 434 9243159; Fax: +1 434 9244338; URL: www.lib.virginia.edu/brown/scilibs/chem-lib.html
24 000 vols 30729

– Claude Moore Health Sciences Library, Univ Va Health System, 1300 Jefferson Park Ave, Charlottesville, VA 22908; P.O. Box 800722, Charlottesville, VA 22908-0722
T: +1 434 9240058; Fax: +1 434 9240379; URL: www.healthsystem.virginia.edu/internet/library
1911; Gretchen Arnold
Kerr L White Health Care Coll, Philip S Hench/Walter Reed Yellow Fever Coll, American Lung Assn of Virginia Arch
134 000 vols 30730

– Clemons Library, PO Box 400710, Charlottesville, VA 22904-4710
T: +1 434 9243684; Fax: +1 434 9247468; E-mail: clemons@virginia.edu; URL: www.lib.virginia.edu/clemons/home.html
Donna Tolson
Film, media studies
124 000 vols 30731

– Colgate Darden Graduate School of Business – Camp Library, Student Services Bldg, 100 Darden Blvd, Charlottesville, VA 22903; P.O. Box 6550, Charlottesville, VA 22906-6500
T: +1 434 9247321; Fax: +1 434 9243533
1955; Karen King
80 000 vols; 2 000 reels of microfilm of per, 200 000 annual rpts on microfiche 30732

– Education Library, Ruffner Hall, 3rd Flr, 405 Emmet St S, Charlottesville, VA 22904-4278; P.O. Box 400278, Charlottesville, VA 22904-4278
T: +1 434 9247040; Fax: +1 434 9243886; E-mail: education.library@virginia.edu; URL: www.lib.virginia.edu/education
Kay A. Buchanan
Education
50 000 vols 30733

1970; N.N.
Frances Benjamin Johnston Photogr Coll of Virginia Architecture; William Morris Libr on Forgery of Works of Art, 15th c to present
152 000 vols; 25 000 microforms; 198 000 architecture slides 30734

– Mathematics Library, 107 Kerchof Hall, Charlottesville, VA 22903; P.O. Box 400140, Charlottesville, VA 22904-4140
T: +1 434 9247806; Fax: +1 434 9243104; URL: www.lib.virginia.edu/science/scilibs/math-lib.html
34 000 vols 30735

– Music Library, Old Cabell Hall, Charlottesville, VA 22904-4175; P.O. Box 400175, Charlottesville, VA 22904-4175
T: +1 434 9247041; Fax: +1 434 9246033; E-mail: musiclib@virginia.edu; URL: www.lib.virginia.edu/MusicLib/index.html
Erin Mayhood
Mackay-Smith Coll (18th c Imprints), Monticello Music Coll, John Powell Coll, 19th-c American Sheet Music
72 000 vols; 20 000 microforms; 340 av-mat; 42 000 sound-rec 30736

– Physics Library, 323 Physics Bldg, Charlottesville, VA 22903-2458; P.O. Box 400714, Charlottesville, VA 22904-4714
T: +1 434 9246589; URL: www.lib.virginia.edu/science/scilibs/phys-lib.html
Vicky Ingram
33 000 vols 30737

University of Minnesota, Andersen Horticultural Library, 3675 Arboretum Dr, mail address: *Chaska*, MN 55318
T: +1 952 4431405; Fax: +1 952 4432521; E-mail: r-isaa@tc.umn.edu; URL: www.arboretum.umn.edu
1970; Richard T. Isaacson
Williams Hosta Coll, Seed and Nursery, Nursery Cat, Wildflowers (Botanical Ills)
12 000 vols; 520 curr per; 1 388 microforms; 12 av-mat; 9 feet of mss 30738

Tennessee Temple University, Cierpke Memorial Library, 1815 Union Ave, *Chattanooga*, TN 37404
T: +1 423 4934250; Fax: +1 423 4934497; E-mail: cierpke@prodigy.net; URL: www.tntemple.edu
1946; Kevin Woodruff
Rare Out of Print Bks, Religious Education Mat
196 000 vols; 50 curr per; 1 000 e-journals 30739

University of Tennessee at Chattanooga, T. Cartter & Margaret Rawlings Lupton Library, 615 McCallie Ave, Dept 6456, *Chattanooga*, TN 37403-2598
T: +1 423 4254510; Fax: +1 423 4254775; URL: www.lib.utc.edu
1872; Theresa Liedtka
Local hist, Southern lit, Civil War
504 000 vols; 1 719 curr per 30740

Eastern Washington University, John F. Kennedy Memorial Library, 816 F St, 100 LIB, *Cheney*, WA 99004-2453
T: +1 509 3592492; Fax: +1 509 3594840; URL: www.ewu.edu/library
1890; Patricia M. Kelley
Northwest Hist, Science Fiction, Religious Architecture
796 000 vols; 5 653 curr per; 9 610 e-journals; 3 000 e-books; 879 297 govt docs; 795 000 microforms; 6 880 av-mat; 19 020 sound-rec; 2 200 digital data carriers; 233 Bks on Deafness & Sign Lang
libr loan 30741

White Pines College, Wadleigh Library, 40 Chester St, *Chester*, NH 03036-4301
T: +1 603 8877454; Fax: +1 603 8871777; URL: www.chestercollege.edu
1965; Marie Anne Lasher
Art, Graphic design, Photography, Writing
26 000 vols; 100 curr per; 4 000 e-books
libr loan 30742

Widener University, Wolfgram Memorial Library, One University Pl, *Chester*, PA 19013-5792
T: +1 610 4994067; Fax: +1 610 4994588; URL: www.widener.edu/libraries
1821; Robert Danford
Lindsay Law Libr, Wolfgram Coll (English & American Lit)
404 000 vols; 2 256 curr per; 150 000 microforms; 552 sound-rec; 86 digital data carriers
libr loan 30743

Logan College of Chiropractic, Learning Resources Center, 1851 Schoettler Rd, *Chesterfield*, MO 63006; P.O. Box 1065, Chesterfield, MO 63006-1065
T: +1 636 2272100; Fax: +1 636 2072448; E-mail: circdesk@logan.edu; URL: www.logan.edu
1960; Bob Snyders
Osseous & Synthetic Human Models; 300 items
13 000 vols; 240 curr per; 150 e-journals; 2 600 diss/theses; 7 microforms; 643 av-mat; 400 sound-rec; 150 digital data carriers; 1 150 slide/tape programs, 544 ACR files
libr loan; CLIBCON, MLNC 30744

Washington College, Clifton M. Miller Library, 300 Washington Ave, *Chestertown*, MD 21620-1197
T: +1 410 7787292; Fax: +1 410 7787288; URL: libraryweb.washcoll.edu
1782; Dr. Ruth C. Shoge
Maryland Coll
172 000 vols; 602 curr per; 15 080 e-journals; 2 000 e-books 30745

Boston College Libraries, 140 Commonwealth Ave, *Chestnut Hill*, MA 02467
T: +1 617 5523195; Fax: +1 617 5520599; URL: www.bc.edu/libraries
1863; Jerome Yavarkovsky
Boston Hist, Balkan Studies, Caribbeana, British Catholic Authors, Irish Coll, Jesuitana, Samuel Beckett Coll, William Butler Yeats Coll, Theodore Dreiser Coll, Flann O'Brien Papers, Graham Greene Libr
2 000 000 vols; 22 000 curr per
libr loan 30746

– Bapst Art Library, 140 Commonwealth Ave, Chestnut Hill, MA 02467
T: +1 617 5523200; Fax: +1 617 5520510; E-mail: adeane.bregman@bc.edu; URL: www.bc.edu/libraries/collections/bapst.html
1994; Adeane Bregman
51 000 vols; 150 curr per; 42 digital data carriers
libr loan; ARLIS/NA, VRA 30747

– Catherine B. O'Connor Library, Weston Observatory, 381 Concord Rd, Weston, MA 02193-1340
T: +1 617 5524450; Fax: +1 617 5528388; URL: www.bc.edu/libraries/collections/weston.html
1947; Enid Karr
Seismology, Geophysics, Geology
8 000 vols; 35 curr per; 160 diss/theses; 10 000 maps; 2 sound-rec
libr loan 30748

– Educational Resource Center, 140 Commonwealth Ave, Chestnut Hill, MA 02467
T: +1 617 5524920; Fax: +1 617 5521769; E-mail: erc@bc.edu; URL: www.bc.edu/libraries/collections/erc.html
Margaret Cohen
37 000 vols; 96 curr per 30749

– Graduate School of Social Work Library, McGuinn Hall 038, 140 Commonwealth Ave, Chestnut Hill, MA 02467-3810
T: +1 617 5520109; Fax: +1 617 5523199; E-mail: swlib@bc.edu; URL: www.bc.edu/libraries/collections/socialwork.html
1936; Jane Morris
Child welfare, Geriatrics & gerontology
41 000 vols; 355 curr per; 2 000 microforms; 3 digital data carriers 30750

– John J. Burns Library of Rare Books & Special Collections, 140 Commonwealth Ave, Chestnut Hill, MA 02167
T: +1 617 552-3282; Fax: +1 617 552-2465; E-mail: robert.oneill.1@bc.edu; URL: www.bc.edu/libraries/collections/burns.html

Robert K. O'Neill
Catholic church, Jesuit, Nursing; Balkan Coll, British Catholic Authors Coll, Burns & Oats Coll, Congregationalism, Congressional Archs, DeFacto School Segregation, Detective Fiction, Ethnology, Evelyn Waugh, Folklore (Jamaica & West Africa), Freemasons, Graham Greene Coll, Irish Colls, Jesuitana, Judaica, Women's Hist, World War I & II
150 000 vols; 32 curr per; 4 000 000 mss 30751

– Law Library, 885 Centre St, Newton Centre, MA 02459
T: +1 617 5524066; Fax: +1 617 5522889; URL: www.bc.edu/lawlibrary
Filippa Marullo Anzalone
St Thomas More Law Coll
468 000 vols; 3 386 curr per; 888 000 microforms; 10 digital data carriers
libr loan 30752

– Thomas P. O'Neill Jr Library (Central Library), 140 Commonwealth Ave, Chestnut Hill, MA 02467
T: +1 617 5528038; Fax: +1 617 5528828
N.N.
2 447 000 vols 30753

Pine Manor College, Annenberg Library & Communications Center, 400 Heath St, *Chestnut Hill*, MA 02467
T: +1 617 7317081; Fax: +1 617 7317045; URL: www.pmc.edu/library/library.html
1911; Marilyn Bregoli
Education, Psychology; First Editions of Noted American Women Authors
63 000 vols; 250 curr per 30754

Cheyney University, Leslie Pinckney Hill Library, 1837 University Circle, *Cheyney*, PA 19319; P.O. Box 200, Cheyney, PA 19319-0200
T: +1 610 3992203; Fax: +1 610 3992491; URL: www.cheyney.edu
1853; Lut R. Nero
Afro-American Studies, Ethnic Coll, Univ Arch
228 000 vols; 1 126 curr per; 257 716 govt docs; 381 000 microforms; 1 279 av-mat; 5 digital data carriers; 118 Journals
libr loan 30755

American Islamic College, Library, 640 W Irving Park Rd, *Chicago*, IL 60613
T: +1 773 2814700; Fax: +1 773 2818552; E-mail: info@aicusa.edu; URL: www.aicusa.edu
Asad Busool
17 000 vols 30756

Chicago School of Professional Psychology Library, 325 N Wells St, 6th Flr, *Chicago*, IL 60610
T: +1 312 3296630; Fax: +1 312 6446075; E-mail: library@thechicagoschool.edu; URL: www.thechicagoschool.edu/content.cfm/library
Indu Aggerwal
Psychotherapy, child psychology; Israel Goldiamond Special Coll
10 000 vols; 220 curr per 30757

Chicago State University, Paul & Emily Douglas Library, 9501 S Martin Luther King Jr Dr, LIB 206-B, *Chicago*, IL 60628-1598
T: +1 773 9952222; Fax: +1 773 8212581; URL: www.csu.edu/library
1867; Rose E. Smith
Education (Learning Mat, Children's Lit)
477 000 vols; 1 005 curr per; 17 000 e-journals; 62 digital data carriers
libr loan 30758

Columbia College, Library, 624 S Michigan Ave, *Chicago*, IL 60605-1996; mail address: 600 S Michigan Ave, Chicago, IL 60605-1996
T: +1 312 3447370; Fax: +1 312 3448062; E-mail: libraryweb@colum.edu; URL: www.lib.colum.edu
1890; R. Conrad Winke
Art, Dance, Film, Photography, Theater, Journalism, Radio & Television; George Lurie Fine Arts Coll; Black Music Research Ctr
283 000 vols; 1 300 curr per; 4 000 e-books; 14 000 av-mat; 8 500 sound-rec; 3 300 digital data carriers; 1 300 Journals, 300 Bks on Deafness & Sign Lang 30759

DePaul University, John T. Richardson Library, 2350 N Kenmore, ***Chicago***, IL 60614
T: +1 773 3253725; Fax: +1 773 3257870; URL: www.lib.depaul.edu
1898; Linda Morrissett
Art Books, Sports Coll, Charles Dickens Coll, Horace Coll, Napoleon Coll
849 000 vols; 5 778 curr per
libr loan 30760

– Law Library, 25 E Jackson Blvd, 5th Flr, Chicago, IL 60604-2287
T: +1 312 3628701; Fax: +1 312 3626908; URL: www.law.depaul.edu/library/library_resources/library
1920; Allen Moye
Jameson Coll (Constitutional Law), Stuart A Weisler Coll (Environmwntal Law), Graduate Health Law Coll, Graduate Taxation Law Coll, Int Human Rights, Nathan Schwartz Coll (Supreme Court Justices' Signatures)
371 000 vols; 5 085 curr per; 3 000 microforms; 9 digital data carriers
libr loan 30761

– Loop, One E Jackson Blvd, 10th Flr, Chicago, IL 60604
T: +1 312 3628432; Fax: +1 312 3626186
250 000 vols; 200 curr per 30762

Illinois College of Optometry, Carl F. Shepard Memorial Library, 3241 S Michigan Ave, ***Chicago***, IL 60616
T: +1 312 9497158; Fax: +1 312 9497337; E-mail: cshepard@ico.edu; URL: library.ico.edu
1955; Gerald Dujsik
Senior Research projects 1421
24 000 vols; 231 curr per; 8 000 e-journals; 2 000 microforms; 638 sound-rec; 125 digital data carriers; 2 129 pamphlets and theses
libr loan; MLA, AVSL 30763

Illinois Institute of Art – Chicago Library, 350 N Orleans St, ***Chicago***, IL 60654-1593
T: +1 312 7778729; Fax: +1 312 7778782; E-mail: lib331@stu.aii.edu; URL: lib331.aisites.com
Juliet Teipel
34 000 vols; 283 curr per 30764

Illinois Institute of Technology, Paul V. Galvin Library, 35 W 33rd St, ***Chicago***, IL 60616
T: +1 312 5673616; Fax: +1 312 5675318; E-mail: library@iit.edu; URL: www.gl.iit.edu
1891; Christopher Stewart
Marvin Camras Coll, David Spaeth Coll
598 000 vols; 9 678 curr per; 257 390 depository docs 30765

– Downtown Campus Libraries, Chicago-Kent College of Law, 565 W Adams St, Chicago, IL 60661
T: +1 312 9065642; Fax: +1 312 9065685; E-mail: library@kentlaw.edu; URL: library.kentlaw.edu; library.stuart.edu
Keith Ann Stiverson
Aging, Int relations, Law; Libr of Int Relations, Stuart Business Libr
575 000 vols; 9 158 curr per 30766

The John Marshall Law School, Louis L. Biro Law Library, 315 S Plymouth Ct, ***Chicago***, IL 60604
T: +1 312 4272737; Fax: +1 312 4278307; URL: www.jmls.edu
1899; Anne Abramson
Chicago Bar Assn Core Coll, CCH Tax Libr Coll, HIS Legislative Hist, Illinois Supreme Court Briefs, US Congressional Publs, US Supreme Court Briefs
392 000 vols; 6 047 curr per; 742 000 microforms; 11 digital data carriers
libr loan 30767

Loyola University Chicago, University Libraries, 6525 N Sheridan Rd, ***Chicago***, IL 60626
T: +1 773 5082641; Fax: +1 773 5082993; URL: libraries.luc.edu
1870; Robert Seal
Jesuitica; Paul Claudel Coll
1 401 000 vols; 8 632 curr per; 368 483 govt docs; 1 628 000 microforms; 9 595 sound-rec; 499 digital data carriers
libr loan; ALA, ACRL, CRL 30768

– Elizabeth M. Cudahy Memorial Library, 6525 N Sheridan Rd, Chicago, IL 60626
T: +1 773 5082654; Fax: +1 773 5082993; URL: libraries.luc.edu/about/cudahy.htm
Chulin Meng
Humanities, Philosophy, Theology
912 000 vols 30769

– Health Sciences Library, Bldg 101, Rm 1717, 2160 S First Ave, Maywood, IL 60153-5585
T: +1 708 2165308; Fax: +1 708 2166777; E-mail: researchservices@lumc.edu; URL: library.luhs.org
1968; Dr. Logan Ludwig
Hist of Medicine Coll
207 000 vols; 4 943 curr per; 1 500 e-journals; 165 e-books; 971 diss/theses; 182 microforms; 5 672 slides
libr loan; MLA, SLA 30770

– Law School Library, 25 E Pearson St, 3rd Flr, Chicago, IL 60611
T: +1 312 9157202; Fax: +1 312 9156797; E-mail: law-library@luc.edu; URL: www.luc.edu/law/library
1909; Julia Wentz
Antitrust law; GPO Depot, Medical Jurisprudence, Child law
181 000 vols; 1 313 curr per; 805 000 microforms; 7 digital data carriers 30771

– Lewis Library, 25 E Pearson St, 6th Flr, Chicago, IL 60611
T: +1 312 9156631; Fax: +1 312 9156637; URL: www.libraries.luc.edu/about/lewis.shtml
Yolande Wersching
Business, Social work, Criminal justice
260 000 vols; 22 digital data carriers
libr loan 30772

Meadville-Lombard Theological School Library, 5701 S Woodlawn Ave, ***Chicago***, IL 60637
T: +1 773 2563000 ext 225; Fax: +1 773 2563007; E-mail: ngerdes@meadville.edu; URL: www.meadville.edu
1844; Rev Neil W. Gerdes
Unitarian universalism, Social ethics, Hist of religions; Jenkin Lloyd Jones, A Powell Davies, Jack Mendelsohn, Vincent Silliman Papers; William Ellery Channing Original Mss
112 000 vols; 130 curr per; 110 microforms 30773

North Park University, Brandel Library, 5114 N Christiana Ave, ***Chicago***, IL 60625; mail address: 3225 W Foster Ave, Chicago, IL 60625
T: +1 773 2445580; Fax: +1 773 2444891; URL: www.northpark.edu/library
1891; Sarah Anderson
Music, Nursing, Religion, Scandinavia; Jenny Lind Coll, Walter Johnson Scandinavian Studies Coll, Paul L. Holmer Coll, G. Anderson Lincoln Coll, Harald Jacobson China Studies Coll, Nils William Olsson Coll
229 000 vols; 961 curr per; 39 e-journals; 6 459 mss; 10 269 music scores; 280 000 microforms; 5 716 sound-rec; 20 digital data carriers; 50 linear feet of vertical file mats
libr loan 30774

Northeastern Illinois University, Ronald Williams Library, 5500 N Saint Louis Ave, ***Chicago***, IL 60625-4699
T: +1 773 4424509; Fax: +1 773 4424530; URL: www.neiu.edu/~neiulib
1961; Sharon Scott
US & Illinois Doc Depositories, Illinois Regional Arch Depository for Chicago and Cook County, African-American Lit, William Gray Reading Coll
711 000 vols; 1 050 e-journals; 38 000 e-books; 465 diss/theses; 196 722 govt docs; 2 074 music scores; 4 670 maps; 869 000 microforms; 4 122 av-mat; 8 290 sound-rec
libr loan; ALA 30775

Northwestern University, ***Chicago***
– Galter Health Sciences Library, Montgomery Ward Bldg, 303 E Chicago Ave, Chicago, IL 60611
T: +1 312 5031908; Fax: +1 312 5031204; E-mail: ghsl-ref@northwestern.edu; URL: www.galter.northwestern.edu
1883; James Shedlock
Dental Hist; Med Classics; Med Hist; Rare Bks
286 000 vols; 849 e-books; 3 incunabula
libr loan 30776

– Joseph Schaffner Library, Wieboldt Hall, 2nd Flr, 339 E Chicago Ave, Chicago, IL 60611
T: +1 312 5038422; Fax: +1 312 5038930; E-mail: schaffner-reference@northwestern.edu; URL: www.library.northwestern.edu/schaffner
1908; Scott Garton
20 000 vols 30777

– School of Law Library, 357 E Chicago Ave, Chicago, IL 60611
T: +1 312 5038451; Fax: +1 312 5039230; URL: www.law.northwestern.edu/lawlibrary/
1859; Jim McMasters
Rare Bk Coll
563 000 vols; 5 452 curr per; 129 000 microforms
libr loan 30778

Passionist Academic Institute Library, 5700 N Harlem Ave, ***Chicago***, IL 60631
T: +1 773 6311686; Fax: +1 773 6318059
1965; Irene Horst
Theology, Philosophy; Cardinal Newman Coll
20 000 vols; 35 curr per
libr loan 30779

Psychiatric Institute UIC, Professional Library, 1601 W Taylor St, Rm 438, ***Chicago***, IL 60612
T: +1 312 4134548; Fax: +1 312 4134556
1959; Margo McClelland
15 000 vols; 15 curr per; 753 av-mat; 505 sound-rec; 1 digital data carriers 30780

Robert Morris College, Library, 401 S State St, ***Chicago***, IL 60605
T: +1 312 9354810; Fax: +1 312 9356253; E-mail: sdutler@robertmorris.edu; URL: www.robertmorris.edu/library
1913; Sue Dutler
116 985 vols; 193 curr per 30781

Roosevelt University, Murray-Green Library, 430 S Michigan Ave, ***Chicago***, IL 60605
T: +1 312 3413649; Fax: +1 312 3412425; URL: www.roosevelt.edu/library
1945; Barbara Schoenfield
Children's bks; American Civilization and English Lit (Libr Resources Microbk Coll), Music Coll
187 000 vols; 1 165 curr per; 3 000 microforms; 900 av-mat; 19 513 sound-rec; 20 digital data carriers
libr loan 30782

– Performing Arts Library, 430 S Michigan Ave, Chicago, IL 60605
T: +1 312 3413651; Fax: +1 312 3412425; E-mail: celias@roosevelt.edu; URL: www.roosevelt.edu/library
1945; Richard Schwegel
History and performance practice of Western art music, contemporary music and jazz studies; Arch of Chicago Musical College
26 000 vols; 100 curr per; 47 000 music scores; 13 700 sound-rec 30783

Saint Xavier University, Byrne Memorial Library, 3700 W 103rd St, ***Chicago***, IL 60655-3105
T: +1 773 2983352; Fax: +1 773 7795231; URL: www.sxu.edu/library
1916; Mark Vargas
172 000 vols; 1 607 curr per 30784

Spertus Institute of Jewish Studies, Norman & Helen Asher Library, 610 S Michigan Ave, ***Chicago***, IL 60605
T: +1 312 3221749; Fax: +1 312 9220455; E-mail: asherlib@spertus.edu; URL: www.spertus.edu
1925; Kathleen Bloch
Chicago Jewish Arch, Chicago Jewish Hist, Jewish Art, Jewish Music (Targ Ctr for Jewish Music)
100 000 vols; 550 curr per; 33 maps; 6 000 microforms; 930 av-mat; 3 800 sound-rec; 18 digital data carriers
libr loan; ALA, AJL 30785

University of Chicago, Library, 1100 E 57th St, ***Chicago***, IL 60637-1502
T: +1 773 7028740; Fax: +1 773 7026623; URL: www.lib.uchicago.edu
1891; Judith Nadler (Director and University Librarian)
Children's Books, Drama, English Bibles, Kentucky History, Judaica, Literature, Balzac, Chaucer, Homer, Italian Women Writers, History, History of Science & Medicine; Special Collections Research Center, Special Collections and Preservation Division
8 830 151 vols; 13 128 curr per; 25 494 e-journals; 831 860 e-books; 438 094 maps; 3 147 978 microforms; 9 367 av-mat; 60 778 sound-rec
libr loan; ALA, ARL, CRL, CNI, CIC, CASE, CLIR, DLF, D-NABG, Hyde Park Cultural Alliance, ILA, IUG, IATUL, Kuali Foundation, OCLC, SPARC 30786

– D'Angelo Law Library, 1121 E 60th St, Chicago, IL 60637-2786
T: +1 773 7029615; Fax: +1 773 7022889; URL: www.lib.uchicago.edu/e/law
1902; Judith Wright
Henry Simons Papers, Karl Llewelyn Papers, US Supreme Court Briefs & Records Depot
636 000 vols; 8 855 curr per; 44 000 microforms; 26 digital data carriers 30787

– Eckhart Library, 1118 E 58th St, Chicago, IL 60637
T: +1 773 7028778; Fax: +1 773 7027535; E-mail: eckhart-library@lib.uchicago.edu; URL: www.lib.uchicago.edu/e/eck
Brenda Rice
Computer science, Mathematics, Statistics
55 000 vols; 520 curr per 30788

– Harper Library, 1116 E 59th St, Chicago, IL 60637
T: +1 773 7027960; E-mail: harper-library@lib.uchicago.edu; URL: www.lib.uchicago.edu/e/harper
Hayden Asahi
67 000 vols; 74 curr per 30789

– John Crerar Library, 5730 S Ellis Ave, Chicago, IL 60637-1434
T: +1 773 7027715; Fax: +1 773 7023317; E-mail: crerar-reference@lib.uchicago.edu; URL: www.lib.uchicago.edu/crerar
1891; Kathleen A. Zar
Astrophysics, Biomedical, Botany, Clinical medicine, Earth sciences, Hist of science & medicine, Oceanography, Physics, Technology, Zoology; Incunabula, Joseph Regenstein Libr, Mss
1 300 000 vols 30790

– Social Service Administration Library, 969 E 60th St, Chicago, IL 60637-2627
T: +1 773 7021199; Fax: +1 773 7020874; URL: www.lib.uchicago.edu/libinfo/libraries/ssad
1965; Eileen Libby
35 000 vols; 141 curr per; 7 000 microforms; 3 800 pamphlets 30791

University of Illinois at Chicago, Richard J. Daley Library, 801 S Morgan St, ***Chicago***, IL 60607; M/C 234, P.O. Box 8198, Chicago, IL 60680-8198
T: +1 312 9964886; Fax: +1 312 4130424; URL: www.uic.edu/depts/lib
1946; Mary Case
Architecture, Chicago Design Arch, Chicago Fairs & Expositions, Chicago Lit, Chicago Photo Arch, Midwest Women's Hist Coll
2 237 000 vols
libr loan 30792

– Library of the Health Sciences, MC 763, 1750 W Polk St, Chicago, IL 60612
T: +1 312 9968966; Fax: +1 312 9969163; E-mail: lib-cref@uic.edu; URL: library.uic.edu/lhs-chicago
1881
Kiefer Coll (Urology), Herbals & Pharmacopoeias, Bailey Coll (Neurology & Psychiatry), Medical Ctr Arch
671 632 vols; 3 682 curr per 30793

– Science Library, 845 W Taylor St, Chicago, IL 60607
T: +1 312 9965396; Fax: +1 312 9967822; E-mail: lib-sci@uic.edu; URL: www.uic.edu/depts/lib/science
1970; Julie M. Hurd
Sadtler Res Laboratories Standard Colls of Spectra, Industry Standards & Codes
144 177 vols; 1 279 curr per; 93 636 tech rpts 30794

University of Science & Arts of Oklahoma, Nash Library, 1901 S 17 St, ***Chickasha***, OK 73018; mail address: 1727 W Alabama Ave, Chickasha, OK 73018
T: +1 405 5741343; Fax: +1 405 5741220; URL: library.usao.edu/home/
1909; Kelly Brown
80 000 vols; 200 curr per; 50 e-books; 504 microforms; 1 975 sound-rec; 300 Bks on Deafness & Sign Lang
libr loan 30795

California State University, Chico, Meriam Library, 400 W First St, ***Chico***, CA 95929-0295
T: +1 530 8986479; Fax: +1 530 8984443; URL: www.csuchico.edu/library
1887; Carolyn Dusenbury
Calif & Rand Corp, Northeast California Coll
949 000 vols; 675 curr per
libr loan 30796

College of Our Lady of the Elms, Alumnae Library, 291 Springfield St, ***Chicopee***, MA 01013-2839
T: +1 413 2652280; Fax: +1 413 5947418; URL: www.elms.edu/departments/library
1928; Patricia Bombardier
Religion, Nursing, Natural science; Ecclesiology, 16th & 17th c editions; 18th c editions of English Authors; Sir Walter Scott Coll, first editions; Fed Depot Coll
12 000 vols; 823 curr per 30797

Athenaeum of Ohio, Eugene H. Maly Memorial Library, 6616 Beechmont Ave, ***Cincinnati***, OH 45230-2091
T: +1 513 2312223; Fax: +1 513 2313254; URL: www.athenaeum.edu
1829; Tracy Koenig
Biblical studies, Roman Catholic theology, Church hist, Pastoral counseling; American Church Hist (Archbishop Purcell Special Coll), Unusual Bibles (Rare Book Coll)
109 000 vols; 408 curr per; 7 000 e-books; 22 mss; 22 incunabula; 400 diss/theses; 1 000 microforms; 250 av-mat; 110 sound-rec; 12 digital data carriers
libr loan; ATLA 30798

College of Mount Saint Joseph, Archbishop Alter Library, 5701 Delhi Rd, ***Cincinnati***, OH 45233-1671
T: +1 513 2444307; Fax: +1 513 2444355; E-mail: library@mail.msj.edu; URL: inside.msj.edu/departments/library
1920; Paul Owen Jenkins
Religion; US Catholic Conference Docs Coll
97 000 vols; 248 curr per; 17 000 e-books; 289 000 microforms; 212 sound-rec; 30 digital data carriers 30799

Hebrew Union College (HUC-JIR) – Jewish Institute of Religion, Klau Library, HUC-JIR, 3101 Clifton Ave, ***Cincinnati***, OH 45220-2488
T: +1 513 4873287; Fax: +1 513 2210519; E-mail: klau@huc.edu; URL: www.huc.edu/libraries
1875; Dr. David J. Gilner
Spinoza Coll, Josephus Coll, Incunabula, Hebrew Mss, Jewish Americana to 1850, 16th c Hebrew Printing, Printed Bibles, Assyriology, Jewish Music, Broadside, Inquisition, Yiddish Theater, Anti-Semitism
460 000 vols; 2 500 curr per; 35 e-journals; 2 300 mss; 32 000 microforms; 158 sound-rec; 160 digital data carriers; 200 Electronic Media & Resources
libr loan; RLG, AJL, ALA 30800

University of Cincinnati Libraries, PO Box 210033, ***Cincinnati***, OH 45221-0033
T: +1 513 5561461; Fax: +1 513 5563141; URL: www.libraries.uc.edu
1819; Victoria A. Montavon
20th c English language poetry, 19th & 20th c astronomy, 18th c British anonymus poetical pamphlets, hist of chemistry, 19th & 20th c German-Americana, American labor hist, D.H. Lawrence mss & Dorothy Brett correspondence, modern Greek Studies, classical studies; Archive and rare bks library
3 418 000 vols; 183 digital data carriers
libr loan; ARL, ALA 30801

– Chemistry-Biology Library, 503 Rieveschl, A-3, Cincinnati, OH 45221; P.O. Box 210151, Cincinnati, OH 45221-0151
T: +1 513 5561498; Fax: +1 513 5561103; URL: www.libraries.uc.edu/libraries/chem-bio
John Tebo
Oesper Coll of Rare Bks & Portraits in Hist of Chemistry
62 000 vols; 66 637 curr per 30802

– Classics Library, 320 Blegen Library, Cincinnati, OH 45221; P.O. Box 210191, Cincinnati, OH 45221-0191
T: +1 513 5561315; Fax: +1 513 5566244; URL: www.libraries.uc.edu/libraries/classics
Jacquelene Riley
187 000 vols; 29 092 curr per 30803

– College of Applied Science Library, 2220 Victory Pkwy, Cincinnati, OH 45206-2839
T: +1 513 5566594; Fax: +1 513 5564217; URL: www.libraries.uc.edu/libraries/cas
1828; Ted Baldwin
45 000 vols; 1 015 diss/theses; 6 000 microforms; 250 av-mat; 561 digital data carriers 30804

– College-Conservatory of Music, Library, 417 Blegen Library, Cincinnati, OH 45221; P.O. Box 210152, Cincinnati, OH 45221-0152
T: +1 513 5561970; Fax: +1 513 5563777; URL: www.libraries.uc.edu/libraries/ccm
Mark Palkovic
31 000 vols; 519 curr per 30805

– Curriculum Resources Center, 400 Teachers College, Cincinnati, OH 45221; P.O. Box 210219, Cincinnati, OH 45221-0219
T: +1 513 5561430; Fax: +1 513 5563006; URL: www.libraries.uc.edu/libraries/cech
1971; Cheryl Ghosh
50 000 vols; 5 751 curr per; 2 760 av-mat; 190 digital data carriers; 317 tests 30806

– Design, Architecture, Art & Planning Library, 5480 Aronoff Ctr, Cincinnati, OH 45221; P.O. Box 210016, Cincinnati, OH 45221-0016
T: +1 513 5561335; Fax: +1 513 5563006; URL: www.libraries.uc.edu/libraries/daap
1929; N.N.
Ladislas Segoe Coll (City Planning)
80 000 vols; 400 curr per; 5 000 microforms; 100 planning rpts 30807

– Engineering Library, 850 Baldwin Hall, Cincinnati, OH 45221; P.O. Box 210018, Cincinnati, OH 45221-0018
T: +1 513 5561550; Fax: +1 513 5562654; URL: www.engrlib.uc.edu
1911; Dorothy Byers
75 000 vols; 53 326 curr per 30808

– Geology-Mathematics- Physics Library, 240 Braunstein Hall, Cincinnati, OH 45221; P.O. Box 210153, Cincinnati, OH 45221-0153
T: +1 513 5561324; Fax: +1 513 5561930; URL: www.libraries.uc.edu/libraries/geo-math-phys
1907; Angela Gooden
SV Hvabar-Exxon Guidebk Coll, Isay Balinkin Color Coll, Observatory Coll
145 000 vols; 700 curr per; 600 diss/theses; 125 243 maps; 413 digital data carriers 30809

University of Cincinnati – Raymond Walters College, Library, Muntz Hall, Rm 115, 9555 Plainfield Rd, ***Cincinnati***, OH 45236-1096
T: +1 513 7455710; Fax: +1 513 7455767; URL: www.libraries.uc.edu/libraries/rwc
1967; Stephena Harmony
Dental Hygiene Coll
50 000 vols; 450 curr per; 15 000 microforms; 831 av-mat; 830 sound-rec; 8 digital data carriers
libr loan 30810

University of Cincinnati – Robert S. Marx Law Library, 2540 Clifton Ave, ***Cincinnati***, OH 45219; P.O. Box 210142, Cincinnati, OH 45221-0142
T: +1 513 5560163; Fax: +1 513 5566265; URL: www.law.uc.edu
1874; N.N.
Cincinnati Legal Hist & Rpts of Various Courts; Early Ohio Legal Coll, hist rpts; Morgan Coll on Human Rights; Land Planning (Robert N Cook Coll); Rare Bk Coll; Church & State Coll
391 000 vols; 2 587 curr per; 624 000 microforms; 884 sound-rec; 8 digital data carriers
libr loan 30811

Xavier University, McDonald Memorial Library, 3800 Victory Pkwy, ***Cincinnati***, OH 45207-5211
T: +1 513 7454822; Fax: +1 513 7451932; E-mail: xulib@xavier.edu; URL: www.xavier.edu/library
1831; Bob Cotter
Bibles, Incunabula & Jesuitica; Catholic Boy's Fiction (Francis Finn, SJ, Coll)
304 000 vols; 1 631 curr per; 8 000 e-books; 137 av-mat; 52 High Interest/Low Vocabulary Bks, 112 Bks on Deafness & Sign Lang, 140 Talking Bks
libr loan 30812

Ohio Christian University, Melvin & Laura Maxwell Library, 1476 Lancaster Pike, ***Circleville***, OH 43113; P.O. Box 460, Circleville, OH 43113-0460
T: +1 740 4748896; Fax: +1 740 4777855; E-mail: library@ohiochristian.edu; URL: ohiochristian.edu/maxwell_library/index.htm
1947; David Tipton
Stout Bible Coll
53 000 vols; 121 curr per
libr loan; ACL 30813

Claremont Colleges Libraries, Honnold-Mudd Library, 800 Dartmouth Ave, ***Claremont***, CA 91711
T: +1 909 6218045; Fax: +1 909 6218681; URL: libraries.claremont.edu
1887; Bonnie J. Clemens
Honnold Libr (Californian & Western Americana), Philbrick Libr of Dramatic Arts, Cartography of the Pacific Coast, Irving Wallace, Northern Europe, Oxford, Renaissance, Water Resources of Southern California, Colleges Arch
2 232 000 vols; 5 958 curr per; 3 000 e-books; 1 200 digital data carriers
libr loan 30814

– Ella Strong Denison Library, Scripps College Campus, 1030 N Columbia Ave, Claremont, CA 91711
T: +1 909 6073941; Fax: +1 909 6071548
1931; Judy Harvey Sahak
Humanities, Fine arts; Perkins & Kirby Coll (Book Hist & Book Arts), Macpherson Coll (Women), Metcalf Coll (Gertrude Stein), Pacific Coast Browning Foundation (Browning), Hanna Coll (Southwest), Miller-Howard Coll (Latin America), Ament Coll (Melville), Louise Seymour Jones Bookplates Coll, Scripps College Arch
120 000 vols; 140 curr per; 850 sound-rec 30815

– Norman F. Sprague Memorial Library, 301 E 12th St, Claremont, CA 91711-5990
T: +1 909 6218920; Fax: +1 909 6077437; URL: libraries.claremont.edu/sprague
1897; Jezmynne Westcott
Science, Hist of science, Engineering; Carruthers Aviation Coll, Herbert Hoover Coll on the Hist of Science & Metallic Arts
50 000 vols; 15 000 microforms; 1 digital data carriers; 27 000 tech rpts, 1 050 pamphlets 30816

– Seeley G. Mudd Science Library, Pomona College Campus, 640 N College Ave, Claremont, CA 91711-6345
T: +1 909 6218924; URL: voxlibris.claremont.edu/libraries/pomona.html
1983; Bruce Taylor
Botany, Zoology, Chemistry, Geology, Mathematics, Physics, Astronomy; AO Woodford Geology Libr
100 000 vols; 30 diss/theses; 5 000 govt docs; 6 000 maps; 4 000 microforms; 600 digital data carriers 30817

Claremont Graduate University, George G. Stone Center for Children's Books, 740 North College Avenue, ***Claremont***, CA 91711
T: +1 909 6073670; URL: www.cgu.edu/pages/3613.asp
1965; Carolyn Angus
35 000 vols; 26 curr per
ALA 30818

Claremont School of Theology Library, 1325 N College Ave, ***Claremont***, CA 91711-3199
T: +1 909 4472589; Fax: +1 909 4476285; URL: www.cst.edu/library
1957; John Dickason
Biblical studies, Methodistica; Kirby Page Mss, Bishop James Baker Mss, Mitchell Hymnology Coll
192 000 vols; 641 curr per; 6 000 microforms; 343 av-mat; 172 sound-rec; 10 digital data carriers
libr loan 30819

Rogers State University, Stratton Taylor Library, 1701 W Will Rogers Blvd, ***Claremore***, OK 74017-3252
T: +1 918 3437720; Fax: +1 918 3437897; E-mail: library@rsu.edu; URL: www.rsu.edu/library
Alan Lawless
72 000 vols; 422 curr per; 38 000 e-books; 5 000 av-mat; 1 300 sound-rec; 60 Electronic Media & Resources 30820

Clarion University of Pennsylvania, Rena M. Carlson Library, 840 Wood St, ***Clarion***, PA 16214
T: +1 814 3931841; Fax: +1 814 3932344; E-mail: libsupport@clarion.edu; URL: www.clarion.edu/library
1867; Dr. Terry S. Latour
British Commonwealth Hist Coll
495 000 vols; 20 591 curr per; 18 000 e-books; 2 496 maps; 1 386 000 microforms; 3 745 sound-rec; 16 digital data carriers; 1 191 Slides
libr loan 30821

Austin Peay State University, Felix G. Woodward Library, 601 E College St, ***Clarksville***, TN 37044; P.O. Box 4595, Clarksville, TN 37040
T: +1 931 2217346; Fax: +1 931 2217296; E-mail: librarian@apsu.edu; URL: library.apsu.edu
1969; Deborah Fetch
350 000 vols; 1 200 curr per; 12 000 e-journals; 55 000 e-books; 110 000 govt docs; 2 014 maps; 620 000 microforms; 4 004 sound-rec; 588 digital data carriers
libr loan 30822

University of the Ozarks, Robson Library, 415 N College Ave, ***Clarksville***, AR 72830
T: +1 479 9791382; Fax: +1 479 9791477; URL: www.ozarks.edu
1891; Stuart P. Stelzer
88 000 vols; 450 curr per; 16 000 e-journals; 38 000 e-books; 50 000 govt docs; 850 av-mat; 2 600 sound-rec
libr loan 30823

Clearwater Christian College, Easter Library, 3400 Gulf-to-Bay Blvd, ***Clearwater***, FL 33759-4595
T: +1 727 7261153; Fax: +1 727 7238566; E-mail: library@clearwater.edu; URL: www.clearwater.edu/library
1966; Elizabeth Werner
Creationism
103 000 vols; 15 000 curr per; 2 881 e-journals; 10 280 e-books; 158 microforms; 1 555 av-mat; 2 645 sound-rec; 100 digital data carriers
libr loan; OCLC, ACL 30824

Clemson University, Cooper Library, Box 343001, ***Clemson***, SC 29634-3001
T: +1 864 6565186; Fax: +1 864 6560758; URL: www.lib.clemson.edu
1893; Wall L. Kay
Edgar A. Brown Papers, James F. Byrnes Papers, John C. Calhoun Letters, 1805-1850, A. Frank Lever Papers, Strom Thurmond Papers, Benjamin R. Tillman Papers

1 259 000 vols; 15 185 curr per; 21 220 maps; 726 000 microforms; 606 sound-rec; 1 678 digital data carriers
libr loan 30825

– Gunnin Architecture Library, 112 Lee Hall, Clemson, SC 29634-0501
T: +1 864 6563932; Fax: +1 864 6563932; E-mail: kathye@clemson.edu; URL: www.lib.clemson.edu/gunnin
Kathy Edwards
36 304 vols; 235 curr per; 2 000 planning docs, 90 000 slides 30826

Case Western Reserve University, Kelvin Smith Library, 11055 Euclid Ave, ***Cleveland***, OH 44106; mail address: 10900 Euclid Ave, Cleveland, OH 44106-7151
T: +1 216 3683517; Fax: +1 216 3686950; E-mail: ulrefer@case.edu; URL: library.case.edu
1967; Dr. Joanne Eustis
American, French, German & English Lit & Hist Colls; Hist of Science & Technology; Fine Arts Colls; Mss Colls; Papers of Clevelanders
1 754 000 vols; 1 877 000 microforms; 150 digital data carriers
libr loan 30827

– Lillian & Milford Harris Library, Mandel School of Applied Social Sciences, 11235 Bellflower Rd, mail address: 10900 Euclid Ave, Cleveland, OH 44106-7164
T: +1 216 3682302; Fax: +1 216 3682106; E-mail: harrisref@case.edu; URL: msass.case.edu/harrislibrary
1916; Samantha C. Skutnik
Social work
40 000 vols; 267 curr per; 544 microforms; 206 sound-rec; 37 linear ft of mss, 136 computer files
libr loan 30828

– School of Law Library, 11075 East Blvd, Cleveland, OH 44106-7148
T: +1 216 3688862; Fax: +1 216 3681002; URL: law.cwru.edu
1893; Kathleen M. Carrick
Anglo-American Common Law, Govt Docs Coll, Legal Clinic, Rare Bk Coll
298 000 vols; 9 962 curr per; 431 000 microforms; 1 986 sound-rec; 214 digital data carriers; 1 958 unbound rpts, pamphlets, docs 30829

Cleveland Health Sciences Library, School of Medicine, Robbins Bldg, 2109 Adelbert Rd, ***Cleveland***, OH 44106-4914
T: +1 216 3686420; Fax: +1 216 3683008; E-mail: hclref@case.edu; URL: www.case.edu/chsl/homepage.htm
1965; Virginia Saha
Hist of Medicine Dittrick Museum of Medical Hist, all media, archival mat; Darwin Coll; Marshall Herbal Coll; Freud Coll; Cole Coll of Venereals
400 000 vols; 1 000 curr per
libr loan 30830

– Dittrick Medical History Center, Library, 11000 Euclid Ave, Cleveland, OH 44106-7130
T: +1 216 3683648; Fax: +1 216 3680165; URL: www.case.edu/artsci/dittrick/site2
James Edmonson
60 000 vols 30831

Cleveland Institute of Art, Jessica Gund Memorial Library, 11141 East Blvd, ***Cleveland***, OH 44106
T: +1 216 4217441; Fax: +1 216 4217439; E-mail: referencehelp@gate.cia.edu; URL: www.cia.edu/library
1882; Cristine C. Rom
Artists' Bks
45 000 vols; 137 curr per; 18 000 microforms; 100 av-mat; 2 000 sound-rec; 550 digital data carriers; 10 000 pictures
libr loan; ALA, ARLIS/NA 30832

Cleveland State University, ***Cleveland***
– Cleveland-Marshall Law Library, Cleveland-Marshall College of Law, 2121 Euclid Ave, LB138, Cleveland, OH 44115-2403
T: +1 216 6876877; Fax: +1 216 6876881; URL: www.law.csuohio.edu/lawlibrary
1897; N.N.
Ohio Supreme Court Briefs & Records, microfrom; US Supreme Court Briefs & Records, microform; CIS Index & US Legislative Hist, microfiche
526 000 vols; 3 141 curr per 30833

– University Library, Rhodes Tower, 2121 Euclid Ave, Cleveland, OH 44115-2214
T: +1 216 6872478; Fax: +1 216 6879380; URL: library.csuohio.edu
1928; Dr. Glenda A. Thornton
Cleveland Colls, Black Hist, Contemporary Poetry Coll, Bridge Eng
1 084 000 vols; 10 944 curr per; 31 000 e-books; 220 digital data carriers; 94 377 slides 30834

David N. Myers College, Library Resource Center, 3921 Chester Ave, ***Cleveland***, OH 44114
T: +1 216 4328990; Fax: +1 216 4269296; E-mail: library@myers.edu; URL: library.myers.edu
1968; Richard Brhel
Business & management, Economy, Law, Public administration; Spencerian Arch of Hist of Business & Business Education
17 000 vols; 125 curr per; 8 digital data carriers
libr loan 30835

Delta State University, Roberts-LaForge Library, Laflore Circle at Fifth Ave, ***Cleveland***, MS 38733-2599
T: +1 662 8464448; Fax: +1 662 8464443; E-mail: refdesk@deltastate.edu; URL: library.deltastate.edu
1925; Jeff M. Slagell
Arch (Walter Sillers Coll), Art Coll, Mississippiana
360 000 vols; 17 368 curr per; 51 000 e-books
libr loan 30836

Siegal College of Judaic Studies, Aaron Garber Library, 26500 Shaker Blvd, ***Cleveland***, OH 44122
T: +1 216 4644050; Fax: +1 216 4645827; E-mail: circulation@siegalcollege.edu; URL: www.siegalcollege.edu
Jean Loeb Lettofsky
Holocaust, Hebrew Lit
34 000 vols; 150 curr per; 10 digital data carriers 30837

Hamilton College, Burke Library, 198 College Hill Rd, ***Clinton***, NY 13323-1299
T: +1 315 8594735; Fax: +1 315 8594578; E-mail: askref@hamilton.edu; URL: www.hamilton.edu/library/
1812; Randall Ericson
Civil War, Contemporary poetry, Feminism, Govt, Hist, Religious studies; Almanacs, Bk Arts, Communal Societies, Ezra Pound Coll, Hamiltonia, Lesser Antilles (Beineke Coll)
615 000 vols; 1 900 curr per
libr loan 30838

Mississippi College, Leland Speed Library, 101 W College St, ***Clinton***, MS 39058; Box 4047, Clinton, MS 39058-4047
T: +1 601 9253232; Fax: +1 601 9253435; E-mail: mbhc@mc.edu; URL: www.mc.edu/campus/library/
1826; Kathleen Hutchison
Mississippi Baptist Hist Coll
242 000 vols; 770 curr per; 205 000 microforms; 4 333 av-mat; 4 931 sound-rec; 14 digital data carriers; 65 Slides 30839

Presbyterian College, James H. Thomason Library, 211 E Maple St, ***Clinton***, SC 29325
T: +1 864 8338299; Fax: +1 864 8338315; E-mail: library@presby.edu; URL: www.presby.edu/library/
1880; Dave Chatham
Caroliniana
137 000 vols; 684 curr per; 14 000 microforms; 3 789 sound-rec; 45 digital data carriers; 2 913 Videotapes
libr loan 30840

Clovis Community College, Clovis Campus Library, 417 Schepps Blvd, ***Clovis***, NM 88101
T: +1 575 7694080; Fax: +1 575 7694190; E-mail: ccclib@clovis.edu; URL: www.clovis.edu
1969; Dr. Deborah McBeth Anderson
Rare Bks, Southwest
70 000 vols; 350 curr per 30841

San Joaquin College of Law, Library, 901 Fifth St, ***Clovis***, CA 93612
T: +1 559 3232100; Fax: +1 559 3235566; URL: www.sjcl.edu/library
1969; Tara M. Crabtree
Water Policy, Land Use & Public Trust Doctrine
56 000 vols; 300 curr per 30842

State University of New York College of Agriculture & Technology, Van Wagenen Library, 142 Schenectady Ave, ***Cobleskill***, NY 12043
T: +1 518 2345841; Fax: +1 518 2555843; URL: www.cobleskill.edu/library
1920; Nancy Van Deusen
County hist, Hist Mat Related to Agriculture
65 000 vols; 200 curr per; 50 maps; 300 microforms; 1 500 sound-rec; 1 700 other items
libr loan; ALA 30843

North Idaho College, Molstead Library, 1000 W Garden Ave, ***Coeur d'Alene***, ID 83814-2199
T: +1 208 7693355; Fax: +1 208 7693428; E-mail: Jim_DeMoss@nic.edu; URL: www.nic.edu
1933
Northwest Coll, Indian Affairs
60 000 vols; 500 curr per 30844

Saint Michael's College, Durick Library, One Winooski Park, Box L, ***Colchester***, VT 05439-2525
T: +1 802 6542405; Fax: +1 802 6542630; URL: www.smcvt.edu/library
1904; John K. Payne
Music (Richard Stoehr Coll), printed music; New England Culinary Institute Coll; Arch of the Society of St Edmund
260 000 vols; 10 000 e-books; 423 Electronic Media & Resources
libr loan 30845

Williams Baptist College, Felix Goodson Library, 91 W Fulbright, ***College City***, AR 72476; P.O. Box 3738 WBC, Walnut Ridge, AR 72476-4669
T: +1 870 7594139; Fax: +1 870 7594135; URL: www.wbcoll.edu
1941; Marilyn Goodwin
Southern Baptist Convention Lab
69 000 vols; 152 curr per
libr loan 30846

University of Maryland, University Libraries, ***College Park***, MD 20742
T: +1 301 4059128; Fax: +1 301 3149408; URL: www.lib.umd.edu/
1813; N.N.
East Asia Coll, Broadcast Pioneer Lib, Katherine Anne Porter Coll, Int Piano Arch at Maryland, Marylandia, National Public Broadcasting Arch
3 083 000 vols; 12 808 curr per; 22 500 e-journals; 107 000 e-books
libr loan 30847

– Architecture Library, College Park, MD 20742-7011
T: +1 301 4059178; Fax: +1 301 3149416; URL: www.lib.umd.edu/ARCH/architecture.html
Joan Stahl
Archit;Landscape archit
41 000 vols; 158 curr per 30848

– Art Library, Art/Sociology Bldg, College Park, MD 20742
T: +1 301 4059061; Fax: +1 301 3149725; URL: www.lib.umd.edu/art/art.html
Louise Greene
Pre-Columbian art;African lang
84 000 vols; 394 curr per 30849

– Broadcast Pioneers Library of American Broadcasting, Hornbake Library, College Park, MD 20742
T: +1 301 4059160; Fax: +1 301 3142634; E-mail: labcast@umd.edu; URL: www.lib.umd.edu/lab/
Charles Howell
Radio and television
10 000 vols; 382 curr per 30850

– Engineering & Physical Sciences Library, College Park, MD 20742-7011
T: +1 301 4059178; Fax: +1 301 4059164; URL: www.lib.umd.edu/engin; www.lib.umd.edu/blogs/engin
Nevenka Zdravkovska
Eng;Phys sci
397 000 vols; 1 450 curr per 30851

– R. Lee Hornbake Undergraduate Library, College Park, MD 20742-7011
T: +1 301 4059257; Fax: +1 301 3149419
Glenn Moreton
205 000 vols; 34 curr per 30852

– Michelle Smith Performing Arts Library, College Park, MD 20742-7011
T: +1 301 4059217; Fax: +1 301 3147170; URL: www.lib.umd.edu/umcp/music/music.html
Yale Fineman
101 000 vols; 345 curr per 30853

– Theodore R. McKeldin Library, College Park, MD 20742-7011
T: +1 301 4059075; Fax: +1 301 3149408; URL: www.lib.umd.edu/MCK/mckeldin.html
Tanner Wray
Rare books
1 975 000 vols; 9 711 curr per 30854

– White Memorial Chemistry Library, 1526 Chemistry Bldg, College Park, MD 20742-7011
T: +1 301 4059178; Fax: +1 301 4059164; URL: www.lib.umd.edu/chem/chemistry.html
Sylvia O'Brien
85 000 vols; 455 curr per 30855

Walla Walla University, Peterson Memorial Library, 104 S College Ave, ***College Place***, WA 99324-1159
T: +1 509 5272133; Fax: +1 509 5272001; E-mail: carolyn.gaskell@wallawalla.edu; URL: www.wallawalla.edu
1892; Carolyn Gaskell
Denominational Hist
2 000 e-books
libr loan 30856

Texas A&M University, University Libraries, Spence St, ***College Station***, TX 77843-5000; mail address: 5000 TAMU, College Station, TX 77843-5000
T: +1 979 8455741; Fax: +1 979 8456238; URL: library.tamu.edu
1876; Colleen Cook
Military Hist, Oceanography
3 018 000 vols; 45 710 curr per 30857

– Cushing Library, College Station, TX 77843-5000
T: +1 979 8451951; Fax: +1 979 8451441; E-mail: cushing-library@tamu.edu; URL: library.tamu.edu/cushing
1930; Steven Escar Smith
Military hist, Science fiction, Texas hist, American illust, Nautical archaeology; Special Colls: Rudyard Kipling, P G Wodehouse, Sea Fiction, Western Fiction, Katherine Ann Porter, Military Hist, Texas Hist, Ranching Hist, Naval Architecture, Naval Science, Ships, Botanicals, Fore Edge Paintings, American Illustrators, Univ Arch, etc
180 000 vols; 56 curr per; 100 incunabula; 30 000 diss/theses; 500 music scores; 10 000 microforms; 150 000 av-mat; 125 sound-rec; 15 digital data carriers; 5 000 000 mss, 60 feet of clippings on TAMU history
ALA, ACRL 30858

– Cushing Library, Science Fiction Research Collection, College Station, TX 77843-5000
T: +1 979 8621840; E-mail: hal-hall@tamu.edu
1973; Steven Smith
Michael Moorcock Coll, George R. R. Martin Coll, Chad Oliver Coll, Otto Binder Coll
40 200 vols; 140 curr per; 200 linear feet of mss, 60 microform science fiction serials, 300 science fiction films and TV series
libr loan 30859

– Cushing Memorial Library & Archives, 5000 TAMU, College Station, TX 77843-5000
T: +1 979 8451951; Fax: +1 979 8451441; E-mail: cushing-library@tamu.edu
Dr. David L. Chapman
180 000 vols 30860

– Medical Sciences Library, University & Olsen Blvd, College Station, TX 77843-4462; mail address: 4462 TAMU, College Station, TX 77843-4462
T: +1 979 8457428; Fax: +1 979 8457493
N.N.
120 000 vols; 1 633 curr per 30861

– Medical Sciences Library, University & Olsen Blvd, 4462 TAMU, College Station, TX 77843-4462
T: +1 979 8457428; Fax: +1 979 8457493; URL: msl.tamu.edu
1940; Martha Bedard
Ethnic Medicine, Veterinary Medicine
119 700 vols; 1 633 curr per; 1 169 microforms
libr loan 30862

– Policy Sciences & Economics Library, 1016 Annenberg Presidential Conference Ctr, College Station, TX 77843-5002
T: +1 979 8623544; Fax: +1 979 8623791; E-mail: bushdesk@lib-gw.tamu.edu
Leslie J. Reynolds
8 000 vols; 80 curr per 30863

Southern Adventist University, McKee Library, 4851 Industrial Dr, ***Collegedale***, TN 37315; P.O. Box 629, Collegedale, TN 37315-0629
T: +1 423 2362794; Fax: +1 423 2361788; URL: library.southern.edu
1890; Genevieve Cottrell
Education, nursing, religion; Lincoln/Civil War mat; Seventh-Day Adventist Coll
155 000 vols; 1 047 curr per; 25 000 e-books; 270 diss/theses; 1 684 govt docs; 1 899 music scores; 643 maps; 449 000 microforms; 1 995 sound-rec; 33 digital data carriers; 225 linear feet of mss, 100 Bks on Deafness & Sign Lang
libr loan; ALA 30864

Saint John's University, Alcuin Library, Alcuin Library Bldg, ***Collegeville***, MN 56321; P.O. Box 2500, Collegeville, MN 56321
T: +1 320 3632125; Fax: +1 320 3632126; URL: www.csbsju.edu/library
1856; Kathleen Parker
Benedictine studies, Theology, Nursing
425 000 vols; 878 curr per
libr loan 30865

Ursinus College, Myrin Library, 601 E Main St, ***Collegeville***, PA 19426; P.O. Box 1000, Collegeville, PA 19426-1000
T: +1 610 4093607; Fax: +1 610 4890634; E-mail: library@ursinus.edu; URL: myrin.ursinus.edu
1870; Charles A. Jamison
Linda Grace Hoyer Updike Literary Papers, Pennsylvania German Studies Arch
375 000 vols; 1 200 curr per
libr loan 30866

Colorado College, Charles Leaming Tutt Library, 1021 N Cascade Ave, ***Colorado Springs***, CO 80903-3252
T: +1 719 3896658; Fax: +1 719 3896082; URL: www.coloradocollege.edu/library
1874; Carol Dickerson
College Coll, Colorado Coll, Rare Bks, Chess Coll, Special Editions, Lincoln Coll
501 240 vols; 1 350 curr per; 3 800 e-journals; 326 617 microforms; 22 720 av-mat; 485 digital data carriers
libr loan 30867

University of Colorado at Colorado Springs, Kraemer Family Library, 1420 Austin Bluffs Pkwy, ***Colorado Springs***, CO 80918; P.O. Box 7150, Colorado Springs, CO 80933-7150
T: +1 719 2623295; Fax: +1 719 5285227; E-mail: refdesk@uccs.edu; URL: web.uccs.edu/library
1965; Teri R. Switzer
Business, Education, Electrical eng, Psychology
383 000 vols; 546 curr per; 21 000 e-journals; 9 573 maps; 290 000 microforms; 3 177 sound-rec; 302 digital data carriers
libr loan 30868

Benedict College, Benjamin F. Payton Learning Resources Ctr Library, 1600 Harden St, ***Columbia***, SC 29204
T: +1 803 7054364; Fax: +1 803 7487539; URL: www.benedict.edu
1870; Darlene Zinnerman-Bethea
African-American studies
118 000 vols; 320 curr per 30869

Columbia College, J. Drake Edens Library, 1301 Columbia College Dr, ***Columbia***, SC 29203-9987
T: +1 803 7863878; Fax: +1 803 7863700; URL: www.columbiacollegesc.edu
1854; John C. Pritchett
Women; Religious Lit for Children, Local Authors
146 000 vols; 323 curr per; 3 190 av-mat; 940 sound-rec; 250 digital data carriers
libr loan 30870

Columbia College, Stafford Library, 1001 Rogers, ***Columbia***, MO 65216
T: +1 573 8757381; Fax: +1 573 8757379; E-mail: reference@ccis.edu; URL: www.ccis.edu/offices/library/about.asp
1989; Janet Caruthers
63 543 vols; 428 curr per; 9 516 microforms; 1 230 av-mat; 2 561 sound-rec; 10 digital data carriers
libr loan; ACRL 30871

Columbia College Library, J. W. & Lois Stafford Library, 1001 Rogers St, ***Columbia***, MO 65216
T: +1 573 8757381; Fax: +1 573 8757379; E-mail: reference@ccis.edu; URL: www.ccis.edu/offices/library/
Janet Caruthers
History, literature, education
63 000 vols; 282 curr per; 10 000 e-books 30872

Columbia International University, G. Allen Fleece Library, 7435 Monticello Rd, ***Columbia***, SC 29230-1599; P.O. Box 3122, Columbia, SC 29203-3122
T: +1 803 8075102; Fax: +1 803 7441391; E-mail: refdesk@ciu.edu; URL: www.ciu.edu/library
1923; Jo Ann Rhodes
Visual Aids for Religious Education, Missionary Curios
128 900 vols; 360 curr per; 7 550 e-books; 211 maps; 575 microforms; 20 925 av-mat; 6 748 sound-rec; 2 660 digital data carriers; 80 Electronic Media & Resources
libr loan 30873

Stephens College, Hugh Stephens Library, 1200 E Broadway, ***Columbia***, MO 62515
T: +1 573 8767181; Fax: +1 573 8767264; E-mail: chutchinson@stephens.edu; URL: www.stephens.edu/library
1833; Corrie Hutchinson
Women's studies; Educational Resources & Childrens Lit, Women's Studies Coll
125 000 vols; 360 curr per; 880 av-mat; 310 sound-rec; 66 Journals
libr loan 30874

University of Missouri-Columbia – Elmer Ellis Library, Ellis Library Bldg, Rm 104, ***Columbia***, MO 65201-5149
T: +1 573 8824581; Fax: +1 573 8828044; E-mail: ellisref@missouri.edu; URL: mulibraries.missouri.edu
1839; James A. Cogswell
American Best Sellers (Frank Luther Mott Coll), Philosophy (Thomas Moore Johnson Coll), Rare Book Coll, World War I & II Poster Coll, Cartoons (John Tinney McCutcheon Coll), Fourth of July Oration Coll, Univ of Missouri Coll, Italian Lit (Anthony C DeBellis Coll), William Peden Short Story Coll, Comic Art Coll, Mary Lago Coll (Edwardian Lit)
692 000 vols; 26 886 curr per; 1 669 853 govt docs; 6 738 000 microforms; 5 571 av-mat; 2 451 Audio Bks
libr loan 30875

– Engineering Library & Technology Commons, W2001 Lafferre Hall, Columbia, MO 65211
T: +1 573 8822379; Fax: +1 573 8844499; URL: mulibraries.missouri.edu/engr
1906; Judy Siebert Maseles
50 000 vols; 736 curr per; 500 e-journals; 4 000 e-books; 3 digital data carriers; 710 CDs 30876

– Geology Library, 201 Geological Sciences, Columbia, MO 65211
T: +1 573 8824860; Fax: +1 573 8825458; URL: mulibraries.missouri.edu/geology
1875; Stephen Stanton
19th C State Geological Survey Publs; 19th C Federal Survey Publs
69 000 vols; 663 curr per; 200 000 maps; 15 000 microforms; 60 av-mat; 200 sound-rec; 298 digital data carriers 30877

– Journalism Library, 27 Neff Annex, Columbia, MO 65211
T: +1 573 8823224; Fax: +1 573 8844963; URL: mulibraries.missouri.edu/journalism
1908; Dorothy Carner
Newspaper Libr
50 000 vols; 200 curr per; 500 diss/theses; 20 digital data carriers
libr loan 30878

– Law Library, 203 Hulston Hall, Columbia, MO 65211-4190
T: +1 573 8824597; Fax: +1 573 8829676; URL: www.law.missouri.edu/library
1872; Randy Diamond
19th Century Trials (John D. Lawson Coll)
343 000 vols; 1 700 curr per; 415 000 microforms; 17 digital data carriers
libr loan 30879

– Mathematical Sciences Library, 206 Math Sciences Bldg, Columbia, MO 65211
T: +1 573 8823224; Fax: +1 573 8840058; URL: mulibraries.missouri.edu/math
1969; William Christopher McCrary III
26 000 vols; 258 curr per
libr loan 30880

– J. Otto Lottes Health Sciences Library, One Hospital Dr, Columbia, MO 65212
T: +1 573 8820467; Fax: +1 573 8825574; URL: www.muhealth.org/~library
1903; Deborah Ward
227 530 vols; 1 750 curr per; 230 microforms; 400 av-mat; 500 sound-rec; 2 000 slides 30881

– Veterinary Medical Library, W-218 Veterinary-Medical Bldg, Columbia, MO 65211
T: +1 573 8822461; Fax: +1 573 8822950; E-mail: vetlib@missouri.edu
1951; Trenton Boyd
50 896 vols; 290 curr per; 747 microforms; 1 000 av-mat; 6 digital data carriers
libr loan 30882

– Western Historical Manuscript Collection – Columbia, 23 Ellis Library, Columbia, MO 65201-5149
T: +1 573 8826028; Fax: +1 573 8840345; E-mail: whmc@umsystem.edu; URL: www.umsystem.edu/whmc
David Moore
19 000 vols; 14 curr per 30883

University of South Carolina, Thomas Cooper Library, 1322 Green St, ***Columbia***, SC 29208-0103
T: +1 803 7774866; Fax: +1 803 7774661; URL: www.sc.edu/library
1801; Paul A. Willis
American & British Hist & Lit Colls, Voyages, Scottish Lit, Early Geology, 19th C Italy, 18th C Botany & Natural Hist, Ornithology, Hist of Books & Printing, Archaeology
3 374 000 vols; 22 844 curr per; 225 174 maps; 5 800 000 microforms; 28 979 sound-rec; 1 910 digital data carriers
libr loan 30884

– Coleman Karesh Law Library, USC Law Ctr, 701 Main St, Columbia, SC 29208
T: +1 803 7775944; Fax: +1 803 7779405; URL: www.law.sc.edu
1923; Duncan Alford
South Carolina Legal Hist Coll
537 000 vols; 3 349 curr per
libr loan 30885

– Elliot White Springs Business Library, Francis M Hipp-William H Close Bldg, 1705 College St, Columbia, SC 29208
T: +1 803 7776032; Fax: +1 803 7776876; E-mail: gardnerd@gwm.sc.edu; URL: www.sc.edu/library/pubserv/business.html
1973; Dwight Gardner
Japan Business Coll
36 000 vols; 350 curr per 30886

– Mathematics Library, LeConte College 3rd Flr, Columbia, SC 29208
T: +1 803 7774741; URL: www.sc.edu/library/math.html
Danley Reed
25 000 vols; 150 curr per 30887

– Music Library, 813 Assembly St, Columbia, SC 29208
T: +1 803 7775139; Fax: +1 803 7771426; URL: www.sc.edu/library/music/index.html
Jennifer Ottervik
90 000 vols; 135 curr per 30888

– School of Medicine Library, 6311 Garners Ferry Rd, Columbia, SC 29208
T: +1 803 7333361; Fax: +1 803 7331509; URL: uscm.med.sc.edu
Ruth Riley
Disabilities
103 000 vols; 700 curr per 30889

– South Caroliniana Library, Columbia, SC 29208-0103
T: +1 803 7773131; Fax: +1 803 7775747
Alan Stokes
101 000 vols; 325 curr per 30890

Beacon University Library, 6003 Veterans Pkwy, ***Columbus***, GA 31909
T: +1 706 3235364; Fax: +1 706 3235891; URL: www.beacon.edu
Joann Lessner
77 000 vols; 116 curr per 30891

Capital University, Blackmore Library, One College & Main, ***Columbus***, OH 43209-2394
T: +1 614 2366351; Fax: +1 614 2366490; E-mail: refdesk@capital.edu; URL: www.capital.edu/library/
1876; Belen C. Fernandez
Juvenile Lit (Lois Lenski Coll), Arch (Univ Arch)
199 000 vols; 504 curr per; 5 650 e-journals; 17 000 e-books; 1 870 Electronic Media & Resources
libr loan 30892

– Law School Library, 303 E Broad St, Columbus, OH 43215
T: +1 614 2366464; Fax: +1 614 2366957; URL: www.law.capital.edu
1903; Donald A. Hughes Jr
258 000 vols; 2 411 curr per; 52 000 microforms
libr loan 30893

Columbus College of Art & Design, Packard Library, 107 N Ninth St, ***Columbus***, OH 43215-3875
T: +1 614 2223273; Fax: +1 614 2226193; URL: www.ccad.edu
1931; Chilin Yu
50 000 vols; 275 curr per; 50 400 slides, 35 500 pictures & prints 30894

Columbus State University, Simon Schwob Memorial Library, 4225 University Ave, ***Columbus***, GA 31907
T: +1 706 5682451; Fax: +1 706 5682084; URL: library.colstate.edu
1961; Callie B. McGinnis
CSU Arch, Chattahoochee Valley Hist Colls, US Representative J Brinkley Coll
376 000 vols; 1 617 curr per; 807 000 microforms; 632 mss
libr loan 30895

Indiana University-Purdue University, Columbus Campus Library, 4555 Central Ave, LC 1600, ***Columbus***, IN 47203
T: +1 812 3148708; Fax: +1 812 3148722; E-mail: colref-l@iupui.edu; URL: www.iupui.edu/library; www.ivytech.edu/library/columbus/index.html
Steven J. Schmidt
40 000 vols; 200 curr per 30896

Mississippi University for Women, John Clayton Fant Memorial Library, P.O. Box W1625, ***Columbus***, MS 39701
T: +1 601 329-7332; Fax: +1 601 329-7348; E-mail: cyoung@muw.edu; URL: www.muw.edu/library
1884
George Eliot First Editions & Criticisms; George Eliot (Blanche Colton Williams' Biography of George Eliot), original mss; Univ hist, bks, micro; Mississippiana; State & Local Histroy (General E T

Sykes Scrapbook Coll), clippings
233 403 vols; 1 619 curr per; 295 009 microforms; 7 digital data carriers
libr loan 30897

Ohio Dominican University Library, 1216 Sunbury Rd, ***Columbus***, OH 43219
T: +1 614 2514637; Fax: +1 614 2522650; E-mail: library@ohiodominican.edu; URL: www.ohiodominican.edu/library/
1924; James E. Layden
Humanities – Theology, Philosophy; Anne O'Hara McCormick Coll, Mary Tetter Zimmerman Coll
105 000 vols; 600 curr per; 6 000 microforms; 12 digital data carriers 30898

Ohio State University, University Libraries, 1858 Neil Ave Mall, ***Columbus***, OH 43210-1286
T: +1 614 2926785; Fax: +1 614 2927859; URL: library.osu.edu/
1873; Joseph Branin
Japanese Coll, Chinese Coll, Rare Books & Mss, Author Colls
5 000 000 vols; 36 020 curr per; 174 000 maps; 4 410 253 microforms of per
libr loan
Currently closed for renovation, essential library services offered by the Sullivant Library, Mat previously stored in the stacks tower at Thompson have been relocated to the Ackerman Library.) Sullivant Library is the new home for current periodicals and newspapers, videos and DVDs, reference books and non-circulating bound journals in the humanities and social sciences 30899

– Biological Sciences & Pharmacy Library, 102 Riffe Bldg, 496 W 12th Ave, Columbus, OH 43210-1214
T: +1 614 2921744; Fax: +1 614 6883123; URL: library.osu.edu/sites/biosci
1994; Natalie Kupferberg
120 000 vols; 1 300 curr per 30900

– Business Library, Raymond E Mason Hall, 250 W Woodruff Ave, Columbus, OH 43210-1395
T: +1 614 2922136; Fax: +1 614 2925559; URL: fisher.osu.edu/library
1925; Charles Popovich
Annual Rpts of Corporations
145 000 vols; 900 curr per; 28 000 microforms 30901

– Cartoon Research Library, 27 W 17th Ave Mall, Columbus, OH 43210-1393
T: +1 614 2920538; Fax: +1 614 2929101; E-mail: cartoons@osu.edu; URL: cartoons.osu.edu
1977; Lucy Shelton Caswell
15 000 vols; 660 curr per; 2 500 mss; 365 000 graphic mat incl 230 000 original cartoons 30902

– Edgar Dale Educational Media & Instructional Materials Laboratory, Library, 260 Ramseyer Hall, 29 W Woodruff Ave, Columbus, OH 43210-1177
T: +1 614 2921177; Fax: +1 614 2927900; E-mail: freeman.5@osu.edu
1978; Dr. Evelyn Freeman
Hist Children's Lit Coll
26 800 vols; 2 000 curr per
OCLC, OHIONET 30903

– English, Theatre & Communication Reading Room, 1858 Neil Ave Mall, Main Library, Rm 200N, Columbus, OH 43210-1286
T: +1 614 2922786; Fax: +1 614 2927859
James Bracken
20 000 vols; 500 curr per 30904

– Fine Arts Library, Wexner Center for the Arts, 1871 N High St, Columbus, OH 43210
T: +1 614 2926184; Fax: +1 614 2924573; URL: library.osu.edu/sites/finearts
1948; Gretchen Donelson
140 000 vols; 400 curr per 30905

– Food, Agricultural & Environmental Sciences Library, 045 Agriculture Administration Bldg, 2120 Fyffe Rd, Columbus, OH 43210-1066
T: +1 614 2929563; Fax: +1 614 2920590
1956; Eboni A. Francis
Agriculture, agronomy, animal sciences, food science and technology, plant pathology, rural sociology, natural resources
89 000 vols; 1 138 curr per; 1 105 diss/theses; 3 000 govt docs; 7 000 microforms; 242 digital data carriers
ALA 30906

– Grant Morrow III MD Library at Children's Hospital, 700 Children's Dr, Rm ED-244, Columbus, OH 43205
T: +1 614 7223200; Fax: +1 614 7223205; URL: library.osu.edu/sites/chi
1953; Linda DeMuro
Pediatrics; Consumer Health in Pediatrics
25 000 vols; 225 curr per; 1 digital data carriers 30907

– John A. Prior Health Sciences Library, 376 W Tenth Ave, Columbus, OH 43210-1240
T: +1 614 2924861; Fax: +1 614 2921920; E-mail: info@library.med.ohio-state.edu; URL: library.med.ohio-state.edu
1849; Susan M. Kroll
Medical Heritage Ctr
216 000 vols; 19 106 curr per; 953 e-books; 1 000 govt docs; 25 000 microforms; 7 digital data carriers
libr loan 30908

– Journalism Library, 100 Journalism Bldg, 242 W 18th Ave, Columbus, OH 43210-1107
T: +1 614 2928747; Fax: +1 614 2476363; E-mail: block.3@osu.edu; URL: www.lib.ohio-state/osu_profile/jouweb
1967; Eleanor Block
33 262 vols; 350 curr per 30909

– Music & Dance Library, 186 Sullivant Hall, 1813 N High St, Columbus, OH 43210-1307
T: +1 614 6880106; Fax: +1 614 2476794; URL: library.osu.edu/sites/music
1947; Alan Green
American Popular Songs; Dance, V-tapes; Medieval Chant, microfilm; Nordic Music Arch; Renaissance Music, microfilms
140 000 vols; 615 curr per; 26 000 sound-rec; 80 digital videodiscs
libr loan; MLA, IAML 30910

– Orton Memorial Library of Geology, 180 Orton Hall, 155 S Oval Mall, Columbus, OH 43210
T: +1 614 2922428; Fax: +1 614 2921496; URL: library.osu.edu/sites/geology
1923; Mary Woods Scott
111 000 vols; 651 curr per; 125 657 maps 30911

– Science & Engineering Library, 175 W 18th Ave, Columbus, OH 43210
T: +1 614 2923022; Fax: +1 614 2923062; URL: library.osu.edu/sites/sel
1993; Martin Jamison
Coll of US Patents
370 000 vols; 2 600 curr per; 632 digital data carriers
libr loan 30912

– Sullivant Library, 1813 N High St, 110 Sullivant Hall, Columbus, OH 43210
T: +1 614 2922075; Fax: +1 614 2928012; E-mail: greenberg.3@osu.edu; URL: library.osu.edu/sites/education/
1998; Gerry Greenberg, Deidra Herring
Education, Psychology, Human Ecology, Social Work, Sports and Recreation, Physical Education; Special colls: ERIC docs on microfiche (459 309), Histc Test Coll, Mental Health Advocacy papers (38 inches)
680 685 vols; 2 400 curr per; 18 000 diss/theses; 90 666 microforms
libr loan; ARL 30913

– Veterinary Medicine Library, 225 Veterinary Medicine Academic Bldg" Columbus, OH 43210
T: +1 614 2926107; Fax: +1 614 2927476; URL: library.osu.edu/sites/vetmed
1929; Sarah A. Murphy
40 000 vols; 630 curr per
libr loan 30914

The Ohio State University – Moritz Law Library, 55 W 12th Ave, ***Columbus***, OH 43210-1391
T: +1 614 2926691; Fax: +1 614 2923202; E-mail: lawlibref@osu.edu; URL: www.moritzlaw.osu.edu/library/
1891; Bruce S. Johnson
Dispute resolution; Ohio Legal Mat
790 000 vols; 7 192 curr per; 947 000 microforms; 14 digital data carriers 30915

Texas A&M University – Commerce, James Gilliam Gee Library, 2600 S Neal St, ***Commerce***, TX 75429; P.O. Box 3011, Commerce, TX 75429-3011
T: +1 903 8865741; Fax: +1 903 8865723; URL: www7.tamu-commerce.edu/library
1894; Dr. Paul Zelhart
Education (Curriculum Libr), Foreign Diplomatic Service (Ambassador Fletcher Warren Papers), Texas Lit (Elithe Hamilton Kirkland Paper), Texas Poetry (Faye Carr Adams Coll, Texas Political Hist
1 090 000 vols; 1 667 curr per; 22 000 e-journals; 43 000 e-books; 332 685 govt docs; 126 maps; 1 162 000 microforms; 4 522 av-mat; 5 336 sound-rec; 206 digital data carriers
libr loan 30916

Barber Scotia College, Sage Memorial Library, 145 Cabarrus Ave W, ***Concord***, NC 28025
T: +1 704 7892953; Fax: +1 704 7892955; E-mail: sagememorial@vnet.net; URL: www.b-sc.edu
1867; Minora Hicks
African-American art
49 000 vols; 102 curr per 30917

Central Baptist College, J.E. Cobb Library, 1501 College Ave, ***Conway***, AR 72034
T: +1 501 3296872; Fax: +1 501 3292941; URL: www.cbc.edu
1952; Anne Clements
Bible, Church hist, Theology; Baptist Missionary Assn of Arkansas Hist
54 000 vols; 310 curr per 30918

Costal Carolina University, Kimbel Library, 755 Hwy 544, ***Conway***, SC 29526; P.O. Box 261954, Conway, SC 29528-6054
T: +1 843 3492400; Fax: +1 843 3492412; URL: www.coastal.edu/library
1954; Michael Lackey
Marine Science Coll
150 000 vols; 500 curr per 30919

Hendrix College, Olin C. Bailey Library, 1600 Washington Ave, ***Conway***, AR 72032
T: +1 501 4501289; Fax: +1 501 4503800; URL: www.hendrix.edu/baileylibrary
1876; Amanda Moore
Arkansas Methodism, Arkansasiana
180 000 vols; 845 curr per
libr loan 30920

University of Central Arkansas, Torreyson Library, 201 Donaghey Ave, ***Conway***, AR 72035
T: +1 501 4505201; Fax: +1 501 4505208; URL: library.uca.edu
1907; Art A. Lichtenstein
Arkansas; Children's Lit (Laboratory Coll), Govt Docs
432 000 vols; 1 057 curr per; 525 000 microforms; 18 digital data carriers; 14 241 Electronic Media & Resources
libr loan 30921

Tennessee Technological University, Volpe Library, 1100 N Peachtree Ave, ***Cookeville***, TN 38505; TTU Campus Box 5066, Cookeville, TN 38505
T: +1 931 3723710; Fax: +1 931 3726112; URL: www.tntech.edu/library
1915; Dr. Winston A. Walden
Engineering; Harding Studio Coll, Joe L Evins Coll, Tennessee Hist Coll, Upper Cumberland Hist Coll
350 000 vols; 1 928 curr per; 68 000 e-books; 990 000 microforms; 1 120 sound-rec; 14 digital data carriers; 140 431 Gov Docs, 21 500 Electronic Media & Resources
libr loan 30922

Central Arizona College, Signal Peak Library, 8470 N Overfield Rd, ***Coolidge***, AZ 85228
T: +1 520 4945286; Fax: +1 520 4945284; URL: www.centralaz.edu/library
1969; Jeffrey Middleton
140 000 vols; 200 curr per 30923

University of Miami, Otto G. Richter Library, 1300 Memorial Dr, ***Coral Gables***, FL 33146; P.O. Box 248214, Coral Gables, FL 33124-0320
T: +1 305 2846102; Fax: +1 305 2844027; URL: www.library.miami.edu
1926; William D. Walker
Univ Hist, Florida, Caribbean Region, Latin America, American Lit, Cuban Heritage Coll, Marine & Atmospheric Sciences
2 600 000 vols; 17 500 curr per
libr loan 30924

– Law Library, 1311 Miller Dr, Coral Gables, FL 33146; P.O. Box 248087, Coral Gables, FL 33124-0247
T: +1 305 2843563; Fax: +1 305 2843554; URL: library.law.miami.edu
1928; Sally H. Wise
421 000 vols; 7 911 curr per; 775 000 microforms; 8 digital data carriers 30925

Texas A&M University – Corpus Christi, Mary & Jeff Bell Library, 6300 Ocean Dr, ***Corpus Christi***, TX 78412-5501
T: +1 361 8252643; Fax: +1 361 8255973; URL: rattler.tamucc.edu
1973; Christine Shupala
Texas Southwest Coll, 19th C Maps & Land Title Papers of South Texas & Northern Mexico, Texas Legislature, Univ Hist, Mexico Archs
433 000 vols; 1 706 curr per
libr loan 30926

State University of New York College at Cortland, Memorial Library, 81 Prospect Terrace, ***Cortland***, NY 13045; P.O. Box 2000, Cortland, NY 13045
T: +1 607 7532526; Fax: +1 607 7535669; E-mail: library@cortland.edu; URL: library.cortland.edu
1868; Gail Wood
Education, Health education, Recreation; Teaching Mat Ctr, College Arch, Rare Bk Coll
417 000 vols; 991 curr per; 96 e-journals; 2 428 maps; 855 000 microforms; 255 sound-rec; 101 digital data carriers; 10 000 pamphlets, 17 000 pictures, 50 VF drawers
libr loan 30927

Oregon State University, The Valley Library, 121 The Valley Library, ***Corvallis***, OR 97331-4501
T: +1 541 7374488; Fax: +1 541 7378224; E-mail: library.web@oregonstate.edu; URL: osulibrary.oregonstate.edu/
1887; Karyle S. Butcher
Northwest Coll, Atomic Energy Coll, Ava Helen & Linus Pauling Papers
1 530 000 vols; 4 556 curr per; 475 655 govt docs; 176 837 maps; 1 827 000 microforms; 727 av-mat; 380 digital data carriers; 58 454 photos, Pictures, Prints
libr loan 30928

Vanguard University of Southern California, O. Cope Budge Library, 55 Fair Dr, ***Costa Mesa***, CA 92626
T: +1 714 5563610; Fax: +1 714 9665478
1920; Alison English
Christian Religion & Judaic Coll, Pentecostal Coll, Drama Coll
180 000 vols; 800 curr per; 37 000 e-books
libr loan 30929

Whittier College, School of Law Library, 3333 Harbor Blvd, ***Costa Mesa***, CA 92626
T: +1 714 4444141; Fax: +1 714 4443609; E-mail: ill@law.whittier.edu; URL: wollfpac.law.whittier.edu
1976; J. Denny Haythorn
360 000 vols; 4 700 curr per
libr loan; SLA 30930

Tulane University, Tulane Regional Primate Research Center Library, 18703 Three Rivers Rd, *Covington*, LA 70433-8915
T: +1 985 8922040 ext 6366; Fax: +1 985 8931352; E-mail: sharon@tpc.tulane.edu; URL: www.tnprc.tulane.edu/index.shtml
1963; Sharon Nastasi
Immunology, Urology
9 540 vols; 50 curr per; 175 diss/theses; 175 microforms
libr loan 30931

Wabash College, Lilly Library, PO Box 352, *Crawfordsville*, IN 47933
T: +1 765 3616376; Fax: +1 765 3616295; URL: www.wabash.edu/library
1832; John Lamborn
438 000 vols; 1 010 curr per; 5 000 e-books; 5 740 Electronic Media & Resources
libr loan 30932

Mount Aloysius College, Library, 7373 Admiral Peary Hwy, *Cresson*, PA 16630-1999
T: +1 814 8866445; Fax: +1 814 8865767; URL: www.mtaloy.edu
1939; Robert Stere
Law Libr, Ecumenical Studies Coll
80 000 vols; 275 curr per 30933

Thomas More College, Library, 333 Thomas More Pkwy, *Crestview Hills*, KY 41017-2599
T: +1 859 3443300; Fax: +1 859 3443342; E-mail: reference@thomasmore.edu; URL: library.thomasmore.edu
1921; James McKellogg
Thomas More Coll
113 000 vols; 601 curr per
libr loan 30934

Doane College, Perkins Library, 1014 Boswell Ave, *Crete*, NE 68333-2421
T: +1 402 8268287; Fax: +1 402 8268199; URL: www.doane.edu
1872; Peggy Brooks Smith
Doane College Arch Coll, Rossman Historiograpy; United Church of Christ Coll, Rall Art Gallery
96 000 vols; 515 curr per; 6 000 e-books 30935

University of Minnesota – Crookston, Media Resources – Kiehle Library, 2900 University Ave, *Crookston*, MN 56716
T: +1 218 2818399; Fax: +1 218 2818080; E-mail: umclib@umn.edu; URL: library.umcrookston.edu
1966; Owen Williams
Agriculture, Horsemanship, Business, Foods, Hospitality, Hotel Management; Equine Resource Ctr
31 785 vols; 766 curr per; 25 006 microforms; 1 900 av-mat
libr loan 30936

Western Carolina University, Hunter Library, 176 Central Dr, *Cullowhee*, NC 28723
T: +1 828 2277465; Fax: +1 828 2277015; URL: www.wcu.edu/library
1922; Dana Sally
Appalachia, Cherokee Indians, Spider Coll, Southern Highlands, Map Room
702 000 vols; 3 330 curr per; 23 000 e-books 30937

North Georgia College & State University, Stewart Library, 238 Georgia Circle, *Dahlonega*, GA 30597-3001
T: +1 706 8641520; Fax: +1 706 8641867; E-mail: refdesk@ngcsu.edu; URL: www.ngcsu.edu
1873; Shawn Tonner
Military hist; Educational Resources Info Ctr Coll
144 000 vols; 500 curr per
libr loan 30938

The Art Institute of Dallas, Mildred M. Kelley Library, Two North Park E, 8080 Park Lane, Ste 100, *Dallas*, TX 75231-5993
T: +1 214 6928080; Fax: +1 214 6928106; URL: www.aidlrc.aiiresources.com
Lisa Casto
Interior design, graphic design, computer art
24 000 vols; 264 curr per 30939

College Misericordia, Francesca McLaughlin Memorial Library, 301 Lake St, *Dallas*, PA 18612-1098
T: +1 570 6746231; Fax: +1 570 6746342; E-mail: library@misericordia.edu; URL: www.misericordia.edu/library
1924; Barbara Burd
Education, Occupational therapy, Nursing; ANA & NLN Coll, publs currently in print
75 000 vols; 597 curr per
libr loan 30940

Criswell College, Wallace Library, 4010 Gaston Ave, *Dallas*, TX 75246
T: +1 214 8181378; Fax: +1 214 8181310; URL: www.criswell.edu
1976; Andrew Streett
Baptist Hist & Theology, 17th-18th C Religions Bks & Tracts
90 000 vols; 428 curr per; 4 000 microforms; 2 000 sound-rec 30941

Dallas Baptist University, Vance Memorial Library, 3000 Mountain Creek Pkwy, *Dallas*, TX 75211-9299
T: +1 214 3335221; Fax: +1 214 3335323; E-mail: lib_ref@dbu.edu; URL: www3.dbu.edu/library/
1898; Debra Collins
ERIC, Libr of American Lit, Libr of English Lit
292 846 vols; 1 119 curr per; 48 641 e-books; 520 430 microforms; 1 940 av-mat; 2 361 sound-rec; 448 digital data carriers
libr loan 30942

Dallas Christian College, C.C. Crawford Memorial Library, 2700 Christian Pkwy, *Dallas*, TX 75234
T: +1 972 2413371; Fax: +1 972 2418021; E-mail: library@dallas.edu; URL: www.dallas.edu/academics/library.cfm
Susan Springer
Biblical studies, Religion, Theology; Hist & Writings of the Restoration Movement
30 000 vols; 425 curr per; 8 000 e-books 30943

Paul Quinn College, Zale Library, 3837 Simpson Stuart Rd, *Dallas*, TX 75241
T: +1 214 3023565; Fax: +1 214 3715889; URL: www.pqc.edu
Clarice Weeks
Afro-American Ethnic & Cultural Coll, AME Church Arch, College Arch
80 000 vols; 163 curr per 30944

Southern Methodist University, Central University Libraries, 6414 Hilltop, University Park, *Dallas*, TX 75205; P.O. Box 750135, Dallas, TX 75275-0135
T: +1 214 7683229; Fax: +1 214 7683815; URL: www.smu.edu/cul
1915; Gillian M. McCombs
2 100 000 vols; 8 614 curr per; 2 417 e-journals; 4 725 mss; 682 603 govt docs; 44 132 music scores; 223 303 maps; 619 000 microforms; 10 762 av-mat; 26 589 sound-rec; 1 800 digital data carriers 30945

– DeGolyer Library of Special Collections, 6404 Hilltop Dr, Dallas, TX 75205; P.O. Box 750396, Dallas, TX 75275
T: +1 214 7683231; Fax: +1 214 7681565; E-mail: degolyer@mail.smu.edu; URL: www.smu.edu/~cul/degolyer
1956; Russell Martin
Travel, Voyages; Western US Hist, Hist of the Spanish Borderlands, Hist of Railroads
100 000 vols; 400 curr per; 3 000 microforms; 4 500 cubic feet of mss & archival colls 30946

– Fondren Library Center, 6414 Hilltop, Dallas, TX 75275; P.O. Box 750135, Dallas, TX 75275-0135
T: +1 214 7683815; Fax: +1 214 7683815; URL: www.smu.edu/cul/flc
Dr. Gillian McCombs
1 200 000 vols; 8 000 curr per 30947

– Hamon Arts Library, 6101 Bishop Blvd, Dallas, TX 75275; P.O. Box 750356, Dallas, TX 75275-0356
T: +1 214 7682894; Fax: +1 214 7681800; URL: www.smu.edu/cul/hamon
Alisa Rata Stutzbach
152 000 vols; 300 curr per 30948

– Institute for Study of Earth & Man Reading Room, N L Heroy Science Hall, Rm 129, 3225 Daniels Ave, Dallas, TX 75275; P.O. Box 750274, Dallas, TX 75275-0274
T: +1 214 7682430; Fax: +1 214 7684289; URL: www.smu.edu
John Phinney
10 000 vols; 50 curr per 30949

– Science-Engineering Library, P.O. Box 750375, Dallas, TX 75275-0375
T: +1 214 7682444; Fax: +1 214 7684236; E-mail: dbicksto@mail.smu.edu; URL: www.smu.edu/cul
1961; Deverett D. Bickston
Earth Sci (DeGolyer & MacNaughton Coll), Maps (Foscoe Coll), Arthur Collins Telecommunications Coll
260 000 vols; 882 curr per; 277 832 govt docs; 213 591 maps 30950

– Underwood Law Library, 6550 Hillcrest Ave, P.O. Box 750354, Dallas, TX 75275-0354
T: +1 214 7683216; Fax: +1 214 7684330; E-mail: gdaly@mail.smu.edu; URL: library.law.smu.edu
1925; Gail M. Daly
Joseph Gold Collection, Dallas (Texas) Independent School District Desegregation Litigation archive
381 561 vols; 4 546 curr per; 2 incunabula; 112 786 diss/theses; 141 028 microforms; 687 av-mat; 3 835 sound-rec
libr loan 30951

Southern Methodist University – Bridwell Library-Perkins School of Theology, 6005 Bishop Blvd, *Dallas*, TX 75205; P.O. Box 750476, Dallas, TX 75275-0476
T: +1 214 7683483; Fax: +1 214 7684295; E-mail: bridadmin@mail.smu.edu; URL: www.smu.edu/bridwell
1915; Dr. James McMillin
Methodist hist, Theology, Cultural hist; Bibles, Fine Bindings, Fine Printing & Private Press, Methodism, 15th C Printing, Reformation, Savonarola Coll
375 000 vols; 1 600 curr per; 129 000 microforms; 12 digital data carriers
libr loan 30952

Texas Woman's University, F.W. & Bessie Dye Memorial Library, 1810 Inwood Rd, *Dallas*, TX 75235-7299
T: +1 214 6896580; Fax: +1 214 6896583; URL: www.twu.edu/library
1966; Oliphant Eula
Health care administration, Nursing, Occupational therapy, Psychology, Physical therapy
12 000 vols; 160 curr per; 225 diss/theses; 12 000 electronic books 30953

University of Texas Southwestern Medical Center at Dallas Library, 5323 Harry Hines Blvd, *Dallas*, TX 75390-9049
T: +1 214 6482001; Fax: +1 214 6482826; URL: www.utsouthwestern.edu/library
1943; Brian Bunnett
Hist of Health Sciences Coll
85 000 vols
libr loan 30954

Western Connecticut State University, Ruth A. Haas Library, 181 White St, *Danbury*, CT 06810
T: +1 203 8379100; Fax: +1 203 8379108; URL: www.wcsu.edu/library
1905; Ralph Holibaugh
Local hist; Fairfield County & Connecticut Hist (Connecticut Room), Instructional Media Ctr, Gov Docs, Young Business Coll, Music Education
200 000 vols; 1 020 curr per; 4 800 music scores; 400 000 microforms; 8 700 av-mat; 7 digital data carriers; 8 700 Audio Bks
libr loan 30955

Averett University, Mary B. Blount Library, 344 W Main St, *Danville*, VA 24541-2849
T: +1 434 7915690; Fax: +1 434 7915637; E-mail: aclib@averett.edu; URL: www.averett.edu/library
1859; Elaine L. Day
Averett Univ Arch, Dan Daniel Arch, Danville Coll
152 000 vols; 11 438 curr per; 45 000 e-books 30956

Centre College of Kentucky, Grace Doherty Library, 600 W Walnut St, *Danville*, KY 40422
T: +1 859 2385275; Fax: +1 859 2367925; URL: www.centre.edu/web/library/homepage.html
1819; Stanley R. Campbell
Dante Coll, LeCompte Davis Coll, Kentucky Coll, Centre College Arch
290 000 vols; 750 curr per; 17 874 e-journals; 31 000 e-books; 3 340 CDs
libr loan 30957

Saint Ambrose University, O'Keefe Library, 518 W Locust St, *Davenport*, IA 52803
T: +1 563 3336246; Fax: +1 563 3336248; URL: library.sau.edu
1882; Mary B. Heinzman
Imprints Coll, Univ Hist Mat Coll, Catholic Messenger
153 000 vols; 671 curr per; 13 000 e-books
libr loan 30958

Davidson College, E. H. Little Library, 209 Ridge Rd, *Davidson*, NC 28036-0001; P.O. Box 7200, Davidson, NC 28035-7200
T: +1 704 8942159; Fax: +1 704 8942625; E-mail: referencedesk@davidson.edu; URL: www.davidson.edu
1837; Jill Gremmels
Robert Burns Coll, W. P. Cumming Map Coll, Davidsoniana Coll, Mecklenburg Declaration of Independence Coll, Peter S. Ney Coll, Bruce Rogers Coll, Woodrow Wilson Coll
637 000 vols; 3 400 Electronic Media & Resources
libr loan 30959

Broward Community College, University-College Library, 3501 SW Davie Rd, *Davie*, FL 33314
T: +1 954 4756648; Fax: +1 954 4236490; URL: ucl.broward.cc.fl.us
1960; Miguel Menendez
228 120 vols; 1 170 curr per; 3 938 av-mat 30960

University of California, Davis – General Library, 100 NW Quad, *Davis*, CA 95616-5292
T: +1 530 7521202; Fax: +1 530 7526899; E-mail: libraryinfo@ucdavis.edu; URL: www.lib.ucdavis.edu
1908; Marilyn J. Sharrow
Rare Books, Women's Hist, Mss, Univ Arch, Agricultural Technology, Apiculture, Californian Art, Ecology, Enology & Viticulture, German Lit, Food Industry & Technology, British Coll
3 354 000 vols; 50 442 curr per; 235 347 maps; 3 100 000 microforms; 12 670 sound-rec; 2 482 digital data carriers
libr loan 30961

– Agricultural & Resource Economics Library, One Shields Ave, Davis, CA 95616-8512
T: +1 530 7521540; Fax: +1 530 7525614; E-mail: arel@ucdavis.edu; URL: arelibrary.ucdavis.edu
1951; Barbara Hegenbart
9 000 vols; 763 curr per; 263 376 pamphlets
ALA 30962

– Law Library, 400 Mrak Hall Dr, Davis, CA 95616
T: +1 530 7523327; Fax: +1 530 7528766; E-mail: lawlibref@ucdavis.edu; URL: law.ucdavis.edu/library
Judy Janes
295 000 vols; 4 662 curr per 30963

– Loren D. Carlson Health Sciences Library, Med Sci B, One Shields Ave, Davis, CA 95616-5291
T: +1 530 7526379; Fax: +1 530 7524718; E-mail: hslref@ucdavis.edu; URL: www.lib.ucdavis.edu/dept/hsl
1966; Terri Malmgren
Veterinary Hist Coll
326 000 vols; 462 curr per; 13 000 microforms; 71 av-mat; 130 sound-rec; 6 digital data carriers; 11 slides
libr loan 30964

– Physical Sciences & Engineering Library, One Shields Ave, Davis, CA 95616-8676
T: +1 530 7520459; Fax: +1 530 7524719; E-mail: pse@ucdavis.edu; URL: www.lib.ucdavis.edu/dept/psel
1971; Karen Andrews
Dept of Energy, National Aeronautics & Space Administration, microfiche rpts coll
353 000 vols; 3 083 curr per; 14 198 maps; 1 200 000 microforms; 1 937 pamphlets 30965

Bryan College, Library, 585 Bryan Dr, ***Dayton***, TN 37321; P.O. Box 7000, Dayton, TN 37321-7000
T: +1 423 7757307; Fax: +1 423 7757309; URL: www.bryan.edu
1930; Laura Kaufmann
113 000 vols 30966

University of Dayton, Roesch Library, 300 College Park Dr, ***Dayton***, OH 45469-1360
T: +1 937 2294221; Fax: +1 937 2294215; E-mail: ref@udayton.edu; URL: library.udayton.edu
1850; Kathleen Marie Webb
Congressman Charles W. Whalen, Jr Coll, public papers; Science Fiction Writers of America Coll
649 000 vols; 9 190 curr per
libr loan 30967

– Marian Library, 300 College Park Dr, Dayton, OH 45469-1390
T: +1 937 2294214; Fax: +1 937 2294258
1943; Brother Thomas A. Thompson
Clugnet Coll (Marian Shrines), Religious Art
100 000 vols; 160 curr per; 4 mss; 30 av-mat; 560 sound-rec; 53 300 clippings, 2 200 slides, 5 000 pictures, 10 000 holy cards 30968

University of Dayton School of Law, Zimmerman Law Library, 300 College Park, ***Dayton***, OH 45469-2780
T: +1 937 2292314; Fax: +1 937 2292555; URL: law.udayton.edu
1974; Thomas L. Hanley
319 000 vols; 4 778 curr per; 640 000 microforms; 50 digital data carriers 30969

Wright State University, University Libraries, 126 Dunbar Library, 3640 Colonel Glenn Hwy, ***Dayton***, OH 45435-0001
T: +1 937 7752925; Fax: +1 937 7754109; URL: www.libraries.wright.edu
1967; Marty Jenkins
Early Aviation, Miami Valley Hist, Univ Arch, Aerospace Medicine, Anthropometry, Children's Lit, Hist of Medicine, Local Hist
871 000 vols; 4 320 curr per; 6 500 e-journals
libr loan 30970

– Fordham Health Sciences Library, 3640 Colonel Glenn Hwy, Dayton, OH 45435
T: +1 937 7752003; Fax: +1 937 7752232; URL: www.libraries.wright.edu
1974; Sheila Shellabarger
Ross A McFarland Coll in Aerospace Medicine & Human Factors Engineering; Aerospace Medical Assn Arch; H.T.E. Hertzburg Coll in Anthropometry
139 309 vols; 2 619 curr per; 2 655 av-mat; 478 computer software programs 30971

Bethune-Cookman College, Carl S. Swisher Library & Learning Resource Center, 640 Mary McLeod Bethune Blvd, ***Daytona Beach***, FL 32114
T: +1 386 4812186; Fax: +1 386 4812182; URL: www.cookman.edu/subpages/about_library.asp
1904; Tasha Lucas-Youmans
Archival Coll, Africa, Art, Children's Coll
175 000 vols; 770 curr per
libr loan 30972

Embry-Riddle Aeronautical University, Jack R. Hunt Memorial Library, 600 S Clyde Morris Blvd, ***Daytona Beach***, FL 32114-3900
T: +1 386 2266592; Fax: +1 386 2266368; URL: amelia.db.erau.edu
1965; Kathleen Citro
Aviation Hist & Aeronautical Engineering Coll
96 000 vols; 840 curr per; 34 150 govt docs; 300 000 microforms; 3 500 av-mat; 370 sound-rec; 17 digital data carriers
libr loan 30973

Saint Norbert College, Todd Wehr Library, 301 Third St, ***De Pere***, WI 54115; mail address: 100 Grant St, De Pere, WI 54115-2002
T: +1 920 4033466; Fax: +1 920 4034064; E-mail: library@snc.edu; URL: www.snc.edu/library
1898; Felice Maciejewski
College Arch, Papers of John F Bennett
212 000 vols; 24 766 curr per; 15 400 e-journals; 12 600 e-books; 1 470 Music Scores 30974

Davenport University, Dearborn Campus Library, 4801 Oakman Blvd, ***Dearborn***, MI 48126
T: +1 313 5814400 ext 272; Fax: +1 313 5814762; E-mail: de_linc@davenport.edu; URL: libraries.davenport.edu
100 000 media items; 467 curr per 30975

University of Michigan Dearborn, Mardigian Library, 4901 Evergreen Rd, ***Dearborn***, MI 48128-2406
T: +1 313 5935445; Fax: +1 313 5935478; URL: library.umd.umich.edu
1959; Timothy F. Richards
300 000 vols; 589 curr per 30976

Agnes Scott College, McCain Library, 141 E College Ave, ***Decatur***, GA 30030-3770
T: +1 404 4716090; Fax: +1 404 4715037; E-mail: library@agnesscott.edu; URL: library.agnesscott.edu
1889; Elizabeth Leslie Bagley
History, Women's studies; Robert Frost Coll, Frontier Religion, Faculty & Student Publs, Catherine Marshall Papers
234 490 vols; 182 curr per; 28 598 e-journals; 50 386 e-books; 33 591 microforms; 23 534 av-mat; 14 666 sound-rec; 160 digital data carriers
libr loan; OCLC, Lyrasis, GALILEO, Oberlin Group 30977

Millikin University, Staley Library, 1184 W Main, ***Decatur***, IL 62522
T: +1 217 4246214; Fax: +1 217 4243992; URL: www.millikin.edu/staley/
1902; Cindy Fuller
Stephen Decatur, Bk Plates, World War I Pamphlets, Alice in Wonderland
212 000 vols; 460 curr per; 2 400 av-mat; 600 sound-rec; 7 digital data carriers
libr loan 30978

Luther College, Preus Library, 700 College Dr, ***Decorah***, IA 52101
T: +1 563 3871166; Fax: +1 563 3871657; E-mail: library@luther.edu; URL: lis.luther.edu/research
1861; Christopher D. Barth
Fine Arts Coll; Rare Books Coll
335 000 vols; 882 curr per; 350 maps; 1 900 av-mat; 8 500 sound-rec 30979

Trinity International University, James E. Rolfing Memorial Library, 2065 Half Day Rd, ***Deerfield***, IL 60015-1241
T: +1 847 3174011; Fax: +1 847 3174012; E-mail: libref@tiu.edu; URL: www.tiu.edu/library
1970; Dr. Robert H. Krapohl
202 000 vols; 1 382 curr per; 546 music scores; 111 000 microforms; 4 012 sound-rec; 10 digital data carriers; 51 Electronic Media & Resources
libr loan; ACRL, ATLA 30980

Defiance College, Pilgrim Library, 201 College Pl, ***Defiance***, OH 43512-1667
T: +1 419 7832481; Fax: +1 419 7832594
Ann Bible
Afro-American Coll; American Hist (Indian Wars of Northwest Ohio, 1785-1815), bk & micro
107 000 vols; 375 curr per 30981

Northern Illinois University, University Libraries, Founders Memorial Library, ***DeKalb***, IL 60115-2868
T: +1 815 7531094; Fax: +1 815 7539803; E-mail: Lib-Admin@niu.edu; URL: www.ulib.niu.edu
1899; Patrick J. Dawson
Donn V Hart Southeast Asian Coll, American & English Lit Colls
2 000 000 vols; 25 789 curr per; 216 734 maps; 3 180 000 microforms; 46 528 sound-rec 30982

– David C. Shapiro Memorial Law Library, DeKalb, IL 60115-2890
T: +1 815 7530519; Fax: +1 815 7539499; URL: law.niu.edu/law/library
1974; Gary L. Vander Meer
257 000 vols; 1 600 curr per 30983

– Music Library, School of Music, Rm 175, DeKalb, IL 60115
T: +1 815 7531426; Fax: +1 815 7539836
Michael Duffy IV
21 000 vols; 115 curr per 30984

Stetson University, DuPont-Ball Library, 421 N Woodland Blvd, Unit 8418, ***DeLand***, FL 32723
T: +1 386 8227175; Fax: +1 386 8227199; URL: www.stetson.edu/library
1883; Betty D. Johnson
Religion (Garwood Baptist Hist Coll)
278 000 vols; 964 curr per; 20 000 e-journals; 3 000 e-books; 2 860 av-mat; 20 200 sound-rec 30985

Ohio Wesleyan University, L.A. Beeghly Library, 43 Rowland, ***Delaware***, OH 43015-2370
T: +1 740 3683225; Fax: +1 740 3683222; E-mail: refdesk@owu.edu; URL: lis.owu.edu
1842; Theresa Byrd
Browning (Gunsaulus Coll), Walt Whitman, (Bayley Coll), Religion (Methodist Hist Coll), Schubert (20th C Imprints Coll)
424 000 vols; 1 083 curr per; 6 750 e-journals 30986

State University of New York – College at Delhi, Louis & Mildred Resnick Library, 2 Main St, Bush Hall, ***Delhi***, NY 13753
T: +1 607 7464635; E-mail: library@delhi.edu; URL: wc.delhi.edu/library
1915; Pamela J. Peters
Travel File, pamphlet
54 670 vols; 421 curr per; 8 852 govt docs; 267 maps; 24 200 microforms; 221 av-mat; 1 571 sound-rec; 8 777 pamphlets, 8 448 slides, 769 overhead transparencies, 43 art reproductions
libr loan 30987

Piedmont College, Arrendale Library, 165 Central Ave, ***Demorest***, GA 30535; P.O. Box 40, Demorest, GA 30535-0040
T: +1 706 7760111; Fax: +1 706 7763338; E-mail: refdept@piedmont.edu; URL: library.piedmont.edu
1897; Bob Glass
116 000 vols; 345 curr per; 900 microforms; 2 digital data carriers 30988

Voorhees College, Wright Potts Library, 5480 Voorhees Rd, ***Denmark***, SC 29042; P.O. Box 678, Denmark, SC 29042-0678
T: +1 803 7933351; Fax: +1 803 7930471; URL: www.voorhees.edu
1935; Marie S. Martin
Hist Papers, Voorhees College Docs, Ten-Year Developmental Study of Episcopal Church Bk of Common Prayer
110 000 vols; 213 curr per 30989

Texas Woman's University, Mary Evelyn Blagg-Huey Library, 1200 Frame St, ***Denton***, TX 76204; P.O. Box 425528 TWU Sta, Denton, TX 76204-5528
T: +1 940 8983748; Fax: +1 940 8983764; URL: www.twu.edu/library
1901; Sherilyn Bird
American fiction by female authors (L.H. Wright Coll), LaVerne Harrell Clark Coll, Hist of Texas Women Coll, Women Airforce Service Pilots (WASP) Coll
534 000 vols; 2 505 curr per; 490 e-journals; 87 000 e-books; 1 514 000 microforms; 4 962 sound-rec; 33 digital data carriers
libr loan 30990

University of North Texas, Willis Library, PO Box 305190, ***Denton***, TX 76203-5190
T: +1 940 5652495; Fax: +1 940 3698760; E-mail: circ@library.unt.edu; URL: www.library.unt.edu
1890; B. Donald Grose
Anson Jones Libr, Music (Duke Ellington, Don Gillis, Lloyd Hibberd, Stan Kenton & Arnold Schoenberg Coll), Source Magazine Arch, WBAP Coll, Weaver Coll, WFAA Coll, Gerontological Film Coll, Texana Coll, Rare Book Room
1 723 000 vols; 30 131 curr per; 109 000 e-books
libr loan 30991

Auraria Library, 1100 Lawrence St, ***Denver***, CO 80204-2095
T: +1 303 5562639; Fax: +1 303 5563528; URL: library.auraria.edu
1976; David Gleim
Architecture; Lit & Literary Criticism (Donald Sutherland Coll), State & Local Policy (Seasongood Libr, National Municipal League, Conservative Think Tanks), Auraria Higher Education Center Arch, Civil Liberties in Colorado
600 000 vols; 3 000 curr per; 20 000 e-journals
libr loan 30992

Education Management Corporation – AiC Library, The Art Institute of Colorado Library, 1200 Lincoln St, ***Denver***, CO 80203-2114
T: +1 303 8244787; Fax: +1 303 8244890; URL: www.aic.artinstitutes.edu/library.asp
Glenn Pflum
30 000 vols; 200 curr per 30993

Iliff School of Theology, Ira J. Taylor Library, 2201 S University Blvd, ***Denver***, CO 80210
T: +1 303 7653172; Fax: +1 303 7770164; URL: www.iliff.edu/research
1892; Dr. Debbie Creamer
Hymnals, Church Hist of the United Methodist Rocky Mountain Conference
206 000 vols; 660 curr per; 449 mss; 14 maps; 57 000 microforms; 86 av-mat; 2 402 sound-rec; 17 digital data carriers; 40 Slides
libr loan; ALA, ATLA 30994

Regis University, Dayton Memorial Library, 3333 Regis Blvd, ***Denver***, CO 80221-1099
T: +1 303 4584030; Fax: +1 303 9645497; URL: www.regis.edu
1877; Ivan Gaetz
Religion, Health Care, Business, Liberal arts; Western Jesuitica, Women's Coll, College Archs
300 000 vols; 25 000 curr per; 10 000 e-books; 107 000 microforms; 770 av-mat; 6 700 sound-rec; 6 digital data carriers; 902 linear feet of mss & archs, 110 000 slides
libr loan 30995

– Colorado Springs Campus Library, 7450 Campus Dr, Ste 100, Colorado Springs, CO 80920
T: +1 719 2647080; Fax: +1 719 2647082; E-mail: cslib@regis.edu; URL: www.regis.edu
1987; Linda Bourgeois
Computer science, Religion
10 000 vols; 25 curr per; 30 000 microforms; 5 digital data carriers 30996

Teikyo Loretto Heights University, University Library, 3001 S Federal Blvd, ***Denver***, CO 80236
T: +1 303 9374246; Fax: +1 303 9374224; URL: www.tlhu.edu
1989; Larry Grieco
East Asian Coll
118 000 vols; 60 curr per; 10 000 microforms; 1 digital data carriers
libr loan 30997

University of Denver, Penrose Library, 2150 E Evans, ***Denver***, CO 80208-2007
T: +1 303 8713441; Fax: +1 303 8712290; URL: library.du.edu
1864; Nancy Allen
Levette J. Davidson Folklore Coll, Margaret Husted Culinary Coll, Judaica, Miller Civil War Coll
909 000 vols; 5 540 curr per; 933 000 microforms; 1 793 digital data carriers
libr loan 30998

– Westminster Law Library, 2255 E Evans Ave, Denver, CO 80208
T: +1 303 8716153; Fax: +1 303 8716999; URL: www.law.du.edu/library/
1892; Gary Alexander
Howard Jenkins Mss, Lowell Thomas Coll
330 000 vols; 4 348 curr per; 280 sound-rec; 50 digital data carriers
libr loan 30999

Drake University, Cowles Library, 2725 University Ave, ***Des Moines***, IA 50311; mail address: 2507 University Ave, Des Moines, IA 50311-4505
T: +1 515 2713993; Fax: +1 515 2713933; E-mail: cowles-ill@drake.edu; URL: www.lib.drake.edu
1881; Rod Neal Henshaw
Religion, Music; Disciples of Christ Hist Coll, Philip Duffield Strong Papers, Gardner (Mike) Cowles Jr Papers, John Cowles Papers, Univ Arch
515 000 vols; 1 447 curr per; 30 000 e-journals; 47 000 e-books; 746 000 microforms; 980 av-mat; 240 sound-rec; 530 digital data carriers
libr loan 31000

– Drake Law Library, Opperman Hall, 27th & Carpenter Ave, Des Moines, IA 50311-4505; mail address: 2507 University Ave, Des Moines, IA 50311-4516
T: +1 515 2713759; Fax: +1 515 2712530; URL: www.law.drake.edu
1865; John D. Edwards
Foreign Law, Agricultural Law, Constitutional Law, Iowa Legal Hist
330 000 vols; 3 247 curr per; 10 000 govt docs; 257 000 microforms; 35 av-mat; 375 sound-rec; 38 digital data carriers
libr loan 31001

Grand View College, Library, 1351 Grandview Ave, ***Des Moines***, IA 50316-1494
T: +1 515 2632878; Fax: +1 515 2632998; E-mail: Library@gvc.edu; URL: library.gvc.edu
1896; Pam Rees
Danish Immigrant Archs
120 000 vols; 400 curr per; 16 000 microforms; 150 sound-rec
libr loan 31002

Mercy College of Health Sciences Library, 928 Sixth Ave, ***Des Moines***, IA 50309-1239
T: +1 515 6436613; Fax: +1 515 6436695; URL: www.mchs.edu
Eileen Hansen
10 000 vols; 111 curr per 31003

Center for Creative Studies Library, Manoogian Visual Resource Ctr, 301 Frederick Douglass Dr, ***Detroit***, MI 48202-4034
T: +1 313 6647803; Fax: +1 313 6647880; URL: www.lib.ccscad.edu
1966; Beth E. Walker
Art History Coll
40 000 vols; 250 curr per 31004

Marygrove College, Library, 8425 W McNichols Rd, ***Detroit***, MI 48221-2599
T: +1 313 9271346; Fax: +1 313 9271366; URL: www.marygrove.edu
1925; Dana Zurawski
99 000 vols; 450 curr per 31005

University of Detroit Mercy, McNichols Campus Library, 4001 W McNichols Rd, ***Detroit***, MI 48221; P.O. Box 19900, Detroit, MI 48219-0900
T: +1 313 9931071; Fax: +1 313 9931780; E-mail: research@udmercy.edu; URL: research.udmercy.edu/
1877; Margaret E. Auer
Architecture, Philosophy, Theology; Marine Hist Coll, Great Lakes Shippping, Mss Colls
484 000 vols; 690 curr per
libr loan 31006

– Kresge Law Library, 651 E Jefferson, Detroit, MI 48226
T: +1 313 5960241; Fax: +1 313 5960245; E-mail: lawlibrary@udmercy.edu; URL: www.law.udmercy.edu/lawlibrary
1912; Byron D. Cooper
Labor, Taxes
347 000 vols; 1 500 curr per; 50 000 microforms; 50 av-mat; 200 sound-rec; 8 digital data carriers 31007

– Outer Drive Campus Library, 8200 W Outer Dr, P.O. Box 19900, Detroit, MI 48219-0900
T: +1 313 9936180; Fax: +1 313 9936329; E-mail: research@udmercy.edu; URL: research.udmercy.edu
1877; La Verne Fant Calloway
159 870 vols; 510 curr per
libr loan 31008

Wayne State University, University Libraries, Office of the Dean, 3100 Undergraduate Library, 5155 Gullen Mall, ***Detroit***, MI 48202
T: +1 313 5774023; Fax: +1 313 5775525; URL: www.lib.wayne.edu/index.php
Sandra G. Yee
3 324 000 vols; 18 643 curr per; 76 500 mss; 411 659 govt docs; 51 775 maps; 3 484 000 microforms; 38 090 sound-rec
libr loan; ALA, IFLA 31009

– Arthur Neef Law Library, 474 Ferry Mall, Detroit, MI 48202
T: +1 313 5773925; Fax: +1 313 5775498; URL: www.lib.wayne.edu/lawlibrary/
1927; Virginia C. Thomas
Michigan Legal Coll; Alwyn V Freeman International Law Coll
408 000 vols; 4 983 curr per; 1 129 000 microforms; 189 av-mat; 256 sound-rec; 117 digital data carriers 31010

– David Adamany Undergraduate Library, 5155 Gullen Mall, Detroit, MI 48202-3962
T: +1 313 5778854; Fax: +1 313 5775265; URL: www.lib.wayne.edu/geninfo/units/ugl.php
Lothar Spang
Undergraduate studies, computer education
62 000 vols; 352 curr per 31011

– Purdy-Kresge Library, 5265 Cass Ave, Detroit, MI 48202
T: +1 313 5774043; Fax: +1 313 5773436
1973; Rhonda McGinnis
Economics & business administration, Education, Humanities, Libr science, Social sciences, Fine arts; Leonard N. Simons Coll (Detroit & Michigan Hist), Eloise Ramsey Coll of Lit for Young People, Mildred Jeffrey Coll for Peace & Conflict Resolution, Arthur L. Johnson Coll for Civil Rights
1 555 000 vols; 5 885 curr per; 1 820 000 microforms; 12 digital data carriers; 360 661 docs, 65 VF drawers of pamphlets & clippings
libr loan 31012

– Science and Engineering, 5048 Gullen Mall, Detroit, MI 48202-3918
T: +1 313 5774066; Fax: +1 313 5773613; URL: www.lib.wayne.edu/geninfo/units/sel.php/
Nancy Wilmes
Natural sciences
616 000 vols; 3 108 curr per 31013

– Vera P. Shiffman Medical Library, Rackham Bldg Rm 044, 60 Farnsworth, Detroit, MI 48202
T: +1 313 5771094; Fax: +1 313 5776668; E-mail: askmed@wayne.edu; URL: www.lib.wayne.edu/shiffman/index.php
1949; Ellen Marks
Community Health Info Services, Pharmacy & Allied Health Learning Resources Ctr Coll, Detroit Community Aids Libr
348 000 vols; 3 829 curr per; 13 000 microforms; 18 digital data carriers
libr loan 31014

– Walter P. Reuther Library of Labor & Urban Affairs, 5401 Cass Ave, Detroit, MI 48202
T: +1 313 5774024; Fax: +1 313 5774300; URL: www.reuther.wayne.edu
1960; Michael Smith
Archs, Mss Coll, Photogr Coll
9 000 vols; 650 curr per 31015

Lake Region State College, Paul Hoghaug Library, 1801 N College Dr, ***Devils Lake***, ND 58301
T: +1 701 6621533; Fax: +1 701 6621570; URL: www.lrsc.nodak.edu
1966; Celeste Ertelt
Law, Paralegal; Irish Hist & Culture
35 000 vols; 200 curr per; 10 000 e-books 31016

Dickinson State University, Stoxen Library, 291 Campus Dr, ***Dickinson***, ND 58601
T: +1 701 4832135; Fax: +1 701 4832006; URL: dickinsonstate.com/library.asp
1918; Lillian Crook
Education, Business; Teddy Roosevelt Coll
159 000 vols; 451 curr per; 10 000 e-books; 400 maps; 11 000 microforms; 800 av-mat; 974 sound-rec; 20 digital data carriers
libr loan 31017

Western Montana College, Lucy Carson Memorial Library, 710 S Atlantic St, ***Dillon***, MT 59725
T: +1 406 6837491; Fax: +1 406 6837493; URL: www.umwestern.edu/library
1897; Michael Schulz
Education; Montana Hist Coll, EPA Educational Libr, NASA Teacher Resource Ctr, State Educational Media Libr
66 000 vols; 358 curr per 31018

Five Towns College, Library, 305 N Service Rd, ***Dix Hills***, NY 11746
T: +1 631 6562138; Fax: +1 631 6562171; URL: www.ftc.edu
1972; Heidi Sanchez
Sheet Music, Songbks
30 000 vols; 500 curr per 31019

Mercy College Libraries, 555 Broadway, ***Dobbs Ferry***, NY 10522
T: +1 914 6747259; Fax: +1 914 6747581; E-mail: libref@mercy.edu; URL: www.mercy.edu/library
1960; Judith Liebman
Libr of American Civilization, Libr of English Lit, Vanderpoel Print Coll, Eric doc, Peter Carl Goldmark Record Coll
134 000 vols; 369 curr per; 6 760 av-mat; 145 digital data carriers
libr loan 31020

Troy University Dothan Library, 502 University Dr, ***Dothan***, AL 36306; P.O. Box 8368, Dothan, AL 36304-0368
T: +1 334 9836556; Fax: +1 334 9836327; URL: dothan.troy.edu/library
1973; Christopher Shaffer
Business and management, Computer science, Criminal law and justice, Education, Hist; Wiregrass Hist Coll
102 000 vols; 410 curr per; 40 000 e-books; 722 maps; 176 000 microforms; 3 700 av-mat; 1 860 sound-rec; 19 digital data carriers 31021

Delaware State University, William C. Jason Library-Learning Center, 1200 N Dupont Hwy, ***Dover***, DE 19901-2277
T: +1 302 8576195; Fax: +1 302 8576177; URL: www.desu.edu/library
1891; Rebecca E. Batson
Education, Nursing, Business, African-American studies; Hist Resource Coll, Delaware Coll
240 000 vols; 1 300 curr per; 18 000 e-journals; 484 e-books; 359 000 microforms; 29 digital data carriers
31022

Wesley College, Robert H. Parker Library, 120 N State St, ***Dover***, DE 19901
T: +1 302 7362413; Fax: +1 302 7362533; URL: www.wesley.edu
1873; Susan Matusak
French Lit (Neves Coll); Delaware Poetry (Edwards Coll), political science, nursing
102 000 vols; 300 curr per; 600 av-mat; 8 digital data carriers 31023

Southwestern Michigan College, Fred L. Mathews Library, 58900 Cherry Grove Rd, ***Dowagiac***, MI 49047
T: +1 269 7821339; E-mail: info@swmich.edu; URL: www.swmich.edu/
1964
Civil War Coll
38 000 vols; 1 500 curr per 31024

Midwestern University, Alumni Memorial Library, 555 31st St, ***Downers Grove***, IL 60515
T: +1 630 5156200; Fax: +1 630 5156195; URL: www.midwestern.edu
1913; Faith Ross
Medicine, Allied health sciences, Osteopathy; Hist of Medicine
11 000 vols; 739 curr per; 1 694 av-mat; 1 468 sound-rec; 29 digital data carriers; 36 479 slides 31025

Delaware Valley College, Joseph Krauskopf Memorial Library, 700 E Butler Ave, ***Doylestown***, PA 18901-2699
T: +1 215 4894968; Fax: +1 215 2302967; URL: www.delval.edu
1896; Peter Kupersmith
Animal and Plant Sciences
49 000 vols; 290 curr per; 12 000 e-journals; 300 e-books; 97 000 microforms; 87 av-mat; 16 digital data carriers
libr loan; OCLC 31026

Pennsylvania State University, Du Bois Commonwealth College Library, College Pl, Hiller Bldg, Rm 113, 301 E DuBois Ave, ***Du Bois***, PA 15801
T: +1 814 3754756; Fax: +1 814 3754784; URL: www.libraries.psu.edu/dubois/
1935; Dr. Janice Norris
Wildlife Technology (Paul A Handwerk & David D Wanless Coll)
43 000 vols; 233 curr per 31027

Clarke College, Nicholas J. Schrup Library, 1550 Clarke Dr, ***Dubuque***, IA 52001
T: +1 563 5886421; Fax: +1 563 5888160; E-mail: library@clarke.edu; URL: www.clarke.edu/page.aspx?id=584
1843; Sue Leibold
BVM Heritage Coll
125 000 vols
libr loan 31028

Loras College, Wahlert Memorial Library, 1450 Alta Vista St, ***Dubuque***, IA 52004-4327; P.O. Box 164, Dubuque, IA 52004-0164
T: +1 563 5887929; Fax: +1 563 5887292; URL: depts.loras.edu/library/
1839; Joyce A. Meldrem
Horace Coll, T. S. Eliot Coll; Center for Dubuque History Collection; Loras College Archives Collection
361 000 vols; 10 390 curr per; 10 390 e-journals; 5 000 e-books; 50 incunabula; 54 000 govt docs; 5 513 maps; 10 000 microforms; 450 sound-rec; 754 digital data carriers
libr loan 31029

University of Dubuque, Charles C. Myers Library, 2000 University Ave, ***Dubuque***, IA 52001
T: +1 563 5893100; Fax: +1 563 5893722; E-mail: libcirc@dbq.edu; URL: www.dbq.edu/library
1852; Mary Anne Knefel
Religion, Native Americans; German Presbyterian Coll, Hist of Iowa & Upper Mississippi Valley
173 000 vols; 484 curr per; 17 030 e-journals; 12 600 e-books; 4 000 microforms 31030

Nichols College, Conant Library, 124 Center Rd, ***Dudley***, MA 01571; P.O. Box 5000, Dudley, MA 01571-5000
T: +1 508 2132222; Fax: +1 508 2132323; E-mail: reference@nichols.edu; URL: www.nichols.edu/library/
1815; Jim Douglas
Management, Advertising, Finance & accounting, Small business, Marketing, Taxation, Economics, Int trade, Humanities; College Hist
43 000 vols; 211 curr per; 6 000 microforms; 1 079 sound-rec; 51 digital data carriers; 675 corporal rpts, 6 slides
31031

College of Saint Scholastica, Library, 1200 Kenwood Ave, ***Duluth***, MN 55811-4199
T: +1 218 7236178; Fax: +1 218 7235948; E-mail: library@css.edu; URL: academics.css.edu/library
1909; Kevin McGrew
American Indian Studies Coll, Children's Coll
125 000 vols; 615 curr per; 8 000 e-books; 3 550 av-mat; 896 Journals
libr loan 31032

University of Minnesota Duluth, Library, 416 Library Dr, ***Duluth***, MN 55812
T: +1 218 7268130; Fax: +1 218 7266205; E-mail: lib@d.umn.edu; URL: www.d.umn.edu/lib
1902; Basil Sozansky
Northeast Minnesota Hist Ctr, Ramseyer-Northern Bible Society Museum Coll, Voyageur Coll

369 000 vols; 1 411 curr per; 25 959 e-journals; 31 000 e-books; 124 254 govt docs; 59 452 music scores; 774 maps; 63 000 microforms; 6 080 av-mat; 10 207 sound-rec; 4 350 digital data carriers; 1 484 periodicals from EBSCO
libr loan 31033

Pennsylvania State University, Worthington Scranton Commonwealth College Library, 120 Ridge View Dr, **Dunmore**, PA 18512-1699
T: +1 570 9632630; Fax: +1 570 9632635; E-mail: wscrant@psulias.psu.edu; URL: www.psu.edu
1923; Richard Fitzsimmons
66 000 vols; 100 curr per 31034

Fort Lewis College, John F. Reed Library, 1000 Rim Dr, **Durango**, CO 81301-3999
T: +1 970 2477250; Fax: +1 970 2477149; URL: library.fortlewis.edu
1911; Chandler Jackson
Center for Southwest Studies
184 000 vols; 460 curr per
libr loan 31035

Southeastern Oklahoma State University, Henry G. Bennett Memorial Library, Sixth St University Blvd, **Durant**, OK 74701; mail address: 1405 N Fourth Ave, PMB 4105, Durant, OK 74701-0609
T: +1 580 7453172; Fax: +1 580 7457463; URL: www.sosu.edu/lib
1913; Sharon Morrison
Curriculum Mat, Native American Coll, Juvenile Lit
183 000 vols; 878 curr per; 8 000 e-books; 2 570 av-mat; 5 080 sound-rec
libr loan 31036

Duke University, William R. Perkins Library, Research Dr, **Durham**, NC 27708; P.O. Box 90193, Durham, NC 27708-0193
T: +1 919 6605800; Fax: +1 919 6605964; E-mail: askref@duke.edu; URL: www.library.duke.edu
1838; Deborah Jakubs
American & English Lit, German Baroque Lit, Advertising, Architecture, mss, Methodist Church, Philippines, Utopias, Wesleyana, Latin American Hist, African Coll, Classics Coll, Drama Coll, Judaica
5 872 000 vols; 47 122 curr per; 143 000 e-books
libr loan 31037

– Divinity School Library, Gray Bldg, 102 Chapel Dr, Durham, NC 27708; P.O. Box 90972, Durham, NC 27708-0972
T: +1 919 6603450; Fax: +1 919 6817594; URL: www.lib.duke.edu/divinity/
Roger Loyd
Biblical studies, Christian theology, Methodism
370 000 vols; 708 curr per; 28 digital data carriers
ATLA 31038

– Duke University Marine Laboratory, Pearce Memorial Library, 135 Duke Marine Lab Rd, Beaufort, NC 28516-9721
T: +1 252 5047510; Fax: +1 252 5047622; E-mail: david.talbert@duke.edu
David Talbert
Biochemistry, Botany, Coastal resource management, Marine biology, Marine biotechnology, Oceanography
25 610 vols 31039

– Ford Library, One Towerview Rd, Durham, NC 27708, Fuqua School of Business, P.O. Box 90122, Durham, NC 27708-0122
T: +1 919 6607870; Fax: +1 919 6607950; URL: library.fuqua.duke.edu/index.html
Meg Trauner
Business
30 000 vols; 300 curr per 31040

– Fuqua School of Business, Ford Library, One Towerview Rd, Durham, NC 27708; P.O. Box 90122, Durham, NC 27708-0122
T: +1 919 6607870; Fax: +1 919 6607950; E-mail: mtrauner@mail.duke.edu; URL: www.lib.duke.edu/fsb/index.htm
1983; Margaret Trauner
28 450 vols; 250 curr per; 2 520 av-mat 31041

– Law School Library, Science Dr & Towerview Rd, Campus Box 90361, Durham, NC 27708
T: +1 919 6137128; Fax: +1 919 6137237; URL: www.law.duke.edu/lib
1868; Melanie Dunshee
442 000 vols; 3 411 curr per; 25 e-journals
libr loan; IALL, AALL, CALI, RLG 31042

– Lilly Library, 1348 Campus Dr, Box 90725, Durham, NC 27708
T: +1 919 6605995; Fax: +1 919 6605999; E-mail: lilly-requests@duke.edu; URL: library.duke.edu/lilly
1927; Kelley Lawton
Art hist, Drama, Philosophy, Visual arts
275 670 vols; 400 curr per; 8 000 av-mat; 70 digital data carriers; 4 000 pamphlets
libr loan 31043

– Medical Center Library, DUMC Box 3702, Ten Bryan-Searle Dr, Durham, NC 27710-0001
T: +1 919 6601150; Fax: +1 919 6601188; E-mail: mclref@mc.duke.edu; URL: www.mclibrary.duke.edu
1930; Patricia Thibodeau
Anesthesia, Early Printed Medical Bks, Mss, Human Sexuality, Military Medicine, Obstetrics, Pediatrics, Vivisection, Yellow Fever, Andreas Vesalius, Benjamin Rush, Benjamin Waterhouse, Josiah C. Trent Coll on Hist of Medicine
296 000 vols; 30 digital data carriers
libr loan 31044

– Music Library, 113 Mary Duke Biddle Bldg, Box 90661, Durham, NC 27708-0661
T: +1 919 6605950; Fax: +1 919 6846556; E-mail: music-requests@duke.edu; URL: library.duke.edu/music
1974; Laura Williams
African-American Music (William Grant Still Coll); Venetian Music (Berdes Papers); Viennese Music (Weinmann Coll); Organ Music (Faxon Coll); Central & Eastern European Hist (Riethus Coll); Robert Ward Arch; Jane L. Berdes Arch for Women in Music
110 000 vols; 400 curr per; 55 diss/theses; 40 000 music scores; 10 000 microforms; 20 digital data carriers; 22 000 sound & video recordings
libr loan 31045

North Carolina Central University, James E. Shepard Memorial Library, 1801 Fayetteville St, **Durham**, NC 27707-3129; P.O. Box 19436, Durham, NC 27707-0019
T: +1 919 5306473; Fax: +1 919 5307612; URL: www.nccu.edu/shepardlibrary
1923; Dr. Theodosia Shields
Local Hist; Martin Coll (African-American)
501 000 vols; 1 928 curr per 31046

– Music Library, 1801 Fayetteville St, Durham, NC 27707
T: +1 919 5306220; Fax: +1 919 5307979; URL: www.nccu.edu/library/music.html
Vernice Faison
12 000 vols 31047

– School of Law Library, 1512 S Alston Ave, Durham, NC 27707
T: +1 919 5306715; Fax: +1 919 5307926; URL: www.nccu.edu/law/library/index.html
1939; Deborah Jefferies
Civil Rights (McKissick Coll)
352 000 vols; 4 500 curr per; 169 000 microforms; 58 digital data carriers 31048

– School of Library & Information Sciences, James E Shepard Memorial Library, 3rd Flr, 1801 Fayetteville St, Durham, NC 27707
T: +1 919 5307323; Fax: +1 919 5306402; URL: www.nccuslis.org/slislib/
Virginia Purefoy Jones
Children's literature, library science
43 000 vols; 470 curr per 31049

University of New Hampshire, University Library, 18 Library Way, **Durham**, NH 03824
T: +1 603 8621535; Fax: +1 603 8620247; URL: www.library.unh.edu
1868; Dr. Claudia Morner
Angling (Milne Coll), Galway Kinnell Coll,

Contra Dance & Folk Music (Ralph Page Coll), Frost Arch, Amy Beach Papers, Senator Norris Cotton Papers, Donald Hall Coll, Senator Thomas McIntyre Papers
1 805 000 vols; 54 691 curr per 31050

– Biological Sciences Library, Kendall Hall, 129 Main St, Durham, NH 03824-3590
T: +1 603 8623718; Fax: +1 603 8622789; E-mail: dml2@cisunix.unh.edu; URL: grinnell.unh.edu
David Lane
75 000 vols; 800 curr per 31051

– Chemistry Library, Parsons Hall, 23 College Rd, Durham, NH 03824-3598
T: +1 603 8621083; Fax: +1 603 8624278; URL: www.library.unh.edu/branches/chemgide.html
Emily LeViness Poworoznek
Materials science
32 000 vols; 129 curr per 31052

– David G. Clark Memorial Physics Library, DeMeritt Hall, Nine Library Way, Durham, NH 03824-3568
T: +1 603 8622348; Fax: +1 603 8622998; URL: www.library.unh.edu/branches/physlib.html
1976; Emily LeViness Poworoznek
30 000 vols; 161 curr per; 130 diss/theses 31053

– Engineering, Mathematics & Computer Science Library, New Hampshire Hall (Rear), 124 Main St, Durham, NH 03824
T: +1 603 8621196; Fax: +1 603 8624112; URL: www.library.unh.edu/branches/engmathcs.html
Emily L. Poworoznek
50 000 vols; 720 curr per 31054

Michigan State University, University Library, 100 Library, **East Lansing**, MI 48824-1048
T: +1 517 4326123; Fax: +1 517 4323532; URL: www.lib.msu.edu
1855; Clifford H. Haka
Cesar E. Chavez Coll, Russel B. Nye Popular Culture Coll, Changing Men Coll, American Radicalism Coll, Veterinary Medicine Historical Coll, Charles Schmitter Fencing Coll, Cookery Coll, Comic Art Coll, Software Coll; Africana; Fine Arts; Gov Docs; Labor & Industrial Relations; Maps; Special Coll; Turfgrass Info Ctr; Vincent Voice Library
4 916 000 vols; 74 177 curr per; 34 572 e-journals; 245 539 maps; 6 486 000 microforms; 2 500 linear feet of mss
libr loan; ALA, CRL, ARL 31055

– Benjamin H. Anibal Engineering Library, 1515 Engineering Bldg, East Lansing, MI 48824
T: +1 517 3558536; Fax: +1 517 3539041; E-mail: volkenin@msu.edu; URL: www.lib.msu.edu/coll/branches/engin
1963; Thomas C. Volkening 31056

– Biomed Library, 1440E Biomedical & Physical Sciences Bldg, East Lansing, MI 48824-5320
T: +1 517 4324900 ext 1990; Fax: +1 517 4324901; E-mail: matthe20@msu.edu; URL: www.lib.msu.edu
Judith Matthews 31057

– Geology Library, 5 Natural Sciences Bldg, East Lansing, MI 48824
T: +1 517 3537988; Fax: +1 517 4323896; E-mail: baclaws2@msu.edu; URL: www.lib.msu.edu
1967; Diane Baclawski 31058

– Gull Lake Library, 3700 East Gull Lake Dr, Hickory Corners, MI 49060
T: +1 269 6712310; Fax: +1 269 6711191; URL: www.lib.msu.edu
Melissa Yost 31059

– Mathematics Library, 101-D Wells Hall, East Lansing, MI 48824
T: +1 517 3538852; Fax: +1 517 3537215; E-mail: flynnhol@mail.lib.msu.edu; URL: www.lib.msu.edu
1967; Holly Flynn 31060

– Planning and Design Library, 212 Urban Planning & Landscape, East Lansing, MI 48824
T: +1 517 3533941; Fax: +1 517 3539888; E-mail: weessie2@msu.edu; URL: www.lib.msu.edu
Kathleen Weessies 31061

– Veterinary Medical Center Library, G-201 Vet Med Center, East Lansing, MI 48824
T: +1 517 3535099; Fax: +1 517 4323797; E-mail: lukasn@msu.edu; URL: www.lib.msu.edu
Nancy Lukas 31062

– William C. Gast Business Library, DCL/Business Libr Bldg, East Lansing, MI 48824
T: +1 517 3553380; Fax: +1 517 3536648; E-mail: lucasn@msu.edu; URL: www.lib.msu.edu
1962; Nancy Lucas 31063

Michigan State University – College of Law Library, 115 Law College Bldg, **East Lansing**, MI 48824-1300
T: +1 517 4326870; Fax: +1 517 4326861; E-mail: reference@law.msu.edu; URL: www.law.msu.edu
1891; Charles Ten Brink
285 000 vols; 712 000 microforms; 220 sound-rec; 4 digital data carriers 31064

Kent State University, East Liverpool Campus Library, 400 E Fourth St, Rm 216, **East Liverpool**, OH 43920-5769
T: +1 330 3827421; Fax: +1 330 3827561; URL: www.kentliv.kent.edu/CurrentStudents/Library/
1968; Susan Weaver
31 000 vols; 44 curr per; 9 maps; 325 av-mat; 5 digital data carriers
libr loan 31065

Atlanta Christian College, James A. Burns Memorial Library, 2605 Ben Hill Rd, **East Point**, GA 30344-1999
T: +1 404 6692097; Fax: +1 404 6694009; URL: www.acc.edu
1937; Michael Bain
Alumni Coll, Libr of American Civilizatiion (Core Coll)
55 000 vols; 210 curr per 31066

East Stroudsburg University, Kemp Library, 216 Normal St, **East Stroudsburg**, PA 18301-2999
T: +1 570 4223914; Fax: +1 570 4223151; URL: www.esu.edu/library
1893; David G. Schappert
Cohn Jazz Coll
460 000 vols; 1 100 curr per; 1 161 000 microforms; 5 796 sound-rec; 2 digital data carriers; 72 362 gov docs 31067

Lafayette College, David Bishop Skillman Library, 710 Sullivan Rd, **Easton**, PA 18042-1797
T: +1 610 3305151; Fax: +1 610 2520370; URL: www.library.lafayette.edu
1826; Neil J. McElroy
American Friends of Lafayette, Stephen Crane Coll, Conahay, Tinsman & Fox Angling Coll, Marquis de Lafayette Coll, Robert & Helen Meyner Coll, Wm E Simon Coll, Howard Chandler Christy Papers, Dixie Cup Co. Coll
510 000 vols; 8 767 curr per; 10 000 e-journals; 4 incunabula; 1 000 microforms; 1 922 cu. feet mss
libr loan 31068

– Kirby Library of Government & Law, Kirby Hall of Civil Rights, 716 Sullivan Rd, Easton, PA 18042-1780
T: +1 610 3305398; Fax: +1 610 3305397
1930; Mercedes Benitez-Sharpless
British Parliamentary Debates
31 000 vols; 126 curr per; 3 VF drawers of pamphlets 31069

Stonehill College, MacPhaidin Library, 320 Washington St, **Easton**, MA 02357-4015
T: +1 508 5651310; Fax: +1 508 5651424; URL: www.stonehill.edu/library
1948; Edward Hynes
Religion, Business & management; Rep Joseph W Martin Jr, Papers & Memorabilia; Michael Novak Papers; Tofias Business Arch
225 000 vols; 928 curr per; 1 290 e-journals; 3 045 av-mat; 3 100 sound-rec; 450 digital data carriers
libr loan 31070

University of Wisconsin – Eau Claire, William D. McIntyre Library, 105 Garfield Ave, ***Eau Claire***, WI 54702-4004
T: +1 715 8365377; Fax: +1 715 8362949; URL: www.uwec.edu/library
1916; John Pollitz
Area Research Ctr, Campus Evolution Records, Chippewa Valley Hist Mss & Local Govt Records
491 000 vols; 1 369 curr per
libr loan 31071

Edinboro University of Pennsylvania, Baron-Forness Library, 200 Tartan Ave, ***Edinboro***, PA 16444
T: +1 814 7322946; Fax: +1 814 7322883; E-mail: library@edinboro.edu
1857; Dr. Donald Dilmore
Edinboro area photos
501 000 vols; 1 523 curr per; 20 790 e-journals; 2 000 e-books; 770 maps; 1 404 000 microforms; 2 450 av-mat; 10 704 sound-rec; 420 linear ft of mss
libr loan; OCLC 31072

University of Texas – Pan American Library, 1201 W University Dr, ***Edinburg***, TX 78541-2999
T: +1 956 3812755; Fax: +1 956 3185396; URL: www.lib.panam.edu
1927; Dr. Farzaneh Razzaghi
Depot for the Texas Regional Hist Resource Depot Program, Cameron, Hidalgo, Jim Hogg, Starr, Webb, Willacy & Zapata Counties; Lower Rio Grande Valley Coll; Rare Books; Univ Arch
598 000 vols; 1 846 curr per; 33 160 e-journals; 46 000 e-books; 5 680 av-mat; 1 110 sound-rec; 210 digital data carriers; 710 Bks on Deafness & Sign Lang
libr loan 31073

Oklahoma Christian University, Tom & Ada Beam Library, 2501 E Memorial Rd, ***Edmond***, OK 73013; P.O. Box 11000, Oklahoma City, OK 73136-1100
T: +1 405 4255322; Fax: +1 405 4255313; E-mail: referencedesk@oc.edu; URL: www.oc.edu/library
1950; Tamie Lyn Willis
Rare Bks & Clipping File Daily Oklahoman (1907-1981), Oklahoma Symphony Orchestra Master Tapes 1950-1984
93 000 vols; 600 curr per 31074

University of Central Oklahoma, Chambers Library, 100 N University, ***Edmond***, OK 73034; P.O. Box 192, Edmond, OK 73034-0192
T: +1 405 9742878; Fax: +1 405 9743874; URL: library.ucok.edu
1890; Habib Tabatabai
Oklahoma Coll, World War II
960 000 vols; 2 147 curr per
libr loan 31075

Southern Illinois University Edwardsville, Elijah P. Lovejoy Library, Campus Box 1063, 30 Hairpin Circle, ***Edwardsville***, IL 62026-1063
T: +1 618 6502711; Fax: +1 618 6502717; URL: www.siue.edu/lovejoy/library
1957; Regina McBride
Documents Coll, Illinois Coll, Music Coll
788 000 vols; 14 371 curr per; 2 000 e-books
libr loan 31076

Christian Heritage College, Library, 2100 Greenfield Dr, ***El Cajon***, CA 92019-1161
T: +1 619 4412200; Fax: +1 619 5902157; URL: www.sdcc.edu/library
1970; Ruth Martin
Books on religion
67 000 vols; 232 curr per; 48 000 e-books; 50 maps; 117 microforms; 135 av-mat; 1 875 sound-rec; 125 digital data carriers
libr loan; OCLC 31077

University of Texas at El Paso, Library, 500 W University Ave, ***El Paso***, TX 79968-0582
T: +1 915 7475683; Fax: +1 915 7475345; E-mail: libraryadmin@utep.edu; URL: www.libraryweb.utep.edu
1919; Carol M. Kelley
Southwest & Border Studies, Rare Books, Western Fiction, Judaica, Mexican Arch, Oral Hist, Military Hist, Chicano Studies
1 274 000 vols; 13 742 curr per; 8 000 e-books
libr loan 31078

Judson College, Benjamin P. Browne Library, 1151 N State St, ***Elgin***, IL 60123
T: +1 847 6282030; Fax: +1 847 6252045; URL: www.judsonu.edu/campuslife/library/
1963; Larry C. Wild
Religion, Music, Art; Edmondson Coll of Contemporary Christian Music; Baptist Hist & Missions
105 000 vols; 350 curr per; 27 000 microforms; 13 000 sound-rec; 20 digital data carriers; 6 300 dramatic plays, 8 200 scores, 400 radio shows
libr loan; ALA, ATLA 31079

Elizabeth City State University, G.R. Little Library, 1704 Weeksville Rd, ***Elizabeth City***, NC 27909
T: +1 252 3353586; Fax: +1 252 3353094; URL: www.ecsu.edu
1892; Juanita Midgette
194 000 vols; 1 735 curr per; 1 687 av-mat; 1 300 sound-rec; 300 Bks on Deafness & Sign Lang 31080

Elizabethtown College, The High Library, One Alpha Dr, ***Elizabethtown***, PA 17022-2227
T: +1 717 3611461; Fax: +1 717 3611167; E-mail: ask-a-librarian@etown.edu; URL: www.etown.edu/HighLibrary.aspx
1899; BethAnn Zambella
Brethren Heritage Coll, College Arch
261 000 vols; 1 150 curr per; 2 000 e-books
libr loan 31081

Davis & Elkins College, Booth Library, 100 Campus Dr, ***Elkins***, WV 26241
T: +1 304 6371233; Fax: +1 304 6371415; E-mail: library@davisandelkins.edu; URL: www.earthhome.net/library
1904; Jacqueline D. Schneider
Traditional Music Arch Coll
116 000 vols; 459 curr per
libr loan 31082

Pennsylvania College of Optometry, Library, 8360 Old York Rd, ***Elkins Park***, PA 19027
T: +1 215 7801260; Fax: +1 215 7801263; URL: www.pco.edu/college/library.htm
1919; Keith Lammers
Antique Eyewear & Ophthalmic Instruments, Old Visual Science Bks
26 000 vols; 316 curr per; 280 diss/theses; 900 sound-rec; 8 digital data carriers; 2 VF drawers, 125 video tapes, 9 500 slides
libr loan 31083

Central Washington University, James E. Brooks Library, 400 E University Way, ***Ellensburg***, WA 98926-7548
T: +1 509 9633682; Fax: +1 509 9633684; URL: www.lib.cwu.edu
1891; Dr. Tolin Philip
522 000 vols; 454 383 govt docs; 84 795 maps; 1 148 000 microforms; 4 272 av-mat; 13 218 sound-rec; 150 digital data carriers; 725 slides, 89 overhead transparencies
libr loan 31084

Elmhurst College, A.C. Buehler Library, 190 Prospect St, ***Elmhurst***, IL 60126
T: +1 630 6173160; Fax: +1 630 6173332; E-mail: ref@elmhurst.edu; URL: www.elmhurst.edu/library
1871; Susan Swords Steffen
Nursing
221 000 vols; 1 100 curr per; 682 av-mat; 6 digital data carriers; 3 382 annual rpts, 3 856 curric bks, 34 552 slides
libr loan 31085

Elmira College, Gannett-Tripp Library, One Park Pl, ***Elmira***, NY 14901
T: +1 607 7351862; Fax: +1 607 7351158; E-mail: gtl@elmira.edu; URL: www.elmira.edu/academics/library
1855; Elizabeth Wavle
Business, Criminal justice, NY State hist, Women's studies; Elmira College Regional Hist, American Music, American Lit (Mark Twain Arch), American & English Rare Books (Lande)
251 500 vols; 855 curr per; 3 700 e-books; 250 000 govt docs; 440 maps; 1 600 000 microforms; 46 090 av-mat; 2 506 sound-rec; 128 digital data carriers
libr loan; ALA 31086

Elon University, Carol Grotnes Belk Library, 308 N O'Kelly Ave, ***Elon***, NC 27244-0187; mail address: 2550 Campus Box, Elon, NC 27244
T: +1 336 2786600; Fax: +1 336 2786637; URL: www.elon.edu/library
1889; Kate Donnelly Hickey
Church Hist Coll (United Church of Christ-Southern Conference Arch); Elon College Arch; North Carolina Authors (Johnson Coll), autographed first editions; Spence Coll; Faculty Publs
278 000 vols; 7 401 curr per; 25 300 e-journals; 30 000 e-books; 2 078 av-mat
libr loan 31087

Principia College, Marshall Brooks Library, One Maybeck Pl, ***Elsah***, IL 62028-9703
T: +1 618 3745235; Fax: +1 618 3745107; URL: www.prin.edu/college/library
1898; Carol D. Stookey
205 000 vols; 800 curr per; 20 000 microforms; 9 000 sound-rec; 50 digital data carriers
libr loan 31088

Mount Saint Mary's University, Hugh J. Phillips Library, 16300 Old Emmitsburg Rd, ***Emmitsburg***, MD 21727-7799
T: +1 301 4475244; Fax: +1 301 4475099; URL: www.msmary.edu/studentsandstaff/library
1961; D. Stephen Rockwood
Theology, Business; Early Catholic Americana, 16th & 17th C Religions
211 000 vols; 910 curr per; 13 000 microforms; 5 550 sound-rec; 38 digital data carriers 31089

Emory & Henry College, Frederick T. Kelly Library, 30480 Armbrister Dr, ***Emory***, VA 24327; P.O. Box 948, Emory, VA 24327-0948
T: +1 276 9446210; Fax: +1 276 9444592; E-mail: askalibrarian@ehc.edu; URL: www.library.ehc.edu
1836; Lorraine Abraham
Liberal arts; Appalachian Oral Hist Coll, Methodist Church Hist (I P Martin Coll), Southwestern Virginiana (Goodrich Wilson Papers)
392 000 vols; 652 curr per; 78 000 e-books
libr loan 31090

Emporia State University, William Allen White Library, 1200 Commercial St, Box 4051, ***Emporia***, KS 66801
T: +1 620 3415208; Fax: +1 620 3416208; URL: library.emporia.edu
1863; Joyce Davis
Children's Lit
650 000 vols; 1 438 curr per; 8 000 e-books
libr loan 31091

Gannon University, Nash Library, 109 University Sq, ***Erie***, PA 16541
T: +1 814 8717559; Fax: +1 814 8715666; URL: www.gannon.edu/library
1925; Ken Brundage
Early American Imprints, microcard; Human Relations Area Files, microfiche; Polish Hist & Polit Sci (K Symmons Coll), curriculum mat
220 000 vols; 606 curr per; 614 av-mat; 620 Audio Bks, 380 CDs
libr loan 31092

Lake Erie College of Osteopathic Medicine, Learning Resource Center, 1858 W Grandview Blvd, ***Erie***, PA 16509-1025
T: +1 814 8668451; Fax: +1 814 8686911; E-mail: library@lecom.edu; URL: www.lecom.edu
Robert Schnick
Osteopathy
8 000 vols; 200 curr per 31093

Mercyhurst College, Hammermill Library, 501 E 38th St, ***Erie***, PA 16546
T: +1 814 8243988; Fax: +1 814 8242219; URL: merlin.mercyhurst.edu
1926; Darci Jones
Ethnic Hist, Northwest Pennsylvania Hist, Women's Hist, Pennsylvania
187 000 vols; 592 curr per 31094

Pennsylvania State Erie, University Libraries, Library, 4951 College Dr, ***Erie***, PA 16563-4115
T: +1 814 8986106; Fax: +1 814 8986350; URL: www.pserie.psu.edu/library/bdindex.htm
1948; Richard L. Hart
127 000 vols; 582 curr per 31095

Northwest Christian College, Edward P. Kellenberger Library, 828 E 11th Ave, ***Eugene***, OR 97401
T: +1 541 6847278; Fax: +1 541 6847307; E-mail: librarian@northwestchristian.edu; URL: www.northwestchristian.edu/library
1895; Steve Silver
Biblical studies; Christian Church Hist (Disciples of Christ, Discipliana), Museum Coll of English Bible (Bushnell Coll), rare bks & Bibles; Museum Coll of African, Oriental & Northwest Pioneer Artifacts Arch
63 000 vols; 261 curr per; 437 maps; 643 microforms; 332 av-mat; 2 951 sound-rec; 8 digital data carriers; 390 cubic feet of mss & unlisted serials
libr loan 31096

University of Oregon Libraries, Knight Library, 1501 Kincaid St, ***Eugene***, OR 97403-1299; mail address: 1299 University of Oregon, Eugene, OR 97403-1299
T: +1 541 3463056; Fax: +1 541 3463485; URL: www.libweb.uoregon.edu
1883; Deborah Carver
American Hist (The American West), American Missions & Missionaries, Children's Lit, Esperanto, Oriental Lit & Art; Politics (20th c American Politics, particularly Conservatism)
2 787 000 vols; 23 186 curr per; 231 000 e-books; 331 000 govt docs; 1 200 000 microforms
libr loan 31097

– Architecture & Allied Arts Library, 200 Lawrence Hall, Eugene, OR 97403; mail address: 1299 University of Oregon, Eugene, OR 97403-1299
T: +1 541 3463637; Fax: +1 541 3462205; E-mail: aaaref@uoregon.edu
1915; Edward Teague
47 000 vols; 270 000 slides, 30 000 prints & photos, 500 architectural drawings, 200 sets of blueprints 31098

– John E. Jaqua Law Library, William W Knight Law Center, 2nd Flr, 1515 Agate St, Eugene, OR 97403-1221; mail address: 1221 University of Oregon, 270 Knight Law Ctr, Eugene, OR 97403-1221
T: +1 541 3463088; Fax: +1 541 3461669; URL: lawlibrary.uoregon.edu
1893; Mary Ann Hyatt
132 000 vols; 3 188 curr per; 744 000 microforms 31099

– Mathematics Library, 210 Fenton Hall, University of Oregon, Eugene, OR 97403; mail address: 1299 University of Oregon, Eugene, OR 97403-1299
T: +1 541 3463023; Fax: +1 541 3463012; URL: www.libweb.uoregon.edu/scilib/mathlib/
1980; Victoria Mitchell
26 000 vols; 350 curr per 31100

– Science Library, Onyx Bridge, Lower Level, University of Oregon, Eugene, OR 97403; mail address: 1299 University of Oregon, Eugene, OR 97403-1299
T: +1 541 3462661; Fax: +1 541 3463012; URL: www.libweb.uoregon.edu/scilib/
1968; Victoria Mitchell
160 000 vols; 2 794 curr per 31101

Louisiana State University-Eunice, LeDoux Library, 2048 Johnson Hwy, ***Eunice***, LA 70535; P.O. Box 1129, Eunice, LA 70535-1129
T: +1 337 5501380; Fax: +1 337 5501455; URL: www.lsue.edu
1967; Gerald Patout
Genealogy, Louisiana
90 000 vols; 218 curr per 31102

Eureka College, Melick Library, 301 E College Ave, ***Eureka***, IL 61530-1563
T: +1 309 4676382; Fax: +1 309 4676386; E-mail: library@eureka.edu; URL: www.eureka.edu/melick/melick.htm
1855; Anthony R. Glass
Christian Church (Disciples of Christ Coll)

78 000 vols; 250 curr per
libr loan   31103

Garrett-Evangelical & Seabury-Western Theological Seminaries, The United Library, 2121 Sheridan Rd, *Evanston*, IL 60201
T: +1 847 8663911; Fax: +1 847 8663957; URL: www.garrett.edu/library
1857; Alva R. Caldwell
Deering-Jackson Methodistica, Keen Bible Coll, Hibbard Egyptian Coll
320 000 vols; 1 700 curr per; 1 000 microforms; 3 digital data carriers
libr loan; ATLA   31104

Northwestern University, University Library, 1970 Campus Dr, *Evanston*, IL 60208-2300
T: +1 847 4917640; Fax: +1 847 4918306; E-mail: refdept@northwestern.edu; URL: www.library.northwestern.edu
1851; Sarah M. Pritchard
Frank Lloyd Wright Coll, Contemporary Music Scores, Africana, European Union, Modern Art & Lit
3 903 000 vols; 31 622 curr per; 2,6 mio microform units
libr loan   31105

– Geology Library, 1850 Campus Dr, Locy Hall, Rm 101, Evanston, IL 60208
T: +1 847 4915525; E-mail: annawu@northwestern.edu; URL: www.library.northwestern.edu/geology
Anna Ren
26 400 vols; 120 curr per   31106

– Melville J. Herskovits Library of African Studies, 1970 Campus Dr, Evanston, IL 60208-2300
T: +1 847 4917684; Fax: +1 847 4671223; E-mail: africana@northwestern.edu; URL: www.library.nwu.edu/africana
1954; David L. Easterbrook
African Language Publs Coll; Arabic/Hausa Mss from Kano; G. M. Carter/T. Karis Coll on South African Politics; African Studies Assn; African Lit Assn
245 000 vols; 2 800 curr per; 10 211 maps; 3 900 feet of vertical files, 11 750 pamphlets, 2 500 posters   31107

– Music Library, 1970 Campus Dr, Evanston, IL 60208-2300
T: +1 847 4913434; Fax: +1 847 4677574; URL: www.library.northwestern.edu/music
D. J. Hoek
Music composed since 1945
46 000 vols; 400 curr per   31108

– Ralph P. Boas Mathematics Library, 2033 Sheridan Rd, Lunt Bldg, Rm 130, Evanston, IL 60208
T: +1 847 4917692; Fax: +1 847 4914655; URL: www.library.northwestern.edu/math
Cunera Buys
34 600 vols; 630 curr per   31109

– Seeley G. Mudd Library for Science & Engineering, 2233 N Campus Dr, Evanston, IL 60208
T: +1 847 4913362; Fax: +1 847 4914655; E-mail: sel@northwestern.edu; URL: www.library.northwestern.edu/SEL
1977; Robert Michaelson
275 000 vols; 1 800 curr per; 75 000 rpts on microfiche   31110

– Transportation Library, 1970 Campus Dr, Evanston, IL 60208
T: +1 847 4918600; Fax: +1 847 4918601; E-mail: trans@northwestern.edu; URL: www.library.northwestern.edu/transportation
Roberto A. Sarmiento
266 000 vols   31111

University of Evansville, University Libraries, 1800 Lincoln Ave, *Evansville*, IN 47722
T: +1 812 4882376; Fax: +1 812 4886996; E-mail: library@evansville.edu; URL: libraries.evansville.edu
1872; William F. Louden
James L Clifford, 18th C Mat; Knecht Cartoons Coll; Law (Kiltz Coll)
281 000 vols; 1 396 curr per; 100 digital data carriers
libr loan   31112

University of Southern Indiana, David L. Rice Library, 8600 University Blvd, *Evansville*, IN 47712
T: +1 812 4641683; Fax: +1 812 4651693; E-mail: libweb@usi.edu; URL: www.usi.edu/library/index.asp
1965; Ruth H. Miller
Ctr for Communal Studies, Children's Lit Coll, Indiana Labor Hist Coll, Mead Johnson Coll, Local Govt Coll, Sun Oil Geology Coll, Univ Arch, Movie Press Kits
222 000 vols; 594 curr per; 3 000 e-books; 4 505 maps; 590 000 microforms; 274 digital data carriers
libr loan   31113

The College of New Jersey, TCNJ Library, 2000 Pennington Rd, *Ewing*, NJ 08628-0718; P.O. Box 7718, Ewing, NJ 08628-0718
T: +1 609 7712417; Fax: +1 609 6375177; URL: www.tcnj.edu/~library
1855; Taras Pavlovsky
Hist of New Jersey, Hist of the American Revolution, Hist of American Education, Hist Textbooks & Hist Children's Bks
580 000 vols; 1 485 curr per; 694 000 microforms; 1 000 digital data carriers
libr loan   31114

University of Alaska Fairbanks, Elmer E. Rasmuson Library, 310 Tanana Dr, *Fairbanks*, AK 99775; P.O. Box 756800, Fairbanks, AK 99775-6800
T: +1 907 4745348; Fax: +1 907 4746841; E-mail: fydir@uaf.edu; URL: www.library.uaf.edu
1917; James Huesmann
Alaska and Polar Regions
809 000 vols; 7 716 curr per; 9 500 av-mat
libr loan; ALA, OCLC, BCR   31115

– BioSciences Library, 186 Arctic Health Research Bldg, P.O. Box 757060, Fairbanks, AK 99775-7060
T: +1 907 4747442; Fax: +1 907 4747820; E-mail: fybmlib@uaf.edu; URL: library.uaf.edu/biosci
1949
55 000 vols; 550 curr per; 5 460 microforms; 584 digital data carriers
libr loan; IAMSLIC   31116

George Mason University, Fenwick Library, 4400 University Dr, MSN 2FL, *Fairfax*, VA 22030-4444
T: +1 703 9932491; Fax: +1 703 9932494; URL: library.gmu.edu
1957; John G. Zenelis
Civil War Colls, Regional Hist, maps, Congressional Papers of Joel T Broyhill, Francis McNamara Papers, American Theater
950 000 vols; 5 857 curr per; 23 000 e-journals; 100 000 e-books
libr loan   31117

– School of Law Library, 3301 N Fairfax Dr, Arlington, VA 22201-4426
T: +1 703 9938120; Fax: +1 703 9938113; URL: www.law.gmu.edu/libtech
1971; Deborah Keene
US Specialized Law & Economics Colls, Patents, Banking Law
253 000 vols; 5 634 curr per; 675 000 microforms; 102 av-mat; 64 digital data carriers   31118

Fairfield University, DiMenna-Nyselius Library, 1073 N Benson Rd, *Fairfield*, CT 06430-5195
T: +1 203 2544000; Fax: +1 203 2544135; E-mail: reference@mail.fairfield.edu; URL: www.fairfield.edu/library
1948; Joan Overfield
Education, Religion, Nursing; Science Hist, American Studies
351 000 vols; 1 424 curr per; 15 e-journals; 43 000 e-books; 5 970 av-mat; 2 100 sound-rec; 1 600 digital data carriers
libr loan   31119

Maharishi University of Management, Library, 1000 N Fourth St, *Fairfield*, IA 52557
T: +1 641 4727000; Fax: +1 641 4721137; E-mail: library@mum.edu; URL: www.mum.edu
1971; Martin Schmidt
Science of Creative Intelligence Coll (Maharishi Mahesh Yogi & M.U.M.

Faculty), Vedic Lit, Maharishi's Vedic Science and Technology
140 000 vols; 115 000 curr per; 10 853 e-journals; 80 diss/theses; 60 000 microforms; 4 419 sound-rec; 80 digital data carriers
libr loan; IPAL, ILA   31120

Miles College, Learning Resources Center, 5500 Myron Massey Blvd, *Fairfield*, AL 35064; P.O. Box 39800, Birmingham, AL 35208-0937
T: +1 205 9291715; Fax: +1 205 9291635; URL: www.miles.edu/lrc.htm
1905; Dr. Geraldine Bell
Afro-American Coll, Instructional Materials Coll
100 000 vols; 400 curr per   31121

Sacred Heart University, Ryan-Matura Library, 5151 Park Ave, *Fairfield*, CT 06825-1000
T: +1 203 3654854; Fax: +1 203 3717833; URL: www.library.sacredheart.edu
1967; Dennis C. Benamati
Religion, Psychology, Business; World Children's Bks, Marian Coll
145 000 vols; 770 microforms; 4 850 sound-rec; 13 digital data carriers; 42 Bks on Deafness & Sign Lang
libr loan   31122

Fairmont State University, Ruth Ann Musick Library, 1201 Locust Ave, *Fairmont*, WV 26554
T: +1 304 3674733; Fax: +1 304 3674677; URL: library.fairmontstate.edu
1867; Thelma Hutchins
218 000 vols; 394 curr per; 27 820 e-journals; 63 000 e-books
libr loan   31123

North Dakota State University, University Libraries, 1201 Albrecht Blvd, *Fargo*, ND 58105-5599
T: +1 701 2318885; Fax: +1 701 2316128; URL: www.lib.ndsu.nodak.edu
1891; James Council
Agriculture; Bonanza Farming Coll, Germans from Russia Heritage Coll, Fred Hultstrand Hist in Pictures Coll, North Dakota Biography Index, North Dakota Pioneer Reminiscences, Senator Milton R Young Photogr Coll
623 000 vols; 8 757 curr per; 1 020 e-journals; 10 000 e-books
libr loan   31124

State University of New York, Thomas D. Greenley Library, 2350 Broadhollow Road, *Farmingdale*, NY 11735-1021
T: +1 631 4202040; Fax: +1 631 4202473; E-mail: knauthmg@farmingdale.edu; URL: www.farmingdale.edu/library/
1912; Michael G. Knauth
125 000 vols; 800 curr per; 120 000 govt docs; 26 500 microforms; 23 000 av-mat; 16 digital data carriers; 22 000 pamphlets
libr loan   31125

University of Maine at Farmington, Mantor Library, 116 South St, *Farmington*, ME 04938-1990
T: +1 207 7787210; Fax: +1 207 7787223; URL: www.umf.maine.edu
1933; Franklin D. Roberts
Univ Arch
99 000 vols; 1 800 curr per
libr loan   31126

Janet D. Greenwood Library, Redford & Race St, *Farmville*, VA 23909; mail address: 201 High St, Farmville, VA 23909
T: +1 434 3952433; Fax: +1 434 3952453; E-mail: libweb@longwood.edu; URL: www.longwood.edu/library/index.htm
1839; Wendell Barbour
275 000 vols; 2 505 curr per   31127

Bevill State Community College, Learning Resources Center, 2631 Temple Ave N, *Fayette*, AL 35555
T: +1 205 9323221 ext 5141; Fax: +1 205 9328821; E-mail: smiddleton@bevillst.cc.al.us; URL: www.bevillst.cc.al.us
1969; Sally Middleton
Albert P. Brewer Coll
40 000 vols   31128

Central Methodist College, Smiley Memorial Library, 411 Central Methodist Sq, *Fayette*, MO 65248
T: +1 660 2486271; Fax: +1 660 2486226; E-mail: library@centralmethodist.edu; URL: www.centralmethodist.edu
1857; Cynthia Dudenhoffer
Religion (Missouri United Methodist Arch)
99 000 vols; 200 curr per; 281 av-mat   31129

Upper Iowa University, Henderson-Wilder Library, 605 Washington St, *Fayette*, IA 52142; P.O. Box 1858, Fayette, IA 52142-1858
T: +1 563 4255217; Fax: +1 563 4255271; E-mail: library@uiu.edu; URL: www.uiu.edu/library
1901; Becky Wadian
NASA Coll
69 000 vols; 290 curr per; 4 000 microforms; 551 av-mat; 1 589 sound-rec; 41 digital data carriers; 1 313 slides
libr loan   31130

Davis Memorial Library, 5400 Ramsey St, *Fayetteville*, NC 28311
T: +1 910 6307587; Fax: +1 910 6307119; E-mail: reference@methodist.edu; URL: www.methodist.edu/library/davis.htm
1960; Tracey Pearson
Catherine Huske Coll, Lafayette Coll
110 000 vols; 470 curr per; 23 000 e-books; 1 803 music scores; 325 maps; 28 000 microforms; 759 av-mat; 2 984 sound-rec; 40 digital data carriers; 14 311 non-books, 5 818 slides, 814 overhead tranparencies, 186 charts, 160 art reproductions
libr loan   31131

Fayetteville State University, Charles W. Chesnutt Library, 1200 Murchison Rd, *Fayetteville*, NC 28301-4298
T: +1 910 6721233; Fax: +1 910 6721312; URL: library.uncfsu.edu
1867; Bobby C. Wynn
African-American, Ethnic studies
292 000 vols; 2 742 curr per; 1 000 e-journals; 13 000 e-books; 5 000 av-mat; 1 120 sound-rec
libr loan   31132

University of Arkansas, University Libraries, 365 N McIlroy Ave, *Fayetteville*, AR 72701-4002
T: +1 479 5756645; Fax: +1 479 5755558; URL: libinfo.uark.edu
1872; Carolyn Henderson Allen
Int Relations, Agriculture, Architecture, Creative writing
1 809 000 vols; 15 189 curr per; 1 393 mss; 222 703 govt docs; 122 116 maps; 3 245 000 microforms; 20 590 sound-rec; 1 366 digital data carriers
libr loan; ACRL, ALA, Greater Western Library Alliance   31133

– Chemistry and Biochemistry Library, 365 N McIlroy Ave, Fayetteville, AR 72701-4002
T: +1 479 5752557; E-mail: lsalisbu@uark.edu; URL: libinfo.uark.edu/chemistry/
1932; Luti Salisbury
Eva Dickson Coll, Harrison Hale Coll
35 000 vols; 170 curr per   31134

– Fine Arts Library, 104 Fine Arts Bldg, Fayetteville, AR 72701; mail address: 365 N McIlroy Ave, Fayetteville, AR 72701-4002
T: +1 479 5754708; URL: libinfo.uark.edu/fal/default.asp
Margaret Boylan
36 000 vols   31135

– Physics Library, 102 Physics, Fayetteville, AR 72701; mail address: 365 N McIlroy Ave, Fayetteville, AR 72701-4002
T: +1 479 5752505
Usha Gupta
19 000 vols   31136

– Robert A. & Vivian Young Law Library, School of Law, Waterman Hall 107, Fayetteville, AR 72701-1201
T: +1 479 5755601; Fax: +1 479 5752053; URL: law.uark.edu/young_law_library
1924; Monika Szakasits
Agricultural Law Coll
158 000 vols; 545 curr per   31137

Ferrum College, Thomas Stanley Library, 150 Wiley Dr, **Ferrum**, VA 24088; P.O. Box 1000, Ferrum, VA 24088-1000
T: +1 540 3654426; Fax: +1 540 3654423; URL: www.ferrumlibrary.net
1913; Cy Dillon
Arch of Governor & Mrs Thomas B Stanley
115 000 vols; 14 600 curr per; 100 000 e-books   31138

The University of Findlay, Shafer Library, 1000 N Main St, **Findlay**, OH 45840-3695
T: +1 419 4344700; Fax: +1 419 4344196; URL: www.findlay.edu/offices/resources/library/default.htm
1882; Robert W. Schirmer
Hist of Churches of God in North America; Congressional Papers (Jackson E Betts Coll & Tennyson Guyer Coll); Wilfred W Black, Coll of Mid-Nineteenth C Americana; College Hist Mat, print & nonprint
136 000 vols; 1 080 curr per
libr loan   31139

Fitchburg State College, Amelia V. Galucci-Cirio Library, 160 Pearl St, **Fitchburg**, MA 01420
T: +1 978 6653194, 6653196; Fax: +1 978 6653069; E-mail: rfoley@fsc.edu; URL: www.fsc.edu/library
1894; Robert Foley
College Arch; Cormier Coll; Moon Coll; Rice Art Coll; Salvatore Coll
2 441 120 vols; 1 343 curr per; 3 600 e-books; 429 950 microforms; 1 850 av-mat; 1 830 sound-rec; 20 digital data carriers; 8 693 pamphlets, 180 art reproductions   31140

Northern Arizona University, Cline Library, Bldg 028, Knoles Dr, **Flagstaff**, AZ 86011; P.O. Box 6022, Flagstaff, AZ 86011-6022
T: +1 928 5236805; Fax: +1 928 5233770; E-mail: library.administration@nau.edu; URL: library.nau.edu
1912; Cynthia Childrey
Arizona Hist (Hist Society Coll), arch, photos; Bruce Babbitt Coll; Harvey Butchart Coll; Colorado Plateau Coll; Fred Harvey Coll; Philip Johnston Coll; Elbert Hubbard-Roycroft Press (Floyd C. Henning Coll); NAU Arch; A. F. Whiting Coll, Emery Kolb Coll
634 000 vols; 30 964 curr per; 37 000 e-books; 36 006 maps; 375 000 microforms; 19 020 sound-rec; 79 digital data carriers
libr loan   31141

Kettering University, University Library, 1700 W Third Ave, **Flint**, MI 48504-4898
T: +1 810 7627814; Fax: +1 810 7629744; URL: www.kettering.edu/library
1928; Dr. Charles Hanson
Engineering, Business & management; SAE & SME Tech Coll, papers; ASTM Standards
132 000 vols; 500 curr per; 35 000 microforms; 131 sound-rec; 12 digital data carriers   31142

University of Michigan – Flint Library, Frances Willson Thompson Library, 303 E Kearsley St, **Flint**, MI 48502-1950
T: +1 810 762340; Fax: +1 810 7623133; URL: lib.umflint.edu
1956; Robert L. Houbeck Jr
Foundation Ctr Regional Coll, Genesee Hist Coll Ctr
235 000 vols; 905 curr per   31143

Francis Marion University, James A. Rogers Library, 4822 Palmetto St, **Florence**, SC 29506; P.O. Box 100547, Florence, SC 29506
T: +1 843 6611311; Fax: +1 843 6611309; URL: www.fmarion.edu/academics/library
1970; Joyce M. Durant
South Caroliniana
391 000 vols; 1 536 curr per; 2 322 e-journals; 28 000 e-books; 341 maps; 417 000 microforms; 26 digital data carriers; 50 332 Journals
libr loan   31144

University of North Alabama, Collier Library, One Harrison Plaza, Box 5028, **Florence**, AL 35632-0001
T: +1 256 7654308; Fax: +1 256 7654438; URL: www2.una.edu/library

1830; Debbie Chaffin
Alabama Hist Coll, Congressman Flippo Coll, Local Hist Arch
380 000 vols; 3 476 curr per; 2 700 e-journals; 56 000 e-books; 7 930 av-mat; 1 400 sound-rec; 150 digital data carriers
libr loan   31145

Wesley College, Library, 166 Hwy 469 N, 111 Wesley Circle, **Florence**, MS 39073; P.O. Box 1070, Florence, MS 39073-1070
T: +1 601 8458562; Fax: +1 601 8452266; URL: www.wesleycollege.edu
Darlene Morgan
English lit, Holiness lit
18 000 vols; 162 curr per   31146

Saint Louis Christian College, Library, 1360 Grandview Dr, **Florissant**, MO 63033
T: +1 314 8376777; Fax: +1 314 8378291; E-mail: librarian@slcconline.edu; URL: www.slcc4ministry.edu
1956; Derek L. Brink
Carl Kitcherside Libr; Libr of American Civilization Coll, micro
30 000 vols; 91 curr per; 16 000 microforms; 2 460 sound-rec; 13 digital data carriers
libr loan; ACL   31147

Queens College – Aaron Copeland School of Music, Queens College Music Library, 65-30 Kissena Blvd, Music Bldg, Room 225, **Flushing**, NY 11367
T: +1 718 9973900; Fax: +1 718 9973928; E-mail: jennifer.oates@qc.cuny.edu; URL: qcpages.qc.cuny.edu/Music_Library/
Dr. Jennifer Oates
Carol Rathaus Archives, K. Robert Schwarz Papers, coll of 18th and 19th c American hymnals, large coll of first edition scores published before World War I
50 000 vols; 20 000 sound-rec   31148

Queens College – Benjamin S. Rosenthal Library, 65-30 Kissena Blvd, **Flushing**, NY 11367-0904
T: +1 718 9973799; Fax: +1 718 9973753; URL: www.qc.edu/Library
1937; N.N.
Louis Armstrong arch, mss, personal papers, photos, recordings, scrapbks, tapes & recordings; New York Stage & Film Coll (through 1950); Theater Coll
780 000 vols; 5 820 curr per; 5 000 e-books
libr loan   31149

Pennsylvania State University – Lehigh Valley, Berks Lehigh Valley College Library-Learning Resources Center, 8380 Mohr Lane, **Fogelsville**, PA 18051-9999
T: +1 610 2855027; Fax: +1 610 2855158; URL: www.lv.psu.edu/library
1912; Dennis J. Phillips
40 000 vols; 172 curr per   31150

Marian College of Fond du Lac, Cardinal Meyer Library, 45 S National Ave, **Fond du Lac**, WI 54935
T: +1 920 9238096; Fax: +1 920 9237154; E-mail: refdesk@marianuniversity.edu; URL: www.marianuniversity.edu
1966; Mary Ellen Gormican
Avie Waxman Jewish Lit, Cardinal Newman Coll, Gromme Bird Coll
91 000 vols; 551 curr per
libr loan   31151

Pacific University, Harvey W. Scott Memorial Library, 2043 College Way, **Forest Grove**, OR 97116
T: +1 503 3521400; Fax: +1 503 3521416; E-mail: reference@pacificu.edu; URL: www.pacificu.edu/library
1849; Marita Kunkel
Optometry, Clinical psychology, Occupational therapy, Physical therapy
209 000 vols; 1 049 curr per
libr loan   31152

Bramson Ort College, Library-Learning Resource Center, 69-30 Austin St, **Forest Hills**, NY 11375
T: +1 718 2615800; Fax: +1 718 5755118; E-mail: rburkos@bramsonort.edu; URL: bramsonort.edu/departments/library/index.html
1977; Rivka Burkos

Business administration, Electronics, English as a second language, Jewish studies; Judaica
13 000 vols; 105 curr per; 800 sound-rec; 300 slides   31153

Colorado State University, William E. Morgan Library, **Fort Collins**, CO 80523-1019
T: +1 970 4911842; Fax: +1 970 4911195; URL: lib.colostate.edu
1870; Catherine Murray-Rust
Agriculture, Agricultural Economics and Engineering, Germans From Russia, Int Poster Coll, Vietnam War Fiction
1 897 000 vols; 21 252 curr per; 63 144 maps; 2 400 000 microforms; 4 788 digital data carriers; 6 737 audio reproductions, 5 493 computer files
libr loan   31154

– Veterinary Teaching Hospital, Library, Colorado State University, Fort Collins, CO 80523
T: +1 970 4911213; Fax: +1 970 4914141; URL: lib.colostate.edu/branches/vet.html
Tom Moothart
12 000 vols; 162 curr per   31155

University of Maine at Fort Kent, Blake Library, 23 University Dr, **Fort Kent**, ME 04743
T: +1 207 8347527; Fax: +1 207 8347518; URL: www.umfk.maine.edu/infoserv/library
1878; Sharon M. Johnson
Saint John Valley Hist
70 000 vols; 362 curr per   31156

Art Institute of Fort Lauderdale, Meinhardt Memorial Library, 1799 SE 17th St, **Fort Lauderdale**, FL 33316; mail address: 1600 SE 17th St, 3rd Flr, Fort Lauderdale, FL 33316
T: +1 954 3082631; Fax: +1 954 4633393; E-mail: aifl_library@aii.edu; URL: aifl.aiiresources.com .
Heather Payne
26 000 vols; 315 curr per   31157

Nova Southeastern University, Alvin Sherman Library, Research & Information Technology Center, 3100 Ray Ferrero Jr Blvd, **Fort Lauderdale**, FL 33314
T: +1 954 2624660; Fax: +1 954 2626830; URL: www.nova.edu/library
1966; Lydia Acosta
405 000 vols; 2 442 curr per; 20 600 e-journals; 18 000 e-books; 253 000 microforms; 200 High Interest/Low Vocabulary Bks, 2100 Bks on Deafness & Sign Lang, 120 Large Print Bks   31158

– Health Professions Division Library, 3200 S University Dr, Fort Lauderdale, FL 33328
T: +1 954 2623106; Fax: +1 954 2621821; URL: www.nova.edu/cwis/hpdlibrary/
Kaye Robertson
Dentistry, optometry, pharmacology, occupational therapy
70 000 vols; 3 388 curr per; 201 e-books   31159

– Oceanography, Library, 8000 N Ocean Dr, Dania Beach, FL 33004
T: +1 954 2623643; Fax: +1 954 2624021; URL: www.nova.edu/ocean/library.html
Kathleen Maxson
10 000 vols; 100 curr per   31160

– Shepard Broad Law Center Library, 3305 College Ave, Fort Lauderdale, FL 33314
T: +1 954 2626100; Fax: +1 954 2623839; E-mail: referencedesk@nsu.law.nova.edu; URL: www.nsulaw.nova.edu
1974
Admiralty, Law & Popular Culture
360 000 vols; 5 466 curr per; 129 000 microforms; 442 av-mat; 729 sound-rec; 208 digital data carriers; 3 490 vols of unbound mats, 459 data files   31161

Florida Gulf Coast University, University Library, 10501 FGCU Blvd S, **Fort Myers**, FL 33965-6501
T: +1 239 5907600; Fax: +1 239 5907609; URL: library.fgcu.edu
1997; Kathleen Miller
187 000 vols; 1 310 curr per; 43 500 e-journals   31162

Fort Valley State University, Henry Alexander Hunt Memorial Library, 1005 State University Dr, **Fort Valley**, GA 31030-4313
T: +1 478 8256343; Fax: +1 478 8256663; URL: www.fvsu.edu
1925; Frank Mahitab
Fort Valley State Univ Coll, Ethnic Heritage Coll
190 000 vols; 805 curr per; 346 000 microforms; 4 654 sound-rec; 45 digital data carriers
libr loan   31163

Indiana University-Purdue University Fort Wayne, Walter E. Helmke Library, 2101 E Coliseum Blvd, **Fort Wayne**, IN 46805-1499
T: +1 260 4816505; Fax: +1 260 4816509; URL: www.lib.ipfw.edu
1964; Cheryl Truesdell
Univ Arch, Bob Englehart Cartoons, Sylvia Bowman Papers
483 000 vols; 814 curr per; 20 190 e-journals; 75 000 e-books; 482 000 microforms; 3 920 sound-rec; 261 digital data carriers   31164

Taylor University – Fort Wayne, S.A. Lehman Memorial Library, 1025 W Rudisill Blvd, **Fort Wayne**, IN 46807
T: +1 260 7448681; URL: fw.taylor.edu/academics/library
1904; Anita Gray
Instructional Mat Ctr
82 000 vols; 450 curr per; 4 877 sound-rec; 106 slides, 222 overhead transparencies
libr loan   31165

University of Saint Francis, University Library, 2701 Spring St, **Fort Wayne**, IN 46808
T: +1 260 3997700; Fax: +1 260 3998166; E-mail: library@sf.edu; URL: www.sf.edu/library
1890; Karla Alexander
ERIC Doc Coll
63 000 vols; 480 curr per   31166

Texas Christian University, Mary Couts Burnett Library, 2913 Lowden St, TCU Box 298400, **Fort Worth**, TX 76129
T: +1 817 2577106; Fax: +1 817 2577282; URL: library.tcu.edu
1873; Kerry Bouchard
European Union Reference Center, International Piano Competition (Cliburn Foundation Arch), Lit (William Luther Lewis Coll), US Hist (James C. Wright Jr Arch)
1 419 000 vols; 51 978 curr per
libr loan   31167

Texas Wesleyan University, Eunice & James L. West Library, 1201 Wesleyan St, **Fort Worth**, TX 76105
T: +1 817 5314821; Fax: +1 817 5314806; URL: ezproxy.txwes.edu
1891; Cindy Potter
Bobby Bragen Baseball Memorabilia, Joe Brown Theatre Coll, Twyla Miranda Juvenile Coll
220 000 vols; 360 curr per; 28 000 e-books; 880 av-mat; 8 170 sound-rec   31168

Framingham State College, Henry Whittemore Library, 100 State St, **Framingham**, MA 01701; P.O. Box 9101, Framingham, MA 01701-9101
T: +1 508 6264651; Fax: +1 508 6264649; URL: www.framingham.edu/wlibrary
1969; Bonnie Mitchell
College & Local Hist Coll, Curriculum Mat, Eric Docs Coll, Faculty Publs, Modern American Poetry
205 000 vols; 400 curr per   31169

Kentucky State University, Paul G. Blazer Library, 400 E Main St, **Frankfort**, KY 40601-2355
T: +1 502 5976880; Fax: +1 502 5975068; URL: www.kysu.edu/academics/library/
1886; Sheila A. Stuckey
Black Studies Coll
316 000 vols; 922 curr per; 17 000 e-books; 420 av-mat; 4 165 sound-rec; 387 digital data carriers; 104 overhead transparencies, 505 art reproductions
libr loan   31170

Dean College, Library, 99 Main St, ***Franklin***, MA 02038-1994
T: +1 508 5411771; Fax: +1 508 5411918; URL: www.dean.edu/administration/Library.cfm
1865; Rick Barr, Judy Tobey
Encyclopedia Britannica Libr of American Civilization & Basic Libr of English Lit, ultra-fiche; New York Times (1851 to date), microfilm
53 000 vols; 150 curr per 31171

Franklin College, B.F. Hamilton Library, 101 Branigin Blvd, ***Franklin***, IN 46131-2623
T: +1 317 7388164; Fax: +1 317 7388787; E-mail: library@franklincollege.edu; URL: www.franklincollege.edu/library
1834; Ronald L. Schuetz
David Demaree Banta Coll of Indiana Hist & Lit, Indiana Baptist Coll, Roger D Branigin Personal Papers
132 000 vols; 273 curr per; 5 740 av-mat
libr loan 31172

Hood College, Beneficial-Hodson Library, 401 Rosemont Ave, ***Frederick***, MD 21701
T: +1 301 6963709; Fax: +1 301 6963796; URL: www.hood.edu/library
1893; Jan Samet O'Leary
Civil War (Landauer Coll), Environ Biology, Sylvia Meagher (Kennedy Assassination Arch)
209 000 vols; 420 curr per; 24 000 e-journals; 4 000 e-books
libr loan; ALA 31173

University of Mary Washington, Simpson Library, 1801 College Ave, ***Fredericksburg***, VA 22401-4665
T: +1 540 6541756; Fax: +1 540 6541067; URL: www.library.umw.edu
1908; LeRoy S. Strohl
Claude Bernard Coll, James Joyce Coll, William Butler Yeats Coll
37 000 vols; 42 000 e-books
libr loan 31174

State University of New York at Fredonia, Daniel A. Reed Library, 280 Central Ave, ***Fredonia***, NY 14063
T: +1 716 6733181; Fax: +1 716 6733185; URL: www.fredonia.edu/library
1826; Randy Gadikian
Chautauqua & Cattaraugus Counties' Hist, Seneca/Iroquois Hist, Stefan Zweig Coll, Holland Land Company Coll
341 000 vols; 283 curr per; 15 000 e-journals; 700 e-books; 32 460 music scores; 1 570 av-mat; 3 720 sound-rec; 170 digital data carriers 31175

Midland Lutheran College, Luther Library, 900 N Clarkson, ***Fremont***, NE 68025
T: +1 402 9416250; Fax: +1 402 7276223; URL: www.mlc.edu/midlandlutheran.aspx?pgid=1303
1883; Thomas Boyle
Biblical Lit Coll
111 000 vols; 450 curr per; 9 000 e-journals; 50 000 e-books; 2 000 av-mat; 1 700 sound-rec; 100 digital data carriers
libr loan 31176

Alliant International University, Kauffman Library, 5130 E Clinton Way, ***Fresno***, CA 93727
T: +1 559 2532265; Fax: +1 559 2532223; URL: library.alliant.edu
1973; Louise A. Colbert-Mar
Psychology; Test Coll
18 000 vols; 35 curr per; 8 000 e-books; 1 000 microforms; 300 av-mat; 315 sound-rec; 7 digital data carriers
libr loan 31177

California Christian College, Library, 4881 E University Ave, ***Fresno***, CA 93703
T: +1 559 2515025; Fax: +1 559 2514231; URL: calchristiancollege.org
1955; Virginia Jolliff
Religion; California Free Will Baptist Hist, Free Will Baptist Denominational Hist
15 000 vols; 56 curr per 31178

California State University, Fresno, Henry Madden Library, 5200 N Barton Ave, Mail Stop ML-34, ***Fresno***, CA 93740-8014
T: +1 559 2782551; Fax: +1 559 2786952; URL: www.lib.csufresno.edu
1911; Peter McDonald
Hist (Roy J. Woodward Memorial Libr of California), Lit (William Saroyan Coll), International Exhibitions, Enology, Credit Foncier Colony, Topolobampo, Sinaloa (Mexico Coll), Arne Nixon Ctr for the Study of Children's Lit
1 070 000 vols; 2 219 curr per; 41 000 e-books; 1 187 000 microforms; 3 449 av-mat; 62 942 sound-rec; 25 digital data carriers 31179

Fresno Pacific University, Hiebert Library, 1717 S Chestnut Ave, ***Fresno***, CA 93702
T: +1 559 4532090; Fax: +1 559 4532124; URL: www.fresno.edu/dept/library
1944; Richard S. Rawls
Religion, Hist; Ctr for Mennonite Brethren Studies
163 000 vols; 1 192 curr per; 245 000 microforms; 6 000 sound-rec; 3 digital data carriers
ATLA 31180

Christendom College, O'Reilly Memorial Library, 263 St Johns Way, ***Front Royal***, VA 22630
T: +1 540 636290; Fax: +1 540 6366569; E-mail: library@christendom.edu
1977; Andrew Armstrong
Hagiography, Rare Bks
60 000 vols; 320 curr per 31181

Frostburg State University, Lewis J. Ort Library, One Stadium Dr, ***Frostburg***, MD 21532
T: +1 301 6874424; Fax: +1 301 6877069; URL: www.frostburg.edu/dept/library
1898; Dr. David M. Gillespie
Selected US Geological Survey Maps, William Price Coll of Railroad photos, George Meyers American Communist Party & Labor Party Mat Coll
272 000 vols; 636 curr per; 39 003 maps; 46 000 microforms; 2 870 av-mat; 7 147 sound-rec; 181 digital data carriers; 19 340 Slides 31182

California State University, Paulina June & George Pollak Library, 800 N State College Blvd, ***Fullerton***, CA 92834; P.O. Box 4150, Fullerton, CA 92834-4150
T: +1 714 2782721; Fax: +1 714 2782439; E-mail: askref@fullerton.edu; URL: www.library.fullerton.edu
1959; Richard C. Pollard
Local Hist Coll, Science Fiction Coll, Press & Fine Printing Coll, Angling Coll, Bowell Coll for the Hist of Cartography, Freedom Center
1 111 000 vols; 2 476 curr per
libr loan 31183

Hope International University, Hugh & Hazel Darling Library, 2500 E Nutwood Ave, ***Fullerton***, CA 92831
T: +1 714 8793901; Fax: +1 714 6817515; E-mail: darlinglibrary@hiu.edu; URL: library.hiu.edu
1928; Robin R. Hartman
Rare book coll, primarily related to hist of Restoration Movement
65 000 vols; 300 curr per; 28 000 e-books; 45 music scores; 191 sound-rec; 423 videorecordings
libr loan 31184

Western State Law Library, 1111 N State College Blvd, ***Fullerton***, CA 92831-3014
T: +1 714 4591113; Fax: +1 714 8714806; URL: www.wsulaw.edu
1966; Cindy Parkhurst
99 000 vols; 76 802 vols in microform 31185

Westminster College – Reeves Memorial Library, 501 Westminster Ave, ***Fulton***, MO 65251-1299
T: +1 573 5925246; Fax: +1 573 6426356; URL: www.westminster-mo.edu/academics/resources/library/pages/default.aspx
1905; Angela Gerling
99 000 vols; 241 curr per; 16 360 e-journals; 7 000 e-books; 1 610 av-mat; 130 sound-rec; 140 digital data carriers
libr loan; ALA 31186

William Woods University, Dulany Memorial Library, One University Ave, ***Fulton***, MO 65251
T: +1 573 5924291; Fax: +1 573 5921159; E-mail: libref@williamwoods.edu; URL: www.williamwoods.edu
1870; Erlene A. Dudley
Education Coll, Equestrian Science Coll
130 000 vols; 419 curr per; 11 000 microforms
libr loan 31187

Limestone College, A.J. Eastwood Library, 1115 College Dr, ***Gaffney***, SC 29340
T: +1 864 4884612; Fax: +1 864 4874613; URL: lib.limestone.edu
1845; Carolyn T. Hayward
Personal Libr of Former Limestone College President (Lee Davis Lodge & Harrison Patillo Griffith)
55 000 vols; 281 curr per; 47 000 e-books 31188

Brenau University, Trustee Library, 625 Academy St NE, ***Gainesville***, GA 30501-3343
T: +1 770 5346113; Fax: +1 770 5346254; URL: www.brenau.edu/library; library.brenau.edu
1878; Marlene Giguere
Education, Music, Nursing; Elson Judaica Coll, Thomas E Watson Coll
80 000 vols; 180 curr per; 50 000 e-books; 266 000 microforms; 1 920 av-mat; 1 860 sound-rec; 240 digital data carriers 31189

University of Florida – George A. Smathers Library, 535 Library West, ***Gainesville***, FL 32611-7000; P.O. Box 117000, Gainesville, FL 32611-7000
T: +1 352 2732505; Fax: +1 352 3927598; URL: www.uflib.ufl.edu
1853; Judith C. Russell
Africana, Florida, Latin America; Belknap-Performing Arts, Rare Bks, Baldwin-Children, New England Authors, Hist of Science, Hist of the Book, Botany, Modern Lit, Christian Theology & Hist
4 022 000 vols; 70 882 curr per
libr loan 31190

– Allen H. Neuharth Journalism and Communications Library, P.O. Box 118400, Gainesville, FL 32611-8400
T: +1 352 3920455; Fax: +1 352 3925809; E-mail: pjr@mail.uflib.ufl.edu; URL: www.uflib.ufl.edu/jour
Patrick Reakes
14 390 vols 31191

– Architecture & Fine Arts Library, 201 Fine Arts Bldg A, P.O. Box 117017, Gainesville, FL 32611
T: +1 352 2732805; Fax: +1 352 8462747; E-mail: annlind@uflib.ufl.edu; URL: www.uflib.ufl.edu/afa/
1965; Ann Lindell
Architectural Preservation, Rare Bks
108 410 vols; 450 curr per; 35 386 microforms; 750 av-mat; 38 digital data carriers; 11 188 architectural drawings, photos, postcards 31192

– Education Library, 1500 Norman Hall, P.O. Box 117016, Gainesville, FL 32611-7016
T: +1 352 2732780; Fax: +1 352 3924789; E-mail: edref@uflib.ufl.edu; URL: www.uflib.ufl.edu/educ
1950; Iona Malanchuk
Hist of Education Coll, ERIC Microfiche Coll
144 630 vols; 729 curr per; 570 414 microforms; 229 av-mat; 119 sound-rec; 178 machine readable data files, 974 graphics 31193

– Governments Documents Department, L120 Marston Science Library, Gainesville, FL 32611-7011
T: +1 352 3920367; Fax: +1 352 3923357; E-mail: janswan@mail.uflib.ufl.edu; URL: docs.uflib.ufl.edu/l/pages
1907; Jan Swanbeck
14 303 vols; 685 804 govt docs; 3 179 maps; 1 888 880 microforms; 25 av-mat; 119 sound-rec; 1 000 digital data carriers; 10 008 tech rpts, 4 063 machine readable data files, 36 211 graphics 31194

– Isser & Rae Price Library of Judaica, George A. Smathers Libraries, 1504 Norman, Gainesville, FL 32611-7051
T: +1 352 3920380; Fax: +1 352 3924789; E-mail: judaica@uflib.ufl.edu; URL: web.uflib.ufl.edu/cm/plj/PLJ.html
1977; Robert Singerman
90 000 vols; 475 curr per; 1 300 microforms
libr loan; CARLJS 31195

– Latin American Collection, 412 Smathers Library, P.O. Box 117001, Gainesville, FL 32611
T: +1 352 3920360; Fax: +1 352 3924787; E-mail: ricphil@uflib.ufl.edu; URL: www.uflib.ufl.edu/lac
1967; Richard Phillips
Caribbean Coll
330 530 vols; 1 100 curr per; 77 605 microforms; 81 av-mat; 11 digital data carriers; 15 000 VF pieces 31196

– Marston Science Library, Bldg 043, Newell Dr, P.O. Box 117011, Gainesville, FL 32611
T: +1 352 3922759; Fax: +1 352 3924787; E-mail: cdrum@ufl.edu; URL: web.uflib.ufl.edu/msl
1987
639 550 vols; 5 600 curr per; 1 186 779 microforms; 175 av-mat; 296 368 docs, 1 547 machine readable data files 31197

– Music Library, 231 Music Bldg, P.O. Box 117900, Gainesville, FL 32611-7900
T: +1 352 2732815; Fax: +1 352 8462748; URL: www.uflib.ufl.edu/music/
Robena Cornwell
Eugene Grissom Trombone Libr, Didier Graffe Mss Scores
23 430 vols 31198

University of Florida – Health Science Center Libraries, 1600 SW Archer Rd, P.O. Box 100206, ***Gainesville***, FL 32610-0206
T: +1 352 273 8408; Fax: +1 352 3922565; E-mail: cecbote@ufl.edu; URL: www.library.health.ufl.edu
1956; Cecilia Botero
Axline Hist Room
329 850 vols; 2 231 curr per; 800 e-journals; 6 191 microforms; 10 731 av-mat; 685 sound-rec; 106 digital data carriers; 2 640 graphics 31199

University of Florida – Levin College of Law, Legal Information Center, 161A Holland Hall, P.O. Box 117628, ***Gainesville***, FL 32611
T: +1 352 2730700; Fax: +1 352 3925093; E-mail: outler@law.ufl.edu; URL: www.law.ufl.edu/lic
1909; Elizabeth Outler
Brasilian Law Coll, British Law Coll
333 850 vols; 6 539 curr per; 284 900 govt docs; 1 555 400 microforms; 1 979 av-mat; 1 145 sound-rec; 932 computer disks, 2 064 machine readable data files 31200

Knox College, Henry W. Seymour Library, 2 E South St, ***Galesburg***, IL 61401
T: +1 309 3417228; Fax: +1 309 3417799; URL: library.knox.edu
1837; Jeffrey A. Douglas
Finley Coll (Northwest Territory Hist), Ray D Smith Coll (Civil War), Preston Player Coll (Mississippi River Valley), Hughes Coll (Hemingway), Strong Coll (American Southwest), Edgar Lee Masters Coll
317 000 vols; 519 curr per
libr loan 31201

Texas A&M University at Galveston, Jack K. Williams Library, 200 Seawolf Pkwy, ***Galveston***, TX 77553; P.O. Box 1675, Galveston, TX 77553-1675
T: +1 409 7404566; Fax: +1 409 7404702; E-mail: library@tamug.edu; URL: www.tamug.edu/library
1972; Natalie H. Wiest
Marine biology, Oceanography, Transportation; Galveston Bay Info Ctr
94 000 vols; 400 curr per; 1 600 maps; 50 000 microforms; 500 sound-rec; 20 digital data carriers
libr loan 31202

University of Texas Medical Branch, Moody Medical Library, 301 University Blvd, **Galveston**, TX 77555-1035
T: +1 409 7722385; Fax: +1 409 7629782; URL: www.library.utmb.edu
1891; Brett Kirkpatrick
Hist of Medicine, Medical Prints & Portraits, Arch
252 000 vols; 5 820 curr per; 31 149 e-journals; 1 393 diss/theses
libr loan; ALA, MLA, TLA, SLA, AAHSLD
31203

Kenyon College, Olin library & Chalmers Memorial Library, 103 College Dr, **Gambier**, OH 43022-9624
T: +1 740 4275571; Fax: +1 740 4275941; URL: lbis.kenyon.edu
1824; Ronald Griggs
Philander Chase Letters, Charles Pettit McIlvaine Letters, Kenyon College Arch, Kenyon Review Arch, Riker Coll of William Butler Yeats Publs, Typography Coll
420 000 vols; 8 185 curr per
libr loan
31204

Adelphi University, Swirbul Library, One South Ave, **Garden City**, NY 11530; P.O. Box 701, Garden City, NY 11530-0701
T: +1 516 8773580; Fax: +1 516 8773673; URL: libraries.adelphi.edu
1896; Charles W. Simpson
Aimee Ornstein Memorial Libr of Banking & Money Management, Christopher Morley Coll, Cuala Press, Expatriate American Writers of the 1920's & 1930's, Gerhard Hauptmann Coll, Musical Instruments, Political & Presidential Letters & Memorabilia, Spanish Civil War Papers, William Blake Coll
486 000 vols; 1 045 curr per; 29 440 e-journals; 747 000 microforms; 5 860 av-mat; 378 digital data carriers
libr loan
31205

George Mercer Jr School of Theology, Mercer Theological Library, 65 Fourth St, **Garden City**, NY 11530
T: +1 516 2484800 ext 39; Fax: +1 516 2484883; E-mail: cegleston@dioceseli.org; URL: www.mercertheoschool.org
1966; Charles Egleston
28 600 vols; 140 curr per; 735 microforms; 250 av-mat; 462 sound-rec; 500 pamphlets
libr loan; NLA
31206

Indiana University Northwest, Library, 3400 Broadway, **Gary**, IN 46408
T: +1 219 9806582; Fax: +1 219 9806558; URL: www.iun.edu/~lib
1940; Timothy Sutherland
Calumet Regional Archs
264 000 vols
31207

State University of New York – College at Geneseo, Milne Library, SUNY Geneseo, One College Circle, **Geneseo**, NY 14454-1498
T: +1 585 2455591; Fax: +1 585 2455769; URL: library.geneseo.edu
1871; Edwin Rivenburgh
Aldous Huxley, Geneseo Valley Hist Coll, Carl F Schmidt Coll in American Architecture, Wadsworth Family Papers,; College Arch
637 000 vols; 828 curr per; 1 200 e-journals; 8 000 e-books; 910 000 microforms; 752 sound-rec; 21 digital data carriers
libr loan
31208

Hobart & William Smith Colleges, Warren Hunting Smith Library, 334 Pulteney St, **Geneva**, NY 14456
T: +1 315 7813552; Fax: +1 315 7813560; URL: academic.hws.edu/library
1824; Vincent W. Boisselle
American & English lit, Behav sciences, Feminism, Hist, Social sciences; College Arch, mss
388 000 vols; 1 053 curr per; 11 105 e-journals; 6 621 maps; 76 000 microforms; 1 797 sound-rec; 25 digital data carriers
libr loan
31209

Georgetown College, Ensor Learning Resource Center, 400 E College St, **Georgetown**, KY 40324
T: +1 502 8638403;
Fax: +1 502 8687740; URL: library.georgetowncollege.edu
1829; Mary Margaret Lowe
Hist, Religion; Christianity (Thompson Coll), Law (Smith Coll), Pre-1660 English Lit, Rankin Civil War Coll
156 000 vols; 551 curr per; 45 000 e-books; 1 500 sound-rec; 750 digital data carriers
31210

Southwestern University, A. Frank Smith Jr Library Center, 1100 E University Ave, **Georgetown**, TX 78626; P.O. Box 770, Georgetown, TX 78627-0770
T: +1 512 8631638; Fax: +1 512 8638198; URL: www.southwestern.edu/library
1840; Lynne M. Brody
Bertha Dobie Papers, Hymnals (Meyer Coll), J Frank Dobie Coll, John G Tower Papers, Rare Bks (Bewick, Bible, Blake), Texana (Clark Coll)
361 000 vols; 2 879 curr per; 33 941 e-journals; 34 000 e-books; 7 digital data carriers
libr loan
31211

Gettysburg College, Musselman Library, 300 N Washington St, **Gettysburg**, PA 17325
T: +1 717 3376604; Fax: +1 717 3377001; URL: www.gettysburg.edu/library/index.dot
1832; Robin Wagner
Asian art, Rare bks, World War II; Civil War Maps Coll, College Hist
349 000 vols; 2 600 curr per
libr loan
31212

Rowan University, Keith & Shirley Campbell Library, 201 Mullica Hill Rd, **Glassboro**, NJ 08028-1701
T: +1 856 2564801; Fax: +1 856 2564924; URL: www.rowan.edu/library
1923; Bruce Alan Whitham
Education, Engineering; US, New Jersey & Delaware Valley Hist (Stewart Coll)
420 000 vols; 11 154 curr per; 2 000 e-books; 5 670 av-mat; 1 690 sound-rec; 770 digital data carriers
31213

Webb Institute, Livingston Library, 298 Crescent Beach Rd, **Glen Cove**, NY 11542
T: +1 866 7089322; E-mail: inquiry@webb-institute.edu; URL: www.webb-institute.edu/library.html
1932
Science & technology; Marine Engineering, Marine Hist, Naval Architecture
45 000 vols; 255 curr per; 466 maps; 1 664 microforms; 1 697 sound-rec; 1 digital data carriers; 5 116 tech rpts
31214

College of Dupage, Library, 425 Fawell Blvd, **Glen Ellyn**, IL 60137-6599
T: +1 630 9422350; Fax: +1 630 8588757; URL: www.cod.edu/library/
1967; Bernard Fradkin
Career & College Info, Vocational & Tech Coll, Art
203 500 vols; 885 curr per; 24 000 av-mat
libr loan
31215

Arizona State University – West, Fletcher Library, Fletcher Library No 317, 4701 W Thunderbird Rd, **Glendale**, AZ 85306; P.O. Box 37100, Phoenix, AZ 85069-7100
T: +1 602 5438567; Fax: +1 602 5438540; E-mail: sondra.brough@asu.edu; URL: library.west.asu.edu/
1984; Sondra Brough
348 000 vols; 27 512 curr per; 698 govt docs; 1 408 000 microforms; 7 346 av-mat; 2 186 sound-rec; 346 digital data carriers
libr loan
31216

Glendale University, College of Law Library, 220 N Glendale Ave, **Glendale**, CA 91206
T: +1 818 2470770; Fax: +1 818 2470872; E-mail: admit@glendalelaw.edu; URL: www.glendalelaw.edu
1967; Judy Greitzer
Tax & Accounting Law, English Common Law Coll, Rare Law Bks
50 000 vols; 125 curr per; 50 digital data carriers
31217

Thunderbird School of Global Management, Merle A. Hinrichs International Business Information Centre, 15249 N 59th Ave, Thunderbird Campus, **Glendale**, AZ 85306-6001
T: +1 602 978-7300; Fax: +1 602 9787762; E-mail: ibicdoc@thunderbird.edu; URL: www.thunderbird.edu/ibic
1946; Carol Hammond
School archives
65 000 vols; 1 800 curr per; 25 000 e-journals; 200 maps; 5 672 microforms; 3 447 av-mat; 143 sound-rec; 50 digital data carriers
libr loan
31218

Joy Memorial Library, 200 College Dr, **Glennallen**, AK 99588; P.O. Box 289, Glennallen, AK 99588-0289
T: +1 907 8223201; Fax: +1 907 8225027
1966; Pam Horst
Religion; Alaska-Arctic Coll
30 000 vols; 102 curr per; 22 microforms
31219

Arcadia University, Bette E. Landman Library, 450 S Easton Rd, **Glenside**, PA 19038-3295
T: +1 215 5722975; Fax: +1 215 5720240; E-mail: librarydesk@arcadia.edu; URL: www.arcadia.edu/library
1963; Eric McCloy
144 000 vols; 832 curr per; 7 000 e-journals; 300 e-books
libr loan
31220

Glenville State College, Robert F. Kidd Library, 100 High St, **Glenville**, WV 26351
T: +1 304 4624109; Fax: +1 304 4624049; E-mail: library@glenville.edu; URL: www.glenville.edu/Resources/RFKLibrary/Default.asp
1930; Thelma Hutchins
Berlin B Chapman – Special Coll Rm
105 000 vols; 130 curr per; 20 000 e-books; 2 720 av-mat
libr loan
31221

College of William & Mary, Virginia Institute of Marine Science Library, Rte 1208, Greate Rd, **Gloucester Point**, VA 23062; P.O. Box 1346, Gloucester Point, VA 23062-1346
T: +1 804 6847115; Fax: +1 804 6847113; URL: www.vims.edu/library
1940; Carol Coughlin
Chesapeake Bay, Coastal zone, Environmental studies, Estuaries, Fisheries, Geology, Marine biology; Expeditions, Sport Fishing & Hunting (Ross H Walker Coll)
80 000 vols; 402 curr per; 4 664 maps; 6 digital data carriers
31222

Colorado School of Mines, Arthur Lakes Library, 1400 Illinois St, **Golden**, CO 80401-1887
T: +1 303 2733690; Fax: +1 303 2733199; URL: www.mines.edu/library/
1874; Joanne V. Lerud-Heck
Mining, Petroleum engineering, Geology, Geophysics, Chemistry, Energy, Environmental studies; Colorado & Mining Hist Coll; Energy, Environmental & Public Policy Coll
691 000 vols; 1 208 curr per; 6 000 e-books; 9 000 diss/theses; 760 000 govt docs; 180 000 maps; 271 000 microforms; 720 digital data carriers
libr loan
31223

Panhandle State University, Marvin E. McKee Library, 323 W Eagle Blvd, **Goodwell**, OK 73939; P.O. Box 370, Goodwell, OK 73939-0370
T: +1 580 3491540; Fax: +1 580 3491541; E-mail: mckeelib@opsu.edu; URL: www.opsu.edu/McKeeLibrary
1909; C. Evlyn Schmidt
Textbook Review Center, Children's Coll & Curriculum Coll
96 000 vols; 261 curr per; 32 000 e-books; 75 av-mat; 100 High Interest/Low Vocabulary Bks, 20 Bks on Deafness & Sign Lang, 25 Large Print Bks
31224

Goshen College – Harold & Wilma Good Library, 1700 S Main, **Goshen**, IN 46526-4794
T: +1 574 5357431; Fax: +1 574 5357438; URL: www.goshen.edu/library
1894; Lisa Guedea Carreno
Religion, Peace; Early American
Hymnody (Jesse Hartzler Coll)
137 000 vols; 487 curr per; 3 000 e-books; 428 Electronic Media & Resources
31225

Florida Baptist Theological College, Ida J. McMillan Library, 5400 College Dr, **Graceville**, FL 32440-1833
T: +1 850 2633261; Fax: +1 850 2635704; URL: www.baptistcollege.edu
1943; John Shaffett
Religion, Social sciences, Music, Humanities; Florida Baptist Historical Coll; College Arch
72 000 vols; 327 curr per
31226

Grambling State University, A.C. Lewis Memorial Library, 403 Main St, **Grambling**, LA 71245; P.O. Box 4256, Grambling, LA 71245-2761
T: +1 318 2743354; Fax: +1 318 2743268; URL: www.gram.edu/library/default.asp
1935; Dr. Rosemary Mokia
Afro-American Rare Bks, fiche; Crime & Juvenile Delinquency, fiche; Education, fiche; English Lit, fiche, Libr of American Civilization, fiche; National Woman's Party Papers, micro; Black Culture, micro
276 000 vols; 43 000 e-books; 253 950 Electronic Media & Resources
31227

University of North Dakota, Chester Fritz Library, 3051 University Ave Stop 9000, **Grand Forks**, ND 58202-9000
T: +1 701 7774629; Fax: +1 701 7773319; E-mail: library@mail.und.edu; URL: www.library.und.edu
1883; Wilbur Stolt
Education, Western Hist; North Dakota Bk Coll, Fred G Aandahl Bk Coll, Genealogy Coll, Orin G. Libby Mss Coll, Univ Arch
1 467 000 vols; 5 622 curr per; 29 000 e-journals; 25 000 e-books; 108 939 maps; 766 000 microforms; 15 900 sound-rec; 1 502 digital data carriers; 15 673 linear feet of mss, 4 460 Electronic Media & Resources, 112 Large Print Bks
libr loan
31228

– Energy & Environmental Research Center Library, 15 North 23rd St Stop 9018, Grand Forks, ND 58202-9018
T: +1 701 7775132; Fax: +1 701 7775181; URL: www.und.nodak.edu/dept/library/resources/eerc/index.jsp
Rosemary Pleva Flynn
18 000 vols; 60 curr per
31229

– Gordon Erickson Music Library, Hughes Fine Arts Ctr 170, 3350 Campus Rd Stop 7125, Grand Forks, ND 58202-7125
T: +1 701 7772817; Fax: +1 701 7773319; URL: www.library.und.edu/coll/music/index.htm
Felicia Clifton
16 000 vols; 41 curr per
31230

– Harley E. French Library of the Health Sciences, School of Medicine and Health Sciences, 501 N Columbia Rd, Stop 9002, Grand Forks, ND 58202-9002
T: +1 701 7772606; Fax: +1 701 7774790; E-mail: hflref@medicine.nodak.edu; URL: undmedlibrary.org
1950; Lila Pedersen
Hist of Medicine (Dr French Coll); James D Barger Coll on Pathology and Quality Control
43 000 vols; 300 curr per; 915 e-books; 2 000 microforms; 133 digital data carriers; 500 pamphlets
libr loan; AAHSL, MLA
31231

– F. D. Holland, Jr Geology Library, 81 Cornell St, Stop 8358, Grand Forks, ND 58202-8358
T: +1 701 7774631; Fax: +1 701 7774449; URL: www.und.edu/dept/library/resources/geology/index.jsp
Darin Buri
North Dakota Geology
55 000 vols; 248 curr per; 300 diss/theses; 10 000 govt docs; 94 100 maps; 12 000 microforms; 324 digital data carriers; 16 500 well logs, 11 630 airphotos
libr loan; Geoscience Information Society
31232

– Thormodsgard Law Library, 215 Centennial Dr, Grand Forks, ND 58202; P.O. Box 9004, Grand Forks, ND 58202-9004
T: +1 701 7773538; Fax: +1 701 7772217; URL: www.law.und.nodak.edu
1899; Rhonda Schwartz
159 000 vols; 2 001 curr per; 653 000 microforms; 481 sound-rec; 137 410 microforms of per 31233

Mesa State College, Tomlinson Library, 1200 College Pl, ***Grand Junction***, CO 81501; mail address: 1100 North Ave, Grand Junction, CO 81501
T: +1 970 2481406; Fax: +1 970 2481930; E-mail: libref@mesastate.edu; URL: www.mesastate.edu/msclibrary/index.htm
1925; Elizabeth W. Brodak
College Arch, Ethridge Indian Pottery Coll, Wayne Aspinall Coll, Walter Walker Memorial Coll
326 000 vols; 917 curr per; 1 510 e-journals; 9 000 e-books
libr loan 31234

Aquinas College, Grace Hauenstein Library, 1607 Robinson Rd SE, ***Grand Rapids***, MI 49506-1799
T: +1 616 6322140; Fax: +1 616 7324534; E-mail: library@aquinas.edu; URL: www.aquinas.edu/library
1936; Francine Paolini & Shellie Jeffries
100 014 vols; 444 curr per; 41 050 e-books; 221 678 microforms; 6 465 av-mat; 4 457 sound-rec; 60 digital data carriers
libr loan 31235

Cornerstone University, Miller Library, 1001 E Beltline Ave NE, ***Grand Rapids***, MI 49525
T: +1 616 2541976; Fax: +1 616 2221405; E-mail: circulation@cornerstone.edu; URL: www.cornerstone.edu/library
1941; Dr. Fred Sweet
Baptist church, Religious education, Theology
127 000 vols; 1 163 curr per; 900 e-journals; 25 000 e-books; 471 diss/theses; 1 maps; 264 000 microforms; 3 927 sound-rec; 400 digital data carriers
libr loan; ALA 31236

Davenport University, Margaret Sneden Library, 6191 Kraft Ave SE, ***Grand Rapids***, MI 49512
T: +1 616 5545612; Fax: +1 616 5545226; E-mail: LibInfoComm@davenport.edu; URL: www.davenport.edu
1866; Julie Gotch
83 000 vols; 307 curr per; 16 000 e-books 31237

Kendall College of Art & Design, Library, 17 Fountain St NW, 2nd Flr, ***Grand Rapids***, MI 49503-3002
T: +1 616 4511868; Fax: +1 616 4519867; URL: www.kcad.edu/student-services/library
1928; Michael J. Kruzich
Res coll documenting the hist of furniture design, ornament, interior design and industrial design
18 000 vols; 110 curr per; 119 electronic databases, 88 auction cat
ALA, ARLIS/NA 31238

Kuyper College, Zondervan Library, 3333 E Beltline NE, ***Grand Rapids***, MI 49525
T: +1 616 2223000; Fax: +1 616 9883608; E-mail: library@kuyper.edu; URL: www.kuyper.edu/Library
1940; Dianne Zandbergen
Religion, theology
56 000 vols; 246 curr per; 11 000 e-journals; 741 microforms; 3 103 sound-rec; 6 digital data carriers; 20 VF drawers
libr loan; ACL 31239

Messiah College, Murray Library, One College Ave, ***Grantham***, PA 17027-0800; P.O. Box 3002, Grantham, PA 17027-9795
T: +1 717 6916006; Fax: +1 717 6916042; URL: www.messiah.edu/murraylibrary
1909; Jonathan D. Lauer
Brethren in Christ Arch, Science & Religion (W Jim Neidhardt Coll), Artists' Bks
296 000 vols; 1 165 curr per
libr loan 31240

New Mexico State University, Grants Branch Library, 1500 N Third St, ***Grants***, NM 87020
T: +1 505 2876639; Fax: +1 505 2876676; E-mail: stafford@nmsu.edu; URL: grants.nmsu.edu/Library%20Services
1968; Cecilia Stafford
New Mexico Hist
34 000 vols; 68 curr per; 6 500 govt docs; 1 500 maps; 38 101 microforms; 2 106 sound-rec; 6 digital data carriers; 25 slides, 20 overhead transparencies
libr loan 31241

Denison University Libraries, William Howard Doane Library, 400 W Loop, ***Granville***, OH 43023; P.O. Box 805, Granville, OH 43023-0805
T: +1 740 5876682; Fax: +1 740 5878280; URL: www.denison.edu/library
1831; Lynn Scott Cochrane
Geology (G K Gilbert Coll)
459 000 vols; 963 curr per; 6 316 e-journals
libr loan 31242

Kentucky Christian College, Young Library, 100 Academic Pkwy, ***Grayson***, KY 41143-2205
T: +1 606 4743240; Fax: +1 606 4743123; URL: library.kcu.edu
1919; Thomas L. Scott
Religion, Education, Biology; Mission Papers (1969-present), Restoration Church Hist
80 000 vols; 11 000 curr per; 315 microforms; 1 985 sound-rec 31243

Bard College at Simon's Rock, Alumni Library, 84 Alford Rd, ***Great Barrington***, MA 01230
T: +1 413 5287370; Fax: +1 413 5287380; E-mail: library@simons-rock.edu; URL: www.simons-rock.edu
1966; Brian Mikesell
American lit, Art, Environmental studies, Liberal arts, Music, Theater; W.E.B. DuBois Coll on the Black Experience, bks; Bernard Krainis Coll of Early Music Scores
66 000 vols; 360 curr per 31244

University of Great Falls, University Library, 1301 20th St S, ***Great Falls***, MT 59405-4948
T: +1 406 7915315; Fax: +1 406 7915395; E-mail: library@ugf.edu; URL: www.ugf.edu/library
1932; David Bibb
Religion, Education, Law; Americana (Microbook Libr of American Civilization
105 000 vols; 560 curr per; 22 000 microforms; 2 954 sound-rec; 5 digital data carriers; 21 669 microforms of per, 1 296 scores 31245

University of Northern Colorado, James A. Michener Library & Skinner Music Library, 501 20th St, ***Greeley***, CO 80639
T: +1 970 3512775; Fax: +1 970 3512963; URL: www.unco.edu/library
1890; Gary M. Pitkin
Education, Music, Business; James A Michener Special Coll, International Gladiolus Hall of Fame Coll, Univ Arch
1 079 000 vols; 1 926 curr per; 20 000 e-journals; 9 000 e-books; 1 040 000 microforms; 1 600 sound-rec; 1 000 digital data carriers
libr loan 31246

– Howard M. Scinner Music Library, 10 th Avenue, Campus Box 68, Greeley, CO 80639
T: +1 970 3512439; E-mail: musiclib@unco.edu; URL: www.unco.edu/music/index.htm
1968; Stephen Luttmann
25 500 vols; 196 curr per; 34 500 music scores; 5 000 microforms; 585 av-mat; 16 500 sound-rec
libr loan; MLA 31247

University of Wisconsin – Green Bay, Cofrin Library, 2420 Nicolet Dr, ***Green Bay***, WI 54311-7001
T: +1 920 4652385; Fax: +1 920 4652136; URL: www.uwgb.edu/library
1967; Leanne Hansen
Belgian-American Ethnic Coll, Local Hist (Area Res Ctr), Univ Arch
295 000 vols; 934 curr per
libr loan 31248

DePauw University, Roy O. West Library, 11 E Larrabee St, ***Greencastle***, IN 46135
T: +1 765 6584420; Fax: +1 765 6584017; E-mail: provine@depauw.edu; URL: www.depauw.edu/library
1837; Rick Provine
Business & management, Economics, Mucic; Bret Harte Libr of First Editions, German Coll, Latin Coll, Pre-Law Coll, Univ Arch
302 900 vols; 2 442 curr per
libr loan 31249

Tusculum College, Library, Hwy 107, 60 Shiloh Rd, ***Greeneville***, TN 37743; P.O. Box 5005, Greenville, TN 37743-0001
T: +1 423 6367320; Fax: +1 423 7878498; E-mail: library@tusculum.edu; URL: library.tusculum.edu
1794; Myron J. Smith Jr
Special Education (Instructional Materials Coll), multi-media; Warren W Hobbie Civic Arts Coll; Coffin Coll
93 000 vols; 206 curr per; 29 390 e-journals; 102 000 e-books; 155 000 microforms; 800 av-mat; 850 sound-rec; 25 digital data carriers 31250

Bennett College, Thomas F. Holgate Library, 900 E Washington St, Campus Box M, ***Greensboro***, NC 27401-3239
T: +1 336 5172139; Fax: +1 336 5172144; E-mail: library@bennett.edu; URL: www.bennett.edu/library
1939; Joan Williams
Black Studies, Women (Afro-American Women's Coll), Bennett College Hist, (Bennett College Arch), Individual Biography, Black (Norris Wright Cuney, Papers), Carnegie Art Coll, Palmer Coll
98 000 vols; 150 curr per 31251

Greensboro College, James Addison Jones Library, 815 W Market St, ***Greensboro***, NC 27401
T: +1 336 2727102;
Fax: +1 336 2177233; URL: library.greensborocollege.edu
1838; Christine A. Whittington
110 000 vols; 302 curr per; 11 High Interest/Low Vocabulary Bks
libr loan 31252

Guilford College, Hege Library, 5800 W Friendly Ave, ***Greensboro***, NC 27410-4175
T: +1 336 3162450; Fax: +1 336 3162950; E-mail: hegelib@guilford.edu; URL: www.guilford.edu/library
1837; Mary Ellen Chijioke
Religious Society of Friends (Quaker)
188 000 vols; 17 618 curr per; 6 290 e-journals; 42 000 e-books; 267 mss; 4 377 diss/theses; 165 music scores; 952 maps; 20 000 microforms; 2 170 av-mat; 2 894 sound-rec; 320 digital data carriers
libr loan; ALA 31253

North Carolina Agricultural and Technical State University, F.D. Bluford Library, 1601 E Market St, ***Greensboro***, NC 27411-0002
T: +1 336 3347158; Fax: +1 336 3347783; E-mail: refemail@ncat.edu; URL: www.library.ncat.edu
1892; Waltrene M. Canada
Afro-American Coll, Chemistry Libr, Univ Arch
580 000 vols; 5 466 curr per; 50 690 e-journals; 44 000 e-books; 981 000 microforms; 800 av-mat; 997 digital data carriers; 800 Audio Bks, 100 780 Electronic Media & Resources
libr loan 31254

University of North Carolina at Greensboro, Walter Clinton Jackson Library, 320 Spring Garden St, ***Greensboro***, NC 27402; P.O. Box 26170, Greensboro, NC 27402-6170
T: +1 336 3344238; Fax: +1 336 3345097; URL: library.uncg.edu
1892; Rosann Bazirjian
Dance Coll; Eugenie Silverman Baizerman Arch; George Herbert Coll; Girls Books in Series Coll; Lois Lenski Coll; Cello Music Colls; Physical Education Hist Coll; Book Arts Coll; Randall Jarrell Coll; Saul Baizerman Arch; Way & Williams Coll; Woman's Coll; Robbie Emily Dunn Coll of American Detective Fiction
2 076 000 vols; 2 879 curr per; 21 720 e-journals; 310 000 e-books; 468 516 govt docs; 17 099 maps; 306 000 microforms; 11 489 sound-rec
libr loan 31255

Seton Hill University, Reeves Memorial Library, One Seton Hill Dr, ***Greensburg***, PA 15601
T: +1 724 8384291; Fax: +1 724 8384203; URL: maura.setonhill.edu/~library
1918; David H. Stanley
Holocaust, Entrepreneurship, Fine Arts
120 000 vols; 350 curr per 31256

University of Pittsburgh at Greensburg, Millstein Library, 150 Finoli Dr, ***Greensburg***, PA 15601-5898
T: +1 724 8369687; Fax: +1 724 8367043; URL: www.pitt.edu/green/millstein.html
1963; Patricia Duck
UPG Arch
78 000 vols; 308 curr per; 2 155 av-mat; 1 620 sound-rec; 540 digital data carriers
libr loan 31257

Bob Jones University, J.S. Mack Library, 1700 Wade Hampton Blvd, ***Greenville***, SC 29614
T: +1 864 2425100; Fax: +1 864 2321729; E-mail: library@bju.edu; URL: www.bju.edu/library/home.html
1927; Joseph Lee Allen Sr
315 486 vols; 1 010 curr per; 30 e-journals; 23 443 e-books; 3 mss; 2 060 diss/theses; 5 490 music scores; 496 520 microforms; 21 292 sound-rec; 5 205 digital data carriers
libr loan 31258

East Carolina University, Joyner Library, E Fifth St, ***Greenville***, NC 27858-4353
T: +1 252 3286677; Fax: +1 252 3286618; E-mail: askref@ecu.edu; URL: www.ecu.edu/lib/
1907; Larry Boyer
Local Hist (North Carolina Coll), Regional Hist (East Carolina Mss Coll)
1 184 000 vols; 6 738 curr per; 865 858 govt docs; 95 766 maps; 1 600 000 microforms; 1 584 av-mat; 13 302 sound-rec; 68 digital data carriers
libr loan 31259

– Music Library, A J Fletcher Music Ctr, Rm A110, Greenville, NC 27858
T: +1 252 3286250; Fax: +1 252 3281243; E-mail: musiclib@ecu.edu; URL: www.ecu.edu/lib/music
1974; David Hursh
23 000 vols; 26 156 music scores; 2 000 microforms; 770 sound-rec; 30 digital data carriers
libr loan; MLA 31260

– William E. Laupus Health Sciences Library, 600 Moye Blvd, Health Sciences Bldg, Greenville, NC 27834
T: +1 252 7442230; Fax: +1 252 7442300; URL: www.hsl.ecu.edu
1969; Dr. Dorothy A. Spencer
Hist of Medicine Special Coll
45 000 vols; 735 curr per; 31 000 microforms; 30 715 microforms of per, 5 536 nonprint mat
libr loan 31261

Furman University, James B. Duke Library, 3300 Poinsett Hwy, ***Greenville***, SC 29613-0600
T: +1 864 2942198; Fax: +1 864 2943004; URL: library.furman.edu
1826; Dr. Janis M. Bandelin
South Carolina Baptist Hist, bks, microfilm & mss; Maxwell Music Library
447 000 vols; 1 400 curr per; 1 233 e-journals; 7 000 e-books; 79 600 govt docs; 62 500 maps; 726 000 microforms; 1 674 sound-rec; 1 120 digital data carriers
libr loan 31262

Greenville College, Ruby E. Dare Library, 315 E College Ave, ***Greenville***, IL 62246
T: +1 618 6646603; Fax: +1 618 6649578; E-mail: libgen@greenville.edu; URL: www.greenville.edu/learningresources/library
1892; Jane L. Hopkins
Free Methodist Church Hist, Greenville College Hist
128 000 vols; 494 curr per; 2 300 av-mat; 300 sound-rec
libr loan 31263

Thiel College, Langenheim Memorial Library, 75 College Ave, ***Greenville***, PA 16125-2183
T: +1 724 5892121; Fax: +1 724 5892122; URL: www.thiel.edu
1866; Allen S. Morrill
187 000 vols; 460 curr per; 124 e-journals; 4 000 e-books; 40 digital data carriers; 480 CDs
libr loan 31264

Heritage Baptist University, University Library, 1301 W County Line Rd, ***Greenwood***, IN 46142
T: +1 317 8822345; Fax: +1 317 8852960; E-mail: info@indianabaptistcollege.com; URL: www.indianabaptistcollege.com
1954; Edna Kehrt
23 000 vols; 33 curr per 31265

Lander University, Larry A. Jackson Library, 320 Stanley Ave, ***Greenwood***, SC 29649-2099
T: +1 864 3888365; Fax: +1 864 3888816; URL: www.lander.edu/library
1872; Ann T. Hare
192 000 vols; 651 curr per; 140 e-journals; 42 000 e-books; 760 av-mat; 720 sound-rec; 130 digital data carriers
libr loan 31266

Grinnell College, Burling Library, 1111 Sixth Ave, ***Grinnell***, IA 50112-1770
T: +1 641 2693351; Fax: +1 641 2694283; E-mail: query@grinnell.edu; URL: www.lib.grinnell.edu
1846; Richard Fyffe
East Asian Coll, James Norman Hall, Iowa Coll, Local Hist & College Arch (Iowa Room Coll), Pinne Coll
769 000 vols; 2 642 curr per; 139 000 e-books; 16 830 Music Scores
libr loan 31267

University of Connecticut at Avery Point, Avery Point Library, 1084 Shennecossett Rd, ***Groton***, CT 06340-6097
T: +1 860 4059146; Fax: +1 860 4059150; E-mail: libadm22@uconnvm.uconn.edu
1967; Barbara Vizoyan
Marine sciences
40 000 vols; 120 curr per; 4 000 govt docs; 1 000 maps; 16 000 microforms; 119 av-mat; 215 sound-rec; 9 digital data carriers
libr loan; IAMSLIC 31268

Grove City College, Henry Buhl Library, 300 Campus Dr, ***Grove City***, PA 16127-2198
T: +1 724 4582047; Fax: +1 724 4582181
1900; Diane Grundy
Mathematics (Locke Coll), bks & pamphlets; Ludwig von Mises Papers, letters, pamphlets & mss
140 000 vols; 300 000 microforms; 1 220 av-mat; 8 digital data carriers
libr loan 31269

Appalachian School of Law Library, 1221 Edgewater Dr, ***Grundy***, VA 24614-7062
T: +1 276 9356688; Fax: +1 276 9357138; URL: www.asl.edu/library
Charles J. Condon
196 000 vols; 3 488 curr per 31270

Stetson University College of Law Library, 1401 61st St S, ***Gulfport***, FL 33707
T: +1 727 5627820; Fax: +1 727 3458973; URL: www.law.stetson.edu/lawlib
1901; Dr. Madison Mosley Jr
398 000 vols; 4 061 curr per 31271

Western State College of Colorado, Leslie J. Savage Library, 600 N Adams St, ***Gunnison***, CO 81231
T: +1 970 9432053; Fax: +1 970 9432042; URL: www.western.edu/lib/Welcome.html
1901; Elizabeth Avery
Colorado Hist (Western Americana); Western Colorado Newspapers
113 191 vols; 428 curr per; 1 000 000 microforms; 2 300 av-mat; 630 sound-rec; 291 digital data carriers; 630 CDs, 2 300 Videos
libr loan 31272

Gwynedd-Mercy College, Lourdes Library, 1325 Sumneytown Pike, ***Gwynedd Valley***, PA 19437; P.O. Box 901, Gwynedd Valley, PA 19437-0901
T: +1 215 6467300; Fax: +1 215 6415596; E-mail: library@gmc.edu; URL: www.gmc.edu
1958; Daniel Schabert
Ireland
104 899 vols; 623 curr per; 140 e-books; 17 795 microforms; 6 430 av-mat; 5 071 sound-rec; 4 digital data carriers
libr loan 31273

Fairleigh Dickinson University, Business Research Library-New Jersey Room, Dickinson Hall, 140 University Plaza Dr, ***Hackensack***, NJ 07601
T: +1 201 6922608; Fax: +1 201 6927048; URL: www.fdu.edu
Maria Kocylowsky
17 000 vols 31274

Centenary College, Taylor Memorial Learning Resource Center, 400 Jefferson St, ***Hackettstown***, NJ 07840
T: +1 908 8521400; Fax: +1 908 8509528; URL: faculty.centenarycollege.edu/library
1867; Nancy Madacsi
Interior design; Centenary College Hist Coll, Lancey Coll on Lincoln
73 000 vols; 75 curr per
libr loan 31275

Paier College of Art, Inc, Library, 20 Gorham Ave, ***Hamden***, CT 06514-3902
T: +1 203 2873023; Fax: +1 203 2873021; E-mail: paierartlibrary@snet.net; URL: www.paierart.com
Beth R. Harris
Picture Reference File (30 000 pictures)
13 000 vols; 142 av-mat; 65 000 slides
ARLIS/NA 31276

Quinnipiac University, Arnold Bernhard Library, 275 Mount Carmel Ave, ***Hamden***, CT 06518
T: +1 203 5828634; Fax: +1 203 5823451; URL: www.quinnipiac.edu
1929; Charles M. Getchell Jr
Holocaust; Albert Schweitzer, Great Hunger (Irish Famine)
167 000 vols; 900 curr per; 12 digital data carriers 31277

– School of Law Library, 275 Mt Carmel Ave, Hamden, CT 06518-1951
T: +1 203 2873300; Fax: +1 203 2873316
1978; Ann M. DeVeaux
250 000 vols; 2 700 curr per 31278

Colgate University, Everett Needham Case Library, 13 Oak Dr, ***Hamilton***, NY 13346-1398
T: +1 315 2287597; Fax: +1 315 2287934; URL: exlibris.colgate.edu
1819; Joanne Schneider
Henry A. Colgate Coll of Joseph Conrad, Richard S. Weiner Coll of George Bernard Shaw, The Powys Family Coll, Orrin E. Dunlap Coll of Radio and Television, Pierrepont Noyes Rhineland Papers, Getrude Stein Coll, James Joyce Coll, T.S. Eliot Coll; George R. Cooley Science Libr
721 000 vols; 1 698 curr per; 29 120 e-journals; 770 e-books; 335 166 govt docs; 475 000 microforms; 7 877 sound-rec
libr loan; ALA 31279

Miami University – Hamilton Campus, Rentschler Library, 1601 University Blvd, ***Hamilton***, OH 45011
T: +1 513 7853235; Fax: +1 513 7853231; URL: www.ham.muohio.edu/library/
1968; Krista McDonald
71 000 vols; 356 curr per
libr loan 31280

Purdue University Calumet Library, 2200 169th St, ***Hammond***, IN 46323-2094
T: +1 219 9892224; Fax: +1 219 9892070; URL: www.calumet.purdue.edu/library
1947; Karen M. Corey
Regional Hist & Institutional Data
266 000 vols; 979 curr per; 729 000 microforms
libr loan 31281

Southeastern Louisiana University, Linus A. Sims Memorial Library, SLU Box 10896, 1211 SGA Dr, ***Hammond***, LA 70402
T: +1 985 5495318; Fax: +1 985 5493490; URL: www.selu.edu/library
1925; Eric Johnson
Genealogy, Education, Mathematics, Music, American Hist; Papers of Congressman James H. Morrison
377 000 vols; 2 381 curr per; 39 000 e-books; 670 000 microforms; 5 920 av-mat; 512 sound-rec; 3 digital data carriers
libr loan 31282

Hampden Sydney College, Eggleston Library, 257 Via Sacra, HSC Box 7, ***Hampden Sydney***, VA 23943
T: +1 434 2236192; Fax: +1 434 2236351; URL: www.hsc.edu
1776; Sharon I. Goad
Humanities, Local hist; Int Video Coll, John Peter Mettauer Coll
251 000 vols; 300 curr per
libr loan 31283

Hampton University, William R. & Norma B. Harvey Library, 130 E Tyler St, ***Hampton***, VA 23668
T: +1 757 7275371; Fax: +1 757 7275952; E-mail: ereference@hamptonu.edu; URL: www.hamptonu.edu/universityservices/library/
1868; Faye Watkins
Black Lit & Hist (George Foster Peabody Coll); Hampton Univ Arch
341 000 vols; 1 150 curr per; 521 000 microforms; 68 digital data carriers 31284

– William H. Moses Jr. Architecture Library, Bemis Laboratory, Room 208, 130 E Tyler St, Hampton, VA 23668
T: +1 757 7286259; URL: set.hamptonu.edu/architecture/facilities.cfm
10 000 vols; 20 000 slides 31285

Finlandia University, Maki Library, 601 Quincy St, ***Hancock***, MI 49930-1882
T: +1 906 4877252; Fax: +1 906 4877297; E-mail: maki.library@finlandia.edu; URL: www.finlandia.edu/maki.html
1896; Yesianne Ramirez
Finnish-American Life & Culture Coll, Upper Peninsula of Michigan
45 000 vols; 217 curr per; 112 diss/theses; 263 maps; 2 000 microforms; 766 sound-rec; 185 digital data carriers
libr loan; ALA, MLA 31286

Hannibal-Lagrange College, L.A. Foster Library, 2800 Palmyra Rd, ***Hannibal***, MO 63401-1999
T: +1 573 2213675; Fax: +1 573 2480294; E-mail: library@hlg.edu; URL: www.hlg.edu/library
1858; Julie Andresen
Missouri Coll, Rare Bks, HLG Arch
101 000 vols; 403 curr per; 11 000 e-books 31287

Dartmouth College, Library, 6025 Baker Berry Library, Rm 115, ***Hanover***, NH 03755-3525
T: +1 603 6462236; Fax: +1 603 6463702; E-mail: dartmouth.college.library@dartmouth.edu; URL: www.dartmouth.edu/~library/
1769; Jeff Horrell
American Calligraphy, Bookplates, Dartmouth Arch, Don Quixote, German & English Plays (Barrett Clark Coll), George Ticknor Libr
2 512 000 vols; 17 533 curr per
libr loan 31288

– Baker-Berry Library, Hinman Box 6025, Hanover, NH 03755-3525
T: +1 603 6462560; Fax: +1 603 6462167; E-mail: baker.library.reference@dartmouth.edu; URL: www.dartmouth.edu/~lbaker/baker.html
Ridie S. Ghezzi
1 696 000 vols; 11 366 curr per 31289

– Biomedical Libraries, Dana Biomedical & Matthews-Fuller Health Sciences Library, 6168 Dana Biomedical Library, Hanover, NH 03755-3880
T: +1 603 6501658, 1656; Fax: +1 603 6501354; E-mail: biomedical.libraries.reference@dartmouth.edu; URL: www.dartmouth.edu/~biomed
1963; William Garitty
Conner Coll of Rare Medical Classics, Henry Kumm Index on Poliomyelitis & Tropical Medicine, Raymond Pearl Longevity Coll, Harry Schroeder Coll of Papers on Trace Elements, Memorabilia of Nathan Smith
249 980 vols; 2 548 curr per; 21 915 microforms; 27 digital data carriers; 19 592 slides 31290

– Feldberg Business Administration & Engineering Library, 6193 Murdough Ctr, Hanover, NH 03755-3560
T: +1 603 6462191; Fax: +1 603 6462384; E-mail: feldberg.reference@dartmouth.edu; URL: www.dartmouth.edu/~feldberg
1973; James R. Fries
Thayer Coll of Military Engineering
116 000 vols; 2 673 curr per; 958 000 microforms; 4 680 digital data carriers 31291

– Kresge Physical Sciences Library & Cook Mathematics Collection, 6115 Fairchild Ctr, Hanover, NH 03755-3571
T: +1 603 6463563; Fax: +1 603 6463681; E-mail: kresge.library.reference@dartmouth.edu; URL: www.dartmouth.edu/~krescook
1974; Jane Quigley
130 000 vols; 1 533 curr per; 53 000 microforms 31292

– Paddock Music Library, 6245 Hopkins Center, Hanover, NH 03755
T: +1 603 6463234; Fax: +1 603 6461219; E-mail: Paddock.Music.Library@Dartmouth.edu; URL: www.dartmouth.edu/~library/paddock
Patricia B. Fisken
96 000 vols; 290 curr per 31293

– Rauner Special Collections Library, Dartmouth College, 6065 Webster Hall, Hanover, NH 03755-3519
T: +1 603 6460538; Fax: +1 603 6460447; E-mail: rauner.special.collections.reference@dartmouth.edu; URL: www.dartmouth.edu/~speccoll/
Jay Satterfield
American Calligraphy, Bookplates, Dartmouth Archs, Don Quixote, George Ticknor Libr, German & English Plays, Horace, New England Early Illust Bks, New Hampshire Hist & Imprints, Stefansson Polar Regions Coll, Chase Streeter Railroads Coll, Robert Frost Coll, Hickmott Shakespeare Coll, Spanish Civilization, Spanish Plays
127 000 vols; 489 curr per 31294

– Sanborn English Library, Dartmouth College, HB6032, Hanover, NH 03755-3525
T: +1 603 6463993; Fax: +1 603 6462159; E-mail: english.department@dartmouth.edu
Darsie Riccio
English literature
9 000 vols; 40 curr per 31295

– Sherman Art Library, Carpentar Hall, Hinman Box 6025, Hanover, NH 03755-3570
T: +1 603 6462305; Fax: +1 603 6461218; E-mail: sherman.library.reference@dartmouth.edu; URL: www.dartmouth.edu/~sherman
1929; Laura Graveline
Architecture, Photography; Artists Bks, Facsimiles of Illuminated Mss
129 000 vols; 642 curr per; 45 000 microforms; 50 digital data carriers 31296

Hanover College, Duggan Library, 121 Scenic Dr, ***Hanover***, IN 47234; P.O. Box 287, Hanover, IN 47243-0287
T: +1 812 8667165; Fax: +1 812 8667172; E-mail: gibson@hanover.edu; URL: www.hanover.edu/Library
1827; Ken Gibson
Church Hist Coll; Civil War Coll; Indiana Hist Coll
533 110 vols; 1 169 curr per
libr loan 31297

Southeastern Illinois College, Melba Patton Library, 3575 College Rd, ***Harrisburg***, IL 62946
T: +1 618 2525400; Fax: +1 618 2522713; URL: www.sic.edu
1960; Gary Jones
40 000 vols; 250 curr per 31298

Widener University – Harrisburg Campus Branch Law Library, 3800 Vartan Way, *Harrisburg*, PA 17110; P.O. Box 69380, Harrisburg, PA 17106-9380
T: +1 717 5413933; Fax: +1 717 5413998; URL: www.law.widener.edu/ LawLibrary
1989; Patricia Fox
202 000 vols   31299

North Arkansas College, Library, 1515 Pioneer Dr, *Harrison*, AR 72601
T: +1 870 3913122; Fax: +1 870 3913245; URL: www.northark.net/ academics/library/index.htm
1974; Jim Robb
25 000 vols; 212 curr per   31300

Eastern Mennonite University, Sadie A. Hartzler Library, 1200 Park Rd, *Harrisonburg*, VA 22802-2462
T: +1 540 4324175; Fax: +1 540 4324977; URL: www.emu.edu/library
1917; Donald D. Smeeton
Anabaptist/Mennonite Hist, Virginiana & Genealogy (Menno Simons Hist Libr & Arch), 16th C bks
168 000 vols; 965 curr per; 16 000 e-books; 606 maps; 858 microforms; 1960 av-mat; 2975 sound-rec; 390 digital data carriers; 260 Charts, 1997 Study Prints
libr loan; ATLA   31301

James Madison University, Carrier Library, 800 S Main St, *Harrisonburg*, VA 22807-0001; MSC 1704, Harrisonburg, VA 22807
T: +1 540 5686691; Fax: +1 540 5686339; URL: www.lib.jmu.edu/
1908; Ralph Alberico
Local hist of Harrisonburg and central Shenandoah Valley; James Madison University hist
464 000 vols; 7173 curr per; 7000 e-journals; 12 000 e-books; 140 mss; 2697 diss/theses; 573 052 govt docs; 19 743 music scores; 32 maps; 1 050 000 microforms; 19 719 sound-rec; 1326 digital data carriers
libr loan   31302

Lincoln Memorial University, Carnegie Vincent Library, Cumberland Gap Pkwy, Box 2012, *Harrogate*, TN 37752
T: +1 423 8696436; Fax: +1 423 8696426; E-mail: library@lmunet.edu; URL: www.lmunet.edu
1897; N.N.
Civil War, Lincoln; Lincoln Memorial Univ Authors, Jesse Stuart Coll
188 000 vols; 119 000 curr per   31303

Rensselaer at Hartford, Robert L. & Sara Marcy Cole Library, 275 Windsor St, *Hartford*, CT 06120-2991
T: +1 860 5482490; Fax: +1 860 2780180; E-mail: lib-info@ewp.rpi.edu; URL: www.ewp.rpi.edu/hartford/library
1955; Mary Dixey
Computer science, Business & management; Computer Science, Corporate Annual Rpts
44 000 vols; 500 curr per
libr loan   31304

Trinity College, Library, 300 Summit St, *Hartford*, CT 06106
T: +1 860 2972255; Fax: +1 860 2972251; URL: www.trincoll.edu/depts/ library/
1823; Richard S. Ross
Education, Hist, Music; Americana, Henry Barnard Coll (American Education, Early American Textbks), Printing, Civil War, Early Voyages, Folklore, Horology, Trumbull-Prime Coll (Incunabula), Witchcraft, Robert Frost Coll, Slavery
905 000 vols; 1407 curr per; 54 000 e-books
libr loan   31305

University of Connecticut, School of Law Library, 39 Elizabeth St, *Hartford*, CT 06105-2287
T: +1 860 5705158; Fax: +1 860 5705116; URL: www.law.uconn.edu
1921; Darcy Kirk
546 000 vols; 982 000 microforms; 515 sound-rec; 25 digital data carriers
libr loan   31306

Coker College, James Lide Coker III Memorial Library, 300 E College Ave, *Hartsville*, SC 29550
T: +1 843 3838125; Fax: +1 843 3838129; E-mail: library@coker.edu; URL: www.coker.edu/library
1908; Alexa Bartel
Arents Tobacco Coll
87 000 vols; 160 curr per; 60 000 e-books
libr loan   31307

Hastings College, Perkins Library, 705 E Seventh St, *Hastings*, NE 68901-7620
T: +1 402 4617330; Fax: +1 402 4617480
1882; Robert Nedderman
Plains & Western Hist (Brown Coll), Holcomb Lewis & Clark Coll
140 000 vols; 540 curr per; 47 000 e-books; 3000 av-mat   31308

University of Southern Mississippi, Joseph Anderson Cook Library, 118 College Dr, No 5053, *Hattiesburg*, MS 39406
T: +1 601 2664249; Fax: +1 601 2666033; URL: www.lib.usm.edu
1912; Carole Kiehl
896 000 vols; 25 748 curr per; 21 259 e-journals; 4587 maps; 1 140 000 microforms   31309

– William David McCain Library & Archives, 118 College Dr, No 5148, Hattiesburg, MS 39406
T: +1 601 2664345; Fax: +1 601 2266269; E-mail: spref@avatar.lib.usm.edu; URL: www.lib.usm.edu
1966; Sherry Laughlin
Editorial Cartoon Coll, Railroads Coll, Children's Lit, Lit Coll, Genealogy Coll, Mississippiana Coll
152 000 vols   31310

William Carey University Libraries, Smith-Rouse Library, 498 Tuscan Ave, WCU Box 5, *Hattiesburg*, MS 39401
T: +1 601 3186170; Fax: +1 601 3186171; URL: library.wmcarey.edu
1911; Furr Patricia
Business, Education, Music, Nursing, Religious studies; Church Music (Clarence Dickinson Coll); Music Listening Libr
95 000 vols; 773 curr per; 18 358 e-journals; 125 diss/theses; 542 music scores; 16 000 microforms; 452 sound-rec
libr loan; CLS, ACRL, ALA, ACL   31311

Haverford College, James P. Magill Library, 370 Lancaster Ave, *Haverford*, PA 19041-1392
T: +1 610 8961175; Fax: +1 610 8961160; E-mail: library@haverford.edu; URL: www.haverford.edu/library
1833; Robert Kieft
Christopher Morley Coll, Cricket Coll, Philips Elizabethan Studies Coll, Maxfield Parrish Coll, Jones Mysticism Coll, Harris Coll of Near Eastern Mss, Quakerism, Roberts Mss Coll, Rufus Jones Writings
588 000 vols; 5475 curr per   31312

– Astronomy Library, Observatory, 370 W Lancaster Ave, Haverford, PA 19041
T: +1 610 8961291; Fax: +1 610 8961102
Dora Wong
Astronomy
46 000 vols; 110 curr per   31313

– Music Library, Union Bldg, 370 W Lancaster Ave, Haverford, PA 19041
T: +1 610 8961005; Fax: +1 610 8961102
Michelle Oswell
31 000 vols; 52 curr per   31314

– White Science Library, 370 W Lancaster Ave, Haversford, PA 19041
T: +1 610 8961291; Fax: +1 610 8961102
Dora Wong
54 000 vols; 101 curr per   31315

Barclay College, Worden Memorial Library, 100 E Cherry St, *Haviland*, KS 67059
T: +1 620 8625274; Fax: +1 620 8625403; E-mail: library@barclaycollege.edu; URL: www.barclaycollege.edu/academics/ library.asp
1979; Emily Harkness

Quaker (Friends) Writings
50 000 vols; 75 curr per; 1000 microforms
libr loan; ACL   31316

Montana State University-Northern, Vande Bogart Library, 300 11th St W, *Havre*, MT 59501; P.O. Box 7751, Havre, MT 59501-7751
T: +1 406 2653706; Fax: +1 406 2653799; URL: www.msun.edu/infotech/ library
1929; Vicki Gist
Education (Educational Resources Info Ctr Coll), micro
141 000 vols; 680 curr per
libr loan   31317

Jarvis Christian College, Olin Library & Communication Center, Hwy 80 E, *Hawkins*, TX 75765; P.O. Box 1470, Hawkins, TX 75765-1470
T: +1 903 7695820; Fax: +1 903 7695822; URL: www.jarvis.edu
1920; Tracy Caradine
Religion, Business, Education; Curriculum Libr, Young Adult Coll
69 000 vols; 126 curr per   31318

Fort Hays State University, Forsyth Library, 600 Park St, *Hays*, KS 67601-4099
T: +1 785 6284351; Fax: +1 785 6284096; URL: www.fhsu.edu/forsyth_lib/
1902; John Ross
Ctr for Ethnic Studies Coll, Western Coll, Univ Arch, Folklore Coll, Children's & Young Adult's Books Coll, Currey Arch of Military Hist, Paschal World War II Hist Coll
350 000 vols; 1800 curr per
libr loan   31319

California State University, East Bay Library, 25800 Carlos Bee Blvd, *Hayward*, CA 94542-3052
T: +1 510 8853765;
Fax: +1 510 8852049; URL: www.library.csueastbay.edu
1957; Myoung-ja Lee Kwon
Bay Area Poetry Coll, Cameos Coll, Early Voyages & Travels Coll, Marco Polo Coll
926 000 vols; 1686 curr per; 13 080 e-journals; 4000 e-books; 5500 av-mat
libr loan   31320

Pennsylvania State University, Hazleton Library, 76 University Dr, *Hazleton*, PA 18202-8025
T: +1 570 4503170; Fax: +1 570 4503128; E-mail: hazelton@psulias.psu.edu; URL: www.hn.psu.edu/department/library
1934; Joseph A. Fennewald
80 000 vols; 230 curr per   31321

Carroll College, Jack & Sallie Corette Library, 1601 N Benton Ave, *Helena*, MT 59625
T: +1 406 4474344; Fax: +1 406 4474525; URL: www.carroll.edu/library
1928; Lois A. Fitzpatrick
98 000 vols; 300 curr per; 897 av-mat; 4088 sound-rec; 10 digital data carriers
libr loan; OCLC, BCR   31322

Phillips Community College of the University of Arkansas, Lewis Library, 1000 Campus Dr, P.O. Box 785, *Helena*, AR 72342
T: +1 870 3386474 ext 1145; E-mail: rpride@pccua.edu; URL: www.pccua.edu
1966
Art, Automotive engineering, Delta blues
43 550 vols; 351 curr per   31323

Hofstra University, *Hempstead*
– Barbara & Maurice A. Deane Law Library, 122 Hofstra University, Hempstead, NY 11549-1220
T: +1 516 4635808; Fax: +1 516 4635129; URL: www.hofstra.edu/libraries/ lawlib/law_library.cfm
Michelle M. Wu
551 000 vols   31324

– Joan & Donald E. Axinn Library, 123 Hofstra University, Hempstead, NY 11549
T: +1 516 4635962; Fax: +1 516 4636387; URL: www.hofstra.edu/library
1935; Dr. Daniel Rubey
Authors Colls of late 19th & early 20th c, William Blake Facsimiles, Coll of Books about Books & Early Printed Books, Georgian Poets, Henry Kroul

Coll of Nazi Culture & Propaganda, New York State/Long Island Hist, Weingrow Coll of Avantgarde Art & Lit, Utopian Communities
1 337 000 vols; 1535 curr per; 34 000 e-books
libr loan   31325

Freed-Hardeman University, Loden-Daniel Library, 158 E Main St, *Henderson*, TN 38340-2399
T: +1 731 9896067; Fax: +1 731 9896065; E-mail: library@fhu.edu; URL: www.fhu.edu/library
1869; Hope Shull
Religion (Restoration Libr Coll), bks & tapes
150 000 vols; 1460 curr per; 14 000 e-books   31326

New England College, H. Raymond Danforth Library, 28 Bridge St, *Henniker*, NH 03242-3298
T: +1 603 4282344; Fax: +1 603 4284273; URL: www.nec.edu
1946; Katherine Van Weelden
Art Coll, New Hampshiriana, Shakespeare Coll (Adams Coll), College Arch
103 000 vols; 300 curr per   31327

Pennsylvania State University, College of Medicine, George T. Harrell Health Sciences Library, Milton S. Hershey Medical Ctr, 500 University Dr, P.O. Box 850, *Hershey*, PA 17033
T: +1 717 5318626; E-mail: crobinson1@hmc.psu.edu; URL: www.hmc.psu.edu/library
1965; Cynthia Robinson
Rare Medical Bks
40 200 vols; 10 000 e-books; 55 sound-rec; 5250 journal titles in all formats   31328

Hesston College, Mary Miller Library, P.O. Box 3000, *Hesston*, KS 67062-3000
T: +1 316 3278245; Fax: +1 316 3278300; E-mail: margaret@hesston.edu; URL: www.hesston.edu/academics/lrc/ mml.htm
1908; Margaret Wiebe
30 000 vols; 225 curr per   31329

Lenoir-Rhyne College, Carl A. Rudisill Library, 625 7th Ave NE, *Hickory*, NC 28601; P.O. Box 7548, Hickory, NC 28603-7548
T: +1 828 3287236; Fax: +1 828 3287338; URL: www.lrc.edu/library/
1891; Virginia Moreland
Martin Luther Works, Quetzalcoatl Coll
150 000 vols; 479 curr per; 19 100 e-journals; 32 000 e-books; 4380 Bks on Deafness & Sign Lang   31330

Michigan State University, Morofsky Memorial Library at Gull Lake, Kellogg Biological Sta, 3700 E Gull Lake Dr, *Hickory Corners*, MI 49060-9516
T: +1 269 6712310; Fax: +1 269 6712309; E-mail: library@kbs.msu.edu; URL: www.kbs.msu.edu/library/index.php
1965
Agriculture, Ecology, Limnology
12 000 vols; 150 curr per   31331

High Point University, Smith Library, 833 Montlieu Ave, *High Point*, NC 27262-4221
T: +1 336 8418102; Fax: +1 336 8415123; E-mail: hpulibrary@highpoint.edu; URL: library.highpoint.edu
1924; David L. Bryden
North Carolina Coll, Home Furnishings Marketing Coll, Methodist Arch
310 000 vols; 25 000 curr per; 54 000 e-books   31332

John Wesley College, Library, 2314 N Centennial St, *High Point*, NC 27265-3197
T: +1 336 8892262; Fax: +1 336 8892221; E-mail: library@johnwesley.edu; URL: www.johnwesley.edu
1936; April Lindsey
Education, Religious studies, Music; John Wesley
38 000 vols; 100 curr per   31333

Northern Kentucky University, W. Frank Steely Library, University Dr, ***Highland Heights***, KY 41099
T: +1 859 5725456; Fax: +1 859 5725390; URL: library.nku.edu
1968; Arne J. Almquist
Kentucky; Caudill Urban Appalachian Coll, Bogardus Ohio River Coll, Gist Hist Society Coll, Morris Garrett Coll, Warren Shonert Coll, Oral Hist Coll, Univ Arch
316 000 vols; 1 579 curr per; 635 e-journals; 35 000 e-books; 2 140 av-mat; 2 270 sound-rec; 8 600 Music Scores
libr loan 31334

Northern Kentucky University – Salomon P. Chase College of Law, Library, Nunn Dr, ***Highland Heights***, KY 41099
T: +1 859 5725394; Fax: +1 859 5726664; URL: chaselaw.nku.edu/library/index.php
Thomas Heard
Siebenthaler Rare Bks Coll
258 000 vols; 2 085 curr per; 113 000 microforms; 24 digital data carriers
libr loan 31335

Jefferson College, Library, 1000 Viking Dr, ***Hillsboro***, MO 63050
T: +1 636 9423000; Fax: +1 636 7893954; URL: www.jeffco.edu/library
1964; Susan Morgan
Jefferson County Hist Ctr
72 000 vols; 130 curr per 31336

Tabor College, Library, 400 S Jefferson St, ***Hillsboro***, KS 67063
T: +1 620 9473121; Fax: +1 620 9472607; E-mail: library@tabor.edu; URL: www.tabor.edu/library
1908; Robin D. Ottoson
Religion; Mennonite Brethren Hist Libr & Archs
83 000 vols; 156 curr per
libr loan 31337

Hillsdale College, Michael Alex Mossey Library, 33 E College St, ***Hillsdale***, MI 49242
T: +1 517 6072404; Fax: +1 517 6072248; URL: www.hillsdale.edu/library
1844; Daniel L. Knoch
Economy & Political Science (Ludwig von Mises Libr), Russell Kirk Libr, Weaver Coll
217 000 vols; 700 curr per; 14 600 e-journals; 28 000 e-books
libr loan 31338

University of Hawaii at Hilo, Edwin H. Mookini Library, 200 W Kawili St, ***Hilo***, HI 96720-4091
T: +1 808 9747343; Fax: +1 808 9747329; E-mail: mookini@hawaii.edu; URL: library.uhh.hawaii.edu
1947; Dr. Linda Marie Golian-Lui
East Asia; Hawaiiana
260 000 vols; 1 200 curr per; 13 000 e-journals; 4 000 e-books; 264 000 microforms
libr loan 31339

Hiram College, Library, 11694 Hayden St, ***Hiram***, OH 44234; P.O. Box 67, Hiram, OH 44234-0067
T: +1 330 5695354; Fax: +1 330 5695491; URL: library.hiram.edu
1850; David Everett
Regional Studies Coll, Institutional Hist, Local Hist Coll, Church Hist, Early Textbook Coll, Juvenile Lit, Hiram Authors, World War I Pamphlets; Rare Bks
215 000 vols; 543 curr per; 6 313 e-journals; 2 048 sound-rec; 6 670 digital data carriers
libr loan 31340

College of the Southwest, Scarborough Memorial Library, 6610 Lovington Hwy, ***Hobbs***, NM 88240
T: +1 505 3926565; Fax: +1 505 3926006; URL: www.csw.edu
1962; John McCance
Southwestern Hist & Art Lit, Southwest Heritage Room (Thelma A. Webber Coll), New Mexico Textbook Evaluation Ctr
76 000 vols; 237 curr per; 1 590 av-mat
libr loan; ALA, ACRL 31341

Stevens Institute of Technology, Samuel C. Williams Library, Castle Point on Hudson, ***Hoboken***, NJ 07030
T: +1 201 2168109; Fax: +1 201 2168319; URL: www.lib.stevens.edu
1870; Ourida Oubraham
F W Taylor Coll (Scientific Management), Lieb Libr of Leonardo da Vinci, Stevens Family Arch, Ironclad Monitor Blueprints
80 000 vols; 120 curr per; 28 193 e-journals; 4 000 e-books; 4 617 diss/theses; 2 maps; 2 000 microforms; 2 736 digital data carriers
libr loan; OCLC 31342

Hope College, Van Wylen Library, 53 Graves Pl, ***Holland***, MI 49422; P.O. Box 9012, Holland, MI 49422-9012
T: +1 616 3957790; Fax: +1 616 3957965; E-mail: libwebteam@hope.edu; URL: www.hope.edu/lib
1866; David P. Jensen
Church Hist (Reformed Church in America), Dutch American Hist, Immigration, Holland Joint Archs; Music dept
367 000 vols; 1 548 curr per; 2 incunabula; 2 065 govt docs; 290 maps; 350 000 microforms; 5 036 sound-rec; 3 825 digital data carriers
libr loan; ALA 31343

Rust College, Leontyne Price Library, 150 E Rust Ave, ***Holly Springs***, MS 38635
T: +1 662 2524661; Fax: +1 662 2528873; E-mail: amoore@rustcollege.edu; URL: www.rustcollege.edu
1866; Anita W. Moore
Roy Wilkins Coll, United Methodist Religious Coll, Int Coll
122 040 vols; 366 curr per 31344

Chaminade University of Honolulu, Sullivan Library, 3140 Waialae Ave, ***Honolulu***, HI 96816-1578
T: +1 808 7394665; Fax: +1 808 7354891; E-mail: library@chaminade.edu; URL: www.chaminade.edu/library
1955; Sharon LePage
Catholic authors, Hawaiiana, Judaica
75 000 vols; 230 curr per; 36 000 e-books; 283 diss/theses; 134 000 microforms; 53 sound-rec; 109 digital data carriers
libr loan 31345

Hawaii Pacific University, Meader Library, 1060 Bishop St, ***Honolulu***, HI 96813-3192
T: +1 808 5440292; Fax: +1 808 5217998; E-mail: lib@hpu.edu; URL: www.hpu.edu/index.cfm?section=llss
1965; Kathleen Chee
Co-operative Education Coll, Corporation Info, Coll, Graduate Professional Paper Coll, Hawaiian-Pacific Coll, English Foundations Program Paperback Coll, Atlas Coll, Foreign Language Coll
110 000 vols; 1 600 curr per; 7 000 e-books; 881 microforms; 27 digital data carriers 31346

University of Hawaii at Manoa, Thomas Hale Hamilton Library, 2550 McCarthy Mall, ***Honolulu***, HI 96822
T: +1 808 9567853; Fax: +1 808 9565968; E-mail: library@hawaii.edu; URL: library.manoa.hawaii.edu
1907; Paula Mochida
Asian Coll, Book Arts, Hawaiian Coll, Pacific Coll, Univ Arch, Jean Charlot Coll, Tsuzaki Reinecke Creole Coll, Hawaii in World War II, Rare Books
3 356 000 vols; 26 605 curr per; 6 096 mss; 249 550 maps; 5 869 000 microforms; 13 023 sound-rec; 65 digital data carriers; 20 708 film & video
libr loan 31347

University of Hawaii at Manoa – School of Medicine, Health Sciences Library, 651 Ilalo St, MEB, ***Honolulu***, HI 96813
T: +1 808 6920810; Fax: +1 808 6921244; E-mail: hslinfo@hawaii.edu; URL: www.hawaii.edu/hslib
2005; Virginia M. Tanji
11 000 vols; 150 curr per; 3 digital data carriers
libr loan; AAHSL 31348

University of Hawaii – William S. Richardson School of Law Library, 2525 Dole St, ***Honolulu***, HI 96822-2328
T: +1 808 9565581; Fax: +1 808 9564615; URL: library.law.hawaii.edu
Leinaala Seeger
Environmental law; Pacific-Asian Law Coll
305 000 vols; 3 748 curr per; 3 digital data carriers 31349

University of Hawaii-College of Education, Western Curriculum Coordination Center, 1776 University Ave UA2-7, ***Honolulu***, HI 96822
T: +1 808 9567834; Fax: +1 808 9563374; E-mail: wccc@hawaii.edu; URL: www.hawaii.edu/wccc/
Lawrence Zane
Curriculum mat, Vocational education
20 000 vols 31350

Houghton College, Willard J. Houghton Memorial Library, One Willard Ave, ***Houghton***, NY 14744
T: +1 585 5679242; Fax: +1 585 5679248; URL: www.houghton.edu/library
1883; Bradley Wilber
Bible, Religion; John Wesley & Methodism, Science & Christian Faith
243 000 vols; 7 598 curr per 31351

Michigan Technological University, J. Robert Van Pelt Library, 1400 Townsend Dr, ***Houghton***, MI 49931-1295
T: +1 906 4872508; Fax: +1 906 4871765; E-mail: reflib@mtu.edu; URL: www.lib.mtu.edu
1887; Phyllis H. Johnson
Copper Country Hist Coll; Foundation Center Regional Coll; Isle Royale (Ben Chynoweth Coll), pamphlets, articles, photos; Spitzbergen (John M. Longyear Coll); Univ Arch; USBM Mine Maps of Michigan
794 000 vols; 5 973 curr per; 3 650 e-journals; 29 000 e-books; 447 257 govt docs; 130 773 maps; 494 000 microforms; 1 414 digital data carriers 31352

Houston Baptist University, Moody Memorial Library, 7502 Fondren Rd, ***Houston***, TX 77074-3298
T: +1 281 6493182; Fax: +1 281 6493489; URL: moody.hbu.edu
1963; Ann A. Noble
Baptist hist, Southwest Texas, Victorian lit; Hist (Jimmy Hicks Memorial Coll), Hist & Lit (Palmer Bradley Coll), Gilbert & Sullivan (Linder Coll)
223 000 vols; 2 964 curr per; 24 000 e-journals; 47 000 e-books; 422 000 microforms; 3 740 av-mat; 1 740 sound-rec; 480 digital data carriers
libr loan 31353

Rice University, Fondren Library, 6100 Main, MS-44, ***Houston***, TX 77005; P.O. Box 1892, Houston, TX 77251-1892
T: +1 713 3484022; Fax: +1 713 3484117; E-mail: reference@rice.edu; URL: library.rice.edu
1912; Geneva Henry
18th-19th c British Maritime & Naval Hist, 18th c British Drama, 19th-20th c Texas, Hist of Aeronautics, Cruikshank Coll, Maximilian & Carlotta Coll, Modern American Lit, US Civil War & Slavery, Hist of Science, Texas Politics, Texas Entrepreneurs/Business; Woodson Research Center, Business Information Center, Electronic Resources Center, Government Publications and Microforms
2 519 000 vols; 53 811 curr per; 11 700 e-journals; 24 000 e-books; 397 mss; 6 incunabula; 104 000 govt docs; 38 849 music scores; 26 867 maps; 3 135 000 microforms; 38 950 sound-rec; 14 107 digital data carriers
libr loan; ARL, RLG, Coalition for Networked Info, Council of Library and Information Resources, Greater Western Library Alliance, Nat'l Initiative for a Networked Cultural Heritage, OCLC, Scholarly Publ & Acad Resources 31354

South Texas College of Law, Parks Law Library, 1303 San Jacinto, ***Houston***, TX 77002-7000
T: +1 713 6461726; Fax: +1 713 6592217; E-mail: stclill@stcl.edu; URL: www.stcl.edu
1924; David G. Cowan
Law School Arch, Rare bks, Texaco-Pennzoil Papers
485 000 vols; 4 326 curr per; 894 000 microforms; 218 digital data carriers 31355

Texas Southern University, Robert James Terry Library, 3100 Cleburne Ave, ***Houston***, TX 77004
T: +1 713 3137148; Fax: +1 713 3131080; URL: www.tsu.edu/library/index.html
1947; Obidike Kamau
Heartman Coll, Barbara Jordan Arch, Traditional African Art Gallery, Univ Arch, Jazz Arch
470 102 vols; 1 715 curr per
libr loan 31356

– Thurgood Marshall School of Law Library, 3100 Cleburne Ave, Houston, TX 77004
T: +1 713 3137125; Fax: +1 713 3134483; URL: www.tsulaw.edu/lawlib
1948; Marguerite Butler
536 000 vols; 2 582 curr per; 20 digital data carriers 31357

University of Houston, M. D. Anderson (Main) Library, 114 University Libraries, ***Houston***, TX 77204-2000
T: +1 713 7439710; Fax: +1 713 7439811; URL: info.lib.uh.edu
1927; Dana Rooks
Texana, Western Hist, Houston Hist, British & American Authors, Hist of the Book & Printing, African-American Authors, Hist of 10th C Physics
2 300 000 vols; 21 840 curr per
libr loan 31358

– Music Library, 220 Moores School of Music Bldg, Houston, TX 77204-4017; mail address: 114 University Libraries, Houston, TX 77204-2000
T: +1 713 7433197; Fax: +1 713 7439918; E-mail: musiclib@uh.edu; URL: info.lib.uh.edu/music
Ericka Patillo
52 000 vols; 220 curr per 31359

– Optometry Library, Rm 2225, Houston, TX 77204-2020
T: +1 713 7431910; Fax: +1 713 7432001; URL: info.lib.uh.edu/local/optometr.htm
1974; Suzanne Ferimer
12 000 vols; 154 curr per; 167 diss/theses; 262 av-mat; 10 sound-rec; 22 digital data carriers; 192 slide sets
libr loan 31360

– The O'Quinn Law Library, 12 Law Library, Houston, TX 77204-6054
T: +1 713 7432331; Fax: +1 713 7432296; URL: www.law.uh.edu/libraries
Spencer Simons
Martitime law, international trade
524 000 vols; 3 326 curr per 31361

– Pharmacy Library, 114 University Libraries, Houston, TX 77204-2000
T: +1 713 7431240; Fax: +1 713 7431233; URL: info.lib.uh.edu/pharmacy
1947; Nelda Cervantes
Hist & Biogr of Pharmacy & Medicine
17 000 vols; 129 curr per; 100 e-books; 10 digital data carriers 31362

– William R. Jenkins Architecture & Art Library, 106 Architecture Bldg, Houston, TX 77204-4000
T: +1 713 7432340; Fax: +1 713 7439917; E-mail: archlib@mail.uh.edu; URL: info.lib.uh.edu/libraries/aa/index.html
1961; Margaret Culbertson
75 000 vols; 225 curr per; 5 digital data carriers 31363

University of Houston – Clear Lake, Neumann Library, 2700 Bay Area Blvd, ***Houston***, TX 77058-1098
T: +1 281 2833930; Fax: +1 281 2833937; URL: nola.cl.uh.edu
1973; Karen Wielhorski
Early English Bks (Pollard & Redgrave Coll)
450 000 vols; 2 000 curr per; 35 000 e-books 31364

University of Houston – Downtown, Dykes Library, One Main St, ***Houston***, TX 77002
T: +1 713 2218467; Fax: +1 713 2218037; URL: www.uhd.edu/library/index.html
1974; Pat Ensor
269 000 vols; 1 122 curr per; 75 000 e-books; 13 000 microforms; 4 306 sound-rec
libr loan 31365

University of Saint Thomas, Robert Pace & Ada Mary Doherty Library, 1100 W Main, ***Houston***, TX 77006; mail address: 3800 Montrose Blvd, Houston, TX 77006
T: +1 713 5256926; Fax: +1 713 5253886; URL: www.stthom.edu/public/index.asp?page_ID=1467
1947; James Piccininni
Philosophy (Thomistic Studies Coll)
215 000 vols; 15 150 curr per; 25 000 e-books; 100 diss/theses; 460 000 microforms; 39 sound-rec; 39 digital data carriers
libr loan; ALA 31366

– Cardinal Beran Library at Saint Mary's Seminary, 9845 Memorial Dr, Houston, TX 77024-3498
T: +1 713 6864345; Fax: +1 713 6817550; E-mail: beran@smseminary.com; URL: www.beran.stthom.edu
1954; Laura Olejnik
64 000 vols; 315 curr per; 1 000 microforms; 2 097 sound-rec; 4 digital data carriers 31367

University of Texas, ***Houston***
– M. D. Anderson Cancer Center, Research Medical Library, Pickens Academic Tower, 21st floor, 1400 Pressler St, Houston, TX 77030-3722
T: +1 713 7922282; Fax: +1 713 5633650; E-mail: RML-Help@mdanderson.org; URL: www3.mdanderson.org/library
1941; Stephanie Fulton
Leland Clayton Barbee Hist of Cancer Coll (Rare Bks & Early Treatises on Cancer)
76 000 vols; 875 curr per; 5 467 e-journals; 408 av-mat; 1 digital data carriers; 100 plus-online databases 31368

– Health Science Center at Houston, Dental Branch Library, 6516 M D Anderson Blvd, Rm 133, Houston, TX 77030; P.O. Box 20068, Houston, TX 77225-0068
T: +1 713 5004094; Fax: +1 713 5004100; URL: www.db.uth.tmc.edu/info-res
1943; Judith Penn
Dentistry Hist Coll
32 000 vols; 204 curr per; 525 av-mat; 20 digital data carriers 31369

– Houston Health Science Center – School of Public Health Library, 1200 Herman Pressler Blvd, Houston, TX 77030-3900; P.O. Box 20186, Houston, TX 77225-0186
T: +1 713 5009131; Fax: +1 713 5009125; URL: www.sph.uth.tmc.edu/library
1969; Helena M. VonVille
Bio-statistics, Community health, Epidemiology, Health promotion, HIV-AIDS, Infectious diseases, Nutrition; Int Census Statistics, Pan American Health Organization, World Health Organization
65 000 vols; 350 curr per 31370

Juniata College, L.A. Beeghly Library, 1815 Moore St, ***Huntingdon***, PA 16652-2120
T: +1 814 6413450; Fax: +1 814 6413435; URL: www.juniata.edu
1876; John Mumford
Early Pennsylvania German Imprints (Abraham Harley Cassel Coll), bks & pamphlets; Church of the Brethren (College Arch), bks, mss; Pennsylvania Folklore (Henry W Shoemaker Coll)
200 000 vols; 1 000 curr per
libr loan 31371

Huntington College, Richlyn Library, 2303 College Ave, ***Huntington***, IN 46750
T: +1 260 3594054; Fax: +1 260 3583698; URL: www.huntington.edu/library
1897; Robert E. Kaehr
College Arch, United Brethren In Christ Church, Curriculum Mat Ctr
180 000 vols 31372

Marshall University, University Libraries, One John Marshall Dr, ***Huntington***, WV 25755-2060
T: +1 304 6962334; Fax: +1 304 6965858; E-mail: library@marshall.edu; URL: www.marshall.edu/library
1929; Barbara A. Winters
Anthropology (Human Relations Area Files); The Rosanna Blake Libr of Confederate Hist; Civil War Newspapers; Congress of Racial Equality Papers; Historic Lit (Pollard, Redgrave & Wing Books, published in England & Scotland, 1400-1700); Hoffman Hist of Medicine Libr; Sociology
428 000 vols; 4 103 curr per; 16 440 e-journals; 6 000 e-books; 6 110 av-mat; 15 420 sound-rec; 80 digital data carriers
libr loan 31373

Marshall University – Health Science Libraries, Joan C. Edwards School of Medicine, 1600 Medical Center Dr, ***Huntington***, WV 25701
T: +1 304 6911752; Fax: +1 304 6911766; E-mail: dzierzak@marshall.edu; URL: musom.marshall.edu/library
1976; Edward Dzierzak
15 620 vols; 423 curr per; 175 av-mat; 45 sound-rec; 3 digital data carriers
libr loan 31374

Alabama Agricultural & Mechanical University, Joseph F. Drake Memorial Learning Resources Center, 4900 Meridian St, ***Huntsville***, AL 35811; P.O. Box 489, Normal, AL 35762-0489
T: +1 256 3724747; Fax: +1 256 8515249; URL: www.aamu.edu
1904; Dr. Clarence Toomer
Archival & Hist Colls, Audio Visual Coll, Black Coll, Carnegie-Mydral Coll, Curriculum Coll, Govt Docs Coll, Int Studies Coll, J. F. Kennedy Memorial Coll, Schomburg Coll, Textbook Coll, YA Coll
254 000 vols; 1 800 curr per; 151 mss; 141 376 govt docs; 46 maps; 32 617 av-mat; 568 sound-rec; 57 digital data carriers; 1 053 college cat, 518 148 ERIC microfiche, 10 979 vertical files
libr loan; ALA 31375

Oakwood College, Eva B. Dykes Library, 7000 Adventist Blvd, ***Huntsville***, AL 35896
T: +1 256 7267248; Fax: +1 256 7267538; URL: www.oakwood.edu/library
1896; Paulette L. Johnson
Black Studies, Seventh-Day Adventist Black Hist, Oakwood College Hist
135 000 vols; 565 curr per
libr loan 31376

Sam Houston State University, Newton Gresham Library, 1830 Bobby K Marks Dr, ***Huntsville***, TX 77340; P.O. Box 2281, Huntsville, TX 77341-2281
T: +1 936 2941614; Fax: +1 936 2943780; URL: library.shsu.edu/
1879; Ann H. Holder
Criminology (Bates, Bennett, Colfield, Eliasburg & McCormirk Coll), Texana & the Southwest Colls, Gertrude Stein Coll, Mark Twain Coll, Col John W Thomason Coll
1 202 000 vols; 4 521 curr per; 24 000 e-books
libr loan 31377

University of Alabama in Huntsville, M. Louis Salmon Library, 301 Sparkman Dr NW, ***Huntsville***, AL 35899
T: +1 256 8246529; Fax: +1 256 8246552; URL: www.uah.edu/library
1967; Dr. Wilson Luquire
Engineering, Science, Technology, Business; Robert E Jones Congressional Papers, Willy Ley Space Coll, Harvie Jone Architectural Coll, Robert Forward Space Coll, Saturn V History Documentation Coll, Skylab Space Station Coll
330 000 vols; 1 044 curr per; 17 500 e-journals; 46 000 e-books; 174 339 govt docs; 962 maps; 505 000 microforms; 1 144 sound-rec; 5 digital data carriers
libr loan; ACRL, NAAL 31378

Bowling Green State University, Firelands College Library, One University Dr, 2nd Flr, ***Huron***, OH 44839-9791
T: +1 419 4335560; Fax: +1 419 4339696; E-mail: firelib@bgnet.bgsu.edu; URL: www.firelands.bgsu.edu/library/index.html
1968; Sharon Britton
Firelands of the Conn Western Reserve
41 000 vols; 249 curr per 31379

Immaculata University, Gabriele Library, 1145 King Rd, ***Immaculata***, PA 19345-0705
T: +1 610 6474400; Fax: +1 610 6405828; URL: library.immaculata.edu
1920; Jeff Rollison
Spanish American & Chicano Lit Coll, Dietetics
142 000 vols; 755 curr per; 510 av-mat
libr loan 31380

Sierra Nevada College, MacLean Library, 999 Tahoe Blvd, ***Incline Village***, NV 89450-9500
T: +1 775 8311314; Fax: +1 775 8326134; E-mail: library@sierranevada.edu; URL: www.sierranevada.edu
1969; Dr. Betts Markle
23 000 vols; 174 curr per 31381

Ohio College of Podiatric Medicine, Library, 6000 Rockside Woods Blvd, ***Independence***, OH 44131
T: +1 216 2313300; Fax: +1 216 4470626; URL: www.ocpm.edu/departments/library
1916; Donna Perzeski
Arch
17 000 vols; 105 curr per; 175 av-mat; 1 500 sound-rec; 50 digital data carriers; 800 reprints, 7 000 slides
libr loan 31382

Indiana University of Pennsylvania, Stapleton Library, 431 S 11th St, ***Indiana***, PA 15705-1096
T: +1 724 3573006; Fax: +1 724 3574891; URL: www.lib.iup.edu
1875; Dr. Phillip Zorich
Education, Liberal arts; Curriculum Mat, Univ School Libr, Charles Darwin, Herman Melville, Mark Twain, Edgar Allan Poe, John Greenleaf Whittier, Norman Mailer, Washington Irving, United Mine Workers Union, Rochester & Pittsburgh Coal Company Papers
722 000 vols; 15 024 curr per; 1 700 000 microforms; 12 digital data carriers
libr loan 31383

– Cogswell Music Library, 101 Cogswell Hall, 422 S 11th St, Indiana, PA 15705-1071
T: +1 724 3572892; Fax: +1 724 3574891; URL: www.lib.iup.edu/depts/musiclib/music.html
1969; Dr. Carl Rahkonen
46 000 vols; 56 curr per 31384

Butler University, Irwin Library, 4600 Sunset Ave, ***Indianapolis***, IN 46208
T: +1 317 9409235; Fax: +1 317 9409711; URL: www.butler.edu
1855; Lewis R. Miller
South Sea Islands Coll, Sibelius Coll, Lincoln Coll, Botanical & Zoological Print Colls, Univ Arch
382 000 vols; 941 curr per; 14 000 e-books
libr loan 31385

– Ruth Lilly Science Library, 740 W 46th St, Indianapolis, IN 46208-3485
T: +1 317 9409401; Fax: +1 317 9409519; E-mail: lmiller@butler.edu; URL: www.butler.edu/libraries
1973; Lewis R. Miller
Chemistry, Mathematics
60 000 vols; 540 curr per; 750 av-mat; 7 digital data carriers 31386

Indiana University, ***Indianapolis***
– Ruth Lilly Medical Library, 975 W Walnut St, IB 100, Indianapolis, IN 46202-5121
T: +1 317 2747182; Fax: +1 317 2782349; E-mail: medlib@iupui.edu; URL: www.medlib.iupui.edu
1908; Julie McGowan
Hist of Medicine Coll
261 000 vols; 310 curr per; 766 diss/theses; 7 000 microforms; 273 digital data carriers
libr loan 31387

– School of Dentistry, Library, 1121 W Michigan St, Rm 128, Indianapolis, IN 46202-5186
T: +1 317 2745203; Fax: +1 317 2781256; E-mail: ds-libry@iupui.edu; URL: www.iusd.iupui.edu/depts/lib/default.aspx
1927; Jan Cox
Arch Coll
27 000 vols; 463 curr per; 40 e-books; 3 000 microforms; 105 sound-rec; 28 digital data carriers
libr loan 31388

– School of Law Library, 530 W New York St, Indianapolis, IN 46202-3225
T: +1 317 2744028; Fax: +1 317 2748825; E-mail: lawlibry@iupui.edu; URL: www.indylaw.indiana.edu/library
1944; Judith Ford Anspach
Commonwealth Coll, Council of Europe, European Communities
318 000 vols; 6 300 curr per; 771 000 microforms; 746 sound-rec
libr loan 31389

Indiana University-Purdue University Indianapolis, University Libraries, 755 W Michigan St, ***Indianapolis***, IN 46202-5195
T: +1 317 2748278; Fax: +1 317 2740492; URL: www.ulib.iupui.edu
1939; David Lewis
Arch Coll, German Americana, Philanthropy, Robert S. Woods Masonic Coll, Herron School of Arts Coll
1 339 000 vols; 4 780 curr per; 21 500 e-journals; 70 000 e-books; 4 800 mss; 2 238 diss/theses; 119 756 govt docs; 50 400 music scores; 1 596 maps; 1 242 000 microforms; 3 146 av-mat; 2 153 sound-rec; 1 235 digital data carriers
libr loan; ARL, CNI 31390

– Herron School of Art Library, 75 W New York St, Rm 117, Indianapolis, IN 46202
T: +1 317 2789484; Fax: +1 317 9202430; URL: www.ulib.iupui.edu/herron
1970; Sonya Staum-Kuniej
24 000 vols; 181 curr per 31391

Marian College, Hackelmeier Memorial Library, 3200 Cold Spring Rd, ***Indianapolis***, IN 46222
T: +1 317 9556090; Fax: +1 317 9556418; E-mail: librarystaff@marian.edu; URL: www.marian.edu/library
1937; Kelley Griffith
Roman Catholic Church, Nursing, Education; Monsignor Doyle Coll, Archbishop Paul Schulte Papers & Bks, American Far West
146 000 vols; 402 curr per
libr loan 31392

University of Indianapolis, Krannert Memorial Library, 1400 E Hanna Ave, ***Indianapolis***, IN 46227-3697
T: +1 317 7886124; Fax: +1 317 7883275; URL: www.kml.uindy.edu
1902; Dr. Philip Young
Evangelical United Brethren Coll, Krannert Coll
154 000 vols; 650 curr per; 30 000 e-books 31393

Simpson College, Dunn Library, 508 North C St, ***Indianola***, IA 50125-1216
T: +1 515 9611519; Fax: +1 515 9611363; URL: www.simpson.edu/library
1860; Cynthia M. Dyer
Avery O. Craven Coll
151 000 vols; 479 curr per; 7 000 e-books; 2 100 av-mat; 3 920 sound-rec; 460 digital data carriers; 23 898 Journals
libr loan 31394

University of West Los Angeles, Law School Library, 9920 S LaCienega Blvd, ***Inglewood***, CA 90301-4423
T: +1 310 3425206; Fax: +1 310 3425298; URL: www.uwla.edu/Academics/library.aspx
1968; Jimmy Rimonte
31 000 vols; 100 curr per; 8 000 microforms; 3 digital data carriers 31395

West Virginia State University, Drain-Jordan Library, Campus Box L17, ***Institute***, WV 25112; P.O. Box 1002, Institute, WV 25112-1002
T: +1 304 7663117; Fax: +1 304 7664103; URL: library.wvstateu.edu
1891; Patrick Hall
African-American hist; John W Davis Papers, College Arch
215 000 vols; 381 curr per; 1 850 av-mat; 340 sound-rec
libr loan 31396

University of Iowa, University Libraries, 125 W Washington St, ***Iowa City***, IA 52242-1420
T: +1 319 3844778; Fax: +1 319 3355830; E-mail: lib-ref@uiowa.edu; URL: www.lib.uiowa.edu
1855; Nancy L. Baker
Brewer Leigh Hunt Coll, Iowa Authors Coll, Bollinger Lincoln Colls, Mabbott Poe Coll, Hist of Hydraulics Coll, Chef Louis

Szathmary Coll of Culinary Arts, Social Docs Colls, Popular Culture Arch
3 788 000 vols; 40 166 curr per
libr loan 31397

– Art Library, 235 Art Bldg West, Iowa City, IA 52242
T: +1 319 3353089; Fax: +1 319 3355900; E-mail: lib-art@uiowa.edu; URL: www.lib.uiowa.edu/art
Rijn Templeton
90 000 vols 31398

– Biological Sciences Library, BSL, Iowa City, IA 52242-1420
T: +1 319 3353083; Fax: +1 319 3352698; E-mail: lib-biology@uiowa.edu; URL: www.lib.uiowa.edu/biology/
Leo Clougherty
47 000 vols 31399

– Chemistry Library, 400 Chemistry Bldg, Iowa City, IA 52242
T: +1 319 3353085; Fax: +1 319 3351193; E-mail: lib-chem@uiowa.edu; URL: www.lib.uiowa.edu/chem/index.html
Leo Clougherty
92 000 vols 31400

– Engineering Library, 2001 Seamans Center, Iowa City, IA 52242-1420
T: +1 319 3356047; Fax: +1 319 3355900; E-mail: lib-engineering@uiowa.edu; URL: www.lib.uiowa.edu/eng/
John W. Forys Jr
110 000 vols 31401

– Geoscience Library, 136 Trowbridge Hall, Iowa City, IA 52242
T: +1 319 3353084; Fax: +1 319 3353419; E-mail: lib-geoscience@uiowa.edu; URL: www.lib.uiowa.edu/geoscience/
Leo Clougherty
55 000 vols 31402

– Hardin Library for the Health Sciences, 100 Hardin Library, Iowa City, IA 52242
T: +1 319 3359871; Fax: +1 319 3359897; E-mail: lib-hardin@uiowa.edu; URL: www.lib.uiowa.edu/hardin
1882; Linda Walton
Hist of medicine
285 000 vols; 25 digital data carriers 31403

– Law Library, 200 Boyd Law Bldg, Iowa City, IA 52242-1166
T: +1 319 3359005; Fax: +1 319 3359039; URL: www.law.uiowa.edu/library/
1868; Mary Ann Nelson
UN Doc (Readex Coll)
1 200 000 vols; 9 500 curr per; 1 265 000 microforms; 129 digital data carriers
libr loan 31404

– Marvin A. Pomerantz Business Library, C320 PBB, Iowa City, IA 52242
T: +1 319 3353077; Fax: +1 319 3353752; E-mail: lib-bus@uiowa.edu; URL: www.lib.uiowa.edu/biz/
1965; J. David Martin
33 000 vols 31405

– Mathematical Sciences Library, 125 MacLean Hall, Iowa City, IA 52242
T: +1 319 3353076; Fax: +1 319 3355900; E-mail: lib-math@uiowa.edu; URL: www.lib.uiowa.edu/math
Lisa McDaniels
57 000 vols 31406

– Physics Library, 350 Van Allen Hall, Iowa City, IA 52242
T: +1 319 3353082; Fax: +1 319 3355900; E-mail: lib-phys@uiowa.edu; URL: www.lib.uiowa.edu/physics
Lisa McDaniels
57 000 vols 31407

– Psychology Library, W202 Seashore Hall, Iowa City, IA 52242
T: +1 319 3353079; Fax: +1 319 3355900; E-mail: lib-psych@uiowa.edu; URL: www.lib.uiowa.edu/psych
Dorothy M. Persson
70 000 vols 31408

– Rita Benton Music Library, 2000 Voxman Music Bldg, Iowa City, IA 52242
T: +1 319 3353086; Fax: +1 319 3352637; E-mail: lib-mus@uiowa.edu; URL: www.lib.uiowa.edu/music/
1957; Ruthann Boles McTyre
99 000 vols 31409

Ohio University – Southern Campus Library, 1804 Liberty Ave, ***Ironton***, OH 45638-2296
T: +1 740 5334622; Fax: +1 740 5334631; E-mail: ouslibrary@mail.southern.ohiou; URL: www.southern.ohiou.edu/library
1956; Mary J. Stout
25 000 vols; 254 curr per 31410

Concordia University, University Library, 1530 Concordia W, ***Irvine***, CA 92612
T: +1 949 8548002; Fax: +1 949 8546893; E-mail: librarian@cui.edu; URL: www.cui.edu/library
1976; Carolina Barton
Dale Hartmann Curriculum Coll, Reformation studies, Robert C Baden Memorial Children's Coll
90 000 vols; 850 curr per; 70 000 microforms; 2 497 sound-rec; 3 digital data carriers; 250 CDs
libr loan; ALA, CLA 31411

University of California, Irvine – Langson Library, P.O. Box 19557, ***Irvine***, CA 92713-9557
T: +1 949 8246836; Fax: +1 949 8243644; URL: www.lib.uci.edu
1965; Gerald Munoff
Meadows Coll of California Hist, Menninger Coll in Horticulture, R Wellek Coll of the Hist of Criticism, Waldmuller Thomas Mann Coll, Contemporary Small Press Poetry Coll, 20th c Political Pamphlets
2 136 232 vols; 18 187 curr per; 8 901 maps; 2 100 000 microforms; 50 962 sound-rec
libr loan 31412

– Ayala Science Library, P.O. Box 19556, Irvine, CA 92713-9556
T: +1 949 8246836; Fax: +1 949 8243114; E-mail: cynthiaj@uci.edu; URL: www.lib.uci.edu
1994; Cynthia Johnson
500 000 vols; 7 000 curr per 31413

University of Dallas, William A. Blakley Library, 1845 E Northgate Dr, ***Irving***, TX 75062-4736
T: +1 972 7215040; Fax: +1 972 7214010; URL: www.udallas.edu/library
1956; Dr. Robert Scott Dupree
Political Philosophy (Kendall Memorial Libr Coll); Theology; Jacques Migne, Patrologiae Cursus Completus, index Thomisticus
209 000 vols; 544 curr per; 2 390 e-journals; 32 000 e-books; 900 av-mat; 1 710 sound-rec; 92 digital data carriers
libr loan 31414

Cornell University, University Library, 201 Olin Library, ***Ithaca***, NY 14853-5301
T: +1 607 2553393; Fax: +1 607 2553609; E-mail: libadmin@cornell.edu; URL: www.library.cornell.edu
1868; Anne R. Kenney
7 830 000 vols; 360 000 e-books
libr loan 31415

– Adelson Library, Laboratory of Ornithology, 159 Sapsucker Woods Rd, Ithaca, NY 14850-1999
T: +1 607 2542165; Fax: +1 607 2542111; E-mail: adelson_lib@cornell.edu; URL: www.birds.cornell.edu
Jacalyn C. Spoon
Ornithology
12 000 vols; 200 curr per 31416

– Albert R. Mann Library, Ithaca, NY 14853-4301
T: +1 607 2553296; Fax: +1 607 2542887; E-mail: mann-ref@cornell.edu; URL: www.mannlib.cornell.edu
1952; Janet McCue
Beekeeping (Everett Franklin Phillips Coll); Lace & Lacemaking (Elizabeth C. Kackenmeister Coll); Language of Flowers; James E. Rice Poultry Libr
764 000 vols; 7 060 curr per; 3 187 maps; 610 000 microforms; 4 016 non-bk
libr loan 31417

– L. H. Bailey Hortorium Library, 412 Mann Bldg, Ithaca, NY 14853
T: +1 607 2550455; Fax: +1 607 2555407; E-mail: pf13@cornell.edu; URL: www.plantbio.cornell.edu/cals/plbio/hortorium/bhortlibr.cfm
1935; P.R. Fraissinet
Horticulture; Worldwide Coll of Seed & Plant Lists & Cats from Botanical Gardens & Commercial Sources; Card File of Sources for Plant Mat
30 000 vols; 200 curr per; 21 500 reprints, 8 000 photogs 31418

– Division of Rare & Manuscript Collections, 2B Carl A Kroch Library, Cornell University, Ithaca, NY 14853
T: +1 607 2553530; Fax: +1 607 2559524; E-mail: rareref@cornell.edu; URL: rmc.library.cornell.edu
Elaine Engst
American hist, Hist of science, Human sexuality; 18th & 19th C French Hist, Anglo-American Lit, Dante Coll, Petrarch, Fiske Icelandic Coll, Witchcraft
270 000 vols 31419

– Edna McConnell Clark Physical Sciences Library, Clark Hall, Ithaca, NY 14853-2501
T: +1 607 2554016; Fax: +1 607 2555288; E-mail: pslref@cornell.edu; URL: www.library.cornell.edu/psl
1965
Indexes of Texas A&M Thermodynamics Res Ctr & Sadtler Res; Special Indexes of Sadtler Res Lab, American Petroleum Institute & Thermodynamics Research Center; X-ray Powder Diffraction Cards of American Society for Testing & Mat
117 000 vols; 202 maps; 26 000 microforms; 152 av-mat 31420

– Engineering Library, Carpenter Hall, Ithaca, NY 14853-2201
T: +1 607 2554318; Fax: +1 607 2550278; E-mail: ENGRanswers@cornell.edu; URL: www.englib.cornell.edu
1937; John Saylor
NASA & DOE tech rpts; NTIS microfiche, 1978-present
380 180 vols; 2 000 curr per; 1 927 923 microforms; 89 av-mat; 65 sound-rec; 4 digital data carriers; 475 computer files 31421

– Fine Arts, Sibley Hall, Ithaca, NY 14853-6701
T: +1 607 2553710; Fax: +1 607 2556718; E-mail: fineartscirc@cornell.edu; URL: library.cornell.edu/finearts
1871; Martha Walker
122 000 vols; 7 000 microforms; 237 av-mat 31422

– Flower-Sprecher Veterinary Library, S2-160 Veterinary Education Ctr, Ithaca, NY 14853-6401
T: +1 607 2533510; Fax: +1 607 2533080; E-mail: vetref@cornell.edu; URL: www.vet.cornell.edu/library
1897; Dr. Erla P. Heyns
102 000 vols; 812 curr per; 26 000 microforms; 28 675 av-mat; 538 sound-rec; 86 digital data carriers
libr loan 31423

– John Henrik Clarke Africana Studies Library, 310 Triphammer Rd, Ithaca, NY 14850
T: +1 607 2553822; Fax: +1 607 2552493; E-mail: afrlib@cornell.edu; URL: www.library.cornell.edu/africana
1969; Eric Kofi Acree
Civil Rights Microfilm Coll
21 000 vols; 81 curr per; 2 000 microforms; 1 000 av-mat; 50 sound-rec; 300 digital data carriers; 930 non-bk 31424

– Johnson Graduate School of Management Library, 101 Sage Hall, Ithaca, NY 14853-3901
T: +1 607 2553389; Fax: +1 607 2558633; URL: www.library.cornell.edu/johnson
1949; Angela Horne
United States & Foreign Corporation Reports
162 000 vols; 600 curr per; 850 000 microforms; 223 non-bks 31425

– Law School Library, Myron Taylor Hall, 524 College Ave, Ithaca, NY 14853-4901
T: +1 607 2559577; Fax: +1 607 2551357; E-mail: lawlib@cornell.edu; URL: www.lawschool.cornell.edu/library
1887; Patricia Court
Benett Coll of Statutory Mat, 19th C Trials, Donovan Coll of Nuremberg Trials, Int & Foreign Law, Rare Bks
721 000 vols; 6 680 curr per; 165 e-journals; 588 e-books; 701 000 microforms; 225 av-mat; 359 sound-rec; 121 digital data carriers
libr loan; AALL, IALL 31426

– Mathematics Library, 420 Malott Hall, Ithaca, NY 14853-4201
T: +1 607 2555076; Fax: +1 607 2545023; E-mail: mathlib@cornell.edu; URL: astech.library.cornell.edu/ast/math/index.cfm
1870; Steven W. Rockey
67 000 vols; 684 curr per; 280 e-journals; 1 000 microforms; 155 av-mat; 1 digital data carriers 31427

– Nestle Hotel School Library, Statler Hall, School of Hotel Administration, Ithaca, NY 14853-6902
T: +1 607 2553673;
Fax: +1 607 2550021; URL: www.hotelschool.cornell.edu/research/library
1921; Donald Schnedeker
Food & Beverage (Herndon & Vehling Coll); Oscar of the Waldorf, bks, menus; American Antiquarian Society Menu Coll; Other Menus
39 000 vols; 500 curr per; 650 diss/theses; 3 000 govt docs; 400 maps; 12 000 microforms; 1 545 av-mat; 138 sound-rec; 164 digital data carriers; 594 non-bks
libr loan; SLA, ALA 31428

– New York State Agricultural Experiment Station, Frank A. Lee Library, Jordan Hall, 630 W North St, Geneva, NY 14456
T: +1 315 7872214; Fax: +1 315 7872276; E-mail: lib@nysaes.cornell.edu; URL: www.nysaes.cornell.edu/library
1882; Martin Schlabach
Wine & Wine Making Coll
51 000 vols; 2 000 microforms; 6 digital data carriers; 349 non-bk 31429

– Olin-Kroch-Uris Library, 201 Olin Library, Ithaca, NY 14853
T: +1 607 2554245; Fax: +1 607 2556788; E-mail: okuref@cornell.edu; URL: www.rmc.library.cornell.edu; www.library.cornell.edu/olinuris; asia.library.cornell.edu
Anne R. Kenney
Govt, Linguistics, Lit, Philosophy, Political science, Religion, Asian studies
3 040 000 vols; 214 682 maps; 2 800 000 microforms; 35 807 mss, 210 288 non-bk 31430

– Population & Development Program Research & Reference Library, Cornell University, B12 Warren Hall, Ithaca, NY 14853-7801
T: +1 607 2554924; Fax: +1 607 2542896; URL: www.einaudi.cornell.edu/pdp
1962; Anne Wilson
10 000 vols; 100 curr per 31431

– School of Industrial & Labor Relations, Martin P. Catherwood Library, Ives Hall, Garden Ave, Ithaca, NY 14853-3901
T: +1 607 2545370; Fax: +1 607 2559641; E-mail: ilrlib@cornell.edu; URL: www.ilr.cornell.edu/library
1945; Gordon Law
Labor Union Journals, Proceedings, & Constitutions; Kheel Ctr for Labor Management Documentation & Arch
233 000 vols; 44 000 microforms; 1 825 av-mat; 2 183 sound-rec; 17 482 cubic feet of mss, 605 motion pictures, 208 computer files 31432

– Sidney Cox Library of Music & Dance, Lincoln Hall, Ithaca, NY 14853-4101
T: +1 607 2554011; Fax: +1 607 2542877; E-mail: musicref@cornell.edu; URL: www.library.cornell.edu/music
Bonna Boettcher
Arch of Field Recordings, Early 16th C Music, 18th C Chamber Music, 18th-21st C American Vocal Music, 19th C Opera, Scarlatti Operas
148 000 vols; 400 curr per; 61 087 sound-rec
libr loan 31433

– Weill Cornell Medical Library, 1300 York Ave, C115, New York, NY 10065-4896
T: +1 212 7466068; Fax: +1 212 7468364; E-mail: infodesk@med.cornell.edu; URL: library.med.cornell.edu
1899; Carolyn Anne Reid
Bio-med, Nursing, Psychiatry; New York

Hospital Arch, Cornell Medical Ctr Arch
179 000 vols; 3 141 curr per; 2 900
e-journals; 13 000 microforms; 751
av-mat; 5 digital data carriers; 26 500
iconographic images, photos, prints
libr loan   31434

Ithaca College, Library, 953 Danby Rd,
Ithaca, NY 14850-7060
T: +1 607 2743206; Fax: +1 607
2741539; E-mail: libweb@ithaca.edu;
URL: www.ithaca.edu/library
1892; Lisabeth Chabot
Music Arch (Gustave Haenschen, Donald
Voorhees & Robert Peters)
308 000 vols; 35 122 curr per; 1 641
e-journals; 44 000 e-books; 62 000
microforms; 2 020 av-mat; 13 000
sound-rec; 4 000 digital data carriers
libr loan   31435

Mississippi Valley State University,
James Herbert White Library, 14000 Hwy
82 W, *Itta Bena*, MS 38941
T: +1 662 2543497; Fax: +1 662
2543499; E-mail: mlhenderson@mvsu.edu;
URL: www.mvsu.edu/library
1950; Mantra Henderson
Martin Luther King Shelf, Mississippi Coll
214 000 vols; 402 curr per; 15 e-journals;
35 000 e-books; 25 Electronic Media &
Resources   31436

Belhaven College, Warren A. Hood
Library, 1500 Peachtree St, *Jackson*,
MS 39202
T: +1 601 9685948; Fax: +1
601 9685968; E-mail:
libcomments@belhaven.edu; URL:
www.belhaven.edu/library
1910; Susan Springer
128 000 vols; 503 curr per; 29 000 e-
books   31437

Information Services Library, 3825
Ridgewood Rd, *Jackson*, MS 39211
T: +1 601 4326313; Fax: +1 601
4326144; URL: sampson.jsums.edu/
screens/informationservices.htm
1971; Melissa Druckrey
Business, Census, Demographic, Social
work
42 000 vols; 500 curr per   31438

Jackson State University, Henry
Thomas Sampson Library, 1325 J R
Lynch St, *Jackson*, MS 39217
T: +1 601 9792123; Fax: +1 601
9792239; URL: sampson.jsums.edu/
screens/libinfo.html
1877; Sandra Nimox
Black Studies (Afro-American), Ayers
Decision Coll, Census & Demographic
Info, Gibbs-Green Coll
1 000 000 vols; 1 600 curr per; 84 maps;
193 av-mat; 2 856 sound-rec; 10 870
Slides, 262 Overhead Transparencies   31439

Lambuth University, Luther L. Gobbel
Library, 705 Lambuth Blvd, *Jackson*,
TN 38301
T: +1 731 4253270; Fax: +1 731
4253200; URL: www.lambuth.edu/
academics/library/library.html
1843; Dr. Pamela Dennis
Civil War, Methodism
113 000 vols; 392 curr per; 42 000 e-
books; 1 230 Music Scores, 270 CDs   31440

Lane College, Library, 545 Lane Ave,
Jackson, TN 38301-4598
T: +1 731 4267654; Fax: +1 731
4267591; URL: www.lanecollege.edu
1882; Lan Wang
Black Studies, AV, bks; Haitian Art;
Juvenile
84 000 vols; 339 curr per   31441

Millsaps College, Millsaps-Wilson
Library, 1701 N State St, *Jackson*,
MS 39210-0001
T: +1 601 9741090; Fax: +1 601
9741082; E-mail: librarian@millsaps.edu;
URL: library.millsaps.edu
1890; Tom Henderson
Lehmann Engel Performing Arts Coll,
Eudora Welty Coll, Paul Ramsey Coll on
Applied Ethics, Harmon Smith Christian
Ethics Coll, Cain Mississippi Methodist
Arch, College Arch
205 000 vols; 615 curr per; 180 000
microforms; 7 296 sound-rec; 12 digital
data carriers
libr loan   31442

Mississippi College, Law Library, 151
E Griffith St, *Jackson*, MS 39201-1391;
P.O. Box 4008, Jackson, MS 39201
T: +1 601 9257120; Fax: +1 601
9257112; E-mail: law@mc.edu; URL:
law.mc.edu/library/index.html
1975; Mary Miller
314 000 vols; 3 500 curr per   31443

Union University, Emma Waters Summar
Library, 1050 Union University Dr,
Jackson, TN 38305-3697
T: +1 731 6615070; Fax: +1 731
6615175; E-mail: library@uu.edu; URL:
www.uu.edu/library
Steve Baker
Bateman Libr Coll, R G Lee Libr Coll,
West Tennessee Baptist Coll
149 000 vols; 713 curr per; 36 000 e-
books
libr loan   31444

Edward Waters College, Centennial
Library, 1658 Kings Rd, *Jacksonville*,
FL 32209
T: +1 904 4708081; E-mail:
vivian.browncarman@ewc.edu; URL:
www.ewc.edu/academics/library
1945; Vivian Brown-Carman
Afro-American Coll
99 000 vols; 150 curr per   31445

Florida Coastal School of Law, Library
& Technology Center, 8787 Baypine Rd,
Jacksonville, FL 32256
T: +1 904 6807612; Fax: +1 904
6807677; URL: www.fcsl.edu/library/
index.asp
Alma Nickell Singleton
139 000 vols; 3 139 curr per   31446

Illinois College, Schewe Library, 245
Park St, *Jacksonville*, IL 62650;
mail address: 1101 W College Ave,
Jacksonville, IL 62650-2299
T: +1 217 2453079; Fax: +1 217
2453082; E-mail: schewe@ic.edu; URL:
www.ic.edu/library.htm
1829; Martin H. Gallas
Civil War Coll, Lincoln Coll
174 000 vols; 608 curr per; 10 000
e-journals; 8 000 microforms; 2 900
sound-rec; 950 dvd's
libr loan; ALA   31447

Jacksonville State University, Houston
Cole Library, 700 Pelham Rd N,
Jacksonville, AL 36265
T: +1 256 7825255; Fax: +1 256
7825812; URL: www.jsu.edu/library
1883; John-Bauer Graham
Alabama Coll; Old & Rare Books
702 000 vols; 5 700 curr per; 12 688
e-journals; 12 000 e-books; 1 406 000
microforms; 142 digital data carriers
libr loan   31448

Jacksonville University, Carl S.
Swisher Library, 2800 University Blvd
N, *Jacksonville*, FL 32211-3394
T: +1 904 2567263; Fax: +1 904
2567259; URL: www.ju.edu/library
1934; David M. Jones
Art, Business, Education, Nursing; Univ
Archs, Delius Coll, Jacksonville Historical
Society, Rare Books
500 000 vols; 350 curr per; 53 000 e-
books; 1 894 maps; 231 000 microforms;
510 av-mat; 25 298 sound-rec; 18 digital
data carriers; 21 294 slides, 47 Bks on
Deafness & Sign Lang
libr loan   31449

MacMurray College, Henry Pfeiffer
Library, 447 E College Ave,
Jacksonville, IL 62650-2510
T: +1 217 4797110; Fax: +1 217
2455214; URL: www.mac.edu/academ/
lib.html
1846; Susan Eilering
Special education, American hist, Nursing;
Abraham Lincoln Coll, Samuel Pepys
(Birdseye Coll), Singing & Voice Culture
(Austin-Ball Coll)
117 000 vols; 80 curr per; 17 000
microforms; 1 053 sound-rec; 150 digital
data carriers; 350 Bks on Deafness &
Sign Lang
libr loan   31450

Trinity Baptist College, Library,
800 Hammond Blvd, *Jacksonville*,
FL 32221-1398
T: +1 904 5962507; Fax: +1 904
5962531; E-mail: mcloss@tbc.edu
Jay Bolan

37 000 vols; 120 curr per; 370 e-books   31451

University of North Florida, Thomas G.
Carpenter Library, Bldg 12-Library, One
UNF Dr, *Jacksonville*, FL 32224-2645
T: +1 904 6202553; Fax: +1 904
6202719; URL: www.unf.edu/library
1970; Dr. Shirley Hallblade
Eartha White Memorial Coll, Arthur N.
Sollee Papers, Senator E. Mathews
Papers, Univ Arch
813 000 vols; 2 350 curr per; 7 860
av-mat; 19 600 sound-rec   31452

Saint John's University, University
Library, 8000 Utopia Pkwy, *Jamaica*,
NY 11439
T: +1 718 9901518; Fax: +1 718
3800353; URL: www.stjohns.edu
1870; James Benson
William M Fischer Tennis Coll, Myer Coll
(Accounting), Baxter Coll (American Lit),
Heller Coll, Irish-American Affairs Colls
1 030 000 vols; 6 108 curr per
libr loan   31453

– Davis Library – Manhattan Campus, 101
Murray St, New York, NY 10007
T: +1 212 2775135; Fax: +1 212
2775140; E-mail: davislibrary@stjohns.edu;
URL: stjohns.campusguides.com/insurance
1870; Ricky Waller
Humanities, Business & management,
Natural science, Pharmacy, Religious
studies; 19th C Catholic Per Coll; Art
Exhibition Cat Coll, vf; Asian Coll; Hugh
Carey Coll; John E Baxter Coll; Northern
Ireland Disturbances, 1970-1977, vf; Paul
O'Dwyer Coll; Saul Heller Coll; James
L Buckley Coll, vf; Tennis (William M
Fischer Lawn Coll), Unidivided Ireland,
1947-1963
130 000 vols; 6 108 curr per   31454

– Kathryn & Shelby Cullom Davis Library,
101 Murray St, 3rd Flr, New York,
NY 10007
T: +1 212 8159263; Fax: +1 212
8159272; E-mail: davislibrary@stjohns.edu
Ismael Rivera
Business and management, natural
science, humanities
110 000 vols; 260 curr per   31455

– Library and Information Science Library,
8000 Utopia Pkwy, Jamaica, NY 11439
T: +1 718 9906024; Fax: +1 718
3800353; E-mail: dealyr@stjohns.edu
Dr. Ross Dealy
36 590 vols; 321 curr per; 1 242
microforms; 3 cabinets of vertical files
and annual rpts   31456

– Rittenberg Law Library, 8000 Utopia
Pkwy, Jamaica, NY 11439
T: +1 718 9906659; Fax: +1 718
9906649; URL: www.law.stjohns.edu
1925; Linda Ryan
St Thomas More Coll, bks & per; New
York State Law Libr
500 000 vols; 5 413 curr per; 1 220 000
microforms; 2 digital data carriers
libr loan   31457

York College, Library, 94-20 Guy R
Brewer Blvd, *Jamaica*, NY 11451
T: +1 718 2622035; Fax: +1 718
2622997; E-mail: library@york.cuny.edu;
URL: york.cuny.edu/library
1966; John A. Drobnicki
Libr of American Civilization,
Anthropology, Papers of the United
Negro College Fund, Geology, Urban
Affairs
176 000 vols; 516 curr per   31458

Jamestown College, Raugust Library,
6070 College Lane, *Jamestown*,
ND 58405-0001
T: +1 701 2523467; Fax: +1 701
2534446; URL: www.jc.edu/Raugust
1971; Phyllis Ann K. Bratton
Children's mat, Western Americana Coll
111 000 vols; 243 curr per; 18 000 e-
journals; 10 000 e-books; 312 music
scores; 285 maps; 10 000 microforms;
559 av-mat; 984 sound-rec; 10 digital
data carriers; 40 linear feet of mss, 530
Audio Bks, 110 Bks on Deafness & Sign
Lang
libr loan   31459

Carson-Newman College, Stephens-
Burnett Memorial Library, 1634 Russell
Ave, *Jefferson City*, TN 37760
T: +1 865 4713335; Fax: +1 865
4713450; URL: library.cn.edu
1851; Bruce Kocour
Baptist mat, Family counseling, Marriage
135 000 vols; 585 curr per
libr loan   31460

Lincoln University, Inman E. Page
Library, 712 Lee Dr, *Jefferson City*,
MO 65101
T: +1 573 6815504; Fax: +1 573
6815511; URL: www.lincolnu.edu/pages/
203.asp
1866; Elizabeth A. Wilson
Lincoln Coll, Pro-Slavery & Antislavery
Tracts, Ethnic Studies Ctr Coll
208 000 vols; 240 curr per; 230
diss/theses; 58 000 microforms; 1 632
av-mat; 1 824 sound-rec; 2 digital data
carriers; 25 VF Drawers
libr loan   31461

New Jersey City University,
Congressman Frank J. Guarini Library,
2039 Kennedy Blvd, *Jersey City*,
NJ 07305-1597
T: +1 201 2002183; Fax: +1 201
2002331
1927; Grace F. Bulaong
Art, Education, Fire science, Hist, Music,
Nursing; Eric Coll, McCarthy Memorial
Coll of Alcohol Lit, Anthropology (Human
Relations Area Files)
245 000 vols; 700 curr per   31462

Saint Peter's College, O'Toole Library,
99 Glenwood Ave, *Jersey City*,
NJ 07306; mail address: 2641 Kennedy
Blvd, Jersey City, NJ 07306
T: +1 201 9159387; Fax: +1 201
4324117; E-mail: libdept@spc.edu; URL:
library.spc.edu
1872; Thomas J. Kenny
Biology, Nursing, Pilosophy, Theology
229 000 vols; 750 curr per   31463

Johnson State College, John Dewey
Library, 337 College Hill, *Johnson*,
VT 05656
T: +1 802 6351274; Fax: +1 802
6351294; URL: library.jsc.vsc.edu;
www.jsc.vsc.edu
1866; Joseph Farara
110 000 vols; 500 curr per; 6 000 e-
books   31464

East Tennessee State University,
Johnson City
– James H. Quillen College of Medicine
Library, Maple St, Bldg 4, Johnson City,
TN 37614; P.O. Box 70693, Johnson
City, TN 37614-0693
T: +1 423 4396252; Fax: +1 423
4397025; E-mail: medref@etsu.edu; URL:
com.etsu.edu/medlib
1975; Biddanda (Suresh) P. Ponnappa
Hist of Medicine Coll, Long Coll
45 000 vols; 618 curr per; 2 000 av-mat;
26 digital data carriers; 1 300 slides
libr loan   31465

– Sherrod Library, Seehorn Dr & Lake St,
Johnson City, TN 37614-0204; P.O. Box
70665, Johnson City, TN 37614-1701
T: +1 423 4394307; Fax: +1 423
4394720; E-mail: refdesk@etsu.edu;
URL: sherrod.etsu.edu
1911; Rita Scher
Arch of Appalachia
493 000 vols; 562 curr per; 52 000 e-
books   31466

University of Pittsburgh, Johnstown
Campus Owen Library, 450 Schoolhouse
Rd, *Johnstown*, PA 15904
T: +1 814 2697289; Fax: +1 814
2697283; URL: www.library.pitt.edu/john/
owen.html
1927; Deborah Rinderknecht
151 000 vols; 360 curr per   31467

University of Saint Francis, Main
Library, 500 Wilcox St, *Joliet*, IL 60435
T: +1 815 7403446; Fax: +1 815
7403364; E-mail: circulation@stfrancis.edu;
URL: www.stfrancis.edu/lib/libindex.htm
1930; Terry Cottrell
Franciscans, Nursing; Contemporary
Business Ethics
110 000 vols; 700 curr per   31468

Arkansas State University, Dean B. Ellis Library, 108 Cooley Dr, ***Jonesboro***, AR 72401; P.O. Box 2040, State University, AR 72467-2040
T: +1 870 9723099; Fax: +1 870 9723199; E-mail: refdesk@astate.edu; URL: www.library.astate.edu
1909; Dr. George C. Grant
Libr Science, Lois Lenski Coll (Children), Cass S. Hough Aeronautical Coll, Judd Hill Coll, Ira F. Twist Jr Coll, Legal Res Coll
613 000 vols; 1 738 curr per
libr loan 31469

Missouri Southern State University, George A. Spiva Library, 3950 E Newman Rd, ***Joplin***, MO 64801-1595
T: +1 417 6259386; Fax: +1 417 6259734; URL: www.mssu.edu/spivalib
1937; Wendy McGrane
Education, Nursing; Arrell Morgan Gibson Coll; Gene Taylor Congressional papers; Tri-State Mining Maps
229 000 vols; 499 curr per
libr loan 31470

Ozark Christian College, Seth Wilson Library, 1111 N Main, ***Joplin***, MO 64801-4804
T: +1 417 6242518; Fax: +1 417 6240090; E-mail: library@occ.edu; URL: occ.edu/campuscommunity/occ.library.aspx
1942
Religion, Archeology, Bible; Christian Restoration Movement
71 000 vols; 370 curr per; 439 av-mat; 7 748 sound-rec; 23 digital data carriers
libr loan; MLNC 31471

University of Alaska Southeast, Juneau, William A. Egan Library, 11120 Glacier Hwy, ***Juneau***, AK 99801-8676
T: +1 907 7966502; Fax: +1 907 7966249; E-mail: egan.library@uas.alaska.edu; URL: www.uas.alaska.edu/library
1956; Carol Hedlin
154 000 vols; 465 curr per; 37 000 e-books
libr loan 31472

Florida Atlantic University, John D. MacArthur Campus Library, 5353 Parkside Dr, ***Jupiter***, FL 33458
T: +1 561 7998530; Fax: +1 561 7998587; URL: www.library.fau.edu/npb/npb.htm
1972; William Miller
80 000 vols; 75 curr per 31473

Kalamazoo College, Upjohn Library, 1200 Academy St, ***Kalamazoo***, MI 49006-3285
T: +1 269 3371748; Fax: +1 269 3377143; URL: www.kzoo.edu/is/library
1850; Stacy Nowicki
Art; Fine Birds Coll, Michigan Baptist Coll, Private Presses, Hist of Bks & Printing, Hist of Science
321 000 vols; 794 curr per
libr loan 31474

Western Michigan University, Dwight B. Waldo Library, Arcadia at Vande Giessen St, ***Kalamazoo***, MI 49008-5080; mail address: 1903 W Michigan Ave, Kalamazoo, MI 49008-5353
T: +1 269 3875156; Fax: +1 269 3875836; URL: www.wmich.edu/library
1903; Joseph Reish
African Studies (Ann Kercher Memorial), American Women's Poetry Coll, Hist (Regional Hist), Ecology (CC Adams Coll), Medieval Studies, (Inst of Cistercian Studies)
1 911 000 vols; 7 743 curr per
libr loan 31475

– Education Library, 3300 Sangren Hall, 3rd Flr, Kalamazoo, MI 49008; mail address: 1903 West Michigan Ave, Kalamazoo, MI 49008-5353
T: +1 269 3875223; Fax: +1 268 3875231
1964; Dennis Strasser
Elementary and high school textbooks
70 000 vols; 600 curr per; 1 400 diss/theses; 450 000 microforms; 100 digital data carriers
libr loan; ALA 31476

– Music & Dance Library, 3006 Dalton Ctr, 3rd Flr, Kalamazoo, MI 49008; mail address: 1093 W Michigan Ave, Kalamazoo, MI 49008-5353
T: +1 269 3875237; Fax: +1 269 3875809
1949; Gregory Fitzgerald
44 000 vols; 512 microforms; 19 339 sound-rec; 5 digital data carriers; 17 608 scores 31477

– Regional History Collection & Archives, 111 East Hall, Kalamazoo, MI 49008; mail address: 1903 W Michigan Ave, Kalamazoo, MI 49008-5353
T: +1 269 3878490; Fax: +1 269 3878484; E-mail: arch_collect@wmich.edu
1957; Sharon Carlson
17 000 vols; 4 000 microforms; 17 000 linear feet of mss 31478

Hawaii Pacific University, Atherton Library, 45-045 Kamehameha Hwy, ***Kaneohe***, HI 96744-5297
T: +1 808 2365805; E-mail: ahollowell@hpu.edu; URL: www.hpu.edu
1967; An K. Hollowell
58 000 vols 31479

Avila College, Hooley-Bundschu Library, 11901 Wornall Rd, ***Kansas City***, MO 64145
T: +1 816 5013621; Fax: +1 816 5012456; URL: www.avila.edu
1916; Kathleen Finegan
Nursing
70 000 vols; 500 curr per 31480

DeVry University-Kansas City, James E. Lovan Library, 11224 Holmes Rd, ***Kansas City***, MO 64131
T: +1 816 9410430; Fax: +1 816 9410896; E-mail: library@kc.devry.edu; URL: library.kc.devry.edu
1931; Jared Rinck
13 500 vols; 1 500 curr per; 435 av-mat; 670 digital data carriers
libr loan; ALA, MLA 31481

Rockhurst University, Greenlease Library, 1100 Rockhurst Rd, ***Kansas City***, MO 64110-2561
T: +1 816 5014142; Fax: +1 816 5014666; URL: www.rockhurst.edu
1917; Laurie Hathman
114 000 vols; 6 000 curr per; 1 524 av-mat; 490 sound-rec
libr loan 31482

Saint Paul School of Theology, Dana Dawson Library, 5123 Truman Rd, ***Kansas City***, MO 64127
T: +1 816 4839600; Fax: +1 816 4839605; URL: www.spst.edu/site/library
1958; Logan S. Wright
Methodistica Coll, Wesleyana Coll
96 000 vols; 650 curr per; 1 161 av-mat; 3 digital data carriers 31483

University of Health Sciences, Library, 1750 Independence Ave, ***Kansas City***, MO 64106-1453
T: +1 816 2832295; Fax: +1 816 2832237; E-mail: library@kcumb.edu; URL: www.kcumb.edu/library
1916; Marilyn J. DeGeus
Osteopathic Medicine
57 000 vols; 452 curr per; 816 e-books; 1 592 av-mat; 126 sound-rec; 462 digital data carriers; 2 952 slide titles, 891 x-ray radiographs
libr loan; MLA 31484

University of Missouri, Kansas City Libraries, 800 E 51st St, ***Kansas City***, MO 64110; mail address: 5100 Rockhill Rd, Kansas City, MO 64110-2446
T: +1 816 2351586; Fax: +1 816 2355531; URL: library.umkc.edu; www.umkc.edu/lib
1933; Sharon Bostick
American Sheet Music Coll, Snyder Coll of Americana, Holocaust Studies, Midwest Ctr for American Music
1 096 000 vols; 5 284 curr per; 3 990 av-mat; 4 000 digital data carriers
libr loan 31485

– Dental Library, 650 E 25th St, Kansas City, MO 64108
T: +1 816 2352030; Fax: +1 816 2356540
1920; Ann Marie Corry
Hist of Dentistry Coll
27 000 vols; 366 curr per 31486

– Health Sciences Library, 2411 Holmes St, Kansas City, MO 64108
T: +1 816 2351880; Fax: +1 816 2356570
1965; Margaret Mullaly-Quijas
79 000 vols; 695 curr per; 41 000 microforms
libr loan 31487

– Leon E. Bloch Law Library, 5100 Rockhill Road, mail address: 5100 Rockhill Rd, Kansas City, MO 64110-2499
T: +1 816 235-1650; Fax: +1 816 235-5274; E-mail: callisterp@umkc.edu; URL: www1.law.umkc.edu/library
1895; Paul Callister
Urban Law Coll
202 810 vols; 1 820 curr per; 504 620 microforms; 379 digital data carriers 31488

University of Nebraska at Kearney, Calvin T. Ryan Library, 2508 11th Ave, ***Kearney***, NE 68849-2240
T: +1 308 8658586; Fax: +1 308 8658452; URL: www.unk.edu/acad/library
1906; Janet Stoeger Wilke
Education, Psychology; Nebraska Hist, Local Hist, Curriculum Coll
401 000 vols; 1 250 curr per; 161 maps; 897 000 microforms; 78 241 av-mat; 6 435 sound-rec; 148 digital data carriers; 22 589 slides, 12 570 overhead transparencies, 1 427 charts, 2 043 art reproductions, 1 031 archs 31489

Antioch New England Graduate School Library, 40 Avon St, ***Keene***, NH 03431-3516
T: +1 603 2832400; Fax: +1 603 3577345; E-mail: library@antiochne.edu; URL: www.antiochne.edu/library
1982; Marcia Leversee
Applied, clinical, & counseling psychology; Ctr for Environmental Education Holdings
38 000 vols; 1 625 curr per; 30 digital data carriers 31490

Keene State College, Wallace E. Mason Library, 229 Main St, ***Keene***, NH 03435-3201
T: +1 603 3582715; Fax: +1 603 3582743; URL: www.keene.edu/library
1909; Irene M. Herold
Education, State hist; Holocaust Resource Ctr, K-12 Curriculum Libr
200 000 vols; 790 curr per; 390 000 microforms; 100 sound-rec; 8 digital data carriers
libr loan 31491

Southwestern Adventist University, Chan Shun Centennial Library, 101 W Magnolia St, ***Keene***, TX 76059
T: +1 817 2026242; Fax: +1 817 5564722; URL: library.swau.edu
1894; Cristina Thomsen
Education, Nursing; Seventh-day Adventist Church Hist
110 000 vols; 460 curr per
libr loan; ALA 31492

Kennesaw State University, Horace W. Sturgis Library, 1000 Chastain Rd, ***Kennesaw***, GA 30144
T: +1 770 4236186; Fax: +1 770 4236185; E-mail: rwilliam@kennesaw.edu; URL: www.kennesaw.edu/library/
1963; Robert Williams
Bentley Rare Book Coll, Difazio Children's Lit Coll, Teen Lit Coll, Western Lit
575 000 vols; 3 800 curr per; 1 000 000 microforms; 7 500 sound-rec; 550 digital data carriers
libr loan 31493

Carthage College, Hedberg Library, 2001 Alford Park Dr, ***Kenosha***, WI 53140-1900
T: +1 262 5515907; Fax: +1 262 5515904; URL: www.carthage.edu
1847; Eugene A. Engeldinger
Civil War (Palumbo Coll), English & American Lit 1890-1950 (Dawe Coll); Historiography (Wilde Coll); Religion & Sociology (Evjen Coll)
145 000 vols; 415 curr per; 12 000 e-books 31494

Gateway Technical College, Learning Resources Center Library, 3520 30th Ave, ***Kenosha***, WI 53144-1690
T: +1 262 5642786; Fax: +1 262 5642787; E-mail: kenoshalrc@gtc.edu; URL: www.gtc.edu/library
1964; Gary Flynn
90 000 vols; 365 curr per; 360 diss/theses; 5 000 microforms; 10 000 av-mat; 12 digital data carriers; 5 200 pamphlets, 6 100 microcomputer programs
libr loan 31495

University of Wisconsin – Parkside Library, 900 Wood Rd, ***Kenosha***, WI 53141; P.O. Box 2000, Kenosha, WI 53141-2000
T: +1 262 5952595; Fax: +1 262 5952545; URL: www.uwp.edu/departments/library/
1967; Vanaja Menon
18th & 19th C American Drama (Teisberg & Perishable Press Coll), Irving Wallace Papers
390 000 vols; 825 curr per; 5 820 av-mat; 4 820 sound-rec; 1 090 digital data carriers
libr loan 31496

Kent State University, University Libraries & Media Services, One Eastway Dr, ***Kent***, OH 44242; P.O. Box 5190, Kent, OH 44242-0001
T: +1 330 6723045; Fax: +1 330 6724811; E-mail: library@kent.edu; URL: www.library.kent.edu
1913; Mark Weber
May 4th Coll, Borowitz True Crime Coll, Open Theater Arch, Saalfeld Publishing Company Arch, American Lit
2 271 000 vols; 11 139 curr per; 40 000 e-books
libr loan 31497

– Architecture Library, 309 Taylor Hall, Kent, OH 44242-0001; P.O. Box 5190, Kent, OH 44242-5190
T: +1 330 6722876
1987; Tom Gates
Historic Town Maps, Manufacturers Mutual Fire Insurance Co Site Maps
15 000 vols; 87 curr per 31498

– Chemistry & Physics Library, 312 Williams Hall, Kent, OH 44242-0001; P.O. Box 5190, Kent, OH 44242-5190
T: +1 330 6722532; Fax: +1 330 6724702; URL: www.library.kent.edu/page/10569
1942; Erica Lilly
60 000 vols; 300 curr per 31499

– Mathematics & Computer Science Library, 333 MSB, Kent, OH 44242-0001
T: +1 330 6722430; Fax: +1 330 6722209
Barbara Schloman
22 000 vols 31500

– Music Library, PO Box 5190, Kent, OH 44242-0001
T: +1 330 6722004; Fax: +1 330 6724482
1967; Daniel Boomhower
Choralist Coll
90 000 vols; 150 curr per; 50 000 music scores; 506 microforms; 600 av-mat; 5 000 sound-rec; 200 digital data carriers 31501

University of Alaska Southeast, Ketchikan Campus Library, 2600 Seventh Ave, ***Ketchikan***, AK 99901
T: +1 907 2254722; Fax: +1 907 2284520; URL: www.ketch.alaska.edu/library
1954; Kathleen Wiechelman
Western Americana, Native people
37 000 vols; 125 curr per
libr loan 31502

Kettering College of Medical Arts, Learning Commons, 3737 Southern Blvd, ***Kettering***, OH 45429-1299
T: +1 937 3958053; Fax: +1 937 3958861; URL: www2.kcma.edu/learningcommons/library
1967; Bev Ervin
26 000 vols; 253 curr per; 50 000 microforms; 822 av-mat; 372 sound-rec; 4 digital data carriers; 319 slides
libr loan 31503

Keuka College, Lightner Library, 141 Central Ave, ***Keuka Park***, NY 14478-0038
T: +1 315 2795224; Fax: +1 315 2795334; E-mail: library@keuka.edu; URL: www.keuka.edu/library.html
1923; Linda Park
90 000 vols; 300 curr per 31504

Potomac State College of West Virginia University, Mary F. Shipper Library & Media Center, 101 Fort Ave 26726, *Keyser*, WV 26726
T: +1 304 7886901; Fax: +1 304 7886946; E-mail: JLGardner@mail.wvu.edu; URL: www.potomacstatecollege.edu
1901; Jill Gardner
Local Hist (Mineral County, Keyser, West Virginia), World War II
45 000 vols; 156 curr per   31505

Central Texas College, Oveta Culp Hobby Memorial Library, 6200 W Central Expressway, *Killeen*, TX 76549; P.O. Box 1800, Killeen, TX 76540-1800
T: +1 254 5261237; Fax: +1 254 5261878; E-mail: mark.plasterer@ctcd.edu; URL: www.ctcd.edu/library/pg-lib.htm
1967; Mark Plasterer
Business & management, Education, Law
80 050 vols; 448 curr per   31506

Tarleton State University, Tarleton Library-Central Texas, 1901 S Clear Creek Rd, *Killeen*, TX 76549
T: +1 254 5261244; Fax: +1 254 5261993; URL: www.tarleton.edu/centraltexas/departments/library/
1973; Melinda Guthrie
NCJRS, Crime & Juvenile Delinquency, microfiche; Solar energy, doc
36 000 vols; 58 curr per; 417 govt docs; 20 000 microforms; 326 av-mat; 32 sound-rec; 1 digital data carriers
libr loan   31507

– Dick Smith Library, 201 Saint Felix, Stephenville, TX 76401; Box T-0450, Stephenville, TX 76402
T: +1 254 9689450; Fax: +1 254 9689467; E-mail: reference@tarleton.edu; URL: www.tarleton.edu/~library
1899; Donna D. Savage
Texana, Agricultural Station Rpts
342 000 vols; 822 curr per; 50 000 e-books   31508

US Merchant Marine Academy, Schuyler Otis Bland Memorial Library, 300 Steamboat Rd, *Kings Point*, NY 11024-1699
T: +1 516 7735864; Fax: +1 516 7735502; URL: www.usmma.edu
1942; Dr. George J. Billy
Merchant Marine; Marad Technical Report Coll, Nuclear Ship Savannah Coll
184 000 vols; 850 curr per; 3 500 maps; 16 000 microforms; 1 600 sound-rec; 30 digital data carriers
libr loan   31509

University of Rhode Island, University Library, 15 Lippitt Rd, *Kingston*, RI 02881
T: +1 401 8742672; Fax: +1 401 8744608; URL: www.uri.edu/library
1892; David Maslyn
Rare Books, Rhode Island Coll, Walt Whitman, Ezra Pound, Fritz Eichenberg Libr
1 380 000 vols; 7 809 curr per; 15 479 maps; 1 398 sound-rec; 35 digital data carriers; 1 566 879 microforms of per, 15 000 electronic media & resources   31510

Texas A&M University – Kingsville, James C. Jernigan Library, 700 University Blvd, MSC 197, *Kingsville*, TX 78363-8202
T: +1 361 5933416; Fax: +1 361 5934093; URL: www.tamuk.edu/library
1925; Dr. Gilda Baeza Ortego
Bilingual studies; Western Americana (McGill Coll); Botany (Runyon Coll)
681 000 vols; 1 400 curr per; 13 700 e-journals; 257 682 govt docs; 299 000 microforms; 815 av-mat; 4 926 sound-rec; 526 digital data carriers   31511

Northwest University, Hurst Library, 5520 108th Ave NE, *Kirkland*, WA 98083-0579; P.O. Box 579, Kirkland, WA 98083-0579
T: +1 425 8895266; Fax: +1 425 8897801; E-mail: library@northwestu.edu; URL: library.northwestu.edu
1934; Charles Diede
Ness Bible Translations Reference Coll, Pentecostal Movement Coll, Pauline Perkins Memorial Libr, Teacher Education Curriculum Libr, Pacific Rim Centre

174 000 vols; 914 curr per; 40 000 microforms; 6 digital data carriers; 13 065 graphic mats, 572 cartographic mats   31512

A. T. Still University of the Health Sciences, A. T. Still Memorial Library, Kirksville Campus, 800 W Jefferson St, *Kirksville*, MO 63501
T: +1 660 6262030; Fax: +1 660 6262333; URL: www.atsu.edu/library/index.htm
1897; Bryant Doug Blansit
69 000 vols; 350 curr per; 33 e-books; 3 digital data carriers
libr loan; MLA, OCLC   31513

Truman State University, Pickler Memorial Library, 100 E Normal, *Kirksville*, MO 63501-4211
T: +1 660 7854008; Fax: +1 660 7854536; URL: library.truman.edu
1867; Richard J. Coughlin
Eugenics Coll (Harry Laughlin), mss; Glenn Frank Coll, mss; Missouriana (Violette McClure Coll); Rare Books Coll; Lincoln (Fred & Ethal Schwengel Coll)
484 000 vols; 7 000 e-books; 39 993 govt docs; 6 269 music scores; 609 maps; 1 514 000 microforms; 13 000 av-mat; 12 975 sound-rec; 2 700 digital data carriers
libr loan   31514

Florida Christian College, Library, 1011 Bill Beck Blvd, *Kissimmee*, FL 34744
T: +1 407 8478966 ext 380; E-mail: linda.stark@fcc.edu; URL: www.fcc.edu/library
1976; Linda Stark
44 000 vols; 240 curr per   31515

Oregon Institute of Technology, Library, 3201 Campus Dr, *Klamath Falls*, OR 97601-8801
T: +1 541 8851772; Fax: +1 541 8851777; E-mail: libtech@oit.edu; URL: www.oit.edu/library
1950; N.N.
Shaw Hist Libr Coll, Western Hist
156 000 vols; 2 110 curr per; 5 219 maps; 58 000 microforms; 1 300 sound-rec; 67 digital data carriers; video tapes 500, slides 400
libr loan   31516

Knoxville College, Alumni Library, 901 Knoxville College Dr, *Knoxville*, TN 37921
T: +1 865 5246553; Fax: +1 865 5246549; URL: www.knoxvillecollege.edu
1876; Carolyn Ashkar
97 000 vols; 205 curr per   31517

University of Tennessee, John C. Hodges University Libraries, 1015 Volunteer Blvd, *Knoxville*, TN 37996-1000
T: +1 865 9744351; Fax: +1 865 9744259; URL: www.lib.utk.edu
1838; Barbara I. Dewey
Cherokee Indians, Congressional Papers, Congreve, Early Imprints, Early Voyages & Travel, Alex Haley Coll, Nineteenth Century American Fiction, Radiation Biology, Tennessee World War II Veterans
2 319 000 vols; 32 099 curr per   31518

– George F. DeVine Music Library, Music Bldg, Rm 301, Knoxville, TN 37996-2600
T: +1 865 9743474; Fax: +1 423 9740564; E-mail: PBAYNE@UTK.EDU; URL: www.lib.utk.edu/music
1971; Pauline S. Bayne
Galston-Busoni Music Coll; Grace Moore Memorabilia; David Van Vactor Coll
37 617 vols; 125 curr per; 22 141 music scores; 2 404 microforms; 1 129 av-mat; 18 929 sound-rec; 45 digital data carriers   31519

– Webster Pendergrass Agriculture & Veterinary Medicine Library, A113 Veterinary Teaching Hospital, Knoxville, TN 37996-4541
T: +1 865 9747338; Fax: +1 865 9744732; E-mail: agvetlib@utk.edu; URL: www.lib.utk.edu/agvet
1880; Sandra Sinsel Leach
130 000 vols; 2 000 curr per
libr loan   31520

University of Tennessee – Joel A. Katz Law Library, Taylor Law Center, 1505 W Cumberland Ave, *Knoxville*, TN 37996-1800
T: +1 865 9743771; Fax: +1 865 9746595; URL: www.law.utk.edu/library/1890; William J. Beintema
Constitutional Law Bks & Mat in Braille
483 000 vols; 6 363 curr per; 32 000 microforms; 30 digital data carriers   31521

Indiana University Kokomo Library, 2300 S Washington St, *Kokomo*, IN 46904; P.O. Box 9003, Kokomo, IN 46904-9003
T: +1 765 4559265; Fax: +1 765 4559276; E-mail: iuklib@iuk.edu; URL: www.iuk.edu/~kolibry
1945; Diane J. Bever
137 000 vols; 1 052 curr per; 324 000 microforms; 2 327 sound-rec
libr loan   31522

Kutztown University, Rohrbach Library, 15200 Kutztown Rd, Bldg 5, *Kutztown*, PA 19530-0735
T: +1 610 6834158; Fax: +1 610 6834747; URL: www.kutztown.edu/library/home.html
1866; Dr. Barbara Darden
Libr Science, bks & per; Russian Culture, bks, per & micro; Pennsylvania, bks, per & micro; Curriculum Mat Ctr
517 000 vols; 795 curr per; 15 050 e-journals; 48 435 maps; 1 199 000 microforms; 2 670 av-mat; 180 sound-rec; 975 digital data carriers; 110 Bks on Deafness & Sign Lang, 390 Large Print Bks
libr loan   31523

University of Wisconsin – La Crosse, Murphy Library Resource Center, 1631 Pine St, *La Crosse*, WI 54601-3748
T: +1 608 7858505; Fax: +1 608 7858639; URL: www.uwlax.edu/murphylibrary
1909; Anita Evans
Contemporary Poetry, Gothic Lit (Arkham House & Skeeters Coll), Inland River Steamboats, Regional Hist Coll, Small Presses Coll
411 000 vols; 1 052 curr per; 8 000 e-books; 194 176 govt docs; 21 801 maps; 1 100 000 microforms; 108 digital data carriers
libr loan   31524

Vitero College, Todd Wehr Memorial Library, 900 Viterbo Dr, *La Crosse*, WI 54601
T: +1 608 7963270; Fax: +1 608 7963275; E-mail: reference@viterbo.edu; URL: www.viterbo.edu/library
1890; Rita Magno
Catholic Hist, Music Scores
109 000 vols; 859 curr per; 8 000 e-books   31525

Eastern Oregon University, Pierce Library, One University Blvd, *La Grande*, OR 97850
T: +1 541 9623579; Fax: +1 541 9623335; URL: pierce.eou.edu
1929; Karen Clay
Genealogy, Local hist, Native American Lit, Oregon
156 000 vols; 854 curr per; 212 e-journals
libr loan   31526

La Grange College, William & Evelyn Banks Library, 601 Broad St, *La Grange*, GA 30240-2999
T: +1 706 8808312; Fax: +1 706 8808040; URL: www.lagrange.edu/library
1836; Loren L. Pinkerman
Hist (Marquis de Lafayette Coll), Lit (Grogan Coll), Methodist Hist
120 000 vols; 704 curr per; 110 000 e-books
libr loan   31527

University of California, San Diego, University Libraries, 9500 Gilman Dr, Mail Code 0175G, *La Jolla*, CA 92093-0175
T: +1 858 5342528; Fax: +1 858 5344970; URL: www.ucsd.edu/library/index.html
1959; Brian E. C. Schottlaender
Renaissance, Spanish Civil War, Pacific Voyages, Melanesian Ethnography, Contemporary American Poetry (Arch for New Poetry), Baja, CA, Science &

Public Policy, Contemporary Music
3 373 000 vols; 7 861 curr per; 18 900 e-journals; 231 000 e-books; 200 273 govt docs; 38 362 music scores; 208 927 maps; 2 880 000 microforms; 3 449 digital data carriers; 43 500 Music Scores, 44 290 CDs
libr loan; ARL, CRL, PSRMLS, SPARC, OCLC, Council of Library and Information Resources, ALA   31528

– The Arts Library, 9500 Gilman Dr, 0175Q, La Jolla, CA 92093-0175
T: +1 858 5348074; Fax: +1 858 5340189; E-mail: aalref@ucsd.edu
Leslie Abrams
173 000 vols   31529

– Biomedical Library, 9500 Gilman Dr, 0699, La Jolla, CA 92093-0699
T: +1 858 5344779; Fax: +1 858 8222219
Alice Witkowski
216 000 vols; 779 curr per   31530

– International Relations & Pacific Studies, University of California, San Diego, 9500 Gilman Dr, 0514, La Jolla, CA 92093-0514
T: +1 858 5341413; Fax: +1 858 5348526
Jim Cheng
139 000 vols; 1 329 curr per   31531

– Medical Center Library, Medical Ctr 8828, 200 W Arbor Dr, San Diego, CA 92103-8828
T: +1 619 5436520; Fax: +1 619 5433289
Alice Witkowski
22 000 vols; 252 curr per   31532

– Science & Engineering, 9500 Gilman Dr, Dept 0175E, La Jolla, CA 92093-0175
T: +1 858 5343258; Fax: +1 858 5345583; E-mail: scilib@ucsd.edu
Mary Linn Bergstrom
291 000 vols; 717 curr per   31533

– Scripps Institution of Oceanography, 9500 Gilman Dr, 0219, La Jolla, CA 92093-0219
T: +1 858 5343274; Fax: +1 858 5345269; E-mail: siocirc@ucsd.edu
Peter Brueggeman
224 000 vols; 1 034 curr per   31534

– Social Science & Humanities Library, 9500 Gilman Dr, Mail Code 0175R, La Jolla, CA 92093-0175
T: +1 858 5346816; Fax: +1 858 5344970
Tammy Dearie
2 245 000 vols; 4 665 curr per   31535

Biola University, University Library, 13800 Biola Ave, *La Mirada*, CA 90639
T: +1 562 9034834; Fax: +1 562 9034840; E-mail: library@biola.edu; URL: www.biola.edu/library
1908; Rodney M. Vliet
Religion, psychology, music; Bible Versions & Translations Coll
260 000 vols; 754 curr per; 37 892 e-journals; 73 469 e-books; 4 475 diss/theses; 7 300 music scores; 8 maps; 214 000 microforms; 6 000 av-mat; 7 000 sound-rec
libr loan; ATLA, ACL   31536

Keystone College, Miller Library, One College Green, *La Plume*, PA 18440-0200
T: +1 570 9453333; Fax: +1 570 9458969; E-mail: millerlibrary@keystone.edu; URL: www.keystone.edu/library
1934; Mari Flynn
Christy Mathewson Coll (Local Hist)
42 000 vols; 283 curr per
libr loan; ALA   31537

University of La Verne, Wilson Library, 2040 Third St, *La Verne*, CA 91750
T: +1 909 5933511; Fax: +1 909 3922733; E-mail: reference@ulv.edu; URL: www.ulv.edu/library
1891; Dr. Taylor Ruhl
Genealogy & Church of the Brethren; Bunnelle of California
149 000 vols; 11 585 curr per; 2 000 e-books; 2 462 av-mat   31538

– College of Law at La Verne, Library, 320 E D St, Ontario, CA 91764
T: +1 909 4602070; Fax: +1 909 4602083; URL: law.ulv.edu/~lawlib
1970; Kenneth Rudolf
Juvenile Law Coll
86 000 vols; 475 curr per; 38 000 microforms; 50 sound-rec; 30 digital data carriers; 4 300 gov docs 31539

Saint Martin's College, O'Grady Library, 5300 Pacific Ave SE, ***Lacey***, WA 98503
T: +1 360 4868802; Fax: +1 360 4868810; E-mail: reference@stmartin.edu; URL: www.stmartin.edu/library
1895; Scot Harrison
Catholic Theology; Hist (Frederick J. Lorden Coll); Children's Lit
97 000 vols; 633 curr per; 7 000 e-books; 8 000 microforms
libr loan; OCLC 31540

University of Louisiana at Lafayette, Edith Garland Dupre Library, 302 E St Mary Blvd, ***Lafayette***, LA 70503; P.O. Box 40199, Lafayette, LA 70504-0199
T: +1 337 4826025; Fax: +1 337 4821176; URL: www.louisiana.edu/infotech/library
1901; Dr. Charles W. Triche III
Southwestern Arch and Mss
1 150 000 vols; 2 236 curr per; 68 210 e-journals; 40 000 e-books; 6 439 maps; 1 320 000 microforms; 1 582 av-mat; 1 930 sound-rec; 240 digital data carriers
libr loan; ALA, LLA 31541

Brigham Young University – Hawaii, Joseph F. Smith Library, 55-220 Kulanui St, BYU-H Box 1966, ***Laie***, HI 96762-1294
T: +1 808 6753878; Fax: +1 808 6753877; URL: lt.byuh.edu/library
1955; Doug Bates
Children's Coll, Mormonism, Pacific Islands
207 000 vols; 1 000 curr per; 16 000 e-journals; 36 000 e-books; 10 500 maps; 802 000 microforms; 5 500 av-mat; 1 444 sound-rec; 620 digital data carriers; 119 slides
libr loan 31542

Citrus Research & Education Center, 700 Experiment Station Rd, ***Lake Alfred***, FL 33850-2299
T: +1 863 9561151; Fax: +1 863 9564631; URL: www.crec.ifas.ufl.edu
1947
Citrus, horticulture, nematology, entomology, food science, postharvest
8 000 vols; 85 curr per; 253 diss/theses; 90 microforms; 12 digital data carriers
FLA, SLA 31543

McNeese State University, Lether E. Frazar Memorial Library, 4205 Ryan St, ***Lake Charles***, LA 70609; P.O. Box 91445, Lake Charles, LA 70609
T: +1 337 4755726; Fax: +1 337 4755719; URL: www.library.mcneese.edu
1939; Nancy L. Khoury
Education; Fore-edge Paintings; Lake Charles, Southwestern Louisiana Arch; 20th C American First Editions; Rosa Hart Little Theater Coll; Gerstner Field Coll; Lake Charles High School Coll; Levingston/Franks Historic Newspaper Coll
383 000 vols; 1 561 curr per; 60 000 e-books; 1 378 000 microforms
libr loan; ALA 31544

Lake Forest College, Donnelley & Lee Library, 555 N Sheridan, ***Lake Forest***, IL 60045
T: +1 847 7355056; Fax: +1 847 7356296; URL: www.library.lakeforest.edu
1857; James Cubit
NY Daily News Coll, Hamill Coll (Humanities, Rare Bks), Stuart Coll (Scotland), Theatre, Printing Hist, Western Americana, Railroad
240 000 vols; 1 000 curr per
libr loan 31545

Warner Southern College, Pontious Learning Resource Center, 13895 Hwy 27, ***Lake Wales***, FL 33859
T: +1 863 6387666; Fax: +1 863 6387675; E-mail: researchhelp@warner.edu; URL: www.warner.edu/lrc
1967; Sherill Lynn Harriger
10 000 vols; 225 curr per; 25 000 e-books 31546

Florida Southern College, Roux Library, 111 Lake Hollingsworth Dr, ***Lakeland***, FL 33801-5698
T: +1 863 6804164; Fax: +1 863 6804126; URL: www.flsouthern.edu/library
1885; Andrew Pearson
Florida United Methodist Coll, James A. Haley Coll, Andy Ireland Coll
182 000 vols; 680 curr per; 27 000 e-books; 2 700 av-mat; 5 170 sound-rec; 1 550 digital data carriers
libr loan 31547

Southeastern College, Steelman Media Center – Mary Stribling Library, 1000 Longfellow Blvd, ***Lakeland***, FL 33801
T: +1 863 6675089; Fax: +1 863 6694160; E-mail: library@seuniversity.edu; URL: www.seuniversity.edu/library.htm
1935; Grace Veach
Religion, Education, Social work, Music, Psychology; Church Builders Alcove, Curriculum Lab, Pentecostal Studies
95 000 vols; 600 curr per 31548

Colorado Christian University, Clifton Fowler Library, 8787 W Alameda Ave, ***Lakewood***, CO 80226
T: +1 303 9633250; Fax: +1 303 9633251; E-mail: ggunderson@ccu.edu; URL: www.ccu.edu/library
1914; Gayle Gunderson
Education, Humanities, Music, Religion
80 000 vols; 400 curr per; 300 000 microforms; 1 400 av-mat; 1 300 sound-rec; 20 digital data carriers
libr loan 31549

Georgian Court College, S. Mary Joseph Cunningham Library, 900 Lakewood Ave, ***Lakewood***, NJ 08701-2697
T: +1 732 9872422; Fax: +1 732 9872017; URL: www.georgian.edu/library/index.html
1908; Dr. Benjamin Willliams
Religious studies, Education; Georgian Court College Arch, Thomas Merton Coll (NASA Resource Ctr)
288 000 vols; 1 000 curr per; 2 512 av-mat; 690 sound-rec; 130 digital data carriers; 170 Audio Bks, 520 CDs, 25 Large Print Bks 31550

Graceland University, Frederick Madison Smith Library, One University Pl, ***Lamoni***, IA 50140
T: +1 641 7845388; Fax: +1 641 7845497; E-mail: library@graceland.edu; URL: www.graceland.edu/library/
1895; Diane Shelton
Health services, Nursing; Mormon Hist Mss, 20th c American Lit
194 000 vols; 721 curr per; 66 e-journals; 7 000 e-books; 77 600 govt docs; 114 000 microforms; 1 540 av-mat; 1 827 sound-rec; 159 digital data carriers; 819 Journals
libr loan; BCR, OCLC 31551

Franklin & Marshall College, Shadek-Fackenthal Library, 450 College Ave, ***Lancaster***, PA 17603-3316; P.O. Box 3003, Lancaster, PA 17603-3003
T: +1 717 291421; Fax: +1 717 2914160; URL: library.fandm.edu
1787; Pamela Snelson
Alexander Corbett Coll of Theatre Memorabilia, German American Imprint Coll, W W Griest Coll of Lincolniana, Anne Figgat Coll of Theatre Arts
502 000 vols; 2 025 curr per; 4 033 av-mat; 2 671 sound-rec; 290 Journals 31552

– Martin Library of the Sciences, PO Box 3003, Lancaster, PA 17604-3003
T: +1 717 2913843; Fax: +1 717 2914088
Pamela Snelson
35 000 vols; 109 curr per 31553

Ohio University – Lancaster Library, McCauley Library, 1570 Granville Pike, ***Lancaster***, OH 43130-1097
T: +1 740 6546711; Fax: +1 740 6879497; E-mail: lancaster@ohio.edu; URL: www.lancaster.ohiou.edu/library
1956; N.N.
Charles Goslin Coll, Herbert M Turner Pioneer Coll
97 000 vols; 267 curr per; 1 000 e-books; 20 av-mat
libr loan 31554

Thaddeus Stevens College of Technology, Kenneth W. Schuler Learning Resources Center, 750 E King St, ***Lancaster***, PA 17602-3198
T: +1 717 2997754; Fax: +1 717 3967186; E-mail: ambruso@stevenscollege.edu; URL: www.stevenstech.org
1976; Diane Ambruso
Thaddeus Stevens Papers
27 000 vols; 600 curr per; 180 000 microforms 31555

Philadelphia Biblical University, Masland Learning Resource Center, 200 Manor Ave, ***Langhorne***, PA 19047
T: +1 215 7024370; Fax: +1 215 7024374; E-mail: library@pbu.edu; URL: www.library.pbu.edu
1913; Timothy Hui
156 120 vols; 829 curr per; 1 200 e-books; 18 276 microforms; 12 700 av-mat; 4 600 sound-rec; 14 digital data carriers; 5 623 Slides 31556

Langston University, G. Lamar Harrison Library, PO Box 1500, ***Langston***, OK 73050-1600
T: +1 405 4663293; Fax: +1 405 4663459; E-mail: brblack@lunet.edu; URL: www.lunet.edu
Bettye Black
60 000 vols; 972 curr per 31557

Great Lakes Christian College, Louis M. Detro Memorial Library, 6211 W Willow Hwy, ***Lansing***, MI 48917
T: +1 517 3210242; Fax: +1 517 3215902; E-mail: library@glcc.edu; URL: www.glcc.edu/library
1949; James Orme
Religion, Bible, Music, Hist; C. S. Lewis Coll
47 000 vols; 239 curr per; 249 maps; 3 000 microforms; 126 av-mat; 1 978 sound-rec; 16 charts
libr loan 31558

Thomas M. Cooley Law School, Brennan Law Library, 300 S Capitol Ave, ***Lansing***, MI 48901; P.O. Box 13038, Lansing, MI 48901-3038
T: +1 517 3715140; Fax: +1 517 3345717; URL: www.cooley.edu/library
1972; Eric Kennedy
Michigan Supreme Court Records & Briefs (1907-present)
535 000 vols; 4 335 curr per; 6 000 e-books; 700 sound-rec; 40 digital data carriers 31559

University of Wyoming, University Libraries, 13th & Ivinson, ***Laramie***, WY 82071; mail address: Dept 3334, 1000 E University Ave, Laramie, WY 82071
T: +1 307 7662070; Fax: +1 307 7662510; URL: www-lib.uwyo.edu
1887; Maggie Farrell
1 413 000 vols; 11 135 curr per; 47 000 e-books
libr loan 31560

– Brinkerhoff Geology Library, SH Knight Bldg, Box 3334, University Sta, Laramie, WY 82070-3334
T: +1 307 7666538; E-mail: jdombrow@uwyo.edu; URL: www.uwyo.edu
1956; Janet Dombrowski
UMI Geology Diss, 1981 to 1993 (microfiche); Wyoming Infrared Photogr; post-Yellowstone Fire Infrared Photogr; Wyoming Geological Survey Publs; Wyoming Black & White Aerial Photogr
189 134 vols; 240 curr per; 10 955 govt docs; 135 000 maps; 23 573 titles in microform 31561

– Science-Technology Library, P.O. Box 3262, University Sta, Laramie, WY 82071-3262
T: +1 307 7665165; Fax: +1 307 7666757; E-mail: lphil@uwyo.edu; URL: www-lib.uwyo.edu
1970; Lori Phillips
329 945 vols; 3 000 curr per; 176 893 microforms; 6 digital data carriers 31562

University of Wyoming – George W. Hopper Law Library, Dept 3035, 1000 E University Ave, ***Laramie***, WY 82071
T: +1 307 7662210; Fax: +1 307 7664044; URL: uwadmnweb.uwyo.edu/lawlib
1920; Tim Kearley
Roman Law (Blume Coll)
143 000 vols; 1 850 curr per; 444 000 microforms; 328 av-mat; 3 digital data carriers; 51 Journals 31563

University of Wyoming – The Learning Resources Center Library, 15th & Lewis Sts, P.O. Box 3374, University Sta, ***Laramie***, WY 82071-3374
T: +1 307 7662527; Fax: +1 307 7662018
1986; Laurn Wilhelm
Practical education; Children/Young Adult Coll, Curriculum Coll, NASA Regional Teachers Resource Ctr
16 000 vols; 15 curr per; 250 microforms; 25 sound-rec; 300 digital data carriers 31564

Texas A&M International University, Sue & Radcliffe Killam Library, 5201 University Blvd, ***Laredo***, TX 78041-1900
T: +1 956 3262138; Fax: +1 956 3262399
1970; Rodney Webb
Int trade, Nursing; Laredo Spanish Arch, microfiche; Raza Unida Papers
221 000 vols; 1 492 curr per; 8 000 e-journals; 11 000 e-books
libr loan 31565

New Mexico State University, University Library, 2911 McFie Circle, ***Las Cruces***, NM 88003; P.O. Box 30006, MSC 3475, Las Cruces, NM 88003-8006
T: +1 575 6461508; Fax: +1 575 6464335; URL: www.lib.nmsu.edu
1888; Dr. Elizabeth A. Titus
Southwest Border Res Ctr Coll, Rio Grande Hist Coll
1 750 000 vols; 2 711 maps; 150 digital data carriers 31566

New Mexico Highlands University, Thomas C. Donnelly Library, Ninth & National Ave, ***Las Vegas***, NM 87701
T: +1 505 4543403; Fax: +1 505 4540026; URL: donnelly.nmhu.edu
1893; Ruben F. Aragon
Fort Union Arch, mss; Govt Docs Coll; Southwest Hist (Arrott Coll)
169 000 vols; 519 curr per; 11 000 e-books
libr loan 31567

Capitol College, John G. & Beverly A. Puente Library, 11301 Springfield Rd, ***Laurel***, MD 20708
T: +1 301 3692800; Fax: +1 301 3692552; URL: www.capitol-college.edu
1966; Rick A. Sample
Electronics, Computer science, Telecommunications, Optoelectronics, Systems management; Data Bk Ref Coll
10 000 vols; 87 curr per; 300 av-mat; 20 digital data carriers; 20 slides 31568

Johns Hopkins University – Applied Physics Laboratory, R.E. Gibson Library & Information Center, 11100 Johns Hopkins Rd, ***Laurel***, MD 20723-6099
T: +1 443 7785151; Fax: +1 443 7785353; URL: lib2.jhuapl.edu
1945; Robert S. Gresehover
30 000 vols; 300 curr per; 43 871 microforms; 8 digital data carriers 31569

Southeastern Baptist College, A.R. Reddin Memorial Library, 4229 Hwy 15 N, ***Laurel***, MS 39440
T: +1 601 4266346; Fax: +1 601 4266347; E-mail: rkitchens@southeasternbaptist.edu; URL: www.southeasternbaptist.edu
1955
Baptist Missionary Assn of America Coll
25 000 vols; 280 curr per 31570

Saint Andrews Presbyterian College, DeTamble Library, 1700 Dogwood Mile, ***Laurinburg***, NC 28352
T: +1 910 2775049; Fax: +1 910 2775050; E-mail: johnsonrd@sapc.edu; URL: www.sapc.edu/library/
1896; Rita D. Johnson
Scottish Coll, Thistle & Shamrock Scottish Music Coll on Tape, Special and Rare Bks
110 105 vols; 358 curr per; 21 443 e-journals; 805 maps; 14 808 microforms; 58 av-mat; 846 sound-rec; 27 digital data carriers
libr loan; ALA, ACRL 31571

Haskell Indian Nations University, Tommaney Library, 155 Indian Ave, ***Lawrence***, KS 66046
T: +1 785 8326661; Fax: +1 785 7498473; URL: www.haskell.edu
1884; Dr. Marilyn L. Russell
Indians of North America Coll
70 000 vols; 250 curr per 31572

University of Kansas, Watson Library, 1425 Jayhawk Blvd, ***Lawrence***, KS 66045-7544
T: +1 785 8643956; Fax: +1 785 8645311; URL: www.lib.ku.edu
1866; Lorraine Haricombe
English & Irish Lit & Hist, Spanish Plays, French Revolution, Botany, Chinese Classics, Modern American Poetry, Ornithology
4 194 000 vols; 48 037 curr per; 313 396 maps; 4 784 digital data carriers; 26 049 recordings
libr loan 31573

– Anshutz Library, 1301 Hoch Auditoria Dr, Lawrence, KS 66045-7537
T: +1 785 8644928; Fax: +1 785 8645705
1989; Lorraine Haricombe
460 650 vols; 5 152 curr per; 21 330 govt docs; 153 638 microforms 31574

– Gorton Music & Dance Library, 1530 Naismith Dr, Lawrence, KS 66045
T: +1 785 8643496; Fax: +1 785 8645310
1953; George Gibbs
Hymnals, Sheet Music, American Music
100 000 vols; 400 curr per 31575

– Murphy Art & Architecture Library, 1301 Mississippi St, Lawrence, KS 66045-8500
T: +1 785 8643020; Fax: +1 785 8644608
1980; Susan V. Craig
Ephemeral Coll of Museum & Gallery Publs
128 000 vols; 600 curr per; pamphlet files on artists 31576

– Spahr Engineering Library, Spahr Library-Learned Hall, 1532 W 15 St, Lawrence, KS 66045
T: +1 785 8643866; Fax: +1 785 8645755; E-mail: spahr-ref@ukans.edu
Jim Neely
92 000 vols; 1 027 curr per 31577

– Wheat Law Library, Green Hall, Rm 200, 1535 W 15th St, Lawrence, KS 66045-7577
T: +1 785 8643025; Fax: +1 785 8643680; URL: www.law.ku.edu/library/
1878; Joyce Pearson
Kansas State Law, Basic Coll of the Law of Great Britain & Canada
400 000 vols; 4 500 curr per; 4 000 microforms; 46 digital data carriers; 52 computer files
libr loan 31578

Rider University, Franklin F. Moore Library, 2083 Lawrenceville Rd, ***Lawrenceville***, NJ 08648-3099
T: +1 609 8965118; Fax: +1 609 8968029; URL: library.rider.edu
1865; F. William Chickering
Delaware Valley Newspapers (from Colonial Times to present), Dispatches of United States Envoys in Britain & France During Civil War Period, Rideriana, Shorthand Mat, Typewriting Hist
1 545 000 vols; 3 031 curr per; 527 000 microforms; 30 digital data carriers
libr loan 31579

– Westminster Choir College, Katherine Houk Talbott Library, Westminster Choir College, 101 Walnut Lane, Princeton, NJ 08540-3899
T: +1 609 9217100; Fax: +1 609 4970243; URL: www.rider.edu/talbott
Mi-Hye Chyun
Choral Music Performance Coll; Tams-Witmark Coll of Choral Music; Hymnology (Routley Coll); Music Education Resource Ct; Organ Hist Society Arch
66 000 vols; 200 curr per; 1 000 microforms; 9 010 sound-rec; 2 digital data carriers; 24 415 scores, 8 000 slides 31580

Saint Paul's College, Russell Memorial Library, 115 College Dr, ***Lawrenceville***, VA 23868-1299
T: +1 434 8481841; Fax: +1 434 8481861; URL: www.saintpauls.edu
Marc Finney
Black Studies (Schomburg Coll), West Indies (Short Coll)
54 000 vols; 245 curr per 31581

Cameron University, University Library, 2800 W Gore Blvd, ***Lawton***, OK 73505-6377
T: +1 580 5812403; Fax: +1 580 5812386; E-mail: library@cameron.edu; URL: www.cameron.edu/library
1908; Dr. Sherry Young
242 000 vols; 371 curr per; 309 000 microforms; 17 digital data carriers
libr loan 31582

University of Saint Mary, De Paul Library, 4100 S Fourth St Trafficway, ***Leavenworth***, KS 66048-5082
T: +1 913 7586163; Fax: +1 913 7586200; URL: www.stmary.edu
1923; Penelope Lonergan
Americana (incl ethnic minorities), Abraham Lincoln, Bible, Music (orchestral scores), Shakespeare
119 000 vols; 205 curr per; 5 000 e-books; 15 av-mat; 15 Audio Bks 31583

Cumberland University, Dorothy & Harry Vise Library, One Cumberland Sq, ***Lebanon***, TN 37087
T: +1 615 5471299; Fax: +1 615 4442569; E-mail: library@cumberland.edu; URL: www.cumberland.edu/library
1842; Eloise Hitchcock
Business, Management, Tennessee hist; Stockton Arch Coll; Nobel Laureate Coll
39 000 vols; 270 curr per; 35 000 e-books
ALA 31584

McKendree College, Holman Library, 701 College Rd, ***Lebanon***, IL 62254-1299
T: +1 618 5376952; Fax: +1 618 5378411; E-mail: libcirculation@mckendree.edu; URL: www.mckendree.edu/library
1828; Stephen Banister
Arch of the Southern Illinois Conference of the United Methodist Church
100 000 vols; 500 curr per
libr loan 31585

Pennsylvania State University – Wilkes-Barre Commonwealth College, Nesbitt Library, PO Box PSU, ***Lehman***, PA 18627-0217
T: +1 570 6759295; Fax: +1 570 6757436; URL: www.wb.psu.edu/
1916; Bruce D. Reid
34 000 vols; 425 curr per 31586

Bucknell University, Ellen Clarke Bertrand Library, 221 Bertrand Library, Information Services & Resources, ***Lewisburg***, PA 17837
T: +1 570 5771462; Fax: +1 570 5773313; URL: www.bucknell.edu/x1263.xml
1846; Nancy S. Dagle
Irish Lit, Fine Presses
778 000 vols; 2 218 curr per; 2 730 e-journals; 42 000 e-books; 320 261 govt docs; 7 764 maps; 715 000 microforms; 12 200 av-mat; 6 824 sound-rec; 13 730 digital data carriers
libr loan 31587

West Virginia School of Osteopathic Medicine, Library, 400 N Lee St, ***Lewisburg***, WV 24901
T: +1 304 6476261; Fax: +1 304 6454443; E-mail: library@wvsom.edu; URL: www.wvsom.edu/library
1973; Mary Frances Bodemuller
24 000 vols; 231 curr per; 15 e-books; 1 500 av-mat; 2 000 sound-rec; 20 digital data carriers; 2 500 slides
libr loan 31588

Bates College, George & Helen Ladd Library, 48 Campus Ave, ***Lewiston***, ME 04240
T: +1 207 7866264; Fax: +1 207 7866055; URL: www.bates.edu/library
1855; Eugene L. Wiemers
Edmund S Muskie Coll, Muskie Oral Hist Project, Batesiana, Maine Small Press Publs, College Arch, Jonathan Stanton Natural Hist Coll, Berent Judaica Coll, Signed 1st Editions
619 000 vols; 2 151 curr per; 23 415 Electronic Media & Resources 31589

Lewis-Clark State College, Library, 500 Eighth Ave, ***Lewiston***, ID 83501
T: +1 208 7922396; Fax: +1 208 7922831; URL: www.lcsc.edu/library
1893; Susan Niewenhous
Pacific Northwest Hist Coll, Curriculum Coll, Children's Lit Coll, ERIC docs
257 000 vols; 22 521 govt docs; 2 458 maps; 314 sound-rec; 239 digital data carriers
libr loan; ALA 31590

Transylvania University, University Library, 300 N Broadway, ***Lexington***, KY 40508
T: +1 859 2338225; Fax: +1 859 2338779; E-mail: library@transy.edu; URL: www.transy.edu/academics/library.htm
1780; Susan M. Brown
Horse, Sporting & Natural Hist, Kentucky Hist, Medicine to 1850 (Transylvania Medical Libr), Univ arch
125 000 vols; 500 curr per; 13 000 microforms; 1 768 av-mat; 100 sound-rec; 3 digital data carriers
libr loan; ALA 31591

University of Kentucky, William T. Young Library, I-85 William T Young Library, 401 Hilltop Ave, ***Lexington***, KY 40506-0456; mail address: 500 S Limestone St, Lexington, KY 40506-0456
T: +1 859 2570500; Fax: +1 859 2570502; E-mail: lib-ts@lsv.uky.edu.; URL: libraries.uky.edu/WTYL
1909; Carol Pitts Diedrichs
English Lit, Anthropology, Medicine, Musicology, Graphic Arts, Printing
3 538 000 vols; 22 900 e-journals; 227 000 e-books; 41 650 av-mat
libr loan 31592

– Center for Applied Energy Research Libraries, 2540 Research Park Dr, Lexington, KY 40511-8410
T: +1 859 2570309; Fax: +1 859 2570302; URL: www.caer.uky.edu
Theresa Wiley
8 000 vols 31593

– Chemistry-Physics Library, 150 Chemistry-Physics Bldg, Lexington, KY 40506-0055
T: +1 859 2575954; Fax: +1 859 3234988; E-mail: jbcarv1@email.uky.edu; URL: www.uky.edu/libraries
1963; Maggie Johnson
87 860 vols; 300 curr per; 51 digital data carriers 31594

– Design Library, 200 Pence Hall, Lexington, VA 40506-0041
T: +1 859 2571533; Fax: +1 859 2574305; E-mail: fharders@uky.edu; URL: www.uky.edu/libraries
1962; Faith Harders
Kentucky Architecture, Le Corbusier, Rare Bks
41 230 vols; 96 curr per; 396 maps; 6 498 microforms; 2 digital data carriers; 608 sets of plans & drawings 31595

– Education Library, 205 Dickey Hall, Lexington, KY 40506-0017
T: +1 859 2571351; Fax: +1 859 3231976; E-mail: jill@email.uky.edu; URL: www.uky.edu/libraries/educ.html
Gillian Buckland
ERIC Coll, microfiche
115 740 vols; 350 curr per; 487 537 microforms; 35 digital data carriers 31596

– Engineering Library, 355 Anderson Hall, Lexington, KY 40506-0046
T: +1 859 2572965; Fax: +1 859 3231911; E-mail: susan.smith@uky.edu; URL: www.uky.edu/libraries/engin.html
1965; Susan Smith
Environmental Rpts, Industry Standards, Robotics
130 000 vols; 533 curr per; 124 016 microforms; 14 digital data carriers 31597

– Geological Sciences Library, 410 King Bldg, Lexington, KY 40506-0039
T: +1 859 2575730; Fax: +1 859 3233225; E-mail: klimrs@uky.edu; URL: www.uky.edu/libraries/geo.htm
1923; Mary Spencer
Kentucky Geology, Map Coll
62 540 vols; 325 curr per; 139 077 maps; 24 530 microforms; 129 digital data carriers 31598

– Law Library, 620 S Limestone St, Lexington, KY 40506-0048
T: +1 859 2578686; Fax: +1 859 3234906; E-mail: helanedavis@uky.edu; URL: www.uky.edu/law/library
1908; Helane Davis
Jurisprudence (Kocourek Coll); US Supreme Court, briefs, records; Human Rights
260 000 vols; 3 910 curr per; 156 104 govt docs; 1 060 330 microforms; 6 digital data carriers 31599

– Lucille Little Fine Arts Library & Learning Center, 160 Patterson Dr, Lexington, KY 40506-0224
T: +1 859 2572800; Fax: +1 859 2574662; E-mail: gail.kennedy@uky.edu; URL: www.uky.edu/libraries
2000; Gail Kennedy
Wilcox Coll of American Music, Alred Cortot Coll, John Jacob Niles Coll, Charles Faber Country Music Sound Recordings
122 250 vols; 500 curr per; 7 300 microforms; 26 350 av-mat; 16 000 sound-rec 31600

– Mathematical Sciences Library, OB-9 Patterson Office Tower, Lexington, KY 40506-0027
T: +1 859 2578365; Fax: +1 859 2578365; E-mail: tom.hecker@uky.edu
Tom Hecker
Programming Languages
47 430 vols; 235 curr per; 150 diss/theses; 50 digital data carriers
libr loan 31601

– Medical Center Library, 800 Rose St, Lexington, KY 40536-0298
T: +1 859 3235300; Fax: +1 859 3231040; E-mail: mclstith@email.uky.edu; URL: www.mc.uky.edu/medlibrary
1957; Janet Stith
236 980 vols; 2 930 curr per; 1 230 e-journals; 600 e-books; 1 275 sound-rec; 17 digital data carriers; 47 210 slides 31602

– Special Collections Library, King Bldg, Lexington, KY 40506-0039
T: +1 859 2578611; Fax: +1 859 2571949; E-mail: gehogg01@uky.edu; URL: www.uky.edu/Libraries/
Gordon Hogg
Kentuckiana, Appalachian Coll, Modern Political Archs
147 680 vols; 64 curr per 31603

Virginia Military Institute, J.T.L. Preston Library, ***Lexington***, VA 24450
T: +1 540 4647228; Fax: +1 540 4647279; URL: www.vmi.edu/library/
1839; Don Samdahl
Civil War, Thomas Stonewall Jackson Coll
216 000 vols; 618 curr per; 20 430 e-journals; 176 253 govt docs; 3 360 av-mat; 1 500 sound-rec; 400 digital data carriers
libr loan 31604

Washington & Lee University, ***Lexington***

– James Graham Leyburn Library, 204 W Washington St, Lexington, VA 24450-0303
T: +1 540 4588644; Fax: +1 540 4588964; URL: library.wlu.edu
Merrily Taylor
702 000 vols; 1 050 curr per; 3 000 e-books 31605

– Wilbur C. Hall Law Library, Lewis Hall, E Denny Circle, Lexington, VA 24450
T: +1 540 4588553; Fax: +1 540 4588967; URL: law.wlu.edu/library
1849; Sarah K. Wiant
439 000 vols; 3 826 curr per; 5 000 e-books; 808 000 microforms; 1 122 av-mat; 1 892 sound-rec; 107 digital data carriers
libr loan; AALL 31606

William Jewell College, Charles F. Curry Library, Campus Box 1097, 500 College Hill, ***Liberty***, MO 64068-1843
T: +1 816 4157610; Fax: +1 816 4155021; URL: campus.jewell.edu/academics/curry/library/default.html
1849; Dr. Hugh G. Stocks
Partee Ctr for Baptist Hist Studies, Children's Lit (Lois Lenski Coll), Missouri Coll, Puritan Lit (Charles Haddon Spurgeon Coll), Western Americana (Settle Coll)

258 000 vols; 613 curr per
libr loan 31607

Ohio State University & Lima Technical College, Lima Campus Library, 4240 Campus Dr, *Lima*, OH 45804
T: +1 419 9958326; Fax: +1 419 9958138; URL: www.lima.ohio-state.edu/library/index.php
1966; Dr. Mohamed Zehery
80 000 vols; 517 curr per 31608

Lincoln College, McKinstry Library, 300 Keokuk, *Lincoln*, IL 62656
T: +1 217 7323155 ext 290; Fax: +1 217 7324465; E-mail: McKinstry@lincolncollege.edu; URL: www.lincolncollege.edu/academics/library.php
1865; June Burke
Lincoln Coll
40 000 vols; 250 curr per 31609

Nebraska Wesleyan University, Cochrane-Woods Library, 50th & St Paul, *Lincoln*, NE 68504
T: +1 402 4652401; Fax: +1 402 4652189; E-mail: library@nebrwesleyan.edu; URL: library.nebrwesleyan.edu
1888; John Montag
Methodism, Religion; Mignon G Eberhart Coll, Publs of Faculty, bks, mss; Rare Bks/College Arch
203 000 vols; 641 curr per 31610

Union College, Library, 3800 S 48th St, *Lincoln*, NE 68506-4386
T: +1 402 4862514; Fax: +1 402 4862678; URL: www.ucollege.edu/library
1891; Sabrina Riley
Genealogy, Seventh-Day Adventists; College Arch, E N Dick Coll, Heritage Room
162 000 vols; 594 curr per; 36 000 e-books; 80 sound-rec
libr loan 31611

University of Nebraska – Lincoln, University Libraries, 13th & R St, *Lincoln*, NE 68588; P.O. Box 884100, Lincoln, NE 68588-04100
T: +1 402 4722526; Fax: +1 402 4725131; E-mail: jgiesecke1@unl.edu; URL: iris.unl.edu
1869; Dr. Joan R. Giesecke
Nebraskana, World Wars I and II, Ethnicity, French Revolution, Railroads, American Folklore, Univ Arch, Russian Coll, Czechoslovakia Colls, Latvian Coll
2 875 000 vols; 42 480 curr per
libr loan 31612

– Architecture Library, 308 Architectural Hall, City Campus 0108, Lincoln, NE 68588-0108
T: +1 402 4721208; Fax: +1 402 4720665; E-mail: archmail@unlnotes.unl.edu; URL: iris.unl.edu
1942; Kay Logan-Peters
WK Kellogg Coll of Rural Community Development Resources
50 000 vols; 300 curr per; 1 885 vols on microfilm, 7 454 vols on microfiche, 10 VF drawers, 50 000 slides 31613

– Engineering Library, Nebraska Hall, Rm W204, City Campus 0516, Lincoln, NE 68588-0516
T: +1 402 4723411; Fax: +1 402 4720663; E-mail: englibm@unlnotes.unl.edu; URL: iris.unl.edu
1973; Virginia Baldwin
315 000 vols; 1 231 curr per; 242 158 microforms 31614

– Geology Library, Bessey Hall, Rm 10, City Campus 0344, Lincoln, NE 68588-0344
T: +1 402 4722653; E-mail: geolmail@unlnotes.unl.edu
Adonna Fleming
52 550 vols; 34 507 curr per 31615

– Marvin & Virginia Schmid Law Library, 40 Fair St, Lincoln, NE 68583; P.O. Box 830902, Ross McCollum Hall, Lincoln, NE 68583-0902
T: +1 402 4723547; Fax: +1 402 4728260; URL: www.law.unl.edu
Richard Leiter
199 000 vols; 2 858 curr per 31616

– Mathematics Library, 14 Avery Hall, Lincoln, NE 68588-0129
T: +1 402 4726900; E-mail: mathmail@unlnotes.unl.edu
Kay Logan-Peters
29 000 vols; 214 curr per 31617

– Music Library, 30 Westbrook Music Bldg, Lincoln, NE 68588-0101
T: +1 402 4726300; Fax: +1 402 4721592; E-mail: abreckbill1@unl.edu
1980; Anita Breckbill
Ruth Etting Coll, Jazz Colls, Pee Wee Erwin Coll, Lowenberg Coll on Emily Dickinson & Music
55 000 vols; 160 curr per; 2 382 microforms; 23 289 sound-rec 31618

– C. Y. Thompson Library, East Campus, 38th & Holdrege Sts, Lincoln, NE 68583-0717
T: +1 402 4724407; Fax: +1 402 4727005; E-mail: cytref@unlnotes.unl.edu; URL: iris.unl.edu
Tracy Bicknell-Holmes
Agriculture, Natural resources, Dentistry, Special education, Home economics, Food science, Textiles/Clothing
374 000 vols; 2 139 curr per; 1 500 diss/theses; 14 000 govt docs; 6 100 maps; 550 microforms; 15 digital data carriers; 20 800 Bks on Deafness & Sign Lang
libr loan 31619

Lincoln University, Langston Hughes Memorial Library, 1570 Old Baltimore Pike, *Lincoln University*, PA 19352; P.O. Box 147, Lincoln University, PA 19352-0147
T: +1 610 9328300 ext 3367; Fax: +1 610 9321206; E-mail: bgizzi@lincoln.edu; URL: www.lincoln.edu/library
1898; Patrick Hall
Liberal arts, Protestant theology, Presbyterian lit, Science; African Studies, Afro-American Studies, Langston Hughes Personal Libr, Therman O'Daniel Personal Libr, Rare Antislavery Pamphlets
163 760 vols; 725 curr per
libr loan 31620

Bethany College, Wallerstedt Library, 235 E Swensson, *Lindsborg*, KS 67456-1896
T: +1 785 2273380; URL: www.bethanylb.edu
1907; Denise Carson
121 000 vols; 60 curr per; 37 000 microforms; 611 av-mat; 4 388 sound-rec; 18 digital data carriers; 51 slides, 24 overhead transparencies
libr loan 31621

Benedictine University, Library, 5700 College Rd, *Lisle*, IL 60532-0900
T: +1 630 8296057; Fax: +1 630 9609451; E-mail: libref@ben.edu; URL: www.ben.edu/library
1887; Jack Fritts
Abraham Lincoln Coll, Autographed Coll, John Erlenborn Papers, College Arch, Rare Bks & Mss
126 000 vols; 676 curr per; 252 000 microforms; 52 digital data carriers; 35 985 govt docs
libr loan 31622

Arkansas Baptist College, Library, 1621 Martin Luther King Dr, *Little Rock*, AR 72202
T: +1 501 2445109; Fax: +1 501 2445102; URL: www.arkansasbaptist.edu
Sonya Locket
30 000 vols; 98 curr per 31623

Philander Smith College, M. L. Harris Library, One Trudie Kibbe Reed Dr, *Little Rock*, AR 72202
T: +1 501 3705262; Fax: +1 501 3705307; URL: www.philander.edu
Teresa Ojezua
African-American hist, Faculty development; Philander Smith College Arch
74 000 vols; 298 curr per; 6 digital data carriers
libr loan 31624

University of Arkansas at Little Rock, Ottenheimer Library, 2801 S University Ave, *Little Rock*, AR 72204
T: +1 501 5693120; Fax: +1 501 5693071; E-mail: LibWebTeam@ualr.edu; URL: library.ualr.edu
1927; Wanda Dole

Arkansas Coll, Architectural Drawings, Myers Shakespeare Coll
503 000 vols; 35 533 curr per; 20 400 e-journals; 18 000 e-books; 3 427 maps; 817 000 microforms; 2 000 av-mat; 11 251 sound-rec; 17 digital data carriers 31625

– Pulaski County Law Library, 1203 McMath Ave, Little Rock, AR 72202-5142
T: +1 501 3249444; Fax: +1 501 3249447; URL: www.law.ualr.edu/library
1965; June Stewart
Arkansas Supreme Court Records and Briefs, 1836-1926
278 000 vols; 3 252 curr per; 440 000 microforms; 200 av-mat; 1 000 sound-rec; 60 digital data carriers
libr loan 31626

University of Arkansas for Medical Sciences, Library, 4301 W Markham St, SLOT 586, *Little Rock*, AR 72205-7186
T: +1 501 6865980; Fax: +1 501 6866745; URL: www.library.uams.edu
1879; Mary Ryan
Pathology (Schlumberger Coll), Hist of Medicine in Arkansas
46 000 vols; 3 100 curr per; 100 e-books; 18 digital data carriers 31627

University of California, Livermore – Lawrence Livermore National Laboratory, Main Library, 7000 East Ave, *Livermore*, CA 94550; P.O. Box 5500, Livermore, CA 94550-5500
T: +1 925 4246105; Fax: +1 925 4242921; E-mail: library-reference@llnl.gov; URL: www.llnl.gov/library
1952; Isom Harrison
Technology
335 000 vols; 7 179 curr per; 100 av-mat; 200 sound-rec; 25 digital data carriers; 245 000 fed rpts
libr loan 31628

University of West Alabama, Julia Tutwiler Library, Station 12, *Livingston*, AL 35470
T: +1 205 6523614; Fax: +1 205 6522332; URL: www.library.uwa.edu
1835; Dr. Monroe C. Snider
Alabama Room; Folklore (Ruby Pickens Tartt Coll), mss; Microfiche Coll
250 000 vols; 200 curr per 31629

Madonna University, University Library, 36600 Schoolcraft Rd, *Livonia*, MI 48150-1173
T: +1 734 4325767; Fax: +1 734 4325687; URL: ww3.madonna.edu/library
1947; Joanne Lumetta
Education, Lit, Nursing; Institutional Archs, Rare Bk Coll
115 000 vols; 600 curr per; 15 400 e-journals
libr loan 31630

Lock Haven University, George B. Stevenson Library, 401 N Fairview Ave, *Lock Haven*, PA 17745-2390
T: +1 570 4842310; Fax: +1 570 4842506; URL: www.lhup.edu/library/home.htm
1870; Tara Lynn Fulton
Eden Phillpotts Coll, bks, micro
353 000 vols; 848 curr per; 1 895 maps; 671 000 microforms; 2 959 sound-rec; 29 digital data carriers
libr loan 31631

Felician College, Library, 262 S Main St, *Lodi*, NJ 07644-2198
T: +1 201 5596071; Fax: +1 973 7773917; URL: www.felician.edu/library
1942; N.N.
A R Ammons' Personal Libr, Children's Bks
116 000 vols; 507 curr per
libr loan; ALA 31632

Utah State University Libraries, Merrill Library, 3000 Old Main Hill, *Logan*, UT 84322-3000
T: +1 435 7972680; Fax: +1 435 7972677; URL: library.usu.edu
1888; Larry Smith
Arch of Society of American Range Management, Czechoslovakia (Masaryk Coll & Spencer Taggart Coll), Fife Folklore Coll, Hand Folklore Coll, Yoder Folklore Coll, Laws Folklore Coll, Medical bks (Robert & Mary Ann Simmons McDill Coll), Mormons & Mormonism (Arrington Coll, Ellsworth Coll, Sonne Coll, Peirce

Coll), Utah State Documents depository, US regional depository, Univ Arch, Jack London Coll, William Lye South African Coll, Environmental Groups, Intermountain West, Western American Lit, Cowboy Poetry, Agriculture
1 574 000 vols; 12 369 curr per; 175 000 e-books; 12 235 diss/theses; 1 756 059 govt docs; music scores; 99 487 maps; 2 433 000 microforms; 5 000 av-mat; 18 000 sound-rec; 4 000 digital data carriers; 456 052 photos, mss (11 993 linear feet)
libr loan; BCR, SPARC 31633

– Ann Carroll Moore Children's Library, 6700 Old Main Hill, Logan, UT 84322
T: +1 435 7973093
Vaughn Larson
22 000 vols 31634

Loma Linda University, Del E. Webb Memorial Library, 11072 Anderson St, *Loma Linda*, CA 92350-0001
T: +1 909 5584925; Fax: +1 909 5584188; URL: www.library.llu.edu
1907; Carlene Drake
Health sciences, Religion; Seventh-Day Adventist (Heritage Coll), Remondino Coll (Hist of Medicine), 19th C Health Reform in America
366 000 vols; 820 curr per; 5 800 e-journals; 63 000 microforms; 8 831 av-mat; 6 265 sound-rec; 20 digital data carriers; 1 160 feet of archival mat
libr loan 31635

National University of Health Sciences, Sordoni-Burich Library, 200 E Roosevelt Rd, Bldg C, *Lombard*, IL 60148-4583
T: +1 630 8896612; Fax: +1 630 4956658; URL: www.nuhs.edu/lrc
1920; Joyce Ellen Whitehead
Alternative healing, Biomedical science; Hist of Chiropractic, Natural Healing
16 000 vols; 375 curr per; 3 000 microforms; 791 av-mat; 738 sound-rec; 2 digital data carriers; 302 slides 31636

California State University, Long Beach, University Library, 1250 Bellflower Blvd, *Long Beach*, CA 90840-1901
T: +1 562 9854047; Fax: +1 562 9858131; URL: www.csulb.edu/library
1949; Roman V. Kochan
Abolition Movement (Dumond Coll), Radical Politics in California (Dorothy Healey Coll), Arts in Southern California Arch, California Hist (Bekeart Coll), Theater Hist (Pasadena Playhouse Coll, playscripts), US Revolutionary War, Art prints & Photo Coll, Univ arch
1 086 000 vols; 821 curr per; 18 682 e-journals; 93 000 e-books; 2 277 sound-rec; 580 Audio Bks
libr loan 31637

University of Southern Mississippi, Gulf Coast, Richard G. Cox Library, 730 E Beach Blvd, *Long Beach*, MS 39560-2698
T: +1 228 8654510; Fax: +1 228 8654544; URL: www.lib.usm.edu
1972; Edward McCormack
Gulf of Mexico Program; Curriculm Mat Ctr
60 000 vols; 630 curr per
libr loan 31638

Fiorello H. Laguardia Community College, 31-10 Thomson Ave, *Long Island City*, NY 11101
T: +1 718 4825421; Fax: +1 718 4825444, 7186092011; URL: libraries.cuny.edu&lib-lg.htm; www.lagcc.cuny.edu/library
1973; Jane Devine
Coop education, Deaf education, Nursing, Nutrition, Occupational therapy, Veterinary medicine; Eric Educational Rpts, Govnt Docs, New York Times on microfilm
117 330 vols; 1 260 curr per; 4 500 av-mat; 1 280 sound-rec; 210 digital data carriers 31639

Bay Path College, Frank & Marian Hatch Library, 539 Longmeadow St, *Longmeadow*, MA 01106; mail address: 588 Longmeadow St, Longmeadow, MA 01106
T: +1 413 5651376; Fax: +1 413 5678345; E-mail: library@baypath.edu; URL: library.baypath.edu

1897; Michael J. Moran
Business, Info technology, Legal, Occupational therapy, Psychology
70 000 vols; 151 curr per; 11 000 e-books
libr loan 31640

Letourneau University, Margaret Estes Library, 2100 S Mobberly Ave, ***Longview***, TX 75602-3524; P.O. Box 7001, Longview, TX 75607-7001
T: +1 903 2333271; Fax: +1 903 2333263; URL: www.letu.edu
1946; Henry Whitlow
Robert G LeTourneau Memorabilia, Abraham Lincoln, Billy Sunday Coll, Harmon General Hospital Coll, Rare Afro Art & Native Antiques
94 000 vols; 527 curr per; 22 550 e-journals; 500 e-books; 170 maps; 84 000 microforms; 12 digital data carriers 31641

Covenant College, Anna Emma Kresge Memorial Library, 14049 Scenic Hwy, ***Lookout Mountain***, GA 30750
T: +1 706 4191438; Fax: +1 706 4191435; E-mail: library@covenant.edu; URL: library.covenant.edu
1955; George A. Mindeman
American Hist Coll, Christian Authors (Kresge Coll), John Bunyan Coll, T. Stanley Soltau Libr
93 000 vols; 600 curr per; 41 000 e-books 31642

Saint Francis University, Pasquerilla Library, 106 Franciscan Way, ***Loretto***, PA 15940; P.O. Box 600, Loretto, PA 15940-0600
T: +1 814 4723163; Fax: +1 814 4723093; E-mail: cirli@francis.edu; URL: library.francis.edu
1847; Sandra A. Balough
Captain Paul Boyton Coll; College Arch; Franciscana; Franciscan Arch; Prince Gallitzin Coll
22 000 vols; 152 curr per; 6 880 e-journals; 5 000 e-books; 30 diss/theses; 6 000 microforms; 2 850 av-mat; 205 sound-rec; 85 digital data carriers
libr loan 31643

California State University, Los Angeles, John F. Kennedy Memorial Library, 5151 State University Dr, ***Los Angeles***, CA 90032-8300
T: +1 323 3434928; Fax: +1 323 3433993; E-mail: tmetcal@calstatela.edu; URL: www.calstatela.edu/library
1947; Alice Kawakami
Perry R. Long Coll of Books on Printing, Joseph Wambaugh Manuscript Coll, Roy Harris & Stan Kenton Music Arch, Eugene List & Carol Glenn Coll of Musical Scores, Otto Klemperer Coll of Musical Scores, Anthony Quinn Coll of Film Scripts, Arthur M. Applebaum Theatre Arts Coll, Public Officials Papers
1 205 000 vols; 205 748 curr per; 3 000 e-books; 910 000 microforms; 2 000 sound-rec; 9 246 digital data carriers; 702 740 Electronic Media & Resources
libr loan 31644

– Roybal Institute for Applied Gerontology Library, 5151 State University Dr, Los Angeles, CA 90032-8903
T: +1 323 3434724; Fax: +1 323 3436410; URL: www.calstatela.edu
1991; Jorge Lambrinos
750 000 vols 31645

Cleveland Chiropractic College, Library, 590 N Vermont Ave, ***Los Angeles***, CA 90004
T: +1 323 6606166; Fax: +1 323 9062092; URL: www.cleveland.edu
1978; Marian Hicks
Hist of Chiropractic
23 000 vols; 1 009 av-mat; 2 703 sound-rec; 7 digital data carriers; 10 021 X-rays, 9 270 slides 31646

Hebrew Union College, Jewish Institute of Religion, Frances-Henry Library, 3077 University Ave, ***Los Angeles***, CA 90007
T: +1 213 7493424; Fax: +1 213 7491937; URL: www.huc.edu/libraries/losangeles
1958; Dr. Yaffa Weisman
West Coast Microfilm Branch of the American Jewish Arch; American Jewish Periodical Center; Rare Hebraica
112 000 vols; 500 curr per; 300 diss/theses; 5 000 microforms; 2 000 sound-rec; 100 digital data carriers; 20 Bks on Deafness & Sing Lang
libr loan; RLG 31647

Loyola Marymount University – Charles von der Ahe Library, One LMU Dr, ***Los Angeles***, CA 90045-8203
T: +1 310 3382788; Fax: +1 310 3384484; E-mail: kbrancol@lmu.edu; URL: www.lib.lmu.edu
1929; Kristine Brancolini
20th C Film & Theatre Coll, Rare Bks, Oliver Goldsmith Coll, Saint Thomas Moore Coll, Arthur P Jacobs Coll
380 000 vols; 3 140 curr per; 4 000 e-books; 40 000 av-mat 31648

Loyola Marymount University – William M. Rains Law Library, 919 S Albany St, ***Los Angeles***, CA 90015-1211
T: +1 213 7361117; Fax: +1 213 4872204; URL: www.lmu.edu/main/submain/library.htm
1920; Daniel W. Martin
Acid rain; Loyola Law School Arch Coll, Frank Gehry Mats, British Trials, O.J. Simpson Trial Mat, Rare Bks on Law
585 000 vols; 7 351 curr per; 1 299 000 microforms; 1 000 sound-rec; 448 digital data carriers
libr loan 31649

Mount Saint Mary's College, Charles Willard Coe Memorial Library, 12001 Chalon Rd, ***Los Angeles***, CA 90049-1599
T: +1 310 9544370; Fax: +1 310 9544379; URL: www.msmc.la.edu
1925; Claudia Reed
Humanities, Nursing, Physical therapy; Cardinal Newman Coll
132 000 vols; 800 curr per 31650

Occidental College, Mary Norton Clapp Library, 1600 Campus Rd, ***Los Angeles***, CA 90041
T: +1 323 2592818; Fax: +1 323 3414991; E-mail: reference@oxy.edu; URL: departments.oxy.edu/library
1887; Laura Serafini
Robinson Jeffers Coll, California & Western Hist Coll, Guymon Mystery & Detective Fiction Coll, William Jennings Bryan Coll, William Henry Coll, Upton Sinclair Coll, Ward Ritchie Press Coll
493 000 vols; 870 curr per; 22 300 e-journals; 4 000 e-books
libr loan 31651

Otis College of Art and Design, The Millard Sheets Library, 9045 Lincoln Blvd, ***Los Angeles***, CA 90045
T: +1 310 6656925; E-mail: mberry@otis.edu; URL: www.otis.edu/life_otis/library/index.html
1917; Sue Maberry
Fine arts, Fashion, Photography; Artists bks
45 000 vols; 139 curr per; 5 100 av-mat; 267 sound-rec; 100 digital data carriers; 62 VF drawers, 100 000 slides, 2 500 art reproductions 31652

Southern California Institute of Architecture, Kappe Library, 960 E Third St, ***Los Angeles***, CA 90013
T: +1 213 6132200; Fax: +1 213 6132260; URL: www.sciarc.edu
1972; Kevin McMahon
50 000 vols; 120 curr per 31653

Southwestern Law School, Leigh H. Taylor Law Library, 3050 Wilshire Blvd, ***Los Angeles***, CA 90010
T: +1 213 7386728; Fax: +1 213 7385792; E-mail: library@swlaw.edu; URL: library.swlaw.edu
Linda A. Whisman
492 000 vols; 4 000 curr per; 22 000 e-books 31654

Southwestern University, Law Library, 675 S Westmoreland Ave, ***Los Angeles***, CA 90005-3992
T: +1 213 7385771; Fax: +1 213 7385792; E-mail: library@swlaw.edu; URL: www.swlaw.edu/main.html
1911; Linda A. Whisman
249 030 vols; 4 607 curr per; 961 937 microforms; 1 300 av-mat; 1 370 sound-rec; 38 digital data carriers
libr loan 31655

University of California, Los Angeles – Ralph J. Bunche Center for African American Studies Library, 135 Haines Hall, Box 951545, ***Los Angeles***, CA 90095-1545
T: +1 310 8256060; Fax: +1 310 8255019; URL: www.bunchecenter.ucla.edu
1969; Dalena E. Hunter
Afro-Americans & Blacks in the Caribbean & Central & South America
8 000 vols; 21 curr per; 100 av-mat; 10 digital data carriers
ALA 31656

University of California, Los Angeles – University Library,Young Research Library, 405 Hilgard, 11334 YRL, ***Los Angeles***, CA 90095; P.O. Box 951575, Los Angeles, CA 90095-1575
T: +1 310 8251201; Fax: +1 310 2064109; URL: www.library.ucla.edu
1919; Gary E. Strong
British Commonwealth Hist, especially Australia & New Zealand, British Hist, California Hist, Contemporary Western Writers, Early Italian Printing, Elmer Belt Libr of Vinciana, Folklore, Hist of Medicine, Latin American Studies
8 157 000 vols; 77 509 curr per; 5 200 000 microforms; 9 322 av-mat; 12 690 sound-rec; 426 digital data carriers; 53 668 mss
libr loan 31657

– The Arts Library, 1400 Public Policy Bldg, Los Angeles, CA 90095; Box 951392, Los Angeles, CA 90095-1392
T: +1 310 8253817; Fax: +1 310 8251303; URL: www2.library.ucla.edu/libraries/arts/index.cfm
Gordon Theil
Elmer Belt Libr of Vinciana
255 000 vols; 1 984 curr per 31658

– College Library, Powell Library Bldg, Los Angeles, CA 90095; Box 951450, Los Angeles, CA 90095-1450
T: +1 310 8255756; Fax: +1 310 2069312
Sarah Watstein
225 000 vols; 3 319 curr per; 4 digital data carriers 31659

– English Reading Room, 1120 Rolfe Hall, Los Angeles, CA 90095
T: +1 310 8254511
1950; Teresa Omidsalar
Josephine Miles Poetry Coll, Modern Contemporary Poetry
31 000 vols; 165 curr per; 6 digital data carriers 31660

– Hugh & Hazel Darling Law Library, 112 Law Bldg, Box 951458,385 Charles E Young Dr E, Los Angeles, CA 90095-1458
T: +1 310 8257826; Fax: +1 310 8251372; URL: www.law.ucla.edu/library
1949; Kevin Gerson
East Asia, Latin America; David Bernard Memorial Aviation Law Libr
550 000 vols; 2 749 curr per; 366 000 microforms; 521 sound-rec; 7 digital data carriers; 1 334 mss 31661

– Louise M. Darling Biomedical Library, 10833 LeConte Ave, 12-077 Center for the Health Sciences, Los Angeles, CA 90095-1798
T: +1 310 8254055; Fax: +1 310 2068675
1947; Judy Consales
Hist of the Health Sciences; Hist of the Life Sciences; Japanese Medical Books & Prints, 17th-19th C; S. Weir Mitchell Coll; Florence Nightingale Coll; Dr M.N. Beigelman Coll (Opthalmology); Near Eastern Medical Mss
651 000 vols; 3 471 curr per; 64 000 microforms; 1 792 av-mat; 3 284 sound-rec; 7 digital data carriers; 336 mss, 6 642 pamphlets, 28 895 slides, 15 105 pictorial items, 1 287 machine readable data files 31662

– Management Library, 110 Westwood Plaza, E-302, Los Angeles, CA 90095
T: +1 310 8253047; Fax: +1 310 8256632; URL: www.anderson.ucla.edu/library.xml
1962; M. Rita Costello
Rare Bks in Business & Economics, Corporate Hist, Goldsmiths-Kress Libr of Economic Lit
177 000 vols; 683 curr per; 498 000 microforms; 2 digital data carriers; 274 235 pamphlets, 85 000 hardcopy Annual & 10K rpts of corporations 31663

– Music Library, 1102 Schoenberg Hall, P.O. Box 951490, Los Angeles, CA 90095-1490
T: +1 310 8251353; Fax: +1 310 2067322; E-mail: gtheil@library.ucla.edu; URL: www.library.ucla.edu/libraries/music/
1942; Gordon Theil
Venetian Libretti, 1637-1674; Film & Television Arch; Jazz Arch; Emigre Musicians Arch; Arch of Popular American Music
156 522 vols; 1 242 curr per; 24 504 microforms; 170 944 sound-rec; 68 000, 150 000 mss, 1 919 slides; 366 pictorial items, 350 283 pamphlets 31664

– Richard C. Rudolph East Asian Library, 21617 Research Library YRL, Los Angeles, CA 90095-1575
T: +1 310 8254836; Fax: +1 310 2064960
1948; Ching Fen (Amy) Tsiang
Chinese Archeology, Japanese Buddhism, Korean Lit, Archival Coll on Democracy and Unification in Korea, Chinese Civil Service Examination Mat, China Democracy Movement and Tiananmen Square Incident Arch
493 000 vols; 2 847 curr per; 13 000 microforms 31665

– Science & Engineering Libraries, 8270 Boelter Hall, Los Angeles, CA 90095
T: +1 310 8253646; Fax: +1 310 2063908
1944; Audrey Jackson
Astronomy; Tech Rpts Coll (incl depot items from NASA)
591 000 vols; 4 498 curr per; 3 976 diss/theses; 68 000 govt docs; 12 286 maps; 1 963 000 microforms; 44 av-mat; 277 digital data carriers; 547 floppy disks
libr loan 31666

– University Elementary School Library, 1017 University Elementary School Library, P.O. Box 951619, Los Angeles, CA 90095-1619
T: +1 310 8254928; Fax: +1 310 1064452
1920; Judith Kantor
Children's Lit; Folk Lit; Poetry Coll
18 775 vols; 29 curr per; 347 bibliographies, 758 pamphlets 31667

– William Andrews Clark Memorial Library, 2520 Cimarron St, Los Angeles, CA 90018
T: +1 323 7318529; Fax: +1 323 7318617; E-mail: whiteman@humnet.ucla.edu; URL: www.humnet.ucla.edu/humnet/clarklib
1926; Peter Reill
17th & early 18th c English Civilization, Modern Fine Printing, Montana Hist, Robert Boyle Coll, John Dryden Coll, Eric Gill Coll, Robert Gibbings Coll, Oscar Wilde & the Nineties
100 000 vols; 186 curr per; 21 000 mss; 45 incunabula; 1 000 music scores; 342 maps; 6 500 microforms 31668

University of Judaism, Ostrow Library, 15600 Mulholland Dr, ***Los Angeles***, CA 90077
T: +1 310 4769777; Fax: +1 310 4765423; URL: libraryuj.edu
Rick Burke
Judaica & Hebraica Coll
115 000 vols; 390 curr per; 750 000 newspaper clippings
libr loan 31669

University of Southern California, Information Services Division Libraries, University Park Campus, 3550 Trousdale Pkwy, ***Los Angeles***, CA 90089-0182
T: +1 213 7404039; Fax: +1 213 7491221; URL: www.usc.edu/libraries
1880; Catherine Quinlan
American Lit Coll (1850-), Cinema Coll, German Lit in Exile Coll (Feuchtwanger & Heinrich Mann), Internat Relations Coll, Philosophy (Gomperz Coll), Latin American (Boeckmann Coll), Printing Arts Coll, East Asian Coll
3 509 000 vols; 37 000 e-journals; 316 000 e-books; 45 627 mss; 262 912 govt docs; 84 490 maps; 6 169 000 microforms;

27 668 sound-rec
libr loan 31670

– Applied Social Sciences Library, Von KleinSmid Ctr, 3518 Trousdale Parkway, Los Angeles, CA 90089-0048
T: +1 213 7401769; Fax: +1 213 7491221; E-mail: vkc@usc.edu; URL: www.usc.edu/isd/locations/international/vkc
1929; Robert Labaree
Int relations, Political science, Public administration; Int Docs Coll; Planning Docs Coll
195 000 vols; 450 curr per; 150 000 microforms; 250 av-mat; 70 digital data carriers 31671

– Asa V. Call Law Library, University Park Campus, MC 0072, 699 Exposition Blvd, No 202, Los Angeles, CA 90089-0072
T: +1 213 7406482; Fax: +1 213 7407179; E-mail: abrecht@law.usc.edu; URL: lawweb.usc.edu
1896; Albert O. Brecht III
292 000 vols; 4 400 curr per; 473 266 microforms; 1 107 sound-rec; 2 digital data carriers
libr loan 31672

– Crocker Business Library, Hoffman Hall HOH 201, Los Angeles, CA 90089-1422
T: +1 213 7408520; Fax: +1 213 7474176; E-mail: library@marshall.usc.edu; URL: www.marshall.usc.edu
1967; John E. Juricek
100 000 vols; 800 curr per; 515 000 microforms; 16 digital data carriers 31673

– East Asian Library, Doheny Memorial Library, 3550 Trousdale Parkway, University Park Campus, Los Angeles, CA 90089-0185
T: +1 213 7401772; URL: www.usc.edu/libraries/locations/east_asian/
1956
Chinese Coll, Japanese Coll, Korean Heritage Libr
82 000 vols; 300 curr per 31674

– Helen Topping Architecture & Fine Arts Library, Watt Hall, 850 Bloom Walk, B-4, Los Angeles, CA 90089-0294
T: +1 213 7401956; Fax: +1 213 7491221; E-mail: afa@usc.edu; URL: www.usc.edu/libraries/locations/arts/
Art & Architecture Ephemera, Artist's Bks, Architectural Drawings
78 000 vols; 260 av-mat; 100 digital data carriers; 275 000 slides 31675

– Hoose Library of Philosophy, Mudd Memorial Hall of Philosophy, 3709 Trousdale Parkway, Los Angeles, CA 90089-0182
T: +1 213 7407434; Fax: +1 213 7491221; E-mail: scimeca@usc.edu; URL: www.usc.edu/libraries/locations/philosophy/
1929; Dr. Ross V. Scimeca
Western European Philosophers, First & Early Editions, 1700-1850
55 000 vols; 150 curr per; 3 digital data carriers 31676

– Jennifer Ann Wilson Dental Library & Learning Center, 925 W 34th St, DEN 21, University Park – MC 0641, Los Angeles, CA 90089-0641
T: +1 213 7408578; Fax: +1 213 7488565; E-mail: wdl@usc.edu; URL: www.usc.edu/hsc/dental/library; www.usc.edu/wdl
1897; John P. Glueckert
Dentistry Rare Bks Coll
19 000 vols; 475 curr per; 57 e-books; 367 diss/theses; 313 sound-rec; 67 digital data carriers; 5 526 slides
libr loan; PSRMLS 31677

– Norris Medical Library, 2003 Zonal Ave, Los Angeles, CA 90089-9130
T: +1 323 4421130; Fax: +1 323 2211235; E-mail: medlib@usc.edu; URL: www.usc.edu/nml
1928; William A. Clintworth
American Indian Ethnopharmacology Coll, Salerni Collegium Hist of Medicine Coll, Far West Medicine Coll
48 000 vols; 2 219 curr per; 2 120 e-journals; 545 e-books; 9 000 microforms; 593 av-mat; 2 634 sound-rec; 54 digital data carriers; 38 650 slides
libr loan 31678

– Science & Engineering Library, 910 Bloom Walk, Los Angeles, CA 90089-0481
T: +1 213 7408507; Fax: +1 213 8214214; E-mail: sci@usc.edu; URL: www.usc.edu/libraries/locations/science/
1970; Sara Tompson
250 000 vols; 2 500 curr per; 252 000 microforms; 27 000 tech rpts 31679

Siena College, J. Spencer & Patricia Standish Library, 515 Loudon Rd, ***Loudonville***, NY 12211-9998
T: +1 518 7832518; Fax: +1 518 7832958; E-mail: sienalibrary@siena.edu; URL: www.siena.edu/library
1937; Gary B. Thompson
Multicultural studies; Franciscana Coll, T.E. Lawrence Coll, Medieval & Early Modern Studies (Convivium Coll)
273 000 vols; 827 curr per; 5 570 e-journals; 4 000 e-books; 3 670 av-mat; 2 160 sound-rec; 620 digital data carriers 31680

Ballarmine University, W.L. Lyons Brown Library, 2001 Newburg Rd, ***Louisville***, KY 40205-0671
T: +1 502 4528317; Fax: +1 502 4528038; URL: www1.bellarmine.edu/library
1950; John Stemmer
Health sciences; Louisville Archdiocesan Coll, Louisville Hist League, Thomas Merton Coll
114 000 vols; 541 curr per
libr loan 31681

Spalding University, University Library, 853 Library Lane, ***Louisville***, KY 40203-9986
T: +1 502 5857130; Fax: +1 502 5857156; E-mail: jlenarz@spalding.edu; URL: www.spalding.edu/library/home1
1920; Jackie Lenarz
Kentucky Coll, Edith Stein Coll, Rare Bks; Curriculum Lab for School of Education
162 560 vols; 489 curr per; 15 541 microforms; 1 263 av-mat; 2 604 sound-rec; 11 digital data carriers; 4 529 slides, 430 art reproductions
libr loan; ACRL, ALA 31682

University of Louisville Libraries, ***Louisville***, KY 40292 31683

– Brandeis School of Law Library, Belknap Campus, Louisville, KY 40292
T: +1 502 8526392; Fax: +1 502 8528906; URL: www.louisville.edu/library/la/html
1926; David Ensign
Correspondence of Justice Brandeis & John M Harlan Sr
239 000 vols; 5 420 curr per 31684

– Dwight Anderson Music Library, Belknap Campus, Louisville, KY 40292
T: +1 502 8525659; Fax: +1 502 8527701
1947; Karen Little
Early American Sheet Music, Kentucky Imprints, Kentucky Composers
119 000 vols; 274 curr per; 912 microforms; 16 500 sound-rec; 30 000 scores, 20 000 uncat items 31685

– Ekstrom Library, Belknap Campus, Louisville, KY 40292
T: +1 502 8526745; Fax: +1 502 8527394; E-mail: h.rader@louisville.edu; URL: library.louisville.edu
1911; Hannelore B. Rader
Burrough Tarzan Colls; Photographic archs, Rare Bks and special colls
2 081 825 vols; 13 860 curr per; 24 071 e-journals; 5 incunabula; 21 150 diss/theses; 44 445 music scores; 22 260 maps; 2 201 300 microforms; 8 832 av-mat; 32 093 sound-rec; 10 digital data carriers; 600 linear ft of mss, 20,994.33 linear ft of archival records, 1,5 mio photos
libr loan; IFLA, ALA, ACRL, ARL, MLA, SLA 31686

– Kornhauser Health Sciences Library, Health Sciences Campus, Louisville, KY 40292
T: +1 502 8525775; Fax: +1 502 8521631; URL: www.louisville.edu/library/kornhauser
Neal Nixon
254 000 vols; 12 957 curr per 31687

– Margaret Bridwell Art Library, Schneider Hall, Belknap Campus, Louisville, KY 40292
T: +1 502 8526741; URL: library.louisville.edu/art
Gail Gilbert
90 000 vols; 337 curr per 31688

University of Massachusetts Lowell, O'Leary Library, 61 Wilder St, ***Lowell***, MA 01854-3098
T: +1 978 9344551; Fax: +1 978 9343020; URL: www.uml.edu/libraries
Patricia Noreau
398 000 vols; 520 curr per; 12 500 e-journals; 4 000 e-books; 286 000 govt docs; 685 000 microforms; 6 170 av-mat; 6 148 sound-rec; 12 digital data carriers
libr loan 31689

Lubbock Christian University, Library, 5601 19th St, ***Lubbock***, TX 79407-2009
T: +1 806 7207326; Fax: +1 806 7207255; E-mail: library@lcu.edu; URL: www.lcu.edu/LCU/cstudent/library/default.htm
1957; Rebecca Vickers
Religious studies
12 000 vols; 550 curr per; 80 000 microforms; 12 sound-rec 31690

Texas Tech University, University Libraries, 18th & Boston Ave, ***Lubbock***, TX 79409-0002
T: +1 806 7422261; Fax: +1 806 7420737; URL: www.library.ttu.edu/ul
1923; Dr. Donald Dyal
Vietnam Arch, Turkish Oral Narrative Arch, CNN World News Rpt Arch, Southwest Coll
2 387 000 vols; 30 788 curr per; 1 625 579 govt docs; 66 092 maps
libr loan 31691

– Southwest Collection Special Collections Library, P.O. Box 41041, Lubbock, TX 79409-1041
T: +1 806 7423676; Fax: +1 806 7420496; E-mail: jason.price@ttu.edu; URL: swco.ttu.edu
Agriculture, Education, Land colonization, Mining, Oil, Politics, Railroads, Ranching, Urban development, Water; Arch of the Vietnam Conflict Coll, Rare Bk Coll, Turkish Oral Hist Coll
60 250 vols 31692

Texas Tech University – School of Law Library, School of Law Bldg, 1802 Hartford Ave, ***Lubbock***, TX 79409
T: +1 806 7423957; Fax: +1 806 7421629; URL: www.law.ttu.edu/lawlibrary/library
1967; Arturo L. Torres
Commercial law
188 000 vols; 2 597 curr per; 150 000 microforms; 100 digital data carriers
libr loan 31693

Angelina College, Library, 3500 S First St, ***Lufkin***, TX 75901-7328; P.O. Box 1768, Lufkin, TX 75902-1768
T: +1 936 6335219; Fax: +1 936 6394299; E-mail: jsublett@angelina.edu; URL: angelina.cc.tx.us/library/index.html
1968; Janet Avery-Sublett
38 641 vols; 193 curr per 31694

Liberty University, A. Pierre Guillermin Library, 1971 University Blvd, ***Lynchburg***, VA 24502
T: +1 434 5822506; Fax: +1 434 5822017; URL: www.liberty.edu/index.cfm?PID=4929; www.liberty.edu/informationservices/ilrc/library/
1971; Dr. David Barnett
Religion; Dr Jerry Falwell Coll
201 000 vols; 766 curr per; 9 000 e-journals; 64 000 e-books 31695

Lynchburg College, Knight-Capron Library, 1501 Lakeside Dr, ***Lynchburg***, VA 24501-3199
T: +1 434 5448441; Fax: +1 434 5448499; URL: www.lynchburg.edu/library
1903; Christopher Millson-Martula
Business & management, Education, Nursing; Iron Industry (Capron); 17th, 18th & 19th C Maps of North America, particularly Virginia (Capron)
228 000 vols; 473 curr per; 27 000 e-books
libr loan 31696

Motlow State Community College, Clayton-Class Library, P.O. Box 8500, ***Lynchburg***, TN 37352-8500
T: +1 931 3931670; E-mail: library@mscc.edu; URL: www.mscc.edu/library
1969
54 970 vols; 114 curr per 31697

Randolph College, Lipscomb Library, 2500 Rivermont Ave, ***Lynchburg***, VA 24503
T: +1 434 9478133; Fax: +1 434 9478134; URL: library.randolphcollege.edu
1891; Theodore J. Hostetler
Classical Culture (Lipscomb Coll); Writings by Virginia Women; Pearl S Buck Coll; Lininger, Children's Lit Coll; Watts Rare Book Room; College Arch
197 000 vols; 21 000 curr per; 40 000 e-books; 5 digital data carriers
libr loan 31698

Virginia University of Lynchburg, Mary Jane Cachelin Library, 2058 Garfield Ave, ***Lynchburg***, VA 24501-6417
T: +1 434 5285276; Fax: +1 434 5284257; URL: vul.edu/facil_library.html
1887; Swannie Thompson
30 000 vols; 50 curr per 31699

Lyndon State College, Samuel Read Hall Library, 1001 College Rd, ***Lyndonville***, VT 05851; P.O. Box 919, Lyndonville, VT 05851
T: +1 802 6266450; Fax: +1 802 6269576; URL: www.lyndonstate.edu/library; www.lsc.vsc.edu/library
1911; Garet Nelson
105 000 vols; 1 132 curr per 31700

University of Maine at Machias, Merrill Library, Nine O'Brien Ave, ***Machias***, ME 04654-1397
T: +1 207 2551234; Fax: +1 207 2551356; URL: www.umm.maine.edu/content/page.php?cat=6&content_id=11
1909; Angelynn King
Maine Coll
91 000 vols; 470 curr per 31701

Western Illinois University Libraries, One University Circle, ***Macomb***, IL 61455
T: +1 309 2982761; Fax: +1 309 2982791; URL: www.wiu.edu/library
1903; Phyllis Self
Ctr for Icarian Studies, West Central Illinois Local Hist Coll, Birds of Prey, Theatre, Politics
998 000 vols; 3 200 curr per; 1 343 000 microforms; 2 005 digital data carriers
libr loan 31702

Mercer University, Jack Tarver Library, 1300 Edgewood Ave, ***Macon***, GA 31207
T: +1 478 3012960; Fax: +1 478 3012111; URL: tarver.mercer.edu
1833; Elizabeth D. Hammond
Baptists, Civil War, Southern hist & culture; Cooperative Baptist Fellowship Arch, Georgia Baptist Hist Coll, Robert Burns Coll, Percy Shelley Coll
250 000 vols; 4 500 curr per; 1 500 av-mat
libr loan 31703

– School of Medicine, Medical Library & LRC, 1550 College St, Macon, GA 31207
T: +1 478 3014056; Fax: +1 478 3012051; URL: medicine.mercer.edu/library_home
1974; Jan H. LaBeause
Southern Hist of Medicine Coll
97 000 vols; 838 curr per; 4 000 microforms; 11 827 gov docs
libr loan 31704

– Walter F. George School of Law Library, 1021 Georgia Ave, Macon, GA 31201-1001; mail address: 1400 Coleman Ave, Macon, GA 31207-0003
T: +1 478 3012334; Fax: +1 478 3012284; URL: www.law.mercer.edu/library
1850; Suzanne L. Cassidy
Griffin B Bell Papers
333 000 vols; 1 100 curr per; 6 digital data carriers 31705

Wesleyan College, Willet Memorial Library, 4760 Forsyth Rd, **Macon**, GA 31210-4462
T: +1 478 7575200; Fax: +1 478 7573898; E-mail: wlibrary@wesleyancollege.edu; URL: www.wesleyancollege.edu
1836; Sybil B. McNeil
Americana (McGregor Coll), Georgiana (Park Coll)
139 000 vols; 612 curr per; 16 000 e-books; 344 av-mat; 2 551 sound-rec; 1 190 Journals   31706

Dakota State University, Karl E. Mundt Library, 820 N Washington Ave, **Madison**, SD 57042-1799
T: +1 605 2565203; Fax: +1 605 2565208; E-mail: reference@dsu.edu; URL: www.departments.dsu.edu/library
1881; Ethelle S. Bean
South Dakota, Senator Karl E Mundt Arch
178 000 vols; 382 curr per; 110 maps; 2 000 microforms; 522 av-mat; 1 223 sound-rec; 280 digital data carriers; 24 Art Repros
libr loan; ALA, AASL, ACRL, LITA, RUSA   31707

Drew University, Library, 36 Madison Ave, **Madison**, NJ 07940
T: +1 973 4083471; Fax: +1 973 4083770; E-mail: drewlib@drew.edu; URL: www.drew.edu/depts/library.aspx
1867; Dr. Andrew D. Scrimgeour
Methodistica, Slavery; Hymnology (David A. Creamer Coll), Methodistica (Tyerman Coll), Tipple & Maser Coll of Wesleyana, Reformation Church Hist (Koehler Coll), Theology Coll, United Nations Coll, Political Journalism, Georges Simenon Coll
577 000 vols; 2 571 curr per; 1 000 e-books; 218 000 microforms; 2 000 av-mat; 2 000 sound-rec; 48 720 Pamphlets, 125 VF Drawers of mss
libr loan   31708

Edgewood College, Oscar Rennebohm Library, 1000 Edgewood College Dr, **Madison**, WI 53711-1997
T: +1 608 6633284; Fax: +1 608 6636778; URL: library.edgewood.edu
1941; Sylvia Contreras
Education, Religion
120 000 vols; 331 curr per; 9 000 e-books   31709

Fairleigh Dickinson University, College at Florham Library, 285 Madison Ave, MLA-003, **Madison**, NJ 07940
T: +1 973 4438532; Fax: +1 973 4438525; URL: library.fdu.edu
1958; Dr. James Marcum
Kahn Coll (Hist of Film & Photography), Black Lit Coll, Florham-Madison Campus and Vanderbilt Estate colls, arch of Fairleigh Dickinson University (FDU) founder Peter Sammartino, New York Cultural Ctr arch
152 000 vols; 406 curr per; 538 e-journals; 40 000 e-books; 2 068 diss/theses; 300 maps; 19 000 microforms; 50 sound-rec; 90 digital data carriers; 75 Electronic Media @ Resources
libr loan; ALA   31710

University of Wisconsin – Madison, General Library System & Memorial Library, 728 State St, **Madison**, WI 53706
T: +1 608 2637360; Fax: +1 608 2652754; URL: www.library.wisc.edu
1850; Edward Van Gemert
Alchemy (Duveen Coll), American Gift Bks & Annuals, Balcanica Coll, Book Plates, Robert Boyle Coll, Brazilian Positivism, Brodhead Mss, Buddhism, Burgess Coll of Children's Lit, C S Lewis Letters, Calvinist Theology & Dutch Hist (Tank Coll), Southeast Asian Colls, South Asian Colls
6 057 000 vols; 39 802 curr per   31711

– Biology Library, 430 Lincoln Dr, Madison, WI 53706
T: +1 608 2622740; Fax: +1 608 2629003; E-mail: biolib@library.wisc.edu; URL: www.library.wisc.edu/libraries/biology/
1907; Elsa Althen
50 000 vols   31712

– Business Library, Grainger Hall, Rm 2200, 975 University Ave, Madison, WI 53706
T: +1 608 2625935; Fax: +1 608 2629001; E-mail: askbusiness@library.wisc.edu; URL: business.library.wisc.edu
1955; Michael G. Enyart
Corporate Annual Rpts, Johnson Foundation Coll (Productivity)
53 000 vols; 315 000 microforms; 120 av-mat; 1 000 master's papers   31713

– Center for Demography Library, 4471 Social Science Bldg, 1180 Observatory Dr, Madison, WI 53706-1393
T: +1 608 2636372; Fax: +1 608 2628400; E-mail: library@ssc.wisc.edu; URL: www.ssc.edu/cde/library/home.htm
1970; John Carlson
18 000 vols; 185 curr per   31714

– Center for Instructional Materials & Computing, 225 N Mills St, Madison, WI 53706
T: +1 608 2634751; Fax: +1 608 2626050; E-mail: askcimc@education.wisc.edu; URL: cimc.education.wisc.edu
1848; Jo Ann Carr
Teacher education; Standardized Tests, Curriculum Guides on microfiche
65 000 vols; 386 curr per; 29 digital data carriers   31715

– Chemistry Library, 1101 University Ave, Madison, WI 53706
T: +1 608 2622942; Fax: +1 608 2629002; URL: www.chemistry.library.wisc.edu
1947
51 000 vols   31716

– College (Undergraduate) Library, 600 N Park St, Madison, WI 53706
T: +1 608 2623245; Fax: +1 608 2624631; E-mail: helenc@library.wisc.edu; URL: college.library.wisc.edu
Carrie Kruse
97 000 vols   31717

– Cooperative Children's Book Center, 4290 Helen C White Hall, 600 N Park St, Madison, WI 53706
T: +1 608 2633720; Fax: +1 608 2624933; E-mail: ccbcinfo@education.wisc.edu; URL: www.education.wisc.edu/ccbc/
1963; Kathleen T. Horning
25 000 vols   31718

– Ebling Library for Health Sciences, 750 Highland Ave, Madison, WI 53705
T: +1 608 2622020; Fax: +1 608 2624732; E-mail: eblinghelp@library.wisc.edu; URL: ebling.library.wisc.edu
1927; Terrance Burton
Hist of Medicine, esp Anatomical Works
372 000 vols; 4 605 pamphlets   31719

– Geography Library, 280 Science Hall, 550 N Park St, Madison, WI 53706
T: +1 608 2621706; E-mail: geoglib@library.wisc.edu; URL: geography.library.wisc.edu
Thomas Tews
75 000 vols   31720

– C. K. Keith Geology & Geophysics Library, 1215 W Dayton St, Madison, WI 53706-1692
T: +1 608 2628956; Fax: +1 608 2620693; E-mail: geolib@library.wisc.edu; URL: www.geology.wisc.edu/library/
1974; Marie Dvorzak
80 000 vols; 700 diss/theses; 2 300 maps; 33 digital data carriers   31721

– Kohler Art Library, 800 University Ave, Madison, WI 53706
T: +1 608 2632258; Fax: +1 608 2632255; E-mail: askart@library.wisc.edu; URL: www.library.wisc.edu/libraries/art/
1970; Lynette Korenic
Artists Bks, Frank Lloyd Wright Coll
175 000 vols; 460 curr per; 20 000 microforms; 28 VF drawers of uncat exhibit catalogs   31722

– Kurt Wendt Engineering Library, 215 N Randall Ave, Madison, WI 53706
T: +1 608 2623493; Fax: +1 608 2624739; E-mail: askwendt@engr.wisc.edu; URL: wendt.library.wisc.edu
Deborah Helman
305 000 vols   31723

– Law School Library, 975 Bascom Mall, Madison, WI 53706
T: +1 608 2621128; Fax: +1 608 2622775; URL: library.law.wisc.edu/
1968; Steven Barkan
404 000 vols; 6 073 curr per; 689 000 microforms   31724

– Mills Music Library, 728 State St, Madison, WI 53706
T: +1 608 2631884; Fax: +1 608 2652754; E-mail: askmusic@library.wisc.edu; URL: music.library.wisc.edu
1939; Jeanette Casey
19th C American Music Imprints, Civil War Band Bks, Wisconsin Music Arch, Stratman-Thomas Coll (Wisconsin Folk Songs), Tams-Witmark Coll (American Musical Theater)
72 000 vols; 1 000 microforms; 130 849 sound-rec   31725

– Physics Library, 4220 Chamberlin Hall, 1150 University Ave, Madison, WI 53706
T: +1 608 2629500; Fax: +1 608 2652754; E-mail: physlib@library.wisc.edu; URL: physics.library.wisc.edu
1972; Kerry L. Kresse
56 000 vols; 4 852 govt docs; 7 000 microforms; 22 sound-rec; 9 digital data carriers
libr loan   31726

– Plant Pathology Memorial Library, 1630 Linden Dr, Rm 584, Madison, WI 53706
T: +1 608 2628698; Fax: +1 608 2632626; E-mail: www.plantpath.wisc.edu/library/
1911; Steve Cloyd
Johnson-Hoggan-Fulton Virus Coll
14 000 vols; 75 curr per; 400 diss/theses; 78 046 reprints, 21 600 abstracts   31727

– F.B Power Pharmaceutical Library, 2202 Rennebohm Hall, 777 Highland Ave, Madison, WI 53705
T: +1 608 2622894; Fax: +1 608 2655889; E-mail: grwanser@facstaff.wisc.edu; URL: www.hsl.wisc.edu/pharmacy/index.cfm
1890; Gregory Higby
38 069 vols   31728

– Primate Center Library, 1220 Capitol Ct, Madison, WI 53715
T: +1 608 2633512; Fax: +1 608 2634031; URL: pin.primate.wisc.edu
1973; Ray Hamel
Primatology Coll
22 000 vols; 172 curr per; 1 060 av-mat; 3 digital data carriers
libr loan   31729

– School of Library & Information Studies, Library, 600 N Park St, Rm 4191, Madison, WI 53706
T: +1 608 2632960; Fax: +1 608 2634849; E-mail: slislib@library.wisc.edu; URL: www.library.wisc.edu/libraries/slislib
1906; Michelle Besant
66 000 vols; 2 000 microforms; 8 800 pamphlets   31730

– Social Science Reference Library, 1180 Observatory Drl, Rm 8432, Madison, WI 53706
T: +1 608 2626195; URL: www.library.wisc.edu/libraries/socialsciref/
Tom Durkin
Industrial Relations, Poverty, Social Systems
15 000 vols   31731

– Social Work Library, 1350 University Ave, Rm 236, Madison, WI 53706
T: +1 608 2633840; Fax: +1 608 2652754; E-mail: socworklib@mail.library.wisc.edu; URL: library.wisc.edu/libraries/socialwork
1972; Jane Linzmeyer
26 000 vols   31732

– Space Science & Engineering Center, Schwerdtfeger Library, 1225 W Dayton St, Madison, WI 53706
T: +1 608 2620987; Fax: +1 608 2625974; E-mail: library@ssec.wisc.edu; URL: library.ssec.wisc.edu
Jean M. Phillips
40 000 vols   31733

– Steenbock Memorial Agricultural Library, 550 Babcock Dr, Madison, WI 53706
T: +1 608 2629635; Fax: +1 608 2633221; E-mail: asksteenbock@library.wisc.edu; URL: steenbock.library.wisc.edu
1969; Jean Gilbertson
Miller Beekeeping Coll, Swanton Cooperative Coll, Levitan Cookbk Coll
225 000 vols; 472 556 govt docs; 4 706 maps; 106 000 microforms; 397 av-mat; 128 sound-rec; 53 digital data carriers; 3 054 slides
libr loan   31734

– Stephen Cole Kleene Mathematics Library, 480 Lincoln Dr, Madison, WI 53706
T: +1 608 2623596; Fax: +1 608 2638891; E-mail: mathlib@library.wisc.edu; URL: math.library.wisc.edu
1963; Travis Warwick
56 000 vols; 1 digital data carriers   31735

– Woodman Astronomical Library, 6521 Sterling Hall, 475 N Charter St, Madison, WI 53706
T: +1 608 2621320; Fax: +1 608 2366386; E-mail: astrolib@library.wisc.edu; URL: astronomy.library.wisc.edu/
Kerry Kresse
19 000 vols   31736

Southern Arkansas University, Magale Library, 100 E University, **Magnolia**, AR 71753-5000; SAU Box 9401, Magnolia, AR 71754-9401
T: +1 870 2354170; Fax: +1 870 2355018; URL: www.saumag.edu/library
1909; Peggy Walters
Social work; Arkansiana
149 000 vols; 612 curr per
libr loan   31737

Ramapo College of New Jersey, George T. Potter Library, 505 Ramapo Valley Rd, **Mahwah**, NJ 07430-1623
T: +1 201 6847574; Fax: +1 201 6847628; URL: library.ramapo.edu/
1968; Judith E. Jeney
178 000 vols; 453 curr per
libr loan   31738

Pepperdine University, Payson Library, 24255 Pacific Coast Hwy, **Malibu**, CA 90263
T: +1 310 5064252; Fax: +1 310 5067225; URL: library.pepperdine.edu
1937; Mark Roosa
French Coll on 19th c Paris (Mlynarsky Coll), Religious Hist (Churches of Christ), Early Children's Lit, T E Lawrence
350 000 vols; 1 515 curr per; 396 000 microforms; 4 078 sound-rec; 108 digital data carriers
libr loan   31739

– School of Law-Jerene Appleby Harnish Law Library, 24255 Pacific Coast Hwy, Malibu, CA 90263
T: +1 310 5064643; Fax: +1 310 5064836; URL: www.law.pepperdine.edu
Daniel Martin
365 000 vols; 3 603 curr per   31740

Penn State University, Great Valley Library, 30 E Swedesford Rd, **Malvern**, PA 19355
T: +1 610 6483215; Fax: +1 610 7255223; URL: www.libraries.psu.edu/greatvalley
1963; Dr. Dolores Fidishun
Education, Computer science, Engineering, Management, Mathematics; Eric Docs, Curriculum, Psychological Tests
40 000 vols; 380 curr per; 10 av-mat; 200 sound-rec; 3 digital data carriers   31741

Hesser College, Library, 3 Sundial Ave, **Manchester**, NH 03103
T: +1 603 2966346; E-mail: library@hesser.edu; URL: www.hesser.edu/Pages/Academic_Services.aspx
Ada Kemp
30 000 vols; 250 curr per   31742

Saint Anselm College, Geisel Library, 100 Saint Anselm Dr, ***Manchester***, NH 03102-1310
T: +1 603 6417300; Fax: +1 603 6417345; URL: www.anselm.edu/library
1889; Joseph W. Constance Jr
New England Hist, Arch of College
238 000 vols; 1 600 curr per; 68 000 microforms; 7 100 sound-rec; 10 digital data carriers 31743

Southern New Hampshire University, Shapiro Library, 2500 N River Rd, ***Manchester***, NH 03106-1045
T: +1 6036459605; Fax: +1 603 6459685; E-mail: circulation@snhu.edu; URL: www.snhu.edu/library.asp
1963; Kathy Growney
Business Hist Coll, micro; Business Teacher Education (BTE Coll); Social Science & Hist (Libr of American Civilization), fiche; AMEX & NYSE 10K & Annual Reports, fiche
94 000 vols; 755 curr per; 312 000 microforms; 1 658 av-mat; 82 sound-rec; 99 digital data carriers; 6 VF drawers, 363 art reproductions 31744

University of New Hampshire at Manchester, Library, 400 Commercial St, ***Manchester***, NH 03101
T: +1 603 6414183; Fax: +1 603 6414124; E-mail: unhm.library@unh.edu; URL: unhm.unh.edu
1967; Ann E. Donahue
Deafness, sign language
32 000 vols; 259 curr per; 6 digital data carriers
libr loan 31745

Kansas State University, Hale Library, 137 Mid-Campus Dr, ***Manhattan***, KS 66506
T: +1 785 5327421; Fax: +1 785 5327415; E-mail: library@ksu.edu; URL: www.lib.k-state.edu
1863; Lori A. Goetsch
Physical Fitness Coll, Human Relations Area Files, Travels in the West & Southwest, Lincoln Coll, Cookery, Hist of American Farming
1 064 000 vols; 13 875 curr per; 680 000 govt docs; 23 792 maps; 3 938 000 microforms; 650 av-mat; 17 469 sound-rec; 1 688 digital data carriers; 29 500 slides, 8 662 scores
libr loan 31746

– Mathematics & Physics Library, Cardwell Hall, Rm 105, Manhattan, NY 66506
T: +1 785 5326827; Fax: +1 785 5326806; E-mail: library@ksu.edu; URL: www.lib.ksu.edu
1963; Barbara Steward
World Meteorological Organization Publs
27 500 vols; 440 curr per 31747

– Paul Weigel Library of Architecture, Planning & Design, Seaton Hall, Rm 323, Manhattan, KS 66506
T: +1 785 5325968; Fax: +1 785 5326722; E-mail: patw@ksu.edu; URL: www.lib.k-state.edu/branches/arch/
1917; Jeff Alger
English & French Architecture prior to World War I
38 810 vols; 200 curr per 31748

– Veterinary Medical Library, Trotter Hall, Rm 408, Veterinary Medical Complex, Manhattan, KS 66506-5614
T: +1 785 5326006; Fax: +1 785 5842838; E-mail: vetlib@vet.ksu.edu; URL: www.vet.ksu.edu/library
1936; Gayle Willard
Faculty Reprints, Animal Nutrition Coll, Animal Welfare Info Coll, German Theses from Univ of Hanover Veterinary School, Human-Animal Relationships, Veterinary Hist, Practice Management Coll
42 220 vols; 950 curr per; 4 digital data carriers 31749

Manhattan Christian College, B.D. Phillips Memorial Library, 1415 Anderson Ave, ***Manhattan***, KS 66502-4081
T: +1 785 5393571; Fax: +1 785 5390832; E-mail: mcclib@mccks.edu; URL: www.mccks.edu/academics/library.html
1927; Mary Ann Buhler
Commentaries
43 000 vols; 113 curr per; 11 218 e-journals; 2 000 e-books; 15 maps; 2 000 microforms; 1 886 sound-rec
libr loan 31750

Silver Lake College, Zigmunt Library, 2406 S Alverno Rd, ***Manitowoc***, WI 54220-9319
T: +1 920 6866134; Fax: +1 920 6847082; E-mail: Ritarose.Stahl@sl.edu; URL: www.sl.edu
1939; Sister Ritarose Stahl
Rare Books; Juvenile Lit; Kodaly Music
60 000 vols; 283 curr per; 1 incunabula; 307 diss/theses; 862 music scores; 121 maps; 414 microforms; 45 av-mat; 7 208 sound-rec; 90 digital data carriers
libr loan; ALA, MVLS, CLA, WCLA 31751

University of Wisconsin – Manitowoc Library, 705 Viebahn St, ***Manitowoc***, WI 54220-6699
T: +1 920 6834718; Fax: +1 920 6834776; E-mail: manlib@uwc.edu; URL: www.manitowoc.uwc.edu/library/
1962; Robert A. Bjerke
28 976 vols; 165 curr per; 2 300 maps; 3 390 microforms; 228 av-mat; 2 480 sound-rec; 9 digital data carriers 31752

Bethany Lutheran College, Memorial Library, 700 Luther Dr, ***Mankato***, MN 56001-4490
T: +1 507 3447350; Fax: +1 507 3447376; E-mail: circulation@blc.edu; URL: www.blc.edu/library
1927; Orrin H. Ausen
67 000 vols; 230 curr per; 5 000 e-books 31753

Minnessota State University Mankato, Library Services, ML3097, ***Mankato***, MN 56001; P.O. Box 8419, Mankato, MN 56002-8419
T: +1 507 3895958; Fax: +1 507 3895155; URL: www.lib.mnsu.edu
1868; Joan Roca
Minnesota Hist (Center for Minnesota Studies Coll), Curriculum Mat Coll, Oral Hist
827 000 vols; 2 195 curr per
libr loan 31754

Mansfield University, North Hall Library, ***Mansfield***, PA 16933
T: +1 570 6624689; Fax: +1 570 6624993; URL: www.mnsfld.edu/depts/
1857; Scott R. DiMarco
Criminal justice, Education, Music; Annual Rpt
240 000 vols; 26 050 curr per
libr loan 31755

Ohio State University – Mansfield Campus, Louis Bromfield Library, 1660 University Dr, ***Mansfield***, OH 44906-1599
T: +1 419 7554324; Fax: +1 419 7554327; URL: library.mansfield.ohio-state.edu
1966; Kay Foltz
Education; Louis Bromfield Papers
50 000 vols; 300 curr per; 50 av-mat 31756

Life University Library, 1269 Barclay Circle, ***Marietta***, GA 30060
T: +1 770 4262688; Fax: +1 770 4262745; E-mail: library@life.edu; URL: www.life.edu/lifes_campus/library.asp
1975; Susan A. Stewart
Chiropractic, health sciences, nutrition, anatomical models
62 000 vols; 126 curr per; 24 763 e-journals; 23 000 e-books; 23392 electronic bks
libr loan; ALA, GLA 31757

Marietta College, Dawes Memorial Library, 220 Fifth St, ***Marietta***, OH 45750
T: +1 740 3764757; Fax: +1 740 3764540; E-mail: library@marietta.edu; URL: library.marietta.edu
1835; Douglas Anderson
Rodney M. Stimson Coll, General Rufus Putnam Papers Coll & Ohio Company of Associates Coll, Charles G Slack Coll, Ohio Hist & Scientific Coll (Samuel P Hildreth Coll), 16th-19th c rare bk coll
246 000 vols; 759 curr per
libr loan 31758

Southern Polytechnic State University, Lawrence V. Johnson Library, 1100 S Marietta Pkwy, ***Marietta***, GA 30060-2896
T: +1 678 9157471; Fax: +1 678 9154944; URL: www.spsu.edu/library/library.html
1948; Dr. Joyce W. Mills
Engineering & technology, Art, Business; Surveying Maps, Architecture
120 000 vols; 1 216 curr per; 48 000 microforms
libr loan 31759

Indiana Wesleyan University, Jackson Library, 4201 S Washington St, ***Marion***, IN 46953
T: +1 765 6772603; Fax: +1 765 6772676; URL: www.indwes.edu/library/
1920; Sheila O. Carlblom
Religion, Education, Nursing; Holiness, Wesleyan Church Hist
147 000 vols; 680 curr per; 5 000 e-books
libr loan 31760

Judson College, Bowling Library, 306 E Dekalb, ***Marion***, AL 36756
T: +1 334 6835184; Fax: +1 334 6835188; URL: www.judson.edu/library.html
1839; Meg Truman
Feminism, Religion; Alabama Women's Hall of Fame Coll
72 000 vols; 429 curr per; 2 000 microforms
libr loan 31761

Ohio State University, Marion Campus Library, 1469 Mount Vernon Ave, ***Marion***, OH 43302
T: +1 740 7256254; Fax: +1 740 7256309; E-mail: marionlibrary@osu.edu; URL: marionlibrary.osu.edu
1957; Betsy L. Blankenship
Warren G. Harding/Norman M. Thomas 'Age of Normalcy' Res Coll
51 000 vols; 251 curr per; 3 diss/theses; 1 004 maps; 3 000 microforms; 1 729 sound-rec; 3 digital data carriers; 2 790 pamphlets, 1 034 video recordings
libr loan; ALA, ACRL
Also libr for Marion Technical College 31762

Marlboro College, Howard & Amy Rice Library, 64 Dalrymple Rd, ***Marlboro***, VT 05344-0300; P.O. Box A, Marlboro, VT 05344-0300
T: +1 802 4517577; Fax: +1 802 4517550; E-mail: library@marlboro.edu; URL: www.marlboro.edu/resources/index.html
1947; Mary H. White
Rudyard Kipling Coll
65 000 vols; 168 curr per; 35 000 e-books; 103 av-mat
libr loan; ALA 31763

Northern Michigan University, Lydia M. Olson Library, 1401 Presque Isle, ***Marquette***, MI 49855
T: +1 906 2272065; Fax: +1 906 2271333; E-mail: Info@nmu.edu; URL: www.nmu.edu/library
1899; Darlene M. Walch
Moses Coit Tyler Coll, Holocaust Coll
615 000 vols; 1 722 curr per; 77 e-journals; 20 000 e-books; 4 580 Electronic Media & Resources 31764

Renfro Library, Harris Media Center, 124 Cascade St, ***Mars Hill***, NC 28754; P.O. Box 220, Mars Hill, NC 28754-0220
T: +1 828 6891454; Fax: +1 828 6891474; URL: library.mhc.edu
1856; Bev Robertson
Education, Music; Bascom Lamar Lunsford Coll (Mountain Music & Dance), Appalachian Photography
90 000 vols; 13 055 curr per; 55 000 e-books
libr loan 31765

East Texas Baptist University, Jarrett Library, 1209 N Grove St, ***Marshall***, TX 75670-1498
T: +1 903 9232262; Fax: +1 903 9353447; URL: www.etbu.edu/library/
1917; Cynthia L. Peterson
Cope Coll of Texana, East Texas Ante-Bellum Hist, Lentz Coll of Texana, Teacher Education Coll
120 000 vols; 150 curr per; 15 000 e-journals; 85 000 e-books; 3 500 music scores; 3 000 microforms; 1 250 av-mat; 2 000 sound-rec; 400 digital data carriers; 30 000 Journals 31766

Missouri Valley College, Murrell Memorial Library, Missouri Valley College, Tech Center Bldg, 500 E College St, ***Marshall***, MO 65340
T: +1 660 8314005; Fax: +1 660 8314068; E-mail: library@moval.edu; URL: www.moval.edu/library1
1889; Pamela K. Reeder
Business, Education; Cumberland Presbyterian Church Arch
59 000 vols; 443 curr per 31767

Southwest Minnesota State University (SMSU), University Library, 1501 State St, ***Marshall***, MN 56258
T: +1 507 5377278; Fax: +1 507 5376200; URL: www.southwestmsu.edu/library/
1967; Sandra Fuhr
Autogr (Z L Begin Coll), Rare Bks Coll, Grants-Scholarship Coll
197 000 vols; 804 curr per; 270 e-journals; 8 000 e-books
libr loan 31768

Wiley College, Thomas Winston Cole Sr Library, 711 Wiley Ave, ***Marshall***, TX 75670-5151
T: +1 903 9273275; Fax: +1 903 9349333; URL: www.wileyc.edu
1873; Dr. Evelyn K. Bonner
Black Studies (Schomburg Coll of Negro Lit & Hist), Wiley College Memorabilia
74 000 vols; 306 curr per 31769

University of Tennessee at Martin, Paul Meek Library, Ten Wayne Fisher Dr, ***Martin***, TN 38238
T: +1 731 8817092; Fax: +1 731 8817074; URL: www.utm.edu/library.php
1900; Mary Vaughan Carpenter
Congressman Ed Jones Papers, Governor Ned Ray McWherter Papers, Harry Harrison Kroll, Holland McCombs, Weekly County Chancery Records
352 000 vols; 1 134 curr per; 27 000 e-books; 4 500 av-mat; 1 700 sound-rec; 420 digital data carriers
libr loan 31770

Marylhurst University, Shoen Library, 17600 Pacific Hwy (Hwy 43), ***Marylhurst***, OR 97036-7036; P.O. Box 261, Marylhurst, OR 97036-0261
T: +1 503 6996261; Fax: +1 503 6361957; E-mail: library@marylhurst.edu; URL: www.marylhurst.edu/shoenlibrary/
1893; Nancy Hoover
Sacred Music, Art Therapy, Northwest United States history
82 000 vols; 304 curr per; 420 e-journals; 7 000 e-books; 124 diss/theses; 3 990 music scores; 216 microforms; 1 260 av-mat; 2 563 sound-rec; 280 digital data carriers
libr loan 31771

Maryville College, Lamar Memorial Library, 502 E Lamar Alexander Pkwy, ***Maryville***, TN 37804-5907
T: +1 865 9818256; Fax: +1 865 9818267; E-mail: angela.quick@maryvillecollege.edu; URL: library.maryvillecollege.edu
1819; Angela M. Quick
19th c Hymnals, College Arch; Fine Arts dept
128 000 vols; 1 118 curr per; 21 000 e-books
libr loan; OCLC 31772

Northwest Missouri State University, B.D. Owens Library, 800 University Dr, ***Maryville***, MO 64468-6001
T: +1 660 5621192; Fax: +1 660 5621049; E-mail: library@nwmissouri.edu; URL: www.nwmissouri.edu/library/index.html
1905; Robert W. Frizzell
Missouri Hist & Govt (Missouriana)
250 000 vols; 688 curr per; 10 500 e-journals; 39 000 e-books 31773

Byrnes-Quanbeck Library at Mayville State University, 330 Third St NE, ***Mayville***, ND 58257
T: +1 701 7884815; Fax: +1 701 7884846; URL: www.mayvillestate.edu
1889; Sarah Batesel
North Dakota Coll
105 000 vols; 497 curr per
libr loan 31774

Pennsylvania State University – McKeesport Commonwealth College, J. Clarence Kelly Library, 4000 University Dr, ***McKeesport***, PA 15132-7698
T: +1 412 6759110; Fax: +1 412 6759113; URL: www.libraries.psu.edu/mckeesport
1948; Kay Ellen Harvey
43 000 vols; 381 curr per 31775

Bethel College, Burroughs Learning Center, 325 Cherry Ave, ***McKenzie***, TN 38201
T: +1 731 3524081; Fax: +1 731 3524070; E-mail: library@bethel-college.edu; URL: www.bethel-college.edu/library/
Harold Kelly
Cumberland Presbyterian Hist Coll
64 000 vols; 340 curr per
libr loan 31776

Linfield College, Nicholson Library, 900 S Baker St, ***McMinnville***, OR 97128
T: +1 503 8832534; Fax: +1 503 8832566; URL: www.linfield.edu/library
1849; Susan Barnes Whyte
Canadiana, Costa Rica, Pacific Northwest, Thomas Hobbes; Baptist Pioneer Hist Coll
183 000 vols; 1 268 curr per
libr loan 31777

– Portland Campus Library, 2255 NW Northrup, Portland, OR 97210; mail address: 1015 NW 22nd Ave, Portland, OR 97210
T: +1 503 4137448; Fax: +1 503 4138016
Patrice O'Donovan
Nursing
13 000 vols; 155 curr per 31778

Central Christian College of Kansas, Briner Library, 1200 S Main, ***McPherson***, KS 67460; P.O. Box 1403, McPherson, KS 67460
T: +1 620 2410723; Fax: +1 620 2416032; URL: www.centralchristian.edu/library.html
1894; Judy Stockstill
Religion, Business; Free Methodist Church Coll
28 000 vols; 136 curr per; 7 000 e-books 31779

McPherson College, Miller Library, 1600 E Euclid, ***McPherson***, KS 67460-3899; P.O. Box 1402, McPherson, KS 67460-1402
T: +1 620 2420490; Fax: +1 620 2418443; E-mail: library@mcpherson.edu; URL: www.mcpherson.edu/library/index.html
1906; Dr. Susan Taylor
Brethren Arch, College Arch, Kansas & McPherson County Mat Coll
95 000 vols; 300 curr per; 31 microforms; 9 digital data carriers
libr loan; ALA 31780

Allegheny College, Lawrence Lee Pelletier Library, 555 N Main St, ***Meadville***, PA 16335; mail address: 520 N Main St, Meadville, PA 16335
T: +1 814 3323790; Fax: +1 814 3375673; URL: library.allegheny.edu
1815; Linda Gail Bills
Original Libr, 1819-23 (Gifts of James Winthrop, Isaiah Thomas & William Bentley); Lincoln, bks, pamphlets; Ida M Tarbell, letters, mss, bks; Arch of the Western Pennsylvania Conference of the United Methodist Church
797 000 vols; 800 curr per; 5 000 e-journals; 150 e-books; 7 500 av-mat; 30 digital data carriers 31781

Tufts University, Tisch Library, 35 Professors Row, ***Medford***, MA 02155-5816; mail address: 35 Packard Ave, Medford, MA 02155
T: +1 617 6273460; Fax: +1 617 6273002; URL: www.library.tufts.edu/tisch
1852; Jo-Ann Michalak
Asa Alfred Tufts Coll, Confederate Arch, Edwin Bolles Coll, Henri Gioiran Coll, Hosea Ballou Coll, John Holmes Coll, P. T. Barnum Coll, Ritter Coll, Ryder Coll, Stearus Coll, William Bentley Sermon Coll, Musicology
885 000 vols; 2 235 curr per; 4 000 e-journals; 1 000 e-books; 210 Electronic Media & Resources, 23 060 CDs
libr loan 31782

– Edwin Ginn Library, Mugar Bldg, 1st Flr, 160 Packard St, Medford, MA 02155-7082
T: +1 617 6273852; Fax: +1 617 6273736; E-mail: ginnref@tufts.edu; URL: www.library.tufts.edu/ginn/index.html
1933; Barbara Boyce
Int law; Ambassador Phillips K. Crowe Papers, Int Labor Office, League of Nations, Ambassador John Moors Cabot Papers, Edward R. Murrow Papers, Permanent Court of Int Justice, United Nations Coll
114 000 vols; 900 curr per; 201 000 microforms; 11 digital data carriers
libr loan 31783

Pennsylvania Institute of Technology Library, 800 Manchester Ave, ***Media***, PA 19063-4098
T: +1 610 8921524; Fax: +1 610 8921523
Lynea Anderman
16 000 vols; 25 curr per; 25 000 e-books 31784

Pennsylvania State University, Delaware County Commonwealth College Library, Brandywine Campus, 25 Yearsley Mill Rd, ***Media***, PA 19063-5596
T: +1 610 8921386; Fax: +1 610 8921359; URL: www.brandywine.psu.edu
1967; Sara Lou Whildin
75 000 vols; 100 curr per 31785

Florida Institute of Technology, Evans Library, 150 W University Blvd, ***Melbourne***, FL 32901-6988
T: +1 321 6747539; Fax: +1 321 7242559
1958; Dr. Celine Lang
Science & technology, Engineering, Aeronautics, Mathematics; Botanical Coll, Edwin A Link Coll (ocean related personal papers), Indian River Lagoon Coll, General John Bruce Medaris Coll (personal papers & memorabilia)
419 000 vols; 17 148 curr per; 193 000 govt docs; 185 000 microforms; 53 av-mat; 688 sound-rec; 100 digital data carriers; 426 readable mats, 2 935 slides, 507 overhead transparencies, 869 charts, 100 art reproductions
libr loan 31786

Gratz College, The Tuttleman Library, 7605 Old York Rd, ***Melrose Park***, PA 19027
T: +1 215 6357300; Fax: +1 215 6357320; URL: www.gratzcollege.edu
1895; Eliezer M. Wise
Arch Coll, Anti-Semitica Coll, Hebraica & Judaica, Mat for Training Teachers of Hebrew Language & Culture
96 000 vols; 181 curr per; 130 av-mat; 400 sound-rec; 9 digital data carriers
AJL 31787

– Abner & Mary Schreiber Jewish Music Library, 7605 Old York Rd, Melrose Park, PA 19027
T: +1 215 6357300; Fax: +1 215 6357320
Eliezer M. Wise
29 000 vols; 68 curr per 31788

Christian Brothers University, Plough Library, 650 East Parkway South, ***Memphis***, TN 38104
T: +1 901 3213432; Fax: +1 901 3213219; E-mail: library@cbu.edu; URL: sun.cbu.edu/cbu/Library/index.htm
1871; Chris Matz
Napoleonic Era Coll; Univ Hist; Higgins Coll (Hist of Bolivia)
97 000 vols; 384 curr per; 58 000 e-books; 765 sound-rec; 4 digital data carriers; 3 000 Slides
libr loan 31789

Crichton College, J.W. & Dorothy Bell Library, 255 N Highland St, ***Memphis***, TN 38111
T: +1 901 3209770; Fax: +1 901 3209785; URL: www.crichton.edu
1947; Pam Walker
50 000 vols; 361 curr per 31790

Harding University Graduate School of Religion, L.M. Graves Memorial Library, 1000 Cherry Rd, ***Memphis***, TN 38117
T: +1 901 7611354; E-mail: hgslib@hugsr.edu; URL: www.hugsr.edu
1958; Don Meredith
Restoration Hist
126 000 vols; 589 curr per; 10 000 microforms; 2 518 sound-rec
OCLC 31791

Le Moyne-Owen College, Hollis F. Price Library, 807 Walker Ave, ***Memphis***, TN 38126
T: +1 901 4351352; Fax: +1 901 4351374; URL: www.loc.edu
1870; Annette C. Berhe-Hunt
African-American; Sweeney Coll
86 000 vols; 420 curr per 31792

Rhodes College, Paul Barret Jr Library, 2000 North Pkwy, ***Memphis***, TN 38112-1694
T: +1 901 8433745; Fax: +1 901 8433404; URL: www.rhodes.edu
1848; Darlene Brooks
Art, Paintings, Objets d'Art; 19th & 20th C English & American Lit (Walter Armstrong Rare Bk Coll), autogr first eds
275 000 vols; 1 200 curr per; 490 e-journals; 34 000 e-books; 392 av-mat; 400 sound-rec
libr loan 31793

The University of Memphis, Cecil C. Humphreys School of Law Library, 3715 Central Ave, ***Memphis***, TN 38152
T: +1 901 6782426; Fax: +1 901 6785293; URL: www.law.memphis.edu
1963; Gregory K. Laughlin
180 000 vols; 2 513 curr per; 84 000 microforms; 26 digital data carriers 31794

University of Memphis, University Libraries, 126 Ned R McWherter Library, ***Memphis***, TN 38152-3250
T: +1 901 6782356; Fax: +1 901 6788218; URL: exlibris.memphis.edu/about/about_ul.html
1914; Dr. Sylverna V. Ford
Memphis Multimedia Project (Race Relations), Theater Coll, Sanitation Strike of Memphis, Robert R Church Family, US Circus Hist, Assassination of Martin Luther King, Mss Colls, Univ Arch
1 311 000 vols; 11 541 curr per
libr loan 31795

– Audiology & Speech Language Pathology, Library, 807 Jefferson St, Rm 110, Memphis, TN 38105-5042
T: +1 901 6785846; Fax: +1 901 6788281; URL: exlibris.memphis.edu/ausp/index.html
1965; John Swearengen
8 000 vols; 86 curr per; 111 microforms; 80 av-mat; 1 digital data carriers 31796

– Chemistry, Library, 316 Smith Chemistry Bldg, Memphis, TN 38152
T: +1 901 6782625; Fax: +1 901 6783447; URL: exlibris.memphis.edu/chemistry/index.html
1966; John Barnett
33 000 vols; 177 curr per; 6 000 microforms 31797

– Earth Sciences, Center for Earthquake Research & Information, Center for Earthquake Research & Information, 3892 Central Ave, Memphis, TN 38152
T: +1 901 6784868; Fax: +1 901 6784734
1991
18 000 vols; 75 curr per; 254 diss/theses; 1 830 govt docs; 3 017 maps; 1 000 microforms; 102 av-mat; 7 sound-rec; 5 digital data carriers; 406 pieces of vertical file
libr loan 31798

– Mathematics, Library, 341 Dunn Hall, Memphis, TN 38152
T: +1 901 6782385; Fax: +1 901 6782480; URL: exlibris.memphis.edu/math/index.html
1965; Carol Washington
31 000 vols; 283 curr per; 83 microforms 31799

– Music, Library, 115 Music Bldg, Memphis, TN 38152
T: +1 901 6782330; Fax: +1 901 6783096; URL: exlibris.memphis.edu/music/index.html
1967; Anna Neal
43 000 vols; 174 curr per; 11 000 microforms; 15 801 sound-rec 31800

University of Wisconsin – Fox Valley Library, 1478 Midway Rd, ***Menasha***, WI 54952-1297
T: +1 920 8322676; Fax: +1 920 8322874; E-mail: ane.carriveau@uwc.edu; URL: www.uwfox.uwc.edu/library/
1937; Ane Carriveau
35 000 vols; 200 curr per
libr loan 31801

University of Wisconsin – Stout, Library Learning Center, 315 Tenth Ave, ***Menomonie***, WI 54751-0790
T: +1 715 2321353; Fax: +1 715 2321783; E-mail: library@uwstout.edu; URL: www.uwstout.edu/lib
1891; Paul R. Roberts
Industrial Technology, Hospitality & Tourism, Human Development & Family Life Education, Home Economics, Business, Vocational Rehabilitation, Early Childhood Education, Manufacturing Engineering, Technology Education
225 000 vols; 1 030 curr per; 250 maps; 1 157 000 microforms; 5 611 sound-rec; 47 digital data carriers; 532 slides, 217 overhead transparencies, 13 charts
libr loan 31802

Concordia University Wisconsin, Rincker Memorial Library, 12800 N Lake Shore Dr, ***Mequon***, WI 53097-2402
T: +1 262 2434420; Fax: +1 262 2434424; URL: www.cuw.edu/tools/library.html
1881; Richard L. Wohlers
German Hymnals, 16th & 17th C Lutheran Theology
114 000 vols; 600 curr per; 1 950 av-mat; 350 sound-rec; 70 digital data carriers; 1 500 slides, 200 overhead transparencies 31803

Thomas More College of Liberal Arts, Warren Memorial Library, Six Manchester St, ***Merrimack***, NH 03054-4805
T: +1 603 8800425;
Fax: +1 603 8809280; URL: www.thomasmorecollege.edu
1978; Thomas W. Syseskey
Humanities, literature, philosophy, political science, religion
40 000 vols; 20 curr per; 250 sound-rec; 12 digital data carriers; 150 video-recordings
libr loan 31804

Barry University, Monsignor William Barry Memorial Library, 11300 NE Second Ave, ***Miami***, FL 33161
T: +1 305 8993760; Fax: +1 305 8994792; URL: www.barry.edu/libraryservices
1940; Estrella Iglesias
Religious studies, Catholicism; Catholic American Coll
980 000 vols; 1 200 curr per; 42 digital data carriers
libr loan 31805

Florida International University, Steven & Dorothea Green Library, 11200 SW Eighth St, ***Miami***, FL 33199
T: +1 305 3482451; Fax: +1 305 3486579; URL: library.fiu.edu
1972; Laura Probst
Urban & Regional Docs, Geological Survey Maps, Latin Am & Caribbean
1 320 000 vols; 2 474 curr per; 5 700 e-journals; 57 000 e-books; 2 540 000 microforms; 13 170 av-mat
libr loan 31806

International Fine Arts College, Daniel M. Stack Memorial Library, 1737 N Bayshore Dr, ***Miami***, FL 33132
T: +1 305 9955011; Fax: +1 305 3740190; E-mail: itomshinsky@ifac.edu; URL: www.ifac.edu
1967; Ida Tomshinsky
Coll of Hist of Fashion; Graphic Design, Interior Design, Film, Fashion Design, Computer Animation
18 200 vols; 205 curr per; 5 incunabula; 3 diss/theses; 1 200 av-mat; 204 digital data carriers; 16 000 slides
libr loan; ALA, ACRL, ARLIS/NA 31807

Miami Dade College – Kendall Campus Library, 11011 SW 104th St, ***Miami***, FL 33176-3393
T: +1 305 2372291; Fax: +1 305 2372923; URL: www.mdc.edu/kendall/library
Estrella Iglesias
140 000 vols; 431 curr per 31808

Miami Dade College – North Campus Library, 11380 NW 27th Ave, ***Miami***, FL 33167
T: +1 305 2371414; Fax: +1 305 2378276; URL: www.mdc.edu/north/library
1960; Nancy Kalikow Maxwell
145 000 vols; 650 curr per 31809

Northeastern Oklahoma A&M College, Learning Resources Center, 200 I NE, **Miami**, OK 74354
T: +1 918 5406250; Fax: +1 918 5427065; E-mail: webmaster@neo.edu; URL: www.neo.edu
1925; S. C. Brown
76 000 vols; 476 curr per 31810

Saint John Vianney College, Seminary Library, 2900 SW 87th Ave, **Miami**, FL 33165
T: +1 305 2234561; Fax: +1 305 2230650
1960; Maria Rodriguez
Religion, Philosophy, Psychology; Bilingual Lit Coll
53 000 vols; 155 curr per 31811

University of Miami, Rosenstiel School of Marine & Atmospheric Science Library, 4600 Rickenbacker Causeway, **Miami**, FL 33149-1098
T: +1 305 4214060; Fax: +1 305 3619306; E-mail: libcirc@rsmas.miami.edu; URL: www.rsmas.miami.edu/support/lib
1943; Elizabeth Fish
Atlases, Expedition Rpts, Marine Science Newsletters Coll, Nautical Charts
70 000 vols; 765 curr per; 2 500 maps; 9 000 microforms; 65 digital data carriers
libr loan; IAMSLIC 31812

– Louis Calder Memorial Library, University Of Miami Miller School Of Medicine, Miller School of Medicine, 1601 NW Tenth Ave, Miami, FL 33136; P.O. Box 016950 (R950), Miami, FL 33101
T: +1 305 2436648; Fax: +1 305 3258853; URL: calder.med.miami.edu
1952; N.N.
Weinstein Coll(Paramedical Sciences); Ophthalmology Coll; History of Medicine Coll; Florida Coll; Consumer Health Coll; Rare Books
231 000 vols; 1 159 curr per; 4 500 e-journals; 8 mss; 596 diss/theses; 41 maps; 1 500 microforms; 614 av-mat; 3 162 sound-rec; 33 digital data carriers; 729 pamphlets; 426 slides; 212 medallions; 226 portraits
libr loan; MLA 31813

– Mary & Edward Norton Library of Ophthalmology, Bascom Palmer Eye Inst, 900 NW 17th St, Miami, FL 33136
T: +1 305 3266078; Fax: +1 305 3266066; URL: www.bascompalmer.org
1962; Cynthia Birch
Hist Coll, AV Coll
18 000 vols; 230 curr per; 25 digital data carriers
libr loan 31814

Florida Memorial University, Nathan W. Collier Library, 15800 NW 42nd Ave, **Miami Gardens**, FL 33054
T: +1 305 6263640; Fax: +1 305 6263625; URL: www.fmuniv.edu/library
1879; Gloria Oswald
Religion, Social sciences; Arch Coll, Black Coll
114 000 vols; 715 curr per 31815

St. Thomas University, University Library, 16401 NW 37th Ave, **Miami Gardens**, FL 33054
T: +1 305 6286668; Fax: +1 305 6286666; URL: www.stu.edu/library
1962; Bryan Cooper
Jackie Gleason Kinescope Arch, Walt Whitman Coll, Black Catholic Arch, Dorothy Day Coll
235 000 vols; 900 curr per
libr loan 31816

– Law Library, 16401 NW 37th Ave, Miami Gardens, FL 33054
T: +1 305 6232330; Fax: +1 305 6232337; URL: www.stu.edu/lawlib
Karl T. Gruben
120 000 vols 31817

Middlebury College, Egbert Starr Library, 110 Storrs Ave, **Middlebury**, VT 05753-6007
T: +1 802 4435498; Fax: +1 802 4435698; URL: web.middlebury.edu/lis/lib/default.htm
1800; Barbara Doyle-Wilch
American Lit, European Languages, Anglo-American Ballad; Abernethy American Lit Coll, Arch of Traditional Music, Flanders Ballad Coll, Vermont Coll, College Arch
711 000 vols; 2 908 curr per; 262 597

govt docs; 81 519 maps; 144 000 microforms; 1 177 av-mat; 25 600 sound-rec; 1 145 digital data carriers
libr loan; NELINET, CLIR, OCLC 31818

– Armstrong Library, McCardell Bicentennial Hall, Middlebury, VT 05753
T: +1 802 4435449; Fax: +1 802 4432016
Carrie M. Macfarlane
108 000 vols; 36 000 microforms; 1 digital data carriers 31819

– Music Library, Center for the Arts, 72 Porter Field Rd, Middlebury, VT 05753-6177
T: +1 802 4435217; Fax: +1 802 4432332; URL: www.middlebury.edu/academics/lis/lib/library_info/music_library
1969; Joy Pile
12 000 vols; 17 823 music scores; 1 maps; 3 000 microforms; 1 254 av-mat; 24 138 sound-rec; 24 digital data carriers
libr loan; MLA, NELINET 31820

Miami University Libraries, Southwest Ohio Regional Depository (SWORD), Middletown Campus, **Middletown**, OH 45042
T: +1 513 7273474; Fax: +1 513 7273478
Sue Berry
1 810 000 vols 31821

Pennsylvania State University – Harrisburg, Library, 351 Olmsted Dr, **Middletown**, PA 17057-4850
T: +1 717 9486071; Fax: +1 717 9486381; E-mail: jcd3@psu.edu; URL: www.hbg.psu.edu/library
1966; Dr. Gregory A. Crawford
Energy Coll, Environment Coll, Women's Hist, Holocaust
280 000 vols; 500 curr per
libr loan 31822

Wesleyan University, Olin Memorial Library, 252 Church St, **Middletown**, CT 06459-3199
T: +1 860 6853844; Fax: +1 860 6852661; URL: www.wesleyan.edu/libr
1831; B. Jones
1 301 000 vols; 1 937 curr per; 4 300 e-journals
libr loan 31823

– Art Library, 301 High St, Middletown, CT 06459-3199
T: +1 860 6853327
Susanne Javorski
24 000 vols 31824

– Science Library, 262 Church St, Middletown, CT 06459-3199
T: +1 860 6852860
Steve Bischof
191 000 vols 31825

– Scores & Recordings Collection, 252 Church St, Middletown, CT 06459-3199
T: +1 860 6853898
Alec McLane
20 000 vols 31826

Northwood University, Strosacker Library, 4000 Whiting Dr, **Midland**, MI 48640-2398
T: +1 989 8374333; Fax: +1 989 8325031; E-mail: milibrary@northwood.edu; URL: www.northwood.edu/mi/academics/strosackerlibrary
1959; Sandra Potts
40 000 vols; 334 curr per; 5 digital data carriers
libr loan; ALA, MLA 31827

Little Memorial Library, 512 E Stephens, **Midway**, KY 40347-9731
T: +1 859 8465316; Fax: +1 859 8465333; E-mail: library@midway.edu; URL: www.midway.edu/library
1847; Cathy Reilender
Horse Industry, Legal Coll for Paralegal Program, Women's Studies
54 000 vols; 450 curr per
libr loan 31828

Georgia College & State University, Library & Information Technology Center, 320 N Wayne St, **Milledgeville**, GA 31061-3397; Campus Box 043, Milledgeville, GA 31061
T: +1 478 4450979; Fax: +1 478 4456847; URL: library.gcsu.edu
1889; Dr. Rachel Schipper
Flannery O'Connor Coll, Cookbk Coll,

Horology Coll
200 000 vols; 415 curr per; 27 000 e-books; 600 000 microforms; 835 av-mat; 3 301 sound-rec; 8 digital data carriers; 20 283 slides, 155 overhead transparencies
libr loan 31829

Millersville University of Pennsylvania, Helen A. Ganser Library, Nine N George St, **Millersville**, PA 17551; P.O. Box 1002, Millersville, PA 17551-0302
T: +1 717 8723611; Fax: +1 717 8723854; URL: library.millersville.edu
1855; Dr. Marjorie Warmkessel
Local & PA hist; Arch of American Industrial Arts Assn, Arch of the Pennsylvania State Modern Language Assn, Arch of Pennsylvania Sociological Assn, The Carl Van Vechten Memorial Coll of Afro American Arts & Letters, Wickersham Coll of 19th c Textbks
565 000 vols; 10 105 curr per; 4 000 e-books
libr loan 31830

Milligan College, P.H. Welshimer Memorial Library, 200 Blowers Blvd, **Milligan College**, TN 37682; P.O. Box 600, Milligan College, TN 37682
T: +1 423 4618703; Fax: +1 423 4618984; URL: www.milligan.edu
1881; Gary F. Daught
Restoration Hist (Restoration of New Testament Christianity)
143 000 vols; 463 curr per; 40 000 e-books 31831

Curry College, Louis R. Levin Memorial Library, 1071 Blue Hill Ave, **Milton**, MA 02186-9984
T: +1 617 3332177; Fax: +1 617 3332164; URL: www.curry.edu/academics/levin+library
1952; Jane Lawless
Learning disabilities
94 000 vols; 725 curr per; 60 av-mat 31832

Alverno College, Library, 3400 S 43rd St, P.O. Box 343922, **Milwaukee**, WI 53234
T: +1 414 3826062; Fax: +1 414 3826354; E-mail: carol.brill@alverno.edu; URL: www.depts.alverno.edu/library
1936; Carol Brill
Children's Lit Coll, Teaching Mat Coll, Faculty Resource Ctr (Higher Education)
89 690 vols; 1 274 curr per
libr loan 31833

Cardinal Stritch University, University Library, 6801 N Yates Rd, **Milwaukee**, WI 53207-3985
T: +1 414 4104261; Fax: +1 414 4104268; E-mail: reference@stritch.edu; URL: library.stritch.edu
1937; David Wineberg-Kinsey
139 000 vols; 4 215 curr per; 12 000 e-books 31834

Marquette University, Raynor Memorial Libraries, 1355 W Wisconsin Ave, **Milwaukee**, WI 53233; P.O. Box 3141, Milwaukee, WI 53201-3141
T: +1 414 2887214; Fax: +1 414 2887813; E-mail: memref@marquette.edu; URL: www.marquette.edu/library
1881; Janice Welburn
Lester W. Olson Lincoln Coll, Jesuitica, Catholic Church
1 015 000 vols; 3 145 curr per; 242 780 e-journals; 247 000 e-books; 548 000 microforms; 300 digital data carriers
libr loan 31835

– Law Library, Sensenbrenner Hall, 1103 W Wisconsin Ave, Milwaukee, WI 53233-2313; P.O. Box 3137, Milwaukee, WI 53201-3137
T: +1 414 2887092; Fax: +1 414 2885914; URL: www.mu.edu/law/library/research.html
1908; Patricia Cervenka
308 000 vols; 3 231 curr per; 638 000 microforms; 11 digital data carriers; 132 linear feet of unbound per 31836

Medical College of Wisconsin, Libraries, Health Research Ctr, Third Flr, 8701 Watertown Plank Rd, **Milwaukee**, WI 53226-0509
T: +1 414 4568302; Fax: +1 414 2668681; E-mail: asklib@mcw.edu; URL: www.lib.mcw.edu/mcw/libraries/aboutus.htm
1913; Mary B. Blackwelder

Medical Hist (Horace Manchester Brown Coll)
250 000 vols; 2 134 curr per; 7 000 microforms; 10 digital data carriers
libr loan 31837

Milwaukee School of Engineering, Walter Schroeder Library, 500 E Kilbourn Ave, mail address: 1025 N Broadway, **Milwaukee**, WI 53202
T: +1 414 2777180; Fax: +1 414 2777186; E-mail: library@msoe.edu; URL: w3.msoe.edu/library/
1903; Gary S. Shimek
56 000 vols; 514 curr per; 3 000 e-books; 7 640 govt docs; 71 000 microforms; 36 sound-rec; 39 digital data carriers; 529 films & AV mats
libr loan; ALA, WILS, OCLC 31838

Mount Mary College, Patrick & Beatrice Haggerty Library, 2900 N Menomonee River Pkwy, **Milwaukee**, WI 53222-4597
T: +1 414 2584810 ext 337; Fax: +1 414 2561205; E-mail: kriegiv@mtmary.edu; URL: www.mtmary.edu/library.htm
1929; Julie Kamikawa
104 610 vols; 207 curr per; 44 460 e-journals; 15 200 e-books; 531 diss/theses; 45 000 microforms; 14 293 av-mat; 2 484 sound-rec; 395 digital data carriers; 10 952 graphics
libr loan 31839

University of Wisconsin – Milwaukee, Golda Meir Library, 2311 E Hartford Ave, **Milwaukee**, WI 53201; P.O. Box 604, Milwaukee, WI 53201-0604
T: +1 414 2296202; Fax: +1 414 2296766; URL: www.uwm.edu/library/
1956; Ewa Barczyk
Albert Camus Arch, Curriculum Coll, George Hardie Aviation & Aerospace Coll, Music Coll, Philip J Hohlweck Civil War Coll, J Max Patrick Coll, Shakespeare Res Coll
1 373 000 vols; 8 000 curr per 31840

– American Geographical Society Library, Golda Meir Library, 2311 E Hartford Ave, Milwaukee, WI 53211; P.O. Box 399, Milwaukee, WI 53201
T: +1 414 2296282; Fax: +1 414 2293624; E-mail: agsl@uwm.edu; URL: www.uwm.edu/Libraries/AGSL/
1851; Christopher Baruth
210 167 vols; 2 000 curr per; 490 191 maps; 1 361 digital data carriers; 207 320 lands at images, 33 707 pamphlets, 158 823 photos, 40 731 35mm slides, 24 084 glass plates, lantern slides, film negatives, 81 rare & special globes
libr loan 31841

Wisconsin Lutheran College, Marvin M. Schwan Library, 8800 W Bluemound Rd, **Milwaukee**, WI 53226
T: +1 414 4438864; Fax: +1 414 4438505; E-mail: library@wlc.edu; URL: www.wlc.edu/library
1978
75 000 vols 31842

Augsburg College, The James G. Lindell Family Library, 2211 Riverside Ave, **Minneapolis**, MN 55454
T: +1 612 3301604; Fax: +1 612 3301436; URL: www.augsburg.edu/library
1869; Jane Ann Nelson
Meridel LeSueur Papers, Modern Scandinavian Music
185 000 vols; 620 curr per; 13 000 microforms; 4 927 sound-rec; 15 digital data carriers; 325 slides
libr loan 31843

College of Saint Catherine, Minneapolis Campus Library & AV Services, 601 25th Ave S, **Minneapolis**, MN 55454
T: +1 651 6907784; Fax: +1 651 6908636; E-mail: library@stkate.edu; URL: www.stkate.edu/library
1964; Carol Johnson
Allied health, Nursing
32 000 vols; 280 curr per 31844

Minneapolis College of Art & Design, Library, 2501 Stevens Ave, **Minneapolis**, MN 55404-3593
T: +1 612 8743791; Fax: +1 612 8743704; E-mail: library@mcad.edu; URL: library.mcad.edu
1886; Suzanne Degler
Contemporary art, Film, Photography, Visual arts; Artists Bks, MSA & MCAD

Arch
55 000 vols; 115 curr per; 1 000
microforms; 600 av-mat; 1 238 sound-rec;
60 digital data carriers; 133 200 slides
31845

North Central University, T.J. Jones
Information Resource Center, 915 E
14th St, *Minneapolis*, MN 55404; mail
address: 910 Elliot Ave, Minneapolis,
MN 55404-1391
T: +1 612 3434490; Fax: +1 612
3438069; URL: www.northcentral.edu
1930; Joy Jewett
Coll of Classical Pentecostal Mat, Islamic
Studies
73 000 vols; 282 curr per; 197 mss; 62
music scores; 9 000 microforms; 293
av-mat; 2 583 sound-rec; 11 digital data
carriers
libr loan; OCLC, ACL, ATLA   31846

**University of Minnesota – Twin
Cities**, Libraries, 499 O Meredith Wilson
Library, 309 19th Ave S, *Minneapolis*,
MN 55455-0414
T: +1 612 6259148; Fax: +1 612
6269353; URL: www.lib.umn.edu
1851; Wendy Pradt Lougee
Philosophy; African-American, Children's
Lit Res Coll, Anderson Horticultural
Lib, Sherlock Holmes Coll, Swedish
Americana
6 587 000 vols; 43 303 curr per; 43 310
mss; 2 661 784 govt docs; 402 471 maps;
5 798 000 microforms
libr loan; ACRL, ALA, ARL, CLIR, CLR,
CRL, IFLA, NAL   31847

– Ames Library of South Asia, 309 19th
Ave S, S-10 Wilson Libr, Minneapolis,
MN 55455
T: +1 612 6244857; Fax: +1 612
6269353; E-mail: d-john4@tc.umn.edu;
URL: www.ames.umn.edu
1961; Donald Clay Johnson
175 000 vols; 750 curr per; 8 924
microforms; mss (115 linear ft)   31848

– Architecture & Landscape Architecture
Library, 210 Rapson Hall, 89 Church St
SE, Minneapolis, MN 55455
T: +1 612 6246383; Fax: +1 612
6255597; URL: www.lib.umn.edu
1913; Joon Mornes
37 192 vols; 175 curr per; 372
microforms   31849

– Bio-Medical Information Services,
505 Essex St SE, 305 Diehl Hall,
Minneapolis, MN 55455
T: +1 612 6263730; Fax: +1 612
6263824; E-mail: enagle@tc.umn.edu;
URL: www.biomed.lib.umn.edu
1892; Carolyn Wahrman
Wangensteen Hist Libr of Biology &
Medicine
467 277 vols; 65 136 microforms; 1 779
av-mat; 6 digital data carriers; 156 209
monogr, 358 computer programs   31850

– Children's Literature Research Collections
(Kerlan & Hess Collections), 222-21st
Ave S, Rm 113, Minneapolis, MN 55455
T: +1 612 6244576; Fax: +1 612
6260377; E-mail: k-hoyl@umn.edu; URL:
special.lib.umn.edu/clrc
1949; Karen Nelson Hoyle
Wanda Giag; Gustaf Tenggren; Paul
Bunyan
118 503 vols; 50 curr per; 6 000 mss; 25
diss/theses; 771 microforms; 65 av-mat;
3 031 mss, 1 913 linear feet of ill, 1 732
posters; 4 000 original ill
libr loan; IFLA, ALA   31851

– East Asian Library, 309-19th Ave S,
Minneapolis, MN 55455
T: +1 612 6249833; Fax: +1 612
6253428; E-mail: suchen@umn.edu;
URL: eastasian.lib.umn.edu/
1965; Su Chen
136 135 vols; 850 curr per; 320 e-
journals; 5 045 microforms; 267 digital
data carriers
libr loan; ALA, ACRL, AAS, CEAL
31852

– Government Publications Library,
309 19th Ave S, 10 Wilson Library,
Minneapolis, MN 55455
T: +1 612 6245073; Fax: +1 612
6269353; E-mail: govref@tc.umn.edu;
URL: govpubs.lib.umn.edu
Amy West
Atomic Energy Commission (to 1968);

Great Britain; OAS; Organization for
Economic Cooperation & Development;
United Nations
290 000 vols; 5 000 curr per; 2 700 000
microforms; 5 000 digital data carriers;
2 000 000 docs
libr loan   31853

– Immigration History Research Center,
College of Liberal Arts – Elmer L.
Anderson Library, Suite 311, 222-21st
Ave S, Minneapolis, MN 55455
T: +1 612 6254800; Fax: +1 612
6260018; E-mail: ihrc@umn.edu; URL:
www.umn.edu/ihrc
1965; Donna Gabaccia
45 000 vols; 150 curr per; 5 200
microforms; 500 sound-rec; 1 000 colls
of mss   31854

– James Ford Bell Library, 309 19th
Ave S, 462 Wilson Libr, Minneapolis,
MN 55455
T: +1 612 6241528; Fax: +1 612
6269353; E-mail: c-urne@tc.umn.edu;
URL: www.bell.lib.umn.edu
1953; Marguerite Ragnow
Hist of Europ Expansion Prior to 1800
20 000 vols; 2 650 maps; 200 microforms
31855

– John R. Borchert Map Library, 309 19th
Ave S, S-76 Wilson Libr, Minneapolis,
MN 55455
T: +1 612 6244549; Fax: +1 612
6269353; E-mail: mapref@tc.umn.edu;
URL: www-map.lib.umn.edu
1940; Kristi Jensen
10 000 vols; 40 curr per; 372 428 govt
docs; 300 000 maps; 55 microforms;
331 000 aerial photos, 450 computer files
31856

– Law Library, 120 Mondale Hall, 229 19th
Ave S, Minneapolis, MN 55455
T: +1 612 6254300; Fax: +1 612
6254378; URL: www.law.umn.edu/library/
home.html
1888; Joan S. Howland
Scandinavian Law; American Indians;
British Commonwealth Legal Mat
libr loan   31857

– Mathematics Library, 310 Vincent
Hall, 206 Church St SE, Minneapolis,
MN 55455
T: +1 612 6246075; Fax: +1 612
6244302; E-mail: library@math.umn.edu;
URL: math.lib.umn.edu
Kristine K. Fowler
40 000 vols; 375 curr per; 1 digital data
carriers   31858

– Music Library, 70 Ferguson Hall, 2106
4th St S, Minneapolis, MN 55455
T: +1 612 6245890; Fax: +1 612
6256994; E-mail: hayco001@tc.umn.edu;
URL: www.lib.umn.edu/music
S. Timothy Maloney
Donald N Ferguson Coll of Rare Bks
& Scores, Operas of 18th & early 19th
c, Latin American Music Scores, Berger
Band Libr, Hubbard Music Coll
60 823 vols; 300 curr per; 1 885
microforms; 42 303 av-mat   31859

– Reference Services (Physical Sciences
& Engineering), 108 Walter Libr, 117
Pleasant St SE, Minneapolis, MN 55455
T: +1 612 6240224; Fax: +1 612
6255583; E-mail: j-jagu@umn.edu; URL:
sciweb.lib.umn.edu
1985; Janice Jaguszewski
Arch for Hist of Quantum Physics,
Helmut Heinrich Parachute Technology
Coll
371 261 vols; 3 126 curr per; 250 000
govt docs; 93 160 maps; 335 394
microforms; 205 av-mat; 3 feet of mss
31860

– Reference Services (Wilson Library), 309
19th Ave S, Minneapolis, MN 55455
T: +1 612 6242227; Fax: +1 612
6269353; E-mail: hawki003@umn.edu
Mary Schoenborn
41 528 vols; 579 877 microforms   31861

– Special Collections and Rare Books, 111
Elmer L. Andersen Library, 222 21st Ave
South, Minneapolis, MN 55455
T: +1 612 6269166; Fax: +1 612
6255525; E-mail: johns976@tc.umn.edu;
URL: special.lib.umn.edu/rare
Timothy J. Johnson
American & English lit (selected

authors), ballooning, African American lit,
English hist (17th c), Sherlock Holmes,
photomechanics, private press bks,
Swedish-Americana, Austrian hist &
culture, August Strindberg
123 915 vols
libr loan   31862

– University Archives, 222 21st Ave S, 218
Andersen Libr, Minneapolis, MN 55455
T: +1 612 6255525; E-mail: p-kros@umn.edu; URL:
www.lib.unm.edu/special.uarch
1928; Beth Kaplan
63 181 vols; 1 383 curr per; 383 maps;
1 014 microforms; 51 000 av-mat; 15 955
mss, 40 VF drawers   31863

Minot State University, Gordon B. Olson
Library, 500 University Ave W, *Minot*,
ND 58707
T: +1 701 8583200; Fax: +1 701
8583581; URL: www.misu.nodak/library/
index1.htm
1913; Larry Greenwood
American West, Geology, Education;
Indians of the North Central States,
Dakota Territory & North Dakota Hist
421 000 vols; 752 curr per; 410 e-
journals
libr loan   31864

Pfeiffer University, G. A. Pfeiffer Library,
48380 US Hwy 52 N, *Misenheimer*,
NC 28109; P.O. Box 930, Misenheimer,
NC 28109-0930
T: +1 704 4633350; Fax: +1 704
4633356; URL: library.pfeiffer.edu
1917; Lara B. Little
Education, English lit, Religious studies,
Music; Pfeiffer Univ Arch Mat
126 000 vols; 230 curr per; 20 000 e-
books
ALA   31865

Bethel College, Bowen Library, 1001 W
McKinley Ave, *Mishawaka*, IN 46545
T: +1 574 2573329; Fax: +1 574
2573499; URL: www.bethel-in.edu
1947; Dr. Clyde R. Root
Religion, Education, Religion; Archs for
Dr Otis Bowen Missionary Church, Inc,
College Archs
133 000 vols; 457 curr per
libr loan; ACL   31866

Mississippi State University, Mitchell
Memorial Library, 395 Hardy Rd,
Mississippi State, MS 39762; P.O. Box
5408, Mississippi State, MS 39762-5408
T: +1 662 3257668; Fax: +1 662
3259344; URL: nt.library.msstate.edu
1881; Frances N. Coleman
Forestry, Agriculture, Engineering; John
C Stennis Coll, Mississippi Politics, David
Bowen Coll, Univ Arch
2 052 000 vols; 18 103 curr per
libr loan   31867

**University of Montana – Maureen &
Mike Mansfield Library**, 32 Campus
Dr, No 9936, *Missoula*, MT 59812-9936
T: +1 406 2436736; Fax: +1 406
2436864; E-mail: ic@mail.lib.umt.edu;
URL: www.lib.umt.edu
1893; Bonnie Allen
Montana Colls, Natural Hist, Whicker Coll
(English & American Lit), Chaucer Coll
1 181 000 vols; 31 614 curr per   31868

**University of Montana School of
Law**, William J. Jameson Law Library,
Missoula, MT 59812-9999
T: +1 406 2432699; Fax: +1
406 2436358; E-mail:
stacey.gordon@umontana.edu; URL:
www.umt.edu/LAW/library/default.htm
1911; Fritz Snyder
Indian Law Coll
19 440 vols; 1 795 curr per   31869

Dakota Wesleyan University, Layne
Library, 1201 McGovern Ave, *Mitchell*,
SD 57301
T: +1 605 9952618; Fax: +1 605
9952893; E-mail: library@dwu.edu; URL:
www.dwu.edu
1885; Kevin J. Kenkel
Senator Francis Case Coll, Senator
George McGovern Coll, Jennewein
Western Libr Coll
78 000 vols; 893 curr per; 350 e-journals;
12 000 e-books; 1 370 av-mat; 545 digital
data carriers
libr loan   31870

Central Christian College, Reese
Resource Center, 911 E Urbandale
Dr, *Moberly*, MO 65270
T: +1 660 2633900; Fax: +1 660
2633936; E-mail: library@cccb.edu; URL:
www.cccb.edu
1957; Patricia A. Agee
Walter S Coble Mission Files Coll,
Preaching Charts of John Hall
72 000 vols; 192 curr per; 173 maps;
6 000 microforms; 100 av-mat; 1 270
sound-rec; 42 digital data carriers; 173
charts   31871

University of Mobile, J.L. Bedsole
Library, 5735 College Pkwy, *Mobile*,
AL 36613-2842
T: +1 251 4422242; Fax: +1 251
4422515; URL: library.umobile.edu
1961; Jeffrey D. Calametti
Alabama, Civil War, Education, Local
hist, Religion; Southern Baptist Hist Coll
71 000 vols; 325 curr per; 37 000 e-
books; 84 000 microforms; 11 digital data
carriers   31872

University of South Alabama, University
Library, 307 University Blvd N, Rm 145,
Mobile, AL 36688
T: +1 251 4607021; Fax: +1
251 4607636; E-mail:
webref@jaguar1.usouthal.edu; URL:
library.southalabama.edu
1964; Dr. Richard J. Wood
Business and management, Education,
Ethnic studies, Geriatrics and gerontology,
hist, medicine; Univ Arch
312 000 vols; 1 794 curr per; 9 000 e-
books; 5 300 av-mat; 587 sound-rec; 859
digital data carriers
libr loan   31873

– Baugh Biomedical Library, 5791 USA
Drive North, Mobile, AL 36688-0002
T: +1 251 4607044; Fax: +1
251 4607638; E-mail:
medlib@bbl.usouthal.edu; URL:
biomedicallibrary.southalabama.edu/library
1972
18 000 vols; 1 200 curr per; 10 632
microforms; 24 digital data carriers
libr loan; MLA   31874

Pennsylvania State University, Beaver
Commonwealth College Library, 100
University Dr, *Monaca*, PA 15061
T: +1 724 7733790; Fax: +1 724
7733793; URL: www.libraries.psu.edu/
beaver
1965; Martin Goldberg
Steel industry; Afro-American Autobiogr
55 000 vols; 142 curr per; 108 av-mat
31875

Monmouth College, Hewes Library, 700
E Broadway, *Monmouth*, IL 61462-1963
T: +1 309 4572303; Fax: +1 309
4572226; E-mail: library@monm.edu;
URL: department.monm.edu/library
1853; John Richard Sayre
Monmouthiana, Govt Docs, James
Christie Shields Coll of Ancient Art &
Antiques
309 000 vols; 392 curr per; 1 060 e-
journals; 8 000 e-books; 2 164 av-mat;
910 sound-rec   31876

Western Oregon University, Hamersly
Library, 345 N Monmouth Ave,
Monmouth, OR 97361-1396
T: +1 503 8388418; Fax: +1 503
8388399; E-mail: refdesk@wou.edu;
URL: www.wou.edu/library
1856; Dr. Allen McKiel
University Arch, Arch for Former Oregon
Governor Robert W. Straub
237 000 vols; 798 curr per; 1 550 e-
journals; 3 000 e-books; 2 480 maps; 185
sound-rec; 107 digital data carriers
libr loan   31877

University of Louisiana at Monroe,
Library, 700 University Ave, *Monroe*,
LA 71209-0720
T: +1 318 3421067; Fax: +1 318
3421075; URL: www.ulm.edu/library
1931; Donald R. Smith
Regional Hist (Otto E Passman Papers),
Civil War, Griffin Photogr Coll
447 000 vols; 153 curr per; 40 000 e-
books; 26 000 e-books
libr loan   31878

Pennsylvania State University, Mont Alto Commonwealth College Library, One Campus Dr, ***Mont Alto***, PA 17237-9703
T: +1 717 7496182; Fax: +1 717 7496059; URL: www.libraries.psu.edu/montalto/
1963; Alica Lisa White
36 000 vols; 180 curr per 31879

Montclair State University, Harry A. Sprague Library, One Normal Ave, ***Montclair***, NJ 07043-1699
T: +1 973 6554301; Fax: +1 973 6557780; URL: library.montclair.edu; www.montclair.edu/library
1908; Dr. Judith Lin Hunt
473 000 vols; 2 955 curr per 31880

Monterey Institute of International Studies, William Tell Coleman Library, 425 Van Buren St, ***Monterey***, CA 93940; mail address: 460 Pierce St, Monterey, CA 93940
T: +1 831 6474135; Fax: +1 831 6473518; URL: monti.miis.edu
1955; Peter Y. Liu
Int, Languages, Trade, Lit, Humanities; Foreign Language, General & Technical Dictionaries (English & foreign languages), MIIS Theses, Int Business
92 000 vols; 600 curr per; 3 000 e-books; 450 maps; 2 digital data carriers; 25 microfiche titles, 6 reels titles
libr loan; RLG 31881

Naval Postgraduate School, Dudley Knox Library, 411 Dyer Rd, ***Monterey***, CA 93943-5101
T: +1 831 6562342; Fax: +1 831 6562050; E-mail: gmarlatt@nps.edu; URL: web.nps.edu/library
1946; Eleanor Uhlinger
Naval Hist & the Sea (Christopher Buckley Jr Coll)
450 000 vols; 1 100 curr per; 1 500 000 microforms; 600 av-mat; 12 digital data carriers; 90 000 tech rpts
libr loan 31882

University of Montevallo, Oliver Cromwell Carmichael Library, Station 6100, ***Montevallo***, AL 35115-6100
T: +1 205 6656100; Fax: +1 205 6656112; E-mail: library@montevallo.edu; URL: www.montevallo.edu/library/
1896; Rosemary H. Arneson
Alabama Hist & Descriptions, Alabama Authors
252 000 vols; 738 curr per 31883

Alabama State University, Levi Watkins Learning Resource Center, 915 S Jackson St, ***Montgomery***, AL 36104; P.O. Box 271, Montgomery, AL 36101-0271
T: +1 334 2294109; Fax: +1 334 2294940; URL: www.lib.alasu.edu
1921; Dr. Janice R. Franklin
Accounting, Biological science, Education, Health sciences; E.D. Nixon Coll, Ollie L. Brown Afro-American Heritage Coll
417 000 vols; 2 082 curr per
libr loan 31884

Auburn University, Montgomery Library, 7440 East Dr, ***Montgomery***, AL 36117; P.O. Box 244023, Montgomery, AL 36124-4023
T: +1 334 2443649; Fax: +1 334 2443720; URL: aumnicat.aum.edu
1969; Rickey Best
Southern Women's lit, genealogy; Local & Regional Studies
372 000 vols; 731 curr per; 240 e-journals; 51 000 e-books; 1 900 av-mat; 314 digital data carriers
libr loan 31885

Faulkner University, Gus Nichols Library, 5345 Atlanta Hwy, ***Montgomery***, AL 36109-3398
T: +1 334 3867299; Fax: +1 334 3867481; URL: www.faulkner.edu/gnl.asp
1944; Barbara Kelly
Churches of Christ Mat
107 000 vols; 504 curr per; 56 maps; 77 000 microforms; 811 sound-rec; 342 digital data carriers; 100 strips
libr loan; ALA, ACRL 31886

– Jones School of Law Library, 5345 Atlanta Hwy, Montgomery, AL 36109
T: +1 334 2606219; Fax: +1 334 2606223; URL: www.faulkner.edu
1928; Judy Hughes
35 000 vols; 100 curr per 31887

Huntingdon College, Houghton Memorial Library, 1500 E Fairview Ave, ***Montgomery***, AL 36106
T: +1 334 8334560; Fax: +1 334 2634465; URL: www.library.huntingdon.edu/library
1854; Eric A. Kidwell
Alabamiana, Arch & Hist of United Methodist Church, Autographed Book Coll, College Arch, Rare Bk Coll
108 000 vols; 261 curr per; 44 000 e-books; 1 540 av-mat; 90 digital data carriers 31888

Southern Christian University, University Library, 1200 Taylor Rd, ***Montgomery***, AL 36117
T: +1 334 3877546; Fax: +1 334 3873878; E-mail: library@southernchristian.edu; URL: www.southernchristian.edu
1967
Theology, Counseling
75 000 vols; 200 curr per; 4 digital data carriers 31889

Troy State University Montgomery, Rosa Parks Library, 252 Montgomery St, ***Montgomery***, AL 36104-3425
T: +1 334 2419576; Fax: +1 334 2419590; E-mail: m01library@troy.edu; URL: montgomery.troy.edu/library/default.htm
1970; Kent E. Snowden
Nursing
26 000 vols; 498 curr per; 42 000 e-books
libr loan 31890

West Virginia University Institute of Technology, Vining Library, 405 Fayette Pike, ***Montgomery***, WV 25136-2436
T: +1 304 4423082; Fax: +1 304 4423091; URL: www.wvit.wvnet.edu/library
1897; Dr. Barbara Crist
West Virginia
150 000 vols; 399 curr per; 12 Electronic Media & Resources 31891

University of Arkansas, Monticello, Taylor Library & Technology Center, 514 University Dr, ***Monticello***, AR 71656; P.O. Box 3599, Monticello, AR 71656-3599
T: +1 870 4601080; Fax: +1 870 4601980; URL: www.uamont.edu/library
1909; Sandra Campbell
Arkansas Coll; Gov Docs; Univ Arch
151 000 vols; 1 000 curr per; 10 diss/theses; 81 000 govt docs; 250 000 microforms
libr loan 31892

Vermont College Division of Norwich University, Gary Memorial Library, 36 College St, ***Montpelier***, VT 05602
T: +1 802 8288747; Fax: +1 802 8288748; E-mail: library@tui.edu; URL: www.tui.edu/library
1883; Matthew Pappathan
50 000 vols; 45 000 e-books 31893

Robert Morris University, Library, 6001 University Blvd, ***Moon Township***, PA 15108-1189
T: +1 412 2628272; Fax: +1 412 2624049; E-mail: librarym@rmu.edu; URL: www.rmu.edu/library
1962; Frances Caplan
Business
119 000 vols; 582 curr per; 2 000 e-books; 63 398 govt docs; 340 000 microforms; 974 sound-rec; 1 267 digital data carriers
libr loan; OCLC, PALINET 31894

Hillsdale Free Will Baptist College, Library, 3701 S I-35, ***Moore***, OK 73160; P.O. Box 7208, Moore, OK 73153-1208
T: +1 405 9129024; Fax: +1 405 9129050; URL: www.library.hc.edu
1968; Nancy Draper III
Free Will Baptist Hist Coll
26 000 vols; 232 curr per; 59 microforms; 540 sound-rec; 30 digital data carriers
libr loan 31895

Concordia College, Carl B. Ylvisaker Library, 901 S Eighth St, ***Moorhead***, MN 56562
T: +1 218 2994640; Fax: +1 218 2994253; URL: library.cord.edu
1891; Sharon Hoverson
Religious studies, Lutheran hist, Philosophy, Int studies, Scandinavian studies
321 000 vols; 1 435 curr per; 31 000 microforms; 570 av-mat; 4 230 sound-rec; 71 digital data carriers
libr loan 31896

Moorhead State University, Livingston Lord Library, 1104 Seventh Ave S, ***Moorhead***, MN 56563
T: +1 218 4772345; Fax: +1 218 4775924; URL: www.mnstate.edu/library
1887; Brittney Goodman
Juvenile coll, Northwest Minnesota hist
570 000 vols; 4 000 curr per; 964 diss/theses; 73 988 govt docs; 4 567 music scores; 9 000 microforms; 1 000 sound-rec
libr loan 31897

Saint Mary's College of California, Saint Albert Hall Library, 1928 Saint Mary's Rd, ***Moraga***, CA 94575; P.O. Box 4290, Moraga, CA 94575-4290
T: +1 925 6314229; Fax: +1 925 3766097; URL: library.stmarys-ca.edu
1863; Tom Carter
Religion, Philosophy, Hist; John Henry Newman and His Times, Libr for Lasallian Studies, College Arch, Byron Bryant Film Coll, v-tapes, California Mathematical Society Archives, Spirituality of 17th & 18th C, Oxford Movement
227 000 vols; 773 curr per; 300 e-journals; 651 e-books 31898

Morehead State University, Camden-Carroll Library, 150 University Blvd, ***Morehead***, KY 40351
T: +1 606 7835107; Fax: +1 606 7835037; E-mail: library@moreheadstate.edu; URL: www.moreheadstate.edu/library
1922; Elsie T. Pritchard
Kentucky & Appalachian Colls, Roger W. Barbour Coll, James Still Coll, Jesse Stuart Coll, Learning Resource Center, Rare Books
524 000 vols; 2 199 curr per; 59 000 e-books; 771 000 microforms; 4 419 av-mat; 10 481 sound-rec; 645 digital data carriers
libr loan 31899

West Virginia University, WVU Libraries, WVU Libraries, 1549 University Ave, ***Morgantown***, WV 26506; P.O. Box 6069, Morgantown, WV 26506-6069
T: +1 304 2930368; Fax: +1 304 2936638; URL: www.libraries.wvu.edu
1867; Frances O'Brien
African Coll; West Virginia Hist; Appalachian Room; Rare Book Room
1 602 000 vols; 34 609 curr per; 16 508 sound-rec; 682 digital data carriers 31900

– George R. Farmer, Jr College of Law Library, One Law Center Dr, Morgantown, WV 26506; P.O. Box 6135, Morgantown, WV 26506-6135
T: +1 304 2935300; Fax: +1 304 2936020; URL: www.wvu.edu/~law/library/index.htm
1878; Camille Riley
340 000 vols; 1 070 curr per; 386 000 microforms; 770 av-mat 31901

– Health Sciences Library, One Medical Ctr Dr, Health Sciences Ctr N, P.O. Box 9801, Morgantown, WV 26506-9801
T: +1 304 2932113; Fax: +1 304 2937319; E-mail: susan.arnold@mail.wvu.edu; URL: www.hsc.wvu.edu/library
1959; Susan Arnold
Medicine in West Virginia, Occupational Respiratory Diseases
259 540 vols; 2 227 curr per; 910 diss/theses; 33 182 microforms; 4 000 av-mat 31902

– Mathematics Library, 421 Armstrong Hall, P.O. Box 6468, Morgantown, WV 26506
T: +1 304 2936011; Fax: +1 304 2933753; URL: www.libraries.wvu.edu/math
Carol Wilkinson
18 590 vols 31903

– West Virginia and Regional History Collection, 1549 University Ave, Morgantown, WV 26506-6069
T: +1 304 2933536; Fax: +1 304 2933981; E-mail: jcuthber@wvu.edu; URL: www.libraries.wvu.edu/wvcollection
1933; John Cuthbert
Archives and mss, Rare Books Coll
39 000 vols; 200 curr per; 5 000 000 mss; 5 500 diss/theses; 4 700 govt docs; 250 music scores; 5 000 maps; 38 000 microforms; 2 000 sound-rec; 15 000 pamphlets, 20 000 linear feet of mss & archives; 60 000 iconographs, 1 200 newspaper titles, 100 000 photos
libr loan 31904

University of Minnesota – Morris, Rodney A. Briggs Library, 600 E Fourth St, ***Morris***, MN 56267
T: +1 320 5896176; Fax: +1 320 5896168; URL: www.morris.umn.edu/library/
1960; LeAnn Lindquist Dean
220 000 vols; 23 500 curr per 31905

College of Saint Elizabeth, Mahoney Library, Two Convent Rd, ***Morristown***, NJ 07960-6989
T: +1 973 2904240; Fax: +1 973 2904226; URL: www.cse.edu
1899; Brother Paul B. Chervenie
Atlases (Phillips Coll), Hist of Chemistry (Florence E Wall Coll), Hist of Women in America (Doris & Yisral Mayer Coll), World War I (Henry C & Ann Fox Wolfe Coll)
110 000 vols; 609 curr per; 38 000 microforms; 570 av-mat; 2 238 sound-rec; 94 digital data carriers
libr loan 31906

Rabbinical College of America, Hoffman Memorial Library, 226 Sussex Ave, ***Morristown***, NJ 07960-3600; P.O. Box 1996, Morristown, NJ 07962-1996
T: +1 973 2679404; Fax: +1 973 2675208; E-mail: rca079@aol.com
Rabbi Israel Gordon
Hebraica, Talmud; Judaica Coll
17 000 vols; 50 curr per 31907

Morrisville State College, The Donald G. Butcher Library, Eaton Street, P.O. Box 902, ***Morrisville***, NY 13408-0902
T: +1 315 6846055; Fax: +1 315 6846115; E-mail: reference@morrisville.edu; URL: library.morrisville.edu
1908; Marion Hildebrand
New York State Hist Coll, Food Service Hist Coll
110 000 vols; 350 curr per; 98 maps; 12 973 microforms; 1 770 av-mat; 2 452 sound-rec; 15 digital data carriers; 160 slides, 1 990 art reproductions
libr loan 31908

Clayton State College & State University, Library, 2000 Clayton State Blvd, ***Morrow***, GA 30260; P.O. Box 285, Morrow, GA 30260
T: +1 678 4664331; Fax: +1 678 4664349; E-mail: reference@clayton.edu; URL: adminservices.clayton.edu/library
1969; Dr. Gordon N. Baker
Civil War (War of the Rebellion); Georgia (Southern Hist)
100 000 vols; 750 curr per 31909

University of Idaho, Library, Rayburn St, ***Moscow***, ID 83844; P.O. Box 442350, Moscow, ID 83844-2350
T: +1 208 8856843; Fax: +1 208 8856817; URL: www.lib.uidaho.edu
1889; Lynn Baird
Idaho and pacific Northwest, Idaho State Publs, Sir Walter Scott Coll, Ezra Pound Coll, Imprints (Caxton Press, Idaho); Western Americana
1 280 000 vols; 6 179 curr per; 35 000 e-journals; 42 000 e-books; 1 481 390 govt docs; 208 613 maps; 108 000 microforms; 2 100 sound-rec; 2 110 digital data carriers
libr loan; ALA, PNLA, ILA 31910

– College of Law, Library, 711 Rayburn St, Moscow, ID 83844; 875 Perimeter Dr, P.O. Box 442324, Moscow, ID 83844-2324
T: +1 208 8852159; Fax: +1 208 8852743; E-mail: lawlib@uidaho.edu; URL: www.law.uidaho.edu/library
Dr. John Hasko
Natural resources
248 000 vols; 4 529 curr per 31911

Berry College, Memorial Library, 2277 Martha Berry Hwy, ***Mount Berry***, GA 30149
T: +1 706 2334056; Fax: +1 706 2387937; E-mail: library@berry.edu; URL: www.berry.edu/library

1926; Rebecca N. Roberts
Educational Resources Info Ctr, College Arch
177 000 vols; 2 100 curr per; 22 900 e-journals; 15 000 e-books; 200 000 microforms
libr loan   31912

Central Michigan University, Charles V. Park Library, Park 407, *Mount Pleasant*, MI 48859; mail address: 300 E Preston, Mount Pleasant, MI 48859
T: +1 989 7743470; Fax: +1 989 7742179; URL: www.lib.cmich.edu
1892; Thomas J. Moore
1 004 410 vols; 3 670 curr per; 5 070 e-journals; 7 200 e-books; 6 540 av-mat; 22 130 sound-rec; 850 digital data carriers
libr loan   31913

– Clarke Historical Library, Mount Pleasant, MI 48859
T: +1 989 7743352; Fax: +1 989 7742160; E-mail: clarke@cmich.edu; URL: clarke.cmich.edu
1954; Frank Boles
Lucile Clarke Memorial Children's Libr; Wilbert Wright Coll Afro-Americana; Reed T Draper Angling Coll; Presidential Campaign Biography Coll; Univ Arch
60 000 vols; 225 curr per; 3 564 music scores; 1 440 maps; 8 072 microforms; 50 sound-rec; 3 274 mss, 1 424 serial titles, 1 100 broadsides, 26 400 photos, 12 000 pieces of ephemera   31914

Iowa Wesleyan College, J. Raymond Chadwick Library, 107 W Broad St, *Mount Pleasant*, IA 52641
T: +1 319 3856318; Fax: +1 319 3856324; E-mail: reference@iwc.edu; URL: chadwick.iwc.edu
1857; Paula Kinney
German-Americanism, Iowa Conference of the United Methodist Church Arch, Iowa Hist
110 000 vols; 441 curr per   31915

Cornell College, Russell D. Cole Library, 620 Third St SW, *Mount Vernon*, IA 52314-1012
T: +1 319 8954260; Fax: +1 319 8955936; E-mail: library@cornellcollege.edu; URL: www.cornellcollege.edu
1853; Laurel Whisler
214 247 vols; 464 curr per; 1 500 e-books; 75 514 microforms; 7 871 av-mat; 1 871 Audio Bks
libr loan   31916

Mount Vernon Nazarene University, Thorne Library/Learning Resource Center, 800 Martinsburg Rd, *Mount Vernon*, OH 43050-9500
T: +1 740 3979000; Fax: +1 740 3978847; URL: library.mvnu.edu
1968; Edythe Feazel
Church of the Nazarene, Doctrine, Hist & Missions
116 000 vols; 541 curr per; 6 900 e-journals; 36 000 e-books; 11 maps; 11 000 microforms; 1 590 sound-rec; 179 digital data carriers; 100 Electronic Media & Resources
libr loan; ACL, LOEX   31917

Ball State University, Alexander M. Bracken Library, 2000 W University Ave, *Muncie*, IN 47306-1099
T: +1 765 2855143; Fax: +1 765 2852644; E-mail: library@bsu.edu; URL: www.bsu.edu/library
1918; Dr. Arthur W. Hafner
Steinbeck Coll, Huxley Coll, Klu Klux Klan Coll, World War I Posters, R. Roller Coll on Glass Manufacturing, Nazi Coll, Contemporary Poetry Coll
1 202 000 vols; 3 408 curr per; 10 300 e-journals; 2 000 e-books
libr loan   31918

– Architecture Library, College of Architecture & Planning Bldg, Rm 116, Muncie, IN 47306
T: +1 765 2855858; Fax: +1 765 2852644; E-mail: wmeyer@gw.bsu.edu; URL: www.bsu.edu/library/thelibraries/units/architecture
1966; Wayne Meyer
38 816 vols; 100 curr per; 550 maps; 6 916 microforms; 111 569 35mm slides
libr loan   31919

– Science-Health Science Library, Cooper Science Bldg, CN16, Muncie, IN 47306
T: +1 765 2855143; Fax: +1 765 2852644; URL: www.bsu.edu/library/collections/shsl
Kevin E. Brooks
63 000 vols; 281 curr per   31920

Chowan University, Whitaker Library, One University Pl, *Murfreesboro*, NC 27855
T: +1 252 3986212; Fax: +1 252 3981301; E-mail: library@chowan.edu; URL: www.chowan.edu/lib
1848; Georgia E. Williams
Baptist Coll
120 000 vols; 900 curr per; 1 419 music scores; 32 000 microforms; 616 av-mat; 3 251 sound-rec; 34 digital data carriers
libr loan; ALA, ACRL   31921

Middle Tennessee State University, Walker Library, MTSU, PO Box 13, *Murfreesboro*, TN 37132
T: +1 615 8982650; Fax: +1 615 9048225; URL: library.mtsu.edu
1911; J. Donald Craig
Tennesseana
749 000 vols; 4 144 curr per; 1 162 000 microforms; 225 digital data carriers
libr loan   31922

Murray State University, Harry Lee Waterfield Library, 205 Waterfield Library, Dean's Office, *Murray*, KY 42071-3307
T: +1 270 8092291; Fax: +1 270 8093736; URL: www.murraystate.edu/msml/msml.htm
1923; Adam Murray
Jesse Stuart Coll, Irvin S Cobb Mss Coll, Forrest C Pogue War & Diplomacy Coll, Regional Politics & Govt, Regional Hist & Culture, Jack London First Editions, NASA Educator Resource Ctr
408 000 vols; 1 418 curr per; 599 e-books; 228 424 govt docs; 21 216 maps; 731 000 microforms; 6 090 av-mat; 2 290 sound-rec; 330 digital data carriers
libr loan   31923

Baker College of Muskegon, Library, 1903 Marquette Ave, *Muskegon*, MI 49442-3404
T: +1 231 7775330; Fax: +1 231 7775334; E-mail: library-mu@baker.edu; URL: www.baker.edu
1885; Gail Powers-Schaub
Business & management, Economics, Accounting, Law
36 000 vols; 150 curr per   31924

Stephen F. Austin State University, Ralph W. Steen Library, PO Box 13055, SFA Sta, *Nacogdoches*, TX 75962-3055
T: +1 936 4684100; Fax: +1 936 4687610; URL: libweb.sfasu.edu
1923; Al Cage
Forestry; East Texas Hist, Business Docs & Papers of Major East Texas Lumber Companies
550 000 vols; 2 458 curr per
libr loan   31925

Northwest Nazarene University, John E. Riley Library, 623 Holly St, *Nampa*, ID 83686; mail address: 623 Holly St, Nampa, ID 83686
T: +1 208 4678606; Fax: +1 208 4678610; E-mail: library@nnu.edu; URL: www.nnu.edu/library
1913; Dr. Sharon I. Bull
John Wesley Coll
175 000 vols; 855 curr per; 2 723 sound-rec; 227 digital data carriers
libr loan   31926

North Central College, Oesterle Library, 320 E School St, *Naperville*, IL 60540
T: +1 630 6375700; Fax: +1 630 6375716; URL: library.noctrl.edu
1861; Carolyn A. Sheehy
Leffler Lincoln Coll, Sang Limited Edition Coll, Sang Jazz Coll, Harris W Fawell Congressional Papers, Tholin Chicagoana Coll
146 000 vols; 3 427 curr per; 1 813 e-journals; 743 diss/theses; 187 000 microforms; 793 sound-rec; 10 digital data carriers
libr loan   31927

International College, Library, 2655 Northbrooke Dr, *Naples*, FL 34119
T: +1 239 9387812; Fax: +1 239 9387886
1990; Jan Edwards
Rare Bks Coll
37 000 vols; 225 curr per; 1 000 e-books   31928

University of Rhode Island – Graduate School of Oceanography, Pell Marine Science Library, Narragansett Bay Campus, South Ferry Rd, *Narragansett*, RI 02882-1197
T: +1 401 8746161; Fax: +1 401 8746101; E-mail: pellib@gso.uri.edu; URL: www.gso.uri.edu/pell/pell.html
1959; Roberta E. Doran
Marine & Polar Expeditionary Rpts, Narragansett Bay, Barge North Cape Oil Spill
70 000 vols; 228 curr per; 2 000 microforms; 89 digital data carriers; 1 656 charts, 16 000 reprints   31929

Daniel Webster College, Anne Bridge Baddour Library, 20 University Dr, *Nashua*, NH 03063-1300
T: +1 603 5776541; Fax: +1 603 5776199; E-mail: librarian@dwc.edu; URL: www.dwc.edu/library
1965; N.N.
Advisory Circulars; Federal Aviation Regulations (FARS)
35 000 vols; 175 curr per; 1 000 e-books; 500 microforms; 780 av-mat; 4 digital data carriers
libr loan   31930

Rivier College, Regina Library, 420 S Main St, *Nashua*, NH 03060-5086
T: +1 603 8978256; Fax: +1 603 8978889; E-mail: libmail@rivier.edu; URL: www.rivier.edu
1933; Daniel Speidel
Franco-American Literary Criticism (Rocheleau-Rouleau Coll), Patristics (Gilbert Coll)
100 000 vols; 450 curr per; 1 500 e-journals; 4 000 e-books; 6 sound-rec; 21 digital data carriers; 129 992 nonprint items   31931

Belmont University, Lila D. Bunch Library, 1900 Belmont Blvd, *Nashville*, TN 37212-3757
T: +1 615 4606033; Fax: +1 615 4605482; URL: library.belmont.edu
1951; Dr. Ernest William Heard
243 000 vols; 1 106 curr per
libr loan   31932

Fisk University, Hohn Hope & Aurelia E. Franklin Library, Fisk University, 1000 17th Ave N, *Nashville*, TN 37208-3051
T: +1 615 3298640; Fax: +1 615 3298761; URL: www.fisk.edu
1866; Dr. Jessie Carney-Smith
Black Lit, Civil Rights & Politics, Music & Musical Lit, Sociology
207 000 vols; 2 566 curr per   31933

Lipscomb University, Beaman Library, One University Park Dr, *Nashville*, TN 37204-3951
T: +1 615 9666037; Fax: +1 615 9661807; URL: library.lipscomb.edu
1891; Carolyn T. Wilson
C E W Dorris Coll, Hymnology Coll (approx 1200 vols), Herald of Truth videotapes
219 000 vols; 919 curr per; 18 000 e-books; 405 000 microforms; 18 000 other items
libr loan; ALA, SLA, TLA, ATLA   31934

Meharry Medical College Library, Kresge Learning Resource Center, 1005 Dr D B Todd Jr Blvd, *Nashville*, TN 37208
T: +1 615 3276728; Fax: +1 615 3276448; URL: www.mmc.edu/library
1940; Fatima Barnes
Black Medical Hist, Meharry Arch
68 000 vols; 2 386 curr per; 1 860 e-journals; 348 e-books; 515 av-mat; 2 sound-rec; 10 digital data carriers; 800 slide/tape sets
libr loan   31935

Scarritt-Bennett Center, Virginia Davis Laskey Research Library, 1008 19th Ave S, *Nashville*, TN 37212-2166
T: +1 615 3407479; Fax: +1 615 3407551; URL: www.scarrittbennett.org/about/library.aspx

1892; Stephen Gateley
Christian education, Church music; Bibles, Church & Community, United Methodist Church
49 000 vols; 77 curr per; 311 diss/theses; 46 maps; 500 av-mat; 509 sound-rec; 9 digital data carriers; 75 curriculum, 32 art repros
ATLA   31936

Tennessee State University, Brown-Daniel Library, 3500 John A Merritt Blvd, *Nashville*, TN 37209-1561
T: +1 615 9635211; Fax: +1 615 9635216; URL: www.tnstate.edu/library
1912; Dr. Yildiz Barlas Binkley
Black Hist, Jazz Recordings, Tennessee Hist, Tennessee State Univ Hist
542 000 vols; 1 706 curr per; 370 e-journals; 300 000 e-books; 2 040 av-mat; 2 000 Talking Bks
libr loan   31937

Trevecca Nazarene University, Waggoner Library, 73 Lester Ave, *Nashville*, TN 37210-4227; mail address: 333 Murfreesboro Rd, Nashville, TN 37210-2834
T: +1 615 2481214; Fax: +1 615 2481471; URL: library.trevecca.edu
1901; Ruth T. Kinnersley
98 000 vols; 627 curr per; 21 000 e-books; 50 bks on Deafness & Sign Lang; 680 Music Scores
libr loan   31938

University of Tennessee College, Social Work Library at Nashville, 193-E Polk Ave, Ste 292, *Nashville*, TN 37210
T: +1 615 2511774; fax: +1 615 7421085; URL: www.lib.utk.edu/~swn
Elsie Pettit
19 000 vols; 60 curr per   31939

Vanderbilt University, Jean and Alexander Heard Library, 419 21st Ave S, *Nashville*, TN 37240-0007
T: +1 615 3227100; Fax: +1 615 3438279; URL: www.library.vanderbilt.edu
1873; Flo Wilson
3 312 000 vols; 44 199 curr per; 103 852 maps; 2 774 000 microforms; 3 640 sound-rec; 150 digital data carriers   31940

– Alyne Queener Massey Law Library, 131 21st Ave S, Nashville, TN 37203
T: +1 615 3222568;
Fax: +1 615 3431265; URL: www.law.vanderbilt.edu/library
1874; Mary C. Miles
James Cullen Looney Medico-Legal Coll
505 000 vols; 6 846 curr per; 636 364 microforms; 15 digital data carriers   31941

– Anne Potter Wilson Music Library, Blair School of Music, 2400 Blakemore Ave, Nashville, TN 37212
T: +1 615 3227696; Fax: +1 615 3430050; URL: www.library.vanderbilt.edu/music
1945; Holling Smith-Borne
Seminar in Piano Teaching
45 000 vols; 170 curr per; 2 000 microforms; 18 694 sound-rec; 12 digital data carriers; 573 videotapes; 248 videodiscs
libr loan   31942

– Annette and Irwin Eskind Biomedical Library, 2209 Garland Ave, Nashville, TN 37232-8340
T: +1 615 9361405; Fax: +1 615 9361407; URL: www.mc.vanderbilt.edu/biolib
1906; Nunzia B. Giuse
Hist of Medicine Coll; Nutrition Hist Colls; Moll Hypnosis Coll
202 000 vols; 4 325 curr per; 3 240 e-journals; 790 e-books; 9 000 microforms; 26 digital data carriers; 8 391 gov publs, 2 974 linear feet of arch mat, 805 computer diskettes   31943

– Central Library, 412 General Library Bldg, 419 21st Ave S, Nashville, TN 37240-0007
T: +1 615 3222407; Fax: +1 615 3437451; URL: www.library.vanderbilt.edu/central
N.N.
Humanities, art
1 375 000 vols; 3 200 curr per   31944

– Divinity Library, 419 21st Ave S, Nashville, TN 37203-2427
T: +1 615 3222865; Fax: +1 615 3432918; URL: divinity.library.vanderbilt.edu
1894; William Hook
Judaica Coll, Kelly Miller Smith Coll
200 000 vols; 827 curr per; 7 digital data carriers 31945

– The Peabody Library, 1210 21st Ave S, Nashville, TN 37203; Peabody No 0135, 230 Appleton Pl, Nashville, TN 37203-5721
T: +1 615 3228866; Fax: +1 615 3437923; URL: www.library.vanderbilt.edu/peabody
1919; Celia S. Walker
Child Study (Peabody Coll of Bks on Children), Curriculum Laboratory, Juvenile Lit Coll
223 000 vols; 372 curr per; 592 000 microforms; 62 digital data carriers 31946

– Science and Engineering Library, 3200 Stevenson Ctr, 419 21st Ave S, Nashville, TN 37240-0007
T: +1 615 3442408; Fax: +1 615 3437249; E-mail: sciren@vanderbilt.edu; URL: www.library.vanderbilt.edu/science
Tracy Primich
349 000 vols; 2 005 curr per 31947

– Special Collections, 419 21st Ave S, Nashville, TN 37240-0007
T: +1 615 3222807; Fax: +1 615 3439832; E-mail: archives@vanderbilt.edu; URL: www.library.vanderbilt.edu/speccol
1965; Juanita Murray
American Lit & Criticism, 1920 to Present (Jesse E. Wills Fugitive-Agrarian Coll); Sevier & Rand Coll; Theatre, Music & Dance (Francis Robinson Coll); 20th c Film (Delbert Mann Coll)
43 000 vols; 2 500 cubic feet of mss, 3 500 cubic feet of arch 31948

– Walker Management Library, Owen Graduate School of Management, 401 21st Ave S, Nashville, TN 37203
T: +1 615 3223635; Fax: +1 615 3430061; URL: www.library.vanderbilt.edu/walker
1970; Hilary A. Craiglow
Career Planning & Placement Resource Coll; Corp Rpts
58 000 vols; 1 102 curr per; 354 000 microforms; 16 digital data carriers 31949

Northwestern State University of Louisiana, Eugene P. Watson Memorial Library, 913 University Pkwy, ***Natchitoches***, LA 71497
T: +1 318 3575465; Fax: +1 318 3574470; URL: www.nsula.edu/watson_library
1884; Fleming A. Thomas
North Louisiana Coll, Carl F. Gauss Coll, Poetry; Cammie Henry Res Ctr
312 000 vols; 1 271 curr per; 23 000 e-books; 1 500 diss/theses; 344 721 govt docs; 115 000 microforms; 679 sound-rec; 12 digital data carriers
libr loan 31950

Cottey College, Blanche Skiff Ross Memorial Library, 1000 W Austin, mail address: 1000 W Austin, ***Nevada***, MO 64772
T: +1 417 6678181 ext 2153; Fax: +1 417 6678103; E-mail: enrollmgt@cottey.edu; URL: www.cottey.edu
1884
Women's Coll
54 530 vols; 205 curr per; 945 music scores; 300 maps; 5 000 microforms; 592 av-mat; 1 572 sound-rec; 8 digital data carriers; Archives (228), Music Coll (7 000), 5 970 slides
libr loan 31951

Indiana University Southeast, Library, 4201 Grant Line Rd, ***New Albany***, IN 47150
T: +1 812 9412262; Fax: +1 812 9412656; E-mail: crosen@ius.edu; URL: www.ius.edu/library
1941; Claude Martin Rosen
Ctr for Cultural Resources, Univ Archs, Ars Femina Musical Scores, William L Simon Sheet Music Coll
250 000 vols; 1 040 curr per
libr loan 31952

Central Connecticut State University, Elihu Burritt Library, 1615 Stanley St, ***New Britain***, CT 06050
T: +1 860 8323408; Fax: +1 860 8322118; URL: library.ccsu.edu
1849; Roy Temple
Connecticut Polish American Arch & Mss Coll, Bookseller's Catalog Coll, Elihu Burrett Coll, Mark Twain Coll, Thomas Hardy Coll, Polish Heritage Coll, Private Press Coll, Gay & Lesbian Coll, Univ Arch
718 000 vols; 2 603 curr per; 17 000 e-journals; 2 000 e-books
libr loan 31953

Rutgers University, University Libraries, 169 College Ave, ***New Brunswick***, NJ 08901-1163
T: +1 732 9327505; Fax: +1 732 9327637; URL: www.libraries.rutgers.edu/
1766; Marianne Gaunt
New Jersey Colls, Rutgers Univ Arch, Women's Arch, Literary Coll, Political Papers Coll, Westerners in Japan Colls, Roebling Family Colls, Social Policy Colls, Edward J Blaustein Dictionnary Coll, Early American Newspapers, Rare Books
3 562 000 vols; 45 533 curr per
libr loan; ARL, RLG 31954

– Archibald Stevens Alexander Library, 169 College Ave, New Brunswick, NJ 08901-1163
T: +1 732 9327509; Fax: +1 732 9321101; URL: www.libraries.rutgers.edu/rul/libs/alex_lib/alex_lib.shtml
Francoise Puniello
1 197 000 vols; 6 325 curr per; 2 000 000 microforms 31955

– Art Library, Voorhees Hall, 71 Hamilton St, New Brunswick, NJ 08901-1248
T: +1 732 9327739; Fax: +1 732 9326743; URL: www.libraries.rutgers.edu/rul/libs/art_lib/art_lib.shtml
Louis E Stern Coll of Modern Art, Bartlett Cowdrey Coll of American Art, Howard Hibbard Coll of Italian Renaissance & Baroque Art
72 000 vols; 89 curr per; 120 VF drawers 31956

– Blanche & Irving Laurie Music Library, Eight Chapel Dr, New Brunswick, NJ 08901-8527
T: +1 732 9329783; Fax: +1 732 9326777; URL: www.libraries.rutgers.edu/rul/libs/music_lib/music_lib.shtml
1982; John Shepard
57 000 vols; 234 curr per; 66 000 music scores; 1 000 microforms; 1 500 av-mat; 250 digital data carriers 31957

– Center of Alcohol Studies Library, Brinkley & Adele Smithers Hall, 607 Allison Rd, Piscataway, NJ 08854-8001
T: +1 732 4454442; Fax: +1 732 4455944; URL: www.libraries.rutgers.edu/rul/libs/alcohol/alcohol.shtml
1940; Patricia Bellanca
Alcohol-drug abuse; Connor Alcohol Res Ref Files (500 survey instruments), McCarthy Memorial Coll of 50,000 alcohol res docs
17 000 vols; 220 curr per; 1 750 diss/theses; 2 000 microforms; 220 av-mat; 150 boxes of archival mat, 20 000 abstracts, 120 bibliographies
libr loan 31958

– Chemistry Library, Wright-Rieman Laboratories, 610 Taylor Rd, Piscataway, NJ 08854-8066
T: +1 732 4452625; Fax: +1 732 4453255; URL: www.libraries.rutgers.edu/rul/libs/chem_lib/chem_lib.shtml
Francoise Puniello
14 000 vols; 153 curr per; 2 000 microforms 31959

– East Asian Library, Alexander Library, 169 College Ave, New Brunswick, NJ 08901-1163
T: +1 732 9327161; Fax: +1 732 9326808; URL: www.libraries.rutgers.edu/rul/libs/east_asia_lib/east_asia_lib.shtml
1970
Microfilm Coll of the Rare Bks in the Nat Central Libr, Taiwan; Prof Ho Kuang-chang's Coll; Prof Ying-shih Yu's Office Coll; Prof Maurice Jansen's Office Coll
126 000 vols; 460 curr per; 4 000 microforms; 5 000 pamphlets
libr loan; RLG, CEAL, AAS 31960

– Kilmer Library, 75 Avenue E, Piscataway, NJ 08854-8040
T: +1 732 4453613; Fax: +1 732 4453472; URL: www.libraries.rutgers.edu/rul/libs/kilmer_lib/kilmer_lib.shtml
1969; Francoise Puniello
Business; Film & Video Coll
144 000 vols; 237 curr per 31961

– Library of Science and Medicine, 165 Bevier Rd, Piscataway, NJ 08854-8009
T: +1 732 4452895; Fax: +1 732 4455703; URL: www.libraries.rutgers.edu/rul/libs/lsm_lib/lsm_lib.shtml
1970; Francoise Puniello
290 000 vols; 2 493 curr per; 18 847 maps; 789 000 microforms; 477 932 gov docs 31962

– Mabel Smith Douglass Library, Eight Chapel Dr, New Brunswick, NJ 08901-8527
T: +1 732 9329407; Fax: +1 732 9326777; URL: www.libraries.rutgers.edu/rul/libs/douglass_lib/douglass_lib.shtml
1918; Francoise Puniello
Agriculture, Environ, Fine arts, Science, Women's studies; Elizabeth Cady Stanton Papers
203 000 vols; 339 curr per 31963

– Mathematical Sciences Library, Hill Center for Mathematical Sciences, 110 Frelinghuysen Rd, Piscataway, NJ 08854-8019
T: +1 732 4453735; Fax: +1 732 4453064; URL: www.libraries.rutgers.edu/rul/libs/math_lib/math_lib.shtml
Mei Ling Lo
45 000 vols; 534 curr per; 5 000 microforms; 3 000 tech rpts 31964

– Physics Library, Serin Physics Laboratory, 136 Frelinghuysen Rd0, Piscataway, NJ 08854-8019
T: +1 732 4452500; Fax: +1 732 4454964; URL: www.libraries.rutgers.edu/rul/libs/physics_lib/physics_lib.shtml
Howard M. Dess
17 000 vols; 146 curr per; 70 636 preprints 31965

– Special Collections and University Archives, 169 College Ave, New Brunswick, NJ 08901-1163
T: +1 732 9327006; Fax: +1 732 9327012; URL: www.libraries.rutgers.edu/rul/libs/scua/scua.shtml
1946; Ronald L. Becker
British & American Lit, Mss Colls, Rare Bk Coll, Hist of labor
153 000 vols; 1 872 curr per 31966

– Stephen & Lucy Chang Science Library, Walter E Foran Hall, 59 Dudley Rd, New Brunswick, NJ 08901-8520
T: +1 732 9320305; Fax: +1 732 9320311; URL: www.libraries.rutgers.edu/rul/libs/chang_lib/chang_lib.shtml
Martin Kesselman
Agriculture, bioengineering, fisheries, food science
13 000 vols; 131 curr per 31967

Wilmington University, Robert C. and Dorothy M. Peoples Library, 320 DuPont Hwy, ***New Castle***, DE 19720
T: +1 302 3289401; Fax: +1 302 3280914; URL: wilmcoll.edu/library
1968; James M. McCloskey
Education, Nursing, Business & management, Aviation
100 000 vols; 500 curr per 31968

Muskingum College, Library, 163 Stormont St, ***New Concord***, OH 43762-1199
T: +1 740 8268152; Fax: +1 740 8268404; E-mail: library@muskingum.edu; URL: www.muskingum.edu/home/library
1837; Dr. Sheila Ellenberger
Arch Coll
205 000 vols; 350 curr per; 25 000 e-books
libr loan 31969

Albertus Magnus College, Library, 700 Prospect St, ***New Haven***, CT 06511
T: +1 203 7738511; Fax: +1 203 7738588; E-mail: refdesk@albertus.edu; URL: www.albertus.edu/library/
1925; Joanne Day
Donald Grant Mitchell Coll, Louis Imogen Guiney Coll, Samuel Bemis (archival mats), Dominicana
104 000 vols; 588 curr per 31970

Southern Connecticut State University, Hilton C. Buley Library, 501 Crescent St, ***New Haven***, CT 06515
T: +1 203 3925750; Fax: +1 203 3925740; URL: www.library.southernct.edu
1895; Ed Harris
Education, Library & info science, Women's studies; African Artifacts; Children's Bks (Caroline Sherwin Bailey Hist Coll); Contemporary Juvenile Coll; Early American Textbks; Connecticut bks, pamphlets, photos, maps, artifacts (Connecticut Room)
600 000 vols; 1 450 curr per
libr loan 31971

Yale University, University Library, 120 High St, ***New Haven***, CT 06520; P.O. Box 208240, New Haven, CT 06520-8240
T: +1 203 4321775; Fax: +1 203 4321294; E-mail: smlref@yale.edu; URL: www.library.yale.edu
1701; Alice Prochaska
12 026 000 vols; 76 022 curr per; 56 375 e-journals; 61 000 av-mat; 109 digital data carriers; 511 866 Rare books, 4,67 Mio Microtext 31972

– American Oriental Society Library, 329 Sterling Memorial Library, New Haven, CT 06520
T: +1 203 4322455; Fax: +1 203 4324087; URL: www.umich.edu/~aos
1942
23 000 vols; 250 curr per 31973

– Art & Architecture Library, 270 Crown St, New Haven, CT 06511-6610; P.O. Box 208318, New Haven, CT 06520-8318
T: +1 203 4322645; Fax: +1 203 4320549; E-mail: art.library@yale.edu; URL: library.yale.edu/art
Christine de Vallet
Bks on Color (Faber Birren Coll)
130 000 vols; 497 curr per 31974

– Astronomy Library, J W Gibbs Lab, Rm 217, 260 Whitney Ave, New Haven, CT 06511-8903; P.O. Box 208101, New Haven, CT 06520-8101
T: +1 203 4323033; Fax: +1 203 4325048; URL: www.astro.yale.edu
1871; Kim Monocchi
Astronomy Slides, Domestic & Foreign Observatory Publs
38 000 vols; 13 digital data carriers 31975

– Babylonian Collection, 130 Wall St, New Haven, CT 06520; P.O. Box 208240, New Haven, CT 06520-8240
T: +1 203 4321837; Fax: +1 203 4327231; E-mail: bab33@pantheon.yale.edu; URL: www.yale.edu/nelc/babylonian.html
1911; Benjamin R. Foster
BB
20 000 vols; 42 curr per; 110 diss/theses; 50 maps; 35 000 cuneiform tablets, 3 000 ancient seals, 2 000 clay and stone artifacts, 4 000 photos of cuneiform tablets
libr loan 31976

– Beinecke Rare Book & Manuscript Library, 121 Wall St, New Haven, CT 06520; P.O. Box 208240, New Haven, CT 06520-8240
T: +1 203 4322972; Fax: +1 203 4324047; E-mail: beinecke.library@yale.edu; URL: www.library.yale.edu/beinecke/
1963; Frank M. Turner
Lit, Children's lit, Alchemy, European hist, Judaica, Theology, Travel; English & American Authors Colls
575 000 vols; 2 250 000 mss 31977

– Classics Library, Phelps Hall, 344 College St, Rm 505, New Haven, CT 06520
T: +1 203 4320854
1892; Carla Lukas
27 000 vols 31978

– Divinity School Library, 409 Prospect St, New Haven, CT 06511-2108
T: +1 203 4326374; Fax: +1 203 4323906; E-mail: divinity.library@yale.edu; URL: www.library.yale.edu/div/
1932; Carolyn Hardin Engelhardt
Hist Libr of Missions
510 000 vols; 1 700 curr per; 175 000 microforms; 12 digital data carriers 31979

– Drama Library, 222 York St, Room 305, P.O. Box 208244, New Haven, CT 06520-8244
T: +1 203 4321554; Fax: +1 203 4321550; E-mail: drama.library@yale.edu; URL: www.library.yale.edu/drama
1925; Pamela C. Jordan
George Pierce Baker Coll, Hist of Costume, Interiors & Furnishings, Rockefeller Prints Coll,; Abel Thomas
30 000 vols; 90 curr per; 150 diss/theses; 100 av-mat; 160 sound-rec; 100 scrapbooks of clippings
libr loan; TLA 31980

– Engineering & Applied Science Library, 15 Prospect St, New Haven, CT 06511; P.O. Box 208284, New Haven, CT 06520-8284
T: +1 203 4322928; Fax: +1 203 4327465; E-mail: engineering.library@yale.edu; URL: www.library.yale.edu/science/library/engineering.html
1969; Andrew Shimp
Yale Univ Tech Rpts
48 000 vols; 6 digital data carriers 31981

– Epidemiology & Public Health Library, Epidemiology & Public Health, 60 College St, New Haven, CT 06520; P.O. Box 208034, New Haven, CT 06520-8034
T: +1 203 7852835; Fax: +1 203 7854998; E-mail: eph.lib@yale.edu; URL: info.med.yale.edu/eph/library
1940; Matthew Wilcox
World Health Organization, Nat Ctr for Health Statistics
25 000 vols; 400 curr per; 6 digital data carriers 31982

– Forestry & Environmental Studies Library, Sage Hall, 205 Prospect St, 4th Flr, New Haven, CT 06511
T: +1 203 4325133; Fax: +1 203 4325942; E-mail: feslcirc@yale.edu; URL: www.library.yale.edu/science/subject/forestry.html
1901; Carla Heister
140 000 vols; 130 curr per; 900 diss/theses; 500 maps; 2 000 microforms; 125 newsletter titles 31983

– Geology Library, Kline Science, 210 Whitney Ave Rm 328, New Haven, CT 06520; P.O. Box 208109, New Haven, CT 06520-8109
T: +1 203 4323157; Fax: +1 203 4323134; E-mail: geocirc@yale.edu; URL: www.library.yale.edu/science/subject/geology.html
1963; David Stern
114 000 vols; 1 395 curr per; 192 756 maps; 42 digital data carriers; 14 000 reprints 31984

– Irving S. Gilmore Music Library, 120 High St, New Haven, CT 06520; mail address: 130 Wall St, Box 208240, New Haven, CT 06520-8240
T: +1 203 4320496; Fax: +1 203 4327339; E-mail: musiclibrary@yale.edu; URL: www.library.yale.edu/musiclib/
1917; Suzanne Eggleston Lovejoy
Horowitz Coll, Hindemith Coll, Kurt Weill Coll, Benny Goodman Coll, German Theoretical Lit 16th-18th C
140 000 vols; 9 000 mss; 5 incunabula; 80 570 music scores; 7 500 maps; 26 000 sound-rec; 3 digital data carriers
libr loan 31985

– Kline Science Library, 219 Prospect St, P.O. Box 208111, New Haven, CT 06520-8111
T: +1 203 4323439; Fax: +1 203 4323441; E-mail: science.reference@yale.edu; URL: www.library.yale.edu/science/library/kline.html
1966; David Stern
Bryology & Lichenology (Evans Coll); Early Science Classics; Various 19th C Expeditions Rpts
346 611 vols; 1 600 curr per; 5 digital data carriers 31986

– Lewis Walpole Library, 154 Main St, Farmington, CT 06032
T: +1 860 6772140; Fax: +1 860 6776369; E-mail: walpole@yale.edu; URL: www.library.yale.edu/walpole
1979; Margaret K. Powell
Horace Walpole Coll, Sir Charles Hanbury Williams Coll, Edward Weston papers, William Hogarth, Thomas Rowlandson Coll, James Gillray Coll
35 000 vols; 32 curr per; 5 diss/theses; 1 digital data carriers; 60 linear feet of mss, 250 microfilm reels, photos & photos, 37 000 aperture cards for the print coll, 40 000 British 18th-c satirical prints, British topographical prints, engraved watercolors, drawings, oil paintings, furnishings
libr loan; American Society for Eighteen-Century Studies, New England Archivists 31987

– Lillian Goldman at Yale Law School Library, 127 Wall St, New Haven, CT 06511; P.O. Box 208215, New Haven, CT 06520-8215
T: +1 203 4321640; Fax: +1 203 4322112; URL: www.yale.edu/law.library
1834; S. Blair Kauffman
American Statute Law (Cole Coll), Blackstone Coll, Int Law, Italian Medieval Statutes, Roman Law (Wheeler Coll)
1 000 000 vols; 10 digital data carriers 31988

– Mathematics Library, Leete Oliver Memorial Hall, 12 Hillhouse Ave, New Haven, CT 06511; P.O. Box 208283, New Haven, CT 06520-8283
T: +1 203 4324179; Fax: +1 203 4327316; URL: www.library.yale.edu/science
1951
21 000 vols; 331 curr per 31989

– Medical Library, Sterling Hall of Medicine, 333 Cedar St, L110 SHM, New Haven, CT 06520; P.O. Box 208014, New Haven, CT 06520-8014
T: +1 203 7855354; Fax: +1 203 7854369; URL: www.med.yale.edu/library
1814; Regina Kenny Marone
Hist of Medicine; Early Ichthyology (George Milton Smith Coll); Med Prints & Drawings (Clements C. Fry Coll); Weights & Measures (Edward Clark Streeter Coll), artifacts
433 000 vols; 2 483 curr per; 325 incunabula; 4 digital data carriers; 50 mss codices before 1600 31990

– Ornithology Library, 21 Sachem St, Environmental Science Ctr, Rm 151, New Haven, CT 06520; Box 208109, New Haven, CT 06520-8109
T: +1 203 4364892; Fax: +1 203 4329816; E-mail: ornithology.library@yale.edu
Jorge de Leon
William R. Coe Coll
80 000 vols
libr loan 31991

– Seeley G. Mudd Library, 38 Mansfield St, New Haven, CT 06511; P.O. Box 208294, New Haven, CT 06520-8294
T: +1 203 4323203; Fax: +1 203 4323214; E-mail: muddcirc@yale.edu
1982; Dana Scott Peterman
1 348 000 vols 31992

– Social Science Libraries & Information Services, 140 Prospect St, New Haven, CT 06511; P.O. Box 208263, New Haven, CT 06520-8263
T: +1 203 4323303; Fax: +1 203 4328979; E-mail: sslref@yale.edu; URL: www.library.yale.edu/socsci
1972; Sandra K. Peterson
Economic Growth Center Coll, Roper Center Arch, Social Science Data Arch
100 000 vols; 15 digital data carriers 31993

– Sterling Chemistry Library, 225 Prospect St, New Haven, CT 06511-8499; P.O. Box 208107, New Haven, CT 06520-8107
T: +1 203 4323960; Fax: +1 203 4323049; URL: www.library.yale.edu/science
1923; David Stern
18 000 vols; 144 curr per; 6 digital data carriers
libr loan 31994

– Sterling Memorial Library, 120 High St, New Haven, CT 06520; P.O. Box 208240, New Haven, CT 06520-8240
T: +1 203 4321818; Fax: +1 203 4329486
1932; Alice Prochaska
American & English Lit & Hist, Judaica, Latin America, Arabic & Sanskrit, Italian Lit & Travel, Journalism, Science & Technology
3 000 000 vols 31995

Colby-Sawyer College, Susan Colgate Cleveland Library & Learning Center, 541 Main St, ***New London***, NH 03257-4648
T: +1 603 5263685; Fax: +1 603 5263777; E-mail: library@colby-sawyer.edu; URL: www.colby-sawyer.edu/information/index.html
1837; Carrie P. Thomas
98 000 vols; 600 curr per
libr loan; OCLC 31996

Connecticut College, Charles E. Shain Library, 270 Mohegan Ave, ***New London***, CT 06320-4196
T: +1 860 4392655; Fax: +1 860 4392871; E-mail: libref@conncoll.edu; URL: www.conncoll.edu/is
1911; W. Lee Hisle
Humanities, Judaica, Chinese language, Art, Natural science, Social science; Children's Lit, Poetry Coll, Printing Hist
585 000 vols; 2 175 curr per
libr loan 31997

– Greer Music Library, 270 Mohegan Ave, Box 5234, New London, CT 06320-4196
T: +1 860 4392711; Fax: +1 860 4392871; URL: www.conncoll.edu/is/info-resources/greer
Carolyn A. Johnson
12 000 vols 31998

US Coast Guard Academy Library, 35 Mohegan Ave, ***New London***, CT 06320-4195
T: +1 860 4448553; Fax: +1 860 4448516
1876; Patricia Daragan
Civil engineering, Electrical engineering, Marine sciences, Technology, Engineering
153 000 vols; 522 curr per; 40 000 microforms; 10 digital data carriers; 90 000 gov docs
libr loan 31999

Dillard University, Will W. Alexander Library, 2601 Gentilly Blvd, ***New Orleans***, LA 70122-3097
T: +1 504 8167486; Fax: +1 504 8164787; E-mail: dulibrary@dillard.edu; URL: books.dillard.edu; www.dillard.edu
1961; Tommy S. Holton
Lit & Architecture (McPherson Memorial Freedom Coll)
106 000 vols; 100 curr per; 88 000 e-books 32000

Louisiana State University Health Sciences Center, John P. Ische Library, 433 Bolivar St, Box B3-1, ***New Orleans***, LA 70112-2223
T: +1 504 5686105; Fax: +1 504 5687720; URL: www.lsuhsc.edu/library/NO
1931; Debbie Sibley
Hist of Medicine in Louisiana, Yellow Fever
255 000 vols; 7 375 curr per; 3 884 av-mat; 19 digital data carriers
libr loan 32001

Loyola University, J. Edgar & Louise S. Monroe Library, 6363 Saint Charles Ave, ***New Orleans***, LA 70118-6195; Campus Box 198, 6363 Saint Charles Ave, New Orleans, LA 70118
T: +1 504 8647144; Fax: +1 504 8647247; E-mail: libref@loyno.edu; URL: library.loyno.edu
1912; Mary Lee Sweat
American hist & lit, English lit, Philosophy, Religion; Hist (Spanish Docs & French Docs); Louisiana Coll; New Orleans Province of Society of Jesus, arch; Univ Arch
343 000 vols; 1 339 curr per; 14 000 e-books
libr loan 32002

– Loyola Law Library, School of Law, 7214 St Charles Ave, New Orleans, LA 70118
T: +1 504 8615545; Fax: +1 504 8615895; URL: law.loyno.edu/library
1914; P. Michael Whipple
US Supreme Court Records & Briefs, GATT
195 000 vols; 3 574 curr per; 600 000 microforms of per 32003

Our Lady of Holy Cross College, Blaine S. Kern Library, 4123 Woodland Dr, ***New Orleans***, LA 70131
T: +1 504 3982103; Fax: +1 504 3912421; E-mail: hfontenot@olhcc.edu; URL: www.olhcc.edu
1916; Sister Helen Fontenot
56 000 vols; 936 curr per; 67 000 e-books; 127 000 microforms; 10 digital data carriers 32004

Southern University in New Orleans, Leonard S. Washington Memorial Library, 6400 Press Dr, ***New Orleans***, LA 70126
T: +1 504 2865224; Fax: +1 504 2845490; URL: www.suno.edu
1959; Dr. Mary Penny
Afro-French Coll
231 600 vols; 512 curr per 32005

Tulane University, Howard-Tilton Memorial Library, 7001 Freret St, ***New Orleans***, LA 70118-5682
T: +1 504 8655131; Fax: +1 504 8656773; URL: library.tulane.edu
1834; Lance Query
Latin America; Southeastern Architectural Arch, Hogan Jazz Arch, Louisiana Hist
2 331 000 vols; 14 141 curr per; 1 280 000 microforms 32006

– Architecture Library, Richardson Memorial Bldg, Rm 202, 6823 St Charles Ave, New Orleans, LA 70118
T: +1 504 8655391; Fax: +1 504 8628966
1948; Alan Velasquez
30 000 vols; 330 curr per 32007

– A. H. Clifford Mathematics Research Library, 431 Gibson Hall, 6823 St Charles Ave, New Orleans, LA 70118
T: +1 504 8623455; Fax: +1 504 8655063; URL: www.math.tulane.edu/Library
Julia Sathler
Edward D. Conway Coll, Frank D. Quigley Coll
35 000 vols; 110 curr per 32008

– Latin American Library, 7001 Freret St, Howard-Tilton Memorial Library, 4th Flr, New Orleans, LA 70118-5549
T: +1 504 8655681; Fax: +1 504 8628970; E-mail: lal@tulane.edu; URL: www.tulane.edu/~latinlib/lalhome.html
1924; Hortensia Calvo
Latin American Photographic Arch; Merle Greene Robertson Rubbings Coll; William E Gates Colls of Mexicana; France V Scholes Coll; Nicolas Leon Coll; Viceregal & Ecclesiastical Mexican Colls
325 000 vols; 475 curr per; 3 400 maps; 775 linear feet of mss, 4 000 pamphlets 32009

– Matas Medical Library, Health Sciences Ctr, 1430 Tulane Ave, SL-86, New Orleans, LA 70112-2632
T: +1 504 9882405; Fax: +1 504 9887417; E-mail: medref@tulane.edu; URL: medlib.tulane.edu
1844; Neville Prendergast
Louisiana Medicine & Medical Biogr
200 000 vols; 25 000 e-books; 3 digital data carriers 32010

– Maxwell Music Library, 7001 Freret St, Howard-Tilton Memorial Library, New Orleans, LA 70118-5682
T: +1 504 8655642; Fax: +1 504 8656773; E-mail: musiclib@pulse.tcs.tulane.edu; URL: www.tulane.edu/~musiclib
1909; Leonard Bertrand
44 424 vols; 166 curr per; 20 932 sound-rec 32011

– Monte M. Lemann Memorial Law Library, 6329 Freret St, New Orleans, LA 70118-5600
T: +1 504 8655952; Fax: +1 504 8655917; URL: library.law.tulane.edu
1847; Trina Robinson
French Civil Law Coll, Maritime Law Coll, Roman Law Coll
275 000 vols; 6 000 curr per; 6 digital data carriers 32012

– Music & Media Library, 7001 Freret St, New Orleans, LA 70118-5682
T: +1 504 8655642; Fax: +1 504 8656773
Leonard Bertrand
22 000 vols; 90 curr per 32013

– Turchin Library, Goldring/Woldenberg Hall I, 3rd Flr, Seven McAlister Dr, New Orleans, LA 70118
T: +1 504 8655376; Fax: +1 504 8628953; URL: www.freeman.tulane.edu/lib-tech/turchin
1926; William Strickland
Finance, Accounting
30 000 vols; 300 curr per   32014

University of New Orleans, Earl K. Long Library, 2000 Lakeshore Dr, *New Orleans*, LA 70148
T: +1 504 2806549; Fax: +1 504 2807277; URL: library.uno.edu
1958; Dr. Sharon Mader
Egyptology (Judge Pierre Crabites Coll), European Community, Louisiana Hist, William Faulkner (Frank A. Von der Haar Coll), Nuclear Regulatory Commission, Orleans Parish School Board, Supreme Court of Louisiana Arch
919 000 vols; 3679 curr per; 31 802 maps; 20 302 sound-rec; 221 digital data carriers; 2 237 879 microforms of per
libr loan   32015

Xavier University of Louisiana, University Library, One Drexel Dr, *New Orleans*, LA 70125-1098
T: +1 504 5207311; Fax: +1 504 5207940; E-mail: library@xula.edu; URL: www.xula.edu/library/index.php
1925; Robert E. Skinner
Black Studies, Southern & Black Catholica; Southern Writers, Western Exploration
255 000 vols; 38 000 e-books; 4 160 sound-rec; 2 300 digital data carriers; 500 Music Scores
libr loan   32016

State University of New York College at New Paltz, Sojourner Truth Library, 300 Hawk Dr, *New Paltz*, NY 12561-2493
T: +1 845 2573719; Fax: +1 845 2573718; URL: lib.newpaltz.edu
1886; Chui-chun Lee
Africa & Asia Colls; New Paltz Coll; College Arch; Early American Imprints; Early English Books
499 000 vols; 999 curr per; 8 000 e-books; 171 digital data carriers
libr loan   32017

Kent State University, Tuscarawas Campus Library, University Dr NE, *New Philadelphia*, OH 44663
T: +1 330 3393391 ext 47456; Fax: +1 330 3397888; E-mail: mkobul@tusc.kent.edu; URL: www.tusc.kent.edu/
1968
Local hist, Nursing; Moravian Coll, Ohio Authors Coll, Olmstead Local Hist Coll
58 000 vols; 250 curr per   32018

College of New Rochelle, Gill Library, 29 Castle Pl, *New Rochelle*, NY 10805-2308
T: +1 914 6545340; Fax: +1 914 6545884; E-mail: gillrefdesk@cnr.edu; URL: cnr.edu/home/library/index.htm
1904; Dr. James T. Schleifer
Early English Text Society, English Lit (Thomas More), James Joyce, Religious Hist (Ursuline Coll)
197 000 vols; 1 100 curr per
libr loan   32019

Iona College, Ryan Library, 715 North Ave, *New Rochelle*, NY 10801-1890
T: +1 914 6332351; Fax: +1 914 6332136; E-mail: library@iona.edu; URL: www.iona.edu/library
1940; Richard L. Palladino
Brother Edmund Rice Coll, Committee on the Art of Teaching Coll, Sean Mc Bride Coll
249 000 vols; 712 curr per; 9 000 e-books   32020

– Helen T. Arrigoni Library-Technology Center, 715 North Ave, New Rochelle, NY 10801-1890
T: +1 914 6372791; Fax: +1 914 6332136
Anissa McEachin
Mass communications, computer science
14 000 vols   32021

Martin Luther College, Library, 1995 Luther Ct, *New Ulm*, MN 56073-3965
T: +1 507 3548221; Fax: +1 507 2339107
1995; David M. Gosdeck
American Civilization Coll, micro, bks
166 000 vols; 308 curr per; 15 300 e-journals; 16 000 e-books; 2 incunabula; 2 021 music scores; 780 maps; 32 000 microforms; 4 197 sound-rec; 419 digital data carriers
libr loan; MINITEX, OCLC   32022

Westminster College, McGill Library, S Market St, *New Wilmington*, PA 16172-0001
T: +1 724 9467330; Fax: +1 724 9466220; URL: www.westminster.edu/library
1852; Molly P. Spinney
Autographed Bks; Bibles in Foreign Languages; James Fenimore Cooper, early eds
247 000 vols; 864 curr per   32023

Audren Cohen College, 431 Canal Street, 12th Fl, *New York*, NY 10013
T: +1 212 3431234; Fax: +1 212 3437398; E-mail: library@metropolitan.edu; URL: www.metropolitan.edu
1966; Lou Acierno
Human services, Management; College Archs
32 000 vols; 400 curr per   32024

Bank Street College of Education, Library, 610 W 112th St, 5th Flr, *New York*, NY 10025
T: +1 212 8754450; Fax: +1 212 8754558; URL: streetcat.bankstreet.edu
1916; Linda Greengrass
Children's lit, Early childhood/elementary education
126 000 vols; 363 curr per; 3 500 diss/theses; 42 maps; 266 000 microforms; 650 av-mat; 1 214 sound-rec; 4 digital data carriers   32025

The Bard Graduate Center for Studies in the Decorative Arts, Design, and Culture, Library, 38 W 86th St, *New York*, NY 10024
T: +1 212 5013035; Fax: +1 212 5013098; E-mail: earnest@bgc.bard.edu; URL: www.bard.edu/academics/libraries/
1992; Greta K. Earnest
Jeremy Cooper Libr
40 000 vols; 200 curr per; 55 diss/theses; 20 microforms; 100 av-mat
libr loan; ALA, ARLIS/NA, NA, CRL, IFLA   32026

Barnard College, Wollman Library, 3009 Broadway, *New York*, NY 10027-6598
T: +1 212 8543946; Fax: +1 212 8543766; E-mail: refdesk@barnard.edu; URL: www.barnard.edu/library
1889; Carol Falcione
American Women Writers (Overbury Coll), Gabriela Mistral Libr
210 000 vols; 419 curr per
libr loan   32027

Baruch College-CUNY, William & Anita Newman Library, 151 E 25 St, Box H-0520, *New York*, NY 10010-2313
T: +1 646 3121610; Fax: +1 646 3121651; URL: newman.baruch.cuny.edu
1968; Dr. Arthur Downing
Business, Economics, Finance, Social studies, Hist
543 000 vols; 4 154 curr per; 2 052 000 microforms; 6 sound-rec; 70 digital data carriers
libr loan   32028

Boricua College, Library & Learning Resources Center, 3755 Broadway, *New York*, NY 10032
T: +1 212 6941000 ext 666 or 667; Fax: +1 212 6941015; E-mail: lrivera@boricuacollege.edu
1974; Liza Rivera
Latin America, Puerto Rico, Maps, Music; Baoillo Papers, Puerto Rican Repository
112 000 vols; 400 curr per   32029

City College of the City University of New York, Morris Raphael Cohen Library, North Academic Ctr, 160 Convent Ave, *New York*, NY 10031
T: +1 212 6507612; Fax: +1 212 6507604; E-mail: ill@ccny.cuny.edu; URL: www1.ccny.cuny.edu/library/
1847; Pamela Gillespie
Humanities, Social sciences, Education;

Costume Coll, English Civil War Pamphlets, 18th & Early 19th C Plays, Astronomy (Newcomb Coll), Socio-Economic Broadsides before 1800 on microfilm (Gitelson Coll)
1 444 000 vols; 2 476 curr per; 29 900 e-journals; 44 000 e-books; 8 200 e-books, 17 700 Music Scores, 17 900 CDs
libr loan   32030

– Dominican Studies Institute Research Library & Archives, NAC 2/204, 160 Convent Ave, New York, NY 10031-0198
T: +1 212 6507170; Fax: +1 212 6507489; E-mail: dsi@ccny.cuny.edu; URL: www1.ccny.cuny.edu/ci/dsi/library.cfm
Sarah Aponte
16 000 vols   32031

City University of New York, Mina Rees Library of Graduate School & University Center, 365 Fifth Ave, *New York*, NY 10016-4309
T: +1 212 8177083; Fax: +1 212 8172982; URL: library.gc.cuny.edu
1964; Julie Cunningham
American Hist (US Presidential Papers), microfilm; City Univ of New York; Old York Libr (Seymour Durst Coll)
285 000 vols; 11 000 curr per; 1 600 music scores; 537 000 microforms; 2 500 sound-rec; 8 digital data carriers; 214 625 slides
libr loan   32032

City University of New York Hunter College Libraries, Jacqueline Grennan Wexler Library, 695 Park Ave, *New York*, NY 10065
T: +1 212 7724192; Fax: +1 212 7724142; URL: library.wexler.hunter.cuny.edu/
1870; Dr. Louise S. Sherby
Early English Novels (Stonehill Coll), Eileen Cowe Hist Textbks, Lenox Hill Neighborhood Coll, Women's City Club of NY Coll
798 000 vols; 962 curr per; 6 000 e-books; 1 600 av-mat
libr loan   32033

– Health Professions Library, Hunter College Brookdale Campus, 425 E 25th St, New York, NY 10010
T: +1 212 4815117; Fax: +1 212 7725116
1909; John Carey
26 000 vols; 330 curr per; 135 000 Envirofiche   32034

– School of Social Work Library, 129 E 79th St, New York, NY 10021
T: +1 212 4527076; Fax: +1 212 4527125; URL: library.hunter.cuny.edu/ssw/index.htm
1969; Philip Swan
Paul Schreiber Coll (Hist of Social Welfare)
47 000 vols; 139 curr per; 139 diss/theses; 3 000 microforms   32035

Columbia University, University Libraries, Butler Library, Rm 517, 535 W 114th St, *New York*, NY 10027
T: +1 212 8542235; Fax: +1 212 8549099; E-mail: lio@columbia.edu; URL: www.columbia.edu/cu/lweb
1761; James G. Neal
7 266 000 vols; 4 995 000 microforms   32036

– African Studies Library, 420 W 118th St, New York, NY 10027
T: +1 212 8548045; Fax: +1 212 8543834; E-mail: africa@libraries.cul.columbia.edu; URL: www.columbia.edu/cu/libraries/indiv/africa
Joseph Caruso
104 600 vols   32037

– Augustus C. Long Health Sciences Library, 701 W 168th St, Lobby Level, New York, NY 10032
T: +1 212 3053605; Fax: +1 212 2340595; E-mail: hs-library@columbia.edu; URL: library.cpmc.columbia.edu/hsl/
1928; Alena Ptak-Danchak
Anatomy (Huntington Coll), Cancer Res, Physiology (Curtis Coll), Plastic Surgery (Jerome P. Webster Coll)
551 000 vols; 4 416 curr per   32038

– Avery Architectural & Fine Arts Library, 1172 Amsterdam Ave, MC 0301, New York, NY 10027
T: +1 212 8548403; Fax: +1 212 8548904; E-mail: avery@libraries.cul.columbia.edu
1890; N.N.
350 000 vols; 2 000 curr per; Architectural Drawings Coll; Archs   32039

– Biological Sciences Library, 601 Fairchild, 1212 Amsterdam Ave, MC 2457, New York, NY 10027
T: +1 212 8544182; Fax: +1 212 8548972
Kathleen Kehoe
60 000 vols   32040

– The Burke Library at Union Theological Seminary, 3041 Broadway, New York, NY 10027
T: +1 212 8515606; Fax: +1 212 8515613; E-mail: refdesk@uts.columbia.edu; URL: www.columbia.edu/cu/lweb/indiv/burke/
1836; N.N.
Bible, Ecumenics, Christian ethics, Church hist, Art; Bonhoeffer Coll, British Hist & Theology (McAlpin Coll), Ecumenics & Church Union (William Adams Brown Coll), Sacred Music Coll, Van Ess Coll, Auburn Coll
604 000 vols; 1 719 curr per; 1 820 music scores; 148 000 microforms; 40 av-mat; 1 770 sound-rec; 13 digital data carriers
libr loan   32041

– Butler Library Reference Department, 301 Butler Library, New York, NY 10027; mail address: 535 W 114th St, New York, NY 10027
T: +1 212 8545477; E-mail: reference@columbia.edu
Jeff Barton
58 000 vols   32042

– Chemistry Library, 454 Chandler, 3010 Broadway, MC 3177, New York, NY 10027
T: +1 212 8545778; Fax: +1 212 8545804
1900; Song Yu
55 000 vols   32043

– Diamond Law Library, 435 W 116th St, New York, NY 10027
T: +1 212 8543743; Fax: +1 212 8543295; URL: www.library.law.columbia.edu
1910; Kent McKeever
League of Nations & UN docs, John Jay & James Kent Coll on US & English Legal Hist, Arch of Telford Taylor
825 000 vols; 7 350 curr per; 618 000 microforms; 47 av-mat; 11 digital data carriers; 1 500 linear feet of mss & archival mat   32044

– Engineering Library, 422 SW Mudd, 500 W 120th St, MC 4707, New York, NY 10027
T: +1 212 8543206; Fax: +1 212 8543323
1883; Ujwal Ranadive
Tech Rpts
50 000 vols; 8 000 microforms   32045

– Gabe M. Wiener Music and Arts Library, 701 Dodge Hall, 2960 Broadway, New York, NY 10027
T: +1 212 8544711; Fax: +1 212 8544748; URL: www.columbia.edu/cu/lweb/indiv/music
Elizabeth Davis
80 000 vols; 50 000 sound-rec   32046

– Geology Library, 601 Schermerhorn, 1190 Amsterdam Ave, New York, NY 10027
T: +1 212 8544713; Fax: +1 212 8544716; E-mail: geology@libraries.cul.columbia.edu
1912; Amanda Bielskas
Geological Survey Coll
96 000 vols   32047

– Lamont-Doherty Geoscience Library, Lamont-Doherty Earth Observatory, 61 Rte 9 W, PO Box 1000, Palisades, NY 10964-8000
T: +1 845 3658808; Fax: +1 845 3658151; E-mail: geology@libraries.cul.columbia.edu; URL: www.columbia.edu/cu/lweb/indiv/geosci/index.html
1960; Amanda Bielskas
30 000 vols; 500 curr per   32048

– Lehman Social Science Library, 300 International Affairs Bldg, 420 West 118th St, New York, NY 10027
T: +1 212 8544170; Fax: +1 212 8542495; E-mail: winland@columbia.edu; URL: www.columbia.edu/cu/lweb/indiv/lehman
1971; Jane Winland
Documents Service Center, Electronic Data Service, Map Coll
1 010 684 vols
libr loan   32049

– Mathematics-Science Library, 303 Mathematics, 2990 Broadway, MC 4702, New York, NY 10027
T: +1 212 8544712; Fax: +1 212 8548849; E-mail: mathsci@libraries.cul.columbia.edu
1934; Vivian Sukenik
126 000 vols   32050

– Middle East and Jewish Studies Library, 303 International Affairs Bldg, 420 W 118th St, New York, NY 10027; mail address: 303 Lehmann Libr, MC 3301, New York, NY 10027
T: +1 212 8543995; Fax: +1 212 8543834; E-mail: mideast@libraries.cul.columbia.edu; URL: www.columbia.edu/cu/libraries/indiv/mideast
Dr. Hossein Kamaly
500 000 vols; 400 curr per; 300 e-journals; 1 000 e-books; 300 av-mat; 300 sound-rec; 100 digital data carriers   32051

– Philip L. Milstein Family College Library, 208 Butler Library, 535 W 114th St, New York, NY 10027
T: +1 212 8545327; Fax: +1 212 8540089; E-mail: undergrad@libraries.cul.columbia.edu; URL: www.columbia.edu/cu/lweb/indiv/under
1998; Anice Mills
92 000 vols   32052

– Physics-Astronomy Library, 810 Pupin, 8th Flr, 550 W 120th St, MC 5251, New York, NY 10027
T: +1 212 8543943; Fax: +1 212 8541364; E-mail: physics@libraries.cul.columbia.edu
1898; Kathleen Kehoe
44 000 vols   32053

– Psychology Library, 409 Schermerhorn, 1190 Amsterdam Ave, MC 5503, New York, NY 10027
T: +1 212 8544714; Fax: +1 212 8545660; E-mail: psychology@libraries.cul.columbia.edu
1912; Vivian Sukenik
37 000 vols; 3 000 microforms   32054

– Rare Book & Manuscript Library, Butler Libr, 6th Fl E, 535 W 114th St, New York, NY 10027
T: +1 212 8542231; Fax: +1 212 8541365; E-mail: rbml@libraries.cul.columbia.edu; URL: www.columbia.edu/cu/lweb/
1930; Michael Ryan
Bakhmeteff Arch of Russian & East European Hist & Culture, Carnegie Corp Arch, Herbert H Lehman Suite & Papers
419 294 vols; 31 000 000 mss; 1 200 incunabula
libr loan   32055

– Russian, Eurasian & European Studies Library, 306 Int Affairs Bldg, 420 W 118th St, New York, NY 10027
T: +1 212 8544701; Fax: +1 212 8543834; E-mail: slavic@libraries.cul.columbia.edu; URL: www.columbia.edu/cu/lweb/indiv/slavic
Jared Ingersoll
945 800 vols   32056

– Social Work Library, Columbia Univ School of Social Work, 1255 Amsterdam Ave, 2nd Flr, New York, NY 10027
T: +1 212 8512194; Fax: +1 212 8512199; E-mail: socwk@libraries.cul.columbia.edu; URL: www.columbia.edu/cu/lweb/indiv/socwk
1898; Alysse Jordan
Agency Coll, Masters Theses
146 921 vols; 646 curr per
libr loan   32057

– South & Southeast Asia Library, 304 Int Affairs Bldg, 420 W 118th St, New York, NY 10027
T: +1 212 8548046; Fax: +1 212 8543834; URL: www.columbia.edu/cu/libraries/indiv/southasia
David Magier
293 500 vols   32058

– C. V. Starr East Asian Library, 300 Kent Hall, MC 3901, 1140 Amsterdam Ave, New York, NY 10027
T: +1 212 8542578; Fax: +1 212 6626286; E-mail: starr@libraries.cul.columbia.edu; URL: www.columbia.edu/lweb/indiv/eastasian
1902; Amy Heinrich
Chinese Local Hist & Genealogies, Japanese Woodblock-printed Bks, Korean Coll of Yi Song-ui
735 000 vols; 4 500 curr per; 31 000 microforms; 297 av-mat   32059

– Watson Library of Business & Economics, 130 Uris Hall, 3022 Broadway, MC 9163, New York, NY 10027
T: +1 212 8547804; Fax: +1 212 8545723
1920; Jill Parchuck
375 000 vols; 746 000 microforms   32060

Columbia University – Teachers College, The Gottesman Libraries of Teachers College, 525 W 120th St, **New York**, NY 10027-6696
T: +1 212 6783023; Fax: +1 212 6783092; URL: www.library.tc.columbia.edu
1887; Dr. Gary J. Natriello
Children's Books (18th & 19th c), Education (Rare Bks of 15th-19th c), Black Hist Series (Educational Coll) films, Learning Technology Services, Hist of Nursing Coll
430 000 vols; 1 425 curr per; 17 000 e-journals; 305 e-books; 545 000 microforms; 3 180 av-mat; 990 sound-rec; 4 770 digital data carriers   32061

Fashion Institute of Technology – SUNY, Gladys Marcus Library, Seventh Ave at 27th St, **New York**, NY 10001-5992
T: +1 212 2174364; Fax: +1 212 2174361; URL: www.w3.fitnyc.edu/library
1944; NJ Wolfe
Fashion Sketches, Fashion Illust Art, Oral Hist, FIT Arch, Hist Mat
147 000 vols; 500 curr per; 1 000 e-journals; 25 mss; 272 diss/theses; 1 007 govt docs; 5 000 microforms; 186 sound-rec; 2 695 digital data carriers; 179 850 picture files
libr loan; OCLC, ARLIS   32062

Fordham University at Lincoln Center, Quinn Library, Leon Lowenstein Bldg, 113 W 60th St, **New York**, NY 10023-7480
T: +1 212 6366085; Fax: +1 212 6366766; URL: www.library.fordham.edu
1969; Dr. James McCabe
Business, Education, Social work; Holocaust Coll, Education (ERIC Docs), microfiche
450 000 vols; 1 500 curr per; 60 digital data carriers; 450 000 microforms of per   32063

Fordham University School of Law, Leo T. Kissam Memorial Library, 140 W 62nd St, **New York**, NY 10023-7485
T: +1 212 6366900; Fax: +1 212 6367357; E-mail: refdesk@law.fordham.edu; URL: law.fordham.edu/library.htm
1905; Robert J. Nissenbaum
European Union Law
613 000 vols; 7 693 curr per; 787 000 microforms; 20 digital data carriers
libr loan   32064

Hebrew Union College – Jewish Institute of Religion (NY), Klau Library, Brookdale Center, One West 4th Street, **New York**, NY 10012-1186
T: +1 212 6745300; Fax: +1 212 3881720; URL: www.huc.edu/libraries/NY/
1922; Dr. Philip E. Miller
145 000 vols; 195 curr per; 13 e-journals; 4 incunabula; 15 digital data carriers
libr loan   32065

John Jay College of Criminal Justice, Lloyd George Sealy Library, 899 Tenth Ave, **New York**, NY 10019
T: +1 212 2378257; Fax: +1 212 2378221; E-mail: libinfo@jjay.cuny.edu; URL: www.lib.jjay.cuny.edu
1965; Dr. Larry E. Sullivan
Flora R Schreiber Papers; NYC Police Dept Blotters, Manhattan 1920-1933; New York Criminal Court Transcripts & Records 1890-1920; Police Department Annual Reports, Sing Sing Prison, Papers of the Warden (Lewis E Lawes Coll); Prison Reform (John Howard Coll)
248 000 vols; 14 374 curr per; 42 000 e-books; 14 digital data carriers   32066

Juilliard School, Lila Acheson Wallace Library, 60 Lincoln Center Plaza, **New York**, NY 10023-6588
T: +1 212 7995000 ext 265; Fax: +1 212 7696421; E-mail: library@juilliard.edu; URL: www.juilliard.edu/libraryarchives/libraryarchives.html
1905; Jane Gottlieb
Dance, Drama, Music; First & Early Editions of Liszt Piano Works, Opera Piano-Vocal Scores, Oboe Coll, Opera Librettos of 19th C, Soulima and Igor Stravinsky Coll
86 140 vols; 220 curr per; 250 mss; 60 000 music scores; 783 av-mat; 33 691 sound-rec
libr loan   32067

Manhattan School of Music, Frances Hall Ballard Library, 120 Claremont Ave, **New York**, NY 10027
T: +1 917 4934511; Fax: +1 212 7495471; E-mail: library@msmnyc.edu; URL: www.msmnyc.edu/libraries/; library.msmnyc.edu
1925; Peter Caleb
115 000 vols; 140 curr per; 1 000 diss/theses; 45 000 music scores; 300 av-mat; 23 158 sound-rec
libr loan; MLA   32068

Mannes College of Music, Harry Scherman Library, 150 W 85th St, **New York**, NY 10024-4499
T: +1 212 5800210; Fax: +1 212 5801738; URL: library.newschool.edu
1954; Ed Scarcelle
Salzedo Harp Coll, Konstantin Ivanow Viola Coll, Sylvia Marlowe Harpsichord Coll
8 000 vols; 74 curr per; 28 014 music scores; 99 av-mat; 8 689 sound-rec; 3 digital data carriers
libr loan; IAML, MLA   32069

Marymount Manhattan College, Thomas J. Shanahan Library, 221 E 71st St, **New York**, NY 10021
T: +1 212 7744808; Fax: +1 212 4528207; URL: www.marymount.mmm.edu
1948; Donna Hurwitz
Communications, Theatre, Women's studies; Geraldine A. Ferraro papers, William Harris Coll
80 000 vols; 650 curr per   32070

Mount Sinai School of Medicine, Gustave L. & Janet W. Levy Library, One Gustave L Levy Pl, **New York**, NY 10029; P.O. Box 1102, New York, NY 10029-6574
T: +1 212 2417795; Fax: +1 212 8312625; URL: www.mssm.edu/library
1968; Lynn Kasner Morgan
Arch; Biomedical Audiovisual & Computer Software
36 000 vols; 24 152 curr per; 6 000 e-books; 2 864 av-mat; 1 377 sound-rec; 36 digital data carriers; 2 348 slides   32071

New York College of Podiatric Medicine, Sidney Druskin Memorial Library, 53-55 E 124th St, **New York**, NY 10035
T: +1 212 4108020; Fax: +1 212 8769426; E-mail: enrollment@nycpm.edu; URL: www.nycpm.edu/library.asp
1911; Thomas Walker
Dermatology, Orthopedics
13 000 vols; 260 curr per
libr loan   32072

New York Institute of Technology, Manhattan Campus, 1855 Broadway, **New York**, NY 10023
T: +1 212 2611526; Fax: +1 212 2611681; URL: www.nyit.edu/library
1958; Elisabete Ferretti
Art, Architecture, Computer science
43 000 vols; 764 curr per   32073

New York Law School Library, 57 Worth St, **New York**, NY 10013
T: +1 212 9658839; URL: www.nyls.edu
1891; Joyce Saltalamachia
435 000 vols; 1 450 curr per; 77 000 microforms; 104 sound-rec; 103 digital data carriers; 78 computer disk titles, 41 VHS video titles
libr loan; ALA, AALL   32074

New York University, Elmer Holmes Bobst Library, 70 Washington Sq S, **New York**, NY 10012-1091
T: +1 212 9982511; Fax: +1 212 9954070; URL: library.nyu.edu
1831; Carol A. Mandel
Alfred C. Berol Lewis Carroll Coll, bks, letters, mss, photos; Robert Frost Libr Coll; Erich Maria Remarque Libr Coll; Rare Judaica and Hebraica; Wiet Coll of Islamic Materials; United Nations Coll
3 657 000 vols; 35 858 curr per
libr loan   32075

– Conservation Center Library, 14 E 78th St, New York, NY 10021-1706
T: +1 212 9925854; Fax: +1 212 9925851; URL: www.nyu.edu/gsas/dept/fineart/cfa/research/libraries.htm
Daniel Biddle
14 000 vols; 202 curr per   32076

– Courant Institute of Mathematical Sciences Library, 251 Mercer St, 12th Flr, New York, NY 10012-1185
T: +1 212 9983315; Fax: +1 212 9954808
1954; Carol Hutchins
Mathematics (Courant Reprints & Bohr Reprints)
66 000 vols; 300 curr per   32077

– Fales Library & Special Collections, Elmer Holmes Bobst Libr, 3rd Flr, 70 Washington Square S, New York, NY 10012
T: +1 212 9982596; Fax: +1 212 9953835; E-mail: fales.library@nyu.edu; URL: www.nyu.edu/library/bobst/research/fales
Marvin Taylor
200 000 vols   32078

– Stephen Chan Library of Fine Arts, Institute of Fine Arts, One E 78th St, New York, NY 10021
T: +1 212 9925825; Fax: +1 212 9925807; E-mail: ifa.library@nyu.edu
1938; Sharon Chickanzeff
200 000 vols; 508 curr per; 690 microforms   32079

– Tamiment Library/Robert F. Wagner Labor Archives, Elmer Holmes Bobst Library, 70 Washington Sq S, 10th Flr, New York, NY 10012; URL: www.nyu.edu/library/bobst/research/tam/
Michael Nash
Labor, Utopianism
61 000 vols   32080

New York University School of Law, Library, 40 Washington Sq S, **New York**, NY 10012-1099
T: +1 212 9986312; Fax: +1 212 9954559; URL: www.law.nyu.edu/library/
1860; Radu D. Popa
1 050 000 vols; 7 021 curr per; 916 000 microforms; 6 digital data carriers   32081

New York University School of Medicine, Frederick L. Ehrman Medical Library, Medical Science Bldg 195, 550 First Ave, **New York**, NY 10016-6450
T: +1 212 2635397; Fax: +1 212 2636534; URL: library.med.nyu.edu
1914; Karen Brewer
Hist of Medicine (Heaton Coll), Environmental Medicine, Arch
202 000 vols; 11 772 curr per; 11 480 e-journals; 9 000 e-books; 116 sound-rec; 248 digital data carriers; 62 Bks on Deafness & Sign Lang   32082

New York University School of Medicine – John & Bertha E. Waldmann Memorial Library, Manhattan VA Medical Center, 423 East 23rd St, 2nd Flr S, **New York**, NY 10010
T: +1 212 9989787; Fax: +1 212

9953529; E-mail: dental.library@nyu.edu; URL: www.nyu.edu/dental/library
Van B. Afes
Dentistry; Hist of Dentistry (Blum, Mestel & Weinburger Coll), rare bks
34 000 vols; 681 curr per; 12 microforms; 234 av-mat; 61 sound-rec; 656 slides, 12 VF drawers of archival mats   32083

Oskar Diethelm Library, 525 E 68th St, Weill Cornell Medical Library, Room F-1212, Baker Pavillon, **New York**, NY 10029
T: +1 212 7463728; Fax: +1 212 4230266; E-mail: der2006@med.cornell.edu; URL: www.cornellpsychiatry.org/history/osk_die_lib/
1936; Paul S. Bunten
Hist of Psychiatry and Behavioral Sciences; Early Medical Doctoral Diss on Psychiatric Topics; Psychiatric Hospital Annual Rpts
34 000 vols; 45 curr per; 500 cubic feet of mss   32084

Pace University, Henry Birnbaum Library, New York Civic Ctr, One Pace Plaza, **New York**, NY 10038-1502
T: +1 212 3461331; Fax: +1 212 3461615; URL: library.pace.edu
1934; Melvin Isaacson
416 000 vols; 750 curr per; 26 000 microforms; 25 digital data carriers; 8 500 pamphlets   32085

Parsons School of Design, New School for Social Research, Adam & Sophie Gimbel Design Library, Two W 13th St, 2nd Flr, **New York**, NY 10011
T: +1 212 2298914; Fax: +1 212 2292806; URL: library.newschool.edu/gimbel/
1896; Amy Schofield
Parsons Arch, Fashion Design (Claire McCardell Coll), Picture Coll (50 000 images), Sketchbks
56 000 vols; 222 curr per; 20 digital data carriers; 85 000 slides
libr loan; ALA, ARLIS/NA, SLA   32086

Rockefeller University, University Library, 1230 York Ave, RU Box 263, **New York**, NY 10065
T: +1 212 3278904; Fax: +1 212 3277840; E-mail: libref@rockefeller.edu; URL: rockefeller.edu/library
1906; Carol Ann Feltes
60 000 vols; 576 curr per
libr loan   32087

School of Visual Arts Library, 380 Second Ave, 2nd Flr, **New York**, NY 10010-3994
T: +1 212 5922660; Fax: +1 212 5922655; URL: www.schoolofvisualarts.edu
1962; Robert Lobe
Alumni Book Coll, Filmscripts, Comics
60 000 vols; 295 curr per; 330 av-mat; 1 250 sound-rec; 3 digital data carriers; 2 000 pamphlets, 152 000 slides, 270 000 pictures

State University of New York – State College of Optometry, Harold Kohn Vision Science Library, 33 W 42nd St, **New York**, NY 10036-8003
T: +1 212 9385693; Fax: +1 212 9385696; URL: www.sunyopt.edu/library/libhome.shtml
1971; Elaine Wells
Optometry, Ophthalmology, Learning disabilities, Optics
38 000 vols; 483 curr per; 1 000 e-books; 543 microforms; 508 av-mat; 1 297 sound-rec; 42 digital data carriers
libr loan   32089

Touro College Libraries, 43 W 23rd St, **New York**, NY 10010
T: +1 212 4630400; Fax: +1 212 6273696; URL: www.touro.edu/library
Dr. Jacqueline A. Maxin
287 000 vols; 550 curr per; 26 510 e-journals; 38 000 e-books; 1 150 av-mat; 1 770 sound-rec; 300 digital data carriers   32090

Yeshiva University, University Libraries, 500 W 185th St, **New York**, NY 10033
T: +1 212 9605363; Fax: +1 212 9600066; E-mail: unilib@yu.edu; URL: library.aecom.yu.edu; yu.edu/libraries; www.cardozo.yu.edu/library/index.asp
1897; Pearl Berger
Medicine, Law, Psychology, Social Work,

Jewish Studies; Rare Bks and Mss; Yeshiva Univ Arch
1 083 000 vols; 175 000 e-books; 1 057 mss; 39 incunabula; 1 415 000 microforms
libr loan; OCLC, RLG, CARLJS, AJL, MLA   32091

– Dr Lillian & Dr Rebecca Chutick Law Library, Benjamin N Cardozo School of Law, 55 Fifth Ave, New York, NY 10003-4301
T: +1 212 7900220; Fax: +1 212 7900236; E-mail: lawlib@yu.edu; URL: www.cardozo.yu.edu/library
1976; Lynn Wishart
Israeli law
531 000 vols; 6 407 curr per
libr loan   32092

– Mendel Gottesman Library of Hebraica-Judaica, 2520 Amsterdam Ave, mail address: 500 W 185th St, New York, NY 10033
T: +1 212 9605379; Fax: +1 212 9600066; URL: www.yu.edu/libraries/page.asp?id=31
1897; Leah Adler
Hebraica Rare Bks & Mss, Sephardic Studies Coll
276 000 vols; 963 curr per; 37 000 e-books; 1 000 mss; 39 incunabula; 500 diss/theses; 95 maps; 12 000 microforms; 2 920 av-mat
libr loan   32093

– Pollack Library/Landowne Bloom Library, 2520 Amsterdam Ave, New York, NY 10033; mail address: Wilf Campus, 500 W 185th St, New York, NY 10033
T: +1 212 9605378; Fax: +1 212 9600066; URL: www.yu.edu/libraries
1938; John Moryl
Social work
308 000 vols; 263 curr per
libr loan   32094

– Stern College for Women, Hedi Steinberg Library, 245 Lexington Ave, New York, NY 10016
T: +1 212 3407785; Fax: +1 212 3407808
1954; Edith Lubetski
Judaica & Hebraica
144 000 vols; 578 curr per
libr loan   32095

Essex County College, Martin Luther King, Jr. Library, 303 University Ave, **Newark**, NJ 07102-1798
T: +1 973 8773238; Fax: +1 973 8771887; E-mail: slaton@essex.edu; URL: www.essex.edu/library
1968; Gwendolyn C. Slaton
94 881 vols; 416 curr per; 31 061 microforms; 3 900 av-mat; 1 549 sound-rec; 250 digital data carriers
libr loan; ALA   32096

New Jersey Institute of Technology, Robert W. Van Houten Library, University Heights, **Newark**, NJ 07102-1982
T: +1 973 5963204; Fax: +1 973 6435601; URL: www.library.njit.edu
1881; Davida Scharf
Electronic theses and diss (www.library.njit.edu/etd); Weston Hist of Science & Technology
140 000 vols; 20 000 curr per; 4 000 diss/theses; 1 100 maps; 5 000 microforms; 2 774 digital data carriers
libr loan   32097

– Barbara & Leonard Littman Architecture Library, 456 Weston Hall, 323 King Blvd, Newark, NJ 07102-1982
T: +1 973 5963083; Fax: +1 973 6435601; URL: www.library.njit.edu/archlib
Maya Gervits
14 000 vols; 100 curr per   32098

Ohio State University at Newark & Central Ohio Technical College, Newark Campus Library, 1179 University Dr, **Newark**, OH 43055-1797
T: +1 740 3669308; Fax: +1 740 3669264; URL: www.newarkcolleges.com; www.newarkcampus.org/library
1957; John D. Crissinger
50 000 vols; 400 curr per; 36 000 e-books   32099

Rutgers, The State University of New Jersey, **Newark**
– Criminal Justice NCCD Collection, 123 Washington St, Ste 350, Newark, NJ 07102
T: +1 973 3533118; Fax: +1 973 3531275; E-mail: pschultz@andromeda.rutgers.edu
1921; Phyllis Schultze
18 600 vols; 275 curr per; 21 720 microforms; 85 av-mat; 3 digital data carriers; 6 738 diss, 49 738 res rpts
libr loan   32100

– Institute of Jazz Studies Library, Dana Library, 185 University Ave, 4th Flr, Newark, NJ 07102
T: +1 973 3535595; Fax: +1 973 3535944; URL: newarkwww.rutgers.edu/ijs; www.libraries.rutgers.edu/rul/libs/jazz/jazz.shtml
1952; Dan Morgenstern
Jazz Arch Coll, Memorabilia, Realia
10 000 vols; 150 curr per; 146 735 sound-rec   32101

– John Cotton Dana Library, 185 University Ave, Newark, NJ 07102
T: +1 973 3535902; Fax: +1 973 3535257; URL: www.libraries.rutgers.edu/rul/librs/dana/dana_lib.shtml
1927; Mark Winston
Humanities, Nursing
346 000 vols; 3 223 curr per   32102

– Library for the Center for Law & Justice, 123 Washington St, Newark, NJ 07102-3094
T: +1 973 3533121; Fax: +1 973 3531356; URL: law-library.rutgers.edu
1946; Carol A. Roehrenbeck
Law Libr of US Supreme Court Justice Bradley (Bradley Coll)
372 000 vols; 618 000 microforms; 1 468 sound-rec; 35 digital data carriers   32103

Seton Hall University, Peter W. Rodino Jr Law Library, One Newark Ctr, **Newark**, NJ 07102
T: +1 973 6428861; Fax: +1 973 6428748; URL: law.shu.edu
1950; Charles A. Sullivan
460 000 vols; 3 367 curr per   32104

University of Delaware, University Library, 181 S College Ave, **Newark**, DE 19717-5267
T: +1 302 8312236; Fax: +1 302 8311046; URL: www.lib.udel.edu
1834; Susan Brynteson
Hist of Papermaking, Irish Lit, 20th c American Lit, Hist of Chemistry, Hist of Horticulture
2 705 000 vols; 12 532 curr per; 430 550 govt docs; 124 956 maps; 3 252 000 microforms; 1 703 sound-rec; 1 197 slides sets, kits and floppy disks, 3 591 linear feet of mss and arch
libr loan; ALA, ARL, CRL, CIRLA, CNI, CLIR, IFLA, SPARC   32105

– College of Education Resource Center Library, 012 Willard Hall Education Bldg, Newark, DE 19716-2940
T: +1 302 8312335; Fax: +1 302 8318404; URL: www.udel.edu/erc
Margaret Phillis Dillner
42 000 vols; 39 curr per   32106

University of Medicine & Dentistry of New Jersey, George F. Smith Library of the Health Sciences, 30 12th Ave, **Newark**, NJ 07103-2706; P.O. Box 1709, Newark, NJ 07101-1709
T: +1 973 9724358; Fax: +1 973 9727474; URL: www.umdnj.edu/librweb/newarklib/library.html
1956; Judith Cohn
Hist Medicine Coll
90 000 vols; 1 359 curr per; 12 831 e-journals; 425 e-books; 22 000 microforms; 4 576 sound-rec; 2 698 Electronic Media & Resources, 168 042 Journals   32107

George Fox University, Murdock Learning Resource Center, 416 N Meridian St, **Newberg**, OR 97132
T: +1 503 5542410; Fax: +1 503 5543599; URL: library.georgefox.edu
1891; Merrill Johnson
Herbert Hoover Coll, Peace Coll, Quaker Coll
117 000 vols; 772 curr per; 5 000 e-books
libr loan   32108

– Portland Center Library, Hampton Plaza, 12753 SW 68th Ave, Portland, OR 97223
T: +1 503 5546130; Fax: +1 503 5546134
1947; Charlie Kamilos
Theology, Counseling
58 000 vols; 293 curr per; 65 diss/theses; 154 maps; 5 000 microforms; 750 av-mat; 1 581 sound-rec; 15 charts, 55 online databases
OCLC, ATLA   32109

Newberry College, Wessels Library, 2100 College St, **Newberry**, SC 29108-2197
T: +1 803 3215229; Fax: +1 803 3215232; URL: www.newberry.edu/wessels
1858; Lawrence E. Ellis
Regional Lutheran Mat; South Caroliniana, Newberry College Mat
81 000 vols; 261 curr per
libr loan   32110

Mount Saint Mary College, Curtin Memorial Library, 330 Powell Ave, **Newburgh**, NY 12550-3494
T: +1 845 5693241; Fax: +1 845 5610999; URL: library.msmc.edu
1959; Carol Wu
100 000 vols; 500 curr per   32111

Oregon State University, Marilyn Potts Guin Library, 2030 Marine Science Dr, **Newport**, OR 97365
T: +1 541 8670249; Fax: +1 541 8670105; E-mail: hmsc.library@oregonstate.edu; URL: osulibrary.oregonstate.edu/guin
1967; Janet G. Webster
Marine science, Marine fisheries, Marine mammals, Aquaculture
35 000 vols; 310 curr per   32112

Salve Regina University, McKillop Library, 100 Ochre Point Ave, **Newport**, RI 02840-4192
T: +1 401 3412330; Fax: +1 401 3412951; URL: library.salve.edu; www.salve.edu
1947; Kathleen Boyd
Jewish Holocaust (Dora & Elias Blumen Libr for the Study of Holocaust Lit), Whiteker Record Coll
120 000 vols; 888 curr per   32113

US Naval War College, Library, 686 Cushing Rd, **Newport**, RI 02841-1207
T: +1 401 8413052; Fax: +1 401 8414804; E-mail: libref@nwc.navy.mil; URL: www.nwc.navy.mil/library
1884; Robert E. Schnare Jr
US Pre-1900 Geography, Mss, Personal Papers & Oral Histories Concerning the Navy & Narragansett Bay; US Pre-1900 Naval & Military Hist, Art & Science
227 000 vols; 750 curr per; 8 000 e-books; 578 000 microforms; 211 av-mat; 55 digital data carriers; 180 Talking Bks
libr loan   32114

Christopher Newport University, Captain John Smith Library, One University Pl, **Newport News**, VA 23606
T: +1 757 5947249; Fax: +1 757 5947772; E-mail: library@cnu.edu; URL: www.cnu.edu/library/libhome.html
1961; Mary Sellen
Nautical Coll (Alexander C. Brown); Josephine Hughes Music Coll; Virginia Authors Coll
401 000 vols; 1 027 curr per; 16 000 e-books, 195 000 microforms, 10 923 sound-rec
libr loan; ALA   32115

Lasell College, Brennan Library, 80 A Maple Ave, **Newton**, MA 02466
T: +1 617 2432244; Fax: +1 617 2432458; URL: www.lasell.edu/studentlife/brennan_library.asp
1851; Marilyn Negip
Lasell Hist Coll
58 000 vols; 193 curr per   32116

Hebrew College, Rae and Joseph Gann Library, 160 Herrick Rd, **Newton Centre**, MA 02459
T: +1 617 5598750; Fax: +1 617 5598751; E-mail: library@hebrewcollege.edu; URL: www.hebrewcollege.edu/library
1918; Dr. Judith Segal
Canadian Jewry; Hassidic & Kabbalistic Lit; Holocaust; Japanese-Judaica; Jewish Education (Herman & Peggy Vershbow Pedagogic Center); Jewish Genealogy;

Jewish Medical Ethics (Harry A. & Beatrice Savitz Coll); Jewish Women's Studies
125 000 vols; 281 curr per; 77 mss; 2 incunabula; 65 diss/theses; 250 music scores; 81 maps; 3 648 microforms; 81 av-mat; 2 125 sound-rec; 22 digital data carriers; 46 Bks on Deafness & Sign Lang
ALA, AJL 32117

Niagara University, University Library, Lewiston Rd, ***Niagara University***, NY 14109
T: +1 716 2868013; Fax: +1 716 2868030; E-mail: reflib@niagara.edu; URL: www.niagara.edu/library
1856; David Schoen
15th-17th c Religious Mat
277 000 vols; 7 500 curr per; 3 000 e-books; 50 Bks on Deafness & Sign Lang
libr loan; OCLC 32118

University of Alaska – Northwest Campus, Learning Resource Center, Pouch 400, ***Nome***, AK 99762
T: +1 907 4438415; Fax: +1 907 4432909; E-mail: smwolf@alaska.edu; URL: www.nwc.uaf.edu
1980; Susan Wolf
12 720 vols; 106 curr per; 57 maps
OCLC 32119

Eastern Virginia Medical School, Edward E. Brickell Medical Sciences Library, 740 W Olney Rd, ***Norfolk***, VA 23501; P.O. Box 1980, Norfolk, VA 23501-1980
T: +1 757 4465851; Fax: +1 757 4465134; E-mail: library@evms.edu; URL: www.evms.edu/evmslib
1972; Judith G. Robinson
90 000 vols; 700 curr per; 2 000 microforms; 25 digital data carriers; 4 186 unbound per, 2 117 media titles 32120

Norfolk State University, Lyman Beecher Brooks Library, 700 Park Ave, ***Norfolk***, VA 23504-8010
T: +1 757 8238481; Fax: +1 757 8232431; URL: library.nsu.edu
1935; Tommy L. Bogger
Black Coll, Juvenile Coll, Theses Coll, Harrison B Wilson Arch, E Woods Museum Libr (African Art)
341 000 vols; 1 395 curr per; 1 250 e-journals; 40 000 e-books; 180 music scores; 74 000 microforms; 200 av-mat; 600 sound-rec; 130 digital data carriers 32121

Old Dominion University, Perry Library, 4427 Hampton Blvd, ***Norfolk***, VA 23529-0256
T: +1 757 6834170; Fax: +1 757 6835035; URL: www.lib.odu.edu
1930; Virginia S. O'Herron
Hist Arch, Scottish hist, Tidewater Hist, Recordings
1 289 000 vols; 14 651 curr per; 433 000 microforms; 50 346 sound-rec; 179 digital data carriers
libr loan 32122

– Elise N. Hofheimer Art Library, Diehn Fine & Performing Arts Ctr, Rm 109, Norfolk, VA 23529
T: +1 757 6834059; URL: www.lib.odu.edu/artlib/index.htm
Clay Lee Vaughan
16 000 vols; 94 curr per 32123

Virginia Wesleyan College, Henry Clay Hofheimer II Library, 1584 Wesleyan Dr, ***Norfolk***, VA 23502-5599
T: +1 757 4552132; Fax: +1 757 4552129; E-mail: library@vwc.edu; URL: www.vwc.edu/library
1966; Jan S. Pace
123 000 vols; 134 curr per; 19 600 e-journals; 52 000 e-books; 330 music scores; 1 230 av-mat; 470 sound-rec 32124

Illinois State University, Milner Library, 201 N School St, ***Normal***, IL 61790-8900; Campus Box 8900, Normal, IL 61790-8900
T: +1 309 4387321; Fax: +1 309 4383676; URL: www.library.ilstu.edu
1890; Cheryl Elzy
19th & 20th c Circus Engravings, Children's Lit, Elementary & Secondary School Texbooks, Lincoln Coll
1 632 000 vols; 4 873 curr per
libr loan 32125

University of Oklahoma, University Libraries, 401 W Brooks, ***Norman***, OK 73019
T: +1 405 3254231; Fax: +1 405 3257550; URL: libraries.ou.edu
1895; Sul H. Lee
4 900 000 vols; 63 000 curr per
libr loan 32126

– Architecture Library, Architecture Library, LLG8, 830 Van Vleet Oval, Norman, OK 73019
T: +1 405 3255521; Fax: +1 405 3256637
1929; Matt Stock
Lt Orville S Witt Memorial Coll
32 000 vols 32127

– Bass Business History Collection, Bass Collection, 507 NW, 401 W Brooks St, Norman, OK 73019
T: +1 405 3253941
1955; Daniel Wren
J & W Seligman & Company Arch, Management Horizons Arch, Sears Roebuck Cats (microfilm), Rare Bks in Economic Hist
23 000 vols; 1 incunabula; 11 000 microforms; 56 av-mat; 14 sound-rec; 1 750 pamphlets 32128

– Chemistry-Mathematics Library, Physical Sciences Center, Chemistry & Mathematics, Rm 207, Norman, OK 73019
T: +1 405 3255628; Fax: +1 405 3257650
1921; Lina Ortega
83 000 vols; 12 000 microforms 32129

– Donald E. Pray Law Library, 300 Timberdell Rd, Norman, OK 73019
T: +1 405 3254311; Fax: +1 405 3256282; URL: www.law.ou.edu/library
1909; Darin K. Fox
Native Peoples Law
207 000 vols; 4 371 curr per; 201 000 microforms; 161 digital data carriers 32130

– Engineering Library, Engineering Library, 222FH, 865 Asp Ave, Norman, OK 73019
T: +1 405 3252941; Fax: +1 405 3250345
James Bierman
79 000 vols; 50 000 microforms 32131

– Fine Arts Library, Fine Arts Library, 20, 500 W Boyd St, Norman, OK 73019
T: +1 405 3254243; Fax: +1 405 3254243
1986; Matt Stock
Bixler files (clippings on theater, film, and dance), Spencer Norton Coll, Joseph Benton Coll, Harrison Kerr Coll
116 000 vols; 10 000 microforms 32132

– Geology Library, Youngblood Energy Library, R220, 100 E Boyd, Norman, OK 73019
T: +1 405 3256451; Fax: +1 405 3256451
1904; Jody Foote
Theses on Oklahoma Geology
106 000 vols; 150 000 maps; 279 700 PI completion cards for Oklahoma 32133

– History of Science Collection, Rm 521 NW, Norman, OK 73019
T: +1 405 3252741
1951; Kerry Magruder
DeGolyer, Klopsteg, Crew, Sally Hall, Nielsen, Roller, Lacy, ADF, & Harlow Colls
92 000 vols 32134

– Physics & Astronomy Library, Physics & Astronomy, 219NH, 440 W Brooks, Norman, OK 73019
T: +1 405 3252887; Fax: +1 405 3253640
1948; Kathryn Caldwell
Survey Plates of the Northern & Southern Skies
38 000 vols 32135

– Western History Collection, Western History Collection, 452 MH, 630 Parrington Oval, Norman, OK 73019
T: +1 405 3253641; Fax: +1 405 3256069
1927; John Lovett
Indians of North America
71 000 vols; 5 500 maps; 20 000 microforms; 2 900 sound-rec; mss (13 000 linear ft) 32136

Massachusetts College of Liberal Arts, Eugene L. Freel Library, 375 Church St, Ste 9250, ***North Adams***, MA 01247
T: +1 413 6625321; Fax: +1 413 6625286; E-mail: library@mcla.edu; URL: www.mcla.edu
1894; Allen S. Morrill
Hoosac Valley Coll for Local Hist, Teacher Resources Coll, College Arch, McFarlin Printing Coll
170 000 vols; 75 curr per; 254 000 microforms; 3 004 av-mat; 1 534 sound-rec; 29 digital data carriers 32137

Merrimack College, McQuade Library, 315 Turnpike St, ***North Andover***, MA 01845
T: +1 978 8375215; Fax: +1 978 8375434; E-mail: merref@noblenet.org; URL: www.noblenet.org/merrimack
1947; Barbara Lachance
Augustinian Studies
87 560 vols; 1 037 curr per; 10 000 microforms; 800 sound-rec; 5 digital data carriers
libr loan 32138

DeVry Institute Library, 630 US Hwy One, ***North Brunswick***, NJ 08902
T: +1 732 7293840; E-mail: nbr-refhelp@devry.edu; URL: www.nj.devry.edu/library.html
Joe Louderback
33 100 vols; 96 curr per 32139

Kent State University, Stark Campus Library, 6000 Frank Ave NW, ***North Canton***, OH 44720-7548
T: +1 330 2443330; Fax: +1 330 4946212; E-mail: starklibrary@listserv.kent.edu; URL: www.stark.kent.edu/library
1967; Rob Kairis
74 000 vols; 258 curr per 32140

Walsh University, Brother Edmond Drouin Library, 2020 E Maple St NW, ***North Canton***, OH 44720-3336
T: +1 330 4907204; Fax: +1 330 4907270; E-mail: library@walsh.edu; URL: library.walsh.edu
1960; Daniel Suvak
138 000 vols; 420 curr per; 28 000 e-journals; 60 000 e-books; 200 sound-rec; 800 digital data carriers
libr loan 32141

Rosalind Franklin University of Medicine & Science, Boxer University Library, 3333 Green Bay Rd, ***North Chicago***, IL 60064
T: +1 847 5783000; Fax: +1 847 5783401; URL: www.rosalindfranklin.edu
1912
119 000 vols; 380 curr per; 1 830 e-journals; 500 diss/theses; 3 000 microforms; 12 drawers of cat 32142

University of Massachusetts Dartmouth, Library Communications Center, 285 Old Westport Rd, ***North Dartmouth***, MA 02747-2300
T: +1 508 9996951; Fax: +1 508 9999240; E-mail: libweb@umassd.edu; URL: www.lib.umassd.edu
1960; Sharon Weiner
Lit, Portuguese language, Technology; American Imprints Coll, Hansard Parliamentary Debates, Robert Kennedy Assassination Arch, Arch of the Ctr for Jewish Culture
468 000 vols; 2 754 curr per; 243 e-books; 695 000 microforms; 8 600 sound-rec; 340 digital data carriers 32143

Manchester College, Funderburg Library, 604 E College Ave, ***North Manchester***, IN 46962
T: +1 260 9825364; Fax: +1 260 9825362; URL: www.manchester.edu/oaa/library
1889; Robin J. Gratz
Church of the Brethren Coll, Peace Studies, College Archs
181 000 vols; 500 curr per 32144

Florida International University, Biscayne Bay Campus Library, 3000 NE 151st St, ***North Miami***, FL 33181-3600
T: +1 305 9195718; Fax: +1 305 9406865; URL: library.fiu.edu
1977; Laura Probst
Curriculum Coll, Holocaust Oral Video Coll
423 000 vols; 1 115 curr per; 5 830 e-journals; 57 000 e-books; 5 400 av-mat 32145

Bethel College, Library, 300 E 27th St, ***North Newton***, KS 67117-0531
T: +1 316 2845361; Fax: +1 316 2845843; URL: www.bethelks.edu/services/library
1891
Mennonite coll
102 000 vols; 291 curr per; 1 000 e-books; 1 370 music scores; 10 000 microforms; 430 av-mat; 830 sound-rec; 40 digital data carriers
libr loan 32146

– Mennonite Library & Archives, 300 E 27th St, North Newton, KS 67117-0531
T: +1 316 2845304; Fax: +1 316 2845843; E-mail: mla@bethelks.edu; URL: www.bethelks.edu/mla
1936; John Thiesen
Baptists, Biblical studies, Reformation, Dutch & German language; Mss Coll, General Conference Church Arch, World War I & II Oral Hist, Rare Anabaptist Bks, HR Voth Coll on Hopi Indians, Rodolphe Petter Coll on Cheyenne Indians, Mennonite Hymnbks
33 000 vols; 280 curr per; 150 maps; 1 000 microforms; 161 710 av-mat; 230 sound-rec
libr loan 32147

Smith College Libraries, William Allan Neilson Library, ***Northampton***, MA 01063
T: +1 413 5852960; Fax: +1 413 5854485; URL: www.smith.edu/libraries
1875; Christopher B. Loring
Women's hist and organizations, hist of the book, college arch, Sophia Smith Coll, Mortimer Rare Book Rm, Sylvia Plath Coll, Virginia Woolf Coll
1 339 000 vols; 2 345 curr per; 97 677 govt docs; 150 562 maps; 107 000 microforms; 5 314 av-mat; 54 821 sound-rec; 691 digital data carriers; 11 270 linear ft of mss, 795 slides 32148

– Anita O'K. & Robert R. Young Science Library, Clark Science Ctr, Bass Hall, Northampton, MA 01063
T: +1 413 585-2950; Fax: +1 413 585-4480; E-mail: sciinfo@smith.edu; URL: www.smith.edu/libraries/libs/young
Josephine Hernandez
Maps, printed, mss, wall, raised relief & gazetteers
154 330 vols; 707 curr per; 5 636 govt docs; 151 300 maps; 22 179 microforms; 97 sound-rec; 24 digital data carriers
libr loan 32149

– Hillyer Art Library, Bell Hall, 45 Round Hill Rd, Northampton, MA 01063
T: +1 413 5852946; Fax: +1 413 5852904; E-mail: hillinfo@smith.edu; URL: www.smith.edu/libraries/libs/hillyer
1918; Barbara Polowy
86 000 vols; 325 curr per; 220 diss/theses; 37 000 microforms; 60 digital data carriers; 10 000 auction cat, 24 500 vertical file items
libr loan; ALSNA 32150

– Mortimer Rare Book Room, Northampton, MA 01063
T: +1 413 5852906; Fax: +1 413 5854486; URL: www.smith.edu/libraries/libs/rarebook
Martin Antonetti
Amerian and British literature, printing
31 000 vols 32151

– Werner Josten Performing Arts Library, Mendenhall Ctr for the Performing Arts, Northampton, MA 01063
T: +1 413 5852930; Fax: +1 413 5853180; URL: www.smith.edu/libraries/libs/josten
1911; Marlene M. Wong
Dance, Music, Theatre; Einstein Coll (music of the 16th & 17th c, copied in score by Alfred Einstein), Music & Correspondence of Werner Josten
89 000 vols; 439 curr per; 2 676 microforms; 600 av-mat; 55 000 sound-rec; 33 digital data carriers; 139 microtext 32152

Carleton College, Laurence McKinley Gould Library, One N College St, **Northfield**, MN 55057-4097
T: +1 507 2224260; Fax: +1 507 2224087; URL: www.library.carleton.edu
1867; Samuel Demas
Polar expeditions, Donald Beaty Bloch Coll of Western Americana, American Studies, art hist, Lucas Jazz Records, photos of famous authors by famous photographers, Warming Orchid Books, Thorsten Veblen's Libr
841 000 vols; 1314 curr per; 5640 e-journals; 300 000 e-books; 400 000 govt docs; 4610 av-mat; 1915 sound-rec; 42 000 digital data carriers; 500 Bks on deafness & Sign Lang   32153

Norwich University, Kreitzberg Library, 23 Harmon Dr, **Northfield**, VT 05663
T: +1 802 4852176; Fax: +1 802 4852173; E-mail: library@norwich.edu; URL: www.norwich.edu/academics/library/
1819; Ellen Hall
Norwichiana Coll
172 000 vols; 813 curr per; 24 180 e-journals; 43 000 e-books   32154

Saint Olaf College – Howard V. & Edna H. Hong Kierkegaard Library, 1510 Saint Olaf Ave, **Northfield**, MN 55057-1097
T: +1 507 6463846; Fax: +1 507 6463858; URL: www.stolaf.edu/collections/kierkegaard
1976; Gordon Marino
11 000 vols; 300 microforms; 10 av-mat; 100 slides   32155

Saint Olaf College – Rolvaag Memorial Library, Glasoe Science Library, Halvorson Music Library, 1510 St Olaf Ave, **Northfield**, MN 55057-1097
T: +1 507 6463634; Fax: +1 507 6463734; URL: www.stolaf.edu/library
1874; Bryn Geffert
Scandinavian lit & hist, Religious studies; Norwegian-American Hist Assn Coll, Pre-1801 Imprints (Vault Coll)
623 420 vols; 1743 curr per   32156

California State University, Northridge, Delmar T. Oviatt Library, University Archives, 18111 Nordhoff St, **Northridge**, CA 91330-8326
T: +1 818 6772285; Fax: +1 818 6774136; E-mail: libref@csun.edu; URL: www.library.csun.edu
1958; Susan C. Curzon
Local Hist, Music; Contemporary 20th Century American Writers (Vern & Bonnie Bullough Coll), 19th Century English & American Playbills, Revolutionary & Political Movements in Russia 1875-1937, Women Music Composers, Human Sexuality, Japanese-American World War II
1 274 000 vols; 18 654 curr per; 5000 e-books; 1 900 000 microforms; 712 av-mat; 16 430 sound-rec; 5514 slides, 282 overhead transparencies, 59 780 art reproductions
libr loan   32157

Wheaton College, Madeleine Clark Wallace Library, 26 E Main St, **Norton**, MA 02766-2322
T: +1 508 2863701; Fax: +1 508 2868275; URL: www.wheatonma.edu/library
1840; Terry Metz
Children's Coll, Lucy Larcom Coll, Private Press Bks
369 000 vols; 10 923 curr per; 9770 e-journals; 10 621 sound-rec; 79 digital data carriers   32158

Saint Mary's College, Cushwa-Leighton Library, **Notre Dame**, IN 46556-5001
T: +1 574 2845280; Fax: +1 574 2844791; URL: www.saintmarys.edu/~library
1855; Janet Fore
Dante Coll
229 000 vols; 584 curr per; 210 av-mat
libr loan   32159

University of Notre Dame, University Libraries, Library Advancement Office, 221 Hesburgh Library, **Notre Dame**, IN 46556
T: +1 574 6315252; Fax: +1 574 6316772; E-mail: blackstead.1@nd.edu; URL: www.nd.edu/~ndlibs

1873; Jennifer A. Younger
Zahm Coll on Dante, Bennett Shaw Coll, Rare Books Coll, Edward L Greene Coll on Botany, Catholic Americana Coll, Hackenbruch Coll of Early American Newspapers, Medieval & Renaissance Mss, Theodore S Weber Coll of Penguin Publs, Durand Coll on Garcilaso de la Vega & the Hist of Peru, Stamp Colls
3 334 000 vols; 9618 curr per; 5000 e-journals; 6660 av-mat; 6400 sound-rec; 630 digital data carriers
libr loan   32160

– Architecture Library, 117 Bond Hall, Notre Dame, IN 46556-5652
T: +1 574 6316654; Fax: +1 574 6319662; E-mail: library.archlib.1@nd.edu; URL: www.nd.edu/~archlib
1931; Deborah Webb
Rare Bk Coll
34 000 vols; 95 curr per; 3000 microforms; 110 av-mat; 3 digital data carriers
libr loan   32161

– Chemistry-Physics Library, 231 Nieuwland Science Hall, Notre Dame, IN 46556
T: +1 574 6317203; Fax: +1 574 6319661; E-mail: chplibr@vma.cc.nd.edu; URL: www.nd.edu/~chemlib
1963; Thurston D. Miller
33 000 vols; 91 curr per; 550 e-journals   32162

– Engineering Library, 149 Fitzpatrick Hall, Notre Dame, IN 46556
T: +1 574 6316665; Fax: +1 574 6319208; E-mail: library.engrlib.1@nd.edu; URL: www.nd.edu/~engrlib
Carol A. Brach
56 000 vols; 134 curr per; 1230 e-journals; 25 000 microforms
libr loan   32163

– Life Sciences Library, B149 Paul V Galvin Life Science Ctr, Notre Dame, IN 46556
T: +1 574 6317209; Fax: +1 574 6319207; E-mail: lifeslib.1@nd.edu; URL: www.nd.edu/~lifeslib
1938; Parker Ladwig
27 000 vols; 252 curr per; 880 e-journals; 1 digital data carriers   32164

– Mathematics Library, 001 Hayes-Healy Ctr, Notre Dame, IN 46556-5641
T: +1 574 6317278; Fax: +1 574 6319660; E-mail: library.mathlib.1@nd.edu; URL: www.nd.edu/~mathlib
1962; Parker Ladwig
Marston Morse Coll
52 000 vols; 91 curr per; 300 e-journals; 1 digital data carriers   32165

– Medieval Institute Library, 715 Hesburgh Library, Notre Dame, IN 46556-5629
T: +1 574 6316603; Fax: +1 574 6318644; URL: www.nd.edu/~medvllib
1948; Derek Webb
Frank M Folsom Ambrosiana Coll, Medieval Education, Intellectual Hist, Medieval Theology, Monasticism, Paleography, Medieval Mss, Milton V Anastos Libr of Byzantine Civilization
90 000 vols; 470 curr per; 16 000 microforms; 4 digital data carriers; 12 VF drawers   32166

University of Notre Dame – Kresge Law Library, Notre Dame Law School, 3113 Eck Hall of Law, **Notre Dame**, IN 46556; P.O. Box 535, Notre Dame, IN 46556-0535
T: +1 574 6317024; Fax: +1 574 6316371; E-mail: lawlib@nd.edu; URL: www.nd.edu/~lawlib
1869; Ed Edmonds
519 000 vols; 181 curr per; 225 digital data carriers   32167

College of Marin, Indian Valley Campus, Library, 1800 Ignacio Blvd, **Novato**, CA 94949
T: +1 415 4578811; Fax: +1 415 8836980; E-mail: WeListen@marin.edu; URL: www.marin.edu
1971; Glade Van Loan
85 000 vols; 45 curr per   32168

Nyack College, Library, One South Blvd, **Nyack**, NY 10960-3698
T: +1 845 3581710 ext 105; Fax: +1 845 3530817; E-mail: Linda.Postan@nyack.edu; URL: www.nyack.edu/library
1882; Linda Poston
Bible, Missions, Theology
96 000 vols; 300 curr per; 11 000 e-journals; 32 400 e-books; 34 022 microforms; 12 036 av-mat
libr loan   32169

Dowling College, Library, 150 Idle Hour Blvd, **Oakdale**, NY 11769-1999
T: +1 631 2443280; Fax: +1 631 2443374; E-mail: reference@dowling.edu; URL: library.dowling.edu
1955; Priscilla Powers
Aviation, Business, Education, Transportation; Long Island Hist, Vanderbilt Family Coll
215 000 vols; 961 curr per; 9000 e-books
libr loan   32170

California College of Arts & Crafts, Meyer Library, 5212 Broadway, **Oakland**, CA 94618
T: +1 510 5943658; E-mail: refdesk@cca.edu; URL: library.cca.edu
1907; Janice Woo
Fine arts, Architecture; Industrial Design (Jo Sinel Coll)
36 000 vols; 124 curr per; 600 diss/theses; 620 av-mat; 120 sound-rec; 20 digital data carriers
libr loan; ALA   32171

Holy Names College, Paul J. Cushing Library, 3500 Mountain Blvd, **Oakland**, CA 94619
T: +1 510 4361332; Fax: +1 510 4361260; URL: www.hnc.edu
1880; Joyce McLean
Religion, Nursing, Music
111 000 vols; 200 curr per; 25 000 e-books; 85 av-mat; 85 Talking Bks
libr loan   32172

Mills College, F.W. Olin Library, 5000 MacArthur Blvd, **Oakland**, CA 94613
T: +1 510 4302180; Fax: +1 510 4302278; URL: www.mills.edu
1852; Janice Braun
Lit, Art, Dance; Albert M Bender Coll, Jane Bourne Parton Coll (American Dance); Elias Olan James Coll; Mills College Arch
224 000 vols; 920 curr per   32173

Patten University, Harry & Dorothy Blumenthal Library, 2433 Coolidge Ave, **Oakland**, CA 94601
T: +1 510 5344344; E-mail: jkadarkwa@patten.edu; URL: www.patten.edu
1960; Joshua Adarkwa
Biblical studies, Church hist, Education, Archeology; Dienstein Criminology Coll
35 500 vols; 210 curr per   32174

Oakland City University, Barger-Richardson LRC, 605 W Columbia St, **Oakland City**, IN 47660
T: +1 812 7491269; Fax: +1 812 7491414; URL: oak.oak.edu
1890; Denise J. Pinnick
115 000 vols; 301 curr per; 30 diss/theses; 95 000 microforms; 1310 sound-rec
libr loan; ALA, ACL, IFLA   32175

Oberlin College, Library, 148 W College St, **Oberlin**, OH 44074
T: +1 440 7758285; Fax: +1 440 7756586; E-mail: reference.desk@oberlin.edu; URL: www.oberlin.edu/library
1833; Ray English
American Dime Novels, Anti-Slavery, Hist of the Book, Oberliniana, Edwin Arlington Robinson, Spanish Drama
1 398 000 vols; 43 246 curr per; 647 digital data carriers; 106 000 Music Scores
libr loan   32176

– Clarence Ward Art Library, Allen Art Bldg, 83 N Main St, Oberlin, OH 44074-1193
T: +1 440 7758635; Fax: +1 440 7755145; URL: www.oberlin.edu/library/libncollect/art/Default.html
1917; Barbara Prior
Thomas Jefferson's Architectural Libr; Artists' books

93 000 vols
libr loan   32177

– Mary M. Vial Music Library, Oberlin Conservatory of Music, 77 W College St, Oberlin, OH 44074-1588
T: +1 440 7758280; Fax: +1 440 7758203; E-mail: library.webmaster@oberlin.edu; URL: www.oberlin.edu/library/
Deborah Campana
172 000 vols   32178

– Science Library, Science Center N174, 119 Woodland St, Oberlin, OH 44074-1083
T: +1 440 7758310; Fax: +1 440 7755152; URL: www.oberlin.edu/library/libncollect/sci/Default.html
1965; Alison Ricker
Biology, Chemistry, Earth Sciences, Medicine, Technology
99 000 vols   32179

University of Texas of the Permian Basin, J. Conrad Dunagan Library, 4901 E University Blvd, **Odessa**, TX 79762
T: +1 432 5522370; Fax: +1 432 5522374; E-mail: shults_c@utpb.edu; URL: www.utpb.edu/library/index
1973; Charlene Shults
Education, Texana; J Frank Dobie Coll, bks & papers; Texas Writers Coll
274 000 vols; 700 curr per   32180

Weber State University, Stewart Library, 2901 University Circle, **Ogden**, UT 84408-2901
T: +1 801 6266384; Fax: +1 801 6268521; URL: library.weber.edu
1924; Joan Hubbard
Art (Paul Bransom Coll), Lit (James A Howell Coll), Mormon Lit (Hyrum & Ruby Wheelwright Coll), Oriental Artifacts (Frank Vercroft Coll), Porcelain (Jeanette McKay Morrell Coll)
534 000 vols; 1711 curr per
libr loan   32181

Clarion University of Pennsylvania, Venango Campus Library, 1801 W First St, **Oil City**, PA 16301
T: +1 814 6766591; Fax: +1 814 6773987; URL: www.clarion.edu/library
1961; Linda Cheresnowski
49 000 vols; 182 curr per
libr loan   32182

Mid-America Christian University, Charles Ewing Brown Library, 3500 SW 119th St, **Oklahoma City**, OK 73170-9797
T: +1 405 6923174; Fax: +1 405 6923165; E-mail: library@macu.edu; URL: library.macu.edu
1953; Elissa Patadal, Michael Foote
Arch of Church of God (Charles Ewing Brown Coll), Wesleyan Holiness Theology (Kenneth E. Jones Coll)
53 861 vols; 250 curr per; 35 diss/theses; 300 maps; 12 304 microforms; 114 av-mat; 1524 sound-rec; 20 digital data carriers; 27 overhead transparencies
libr loan; ALA, OLA, ACRL, ACL   32183

Oklahoma City University, **Oklahoma City**
– Dulaney-Browne Library, 2501 N Blackwelder, Oklahoma City, OK 73106
T: +1 405 2085874; Fax: +1 405 2085291; E-mail: askalibrarian@okcu.edu; URL: www.okcu.edu/library
1904; Dr. Victoria Swinney
Alexander the Third & Chinese Communism, micro; Methodist Hist; Oklahoma Hist (George H Shirk Hist Ctr Coll); Russian Revolution Coll, micro
184 000 vols; 528 curr per; 36 000 e-books; 10 digital data carriers   32184

– Law Library, 2501 N Blackwelder, Oklahoma City, OK 73106
T: +1 405 2085271; Fax: +1 405 2085172; E-mail: askalibrarian@okcu.edu; URL: www.okcu.edu/law/lawlib
1922; Judith Morgan
317 000 vols   32185

Mid-America Nazarene University, Mabee Library & Learning Resource Center, 2030 College Way, **Olathe**, KS 66062
T: +1 913 7913485; Fax: +1 913 7913285; E-mail: rmorriso@mnu.edu; URL: www.mnu.edu/academics/mabee
1968; Dr. Ray L. Morrison

Americana, Church of the Nazarene Publs
995 110 vols; 225 curr per; 360 000 av-mat; 1 150 sound-rec; 70 digital data carriers
32186

New York Institute of Technology, Wisser Library, PO Box 8000, *Old Westbury*, NY 11568-8000
T: +1 516 6867657; Fax: +1 516 6861320; URL: www.nyit.edu/library
1955; Gerri Flanzraich
Architecture, Culinary Arts, Engineering; Ctr for Prejudice Reduction
102 000 vols; 432 curr per; 15 000 e-books; 738 000 microforms; 66 digital data carriers
libr loan; ALA, LILRC, OCLC 32187

– Education Hall Library Art & Architecture Collection, PO Box 8000, Old Westbury, NY 11568
T: +1 516 6867422; Fax: +1 516 6867814; URL: iris.nyit.edu/library/campus/edhall/index.html
Linda Heslin
26 000 vols; 215 curr per 32188

State University of New York – College at Old Westbury, Library, 223 Store Hill Rd, *Old Westbury*, NY 11568; P.O. Box 229, Old Westbury, NY 11568-0229
T: +1 516 8763151; Fax: +1 516 8763325; URL: www.oldwestbury.edu
1967; Barbara Walsh
Ethnic studies, Feminism, Social sciences, Behav sciences; Black Poetry & Lit Coll; Slavery Source Mat Coll, (micro); Underground Press Coll, (micro); Women's Studies Coll
212 000 vols; 900 curr per; 163 maps; 210 000 microforms; 2216 av-mat; 1307 sound-rec; 3 digital data carriers; 4515 slides
libr loan 32189

Olivet College, Burrage Library, 333 S Main St, *Olivet*, MI 49076-9730
T: +1 269 7497608; Fax: +1 269 7497121; URL: www.olivetcollege.edu
1844; Jane Reiter
Arctic, Education; Arctic Coll
90 000 vols; 180 curr per
libr loan 32190

Evergreen State College, Daniel J. Evans Library, Library Bldg, Rm 2300, 2700 Evergreen Pkwy NW, *Olympia*, WA 98505-0002
T: +1 360 8676250; Fax: +1 360 8676688; URL: www.evergreen.edu/library
1969; Gregg Sapp
College Arch, Nisqually Delta Assn Arch, Washington State Folklife Council Project Arch, Japanese Culture, Chicano Arch, Peoples of Washington Coll, Washington Worm Growers Assn Coll
467 000 vols; 459 curr per 32191

Washington State University Extension, Energy Program Library, 905 Plum St SE, *Olympia*, WA 98504; P.O. Box 43169, Olympia, WA 98504-3169
T: +1 360 9562076; Fax: +1 360 2362076; E-mail: library@energy.wsu.edu; URL: www.energy.wsu.edu/library
Angela Santamaria
17 000 vols; 350 curr per
libr loan 32192

College of Saint Mary, Library, 7000 Mercy Rd, *Omaha*, NE 68106-2606
T: +1 402 3992631; Fax: +1 402 3992686; E-mail: csmlibrary@csm.edu; URL: www.csm.edu/academics_majors/library
1923; Faye Couture
Business, Education, Nursing
83 000 vols; 369 curr per; 15 000 e-books
libr loan 32193

Creighton University, Reinert Alumni Memorial Library, 2500 California Plaza, *Omaha*, NE 68178-0209
T: +1 402 2802706; Fax: +1 402 2802435; E-mail: askus@creighton.edu; URL: reinert.creighton.edu
1878; Michael J. LaCroix
Physics, Theology; Early Christian Writings, Fables of Aesop & La Fontaine
428 000 vols; 1594 curr per; 222 230 e-journals; 3 000 e-books; 685 000 microforms; 4 000 av-mat; 1740 sound-rec; 260 digital data carriers
32194

– Health Sciences Library, Learning Resources Center, 2770 Webster St., Omaha, NE 68178
T: +1 402 280-5108; Fax: +1 402 280-5134; E-mail: jbothmer@creighton.edu; URL: hsl.creighton.edu
1977; A. James Bothmer
National Football League Coll, Autism Coll
239 600 vols; 1093 curr per; 107 535 microforms; 11 940 av-mat; 127 digital data carriers; 227 331 physical units
32195

– Klutznick Law Library, School of Law, 2500 California Plaza, Omaha, NE 68178-0340
T: +1 402 2802875; Fax: +1 402 2802244; URL: law.creighton.edu/library
1904; Kay L. Andrus
378 000 vols; 41 000 e-books; 965 000 microforms; 47 digital data carriers
libr loan; AALL 32196

Grace University, University Library, 823 Worthington, *Omaha*, NE 68108-3642; mail address: 1311 S Ninth St, Omaha, NE 68108-3629
T: +1 402 4492893; Fax: +1 402 4492919; URL: www.graceuniversity.edu/libraryindex.cfm
1943; Stanley Udd
Bible & Theology, Counseling
54 000 vols; 233 curr per; 18 000 e-books; 202 microforms; 80 av-mat; 530 sound-rec; 7 digital data carriers; 399 slides 32197

University of Nebraska at Omaha, Dr C.C. & mabel L. Criss Library, 6001 Dodge St, *Omaha*, NE 68182-0237
T: +1 402 5542661; Fax: +1 402 5543593; URL: library.unomaha.edu/
1908; Stephen R. Shorb
Arthur Paul Afghanistan Coll, Mary L Richmond – Cummings Press Coll, Icarian Community Coll, Nebraska Authors & Hist, Omaha Federal Writers Project Papers (WPA)
750 000 vols
libr loan 32198

Hartwick College, Stevens-German Library, One Hartwick Dr, *Oneonta*, NY 13820
T: +1 607 4314448; Fax: +1 607 4314457; URL: hartwick.edu/library/homepage.html
1928; Elizabeth Orgeron
Judge William Cooper Papers, North American Indians (Yager Coll)
315 000 vols; 2 300 curr per
libr loan 32199

State University of New York – College at Oneonta, James M. Milne Library, *Oneonta*, NY 13820-4014
T: +1 607 4362720; Fax: +1 607 4363081; URL: www.oneonta.edu/library
1889; Janet L. Potter
Education, Home economics; Early Textbks & Early Educational Theory, New York State Hist Coll, New York State Verse Coll, 19th & Early 20th C Popular Fiction, James Fenimore Cooper
558 000 vols; 1063 curr per; 18 000 e-journals; 4 000 e-books; 3 290 music scores; 2450 av-mat; 2 310 sound-rec; 290 digital data carriers
libr loan 32200

Chapman University, Leatherby Libraries, One University Dr, *Orange*, CA 92866-1099
T: +1 714 5327714; Fax: +1 714 5327743; URL: www1.chapman.edu/library/
1923; Charlene Baldwin
County hist; Disciple of Christ Church Hist, Charles C. Chapman Rare Bk Coll
196 000 vols; 1802 curr per
32201

Chapman University School of Law, Harry and Diane Rinker Law Library, 370 N Glassell St, Rm 325, *Orange*, CA 92866
T: +1 714 6282595; URL: www.chapman.edu/law/library
Sheryl Summers-Kramer
280 000 vols 32202

Santiago Canyon College, Library, 8045 Chapman Ave, *Orange*, CA 92869
T: +1 714 6285000; E-mail: Banderas_Justin@sccollege.edu; URL: www.sccollege.edu/Library/Pages/default.aspx
50 000 vols; 64 curr per; 14 000 e-books; 1 000 av-mat; 1200 sound-rec 32203

Northwestern College, Ramaker Library & Learning Resource Center, 101 Seventh St SW, *Orange City*, IA 51041-1996
T: +1 712 7077234; Fax: +1 712 7077247; URL: www.nwciowa.edu
1882; Daniel Daily
Religion, Hist, Lit; Dutch Related-Reformed Church, Congressman Hoeven Coll
108 000 vols; 530 curr per
libr loan 32204

Claflin University, H.V. Manning Library, 400 Magnolia St, *Orangeburg*, SC 29115
T: +1 803 5355308; Fax: +1 803 5355091; E-mail: library@claflin.edu; URL: www.claflin.edu/library/index.asp
Marilyn Y. Gibbs
Hist, Music, Religious studies; Black Life & Hist, bks, microfilm
162 000 vols; 450 curr per 32205

South Carolina State College, Miller F. Whittaker Library, 300 College St NE, *Orangeburg*, SC 29115-4427; P.O. Box 7491, Orangeburg, SC 29117
T: +1 803 5367045; Fax: +1 803 5368902; URL: library.scsu.edu
1913; Adrienne Webber
Black Hist, South Carolina State College Hist Coll, South Carolina State Data Ctr
304 000 vols; 974 curr per
libr loan 32206

Saint Mary's College, Alumni Memorial Library, 3535 Indian Trail, *Orchard Lake*, MI 48324
T: +1 248 7064211; Fax: +1 248 6830526; E-mail: library@sscms.edu; URL: www.sscms.edu/library/index.html
1885; Caryn Noel
Audiovisual, Polish Language Coll, Rare Bks
90 000 vols; 150 curr per; 22 000 microforms; 1584 sound-rec; 18 digital data carriers; 525 videotapes, 106 slides
libr loan; ALA 32207

Utah Valley State College, Library, 800 W University Parkway, *Orem*, UT 84058-5999
T: +1 801 8638840; Fax: +1 801 8637065; URL: www.uvsc.edu/library
1978; Michael J. Freeman
LDS Religion Collection, education
186 000 vols; 800 curr per; 27 000 e-journals; 6 000 e-books; 274 music scores; 2 969 maps; 10 microforms; 11 500 av-mat; 1 735 sound-rec; 4 000 digital data carriers
libr loan; OCLC, SPARC 32208

Florida Agricultural & Mechanical University, Law Library, 201 Beggs Ave, *Orlando*, FL 32801
T: +1 407 2543263; Fax: +1 407 2543273; URL: www.famu.edu
Grace Mills
350 000 vols 32209

Florida Hospital College of Health Sciences, Robert Arthur Wiliams Library, 671 Winyah Dr, *Orlando*, FL 32803
T: +1 407 3031851; Fax: +1 407 3039622; URL: www.fhchs.edu
1992; Deanna L. Stevens
Nursing Coll, Seventh-da Adventist Mat
16 000 vols; 174 curr per; 108 av-mat; 118 sound-rec; 8 digital data carriers; 303 indexes
libr loan; ASDAL, SOLINET 32210

University of Central Florida, University Libraries, 4000 Central Florida Blvd, Bldg 2, *Orlando*, FL 32816-2666; P.O. Box 162666, Orlando, FL 32816-2666
T: +1 407 8232580; Fax: +1 407 8235865; URL: library.ucf.edu
1963; Barry B. Baker
Bryant West Indies Coll, H. Eves Mathematics Coll, Finney Accounting Coll, Solar Energy Coll, Malkoff Book Arts Coll
1 587 000 vols; 16 074 curr per; 7 500 e-journals; 197 000 e-books
libr loan 32211

Valencia Community College – Raymer Maguire Jr Learning Resources Center, West Campus, 1800 S Kirkman Rd, *Orlando*, FL 32811
T: +1 407 5821210; Fax: +1 407 5821686; URL: www.valenciacc.edu/library/west
1967; Karen Blondeau
Education, Nursing, Hotel administration, Horticulture
80 000 vols; 131 curr per
libr loan 32212

University of Maine, Raymond H. Fogler Library, 5729 Fogler Library, *Orono*, ME 04469-5729
T: +1 207 5811674; Fax: +1 207 5811653; URL: www.library.umaine.edu
1865; Joyce Rumery
Abolition & Anti-slavery (O'Brien Coll), Cole Maritime Coll; Folklore, Indian Lore & Etymology (Eckstorm Coll); Hannibal Hamlin Family Papers, State of Maine Coll, Maine State Docs Coll, Philip H Taylor Coll of Modern Hist, War, and Diplomacy, William S. Cohen Papers, Univ Arch
1 088 000 vols; 5 438 curr per; 2 220 254 govt docs; 67 000 maps; 1 584 000 microforms; 10 000 sound-rec; 550 digital data carriers
libr loan 32213

University of Wisconsin – Oshkosh, Forrest R. Polk Library, 801 Elmwood Ave, *Oshkosh*, WI 54901
T: +1 920 4244333; Fax: +1 920 4247338; E-mail: infodesk@uwosh.edu; URL: www.uwosh.edu/library
1871; Patrick J. Wilkinson
Local Hist (Wisconsin Area Research Ctr); Pare Lorenz Coll, flm, mss, monographs
608 000 vols; 900 curr per; 15 000 e-journals; 675 000 govt docs; 32 500 maps; 2 300 000 microforms; 1 500 digital data carriers
libr loan 32214

William Penn University, Wilcox Library, 201 Trueblood Ave, *Oskaloosa*, IA 52577
T: +1 641 6731096; Fax: +1 641 6731098; URL: www.wmpenn.edu/library.html
1873; Julie E. Hansen
Quaker Coll
68 000 vols; 349 curr per; 1219 av-mat
libr loan 32215

State University of New York at Oswego, Penfield Library, SUNY Oswego, *Oswego*, NY 13126-3514
T: +1 315 3124267; Fax: +1 315 3123194; E-mail: refdesk@oswego.edu; URL: www.oswego.edu/library
1861; Mary Beth Bell
College Arch, Local Hist (Marshall Family Coll), Local & State Hist (Safe Haven Coll), Presidential Papers (Millard Fillmore Coll)
474 000 vols; 596 curr per; 4 308 maps; 1 395 000 microforms; 1 437 av-mat; 8 359 sound-rec; 61 digital data carriers; 1 801 495 microforms of per, 28 719 slides
libr loan 32216

Ottawa University, Myers Library, 1001 S Cedar, *Ottawa*, KS 66067-3399
T: +1 785 2425200; Fax: +1 785 2291012; E-mail: library@ottawa.edu; URL: www.ottawa.edu/residential/content/view/330
1865; Gloria Creed-Dikeogu
Chinese Art & Related Asiatic Studies
71 000 vols; 84 curr per; 2 000 e-books; 212 maps; 180 000 microforms; 68 av-mat; 1 171 sound-rec; 959 digital data carriers; 68 Audio Bks
libr loan 32217

Brescia University, Father Leonard Alvey Library, 717 Frederica St, *Owensboro*, KY 42301
T: +1 270 6864212; Fax: +1 270 6864266; E-mail: library@brescia.edu; URL: www.brescia.edu/bulibrary
1950; Sister Judith N. Riney
Lit, Religion, Music; Kentuckiana, Contemporary Woman
121 000 vols; 19 942 e-journals; 30 000 e-books
libr loan; ALA, OCLC, SOLINET 32218

Kentucky Wesleyan College, Library Learning Center, 3000 Frederica St, ***Owensboro***, KY 42301; P.O. Box 1039, Owensboro, KY 42302-1039
T: +1 270 8523258; Fax: +1 270 9263196; URL: www.kwc.edu/library
1858; Pat McFarling
First Editions-American & English Lit, Kentuckiana, Kentucky United Methodist Heritage Ctr Coll, Matsumoto Memorial Libr of Japanese Culture, Dan M. King Architecture Coll
98 000 vols; 305 curr per
libr loan 32219

Baker College of Owosso, Library, 1020 S Washington, ***Owosso***, MI 48867-4400
T: +1 989 7293325; Fax: +1 989 7293429; E-mail: library-ow@baker.edu; URL: www.baker.edu
1984; Brian Ryckman
35 000 vols; 200 curr per 32220

Miami University, Edgar W. King Library, 225 King Library, ***Oxford***, OH 45056
T: +1 513 5296147; Fax: +1 513 5291682; URL: www.lib.muohio.edu
1809; Judith A. Sessions
Botanical Medicine, Shakespeare's first four folios, Ibsen, Strindberg, Orwell, Shaftesbury, Whitman, Literary Society Libs, Russian Military Hist, King Coll of Early Juvenile Books & Periodicals
1 999 000 vols; 34 065 curr per; 596 769 govt docs; 104 760 maps; 2 836 000 microforms; 879 av-mat; 17 600 sound-rec; 194 digital data carriers; 375 slides, 454 overhead transparencies
libr loan 32221

– Amos Music Library, Center for Performing Arts, Oxford, OH 45056
T: +1 513 5292299; Fax: +1 513 5291378
1969; Barry Zaslow
38 000 vols; 97 curr per; 16 000 sound-rec; 26 digital data carriers; 19 800 music scores 32222

– Brill Science Library, Hughes Laboratory Bldg, Oxford, OH 45056
T: +1 513 5296886; Fax: +1 513 5291736
1978; Jerome Conley
Paper Science Coll, Kuchler Vegetation Maps Coll
169 000 vols; 666 curr per; 101 831 maps; 359 000 microforms; 14 VF drawers of pamphlets 32223

– Wertz Art-Architecture Library, Alumni Hall, Oxford, OH 45056
T: +1 513 5296638; Fax: +1 513 5294159
1952; Stacy Brinkman
60 000 vols; 203 curr per; 5 digital data carriers 32224

Salish Kootenai College, D'Arcy McNickle Library, 58138 US Hwy 93, ***Pablo***, MT 59855; P.O. Box 70, Pablo, MT 59855-0070
T: +1 406 2754874; Fax: +1 406 2754812; URL: skclibrary.skc.edu
1979; Carlene Engstrom
Environmental studies, Native American studies, Nursing; Confederated Salish & Kootenai Tribal Hist
47 000 vols; 200 curr per; 789 av-mat; 3 digital data carriers
libr loan 32225

Lake Erie College, James F. Lincoln Library, 391 W Washington St, ***Painesville***, OH 44077-3309
T: +1 440 3757400; Fax: +1 440 3757404; URL: www.lec.edu/library
1856; Christopher Bennett
Equestrian studies; Thomas Harvey Coll
80 000 vols; 9 000 curr per 32226

University of Alaska Anchorage Matanuska-Susitna College, Alvin S. Okeson Library, Mile Two Trunk Rd, P.O. Box 5001, ***Palmer***, AK 99645-5001
T: +1 907 7459740; Fax: +1 907 7459777; E-mail: cballain@matsu.alaska.edu; URL: matsu.alaska.edu/office/library
1961; Craig Ballain
Agriculture, Computer science, Electronics; Local Hist
50 000 vols; 200 curr per
libr loan 32227

Pacific Graduate School of Psychology, Research Library, 935 E Meadow Dr, ***Palo Alto***, CA 94303-4233
T: +1 650 8433555; Fax: +1 650 4936147; URL: www.pgsp.edu/
1975; Christine Kidd
8 000 vols; 1 200 curr per; 15 000 e-books; 2 000 microforms; 150 av-mat; 350 sound-rec; 2 digital data carriers; 500 slides 32228

Nebraska Christian College, Loren T. & Melva M. Swedburg Library, 12550 S 114th St, ***Papillon***, NE 68046
T: +1 402 9359440; Fax: +1 402 9359500; E-mail: llloyd@nechristian.edu; URL: www.nechristian.edu
1945; Linda Lu Lloyd
Gunderson Rare Bibles Coll, Guy B. Dunning Libr (Religious Studies)
25 490 vols; 173 curr per; 1 500 archival items 32229

Park University Library, 8700 NW River Park Dr, ***Parkville***, MO 64152
T: +1 816 5846704; Fax: +1 816 7414911; URL: www.park.edu/library
1875; Ann Schultis
Hist (Platte County Hist Society Arch), Park College Hist
153 000 vols; 591 curr per; 1 240 av-mat; 60 digital data carriers 32230

Art Center College of Design, James Lemont Fogg Memorial Library, 1700 Lida St, ***Pasadena***, CA 91103
T: +1 626 3962237; Fax: +1 626 5680428; URL: www.artcenter.edu/library
1930; Elizabeth Galloway
100 000 vols; 400 curr per; 6 000 av-mat; 2 000 digital data carriers; 2 086 exhibit cat, 1 900 annual rpts, 32 VF drawers, 95 000 slides, 400 Special Interest Per Sub 32231

California Institute of Technology – Caltech Library System 1-32, M/C 1-32, M/C 1-32, 1201 E California Blvd, ***Pasadena***, CA 91125
T: +1 626 3956401; Fax: +1 626 7927540; E-mail: library@caltech.edu; URL: www.library.caltech.edu
1891; Kimberly Douglas
Enginering, Hist of science, Social sciences, Technology; NACA/NASA Technical Reports
549 000 vols; 3 144 curr per; 586 000 microforms; 3 026 sound-rec; 478 digital data carriers; 2 051 linear feet of docs
libr loan 32232

– Astrophysics Library, 1201 E California Blvd, M/C 105-24, Pasadena, CA 91125
T: +1 626 3954008; Fax: +1 626 5689352
Caroline Smith
12 000 vols; 35 curr per 32233

– Earthquake Engineering Research Library, 1200 E California Blvd, M/C 104-44, Pasadena, CA 91125
T: +1 626 3954227; Fax: +1 626 5682719; E-mail: eerllib@caltech.edu; URL: library.caltech.edu/collections/earthquake.htm
Jim O'Donnell
15 000 vols; 25 curr per 32234

– Sherman Fairchild Library of Engineering & Applied Science, Fairchild Library I-43, Pasadena, CA 91125
T: +1 626 3953404; Fax: +1 626 4312681
George Porter
75 000 vols 32235

Pacific Oaks College, Andrew Norman Library, 55 Eureka, Ste 145, ***Pasadena***, CA 91103
T: +1 626 3971360, 1354; Fax: +1 626 3971356; E-mail: library@pacificoaks.edu; URL: www.pacificoaks.edu/library
1945; Diane Gray-Reed
Society of Friends; Hist Development of Children's Lit (Critical), Children's Coll
20 000 vols; 110 curr per
libr loan 32236

Texas Chiropractic College, Mae Hilty Memorial Library, 5912 Spencer Hwy, ***Pasadena***, TX 77505
T: +1 281 9986049; Fax: +1 281 4874168; URL: www.txchiro.edu
Caroline Webb
Medicine, chiropractic
14 000 vols; 175 curr per 32237

Callahan Library, Saint Joseph's College, 25 Audubon Ave, ***Patchogue***, NY 11772-2399
T: +1 631 4473226; Fax: +1 631 6543255; E-mail: callahan@sjcny.edu; URL: libraries.sjcny.edu
1972; Terri Corbin-Hutchinson
Business, Child studies, Education, Health sciences; Long Island Hist
118 000 vols; 258 curr per; 2 000 e-books 32238

Anna Maria College, Mondor-Eagen Library, 50 Sunset Lane, ***Paxton***, MA 01612-1198
T: +1 508 8493473; Fax: +1 508 8493408; URL: www.annamaria.edu/library/index.php
1946; Ruth Pyne
American hist & social work, Business, Criminal justice, English & French lit, Nursing, Religion
75 000 vols; 290 curr per; 283 maps; 483 av-mat; 3 285 sound-rec; 5 digital data carriers; 6 826 Slides, 51 Charts
libr loan; NELINET, WACL, OCLC 32239

Central College, Geisler Library, Campus Box 6500, 812 University St, ***Pella***, IA 50219-1999
T: +1 641 6285193; Fax: +1 641 6285327; URL: www.central.edu/library/libhome.htm
1853; Natalie Hutchinson
Scholte Coll (Dutch in America & Iowa), Helen Van Dyke Miniature Bks Coll, Local Hist, Coll of Rumanian Music
234 000 vols; 1 150 curr per; 50 e-journals; 11 000 e-books; 4 500 music scores; 200 maps; 56 000 microforms; 104 740 av-mat; 8 840 sound-rec; 12 670 digital data carriers; 150 Bks on Deafness & Sign Lang
libr loan 32240

University of North Carolina at Pembroke, Sampson-Livermore Library, Faculty Row, ***Pembroke***, NC 28372; P.O. Box 1510, Pembroke, NC 28372-1510
T: +1 910 5216516; Fax: +1 910 5216547; URL: www.uncp.edu/library
1887; Elinor Foster
Lumbee Indian Hist Coll
368 000 vols; 30 199 curr per; 37 000 e-books; 260 Electronic Media & Resources
libr loan; ALA 32241

University of West Florida, John C. Pace Library, 11000 University Pkwy, ***Pensacola***, FL 32514-5750
T: +1 850 4742414; Fax: +1 850 4743338; URL: library.uwf.edu
1966; Helen Wigersma
West Florida History
724 000 vols; 5 019 curr per; 1 735 e-journals; 91 000 e-books; 839 970 mss; 1 incunabula; 922 diss/theses; 713 326 govt docs; 3 212 music scores; 3 931 maps; 1 156 000 microforms; 3 134 av-mat; 4 200 sound-rec; 1 376 digital data carriers
libr loan; SOLINET, OCLC
The University of West Florida Libraries include John C. Pace Library (the main library), the Curriculum Materials Library, the Music Library and the Fort Walton Beach Campus Library. 32242

Bradley University, Cullom-Davis Library, 1501 W Bradley Ave, ***Peoria***, IL 61625
T: +1 309 6772837; Fax: +1 309 6772558; URL: library.bradley.edu
1897; Barbara A. Galik
Abraham Lincoln & Civil War (Martin L. Howser Coll), Public Safety Communications Hist, Industrial Arts Hist
435 000 vols; 1 488 curr per
libr loan 32243

Ursuline College, Ralph M. Besse Library, 2550 Lander Rd, ***Pepper Pike***, OH 44124-4398
T: +1 440 4494202; Fax: +1 440 4493180; URL: www.ursuline.edu/library
1871; Betsey Belkin
Besse Rivers Coll, Global Studies, Picture Arch Coll
134 000 vols; 305 curr per; 12 990 e-journals; 25 000 e-books 32244

Northwest Ohio Regional Book Depository, 1655 N Wilkinson Way, ***Perrysburg***, OH 43551
T: +1 419 8744891; Fax: +1 419 8744385
Michael McHugh
1 100 000 vols 32245

Peru State College, Library, 600 Hoyt St, ***Peru***, NE 68421
T: +1 402 8722218; Fax: +1 402 8722298; E-mail: library@oakmail.peru.edu; URL: www.peru.edu
1867
Marion Marsh Brown Coll
106 000 vols; 8 000 e-books; 3 500 High Interest/Low Vocabulary Bks, 300 Special Interest Per Sub, 3 000 Large Print Bks
libr loan 32246

Virginia State University, Johnston Memorial Library, One Hayden Dr, ***Petersburg***, VA 23806-0001; P.O. Box 9406, Petersburg, VA 23806-9406
T: +1 804 5245582; Fax: +1 804 5245482; URL: library.vsu.edu/
1882; Elsie Wetherington
Education, African-American studies
311 000 vols; 4 437 curr per; 2 830 e-journals; 15 000 e-books; 700 music scores; 1 560 av-mat; 470 sound-rec; 270 digital data carriers
libr loan 32247

North Central Michigan College, Library, 1515 Howard St, ***Petoskey***, MI 49770
T: +1 231 3486615; Fax: +1 231 3486629; E-mail: library@ncmich.edu; URL: library.ncmich.edu
1958; Eunice Teel
Thomas F. Schweigert Leadership Exhibit, Nuclear Docs
30 000 vols; 300 curr per; 10 000 govt docs; 500 maps; 5 500 microforms; 600 av-mat; 2 000 sound-rec
libr loan 32248

Northwest-Shoals Community College, Library / Learning Resources Center, 2080 College Rd, ***Phil Campbell***, AL 35581
T: +1 256 3316271; Fax: +1 256 3316272; E-mail: colvint@nwscc.edu; URL: nwscc.cc.al.us/index4.html
1963
Alabama Room Coll, Nursing Coll, Children's Coll
31 000 vols; 210 curr per 32249

Center for Judaic Studies Library, 420 Walnut St, ***Philadelphia***, PA 19106-3703
T: +1 215 2381290; Fax: +1 215 2381540; URL: www.library.upenn.edu/cjs
1907; Dr. Arthur Kiron
Judaica, Biblical studies, Near Eastern studies; American-Jewish Hist, Arch; Arabica (Prof Skoss Coll); Bible (Prof Max Margolis Coll); Genizah Fragments; Hist of Philadelphia; Hist of Jewish & Oriental Studies; Judaica Rare Bks (Mayer Sulzberger & Isaac Leeser Coll); Hebrew Mss; USSR Coll; Poland & Hungary Coll; Oriental Mss
180 000 vols; 92 curr per; 450 mss; 30 incunabula; 200 diss/theses; 10 digital data carriers
libr loan 32250

Chestnut Hill College, Logue Library, 9601 Germantown Ave, ***Philadelphia***, PA 19118-2695
T: +1 215 2487050; Fax: +1 215 2487056; URL: www.chc.edu/library
1924; Mary Jo Larkin
Education, Liberal arts, Religious studies; Catholic Church Music (Montani Coll), Irish Hist & Lit
145 000 vols; 543 curr per
libr loan 32251

College of Physicians of Philadelphia, Historical Library, 19 S 22nd St, ***Philadelphia***, PA 19103-3097
T: +1 215 5633737; Fax: +1 215 5616477; E-mail: histref@collphyphil.org; URL: www.collphyphil.org
1788; Andrea Lee Kenyon
Hist of medicine; Arch of the College of Physicians, Samuel Gross Libr of Surgery, Gerontology (Joseph T Freeman Coll), William Harvey, Helfand-Radbill Medical Bookplate Coll, Samuel Lewis

Curio Coll, Medical Autogr Coll, Medical Portraits Coll
350 000 vols; 12 curr per; 450 incunabula; 34 000 diss/theses; 1 mio mss items, 10 000 autographs, 18 000 portraits, engravings, pictures
libr loan   32252

Drexel University, W.W. Hagerty Library, 33rd & Market Sts, *Philadelphia*, PA 19104-2875; mail address: 3141 Chestnut St, Philadelphia, PA 19104-2875
T: +1 215 8952769; Fax: +1 215 8952070; E-mail: qmlib@drexel.edu;
URL: www.library.drexel.edu
1891; Jane Bryan
Charles Lukens Huston Ethics Coll, Hist of the Bk
375 000 vols; 8 950 curr per; 10 000 e-books, 781 000 microforms; 414 sound-rec; 1 digital data carriers   32253

– Queen Lane Library, 2900 Queen Lane, Philadelphia, PA 19129
T: +1 215 9918740; Fax: +1 215 8430840; URL: www.library.drexel.edu
Martha Kirby
Anatomy, immunology, pharmacology, biochemistry
10 000 vols; 8 curr per; 80 000 e-books   32254

Holy Family University, Library, 9801 Frankford Ave, *Philadelphia*, PA 19114
T: +1 267 3413584; Fax: +1 215 6328067; E-mail: reference@holyfamily.edu; URL: www.holyfamily.edu/library
1954; Lori A. Schwabenbauer
Curriculum Libr Arch, Newtown Learning Resource Ctr
107 000 vols; 1 008 curr per; 5 000 e-books; 4 000 microforms; 400 sound-rec; 142 digital data carriers; 7 online database subscriptions
libr loan; ALA, CLA, PALINET   32255

La Salle University, Connelly Library, 1900 W Olney Ave, *Philadelphia*, PA 19141-1199
T: +1 215 9511287; Fax: +1 215 9511595; URL: www.lasalle.edu/library
1863; John S. Baky
Germantowniana Coll, Graham Green, Japanese Tea Ceremony, Lasalliana, Katherine Ann Porter Coll, Walker Percy Coll, Charles Willson Peale Coll, Vietnam War, Fiction; Imaginative Representations of the Vietnam War
365 000 vols; 4 789 curr per; 1 088 av-mat; 600 sound-rec; 6 digital data carriers; 216 Journals
libr loan; OCLC, PALINET   32256

Moore College of Art & Design, Library, 20th St & The Parkway, *Philadelphia*, PA 19103-1179
T: +1 215 9654054; Fax: +1 215 9658544; URL: library.moore.edu
1848; Sharon Watson-Mauro
Bookworks Coll, artists' bks; Joseph Moore Jr Coll; John Sartain Coll (engravings & prints); Sartain Family Correspondence; Philadelphia School of Design for Women Arch
40 000 vols; 110 curr per; 562 av-mat; 1 400 sound-rec; 7 digital data carriers; 400 folios, 98 VF drawers, 117 000 slides   32257

Peirce College, Library, 1420 Pine St, *Philadelphia*, PA 19102-4699
T: +1 215 6709269; Fax: +1 215 6709338; E-mail: library@peirce.edu; URL: library.peirce.edu
1963; Debra S. Schrammel
Business; Law Coll, Pre-1900 Business Textbks
44 000 vols; 161 curr per
libr loan   32258

Philadelphia College of Osteopathic Medicine, O.J. Snyder Memorial Library, 4170 City Ave, *Philadelphia*, PA 19131-1694
T: +1 215 8716470; Fax: +1 215 8716478; E-mail: library@pcom.edu; URL: www.pcom.edu/library
1899; Etheldra Templeton
Archival Hist of Medicine; First Editions in Osteopathy; Osteopathic Periodicals
39 000 vols; 525 curr per; 323 microforms; 1 775 av-mat; 3 809 sound-rec; 8 780 slides
libr loan   32259

Philadelphia University, Paul J. Gutman Library, School House Lane & Henry Ave, *Philadelphia*, PA 19144-5497; mail address: 4201 Henry Ave, Philadelphia, PA 19144
T: +1 215 9512848; Fax: +1 215 9512574; URL: www.philau.edu/library
1884; Kathy Mulroy
Textiles, Fashion, Design; Textile Hist Coll
110 000 vols; 950 curr per; 10 000 microforms; 780 sound-rec; 9 digital data carriers   32260

Saint Joseph's University, Francis A. Drexel Library, 5600 City Ave, *Philadelphia*, PA 19131-1395
T: +1 610 6601905; Fax: +1 610 6601916; URL: www.sju.edu/libraries/drexel
1851; Evelyn Minick
Religion & theology, Hist; SJU Publs Coll
356 000 vols; 1 418 curr per; 7 480 e-journals; 816 000 microforms; 75 digital data carriers
libr loan   32261

Temple University, Samuel Paley Library, Paley Library (017-00), 1210 W Berks St, *Philadelphia*, PA 19122-6088
T: +1 215 2040744; Fax: +1 215 2045201; URL: library.temple.edu
1892; Larry Alford
Blockson African American Coll, Contemporary Culture Coll, Conwellana-Templana Coll, Rare Book & Mss Coll, Philadelphia Urban Colls
2 972 000 vols; 23 567 curr per
libr loan; PALINET, OCLC, RLG, CNI, ARL   32262

– Ambler Library, 580 Meetinghouse Rd, Ambler, PA 19002
T: +1 267 4688640; Fax: +1 267 4688641
Sandra Thompson
90 000 vols; 446 curr per   32263

– Health Science Center Libraries, 3440 N Broad St, Philadelphia, PA 19140
T: +1 215 7072850; Fax: +1 215 7074135; URL: eclipse.hsclib.temple.edu
1901; Mark Allen Taylor
Medical Hist Coll
131 000 vols; 1 410 curr per
libr loan   32264

– Law Library, Charles Klein Law Bldg, 1719 N Broad St, Philadelphia, PA 19124
T: +1 215 2047891; Fax: +1 215 2041785; URL: www2.law.temple.edu
1897; John Necci
Hist Trials (Temple Univ Trials Coll)
550 000 vols; 850 curr per; 647 000 microforms; 18 digital data carriers
libr loan; AALL, GPLAA, East Asian Libraries Association   32265

– Science, Engineering & Architecture Library, College of Engineering & Architecture, 1947 N 12th St, Room 201, Philadelphia, PA 19122
T: +1 215 2047828; Fax: +1 215 2047720; E-mail: gsneff@temple.edu; URL: www.library.temple.edu/seilib/engr.htm
1921; Gretchen Sneff
26 000 vols; 306 curr per; 200 av-mat   32266

– Tyler School of Art Library, 7725 Penrose Ave, Elkins Park, PA 19027
T: +1 215 7822849; Fax: +1 215 7822799
1935
34 000 vols; 118 curr per; 6 VF drawers of pictures   32267

Temple University School of Podiatric Medicine, Charles E. Krausz Library, 8th and Race Streets, 6th Floor, *Philadelphia*, PA 19107
T: +1 215 6290300; Fax: +1 215 6291622; URL: library.temple.edu/help/numbers/?bhcp=1
1962
Anthony Sabatella Coll; Ctr for the Hist of Foot Care; Stewart E Reed Coll, bks, prints, per, monogr
10 500 vols; 350 curr per; 50 microforms; 1 000 av-mat; 2 digital data carriers; 2 500 pamphlets, 16 VF drawers
libr loan; MLA, SLA   32268

Thomas Jefferson University, Scott Memorial Library, 1020 Walnut St, *Philadelphia*, PA 19107
T: +1 215 5036773; Fax: +1 215 9233203; URL: jeffline.jefferson.edu
1896; Edward W. Tawyea
Medicine & allied health sciences; Obstetrics & Gynecology (Bland Coll)
204 000 vols; 1 872 curr per; 11 000 microforms; 763 av-mat; 221 sound-rec; 26 digital data carriers; 446 Slides
libr loan   32269

University of Pennsylvania, University Libraries, 3420 Walnut St, *Philadelphia*, PA 19104-6206
T: +1 215 8987555; Fax: +1 215 8980559; E-mail: library@pobox.upenn.edu; URL: www.library.upenn.edu
1750; Carton Rogers
Church Hist, Spanish Inquisition, Canon Law & Witchcraft (Henry Charles Lea Libr), Shakespeariana, Tudor & Stuart Drama (Horace Howard Furness Libr), Hist of Alchemy & Chemistry
5 971 000 vols; 13 870 e-journals; 285 000 e-books; 12 170 av-mat   32270

– Annenberg School of Communication, 3620 Walnut, Philadelphia, PA 19104-6220
T: +1 215 8987027; Fax: +1 215 8985388
Sharon Black
Mass media, history and technology of communication, attitude and opinion research
10 000 vols; 305 curr per   32271

– Biddle Law Library, Tanenbaum Hall, Flrs 3-5, 3443 Sansom St, Philadelphia, PA 19104; mail address: 3460 Chestnut St, Philadelphia, PA 19104-3406
T: +1 215 8987488; Fax: +1 215 8986619; URL: www.law.upenn.edu/bll/
1886; Paul M. George
Papers of Judge David L Bazelon, Trent Coll on the Black Lawyer in America, American Law Institute Arch, Nat Conference of Commissioners on Uniform State Laws Arch, 16 C English Yearbks
601 000 vols; 819 000 microforms   32272

– Biomedical Library, Johnson Pavilion, 3610 Hamilton Walk, Philadelphia, PA 19104-6060
T: +1 215 8985817; Fax: +1 215 5734143; E-mail: penav@mail.med.upenn.edu; URL: www.library.upenn.edu/biomed
1931; Valerie Pena
200 000 vols; 2 928 curr per; 351 av-mat   32273

– Chemistry Library, 3301 Spruce St, 5th Flr, Philadelphia, PA 19104-6323
T: +1 215 8982177; Fax: +1 215 8980741; E-mail: chemlib@pobox.upenn.edu
1967; Judith Currano
39 000 vols   32274

– Dental Library, 240 S 40th St, Philadelphia, PA 19104-6030
T: +1 215 8988969; Fax: +1 215 8987985
Pat Heller
Dental Patents, Thomas W Evans Hist Docs, Dental Cat, Foreign Dental Diss, Rare Dental Bks
59 000 vols; 1 digital data carriers
libr loan   32275

– Engineering & Applied Science Library, 220 S 33rd St, Philadelphia, PA 19104-6315
T: +1 215 8987266; Fax: +1 215 5732011
1947; Douglas McGee
Fluid Mechanics, Heat Transfer, NASA Rpts, Robotics
44 000 vols; 100 av-mat   32276

– Fisher Fine Arts Library, Furness Bldg, 220 S 34th St, Philadelphia, PA 19104-6308
T: +1 215 8988325; Fax: +1 215 5732066
1890; William Keller
Rare Architectural Bks, 16th to 20th c
160 000 vols; 59 000 mounted photos, 450 000 35mm slides, 30 000 digitized images   32277

– High Density Storage, 3001 Market St, Ste 10, Philadelphia, PA 19104-6316
T: +1 215 5735662; Fax: +1 215 5735660; E-mail: storage@pobox.upenn.edu
Andrea Loigman
1 200 000 vols   32278

– Lippincott-Wharton School, Library, 3420 Walnut St, Philadelphia, PA 19104-6207
T: +1 215 8985924; Fax: +1 215 8982261; E-mail: lippinco@wharton.upenn.edu
1927; Michael Halperin
Business & finance, Marketing; Corp Annual Rpts; New York Stock Exchange & American Stock Exchange, microfiche; Financial, Investment Sources; Lipman Criminology Coll
189 000 vols; 228 000 microforms   32279

– Math-Physics-Astronomy Library, David Rittenhouse Lab, 209 S 33rd St, Philadelphia, PA 19104-6317
T: +1 215 8988173; Fax: +1 215 5732009; E-mail: mpalib@pobox.upenn.edu
1948; Doug McGee
53 000 vols   32280

– Museum Library, 3260 South St, Philadelphia, PA 19104-6324
T: +1 215 8984021; Fax: +1 215 5732008; E-mail: muselib@pobox.upenn.edu
1898; John Weeks
Anthropology, Archaeology, Ethnology; Egyptology, Aboriginal American Linguistics & Ethnology (Daniel Garrison Brinton Coll)
140 000 vols; 100 mss; 125 av-mat
libr loan   32281

– Rare Book & Manuscript Library, 3420 Walnut St, Philadelphia, PA 19104
T: +1 215 8987088; E-mail: rbml@pobox.upenn.edu; URL: www.library.upenn.edu/collections/rbm
David McKnight
250 000 vols   32282

– Veterinary Library, 380 S University Ave, Philadelphia, PA 19104-6008
T: +1 215 8988895; Fax: +1 215 5732007; E-mail: vetlib@pobox.upenn.edu; URL: www.library.upenn.edu/vet
1908; Barbara Cavanaugh
Fairman Rogers Coll on Equitation & Horsemanship
34 000 vols; 18 VF drawers
libr loan; AMLA   32283

University of the Arts, University Libraries, Anderson Hall, 1st Flr, 333 S Broad St, *Philadelphia*, PA 19102; mail address: 320 S Broad St, Philadelphia, PA 19102-4994
T: +1 215 7176283; Fax: +1 215 7176287; URL: library.uarts.edu
1876; Carol Graney
Visual and Performing Arts, Book Arts Coll, Textiles Coll; Music Library
154 000 vols; 382 curr per; 14 978 music scores; 461 microforms; 17 930 sound-rec; 100 digital data carriers; 117 429 Pictures
libr loan; PALINET   32284

University of the Sciences in Philadelphia, Joseph W. England Library, 4200 Woodland Ave, *Philadelphia*, PA 19104-4491
T: +1 215 5968960; Fax: +1 215 5968760; URL: www.usip.edu/library
1822; Charles J. Myers
College Arch; Rare Books
84 000 vols; 318 curr per; 40 e-books; 34 000 microforms; 109 digital data carriers; videos
libr loan   32285

Alderson-Broaddus College, Pickett Library, College Hill Rd, *Philippi*, WV 26416
T: +1 304 4576229; Fax: +1 304 4576239; URL: www.ab.edu
David E. Hoxie
Education, Health sciences; Baptist Arch Coll, Civil War Coll
60 000 vols; 89 curr per; 50 000 e-books; 9 000 Electronic Media & Resources   32286

Grand Canyon University, Fleming Library, 3300 W Camelback Rd, ***Phoenix***, AZ 85017-3030; P.O. Box 11097, Phoenix, AZ 85061-1097
T: +1 602 5892420; Fax: +1 602 5892895; E-mail: library@gcu.edu; URL: library.gcu.edu
1949; Suella H. Baird
Business, Education, Religion, Nursing; Children's Lit (Vera Butler Coll), Music Rec (Brantner Coll), Rare Books (William Schattner Coll)
87000 vols; 1174 curr per; 18000 e-books; 49860 govt docs; 80720 microforms; 5287 av-mat; 4000 sound-rec; 1208 digital data carriers
libr loan; CLC 32287

Southwestern College, Dr R. S. Beal Sr Library, 2625 E Cactus Rd, ***Phoenix***, AZ 85032-7097
T: +1 602 9926101; Fax: +1 602 4042159; E-mail: library@swcaz.edu; URL: www.library.swcaz.edu
1960; Alice Eickmeyer
27000 vols; 130 curr per; 19000 e-books; 20000 microforms; 566 av-mat; 16 digital data carriers
libr loan 32288

J. Robert Ashcroft Memorial Library, 1401 Charlestown Rd, ***Phoenixville***, PA 19460
T: +1 610 9172001; Fax: +1 610 9172008; E-mail: research@vfcc.edu; URL: www.vfcc.edu
1939; Paul James Mathias III
Bible, Religion; Pentecostalism Coll
67000 vols; 175 curr per; 3 digital data carriers 32289

Pikeville College, Frank M. Allara Library, 147 Sycamore St, ***Pikeville***, KY 41501-9118
T: +1 606 2185605; Fax: +1 606 2185613; URL: library.pc.edu
1920; Karen S. Chafin
Material about Pikeville College, Kentucky, Appalachia; 213 linear feet of mss and archives
71000 vols; 357 curr per; 31000 microforms; 6 digital data carriers
libr loan; ALA 32290

Melville Library, UAMS-AHEC, Pine Bluff, 4010 S Mulberry St, ***Pine Bluff***, AR 71603
T: +1 870 5417629; Fax: +1 870 5417628
Julie Dobbins
Medicine, Nursing
185000 vols 32291

University of Arkansas, Pine Bluff, Watson Memorial Library-Learning & Instructional Resources Centers, 1200 N University Dr, ***Pine Bluff***, AR 71601
T: +1 870 5758411; Fax: +1 870 5754651; URL: www.uapb.edu
1938; Edward J. Fontenette
Industrial arts, Nursing; Afro-American Coll, Literature (Rare Bks Coll), Know Nelson Coll
351000 vols; 34301 goc docs
libr loan 32292

Louisiana College, Richard W. Norton Memorial Library, 1140 College Blvd, ***Pineville***, LA 71359
T: +1 318 4877184; Fax: +1 318 4877143; URL: norton.lacollege.edu/lacollege/default.asp
1906; Terry Martin
Baptist Hist Coll
165000 vols; 502 curr per 32293

Pittsburg State University, Leonard H. Axe Library, 1605 S Joplin St, ***Pittsburg***, KS 66762-5889; mail address: 1701 S Broadway, Pittsburg, KS 66762-5876
T: +1 620 2354890; Fax: +1 620 2354090; E-mail: libr@pittstate.edu; URL: library.pittstate.edu
1903; Robert A. Walter
Southeast Kansas Coll, Univ Arch, Haldemann-Julius Coll
498000 vols; 3425 curr per; 2006 digital data carriers; 145 Bks on Deafness & Sign Lang
libr loan 32294

Art Institute of Pittsburgh, 420 Boulevard of the Allies, ***Pittsburgh***, PA 15219-1328
T: +1 412 2916357; Fax: +1 412 2633715; URL: www.aip.aii.edu; www.aip.artinstitutes.edu/pittsburgh
Kathleen S. Ober
28000 vols; 185 curr per 32295

Carlow College, Grace Library, 3333 Fifth Ave, ***Pittsburgh***, PA 15213
T: +1 412 5786137; Fax: +1 412 5786242; E-mail: gracelibrary@carlow.edu; URL: library.carlow.edu
1929; Elaine J. Misko
American lit, Education, Theology, English, Irish; Black Studies, Career Resources, Peace Studies
106000 vols; 363 curr per
libr loan 32296

Carnegie Mellon University, University Libraries, Hunt Library, 4909 Frew St, ***Pittsburgh***, PA 15213-3890
T: +1 412 2682446; Fax: +1 412 2682793; E-mail: huntref@andrew.cmu.edu; URL: www.library.cmu.edu
1920; Dr. Gloriana St Clair
Bookbindings (incl Edwards of Halifax), Early Scientific Works, Important Early Printers (Aldus, Plantin & Estienne Coll), Private Presses (Kimscott & Doves Coll), Architecture Arch
1021000 vols; 2945 curr per; 639000 microforms
libr loan 32297

– Software Engineering Institute Library, 4500 Fifth Ave, Pittsburgh, PA 15213-2612
T: +1 412 2687733; Fax: +1 412 2685758; E-mail: library@sei.cmu.edu
Sheila Rosenthal
13000 vols; 245 curr per 32298

Chatham College, Jennie King Mellon Library, Woodland Rd, ***Pittsburgh***, PA 15232
T: +1 412 3651245; Fax: +1 412 3651465; URL: www.chatham.edu/academics/library.cfm
1869; Jill Ausel
Women's studies; African-American (Wray Coll), Mayan Art & Civilization (Snowdon Coll)
97000 vols; 365 curr per 32299

Duquesne University, Gumberg Library, 600 Forbes Ave, ***Pittsburgh***, PA 15282
T: +1 412 3966133; Fax: +1 412 3961800; E-mail: saunders@duq.edu; URL: www.duq.edu/library/
1928; Dr. Laverna Saunders
Medieval Judaic-Christian Relations (Rabbi Herman Hailperin), Simon Silverman Phenomenology Coll, Cardinal John Wright Coll, Judge Michael Musmanno Coll
601000 vols; 552 curr per; 1370 e-journals; 707 e-books; 38940 Electronic Media & Resources, 102 High Interest/Low Vocabulary Bks, 190 Bks on Deafness & Sign Lang 32300

– Duquesne University School of Law, Duquesne University Center for Legal Information, 600 Forbes Ave, Pittsburgh, PA 15282; mail address: 900 Locust St, Pittsburgh, PA 15282
T: +1 412 3965017; Fax: +1 412 3966294; URL: www.lawlib.duq.edu
1911; Frank Y. Liu
266000 vols; 5126 curr per; 58000 microforms; 61 digital data carriers
AALL 32301

La Roche College, John J. Wright Library, 9000 Babcock Blvd, ***Pittsburgh***, PA 15237
T: +1 412 5361063; Fax: +1 412 5361062; E-mail: laverne.collins@laroche.edu; URL: intranet.laroche.edu/Library
1963; LaVerne P. Collins
61000 vols; 604 curr per
libr loan 32302

Point Park University Library, 414 Wood St, ***Pittsburgh***, PA 15222
T: +1 412 3923171; Fax: +1 412 3923168; E-mail: library@pointpark.edu; URL: www.pointpark.edu
Elizabeth Evans
95000 vols; 270 curr per; 30000 e-books 32303

University of Pittsburgh, Oakland Campus Libraries, ***Pittsburgh***, PA 15260
URL: www.library.pitt.edu
1873; Rush Miller

– Bevier Engineering Library, 126 Benedum Hall, Pittsburg, PA 15261
T: +1 412 6249620; Fax: +1 412 6248103; E-mail: engineering@library.pitt.edu; URL: www.library.pitt.edu/libraries/engineering/engineer.html
1956; Kate Thomes
35000 vols; 1012 curr per; 3926 govt docs; 84409 microforms 32304

– Business Library, 118 Mervis Hall, Pittsburgh, PA 15260
T: +1 412 6481669; Fax: +1 412 6481586; E-mail: business@library.pitt.edu; URL: www.library.pitt.edu/libraries/business/business.html
1961; Eve Wider
15000 vols; 600 curr per; 211000 microforms 32305

– Chemistry Library, 200 Eberly Hall, Pittsburg, PA 15261
T: +1 412 6248294; Fax: +1 412 6248296; E-mail: chemistry@library.pitt.edu; URL: www.library.pitt.edu/libraries/chemcomp/chem_comp.html
1975; Margarete Bower
50000 vols; 350 curr per; 2605 microforms 32306

– Graduate School of Public & International Affairs/Economics, Library, 1G12 Wesley W Posvar Hall, Pittsburg, PA 15260
T: +1 412 6487575; Fax: +1 412 6487569; E-mail: sgn@pitt.edu; URL: www.library.pitt.edu/libraries/gspia/economics.html
1958; Susan Neumann
135051 vols; 750 curr per; 3505 microforms 32307

– Henry Clay Frick Fine Arts Library, Frick Fine Arts Bldg, 1rst Flr, Pittsburg, PA 15260
T: +1 412 6482410; Fax: +1 412 6487568; E-mail: frickart@pitt.edu; URL: www.library.pitt.edu/libraries/frick/fine_arts.html
1927; James P. Cassaro
Medieval Illuminated Mss Facsimiles
83136 vols; 334 curr per; 22781 microforms; 1518 pamphlets 32308

– Hillman Library, 3960 Forbes Ave, Pittsburg, PA 15260
T: +1 412 6483330; Fax: +1 412 6487887; E-mail: rgmiller@pitt.edu; URL: www.library.pitt.edu/libraries/hillman/hillman.html
Rush Miller
Rare Books and General Mss Colls, Arch of Popular Culture, Arch of Scientific Philosophy in the Twentieth-First Century, Bernard S Horne Memorial Coll, Fidelis Zitterbart Coll, Flora and Norman Winkler Coll, Ford E and Harriet R Curtis Theatre Coll, Hervey Allen Coll, Lawrence Lee Coll, Mary Roberts Rinehart Coll, Pavlova-Herinrich Dance Coll, Ramon Gomez de la Serna Coll, Ripon England Docs, Robert Watson Coll, William Steinberg Coll, World War II Picture Coll; African American Coll, Alldred Coll, Alliance College Polish Coll, Buhl Social Work Coll, East Asian Libr, Eduardo Lozano Latin American Coll, Gertrude and Philip Hofman Judaic Coll, Geographic Info Service, Govnt Publs Coll, Japan Information Centre, Map Coll, Media Resources, Microforms Coll, Periodicals Coll, Special Colls
1500000 vols; 2200000 microforms 32309

– Information Sciences Library, 135 N Bellefield Ave, Pittsburg, PA 15260
T: +1 412 6244710; Fax: +1 412 6244062; E-mail: information-sciences@library.pitt.edu; URL: www.library.pitt.edu/libraries/is/info_sci.html
1966; Elizabeth T. Mahoney
Elizabeth Nesbitt Room (Hist Children's Lit); Clifton Fadiman Coll of 20th C Children's Lit; Children's Television Arch; Mr Roger's Neighborhood Arch
87912 vols; 833 curr per; 22703 microforms 32310

– Langley Library, 217 Langley Hall, Pittsburg, PA 15260
T: +1 412 6244490; Fax: +1 412 6241809; E-mail: langley@library.pitt.edu; URL: www.library.pitt.edu/libraries/langley/langley.html
1961; Laura McIntyre
Neuroscience, biological sciences, psychology, life sciences
70000 vols; 650 curr per; 4300 microforms 32311

– Mathematics Library, 430 Thackeray Hall, Pittsburg, PA 15260
T: +1 412 6248205; Fax: +1 412 6243809; E-mail: mathematics@library.pitt.edu; URL: www.library.pitt.edu/libraries/math/math.html
1963; Michael Ford
25000 vols; 285 curr per; 1000 microforms 32312

– Theodore M. Finney Music Library, B-28 Music Bldg, Pittsburg, PA 15260
T: +1 412 6244130; Fax: +1 412 6244180; E-mail: music@library.pitt.edu; URL: www.library.pitt.edu/libraries/music/music.html
1966; James P. Cassaro
65000 vols; 150 curr per; 130 diss/theses; 1500 microforms; 500 av-mat; 25000 sound-rec; 17 digital data carriers
libr loan 32313

– Western Psychiatric Institute & Clinic Library, Health Sciences Library System, Thomas Detre Hall, 3811 O'Hara St, Pittsburg, PA 15213
T: +1 412 6242378; Fax: +1 412 2465490; E-mail: wpicref@pitt.edu; URL: www.hsls.pitt.edu
1942; Jim Fischerkeller
30000 vols; 100 curr per; 70000 mss; 900 av-mat; 250 sound-rec
libr loan; AMHL, MLA, SLA 32314

University of Pittsburgh – Barco Law Library, Law Bldg, 3900 Forbes Ave, 4th Flr, ***Pittsburgh***, PA 15260
T: +1 412 6481330; Fax: +1 412 6481352; URL: www.law.pitt.edu/library
1915; George Pike
185000 vols; 4674 curr per; 1147000 microforms; 25 digital data carriers 32315

University of Pittsburgh – Falk Library of the Health Sciences, 200 Scaife Hall, DeSoto & Terrace Sts, ***Pittsburgh***, PA 15261
T: +1 412 6482037; Fax: +1 412 6489020; E-mail: medlibq+@pitt.edu; URL: www.hsls.pitt.edu
1957; Barbara A. Epstein
Hist of Medicine Coll; Consumer Health Coll
447000 vols; 3182 curr per; 1700 e-journals; 1000 av-mat; 500 sound-rec; 244 digital data carriers 32316

Goddard College, Eliot D. Pratt Library, 123 Pitkin Rd, ***Plainfield***, VT 05667
T: +1 802 4548311; Fax: +1 802 4541451; E-mail: library@goddard.edu; URL: www.goddard.edu
1938; Clara Bruns
BA, MA & MFA Theses, Goddard Authors Coll, Goddard Arch
50000 vols; 31000 e-books 32317

Wayland Baptist University, Mabee Learning Resources Center, 1900 W Seventh, ***Plainview***, TX 79072-6957
T: +1 806 2913700; Fax: +1 806 2911964; URL: www.wbu.edu/lrc/library.asp
1908; Dr. Polly Lackey
119000 vols; 541 curr per; 148000 microforms; 4309 sound-rec; 15 digital data carriers; 3869 Slides, 348 Overhead Transparencies
libr loan 32318

University of Wisconsin – Platteville, Elton S. Karrmann Library, One University Plaza, ***Platteville***, WI 53818-3099
T: +1 608 3421668; Fax: +1 608 3421645; E-mail: reference@uwplatt.edu; URL: www.uwplatt.edu/library
1866; John Krogman
Arch (Area Res Ctr), mss; Regional Hist
241000 vols; 927 curr per; 13 e-books; 11500 av-mat
libr loan 32319

State University of New York College at Plattsburgh, Benjamin F. Feinberg Library, Two Draper Ave, *Plattsburgh*, NY 12901-2697
T: +1 518 5645182; Fax: +1 518 5645295; URL: www.plattsburgh.edu/library
1889; Cerise G. Oberman
Hist of Northern New York, M Lansing Porter Folklore Coll, Rockwell Kent Coll
367 000 vols; 4 052 curr per; 2 520 e-journals; 7 000 e-books; 24 000 av-mat; 240 digital data carriers
libr loan 32320

John F. Kennedy University, Robert M. Fisher Library, 100 Ellinwood Way, *Pleasant Hill*, CA 94523
T: +1 925 9693100; Fax: +1 925 9693101; URL: library.jfku.edu
1964; Claudia Chester
Business and management, Education, Holistic studies, Liberal arts, Museology, Psychology
551 000 vols; 438 curr per; 1 000 e-books; 1 000 av-mat; 600 sound-rec 32321

– Law Library, 547 Ygnacio Valley Rd, Walnut Creek, CA 94596
T: +1 925 9693100; E-mail: reference@jfku.edu; URL: library2.jfku.edu/Law_Library.html
25 000 vols; 150 curr per 32322

Pace University, Edward & Doris Mortola Library, 861 Bedford Rd, *Pleasantville*, NY 10570-2799
T: +1 914 7733381; Fax: +1 914 7733508; URL: library.pace.edu/
1963; William J. Murdock
Saint Joan of Arc, Rene Dubos
215 000 vols; 111 curr per; 20 180 e-journals; 10 000 e-books; 17 000 microforms; 280 av-mat; 580 digital data carriers 32323

Plymouth State University, Herbert H. Lamson Library, Highland St, *Plymouth*, NH 03264-1595
T: +1 603 5352455; Fax: +1 603 5352445; URL: www.plymouth.edu/library
David Berona
350 000 vols; 1 100 curr per 32324

Idaho State University, E.M. Oboler Library, Idaho State University, 850 S Ninth Ave, *Pocatello*, ID 83209-8089; mail address: 921 S Eighth Ave, Stop 8089, Pocatello, ID 83209-8089
T: +1 208 2823248; Fax: +1 208 2825847; E-mail: refdesk@isu.edu; URL: www.isu.edu/library/
1902; Kay A. Flowers
Health Sciences; Intermountain West Coll, Richard Stallings Papers, Samuel Johnson Coll
517 000 vols; 4 711 curr per; 5 900 e-journals; 78 e-books; 42 884 maps; 1 700 000 microforms; 723 digital data carriers 32325

College of the Ozarks, Lyons Memorial Library, One Opportunity Ave, *Point Lookout*, MO 65726; P.O. Box 17, Point Lookout, MO 65726-0017
T: +1 417 3346411 ext 3411; Fax: +1 417 3343085; URL: www.cofo.edu/library/default.asp
1906; Judy Holmes
Ozarkiana Coll
121 000 vols; 715 curr per 32326

California State Polytechnic University Library, 3801 W Temple Ave, Bldg 15, *Pomona*, CA 91768
T: +1 909 8693111; Fax: +1 909 8694375; URL: www.csupomona.edu/~library
1938; Harold B. Schleifer
W.K. Kellogg Libr, English and American Lit
760 000 vols; 2 384 curr per; 3 460 e-journals; 12 000 e-books; 12 313 maps; 2 900 000 microforms; 4 460 sound-rec; 8 digital data carriers; 6 377 Electronic Media & Resources
libr loan 32327

Richard Stockton College of New Jersey, Library, Jim Leeds Rd, *Pomona*, NJ 08240; P.O. Box 195, Pomona, NJ 08240-0195
T: +1 609 6524343; Fax: +1 609 6524964; E-mail: librarian@stockton.edu; URL: library.stockton.edu

1971; David Pinto
New Jersey Pine Barrens Coll, Holocaust Coll
237 000 vols; 1 065 curr per
libr loan 32328

Western University of Health Sciences, Health Sciences Library, 287 E Third St, *Pomona*, CA 91766-1854; mail address: 309 E Second St, Pomona, CA 91766
T: +1 909 4695320; Fax: +1 909 4695486; E-mail: reference@westernu.edu; URL: www.westernu.edu/library/index.html
1977; Patricia A. Vader
Medicine; Osteopathic Lit
14 000 vols; 374 curr per 32329

University of Texas at Austin – Marine Science Library, Marine Science Institute, 750 Channelview Dr, *Port Aransas*, TX 78373-5015
T: +1 361 7496723; Fax: +1 361 7496725; URL: www.lib.utexas.edu/msl
1941; Liz DeHart
Gulf of Mexico
40 000 vols; 100 curr per; 5 000 microforms; 30 000 rpts
libr loan; IAMSLIC, SAIL 32330

Lamar State College, Gates Memorial Library, 317 Stilwell Blvd, *Port Arthur*, TX 77640; P.O. Box 310, Port Arthur, TX 77641-0310
T: +1 409 9846220; Fax: +1 409 9846080; URL: library.lamarpa.edu
1909; Peter B. Kaatrude
Texana Coll
60 000 vols; 200 curr per; 28 000 e-books 32331

Florida Atlantic University – Port Saint Lucie Branch, Library, 180 SW Prima Vista Blvd, *Port Saint Lucie*, FL 34983
T: +1 561 8715450; Fax: +1 561 8715454; URL: www.stlucieco.gov/library/psl_branch.htm
1970
46 460 vols 32332

Eastern New Mexico University, Golden Library, 1500 S Ave K, Sta 32, *Portales*, NM 88130-7402; Sta 32, Portales, NM 88130
T: +1 505 5622624; Fax: +1 505 5622647; E-mail: Golden.Library@enmu.edu; URL: www.enmu.edu/academics/library/index.shtml
1934; Melveta Walker
Business, Education, Art; Harold Runnels Coll, Textbk Review Ctr, Williamson Science Fiction Coll
270 940 vols; 15 264 curr per; 11 000 e-books; 23 818 av-mat; 540 Talking Bks
libr loan 32333

Art Institute of Portland Library, 1122 NW Davis St, *Portland*, OR 97209-2911
T: +1 503 2286528; Fax: +1 503 2282895; URL: www.aii.edu
1966; Nancy Thurston
Design, Fashion hist, Furniture; Fashion Clip File (1965-1995)
24 000 vols; 200 curr per; 236 digital data carriers
libr loan; ALA, ARLIS/NA, SLA 32334

Concordia University, University Library, 2811 NE Holman St, *Portland*, OR 97211-6067
T: +1 503 2808507; Fax: +1 503 2808697; E-mail: library@cu-portland.edu; URL: www.cu-portland.edu/library
1905; Brent Mai
Religious Hist (Luther & Reformation Res Coll)
88 000 vols; 250 curr per; 23 000 e-journals; 73 000 e-books; 150 diss/theses; 200 000 microforms
libr loan; ALA, SLA 32335

Lewis & Clark College, *Portland*
– Aubrey R. Watzek Library, 0615 SW Palatine Hill Rd, Portland, OR 97219-7899
T: +1 503 7687274; Fax: +1 503 7687282; URL: library.lclark.edu
1867; Jim Kopp
Gender studies, Pacific Northwest hist; Lewis & Clark Expedition Coll, North American Indians Coll
290 000 vols; 292 000 microforms; 733 sound-rec; 158 digital data carriers 32336

– Paul L. Boley Law Library, Lewis & Clark Law School, 10015 SW Terwilliger Blvd, Portland, OR 97219
T: +1 503 7686776; Fax: +1 503 7686760; E-mail: lawlib@lclark.edu; URL: lawlib.lclark.edu
1884; Peter S. Nycum
Milton S Pearl Environmental Law Libr, Patent Law Coll, Samuel S Johnson Public Land Law Review Commission Coll
214 000 vols; 4 840 curr per 32337

E. W. McMillan Library at Cascade College, 9101 E Burnside, *Portland*, OR 97216-1515
T: +1 503 2571360; Fax: +1 503 2571222; E-mail: library@cascade.edu; URL: www.cascade.edu/lib
1957; Michael Clark
Bible, Business, Education, Environmental studies; First C Christianity to the Restoration Movement
34 000 vols; 245 curr per; 26 000 microforms; 2 digital data carriers 32338

Oregon Health & Science University, Library, 3181 SW Sam Jackson Park Rd, *Portland*, OR 97239-3098; P.O. Box 573, Portland, OR 97207-0573
T: +1 503 4943460; Fax: +1 503 4943227; URL: www.ohsu.edu/library
1919; Chris Shaffer
Hist of Medicine (Pacific Northwest & Rare Bk); Oregon Memorial Libr for Bereaved Parents
221 000 vols; 2 429 curr per; 2 digital data carriers; 973 Videos, 14 516 Slides 32339

Pacific Northwest College of Art, Charles Voorhies Fine Art Library, 1241 NW Johnson St, *Portland*, OR 97209
T: +1 503 8218966; URL: library.pnca.edu
Claire Rivers
18 000 vols; 220 curr per 32340

Portland State University, Branford Price Millar Library, 1875 SW Park Ave, *Portland*, OR 97201-3220; P.O. Box 1151, Portland, OR 97207-1151
T: +1 503 7254616; Fax: +1 503 7254524; URL: library.pdx.edu
1946; Helen H. Spalding
Middle East Studies
1 303 000 vols; 13 000 curr per; 35 530 e-journals; 477 622 govt docs; 55 635 maps; 2 289 000 microforms; 24 307 sound-rec; 3 370 digital data carriers
libr loan 32341

Reed College, Eric V. Hauser Memorial Library, 3203 SE Woodstock Blvd, *Portland*, OR 97202-8199
T: +1 503 7777750; Fax: +1 503 7777786; URL: www.library.reed.edu
1912; Victoria L. Hanawalt
592 000 vols
libr loan 32342

University of Maine School of Law, Donald L. Garbrecht Law Library, 246 Deering Ave, *Portland*, ME 04102
T: +1 207 7804829; Fax: +1 207 7804913; E-mail: lawlib@usm.maine.edu; URL: www.mainelaw.maine.edu/library
1962; Christopher A. Knott
325 000 vols; 1 175 curr per; 50 000 microforms of per 32343

University of New England – Westbrook College Campus, Josephine S. Abplanalp Library, Westbrook College Campus, 716 Stevens Ave, mail address: 11 Hills Beach Rd, *Portland*, ME 04103
T: +1 207 2214363; Fax: +1 207 2214893
1831; Andrew J. Golub
Maine Women Writers Coll
150 000 vols; 40 000 curr per; 6 000 e-books
libr loan 32344

University of Portland, Wilson W. Clark Memorial Library, 5000 N Willamette Blvd, *Portland*, OR 97203-5743
T: +1 503 9437111; Fax: +1 503 9437491; E-mail: library@up.edu; URL: library.up.edu
1901; Drew Harrington
Forestry (Daniel D Buckley Coll), Salvador J Macias Coll of Spanish Lit, David W Hazen Coll in American Hist, Anthony Juliano Drama Coll

360 000 vols; 1 600 curr per; 231 000 microforms; 4 357 sound-rec; 30 digital data carriers
libr loan 32345

University of Southern Maine, Glickman Family Library, 314 Forest Ave, *Portland*, ME 04104; P.O. Box 9301, Portland, ME 04104-9301
T: +1 207 7804272; Fax: +1 207 7804042; URL: library.usm.maine.edu
1878; David J. Nutty
Fine Printing Coll, Early Textbk Coll, Rice-Children's Coll, African American Coll, Lesbian & Gay Coll, Judaic Coll, Franco-American Heritage Coll, J Byers Simpson Ctr for Diversity in Maine Coll, Univ Arch, Antique Cartogr Mat
369 000 vols; 2 649 curr per; 650 sound-rec; 235 digital data carriers; 1 023 581 microforms of per
libr loan 32346

Warner Pacific College, Otto F. Linn Library, 2219 SE 68th Ave, *Portland*, OR 97215-4099
T: +1 503 5171102; Fax: +1 503 5171351; URL: www.warnerpacific.edu/library
1937; Sue Kopp
Church of God Arch
59 000 vols; 315 curr per; 40 diss/theses; 313 maps; 1 000 av-mat; 1 068 sound-rec; 25 digital data carriers; 300 non-print
libr loan; ALA, ACRL 32347

Western States Chiropractic College, W.A. Budden Library, 2900 NE 132nd Ave, *Portland*, OR 97230-3099
T: +1 503 2515752; Fax: +1 503 2515759; E-mail: librarian@wschiro.edu; URL: www.wschiro.edu
1909; Janet Tapper
Chiropractic History
14 000 vols; 300 curr per; 125 diss/theses
libr loan; MLA, SLA 32348

Shawnee State University, Clark Memorial Library, 940 Second St, *Portsmouth*, OH 45662-4344
T: +1 740 3513519; Fax: +1 740 3513432; URL: www.shawnee.edu/off/cml/index.html
1967; Connie Salyers Stoner
Louis A Brennan Coll, Albert Parry Coll, Jessie Stuart Coll, Southern Ohio Valley Writers, Vernal G Riffe Memorabilia
137 000 vols; 811 curr per
libr loan 32349

Clarkson University, Harriet Call Burnap Library, Andrew S. Schuler Educational Resources Center, Andrew S Schuler Educational Resources Center, Eight Clarkson Ave, *Potsdam*, NY 13699; P.O. Box 5590, Potsdam, NY 13699-5590
T: +1 315 2682297; Fax: +1 315 2687655; E-mail: refdesk@clarkson.edu; URL: www.clarkson.edu/library
1896; J. Natalia Stahl
142 000 vols; 100 curr per; 1 680 e-journals; 18 000 e-books; 1 316 diss/theses; 421 maps; 272 000 microforms; 700 av-mat; 2 500 sound-rec; 80 digital data carriers
libr loan 32350

State University of New York – College at Potsdam, Frederick W. Crumb Memorial Library, 44 Pierrepont Ave, *Potsdam*, NY 13676-2294
T: +1 315 2672485; Fax: +1 315 2672744; E-mail: libcrumb@potsdam.edu; URL: www.potsdam.edu/library.html
1816; J. Rebecca Thompson
Architecture, Art; Archs, Education, St Lawrence Seaway (Bertrand H Snell Papers)
238 480 vols; 500 curr per; 5 717 maps; 668 332 microforms; 64 001 800 av-mat; 2 764 sound-rec; 27 digital data carriers
libr loan 32351

– Julia E. Crane Memorial Library, Crane School of Music, Potsdam, NY 13676-2294
T: +1 315 2672451; Fax: +1 315 2672744; E-mail: libcrumb@potsdam.edu; URL: www.potsdam.edu/library.html
1925; Nancy Alzo
Music education, Performance; Music/Music Education (Julia E. Crane

Papers); Helen Hosmer Papers
45 000 vols; 70 curr per; 30 000 music scores; 1 552 microforms; 40 094 av-mat; 9 000 sound-rec; 1 digital data carriers
libr loan; MLA, IAML   32352

Marist College, James A. Cannavino Library, 3399 North Rd, *Poughkeepsie*, NY 12601-1387
T: +1 845 5753199; Fax: +1 845 5753150; URL: www.library.marist.edu
1929; Verne Newton
Hudson Valley Regional studies; Hudson River Environmental Society Coll, Maristiana, Lowell Thomas Coll, John Tillman Newscasts, Rick Whitsell R&B Rec Coll
197 000 vols; 629 curr per
libr loan   32353

Vassar College, Library, 124 Raymond Ave, Maildrop 20, *Poughkeepsie*, NY 12604-0020
T: +1 845 4375760; Fax: +1 845 4375864; URL: library.vassar.edu
1861; Sabrina L. Pape
College Hist & Arch, Early Atlases & Maps, Incunabula, Elizabeth Bishop Papers, John Burroughs Journals, Mark Twain Coll, Robert Owens Coll
890 000 vols
libr loan   32354

– Art Library, 124 Raymond Ave, P.O. Box 512, Poughkeepsie, NY 12604
T: +1 845 4375790; Fax: +1 845 4375864; E-mail: researchhelp@vassar.edu; URL: artlibrary.vassar.edu
1864
55 000 vols; 200 curr per   32355

Green Mountain College, Griswold Library, One College Circle, *Poultney*, VT 05764-1199
T: +1 802 2878225; Fax: +1 802 2878222
1834; Paul Millette
Environmental liberal arts; Libr of American Civilization, ultrafiche; Early American Decoration (Ramsey Coll); Welsh Coll, ultrafiche, bks, patterns & art objects
90 000 vols; 234 curr per
libr loan; OCLC, Nelinet   32356

Prairie View A&M University, John B. Coleman Library, P.O. Box 519, *Prairie View*, TX 77446-0519
T: +1 936 8572012; Fax: +1 936 8572755; URL: www.tamu.edu/pvamu/library
1878; Helen Yeh
African-American Heritage of the West Coll, African-Americans in the Military Coll, Black Lit (W D Lawless Coll)
333 060 vols; 872 curr per
libr loan   32357

Embry-Riddle Aeronautical University, Prescott Campus Library, 3700 Willow Creek Rd, *Prescott*, AZ 86301-3720
T: +1 928 7773761; Fax: +1 928 7776988; E-mail: prlib@erau.edu; URL: library.pr.erau.edu
1978; Sarah K. Thomas
Aircraft Accident Investigation, Kalusa Miniature Aircraft Coll
41 000 vols; 668 curr per; 8 000 e-books
libr loan   32358

Yavapai College, Library, 1100 E Sheldon St, Bldg 19, *Prescott*, AZ 86301
T: +1 928 7762260; Fax: +1 928 7762275; E-mail: library@yc.edu; URL: www.yc.edu/library
1969; Lisa Griest
Southwest; College Arch
114 000 vols; 557 curr per   32359

University of Maine at Presque Isle, Library, 181 Main St, *Presque Isle*, ME 04769-2888
T: +1 207 7689599; Fax: +1 207 7689644; URL: www.umpi.maine.edu/library
1903; Gregory Curtis
Education, Art; Aroostook County Hist Coll, Maine Coll, Rare bks
162 000 vols; 300 curr per
libr loan   32360

College of Eastern Utah, Library, 451 E & 400 N, *Price*, UT 84501
T: +1 435 613-5209; Fax: +1 435 613-5863; E-mail: Barbara.Steffee@ceu.edu; URL: library.ceu.edu/
1938; Barbara Steffee
47 500 vols; 170 curr per
libr loan   32361

University of Maryland – Eastern Shore, Frederick Douglass Library, Backbone Rd, *Princess Anne*, MD 21853
T: +1 410 6512200 ext 6621; Fax: +1 410 6516269; URL: www.fdl.umes.umd.edu
1968; Dr. Theodosia Shields
Physical therapy; Juvenile Coll, Maryland Coll
174 000 vols; 937 curr per   32362

Institute for Advanced Study, Mathematics-Natural Sciences Library, Einstein Dr, *Princeton*, NJ 08540
T: +1 609 7348181; Fax: +1 609 9248399; E-mail: mnlib@ias.edu; URL: www.admin.ias.edu/hslib/ls.htm
1940; Momota Ganguli
Art, Classical studies, Hist, Mathematics, Natural science, Social science; Hist of Science (Rosenwald Coll)
130 000 vols; 1 000 curr per
libr loan   32363

Princeton University, Firestone Library, One Washington Rd, *Princeton*, NJ 08544-2098
T: +1 609 2583272; Fax: +1 609 2584105; URL: libweb.princeton.edu
Karen Trainer
Ainsworth, Barrie, the Brontes, Bulwer-Lytton, Collins, Mrs Craik, Dickens, Disraeli, Dogson, George Eliot, Mrs Gaskell, Hardy, Hughes, Kingsley, Lever, Reade, Stevenson, Thackeray, Trollope (Parrish Coll of Victorian Novelists), Aeronautics Coll, Americana Coll, American Hist Mss Coll, European Legal Docs, Chess Coll, Mss, Vergil Coll, Theater Coll
5 315 000 vols; 37 629 curr per; 1 750 000 microforms
libr loan   32364

– Astrophysics Library, Peyton Hall, Ivy Lane, Princeton, NJ 08544-1001
T: +1 609 2583820; Fax: +1 609 2581020; E-mail: library@astro.princeton.edu; URL: www.astro.princeton.edu/library
Jane Holmquist
16 600 vols; 300 curr per; 500 slides
                                       32365

– Biology Library, Fine Hall, MCDonnell Hall, One Washington Ave, Princeton, NJ 08544
T: +1 609 2583235; Fax: +1 609 2582627; E-mail: biolib@princeton.edu; URL: biolib.princeton.edu
Steven Adams
59 000 vols; 800 curr per; 500 diss/theses   32366

– Chemistry Library, Frick Laboratory, One Washington Rd, Princeton, NJ 08544
T: +1 609 2583238; Fax: +1 609 2586746; E-mail: chemlib@princeton.edu; URL: chemlib.princeton.edu
1929; Juliette Arnheim
54 840 vols; 400 curr per; 1 400 diss/theses; 9 200 microforms   32367

– Department of Rare Books, One Washington Rd, Princeton, NJ 08544
T: +1 609 2583184; E-mail: rbsc@princeton.edu; URL: www.princeton.edu/~rbsc
Mark Farrell
200 000 vols   32368

– Donald E. Stokes Library – Public & International Affairs & Population Research, Wallace Hall, Princeton, NJ 08544
T: +1 609 2585455; Fax: +1 609 2586844; E-mail: piaprlib@princeton.edu; URL: stokeslib.princeton.edu/main.htm
Jacqueline Druery
55 000 vols; 500 curr per   32369

– East Asian Library & Gest Collection, 33 Frist Campus Ctr, Rm 317, Princeton, NJ 08544-1100
T: +1 609 2583182; Fax: +1 609 2584573; E-mail: gest@princeton.edu; URL: eastasianlib.princeton.edu
1937; Tai-Loi Ma
Buddhist Sutras, Sung & Yuan Editions, Ming Editions, Hishi Copies, Ming Works reproduced in Japan, Chinese Medicine & Materia Medica, Go Coll, Rare Bks
396 600 vols; 3 300 curr per; 3 000 mss; 25 978 microforms   32370

– Engineering Library, One Washington Rd, Princeton, NJ 08544
T: +1 609 2583200; Fax: +1 609 2587366; E-mail: englib@princeton.edu; URL: englib.princeton.edu
1963; Juliette Arnheim
DOE & NASA Rpts; Society of Automotive Engineers & American Institute of Aeronautics & Astronautics Conference Papers; IEEE, ACM, & ASME Conference Proceedings
124 550 vols; 1 400 curr per; 650 000 tech rpts & gov docs   32371

– Fine Hall Library-Mathematics, Physics & Statistics, One Washington Rd, Princeton, NJ 08544-0001
T: +1 609 2583187; Fax: +1 609 2582627; E-mail: finelib@princeton.edu; URL: www.priceton.edu/~finelib
1920; Ateven Adams
18th C German Mathematics Diss, High Energy & Statistics Preprints, Rare Mathematic Texts
124 770 vols; 692 curr per; 5 000 pamphlets, 10 VF drawers of theses, 16 VF drawers of undergraduate theses, 259 VF drawers of uncat pamphlets
                                       32372

– Geosciences & Maps Library, 1 Washington Rd, Fine Hall (B-Level), Princeton, NJ 08544
T: +1 609 2583267; Fax: +1 609 2584607; E-mail: geolib@princeton.edu; URL: www.princeton.edu/~geolib
1909; Patricia Gaspari-Bridges
76 000 vols; 1 400 curr per; 2 000 diss/theses; 297 000 maps; 850 tech rpts
libr loan   32373

– Graphic Arts Collection, One Washington Rd, Princeton, NJ 08544-2098
T: +1 609 2583197; Fax: +1 609 2582324; E-mail: davidson@princeton.edu; URL: www.princeton.edu/~rbsc/department/graphicarts
1940; Julie Mellby
Bookbinding & illust, Calligraphy, Hist of bks, Printmaking, Typography
30 000 vols   32374

– Harold P. Furth Library, Forrestal Campus, C-Site A 108, P.O. Box 451, Princeton, NJ 08543-0451
T: +1 609 2433565; Fax: +1 609 2432299; E-mail: ppllib@princeton.edu; URL: www.pppl.gov/library
1951; Adriana Popescu
18 010 vols; 60 curr per; 40 000 microforms, 3 000 project rpts, 20 000 tech rpts & reprints   32375

– Industrial Relations Library, One Washington Rd, Social Ref Ctr, Princeton, NJ 08544
T: +1 609 2584936; Fax: +1 609 2582907; E-mail: ssrc@library.princeton.edu; URL: www.princeton.edu/~ssrc/lrmain.html
1922; Kevin Barry
8 000 vols; 650 curr per; 105 VF drawers, 100 000 pamphlets, labor union docs   32376

– Marquand Library of Art & Archaeology, McCormick Hall, Princeton, NJ 08544
T: +1 609 2583783; Fax: +1 609 2587650
1908; Janice Powell
260 000 vols; 700 curr per   32377

– Near East Collections, Princeton, NJ 08544
T: +1 609 2583266; E-mail: jwwein@princeton.edu; URL: www.princeton.edu/~nes/resources.html
Dr. James Weinberger
370 000 vols; 1 400 curr per; 15 000 vols of mss   32378

– Psychology Library, Green Hall, Princeton, NJ 08544-1010
T: +1 609 2583239; Fax: +1 609 2581113; E-mail: psychlib@princeton.edu; URL: psychlib.princeton.edu
1893; Mary Chaikin
Psychologie, cognitive neuroscience, artificial intelligence, neuroscience
38 000 vols; 300 curr per; 400 e-journals; 700 govt docs; 3 500 microforms
libr loan; SLA, ALA, AMHL   32379

– Public Administration Collection, One Washington Rd, Firestone Libr, A-17-J-1, Princeton, NJ 08544
T: +1 609 2583209; Fax: +1 609 2586154; E-mail: rosemary@pucc.princeton.edu; URL: www.princeton.edu/~pubadmin
1930; Rosemary Little
Codebks for machine readable data files
20 070 vols; 18 000 pamphlets, 28 VF drawers of clippings   32380

– School of Architecture Library, Architecture Bldg, 2nd Flr, S-204, One Washington Rd, Princeton, NJ 08544
T: +1 609 2583256; E-mail: fmchen@Princeton.edu
1967; Frances Chen
32 000 vols; 300 curr per   32381

Brown University, John D. Rockefeller Jr Library, Ten Prospect St, Box A, *Providence*, RI 02912
T: +1 401 8632165; Fax: +1 401 8631272; E-mail: rock@brown.edu; URL: www.brown.edu/library
1767; Florence K. Doksansky
American Poetry and Lit, Sermons, Abraham Lincoln Coll, Dante, Children's Bks, East Asia, Hist of Medicine and Science, South America, Military Hist
3 257 000 vols; 21 257 curr per; 9 220 sound-rec; 1 378 digital data carriers; 1 618 582 microforms of per
libr loan   32382

– John Carter Brown Library, P.O. Box 1894, Providence, RI 02912
T: +1 401 8632725; Fax: +1 401 8633477; E-mail: JCBL_Information@Brown.edu; URL: www.jcbl.org
1846; Norman Fiering
Colonial Hist of the Americas (North & South) 1492-1835
54 500 vols; 13 curr per; 1 000 000 mss; 1 200 maps; 2 150 microforms; 40 av-mat; 100 digital data carriers
                                       32383

– Orwig Music Library, Orwig Music Bldg, One Young Orchard Ave, Providence, RI 02912; Box A, Providence, RI 02912
T: +1 401 8633759
Ned Quist
21 000 vols; 150 curr per   32384

– Sciences Library, 201 Thayer St, Providence, RI 02912; P.O. Box I, Providence, RI 02912
T: +1 401 8633333; Fax: +1 401 8639639
Florence Doksansky
625 000 vols   32385

Johnson & Wales University, University Library, 111 Dorrance St, *Providence*, RI 02903; mail address: 8 Abbott Park Pl, Providence, RI 02903
T: +1 401 5981121; Fax: +1 401 5981834; E-mail: library@jwu.edu; URL: jwu-ri.libguides.com/
1914; Rosita Hopper
Fritzche Cookbook Coll (7000 vols of US & foreign cookbks)
83 000 vols; 600 curr per; 380 159 microforms; 54 000 av-mat; 396 sound-rec; 18 digital data carriers   32386

Providence College, Phillips Memorial Library, 549 River Ave, *Providence*, RI 02918
T: +1 401 8652242; Fax: +1 401 8652823; URL: www.providence.edu/academics/phillips+memorial+library/
1917; Donald Russel
Blackfriars' Guild Coll, Genealogy Coll, English & Colonial 18th C Trade Statistics Coll, Irish Lit, Rhode Island Colls
2 267 000 vols; 1 775 curr per; 9 490 e-journals; 8 000 e-books
libr loan; ALA   32387

Rhode Island College, James P. Adams Library, 600 Mt Pleasant Ave, *Providence*, RI 02908-1924
T: +1 401 4568126; Fax: +1 401 4569646; URL: www.ric.edu
1854; Tjalda Nauta
Education, Social work; Children's Lit (Amy Thompson Coll)
643 000 vols; 1 448 curr per   32388

Rhode Island School of Design, Fleet Library, 15 Westminster St *Providence*, RI 02903; mail address: 2 College St, Providence, RI 02903
T: +1 401 7095902; Fax: +1 401 7095932; E-mail: risdlib@risd.edu; URL: library.risd.edu/
1878; Carol S. Terry
Artists' Bks, Lowthorpe Coll of Landscape Architecture, Gorham Libr of Design; RISD Archives, Special Colls
150 000 vols; 375 curr per; 5 200 e-journals; 312 diss/theses; 1 927 maps; 16 microforms; 429 sound-rec; 192 digital data carriers
libr loan; ALA, ARLIS/NA, VRA   32389

University of Rhode Island – Providence Campus, College of Continuing Education Library, 80 Washington St, *Providence*, RI 02903-1803
T: +1 401 2775130; Fax: +1 401 2775148; URL: www.uri.edu/library
1964; Joanna M. Burkhardt
26 000 vols; 325 curr per   32390

Brigham Young University, *Provo*
– Harold B. Lee Library, 2060 HBLL, Provo, UT 84602
T: +1 801 4227652; Fax: +1 801 4220466; URL: www.library.byu.edu
1875; Randy J. Olsen
Middle American Linguistics (William Gates Coll), Welsh Languages & Lit, Herman Melville, William Wordsworth, Robert Burns; Mormon & Western Americana, Walt Whitman, Children's Lit, Victorian Lit, Modern Fine Press, Incunabula
3 121 000 vols; 22 714 curr per
libr loan   32391

– Howard W. Hunter Law Library, 256 JRCB, Provo, UT 84602-8000
T: +1 801 4225481; Fax: +1 801 4220404; URL: www.law2.byu.edu
1972; Kory Staheli
345 000 vols; 4 121 curr per; 3 000 e-books
libr loan   32392

Colorado State University, Pueblo Library, 2200 Bonforte Blvd, *Pueblo*, CO 81001-4901
T: +1 719 5492333; Fax: +1 719 5492738; URL: library.colostate-pueblo.edu
1933; Rhonda Gonzales
US Western Hist, Univ Arch
187 000 vols; 3 121 curr per; 2 500 e-journals; 8 000 e-books; 15 702 maps; 29 000 microforms; 5 100 av-mat; 12 999 sound-rec; 449 digital data carriers
libr loan   32393

Martin Methodist College, Warden Memorial Library, 433 W Madison St, *Pulaski*, TN 38478-2799
T: +1 931 3639844; Fax: +1 931 3639844; E-mail: library@martinmethodist.edu; URL: www.martinmethodist.edu/library
1870; Richard Madden
Psychology (William Fitts Coll); Methodist Hist; Glatzer/Zimmerman Judaica; Gregory McDonald mss; Senator Ross Bass Arch
84 000 vols; 670 curr per; 46 000 e-books; 2 mss; 100 diss/theses; 2 000 microforms; 550 av-mat; 200 sound-rec; 14 digital data carriers
libr loan; SOLINET, TLA   32394

Washington State University, University Libraries, 100 Dairy Rd, *Pullman*, WA 99164-5610; P.O. Box 645610, Pullman, WA 99164-5610
T: +1 509 3354558; Fax: +1 509 3350934; URL: www.wsulibs.wsu.edu
1892; Jay Starratt
Leonard & Virginia Woolf Libr, Bloomsbury Authors, Pacific Northwest Agricultural Hist Arch, Veterinary Hist, Pacific Northwest Publishers' Arch, 20th C Music Arch, Angling, Germans from Russia, Comix Colls, Hansen Beatles

Coll, Univ Arch
2 206 000 vols; 29 327 curr per; 3 840 000 microforms; 7 303 sound-rec; 22 866 Videos
libr loan   32395

– George B. Brain Education Library, 130 Cleveland Hall, P.O. Box 642112, Pullman, WA 99164-2112
T: +1 509 3351492; E-mail: educref@mail.wsu.edu; URL: www.wsulibs.wsu.edu/educ/brain.htm
1963; Christy Zlatos
74 800 vols; 320 curr per; 31 cabinets of microforms, 9 VF cabinets of proprietary tests, 765 kits & records   32396

– George W. Fischer Agricultural Sciences Library, Johnson Annex C-2, Pullman, WA 99164-7150
T: +1 509 3352266; Fax: +1 509 3356782; E-mail: agsciref@wsu.edu
Cindy Stewart Kaag
26 000 vols   32397

– Owen Science & Engineering Library, P.O. Box 643200, Pullman, WA 99164-3200
T: +1 509 3352672; Fax: +1 509 3352534; E-mail: lcrook@wsu.edu; URL: www.wsulibs.wsu.edu/science/owen.htm
1892
500 000 vols; 2 560 curr per; 1 500 e-journals; 80 000 maps; 120 cabinets of microforms   32398

Indiana University of Pennsylvania, Punxsutawney Campus Library, 1012 Winslow St, *Punxsutawney*, PA 15767
T: +1 814 9384870; Fax: +1 814 9385900; URL: www.lib.iup.edu
1962; Carol Asamoah
20 000 vols; 82 curr per
libr loan; OCLC, PALINET   32399

Manhattanville College, Library, 2900 Purchase St, *Purchase*, NY 10577
T: +1 914 3235275; Fax: +1 914 3238139; E-mail: library@mville.edu; URL: www.mville.edu/library
1841; Rhonna A. Goodman
Allain Biogr Coll, Buddhism & Hinduism (Zigmund Cerbu Coll); Alexander Stephens, letter, mss
250 209 vols; 351 curr per; 40 156 e-journals; 531 109 microforms; 5 312 av-mat; 700 sound-rec
libr loan   32400

State University of New York – Purchase College, Library, 735 Anderson Hill Rd, *Purchase*, NY 10577-1400
T: +1 914 2516410; Fax: +1 914 2516437; E-mail: lib.reference@purchase.edu; URL: www.purchase.edu/library
1967; Patrick F. Callahan
Art, Music, Performing arts
247 000 vols; 470 curr per; 24 130 e-journals; 4 000 e-books; 1 600 av-mat; 1 040 digital data carriers
libr loan   32401

US Marine Corps, Library, 2040 Broadway St, Gray Research Ctr, *Quantico*, VA 22134-5207
T: +1 703 7844400; Fax: +1 703 7844306; E-mail: ramkeyce@grc.usmcu.edu; URL: www.mcu.usmc.mil
1928; Carol E. Ramkey
Military science & hist, Int relations & studies; Marine Corps Hist, Military Operations
160 000 vols; 300 curr per   32402

Eastern Nazarene College, Nease Library, 23 E Elm Ave, *Quincy*, MA 02170
T: +1 617 7453850; Fax: +1 617 7453913; E-mail: askalibrarian@enc.edu; URL: library.enc.edu
1900; Susan J. Watkins
Theology Coll
137 000 vols; 500 curr per; 25 diss/theses; 295 music scores; 78 maps; 57 000 microforms; 771 sound-rec
libr loan   32403

Quincy College, Anselmo Library, Newport Hall, 150 Newport Ave, *Quincy*, MA 02171
T: +1 617 9841680; E-mail: jlanigan@quincycollege.edu; URL: www.quincycollegelibrary.org
1958; Janet Lanigan
50 000 vols; 500 curr per   32404

Quincy University, Brenner Library, 1800 College Ave, *Quincy*, IL 62301-2699
T: +1 217 2285432; Fax: +1 217 2285354; URL: www.quincy.edu/Library/index.php
1860; Patricia Tomczak
English & American lit, American hist, Theology; Early Christian & Medieval (Bonaventure Coll), Local Hist, Spanish-American Hist (Biblioteca Fraborese), Rare Book Libr, East Asian Coll
198 000 vols; 455 curr per; 15 e-journals; 10 e-books; 1 350 music scores; 2 500 av-mat; 870 sound-rec; 600 digital data carriers
libr loan   32405

Radford University, John Preston McConnell Library, 801 E Main St, *Radford*, VA 24142-0001; P.O. Box 6881, Radford, VA 24142-6881
T: +1 540 8316624; Fax: +1 540 8316138; E-mail: refdesk@radford.edu; URL: lib.radford.edu
1913; David Hayes
Southwestern Virginia Regional Hist
389 000 vols; 8 029 curr per; 13 660 e-journals; 47 000 e-books; 1 433 digital data carriers; 1 483 CDs
libr loan   32406

Cabrini College, Library, 610 King of Prussia Rd, *Radnor*, PA 19087-3698
T: +1 610 9028538; Fax: +1 610 9028539; URL: www.cabrini.edu/library
1957; Dr. Roberta C. Jacquet
Education, Immigration; Franklin Delano Roosevelt Coll
85 000 vols; 340 curr per; 600 e-books; 54 000 microforms; 300 digital data carriers
libr loan; ACRL   32407

Meredith College, Carlyle Campbell Library, 3800 Hillsborough St, *Raleigh*, NC 27607-5298
T: +1 919 7608531; Fax: +1 919 7602830; E-mail: library@meredith.edu; URL: www.meredith.edu/library
1899; Laura Davidson
Art & architecture, British lit, Music, Women's studies; American Hist (Libr of American Civilization Core Coll), micro; Anthropology (Human Relations Resource Files), micro
194 010 vols; 2 290 curr per; 41 200 e-books; 15 000 av-mat; 8 160 Music Scores
libr loan   32408

North Carolina State University, University Libraries, Two Broughton Dr, *Raleigh*, NC 27695-7111; NC State University, Campus Box 7111, Raleigh, NC 27695-7111
T: +1 919 5153364; Fax: +1 919 5153628; URL: www.lib.ncsu.edu
1887; Carolyn Argentati
Metcalf & Tippmann Colls (Entomology), Tom Regan Coll (Animal Rights), Forestry, Textiles, Visual Design, Architectural Records, Technical Rpts
3 800 000 vols; 53 000 curr per; 35 036 maps; 6 000 av-mat; 3 000 linear feet of mss & archives, 3 328 machine-readable mat
libr loan   32409

– Burlington Textiles Library, North Carolina State University Libraries, 4411 College of Textiles, Campus Box 8301, Raleigh, NC 27695-8301
T: +1 919 5153043; Fax: +1 919 5153926; E-mail: tx-lib@tx.ncsu.edu
1945; Honora Eskridge
Speizman Hosiery coll, Harris coll of Modern Fabrics, Barnhardt Leadership coll
50 000 vols; 200 curr per; 1 000 microforms; 400 av-mat; 44 digital data carriers; 19 VF drawers of pamphlets and clippings, 7 VF drawers of textile machinery trade cat
libr loan   32410

– Harrye B. Lyons Design Library, 209 Brooks Hall, Campus Box 7701, Raleigh, NC 27695-7701
T: +1 919 5152207; Fax: +1 919 5157330; URL: www.lib.ncsu.edu/design
1942; Karen De Witt
30 000 vols; 467 maps; 1 500 vertical file mat, 125 videos, 64 000 slides   32411

– Learning Resources Library, 400 Poe Hall, Campus Box 7801, Raleigh, NC 27695-7801
T: +1 919 5153191; Fax: +1 919 5157634
1964; Ann Akers
Education, Psychology; North Carolina state adopted textbooks; Standardized Test Libr
12 000 vols; 80 curr per; 250 maps; 7 000 microforms; 1 100 av-mat; 20 digital data carriers   32412

– Natural Resources Library, Jordan Hall, Rm 1102, 2800 Faucette Dr, Campus Box 7114, Raleigh, NC 27695-7114
T: +1 919 5152306; Fax: +1 919 5157802
1970; Karen Ciccone
US Forest Service Rpts, North Carolina Topographic Maps
22 000 vols; 150 curr per; 1 000 microforms; 5 digital data carriers; 29 193 rpts   32413

– Veterinary Medical Library, 4700 Hillsborough St, Campus Box 8401, Raleigh, NC 27606
T: +1 919 5136218; Fax: +1 919 5136400
1982; Kristine Alpi
17 000 vols; 7 digital data carriers
libr loan; ARL   32414

Saint Augustine's College, The Prezell R. Robinson Library, 1315 Oakwood Ave, *Raleigh*, NC 27610-2298
T: +1 919 5164148; Fax: +1 919 5164758; E-mail: askthelibrarian@st-aug.edu; URL: www.st-aug.edu/library/library.htm
Linda Simmons-Henry
Curriculum Mat, St Agnes Coll
100 000 vols; 200 curr per   32415

Shaw University, James E. Cheek Learning Resources Center, 118 E South St, *Raleigh*, NC 27601
T: +1 919 5468450; Fax: +1 919 8311161; URL: www.shawuniversity.edu
1865; Musette McKelvey
Oral hist; Black Coll, African & Afro-American Hist (Yergan Coll), John W Fleming Afro-American Coll, North American Indian Coll, Art Hist Coll, Religion & Philosophy (Gilmour Coll), Shaw Univ (Arch Coll), Schomburg Coll
140 000 vols; 687 curr per   32416

South Dakota School of Mines & Technology, Devereaux Library, 501 E Saint Joseph St, *Rapid City*, SD 57701-3995
T: +1 605 3941255; Fax: +1 605 3941256; E-mail: library.reference@sdsmt.edu; URL: library.sdsmt.edu
1885; Patricia M. Andersen
Black Hills & Western South Dakota Hist; Mining Hist of South Dakota & Adjacent Areas
212 000 vols; 527 curr per; 2 256 diss/theses; 14 874 maps; 405 000 microforms; 300 av-mat; 420 digital data carriers; 2 279 Vols of docs, 461 VF Items
libr loan   32417

Albright College, F. Wilbur Gingrich Library, 13th & Exeter Sts, *Reading*, PA 19604; P.O. Box 15234, Reading, PA 19612-5234
T: +1 610 9217209; Fax: +1 610 9217509; E-mail: libraryref@alb.edu; URL: www.albright.edu/library/
1856; Rosemary L. Deegan
Albrightiana, Evangelical United Brethren Church (Eastern PA); Byron Vazakas Coll, Dick Coll, J. Bennett Nolan Coll, Pennsylvania Dutch Coll
218 000 vols; 700 curr per; 15 000 microforms; 60 Bks on Deafness & Sign Lang
libr loan   32418

Alvernia College, Dr Frank A. Franco Library, 400 St Bernardine St, ***Reading***, PA 19607-1737
T: +1 610 7968223; Fax: +1 610 7968347; URL: www.alvernia.edu/library
1958; Sharon Neal
Italian-American Cultural Ctr, Polish Coll
70 000 vols; 323 curr per; 12 000 e-books; 1 600 av-mat; 450 sound-rec; 77 digital data carriers; 909 Journals
libr loan; ALA, PALINET 32419

Pennsylvania State University, Berks-Lehigh Valley College Thun Library, Berks Campus, Tulpehocken Rd, ***Reading***, PA 19610; P.O. Box 7009, Reading, PA 19610-7009
T: +1 610 3966240; Fax: +1 610 3966249; URL: www.libraries.psu.edu/berks/
1958; Deena J. Morganti
50 000 vols; 130 curr per; 50 av-mat 32420

Simpson College, Start-Kilgour Memorial Library, 2211 College View Dr, ***Redding***, CA 96003-8606
T: +1 530 2264116; Fax: +1 530 2264858; E-mail: library@simpsonuniversity.edu; URL: library.simpsonuniversity.edu
1921; Larry L. Haight
Religion, Education; Christian & Missionary Alliance Denominational Hist (A B Simpson Memorial)
96 000 vols; 202 curr per; 39 000 e-books
libr loan 32421

University of Redlands, George & Verda Armacost Library, 1249 E Colton Ave, ***Redlands***, CA 92374-3758
T: +1 909 7484724; Fax: +1 909 3355392; URL: www.redlands.edu/library.xml
1907; William Kennedy
Harley Farnsworth & Florence Ayscough McNair Far Eastern Libr, James Irvine Foundation Map Coll, Californiana & the Great Southwest (Vernon & Helen Farquhar Coll)
212 000 vols; 1 800 curr per; 13 000 e-journals; 166 641 govt docs; 19 313 maps; 291 000 microforms; 1 300 av-mat; 8 500 sound-rec
libr loan 32422

Desert Research Institute, Patrick Squires Library, 2215 Raggio Pkwy, ***Reno***, NV 89512-1095
T: +1 775 6747042; Fax: +1 775 6747183; URL: www.dri.edu/library
1977; Melanie Scott
Environmental science, Atmospheric science
13 000 vols; 21 371 govt docs; 5 000 microforms; 5 digital data carriers; 973 tech rpts 32423

National Judicial College, Law Library, Univ of Nev, Judicial College Bldg/MS 358, ***Reno***, NV 89557
T: +1 775 7846747; Fax: +1 775 7841253; E-mail: info@judges.org; URL: www.judges.org
1966; Randall J. Snyder
Benchbks, Court Administration, Dispute Resolution, Evidence, Jury Instructions, Sentencing, SJI Products (Judicial Education)
83 000 vols; 650 curr per; 235 digital data carriers 32424

University of Nevada-Reno, Noble H. Getchell Library, 1664 N Virginia St, Mailstop 0322, ***Reno***, NV 89557-0322
T: +1 775 6825625; Fax: +1 775 7844529; URL: www.library.unr.edu
1886; Steven D. Zink
Basque Studies, Nevada & the Great Basin, Modern English & American Authors, Women in the West
1 150 000 vols; 10 499 curr per; 137 814 maps; 8 200 av-mat; 40 460 sound-rec; 1 330 digital data carriers; 3 311 040 microforms of per
libr loan; ALA 32425

– DeLaMare Library, 1664 N Virginia St, MS 262, Reno, NV 89557-0262
T: +1 775 7846945; Fax: +1 775 7846949
Brenda Mathenia
Computer science, engineering
37 000 vols; 1 300 curr per 32426

– Life & Health Sciences Library, Fleischman Agriculture, Rm 300, Mailstop 206, Reno, NV 89557
T: +1 775 7846616; Fax: +1 775 7841046; E-mail: ashannon@unr.edu; URL: www.library.unr.edu/~lhsl
1958; Amy Shannon
44 000 vols; 800 curr per; 21 977 microforms; 8 251 bound govt rpts, 68 451 unbound govt rpts 32427

– Savitt Medical Library & IT Department, Pennington Medical Education Bldg, 1664 N Virginia St, Mail Stop 306, Reno, NV 89557
T: +1 775 7844625; Fax: +1 775 7844489; URL: www.med.unr.edu/medlib
James Allen Curtis
Consumer health, public health, speech pathology
72 000 vols; 3 443 curr per; 41 e-books 32428

Saint Joseph's College, Keith & Kate Robinson Memorial Library, Hwy 231 S, ***Rensselaer***, IN 47978; P.O. Box 990, Rensselaer, IN 47978-0990
T: +1 219 8666209; Fax: +1 219 8666135; URL: www.saintjoe.edu/library
1892; Catherine Salyers
132 000 vols; 425 curr per
libr loan 32429

Brigham Young University-Idaho, David O. McKay Library, 525 S Center St, ***Rexburg***, ID 83440-0405
T: +1 208 4962351; Fax: +1 208 4962390; URL: www.lib.byui.edu
1888; Martin Raish
Idaho Hist; Mormon Church Coll
190 000 vols; 525 curr per; 29 300 e-journals; 600 000 e-books; 779 mss; 50 669 govt docs; 140 000 music scores; 14 317 maps; 10 000 microforms; 17 000 av-mat; 5 800 sound-rec; 650 digital data carriers; 46 art reproductions, 2 820 posters, 109 flags
libr loan 32430

University of Wisconsin – Barron County Library, 1800 College Dr, ***Rice Lake***, WI 54868-2497
T: +1 715 2348176 ext 5448; Fax: +1 715 2341975; E-mail: zsampson@uwc.edu; URL: www.barron.uwc.edu/library/facilities/index.html
1966; Zora Sampson
Music, business, hist, art
30 984 vols; 88 curr per; 5 000 govt docs; 1 956 maps; 9 309 microforms; 4 000 av-mat; 40 sound-rec; 5 digital data carriers; ser sub 102
libr loan; ALA, ACRL 32431

University of Texas at Dallas, Eugene McDermott Library, 800 W Campbell Rd, ***Richardson***, TX 75080; P.O. Box 830643, MC33, Richardson, TX 75083-0643
T: +1 972 8832900; Fax: +1 972 8832473; URL: www.utdallas.edu/library
1964; Dr. Larry D. Sall
Hist of Aviation Coll, Wineburgh Philatelic Research Libr, Belsterling Botanical Libr, Wineburgh Philatelic Res Libr
1 343 000 vols; 1 100 curr per; 14 660 e-journals; 400 000 e-books; 2 000 diss/theses; 300 000 govt docs; 12 300 maps; 1 855 000 microforms; 4 200 av-mat; 4 500 sound-rec; 1 500 digital data carriers
libr loan 32432

Washington State University Tri-Cities, Consolidated Information Center, 2770 University Dr, ***Richland***, WA 99354; mail address: 2710 University Dr, Richland, WA 99354
T: +1 509 3727430; Fax: +1 509 3727281; E-mail: lib-req@tricity.wsu.edu; URL: www.tricity.wsu.edu
1990; Joseph Judy
34 000 vols; 20 000 e-books 32433

Earlham College, Lilly Library, 801 National Rd W, ***Richmond***, IN 47374-4095
T: +1 765 9831360; Fax: +1 765 9831304; URL: www.earlham.edu/~libr
1847; Thomas G. Kirk Jr
East Asian Mats, Religious Society of Friends
406 000 vols; 22 439 curr per; 15 501 maps; 224 000 microforms; 1 972 sound-rec; 30 945 slides
libr loan 32434

Eastern Kentucky University, University Libraries, 521 Lancaster Ave, ***Richmond***, KY 40475-3102
T: +1 859 6221778; Fax: +1 859 6221174; URL: www.library.eku.edu
1906; Carrie Cooper
Kentuckiana, Learning Resources Ctr
659 000 vols; 1 566 curr per; 13 130 e-journals; 4 000 e-books; 9 100 music scores; 6 500 av-mat; 4 500 sound-rec; 290 digital data carriers 32435

– Justice & Safety Library, Stratton Bldg, Richmond, KY 40475
T: +1 859 6221798; Fax: +1 859 6228028; E-mail: web.library@eku.edu; URL: www.library.eku.edu/branches/justice/
1975; Nicole M. Montgomery
35 000 vols; 100 curr per; 15 000 microforms; 43 digital data carriers
libr loan 32436

– Music Library, Foster Bldg, 521 Lancaster Ave, Richmond, KY 40475
T: +1 859 6221795; Fax: +1 859 6221174; URL: www.library.eku.edu/branches/music
1969; Dr. Greg Engstrom
12 000 vols; 129 curr per; 9 928 music scores; 7 000 microforms; 220 av-mat; 10 714 sound-rec 32437

Indiana University East, Library Services, 2325 Chester Blvd, ***Richmond***, IN 47374
T: +1 765 9738309; Fax: +1 765 9738315; URL: www.iue.edu/library/
1975; Julianne Stout
71 000 vols; 488 curr per
libr loan 32438

University of Richmond, Boatwright Memorial Library, 28 Westhampton Way, ***Richmond***, VA 23173
T: +1 804 2898672; Fax: +1 804 2898757; URL: library.richmond.edu
1830; James R. Rettig
Virginia Baptists, 19th-20th C American Lit, Virginia Hist, Civil War Confederate Imprints
479 000 vols; 2 232 curr per; 57 000 e-books; 3 incunabula; 678 diss/theses; 218 000 govt docs; 11 255 music scores; 1 670 maps; 363 000 microforms; 1 836 av-mat; 12 172 sound-rec; 4 880 digital data carriers; 16 250 Music Scores, 8 820 CDs
libr loan 32439

– Parsons Music Library, Modlin Center for the Arts, Webb Tower, University of Richmond, VA 23173
T: +1 804 2898286; Fax: +1 804 2876899
Dr. Linda B. Fairtile
Music performance, music theory, popular music, Jazz, musicology, ethnomusicology
9 000 vols; 44 curr per 32440

– William T. Muse Law Library, 28 Westhampton Way, Richmond, VA 23173
T: +1 804 2898685; Fax: +1 804 2871845; E-mail: lawrefdesk@richmond.edu; URL: law.richmond.edu/librarytech/index.php
Joyce Manna Janto
383 000 vols; 4 573 curr per 32441

Virginia Commonwealth University, University Library Services, 901 Park Ave, ***Richmond***, VA 23284-2033; P.O. Box 2033, Richmond, VA 23284-2033
T: +1 804 8281105; Fax: +1 804 8280151; URL: www.library.vcu.edu
1838; John E. Ulmschneider
1 863 000 vols; 23 863 curr per; 26 500 e-journals; 140 000 e-books 32442

– James Cabell Branch Library, 901 Park Ave, Academic Campus, Richmond, VA 23284-2033
T: +1 804 8281110; Fax: +1 804 8280151; E-mail: ulsjbcref@gems.vcu.edu; URL: www.library.vcu.edu/jbc
N.N.
Contemporary Virginia Authors Coll; 20th C American Cartoons; Comic Books; Animation; Book Art
1 311 020 vols; 15 581 curr per 32443

– Tompkins-McCaw Library, Medical College of Virginia Campus, 509 N 12th St, P.O. Box 980582, Richmond, VA 23298-0582
T: +1 804 8280636; Fax: +1 804 8286089; URL: www.library.vcu.edu/tml
1897; Jean P. Shipman
19th C Medical Hist Coll, Nursing Hist Coll, Medical College of Virginia Arch, Virginia Health Sciences Arch, Medical Artifacts Coll
313 160 vols; 3 857 curr per; 82 818 microforms; 5 179 av-mat; 58 digital data carriers 32444

Virginia Union University, L. Douglas Wilder Library & Learning Resource Center, 1500 N Lombardy St, ***Richmond***, VA 23220
T: +1 804 2575822; Fax: +1 804 2575818; E-mail: reference@vuu.edu; URL: www.vuu.edu/library/home.htm
1997; Dr. Delores Pretlow
Black Studies (Schomberg Coll, L D Wilder Coll)
167 000 vols; 224 curr per; 43 000 e-books; 750 av-mat 32445

Franklin Pierce College, Library, 20 College Rd, ***Rindge***, NH 03461-3114; P.O. Box 60, Rindge, NH 03461
T: +1 603 8994140; Fax: +1 603 8994375; URL: library.fpc.edu
1962; Mary Ledoux
Ecology, Graphic arts, Mass communications, New Hampshire
130 000 vols
libr loan; ALA, ACRL 32446

University of Rio Grande, Jeanette Albiez Davis Library, 585 E College Ave, ***Rio Grande***, OH 45674
T: +1 740 2457322; Fax: +1 740 2457096; E-mail: refdesk@rio.edu; URL: library.rio.edu
1876; J. David Mauer
91 000 vols; 788 curr per
libr loan 32447

Ripon College, Lane Library, 300 Seward St, ***Ripon***, WI 54971; P.O. Box 248, Ripon, WI 54971-0248
T: +1 920 7488175; Fax: +1 920 7487243; E-mail: askref@ripon.edu; URL: www.ripon.edu/library
1851; Sharon Wielgus
Liberal arts; Local Hist (Pedrick Coll)
165 000 vols; 794 curr per
libr loan 32448

University of Wisconsin – River Falls, Chalmer Davee Library, 410 S Third St, ***River Falls***, WI 54022
T: +1 715 4253222; Fax: +1 715 4250609; URL: www.uwrf.edu/library
1875; Valerie I. Malzacher
Education, Agriculture, Business; Western Americana & Frontier Hist, Pierce & St Croix County Hist
308 000 vols; 1 143 curr per; 11 000 e-books; 161 060 govt docs; 539 000 microforms; 541 digital data carriers
libr loan 32449

Concordia University, Klinck Memorial Library, 7400 Augusta St, ***River Forest***, IL 60305-1499
T: +1 708 2093050; Fax: +1 708 2093175; E-mail: library@CUChicago.edu; URL: www.cuchicago.edu/library
1864; Yana V. Serdyuk
Education, Religion, Music; Curriculum Libr, Educational Resources Info Ctr
170 000 vols; 534 curr per; 413 000 microforms; 5 800 av-mat; 4 040 sound-rec; 15 digital data carriers
libr loan; OCLC 32450

Dominican University, Rebecca Crown Library, 7900 W Division St, ***River Forest***, IL 60305-1066
T: +1 708 5246877; Fax: +1 708 3665360; URL: www.domweb.dom.edu/library/crown
1918; Dr. Bella Karr Gerlich
Libr & info science; 18th & 19th C British Culture, American Fiction (1774-1900), Hist, Small Press Coll, Western Americana
300 000 vols; 16 000 curr per
libr loan 32451

Manhattan College, O'Malley Library, 4513 Manhattan College Pkwy, ***Riverdale***, NY 10471
T: +1 718 8627295; Fax: +1 718 8628028; E-mail: stacy.pober@manhattan.edu; URL: www.manhattan.edu/library
1853; Maire Duchon
American Culture, English Lit, First Editions (Fales Coll); Lydia Cabrera Coll (Spanish)
226 000 vols; 1 160 curr per 32452

California Baptist University, Annie Gabriel Library, 8432 Magnolia Ave, ***Riverside***, CA 92504
T: +1 909 3434228; Fax: +1 909 3434523; E-mail: library@calbaptist.edu; URL: www.calbaptist.edu/library
1950; Erica McLaughlin
Religion, Education, Hymnology, Holocaust, Psychology; Hymnology (P Boyd Smith Coll); Wallace Coll; Nie Wieder Coll; Virginia Hyatt Coll
152 000 vols; 500 curr per; 52 000 microforms; 357 sound-rec; 19 digital data carriers 32453

La Sierra University, University Library, 4500 Riverwalk Pkwy, ***Riverside***, CA 92505-3344
T: +1 951 7852396; Fax: +1 951 7852445; URL: www.lasierra.edu/library
1927; Kitty Simmons
Libr of American Civilization; Reformation Hist (William M. Landeen Coll); Far Eastern, Hist (W. A. Scharffenberg Coll); Libr of English Lit; Seventh Day Adventist Coll
214 000 vols; 1 016 curr per; 32 000 e-books
libr loan 32454

University of California, Riverside, University Library / Tomás Rivera Library, 3401 Watkinds Dr, ***Riverside***, CA 92521; PO Box 5900, Riverside, CA 92517-5900
T: +1 951 8274392; Fax: +1 951 8272255; URL: library.ucr.edu
1954; Ruth Jackson
20th c English Lit, Victorian & Edwardian Lits, Ezra Pound, Eaton Coll of Fantasy & Science Fiction, Hist Coll of Boys Series Books, Skinner-Ropes Coll, Jack Hirschmann Poetry, William Blake Coll, B. Traven Coll, German National Socialism, Hist of the Panama Canal, Women, Paraguay
2 483 000 vols; 29 941 curr per; 87 055 maps; 1 500 000 microforms; 14 000 av-mat; 13 911 sound-rec; 2 305 digital data carriers
libr loan 32455

– Music Library, Riverside, CA 92521
T: +1 951 8272268; E-mail: caitlins@ucr.edu; URL: library.ucr.edu
1963
Bks on Bells & Carillons; Niels Wilhelm Gade Coll; Oswald Jonas Memorial Arch, incorporating the Heinrich Schenker Arch
35 000 vols; 200 curr per; 387 microforms; 15 740 sound-rec; 26 073 scores 32456

– Raymond L. Orbach Science Library, 3401 Watkins Dr, Riverside, CA 92521; P.O. Box 5900, Riverside, CA 92517-5900
T: +1 951 8273316; URL: library.ucr.edu
1961
Citrus Coll; Desert Ecology; Arid-Land Research; Colls on Jojoba, Guayule; Geologic Maps
533 000 vols; 3 025 curr per; 75 217 microforms 32457

Hollins University, Wyndham Robertson Library, 7950 E Campus Dr, ***Roanoke***, VA 24020-1000; P.O. Box 9000, Roanoke, VA 24020-1000
T: +1 540 3626591; Fax: +1 540 3626756; E-mail: askref@hollins.edu; URL: www.hollins.edu/library
1842; Joan Ruelle
Canadiana, Children's Lit (Margaret Wise Brown Coll), Benjamin Franklin Coll, French Symbolist Lit (Enid Starkie Coll), Robert Frost (McVitty Coll), Hollins Authors, Incunabula, Paper-Making, Printing, Private Presses
195 000 vols; 14 000 e-books 32458

Nazareth College of Rochester, Lorette Wilmot Library, 4245 East Ave, ***Rochester***, NY 14618-3790; P.O. Box 18950, Rochester, NY 14618-0950
T: +1 585 3892122; Fax: +1 585 3892145; E-mail: library@naz.edu; URL: www.naz.edu/library
1924; Catherine Doyle
Maurice Baring Coll, Hilaire Belloc Coll, Sitwells Coll, Chesterton Coll, Byrne Coll, Hendrick Papers, Thomas Merton Coll
235 160 vols; 1 619 curr per; 375 788 microforms; 5 730 av-mat; 12 293 sound-rec; 30 digital data carriers
libr loan; OCLC 32459

Oakland University, Kresge Library, 2200 N Squirrel Rd, ***Rochester***, MI 48309-4484
T: +1 248 3704425; Fax: +1 248 3702474; E-mail: ref@oakland.edu; URL: www.kl.oakland.edu
1959; Julie Voelck
Biology; Rare Book Room Coll, Folklore (James Coll), Lincolniana (William Springer Coll), Women in Lit, 17th-19th c (Hicks Coll), Underground Press Coll
424 900 vols; 1 315 curr per; 11 330 e-journals; 10 200 e-books; 10 190 av-mat 32460

Rochester Institute of Technology, Wallace Library, 90 Lomb Memorial Dr, ***Rochester***, NY 14623-5604
T: +1 585 4752562; Fax: +1 585 4757007; URL: wally.rit.edu
1829; Chandra McKenzie
RIT Arch, Melbert B Cary Graphic Arts Libr, New Design Arch, Deafness Coll, Polish Posters, AMICO Coll
272 000 vols; 3 217 curr per; 2 696 diss/theses; 309 000 microforms; 2 447 sound-rec; 57 digital data carriers; 7 026 videos, 140 VF drawers, 69 850 slides 32461

Saint John Fisher College, Lavery Library, 3690 East Ave, ***Rochester***, NY 14618-3599
T: +1 585 3857340; Fax: +1 585 3858445; URL: library.sjfc.edu
1951; Melissa Jadlos
Big Band Recordings (Bill Givens), Book Plates, Early Genesee Country Newspapers, Dime Novels, Frederick Douglass Anti-Slavery Mat, Post Cards, Grand Army of the Republic Coll
179 000 vols; 861 curr per; 9 000 e-books; 50 av-mat; 50 Audio Bks 32462

University of Rochester, Rush Rhees Library, Wilson Blvd, ***Rochester***, NY 14627-0055
T: +1 585 2754461; Fax: +1 585 2735309; URL: www.library.rochester.edu
1850; Susan Gibbons
Rare Books, Literacy & Historical Mss, Local Hist, Restoration & 19th C British Theater & Drama, Hist of Law & Political Theory, 19th c American Political Hist, American & English Lit, Social & Natural Hist Colls
3 185 000 vols; 21 659 curr per; 100 digital data carriers; 4 126 896 microforms of per
libr loan 32463

– Art/Music Library, Rush Rhees Library, Rochester, NY 14627
T: +1 585 2754476; Fax: +1 585 2731032; E-mail: artlib@library.rochester.edu
1955; Ronald Dow
Robert MacCameron Coll, papers & photos; Sotheby's Auction Cats
65 000 vols; 200 curr per; 2 000 av-mat; 200 sound-rec; 500 digital data carriers 32464

– Memorial Art Gallery, Charlotte Whitney Allen Library, Memorial Art Gallery, 500 University Ave, Rochester, NY 14607
T: +1 585 2768999; Fax: +1 585 4736266; E-mail: maglibinfo@mag.rochester.edu; URL: mag.rochester.edu/library
1913; Lu Harper
Fritz Trautman Coll; Memorial Art Gallery Arch; Visual Resources Coll
43 000 vols; 60 curr per; 150 av-mat; 60 digital data carriers; 24 VF drawers of archivak mat
libr loan 32465

– Physics-Optics-Astronomy Library, 374 Bausch & Lomb Hall, Rochester, NY 14627-0171
T: +1 585 2754469; Fax: +1 585 2735321; E-mail: psulouff@rcl.lib.rochester.edu
Ronald Dow
Hist of Optics; Lens Design Patents
34 035 vols; 225 curr per; 1 318 diss/theses 32466

– River Campus Libraries – Science & Engineering Library, Carlson Libr, 160 Trustee Rd, Rochester, NY 14627-0236
T: +1 585 2754488; Fax: +1 585 2734656
1972; Ronald Dow
Biology, Chemistry, Computer science, Engineering, Geology, Mathematics, Statistics; USGS Topographic Maps
165 000 vols; 1 300 curr per; 14 digital data carriers; 14 000 tech rpts 32467

– Rossell Hope Robbins Library, Rush Rhees Library, Rochester, NY 14627
T: +1 716 2750110; E-mail: alupak@rcl.lib.rochester.edu; URL: www.lib.rochester.edu/camelot/robhome.stm
1987; Alan Lupack
Medieval English lit, Medieval studies; Medieval Lit (Offprint Coll)
15 400 vols; 35 curr per; 19 digital data carriers; 5 000 Offprints 32468

– Sibley Music Library, 27 Gibbs St, Rochester, NY 14604-2596
T: +1 585 2741300; Fax: +1 585 2741380; URL: www.esm.rochester.edu/sibley
1904; James Farrington
Oscar Sonneck Libr; Chamber Music; Folk Music (Krehbiel Coll); 17th c Sacred Music (Olschki Coll); Music Biography, Theatre & Librettos (Pougin Coll); Music Mss; Music Publishing (Sengstack Arch); Performers' Coll (Malcolm Frager, Jan DeGaetani & Jacques Gordon Colls)
359 000 vols; 620 curr per; 16 000 microforms
libr loan 32469

Winthrop University, Ida Jane Dacus Library, 824 Oakland Ave, ***Rock Hill***, SC 29733
T: +1 803 3232274; Fax: +1 803 3232215; URL: www.winthrop.edu/dacus
1895; Dr. Mark Y. Herring
Catawba Indians, mss; Education Resources Info Ctr Rpts, microfiche; Libr of English Lit, ultrafiche; Draper Mss Coll, microfilm; Libr of American Civilization, ultrafiche; South Carolina Hist, mss; UN Publs; Winthrop College Hist
590 000 vols; 1 447 curr per; 14 000 e-books; 2 420 maps; 1 099 000 microforms; 1 783 sound-rec; 117 digital data carriers
libr loan 32470

Augustana College, Thomas Tredway Library, 3435 9 1/2 Ave, ***Rock Island***, IL 61201-2296; mail address: 639 38th St, Rock Island, IL 61201-2296
T: +1 309 7947585; Fax: +1 309 7947230; URL: www.augustana.edu/library
1860; Carla Tracy
French Revolution (Charles XV Coll); Upper Mississippi Valley (Hauberg Coll); English Lit, 17th-19th c
214 000 vols; 541 curr per
libr loan 32471

Rockford College, Howard Colman Library, 5050 E State St, ***Rockford***, IL 61108-2393
T: +1 815 2264165; Fax: +1 815 2264084; URL: www.rockford.edu/library; 65.36.189.110/library/libinfo.asp
1847; Stephanie Quinn
Jane Addams Papers, Julia Lathrop Papers, Holbrook ABC Coll (Children's Lit)
141 000 vols; 530 curr per
libr loan 32472

Sierra Joint Community College District, Learning Resource Center, 5000 Rocklin Rd, ***Rocklin***, CA 95677
T: +1 916 6607230; Fax: +1 916 6607232; E-mail: ReferenceDesk@sierracollege.edu; URL: lrc.sierra.cc.ca.us/library.html
1914
California, Mining
66 000 vols; 190 curr per; 1 000 maps; 1 363 microforms; 2 704 sound-rec; 21 828 film & videotapes, 169 rare bks, 60 realia
libr loan 32473

William Jessup University, Library, 333 Sunset Blvd, ***Rocklin***, CA 95765
T: +1 916 5772288; Fax: + 916 5772290; E-mail: library@jessup.edu; URL: www.jessup.edu/library
1939; Kevin Pischke
Bible Commentaries; Restoration Hist
62 000 vols; 195 curr per; 550 av-mat; 932 sound-rec; 1 digital data carriers; 234 kits, 21 860 stored journal issues 32474

Molloy College, James Edward Tobin Library, 1000 Hempstead Ave, ***Rockville Centre***, NY 11571; P.O. Box 5002, Rockville Centre, NY 11571-5002
T: +1 516 6785000; Fax: +1 516 6788908; URL: www.molloy.edu/library1
1955; Robert D. Martin
110 000 vols; 695 curr per; 550 av-mat; 3 000 Microbks of English Lit
libr loan 32475

North Carolina Wesleyan College, Pearsall Library, 3400 N Wesleyan Blvd, ***Rocky Mount***, NC 27804
T: +1 252 9855350; Fax: +1 252 9855235; E-mail: reference@ncwc.edu; URL: library.ncwc.edu
1960; Katherine R. Winslow
Black Mountain College Coll, Music Coll, United Methodist Church & North Caroliniana (Hardee-Rives Coll), rare bks, fine eds
80 000 vols; 210 curr per; 25 000 e-books 32476

Sonoma State University, Jean & Charles Schulz Information Center, 1801 E Cotati Ave, ***Rohnert Park***, CA 94928-3609
T: +1 707 6642397; Fax: +1 707 6642090; URL: www.libweb.sonoma.edu
1961; Barbara Butler
Celtic Coll, Small California Press
560 000 vols; 1 400 000 microforms; 6 197 sound-rec; 42 digital data carriers; 17 802 slides
libr loan 32477

University of Missouri – Rolla, Curtis Laws Wilson Library, 400 W 14th St, ***Rolla***, MO 65409-0060
T: +1 573 3414008; Fax: +1 573 3414233; E-mail: library@mst.edu; URL: library.mst.edu
1871; J. Andrew Stewart
Mining, Mat science, Metallurgy, Engineering, Geology, Earth sciences
551 000 vols; 630 curr per; 121 000 e-books; 72 780 maps; 248 digital data carriers; 566 064 microforms of per 32478

Shorter College, Livingston Library, 315 Shorter Ave, ***Rome***, GA 30165
T: +1 706 2912121; Fax: +1 706 2361512; URL: www.shorter.edu
1873; Kimmetha Herndon
Religion, Music; Georgia Baptist Hist, Baptist Convention & Assn Minutes
139 000 vols; 832 curr per 32479

Lewis University, University Library, One University Pkwy, ***Romeoville***, IL 60446-2200
T: +1 815 8380500; Fax: +1 815 8389456; URL: www.lewisu.edu/library/
1952; Mary Hollerich
Business & management, Nursing, Religion; Contemporary Print Arch, Libr of English Lit, Libr of American Civilization
176 000 vols; 602 curr per; 163 800 govt docs; 31 000 microforms; 2 000 av-mat
libr loan 32480

Northeastern Ohio Universities Colleges of Medicine and Pharmacy, Ocasek Medical Library, 4209 SR 44, ***Rootstown***, OH 44272; P.O. Box 95, Rootstown, OH 44272-0095
T: +1 330 3256604; Fax: +1 330 3250522; URL: www.neoucom.edu/audience/library
1974; N.N.
Archival Coll, Audiovisual Coll, Medical Implements, 19th C Medical Bks
123 000 vols; 2 693 curr per; 4 digital data carriers; 2 758 charts, 270 linear feet of arch mat & mss
libr loan 32481

University of the West Library, 1409 N Walnut Grove Ave, ***Rosemead***, CA 91770
T: +1 626 5718811; Fax: +1 626 5711413; E-mail: library@uwest.edu; URL: library.uwest.edu
1991; Ling-Ling Kuo
Buddhist Lit, Far Eastern Languages, East and West; Buddhism Coll
60 000 vols; 200 curr per; 10 000 e-journals
libr loan; OCLC 32482

Rosemont College, Gertrude Kistler Memorial Library, 1400 Montgomery Ave, ***Rosemont***, PA 19010-1631
T: +1 610 5270200; Fax: +1 610 5252930; URL: trellis.rosemont.edu
1921; Kathleen Deeming
Liberal arts; Rosemont Coll, Yvonne Chismpeace Women's Poetry Coll, Publisher's Binding Coll
170 000 vols; 459 curr per; 650 e-journals; 500 e-books; 50 sound-rec; 47 Bks on Deafness & Sign Lang
libr loan 32483

Arkansas Tech University, Ross Pendergraft Library & Technology Center, 305 West Q St, ***Russellville***, AR 72801-2222
T: +1 479 9640568; Fax: +1 479 9640559; URL: library.atu.edu
1909; Bill Parton
Engineering, Humanities, Music, Nursing; Parks & Recreation Administration; Music dept
295 000 vols; 929 curr per; 115 000 govt docs; 890 000 microforms; 4 000 sound-rec; 400 digital data carriers
libr loan; ALA 32484

Louisiana Tech University, Prescott Memorial Library, Everett St at The Columns, ***Ruston***, LA 71272; P.O. Box 10408, Ruston, LA 71272-0046
T: +1 318 2572577; Fax: +1 318 2572447; URL: www.latech.edu/library
1895; Michael A. DiCarlo
456 000 vols; 2 654 curr per; 57 000 e-books; 30 442 maps; 1 728 000 microforms; 1 217 sound-rec; 12 digital data carriers
libr loan 32485

College of Saint Joseph, Library, 71 Clement Rd, ***Rutland***, VT 05701
T: +1 802 7735900; Fax: +1 802 7765258; URL: www.csj.edu/library.html
1950; Doreen J. McCullough
Special education; Kyran McGrath Irish Studies Coll, Sister St George Vermont Coll
56 000 vols; 84 curr per
libr loan 32486

California State University, Sacramento Library, 2000 State University Dr E, ***Sacramento***, CA 95819-6039
T: +1 916 2786395; Fax: +1 916 2785917; URL: www.csus.edu/csuslibr
1947; Terry D. Webb
Business, Education, Engineering, Humanities, Social sciences; Classic Radio Coll, Congressman John Moss Papers, Dissent & Social Change Coll, Florin Japanese-American Citizens League Oral Hist Project, Charles M Goethe Coll, Phillip Isenberg Papers, Japanese American Archival Coll, OOVR-TV News Historic Film Coll, Port of Sacramento Coll, State Senator Albert Rodda Papers, Mary Tsukamoto Japanese Coll
1 328 000 vols; 3 143 curr per; 4 000 e-books; 14 500 av-mat
libr loan 32487

– Tsakopoulos Hellenic Collection, 2000 State University Dr East, Sacramento, CA 95819-6039
T: +1 916 2784361; Fax: +1 916 2785917; E-mail: paganelis@csus.edu; URL: www.library.csus.edu/tsakopoulos
2003; George I. Paganelis
Geographically covering Greece, Turkey, the Balkans, the Near and Middle East; chronologically spanning the ancient world to the present. Encompasses all major disciplines within the social sciences and humanities
70 000 vols; 75 curr per; 50 e-journals; 450 diss/theses; 2 500 govt docs; 30 music scores; 475 microforms; 200 av-mat; 40 sound-rec; 40 digital data carriers; 270 linear feet of the Dr Basil Vlavianos arch; pamphlet coll
libr loan; ALA, ACRL, Modern Greek Studies Association (MGSA) 32488

University of the Pacific – McGeorge School of Law, Gordon D. Schaber Law Library, 3282 Fifth Ave, ***Sacramento***, CA 95817
T: +1 916 7397164; Fax: +1 916 7397273; URL: www.mcgeorge.edu
1924
509 000 vols; 4 572 curr per; 50 digital data carriers
libr loan 32489

Flagler College, William L. Proctor Library, 74 King St, ***Saint Augustine***, FL 32084-4302; P.O. Box 1027, Saint Augustine, FL 32085-1027
T: +1 904 8196206; Fax: +1 904 8238511; E-mail: library@flagler.edu; URL: www.flagler.edu/library
1968; Michael A. Gallen
90 000 vols; 488 curr per; 83 000 e-books
libr loan 32490

Mount Angel Abbey Library, One Abbey Dr, ***Saint Benedict***, OR 97373
T: +1 503 8453303; Fax: +1 503 8453500; URL: www.mtangel.edu/library
1882; Victoria Ertelt
Theology, Philosophy, Humanities; Chicano/Chicana Lit, Civil War, William Redman Duggan Coll (Africa), Patristic & Latin Christian Studies, Spanish-American Lit
230 000 vols; 380 curr per; 40 000 microforms; 6 500 sound-rec; 350 digital data carriers; 2 000 slides, 31 000 postcards
libr loan 32491

Saint Bonaventure University, Friedsam Memorial Library, Rte 417, ***Saint Bonaventure***, NY 14778
T: +1 716 3752323; Fax: +1 716 3752389; URL: www.sbu.edu/friedsam
1858; Paul J. Spaeth
Jim Bishop Coll, mss; Thomas Merton Coll; Franciscan Inst; Robert Lax Coll
285 000 vols; 1 200 curr per 32492

Crown College, Peter Watne Memorial Library, 8700 College View Dr, ***Saint Bonifacius***, MN 55375-9002
T: +1 952 4464241; Fax: +1 952 4464149; URL: www.crown.edu/library/
1916; Dr. Dennis Ingolfsland
Evans Coll, microprint; early American imprints; early American newspapers
96 000 vols; 200 curr per; 15 000 e-journals; 30 000 e-books; 1 100 av-mat; 37 digital data carriers 32493

Lindenwood University, Margaret L. Butler Library, 209 S Kingshighway, ***Saint Charles***, MO 63301
T: +1 636 9494144; Fax: +1 636 9494822; E-mail: library@lindenwood.edu; URL: www.lindenwood.edu/library
1827; MacDonald Elizabeth
McKissack Ctr for Black Children's Lit, Japanese Culture
130 000 vols; 578 curr per; 7 000 e-books 32494

Belmont Technical College, Learning Resource Center, 120 Fox-Shannon Pl, ***Saint Clairsville***, OH 43950-9735
T: +1 740 6959500; Fax: +1 740 6952247; E-mail: refdesk@btc.edu; URL: www.btc.edu
Joyce Baker
13 000 vols; 341 curr per 32495

Ohio University Eastern, Library, Shannon Hall, 1st Flr, 45425 National Rd, ***Saint Clairsville***, OH 43950-9724
T: +1 740 6992344; Fax: +1 740 6957075; URL: www.eastern.ohiou.edu/directory/library/library.htm
1957; Patricia Murphy
Contemporary American Poetry Coll, James Wright Poetry Festival mat
73 000 vols; 774 curr per; 1 701 govt docs; 45 000 microforms; 4 552 sound-rec; 25 digital data carriers; 2 189 other items
libr loan; OCLC, OHIONET 32496

St. Cloud State University, Learning Resources & Technology Services, 720 4th Ave S, 112 Miller Ctr, ***Saint Cloud***, MN 56301-4498
T: +1 320 3082022; Fax: +1 320 3084778; E-mail: lrtsinfo@stcloudstate.edu; URL: lrs.stcloudstate.edu
1869; Kristi M. Tornquist
State Author Mss Coll, William Lindgren Asian Coll, Don Boros Theater Coll
631 345 vols; 1 454 curr per; 1 274 307 govt docs; 61 347 maps; 671 435 microforms; 20 000 av-mat; 12 462 sound-rec; 262 digital data carriers 32497

Eastern University, Warner Memorial Library, 1300 Eagle Rd, ***Saint Davids***, PA 19087
T: +1 610 3411777; Fax: +1 610 3411375; E-mail: reference1777@eastern.edu; URL: www.eastern.edu/library/
1952; James L. Sauer
Harry C Goebel Coll on Fine Printing; Bruce Rogers Coll, Marcus Aurelius
167 000 vols; 650 curr per 32498

College of Saint Benedict, Clemens Library, 37 S College Ave, ***Saint Joseph***, MN 56374
T: +1 320 3632119; Fax: +1 320 3635197; URL: www.csbsju.edu/library
Kathleen Parker
Nursing, women's studies
204 000 vols; 562 curr per 32499

Missouri Western State University, Hearnes Center, 4525 Downs Dr, ***Saint Joseph***, MO 64507-2294
T: +1 816 2714573; Fax: +1 816 2714574; E-mail: refdesk@missouriwestern.edu
1915; Julia Schneider
Women Writers Along the Rivers
239 000 vols; 1 481 curr per
libr loan 32500

Saint Leo University, Cannon Memorial Library, 33701 State Rd 52, ***Saint Leo***, FL 33574; P.O. Box 6665, MC2128, Saint Leo, FL 33574-6665
T: +1 352 5888258; Fax: +1 352 5888484; URL: www.saintleo.edu
1959; Brent Short
Church hist, Monasticism
159 000 vols; 750 curr per; 46 000 e-books; 29 000 microforms; 1 125 sound-rec; 16 digital data carriers; 237 pamphlets, 2 realia, 17 computerdiscs, 1 chart, 313 filmstrips, 2 823 slides
libr loan; ALA, CLA 32501

Fontbonne University, Library, 6800 Wydown Blvd, ***Saint Louis***, MO 63105
T: +1 314 8891417; Fax: +1 314 7198040; URL: www.fontbonne.edu/library
1923; Sharon McCaslin
86 000 vols; 260 curr per; 19 300 e-journals; 9 000 e-books; 1 600 av-mat; 720 sound-rec
libr loan 32502

Harris-Stowe State College, Library, 3026 Laclede Ave, ***Saint Louis***, MO 63103-2199
T: +1 314 3403621; Fax: +1 314 3403630; URL: www.hssu.edu
1857; Barbara N. Noble
Education; Elementary Education; Education of Exceptional Children; Black Studies; Juvenile Lit; Civil Rights; St Louis Public School Arch
94 000 vols; 333 curr per; 2 000 microforms
libr loan 32503

Maryville University, University Library, 650 Maryville University Dr, ***Saint Louis***, MO 63141
T: +1 314 5299595; Fax: +1 314 5299941; URL: www.maryville.edu/library
1872; Dr. Genie V. McKee
Education, Music therapy; Curriculum Mat, Papers of Edward S Dowling, Maryville Arch, Murphy-Meisgeier Coll
154 000 vols; 668 curr per; 38 000 e-books; 545 maps; 304 000 microforms; 1 100 av-mat; 130 sound-rec; 140 digital data carriers 32504

Missouri Baptist College, Library, One College Park Dr, ***Saint Louis***, MO 63141-8698
T: +1 314 4341115; Fax: +1 314 3922343; E-mail: circdsk@mobap.edu; URL: www.mobap.edu/library
1968; Nitsa Hindeleh
Southern Baptist Convention Curriculum
91 000 vols; 375 curr per; 391 av-mat 32505

Saint Louis College of Pharmacy, O. J. Cloughly Alumni Library, 4588 Parkview Pl, ***Saint Louis***, MO 63110
T: +1 314 4468361; Fax: +1 314 4468360; E-mail: library@stlcop.edu; URL: www.stlcop.edu/library
1948; Jill Nissen
Arch
54 000 vols; 400 curr per; 500 sound-rec; 5 digital data carriers 32506

Saint Louis University, Pius XII Memorial Library, 3650 Lindell Blvd, ***Saint Louis***, MO 63108-3302
T: +1 314 9773580; Fax: +1 314 9773108; E-mail: piusweb@slu.edu; URL: www.slu.edu/libraries/pius
1818; David Cassens
Medieval Mss, Western Americana
1 423 000 vols; 11 847 curr per
libr loan 32507

– Health Sciences Center Library, 1402 S Grand Blvd, Saint Louis, MO 63104
T: +1 314 9778803; Fax: +1 314 9775573; URL: www.slu.edu/libraries/hsc
1890; Patrick McCarthy
129 000 vols; 841 curr per
libr loan; MLA, ARL 32508

– Omer Poos Law Library, Morrissey Hall, 3700 Lindell Blvd, Saint Louis, MO 63108-3478
T: +1 314 9773991; Fax: +1 314 9773966; URL: law.slu.edu/library
1842; Mark P. Bernstein
US & Missouri Govt Docs, Congressman Leonor Sullivan Papers, Father Leo Brown Labor Law Arch, Smurfit Irish Law Ctr, Jewish Law Ctr
392 000 vols; 3 565 curr per; 24 000 e-books; 906 000 microforms
libr loan 32509

University of Missouri-Saint Louis, Thomas Jefferson Library, One University Blvd, ***Saint Louis***, MO 63121
T: +1 314 5165060; Fax: +1 314 5165853; URL: www.umsl.edu
1963; Marilyn Rodgers
Colonial Latin American Hist, Mercantile, Utopian Lit & Science Fiction
1 052 000 vols; 3 614 curr per; 1 200 av-mat
libr loan 32510

– Ward E. Barnes Library, 8001 Natural Bridge Rd, Saint Louis, MO 63121
T: +1 314 5165576; Fax: +1 314 5166468; E-mail: cann@umsl.edu; URL: www.umsl.edu/services/scampus/
1978; Cheryle Cann
Education, Nursing, Optometry
30 000 vols; 350 curr per; 30 000 mss; 100 diss/theses; 2 500 govt docs; 550 000 microforms; 999 av-mat; 1 digital data carriers
libr loan; ALA, MLA 32511

Washington University, University Libraries, One Brookings Dr, Campus Box 1061, ***Saint Louis***, MO 63130-4862
T: +1 314 9355442; Fax: +1 314 9354045; URL: www.library.wustl.edu
1853; Stephanie Atkins
19th & 20th c English & American Lit, Hist of Printing & Book Arts, Western Americana, St Louis Political & Social Welfare Hist
3 746 000 vols; 44 806 curr per; 198 000 e-books; 24 000 av-mat; 6 200 sound-rec; 1 980 digital data carriers 32512

– Art and Architecture Library, One Brookings Dr, Campus Box 1061, Saint Louis, MO 63130-4862
T: +1 314 9355268; Fax: +1 314 9354362
Rina Vecchiola
106 000 vols; 310 curr per 32513

– Biology Library, 1 Brookings Dr, Campus Box 1061, Saint Louis, MO 63130
T: +1 314 9355405; E-mail: biology@wustl.edu; URL: library.wustl.edu/units/biology
Ruth Lewis
70 650 vols; 476 curr per; 3 003 microforms; 130 av-mat   32514

– Chemistry Library, 549 Louderman Hall, Campus Box 1134, Saint Louis, MO 63130
T: +1 314 9354818; Fax: +1 314 9354778; E-mail: chem@wumail.wustl.edu
Dr. Robert McFarland
46 000 vols; 321 curr per   32515

– Earth and Planetary Science Library, One Brookings Dr, Third Flr, Campus Box 1061, Saint Louis, MO 63130-4862
T: +1 314 9355406; Fax: +1 314 9354800; E-mail: eps@wulib.wustl.edu
Clara McLeod
40 000 vols; 215 curr per   32516

– East Asian Library, One Brookings Dr, Campus Box 1061, Saint Louis, MO 63130-4862
T: +1 314 9355525; Fax: +1 314 9357505; E-mail: ea@wumail.wustl.edu; URL: library.wustl.edu/units/ea
Tony Chang
141 000 vols; 272 curr per   32517

– Gaylord Music Library, Gaylord Hall, 6500 Forsyth Blvd, Saint Louis, MO 63105; One Brookings Dr, Campus Box 1061, Saint Louis, MO 63130-4862
T: +1 314 9355563; Fax: +1 314 9354263; E-mail: music@wulib.wustl.edu; URL: libguides.wustl.edu/music
Bradley Short
63 000 vols; 357 curr per   32518

– GWB School of Social Work, Library & Learning Resources Center, One Brookings Dr, Campus Box 1196, Saint Louis, MO 63130-4862
T: +1 314 9356600; Fax: +1 314 9358511; URL: gwbweb.wustl.edu
Sylvia Toombs
Wortman Coll (Family Therapy)
51 000 vols; 494 curr per; 250 diss/theses; 113 microforms; 350 av-mat   32519

– Kopolow Business Library, One Brookings Dr, Campus Box 1133, Saint Louis, MO 63130-4899
T: +1 314 9356963; Fax: +1 314 9354970; URL: www.olin.wustl.edu/library
1925; Ron Allen
32 000 vols; 460 curr per; 308 diss/theses; 386 000 microforms; 2 693 digital data carriers; 1 990 working papers   32520

– Law Library, Washington Univ Sch Law, Anheuser-Busch Hall, One Brookings Dr, Campus Box 1171, Saint Louis, MO 63130
T: +1 314 9356400; Fax: +1 314 9357125; URL: law.wustl.edu/library
Phillip Berwick
676 000 vols; 6 109 curr per   32521

– Mathematics Library, One Brookings Dr, Campus Box 1061, Saint Louis, MO 63130-4899
T: +1 314 9355048; Fax: +1 9354045; E-mail: rlewis@ wustl.edu; URL: library.wustl.edu/units/math/
Ruth Lewis
12 500 vols; 60 curr per   32522

– Pfeiffer Physics Library, One Brookings Dr, 340 Compton Lab, Saint Louis, MO 63130
T: +1 314 9356215; Fax: +1 314 9356219; URL: libguides.wustl.edu/physics
Alison Verbeck
49 000 vols; 233 curr per   32523

– School of Medicine, Bernard Becker Medical Library, 660 S Euclid Ave, Campus Box 8132, Saint Louis, MO 63110
T: +1 314 7470029; Fax: +1 314 3629630; URL: becker.wustl.edu
1837; Paul Schoening
Bernard Becker Coll in Ophthalmology & Optics; Max Goldstein-C.I.D. Coll in Otology & Deaf Education; Henry J McKellops Coll in Dental Medicine; Robert E Schlueter Paracelsus Coll; Archives of Rare Books Dept

296 000 vols; 2 595 curr per; 176 e-books; 1 812 sound-rec; 2 273 feet of archives, 4 113 nonprint items
libr loan; MLA   32524

Saint Mary-of-the-Woods College, Library, 3301 Saint Mary's Rd, *Saint Mary-of-the-Woods*, IN 47876
T: +1 812 5355223; Fax: +1 812 5355127; E-mail: library@smwc.edu; URL: library.smwc.edu
1840; Judith Tribble
Catholic Americana, 17th & 18th c French Religious Bks; Law Libr, Women in Small Business Ctr, Curriculum Media Ctr
133 000 vols; 150 curr per
libr loan; INCOLSA, OCLC   32525

Saint Mary's College of Maryland, Library, 18952 E Fisher Rd, *Saint Mary's City*, MD 20686-3001
T: +1 240 8954256; Fax: +1 240 8954914; E-mail: cerabinowitz@smcm.edu; URL: www.smcm.edu/library
1840; Dr. Celia E. Rabinowitz
Marylandiana, St Mary's County Hist
157 080 vols; 2 829 curr per; 1 600 music scores; 3 100 av-mat; 1 100 sound-rec; 600 digital data carriers   32526

Saint Meinrad Archabbey & School of Theology, Archabbey Library, 200 Hill Dr, *Saint Meinrad*, IN 47577
T: +1 812 3576717; Fax: +1 812 3576398; E-mail: library@saintmeinrad.edu; URL: www.saintmeinrad.edu
1854; Dan Kolb
175 000 vols; 333 curr per; 412 e-journals
libr loan   32527

Bethel University, Library, 3900 Bethel Dr, *Saint Paul*, MN 55112
T: +1 651 6386222; Fax: +1 651 6386001; URL: library.bethel.edu
1871; Robert C. Suderman
Education, Hist
174 000 vols; 842 curr per; 21 000 e-journals; 13 500 av-mat
libr loan   32528

College of Saint Catherine, Libraries, Media Services & Archives, Library, 2004 Randolph Ave, Mail F-10, *Saint Paul*, MN 55105-1794
T: +1 651 6906652; Fax: +1 651 6908636; URL: library.stkate.edu
1905; Carol P. Johnson
Ruth Sawyer Coll (Children's Lit), Autogr & Mss (Mother Antonia McHugh Coll), Liturgical Art (Ade Bethune Coll), Printing (Muellerleihe Coll), Rare Bks
247 000 vols; 560 curr per; 31 400 e-journals; 17 000 e-books; 4 100 av-mat; 9 000 sound-rec; 210 digital data carriers; 300 Bks on Deafness & Sign Lang
libr loan   32529

Concordia University, Library Technology Center, 1282 Concordia Ave, *Saint Paul*, MN 55104; mail address: 275 N Syndicate St, Saint Paul, MN 55104
T: +1 651 6418812; Fax: +1 651 6418782; E-mail: reference@csp.edu; URL: concordia.csp.edu/library
1893; Dr. Charlotte M. Knoche
Education (Children's Coll; Curriculum Coll); Hist Textbks; Hymnbook Coll; 16-19th C Coll
127 000 vols; 450 curr per; 6 000 microforms; 4 000 sound-rec; 385 digital data carriers   32530

Hamline University, *Saint Paul*
– Bush Memorial Library, 1536 Hewitt, Saint Paul, MN 55104
T: +1 651 5232375; Fax: +1 651 5232199; E-mail: bush_reference_email@gw.hamline.edu; URL: www.hamline.edu/bushlibrary/index.html
1854
Brass Rubbing Coll, Methodism
154 000 vols; 1 148 curr per
libr loan   32531

– School of Law Library, 1536 Hewitt Ave, Saint Paul, MN 55104-1237
T: +1 651 5232379; Fax: +1 651 5232863; URL: www.hamline.edu/law/library/law_library.html
1972; Connie Lenz
274 000 vols; 76 000 microforms; 3 digital data carriers
libr loan   32532

Macalester College, DeWitt Wallace Library, 1600 Grand Ave, *Saint Paul*, MN 55105-1899
T: +1 651 6966530; Fax: +1 651 6966617; URL: www.macalester.edu/library
1874; Teresa A. Fishel
Early Minnesota, Sinclair Lewis Coll
440 000 vols; 2 038 curr per; 5 000 e-books
libr loan   32533

Northwestern College, Berntsen Library, 3003 Snelling Ave N, *Saint Paul*, MN 55113
T: +1 651 2867708; Fax: +1 651 6315598; URL: nwc.edu/library
1902; Ruth McGuire
W.B. Riley Coll
100 000 vols; 965 curr per; 300 maps; 77 000 microforms; 3 700 sound-rec; 12 digital data carriers; 484 slides
libr loan; ACL   32534

University of Minnesota – Magrath Library, Saint Paul Campus Libraries, 1984 Buford Ave, *Saint Paul*, MN 55108
T: +1 612 6242233; Fax: +1 612 6249245, 6253134;
E-mail: stpcirc@umn.edu; URL: magrath.lib.umn.edu
1890
398 600 vols; 3 307 curr per; 55 098 microforms; 352 av-mat; 933 digital data carriers; 436 578 docs
libr loan   32535

– Entomology, Fisheries & Wildlife Library, 375 Hodson Hall, 1980 Folwell Ave, Saint Paul, MN 55108
T: +1 612 6249288; Fax: +1 612 6240719; URL: efw.lib.umn.edu
1905; Margaret Borg
Bee Coll; Early Zoology Imprint Coll
75 000 vols; 781 curr per; 23 313 govt docs; 105 maps; 12 022 microforms; 700 linear feet gov docs
libr loan   32536

– Forestry Library, B50 Skok Hall, 2003 Upper Buford Circle, Saint Paul, MN 55108
T: +1 612 6243222; Fax: +1 612 6243733; E-mail: forlib@umn.edu; URL: forestry.lib.umn.edu
1899; Linda Eells
54 295 vols; 600 curr per; 67 670 govt docs; 4 434 maps; 3 670 microforms; 675 av-mat; 3 digital data carriers; 45 items   32537

University of Saint Thomas, O'Shaughnessy-Frey Library, 2115 Summit Ave, Mail Box 5004, *Saint Paul*, MN 55105
Fax: +1 651 9625406; URL: www.stthomas.edu/libraries/
1885; Daniel Ross Gjelten
Belloc-Chesterton, Celtic Coll, French Memoir Coll, Luxembourgian Coll, Univ Arch
349 000 vols; 1 807 curr per; 32 170 e-journals; 47 000 e-books; 32 170 av-mat; 3 700 sound-rec; 8 770 digital data carriers; 200 Electronic Media & Resources, 61 500 Journals   32538

– Archbishop Ireland Memorial Library, 2260 Summit Ave, Mail No IRL, Saint Paul, MN 55105
T: +1 651 9625453; Fax: +1 651 9625460; URL: www.stthomas.edu/libraries/ireland
1894; N. Curtis Lemay
Theology
110 000 vols; 595 curr per; 32 170 e-journals; 47 000 e-books   32539

– Charles J. Keffer Library, 1000 LaSalle Ave, MOH 206, Minneapolis, MN 55403
T: +1 651 9624642; Fax: +1 651 9624648; URL: www.stthomas.edu/libraries/keffer
1992; Janice Kragness
Education, Business, Psychology
32 000 vols; 340 curr per; 32 170 e-journals; 47 000 e-books   32540

– Schoenecker Law Library, 1101 Harmon Pl, Minneapolis, MN 55403; mail address: 1000 LaSalle Ave, MSL 112, Minneapolis, MN 55403-2015
T: +1 651 9624809; Fax: +1 651 9624946; E-mail: lawlibrary@stthomas.edu; URL: www.stthomas.edu/law/library/

Ann L. Bateson
55 000 vols; 1 380 curr per; 36 000 e-books   32541

William Mitchell College of Law, Warren E. Burger Library, 871 Summit Ave, *Saint Paul*, MN 55105
T: +1 651 2906424; Fax: +1 651 2906318; E-mail: reference@wmitchell.edu; URL: www.wmitchell.edu/library/
1958; Ann L. Bateson
319 000 vols; 4 074 curr per; 697 000 microforms; 7 digital data carriers   32542

Gustavus Adolphus College, Folke Bernadotte Memorial Library, 800 W College Ave, *Saint Peter*, MN 56082
T: +1 507 9337556; Fax: +1 507 9336292; E-mail: folke@gustavus.edu; URL: www.gac.edu/academics/library/
1862; Barbara Fister
Gene Basset Cartoons; Selma Lagerloff (Nils Sahil Coll), bks, pamphlets; Mettetal Record Coll; Swedish American (Hist Arch Coll); Korean Porcelain & Japanese Prints (Oriental Art Coll), artifacts
306 000 vols; 906 curr per   32543

Eckerd College, Peter H. Armacost Library, 4200 54th Ave S, *Saint Petersburg*, FL 33711
T: +1 727 8648475; Fax: +1 727 8648997; URL: www.eckerd.edu/library
1959; David W. Henderson
165 000 vols; 821 curr per
libr loan   32544

University of South Florida Saint Petersburg, Nelson Poynter Memorial Library, 140 Seventh Ave S, *Saint Petersburg*, FL 33701
T: +1 727 8734401; Fax: +1 727 8734196; URL: www.nelson.usf.edu
1968; N.N.
D Hubbell Mark Twain Coll; Oral Hist of Modern America; J. Briggs Ichthyology Coll; Papers of Nelson Poynter
232 000 vols; 561 curr per; 37 mss; 378 maps; 841 000 microforms; 5 000 av-mat; 2 149 sound-rec; 68 digital data carriers
libr loan   32545

Allegheny Wesleyan College Library, 2161 Woodsdale Rd, *Salem*, OH 44460
T: +1 330 3376403; Fax: +1 330 3376255
1973; Angela Reynolds
18 000 vols; 118 curr per   32546

Corban College, Library, 5000 Deer Park Dr SE, *Salem*, OR 97317-9392
T: +1 503 3757016; Fax: +1 503 3757196; E-mail: library@corban.edu; URL: www.corban.edu/library
1946; Floyd M. Votaw
Bible, Theology, Missions; Museum in Middle Eastern Archaeology
99 000 vols; 585 curr per; 8 000 e-books
libr loan; ACL, NAPCU   32547

Kent State University, Salem Campus Library, 2491 State Rte 45-S, *Salem*, OH 44460-9412
T: +1 330 3374211; Fax: +1 330 3374144
1962; Lilith R. Kunkel
22 000 vols; 98 curr per   32548

Roanoke College, Fintel Library, 220 High St, *Salem*, VA 24153
T: +1 540 3752294; E-mail: library@roanoke.edu; URL: www.roanoke.edu/library
1842; Stan Umberger
Henry F Fowler Coll, Roanoke College Coll
173 000 vols; 683 curr per; 42 000 e-books   32549

Salem International University, Benedum Library, KD Hurley Blvd, *Salem*, WV 26426; P.O. Box 500, Salem, WV 26426-0500
T: +1 304 3261390; Fax: +1 304 3261240; E-mail: library@salemu.edu; URL: library.salemu.edu
1888; Dr. Phyllis D. Freedman
Equine Mat, Seventh-Day Baptists
112 000 vols; 450 curr per; 297 diss/theses; 9 333 govt docs; 242 000 microforms; 700 av-mat; 25 digital data carriers; 17 Special Interest Per Sub
libr loan; ALA   32550

Salem State College, Library, 352 Lafayette St, ***Salem***, MA 01970-4589
T: +1 978 5426230; Fax: +1 978 5422132; URL: www.salemstate.edu/library
1854; Susan E. Cirillo
Cartography, Education, Nursing; Annual Reports & 10 K's, Education (19th C Normal School Texts Coll of the College), Eric 1968 to date, Federal Govt (Representative William H. Bates Memorial Arch & Representative Michael Harrington Papers), US Geological Survey Topographical Map Coll, Beat poets
319 000 vols; 692 curr per; 30 700 e-journals; 4 000 e-books; 75 200 maps; 214 000 microforms; 25 digital data carriers; 110 Electronic Media & Resources
libr loan 32551

Willamette University, Mark O. Hatfield Library, 900 State St, ***Salem***, OR 97301
T: +1 503 3706610; Fax: +1 503 3706141; E-mail: library@willamette.edu; URL: library.willamette.edu
1843; Deborah Dancik
Pacific Northwest, Gender studies
245 000 vols; 2 485 curr per; 5 000 e-books; 6 640 Music Scores
libr loan 32552

– J. W. Long Law Library, 245 Winter St SE, Salem, OR 97301
T: +1 503 3706386; Fax: +1 503 3755426; URL: www.willamette.edu/law/longlib
1842; Richard Breen
297 000 vols; 1 357 curr per; 146 000 microforms; 170 av-mat; 6 digital data carriers
libr loan 32553

Kansas State University – Salina College of Technology & Aviation, Library, Technology Center Bldg, 2310 Centennial Rd, Rm 111, ***Salina***, KS 67401
T: +1 785 8262637; Fax: +1 785 8262937; URL: www.sal.ksu.edu/library/
1966; Alysia Starkey
Computer Programming, Management, Applied Engineering
30 000 vols; 100 curr per; 200 av-mat; 103 sound-rec; 7 digital data carriers; 100 art reproductions 32554

Kansas Wesleyan University, Memorial Library, 100 E Claflin, ***Salina***, KS 67401-6100
T: +1 785 8275541 ext 4120; Fax: +1 785 8270927; E-mail: library@kwu.edu; URL: www.kwu.edu/library
1886; Angela A. Allen
Curriculum Mat Coll
78 500 vols; 120 curr per; 7 500 e-books; 193 microforms 32555

Catawba College, Corriher-Linn-Black Library, 2300 W Innes St, ***Salisbury***, NC 28144-2488
T: +1 704 6374448; Fax: +1 704 6374304; E-mail: ill@catawba.edu; URL: www.lib.catawba.edu
1851; Steve McKinzie
Poetry Council of North Carolina Coll, United Church of Christ Hist
167 000 vols; 594 curr per; 450 music scores; 1 200 av-mat; 40 sound-rec; 80 digital data carriers
libr loan 32556

Livingstone College, Andrew Carnegie Library, 701 W Monroe St, ***Salisbury***, NC 28144
T: +1 704 2166033; Fax: +1 704 2166798; URL: www.livingstone.edu/lib/index.html
1879; Dr. G. Peart
African-American hist; Black Lit & Hist (American Black & African Studies), African Methodist Episcopal Zion Church, Ecumenical Methodist Conference
75 000 vols; 78 000 microforms; 1 300 av-mat; 835 sound-rec; 412 digital data carriers 32557

– Hood Theological Seminary, Library, 1810 Lutheran Synod Dr, Salisbury, NC 28144
T: +1 704 6366840; Fax: +1 704 6367699; URL: www.hoodseminary.edu
1885; Cynthia D. Keever
AME Zion Coll
32 000 vols; 360 curr per; 43 microforms; 7 digital data carriers; 115 films & tapes 32558

Salisbury University, Blackwell Library, 1101 Camden Ave, ***Salisbury***, MD 21801-6863
T: +1 410 5485988; Fax: +1 410 5436203; URL: www.salisbury.edu/library
1925; Dr. Alice H. Bahr
Civil War (Les Callette Memorial Coll), Maryland Room, Teacher Education (Educational Resources Coll), AV, micro, Federal & Maryland State Docs
271 000 vols; 1 153 curr per; 196 955 govt docs; 990 maps; 684 000 microforms; 3 700 av-mat; 885 sound-rec; 300 digital data carriers
libr loan; PALINET, OCLC 32559

University of Utah, ***Salt Lake City***
– Marriott Library, 295 S 1500 East, Salt Lake City, UT 84112-0860
T: +1 801 5816010; Fax: +1 801 5857185; URL: www.lib.utah.edu
1850; Joyce Ogburn
Fine arts, Mathematics, Rare bks, Western Americana; 2002 Winter Olympics, Archs
2 900 000 vols; 29 645 curr per
libr loan 32560

– S. J. Quinney Law Library, 332 S 1400 East, Salt Lake City, UT 84112-0731
T: +1 801 5816594; Fax: +1 801 5853033; URL: www.law.utah.edu/sjqlibrary
1923; Rita T. Reusch
320 000 vols; 2 300 curr per; 531 000 microforms; 20 digital data carriers 32561

– Spencer S. Eccles Health Sciences Library, Bldg 589, 10 N 1900 E, Salt lake City, UT 84112
T: +1 801 5815534; Fax: +1 801 5813632; URL: library.med.utah.edu
1966
Hist of Medicine Coll
59 500 vols; 1 789 curr per
libr loan 32562

Westminster College, Giovale Library, 1840 S 1300 East, ***Salt Lake City***, UT 84105-3697
T: +1 801 8322256; Fax: +1 801 8323109; URL: www.westminstercollege.edu/library
1875; David A. Hales
Archival Mat (Early Hist of the College & Early Hist of the Presbyterian Church in Utah)
126 000 vols; 312 curr per; 21 250 e-journals; 46 000 e-books; 95 Electronic Media & Resources
libr loan; LOEX 32563

Angelo State University, Porter Henderson Library, 2025 S Johnson, ***San Angelo***, TX 76904-5079; ASU Station No 11013, San Angelo, TX 76909-1013
T: +1 325 9422154; Fax: +1 325 9422198; E-mail: library@angelo.edu; URL: www.angelo.edu/services/library
1928; Dr. Maurice G. Fortin
492 000 vols; 1 550 curr per; 16 450 e-journals; 28 000 e-books; 8 500 av-mat; 760 sound-rec 32564

Donald E. O'Shaughnessy Library, Oblate School of Theology, 285 Oblate Dr, ***San Antonio***, TX 78216-6693
T: +1 210 3411368; Fax: +1 210 3414519; E-mail: library@ost.edu; URL: www.ost.edu
1903; Father Donald J. Joyce
Religious studies, Mexican-American studies; Faculty Diss, Mission Docs
85 000 vols; 425 curr per; 1 000 microforms; 9 digital data carriers; 700 pamphlets, 400 charts 32565

Our Lady of the Lake University, Sueltenfuss Library, 411 SW 24th St, ***San Antonio***, TX 78207-4689
T: +1 210 4346711; Fax: +1 210 4361616; URL: www.ollusa.edu; library.ollusa.edu
1896; Dr. Paul Frisch
Hist of the Southwest (Texana)
144 000 vols; 724 curr per; 21 000 e-books; 15 av-mat; 52 Bks on Deafness & Sign Lang, 15 Talking Bks 32566

– Worden School of Social Service Library, 411 SW 24th St, San Antonio, TX 78207-4689
T: +1 210 4346711; Fax: +1 210 4314028; E-mail: delga@lake.ollusa.edu; URL: lib.ollusa.edu
1942; Ana Leyba Delgado
262 239 vols; 88 curr per; 259 diss/theses; 325 case records 32567

Saint Mary's University, ***San Antonio***
– Academic Library, One Camino Santa Maria, San Antonio, TX 78228-8608
T: +1 210 4363441; Fax: +1 210 4363782; URL: www.stmarytx.edu/acadlib
1852; Dr. H. Palmer Hall
Hilaire Belloc Coll & G K Chesterton Coll (complete sets of 1st editions), Peninsular Wars Coll; Political Buttons Coll; Spanish Arch of Laredo, Texas
270 000 vols; 1 100 curr per
libr loan 32568

– Sarita Kenedy East Law Library, One Camino Santa Maria, San Antonio, TX 78228-8605
T: +1 210 4363435; Fax: +1 210 4363240; URL: www.stmarytx.edu
1927; Robert H. Hu
Early Spanish Law Coll
339 000 vols; 3 474 curr per; 681 000 microforms; 1 668 sound-rec; 5 digital data carriers 32569

San Antonio College, Library Department, 1001 Howard St, ***San Antonio***, TX 78212
T: +1 210 7856201; URL: www.accd.edu/sac/lrc
1926; Dr. Alice Johnson
18th c British Lit (Morrison Coll), Los Pastores, McAllister Coll, Texana, Western Coll
217 470 vols; 120 curr per; 17 430 e-journals; 24 300 e-books; 7 500 av-mat; 330 sound-rec; 120 Electronic Media & Resources, 490 Bks on Deafness & Sign Lang
libr loan 32570

Trinity University, Coates Library, One Trinity Pl, ***San Antonio***, TX 78212-7200
T: +1 210 9998127; Fax: +1 210 9998021; URL: www.trinity.edu/departments/library/library.html
1869; Diane J. Graves
American Lit (Helen Miller Jones Coll), Beretta Texana, Mexican Arch of Monterrey, Mexico & the State of Nuevo Leon, Space Exploration & Space Medicine (Paul A. Campbell Man & Space, Jim Maloney)800= 201
933 000 vols; 2 408 curr per; 122 000 e-books; 197 256 govt docs; 6 129 maps; 1 239 000 microforms; 17 541 sound-rec; 776 digital data carriers; 2 954 videos, 27 480 slides
libr loan; ALA 32571

University of Texas at San Antonio, Library, One UTSA Circle, ***San Antonio***, TX 78249-0671
T: +1 210 4584573; Fax: +1 210 4584571; E-mail: librarycolldev@utsa.edu; URL: www.lib.utsa.edu
1972; Dr. David R. Johnson
Texana, especially San Antonio & South Texas
668 000 vols; 2 000 curr per; 35 000 e-journals
libr loan 32572

University of the Incarnate Word, J.E. & L.E. Mabee Library, 4301 Broadway, UPO Box 297, ***San Antonio***, TX 78209-6397
T: +1 210 8296010; Fax: +1 210 8296041; URL: www.uiwtx.edu/~libweb/mabee.html
1897; Basil A. Aivaliotis
Texana, Ezra Pound, Women's studies; Unique Editions & Rare Bks
269 000 vols; 17 018 curr per; 24 000 e-books; 500 Bks on Deafness & Sign Lang
libr loan 32573

California State University, San Bernardino Library, John M Pfau Library, 5500 University Pkwy, ***San Bernardino***, CA 92407-2397
T: +1 909 5375091; Fax: +1 909 5377048; URL: www.lib.csusb.edu
1963; Johnnie Ann Ralph
685 000 vols; 1 504 curr per; 12 000 e-books; 5 200 av-mat
libr loan 32574

Alliant International University, Walter Library, 10455 Pomerado Rd, ***San Diego***, CA 92131-1799
T: +1 858 6354511; Fax: +1 858 6354599; E-mail: wlibrary@alliant.edu; URL: library.alliant.edu
1952; Scott Zimmer
Education, Business, Psychology
75 000 vols; 193 curr per; 7 000 e-books; 1 000 microforms; 600 av-mat; 15 digital data carriers
libr loan 32575

California Western School of Law Library, 290 Cedar St, ***San Diego***, CA 92101; mail address: 225 Cedar St, San Diego, CA 92101
T: +1 619 5251437; Fax: +1 619 6852918; URL: www.cwsl.edu/library
1958; Phyllis Marion
Congressional Info Service (US Congress Coll); US Supreme Court Records & Briefs
341 000 vols; 3 957 curr per; 213 e-journals; 89 000 microforms; 1 177 sound-rec; 315 digital data carriers
libr loan; AALL 32576

National University, University Library, 9393 Lightwave Ave, ***San Diego***, CA 92123-1447
T: +1 858 5417923; Fax: +1 858 5417994; E-mail: refdesk@nu.edu; URL: www.nu.edu/library
1975; Anne Marie Secord
Adult Learners Coll
274 000 vols; 1 188 curr per; 17 000 e-journals; 89 000 e-books; 6 500 av-mat; 102 sound-rec; 1 700 digital data carriers; 311 Bks on Deafness & Sign Lang 32577

Point Loma Nazarene University, Ryan Library, 3900 Lomaland Dr, ***San Diego***, CA 92106-2899
T: +1 619 8492338; Fax: +1 619 2220711; E-mail: reflib@pointloma.edu; URL: www.pointloma.edu/RyanLibrary
1902; Frank Quinn
Religion (Holiness Authors), Armenian-Wesleyan Theological Libr
165 000 vols; 613 curr per; 5 000 microforms; 900 av-mat; 1 600 sound-rec; 12 digital data carriers
libr loan 32578

San Diego Miramar College, Library / Learning Resources Center, 10440 Black Mountain Rd, ***San Diego***, CA 92126-2999
T: +1 619 3887310; Fax: +1 619 3887918; E-mail: spesce@sdccd.net; URL: www.miramar.sdccd.net/depts/library
1973; Glen Magpuri
Automotive, Emergency medicine, Fire science, Transportation; Law Libr
22 000 vols; 102 curr per; 1 000 av-mat 32579

San Diego State University, University Library, 5500 Campanile Dr, ***San Diego***, CA 92182-8050
T: +1 619 5946730; Fax: +1 619 5943270; URL: infodome.sdsu.edu
1897; Connie V. Dowell
Hist of Astronomy & Sciences, Entomology & the Hist of Biology, Orchids, Miniature Bks, Artist Bks, Edward Gorey, San Diego Hist
1 649 000 vols; 4 177 curr per; 24 371 e-journals; 109 000 e-books; 13 incunabula; 627 718 diss/theses; 141 503 maps; 157 microforms; 6 000 av-mat; 15 158 sound-rec; 129 digital data carriers; 5 924 linear ft of archs & mss
libr loan 32580

University of San Diego, Helen K. & James Copley Library, 5998 Alcala Park, ***San Diego***, CA 92110
T: +1 619 2604799; Fax: +1 619 2604617; URL: marian.sandiego.edu
1949; Edward D. Starkey
Catholicism, Hist, Literature, Philosophie, Religion, Humanities
500 000 vols; 2 200 curr per; 400 maps; 150 000 microforms; 2 000 sound-rec; 70 digital data carriers 32581

– Legal Research Center Library, 5998 Alcala Park, San Diego, CA 92110-2492
T: +1 619 2604542; Fax: +1 619 2604616; URL: www.sandiego.edu/lrc
1954; L. Ruth Levor
535 000 vols; 4 760 curr per; 894 000 microforms; 15 av-mat; 2 digital data carriers
libr loan   32582

Alliant International University, Rudolph Hurwich Library, One Beach St, **San Francisco**, CA 94133
T: +1 415 9552157; Fax: +1 415 9552180; URL: www.alliant.edu/library
1969; Deanna Gaige
Social sciences; Psychological Assessment Coll,Education Coll
24 000 vols; 100 curr per; 1 790 diss/theses; 499 sound-rec; 1 digital data carriers
libr loan   32583

California College of the Arts Libraries, Simpson Library, 1111 Eighth St, **San Francisco**, CA 94107
T: +1 415 7039574
Michael Lordi
Architecture, design
19 000 vols; 154 curr per   32584

California Institute of Integral Studies Library, Laurance S. Rockefeller Library, 1453 Mission St, 3rd Fl, **San Francisco**, CA 94103
T: +1 415 5756185; Fax: +1 415 5751264; E-mail: library@ciis.edu; URL: library.ciis.edu
1968; Lise M. Dyckman
Psychology, Philosophy, Religion, Cultural anthropology
33 000 vols; 201 curr per; 800 diss/theses; 100 maps; 120 av-mat; 500 sound-rec; 7 digital data carriers
libr loan   32585

California Pacific Medical Center-University of the Pacific School of Dentistry, Health Sciences Library, 2395 Sacramento St, **San Francisco**, CA 94115-2328; P.O. Box 7999, San Francisco, CA 94120-7999
T: +1 415 9233240; Fax: +1 415 9236597; E-mail: cpmclib@sutterhealth.org; URL: www.cpmc.org/hslibrary
1970; Anne Shew
99 000 vols; 370 curr per; 350 av-mat; 1 024 sound-rec; 1 digital data carriers   32586

Golden Gate University, University Library, 536 Mission St, **San Francisco**, CA 94105-2967
T: +1 415 4427256; Fax: +1 415 5436779; E-mail: askalibrarian@ggu.edu; URL: www.ggu.edu/library; library.ggu.edu
1851; Janice Carter
Business and Film (AV mat, VHS and DVD)
300 000 vols; 1 500 curr per; 83 e-journals; 322 diss/theses; 199 digital data carriers   32587

– School of Law Library, 536 Mission St, San Francisco, CA 94105
T: +1 415 4426680; Fax: +1 415 5129395; URL: www.ggu.edu/lawlibrary
1901; Margaret G. Arnold
250 000 vols; 3 424 curr per; 69 digital data carriers
libr loan   32588

New College of California – Humanities Library, 777 Valencia St, **San Francisco**, CA 94110
T: +1 415 4373455
Gary Tombleson
25 000 vols   32589

New College of California – Law Library, 50 Fell St, **San Francisco**, CA 94102
T: +1 415 2411300; Fax: +1 415 2411353; URL: www.newcollege.edu/lib/library.html
1973; Lois Schwartz
Modern American Poets (Poetics Coll), tapes, journals
10 000 vols; 10 curr per; 170 sound-rec; 10 digital data carriers
libr loan   32590

San Francisco Conservatory of Music, Bothin Library, 1201 Ortega St, **San Francisco**, CA 94122
T: +1 415 5648086; Fax: +1 415 7593499; URL: www.sfcm.edu
1957; Deborah Smith
Performance Mat
75 000 vols; 73 curr per; 19 678 sound-rec; 2 digital data carriers; 33 474 musical scores, 470 slides   32591

San Francisco State University, J. Paul Leonard Library, 1630 Holloway Ave, **San Francisco**, CA 94132-4030
T: +1 415 3381854; Fax: +1 415 3381504; E-mail: rcolunga@sfsu.edu; URL: www.library.sfsu.edu
1899; Russell Colunga
Art, Architecture, Business and management, Economics, Education, Hist; Archer Children's Book Coll, Bay Area TV Archives, Frank V de Bellis Coll, Labor Archives & Research Center
841 000 vols; 16 000 curr per; 10 000 e-books
libr loan   32592

University of California – Hastings College of the Law Library, 200 McAllister St, **San Francisco**, CA 94102-4978
T: +1 415 5654751; Fax: +1 415 6214859; URL: www.uchastings.edu/library
1878; Jenni Parrish
Criminal law and justice
438 000 vols; 7 857 curr per; 72 000 microforms; 500 av-mat; 2 372 sound-rec; 37 digital data carriers
libr loan   32593

University of California, San Francisco – Paul & Lydia Kalmanovitz Library & The Center for Knowledge Management, 530 Parnassus Ave, **San Francisco**, CA 94143-0840
T: +1 415 4762337; Fax: +1 415 4764653; URL: www.library.ucsf.edu
1864; Peggy Tahir
Health Sciences, Tobacco; California Medicine (communicable diseases, high altitude physiology, industrial/organizational medicine), East Asian Medicine, Hist of Health Sciences, Homeopathy, Univ Arch
812 000 vols; 167 000 microforms; 10 digital data carriers   32594

University of San Francisco, Richard A. Gleeson Library – Charles & Nancy Gieschke Resource Center, 2130 Fulton St, **San Francisco**, CA 94117-1080
T: +1 415 4222039; Fax: +1 415 4225949; URL: www.usfca.edu/library
1855; Tyrone Cannon
1890's English Lit, Fine Printing, Sir Thomas Moore & Contemporaries
710 000 vols; 2 104 curr per; 569 e-journals; 28 000 e-books; 2 000 av-mat; 1 500 sound-rec; 370 Bks on Deafness & Sign Lang
libr loan   32595

– Zief Law Library, 2101 Fulton St, San Francisco, CA 94117-1004; mail address: 2130 Fulton St, San Francisco, CA 94117-1080
T: +1 415 4226679; Fax: +1 415 4222345; URL: www.usfca.edu/law_library/
1912; Shannon S. Burchard
355 000 vols; 3 293 curr per; 996 000 microforms; 62 digital data carriers   32596

Lincoln Law School of San Jose, James F. Boccardo Law Library, One N First St, **San Jose**, CA 95113
T: +1 408 9777227; Fax: +1 408 9777228; URL: www.lincolnlaw.sj.edu
12 000 vols; 15 curr per   32597

San Jose State University, King Library, One Washington Sq, **San Jose**, CA 95192-0028
T: +1 408 8082304; Fax: +1 408 8082082; URL: www.sjlibrary.org/
1857; Ruth E. Kifer
Beethoven Studies, World War II Diplomatic & Military Hist, John Steinbeck Coll, California & Santa Clara County Hist, Gay/Lesbian Community in San Francisco (Ted Sahl Arch), John Gordon Coll of Photography
1 283 000 vols; 465 curr per; 5 000 e-books; 19 631 diss/theses; 318 790 govt

docs; 10 367 maps; 1 500 000 microforms; 21 056 sound-rec; 30 digital data carriers
libr loan; ALA   32598

California Polytechnic State University, Robert E. Kennedy Library, One Grand Ave, **San Luis Obispo**, CA 93407
T: +1 805 7562345; Fax: +1 805 7562346; E-mail: lib-admin@calpoly.edu; URL: lib.calpoly.edu
1901; Anna K. Gold
Agriculture, Art, Architecture; Barton Coll of Landscape Architecture, Julia Morgan Papers, Hearst Castle Coll, Diablo Canyon Nuclear Plant Depository, California Promotional & Travel Lit, Western Fairs Coll, Upton Sinclair Coll, William F. Cody Coll
804 360 vols; 737 curr per; 32 219 e-journals; 3 236 e-books; 196 mss; 3 222 diss/theses; 5 320 govt docs; 24 584 maps; 2 157 216 microforms; 8 486 av-mat; 2 308 sound-rec; 14 409 digital data carriers; 4 832 linear feet of manuscripts, 37 053 graphic materials
libr loan; WEST, CNI, USAIN   32599

California State University, San Marcos Library, 333 S Twin Oaks Valley Rd, **San Marcos**, CA 92096-0001
T: +1 760 7504340; Fax: +1 760 7503287; URL: library.csusm.edu
1989; Marion T. Reid
Spanish Lit Coll
269 000 vols; 665 curr per; 15 100 e-journals; 29 000 e-books; 58 diss/theses; 819 000 microforms; 3 000 sound-rec; 1 705 digital data carriers
libr loan   32600

Texas State University-San Marcos, Albert B. Alkek Library, Wood & Talbot St, **San Marcos**, TX 78666-4604; mail address: 601 University Dr, San Marcos, TX 78666-4604
T: +1 512 2452685; Fax: +1 512 2453002; URL: www.library.txstate.edu
1899; Connie Todd
Early Textbooks (DAR), John Wesley Hardin, Lyndon B. Johnson, T. E. Lawrence, Southwestern Writers Coll (J. Frank Dobie, John Graves, Preston Jones, Larry L. King, Russell Lee, Edwin Shrake, Texana (Elliott Coll)
1 410 000 vols; 6 100 curr per; 32 723 e-journals
libr loan   32601

Dominican University of California, Archbishop Alemany Library, 50 Acacia Ave, **San Rafael**, CA 94901-2298
T: +1 415 4853251; Fax: +1 415 4592300; E-mail: circdesk@dominican.edu; URL: www.dominican.edu/academics/resources/library
1917; Gary Gorka
Adm Chester Nimitz Coll, Ansel Adams Coll
91 800 vols; 450 curr per; 400 e-books; 2 700 microforms; 400 av-mat; 7 digital data carriers; 18 000 Electronic Media & Resources, 1 244 Bks-By-Mail
libr loan; ALA   32602

Trinity International University, Law Library & Information Center, 2200 N Grand Ave, **Santa Ana**, CA 92705-7016
T: +1 714 7967172; Fax: +1 714 7967190; URL: www.tiu.edu/law
Kirk JD Womack
Human rights
50 000 vols; 56 curr per   32603

Brooks Institute, School of Photography Library, 27 E Cota St, **Santa Barbara**, CA 93101
T: +1 805 9663888 ext 3027; Fax: +1 805 5641475; E-mail: library@brooks.edu; URL: www.brooks.edu
1972; Susan Shiras
8 000 vols; 100 curr per; 106 diss/theses; 200 av-mat; 28 digital data carriers; 25 000 unbound periodicals, 2 500 pamphlets   32604

University of California, Santa Barbara, Davidson Library, U C S B, **Santa Barbara**, CA 93106-9010
T: +1 805 8933491; Fax: +1 805 8934676; URL: www.library.ucsb.edu
1909; Cathy Chiu
American Religions Coll, Bernath Memorial Coll, California Ethnic & Multicultural Arch, Darwin/Evolution

Coll, Morris Ernst Banned Book Coll, Humanistic Psychology Arch, Performing Art Colls, Rare Books, L. B. Romaine Trade Card Coll, Santa Barbara Hist Colls, Skofield Printers Coll, Wyles Coll, Univ Arch
2 949 000 vols; 23 218 curr per; 405 000 maps; 3 800 000 microforms; 8 000 av-mat; 65 000 sound-rec; 150 digital data carriers
libr loan   32605

– Arts Library, Davidson Library, Santa Barbara, CA 93106-9010
T: +1 805 8932850; Fax: +1 805 8935879; URL: www.library.ucsb.edu/depts/arts/
Susan Moon
320 000 vols; 534 curr per   32606

Westmont College, Roger John Voskuyl Library, 955 La Paz Rd, **Santa Barbara**, CA 93108-1099
T: +1 805 5656147; Fax: +1 805 5656220; URL: www.library.westmont.edu
1940; John D. Murray
Christ & Culture Coll
123 000 vols; 300 curr per; 13 000 microforms; 4 392 sound-rec; 6 digital data carriers; 11 slides
libr loan   32607

Santa Clara University, Library, 500 El Camino Real, **Santa Clara**, CA 95053-0500
T: +1 408 5544415; Fax: +1 408 5546827; E-mail: libraryreference@scu.edu; URL: www.scu.edu/library
1851
Californiana, Labor Relations in California, Denise Levertov, Jose Antonio Villarreal
3 800 vols; 2 630 curr per; 4 046 sound-rec; 1 330 scores   32608

– Heafey Law Library, School of Law, 500 El Camino Real, Santa Clara, CA 95053-0430
T: +1 408 5544072; Fax: +1 408 5545318; E-mail: memery@scu.edu; URL: www.scu.edu/law/library/index.html
1912; Mary B. Emery
Proceedings of the House Judiciary Committee on the Watergate Hearings
169 000 vols; 4 153 curr per; 744 000 microforms; 10 digital data carriers
libr loan   32609

The Master's College, Robert L. Powell Library, 21726 W Placerita Canyon Rd, **Santa Clarita**, CA 91321-1200
T: +1 661 2593540; Fax: +1 661 2229159; URL: www.masters.edu/deptpagenew.asp?pageid=22
1927; John W. Stone
Biblical studies
151 000 vols; 494 curr per; 18 500 e-journals; 10 000 e-books; 73 digital data carriers   32610

University of California, Santa Cruz, University Library, UCSC Library, 1156 High St, **Santa Cruz**, CA 95064
T: +1 831 4592076; Fax: +1 831 4598206; URL: library.ucsc.edu
1965; Virginia Steel
Gregory Bateson Coll, Kenneth Patchen Arch, Trianon Press Arch, Santa Cruz County Hist Coll, Thomas Carlyle Coll
1 613 000 vols; 31 047 curr per; 167 776 maps; 731 000 microforms; 25 027 sound-rec; 1 336 digital data carriers; 263 666 slides
libr loan   32611

– Science & Engineering Library, 1156 High St, Santa Cruz, CA 95064
T: +1 831 4592886; Fax: +1 831 4592797; URL: www.library.ucsc.edu/science
Catherine Soehner
400 000 vols   32612

College of Santa Fe, Fogelson Library Center, 1600 St Michael's Dr, **Santa Fe**, NM 87505-7634
T: +1 505 4736569; Fax: +1 505 4736593; E-mail: reference@santafeuniversity.edu; URL: library.santafeuniversity.edu/
1874; Valerie Nye
Southwest & New Mexico Hist, Long Playing Record Albums, Hist of Photography, Chase Art Hist Libr
222 000 vols; 15 500 curr per; 1 300

music scores; 1 500 av-mat
libr loan 32613

Saint John's College, Meem Library, 1160 Camino Cruz Blanca, *Santa Fe*, NM 87505
T: +1 505 9846042; Fax: +1 505 9846004; URL: www.stjohnscollege.edu
1964; Jennifer Sprague
Music (Grumman, Holzman, Schmidt & White), Hunt Coll
70 000 vols; 120 curr per
libr loan 32614

Southwestern College, Quimby Memorial Library, San Felipe at Airport Rd., Rte 20, Box 29-D, *Santa Fe*, NM 87501; P.O. Box 4788, Santa Fe, NM 87502
T: +1 505 467-6825 ext 23; Fax: +1 505 467-6826; E-mail: library@swc.edu; URL: www.swc.edu
Sandra Hareld
18 500 vols; 57 curr per; 10 diss/theses; 260 av-mat; 62 sound-rec
libr loan; ALA 32615

Thomas Aquinas College, Saint Bernadine Library, 10000 N Ojai Rd, *Santa Paula*, CA 93060-9980
T: +1 805 5254417; Fax: +1 805 5259342; URL: www.thomasaquinas.edu
1971; V. A. Jatulis
Theology, Philosophy
61 000 vols; 67 curr per; 400 av-mat 32616

New College of Florida – University of South Florida, Sarasota-Manatee, Jane Bancroft Cook Library, 5800 Bay Shore Rd, *Sarasota*, FL 34243-2109
T: +1 941 4874301; Fax: +1 941 4874307; URL: www.ncf.edu/library
1962; Paul Gherman
Education, Humanities; Hagberg Mss
286 000 vols; 855 curr per; 170 000 e-books 32617

Ringling College of Art & Design, Verman Kimbrough Memorial Library, 2700 N Tamiami Trail, *Sarasota*, FL 34234
T: +1 941 3597587; Fax: +1 941 3597632; E-mail: library@ringling.edu; URL: www.lib.ringling.edu
1931; Kathleen L. List
Artists books; art catalogues raisonnés
55 000 vols; 360 curr per; 4 000 e-journals; 33 000 e-books; 928 sound-rec; 3 230 digital data carriers
libr loan; ARLIS/NA, TBLC 32618

Skidmore College, Lucy Scribner Library, 815 N Broadway, *Saratoga Springs*, NY 12866
T: +1 518 5805000; Fax: +1 518 5805541; E-mail: illdesk@skidmore.edu; URL: www.skidmore.edu
1911; Ruth Copans
Art Coll, Autographed First Editions (Frances Steloff Coll), Hebraica-Judaica (Leo Usdan Coll), Saratogiana (Anita P Yates Coll)
400 000 vols; 112 000 govt docs; 10 904 maps; 253 000 microforms; 1 155 sound-rec; 32 digital data carriers; 115 266 slides
libr loan 32619

Lake Superior State University, Kenneth J. Shouldice Library, 906 Ryan Ave, *Sault Ste. Marie*, MI 49783
T: +1 906 6352815; Fax: +1 906 6352193; URL: www.lssu.edu/library/
1946; Fredrick Michels
Michigan Hist (Michigan Room); Great Lakes (Marine-Laker)
111 000 vols; 811 curr per
libr loan 32620

Armstrong Atlantic State University, Lane Library, 11935 Abercorn St, *Savannah*, GA 31419
T: +1 912 9275332; Fax: +1 912 9275387; E-mail: fraziedo@mail.armstrong.edu; URL: www.library.armstrong.edu
1935; Doug Frazier
Georgia; Educational Resources Info Ctr Coll, Libr of American Civilization, Libr of English Lit
195 000 vols; 1 100 curr per; 118 maps; 466 425 microforms; 5 700 av-mat; 1 835 sound-rec; 23 digital data carriers
libr loan 32621

Savannah State University, Asa H. Gordon Library, 2200 Tompkins Rd, *Savannah*, GA 31404; P.O. Box 20394, Savannah, GA 31404-0705
T: +1 912 3562932; Fax: +1 912 3562874; URL: www.library/savstate.edu
1891; Mary Jo Fayoyin
Black studies; Educational Resources Info Ctr
189 000 vols; 566 curr per 32622

Illinois Institute of Art, Learning Resource Center, 1000 Plaza Dr, Ste 100, *Schaumburg*, IL 60173-4990
T: +1 847 6193450; Fax: +1 847 6193064; URL: www.ilia.aii.edu
1986; Rich Wilson
Applied arts
9 000 vols; 175 curr per; 3 700 av-mat; 3 585 sound-rec; 100 digital data carriers; paper samples
libr loan; OCLC 32623

Union College, Schaffer Library, 807 Union St, *Schenectady*, NY 12308
T: +1 518 3886282; Fax: +1 518 3886619; URL: www.union.edu/public/library
1795; Thomas G. McFadden
French Civilization to end of 19th c (John Bigelow Libr Coll), Local Hist (Schenectady Coll), 19th c American Wit & Humor (Bailey Coll), Schenectady Arch of Science & Technology Coll
614 000 vols; 6 248 curr per
libr loan 32624

Pennsylvania State University, Schuylkill Campus, Capital College, Ciletti Memorial Library, 240 University Dr, *Schuylkill Haven*, PA 17972-2210
T: +1 570 3856234; Fax: +1 570 3856232; E-mail: mwl2@psu.edu; URL: www.hbg.psu.edu/library/ciletti/index.html
1934; Michael W. Loder
County Coll, O'Hara Coll, Richter Coll, Treasure Coll
44 000 vols; 170 curr per 32625

Bethany College, Wilson Library, 800 Bethany Dr, *Scotts Valley*, CA 95066-2898
T: +1 831 4383800; Fax: +1 831 4384517; URL: www.bethany.edu/library
1919; Anna Temple
Religion, Liberal Arts, Bible, Theology; Pentecostalism; Education; Curriculum
64 000 vols; 650 curr per; 30 diss/theses; 847 music scores; 200 maps; 5 000 microforms; 880 sound-rec; 28 digital data carriers 32626

Marywood University, Learning Resources Center, 2300 Adams Ave, *Scranton*, PA 18509-1598
T: +1 570 3486205; Fax: +1 570 9614769; E-mail: libraryhelp@marywood.edu; URL: www.marywood.edu/library
1915; Cathy Schappert
220 000 vols; 476 curr per; 13 140 e-journals; 3 060 sound-rec; 10 digital data carriers; 6 840 Videos, 22 390 Slides, 258 art Reproductions
libr loan 32627

University of Scranton, Harry & Jeanette Weinberg Memorial Library, Monroe & Linden, *Scranton*, PA 18510-4634
T: +1 570 9417524; Fax: +1 570 9414002; URL: www.scranton.edu/library
1888; Charles E. Kratz
Early Printed Bks & Mss (William W. Scranton Coll), Univ Arch Coll
493 000 vols; 1 208 curr per; 16 000 e-journals; 14 000 e-books; 220 Bks on Deafness & Sign Lang
libr loan 32628

Harding University, Brackett Library, 915 E Market St, *Searcy*, AR 72149-2267
T: +1 501 2794354; URL: quest.harding.edu
1924; Ann Dixon
Williams-Miles Science Hist Coll, LAC
596 000 vols; 1 206 curr per; 4 572 govt docs; 580 digital data carriers
libr loan 32629

California State University, Monterey Bay Library, 100 Campus Ctr, Bldg 12, *Seaside*, CA 93955
T: +1 831 5823733; Fax: +1 831 5823875; E-mail: library_circulation@csumb.edu; URL:

library.csumb.edu
1995; Bill Robnett
Environmental Clean-up (Ford Ord Reuse Coll)
69 000 vols; 496 curr per; 4 000 e-books 32630

Cornish College of the Arts, Library, 1000 Lenora St, mail address: 710 E Roy St, *Seattle*, WA 98121
T: +1 206 7265041; Fax: +1 206 3155811; E-mail: libraryref@cornish.edu; URL: www.cornish.edu/cornish_library
1964; Hollis P. Near
Applied arts
20 000 vols; 154 curr per; 3 000 e-books; 3 500 music scores; 4 500 sound-rec; 30 000 slides, 500 videos 32631

Seattle Pacific University, University Library, 3307 Third Ave W, *Seattle*, WA 98119
T: +1 206 2812154; Fax: +1 206 2812936; URL: www.spu.edu/depts/library
1891; Bryce Nelson
Free Methodism Coll
192 000 vols; 1 230 curr per; 3 000 e-books 32632

Seattle University, A.A. Lemieux Library, 901 12th Ave, *Seattle*, WA 98122-4411; P.O. Box 222000, Seattle, WA 98122-1090
T: +1 206 2966222; Fax: +1 206 2966224; URL: www.seattleu.edu/lemlib
John Popko
217 000 vols; 1 604 curr per; 483 maps; 490 000 microforms; 2 689 sound-rec; 10 digital data carriers
libr loan; OCLC 32633

Seattle University School of Law Library, Sullivan Hall, 901 12th Ave, *Seattle*, WA 98122-4411; P.O. Box 222000, Seattle, WA 98122-1090
T: +1 206 3984227; Fax: +1 206 3984194; URL: www.law.seattleu.edu/library
1972; Kristin Cheney
162 000 vols; 930 000 microforms; 19 digital data carriers
libr loan 32634

University of Washington, University Libraries, Allen Library, 4th Flr, Rm 482, Box 352900, Seattle, WA 98195-2900
T: +1 206 5431878; Fax: +1 206 6858049; URL: www.lib.washington.edu
1862; Lizabeth A. Wilson
Pacific Northwest, 19th C American Lit, Historic Maps, Puget Sound Architectural Drawings, Book Arts, Travel & Exploration, Historical Children's Lit, Hans Christian Andersen, William Blake, William Butler Yeats, Rare Bks
6 000 000 vols; 50 245 curr per; 335 897 maps; 6 870 000 microforms; 67 260 sound-rec; 935 136 graphic 32635

– Architecture-Urban Planning Library, 334 Gould Hall, Box 355730, Seattle, WA 98195-5730
T: +1 206 5434067; E-mail: arch@lib.washington.edu
1923; Alan R. Michelson
41 000 vols; 8 000 microforms; 2 355 rpts 32636

– Art Library, 101 Art Bldg, Box 353440, Seattle, WA 98195
T: +1 206 5430648; E-mail: art@lib.washington.edu; URL: www.lib.washington.edu
Connie T. Okada
44 000 vols 32637

– Chemistry Library, 60 Chemistry Libr Bldg, Box 351700, Seattle, WA 98195-1700
T: +1 206 5431603; Fax: +1 206 5433863; E-mail: chemlib@u.washington.edu
Susanne J. Redalje
62 000 vols; 13 000 microforms 32638

– Drama Library, Hutchinson Hall, Rm 145, Box 353950, Seattle, WA 98195-3950
T: +1 206 5435148; Fax: +1 206 5438512; E-mail: drama@lib.washington.edu
1931; Angela Weaver
19th c Acting Editions
30 000 vols; 3 000 microforms; 5 900 av-mat; 657 sound-rec 32639

– East Asia Library, 322 Gowen Hall, 3rd Flr, Box 353527, Seattle, WA 98195-3527
T: +1 206 5434490; Fax: +1 206 2215298; E-mail: ealcirc@lib.washington.edu
1937; Zhijia Shen
Works in Chinese, Japanese, Korean, Tibetan, Vietnamese, Mongolian, & Manchu
451 000 vols; 2 728 curr per; 48 000 microforms; 4 589 pamphlets 32640

– Engineering Library, Engineering Library Bldg, Box 352170, Seattle, WA 98195-2170
T: +1 206 5430740; Fax: +1 206 5433305; E-mail: englib@u.washington.edu; URL: www.lib.washington.edu/engineering
Mel DeSart
165 000 vols; 3 136 curr per 32641

– Fisheries-Oceanography Library, 151 Oceanography Bldg, Box 357952, Seattle, WA 98195-7952
T: +1 206 5434279; Fax: +1 206 5434909; E-mail: fishlib@u.washington.edu; URL: www.lib.washington.edu/fish
1950; Louise M. Richards
Pacific Salmon Lit Compilation; Canadian Translations of Fisheries & Aquatic Sciences (microfiche)
67 000 vols; 500 maps; 28 000 microforms 32642

– Foster Business Library, Bank of America Exec Educ Ctr, Rm 013, Box 353224, Seattle, WA 98195-3224
T: +1 206 5434360; Fax: +1 206 6166430; E-mail: buslib@u.washington.edu
1951; Gordon Aamot
76 000 vols; 238 000 microforms 32643

– Friday Harbor Laboratories Library, 620 University Rd, Box 352900, Friday Harbor, WA 98250-2900
T: +1 206 6160758; Fax: +1 206 5431273; E-mail: frihar@u.washington.edu; URL: www.lib.washington.edu/fhl
Maureen D. Nolan
18 000 vols 32644

– Health Sciences Libraries, T-334 Health Sciences Bldg, Box 357155, 1959 NE Pacific St, Seattle, WA 98195-7155
T: +1 206 5433441; Fax: +1 206 5438066; E-mail: hsl@u.washington.edu; URL: healthlinks.washington.edu
1949; Sherrilynne S. Fuller
History of Medicine Coll
352 000 vols; 3 856 curr per; 80 000 microforms; 25 digital data carriers
libr loan 32645

– Marian Gould Gallagher Law Library, William H. Gates Hall, Box 353025, Seattle, WA 98195-3025
T: +1 206 5434086; Fax: +1 206 6852165; URL: lib.law.washington.edu
1899; Penny A. Hazelton
Chinese, Korean and Japanese legal coll; East Asia Law Dept
605 991 vols; 5 581 curr per; 1 223 e-journals; 2 incunabula; 42 427 microforms; 380 av-mat; 159 sound-rec; 11 digital data carriers
libr loan; AALL 32646

– Mathematics Research Library, C-306 Padelford Hall, Box 354350, Seattle, WA 98195-4350
T: +1 206 5437296; E-mail: mathlib@u.washington.edu; URL: www.lib.washington.edu/math
Martha A. Tucker
61 000 vols 32647

– Music Library, 113 Music Bldg, Box 353450, Seattle, WA 98195-3450
T: +1 206 5431168; E-mail: musiclib@u.washington.edu; URL: www.lib.washington.edu/music
Judy Tsou
Offenbacher Mozart Coll, Harris wind instruments recordingcoll, American Music Center
60 000 vols; 40 mss; 530 diss/theses; 41 730 music scores; 4 000 microforms; 3 800 av-mat; 45 913 sound-rec; 25 digital data carriers
libr loan; MLA, IAML, IASA 32648

– Natural Sciences Library, Allen Library S, Ground & First Flrs, Box 352900, Seattle, WA 98195-2900
T: +1 206 5431243; Fax: +1 206 6851665; E-mail: natsci@u.washington.edu; URL: www.lib.washington.edu/natsci
Maureen Nolan
206 000 vols 32649

– Odegaard Undergraduate Library, Box 353080, Seattle, WA 98195-3080
T: +1 206 6853752; Fax: +1 206 6858485; URL: www.lib.washington.edu/ougl
Jill M. McKinstry
180 000 vols 32650

– Physics-Astronomy Library, C-620 Physics Astronomy Bldg, 3910 15th Ave NE, Box 351560, Seattle, WA 98195-1560
T: +1 206 5432988; Fax: +1 206 6850635; E-mail: phylib@u.washington.edu; URL: www.lib.washington.edu/physics
1935; Pamela F. Yorks
29 000 vols; 113 digital data carriers; 3 272 sky atlases 32651

– Social Work Library, 252 Social Work Bldg, Box 354900, 15th Ave NE, Seattle, WA 98195-4900
T: +1 206 6852180; Fax: +1 206 6857647; E-mail: swl@u.washington.edu; URL: healthlinks.washington.edu/hs/
1954; Angela Lee
40 000 vols; 247 curr per; 29 sound-rec; 1 digital data carriers 32652

– Tacoma Branch, 1900 Commerce St, P.O. Box 358460, Tacoma, WA 98402
T: +1 253 6924440; Fax: +1 253 6924445; E-mail: taclib@uwashington.edu; URL: www.tacoma.washington.edu/library
Charles Lord
60 000 vols; 300 curr per 32653

Texas Lutheran University, Blumberg Memorial Library, 1000 W Court St, **Seguin**, TX 78155-5978
T: +1 830 3728100; Fax: +1 830 3728156; URL: www.tlu.edu
1891; Martha Rinn
Liberal arts, Religious studies; American Lutheran Church, German Lit & Culture, Rundell Rare Bk Coll
102 000 vols; 597 curr per 32654

Susquehanna University, Blough-Weis Library, 514 University Ave, **Selinsgrove**, PA 17870-1050
T: +1 570 3724022; Fax: +1 570 3724310; URL: www.susqu.edu/library
1858; Kathleen Gunning
Business, Environmental studies, Music, Theater; Jane Apple Shakespeare Coll, Wilt Music Coll
282 000 vols; 15 766 curr per
libr loan; ALA 32655

Concordia College, Ellwanger-Hunt Learning Resource Center, 1804 Green St, **Selma**, AL 36701; P.O. Box 1329, Selma, AL 36702-1329
T: +1 334 8745700; Fax: +1 334 8745755; URL: www.concordiaselma.edu
Millie MacMillan
60 000 vols 32656

New York Chiropractic College, Library, 2360 State Rte 89, **Seneca Falls**, NY 13148-9460; P.O. Box 800, Seneca Falls, NY 13148-0800
T: +1 315 5683249; Fax: +1 315 5683119; URL: www.nycc.edu/library
1919; Daniel Kanaley
Chiropractic Hist
46 000 vols; 306 curr per; 1 digital data carriers 32657

University of the South, Jessie Ball duPont Library, 735 University Ave, **Sewanee**, TN 37383-1000
T: +1 931 5981664; Fax: +1 931 5981702; URL: library.sewanee.edu
1858; Todd D. Kelley
Liberal arts, Theology; Anglican Prayer Book, Anglican Studies, Ayres Architecture, Episcopal Church in Southeast Hist, Limited Editions Club Publs, Ward Ritchie Coll, Hudson Stuck Coll, Allen Tate Coll, Sewaneena, Southern Lit & Hist
710 000 vols; 2 531 curr per; 837 e-journals; 148 311 govt docs; 11 163 music scores; 107 000 microforms; 7 316

sound-rec; 428 digital data carriers
libr loan 32658

Concordia University, Link Library, 800 N Columbia Ave, **Seward**, NE 68434-1595
T: +1 402 6437254; Fax: +1 402 6434218; E-mail: library@cune.edu; URL: www.cune.edu/academics/library/
1912; Philip Hendrickson
Education & Religion; Koschman Memorial Coll of Children's Lit
libr loan 32659

Pennsylvania State University, Shenango Commonwealth College, Library, 177 Vine Ave, **Sharon**, PA 16146
T: +1 724 9832876; Fax: +1 724 9832881; URL: www.libraries.psu.edu/shenango
1965; Matthew Ciszek
35 000 vols; 319 curr per 32660

James Kelly Library at Saint Gregory's University, 1900 W MacArthur Dr, **Shawnee**, OK 74804
T: +1 405 8785111; Fax: +1 405 8785198; E-mail: cjbuckley@stgregorys.edu; URL: intranet.stgregorys.edu/places/library/
1915; Anita Semtner
74 820 vols; 106 curr per; 4 000 e-books; 1 240 av-mat; 690 Talking Bks 32661

Oklahoma Baptist University, Mabee Learning Center, 500 W University, OBU Box 61310, **Shawnee**, OK 74804-2504
T: +1 405 8782269; Fax: +1 405 8782256; URL: www.okbu.edu/library/
1911; Richard O. Cheek
Baptist Resource Ctr, Hershel Hobbs Coll
225 000 vols; 450 curr per
libr loan 32662

Lakeland College, John Esch Library, PO Box 359, **Sheboygan**, WI 53082-0359
T: +1 920 5651238; Fax: +1 920 5651206
1940; Ann Penke
65 000 vols; 310 curr per; 12 000 e-books; 30 000 microforms; 470 av-mat; 336 sound-rec; 4 digital data carriers
libr loan; WLA 32663

Shepherd University, Scarborough Library, 301 N King St, **Shepherdstown**, WV 25443; P.O. Box 5001, Shepherdstown, WV 25443-5001
T: +1 304 8765217; Fax: +1 304 8760731; URL: www.shepherd.edu/libweb
1872; John B. Sheridan
West Virginia Coll
170 000 vols; 426 curr per 32664

Austin College, George T. & Gladys H. Abell Library Center, 900 N Grand Ave, Ste 6L, **Sherman**, TX 75090-4402
T: +1 903 8132556; Fax: +1 903 8132287; URL: abell.austincollege.edu/Abell/
1849; John R. West
Alexander the Great (Berzunza Coll), Texana (Pate Texana & Hoard Texas Coll)
222 000 vols; 27 000 e-books; 271 249 govt docs; 99 000 microforms; 746 sound-rec; 12 digital data carriers
libr loan; ALA, ACRL 32665

Shippensburg University, Ezra Lehman Memorial Library, 1871 Old Main Dr, **Shippensburg**, PA 17257-2299
T: +1 717 4771462; Fax: +1 717 4771389; E-mail: libref@ship.edu; URL: www.ship.edu/~library
1871; Dr. Marian Schultz
Pennsylvania Coll, Rare Bks, Univ Arch, Media/Curricular Ctr
449 000 vols; 1 243 curr per; 7 971 maps; 1 797 000 microforms; 4 400 av-mat; 7 790 sound-rec; 1 673 Computer Files
libr loan 32666

Centenary College of Louisiana, John F. Magale Memorial Library, 2834 Woodlawn St, **Shreveport**, LA 71104-3335; P.O. Box 41188, Shreveport, LA 71134-1188
T: +1 318 8695171; Fax: +1 318 8695004; URL: www.centenary.edu/library
1825; Christy Jordan Wrenn
Geology, American Lit, Music education; Centenary College Arch, Jack London Papers, North Louisiana Hist Assn Arch,

United Methodist Church Arch, Sam Peters Res Ctr
208 000 vols; 973 curr per; 62 e-journals; 686 e-books; 12 000 music scores; 1 510 maps; 362 000 microforms; 1 930 sound-rec; 440 digital data carriers; 55 960 Electronic Media & Resources, 2 500 Bks-By-Mail, 350 Special Interest Per Sub, 2 880 Videos
libr loan 32667

Louisiana State University Health Sciences Center – Shreveport, Library, 1501 Kings Hwy, **Shreveport**, LA 71130; P.O. Box 33932, Shreveport, LA 71130-3932
T: +1 318 6755445; Fax: +1 318 6755442; URL: lib-sh.lsuhsc.edu
1968; Marianne L. Comegys
Medical Fiction, Hist of Medicine
191 000 vols; 325 curr per; 704 microforms; 3 604 sound-rec; 528 digital data carriers; 2 558 slides
libr loan; MLA 32668

Louisiana State University in Shreveport, Noel Memorial Library, One University Pl, **Shreveport**, LA 71115-2399
T: +1 318 7975225; Fax: +1 318 7984138; URL: www.lsus.edu/library
1966; Dr. Alan D. Gabehart
James Smith Noel Coll
324 000 vols; 428 curr per; 48 000 e-books; 1 046 maps; 365 000 microforms; 894 digital data carriers
libr loan; OCLC 32669

Southern University, Shreveport-Bossier City Campus Library, 3050 Martin Luther King Dr, **Shreveport**, LA 71107
T: +1 318 6743400; Fax: +1 318 6743403; E-mail: joriley@susla.edu; URL: web.susla.edu/library/Pages/Library-home.aspx
1967; Jane O'Riley
Natural science, Ethnic studies; Black Studies, Louisiana Coll
49 720 vols; 208 curr per 32670

John Brown University, Arutunoff Library, 2000 W University, **Siloam Springs**, AR 72761
T: +1 479 5247202; Fax: +1 479 5247335; E-mail: library@jbu.edu; URL: www.jbu.edu/library
1956; Mary E. Habermas
John Brown University Archives, McGee Biblical Studies Coll, Romig Juvenile Collection, Oliver Marriage & Family Coll; Music Library, Soderquist Business Library, Engineering & Construction Management Library
107 000 vols; 450 curr per; 6 000 e-journals; 11 000 e-books
libr loan; ALA, ACRL, ACL 32671

Western New Mexico University, Miller Library, 1000 W College, **Silver City**, NM 88061; P.O. Box 680, Silver City, NM 88062-0680
T: +1 505 5386350; Fax: +1 505 5386178; URL: voyager.wnmu.edu
1893; Gilda Ortego
Culture, Hist, Southwest; Education (ERIC), Hist (Libr of American Civilization, Indian Claims Commission, Contemporary Newspapers of the North American Indian & Western Americana Hist), Music (Musicache Coll), Local Newspapers from 1886
140 000 vols; 950 curr per; 2 500 maps; 5 digital data carriers; 88 Bks on Deafness & Sign Lang
libr loan 32672

School of Art & Design at Montgomery College, Library, 10500 Georgia Ave, **Silver Spring**, MD 20902
T: +1 301 6494454; Fax: +1 301 6492940
1977; Kate Cooper
11 000 vols; 42 curr per; 30 000 slides 32673

Dordt College, John & Louise Hulst Library, 498 Fourth Ave NE, **Sioux Center**, IA 51250
T: +1 712 7226042; Fax: +1 712 7221198; E-mail: library@dordt.edu; URL: www.dordt.edu/academics/library
1955; Sheryl Sheeres Taylor
Religion, Education, Hist; Arch & Dutch Memorial Coll
303 000 vols; 650 curr per 32674

Briar Cliff University, Mueller Library, 3303 Rebecca St, **Sioux City**, IA 51104-2324
T: +1 712 2795449; Fax: +1 712 2791723; E-mail: library@briarcliff.edu; URL: www.briarcliff.edu/library
1930; Rachel Crowley
Nursing, social work, theology
100 000 vols; 1 800 curr per; 2 000 e-books
libr loan 32675

Morningside College, Hickman-Johnson-Furrow Library Center, 1601 Morningside Ave, **Sioux City**, IA 51106
T: +1 712 2745193; Fax: +1 712 2745224; URL: library.morningside.edu
1894; Daria L. Bossman
Native American studies
99 000 vols; 443 curr per; 230 000 microforms; 5 216 sound-rec; 4 digital data carriers
libr loan; ALA 32676

Augustana College, Mikkelsen Library, 2001 S Summit Ave, **Sioux Falls**, SD 57197-0001
T: +1 605 2744921; Fax: +1 605 2745447; URL: www.augie.edu/library
1860; Ronelle Thompson
Norwegian Coll, Krause Coll, Ctr for Western Studies
210 000 vols; 400 curr per; 13 000 e-books; 1 232 maps; 109 000 microforms; 1 500 av-mat; 3 776 sound-rec; 1 248 digital data carriers
libr loan 32677

– Center for Western Studies Library, 2201 S Summit Ave, Sioux Falls, SD 57197; P.O. Box 727, Sioux Falls, SD 57197-0727
T: +1 605 2744007; Fax: +1 605 2744999; E-mail: cws@augie.edu
1970; Arthur R. Huseboe
Upper Great Plains, Episcopal Diocese of SD Arch, United Church of Christ SD Conference Arch
35 000 vols; 30 curr per; 4 000 linear feet of mss, photos, artifacts 32678

University of Sioux Falls, Mears Library, 1101 W 22nd St, **Sioux Falls**, SD 57105-1699
T: +1 605 3316664; URL: www.usiouxfalls.edu
1883; Judy Clauson Krull
Baptist, South Dakota
88 000 vols; 378 curr per
libr loan 32679

Sheldon Jackson College, Stratton Library, 801 Lincoln St, **Sitka**, AK 99835
T: +1 907 7475259; Fax: +1 907 7475237; URL: www.sheldonjackson.edu
1944; Ginny Norris Blackson
Alaska Reference Room; Rare Books & Arch Room
40 000 vols; 60 curr per; 5 000 microforms; 35 digital data carriers; 5 000 slides
libr loan 32680

Hebrew Theological College, Saul Silber Memorial Library, 7135 N Carpenter Rd, **Skokie**, IL 60077-3263
T: +1 847 9822500; Fax: +1 847 6746381; URL: www.htc.edu
1922; Alan Kagan
Jewish hist & lit, Biblical studies; J. Rapoport, R. Farber; Saul Silber, Max Shulman Zionist Libr, Rev Newman Hebrew Periodical Coll, Rabbi Simon H Album Halakha Coll, Rabbi Leonard C Mishkin Holocaust Coll, Moses Wolfe Coll of Women in Judaism, Lazar Holocaust Memorial Coll
70 000 vols; 182 curr per; 49 sound-rec; 10 digital data carriers
libr loan 32681

Slippery Rock University of Pennsylvania, Bailey Library, **Slippery Rock**, PA 16057-9989
T: +1 724 7382058; Fax: +1 724 7382661; URL: academics.sru.edu/library/new/index.htm
1889; Philip Tramdack
Japan, Italy, Physical Education, Recreation & Sports
600 000 vols; 820 curr per 32682

Bryant Universtiy, Douglas & Judith Krupp Library, 1150 Douglas Pike, ***Smithfield***, RI 02917-1284
T: +1 401 2326299; Fax: +1 401 2326869; E-mail: library@bryant.edu; URL: library.bryant.edu
1955; Mary Moroney
Finance, Small business, Taxation
144 000 vols; 400 curr per; 20 520 e-journals; 8 000 e-books; 13 000 microforms; 300 sound-rec; 20 digital data carriers
libr loan 32683

University of Maryland Center for Environmental Science, Chesapeake Biological Laboratory Library, One Williams St, ***Solomons***, MD 20688; P.O. Box 38, Solomons, MD 20688-0038
T: +1 410 3267223; Fax: +1 410 3267430; E-mail: ill@cbl.umces.edu; URL: www.cbl.umces.edu/library/index.html
Kathy Heil
Ecology, marine biology
12 000 vols; 500 curr per 32684

Indiana University South Bend, Franklin D. Schurz Library, 1700 Mishawaka Ave, ***South Bend***, IN 46615; P.O. Box 7111, South Bend, IN 46634-7111
T: +1 574 5204449; Fax: +1 574 5204472; URL: www.iusb.edu/~libg
1940; Michele C. Russo
James Lewis Casaday Theatre Coll, Lincoln Coll
315 000 vols; 1 599 curr per
libr loan 32685

Notre Dame College, Clara Fritzsche Library, 4545 College Rd, ***South Euclid***, OH 44121
T: +1 216 3735267; Fax: +1 216 3813227; URL: www.notredamecollege.edu/library
1922; Karen Zoller
Tolerance Resource Center; Eastern Church Resource Center; Le Cercle des Conferences Francaises French Coll; Curriculum Libr and Juvenile Coll for Education Dept
88 000 vols; 244 curr per; 413 maps; 17 000 microforms; 1 429 sound-rec; 13 digital data carriers
libr loan; ALA 32686

Mount Holyoke College Library, Information & Technology Services, 50 College St, ***South Hadley***, MA 01075-1423
T: +1 413 5382423; Fax: +1 413 5382370; URL: www.mtholyoke.edu/lits
1837; Charlotte Slocum Patriquin
Economics, Feminism, Hist, Natural sciences; Alumnae Letters & Diaries, Faculty Papers, Illust Editions of Dante's Divine Comedy (Giamatti Dante Coll), Women's Education 1920-
721 000 vols; 3 805 curr per; 1 540 e-journals; 5 000 e-books
libr loan 32687

Atlantic Union College, G. Eric Jones Library, 138 Main St, ***South Lancaster***, MA 01561; P.O. Box 1209, South Lancaster, MA 01561-1209
T: +1 978 3682455; Fax: +1 978 3682456; URL: www.auc.edu/
1882; Monica K. McCarter
Religion, Theology, Literary criticism; George H. Reavis Education Mat, Seventh-Day Adventist Coll, 20th C British & American Poets (Stafford Poetry Coll), Career Ref
110 000 vols; 533 curr per; 12 000 microforms; 19 digital data carriers
libr loan 32688

Seton Hall University, Walsh Library, Walsh Library Bldg, 400 S Orange Ave, ***South Orange***, NJ 07079
T: +1 973 7619441; Fax: +1 973 7619432; E-mail: library@shu.edu; URL: library.shu.edu
1856; Howard F. McGinn
Civil War (Gerald Murphy Coll), Irish Lit & Hist (McManus Coll), Seton Hall Univ & Archdiocesan Arch, Complete Works of Liam O'Flaherty, Classical Studies (Steciuk Coll) Autogr Coll
574 000 vols; 980 curr per
libr loan 32689

Vermont Law School, Julien & Virginia Cornell Library, 68 North Windsor, ***South Royalton***, VT 05068; P.O. Box 68, South Royalton, VT 05068
T: +1 802 8311441; Fax: +1 802 7637159; URL: www.vermontlaw.edu/library/index.cfm
1973; Carl A. Yirka
Environmental & Historic Preservation Coll
230 000 vols; 2 420 curr per; 29 775 e-journals; 920 microforms; 306 sound-rec; 23 digital data carriers
libr loan; NELLCO, AALL, LLNE, SLA, ALA, ACRL 32690

Lawrence Technological University, Library, 21000 W Ten Mile Rd, ***Southfield***, MI 48075-1058
T: +1 248 2043000; Fax: +1 248 2043005; E-mail: refdesk@ltu.edu; URL: library.ltu.edu
1932; Gary R. Cocozzoli
Architecture, Engineering, Management; Architectural Mat (Albert F. Kahn Coll)
113 000 vols; 1 000 curr per; 27 000 e-journals; 24 000 e-books; 16 digital data carriers
libr loan 32691

Saint Thomas Aquinas College, Lougheed Library, 125 Rte 340, ***Sparkill***, NY 10976
T: +1 845 3984219; Fax: +1 845 3599537; URL: www.stac.edu
1952; Mary Anne Lenk
110 000 vols; 548 curr per; 273 000 microforms; 2 000 av-mat
libr loan; ALA 32692

Converse College, Mickel Library, 580 E Main St, ***Spartanburg***, SC 29302-0006
T: +1 864 5969072; Fax: +1 864 5969075; URL: www.converse.edu/library
1890; Wade Woodward
Education, Music; School Prize Texts (A B Taylor)
151 000 vols; 591 curr per 32693

University of South Carolina – Upstate-Spartanburg, Library, 800 University Way, ***Spartanburg***, SC 29303
T: +1 864 5035638; Fax: +1 864 5035601; URL: www.uscupstate.edu/library/default.aspx
1967; Frieda Davison
172 000 vols; 638 curr per; 30 530 e-journals; 27 000 e-books; 520 sound-rec; 390 digital data carriers 32694

Wofford College, Sandor Teszler Library, 429 N Church St, ***Spartanburg***, SC 29303-3663
T: +1 864 5974300; Fax: +1 864 5974329; URL: www.wofford.edu/sandorteszlerlibrary
1854; Oakley H. Coburn
Book Arts, Geography & Travel, Hymns & Hymnody, Press Bks, 16th & 17th C Books, South Caroliniana, Hist Coll of Mat Related to the SC Conference of the United Methodist Church
196 000 vols; 532 curr per; 9 520 e-journals; 38 000 e-books 32695

Black Hills State University, E. Y. Berry Library-Learning Center, 1200 University St, Unit 9676, ***Spearfish***, SD 57799-9676
T: +1 605 6426833; Fax: +1 605 6426298; E-mail: librarydirector@bhsu.edu; URL: iis.bhsu.edu/lis/index.cfm
1883; Rajeev Bukralia
Congressman E Y Berry Papers; Western Hist Studies; Arrow, Inc Coll
206 000 vols; 2 621 curr per; 17 000 e-journals; 13 000 e-books
libr loan 32696

Cooperative Academic Library Services (CALS), Washington State University and Eastern Washington University, 668 N Riverpoint Blvd, Box C, ***Spokane***, WA 99201-1677
T: +1 509 3587930; Fax: +1 509 3587928; E-mail: spoklib@mail.wsu.edu; URL: www.spokane.wsu.edu/libr.html
1992; David Buxton
12 000 vols; 570 curr per; 200 diss/theses; 200 references
libr loan 32697

Gonzaga University, Foley Library, 502 E Boone Ave, ***Spokane***, WA 99258-0095
T: +1 509 3236532; Fax: +1 509 3235904; URL: www.foley.gonzaga.edu
1992; Eileen Bell-Garrison
Gerard Manley Hopkins Coll, Bing Crosby Memorabilia, Hanford Health & Info Arch, Gonzaga Univ Arch, Jesuitica, Labor Unions (Jay Fox Coll)
306 000 vols; 1 169 curr per; 18 000 e-journals; 500 e-books; 712 maps; 501 000 microforms; 1 935 sound-rec; 349 digital data carriers
libr loan 32698

– Chastek Library, 721 N Cincinnati St, Spokane, WA 99202; P.O. Box 3528, Spokane, WA 99220-3528
T: +1 509 3233755; Fax: +1 509 3235534; E-mail: reference@lawschool.gonzaga.edu; URL: www.law.gonzaga.edu/library/
1912; Elizabeth Thweatt
Canon Law Mat, Federal Legislative Histories, Heins American Law Institution Publs, 19th C Treatises
153 000 vols; 2 585 curr per; 108 000 microforms
libr loan 32699

Whitworth University, Harriet Cheney Cowles Memorial Library, 300 W Hawthorne Rd, MS 0901, ***Spokane***, WA 99251-0901
T: +1 509 7774482; Fax: +1 509 7773221; URL: www.whitworth.edu/library
1890; Dr. Hans E. Bynagle
Daniel Photogr Coll, Protestantism in Pacific NW, Presbyterianism, Whitworth Coll
201 000 vols; 1 478 curr per; 862 e-journals; 3 000 e-books; 20 000 mss; 6 301 music scores; 67 000 microforms; 3 750 sound-rec; 488 digital data carriers
libr loan 32700

Spring Arbor University, Hugh A. & Edna C. White Library, 106 E Main St, ***Spring Arbor***, MI 49283
T: +1 517 7506439; Fax: +1 517 7502108; URL: www.arbor.edu/whitelibrary
1873; Roy Meador
102 000 vols; 572 curr per
libr loan 32701

American International College, James J. Shea Sr Memorial Library, 1000 State St, ***Springfield***, MA 01109
T: +1 413 2053225; Fax: +1 413 2053904; URL: www.aic.edu/library
1885; Dr. F. Knowlton Utley
Education, Health science, Psychology; Curriculum Libr, rare bks; Oral Hist Ctr
75 000 vols; 525 curr per 32702

Drury University, F.W. Olin Library, 900 N Benton Ave, ***Springfield***, MO 65802
T: +1 417 8737338; Fax: +1 417 8737432; URL: library.drury.edu
1873; Stephen K. Stoan
Architecture, Women's studies; Claude Thornhill Music Coll, John F Kennedy Memorabilia
176 000 vols; 993 curr per; 590 e-journals; 16 000 e-books; 3 500 av-mat; 1 300 sound-rec
libr loan 32703

Evangel University, Klaude Kendrick Library, 1111 N Glenstone Ave, ***Springfield***, MO 65802
T: +1 417 8652815; URL: www.evangel.edu/library/index.asp
1955; Woodvall R. Moore
Civil War, Indian hist; Libr of American Civilization, O'Reilly Hospital
130 000 vols; 700 curr per; 12 000 e-journals; 7 000 e-books; 500 av-mat; 1 250 sound-rec; 400 digital data carriers
libr loan 32704

Missouri State University, Duane G. Meyer Library, 850 S John Q Hammons Pkwy, ***Springfield***, MO 65807; mail address: 901 S National, Springfield, MO 65897-0001
T: +1 417 8364525; Fax: +1 417 8364538; URL: www.library.missouristate.edu
1907; Neosha A. Mackey
Michel Butor, Jean Arthur Rimbaud (William Jack Jones Coll), Lena Wills Genealogical Coll, Robert Wallace Coll
1 739 000 vols; 4 200 curr per
libr loan 32705

– SMSU-WP Garnett, 304 W Trish Knight St, West Plains, MO 65775
T: +1 417 2557945; Fax: +1 417 2557944
Rose Scarlet
Nursing, Rare bks; Local Hist Audiocassette Coll
40 000 vols; 200 curr per 32706

School of Law Library, 1215 Wilbraham Rd, ***Springfield***, MA 01119
T: +1 413 7821457; Fax: +1 413 7821745; E-mail: bwest@law.wnec.edu; URL: www.law.wnec.edu/library
1973; Barbara West
Labor, Tax; Govt Docs, Massachusetts Continuing Legal Education Mat
367 540 vols; 4 984 curr per; 177 822 microforms; 408 av-mat; 1 526 sound-rec; 35 digital data carriers; 194 software
libr loan; AALL, LLNE 32707

Southern Illinois University – School of Medicine Library, 801 N Rutledge, ***Springfield***, IL 62702; P.O. Box 19625, Springfield, IL 62794-9625
T: +1 217 5452658; Fax: +1 217 5450988; E-mail: reference@siumed.edu; URL: www.siumed.edu/lib
1970; Connie Poole
Hist of Medicine
84 000 vols; 1 182 curr per; 529 e-journals; 763 e-books; 1 020 govt docs; 6 000 microforms; 865 digital data carriers
libr loan; OCLC 32708

Springfield College, Babson Library, 263 Alden St, ***Springfield***, MA 01109-3797
T: +1 413 7483315; Fax: +1 413 7483631; E-mail: ataupier@spfldcol.edu; URL: www.spfldcol.edu/library
1877; Anderea S. Taupier
Sports Rules, US Volleyball Assn Mat
197 320 vols; 1 155 curr per 32709

University of Illinois at Springfield, Norris L. Brookens Library, 1 University Plaza, MS BRK-140, ***Springfield***, IL 62703-5407
T: +1 217 2066597; Fax: +1 217 2066354; E-mail: jtrea1@uis.edu; URL: library.uis.edu
1970; Jane Treadwell
Illinois Regional Arch Depot, Handy Writers Colony Coll, Univ Arch, Oral Hist Coll, Walt Whitman Coll
566 942 vols; 74 179 curr per; 47 377 e-journals; 38 657 e-books; 5 356 diss/theses; 111 063 govt docs; 5 046 maps; 1 863 225 microforms; 26 686 av-mat; 10 269 sound-rec; 3 125 lin ft. mss
libr loan 32710

Western New England College – D'Amour Library, 1215 Wilbraham Rd, ***Springfield***, MA 01119
T: +1 413 7821654; Fax: +1 413 7962011; URL: libraries.wnec.edu
1983; Priscilla L. Perkins
John F. Kennedy, SAEX Judaica Resource Ctr
120 000 vols; 60 curr per; 4 000 e-books; 353 000 microforms; 562 sound-rec; 7 digital data carriers 32711

Wittenberg University, Thomas Library, 807 Woodlawn Ave, ***Springfield***, OH 45504; P.O. Box 7207, Springfield, OH 45501-7207
T: +1 937 3277511; Fax: +1 937 3276139; E-mail: refdesk@wittenberg.edu; URL: www6.wittenberg.edu/lib
1845; Douglas K. Lehman
East Asian studies, Music; Dos Passos Entomological Libr, Hymn Book Coll, Japan (Matsumoto Coll), Martin Luther Reformation
424 000 vols; 14 551 curr per; 11 519 e-journals; 21 000 e-books; 9 640 maps; 68 000 microforms; 17 713 sound-rec; 207 digital data carriers
libr loan; OCLC 32712

Saint Basil College Library, 39 Clovelly Rd, ***Stamford***, CT 06902-3004; mail address: 195 Glenbrook Rd, Stamford, CT 06902-3099
T: +1 203 3277899; Fax: +1 203 9679948; E-mail: ukrmulrec@optonline.net; URL: www.umlsct.org
John Terlecky
131 000 vols 32713

University of Connecticut at Stamford, Jeremy Richard Library, One University Pl, ***Stamford***, CT 06901-2315
T: +1 203 2519599; Fax: +1 203 2518501; E-mail: nancy.gillies@uconn.edu; URL: www.lib.uconn.edu
1962; Nancy Gillies
Economics, Social sciences, Lit
90 000 vols; 200 curr per
libr loan 32714

Saint Joseph's College, Wellehan Library, 278 Whites Bridge Rd, ***Standish***, ME 04084-5263
T: +1 207 8937725; Fax: +1 207 8937883; E-mail: library@sjcme.edu; URL: www.sjcme.edu/library
1912; Shelly Davis
Thomas Merton Coll
100 000 vols; 275 curr per 32715

Stanford University, Cecil H. Green Library, 557 Escondido Mall, ***Stanford***, CA 94305-6004
T: +1 650 7231493; E-mail: infocenter@stanford.edu; URL: www.sul.stanford.edu
1892; Michael A. Keller
C Ashley Felton Memorial Libr, Gunst Memorial Libr of the Book Arts, Barchas Coll on the Hist of Science & Ideas, Brasch Coll on Newton & the Hist of Scientific Thought, Memorial Libr of Music, Antoine Borel Coll, Healey Coll of Irish Lit, Steinbeck Coll, Hemingway Coll, Saroyan Coll, Levertov Coll, Creeley Coll, Ginsberg Coll, Taube-Baron Coll of Jewish Hist & Culture
2 871 000 vols; 7 677 curr per; 220 000 maps; 1 804 000 microforms; 1 800 av-mat; 2 573 sound-rec; 23 689 750 mss & archives, 15 780 tech rpts 32716

– Art & Architecture Library, 102 Cummings Art Bldg, Main Flr, Stanford, CA 94305-2018
T: +1 650 7251037; Fax: +1 650 7250140; E-mail: artlibrary@stanford.edu
1969; Peter Blank
Hist of world art and architecture, with special emphasis on the US, Western Europe, and the Far East
187 000 vols; 501 curr per; 53 000 microforms; 105 digital data carriers
libr loan 32717

– Branner Earth Sciences & Map Collections, Mitchell Bldg, 2nd Flr, Stanford, CA 94305-2174
T: +1 650 7251103; Fax: +1 650 7252534
Julie Sweetkind-Singer
140 000 vols; 1 976 curr per 32718

– Cubberley Education Library, Education Bldg, Rm 202-205, Stanford, CA 94305-3096
T: +1 650 7232121; Fax: +1 650 7360536; E-mail: cubberley@stanford.edu
Kathy Kerns
170 000 vols; 1 195 curr per 32719

– East Asia Library, Stanford, CA 94305-6004
T: +1 650 7253435; Fax: +1 650 7242028; E-mail: eastasialibrary@stanford.edu
Dongfang Shao
Korean, Japanese, Chinese
546 000 vols; 2 809 curr per 32720

– Engineering Library, Terman Engineering Ctr, 2nd Flr, Stanford, CA 94305-4029
T: +1 650 7230001; Fax: +1 650 7251096; E-mail: englibrary@stanford.edu
Helen Josephine
112 000 vols; 1 265 curr per 32721

– Falconer Biology Library, Herrin Hall, 3rd Flr, Stanford, CA 94305-5020
T: +1 650 7231528; Fax: +1 650 7257712
Michael Newman
102 000 vols; 906 curr per 32722

– J. Henry Meyer Memorial Library, 560 Escondido Mall, Stanford, CA 94305
T: +1 650 7245600, 7232434; Fax: +1 650 7258495
Ed McGuigan
Foreign Language Instruction, Instructional Maps, Media Coll, Snowglobes
85 000 vols; 29 curr per 32723

– Hoover Institution on War, Revolution & Peace Library, Stanford, CA 94305-6004
T: +1 650 7231754; Fax: +1 650 7231687
Paul Thomas
915 000 vols; 9 curr per 32724

– J. Hugh Jackson Library, Graduate School of Business, 350 Memorial Way, Stanford, CA 94305-5016
T: +1 650 7232162; Fax: +1 650 7230281; E-mail: jacksonlibrary@gsb.stanford.edu; URL: www-gsb.stanford.edu/services/library/
1933; Kathy Long
Economics; Pacific Northwest Economic Hist (Favre Coll)
205 000 vols; 1 875 curr per; 1 147 000 microforms; 661 digital data carriers; 184 809 corporate rpts, 32 034 tech rpts
libr loan 32725

– Lane Medical Library, Stanford University Medical Ctr, 300 Pasteur Dr, Rm L109, Stanford, CA 94305-5123
T: +1 650 7254584; Fax: +1 650 7257471; URL: www.lane.stanford.edu
1906; Heidi Heilemann
Fleischmann Learning, Hist of Medicine
370 000 vols; 4 568 curr per; 6 000 microforms; 1 004 598 mss & archs, 402 computer mat
libr loan 32726

– Mathematics & Computer Sciences, Sloan Mathematics Ctr, Bldg 380, 4th Flr, Stanford, CA 94305-2125
T: +1 650 7234672; Fax: +1 650 7258998; E-mail: mathcslib@stanford.edu
Linda Yamamoto
124 000 vols; 1 331 curr per 32727

– Miller Library at Hopkins Marine Station, Hopkins Marine Sta, Pacific Grove, CA 93950-3094
T: +1 831 6556228; Fax: +1 831 3737859
1920; Joseph G. Wible
MacFarland Opisthobranchiate Molluscan Coll; G M Smith Algae Reprint Coll
43 000 vols; 300 curr per; 150 maps; 111 microforms
libr loan 32728

– Music Library, Braun Music Ctr, 541 Lasuen Mall, Stanford, CA 94305-3076
T: +1 650 7231212; Fax: +1 650 7251145; E-mail: muslibcirc@stanford.edu
Jerry McBride
125 000 vols; 307 curr per 32729

– Physics Library, Varian Bldg, 3rd Flr, Rm 300, 328 Via Pueblo Mall, Stanford, CA 94305-4060
T: +1 650 7234342; Fax: +1 650 7252079
Stella Ota
58 000 vols; 253 curr per 32730

– Robert Crown Law Library, Stanford Law School, 559 Nathan Abbott Way, Stanford, CA 94305-8610
T: +1 650 7231932; Fax: +1 650 7231933; E-mail: reference@law.stanford.edu; URL: www.law.stanford.edu/library/
1897; Paul Lomio
415 000 vols; 6 569 curr per; 450 000 microforms
libr loan; AALL 32731

– Stanford Auxiliary Library, 691 Pampas Lane, Stanford, CA 94305
T: +1 650 7239201
Regina Wallen
1 982 000 vols; 68 curr per 32732

– Stanford Linear Accelerator Center Research Library, 2575 Sand Hill Rd, MS82, Menlo Park, CA 94025
T: +1 650 9262411; Fax: +1 650 9264905; E-mail: library@slac.stanford.edu; URL: www.slac.stanford.edu/library
1962; Ann Redfield
Arch of the Stanford Linear Accelerator Ctr, & Professional Papers of Key Scientific & Tech Staff (2 072 cu ft of records, 9 000 photos)
19 000 vols; 483 curr per; 60 000 microforms; 8 digital data carriers; 400 000 tech rpts
libr loan 32733

– Swain Library of Chemistry & Chemical Engineering, Organic Chemistry Bldg, 364 Lomita Dr, Stanford, CA 94305-5080
T: +1 650 7239237; Fax: +1 650 7252274; E-mail: swainlibrary@stanford.edu
Grace Baysinger
43 000 vols; 307 curr per 32734

College of Staten Island, Library, 2800 Victory Blvd, 1L, ***Staten Island***, NY 10314-6609
T: +1 718 9824011; Fax: +1 718 9824002; E-mail: library@mail.csi.cuny.edu; URL: www.library.csi.cuny.edu
1976; Wilma L. Jones
Staten Island, Senator John Marchi
233 000 vols; 8 215 curr per; 15 000 e-journals; 3 000 e-books; 523 000 microforms; 4 848 sound-rec; 16 digital data carriers; 77 Electronic Media & Resources
libr loan 32735

Saint John's University, Loretto Memorial Library, Staten Island Campus, 300 Howard Ave, ***Staten Island***, NY 10301
T: +1 718 3904456; Fax: +1 718 3904290; URL: www.stjohns.edu/academics/libraries
1972; Mark Meng
154 000 vols; 337 curr per; 19 000 e-books; 317 000 microforms; 6 digital data carriers
libr loan 32736

Wagner College, Horrmann Library, One Campus Rd, ***Staten Island***, NY 10301-4495
T: +1 718 3903401; Fax: +1 718 4204218; E-mail: library@wagner.edu; URL: www.wagner.edu/library
1889; Dorothy Davison
Lit (Edwin Markham Coll)
209 000 vols; 548 curr per; 11 880 e-journals; 1 800 av-mat; 2 200 sound-rec; 37 000 journals
libr loan 32737

Georgia Southern University, Zach S. Henderson Library, One Lake Dr, ***Statesboro***, GA 30460-8074; P.O. Box 8074, Statesboro, GA 30460
T: +1 912 6815647; Fax: +1 912 6810093; URL: library.georgiasouthern.edu
1906; Dr. W. Bede Mitchell
Commander William M Rigdon Coll, Zachert Coll of Private Press Bks
577 000 vols; 2 687 curr per 32738

Mary Baldwin College, Martha S. Grafton Library, 109 E Frederick St, ***Staunton***, VA 24401
T: +1 540 8877317; Fax: +1 540 8877137; E-mail: ask@mbc.edu; URL: library.mbc.edu
1842; Carol Creager
Women's studies; Mary Baldwin Coll, College Hist
148 000 vols; 337 curr per; 17 000 e-journals; 22 000 e-books 32739

Sterling College, Mabee Library, 125 W Cooper, ***Sterling***, KS 67579; P.O. Box 98, Sterling, KS 67579
T: +1 620 2784234; Fax: +1 620 2784414; E-mail: vstarr@sterling.edu; URL: www.sterling.edu/academics/mabee-library
1887/1995; Charles T. Kendall
85 484 vols; 450 curr per; 1 413 microforms; 46 digital data carriers; 900 slides
libr loan; ALA 32740

Franciscan University of Steubenville, John Paul II Library, 1235 University Blvd, ***Steubenville***, OH 43952-1763
T: +1 740 2836208; Fax: +1 740 2847239; URL: www.franciscan.edu
1946; William Jakub
Franciscana, Mulloy, Kirk
201 000 vols; 728 curr per; 155 000 microforms; 8 digital data carriers
libr loan 32741

University of Wisconsin – Stevens Point, University Library, 900 Reserve St, ***Stevens Point***, WI 54481-1985
T: +1 715 3462540; Fax: +1 715 3463857; E-mail: lrcsec@uwsp.edu; URL: library.uwsp.edu
1894; Dr. Kathy Davis
Censorship, John F Kennedy Assassination, Native Americans
321 000 vols; 8 470 curr per; 587 789 govt docs; 4 482 maps; 854 000 microforms; 623 digital data carriers
libr loan 32742

Villa Julie College, Library, 1525 Greenspring Valley Rd, ***Stevenson***, MD 21153
T: +1 410 4867000; Fax: +1 410 4867329; URL: web.vjc.edu/library
1953; Maureen Anne Beck
Education, Nursing, Paralegal
75 000 vols; 689 curr per 32743

Oklahoma State University, University Library, 216 Library, ***Stillwater***, OK 74078-1071
T: +1 405 7449729; Fax: +1 405 7445183; URL: www.library.okstate.edu
1894; Sheila G. Johnson
Architecture, Veterinary Medicine, US Patents and Trademarks
2 427 870 vols; 38 750 curr per; 5 388 mss; 279 000 maps; 3 486 600 microforms; 70 000 av-mat; 3 780 sound-rec
libr loan; OCLC 32744

Humphreys College, Library, 6505 Inglewood Ave, ***Stockton***, CA 95207
T: +1 209 4780800; Fax: +1 209 4788721; URL: www.humphreys.edu
Dr. Stanislav Perkner
Law
22 000 vols; 113 curr per 32745

University of the Pacific, Library, 3601 Pacific Ave, ***Stockton***, CA 95211
T: +1 209 9462434; Fax: +1 209 9462805; E-mail: jpurnell@uop.edu; URL: library.pacific.edu
1851; Jean M. Purnell
Folk Dance Coll, Methodist Hist Coll, John Muir Papers, Dave Brubeck Coll
373 750 vols; 1 830 curr per; 2 027 maps; 9 100 av-mat; 13 621 sound-rec; 12 digital data carriers 32746

Mississippi State University Agricultural & Forestry Experiment Station, Delta Research & Extension Center Library, Bldg 1532, 82 Stoneville Rd, ***Stoneville***, MS 38776; P.O. Box 197, Stoneville, MS 38776-0197
T: +1 662 6863261; Fax: +1 662 6863342; URL: msucares.com/drec/index.html
1966; Rhonda Holman Watson
Publs of State Experiment Stations and the USDA
25 000 vols; 6 digital data carriers; 50 000 other items
libr loan 32747

State University of New York at Stony Brook, Frank Melville Jr Memorial Library, W-1502 Melville Library, John S Toll Rd, ***Stony Brook***, NY 11794-3300
T: +1 631 6327115; Fax: +1 631 6327116; URL: www.stonybrook.edu/library
1957; Christian Filstrup
Conrad Potter Aiken, Jorge Carrera Andrade, Children's Lit, Chilean Theater Pamphlets, Roberts Creeley, Fortune Press, London, Latin American Pamphlets, Pablo Neruda, Robert Payne, Ezra Pound, W. B. Yeats Mss Coll, Printing & Publishing Coll, Long Island Fiction
1 976 000 vols; 35 222 curr per; 2 000 e-books; 229 695 govt docs; 122 237 maps; 3 700 000 microforms; 30 998 sound-rec; 5 450 linear feet of mss
libr loan; ARL, RLG, CRL 32748

– Chemistry Library, Chemistry Bldg, C-215, Stony Brook, NY 11794-3425
T: +1 631 6327150; Fax: +1 631 6319191; URL: ws.cc.sunysb.edu/chemlib
1965; Dana Antonucci-Durgan
68 000 vols; 200 curr per; 400 microforms
libr loan; ARL 32749

– Computer Science Library, 2120 Computer Science Bldg, Stony Brook, NY 11794-3855
T: +1 631 6327628; Fax: +1 631 6327401; E-mail: library@cs.sunysb.edu
1987; Karen Kostner
15 669 vols; 230 curr per; 3 000 rpts 32750

– Health Sciences Center Library, HSC Level 3, Rm 136, Stony Brook, NY 11794-8034
T: +1 631 4442512; Fax: +1 631 4446649
1969; Spencer Marsh
Hist of Medicine, Dentistry & Nursing
282 000 vols; 2 175 curr per; 17 000 microforms; 224 sound-rec; 11 digital data carriers
libr loan 32751

– Mathematics-Physics-Astronomy Library, Physics Bldg, C-124, Stony Brook, NY 11794-3333
T: +1 631 6327145; Fax: +1 631 6329192; E-mail: sherrychang@cc.mail.sunysb.edu; URL: sunysb.edu/library/math/
1964; Sherry Chang
91 000 vols; 400 curr per; 1 000 diss/theses; 1 300 microforms; 800 unbound lecture notes 32752

– Music Library, Melville Libr, Rm W1530, Stony Brook, NY 11794-3333
T: +1 631 6327097; Fax: +1 631 6327116; E-mail: gglover@notes.cc.sunysb.edu; URL: sunysb.edu/library/music/
1974; Gisele Ira Schierhorst
66 000 vols; 360 curr per; 10 000 microforms; 28 000 sound-rec 32753

Buena Vista University, Ballou Library, 610 W Fourth St, ***Storm Lake***, IA 50588
T: +1 712 7492096; Fax: +1 712 7492059; E-mail: library@bvu.edu; URL: www.bvu.edu/library/
1891; James R. Kennedy
Iowa, Hist
137 000 vols; 623 curr per; 40 e-journals; 10 000 e-books; 3 500 av-mat; 1 140 sound-rec; 230 digital data carriers 32754

University of Connecticut, University Library, 369 Fairfield Rd, ***Storrs***, CT 06069-1005
T: +1 860 4862219; Fax: +1 860 4860584; URL: www.lib.uconn.edu
1881; Brinley Franklin
Childrens's lit, Latin America, Connecticut Hist, Belgium, French & Italian Hist, Chilean Hist, Labor Hist; Alternative Press, Belgium History, Black Mountain Poets, Bookplates, Charles Olson Coll
3 168 620 vols; 42 060 curr per; 167 447 maps; 2 440 000 microforms; 20 167 sound-rec; 246 digital data carriers
libr loan 32755

– Music & Dramatic Arts Library, 1295 Storrs Rd, Unit 1153, 69 North Eagleville Rd, Rm 228, Storrs, CT 06269-1153
T: +1 860 4862502; Fax: +1 860 4865551; URL: www.lib.uconn.edu/music
Tracey Rudnick
75 000 vols; 150 curr per 32756

– Pharmacy Library, Pharmacy/Biology Bldg, Storrs, CT 06269-3092
T: +1 860 4862218; Fax: +1 860 4864998; URL: www.lib.uconn.edu/online/research/speclib/pharmacy/
Sharon Giovenale
28 000 vols; 50 curr per 32757

– Thomas J. Dodd Research Center, Unit 1205, 405 Babbidge Rd, Storrs, CT 06269-1205
T: +1 860 4862524; Fax: +1 860 4864521; E-mail: dodref@lib.uconn.edu; URL: doddcenter.uconn.edu/about/about.htm
1964; Thomas Wilsted
Alternative Politics & Culture; American and English Lit; Americana; Children's Lit; Connecticut Business & Enterprise; Connecticut Historic Preservation; Connecticut Labor; Connecticut Politics & Public Affairs; Ethnic Heritage & Immigration; Graphic & Book Arts; Hispanic Hist & Culture; Hist of Nursing; Human Rights; Natural Hist; Univ Arch
50 000 vols; 150 000 per 32758

University of Medicine & Dentistry of New Jersey, Health Science Library, Academic Ctr, One Medical Center Dr, ***Stratford***, NJ 08084
T: +1 856 5666800; Fax: +1 856 5666380; URL: www3.umdnj.edu/stlibweb
1970; Jan Skica
Osteopathy Coll
30 000 vols; 450 curr per; 200 microforms; 18 digital data carriers; 3 000 monogr 32759

International College of Hospitality Management Library, 1760 Mapleton Ave, ***Suffield***, CT 06078-1463
T: +1 860 6683515; Fax: +1 860 6687369; URL: ichm.edu
Eileen Roehl
10 000 vols; 40 curr per 32760

Morris College, Richardson-Johnson Learning Resources Center, 100 W College St, ***Sumter***, SC 29150-3599
T: +1 803 9343230; Fax: +1 803 7782923; URL: www.morris.edu
1920; Margaret N. Mukooza
African American Coll
75 000 vols; 339 curr per; 1 600 av-mat; 1 000 High Interest/Low Vocabulary Bks
libr loan 32761

University of South Carolina at Sumter, Library, 200 Miller Rd, ***Sumter***, SC 29150-2498
T: +1 803 9383736; Fax: +1 803 9383811; URL: www.uscsumter.edu/library/index.shtml; www.uscsumter.edu
1966; Sharon Chapman
100 000 vols; 125 curr per 32762

Cogswell Polytechnical College, Library, 1175 Bordeaux Dr, ***Sunnyvale***, CA 94089
T: +1 408 5410100 ext 144; Fax: +1 408 7470764; E-mail: library@cogswell.edu; URL: www.cogswell.edu
1887
Electrical engineering; Manufacturers' Data Manuals
13 000 vols; 100 curr per 32763

University of Wisconsin – Superior, Jim Dan Hill Library, PO Box 2000, Belknap & Catlin, ***Superior***, WI 54880-2000
T: +1 715 3948341; Fax: +1 715 3948462; URL: www.uwsuper.edu
1896; Felix Unaeze
Popular Lit (John W R Beecroft), Regional Hist, Lake Superior Marine Museum Assn Coll
250 000 vols; 1 894 curr per
libr loan; ALA 32764

Lassen Community College, Library & Learning Center, 478-200 Hwy 139, P.O. Box 3000, ***Susanville***, CA 96130
T: +1 916 2518830; Fax: +1 916 2578964; E-mail: rbrown@lassencollege.edu; URL: www.lassencollege.edu/cp_library.html
1926; Rosanna Brown
Criminal law; North American Indian Coll, California Coll
21 000 vols; 217 curr per 32765

Warren Wilson College, Pew Learning Center & Ellison Library, 701 Warren Wilson Rd, ***Swannanoa***, NC 28778; Campus Box 6358, P.O. Box 9000, Asheville, NC 28815-9000
T: +1 828 7713058; Fax: +1 828 7717085; URL: www.warren-wilson.edu/~library/
1894; Chris Nugent
Arch, James McClure Clarke Papers, Arthur S Link Libr of American Hist
107 000 vols; 11 076 curr per; 50 000 e-books; 34 000 microforms; 130 sound-rec; 48 Special Interest Per Sub, 92 linear feet mss
libr loan 32766

Friends Historical Library of Swarthmore College, 500 College Ave, ***Swarthmore***, PA 19081-1399
T: +1 610 3288496; Fax: +1 610 6905728; E-mail: friends@swarthmore.edu; URL: www.swarthmore.edu/library/friends
1871; Christopher Densmore
Friends Meeting Records (Record Group 2), mss & arch; Lucretia Mott, mss; John G Whittier, (Whittier Coll), bks, mss & pictures, Swarthmore College Arch
44 000 vols; 212 curr per; 300 maps; 3 000 microforms; 380 sound-rec; 4 digital data carriers; 10 121 mss & archives 32767

Swarthmore College, McCabe Library, 500 College Ave, ***Swarthmore***, PA 19081-1081
T: +1 610 3288477; Fax: +1 610 3287329; E-mail: librarian@swarthmore.edu; URL: www.swarthmore.edu/library
1864; Pam Harris
W H Auden Coll, British Writings on Travel in America, Hist of Technology (Bathe Coll), Private Press (Charles B Shaw Coll), Recorded Lit (Potter Coll), Romantic Poetry (Wells Wordsworth & Thomson Coll), Peace Coll
740 000 vols; 1 900 curr per; 297 615 govt docs; 70 000 microforms; 10 000 av-mat; 13 728 sound-rec; 1 300 digital data carriers
libr loan; ALA, ACRL 32768

– Cornell Science & Engineering Library, 500 College Ave, Swarthmore, PA 19081-1399
T: +1 610 3287685; Fax: +1 610 6905776
Meg E. Spencer
62 000 vols; 500 curr per 32769

– Daniel Underhill Music & Dance Library, 500 College Ave, Swarthmore, PA 19081-1399
T: +1 610 3288232
Donna Fournier
19 000 vols; 32 curr per 32770

Sweet Briar College, Mary Helen Cochran Library, 134 Chapel Rd, ***Sweet Briar***, VA 24595-1200; P.O. Box 1200, Sweet Briar, VA 24595-1200
T: +1 434 3816138; Fax: +1 434 3816173; URL: www.cochran.sbc.edu
1901; John G. Jaffe
Wystan Hugh Auden Coll, Incunabula, Kellogg Childrens Coll, George Meredith Coll, Evelyn D Mullen, T E Lawrence Coll, Fletcher Williams Founders Coll, Virginia Woolf Coll
223 000 vols; 982 curr per; 350 000 microforms; 7 757 sound-rec; 23 digital data carriers
libr loan 32771

– Junius P. Fishburn Music Library, Babcock Fine Arts Bldg, Sweet Briar, VA 24595
T: +1 434 3816250; Fax: +1 434 3816173
1961
11 000 vols; 3 700 music scores; 3 086 sound-rec
libr loan 32772

– Martin C. Shallenberger Art Library, Anne Gary Pannell Ctr, Sweet Briar, VA 24595
T: +1 434 3816294
1961
18 000 vols
libr loan 32773

Lourdes College, Duns Scotus Library, 6832 Convent Blvd, ***Sylvania***, OH 43560
T: +1 419 8243761; Fax: +1 419 8243511; URL: www.lourdes.edu
1950; Sister Sandra Rutkowski
Bible Coll, Franciscan Order
70 000 vols; 448 curr per; 250 maps; 14 000 microforms; 1 200 av-mat; 224 sound-rec; 11 digital data carriers; 350 art reproductions
libr loan; OPAL, OhioLINK, ALA, ACRL, CLA 32774

Le Moyne College, Noreen Reale Falcone Library, 1419 Salt Springs Rd, ***Syracuse***, NY 13214-1301
T: +1 315 4454153; Fax: +1 315 4454642; URL: www.lemoyne.edu/library
1946; James J. Simonis
Philosophy, Religion; Danny Biasone Syracuse Nationals Coll, Arch, Jesuitica (Jesuit Hist), Irish Lit (Father William Noon, SJ Coll), McGrath Music Coll
217 000 vols; 34 415 curr per; 10 000 e-books; 14 microforms; 3 999 sound-rec
libr loan 32775

State University of New York – College of Environmental Science & Forestry, F. Franklin Moon Library, One Forestry Dr, ***Syracuse***, NY 13210
T: +1 315 4706711; Fax: +1 315 4706512; URL: www.esf.edu/moonlib
1919; Elizabeth A. Elkins
Landscape Architect Fletcher Steele Mat
135 000 vols; 2 022 curr per; 5 376 diss/theses; 154 000 microforms
libr loan 32776

State University of New York Educational Opportunity Center, Paul Robeson Library, 100 New St, ***Syracuse***, NY 13202
T: +1 315 4720130 ext 30; Fax: +1 315 4721241; E-mail: wallam@morrisville.edu; URL: www.syracuseeoc.com
1969
African-American Bks (Frazier Libr Coll), National Arch Coll of Afro-American Artists
12 000 vols; 100 curr per; 200 av-mat; 575 sound-rec; 40 VF drawers 32777

Suny Upstate Medical University, Health Sciences Library, 766 Irving Ave, ***Syracuse***, NY 13210-1602
T: +1 315 4644581; Fax: +1 315 4644584; E-mail: library@upstate.edu; URL: www.upstate.edu/library/
1912; Cristina Pope
Medicine (Americana Coll), Geneva Coll, Stephen Smith Coll, Rare Books Coll
216 000 vols; 1 800 curr per; 72 e-books
libr loan 32778

Syracuse University, Bird Library, 222 Waverly Ave, ***Syracuse***, NY 13244-2010
T: +1 315 4432573; Fax: +1 315 4432060; URL: library.syr.edu
1871; Suzanne Elizabeth Thorin
Stephen Crane First Eds & Mss, Spire Coll on Loyalists in the American Revolution, Novotny Libr of Economic Hist, William Hobart-Royce Balzac Coll, Sol Feinstone Libr, Leopold von Ranke Libr, Shaker Coll, Rare Books, Albert Schweitzer Coll, Business Hist, Marcel Breuer Coll
3 180 000 vols; 22 865 curr per; 311 200 govt docs; 190 600 maps; 4 650 000 microforms
libr loan 32779

– H. Douglas Barclay Law Library, College of Law, E I White Hall, Syracuse, NY 13244-1030
T: +1 315 4439570; Fax: +1 315 4439567; E-mail: library@law.syr.edu
1899; Thomas R. French
194 000 vols; 3 261 curr per; 202 000 microforms; 25 digital data carriers
libr loan 32780

– Geology Library, 300 Heroy Geology Laboratory, Syracuse, NY 13244-1070
T: +1 315 4433337; Fax: +1 315 4433363; E-mail: eawallac@syr.edu; URL: library.syr.edu/information/geology
Elizabeth Wallace
45 000 vols; 75 curr per; 300 diss/theses; 2 500 govt docs; 500 maps; 600 microforms; 70 av-mat; 50 digital data carriers
libr loan 32781

– Mathematics Library, 308 Carnegie Bldg, Syracuse, NY 13244-1150
T: +1 315 4432092; Fax: +1 315 4435539; URL: library.syr.edu/information/math/
Mary DeCarlo
50 000 vols 32782

– Physics Library, 208 Physics Bldg, Syracuse, NY 13244-1130
T: +1 315 4432692; Fax: +1 315 4435549; E-mail: jlpease@syr.edu; URL: library.syr.edu/information/physics
Janet Pease
45 000 vols; 89 curr per 32783

– Science & Technology Library, Carnegie Libr Bldg, Syracuse, NY 13244-2010
T: +1 315 4432160; Fax: +1 315 4435549; E-mail: mmdecarl@library.syr.edu; URL: library.syr.edu/information/scitechlib/
1971; Mary DeCarlo
Rpts from Atomic Energy Commission, Energy Res & Development Administration, DOE, NASA, Society of Automotive Engineers
400 000 vols; 1 900 curr per; 2 000 000 microforms 32784

Pacific Lutheran University, Robert A. L. Mortvedt Library, 12180 Park Ave S, ***Tacoma***, WA 98447-0001
T: +1 253 5357507; Fax: +1 253 5365110; E-mail: libr@plu.edu; URL: www.plu.edu/~libr
1894; Francesca Lane Rasmus
Scandinavian Immigrant Experience Coll
345 000 vols; 4 474 curr per
libr loan 32785

University of Puget Sound, Collins Memorial Library, 1500 N Warner St, Campus Mail Box 1021, **Tacoma**, WA 98416-1021
T: +1 253 8793257; Fax: +1 253 8793670; E-mail: libref@ups.edu; URL: www.library.ups.edu
1888; Jane A. Carlin
Liberal arts; Music Recordings & Scores Coll
475 000 vols; 1 426 curr per; 24 000 e-books; 566 000 microforms; 8 822 sound-rec
libr loan   32786

Northeastern State University, John Vaughan Library-Learning Resource Center, 711 N Grand Ave, **Tahlequah**, OK 74464-2333
T: +1 918 4565511; Fax: +1 918 4582197; URL: library.nsuok.edu
1909; Dr. Shiela Collins
Native American and Local Hist Coll; Archives and Genealogy Dept
415 000 vols; 5 685 curr per; 75 000 e-books; 85 govt docs; 3 868 maps; 725 000 microforms; 3 500 av-mat; 3 926 sound-rec; 1,349.85 cubic feet of mss, 2 780 computer files, 278 other items
libr loan   32787

Columbia Union College, Weis Library, 7600 Flower Ave, **Takoma Park**, MD 20912-7796
T: +1 301 8914222; Fax: +1 301 8914204; URL: www.cuc.edu/library/
1904; Lee Marie Wisel
Seventh-day Adventists Hist & Publs
130 000 vols; 369 curr per; 12 digital data carriers; 34 VF drawers of rpts, mss, clippings
libr loan   32788

Talladega College, Savery Library, 627 W Battle St, **Talladega**, AL 35160
T: +1 256 7616279; Fax: +1 256 7619206; URL: www.talladega.edu
1939; Juliette S. Smith
Amistad Mutiny – Murals, Black Studies
120 700 vols; 362 curr per   32789

Florida Agricultural & Mechanical University, Samuel H. Coleman Memorial Library, 1500 S Martin Luther King Blvd, **Tallahassee**, FL 32307-4700
T: +1 850 5993370; Fax: +1 850 5612293; URL: www.famu.edu/library
1887; Dr. Lauren B. Sapp
Afro-American Culture & Hist
807 000 vols; 19 735 curr per; 24 490 e-journals; 55 000 e-books; 1 200 av-mat; 537 Electronic Media & Resources, 10 090 Journals
libr loan   32790

– Architecture Library, 1938 S Martin Luther King Jr Blvd, Tallahassee, FL 32307
T: +1 850 5998776; Fax: +1 850 5993535; URL: www.famu.edu/oldsite/acad/coleman/architecture.html
Jeneice Williams-Smith
Landscape architecture, construction technology
25 000 vols; 100 curr per   32791

– Frederic S. Humphries Science Research Center, 307 Pershint St, Tallahassee, FL 32309
T: +1 850 5993393; Fax: +1 850 5993422; URL: www.famu.edu/oldsite/acad/coleman/science.html
Pauline Hicks
24 000 vols; 915 curr per   32792

Florida State University, Tallahassee
– Center for Demography and Population Health, Library, 601Bellamy Bldg, 113 Collegiate Loop, Tallahassee, FL 32306-2240
T: +1 850 6441762; Fax: +1 850 6448818; E-mail: popctr@fsu.edu; URL: www.fsu.edu/~popctr
1973; Judy Kirk
World Fertility Survey Comparitive Studies, Scientific Rpts & Occasional Papers; Aids; US Population & Housing Census; Charles M Grigg Memorial Coll; Soviet Population Materials (Galina Selegan Coll)
13 500 vols; 50 curr per; 80 digital data carriers; 3 200 vertical files   32793

– Harold Goldstein Library, College of Information, 142 Collegiate Way, Tallahassee, FL 32306-2100
T: +1 850 6441803; Fax: +1 850 6440460; E-mail: ci-goldstein@admin.fsu.edu; URL: goldstein.ci.fsu.edu
Pam Doffek
62 000 vols; 127 curr per; 2 000 e-books   32794

– Law Library, 425 W Jefferson St, Tallahassee, FL 32306; P.O. Box 1600, Tallahassee, FL 32306-1600
T: +1 850 6444578; Fax: +1 850 6445216; URL: www.law.fsu.edu/library/
1966; Faye Jones
Florida Supreme Court Briefs (microfiche) & Oral Arguments (videos); Selective US Gov Doc Depot
511 000 vols; 3 580 curr per; 901 000 microforms; 3 000 av-mat; 1 338 sound-rec; 380 digital data carriers
libr loan   32795

– Paul A. M. Dirac Science Library, Tallahassee, FL 32306-4140
T: +1 850 6445534; Fax: +1 850 6440025; URL: www.fsu.edu/~library
1988; Sharon Schwerzel
550 000 vols; 3 500 curr per; 11 000 e-journals   32796

– Robert Manning Strozier Library, Strozier Library Bldg, 105 Dogwood Way, Tallahassee, FL 32306-2047
T: +1 850 6442706; Fax: +1 850 6444702; URL: www.fsu.edu
Dr. F. William Summers
2 890 000 vols; 42 076 curr per   32797

University of South Florida, Tampa
– Louis de la Parte Florida Mental Health Institute, Research Library, 13301 Bruce B. Downs Blvd, Tampa, FL 33612-3899
T: +1 813 9744471; Fax: +1 813 9747242; E-mail: hanson@fmhi.usf.edu; URL: lib.fmhi.usf.edu/
1974; Ardis Hanson
Institute Arch; Aids/HIV Coll; Children with Special Needs; Multi-cultural Coll; Attention-Deficit Disorder Coll & the Autism Coll; Managed Behavioral Health Care
24 390 vols; 210 curr per; 450 govt docs; 259 microforms; 280 av-mat; 7 digital data carriers; 3 400 unbound per, 74 kits   32798

– Shimberg Health Sciences Library, 12901 Bruce B. Downs Blvd, MDC Box 31, Tampa, FL 33612
T: +1 813 9742243; Fax: +1 813 9744840; E-mail: bshattuc@hsc.usf.edu; URL: library.hsc.usf.edu
1971; Beverly A. Shattuck
29 870 vols; 1 373 curr per; 5 digital data carriers   32799

– Tampa Campus Library, 4202 E Fowler Ave, LIB 122, Tampa, FL 33620-5400
T: +1 813 9741611
1960; Phyllis Ruscella
19th C American Playscript Coll, 19th C American Ephemera Coll, Songbk Coll, Almanacs, Toybks, Arch & Mss Coll, Cigar Art Coll, Dime Novel Coll, Dobkin Coll of American 19th C Lit, Early American Textbks, Florida Sheet Music Coll, Rare Bks, Rare Maps, Floridiana
2 324 000 vols; 32 400 curr per
libr loan   32800

University of Tampa, McDonald-Kelce Library, 401 W Kennedy Blvd, **Tampa**, FL 33606-1490
T: +1 813 2536231; Fax: +1 813 2587426; URL: utopia.ut.edu
1931; Marlyn Pethe
Blanche Yurka Drama Coll, Stanley Kimmel Coll (John Wilkes Booth), Florida Military, Local Hist, Univ Arch
275 000 vols; 10 500 curr per; 295 maps; 2 233 sound-rec; 63 digital data carriers; 30 Electronic Media & Resources
libr loan   32801

Fairleigh Dickinson University, Weiner Library, 1000 River Rd, **Teaneck**, NJ 07666-1914
T: +1 201 6922278; Fax: +1 201 6929815
1954; Dr. James W. Marcum
Lincoln, Mf Coll Presidential Papers
203 000 vols; 756 curr per; 40 000 e-books   32802

Arizona State University, University Libraries, 300 E Orange Mall Dr, **Tempe**, AZ 85287-1006; P.O. Box 871006, Tempe, AZ 85287-1006
T: +1 480 9656164; Fax: +1 480 9659169; URL: lib.asu.edu
1891; Sherrie Schmidt
Pre-Rapaelite Brotherhood Coll, Patten Coll of Early Herbals & Gardening Books; Child Drama Coll, Laos Research Coll, William Burroughs Arch, Mexican Numismatics, Arizona Hist
4 342 000 vols; 65 586 curr per   32803

– Architecture & Environmental Design Library, P.O. Box 871705, Tempe, AZ 85287-1705
T: +1 480 965-6400; Fax: +1 480 7276965; E-mail: Deborah.Koshinsky@asu.edu; URL: lib.asu.edu/architecture
1960; Debra Koshinsky
Paolo Soleri Special Res Coll, Frank Lloyd Wright Special Res Coll
39 000 vols; 150 sound-rec   32804

– Daniel E. Noble Science & Engineering Library, 601 E Tyler Mall, Tempe, AZ 85281; P.O. Box 871006, Tempe, AZ 85287-1006
T: +1 480 9657609; Fax: +1 480 9650883; URL: www.lib.asu.edu/noble
1983
450 000 vols; 2 700 curr per; 200 000 maps; 410 000 microforms   32805

– Music Library, Music Bldg, Tempe, AZ 85287; P.O. Box 870505, Tempe, AZ 85287-0505
T: +1 480 9653513; Fax: +1 480 9659598
1965; Dr. Christopher Mehrens
Pablo Casals Int Cello Libr; Wayne King Coll of Popular Music
69 000 vols; 144 curr per; 9 000 music scores; 3 000 microforms; 37 000 sound-rec   32806

– John J. Ross – William C. Blakley Law Library, PO Box 877806, Tempe, AZ 85287-7806
T: +1 480 9656141; Fax: +1 480 9654283; URL: www.lawlib.asu.edu
1966; Victoria K. Trotta
Native Americans
407 000 vols; 3 901 curr per; 541 000 microforms; 55 digital data carriers
libr loan   32807

Indiana State University, Cunningham Memorial Library, 650 Sycamore, **Terre Haute**, IN 47809
T: +1 812 2373700; Fax: +1 812 2373376; URL: library.indstate.edu
1870; Alberta Davis Comer
American Education, American Labour Movement, Indian Culture, Indiana Federal Writers' Materials, Indiana Hist, Music, Publ Hist & Culture, Rare Books Coll; Career Center Libr
1 336 000 vols; 3 000 e-books; 603 Audio Bks, 453 Bks on Deafness & Sign Lang, 82 Large Print Bks
libr loan   32808

Rose-Hulman Institute of Technology, John A. Logan Library, 5500 Wabash Ave, **Terre Haute**, IN 47803
T: +1 812 8778200; Fax: +1 812 8778579; URL: www.rose-hulman.edu/Library
1874; Rachel Crowley
78 000 vols; 480 curr per; 675 microforms; 5 digital data carriers; 2 100 docs, 1 200 archival vols
libr loan   32809

Texarkana College, Palmer Memorial Library – John F. Moss Library, 1024 Tucker St, P.O. Box 9150, **Texarkana**, TX 75599
T: +1 903 2233088; Fax: +1 903 8317429; E-mail: teri.stover@texarkanacollege.edu; URL: www.tc.cc.tx.us
1927
Nursing; Interstate Commerce (Transportation Coll), Rare Bks
40 550 vols; 561 curr per
libr loan   32810

Texas A&M University – Texarkana, John F. Moss Library, 1024 Tucker St, **Texarkana**, TX 75501; P.O. Box 6187, Texarkana, TX 75505
T: +1 903 2233088; Fax: +1 903 8317429; URL: library.tamut.edu
1971; Jimmie Sue Simmons
136 000 vols; 404 curr per; 51 000 e-books   32811

College of the Mainland, Library, 1200 Amburn Rd, **Texas City**, TX 77591-2499
T: +1 409 9338448; Fax: +1 409 9388918; E-mail: dl-library-reference@com.edu; URL: library.com.edu
1967
Texana; Black Studies, Mexican Americans & American Indians (Ethnic Coll)
50 410 vols; 247 curr per   32812

Nicholls State University, Allen J. Ellender Memorial Library, 906 E First St, **Thibodaux**, LA 70310; P.O. Box 2028, Thibodaux, LA 70310
T: +1 985 4484646; Fax: +1 985 4484925; URL: www.nicholls.edu/library
1948; Carol Mathias
Local Hist Coll, Cajun & Zydeco Music Heritage, Ctr for Traditional Louisiana Boat Building, Sugar Cane Plantations, US Senators' Papers
533 000 vols; 1 650 curr per; 9 200 av-mat; 1 322 sound-rec; 385 662 microforms of per, 9 320 art reproductions
libr loan   32813

Alabama Southern Community College, Library, 30755 Hwy 43, P.O. Box 2000, **Thomasville**, AL 36784
T: +1 334 6369642; Fax: +1 334 6361478; E-mail: drankins@ascc.edu; URL: www.ascc.edu
Angela Roberts
43 000 vols; 250 curr per   32814

Thomas University, University Library, 1501 Millpond Rd, **Thomasville**, GA 31792
T: +1 229 2261621; Fax: +1 229 2261679
1950; Gary Cooper
85 000 vols; 412 curr per   32815

California Lutheran University, Pearson Library, 60 W Olsen Rd, MC 5100, **Thousand Oaks**, CA 91360-2787
T: +1 805 4933250; Fax: +1 805 4933842; E-mail: libcirc@clunet.edu; URL: www.clunet.edu
1961; Pat Hilker
Counseling, Marriage, Pacific Islands; Scandinavian Lutheranism & Hist
135 000 vols; 425 curr per
libr loan   32816

Heidelberg College, Beeghly Library, Ten Greenfield St, **Tiffin**, OH 44883-2420
T: +1 419 4482104; Fax: +1 419 4482578; URL: www.heidelberg.edu/beeghlylibrary
1850; Edward Krakora
Religious studies; Ballet & Dance (Pohlable Coll)
150 000 vols; 392 curr per
libr loan   32817

Tiffin University, Pfeiffer Library, 139 Miami St, **Tiffin**, OH 44883-2162
T: +1 419 4483435; Fax: +1 419 4485013; E-mail: ffleet@tiffin.edu; URL: www.tiffin.edu/library
1956; Frances A. Fleet
Business; NCJRS Doc Microfiche Coll
29 700 vols; 255 curr per; 563 maps; 20 digital data carriers   32818

University of Georgia College of Agriculture & Environmental Sciences, Tifton Campus Library, 4601 Research Way, **Tifton**, GA 31793; P.O. Box 748, Tifton, GA 31793-0748
T: +1 912 3863447; Fax: +1 912 3912501; E-mail: librtif@uga.edu
1924; Duncan McClusky
8 000 vols; 150 curr per
GLA, IAMSLIC, SLA   32819

North Greenville College, Hester Memorial Library, PO Box 1892, **Tigerville**, SC 29688-1892
T: +1 864 9777093; Fax: +1 864 9772126; URL: www.ngu.edu
1892; Jonathan Bradsher
Edith Duff Miller Bible Museum Coll
52 000 vols; 426 curr per   32820

Toccoa Falls College, Seby Jones Library, PO Box 800749, ***Toccoa Falls***, GA 30598
T: +1 706 8866831; Fax: +1 706 2826010; URL: www.tfc.edu/library
1911; Patricia Fisher
Religion, Education, Music; Religion (R. A. Forrest Coll-Founder & First President of College)
146 000 vols; 299 curr per; 23 000 microforms; 5 digital data carriers; 563 vertical files
libr loan 32821

Medical College of Ohio, Raymon H. Mulford Library, Health Science Campus, Mail Stop 1061, 3000 Arlington Ave, ***Toledo***, OH 43614-5805
T: +1 419 3834223; Fax: +1 419 3836146; URL: www.utoledo.edu/library/mulford
1967; David W. Boilard
154 000 vols; 2 376 curr per; 3 000 microforms; 200 av-mat; 8 digital data carriers 32822

Mercy College of Northwest Ohio, Health Sciences Library, 2221 Madison Ave, ***Toledo***, OH 43624-1120
T: +1 419 2511700; Fax: +1 419 2511730; E-mail: library@mercycollege.edu; URL: www.mercycollege.edu
1973
23 000 vols; 175 curr per; 25 govt docs; 138 digital data carriers
libr loan; ALA, HSLNO, OLA, ACRL 32823

University of Toledo, William S. Carlson Library, 2801 W Bancroft St, Mail Stop 509, ***Toledo***, OH 43606-3390
T: +1 419 5302324; Fax: +1 419 5302726; URL: www.library.utoledo.edu
1917; Marcia Suter
Afro-American Lit, American Women's Social Hist, Glass Manufacturing, Henry David Thoreau Coll, Southern Authors
1 600 000 vols; 8 154 curr per
libr loan 32824

– Law Library, 2801 W Bancroft St, MS 508, Toledo, OH 43606-3399
T: +1 419 5302733; Fax: +1 419 5302821; URL: www.utlaw.edu/students/lawlibrary/index.htm
Rick Goheen
Int & Comparative Law Coll
216 000 vols; 3 371 curr per; 640 000 microforms; 300 av-mat; 20 digital data carriers 32825

Washburn University, Mabee Library, 1700 SW College Ave, ***Topeka***, KS 66621
T: +1 785 6701483; Fax: +1 785 6703223; URL: www.washburn.edu/mabee
1865; Alan Bearman
College & Univ Hist (Washburn Arch); Bradbury Thompson Mat; Curriculum Resources Ctr, William I. Koch Art Hist Coll
346 000 vols; 1 672 curr per; 21 260 e-journals; 10 000 e-books; 2 100 av-mat; 1 100 sound-rec
libr loan 32826

– School of Law Library, 1700 SW College Ave., Topeka, KS 66621
T: +1 785 6701088; Fax: +1 785 6703194; E-mail: lawlibrary@washburnlaw.edu; URL: www.washburnlaw.edu/library
1903; John E. Christensen
Brown v. Board Oral Hist Collection, Kansas Supreme Court Briefs, U.S. Supreme Court Autographs Collection, Federal and State of Kansas Federal Selective Depository Collections
178 000 vols; 4 064 curr per; 2 000 e-journals; 25 000 e-books; 211 000 microforms
libr loan; AALL 32827

Heritage College, Don North Library, 3240 Fort Rd, ***Toppenish***, WA 98948
T: +1 888 2726190; URL: www.heritage.edu/CurrentStudents/Library/tabid/307/Default.aspx
1982
Bilingual education, Social work
32 000 vols; 200 curr per 32828

Eastern Wyoming College, Library, 3200 West C, ***Torrington***, WY 82240
T: +1 307 5328210; Fax: +1 307 5328225; URL: www.ewc.wy.edu
1948; Marilyn Miller
34 000 vols; 121 curr per; 3 000 microforms; 35 digital data carriers
libr loan 32829

Tougaloo College, L. Zenobia Coleman Library, Tougaloo College, 500 W County Line Rd, ***Tougaloo***, MS 39174-9799
T: +1 601 9777706; Fax: +1 601 9777714; URL: www.tougaloo.edu/library
1869; Dr. Dorothy Burnett
African Mat (Ross Coll), Civil Rights & Liberties (Charles Horowitz Papers), Civil Rights Movement (Tracy Sugarman Print Coll of 1964), Mississippi Civil Rights Lawsuits of the 1960's, Music (B. B. King Coll), Radical Papers (Kudzu File)
139 000 vols; 389 curr per 32830

Towson University, Albert S. Cook Library, 8000 York Rd, ***Towson***, MD 21252-0001
T: +1 410 7042456; Fax: +1 410 7043829; URL: cooklibrary.towson.edu
1866; Deborah A. Nolan
Educational Resources Information Center, Libr of American Civilization, Libr of English Lit
580 000 vols; 1 240 curr per; 3 240 e-journals; 173 000 e-books; 8 500 av-mat; 2 520 sound-rec; 730 digital data carriers 32831

Northwestern Michigan College, Mark & Helen Osterlin Library, 1701 E Front St, ***Traverse City***, MI 49686-3061
T: +1 231 9951060; Fax: +1 231 9951056; E-mail: library@nmc.edu; URL: www.nmc.edu/library
1951; Tina Ulrich
American Culture Series, microfilm; American Per Series, microfilm
92 000 vols; 728 curr per; 2 100 maps; 30 250 microforms; 1 200 av-mat; 2 980 sound-rec; 70 digital data carriers
libr loan 32832

Trinity College, Raymond H. Center Library, 2430 Welbilt Blvd, ***Trinity***, FL 34655
T: +1 727 3766911; Fax: +1 727 3760781; URL: www.trinitycollege.edu
Janet Kuehne
45 000 vols; 100 curr per 32833

Rensselaer Research Libraries, Folsom Library, Rensselaer Polytechnic Inst, 110 Eighth St, ***Troy***, NY 12180-3590
T: +1 518 2768300; Fax: +1 518 2762044; URL: library.rpi.edu
1824; Bob Mayo
Hist of Science & Technology; Technical rpts; Geological Survey Quadrangle Maps
387 000 vols; 42 000 e-journals; 52 000 e-books; 800 000 govt docs; 62 438 maps; 4 288 sound-rec; 84 000 slides
libr loan 32834

The Sage Colleges, James Wheelock Clark Library, 45 Ferry St, ***Troy***, NY 12180
T: +1 518 2442249; Fax: +1 518 2442400; E-mail: libref@sage.edu; URL: www.sage.edu
1916; Kingsley W. Greene
Allied health, Women's studies; 20th C Poetry (Carol Ann Donahue Memorial Coll)
154 000 vols; 284 curr per; 10 000 e-books; 29 000 microforms; 2 879 sound-rec; 36 digital data carriers; 4510 Audio Bks
libr loan 32835

Troy State University, University Library, Wallace Hall, ***Troy***, AL 36082
T: +1 334 6703470; Fax: +1 334 6703955; URL: library.troy.edu
1887; Dr. Henry R. Stewart
Education, Indians
556 000 vols; 3 332 curr per; 49 000 e-books; 15 av-mat; 75 digital data carriers; 160 Bks on Deafness & Sign Lang, 15 Talking Bks
libr loan 32836

Walsh College of Accountancy & Business Administration, Troy Campus Library, 3838 Livernois Rd, ***Troy***, MI 48083-5066; P.O. Box 7006, Troy, MI 48007-7006
T: +1 248 8231337; Fax: +1 248 6899066; E-mail: librarian@walshcollege.edu; URL: www.walshcollege.edu
1928; Dr. Jonathan Campbell
annual rpts of 1 300 companies
30 000 vols; 215 curr per; 2 000 microforms; 7 digital data carriers; 500 pamphlets 32837

University of Arizona, Main Library, 1510 E University Blvd, ***Tucson***, AZ 85721; P.O. Box 210055, Tucson, AZ 85721-0055
T: +1 520 6212101; Fax: +1 520 6219733; URL: www.library.arizona.edu
1891; Carla Stoffle
Fiction, Hist of science, Science; Photography as an Art Form, Fine Arts (Hanley Coll), Drama (W Stevens Coll), Southwestern Americana, Mexican Colonial Hist Coll, Private Press (Frank Holme Coll)
5 050 000 vols; 23 280 curr per
libr loan 32838

– Arizona Health Sciences Library, 1501 N Campbell Ave, Tucson, AZ 85724; P.O. Box 245079, Tucson, AZ 85724-5079
T: +1 520 6266125; Fax: +1 520 6262922; E-mail: refdesk@ahsl.arizona.edu; URL: ahsl.arizona.edu
1965; Gary A. Freiburger
Arizona Health Sciences Ctr & Hist of Health Care & The Healing Arts in the Southwest, Arch
219 000 vols; 2 194 curr per 32839

– Center for Creative Photography Library, 1030 N Olive Rd, Tucson, AZ 85721-0001; P.O. Box 210103, Tucson, AZ 85721-0103
T: +1 520 6217968; Fax: +1 520 6219444; E-mail: oncenter@ccp.library.arizona.edu; URL: www.creativephotography.org
1975; Laura Earles
Photogr Arch
22 000 vols; 100 curr per; 600 microforms; 600 av-mat; 40 feet of biogr files 32840

– College of Agriculture & Life Sciences Arid Lands Information Center, 1955 E Sixth St, Tucson, AZ 85719-5224
T: +1 520 6218571; Fax: +1 520 6213816; URL: www.arid.arizona.edu/Divisions/division.asp?div=ALIC
1968; Carla Long Casler
Coll of Jojoba, Guayule; Republik of Niger; Developing Country Profiles; Desert Slide Coll
35 000 vols; 15 curr per; 10 digital data carriers; 100 environmental impact statements 32841

– East Asian Collection, 1510 E University Blvd, Tucson, AZ 85720; P.O. Box 210055, Tucson, AZ 85721-0055
T: +1 520 6216384; Fax: +1 520 6213655; E-mail: askref@u.library.arizona.edu; URL: www.library.arizona.edu
Ping Situ
Japanese studies, Chinese
186 000 vols; 239 curr per 32842

– James E. Rogers College of Law Library, PO Box 210176, Tucson, AZ 85721-0176
T: +1 520 6211413; Fax: +1 520 6213138; URL: www.law.arizona.edu/library/
Michael G. Chiorazzi
Natural resources
400 000 vols; 4 500 curr per 32843

– Music Collection, 1510 E University Blvd, Rm 233, P.O. Box 210055, Tucson, AZ 85721-0055
T: +1 520 6217010; Fax: +1 520 6261630; E-mail: jmarley@bird.library.arizona.edu; URL: www.library.Arizona.EDU/library/type1/branches/data/Music-Library.html
1959; Bob Diaz
Nat Flute Assn Music Libr, Int Trombone Assn Resource Libr, Arizona and Southwest, Hist Popular Sheet Music
60 000 vols; 280 curr per; 37 400 music scores; 21 875 sound-rec; 55 000 scores 32844

– Poetry Center in the College of Humanities, 1508 E Helen St, Tucson, AZ 85721-0129
T: +1 520 6263765; E-mail: poetry@email.arizona.edu
1960; Gail Browne
45 000 vols; 200 curr per; 20 mss; 100 av-mat; 1 050 sound-rec
PALM (Promoting Archives, Libraries and Museums) 32845

– Science-Engineering Library, 744 N Highland, Bldg 54, Tucson, AZ 85721; P.O. Box 210054, Tucson, AZ 85721-0054
T: +1 520 6216394; Fax: +1 520 6213655; URL: www.library.arizona.edu
1963; Jeanne Pfander
Arid Lands
520 000 vols; 10 000 curr per; 27 000 govt docs; 1 400 000 microforms 32846

University of Tennessee Space Institute Library, Library, MS-25, 411 B H Goethert Pkwy, ***Tullahoma***, TN 37388-9700
T: +1 931 3937315; Fax: +1 931 3937518; E-mail: library@utsi.edu; URL: www.utsi.edu/library
1965; Emily S. Moore
Herman Diederich Memorial Coll (Propulsion)
25 000 vols; 158 curr per; 1 194 diss/theses; 209 000 microforms; 64 digital data carriers; 52 327 techn rpts
libr loan 32847

Oklahoma State University – College of Osteopathic Medicine, Medical Library, 1111 W 17th St, ***Tulsa***, OK 74107-1898; mail address: 1117 W 17th St, Tulsa, OK 74107
T: +1 918 5611119;
Fax: +1 918 5618412; URL: www.healthsciences.okstate.edu/medlibrary/index.html
1974; Beth Anne Freeman
Biomed; Anatomical Models & Realia Coll, Case Histories for Massachusetts General Hospital, College Arch, Nat Libr of Medicine Lit Searches Coll, Osteopathy Coll
58 000 vols; 356 curr per; 170 e-books; 41 000 microforms; 3 296 sound-rec; 1 digital data carriers; 2 156 videos, 120 799 slides, 89 charts
libr loan 32848

Oklahoma State University – Tulsa Library, 700 N Greenwood Ave, ***Tulsa***, OK 74106-0700
T: +1 918 5948132; Fax: +1 918 5948145; URL: www.osu-tulsa.okstate.edu/library/
1986; Beth Anne Freeman
Eric Coll (from 1980), ETS Test Coll
110 000 vols; 9 495 curr per; 24 016 e-journals; 50 000 e-books; 496 govt docs; 512 000 microforms; 2 000 av-mat; 1 000 sound-rec; 615 digital data carriers; 360 Audio Bks, 150 Electronic Media & Resources, 260 Bks on Deafness & Sign Lang, 31 Large Print Bks
libr loan; OCLC 32849

Oral Roberts University, John D. Messick Learning Resources Center, 7777 South Lewis Ave, ***Tulsa***, OK 74171
T: +1 918 4956391; Fax: +1 918 4956893; E-mail: libref@oru.edu; URL: www.oru.edu/university/library
1965; Dr. William W. Jernigan
Holy Spirit Research Center, Oral Roberts Ministry Arch, Oral Hist Program
274 000 vols; 434 curr per; 23 290 e-journals; 43 000 e-books
libr loan 32850

– Holy Spirit Research Center Library, 7777 S Levis Ave, Tulsa, OK 74171
T: +1 918 4956391; Fax: +1 918 4956662; E-mail: hsrc@oru.edu; URL: www.oru.edu/university/library
1962; Mark E. Roberts
Edward Irving Coll
16 000 vols; 484 curr per; 7 mss; 45 music scores; 37 microforms; 410 av-mat; 8 300 sound-rec; 55 VF drawers of pamphlets
CL, ACRL, ALA, ATLA, MLA, OCLC 32851

Spartan College of Aeronautics & Technology Library, 8820 E Pine St, ***Tulsa***, OK 74115
T: +1 918 8366886; Fax: +1 918 8315245; URL: www.spartan.edu
Melody Watts
13 000 vols 32852

University of Tulsa, McFarlin Library, 2933 E Sixth St, ***Tulsa***, OK 74104-3123
T: +1 918 6312880; Fax: +1 918 6313791; URL: www.lib.utulsa.edu
1984; Francine Fisk
American Indian law & hist (Robertson-Shleppey-Milam Coll), Modernist Lit incl libs of Edmond Wilson & Cyril Connelly, strong holdings of James Joyce, Robert Graves, D. H. Lawrence
794 000 vols; 2 258 curr per
libr loan 32853

– University of Tulsa College of Law, Mabee Legal Information Center, Library, 3120 E Fourth Pl, Tulsa, OK 74104-3189
T: +1 918 6312404; Fax: +1 918 6313556; URL: www.utulsa.edu/law/library
1923; Richard E. Ducey
Alternative Dispute Resolution, American Indian Law, Energy & Environmental Law
375 000 vols; 3 253 curr per; 119 000 microforms; 70 av-mat; 253 digital data carriers
libr loan; AALL, SWALL, MALSC 32854

California State University, Stanislaus University Library, One University Circle, ***Turlock***, CA 95382
T: +1 209 6673234; Fax: +1 209 6673164; E-mail: library@library.csustan.edu; URL: library.csustan.edu
1960; Carl E. Bengston
Assyriana (Eshoo Paul Sayad Coll), North San Joaquin Valley Local Hist, Stanislaus County Hist Docs & Photo Coll, Portuguese Cultural Photo Coll
373 000 vols; 841 curr per; 20 917 e-journals; 6 000 e-books; 1 781 diss/theses; 116 557 govt docs; 1 642 music scores; 10 095 maps; 1 138 000 microforms; 2 010 sound-rec; 187 digital data carriers
libr loan 32855

Stillman College, William H. Sheppard Library, 3601 Stillman Blvd, ***Tuscaloosa***, AL 35403; P.O. Box 1430, Tuscaloosa, AL 35403-1430
T: +1 205 3668851; Fax: +1 205 2478042; URL: www.stillman.edu
1876; Robert Heath
Relig studies; Afro-American Coll, Black hist & lit (19th c & early 20th c)
118 000 vols; 6 000 microforms; 748 sound-rec; 12 digital data carriers; 816 slides
libr loan 32856

University of Alabama, University Libraries, University of Alabama Campus, Capstone Dr, ***Tuscaloosa***, AL 35487; Box 870266, Tuscaloosa, AL 35487-0266
T: +1 205 3486047; Fax: +1 205 3489564; URL: www.lib.ua.edu
1831; Dr. Louis A. Pitschmann
Cartography Coll, First Editions, Alabamiana, Southern Authors
2 176 000 vols; 29 374 curr per; 73 500 e-journals; 40 000 e-books; 23 435 maps; 2 862 000 microforms; mss (26 569 linear ft)
libr loan; ARL, CRL, ASERL, SOLINET, NAAL, CNI 32857

– Angelo Bruno Business Library, P.O. Box 870266, Tuscaloosa, AL 35487-0266
T: +1 205 3481080; Fax: +1 205 3480803; E-mail: kchapman@bruno.cba.ua.edu; URL: brunolib.cba.ua.edu
1925; Lee E. Pike
165 912 vols; 1 125 curr per; 352 337 microforms; 3 000 bound annual report vols 32858

– McLure Education Library, P.O. Box 870266, Tuscaloosa, AL 35487-0266
T: +1 205 3486055; Fax: +1 205 3486602; E-mail: hvissche@bama.ua.edu; URL: www.lib.ua.edu/libraries/mclure/
1954; Helga B. Visscher
Curriculum Mats Center; Children's Coll; Hist of Education in America; School Libr, Curriculum Materials Coll
168 844 vols; 901 curr per; 703 037 microforms; 8 vertical file drawers
libr loan; ALA, Alabama Virtual Library 32859

– Rodgers Library for Science and Engineering, P.O. Box 870266, Tuscaloosa, AL 35487-0266
T: +1 205 3482100; Fax: +1 205 3482113; E-mail: jsandy@bama.ua.edu; URL: www.lib.ua.edu/rodgers/home.htm
1990; John H. Sandy
226 800 vols; 1 467 curr per; 115 630 microforms 32860

– School of Law Library, 101 Paul Bryant Dr, Tuscaloosa, AL 35487; Box 870383, Tuscaloosa, AL 35487-0383
T: +1 205 3485925; Fax: +1 205 3481112; URL: www.law.ua.edu
1872; James Leonard
Former US Supreme Court Justice Hugo L. Black, Former US Senator Howell Heflin
438 000 vols; 3 368 curr per 32861

– William Stanley Hoole Special Collections Library, P.O. Box 870266, Tuscaloosa, AL 35487-0266
T: +1 205 3480500; Fax: +1 205 3481699; URL: www.lib.ua.edu/hoole
1947; Clark E. Center
Alabamiana; Black Folk Music; Oral Hist; Rare books; Mss; Univ Arch
78 144 vols; 135 curr per; 12 500 diss/theses; 5 000 music scores; 25 000 maps; 10 000 sound-rec 32862

University of Alabama – Health Sciences Library, University Medical Center, 850 Fifth Ave E, ***Tuscaloosa***, AL 35401; Box 870378, Tuscaloosa, AL 35487-0378
T: +1 205 3481360; Fax: +1 205 3489563; URL: www.cchs.ua.edu/hsl/ind.cfm
1973; Nell Williams
AV Programs on Clinical Medicine
18 000 vols; 295 curr per; 1 500 av-mat; 125 sound-rec; 1 digital data carriers 32863

Tuskegee University, Ford Motor Company Library-Learning Resource Center, Hollis Burke Frissell Library Bldg, ***Tuskegee***, AL 36088
T: +1 334 7278894; Fax: +1 334 7279282; URL: www.tuskegee.edu
1881; Juanita Roberts
Blacks (Washington Coll)
310 000 vols; 1 500 curr per 32864

– Veterinary Medical Library, Patterson Hall, Tuskegee, AL 36088
T: +1 334 7278780; Fax: +1 334 7278442
Margaret Alexander
19 000 vols; 300 curr per 32865

College of Southern Idaho, Library, 315 Falls Ave, ***Twin Falls***, ID 83303-1238
T: +1 208 7326500; Fax: +1 208 7363087; URL: www.library.csi.edu
1965; Teri Fattig
69 000 vols; 383 curr per 32866

Texas College, D.R. Glass Library, 2404 N Grand, ***Tyler***, TX 75702-4500
T: +1 903 5938311; Fax: +1 903 5264426; URL: www.texascollege.edu
1894; Cynthia Charles
African-American studies
79 000 vols; 150 curr per 32867

University of Texas at Tyler, Robert R. Muntz Library, 3900 University Blvd, ***Tyler***, TX 75799
T: +1 903 5667343; Fax: +1 903 5655562; E-mail: library@uttyler.edu; URL: library.uttyler.edu
1973; Jeanne Pyle
203 000 vols; 423 curr per; 166 000 e-books; 3 000 av-mat; 1 700 sound-rec; 13 digital data carriers; 470 Music Scores
libr loan 32868

Kean University, Nancy Thompson Library, 1000 Morris Ave, ***Union***, NJ 07083
T: +1 908 7374600; Fax: +1 908 7374620; URL: www.library.kean.edu
1914; Dr. Barbara Simpson Darden
Hist (New Jerseyiana), Political Science (Dwyer Papers)
280 000 vols; 16 000 curr per; 187 000 microforms; 110 VF drawers
libr loan 32869

Pennsylvania State University, Fayette Commonwealth College, Library, One University Dr, ***Uniontown***, PA 15401; P.O. Box 519, Uniontown, PA 15401-0519
T: +1 724 4304155; Fax: +1 724 4304152; URL: www.libraries.psu.edu/fayette/
1965; John Riddle
50 000 vols; 125 curr per; 6 000 microforms; 5 007 sound-rec; 4 365 pamphlets 32870

University of Mississippi, John Davis Williams Library, One Library Loop, ***University***, MS 38677; P.O. Box 1848, University, MS 38677-1848
T: +1 662 9155867; Fax: +1 662 9155453; URL: www.olemiss.edu/depts/general_library
1848; Julia M. Rholes
Blues Arch, William Faulkner Coll, Lumber Arch, Mississippiana, Southern Culture, Space Law, Stark Young Coll
1 268 000 vols; 8 500 curr per
libr loan 32871

– Science Library, 1031 Natural Products Ctr, University, MS 38677
T: +1 662 9157381; Fax: +1 662 9157549; E-mail: scireply@olemiss.edu; URL: www.olemiss.edu/depts/general_library/files/science/index.html
1997; Elizabeth M. Choinski
70 000 vols; 450 curr per 32872

University of Mississippi – Law Library, Three Grove Loop, ***University***, MS 38677; P.O. Box 1848, University, MS 38677-1848
T: +1 662 9156812; Fax: +1 662 9157731; URL: library.law.olemiss.edu
1854; Kris Gilliland
Space law, Tax law; Senator James O Eastland Papers Coll
336 000 vols; 24 000 e-books; 143 000 microforms; 52 digital data carriers
libr loan 32873

Saginaw Valley State University, Melvin J. Zahnow Library, 7400 Bay Rd, ***University Center***, MI 48710
T: +1 989 9644242; Fax: +1 989 9642003; E-mail: library@svsu.edu; URL: www.svsu.edu/library
1963; Linda Farynk
Local Hist Coll; Cramton Jazz Coll, rec; Univ Arch
226 000 vols; 9 000 e-books; 275 000 microforms; 17 414 sound-rec; 6 digital data carriers
libr loan 32874

John Carroll University, Grasselli Library & Breen Learning Center, 20700 N Park Blvd, ***University Heights***, OH 44118
T: +1 216 3974233; Fax: +1 216 3974256; URL: library.jcu.edu
1886; Jeanne Somers
G K Chesterton (John R Bayer Chesterton Coll), bks, micro; Far East (Daniel A Hill Far Eastern, Coll)
741 000 vols; 1 309 curr per; 8 517 sound-rec; 75 digital data carriers; 680 005 microforms of per
libr loan 32875

Governors State University, University Library, One University Pkwy, ***University Park***, IL 60466-0975
T: +1 708 2357508; Fax: +1 708 5344564; URL: www.govst.edu/library
1969; Diane Dates Casey
Afro-American Lit (Schomberg Coll), Art Slides Coll, ERIC Docs Coll
264 000 vols; 1 917 curr per; 14 303 e-journals; 2 000 e-books
libr loan 32876

Pennsylvania State University Libraries, Pattee Library & Paterno Library, 510 Paterno Library, ***University Park***, PA 16802; mail address: 515 Paterno Library, University Park, PA 16802-1812
T: +1 814 8650401; Fax: +1 814 8653665; E-mail: slw1@psulias.psu.edu; URL: www.libraries.psu.edu
1857; Nancy L. Eaton
American Lit, Australiana, Bible Coll, German-American Lit, Television Hist Coll
5 031 000 vols; 68 445 curr per; 447 000 maps; 3 800 000 microforms; 21 671 linear feet of mss
libr loan 32877

– Eberly Family Special Collections, 104 Paterno Library, University Park, PA 16802-1808
T: +1 814 8651793, 8657931; Fax: +1 814 8635318; E-mail: spcolref@psulias.psu.edu; URL: www.libraries.psu.edu/speccolls
1904; William L. Joyce
Mss, Rare bks, Arch
180 000 vols 32878

– Fletcher L. Byrom Earth & Mineral Sciences Library, 105 Deike Bldg, University Park, PA 16802
T: +1 814 8659517; Fax: +1 814 8651379; E-mail: ems@psulias.psu.edu; URL: www.libraries.psu.edu/emsl/
1931; Linda Musser
120 000 vols 32879

– George & Sherry Middlemas Arts & Humanities Library, 202 Pattee Library, University Park, PA 16802-1801; mail address: W 337 Pattee Libr, University Park, PA 16802-1803
T: +1 814 8656481; Fax: +1 814 8637502; E-mail: alm8@PSULias.psu.edu; URL: www.libraries.psu.edu/artshumanities
1964; Daniel Mack
Warren Mack Memorial Coll
128 000 vols; 850 curr per; 28 100 sound-rec; 22 000 scores & parts 32880

– Physical & Mathematical Sciences Library, 201 Davey Lab, University Park, PA 16802-6301
T: +1 814 8657617; Fax: +1 814 8652565; E-mail: math@psulias.psu.edu; URL: www.libraries.psu.edu/pams
2005; Nancy J. Butkovich
70 000 vols; 300 curr per 32881

Taylor University, Zondervan Library, 236 W Reade Ave, ***Upland***, IN 46989-1001
T: +1 765 9984357; Fax: +1 765 9985569; E-mail: zonlib@taylor.edu; URL: www.taylor.edu/academics/library
1846; Daniel J. Bowell
Old & Scarce Bks (Ayres Coll)
186 000 vols; 730 curr per; 8 000 microforms
libr loan 32882

Pennsylvania State University, New Kensington Commonwealth College, Elisabeth S. Blissell Library, 3550 Seventh St Rd, Rte 780, ***Upper Burrell***, PA 15068-1798
T: +1 724 3346071; Fax: +1 724 3346113; URL: www.libraries.psu.edu/newken/
1958; Jennifer R. Gilley
36 000 vols; 100 curr per 32883

University of Illinois at Urbana-Champaign, Library, 230 Main Library, 1408 W Gregory Dr, ***Urbana***, IL 61801
T: +1 217 3332290; Fax: +1 217 2444358; URL: www.library.uiuc.edu
1868; Barbara J. Ford
American Humor & Folklore, Baskette Coll on Freedom of Expression, Confederate Imprints, Incunabula, Hollander Libr of Economic Hist, Abraham Lincoln, John Milton, 17th c Newsletters, 18th c English Lit, 17th c Political & Religious Pamphlets, 19th c Publishing, T W Baldwin Elizabethan Libr, Proust Correspondence, Ingold Shakespeare Coll, Carl Sandburg, Shana Alexander, James Scotty Reston
10 000 000 vols; 90 962 curr per; 1 000 incunabula; 655 720 music scores; 9 201 000 microforms; 150 894 sound-rec; 1 349 547 other items
libr loan; ALA, IFLA, SLA, CRL, CLIR, SPARC, ILA 32884

– Africana Library, 328 Main Library, 1408 W Gregory Dr, Urbana, IL 61801
T: +1 217 2441903; Fax: +1 217 3332214; URL: library.afrst.uiuc.edu
Alfred Kagan
180 000 vols 32885

– Applied Life Studies Library, 146 Main Library, 1408 W Gregory Dr, Urbana, IL 61801
T: +1 217 3333615; Fax: +1 217 3338384; URL: www.library.uiuc.edu/ahs
Mary Beth Allen
25 000 vols 32886

– Architecture & Art Library, 208 Architecture, 608 E Lorado Taft Dr, Urbana, IL 61801
T: +1 217 3330224; Fax: +1 217 2445169; E-mail: rickerlibrary@library.uiuc.edu; URL: www.library.illinois.edu/arx
1873; Jane Block
58 000 vols   32887

– Asian Library, 325 Main Library, 1408 W Gregory Dr, Urbana, IL 61801
T: +1 217 3331501; Fax: +1 217 3332214; URL: www.library.illinois.edu/asx
1965; Karen Wei
420 000 vols; 1 250 curr per   32888

– Biology Library, 101 Burrill Hall, 407 S Goodwin Ave, Urbana, IL 61801
T: +1 217 2443591; Fax: +1 217 3333662; E-mail: biolib@uiuc.edu; URL: www.library.illinois.edu/bix
1884; Diane Schmidt
Microfiche Coll of Vascular Plant Types
137 000 vols; 1 169 curr per   32889

– Business & Economics Library, 101 Main Library, 1408 W Gregory, Urbana, IL 61801
T: +1 217 3333619; Fax: +1 217 2441931; URL: library.illinois.edu/bel
1908; Rebecca Smith
65 000 vols; 1 200 curr per; 3 700 000 company annual rpts & SEC filings on microfiche   32890

– Chemistry Library, 170 Noyes Lab, MC-712, 505 S Matthews, Urbana, IL 61801
T: +1 217 3333737; URL: www.library.uiuc.edu/chx
1892; Tina Chrazstowski
70 000 vols; 925 films & microfiche   32891

– City Planning & Landscape Architecture Library, 203 Mumford Hall, Urbana, IL 61801
T: +1 217 3330424; Fax: +1 217 2656241; E-mail: pcyu@uiuc.edu; URL: www.library.uiuc.edu/cpx/
Priscilla Yu
24 440 vols   32892

– Classics Library, 419A Main Library, 1408 W Gregory Dr, Urbana, IL 61801
T: +1 217 2441872; Fax: +1 217 3332214; E-mail: b-swann@uiuc.edu; URL: www.library.uiuc.edu/clx
1907; Bruce W. Swann
Dittenberger-Vahlen Coll
54 000 vols; 395 curr per
libr loan   32893

– Communications Library, 122 Gregory Hall, Urbana, IL 61801
T: +1 217 3332216; URL: www.library.illinois.edu/cmx
Lisa Romero
17 000 vols   32894

– Education & Social Science Library, 100 Main Library, MC-522" Urbana, IL 61801
T: +1 217 2441864; Fax: +1 217 3332214; E-mail: educlib@library.uiuc.edu; URL: www.library.uiuc.edu/edx
1928
Arms Control Coll, CW Odell Test Coll, Children's Lit Coll, Human Relations Area Files, Mandeville Coll of Parapsychology & the Occult, Curriculum Coll
350 000 vols; 1 151 curr per; 556 000 microforms; 32 sound-rec; 98 digital data carriers
libr loan   32895

– English Library, 321 Main Libr" Urbana, IL 61801
T: +1 217 3332220; Fax: +1 217 3332214; URL: www.library.illinois.edu/egx
1911; Kathleen M. Kluegel
35 000 vols   32896

– Funk Agricultural, Consumer & Environmental Sciences Library, 1101 S Goodwin, MC-633, Urbana, IL 61801
T: +1 217 3332416; Fax: +1 217 3330558; URL: www.library.illinois.edu/agx
1915; Joseph Zumalt
State Experiment Station Publs
120 000 vols; 12 000 microforms   32897

– Geology Library, 223 Natural History Bldg, 1301 W Green St, Urbana, IL 61801
T: +1 217 3331266; Fax: +1 217 2444319; URL: www.library.illinois.edu/gex
Lura Joseph
100 000 vols   32898

– Government Documents Library, 200-D Main Libr, 1408 W Gregory Dr, Urbana, IL 61801
T: +1 217 2446445; Fax: +1 217 3332214; E-mail: gdoclib@library.uiuc.edu; URL: www.library.illinois.edu/doc
1980; Mary Mallory
Statistics, Legislation, Regulation, Executive, Historical Bibliography; US Government, Illinois State & United Nations Colls
230 000 vols; 1 200 000 microforms   32899

– Grainger Engineering Library Information Center, 1301 W Springfield, Urbana, IL 61801
T: +1 217 3333576; Fax: +1 217 2447764; E-mail: enginlib@uiuc.edu; URL: www.library.illinois.edu/grainger
1916; William Mischo
272 000 vols; 14 000 vols of college docs   32900

– History, Philosophy & Newspaper Library, 246 Main Library, 1408 W Gregory Dr, Urbana, IL 61801
T: +1 217 3331509; URL: www.library.illinois.edu/hpnl
1918; Mary Stuart
Horner Lincoln Coll
41 000 vols   32901

– Illinois History & Lincoln Collections, 422 Main Library, 1408 W Gregory Dr, Urbana, IL 61801
T: +1 217 3331777; URL: www.library.illinois.edu/ihx
2006; John Hoffmann
25 000 vols   32902

– Illinois Natural History Survey Library, 1816 South Oak St, MC-652, Champaign, IL 61820-6970
T: +1 217 3336892; Fax: +1 217 3334949; URL: www.library.uiuc.edu/nhx
1858; Beth Wohlgemuth
44 000 vols; 550 curr per   32903

– Latin American & Caribbean Library, 324 Main Libr, MC 522, 1408 W Gregory Dr, Urbana, IL 61801
T: +1 217 3332786; Fax: +1 217 3332214; E-mail: ngonzale@uiuc.edu; URL: www.library.uiuc.edu/lat
1904; Nelly S. Gonzalez
Gabriel Garcia Marquez Coll
415 000 vols; 212 curr per   32904

– Law Library, 142 Law Bldg, 504 E Pennsylvania Ave, Champaign, IL 61820
T: +1 217 3330931; Fax: +1 217 2448500; URL: www.law.uiuc.edu/library
1897; Janis Johnston
Foreign law; Arch of the American Assn of Law Libraries, European Economic Community
762 000 vols; 8 800 curr per; 810 000 microforms; 209 digital data carriers; 2 957 pamphlets   32905

– Library Science & Information Science Library, 306 Main Library, 1408 W Gregory Dr, Urbana, IL 61801
T: +1 217 3333804; E-mail: lislib@library.uiuc.edu; URL: www.library.illinois.edu/lsx
Susan Searing
25 000 vols; 600 curr per   32906

– Map & Geography Library, 1408 W Gregory Dr, Urbana, IL 61801
T: +1 217 3330827; Fax: +1 217 3332214; URL: www.library.uiuc.edu/max
Jenny Marie Johnson
591 000 vols   32907

– Mathematics Library, 1409 W Green St, Urbana, IL 61801
T: +1 217 3330258; E-mail: math@library.uiuc.edu; URL: www.library.illinois.edu/mtx
1906; Timothy Cole
100 000 vols; 1 000 microforms   32908

– Modern Languages & Linguistics Library, 425 Main Library, MC-522, 1408 W Gregory Dr, Urbana, IL 61801
T: +1 217 3330076; Fax: +1 217 3332214; URL: www.library.illinois.edu/mdx
Bruce Swan
17 000 vols   32909

– Music & Performing Arts Library, 2146 Music Bldg, MC-056, 1114 W Nevada St, Urbana, IL 61801-3859
T: +1 217 3331173; Fax: +1 217 2449097; E-mail: musiclib@library.uiuc.edu; URL: library.illinois.edu/mux
John Wagstaff
311 000 vols   32910

– Physics-Astronomy Library, 204 Loomis Lab, 110 W Green St, Urbana, IL 61801
T: +1 217 2448530; URL: www.library.illinois.edu/phx
1948; Mary Schlembach
47 000 vols   32911

– Rare Book & Manuscript Library, 346 Main Library, MC-522, 1408 W Gregory Dr, Urbana, IL 61801
T: +1 217 3333777; Fax: +1 217 2441755; URL: www.library.illinois.edu/rbx
Valerie Hotchkiss
173 000 vols   32912

– Slavic and East European Library, 225 Main Library, 1408 W Gregory Dr, Urbana, IL 61801
T: +1 217 3331349; Fax: +1 217 2448976; E-mail: srscite@cliff.library.uiuc.edu; URL: gateway.library.uiuc.edu/spx
1959; Miranda Beaven Remnek
17 000 vols; 92 000 vols in microform   32913

– Undergraduate Library, 1402 W Gregory Dr, Urbana, IL 61801
T: +1 217 3338589; Fax: +1 217 2650936; E-mail: uglcirc@library.uiuc.edu; URL: www.library.illinois.edu/ugl
Lisa Hinchliffe
250 000 vols   32914

– University Laboratory High School, Library, 1212 W Springfield Ave, Rm 201, Urbana, IL 61801
T: +1 217 3331589; Fax: +1 217 3334064; URL: www.uni.uiuc.edu/library
Frances Jacobsen Harris
14 000 vols   32915

– Veterinary Medicine Library, 1257 Veterinary Med Basic Science Bldg, 2001 S Lincoln Ave, Urbana, IL 61802
T: +1 217 3338778; Fax: +1 217 3332286; URL: www.library.illinois.edu/vex
Greg Youngun
52 000 vols   32916

Urbana University, Swedenborg Memorial Library, 579 College Way, **Urbana**, OH 43078-2091
T: +1 937 4841298; Fax: +1 937 6538551; E-mail: library@urbana.edu; URL: www.urbana.edu
1850; Barbara M. Macke
Children's Lit (19th century); Emanuel Swedenborg Coll; Church of the New Jerusalem
77 000 vols; 118 curr per; 40 diss/theses; 8 000 microforms; 10 digital data carriers
libr loan; OCLC   32917

State University of New York – Institute of Technology, Peter J. Cayan Library, Rte 12 N & Horatio St, **Utica**, NY 13502; P.O. Box 3051, Utica, NY 13504-3051
T: +1 315 7927245; Fax: +1 315 7927517; URL: www.sunyit.edu/library
1969; Daniel Schabert
Nursing, technology, computer science
192 000 vols; 545 curr per; 60 133 govt docs; 65 000 microforms; 800 av-mat; 2 582 sound-rec; 113 digital data carriers
libr loan   32918

Utica College, Frank E. Gannett Memorial Library, 1600 Burrstone Rd, **Utica**, NY 13502-4892
T: +1 315 7923262; Fax: +1 315 7923361; URL: www.utica.edu
1946; Beverly Marcoline
Welsh Language Imprints of New York State, Fiction-Scene in Upstate New York Since 1929
180 000 vols; 1 400 curr per
libr loan   32919

Valdosta State University, Odum Library, 1500 N Patterson St, **Valdosta**, GA 31698-0150
T: +1 229 3335860; Fax: +1 229 2595055; URL: www.valdosta.edu/library
1913; George R. Gaumond
Emily Hendree Park Memorial Coll, US Maps Coll, Arch of Contemporary South Georgia Hist
529 000 vols; 2 776 curr per; 65 digital data carriers
libr loan   32920

California Institute of the Arts, Division of Library & Information Resources, 24700 McBean Pkwy, **Valencia**, CA 91355
T: +1 661 2537889; Fax: +1 661 2544561; E-mail: libref@calarts.edu; URL: www.calarts.edu/library
1968; Jeff Gatten
Artists Bks, Film Hist, Viola Hegyi Swisher Coll of Dance & Theatre Papers
194 000 vols; 463 curr per; 205 e-books; 10 000 microforms; 16 378 sound-rec; 102 digital data carriers; 115 748 other items   32921

New York Medical College, Health Sciences Library, Basic Science Bldg, 15 Dana Road, **Valhalla**, NY 10595
T: +1 914 5944200; Fax: +1 914 5943171; E-mail: hsl_nymc@nymc.edu; URL: library.nymc.edu
1976; Diana Cunningham
Hist of Medicine & Homeopathy (Hist Coll), Hist of Orthopedics (Alfred Haas Coll), Rare Books (J. Alexander van Heuven Coll)
208 535 vols; 374 curr per; 14 597 e-journals; 810 e-books; 1 000 microforms; 17 digital data carriers
libr loan; MLA, National Network of Libraries of Medecine (NYLINK),), NYSHEI, NERL, OCLC, METRO   32922

California Maritime Academy Library, California State University, 200 Maritime Academy Dr, **Vallejo**, CA 94590
T: +1 707 6541092; Fax: +1 707 6541094; E-mail: library@csum.edu; URL: library.csum.edu
1959; Carl D. Phillips
California Maritime Academy Arch; Ship's Library
40 000 vols; 276 curr per; 300 govt docs; 150 maps; 21 000 microforms; 220 av-mat; 26 digital data carriers
libr loan   32923

Valley City State University, Allen Memorial Library, 101 College St SW, **Valley City**, ND 58072-4098
T: +1 701 8457279; Fax: +1 701 8457437; E-mail: library.office@vcsu.edu; URL: library.vcsu.edu
1890; Donna James
Education; North Dakota Coll, Valley City State Univ Hist Coll, Woiwode Coll
95 000 vols; 246 curr per; 15 000 e-books; 55 digital data carriers
libr loan   32924

Valparaiso University, Christopher Center for Library & Information Resources, 1410 Chapel Dr, **Valparaiso**, IN 46383-6493
T: +1 219 4646890; URL: www.valpo.edu/library
1859; Richard A. AmRhein
Theology; Univ Arch Coll
348 000 vols; 2 600 curr per
libr loan   32925

– School of Law Library, 656 S Greenwich St, Valparaiso, IN 46383
T: +1 219 4657827; Fax: +1 219 4657808; URL: www.valpo.edu/law
1879; Gail Hartzell
Supreme Court Records & Briefs Coll
320 000 vols; 2 059 curr per; 800 000 microforms; 80 digital data carriers
libr loan   32926

University of South Dakota, I.D. Weeks Library, 414 E Clark St, **Vermillion**, SD 57069
T: +1 605 6775371; Fax: +1 605 6775488; E-mail: library@usd.edu; URL: www.usd.edu/library
1882; Anne Moore

Herman P Chilson Western Americana Coll, Country Schools Survey, South Dakota Hist & Hist Maps (Richardson Arch)
551 000 vols; 1 509 curr per; 6 200 av-mat; 400 digital data carriers
libr loan 32927

– McKusick Law Library, 414 E Clark St, Vermillion, SD 57069-2390
T: +1 605 6776354; Fax: +1 605 6775417; E-mail: lllibrary@usd.edu; URL: www.usd.edu/lawlib
1901; John F. Hagemann
Indian Law, Professional Responsibility Law, Arts & the Law
200 000 vols; 20 digital data carriers; 8 VF drawers of pamphlets, 4 VF drawers of arch 32928

University of South Dakota – Christian P. Lommen Health Sciences Library, Sanford School of Medicine, 414 E Clark, *Vermillion*, SD 57069-2390
T: +1 605 6775121; Fax: +1 605 6775124; URL: www.usd.edu/lhsl/
1907; David Hulkonen
Hist of Medicine Arch, Rare Books, Medical School Arch
91 000 vols; 677 curr per; 580 microforms; 750 av-mat; 8 digital data carriers
libr loan; MLA 32929

University of Houston, Victoria College Library, 2602 N Ben Jordan St, *Victoria*, TX 77901-5699
T: +1 361 5704177; Fax: +1 361 5704155; URL: www.vcuhvlibrary.uhv.edu
1925; Dr. Joe F. Dahlstrom
287 000 vols; 285 curr per
libr loan 32930

Ohio Valley College, Icy Belle Library, One Campus View Dr, *Vienna*, WV 26105-8000
T: +1 304 8656112; Fax: +1 304 8656001; URL: www.ovc.edu/library
1960; John H. Foust
Religious studies
34 000 vols; 142 curr per; 94 000 e-books 32931

Villanova University, Falvey Memorial Library, 800 Lancaster Ave, *Villanova*, PA 19085
T: +1 610 5194274; Fax: +1 610 5195018; URL: www.library.villanova.edu
1842; Joseph Lucia
Irish American Hist & Lit (Joseph McGarrity Coll), Saint Augustine Coll
683 000 vols; 12 000 curr per; 2 000 e-books; 93 av-mat
libr loan 32932

– Law Library, Garey Hall, 299 N Spring Mill Rd, Villanova, PA 19085
T: +1 610 5197024; Fax: +1 610 5197033; URL: www.law.villanova.edu/library
1953; William James
Tax law; Church & State Coll
337 000 vols; 3 703 curr per; 736 000 microforms; 79 digital data carriers 32933

Vincennes University, Shake Learning Resources Library, 1002 N First St, *Vincennes*, IN 47591
T: +1 812 8885130; Fax: +1 812 8885471; E-mail: libref@vinu.edu; URL: www.vinu.edu/cms/opencms/academic_resources/library/
1959; Robert A. Slayton
Lewis Hist Libr Coll
89 000 vols; 433 curr per; 7 000 e-books; 550 av-mat
libr loan 32934

ECPI College of Technology, Virginia Beach Main Campus Library, 5555 Greenwich Rd, *Virginia Beach*, VA 23462
T: +1 757 4909090; URL: www.ecpi.edu
1984
Computer science, Data processing; Computer Technology
12 000 vols; 171 curr per
libr loan; ALA, VLA, NCLA, SLA 32935

Regent University, University Library, 1000 Regent University Dr, *Virginia Beach*, VA 23464
T: +1 757 2264150; Fax: +1 757 2264167; E-mail: refer@regent.edu; URL: www.regent.edu/general/library
1978; Sara Baron
Animated Films, Baptista Film Mission Arch, Christian Films Research Coll, Clark Hymnology Coll, Scott Ross Cultural Coll
280 000 vols; 1 899 curr per; 160 e-journals; 100 000 e-books; 81 Bks on Deafness & Sign Lang
libr loan 32936

– Law Library, 1000 Regent University Dr, Virginia Beach, VA 23464-9800
T: +1 757 2264145; Fax: +1 757 2264167; E-mail: lawref@regent.edu; URL: law.regent.edu/library
1986; Charles H. Oates
Ralph Bunche Coll, Founders Coll, Roscoe Bound Papers, John Brabner-Smith Libr & Papers, First Amendment & Civil Rights Coll
374 000 vols; 4 208 curr per; 801 000 microforms; 15 av-mat; 958 sound-rec; 101 digital data carriers 32937

Baylor University, University Libraries, 1312 S Third, *Waco*, TX 76798; mail address: One Bear Pl, No 97148, Waco, TX 76798-7148
T: +1 254 7102340; Fax: +1 254 7103116; URL: www.baylor.edu/library
1845; Pattie Orr
Robert Browning, Legislative Coll, Texas Hist
2 378 000 vols; 18 910 e-journals; 347 000 e-books
libr loan 32938

– Armstrong Browning Library, 710 Speight Ave, Waco, TX 76798-7152; mail address: One Bear Pl, No 97152, Waco, TX 76798-7152
T: +1 254 7103566; Fax: +1 254 7103552; URL: www.browninglibrary.org
1918; N.N.
19th c American & British Lit
25 000 vols; 22 curr per; 673 microforms; 34 sound-rec; 5 059 mss & letters, 21 vols of clippings, 6 303 slides & taped lectures, 1 500 music scores 32939

– Baylor Collections of Political Materials, 201 Baylor Ave, Waco, TX 76706; mail address: One Bear Pl, No 97153, Waco, TX 76798-7153
T: +1 254 7103540; Fax: +1 254 7103059; E-mail: poage_library@baylor.edu; URL: www.baylor.edu/lib/poage
1979; Ben Rogers
20th c politics, terrorism, JFK assassination
17 000 vols; 759 maps; 2 285 linear feet of personal & legislative papers & other docs 32940

– Crouch Fine Arts Library, 1312 S Third St, mail address: One Bear Pl, No 97148, Waco, TX 76798-7148
T: +1 254 7102164; Fax: +1 254 7103116; E-mail: sha_towers@baylor.edu; URL: www.baylor.edu/library
1929; Sha Towers
18th & 19th C English Bks on Music & Scores, 18th C Eds of Ensemble Music, Mrs J W Jennings Coll (Medieval Music Mss & Early Printed Music), Travis Johnson Coll (Early American Songbks), Francis G Spencer Coll of American Printed Music
34 300 vols; 200 curr per; 650 microforms; 2 500 av-mat; 37 756 sound-rec; 75 000 musical items, DVDs, laser discs 32941

– J. M. Dawson Church-State Research Center Library, Carroll Library Bldg, 1311 S Fifth St, Waco, TX 76798-7308; mail address: One Bear Pl, No 97308, Waco, TX 76798-7308
T: +1 254 7101510; Fax: +1 254 7101571
1957; Christopher Marsh
Church & State Coll; Joseph Martin Dawson Coll; E S James Coll, microforms & doc; Leo Pfeffer Coll
16 000 vols; 235 curr per; 182 microforms 32942

– Jesse H. Jones Library, 1312 S Third, mail address: One Bear Pl, No 97148, Waco, TX 76798-7148
T: +1 254 7102112; Fax: +1 254 7103116; E-mail: bill_hair@baylor.edu; URL: www.baylor.edu/Library
1845; William B. Hair
2 118 851 vols; 8 816 curr per
libr loan 32943

– Sheridan & John Eddie Williams Legal Research & Technology Center, Library, 1114 S University Parks Dr, One Bear Pl, No 97128, Waco, TX 76798-7128
T: +1 254 7102168; Fax: +1 254 7102294; URL: law.baylor.edu/library/main.htm
1857; Brandon Quarles
Frank M Wilson Rare Bk Coll
108 000 vols; 2 206 curr per; 361 000 microforms; 34 digital data carriers 32944

– Texas Collection, 1429 S Fifth St, Waco, TX 76706; mail address: One Bear Pl, No 97142, Waco, TX 76798-7142
T: +1 254 7101268; Fax: +1 254 7101368; E-mail: txcoll@baylor.edu; URL: www3.baylor.edu/Library/Texas
1923; Dr. Thomas L. Charlton
Regional Hist Resource Depot; Depot for Texas State pubs & docs; newsfilm arch, KWTX-TV, Texas
110 000 vols; 1 101 curr per; 9 650 maps; 3 450 sound-rec; 2 910 mss, 2 026 videos, 6 473 slides, 53 000 photogs 32945

Whitman College, Penrose Library, 345 Boyer Ave, *Walla Walla*, WA 99362
T: +1 509 5275191; Fax: +1 509 5275900; URL: www.whitman.edu/penrose
1882; Dalia Rogan
Dogwood Press Coll; McFarlane Coll, early illustrated bks; Stuart Napoleon Coll
387 000 vols; 21 024 curr per
libr loan; ALA 32946

University of Maine, Darling Marine Center Library, 193 Clarks Cove Rd, *Walpole*, ME 04573
T: +1 207 5633146 ext 226; Fax: +1 207 5633119; E-mail: randy.lackovic@umit.maine.edu; URL: www.dmc.maine.edu/library.html
1966; Randy Lackovic
9 000 vols; 160 curr per; 1 800 microforms
ALA 32947

Bentley College, Solomon R. Baker Library, 175 Forest St, *Waltham*, MA 02452-4705
T: +1 781 8912300; Fax: +1 781 8912830; E-mail: library@bentley.edu; URL: ecampus.bentley.edu/dept/li
1917; Phillip Knutel
Accounting, Business & management, Economics, Finance; Accounting, 17th-19th c (rare book), Business Hist
225 000 vols; 624 curr per; 225 000 microforms; 624 sound-rec; 12 digital data carriers 32948

Brandeis University, University Libraries, 415 South St, Mailstop 045, *Waltham*, MA 02454-9110
T: +1 781 7367777; Fax: +1 781 7364719; URL: lts.brandeis.edu
1948; Susan V. Wawrzaszek
Judaica; Shakespeare Coll
1 207 000 vols; 23 130 e-journals; 880 000 microforms
libr loan; RLG, CRL 32949

Kent State University, Trumbull Campus Library, 4314 Mahoning Ave NW, *Warren*, OH 44483-1998
T: +1 330 6758865; Fax: +1 330 6758825; URL: www.trumbull.kent.edu/library/
Rose Guerrieri
75 000 vols; 235 curr per 32950

University of Central Missouri, James C. Kirkpatrick Library, 601 S Missouri, *Warrensburg*, MO 64093
T: +1 660 5434283; Fax: +1 660 5438001; URL: library.ucmo.edu
1872; Mollie Dinwiddie
Civil War (Personal Narratives & Unit Hist), Lit (Izaac Walton's Compleat Angler), Geography (Missouri & Int Speleology), Missouri Coll, Res Coll in Children's Lit
863 000 vols; 410 curr per; 450 av-mat 32951

New England Institute of Technology Library, 2500 Post Rd, *Warwick*, RI 02886-2266
T: +1 401 7395000; Fax: +1 401 7384061; URL: library.neit.edu
Sharon J. Charette
48 000 vols; 15 744 curr per; 17 309 e-journals; 6 000 e-books
libr loan; ALA, ACRL 32952

American College of Obstetricians & Gynecologists, Resource Center, 409 12th St SW, *Washington*, DC 20024-2188; P.O. Box 96920, Washington, DC 20090-6920
T: +1 202 8632518; Fax: +1 202 4841595; E-mail: resources@acog.org; URL: www.acog.org
1969; Mary A. Hyde
Hist Ob-Gyn; Archs
12 000 vols; 450 curr per; 6 digital data carriers
libr loan; MLA, SLA, AMIA 32953

The American University, Jack I. & Dorothy G. Bender Library & Learning Resources Center, 4400 Massachusetts Ave NW, *Washington*, DC 20016-8046
T: +1 202 8853238; Fax: +1 202 8853226; E-mail: librarymail@american.edu; URL: www.library.american.edu
1892; Bill Mayer
John Hickman Coll, Drew Pearson Coll, Mathematics (Artemas Martin Coll), Asia & the East, Japanese Culture (Spinks Coll), The Papers of the National Commission on the Public Service
1 000 000 vols; 4 200 curr per; 14 400 e-journals; 193 000 e-books; 12 345 diss/theses; 13 170 music scores; 1 121 000 microforms; 10 000 av-mat; 36 746 sound-rec; 1 400 digital data carriers
libr loan 32954

American University – Washington College of Law, Pence Law Library, 4801 Massachusetts Ave NW, *Washington*, DC 20016-8182
T: +1 202 2744300; Fax: +1 202 2744365; E-mail: reflib@wcl.american.edu; URL: library.wcl.american.edu
1896; Billie Jo Kaufman
Administrative Conference of the US coll; Goodman Coll of Rare Law Bks; Peter M. Cicchino Collection; Richard R. Baxter Coll.; Working Papers of the Bankruptcy Review Commission
598 029 vols; 1 294 curr per; 47 260 e-journals; 21 800 e-books; 8 diss/theses; 55 360 govt docs; 95 956 microforms; 119 sound-rec; 209 digital data carriers
AALL 32955

Capuchin College, Library, 4121 Harewood Rd NE, *Washington*, DC 20017
T: +1 202 5292188; Fax: +1 202 5266664
1917; Sonia Bernardo
Capuchin hist, Franciscan hist & theology, Roman Catholic theology
75 000 vols; 50 curr per; 2 mss; 2 incunabula; 50 diss/theses 32956

Catholic University of America, John K. Mullen of Denver Memorial Library, 620 Michigan Ave NE, 315 Mullen Library, *Washington*, DC 20064
T: +1 202 3195055; Fax: +1 202 3194735; URL: libraries.cua.edu
1889; Steve Connaghan
Church Hist, Greek & Latin, Patristics, Medieval Studies, Canon Law, Labor, Immigration; Catholic Americana, Celtic, Church Hist, Knights of Malta, Libr of Pope Clement XI (Clementine Libr); Luso-Brazilian Studies (Lima Libr)
1 373 000 vols; 5 203 curr per; 13 370 e-journals; 2 000 e-books; 5 600 av-mat; 150 digital data carriers; 30 000 Electronic Media & Resources
libr loan 32957

– Engineering-Architecture & Mathematics Library, 620 Michigan Ave NE, 200 Pangborn Hall, Washington, DC 20064
T: +1 202 3195167; Fax: +1 202 3194485; E-mail: ogrady@cua.edu; URL: libraries.cua.edu/engcoll.html
Christina O'Grady
38 259 vols; 413 curr per 32958

– Judge Kathryn J. DuFour Law Library, 3600 John McCormack Rd NE, Washington, DC 20064-8206
T: +1 202 3195156; Fax: +1 202 3195581; URL: law.cua.edu/library
1898; Stephen Margeton
Religion & the Law Coll
400 000 vols; 4 853 curr per; 26 digital data carriers   32959

– Music Library, 101 Ward Hall, 620 Michigan Ave NE, Washington, DC 20064
T: +1 202 3195424; Fax: +1 202 3196280; E-mail: saylor@cua.edu; URL: libraries.cua.edu/musicoll.html
1952; Maurice Saylor
Grentzer Spivacke Coll (Music Education), Latin American Music Coll
25 320 vols; 113 curr per; 19 805 sound-rec; 12 005 pieces of music   32960

– Nursing-Biology Library, 212 Gowan Hall, 620 Michigan Ave NE, Washington, DC 20064
T: +1 202 3195411; Fax: +1 202 3195410; E-mail: womack@cua.edu; URL: libraries.cua.edu/nurscoll.html
1932; Kristina Womack
Nursing Hist Coll
56 260 vols; 435 curr per; 2 727 diss/theses; 3 295 microforms; 1 digital data carriers   32961

– Oliveira Lima Library, 22 Mullen Library, 620 Michigan Ave NE, Washington, DC 20064
T: +1 202 3195059; Fax: +1 202 3194735; URL: libraries.cua.edu/ limacoll.html
1916; Maria Angela Leal
Oliveira Lima Family Papers, Tracts on Portuguese Inquisition, 19th C Portuguese Liberalism Pamphlets Coll, Dutch Pamphlets on 17th C Brazil, Portuguese Restoration, Society of Jesus
59 000 vols   32962

– Rare Books / Special Collections, 214 Mullen Library, 620 Michigan Ave NE, Washington, DC 20064
T: +1 202 3195091; Fax: +1 202 3194735; E-mail: rouse@cua.edu; URL: www.libraries.cua.edu/www/rarecoll
Lenore Rouse
40 000 vols   32963

– Reference & Instructional Services Division, 620 Michigan Ave NE, 124 Mullen Library, Washington, DC 20064
T: +1 202 3195070; Fax: +1 202 3196054; E-mail: lesher@cua.edu; URL: libraries.cua.edu/refcoll.html
Anne Lesher
20 400 vols   32964

– Semitics/ICOR Library, 620 Michigan Ave NE, 18 Mullen Library, Washington, DC 20064
T: +1 202 3195084; Fax: +1 202 3194735; URL: libraries.cua.edu/ semicoll.html
1895; Dr. Monica Blanchard
Ostraca & Papiri Colls, Syriac Digital Libr
46 000 vols; 413 curr per; 150 mss   32965

Gallaudet University, University Library, 800 Florida Ave NE, *Washington*, DC 20002
T: +1 202 6515217; Fax: +1 202 6515213; E-mail: library.help@gallaudet.edu; URL: library.gallaudet.edu
1876; Leida Torres
Deafness, Audiology & Hearing (Mats Relating to Deafness), bks, micro, flm, v-tapes, Arch Mats
242 630 vols; 1 259 curr per; 50 e-books; 436 665 microforms; 7 237 av-mat; 302 Electronic Media & Resources, 70 380 Bks on Deafness & Sign Lang
libr loan   32966

George Washington University, Melvin Gelman Library, 2130 H St NW, Ste 201, *Washington*, DC 20052
T: +1 202 9946455; Fax: +1 202 9941340; URL: www.gwu.edu/gelman
1821
Economic, Social, Political & Cultural Hist of Washington DC, Jewish Community Council of Greater Washington, Washington Writing Arch, The Univ Arch, Oral Hist Tapes & Transcripts, Frederick

Kuh, Samual Shaffer, Sports & Special Events Management Coll, Hist of Printing Coll
1 766 000 vols; 42 098 curr per
libr loan   32967

– Eckles Library, 2100 Foxhall Rd NW, Washington, DC 20007-1199
T: +1 202 2426666; Fax: +1 202 2426822; E-mail: eckles@gwu.edu; URL: www.gwu.edu/gelman/eckles
1875; John Danneker
Art & Art hist, Interior design, Women's studies; Culinary Bk Coll, Walter Beach Archs of the American Political Science Assn
63 000 vols; 154 curr per   32968

– Jacob Burns Law Library, 716 20th St NW, Washington, DC 20052
T: +1 202 9944156; Fax: +1 202 9942874; URL: www.law.gwu.edu/burns/
1865; Scott B. Pagel
621 000 vols; 4 304 curr per; 512 000 microforms; 630 sound-rec; 101 digital data carriers
libr loan   32969

– Paul Himmelfarb Health Sciences Library, 2300 I St NW, Medical Ctr, Washington, DC 20037
T: +1 202 9942850; Fax: +1 202 2233691; E-mail: library@gwumc.edu; URL: www.gwumc.edu/library
1857; Anne Linton
Interviews with George Washington Univ VIPs from 1930-50's
121 024 vols; 1 550 curr per; 1 500 av-mat; 16 digital data carriers
libr loan   32970

George Washington University – National Clearinghouse for English Language Aquisition & Language Instruction Educational Programs, Library, 2121 K St NW, Ste 260, *Washington*, DC 20037
T: +1 202 4670867; Fax: +1 202 4674283; E-mail: askncela@ncela.gwu.edu; URL: www.ncela.gwu.edu/
Minerva Gorena
Mats Produced by Title VII Programs
19 000 vols; 22 curr per   32971

Georgetown University, Joseph Mark Lauinger Library, 37th & N St NW, *Washington*, DC 20057-1174
T: +1 202 6877425;
Fax: +1 202 6871215; URL: www.library.georgetown.edu
1789; Artemis Kirk
Arch of Dag Hammarskjold College, Arch of Maryland Province, Society of Jesus, Arch of the American Political Science Assn, Arch of Woodstock College, Catholic History, Dipolomacy & Foreign Affairs, Political Science, Unitec States-American & English Literature, University Arch, Woodstock Thelogical Libr, Shea Coll (Americana), Parsons Coll (Early Catholic Americana), Endicott Coll (Panama & the Canal), Bowen Coll on Intelligence & Covert Activities
1 581 000 vols; 16 025 curr per
libr loan   32972

– Edward Bennett Williams Law Library, 111 G St NW, Washington, DC 20001
T: +1 202 6629140; Fax: +1 202 6629168; URL: www.ll.georgetown.edu
1870; N.N.
Federal Legislative Hist, UN Docs
1 000 000 vols; 2 300 000 microforms
libr loan   32973

– John Vinton Dahlgren Memorial Library, Preclinical Science Bldg GM-7, 3900 Reservoir Rd NW, Washington, DC 20007; P.O. Box 571420, Washington, DC 20057-1420
T: +1 202 6871448; Fax: +1 202 6871862; URL: www.georgetown.edu/ dml
1912; Jeff McCann
Hist of Ophthalmology (Julius Hirshberg Coll), Medicine (Alexis Carrel Coll)
30 000 vols; 786 curr per; 1 295 diss/theses; 1 000 microforms; 8 digital data carriers; 375 software titles
libr loan   32974

– National Center for Education in Maternal & Child Health, Library, 2115 Wisconsin Ave NW, Ste 601, Washington, DC 20007; P.O. Box 571272, Washington, DC 20057-1272
T: +1 202 7849770; Fax: +1 202 7849777; E-mail: mchgroup@georgetown.edu; URL: www.mchlibrary.info
1982; Olivia K. Pickett
Final Rpts of Projects Funded by Maternal & Child Health Bureau, Maternal Health Rpts of State Title V Coll
23 000 vols; 50 curr per   32975

Howard University Libraries, 500 Howard Pl NW, *Washington*, DC 20059
T: +1 202 8065064; Fax: +1 202 8065903; E-mail: refdept@howard.edu; URL: www.howard.edu/library
1867; Mohamed Mekkawi
Channing Pollock Theatre Coll, Moorland Springarn Res Ctr for Oral Hist
2 507 000 vols; 12 216 curr per; 3 680 000 microforms; 36 digital data carriers
libr loan   32976

– Architecture Library, 2366 Sixth St NW, Washington, DC 20059
T: +1 202 8067774; Fax: +1 202 8064441
1971; Alliah Humber
Art & architecture, Interior design, Historic preservation
34 000 vols; 62 curr per; 428 maps; 2 000 microforms; 33 921 slides, films, lantern frames   32977

– Business Library, 2600 Sixth St NW, Washington, DC 20059
T: +1 202 8061561; Fax: +1 202 7976393
1970; Lucille Smiley
94 000 vols; 1 100 curr per; 7 digital data carriers; 25 000 reels of microfilm, 500 000 10K rpts on microfiche   32978

– Department of Afro-American Studies Resource Center, 500 Howard Pl NW, Rm 300, Washington, DC 20059
T: +1 202 8067242; Fax: +1 202 9860538
1969; E. Ethelbert Miller
32 000 vols   32979

– Divinity Library, 1400 Shepherd St NE, Washington, DC 20017
T: +1 202 8060760; Fax: +1 202 8060711
1935; Carrie Hackney
Reformed theology, Biblical studies, Church hist; African Heritage Coll
121 000 vols; 240 curr per   32980

– Founders & Undergraduate Library, 500 Howard Pl NW, Washington, DC 20059
T: +1 202 8065716; Fax: +1 202 8064622
Dr. Arthuree Wright
Engineering, mathematics, humanities
1 173 000 vols; 6 964 curr per   32981

– Law Library, 2929 Van Ness St NW, Washington, DC 20008
T: +1 202 8068208; Fax: +1 202 8068400; URL: www.law.howard.edu/ library/
1868; Rhea Ballard-Thrower
Civil Rights Project Coll, Indritz Papers
253 000 vols; 621 curr per; 59 000 microforms; 600 av-mat; 269 sound-rec; 14 digital data carriers   32982

– Louis Stokes Health Sciences Library, 501 W St NW, Washington, DC 20059
T: +1 202 8841730; Fax: +1 202 8841506; URL: hsl.howard.edu
1927; Ellis B. Beteck
Biographical files on Blacks in Medicine, Dentistry and Nursing; Local Hist, Howard University, Colleges of Medicine, Dentistry, Nursing & Allied Health, Sickle Cell Anemia
134 000 vols; 1 089 curr per; 21 e-journals; 550 microforms; 40 digital data carriers; 515 bibiogr, 121 shelves of AV progr, 26 VF drawers, 115 drawers of microfilm
libr loan; ALA, ARL   32983

– Moorland-Spingarn Research Center Library, 500 Howard Pl NW, Washington, DC 20059
T: +1 202 8064237; Fax: +1 202 8066405
1914; Dr. Thomas C. Battle

Africa; Afro-American & Afro-Latin Authors (Spingarn Coll); Civil Rights (Ralph S. Bunche Oral Hist Coll); Journalism (Documentary Series on the Black Press), Arthur B. Spingarn Music Coll, sheet music; Howardiana
200 000 vols; 632 curr per; 6 600 mss; 7 000 diss/theses; 40 000 microforms; 100 000 images, 22 000 vertical file folders, 17 000 linear ft mss & arch colls   32984

– Social Work Library, 601 Howard Pl NW, Washington, DC 20059
T: +1 202 8067316; URL: www.howard.edu/library/Social_Work_Library
1971; Audrey M. Thompson
40 000 vols; 344 curr per   32985

Inter-American Defense College, Colegio Interamericano de Defensa, Library, Fort McNair, Bldg 52, *Washington*, DC 20319-6100
T: +1 202 6461337; Fax: +1 202 6461340; E-mail: Personnel@jid.org; URL: www.jid.org
1962
25 000 vols; 207 curr per; 11 000 docs & pamphlets   32986

Johns Hopkins University School of Advanced International Studies, Sydney R. & Elsa W. Mason Library, 1740 Massachusetts Ave NW, *Washington*, DC 20036
T: +1 202 6635901; Fax: +1 202 6635916; E-mail: saislibrary@jhu.edu; URL: www.sais-jhu.edu/library
1943; Sheila Thalhimer
Int Affairs since 1945, Int Economics & Law, Hist, Politics, Sociology
110 000 vols; 950 curr per; 46 000 microforms
libr loan   32987

Saint Paul's College, Library, 3015 Fourth St NE, *Washington*, DC 20017
T: +1 202 8326262; Fax: +1 202 2692507
1889; Denise Eggers
Theology, Catholicism; Isaac T Hecker Arch Coll, Paulist Fathers Arch, 17th & 18th C Works
45 000 vols; 80 curr per; 90 microforms; 2 500 pamphlets   32988

Strayer University, Wilkes Library, 1133 15th St NW, *Washington*, DC 20005
T: +1 202 4190483; Fax: +1 202 8330528; URL: icampus.strayer.edu/lrc/ about
1965; David Moulton
Business administration, Data processing, Accounting, Economics, Int business
75 000 vols; 600 curr per; 500 av-mat; 25 sound-rec; 7 digital data carriers   32989

Trinity College, Sister Helen Sheehan Library, 125 Michigan Ave NE, *Washington*, DC 20017
T: +1 202 8849350; Fax: +1 202 8849362; E-mail: trinitylibrary@trinitydc.edu; URL: library.trinitydc.edu
1897; Kaye Gapen
Hist, Lit, Women's studies
240 000 vols; 550 curr per; 40 e-journals; 20 200 av-mat; 100 digital data carriers; 20 000 slides, 200 videos   32990

University of the District of Columbia, Learning Resources Division, 4200 Connecticut Ave NW, *Washington*, DC 20008
T: +1 202 2746122; Fax: +1 202 2746012; URL: www.lrdudc.wrlc.org
1976; Albert J. Casciero
Atlanta Univ Black Culture Coll, Slavery Source Mat, Schomburg Clipping File
550 000 vols; 594 curr per; 606 000 microforms; 19 178 sound-rec; 16 digital data carriers   32991

– David A. Clarke School of Law, Charles N. & Hilda H. M. Mason Law Library, Bldg 39, Rm B-16, 4200 Connecticut Ave NW, Washington, DC 20008
T: +1 202 2747310; Fax: +1 202 2747311; E-mail: lawlibrary@udc.edu; URL: www.law.udc.edu; catalog.law.udc.edu
Roy Balleste
255 000 vols   32992

US Department of Defense – National Defense University, University Library, Fort McNair, Bldg 62 Marshall Hall, 300 Fifth Ave, ***Washington***, DC 20319-5066
T: +1 202 6853964; Fax: +1 202 6853733; URL: www.ndu.edu/library
1976; Meg Tulloch
Personal Papers of C. Powell, J. Shalikashvili, G. Joulwar, J. Galvon, M. D. Taylor, L. L. Lemnitzer, A. J. Goodpaster, F. S. Besson Jr; Speeches by J. Carlton Ward; Hudson Institute Arch; Libraries of Hoffman Nickerson & Ralph L. Powell; Arch form the Final Report to Congress – Conduct of the Persian Gulf War
675 000 vols; 1 200 curr per; 18 000 e-journals; 42 000 maps; 500 000 microforms; 1 050 sound-rec; 12 digital data carriers; 60 Electronic Media & Resources
libr loan 32993

Washington & Jefferson College Library, U. Grant Miller Library, 60 S Lincoln St, ***Washington***, PA 15301
T: +1 724 2236072; Fax: +1 724 2235272; URL: www.washjeff.edu/library
1781; Rachael Bolden
College Hist; Washington County (Hist); Western Pennsylvania & Upper Ohio Valley, mss
161 000 vols; 522 curr per; 13 000 e-books
libr loan 32994

Post University, Traurig Library & Learning Resources Center, 800 Country Club Rd, ***Waterbury***, CT 06723-2540
T: +1 203 5964560; Fax: +1 203 5759691; E-mail: library@post.edu; URL: www.post.edu/maincampus/library.shtml
1890; Tracy Ralston
Africanus Coll, University Archival Coll, Zwicker Tax Institute Coll,
84 000 vols; 500 curr per; 5 000 microforms; 800 av-mat; 650 sound-rec; 15 digital data carriers 32995

University of Connecticut, Waterbury Regional Campus Library, 99 E Main St, ***Waterbury***, CT 06702-2311
T: +1 203 2369902; Fax: +1 203 2369905; URL: www.lib.uconn.edu
1946; Shelley Roseman
Philemon J. Hewitt, Jr. Apicultural Coll
35 000 vols; 100 curr per; 2 000 microforms; 2 digital data carriers
libr loan 32996

Colby College Libraries, 5100 Mayflower Hill, ***Waterville***, ME 04901
T: +1 207 8595100; Fax: +1 207 8595105; E-mail: circmill@colby.edu; URL: www.colby.edu/academics_cs/library/
1813; Clem P. Guthro
American Regional Lit Coll, Irish Lit Coll, Bern Porter Coll of Contemporary Letters, Thomas Hardy Coll, Thomas Mann Coll
492 380 vols; 1 145 curr per; 17 030 e-journals; 115 000 e-books; 9 000 music scores; 30 000 av-mat; 300 sound-rec; 6 500 digital data carriers 32997

– Bixler Art & Music Library, 5660 Mayflower Hill Dr, Waterville, ME 04901
T: +1 207 8595660; Fax: +1 207 8595105
Margaret D. Ericson
53 000 vols; 100 curr per 32998

– Science Library, 5890 Mayflower Hill, Waterville, ME 04901
T: +1 207 8595791; Fax: +1 207 8595105; URL: www.colby.edu/academics_cs/library/science/index.cfm
Susan W. Cole
46 000 vols 32999

Thomas College, Marriner Library, 180 W River Rd, ***Waterville***, ME 04901
T: +1 207 8591204; URL: www.thomas.edu/library
1894; Christopher Rhoda
Business & management, Education
24 000 vols; 1 digital data carriers 33000

Carroll College, Todd Wehr Memorial Library, 100 N East Ave, ***Waukesha***, WI 53186
T: +1 262 6504892; Fax: +1 262 5247377; URL: divisions.cc.edu/library/
1846; Dr. Lelan McLemore
Religious studies, Hist; English & Scottish 19th c Lit, Hist (W Norman FitzGerald Civil War Coll), Welsh Lit & Language
150 000 vols; 341 curr per; 14 000 e-journals; 7 000 e-books
libr loan 33001

University of Wisconsin Colleges, Waukesha Library & Media Services, 1500 University Dr, ***Waukesha***, WI 53188
T: +1 262 5215473; E-mail: scott.silet@uwc.edu; URL: waukesha.uwc.edu/Faculty—Staff/Departments/Library.aspx
1966; Scott Silet
61 000 vols; 2 750 av-mat 33002

University of Wisconsin Center – Marathon County Library, 518 S Seventh Ave, ***Wausau***, WI 54401-5396
T: +1 715 2616220; Fax: +1 715 2616330; URL: www.uwmc.uwc.edu
1938; Judy Palmateer
42 000 vols; 101 curr per 33003

Wartburg College, Vogel Library, 100 Wartburg Blvd, ***Waverly***, IA 50677-0903
T: +1 319 3528506; Fax: +1 319 3528312; E-mail: asklibrarian@wartburg.edu; URL: www.wartburg.edu/library/
1852; Eileen Myers
College Arch, Koob Coll, Namibia Coll, Archs of Iowa Broadcasting Hist
184 000 vols; 801 curr per; 11 000 e-books; 4 202 Electronic Media & Resoruces
libr loan 33004

Southwestern Assemblies of God University, P.C. Nelson Memorial Library, 1200 Sycamore, ***Waxahachie***, TX 75165-2342
T: +1 972 8254761; Fax: +1 972 9230488; E-mail: library@sagu.edu; URL: www.sagu.edu/library
1927; Eugene Holder
Religious studies, Education; Charismatic Authors, Hist & Mat (Pentecostal Alcove)
100 000 vols; 609 curr per; 72 mss; 15 diss/theses; 133 maps; 84 000 microforms; 2 838 sound-rec; 20 digital data carriers; 9 959 pamphlets, 1 608 slides, 245 overhead transparencies
CLA 33005

Wayne State College, Conn Library, 1111 Main St, ***Wayne***, NE 68787
T: +1 402 3757258; Fax: +1 402 3757538; E-mail: library@wsc.edu; URL: academic.wsc.edu/conn_library
1892; David Graber
Instructional Resources, Juvenile & Young Adults, Kessler Art Coll, Val Peterson Arch
886 000 vols; 2 180 curr per; 15 300 e-journals; 65 000 govt docs; 600 000 microforms; 6 700 av-mat; 1 900 sound-rec; 827 digital data carriers; 380 Electronic Media & Resources, 4 500 Music Scores, 5 110 High Interest/Low Vocabulary Bks 33006

William Paterson University of New Jersey, David & Lorraine Cheng Library, 300 Pompton Rd, ***Wayne***, NJ 07470
T: +1 973 7203180; Fax: +1 973 7203171; E-mail: refdesk@wpunj.edu; URL: www.wpunj.edu/library
1924; Anne Ciliberti
Business, Education, Law, Psychology; Professional Papers of William Paterson (1745-1806), First & Limited Editions of 19th & 20th c American & British Authors, New Jerseyiana
382 000 vols; 1 400 curr per; 4 000 e-books; 5 500 av-mat; 2 800 sound-rec; 2 000 Music Scores
ALA, NJLA, ACRL 33007

Waynesburg College, Eberly Library, 93 Locust Ave, ***Waynesburg***, PA 15370-1242
T: +1 724 8523419; Fax: +1 724 6274188; E-mail: library@waynesburg.edu; URL: www.waynesburg.edu/depts/eberly
1849; Rea Andrew Redd
Business, Education, Sports medicine, Nursing; Western Pennsylvania Hist (Trans-Appalachian Coll)
87 000 vols; 395 curr per; 582 av-mat
libr loan 33008

Southwestern Oklahoma State University, Al Harris Library, 100 Campus Dr, ***Weatherford***, OK 73096-3002
T: +1 580 7743130; Fax: +1 580 7743112; URL: www.swosu.edu/library/
1902; Dr. Jonathan Sparks
Education, Pharmacy; SWOSU Univ Publ
230 000 vols; 300 curr per; 62 000 e-books; 42 865 govt docs; 194 maps; 1 207 000 microforms; 886 digital data carriers
libr loan; ALA 33009

Wellesley College, Margaret Clapp Library, 106 Central St, ***Wellesley***, MA 02481-8275
T: +1 781 2832096; Fax: +1 781 2833690; URL: www.wellesley.edu/library
1875; Micheline Jedry
First & Rare Editions of English & American Poetry, Italian Renaissance (Plimpton Coll), Ruskin, Book Arts, Slavery (Elbert Coll)
100 000 vols; 2 305 curr per
libr loan 33010

– Art Library, Jewett Arts Ctr, 106 Central St, Wellesley, MA 02481
T: +1 781 2832944; E-mail: jhablani@wellesley.edu; URL: web.wellesley.edu/web/Dept/LT/Collections/artlib.psml
1875; Brooke Henderson
57 259 vols; 172 curr per 33011

– Music Library, Jewett Arts Center, Wellesley, MA 02481
T: +1 781 2832075; Fax: +1 781 2833687; E-mail: pbristah@wellesley.edu; URL: www.wellesley.edu/library/Music/musiclib.html
1904; Pamela Bristah
36 000 vols; 100 curr per; 14 490 sound-rec
libr loan 33012

– Science Library, Science Ctr, 106 Central St, Wellesley, MA 02481
T: +1 781 2833085; Fax: +1 781 2833642
1976
105 000 vols; 625 curr per 33013

Gordon College, Jenks Library, 255 Grapevine Rd, ***Wenham***, MA 01984-1899
T: +1 978 8674339; Fax: +1 978 8674660; E-mail: library@gordon.edu; URL: www.gordon.edu/library/
1889; Dr. Myron Schirer-Suter
Edward Payson Vining Coll
150 000 vols; 327 curr per; 41 000 e-journals; 800 e-books; 19 000 microforms; 1 560 sound-rec; 72 digital data carriers
libr loan 33014

West Chester University, Francis Harvey Green Library, 25 W Rosedale Ave, ***West Chester***, PA 19383
T: +1 610 4362747; Fax: +1 610 7380554; E-mail: refdesk@wcupa.edu; URL: www.wcupa.edu/library.fhg
1871; Richard H. Swain
William Darlington Libr (Rare Scientific & Botanical Mat), Chester County Cabinet Libr, Philips Autograph Libr, Shakespeare Folios, Ehinger Libr (Hist Mat on Physical Education), Weintraub Mss & Res Coll, Chester County Coll, Holocaust Studies
784 000 vols; 8 850 curr per; 1 530 e-journals; 60 000 e-books; 2 650 Music Scores
libr loan 33015

Fordham University Library at Marymount, Gloria Gaines Memorial Library, 400 Westchester Ave, ***West Harrison***, NY 10604
T: +1 914 3673061; URL: www.library.fordham.edu
1920; Diane Deery
Catholic News, 1888-1965 (micro); Mussolini Papers (micro); Thomas More; Washington Irving; Women
23 000 vols; 2 incunabula; 14 000 microforms; 123 sound-rec; 100 digital data carriers
libr loan; ALA 33016

Saint Joseph College, Pope Pius XII Library, 1678 Asylum Ave, ***West Hartford***, CT 06117-2791
T: +1 860 2315208; Fax: +1 860 5234356; URL: www.sjc.edu/library
1932; Linda Geffner
Catholicism, Counseling, Nursing, Child studies, Gerontology
133 500 vols; 624 curr per; 1 724 sound-rec; 31 digital data carriers; 7 000 slides
libr loan 33017

University of Connecticut – Greater Hartford Campus – School of Social Work, Harleigh B. Trecker Library, 1800 Asylum Ave, ***West Hartford***, CT 06117
T: +1 860 5709024; Fax: +1 860 5709027; E-mail: treckref@uconn.edu; URL: www.lib.uconn.edu
1985; William Uricchio
Social sciences, Business & management; Social Work Theses
100 000 vols; 200 curr per; 4 000 microforms; 14 digital data carriers; 70 drawers of annual rpts & company hist
libr loan 33018

University of Hartford Libraries and Learning Resources, W.H. Mortensen Library, 200 Bloomfield Ave, ***West Hartford***, CT 06117
T: +1 860 7684142; Fax: +1 860 7684274; URL: library.hartford.edu
1957; Randi Lynn Ashton-Pritting
Black Lit Coll, Millie & Irving Bercowetz Family Coll
450 000 vols; 2 089 curr per; 2 000 av-mat
libr loan 33019

– University of Hartford, Allen Library / Mortensen Library, 200 Bloomfield Ave, West Hartford, CT 06117-0395
T: +1 860 7684491; Fax: +1 860 7685295; URL: library.hartford.edu/allen.allenhome.html
1938; Linda Solow Blotner
Kalmen Opperman Clarinet Coll; Stuart Smith Coll, mss, writings & published works
12 000 vols; 211 curr per; 271 diss/theses; 35 241 music scores; 65 microforms; 18 000 sound-rec; 18 000 mixed media
libr loan; MLA, IAML 33020

University of New Haven, Marvin K. Peterson Library, 300 Boston Post Rd, ***West Haven***, CT 06516
T: +1 203 9327190; Fax: +1 203 9321469; URL: library.newhaven.edu
1920; Hanko H. Dobi
Forensic Science, Criminal Justice, Engineering
300 000 vols; 1 000 curr per; 14 796 e-journals; 455 diss/theses; 162 863 govt docs; 413 maps; 551 000 microforms; 1 325 sound-rec; 279 digital data carriers; 4451 pamphlets; 96 pieces of art
libr loan; NELINET, OCLC 33021

Purdue University, University Libraries, 504 W State St, ***West Lafayette***, IN 47907-2058
T: +1 765 4942900; Fax: +1 765 4940156; URL: www.lib.purdue.edu
1874; Dr. James L. Mullins
Papers of aviator Amelia Earhart, playwright George Ade, cartoonist John T. McCutcheon, Nobel-prize winning chemist Herbert C. Brown, author Charles Major, designer of the Golden Gate Bridge Charles Ellis, and time and motion study pioneers Frank and Lillian Gilbreth, Krannert Special Coll on the hist of economics and economic thought, the W.F.M. Goss Library of engineering hist, and the Anna Embree Baker Coll of bks designed and printed by renowned typographer and book designer Bruce Rogers, the university archives, which document the origin and hist of Purdue University
2 505 000 vols; 40 073 curr per; 32 000 e-journals; 3 100 000 microforms
libr loan 33022

– Earth & Atmospheric Sciences Library, 2215 Civil Engineering Bldg, West Lafayette, IN 47907-1210
T: +1 765 4943264; Fax: +1 765 4961210; E-mail: easlib@purdue.edu; URL: www.lib.purdue.edu/eas/
1970; Michael Fosmire
29 822 vols; 251 curr per; 170 000 maps; 30 686 microforms; 10 000 aerial photos 33023

– Humanities, Social Science, and Education Library, Stewart Ctr, Rm 150, West Lafayette, IN 47907-1530
T: +1 765 4942831; Fax: +1 765 4949007; E-mail: hsselib@purdue.edu; URL: www.lib.purdue.edu/hsse/
1874; Judith M. Nixon
788 987 vols; 3 251 curr per; 248 553 microforms
libr loan
33024

– Life Science Library, Lilly Hall, Rm 2-400, West Lafayette, IN 47907-1323
T: +1 765 4942910; E-mail: lifelib@purdue.edu; URL: www.lib.purdue.edu/life/
1959; Vicki Killion
87 085 vols; 1 211 curr per; 76 914 microforms
33025

– Management and Economics Library, Krannert Bldg, 2nd & 3rd Flrs, West Lafayette, IN 47907-1340
T: +1 765 4942920; Fax: +1 765 4949658; E-mail: kranlib@purdue.edu; URL: www.lib.purdue.edu/mel/
1959; Tomalee Doan
Estey Coll (Business Cycles); Rare Bks in Economics & Business Hist, 16th-19th c
144 641 vols; 1 416 curr per; 103 304 microforms; 5 067 bound annual rpts
33026

– Mathematical Sciences Library, Mathematical Sciences Bldg, Rm 311, West Lafayette, IN 47907-1385
T: +1 765 4942855; E-mail: mathlib@purdue.edu; URL: www.lib.purdue.edu/math/
1910; Michael Fosmire
59 199 vols; 547 curr per; 1 291 microforms; 1 200 tech rpts
33027

– M. G. Mellon Library of Chemistry, Wetherhill Lab of Chemistry Bldg, Rm 301, West Lafayette, IN 47907-1333
T: +1 765 4942862; Fax: +1 765 4940239; E-mail: chemlib@purdue.edu; URL: www.lib.purdue.edu/chem/
1874; F. Bartow Culp
Arch of Herbert C Brown
53 526 vols; 308 curr per; 13 220 microforms
33028

– Physics Library, Physics Bldg, Rm 290, West Lafayette, IN 47907-1321
T: +1 765 4942858; Fax: +1 765 4940706; E-mail: physlib@purdue.edu; URL: www.lib.purdue.edu/phys/
1905; Michael Fosmire
57 237 vols; 273 curr per
33029

– Siegesmund Engineering Library, Potter Engineering Ctr, Rm 160, West Lafayette, IN 47907-1250
T: +1 765 4942873; Fax: +1 765 4940156; E-mail: enginlib@purdue.edu
1977; Michael Fosmire
Goss Hist of Engineering Libr
374 527 vols; 1 764 curr per; 930 648 microforms
33030

– Veterinary Medical Library, Lynn Hall, Rm 1133, West Lafayette, IN 47907-1537
T: +1 765 4942853; Fax: +1 765 4940781; E-mail: vetmlib@purdue.edu; URL: www.lib.purdue.edu/vetmed/
1960; Gretchen Stephens
35 522 vols; 544 curr per
33031

West Liberty State College, Paul N. Elbin Library, CSC No 135, **West Liberty**, WV 26074; P.O. Box 295, West Liberty, WV 26074-0295
T: +1 304 3368261; Fax: +1 304 3368186; URL: www.westliberty.edu
1832; Cheryl Harshman
Education, Music; Nelle Krise Rare Bk Room, College Arch
195 000 vols; 125 curr per
libr loan
33032

Monmouth University, Guggenheim Memorial Library, 400 Cedar Ave, **West Long Branch**, NJ 07764
T: +1 732 5713438; Fax: +1 732 2635124; URL: library.monmouth.edu
1933; Dr. Ravindra Sharma
New Jersey Hist Coll, Lewis Mumford Coll
234 000 vols; 2 726 curr per
33033

Northwood University, Dr & Mrs Peter C. Cook Library, 2600 N Military Trail, **West Palm Beach**, FL 33409-2911
T: +1 561 4785537; Fax: +1 561 6973138; E-mail: fl.library@northwood.edu; URL: www.northwood.edu
1984; Sue Ann Berard
Business
20 000 vols; 150 curr per
libr loan
33034

Warren Library, Palm Beach Atlantic University, 300 Pembroke Pl, **West Palm Beach**, FL 33401-6503; P.O. Box 24708, West Palm Beach, FL 33416-4708
T: +1 561 8032226; Fax: +1 561 8032235; E-mail: library_reference@pba.edu; URL: www.pba.edu/library/
1968; Steven Baker
122 438 vols; 4 047 curr per; 24 873 e-journals; 82 445 e-books; 1 678 music scores; 79 097 microforms; 3 801 av-mat; 1 525 sound-rec
libr loan; Lyrasis, ACL
33035

US Military Academy, Library, Bldg 757, **West Point**, NY 10996-1799
T: +1 845 9388325; Fax: +1 845 9383752; URL: www.library.usma.edu/screens/libinfo.asp
1802; Joseph Barth
US Army Hist, West Pointiana, Hudson Highlands Hist, Military Arts & Sciences, Cadet Textbks, Chess Coll, Omar N Bradley Papers, Orientalia
457 000 vols; 900 curr per
libr loan
33036

Otterbein College, Courtright Memorial Library, 138 W Main St, mail address: One Otterbein College, **Westerville**, OH 43081
T: +1 614 8231215; Fax: +1 614 8231921; E-mail: library@otterbein.edu; URL: library.otterbein.edu
1847; Lois F. Szudy
Americana (J. Burr & Jessie M. Hughes Memorial), Classics (Marshall B. & Mary M. Fanning Fund), Ethnics & Political Science (Lewis E. Myers Memorial), Science (Elvin & Ruth Warrick Fund), Humanities (NEH Fund)
247 270 vols; 1 056 curr per
libr loan; OHIONET
33037

Westfield State College, Ely Library, 577 Western Ave, **Westfield**, MA 01085-2580; P.O. Box 1630, Westfield, MA 01086-1630
T: +1 413 5725234; Fax: +1 413 5725520; URL: www.lib.wsc.ma.edu
1839; Corinne Ebbs
Education, Criminal law & justice
159 000 vols; 638 curr per
33038

McDaniel College, Hoover Library, Two College Hill, **Westminster**, MD 21157-4390
T: +1 410 8572282; Fax: +1 410 8572748; URL: hoover.mcdaniel.edu
1867; Michele M. Reid
ERIC Coll, microfiche
275 000 vols; 1 000 curr per
libr loan
33039

Regis College, Library, 235 Wellesley St, **Weston**, MA 02493
T: +1 781 7687300; Fax: +1 781 7687323; E-mail: library@regiscollege.edu; URL: www.regiscollege.edu
1927; Lynn Triplett
Cardinal Newman Coll, Madeleine Doran Coll
135 000 vols; 607 curr per
33040

Purdue University North Central, Library, Library-Student-Faculty (LSF) Bldg, 2nd Flr, 1401 S US Hwy 421, **Westville**, IN 46391
T: +1 219 7855249; Fax: +1 219 7855501; URL: www.pnc.edu/ls
1967; K. R. Johnson
85 000 vols; 390 curr per
libr loan; ALA, INCOLSA
33041

College Church in Wheaton, Library, 332 E Seminary Ave, **Wheaton**, IL 60187
T: +1 630 6680878; Fax: +1 630 6680984; URL: www.college-church.org
Lisa Kern
Biblical studies, Missions
13 000 vols; 10 curr per
33042

Wheaton College – Billy Graham Center, Evangelism & Missions Collection, Wheaton College Archives & Special Collections, 500 E College Ave, **Wheaton**, IL 60187-5534
T: +1 630 7525705; E-mail: special.collection@wheaton.edu; URL: www.billygrahamcenter.com
1976; Ferne L. Weimer
Evangelicalism, Missions & missionaries; Billy Graham Coll, MK (Missionary Children) Coll, Joint IMC/CBMS Missionary Arch (microfiche), Council for World Mission Arch (microfiche), Moravian Missions to the Indians (microfilm), Early American Imprints, Pamphlets in American Hist, Human Relations Area Files (microfiche)
74 020 vols; 603 curr per; 1 000 diss/theses
33043

Wheaton College – Buswell Memorial Library, 510 Irving Ave, **Wheaton**, IL 60187-5593; mail address: 501 College Ave, Wheaton, IL 60187-5593
T: +1 630 7525101; Fax: +1 630 7525855; E-mail: reference@wheaton.edu; URL: library.wheaton.edu
1860; Lisa Richmond
American hist, Religion, Theology, American & English lit, Science, Education; Nutting, Mormon, Hymnals, Samuel Johnson, James Boswell, John Bunyan, Landon/Southeast Asia, Anti-Secretism, David Aikman, American Scientific Affiliation, Batson/Shakespeare, Devotional Lit
367 000 vols; 1 734 curr per; 2 000 e-books; 11 655 music scores; 472 000 microforms; 15 654 sound-rec; 17 digital data carriers
libr loan
33044

Wheaton College – Marion E. Wade Center Library, 351 E Lincoln, **Wheaton**, IL 60187-4213; mail address: 501 College Ave, Wheaton, IL 60187-5501
T: +1 630 7525908; Fax: +1 630 7525459; E-mail: wade@wheaton.edu; URL: www.wheaton.edu/learnres/wade
Dr. Christopher Mitchell
Children's literature, science fiction, fantasy, theology
13 000 vols; 42 curr per
33045

National-Louis University, University Library, 1000 Capital Dr, **Wheeling**, IL 60090-7201
T: +1 847 9475503; Fax: +1 847 4655659; E-mail: libref@nl.edu; URL: www.nl.edu/library
1920; Kathleen Walsh
E. Harrison Early Childhood Education Archs, Children's Lit, Adult Learning Mat
138 000 vols; 100 curr per; 400 e-journals; 2 000 e-books; 6 500 av-mat; 70 Electronic Media & Resources
33046

– Evanston Campus Library, 2840 Sheridan Rd, Evanston, IL 60201-1796
T: +1 847 4751100 ext 2288; Fax: +1 847 2565172; URL: nlu.nl.edu/ulibrary
1920; Jerry Dachs
Business, Management, Education, Psychology
272 920 vols; 1 266 curr per; 1 240 000 microforms
33047

Wheeling Jesuit University, Bishop Hodges Library, 316 Washington Ave, **Wheeling**, WV 26003-6295
T: +1 304 2432226; Fax: +1 304 2432466; E-mail: library@wju.edu; URL: www.wju.edu/library
1955; Kelly Mummert
132 000 vols; 382 curr per; 9 600 e-journals; 61 000 e-books
libr loan
33048

Mercy College – White Plains Campus Library, 277 Martine Ave & S Broadway, **White Plains**, NY 10601
T: +1 914 9483666; Fax: +1 914 6861858; E-mail: libwp@mercy.edu
Srivalli Rao
16 000 vols
33049

Pace University – School of Law Library, 78 N Broadway, **White Plains**, NY 10603
T: +1 914 4224120; Fax: +1 914 4224139; E-mail: reference@law.pace.edu; URL: web.pace.edu/page.cfm?doc_id=29541

1976; Marie S. Newman
David Sive Archives, Pace Law School Archives
187 069 vols; 1 656 curr per; 825 e-journals; 7 diss/theses; 62 296 microforms; 551 av-mat; 125 sound-rec; 462 digital data carriers
libr loan; AALL
33050

University of Wisconsin – Whitewater, Andersen Library, 800 W Main St, **Whitewater**, WI 53190; P.O. Box 900, Whitewater, WI 53190-0900
T: +1 262 4721032; Fax: +1 262 4725727; E-mail: library@uww.edu; URL: library.uww.edu
1868; Joyce Huang
Business & finance, Education; George A Custer (Kenneth Hammer Coll); Criminology (Steinmetz Coll); Local Hist (Area Res Ctr Coll), bks, mss & arch
672 000 vols; 4 816 curr per; 1 170 e-journals; 31 000 e-books; 170 Electronic Media & Resources
libr loan
33051

Calumet College of Saint Joseph, Speckler Library, 2400 New York Ave, **Whiting**, IN 46394
T: +1 219 4734373; Fax: +1 219 4734259; E-mail: library@ccsj.edu; URL: www.ccsj.edu/library
1963; Virginia Rodes
Theology
94 000 vols; 3 000 microforms; 46 digital data carriers
libr loan
33052

Southern California University of Health Sciences, Learning Resource Center, 16200 E Amber Valley Dr, **Whittier**, CA 90604-4098; P.O. Box 1166, Whittier, CA 90609-1166
T: +1 562 9023368; Fax: +1 562 9023323; URL: www.scuhs.edu
1911; Valerie Fernandez
Chiropractic Hist, Nutrition & Natural Therapeutics
25 000 vols; 222 curr per; 63 000 microforms; 1 679 sound-rec; 100 digital data carriers
libr loan
33053

Whittier College – Bonnie Bell Wardman Library, 7031 Founders Hill Rd, **Whittier**, CA 90608-9984
T: +1 562 9074235; Fax: +1 562 6987168; URL: web.whittier.edu/academic/library
1901; Mary Ellen Vick
Society of Friends (Clifford & Susan Johnson Libr of Quaker Lit); Colls of Jessamyn West, Richard Nixon, Jan de Hartog, John Greenleaf Whittier
302 000 vols; 715 curr per; 10 300 e-journals; 2 000 e-books; 125 mss; 2 incunabula; 653 diss/theses; 90 364 govt docs; 139 000 microforms; 100 av-mat; 200 sound-rec; 61 digital data carriers
libr loan; ALA, ACRL, Oberlin Group
33054

Friends University, Edmund Stanley Library, 2100 W University St, **Wichita**, KS 67213-3397
T: +1 316 2955880; Fax: +1 316 2955080; E-mail: askmax@friends.edu; URL: www.friends.edu/library
1898; Max M. Burson
Quaker Coll
63 000 vols; 350 curr per
33055

Newman University, Ryan Library, 3100 McCormick Ave, **Wichita**, KS 67213-2097
T: +1 316 9424291; Fax: +1 316 9421747; URL: www.newmanu.edu
1933; Joseph E. Forte
Catholicism, Allied health, Nursing, Social work; Cardinal Newman Coll, Chrysostom Coll, College Archs
126 000 vols; 345 curr per
33056

Wichita State University, Ablah Library, Ablah Library, 1845 Fairmount, **Wichita**, KS 67260-0068
T: +1 316 9783586; Fax: +1 316 9783727; URL: library.wichita.edu
1895; Dr. Pal V. Rao
Robert T. Aitchison Coll (Hist of Printing), Tinterow Coll (Hist of Hypnotism), Merrill Coll of W. L. Garrison Papers, W. H. Auden Coll, Kantor Coll of Sanitary Mat, Wichita State Univ Arch
1 711 000 vols; 13 000 curr per; 13 000

e-journals; 13 000 e-books; 34 000 music scores; 11 000 av-mat; 18 500 sound-rec; 4 680 digital data carriers 33057

Midwestern State University, George Moffett Library, 3410 Taft Ave, ***Wichita Falls***, TX 76308-2099
T: +1 940 3974204; Fax: +1 940 3974689; E-mail: ryan.samuelson@mwsu.edu; URL: library.mwsu.edu
1922; Clara M. Latham
Americana (Libr of American Civilization), English Lit, Missouri-Kansas-Texas Railroad Map Coll, Nolan A Moore III Heritage of Print Coll
480 000 vols; 1 694 curr per; 45 000 e-books
libr loan 33058

Central State University, Hallie Q. Brown Memorial Library, 1400 Brush Row Rd, ***Wilberforce***, OH 45384; P.O. Box 1006, Wilberforce, OH 45384-1006
T: +1 937 3766106; Fax: +1 937 3766132; URL: hallie.ces.edu
1948; Johnny W. Jackson
Afro-American Coll
197 000 vols; 386 curr per; 21 000 e-books 33059

Wilberforce University, Rembert Stokes Learning Resources Center Library, 1055 N Bicket Rd, ***Wilberforce***, OH 45384; P.O. Box 1003, Wilberforce, OH 45384-1003
T: +1 937 7085630; Fax: +1 937 7085771; E-mail: library@wilberforce.edu; URL: www.wilberforce.edu
1856; Mark E. Mattheis
Afro-American Hist (Arnett-Coppin & Payne), scrapbks, newspaper clippings, handbills & some correspondence; Hist of African Methodist Episcopal Church
62 000 vols; 10 curr per; 200 e-journals; 12 000 e-books; 200 av-mat 33060

Eastern Oklahoma State College, Media Center, 1301 W Main St, ***Wilburton***, OK 74578
T: +1 918 4651779; Fax: +1 918 4650112; URL: www.eosc.edu/library
1919; Mary Edith Butler
Native American Coll
41 000 vols; 200 curr per 33061

King's College, D. Leonard Corgan Library, 14 W Jackson St, ***Wilkes-Barre***, PA 18711-0850
T: +1 570 2085840; Fax: +1 570 2086022; URL: www.kings.edu
1946; Dr. Terrence Mech
George Korson Folklore Coll; Papers of Honorable Daniel J Flood, MC
178 000 vols; 490 curr per; 523 000 microforms; 1 800 av-mat; 2 978 sound-rec; 9 digital data carriers
libr loan 33062

Wilkes University, E.S. Farley Library, 187 S Franklin St, ***Wilkes-Barre***, PA 18766-0998; mail address: 84 W South St, Wilkes-Barre, PA 18766
T: +1 570 4084254; Fax: +1 570 4087823; E-mail: library@wilkes.edu; URL: www.wilkes.edu/library
1933; Robert Shaddi
Poland, Culture & Hist (Polish Rm), Northeast Pennsylvania Hist
202 000 vols; 522 curr per; 1 000 maps; 750 000 microforms; 1 000 av-mat; 2 059 sound-rec; 25 digital data carriers
libr loan; OCLC, PALINET 33063

University of Chicago, Yerkes Observatory Library, 373 W Geneva St, ***Williams Bay***, WI 53191-9603
T: +1 262 2455555; Fax: +1 262 2459805; E-mail: yerkes@lib.uchicago.edu; URL: www.lib.uchicago.edu/e/yerkes
1897; Judith A. Bausch
Astronomy, Astrophysics; Star Atlases & Charts
25 000 vols; 100 microforms 33064

College of William and Mary in Virginia – Earl Gregg Swem Library, One Landrum Dr, ***Williamsburg***, VA 23187; P.O. Box 8794, Williamsburg, VA 23187-8794
T: +1 757 2213067; Fax: +1 757 2212635; URL: www.swem.wm.edu
1693; Connie K. McCarthy
Virginian Early American Hist, Papermakers; Music, Physics, Geology, Chemistry
1 846 000 vols; 3 000 000 mss; 10 incunabula; 581 943 govt docs; 22 196 maps; 1 311 000 microforms; 20 090 sound-rec; 660 digital data carriers
libr loan 33065

College of William & Mary in Virginia – Marshall-Wythe Law Library, 613 S Henry St, ***Williamsburg***, VA 23187; P.O. Box 8795, Williamsburg, VA 23187-8795
T: +1 757 2213257; Fax: +1 757 2213051; URL: www.wm.edu/law/lawlibrary
1789; James Heller
Thomas Jefferson Law Bks
390 000 vols; 4 000 curr per; 803 000 microforms; 40 digital data carriers
libr loan 33066

Unibversity of the Cumberlands/Cumberland College, Norma Perkins Hagan Memorial Library, 821 Walnut St, ***Williamsburg***, KY 40769
T: +1 606 5394329; Fax: +1 606 5394317; URL: www.ucumberlands.edu/library
1889; Jan Wren
Religion, Education, Appalachia; Children's Lit, Kentucky, US Govt
203 000 vols; 24 683 curr per; 100 000 e-books
libr loan 33067

Lycoming College, John G. Snowden Memorial Library, 700 College Pl, ***Williamsport***, PA 17701-5192
T: +1 570 3214091; Fax: +1 570 3214090; URL: www.lycoming.edu/library
1812; Janet M. Hurlbert
Psychology, Sociology, Religious studies, Business; Central Pennsylvania Conference of the United Methodist Church Arch
186 000 vols; 1 140 curr per; 2 000 e-books 33068

Pennsylvania College of Technology, Library, 999 Hagan Way, ***Williamsport***, PA 17701; One College Ave, DIF No 69, Williamsport, PA 17701
T: +1 570 3202409; Fax: +1 570 3274503; E-mail: library@pct.edu; URL: www.pct.edu/library
1965; Lisette N. Ormsbee
Sloan Art Coll
108 000 vols; 1 147 curr per; 16 000 e-books
libr loan 33069

Williams College – Chapin Library, 26 Hopkins Hall Dr, ***Williamstown***, MA 01267-2560; P.O. Box 426, Williamstown, MA 01267-0426
T: +1 413 5972462; Fax: +1 413 5972929; E-mail: chapin.library@williams.edu; URL: www.williams.edu/resources/chapin
1923; Robert L. Volz
Aldine Press; Bibles & Liturgical Books; Sporting Books; American Cookbooks
53 000 vols; 25 curr per; 40 000 mss; 550 incunabula; 44 microforms 33070

Williams College – Sawyer Library, 55 Sawyer Library Dr, ***Williamstown***, MA 01267
T: +1 413 5972504; Fax: +1 413 5972478; URL: www.library.williams.edu
1793; David M. Pilachowski
William Cullen Bryant Coll, Shaker, Paul Whiteman Coll, Williamsiana Coll
884 000 vols; 1 304 curr per; 13 180 e-journals; 95 000 e-books 33071

Eastern Connecticut State University, J. Eugene Smith Library, 83 Windham St, ***Willimantic***, CT 06226-2295
T: +1 860 4654462; Fax: +1 860 4655522; URL: www.easternct.edu/smithlibrary
1889; Patricia S. Banach
Connecticut Hist Coll, Career Info Ctr
360 000 vols; 1 735 curr per; 5 000 e-books; 3 000 av-mat; 1 300 sound-rec; 582 digital data carriers; 400 Audio Bks, 120 Electronic Media & Resources, 590 Music Scores, 3 784 Journals, 400 Bks on Deafness & Sign Lang
libr loan 33072

University of North Carolina at Wilmington, William Madison Randall Library, 601 S College Rd, ***Wilmington***, NC 28403-5616
T: +1 910 9623272; Fax: +1 910 9623078; URL: library.uncwil.edu
1947; Sherman Hayes
Education, Hist, Marine biology; Audiovisuals
1 024 000 vols; 24 798 curr per
libr loan 33073

Widener University – School of Law Library, 4601 Concord Pike, ***Wilmington***, DE 19803; P.O. Box 7475, Wilmington, DE 19086
T: +1 302 4772244; Fax: +1 302 4772240; E-mail: law.libref@law.widener.edu; URL: law.widener.edu/lawlibrary
1973; Michael J. Slinger
Delaware law, Corporate law, Health law, Tax law
610 000 vols; 8 400 curr per; 27 digital data carriers; 244 700 vols in microform
libr loan 33074

Wilmington College, Sheppard Arthur Watson Library, Pyle Ctr 1227, 1870 Quaker Way, ***Wilmington***, OH 45177-2473
T: +1 937 3826661; Fax: +1 937 3838571; URL: www.watsonlibrary.org
1870; Jean Mulhern
Quakers-Quakerism Coll; Peace Resources Ctr-Hiroshima & Nagasaki Memorial Coll
110 000 vols; 7 000 curr per
libr loan 33075

Asbury College, Kinlaw Library, One Macklem Dr, ***Wilmore***, KY 40390-1198
T: +1 859 8583511 ext 2126; Fax: +1 859 8583921; E-mail: library@asbury.edu; URL: www.asbury.edu
1890; Douglas Butler
College Archs, Missionary Coll, College Publs
195 010 vols; 517 curr per
libr loan 33076

Barton College, Hackney Library, 400 Atlantic Christian College Dr NE, ***Wilson***, NC 27893; P.O. Box 5000, Wilson, NC 27893-7000
T: +1 252 3996502; Fax: +1 252 3996571; E-mail: reference@barton.edu; URL: library.barton.edu
1902; Rodney Lippard
Deaf education; Disciplinana Coll
192 000 vols; 13 437 curr per; 23 000 e-books 33077

Shenandoah University, Alson H. Smith Jr Library, 1460 University Dr, ***Winchester***, VA 22601
T: +1 540 6655424; Fax: +1 540 6654609; E-mail: library@su.edu; URL: www.su.edu/library
1875; Christopher A. Bean
Religion (Evangelical United Brethren Church Hist Room), Shenandoah Univ Arch, Shenandoah Valley Hist
127 000 vols; 1 060 curr per; 430 diss/theses; 15 000 music scores; 123 000 microforms; 3 700 av-mat; 16 000 sound-rec; 10 digital data carriers; 72 Electronic Media & Resources, 16 200 Music Scores
libr loan; ALA 33078

Southwestern College, Memorial Library, 100 College St, ***Winfield***, KS 67156-2498
T: +1 620 2296127; Fax: +1 620 2296382; URL: www.sckans.edu/library
1885; Veronica Mc Asey
Black Hist & Lit, Watmull Coll of Indian Studies
55 000 vols; 106 curr per 33079

Wingate University, Ethel K. Smith Library, PO Box 219, ***Wingate***, NC 28174-1202
T: +1 704 2338097; Fax: +1 704 2338254; E-mail: library_info@wingate.edu; URL: library.wingate.edu
1896; Amee Huneycutt Odom
Children's Coll, Charles A Cannon Family Papers, Wingate University Arch
107 000 vols; 159 curr per; 192 000 microforms; 200 av-mat; 5 000 sound-rec; 60 digital data carriers
libr loan 33080

St. Mary's University of Minnesota, Fitzgerald Library, 700 Terrace Heights, No 26, ***Winona***, MN 55987-1399
T: +1 507 4571489; Fax: +1 507 4571565; URL: www.smumn.edu/sitepages/pid2571.php
1925; Mary J. Moxness
194 000 vols; 588 curr per; 13 820 e-journals; 5 000 e-books; 181 maps; 35 microforms; 89 av-mat; 4 122 sound-rec; 5 digital data carriers; 84 Bks on Deafness & Sign Lang
libr loan 33081

Winona State University, Darrell W. Krueger Library, 176 W Mark St, ***Winona***, MN 55987-5838; P.O. Box 5838, Winona, MN 55987-5838
T: +1 507 4575367; Fax: +1 507 4575594; URL: www.winona.edu/library
1860; Larry Hardesty
WSU Arch; Eric Coll, micro
218 000 vols; 22 000 e-books 33082

North Carolina School of the Arts, Semans Library, 1533 S Main St, ***Winston-Salem***, NC 27127
T: +1 336 7703257; Fax: +1 336 7703271; URL: www.ncarts.edu
1965; Vicki Weavil
School Arch
110 000 vols; 450 curr per; 35 diss/theses; 44 000 music scores; 5 microforms; 4 000 av-mat; 17 000 sound-rec; 1 000 digital data carriers; 49 000 Music Scores
libr loan; SOLINET, ACRL, ALA, TLA, MLA 33083

Salem College, Dale H. Gramley Library, 626 S Church St, ***Winston-Salem***, NC 27108; P.O. Box 10548, Winston-Salem, NC 27108
T: +1 336 7212649; Fax: +1 336 9175339; URL: www.salem.edu
1772; Dr. Rose Simon
Lit, Women's hist; Moravian Church Coll, Salem Academy & College Coll
130 000 vols; 5 500 curr per
libr loan 33084

Wake Forest University, Z. Smith Reynolds Library, PO Box 7777, ***Winston-Salem***, NC 27109-7777
T: +1 336 7584931; Fax: +1 336 7588831; URL: zsr.wfu.edu
1879; Lynn Sutton
Anglo-Irish Lit, Hist Books & Printing, Holocaust Coll, North Carolina Baptist Hist, Selected English & American Authors of the 20th c, Gertrude Stein Coll, Mark Twain Coll, W.J. Cash Mss, Wayne Oates Mss, Harold Hyes Mss, Dolman Press Arch, Giuseppe De Santis Film Arch, Joseph E. Smith Music Coll, Ronald Watkins Libr & Personal Papers
1 949 000 vols; 5 532 curr per; 313 000 e-books; 1 046 000 microforms; 11 000 av-mat; 1 000 linear feet of mss, 6 940 Journals
libr loan; ALA, ACRL, ASERL 33085

– Coy C. Carpenter School of Medicine Library, Medical Center Blvd, Winston-Salem, NC 27157-1069
T: +1 336 7137100; Fax: +1 336 7162186; URL: www.wfubmc.edu/library
1941; Parks Welch
Arts in Medicine, Hist of Medicine & Neurology (Rare Book Coll), Samuel Johnson Coll
153 000 vols; 2 156 curr per; 134 e-books; 3 000 microforms; 480 digital data carriers
libr loan; MLA, AAHSLD, NCLA 33086

– Professional Center Library, Worrell Professional Ctr for Law & Management, 1834 Wake Forest Rd, Winston-Salem, NC 27106; P.O. Box 7206, Winston-Salem, NC 27109-7206
T: +1 336 7584520; Fax: +1 336 7586077; URL: www.pcl.wfu.edu
1894; Marian F. Parker
Law, Management
217 000 vols; 880 000 microforms; 1 009 sound-rec; 107 digital data carriers; 834 809 microforms of per
libr loan 33087

Winston-Salem State University, C.G. O'Kelly Library, 601 Martin Luther King Dr, ***Winston-Salem***, NC 27110; mail address: 227 O'Kelly Library, Winston-Salem, NC 27110
T: +1 336 7502454; Fax: +1 336 7502459; URL: www.wssu.edu/library
1920; Mae L. Rodney
Black Studies (Curriculum Mat Ctr)
203 000 vols; 1 697 curr per; 14 000 e-books
libr loan 33088

Rollins College, Olin Library, 1000 Holt Ave, Campus Box 2744, *Winter Park*, FL 32789-2744
T: +1 407 6462507; Fax: +1 407 6461515; URL: www.rollins.edu/olin/
1885; Jonathan Miller
Jessie B Rittenhouse Coll (Poetry), Florida Hist Coll, Hamilton Holt Coll, William Sloane Kennedy/Whitman Coll, Shiel Coll
301 000 vols; 1 239 curr per; 11 000 e-books; 2 100 av-mat; 500 sound-rec; 750 digital data carriers   33089

University of Virginia's College at Wise, John Cook Wyllie Library, One College Ave, *Wise*, VA 24293
T: +1 276 3280150; Fax: +1 276 3280105; E-mail: vlf3z@uvawise.edu; URL: lib.uvawise.edu
1954; Robin P. Benke
Southwest Virginia (Arch of Southwest Hist Society), James Taylor Adams Papers, Beaty-Flannary Papers, Bruce Crawford Papers, Trigg Floyd Papers, Emory L Hamilton Papers, John Warfield Johnston Papers, Elihu Jasper Sutherland Papers, Virginia Coal Operators
143 000 vols; 560 curr per; 9 270 e-journals; 432 e-books; 62 000 microforms; 1 500 av-mat; 490 sound-rec; 110 digital data carriers
libr loan; ALA   33090

The College of Wooster Libraries, 1140 Beall Ave, *Wooster*, OH 44691-2364
T: +1 330 2632136; Fax: +1 330 2632253; URL: www.wooster.edu/library/
1866; Mark A. Christel
American Politics (Paul O Peters Coll); Drama & Theatre (Gregg D Wolfe Memorial Libr of the Theatre); 17th c British Studies (Wallace Notestein Coll)
622 000 vols; 1 195 curr per
libr loan   33091

Ohio State University, Agricultural Technical Institute Library, Halterman Hall, 1328 Dover Rd, *Wooster*, OH 44691-4000
T: +1 330 2871294; Fax: +1 330 2627634; E-mail: atilibrary@osu.edu; URL: library.osu.edu
1972; Sharon Edwards
Beekeeping Journals, Ohio County Soil Surveys
20 000 vols; 630 curr per; 29 000 microforms; 10 VF drawers of pamphlets, 5 VF drawers of Ohio soil surveys   33092

Assumption College, Emmanuel d'Alzon Library, 500 Salisbury St, *Worcester*, MA 01609
T: +1 508 7677135; Fax: +1 508 7677374; E-mail: library@assumption.edu; URL: www.assumption.edu/dept/library
1904; Dr. Dawn Thistle
Theology, Hist, Lit, Franco-Americans
158 000 vols; 1 100 curr per; 8 000 e-books; 18 000 microforms; 9 digital data carriers
libr loan   33093

Clark University, Robert Hutchings Goddard Library, 950 Main St, *Worcester*, MA 01610-1477
T: +1 508 7937711; Fax: +1 508 7938871; URL: www.clarku.edu/research/goddard
1889; Gwen Arthur
Rare Bks (Robert H. Goddard Coll & G. Stanley Hall Papers); Map Dept, Science Dept
613 000 vols; 1 303 curr per; 90 digital data carriers
libr loan   33094

– Special Collections-Archives, Worcester, MA 01610-1477
T: +1 508 7937572; Fax: +1 508 7938881; URL: www.clarku.edu/research/goddard/
Mott Linn
10 000 vols   33095

College of the Holy Cross – Dinand Library, One College St, *Worcester*, MA 01610
T: +1 508 7932639; Fax: +1 508 7932372; URL: www.holycross.edu/libraries
1843; James E. Hogan
Jesuitiana; Americana; Guiney Coll (Lit);

Walsh Coll; Curley Coll
615 000 vols; 4 818 curr per; 8 000 e-books; 18 000 microforms; 2 069 sound-rec; 11 digital data carriers   33096

College of the Holy Cross – O'Callahan Science Library, Swords Bldg, *Worcester*, MA 01610
T: +1 508 7932739
1958; Barbara Merolli
99 000 vols; 2 073 curr per; 40 000 monographs
libr loan; OCLC, ALA, ACRL, SLA   33097

Worcester Polytechnic Institute, George C. Gordon Library, 100 Institute Rd, *Worcester*, MA 01609-2280
T: +1 508 8315414; Fax: +1 508 8315829; URL: www.wpi.edu/academics/library
1867; Matthew Hall
Environmental studies, Safety; Charles Dickens Coll, WPI Arch, Hist of Engineering
270 000 vols; 900 curr per; 38 000 e-books; 2 099 maps; 800 000 microforms; 4 800 av-mat; 4 201 sound-rec; 14 digital data carriers; 17 166 tech rpts, 8 VF drawers, motion pictures, slides, 997 art reproductions
libr loan   33098

Worcester State College, Learning Resources Center, 486 Chandler St, *Worcester*, MA 01602-2597
T: +1 508 9298027; Fax: +1 508 9298198; E-mail: library@worcester.edu; URL: www.worcester.edu/library
1874; Dr. Donald Hochstetler
Education, Allied health; Education Resources Coll, Children's Coll
190 000 vols; 1 000 curr per; 16 000 e-books; 54 000 microforms; 3 950 sound-rec; 6 digital data carriers   33099

Chesapeake College, Learning Resource Center, PO Box 8, *Wye Mills*, MD 21679
T: +1 410 8275860; Fax: +1 410 8275257; E-mail: lrcdesk@chesapeake.edu; URL: www.chesapeake.edu/library
1967; Pat Cheek
Eastern Shore Lit (Chesapeake Rm)
45 000 vols; 232 curr per   33100

Reconstructionist Rabbinical College, Mordecai M. Kaplan Library, 1299 Church Rd, *Wyncote*, PA 19095
T: +1 215 5760800; Fax: +1 215 5766143; E-mail: kaplanlibrary@rrc.edu; URL: www.rrc.edu
1968; Deborah Stern
Judaica (Mordecai M Kaplan Coll), Creative Liturgy File
47 000 vols; 120 curr per   33101

Mount Marty College, Library, 1105 W Eighth St, *Yankton*, SD 57078-3724
T: +1 605 6681555; Fax: +1 605 6681357; URL: www.mtmc.edu
1936; Sandra Brown
Religious studies
78 000 vols; 327 curr per   33102

Antioch College, Olive Kettering Library, 795 Livermore St, *Yellow Springs*, OH 45387-1694
T: +1 937 7691240; Fax: +1 937 7691239; URL: www.antioch-college.edu/library
1852; Richard Kerns
Arthur E. Morgan Papers, Robert L. Straker Coll of Horace Mann Papers, Herbert Gardner Papers
295 000 vols; 403 curr per; 12 318 e-journals; 18 000 e-books; 1 incunabula; 1 300 diss/theses; 49 000 microforms; 4 773 sound-rec; 34 digital data carriers; 20 Bks on Deafness & Sign Lang
libr loan   33103

Pennsylvania State University, York Commonwealth College, Library, 1031 Edgecomb Ave, *York*, PA 17403-3398
T: +1 717 7714020; Fax: +1 717 7714022; URL: www.libraries.psu.edu/york
David Van de Streek
60 000 vols; 175 curr per   33104

York College, Levitt Library, 1125 E Eighth, *York*, NE 68467-2699
T: +1 402 3635704; Fax: +1 402 3635685; URL: www.york.edu/library/
1956; Ken Gunselman
Church Hist (Restoration Movement); Missions; Yorkana
131 000 vols; 292 curr per; 24 000 e-books; 20 000 microforms; 2 163 sound-rec; 20 digital data carriers; videos, art reproductions
libr loan   33105

York College of Pennsylvania, Schmidt Library, 441 Country Club Rd, *York*, PA 17405-7199
T: +1 717 8151345; Fax: +1 717 8491608; E-mail: library@ycp.edu; URL: www.ycp.edu/library
1787; Susan Campbell
Business, Management, Nursing
186 000 vols; 15 000 curr per
libr loan   33106

Mercy College – Yorktown Campus, 2651 Strang Blvd, *Yorktown Heights*, NY 10598
T: +1 914 2456100; Fax: +1 914 9621042; E-mail: libyktn@mercy.edu
Agnes Cameron
42 000 vols   33107

Youngstown State University, William F. Maag Library, One University Plaza, *Youngstown*, OH 44555-0001
T: +1 330 9413675; Fax: +1 330 9413734; E-mail: library@cc.ysu.edu; URL: www.maag.ysu.edu
1931; Paul Kobulnicky
Early Americana
793 000 vols; 1 344 curr per   33108

Eastern Michigan University, Bruce T. Halle Library, 955 W Circle Dr Library, Rm 200, *Ypsilanti*, MI 48197
T: +1 734 4870020; Fax: +1 734 4841151; URL: www.emich.edu/halle
1849; Rachel Cheng
Education, African-Am
632 000 vols; 12 700 curr per; 26 000 e-books
libr loan   33109

Ohio University – Zanesville / Zane State College, Zanesville Campus Library, 1425 Newark Rd, *Zanesville*, OH 43701
T: +1 740 5881404; Fax: +1 740 4530706; E-mail: fair@ohio.edu; URL: www.zanesville.ohiou.edu/zcl
Shana Fair
Muskingum County Hist (Zanesville Heritage Coll)
62 000 vols; 18 000 curr per   33110

School Libraries

Grays Harbor College, John Spellman Library, 1620 Edward P Smith Dr, *Aberdeen*, WA 98520-7599
T: +1 360 5384050; Fax: +1 360 5384294; E-mail: lib_ref@ghc.ctc.edu; URL: ghc.ctc.edu/library/home/index.htm
1930; Stanley Horton
Careers; Pacific Northwest, Small Business, Water/Fisheries
38 560 vols; 241 curr per   33111

Presentation College, Library, 1500 N Main, *Aberdeen*, SD 57401-1299
T: +1 605 2298468; Fax: +1 605 2298430; E-mail: pclibrary@presentation.edu; URL: www.presentation.edu
1951; Arvyce Burns
Nursing, Religious studies
34 830 vols; 351 curr per; 2 830 sound-rec; 116 digital data carriers   33112

South Dakota School for the Blind & Visually Impaired, Library Media Center, 423 17th Ave SE, *Aberdeen*, SD 57401-7699
T: +1 605 6262675; Fax: +1 605 6262607; URL: www.sdsbvi.sdbor.edu
1900; Pat Geditz
Local School Arch
18 000 vols; 50 curr per   33113

Virginia Highlands Community College, Library, 140 Old Jonesboro Rd, P.O. Box 828, *Abingdon*, VA 24210
T: +1 276 7392512; Fax: +1 276 7392593; E-mail: phunter@vh.vccs.edu; URL: www.vhcc.edu/lis
1969; Dr. Patricia A. Hunter
VIVA (Virtual Libr of Virginia)
30 930 vols; 217 curr per   33114

Roanoke-Chowan Community College, Learning Resources Center, 109 Community College Rd, *Ahoskie*, NC 27910
T: +1 252 8621223; Fax: +1 252 8621358; E-mail: lrc@roanoke.cc.nc.us; URL: www.roanoke.cc.nc.us
1967; Margaret S. Lefler
Audiovisual Prof Learning Lab
31 550 vols; 172 curr per   33115

College of Alameda, Library & Learning Resources Center, 555 Atlantic Ave, *Alameda*, CA 94501
T: +1 510 7482365; Fax: +1 510 7482380; E-mail: mholland@peralta.cc.ca.us
1970; Dr. Mary K. Holland
32 000 vols; 200 curr per   33116

New Mexico School for the Visually Handicapped Library, 1900 N White Sands Blvd, *Alamogordo*, NM 88310
T: +1 505 4373505; Fax: +1 505 4394411; E-mail: kflanary@nmsvh.k12.nm.us
Judy Bates
Blindness, Professional Shelf, Visually Impaired & Multi Handicapped
16 500 vols; 100 curr per; 5 370 sound-rec; 10 digital data carriers   33117

Darton College, Harold B. Wetherbee Library, 2400 Gillionville Rd, *Albany*, GA 31707
T: +1 229 4306760; Fax: +1 229 4306794; E-mail: lowry@mail.dartnet.peachnet.edu; URL: www.darton.edu/~dclib/
1966; Kay Loway
Allied health, Nursing; American Enterprise Institute Coll
88 520 vols; 583 curr per   33118

Linn-Benton Community College, Library, 6500 SW Pacific Blvd, *Albany*, OR 97321-3799
T: +1 541 9174638; Fax: +1 541 9174659; URL: lib.linnbenton.edu/library/portal2.html
1969; Diane Watson
50 000 vols; 150 curr per
libr loan   33119

Maria College of Albany, Learning Resource Center, 700 New Scotland Ave, *Albany*, NY 12208
T: +1 518 4383111 ext 221; Fax: +1 518 4387170; E-mail: ltobin@mariacollege.edu; URL: mariacollege.edu
1958; Lisa Tobin
Allied health, Early childhood
55 000 vols; 225 curr per   33120

The Sage Colleges, Albany Campus Library, 140 New Scotland Ave, *Albany*, NY 12208
T: +1 518 2921742; Fax: +1 518 2921904; E-mail: libref@sage.edu; URL: www.sage.edu
1957; Kingsley W. Greene
Graphic arts
63 000 vols; 119 curr per; 10 000 e-books
libr loan   33121

Southside Virginia Community College, College Libraries, 109 Campus Dr, *Alberta*, VA 23821
T: +1 804 9491000; Fax: +1 804 9490013; E-mail: jack.ancell@sv.cc.va.us; URL: www.sv.vccs.edu/lrs
1970; Jack Ancell
30 000 vols; 270 curr per   33122

Southwestern Indian Polytechnic Institute Libraries, 9169 Coors Rd NW, *Albuquerque*, NM 87120
T: +1 505 3462352; Fax: +1 505 3462381; E-mail: library@sipi.bia.edu
Paula M. Smith
Indian Coll
30 000 vols; 150 curr per   33123

Central Alabama Community College, Thomas D. Russell Library, 1675 Cherokee Rd, *Alexander City*, AL 35010; P.O. Box 699, Alexander City, AL 35011
T: +1 256 2154290; Fax: +1 256 2340384
1965; Gerson Milles
30483 vols; 158 curr per; 2032 microforms; 130 av-mat; 1745 sound-rec
libr loan   33124

Alpena Community College, Stephen H. Fletcher Library, 665 Johnson St, *Alpena*, MI 49707-1495
T: +1 989 3587252; Fax: +1 989 3587556; E-mail: admisreq@alpenacc.edu; URL: www.alpena.edu
1952; Charles E. Tetzlaff
36000 vols; 210 curr per   33125

Amarillo College, Lynn Library Learning Resource Center, 2201 S Washington, P.O. Box 447, *Amarillo*, TX 79178-0001
T: +1 806 3715400; Fax: +1 806 3715470; E-mail: ruddymk@actx.edu; URL: www.actx.edu/library/?&MMN_position=45:45
1929; M. Karen Ruddy
84480 vols; 380 curr per
libr loan   33126

Des Moines Area Community College, Ankeny Campus Library, 2006 S Ankeny Blvd, *Ankeny*, IA 50021
T: +1 515 9646317; Fax: +1 515 9657126; URL: www.dmacc.cc.ia.us
Lisa Stock
37000 vols; 600 curr per   33127

Washtenaw Community College, Richard W. Bailey Library, 4800 E Huron River Dr, *Ann Arbor*, MI 48105-4800
T: +1 734 9733379; Fax: +1 734 6772220; URL: www4.wccnet.edu/resources/library/
1966; Victor Liu
College Arch; Prof Coll, bks, rpts on higher educ
68000 vols; 400 curr per; 16000 e-books   33128

Northern Virginia Community College Libraries, 8333 Little River Turnpike, *Annandale*, VA 22003
T: +1 703 3233096; Fax: +1 703 3233831; E-mail: sbeeson@nvcc.edu; URL: www.nvcc.edu/library
1965; Sandra J. Beeson
Allied Health Mat, Judith DiStephano Women's Hist Coll, College Arch, Northern Virginia Historical Society Arch
255445 vols; 1164 curr per; 7100 e-journals; 1560 e-books; 28500 av-mat; 3690 sound-rec; 104 digital data carriers   33129

Fox Valley Technical College, William M. Sirek Educational Resource Center, 1825 N Bluemound Dr, P.O. Box 2277, *Appleton*, WI 54913-2277
T: +1 920 7355600; Fax: +1 920 7354870; E-mail: parson@foxvalleytec.com; URL: foxvalley.tec.wi.us/library
1967; Karen Parson
52320 vols; 303 curr per   33130

Cabrillo College, Robert E. Swenson Library / Learning Resources Center, 6500 Soquel Dr, *Aptos*, CA 95003-3198
T: +1 831 479-6537; Fax: +1 831 4796500; E-mail: geromero@cabrillo.edu; URL: libwww.cabrillo.edu/
1959; Georg L. Romero
63530 vols; 400 curr per   33131

Anne Arundel Community College, Andrew G. Truxal Library, 101 College Pkwy, *Arnold*, MD 21012-1895
T: +1 410 7772211; Fax: +1 410 7772652; E-mail: library@aacc.edu; URL: www.aacc.edu/library
1961; Cynthia Steinhoff
140000 vols; 375 curr per; 8000 av-mat   33132

Randolph Community College, Learning Resources Center, 629 Industrial Park Ave, P.O. Box 1009, *Asheboro*, NC 27204-1009
T: +1 336 6330204; Fax: +1 336 6294695; E-mail: dsluck@randolph.cc.nc.us; URL: library.randolph.edu/
1963; Deborah Luck
32000 vols; 240 curr per   33133

Asheville-Buncombe Technical Community College, Holly Library, 340 Victoria Rd, *Asheville*, NC 28801
T: +1 828 2541921 ext 300; Fax: +1 828 2516074; E-mail: admissions@abtech.edu; URL: www1.abtech.edu/holly-library
1959; Shirley McLaughlin
40000 vols; 210 curr per; 2255 microforms; 4 digital data carriers   33134

Ashland Community & Technical College, Mansbach Memorial Library, 1400 College Dr, *Ashland*, KY 41101
T: +1 606 3262169; Fax: +1 606 3262186; E-mail: AS_Reference@kctcs.edu; URL: www.ashland.kctcs.edu/library/
1938; Matthew Onion
Jesse Stuart, Kentucky Authors
45660 vols; 353 curr per   33135

Clatsop Community College, Library, 1680 Lexington, *Astoria*, OR 97103
T: +1 503 3382462; Fax: +1 503 3382387; URL: www.clatsopcollege.com/
1962; Sara Campbell
Marine tech, Local & regional hist
42307 vols; 670 curr per   33136

Athens Technical College Library, 700 Hwy 29 N, *Athens*, GA 30601-1500
T: +1 706 3555020; Fax: +1 706 3555162; E-mail: alibrary@damin1.athens.tec.ga.us; URL: www.athenstech.edu/AcademicAffairs/LibraryServices/
1984; Metta L. Nicewarner
Tech education, Allied health
49120 vols; 453 curr per   33137

Atlanta Metropolitan College, Library, 1630 Metropolitan Pkwy, *Atlanta*, GA 30310
T: +1 404 7564010; Fax: +1 404 7565613; E-mail: library@atlm.edu; URL: www.atlm.edu/current_students/library.html
1974; Wanda L. Crenshaw
40000 vols; 320 curr per   33138

Cayuga Community College, Norman F. Bourke Memorial Library/LRC, 197 Franklin St, *Auburn*, NY 13021
T: +1 315 2949019; Fax: +1 315 5925055; E-mail: cay_ref@cayuga-cc.edu; URL: www.cayuga-cc.edu/library
1953; Douglas O. Michael
Criminal justice, Local hist, Telecommunications; Auburn Imprints
79790 vols; 502 curr per; 12495 microforms; 12780 av-mat; 1272 sound-rec; 12 digital data carriers; 2464 other items
libr loan   33139

Green River Community College, Holman Library, 12401 SE 320th St, *Auburn*, WA 98092-3699
T: +1 253 8339111; Fax: +1 253 2883436; URL: www.greenriver.edu/library
1965; Kimberly Nakano
40000 vols; 300 curr per   33140

Oakland Community College – Auburn Hills Campus Library, 2900 Featherstone Rd, *Auburn Hills*, MI 48350
T: +1 248 2324125; Fax: +1 248 2324135; E-mail: masheble@occ.cc.mi.us; URL: www.occ.cc.mi.us/library
1965; Mary Ann Sheble
47750 vols; 242 curr per
libr loan   33141

Marmion Academy Library, 1000 Butterfield Rd, *Aurora*, IL 60504-9742
T: +1 630 8976936; Fax: +1 630 8977086
Mario Pedi
10320 vols   33142

Austin Community College, Library Services, 1212 Rio Grande, *Austin*, TX 78701
T: +1 512 2233066, 2233085; Fax: +1 512 2233431; E-mail: library@austincc.edu; URL: library.austincc.edu
1973; Julie Todaro
Health sciences; Multicultural Coll
120665 vols; 1387 curr per; 24900 e-books; 15650 av-mat
libr loan   33143

Texas School for the Blind, Learning Resource Center Library, 1100 W 45th St, *Austin*, TX 78756
T: +1 512 4548631; Fax: +1 512 2069450; URL: www.tsbvi.edu
Diane Nousanen
Special education for visually handicapped
11000 vols; 50 curr per   33144

South Florida Community College, Library, 600 W College Dr, *Avon Park*, FL 33825-9399
T: +1 863 7847306; Fax: +1 863 4526042; E-mail: lrc@sfcc.fl.us; URL: www.sfcc.cc.fl.us
1966; Lena D. Phelps-Ellerker
42626 vols; 96 curr per; 428 microforms; 1770 av-mat; 382 sound-rec; 5 digital data carriers
libr loan   33145

Bainbridge College, Library, 2500 E Shotwell St, P.O. Box 953, *Bainbridge*, GA 01717-0953
T: +1 229 248-2590; Fax: +1 229 248-2589; E-mail: sralph@bainbridge.edu; URL: www.bainbridge.edu/bclib/sta_inf.htm
1973; Thomas J. Frieling
Apollo Lunar Surface EVA Video Coll
35190 vols; 428 curr per   33146

Bakersfield College, Grace Van Dyke Bird Library, 1801 Panorama Dr, *Bakersfield*, CA 93305-1298
T: +1 661 3954461; Fax: +1 661 3954397; E-mail: aagenjo@bc.cc.ca.us; URL: www.bc.cc.ca.us/library
1913; Anna Agenjo
California; Bell British Plays, 1776-1795; Grove Plays of the Bohemian Club of San Francisco, 1911-1958
57000 vols; 319 curr per   33147

Baltimore City Community College, Bard Library, 2901 Liberty Heights Ave, *Baltimore*, MD 21215
T: +1 410 4628245; Fax: +1 410 4628233; E-mail: Libraryhelp@bccc.edu; URL: www.bccc.edu
1947; Stephanie Reidy
Multicultural Coll, Beacon Grant
72000 vols; 1500 curr per; 14600 microforms; 4000 sound-rec; 8 digital data carriers; 3100 slides   33148

Baltimore International College, George A. Piendak Library, 17 Commerce St, *Baltimore*, MD 21202-3230
T: +1 410 7524710; Fax: +1 410 7526720
1987; Gwendolyn Baker
17000 vols; 150 curr per; 900 av-mat
libr loan   33149

Community College of Baltimore County, Dundalk Campus Library, 7200 Sollers Point Rd, *Baltimore*, MD 21222
T: +1 410 2859640; Fax: +1 410 2859559; E-mail: mlandry@ccbc.cc.md.us
Mary Landry
Counseling, forestry, photography
32000 vols; 140 curr per   33150

Community College of Baltimore County – Essex Campus, James A. Newpher Library, 7201 Rossville Blvd, *Baltimore*, MD 21237-3899
T: +1 410 7806426; Fax: +1 410 3912642; URL: www.ccbcmd.edu
1957; Taylor Ruhl
Allied health, Am & English lit
95528 vols; 455 curr per   33151

Lees-McRae College, James H. Carson Library, 191 Main St W, *Banner Elk*, NC 28604-9238; P.O. Box 128, Banner Elk, NC 28604-0128
T: +1 828 8988770; Fax: +1 828 8988710; URL: www.lmc.edu/sites/library
1900; Russell Taylor
Southern Appalachian Region (A. B. Stirling Coll)
98000 vols; 343 curr per   33152

Gordon College, Library, 419 College Dr, *Barnesville*, GA 30204
T: +1 770 3585078; Fax: +1 770 3585240; E-mail: librarysupport@gdn.edu; URL: www.gdn.edu/library
1939; Nancy D. Anderson
Georgia
95501 vols; 216 curr per; 8673 microforms; 5557 av-mat; 27417 digital data carriers   33153

Barstow College, Thomas Kimball Library, 2700 Barstow Rd, *Barstow*, CA 92311
T: +1 760 2522411 ext 7270; Fax: +1 760 2526725; E-mail: library@gw.barstow.cc.ca.us
1960; Joseph A. Clark
41000 vols; 250 curr per   33154

Genesee Community College, Alfred C. O'Connell Library, One College Rd, *Batavia*, NY 14020-9704
T: +1 585 3456834; Fax: +1 585 3430433; E-mail: NTWarren@genesee.edu; URL: www.genesee.edu/library
1966; Nina Warren
71630 vols; 332 curr per   33155

Kellogg Community College, Emory W. Morris Learning Resource Center, 450 North Ave, *Battle Creek*, MI 49017-3397
T: +1 616 9654122; Fax: +1 616 9654133; E-mail: stilwelm@kellogg.cc.mi.us; URL: www.kellogg.cc.mi.us
1956; Martha Johnson Stilwell
Law Ref Coll
51500 vols; 273 curr per   33156

Faulkner State Community College, Austin R. Meadows Library, 1900 Hwy 31 S, *Bay Minette*, AL 36507
T: +1 251 5802145; Fax: +1 251 9375140; E-mail: relmore@faulkner.cc.al.us
1965; Rheena Elmore
57000 vols; 200 curr per   33157

Queensborough Community College, Kurt R. Schmeller Library, 222-05 56th Ave, *Bayside*, NY 11364-1497
T: +1 718 6316226; Fax: +1 718 2815012; E-mail: kkim@qcc.cuny.edu; URL: www.qcc.cuny.edu/library/
1960; Devin Feldman
152800 vols; 548 curr per; 10000 microforms; 9830 av-mat; 3624 sound-rec; 2 digital data carriers; 3592 Slides
libr loan   33158

Lee College, Erma Wood Carlson Learning Resources Center, 511 S Whiting St, P.O. Box 818, *Baytown*, TX 77522
T: +1 281 4256380; Fax: +1 281 4256557; URL: www.lee.edu/library
1935; Paul Arrigo
Texas gulf coast; Law Libr Coll, Lee College Arch
95000 vols; 600 curr per; 1198 maps; 25559 microforms; 3624 sound-rec; 15 digital data carriers; video tapes 539, slides 336, overhead transparencies 20   33159

Middlesex Community College, Academic Resources Division, Springs Rd, Bldg 1-ARC, mail address: 591 Springs Rd, *Bedford*, MA 01730
T: +1 978 6563370; E-mail: middlesex@middlesex.mass.edu; URL: www.middlesex.mass.edu
1970; Mary Ann Niles
52960 vols; 538 curr per
libr loan   33160

Grady C. Hogue Learning Resource Center Library, 3800 Charco Rd, *Beeville*, TX 78102-9985
T: +1 361 3542740; Fax: +1 361 3542719; E-mail: sarahm@cbc.cc.tx.us
1967; Sarah Milnarich
Texana Coll
40000 vols; 237 curr per   33161

Harford Community College, Library, 401 Thomas Run Rd, *Bel Air*, MD 21015-1698
T: +1 410 8364316; Fax: +1 410 8364198; URL: www.harford.edu/Library/About/Library.asp?FA=Library
1957; Geraldine Yeager
Art; Maryland Hist, Maryland Constitutional Convention File, Rosenburg Rpt
63000 vols; 428 curr per   33162

Southwestern Illinois College, Library, 2500 Carlyle Ave, *Belleville*, IL 62221
T: +1 618 2352700 ext 5597; E-mail: Jennifer.Bone@swic.edu; URL: www.southwestern.cc.il.us/library/
1946
62920 vols; 380 curr per   33163

Bellevue Community College, Library Media Center, 3000 Landerholm Circle SE, *Bellevue*, WA 98007-6484
T: +1 425 5642252; Fax: +1 425 5646186; E-mail: zzreference@bcc.ctc.edu; URL: bellevuecollege.edu/lmc/info.html
1966; Myra Van Vactor
45 000 vols; 200 curr per; 300 microforms; 1 300 av-mat; 42 digital data carriers
libr loan; ALA   33164

Central Oregon Community College, Barber Library, 2600 NW College Way, *Bend*, OR 97701-5998
T: +1 541 383-7560; Fax: +1 541 383-7507; E-mail: dbilyeu@cocc.edu; URL: campuslibrary.cocc.edu/
1950; David Bilyeu
73 000 vols; 364 curr per; 3 535 maps; 1 560 av-mat; 797 sound-rec; 191 digital data carriers
libr loan   33165

Southwestern College of Christian Ministries, Library, C H Springer Bldg, 7210 NW 39th Expressway, *Bethany*, OK 73008; P.O. Box 340, Bethany, OK 73008-0340
T: +1 405 7897661; Fax: +1 405 4950078; E-mail: scu.library@swcu.edu; URL: www.freewebs.com/sculib; www.swcu.edu
1946; Marilyn A. Hudson
Pentecostal Resource Ctr
30 000 vols; 54 curr per; 12 av-mat
                                       33166

Northampton Community College, Paul & Harriett Mack Library, 3835 Green Pond Rd, *Bethlehem*, PA 18020-7599
T: +1 610 8615360; Fax: +1 610 8615373; E-mail: referencedesk@northampton.edu; URL: www.northampton.edu/library
1967
70 270 vols; 376 curr per; 2 500 av-mat; 3 500 sound-rec; 6 digital data carriers
libr loan   33167

Endicott College, Diane M. Halle Library, 376 Hale St, *Beverly*, MA 01915
T: +1 978 2322268; Fax: +1 978 2322700; E-mail: end@noblenet.org; URL: www.endicott.edu
1939; Brian Courtemanche
118 000 vols; 95 curr per; 41 270
Electronic Media & Resources   33168

North Shore Community College, Learning Resource Center, 112 Sohier Rd, *Beverly*, MA 01915-5534; P.O. Box 3340, Danvers, MA 01923-3340
T: +1 978 7624000 ext 5526; Fax: +1 978 9225165; URL: www.northshore.edu
1965; Anne Tullson-Johnsen
65 180 vols; 423 curr per
libr loan   33169

Howard College, Big Spring Campus Library, 1001 Birdwell Lane, *Big Spring*, TX 79720
T: +1 432 2645091; E-mail: wkincade@howardcollege.edu; URL: www.howardcollege.edu
1945; William Luis Kincade
35 400 vols; 344 curr per   33170

Mountain Empire Community College, Wampler Library, PO Drawer 700, *Big Stone Gap*, VA 24219
T: +1 540 5232400 ext 304; Fax: +1 540 5238220; E-mail: jcotham@me.vccs.edu; URL: www.me.cc.va.us/melrc.htm
1972; John M. Cotham
Criminal justice, Nursing, Respiratory therapy
34 270 vols; 220 curr per   33171

Broome Community College, Cecil C. Tyrell Learning Resources Center, P.O. Box 1017, *Binghamton*, NY 13902
T: +1 607 7785020; Fax: +1 607 7785108; E-mail: petrus_r@sunybroome.edu; URL: www.sunybroome.edu
1947; Robin Petrus
Community College Education Coll
67 000 vols; 298 curr per   33172

Jefferson State Community College, James B. Allen Library, 2601 Carson Rd, *Birmingham*, AL 35215-3098
T: +1 205 8568512; Fax: +1 205 8530340; E-mail: cleckler@jeffstateonline.com; URL: www.jeffstateonline.com
1965
65 500 vols; 245 curr per   33173

Bismarck State College, Library, 1500 Edwards Ave, *Bismarck*, ND 58501-1299; P.O. Box 5587, Bismarck, ND 58506-5587
T: +1 701 2245450; Fax: +1 701 2245551; E-mail: BSC.Library.Department@bsc.nodak.edu; URL: www.bismarckstate.edu/library
1955; Marlene Anderson
70 000 vols; 415 curr per
libr loan   33174

Camden County College Library, Charles Wolverton Learning Resource Center, College Drive, P.O. Box 200, *Blackwood*, NJ 08012-0200
T: +1 856 2777200 ext 4408; Fax: +1 856 3744897; E-mail: Blaynor@camdencc.edu; URL: www.camdencc.edu/library
1967
Art, Judaica, Real estate
99 000 vols; 400 curr per   33175

Normandale Community College, Learning Resources Center, 9700 France Ave S, *Bloomington*, MN 55431
T: +1 952 4878290; Fax: +1 952 4878101; E-mail: Carol.Johnson@normandale.edu; URL: www.normandale.edu
1968; Carol Johnson
Career & Academic Planning Ctr Coll, College Success Ctr Coll, Map Libr, Minnesota Authors Coll
90 380 vols; 651 curr per   33176

Northeast State Technical Community College, Wayne G. Basler Library, 2425 Highway 75, *Blountville*, TN 37617-0246; P.O. Box 246, Blountville, TN 37617-0246
T: +1 423 3233191; Fax: +1 423 3230254; E-mail: daparsons@NortheastState.edu; URL: www.NortheastState.edu/library
1966; Duncan A. Parsons
35 000 vols; 411 curr per; 108 maps; 45 000 microforms; 7 500 av-mat; 564 sound-rec; 170 digital data carriers
libr loan; ALA   33177

Montgomery County Community College, Central Campus, Brendinger Library, 340 DeKalb Pike, *Blue Bell*, PA 19422-0796; P.O. Box 440, Blue Bell, PA 19422-0440
T: +1 215 6416594; E-mail: refdesk@mc3.edu; URL: www.mc3.edu
1966
College Arch
95 000 vols; 409 curr per   33178

Northeast Mississippi Community College, Eula Dees Memorial Library, Cunningham Blvd, *Booneville*, MS 38829
T: +1 662 7287751, 7207408; Fax: +1 662 7282428; E-mail: ckillou@necc.cc.ms.us; URL: www.necc.cc.ms.us
1948; Carol W. Killough
Mississippi Coll
45 260 vols; 283 curr per   33179

Bossier Parish Community College, Library, 2719 Airline Dr, *Bossier City*, LA 71111
T: +1 318 7469851 ext 357; Fax: +1 318 7411498; E-mail: gbryan@bpcc.cc.la.us; URL: www.bpcc.cc.la.us/boss
1968; Virginia Bryan
30 000 vols; 350 curr per   33180

Bunker Hill Community College, Library, 250 New Rutherford Ave, *Boston*, MA 02129-2925
T: +1 617 2282213; Fax: +1 617 2283288; E-mail: BHCCLibrary@bhcc.mass.edu; URL: www.noblenet.org/bhcc/
Diane Smith
Europe, Middle East, Nursing
74 000 vols; 30 000 e-journals; 22 000 e-books   33181

Roxbury Community College, Learning Resources Center Library, 1234 Columbus Ave, *Boston*, MA 02120-3400
T: +1 617 5410339; Fax: +1 617 5410339; E-mail: rschwehm@rcc.mass.edu; URL: www.rcc.mass.edu/lib
1973; Roblyn W. Honeysucker
Black United Front Archives
45 911 vols; 162 curr per; 673 av-mat; 354 sound-rec
libr loan   33182

State College of Florida, Manatee-Sarasota – Bradenton Campus, Library, 5840 26th St W, *Bradenton*, FL 34207
T: +1 941 7525305; Fax: +1 941 7525308; E-mail: elliott@scf.edu; URL: www.scf.edu
1958; Tracy Elliott
Lit, Nursing, Paramedics
55 160 vols; 315 curr per; 209 microforms; 1 848 av-mat; 398 sound-rec; 10 digital data carriers
libr loan   33183

Raritan Valley Community College, Evelyn S. Field Library, 118 Lamington Rd, *Brandenburg*, NJ 08876
T: +1 908 2188865; E-mail: bnebeker@raritanval.edu; URL: library.raritanval.edu
1968; Birte Nebeker
Art Coll, slides
80 000 vols; 550 curr per
libr loan   33184

Haselwood Library, Olympic College, 1600 Chester Ave, *Bremerton*, WA 98337
T: +1 360 4757250; Fax: +1 360 4757261; URL: oc.ctc.edu/instruction/lrc/index.shtml
1946; Ruth M Ross
Mountaineering & Outdoor Lit (George W. Martin Coll)
59 116 vols; 386 curr per; 386 microforms; 2 500 av-mat; 700 sound-rec; 20 digital data carriers; 481 overhead transparencies
libr loan; ALA   33185

Blinn College, W.L. Moody Jr Library, 800 Blinn Blvd, *Brenham*, TX 77833; mail address: 902 College Ave, Brenham, TX 77833
T: +1 979 8304250; Fax: +1 979 8304222; E-mail: lflynn@blinn.edu; URL: www.blinn.edu/library/
1883; Linda C. Flynn
College Arch, Local Hist Coll
170 000 vols; 2 000 curr per; 10 800 av-mat   33186

Suffolk County Community College, Western Campus Library, 1001 Crooked Hill Rd, *Brentwood*, NY 11717
T: +1 631 8516740; Fax: +1 631 8516509; E-mail: gabriej@sunysuffolk.edu; URL: www3.sunysuffolk.edu/Library/index.asp
1974; David J. Quinn
45 171 vols; 300 curr per; 6 746 microforms; 9 digital data carriers
libr loan; ALA   33187

Jefferson Davis Community College, Leigh Library, 220 Alco Dr, *Brewton*, AL 36426-2116; P.O. Box 958, Brewton, AL 36427
T: +1 251 8091584; Fax: +1 251 8677399; E-mail: jfaust@acet.net; URL: www.jdcc.net/library.htm
1965; Jeffrey B. Faust
Alabama Coll, College & Career Section, House Escambia County Hist Society mat (local hist & genealogy)
31 010 vols   33188

Housatonic Community College, Library, 900 Lafayette Blvd, *Bridgeport*, CT 06604
T: +1 203 3325070; Fax: +1 203 3325252; E-mail: ho_harvey@commnet.edu; URL: www.hctc.commnet.edu/library/index.html
1967; Bruce Harvey
30 000 vols; 245 curr per   33189

Massasoit Community College, Library, One Massasoit Blvd, *Brockton*, MA 02402
T: +1 508 5889100 ext 1940; Fax: +1 508 4271255; E-mail: jjones@massasoit.mass.edu; URL: www.massasoit.mass.edu

1966; Joanne E. Jones
Allied Health Resources, New York Times, 1851-date
74 100 vols; 339 curr per   33190

Bronx Community College, Library & Learning Center, W 181st St & University Ave, Meister Hall, *Bronx*, NY 10453
T: +1 718 2895439; Fax: +1 718 2896063; E-mail: teresa.mcmanus@bcc.cuny.edu; URL: www.bcc.cuny.edu/library/
1958; Theresa McManus
106 760 vols; 355 curr per; 600 e-books; 3 684 av-mat
libr loan   33191

Hostos Community College, Library, 475 Grand Concourse, *Bronx*, NY 10451
T: +1 718 5184208; Fax: +1 718 5184206; E-mail: dadho@mail.hostos.cuny.edu; URL: www.hostos.cuny.edu/library/index.htm
1968; Dr. Lucinda R. Zoe
Ethnic studies, Science & tech; Allied Health, Black Studies, Spanish American Lit, College Arch
54 451 vols   33192

Kingsborough Community College, Robert J. Kibbee Library, 2001 Oriental Blvd, *Brooklyn*, NY 11235
T: +1 718 3685632; Fax: +1 718 3685482; E-mail: jmurphy@kbcc.cuny.edu; URL: www.kbcc.cuny.edu/kcclibrary
1964; Michael Rossen
Broadcasting tech, Fisheries & marine tech, Judaica, Puppetry, Travel; Coney Island Chamber of Commerce Coll; Manhattan Beach (Herman Field Coll), photos
154 640 vols; 441 curr per   33193

New York City Technical College, Ursula C. Schwerin Library, 300 Jay St, *Brooklyn*, NY 11201
T: +1 718 2605470; Fax: +1 718 2605631; URL: library.nycts.cuny.edu
1946; Darrow Wood
College Arch, Hotel & Restaurant Management (Menu File)
186 000 vols; 600 curr per; 11 941 microforms; 200 av-mat; 3 254 sound-rec; 7 digital data carriers; 162 VF drawers   33194

North Hennepin Community College, Library, 7411 85th Ave N, *Brooklyn Park*, MN 55445-2298
T: +1 763 4240732, 0733; Fax: +1 763 4930569; E-mail: theresa.crosby@nhcc.mnscu.edu; URL: www.nh.cc.mn.us/college.services/library
1966; Theresa Crosby
Genealogy
43 000 vols   33195

Coastal Georgia Community College, Gould Memorial Library, 3700 Altama Ave, *Brunswick*, GA 31520-3644
T: +1 912 2647270; Fax: +1 912 2647274; E-mail: calverr@bc9000.bc.peachnet.edu
1961; Raymond J. calvert
Coastal Georgia Hist
65 000 vols   33196

Saint Mary's School for the Deaf, Library Information Center, 2253 Main St, *Buffalo*, NY 14214
T: +1 716 8347200 ext 152; Fax: +1 716 8372080; E-mail: jmodien@smsdk12.org
1964; Jean Odien
18 000 vols; 90 curr per; 414 microforms
                                       33197

Villa Maria College, Library, 240 Pine Ridge Rd, *Buffalo*, NY 14225-3999
T: +1 716 8960700; Fax: +1 716 8960705; E-mail: library@villa.edu; URL: www.villa.edu
1961; Sr M. Anna Falbo
Polish Coll
40 000 vols; 200 curr per; 16 maps; 17 700 microforms; 706 av-mat; 5 624 sound-rec; 23 digital data carriers; 10 995 slides, 50 overhead transparencies
libr loan; ALA, CLA   33198

Champlain College, Library, 83 Summit St, ***Burlington***, VT 05401-0670; mail address: 163 S Willard St, Burlington, VT 05401-3420
T: +1 802 6515987; Fax: +1 802 8602782; URL: champlain.edu/library
1878; Janet Cottrell
Art Book Coll, Vermontiana, College Arch
40 000 vols; 187 curr per; 40 000 e-books
libr loan 33199

Butler County Community College, John A. Beck Jr Library, College Dr, Oak Hills, P.O. Box 1203, ***Butler***, PA 16003
T: +1 724 2878711 ext 8299; Fax: +1 724 2856047; E-mail: smj1652@bc3.cc.pa.us; URL: www.bc3.org/lib
1966; Stephen Joseph
45 130 vols; 255 curr per 33200

Finger Lakes Community College, Charles J. Meder Library, 3325 Marvin Sands Drive, ***Canandaigua***, NY 14424
T: +1 585 7851432; Fax: +1 585 3948826; E-mail: queenefr@flcc.edu; URL: library.flcc.edu
1968; Frank Queener
Environment conservation, horticulture, nursing, travel; Canandaigua Lake Pure Waters Assn Arch
79 000 vols; 398 curr per
libr loan 33201

Spoon River College, Learning Resource Center, 23235 N County Rd 22, ***Canton***, IL 61520
T: +1 309 6496222; Fax: +1 309 6496235; E-mail: reference@src.cc.il.us; URL: spoonrivercollege.edu
1951; Kathleen A. Menanteau
38 340 vols; 900 curr per
libr loan 33202

Western Nevada Community College, Library & Media Services, 2201 W College Pkwy, ***Carson City***, NV 89703
T: +1 775 4453228; Fax: +1 775 8873087; E-mail: wnclrc@scs.unr.edu; URL: www.library.wncc.nevada.edu
1972; Ken Sullivan
36 100 vols; 180 curr per 33203

Community College of Baltimore County Catonsville, Library Services, 800 S Rolling Rd, ***Catonsville***, MD 21228
T: +1 410 4554219; Fax: +1 410 4556106; E-mail: pcurtis@ccbc.cc.md.us; URL: www.ccbcmd.edu/libraries/cat/index.html
1957; Mary Landry
Criminal law & justice, Mortuary science, Nursing
110 480 vols; 450 curr per; 190 e-books; 56 485 microforms; 6 878 sound-rec; 7 digital data carriers
libr loan 33204

Cazenovia College, Daniel W. Terry Library, Lincklaen St, ***Cazenovia***, NY 13035
T: +1 315 6557132; Fax: +1 315 6558675; URL: www.cazenovia.edu
1824; Stanley J. Kozackza
Women's studies
73 000 vols; 430 curr per 33205

Kirkwood Community College, Library, 6301 Kirkwood Blvd SW, P.O. Box 2068, ***Cedar Rapids***, IA 52406-2068
T: +1 319 3985553; Fax: +1 319 3984908; URL: www.kirkwood.cc.ia.us/library
1967; Jerrie Bourgo
Carl Van Vechten Coll
65 609 vols; 420 curr per; 40 377 microforms; 2 477 av-mat; 1 660 sound-rec; 37 digital data carriers; 14 slides, 80 art reproductions
libr loan 33206

Centralia College, Library Media Center, 600 W Locust St, ***Centralia***, WA 98531
T: +1 360 7369391 ext 241; Fax: +1 360 3307509; E-mail: reference@centralia.ctc.edu; URL: www.library.centralia.ctc.edu
1925; Philip Meany
Centralia Massacre Coll
34 160 vols; 1 868 curr per
libr loan 33207

Glen Oaks Community College, E.J. Shaheen Library, 62249 Shimmel Rd, ***Centreville***, MI 49032-9719
T: +1 616 4679945; Fax: +1 616 4674114; E-mail: bmorgan@glenoaks.cc.mi.us; URL: www.glenoaks.cc.mi.us
1966; Betsy Susan Morgan
Amish Religion Coll
40 140 vols; 289 curr per
libr loan; MLA, ALA 33208

Nunez Community College, Library, 3710 Paris Rd, ***Chalmette***, LA 70043
T: +1 504 6802602; Fax: +1 504 6802584; E-mail: library@nunez.cc.la.us; URL: www.nunez.ca.la.us/library/ncclib.htm
1992; Albert Tate
30 000 vols; 300 curr per 33209

Parkland College, Library, 2400 W Bradley Ave, ***Champaign***, IL 61821-1899
T: +1 217 3512200; Fax: +1 217 3512581; E-mail: amwatkin@parkland.edu; URL: www.parkland.edu/library
1967; Anna Maria Watkin
Education, Nursing
122 330 vols; 331 curr per; 1 400 e-books; 9 047 av-mat; 100 sound-rec 33210

Neosho County Community College, Chapman Library, 800 W. 14th, ***Chanute***, KS 66720-2699
T: +1 620 4312820; Fax: +1 620 4329841; E-mail: ncctlb@ink.org; URL: www.neosho.cc.ks.us
1936; Susan Weisenberger
35 000 vols; 128 curr per
libr loan 33211

Trident Technical College, Main Campus Learning Resources Center, LR-M, P.O. Box 118067, ***Charleston***, SC 29423-8067
T: +1 843 5746089; Fax: +1 843 5746484; E-mail: charnette.singleton@tridenttech.edu; URL: www.tridenttech.edu/library.htm
1964; Charmette Singleton
Sams Photofact Coll (Electronics), Law Libr, Engineering & Technical Bks
70 000 vols; 152 316 microforms; 4 299 av-mat; 1 383 sound-rec; 57 403 slides, 1 242 overhead transparencies, 88 laserdiscs 33212

Central Piedmont Community College, Central Campus Library, 1201 Elizabeth Ave, P.O. Box 35009, ***Charlotte***, NC 28235
T: +1 704 3306885; E-mail: Jennifer.Arnold@cpcc.edu; URL: www.cpcc.edu/library/about-the-library/campus-libraries
1963; Jennifer Arnold
100 000 vols; 700 curr per; 13 000 non-print materials, over 700 telecourse DVDs 33213

Piedmont Virginia Community College, Betty Sue Jessup Library, 501 College Dr, ***Charlottesville***, VA 22902
T: +1 434 9615308; Fax: +1 434 9776842; E-mail: ljc2d@jade.pvcc.cc.va.us; URL: www.pvcc.cc.va.us/library
1972; Linda Cahill
32 660 vols; 206 curr per 33214

Chattanooga State Community College, Augusta R. Kolwyck Library, 4501 Amnicola Hwy, ***Chattanooga***, TN 37406-1097
T: +1 423 6974436; Fax: +1 423 6974409; E-mail: reference@chattanoogastate.edu; URL: library.chattanoogastate.edu
1965; Victoria P. Leather
62 000 vols; 540 curr per 33215

Tidewater Community College, Chesapeake Campus Library, 1428 Cedar Rd, ***Chesapeake***, VA 23322
T: +1 757 8225160; Fax: +1 757 8225173; E-mail: tcthomp@tc.cc.va.us; URL: www.tcc.vccs.edu/lrc
1973; Patricia U. Thomas
43 040 vols; 239 curr per; 43 178 microforms; 10 686 av-mat; 5 digital data carriers
libr loan; OCLC, SOLINET 33216

John Tyler Community College, Chester Campus Library, Moyar Hall, M216, ***Chester***, VA 23831-5399
T: +1 804 706-5196; Fax: +1 804 796-4238; E-mail: amckinney@jt.cc.va.us; URL: www.jt.cc.va.us/lrtc/libweb/libweb1.htm
1967; Arthur W. McKinney Jr
39 000 vols; 94 curr per 33217

Laramie County Community College, Instructional Resources Center, 1400 E College Dr, ***Cheyenne***, WY 82007
T: +1 307 7781205; Fax: +1 307 7781399; E-mail: libref@lccc.cc.wy.us; URL: www.lccc.cc.wy.us/library
1969
Children's Lit; Foundation Center Materials; Wyoming Hist Coll
49 000 vols; 398 curr per; 10 digital data carriers
libr loan 33218

DeVry University, Chicago Campus, Library, 3300 N Campbell Ave, ***Chicago***, IL 60618
T: +1 773 9298500; Fax: +1 773 6972714; URL: www.chi.devry.edu/library
1967; Catherine J. Carter
20 000 vols; 30 curr per; 70 digital data carriers
libr loan; ALA, SLA, ILA 33219

Harold Washington College, City Colleges of Chicago, Library, 30 E Lake St, ***Chicago***, IL 60601-2449
T: +1 312 5535760; Fax: +1 312 5535783; E-mail: hwc-library@ccc.edu; URL: hwclibrary.ccc.edu
1962; Sherry Ledbetter
74 000 vols; 300 curr per; 1 500 av-mat; 2 500 sound-rec; 12 digital data carriers
libr loan 33220

International Graphoanalysis Society, IGAS Library, 111 N Canal St, Ste 995, ***Chicago***, IL 60606
T: +1 312 9309446; Fax: +1 312 9305903; URL: www.igas.com
1929; Anthony Hibbs, Richard Meade
Behavioral sciences, Social sciences
25 000 vols 33221

Kennedy-King College, City Colleges of Chicago Library, 6301 South Halsted St, ***Chicago***, IL 60621
T: +1 773 6025000; Fax: +1 773 6025463; E-mail: kkc-library@ccc.edu; URL: kennedyking.ccc.edu/
1934; Linda King
Nursing
75 000 vols; 200 curr per; 6 800 microforms
libr loan 33222

Malcolm X College, Library, 1900 W Van Buren St, City Colleges of Chicago, ***Chicago***, IL 60612
T: +1 312 8507250; Fax: +1 312 8507187; URL: www.xx.edu/malcolmx/malcolmx/library
1934; Ellen Garrett
45 200 vols; 350 curr per 33223

Olive-Harvey College, City Colleges of Chicago, Learning Resource Center, 10001 S Woodlawn Ave, ***Chicago***, IL 60628
T: +1 773 291-6477/-6354; Fax: +1 773 2916463; E-mail: ohc-ask-a-librarian@ccc.edu; URL: oliveharvey.ccc.edu/library.shtml
1957; Willa Lyn Fox
Spanish language
57 100 vols; 260 curr per 33224

Richard J. Daley College, Kelly Memorial Library, 7500 S Pulaski Rd, ***Chicago***, IL 60652-1200
T: +1 773 8387669; Fax: +1 773 8387670; E-mail: dkoss@ccc.edu; URL: daley.ccc.edu/library/default.aspx
1965; Donald R. Koss
Local hist
63 830 vols; 225 curr per 33225

School of the Art Institute of Chicago, John M. Flaxman Library, 37 S Wabash Ave, ***Chicago***, IL 60603-3103
T: +1 312 8995096; Fax: +1 312 8991851; E-mail: flaxman.refdesk@saic.edu; URL: digital-libraries.saic.edu; www.saic.edu/library
1968; Claire Eike
Joan Flasch Artists Bks Coll, Film Study Coll
103 000 vols; 409 curr per; 164 microforms; 1 000 av-mat; 2 647 sound-rec; 8 digital data carriers; 53 linear feet of vertical files, 1 200 audio items
libr loan 33226

Truman College, Cosgrove Library, 1145 W Wilson Ave, ***Chicago***, IL 60640-5691
T: +1 773 9074865; Fax: +1 773 9076803; URL: www.trumancollege.cc/library/
1956; Leone McDermott
66 800 vols; 251 curr per; 7 925 microforms; 850 av-mat; 3 500 sound-rec; 1 000 slides 33227

Wright College, Main Campus, Library, 4300 N Narragansett Ave, ***Chicago***, IL 60634-1500
T: +1 773 4818408; E-mail: rbazile@ccc.edu; URL: wright.ccc.edu
1934; Bazile Richard
52 000 vols; 245 curr per 33228

Prairie State College, Library, 202 S Halsted, ***Chicago Heights***, IL 60411-8200
T: +1 708 7093552; Fax: +1 708 7093940; E-mail: library@prairie.cc.il.us; URL: www.prairie.cc.il.us/lrc
1958; Suzanne Dubsky
43 000 vols; 525 curr per
libr loan 33229

Southwestern College, Library, 900 Otay Lakes Rd, ***Chula Vista***, CA 91910-7299
T: +1 619 4216700; E-mail: amcgee@swccd.edu; URL: www.swccd.edu/~library/
1961; Greg Sandoval
84 345 vols; 424 curr per 33230

Morton College, Learning Resources Center, 3801 S Central Ave, ***Cicero***, IL 60804
T: +1 708 6568000 ext 321; Fax: +1 708 6563297; E-mail: library@morton.edu; URL: www.morton.edu
1924; Jennifer Butler
Spanish language; Adult New Readers
50 560 vols; 347 curr per 33231

God's Bible School & College, Library, 513 Ringgold St, ***Cincinnati***, OH 45202; mail address: 1810 Young St, Cincinnati, OH 45202-6838
T: +1 513 7217944; E-mail: library@gbs.edu
1901; Elisabeth A. Tyler
45 000 vols; 210 curr per; 30 av-mat 33232

Coahoma Community College, Dickerson Johnson Library, 3240 Frairs Point Rd, ***Clarksdale***, MS 38614
T: +1 662 6272571 ext 461; Fax: +1 662 6279530; E-mail: ystanford@ccc.cc.ms.us; URL: www.ccc.cc.ms.us
1949; Yvonne Stanford
Business & management, Child growth & development, Education; Special Black Studies Coll
30 000 vols; 126 curr per 33233

Georgia Perimeter College, Jim Cherry Learning Resource Center, 555 N Indian Creek Dr, ***Clarkston***, GA 30021-2396
T: +1 678 8913635; Fax: +1 404 2984919; E-mail: elautema@gpc.edu; URL: www.gpc.edu/~clalib/
1964; Eva Lautemann
232 680 vols; 1 073 curr per
libr loan 33234

Cleveland State Community College, Library, 3535 Adkisson Dr, ***Cleveland***, TN 37312-2813; P.O. Box 3570, Cleveland, TN 37320-3570
T: +1 423 4786209, 4727141 ext 209; Fax: +1 423 4786255; E-mail: clscc-info@clscc.cc.tn.us; URL: www.clscc.cc.tn.us/library/index.html
1967; Mary Evelyn Lynn
Bradley County Hist, bks, flm; Polk County Hist
53 482 vols; 368 curr per 33235

Cuyahoga Community College – Eastern Campus Library, 4250 Richmond Rd, Highland Hill Village, ***Cleveland***, OH 44122-6195
T: +1 216 9872085, 2086; Fax: +1 216 9872054
1971; Terry Hancox
47 010 vols; 410 curr per 33236

Cuyahoga Community College – Metropolitan Campus, Library, 2900 Community College Ave, *Cleveland*, OH 44115
T: +1 216 9874294; Fax: +1 216 9874404; E-mail: lawrencecole@tri.cc.oh.us
1968; Lawrence Cole
58 780 vols; 484 curr per   33237

Cuyahoga Community College – Western Campus Library, Western Campus Library, 11000 Pleasant Valley Rd, Parma, *Cleveland*, OH 44130
T: +1 216 9875410; Fax: +1 216 9875050; E-mail: albert.zavar@tri-c.edu; URL: www.tri-c.edu/library/Pages/default.aspx
1966; Al Zavar
64 090 vols; 481 curr per   33238

Truett-McConnell College, Cofer Library, 100 Alumni Dr, *Cleveland*, GA 30528-9799
T: +1 706 8652136; Fax: +1 706 8655130; URL: library.truett.edu/truett/default.asp
1946; Janice E. Wilson
Religion (George W. Truett Coll)
42 000 vols; 140 curr per   33239

Dabney S. Lancaster Community College, Library/Learning Resource Center, 1000 Dabney Dr, *Clifton Forge*, VA 24422
T: +1 540 8632800; Fax: +1 540 8632915; E-mail: dlreidl@dl.cc.va.us; URL: www.dslcc.edu/VPISS/learningresources/index.htm
1964; Laurel Reid
38 980 vols; 376 curr per   33240

Mount Saint Clare College, Library, 400 N Bluff Blvd, *Clinton*, IA 52732
T: +1 563 2424023; Fax: +1 563 2422003; URL: www.ashford.edu
1918; Flora S. Lowe
St Clare & the Medieval Woman Autogr Coll
105 000 vols; 500 curr per   33241

Sampson Community College, Library, P.O. Box 318, *Clinton*, NC 28329-0318
T: +1 910 5928081; Fax: +1 910 5928048; URL: www.sampson.cc.nc.us
1966; Mark Rushing
32 000 vols; 250 curr per   33242

Macomb Community College Libraries – Center Campus, 44575 Garfield Rd, *Clinton Township*, MI 48038-1139
T: +1 586 2862104; Fax: +1 586 2862002; URL: www.macomb.edu/library
1968; Gerald Bosler
86 000 vols; 98 curr per; 10 000 e-books; 5 173 av-mat   33243

West Hills Community College, Fitch Library, 300 Cherry Lane, *Coalinga*, CA 93210
T: +1 559 9350801 ext 3247; Fax: +1 559 9353312; E-mail: nolandj@whccd.cc.ca.us; URL: www.westhills.cc.ca.us
1956; Jon F. Noland
Chicano studies; Critical Thinking
44 000 vols; 1 800 curr per   33244

Middle Georgia College, Roberts Memorial Library, 1100 Second St SE, *Cochran*, GA 31014-1599
T: +1 478 9343074; Fax: +1 478 9343378; E-mail: probards@mgc.edu; URL: www.mgc.edu/academics/library-resources/
1928; Paul Robards
Georgianna Genealogy, County Hist
90 000 vols; 428 curr per
libr loan   33245

BCC-UCF Joint Use Library, Learning Resources Center – Cocoa Campus, 1519 Clearlake Rd, *Cocoa*, FL 32922
T: +1 321 4337662; Fax: +1 321 4337678; E-mail: libraryb@brevardcc.edu; URL: www.brevardcc.edu/library
1960; Dr. Mem Stahley
Genealogy
196 700 vols; 1 300 curr per; 20 000 e-books; 80 Electronic Media & Resources
libr loan   33246

Colorado School for the Deaf & the Blind, Media Center, 33 N Institute St, *Colorado Springs*, CO 80903-3599
T: +1 719 5782206; Fax: +1 719 5782239; URL: www.csdb.org
1874; Mary Rupp
Professional Bks on Deafness & Blindness & Other Handicaps, Braille & Large Print Bks for Children, Videotapes on Sign Language
10 000 vols; 38 curr per   33247

Pikes Peak Community College, Learning Resources Center, 5675 S Academy Blvd, *Colorado Springs*, CO 80906-5498
T: +1 719 5767711; Fax: +1 719 5407523; E-mail: joberger@ppcc.colorado.edu
1969; Jo Berger
PPCC Arch
54 212 vols; 325 curr per; 278 maps; 71 187 microforms; 1 608 av-mat; 2 785 sound-rec; 116 digital data carriers; 318 slides, 44 overhead transparencies
libr loan   33248

Columbia State Community College, Finney Memorial Library, 1665 Hampshire Pike, *Columbia*, TN 38401; P.O. Box 1315, Columbia, TN 38402-1315
T: +1 931 5402560; Fax: +1 931 5402565; E-mail: breeden@coscc.cc.tn.us; URL: www.coscc.cc.tn.us/lrc
1966; Kathy Breeden
South-Central Tennessee Hist, Arch
61 090 vols; 320 curr per   33249

Howard Community College, Library, 10901 Little Patuxent Pkwy, *Columbia*, MD 21044
T: +1 410 7724812; Fax: +1 410 7724993; E-mail: lgardner@howardcc.edu; URL: www.howardcc.edu
1970; Lucy Gardner
Art, Nursing
48 050 vols; 160 curr per   33250

Lindsey Wilson College, Katie Murrell Library, 210 Lindsey Wilson St, *Columbia*, KY 42728; mail address: 210 Lindsey Wilson St, Columbia, KY 42728
T: +1 270 3848250; Fax: +1 270 3844188; E-mail: library@lindsey.edu; URL: www.lindsey.edu/library
1903; Philip Hanna
Mag Coll
61 000 vols; 200 curr per; 110 000 e-books   33251

Columbus State Community College, Educational Resources Center, 550 E Spring St, P.O. Box 1609, *Columbus*, OH 43215
T: +1 614 2872465; Fax: +1 614 2872457; URL: www.cscc.edu
1965; Claire Fohl
36 254 vols; 551 curr per; 6 105 microforms; 6574 av-mat; 2 290 sound-rec; 7 digital data carriers; 29 pamphlets, 22 729 slides
libr loan; ALA   33252

DeVry Institute of Technology, Library, 1350 Alum Creek Dr, *Columbus*, OH 43209
T: +1 614 2537291; Fax: +1 614 2524108; E-mail: library@devrycols.edu; URL: www.devrycols.edu/library
1970; Bruce Weaver
Electronics, Computer science, Business; SAM's Photofact Series; Electronics Databooks
25 000 vols; 70 curr per; 13 000 e-books; 6 digital data carriers   33253

Ohio School for the Deaf Library, 500 Morse Rd, *Columbus*, OH 43214
T: +1 614 7281414; Fax: +1 614 7284060; URL: www.ohioschoolforthedeaf.org
Ada G. Kent
Ohio Chronicle 1868-1999
10 000 vols; 45 curr per   33254

Compton Community College, Emily B. Hart Holifield Library, 1111 E Artesia Blvd, *Compton*, CA 90221-5393
T: +1 310 9001648; Fax: +1 310 9001693; E-mail: stevens_a@compton.cc.ca.us; URL: www.compton.cc.us/library
1927; Roberta Hawkins
Black history
40 890 vols; 32 511 microforms; 9 digital data carriers   33255

Horry – Georgetown Technical College, Conway Campus Library, 2050 Hwy 501 E, *Conway*, SC 29526-9521; P.O. Box 261966, Conway, SC 29528-6066
T: +1 843 3495268; Fax: +1 843 3470352; URL: www.hgtc.edu/library
1966; Peggy E. Smith
61 000 vols; 360 curr per; 57 000 e-books   33256

Anoka-Ramsey Community College, Library, 11200 Mississippi Blvd NW, *Coon Rapids*, MN 55433
T: +1 763 4331100; URL: www.an.cc.mn.us
1965; Al Mamaril, Gina Pancerella-Willis
40 481 vols; 250 curr per   33257

Southwestern Oregon Community College, Library, 1988 Newmark, *Coos Bay*, OR 97420-2956
T: +1 541 8887431; Fax: +1 541 8887605; E-mail: mstricke@southwestern.cc.or.us; URL: www.southwestern.cc.or.us
1962; Mary Stricker
39 300 vols; 260 curr per
libr loan   33258

Corning Community College, Arthur A. Houghton, Jr Library, One Academic Dr, *Corning*, NY 14830
T: +1 607 9629251; Fax: +1 607 9629466; E-mail: lockarbh@corning-cc.edu; URL: www.corning-cc.edu/visitors/library/
1957; Barbara Hornick-Lockard
Rare bks, Art, Local hist
70 400 vols; 958 curr per   33259

Del Mar College, William F. White Jr Library, 101 Baldwin, *Corpus Christi*, TX 78404
T: +1 361 6981310; Fax: +1 361 6981182; E-mail: chris@delmar.edu; URL: library.delmar.edu
1937; Christine Tetzlaff-Belhasen
185 210 vols; 760 curr per; 28 500 e-books; 14 300 av-mat   33260

Navarro College, Richard M. Sanchez Library, 3200 W Seventh Ave, *Corsicana*, TX 75110
T: +1 903 8757442; Fax: +1 903 8757449; E-mail: tim.kevil@navarrocollege.edu; URL: www.navarrocollege.edu
1946; Tim Kevil
Indian Artifacts (R S Reading Coll), Pearce Civil War Docs Coll, Pearce Western Art Coll, Samuels Hobbit Coll
55 400 vols; 290 curr per   33261

Orange Coast College, Library, 2701 Fairview Rd, P.O. Box 5005, *Costa Mesa*, CA 92628-5005
T: +1 714 4325885; Fax: +1 714 4326850; E-mail: jposhek@occ.cccd.edu; URL: www.orangecoastcollege.edu/academics/librar
1948; Joe Poshek
Horticulture, photography
100 000 media items; 453 curr per
libr loan   33262

Iowa Western Community College, Learning Resource Center, 2300 College Rd, P.O. Box 4-C, *Council Bluffs*, IA 51502
T: +1 712 3253247; Fax: +1 712 3253244; URL: www.iwcc.cc.ia.us
1966; Ellen VanWaart
56 240 vols   33263

Union County College, MacKay Library, 1033 Springfield Ave, *Cranford*, NJ 07016
T: +1 908 7097623; Fax: +1 908 7097589; URL: www.ucc.edu/library
1933; Andrea MacRitchie
English as a second language, nursing
132 000 vols; 913 curr per
libr loan   33264

McHenry County College, Library, 8900 US Hwy 14, *Crystal Lake*, IL 60012
T: +1 815 4558533; Fax: +1 815 4553999; E-mail: czange@mchenry.cc.il.us; URL: www.mchenry.cc.il.us/library/library.html
1968; Cathy Zange
Horticulture, Agriculture
36 390 vols; 220 curr per   33265

West Los Angeles College, Library, 9000 Overland Ave, *Culver City*, CA 90230
T: +1 310 2874408; Fax: +1 310 8362867
1969; Ken Lee
Law Coll (support paralegal program)
80 000 vols; 225 curr per; 18 600 microforms; 1 500 av-mat; 10 sound-rec; 8 digital data carriers
libr loan; ALA   33266

Allegany College of Maryland, Library, 12401 Willowbrook Rd SE, *Cumberland*, MD 21502-2596
T: +1 301 7845268; Fax: +1 301 7845017; URL: www.ac.cc.md.us/library/
1961; Mary Huebner
Allied health, Social sciences, Criminal law; Local Hist (Appalachian Coll)
57 890 vols; 440 curr per; 25 394 microforms; 1 226 av-mat; 2 017 sound-rec; 13 digital data carriers; 293 slides
libr loan   33267

Southeast Community College, Library, 207 Chrisman Hall, 700 College Rd, *Cumberland*, KY 40823
T: +1 606 5892145 ext 2086; Fax: +1 606 5894941; URL: www.secc.kctcs.net/library
1960; Warren F. Gray
Kentucky Authors Coll
35 620 vols; 207 curr per
libr loan   33268

De Anza College, A. Robert DeHart Learning Center, 21250 Stevens Creek Blvd, *Cupertino*, CA 95014-5793
T: +1 408 8648313; Fax: +1 408 8648603; E-mail: mccarthyjames@fhda.edu; URL: www.deanza.edu/library/
1967; James M. McCarthy
Hist, Art & architecture, Special education; De Cillis Coll (Vietnam Conflict)
80 700 vols; 1 125 curr per   33269

Andrew College, Pitts Library, 413 College St, *Cuthbert*, GA 31740
T: +1 229 7325944; Fax: +1 229 7325957; E-mail: karanpittman@andrewcollege.edu
1854; Karan Ann Berryman-Pittman
Andrew College Arch, Georgia & Genealogy Coll, Methodist & Religious Hist Coll
33 000 vols; 237 curr per
libr loan; GLA, ALA, SOLINET, OCLC   33270

Cypress College, Library, 9200 Valley View St, *Cypress*, CA 90630
T: +1 714 4847125; Fax: +1 714 8266723; URL: www.cypresscollege.edu/~library/index.htm
1966; Dr. Kay Bruce
68 610 vols; 252 curr per   33271

Pasco-Hernando Community College, Charles E. Conger Library, East Campus, 36727 Blanton Rd, *Dade City*, FL 33523-7599
T: +1 352 5181306; Fax: +1 352 5181350; E-mail: butlerd@phcc.edu; URL: www.phcc.edu
1972
Electronics, Nursing, Agriculture
58 000 vols; 292 curr per   33272

Gaston College, Morris Library & Media Center, 201 Hwy 321 S, *Dallas*, NC 28034
T: +1 704 9226359; Fax: +1 704 9226363; URL: www.gaston.cc.nc.us
1964; Dr. Harry Cooke
Local hist; Civil War Coll
51 000 vols; 450 curr per   33273

Mountain View College, Learning Resources Center, 4849 W Illinois, *Dallas*, TX 75211-6599
T: +1 214 8608669; Fax: +1 214 8608667; E-mail: glo6600@dcccd.edu; URL: www.dcccd.edu/library
1970; Dr. Gwendolyn Oliver
Aviation
49 000 vols; 200 curr per
libr loan   33274

Richland College, Library, 12800 Abrams Rd, **Dallas**, TX 75243-2199
T: +1 972 2386081; Fax: +1 972 2386963; E-mail: richlandlibrary@dcccd.edu; URL: www.richlandcollege.edu/library
1972; Lennjo Henderson
85 000 vols
libr loan   33275

Dalton State College, Derrell C. Roberts Library, 650 College Dr, **Dalton**, GA 30720-3778
T: +1 706 2724575; Fax: +1 706 2724511; URL: www.daltonstate.edu/library
1967; Lydia F. Knight
Arch Georgia (Dalton Room)
127 000 vols; 339 curr per; 57 000 e-books   33276

Quinebaug Valley Community College, Library, 742 Upper Maple St, **Danielson**, CT 06239
T: +1 860 7745967; Fax: +1 860 7793287; E-mail: library@qvcc.commnet.edu; URL: www.qvct.commnet.edu/llc/library.html
1971; Hyunyong C. Kim
30 000 vols; 135 curr per   33277

North Shore Community College, Learning Resource Center, One Ferncroft Rd, **Danvers**, MA 01923
T: +1 978 7624000 ext 5524; Fax: +1 978 9225165; URL: www.northshore.edu
75 960 vols; 592 curr per
libr loan   33278

Danville Area Community College, Learning Resources Center, 2000 E Main St, **Danville**, IL 61832
T: +1 217 4438734; Fax: +1 217 4438147; E-mail: library@dacc.cc.il.us; URL: www.dacc.cc.il.us/library
1962; Sally Duchow
45 500 vols; 196 curr per
libr loan   33279

Danville Community College, Whittington W. Clement Learning Resources Center, 1008 S Main St, **Danville**, VA 24541-4004
T: +1 804 7978453; E-mail: wdey@dc.cc.va.us; URL: www.dc.cc.va.us/lrc2/lrc/lrc2.htm
1968; William L. Dey
Local Hist
47 150 vols; 225 curr per
libr loan   33280

Palmer College of Chiropractic, David D. Palmer Health Sciences Library, 1000 Brady St, **Davenport**, IA 52803-5287
T: +1 563 8845896; Fax: +1 563 8845897; E-mail: refdesk@palmer.edu; URL: www.palmer.edu
1897; Dennis Peterson
BJ Palmer Papers, Lyndon Lee Papers, Kenneth Cronk Papers, Palmer College Arch, BJ Palmer Osteological Coll, Chiropractic Hist & Res, Conservative Health Care
60 000 vols; 151 curr per; 11 000 e-books; 10 000 microforms; 1 493 av-mat; 2 030 sound-rec; 2 digital data carriers; 21 657 slides, 1 653 x-ray sets, 910 biological specimen & models
libr loan   33281

Sinclair Community College, Learning Resources Center, 444 W Third St, **Dayton**, OH 45402-1460
T: +1 937 5122855; Fax: +1 937 5124564; URL: library.sinclair.edu
1887; Douglas Kaylor
147 615 vols; 509 curr per
libr loan   33282

Daytona Beach Community College, Library, 1200 W International Speedway Blvd, **Daytona Beach**, FL 32114; P.O. Box 2811, Daytona Beach, FL 32120-2811
T: +1 386 5063055; Fax: +1 386 5063008
Mercedes Clement
Harding Political Memorabilia Coll, Lapensohn Newspaper Coll
100 000 vols; 752 curr per   33283

Henry Ford Community College, Eshleman Library, 5101 Evergreen Rd, **Dearborn**, MI 48128-1495
T: +1 313 8456375; Fax: +1 313 2715868; E-mail: bluka@hfcc.edu; URL: clara.hfcc.edu/
1938; Barbara Lukasiewicz
Law, Nursing, Performing arts
79 010 vols; 590 curr per   33284

Calhoun Community College, Albert P. Brewer Library, Hwy 31 N, P.O. Box 2216, **Decatur**, AL 35609
T: +1 256 3062775; Fax: +1 256 3062780; URL: lib.calhoun.cc.al.us/lib
1965; Lucinda M. Beddow
Human resources, Alabama Coll, Ctr for the Study of Southern Political Culture
52 000 vols; 150 curr per
libr loan   33285

DeVry Institute of Technology, Library, 250 N Arcadia Ave, **Decatur**, GA 30030
T: +1 404 2927900; URL: www.atl.devry.edu
1969; Mary Antoine
23 000 vols; 75 curr per   33286

East Central Community College, Burton Library, 275 W Broad St, P.O. Box 129, **Decatur**, MS 39327
T: +1 601 6352111 ext 219; Fax: +1 601 6352150; E-mail: burton_library@eccc.cc.ms.us; URL: ecc.cc.ms.us
1977; Gloria Johnson
38 000 vols; 237 curr per   33287

Richland Community College, Kitty Lindsay Learning Resources Center, One College Park, **Decatur**, IL 62521
T: +1 217 8757200 ext 296; Fax: +1 217 8756961; E-mail: dzindel@richland.cc.il.us; URL: www.richland.cc.il.us
1972; David Zindel
38 000 vols; 281 curr per   33288

Wisconsin School for the Deaf, John R. Gant Library, 309 W Walworth Ave, **Delavan**, WI 53115
T: +1 262 7287127; Fax: +1 262 7287129; URL: www.wsd.k12.wi.us/
1930; Shelly McDowell
Depot for Captioned Films for the Deaf; Professional Libr – emphasis on Deafness
11 000 vols; 1 500 filmstrips
libr loan   33289

Methodist Theological School in Ohio Library, 3081 Columbus Pike, **Delaware**, OH 43015-8004; P.O. Box 8004, Delaware, OH 43015-8004
T: +1 740 3631146; Fax: +1 740 3623456; URL: www.tcgc.capital.edu
1960; Dr. Paul Schrodt
Denominational Coll of the United Methodist Church & its Predecessor Bodies, Philip Gatch Mss
128 490 vols; 393 curr per; 485 microforms; 596 av-mat; 3 396 sound-rec; 598 digital data carriers   33290

Grayson County College, Library, 6101 Grayson Dr, **Denison**, TX 75020-8299
T: +1 903 4638637; Fax: +1 903 4654123; E-mail: paikowski@grayson.edu; URL: www.grayson.edu
1965; Gary F. Paikowski
60 000 vols; 100 curr per   33291

Highline Community College, Library, 2400 S 240th St, MS 25-4, **Des Moines**, WA 98198; P.O. Box 98000, Des Moines, WA 98198-9800
T: +1 206 8783710; Fax: +1 206 8703776; URL: flightline.highline.edu/library
1961; Monica Luce
96 000 vols; 697 curr per   33292

Oakton Community College, Library, 1600 E Golf Rd, **Des Plaines**, IL 60016
T: +1 847 6351642; E-mail: libraryask@oakton.edu; URL: www.oakton.edu/library/index.php
1970; Gary Newhouse
92 000 vols; 586 curr per   33293

Wayne County Community College, Arthur Cartwright LRC Library, 1001 W Fort St, **Detroit**, MI 48226
T: +1 313 4962063; Fax: +1 313 4964731; E-mail: scoffer1@wcccd.edu; URL: www.wccc.edu
1974; Stephanie A. Coffer
51 610 vols; 559 curr per   33294

Sauk Valley Community College, Learning Resource Center, 173 IL Rte 2, **Dixon**, IL 61021-9112
T: +1 815 2885511 ext 306; Fax: +1 815 2885651; E-mail: thomasr@svcc.edu; URL: www.svcc.edu
1966; Robert Thomas
Popular Culture, State, Regional & Local Hist Coll
56 000 vols; 275 curr per; 3 000 av-mat; 6 500 sound-rec
libr loan; ALA, ILA   33295

Surry Community College, Learning Resources Center, 630 S Main St, P.O. Box 304, **Dobson**, NC 27017
T: +1 336 3863252; Fax: +1 336 3863692; E-mail: cores@surry.cc.nc.us
1965; Sheila Core
48 419 vols; 334 curr per; 10 281 microforms; 3 270 av-mat; 972 sound-rec; 417 slides, 394 overhead transparencies
libr loan   33296

Dodge City Community College, Learning Resources Center-Library Services, 2501 N 14th, **Dodge City**, KS 67801
T: +1 620 2251321 ext 287; Fax: +1 620 2250918; E-mail: library@dccc.cc.ks.us; URL: www.dccc.ks.us
1935; Cathy L. Reeves
Agriculture, Nursing
30 950 vols; 220 curr per   33297

George C. Wallace State Community College, Learning Resources Center, Rte 6, Box 62, **Dothan**, AL 36303
T: +1 334 9833521 ext 225; Fax: +1 334 9833650; E-mail: mjohnson@wcc.cc.al.us; URL: www.wcc.cc.al.us
1965; Megan Johnson
40 000 vols; 350 curr per   33298

Cochise College, Charles Di Peso Library, 4190 W Hwy 80, **Douglas**, AZ 85607
T: +1 520 5155320; E-mail: IMS@cochise.edu; URL: padme.cochise.edu/wordpress/
1964; Patricia Hotchkiss
Aviation, Hist, Nursing
62 850 vols; 318 curr per   33299

South Georgia College, William S. Smith Library, 100 W College Park Dr, **Douglas**, GA 31533-5098
T: +1 912 3894510; Fax: +1 912 3894469; E-mail: jacqueline.vickers@sgc.edu; URL: www.sgc.edu/library/index.html
1907; Jacqueline Vickers
US Geological Survey Maps
80 300 vols; 334 curr per
libr loan   33300

Tompkins-Cortland Community College, Gerald A. Barry Memorial Library, 170 North St, P.O. Box 139, **Dryden**, NY 13053-0139
T: +1 607 8448222; Fax: +1 607 8446540; E-mail: demow@sunytccc.edu; URL: www.sunytccc.edu/library.htm
1968; William Demo
41 282 vols; 515 curr per   33301

New River Community College, Learning Resource Center, Rte 100, PO Drawer 1127, **Dublin**, VA 24084-1127
T: +1 540 6743600; Fax: +1 540 6763626; URL: nr.cc.va.us/delr/lis/lishome.htm
1968; Gary Bryant
34 500 vols; 406 curr per; 596 maps; 40 000 microforms; 3 300 av-mat; 586 sound-rec; 18 digital data carriers   33302

Georgia Perimeter College – Dunwoody Campus Library, 2101 Womack Rd, **Dunwoody**, GA 30338
T: +1 770 274-5085; Fax: +1 770 5513201; E-mail: jbarnes@gpc.peachnet.edu; URL: www.gpc.edu/~dunlib/
1964; Joseph W. Barnes
70 000 vols; 200 curr per   33303

Durham Technical Community College, 1637 Lawson St, P.O. Box 11307, **Durham**, NC 27703
T: +1 919 6863969; Fax: +1 919 6863471; E-mail: laubei@gwmail.dtce.ec.nc.us
1961; Irene H. Laube
36 120 vols; 262 curr per   33304

Dyersburg State Community College, Learning Resource Center, 1510 Lake Rd, **Dyersburg**, TN 38024
T: +1 901 2863225; Fax: +1 901 2863228; E-mail: lrc@dscclan.dscc.cc.tn.us; URL: www.dscc.cc.tn.us/lrc
1967; Bob Lhota
43 240 vols; 108 curr per   33305

College of Aeronautics, Library, 8601 23rd Ave, **East Elmhurst**, NY 11369
T: +1 718 4296600 ext 184; Fax: +1 718 4787066; E-mail: joannj@aero.edu; URL: www.aero.edu
1932; JoAnn Jayne
Aircraft Maintenance Manuals, NACA & other annual rpts, bound vols, NASA rpts, SAE rpts
34 000 vols; 314 curr per; 138 124 microforms; 1 109 av-mat   33306

Skowhegan School of Painting & Sculpture, Robert Lehman Library, Art School Rd, **East Madison**, ME 04950; mail address: 200 Park Ave S, Ste 1116, New York, NY 10003
T: +1 212 5290505; Fax: +1 212 4731342; URL: www.skoweganart.org
Linda Earle
16 000 vols   33307

Illinois Central College, Learning Resources Center, One College Dr, **East Peoria**, IL 61635-0001
T: +1 309 6945461; Fax: +1 309 6945473; E-mail: Cathryne.Parish@icc.edu; URL: apps.icc.edu/library/
1967; Cate Parish
92 000 vols; 3 500 curr per
libr loan   33308

Chippewa Valley Technical College, Library, 620 W Clairemont Ave, **Eau Claire**, WI 54701-6162
T: +1 715 8336285; Fax: +1 715 8336470; E-mail: eemberson@mail.chippewa.tec.wi.us; URL: www.chippewa.tec.wi.us/library
1965; Ronald Edwards
36 430 vols; 681 curr per; 40 digital data carriers
libr loan   33309

Middlesex County College, Library, 2600 Woodbridge Ave, **Edison**, NJ 08818
T: +1 732 9064252; E-mail: MThompson@middlesexcc.edu; URL: www.middlesex.cc.nj.us
1967; Mark Thompson
Business, Paralegal studies; College Arch
82 644 vols; 765 curr per   33310

Cuyamaca College, Library, 900 Rancho San Diego Pkwy, **El Cajon**, CA 92019-4304
T: +1 619 6604416; Fax: +1 619 6604493; URL: www.cuyamaca.net
1978; Pei-Hua Chou
32 000 vols; 283 curr per   33311

Grossmont College, Lewis F. Smith Learning & Technology Resource Center Library, 8800 Grossmont College Dr, **El Cajon**, CA 92020-1799
T: +1 619 6447364; Fax: +1 619 6447054; E-mail: curtis.stevens@gcccd.net; URL: www.grossmont.net/library
1961; Dr. Curtis L. Stevens
Career Information, Reference Resources
106 150 vols; 431 curr per   33312

Butler County Community College, L.W. Nixon Library, 901 S Haverhill Rd, **El Dorado**, KS 67042-3280
T: +1 316 3223234; Fax: +1 316 3223315; E-mail: bbeattie@butler.buccc.cc.ks.us; URL: www.buccc.cc.ks.us
1927; Brian Beattie
41 000 vols
libr loan; ALA   33313

El Paso Community College, Rio Grande Library, 1111 N Oregon, P.O. Box 20500, *El Paso*, TX 79998-0500
T: +1 915 8314018; Fax: +1 915 8314626; URL: www.epcc.edu/library/library.htm
1971; Charlotte Hollis
54 000 vols; 206 curr per   33314

Elgin Community College, Renner Learning Resources Center, 1700 Spartan Dr, *Elgin*, IL 60123
T: +1 847 2147337; Fax: +1 847 8887995; E-mail: libref@elgin.edu; URL: www.elgin.edu/students.aspx?id=694
1949
70 000 vols; 425 curr per   33315

College of the Albemarle, Elizabeth City Campus Library, Hwy 17 N, P.O. Box 2327, *Elizabeth City*, NC 27909
T: +1 252 3350821 ext 2270; Fax: +1 252 3350649; E-mail: rschenck@albemarle.cc.nc.us; URL: www.albemarle.cc.nc.us/library.php?cat=414
1961; Robert Schenck
North Caroliniana
34 400 vols; 232 curr per; 125 av-mat; 100 sound-rec; 15 digital data carriers   33316

Elizabethtown Community College, Media Center Library, 600 College St, *Elizabethtown*, KY 42701
T: +1 270 7692371 ext 240; Fax: +1 270 7691618; E-mail: ecclbrat@pop.uky.edu; URL: www.elizabethtowncc.com/Resources/Library
1964; Jimmy Bruce
Kentucky; Hardin County Genealogical Coll, Local Newspapers
43 000 vols; 245 curr per
libr loan   33317

Great Basin College, Learning Resources Center, 1500 College Pky, *Elko*, NV 89801
T: +1 775 7532222; Fax: +1 775 7532296; E-mail: karr@gbcnv.edu; URL: www.gbcnv.edu/library
1967; Juanita R. Karr
Juvenile lit; American Indian, Basque, Nevada
30 310 vols; 250 curr per   33318

Jones County Junior College, T. Terrell Tisdale Library, 900 S Court St, *Ellisville*, MS 39437
T: +1 601 4774055; Fax: +1 601 4772600; E-mail: andrew.sharp@jcjc.edu; URL: www.jcjc.edu
1924; Andrew Sharp
Literary Criticism on William Faulkner, James Joyce, Katherine Anne Porter & Eudora Welty, bks & flm; Mississippiana
68 000 vols; 500 curr per   33319

Vermilion Community College, Library, 1900 E Camp St, *Ely*, MN 55731
T: +1 218 3657226; Fax: +1 218 3657218; URL: www.vcc.mnscu.edu/info_sru/library/libmain.htm
1922
Natural resources; Ojibway Native American Coll
30 000 vols; 200 curr per   33320

Asnuntuck Community College, LRC Library, 170 Elm St, *Enfield*, CT 06082-0068
T: +1 860 2533174; Fax: +1 860 2533176; E-mail: aslrcref@acc.commnet.edu; URL: www.acc.commnet.edu/lrc/index.htm
1972; Ravil Veli
Copernicus Coll (Polish Hist & Culture), Literacy Volunteers
31 200 vols; 291 curr per
libr loan   33321

Enterprise State Junior College, Learning Resource Center, 600 Plaza Dr, P.O. Box 1300, *Enterprise*, AL 36331
T: +1 334 3472623 ext 271; Fax: +1 334 3472623 ext 306; E-mail: esjc@cc.al.us; URL: www.esjc.cc.al.us/lrc_hp.htm
1966; Susan Sumblin
Genealogy
40 000 vols; 333 curr per   33322

Snow College, Lucy A. Phillips Library, 255 E Center, P.O. Box 2008, 150 E College Ave, *Ephraim*, UT 84627
T: +1 435 2837363; Fax: +1 435 2837369; E-mail: library@snow.edu; URL: www.library.snow.edu
1888; Russell W. Dean
Utah & Sanpete County Hist, Map Coll, Children's Lit Coll
39 460 vols; 278 curr per   33323

Bay De Noc Community College, Learning Resources Center, 2001 N Lincoln Rd, *Escanaba*, MI 49829-2511
T: +1 906 7865802 ext 1229; Fax: +1 906 7896912; E-mail: holmesc@baydenoc.cc.mi.us; URL: www.lrcweb.baydenoc.cc.mi.us
1963; Christian Holmes
American Welding Society Coll, Delta County Oral Hist, Finnish Language Coll, Fire-fighting training videos
45 000 vols; 230 curr per   33324

Iowa Lakes Community College, Library, 300 S 18th St, *Estherville*, IA 51334
T: +1 712 3627985; Fax: +1 712 3625970; E-mail: info@iowalakes.edu; URL: www.iowalakes.edu
1924; Brenda Colegrove
38 500 vols; 850 curr per; 2 000 av-mat
libr loan   33325

Lane Community College, Library, 4000 E 30th Ave, *Eugene*, OR 97405-0640
T: +1 541 7262220; Fax: +1 541 7444150; E-mail: library@lanecc.edu; URL: www.lanecc.edu/library/
1964; Nadine Williams
66 720 vols; 350 curr per   33326

College of the Redwoods, Library, 7351 Tompkins Hill Rd, *Eureka*, CA 95501
T: +1 707 4764260; Fax: +1 707 4764432; URL: www.redwoods.edu/Eureka/Library
1965; Ruth Coughlin
Nursing; Civil War Coll
67 710 vols; 91 curr per
libr loan   33327

Kendall College, Library, 2408 Orrington Ave, *Evanston*, IL 60201
T: +1 847 8661300 ext 1287; Fax: +1 847 8661320; URL: www.kendall.edu
1934; Iva M. Freeman
Native American Coll, Culinary Coll
35 000 vols; 210 curr per; 2 digital data carriers   33328

Everett Community College, John N. Terrey Library – Media Center, 2000 Tower, *Everett*, WA 98201-1352
T: +1 425 3889501; Fax: +1 425 3889144; E-mail: jleader@ctc.edu; URL: www.evcc.ctc.edu/library
1948; Jeanne Leader
45 490 vols   33329

Bristol Community College, Farley Learning Resources Center, 777 Elsbree St, *Fall River*, MA 02720
T: +1 508 6782811 ext 2500; Fax: +1 508 7303270; E-mail: Lisa.Richter@bristolcc.edu; URL: www.bristolcc.edu/academics/library
1968
Allied health
62 270 vols; 352 curr per   33330

Brookhaven College, Learning Resources Center, 3939 Valley View, *Farmers Branch*, TX 75244-4997
T: +1 972 8604854; Fax: +1 972 8604675; E-mail: sferguson@dcccd.edu; URL: www.dcccd.edu/bhc/campserv/library
1978; Sarah Ferguson
Plotkin Holocaust Coll
55 890 vols; 163 curr per   33331

San Juan College, Library, 4601 College Blvd, *Farmington*, NM 87402
T: +1 505 566-3249; Fax: +1 505 5663381; E-mail: boltonl@sjc.cc.nm.us; URL: www.sanjuancollege.edu/lib
1964; Louise Bolton
Tom Carter Petroleum Geological Coll; Southwestern Americana, bks, maps, rpts
50 000 vols; 475 curr per; 2 000 govt docs; 917 maps; 1 200 av-mat; 6 digital data carriers
libr loan; ALA, NMLA   33332

Tunxis Community College, Library, 271 Scott Swamp Rd, *Farmington*, CT 06032
T: +1 860 6799544; Fax: +1 860 6760021; URL: tunxis.comment.edu/library
1970; Judith Markiewicz
American Lit Coll, Art Coll, Criminal Justice Coll
42 000 vols; 251 curr per
libr loan   33333

Oakland Community College, King Library, 27055 Orchard Lake Rd, Orchard Ridge Campus, *Farmington Hills*, MI 48334-4579
T: +1 248 5223525; Fax: +1 248 5223530; E-mail: nmspringe@oaklandcc.edu; URL: www.oaklandcc.edu/library
1967; Nadja Springer-Ali
90 200 vols; 411 curr per; 3 050 av-mat; 16 500 ebks   33334

Fayetteville Technical Community College, Paul H. Thompson Library, 2201 Hull Rd, P.O. Box 35236, *Fayetteville*, NC 28303
T: +1 910 6788247; Fax: +1 910 6788401; E-mail: wilkinss@faytechcc.edu; URL: www.faytechcc.edu/library/default.asp
1961; Susan Rose
66 250 vols; 370 curr per   33335

Fergus Falls Community College, Library, 1414 College Way, *Fergus Falls*, MN 56537-1000
T: +1 218 7397531; Fax: +1 218 7397475; E-mail: dkelman@mail.ff.cc.mn.us; URL: www.ff.cc.mn.us
1960; Deb Kelman
Environment, Womens' studies
31 000 vols; 150 curr per   33336

Saint Louis Community College – Florissant Valley Campus, Library, 3400 Pershall Rd, *Ferguson*, MO 63135-1499
T: +1 314 5134511; Fax: +1 314 5134053; E-mail: pwilliams@stlcc.cc.mo.us; URL: www.stlcc.edu
1963; Patricia Williams
102 440 vols; 584 curr per; 1 260 av-mat   33337

Blue Ridge Community College, Library, College Dr, *Flat Rock*, NC 28731
T: +1 828 6922572 ext 272; Fax: +1 828 6941692; E-mail: susanw@blueridge.cc.nc.us
1969; Marianne Campbell & Susan D. Williams
30 000 vols   33338

Baker College System Libraries, 1050 W Bristol Rd, *Flint*, MI 48507-5508
T: +1 810 7664237; Fax: +1 810 7662013; E-mail: library-fl@baker.edu; URL: www.baker.edu
1912; Eric Palmer
50 000 vols; 225 curr per   33339

Mott Community College, C.S. Mott Library, 1401 E Court St, *Flint*, MI 48503
T: +1 810 7620403; Fax: +1 810 7620407; E-mail: library@mcc.edu; URL: www.library.mcc.edu
1960; Denise Hooks
Children's Lit, Law Ref
95 630 vols; 334 curr per   33340

Florence-Darlington Technical College Wellman, Inc, Library, P.O. Box 100548, *Florence*, SC 29501
T: +1 843 6618032; Fax: +1 843 6618266; E-mail: bradleyj@flo.tec.sc.us; URL: www.flo.tec.sc.us/library
1964; Jeronell W. Bradley
50 320 vols; 300 curr per; 26 351 microforms; 5 digital data carriers
libr loan   33341

Moraine Park Technical College, Library, 235 N National Ave, P.O. Box 1940, *Fond du Lac*, WI 54936-1940
T: +1 920 9292470; E-mail: hbaierl@morainepark.edu; URL: www.morainepark.edu/services/library
1965; Hans Baierl
65 355 vols; 60 curr per; 2 music scores; 65 maps; 5 microforms; 23 683 av-mat; 1 244 sound-rec; 835 digital data carriers; 842 slides; 270 overhead

transparencies, 2 288 pamphlets, 117 kits, 484 filmstrips, 1 689 other items
libr loan; ALA, WLA   33342

Waldorf College, Voss Memorial Library, 106 S Sixth St, *Forest City*, IA 50436
T: +1 641 5858110; Fax: +1 641 5858111; URL: www.waldorf.edu
1903; Jim M. Kapoun
Bible Coll
41 000 vols; 320 curr per   33343

Iowa Central Community College, Fort Dodge Center Library, 330 Avenue M, *Fort Dodge*, IA 50501
T: +1 515 5767201 ext 2618; Fax: +1 515 5760099 ext 2631; E-mail: schiefelbein@triton.iccc.cc.ia.us; URL: www.iccc.cc.ia.us/libraries
1967; Dan Schiefelbein
Education, Iowa, Local hist
55 000 vols; 350 curr per   33344

Keiser College, Jim Bishop Memorial Library, 1500 NW 49th St, *Fort Lauderdale*, FL 33309
T: +1 954 3514035; Fax: +1 954 3514051; URL: www.keiseruniversity.edu
1976; Benjamin Williams
85 000 vols   33345

Edison State College, Richard H. Rush Library, 8099 College Pkwy SW, Bldg I, *Fort Myers*, FL 33919
T: +1 941 4899300; Fax: +1 941 4899095; E-mail: wshuluk@edison.edu; URL: www.edison.edu/library/
1962; William Shuluk
81 480 vols; 400 curr per
libr loan   33346

Indian River Community College, Charles S. Miley Learning Resources Center, 3209 Virginia Ave, *Fort Pierce*, FL 34981-5599
T: +1 772 4627600; E-mail: library@irsc.edu; URL: www.ircc.cc.fl.us/libraries/index.aspx
1960; Rudolph P. Widman
Area Hist Coll, Florida Power & Light Coll
68 000 vols; 442 curr per; 350 854 microforms; 1 110 sound-rec; 80 digital data carriers; 47 020 slides, 64 overhead transparencies, 53 art reproductions   33347

Boreham Library, 5210 Grand Ave, *Fort Smith*, AR 72903; P.O. Box 3649, Fort Smith, AR 72913-3649
T: +1 479 7887204; Fax: +1 479 7887209; URL: www.uafortsmith.edu
1928; Wilma Cunningham
Grantsmanship Coll, Wilder Hist Coll
81 000 vols; 600 curr per; 18 000 e-books
libr loan   33348

Tarrant County Junior College, South Campus Library, 5301 S Campus Dr, *Fort Worth*, TX 76119-5926
T: +1 817 5154521; Fax: +1 817 5154436; E-mail: ask.librarian@tccd.edu
1967
69 460 vols; 567 curr per   33349

Tarrant County Junior College – Northwest Campus Walsh Library, 4801 Marine Creek Pkwy, *Fort Worth*, TX 76179
T: +1 817 5157725, 5157765; Fax: +1 817 5157720; E-mail: anna.holzer@tccd.net; URL: library.tccd.net
1975; Anna Holzer
41 030 vols; 250 curr per   33350

Emmanuel College, Library, 2261 W Main St, *Franklin Springs*, GA 30639; P.O. Box 69, Franklin Springs, GA 30639-0069
T: +1 706 2457226; Fax: +1 706 2454424; URL: www.emmanuelcollege.edu/academic/library
1935; Joye D. P. Slife
Pentecostal Holiness Archs
90 000 vols; 76 curr per   33351

Frederick Community College, Library, 7932 Opossumtown Pike, *Frederick*, MD 21702
T: +1 301 8462444; Fax: +1 301 8462498; E-mail: moleary@fcc.cc.md.us; URL: www.fcc.cc.md.us/library
1957; Mick O'Leary
Nursing
37 000 vols; 100 curr per   33352

Highland Community College, Library, 2998 W Pearl City Rd, **Freeport**, IL 61032-9341
T: +1 815 2356121 ext 3539; Fax: +1 815 2351366; E-mail: ewelch@admin.highland.cc.il.us; URL: www.hcclibrary.net
1962; Eric C. Welch
Local Authors Coll
45 000 vols; 250 curr per   33353

Ohlone College, Blanchard Learning Resources Center, 43600 Mission Blvd, **Fremont**, CA 94539
T: +1 510 6596160; Fax: +1 510 6596265; E-mail: librarians@ohlone.edu; URL: www2.ohlone.edu/org/library/
1967; Dr. Shirley Peck
Hist, Art & architecture, Indians, Law
68 920 vols; 217 curr per   33354

Queen of the Holy Rosary College, Library, 43326 Mission Blvd, P.O. Box 3908, **Fremont**, CA 94539-0391
T: +1 510 6572468; Fax: +1 510 6571734; E-mail: mparker@infalane.com; URL: www.qhrc.org
1908; Mary Ellen D. Parker
Religion, Theology, Philosophy; Art (Dresden Coll), slides
25 680 vols; 150 curr per   33355

Fresno City College, Learning Resources Center, 1101 E University Ave, **Fresno**, CA 93741
T: +1 559 4428206; Fax: +1 559 2655708; E-mail: nancy.almand@fresnocitycollege.edu; URL: www.fresnocitycollege.edu
1910; Jannett Jackson
Law Coll, State Federal Rpts Codes
75 000 vols; 540 curr per
ALA, CLA   33356

Fullerton College, William T. Boyce Library, 321 E Chapman Ave, **Fullerton**, CA 92832-2095
T: +1 714 9927061; Fax: +1 714 9929961; E-mail: jayala@fullcoll.edu; URL: www.fullcoll.edu
1913; Jackie Boll
Local hist, Library & information science, Literary criticism, Music; Fullerton College Oral Hist Program, Orange County Hist, Topographic Maps of California (US Geological Survey Coll), Arch of Fullerton College
103 000 vols; 667 curr per; 569 diss/theses; 4 605 maps; 87 726 microforms; 1 268 av-mat; 2 277 sound-rec; 1 digital data carriers; 14 336 pamphlets, 145 linear feet of papers, oral hist tapes, transcripts, 3 194 photos, 1 135 slides
libr loan   33357

Itawamba Community College, Learning Resource Center, 602 W Hill, **Fulton**, MS 38843
T: +1 662 8628000 ext 237; Fax: +1 662 8628410; URL: www.icc.cc.ms.us/campus
Dr. Glenda Segars
Mississippi Coll
38 000 vols; 254 curr per   33358

Missouri School for the Deaf, Grover C. Farquhar Library, 505 E Fifth St, **Fulton**, MO 65251-1703
T: +1 573 5922513; Fax: +1 573 5922570
1851; Virginia Johns
14 000 vols; 33 curr per; 20 digital carriers   33359

Gadsden State Community College, Meadows Library, 1001 George Wallace Dr, **Gadsden**, 35902; P.O. Box 227, Gadsden, AL 35902-0227
T: +1 256 5498333; Fax: +1 256 5498401; URL: www.gadsdenstate.edu/library
1965; Dr. Jeff Luzius
Alabama, Law, Southern lit; Gadsden State Arch, Mary Cooper Coll
110 000 vols; 282 curr per   33360

North Central Texas College, Library, 1525 W California St, **Gainesville**, TX 76240-0815
T: +1 940 6684283, 6687731; Fax: +1 940 6684871; E-mail: p.wilson@nctc.cc.tx.us
1926; Patsy Wilson
48 000 vols; 280 curr per   33361

Santa Fe College, Lawrence W. Tyree Library, 3000 NW 83rd St, **Gainesville**, FL 32606
T: +1 352 3955150; Fax: +1 352 3957326; E-mail: myra.sterrett@sfcollege.edu; URL: dept.sfcollege.edu/library
1966; Myra Sterrett
89 030 vols; 550 curr per   33362

Carl Sandburg College, Learning Resources Center, 2400 Tom L Wilson Blvd, **Galesburg**, IL 61401
T: +1 309 3415290; Fax: +1 309 3443526; E-mail: sand@darkstar.rsa.lib.il.us; URL: www.csc.cc.il.us
1967; Michael Eugene Walters
Vocational education; Bill Campbell Cartoon Art Coll, original graphic art; Carl Sandburg & Institutional Arch; Science Fiction Paperback Coll
40 780 vols; 420 curr per
libr loan   33363

Volunteer State Community College, Library-Learning Resources Center, 1480 Nashville Pike, **Gallatin**, TN 37066-3188
T: +1 615 2303400; Fax: +1 615 2303410; E-mail: louise.kelly@volstate.edu; URL: www.volstate.edu/library
1971; Louise Kelly
48 000 vols; 350 curr per; 47 maps; 65 400 microforms; 2 179 av-mat; 8 sound-rec; 8 digital data carriers; 33 overhead transparencies
libr loan   33364

Galveston College, David Glenn Hunt Memorial Library, 4015 Avenue Q, **Galveston**, TX 77550
T: +1 409 7636551 ext 240; Fax: +1 409 7629367; E-mail: gwilson@tusk.gc.edu; URL: www.gc.edu/library
1967; Gary E. Wilson
42 000 vols; 205 curr per   33365

Garden City Community College, Thomas F. Saffell Library, 801 Campus Dr, **Garden City**, KS 67846
T: +1 620 2769511; Fax: +1 620 2769630; E-mail: library@gccc.cc.ks.us; URL: www.gccc.cc.ks.us
1919; William Utz
42 000 vols; 200 curr per
libr loan   33366

Nassau Community College, A. Holly Patterson Library, One Education Dr, **Garden City**, NY 11530-6793
T: +1 516 5727400; Fax: +1 516 5727846; URL: library.ncc.edu
1959; Nancy Williamson
Dozenal Society, G Wilson Knight Interdisciplinary Society
179 000 vols; 500 curr per; 17 875 microforms; 5 804 av-mat; 12 026 sound-rec; 40 digital data carriers; 53 080 Slides, 1 052 Overhead Transparencies   33367

Mount Wachusett Community College, Library, 444 Green St, **Gardner**, MA 01440
T: +1 978 6326600 ext 125; Fax: +1 978 6321210; E-mail: l.oldach@mwcc.mass.edu; URL: www.mwcc.edu/Html/Library/index.html
1964
56 340 vols; 290 curr per   33368

Mississippi Gulf Coast Community College, Jackson County Campus Library, 2300 Hwy 90, P.O. Box 100, **Gautier**, MS 39553
T: +1 228 4977716; Fax: +1 228 4977643; E-mail: pamela.ladner@mgccc.cc.ms.us; URL: www.mgccc.cc.ms.us
1965; Pamela Ladner
30 000 vols; 349 curr per   33369

Delaware Technical & Community College – Stephen J. Betze Library, P.O. Box 630, **Georgetown**, DE 19947
T: +1 302 8569033; Fax: +1 302 8585462; E-mail: jpainter@college.dtcc.edu; URL: www.library.dtcc.edu
1967; John C Painter
Delaware Coll
54 890 vols; 529 curr per; 6 429 microforms; 257 av-mat; 2 743 sound-rec; 5 digital data carriers
libr loan   33370

Montgomery College – Germantown Campus, Library, 20200 Observation Dr, **Germantown**, MD 20876
T: +1 301 3537849; Fax: +1 301 3537859; E-mail: diane.cockrell@montgomerycollege.edu; URL: www.montgomerycollege.edu
1978
50 000 vols; 447 curr per   33371

Gavilan College, Library, 5055 Santa Teresa Blvd, **Gilroy**, CA 95020
T: +1 408 8484812; Fax: +1 408 8464927; E-mail: sauyeung@gavilan.cc.ca.us; URL: www.gavilan.cc.ca.us/library
1963; Shukchun Auyeung
Hispanic Resources Coll
46 720 vols; 2 513 curr per   33372

Glendale Community College, Library Media Center, 6000 W Olive Ave, **Glendale**, AZ 85302
T: +1 623 8453101; Fax: +1 623 8453103; E-mail: stefanie.macias@gcmail.maricopa.edu; URL: lib.gccaz.edu/lmc/
1965; David M. Rodriguez
76 570 vols; 439 curr per
libr loan   33373

Glendale Community College, Library, 1500 N Verdugo Rd, **Glendale**, CA 91208-2894
T: +1 818 2401000 ext 5574; Fax: +1 818 2465107; E-mail: lwinters@glendale.edu; URL: www.glendale.edu/library
1927; Dr. Linda S. Winters
123 700 vols; 270 curr per
libr loan; ALA, CLA   33374

Citrus College, Hayden Memorial Library, 1000 W Foothill Blvd, **Glendora**, CA 91741-1899
T: +1 626 9148640; Fax: +1 626 9632531; URL: www.citruscollege.edu
1915; John R. Thompson
Schlesinger Coll (Astronomy)
38 610 vols; 125 curr per
libr loan   33375

Lewis & Clark Community College, Reid Memorial Library, 5800 Godfrey Rd, **Godfrey**, IL 62035
T: +1 618 4684301; Fax: +1 618 4661294; E-mail: lce.lxe@lcls.lib.il.us; URL: www.lc.cc.il.us/libweb.nsf
1970; Brett Reinert
Monticello College Hist
30 000 vols; 36 curr per   33376

Wayne Community College, Library, 3000 Wayne Memorial Dr, P.O. Box 8002, **Goldsboro**, NC 27533-8002
T: +1 919 7355151 ext 264; Fax: +1 919 7363204; E-mail: sj@wcc.wayne.cc.nc.us; URL: www.wayne.cc.nc.us/library/wcclib.htm
1965; Dr. Shirley T. Jones
Local Genealogy Coll
42 000 vols; 341 curr per; 50 516 microforms; 3 639 av-mat   33377

Idaho School for the Deaf & Blind Library, 1450 Main St, **Gooding**, ID 83330
T: +1 208 9344457; Fax: +1 208 9348352; E-mail: scobble@isdh.state.id.us
Shirley Cobble
30 000 vols; 2000 bks on deafness & sign language   33378

Idaho School for the Deaf & Blind Library, 1450 Main St, **Gooding**, ID 83330
T: +1 208 9344457; Fax: +1 208 9348352; URL: www.isdb.idaho.gov
Shirley Cobble
30 000 vols   33379

Holmes Community College, McMorrough Library, P.O. Box 439, **Goodman**, MS 39079
T: +1 662 4722312 ext 1049; Fax: +1 662 4729155; E-mail: kboggan@holmes.cc.ms.us; URL: www.holmes.cc.ms.us
1928; Kay Bouggan
Mississippi, Shakespeare
51 000 vols; 373 curr per   33380

Grand Rapids Community College, Arthur Andrews Memorial Library, 140 Ransom Ave NE, **Grand Rapids**, MI 49503; mail address: 143 Bostwick Ave NE, Grand Rapids, MI 49503
T: +1 616 2343749; Fax: +1 616 2343889; URL: www.grcc.edu/library
1914; Anita Cook
75 000 vols; 446 curr per; 26 000 e-books; 23 av-mat   33381

Aiken Technical College, ATC Library, 2276 J. Davis Hwy, **Graniteville**, SC 29829; PO Drawer 696, Aiken, SC 29802-0696
T: +1 803 5939231 ext 1330; Fax: +1 803 5932169; E-mail: geisena@atc.edu; URL: www.atc.edu/p35.aspx
1973; Katie Miller
31 900 vols; 221 curr per
libr loan   33382

Rogue Community College, Library, 3345 Redwood Hwy, **Grants Pass**, OR 97527
T: +1 541 9567500; Fax: +1 541 4713588; E-mail: lkettler@rogue.cc.or.us; URL: ch.rogue.or.us/departments/library.html
1971; Lynda Kettler
Allied health, Nursing, Business; Oregon Outdoors
33 600 vols; 390 curr per
libr loan   33383

College of Lake County, Learning Resource Center, 19351 W Washington St, **Grayslake**, IL 60030
T: +1 847 5432619; Fax: +1 847 2237690; E-mail: cbakker@clcillinois.edu; URL: library.clcillinois.edu
1970; Connie Bakker
ERIC Docs Coll
121 970 vols; 550 curr per; 7 270 av-mat; 840 digital data carriers   33384

Barton County Community College, Library, 245 NE 30 Rd, **Great Bend**, KS 67530
T: +1 620 7929362; Fax: +1 620 7923238; URL: www.barton.cc.ks.us/library
1969; Carol Barta
Nursing; Rural Gerontology Grant Coll
33 000 vols; 13 digital data carriers   33385

Aims Community College, Kiefer Library, 5401 W 20th St, **Greeley**, CO 80634-3002; P.O. Box 69, Greeley, CO 80632-0069
T: +1 970 3308008 ext 6618; Fax: +1 970 3396568; E-mail: info@aims.edu; URL: www.aims.edu
1970; Jean Warnke
Colorado Hist Coll
39 800 vols; 257 curr per   33386

Greenfield Community College, Library, One College Dr, **Greenfield**, MA 01301-9739
T: +1 413 7751830; Fax: +1 413 7751838; E-mail: cletson@gcc.mass.edu; URL: www.gcc.mass.edu
1962; Carol Letson
Archibald MacLeish Coll, Massachusetts Census Data Ctr, Pioneer Valley Resource Ctr, Yankee-Rowe Local Public Doc Coll
50 310 vols; 351 curr per   33387

Greenville Technical College, Library, 506 S Pleasantburg Dr, P.O. Box 5539, **Greenville**, SC 29606
T: +1 864 2508319; Fax: +1 864 2508506; E-mail: elliottlge@gvltec.edu; URL: www.greenvilletech.com
1962; Dr. L. Gene Elliott
57 220 vols; 730 curr per; 21 085 microforms; 1 940 sound-rec; 21 digital data carriers   33388

Pitt Community College, Learning Resources Center, Hwy 11 S, PO Drawer 7007, **Greenville**, NC 27835
T: +1 252 3214357; Fax: +1 252 3214404; E-mail: pittlrc@pcc.pitt.cc.nc.us; URL: www.pitt.cc.nc.us/lrc/lrc.htm
1964; Lisa C. Driver
43 830 vols; 575 curr per   33389

Piedmont Technical College, Library, 620 N Emerald Rd, PO Drawer 1467, ***Greenwood***, SC 29648
T: +1 864 9418440; Fax: +1 864 9418558; E-mail: nicholson_r@piedmont.tec.sc.us; URL: www.piedmont.tec.sc.us/library
1966; Ruth Nicholson
31 000 vols; 330 curr per; 3 500 microforms; 3 028 av-mat; 473 sound-rec; 8 digital data carriers; 600 slides with A-tapes 33390

Mount Hood Community College, Library, 26000 SE Stark St, ***Gresham***, OR 97030
T: +1 503 4917161; Fax: +1 503 4917389; E-mail: reference@mhcc.edu; URL: www.mhcc.edu/library
1965; Jeff Ring
60 000 vols; 475 curr per
libr loan 33391

Mississippi Gulf Coast Community College, Jefferson Davis Campus Learning Resource Center, 2226 Switzer Rd, ***Gulfport***, MS 39507
T: +1 228 8962525; Fax: +1 228 8962521; E-mail: foster.flint@mgccc.cc.ms.us; URL: www.mgccc.cc.ms.us
Foster Flint
McNaughton
41 100 vols; 232 curr per 33392

Hagerstown Community College, Library, 11400 Robinwood Dr, ***Hagerstown***, MD 21742-6590
T: +1 301 7902800 ext 237; Fax: +1 301 3933681; E-mail: library@hcc.cc.md.us; URL: www.hcc.cc.md.us/library
1946; James R. Feagin
48 000 vols; 3 digital data carriers 33393

Hilbert College, McGrath Library, 5200 S Park Ave, ***Hamburg***, NY 14075
T: +1 716 6497900; Fax: +1 716 6486530; URL: www.hilbert.edu/library2.asp
1955; Wil Prout
Polish Language
36 000 vols; 337 curr per; 102 av-mat
libr loan 33394

Sacred Heart Academy, Mary & James Dimeo Library, c/o Sacred Heart Academy, 265 Benham St, ***Hamden***, CT 06514
T: +1 203 2882309; Fax: +1 203 2309680; E-mail: library@sha-excelsior.org; URL: sha-excelsior.org
1946; Sister Mary Matthew Papallo, Mary Jo Lee
40 000 vols; 150 curr per; 20 maps; 3 500 microforms; 355 av-mat; 640 sound-rec; 22 digital data carriers; 200 pamphlets 33395

Richmond Community College, Library, Hwy 74, P.O. Box 1189, ***Hamlet***, NC 28345
T: +1 910 5827000, 7043; Fax: +1 910 5827045; E-mail: ehartzell@richmond.cc.nc.us; URL: www.richmond.cc.nc.us
1964; Emily U. Hartzell
Technology, Vocational education; North Carolina
30 000 vols; 200 curr per 33396

Thomas Nelson Community College, Library, Wythe Hall 228, 99 Thomas Nelson Dr, ***Hampton***, VA 23666; P.O. Box 9407, Hampton, VA 23670-0407
T: +1 757 8252876; Fax: +1 757 8252870; URL: www.tncc.edu
1968; Aileen Schweitzer
59 000 vols; 346 curr per 33397

Roane State Community College, Roane County Main Campus, Library, 276 Patton Lane, ***Harriman***, TN 37748
T: +1 865 8824553; Fax: +1 865 8824646; E-mail: librarystaff@roanestate.edu; URL: www.rscc.cc.tn.us/library
1971
52 500 vols; 560 curr per 33398

Harrisburg Area Community College, McCormick Library, One HACC Dr, ***Harrisburg***, PA 17110-2999
T: +1 717 7802460; Fax: +1 717 7802462; E-mail: allubrec@hacc.edu; URL: lib2.hacc.edu
1964; Alice L. Lubrecht
98 685 vols; 597 curr per; 105 e-books; 6 629 av-mat; 17 Bks on Deafness & Sign Lang 33399

University of Hartford, Hartford College for Women, Bess Graham Library, 1265 Asylum Ave, ***Hartford***, CT 06105
T: +1 860 7685647; Fax: +1 860 7685693; E-mail: metcalfe@mail.hartford.edu; URL: libaxp.hartford.edu/llr/online.htm
1939; Sara Metcalfe
Humanities, Legal, Women's studies, Lit
40 000 vols; 135 curr per 33400

Northern Essex Community College, Bentley Library, 100 Elliott St, ***Haverhill***, MA 01830
T: +1 978 5563400; Fax: +1 978 5563738; E-mail: lshea@necc.mass.edu; URL: www.necc.mass.edu/departments/library
1961; Linda Hummel-Shea
62 620 vols; 302 curr per
libr loan 33401

Chabot College, Learning Resource Center, 25555 Hesperian Blvd, ***Hayward***, CA 94545
T: +1 510 7236778; Fax: +1 510 7237005; E-mail: jmatthews@chabotcollege.edu; URL: www.chabotcollege.edu/library/
1961; Kim Morrison
Fire science, Nursing; California Hist Coll
85 460 vols; 295 curr per 33402

Life Chiropractic College – West Library, 25001 Industrial Blvd, ***Hayward***, CA 94545
T: +1 510 780-4507/-4599 ext 2930; Fax: +1 510 7804525; E-mail: aosenga@lifewest.edu; URL: www.lifewest.edu
1971; Annette Osenga
College arch; Chiropractic (rare bk coll)
15 000 vols; 250 curr per; 2 500 av-mat; 750 sound-rec; 100 digital data carriers; videos, charts, models, x-rays, equipment, video discs, slides, transparencies, instructional software
libr loan; MLA 33403

Hazard Community College, Library, One Community College Dr, ***Hazard***, KY 41701
T: +1 606 4365721 ext 347; Fax: +1 606 4391657; E-mail: eileenhaddix@kctcs.net; URL: www.hazcc.kctcs.net/library/library.htm
1968; Eileen C. Haddix
Genealogy, Allied health, Local hist
42 000 vols; 165 curr per 33404

Henderson Community College, Hartfield Library, 2660 S Green St, ***Henderson***, KY 42420-4699
T: +1 270 8271867, 8305267; Fax: +1 270 8278635; URL: www.hencc.kctcs.net/library
1960; Michael Knecht
30 210 vols; 225 curr per 33405

Vance-Granville Community College, Learning Resources Center, P.O. Box 917, ***Henderson***, NC 27536
T: +1 252 4922061; Fax: +1 919 4300460; E-mail: oakley@vgcc.cc.nc.us; URL: www.vgcc.cc.nc.us
1970; Sondra Oakley
37 260 vols; 482 curr per 33406

Herkimer County Community College, Ronald F. Williams Library, Reservoir Rd, ***Herkimer***, NY 13350
T: +1 315 8660300 ext 270; Fax: +1 315 8661806; E-mail: library@herkimer.edu; URL: www.herkimer.edu/library
1967; Scott DiMarco
80 000 vols; 294 curr per
libr loan 33407

Catawba Valley Community College, Learning Resource Center, 2550 Hwy 70 SE, ***Hickory***, NC 28602-9699
T: +1 828 3277000 ext 4229; Fax: +1 828 3245130; E-mail: jpayne@cvcc.cc.nc.us; URL: www.cvcc.cc.nc.us
1960; James M. Payne
Furniture, Decoration
46 000 vols; 270 curr per 33408

Hill College, Library, 112 Lamar, P.O. Box 619, ***Hillsboro***, TX 76645-0619
T: +1 254 5822555 ext 240; Fax: +1 254 5827591; E-mail: library@hill-college.cc.tx.us; URL: www.hill-college.cc.tx.us/library
1962; Joe Shaughnessy
Civil War Res Ctr
32 600 vols; 226 curr per 33409

Southern State Community College, Learning Resources Center, 100 Hobart Dr, ***Hillsboro***, OH 45133-9487
T: +1 937 3933431 ext 2681; Fax: +1 937 3939370; E-mail: lmays@soucc.southern.cc.oh.us; URL: lrc.southern.cc.oh.us
1975; Louis Mays
Southern Ohio Genealogical Society Coll
42 100 vols; 2 349 curr per 33410

New Mexico Junior College, Pannell Library, 1 Thunderbird Circle, ***Hobbs***, NM 88240
T: +1 575 3925473; Fax: +1 505 3923668; E-mail: library@nmjc.edu; URL: www.nmjc.edu
1965; Sharon D. Jenkins
Waste Isolation Pilot Plant Depot
88 000 vols; 343 curr per 33411

Northland Pioneer College, Libraries, P.O. Box 610, ***Holbrook***, AZ 86025
T: +1 928 5247324; Fax: +1 928 5247321; URL: www.npc.edu/lib
1975; Trudy Bender
31 800 vols; 252 curr per; 12 647 govt docs; 86 maps; 6 801 microforms; 705 av-mat; 128 sound-rec; 69 digital data carriers 33412

Holyoke Community College, Library, 303 Homestead Ave, ***Holyoke***, MA 01040-1099
T: +1 413 5522372; E-mail: rstoddard@hcc.edu; URL: www.hcc.edu/campus/library.html
1946
Libr of English Lit, ultrafiche
75 000 vols; 150 curr per 33413

Honolulu Community College, Library, 874 Dillingham Blvd, ***Honolulu***, HI 96817-4598
T: +1 808 8459199; Fax: +1 808 8453618; E-mail: library@hcc.hawaii.edu; URL: www.honolulu.hawaii.edu/library
1965; Irene Mesina
Liberal arts, Occupational; Automotive Tech Coll, Hawaii & Pacific Coll
55 000 vols; 140 curr per 33414

Kapi'olani Community College, Library, 4303 Diamond Head Rd, ***Honolulu***, HI 96816
T: +1 808 7349259; Fax: +1 808 7349453; E-mail: smurata@hawaii.edu; URL: library.kcc.hawaii.edu/error.php
1966; Susan Murata
Hawaii, Japan; Char Coll (Chinese Hist & Culture)
71 570 vols; 400 curr per 33415

Hopkinsville Community College, Library, 720 North Dr, ***Hopkinsville***, KY 42240; P.O. Box 2100, Hopkinsville, KY 42241-2100
T: +1 270 8863921; Fax: +1 270 8856048; E-mail: cynthia.atkins@kctcs.net; URL: www.hopkinsville.kctcs.edu/library/
1965; Cynthia Atkins
Kentucky
43 370 vols; 156 curr per 33416

Houston Community College, Central College Library, 1300 Holman, ***Houston***, TX 77004
T: +1 713 7186133; Fax: +1 713 7186154; E-mail: aska.librarian@hccs.edu; URL: learning.cc.hccs.edu/Library
1972; Ronald Homick
56 925 vols; 97 curr per; 19 000 e-books; 4 161 av-mat 33417

North Harris Montgomery Community College District – North Harris College, Library, 2700 W Thorne Dr, ***Houston***, TX 77073
T: +1 281 6185491; Fax: +1 281 6185695
1973; Maryann Readal
ERIC Junior College Fiche Coll
101 000 vols; 1 041 curr per; 11 870 av-mat 33418

San Jacinto College North, Doctor Edwin E. Lehr Library, 5800 Uvalde Rd, ***Houston***, TX 77049-4589
T: +1 281 4597116; Fax: +1 281 4597166; URL: www.sjcd.edu/library_5225.html
1974; Jan C. Crenshaw
Texana Coll
62 000 vols; 349 curr per 33419

San Jacinto College South, Parker Williams Library, 13735 Beamer Rd, ***Houston***, TX 77089-6099
T: +1 281 9223416; E-mail: richard.mckay@sjcd.edu; URL: www.sanjac.edu/current-students/welcome/libraries
1979; Richard McKay
American hist, American & British lit, Health sciences; Texana
60 000 vols; 300 curr per 33420

Caldwell Community College & Technical Institute, Broyhill Center for Learning Resources, 2855 Hickory Blvd, ***Hudson***, NC 28638
T: +1 828 7262309; Fax: +1 704 7262603; E-mail: mcooke@caldwell.cc.nc.us
1966; Marischa B. Cooke
58 000 vols; 398 curr per 33421

Columbia-Greene Community College, Library, 4400 Rte 23, ***Hudson***, NY 12534
T: +1 518 8284181; Fax: +1 518 8284396; E-mail: erceg@sunycgcc.edu; URL: albweb2.sunyconnect.suny.edu/golgr
1969; Lynn Erceg
Ettelt Children's Coll, Map Coll
54 900 vols; 400 curr per 33422

Golden West College, R. Dudley Boyce Library & Learning Center, 15744 Golden West St, ***Huntington Beach***, CA 92647
T: +1 714 8958741; Fax: +1 714 8952926; URL: goldenwestcollege.edu/library/
1966; Douglas Larson
90 020 vols; 290 curr per
libr loan 33423

Tarrant County College, Northeast Library, 828 W Harwood Rd, ***Hurst***, TX 76054
T: +1 817 5156627; E-mail: ask.librarian@tccd.edu; URL: lib-serv.tccd.edu/libraries/northeast
1968
68 500 vols; 395 curr per
libr loan 33424

Hutchinson Community College, John F. Kennedy Library, 1300 N Plum, ***Hutchinson***, KS 67501
T: +1 316 665-3548; Fax: +1 316 665-3392; E-mail: kellyr@hutchcc.edu; URL: wwwcms.hutchcc.edu/
1928; Pat Vierthaler
FAA Resource Ctr contains info on flying
44 130 vols; 352 curr per
libr loan 33425

Imperial Valley College, Spencer Library Media Center, 380 E Ira Aten Rd, P.O. Box 158, ***Imperial***, CA 92251-0158
T: +1 760 3556378; Fax: +1 760 3551090; E-mail: toni.gamboa@imperial.edu; URL: www.imperial.edu/index.php?pid=790
1922
Local hist
60 680 vols; 463 curr per
libr loan 33426

Rend Lake College, Learning Resource Center, 468 N Ken Gray Pkwy, ***Ina***, IL 62846
T: +1 618 4375321; Fax: +1 618 4375598; E-mail: reference@rlc.cc.il.us; URL: www.rlc.cc.il.us/lrc/lrchome.htm
1956; Andrea Whitoff
35 080 vols; 278 curr per 33427

Inver Hills Community College, Library, 2500 80th St E, *Inver Grove Heights*, MN 55076-3209
T: +1 651 4508625; Fax: +1 651 4508679; E-mail: jbenolk@ih.cc.mn.us; URL: www.inverhills.mnscu.edu/library
1970; Julie Benolken, Ann Schroder
43 000 vols; 250 curr per   33428

Allen Community College, Library, 1801 N Cottonwood, *Iola*, KS 66749-1698
T: +1 620 3655116 ext 208; Fax: +1 620 3653284; E-mail: admissions@ allencc.edu; URL: www.allen.cc.ks.us/web/index.htm
1970; Steven Wells Anderson
Local genealogy
38 650 vols; 94 curr per   33429

Irvine Valley College Library, 5500 Irvine Ctr Dr, *Irvine*, CA 92618
T: +1 949 4515761; Fax: +1 949 4515796; E-mail: fforbes@ivc.cc.ca.us; URL: www.ivc.cc.ca.us/infoserv/library/library.html
1979; Fred Forbes
45 000 vols; 240 curr per   33430

North Lake College, Library, 5001 N MacArthur Blvd, *Irving*, TX 75038-3899
T: +1 972 2733400; Fax: +1 972 2733431; E-mail: ekc7610@dcccd.edu
Enrique Chamberlain
45 000 vols; 325 curr per   33431

Hazard Community College, Lees College Campus Library, 601 Jefferson Ave, *Jackson*, KY 41339
T: +1 606 6667521 ext 653; Fax: +1 606 6668910; URL: www.leecc.uky.edu/lees-library/index.htm
1883; Robert Hilton
Local hist, Religion; Appalachia & Kentuckiana
40 000 vols; 200 curr per   33432

Jackson Community College, Atkinson Learning Resources Center, 2111 Emmons Rd, *Jackson*, MI 49201
T: +1 517 7878622; Fax: +1 517 7968623; URL: www.jackson.cc.mi.us
1928; Cliff Taylor
51 940 vols; 325 curr per; 314 av-mat; 3 246 sound-rec; 1 800 slides   33433

Jackson State Community College, Library, 2046 N Parkway, *Jackson*, TN 38301
T: +1 731 4252609; Fax: +1 731 4252625; E-mail: scohen@jscc.cc.tn.us; URL: www.jscc.cc.tn.us/library
1967; Scott Cohen
59 000 vols; 272 curr per   33434

Lexington School for the Deaf, Library Computer Center, 30th Ave & 75th St, *Jackson Heights*, NY 11370
T: +1 718 3503126; Fax: +1 718 8999846; URL: www.lexnyc.org
Marie-Ann Marchese
African-American, Asian studies, Hispanic, Multicultural, Native American; Arch Related to Deafness, Education of the Deaf & Hist of Lexington School
18 840 vols; 46 curr per; 1 392 av-mat; 20 digital data carriers   33435

Coastal Carolina Community College, Learning Resources Center, 444 Western Blvd, *Jacksonville*, NC 28546-6877
T: +1 910 9386237; Fax: +1 910 4457027; E-mail: asklibrarian@coastal.cc.nc.us; URL: www.costal.cc.nc.us
1965; Linda Muir
42 500 vols; 250 curr per   33436

Florida Community College at Jacksonville, Kent Campus Library, 3939 Roosevelt Blvd, *Jacksonville*, FL 32210
T: +1 904 3813522; Fax: +1 904 3813579; URL: www.fccj.org/library/
1966; Ken Puckett
86 000 vols; 453 curr per   33437

Illinois School for the Deaf, Library for the Deaf, 125 Webster, *Jacksonville*, IL 62650
T: +1 217 4794240; Fax: +1 217 4794244; URL: www.morgan.k12.il.us/isd
Nancy Kelly-Jones
Deafness, Deaf education, Sign Language; captioned videos
15 000 vols; 50 curr per; 25 digital data carriers; 9 000 other items
libr loan   33438

Illinois School for the Visually Impaired Library, 658 E State St, *Jacksonville*, IL 62650-2184
T: +1 217 4794400 ext 471; Fax: +1 217 4794479; URL: www.morgan.k12.il.us/isvi
27 000 vols; 40 curr per   33439

Guilford Technical Community College, Mertys W. Bell Library, 601 High Point Rd, P.O. Box 309, *Jamestown*, NC 27282-0309
T: +1 336 3344822 ext 2292; Fax: +1 336 8414350; E-mail: eref@gtcc.edu; URL: www.gtcc.edu/lib/
1958; A. Beverly Gass
79 110 vols; 372 curr per; 32 483 microforms; 7 286 av-mat; 19 digital data carriers; 8 VF drawers of clippings & pamphlets; 8 VF drawers of annual rpts, 12 VF drawers of archives   33440

Jamestown Community College, Hultquist Library, 525 Falconer St, *Jamestown*, NY 14701
T: +1 716 6655220; Fax: +1 716 6655518; E-mail: DennisBenamat@mail.sunyjcc.edu; URL: www.sunyjcc.edu/hultquist
1950; Dennis C. Benamati
Scandinavian Studies; Chautauqua County, NY
65 000 vols; 450 curr per; 270 maps; 66 497 microforms; 1 749 av-mat; 4 334 sound-rec; 20 digital data carriers; 8 563 slides, 58 overHead transparencies, 112 art repros
libr loan; OCLC, ALA   33441

Wisconsin Center for the Blind & Visually Impaired, Wisconsin School for the Visually Handicapped Library, 1700 W State St, *Janesville*, WI 53546
T: +1 608 7586118; Fax: +1 608 7586161
Michelle Rueckert
12 000 vols; 90 curr per; braille, talking & large print bks
libr loan   33442

Manor College, Basileiad Library, 700 Fox Chase Rd, *Jenkintown*, PA 19046-3399
T: +1 215 8852360 ext 238; Fax: +1 215 5766564; E-mail: jholst@manor.edu; URL: library.manor.edu
1947; Jerome A. Holst
Lit, Ukrainian hist; Ukrainian Language Bk Coll (3 000 vols)
46 000 vols; 212 curr per; 178 microforms; 110 av-mat; 580 sound-rec; 35 digital data carriers; 1 420 slides
ALA, CLA   33443

Fulton-Montgomery Community College, Evans Library, 2805 State Hwy 67, *Johnstown*, NY 12095-3790
T: +1 518 7624651 ext 5600; Fax: +1 518 7623834; E-mail: mdonohue@fmcc.suny.edu; URL: fmcc.suny.edu/library
1964; Mary Donohue
Applied sciences, Human servs, Regional hist
53 000 vols; 150 curr per   33444

Joliet Junior College, Library, Learning Resource Center, 1216 Houbolt Rd, *Joliet*, IL 60436
T: +1 815 7299020 ext 2350; Fax: +1 815 7442465; URL: www.jjc.cc.il.us/dept/library.html
1902; Dr. Denis Wright
Education, Nursing, Horticulture; Soil Surveys, Children's Bks
70 000 vols; 675 curr per   33445

Maui Community College, Library, 310 Kaahumanu Ave, *Kahului*, HI 96732
T: +1 808 9843233; Fax: +1 808 2449644; E-mail: mcclib@hawaii.edu; URL: www.hawaii.edu/maui/library
1970; Dorothy Tolliver
Hawaii
62 000 vols; 298 curr per   33446

Kalamazoo Valley Community College, Texas Township Campus, Library, 6767 W O Ave, P.O. Box 4070, *Kalamazoo*, MI 49003-4070
T: +1 269 4884328; Fax: +1 269 4884488; E-mail: jalm@kvcc.edu; URL: www.kvcc.edu
1968; Janet Alm
Career, Sign language; Alva Dorn Photogr Coll, Mary Mace Spradling

African-American Coll, Ned Rubinstein Memorial Coll (Michigan Hist)
92 810 vols; 329 curr per   33447

Windward Community College, Library, 45-720 Keaahala Rd, *Kaneohe*, HI 96744
T: +1 808 2357436; Fax: +1 808 2357344; E-mail: heu@hawaii.edu; URL: www.library.wcc.hawaii.edu
1972; Nancy Heu
Hawaiian coll
46 780 vols; 195 curr per; 390 maps; 46 microforms; 1 368 av-mat; 1 983 sound-rec; 13 digital data carriers; 216 art reproductions, 3 264 pamphlets
libr loan   33448

Kankakee Community College, Learning Resource Center, River Rd, P.O. Box 888, *Kankakee*, IL 60901
T: +1 815 9330260; Fax: +1 815 9330217; E-mail: dsmith@kcc.cc.il.us; URL: www.kcc.cc.il.us
1966; Donna Smith
Gordon Graves Environmental Coll, Reece L Ayers Soil & Water Conservation Coll
45 160 vols; 392 curr per   33449

Donnelly College, Trant Memorial Library, 608 N 18th St, *Kansas City*, KS 66102
T: +1 913 6218735; Fax: +1 913 6218719; URL: www.donnelly.edu
1949; Tom Brown
Biblical studies, Women's studies; Black Hist (Roe Coll)
30 000 vols; 115 curr per   33450

Kansas City Kansas Community College, Library, 7250 State Ave, *Kansas City*, KS 66112-3098
T: +1 913 3341100 ext 7650; Fax: +1 913 5969606; E-mail: cheryl@toto.net; URL: kckcc.cc.ks.us/college-support-services/institutional-services/library
1923; Cheryl Postlewait
Education, Nursing, Mortuary science; Houston Gray Film Memorabilia
60 000 vols; 400 curr per   33451

Metropolitan Community Colleges – Maple Woods Community College, Library, 2601 NE Barry Rd, *Kansas City*, MO 64156
T: +1 816 4373080; Fax: +1 816 4373082
1969; Linda Wilson
Veterinary Medicine Coll
30 000 vols; 273 curr per   33452

Metropolitan Community Colleges – Penn Valley Community College, Library, 3200 Pennsylvania Ave, mail address: 3201 SW Trafficway, *Kansas City*, MO 64111
T: +1 816 7594082; Fax: +1 816 7594374; URL: mcckc.edu/main.asp?L=DPPennValley
1969
Kansas City Hist
84 260 vols; 184 curr per
libr loan   33453

College of Marin, Library, 835 College Ave, *Kentfield*, CA 94904
T: +1 415 4578811; Fax: +1 415 4575395; E-mail: ccox@marin.cc.ca.us; URL: www.marin.cc.us
1926; Carl Cox
137 000 vols; 331 curr per   33454

Schreiner College, W.M. Logan Library, 2100 Memorial Blvd, *Kerrville*, TX 78028-5697
T: +1 830 7927312; Fax: +1 830 7927448; URL: library.schreiner.edu
1967; Dr. Candice Scott
Liberal arts; Texas Hill Country Coll
130 000 vols; 260 curr per
libr loan; OCLC   33455

Florida Keys Community College, Library, 5901 College Rd, *Key West*, FL 33040
T: +1 305 2969081 ext 210; Fax: +1 305 2925162; E-mail: soule_m@firn.edu; URL: www.fkcc.edu/
1965; Maria J. Soule
30 400 vols; 150 curr per
libr loan   33456

Kilgore College, Randolph C. Watson Library, 1100 Broadway, *Kilgore*, TX 75662
T: +1 903 9838237; Fax: +1 903 9838638; E-mail: kfair@kilgore.edu; URL: www.kilgore.edu/library.asp
1935; Kathy Fair
Habenicht Texana Coll; Hill Texana Coll; Spear Coll (American & English Lit)
85 000 vols   33457

Mohave Community College Library System, 1971 Jagerson Ave, *Kingman*, AZ 86401
T: +1 520 7570883; Fax: +1 520 7570871; E-mail: robshu@mohave.edu; URL: www.mohave.edu
1971; Robert Shupe
55 915 vols; 523 curr per; 436 maps; 682 av-mat; 2 digital data carriers; 247 art reproductions   33458

North Harris Montgomery Community College District – Kingwood College, Library, 20000 Kingwood Dr, *Kingwood*, TX 77339
T: +1 281 3121691; Fax: +1 281 3121456; URL: www.kingwoodcollegelibrary.com
1984; Peggy Whitley
36 600 vols; 441 curr per   33459

Lenoir Community College, Learning Resources Center, 231 Hwy 58 S, *Kinston*, NC 28504-6836; P.O. Box 188, Kinston, NC 28502-0188
T: +1 252 5276223; Fax: +1 252 5270192; E-mail: lrcinfo@lenoircc.edu; URL: www.lenoircc.edu
1964; Stephen N. Hawkins
Local Eastern North Carolina
50 000 vols; 300 curr per   33460

Lakeland Community College, Library, 7700 Clocktower Dr, *Kirtland*, OH 44094
T: +1 440 9537069; Fax: +1 440 9539710; E-mail: lakelandlibrary@lakelandcc.edu; URL: library.lakelandcc.edu
1967
63 000 vols; 1 784 curr per   33461

Pellissippi State Technical Community College, Library Services, 100915 Hardin Valley Rd, P.O. Box 22990, *Knoxville*, TN 37933-0990
T: +1 865 6946516; Fax: +1 865 6946625; E-mail: pnerzak@pstcc.cc.tn.us; URL: www.pstcc.cc.tn.us/library
1975; Peter Nerzak
45 000 vols; 500 curr per; 2 000 av-mat; 50 sound-rec; 6 digital data carriers; 83 121 cubic feet of microfiche, 1 064 reels of microfilm
libr loan; TLA   33462

Western Wisconsin Technical College, Library, 400 N Seventh St, *La Crosse*, WI 54602-0908
T: +1 608 7859142; Fax: +1 608 7896212; E-mail: library@western.tec.wi.us; URL: www.western.tec.wi.us/library
1967; Patrick J. Brunet
31 000 vols; 325 curr per; 11 digital data carriers; 8 VF drawers of pamphlets   33463

Otero Junior College, Wheeler Library, 20 Pinon Ave, *La Junta*, CO 81050-3347
T: +1 719 3846882; Fax: +1 719 3846883; URL: www.ojc.cccoes.edu/library/main.html
1941; Kendra Schwindt Swope
33
34 310 vols; 230 curr per; 3 digital data carriers
libr loan; ALA, CLA   33464

College of Southern Maryland, Library, 8730 Mitchell Rd, P.O. Box 910, *La Plata*, MD 20646-0910
T: +1 301 9342251 ext 7626; Fax: +1 301 9347699; E-mail: tomr@charlescc.md.us
1958; Thomas Repenning
Local hist; Southern Maryland Mss & Genealogy
40 790 vols; 1 112 curr per   33465

Lake City Community College, G.T. Melton Learning Resources Center, Rte 19, Box 1030, *Lake City*, FL 32025
T: +1 386 7544337; Fax: +1 386 7552686; E-mail: morrisj@mail.lakecity.cc.fl.us; URL: www.lakecity.cc.fl.us
1962; Jim Morris
Art, Allied health, Forestry
47 700 vols; 90 curr per   33466

Brazosport College, Library, 500 College Dr, *Lake Jackson*, TX 77566
T: +1 979 2303310; Fax: +1 979 2303443; URL: www.brazosport.edu
1968; Tami Wisofsky
Children's Coll, Small Business Development Coll, New Bk Coll
63 750 vols; 350 curr per   33467

Palm Beach Community College, Harold C. Manor Library, 4200 Congress Ave, *Lake Worth*, FL 33461
T: +1 561 8683800; Fax: +1 561 8683708; E-mail: kelleyb@pbcc.edu; URL: www.pbcc.cc.fl.us/llrc
1933; Brian C. Kelley
Civil War; Finnish Coll
210 600 vols; 1870 curr per; 19 000 e-journals; 18 000 e-books; 6 500 av-mat   33468

Pierce College, Library, 9401 Farwest Dr SW, *Lakewood*, WA 98498
T: +1 253 9646547; Fax: +1 253 9646713; E-mail: dgilchrist@pierce.ctc.edu; URL: www.pierce.ctc.edu/library/index.html
1967; Debra Gilchrist
120 000 vols; 525 curr per   33469

Red Rocks Community College – Marvin Buckels Library, 13300 W Sixth Ave, *Lakewood*, CO 80228-1255
T: +1 303 9146740; Fax: +1 303 9146741; E-mail: library@rrcc.edu; URL: www.rrcc.edu/library
1969; Joseph Sanchez
50 000 vols; 320 curr per; 14 000 microforms; 5 000 av-mat; 12 digital data carriers; 15 000 slides
libr loan   33470

Antelope Valley College, Library, 3041 W Ave K, *Lancaster*, CA 93536-5426
T: +1 661 7226533; Fax: +1 661 7226456; E-mail: library@avc.edu; URL: www.avc.edu/studentservices/library/index.htm
1962; Diana Gonzalez
American Lit (Evelyn Foley)
45 000 vols; 150 curr per
libr loan   33471

Cedar Valley College, Library, 3030 N Dallas Ave, *Lancaster*, TX 75134-3799
T: +1 972 8608140; Fax: +1 972 8608221; E-mail: ewhite@dcccd.edu; URL: www.dcccd.edu/cvc/cvc.htm
1977; Edna White
43 490 vols; 185 curr per; 363 music scores; 93 152 microforms; 7 116 av-mat; 1 719 sound-rec; 17 digital data carriers; 7 116 slides, 127 overhead transparencies, 5 charts
libr loan; ALA   33472

Prince George's County Public Schools, Professional Library, 8437 Landover Rd, *Landover*, MD 20785-3599
T: +1 301 3861595; Fax: +1 301 3861601; E-mail: dbrady1@pgcps.pg.k12.md.us
Diane Brady
Multicultural Education Coll
10 000 vols; 242 curr per   33473

Prince George's County Public Schools – Professional Library, 8437 Landover Rd, *Landover*, MD 20785-3599
T: +1 301 3861595; Fax: +1 301 3861601
Catherine Francover
Multicultural Education Coll
17 000 vols; 242 curr per   33474

Lansing Community College, Library, 200 Technology & Learning Ctr, 400 N Capitol Ave, *Lansing*, MI 48933; mail address: 1510 Library, P.O. Box 40010, Lansing, MI 48901-7210
T: +1 517 4831657; Fax: +1 517 4835300; URL: www.llc.edu/library
1959; Elenka Raschkow
Career Coll

109 300 vols; 413 curr per; 24 500 e-books; 5 980 av-mat; 100 Electronic Media & Resources   33475

Laredo Community College, Harold R. Yeary Library, West End Washington St, *Laredo*, TX 78040
T: +1 956 7215816; Fax: +1 9567215447; E-mail: tlafleur@laredo.cc.tx.us; URL: www.laredo.edu/library
1947; Thomas LaFleur
Laredo Arch, Old Fort MacIntosh Records
86 900 vols; 360 curr per
libr loan   33476

Prince George's Community College, Media Center, 301 Largo Rd, *Largo*, MD 20774-2199
T: +1 301 3220476; Fax: +1 301 8088847; E-mail: LibRefDesk@pgcc.edu; URL: www.pgcc.edu
1958; Dr. Lynda Byrd Logan
93 500 vols; 450 curr per   33477

Community College of Southern Nevada, Department of Libraries, 6375 W Charleston Blvd, *Las Vegas*, NV 89146
T: +1 702 6515716; Fax: +1 702 6515718; E-mail: irving@nevada.edu; URL: www.ccsn.nevada.edu/library
1972; Marcia Arado
58 000 vols; 540 curr per   33478

Metropolitan Community Colleges – Longview Community College, Library, 500 SW Longview Rd, *Lee's Summit*, MO 64081-2105
T: +1 816 6722266, 2080; Fax: +1 816 6722087; E-mail: swalls@longview.cc.mo.us; URL: wilo.Missouri.edu/search~S3
1969; Scarlett Swall
46 000 vols; 260 curr per   33479

Lake-Sumter Community College, Learning Resources Center, 9501 US Hwy 441, *Leesburg*, FL 34788-8751
T: +1 352 3653541; Fax: +1 352 3653590; E-mail: englishd@lscc.cc.fl.us; URL: www.lscc.cc.fl.us/library
1962; Denise K. English
57 800 vols; 350 curr per   33480

Becker College, Paul Swan Library, 13 Washburn Sq, *Leicester*, MA 01524
T: +1 774 3540065; Fax: +1 508 8927472; URL: www.beckercollege.edu
1941; Jean Collins
Psychology, Philosophy, Veterinary science & technology; Samuel May, 19th C Religion & Philosophy
31 000 vols; 104 curr per   33481

South Plains College, Library, 1401 College Ave, *Levelland*, TX 79336
T: +1 806 8949611 ext 2302; Fax: +1 806 8945274; E-mail: info@southplainscollege.edu; URL: www.southplainscollege.edu
1958
75 000 vols; 570 curr per   33482

Davidson County Community College, Grady E. Love Learning Resources Center, 297 DCCC Rd, *Lexington*, NC 27295; P.O. Box 1287, Lexington, NC 27293-1287
T: +1 336 2498186 ext 270; Fax: +1 336 2488531; E-mail: lburke@davidson.cc.nc.us; URL: www.davidson.cc.nc.us/lrc
1963; Brenda Farmer, Linda Burke
American lit, Art, Architecture, Automotive, Law; Furniture Design & Decoration
57 000 vols; 236 curr per; 100 maps; 44 569 microforms; 3 519 av-mat; 665 sound-rec; 7 digital data carriers; 248 slides, 73 overhead transparencies, 30 art reproductions   33483

Lexington Community College, Learning Resource Center, Cooper Dr, 220 Oswald Bldg, *Lexington*, KY 40506-0235
T: +1 859 2574872; Fax: +1 859 3231091; E-mail: crjames@pop.uky.edu; URL: www.uky.edu/LCC/lib
1976; Charles James
36 000 vols; 218 curr per   33484

Seward County Community College, Learning Resource Center, 1801 N Kansas, *Liberal*, KS 67901-2054; P.O. Box 1137, Liberal, KS 67905-1137
T: +1 620 6292656; Fax: +1 620 6292725; E-mail: dgammell@sccc.net
1967; Denyce G. Gammell
38 500 vols; 300 curr per   33485

Kauai Community College, S.W. Wilcox II Learning Resource Center, 3-1901 Kaumualii Hwy, *Lihue*, HI 96766
T: +1 808 2458322; Fax: +1 808 2458294; URL: kauai.hawaii.edu/library/
1967; Robert M. Kajiwara
Allied health, Nursing; Hawaii & Pacific, videotapes
60 000 vols; 169 curr per
libr loan   33486

Community College of Rhode Island, William F. Flanagan Campus, 1762 Louisquisset Pike, *Lincoln*, RI 02865-4585
T: +1 401 3337058; Fax: +1 401 3337115
1972; Charles D'Arezzo
43 650 vols; 400 curr per   33487

Brookdale Community College, Bankier Library, 765 Newman Springs Rd, *Lincroft*, NJ 07738-1597
T: +1 732 2242706; Fax: +1 732 2242982; URL: bcc-library.brookdale.cc.nj.us
1969; David Murray
77 770 vols; 886 curr per   33488

Arkansas School for the Deaf Library, 2400 W Markham St, *Little Rock*, AR 72205; P.O. Box 3811, Little Rock, AR 72203-3811
T: +1 501 3249515; Fax: +1 501 3249553
Fran Miller
10 000 vols; 75 curr per   33489

Arapahoe Community College, Library, 5900 Santa Fe Dr, P.O. Box 9002, *Littleton*, CO 80160-9002
T: +1 303 7975090; E-mail: librarians@arapahoe.edu; URL: www.arapahoe.edu/student-resources/library
1966; Malcolm Brantz
43 500 vols; 404 curr per
libr loan   33490

Las Positas Community College, Learning Resource Center, 3033 Collier Canyon Rd, *Livermore*, CA 94550-7650
T: +1 925 3734950; Fax: +1 925 6067249; E-mail: llucas@clpccd.cc.ca.us; URL: www.lp1.clccd.cc.ca.us/lpc/lrc
1975; Pamela Luster
32 000 vols; 226 curr per   33491

Schoolcraft College, Eric J. Bradner Library, 18600 Haggerty Rd, *Livonia*, MI 48152-2696
T: +1 734 4624400 ext 5440; Fax: +1 734 4624495; E-mail: library@schoolcraft.edu; URL: www.schoolcraft.edu/library
1964
95 000 vols; 650 curr per   33492

Sullivan County Community College, Hermann Memorial Library, 112 College Rd, *Loch Sheldrake*, NY 12759-5108
T: +1 845 4345750 ext 4389; E-mail: library@sullivan.suny.edu; URL: www.sunysullivan.edu/library
1964
Hotel tech
68 000 vols; 400 curr per; 78 e-journals; 39 000 e-books; 5 000 microforms   33493

Germanna Community College, Locust Grove Library, 2130 Germanna Hwy, *Locust Grove*, VA 22508
T: +1 540 7273120; Fax: +1 540 7273210; URL: www.gcc.vccs.edu
1970; Karen Bowers
Genealogy & Local Hist (Germanna Foundation Society Mat & Genealogies)
31 000 vols; 175 curr per   33494

Long Beach City College, Liberal Arts Campus & Pacific Coast Campus Libraries, 4901 E Carson St, *Long Beach*, CA 90808
T: +1 562 9384025; Fax: +1 562 9384777; E-mail: nbuena@lbcc.cc.ca.us; URL: www.lib.lbcc.edu
1927; Nenita Buenaventura
145 640 vols; 466 curr per
libr loan   33495

Lower Columbia College, Alan Thompson Library, 1600 Maple St, P.O. Box 3010, *Longview*, WA 98632
T: +1 360 5772310; Fax: +1 360 5781400; E-mail: info@lowercolumbia.edu; URL: lcc.ctc.edu/info/contactus.xtm
1934; Mike Gabriel
43 000 vols; 150 curr per   33496

Foothill College, Hubert H. Semans Library, 12345 El Monte Rd, *Los Altos Hills*, CA 94022-4599
T: +1 650 9497390; Fax: +1 650 9497123; E-mail: patz@admin.fhda.edu; URL: www.fhda.edu/foothill
1958; Dr. Penny Patz
Art, Lit, Religion, Philosophy
82 760 vols; 477 curr per   33497

Los Angeles City College, Library, 855 N Vermont Ave, *Los Angeles*, CA 90029
T: +1 323 9534000; Fax: +1 323 9534013; URL: www.lacitycollege.edu/resource/library
1929; Barbara J. Vasquez
146 900 vols; 159 curr per   33498

Los Angeles Southwest College, Library, 1600 W Imperial Hwy, *Los Angeles*, CA 90047
T: +1 323 2415235; Fax: +1 323 2415221
1967; Patricia H. McCollum
African-Am, Hispanic
61 000 vols; 360 curr per   33499

Los Angeles Trade Technical College, Library, 400 W Washington Blvd, *Los Angeles*, CA 90015
T: +1 213 7449025; Fax: +1 213 7490002; E-mail: joyce_livingston@laccd.cc.ca.us; URL: www.lattc.cc.ca.us/dept/tlib.htm
1927; Joyce Livingston
Fashion
73 440 vols; 320 curr per; 400 av-mat; 600 sound-rec; 10 digital data carriers; 150 slides, 40 overhead transparencies   33500

Sinai Temple, Blumenthal Library, 10400 Wilshire Blvd, *Los Angeles*, CA 90024
T: +1 310 4813218; Fax: +1 310 4746801; E-mail: info@sinaitemple.org; URL: www.sinaitemple.org
1969; Lisa Silverman
Judaica, Lit, Children's lit; Sinai Akiba Day School Coll for General Studies, William R Blumenthal Rare Book Coll, Haggadot Coll, Parenting Coll
30 000 vols; 38 curr per; 30 maps; 400 av-mat; 375 sound-rec; 50 digital data carriers; 750 slides, 200 charts
AJL, ALA   33501

Louisburg College, C.W. Robbins Library, 502 N Main St, *Louisburg*, NC 27549-7704
T: +1 919 4973269; Fax: +1 919 4965444; E-mail: admission@louisburg.edu; URL: www.louisburg.edu
1890; Pat Hinton
Religion, North Carolina; Louisburg College & Town Arch, mixed
51 000 vols; 120 curr per; 2 325 microforms; 1 625 sound-rec; 110 overhead transparencies   33502

Jefferson Community College, John T. Smith Learning Resources Center, 109 E Broadway, *Louisville*, KY 40202
T: +1 502 2132157; Fax: +1 502 5854425; E-mail: ccljoill@ukcc.uky.edu; URL: www.jcc.kctcs.net/libraries
1968; Sheree Huber-Williams
52 850 vols; 303 curr per
libr loan   33503

Jefferson Community College – Southwest Campus, Library, 1000 Community College Dr, *Louisville*, KY 40272
T: +1 502 9359840 ext 3210; Fax: +1 502 9358653; URL: www.jcc.uky.edu
1972; Larry Rees
32 500 vols; 190 curr per   33504

Kentucky School for the Blind, Library, 1867 Frankfort Ave, *Louisville*, KY 40206
T: +1 502 8971583 ext 264; Fax: +1 502 8971581; E-mail: chicks@ksb.k12.ky.us
1842; Cathy Hicks

Physical handicaps, Recreational reading; Braille, Talking Bks
18 373 vols; 25 curr per; 47 maps; 642 av-mat; 1 716 sound-rec; 6 digital data carriers; 5 549 braille vols   33505

Robeson Community College, Library, Hwy 301, P.O. Box 1420, *Lumberton*, NC 28359
T: +1 910 7387101 ext 231; Fax: +1 910 6185685; E-mail: robeson@cc.nc.us; URL: www.robeson.cc.nc.us
1965; Marilyn Locklear-Hunt
Vocational tech
39 000 vols; 140 curr per   33506

Central Virginia Community College, Library, 3506 Wards Rd, *Lynchburg*, VA 24502-2498
T: +1 434 8327750; Fax: +1 434 3864677; E-mail: Library@cvcc.vccs.edu; URL: www.cvcc.vccs.edu/Library
1967; Michael T. Fein
32 500 vols; 200 curr per; 114 maps; 200 sound-rec; 7 digital data carriers
libr loan   33507

Edmonds Community College, Library, 20000 68th Ave W, *Lynnwood*, WA 98036
T: +1 425 6401529; Fax: +1 425 7750690; E-mail: Knakano@edcc.edu; URL: www.edcc.edu/library
1967; Lauri Kram, Monica Tobin
46 000 vols; 385 curr per
libr loan; ALA, PNLA   33508

Macon State College, Library, 100 College Station Dr, *Macon*, GA 31206
T: +1 478 4712709; Fax: +1 478 4712869; E-mail: it@maconstate.edu; URL: www.maconstate.edu
1968; Patricia Borck
Health sciences, Health tech; College Arch, Horticulture Coll
85 280 vols; 327 curr per   33509

Madison Area Technical College, Information Resource Center, 3550 Anderson St, *Madison*, WI 53704-2599
T: +1 608 2466640; Fax: +1 608 2466880; E-mail: jgores@matcmadison.edu; URL: library.matcmadison.edu
1965; Julie Gores
62 000 vols; 900 curr per; 203 maps; 253 500 microforms; 5 000 av-mat; 4 000 sound-rec; 2 000 pamphlets; 2 400 slide & transparency series
libr loan   33510

North Florida Community College, Marshall W. Hamilton Library, 1000 Turner Davis Dr, *Madison*, FL 32340-1699
T: +1 850 9731624; Fax: +1 850 9731698; URL: www.nfcc.cc/library
1958; Sheila Hiss
Opera Coll, Video Coll, Florida Mat
32 000 vols; 125 curr per; 10 diss/theses; 1 maps; 1 503 microforms; 1 769 av-mat; 975 sound-rec; 3 digital data carriers
libr loan; SOLINET   33511

Hiwassee College, Hardwick Johnston Memorial Library, 225 Hiwassee College Dr, *Madisonville*, TN 37354
T: +1 423 440-1223; E-mail: library@hiwassee.edu; URL: hiwasseecollege.org/ac/library/
1955; Adele Miller
40 000 vols; 230 curr per   33512

Kishwaukee College, Learning Resources Center, 21198 Malta Rd, *Malta*, IL 60150-9699
T: +1 815 8252086 ext 225; Fax: +1 815 8252072; E-mail: judie.kish.cc.il.us
1968; Anne-Marie Eggleston
44 400 vols; 248 curr per   33513

Manchester Community College, Library, Great Path, P.O. Box 1046, *Manchester*, CT 06045-1046
T: +1 860 5123420, 3413; Fax: +1 860 5123411; E-mail: maxrsf@commnet.edu; URL: www.mcc.commnet.edu/library/new/library.htm
1963; Randy Fournier
42 053 vols; 461 curr per   33514

Chipola Junior College, Library, 3094 Indian Circle, *Marianna*, FL 32446
T: +1 850 7182274; Fax: +1 850 7182349; E-mail: poolec@chipola.cc.fl.us; URL: www.chipola.cc.fl.us
1948; Dr. Carolyn E. Poole
Florida Coll
35 000 vols; 220 curr per   33515

Marion Military Institute, Baer Memorial Library, 1101 Washington St, *Marion*, AL 36756
T: +1 334 6832372; Fax: +1 334 6832380
1930; Audra Westbrook
H. O. Murfee Coll, US Hist (Thomas Perkins Abernethy Coll)
32 130 vols; 76 curr per   33516

Marshalltown Community College, B.J. Harrison Library, 3700 S Center St, *Marshalltown*, IA 50158
T: +1 641 8445690; Fax: +1 641 7541442; E-mail: MCCLibrary@iavalley.edu; URL: www.iavalley.cc.ia.us/mcc/index.html
1927
58 000 vols; 1 224 curr per   33517

University of Wisconsin, Marshfield-Wood County Library, 2000 W Fifth St, *Marshfield*, WI 54449-3310
T: +1 715 3896531; Fax: +1 715 3896539; E-mail: relderbr@uwc.edu; URL: www.marshfield.edu
1964; Ruth Elberbrook
31 960 vols; 154 curr per   33518

Patrick Henry Community College, Lester Library, 645 Patriot Ave, P.O. Box 5311, *Martinsville*, VA 24115
T: +1 540 6388777 ext 228; Fax: +1 540 6560327; URL: www.ph.vccs.edu
1962; Carolyn Byrd
Lit (Thomas H. Carter Memorial), Southern Hist (William F. Stone Memorial)
31 800 vols; 231 curr per   33519

Yuba Community College, Learning Resources Center, 2088 N Beale Rd, *Marysville*, CA 95901
T: +1 530 7416757; Fax: +1 530 7416824; URL: yc.yccd.edu/academics/library.aspx
1927; Stephen Cato
65 000 vols; 250 curr per   33520

Lake Land College, Virgil H. Judge Learning Resource Center, 5001 Lake Land Blvd, *Mattoon*, IL 61938
T: +1 217 2345367; Fax: +1 217 2586459; E-mail: dsilvers@lakeland.cc.il.us; URL: www.lakeland.cc.il.us/library
1968; Scott Drone-Silvers
36 310 vols; 223 curr per   33521

Atlantic Cape Community College, William Spangler Library, 5100 Blackhorse Pike, *Mays Landing*, NJ 08330
T: +1 609 3434952; Fax: +1 609 3434957; E-mail: wilinski@atlantic.edu; URL: www.atlantic.edu/library/index.htm
1966; Grant Wilinski
Culinary arts; Southern New Jersey Hist
78 000 vols; 187 curr per; 2 digital data carriers   33522

Maysville Community College, Library, 1755 US 68, *Maysville*, KY 41056
T: +1 606 7597141 ext 206; Fax: +1 606 7597176; E-mail: sonja.eads@kctcs.net; URL: www.maycc.kctcs.net/library/index.html
1968; Sonja R. Eads
Kentucky, Hist, Nursing; Buffalo Trace Region Coll, Kentucky Hist Coll
32 458 vols; 288 curr per   33523

South Texas Community College, Library, Pecan Campus, 3201 W Pecan Blvd, *McAllen*, TX 78501-6661
T: +1 956 6188330; Fax: +1 956 6188398; E-mail: sesin@stcc.cc.tx.us; URL: www.stcc.cc.tx.us/lrc/library/libfront.html
1984; Armandina A. Sesin
42 230 vols; 300 curr per   33524

Delaware County Community College, Library, 901 S Media Line Rd, *Media*, PA 19063-1094
T: +1 610 3595149, 5326; Fax: +1 610 3595272; E-mail: wecare@dccc.edu; URL: www.dccc.edu
1967; Pedro Navarro
70 900 vols; 525 curr per
libr loan   33525

Brevard Community College, Philip F. Nohrr Learning Resource Center, 3865 N Wickham Rd, *Melbourne*, FL 32925-2399
T: +1 321 433-7662; Fax: +1 321 433-7678; E-mail: RegisN@brevard.cc.fl.us; URL: www.brevard.cc.fl.us
1968; Nina Regis
50 000 vols; 275 curr per; 27 000 microforms; 2 500 av-mat; 25 digital data carriers; 65 slides, 29 overhead transparencies
libr loan; ALA   33526

Memphis College of Art, G. Pillow Lewis Memorial Library, 1930 Poplar Ave, *Memphis*, TN 38104
T: +1 901 2725130, 2725131; Fax: +1 901 2726851; E-mail: library@mca.edu
1936; Maxine Strawder
16 000 vols; 120 curr per; 100 av-mat; 24 VF drawers of clipping & pamphlet mat, 1 000 matted prints, 7 700 mounted prints, 36 000 slides   33527

Southern College of Optometry, Library, 1245 Madison Ave, *Memphis*, TN 38104
T: +1 901 7223237; Fax: +1 901 7223292
1938; Dr. Sharon E. Tabachnick
Visionet Database of Journal Citations for Optometry
16 000 vols; 162 curr per; 529 microforms; 245 av-mat; 743 sound-rec; 16 digital data carriers; 13 100 slides
libr loan; MLA, SLA   33528

Southwest Tennessee Community College, Library, P.O. Box 40568, *Memphis*, TN 38174-0568
T: +1 901 3335135; Fax: +1 901 3335141; E-mail: vstewart@sscc.tn.us; URL: www.stcc.cc.tn.library
1972; Vivian Stewart
49 820 vols; 426 curr per   33529

Southwest Tennessee Community College – George E. Freeman Library, 5983 Macon Cove, *Memphis*, TN 38134
T: +1 901 3334105; Fax: +1 901 3334566; E-mail: rburnett@mail.stim.tec.tn.us; URL: stcc.cc.tn.us/library
1967; Rosa S. Burnett
Computer, Engineering; Computer Literacy Coll, Olin F Morris Coll, Tennessee Coll
42 040 vols; 185 curr per   33530

Merced College, Lesher Library, 3600 M St, *Merced*, CA 95348-2898
T: +1 209 3846082; Fax: +1 209 3846084; E-mail: lrcstaff@elite.net; URL: www.merced.cc.ca.us/lrc
1972; Dr. Susan Walsh
40 890 vols; 253 curr per   33531

Meridian Community College, L.O. Todd Library, 910 Hwy 19 N, *Meridian*, MS 39307
T: +1 601 4848760; Fax: +1 601 4823936; E-mail: swelden@meridiance.edu; URL: www.mcc.ms.us
1937; Billy C. Beal
65 000 vols   33532

Mesa Community College, Paul A. Elsner Library & High Technology Complex, 1833 W Southern Ave, *Mesa*, AZ 85202
T: +1 480 4617671; Fax: +1 480 4617681; E-mail: lorna.peralta@mcmail.maricopa.edu; URL: www.mc.maricopa.edu/library
1963; Lorna Peralta
90 400 vols; 500 curr per   33533

Eastfield College, Eastfield Learning Resource Center, 3737 Motley Dr, *Mesquite*, TX 75150-2033
T: +1 972 8607168; Fax: +1 972 8608537; E-mail: karla_greer@dcccd.edu; URL: www.efc.dcccd.edu
1970; Karla Greer

Jazz (Don Ellis Coll)
52 110 vols; 535 curr per; 5 868 microforms
libr loan   33534

Miami Dade Community College – Mitchell Wolfson New World Center Campus Learning Resources, 300 NE Second Ave, *Miami*, FL 33132
T: +1 305 2373451; Fax: +1 305 2373707; URL: www.mdc.edu
1972; Zenaida Fernandez
45 000 vols; 170 curr per; 236 av-mat   33535

Lord Fairfax Community College, Middletown Campus, Paul Wolk Library, 173 Skirmisher Lane, P.O. Box 47, *Middletown*, VA 22645
T: +1 540 8687170; Fax: +1 540 8687171; E-mail: lfci@lfcc.edu; URL: www.lf.vccs.edu/library
1970
57 000 vols; 300 curr per   33536

Middlesex Community College, Jean Burr Smith Library, 100 Training Hill Rd, *Middletown*, CT 06457-4889
T: +1 860 3435830; Fax: +1 860 3435874; E-mail: mx_einsohn@commnet.edu; URL: www.mxctc.commnet.edu/rsrcs/library/library.htm
1966; Howard Einsohn
46 000 vols; 166 curr per   33537

Orange County Community College, SUNY Orange Library, 115 South St, *Middletown*, NY 10940
T: +1 845 3414855; Fax: +1 845 3414424; E-mail: sparryjo@sunyorange.edu; URL: www.sunyorange.edu/lrc
1950; Susan Parry
Orange County Hist & Heritage
98 170 vols; 250 curr per   33538

Midland College, Murray Fasken Learning Resource Center, 3600 N Garfield, *Midland*, TX 79705
T: +1 915 6854560; Fax: +1 915 6854721; E-mail: libraryqa@midland.edu; URL: www.midland.edu
1973
Health Science Libr, Law
58 450 vols; 311 curr per   33539

Rose State College, Learning Resources Center, 6420 SE 15, *Midwest City*, OK 73110
T: +1 405 7337323; Fax: +1 405 7360260; URL: www.rose.edu
1970; Sharon Saulmon
Local hist coll
94 000 vols; 422 curr per; 2 maps; 7 200 microforms; 11 264 av-mat; 1 972 sound-rec
libr loan   33540

Georgia Military College, Sibley-Cone Memorial Library, 201 E Greene St, *Milledgeville*, GA 31061
T: +1 478 4452718; Fax: +1 478 4455592; E-mail: jsimpson@gmc.cc.ga.us; URL: www.gmc.cc.ga.us/milledgeville/library.html
1879; Jane Simpson
Georgia hist, Local hist, Military science
30 000 vols; 200 curr per; 1 800 av-mat   33541

Milwaukee Area Technical College, Rasche Memorial Library, 700 W State St, *Milwaukee*, WI 53233-1443
T: +1 414 2977030; E-mail: library@matc.edu; URL: www.matc.edu
1935; Jeff Jackson
Voigt Graphic Arts Coll
70 000 vols; 380 curr per; 1 941 av-mat; 12 digital data carriers   33542

Minneapolis Community & Technical College, Library, 1501 Hennepin Ave, *Minneapolis*, MN 55403
T: +1 612 3417087; Fax: +1 612 3417480; URL: www.mctc.mnsu.edu/library/index.html
1965; Tom Eland
40 300 vols; 400 curr per; 600 av-mat; 20 digital data carriers
libr loan   33543

Saddleback College, James B. Utt Library, 28000 Marguerite Pkwy, ***Mission Viejo***, CA 92692
T: +1 949 5824523; Fax: +1 949 3640284; E-mail: koconnor@saddleback.edu; URL: www.saddelback.edu
1968; Dr. Kevin O'Connor
101 440 vols; 283 curr per; 9 000 av-mat; 65 Bks on Deafness & Sign Lang
libr loan 33544

Bishop State Community College, Minnie Slade Bishop Library, 351 N Broad St, ***Mobile***, AL 36603-5898
T: +1 334 6906866; Fax: +1 334 4382463; E-mail: rparker@bscc.cc.al.us; URL: www.bscc.cc.al.us
1943; Robert L. Parker
Black Coll
62 715 vols; 236 curr per 33545

Spring Hill College, Burke Memorial Library, 4000 Dauphin St, ***Mobile***, AL 36608
T: +1 251 3803870; Fax: +1 251 4602107; URL: www.shc.edu; shclibrary.shc.edu
1830; Theresa McGonogle Crider
Rare Bks; Mobiliana Coll, Jesuitica Coll
173 000 vols; 543 curr per
libr loan 33546

Modesto Junior College, Library, 435 College Ave, ***Modesto***, CA 95350
T: +1 209 5756868; Fax: +1 209 5756669; E-mail: clarket@mjc.edu; URL: virtual.mjc.edu/mjclibrary
1920; Tobin Clarke
College Arch
69 865 vols 33547

Black Hawk College, Library, 6600 34th Ave, ***Moline***, IL 61265
T: +1 309 796-5700; Fax: +1 309 796-0393; E-mail: keyc@bhcl.bhc.edu; URL: www.bhc.edu/index.asp?NID=358
1946; Charlet Key
62 000 vols; 510 curr per; 2 400 govt docs; 19 500 microforms; 10 digital data carriers
libr loan 33548

Community College of Beaver County, Library, One Campus Dr, ***Monaca***, PA 15061-2588
T: +1 724 7758561 ext 116; Fax: +1 724 7748995; E-mail: kristen.darke@ccbc.cc.pa.us; URL: library.ccbc.cc.pa.us
1967; Kristen Darke
53 000 vols; 340 curr per 33549

Monroe County Community College, Learning Resource Center, 1555 S Raisinville Rd, ***Monroe***, MI 48161-9746
T: +1 734 3844204; Fax: +1 734 3844160; E-mail: cyonovich@monroeccc.edu; URL: www.monroeccc.edu/library
1966; Barbara McNamee
Professional Libr
50 000 vols; 343 curr per; 20 maps; 500 microforms; 9 digital data carriers
libr loan; ALA, ACRL 33550

Alabama Southern Community College, John Dennis Forte Library, Hwy 21 S, P.O. Box 2000, ***Monroeville***, AL 36461-2000
T: +1 251 5753156; Fax: +1 251 5755116; E-mail: lhollinger@ascc.edu; URL: www.ascc.edu
1965; Angela Roberts
Alabamiana Coll, Forestry Coll, Law Enforcement Coll
48 000 vols; 250 curr per 33551

Community College of Allegheny County, Boyce Campus Library, 595 Beatty Rd, ***Monroeville***, PA 15146
T: +1 724 3256712; Fax: +1 724 3256696; URL: ccac.edu/library/boyce
1966; Raymond Martin
Paralegal Coll
63 410 vols; 337 curr per 33552

Monterey Peninsula College, Library, 980 Fremont Blvd, ***Monterey***, CA 93940-4704
T: +1 831 6464095; Fax: +1 831 6461308; URL: www.mpc.edu/library
1947; Mary Anne Teed
55 000 vols; 288 curr per
libr loan 33553

East Los Angeles College, Helen Miller Bailey Library, 1301 Avenida Cesar Chaves, ***Monterey Park***, CA 91754
T: +1 323 2658758; Fax: +1 323 2673714; URL: www.elac.edu/departments/library/library
1946; Choonhee Rhim
Afro-American, Mexican-American
95 000 vols; 134 curr per 33554

Montreat College, L. Nelson Bell Library, 310 Gaither Circle, ***Montreat***, NC 28757; P.O. Box 1267, Montreat, NC 28757
T: +1 828 6698012; Fax: +1 828 3502083; URL: www.montreat.edu
1898; Elizabeth Pearson
Social Studies (Howard Kester Papers), micro; Crosby Adams Music Coll, bks, clippings, music
80 000 vols; 485 curr per; 200 microforms; 478 av-mat; 1 672 sound-rec; 39 digital data carriers; 206 slides, 25 overhead transparencies
libr loan 33555

Mississippi Delta Community College, Stanny Sanders Library, P.O. Box 668, ***Moorhead***, MS 38761
T: +1 662 2466375; Fax: +1 662 2468627; E-mail: mdcc_library@mdcc.cc.ms.us; URL: mdcc.cc.ms.us
Beverly B. Nobile
35 000 vols; 249 curr per 33556

Moorpark College, Library, 7075 Campus Rd, ***Moorpark***, CA 93021-1695
T: +1 805 3781450; Fax: +1 805 3781470; E-mail: inicklas@vcccd.edu; URL: www.moorparkcollege.edu
1967; Nicklas Inajane
77 000 vols; 293 curr per 33557

Western Piedmont Community College, Phifer Learning Resources Center, 1001 Burkemont Ave, ***Morganton***, NC 28655-4504
T: +1 828 4486036; Fax: +1 828 4486173; E-mail: library@wpcc.edu; URL: www.wpcc.edu/lrc/lrc.htm
1966; Dr. Daniel R. Smith
Grace DiSanto Poetry Coll, Senator Sam J. Ervin, Jr Coll, Mark Twain (Ervin Coll), Small Business Ctr Coll
42 000 vols; 167 curr per; 2 000 e-books 33558

Walters State Community College, Learning Resource Center, 500 S Davy Crockett Pkwy, ***Morristown***, TN 37813-6899
T: +1 423 5856902; Fax: +1 423 5856959; E-mail: doug.cross@wscc.cc.tn.us; URL: www.wscc.cc.tn.us/library
1970; Douglas Cross
47 560 vols; 189 curr per 33559

Big Bend Community College, Library, 7662 Chanute St, ***Moses Lake***, WA 98837
T: +1 509 7626246; Fax: +1 509 7622402; URL: library.bb.cc.ctc.edu
1963; Tim Fuhrman
Chinese Art Coll
33 330 vols; 225 curr per 33560

Wabash Valley College, Bauer Media Center, 2200 College Dr, ***Mount Carmel***, IL 62863
T: +1 618 2628641 ext 3400; Fax: +1 618 2628962
1961; Sandra Craig
Children's Book Coll
32 000 vols; 85 curr per 33561

Southern West Virginia Community & Technical College, Harless Library, Dempsey Branch Rd, P.O. Box 2900, ***Mount Gay***, WV 25637
T: +1 304 7927098 ext 202; Fax: +1 304 7522837; E-mail: kimm@southern.wvnet.edu; URL: www.southern.wvnet.edu
1971; Kimberly Maynard
West Virginia Coll
34 810 vols; 145 curr per 33562

Mount Olive College, Moye Library, 634 Henderson St, ***Mount Olive***, NC 28365-1699; mail address: 644 Henderson St, Mount Olive, NC 28365
T: +1 919 6587869; Fax: +1 919 6588934; URL: www.moc.edu
1955; Pamela R. Wood
Free Will Baptist Hist
66 000 vols; 3 755 curr per; 11 000 e-books; 1 700 sound-rec; 35 digital data carriers
libr loan 33563

Brewton-Parker College, Fountain-New Library, 201 David-Eliza Fountain Circle, ***Mount Vernon***, GA 30445; P.O. Box 197, Mount Vernon, GA 30445
T: +1 912 5833235; Fax: +1 912 5833236; URL: www.bpc.edu/academics/library/library.htm
1988; Ann C. Turner
Music, Education, Christianity; Brewton-Parker College Hist Coll
81 000 vols; 410 curr per 33564

Skagit Valley College, Norwood Cole Library, 2405 E College Way, ***Mount Vernon***, WA 98273
T: +1 360 4167850; Fax: +1 360 4167698; E-mail: mv.library@skagit.edu; URL: library.skagit.edu
1926
75 950 vols; 250 curr per
libr loan 33565

Muskegon Community College, Hendrik Meijer Library, 221 S Quarterline Rd, ***Muskegon***, MI 49442
T: +1 231 7739131; E-mail: carol.briggs-erickson@muskegoncc.edu; URL: library.muskegon.cc.mi.us
1926; Carol Briggs-Erickson
Careers, Children's lit; Michigan Authors
53 000 vols; 230 curr per 33566

Bacone College, Library, 2299 Old Bacone Rd, ***Muskogee***, OK 74403
T: +1 918 7817263; URL: www.bacone.edu/academics/library-online.html
1880; Frances Donelson
31 000 vols; 125 curr per; 20 av-mat; 5 digital data carriers; 10 drawers of microfiche, 2 cabinets of microfilm 33567

Luzerne County Community College, Library, 1333 South Prospect St, ***Nanticoke***, PA 18634
T: +1 570 7400415; Fax: +1 570 7356130; E-mail: mbassham@luzerne.edu; URL: depts.luzerne.edu/library/
1966; Mia Bassham
Criminal Justice, bks, flm; Fire Science Technology, bks, flm
59 650 vols; 317 curr per 33568

Napa Valley College, Library and Learning Resource Center, 2277 Napa Vallejo Hwy, ***Napa***, CA 94558-6236
T: +1 707 2567417; E-mail: MSaepharn@napavalley.edu; URL: www.napavalley.edu/llrc/Pages/default.aspx
1942; Rebecca Scott
Jessayman West Coll
56 500 vols; 203 curr per
libr loan 33569

Nashville School of Law Library, 4013 Armory Oaks Dr, ***Nashville***, TN 37204
T: +1 615 2563684; Fax: +1 615 2442383; URL: www.nashvilleschooloflaw.net
Amy Hennemann
Judge Shriver Coll
16 000 vols; 15 curr per 33570

Nashville State Technical Institute, Educational Resource Center, 120 White Bridge Rd, ***Nashville***, TN 37209-4515
T: +1 615 3533555; Fax: +1 615 3533558; URL: library.nsti.tec.tn.us
1969; Charles May
45 000 vols; 325 curr per; 9 540 microforms; 6 555 av-mat; 1 006 sound-rec; 86 digital data carriers 33571

Watkins College of Art & Design Library, 2298 Rosa L Parks Blvd (MetroCenter), ***Nashville***, TN 37228
T: +1 615 2777427; Fax: +1 615 3834849; URL: www.watkins.edu
Beverly Stark
20 000 vols; 60 curr per 33572

Crowder College, Learning Resources Center, 601 Laclede Ave, ***Neosho***, MO 64850
T: +1 417 4513223; Fax: +1 417 4514280; E-mail: circ@crowder.edu; URL: www.crowder.edu
1964; Barbara Schade
35 503 vols; 167 curr per; 150 989 microforms; 2 950 sound-rec; 21 digital data carriers; 50 slides, 2 524 art repros 33573

Gateway Community College, Long Wharf Campus Library, 60 Sargent Dr, ***New Haven***, CT 06511-5970
T: +1 203 2852057; Fax: +1 203 2852055; URL: www.gwcc.comment.edu/libwebpage/libwebpage.html
1968; Michele Cone
53 000 vols; 624 curr per 33574

Mitchell College, Library, 437 Pequot Ave, ***New London***, CT 06320-4498
T: +1 860 7017091; Fax: +1 860 7015099; URL: campus.mitchell.edu
1939; Suzanne M. Risley
Art, Art hist; Robert Penn Warren Letters
48 000 vols; 28 curr per 33575

Delgado Community College, Moss Memorial Library, 615 City Park Ave, ***New Orleans***, LA 70119
T: +1 504 4834119; Fax: +1 504 4831939; E-mail: cvarna@dcc.edu; URL: www.dcc.edu/library
1921; Constance Varnado
Louisiana
123 000 vols; 981 curr per 33576

Borough of Manhattan Community College, A. Philip Randolph Memorial Library, 199 Chambers St, ***New York***, NY 10007
T: +1 212 2201451; E-mail: jgonzalez@bmcc.cuny.edu; URL: lib1.bmcc.cuny.edu
1964; Sidney Eng
96 000 vols; 714 curr per 33577

The New School – Raymond Fogelman Library, 65 Fifth Ave, Lower Level, ***New York***, NY 10003
T: +1 212 2295307; Fax: +1 212 2295306; URL: www.library.newschool.edu
1919; Gary A. Wasdin
Husserl Arch; NACLA Arch of Latin Americana; New School Arch; Hannah Arendt digital arch; New School Dramatic Workshop arch
194 000 vols; 220 curr per; 3 500 diss/theses; 1 260 govt docs; 300 microforms; 60 subscription databases
libr loan; ALA 33578

New York School of Interior Design Library, 170 E 70th St, ***New York***, NY 10021
T: +1 212 4721500; Fax: +1 212 4728175; URL: nysid.net/library
Dr. Eric Wolf
Interior design, architecture
14 000 vols; 104 curr per 33579

Delaware Technical & Community College – Stanton Campus Library, 400 Stanton-Christiana Rd, ***Newark***, DE 19713-2197
T: +1 302 4543939; Fax: +1 302 4533079; URL: /www.library.dtcc.edu/stantlib/index.html
1968; Regina Wells
Social sciences, Fire science, Careers
32 000 vols; 364 curr per; 3 000 microforms
libr loan 33580

Sussex County Community College, Learning Resource Center, College Hill, ***Newton***, NJ 07860
T: +1 973 3002162; Fax: +1 973 3002276; E-mail: aamack@www.sussex.cc.nj.us; URL: www.sussex.cc.nj.us
Dr. Peter Panos
Juvenile, Law
33 000 vols; 205 curr per 33581

Mount Ida College, Learning Resource Center, 777 Dedham St, ***Newton Center***, MA 02159
T: +1 617 9284552; Fax: +1 617 9284038; E-mail: lrcref@tiac.net
1939; Marge Lippincott
Nat Ctr for Death Education Libr
46 000 vols; 530 curr per 33582

Bucks County Community College, Library, 275 Swamp Rd, ***Newtown***, PA 18940-0999
T: +1 215 9688009; Fax: +1 215 9688142; URL: www.bucks.edu/library
1965; Brandi Porter
College Hist, Bucks County Hist
203 110 vols; 439 curr per; 400 maps; 177 av-mat; 5 387 sound-rec; 200 digital data carriers; 60 Art Reproductions 33583

Northwest Florida State College – Niceville Campus, Learning Resources Center, 100 College Blvd, **Niceville**, FL 32578
T: +1 850 7295318; Fax: +1 850 7295295; E-mail: hendersj@nwfsc.edu; URL: lrc.nwfsc.edu
1964; Janice Henderson
Florida & Works of Floridians Coll
84 610 vols; 364 curr per   33584

Cecil Community College, Veterans Memorial Library, One Seahawk Dr, **North East**, MD 21901-1904
T: +1 410 2876060 ext 562; Fax: +1 410 2871026; URL: www.cecil.cc.md.us/library
1968; Shelley Gardner
Genealogy, Maryland
31 000 vols; 150 curr per   33585

Lorain County Community College, Library, 1005 Abbe Rd N, **North Elyria**, OH 44035-1691
T: +1 440 3664026; Fax: +1 440 3664127; URL: library.lorainccc.edu
1964; Keith E. Washburn
115 000 vols; 641 curr per   33586

Cerritos College, Library, 11110 Alondra Blvd, **Norwalk**, CA 90650
T: +1 562 8602451 ext 2412; Fax: +1 562 4675002; E-mail: refuser@cerritos.edu; URL: www.cerritos.edu
1957; John McGinnis
97 710 vols; 299 curr per   33587

Gibbs College, Library, 142 East Ave, **Norwalk**, CT 06851
T: +1 203 8384173 ext 419; Fax: +1 203 8381501; E-mail: jshoup@gibbsnorwalk.com
John Shoup
41 000 vols; 55 curr per   33588

Norwalk Community College, Everett I. L. Baker Library, 188 Richards Ave, **Norwalk**, CT 06854
T: +1 203 8577200; Fax: +1 203 8577380; E-mail: nk_sahlin@apollo.commnet.edu; URL: www.nctc.commnet.edu/library/
1962; Linda Petersen
American Civilization, microfiche; English Lit, microfiche; Human Relations Area Files, microfiche
55 000 vols; 225 curr per; 1 616 av-mat; 1 689 sound-rec; 8 digital data carriers
libr loan   33589

Three Rivers Community College, Mohegan Campus Learning Resources Center, Mahan Dr, **Norwich**, CT 06360
T: +1 860 3835289; Fax: +1 860 8860691; URL: www.trcc.commnet.edu/library
1970; Mildred Hodge
37 000 vols; 415 curr per   33590

Laney College, Peralta Colleges, Library, 900 Fallon St, **Oakland**, CA 94607
T: +1 510 4643497; Fax: +1 510 4643264; E-mail: scoaston@peralta.edu; URL: www.laney.edu/wp/library
1956; Shirley Coaston
78 885 vols; 285 curr per   33591

Merritt College, Library, 12500 Campus Dr, Bldg A, Room 129, **Oakland**, CA 94619-3196
T: +1 510 4362457; Fax: +1 510 5314690; E-mail: merrittlib@peralta.edu; URL: www.merritt.edu/student-services/library
1954
60 000 vols; 120 curr per   33592

Samuel Merritt College, John A. Graziano Memorial Library, 400 Hawthorne Ave, **Oakland**, CA 94609
T: +1 510 8698900; Fax: +1 510 8696633; URL: www.samuelmerritt.edu
1909; Barbara Ryken
Health sciences, Nursing, Occupational therapy, Physical therapy
16 000 vols; 437 curr per; 900 av-mat; 12 digital data carriers; 10 VF drawers of pamphlets   33593

Gainesville State College, Gainesville Campus Library, 3820 Mundy's Mill Rd, **Oakwood**, GA 30566; P.O. Box 1358, Gainesville, GA 30503
T: +1 678 7173639; E-mail: dprosser@gsc.edu; URL: www.gsc.edu
1964; Deborah Prosser
Ed Dodd Coll
86 000 vols; 157 curr per
libr loan   33594

Central Florida Community College, Learning Resources Center, Ocala Campus, Bldg 3, P.O. Box 1388, **Ocala**, FL 34474
T: +1 352 8735805; E-mail: library@cf.edu
1958
Equine Coll, Wisdom Traditions, Women's Hist
61 000 vols; 294 curr per; 5 829 microforms; 1 859 av-mat; 114 sound-rec; 334 digital data carriers
libr loan   33595

Miracosta College, Library, One Barnard Dr, **Oceanside**, CA 92056-3899
T: +1 760 7956715; Fax: +1 760 7956723; E-mail: lnolte@miracosta.edu; URL: www.miracosta.edu
1934; Joseph A. Moreau
60 960 vols; 205 curr per   33596

Odessa College, Murry H. Fly Learning Resources Center, 201 W University, **Odessa**, TX 79764
T: +1 915 3356641; Fax: +1 915 3356610; E-mail: pquintero@odessa.edu; URL: www.odessa.edu/dept/library
1946
75 130 vols; 388 curr per   33597

Utah School for the Deaf & Blind, Educational Center, 742 Harrison Blvd, **Ogden**, UT 84404
T: +1 801 6294817; Fax: +1 801 6294896; URL: www.usdb.org
Lorri Quigley
37 000 vols; 20 curr per   33598

Illinois Valley Community College, Jacobs Memorial Library, 815 Orlando Smith Ave, **Oglesby**, IL 61348-9692
T: +1 815 2240306; Fax: +1 815 2249147; E-mail: jacobs@ivcc.edu; URL: www.ivcc.edu/library
1968; Frances Whaley
Local hist, Nursing
48 000 vols; 967 curr per; 9 200 e-books; 77 025 govt docs; 484 maps; 24 918 microforms; 4 360 av-mat; 35 sound-rec; 916 digital data carriers; 52 slides, 300 overhead transparencies   33599

Oklahoma City Community College, Library, 7777 S May Ave, **Oklahoma City**, OK 73159
T: +1 405 6821611 ext 7251; Fax: +1 405 6827585; E-mail: lking@occc.edu; URL: www.occc.edu/Library
1972; Barbara King
60 000 vols; 500 curr per
libr loan   33600

OSU-OKC Library, 900 N Portland, **Oklahoma City**, OK 73107-6187
T: +1 405 9453251; Fax: +1 405 9453289; E-mail: okcisok@okway.okstate.edu; URL: www.osuokc.edu/library.html
1961; Marla Roberson
Nursing, Police science, Horticulture, Fire protection
14 000 vols; 245 curr per; 30 digital data carriers
libr loan   33601

South Puget Sound Community College, Library, 2011 Mottman Rd SW, **Olympia**, WA 98512
T: +1 360 7547711; Fax: +1 360 6640780
1972; W. Russell Rose
35 000 vols; 300 curr per   33602

Metropolitan Community College, Sonny Foster Library, 30th & Fort Sts, **Omaha**, NE 68111; P.O. Box 3777, Omaha, NE 68103-0777
T: +1 402 4572630; Fax: +1 402 4572768; E-mail: awills@mccneb.edu; URL: www.mccneb.edu/library
1974; Ann Wills
56 350 vols; 772 curr per
libr loan   33603

Treasure Valley Community College, Library, 650 College Blvd, **Ontario**, OR 97914
T: +1 541 8896493 ext 248; Fax: +1 541 8812724; E-mail: dalee@mailman.tvcc.cc.or.us; URL: www.tvcc.or.us/library/index.htm
1963; Dale L. Edwards
Indian Artifact (Horace & Roa Arment)
33 000 vols; 165 curr per; 10 313 microforms; 300 av-mat; 2 500 sound-rec; 18 digital data carriers   33604

Erie Community College – South Campus, Library Resources Center, S-4041 Southwestern Blvd, **Orchard Park**, NY 14127
T: +1 716 8511772; Fax: +1 716 8511778; E-mail: geer@ecc.edu
1974; Judith Geer
Henry Louis Mencken; Women Writers of Western New York
55 765 vols; 298 curr per; 9 govt docs; 136 maps; 16 234 microforms; 902 av-mat; 546 sound-rec; 9 digital data carriers
libr loan   33605

Clackamas Community College, Marshall N. Dana Memorial Library, 19600 S Molalla Ave, **Oregon City**, OR 97045
T: +1 503 657-6958 ext 2417; Fax: +1 503 6558925; E-mail: reference@clackamas.edu; URL: www2.clackamas.edu/library/index.htm
1967; Cynthia Andrews
Oregon Coll
48 890 vols; 294 curr per
libr loan   33606

Valencia Community College – East Campus Learning Resource Center, 701 N Econlockhatchee Trail, **Orlando**, FL 32825
T: +1 407 5822456; Fax: +1 407 5828914; URL: eastlrc.valenciacc.edu
1975; Dr. Dennis Weeks
Law Coll, College Arch
58 000 vols; 385 curr per; 1 diss/theses; 690 microforms; 22 digital data carriers
libr loan; ALA   33607

Butte Community College, Frederick S. Momtgomery Library, 3536 Butte Campus Dr, **Oroville**, CA 95965
T: +1 530 8952596, 8952511; Fax: +1 530 8952924; E-mail: dosuta@butte.cc.ca.us; URL: www.butte.cc.ca.us/library
1967; Tabzeera Dosu
Art, Lit
53 000 vols; 184 curr per; 29 000 art reproductions   33608

Indian Hills Community College, Special Programs-Learning Resource Center, 525 Grandview Ave, Bldg 10, **Ottumwa**, IA 52501-1398
T: +1 641 6835199; Fax: +1 641 6835184; E-mail: library@ihcc.cc.ia.us; URL: www.ihcc.cc.ia.us
1960; Mary Stewart
30 000 vols; 270 curr per   33609

Johnson County Community College, Billington Library, 12345 College Blvd, **Overland Park**, KS 66210
T: +1 913 4693871; Fax: +1 913 4693816; URL: library.jccc.net
1969; Mark Daganaar
Education (ERIC Coll)
100 000 vols; 600 curr per   33610

Oxnard College, Library, 4000 S Rose Ave, **Oxnard**, CA 93033-6699
T: +1 805 9865819; Fax: +1 805 9865880; URL: www.oxnard.cc.ca.us/library/libmain.htm
1975; Delois J. Flowers
Social sciences
31 000 vols; 212 curr per   33611

William Rainey Harper College Library, Resources for Learning, 1200 W Algonquin Rd, **Palatine**, IL 60067
T: +1 847 9256584; Fax: +1 847 9256164; E-mail: refdesk@harpercollege.edu; URL: harpercollege.edu/library
1967; Njambi Kamoche
Harper College Archs
127 580 vols; 279 curr per; 2 000 e-books; 21 000 av-mat   33612

St. John's River State College, B.C. Pearce Learning Resources Center, 5001 St Johns Ave, **Palatka**, FL 32177-3897
T: +1 386 3124150; E-mail: carmencummings@sjrstate.edu; URL: www.sjrstate.edu/libraries/palatka.htm
1958; Carmen M. Cummings
Florida Coll, Civil War Coll
70 000 vols; 421 curr per; 91 011 microforms; 2 025 sound-rec; 17 digital data carriers; 1 047 videotapes   33613

College of the Desert, Library, 43-500 Monterey, **Palm Desert**, CA 92260
T: +1 760 7760105, 7732563; Fax: +1 760 7760134; E-mail: cwhitaker@dccd.cc.ca.us; URL: www.desert.cc.ca.us
1962; Char Whitaker
Winston S Churchill Coll; Desert Coll, rare bks
45 000 vols; 240 curr per; 88 maps; 5 454 microforms; 4 digital data carriers
libr loan   33614

Trinity Christian College, Jennie Huizenga Memorial Library, 6601 W College Dr, **Palos Heights**, IL 60463
T: +1 708 5973000; Fax: +1 708 3855665; URL: www.trnty.edu/library
1959; Marcille Frederick
77 000 vols; 300 curr per; 19 000 microforms; 900 sound-rec
libr loan   33615

Moraine Valley Community College, Robert E. Turner Learning Resources Ctr, 10900 S. 88th Ave., **Palos Hills**, IL 60465-0937
T: +1 708 9745235; Fax: +1 708 9741184; URL: www.morainevalley.edu/lrc
1967; Sylvia Jenkins
77 517 vols; 552 curr per; 847 maps; 204 microforms; 6 442 av-mat; 752 sound-rec; 97 digital data carriers
libr loan   33616

Gulf Coast Community College, Library, 5230 W Hwy 98, **Panama City**, FL 32401
T: +1 850 8723893; Fax: +1 850 8723861; E-mail: ldriscoll@gulfcoast.edu; URL: www.gulfcoast.edu/library/
1957; Lori Driscoll
Smithsonian Institution Annual Rpts 1849-1964, Bureau of Ethnology Annual Rpts 1879-1931
80 000 vols; 800 curr per   33617

Bergen Community College, Sidney Silverman Library & Learning Resource Center, 400 Paramus Rd, No L216A, **Paramus**, NJ 07652-1595
T: +1 201 4477130; Fax: +1 201 4938167; E-mail: pdenholm@bergen.edu; URL: www.bergen.edu/library/
1965; Patricia Denholm
136 062 vols; 677 curr per; 8 994 microforms; 7 150 av-mat; 371 sound-rec; 361 digital data carriers
libr loan   33618

Paris Junior College, Learning Center, 2400 Clarksville St, **Paris**, TX 75460
T: +1 903 7820415, 7820215; Fax: +1 903 7820356; E-mail: ccovert@parisjc.edu; URL: www.parisjc.edu
1924; Carl Covert
A M Aikin Archs
55 000 vols; 340 curr per   33619

Mineral Area College, C.H. Cozean Library, 5270 Flat River rd, **Park Hills**, MO 63601; P.O. Box 1000, Park Hills, MO 63601-1000
T: +1 573 5182141; Fax: +1 573 5182162; E-mail: chris@mail.mac.cc.mo.us; URL: www.mineralarea.edu/library
1969; Carol Harper
52 490 vols; 203 curr per   33620

Pasadena City College, Shatford Library, 1570 E Colorado Blvd, **Pasadena**, CA 91106-2003
T: +1 626 5857360; Fax: +1 626 5857913; E-mail: malaun@pasadena.edu; URL: www.pasadena.edu
1924; Mary Ann Laun
Education, Hist, Local hist
135 670 vols; 350 curr per
libr loan   33621

San Jacinto College, Lee Davis Library, 8060 Spencer Hwy, *Pasadena*, TX 77505
T: +1 281 4761850; Fax: +1 281 4782734; URL: www.sjcd.edu
1961; Jay B. Clark
Pomeroy Archs on Area Hist
157 000 vols; 300 curr per   33622

Columbia Basin College, Library Services, 2600 N 20th Ave, *Pasco*, WA 99301
T: +1 509 5470511 ext 2287; Fax: +1 509 5460401; E-mail: kfoley@ctc.ctc.edu; URL: www.cbc2.org
1955; Katie Foley
Benton-Franklin County Law Libr
53 890 vols; 259 curr per   33623

Passaic County Community College, Library & Learning Resources Center, One College Blvd, *Paterson*, NJ 07505
T: +1 973 6845896; Fax: +1 973 6846675; E-mail: gfallon@pccc.cc.nj.us; URL: www.pccc.cc.nj.us/library
1971; Greg Fallon
Poetry Coll
47 000 vols; 263 curr per; 3 821 microforms; 775 sound-rec; 2 digital data carriers; 595 slides, 309 overhead transparencies   33624

Paul Smiths College of Arts & Sciences, Frank L. Cubley Library, Rte's 30 & 86, *Paul Smiths*, NY 12970
T: +1 518 3276313; Fax: +1 518 3276350; URL: www.paulsmiths.edu
1946; Susan Mitchell
62 000 vols; 460 curr per; 26 550 govt docs; 359 maps; 581 microforms; 405 av-mat; 81 sound-rec; 28 000 forestry pamphlets, 6 923 slides   33625

Leeward Community College, Library, 96-045 Ala Ike, *Pearl City*, HI 96782-3393
T: +1 808 4550379; Fax: +1 808 4536729; E-mail: rtoyama@hawaii.edu; URL: www.leeward.hawaii.edu/library/
1968; Ralph Toyama
Hawaiiana Coll
76 170 vols; 249 curr per   33626

Burlington County College, Library, County Rte 530, *Pemberton*, NJ 08068
T: +1 609 8949311 ext 7222; Fax: +1 609 8944189; E-mail: library@bcc.edu; URL: staff.bcc.edu/library/
1969
Geneology & Local Hist (Pinelands Coll), Cinema Coll
92 000 vols; 340 curr per   33627

Blue Mountain Community College, Library, 2411 NW Carden Ave, *Pendleton*, OR 97801
T: +1 541 2785916; Fax: +1 541 2766119; E-mail: shannon.vankirk@bluecc.edu; URL: www.bluecc.edu/library/about/staff.php
1963; Shannon Van Kirk
World War II; Affiliate Data Ctr (census mat)
38 000 vols; 253 curr per   33628

Tri-County Technical College, Learning Resource Center, P.O. Box 587, *Pendleton*, SC 29670
T: +1 864 6468361 ext 2254; Fax: +1 864 6468256
1963; Caroline Mills
37 290 vols; 276 curr per; 550 av-mat; 8 digital data carriers
libr loan   33629

Pensacola Junior College, Learning Resource Center, 1000 College Blvd, *Pensacola*, FL 32504-8998
T: +1 850 4842002; Fax: +1 850 4841991; E-mail: wbradley@pjc.edu; URL: www.lrc.pjc.edu
1948; Winifred Bradley
151 500 vols; 320 curr per; 13 200 e-books; 5 187 av-mat; 370 sound-rec; 1 100 digital data carriers; 93 High Interest/Low Vocabulary Bks   33630

Mississippi Gulf Coast Community College, Perkinston Campus Learning Resource Center, P.O. Box 849, *Perkinston*, MS 39573
T: +1 601 9285211 ext 6380; Fax: +1 601 9286359; E-mail: elizabeth.mixon@mgccc.cc.ms.us; URL: www.mgccc.cc.ms.us
1925; Liz Mixon
31 000 vols; 212 curr per   33631

Owens Community College, Library, 30335 Oregon Rd, *Perrysburg*, OH 43551; P.O. Box 10000, Toledo, OH 43699-1947
T: +1 419 6617221; Fax: +1 419 6617021; E-mail: libhelp@owens.cc.oh.us; URL: www.owens.cc.oh.us/Library
1966; Thomas Sink
49 800 vols; 530 curr per; 64 music scores; 15 maps; 70 500 microforms; 2 665 av-mat; 859 sound-rec; 14 digital data carriers; 1 682 videocassettes/dvd
libr loan; ALA   33632

Richard Bland College, Library, 11301 Johnson Rd, *Petersburg*, VA 23805
T: +1 804 8626208; Fax: +1 804 8626125; E-mail: library@rbc.edu; URL: www.rbc.edu/library/
1960; Dr. Virginia Rose Cherry
70 000 vols; 125 curr per; 47 000 e-books
libr loan; ALA   33633

Community College of Philadelphia, Library, 1700 Spring Garden St, *Philadelphia*, PA 19130
T: +1 215 7518000; Fax: +1 215 7518762; E-mail: jjohnson@ccp.edu; URL: www.ccp.edu/site/academic/library.php
1964; Joan Johnson
100 000 vols; 450 curr per; 7 360 microforms; 10 digital data carriers   33634

Pennsylvania Academy of Fine Arts Library, 1301 Cherry St, *Philadelphia*, PA 19107
T: +1 215 9727600 ext 2030; Fax: +1 215 5690153; E-mail: pafine@hslc.org
1805; Aurora Deshauteurs
14 500 vols; 89 curr per; 96 VF drawers   33635

Pennsylvania School for the Deaf Library, 100 W School House Lane, *Philadelphia*, PA 19144
T: +1 215 9514743; Fax: +1 215 9514708; E-mail: library@psd.org
Janice VanGorden
18 000 vols; 41 curr per   33636

School District of Philadelphia, Pedagogical Library, 2120 Winter St, *Philadelphia*, PA 19103
T: +1 215 2992543; Fax: +1 215 2992540; E-mail: beryrich@phila.k12.pa.us; URL: www.libraries.phila.k12.pa.us
1883; Berry Richards
District Arch
30 000 vols; 60 curr per; 650 000 microforms; 10 digital data carriers   33637

DeVry Institute of Technology, Library, 2149 W Dunlap Ave, *Phoenix*, AZ 85021-2982
T: +1 602 8709222; Fax: +1 602 7341999; E-mail: spritchard@devry-phx.edu; URL: www.devry.phx.edu
1931; Susan V. Pritchard
Electronics, Computer languages, Business operations, Accounting, Telecommunications; Career Services Directories Coll
20 285 vols; 125 curr per; 250 microforms; 1 400 av-mat; 150 sound-rec; 300 digital data carriers; 250 computer disks
libr loan   33638

Gateway Community College, Library Center, 108 N 40th St, *Phoenix*, AZ 85034-1704
T: +1 602 3925147; Fax: +1 602 3925300; E-mail: garcia_j@gwc.maricopa.edu; URL: www.gwc.maricopa.edus/lib/biblio
1968; Josefa Garcia
34 000 vols; 175 curr per; 245 microforms; 243 sound-rec; 9 digital data carriers; 600 programmed mat & media kits
libr loan   33639

Paradise Valley Community College, Library, 18401 N 32nd St, *Phoenix*, AZ 85032-1200
T: +1 602 7877200; Fax: +1 602 7877205; E-mail: j.chavez@pvmail.maricopa.edu; URL: www.pvc.maricopa.edu/library
John U. Chavez
30 000 vols; 400 curr per   33640

Phoenix College, Fannin Library, 1202 W Thomas Rd, *Phoenix*, AZ 85013
T: +1 602 2857473; Fax: +1 602 2857368; E-mail: info@pcmail.maricopa.edu; URL: www.pc.maricopa.edu/departments/library
1925
Arizona
78 120 vols; 330 curr per   33641

South Mountain Community College, Library, 7050 S 24th St, *Phoenix*, AZ 85042
T: +1 602 2438187; Fax: +1 602 2438180; E-mail: cate.mcnamara@smcmail.maricopa.edu; URL: libguides.southmountaincc.edu/home
1980
52 000 vols   33642

Sandhills Community College, Katharine L. Boyd Library, 3395 Airport Rd, *Pinehurst*, NC 28374
T: +1 910 6953819; Fax: +1 910 6953947; E-mail: staceyj@sandhills.edu; URL: www.sandhills.edu/library/
1965; Dr. John Stacey
88 000 vols; 358 curr per   33643

Saint Petersburg College, M.M. Bennett Libraries, 7200 66th St N, *Pinellas Park*, FL 33781; P.O. Box 13489, Saint Petersburg, FL 33733-3489
T: +1 727 3413719; Fax: +1 727 3413658; URL: www.spcollege.edu
1927; Dr. Susan Anderson
225 000 vols; 1 510 curr per
libr loan   33644

Alice Lloyd College, McGaw Library & Learning Center, 100 Purpose Rd, *Pippa Passes*, KY 41844
T: +1 606 3686117; Fax: +1 606 3686212; E-mail: mcgaw_library@hotmail.com; URL: www.alc.edu
1923; Andrew Busroe
Appalachian Oral Hist, Appalachian Photogr Coll, Children's Lit Coll
74 000 vols; 219 curr per; 120 000 e-books   33645

Edison State Community College, Library, 1973 Edison Dr, *Piqua*, OH 45356
T: +1 937 7788600; Fax: +1 937 7781920; E-mail: library@edison.cc.oh.us; URL: elink.edison.cc.oh.us
1973; Mary Beth Aust-Keefer
Hist, Nursing
33 530 vols; 499 curr per   33646

Community College of Allegheny County, Allegheny Campus Library, 808 Ridge Ave, *Pittsburgh*, PA 15212-6097
T: +1 412 2372585; Fax: +1 412 2376563; URL: www.acd.ccac
1966; Dennis Hennessy
82 000 vols; 366 curr per
libr loan   33647

Berkshire Community College, Jonathan Edwards Library, 1350 West St, *Pittsfield*, MA 01201
T: +1 413 4994660 ext 201; Fax: +1 413 4482700; URL: www.cc.berkshire.org
1960; Nancy A. Walker
Art, Environmental studies, Social sciences, Technology
56 000 vols; 351 curr per   33648

Collin County Community College District, Richard L. Ducote Learning Resource Center, Spring Creek Campus, 2800 E Spring Creek Pkwy, *Plano*, TX 75074
T: +1 972 8815860; Fax: +1 972 8815911; URL: www.cccd.edu
1985; Martha C. Adamson
125 650 vols; 940 curr per; 3 797 microforms; 189 470 av-mat; 3 519 sound-rec; 281 digital data carriers
libr loan   33649

Clinton Community College, LeRoy M. Douglas Learning Resource Center, 136 Clinton Point Dr, *Plattsburgh*, NY 12901-5690
T: +1 518 5624143; Fax: +1 518 5624158; URL: clintoncc.suny.edu
1969; Andrew Hersh-Tudor
Adirondack Coll
47 080 vols; 300 curr per   33650

Diablo Valley College, Library, 321 Golf Club Rd, *Pleasant Hill*, CA 94523-1576
T: +1 925 6851230 ext 241; Fax: +1 925 7983588; E-mail: mdolven@dvc.edu; URL: www.dvc.edu/library
1950; Mary Dolven
Ethnic studies, Art & architecture; Californiana, Topographic Maps, Rare Bks
88 300 vols; 300 curr per
libr loan   33651

Three Rivers Community College, Rutland Library, 2080 Three Rivers Blvd, *Poplar Bluff*, MO 63901
T: +1 573 8409695; Fax: +1 573 8409659; E-mail: library@trcc.cc.mo.us; URL: www.trcc.cc.mo.us/library
1966; Gordon Johnston
Civil War, Missouri Hist, Ozarks Hist
34 000 vols; 450 curr per   33652

Pearl River Community College, Garvin H. Johnston Library, 101 Hwy 11 N, P.O. Box 5660, *Poplarville*, MS 39470-5660
T: +1 601 4031332; Fax: +1 601 4031135; E-mail: library@prcc.edu; URL: www.prcc.cc.ms.us
1926; Jeanne E. Dyar
Mississippi Coll
53 000 vols; 326 curr per   33653

Peninsula Community College, John D. Glann Library, 1502 E Lauridsen Blvd, *Port Angeles*, WA 98362-6698
T: +1 360 4529277; Fax: +1 360 4176295; E-mail: pdohrt@ctc.edu; URL: www.pc.ctc.edu/lmc
1961; Paula B. Doherty
Chamber Music
34 100 vols; 250 curr per   33654

St. Clair County Community College, Learning Resources Center, 323 Erie St, P.O. Box 5015, *Port Huron*, MI 48061-5015
T: +1 810 9895640; Fax: +1 810 9842852; E-mail: library@sc4.edu; URL: www.sc4.edu
1923
60 070 vols; 575 curr per   33655

Porterville College, Library, 100 E College Ave, *Porterville*, CA 93257-5901
T: +1 559 7912318; Fax: +1 559 7912289; E-mail: jspalsbu@pc.cc.ca.us; URL: www.pc.cc.ca.us
1927; Jeff R. Spalsbury
Anthropology Libr, Valley Writers Coll
34 050 vols; 290 curr per   33656

Maine College of Art, Joanne Waxman Library, 522 Congress St, *Portland*, ME 04101
T: +1 207 7755153; Fax: +1 207 7725069; E-mail: library@meca.edu; URL: library.meca.edu
1974; Moira Steven
Rare Book Coll, Artists Books
30 000 vols; 100 curr per; 45 000 slides, 150 videos
libr loan; ALA, ARLIS/NA, MLA   33657

Oregon College of Art & Craft Library, 8245 SW Barnes Rd, *Portland*, OR 97225
T: +1 503 2975544; Fax: +1 503 2979651; URL: www.ocac.edu
Lori Johnson
Craft history, design
10 000 vols; 90 curr per   33658

Portland Community College Libraries, P.O. Box 19000, *Portland*, OR 97280-0990
T: +1 503 9774498; Fax: +1 503 9774977; E-mail: lriester@pcc.edu; URL: www.pcc.edu/library
1964; Berniece Owen
261 000 vols; 1 400 curr per; 380 e-books; 8 400 av-mat; 560 High Interest/Low Vocabulary Bks, 212 Bks on Deafness & Sign Lang, 1 680 Talking Bks   33659

Walla Walla College, Portland Campus, School of Nursing Library, 10345 SE Market St, *Portland*, OR 97216
T: +1 503 2516115; Fax: +1 503 2516249; URL: library.wallawalla.edu
Bruce McClay
8 000 vols; 100 curr per; 200 av-mat; 8 digital data carriers
libr loan   33660

Tidewater Community College, Portsmouth Campus Library, 7000 College Dr, **Portsmouth**, VA 23703 T: +1 757 8222130; Fax: +1 757 8222149; E-mail: mglanzer@tcc.edu; URL: www.tcc.vccs.edu/lrc 1961; Mary Anne Glanzer 64 070 vols; 188 curr per   33661

Dutchess Community College, Library, 53 Pendell Rd, **Poughkeepsie**, NY 12601-1595 T: +1 845 4318630; Fax: +1 845 4318995; E-mail: libinfo@sunydutchess.edu; URL: www.sunydutchess.edu/academics/library/ 1957 89 740 vols; 339 curr per   33662

Northwest College, Hinckley Library, 231 W Sixth St, **Powell**, WY 82435 T: +1 307 7546207; Fax: +1 307 7546010; E-mail: carlsonk@nwc.cc.wy.us; URL: www.nwc.cc.wy.us/library/ reference.html 1948; Kay Carlson 45 000 vols; 396 curr per libr loan   33663

Prestonsburg Community College, Library, One Bert Combs Dr, **Prestonsburg**, KY 41653 T: +1 606 8863863 ext 221; Fax: +1 606 8868683; E-mail: sandra.robertson@kctcs.net; URL: www.prestonsburgcc.com 1964; Sandra D. Robertson Eastern Kentucky Hist Coll 38 000 vols; 225 curr per   33664

Landmark College, Library, River Rd, **Putney**, VT 05346 T: +1 802 3876764 1985; Lisa Griest Learning Disabilities Coll 32 720 vols; 187 curr per libr loan   33665

Adirondack Community College, Library, 640 Bay Rd, Scoville Learning Ctr, **Queensbury**, NY 12804 T: +1 518 7432260; Fax: +1 518 7451442; URL: www.sunyacc.edu 1961; Teresa Ronning Criminal justice, Liberal arts, Nursing; Local Hist (Hill Coll) 58 000 vols; 350 curr per   33666

Northeast Alabama Community College, Cecil B. Word Learning Resources Center, P.O. Box 159, **Rainsville**, AL 35986 T: +1 256 2286001 ext 328; Fax: +1 256 2284350; E-mail: NAC_circ@lmn.lib.al.us 1965; Arlene E. Grede 51 385 vols; 112 curr per; 1752 microforms; 2 digital data carriers libr loan   33667

Peace College, Lucy Cooper Finch Library, 15 E Peace St, **Raleigh**, NC 27604-1194 T: +1 919 5082302; Fax: +1 919 5082769; E-mail: circdesk@peace.edu; URL: faculty.peace.edu/library 1872; Paul F. King Liberal arts 53 000 vols; 4 900 curr per; 6 000 e-books   33668

Saint Mary's College, Sarah Graham Kenan Library, 900 Hillsborough St, **Raleigh**, NC 27603-1689 T: +1 919 4244040; Fax: +1 919 4244134; E-mail: jlmclean@saint-marys.edu; URL: www.saint-marys.edu 1842; Jennifer McLean 46 500 vols; 196 curr per; 4 484 microforms; 414 sound-rec; 24 digital data carriers   33669

Wake Technical Community College, Main Campus, Library, 9101 Fayetteville Rd, **Raleigh**, NC 27603 T: +1 919 8665644; Fax: +1 919 6623575; E-mail: nshebert@waketech.edu; URL: library.waketech.edu 1962; Jackie Case 70 617 vols; 540 curr per   33670

Chaffey College, Library, 5885 Haven Ave, **Rancho Cucamonga**, CA 91737-3002 T: +1 909 6526800; E-mail: library@chaffey.edu; URL: www.chaffey.edu/library/index.shtml 1916; Frank R. Pinkerton, Priscilla Fernandez 73 000 vols; 250 curr per   33671

County College of Morris, Sherman H. Masten Learning Resource Center, 214 Center Grove Rd, **Randolph**, NJ 07869-2086 T: +1 973 3285282; Fax: +1 973 3282370; E-mail: jcohn@ccm.edu; URL: www.ccm.edu/library 1968; John M. Cohn 97 885 vols; 819 curr per libr loan   33672

Vermont Technical College, Hartness Library, **Randolph Center**, VT 05061 T: +1 802 7281237; Fax: +1 802 7281506; E-mail: library@vtc.vsc.edu; URL: www.vtc.edu 1866; David Sturges 60 000 vols; 396 curr per   33673

Ranger College, Golemon Library, College Circle, **Ranger**, TX 76470-3298 T: +1 254 6471414; Fax: +1 254 6471656; E-mail: ranger@cc.tx.us; URL: www.ranger.cc.tx.us/library/library.html 1926; Courtney Hansen Robert E. Howard Coll 32 000 vols; 115 curr per   33674

Hinds Community College District, McLendon Library & Learning Resources Center, Raymond Campus, **Raymond**, MS 39154; P.O. Box 1100, PMB 11258, Raymond, MS 39154-1100 T: +1 601 8573253; Fax: +1 601 8573293; URL: www.lrc.hindscc.edu 1922; Dr. Helen J. Flanders Black Heritage Coll, Govt (John Bell Williams Coll) 173 680 vols; 1 180 curr per   33675

Reading Area Community College, The Yocum Library, P.O. Box 1706, **Reading**, PA 19603-1706 T: +1 610 3724721; Fax: +1 610 6076254; E-mail: library@email.racc.cc.pa.us; URL: www.racc.cc.pa.us 1971; Mary Ellen G. Heckman Film, Nursing, Science fiction; Comic Bks, Music Score Coll, Schuylkill Navigation Co Maps 35 000 vols; 425 curr per   33676

Shasta College, Learning Resources Center Library, 11555 N Old Oregon Trail, P.O. Box 496006, **Redding**, CA 96049 T: +1 530 2427550; Fax: +1 530 2254830; E-mail: jalbright@shastacollege.edu; URL: www.shastacollege.edu/library 1948; Janet Albright Local hist 63 452 vols; 180 curr per; 48 331 microforms; 1 100 av-mat; 1 615 sound-rec   33677

Canada College, Library, 4200 Farm Hill Blvd, **Redwood City**, CA 94061-1099 T: +1 650 3063267; Fax: +1 650 3063434; E-mail: hayward@smccd.net; URL: canadacollege.net/library 1968; Marilyn Hayward Ctr for the American Music Libr, Arch 53 000 vols; 140 curr per   33678

Kings River Community College, Reedley College, Library, 995 N Reed Ave, **Reedley**, CA 93654-2099 T: +1 559 6380352; Fax: +1 559 6380384; E-mail: stephanie.curry@scccd.com; URL: www.reedleycollege.com 1956; Wilifred Louise Alire/Stephanie Curry National College Catalog sets (10 yrs), microfiche 34 590 vols; 185 curr per; 3 167 microforms; 750 av-mat; 500 sound-rec; 10 digital data carriers; 750 videos libr loan   33679

Truckee Meadows Community College, Elizabeth Sturm Libary, 7000 Dandini Blvd, **Reno**, NV 89512-3999 T: +1 775 6737000; Fax: +1 775 6737273; E-mail: tmclrc@unssu.nevada.edu; URL: www.tmcc.edu/library Lorin Lindsay Folklore, Career education, Health education, Industry 39 450 vols; 105 curr per   33680

Nicolet Area Technical College, Richard J. Brown Library, Lake Julia Campus, Hwy G, P.O. Box 518, **Rhinelander**, WI 54501 T: +1 715 3654429; Fax: +1 715 3654404; E-mail: amussehl@nicolet.tec.wi.us; URL: www.nicolet.tec.wi.us/libry/index.html 1969; Allan A. Mussehl 40 000 vols; 350 curr per libr loan   33681

Southwest Virginia Community College, Library, P.O. Box SVCC, **Richlands**, VA 24641 T: +1 540 9642555; Fax: +1 540 9647259; E-mail: su_captain@sw.va.us; URL: www.sw.cc.va.us/library 1968; Sue W. Captain 38 000 vols; 225 curr per   33682

J. Sargeant Reynolds Community College – Downtown Campus Learning Resources Center, 700 E Jackson St, **Richmond**, VA 23219-1543; P.O. Box 85622, Richmond, VA 23285-5622 T: +1 804 7865211; Fax: +1 804 7866200; E-mail: amiah@jsr.cc.va.us 1973; Dr. Abdul J. Miah Business Administration, AV, bks; Health Science, AV, bks 33 830 vols; 215 curr per   33683

J. Sargeant Reynolds Community College – Parham Campus Learning Resources Center, 1651 E Parham Rd, **Richmond**, VA 23228; P.O. Box 85622, Richmond, VA 23285-5622 T: +1 804 3713220; Fax: +1 804 3713086; E-mail: twilliams@jsr.cc.va.us; URL: www.jsr.cc.va.us/library/default.htm 1974; Charles Lewis & Tim Williams 56 000 vols; 230 curr per   33684

Triton College, Library, 2000 N Fifth Ave, **River Grove**, IL 60171 T: +1 708 4560300 ext 3215; E-mail: library@triton.edu; URL: www.triton.edu/library 1964 Science & technology, Nursing 76 650 vols; 470 curr per libr loan   33685

Suffolk County Community College, Riverhead Campus Library, 121 Speonk Riverhead Rd, **Riverhead**, NY 11901-9990 T: +1 631 5482536; Fax: +1 631 3692641; E-mail: millerma@sunysuffolk.edu; URL: www.sunysuffolk.edu 1977; Mary Ann Miller 35 912 vols; 402 curr per   33686

California School for the Deaf Library, 3044 Horace St, **Riverside**, CA 92506 T: +1 951 7826500; URL: www.csdr-cde.ca.gov/ Mary Hanlon American Annals for the Deaf, Volta Review 13 000 vols   33687

Riverside Community College District, Digital Library & Learning Resource Center, 4800 Magnolia Ave, **Riverside**, CA 92506-1299 T: +1 951 2228651; Fax: +1 951 3283679; E-mail: cwong@rccd.cc.ca.us; URL: library.rcc.edu 1919; Cecilia Wong 154 000 vols; 890 curr per; 7 733 av-mat; 183 Bks on Deafness & Sign Lang, 30 Talkings Bks   33688

Central Wyoming College, Library, 2660 Peck Ave, **Riverton**, WY 82501 T: +1 307 8552141; Fax: +1 307 8552094; E-mail: cdeering@cwc.cc.wy.us 1967; Carol Deering American Indian Coll, Wyoming Coll 40 000 vols; 348 curr per; 1 900 maps;

61 192 microforms libr loan   33689

Virginia Western Community College, Brown Library, 3095 Colonial Ave SW, **Roanoke**, VA 24015-4705; P.O. Box 40012, Roanoke, VA 24022-0012 T: +1 703 8576445; Fax: +1 703 8577544; E-mail: lhurt@virginiawestern.edu; URL: www.virginiawestern.edu/library 1966; Lynn Hurt 69 622 vols; 245 curr per   33690

Monroe Community College, LeRoy V. Good Library, 1000 E Henrietta Rd, **Rochester**, NY 14692; P.O. Box 92810, Rochester, NY 14692-8910 T: +1 585 2922090; Fax: +1 585 4241402; E-mail: shr.ill@monroecc.edu; URL: www.monroecc.edu/go/library 1962; Peter Genovese AIDS Resource Libr, Genocide/Holocaust Resource Ctr 108 000 vols; 515 curr per libr loan   33691

Rochester Community & Technical College, Goddard Library, 851 30 Ave SE, **Rochester**, MN 55904 +1 507 2857233; Fax: +1 507 2817772; E-mail: library@ucr.roch.edu; URL: www.roch.edu/library 1915; Dale Pedersen 110 000 vols; 350 curr per libr loan   33692

Rochester College, Muirhead Library, 800 W Avon Rd, **Rochester Hills**, MI 48307 T: +1 248 2182266; Fax: +1 248 2182265; E-mail: library@rc.edu; URL: www.rc.edu/lib 1959; Allie Keller 45 000 vols; 130 curr per; 12 000 e-books; 52 av-mat   33693

Western Wyoming Community College, Hay Library, 2500 College Dr, **Rock Springs**, WY 82901-3402; P.O. Box 428, Rock Springs, WY 82902-0428 T: +1 307 3821701; Fax: +1 307 3827665; E-mail: cabrown@wwcc.wy.edu; URL: www.wwcc.cc.wy.us/library 1959; Carol Brown Local Newspapers (Rock Springs Rocket, Green River Star), micro 60 000 vols; 300 curr per libr loan   33694

Rock Valley College, Estelle M. Black Library, 3301 N Mulford Rd, **Rockford**, IL 61114 T: +1 815 9214626; Fax: +1 815 9214629; E-mail: rvc-libref@rockvalleycollege.edu; URL: www.rockvalleycollege.edu 1965; Dr. Hsiao-Hung Lee 75 000 vols; 515 curr per   33695

Montgomery College – Rockville Campus, Library, 51 Mannakee St, **Rockville**, MD 20850 T: +1 301 2517117, 2795067; Fax: +1 301 2517153; URL: www.montgomerycollege.org 1965; Patricia Mehok 149 240 vols; 780 curr per   33696

Montgomery County Public Schools Professional Library, MCPS Professional Library, 850 Hungerford Dr, Rm 50, **Rockville**, MD 20850-1718 T: +1 301 2793227; Fax: +1 301 2793072; E-mail: professional_library@mcpsmd.org; URL: www.mcps.k12.md.us/departments/media/ prof.html 1961; Janet Biggs Education; Montgomery County Public Schools' Curriculum Guides, Diss by MCPS Employees 21 000 vols; 120 curr per; 17 digital data carriers   33697

Nash Community College, Library, Old Carriage Rd, P.O. Box 7488, **Rocky Mount**, NC 27804 T: +1 252 4434011 ext 244; Fax: +1 252 4430828; URL: www.nash.cc.nc.us 1968; Lynette Finch 40 000 vols; 130 curr per   33698

Rollinsford School Library, 487 Locust St, *Rollinsford*, NH 03869
T: +1 603 7422351; Fax: +1 603 7495629; E-mail: dmnt@comcast.net
Mary Stumhofer
12 000 vols; 26 curr per   33699

Floyd College, Library, 3175 Cedartown Hwy SE, *Rome*, GA 30161; P.O. Box 1864, Rome, GA 30162-1864
T: +1 706 2956318; Fax: +1 706 2956365
1970; Debbie Holmes
Paramedics
62 273 vols; 288 curr per; 329 maps; 19 029 microforms; 7 328 av-mat; 3 046 sound-rec; 8 digital data carriers
libr loan; ALA   33700

West Virginia School for the Deaf & Blind Library, 301 E Main St, *Romney*, WV 26757
T: +1 304 8224860; Fax: +1 304 8224870
Marty Blakely
10 000 vols; 40 curr per   33701

Kirtland Community College, Library, 10775 N St Helen Rd, *Roscommon*, MI 48653
T: +1 989 2755000 ext 246; Fax: +1 989 2758510; E-mail: library@kirtland.cc.mi.us; URL: www.kirtland.cc.mi.us/library
1968; Louise Bucco
30 000 vols; 250 curr per   33702

Umpqua Community College, Library, 1140 College Rd, P.O. Box 967, *Roseburg*, OR 97470-0226
T: +1 541 4404640; Fax: +1 541 4404637; E-mail: hutchid@umpqua.cc.or.us; URL: www.umpqua.cc.or.us/library/libhome.htm
1964; David Hutchinson
Coll of Fire Science, Career, Grant
38 950 vols; 188 curr per; 106 maps; 743 microforms; 1 100 av-mat; 1 459 sound-rec; 8 digital data carriers
libr loan; ALA   33703

Don Bosco Technical Institute, Lee Memorial Library, 1151 San Gabriel Blvd, *Rosemead*, CA 91770
T: +1 626 9402035; Fax: +1 626 9402001; E-mail: GeneralInfo@boscotech.edu; URL: www.boscotech.edu
1957; Margaret Pedregon
Technology, Art, Religion
95 000 vols; 30 curr per   33704

New Mexico Military Institute, Paul Horgan Library – Tole's Learning Center, 101 W College, *Roswell*, NM 88201-5173
T: +1 505 6248385; Fax: +1 505 6248390; E-mail: klopfer@nmmi.edu; URL: www.nmmi.edu/library/index.htm
1902; Jerry Klopfer
Humanities, Natural science; Henry David Thoreau, Paul Horgan Writings, Military Hist, Napoleonic Hist, Southwestern Hist
60 000 vols; 231 curr per; 18 413 microforms; 30 VF drawers   33705

Oakland Community College, Royal Oak Campus Library, 739 S Washington Ave, *Royal Oak*, MI 48067
T: +1 248 2322525; Fax: +1 248 2462520; E-mail: ctbenson@occ.cc.mi.us; URL: www.occ.cc.mi.us/library
1917; Carol Benson, Thomas Lewandowski
Careers
43 830 vols; 800 curr per
libr loan   33706

American River College, Library, 4700 College Oak Dr, *Sacramento*, CA 95841
T: +1 916 4848457; Fax: +1 916 4848657; E-mail: info@arc.losrios.edu; URL: www.arc.losrios.edu
1955; Bruce Kinghorn
California Hist Coll, Vocations-Careers Coll, Weaving Coll, Professional Coll
90 000 vols
libr loan   33707

Cosumnes River College, Library & Learning Resources Center, 8401 Center Pkwy, *Sacramento*, CA 95823-5799
T: +1 916 6917266; Fax: +1 916 6917349; E-mail: info@crc.losrios.edu; URL: www.crc.losrios.edu
1970; Michael Mayer-Kielmann
California Hist, Campus Oral Hist
65 500 vols; 380 curr per
libr loan   33708

Sacramento City College, Learning Resources Division Library, 3835 Freeport Blvd, *Sacramento*, CA 95822
T: +1 916 5582461; Fax: +1 916 5582114; E-mail: mcdonas@scc.losrios.edu; URL: wserver.scc.losrios.edu/~library/
1916
70 000 vols; 410 curr per   33709

Saint Catharine College, Library, 2735 Bardstown Rd, *Saint Catharine*, KY 40061
T: +1 859 3365082 ext 260; Fax: +1 859 3365087; E-mail: library@sccky.edu; URL: Aquinas.sccky.edu
1930; Ilona Burdette
Elizabeth Madox Roberts Coll
40 500 vols; 120 curr per   33710

Dixie State College of Utah, Val A. Browning Library, 225 S 700 E, *Saint George*, UT 84770
T: +1 435 6527714; Fax: +1 435 6564169; URL: library.dixie.edu
1912; William Baer
Mormon & Southwest Hist
100 000 vols; 314 curr per; 12 diss/theses; 374 govt docs; 567 maps; 20 000 microforms; 1 300 av-mat; 1 888 sound-rec; 56 digital data carriers; 1 991 pamphlets, 4 358 vertical files
libr loan; ALA   33711

Missouri School for the Blind Library, 3815 Magnolia Ave, *Saint Louis*, MO 63110
T: +1 314 7764320; Fax: +1 314 7733762; URL: www.msb.k12.mo.us
Mary Dingus
Blind Education
24 000 vols; 115 curr per; 91 000 av-mat; 6 460 sound-rec; 4 digital data carriers   33712

St. Louis Community College – Forest Park Campus Library, 5600 Oakland Ave, *Saint Louis*, MO 63110-1393
T: +1 314 6449210; Fax: +1 314 6449240; E-mail: jswilliams@stlcc.edu; URL: www.stlcc.edu/Libraries/Index.html
1965; June S. Williams
Afro-American Coll
80 710 vols; 457 curr per; 5 sound-rec   33713

St. Louis Community College – Meramec Campus Library, 11333 Big Bend Blvd, *Saint Louis*, MO 63122
T: +1 314 9847797; Fax: +1 314 9847225; E-mail: bsanguinet@stlcc.edu; URL: www.stlcc.edu/libraries/Contact/#MC
1963; Bonnie Sanguinet
93 000 vols; 547 curr per   33714

Saint Charles Community College, Library, 4601 Mid Rivers Mall Dr, *Saint Peters*, MO 63376
T: +1 636 9228434; Fax: +1 636 9228433; E-mail: jclarke@stchas.edu; URL: www.stchas.edu/library
1987; Joan Clarke
47 860 vols; 325 curr per
libr loan; ALA, ACRL, MLA   33715

Chemeketa Community College, Learning Resource Center, 4000 Lancaster Dr NE, *Salem*, OR 97305-1500; P.O. Box 14007, Salem, OR 97309-7070
T: +1 503 3995043 ext 5105; Fax: +1 503 3995214; URL: library.chemeketa.edu
1962; Linda Cochrane
62 000 vols; 725 curr per
libr loan   33716

National College of Business and Technology, Roanoke Valley Campus Library, 1813 E Main St, *Salem*, VA 24153-4598
T: +1 540 9861800; Fax: +1 540 4444195; URL: www.national-college.edu
Betty W. Johnson

23 000 vols; 50 maps
libr loan   33717

Oregon School for the Deaf Library, 999 Locust St NE, *Salem*, OR 97301-0954
T: +1 503 3783825; Fax: +1 503 3783378; URL: www.osd.k12.or.us
1970; Charleen Hoiland
American Annals of the Deaf, Volta, Proceedings of Convention of American Instructors of the Deaf
14 000 vols; 30 curr per; 300 av-mat   33718

Hartnell Community College, Library, 156 Homestead Ave, *Salinas*, CA 93901
T: +1 831 7556872; Fax: +1 831 7596084; E-mail: ghughes@hartnell.cc.ca.us; URL: www.hartnell.cc.ca.us
1936; Gary Hughes
Ornithology (O. P. Silliman Memorial Libr of Natural Hist)
58 000 vols; 260 curr per   33719

Rowan-Cabarrus Community College, Learning Resource Center-North Campus, 1333 Jake Alexander Blvd, P.O. Box 1595, *Salisbury*, NC 28145-1595
T: +1 704 6370760 ext 253; Fax: +1 704 6376642
1964; Sheila Bailey
40 000 vols; 425 curr per   33720

Salt Lake Community College, Library, 4600 S Redwood Rd, *Salt Lake City*, UT 84123-3197; P.O. Box 30808, Salt Lake City, UT 84130-0808
T: +1 801 9574195; Fax: +1 801 9574141; URL: www.slcc.edu/library
1948; Dr. Larry Dorrell
Education, Nursing, Business, Science & technology; Automotive Manuals, Contractors Exams
91 570 vols; 721 curr per; 6 300 e-books; 28 682 av-mat; 219 High Interest/Low Vocabulary Bks, 204 Bks on Deafness & Sign Lang
libr loan; ALA, ULA, BCR, OCLC   33721

Palo Alto College, George Ozuna, Jr Learning Resources Center, 1400 W Villaret, *San Antonio*, TX 78224-2499
T: +1 210 9215100; Fax: +1 210 9215461, +1 210 9215065; E-mail: hilario@accd.edu; URL: www.accd.edu/pac/lrc/index.htm
1985; Gloria E. Hilario
73 810 vols; 455 curr per   33722

St. Philip's College, Learning Resources Center, 1801 Martin Luther King Dr, *San Antonio*, TX 78203-2098
T: +1 210 5313359; Fax: +1 210 5313331; E-mail: pdolat@accd.edu; URL: www.accd.edu/spc/admin/lrc/default.aspx
1927; Dr. Adele S. Dendy
African-American Lit
90 767 vols; 1 027 curr per; 778 microforms; 14 000 av-mat; 2 958 sound-rec
libr loan   33723

San Bernardino Valley College, Samuel E. Andrews Memorial Library, 701 S Mt Vernon Ave, *San Bernardino*, CA 92410
T: +1 909 3844448; Fax: +1 909 3832802; E-mail: mmestas@valleycollege.edu; URL: lr.valley.sbccd.cc.ca.us/libhome.htm
1928; Marie Mestas
Humanities, Social science
101 043 vols; 305 curr per   33724

Skyline College, Library, 3300 College Dr, *San Bruno*, CA 94066-1698
T: +1 650 7384311; Fax: +1 650 7384149; URL: www.pls.lib.ca.us/pls/smcccd/sky/sky.html
1969; Thomas Hewitt
Feminism
45 000 vols; 240 curr per   33725

San Diego City College, Learning Resource Center, 1313 12th Ave, *San Diego*, CA 92101-4712
T: +1 619 3883421; Fax: +1 619 3883410; E-mail: citylib@sdccd.edu; URL: www.sdcity.edu/CollegeServices/LRCLibrar
1916; Robbi Ewell
64 000 vols; 425 curr per   33726

San Diego Mesa College, Library, 7250 Mesa College Dr, *San Diego*, CA 92111-4998
T: +1 619 6272695; Fax: +1 619 6272922; E-mail: jforman@sdccd.net; URL: intergate.sdmesa.sdccd.cc.ca.us/lrc/index.html
1963; Jack Forman
Architecture
95 000 vols; 500 curr per; 29 maps; 96 215 microforms; 2 500 av-mat; 7 digital data carriers
libr loan   33727

Academy of Art College, Library, 180 New Montogomery, 6th Flr, *San Francisco*, CA 94105
T: +1 415 2742270; Fax: +1 415 2638803; E-mail: library@academyart.edu; URL: www.academyart.edu; library.academyart.edu
1977; John Winsor
26 000 vols; 300 curr per; 439 diss/theses; 231 digital data carriers; 1 200 videos, 100 000 slides, 100 picture files
ARLIS/NA, VRA   33728

City College of San Francisco, Rosenberg Library, 50 Phelan Ave, R501, *San Francisco*, CA 94112
T: +1 415 4525454; Fax: +1 415 4525588; E-mail: rjones@ccsf.edu; URL: www.ccsf.edu/lib
1935; Rita W. Jones
Hotel-Restaurant (Alice Statler Coll)
141 000 vols; 680 curr per; 500 e-books; 150 av-mat
libr loan   33729

Mount San Jacinto College, Milo P. Johnson Library, 1499 N State St, *San Jacinto*, CA 92583-2399
T: +1 909 4876752 ext 1580; Fax: +1 909 6548387; E-mail: ddking@msjc.cc.ca.us; URL: www.msjc.cc.ca.us/sjclibrary
1963; David King
36 000 vols; 330 curr per   33730

Evergreen Valley College, Library, 3095 Yerba Buena Rd, *San Jose*, CA 95135-1598
T: +1 408 2706433; Fax: +1 408 5321925; E-mail: victoria.atherton@sjeccd.cc.ca.us; URL: www.library.sjeccd.cc.ca.us
1975; Victoria Atherton
46 000 vols; 160 curr per
libr loan   33731

Palmer College of Chiropractic-West Campus Library, 90 E Tasman Dr, *San Jose*, CA 95134
T: +1 408 9446142; Fax: +1 408 9446181; URL: www.palmer.edu
Wendy Kubow
15 000 vols; 140 curr per   33732

San Jose City College, Library, 2100 Moorpark Ave, *San Jose*, CA 95128-2799
T: +1 408 2883775; Fax: +1 408 2934728; E-mail: joseph.king@sjeccd.cc.ca.us; URL: www.sjc.edu/library
1953; Joseph King
60 239 vols; 232 curr per; 5 000 incunabula; 500 govt docs; 1 000 music scores; 226 maps; 3 514 microforms; 700 av-mat
libr loan; Goldengate Library Network, National Library of Medicine, Chief Council of Librarians California Academic and Research Libraries   33733

Cuesta College, Library Learning Resources Center, Hwy 1, *San Luis Obispo*, CA 93401; P.O. Box 8106, San Luis Obispo, CA 93403-8106
T: +1 805 5463155; Fax: +1 805 5463109; E-mail: ddowell@cuesta.org; URL: library.cuesta.org
1965; Dr. David Dowell
Career Transfer Ctr, Health Coll, Morro Bay Coll
67 000 vols; 280 curr per
libr loan   33734

Palomar College Library – Media Center, 1140 W Mission Rd, *San Marcos*, CA 92069-1487
T: +1 760 7441150 ext 2848; Fax: +1 760 7613500; E-mail: library@palomar.edu; URL: www.palomar.edu/library

1946; Dr. George Mozes
Architecture, ethnic studies, Fine Arts;
California Hist Coll, Early California &
Iowa Frontier, newspapers on microfilm,
Iceland, Indians of North America, World
War I posters
108 400 vols; 900 curr per; 11 826
microforms   33735

College of San Mateo, Library,
1700 W Hillsdale Blvd, *San Mateo*,
CA 94402-3795
T: +1 650 5746100; Fax: +1 650
3586797; E-mail: csmlibrary@smccd.net;
URL: collegeofsanmateo.edu/library
1923; Lorrita E Ford
Art, Architecture, Ethnic studies,
Feminism, Music, Natural science,
Nursing, Social sciences and issues,
Technology; American Hist (LAC, Libr of
American Civilization)
90 930 vols; 265 curr per   33736

Contra Costa College, Library,
2600 Mission Bell Dr, *San Pablo*,
CA 94806-3195
T: +1 510 2357800 ext 4420;
Fax: +1 510 2348161; E-mail:
bcarlton@contracosta.cc.ca.us; URL:
www.contracosta.cc.ca.us/library/index.htm
1950; Bruce Carlton
57 863 vols; 220 curr per
libr loan   33737

Niagara County Community College,
Library Learning Center, 3111 Saunders
Settlement Rd, *Sanborn*, NY 14132
T: +1 716 6146791; E-mail:
melcher@niagaracc.suny.edu; URL:
www.niagaracc.suny.edu
1963; Tamara Anderson
86 770 vols   33738

Central Carolina Community College,
Library Learning Resources, 1105 Kelly
Dr, *Sanford*, NC 27330
T: +1 919 7187244; Fax: +1 919
7187378; E-mail: tguthrie@cccc.edu;
URL: www.cccc.edu/library/
1962; Tara Guthrie
Veterinary Medical Technology, Law Libr
Coll, Nursing & Medical Libr
52 211 vols; 200 curr per; 5 352 av-mat;
32 digital data carriers   33739

Seminole State College, Sanford /
Lake Mary Campus, 100 Weldon Blvd,
Sanford, FL 32773-6199
T: +1 407 7084722; E-mail:
jenningm@seminolestate.edu; URL:
www.seminolestate.edu/slm/
1966; Patricia DeSalvo
83 500 vols; 500 curr per
libr loan   33740

Santa Ana College, Nealley Library, 1530
W 17th St, *Santa Ana*, CA 92706-3398
T: +1 714 5646700; Fax: +1 714
5646729; URL: www.sac.edu/students/
library/nealley
1915; Carolyn Breeden
995 000 vols; 405 curr per   33741

Santa Barbara City College, Eli Luria
Library, 721 Cliff Dr, *Santa Barbara*,
CA 93109-2394
T: +1 805 9650581 ext 2638; Fax: +1
805 9650771; E-mail: neufeld@sbcc.edu;
URL: library.sbcc.edu
1909; Kenley Neufeld
10 000 e-books
123 530 vols; 324 curr per; 3 040 e-
journals; 238 microforms
libr loan   33742

Mission College, Library, 3000
Mission College Blvd, *Santa Clara*,
CA 95054-1897
T: +1 408 8555150; Fax: +1
408 8555462; E-mail:
tim_karas@wvmccd.cc.ca.us; URL:
www.missioncollege.org/lib/lib.html
1975; Timothy Karas
Fire science, Allied health
49 030 vols; 175 curr per; 25 microforms;
3 105 av-mat; 823 sound-rec; 7 digital
data carriers; index (32), 10 newspapers
libr loan   33743

College of the Canyons, William G.
Bonelli Instructional Resource Center,
26455 N Rockwell Canyon Rd, *Santa
Clarita*, CA 91355-1899
T: +1 661 2597800 ext 3330;
Fax: +1 661 2531845; E-mail:
keller_j@mail.coc.cc.ca.us; URL:
www.canyons.edu/offices/library/
CanyonCountryInfo.asp
1969; Jan K. Keller
50 000 vols; 250 curr per   33744

**New Mexico School for the Deaf
Library**, 1060 Cerrillos Rd, *Santa Fe*,
NM 87505
T: +1 505 4766379; Fax: +1 505
4766376; URL: www.nmsd.k12.nm.us
Hedy Udkovich-Stern
Deaf Culture; Deaf Education
16 000 vols; 40 curr per   33745

Allan Hancock College, Learning
Resources Center, 800 S College Dr,
Santa Maria, CA 93454
T: +1 805 9226966 ext 3224;
Fax: +1 805 9223763; E-mail:
library@hancockcollege.edu;
URL: www.hancock.cc.ca.us/
Default.asp?Page=103
1920
Theatre Coll
44 000 vols; 340 curr per   33746

Santa Monica College, Library,
1900 W Pico Blvd, *Santa Monica*,
CA 90405-1628
T: +1 310 4344334; Fax: +1 310
4344387; URL: library.smc.edu
1929; Mona Martin
105 000 vols; 475 curr per   33747

Santa Rosa Junior College, Frank P.
Doyle Library, 1501 Mendocino Ave,
Santa Rosa, CA 95401
T: +1 707 5274261; Fax: +1 707
5274545; E-mail: wbaty@santarosa.edu;
URL: www.santarosa.edu/library
1918; Will Baty
Computer science, Dentistry, Horticulture,
Nutrition, Patents, Theater
122 700 vols; 393 curr per
libr loan   33748

North Country Community College,
Library, 20 Winona Ave, P.O. Box 89,
Saranac Lake, NY 12983
T: +1 518 8912915 ext 218
1967; Patrick F McIntyre
Nettie Marie Jones Fine Arts Coll,
Adirondack Coll
57 010 vols; 177 curr per; 12 445
microforms; 710 av-mat; 554 sound-rec
libr loan; OCLC   33749

West Valley Community College,
Library, 14000 Fruitvale Ave, *Saratoga*,
CA 95070-5698
T: +1 408 7412140; Fax: +1
408 7412134; E-mail: dave-
fishbaugh@wvmccd.cc.ca.us
1964; Dave Fishbaugh
93 484 vols; 400 curr per; 1 700 av-mat;
5 843 sound-rec; 7 digital data carriers
libr loan   33750

**Schenectady County Community
College**, Begley Library & Instructional
Technology Center, 78 Washington Ave,
Schenectady, NY 12305
T: +1 518 3811240; Fax: +1 518
3704659; E-mail: kinglo@sunysccc.edu;
URL: www.sunysccc.edu/library/
1968; Lynne King
College Memorabilia Coll
81 000 vols; 400 curr per   33751

Lehigh Carbon Community College,
Rothrock Library, 4750 Orchard Rd,
Schnecksville, PA 18078
T: +1 610 7991150; Fax: +1 610
7791159; E-mail: rothrock@lccc.edu;
URL: www.lccc.edu/library
1967
Allied health, Criminal justice; College
Arch; New York Times, 1851-present,
micro
51 000 vols; 490 curr per; 6 200 av-mat
libr loan   33752

**Western Nebraska Community
College**, Information Services, 1601
E 27th NE, *Scottsbluff*, NE 69361-1899
T: +1 308 6356040; Fax: +1
308 6356086; E-mail:
vschneid@hannibal.wncc.cc.ne.us; URL:
www.wncc.net/library

1926; Valetta Schneider
34 000 vols   33753

Scottsdale Community College, Library,
9000 E Chaparral Rd, *Scottsdale*,
AZ 85256
T: +1 480 4236651; Fax: +1
480 4236666; E-mail:
scc.library@sccmail.maricopa.edu; URL:
www.sc.maricopa.edu/library
1971
Foreign languages, Careers; Indians of
the Southwest Coll
50 000 vols; 350 curr per   33754

North Seattle Community College,
Library Media Center, 9600 College Way
N, *Seattle*, WA 98103
T: +1 206 5273607; Fax: +1 206
5273614; URL: www.sccd.ctc.edu/~library
1970
46 000 vols; 491 curr per   33755

Seattle Central Community College,
Instructional Resource Services, 1701
Broadway, *Seattle*, WA 98122
T: +1 206 5875420, 5874050;
Fax: +1 206 5873878; E-mail:
BHarding@sccd.ctc.edu; URL:
www.seattlecentral.org/library/index.php
1966; Wai-Fong Lee
60 580 vols; 595 curr per   33756

Shoreline Community College, Ray
W. Howard Library Media Center,
16101 Greenwood Ave N, *Seattle*,
WA 98133-5696
T: +1 206 5466939; E-mail: jbetz-
zall@shoreline.edu; URL: oscar.ctc.edu/
library
1963; Tom Moran
70 000 vols; 360 curr per
libr loan   33757

South Seattle Community College,
Instructional Resources Center, 6000 16th
Ave SW, *Seattle*, WA 98106
T: +1 206 7645394;
Fax: +1 206 7635155; URL:
www.dept.seattlecolleges.com/sslib
1971; Mary Jo White
Landscape-Horticulture Coll
33 090 vols; 650 curr per   33758

Suffolk County Community College,
Ammerman Campus Library, 533 College
Rd, *Selden*, NY 11784-2899
T: +1 631 4514177; Fax: +1 631
4514697; URL: www.sunysuffolk.edu/
library
1960; Hedi Ben Aicha
Allied health, Business, Nursing; Long
Island Coll
124 610 vols; 712 curr per; 9 453
microforms; 1 900 av-mat; 1 418
sound-rec; 6 digital data carriers
libr loan   33759

Seminole State College, David L. Boren
Library, Junction Hwy 9 & David L
Boren Blvd, P.O. Box 351, *Seminole*,
OK 74818-0351
T: +1 405 3829950 ext 243; Fax: +1
405 3829511
1970; Debbie Kensey
Libr of American Civilization (American
Hist Coll), ultrafiche
31 000 vols; 301 curr per   33760

**Northwest Mississippi Community
College**, R.C. Pugh Library, 4975 Hwy
51 N, P.O. Box L-NWCC, *Senatobia*,
MS 38668
T: +1 662 5623277; Fax: +1 662
5623280; E-mail: mrogers@nwcc.cc.ms.us;
URL: www.nwcc.cc.ms.us
1926; Margaret N. Rogers
46 300 vols; 286 curr per   33761

Gloucester County College, Library-
Media Center, 1400 Tanyard Rd, *Sewell*,
NJ 08080
T: +1 856 4685000 ext 2250; Fax: +1
856 4641695; URL: gccnj.edu/library/
services/index.cfm
1968; Jane Lopes-Crocker
Art, Business, Nursing
50 000 vols; 479 curr per   33762

Montcalm Community College Library,
2800 College Dr SW, P.O. Box 300,
Sidney, MI 48885-0300
T: +1 989 3282411; Fax: +1
989 3282950; E-mail:
rickp@montcalm.cc.mi.us; URL:
www.momtcalm.cc.mi.us/library
1966; Richard L. Parker
30 000 vols; 220 curr per   33763

Johnston Community College, Library,
245 College Rd, P.O. Box 2350,
Smithfield, NC 27577
T: +1 919 2092101; Fax: +1
919 2092186; E-mail:
ramseylinda@novell.johnston.cc.nc.us;
URL: www.johnston.cc.nc.us
1969; Linda Ramsey
31 140 vols; 350 curr per   33764

Western Texas College, Learning
Resource Center, 6200 College Ave,
Snyder, TX 79549
T: +1 915 5738511 ext 303; Fax: +1
915 5739321; E-mail: zirons@wtc.cc.tx.us;
URL: www.wtc.cc.tx.us
1971; Zelma A. Irons
30 000 vols; 78 curr per   33765

Somerset Community College, Harold
B. Strunk Learning Resource Center, 808
Monticello Rd, *Somerset*, KY 42501
T: +1 606 6798501 ext 3410; Fax: +1
606 6795139; E-mail: shelley.burgett@
kctcs.net; URL: www.somcc.kctcs.net/
library
1965; Shelley Wood Burgett
Local hist, Japan; Local Newpaper,
microfilm
45 020 vols; 186 curr per   33766

Columbia College, Library, 11600
Columbia College Dr, *Sonora*,
CA 95370-8581
T: +1 209 5885118; Fax: +1 209
5885121; URL: columbia.yosemite.cc.ca.us
1968; Larry R. Steuben
Forestry; Hist of the Mother Lode
34 307 vols; 5 187 sound-rec; 1 digital
data carriers; 113 slides, 118 art
reproductions   33767

South Suburban College, Library, 15800
S State, *South Holland*, IL 60473-9998
T: +1 708 2105751; Fax: +1 708
2105755; E-mail: ssclib@cedar.cic.net;
URL: www.southsuburbancollege.edu
1927; Bob Lhota
38 000 vols; 305 curr per; 20 maps; 333
sound-rec; 78 digital data carriers; video
tapes 804, slides 276   33768

Lake Tahoe Community College,
Library & Media Services, One College
Dr, *South Lake Tahoe*, CA 96150-4524
T: +1 530 5414660; Fax: +1 530
5417852; E-mail: library@ltcc.cc.ca.us;
URL: freel.ltcc.cc.ca.us
1976; Phil Roché
Music Dept Libr
39 631 vols; 200 curr per; 3 523
microforms; 1 200 av-mat; 2 717
sound-rec
libr loan   33769

Spartanburg Methodist College, Marie
Blair Burgess Library, 1200 Textile Dr,
Spartanburg, SC 29301
T: +1 864 5874208; Fax: +1 864
5874352; E-mail: hallerje@smcsc.edu
1911; James Haller
44 000 vols; 200 curr per   33770

Spartanburg Technical College,
Library, PO Drawer 4386, *Spartanburg*,
SC 29305
T: +1 864 5913615; Fax: +1 864
5913762; E-mail: greenm@spt.tec.sc.us;
URL: library.spt.tec.sc.us
Margaret Green
Horiculture, Sensory impaired
37 090 vols; 291 curr per   33771

Isothermal Community College, Library,
286 ICC Loop Rd, P.O. Box 804,
Spindale, NC 28160-0804
T: +1 828 2863636; Fax: +1
828 2868208; E-mail:
svaughan@isothermal.cc.nc.us; URL:
www.isothermal.cc.nc.us/library
1965; Susan Vaughan
Local Hist Coll
39 073 vols; 243 curr per; 3 341
microforms; 2 200 av-mat; 1 065
sound-rec; 34 digital data carriers
libr loan; ALA   33772

Spokane Community College, Library,
N 1810 Greene St, MS 2160, *Spokane*,
WA 99217-5399
T: +1 509 5337045; Fax: +1
509 5337276; E-mail:
mcarr@scc.spokane.cc.wa.us; URL:
www.scc.spokane.cc.wa.us/lrc/library
1963; Mary M. Carr
Career Ctr
45 000 vols; 340 curr per   33773

Spokane Falls Community College, Library, W 3410 Ft George Wright Dr M/S 3020, ***Spokane***, WA 99224-5288
T: +1 509 5333800; Fax: +1 509 5333144
1967; Mary Ann Lund Goodwin
58 827 vols; 400 curr per; 36 051 microforms; 3 540 av-mat; 2 510 sound-rec
libr loan; OCLC 33774

Clark State Community College, Library, 570 E Leffel Lane, P.O. Box 570, ***Springfield***, OH 45501
T: +1 937 3286022; Fax: +1 937 3286133; URL: lib2.clark.cc.oh.us/library/library.html
1966; Nancy Schwerner
35 000 vols; 375 curr per
libr loan 33775

Lincoln Land Community College, Library, 5250 Shepherd Rd, P.O. Box 19256, ***Springfield***, IL 62794-9256
T: +1 217 7862354; Fax: +1 217 7862251; E-mail: jill.campbell@llcc.edu; URL: www.llcc.edu
1968
Nursing
64 000 vols; 265 curr per 33776

Springfield Technical Community College, Library, One Armory Sq, ***Springfield***, MA 01101-9000
T: +1 413 7554845; Fax: +1 413 7556315; E-mail: abeth@stcc.edu; URL: library.stcc.edu
1969; Amy Beth
65 500 vols; 440 curr per 33777

Mitchell Community College, Mildred & J. P. Huskins Library, 500 W Broad St, ***Statesville***, NC 28677
T: +1 704 8783271; Fax: +1 704 8783206; E-mail: rklett@mitchell.cc.nc.us; URL: www.mitchell.cc.nc.us
1852; Rex E. Klett
39 230 vols; 201 curr per 33778

Northeastern Junior College, Monahan Library, 100 N College Dr, ***Sterling***, CO 80751-2399
T: +1 970 5216600 ext 6612; Fax: +1 970 5216759; E-mail: Candace.havely@njc.cccoes.edu; URL: nejc.cc.co.us/library/library.html
1941; Candace Havely
Colorado
33 691 vols; 326 curr per 33779

San Joaquin Delta College, Irving Goleman Library, 5151 Pacific Ave, ***Stockton***, CA 95207-6370
T: +1 209 4745151; Fax: +1 209 4745691; E-mail: cmooney@deltacollege.edu; URL: www.deltacollege.edu/dept/library/
1948
93 100 vols; 629 curr per
libr loan 33780

Ulster County Community College, MacDonald DeWitt Library, Cottekill Rd, ***Stone Ridge***, NY 12484
T: +1 845 6875000; Fax: +1 800 7240833; URL: www.sunyulster.edu/resources/library.asp
1963; Lawrence Berk
Local hist, Psychology
78 280 vols; 440 curr per 33781

Rockland Community College, Library, 145 College Rd, ***Suffern***, NY 10901
T: +1 845 5744000; Fax: +1 845 5744424; E-mail: library@sunyrockland.edu; URL: www.sunyrockland.edu
1959; Dr. Xi Shi
Art, Elaine Magid Mystery & Detective Coll, Judaica, Rockland County Hist
123 000 vols; 300 curr per; 1 000 maps; 26 800 microforms; 11 digital data carriers
libr loan 33782

Waubonsee Community College, Sugar Grove Campus, Todd Library, Rte 47 at Waubonsee Drive, ***Sugar Grove***, IL 60554-9454
T: +1 630 4667900 ext 2400; Fax: +1 630 4667799; URL: www.waubonsee.edu
1967
Deaf
57 210 vols; 530 curr per 33783

Solano Community College, Library, 4000 Suisun Valley Rd, ***Suisun City***, CA 94585
T: +1 707 8647132; Fax: +1 707 8647231; E-mail: qcarter@solano.cc.ca.us; URL: www.solano.cc.ca.us/lrc/lib/scclib.html
1965; Quentin Carter
36 500 vols; 165 curr per; 4 digital data carriers 33784

Southwest Mississippi Community College, Library-Learning Resources Center, College Dr, ***Summit***, MS 39666
T: +1 601 2762004; Fax: +1 601 2763748
1977; Jo Ann Young
Mississippi Coll
34 000 vols; 191 curr per 33785

East Georgia College, Library, 131 College Circle, ***Swainsboro***, GA 30401-2699
T: +1 478 2892083; Fax: +1 478 2892089; E-mail: cbray@ega.peachnet.edu; URL: www.ega.peachnet.edu/library/default.htm
1973; Carol L. Bray
Ehrlich Military Hist Coll, Southeast Georgia Arch Coll
41 000 vols; 125 curr per 33786

Los Angeles Mission College, Library, 13356 Eldridge Ave, ***Sylmar***, CA 91342-3200
T: +1 818 3647750; Fax: +1 818 3647749; E-mail: lamission@cc.ca.us; URL: www.lamission.cc.ca.us
1975; Sandy Thomsen
Los Angeles Mission College Arch
44 000 vols; 193 curr per 33787

Onondaga Community College, Sidney B. Coulter Library, 4585 West Seneca Turnpike, ***Syracuse***, NY 13215
T: +1 315 4982334; E-mail: library@sunyocc.edu; URL: library.sunyocc.edu
1962; Jeff Harr
Berrigan Brothers Coll, Central New York Hist, Faculty Authors, Local Hist Coll, Onondaga Community College Arch
89 110 vols; 400 curr per; 61 138 microforms; 5 200 av-mat; 7 699 sound-rec; 30 digital data carriers; 50 500 pamphlets, 730 mss (linear feet)
libr loan 33788

Tacoma Community College, Library, 6501 S 19th St, ***Tacoma***, WA 98466
T: +1 253 5665087; Fax: +1 253 5665398; E-mail: bsproat@tcc.tacoma.ctc.edu; URL: www.tacoma.ctc.edu/library
1965; Rebecca Sproat
60 000 vols; 1 200 curr per 33789

Taft College, Library, 29 Emmons Park Dr, ***Taft***, CA 93268-2317
T: +1 661 7637707; Fax: +1 661 7637778; E-mail: llrc@taft.org; URL: www.taft.cc.ca.us/library/renovation/firstpage.asp
1952; Dr. Mimi Collins
32 000 vols; 150 curr per 33790

Montgomery College – Takoma Park Campus, Library, 7600 Takoma Ave, ***Takoma Park***, MD 20912
T: +1 240 5671540; E-mail: kathy.swanson@montgomerycollege.edu; URL: www.montgomerycollege.edu
1946; Sarah Fisher
70 000 vols; 385 curr per 33791

Tallahassee Community College, Library, 444 Appleyard Dr, ***Tallahassee***, FL 32304-2895
T: +1 850 2018396; Fax: +1 850 2018380; E-mail: library@tcc.fl.edu; URL: www.tcc.fl.edu/dept/library/index.htm
1966; Cherry Alexander
Paramedics; Florida (Beatrice Shaw Coll)
121 470 vols; 782 curr per
libr loan 33792

Hillsborough Community College, District Library Technical Services, 4001 Tampa Bay Blvd, ***Tampa***, FL 33614; P.O. Box 30030, Tampa, FL 33630-3030
T: +1 813 2596059; Fax: +1 813 2537510; URL: www.hccfl.edu
1968; Patricia Manack
Allied health, Nursing, Music; Art Slides Coll, Literary Criticism File
109 650 vols; 1 008 curr per 33793

Edgecombe Community College, Learning Resources Center, 2009 W Wilson St, ***Tarboro***, NC 27886
T: +1 252 8235166; Fax: +1 252 8236817; URL: www.edgecombe.cc.nc.us
1968; Daniel F. Swartant
39 660 vols; 169 curr per 33794

Cook School for Christian Leadership, Mary M. McCarthy Library, 708 S Lindon Lane, ***Tempe***, AZ 85281
T: +1 480 9689354; Fax: +1 480 9689357
1943; Mark E. Thomas
Religion, Native Americans; Indian Coll, Arch
15 000 vols; 25 curr per; 10 av-mat; 40 sound-rec; 8 VF drawers 33795

Temple College, Hubert M. Dawson Library, 2600 S First St, ***Temple***, TX 76504-7435
T: +1 254 2988426; Fax: +1 254 2988430; E-mail: library@templejc.edu; URL: www.templejc.edu/library/library.html
1926; Kathleen Fulton
54 340 vols; 349 curr per 33796

Florida College, Chatlos Library, 119 N Glen Arven Ave, ***Temple Terrace***, FL 33617-5578
T: +1 813 8496784; Fax: +1 813 8996828; URL: www.floridacollege.edu/library
1946; Wanda D. Dickey
Religion, Hist, Music
112 000 vols; 290 curr per; 29 000 e-books
libr loan 33797

Eastern Arizona College, Alumni Library, 600 Church St, ***Thatcher***, AZ 85552
T: +1 520 4288304; Fax: +1 520 4287462; E-mail: melton@eac.cc.az.us; URL: www.eac.cc.az.us/library
1888; Sue Kramer
Arizona Coll, Indians of North America Coll, Comic Books, TV Social Issues
48 000 vols; 100 curr per 33798

Abraham Baldwin Agricultural College, Baldwin Library, Dormitory Dr, 2802 Moore Hwy, ***Tifton***, GA 31794
T: +1 229 3863223; Fax: +1 229 3867471; E-mail: bsellers@abac.peachnet.edu; URL: www.abac.edu/library/
1933; Brenda A. Sellers
Dorothy King Children's Classic Coll, Georgiana Coll
73 550 vols; 513 curr per; 6 digital data carriers
libr loan 33799

North Harris Montgomery Community College District – Tomball College, Library, 30555 Tomball Pkwy, ***Tomball***, TX 77375-4036
T: +1 281 3513391; Fax: +1 281 3573786; URL: wwwtc.nhmccd.cc.tx.us/bluebonnet/education/lrc
1988; Mary Jean Webster
32 470 vols; 486 curr per 33800

Ocean County College, Learning Resources Center, College Dr, P.O. Box 2001, ***Toms River***, NJ 08754
T: +1 732 2550400; Fax: +1 732 2550421; E-mail: jtoth@ocean.edu; URL: www.ocean.edu/library/welcome.html
1966; Joe Toth
80 790 vols; 491 curr per
libr loan 33801

Northern Oklahoma College, Vineyard Library, 1220 E Grand Ave, P.O. Box 310, ***Tonkawa***, OK 74653-0310
T: +1 580 6286250; Fax: +1 580 6286209; E-mail: bhainlin@nocaxp.north.ok.edu; URL: www.north-ok.edu
1901; Ben K. Hainline
Native American hist; Oklahoma Coll
41 000 vols; 227 curr per 33802

El Camino College, Instructional Services, 16007 S Crenshaw Blvd, ***Torrance***, CA 90506
T: +1 310 6603525; Fax: +1 310 6603513; URL: www.elcamino.edu/library/
1948; Alice B. Grigsby
119 380 vols; 414 curr per; 219 High Interest/Low Vocabulary Bks, 76 Bks on Deafness & Sign Lang 33803

Mercer County Community College, Library, 1200 Old Trenton Rd, ***Trenton***, NJ 08690
T: +1 609 5864800 ext 3554; Fax: +1 609 5883778; E-mail: library@mccc.edu; URL: www.mccc.edu
1947
Mortuary Science Coll
54 660 vols; 537 curr per
libr loan 33804

Trinidad State Junior College, Samuel Freudenthal Memorial Library, 600 Prospect St, ***Trinidad***, CO 81082
T: +1 719 8465593; Fax: +1 719 8467343; E-mail: craig.larson@tsjc.cccoes.edu; URL: www.trinidadstate.edu/library/index.htm
1925; Craig Larson
50 695 vols; 154 curr per; 1 424 incunabula; 25 maps; 16 762 microforms; 1 564 av-mat; 772 sound-rec; 5 digital data carriers
libr loan; ALA 33805

Hudson Valley Community College, Dwight Marvin Learning Resources Center, 80 Vandenburgh Ave, ***Troy***, NY 12180
T: +1 518 6297336; Fax: +1 518 6297509; E-mail: erefdesk@hvcc.edu; URL: www.hvcc.edu
1953; Brenda Twiggs
Hist (Microbook Libr Journal of American Civilization Coll), micro; Hist of Western Civilization (Microbook); Video Encyclopedia of the 20th c; automotive technology
97 630 vols; 240 curr per; 5 776 av-mat 33806

Dine College, Tsaile Campus Library, ***Tsaile***, AZ 86556
T: +1 520 7246757; Fax: +1 520 7243349; URL: www.crystal.ncc.cc.nm.us
1969; Cindy Slivers
North American Indians (Moses Donner Coll)
55 000 vols; 200 curr per 33807

Arizona State Schools for the Deaf & the Blind Library, 1200 W Speedway, ***Tucson***, AZ 85745; P.O. Box 85000, Tucson, AZ 85754-5000
T: +1 520 7703671; Fax: +1 520 7703752
Peg Hartman
Braille, Deaf & Visually Impaired (Professional Coll), Captioned Film Depository, Deaf Studies, Parent Lending Libr
23 000 vols; 40 curr per 33808

Pima Community College, West Campus Library, 2202 W Anklam Rd, ***Tucson***, AZ 85709-0001
T: +1 520 2066821; Fax: +1 520 2063059; URL: www.library.pima.edu
1970; Joseph Labula
Allied health, Crimal law, Ethnic studies; Spanish Language Coll, Children's Coll, SAMS Photofacts
170 000 vols; 600 curr per; 3 215 microforms
libr loan 33809

Tulsa Community College, Metro Campus Learning Resource Center, 909 S Boston, ***Tulsa***, OK 74119-2095
T: +1 918 5957172; Fax: +1 918 5957179; URL: www.tulsacc.edu
1970; Michael David Rusk
50 000 vols; 900 curr per 33810

Shelton State Community College, Brooks-Cork Library, 9500 Old Greensboro Rd, ***Tuscaloosa***, AL 35405
T: +1 205 3913925; Fax: +1 205 3913926; E-mail: library@shelton.cc.al.us; URL: www.shelton.cc.al.us/library
1979; Deborah Grimes
41 000 vols; 400 curr per 33811

Tyler Junior College, Vaughn Library & Learning Resources Center, 1400 E Fifth St, ***Tyler***, TX 75701; P.O. Box 9020, Tyler, TX 75711-9020
T: +1 903 5102503; Fax: +1 903 5102639; E-mail: mjac@tjc.edu
1926; Marian D. Jackson
Allied Health Sciences; Legal Assistants; Texas Hist; Dr Tom Smith Indian Coll (artwork & Edward S. Curtis photogravure set); Al Herrington Native American Coll (primarily Caddo and Mississipian)
95 958 vols; 325 curr per; 4 700

microforms; 23 digital data carriers
libr loan 33812

Mendocino College, Library, 1000 Hensley Creek Rd, P.O. Box 3000, **Ukiah**, CA 95482
T: +1 707 4683053; Fax: +1 707 4683056; E-mail: ysligh@mendocino.cc.ca.us; URL: www.mendocino.cc.ca.us
1973; Yvonne Sligh
30 000 vols; 177 curr per 33813

Shawnee College, Library, 8364 College Rd, **Ullin**, IL 62992
T: +1 618 6343200 ext 271; Fax: +1 618 6343215; E-mail: juliaj@shawnee.cc.il.us; URL: www.shawnee.cc.il.us
1969; Julia Johnson
38 000 vols; 175 curr per 33814

East Central College, Library Services, 1964 Prairie Dell Rd, P.O. Box 529, **Union**, MO 63084-0529
T: +1 636 5835193 ext 2245; Fax: +1 636 5831897; E-mail: library@ecmail.ecc.cc.mo.us; URL: www.ecc.cc.mo.us/ecc/library
1969; Jennifer Dodillet
Lincoln Coll
35 000 vols; 300 curr per
libr loan; ALA 33815

Delta College, Library Learning Information Center, 1961 Delta Rd, **University Center**, MI 48710
T: +1 989 6869016; Fax: +1 989 6864131; E-mail: jackwood@delta.edu; URL: www.delta.edu/llic.aspx
1961; Jack G. Wood
90 000 vols; 375 curr per 33816

Mohawk Valley Community College, Library, 1101 Sherman Dr, **Utica**, NY 13501-5394
T: +1 315 7925408; Fax: +1 315 7925666; E-mail: sfrisbee@mvcc.edu; URL: www.mvcc.edu/library
1946; Stephen Frisbee
Career Ctr
92 000 vols; 650 curr per; 7 629 microforms; 4 400 av-mat; 560 sound-rec; 43 digital data carriers
libr loan 33817

Southwest Texas Junior College, Will C. Miller Memorial Library, 2401 Garner Field Rd, **Uvalde**, TX 78801-6297
T: +1 830 5917254; Fax: +1 830 5914186; E-mail: glenda.swink@swtjc.cc.tx.us; URL: library.swtjc.cc.tx.us
1946; Glenda Swink
Arch; Education Curriculum; Texana
38 000 vols; 390 curr per; 16 digital data carriers
libr loan; TLA 33818

Suny Westchester Community College, Harold L. Drimmer Library & Learning Resource Center, 75 Grasslands Rd, **Valhalla**, NY 10595-1693
T: +1 914 7856960; Fax: +1 914 7856531; E-mail: una.shih@sunywcc.edu; URL: www.sunywcc.edu/library/homepage.htm
1946; Una Shih
College & Career Coll, Legal Coll New York State
138 000 vols; 362 curr per; 50 000 microforms; 4 100 av-mat; 389 sound-rec; 40 digital data carriers
libr loan 33819

Los Angeles Valley College, Library, 5800 Fulton Ave, **Valley Glen**, CA 91401-4096
T: +1 818 9472425; Fax: +1 818 9472751; E-mail: maydg@lavc.edu; URL: www.lavc.edu/library/library.html
1949; David G. May
Ethnic studies, Humanities, Social sciences
123 000 vols; 375 curr per 33820

Clark College, Lewis D. Cannell Library, 1800 E McLoughlin Blvd, **Vancouver**, WA 98663-3598
T: +1 360 9922329; Fax: +1 360 9922869; E-mail: mbagley@clark.edu; URL: www.clark.edu/Library
1933; Michelle Bagley
61 153 vols; 523 curr per; 1931 microforms; 7 digital data carriers
libr loan 33821

Washington School for the Deaf, Learning Resource Center, McGill Library, 611 Grand Blvd, **Vancouver**, WA 98661-4498
T: +1 360 6966525; Fax: +1 360 4180418; URL: www.wsd.wa.gov
Professional Coll for Educators of the Deaf
11 000 vols; 73 curr per 33822

Ventura College, D.R. Henry Library, 4667 Telegraph Rd, **Ventura**, CA 93003-3889
T: +1 805 6546482; Fax: +1 805 6488900; E-mail: kscott@vcccd.edu; URL: www.venturacollege.edu
1926; Kathy Scott
Art, Geology, Ethnic studies, Feminism
66 500 vols; 494 curr per 33823

Vernon Regional Junior College, Wright Library, 4400 College Dr, **Vernon**, TX 76384
T: +1 940 5526291 ext 2220; Fax: +1 940 5529423; E-mail: librarian@vrjc.cc.tx.us; URL: www.vernoncollege.edu/libnew
1972; Marian Grona
36 000 vols; 150 curr per 33824

Victor Valley Community College, Library, 18422 Bear Valley Rd, **Victorville**, CA 92392-9699
T: +1 760 2454271 ext 2262; Fax: +1 760 2454373; E-mail: akinsj@vvc.edu; URL: www.vvc.edu/library
1961
Mojave Desert Coll
54 000 vols; 492 curr per 33825

Cumberland County College, Library, 3322 College Dr, **Vineland**, NJ 08360
T: +1 856 6918600 ext 261; Fax: +1 856 6911969; E-mail: paschmid@cccnj.net; URL: www.cccnj.net
1966; Patti Schmid
Holistic health, Law, Mythology, Nursing; Jerseyanna
37 000 vols; 93 curr per
libr loan 33826

Tidewater Community College, Virginia Beach Campus Learning Resources Center, 1700 College Crescent, **Virginia Beach**, VA 23456
T: +1 804 4277157; Fax: +1 804 4270327
1971; Dr. Janice Sims Johnson
41 805 vols; 206 curr per; 52 105 microforms 33827

College of the Sequoias, Visalia Campus, Library, 915 S Mooney Blvd, **Visalia**, CA 93277
T: +1 559 7303824; Fax: +1 559 7374835; URL: giant.sequoias.cc.ca.us/academic/library/homepage/library_homepage.htm
1926; Nancy Finney
78 000 vols; 430 curr per 33828

McLennan Community College, Library, 1400 College Dr, **Waco**, TX 76708-1498
T: +1 254 2998000; Fax: +1 254 2998062; E-mail: dmartinsen@mclennan.edu; URL: www.mclennan.edu/library
1968; Dan Martinsen
Law
78 000 vols; 350 curr per 33829

Texas State Technical College, Waco Library, 3801 Campus Dr, **Waco**, TX 76705
T: +1 254 8674846; Fax: +1 254 8672339; E-mail: lkoepf@tstc.edu; URL: walib.tstc.edu
1967; Linda S. Koepf
Industrial Standards Coll, Deaf & Sign Language Coll
60 000 vols; 744 maps; 463 345 microforms; 1 808 sound-rec; 24 digital data carriers; 1 657 videos, 6 790 VF items, 17 335 slides, 220 overhead transparencies, 148 charts, 135 art reproductions, 2 442 archival clippings 33830

Southern Union State Community College, McClintock-Ensminger Library, Robert St, P.O. Box 1000, **Wadley**, AL 36276
T: +1 256 3952211 ext 5130; Fax: +1 256 3952215; E-mail: kreynolds@suscc.edu; URL: www.suscc.edu/

frmLearningResoursces.aspx
1922; Kathy E. Reynolds
92 000 vols; 408 curr per
ALA 33831

North Dakota State College of Science, Mildred Johnson Library, 800 N Sixth St, **Wahpeton**, ND 58076-0002
T: +1 701 6712298; Fax: +1 701 6712674; E-mail: Karen.Chobot@ndscs.edu; URL: www.ndscs.nodak.edu
1903; Karen Chobot
75 000 vols; 650 curr per; 435 maps; 23 700 microforms; 2 500 av-mat; 3 150 sound-rec; 12 VF drawers of pamphlets
libr loan 33832

Reinhardt College, Hill Freeman Library, 7300 Reinhardt College Circle, **Waleska**, GA 30183
T: +1 770 7209120; Fax: +1 770 7205602; E-mail: library@reinhardt.edu; URL: www.reinhardt.edu; library.reinhardt.edu/default.htm
1883; Shawn Tonner
46 000 vols; 300 curr per 33833

Walla Walla Community College, Library, 500 Tausick Way, **Walla Walla**, WA 99362-9267
T: +1 509 5274294; Fax: +1 509 5274480; E-mail: library@wwcc.ctc.edu; URL: www.wallawalla.cc/library/wwcclib.html
1967; Ann Daly
42 450 vols; 787 curr per
libr loan 33834

Mount San Antonio College, Library & Learning Resources, 1100 N Grand Ave, **Walnut**, CA 91789
T: +1 909 5945611 ext 4260; Fax: +1 909 4684011; E-mail: mchen@mtsac.edu; URL: www.mtsac.edu
1946; Meghan Chen
77 300 vols; 747 curr per
libr loan 33835

Connors State College, Library, Rte 1, Box 1000, **Warner**, OK 74469-9700
T: +1 918 4632931; Fax: +1 918 4636314; E-mail: idkerley@connors.cc.ok.us; URL: www.connors.cc.ok.us
1950; Margaret Rigney, Izoro Daphane Kerley
33 400 vols; 300 curr per 33836

Macomb Community College Libraries – South Campus, 14500 E 12 Mile Rd, **Warren**, MI 48088-3896
T: +1 586 4457401; Fax: +1 586 4457157; URL: www.macomb.edu
1954; Gerald Bosler
Law
102 500 vols; 160 curr per; 10 000 e-books; 5 290 av-mat; 250 sound-rec; 840 digital data carriers 33837

Rappahannock Community College, Library Center, 52 Campus Dr, **Warsaw**, VA 22572
T: +1 804 3336700; Fax: +1 804 3330589; E-mail: telliott@rcc.vccs.edu; URL: www.rcc.vccs.edu
1973; Tracy A. Elliott
Children's Libr, Cooperative Law Libr, Virginiana
30 000 vols; 203 curr per 33838

Community College of Rhode Island, Learning Resources Center, Knight Campus, 400 East Ave, **Warwick**, RI 02886-1807
T: +1 401 8252216; Fax: +1 401 8252215; URL: www.ccri.edu/
1964; Prof. Brenda Andrade
College Cat Coll, Career & College Info CLSI
100 000 vols; 600 curr per 33839

Warren County Community College, Library-Learning Resources Center, 475 Rte 57 W, **Washington**, NJ 07882-4343
T: +1 908 835-2336/-2339; Fax: +1 908 8351283; E-mail: dorenstein@warren.edu; URL: www.warren.edu
1984; David Orenstein
Humanities, Law, Nursing; Arch, Small Business, New Reader
34 000 vols; 442 curr per 33840

Max R. Traurig Learning Resource Center Library, 750 Chase Pkwy, **Waterbury**, CT 06708
T: +1 203 5758024; Fax: +1 203 5758062; URL: www.nv2.commnet.edu/library
1964; Rosalie C. Levinson
40 000 vols; 518 curr per 33841

Hawkeye Community College, Library, 1501 E Orange Rd, **Waterloo**, IA 50704
T: +1 319 2964006; Fax: +1 319 2969140
1970; Robert Chittenden
Vocational education
33 070 vols; 518 curr per 33842

Perkins School for the Blind, Braille & Talking Book Library, 175 N Beacon St, **Watertown**, MA 02472
T: +1 617 9727240; Fax: +1 617 9727363; E-mail: Library@Perkins.org; URL: www.perkins.org/
1829; Kim L. Charlson
Foreign Language, cassettes; Ref Mat on Blindness & Other Physical Handicaps
44 000 vols; 35 000 sound-rec; 10 000 braille bks 33843

Perkins School for the Blind – Samuel P. Hayes Research Library, 175 N Beacon St, **Watertown**, MA 02472
T: +1 617 9727250; Fax: +1 617 9238076; E-mail: hayeslibrary@perkins.org; URL: www.perkins.org/researchlibrary/
Jan Seymour-Ford
30 000 vols; 150 curr per 33844

State University of New York – Jefferson Community College, Melvil Dewey Library, 1220 Coffeen St, **Watertown**, NY 13601-1897
T: +1 315 7862225; Fax: +1 315 7880716; E-mail: shirley_ellsworth@ccmgate.sunyjefferson.edu; URL: www.sunyjefferson.edu/
1963; Shirley Ellsworth
Jefferson County Local Hist
55 750 vols; 322 curr per
libr loan 33845

Northcentral Technical College, Library, 1000 W Campus Dr, **Wausau**, WI 54401
T: +1 715 6753331; Fax: +1 715 6759776; E-mail: lib@ntc.edu; URL: ntc.edu/ntclibrary
1969; Gail Piotrowski
American Sign Language
30 000 vols; 250 curr per; 30 digital data carriers
libr loan 33846

Waycross College, Library, 2001 S Georgia Pkwy, **Waycross**, GA 31503
T: +1 912 2856136; Fax: +1 912 2842990; E-mail: lkelly@waycross.edu; URL: www.waycross.edu
1976; Sharon Lynn Kelly
Okefenokee Swamp Coll
31 000 vols; 230 curr per
libr loan 33847

Valley Forge Military Academy & College, May H. Baker Memorial Library, 1001 Eagle Rd, **Wayne**, PA 19087-3695
T: +1 610 9891200; Fax: +1 610 9759642; E-mail: jgosh@vfmac.edu; URL: www.vfmac.edu/
1928; Jean Smith
Military science, Classical lit, Govt, Hist; Military & Naval Hist Coll
60 000 vols; 500 curr per; 500 av-mat; 17 Electronic Media & Resources, 50 Per Subs, 200 High Interest/Low Vocabulary Bks 33848

Weatherford College, Library, 225 College Park Dr, **Weatherford**, TX 76086
T: +1 817 5945471 ext 252; Fax: +1 817 5999305, +1 817 5986369; E-mail: mtandy@wc.edu; URL: library.wc.edu
1869; Martha Tandy
Ranching in the Southwest, Texas Counties
60 000 vols; 421 curr per; 20 diss/theses; 160 915 microforms; 1938 sound-rec; 5 digital data carriers; 820 vertical files
libr loan; ALA, TLA 33849

College of the Siskiyous, Library, 800 College Ave, **Weed**, CA 96094
T: +1 530 9385331; Fax: +1 530 9385226; E-mail: library@siskiyous.edu;
URL: www.siskiyous.edu/library
1957; Dennis R. Freeman
Local hist
42 700 vols; 145 curr per
libr loan   33850

West Virginia Northern Community College – Weirton Campus, Library, 150 Park Ave, **Weirton**, WV 26062
T: +1 304 7232210; Fax: +1 304 7236704
Patricia Stroud
35 000 vols; 254 curr per   33851

Halifax Community College, Library, Hwy 158 & College Dr, PO Drawer 809, **Weldon**, NC 27890-0700
T: +1 919 5367237; Fax: +1 919 5360474; E-mail: cooperm@halifaxcc.edu;
URL: www.halifaxcc.edu/
1968; Mary Gail Cooper
34 480 vols; 167 curr per
libr loan   33852

Massachusetts Bay Community College, Perkins Library, 50 Oakland St, **Wellesley**, MA 02181
T: +1 617 2392610; Fax: +1 617 2393621; E-mail: info@massbay.edu;
URL: www.massbay.edu/
1961; Catherine Lee
Allied health
49 000 vols; 400 curr per   33853

Wenatchee Valley College, John A. Brown Library Media Center, 1300 Fifth St, **Wenatchee**, WA 98801
T: +1 509 6642520; Fax: +1 509 6642542
1939; Bruce P. Swenson
Northwest Indian Hist, Pacific Northwest Hist
33 500 vols; 290 curr per
libr loan   33854

Copiah-Lincoln Community College, Evelyn W. Oswalt Library, 1029 J C Redd Dr, P.O. Box 649, **Wesson**, MS 39191
T: +1 601 6438365; Fax: +1 601 6438212; E-mail: ken.chapman@colin.edu;
URL: www.colin.edu/
1928; Kendall P. Chapman
65 000 vols; 28 000 electronic books, current periodicals, and online databases.
33855

Cape Cod Community College, Library-Learning Resource Center, 2240 Iyahough Rd, **West Barnstable**, MA 02668-1599
T: +1 508 3622131; Fax: +1 508 3754020; E-mail:
redesk@capecod.mass.edu; URL:
www.capecod.mass.edu/library
1961; Greg Masterson
Cape Cod Hist Coll, Health Resources Coll, Business Info Ctr
54 000 vols; 483 curr per   33856

Southeastern Community College Library – North Campus, Yohe Memorial Library, 1500 W Agency Rd, **West Burlington**, IA 52655; P.O. Box 180, West Burlington, IA 52655-0180
T: +1 319 7522731; Fax: +1 319 7530322; URL: www.scciowa.edu
1920; Brian P. McAtee
Art, Anthropology, Environmental studies
30 000 vols; 300 curr per   33857

Midlands Technical College, Library, 1260 Lexington Dr, **West Columbia**, SC 29170; P.O. Box 2408, Columbia, SC 29202
T: +1 803 8223530; Fax: +1 803 8223670; E-mail:
libstaff@midlandstech.com; URL:
www.lib.midlandstech.edu
1963
77 000 vols; 500 curr per   33858

Community College of Allegheny County, South Campus Library, ER Crawford, 1750 Clairton Rd, **West Mifflin**, PA 15122-3097
T: +1 412 4696295; Fax: +1 412 4696370; URL: www.ccac.edu
1967; Barbara Chandler
61 000 vols; 437 curr per
libr loan   33859

Berkeley College, Walter A. Brower Library, 44 Rifle Camp Rd, **West Paterson**, NJ 07424
T: +1 973 2785400 ext 230; Fax: +1 973 2782242; E-mail: md@berkeley.org;
URL: www.berkleycollege.edu/
1931; Nancy Weiner
54 000 vols; 196 curr per   33860

Houghton College at West Seneca, Ada M. Kidder Memorial Library, 810 Union Rd, **West Seneca**, NY 14224
T: +1 716 6746363 ext 8766; Fax: +1 716 6740250; URL: www.houghton.edu
1969; George E. Bennett
Theology, Missions, Christian education, Early childhood education
30 000 vols; 119 curr per; 95 microforms; 825 sound-rec
libr loan   33861

Carroll Community College, Library & Media Center, 1601 Washington Rd, **Westminster**, MD 21157-6944
T: +1 410 3868090; Fax: +1 410 8400398; E-mail:
abogage@carroll.cc.md.us; URL:
www.carrollcc.edu/
1976; Alan Bogage
38 330 vols; 305 curr per   33862

Front Range Community College, College Hill Library & Media Center, 3705 W 112th Ave, **Westminster**, CO 80031-2140; mail address: 3645 W 112th Ave, Campus 12, Westminster, CO 80031
T: +1 303 4045500; Fax: +1 303 4045135; E-mail:
clara.burns@frontrange.edu; URL:
www.frontrange.edu/
1968; Clara Burns
Rocky Flats Reading Room
48 000 vols; 251 curr per
libr loan   33863

Blue Ridge Community College, Houff Library, One College Lane, P.O. Box 80, **Weyers Cave**, VA 24486-0080
T: +1 540 4532247; Fax: +1 540 2349598; E-mail: bralexd@br.cc.va.us;
URL: www.br.cc.va.us
1967; Frank Moran
Veterinary tech, Nursing; Virginia Regional Hist Coll
48 970 vols; 233 curr per   33864

Wharton County Junior College, J.M. Hodges Library, 911 Boling Hwy, **Wharton**, TX 77488-3252
T: +1 409 5324560; Fax: +1 409 5322216; E-mail: KweiH@wcjc.cc.tx.us;
URL: www.wcjc.cc.tx.us
1946; Kwei-Feng Hsu
66 820 vols; 512 curr per   33865

West Virginia Northern Community College, Learning Resources Center, 1704 Market St, **Wheeling**, WV 26003-3699
T: +1 304 2335900 ext 4252; Fax: +1 304 2320965; URL:
www.northern.wvnet.edu/~library
1972; Mary Salony
37 120 vols; 170 curr per   33866

Century Community & Technical College, West Campus Library, 3300 Century Ave N, **White Bear Lake**, MN 55110
T: +1 651 7793263; Fax: +1 651 7793464; E-mail:
r.murray@century.mnscu.edu; URL:
www.century.mnscu.edu/library
1967; Raymond Murray
56 890 vols; 486 curr per   33867

Southeastern Community College, Library, P.O. Box 151, **Whiteville**, NC 28472-0151
T: +1 910 642-7141 ext 386; Fax: +1 910 642-5658; E-mail:
mhester@sccnc.edu; URL:
www.sccnc.edu/
1965; Marian Williams
North Carolina Colonial Records 1662-1789; North Carolina Genealogy & Hist Coll; Official Records of the Union & Confederate Armies; Southeastern North Carolina Records, micro
61 410 vols; 236 curr per   33868

Rio Hondo Community College, Library, 3600 Workman Mill Rd, **Whittier**, CA 90601
T: +1 562 9083417; Fax: +1 562 6929948; E-mail: library@rh.cc.ca.us;
URL: library.riohondo.edu/
1965; Jan Coe
82 600 vols; 206 curr per   33869

Wilkes Community College, Learning Resources Center / Pardue Library, 1328 Collegiate Dr, P.O. Box 120, **Wilkesboro**, NC 28697-0120
T: +1 336 8386114; Fax: +1 336 8386515; E-mail: fay.byrd@wilkescc.edu;
URL: www.wilkescc.edu/
default2.aspx?id=430
1966; Dr. Fay Byrd
James Larkin Pearson Coll, Wilkes County Coll
60 000 vols; 122 curr per; 839 microforms; 7 000 av-mat; 1 981 sound-rec; 6 digital data carriers   33870

Martin Community College, Library, 1161 Kehukee Park Rd, **Williamston**, NC 27892-9988
T: +1 252 7921521; Fax: +1 252 7924425; E-mail: pcherry@martin.cc.nc.us;
URL: www.martin.cc.nc.us
1968; Peggy Cherry
Martin County Hist Coll
34 500 vols; 228 curr per   33871

Erie Community College – North Campus, Library Resources Center, 6205 Main St, **Williamsville**, NY 14221-7095
T: +1 716 8511273; Fax: +1 716 8511277; E-mail: mende@ecc.edu; URL:
www.ecc.edu
1946; Lynnette Mende
Allied health, Technology; College Arch
78 250 vols; 380 curr per   33872

Ridgewater College, Library, Willmar Campus, 2101 15th Ave NW, **Willmar**, MN 56201
T: +1 320 234-8565; Fax: +1 320 234-8640; E-mail:
ckelleher@ridgewater.mnscu.edu; URL:
www.ridgewater.mnscu.edu/Current-Students/Library/index.cfm
1962; Carolyn Kelleher
Lawson Libr
30 430 vols; 252 curr per   33873

Cape Fear Community College, Learning Resource Center, 415 N Second St, mail address: 411 N Front St, **Wilmington**, NC 28401-3993
T: +1 910 2515130; Fax: +1 910 2515105; E-mail:
coakley@capefear.cc.nc.us; URL: cfcc.net
1964; Carolyn C. Oakley
Boat building, Fishing, Marine tech
34 880 vols; 782 curr per   33874

Delaware Technical & Community College, Wilmington Campus, John Eugene Derrickson Memorial Library, 333 Shipley St, **Wilmington**, DE 19801
T: +1 302 5735432; Fax: +1 302 5772038; URL: www.dtcc.edu/wilmington/
library/station.html
1973; Donna M. Abed
Business & management, Allied health
34 000 vols; 460 curr per   33875

Goldey Beacom College, J. Wilbur Hirons Library, 4701 Limestone Rd, **Wilmington**, DE 19808
T: +1 302 2256247; Fax: +1 302 9986189; URL: www.gbc.edu/library
1969; Pamela Stewart
Business, Management, Accounting, Business administration, Marketing, Computer science; Delaware Business, College Arch
45 000 vols; 1 051 curr per; 13 maps;
46 000 microforms; 527 sound-rec; 13 digital data carriers; 700 corp annual rpts   33876

Los Angeles Harbor College, Baxter Library, 1111 Figueroa Pl, **Wilmington**, CA 90744-2397
T: +1 310 5228292; Fax: +1 310 5228435; E-mail: gogins@lahc.cc.ca.us;
URL: www.lahc.cc.ca.us/library
1949; Sally Gogin
Local Hist
83 500 vols; 206 curr per
libr loan   33877

Wilson Technical Community College, Learning Resource Center, 902 Herring Ave, P.O. Box 4305, **Wilson**, NC 27893-4305
T: +1 252 2461235; Fax: +1 252 2437148; E-mail:
library_support@email.wilsontech.cc.nc.us;
URL: www.wilsontech.cc.nc.us
1958; Gerry O'Neill
33 000 vols; 420 curr per   33878

Northwestern Connecticut Community College, Library, 100 S Main St, **Winsted**, CT 06098
T: +1 860 7386480; Fax: +1 860 3794995; E-mail:
adodge@nwcc.commnet.edu; URL:
www.nwcc.commnet.edu
1965; Anne Dodge
Art, Deaf, Recreation; Deaf Education; World War I & II; Jazz, compact discs
41 860 vols; 265 curr per; 2 801 microforms; 1 700 sound-rec; 6 digital data carriers
libr loan   33879

Forsyth Technical Community College, Library, 2100 Silas Creek Pkwy, **Winston-Salem**, NC 27103
T: +1 336 7230371 ext 7216; Fax: +1 336 7612399; E-mail:
rcandelaria@forsyth.cc.nc.us; URL:
www.forsyth.tec.nc.us
1964; J. Randel Candelaria
Nursing Allied Health Coll
34 000 vols; 300 curr per; 4 645 microforms; 2 014 av-mat; 3 digital data carriers
libr loan   33880

Polk Community College, James W. Dowdy Memorial Library, 999 Avenue H NE, **Winter Haven**, FL 33881-4299
T: +1 863 2971040; Fax: +1 863 2971065; URL: www.polk.edu/
1965; Jim Horton
Florida, State hist
93 000 vols; 350 curr per; 12 960 microforms; 2 500 av-mat; 1 143 sound-rec; 19 digital data carriers; 570 art reproductions   33881

Los Angeles Pierce College, Library, 6201 Winnetka Ave, **Woodland Hills**, CA 91371
T: +1 818 7196409; Fax: +1 818 7199058; E-mail:
robinfk@piercecollege.edu; URL:
www.piercecollege.edu/library
1947; Florence Robin
Agriculture; Beachy Libr (San Fernando Valley Coll)
112 000 vols; 297 curr per; 4 000 e-books   33882

Becker College, William F. Ruska Library, 61 Sever St, **Worcester**, MA 01615-0071
T: +1 508 7939710; E-mail:
donna.sibley@becker.edu; URL:
www.beckercollege.edu
1887
65 390 vols; 245 curr per; 5 500 av-mat; 1 000 sound-rec; 47 digital data carriers
libr loan; ALA, ACRL   33883

Quinsigamond Community College, Alden Library, 670 W Boylston St, **Worcester**, MA 01606-2092
T: +1 508 8544366; Fax: +1 508 8526943; E-mail: mbejune@qcc.mass.edu;
URL: www.qcc.mass.edu/library
1963
63 500 vols; 320 curr per   33884

Minnesota West Community & Technical College, Library and Academic Resource Center (LARC), 1450 College Way, **Worthington**, MN 56187-3024
T: +1 507 8256817; E-mail:
chris.schafer@mnwest.edu; URL:
www.mnwest.edu/
1936; Chris Schafer
80 000 vols; 200 curr per
libr loan   33885

Yakima Valley Community College, Library, 16th Ave at Nob Hill Blvd, P.O. Box 22520, **Yakima**, WA 98907-2520
T: +1 509 5744991; Fax: +1 509 5744989
1929; Joan L. Weber
32 000 vols; 80 curr per   33886

Young Harris College, Duckworth Library, One College St, P.O. Box 39, **Young Harris**, GA 30582
T: +1 706 3794313; Fax: +1 706 3794314; E-mail: dlamade@yhc.edu; URL: www.yhc.edu
1886; Dawn A. Lamade
Humanities, Religion, Music; Byron Herbert Reece & J A Sharp Coll; Merle Mann Indian Artifacts
60 000 vols; 237 curr per
libr loan; OCLC   33887

Westmoreland County Community College, Learning Resources Center, 400 Armbrust Rd, **Youngwood**, PA 15697-1895
T: +1 724 9254100; Fax: +1 724 9251150; E-mail: keefek@wccc.edu; URL: www.wccc.edu
1970
66 720 vols; 500 curr per   33888

Crafton Hills College, Library, 11711 Sand Canyon Rd, **Yucaipa**, CA 92399
T: +1 909 389-3323, -3322; Fax: +1 909 7949524; E-mail: bbyron@sbccd.cc.ca.us; URL: www.craftonhills.edu/
1972; Betty Byron
67 600 vols; 414 curr per   33889

Arizona Western College & Nau-Yuma, Academic Library, 9500 S Ave 8E, **Yuma**, AZ 85365; P.O. Box 929, Yuma, AZ 85366-0929
T: +1 928 3447777; Fax: +1 928 3447751; URL: www.azwestern.edu
1963; Tim Shove
Southwest Arizona & Lower Colorado River Coll
54 640 vols; 440 curr per
libr loan   33890

Government Libraries

US Army – Center for Health Promotion & Preventive Medicine Library, Bldg E-1570, **Aberdeen Proving Ground**, MD 21010-5403
T: +1 410 4364236; Fax: +1 410 4367766
1945
Medicine, Occupational safety, Public health, Toxicology, Chemistry, Engineering
21 000 vols; 200 curr per
libr loan   33891

US Army – Ordnance Center & School Library, Bldg 3071, **Aberdeen Proving Ground**, MD 21005-5201
T: +1 410 2785615; Fax: +1 410 2788882
1940; Linda Taylor
Dept of the Army Publs
25 000 vols; 200 curr per   33892

US Army – Post Library, Bldg 3320, **Aberdeen Proving Ground**, MD 21005-5001
T: +1 410 2783417; Fax: +1 410 2789537; E-mail: library-aa@mwr.apg.army.mil
Daniel Norum
Military hist, World War II
40 000 vols; 300 curr per
libr loan   33893

National Archives & Records Administration, Dwight D. Eisenhower Library, 200 SE Fourth St, **Abilene**, KS 67410-2900
T: +1 785 2636700; Fax: +1 785 2636715; E-mail: eisenhower.library@nara.gov; URL: eisenhower.archives.gov
1962; Daniel D. Holt
World War II & Eisenhower Administration Manuscript Coll
27 000 vols; 75 curr per; 4 000 microforms; 1 108 sound-rec   33894

US Army Research Laboratory, Technical Library, 2800 Powder Mill Rd, **Adelphi**, MD 20783-1197
T: +1 301 3942536; Fax: +1 301 3941465; E-mail: libraryALC@arl.army.mil
Kathleen Mary Mako
Chemistry, Economics, Electrical engineering, Electronics, Mechanical engineering
36 000 vols; 800 curr per   33895

US Coast Guard Base Library – Otis, Bldg 5205, **Air Station Cape Cod**, MA 02542-5017
T: +1 508 9686456; Fax: +1 508 9686686
Kathy Dill
World War II
40 000 vols; 55 curr per   33896

New York State Court of Appeals Library, 20 Eagle St, **Albany**, NY 12207-1905
T: +1 518 4557770; URL: www.courts.state.ny.us
1870; Elizabeth F. Murray
80 000 vols; 100 curr per   33897

New York State Department of Economic Development, Empire State Development Library, 30 S Pearl St, **Albany**, NY 12245
T: +1 518 2925238; Fax: +1 518 2925805; E-mail: bbeverley@empire.state.ny.us
1944; Barbara S. Beverley
20 000 vols; 800 curr per
libr loan; SLA   33898

New York State Department of Health, Dickerman Library, Empire State Plaza, **Albany**, NY 12201; P.O. Box 509, Albany, NY 12201-0509
T: +1 518 4746172; Fax: +1 518 4743933; E-mail: dohlib@health.state.ny.us; URL: dickerman.wadsworth.org
1914; Thomas Flynn
40 000 vols; 1 100 curr per
libr loan   33899

New York State Department of Law Library, The Capitol, **Albany**, NY 12224
T: +1 518 4743840; Fax: +1 518 4731822; E-mail: albany.library@oag.state.ny.us
1944; Sarah P. Browne
New York State Law Dept Records & Briefs
125 000 vols; 201 curr per   33900

New York State Department of Taxation & Finance Library, Bureau of Tax Policy Analysis, W A Harriman Campus, Bldg 9, Rm 280, **Albany**, NY 12227
T: +1 518 4573512; Fax: +1 518 4851365
1946; Michael Craig
14 500 vols; 60 curr per   33901

New York State Legislative Library, State Capitol, Rm 337, **Albany**, NY 12224-0345
T: +1 518 4552468; Fax: +1 518 4266901
Hist of New York Laws, Legislative Rpts, New York State Agency Rpts
100 000 vols; 100 curr per   33902

New York Supreme Court Appellate Division, Third Department Library, Justice Bldg, Empire State Plaza, **Albany**, NY 12223
T: +1 518 4714777; Fax: +1 518 4714750; URL: www.nycourts.gov/ad3
Helen Rizzo
22 000 vols   33903

US Department of Energy, Albany Research Center Library, 1450 Queen Ave SW, **Albany**, OR 97321-2152
T: +1 541 9675864; Fax: +1 541 9675936
1943; Cathy Wright
30 000 vols; 25 curr per   33904

US Navy – Marine Corps Logistics Base Library, Attn Base Libr, Bldg 7458, 814 Radford Blvd, Ste 20311, **Albany**, GA 31704-0311
T: +1 229 6395242; Fax: +1 229 6395197
Amos Tookes
26 000 vols; 39 curr per   33905

Orleans Correctional Facility Library, 3531 Gaines Basin Rd, **Albion**, NY 14411
T: +1 585 5896820; Fax: +1 585 5896820 ext 3199
Douglas Bauer
14 000 vols; 80 curr per   33906

State Correctional Institution, Albion Library, 10745 Rte 18, **Albion**, PA 16475-0001
T: +1 814 7565778; Fax: +1 814 7569735
Gene Zarnick
25 000 vols; 75 curr per   33907

United States Courts Library, 333 Lomas Blvd NW, Ste 360, **Albuquerque**, NM 87102
T: +1 505 3482135; Fax: +1 505 3482795
Gregory L. Townsend
19 000 vols; 50 curr per   33908

Wende Correctional Facility Library, 3622 Wende Rd, **Alden**, NY 14004; P.O. Box 1187, Alden, NY 14004-1187
T: +1 716 9374000; Fax: +1 716 9370206
William Smith
12 000 vols; 50 curr per   33909

US Army – Headquarters Army Material Command Technical Library, AMCIO-I-L, 5001 Eisenhower Ave, **Alexandria**, VA 22333-0001
T: +1 703 6174115; Fax: +1 703 6175588
1973; Valerie Hicks
Army & Department of Defense Publs
28 000 vols; 300 curr per
libr loan; OCLC   33910

US Department of the Army – Office of the Chief Engineers Library, CEHEC-ZL Casey Bldg, 7701 Telegraph Rd, **Alexandria**, VA 22315-3860
T: +1 703 4286388; Fax: +1 703 4286310; E-mail: cehec-im-l@usace.army.mil; URL: www.hecsa.usace.army.mil/hxlibrary
1942; Connie J. Wiley
Civil & Environmental Engineering; Congressional Mats; Corps of Engineers Hist & Activities, 1776-present; Management Coll
152 000 vols; 427 curr per
libr loan   33911

US Department of the Army – Office of the Chief of Staff, Operational Test & Evaluation Command (OPTEC), Headley Technical Research Center, 4501 Ford Ave, **Alexandria**, VA 22302-1458
T: +1 703 6819234; Fax: +1 703 6814973; E-mail: @optec.army.mil
1972; Marjorie Rust
US Dept of Army Military Publs (Administration, Organization, Training)
10 000 vols; 300 curr per   33912

US Patent & Trademark Office, Scientific & Technical Information Center, Madison West Bldg, Rm 1C35, 600 Dulany St, **Alexandria**, VA 22314
T: +1 571 2723547; Fax: +1 571 2730048; URL: www.uspto.gov/web/patents/sticlibinfo.htm
1836; Kristin Vajs
Biochemistry, Biomedicine, Biotechnology, Chemical & mechanical engineering design, Computer science, Copyright law, Electrical, Federal law; Foreign Patents
129 000 vols; 300 curr per; 17 000 e-journals   33913

Lehigh County Law Library, County Court House, 455 W Hamilton St, **Allentown**, PA 18101-1614
T: +1 610 7823385; Fax: +1 610 8203311; URL: www.lccpa.org
1869; Lorelei A. Broskey
Pennsylvania Law, Local Municipal & Legislative Hist; Ordinances Coll
21 000 vols; 35 curr per   33914

Georgia Department of Corrections, Office of Library Services, Arrendale State Prison, 2023 Gainesville Hwy S, **Alto**, GA 30510; P.O. Box 709, Alto, GA 30510-0709
T: +1 706 7764700; Fax: +1 706 7764710; URL: www.dcor.state.ga.us
Regina Bell
10 000 vols   33915

US Air Force – Altus Air Force Base Library FL4419, 97 FSS/FSDL, 109 E Ave, Bldg 65, **Altus AFB**, OK 73525-5134
T: +1 580 4816302; Fax: +1 580 4820469
1953; Sharon McKenna

Aeronautics, Political science
22 000 vols; 185 curr per   33916

Potter County Law Library, 501 S Filmore, Ste 2B, **Amarillo**, TX 79101
T: +1 806 3792347
1911; Susan Montgomery
State Reporters, 1800-1900
50 000 vols   33917

Siemens Water Technology, Lawlor Library, 600 Arrasmith Trail, **Ames**, IA 50010
T: +1 515 2688400; Fax: +1 515 2688500; URL: www.usfilter.com
Marlys A. Cooper
115 000 vols   33918

Alaska State Court Law Library, 303 K St, mail address: 820 W Fourth Ave, **Anchorage**, AK 99501
T: +1 907 2640585; Fax: +1 907 2640733; E-mail: library@courts.state.ak.us; URL: www.state.ak.us/courts/library.htm
1959; Catherine Lemann
364 000 vols; 328 curr per
libr loan   33919

US Courts Library, 222 W Seventh Ave, Rm 181, **Anchorage**, AK 99513-7586
T: +1 907 2715655; Fax: +1 907 2715640
Catherine A. Davidson
Alaska National Interest Lands Conservation Act, Alaska Native Claims Settlement Act & other Alaska Titles
26 000 vols; 12 curr per   33920

US Air Force – Andrews Air Force Base Library, 89 SVS/SVMG, Brookley & D St, Bldg 1642, **Andrews AFB**, MD 20762
T: +1 301 9816454; Fax: +1 301 9814231
Anette Powell
Political science, Business; Air War College, Children's Coll
31 000 vols; 140 curr per   33921

US Air Force – Malcolm Grow Medical Center Library, **Andrews AFB**, MD 20762-6600
T: +1 240 8572354; Fax: +1 240 8578608
1958; Tina Pinnix-Broome
8 000 vols; 285 curr per; 500 av-mat; 125 digital data carriers
libr loan   33922

Washington Correctional Institute Library, 27268 Hwy 21, **Angie**, LA 70426
T: +1 985 9865064; Fax: +1 985 9865060
Deborah Cook
12 000 vols; 26 curr per   33923

Brazoria County Law Library, 111 E Locust St, Ste 315-A, **Angleton**, TX 77515
T: +1 979 8641225; Fax: +1 979 8641226
Angela Wollam
10 000 vols; 20 curr per   33924

Louisiana State Penitentiary Library, Main Prison Library, A Bldg, **Angola**, LA 70712
T: +1 225 6552600; Fax: +1 225 6552393; URL: www.corrections.state.la.us
1968; Linda Holmes
Criminal Justice, Science Fiction
32 000 vols; 125 curr per   33925

US Geological Survey, John Van Oosten Library, Great Lakes Science Center, 1451 Green Rd, **Ann Arbor**, MI 48105-2807
T: +1 734 214-7210; Fax: +1 734 994-8780; URL: www.glsc.usgs.gov
1965; Christine Schmuckal
20 000 vols; 95 curr per; 450 diss/theses; 500 microforms; 20 digital data carriers
libr loan   33926

Anne Arundel County Circuit Court, Law Library, Seven Church Circle, Ste 303, **Annapolis**, MD 21401; P.O. Box 2395, Annapolis, MD 21404-2395
T: +1 410 2221387; Fax: +1 410 2689762; E-mail: library@circuitcourt.org
Joan Bellistri
Maryland Law Coll
20 000 vols; 30 curr per   33927

Maryland Department of Legislative Services Library, 90 State Circle, **Annapolis**, MD 21401
T: +1 410 9465400; Fax: +1 410 9465405; E-mail: libr@mlis.state.md.us
1966; Johanne Greer
State govt; Codes of 50 States, Maryland County & City Codes, State Publs, Maryland Hist
100 000 vols; 150 curr per   33928

Maryland State Law Library, Courts of Appeal Bldg, 361 Rowe Blvd, **Annapolis**, MD 21401-1697
T: +1 410 2601430; Fax: +1 410 9742063; E-mail: mdlaw.library@courts.state.md.us; URL: www.lawlib.state.md.us
1826; Steven Paul Anderson
State govt, Maryland, Genealogy; Audubon's Birds of America Coll, Early Maryland Newspapers-1745 to date
400 000 vols; 700 curr per   33929

Anoka County Law Library, 325 E Main St, **Anoka**, MN 55303
T: +1 763 4227487; Fax: +1 763 4227453; URL: www.co.anoka.mn.us
Gene Myers
Local Municipal Ordinances
42 000 vols
AALL   33930

Florida Department of Corrections, DeSoto Correctional Institution Library, 13617 SE Hwy 70, **Arcadia**, FL 34266-7800
T: +1 863 4943727; Fax: +1 863 4940971
Lottie Rembert
23 000 vols; 41 curr per   33931

BNA, Inc, Library, 1801 S Bell St, Rm 3200, **Arlington**, VA 22202
T: +1 703 3413317; Fax: +1 703 3411636; E-mail: library@bna.com; URL: www.bna.com
Marilyn M. Bromley
State Regulatory Codes
20 000 vols; 900 curr per; 15 digital data carriers
libr loan; AALL, SLA   33932

US Agency for International Development, Environmental Health Project Information Center, 1611 Kent St, Rm 300, **Arlington**, VA 22209
T: +1 703 2478730; Fax: +1 703 2478610; E-mail: info@ehproject.org; URL: www.ehproject.org
Dan Campbell
Water resources
10 000 vols; 75 curr per   33933

US Air Force – Air Force Legal Services Agency-JAC Library, 1501 Wilson Blvd, Ste 617, **Arlington**, VA 22209
T: +1 703 6968508; Fax: +1 703 6969084
Doneva M. Jones
30 000 vols   33934

US Department of the Interior, Office of Hearings & Appeals Library, 801 N Quincy St, **Arlington**, VA 22203
T: +1 703 2353804; Fax: +1 703 2359014
1970; Theodore Richardson
Environmental studies, Federal law, Indian probate, Pub lands, Surface mining; Congressional Record, vol 115-1969 to vol 121-1975, micro; Federal Register vol 1-1936 to vol 38-1973, micro, vol 39-1974 to present
20 000 vols   33935

US Air Force – Arnold Engineering Development Center Technical Library, FL 2804, 100 Kindel Dr, Ste C212, **Arnold AFB**, TN 37389-3212
T: +1 931 4544430; Fax: +1 931 4545421
1952; Gay D. Goethert
Aerospace science, Engineering, Astronomy, Chemistry, Physics, Mathematics; NACA Tech Rpts, NACA Wartime Rpts
27 000 vols; 273 curr per
libr loan   33936

National Oceanic & Atmospheric Administration, National Climatic Data Center, Library, 151 Patton Ave, Rm 400, **Asheville**, NC 28801-5001
T: +1 828 2714677; Fax: +1 828 2714009; E-mail: linda.d.preston@noaa.gov
1961; Linda D. Preston
10 000 vols; 400 curr per   33937

US Air Force – Air Force Weather Technical Library, AFCCC/DOR, 151 Patton Ave, Rm 120, **Asheville**, NC 28801-5002
T: +1 828 2714277; Fax: +1 828 2714321; URL: www.afccc.af.mil
1975; John Gray
Meteorology, Geology
250 000 vols; 150 curr per
libr loan   33938

Athens County Law Library, Court House, 4th Flr, **Athens**, OH 45701
T: +1 740 5938893; Fax: +1 740 5923282; E-mail: aclawlib@frognet.net
Edward Kruse
10 000 vols; 10 curr per   33939

Centers for Disease Control & Prevention Information Center, MS C-04, 1600 Clifton Rd NE, **Atlanta**, GA 30333
T: +1 404 6391717; Fax: +1 404 6391160
1947; Dr. Jocelyn Rankin
25 000 vols; 1 531 curr per; 6 digital data carriers
libr loan; MLA   33940

Court of Appeals Eleventh Circuit Library, 56 Forsyth St NW, **Atlanta**, GA 30303
T: +1 404 3356500; Fax: +1 404 3356510; URL: www.ca11.uscourts.gov/library
1981; Elaine Fenton
50 000 vols; 300 curr per   33941

Fulton County Law Library, 185 Central Ave, Justice Ctr Tower, 7th Fl, **Atlanta**, GA 30303
T: +1 404 7304544; Fax: +1 404 7304565; E-mail: Jeannie.Ashley@fultoncourt.org; URL: www.fultoncourt.org/sca200807/offices/law-library.html
Jeannie Ashley
19 500 vols; 20 curr per   33942

US Environmental Protection Agency, Region 4 Library, 61 Forsyth St SW, Atlanta Federal Ctr, **Atlanta**, GA 30303-3104
T: +1 404 5628190; Fax: +1 404 5628114; E-mail: r4-library@epa.gov; URL: www.epa.gov/libraries/
1973; Kathy Piselli
Pollution, Environmental law; Superfund Docket, EPA Rpts, Public Display USGS Rpts
27 000 vols; 60 curr per; 15 000 maps; 300 000 microforms; 50 digital data carriers   33943

US Navy – Navy Supply Corps School Station Library, 1425 Prince Ave, Code 0333, **Atlanta**, GA 30606-2205
T: +1 706 3547183; Fax: +1 706 3547318
Steve Toepper
14 000 vols; 50 curr per   33944

Attica Correctional Facility School Library, Exchange St, **Attica**, NY 14011; P.O. Box 149, Attica, NY 14011-0149
T: +1 585 5912000
Edward O'Mara
14 000 vols; 67 curr per   33945

Wyoming Correctional Facility General Library, PO Box 501, **Attica**, NY 14011
T: +1 585 5911010
Brian Martin
15 000 vols; 70 curr per   33946

US Forest Service, Forest Engineering Research Project Library, George W. Andrews Forestry Sciences Lab, 520 Devall Dr, **Auburn**, AL 36849-5418
T: +1 334 8268700; Fax: +1 334 8210037
Janice Jordan
Forestry; Foreign Publications, Forest Engineering
12 000 vols   33947

Maine State Law and Legislative Reference Library, 43 State House Sta, **Augusta**, ME 04333-0043
T: +1 207 2871600; Fax: +1 207 2876467; URL: www.state.me.us/legis/lawlib/homepage.htm
1971; Lynn E. Randall
Legislative Committee master files (111th legislature to present)
114 000 vols; 480 curr per; 162 000 microforms; 219 digital data carriers
libr loan; AALL   33948

Public Utility Commission of Texas Library, 1701 N Congress, 12th Fl, **Austin**, TX 78701; P.O. Box 13326, Austin, TX 78711-3326
T: +1 512 9367080; Fax: +1 512 9367079; E-mail: clements@puc.state.tx.us; URL: www.puc.state.tx.us
1975; Helen P. Clements
Electric Power Res Institute Res Rpts
16 500 vols; 300 curr per   33949

Railroad Commission of Texas Library, Administration Division, Oil & Gas Central Records Dept, William B Travis Bldg, 1701 N Congress, Rm 10-100, **Austin**, TX 78701; P.O. Box 12967, Austin, TX 78711-2967
T: +1 512 4636882; Fax: +1 512 4637200
Susan Bains Rhyne
Energy
28 500 000 vols   33950

Texas Department of State Health Services, Library & Information Services Program, 1100 W 49th St, **Austin**, TX 78756-3199; P.O. Box 149347, Mailcode 1955, Austin, TX 78714-9347
T: +1 512 4587559; Fax: +1 512 4587683; E-mail: library@dshs.state.tx; URL: www.dshs.state.tx.us/library/default.shtm
1958; David McLellan
Public health; Texas Health Statistics
20 000 vols; 250 curr per
libr loan   33951

Texas Department of Transportation-Center for Transportation Research, Research & Tech Transfer Office Library, 3208 Red River, Ste 200, **Austin**, TX 78705
T: +1 512 2323126; Fax: +1 512 2323088; E-mail: ctrlib@uts.cc.utexas.edu; URL: library.ctr.utexas.edu/index.htm
Louise Rosenzweig
Energy, Safety
28 000 vols; 10 curr per   33952

Texas Education Agency, Infrastructure Library, 1701 N Congress Ave, **Austin**, TX 78701-1494
T: +1 512 4639050; Fax: +1 512 4753447
1968; Linda Kemp
Public school education; ERIC Coll, micro
20 000 vols; 250 curr per
libr loan   33953

Texas Legislative Reference Library, State Capitol Bldg, 1100 N Congress Ave, Rm 2N-3, **Austin**, TX 78701; P.O. Box 12488, Austin, TX 78711-2488
T: +1 512 4631252; Fax: +1 512 4754626; E-mail: lrl.service@lrl.state.tx.us; URL: www.lrl.state.tx.us
1969; Dale Propp
Texas Legislative Bills 1973-Present, Texas State Docs
50 000 vols   33954

Texas Natural Resource Conservation Commission Library, Mc-196, 12100 Park 35 Circle, **Austin**, TX 78753; P.O. Box 13087, Austin, TX 78711-3087
T: +1 512 2390020; Fax: +1 512 2390022; URL: www.tceq.state.tx.us/library
1964; Sylvia Von Fange
Geology (US Geological Survey Papers), Civil Engineering (US Corps of, Engineers & US Bureau of Reclamation Papers), Water Resources Development (Texas Water, Development Board Publs)
66 000 vols; 235 curr per; 2 000 microforms
libr loan   33955

Texas State Law Library, Tom C Clark Bldg" **Austin**, TX 78701-1614; P.O. Box 12367, Austin, TX 78711-2367
T: +1 512 4631722; Fax: +1 512 4631728; URL: www.sll.state.tx.us
1971; Marcelino A. Estrada
130 000 vols; 275 curr per   33956

Travis County Law Library, Travis City Admin Bldg, 314 W 11th, Rm 140, **Austin**, TX 78701-2112
T: +1 512 8544569; Fax: +1 512 8549887; URL: www.traviscountylawlibrary.org
Texas and Federal Law Coll
12 000 vols; 20 curr per   33957

California Department of Corrections Library System, Avenal State Prison, One Kings Way, **Avenal**, CA 93204; P.O. Box 8, Avenal, CA 93204-0008
T: +1 559 3860587; Fax: +1 559 3866068
Karen Henry
30 000 vols   33958

New Jersey Department of Corrections, Adult Diagnostic & Treatment Center Library, Eight Production Way, **Avenel**, NJ 07001
T: +1 732 5742250; Fax: +1 732 3828912; URL: www.doc.state.nj.us
Mark Yaros
15 000 vols; 38 curr per   33959

Avon Park Correctional Facility, Main Unit Library, State Rd 64 E, **Avon Park**, FL 33825; P.O. Box 1100, Avon Park, FL 33826-1100
T: +1 863 4533174
1958; Colby Joyner
Fiction; Spanish Coll, large print, easy
12 000 vols; 38 curr per   33960

Kern County Law Library, 1415 Truxtun Ave, Rm 301, **Bakersfield**, CA 93301
T: +1 661 8685320; Fax: +1 661 8685368; URL: www.kerncountylawlibrary.org
1891; Annette Heath
25 000 vols; 56 curr per; 948 sound-rec   33961

Attorney General's Office, Law Library, 200 Saint Paul Pl, 18th Flr, **Baltimore**, MD 21202
T: +1 410 5766400; Fax: +1 410 5767002; URL: www.oag.state.md.us
1917; Beverly Rubenstein
25 000 vols; 25 curr per; 4 000 microforms
libr loan; AALL, SLA   33962

Maryland Department of Planning Library, 301 W Preston St, Rm 1101, **Baltimore**, MD 21201-2305
T: +1 410 7674500, 4458; Fax: +1 410 7674480; E-mail: hjeng@mdp.state.md.us; URL: www.mdp.state.md.us/
1959; Helene W. Jeng
Census, Housing, Land use, Planning, Zoning; Maryland Counties & Municipalities Central Depot of Plans
9 550 vols; 140 curr per   33963

Social Security Administration Library, 6401 Security Blvd, Altmeyer Bldg, Rm 571, P.O. Box 17728, **Baltimore**, MD 21235
T: +1 410 9656114; Fax: +1 410 9662027; E-mail: bill.vitek@ssa.gov; URL: www.ssa.gov
1942; Bill Vitek
Social sciences, Personnel management, Business & management, Law, Public health, Medicine, Hist
80 000 vols; 210 curr per   33964

United States Courts Library, US Courthouse, Rm 3625, 101 W Lombard St, **Baltimore**, MD 21201
T: +1 410 9620997; Fax: +1 410 9629313
Charmaine Metallo
38 000 vols; 125 curr per   33965

USACE Baltimore District Library, Ten S Howard St, **Baltimore**, MD 21201; P.O. Box 1715, Baltimore, MD 21203-1715
T: +1 410 9623423; Fax: +1 410 9621889
1974; Stephen L. Brooks
Civil engineering, Soil mechanics, Water resources, Environmental studies, Law; American Society of Civil Engineers

Proceedings (1083-), Corps of Engineers Annual Rpts (1800-), Industrial Standards, Waterway Legislation
20000 vols; 150 curr per
libr loan 33966

Penobscot County Law Library, 97 Hammond St, ***Bangor***, ME 04401
T: +1 207 5612329
Judith Bennett
15000 vols 33967

US Air Force – Barksdale Air Force Base Library, 744 Douhet Dr, Bldg 4244, ***Barksdale AFB***, LA 71110
T: +1 318 4564182; Fax: +1 318 7520509; E-mail: barksdale.library2@barksdale.af.mil; URL: www.barksdalelibrary.org
1933; C. Frances Morris
Military hist; Project Warrior Coll, Louisiana Hist & Culture
38000 vols; 79 curr per 33968

Polk County Law Library, Courthouse, Rm 3076, 255 N Broadway, ***Bartow***, FL 33830
T: +1 863 5344013; Fax: +1 863 5347443
1956; Irene Morris
Legal Mats Coll
25000 vols; 60 curr per 33969

Clermont County Law Library Association, 270 Main St, ***Batavia***, OH 45103
T: +1 513 7327109; Fax: +1 513 7320974; E-mail: cclaw@cclla.org
1933; Carol A. Suhre
14200 vols; 88 curr per; 10000 microforms 33970

Louisiana Department of Environmental Quality, Environmental Assistance Library, 7290 Bluebonnet Blvd, 2nd Fl, ***Baton Rouge***, LA 70810
T: +1 225 7650169; Fax: +1 225 7650222; E-mail: assist@deq.state.la.us; URL: www.deq.state.la.us
1992; Elizabeth Santa
Air quality, Environ regulations, Radiation protection, Water quality; Chemical Risk/Hazard Data
10000 vols; 75 curr per 33971

Louisiana House of Representatives, David R. Poynter Legislative Research Library, P.O. Box 94062, ***Baton Rouge***, LA 70804-9012
T: +1 225 3426945, 2434; Fax: +1 225 3422431
1952; Suzanne Hughes
Legislative Arch Coll; Legislative Docs & Res Publs, per, clippings
10000 vols; 250 curr per 33972

Minnesota Correctional Facility, Education Library, 970 Pickett St N, ***Bayport***, MN 55003-1490
T: +1 651 7792700; Fax: +1 651 3513602
Deborah Garbison
Law
10000 vols; 53 curr per 33973

Fishkill Correctional Facility Library, Bldg 13, ***Beacon***, NY 12508; P.O. Box 307, Beacon, NY 12508-0307
T: +1 845 8314800; Fax: +1 845 8313199
Nancy Maicovski
African-American studies, Criminology, Spanish language
15000 vols; 35 curr per 33974

US Air Force – Beale Air Force Base Library, Nine SVS/SVMG, 17849 16th St, Bldg 25219, ***Beale AFB***, CA 95903-1611
T: +1 530 6342314; Fax: +1 530 6342032
Bonnie Williams
Aeronautics & California Colls
44000 vols; 150 curr per 33975

US Marine Corps – Air Station Library, P.O. Box 55018, ***Beaufort***, SC 29904-5018
T: +1 843 2287682
1957
Military science; Childrens' Stories, Fairytales, Classics, General Fiction & Non-fiction Coll
27752 vols; 100 curr per 33976

Beaver County Law Library, Court House, 810 Third St, ***Beaver***, PA 15009
T: +1 724 7283934; Fax: +1 724 7284133; URL: www.co.beaver.pa.us/lawlibrary
1972; Bette Sue Dengel
Local Court Opinions & Local Ordinances
28000 vols; 50 curr per 33977

US Department of Labor, Mine Safety & Health Administration Technical Information Center Library, 1301 Airport Rd, ***Beaver***, WV 25813-9426
T: +1 304 2563266; Fax: +1 304 2563372; E-mail: mshalibrary@dol.gov; URL: www.msha.gov/training/library/library.htm
1976; Yvonne S. Farley
Bureau of Mines (1910-present), Mine Accident Rpts (Middle 1800-present)
40000 vols; 150 curr per 33978

New York Department of Correctional Services, Bedford Hills Correctional Facility Library, 247 Harris Rd, ***Bedford Hills***, NY 10507-2499
T: +1 914 2413100; Fax: +1 914 2413100 ext 3199
Anthony Litwinowicz
13000 vols; 40 curr per 33979

Florida Department of Corrections, Glades Correctional Institution Library, 500 Orange Ave Circle, ***Belle Glade***, FL 33430-5222
T: +1 561 8291400; Fax: +1 561 9921322
1968; Edith Tucker
14000 vols; 41 curr per 33980

Centre County Law Library, Court House, 3rd Flr, ***Bellefonte***, PA 16823
T: +1 814 3556754; Fax: +1 814 3556707
Barbara Gallo
20000 vols; 15 curr per 33981

State Correctional Institution, Rockview Library, Box A, Rte 26, ***Bellefonte***, PA 16823-0820
T: +1 814 3554874
Alan Riggall
14000 vols; 50 curr per 33982

Whatcom County Law Library, Courthouse, Ste B-03, 311 Grand Ave, ***Bellingham***, WA 98225
T: +1 360 6766556; E-mail: lawlib@co.whatcom.wa.us
Virginia Tucker
20000 vols 33983

US Department of Agriculture, National Agricultural Library, 10301 Baltimore Ave, ***Beltsville***, MD 20705-2351
T: +1 301 5046709; Fax: +1 301 5045472; URL: www.nal.usda.gov
1862; Peter R. Young
Flock, Herd & Stud; Food & Nutrition, av; Foreign & Domestic Nursery & Seed Trade, cat; Forest Service Photo Coll; Poster Coll; Poultry (James M. Gwin Coll); Milk Sanitation (Charles E. North Coll), mss; Rare Bk Coll
2365000 vols; 19000 curr per
libr loan 33984

New York State Supreme Court Library, Broome County Courthouse, Rm 107, 92 Court St, ***Binghamton***, NY 13901-3301
T: +1 607 7782119; Fax: +1 607 7728331; E-mail: binglawlib@courts.state.ny.us; URL: www.nycourts.gov/6jd/countymaps/law/brmlaw/brmlaw.html
1859; Judy A. Lauer
60000 vols; 270 curr per 33985

Jefferson County Law Library, 2121 Eighth Ave N, Ste 500, ***Birmingham***, AL 35203-0072
T: +1 205 3255628; Fax: +1 205 3225915; E-mail: hand@jcc.co.jefferson.al.us; URL: www.jclawlib.org
1885; Linda M. Hand
55148 vols; 78 curr per; 62 digital data carriers 33986

Cochise County Law Library, PO Drawer P, ***Bisbee***, AZ 85603-0050
T: +1 520 4329339; Fax: +1 520 4329293; E-mail: lventura@sp.state.courts.az.us
1930; Lucia Ventura
27000 vols 33987

North Dakota Legislative Council Library, 600 E Boulevard Ave, ***Bismarck***, ND 58505-0660
T: +1 701 3284900; Fax: +1 701 3283615
Marilyn Johnson
20000 vols; 120 curr per 33988

North Dakota Supreme Court, Law Library, Judicial Wing, 2nd Flr, 600 E Boulevard Ave, Dept 182, ***Bismarck***, ND 58505-0540
T: +1 701 3284594; Fax: +1 701 3283609; URL: www.court.state.nd.us/lawlib/www6.htm
1889; Ted Smith
North Dakota Legal Mat
72000 vols 33989

Bland Correctional Center, Department of Correctional Education Library, 256 Bland Farm, ***Bland***, VA 24315-9615
T: +1 276 6883341
Harmie E. Wiley
12000 vols; 40 curr per 33990

Calhoun Correction Institution Library, 19562 SE Institution Dr, Unit 1, ***Blountstown***, FL 32424
T: +1 850 2376500
Dan Nolen
10000 vols; 40 curr per 33991

California Department of Corrections Library System, Chuckawalla Valley State Prison, 19025 Wiley's Well Rd, ***Blythe***, CA 92225; P.O. Box 2289, Blythe, CA 92226-2289
T: +1 760 9225300; Fax: +1 760 9229755
Matt Hartman
39000 vols 33992

California Department of Corrections Library System, Ironwood State Prison, 19005 Wilegis Well Rd, ***Blythe***, CA 92225; P.O. Box 2229, Blythe, CA 92226-2229
T: +1 760 9213000; Fax: +1 760 9214329
Linda Oyas
41000 vols 33993

Idaho State Law Library, 451 W State St, ***Boise***, ID 83720; P.O. Box 83720, Boise, ID 83720-0051
T: +1 208 3343316; Fax: +1 208 3344019; E-mail: lawlibrary@idcourts.net; URL: www.isll.idaho.gov
1869; Kristin Quigley
Law, Legislation; Idaho Apellate Court Briefs
150000 vols; 27 curr per 33994

John H. Lilley Correctional Center, Leisure Library, PO Box 1908, ***Boley***, OK 74829-1908
T: +1 918 6674246; Fax: +1 918 6674245
Melody Sharver
8000 vols; 10 curr per 33995

Albert C. Wagner Youth Correctional Facility Library, 500 Ward Ave, P.O. Box 500, ***Bordentown***, NJ 08505
T: +1 609 2980500 ext 1535; Fax: +1 609 2983639
Joseph Butler
14300 vols 33996

Commonwealth of Massachusetts, Office of the Attorney General Library, One Ashburton Pl, ***Boston***, MA 02108-1698
T: +1 617 7272200 ext 2098; Fax: +1 617 7275768
1975; Karin Thurman
23500 vols; 200 curr per 33997

John Joseph Moakley United States Courthouse, Library, John Joseph Moakley US Courthouse, Ste 9400, One Courthouse Way, ***Boston***, MA 02210
T: +1 617 7489044; Fax: +1 617 7489358
Karen M. Moss
Govnt Docs Selective Depository
40000 vols; 170 curr per; 95000 microforms; 5 digital data carriers 33998

National Archives & Records Administration, John F. Kennedy Presidential Library & Museum, Columbia Point, ***Boston***, MA 02125
T: +1 617 5141600; Fax: +1 617 5141593; URL: www.jfklibrary.org
1963; Thomas Putnam
Ernest Hemingway Coll, mss, photography; The Life & Times of John F. Kennedy; Mid-twentieth C American Politics & Govt (oral hist)
35000 vols; 10 curr per; 6000 microforms; 7313 sound-rec; archives (31 mio pages); still (181000), 305 art reproductions
libr loan 33999

National Center for Atmospheric, Research Library, 1850 Table Mesa Dr, ***Boulder***, CO 80305; P.O. Box 3000, Boulder, CO 80307-3000
T: +1 303 4971180; Fax: +1 303 4971170; E-mail: ncarref@ucar.edu; URL: www.ucar.edu/library
1962; Gayl Gray
Meteorology, Physics, Computer science, Chemistry, Electronics, Oceanography; Meteorological Atlases & Data
35000 vols; 642 curr per; 700 maps; 75375 microforms; 10 digital data carriers; 22000 hard copy, 270 meteorological data titles 34000

Florida Department of Corrections, Hardee Correctional Institution Library, 6901 State Rd 62, ***Bowling Green***, FL 33834
T: +1 863 7674500
Maurice Williams
14000 vols; 41 curr per 34001

Wood County Law Library, 168 S Main St, ***Bowling Green***, OH 43402
T: +1 419 3533921; Fax: +1 419 3529269; E-mail: jgill@wcnet.org
Judith Gill
24000 vols; 40 curr per 34002

US Navy – Engineering Library, Code 1143, Puget Sound Naval Shipyard, ***Bremerton***, WA 98314-5001
T: +1 360 4762767; Fax: +1 360 4764667; E-mail: engineeringlibrary@psns.navy.mil
1936; Julie Brummond
20000 vols; 150 curr per 34003

Connecticut Judicial Branch, Law Library at Bridgeport, Court House, 7th Flr, 1061 Main St, ***Bridgeport***, CT 06604
T: +1 203 5797244; URL: www.jud.ct.gov/lawlib
Willie E. Jackson
38000 vols; 92 curr per 34004

Massachusetts Department of Corrections, Institutional Library at Massachusetts Treatment Center, Administration Rd, ***Bridgewater***, MA 02324
T: +1 508 2798100
Natalya Pushkina
9000 vols; 31 curr per 34005

Florida Department of Corrections, Liberty Correctional Institution Library, 11064 NW Dempsey Barron Rd, ***Bristol***, FL 32321-9711
T: +1 850 6439400; Fax: +1 850 6439412
15000 vols; 41 curr per 34006

Criminal Court of the City of New York Library, 215 E 161st St, Rm 9-5A, ***Bronx***, NY 10451; mail address: 100 Centre St, Rm 316, New York, NY 10013
T: +1 718 5902931; Fax: +1 718 5907297; E-mail: dstephen@courts.state.ny.us
Douglas B. Stephens
20000 vols; 106 curr per
libr loan 34007

Brooklyn Correctional Institution Library, 59 Hartford Rd, ***Brooklyn***, CT 06234
T: +1 860 7792600; Fax: +1 860 7794579
Jean Hansen
12000 vols; 20 curr per 34008

New York State Supreme Court, Appellate Division Second Department Library, 45 Monroe Pl, *Brooklyn*, NY 11201
T: +1 718 7226356; Fax: +1 718 7226302; URL: www.courts.state.ny.us/courts/ad2
Bruce Bosso
Law Coll
35 000 vols   34009

New York State Supreme Court Library, Brooklyn, Supreme Court Bldg, Rm 349, 360 Adams St, *Brooklyn*, NY 11201-3782
T: +1 347 2961144; Fax: +1 718 6432412
1850; Paul Monical
Records & Briefs of the Four Appellate Courts, the Court of Appeals of the State of New York
250 000 vols; 56 000 microforms   34010

US Army – Post Library, Fort Hamilton, 404 Pershing Loop, Fort Hamilton, *Brooklyn*, NY 11252-5100
T: +1 718 6304875; Fax: +1 718 6304038
1942; Michael Gerard Plumley
Military hist & science; Black Hist, New York City & State Hist (New Yorkana Coll)
22 000 vols; 40 curr per   34011

US Courts Library, Eastern District of New York Library, 225 Cadman Plaza E, *Brooklyn*, NY 11201
T: +1 718 2602320; Fax: +1 718 3301754
1865; John Saiz
39 000 vols; 110 curr per   34012

US Air Force – The Aeromedical Library, 2511 Kennedy Circle, *Brooks AFB*, TX 78235-5116
T: +1 210 5363321; Fax: +1 210 5363239
1918; Joseph J. Franzello
Hist of Aerospace Medicine Coll
43 000 vols; 431 curr per; 117 000 microforms; 1 050 sound-rec; 17 digital data carriers; 163 971 tech rpts
libr loan   34013

US Air Force – Brooks Air Force Base Library, FL 2857, 311 ABG/SVMG, 8103 Outer Circle Rd, Ste 1, *Brooks AFB*, TX 78235-5234
T: +1 210 5362634; Fax: +1 210 5366625; E-mail: joanna.hansen@brooks.af.mil
Joanna Hansen
19 380 vols; 164 curr per   34014

Williams County Law Library Association, Court House, 3rd Fl, *Bryan*, OH 43506
T: +1 419 6364600; Fax: +1 419 6369886
John Shaffer
10 000 vols   34015

Arizona Department of Corrections – Adult Institutions, Arizona State Prison Complex – Lewis Library, 26700 S Hwy 85, *Buckeye*, AZ 85326; P.O. Box 70, Buckeye, AZ 85326-0070
T: +1 623 3866160; Fax: +1 623 3866160
Ruby Padilla
30 000 vols   34016

Buena Vista Correctional Complex Library, 15125 Hwys 24 & 285, *Buena Vista*, CO 81211; P.O. Box 2017, Buena Vista, CO 81211-2017
T: +1 719 3957254; Fax: +1 719 3957214
Diana Reese
10 000 vols; 48 curr per   34017

New York Supreme Court, Eighth Judicial District Library, 77 W Eagle St, *Buffalo*, NY 14202
T: +1 716 8459400; Fax: +1 716 8523454; E-mail: sclbuff@courts.state.ny.us; URL: www.courts.ny.us/8jd/nyssclawlib/nysscbuff.htm
1863; James R. Sahlem
Court of Appeals Records & Briefs, Law Rpts (Old English & Canadian Rpts Coll), New York Nominatives, NYCRR Backfile
350 000 vols; 226 curr per   34018

US Army – Corps of Engineers, Buffalo District Technical Library, 1776 Niagara St, *Buffalo*, NY 14207-3199
T: +1 716 8794114; Fax: +1 716 8794310; E-mail: thomas.l.vanwart@usace.army.mil
1976; Thomas Van Wart
Aerial photography, govt docs, Great Lakes res, nuclear waste disposal, Microcomputer Software Bank, radioactive waste isolation
15 000 vols; 70 curr per   34019

Miami Correctional Facility, Phase I. Library, 3038 W 850 S, *Bunker Hill*, IN 46914; P.O. Box 900, Bunker Hill, IN 46914-0900
T: +1 765 6898920; Fax: +1 765 6895964
Robert Moore
10 000 vols; 8 curr per   34020

Florida Department of Corrections, Sumter Correctional Institution Library, 9544 CR 476B, *Bushnell*, FL 33513
T: +1 352 7932525; Fax: +1 352 7933542
Rosanne Powers
19 000 vols; 41 curr per   34021

Florida Division of Adult Corrections, Sumter Correctional Institution Library, Hwy 476 B, P.O. Box 667, *Bushnell*, FL 33513-0667
T: +1 352 7932525 ext 357; Fax: +1 352 7933542
1968; Jeannie Chancellor
Spanish language, Science fiction; Black Interests, Easy-Reading, Paperbacks
15 560 vols; 98 curr per   34022

Butler County Law Library, 124 W Diamond St, Courthouse, *Butler*, PA 16001
T: +1 724 2845206; Fax: +1 724 2845210; E-mail: smegarry@co.butler.pa.us
Susan Megarry
20 000 vols; 60 curr per   34023

Massachusetts Maritime Academy, Captain Charles H. Hurley Library, 101 Academy Dr, *Buzzards Bay*, MA 02532
T: +1 508 8305034; Fax: +1 508 8305071; URL: library.maritime.edu
1970; Susan Berteaux
Oceanography, Ocean engineering, Nautical hist, Law; Rare Bks on Maritime Hist, Cape Cod Canal Coll
54 000 vols; 50 curr per; 25 000 microforms   34024

State of Ohio Department of Corrections, Noble Correctional Institution Library, 15708 Collinsville Rd, *Caldwell*, OH 43724
T: +1 740 7485188; Fax: +1 740 7485010
Julie Hupp
12 000 vols; 60 curr per   34025

California Department of Corrections Library System, Calipatria State Prison, 7018 Blair Rd, *Calipatria*, CA 92233; P.O. Box 5001, Calipatria, CA 92233-5001
T: +1 760 3487000; Fax: +1 760 3486041
Jeff Schaller
20 000 vols   34026

Guernsey County Law Library, Guernsey County Court House, 801 Wheeling Ave, Rm D 301, *Cambridge*, OH 43725
T: +1 740 4329258
Richard A. Baker
Ohio law, federal law
15 000 vols   34027

John A. Volpe National Transportation Systems Center, Technical Reference Center, Library, RTV-9L, 55 Broadway, Bldg 4, 1st Flr, Rm 117, *Cambridge*, MA 02142-1093
T: +1 617 4942306; Fax: +1 617 4943125; E-mail: volpetrc@volpe.dot.gov; URL: www.volpe.dot.gov/library/index.html
1970; Susan C. Dresley
Fed Aviation Administration Documents; Transportation Statistics Documents
32 000 vols; 250 curr per; 350 000 microforms; 100 digital data carriers
libr loan; OCLC   34028

Marine Corps Base Hawaii Libraries, Camp Smith Branch, Bldg 1, Rm 201, *Camp H M Smith*, HI 96861-4123; Box 64123, Camp H M Smith, HI 96861-4123
T: +1 808 4776348; Fax: +1 808 4772452
Polly Chan
Commandants Reading List
14 000 vols; 72 curr per   34029

State Correctional Institution, Camp Hill Library, PO Box 8837, *Camp Hill*, PA 17011-8837
T: +1 717 7374531; Fax: +1 717 7372202
Lizhu Zhong
12 000 vols; 40 curr per   34030

United States Marine Corps, Harriotte B. Smith Library, 1401 West Rd, Bldg 1220, *Camp Lejeune*, NC 28547-2539
T: +1 910 4515724; Fax: +1 910 4511871; URL: www.mccslejeune.com/library.html; library.usmc-mccs.org
1942; Janice Woodward
Military Hist Coll
150 000 vols; 308 curr per; 75 maps; 78 000 microforms; 9 300 av-mat; 10 748 sound-rec; 300 digital data carriers
libr loan   34031

United States Marine Corps – Library Services, Bldg 1146, *Camp Pendleton*, CA 92055; P.O. Box 555005, Camp Pendleton, CA 92055-5005
T: +1 760 7255669; Fax: +1 760 7256569; URL: library.usmc-mccs.org
1950; Sandra Jensen
Military hist
16 000 vols; 27 000 microforms; 50 digital data carriers   34032

United States Marine Corps – Seaside Square Library, San Onofre, Bldg 51093, *Camp Pendleton*, CA 92055
T: +1 760 7257325; Fax: +1 760 7631360
Geraldine Hagen
10 000 vols; 70 curr per   34033

US Marine Corps – Seaside Square Library, San Onofre, Bldg 51093, *Camp Pendleton*, CA 92055
T: +1 760 7257325; Fax: +1 760 7257325; URL: www.library.usmc-nccs.org/pendleton/index.htm
Geraldine Hagen
10 130 vols; 70 curr per   34034

US Navy – Crews Library, Naval Hospital, Box 555191, *Camp Pendleton*, CA 92055-5191
T: +1 760 7251229; Fax: +1 760 7251229; E-mail: pen1kdt@pen10.med.navy.mil
Kathleen Dunning-Torbert
10 000 vols; 87 curr per   34035

US Navy – Mobile Library, Bldg 1122, E St, *Camp Pendleton*, CA 92055-5005
T: +1 760 7255104; Fax: +1 760 7256177
Patrick J. Carney
8 400 vols   34036

US Navy – South Mesa Library, Bldg 200090, Oceanside, *Camp Pendleton*, CA 92055
T: +1 760 7252032; Fax: +1 760 7252041
Aurora M. Ulatan
23 000 vols; 72 curr per   34037

US Air Force – Cannon Air Force Base Library, 27 SVS/SVMG, 107 Trident Ave, Bldg 75, *Cannon AFB*, NM 88103-5211
T: +1 505 7842786; Fax: +1 505 7846929
Melissa Haraughty
35 000 vols; 160 curr per
libr loan   34038

Colorado Department of Corrections – Centennial Correctional Facility Library, PO Box 600, *Canon City*, CO 81215-0600
T: +1 719 2695546; Fax: +1 719 2695545
Jean Frost
8 000 vols; 28 curr per   34039

Colorado Department of Corrections – Colorado State Penitentiary Library, PO Box 777, *Canon City*, CO 81215-0777
T: +1 719 2695268; Fax: +1 719 2695125; URL: www.doc.state.co.us
Ladell Marta
12 000 vols   34040

Colorado Department of Corrections – Four Mile Correctional Center Library, PO Box 200, *Canon City*, CO 81215-0200
T: +1 719 2695601; Fax: +1 719 2695364; URL: www.doc.state.co.us
Blair Austin
10 000 vols; 25 curr per   34041

Colorado Department of Corrections – Fremont Correctional Facility Library, US Hwy 50, Evans Blvd, *Canon City*, CO 81215; P.O. Box 999, Canon City, CO 81215-0999
T: +1 719 2695002; Fax: +1 719 2695048; URL: www.doc.state.co.us
Darlene M. Cole
18 000 vols; 52 curr per   34042

Colorado Territorial Correctional Facility Library, PO Box 1010, *Canon City*, CO 81215-1010
T: +1 719 2754181; Fax: +1 719 2694115; E-mail: ctcflib@my.amigo.net
Linda Hyatt
Poetry
13 000 vols; 42 curr per   34043

Illinois Department of Corrections, Illinois River Correctional Center Library, 1300 W Locust St, *Canton*, IL 61520-8791; P.O. Box 999, Canton, IL 61520
T: +1 309 6477030; Fax: +1 309 6470353
Don Burkhart
Law
19 000 vols; 27 curr per   34044

Deerfield Correctional Center, DCE Library, 21360 Deerfield Dr, *Capron*, VA 23829
T: +1 434 6584368; Fax: +1 434 6584371
1994; Susan Gillette
15 000 vols; 30 curr per   34045

Cumberland County Law Library, One Courthouse Sq, *Carlisle*, PA 17013-3387
T: +1 717 2406200; Fax: +1 717 2406462; URL: www.ccpa.net
Steve Lipson
17 000 vols; 150 curr per   34046

US Army Military Heritage & Education Center US, 950 Soldiers Dr, *Carlisle*, PA 17013-5021
T: +1 717 2453130; Fax: +1 717 2453067; URL: www.carlisle.army.mil/ahec
1967; Dr. Conrad C. Crane
Mss Arch, Photogr Arch
284 000 vols; 185 curr per
libr loan   34047

Michigan Department of Corrections, Boyer Road Correctional Facility Library, PO Box 5000, *Carson City*, MI 48811-5000
T: +1 989 5843941; Fax: +1 989 5846535
Terri Halfman
10 000 vols   34048

Michigan Department of Corrections, Carson City Correctional Facility Library, PO Box 5000, *Carson City*, MI 48811-5000
T: +1 989 5843941; Fax: +1 989 5846535
Elisha Johnson
20 000 vols; 15 curr per   34049

Nevada Supreme Court Library, Supreme Court Bldg, 201 S Carson St, Ste 100, *Carson City*, NV 89701-4702
T: +1 775 6841640; Fax: +1 775 6841662; URL: lawlibrary.nvsupremecourt.us
1973; Kathleen Harrington
American & English Law
132 000 vols; 330 curr per   34050

New York State Supreme Court Library, Emory A. Chase Memorial Law Library, Greene County Courthouse, 320 Main St, *Catskill*, NY 12414
T: +1 518 9433130; Fax: +1 518 9437763
1908; Angelina Knott
New York state law; Van Orden Survey Coll
14 000 vols 34051

Linn County Law Library, Linn County Courthouse, *Cedar Rapids*, IA 52401
T: +1 319 3983920 ext 300
1925; Karan Ruzicka
16 000 vols; 2 curr per 34052

Centralia Correctional Center Library, Shattuc Rd, *Centralia*, IL 62801; P.O. Box 1266, Centralia, IL 62801-1266
T: +1 618 5334111
1980; Larry Taylor
18 000 vols; 66 curr per 34053

Century Correctional Institution Library, 400 Tedder Rd, *Century*, FL 32535
T: +1 850 2560701
Jeff Wheelahan
13 000 vols; 43 curr per 34054

Franklin County Law Library, Courthouse, 157 Lincoln Way E, *Chambersburg*, PA 17201-0459
T: +1 717 2634809; Fax: +1 717 2641992; E-mail: fcll@innernet.net
1840; Paula S. Rabinowitz
21 300 vols; 10 curr per 34055

US Army – Construction Engineering Research Laboratories, 2902 Newmark Dr, *Champaign*, IL 61822; P.O. Box 9005, Champaign, IL 61826-9005
T: +1 217 3737217; Fax: +1 217 3737222; E-mail: m_wakefield@cecer.army.mil
1969; Lacey Pat
19 000 vols; 500 curr per
libr loan 34056

United States Courts Library, US Court House, 300 Virginia St E, Rm 7400, *Charleston*, WV 25301
T: +1 304 3473420; Fax: +1 304 3473423; URL: www.wvsd.uscourts.gov
Marjorie Price
25 000 vols; 50 curr per 34057

West Virginia State Supreme Court of Appeals, State Law Library, Bldg 1, Rm E-404,1900 Kanawha Blvd E, *Charleston*, WV 25305-0833
T: +1 304 5583637; Fax: +1 304 5583673; URL: www.state.wv.us/wvsca/library/menu.htm
1867; Kaye L. Maerz
155 000 vols; 290 curr per; 10 digital data carriers 34058

US Air Force – Charleston Air Force Base Library, 437 SVS/SVMG, 106 W McCaw St, Bldg 215, *Charleston AFB*, SC 29404-4700
T: +1 843 9633320; Fax: +1 843 9633840
1953; Angela Aschenbrenner
Military hist, Total quality management
32 000 vols; 174 curr per
libr loan 34059

Mecklenburg County Law & Government Library, 700 E Trade St, *Charlotte*, NC 28202-3076
T: +1 704 3367359; Fax: +1 704 3367935; E-mail: law@plmc.lib.nc.us; URL: www.plcmc.lib.nc.us/libloc/branchlaw&gov
1991; Joyce Reimann
30 000 vols; 28 curr per 34060

Hamilton County Governmental Law Library, City County Courts Bldg, 600 Market St, Rm 305, *Chattanooga*, TN 37402
T: +1 423 2097595; Fax: +1 423 2097596
1953; Martha Wilson
Federal & State Codes Coll, West Reporter Coll
13 000 vols 34061

Tennessee Valley Authority, Corporate Library-Chattanooga, 1101 Market St, LP 4A-C, *Chattanooga*, TN 37402
T: +1 423 7514913; Fax: +1 423 7514914; E-mail: corplibchatt@tva.gov
1957; Ann Holder
Electric Power, Nuclear energy, Environmental studies; Electric Power Res Institute Rpts
35 000 vols; 20 curr per 34062

Lewis County Law Library, 345 Main St, *Chehalis*, WA 98532
T: +1 360 7489121
Gene Butler
12 000 vols 34063

United States Marine Corps, Air Station Library, Bldg 298, Marine Corps Air Station, *Cherry Point*, NC 28533-0009; PSC Box 8009, Cherry Point, NC 28533-8009
T: +1 252 4663552; Fax: +1 252 4662476; URL: library.usmc-mccs.org
1942; Suzanne L. Shell
Military science & hist; Libr of American Civilization
67 000 vols; 171 curr per 34064

Georgia Department of Corrections, Office of Library Services, Dodge State Prison, 2971 Old Bethel Rd, *Chester*, GA 31012; P.O. Box 276, Chester, GA 31012-0276
T: +1 478 3587200; Fax: +1 478 3587303; URL: www.dcor.state.ga.us
Tina Sanders
10 000 vols 34065

Wyoming Supreme Court, Wyoming State Law Library, Supreme Court Bldg, 2301 Capitol Ave, *Cheyenne*, WY 82002-0450
T: +1 307 7777509; Fax: +1 307 7777240; URL: www.courts.state.wy.us/lawlibrary
1871; Kathleen Carlson
113 000 vols; 240 curr per 34066

Chicago Transit Authority-Law Library, 567 W Lake St, *Chicago*, IL 60661-1498; P.O. Box 7564, Chicago, IL 60680-7564
T: +1 312 6812778; Fax: +1 312 6812795; URL: www.transitchicago.com
1969; Bart Lind Smith
9 000 vols; 62 e-books
libr loan 34067

City of Chicago, Department of Law Library, 600 City Hall, *Chicago*, IL 60602
T: +1 312 7440200, 7632; Fax: +1 312 7441974
Scott Burgh
16 000 vols 34068

Cook County Law Library, 2900 Richard J Daley Ctr, 50 W Washington, *Chicago*, IL 60602
T: +1 312 6035423; Fax: +1 312 6034716
1966; Bennie Martin
246 000 vols; 1 020 curr per 34069

United States Railroad Retirement Board Library, 844 N Rush St, *Chicago*, IL 60611-2031
T: +1 312 7514926; Fax: +1 312 7514924; E-mail: library@rrb.gov; URL: usrrb.gov
1940; Katherine Tsang
Law
45 000 vols; 162 curr per 34070

US Environmental Protection Agency, Region V Library, 77 W Jackson Blvd, PL-12J, *Chicago*, IL 60604-3590
T: +1 312 3532022; Fax: +1 312 3532001; E-mail: patricia.krause@epa.gov; URL: www.epa.gov/region5/library
1972; Patricia Krause
Air Pollution, Great Lakes, Radiation, Environmental law; EPA Rpts, Microfiche Federal Registers
17 500 vols; 300 curr per
libr loan 34071

William J. Campbell Library of the US Courts, 219 S Dearborn St, Rm 1637, *Chicago*, IL 60604-1769
T: +1 312 4355660; Fax: +1 312 4085031; URL: www.lb7.uscourts.gov
Jerry Lewis
Govt publs
30 000 vols 34072

Chillicothe Correctional Institution Library, 15802 State Rte 104 N, *Chillicothe*, OH 45601; P.O. Box 5500, Chillicothe, OH 45601-5500
T: +1 740 7747080; Fax: +1 740 7747082
Cathy Pummill
15 000 vols; 100 curr per 34073

US Navy – Naval Air Warfare Center Weapons Division Technical Library, One Administration Circle, Stop 6203, *China Lake*, CA 93555-6104
T: +1 760 9393389; Fax: +1 760 9392431; E-mail: nwtechlib@navy.mil
1946; Barbara Lupei
41 000 vols; 400 curr per; 1 mio tech rpts
libr loan 34074

US Navy – Naval Air Warfare Station Library, 861000D, Bldg 17 Naws, *China Lake*, CA 93555-6001
T: +1 760 9392595; Fax: +1 760 9393013
1945; Doreen Hobson
Military Aviation Coll
40 000 vols; 116 curr per 34075

California Department of Corrections Library System, California Institution for Men, 14901 S Central Ave, *Chino*, CA 91710; P.O. Box 128, Chino, CA 91708-0128
T: +1 909 5971821; +1 909 6067012
Brian Maughan
15 000 vols 34076

Division of Juvenile Justice & Department of Corrections & Rehabilitation, Herman G. Stark Youth Correctional Facility Library, 15180 Euclid Ave, *Chino*, CA 91710
T: +1 909 6065000; Fax: +1 909 6065063
Rose Lepley
10 000 vols 34077

California Department of Corrections Library System, Central California Women's Facility, 23370 Rd 22, *Chowchilla*, CA 93610; P.O. Box 1501, Chowchilla, CA 93610-1501
T: +1 559 6655531; Fax: +1 559 6656037
Phil Renteria
25 000 vols 34078

California Department of Corrections Library System, Valley State Prison for Women, 21633 Avenue 24, *Chowchilla*, CA 93610; P.O. Box 92, Chowchilla, CA 93610-0092
T: +1 559 6656100; Fax: +1 559 6656113
Diane Johnson
40 000 vols 34079

San Diego County Law Library – South Bay, 500 Third Ave, *Chula Vista*, CA 91910-5617
T: +1 619 6914929; Fax: +1 619 4277521; URL: www.sdccll.org
Edna Thiel
16 500 vols 34080

US Court of Appeals for the Sixth Circuit Library, 312 Potter Stewart US Courthouse, *Cincinnati*, OH 45202; mail address: 317 Potter Stewart US Courthouse, Cincinnati, OH 45202
T: +1 513 5647321; Fax: +1 513 5647329; URL: www.ca6.uscourts.gov
1895; Kathy Welker
Primarily Anglo-American Legal Coll
250 000 vols 34081

US Department of Health & Human Services, National Institute for Occupational Safety & Health Library, 4676 Columbia Pkwy, *Cincinnati*, OH 45226
T: +1 513 5338321; Fax: +1 513 5338382
Lawrence Foster
10 000 vols; 100 curr per 34082

Saint Louis County Law Library, Courts Bldg, Ste 536, 7900 Carondelet Ave, *Clayton*, MO 63105
T: +1 314 6154726
Mary C. Dahm
31 000 vols; 20 curr per 34083

Clearfield County Law Library, Courthouse, 2nd Flr, Ste 228, 230 E Market St, *Clearfield*, PA 16830
T: +1 814 7652641; Fax: +1 814 7657649
Carol Mease
14 000 vols; 71 curr per 34084

Pinellas County Law Library, 324 S Ft Harrison Ave, *Clearwater*, FL 33756-5165
T: +1 727 4643411; Fax: +1 727 4644571; URL: www.jud6.org/lawlibraries
1950; Donna L. Haverkamp
Laws of Florida
34 000 vols 34085

Florida Department of Corrections, Lake Correctional Institution Library, 19225 Hwy 27, *Clermont*, FL 34711
T: +1 352 3946146 ext 295
1974; Guy Kovacs
Vocational education
12 200 vols; 50 curr per 34086

NASA – John H. Glenn Research Center at Lewis Field, Technical Library, 21000 Brookpark Rd, MS 60-3, *Cleveland*, OH 44135
T: +1 216 4335761; Fax: +1 216 4338192; E-mail: library@grc.nasa.gov; URL: grctechlib.grc.nasa.gov
1941; Susan Oberc
Project Doc Ctr
80 000 vols; 300 curr per 34087

Correctional Institution for Women, Edna Mahan Hall Library, PO Box 4004, *Clinton*, NJ 08809-4004
T: +1 908 7357111; Fax: +1 908 7350108
Kelly Scott
Law; Criminal Law, Women in Prison
9 000 vols; 40 curr per 34088

South Carolina Department of Disabilities & Special Needs, Whitten Center Library & Media Resource Services, P.O. Box 239, *Clinton*, SC 29325
T: +1 864 9383331; Fax: +1 864 9383179; E-mail: nwells@ddsn.state.sc.us
1965; Nancy Wells
American Journal of Mental Deficiency, (1922 to present), per; Annual Report of South Carolina State, Training School for the Feebleminded (1918-1968); Bibliography of Professional Materials, on Mental Retardation & Reference, (1972, 1975)
16 500 vols; 25 curr per 34089

Macomb Intermediate School District, Beal Library, 44001 Garfield Rd, *Clinton Township*, MI 48038-1100
T: +1 810 2283400; Fax: +1 810 2861523; E-mail: richard.palmer@mda.net; URL: www.misd.net
1971; Richard J Palmer
Education; Curriculum Guides on Microfilm
40 000 vols; 475 curr per; 800 250 microforms; 600 av-mat; 20 digital data carriers 34090

California Department of Corrections Library System, Pleasant Valley State Prison, 24863 W Jayne Ave, *Coalinga*, CA 93210; P.O. Box 8500, Coalinga, CA 93210-1135
T: +1 559 9354900
44 000 vols; 54 curr per 34091

Florence Crane Correctional Facility Library, 38 Fourth St, *Coldwater*, MI 49036; P.O. Box 307, Coldwater, MI 49036
T: +1 517 2799165; Fax: +1 517 2788265
Mary Gray
10 000 vols; 20 curr per 34092

Collins Correctional Facility Library, PO Box 490, *Collins*, NY 14034-0490
T: +1 716 5324588
24 000 vols; 66 curr per 34093

El Paso County Law Library, 20 N Cascade Ave, *Colorado Springs*, CO 80903
T: +1 719 5316333; URL: ppld.org/aboutyourlibrary/services/lawlibrary/default.asp
1973; Sara Goroski
22 000 vols 34094

Columbia Environmental Research Center Library, US Geological Survey, 4200 New Haven Rd, *Columbia*, MO 65201
T: +1 573 8761853; Fax: +1 573 8761833; E-mail: cerc.library@usgs.gov
1970; Julia Towns
Biological toxicity, Chemical pesticides, Environmental contaminants, Missouri River, Water quality; Pesticides (Reprint Coll)
8 500 vols; 80 curr per
libr loan   34095

South Carolina Attorney General's Office, Daniel R. McLeod Law Library, 1000 Assembly St, Ste 742, *Columbia*, SC 29201-3117; P.O. Box 11549, Columbia, SC 29211-1549
T: +1 803 7343769; Fax: +1 803 2536283; E-mail: agshusman@ag.state.sc.us; URL: www.scattorneygeneral.org
1974; Susan Husman
Attorney General's Opinions
20 000 vols; 20 curr per; 13 av-mat; 10 sound-rec; 5 digital data carriers   34096

South Carolina Department of Corrections, Library Services, 4444 Broad River Rd, *Columbia*, SC 29210-4000; P.O. Box 21787, Columbia, SC 29221-1787
T: +1 803 8961567; Fax: +1 803 8961513
1968; Daisy Lindler
94 713 vols; 28 curr per   34097

South Carolina Supreme Court Library, 1231 Gervais St, *Columbia*, SC 29201-3206; P.O. Box 11330, Columbia, SC 29211-1330
T: +1 803 7341080; Fax: +1 803 7340519; URL: www.sccourts.org
1871; Janet F. Meyer
Court Cases (1918 to present), micro
50 000 vols; 115 curr per   34098

Lowndes County Law Library, County Court House, Rm 210, P.O. Box 1364, *Columbus*, MS 39703-1364
T: +1 662 3295889; Fax: +1 662 3295870
Lena Mae Duncan
12 000 vols; 80 curr per   34099

Ohio Attorney General, Law Library, 30 E Broad St, 15th Flr, *Columbus*, OH 43215
T: +1 614 4662465; Fax: +1 614 7529867
1846; Madelaine A. Gordon
22 000 vols; 25 curr per   34100

Ohio Department of Transportation Library, 1980 W Broad St, *Columbus*, OH 43223
T: +1 614 4667680; Fax: +1 614 7286694; E-mail: janet.bix@dot.state.oh.us; URL: slonet.state.oh.us
1977; Janet Bix
24 000 vols; 250 curr per; Dept of Ohio Transportation Publs
libr loan   34101

Ohio Legislative Service Commission Library, 77 S High St, 9th Fl, *Columbus*, OH 43266-0342
T: +1 614 4665312; Fax: +1 614 6441721
1953; Debbie Tavenner
Bulletins of the Ohio General Assembly, Journals of the Ohio House & Senate, Laws of Ohio
15 000 vols; 250 curr per; 3 digital data carriers   34102

Supreme Court of Ohio, Law Library, 65 S Front St, 11th Flr, *Columbus*, OH 43215-3431
T: +1 614 3879650; Fax: +1 614 3879689; E-mail: libref@sconet.state.oh.us; URL: www.sconet.state.oh.us/lawlibrary
1860; Ken Kozlowski
Old Ohio & Anglo-American Legal Treatises, Cases & Statutes
400 000 vols; 1 740 curr per   34103

US Air Force – Columbus Air Force Base Library, FL 3022 (AETC), *Columbus AFB*, MS 39710-5102
T: +1 662 4342934; Fax: +1 662 4346291
Aviation, Military strategy
16 640 vols; 87 curr per   34104

Great Meadow Correctional Facility Library, 11739 State Rte 22, *Comstock*, NY 12821; P.O. Box 51, Comstock, NY 12821
T: +1 518 6395516; Fax: +1 518 6395816 ext 2099
Heather Larrow
Spanish language, African-American studies
12 000 vols; 50 curr per   34105

Washington Correctional Facility Library, PO Box 180, *Comstock*, NY 12821-0180
T: +1 518 6394486; Fax: +1 518 6393299
D. Cartmell
10 000 vols   34106

Massachusetts Correctional Institution-Concord Library, 965 Elm St, *Concord*, MA 01742-2119; P.O. Box 9106, Concord, MA 01742-9106
T: +1 978 3693220; Fax: +1 978 4056108
Margaret Mubiru-Musoke
Law; Spanish Coll
8 000 vols; 20 curr per   34107

New Hampshire Department of Justice, Office of the Attorney General Library, 33 Capitol St, *Concord*, NH 03301-6397
T: +1 603 2713658; Fax: +1 603 2712110
Ellen Webb
10 000 vols   34108

New Hampshire Law Library, Supreme Court Bldg, One Noble Dr, *Concord*, NH 03301-6160
T: +1 603 2713777; Fax: +1 603 2712168; URL: www.courts.state.nh.us/lawlibrary
1819; Mary Searles
Laws & Court Rpts, New Hampshire Legal Mat
100 000 vols; 110 curr per
libr loan   34109

New Hampshire State Prison Library, 281 N State St, *Concord*, NH 03301; P.O. Box 14, Concord, NH 03302-0014
T: +1 603 2711929; Fax: +1 603 2710401
1918; Becky Harding
Law materials, Lois Law on CD-ROM, 13 U.S. Circuit Courts of Appeal, Supreme Court Reports
18 000 vols; 12 curr per; 4 digital data carriers; Newspapers 12
libr loan; OCLC   34110

US Army – Corps of Engineers, New England District Library, 696 Virginia Rd, *Concord*, MA 01742-2751
T: +1 978 3188349; Fax: +1 978 3188693; E-mail: timothy.p.hays@usace.army.mil
1947; Timothy Hays
Civil engineering, Structural engineering, Ecology, Natural science, Soil science, Water resources
21 000 vols   34111

California Department of Corrections Library System, California Substance Abuse Treatment Facility & State Prison, 900 Quebec Ave, *Corcoran*, CA 93212; P.O. Box 7100, Corcoran, CA 93212-7100
T: +1 559 9927100; Fax: +1 559 9927182
Ross Zimmerman
Law reference books
50 000 vols; 72 curr per   34112

US Navy – Naval Air Station Library, Bldg 5, *Corpus Christi*, TX 78419
T: +1 361 9613574
1941; Sharon Faith Scott
Aviation Coll, World War II
34 000 vols; 72 curr per; 396 av-mat
                                              34113

Environmental Protection Agency, NHEERL Western Ecology Division Library, 200 SW 35th St, *Corvallis*, OR 97333
T: +1 541 7544731; Fax: +1 541 7544799; E-mail: libary@mail.cor.epa.gov
1966; Mary O'Brien
8 000 vols; 115 curr per
libr loan; SLA   34114

Avoyelles Correctional Center Library, 1630 Prison Rd, *Cottonport*, LA 71327
T: +1 318 8762891; Fax: +1 318 8764220
Russell Clayton
12 000 vols; 11 curr per   34115

California Department of Corrections Library System, Pelican Bay State Prison, 5905 Lake Earl Dr, *Crescent City*, CA 95532; P.O. Box 7000, Crescent City, CA 95531-7000
T: +1 707 4651000; Fax: +1 707 4659120
Charles Cox
10 000 vols   34116

Florida Department of Corrections, Okaloosa Correctional Institution Library, 3189 Little Silver Rd, *Crestview*, FL 32539
T: +1 850 6820931; Fax: +1 850 6897803; E-mail: okaloosaci@mail.dc.state.fl.us; URL: www.dc.state.fl.us
Dana Quillin
15 000 vols; 41 curr per   34117

Cross City Correctional Institution Library, Old Airforce Radar Rd, *Cross City*, FL 32628; P.O. Box 1500, Cross City, FL 32628-1500
T: +1 352 4984444; Fax: +1 352 4981275; URL: www.dc.state.fl.us
1974; Ttonya Hudson
Science fiction, Western, Mysteries, Hispanic, African-American
10 000 vols; 26 curr per   34118

Colorado Department of Corrections, Arkansas Valley Correctional Facility Library, 12750 Hwy 96, Lane 13, *Crowley*, CO 81034; P.O. Box 1000, Crowley, CO 81034
T: +1 719 2673520; Fax: +1 719 2675024; E-mail: linda.hollis@doc.state.co.us
Diane Walden
16 000 vols; 49 curr per; 27 av-mat
                                              34119

Allegany County Circuit Court Law Library, Court House, 30 Washington St, *Cumberland*, MD 21502
T: +1 301 7775925; Fax: +1 301 7772055
Anne M. SanGiovanni
Maryland County Charters, Maryland Court of Appeals Records & Briefs, 1893-1931
16 000 vols   34120

Western Correctional Institution Library, 13800 McMullen Hwy SW, *Cumberland*, MD 21502
T: +1 301 7297184; Fax: +1 301 7297150
Charles Albright
14 000 vols; 34 curr per   34121

US Navy – Dahlgren Laboratory, General Library, Naval Surface Warfare Ctr, *Dahlgren*, VA 22448-5000
T: +1 540 6537474
1954; Carolyn Bradley
18 000 vols; 36 curr per   34122

Dallas County Law Library – Civil Law Collection, George Allen Courts Bldg, 600 Commerce St, Ste 292, *Dallas*, TX 75202-4606
T: +1 214 6537481; Fax: +1 214 6536103; E-mail: lawlibrary@dallascounty.org; URL: www.dallascounty.org
Mary Rankin
1 branch libr
83 000 vols   34123

Dallas County Law Library – Criminal Law Collection, Frank Crowley Courts Bldg, 133 N Industrial, Rm A-9, *Dallas*, TX 75207
T: +1 214 6535990; Fax: +1 214 6535993; URL: www.dallascounty.org
Mary Rankin
20 000 vols   34124

Connecticut Judicial Branch, Danbury Law Library, Courthouse, 146 White St, *Danbury*, CT 06810
T: +1 203 2078625; Fax: +1 203 2078627
Linda Mellick
Connecticut Legislative Hist, Hist Info (Connecticut Law)
28 000 vols; 30 curr per   34125

US Air Force – Davis-Monthan Air Force Base Library, FL4877, 355 SVS/SVMG, 355 SVS/SVMG, 5427 E Madera St, Bldg 4339, *Davis Monthan AFB*, AZ 85707-4930
T: +1 520 2287148; Fax: +1 520 2283998; URL: dmlibrary.org
Pipei Guo
41 000 vols; 135 curr per   34126

Florida Department of Corrections, Tomoka Correctional Institution Library, 3950 Tiger Bay Rd, *Daytona Beach*, FL 32124-1098
T: +1 386 3231195; Fax: +1 386 3231168
Samy Ghaly
15 000 vols; 93 curr per   34127

Volusia County Law Library, Courthouse Annex, Rm 208, 125 E Orange Ave, *Daytona Beach*, FL 32114
T: +1 386 2576041; Fax: +1 386 2576052
Deborah Patterson
25 000 vols   34128

Montana State Prison Library, Conley Lake Rd, *Deer Lodge*, MT 59722
T: +1 406 8461320 ext 2410; Fax: +1 406 8462951
Dave Beatty
Law, College ref works
8 000 vols; 30 curr per   34129

Defiance County Law Library, 510 Court St, *Defiance*, OH 43512
T: +1 419 7821186; Fax: +1 419 7822437
Jill S. Pixler
20 000 vols; 1 000 curr per   34130

Florida Department of Corrections, Walton Correctional Institution Library, 691 WWII Veterans Lane, *DeFuniak Springs*, FL 32433
T: +1 850 9511437; Fax: +1 850 8923691
John Kilhefner
10 000 vols; 38 curr per   34131

California Department of Corrections Library System, North Kern State Prison, 2737 W Cecil Ave, *Delano*, CA 93215; P.O. Box 567, Delano, CA 93216-0567
T: +1 661 7212345
Brooke Lilly
State law, federal law
24 000 vols; 22 curr per   34132

Delaware County Supreme Court Library, 3 Court St, *Delhi*, NY 13753-9990
T: +1 607 7463959; Fax: +1 607 7468198; URL: www.courts.state.ny.us
1880; Charle Cash
8 500 vols   34133

Bureau of Land Management Library, Bldg 50, RS 150 A, Denver Federal Ctr, P.O. Box 25047, *Denver*, CO 80225-0047
T: +1 303 2366650; Fax: +1 303 2364810; E-mail: BLM-LIBRARY@sc.blm.gov
1971; Elizabeth Araki
40 000 vols; 225 curr per
libr loan   34134

Colorado Department of Corrections, Denver Women's Correctional Facility Library, 3600 Havana St, *Denver*, CO 80239
T: +1 303 3072500; Fax: +1 303 3072674
Irene A. Betin
10 000 vols; 40 curr per   34135

Colorado Department of Transportation Library, 4201 E Arkansas Ave, *Denver*, CO 80222
T: +1 303 7579972; Fax: +1 303 7579242; E-mail: joan.pinamont@dot.state.co.us; URL: www.dot.state.co.us/business/library/index.htm
1949; Joan Pinamont
18 000 vols; 12 curr per
SLA   34136

Colorado Supreme Court Library, B112 State Judicial Bldg, Two E 14th Ave, *Denver*, CO 80203-2116
T: +1 303 8373720; Fax: +1 303 8644510; URL: www.state.co.us/courts/sctlib
1874; Dan Cordova
80 000 vols; 300 curr per   34137

US Bank, Law Library, 950 17th St, US Bank Tower Bldg, **Denver**, CO 80202
T: +1 303 8258400; Fax: +1 303 8256525; E-mail: awills@ojrnr.com
1975; Ann Marie Wills
33 000 vols; 96 curr per
libr loan   34138

US Courts Library, Tenth Circuit Court of Appeals, Byron Rogers Courthouse, 1929 Stout St, Rm 430, **Denver**, CO 80294
T: +1 303 8443591; Fax: +1 303 8443558; E-mail: 10th_circuit_library@ca10.uscourts.gov
1929; J. Terry Hemming
36 000 vols   34139

US Department of Defense, Defense Finance & Accounting Service – Denver Center – Learning Center, 6760 E Irvington Pl, **Denver**, CO 80279-8000
T: +1 303 6767566; Fax: +1 303 6766003
1951; Harvey Reynolds
Accounting, Computer science
8 000 vols   34140

US Department of the Interior – Bureau of Reclamation, Reclamation Denver Office Library, Denver Fed Ctr D7925, P.O. Box 25007, **Denver**, CO 80225
T: +1 303 4452072; Fax: +1 303 4456303; E-mail: library@do.usbr.gov; URL: library.usbr.gov
1945; Dianne Powell
Construction Specifications
90 000 vols; 1 000 curr per; 10 000 microforms; 150 digital data carriers; 250 000 rpts
libr loan; ALA, SLA, CLA   34141

US Geological Survey Library, Denver Federal Ctr, Bldg 20, Rm C2002, **Denver**, CO 80225-5046, Stop 914, Denver, CO 80225-0046
T: +1 303 2361000; Fax: +1 303 2360015; E-mail: den_lib@usgs.gov; URL: library.usgs.gov
1948; Tommie Ann Gard
Earth science; Field Records Libr, GEO Ctr, Photo Libr
230 000 vols; 700 curr per   34142

Iowa Department of Human Services Library, Hoover Bldg, 1305 E. Walnut, **Des Moines**, IA 50319
T: +1 515 2816033; Fax: +1 515 2814243; E-mail: kelliot@dhs.state.ia.us
1968; Kay M. Elliott
9 000 vols; 100 curr per; 290 av-mat; 6 digital data carriers
libr loan   34143

State Library of Iowa – Iowa State Law Library, State Capitol Bldg, 1007 E Grand Ave, **Des Moines**, IA 50319
T: +1 515 2815124; Fax: +1 515 2815405; E-mail: law@lib.state.ia.us; URL: www.silo.lib.ia.us/lawlib.html
1835; Linda Robertson
English Law Coll, Iowa Law
160 000 vols   34144

US Court of Appeals, Branch Library, 110 E Court Ave, Ste 358, **Des Moines**, IA 50309
T: +1 515 2846228; Fax: +1 515 2846451
Cheryl Gritton
15 000 vols; 50 curr per   34145

City of Detroit, Law Department Library, First National Bldg, 660 Woodward Ave, Ste 1650, **Detroit**, MI 48226
T: +1 313 2244550; Fax: +1 313 2245505
1935; Thomas R. Killian
35 000 vols; 30 curr per   34146

Ralph M. Freeman Memorial Library for the US Courts, 436 US Courthouse, **Detroit**, MI 48226-2719
T: +1 313 2345255; Fax: +1 313 2345383
1975; Linda D. Smith
Federal Judicial Ctr Publs Coll
43 000 vols; 80 curr per   34147

Thirty Sixth District Court Law Library, 421 Madison Ave, **Detroit**, MI 48226
T: +1 313 9652792; Fax: +1 313 9654057
Theodosia Clemons
30 000 vols; 75 curr per   34148

Buckingham Correctional Center, DCE Library, 1349 Correctional Center Rd, **Dillwyn**, VA 23936; P.O. Box 430, Dillwyn, VA 23936-0430
T: +1 434 3915980; Fax: +1 434 9831296
David Jobe
14 000 vols; 43 curr per   34149

Dillwyn Correctional Center, DCE Library, Rte 20, P.O. Box 670, **Dillwyn**, VA 23936-0670
T: +1 804 9835034 ext 1733, 1713; Fax: +1 804 9831821
1994; Steve Sarazin
10 000 vols; 31 curr per   34150

Dixon Correctional Center Library, 2600 N Brinton Ave, **Dixon**, IL 61021-9524
T: +1 815 2885561
Carole O'Neal
Federal and Illinois Law
10 000 vols   34151

Arizona Department of Corrections – Adult Institutions, Arizona State Prison Complex – Douglas Library, 6911 N BDI Blvd, **Douglas**, AZ 85608; P.O. Drawer 3867, Douglas, AZ 85608-3867
T: +1 520 3647521; Fax: +1 520 8055971
Kathleen Fry
27 000 vols   34152

Delaware Department of Transportation Library, 800 Bay Rd, P.O. Box 778, **Dover**, DE 19903
T: +1 302 7602104; Fax: +1 302 7396371; E-mail: jcheng@mail.dot.state.de.us
Julianna Cheng
13 000 vols; 12 curr per   34153

Delaware State Law Library in Kent County, Thomas Collins Bldg, 540 S DuPont Hwy, Ste 3, **Dover**, DE 19901-3615
T: +1 302 7395467; Fax: +1 302 7396721
Karen Parrott
Early & Unusual Law Books
35 000 vols; 30 curr per   34154

US Air Force – Dover Air Force Base Library, 436 SVS/SVMG, 262 Chad St, **Dover AFB**, DE 19902-7235
T: +1 302 6773995; Fax: +1 302 6775490
Richard Krueger
Quality Air Force, Transition Assistance
38 000 vols; 150 curr per   34155

Bucks County Law Library, Court House, 55 E Court St, First Flr, **Doylestown**, PA 18901
T: +1 215 3436023; Fax: +1 215 3486827; URL: www.buckscounty.org
Donald W. Jacobs
Fed Statutes, State Statutes (PA,NJ,NY,DE,FL,MD), National Reporter & system, Law Encyclopedias, Treaties, Dictionaries, Per (law-related), Form Bks, Court Rules, Digests, Case Law
35 000 vols; 100 curr per   34156

US Army – Post Library, 5124 Kister Ave, IMWE-DUG-MWL MS3, **Dugway**, UT 84022-5003
T: +1 435 8312633; Fax: +1 435 8313543; URL: www.dugwaymwr.com/lib/index.htm
Michael Beier
38 000 vols; 50 curr per; 800 av-mat   34157

US Army – West Desert Technical Information Center, Dugway Proving Ground, Kuddes Bldg 4531, Rm 116, **Dugway**, UT 84022; CSTE-DTC-DP-WD-JC-L, Dugway, UT 84022
T: +1 435 8313822; Fax: +1 435 8313931; E-mail: dpg@army.mil; URL: www.atc.army.mil
1950; Steven B. Christensen
8 000 vols   34158

Saint Louis County Law Library, 100 N Fifth Ave W, Rm 515, **Duluth**, MN 55802
T: +1 218 7262611; Fax: +1 218 7262612
Michele Des Rosier
20 000 vols; 15 curr per   34159

US Environmental Protection Agency, Mid-Continent Ecology Division Library, 6201 Congdon Blvd, **Duluth**, MN 55804-2558
T: +1 218 5295000; Fax: +1 218 5295003; E-mail: bankson.john@epamail.epa.gov; URL: www.epa.gov/med
1967; John Bankson
Effluent testing, Freshwater toxicology
12 000 vols; 145 curr per; 10 digital data carriers   34160

Illinois Department of Corrections, Dwight Correctional Center Library, 23813 E 3200 N Rd, **Dwight**, IL 60420
T: +1 815 5842806; Fax: +1 815 8541432
Beatrice Stanley
10 000 vols   34161

Alaska State Department of Corrections, Hiland Mountain Correctional Center Library, P.O. Box 600, **Eagle River**, AK 99577-0600
T: +1 907 6949511 ext 139; Fax: +1 907 6944507
1974; Stanley M- Reed
8 719 vols; 27 curr per; 30 digital data carriers   34162

Florida Department of Corrections, Putnam Correctional Institution Library, 128 Yelvington Rd, **East Palatka**, FL 32131-2100
T: +1 386 3266800; Fax: +1 386 3122219
Dale Brown
8 000 vols; 41 curr per   34163

Northampton County Law Library, 669 Washington St, **Easton**, PA 18042-7468
T: +1 610 5596751; Fax: +1 610 5596750; E-mail: ncll@nccpa.org; URL: www.nccpa.org
1860; Anita L. DeBona
Typical Coll for Pa County Law Libr
23 399 vols; 18 digital data carriers   34164

Cambria County Free Law Library, Court House, S Center St, **Ebensburg**, PA 15931
T: +1 814 4721501; Fax: +1 814 4724799
1920; Sandy Papcunik
40 000 vols; 58 curr per   34165

Hidalgo County Law Library, Courthouse, 100 Closner, **Edinburg**, TX 78539
T: +1 956 3182155; Fax: +1 956 3814269
Angie Chapa
22 000 vols   34166

US Air Force – Air Force Flight Test Center Technical Library, 812 TSS/ENTL,307 E Popson Ave, Bldg 1400, Rm 106, **Edwards AFB**, CA 93524-6630
T: +1 661 2773606; Fax: +1 661 2776451
1955; Marie L. Nelson
Aeronautics, Chemistry, Physics, Electrical eng, Mathematics
32 000 vols; 122 000 microforms; 376 av-mat; 10 digital data carriers; 375 391 tech rpts
libr loan   34167

US Air Force – Edwards Air Force Base Library, 95th FSS/FSDL, Five W Yeager Blvd, Bldg 2665, **Edwards AFB**, CA 93524-1295
T: +1 661 2752665; Fax: +1 661 2776100
1942; Dr. Tatiana Verren
Education, Recreation
43 000 vols; 150 curr per   34168

Air Force Research Laboratory, Munitions Directorate Technical Library, 203 W Eglin Blvd, Ste 300, **Eglin AFB**, FL 32542-6843
T: +1 850 8825586; Fax: +1 850 8824476
Cheryl Mack
12 000 vols; 200 curr per   34169

US Air Force – Eglin Air Force Base Library, 305 W F St, Bldg 278, **Eglin AFB**, FL 32542-6842
T: +1 850 8823462; Fax: +1 850 8822621; URL: www.youseemore.com/eglin
1942; Vicky Stever
Aeronautics, Military hist, Business & management
43 000 vols; 198 curr per; 8 000 e-books; 1 200 av-mat   34170

US Air Force – Technical Library, 203 W Eglin Blvd, Ste 300, **Eglin AFB**, FL 32542-6843
T: +1 850 8823212, 5586; Fax: +1 850 8823214
1955; Jim Elkins
Aeronautics, Electronics, Physics, Mathematics, Chemistry, Biology
13 000 vols; 450 curr per
libr loan   34171

US Air Force – Eielson Air Force Base Library, 2555 Coman Ave, **Eielson AFB**, AK 99702
T: +1 907 3773174; Fax: +1 907 3771683
1946; Cathy Rasmussen
Arctic Coll, Total Quality Management (TQM), Professional Military Education
30 000 vols; 63 curr per; 11 e-books   34172

San Diego County Law Library – East County, 250 E Main, **El Cajon**, CA 92020-3941
T: +1 619 4414444; Fax: +1 619 4410235; URL: www.sdcll.org
Carolyn Dulude
10 900 vols   34173

Imperial County Law Library, Courthouse, 939 W Main St, **El Centro**, CA 92243
T: +1 760 4824374; Fax: +1 760 3523184
Ramona Gieck
12 100 vols   34174

US Navy – Naval Air General Library, 2003 D St, Bldg 318, MWR Facility, **El Centro**, CA 92243
T: +1 760 3392470; Fax: +1 760 3392470
Letti Earle
8 000 vols; 14 curr per   34175

Kansas Department of Corrections, El Dorado Correctional Facility Library, 1737 SE Hwy 54, **El Dorado**, KS 67042; P.O. Box 311, El Dorado, KS 67042
T: +1 316 3217284; Fax: +1 316 3222018
Margaret Adamson
10 000 vols; 20 curr per   34176

El Paso County Law Library, 500 E San Antonio St, Rm 1202, **El Paso**, TX 79901
T: +1 915 5462245; Fax: +1 915 5420440; E-mail: lsanchez@co.el-paso.tx.us; URL: www.co.el-paso.tx.us
Lynn Sanchez
30 000 vols; 49 curr per   34177

Bureau of Prisons, Federal Correctional Institution Library, P.O. Box 1000, **El Reno**, OK 73036
T: +1 405 2624875; Fax: +1 405 2626266
1960; Wayne Huffman
12 000 vols; 14 curr per   34178

Union County Law Library, 2 Broad St, Union County Courthouse, **Elizabeth**, NJ 07207
T: +1 908 6594166
Nicholas Tomich
40 000 vols   34179

North Carolina Department of Corrections, Pasquotank Correctional Institution Library, 527 Commerce Dr, Caller No 5005, **Elizabeth City**, NC 27906-5005
T: +1 252 3314881; Fax: +1 252 3314867
Andre Williams
12 000 vols; 20 curr per   34180

Hancock County Law Library, 60 State St, **Ellsworth**, ME 04605; mail address: 97 Hammond St, Bangor, ME 04401
Judith Bennett
10 000 vols   34181

Kansas Department of Corrections, Ellsworth Correctional Facility Library, PO Box 107, **Ellsworth**, KS 67439
T: +1 785 4725501; Fax: +1 785 4724032
Francis J. Devadason
10 000 vols; 10 curr per   34182

US Air Force – Ellsworth Air Force Base Library, Holbrook Library FL 4690, 28 SVS/SVMG, 2650 Doolittle Dr, Bldg 3910, **Ellsworth AFB**, SD 57706-4820
T: +1 605 3851688; Fax: +1 605 3854467; URL: www.ellsworthservices.com
Jeanne Stoltenburg
42 000 vols; 175 curr per; 6 000 e-books; 1 515 av-mat; 690 sound-rec; 63 digital data carriers
libr loan   34183

US Air Force – Elmendorf Air Force Base Library, FL 5000, 10480 22nd St, **Elmendorf AFB**, AK 99506-2530
T: +1 907 5523787; Fax: +1 907 5521707; E-mail: ill@gci.net; URL: www.elmendorf.af.mil/library/
1950; Martha K. Sumpter
Air War College Coll, Alaskana Coll
49 905 vols; 237 curr per   34184

Charles B. Swartwood Supreme Court Library, 203-205 Lake St, **Elmira**, NY 14901
T: +1 607 7372983; Fax: +1 607 7339863; E-mail: lhubbard@courts.state.ny.us
1895; Laurie A. Hubbard
15 000 vols   34185

New York State Department of Correctional Services, Elmira Correctional Facility Library, 1879 Davis St, **Elmira**, NY 14901-1042; P.O. Box 500, Elmira, NY 14902-0500
T: +1 607 7343901
Greg Harris
Vocational Guidance Coll
11 000 vols; 90 curr per   34186

Nevada Department of Corrections, Ely State Prison Library, 4569 N State Rte 490, **Ely**, NV 89301; Mountain High School, Ave C, Ely, NV 89301
T: +1 775 2898800; Fax: +1 775 2891273; URL: www.doc.nv.gov
Bill Thompson
14 000 vols; 10 curr per   34187

Lorain County Law Library, 226 Middle Ave, **Elyria**, OH 44035
T: +1 440 3295567; Fax: +1 440 3221724
1889; Mary Kovacs
25 000 vols; 25 curr per; 20 000 microforms; 250 sound-rec   34188

National Emergency Training Center, Learning Resource Center, 16825 S Seton Ave, **Emmitsburg**, MD 21727
T: +1 301 4471030; Fax: +1 301 4473217; E-mail: netclrc@dhs.gov; URL: www.lrc.dhs.gov
Edward J. Metz
85 000 vols; 300 curr per   34189

Cameron County Law Library, Court House, 20 E Fifth St, **Emporium**, PA 15834
T: +1 814 4862315; Fax: +1 814 4860464; URL: www.cameroncountypa.com
Brenda Munz
20 000 vols; 57 curr per   34190

Carl Robinson Correctional Institution Library, 285 Shaker Rd, **Enfield**, CT 06083; P.O. Box 1400, Enfield, CT 06083-1400
T: +1 860 7636230; Fax: +1 860 7636345
Nafi Donat
30 000 vols; 58 curr per   34191

Willard-Cybulski Correctional Institution Library, 391 Shaker Rd, **Enfield**, CT 06082
T: +1 860 7636590; Fax: +1 860 7636517
Gay Maiolo
10 000 vols; 15 curr per   34192

Erie County Law Library, Court House, Room 01, **Erie**, PA 16501-1032
T: +1 814 4516319; Fax: +1 814 4516320; E-mail: mpeaster@eriecountygov.org
1876; Max C Peaster
14 000 vols; 10 curr per; 7 695 microforms
libr loan   34193

Humboldt County Law Library, Courthouse, 825 Fourth St, RM 812, **Eureka**, CA 95501
T: +1 707 4762356; Fax: +1 707 4456297; E-mail: redgar@co.humboldt.ca.us
1907; Richard Edgar
21 000 vols   34194

Snohomish County Law Library, M/S 703, Rm 139, 3000 Rockefeller Ave, **Everett**, WA 98201
T: +1 425 3883010; Fax: +1 425 3883020; URL: www.co.snohomish.wa.us
1932; Lettice Parker
20 000 vols   34195

US Air Force – Base Library/FL 4620, Two W Castle St, **Fairchild AFB**, WA 99011-8532
T: +1 509 2475556; Fax: +1 509 2473365
1950; Sherry Ann Hokanson
Military Hist & Science; Northwest
28 000 vols; 87 curr per; 10 000 microforms; 700 av-mat; 1 200 sound-rec; 33 digital data carriers
libr loan; OCLC   34196

Fairfax Law Library, 4110 Chain Bridge Rd, Rm 308, **Fairfax**, VA 22030
T: +1 703 2462174; Fax: +1 703 5910310; E-mail: liblawlibrary@fairfaxcounty.gov; URL: www.fairfaxcounty.gov/courts/lawlib/
1970; Anna Christine Hall
Virginia Law
36 000 vols; 25 curr per   34197

Solano County Law Library, Hall of Justice, 600 Union Ave, **Fairfield**, CA 94533
T: +1 707 4216520; Fax: +1 707 4216516; E-mail: mmoore@solanocounty.com
1911; Marianna G. Moore
California Statutes, Codes of Solano Co & Cities therein
24 000 vols; 99 curr per   34198

US Armed Forces – Office of the Army Surgeon General Medical Library, 5109 Leesburg Pike, Skyline 6, Rm 670, **Falls Church**, VA 22041-3258
T: +1 703 6818028; Fax: +1 703 6818034; URL: www.tricare.osd.mil/afml
1969; Diane Zehnpfennig
Hospital administration; Annual Rpt of the Surgeon General 1818-present
12 000 vols; 430 curr per   34199

US Courts Branch Library, 655 First Ave N, Ste 310, **Fargo**, ND 58102
T: +1 701 2977280; Fax: +1 701 2977285; E-mail: suzanne_morrison@ca8.uscourts.gov
1986; Suzanne Morrison
16 000 vols; 150 curr per
libr loan   34200

Missouri Department of Corrections, Farmington Correctional Center, 1012 W Columbia St, **Farmington**, MO 63640-2902
T: +1 573 2187100; Fax: +1 573 2187106
17 000 vols; 96 curr per   34201

Hancock County Law Library Association, 300 S Main St, 4th Flr, **Findlay**, OH 45840
T: +1 419 4247077; Fax: +1 419 4254136
1903; Deborah L. Ward
24 000 vols; 6 000 microforms; 10 digital data carriers   34202

Massachusetts Trial Court, Fitchburg Law Library, Superior Court House, 84 Elm St, **Fitchburg**, MA 01420-3296
T: +1 978 3456726; Fax: +1 978 3457334; E-mail: fll@ma.ultranet.com; URL: www.lawlib.state.ma.us
1871; Peter Anderegg
26 700 vols   34203

Coconino County Superior Court, Coconino County Law Library and Self-Help Center, 222 E Birch Ave, **Flagstaff**, AZ 86001
T: +1 520 7796535; Fax: +1 520 7796655; E-mail: cocolawlib@yahoo.com
Vicki Vega
15 000 vols; 154 curr per   34204

US Department of the Interior & the National Park Service, Carl Sandburg Home National Historic Site, Library, 1928 Little River Rd, **Flat Rock**, NC 28731
T: +1 828 6934178; Fax: +1 828 6934179; E-mail: CARL_Administration@nps.gov; URL: www.nps.gov/Carl
1972; Connie Hudson Backlund
American Lit, Carl Sandburg coll, National Park Service, Natural Hist
12 000 vols; 4 curr per; 40 000 govt docs; 50 maps; 550 av-mat; 2 006 sound-rec; 15 scrapbooks, 15 000 photos, 100 architectural drawings, 15 aerial photos   34205

Arizona Department of Corrections – Adult Institutions, Arizona State Prison Complex – Eyman Library, 4374 Butte Ave, **Florence**, AZ 85232; P.O. Box 3500, Florence, AZ 85232-3500
T: +1 520 8680201; Fax: +1 520 8688556
29 000 vols   34206

Arizona Department of Corrections – Adult Institutions, Arizona State Prison Complex – Eyman Library, 4374 Butte Ave, **Florence**, AZ 85232; P.O. Box 3500, Florence, AZ 85232-3500
T: +1 520 8680201; Fax: +1 520 8688556
I. Delfine
29 000 vols   34207

Fond Du Lac Circuit Court, Law Library, P.O. Box 1355, **Fond Du Lac**, WI 54935-1355
T: +1 920 9064748; Fax: +1 920 9299333
Carol Marx
13 000 vols   34208

Taycheedah Correctional Institution Library, 751 Hwy K, **Fond du Lac**, WI 54935-9099
T: +1 920 9293800; Fax: +1 920 9297899
1967; Mary Wood
Legal Coll – Women's Issues
14 000 vols; 45 curr per   34209

Georgia Department of Transportation, Research Library, 15 Kennedy Dr, **Forest Park**, GA 30297-2599
T: +1 404 3637540; Fax: +1 404 3637684
1974; Stardina Wyche
Govt Publs, Hazardous Waste, TRB Coll
30 000 vols; 30 curr per; 7 100 govt docs; 50 sound-rec; 25 digital data carriers
libr loan   34210

US Army – The Institute of Heraldry Library, 9325 Gunston Rd, Ste S116, **Fort Belvoir**, VA 22060-5579
T: +1 703 8064967; Fax: +1 703 8064964; URL: www.tioh.hqda.pentagon.mil
1961; Roy Ellis Cornwell
26 000 vols; 26 curr per   34211

US Army – Van Noy Library, Bldg 1024, Ste 101, 5966 12th St, **Fort Belvoir**, VA 22060-5554
T: +1 703 8060093; Fax: +1 703 8060091; E-mail: fbdmwr.library@conus.army.mil; URL: belvoirmwr.com/vannoy/index.html
1939; Donna Epps Ramsey
67 000 vols; 300 curr per; 1 058 av-mat   34212

US Army – Donovan Research Library, 6751 Constitution Loop, Rm 101, **Fort Benning**, GA 31905
T: +1 706 5455661; Fax: +1 706 5458590
1919; Erika Loze-Hudson
Military hist; Army Unit Hist, After Action Rpts, Rare Bks
58 000 vols; 84 curr per
libr loan   34213

US Army – Martin Army Community Hospital Professional Library, Bldg 9200, Rm 10 MCXB-CSL, **Fort Benning**, GA 31905-6100
T: +1 706 5442041; Fax: +1 706 5443532
Susan Waldrop
16 677 vols; 85 curr per   34214

US Army – Sayers Memorial Library, Bldg 93, **Fort Benning**, GA 31905
T: +1 706 5457141; Fax: +1 706 5456363; URL: www.benningmwr.com/library31905.htm
1942; Sherrie Floyd
Hist, Business, Political science, Music
55 000 vols; 227 curr per   34215

Western Hemisphere Institute for Security Cooperation, John B. Amos Library, Bldg 35, Rm 257, **Fort Benning**, GA 31905-6245
T: +1 706 5451247; Fax: +1 706 5451064; URL: www.benning.army.mil/whinsec
Yamill Collazo
20 000 vols; 75 curr per   34216

US Army – Mickelsen Library, Mickelsen Library, Bldg 2E Sheridan Rd, **Fort Bliss**, TX 79916-3802
T: +1 915 5683089; Fax: +1 915 5685754
1942; Kathryn P. Thomson
35 000 vols; 65 curr per; 100 000 govt docs; 16 000 microforms; 336 sound-rec; 2 digital data carriers; Military science, hist of battles
libr loan   34217

US Army – Othon O. Valent Learning Resources Center, 11291 SGT E Churchill St, **Fort Bliss**, TX 79918-8002; Commandant USASMA, ATTN: ATSS-SL, Fort Bliss, TX 79918-8002
T: +1 915 5688451; Fax: +1 915 5688484; E-mail: atss-sl@bliss.army.mil; URL: www.bliss.army.mil/usasma
1972; Angelica Garcia
Army Unit Hist
40 000 vols; 249 curr per
libr loan   34218

US Army – John L. Throckmorton Library, IMSE-BRG-MWR-L Bldg 1-3346, Randolph St, **Fort Bragg**, NC 28310-5000
T: +1 910 3962665; Fax: +1 910 9072274; URL: www.fortbraggmwr.com/library.php
1941; Philip Quinones
Military science, Current affairs, Int relations, Hist
125 000 vols; 200 curr per   34219

US Army – Marquat Memorial Library, Bank Hall, Bldg D-3915, 3004 Ardennes St, **Fort Bragg**, NC 28310-9610
T: +1 910 3965370; Fax: +1 910 4327788
1952; Margaret Harrison
Geopolitics, Int relations, Languages, Political science, Special forces & operations, Terrorism, Unconventional warfare
33 000 vols; 126 curr per
libr loan   34220

US Army – R. F. Sink Memorial Library, 38 Screaming Eagle Blvd, **Fort Campbell**, KY 42223-5342
T: +1 270 7981217; Fax: +1 270 7980369; URL: www.fortcampbellmwr.com/library
1941; James Moore
Military hist; Local hist on fiche, World War II; Official Records of the Civil War
88 000 vols; 315 curr per; 2 414 av-mat   34221

US Army – Grant Library, 4950 Flint St, **Fort Carson**, CO 80913-4105
T: +1 719 5262350; Fax: +1 719 5268139; E-mail: harrisr@carson.army.mil
1942; Rebecca Harris
Military Arts & Sciences, Colorado, World War II
52 000 vols; 250 curr per   34222

National Forest Service Library, 240 W Prospect Rd, **Fort Collins**, CO 80526
T: +1 970 4981207; Fax: +1 970 4981059; E-mail: fslibrary@fs.fed.us; URL: fsinfo.fs.fed.us/cgi-bin/gw/chameleon
1966; Carol A. Ayer
Natural resources, wildlife, recreation

50 000 vols; 200 curr per; 500 e-journals; 20 e-books; 5 000 mss; 1 000 diss/theses; 100 000 govt docs; 100 maps; 100 microforms; 200 av-mat; 30 digital data carriers; monographic serials, technical reports (50 000)
libr loan; OCLC, Fedlink   34223

US Centers for Disease Control & Prevention, Division of Vector-Borne Infectious Diseases Library, US Public Health Serv, Foothill Campus, P.O. Box 2087, *Fort Collins*, CO 80522-2087
T: +1 970 2216400; Fax: +1 970 2216476
Diana White
10 000 vols; 100 curr per   34224

US Forest Service, Rocky Mountain Research Station, 240 W Prospect, *Fort Collins*, CO 80526
T: +1 970 4981268; Fax: +1 970 4981059; E-mail: sdunphy@fs.fed.us
Carin Batt
Rpts, pioneer res 1910
20 000 vols; 300 curr per   34225

US Geological Survey – Biological Resources Division, Mid-Continent Ecological Science Center Library, 4512 McMurry Ave, *Fort Collins*, CO 80525-3400
T: +1 970 2269403; Fax: +1 970 2269230; E-mail: liz.Lucke@usgs.gov; URL: www.mesc.usgs.gov
Liz Lucke
Biology, Ecology, Wildlife, Fisheries
12 000 vols; 89 curr per; 2 digital data carriers   34226

US Army at Fort Dix, General Library, Bldg 5403, First & Delaware St, *Fort Dix*, NJ 08640-5332
T: +1 609 5625228; Fax: +1 609 5623554
1917; Wanda James
30 000 vols; 14 curr per   34227

US Army – Robert C. McEwen Library, 4300 Camp Hale Rd, *Fort Drum*, NY 13602-5284
T: +1 315 7726005; Fax: +1 315 7724912
1941; Marian O. Nance
Military, Job hunting, Self defense, New York State
55 000 vols; 200 curr per; 6 011 av-mat
libr loan   34228

US Army – Groninger Library, Army Transportation Ctr, Bldg 1313, *Fort Eustis*, VA 23604-5107
T: +1 757 8785017; Fax: +1 757 8781024; E-mail: levyc@eustis.emh1.army.mil
Claudia L. Levy
60 000 vols; 70 curr per
libr loan   34229

US Army – Transportation School Library, Bldg 705, Rm 36, *Fort Eustis*, VA 23604-5450
T: +1 757 8785563; Fax: +1 757 8786256; E-mail: renickt@eustis.army.mil; URL: www.lic.eustis.army.mil
1944; Tim Renick
32 450 vols; 550 curr per
libr loan   34230

US Army – Post Library, Bldg 4418, Llewellyn Ave, *Fort George G. Meade*, MD 20755-5068
T: +1 301 6773594; Fax: +1 301 6772694; URL: www.ftmeademwr.com/library.htm
Karen L. Hayward
Foreign language
27 000 vols; 144 curr per   34231

US Army – Eisenhower Army Medical Center, Health Sciences Library, DDEAMC, *Fort Gordon*, GA 30905-5650
T: +1 706 7876765; Fax: +1 706 7872327
Judy M. Krivanek
16 000 vols; 500 curr per   34232

Arizona Department of Corrections, Arizona State Prison Complex-Safford, Fort Grant Library, NW Sulphur Springs Valley, Graham County, N of Bonita at end of SR 266 Spur Rd, *Fort Grant*, AZ 85643; mail address: 896 S Cook Rd, Safford, AZ 85546
T: +1 928 8283393
Michael Kaliher

Addiction & Recovery Coll, Southwestern States History Coll, Hispanic Coll, Man-Woman Relationships Coll, African American Coll, Native American Coll
8 000 vols; 30 curr per   34233

Veterans Administration Center, Fort Harrison, *Fort Harrison*, MT 59636
T: +1 406 4426410 ext 7345; Fax: +1 406 4477948
1930; Charles Grasmick
17 000 vols   34234

US Army – Casey Memorial Library, 761 Tank Battalion, *Fort Hood*, TX 76544
T: +1 254 2870025; Fax: +1 254 2884029; E-mail: sheltonp@hood-emh3.army.mil; URL: www.hoodmwr.com/casey_library.htm
1942; Pamela Shelton
Military science, General reading, Grad & undergrad studies
90 000 vols; 285 curr per   34235

US Army – Army Intelligence Center & School Library, Bldg 52065, *Fort Huachuca*, AZ 85613-6000
T: +1 520 5334101; Fax: +1 520 5382119
1955; Chris Hurd
Military hist; Apache Indian Wars, Battle of Gettysburg
35 000 vols; 110 curr per
libr loan   34236

US Army – Fort Huachuca Library Branch, Bldg 52065 ATZS-ESD-L, *Fort Huachuca*, AZ 85613-6000
T: +1 520 5333041; Fax: +1 520 5382119
Chris Hurd
Military Affairs, Southwest Coll
65 000 vols; 400 curr per
libr loan   34237

US Army – Fort Irwin Post Library, National Training Ctr, Bldg 331, Second St & F Ave, *Fort Irwin*, CA 92310; P.O. Box 105091, Fort Irwin, CA 92310-5091
T: +1 760 3804337; Fax: +1 760 3805071; E-mail: irwin-dmwr-library@conus.army.mil; URL: www.fortirwinmwr.com
1981; Joan Leopold
Military science & hist; California
31 000 vols; 94 curr per; 859 av-mat   34238

US Army – Fort Jackson Main Post Library, Thomas Lee Hall Main Post Library, Bldg 4679, *Fort Jackson*, SC 29207
T: +1 803 7514816; Fax: +1 803 7511065
1946; John Anthony Vassallo
Military science, Hist, Business
70 000 vols; 50 curr per   34239

US Army – Armor School Library, Rockenbach Hall Bldg 2020, 159 Blackhorse Regt Ave, *Fort Knox*, KY 40121-4130
T: +1 502 6245449; Fax: +1 502 6243365; E-mail: knox.armor.library@knox.army.mil; URL: www.knox.army.mil
1941; William H. Hansen
Military hist, Civil War; Student Papers, After Action Rpts
25 000 vols; 301 curr per; 2 500 govt docs; 500 maps; 1 003 000 microforms; 600 av-mat; 250 digital data carriers
libr loan; FEDLINK   34240

US Army – Barr Memorial Library, 400 Quartermaster St, *Fort Knox*, KY 40121
T: +1 502 6241232; Fax: +1 502 6247528; E-mail: curtissc@ftknox-emk3.army.mil
Carmen Curtiss
Military science, Kentucky
84 000 vols; 250 curr per; 1 100 av-mat
libr loan; OCLC   34241

US Army – Patton Museum of Cavalry & Armor, Emert L. Davis Memorial Library, 4554 Fayette Ave, *Fort Knox*, KY 40121-0208
T: +1 502 6243812; Fax: +1 502 6242364; E-mail: museum@knox.army.mil
1975; Candace L. Fuller
Robert J Icks' Photo & Mss Coll on Armored Equipment, bks, maps, photogs;

General George S Patton, Jr Coll, photos
9 000 vols; 20 curr per   34242

Broward County Law Library, 1800 Broward County Judicial Complex, 201 SE Sixth St, *Fort Lauderdale*, FL 33301
T: +1 954 8316226
1956; Jeanne Underhill
50 000 vols; online subscriptions and CDs available in library only, relating to legal research
AALL   34243

United States Army – Combined Arms Research Library, US Army Command & General Staff College, Eisenhower Hall, 250 Gibbon Ave, *Fort Leavenworth*, KS 66027-2314
T: +1 913 7583033; Fax: +1 913 7583014; URL: cgsc.leavenworth.army.mil/carl
1882; Edwin B. Burgess
Combined Arms & Fort Leavenworth Archs
285 000 vols; 642 curr per; 255 digital data carriers; 342 226 docs, 1 444 885 microforms
libr loan; ALA, SLA   34244

US Army – US Disciplinary Barracks Library, 300 McPherson Ave, *Fort Leavenworth*, KS 66027-7140
T: +1 913 6844313; Fax: +1 913 6517108
Sally Quinn
22 500 vols; 84 curr per   34245

US Army – The Army Logistics Library, US Army Logistics Mgmt College, Bldg 12500, 2401 Quarters Rd, *Fort Lee*, VA 23801-1705
T: +1 804 7654176; Fax: +1 804 7654660; E-mail: armyloglib@lee.army.mil; URL: www.almc.army.mil
1971; Tim Renick
US Govnt Publs
40 000 vols; 200 curr per
libr loan   34246

US Army – Bruce C. Clarke Community Library, 597 Manscen Loop, Ste 100, *Fort Leonard Wood*, MO 65473-8928
T: +1 573 5634113; Fax: +1 573 5634118; URL: www.wood.army.mil/isd/comlib.htm; www.wood.army.mil/ttic
1941; Joyce Waybright
Military art & science, Business, Foreign language; Children's Libr
68 000 vols; 120 curr per   34247

US Army – Manscen Academic Library, Bldg 3202, 597 Manscen Loop, Ste 200, *Fort Leonard Wood*, MO 65473-8928
T: +1 573 5634109; Fax: +1 573 5634118; URL: www.library.wood.army.mil
1935; Claretta Crawford
Military engineering, Miltary hist; Civil War Era (Rare Bk Coll)
64 000 vols; 45 curr per   34248

US Army – Fort Lewis Library System, Bldg 2109, *Fort Lewis*, WA 98433-9500; Box 339500, Fort Lewis, WA 98433-9500
T: +1 253 9661300; Fax: +1 253 9673922; E-mail: lewisdmwrflls@conus.army.mil; URL: www.lewis.army.mil/dpca/library
1942; Bonnie Tucker
German, Korean, Spanish Lang, Military Science
130 000 vols; 100 curr per   34249

Iowa State Penitentiary, Library, Three John Bennett Dr, *Fort Madison*, IA 52627
T: +1 319 3725432; Fax: +1 319 3726967
Bob Ensminger
32 000 vols; 10 curr per   34250

Iowa State Penitentiary – John Bennett Correctional Center Library, Three John Bennett Dr, *Fort Madison*, IA 52627
T: +1 319 3725432
Mike Brixius
12 000 vols; 19 curr per   34251

US Army – Fort McPherson Library, 1794 Walker Ave SW, *Fort McPherson*, GA 30330-1013
T: +1 404 4642665; Fax: +1 404 4643801; E-mail: blackburnc@forscom.army.mil
1941; Clayton E. Blackburn
Military science; Dept of Army Publs, Dept of Defense Docs & Publs
42 000 vols; 300 curr per
libr loan   34252

US Army – Communications-Electronics Command R&D Technical Library, AMSEL-IM-BM-I-L-R, Bldg 2700, *Fort Monmouth*, NJ 07703
T: +1 723 4272235; Fax: +1 723 4271551; E-mail: amsel-im-bm-i-l-r@mail1.monmouth.army.mil
1941; Henry McAteer
43 340 vols; 300 curr per   34253

US Army – Van Deusen Post Library, Bldg 502, *Fort Monmouth*, NJ 07703-5117
T: +1 732 5323172; Fax: +1 732 5324766
1917; Heather Goyette
Military art, Military science
30 000 vols; 98 curr per
libr loan   34254

US Army – Fort Monroe General Library, Bldg 7, *Fort Monroe*, VA 23651-5123
T: +1 757 7272909; Fax: +1 757 7273589; E-mail: abellr@monroe.army.mil
1824; Raymond Abell
32 560 vols; 109 curr per
libr loan   34255

US Army – Fort Myer Post Library, 210 McNair Rd, Bldg 469, *Fort Myer*, VA 22211-1101
T: +1 703 6963555; Fax: +1 703 6968587; URL: 69.63.217.2/f10062staff/opac/index.asp; www.fmmcmwr.com/librarymyer.htm
Cynthia D. Earman
Popular Coll
45 000 vols; 250 curr per   34256

Lee County Law Library, Lynn Gerald Law Library, 1700 Monroe St, Lee County Justice Ctr, *Fort Myers*, FL 33901
T: +1 941 3352230; Fax: +1 941 3352598
1959; Virginia Groth
17 500 vols   34257

Saint Lucie County Law Library, 218 S Second St, Rm 102, *Fort Pierce*, FL 34950
T: +1 561 4622370; Fax: +1 561 4622145
Delana Cook, Melene Phillips
15 000 vols   34258

US Army – Allen Memorial Library, Bldg 660, 7460 Colorado Ave, *Fort Polk*, LA 71459
T: +1 337 5312665; Fax: +1 337 5316687
Kelly Herbert
Military hist, Louisiana
70 000 vols; 260 curr per   34259

US Army – Fort Riley Post Library, Bldg 5306, Hood Dr, *Fort Riley*, KS 66442-6416
T: +1 785 2395305; Fax: +1 785 2394422
Terri Seaman
Military hist
22 000 vols; 60 curr per; 320 av-mat   34260

US Army – Aeromedical Research Laboratory Science Support Center Library, P.O. Box 620577, *Fort Rucker*, AL 36362-0577
T: +1 334 2556907; Fax: +1 334 2556067
1963; Diana L. Hemphill
15 000 vols; 170 curr per   34261

US Army – Aviation Center Library, Bldg 212, *Fort Rucker*, AL 36362
T: +1 334 2553695; Fax: +1 334 2551567
1954; Leslie R. Waltman Jr
German (Language)
65 000 vols; 112 curr per   34262

US Army – Aviation Technical Library, Bldg 9204, Fifth Ave, *Fort Rucker*, AL 36362
T: +1 334 2553177
1955; Jill Redington
DTIC Tech Rpts, Army Regulations, FFA Regulations
41 000 vols; 431 curr per   34263

US Army – Medical Library, USAAMC, Lyster US Army Community Hospital, Aeromedical Ctr, Bldg 301, *Fort Rucker*, AL 36362-5333
T: +1 334 2557350; Fax: +1 334 2057714; E-mail: prott1@ns.awanet.com; URL: www.rucker.amedd.army.mil/lindex.htm
Mary Fran Prottsman
12 674 vols; 148 curr per   34264

US Army – Brooke Army Medical Center Library, 3851 Roger Brooke Dr, Bldg 3600, Rm 330-19, Medical Library MCHE-CSL, *Fort Sam Houston*, TX 78234-6200
T: +1 210 9161119; Fax: +1 210 9165709; E-mail: medicallibrary.bamc@cen.amedd.army.mil; URL: www.gprmc.amedd.army.mil
1914; Beverly Rakowitz
19 000 vols; 656 curr per
libr loan   34265

US Army – Fort Sam Houston Library, 2601 Harney Rd, Ste 29, *Fort Sam Houston*, TX 78234-5029
T: +1 210 2214702; Fax: +1 210 2275921
1918; Alfonso Butcher
98 000 vols; 245 curr per; 10 000 microforms; 826 av-mat; 3 120 sound-rec; 15 digital data carriers
libr loan   34266

US Army – Stimson Library, Medical Department Center & School, 2250 Stanley Rd, Bldg 2840, Ste 106, *Fort Sam Houston*, TX 78234-6160
T: +1 210 2216230; Fax: +1 210 2218264; E-mail: stimson.library@amedd.army.mil; URL: digitallib.amedd.army.mil; ameddlib.amedd.army.mil; www.cs.amedd.army.mill/stimlib
1932; Norma L. Sellers
60 000 vols; 600 curr per   34267

US Army – Fort Shafter Library, Bldg 650, *Fort Shafter*, HI 96858-5005
T: +1 808 4389521; Fax: +1 808 4383100
1943; Janet Howard
Hawaii
18 000 vols; 86 curr per   34268

US Army – Morris Swett Technical Library, Snow Hall, Bldg 730, *Fort Sill*, OK 73503-0312
T: +1 580 4424525; Fax: +1 580 4427300
1911; Jo Ann Knight
Weapons, Ballistics, Military hist & science; Janes Series; Military Per Analytical Index File, VF; Unit Histories-Field Artillery; Rare Bks Coll; Special Bibliogr
79 000 vols; 65 curr per   34269

US Army – Nye Library, 2404 NW Randolph Rd, *Fort Sill*, OK 73503-9022
T: +1 580 4425111; Fax: +1 580 4427347
1953; Joan Ransom
Adult education, Consumer education, Family life, Military science; Bks for College Libraries, microfiche; College Cat, microfiche; Webster Univ Deposit Coll
83 000 vols; 77 curr per; 550 av-mat   34270

Sebastian County Law Library, 623 Garrison Ave, *Fort Smith*, AR 72901
T: +1 501 7834730
Jeri Kizer
15 000 vols   34271

US Army – Fort Stewart Main Post Library, Bldg 411, 316 Lindquist Rd, *Fort Stewart*, GA 31314-5126
T: +1 912 7672260;
Fax: +1 912 7673794; URL: www.stewlib3.stewart.army.mil
1942; Marilyn Mancuso
Military hist; Bks for College Libs, Libr of American Civilization, Libr of English Lit
68 000 vols; 250 curr per
libr loan   34272

US Army – Fort Story Library, Solomons Rd, Bldg T-530, *Fort Story*, VA 23459-5067
T: +1 757 4227548; Fax: +1 757 4227773; E-mail: fashionv@eustis.army.mil
Valerie D. Fashion
34 000 vols; 60 curr per   34273

US Army – Post Library, Santiago Ave, Bldg 3700, *Fort Wainwright*, AK 99703-6600; mail address: 1060 Gaffney Rd, Fort Wainwright, AK 99703-6600
T: +1 907 3532642; Fax: +1 907 3532609; E-mail: ftwainwrightlibrary@wainwright.army.mil
1951; Alfred L. Preston
Alaska
41 000 vols; 200 curr per; 3 087 av-mat; 4 843 sound-rec; 1 digital data carriers
libr loan   34274

Tarrant County Law Library, 100 W Weatherford, Rm 420, *Fort Worth*, TX 76196-0800
T: +1 817 8841481; Fax: +1 817 8841509; URL: www.tarrantcounty.com/law_library/
1942; Sharon Wayland
55 000 vols; 270 curr per   34275

US Army – Corps of Engineers, Fort Worth District Technical Library, 819 Taylor St, P.O. Box 17300, *Fort Worth*, TX 76102-0300
T: +1 817 8861326; Fax: +1 817 9782472
1950; Carolyn Solomon
Civil & Military Design Rpts; Law (Federal, Texas, Louisiana & New Mexico); State of Texas Water Resources Rpts
35 000 vols; 500 curr per
libr loan   34276

US Navy – Naval Air Station Library, 1802 Doolittle Ave, NAS Fort Worth JRB, *Fort Worth*, TX 76127
T: +1 817 7827735; Fax: +1 817 7827219
Patricia Elaine Tellman
Military hist
31 000 vols; 15 curr per   34277

Fox Lake Correctional Institution Library, PO Box 147, *Fox Lake*, WI 53933-0147
T: +1 920 9283151; Fax: +1 920 9286229
1962; Bob Zabkowicz
15 000 vols; 50 curr per   34278

State Correctional Institution, Frackville Library, 1111 Altamont Blvd, *Frackville*, PA 17931
T: +1 570 8744516; Fax: +1 570 7942013
Suzanne Domalakes
12 000 vols; 40 curr per   34279

Kentucky Department for Environmental Protection, EPIC Library, 14 Reilly Rd, *Frankfort*, KY 40601
T: +1 502 5642150 ext 134; Fax: +1 502 5644245; E-mail: snodgrass-b@nrdep.nr.state.ky.us; URL: www.nr.state.ky.us
Billie Ann Snodgrass
State of Kentucky Environmental Rpts, EPA
16 000 vols; 250 curr per; 2 digital data carriers   34280

State Law Library, 700 Capital Ave, Ste 200, *Frankfort*, KY 40601-3489
T: +1 502 5644848; Fax: +1 502 5645041; E-mail: statelawlibrary@kycourts.net; URL: www.kycourts.net
1954; Jennifer Frazier
Anglo-American law; Kentucky Law Coll
134 000 vols; 304 e-journals   34281

Venango County Law Library, Venango County Court House, Liberty St, P.O. Box 831, *Franklin*, PA 16323-0831
T: +1 814 4329500 ext 243; Fax: +1 814 4329615
Sandy Baker
Law cases from 25 reporter systs
22 000 vols; 60 curr per   34282

Frederick County Law Library, Courthouse, 100 W Patrick St, *Frederick*, MD 21701
T: +1 301 6941993
Jacob Kiser, Betty Tucker
27 000 vols   34283

US Army – Fort Detrick Post Library, Fort Detrick, 1520 Freedman Dr, *Frederick*, MD 21702
T: +1 301 6197519; Fax: +1 301 6192884
33 000 vols; 90 curr per   34284

US Army – Medical Research Institute of Infectious Diseases Library, Fort Detrick, 1425 Porter, *Frederick*, MD 21702-5011
T: +1 301 6192717; Fax: +1 301 6196059
Denise Lupp
10 000 vols; 220 curr per   34285

Michigan Department of Corrections, Saginaw Correctional Facility Library, 9625 Pierce Rd, *Freeland*, MI 48623
T: +1 989 6959880; Fax: +1 989 6956345
Ervin Bell
20 000 vols; 50 curr per   34286

California Court of Appeal Fifth Appellate District Library, 2424 Ventura St, *Fresno*, CA 93721
T: +1 559 4455686; Fax: +1 559 4456684
Tara Crabtree
25 000 vols; 250 curr per   34287

Fresno County Law Library, Fresno County Courthouse, Rm 600, 1100 Van Ness Ave, *Fresno*, CA 93721-2069
T: +1 559 2372227; Fax: +1 559 4424960; URL: www.fresnolawlibrary.org
1891; Sharon E. Borbon
Philosophy; Frank E Wells Memorial Coll
45 000 vols; 233 curr per   34288

Fresno County Office of Education, School Library Media Services, 1111 Van Ness Ave, *Fresno*, CA 93721
T: +1 559 2653091; Fax: +1 559 2373525; E-mail: jmisakian@fcoe.k12.ca.us; URL: www.fcoe.k12.ca.us/eduscrvc/librarymedia.htm
1945; Jo Ellen Misakian
Adult Basic Education-English as a Second Language; Best of the Best, Children's Coll; Instructional Mat Display; Lit Sets; Microcomputer/Software Evaluation Coll; Publishers' Samples; Teachers Professional Libr
12 000 vols; 10 curr per
libr loan; ALA, AASL, CSLA   34289

United States Courts Library, 2500 Tulare St, Ste 2401, *Fresno*, CA 93721
T: +1 559 4995615
Martin Schwartz
17 000 vols   34290

Florida Department of Agriculture & Consumer Services, Division of Plant Industry Library, 1911 SW 34th St, *Gainesville*, FL 32608; P.O. Box 147100, Gainesville, FL 32614-7100
T: +1 352 3723505 ext 131; Fax: +1 352 9552301; E-mail: dpilib@doacs.state.fl.us; URL: www.neslin.org/dpi
1915; Beverly Pope
Entomology
17 000 vols; 590 curr per   34291

National Institute of Standards & Technology, Information Services Division, Research Library & Information Group, Bureau Dr, Stop 2500, *Gaithersburg*, MD 20899-2500
T: +1 301 9752784; Fax: +1 301 8698071; E-mail: library@nist.gov; URL: nvl.nist.gov
1901; Rosa Liu
Artifacts of the Nat Bureau of Standards (Hist Museum Coll), Nat Bureau of Standards Personalities, Old & Rare 17th & 18th C Scientific Meteorology Treaties (Hist Coll), Significant Compilations of Atomic & Molecular Properties, Chemical Kinetics (Colloid & Surface)
200 000 vols; 1 500 curr per; 59 000 microforms; 12 digital data carriers
libr loan   34292

Hill Correctional Center Library, 600 Linwood Rd, P.O. Box 1327, *Galesburg*, IL 61401
T: +1 309 3434212 ext 360; Fax: +1 309 3434212 ext 123
1986; Camilla A. Willmart
Law Libr Coll (8500 vol)
12 000 vols; 10 curr per   34293

US Army – Galveston District Corps of Engineers Library, 2000 Fort Point Rd, Rm 308, P.O. Box 1229, *Galveston*, TX 77553
T: +1 409 7663196; Fax: +1 409 7663905; E-mail: clark.bartee@usace.army.mil; URL: www.swg.usace.army.mil/library.htm
Clark Bartee
Dredging, Ecology, Flood control, Law, Navigation, Shoreline studies, Soils, Water resources; Archaeological Rpts, Civil Works Rpts, Congressional Docs, Environmental Impact Statements & Assessments
10 620 vols; 136 curr per   34294

US National Park Service, Hubbell Trading Post National Historic Site Library, P.O. Box 150, *Ganado*, AZ 86505-0150
T: +1 928 7553475; Fax: +1 520 7553405; URL: www.nps.gov/hutr
Yolanda Lincoln
15 000 vols; 40 av-mat; 300 sound-rec   34295

Georgia Department of Corrections, Office of Library Services, Coastal State Prison, 200 Gulfstream Rd, *Garden City*, GA 31418; P.O. Box 7150, Garden City, GA 31418-7150
T: +1 912 9656330; Fax: +1 912 9666799; URL: www.dcor.state.ga.us
Sadie Jenkins
15 000 vols; 41 curr per   34296

Lake County Central Law Library, Indiana University Northwest, 3400 Broadway, *Gary*, IN 46408
T: +1 219 9806946; Fax: +1 219 9806558
Timothy Sutherland
Indiana Law, Pre-National Reporter Coll of State Rpts
33 000 vols   34297

Adams County Law Library, Court House, 111-117 Baltimore St, *Gettysburg*, PA 17325
T: +1 717 3379812; Fax: +1 717 3341625
Cecelia Brown
50 000 vols; 10 curr per   34298

Virginia Department of Environmental Quality Library, 4949 A Cox Rd, *Glen Allen*, VA 23060
T: +1 804 5275020; Fax: +1 804 5275106
Zelda Hardy
Agency publs
15 000 vols; 50 curr per; 14 av-mat; 25 slides   34299

Georgia Department of Corrections, Office of Library Services, Smith State Prison, 9676 Hwy 301 N, *Glennville*, GA 30427; P.O. Box 726, Glennville, GA 30427-0726
T: +1 912 6545090; Fax: +1 912 6545131; URL: www.dcor.state.ga.us
Julia Steele
11 000 vols; 40 curr per; 10 av-mat   34300

Gila County Law Library, 1400 E Ash St, *Globe*, AZ 85501
T: +1 928 4253231 ext 275; Fax: +1 928 4250319; E-mail: bkandt@gilacold.lib.az.us
1908; Rosemary Bigando
30 000 vols; 20 curr per   34301

Federal Law Enforcement Training Center Library, Bldg 262, 1131 Chapel Crossing Rd, *Glynco*, GA 31524
T: +1 912 2672320
1975
40 000 vols   34302

US Air Force – FL 3030, Base Library, Bldg 712, 17 MSG/SVMG, 271 Fort Phantom Hill Ave, *Goodfellow AFB*, TX 76908-4711
T: +1 325 6543045; Fax: +1 325 6544731
1942; Pat Rossi
28 000 vols; 129 curr per   34303

**Arizona Department of Corrections –
Adult Institutions**, Arizona State Prison
Complex – Perryville Library, 2014 N
Citrus Rd, *Goodyear*, AZ 85338; P.O.
Box 3000, Goodyear, AZ 85338-3000
T: +1 623 8530304; Fax: +1 623
8530304
Joyce Kelly
16 000 vols   34304

US Navy – Naval Weapons Stations,
CHASN Library, Bldg 732, 2316
Red Bank Rd, *Goose Creek*,
SC 29445-8601
T: +1 843 7647900; Fax: +1 843
7644054; URL: www.nwschs.navy.mil
1966; Vivian Skipworth
Bks on cassettes, Children's films, Educ
films, Large print, Self development; CD-
ROM Coll, Help Wanted USA Claasifies
Ads (microfiche)
27 000 vols   34305

NYS Supreme Court, Law Library of
Orange County, Orange County Govt Ctr,
255-275 Main St, *Goshen*, NY 10924
T: +1 845 2913138; Fax: +1 845
2912595
Margaret S. O'Loughlin
24 000 vols; 10 curr per; 14 microforms;
16 digital data carriers   34306

Gouverneur Correctional Facility,
General Library, Scott Settlement Rd,
Gouverneur, NY 13642
T: +1 315 2877351; Fax: +1 315
2877351
Lynne H. Matott
20 000 vols   34307

Gowanda Correctional Facility Library,
PO Box 350, *Gowanda*, NY 14070-0350
T: +1 716 5320177; Fax: +1 716
5320177
Corinne Leone
24 000 vols; 80 curr per   34308

**US Air Force – Grand Forks Air
Force Base Library**, 319 SVS/SVMG,
511 Holzapple St, Bldg 201, *Grand
Forks AFB*, ND 58205
T: +1 701 7473046; Fax: +1 701
7474584; URL: www.gf-services.com/
new/library-events.htm
Arlene Ott
Air War College; Air Univ; Quality
Improvement; Chief of Staff Reading
List
40 000 vols; 61 curr per; 8 000 e-books;
1 600 av-mat; 2 000 sound-rec; 1 068
digital data carriers
libr loan; ALA   34309

US Department of Justice, US
Attorney's Office Library, 330 Ionia, Ste
501, *Grand Rapids*, MI 49503; P.O.
Box 208, Grand Rapids, MI 49501-0208
T: +1 616 4562462; Fax: +1
616 4562408; E-mail:
june.vanwingen@uddoj.gov; URL:
www.usdoj.gov/usao/miw
June L. Van Wingen
10 000 vols; 20 curr per   34310

Oklahoma State Reformatory Library,
1700 E First St, *Granite*, OK 73547;
P.O. Box 514, Granite, OK 73547-0514
T: +1 580 4803700; Fax: +1 580
4803989
John Slater
12 000 vols; 50 curr per   34311

Josephine County Law Library, County
Courthouse, 500 NW Sixth St, *Grants
Pass*, OR 97526
T: +1 541 4745488; Fax: +1 541
4745223
Beecher Ellison
15 000 vols   34312

US Navy – Medical Library, 3001A
Sixth St, Naval Hospital, *Great Lakes*,
IL 60088-5230
T: +1 847 6886969; Fax: +1
847 6886701; E-mail:
grl1cjs@grl10.med.navy.mil
1945; Carol Struck
13 000 vols; 240 curr per
libr loan   34313

**US Navy – Naval Training Center
Library**, Bldg 160, 2601E Paul Jones
St, *Great Lakes*, IL 60088-2845
T: +1 847 6884617; Fax: +1 847
6883602; E-mail: library@mwrgl.com;
URL: mwrgl.cnic.navy.mil/recreation/library/
library.htm
1912; Kevin R. Jones
22 000 vols; 95 curr per; 450 av-mat
                                     34314

Weld County District Court, Law
Library, 915 Tenth St, P.O. Box C,
Greeley, CO 80632-0138
T: +1 970 3517300 ext 4515; Fax: +1
970 3564356
1918; Shairan Whitman
18 000 vols   34315

**Green Bay Correctional Institution
Library**, PO Box 19033, *Green Bay*,
WI 54307-9033
T: +1 920 4324877; Fax: +1 920
4325388
Mark Kulieke
8 000 vols; 50 curr per   34316

NASA Goddard Space Flight Center,
The Goddard Library, Homer E. Newell
Memorial Library, Library, Bldg 21, Code
272, *Greenbelt*, MD 20771
T: +1 301 2867217; Fax: +1
301 2861755; E-mail:
library@listserv.gsfc.nasa.gov; URL:
library.gsfc.nasa.gov
1959; Robin Miller Dixon
104 000 vols; 2 371 curr per   34317

Putnamville Correctional Facility,
Learning Resource Center, 1946 W US
40, *Greencastle*, IN 46135-9275
T: +1 765 6538441; Fax: +1 765
6534157
Jimmie Bowman
11 000 vols; 80 curr per   34318

Westmoreland County Law Library,
Two N Main St, Ste 202, *Greensburg*,
PA 15601
T: +1 724 8303266;
Fax: +1 724 8303042; URL:
www.co.westmoreland.pa.us
Betty Ward
24 000 vols; 35 curr per; 6 digital data
carriers
libr loan; AALL   34319

Greenville Law Library, 124 W Fifth St,
Greenville, OH 45331
T: +1 937 5479741; Fax: +1
937 5485730; E-mail:
darkelawlibrary@who.rr.com; URL:
www.npo.countystart.com/~lawlibrary
Eileen Litchfield
8 500 vols   34320

**US Coast Guard Research &
Development Center**, Technical
Library, 1082 Shennecossett Rd, *Groton*,
CT 06340-6096
T: +1 860 4412648; Fax: +1 860
4412792; E-mail: mkendall@rdc.uscg.mil;
URL: www.rdc.uscg.mil
Sandra A. Brown
Ocean engineering; Coast Guard Rpts
10 000 vols; 100 curr per; 20 digital data
carriers   34321

US Navy – Base Library, Naval
Submarine Base New London, Bldg 164,
P.O. Box 15, *Groton*, CT 06349-5014
T: +1 860 6943723; Fax: +1
860 6945056; E-mail:
parkerm@subasenlon.navy.mil
1942; Maree Parker
20 000 vols; 70 curr per   34322

**Georgia Department of Corrections,
Office of Library Services**, Augusta
State Medical Prison, 3001 Gordon Hwy,
Grovetown, GA 30813
T: +1 706 8554882; Fax: +1 706
8554924; URL: www.dcor.state.ga.us
Saundra Hood
14 000 vols; 27 curr per   34323

Harrison County Law Library, 1801
23rd Ave, *Gulfport*, MS 39501
T: +1 228 8654068; Fax: +1 228
8654067
1967; Fran J. Perry
24 000 vols   34324

US Navy, Naval Construction Battalion
Center Base Library, MWR CBC Libr
Code 300, Rm 201, 1800 Dong Xoai
Ave, *Gulfport*, MS 39501
T: +1 228 8712409; Fax: +1 228
8712539
1966; Tina Dahl
12 600 vols; 45 curr per   34325

Utah Department of Corrections,
Central Utah Correctional Facility Library,
PO Box 898, *Gunnison*, UT 84634-0898
T: +1 435 5286000; Fax: +1 435
5286234
Alvin Hatch
20 000 vols; 10 curr per   34326

Bergen County Law Library, Justice
Center Bldg, Ten Main St, Rm 123,
Hackensack, NJ 07601
T: +1 201 5272274; Fax: +1 201
3711121
Henry Gozdz
36 000 vols; 10 curr per   34327

**Maryland Correctional Institution-
Hagestown Library**, 18601 Roxbury
Rd, *Hagerstown*, MD 21746
T: +1 240 4201000; Fax: +1 301
7978448
Ken R. Baker
8 000 vols; 39 curr per   34328

**Roxbury Correctional Institution
Library**, 18701 Roxbury Rd,
Hagerstown, MD 21746
T: +1 240 4203000; Fax: +1 301
7332672
Addis Kambule
10 000 vols; 39 curr per   34329

**Washington County Board of
Education**, Instructional Resource
Library, 820 Commonwealth Ave,
Hagerstown, MD 21740
T: +1 301 7662941; Fax: +1 301
7662942; E-mail: rfisher@umd5.umd.edu;
URL: www.wcboe.k12.md.us
1967; Roseanne W. Fisher
India, artifacts, slides, flm
20 000 vols; 120 curr per   34330

Washington County Law Library,
Circuit Court House, 95 W Washington
St, *Hagerstown*, MD 21740
T: +1 240 3132570; URL:
www.courts.state.md.us/
20 000 vols; 10 curr per   34331

NASA, Langley Research Center Technical
Library, 2 W Durand St, MS 185,
Hampton, VA 23681-2199
T: +1 757 8642356; Fax: +1
757 8642375; E-mail: tech-
library@larc.nasa.gov; URL:
library.larc.nasa.gov
1920; Mary McCaskill
80 000 vols; 800 curr per   34332

Kings County Law Library, Kings
County Government Ctr, 1400 W Lacey
Blvd, *Hanford*, CA 93230
T: +1 559 5823211 ext 4430
1898; Isabelle Baltazar
13 000 vols; 16 curr per   34333

Hanover Juvenile Correctional Center,
Dept of Correctional Education Library,
Three Broad Neck Rd, P.O. Box 507,
Hanover, VA 23069-0527
T: +1 804 5375602 ext 405; Fax: +1
804 5375491
James Adams
64 000 vols; 30 curr per   34334

**US Army – Cold Regions Research
and Engineering Laboratory Library**,
72 Lyme Rd, *Hanover*, NH 03755-1290
T: +1 603 6464221; Fax: +1
603 6464712; E-mail:
nliston@crrel.usace.army.mil; URL:
www.crrel.usace.army.mil/library
1952; Nancy C. Liston
Arctic science & engineering bibliography
21 000 vols; 400 curr per; 1 500
diss/theses; 25 000 govt docs; 30 000
maps; 150 000 microforms; 100 digital
data carriers
libr loan   34335

**US Air Force – Air Force Research
Laboratory Library**, Five Wright St,
Hanscom AFB, MA 01731-3004
T: +1 781 3774742; Fax: +1 781
3774896
1945; John F. Griffin
Air Flight Coll, Early Ballooning &
Aeronautics, Oriental Science Libr,
Science Mss of Lords Rayleigh III &
IV, Geophysics of Space Coll
253 000 vols; 425 curr per
libr loan   34336

**US Air Force – Hanscom Air Force
Base Library**, 66 SVS/SVMG, 98
Barksdale St, Bldg 1530, *Hanscom
AFB*, MA 01731-1807
T: +1 781 3772177;
Fax: +1 781 3774482; URL:
www.hanscomservices.com/library.html
1952; Teresa Hathaway
Military hist; Air War Coll, Project Warrior
21 000 vols; 110 curr per   34337

National Park Service, Harpers Ferry
Center Library, 68 Mather Pl, *Harpers
Ferry*, WV 25425; P.O. Box 50, Harpers
Ferry, WV 25425-0050
T: +1 304 5356262; Fax: +1 304
5356492; URL: www.nps.gov/hfc
1973; David Nathanson
Cultural Resources Bibliogr Coll, Evelyn
Wallace Domestic Economy Coll, Harold
L Peterson Military Art & Science Coll,
Nat Park Service Hist Coll, Rare Bk Coll,
Vera Craig Pictorial Arch of Interiors
36 000 vols; 200 curr per
libr loan   34338

Dauphin County Law Library, Dauphin
County Courthouse, Front & Market Sts,
4th Flr, *Harrisburg*, PA 17101
T: +1 717 7806605; Fax: +1 717
7806481; URL: www.dauphinc.org
1865; Tracey McCall
36 000 vols; 9 curr per; 6 microforms; 4
digital data carriers
AALL   34339

**Pennsylvania Department of
Conservation & Natural Resources**,
Bureau of Topographic & Geologic
Survey Library, P.O. Box 8453,
Harrisburg, PA 17105-8453
T: +1 717 7838077; E-mail:
rkeen@state.pa.us
1850; Richard C. Keen
Earth science, Geology, Geography;
Maps (US Geology Survey Seven
& One Half Minute Maps), Aerial
Photography
10 000 vols; 70 curr per   34340

**Pennsylvania Legislative Reference
Bureau Library**, Main Capitol, Rm 641,
Harrisburg, PA 17120-0033
T: +1 717 7874816; Fax: +1
717 7832396; E-mail:
lrblibrary@legis.state.pa.us
1909; Susan K. Zavacky
12 000 vols; 25 curr per   34341

**Pennsylvania Office of Attorney
General**, Law Library, 1525 Strawberry
Sq, *Harrisburg*, PA 17120
T: +1 717 7873176; Fax: +1 717
7724526
1873; Bob Robitaille
26 000 vols   34342

Senate Library of Pennsylvania, Main
Capitol Bldg, Rm 157, *Harrisburg*,
PA 17120-0030
T: +1 717 7876120; Fax: +1
717 7835021; E-mail:
eandrews@os.pasen.gov; URL:
www.pasen.gov/senate_library.html
Evelyn F. Andrews
Histories of Legislation for Senate
& House of Representatives of
Pennsylvania, Legislative Journals,
Transcripts of Hearings
10 000 vols; 15 curr per   34343

Connecticut Judicial Department,
Hartford Law Library, Superior Court,
Rm 201, 95 Washington St, *Hartford*,
CT 06106
T: +1 860 5482866; Fax: +1 860
5482868; URL: www.jud.ct.gov
Sandra Phillips
40 000 vols   34344

Connecticut Legislative Library, Legislative Office Bldg, Rm 5400, ***Hartford***, CT 06106-1591
T: +1 860 2408888; Fax: +1 860 2408881; E-mail: legislative.library@cga.ct.gov
Susan M. Southworth
10 000 vols 34345

Haynesville Correctional Center, Department of Correctional Education Library, 241 Barnfield Rd, ***Haynesville***, VA 22472; P.O. Box 129, Haynesville, VA 22472-0129
T: +1 804 3333577; Fax: +1 804 3331295
Edie Hudgins
9 000 vols; 36 curr per 34346

Bernard E. Witkin Alameda County Law Library – South County Branch, 224 W Winton Ave, Rm 162, ***Hayward***, CA 94544
T: +1 510 6705230; Fax: +1 510 6705292
1967; Mrs Cossette T. Sun
Local Municipal Codes & City Ordinances
30 000 vols 34347

Montana Legislative Reference Center, State Capitol Rm 110, 1301 E. Sixth Ave, ***Helena***, MT 59620-1706; P.O. Box 201706, Helena, MT 59620-1706
T: +1 406 4444043; Fax: +1 406 4442588; E-mail: leglib@state.mt.us; URL: leg.state.mt.us
1975; Lisa Mecklenberg Jackson
Montana Legislature Interim Study Arch
8 200 vols; 507 curr per; 40 maps; 310 microforms; 7 627 av-mat; 22 sound-rec; 20 digital data carriers; 10 507 clippings
libr loan 34348

Montana State Department of Natural Resources & Conservation, Research & Information Center, 48 N Last Chance Gulch, P.O. Box 201601, ***Helena***, MT 59620-1601
T: +1 406 4446603; Fax: +1 406 4445918
Bud Clinch
Energy Planning & Development; Environmental Impact Statements; Montana Department of Natural Resources Publications & Water Planning, (incl Columbia, Missouri & Yellowstone River basin studies & general resource planning)
9 000 vols; 91 curr per 34349

State Law Library of Montana, 215 N Sanders, ***Helena***, MT 59601-4522; P.O. Box 203004, Helena, MT 59620-3004
T: +1 406 4443660; Fax: +1 406 4443603; E-mail: mtlawlibrary@mt.gov; URL: www.lawlibrary.mt.gov
1866; Judith Meadows
Legal hist, Legislative hist; State Justice Institute Depot
180 000 vols; 550 curr per
libr loan; AALL, MLA 34350

Herkimer County Law Library, 301 N Washington St, Ste 5511, P.O. Box 916, ***Herkimer***, NY 13350-1299
T: +1 315 8671172; Fax: +1 315 8667991; E-mail: herklaw@pppmail.appliedtheory.com
1941; Deborah E. Melnick
Legal Ref Coll
14 000 vols 34351

US Air Force – Hickam Air Force Base Library, 15 SVS/SVMG, 990 Mills Blvd, Bldg 595, ***Hickam AFB***, HI 96853-5316
T: +1 808 4498299; Fax: +1 808 4498298; E-mail: 15svs.svmg@hickam.af.mil; URL: www.hickamservices.com/library.asp
1957; Phyllis Frenzel
Military hist, Hawaii
36 000 vols; 75 curr per 34352

US Air Force – Gerrity Memorial Library, 75 MSG/SVMG, Bldg 440, 7415 Eighth St, ***Hill AFB***, UT 84056-5006
T: +1 801 7773833; Fax: +1 801 7776667
1941; Rose Burton
40 000 vols; 107 curr per
libr loan 34353

Washington County Law Library, 111 NE Lincoln St, ***Hillsboro***, OR 97124
T: +1 503 6488880; Fax: +1 503 6403515
1926; Ann Karlen
17 000 vols 34354

Kulani Correctional Facility Library, HC-01 Stainback Hwy, ***Hilo***, HI 96720
T: +1 808 9359268; Fax: +1 808 9359268; E-mail: kcf@lava.net
Robert Li
23 000 vols; 60 av-mat 34355

State Supreme Court, Third Circuit Court-Law Library, State Office Bldg, Rm 206, 75 Aupuni St, ***Hilo***, HI 96720
T: +1 808 9617438; Fax: +1 808 9617416
Debra Kaido
30 000 vols 34356

Blair County Courthouse, Law Library, 423 Allegheny St Ste 227, ***Hollidaysburg***, PA 16648
T: +1 814 6933090; Fax: +1 814 6933289; E-mail: lawlibrary@blairco.org
1900; Lucille H. Wolf
9 000 vols; 30 curr per; Court of Common Pleas, State superior Court and State Supreme Court Opinions, Various Pennsylvania bar Assoc Mat 34357

San Benito County Law Library, Courthouse, Rm 206, 440 Fifth St, ***Hollister***, CA 95023
Fax: +1 831 6364044
Maria Alfaro
12 540 vols 34358

US Air Force – Holloman Air Force Base Library, FL 4801, 596 Fourth St, ***Holloman***, NM 88330-8038
T: +1 505 5723939; Fax: +1 505 5725340; E-mail: marie.ludwig@holloman.af.mil
Marie Ludwig
Military Studies, Foreign Relations & Management
45 000 vols; 170 curr per; 7 000 e-books 34359

David Wade Correctional Center, Wade Library, 670 Bell Hill Rd, ***Homer***, LA 71040
T: +1 318 9270424; Fax: +1 318 9270423
Jamie Fussell
10 000 vols 34360

Legislative Reference Bureau Library, State Capitol, Rm 005, ***Honolulu***, HI 96813
T: +1 808 5870690; Fax: +1 808 5870699; URL: www.hawaii.gov/lrb
1943
55 000 vols; 300 curr per 34361

State of Hawaiin Department of Business, Economic Development & Tourism, DBEDT Library, 250 S Hotel St, 4th Fl, ***Honolulu***, HI 96813; P.O. Box 2359, Honolulu, HI 96804-2359
T: +1 808 5862424; Fax: +1 808 5872790; E-mail: library@dbedt.hawaii.gov; URL: www.hawaii.gov/dbedt
1962; Debra Miyashiro
6 000 vols; 305 curr per; 20 maps; 109 digital data carriers; government docs 34362

Supreme Court Law Library, 417 S King St, Rm 115, ***Honolulu***, HI 96813; mail address: 417 S King St, Rm 119, Honolulu, HI 96813
T: +1 808 5394964; Fax: +1 808 5394974; E-mail: lawlibrary@courts.state.hi.us
1851; Mark Skrimstad
155 000 vols; 206 curr per; 289 000 microforms; 405 sound-rec; 13 digital data carriers 34363

US Army – Aliamanu Library, Aliamanu Military Reservation, 1782 Bougainvillea Loop, ***Honolulu***, HI 96818-1311
T: +1 808 8334851; Fax: +1 808 8333714
1979; Jeeaam Hong
Children's & Juvenile lit
89 000 vols; 30 curr per 34364

US Courts Library, 300 Ala Moana Blvd C-341, ***Honolulu***, HI 96850
T: +1 808 5411797; Fax: +1 808 5413667
Patricia L. Butson
35 000 vols; 50 curr per 34365

City of Houston – Legal Department Library, 900 Bagby, 4th Fl, P.O. Box 1562, ***Houston***, TX 77251
T: +1 713 2471296; Fax: +1 713 2471017
1907; Evangeline Bell
17 200 vols; 15 curr per 34366

Harris County Law Library, Congress Plaza, 1019 Congress, 17th Flr, ***Houston***, TX 77002
T: +1 713 7555183; URL: www.co.harris.tx.us/hcll
1913; John R. Eichstadt
100 000 vols 34367

NASA – Johnson Space Center Scientific & Technical Information Center, 2101 NASA Rd One, Mail Code GP23, ***Houston***, TX 77058-3696
T: +1 281 4834245; Fax: +1 281 4832527; E-mail: sticnter@ems.jsc.nasa.gov; URL: sticjsc.nasa.gov
1962; Jane Hultberg
68 900 vols; 170 curr per
libr loan 34368

United States Courts Library, 515 Rusk Ave, Rm 6311, ***Houston***, TX 77002
T: +1 713 2505696; Fax: +1 713 2505091; URL: www.lb5.uscourts.gov
45 000 vols; 101 curr per 34369

State Correctional Institution, Retreat Library, 660 State Rte 11, ***Hunlock Creek***, PA 18621
T: +1 570 7358754; Fax: +1 570 7402406
Karen Stroup
9 000 vols; 25 curr per 34370

State Correctional Institution, Smithfield Library, 1120 Pike St, ***Huntingdon***, PA 16652; P.O. Box 999, Huntingdon, PA 16652-0999
T: +1 814 6436520; Fax: +1 814 5061022
Renee Lubert
19 000 vols; 58 curr per 34371

Texas Department of Criminal Justice Library, 804 Bldg A, FM 2821 W, ***Huntsville***, TX 77340; P.O. Box 40, Huntsville, TX 77342-0040
T: +1 936 2915384; Fax: +1 936 2914656
1970; Janice Warren
573 000 vols 34372

US Army – Ordnance Missile & Munitions Center & School Technical Library, 3323 Redeye Rd, ***Huntsville***, AL 35898
T: +1 256 8764741; Fax: +1 256 8763949; URL: www.redstonemwr.com
1952; Gail M. Alden
Electronics, Military hist, Education
30 000 vols; 75 curr per; 1 000 microforms; 2 000 av-mat; 25 digital data carriers; 20 000 rpts
libr loan 34373

US Air Force – Hurlburt Base Library, Base Library, 16SVS/SVMG, ***Hurlburt Field***, FL 32544
T: +1 850 8846947; Fax: +1 850 8846050; URL: commandolibrary.com
1955; Kelly B. Desmond
Military Art & Science, Aeronautics; Military Special Operations
35 000 vols; 170 curr per; 44 000 microforms; 80 digital data carriers
libr loan 34374

Florida Department of Corrections, Hendry Correctional Institution Library, 12551 Wainwright Dr, ***Immokalee***, FL 34142-9747
T: +1 239 8672189
8 000 vols; 25 curr per 34375

California Department of Corrections Library System, Centinela State Prison, 2302 Brown Rd, ***Imperial***, CA 92251; P.O. Box 731, Imperial, CA 92251-0731
T: +1 760 3377900; Fax: +1 760 3377631
Rosa Pitones
35 000 vols 34376

Illinois Department of Corrections, Big Muddy River Correctional Center Library, 251 N Illinois Hwy 37, ***Ina***, IL 62846; P.O. Box 1000, Ina, IL 62846-1000
T: +1 618 4375300; Fax: +1 618 4375627
Jennifer Wilson
32 000 vols; 54 curr per 34377

National Archives & Records Administration, Harry S. Truman Presidential Library & Museum, 500 W US Hwy 24, ***Independence***, MO 64050-1798
T: +1 816 2688200; Fax: +1 816 2688295; E-mail: truman.library@nara.gov; URL: www.trumanlibrary.org
1957; Michael Devine
Admin of President Harry S. Truman, Career; Papers of Harry S. Truman & 500 other individuals
40 000 vols; 39 curr per
libr loan 34378

US Navy – Albert T. Camp Library, Naval Surface Warfare Ctr, Bldg 299, ***Indian Head***, MD 20640
T: +1 301 7444742; Fax: +1 301 7444192; E-mail: sikoramp@ih.navy.mil
Mike Sikora
Chemistry, Engineering; Ordnance, Research, Rocketry, Test & Evaluation Rpts
12 150 vols; 417 curr per
libr loan 34379

Indiana County Law Library, County Court House, ***Indiana***, PA 15701
T: +1 724 4653956; Fax: +1 724 4653152
Wayne Kablack
15 000 vols 34380

Indiana Supreme Court Law Library, 316 State House, 200 W Washington St, ***Indianapolis***, IN 46204-2788
T: +1 317 2322557; Fax: +1 317 2328372
1867; Teri Ross
83 000 vols; 198 curr per 34381

Indiana Women's Prison Library, 401 N Randolph St, ***Indianapolis***, IN 46201
T: +1 317 6392671; Fax: +1 317 6849643
Judith Richey
14 000 vols; 44 curr per 34382

Marion County Law Library, 200 E Washington St, Ste T-360, ***Indianapolis***, IN 46204
T: +1 317 3275499; Fax: +1 317 3275461
Terri Lea Ross
14 500 vols; 15 curr per 34383

Martin Correctional Institution Library, 1150 SW Allapattah Rd, ***Indiantown***, FL 34956-4310
T: +1 772 5973705; Fax: +1 772 5974529
Teddi Greaves
19 000 vols; 79 curr per 34384

Riverside County Law Library, Indio Branch, Larson Justice Ctr, 46-200 Oasis St, ***Indio***, CA 92201
T: +1 760 8638316; Fax: +1 760 3422581
Patricia Stewart
23 000 vols; 14 curr per 34385

Michigan Department of Corrections, Parnall Correctional Facility Library, 1790 E Parnall Rd, ***Jackson***, MI 49201-9037
T: +1 517 7806000; Fax: +1 517 7806399
Louis Yonke
10 000 vols 34386

Mississippi Department of Environmental Quality Library, Southport Ctr, 2380 Hwy 80 W, ***Jackson***, MS 39204; P.O. Box 20305, Jackson, MS 39289-1305
T: +1 601 9615650; Fax: +1 601 9615521; URL: www.deq.state.ms.us/mdeq.nsf/page/main_home?opendocument
1850; Michael B. Bograd
Canadian & British Geological Survey Publs, Environmental Protection Agency Rpts, Topographic Maps, US & State Geological Survey Publs
48 000 vols; 126 curr per 34387

Mississippi State Department of Archives & History, Archives & Library Division, 200 North St, ***Jackson***, MS 39201; P.O. Box 571, Jackson, MS 39205
T: +1 601 5766876; Fax: +1 601 5766964; E-mail: refdesk@mdah.state.ms.us; URL: www.mdah.state.ms.us
1902; Julia Marks Young

County Records, micro; Federal Govt Records Pertaining to Mississippi; Map Coll; Mississippi Businesses & Organizations, private papers, mss; Mississippi Coll, newspapers; newsfilm; Mississippi Confederate Records; Photogr Coll; State, Territorial & Provincial Govt, arch
68 000 vols; 300 curr per   34388

G. Robert Cotton Regional Correctional Facility Library, 3500 N Elm Rd, *Jackson*, MI 49201
T: +1 517 7805172; Fax: +1 517 7805100
Hatatu Elum
10 000 vols; 14 curr per   34389

Tennessee State Law Library, Supreme Court Bldg, 6 Hwy 45 By-Pass, *Jackson*, TN 38301
T: +1 731 4235849
Debbie Durham
30 000 vols; 30 curr per   34390

Duval County Law Library, 330 E Bay St, Rm 102, *Jacksonville*, FL 32202
T: +1 904 6302560; Fax: +1 904 6302309
1939; Bud Mauer
47 000 vols   34391

Illinois Department of Corrections, Jacksonville Correctional Center Library, 2268 E Morton Rd, *Jacksonville*, IL 62650-9347
T: +1 217 2451481 ext 334; Fax: +1 217 2451481 ext 324; E-mail: alsq@darkstar.rsa.lib.il.us
1984; Susan Crowfoot
Law Coll, Self-Help Legal Bks
8 800 vols; 26 curr per
libr loan   34392

US Army – Corps of Engineers, Technical Library, 701 San Marco Blvd, Rm 430-W, *Jacksonville*, FL 32207; P.O. Box 4970, Jacksonville, FL 32232-0019
T: +1 904 2323643; Fax: +1 904 2321838
1978; Oriana Brown West
Engineering; Cross Florida Barge Canal Study
27 000 vols; 400 curr per; 5 000 techn rpts, 17 000 trade bks
libr loan   34393

US Marine Corps – MCAS Station Library, New River Air Sta, Bldg 213, *Jacksonville*, NC 28545-5001
T: +1 910 4516715; Fax: +1 910 4516037
1956; Ian R. Smith
Alcohol-abuse control, Drug, Automobile maintenance repair, Human relations, Military preparedness, Sports, Vietnam, World War II
21 000 vols; 74 curr per   34394

Queens County Supreme Court Library, General Court House, 88-11 Sutphin Blvd, *Jamaica*, NY 11435
T: +1 718 2981206; Fax: +1 718 5203589; E-mail: law.library_queens@courts.state.ny.us; URL: www.nycourts.gov/library/queens
1911; Andrew Tschinkel
State law, New York City
125 000 vols; 100 curr per; 172 000 microforms; 26 digital data carriers
34395

California Department of Corrections Library System, Sierra Conservation Center, 5100 O'Byrnes Ferry Rd, *Jamestown*, CA 95327
T: +1 209 9845291; Fax: +1 209 9844563
Dennis F. Ward
Federal & California Law
43 000 vols   34396

Missouri Department of Corrections, Offender Libraries, 2729 Plaza Dr, *Jefferson City*, MO 65109-1146; P.O. Box 236, Jefferson City, MO 65102-0236
T: +1 573 5221928; Fax: +1 573 7514099
1970; Margaret Booker
Civil rights, Law
145 000 vols   34397

Missouri Supreme Court Library, Supreme Court Bldg, 207 W High St, *Jefferson City*, MO 65101
T: +1 573 7512636; Fax: +1 573 7512573
1820; Tyronne M. Allen
110 000 vols; 180 curr per   34398

Hudson County Law Library, 595 Newark Ave, Hudson County Admin Bldg, *Jersey City*, NJ 07306
T: +1 201 7956629; Fax: +1 201 7956603
Theresa Banks
19 000 vols   34399

Maryland House of Correction-Annex Library, Rte 175, Box 534, *Jessup*, MD 20794
T: +1 410 7990100; Fax: +1 410 7994487
Penny Cully
10 000 vols; 39 curr per   34400

Carbon County Law Library, Carbon County Courthouse, P.O. Box 207, *Jim Thorpe*, PA 18229-0207
T: +1 570 3253111
Mary Alice Herman
14 143 vols; 30 curr per   34401

Stateville Correctional Center Libraries, PO Box 112, *Joliet*, IL 60434-0112
T: +1 815 7273607
Phyllis Baker
Law, criminal law and justice
39 000 vols; 140 curr per   34402

Marine Corps Base Hawaii Libraries, MCCS Hawaii Libraries, Bldg 219, Marine Corps Base Hawaii, *Kaneohe Bay*, HI 96863-3073; P.O. Box 63073, Marine Corps Base Hawaii, Kaneohe Bay, HI 96863-3073
T: +1 808 2547623; Fax: +1 808 2547623; URL: www.mccshawaii.com/library.htm; www.usmc-mccs.org/library
1953; Murray R. Visser
Children's Coll, CMC Reading Program, Hawaiiana, MCI Courses, US Marine Corps Coll
58 000 vols; 239 curr per; 303 maps; 34 000 microforms; 841 sound-rec; 17 digital data carriers   34403

Jackson County Law Library, 1125 Grand Blvd, Ste 1050, *Kansas City*, MO 64106
T: +1 816 2212221; Fax: +1 816 2216607; E-mail: info@jcll.org; URL: www.jcll.org
1871; Jan D. Medved
47 000 vols; 31 curr per
libr loan; AALL, SLA   34404

Missouri Court of Appeals Library, Western District, 1300 Oak St, *Kansas City*, MO 64106-2910
T: +1 816 8893639; Fax: +1 816 8893668; URL: www.courts.mo.gov
1885; Janine Estrada-Lopez
32 000 vols; 150 curr per   34405

National Archives & Records Administration, Central Plains Region (Kansas City), 2312 E Bannister Rd, *Kansas City*, MO 64131-3011
T: +1 816 2688000; Fax: +1 816 2688037; E-mail: kansascity.archives@nara.gov; URL: www.archives.gov/central-plains/kansas-city
Reed Whittaker
Economics, ethnology, history, political science, public admiistration
38 000 vols; 41 curr per   34406

US Army – Corps of Engineers, District Library, 601 E 12th St, Rm 747, *Kansas City*, MO 64106-2896
T: +1 816 9833421; Fax: +1 816 4262730; URL: www.nwk.usace.army.mil/library.html
Pat Adrian
10 000 vols; 150 curr per   34407

US Courts, Branch Library, 9440 Charles Evans Whittaker US Courthouse, 400 E Ninth St, *Kansas City*, MO 64106
T: +1 816 5125790; Fax: +1 816 5125799
Deborah Showalter-Johnson
40 000 vols   34408

US Courts – Kansas City Branch Library, 532 US Courthouse, 500 State Ave, *Kansas City*, KS 66101-2441
T: +1 913 5515648; Fax: +1 913 5516547; E-mail: sharon_hom@ksd.uscourts.gov
1994; Sharon L. Hom
10 000 vols; 12 curr per   34409

Wyandotte County Law Library, Court House, 710 N Seventh St, 5th Flr, *Kansas City*, KS 66101-3999
T: +1 913 5732899; Fax: +1 913 5732892
1925; Brenda Eaton
50 000 vols   34410

US Geological Survey – Leetown Science Center, Technical Information Services, 1700 Leetown Rd, *Kearneysville*, WV 25430
T: +1 304 7244447; Fax: +1 304 7244435; E-mail: vi_catrow@usgs.gov; URL: www.lsc-tis.library.net
1959; Vi Catrow
Aquaculture, Diseases, Fish bacteriology, Genetics, Histology, Immunology, Nutrition, Parasitology, Physiology, Virology
22 000 vols; 200 curr per
libr loan   34411

US Air Force – McBride Library, 512 Larcher Blvd, Bld 2222, *Keesler AFB*, MS 39534-2345
T: +1 228 3772181; Fax: +1 228 4350203; URL: www.keesler.af.mil
1942; William R. Province
36 500 vols; 300 curr per   34412

NASA – John F. Kennedy Space Center Library, *Kennedy Space Center*, FL 32899
T: +1 321 8673600; Fax: +1 321 8674534; URL: www-pao.ksc.nasa.gov/media/referlib.htm
1962; Wlliam G. Cooper
Aerospace science; Arch Coll, Kennedy Space Ctr Hist Coll, photos
18 000 vols; 3 200 curr per; 15 digital data carriers   34413

Queens County Supreme Court Library, Kew Gardens Branch, 125-01 Queens Blvd, 7th Flr, *Kew Gardens*, NY 11415
T: +1 718 2981327; Fax: +1 718 5204661
Andrew Tschinkel
Criminal law
25 000 vols   34414

Michigan Department of Corrections, Pugsley Correctional Facility Library, 7401 Walton Rd, *Kingsley*, MI 49649
T: +1 231 2635253; Fax: +1 231 2637606
Denise Bearre
10 000 vols; 20 curr per   34415

New York State Supreme Court, Third Judicial District Law Library, 285 Wall St, *Kingston*, NY 12401
T: +1 914 3403053; Fax: +1 914 3403773
Michael Birzenieks
17 000 vols   34416

US Navy – Naval Air Station Library, 904 Dealy Ave, Ste 110, *Kingsville*, TX 78363-5034
T: +1 361 5166271; Fax: +1 361 5166971; E-mail: naslib@intcomm.com
1943; Vicky Jacobson
Military, Military hist
14 000 vols; 44 curr per   34417

Kirtland Air Force Base Library, 2050-B Second St SE, Bldg 20204, 377 SVS/SVMG, *Kirtland AFB*, NM 87117-5525
T: +1 505 8461071; Fax: +1 505 8466674
1945; Amie Stone
United States Hist (Southwest Coll); Military Art & Science
42 000 vols; 280 curr per; 2 000 av-mat; 1 874 sound-rec; 887 digital data carriers
34418

US Air Force – Phillips Site Technical Library FL2809, AFRL/VSII, AFRL/RVIL, 3550 Aberdeen Ave SE Bldg 419, *Kirtland AFB*, NM 87117-5776
T: +1 505 8464767; Fax: +1 505 8464790
1947; Becky Smith
Engineering, Physical science
38 000 vols; 800 curr per; 678 367 govt docs; 1 149 000 microforms; 815 av-mat; 27 sound-rec   34419

Knox County Governmental Library, M-47 City County Bldg, 400 Main St, *Knoxville*, TN 37902
T: +1 865 2152368; Fax: +1 865 2152920
Katherine Meredith Douglas
60 000 vols   34420

Knoxville-Knox County Metropolitan Planning Commission Library, 400 Main St, City & County Bldg, Ste 403, *Knoxville*, TN 37902-2476
T: +1 865 2152500; Fax: +1 865 2152068; E-mail: contact@knoxmpc.org; URL: www.knoxmpc.org
1975; Gretchen F. Beal
Urban affairs
10 000 vols; 93 curr per   34421

Municipal Technical Advisory Service Library, 600 Henley St, Univ Tennessee, UT Conference Ctr Bldg, Ste 120, *Knoxville*, TN 37996-4105
T: +1 865 9740411; Fax: +1 865 9740423; E-mail: adams-obrien@utk.edu; URL: www.mtas.utk.edu
1950; Frances Adams-O'Brien
City Ordinances, Municipal Law
14 600 vols; 250 curr per; 10 000 pamphlets, 3 000 sample ordinances, 3 000 indexed articles   34422

Tennessee Supreme Court Law Library, 505 W Main St, Ste 200, *Knoxville*, TN 37902
T: +1 865 5946128
1937; Susan Delp
30 000 vols   34423

Kentucky State Reformatory Library, 3001 W Hwy 146, *La Grange*, KY 40032
T: +1 502 2229441; Fax: +1 502 2229022
1938; Linda Goble
Fiction
23 000 vols; 81 curr per; 662 av-mat   34424

Luther Luckett Correctional Complex Library, 1612 Dawkins Rd, *La Grange*, KY 40031; P.O. Box 6, La Grange, KY 40031-0006
T: +1 502 2220363
Mary Morgan
9 000 vols; 85 curr per   34425

Charles County Circuit Court, Charles County Public Law Library, 200 Charles St, *La Plata*, MD 20646; P.O. Box 3060, La Plata, MD 20646-3060
T: +1 301 9323322; Fax: +1 301 9323324; URL: www.charlescountylawlibrary.com/index.html
Mary Rice
8 000 vols   34426

New Hampshire Department of Corrections, Lakes Region Facility Library, One Right Way Path, *Laconia*, NH 03246
T: +1 603 5289238; URL: www.state.nh.us/nhdoc
Rebecca Frame
9 000 vols   34427

US Geological Survey, National Wetlands Research Center, 700 Cajundome Blvd, *Lafayette*, LA 70506-3152
T: +1 337 2668692; Fax: +1 337 2668513; E-mail: judy_buys@usgs.gov; URL: www.nwrc.gov/library.html
1978; Judy Buys
Forestry, Global climate change, Wetland ecology; Arch of Ctr Publs
13 000 vols; 48 curr per; 185 diss/theses; 3 000 govt docs; 1 000 maps; 1 080 microforms; 385 av-mat; 342 sound-rec; 176 digital data carriers; 5 000 slides
libr loan   34428

Florida Department of Corrections, Reception & Medical Center Library (Main Unit), 7765 S CR 231, *Lake Butler*, FL 32054; P.O. Box 628, Lake Butler, FL 32054-0628
T: +1 386 4966000; Fax: +1 386 4963287
Criminal law, Large type print; Hi-Lo Coll, Popular Music, Vocational Material Coll
17 000 vols; 41 curr per   34429

Supreme Court Library, Warren County Municipal Ctr, 1340 State Rte 9, *Lake George*, NY 12845-9803
T: +1 518 7616442; Fax: +1 518 7616586
Vickie Cady
30 000 vols; 19 curr per   34430

Second District Court of Appeals, Law Library, 1005 E Memorial Blvd, *Lakeland*, FL 33801; P.O. Box 327, Lakeland, FL 33802
T: +1 863 4992290; Fax: +1 863 4992277
Helen Jensen
21 000 vols   34431

California Department of Corrections Library System, California State Prison, Los Angeles County, 44750 60th St W, *Lancaster*, CA 93536-7620
T: +1 661 7292000; Fax: +1 661 7296993
Linda Rowe
27 000 vols   34432

Lancaster County Law Library, 50 N Duke St, *Lancaster*, PA 17602; P.O. Box 83480, Lancaster, PA 17608-3480
T: +1 717 2998090; Fax: +1 717 2952509; URL: www.co.lancaster.pa.us
1867; Eleanor Gerlott
27 000 vols; 25 curr per   34433

US Air Force – Langley Air Force Base Library, 42 Ash Ave, *Langley AFB*, VA 23665
T: +1 757 7642906; Fax: +1 757 7643315; E-mail: 1svs.svmg@langley.af.mil; URL: www.langleylibrary.org
1942; Leslie Smail
35 000 vols; 80 curr per   34434

Kansas Department of Corrections, Lansing Correctional Facility Library, 801 E Kansas Hwy, *Lansing*, KS 66043; P.O. Box 2, Lansing, KS 66043-0002
T: +1 913 7273235; Fax: +1 913 7272997
Mark York
20 000 vols; 18 curr per   34435

Michigan Department of Transportation Information Services, 425 W Ottawa, P.O. Box 30050, *Lansing*, MI 48909
T: +1 517 2414140; Fax: +1 517 3730168; E-mail: midot.michtrans6.thomasj@state.mi.us
1964; Tricia Aper
Transportation Res Board
18 000 vols; 210 curr per   34436

Michigan State Legislative Service Bureau, Michigan Nat Tower, 4th Fl, P.O. Box 30036, *Lansing*, MI 48909-7536
T: +1 517 3730472; Fax: +1 517 3730171
1941; Leo Kennedy
Michigan Law (bks), Legislative Rpts (docs)
17 350 vols; 115 curr per   34437

Wyoming State Geological Survey Library, 11th & Lewis St, Geological Survey Bldg, P.O. Box 3008, *Laramie*, WY 82071-3008
T: +1 307 7662286; Fax: +1 307 7662605; E-mail: sales@wsgs.uwyo.edu; URL: www.wsgs.uwyo.edu
1933
Wyoming Electric Log Files Coll, Wyoming Petroleum Info Cards Coll, US Geological Survey Open-File Rpts for Wyoming
25 000 vols; 10 curr per   34438

New Mexico Corrections Department, Southern New Mexico Correctional Facility Library, 1983 Joe R Silva Blvd, *Las Cruces*, NM 88004; P.O. Box 639, Las Cruces, NM 88004-0639
T: +1 505 5233200; Fax: +1 505 5233337
Dinah Jentgen
10 000 vols; 40 curr per   34439

Clark County Law Library, 309 S Third St, Ste 400, *Las Vegas*, NV 89155; P.O. Box 557340, Las Vegas, NV 89155-7340
T: +1 702 4554696; Fax: +1 702 4555120; E-mail: askccll@co.clark.nv.us; URL: www.co.clark.nv.us/law_library/home_page.htm
1923; Kevin Clanton
Nevada Coll, Statutes of each state
68 000 vols; 420 curr per   34440

US Environmental Protection Agency, Environmental Sciences Division (ESD) Library, 944 E Harmon Ave, *Las Vegas*, NV 89119; P.O. Box 93478, Las Vegas, NV 89193
T: +1 702 7982648; Fax: +1 702 7982622; E-mail: library-lvgroup@epa.gov; URL: www.epa.gov/libraries/esd.html
1966; Jessie Choate
Radiation effects; Nat Estuary Surveys, OSWER Directives
16 140 vols; 60 curr per   34441

US Air Force – Laughlin Air Force Base Library, 427 Fourth St, Bldg 257, *Laughlin AFB*, TX 78843-5125
T: +1 830 2985119; Fax: +1 830 2987903; E-mail: lafblibrary@bizstx.rr.com; URL: www.laughlinlibrary.com
Sue A. Blankemeyer
Career Transition Assistance
22 000 vols; 100 curr per   34442

US Geological Survey, Biological Resources Division, Patuxent Wildlife Research Center, 12100 Beech Forest Rd, *Laurel*, MD 20708-4030
T: +1 301 4975550; Fax: +1 301 4975545; E-mail: Lynda_Garrett@usgs.gov; URL: www.pwrc.usgs.gov/library
1942; Lynda Garrett
Pesticides & Pollution, reprints
9 000 vols; 350 curr per; 1 digital data carriers
libr loan; OCLC   34443

Kansas Geological Survey Library, University of Kansas, 1930 Constant Ave, *Lawrence*, KS 66047-3726
T: +1 785 8642098; Fax: +1 785 8645317; E-mail: sorensen@kgs.ukans.edu; URL: www.kgs.ukans.edu
1973; Janice H. Sorensen
17 000 vols; 55 curr per; 200 microforms; 25 digital data carriers; 1 000 open file rpts, 200 computer contributions, 250 overhead transparencies, 1 000 slides & photos   34444

Brunswick Correctional Center, DCE Library, 1147 Planters Rd, *Lawrenceville*, VA 23868-3499
T: +1 434 8484131; Fax: +1 434 8480971
Mary Geist
12 000 vols; 45 curr per   34445

Gwinnett County Law Library, 75 Langley Dr, *Lawrenceville*, GA 30045
T: +1 770 8228575; Fax: +1 770 8228570; URL: www.gcll.org
Grace Holloway
13 000 vols; 25 curr per   34446

Florida Department of Corrections, Lawtey Correctional Institution Library, 22298 NE CR200B, *Lawtey*, FL 32058; mail address: 7819 NW 228th St, Raiford, FL 32026
T: +1 904 7822000; Fax: +1 904 7822005
Richard Rogers
Criminal law
11 000 vols; 41 curr per   34447

Mississippi Department of Corrections, South Mississippi Correctional Institution Library, Hwy 63 N, *Leakesville*, MS 39451; P.O. Box 1419, Leakesville, MS 39451-1419
T: +1 601 3945600; Fax: +1 601 3945600 ext 1182; URL: www.mdoc.state.ms.us

Alvin Moody
9 000 vols   34448

State of Ohio Department of Corrections, Trumbull Correctional Institution Library, 5701 Burnette Rd, *Leavittsburg*, OH 44430; P.O. Box 901, Leavittsburg, OH 44430-0901
T: +1 330 8980820; Fax: +1 330 8982011
Gene DeCapua
10 000 vols; 66 curr per   34449

Lebanon Correctional Institution Library, 3791 SR 63, P.O. Box 56, *Lebanon*, OH 45036
T: +1 513 9321211; Fax: +1 513 9325803
1961; Billy Bailey
12 000 vols   34450

Warren County Law Library Association, 500 Justice Dr, *Lebanon*, OH 45036
T: +1 513 9324040 ext 1381; Fax: +1 513 6952947
Kristina Carter
16 000 vols   34451

Bayside State Prison Library, mail address: *Leesburg*, NJ 08327
T: +1 856 7850040 ext 5408; Fax: +1 856 7852463
Isac George
42 000 vols; 37 curr per   34452

Nez Perce County Law Library, Court House, 1230 Main, P.O. Box 896, *Lewiston*, ID 83501
T: +1 208 7993040; Fax: +1 208 7993058; E-mail: lawclerk@co.nezperce.id.us; URL: www.co.nezperce.id.us
Steve Carlor
10 000 vols   34453

Council of State Governments, States Information Center, 2760 Research Park Dr, P.O. Box 11910, *Lexington*, KY 40578
T: +1 859 2448253; Fax: +1 859 2448001; E-mail: sic@csg.org; URL: www.csg.org
1947; Don Hunter
Legislation; State Budgets, State Govt, bluebks & related mats
23 220 vols; 300 curr per   34454

Allen Correctional Institution Library, 2338 N West St, *Lima*, OH 45801; P.O. Box 4501, Lima, OH 45802
T: +1 419 2248000 ext 3006; Fax: +1 419 9985618
1987; Elizabeth Osborne
Law Libr
15 685 vols; 61 curr per   34455

Allen County Law Library, Court of Appeals, Rm 102, 204 N Main St, mail address: 3233 Spencerville Rd, *Lima*, OH 45801-4456
T: +1 419 2231426; E-mail: lawbooks@bright.net
Bonnie Everett
24 000 vols; 18 curr per   34456

Oakwood Correctional Facility Library, 3200 N West St, *Lima*, OH 45801-2048
T: +1 419 2258052; Fax: +1 419 2258000
1961; Ronald A. Hawkins
Philosophy & psychology, Forensic psychiatry
12 000 vols; 22 curr per; 516 av-mat; 720 sound-rec; 13 digital data carriers; 20 slides
libr loan   34457

Colorado Department of Corrections, Limon Correctional Facility Library-General, 49030 State Hwy 71, *Limon*, CO 80826
T: +1 719 7759221; Fax: +1 719 7757676; URL: www.doc.state.co.us
Phyllis Wilson
12 000 vols; 55 curr per   34458

Lincoln Correctional Center Library, RR 3, P.O. Box 549, *Lincoln*, IL 62656
T: +1 217 7355411 ext 368; Fax: +1 217 73513361
Lana Wildman
8 500 vols   34459

Lincoln Correctional Center Library, 3216 W Van Dorn St, *Lincoln*, NE 68522; P.O. Box 22800, Lincoln, NE 68542-2800
T: +1 402 4712861; Fax: +1 402 4796100
Sandra Elton
10 000 vols; 44 curr per   34460

Nebraska Department of Roads, Resource Library, 1500 Nebraska Hwy 2, *Lincoln*, NE 68502-5480; P.O. Box 94759, Lincoln, NE 68509-4759
T: +1 402 4794316; Fax: +1 402 4793989; URL: www.nebraskatransportation.org
Mary Neben
Transportation
20 000 vols; 300 curr per
SLA   34461

Nebraska Legislative Council, Legislative Reference Library, State Capitol, Rm 1201, P.O. Box 94945, *Lincoln*, NE 68509
T: +1 402 4712221; Fax: +1 402 4790967
1980; Peg Jones & Mary Rasmussen
11 500 vols; 250 curr per   34462

Nebraska State Penitentiary Library – Department of Corrections, PO Box 2500, Sta B, *Lincoln*, NE 68502
T: +1 402 4713161; Fax: +1 402 4795819
Sam Shaw
15 000 vols; 79 curr per   34463

Minnesota Department of Corrections, Minnesota Correctional Facility – Lino Lakes, 7525 Fourth Ave, *Lino Lakes*, MN 55014
T: +1 651 7176684; Fax: +1 651 7176598; URL: www.doc.state.mn.us
Sandra Vadnais
10 000 vols   34464

Columbiana County Law Library, 32 N Park Ave, *Lisbon*, OH 44432
T: +1 330 4203662; Fax: +1 330 4247902; E-mail: lcolumbianalaw@neo.rr.com; URL: www.columbianacountylawlibrary.com
Jan Morley
15 000 vols   34465

Arkansas Supreme Court Library, 625 Marshall St, Ste 1500, *Little Rock*, AR 72201
T: +1 501 6822147; Fax: +1 501 6826877; URL: courts.state.ar.us/courts/sc_library.html
1851; Ava M. Hicks
90 000 vols   34466

US Court of Appeals, Branch Library, 600 W Capitol Ave, Rm 224, *Little Rock*, AR 72201
T: +1 501 6045215; Fax: +1 501 6045217; E-mail: crata_castleberry@ca8.uscourts.gov; URL: www.ca8.uscourts.gov/library/library.html
1981; Crata Castleberry
16 000 vols; 65 curr per   34467

US Air Force – Little Rock Air Force Base Library, 976 Cannon Dr, FL 4460, *Little Rock AFB*, AR 72099-5289
T: +1 501 9876979
1956; Bethry J. Becker
Aeronautics, Business & management
21 000 vols   34468

Madison County Law Library, One N Main, Rm 205, *London*, OH 43140-1068
T: +1 740 8529515; Fax: +1 740 8527144; E-mail: jkronk@ix.netcom.com
1903; Janet C. Kronk
20 000 vols   34469

State of Ohio Department of Corrections, London Correctional Institute Library, 1580 State Rte 56 SW, *London*, OH 43140; P.O. Box 69, London, OH 43140-0069
T: +1 740 8522454; Fax: +1 740 8521591
1970; Gilbert Arthur Hurwood
Westerns
17 000 vols; 45 curr per   34470

937

US Department of Justice, National Institute of Corrections Information Center, 1860 Industrial Circle, Ste A, **Longmont**, CO 80501
T: +1 303 6820213; Fax: +1 303 6820558; E-mail: asknicic@nicic.org; URL: www.nicic.org
Eileen Conway
Prisons
15 000 vols; 200 curr per   34471

California Department of Justice Library, North Tower, 7th Flr, 300 S Spring St, **Los Angeles**, CA 90013
T: +1 213 8972342; Fax: +1 213 8972335
Janet Raffalow
35 000 vols   34472

California Second District Court of Appeals, 300 S Spring St, Rm 3547, **Los Angeles**, CA 90013
T: +1 213 8307241; Fax: +1 213 8972429
1967; Carol David Ebbinghouse
California
65 000 vols   34473

California State Department of Corporations Libary, 320 W 4th St, Ste 750, **Los Angeles**, CA 90013
T: +1 213 5767667; Fax: +1 213 5767182; E-mail: sakey@corp.ca.gov; URL: www.corp.ca.gov
1980; Sharon Akey
Corporate securities
12 000 vols; 100 curr per
AALL   34474

Los Angeles County Counsel Law Library, 500 W Temple St, Rm 610, **Los Angeles**, CA 90012
T: +1 213 9741982; Fax: +1 213 6267446; URL: www.lacounty.info
Eugene S. Drayton
15 000 vols
libr loan; AALL   34475

Los Angeles County Law Library, 301 W First St, **Los Angeles**, CA 90012-3100
T: +1 213 7852513; Fax: +1 213 6131329; URL: www.lalaw.lib.ca.us/
1891; Marcia J. Koslow
8 branch libs
992 000 vols; 11 876 curr per; 673 000 microforms; 3 081 sound-rec; 44 digital data carriers
libr loan   34476

Los Angeles County Metropolitan Transportation Authority, Dorothy Peyton Gray Transportation Library-Research Center, One Gateway Plaza, 15th Flr, Mail Stop 99-15-1, **Los Angeles**, CA 90012-2952
T: +1 213 9224859; Fax: +1 213 9227955; E-mail: library@metro.net; URL: www.metro.net/library
1971; Matthew Barrett
Deeds, local transit, urban planning, photography (early 1900 to present)
200 000 vols; 165 curr per; 1 mss; 15 000 microforms; 1 100 av-mat; 300 sound-rec; 5 digital data carriers   34477

Los Angeles County Superior Court, Law Library, 111 N Hill St, Rm 344, **Los Angeles**, CA 90012
T: +1 213 9744867; Fax: +1 213 6805814
Ramon Zamudio
200 000 vols   34478

US Courts Library, 255 E Temple St, Rm 132, **Los Angeles**, CA 90012
T: +1 213 8948900; Fax: +1 213 8948906
Zora Maynard
Law
30 000 vols   34479

US Department of Justice, US Attorney Central District of California Library, 1214 US Courthouse, 312 N Spnring St, **Los Angeles**, CA 90012
T: +1 213 8942419; Fax: +1 213 8941381
1888; Cornell Winston
15 000 vols   34480

Southern Ohio Correctional Facility Library, P.O. Box 45699, **Lucasville**, OH 45699-5699
T: +1 740 2595544 ext 3592
1972; Joyce Brady
13 602 vols; 50 curr per   34481

US Air Force – Luke Air Force Base Library, Bldg 219, 7424 N Homer Dr, 56 SVS/SVMG FL 4887, **Luke AFB**, AZ 85309-1220
T: +1 623 8567191; Fax: +1 623 9352023
1951; Katherine Gillen
Aeronautics, Military hist; Arizona, Mission Support
50 000 vols; 225 curr per; 450 e-books; 4 500 av-mat
libr loan   34482

US Air Force – MacDill Air Force Base Library, 8102 Condor St, **MacDill AFB**, FL 33621-5408
T: +1 813 8283607; Fax: +1 813 8284416; E-mail: 6svslibrary@macdill.af.mil; URL: amccrooms.sirsi.net
1940; Kathleen M. Brady
Military hist, Middle East
21 000 vols; 129 curr per   34483

Madera County Law Library, County Government Ctr, 209 W Yosemite Ave, **Madera**, CA 93637-3596
T: +1 559 6730378
1909; Darla Hix
8 120 vols   34484

Florida Department of Corrections, Madison Correctional Institution Library, 382 SW MCI Way, **Madison**, FL 32340
T: +1 850 9735429
Becki Johnson
21 000 vols; 41 curr per   34485

Forest Products Laboratory Library, USDA Forest Service, One Gifford Pinchot Dr, **Madison**, WI 53726-2398
T: +1 608 2319491; Fax: +1 608 2319311; E-mail: pdl_fpl_library@fs.fed.us; URL: www.fpl.fs.fed.us/library; maple.cybertoolsforlibraries.com/cgi-bin/CyberHTML?USDAFSHO
1910; Julie Blankenburg
99 000 vols; 340 curr per; 6 000 microforms; 130 sound-rec; 50 digital data carriers
libr loan; ALA, SLA, FSLN   34486

Wisconsin Department of Justice, Law Library, 17 W Main St, **Madison**, WI 53703-3305; P.O. Box 7857, Madison, WI 53707-7857
T: +1 608 2660325; Fax: +1 608 2942958
1969; Sara J. Paul
Wisconsin Attorney General Mat, Consumer Protection Mat, Forensics
40 000 vols; 30 curr per   34487

Wisconsin Department of Public Instruction, Interlibrary Loan & Resource Sharing, 2109 S Stoughton Rd, **Madison**, WI 53716-2899
T: +1 608 2246167; Fax: +1 608 2246178; URL: dpi.wi.gov/rll/index.html
Sally Drew
151 000 vols; 60 curr per
libr loan   34488

Wisconsin Legislative Reference Bureau, Reference & Library Sections, One E Main St, **Madison**, WI 53703-3373; P.O. Box 2037, Madison, WI 53701-2037
T: +1 608 2660341; Fax: +1 608 2665648
1901; Marian Rogers
Bill Drafting Records, files, micro; Bill Index, 1897-present; Clipping Coll, 1900-present; Wisconsin Bills, 1897-present; Wisconsin State Govt Docs
100 000 vols   34489

Wisconsin State Department of Natural Resources – Department Library, 101 S Webster St, **Madison**, WI 53702-0005; P.O. Box 7921, Madison, WI 53707-7921
T: +1 608 2668933; Fax: +1 608 2668943; E-mail: dnrlibrary@dnr.state.wi.us; URL: www.dnr.state.wi.us
1978; Dreux J. Watermolen
33 000 vols; 300 curr per   34490

Wisconsin State Department of Transportation Library, 4802 Sheboygan Ave, Rm 803, **Madison**, WI 53707; P.O. Box 7957, Madison, WI 53707-7957
T: +1 608 2660724; Fax: +1 608 2616306; E-mail: library@dot.state.wi.us;
URL: www.dot.wisconsin.gov/library
1968; John Cherney
Regional Planning Commission Rpts, Transportation (Transportation Res Board)
35 000 vols; 350 curr per   34491

Wisconsin State Law Library, 120 Martin Luther King Jr Blvd, 2nd Flr, P.O. Box 7881, **Madison**, WI 53707-7881
T: +1 608 2679696; Fax: +1 608 2672319; E-mail: wsll.ref@wicourts.gov; URL: wsll.state.wi.us
1836; Jane E. Colwin
Wisconsin Appendices & Briefs, 1836-present; Wisconsin Court of Appeals Unpublished Opinions, 1978-present; 2 branch libs
150 000 vols; 1 000 curr per   34492

US Air Force – Malmstrom Air Force Base Arden G. Hill Memorial Library, 341 SVS/SVMG, 7356 Fourth Ave N, **Malmstrom AFB**, MT 59402-7506
T: +1 406 7314638; Fax: +1 406 7313667; URL: www.341services.com
1953
31 000 vols; 125 curr per   34493

Bare Hill Correctional Facility Library, 181 Brand Rd, **Malone**, NY 12953
T: +1 518 4838411; Fax: +1 518 4838411
Gwen Eagan
11 000 vols; 39 curr per   34494

Florida Department of Corrections, Jackson Correctional Institution Library, 5563 Tenth St, **Malone**, FL 32445
T: +1 850 5695260
Dr. Teri Maggio
14 000 vols; 27 curr per   34495

Prince William County Circuit Court, Law Library, 9311 Lee Ave, JU-170, **Manassas**, VA 22110
T: +1 703 7926262; Fax: +1 703 7926298; E-mail: rldavis@pwcgov.org; URL: www.pwcgov.org/ccourt
Robert Davis
10 000 vols   34496

Hillsborough County Law Library, 300 Chestnut St, **Manchester**, NH 03101
T: +1 603 6275605
Pam Boardman
11 000 vols   34497

Michigan Department of Corrections, Oaks Correctional Facility Library, 1500 Caberfae Way, **Manistee**, MI 49660-0038
T: +1 231 7238272; Fax: +1 231 7238430
Danielle Straubel
15 000 vols   34498

Richland Correctional Institution Library, 1001 Olivesburg Rd, **Mansfield**, OH 44905-1228; P.O. Box 8107, Mansfield, OH 44901-8107
T: +1 419 5262100; Fax: +1 419 5212814
Rebecca Williams
Law
13 000 vols; 70 curr per   34499

Richland County Law Library, 50 Park Ave E, **Mansfield**, OH 44902
T: +1 419 7745595
Traycee Conner
19 000 vols; 15 curr per   34500

State of Ohio Department of Corrections, Mansfield Correctional Institution Library, State Rd 545 N, 1150 N Main St, **Mansfield**, OH 44903; P.O. Box 788, Mansfield, OH 44901-0788
T: +1 419 5262000; Fax: +1 419 5261763
John Babajide
Criminal law
8 000 vols; 160 curr per   34501

North Carolina Division of Archives & History, Roanoke Island, Festival Park Complex, **Manteo**, NC 27954; P.O. Box 250, Manteo, NC 27954-0250
T: +1 252 4732655; Fax: +1 252 4731483; E-mail: obhc@ncmail.net; URL: www.obhistorycenter.net
1988; Kaeli Spiers
Local hist, Genealogy, Civil War, Maritime hist, Oral hist; Colonial American Hist, Early Exploration, Fishing Industry, US Life Saving Services
30 000 vols; 125 curr per   34502

Arizona Department of Corrections – Adult Institutions, Marana Community Correctional Facility Library, 12610 W Silverbell Rd, **Marana**, AZ 85653
T: +1 520 6822077; Fax: +1 520 6824080
William Poe
25 000 vols; 24 curr per   34503

Mid-State Correctional Facility Library, PO Box 216, **Marcy**, NY 13403-0216
T: +1 315 7688581; Fax: +1 315 7688581 ext 2099
Charles L. Youmans III
9 000 vols; 61 curr per   34504

Washington County Law Library, 205 Putnam St, **Marietta**, OH 45750-3017
T: +1 740 3736623 ext 214; Fax: +1 740 3732085; E-mail: timgaa@sys1.openohio.com
Juanita Hennigar
Ohio Law Cases 1800's Through Current
15 050 vols; 10 curr per   34505

Marion County Law Library, 258 W Center St, **Marion**, OH 43302
T: +1 740 2234170; E-mail: lawlib@on-ramp.net
Heather Ebert
17 000 vols; 18 curr per   34506

Contra Costa County Law Library, 1020 Ward St, 1st Flr, **Martinez**, CA 94553-1360
T: +1 925 6462783; Fax: +1 925 6462438; E-mail: libinfo@ll.cccounty.us; URL: www.cccpllib.org
Naomi Little
36 000 vols; 15 curr per   34507

North Carolina Department of Corrections, Eastern Correctional Institution Library, PO Box 215, **Maury**, NC 28554-0215
T: +1 252 7478101; Fax: +1 252 7475697
Gloria Farmer
8 000 vols; 36 curr per   34508

US Air Force – Air University Library, Air University Library, 600 Chennault Circle, **Maxwell AFB**, AL 36112-6010
T: +1 334 9532888; URL: www.au.af.mil/au/aul/aul.htm
1946; Dr. Jeff Luzius
Business and management, Economics, Education, Foreing affairs, Hist, Military science; Sir University Coll, rare bks & periodicals related to flight
328 000 vols; 1 897 curr per; 2 883 av-mat; 345 sound-rec
libr loan   34509

US Air Force – Maxwell Air Force Base Library, FL 3300, Bldg 28, 355 Kirkpatrick Ave E, **Maxwell AFB**, AL 36112
T: +1 334 9536484; Fax: +1 334 9537643
1958
68 000 vols; 275 curr per; 1 700 av-mat; 1 445 sound-rec; 7 digital data carriers   34510

US Air Force – Maxwell Gunter Community Library System, MSD/MSEL, 481 Williamson St, Bldg 1110 Gunter Annex, **Maxwell AFB**, AL 36114
T: +1 334 4163179; Fax: +1 334 4162949
Bernadette Roche
41 000 vols   34511

Florida Department of Corrections, Mayo Correctional Institution Library, 8784 US Hwy 27 W, **Mayo**, FL 32066
T: +1 386 2944500
15 000 vols   34512

US Navy – Mayport Naval Station Library, Naval Station Mayport, **Mayport**, FL 32228-0037
T: +1 904 2705393; Fax: +1 904 2705547
Brent Brown
Military science, Naval hist
13 400 vols; 60 curr per   34513

Chautauqua County Law Library, 3 N Erie St, Gerace Office Bldg, **Mayville**, NY 14757
T: +1 716 7534247; Fax: +1 716 7534129
Stephen Abdella
Air pollution, Education, govt, Labor, Real estate; Chautauqua County Hist
12 800 vols; 8 curr per   34514

Oklahoma Department of Corrections & the Oklahoma Department of Libraries, State Penitentiary Library, PO Box 97, *McAlester*, OK 74502-0097
T: +1 918 4234700; Fax: +1 918 4233862
1975; Wayne Brackensick
12 000 vols; 30 curr per   34515

US Army – John L. Byrd Jr Technical Library for Explosives Safety, Bldg 35, One C Tree Rd, *McAlester*, OK 74501
T: +1 918 4208707; Fax: +1 918 4208473; URL: www.dac.army.mil
Christine L. Holiday
27 000 vols; 16 curr per   34516

US Air Force – McChord Air Force Base Library, FL4479 62 SVS/SVMG, 851 Lincoln Blvd, *McChord AFB*, WA 98438-1317
T: +1 253 9843454; Fax: +1 253 9843944; E-mail: david.english@mcchord.af.mil
1941; David English
Aviation, Military hist
30 000 vols; 193 curr per   34517

US Air Force – McGuire Air Force Base Library, 305 SVS/SVMG, 2603 Tuskegee Airmen Ave, *McGuire AFB*, NJ 08641-5016
T: +1 609 7545159; Fax: +1 609 7545108; URL: www.305services.com
Kathleen Hanselmann
Aviation, Military hist, Quality management; Air Force Hist Coll
36 000 vols; 200 curr per   34518

Collin County Law Library, Courthouse, Ste 203, 210 S McDonald St, *McKinney*, TX 75069
T: +1 972 5484255; Fax: +1 972 5475734; URL: www.co.collin.tx.us
Judith A. McCullough
25 000 vols; 15 curr per   34519

Farm Credit Administration Information Center, 1501 Farm Credit Dr, *McLean*, VA 22102-5090
T: +1 703 8834296; Fax: +1 703 7341950
1984
10 000 vols; 150 curr per   34520

Crawford County Law Library, Courthouse, 2nd Flr, 903 Diamond Park, *Meadville*, PA 16335
T: +1 814 3338157; Fax: +1 814 3337489
Jenny Kightlinger
26 000 vols; 127 curr per   34521

Jackson County Law Library, Justice Bldg, 100 S Oakdale, *Medford*, OR 97501
T: +1 541 7746436; Fax: +1 541 7746767; E-mail: lawlib@jcls.org
Pam Pfeil
15 000 vols; 10 curr per   34522

Judge Francis J. Catana Law Library, Court House, 201 W Front St, *Media*, PA 19063
T: +1 610 8914462; Fax: +1 610 8914480; URL: www.delcolibraries.org/lawlibrary/
1902; N.N.
Pennsylvania Law
30 000 vols; 30 curr per   34523

Medina County Law Library Association, 93 Public Sq, *Medina*, OH 44256
T: +1 330 7259744; Fax: +1 330 7239608
1899; MaryAnn Lapina
20 000 vols; 33 000 microforms; 10 digital data carriers
AALL   34524

Shelby County Law Library, Shelby County Courthouse, 140 Adams Ave, Rm 334, *Memphis*, TN 38103
T: +1 901 5278498; Fax: +1 901 5228935; E-mail: lawlibrary@bellsouth.net; URL: www.shelbycountylawlibrary.com
1874; Gary L. Johnson
Early English Law, Early Laws of North Carolina, Statutes at Large for First Congress of the US
51 000 vols   34525

East Central Wisconsin Regional Planning Commission Library, 132 Main St, *Menasha*, WI 54952
T: +1 920 7514770; Fax: +1 920 7514771
Vicky Johnson
US Census Coll, bks, micro
10 000 vols   34526

Mercer County Law Library, 305 Courthouse, *Mercer*, PA 16137-0123
T: +1 724 6623800; Fax: +1 724 6620620; E-mail: kdeniker@mcc.co.mercer.pa.us; URL: www.mcc.co.mercer.pa.us/library.default
1913; Kim Deniker
17 500 vols; 1 200 curr per   34527

Florida Department of Corrections, South Florida Reception Center Library, 14000 NW 41 St, *Miami*, FL 33178-3003
T: +1 305 5929567; Fax: +1 305 4705765
Leroy Wilson
12 000 vols; 41 curr per   34528

Miami-Dade County Law Library, County Courthouse, Rm 321A, 73 W Flagler St, *Miami*, FL 33130
T: +1 305 3497548; Fax: +1 305 3497552; E-mail: refdesk@mdcll.org; URL: www.mdcll.org
1937; Johanna Porpiglia
127 000 vols; 30 curr per   34529

Third District Court of Appeals, Law Library, 2001 SW 117th Ave, *Miami*, FL 33175
T: +1 305 2293200
1957; Joanne Sargent
25 000 vols; 65 curr per   34530

US National Oceanic & Atmospheric Administration – Miami Regional Library, 4301 Rickenbacker Causeway, *Miami*, FL 33149
T: +1 305 3614428; Fax: +1 305 3614448; E-mail: aoml.library@noaa.gov; URL: www.aoml.noaa.gov/general/lib/
1956; Linda L. Pikula
Meteorology; Technical Reports Coll, Tropical Typhoons & Cyclones
21 000 vols; 187 curr per   34531

US National Oceanic & Atmospheric Administration – Miami Regional Library at AOML, 4301 Rickenbacker Causeway, *Miami*, FL 33149-1097
T: +1 305 3614428, 4429; Fax: +1 305 3614429; E-mail: pikula@aoml.noaa.gov; URL: www.aoml.noaa.gov/general/lib
1972; Linda L. Pikula
Oceanography, Ocean engineering, Mathematics; US Coast & Geodetic Survey Rpt 1851-1928
20 000 vols; 200 curr per
libr loan   34532

Indiana State Prison, Michael S. Thomas Learning Resource Center, One Park Row St, *Michigan City*, IN 4630-6597
T: +1 219 8747256; Fax: +1 219 8740335
1969; Kenneth J. Boyle
Fiction, Careers
15 000 vols; 27 curr per   34533

Federal Correctional Institution Library, E Arkona Rd, *Milan*, MI 48160; P.O. Box 49999, Milan, MI 48160
T: +1 734 4391511; Fax: +1 734 4393608
Bureau of Prison Program Statements & Institutions Supplements, Federal Law Bks & Statutes, Ref Works
20 000 vols; 35 curr per   34534

US Navy – Navy Personel Research Studies & Technology, Library, PERS 102, 5720 Integrity Dr, *Millington*, CA 38055
T: +1 901 8742116; E-mail: library@nprdc.navy.mil
1953
Statistics
15 000 vols; 330 curr per   34535

Florida Department of Corrections, Santa Rosa Correctional Institution Library (Main Unit), 5850 E Milton Rd, *Milton*, FL 32583
T: +1 850 9835800; Fax: +1 850 9835907
Michael Blue
12 000 vols; 41 curr per   34536

US Navy – Naval Air Station, Whiting Field Library, 7180 Langley St, Bldg 1417, *Milton*, FL 2570-5000
T: +1 850 6237502 ext 28; Fax: +1 850 6237561; URL: ci.mount-dora.fl.us/departments/library.htm
1969; Lisa J. Erhardt
Aeronautics, Meteorology, Electronics, Mathematics
13 500 vols; 36 curr per   34537

Legislative Reference Bureau, City Hall, Rm B-11, 200 E Wells St, *Milwaukee*, WI 53202-3567
T: +1 414 2862297; Fax: +1 414 2863004; URL: www.ci.milwaukee.wi.us/legislativereference1136.htm
1908; Eileen Lipinski
Municipal administration, Urban affairs
50 000 vols; 180 curr per   34538

United States Courts Library, 517 E Wisconsin Ave, Rm 516, *Milwaukee*, WI 53202
T: +1 414 2971698; Fax: +1 414 2971695
Mary B. Jones
20 000 vols   34539

Nassau County Supreme Court, Law Library, 100 Supreme Court Dr, *Mineola*, NY 11501
T: +1 516 5713883; Fax: +1 516 5710752
1902; Robert F. Cambridge
409 000 vols; 2 016 curr per   34540

Hennepin County Law Library, C-2451 Government Ctr, 300 S Sixth St, *Minneapolis*, MN 55487
T: +1 612 3483022; Fax: +1 612 3484230; E-mail: ll.reference@co.hennepin.mn.us; URL: hclaw.co.hennepin.mn.us
1883; Edward Carroll (Acting Director)
108 000 vols
libr loan   34541

US Court of Appeals, Branch Library, 1102 US Courthouse, 300 Fourth St, *Minneapolis*, MN 55415
T: +1 612 6645830; Fax: +1 612 6645835; E-mail: joyce_larson_schampel@mnd.uscourts.gov; URL: www.c28.uscourts.gov/library/library.html
Joyce Larson Schampel
18 000 vols   34542

US Air Force – Minot Air Force- Base Library, 5 FSS/FSDL, 210 Missile Ave, Bldg 156 Unit 1, *Minot AFB*, ND 58705-5026
T: +1 701 7233344; Fax: +1 701 7279850; URL: library.minot.accqolnet.org
1961; Lindsey Guderjahn
Air War College CSAF Reading List, McNaughton Rental Coll
27 000 vols; 92 curr per; 1 300 av-mat; 523 sound-rec; 44 digital data carriers
libr loan   34543

Tomorrow Center Library, Iowa Correctional Institution for Women, 300 Elm Ave SW, *Mitchellville*, IA 50169
T: +1 515 9674236; Fax: +1 515 9675347
Nancy Ellingson
Feminism, Ethnic studies
20 000 vols; 20 curr per   34544

Missouri Department of Corrections, Moberly Correctional Center, Bus Rte 63 S, 5201 S Morley, *Moberly*, MO 65270; P.O. Box 7, Moberly, MO 65270-0007
T: +1 660 2633778; Fax: +1 660 2631730
Jennifer Cook
Law
8 000 vols; 64 curr per   34545

Mobile County Public Law Library, Mobile Government Plaza, 205 Government St, *Mobile*, AL 36644-2308
T: +1 251 5748436; Fax: +1 251 5744757
Jacquelyn Carson
60 000 vols; 59 curr per   34546

Stanislaus County Law Library, 1101 13th St, *Modesto*, CA 95354
T: +1 209 5587759; Fax: +1 209 5588284; E-mail: lawlibrary@arrival.net; URL: www.stanislauslawlibrary.arrival.net
Janice K. Milliken
Local Municipal & County Codes
21 000 vols; 51 curr per
AALL   34547

NASA Armes Research Center, Life Sciences Library, Mail Stop 239-13, *Moffett Field*, CA 94035-1000
T: +1 650 6045387; Fax: +1 650 6047741; URL: www.library.arc.nasa.gov
1965; Gerri Stephenson
Aerospace Biology & Medicine, Biogenesis Coll, Biochemistry (Origin of Life Coll), Evolution Genetics
20 000 vols; 112 curr per   34548

NASA – Armes Research Center, Technical Library, Mail Stop 202-3, *Moffett Field*, CA 94035
T: +1 650 6046325; Fax: +1 650 6044988; URL: ameslib.arc.nasa.gov
1940; Mary Walsh
NASA & NACA Reports
87 000 vols; 300 curr per; 980 000 microforms; 18 digital data carriers
libr loan   34549

Wisconsin State Department of Natural Resources – Research Library, 1350 Femrite Dr, *Monona*, WI 53716
T: +1 608 2216325; Fax: +1 608 2216353; E-mail: dnrlibrary@mail01.dnr.state.wi.us
1964; Sonia Slemrod
Fish (Dr Schneberger, Lyle Christensen & Warren Churchill Coll), Richard Hunt and Carroll D Besadny Coll, Uncat Mat
25 000 vols; 350 curr per
libr loan   34550

Monroe County Law Library, 106 E First St, Court House, *Monroe*, MI 48161
T: +1 734 2407070; Fax: +1 734 2407056
William F. LaVoy
10 000 vols   34551

Grays Harbor County, Law Library, 102 W Broadway, Rm 203, *Montesano*, WA 98563
T: +1 360 2495311; Fax: +1 360 2496391
Elaine Urquhart
15 000 vols   34552

Alabama Supreme Court & State Law Library, Heflin-Torbert Judicial Bldg, 300 Dexter Ave, *Montgomery*, AL 36104
T: +1 334 2290581; Fax: +1 334 2290544; E-mail: reference@alalinc.net; URL: www.alalinc.net
1828; Timothy A. Lewis
Alabama Law
236 000 vols; 776 curr per   34553

Montgomery County Law Library, 251 S Lawrence St, *Montgomery*, AL 36104; P.O. Box 172, Montgomery, AL 36101-0172
T: +1 334 8321394; Fax: +1 334 2659536
Tiffany McCord
13 500 vols   34554

Florida Department of Corrections, Jefferson Correctional Institution Library, 1050 Big Joe Rd, *Monticello*, FL 32344-0430
T: +1 850 3420422
Dan Hamedani
10 000 vols; 53 curr per   34555

Susquehanna County Law Library, Court House, P.O. Box 218, *Montrose*, PA 18801-0218
T: +1 570 2784600
Mary Foster
12 500 vols   34556

US Air Force – Moody Air Force Base Library, 347 SVS/SVMG, 5107 Austin Ellipse, Ste A Bldg 103, *Moody AFB*, GA 31699-1594
T: +1 229 2573018; Fax: +1 229 2574119
1952; Elizabeth Britt Mary
Military hist
40 000 vols; 250 curr per   34557

Minnesota Department of Corrections, Minnesota Correctional Facility – Willow River/Moose Lake, 1000 Lake Shore Dr, *Moose Lake*, MN 55767
T: +1 218 4855000; Fax: +1 218 4855113; URL: www.doc.state.mn.us
Becky Pemberton
10 000 vols; 68 curr per   34558

North Carolina Department of Corrections, Western Youth Institution Library, 5155 Western Ave, *Morganton*, NC 28655-9696; P.O. Box 1439, Morganton, NC 28680-1439
T: +1 828 4386037; Fax: +1 828 4386076
Bill Smith
13 000 vols; 161 curr per   34559

Federal Correctional Institution – Morgantown Library, PO Box 1000, *Morgantown*, WV 26507-1000
T: +1 304 2964416; Fax: +1 304 2843622
Lisa Luzier
Prisoner Rights
11 000 vols; 62 curr per   34560

US Department of Energy, Federal Energy Technology Laboratory Library, 3610 Collins Ferry Rd, P.O. Box 880, *Morgantown*, WV 26507-0880
T: +1 304 2854764, 2854184; Fax: +1 304 2854188; E-mail: mmarst@fetc.doe.gov; URL: www.netl.doe.gov
1953; JoAnn Yuill
Mining (US Bureau of Mines, Rpts of Investigations, Info Circulars); Mining (US Bureau of, Mines Open File Rpts); Energy (US Dept of Energy); DOE Fossil Energy Rpts
30 000 vols; 150 curr per   34561

Morris County Law Library, Court House, Eight Ann St, *Morristown*, NJ 07963-0900; P.O. Box 910, Morristown, NJ 07963-0910
T: +1 973 6563917; Fax: +1 973 6563949
1970; Peter DeLucia
Pilch Libr (Henry Pilch Family Coll); West's National Reporter System Coll, ultrafiche
32 000 vols; 35 curr per   34562

Georgia Department of Archives & History, Public Services Division Library, 5800 Jonesboro Rd, *Morrow*, GA 30260
T: +1 678 3643710; Fax: +1 678 3643856; URL: www.georgiaarchives.org
1918; Kayla Barrett
Local genealogy & hist
20 000 vols; 193 curr per; 10 500 maps; 30 000 microforms; 200 charts   34563

Mount Pleasant Correctional Facility Library, 1200 E Washington, *Mount Pleasant*, IA 52641
T: +1 319 3859511; Fax: +1 319 3858465
Georgia Houseman
8 000 vols; 40 curr per; 50 av-mat
                                                            34564

Illinois Appellate Court, Fifth District, Law Library, 14th & Main Sts, P.O. Box 867, *Mount Vernon*, IL 62864-0018
T: +1 618 2426414; Fax: +1 618 2429133
1857; Janet Girvan
13 000 vols; 20 curr per; 6 752 microfilms; 2 080 digital data carriers
                                                            34565

US Air Force – Mountain Home Air Force Base Library, 366 SVS/SVMG, 520 Phantom Ave Bldg 2427, *Mountain Home AFB*, ID 83648-5000
T: +1 208 8282326; Fax: +1 208 8329840; E-mail: david.patterson@mountainhome.af.mil; URL: library.mountainhome.accqolnet.org/library
1952; David R. Patterson
Military hist, Aviation, Idaho
40 000 vols   34566

Michigan Department of Corrections, Alger Maximum Correctional Facility Library, Industrial Park Dr, P.O. Box 600, *Munising*, MI 49862
T: +1 906 3875000; Fax: +1 906 3875033
1990; Janice Yoak
ABE, preGED & GED Prepatory Mat; Native American, African American & Hispanic Hist
8 000 vols   34567

Michigan Department of Corrections, Muskegon Correctional Facility Library, 2400 S Sheridan Dr, *Muskegon*, MI 49442
T: +1 231 7733201; Fax: +1 616 7733657
David Amo
13 000 vols; 20 curr per   34568

Muskogee Law Library Association, Muskogee County Court House, *Muskogee*, OK 74401
T: +1 918 6827873
Paula Sexton
10 000 vols   34569

Napa County Law Library, Old Courthouse, Rm 132, 825 Brown St, *Napa*, CA 94559
T: +1 707 2591201; Fax: +1 707 2534229
Maxine C. Oellien
14 700 vols   34570

Eastern Correctional Facility Library, PO Box 338, *Napanoch*, NY 12458-0338
T: +1 845 6477400; Fax: +1 845 6477400 ext 5099
Christine M. Nielsen
11 000 vols; 88 curr per   34571

Tennessee Department of Corrections, Tennessee Prison for Women Library, 3881 Stewarts Lane, *Nashville*, TN 37218-5256
T: +1 615 7411255; Fax: +1 615 2536323
Earlene Guida
17 000 vols; 10 curr per   34572

Tennessee Department of Transportation Library, James K Polk Bldg, Ste 300, *Nashville*, TN 37243-0345
T: +1 615 7412330; Fax: +1 615 7411791; E-mail: rletson@mail.state.tn.us
1973; Ruth S. Letson
Photo Log
14 000 vols; 100 curr per; 304 microforms; 13 sound-rec   34573

Tennessee State Law Library, Law Library, Supreme Ct Bldg, 401 Seventh Ave N, *Nashville*, TN 37219-1407
T: +1 615 7412016; Fax: +1 615 7417186
1938; Stephen M. Jackson
55 000 vols; 30 curr per   34574

US Court of Appeals for the Sixth Circuit, Harry Phillips Memorial Library, United States Courthouse, Rm A-830, 110 Ninth Ave S, *Nashville*, TN 37203
T: +1 615 7367492; Fax: +1 615 7362045
Joe D. McClure
27 000 vols   34575

US Army – Technical Library, AMSSB-ROC-T(N), AMSRD-NSC-OC-T, Kansas St, *Natick*, MA 01760-5056
T: +1 508 2334542; Fax: +1 508 2334248; E-mail: nati-amsrd-nsc-ad-b@conus.army.mil; URL: www.ssc.army.mil
1946; Denice M. Czedik
Engineering, Biology, Chemistry, Mathematics, Medicine, Psychology, Social sciences, Textiles
48 000 vols; 100 curr per
libr loan   34576

General Dynamics Communication System Library, 77 A St, *Needham Heights*, MA 02194-2806
T: +1 781 4554461; Fax: +1 781 4554460; E-mail: harriet.randall@gd-cs.com
Harriet Randall
Telecommunications, Data processing, Computer science, Electronics
15 000 vols; 200 curr per; 300 000 microforms   34577

US Air Force – Nellis Air Force Base Library, 99 SVS/SVMG, 4311 N Washington Blvd, Bldg 312, Ste 101, *Nellis AFB*, NV 89191-7064
T: +1 702 6527704; Fax: +1 702 6529224; E-mail: 99svs.svmg@nellis.af.mil; URL: www.nellis.af.mil/facilities5.htm
1942; Jo Davis
Aeronautics, Military hist, Business & management
38 000 vols; 60 curr per   34578

Middlesex County Law Library, One Kennedy Sq, *New Brunswick*, NJ 08901
T: +1 732 9813099; Fax: +1 732 9813233
Betty Agin
22 500 vols   34579

Lawrence County Law Library, 430 Court St, *New Castle*, PA 16101
T: +1 724 6562136; Fax: +1 724 6584489
Colleen Mojock
25 000 vols; 15 curr per   34580

Connecticut Judicial Branch, Law Library at New Haven, County Courthouse, 235 Church St, *New Haven*, CT 06510
T: +1 203 5036828; Fax: +1 203 7896499; URL: www.jud.ct.gov/lawlib
Ann Doherty
55 000 vols; 75 curr per   34581

New Lisbon Correctional Institution Library, 2000 Progress Rd, *New Lisbon*, WI 53950
T: +1 608 5627375; Fax: +1 608 5626410
Stephanie Flint
8 000 vols; 45 curr per   34582

Law Library of Louisiana, Supreme Court Bldg, 2nd Flr, 400 Royal St, *New Orleans*, LA 70130-2104
T: +1 504 3102515; Fax: +1 504 3102419; E-mail: library@lasc.org; URL: www.lasc.org
1855; Georgia Chadwick
19th C American, British & French Law, Louisiana Law
175 000 vols; 600 curr per
libr loan   34583

Louisiana Departments of Army & Air Force Headquarters, Louisiana National Guard, Jackson Barracks Military Library, Jackson Barracks, Bldg 53, *New Orleans*, LA 70146-0030
T: +1 504 2788241; Fax: +1 504 2788751; E-mail: pughs@la-army.ngb.army.mil
Sherrie S. Pugh
Civil War hist, Hist, Spanish American
25 000 vols   34584

US Army – Corps of Engineers, Technical Library, Rm 389, P.O. Box 60267, *New Orleans*, LA 70160-0267
T: +1 504 8622558; Fax: +1 504 8621721; E-mail: batesdan@smtp.1mn.usace.army.mil
1974; Daniel Bates
Engineering, Environmental studies, Flood control, Navigation; Environmental Impact Statements, Govt Rpts, Standards & Specifications
14 000 vols; 500 curr per
libr loan   34585

US Court of Appeals, Fifth Circuit Library, 600 Camp St, Rm 106, *New Orleans*, LA 70130
T: +1 504 5896510; Fax: +1 504 5896517; URL: www.ca5.uscourts.gov
1920; Kay Guillot
Govnt Docs Coll
75 000 vols; 30 000 microforms; 50 digital data carriers   34586

US Department of Agriculture, Southern Regional Research Center, 1100 Robert E Lee Blvd, *New Orleans*, LA 70124; P.O. Box 19687, New Orleans, LA 70179-0687
T: +1 504 2864288; Fax: +1 504 2864396; URL: www.ars.usda.gov
1941; Suhad Wojkowski
Bacteriology, Chemistry, Entomology, Food science, Microbiology, Textiles
25 000 vols; 500 curr per   34587

US Navy – Naval Support Activity Library, 2300 General Meyer Ave, Bldg 24, *New Orleans*, LA 70142-5007
T: +1 504 6782727; E-mail: m.g.Cantrell@hotmail.com
Martha G. Cantrell
Naval Hist Coll, Louisiana
9 090 vols; 56 curr per   34588

Tuscarawas County Law Library Association, 101 E High Ave, *New Philadelphia*, OH 44663-2599
T: +1 330 3643703; Fax: +1 330 3435509; E-mail: tusclawl@adelphia.net
Kathy Moreland
17 000 vols   34589

Civil Court of the City of New York Library, 111 Centre St, Rm 1034, *New York*, NY 10013
T: +1 646 3865430; Fax: +1 212 7485171
1962; Deborah Melnick
Commercial law, Procedures, Real property
60 000 vols   34590

Environmental Protection Agency, Region 2 Library, 290 Broadway, Rm 1660, *New York*, NY 10007
T: +1 212 6373185; Fax: +1 212 6373086; E-mail: library.region2@epa.gov; URL: www.epa.gov/region2/library
1965; Rebecca Garvin
Hazardous Waste Coll
15 000 vols; 100 curr per   34591

NASA – Goddard Institute for Space Studies Library, 2880 Broadway, Rm 710, *New York*, NY 10025
T: +1 212 6785613; Fax: +1 212 6785552
1961; Zoe Wai
10 000 vols   34592

New York City Law Department, Office of Corporation Counsels Law Library, 100 Church St, Rm 6-310, *New York*, NY 10007
T: +1 212 7881609; Fax: +1 212 7881239; URL: www.nyc.law.gov
1856; Catherine Fitzgerald
71 600 vols; 3 000 curr per   34593

New York City Police Department, Training Resource Unit Library, 235 E 20th St, Rm 639, *New York*, NY 10003
T: +1 212 4779723; Fax: +1 212 4779270
Camilla Browder
11 000 vols; 70 curr per   34594

New York County District Attorney's Office Library, One Hogan Pl, *New York*, NY 10013
T: +1 212 3354292; Fax: +1 212 3859789; E-mail: matuszak@interport.net
Mary E. Matuszak
26 970 vols; 56 curr per   34595

New York State Department of Law Library, 120 Broadway, 25th Fl, *New York*, NY 10271
T: +1 212 4168002; Fax: +1 212 4166130
Fran Sheinwald
25 000 vols; 167 curr per   34596

New York State Division of Housing & Community Renewal, Reference Room, 25 Beaver St, 7th Fl, *New York*, NY 10004
T: +1 212 4807403; Fax: +1 212 4807416
Robert Muniz
11 000 vols   34597

New York State Psychiatric Institute, Research Library, 1051 Riverside Dr, *New York*, NY 10032
T: +1 212 5435675; Fax: +1 212 5435092; URL: nyspi.org/library
1896; David T. Lane
30 000 vols; 400 curr per
libr loan   34598

New York State Supreme Court, First Judicial District Civil Law Library, 60 Centre St, *New York*, NY 10007
T: +1 646 3863670; Fax: +1 212 3748159
Julie Gick
New York City Codes; New York State Statutes; Records & Briefs NY Court of Appeals & Appellate Divisions, 1984-present, micro
75 000 vols   34599

New York State Supreme Court – First Judicial District Criminal Law Library, 100 Centre St, 17th Flr, *New York*, NY 10013
Fax: +1 212 7487908; E-mail: reflibny@courts.state.ny.us; URL: www.nycourts.gov/library/nyc_criminal/; www.bloglines.com/blog/pll
David G. Badertscher
Trial Transcripts for New York State, First JD Supreme Court-Criminal Branch
107 000 vols   34600

Supreme Court, Appellate Division, First Dept Law Library, 27 Madison Ave, *New York*, NY 10010
T: +1 212 3400478
1901; Gene Preudhomme
70 000 vols; 60 curr per   34601

US Court of International Trade, Court Library, One Federal Plaza, *New York*, NY 10278
T: +1 212 2642816; Fax: +1 212 2643242; URL: www.cit.uscourts.gov
Christina Rattiner
Customs law, Law
50 000 vols; 100 curr per   34602

US Department of Energy – Environmental Measurements Laboratorium Library, 201 Varick St, *New York*, NY 10014
T: +1 212 6207865; Fax: +1 212 6203600; E-mail: fdipasq@eml.doe.gov; URL: www.eml.doe.gov
1947; Frances DiPasqua
13 500 vols   34603

Essex County Law Library, 50 W Market St, Rm 512, *Newark*, NJ 07102
T: +1 973 6935723
1907; Debra Womeck
American & English Law Rpts, Legal Per, Legal Treatises & Texts
38 000 vols; 25 curr per   34604

US Court of Appeals Library, King Courthouse, Rm 5007, 50 Walnut St, *Newark*, NJ 07102; P.O. Box 1068, Newark, NJ 07101-1068
T: +1 973 6453034
Andrea Battel
22 000 vols; 30 curr per   34605

US Navy – Academic Resources Information Center, 440 Meyerkord Rd, *Newport*, RI 02841
T: +1 401 8414352; Fax: +1 401 8412805
1917; James F. Aylward
Academic, Adult education, Leadership management; Military Docs, fiche; Naval Training Manuals
100 000 vols; 275 curr per   34606

US Navy – Naval Undersea Warfare Center Division, Newport Technical Library, 5141, Bldg 103, *Newport*, RI 02841
T: +1 401 8323124; Fax: +1 401 8323699
1970
Engineering, Computers, Underwater acoustics, Underwater ordnance
20 000 vols; 150 curr per   34607

US Navy – Officer Indoctrination School Library, 291 Kollmeyer Ave, *Newport*, RI 02841-1524
T: +1 401 8414310; Fax: +1 401 8413323
1972; Commander Walker
Leadership, Ocean science; Naval Manuals
9 500 vols; 72 curr per   34608

Gates Correctional Institution Library, 131 N Bridebrook Rd, *Niantic*, CT 06357
T: +1 860 6914772; Fax: +1 860 6914769
Black Heritage, Spanish Heritage
16 000 vols; 22 curr per   34609

York Correctional Institution Library, 201 W Main St, *Niantic*, CT 06357
T: +1 860 6916810; Fax: +1 860 6916864
Joe Lea
Law
10 000 vols; 20 curr per   34610

Anvil Mountain Correctional Center Library, PO Box 730, *Nome*, AK 99762-0730
T: +1 907 4432241; Fax: +1 907 4435195
Howard Appel
60 000 vols   34611

California Department of Corrections Library System, California Rehabilitation Center, Fifth St & Western, *Norco*, CA 92860; P.O. Box 1841, Norco, CA 91760-1841
T: +1 951 7372683
Pat Osby
25 000 vols   34612

Joint Forces Staff College, Library, Ike Skelton Library, 7800 Hampton Blvd, *Norfolk*, VA 23511-1702
T: +1 757 4436401; Fax: +1 757 4436047; E-mail: reference@jfsc.ndu.edu; URL: www.jfsc.ndu.edu/library
1947; Dr. Janet Gail Nicula
JFSC Arch 1946-current
136 000 vols; 460 curr per; 150 digital data carriers
libr loan; FEDLINK, OCLC   34613

United States Courts Library, Walter E Hoffman US Courthouse, 600 Granby St, Rm 319, *Norfolk*, VA 23510
T: +1 757 2227044; Fax: +1 757 2227047
1986; Karen J. Johnson
17 000 vols
AALL   34614

US Navy – Naval Amphibious Base Library, 1481 D St, Bldg 3016, *Norfolk*, VA 23521
T: +1 757 4627691; Fax: +1 757 4624950
1942; Pauline Reimold
Naval hist; CLEP Preparation Videos
35 000 vols; 85 curr per   34615

Law Library of Montgomery County, Court House, Swede & Airy Streets, *Norristown*, PA 19404; P.O. Box 311, Norristown, PA 19404-0311
T: +1 610 2783806; Fax: +1 610 2785998; URL: llmc.montcopa.org
1869; Bruce S. Piscadlo
70 000 vols; 100 curr per   34616

Middlesex County Adult Correction Center Library, Rte 130, Apple Orchard Lane, *North Brunswick*, NJ 08902; P.O. Box 266, New Brunswick, NJ 08903-0266
T: +1 732 2973636
Jerry Berkowitz
18 000 vols; 25 curr per   34617

Commonwealth of Massachusetts Trial Court, Hampshire Law Library, Courthouse, 99 Main St, *Northampton*, MA 01060
T: +1 413 5862297; Fax: +1 413 5840870; URL: www.lawlib.state.ma.us
1894; Barbara Fell-Johnson
Statutes of New England States & New York
23 000 vols; 40 curr per
libr loan   34618

Huron County Law Library Association, Court House, 3rd Fl, 2 E Main St, *Norwalk*, OH 44857
T: +1 419 6685127; Fax: +1 419 6635026; E-mail: hclawlib@accnorwalk.com
Erin Gail Bartle
13 000 vols   34619

Connecticut Judicial Branch, Law Library at Norwich, One Court House Sq, *Norwich*, CT 06360
T: +1 860 8872398; Fax: +1 860 8231752; URL: www.jud.ct.gov/lawlib
Lawrence Cheeseman
Connecticut & Federal Law
23 000 vols; 64 curr per   34620

New York State Supreme Court Law Library, David L. Follett Memorial Library, 5 W Main St, *Norwich*, NY 13815-1899
T: +1 607 3349463; Fax: +1 607 3349236
1902; Lorraine Knapp
13 000 vols   34621

US Department of Energy – Office of Scientific & Technical Information, One Science.gov Way, *Oak Ridge*, TN 37830; P.O. Box 62, Oak Ridge, TN 37831-0062
T: +1 865 5761188; Fax: +1 865 5763609; URL: www.osti.gov
Walter L. Warnick
US Dept of Energy Res & Development Rpts & Monogr
4 700 000 vols; 50 curr per   34622

Bernard E. Witkin Alameda Country Law Library, 12th & Oak St Bldg, 125 Twelfth Street, *Oakland*, CA 94607-4912
T: +1 510 2084832; Fax: +1 510 2084823; E-mail: lawlib@acgov.org; URL: www.acgov.org/law
1891; Mrs Cossette T. Sun
Alameda County Ordinance Code, Administrative Code, Budgets & Other Special Publications, California Supreme Court & Appellate Briefs, microfiche, Federal & State Statutes, Indexes & Digests, Legal Periodical Coll, California & US Law Schools, Legal Practice Material, California & Federal, Local Municipal Codes
75 000 vols; 1 898 curr per; 38 av-mat; 80 sound-rec
libr loan   34623

Keen Mountain Correctional Center, Department of Correctional Education Library, State Rd 629, *Oakwood*, VA 24631; P.O. Box 860, Oakwood, VA 24631-0860
T: +1 276 4987411; Fax: +1 276 4987341
1990; Mary Marshall
19 000 vols; 25 curr per
libr loan   34624

Florida Department of Corrections, Florida Correctional Institution Library, 11120 Gainesville Rd, *Ocala*, FL 34482; P.O. Box 147, Ocala, FL 32663
T: +1 352 6225151 ext 379; Fax: +1 352 3692159
1967; M. Miller
Law
11 290 vols; 100 curr per   34625

US Air Force – Offutt Air Force Base Library, 55 SVS/SVMG, Bldg 73, 510 Custer Dr, *Offutt AFB*, NE 68113-2150
T: +1 402 2945823; Fax: +1 402 2947114; E-mail: 55fss.fsdl@acc.af.mil; URL: www.offuttbaselibrary.org; www.offutt55fss.com/amenities/library.html
Rebecca Sims
70 000 vols; 50 curr per; 1 500 av-mat   34626

New York State Department of Correctional Services, Ogdensburg Correctional Facility General Library, One Correction Way, *Ogdensburg*, NY 13669-2288
T: +1 315 3930281; Fax: +1 315 3930281 ext 3299
1982; Thomas E. Lawrence
Black culture Hist & Spanish Language
19 000 vols; 65 curr per; 234 maps; 362 microfilms; 365 av-mat; 739 sound-rec; 4 digital data carriers; 178 equipment
libr loan   34627

Florida Department of Corrections, Okeechobee Correctional Institution Library, 3420 NE 168th St, *Okeechobee*, FL 34972
T: +1 863 4625400; Fax: +1 863 4625612
Timothy Haselden
9 000 vols; 41 curr per   34628

Federal Aviation Administration, Mike Monroney Aeronautical Center Library, Academy Bldg, Rm 114, 6500 S MacArthur Blvd, *Oklahoma City*, OK 73169; AMA-300A, P.O. Box 25082, Oklahoma City, OK 73125-0082
T: +1 405 9542665; Fax: +1 405 9544742; URL: www.academy.faa.gov/library
1962; Daniel Spencer
Aeronautics, aviation; Coll of Federal Aviation Administration publs
28 000 vols; 65 curr per; 2 000 microforms; 133 av-mat; 3 sound-rec; 10 digital data carriers; 10 000 technical rpts – 10 000 in paper and 80 000 in microfiche
libr loan; OCLC, FEDLINK   34629

Oklahoma County Law Library, 321 Park Ave, Rm 247, *Oklahoma City*, OK 73102-3695
T: +1 405 7131353; Fax: +1 405 7131852; URL: www.oklahomacounty.org
Venita L. Hoover
33 000 vols; 25 curr per   34630

US Courts Library, 2305 US Courthouse, 200 NW Fourth, *Oklahoma City*, OK 73102
T: +1 405 6095460; Fax: +1 405 6095461
1990; Jerry E. Stephens
30 000 vols; 450 curr per   34631

Johnson County Law Library, Courthouse, Rm 101, 100 N Kansas Ave, *Olathe*, KS 66061
T: +1 913 7154154; Fax: +1 913 7154152; URL: lawlibrary.jocogov.org
1954; John Pickett
23 000 vols; 50 curr per   34632

Attorney General's Office, Law Library, 1101 Washington St SE, Ste 260, P.O. Box 40115, *Olympia*, WA 98504-0115
T: +1 360 7532681; Fax: +1 360 7533490
Jane Halligan
11 000 vols   34633

Washington State Department of Natural Resources, Division of Geology & Earth Resources Library, P.O. Box 47007, *Olympia*, WA 98504-7007
T: +1 360 9021472; Fax: +1 360 9021785; E-mail: connie.manson@wadnr.gov; URL: www.wa.gov/dnr/htdocs/ger/library.htm
1935; Connie J Manson
Theses; geologic maps
42 000 vols; 100 curr per; 2 000 diss/theses; 10 000 govt docs; 2 000 maps; 220 microforms; 50 digital data carriers   34634

Washington State Law Library, Temple of Justice, *Olympia*, WA 98504; P.O. Box 40751, Olympia, WA 98504-0751
T: +1 360 3572136; Fax: +1 360 3572143; E-mail: law.library@courts.wa.gov; URL: www.courts.wa.gov/library
1889; Kay Newman
Legal res; Legal Per
425 000 vols; 915 curr per
libr loan   34635

Douglas County District Court, Law Library, Hall of Justice, 1701 Farnam St, 1st Flr, *Omaha*, NE 68183
T: +1 402 4447174; Fax: +1 402 4443927
1905; Ann Borer
25 000 vols; 12 curr per   34636

Omaha Correctional Center Library, 2323 East Ave J, *Omaha*, NE 68111
T: +1 402 5953964; Fax: +1 402 5952227
Denise Morton
8 000 vols; 25 curr per   34637

US Army – Corps of Engineers, Omaha District, 106 S 15th St, *Omaha*, NE 68102-1618
T: +1 402 2213230; Fax: +1 402 2214198; E-mail: cenwo.library@usace.army.mil; URL: www.nwo.usace.army.mil/html/im-c/lib/hp.htm
Karen L. Stefero
Engineering Design Memo, Sediment Series
13 000 vols; 131 curr per   34638

US Court of Appeals, Branch Library, 111 S 18th Plaza, Ste 4104, *Omaha*, NE 68102-1322
T: +1 402 6617590; Fax: +1 402 6617591
Angela Lange
14 000 vols   34639

Orange County Law Library, 101 E Central Blvd, *Orlando*, FL 32801
T: +1 407 8357323; Fax: +1 407 8357646
1959; Gregg Gronlund
50 000 vols; 97 curr per   34640

Butte County Law Library, 1675 Montgomery St, *Oroville*, CA 95965
T: +1 530 5387122; Fax: +1 530 5341499; E-mail: publiclawlibrarian@gmail.com; URL: home.surewest.net/buttelaw
1907; John A. Zorbas
21 000 vols; 31 curr per   34641

Oshkosh Correctional Institution Library, 1730 W Snell Rd, *Oshkosh*, WI 54903; P.O. Box 3530, Oshkosh, WI 54903-3530
T: +1 920 2314010; Fax: +1 920 2362626
Cassandra Chaney
30 000 vols; 40 curr per   34642

Winnebago County Law Library, Court House, 415 Jackson St, P.O. Box 2808, *Oshkosh*, WI 54903
T: +1 920 2364808; Fax: +1 920 3034783
Dana Hartel
22 150 vols   34643

New York State Department of Correctional Services, Sing Sing Correctional Facility Library, 354 Hunter St, *Ossining*, NY 10562
T: +1 914 9410108; Fax: +1 914 9416583
Robert Richter
18 000 vols; 71 curr per   34644

Third District Appellate Court Library, 1004 Columbus St, *Ottawa*, IL 61350
T: +1 815 4345050
Sharon M. Smith
15 000 vols   34645

USDA-ARS, National Sedimentation Laboratory Library, 598 McElroy Dr, *Oxford*, MS 38655-2117; P.O. Box 1157, Oxford, MS 38655-1157
T: +1 662 2322996; Fax: +1 662 2322920
Dr. Mathias Romkens
20 000 vols; 15 curr per   34646

Missouri Department of Corrections, Missouri Eastern Correctional Center, 18701 US Hwy 66, *Pacific*, MO 63069-3525
T: +1 636 2573322; Fax: +1 636 2575296
Afro-American Coll
21 000 vols; 60 curr per   34647

US Navy – Naval Coastal Systems Station, 6703 W Hwy 98, Naval Surface Warfare Ctr, *Panama City*, FL 32407-7001
T: +1 850 2344848; Fax: +1 850 2354844; E-mail: whatleyah@ncsc.navy.mil
1948; Angelia Whatley
16 000 vols; 125 curr per   34648

United States Marine Corps, Recruit Depot Station Library, PO Box 5070, *Parris Island*, SC 29905-0070
T: +1 843 2283302; Fax: +1 843 2283840; URL: library.usmc-mccs.org/parris/index.htm
1940; Grace Cummings
27 000 vols; 83 curr per   34649

US Court of Appeals, Ninth Circuit Library, 125 S Grand Ave, *Pasadena*, CA 91105
T: +1 626 2297190; Fax: +1 626 2297460
1985; Kathryn A. Way
27 000 vols; 90 curr per   34650

Passaic Vicinage Law Library, 77 Hamilton St, 2nd Flr, *Paterson*, NJ 07505-2096
T: +1 973 2478013; URL: www.judiciary.state.nj.us
1930; Tami Lowe
57 000 vols   34651

US Air Force – Patrick Air Force Base Library, Bldg 722B, 842 Falcon Ave, *Patrick AFB*, FL 32925-3439
T: +1 321 4944801; Fax: +1 321 4944190; E-mail: base.library@patrick.af.mil
1950; Marta Demopoulos
Military hist; Air War College, Quality Management Coll
40 000 vols; 90 curr per   34652

US Air Force – 45th Space Wing Technical Library, 1030 S Hwy A1A, *Patrick AFB*, FL 32925-3002; P.O. Box 4127, Patrick AFB, FL 32925-0127
T: +1 321 4946636, 6638; Fax: +1 321 4946638
1952; Valerie G. Mutter
Science & technology, Aeronautics, Mathematics, Engineering
10 362 vols; 864 curr per
libr loan   34653

US Navy – Patuxent River Central Library, 22269 Cedar Point Rd B407, *Patuxent River*, MD 20670-1120
T: +1 301 3425348; Fax: +1 301 3421933; E-mail: technical.library.fct@navy.mil
Carolyn K. Eaton
Engineering, Aeronautics, Computer science, US govt
35 000 vols; 100 curr per   34654

Pendleton Correctional Facility, Law, 4490 W Reformatory Rd, *Pendleton*, IN 46064; P.O. Box 28, Pendleton, IN 46064-0028
T: +1 765 7782107; Fax: +1 765 7783395
Larry Fowler
12 000 vols; 41 curr per   34655

Pendleton Correctional Facility, Fender, 4490 W Reformatory Rd, *Pendleton*, IN 46064; P.O. Box 28, Pendleton, IN 46064-0028
T: +1 765 7782107; Fax: +1 765 7788166
Larry Fowler
Science fiction
17 000 vols; 48 curr per   34656

Escambia County Law Library, 190 Governmental Ctr, *Pensacola*, FL 32502
T: +1 850 5954468; Fax: +1 850 5954470
Susan Dobinson
27 000 vols; 10 curr per   34657

US Navy – Naval Air Station Library, Commanding Officer Sta Libr, 190 Radford Blvd, Bldg 624, *Pensacola*, FL 32508-5217
T: +1 850 4524362; Fax: +1 850 4532028; E-mail: judith.walker@cnet.navy.mil
1914; Judith Walker
Navy hist, Aviation, US hist
28 000 vols; 150 curr per   34658

Peoria County Law Library, Peoria County Court House – 324 Main, Rm 211, *Peoria*, IL 61602
T: +1 309 6726084
Vicky Mundwiler
14 000 vols   34659

US Department of Agriculture, National Center for Agricultural Utilization Research, Agricultural Research Service, Research Library, 1815 N University St, *Peoria*, IL 61604
T: +1 309 6816526; Fax: +1 309 6816681; E-mail: blumenj@ncaur.usda.gov; URL: www.ncaur.usda.gov
1941; Joyce Blumenshine
22 000 vols; 162 curr per; 300 microforms
libr loan; SLA, ALA   34660

Florida Department of Corrections, Taylor Correctional Institution Library, 8515 Hampton Springs Rd, *Perry*, FL 32348
T: +1 850 8384080
Jerry Wells
13 000 vols; 41 curr per   34661

State of Ohio Department of Corrections, Mohican Juvenile Correctional Facility Library, 1012 ODNR Mohican S, *Perrysville*, OH 44864; Box 150 State Park Rd, Loudonville, OH 44824
T: +1 419 9944127; Fax: +1 419 9943441
Susan Ridgeway
8 000 vols   34662

US Air Force – Peterson Air Force Base Library FL 2500, 21 SVS/SVRL, 21 SVS/SVMG, 201 W Stewart Ave, Bldg 1171, *Peterson AFB*, CO 80914-1600
T: +1 719 5567463; Fax: +1 719 5566752
1950; Kathleen Kucharski
Aeronautics, Space science, Business & management
35 000 vols; 150 curr per   34663

Philadelphia Common Pleas & Municipal Court Law Library, Rm 600, City Hall, *Philadelphia*, PA 19107
T: +1 215 6863799; Fax: +1 215 6863737
1970; James M. Clark
59 000 vols; 132 curr per   34664

US Court of Appeals, William HAC Library, Burn Courthouse, 1st Flr, 601 Market St, *Philadelphia*, PA 19106
T: +1 267 2994300; Fax: +1 267 2994328
Susan B. English
100 000 vols; 150 curr per   34665

US Environmental Protection Agency, Region 3, Regional Center for Environmental Information, 1650 Arch St, *Philadelphia*, PA 19103
T: +1 215 814-5254; Fax: +1 215 814-5253; E-mail: library-reg3@epa.gov; URL: www.epa.gov/libraries/region3.html
1972; Domenico Lazzaro
14 000 vols; 200 curr per   34666

US Navy – Naval Ship Systems Engineering Station Library, Philadelphia Naval Surface Warfare Ctr-Carderock Div, Bldg 4, *Philadelphia*, PA 19112-5083
T: +1 215 8977816; Fax: +1 215 8978966
1909; Kathleen Schollenberger
Navy Tech Manuals Coll
25 000 vols; 25 curr per   34667

Arizona Department of Corrections – Adult Institutions, Arizona State Prison Complex – Phoenix Library, 2500 E Van Buren St, *Phoenix*, AZ 85008; P.O. Box 52109, Phoenix, AZ 85072-2109
T: +1 602 6853100; Fax: +1 602 6853129
Ron Brugman
15 000 vols   34668

Arizona Department of Economic Security, DES Authority Library, 1789 W Jefferson, *Phoenix*, AZ 85007
T: +1 602 5424777; Fax: +1 602 5424496
1977; Diane Baker
Federal & State Regulations Internal Docs & Codes
10 000 vols; 103 curr per
libr loan   34669

Arizona Department of Education, Library/Educational Information Center, 1535 W. Jefferson St Bin 48, *Phoenix*, AZ 85007
T: +1 602 5425416; Fax: +1 602 5420520
Linda Edgington
Arizona Education
10 000 vols; 130 curr per   34670

Arizona Department of Environmental Quality, ADEQ Library, 1110 W Washington St, *Phoenix*, AZ 85007
T: +1 602 771-2217; Fax: +1 602 2074872; E-mail: cona.lorraine@azdeq.gov; URL: www.azdeq.gov/function/assistance/library.html
Lorraine E. Akey
10 000 vols   34671

Arizona Department of Transportation Library, 2739 E Washington, *Phoenix*, AZ 85034; mail address: 206 S 17th Ave, No 075R, Phoenix, AZ 85007
T: +1 602 7123138; Fax: +1 602 7123140; E-mail: library@azdot.gov; URL: www.azdot.gov/TPD/ATRC/library/index.asp
1990; Dale Steele
35 000 vols; 85 curr per; 350 av-mat; 24 digital data carriers; 10 slides
libr loan; SLA   34672

Maricopa County Jail Library, 3150 W Lower Buckeye Rd, *Phoenix*, AZ 85009
T: +1 602 8765638; Fax: +1 602 3531546
Ed Van Winkle
20 000 vols   34673

Superior Court Law Library, 101 W Jefferson, *Phoenix*, AZ 85003
T: +1 602 5063461; Fax: +1 602 5063677; E-mail: services@scll.maricopa.gov; URL: www.superiorcourt.maricopa.gov/lawlibrary
1913; Richard Tennstra
140 000 vols   34674

United States Courts Library, Sandra Day O'Connor United States Courthouse, Ste 410, 401 W Washington St, SPC16, *Phoenix*, AZ 85003-2135
T: +1 602 3227295; Fax: +1 602 3227299
Tim Blake
34 000 vols   34675

South Dakota Supreme Court, Law Library, 500 E Capitol Ave, *Pierre*, SD 57501-5070
T: +1 605 7734898; Fax: +1 605 7736128
Sheridan Cash Anderson
30 000 vols   34676

Allegheny County Court of Common Pleas Library, 921 City County Bldg, 9th Fl, 414 Grant St, *Pittsburgh*, PA 15219-2465
T: +1 412 3505353; Fax: +1 412 3505889
Joel Fishman
24 000 vols; 424 curr per   34677

Allegheny County Health Department Library, 301 39th St, *Pittsburgh*, PA 15201
T: +1 412 5788028; Fax: +1 412 5788144; E-mail: jacksonl@iaa2.org
1973; N.N.
13 230 vols; 133 curr per   34678

Allegheny County Law Library, 921 City-County Bldg, 414 Grant St, *Pittsburgh*, PA 15219-2143
T: +1 412 3505353; Fax: +1 412 3505889; URL: www.duq.edu/law/lawlib/lawlib.html
1867; Frank Y. Liv
US Supreme Court Records & Briefs, 1912 to present; Alaskan Boundary Dispute 1893; Mss of D T Watson
110 000 vols; 109 curr per; 22 Journals
libr loan   34679

US Court of Appeals, Pittsburgh Branch Library, 512 US Courthouse, *Pittsburgh*, PA 15260
T: +1 412 6446485; Fax: +1 412 6445975
Linda Schneider
43 000 vols; 130 curr per   34680

El Dorado County Law Library, 550 Main St, Ste A, *Placerville*, CA 95667-5699
T: +1 530 6216423; Fax: +1 530 6261932
Marcus N. Prenk
California; Californian & Federal Legal Mats
10 300 vols   34681

New York State Supreme Court Fourth District, Law Library, 72 Clinton St, *Plattsburgh*, NY 12901
T: +1 518 5654808; Fax: +1 518 5621193; URL: www.courts.state.ny.us
Lynn Bezio
12 000 vols   34682

Kettle Moraine Correctional Institution Library, P.O. Box 31, *Plymouth*, WI 53073
T: +1 920 5263244 ext 309; Fax: +1 920 5263989
Conrad Reedy
15 000 vols; 45 curr per   34683

Michigan Department of Corrections, Robert Scott Correctional Facility Library, 47500 Five Mile Rd, *Plymouth*, MI 48170
T: +1 734 4597400; Fax: +1 734 4596148
Gayle Leach
10 000 vols   34684

Idaho State Correctional Institution Library, Pocatello Women's Correctional Center Library, 1451 Fore Rd, *Pocatello*, ID 83204
T: +1 208 2366360; Fax: +1 208 2366362
Dr. Betty Jo Willey
13 000 vols; 19 curr per   34685

US Navy – Naval Base Ventura Library/Resource Center, Bldg 221, Code N92V1, Point Mugu Sta, *Point Mugu*, CA 93042-5000; mail address: 10000 23rd Ave, Code MW400, Port Hueneme, CA 94043-4301
T: +1 805 9897771; Fax: +1 805 9898674
1946; Diana Hoslett
Military hist, Aeronautics
30 000 vols; 100 curr per   34686

Florida Department of Corrections, Polk Correctional Institution Library, 10800 Evans Rd, *Polk City*, FL 33868
T: +1 863 9842273; Fax: +1 863 9843072
Henry P. Ziegler Jr
14 000 vols; 41 curr per   34687

Adams-Pratt Oakland County Law Library, 1200 N Telegraph Rd, Bldg 14 E, *Pontiac*, MI 48341-0481
T: +1 248 8580012; Fax: +1 248 4529145; URL: www.oakgov.com/lawlib
1904; Laura Mancini
State law; Michigan Law (Supreme Court Records & Briefs, 1976 to date, House & Senate Bills, 1973 to date, Michigan Statutes, 1838 to date), micro, bk, binders
65 000 vols; 400 curr per   34688

Oakland County Jail Library, 1200 N Telegraph Rd, *Pontiac*, MI 48341-0450
T: +1 248 8580012; Fax: +1 248 4529145
David O. Conklin
25 000 vols; 18 curr per   34689

Oakland County Research Library, 1200 N Telegraph Rd, Bldg 14 E, *Pontiac*, MI 48341-0453
T: +1 248 8580738; Fax: +1 248 4529145; URL: www.oakgov.com/reslib
1973; N.N.
Business & management, Census, Local govt, Personnel management, Planning, Sewage, Sociology, Transportation; Census, Local Docs
22 000 vols; 130 curr per   34690

Pontiac Correctional Center Library, 700 W Lincoln St, *Pontiac*, IL 61764-2323; P.O. Box 99, Pontiac, IL 61764-0099
T: +1 815 8422816; Fax: +1 815 8423051
Malini Patel
Law
32 000 vols; 14 curr per   34691

US Air Force – Pope Air Force Base Library, FL 4488, 43 SVS/SVMG, 396 Sonic St Bldg 373, *Pope AFB*, NC 28308-5225
T: +1 910 3942195; Fax: +1 910 3941172; E-mail: robin.demark@pope.af.mil; URL: www.pope.af.mil
Faye Couture
Aeronautics, Military hist
30 000 vols   34692

US Navy – Naval Construction Battalion Center Library – The Resource Center, 1000 23rd Ave, Code 193, Bldg 1180, *Port Hueneme*, CA 93043-4301
T: +1 805 9824746; Fax: +1 805 9823514; E-mail: djhoslett@cbeph.navy.mil
1947; Diana Hoslett
World War II
40 000 vols; 53 curr per   34693

US Navy – Naval School, Civil Engineer Corps Officers, Moreell Library, 3502 Goodspeed St, Ste 1, *Port Hueneme*, CA 93043-4336
T: +1 805 9822826; Fax: +1 805 9822918; URL: www.cecos.navy.mil
1946; Deborah M. Gunia
Civil engineering, Construction; Admiral Ben Moreell Coll, Admiral Robert E. Peary Coll
34 000 vols; 46 curr per
libr loan; FEDLINK, OCLC   34694

US Navy – NFESC Technical Information Center, Technical Library, 1100 23rd Ave, *Port Hueneme*, CA 93043-4370
T: +1 805 9821124
1948; Bryan Thompson
Civil engineering, Construction, Soil mechanics, Ocean engineering
70 000 vols; 50 curr per
libr loan   34695

Kitsap County Law Library, 614 Division St, *Port Orchard*, WA 98366
T: +1 360 3375788; Fax: +1 360 3375789; E-mail: library@kitsapbar.com; URL: www.kitsapbar.com/library
Paul Fjelstad
10 000 vols   34696

Columbia Correctional Institution Library, 2925 Columbia Dr, *Portage*, WI 53901; P.O. Box 950, Portage, WI 53901-0950
T: +1 608 7429100; Fax: +1 608 7429111
1986; Glen W. Singer
Law Libr
24 000 vols; 38 curr per; 50 govt docs; 200 av-mat; 1 digital data carriers
libr loan   34697

Oregon Department of Environmental Quality Library, 811 SW Sixth Ave, *Portland*, OR 97204
T: +1 503 2296854; Fax: +1 503 2295850
1988; Olivia Jonason
10 000 vols; 315 curr per
libr loan   34698

Oregon Department of Geology & Mineral Industries Library, Ste 965, No 28, 800 NE Oregon St, *Portland*, OR 97232-2162
T: +1 971 6731555; Fax: +1 971 6731562; URL: www.oregongeology.com
1937; Margaret D. Jenks
Theses & diss on Geology of Oregon, Unpubl Data & Rpts on Geology of Oregon incl site-specific seismic studies
30 000 vols; 1000 diss/theses; 20 000 govt docs; 8000 maps; 16 000 slides & photos
libr loan; WAML   34699

US Army – Corps of Engineers, Technical Library, Portland District & North Pacific Division, CENWP-RL, 333 SW First Ave, *Portland*, OR 97204; P.O. Box 2946, Portland, OR 97208-2946
T: +1 503 8085140; Fax: +1 503 8085142; E-mail: cenwp.library@usace.army.mil; URL: www.nwd.usace.army.mil/im/library/home.asp
1938; N.N.
Civil engineering, Dredging, Water res planning; Portland District & N Pac Div Rpts
23 000 vols; 40 curr per
libr loan   34700

US Court of Appeals, Ninth Circuit Library, Pioneer Courthouse, 555 SW Yamhill, *Portland*, OR 97204-1494
T: +1 503 3262124; Fax: +1 503 3267047; E-mail: Scott_McCordy@lb9.uscourts.gov
Scott McCurdy
14 000 vols; 50 curr per
libr loan   34701

US Department of Energy, Bonneville Power Administration Library – CILL-1, 905 NE 11th Ave, *Portland*, OR 97232; P.O. Box 3621, Portland, OR 97208-3621
T: +1 503 2304171; Fax: +1 503 2305911
1939
BPA Coll of conservation, environment, fish and wildlife Reports
50 000 vols; 200 curr per; 500 microforms; 100 digital data carriers
libr loan   34702

US District Court, Central Library, 7A40 Mark O Hatfield US Courthouse, 1000 SW Third Ave, *Portland*, OR 97204
T: +1 503 3268140; Fax: +1 503 3268144
1975; Scott McCurdy
17 000 vols; 150 curr per   34703

US Navy – Naval Regional Medical Center, Bldg 1, 4th Flr, 620 John Paul Jones Circle, *Portsmouth*, VA 23708-2197
T: +1 757 9535383; Fax: +1 757 9537533; E-mail: nmcp-libraryservices@mar.med.navy.mil; URL: www.nmcp.med.navy.mil/library
Jane Pellegrino
32 000 vols; 390 curr per   34704

US Navy – Norfolk Naval Shipyard, Technical Library, Eng & Planning Dept, Code 202-3, Bldg 29, 2nd Fl, *Portsmouth*, VA 23709-5000
T: +1 757 3965674; Fax: +1 757 3964103
Pamela L. Stevens
Technical Manuals Coll
57 000 vols; 25 curr per   34705

Schuylkill County Law Library, Schuylkill County Court House, 401 N Second St, *Pottsville*, PA 17901
T: +1 570 6281235; Fax: +1 570 6281017
1888; D. Susan Kost
30 000 vols; 20 curr per   34706

New York State Supreme Court, Law Library, Court House, Ninth Judicial District, 50 Market St, *Poughkeepsie*, NY 12601-3203
T: +1 845 4862215; Fax: +1 845 4862216; E-mail: mboback@pppmail.appliedtheory.com
1904; Mary Boback
Legal Ref Mat
14 500 vols   34707

Yavapai County Law Library, Yavapai County Courthouse, Rm 201, *Prescott*, AZ 86301
T: +1 520 7713309; Fax: +1 520 7713389
Laura Baxter
12 275 vols; 44 curr per; 5 digital data carriers   34708

National Oceanic & Atmospheric Administration, Geophysical Fluid Dynamics Laboratory Library, Forrestal Campus, US Rte 1, P.O. Box 308, *Princeton*, NJ 08542
T: +1 609 4526550; Fax: +1 609 9875063; E-mail: gth@gfdl.gov; URL: www.gfdl.gov
1968; Gail T. Haller
Applied mathematics, Meteorology, Oceanography, Science & technology
10 000 vols; 130 curr per; 10 digital data carriers
libr loan   34709

Rhode Island State Law Library, Frank Licht Judical Complex, Frank Licht Judical Complex, 250 Benefit St, *Providence*, RI 02903
T: +1 401 2223275; Fax: +1 401 2223865
1827; Karen Quinn
Rhode Island Colonial Laws, Old English Law
100 000 vols; 366 curr per; 1 incunabula; 350 govt docs; 350 maps; 20 000 microforms; 40 sound-rec; 43 digital data carriers
ALA, AALL   34710

US Court of Appeals, First Circuit Satellite Library, One Exchange Terrace, Rm 430, *Providence*, RI 02903-1746
T: +1 401 7527240; Fax: +1 401 7527245; E-mail: stephanie_mutty@ca1.uscourts.gov
1990; Stephanie S. Mutty
10 000 vols; 20 curr per
libr loan   34711

Colorado Department of Corrections, LaVista Correctional Facility Library, PO Box 3, *Pueblo*, CO 81002
T: +1 719 5444800; Fax: +1 719 5835755
Laura Heberly
8 000 vols; 35 curr per   34712

Florida Department of Corrections, Charlotte Correctional Institution Library, 33123 Oil Well Rd, *Punta Gorda*, FL 33955
T: +1 941 8332344; Fax: +1 941 5755763
Willie Davenport
14 000 vols; 41 curr per   34713

Connecticut Judicial Branch, Putnam Law Library, Court House, 155 Church St, *Putnam*, CT 06260
T: +1 860 9283716; Fax: +1 860 9637531; E-mail: donna.izbicki@jud.state.ct.us; URL: www.jud.state.ct.us/lawlib/index.html
Donna R. Izbicki
17 500 vols   34714

Federal Bureau of Investigation, Academy Library, MCB No 4, *Quantico*, VA 22135
T: +1 703 6323200; Fax: +1 703 6323214; E-mail: librarian@fbiacademy.edu; URL: fbilibrary.fbiacademy.edu
1972; Eugenia B. Ryner
Criminal justice, Law enforcement, Police
45 000 vols; 327 curr per
libr loan; ALA   34715

US Marine Corps – Quantico Family Library, 2040 Broadway St, *Quantico*, VA 22134-5107
T: +1 703 7844353; Fax: +1 703 7844306; URL: www.mcu.usmc.mil/mcrcweb/library6.html
Carol E. Ramkey
30 000 vols; 100 curr per   34716

Gadsen Correctional Institution Library, 6044 Greensboro Hwy, *Quincy*, FL 32351
T: +1 850 8759701; Fax: +1 850 8759710
Gary Kitch
12 000 vols; 71 curr per   34717

Racine County Law Library, 730 Wisconsin Ave, *Racine*, WI 53403-1247
T: +1 262 6363773; Fax: +1 262 6363341
Theresa Wheary
16 000 vols   34718

East Jersey State Prison Library, Lock Bag R, Woodbridge Ave, *Rahway*, NJ 07065
T: +1 732 3962695; Fax: +1 732 4995023
Rick Liss
10 000 vols   34719

Florida Department of Corrections, New River East Correctional Institution Library, 7819 NW 228th St, *Raiford*, FL 32026; P.O. Box 3040, Raiford, FL 32026-3040
T: +1 904 3683000; Fax: +1 904 3683283; URL: www.myflorida.com
Ketura Williams
8 000 vols; 41 curr per   34720

Florida State Prison Library, Main Unit, 7819 NW 228 St, *Raiford*, FL 32026
T: +1 904 3682583
1968; Simeon Cerdan
Records, Legal
14 000 vols; 41 curr per   34721

Union Correctional Institution Library, State Rd 16, P.O. Box 221, *Raiford*, FL 32083-0221
T: +1 386 4312000 ext 2173; Fax: +1 386 4311023
1968; M. Peterson
Lit, Science fiction, African-American, Hispanic; Law Coll
10 000 vols; 85 curr per
libr loan   34722

North Carolina Department of Environment & Natural Resources Library, 512 N Salisbury St, Rm 719, *Raleigh*, NC 27604-1170; mail address: 1610 Mail Service Ctr, Raleigh, NC 27699-1610
T: +1 919 715-4160; Fax: +1 919 733-2622; E-mail: mary.tucker@ncmail.net; URL: www.eenorthcarolina.org/library/about.htm
1987; Mary Tucker
20 000 vols; 50 curr per
libr loan; ALA, SLA, NCLA   34723

North Carolina Legislative Library, 500 Legislative Office Bldg, 300 N Salisbury St, *Raleigh*, NC 27603-5925
T: +1 919 7339390; Fax: +1 919 7155460; URL: www.ncleg.net/leglibrary
1967; Cathy L. Martin
Legislation (Standing & Study Committee Notebks)
20 000 vols   34724

North Carolina Supreme Court Library, 500 Justice Bldg, Two E Morgan St, *Raleigh*, NC 27601-1428 T: +1 919 8315709; Fax: +1 919 7330486; E-mail: tpd@sc.state.nc.us; URL: www.aoc.state.nc.us/www/public/html/sc_library.htm
1812; Thomas Davis
143 000 vols; 319 curr per; 117 000 microforms; 18 sound-rec; 35 digital data carriers 34725

Crestmont College Salvation Army, Elftman Memorial Library, 30840 Hawthorne Blvd, *Rancho Palos Verdes*, CA 90275 T: +1 310 5446475; Fax: +1 310 2656514; E-mail: mukbil1@msn.com
1923; Misty Jesse
Religion, Psychology; Salvation Army Publs
36 000 vols; 153 curr per; 40 digital data carriers; 75 slides, 100 overhead transparencies
libr loan; ALA 34726

US Air Force Air Education & Training Command, Library Program, HQ AETC/SVP, 100 H St W, *Randolph AFB*, TX 78154 T: +1 210 6522515; Fax: +1 210 6526683; E-mail: darlene.price@randolph.af.mil
Darlene Price-Boiser
1 000 000 vols; 7 300 curr per; 223 000 microforms; 1 800 000 tech rpts & doc 34727

US Air Force – Randolph Air Force Base Library, 12 SPTG/SVMG, Bldg 598, Fifth St E, *Randolph AFB*, TX 78150-4405 T: +1 210 6522617, +1 210 6525578; Fax: +1 210 6523261; E-mail: david.ince@randolph.af.mil; URL: rafblibrary.org
1933; David Ince
Military hist, Management, Professional development; Air War College, Project Warrior, Texana
36 800 vols; 125 curr per 34728

Portage County Law Library, 241 S Chestnut St, *Ravenna*, OH 44266 T: +1 330 2973661
Mary Alice Law
17 000 vols; 20 curr per 34729

Berks County Law Library, Courthouse, Tenth Flr, 633 Court St, *Reading*, PA 19601-4302 T: +1 610 4783370; Fax: +1 610 4786375; E-mail: lawlibrary@countyofberks.com; URL: www.co.berks.pa.us/lawlibrary
1859; Melanie R. Solon
34 000 vols; 1 300 curr per 34730

Shasta County Law Library, 1500 Court St, Court House, B-7, *Redding*, CA 96001 T: +1 530 2456243; Fax: +1 530 2456966; E-mail: lawlib@snowcrest.com
Jeanne Capell
13 500 vols; 12 curr per; 1 500 microforms
libr loan 34731

US Army – Aviation & Missile Command Redstone Scientific Information Center, AMSAM-RD-AS-I-RSIC, AMSAM-RD-AS-I-RSIC, Bldg 4484, *Redstone Arsenal*, AL 35898-5000 T: +1 256 8769309; Fax: +1 256 8427415; E-mail: rsic@amrdec.army.mil; URL: rsic.redstone.army.mil
1949; Martha Knott
Aeronautics, Astronautics, Chemistry; Guidance & Control; Helicopter, Metrication, Peenemuende Papers, Rocket Technology, Space Defense
240 000 vols; 350 curr per 34732

San Mateo County Law Library, 710 Hamilton St, *Redwood City*, CA 94063 T: +1 650 3634913; Fax: +1 650 3678040; URL: www.smcll.org
Karen M. Lutke
36 000 vols; 20 curr per 34733

Georgia Department of Corrections, Office of Library Services, Rogers State Prison, 1978 Georgia Hwy 147, *Reidsville*, GA 30453 T: +1 912 5577771; Fax: +1 912 5577051; URL: www.dcor.state.ga.us
14 000 vols; 30 curr per 34734

Library of the US Courts, 400 S Virginia St, Rm 1001, *Reno*, NV 89501 T: +1 775 6865776; Fax: +1 775 6865779
Cheryl Bjerke
22 000 vols 34735

Washoe County Law Library, Courthouse, PO Box 30083, *Reno*, NV 89520-0083 T: +1 775 3283250; Fax: +1 775 3283441; E-mail: lawlib@washoecounty.us; URL: www.washoecounty.us/lawlib
1915; Sandra Marz
General Law Libr (Nevada Law, California Law Coll)
53 000 vols; 315 curr per 34736

California Department of Corrections Library System, California State Prison, Sacramento Inmate Library, One Prison Rd, *Represa*, CA 95671; P.O. Box 290002, Represa, CA 95671-0002 T: +1 916 9858610; Fax: +1 916 9853407
J. Salter
Genealogy, law
28 000 vols 34737

California Department of Corrections Library System, Folsom State Prison, 300 Prison Rd, *Represa*, CA 95671; P.O. Box 71, Represa, CA 95671-4071 T: +1 916 9852561; Fax: +1 916 3513002
Robert Morris
Mental health, law, literacy
70 000 vols 34738

Folsom State Prison Library, Prison Rd, P.O. Box 71, *Represa*, CA 95671 T: +1 916 9852561 ext 4236; Fax: +1 916 3513002
1949; Robert Morris
Law, Literacy, Mental Health, African-American, American Indians, Selfhelp, Spanish; Gillmore/Lynch Law coll
40 000 vols; 65 curr per
libr loan; ALA, CLA 34739

US Geological Survey Library, Clarencê King Library, National Ctr, Rm 1D100, 12201 Sunrise Valley Dr, *Reston*, VA 20192 T: +1 703 6486243; Fax: +1 703 6486373; E-mail: library@usgs.gov; URL: library.usgs.gov
1882; Robert A. Bier Jr
Gems & Minerals (George F Kunz Coll), Russian Geology Bks (Alverson)
1 000 000 vols; 700 curr per; 500 e-journals; 1 000 e-books; 330 000 maps; 387 000 microforms; 275 sound-rec; 125 digital data carriers; 273 000 pamphlets
libr loan 34740

US Court of Appeals, Fourth Circuit Library, United States Courthouse, 1000 E Main St, *Richmond*, VA 23219-3517 T: +1 804 9162319
Aly McClure
100 000 vols; 300 curr per 34741

Virginia State Law Library, Supreme Court Bldg, 2nd Flr, 100 N Ninth St, *Richmond*, VA 23219-2335 T: +1 804 7862075; Fax: +1 804 7864542; E-mail: lawlibrary@courts.state.va.us; URL: www.courts.state.va.us/library
1899; Gail Warren
English Rpts (Nominative Rpts, Mostly Originals), 18th C Treatises
100 000 vols; 336 curr per 34742

Tenth Judicial District Supreme Court Law Library, Arthur M Cromarty Court Complex, 220 Center Dr, *Riverhead*, NY 11901-3312 T: +1 631 8521887; Fax: +1 631 8521782
1909; Lynn L. Fullshire
2 branch libs
110 000 vols; 150 curr per 34743

California State Court of Appeal, Fourth Appellate District, Division Two Library, 3389 12th St, *Riverside*, CA 92501 T: +1 909 2480220; Fax: +1 909 2480235
Terry R. Lynch
30 000 vols 34744

Riverside County Law Library, 3989 Lemon St, *Riverside*, CA 92501-3674 T: +1 951 9556397; Fax: +1 951 9556394; E-mail: lawlib-r@co.riverside.ca.us; URL: www.lawlibrary.co.riverside.ca.us
Gayle E. Webb
Municipal Codes Coll
103 000 vols 34745

New York State Judicial Department, Appellate Division Law Library, M Dolores Denman Courthouse, 50 East Ave, Ste 100, *Rochester*, NY 14604-2214 T: +1 585 5303251; Fax: +1 585 5303270; URL: www.courts.state.ny.us/ad4/lib/
1849; David Voisinet
Pre-1850 Law bks, Regional Directories & Plat Bks
310 000 vols; 650 curr per 34746

Rock Island County Law Library, 210 5th St, Courthouse, *Rock Island*, IL 61201 T: +1 309 7864451 ext 3255; Fax: +1 309 5583263
Evelyn Schafer
10 000 vols 34747

US Army – Corps of Engineers, Rock Island District Technical Library, Rodman Ave, Clock Tower Bldg, P.O. Box 2004, *Rock Island*, IL 61204-2004 T: +1 309 7945576; Fax: +1 309 7945807; URL: iepacl.brodart.com/search/um; www.mvr.usace.army.mil/library
1975; Bob Romic
Civil engineering, Soil mechanics; Corps of Engineers Hist Coll, Environmental Resources, Hydraulics
11 000 vols; 100 curr per 34748

Winnebago County Law Library, Courthouse Bldg, Ste 301, *Rockford*, IL 61101-1221 T: +1 815 3194967; Fax: +1 815 3194801
1975; Brian L. Buzard
Illinois Law, Illinois Law School Law Reviews
20 000 vols; 20 curr per 34749

Connecticut Judicial Branch, Law Library at Rockville, 69 Brooklyn St, *Rockville*, CT 06066 T: +1 860 8964955; Fax: +1 860 8753213; URL: www.jud.ct.gov/lawlib
Lawrence Cheeseman
25 000 vols 34750

Montgomery County Circuit Court, Law Library, Judicial Ctr, 50 Maryland Ave, Ste 326, *Rockville*, MD 20850 T: +1 240 7779120; Fax: +1 240 7779126; URL: www.courts.state.md.us/montgomery
Janet Camillo
Continuing education, Legal
80 000 vols; 110 curr per 34751

US Department of Health & Human Services – FDA Center for Devices & Radiological Health Library, 9200 Corporate Blvd, Rm 030, *Rockville*, MD 20850 T: +1 301 8276901; Fax: +1 301 5942352; URL: www.cdrh.fda.gov
1976; Harriet Albersheim
10 000 vols; 600 curr per 34752

US Nuclear Regulatory Commission Library, 11545 Rockville Pike, *Rockville*, MD 20852-2738; T2C8, Washington, DC 20555 T: +1 301 4157204; Fax: +1 301 4155365; E-mail: library@nrc.gov
Thomas E. Smith
Int Atomic Energy Agency Publs
20 000 vols; 400 curr per; 450 000 microforms 34753

Missouri Department of Natural Resources, Division of Geology & Land Survey Library, 111 Fairgrounds Rd, *Rolla*, MO 65401-2909; P.O. Box 250, Rolla, MO 65401-2250 T: +1 573 3682100; Fax: +1 573 3682111; E-mail: geology@dnr.mo.gov
1853; Carolyn Ellis
Missouri Geology (Fowler File & Ed Clark Museum)
58 000 vols; 37 curr per 34754

Oneida Correctional Facility Library, 6100 School Rd, *Rome*, NY 13440 T: +1 315 3396880; Fax: +1 315 3396880 ext 3299
Bruce Thompson
8 000 vols; 65 curr per 34755

US Air Force – Air Force Research Laboratory-Information Directorate Technical Library, FL 2810, AFRL/IFOIL, 26 Electronic Pky, *Rome*, NY 13441-4514 T: +1 315 3307600; Fax: +1 315 3303086; E-mail: IFOIL@RL.AF.MIL; URL: www.rl.af.mil
1952; Mike Heines
Computer science, Electronics, Mathematics, Radar
21 000 vols; 358 curr per
libr loan; OCLC 34756

Douglas County Law Library, Justice Bldg, Rm 305, *Roseburg*, OR 97470 T: +1 541 4404341; E-mail: dclawlib@co.douglas.or.us
Diana L. Hadley
11 000 vols; 10 curr per 34757

Minnesota Department of Corrections, Minnesota Correctional Facility – Rush City, 7600 – 525th St, *Rush City*, MN 55069 T: +1 320 3580400; Fax: +1 763 6897555; URL: www.doc.state.mn.us
Jonathan P. Chapman
17 000 vols; 60 curr per 34758

California Department of Justice, Attorney General's Law Library, 1300 I St, P.O. Box 944255, *Sacramento*, CA 95814 T: +1 916 3245314; Fax: +1 916 3235342
John Hoffman
California
32 000 vols 34759

California Department of Pesticide Regulation Library, 1001 I St, *Sacramento*, CA 95814; P.O. Box 4015, Sacramento, CA 95814-2828 T: +1 916 3243548; Fax: +1 916 3241719; E-mail: librarydesk@cdpr.ca.gov
1981; Chizuko Kawamoto
60 000 vols; 78 curr per 34760

California State Court of Appeal, Law Library, 914 Capitol Mall, *Sacramento*, CA 95814; mail address: 900 N St, Ste 400, Sacramento, CA 95814 T: +1 916 6530207; Fax: +1 916 6530324
Linda Wallihan
California Law Coll
25 000 vols; 50 curr per 34761

California State Department of Food & Agriculture, Plant Test Diagnostic Library, 3294 Meadowview Rd, *Sacramento*, CA 95832-1448 Fax: +1 916 2621191; URL: www.cdfa.ca.gov
Ramona Randolph
30 000 vols; 300 curr per 34762

California State Department of Transportation, Law Library, 1120 N St, P.O. Box 1438, *Sacramento*, CA 95812-1438 T: +1 916 6542630 ext 3179; Fax: +1 916 6546128
Arno Nappi
17 000 vols; 183 curr per 34763

California State Department of Water Resources, Law Library, 1416 Ninth St, Rm 1118-13, *Sacramento*, CA 95814; P.O. Box 942836, Sacramento, CA 94236-0001 T: +1 916 6538001; Fax: +1 916 6530952
1966; Mary Ann Parker
20 000 vols; 30 curr per 34764

California State Legislative Counsel, Law Library, 925 L St, Lower Level, *Sacramento*, CA 95814-3772 T: +1 916 3418036; E-mail: lcb.library@lc.ca.gov; URL: www.leginfo.ca.gov
1951; Linda Heatherly
Legal Coll
45 000 vols; 152 curr per 34765

Sacramento County Public Law Library, 813 Sixth St, 1st Flr, **Sacramento**, CA 95814-2403
T: +1 916 8748541; Fax: +1 916 8745691; URL: www.saclaw.lib.ca.us
Shirley H. David
79 000 vols; 2 250 curr per   34766

State Public Defender, Law Library, 801 K St, Ste 1100, **Sacramento**, CA 95814
T: +1 916 3222676; Fax: +1 916 3249792
Moniquette Kaduk
Brief Bank
25 000 vols   34767

US Department of the Interior, Bureau of Reclamation, Mid-Pacific Regional Library, 2800 Cottage Way, Rm W-1825, **Sacramento**, CA 95825-1898
T: +1 916 9785593; Fax: +1 916 9785599
1946; Stephen Jones
Water resources; California, Nevada & Oregon Climatological Data, microfiche, San Joaquin Valley Drainage Program
21 000 vols; 120 curr per
libr loan   34768

Arizona Department of Corrections – Adult Institutions, Arizona State Prison Complex – Safford Library, 896 S Cook Rd, **Safford**, AZ 85546
T: +1 928 4284698
Tamara Swerline
9 000 vols   34769

Saginaw County Law Library, 111 S Michigan Ave, Rm LL007, **Saginaw**, MI 48602
T: +1 989 7905533; Fax: +1 989 7905248
Patricia Becker
20 000 vols; 20 digital data carriers   34770

Juvenile Correction Center Library, 2220 E 600 North, **Saint Anthony**, ID 83445; P.O. Box 40, Saint Anthony, ID 83445-0040
T: +1 208 6243462; Fax: +1 208 6240973
Kate Jensen
9 000 vols   34771

Belmont County Law Library, Court House, 101 W Main St, **Saint Clairsville**, OH 43950
T: +1 740 6952121; Fax: +1 740 6954968
Kate Subasic
25 000 vols   34772

Minnesota Department of Corrections, Minnesota Correctional Facility – St Cloud, 2305 Minnesota Blvd SE, **Saint Cloud**, MN 56304
T: +1 320 2403071; URL: www.doc.state.mn.us
Teri Hams
13 000 vols; 56 curr per   34773

Northeast Regional Correctional Facility Library, 1270 US Rte 5 S, **Saint Johnsbury**, VT 05819
T: +1 802 7511410; Fax: +1 802 7481482
Anne Cote
Literacy for adults
8 000 vols; 30 curr per   34774

Mid-Michigan Correctional Facility Library, 8201 N Croswell Rd, **Saint Louis**, MI 48880
T: +1 989 6814361; Fax: +1 989 6814203
1990
Hist; Law
15 000 vols   34775

Missouri Court of Appeals-Eastern District Library, One Old Post Office Sq, Rm 304, 815 Olive St, **Saint Louis**, MO 63101
T: +1 314 5394300; Fax: +1 314 5394324
Maureen Jacquot
Regional Rpts Appellate; US Supreme Court Cases
29 000 vols; 12 curr per   34776

US Court of Appeals, 8th Circuit Library, Thomas F Eagleton US Courthouse, 111 S Tenth St, Rm 22-300, **Saint Louis**, MO 63102
T: +1 314 2442665; Fax: +1 314 2442675; E-mail: library8th@ca8.uscourts.gov; URL: www.lb8.uscourts.gov
Ann T. Fessenden
64 000 vols; 58 000 microforms; 299 digital data carriers   34777

US Department of Defense – National Imagery & Mapping Agency Reference Library, 3200 South Second St, **Saint Louis**, MO 63118; mail address: 3838 Vogel Rd, L-37/ISDRS, Arnold, MO 63010-6238
T: +1 314 2634266; Fax: +1 314 2634441
1943; Paige Shoger
32 000 vols; 450 curr per   34778

Metropolitan Council Library, 230 E Fifth St, **Saint Paul**, MN 55101
T: +1 651 6021310; Fax: +1 651 6021464; E-mail: jan.price@metc.state.mn.us; URL: www.metrocouncil.org
1967; Jan Price
Housing, Regional planning, Transportation; Local Govt Comprehensive Plans
8 400 vols; 300 curr per; 3 426 microforms; 42 digital data carriers
libr loan   34779

Minnesota Attorney General Library, Bremer Tower, Ste 1050, 445 Minnesota St, **Saint Paul**, MN 55101-2109
T: +1 651 2968152; Fax: +1 651 2967000; URL: www.ag.state.mn.us
Anita Anderson
Minnesota Attorney General Opinions Coll, micro, VF
21 000 vols; 600 curr per   34780

Minnesota Department of Transportation Library, 395 John Ireland Blvd, MS 155, **Saint Paul**, MN 55155
T: +1 651 3663791; Fax: +1 651 3663789; E-mail: library@dot.state.mn.us; URL: www.dot.state.mn.us/library
1957; Jerome C. Baldwin
28 000 vols; 300 curr per   34781

Minnesota Legislative Reference Library, 100 Constitution Ave, 645 State Off Bldg, **Saint Paul**, MN 55155
T: +1 651 2963398; Fax: +1 651 2969731; E-mail: refdesk@library.leg.state.mn.us; URL: www.library.leg.state.mn.us
1969; Robbie LaFleur
Bills Introduced for Ten Years; Interim Committee Rpts; Minnesota Docs Coll; Minnesota Govt Manual, 1887-present; Minnesota Govt Publs, on fiche; Senate & House Journals, tapes, committee minutes
39 000 vols; 600 curr per
libr loan   34782

Minnesota Pollution Control Agency Library, 520 Lafayette Rd, **Saint Paul**, MN 55155-4194
T: +1 651 2967719; Fax: +1 651 2825446; URL: www.pca.state.mn.us
Kathleen Malec
EPA Coll, micro
25 000 vols; 200 curr per   34783

Minnesota State Law Library, Minnesota Judicial Ctr, Rm G25, 25 Rev Dr Martin Luther King Jr Blvd, **Saint Paul**, MN 55155
T: +1 651 2962775; Fax: +1 651 2966740; E-mail: askalibrarian@courts.state.mn.us; URL: www.lawlibrary.state.mn.us
1849; Barbara Golden
Minnesota Legal Per Index, MN Trial Coll, Program to Collect Prof Papers of Retired Justices of Minnesota Supreme Court
260 000 vols; 600 curr per; 190 000 govt docs; 24 000 microforms; 25 digital data carriers   34784

Ramsey County Law Library, 1815 Court House, **Saint Paul**, MN 55102
T: +1 651 2668391; Fax: +1 651 2668399; E-mail: florin@co.ramsey.mn.us
1936; Carol Florin
28 000 vols; 30 curr per   34785

US Courts Library, Eighth Circuit Library, 541 Federal Court Bldg, 316 N Robert St, **Saint Paul**, MN 55101
T: +1 651 8481320; Fax: +1 651 8481325; URL: www.ca8.uscourts.gov
Nancee Halling
20 000 vols; 30 curr per   34786

Pinellas County Law Library, Saint Petersburg Branch, 545 First Ave N, Rm 500, **Saint Petersburg**, FL 33701-3769
T: +1 727 5827875; Fax: +1 727 5827874; E-mail: pcll@intnet.net
1949; Rebecca A. Frank
20 000 vols; 75 curr per   34787

Marion County Law Library, 555 Court St, **Salem**, OR 97309; P.O. Box 14500, Salem, OR 97309
T: +1 503 5885090; Fax: +1 503 3734386; E-mail: lawlibrary@co.marion.or.us; URL: legal.co.marion.or.us/lawlibrary
Martha Renick
10 000 vols; 50 curr per   34788

Oregon Department of Transportation, ODOT Library & Information Resource Center, 355 Capitol St NE, Rm 22, **Salem**, OR 97301-3871
T: +1 503 9863280; Fax: +1 503 9864025; E-mail: garnet.k.elliott@odot.state.or.us; URL: www.odot.state.or.us/ssbbsrvcpublic/rm/library.htm
1937; Garnet Elliott
15 000 vols; 90 curr per; 9 000 govt docs; 2 000 maps; 700 av-mat; 100 digital data carriers
libr loan; SLA   34789

Oregon Legislative Library, 347 State Capitol, 900 Court St, **Salem**, OR 97301
T: +1 503 9861668; Fax: +1 503 9861005; E-mail: david.b.harrell@state.or.us; URL: www.leg.state.or.us/comm/commsrvs/home.html
David Harrell
10 000 vols; 160 curr per
libr loan   34790

Oregon State Correctional Institution Library, 3405 Deer Park Dr SE, **Salem**, OR 97310-3985
T: +1 503 3737523; Fax: +1 503 3788919
1959; Greg Hunter
40 000 vols   34791

Oregon State Penitentiary Library, OSP Maximum, 2605 State St, **Salem**, OR 97310
T: +1 503 3782081
1953; M. Davidson
Agriculture & Horticulture, Art & architecture, Philosophy; Classical, Contemporary, Country & Western & Rock Music, LPs & tapes; Recordings of Classical Radio Broadcasts; Paintings by Inmate Artists, Indian Hist
20 000 vols   34792

Oregon State Penitentiary Library – Oregon Women's Correctional Center, 2809 State St, **Salem**, OR 97310
T: +1 503 3731911; Fax: +1 503 3788335
Coleen McMullen
10 000 vols   34793

State of Oregon Law Library, Supreme Court Bldg, 1163 State St, **Salem**, OR 97301-2563
T: +1 503 9865640; Fax: +1 503 9865623; E-mail: state.law.library@ojd.state.or.us; URL: ojd.state.or.us/osca/acs/lawlibrary/index.htm; egov.oregon.gov/soll/index.shtml
1848; Joe Stephens
100 000 vols; 200 curr per   34794

North Carolina Justice Academy, Learning Resource Center, 204 Faculty Ave, **Salemburg**, NC 28385; P.O. Box 99, Salemburg, NC 28385-0099
T: +1 910 5254151; Fax: +1 910 5254491; URL: www.jus.state.nc.us/ncja
1975; Alana Fisher
23 000 vols; 150 curr per   34795

Monterey County Law Library, 100 W Alisal, Ste 144, **Salinas**, CA 93901
T: +1 831 7555046; Fax: +1 831 4229593; E-mail: mcolawlib@redshift.com; URL: fp.redshift.com/mcolawlib
Joseph Wyatt
California Legal
15 800 vols   34796

United States Courts Library, US Courthouse, Rm 201,350 S Main St, **Salt Lake City**, UT 84101
T: +1 801 5243505; Fax: +1 801 5245375
Patricia Hummel
18 000 vols   34797

Utah State Law Library, 450 S State St, W-13, **Salt Lake City**, UT 84111-3101; P.O. Box 140220, Salt Lake City, UT 84114-0220
T: +1 801 2387990; Fax: +1 801 2387993; URL: www.utcourts.gov/lawlibrary
Jessica Van Buren
50 000 vols   34798

Calaveras County Law Library, County Counsel, Government Ctr, **San Andreas**, CA 95249
T: +1 209 7546314
Mike Ibold
8 000 vols   34799

Bexar County Law Library, Bexar County Courthouse, 5th Flr, 100 Dolorosa, **San Antonio**, TX 78205
T: +1 210 2278822; Fax: +1 210 2719614
James M. Allison
77 000 vols; 86 curr per   34800

US Air Force – AFIWC Library, 102 Hall Blvd, Ste 315, **San Antonio**, TX 78243-7078
T: +1 210 9772804; Fax: +1 210 9776621
1994; Emily Mardis
Electronic warfare, Engineering; Air War College, Project Warrior, Total Quality Management & TAPS (Transition Assistance Program)
20 000 vols; 20 curr per   34801

US Air Force – Kelly Air Force Base Library, 250 Goodrich Dr, Ste E6, **San Antonio**, TX 78241-5806; mail address: 37 MSG/SVMGK, 2261 Hughes Ave, Ste 115, San Antonio, TX 78236-9813
T: +1 210 9253214; Fax: +1 210 9257421
1917; JoAnn Castilleja
Military education, Job search, Transition; Hispanic Culture Coll, Military Hist, Texana
22 000 vols; 49 curr per   34802

US Department of the Air Force, HQ Air Force Services Agency, Directorate of Programs, Air Force Libr & Info Syst, HQ-AFSVA/SVPAL, 10100 Reunion Pl, Ste 502, **San Antonio**, TX 78216-4138
T: +1 210 6524589, +1 210 6523037; Fax: +1 210 6527039; E-mail: barabara.wrinkleb@agency.afsv.af.mil
Barbara Wrinkle
Aeronautics, Military hist
47 200 curr per   34803

Law Library for San Bernardino County, 402 North D St, **San Bernardino**, CA 92401
T: +1 909 8853020; URL: www.sblawlibrary.com; www.sblawlibrary.org
1891; Lawrence R. Meyer
American Law & Foreign Law; 2 branch libs
105 000 vols   34804

California Court of Appeal, Fourth Appellate District-Division One Law Library, 750 B St, Ste 300, **San Diego**, CA 92101-8173
T: +1 619 6452833; Fax: +1 619 6453009
1929; Nanna Frye
California law
40 000 vols   34805

California Department of Justice Library, 110 West A St, Ste 1311, **San Diego**, CA 92101
T: +1 619 6452162; E-mail: henexson@class.org
1972; Fay Henexson
26 000 vols; 50 curr per   34806

Marine Corps Recruit Depot Library, 3800 Chosin Ave, Bldg 7 W, *San Diego*, CA 92140-5196
T: +1 619 5241849; Fax: +1 619 5248243
Dan Cisco
Marine Corps Professional Reading Program
31 000 vols; 72 curr per   34807

San Diego City Attorney Library, 1200 Third Ave, Ste 1307, *San Diego*, CA 92101
T: +1 619 2366078; Fax: +1 619 2367215; E-mail: mlhyde@sandiego.gov
Mary Lynn Hyde
15 000 vols; 75 curr per   34808

San Diego County Public Law Library, Main Law Library, 1105 Front St, *San Diego*, CA 92101-3904
T: +1 619 6856552; Fax: +1 619 2387716; URL: www.sdcpll.org
1891; Charles R. Dyer
Local Legal Hist, California Appellate Court Briefs; 3 branch libs
222 000 vols; 298 curr per; 57 000 microforms; 190 av-mat; 3 719 sound-rec; 29 digital data carriers   34809

US Courts Library, 940 Front St, Rm 3185, *San Diego*, CA 92101-8920
T: +1 619 5575066; Fax: +1 619 5575077
Valerie A. Railey
52 000 vols   34810

US Navy – The Command Library, Fleet Anti-Sub Warfare Training Ctr, 32444 Echo Lane, Ste 100, *San Diego*, CA 92147-5199
T: +1 619 5241908; Fax: +1 619 5246875
1967; Donna Murico
18 000 vols; 30 curr per   34811

US Navy – Marine Corps Recruit Depot Library, W Chosin Ave, Bldg 7, *San Diego*, CA 92140-5196
T: +1 619 5241849, 1850; Fax: +1 619 5248243
1927; Rebecca B. Young
Marine Corps Hist Coll
36 000 vols; 75 curr per   34812

US Navy – Naval Base Coronado Library, MWR Base Library, MWR Base Library, 2478 Munda Rd, *San Diego*, CA 92155-5396
T: +1 619 4373026; Fax: +1 619 4373891
1944; Barbara Siemer
Vietnam
22 000 vols; 124 curr per; 1 800 av-mat; 300 sound-rec; 20 digital data carriers
libr loan; ALA, CLA   34813

US Navy – Naval Medical Center, Naval Med Ctr, Libr Bldg 5-2, *San Diego*, CA 92134-5200
T: +1 619 5327950; Fax: +1 619 5329293; E-mail: sndlmws@snd10.med.navy.mil
1922; Jan B. Dempsey
Layman Health Info
35 868 vols; 950 curr per; 8 digital data carriers   34814

US Navy – Spawar Systems Center San Diego Technical Library, Code 21512, 53560 Hull St, *San Diego*, CA 92152-5001
T: +1 619 5534890; Fax: +1 619 5534893
1949; Barbara Busch
Engineering, Electronics, Physics, Marine biology; War Res, Navy Radio & Sound Lab Rpts; National Defense Research Committee Coll, University of California Division of War Research, Navy Radio & Sound Laboratory Reports
225 000 vols; 700 curr per
libr loan   34815

US Navy – Supervisor of Shipbuilding Conversion & Repair Library, P.O. Box 368119, *San Diego*, CA 92136-5066
T: +1 619 5563385; Fax: +1 619 5563456; E-mail: parkerpat@exchange.sssd.navy.mil
1950; Pat Parker
20 000 vols   34816

US Navy – Wilkins Biomedical Library, Naval Health Research Center, P.O. Box 85122, *San Diego*, CA 92186-5122
T: +1 619 5538425; Fax: +1 619 5530213; E-mail: aldous@nhrc.navy.mil
1959; Mary Aldous
Medicine, Biochemistry, Immunology; Naval Medicine Coll, Prisoner of War Studies Publs
15 700 vols; 354 curr per; 310 microforms; 4 digital data carriers; 6 986 tech rpts   34817

City & County of San Francisco – Law Library, San Francisco City Attorney, 1390 Market St, Ste 500, *San Francisco*, CA 94102
T: +1 415 5543910; Fax: +1 415 5543837
Maria Protti
Municipal law; Municipal Rpts 1862 to date; San Francisco City Attorney Opinions 1893 to date; Statutes & Amendments to Codes of California Coll
24 250 vols; 26 curr per   34818

San Francisco Law Library, 401 Van Ness Ave, Rm 400, *San Francisco*, CA 94102-4672
T: +1 415 5546821; Fax: +1 415 5546820; URL: www.ci.sf.ca.us/sfll
1870; Marcia R. Bell
2 branch libs
308 000 vols; 492 curr per   34819

San Francisco Law Library – Market Street, 685 Market St, Ste 420, *San Francisco*, CA 94105
T: +1 415 8829310; Fax: +1 415 8829594
Marcia Bell
46 000 vols; 53 curr per   34820

United States Courts for the Ninth Circuit – Library, 95 Seventh St, *San Francisco*, CA 94103; P.O. Box 193939, San Francisco, CA 94119-3939
T: +1 415 5569500; Fax: +1 415 5569927
1891; Eric D. Wade
100 000 vols; 2 500 curr per; 406 000 microforms; 370 sound-rec; 7 digital data carriers   34821

US Army – Corps of Engineers, San Francisco Area Technical Library, 333 Market St, Rm 925, *San Francisco*, CA 94105-2197
T: +1 415 9778601; Fax: +1 415 9778423; E-mail: fconway@smtp.spd.usace.army.mil; URL: www.usace.mil:80/library.html
Frank Conway
CE Projects (Corps of Engineers Annual Rpts), San Francisco Bay-Delta, Wetlands
14 000 vols; 300 curr per; 38 digital data carriers   34822

US District Court, Law Library, 450 Golden Gate Ave, Box 36060, *San Francisco*, CA 94102
T: +1 415 4368130; Fax: +1 415 4368134
1960; Susan Wong Caulder
37 000 vols; 1 145 curr per   34823

US Environmental Protection Agency, Region IX Library, 75 Hawthorne St, *San Francisco*, CA 94105
T: +1 415 9723671; Fax: +1 415 7441474; E-mail: library-reg9@epamail.epa.gov; URL: www.epa.gov
1970; Colette Myles
Hazardous Waste, Nat Priorities List Docs, Pollution Prevention, Topographic Maps
8 000 vols; 200 curr per
libr loan   34824

Santa Clara County Law Library, 360 N First St, *San Jose*, CA 95113
T: +1 408 2993568; Fax: +1 408 2869283; URL: www.sccll.org
1874; Elaine Taranto
70 000 vols; 200 curr per   34825

Santa Clara Valley Water District Library, 160 Great Oaks Blvd, *San Jose*, CA 95119; mail address: 5750 Almaden Expressway, San Jose, CA 95118
T: +1 408 2652600; Fax: +1 408 9795693; URL: www.valleywater.org

Robert J. Teeter
Water, engineering, environment
20 000 vols; 400 curr per; 100 diss/theses; 8 000 govt docs; 100 maps; 400 av-mat; 100 sound-rec; 100 digital data carriers
libr loan; SLA, OCLC   34826

United States District Court Library, 280 S First St, *San Jose*, CA 95113
T: +1 408 5355323; Fax: +1 408 5355322
Nancy B. Selan
20 000 vols; 20 curr per   34827

California Department of Corrections Library System, California Men's Colony-West, Hwy 1, *San Luis Obispo*, CA 93409; P.O. Box 8101, San Luis Obispo, CA 93403-8101
T: +1 805 5477900; Fax: +1 805 5477792
Patrick Moloney
25 000 vols; 65 curr per   34828

California Department of Corrections Library System, California Men's Colony-East, Hwy 1 Drawer B, *San Luis Obispo*, CA 93409; P.O. Box 8101, San Luis Obispo, CA 93409
T: +1 805 5477900; Fax: +1 805 5477517
Joan Lienemann
40 000 vols; 125 curr per   34829

San Luis Obispo County Law Library, County Government Ctr, Rm 125, 1050 Monterey St, *San Luis Obispo*, CA 93408
T: +1 805 7815855; Fax: +1 805 7814172; E-mail: slolawli@rain.org; URL: www.thegrid.net/slocll
1896; Kathleen Boyd
California
26 035 vols; 60 curr per   34830

California Department of Corrections Library System, San Quentin State Prison, *San Quentin*, CA 94964
T: +1 415 4541460; Fax: +1 415 4555049
Doug Jeffrey
40 000 vols   34831

Marin County Law Library, 20 N San Pedro Rd, Ste 2015, *San Rafael*, CA 94903
T: +1 415 4996356; Fax: +1 415 4996837
1891; Hal Aigner
Continuing Education of the Bar Coll
29 000 vols   34832

Florida Department of Corrections, Baker Correctional Institution Library, 20706 US Hwy 90 W, *Sanderson*, FL 32087
T: +1 386 7194614; Fax: +1 386 7194516
Evan Brown
15 000 vols; 35 curr per   34833

Sandusky Bay Law Library Association, 323 Columbus Ave, *Sandusky*, OH 44870
T: +1 419 6264823; Fax: +1 419 6264826
1890; Kelly Del Vecchio
14 450 vols   34834

Orange County Public Law Library, 515 N Flower St, *Santa Ana*, CA 92703-2354
T: +1 714 8343397; Fax: +1 714 8344375; URL: www.ocpll.org
1891; Maryruth Storer
California SC, USSC
351 000 vols; 942 curr per; 904 000 microforms; 1 961 sound-rec; 153 digital data carriers
AALL, IALL   34835

McMahon Law Library of Santa Barbara County, County Court House, *Santa Barbara*, CA 93101
T: +1 805 5682296; Fax: +1 805 5682299; URL: www.countylawlibrary.org
1891; Raymond MacGregor
33 300 vols   34836

Santa Cruz County Law Library, 701 Ocean St, Rm 070, *Santa Cruz*, CA 95060
T: +1 831 4202205; Fax: +1 831 4572255; E-mail: scclawlib@yahoo.com; URL: www.lawlibrary.org
1896; Renee J. Fleming
18 000 vols   34837

New Mexico Corrections Department, Education Bureau, 4337 State Rd 14, *Santa Fe*, NM 87508; P.O. Box 27116, Santa Fe, NM 87505
T: +1 505 8278503; Fax: +1 505 8278548; URL: corrections.state.nm.us
Gail Oliver
Arizona Legal Access Model
115 000 vols; 210 curr per
libr loan   34838

New Mexico Supreme Court, Law Library, 237 Don Gaspar, *Santa Fe*, NM 87501; P.O. Drawer L, Santa Fe, NM 87504-0318
T: +1 505 8274850; Fax: +1 505 8274852; E-mail: libref@nmcourts.com; URL: www.supremecourtlawlibrary.org
1853; Michael Poulson
Anglo-American law, Pre-1850 Mexican law
185 000 vols; 412 curr per; 65 000 govt docs; 15 000 microforms; 20 digital data carriers   34839

Santa Barbara County Law Library, Santa Maria Branch, 312-E Cook St, *Santa Maria*, CA 93454
T: +1 805 3467548; Fax: +1 805 3467692; E-mail: smlaw@rain.org; URL: www.rain.org/~sblaw/santamaria.htm
Ray McGregor
16 000 vols   34840

Sonoma County Law Library, Hall of Justice, Rm 213-J, 600 Administration Dr, *Santa Rosa*, CA 95403-2879
T: +1 707 5652668; Fax: +1 707 5651126; URL: www.sonomacountylawlibrary.org
1891; Kim Tucker
29 000 vols   34841

N.Y.S. Supreme Court Library, Fourth Judicial District, 474 Broadway, Ste 10, *Saratoga Springs*, NY 12866-2297
T: +1 518 5844862; Fax: +1 518 5810966
1866; Vicki Cady
Directories of Saratoga Springs (1884-present)
20 000 vols; 12 digital data carriers
AALL   34842

City of Savannah, Municipal Research Library, City Hall, Rm 103, Two E Bay St at Bull St, *Savannah*, GA 31401; P.O. Box 1027, Savannah, GA 31402
T: +1 912 6516412; Fax: +1 912 2331992; URL: www.savannahga.gov
1974; Glenda E. Anderson
Savannah Area Local Docs
42 000 vols; 100 curr per   34843

Joseph F. Egan Memorial Supreme Court Library, Schenectady County Judicial Bldg, 612 State St, *Schenectady*, NY 12305
T: +1 518 2858518; Fax: +1 518 3775909
Patricia L. Schultz
28 000 vols   34844

US Army – Sergeant Rodney J. Yano Main Library, Bldg 560, *Schofield Barracks*, HI 96857-5000
T: +1 808 6558002; Fax: +1 808 6556375; URL: www.mwrarmyhawaii.com/leisureactivities/librariesla.asp
1915; Bonnie Dong
Military hist, Health; Hawaiiana Coll
80 000 vols; 200 curr per; 125 maps; 1 800 av-mat; 590 sound-rec; 350 digital data carriers
libr loan; ALA, HLA, OCLC
Military Post Libr. Main Libr for Army Libs (general) on O'ahu. For use by active duty and military retired, military dependents, and DOD civilians only   34845

US Air Force – Scott Air Force Base Library, 375 SVS/SVMG, 510 Ward Dr, *Scott AFB*, IL 62225-5360
T: +1 618 2565100; Fax: +1 618 2564558; URL: www.375services.com/scott_library.htm
1926; Sandra Koontz
Military hist; Education Office Coll, TQM
40 000 vols; 281 curr per   34846

New Jersey State Police Training Bureau Library, Training Center, *Sea Girt*, NJ 08750
T: +1 732 4495200 ext 5202; Fax: +1 732 4498763; URL: www.njsp.org
Linda DuBois
18 000 vols   34847

King County Department of Natural Resources & Parks, Technical Document & Research Center, 201 South Jackson St, Ste 190, *Seattle*, WA 98104
T: +1 206 6841129; Fax: +1 206 2960196; E-mail: research.center@metroke.gov
1975; Dawn Duddleson
Natural resources, Tranportation
15 000 vols; 50 curr per   34848

King County Law Library, W 621 King County Courthouse, 516 Third Ave, *Seattle*, WA 98104
T: +1 206 2960940; Fax: +1 206 2050513; E-mail: kcll@metrokc.gov; URL: www.kcll.org
Marcus L. Hochstetler
90 000 vols; 200 curr per   34849

National Environmental Satellite Data & Information Services, NOAA Library Seattle, 7600 Sand Point Way NE, Bldg 3, *Seattle*, WA 98115
T: +1 206 5266241; Fax: +1 206 5264535; E-mail: seattle.library@noaa.gov; URL: www.wrclib.noaa.gov
Brian Voss & Maureen Woods
Meteorology, Oceanography
15 000 vols   34850

US Army – Corps of Engineers, Seattle District Library, 4735 E Marginal Way, *Seattle*, WA 98134; P.O. Box 3755, Seattle, WA 98124-3755
T: +1 206 7643728; Fax: +1 206 7666444; E-mail: cenws.library@nws02.usace.army.mil; URL: www.nws.usace.army.mil
1940; Shelly Trulson
25 000 vols; 300 curr per; 50 000 microforms; 500 av-mat; 45 digital data carriers; 20 000 rpts & docs
libr loan; SLA   34851

US Courts Library, 700 Stewart St, Rm 19105, *Seattle*, WA 98101
T: +1 206 3708975; Fax: +1 206 3708976
1939; Timothy Sheehy
42 000 vols; 1 400 curr per   34852

US Army – Garrison-Selfridge Library, Bldg 780 W, *Selfridge Air National Guard Base*, MI 48045-5016
T: +1 810 3075238
1972; Jo Ann Bonnett
Military hist
26 390 vols; 331 curr per   34853

US Air Force – Seymour Johnson Air Force Base Library, 4 SVS/SVMG, 1520 Goodson St, *Seymour Johnson AFB*, NC 27531
T: +1 919 7225825; Fax: +1 919 7225835; URL: library.seymourjohnson.accqolnet.org
1956; Kim Huskins Webb
41 000 vols; 50 curr per   34854

Minnesota Department of Corrections, Minnesota Correctional Facility – Shakopee, 1010 W Sixth Ave, *Shakopee*, MN 55379-2213
T: +1 952 4964916; Fax: +1 952 4964460; URL: www.doc.state.mn.us
Andrea Smith
10 000 vols; 45 curr per   34855

US Air Force – Shaw Air Force Base Library, FL 4803, 451 Johnson St, Bldg 405, *Shaw AFB*, SC 29152
T: +1 803 8959810; Fax: +1 803 8959816; URL: www.20thservices.com/library.html
Janet Price
Transition Assistant Program-Military
40 000 vols; 76 curr per   34856

US Air Force – Sheppard Air Force Base Library, 82 SVS/SVMG, 425 Third Ave Bldg 312, *Sheppard AFB*, TX 76311-3043
T: +1 940 6766152; Fax: +1 940 8558854
1949
25 000 vols; 140 curr per   34857

US Air Force – 882nd Training Group Academic Library, 939 Missile Rd, Bldg 1900, *Sheppard AFB*, TX 76311-2245
T: +1 940 6762736; Fax: +1 940 6764025
1958; Patricia Boyd
Medicine, Dentistry, Nursing
10 000 vols; 150 curr per   34858

Sheridan Correctional Center Library, 4017 E 2603 Rd, P.O. Box 38, *Sheridan*, IL 60551
T: +1 815 4962311
Ramona Witte
10 500 vols; 110 curr per   34859

United States Courts Library, 300 Fannin St, Rm 5012, *Shreveport*, LA 71101-6305
T: +1 318 6763230; Fax: +1 318 9344866; E-mail: 5satlib-shreveport@ca5.uscourts.gov
Marian Drey
10 000 vols; 40 curr per   34860

Walter Reed Army Institute of Research, Gorgas Memorial Library, 503 Robert Grant Ave, *Silver Spring*, MD 20910-7500
T: +1 301 3199555; Fax: +1 301 3199402; E-mail: judy.kessenich@na.amedd.army.mil; URL: wrair-www.army.mil
1943; Judy H. Kessenich
Biomedical res, especially tropical infectious diseases
14 000 vols; 450 curr per; 16 000 microforms; 5 digital data carriers; 13 000 monogr & serials (bound), 15 000 bound journals vols
libr loan; MLA, SLA   34861

Woodbury County Law Library, Woodbury County Court House, 620 Douglas St, 6th Fl, *Sioux City*, IA 51102
T: +1 712 2796609; Fax: +1 712 2796577
1918; Richard P. Hustig
13 000 vols; 6 curr per   34862

South Dakota State Penitentiary, Donald M. Cole & Jameson Annex Library, 1600 N Dr, *Sioux Falls*, SD 57104-0915; P.O. Box 5911, Sioux Falls, SD 57117-5911
T: +1 605 3675171
Michelle Wysuph
8 000 vols; 40 curr per   34863

US Geological Survey, Earth Resources Observation Systems, EROS Data Center Library, EROS Data Ctr, *Sioux Falls*, SD 57198
T: +1 605 5946565; Fax: +1 605 5946529; E-mail: ortega@edcmail.cr.usgs.gov
1973; Laurie Ortega
10 000 vols; 75 curr per
libr loan; ALA, SLA   34864

Florida Department of Corrections, Apalachee Correctional Institution-East Unit Library, 35 Apalachee Dr, *Sneads*, FL 32460
T: +1 850 5936431; Fax: +1 850 5936445
1968; Judith Varn
Florida Legal Coll
13 000 vols; 50 curr per   34865

Osborn Correctional Institution, Osborn CI Library, 335 Bilton Rd, *Somers*, CT 06071; P.O. Box 100, Somers, CT 06071-0100
T: +1 860 5667500; Fax: +1 860 7633157
Darrell Harrison
Law
18 000 vols   34866

Somerset County Law Library, Court House, 111 E Union St, Ste 60, *Somerset*, PA 15501
T: +1 814 4451508; Fax: +1 814 4451455
Daniel Stants
22 000 vols; 75 curr per   34867

State Correctional Institution, Laurel Highlands Library, 5706 Glades Pine, *Somerset*, PA 15501; P.O. Box 631, Somerset, PA 15501-0631
T: +1 814 4456501; Fax: +1 814 4430269
Eleanor Silvis
10 000 vols   34868

State Correctional Institution, Somerset Library, 1590 Walters Mill Rd, *Somerset*, PA 15510-0001
T: +1 814 4438100; Fax: +1 814 4438157
Marcia Roman
15 000 vols; 70 curr per   34869

Somerset County Law Library, 20 N Bridge St, *Somerville*, NJ 08876; P.O. Box 3000, Somerville, NJ 08876-1262
T: +1 908 2317612; Fax: +1 908 2538590
Brad Small
20 000 vols   34870

Tuolumne County Law Library, 68 N Washington St, *Sonora*, CA 95370
T: +1 209 5360308; Fax: +1 209 5360718; E-mail: lawlib@mlode.com
David Holstrom
Statutes & Amendments to the Codes 1851-1993, California
8 500 vols   34871

Livingston Correctional Facility Library, PO Box 49, *Sonyea*, NY 14556
T: +1 585 6583710; Fax: +1 585 6584841
Amy Kelley
10 000 vols; 45 curr per   34872

Library of the US Courts, 204 S Main St, Robert A Grant Courthouse, Rm 316, *South Bend*, IN 46601
T: +1 219 2468050; Fax: +1 219 2468002; E-mail: law.library@lb7.uscourts.gov; URL: www.lb7.uscourts.gov
1986; Patricia Piasecki
14 000 vols; 135 curr per   34873

Saint Joseph County Law Library, Court House, 101 S Main St, *South Bend*, IN 46601
T: +1 219 2359657; Fax: +1 219 2359905; E-mail: dford@sjcba.org; URL: www.sjcba.org
Diane Ford
Law Books Coll, Reporters Coll, Statutes Coll
18 350 vols   34874

Massachusetts Department of Corrections, Institution Library at MCI Cedar Junction, PO Box 100, *South Walpole*, MA 02071-0100
T: +1 508 6682100
1956; Beverly Veglas
Law Libr Coll
10 000 vols   34875

US Army – Fort McCoy Post Library, Fort McCoy, B2000 11th Ave, *Sparta*, WI 54656-5161
T: +1 608 3882410; Fax: +1 608 3882690; E-mail: david.onstad@emh2.mccoy.army.mil
David Onstad
Military hist; Encyclopaedia Britannica Coll, ninth ed ca 1892; Military Insignia & Memorabilia & Patch Coll
10 000 vols; 20 curr per   34876

Spokane County Law Library, 421 Riverside Ave, Ste 1020, *Spokane*, WA 99201
T: +1 509 4773680; Fax: +1 509 4774722; URL: www.spokanecounty.org
1920
25 000 vols; 50 curr per   34877

US Department of the Interior – US Geological Survey Mineral Resources Library, W 904 Riverside Ave, Rm 202, *Spokane*, WA 99201
T: +1 509 3683101; Fax: +1 509 3683199
1946; Dave Frank
US Geological Survey Publs
11 000 vols; 25 curr per   34878

Bureau of Land Management, Eastern States Office Library, 7450 Boston Blvd, *Springfield*, VA 22153-3121
T: +1 703 4401561, 4401562; Fax: +1 703 4401570
Vernadine White
10 000 vols   34879

Illinois Environmental Protection Agency Library, 1021 N Grand Ave E, P.O. Box 19276, *Springfield*, IL 62794-9276
T: +1 217 7829691; Fax: +1 217 5244916; E-mail: nancy.simpson@epa.state.il.us; URL: www.epa.state.il.us
1970; Nancy R. Simpson
30 100 vols; 300 curr per
libr loan; SLA   34880

Illinois State Department of Transport, Technical Reference Library, 2300 S Dirksen Pkwy, 320 Admin Bldg, *Springfield*, IL 62764-0001
T: +1 217 7826680; Fax: +1 217 5243834; E-mail: motzkusgh@nt.dot.state.il.us
1965; Gisela Motzkus
25 000 vols; 293 curr per; 5 000 govt docs; 13 428 microforms; 167 digital data carriers
libr loan; SLA   34881

Legislative Reference Bureau Law Library, State Capitol, Rm 112, *Springfield*, IL 62706
T: +1 217 7826625; Fax: +1 217 7854583; URL: www.ilga.gov
1913; Mike Trudeau
Annotated Statutes for all Fifty States, Legislative Synopsis & Digests; Illinois Laws (since 1840)
20 000 vols; 29 curr per   34882

Massachusetts Trial Court, Hampden Law Library, 50 State St, *Springfield*, MA 01103-2021; P.O. Box 559, Springfield, MA 01102-0559
T: +1 413 7487923; Fax: +1 413 7342973; E-mail: hampdenlawlibrary@yahoo.com; URL: masslaw.library.net
1890; Kathleen M. Flynn
Fed law; Massachusetts Law
61 000 vols; 389 curr per   34883

Mike Durfee State Prison, Carl G. Lawrence Library, 1412 Wood St, *Springfield*, SD 57062
T: +1 605 3692201; Fax: +1 604 3692813
Mark Stoebner
30 000 vols; 50 curr per   34884

Missouri State Court of Appeals, Southern District Law Library, 300 Hammons Pkwy, University Plaza, *Springfield*, MO 65806
T: +1 417 8956813; Fax: +1 417 8956817
Beverly Heist
16 090 vols; 30 curr per   34885

Supreme Court of Illinois Library, Supreme Court Bldg, 200 E Capital Ave, *Springfield*, IL 62701-1791
T: +1 217 7822424; Fax: +1 217 7825287
1839; Brenda Larison
Law, Illinois; SJI Depot
100 000 vols; 500 curr per   34886

United States Courts Library, 600 E Monroe St, Rm 305, *Springfield*, IL 62701
T: +1 217 4924191; Fax: +1 217 4924192
Martha Doyle
8 000 vols; 35 curr per   34887

Connecticut Judicial Branch, Law Library at Stamford, Court House, 123 Hoyt St, *Stamford*, CT 06905
T: +1 203 9655250; Fax: +1 203 9655784; URL: www.jud.ct.gov/lawlib
Jonathan C. Stock
36 000 vols; 44 curr per   34888

Stanley Correctional Institution Library, 100 Corrections Dr, *Stanley*, WI 54768
T: +1 715 6442960; Fax: +1 715 6443777
Stacey Birch
14 000 vols; 95 curr per   34889

Department of Correctional Education, Powhatan, James River & Deep Meadow Correctional Center, 1954 State Farm Rd, *State Farm*, VA 23160-9998
T: +1 804 7843551; Fax: +1 804 7842480
1968; Abiodun Solanke
Black Studies (Paul Robeson Memorial Coll)
16 000 vols; 350 curr per   34890

Supreme Court Library, Richmond County Court House, 130 Stuyvesant Pl, Rm 300, **Staten Island**, NY 10301-1968
T: +1 718 3905291; Fax: +1 718 7274106
1920; Philip A. Klingle
75 000 vols; 281 curr per   34891

US Navy – Matthew Fontaine Maury Oceanographic Library, 1002 Balch Blvd, Bldg 1003, **Stennis Space Center**, MS 39522-5001
T: +1 228 6884597; Fax: +1 228 6884191; E-mail: library@navo.navy.mil; URL: www.navo.navy.mil
1871; Dr. Jack Breyer
Hydrographic Office Publs Coll; Oceanographic Expeditions Coll
400 000 vols; 250 curr per; 60 e-journals; 30 000 tech rpts
libr loan; Consortium of Naval Libraries, FEDLINK, OCLC   34892

Colorado Department of Corrections – Sterling Correctional Facility Library-East Side, 12101 Hwy 61, **Sterling**, CO 80751; P.O. Box 6000, Sterling, CO 80751
T: +1 970 5215010; Fax: +1 970 5218905; URL: www.doc.state.co.us
Jean Heverly
14 000 vols; 50 curr per   34893

Colorado Department of Corrections – Sterling Correctional Facility Library-West Side, 12101 Hwy 61, **Sterling**, CO 80751; P.O. Box 6000, Sterling, CO 80751
T: +1 970 5215010; Fax: +1 970 5218905; URL: www.doc.state.co.us
Jean Heverly
17 000 vols; 50 curr per   34894

Jefferson County Law Library Association, 301 Market St, **Steubenville**, OH 43952
T: +1 740 2838553; Fax: +1 740 2838629; E-mail: law_library@jeffcch.com
Ardis J. Stein
20 000 vols; 30 curr per   34895

Oklahoma Department of Career & Technology Education, Resource Center, 1500 W Seventh Ave, **Stillwater**, OK 74074-4364
T: +1 405 3772000; Fax: +1 405 7435142; URL: www.okcareertech.org/informationcommons/
1970; Denise Christy
20 000 vols   34896

California Youth Authority – N. A. Chaderjian Youth Correctional Facility Library, 7650 S Newcastle Rd, **Stockton**, CA 95213; P.O. Box 213014, Stockton, CA 95213-9014
T: +1 209 9446444; Fax: +1 209 9446167
Mark Huston
10 000 vols; 31 curr per   34897

California Youth Authority – O. H. Close Youth Correctional Facility Library, 7650 S Newcastle Rd, **Stockton**, CA 95213-9001; P.O. Box 213001, Stockton, CA 95213-9001
T: +1 209 9446346; Fax: +1 209 9446136
Leonardo Cortez
10 000 vols   34898

San Joaquin County Law Library, Kress Legal Center, 20 N Sutter St, **Stockton**, CA 95202
T: +1 209 4683920; Fax: +1 209 4689968; E-mail: info@sjclawlib.org; URL: www.sjclawlib.org
Barbara M. Zaruba
30 000 vols
AALL   34899

Green Haven Correctional Facility Library, Rte 216, **Stormville**, NY 12582
T: +1 845 2212711; Fax: +1 914 2212711 ext 2099
1942; Susan Blase
African-American studies, Coping skills mat, Hispanic studies, Occult, Yoga; Black Hist, Spanish
24 940 vols; 75 curr per   34900

Racine Correctional Institution Library, 2019 Wisconsin St **Sturtevant**, WI 53177; P.O. Box 900, Sturtevant, WI 53177-0900
T: +1 262 8863214; Fax: +1 262 8863514
Wendy Cramer
12 000 vols; 96 curr per   34901

US Census Bureau Library, 4600 Silver Hill Rd, Rm 1L1001, **Suitland**, MD 20746
T: +1 301 7632511; Fax: +1 301 4572194; E-mail: library@census.gov
1952; Catherine Mulhearn Earles
Statistics, Govt, Economics; Census Authors Papers, Foreign Statistics (Int Statistical Coll), US Census Coll
350 000 vols; 270 curr per
libr loan   34902

Northumberland County Law Library, 201 Market St, Court House, **Sunbury**, PA 17801-3471
T: +1 570 9884162; Fax: +1 570 9884497
1886; Catherine L. Kroh
10 550 vols   34903

Douglas County Law Library, 1313 Belknap St, **Superior**, WI 54880
T: +1 715 3951207
Sandy Picard
10 500 vols   34904

California Department of Corrections Library System, California Correctional Center, 711-045 Center Rd, **Susanville**, CA 96127; P.O. Box 790, Susanville, CA 96130-0790
T: +1 530 2572181; Fax: +1 530 2523020
Howard Cron
11 000 vols   34905

New York State Unified Court System, Syracuse Supreme Court Law Library, 401 Montgomery St, **Syracuse**, NY 13202
T: +1 315 6711143; Fax: +1 315 6711160
1849; Cynthia Kesler
Native American Law Mat Coll
165 000 vols; 1 200 curr per   34906

Pierce County Law Library, County-City Bldg, 930 Tacoma Ave S, Rm 1A – 105, **Tacoma**, WA 98402-2174
T: +1 253 7987494; Fax: +1 253 7982989; URL: www.piercecountywa.org/pc/services/lawjust/library/home.htm
Laurie B. Miller
35 000 vols; 20 curr per   34907

US Army – Madigan Army Medical Center, Medical Library, Bldg 9040 Fitzsimmons Dr, 2nd Flr, **Tacoma**, WA 98431; Attn: MCHJ-EDML (Medical Library), Tacoma, WA 98431-1100
T: +1 253 9680118; Fax: +1 253 9680958; E-mail: mamcmedlib@amedd.army.mil; URL: www.mamc.amedd.army.mil/medlib/ml_hmeinfo.htm
1944; Edean Berglund
27 000 vols; 450 curr per   34908

Talladega County Law Library, Talladega County Judicial Bldg, Northeast St, **Talladega**, AL 35161; P.O. Box 459, Talladega, AL 35161-0459
T: +1 256 7612116; Fax: +1 256 4805293
1955
35 000 vols; 100 curr per   34909

Department of State Library of Florida, The Capitol, Rm 701, **Tallahassee**, FL 32399
T: +1 850 4882812; Fax: +1 850 4889879; URL: dlis.dos.state.fl.us/leglib
1949; Cheri Ellison
22 000 vols; 410 curr per   34910

Federal Correctional Institution Library, 501 Capital Circle NE, **Tallahassee**, FL 32301-3572
T: +1 850 8782173; Fax: +1 850 6716121
1947; Jenny Warfield
11 000 vols; 90 curr per   34911

First District Court of Appeal Library, 301 Martin Luther King Jr Blvd, **Tallahassee**, FL 32399
T: +1 850 4888136; Fax: +1 850 4887989
Janet McPherson
25 000 vols   34912

Florida Attorney General's Law Library, Colins Bldg, 107 W Gaines St, Rm 437, **Tallahassee**, FL 32399-1050; mail address: The Capitol, PL 01, Tallahassee, FL 32301
T: +1 850 4143300; Fax: +1 850 9215784; E-mail: library@oag.state.fl.us
Betsy L. Stupski
50 000 vols; 20 curr per
libr loan   34913

Florida Department of Environmental Protection, Florida Geological Survey Library, 903 W Tennessee St, **Tallahassee**, FL 32304-7700
T: +1 850 4889380; Fax: +1 850 4888086; E-mail: carol.armstrong@dep.state.fl.us
1908; Carol J. Armstrong
Florida, Geology; Florida Aeral photos, Florida Sinkhole Res Institute Arch, Photo Arch Coll
33 000 vols; 40 curr per; 200 diss/theses; 13 000 govt docs; 16 000 maps; 200 microforms; 30 digital data carriers
libr loan   34914

Florida Department of Environmental Protection Library, 2600 Blair Stone Rd, Rm 176, **Tallahassee**, FL 32399-2400
T: +1 850 4880890; Fax: +1 850 9226661; E-mail: bruny.betancourt@dep.state.fl.us
1973; Brunilda Betancourt
Ecology, Natural resources, Pollution, Toxicology
18 000 vols; 150 curr per   34915

Florida Department of Transportation, Research Management Library, 605 Suwannee St, Burns Bldg, MS 30, **Tallahassee**, FL 32399
T: +1 850 4144615; Fax: +1 850 4130657; URL: myflorida.com
1967; Richard C. Long
Air transportation; Hist DOT Coll, HRD Mat, Transportation Res-Related Rpts
15 000 vols; 20 curr per   34916

Florida Public Service Commission, Resource Center, 2540 Shumard Oak Blvd, Easley Bldg, Rm 204, **Tallahassee**, FL 32399-0850
T: +1 850 4136860; Fax: +1 850 4136861
1977; Brenda B. Monroe
Law
15 000 vols; 400 curr per   34917

Florida Supreme Court Library, 500 S Duval St, **Tallahassee**, FL 32399-1926
T: +1 850 4888919; Fax: +1 850 9225219; URL: library.flcourts.org
1845; Billie Blaine
Florida Supreme Court Hist Society
115 000 vols; 1 476 curr per   34918

Hillsborough County Law Library, 701 E Twiggs St, **Tampa**, FL 33602
T: +1 813 2725818; Fax: +1 813 2725226
Norma Wise
40 000 vols; 75 curr per   34919

Lake County Law Library, 202 N Sinclair Ave, P.O. Box 7800, **Tavares**, FL 32778
T: +1 352 7424161; Fax: +1 352 7424190; E-mail: lawlib@lakecounty.clerk.org; URL: www.clerk.lake.fl.us
Faye Osebold
16 500 vols; 8 curr per   34920

Nebraska Department of Corrections, Tecumseh State Correctional Institution Library, 2725 N Hwy 50, **Tecumseh**, NE 68450
T: +1 402 3355998; Fax: +1 402 3355115
Jody Herd
9 000 vols; 59 curr per; 10 av-mat   34921

Court of Appeals for Sixth District of Texas, Law Library, 100 N State Line Ave, No 20, **Texarkana**, TX 75501
T: +1 903 7983046; Fax: +1 903 7983034; E-mail: 6thcourt@gte.net
1907; Linda Rogers
15 530 vols   34922

US Army – AM School of Engineering & Logistics, Library, Red River Army Depot, **Texarkana**, TX 75507-5000
T: +1 903 3342141; Fax: +1 903 3343696
1970
Govt Publs
22 150 vols; 200 curr per   34923

Seneca County Law Library, Seneca County Courthouse Annex, 117 E Market St, Ste 4303, **Tiffin**, OH 44883
T: +1 567 2300204
Lisa Russell
17 000 vols; 20 curr per   34924

US Air Force – Tinker Air Force Base Library, FL 2030 72 SPTG/SVMG, Bldg 1045D, Rapcon St, **Tinker AFB**, OK 73145-8101
T: +1 405 7342626; Fax: +1 405 7333655; E-mail: daver.bruce@tinker.af.mil
1943; Dave Moinett
Aeronautical engineering; Total Quality Management
9 500 vols; 115 curr per   34925

US Army – Tobyhanna Army Depot Post Library, 11 Hap Arnold Blvd, **Tobyhanna**, PA 18466-5099
T: +1 570 8957316; Fax: +1 570 8957419; E-mail: Jeffrey.davis@tobyhanna.army.mil
1959; Jeff Davis
14 000 vols; 26 curr per   34926

State of Ohio Department of Corrections, Toledo Correctional Institution Library, 2001 E Central Ave, **Toledo**, OH 43608
T: +1 419 7267977; Fax: +1 419 7267158
Rose Kuei-Hsiang Shaddy
16 000 vols; 63 curr per   34927

United States Courts, Law Library, 418 US Courthouse, 1716 Spielbusch Ave, **Toledo**, OH 43604
T: +1 419 2135655; Fax: +1 419 2135660
Marianne C. Mussett
16 000 vols; 30 curr per   34928

Ocean Vicinage Law Library, Ocean County Courthouse, Rm 103, 118 Washington St, **Toms River**, NJ 08753
T: +1 732 5065026
Dr. Barbara J. Woods
            34929

Nye County Law Library, 101 Radar Rd, **Tonopah**, NV 89049; P.O. Box 393, Tonopah, NV 89049
T: +1 775 4828103; Fax: +1 775 4828198
Jerie Clifford
10 000 vols   34930

Baltimore County Circuit Court, Law Library, 401 Bosley Ave, **Towson**, MD 21204
T: +1 410 8873086;
Fax: +1 410 8874807; URL: www.baltimorecountymd.gov/go/lawlibrary
Stephanie Levasseur
CA & CSA Briefs, microfiche; Micpels
100 000 vols; 60 curr per   34931

California Department of Corrections – Deuel Vocational Institution Library, 23500 Kasson Rd, P.O. Box 378001, **Tracy**, CA 95378-8001
T: +1 209 4668055; Fax: +1 209 8364144
Warren Schlatter
Law, state & federal codes
14 000 vols; 10 curr per   34932

US Air Force – David Grant Medical Center Library, 101 Bodin Circle, **Travis AFB**, CA 94535-1800
T: +1 707 4235344; Fax: +1 707 4237965; E-mail: regina.rowell@60mdg.travis.af.mil
Regina Ann Rowell, John Sims
8 600 vols; 425 curr per   34933

US Air Force – Travis Air Force Base Library FL4427, 60 SVS/SVMG, 510 Travis Ave, **Travis AFB**, CA 94535-2168
T: +1 707 4243279; Fax: +1 707 4243809
1943; Susan D. Gilroy
40 000 vols; 242 curr per   34934

Florida Department of Corrections, Lancaster Correctional Institution Library, 3449 SW SR 26, **Trenton**, FL 32693
T: +1 352 4634100; Fax: +1 352 4633476; E-mail: lancasterci@mail.dc.state.fl.us
9 000 vols; 56 curr per   34935

New Jersey State Department of Law & Public Safety, Attorney General's Library, 25 Market St, West Wing, 6th Flr, **Trenton**, NJ 08625-2148; P.O. Box 115, Trenton, NJ 08625-0115
T: +1 609 2924958; Fax: +1 609 6336555; E-mail: lpalibr@lps.state.nj.us
Marilyn Maney
New Jersey Legislative Mat Coll, Legal Memoranda Coll
50 000 vols; 100 curr per   34936

Georgia Department of Corrections, Office of Library Services, Hays State Prison, 777 Underwood Rd, **Trion**, GA 30753; P.O. Box 668, Trion, GA 30753-0668
T: +1 706 8570400; Fax: +1 706 8570551; URL: www.dcor.state.ga.us
Carole Farr
13 000 vols; 150 curr per   34937

US Army – Medical Library, One Jarrett White Rd, **Tripler AMC**, HI 96859-5000
T: +1 808 4336391; Fax: +1 808 4334892
1946; Linda Requena
11 500 vols; 750 curr per   34938

Miami County Law Library, 201 W Main St, **Troy**, OH 45373
T: +1 937 4405994
Carolyn Bolin
11 000 vols; 10 curr per   34939

New York State Supreme Court Library, Court House Second St Annex, 86 Second St, **Troy**, NY 12180-4098
T: +1 518 2703717; Fax: +1 518 2740590; URL: www.nycourts.gov
1909; Karlye Ann Pillai
43 000 vols   34940

Arizona Department of Corrections – Adult, Arizona State Prison Complex – Tucson Library, 10000 S Wilmot Rd, P.O. Box 24400, **Tucson**, AZ 85734-4400
T: +1 520 5740024; Fax: +1 520 5747308
Sally Berg & Sue Boers
46 230 vols   34941

Arizona Department of Corrections – Adult Institutions, Arizona State Prison Complex – Tucson Library, 10000 S Wilmot Rd, **Tucson**, AZ 85734; P.O. Box 24400, Tucson, AZ 85734-4400
T: +1 520 5740024; Fax: +1 520 5747308
Sally Berg
46 000 vols   34942

Pima County Law Library, Superior Court Bldg, 110 W Congress, Rm 256, **Tucson**, AZ 85701-1317
T: +1 520 7408456; Fax: +1 520 7919122; E-mail: pcll@sc.pima.gov; URL: www.sc.co.pima.gov/lawlib/
1915; Brenda B. Kelley
35 000 vols   34943

Tulsa County Law Library, 500 S Denver Ave, **Tulsa**, OK 74103
T: +1 918 5965404; Fax: +1 918 5964509; E-mail: lawlibrary@tulsacounty.org; URL: www.tulsalawlib.com
1949; Joyce M. Pacenza
30 000 vols   34944

United States Marine Corps, MCCS Lifelong Learning Library, MCAGCC, Box 788150, **Twentynine Palms**, CA 92278-8150
T: +1 760 8306875; Fax: +1 760 8304497; URL: www.librarylink.info/29palms
1957; Claire Stults
Military hist
31 000 vols; 84 curr per   34945

US Air Force – Air Base & Environmental Tech Library, 139 Barnes Dr, Ste 2, **Tyndall AFB**, FL 32403-5323
T: +1 850 2836285; Fax: +1 850 2836500; E-mail: poulis@afcesa1.af.mil
1968; Andrew D. Poulis
Engineering; Air-Bird Strikes, Hazardous Mats, Geotech Centrifuges, Rapid Runway Repair, Sonic Boom Res
12 500 vols; 400 curr per
libr loan   34946

US Air Force – Tyndall Air Force Base Library, 325 SVS/SVMG/45, 640 Suwanee Rd, Bldg 916, **Tyndall AFB**, FL 32403-5531
T: +1 850 2834287; Fax: +1 850 2834994
William Niblick
Military hist, Aeronautics
27 000 vols; 300 curr per   34947

Mendocino County Law Library, 100 N State St, Courthouse, Rm 307, **Ukiah**, CA 95482
T: +1 707 4634201; Fax: +1 707 4683459; E-mail: lawlib@pacific.net; URL: www.pacificsites.com/~lawlib
Tom Anderson
8 000 vols   34948

Two Rivers Correctional Institute Library, 82911 Beach Access Rd, **Umatilla**, OR 97882
T: +1 541 9222177
Sharon Justus
14 000 vols   34949

Fayette County Law Library, Court House, 61 E Main St, **Uniontown**, PA 15401
T: +1 724 4301228; Fax: +1 724 4304886; E-mail: falawlib@atlanticbbn.net
1927; Barry Richard Blaine
25 000 vols   34950

Prince George's County Circuit Court, Law Library, 14735 Main St, P.O. Box 1696, **Upper Marlboro**, MD 20773
T: +1 301 9523438; Fax: +1 301 9522770; URL: www.co.pg.md.us
1900; Pamela J. Gregory
37 000 vols; 67 curr per   34951

US Air Force Academy – Community Center, 5136 Red Tail Dr, Ste H103, **USAF Academy**, CO 80840-2600
T: +1 719 3333198; Fax: +1 719 3339491
N.N.
31 000 vols; 102 curr per   34952

US Air Force Academy Library, McDermott Library, 2354 Fairchild Dr, Ste 3A10, **USAF Academy**, CO 80840-6214
T: +1 719 3334406; Fax: +1 719 3334754; E-mail: mail.dflib@usafa.af.mil
1955; Dr. Edward A. Scott
Aviation, Military hist; Aeronautics Hist before 1910 (Col Richard Gimbel Coll), US Air Force Academy Arch, Falconry
542 150 vols; 1 590 curr per; 5 500 e-books; 2 142 maps; 658 500 microforms; 3 640 av-mat; 4 900 sound-rec; 74 digital data carriers; 195 000 gov docs
libr loan   34953

Oneida County Supreme Court, New York State United Court System, Law Library, 235 Elizabeth St, **Utica**, NY 13501
T: +1 315 7985703; Fax: +1 315 7986470; URL: www.courts.state.ny.us/library/oneida/index.shtml
Paula J. Eannace
Legal Treatises for Practitioners, 18th & 19th C
50 000 vols; 300 curr per   34954

Georgia Department of Corrections, Office of Library Services, Valdosta State Prison, PO Box 310, **Valdosta**, GA 31603
T: +1 229 3337900; Fax: +1 229 3335387
Zebedee Moore
11 000 vols; 24 curr per   34955

US National Park Service, Valley Forge National Historic Park – Horace Willcox Memorial Library, P.O. Box 953, **Valley Forge**, PA 19482-0953
T: +1 610 2962593; Fax: +1 610 2964834; URL: www.nps.gov
Lee Boyle
Revolutionary war hist
65 000 vols; 25 curr per   34956

US Air Force – Vance Air Force Base Library, 71 FTW/CSC-CSSL, 446 McAffrey Ave, Bldg 314, Ste 24, **Vance AFB**, OK 73705-5710
T: +1 580 2137368; Fax: +1 580 2378100; URL: www.youseemore.com/vanceafb
1941; Mary Arthur
Aviation, Flying
20 000 vols; 50 curr per   34957

Clark County Law Library, 1200 Franklin St, **Vancouver**, WA 98660; P.O. Box 5000, Vancouver, WA 98666-5000
T: +1 360 3972268; E-mail: lawlibrary@clark.wa.gov; URL: www.clark.wa.gov/law-library/index.html
Maria Sosnowski
Washington Law
19 000 vols   34958

Vandalia Correctional Center Library, PO Box 500, **Vandalia**, IL 62471
T: +1 618 2834170
Bruce Griggs
11 000 vols; 31 curr per   34959

US Air Force – Technical Library, 30th Space Wing, 806 13th St, Bldg 7015, Ste A, **Vandenberg AFB**, CA 93437-5223
T: +1 805 6061110 ext 69745; Fax: +1 805 7348232 ext 58941; E-mail: techlib@plansvaft.af.mil
Michele Knight
11 500 vols; 480 curr per; 3 100 maps; 17 digital data carriers; 75 000 tech rpts   34960

US Air Force – Vandenberg Air Force Base Library, 100 Community Loop Bldg 10343-A, **Vandenberg AFB**, CA 93437-6111
T: +1 805 6061110 ext 66414; Fax: +1 805 7341201; E-mail: spaceprt@rain.org
1959; Dixie Paronto
Aerospace science, California; Air Force Hist (Air War College Coll)
48 500 vols; 133 curr per   34961

Ventura County Law Library, 800 S Victoria Ave, **Ventura**, CA 93009-2020
T: +1 805 6428982; Fax: +1 805 6427177; E-mail: vcll@rain.org; URL: www.vencolawlib.com
1891; Jane G. Meyer
81 000 vols; 328 curr per   34962

US Army – Engineer Research & Development Center Library, 3909 Halls Ferry Rd, **Vicksburg**, MS 39180-6199
T: +1 601 6342355; Fax: +1 601 6342542; URL: itl.erdc.usace.army.mil/library; libweb.wes.army.mil/index.htm
Carol McMillin
25 000 vols; 1 500 curr per   34963

Shawnee Correctional Center Library, 6665 State Rte 146E, **Vienna**, IL 62995
T: +1 618 6588331 ext 2120
1985; Leanne Pate
Federal & Illinois Law
15 000 vols; 38 curr per   34964

Vienna Correctional Center Library, P.O. Box 200, **Vienna**, IL 62995
T: +1 618 6588371 ext 270; Fax: +1 618 6583609
1972; J. Karen Jones
Law; Criminology Coll, Law Libr
17 000 vols; 40 curr per   34965

Tulare County Law Library, 221 S Mooney Blvd, County Courthouse, Rm 1, **Visalia**, CA 93291
T: +1 559 7336395; Fax: +1 559 7302613; E-mail: abernard@co.tulare.ca.us; URL: www.co.tulare.ca.us
1892; Anne R. Bernardo
18 500 vols; 105 curr per   34966

Tulare County Office of Education, Educational Resource Services, 7000 Doe Ave, Ste A, **Visalia**, CA 93291
T: +1 559 6513031; Fax: +1 559 6511012; URL: www.erslibrary.org
1927; Elainea Scott
Instructional Mats Display Ctr of Textbooks & Media
250 000 vols; 50 curr per   34967

San Diego County Law Library – North County, 325 S Melrose, Ste 300, **Vista**, CA 92081-6697
T: +1 760 9404386; Fax: +1 760 7247694
20 000 vols   34968

McLennan County Law Library, 500 Washington, **Waco**, TX 76701
T: +1 254 7575191
Janet Gomez
11 200 vols   34969

Hawaii State Circuit Court-Second Circuit, Law Library, 2145 Main St, **Wailuku**, HI 96793
T: +1 808 2442959; Fax: +1 808 2442932
Morris Haole
17 970 vols   34970

Maui Correctional Center Library, 600 Waiale Dr, **Wailuku**, HI 96753
T: +1 808 2435855; Fax: +1 808 2427867
Sandra Wada
52 000 vols; 25 curr per   34971

US Army – Corps of Engineers, Walla Walla District Technical Library, 201 N Third, **Walla Walla**, WA 99362-1876
T: +1 509 5277427; Fax: +1 509 5277816; E-mail: cenww-im-sl@usace.army.mil; URL: www.nww.usace.army.mil
1948; Barbara Hackett
Law Coll
19 000 vols; 300 curr per
libr loan   34972

Walla Walla County Law Library, County Courthouse, 315 W Main St, **Walla Walla**, WA 99362
T: +1 509 5273229; Fax: +1 509 5273214; E-mail: tdriver@co.walla-walla.wa.us
Tina Driver
14 530 vols   34973

Auglaize County Law Library, County Courthouse, 201 Willipie St, Ste 207, **Wapakoneta**, OH 45895
T: +1 419 7383124; Fax: +1 419 7384713; E-mail: lawlib@bright.net
1898; Bridget Weller
20 000 vols   34974

Trumbull County Law Library, 120 High St NW, **Warren**, OH 44481
T: +1 330 6752525; Fax: +1 330 6752527; URL: www.tclla.org
George Baker
32 000 vols; 56 curr per   34975

California Department of Corrections Library System, Wasco State Prison, 701 Scofield Ave, **Wasco**, CA 93280; P.O. Box 8800, Wasco, CA 93280-8800
T: +1 661 7588400; Fax: +1 661 7587049
Gabrielle McClure
44 000 vols   34976

Administrative Office of the United States Courts Library, Law Library & Research Room, One Columbus Circle NE, Ste 4-400, **Washington**, DC 20544
T: +1 202 5021237; Fax: +1 202 5022833; URL: www.uscourts.gov/adminoff.html
Elizabeth Stroup Endicott
Criminal law, Civil law, US courts
22 000 vols; 80 curr per; 24 e-books   34977

Armed Forces Institute of Pathology, Ash Library, Bldg 54, Rm 4077, WRAMC, **Washington**, DC 20306-6000
T: +1 202 7821830; Fax: +1 202 7826403; E-mail: ashlib@afip.osd.mil; URL: www.afip.org
1951; Ruth Li
19 000 vols   34978

Board of Governors of the Federal Reserve System – Law Library, 20th & C St NW, MS 7, **Washington**, DC 20551
T: +1 202 4523040; Fax: +1 202 4523101; URL: www.federalreserve.gov
1975; Scott Finet
Administrative law; Congressional (Legislative Hist Coll of Banking-Related Statutes of the US), micro
30 000 vols; 500 curr per   34979

Board of Governors of the Federal Reserve System – Research Library, 20th & C St NW, MS 102, *Washington*, DC 20551
T: +1 202 4523336; Fax: +1 202 5306222
1914; Kristen Vats
Federal Reserve System; Foreign Central Bank publs
62 000 vols; 1 200 curr per; 2 000 microforms; 20 digital data carriers
SLA   34980

Central Intelligence Agency, Office of General Counsel Law Library, NHB, Rm 6U25, *Washington*, DC 20505
T: +1 703 8743188; Fax: +1 703 8743208
Claudette McLeod
20 000 vols   34981

Congressional Budget Office Library, Second & D St SW, Rm 472, Ford House Office Bldg, *Washington*, DC 20515
T: +1 202 2262635; Fax: +1 202 2251484
Dr. Majid S. Moghaddam
19 000 vols; 842 curr per; 7 digital data carriers   34982

DC Court of Appeals Library, 500 Indiana Ave NW, 6th Flr, Rm 6000, *Washington*, DC 20001
T: +1 202 8792767; Fax: +1 202 6380383
1977; Letty Limbach
20 000 vols   34983

Defense Intelligence Agency, Analytic Support Division, Departement of Defense, 2000 MacDill Blvd, Bolling Air Force Base IDD-2, *Washington*, DC 20340
T: +1 202 2314444; Fax: +1 202 2313231
1963; Gloria Fuller
Doc Coll
227 000 vols; 1 181 curr per; 2 000 000 microforms; 1 500 av-mat   34984

Defense Intelligence Agency – Joint Military Intelligence College, John T. Hughes Library, Defense Intelligence Agency, John T Hughes Library (MCA-4), Bldg 6000, Bolling Air Force Base, *Washington*, DC 20340-5100
T: +1 202 2313777; Fax: +1 202 2313838
George Jupin
98 000 vols; 854 curr per; 2 200 diss/theses; 12 000 maps; 230 000 microforms; 3 500 av-mat; 500 sound-rec; 15 digital data carriers
libr loan   34985

District of Columbia Office of the Corporation Counsel, Law Library, 441 Fourth St NW, 1C-S010, *Washington*, DC 20001
T: +1 202 4429894; Fax: +1 202 4424160; E-mail: oag@dc.gov; URL: www.oag.dc.gov
1932; Bonnie Morgan
District of Columbia Coll
30 000 vols   34986

District of Columbia Superior Court Judges Library, 500 Indiana Ave NW, Rm 5400, *Washington*, DC 20001-2131
T: +1 202 8791435; URL: www.dccourts.gov
Yousuf Galeel
30 000 vols; 30 curr per   34987

Executive Office of the President Libraries, New Exec Off Bldg, Rm G-007, 725 17th St NW, *Washington*, DC 20503
T: +1 202 3954690; Fax: +1 202 3956137
Lea Uhre
US Presidents, Economics; Congressional Appropriations Legislation Coll, Federal Budget Docs Coll, Legislative Hist of Federal Govt Reorganization Plans Coll, The Presidency, Legal Coll
100 000 vols; 670 curr per; 698 000 microforms; 100 digital data carriers
34988

Federal Bureau of Investigation, Forensic Science Information Resource System, 935 Pennsylvania Ave NW, Rm 3865, *Washington*, DC 20535-0001
T: +1 202 3244384; Fax: +1 202 3244323; E-mail: labfsirs@fbi.gov; URL: www.fbi.gov
1985; Colleen Wade
10 000 vols; 400 curr per   34989

Federal Communications Commission Library, 445 12th St SW, *Washington*, DC 20024-2101
T: +1 202 4180450; Fax: +1 202 4182805
Gloria Jean Thomas
Telecommunications, Computer science, Law, Economics; Legislative Hist, Communications Act of 1934
32 000 vols; 432 curr per   34990

Federal Election Commission, Law Library, 999 E St NW, Rm 801, *Washington*, DC 20463
T: +1 202 6941660; Fax: +1 202 2083579; URL: www.fec.gov
Leta Holley
10 000 vols; 180 curr per   34991

Federal Emergency Management Agency Library, 500 C St SW, Rm 123, *Washington*, DC 20472
T: +1 202 6463771; Fax: +1 202 6464295; URL: www.fema.gov
1980; Mercedes Lopez Emperado
Emergency Preparedness Coll, Security Management Coll
100 000 vols; 120 curr per; 3 000 govt docs; 50 av-mat; 3 digital data carriers
libr loan; ALA, SLA   34992

Federal Energy Regulatory Commission Library, 888 First St NE, Rm 95-01, *Washington*, DC 20426
T: +1 202 2082179; Fax: +1 202 5012870
Robert F. Kimberlin
Electric power, Natural gas; Federal Power Commission/FERC Publs
40 000 vols; 800 curr per   34993

Federal Judicial Center, Information Services Office, One Columbus Circle NE, *Washington*, DC 20878
T: +1 202 5024153; Fax: +1 202 5024077; URL: www.fjc.gov
Roger Karr
14 000 vols; 450 curr per
libr loan; FEDLINK   34994

Federal Maritime Commission Library, 800 North Capitol St NW, *Washington*, DC 20573
T: +1 202 5235762; Fax: +1 202 5235738; E-mail: davev@fmc.gov
1961; David J. Vespa
11 300 vols; 88 curr per; 17 digital data carriers
libr loan   34995

Federal Trade Commission Library, 600 Pennsylvania Ave NW, Rm 630, *Washington*, DC 20580
T: +1 202 3262395; Fax: +1 202 3262732
1914; Margie Knott
Economics, Law; Arch, Legislative Hist
117 000 vols; 350 curr per; 200 000 microforms; 10 digital data carriers
libr loan   34996

General Services Administration Library, 1800 F St NW, Rm 1033, *Washington*, DC 20405
T: +1 202 5010788; Fax: +1 202 5014452; E-mail: emily.mosery@gsa.gov; URL: www.gsa.gov
1961; Emily Moser
Telecommunications, Architecture; FAI Procurement Coll, GPO Depot, Karel Yasko Coll (hist docs on US public buildings), Legislative Hist of Public Laws
78 000 vols; 30 curr per; 20 000 govt docs; 10 000 microforms; 130 av-mat; 100 sound-rec; 50 digital data carriers
libr loan   34997

Health & Human Services Department, Office of the General Counsel Library, 330 Independance Ave SW, Cohen Bldg, G-400, *Washington*, DC 20201
T: +1 202 6190190; Fax: +1 202 6193719; E-mail: llibrary@os.dhhs.gov
Carl R. Kessler
25 000 vols; 190 curr per   34998

House Library, Cannon Bldg, B-106, Legislative Resource Ctr, *Washington*, DC 20515-6606
T: +1 202 2265200; Fax: +1 202 2265207; E-mail: info.clerkweb@mail.house.gov
1792; Rae Ellen Best
200 000 vols; Printed Docs & Mats Produced by the House of Representatives   34999

NASA Headquarters Library, 300 E St SW, Code CO-2, *Washington*, DC 20546-0001
T: +1 202 3580168; Fax: +1 202 3583469; E-mail: library@hq.nasa.gov; URL: www.hq.nasa.gov/office/hqlibrary
1958; Jannie Pratte
Aerospace science; NACA Docs
16 000 vols; 200 curr per; 1 050 000 microforms   35000

National Endowment for the Humanities Library, 1100 Pennsylvania Ave NW, *Washington*, DC 20506
T: +1 202 6068244; Fax: +1 202 6068457; E-mail: library@neh.gov
1965; Donna McClish
Jefferson Lectures; NEH Rpts
40 000 vols; 64 curr per   35001

National Guard Memorial Library, One Massachusetts Ave NW, *Washington*, DC 20001
T: +1 202 4085890; Fax: +1 202 6829358; URL: www.ngef.org
1957; John Sterne
State Hist, bks, corr, clippings; Aircraft, photos
32 000 vols; 75 curr per   35002

National Library of Education, 400 Maryland Ave NW, BE-101, *Washington*, DC 20202-5523
T: +1 202 2055019; Fax: +1 202 2607364; E-mail: library@ed.gov; URL: www.ed.gov/NLE
1994; Christina Jordan Dunn
ERIC Database, Early American Textbks, Rare Bk Coll, Arch Colls
80 000 vols; 672 curr per; 1 000 000 microforms; 50 digital data carriers
libr loan   35003

Navy Department Library, 805 Kidder-Breese St SE, *Washington*, DC 20374-5060
T: +1 202 4334132; Fax: +1 202 4339553; E-mail: navylibrary@navy.mil; URL: www.history.navy.mil/library
1800; Glenn Edward Helm
Naval and Maritime Hist; Special Colls: Administrative Hist of World War II, Naval Registers, Regulations, Signal Bks, Cruisebks, Diss on Naval & Military Hist, Rare Bks & Mss
250 000 vols; 375 curr per; 700 e-journals; 4 514 govt docs; 500 maps; 11 000 microforms; 60 digital data carriers
libr loan; ALA   35004

Office of Thrift Supervision Library, Columbus Memorial Bldg, 1700 G St NW, *Washington*, DC 20552
T: +1 202 9066470; Fax: +1 202 9067591; URL: www.ots.treas.gov
1935; Joseph Thornton
Annotated Compilations of Court Decisions, US Supreme Court Rpts, State Rpts, Compiled Legislative Hist Coll; Archs, Docs Ctr
80 000 vols; 800 music scores; 1 500 maps; 6 000 microforms; 100 sound-rec; coll of mss, diss, 241 401 OAS docs, 45 000 photos
libr loan   35005

Osha, Technical Data Center, Rm N-2625, 200 Constitution Ave NW, *Washington*, DC 20210-2001
T: +1 202 6932350; Fax: +1 202 6931648
Chris Aaron
Engineering, Medicine; OSHA Rulemaking Records 1971-present
12 500 vols; 206 curr per; 300 000 microforms; 47 digital data carriers
libr loan   35006

Overseas Private Investment Corporation Library, 1100 New York Ave NW, 11th Fl, *Washington*, DC 20527
T: +1 202 3368565; Fax: +1 202 4089860; E-mail: lpaul@opic.gov; URL: www.opic.gov
1974; Lena Paulsen
Economics, Finance; Country File, Foreign Assistance & Int Development (Legislative Hist Coll)
10 000 vols; 150 curr per   35007

Supreme Court of the United States Library, One First St NE, *Washington*, DC 20543; mail address: 3035 V St NE, Washington, DC 20018
T: +1 202 4793177; Fax: +1 202 4793477
1887; Judith A. Gaskell
Supreme Court Legislative Coll
500 000 vols; 5 500 curr per; 14 digital data carriers   35008

US Aid Library, US Agency for International Development, 1300 Pennsylvania Ave NW, Ronald Reagan Bldg, Rm M01-010, *Washington*, DC 20523-1000
T: +1 202 7120579; Fax: +1 202 2163515
1967; David Wolfe
Project Doc Arch 1958-1974, US Agency for Int Development Program Documentation 1950's to Present
10 000 vols; 400 curr per   35009

US Air Force – Bolling Air Force Base Library, FL 7054 HQ 11 MSG/SVMG, 410 Tinker St Bolling AFB, *Washington*, DC 20032-0703
T: +1 202 7675578; Fax: +1 202 4048526
1931; Shirley Foster
Military hist
28 000 vols; 33 curr per   35010

US Court of Appeals for the Armed Forces Library, 450 E St NW, *Washington*, DC 20442-0001
T: +1 202 7611466
1951; Agnes Kiang
Criminal law & justice
20 000 vols; 40 curr per   35011

US Court of Appeals for the District of Columbia, Judges' Library, US Court House, 333 Constitution Ave NW, Rm 5518, *Washington*, DC 20001
T: +1 202 2167396; URL: www.cadc.uscourts.gov
Patricia Michalowskij
90 000 vols; 160 curr per   35012

US Court of Appeals for the Federal Circuit Library, Howard T Markey National Courts Bldg, 717 Madison Pl NW, Ste 218, *Washington*, DC 20439
T: +1 202 3125500; Fax: +1 202 7867015
1967; Patricia M. McDermott
Legislative Hist
48 000 vols; 75 curr per   35013

US Customs Service, Informaion Resources Center, Mint Annex, Tenth Flr, 799 Nineth St NW, *Washington*, DC 20004; Mint Annex, Tenth Flr, 1300 Pennsylvania Ave, NW, Washington, DC 20229-0001
T: +1 202 3250172; Fax: +1 202 3250170; E-mail: cbp.library@dhs.gov; URL: 207.67.203.70/u40009staff/opac
1975; Linda Cullen
Int trade, Economics, Drug abuse, Law enforcement
59 000 vols; 856 curr per; 7 000 e-books
35014

US Department of Agriculture, Economic Research Service Reference Center, 1800 M St, Rm N-3050, *Washington*, DC 20036
T: +1 202 6945065; Fax: +1 202 6945689; E-mail: mgraham@econ.ag.gov
1978; Marilynn Graham
Agricultural economics; ERS/ESCS Publs, micro
20 000 vols; 289 curr per   35015

US Department of Agriculture, Office of the General Counsel Law Library, 1400 Independence Ave SW, Rm 0325-South, *Washington*, DC 20250-1400
T: +1 202 7207751; Fax: +1 202 6900682
Peter MacHare
Agricultural law
90 000 vols; 50 curr per   35016

US Department of Commerce, Office of the Library & Information Services, HCHB, Rm 7047, 1401 Constitution Ave NW, *Washington*, DC 20230
T: +1 202 4825511; Fax: +1 202 4825685; URL: www.doc.gov/lib
1913
Diversity Resources Ctr
150 000 vols; 200 curr per; 250 000 microforms; 140 digital data carriers
libr loan   35017

US Department of Commerce – Law Library, 14th & Constitution Ave NW, Rm 1894, *Washington*, DC 20230
T: +1 202 4821154; Fax: +1 202 4820221; E-mail: lawlibrary@doc.gov; URL: www.osec.doc.gov/lib
1973; Jane Sessa
Int trade; Legislative Coll
190 000 vols; 80 curr per   35018

US Department of Energy – Energy Library, 1000 Independence Ave SW, Rm GA-138, *Washington*, DC 20585
T: +1 202 5869534; Fax: +1 202 5860573; E-mail: forrestal.library@hq.doe.gov
1948; Denise Diggin
Atomic Energy Commission, Dept of Energy, Energy Research & Development Administration, Federal Energy Administration Reports
15 000 vols; 2 160 curr per; 5 030 microforms
libr loan   35019

US Department of Energy – Law Library, GC Law Library, 6A-156, 1000 Independence Ave SW, *Washington*, DC 20585-0103
T: +1 202 5864849; Fax: +1 202 5860865
1942; N.N.
30 000 vols; 75 curr per   35020

US Department of Housing and Urban Development, HUD Library, 451 Seventh St SW, Rm 8141, *Washington*, DC 20410
T: +1 202 7082370; Fax: +1 202 7081485; E-mail: webmanager@hud.gov; URL: www.hud.gov
1934; Robin Lewis
American housing, Architecture, Building construction, Regional planning, Economics; Comprehensive Housing Affordability Strategy Rpts; Housing in the 70's Background Papers; Management Evaluation Reports
680 000 vols; 2 200 curr per
libr loan   35021

US Department of Justice – Justice Libraries, 950 Pennsylvania Ave, Ste 5313, *Washington*, DC 20530
T: +1 202 5143775; Fax: +1 202 5143546; URL: www.usdoj.gov
1831; Blane K. Dessy
American, Canadian & British Law, Dept of Justice Publs; US Supreme Court Records & Briefs
300 000 vols; 1 700 curr per   35022

US Department of Labor, Wirtz Labor Library, 200 Constitution Ave NW, Rm N-2445, *Washington*, DC 20210-0002
T: +1 202 6936600; Fax: +1 202 6936644; URL: library.dol.gov
1917; Jean M. Bowers
Labor Unions, docs, mat
570 000 vols; 500 curr per; 9 500 microforms; 11 digital data carriers
libr loan   35023

US Department of State, Ralph J. Bunche Library, 2201 C St NW, Rm 3239, *Washington*, DC 20520-2442
T: +1 202 6471099; Fax: +1 202 6472971; E-mail: library@state.gov; URL: www.state.gov/m/a/ls
1789; C. Elaine Cline
Int law; Dept of State Publs, Diplomatic Lists
400 000 vols; 800 curr per; 338 e-journals; 7 000 e-books; 406 000 microforms; 25 digital data carriers; 51 Electronic Media & Resources   35024

US Department of the Army – The Pentagon Library, 6605 Army Pentagon, *Washington*, DC 20310-6605
T: +1 703 6952006; Fax: +1 703 6954009; URL: www.hqda.army.mil/library
1944; Mena Whitmore
Army studies, Law, Military hist;

Regulatory Publs
200 000 vols; 975 curr per   35025

US Department of the Interior Library, 1849 C St NW, Rm 1151, *Washington*, DC 20240
T: +1 202 2085815; Fax: +1 202 2086771; E-mail: library@nbc.gov; URL: library.doi.gov
1949; George Franchois
Natural resources, Energy, Mining; Conservation & Natural Resources, (Diss)
1 000 000 vols; 2 000 curr per; 560 000 microforms; 20 digital data carriers
                                               35026

US Department of the Navy – Office of the General Counsel, Law Library, Bldg 36, Rm 213, 720 Kennon St SE, *Washington*, DC 20374
T: +1 202 6856944; Fax: +1 202 6856959
1949
Legislative Histories Coll
30 000 vols   35027

US Department of the Navy – Office of the Judge Advocate General Law Library, Washington Navy Yard, 1322 Patterson Ave SE, Ste 3000, *Washington*, DC 20374-5066
T: +1 202 6855270; Fax: +1 202 6855171; E-mail: roachss@jag.navy.mil
1939; Sue Roach
Naval Legal Mat
45 000 vols   35028

US Department of the Treasury – Comptroller of the Currency, Administrator of National Banks Library, 250 E St SW, *Washington*, DC 20219
T: +1 202 8744720; Fax: +1 202 8745138; URL: www.occ.treas.gov
1974; Laura Keen
Economics, Law
13 500 vols; 500 curr per   35029

US Department of the Treasury – Treasury Library, 1500 Pennsylvania Ave NW, Rm 1428 MT, *Washington*, DC 20220
T: +1 202 6220990; Fax: +1 202 6220018; E-mail: library.reference@do.treas.gov
1817; Judy Lim-Sharpe
Economics, Law, Tax; Administrative Hist of World War II Civilian Agencies; Congressional Record, Predecessors & Congressional Serial set from 1789-present; League of Nations Publs; Legislative Compilations Dealing with Federal Taxes; Treasury Hist
73 000 vols; 600 curr per   35030

US Department of Transportation, TASC Library, 400 7th St SW, *Washington*, DC 20590
T: +1 202 3660745; Fax: +1 202 3667779; E-mail: library@dot.gov; URL: dotlibrary.dot.gov
1969; Phyllis Bell
200 000 vols; 600 curr per
libr loan   35031

US Department of Transportation – Federal Highway Administration, Chief Counsel's Law Library, 400 Seventh St SW, *Washington*, DC 20590
T: +1 202 3661388; Fax: +1 202 3661380; E-mail: sherie.abbasi@fhwa.dot.gov; URL: www.fhwa.dot.gov/legsregs/legislat.html
Sherie Abbasi
Highway legislation; Special coll: Federal Highway Administration (FHWA) legislative histories 1893-present time
10 000 vols   35032

US International Trade Commission – Law Library, 500 E St SW, Rm 614, *Washington*, DC 20436
T: +1 202 2053287; Fax: +1 202 2053111
1972
Legislative Hist dealing with Trade & Tariff Acts
100 000 vols; 75 curr per   35033

US International Trade Commission – National Library of International Trade, 500 E St SW, Rm 300, *Washington*, DC 20436
T: +1 202 2052630; Fax: +1 202 2052316; URL: www.usitc.gov
1916; Wendy Willis
100 000 vols; 2 500 curr per; 165 000 microforms; 100 sound-rec; 40 digital data carriers
libr loan; IFLA   35034

US Marine Corps – Marine Corps Historical Center Library, Washington Navy Yard, 901 M St SE, Code HDS-3, Bldg 58, *Washington*, DC 20374-5040
T: +1 202 4333447; Fax: +1 202 4337265; E-mail: eengland@notes.hqi.usmc.mil
1843; Evelyn A. Englander
Military hist; Biogr of US Marines, Foreign Marine Corps, Hist of Marine Corps & Amphibious Warfare, General Naval & Military Hist, Marine Corps Operational Docs & After Action Rpts, Muster Rolls & Unit Diaries
41 000 vols; 12 curr per   35035

US Merit Systems Protection Board Library, 1120 Vermont Ave NW, Rm 828, *Washington*, DC 20419
T: +1 202 6537132; Fax: +1 202 6536182
Kathleen O'Sullivan
8 000 vols; 75 curr per   35036

US Naval Observatory, James Melville Gilliss Library, 3450 Massachusetts Ave NW, *Washington*, DC 20392-5420
T: +1 202 7621463; Fax: +1 202 7621083; URL: www.usno.navy.mil/usno/library
1830; Sally Bosken
Pre-19th C Bks Covering Subjects of Astronomy, Mathematics, Physics & Navigation
85 000 vols; 200 curr per   35037

US Office of Personnel Management, Resource Center, 1900 E St NW, *Washington*, DC 20415-7740
T: +1 202 6061381; Fax: +1 202 6060909; URL: www.opm.gov
1941; Leon Brody
Civil serv
15 000 vols; 30 curr per   35038

US Postal Service Library, 475 L'Enfant Plaza SW, Rm 11800, *Washington*, DC 20260-1540
T: +1 202 2682904; Fax: +1 202 2684423; URL: www.usps.gov
1955; Carolina Menendez
Economics, Marketing, Law, Data processing; Congressional Rpts (US Congressional Serial Doc Set)
115 000 vols; 850 curr per; 30 digital data carriers   35039

US Securities & Exchange Commission Library, 100 F St NE, Rm 1500, *Washington*, DC 20549-0002
T: +1 202 5515450; Fax: +1 202 7729326; E-mail: library@sec.gov
1934; Cindy Plisch
Securities, Economics, Finance, Law; Legislative Hist of Statutes Administered by Agency
51 000 vols; 392 curr per
libr loan   35040

US Senate Library, FRB-15, Senate Russell Bldg, *Washington*, DC 20510-7112
T: +1 202 2247106; Fax: +1 202 2240879
1871; Mary Cornaby
Congressional Hearings, Congressional Record, Bills & Resolutions
200 000 vols; 115 curr per; 1 000 000 microforms   35041

US Tax Court Library, 400 Second St NW, *Washington*, DC 20217
T: +1 202 6068707; Fax: +1 202 2193794; E-mail: tclib@ustaxcourt.gov
1924; Elsa B. Silverman
Tax Laws
36 000 vols; 1 500 curr per   35042

Walter Reed Army Medical Center – Medical Library Service, US Army Medical Command, Bldg 2, Rm 2G, *Washington*, DC 20307-5001
T: +1 202 7826238; Fax: +1 202 7826803; URL: www.wramc.amedd.army.mil
Robert Mohrmann, Kevin Canning
Clinical medicine; Hist of Military Medicine (Fred C Ainsworth Endowment Libr), Patient Education
31 830 vols; 880 curr per; 5 digital data carriers
libr loan; ALA, SLA, MLA   35043

Washington County Law Library, One S Main St, Ste G004, *Washington*, PA 15301-6813
T: +1 724 2286747
1867; Nancy Weiss
22 000 vols; 20 curr per   35044

Fayette County Law Library, 110 E Court House, *Washington Court House*, OH 43160-1355
T: +1 740 3353608; Fax: +1 740 3333530
Rollo Marchant
Ohio Laws
14 000 vols   35045

Connecticut State Judicial Department, Law Library at Waterbury, Court House, 300 Grand St, *Waterbury*, CT 06702
T: +1 203 5913338; Fax: +1 203 5964137; URL: www.jud.ct.gov/lawlib
Mary Fuller
40 000 vols; 35 curr per   35046

New York State Supreme Court, Law Library at Watertown, Court House, 163 Arsenal St, *Watertown*, NY 13601
T: +1 315 7853064; Fax: +1 315 7853330
1944; Deanna Morse
45 000 vols; 34 curr per   35047

William D. Block Memorial Law Library, 18 N County St, *Waukegan*, IL 60085-4339
T: +1 847 3772800; URL: 19thcircuitcourt.state.il.us/library/libr_toc.htm
1845; Peggy Grady
Illinois Appellate Court Briefs of the Second Judicial District, 1964-present; Law Books on the State of Illinois
25 000 vols; 50 curr per   35048

Waupun Correctional Institution Library, 200 S Madison St, *Waupun*, WI 53963-2069; P.O. Box 351, Waupun, WI 53963-0351
T: +1 920 3245571; Fax: +1 920 3247250
Nevin B. Webster
13 000 vols; 90 curr per   35049

Georgia Department of Corrections, Office of Library Services, Ware State Prison, 3620 Harris Rd, *Waycross*, GA 31503
T: +1 912 2856400; Fax: +1 912 2876520; URL: www.dcor.state.ga.us
Patti DeMarco
15 000 vols; 38 curr per   35050

Greene County Law Library, Court House, *Waynesburg*, PA 15370
T: +1 412 8525237; Fax: +1 412 6274716
Audrey Szoyka
12 500 vols   35051

Chelan County Law Library, County Courthouse, 401 Washington St, *Wenatchee*, WA 98801
T: +1 509 6645213; Fax: +1 509 6645588
Mark DeForest
10 000 vols; 10 curr per   35052

Naval Surface Warfare Center, Carderock Division Technical Information Center, 9500 MacArthur Blvd, *West Bethesda*, MD 20817-5700
T: +1 301 2271433; Fax: +1 301 2275307
David Glenn
38 000 vols; 10 curr per   35053

Worcester County Jail & House of Correction Library, Five Paul X Tivnan Dr, *West Boylston*, MA 01583
T: +1 508 8541800; Fax: +1 508 8528754
Christina Moore
8 000 vols; 17 curr per   35054

Chester County Law Library, Bar Association Bldg, 15 W Gay St, *West Chester*, PA 19380-3014
T: +1 610 3446166; Fax: +1 610 3446994; E-mail: lawlibrary@chesco.org; URL: dsf.chesco.org
1862; Jeannie Naftzger
33 000 vols; 40 curr per   35055

Eastern Kentucky Correctional Complex Library, 200 Road to Justice, *West Liberty*, KY 41472
T: +1 606 7432800; Fax: +1 606 7432811
Darryl Thompson
18 000 vols; 35 curr per   35056

Palm Beach County Law Library, County Courthouse, Rm 12200, 205 N Dixie Hwy, *West Palm Beach*, FL 33401
T: +1 561 3552928; Fax: +1 561 3551654
1947; Linda Sims
30 000 vols; 30 curr per   35057

US Army – West Point – Post Library, 622 Swift Rd, Ste 3, *West Point*, NY 10996-1981
T: +1 845 9382974; Fax: +1 845 9383019
Suzanne Moskala
Military hist
37 000 vols; 65 curr per   35058

National Ground Water Association, Ground Water Information Center, 601 Dempsey Rd, *Westerville*, OH 43081-8978
T: +1 614 8987791; Fax: +1 614 8987786; URL: www.ngwa.org/gwonline/gwol.cfm
Thad Plumley
24 000 vols; 100 curr per; 100 maps; 12 000 microforms; 50 av-mat; 10 digital data carriers   35059

Carroll County Court Library, Court & Willis Sts, *Westminster*, MD 21157
T: +1 410 3862672; Fax: +1 410 7515240; URL: www.lawlib.state.md.us
Florence Green
8 000 vols; 10 curr per   35060

Eastern Correctional Institution, West Library, 30420 Revells Neck Rd, *Westover*, MD 21890-3358
T: +1 410 8454000; Fax: +1 410 8454206
June Brittingham
9 000 vols; 35 curr per; 20 av-mat   35061

Westville Correctional Facility, Resident Library, 5501 S 1100 W, *Westville*, IN 46391
T: +1 219 7852511; Fax: +1 219 7854864; URL: www.in.gov/idoc/2401.htm
Brad Stigler
29 000 vols; 90 curr per   35062

Gulf Correctional Institution Library, 500 Ike Steel Rd, *Wewahitchka*, FL 32465
T: +1 850 6391480; Fax: +1 850 6391182
N.N.
13 000 vols; 72 curr per   35063

Ohio County Law Library, City-County Bldg, Rm 406, 1500 Chapline St, *Wheeling*, WV 26003
T: +1 304 2343780; Fax: +1 304 2346437; URL: www.state.wv.us/wvsca
1919; Nancy Chatlak
Supreme Court Records & Briefs
37 000 vols   35064

NYS Supreme Court Library, Ninth Judicial District, 9th Flr, 111 Dr Martin Luther King Blvd, *White Plains*, NY 10601
T: +1 914 8245660
1908; Sonja Davis
Records on Appeal (Four Appellate Divisions & Court of Appeals)
300 000 vols; 45 curr per   35065

US Army – Consolidated Library, Bldg 465, Rm 113, *White Sands Missile Range*, NM 88002-5039
T: +1 575 6785820; Fax: +1 575 6782270
1950; Mac Odom
Engineering, Southwest, US military; Military Science, Southwest Coll
55 000 vols; 45 curr per; 700 e-books; 300 av-mat
libr loan   35066

US Air Force – Whiteman Air Force Base Library, FL 4625, 509 SVS/SVMG, 750 Arnold Ave Bldg 527, *Whiteman AFB*, MO 65305-5019
T: +1 660 6875614; Fax: +1 660 6876240
Selvin Royal
Military, Political science, Business
23 120 vols; 115 curr per   35067

Sedgwick County Law Library, 225 N Market St, Ste 210, *Wichita*, KS 67202-2023
T: +1 316 2632251; Fax: +1 316 2630629; URL: www.wichitabar.org
John Lewallen
30 000 vols   35068

US Air Force – Air Mobility Command, McConnell Air Force Base Library, McConnell AFB, 53476 Wichita St, Bldg 412, *Wichita*, KS 67221
T: +1 316 7594207; Fax: +1 316 7594254; URL: www.amccrooms.sirsi.net
1953; Darla Cooper
Air War College Coll, Kansas Coll, Military Aviation Hist Coll, Project Warrior
40 000 vols; 50 curr per   35069

US Courts Library, B55 US Courthouse, 401 N Market St, *Wichita*, KS 67202-2011
T: +1 316 2696162; Fax: +1 316 2696168
Linda Miller
10 000 vols; 12 curr per   35070

Lycoming County Law Library, Court House, 48 W Third St, Basement, *Williamsport*, PA 17701
T: +1 570 3272475; Fax: +1 570 3273288
1927; Deborah C. Smith
20 500 vols; 15 curr per   35071

Clinton County Law Library, 46 S South St, *Wilmington*, OH 45177
T: +1 937 3822428; Fax: +1 937 3827632; E-mail: cclibr1@erinet.com
1905; Judy A. Gano
25 000 vols   35072

New Castle County Law Library, 500 King St, Ste 2500, *Wilmington*, DE 19801
T: +1 302 2550847; Fax: +1 302 2552223
1911; Alda Monsen
25 000 vols   35073

US Court of Appeals, Branch Library, 844 King St, US Courthouse, Lock Box 43, *Wilmington*, DE 19801
T: +1 302 5736178; Fax: +1 302 5736430
1974; Lesley Lawrence
14 000 vols; 25 curr per   35074

Randolph Circuit Court, Law Library, 100 S Main St, *Winchester*, IN 47394
T: +1 765 5847070 ext 230; Fax: +1 765 5842958
Mike O'Neal
14 000 vols   35075

Maine Department of Corrections, Maine Correctional Center Library, 17 Mallison Falls Rd, *Windham*, ME 04062
T: +1 207 8937000; Fax: +1 207 8937001
Francine Bowden
12 000 vols; 30 curr per; 25 av-mat   35076

Arizona Department of Corrections – Adult Institutions, Arizona State Prison Complex – Winslow Library, 2100 S Hwy 87, *Winslow*, AZ 86047
T: +1 928 2899551; Fax: +1 928 2899551
Gary D. Calhoun
15 000 vols   35077

Lincoln County Law Library, High St, P.O. Box 249, *Wiscasset*, ME 04578-0249
T: +1 207 8827517
10 000 vols   35078

Yolo County Law Library, 625 Court St, No B07, *Woodland*, CA 95695
T: +1 530 6668918; Fax: +1 530 6668618; E-mail: lawlib@cndcn.davis.ca.us
Ramona Margherio
13 400 vols   35079

Marine Biological Laboratory, MBL/WHOI Library, 7 MBL St, *Woods Hole*, MA 02543-1026
T: +1 508 2897002; Fax: +1 508 5406902; E-mail: library@mbl.edu; URL: www.mblwhoilibrary.org
1888; Cathy Norton
Institutional archives; scientific data; oceanographic instruments
250 000 vols; 2 000 curr per; 2 215 e-journals; 200 diss/theses; 1 000 maps; Maps, Nautical Charts, Rare Books, Scientific data
libr loan; IAMSLIC, Nelinet, OCLC   35080

McHenry County Law Library, 2200 N Seminary Ave, *Woodstock*, IL 60098
T: +1 815 3344166; Fax: +1 815 3341005; URL: www.co.mchenry.il.us
Louise Joseph
12 000 vols   35081

Massachusetts Trial Court, Worcester Law Library, Two Main St, *Worcester*, MA 01608
T: +1 508 7701899; Fax: +1 508 7549933; E-mail: worcesterlaw@yahoo.com; URL: www.lawlib.state.ma.us
1842; Suzanne M. Hoey
Early Law Rpts (American Coll); Legal Textbks (Major Coll on General Law), Hist of Worcester County & its Cities & Towns
103 000 vols; 110 curr per   35082

US Air Force – Air Force Institute of Technology, Academic Library FL3319, AFIT/ENWL, 2950 Hobson Way, Bldg 642, *Wright-Patterson AFB*, OH 45433-7765
T: +1 937 2553005; Fax: +1 937 6567746; URL: library.afit.edu
1946; James T. Helling
130 000 vols; 1 500 curr per; 18 000 e-journals; 1 200 000 microforms; 25 digital data carriers
libr loan   35083

US Air Force – Air Force Research Lab, Wright Research Site Technical Library, Det 1 AFRL/WSC, Bldg 642, Rm 1300, 2950 Hobson Way, *Wright-Patterson AFB*, OH 45433-7765
T: +1 937 2555511; Fax: +1 937 6567746; E-mail: afrl.wsc.library@wpafb.af.mil; URL: www.afrl.af.mil/wrslibrary
1919; Ann Lewis
Lahm-Chandler Coll of Aeronautica
84 000 vols; 170 curr per
libr loan   35084

US Air Force – National Air Intelligence Center Information Center, NAIC/DXOA, 4180 Watson Way, *Wright-Patterson AFB*, OH 45433-5648
T: +1 937 2572452; Fax: +1 937 2570122
Janet Burke
17 000 vols; 800 curr per   35085

US Air Force – Wright-Patterson Air Force Base Library, FL2300, 88 SPTG/SVMG, Bldg 1044, 5651 Fir St, Ste 3, *Wright-Patterson AFB*, OH 45433-5428
T: +1 937 2574815; Fax: +1 937 6561776; E-mail: mary.rinas@88abw.wpafb.af.mil
1942; Patrick Colucci
Military hist; Total Quality Management Coll
55 000 vols; 500 curr per   35086

US Air Force – Wright-Patterson Medical Center Library, 74 MDSS/SGSFL, Bldg 830, 4881 Sugar Maple Dr, *Wright-Patterson AFB*, OH 45433-5529
T: +1 937 2574506; Fax: +1 937 2570741; E-mail: mary.auer@wpafb.af.mil
1942; Mary A. Auer
23 000 vols; 758 curr per
libr loan   35087

US Department of Defense, Defense Institute of Security Assistance Management, Library, Bldg 125, 2335 Seventh St, Rm 2317, *Wright-Patterson AFB*, OH 45433-7803
T: +1 937 2555567, 9211; Fax: +1 937 2558258; E-mail: patricia.white@disam.dsca.osd.mil; URL: www.disam.osd.mil/
1977; Patricia White
Political science; Human Rights, Regional Studies, Security Assistance
12 700 vols; 221 curr per   35088

Mid State Correctional Facility Library, Range Rd, P.O. Box 866, *Wrightstown*, NJ 08562-0466
T: +1 609 7234221 ext 8415; Fax: +1 609 7238271
Emma Pervall
8 701 vols; 62 curr per   35089

US Department of Agriculture, Agricultural Research Service, Eastern Regional Research Center, Scientific Information Resources, 600 E Mermaid Lane, *Wyndmoor*, PA 19038-8598
T: +1 215 2336602, 6672; Fax: +1 215 2336606; E-mail: wkramer@arserrc.gov
1940; Wendy H. Kramer
31 000 vols; 270 curr per   35090

Greene County Law Library, Court House, Rm 309, *Xenia*, OH 45385
T: +1 937 3765115; Fax: +1 937 3765116
Nancy Hedges
24 000 vols; 50 curr per   35091

Yakima County Law Library, County Courthouse, *Yakima*, WA 98901
T: +1 509 5742692
1923; Letha Hammer
20 000 vols   35092

Garden State Youth Correctional Facility Library, PO Box 11401, *Yardville*, NJ 08620-1401
T: +1 609 2986300; Fax: +1 609 2988682
Vera Morris
15 000 vols; 48 curr per   35093

York County Law Library, Court House, *York*, PA 17401-1583
T: +1 717 8540754; Fax: +1 717 8437394; E-mail: yorkcolawlibrary@netscape.net
Susan Hedge
17 768 vols   35094

Post Library, Yuma Proving Ground, AZ, U.S. Army, Bldg 530, 301 C St, *Yuma*, AZ 85365-9848
T: +1 928 3283068; Fax: +1 928 3283055
Carol A. Cowperthwaite
43 000 vols; 24 curr per; 1 600 av-mat; 800 sound-rec; 10 digital data carriers
libr loan; ALA   35095

United States Marine Corps, Station Library, Bldg 633, *Yuma*, AZ 85369; P.O. Box 99119, Yuma, AZ 85369-9119
T: +1 928 2692785; Fax: +1 928 3445592
1948; Bonnie Miller
Aviation
20 000 vols; 93 curr per   35096

Yuma County Law Library, 219 W Second St, *Yuma*, AZ 85364
T: +1 928 3292255
Beth Roe
16 000 vols   35097

Muskingum County Law Library, Court House, 401 Main St, *Zanesville*, OH 43701
T: +1 740 4557154; Fax: +1 740 4557177
Donna Herron
13 000 vols   35098

Florida Department of Corrections, Zephyrhills Correctional Institution Library, 2739 Gall Blvd, *Zephyrhills*, FL 33541
T: +1 813 7825221; Fax: +1 813 7800123
Lois Sadd
8 000 vols; 41 curr per   35099

Ecclesiastical Libraries

First Baptist Church, Media Library, 11333 N 3rd St, P.O. Box 85, *Abilene*, TX 79601
T: +1 915 6735031; Fax: +1 915 6758189
1948; Karin Richardson
18 000 vols; 12 curr per; 680 sound-rec; 4 VF drawers, 185 slides   35100

High Street Christian Church, H.A. Valentine Memorial Library, 131 S High St, **Akron**, OH 44308
T: +1 330 4341039; Fax: +1 330 4347271; E-mail: office@highstcc.org; URL: www.highstcc.org
1965
Church Arch
9000 vols; 12 curr per; 44 sound-rec
35101

First Baptist Church Library, 101 Broadway NE, **Albuquerque**, NM 87102
T: +1 505 2473611; Fax: +1 505 2470345
1937; Gloria Potts
11000 vols
35102

Virginia Theological Seminary, Bishop Payne Library, 3737 Seminary Rd, **Alexandria**, VA 22304-5201
T: +1 703 4611731; Fax: +1 703 3700935; E-mail: paynelib@vts.edu; URL: www.vts.edu/library
1823; Mitzi J. Budde
Anglicanism & Episcopal Church in the USA
192000 vols; 1050 curr per; 147 music scores; 69 maps; 7000 microforms; 2960 sound-rec; 40 digital data carriers; 550 linear feet mss
libr loan; OCLC
35103

Detroit Baptist Theological Seminary, Library, 4801 Allen Rd, **Allen Park**, MI 48101
T: +1 313 3810111; Fax: +1 313 3810798; E-mail: library@dbts.edu; URL: library.dbts.edu
1976; Mark A. Snoeberger
45000 vols; 300 curr per
OCLC, MLC
35104

Trinity Episcopal School for Ministry Library, 311 11th St, **Ambridge**, PA 15003
T: +1 724 2663838; Fax: +1 724 2664617
1975; Susanah Hanson
85000 vols; 440 curr per
35105

Temple Beth Zion, Suburban Library, 700 Sweet Home Rd, **Amherst**, NY 14226
T: +1 716 8867150; Fax: +1 716 8867152
1915; Madeline Davis
Reform Judaism; Jewish Beliefs & Practices; American Jewish Hist
13000 vols; 18 curr per; 200 sound-rec; 250 filmstrips & slides
35106

Temple Shaarey Zedek, Rabbi Isaac Klein Library, 621 Getzville Rd, **Amherst**, NY 14226
T: +1 716 8383232; Fax: +1 716 8356154
1959; Judy Carrel
Children's Judaica
9000 vols; 13 curr per
35107

Church of the Redeemer, King Library, 1065 Bristol Pike, **Andalusia**, PA 19020
T: +1 215 6394387
1882; Dr. John S. Keefer
Key Coll
12415 vols; 1 curr per; 40 av-mat; 800 sound-rec; 3 digital data carriers; 40 videos
35108

Faith Baptist Bible College & Theological Seminary, John L. Patten Library, 1900 NW Fourth St, **Ankeny**, IA 50023
T: +1 515 9640601; Fax: +1 515 9641638; URL: www.faith.edu
1921; Karen Houghton
68000 vols; 438 curr per
ATLA, ACL
35109

Arlington Baptist College, Earl K. Oldham Library, 3001 W Division, **Arlington**, TX 76012-3425
T: +1 817 4618741; Fax: +1 817 2741138; E-mail: abclibrary@abconline.org; URL: www.abconline.edu
Denise Watts
31000 vols; 280 curr per
35110

Ashland Theological Seminary, Roger E. Darling Memorial Library, 910 Center St, **Ashland**, OH 44805
T: +1 419 2895434; Fax: +1 419 2895969; URL: library.ashland.edu
1930; Sylvia L. Locher
Religious Debates (Darling Debate); Mary, Queen of Scots (Ronk Coll); Artifacts
82000 vols; 480 curr per; 1972 sound-rec
35111

American Baptist Churches-USA, Historical Society-Research Library & Archives, 2930 Flowers Rd S, **Atlanta**, GA 30341; mail address: 3001 Mercer University Dr, Atlanta, GA 30341
T: +1 678 5476680
1854; Dr. Deborah Van Broekhoven
American Baptist Church/USA Nat Board & Baptist World Alliance, Mission Artifacts
84000 vols; 91 curr per
35112

American Baptist Historical Society, Library, 3001 Mercer University Drive, **Atlanta**, GA 30341; URL: www.abhsarchives.org
1853; Betsy Dunbar
Danish-Norwegian Baptist General Conference of American Arch, 17th & 18th C English Baptist Hist (Henry S Burrage), American Bible Union Archs, Samuel Colgate Hist Coll
60000 vols; 300 curr per; 175 sound-rec; 25000 unbd serials, 8000 pamphlets, 800 slides
libr loan; ATLA
35113

Beulah Heights Bible College, Barth Memorial Library, 892 Berne St SE, **Atlanta**, GA 30316; P.O. Box 18145, Atlanta, GA 30316-0145
T: +1 404 6272681; Fax: +1 404 6270702; URL: www.beulah.org
1918; Pradeep Das
Leadership, World Missions, Urban Ministry, Black Church and Black Hist
46000 vols; 284 curr per; 6 diss/theses; 5 maps; 100 microforms; 210 av-mat; 920 sound-rec; 10 digital data carriers
libr loan; ALA, SLA, ACL, GLA
35114

Carver Bible College, Library, 3870 Cascade Rd, **Atlanta**, GA 30331
T: +1 404 5274520; Fax: +1 404 5274524; URL: www.carver.edu
1943; L
16000 vols; 60 curr per
35115

Mount Paran Church of God, Ruth Holt Library, 2055 Mount Paran Rd NW, **Atlanta**, GA 30327
T: +1 404 2333949; Fax: +1 404 2399460; URL: www.psy.edu
1977; John Hughes
14000 vols; 39 curr per; 50 boxes of arch mats
35116

Peachtree Presbyterian Church, Pattillo Library, 3434 Roswell Rd NW, **Atlanta**, GA 30305
T: +1 404 8425813; Fax: +1 404 8425858; E-mail: info@peachtreepres.org; URL: www.peachtreepres.org
1960
10000 vols
35117

Austin Presbyterian Theological Seminary, David L. and Jane Stitt Library, 100 E 27th St, **Austin**, TX 78705-5797
T: +1 512 4726736; Fax: +1 512 3220901; E-mail: library@austinseminary.edu; URL: www.austinseminary.edu
1902; Timothy D. Lincoln
Rumble Communion Token Coll
157000 vols; 590 curr per; 50 diss/theses; 3000 microforms; 10 digital data carriers
libr loan; ATLA, TLA
35118

Episcopal Theological Seminary of the Southwest, Booher Library, 606 Rathervue Pl, **Austin**, TX 78705; P.O. Box 2247, Austin, TX 78768-2247
T: +1 512 4724133; Fax: +1 512 4724620; URL: www.etss.edu/
1951; Donald E. Keeney
19th C English Lit-Fine Editions (Charles L Black Coll); Hist, Lit & Culture of Latin America (Sophy H Winterbotham Coll)
123000 vols; 260 curr per; 1000 microforms; 1490 sound-rec
libr loan; ATLA
35119

LSPS/Seminex Library, Lutheran Seminary Program in the SW, 607 Rathervue Pl, P.O. Box 4790, **Austin**, TX 78765
T: +1 512 4785212; Fax: +1 512 4776693; E-mail: lsps@lsps.edu
1974; Lucille Hager
39668 vols; 118 curr per; 13236 microforms
libr loan
35120

Baltimore Hebrew Congregation, Julius & Doris Myerberg Library, 7401 Park Heights Ave, **Baltimore**, MD 21208
T: +1 410 7641587; Fax: +1 410 7647948; URL: bhcong.org
Susan F. Berman
Judaica
15000 vols; 30 curr per; 50 av-mat
35121

Carmelite Monastery, Library & Archives, 1318 Dulaney Valley Rd, **Baltimore**, MD 21286-1399
T: +1 410 8237415; Fax: +1 410 8237418; E-mail: carmelit@starpower.net; URL: www.karmel.at/eng/res.htm
1790; Sr Constance FitzGerald
Theology; Works of St Teresa of Avila & St John of the Cross; Arch of the First Community of Roman Catholic Nuns in the 13 Original States
30000 vols; 55 curr per
35122

Faith Theological Seminary, Library, 529-531 Walker Ave, **Baltimore**, MD 21212-2624
T: +1 410 3236211; Fax: +1 410 3236331; E-mail: fts@faiththeological.org; URL: www.faiththeological.org
1937; Katherine Finnegan
28000 vols; 20 curr per
35123

Saint Mary's Seminary & University, The Knott Library, 5400 Roland Ave, **Baltimore**, MD 21210-1994
T: +1 410 8643621; Fax: +1 410 4358571; URL: www.stmarys.edu
1791; David Carter
Religion, Theology; Early Catholic Americana, Scripture/Orientalia, Semitics
120000 vols; 325 curr per
35124

Talmudical Academy of Baltimore Library, 4445 Old Court Rd, **Baltimore**, MD 21208
T: +1 410 4846600; Fax: +1 410 4845717
1947
12000 vols; 12 curr per
35125

Bangor Theological Seminary, Moulton Library, PO Box 411, **Bangor**, ME 04402-0411
T: +1 207 9426781; Fax: +1 207 9901267; E-mail: library@bts.edu; URL: www.bts.edu
1814; Beth Bidlack
Coll of Sermons, 19th c Pamphlets (2500)
137000 vols; 431 curr per; 936 microforms; 364 sound-rec
libr loan
35126

Bangor Theological Seminary, General Theological Library, c/o Husson University, One College Circle, **Bangor**, ME 04401
T: +1 207 9417188; E-mail: admit@husson.edu; URL: www.bts.edu/current/library/index.html
1860
35000 vols; 15 curr per
35127

Zion Bible College, Library, 27 Middle Hwy, **Barrington**, RI 02806
T: +1 401 6282116; Fax: +1 401 2460906
1956; Ginger McDonald
41000 vols; 100 curr per; 2600 sound-rec
ACL
35128

The Temple – Tifereth Israel, Lee & Dolores Hartzmark Library, 26000 Shaker Blvd, **Beachwood**, OH 44122
T: +1 216 8313233 ext 119; Fax: +1 216 8314216; E-mail: adavidson@ttti.org; URL: www.ttti.org
1896; Andrea Davidson
Judaica; Abba Hillel Silver Arch
25000 media items; 75 curr per
ALA
35129

Carmelite Monastery Library, 139 Depuyster Ave, **Beacon**, NY 12508-3599
1950; Sr Michalene
Works of St John of the Cross & St Teresa of Jesus
8400 vols; 40 curr per; 6 VF drawers of rpts
35130

Saint Benedict's Abbey, Benet Library, 12605 224th Ave, **Benet Lake**, WI 53102-0333
T: +1 262 3964311; Fax: +1 262 3964365; E-mail: info@benetlake.org; URL: benetlake.org
1945
19000 vols; 45 curr per
35131

Academy of American Franciscan History Library, 1712 Euclid Ave, **Berkeley**, CA 94709-1208
T: +1 510 5481755; Fax: +1 510 5499466; E-mail: acadafh@aol.com; URL: www.aafh.org
1944; John F. Schwaller
Hist of Southwest (Mary Coleman Powell Coll), 700 vols; Arch of the Indies (Seville Coll), Franciscan related docs; Archivo Historico Nacional de Mexico, docs
20000 vols; 100 curr per
35132

Swedenborgian Library & Archives, 1798 Scenic Ave, **Berkeley**, CA 94709
T: +1 510 8498228; Fax: +1 510 8498296
1866; Michael Yockey
Church hist, Theology; Church of the New Jerusalem; Emanuel Swedenborg, arch, letters; New Church; Swedenborgian Church
27000 vols; 20 curr per
35133

Holy Cross Abbey Library, 901 Cool Spring Lane, **Berryville**, VA 22611-2700
T: +1 540 9551425; Fax: +1 540 9551356
1950; Father Vincent Collins
25000 vols; 59 curr per
35134

Abbey of Regina Laudis Library, Flanders Rd, **Bethlehem**, CT 06751
T: +1 203 2667727; URL: www.abbeyofreginalaudis.com
1947; Mother Lucia Kuppens
Patrology, Scripture, Ecclesiastical hist, Art, Social sciences, Musicology, New England hist, Natural sciences, Lit; Patristic Writings, Newman's Writings, Medieval Mystics
24000 vols; 12 curr per; 500 maps; 600 microforms; 500 av-mat; 3500 sound-rec; 3000 slides, 100 charts, 1600 pamphlets
libr loan
35135

Moravian College & Moravian Theological Seminary, Reeves Library, 1200 Main St, **Bethlehem**, PA 18018-6650
T: +1 610 8611540; Fax: +1 610 8611577; URL: www.moravian.edu/public/reeves/index.htm
1742; Rita Berk
Liberal arts, Theology; Hist of the Moravian Church Coll, Map Coll, Moravian Theological Seminary Coll, Music dept contains the (Herman Adler Coll) of 4500 Baroque & predominantly German Classic & Romantic recordings
263000 vols; 1037 curr per; 2240 e-journals; 26000 e-books; 8809 music scores; 9000 microforms; 2410 av-mat; 16013 sound-rec; 370 digital data carriers; 103 linear feet of mss and archives
libr loan; ALA, ATLA
35136

Trinity Lutheran Seminary, Hamma Library, 2199 E Main St, **Bexley**, OH 43209-2334
T: +1 614 3844645; Fax: +1 614 2380263; E-mail: tlslibrary@trinitylutheranseminary.edu; URL: www.trinitylutheranseminary.edu
1830; Ray A. Olson
New Testament Studies (Lenski Memorial Coll); Hymnals; Catechisms
137000 vols; 430 curr per; 2000 microforms; 4510 sound-rec; 125 digital data carriers
libr loan; ATLA
35137

Southeastern Bible College, Gannett-Estes Library, 2545 Valleydale Rd, **Birmingham**, AL 35244
T: +1 205 9709233; Fax: +1 205 9709207; E-mail: library@sebc.edu; URL: www.sebc.edu
1935; Paul A. Roberts
60 000 vols; 151 curr per; 30 000 e-books; 156 maps; 1 400 CDs   35138

Saint Thomas Seminary-Hartford Archdiocese, Archbishop Henry J. O'Brien Library, 467 Bloomfield Ave, **Bloomfield**, CT 06002
T: +1 860 2425573; Fax: +1 860 2424981; URL: www.stseminary.org/obrienlibrary.htm
1897; Karen Lesiak
Bibles, Early Catholic Americana Coll
30 000 vols; 87 curr per; 78 av-mat
libr loan; CLA, ALA   35139

Reform Episcopal Seminary, Fred C. Kuehner Memorial Library, 826 Second Ave, **Blue Bell**, PA 19422
T: +1 610 2929852; Fax: +1 610 2929853; E-mail: library@reseminary.edu; URL: www.reseminary.edu
1886
23 000 vols; 54 curr per
libr loan; ATLA   35140

Reformed Episcopal Seminary, Fred C. Kuehner Memorial Library, 826 Second Ave, **Blue Bell**, PA 19422
T: +1 610 2929852; Fax: +1 610 2929853; URL: library.lts.org/septla/pts.htm
Jonathan S. Riches
Reformed Episcopal church, Puritans, British church history
21 000 vols   35141

Boise Bible College, Library, 8695 W Marigold St, **Boise**, ID 83714-1220
T: +1 208 3767731; Fax: +1 208 3767743; E-mail: boisebible@boisebible.edu; URL: www.boisebible.edu
1945; Glennis Thomas
Restoration Movement Mat
35 000 vols; 77 curr per; 34 maps; 1 000 microforms; 620 sound-rec; 15 digital data carriers
libr loan; ACL   35142

American Congregational Association, Congregational Library, 14 Beacon St, **Boston**, MA 02108-9999
T: +1 617 5230470;
Fax: +1 617 5230491; URL: www.congregationallibrary.org
1853; Dr. Margaret L. Bendroth
Religion, Theology, Missions, Philosophy, Lit, Psychology, Natural science; Congregationalism (National & General Council Congregational Christian Churches), Church Records, Educational & Missionary Societies, State & Intrastate Clerical Organizations, Mather Family Coll
255 000 vols; 40 curr per; 1 260 microforms; 100 000 pamphlets & per   35143

Temple Israel, Dr. Arnold L. Segel Library Center, 477 Longwood Ave, **Boston**, MA 02215
T: +1 617 5663960; Fax: +1 617 7313711; E-mail: aabrams@tisrael.org; URL: tisrael.org/librarymainpage.asp
Ann Abrams
Judaica
10 000 vols; 12 curr per   35144

St. Vincent de Paul Regional Seminary, Library, 10701 S Military Trail, **Boynton Beach**, FL 33436-4811
T: +1 561 7324424; Fax: +1 561 7372205; URL: www.svdp.edu
1963; Arthur G. Quinn
Theology, Philosophy, Latin America; Spanish Language Coll, Loeb Series, Sources Chretiennes Series
67 000 vols; 380 curr per; 774 microforms; 884 sound-rec; 2 digital data carriers   35145

Saint John's Seminary, Library, 99 Lake St, **Brighton**, MA 02135-3846
T: +1 617 7465426; Fax: +1 617 7465495
1884; Lawrence McGrath
152 500 vols; 350 curr per   ∗ 35146

Yeshiva Torah Vodaath and Mesifta, Torah Vodaath Library, 425 E 9th St, **Brooklyn**, NY 11218
T: +1 718 9418000; Fax: +1 718 9418032
1918; Mr Gany
24 075 vols; 65 curr per; 600 unbound periodicals, 280 unbound pamphlets   35147

Temple Beth Zion Library, 805 Delaware Ave, **Buffalo**, NY 14209
T: +1 716 8366565
1915; Madeline Davis
Children's Judaica (Religious School Libr)
20 000 vols; 22 curr per   35148

St. John's Seminary – Carrie Estelle Doheny Memorial Library, 5012 Seminary Rd, **Camarillo**, CA 93012
T: +1 805 4822755 ext 2043; Fax: +1 805 9870885; E-mail: plyons@stjohnsem.edu; URL: library.stjohnsem.edu
1962; Dr. Patricia Lyons
Humanities, Philosophy
75 000 vols; 200 curr per
OCLC   35149

St. John's Seminary – Edward Laurence Doheny Library, 5012 Seminary Rd, **Camarillo**, CA 93012
T: +1 805 4822755; Fax: +1 805 9870885; E-mail: plyons@stjohnsem.edu; URL: library.stjohnsem.edu/
1940; Dr. Patricia Lyons
Greek & Latin Patrology
69 000 vols; 200 curr per; 2 094 microforms; 426 av-mat; 387 sound-rec; 7 digital data carriers
libr loan   35150

Episcopal Divinity School & Weston Jesuit School of Theology, Library, 99 Brattle St, **Cambridge**, MA 02138
T: +1 617 3493602; Fax: +1 617 3493603; URL: www.edswjst.org
1867; Esther A. Griswold
220 000 vols; 1 163 curr per; 150 diss/theses; 1 000 microforms; 494 av-mat; 13 digital data carriers
libr loan; ATLA   35151

Tabernacle Baptist Church, Library, 150 Tabernacle Dr, **Carrollton**, GA 30117
T: +1 770 8327063; Fax: +1 770 8342777; URL: www.tabernacle.org
8 500 vols   35152

First Presbyterian Church of Charleston, Library, 16 Leon Sullivan Way, **Charleston**, WV 25301
T: +1 304 3438961; Fax: +1 304 3438970; E-mail: firstpresby@firstpresby.com; URL: www.firstpresby.com
1941
9 000 vols; 12 curr per; 150 av-mat; 150 sound-rec   35153

AMG International Library, 6815 Shallowford Rd, **Chattanooga**, TN 37421-1755
T: +1 423 8946060; Fax: +1 423 5108074
Dale Anderson
Theology, Commentaries, Sermons, Church hist, Missions, Biogr
32 000 vols; 23 000 microforms   35154

Catholic Theological Union, Paul Bechtold Library, 5416 S Cornell Ave, **Chicago**, IL 60615-5698
T: +1 773 3715464; Fax: +1 773 7535340; URL: www.ctu.lib.il.us
1968; Melody Layton McMahon
Franciscan Order, Roman Catholic theology; Christian Art, Canon Law, Missiology
152 000 vols; 500 curr per; 46 microforms; 688 sound-rec
libr loan   35155

Chicago Theological Seminary, Hammond Library, 5757 S University Ave, **Chicago**, IL 60637-1507
T: +1 773 3220225; Fax: +1 773 7527194; E-mail: library@ctschicago.edu; URL: www.ctschicago.edu
1855; Neil Gerdes
Congregational Church hist (Lowenbach Coll); Psychiatry, Psychology & Religion (Anton Boisen Coll)
117 000 vols; 144 curr per; 4 incunabula; 135 diss/theses; 795 microforms; 12 digital data carriers
libr loan   35156

Lutheran School of Theology at Chicago & McCormick Theological Seminary, Jesuit-Krauss-McCormick Library, 1100 E 55th St, **Chicago**, IL 60615-5199
T: +1 773 2560739; Fax: +1 773 2560737; E-mail: library@jkmlibrary.org
1975; Christine Wenderoth
Biblical Mss, 9th to 16th c; Church Ministry Resources
356 000 vols; 597 curr per; 505 av-mat; 494 sound-rec; 12 digital data carriers
libr loan   35157

Moody Bible Institute, Crowell Library, 820 N La Salle Blvd, **Chicago**, IL 60610-3284
T: +1 312 3294136; Fax: +1 312 3298959; E-mail: library@moody.edu; URL: library.moody.edu
1889; Jim Preston
Religion; Moodyana Coll, Music Libr
171 000 vols; 349 curr per; 16 030 e-journals; 2 000 e-books; 4 740 av-mat; 17 240 sound-rec; 128 Braille Vols
libr loan   35158

Cincinnati Christian University, George Mark Elliott Library, 2700 Glenway Ave, **Cincinnati**, OH 45204-3200
T: +1 513 2448197; Fax: +1 513 2448434; E-mail: library@ccuniversity.edu; URL: library.ccuniversity.edu
1924; James H. Lloyd
Restoration Movement Coll
115 000 vols; 625 curr per; 10 500 diss/theses; 80 000 music scores; 500 maps; 5 000 microforms; 12 148 sound-rec; 25 digital data carriers
libr loan; ATLA, ACL   35159

Isaac M. Wise Temple Library, 8329 Ridge Rd, **Cincinnati**, OH 45236
T: +1 513 7932556; Fax: +1 513 7933322; URL: wisetemple.org
1931; Andrea Rapp
Judaica, Holocaust
20 000 vols; 35 curr per; 150 av-mat; 504 sound-rec; 15 overhead transparencies
AJL   35160

Rockdale Temple, Sidney G. Rose Library, 8501 Ridge Rd, **Cincinnati**, OH 45236
T: +1 513 8919900; Fax: +1 513 8910515; E-mail: mail@rockdaletemple.org; URL: rockdaletemple.club.officelive.com/default.aspx
Judaica; Interfaith Marriages Coll, Holocaust Coll, Judaica Art
9 000 vols; 11 curr per   35161

Baptist Bible College and Seminary, Richard J. Murphy Memorial Library, 538 Venard Rd, **Clarks Summit**, PA 18411-1250
T: +1 570 5859282; Fax: +1 570 5859244; URL: www.bbc.edu
1932; Joshua Michael
104 000 vols; 356 curr per; 10 000 microforms; 11 136 sound-rec; 21 digital data carriers; 39 VF drawers of pamphlets & clippings, 1 697 film & video mat   35162

Lee University – Church of God Theological Seminary, William G. Squires Library, 260 11th St NE, **Cleveland**, TN 37311
T: +1 423 6148551; Fax: +1 423 6148555; E-mail: library@leeuniversity.edu; URL: library.leeuniversity.edu
1941; Barbara McCullough
Pentecostal Research Ctr Coll, multi-media
165 000 vols; 630 curr per; 54 000 microforms; 1 424 av-mat; 12 000 sound-rec; 65 digital data carriers
libr loan; SOLINET   35163

St Andrew's Abbey, Slovak Institute Library & Museum, 10510 Buckeye Rd, **Cleveland**, OH 44104
T: +1 216 7215300; Fax: +1 216 7918268; E-mail: slovakinstitute@csnmail.net
Andrew F. Hudak Jr
11 000 vols; 200 curr per   35164

Park Synagogue, Kravitz Memorial Library, 3300 Mayfield Rd, **Cleveland Heights**, OH 44118-1899
T: +1 216 3712244; Fax: +1 216 3210639; URL: www.parksyn.org/library.htm
1923; Sarajane Dolinsky
Holocaust, Israel; Judaica (adult, juvenile, Hebrew, Yiddish & English fiction)
20 000 vols; 35 curr per   35165

Nazarene Bible College, Trimble Library, Loop Park, 1111 Academy, **Colorado Springs**, CO 80910-3704
T: +1 719 8845071; Fax: +1 719 8845199; E-mail: library@nbc.edu
1967; Vernell Posey
Wesleyana Coll
68 592 vols; 286 curr per; 215 av-mat; 2 189 sound-rec; 8 VF drawers of pamphlets, 67 art reproductions
libr loan   35166

Lutheran Theological Southern Seminary, Lineberger Memorial Library, 4201 N Main St, **Columbia**, SC 29203
T: +1 803 7865150; Fax: +1 803 4613278; URL: www.ltss.edu
1830; Dr. Lynn A. Feider
Religious studies; 16th-18th C German Pietism, German Hymnbooks & Catechisms
100 000 vols; 474 curr per; 8 000 microforms; 2 464 sound-rec; 3 digital data carriers; 451 linear feet of arch   35167

Pontifical College Josephinum, A. T. Wehrle Memorial Library, 7625 N High St, **Columbus**, OH 43235-1498
T: +1 614 9852295; Fax: +1 614 8852307; E-mail: libreqs@pcj.edu; URL: www.pcj.edu/pcjlibrary/homepage.htm
1888; Peter G. Veracka
Biblical Studies Coll; Books on the Catacombs
145 000 vols; 403 curr per; 18 000 e-books; 2 mss; 11 incunabula; 2 000 microforms; 786 av-mat; 2 022 sound-rec; 9 digital data carriers
libr loan; ATLA, CLA   35168

Conception Abbey & Seminary, Library, 37174 State Hwy W, **Conception**, MO 64433; P.O. Box 501, Conception, MO 64433-0501
T: +1 660 9442803; Fax: +1 660 9442833; URL: library.conception.edu
1873; Brother Thomas Sullivan
Roman Catholic religion, Religious studies, Philosophy, Medieval European hist, Art; 17th-19th c Catholic Theology Coll, American Catholic Church Hist Coll
121 000 vols; 364 curr per; 1 000 microforms; 110 av-mat; 6 901 sound-rec; 3 digital data carriers; 3 500 Rare Books & Incunabula, 12 950 slides, 60 Journals   35169

Mid-America Baptist Theological Seminary, Ora Byram Allison Memorial Library, 2095 Appling Rd, **Cordova**, TN 38016
T: +1 901 7513007; Fax: +1 901 7518454
1972; Terrence Neal Brown
Missions, Semitic languages
150 000 vols; 972 curr per   35170

Monastery of Gertrude, Library, 465 Keuterville Rd, **Cottonwood**, ID 83522
T: +1 208 9623224; Fax: +1 208 9627212; E-mail: monastery@stgertrudes.org; URL: www.stgertrudes.org
40 000 vols   35171

St Francis Library, 11414 W Hwy 33, **Coyle**, OK 73027; P.O. Box 400, Coyle, OK 73027-0400
T: +1 405 4663774; Fax: +1 405 4663722
1982; Chris Contreras
Eastern Orthodox Theology, Patristics, Biblical Theology, Liturgics, Byzantine Music, Philosophy, Homiletics, Roman Catholic Theology, Anglican Theology, Alternative Lifestyles, Holistic Medicine; Monumentae Musicae Byzantinae, Carl G. Jung, Sustainable Agriculture, St Francis of Asissi
21 000 vols; 300 microforms; 1 500 unbound periodicals, 1 000 leaflets, pamphlets   35172

Holy Apostles College and Seminary, Library, 33 Prospect Hill Rd, *Cromwell*, CT 06416-0903
T: +1 860 6323009; Fax: +1 860 6323090; E-mail: info@holyapostles.edu; URL: www.holyapostles.edu
1957; Clare Adamo
Religion, theology, philosophy
61 400 vols; 220 curr per; 15 digital data carriers; 230 reels
ATLA, CLA   35173

All Saints Catholic Church, Parish Resource Library, 5231 Meadowcreek at Arapaho, *Dallas*, TX 75248-4046
T: +1 972 7780327; Fax: +1 972 2335401; E-mail: library@allsaintsdallas.org; URL: www.allsaintsdallas.org/189537.ihtml
1979; Maria Isabel Bellavance
Parish Arch, Antique Bibles, Antique Bks
13 000 vols; 35 curr per; 211 maps; 451 microforms; 862 av-mat; 1 200 slides
                                            35174

Church of the Incarnation, Marmion Library, 3966 McKinney Ave, *Dallas*, TX 75204-2099
T: +1 214 5215101 ext 24; Fax: +1 214 5287209; E-mail: info@incarnation.org; URL: www.incarnation.org
1955
14 700 vols; 25 curr per   35175

Dallas Theological Seminary, Turpin Library, 3909 Swiss Ave, *Dallas*, TX 75204
T: +1 214 8413750; Fax: +1 214 8413745; URL: library.dts.edu
1924; Robert D. Ibach
209 000 vols; 859 curr per; 13 e-journals; 11 000 e-books; 1 incunabula; 76 maps; 46 000 microforms; 1 920 av-mat; 16 677 sound-rec; 500 digital data carriers; 130 Braille vols
libr loan; ALA, ATLA   35176

First Baptist Church of Dallas, Truett Memorial Library, 1707 San Jacinto, *Dallas*, TX 75201
T: +1 214 9692442; Fax: +1 214 9692471; E-mail: truettlibrary@firstdallas.org; URL: www.firstdallas.org
1898; Ruth Turner
Hymnals
61 000 vols; 210 curr per; 2 digital data carriers   35177

Highland Park Presbyterian Church, Meyercord Library, 3821 University Blvd, *Dallas*, TX 75205
T: +1 214 5267457; Fax: +1 214 5595311; URL: www.hppc.org
1953; Diana Wehrmann
Religion; Birds, Antique Bibles
17 000 vols   35178

Highland Park United Methodist Church Library, 3300 Mockingbird Lane, *Dallas*, TX 75205
T: +1 214 5232273; E-mail: williama@hpumc.org; URL: www.hpumc.org/pages/09_Library
1950; Ann L. Williams
Large Print Bks; Nat Geographic, 1911-1987; Texas Libr
18 000 vols; 14 curr per; 780 sound-rec
                                            35179

Park Cities Baptist Church, Media Library, 3933 Northwest Pkwy, *Dallas*, TX 75225-3333; P.O. Box 12068, Dallas, TX 75225-0068
T: +1 214 8601500; Fax: +1 214 8601538; URL: www.pcbc.org
1944; Beth Andrews
20 000 vols; 35 curr per   35180

Temple Emanu-El, Alex F. Weisberg Library, 8500 Hillcrest Rd, *Dallas*, TX 75225
T: +1 214 7060000; Fax: +1 214 7060025; E-mail: information@tedallas.org; URL: www.tedallas.org
1957
12 000 vols; 30 curr per   35181

Saint Norbert Abbey, Augustine Library, 1016 N Broadway, *De Pere*, WI 54115-2697
T: +1 920 3374354
Karen E. Mand
Canon law, church history, philosophy
11 000 vols   35182

Columbia Theological Seminary, John Bulow Campbell Library, 701 Columbia Dr, *Decatur*, GA 30030
T: +1 404 6874549; Fax: +1 404 6874687; E-mail: ref-desk@ctsnet.edu; URL: www.ctsnet.edu/library/libintro.asp
1828; Sara Myers
Religion, Philosophy
200 000 vols; 862 curr per   35183

Archbishop Vehr Theological Library, 1300 S Steele St, *Denver*, CO 80210
T: +1 303 7153146; Fax: +1 303 7152037
1907; Dr. Michael Woodward
Theology, Bible, Social problems; Anglican Studies, Hispanic Pastoral Ministry Coll, Catholic Theology, Church Hist, Religious Art
150 000 vols; 300 curr per   35184

Congregation Emanuel Library, Brenner Library, 51 Grape St, *Denver*, CO 80220
T: +1 303 3884013;
Fax: +1 303 3886328; URL: www.congregationemanuel.com
Katherine Steinberg
Judaism
10 000 vols; 30 curr per; 40 av-mat
                                            35185

Ecumenical Theological Seminary, John E. Biersdorf Library, 2930 Woodward Ave, *Detroit*, MI 48201
T: +1 313 8315200; Fax: +1 313 8311353
Jonathan Chad Roach
31 000 vols; 68 curr per   35186

Sacred Heart Major Seminary, Cardinal Edmund Szoka Library, 2701 Chicago Blvd, *Detroit*, MI 48206
T: +1 313 8838654; Fax: +1 313 8688594; E-mail: szokalibrary@shms.edu; URL: www.shms.edu
1921; Dr. Herman A. Peterson
Theology, Philosophy; Early Michigan (Gabriel Richard Coll), Church Hist
123 000 vols; 500 curr per; 75 diss/theses; 3 000 microforms; 169 av-mat; 2 492 sound-rec
libr loan; MLA, ALA   35187

Emmaus Bible College, Library, 2570 Asbury Rd, *Dubuque*, IA 52001-3096
T: +1 563 5888000; Fax: +1 563 5881216
1941; John Rush
Plymouth Brethren Writings
110 000 vols; 310 curr per   35188

Wartburg Theological Seminary, Reu Memorial Library, 333 Wartburg Pl, *Dubuque*, IA 52003
T: +1 563 5890266; Fax: +1 563 5890333; E-mail: library@wartburgseminary.edu; URL: www.wartburgseminary.edu/campus/library.htm
1853; Susan J. Ebertz
85 000 vols; 246 curr per
libr loan   35189

Erskine College & Theological Seminary, McCain Library, One Depot St, *Due West*, SC 29639; P.O. Box 188, Due West, SC 29639-0188
T: +1 864 3798714; Fax: +1 864 3792900; E-mail: library@erskine.edu; URL: www.erskine.edu/library
1837; John F. Kennerly
Religion, genealogy, local hist, hist; Associate Reformed Presbyterian Church records
178 000 vols; 1 333 curr per; 360 e-journals; 36 000 e-books   35190

Northland Baptist Bible College, Library, W10085 Pike Plains Rd, *Dunbar*, WI 54119
T: +1 715 3246900; Fax: +1 715 3246133; E-mail: library@nbbc.edu; URL: www.nbbc.edu
Van Carpenter
Rare Theological Books
47 000 vols; 220 curr per   35191

Mid-America Reformed Seminary, Library, 229 Seminary Dr, *Dyer*, IN 46311
T: +1 219 8642400; Fax: +1 219 8642410; URL: www.midamerica.edu
Alan D. Strange
40 000 vols; 150 curr per   35192

Christ the King Seminary, Library, 711 Knox Rd, *East Aurora*, NY 14052; P.O. Box 607, East Aurora, NY 14052-0607
T: +1 716 6528959; Fax: +1 716 6528903
1951; Bonaventure Hayes
Early French Canadian & Niagara Frontier Hist (Msgr James Bray Coll)
176 000 vols; 437 curr per; 3 508 microforms; 1 660 av-mat; 1 280 sound-rec; 3 digital data carriers
libr loan; ATLA, CLA   35193

Rio Grande Bible Institute & Language School, Richard Wade & Glen Vyck McKinney Library, 4300 S Business 281, *Edinburg*, TX 78539
T: +1 956 3808138; Fax: +1 956 3808256; E-mail: biblioteca@riogrande.edu; URL: www.riogrande.edu
Laura G. Coon
Missions, Christian theological works in Spanish
25 000 vols; 86 curr per   35194

Roanoke Bible College, Watson-Griffith Library, 715 N Poindexter St, *Elizabeth City*, NC 27909-4054
T: +1 252 3342046; Fax: +1 252 3342071; URL: www.roanokebible.edu/page/view/31
1948; Frank L. Dodson
Creationism Coll, Deaf Coll, Discipliana Coll
30 000 vols; 188 curr per   35195

Associated Mennonite Biblical Seminary, Library, 3003 Benham Ave, *Elkhart*, IN 46517
T: +1 574 2966253; Fax: +1 574 2950092; URL: www.ambs.edu/library
1945; Eileen K. Saner
Religion; Studer Bible Coll
115 000 vols; 486 curr per; 359 maps; 1 000 microforms; 308 av-mat; 899 sound-rec; 36 digital data carriers
libr loan; ATLA, ALA   35196

Trinity Bible College, Fred J. Graham Library, 50 Sixth Ave S, *Ellendale*, ND 58436-7150
T: +1 701 3495407; Fax: +1 701 3495443; E-mail: tbclibrary@trinitybiblecollege.edu; URL: www.trinitybiblecollege.edu
1948; Diane Olson
College & Denomination Arch, Graham Coll (NDak Hist); Curriculum Lab
85 000 vols; 215 curr per; 12 000 e-books; 35 diss/theses; 100 music scores; 36 maps; 7 000 microforms; 2 400 sound-rec; 200 digital data carriers; 10 000 pamphlets & clippings, 16 VF drawers
libr loan   35197

Divine Word College, Matthew Jacoby Memorial Library, 102 Jacoby Dr SW, *Epworth*, IA 52045-0380
T: +1 563 8763353; Fax: +1 563 8763407; URL: www.dwci.edu/library
1915; Daniel Boice
Religion, Philosophy, Cross-cultural studies; LEP, Vietnamese Lit
94 000 vols; 310 curr per; 135 av-mat
libr loan   35198

Westminster Seminary California, Library, 1725 Bear Valley Pkwy, *Escondido*, CA 92027
T: +1 760 4808474; Fax: +1 760 4800252; URL: www.wscal.edu
1980; Elizabeth Mehne
Theology; 16th/17th C Reformation & Puritanism (microforms)
63 000 vols; 260 curr per; 2 000 microforms; 2 034 sound-rec; 12 digital data carriers   35199

Eugene Bible College, Flint Memorial Library, 2155 Bailey Hill Rd, *Eugene*, OR 97405
T: +1 541 4851780; Fax: +1 541 3435801
1925; Priscilla Cameron
Flint Coll
34 000 vols; 250 curr per
libr loan   35200

Beth Emet Synagogue, Bruce Gordon Memorial Library, 1224 Dempster St, *Evanston*, IL 60202
T: +1 847 8697830; Fax: +1 847 8697830; E-mail: office@bethemet.org; URL: www.bethemet.org

1950
8 000 vols; 40 curr per   35201

Trinity Lutheran College, Library, 2802 Wetmore Ave, *Everett*, WA 98201; P.O. Box 870, Everett, WA 98206
T: +1 425 3920400; Fax: +1 425 3920404; E-mail: library@tlc.edu; URL: www.tlc.edu
1944; Elliott Ohannes
Reformation Works
30 000 vols; 216 curr per; 200 av-mat; 465 sound-rec; 30 digital data carriers; 27 kits, 12 VF drawers   35202

Adat Shalom Synagogue, Jacob E. Siegel Library, 29901 Middlebelt Rd, *Farmington Hills*, MI 48334
T: +1 248 8515100; Fax: +1 248 8513190; E-mail: info@adatshalom.org; URL: www.adatshalom.org
1960; Michael Jaron
Judaica, Israel, Archaeology, Religion; Comparative Religion (I. Leeman Memorial Coll), Gendein Yiddish Coll, I. Elpern Modern Hebrew Lit Coll, Zionism
12 000 vols; 15 curr per   35203

Sisters of St. Benedict, St. Benedict Library, 802 E Tenth St, *Ferdinand*, IN 47532-9154
T: +1 812 3671411 ext 2110; Fax: +1 812 3672313; E-mail: sisters@thedome.org; URL: www.thedome.org
1890
30 000 vols; 75 curr per   35204

Lutheran Brethren Seminary, Christiansen Memorial Library, 815 W Vernon Ave, *Fergus Falls*, MN 56337
T: +1 218 7393375; Fax: +1 218 7391259; URL: www.lbs.edu
18 000 vols   35205

Winebrenner Theological Seminary, Library, 950 N Main St, *Findlay*, OH 45840-3652
T: +1 419 4344200; Fax: +1 419 4344267; E-mail: library@winebrenner.edu; URL: www.winebrenner.edu
1942; Margaret Hirschy
Rosenberry Family coll
44 000 vols; 110 curr per; 12 maps; 373 microforms; 425 sound-rec; 7 digital data carriers
libr loan; ATLA   35206

Concordia Theological Seminary, Walther Library, 6600 N Clinton St, *Fort Wayne*, IN 46825
T: +1 260 4523145; Fax: +1 260 4522126; E-mail: library@ctsfw.edu; URL: www.ctsfw.edu/library
1846; Robert V. Roethemeyer
Missions; 16th and 17th C Lutheran Orthodoxy
170 000 vols; 719 curr per; 5 000 e-books; 12 incunabula; 2 174 diss/theses; 11 000 microforms; 6 138 sound-rec; 32 digital data carriers
libr loan; OCLC, ATLA, INCOLSA, PALNI
                                            35207

Southwestern Baptist Theological Seminary, Roberts Library, 2001 W Seminary Dr, *Fort Worth*, TX 76115-2157; P.O. Box 22490, Fort Worth, TX 76122-0490
T: +1 817 9231921; Fax: +1 817 9218765; E-mail: rlcirc@swbts.edu; URL: www.swbts.edu/libraries
1908; Dr. Berry Driver
Religious education, Theology, Sacred music; J. T. and Zelma Luther Arch, Baptist Hist (James M Carroll, George W Truett & M E Dodd Coll), mss files; Texas Baptist Hist, bks, & mss files; Hymnals
387 000 vols; 3 359 curr per; 128 maps; 12 000 microforms; 6 461 av-mat; 37 181 sound-rec; 500 digital data carriers; 74 386 annuals, 540 000 mss, 410 slides, 34 overhead transparencies
libr loan   35208

Sacred Heart School of Theology, Leo Dehon Library, 7335 S Hwy 100, *Franklin*, WI 53132; P.O. Box 429, Hales Corners, WI 53130-0429
T: +1 414 4258300; Fax: +1 414 5296992; E-mail: library@shst.edu
1932; Kathleen Harty
Religious Americana, Canon Law; Sacred Heart Coll
92 000 vols; 434 curr per; 2 000 e-books;

210 maps; 1 000 microforms; 950 av-mat; 15 455 sound-rec; 6 digital data carriers
libr loan; ATLA, ACRL, CLA   35209

Roman Catholic Diocese of Fresno Library, 1550 N Fresno St, *Fresno*, CA 93703-3788
T: +1 559 4887400; Fax: +1 559 4887464
1934
Central California mats
20 000 vols; 75 VF drawers   35210

Graymoor Friary Library, 1350 Route 9, P.O. Box 300, *Garrison*, NY 10524-0301
T: +1 845 4243671; Fax: +1 845 4242168; E-mail: graymoorcenter@atonementfriars.org; URL: www.atonementfriars.org
1960; Jim Gardner
Anglicanism, English reformation, North American Ecumenical hist & theology, Spirituality; Hist of the Franciscan Friars of the Atonement (Paul Watson Res Ctr)
88 500 vols; 88 curr per; 450 microforms   35211

Flint-Groves Baptist Church, Library, 2017 E Ozark, *Gastonia*, NC 28054
T: +1 704 8654068; Fax: +1 704 8658008; URL: www.flintgroves.org
17 525 vols   35212

Summit Christian College, Library, 2025 21st Street, *Gering*, NE 69341
T: +1 308 6326933; URL: www.summitcc.net
1952
17 000 vols; 25 curr per; 100 microforms   35213

Lutheran Theological Seminary, A.R. Wentz Library, 66 Seminary Ridge, *Gettysburg*, PA 17325
T: +1 717 3383014; Fax: +1 717 3371611; E-mail: library@ltsg.edu; URL: www.ltsg.edu; ecco.easterncluster.com
1826; Dr. B. Bohleke
Lutheran Church Hist in America
178 000 vols; 450 curr per; 6 000 microforms   35214

North Shore Congregation Israel, Romanek Library, 1185 Sheridan Rd, *Glencoe*, IL 60022
T: +1 847 8350724; Fax: +1 847 8355613; URL: www.nsci.org
1952; Janice B. Footlik
Archeology; Jewish Art Bks
20 000 vols; 24 curr per   35215

Westminster Theological Seminary, Montgomery Library, 2960 W Church Rd, *Glenside*, PA 19038; P.O. Box 27009, Glenside, PA 19118-7009
T: +1 215 9353880; Fax: +1 215 8873412; E-mail: library@wts.edu; URL: www.wts.edu/library/
1929; Sandy Finlayson
Bible Texts & Versions; Early Reformed Theology
141 000 vols; 685 curr per; 1 incunabula; 14 000 microforms; 3 530 sound-rec; 5 digital data carriers
libr loan; ATLA   35216

Immanuel Church Library, 74 Park Dr, *Glenview*, IL 60025
T: +1 847 7241080; Fax: +1 847 7243042
25 000 vols; 10 curr per   35217

Calvin College & Theological Seminary, Heckman Library, 1855 Knollcrest Circle SE, *Grand Rapids*, MI 49546-4402
T: +1 616 5266072; Fax: +1 616 5266146; URL: library.calvin.edu
1892; Glenn A. Remelts
Arch of Christian Reformed Church (Heritage Hall Arch), bk & microfilm; H Henry Meeter Calvinism Research Coll, bk & microfilm
765 000 vols; 2 658 curr per; 150 000 govt docs; 4 582 maps; 700 000 microforms; 18 687 sound-rec; 700 digital data carriers
libr loan   35218

Grace Bible College, Bultema Memorial Library, 1011 Aldon St SW, *Grand Rapids*, MI 49509
T: +1 616 2618575; Fax: +1 616 5380599
1945; Kathy Molenkamp
Dispensational Theology
40 000 vols; 150 curr per; 800 microforms; 307 sound-rec; 6 digital data carriers   35219

Temple Beth El, Arnold & Marie Schwartz & Hattie & Albert Grauer Library, 5 Old Mill Rd, *Great Neck*, NY 11023
T: +1 516 4870900 ext 45; Fax: +1 516 4876941
1950; MCH Feinsilver
Children's Coll
15 000 vols; 25 curr per   35220

Christian Union Bible College Library, 1190 N Washington St, *Greenfield*, OH 45123-9719; P.O. Box 27, Greenfield, OH 45123-0027
T: +1 937 9812897; Fax: +1 937 9812897; URL: www.christianunionbiblecollege.com
Neal Reid
Religion, philosophy, psychology
11 000 vols   35221

Currie Library, 617 N Elm St, *Greensboro*, NC 27401
T: +1 336 4784731; E-mail: nfuller@fpcgreensboro.org; URL: www.fpcgreensboro.org
1925; Nancy Fuller
Music
18 000 vols; 21 curr per   35222

First Baptist Church Library, 1000 W Friendly Ave, P.O. Box 5443, *Greensboro*, NC 27403
T: +1 336 2743086; Fax: +1 336 2743288; E-mail: media@fbcgso.org; URL: www.fbcgso.org
1947; Kelly Stephens
17 200 vols; 39 curr per; 340 av-mat; 1 600 sound-rec; 4 VF drawers, 1 000 slides   35223

The Archdiocesan Center at St. Thomas Seminary – Archbishop O'Brien Library, 467 Bloomfield Ave, *Hartford*, CT 06002
T: +1 860 2425573 ext 2608/2609; E-mail: klesiak@stseminary.org; URL: www.stseminary.org/obrienlibrary.htm
1935; Karen Lesiak
10 000 vols   35224

Hartford Seminary, Library, 77 Sherman St, *Hartford*, CT 06105-2260
T: +1 860 5099500; Fax: +1 860 5099509; E-mail: library@hartsem.edu; URL: www.library.hartsem.edu
1834; Steven Peter Blackburn Doctor of Philosophy
Islam, Theology; Arabian Nights (Duncan Black MacDonald Coll), Arabic Mss, Illuminated Qu'rans, Armenian Mss, New England Theologians, Papers & Sermons
92 000 vols; 308 curr per; 55 e-journals; 8 000 e-books; 610 diss/theses; 6 000 microforms; 390 av-mat; 175 sound-rec; 12 digital data carriers; 100 000 archival mat   35225

Biblical Theological Seminary, Library, 200 N Main St, *Hatfield*, PA 19440-2499
T: +1 215 3685000; Fax: +1 215 3686906; E-mail: library@biblical.edu; URL: library.biblical.edu
1971; Daniel LaValla
Biblical Seminary Theses (New York Coll)
38 000 vols; 374 curr per; 4 000 microforms; 43 av-mat; 1276 sound-rec; 6 digital data carriers; 1 072 pamphlets   35226

First Baptist Church, D.H. Reed Memorial Library, 510 W Pine St, *Hattiesburg*, MS 39401
T: +1 601 5440100; Fax: +1 601 5848592
1953; Alfred Foy
12 000 vols; 11 curr per; 400 av-mat; 150 sound-rec; 50 vertical files of clippings   35227

Fruitland Baptist Bible Institute, Randy Kilby Memorial Library, 1455 Gillaim Rd, *Hendersonville*, NC 28792
T: +1 828 6858886; Fax: +1 828 6858888; E-mail: fruitland@mchsi.com; URL: www.fruitland.edu
1949; Scott Thompson
27 000 vols; 50 curr per   35228

Congregation Solel Library, 1301 Clavey Rd, *Highland Park*, IL 60035
T: +1 847 4333555; Fax: +1 847 4333573; E-mail: soleloffice@solel.org; URL: www.solel.org/learning/library
1963
Judaica (art, music), Hagaddot, Holocaust
12 000 vols; 20 curr per   35229

North Suburban Synagogue Beth El, Maxwell Abbell Library, 1175 Sheridan Rd, *Highland Park*, IL 60035
T: +1 847 4328900; Fax: +1 847 4329242; E-mail: nssbe@nssbethel.org; URL: www.nssbethel.org
1959
Holocaust, Yiddish Coll
20 000 vols; 85 curr per; 12 maps; 700 av-mat; 650 sound-rec; 4 digital data carriers; 3 VF drawers of pamphlets   35230

Hobe Sound Bible College, Library, 11440 SE Gomez Ave, *Hobe Sound*, FL 33455-3378; P.O. Box 1065, Hobe Sound, FL 33475-1065
T: +1 772 5451400; Fax: +1 772 5451422; URL: www.hsbc.edu/library
1960; William Snider
Humanities, Music; Child Evangelism Coll
33 000 vols; 220 curr per; 770 microforms; 12 VF drawers   35231

Western Theological Seminary, Beardslee Library, 101 E 13th St, *Holland*, MI 49423
T: +1 616 3928555; Fax: +1 616 3928889; URL: www.westernsem.edu/resources/library
1895; Paul M. Smith
Hist of Reformed Church in America (Kolkman Memorial Arch), mss mat; 15th-18th c theology (rare bks), bound vols
111 000 vols; 442 curr per; 400 microforms; 429 sound-rec; 8 digital data carriers; 3 200 slides
libr loan; ALA, ATLA   35232

Chapelwood United Methodist Church, Carey B. Sayers Memorial Library, 11140 Greenbay Dr, *Houston*, TX 77024-6798
T: +1 713 4653467; Fax: +1 713 3652808; URL: www.chapelwood.org
Dona Badgett
13 000 vols; 12 curr per; 75 av-mat   35233

Congregation Beth Yeshurun, Cantor Rubin Kaplan Memorial Library, 4525 Beechnut Blvd, *Houston*, TX 77096
T: +1 713 6661884; Fax: +1 713 6662924
Monica Woolf
Judaica, Holocaust
50 000 vols   35234

Resurrection Metropolitan Community Church, Botts Memorial Library & Archives, 2025 W 11th St, *Houston*, TX 77008
T: +1 713 8619149; Fax: +1 713 8612520; URL: www.resurrectionmcc.org
1979; Melissa Rogers
Religion, Homosexuality
20 000 vols; 75 000 curr per   35235

Seminary of the Immaculate Conception, Library, 440 W Neck Rd, *Huntington*, NY 11743
T: +1 631 4230483; Fax: +1 631 4232346; URL: www.icseminary.edu
1930; Elyse Hayes
46 000 vols; 368 curr per   35236

Community of Christ, Library, 201 S River, *Independence*, MO 64050
T: +1 816 8331000 ext 2400; Fax: +1 816 5213087; E-mail: rmeisinger@CofChrist.org; URL: www.cofchrist.org/library
1865; Rachel Killebrew
Christianity, Mormon hist & theology, Peace studies
20 000 vols; 150 curr per; 25 maps; 3 500 microforms; 400 av-mat; 1 000 sound-rec
libr loan   35237

Christian Theological Seminary, Library, 1000 W 42nd St, *Indianapolis*, IN 46208
T: +1 317 9312361; Fax: +1 317 9312363; E-mail: renew@cts.edu; URL: www.cts.edu
1942; Dr. Lorna A. Shoemaker
Religion, History, Social sciences, Music; Disciples of Christ Historical Materials; Congregational Resource Center
225 000 vols; 1 420 curr per; 700 e-books; 325 digital data carriers
libr loan; Association of Theological Libraries in America   35238

Crossroads Bible College, Library, 601 N Shortridge Rd, *Indianapolis*, IN 46219-4912
T: +1 317 3528736; Fax: +1 317 3529145; URL: www.crossroads.edu
1980; Liz LeMond
Afro-American Hist & the Black Church Coll
34 000 vols; 90 curr per
libr loan; INCOLSA, OCLC   35239

Free Methodist Church of North America, Marston Memorial Historical Center-White Memorial Library, 770 N High School Rd, P.O. Box 535002, *Indianapolis*, IN 46253-5002
T: +1 317 2443660; Fax: +1 317 2441503; E-mail: hrdept@fmcna.org; URL: www.freemethodistchurch.org
1969
12 500 vols; 10 curr per   35240

Free Methodist Church of North America, Marston Memorial Historical Center & Archives, 770 N High School Rd, *Indianapolis*, IN 46214; P.O. Box 535002, Indianapolis, IN 46253-5002
T: +1 317 2443660; Fax: +1 317 2441247; E-mail: history@fmcna.org; URL: www.marston.freemethodistchurch.org
Cathy Fortner
Methodism (John Wesley Coll), Wesleyana, Free Methodist Memoribilia
12 000 vols; 10 curr per   35241

Kentucky Mountain Bible College, Gibson Library, 855 Hwy 541, *Jackson*, KY 41339; P.O. Box 10, Vancleve, KY 41385-0010
T: +1 606 6665000 ext 100; Fax: +1 606 6667744; URL: www.kmbc.edu/?q=Library
1931
Holiness, Missionary, Religious Biogr
30 000 media items; 191 curr per   35242

Reformed Theological Seminary, Library, 5422 Clinton Blvd, *Jackson*, MS 39209-3099
T: +1 601 9231623; Fax: +1 601 9231621; E-mail: library.jackson@rts.edu; URL: www.rts.edu/libraries
1965; Kenneth R. Elliott
Southern Presbyterianism (Blackburn Coll), Roger G Stribling Rare Bk Room
150 000 vols; 640 curr per; 42 000 microforms; 7 800 sound-rec; 1 digital data carriers; 26 slides
libr loan   35243

Wesley Biblical Seminary, Library, 787 E Northside Dr, *Jackson*, MS 39206; P.O. Box 9938, Jackson, MS 39286-0938
T: +1 601 3668880; Fax: +1 601 3668832; URL: www.wbs.edu
Dr. Dan Burnett
Religion
62 000 vols; 270 curr per; 35 diss/theses; 950 sound-rec; 25 digital data carriers
libr loan; ATLA   35244

Baptist Missionary Association Theological Seminary, Kellar Library, 1530 E Pine St, *Jacksonville*, TX 75766-5407
T: +1 903 5862501; Fax: +1 903 5860378; E-mail: bmatsem@bmats.edu; URL: www.bmats.edu
1957; James C. Blaylock
Annuals of Baptist Yearly Meetings, bk, micro
70 000 vols; 328 curr per; 28 000 e-books; 200 maps; 958 microforms; 6 372 sound-rec; 31 digital data carriers; 1 073 slides; 838 video tapes
libr loan; ATLA, TLA, ALA   35245

Emmanuel School of Religion Library, One Walker Dr, *Johnson City*, TN 37601-9438
T: +1 423 4611543; Fax: +1 423 9266198; E-mail: library@esr.edu; URL: www.esr.edu/library.htm
1965; Thomas E. Stokes
Discipliana Coll, hist materials (Christian Churches & Churches of Christ)
116 000 vols; 712 curr per; 1 incunabula; 634 diss/theses; 47 maps; 28 000 microforms; 661 av-mat; 752 sound-rec; 20 digital data carriers; 70 Slides
libr loan; OCLC, SOLINET   35246

Holy Trinity Orthodox Seminary & Monastery, Library, 1407 Robinson Rd, *Jordanville*, NY 13361-0036; P.O. Box 36, Jordanville, NY 13361-0036
T: +1 315 8580945; Fax: +1 315 8580945; E-mail: library@hts.edu; URL: www.hts.edu
Vladimir Tsurikov
45 000 vols; 75 curr per   35247

Calvary Bible College Theological Seminary, Hilda Kroeker Library, 15800 Calvary Rd, *Kansas City*, MO 64147-1341
T: +1 816 3220110; Fax: +1 816 3314474; URL: www.calvary.edu
1932; Hannah Bitner
59 000 vols; 70 curr per; 397 e-books; 163 diss/theses; 3 000 microforms; 54 VF drawers   35248

Midwestern Baptist Theological Seminary, Library, 5001 N Oak St, Trafficway, *Kansas City*, MO 64118-4620
T: +1 816 4143730; Fax: +1 816 4143790; URL: www.mbts.edu/library/index.html
1958; Craig Kubic
Biblical studies; Baptist Denominational Coll
105 000 vols; 371 curr per; 250 microforms; 300 sermon tapes
libr loan; ALA, OCLC   35249

Nazarene Theological Seminary, Broadhurst Library, 1700 E Meyer Blvd, *Kansas City*, MO 64131
T: +1 816 2685471; Fax: +1 816 8229025; E-mail: library@nts.edu; URL: www.nts.edu
1945; Debra Bradshaw
Methodistica-Wesleyana, Hist of the Holiness Movement
141 000 vols; 526 curr per; 22 000 microforms; 7 digital data carriers; 5 096 pamphlets
libr loan   35250

Western Baptist Bible College, Memorial Library, 2119 Tracy, *Kansas City*, MO 64108
T: +1 816 8424195; Fax: +1 816 8423050; E-mail: wbbible@sbcglobal.net; URL: www.westernbaptistkc.org
1989
13 020 vols   35251

Wyoming Seminary, Kirby Library, 201 N Sprague Ave, *Kingston*, PA 18704-3593
T: +1 570 2702169; Fax: +1 570 2702178; URL: www.kirbylibrary.org
1976; Courtney Lewis
Alinikoff Middle East Hist Coll, Leroy Bugbee Memorial Hist Coll, Robert E Shortz Chess Coll
20 000 vols; 125 curr per   35252

Johnson Bible College, Glass Memorial Library, 7902 Eubanks Dr, *Knoxville*, TN 37998; mail address: 7900 Johnson Dr, Knoxville, TN 37998
T: +1 865 2512277; Fax: +1 865 2512278; E-mail: library@jbc.edu; URL: www.jbc.edu/library
1893; Carrie Beth Lowe
Restoration Movement Hist Coll
108 000 vols; 395 curr per; 31 000 e-books; 15 000 microforms; 8 289 sound-rec; 9 digital data carriers; 14 VF drawers of pamphlets, 6 VF drawers of letters
libr loan   35253

Southside Baptist Church, Hollis-Hays Library, 310 McDonald St, *Lakeland*, FL 33803
T: +1 941 6828764; Fax: +1 941 6825849
1944; Evelyn Winter
10 000 vols; 29 curr per; 589 av-mat;

340 sound-rec; 59 digital data carriers   35254

Lancaster Bible College, Library, 901 Eden Rd, *Lancaster*, PA 17601-5036
T: +1 717 5608250; Fax: +1 717 5608265; E-mail: library@lbc.edu; URL: library.lbc.edu
1933; Gerald Lincoln
Lloyd M Perry Coll (Pastoral Theology), LBC Coll
198 000 vols; 388 curr per; 8 000 e-books   35255

Lancaster Theological Seminary, Philip Schaff Library, 555 W James St, *Lancaster*, PA 17603-9967
T: +1 717 2908707; Fax: +1 717 3934254; E-mail: library@lancasterseminary.edu; URL: www.lancasterseminary.edu
1825; Richard R. Berg
Albright Coll in Church Hist & Liturgics, Rare Bks, Lancastriana
126 000 vols; 283 curr per; 1 incunabula; 6 000 microforms; 300 sound-rec; 8 digital data carriers
libr loan; ATLA   35256

Washington Bible College – Capital Bible Seminary, Oyer Memorial Library, 6511 Princess Garden Pkwy, *Lanham*, MD 20706
T: +1 301 5521400 ext 231; Fax: +1 301 5522775; E-mail: wbc@bible.edu
1937; Marseille Pride
Rare Books
75 228 vols; 525 curr per; 356 diss/theses; 4 752 microforms; 730 av-mat; 5 969 sound-rec; 25 digital data carriers; 2 500 slides, 1 512 rare books
libr loan; ALA, ACL   35257

Calvary Baptist Theological Seminary, Library, 1380 S Valley Forge Rd, *Lansdale*, PA 19446-4797
T: +1 215 3684444; Fax: +1 215 3681003; E-mail: librarian@cbs.edu; URL: seminary.cbs.edu/content/library/index.htm; library.cbs.edu
1975; George Coone
Biblical studies; Baptist Coll
76 000 vols; 340 curr per; 55 000 microforms; 1 528 sound-rec; 3 digital data carriers
libr loan; ATLA   35258

Salvation Army Church Library, 2900 Palamino Lane, *Las Vegas*, NV 89107
T: +1 702 8704430; Fax: +1 702 8704087; E-mail: LasVegasInfo@usw.salvationarmy.org; URL: www1.usw.salvationarmy.org
1970
Christian lit; Salvation Army Lit
20 000 vols   35259

Saint Vincent College & Seminary, Library, 300 Fraser Purchase Rd, *Latrobe*, PA 15650-2690
T: +1 724 8052310; Fax: +1 724 8052905; E-mail: library@stvincent.edu; URL: www.stvincent.edu
1846; Brother David Kelly
Benedictina; Ecclesiastical Hist; Incunabula; Medievalia; Patrology; Pennsylvaniana; Theology
378 000 vols; 528 curr per; 10 905 e-journals; 2 406 diss/theses; 99 000 microforms; 2 006 av-mat; 2 468 sound-rec; 323 digital data carriers
libr loan; PALINET, OCLC, ATLA   35260

Catholic Newman Center, Library, 320 Rose Lane, *Lexington*, KY 40508
T: +1 859 2558566; Fax: +1 859 2547519; E-mail: newmancenter@insightbb.com; URL: newmancenter.home.insightbb.com
1965
35 000 vols; 12 curr per   35261

Lexington Theological Seminary, Bosworth Memorial Library, 631 S Limestone, *Lexington*, KY 40508
T: +1 859 2801229; Fax: +1 859 2816042; URL: www.lextheo.edu
1865; Barbara Pfeifle
John Mason Neale Coll, Disciples of Christ Coll
155 000 vols; 1 000 curr per; 9 000 microforms
libr loan; ATLA   35262

Lincoln Christian College & Seminary, Jessie C. Eury Library, 100 Campus View Dr, *Lincoln*, IL 62656
T: +1 217 7327788; Fax: +1 217 7323785; URL: www.lccs.edu/library
1944; Nancy J. Olson
Theology, Education; Enos E Dowling Rare Bk Room, Chi Lambda Cult/Occult Res Ctr
105 000 vols; 442 curr per; 150 maps; 17 000 microforms; 2 662 av-mat; 70 digital data carriers   35263

Luther Rice Seminary & University, Bertha Smith Library, 3038 Evans Mill Rd, *Lithonia*, GA 30038
T: +1 770 4841204 ext 222; Fax: +1 770 4841155; E-mail: library@LRU.edu; URL: www.LRU.edu
1972; Hal Haller
Christian education; Religion (Christian Ministry), Diss
45 000 vols; 100 curr per; 81 microforms; 3 527 av-mat; 3 390 sound-rec   35264

Denver Seminary, Carey S. Thomas Library, 6399 S Santa Fe Dr, *Littleton*, CO 80120-2912
T: +1 303 7612482 ext 1403; Fax: +1 303 7618060; E-mail: library@denverseminary.edu; URL: www.denverseminary.edu/library
1950; Keith Wells
Biblical stucies, Theology
170 000 vols; 550 curr per; 5 digital data carriers   35265

Northern Baptist Theological Seminary, Brimson Grow Library, 680 E Butterfield Rd, *Lombard*, IL 60148
T: +1 630 6202104; Fax: +1 630 6202170; E-mail: library@seminary.edu; URL: www.seminary.edu/bgl
1913; Blake Walter
Baptist Hist Bks, Seminary Arch
53 000 vols; 296 curr per; 8 000 e-books
libr loan   35266

First Baptist Church, John L. Whorton Media Center Library, 209 E South, *Longview*, TX 75601
T: +1 903 7580681; Fax: +1 903 7530936; URL: www.fbcl.org
1942; Lisa Brown
First Baptist Church Hist Coll
30 000 vols; 14 curr per; 300 sound-rec   35267

Wilshire Boulevard Temple – Libraries, 3663 Wilshire Blvd, *Los Angeles*, CA 90010-2798
T: +1 213 3882401; Fax: +1 213 3882595; URL: www.wilshireboulevardtemple.org
1929
Judaica, World War II (Trials of the Major War Criminals)
17 000 vols; 100 pamphlets, 12 VF drawers of uncataloged pamphlets 35268

California Province of the Society of Jesus, Jesuit Center Library, 300 College Ave, *Los Gatos*, CA 95030; P.O. Box 128, Los Gatos, CA 95031-0128
T: +1 408 8841700; Fax: +1 408 8841701
1851; Father Burk
Jesuitica
50 000 vols; 22 curr per   35269

Louisville Presbyterian Theological Seminary, Ernest Miller White Library, 1044 Alta Vista Rd, *Louisville*, KY 40205-1798
T: +1 502 8953411; Fax: +1 502 8951096; URL: www.lpts.edu/academic_resources/emwhite_library.asp
1853; Dr. Douglas L. Gragg
Religion, Social sciences; Presbyterian Church Hist
165 000 vols; 602 curr per; 8 000 microforms; 5 568 sound-rec; 3 digital data carriers
libr loan   35270

Southern Baptist Theological Seminary, James P. Boyce Centennial Library, 2825 Lexington Rd, *Louisville*, KY 40280-0294
T: +1 502 8974553; Fax: +1 502 8974600; URL: www.sbts.edu
1859; Bruce L. Keisling
Baptist Hist Coll, Billy Graham Room, William F Albright Archaeology Coll, Everett Helm Music Coll, Missions

385 000 vols; 1 419 curr per; 184 738 music scores; 70 000 microforms; 39 678 sound-rec; 7 862 digital data carriers; 125 032 pamphlets
libr loan; ALA, ATLA, KLA, Solinet
                    35271

Southern Baptist Theological Seminary – Music and Audiovisual Library, 2825 Lexington Rd, *Louisville*, KY 40280-0294
T: +1 502 8974055; Fax: +1 502 8974600; E-mail: dgregory@lib.sbts.edu; URL: www.sbts.edu
1944; David Louis Gregory
Hymnology, church music
27 000 vols; 136 curr per; 27 171 sound-rec; 4 749 digital data carriers
                    35272

Walnut Street Baptist Church, Media Library, 1143 S Third St, Ste A, *Louisville*, KY 40203
T: +1 502 5895290; Fax: +1 502 5899323; E-mail: info@walnutstreet.org; URL: walnutstreet.org
1943
13 000 vols; 10 curr per   35273

United Methodist Church, General Commission on Archives & History Library, 36 Madison Ave, *Madison*, NJ 07940
T: +1 973 4083590; Fax: +1 973 4083836; URL: www.depts.drew.edu/lib/uma.html; www.gcah.org
1968; Jennifer Woodruff Tait
100 000 feet of microfilm, 100 tubes of blueprints, 4 000 000 arch items
50 000 vols; 600 curr per   35274

New Orleans Baptist Theological Seminary, North Georgia Campus Library, 100 Johnson Ferry Rd, Ste C115, *Marietta*, GA 30068
T: +1 770 3211606; Fax: +1 770 3215363; URL: www.nobts.edu
Jeff Griffin
21 000 vols; 25 curr per   35275

Blue Cloud Abbey, Library, 46561 147th St, P.O. Box 98, *Marvin*, SD 57251-0098
T: +1 605 3989200; Fax: +1 605 3989201; E-mail: webmonk@bluecloud.org; URL: www.bluecloud.org
1950
Theology, Scriptures, Hist, Indian studies
35 000 vols; 100 curr per
libr loan   35276

Mid-Continent University, Anne P. Markham Library, 99 E Powell Rd, *Mayfield*, KY 42066
T: +1 270 2478521 ext 275; Fax: +1 270 2473115; E-mail: mcc@midcontinent.edu; URL: www.midcontinent.edu/traditional/library
1968
30 000 vols; 200 curr per; 1 600 av-mat
                    35277

Memphis Theological Seminary, Cumberland Presbyterian Church, Library, 168 E Parkway S, *Memphis*, TN 38104
T: +1 901 3345814; Fax: +1 901 4524051; E-mail: library@memphisseminary.edu; URL: www.memphisseminary.edu/library
1964; Steven R. Edscorn
Christian Missions (R. Pierce Beaver Coll); Cumberland Presbyterian Hist
72 000 vols; 382 curr per; 4 000 microforms; 24 sound-rec; 15 digital data carriers
libr loan; ATLA, TLA, ALA   35278

Saint Patrick's Seminary, Carl and Celia Berta Gellert Foundation Memorial Library, 320 Middlefield Rd, *Menlo Park*, CA 94025
T: +1 650 3215655; Fax: +1 650 3235447; E-mail: stpats@ix.netcom.com; URL: www.catalogue.org/stp/
1898; Dr. Cecil R. White
Theology; Libr of Archbishop Alemany (First Archbishop of San Francisco Coll), Western Americana
125 036 vols; 300 curr per; 2 174 microforms; 479 av-mat; 2 484 sound-rec
libr loan; ATLA, CLA   35279

957

Wisconsin Lutheran Seminary, Library, 6633 W Wartburg Circle, **Mequon**, WI 53092-1530
T: +1 262 2428113; Fax: +1 262 2428118; E-mail: library@wls.wels.net; URL: www.wls.wels.net
1878; John P. Hartwig
57 000 vols; 400 curr per   35280

Church of Jesus Christ of Latter-Day Saints, Mesa Regional Family History Center, 41 S Hobson, **Mesa**, AZ 85204
T: +1 480 9641200; Fax: +1 480 9647137; E-mail: admin@mesarfhc.org; URL: www.mesarfhc.org
1930
Genealogy; Family Histories & Biographies, Int Genealogical Index; US Census Rec
20 000 vols   35281

Memorial Presbyterian Church, Greenhoe Library and Rainbow Childrens Library, 1310 Ashman St, **Midland**, MI 48640
T: +1 989 8356759
1945; Esther Frost
Children's Libr
8 000 vols; 12 curr per; 1 351 av-mat; 636 sound-rec
CSLA   35282

Golden Gate Baptist Theological Seminary, Library, 201 Seminary Dr, Box 37, **Mill Valley**, CA 94941-3197
T: +1 415 3801660; Fax: +1 415 3801652; E-mail: library@ggbts.edu; URL: www.ggbts.edu
1944; Kelly Campbell
Musik; Religion; Baptist Hist, Old & Rare Hymnals
156 070 vols; 797 curr per; 4 441 microforms; 1 251 av-mat; 9 083 sound-rec; 28 digital data carriers; 8 VF drawers of pamphlets, 3 965 slides, 40 Bks on Deafness & Sign Lang   35283

Ascension Lutheran Church Library, 1236 S Layton Blvd, **Milwaukee**, WI 53215-1653
T: +1 414 6452933; URL: www.ascension-mke.org
1954
The Dewey 200's, Movie Bks (Richard Krueger Coll), Old Bibles, Sheet Music Coll (organ)
13 000 vols; 2 curr per; 40 av-mat; 600 sound-rec; 4 VF drawers   35284

Congregation Emanu-el B'ne Jeshurun Library, Rabbi Dudley Weinberg Library, 2020 W Brown Deer Rd, **Milwaukee**, WI 53217
T: +1 414 2287545; Fax: +1 414 2287884; URL: www.ceebj.org
1929
Jewish Children's Bks, Out-of-Print-Bks
8 500 vols; 30 curr per; 180 av-mat; 200 sound-rec   35285

Saint Lawrence Catholic Church, Newman Center Library, 1203 Fifth St SE, **Minneapolis**, MN 55414
T: +1 612 3317941; Fax: +1 612 3781771; URL: www.stlawrencenewman.org
1946; John Behnke
12 000 vols; 15 curr per   35286

Temple Israel Library, 2324 Emerson Ave S, **Minneapolis**, MN 55405-2695
T: +1 612 3778680; Fax: +1 612 3776630; E-mail: gkalman@templeisrael.com
1929; Georgia Kalman
Holocaust, Hist, Biography, Art
15 000 vols; 46 curr per
AAJL   35287

Roman Catholic Archdiocese of Los Angeles, Archival Center Library & Historical Museum, 15151 San Fernando Mission Blvd, **Mission Hills**, CA 91345
T: +1 818 3651501; Fax: +1 818 3613276; URL: www.archivalcenter.org
1962; Francis J. Weber
Robert G. Cowan Coll, Estelle Doheny Coll of Californiana
15 000 vols; 350 000 mss, docs, memorabilia   35288

IHM Library/Resource Center, Congregational Library, 610 W Elm Ave, **Monroe**, MI 48162-7909
T: +1 734 2409678; Fax: +1 734 2408347; E-mail: library@ihmsisters.org; URL: www.ihmsisters.org
1927; Sister Marie Gabriel Hungerman
Theology, Religion, Art, Ecology, Fiction, Large type print
30 000 vols; 100 curr per   35289

Presbyterian Heritage Center at Montreat, Research Library, 318 Georgia Terr, P.O. Box 207, **Montreat**, NC 28757
T: +1 828 6696556; Fax: +1 828 6695369; E-mail: info@phcmontreat.org; URL: www.phcmontreat.org/library.htm
1927
Printed Records of the Presbyterian Church and Minutes of the General Assembly, from 1706; Rare denominational & other periodicals; Reference books into Presbyterian history
40 000 vols   35290

John Van Puffelen Library of the Appalachian Bible College, 161 College Dr, **Mount Hope**, WV 25880-1040; P.O. Box ABC, Bradley, WV 25818-1353
T: +1 304 8776428; Fax: +1 304 8775983; URL: www.abc.edu
1950; Ed Arnold Chesley
Missions, Theology, West Virginia; Judaica (Patteson Coll)
50 000 vols; 250 curr per; 985 microforms; 384 av-mat; 2 986 sound-rec; 187 digital data carriers
libr loan; ALA, ACL   35291

University of Saint Mary of The Lake – Mundelein Seminary, Feehan Memorial Library, 1000 E Maple Ave, **Mundelein**, IL 60060
T: +1 847 9704820; Fax: +1 847 5665229
1929; Lorraine H. Olley
Ancient Christian lit, Medieval theology, Catholic theology; Irish Hist, Language & Lit Coll
199 000 vols; 430 curr per; 1 000 microforms; 6 digital data carriers   35292

First Baptist Church Library, 200 E Main St, **Murfreesboro**, TN 37130
T: +1 615 8932514; Fax: +1 615 8955804; URL: www.firstbaptistmurfreesboro.org
1943
Church curriculum, Religious studies; Children's Coll
18 400 vols; 31 curr per   35293

Evangelical School of Theology, Rostad Library, 121 S College St, **Myerstown**, PA 17067
T: +1 717 8665775; Fax: +1 717 8664667; URL: www.evangelical.edu
1954; Dr. Terry M. Heisey
Biblical studies; Evangelical Assn, Evangelical Congregational Church Arch, Pietism
77 000 vols; 550 curr per; 600 microforms; 8 digital data carriers   35294

American Baptist College of the American Baptist Theological Seminary, T.L. Holcomb Library, 1800 Baptist World Center Dr, **Nashville**, TN 37207-4994
T: +1 615 6876904; E-mail: frontdesk@abcnash.edu; URL: www.abcnash.edu
1924; Jean-Marie Cherisna
Religious studies, Ethnic studies; Gandhi Coll
40 000 vols; 200 curr per; 1 078 av-mat; 2 576 vertical file mat   35295

Free Will Baptist Bible College, Welch Library, 3630 W End Ave, **Nashville**, TN 37205; mail address: 3606 W End Ave, Nashville, TN 37205
T: +1 615 8445274; Fax: +1 615 2696028
1942; Margaret Evans Hampton
Free Will Baptist Hist Coll
64 000 vols; 300 curr per; 3 000 e-books   35296

Southern Baptist Historical Library & Archives, 901 Commerce St, Ste 400, **Nashville**, TN 37203-3630
T: +1 615 2440344; Fax: +1 615 7824821; URL: www.sbhla.org
Bill Sumners
Baptist history; Southern Baptist Convention Leaders' Papers, Southern Baptist Convention Archives Depository
31 000 vols; 330 curr per   35297

West End Synagogue, Library, 3810 W End Ave, **Nashville**, TN 37205
T: +1 615 2694592; Fax: +1 615 2694695; E-mail: office@westendsyn.org; URL: westendsyn.org
15 000 vols; 5 curr per
libr loan   35298

United Theological Seminary of the Twin Cities, The Spencer Library, 3000 Fifth St NW, **New Brighton**, MN 55112-2598
T: +1 651 6334311; Fax: +1 651 6334315; E-mail: library@unitedseminary-mn.org; URL: www.spencerlibrary.org
1962; Susan K. Ebbers
81 200 vols; 278 curr per; 101 av-mat; 982 sound-rec; 17 slides
libr loan; ATLA   35299

New Brunswick Theological Seminary, Gardner A. Sage Library, 21 Seminary Pl, **New Brunswick**, NJ 08901-1159
T: +1 732 2475243; Fax: +1 732 2471356; E-mail: sage@nbts.edu; URL: www.nbts.edu
1784; Renee House
Arch of Reformed Church in America, Leiby Coll
155 000 vols; 310 curr per   35300

New Orleans Baptist Theological Seminary – John T. Christian Library, 4110 Seminary Pl, **New Orleans**, LA 70126
T: +1 504 8168018; Fax: +1 504 8168429; E-mail: library@nobts.edu; URL: www.nobts.edu/library
1918; Dr. Griffin Jeff
Keith Coll (Church Music), William Carey Libr, Evangelism & Church Growth Coll, NOBTS Arch
300 000 vols; 1 020 curr per; 17 119 microforms; 25 402 av-mat; 12 digital data carriers; 2 873 pamphlets, 17 881 music scores & anthems, 6 043 VF mats, 730 curriculum mats   35301

New Orleans Baptist Theological Seminary – Martin Music Library, 4110 Seminary Pl, **New Orleans**, LA 70126
T: +1 504 8168018; Fax: +1 504 8168429; E-mail: musiclibrary@nobts.edu; URL: www.nobts.edu/library/martin/default.html
Eric Benoy
21 000 vols; 100 curr per   35302

Notre Dame Seminary, Library, 2901 S Carrollton Ave, **New Orleans**, LA 70118-4391
T: +1 504 8667426; Fax: +1 504 8666260; E-mail: librarian@nds.edu; URL: www.nds.edu/library.htm
1923; George Dansker
Theology, Philosophy
92 000 vols; 165 curr per   35303

Armenian Apostolic Church of America, Saint Nerses Shnorhali Library, 138 E 39th St, **New York**, NY 10016
T: +1 212 6897810; Fax: +1 212 6897168
1975; Houri Ghougassian
Armenian studies; Classical & Modern Armenian Lit Coll
12 000 vols   35304

Cathedral School of Saint John the Divine Library, 1047 Amsterdam Ave, **New York**, NY 10025
T: +1 212 3167495
1957; Sharon Owens
William Reed Huntington Coll; Canon Edward West Coll; Bibles
20 000 vols   35305

Congregation Emanu-El, Ivan M. Stettenheim Library, 1 E 65th St, **New York**, NY 10021-6596
T: +1 212 7441400; Fax: +1 212 5700826; E-mail: establer@emanuelnyc.org; URL: www.emanuelnyc.org

1906; Elizabeth F. Stabler
10 000 vols; 55 curr per; 200 av-mat; 120 sound-rec; 40 digital data carriers
AJL, SLA, ALA   35306

General Theological Seminary, Saint Mark's Library, 175 Ninth Ave, **New York**, NY 10011
T: +1 212 2435150; Fax: +1 212 9246304; E-mail: library@gts.edu; URL: library.gts.edu
1819; Andrew Kadel
Clement Clarke Moore Coll, Early English Theology Coll, Episcopal Church Hist, Latin Bible Coll, Liturgics, T.S. Eliot Coll
255 000 vols; 680 curr per; 108 microforms; 92 sound-rec; 6 digital data carriers   35307

Interchurch Center, Ecumenical Library, 475 Riverside Dr, Rm 900, **New York**, NY 10115
T: +1 212 8703804; Fax: +1 212 8702440; E-mail: tdelduca@interchurch-center.org; URL: interchurch-center.library.net
1978
Religious Res Projects (H Paul Douglass Coll), Denominational Yearbks, Methodist Missionary Records
14 000 vols; 100 curr per; 2 900 microforms; 1 digital data carriers; 35 file drawers of pamphlets & clippings   35308

Jewish Theological Seminary, Library, 3080 Broadway, **New York**, NY 10027
T: +1 212 6788075; Fax: +1 212 6788891; E-mail: library@jtsa.edu; URL: www.jtsa.edu/library
1903; Naomi M. Steinberger
Bible; Liturgical works; Hebrew Incunabula, arch; rare bks; Hebrew Mss; Rabbinics
400 000 vols; 788 curr per; 100 e-journals; 250 e-books; 10 000 mss; 145 incunabula; 3 000 diss/theses; 3 000 music scores; 250 maps; 12 000 microforms; 1 250 av-mat; 13 150 sound-rec; 45 digital data carriers
libr loan; ALA, AJL   35309

New York Theological Seminary, Library, 475 Riverside Dr, **New York**, NY 10115
T: +1 212 8701211; Fax: +1 212 8701236; E-mail: online@myts.edu; URL: www.nyts.edu/library
1902
25 000 vols; 40 curr per; 780 sound-rec   35310

Park Avenue Synagogue, Edmond de Rothschild Library, 50 E 87th St, **New York**, NY 10128
T: +1 212 3692600 ext 137
1956; Ilana Abend-David
Judaica Picture Bks
9 000 vols; 20 curr per; 120 av-mat   35311

American Jewish Historical Society, Wyner Center Library, 160 Herrick Rd, **Newton Centre**, MA 02459
T: +1 617 5598880; Fax: +1 617 5598881; E-mail: ajhs@ajhs.org; URL: www.ajhs.org
1892; Michael Feldberg
Jewish hist & lit; Yiddish Theatre (Kanof-Pascher), posters; Yiddish Sheet Music; Rutenberg & Everett Yiddish Film Coll; The Sang Coll of American Judaica; Eleanor & Morris Soble Coll of American Judaica; Rosenbach Coll of American Judaica
60 000 vols; 750 curr per; 350 microforms; 500 000 unbound per, 7 000 linear feet of mss   35312

Andover Newton Theological School, Franklin Trask Library, 169 Herrick Rd, **Newton Centre**, MA 02459
T: +1 617 9641100; Fax: +1 617 4673051; URL: www.ants.edu/library
1807; Diana Yount
Isaac Backus Coll; Baptist & Congregational Church Records Coll; Jonathan Edwards Coll; New England Baptist Hist Coll
260 000 vols; 357 curr per   35313

Mennonite Church, Conference Resource Library, P.O. Box 306, ***North Newton***, KS 67117-0306
T: +1 316 2836300; Fax: +1 316 2830620; E-mail: crlib@mennowdc.org; URL: www.mennowdc.org
1936; Marlene H. Bogard
Religion; Children's Coll, Sunday School Curriculums
12 000 vols 35314

Congregation Beth Shalom, Rabbi Mordecai S. Halpern Memorial Library, 14601 W Lincoln Blvd, ***Oak Park***, MI 48237
T: +1 248 5477970; Fax: +1 248 5470421; E-mail: cbs@congbethshalom.org
1965; Judy Goldsmith
8 000 vols; 5 curr per; 100 av-mat; 200 sound-rec; 100 digital data carriers
AJL 35315

Michigan Jewish Institute Library, 25401 Coolidge Hwy, ***Oak Park***, MI 48237
T: +1 248 4146900; Fax: +1 248 4146907; E-mail: info@mji.edu; URL: www.mji.edu
Karen Robertson-Henry
Fineberg Coll
25 000 vols; 50 curr per 35316

First United Methodist Church, Jones Memorial Library, 1350 Oak Ridge Tpk, ***Oak Ridge***, TN 37831
T: +1 865 4834357; Fax: +1 865 4839011; URL: www.fumcor.org
1967
Christians & Jews Studies (Munz Coll)
9 000 vols; 25 curr per 35317

Kripke Jewish Federation Library, 333 South 132nd St, ***Omaha***, NE 68154
T: +1 402 3346461; Fax: +1 402 3346464; E-mail: jewishlibrary@jewishomaha.org; URL: www.jewishomaha.org
1945
Judaica, Comparative religion, Holocaust, Israel, Archeology
39 000 vols; 75 curr per; 300 av-mat; 5 600 sound-rec; 1 000 pamphlets, 70 slides 35318

Southern Methodist College, Lynn Corbett Library, 541 Broughton St, ***Orangeburg***, SC 29115; P.O. Box 1027, Orangeburg, SC 29116-1027
T: +1 803 2681331; Fax: +1 803 5347827; E-mail: lcl@smcollege.edu; URL: www.smcollege.edu/library.php
Mary Jo Simmons
Southern Methodist, journals, publs
20 000 vols; 90 curr per 35319

Concregation Beth Shalom, Blanche & Ira Rosenblum Memorial Library, 14200 Lamar Ave, ***Overland Park***, KS 66223
T: +1 913 6477279; E-mail: info@bethshalomkc.org; URL: www.bethshalomkc.org
1948; Frances Wolf
95 000 vols; 20 curr per; 300 av-mat 35320

Reformed Theological Seminary, Orlando Campus, Library, 1231 Reformation Dr, ***Oviedo***, FL 32765
T: +1 407 3669493; Fax: +1 407 3669425; E-mail: library.orlando@rts.edu; URL: library.rts.edu
1989; Dr. John R. Muether
100 000 vols; 300 curr per 35321

Fuller Theological Seminary, McAlister Library, 135 N Oakland Ave, ***Pasadena***, CA 91182
T: +1 626 5845219; Fax: +1 626 5845613; URL: www.fuller.edu/library
1948; David Bundy
Biblical studies, Church hist, Feminism, Philosophy, Psychology, Religion, Theology
227 000 vols; 960 curr per; 19 000 microforms; 3 digital data carriers
libr loan 35322

Watchtower Bible School of Gilead Library, 100 Watchtower Dr, ***Patterson***, NY 12563-9204; mail address: 25 Columbia Heights, Brooklyn, NY 11201-1300
T: +1 718 5605000; URL: www.watchtower.org
1943; Sarah Hall
26 000 vols; 30 curr per 35323

Har Zion Temple, Ida & Matthew Rudofker Library, 1500 Hagys Ford Rd, ***Penn Valley***, PA 19072
T: +1 610 6675000; Fax: +1 610 6672032; E-mail: hzt@harziontemple.org; URL: www.harziontemple.org
Bill Moody
9 000 vols; 23 curr per 35324

New Melleray Library, 6632 Melleray Circle, ***Peosta***, IA 52068
T: +1 563 5882319; E-mail: frsteve@newmelleray.org; URL: www.newmelleray.org
1849
Theology, Comperative religion, Scripture, Hist, Psychology
28 000 vols; 40 curr per 35325

Congregation Rodeph Shalom, Philadelphia & Elkins Park Suburban Center Library, 615 N Broad St, ***Philadelphia***, PA 19123
T: +1 215 6276747; Fax: +1 215 6271313; E-mail: info@rodephshalom.org; URL: www.rodephshalom.org
1802
Judaica, Religion, Hist; Family Life (Sadie Goldberg Memorial), Children's Bks, Jewish Music (Roberta Lee Magaziner Memorial)
10 300 vols; 22 curr per; 8 VF drawers of clippings & pamphlets 35326

Lutheran Theological Seminary, Krauth Memorial Library, 7301 Germantown Ave, ***Philadelphia***, PA 19119-1794
T: +1 215 2486329; Fax: +1 215 2486327; E-mail: request@ltsp.edu; URL: ltsp.edu/library
1864; Karl Krueger
Coll of Lutheran Arch Region 7, Evangelical Lutheran Church in America
188 265 vols; 470 curr per; 13 incunabula; 50 maps; 23 586 microforms; 2 128 av-mat; 3 700 sound-rec; 27 digital data carriers; 5 727 Slides, 130 Overhead Transparencies
libr loan; OCLC, ATLA 35327

Philadelphia Yearly Meeting Library, Religious Society of Friends, 1515 Cherry St, ***Philadelphia***, PA 19102
T: +1 215 2417220 ext 7219; Fax: +1 215 5672096; E-mail: library@pym.org; URL: www.pym.org/pm/lib.php
1961; Rita Varley
Non Violent Alternatives; Peace Education Resources; Quakerism & Quaker Hist; Religion & Psychology (Dora Willson Coll), bks & papers
20 000 vols; 45 curr per; 229 av-mat; 350 sound-rec
CSLA 35328

Presbyterian Church (USA), Presbyterian Historical Society, Library, 425 Lombard St, ***Philadelphia***, PA 19147-1516
T: +1 215 6271852; Fax: +1 215 6270509; E-mail: refdesk@history.pcusa.org; URL: www.history.pcusa.org
1852; Frederick J. Heuser Jr
Nat Arch of the Presbyterian Church (U.S.A.) and its predecessors; American Sunday School Union records; National Council of Churches of Christ in the U.S.A. records; Alaska; 1 branch libr
180 000 vols; 150 curr per; 200 maps; 3 000 microforms; 25 000 sound-rec; 3 digital data carriers; 16 000 cubic feet of archival holdings 35329

American Indian College, Dorothy Cummings Memorial Library, 10020 N 15th Ave, ***Phoenix***, AZ 85021-2199
T: +1 602 9443335; URL: www.aicag.edu
John S. Rose
Religion, theology
23 000 vols; 100 curr per 35330

Diocese of Phoenix, Kino Institute Library, 400 E Monroe, ***Phoenix***, AZ 85004
T: +1 602 3542300; E-mail: kinoinstitute@diocesephoenix.org; URL: www.diocesephoenix.org
1975; Sr. Darcy Peletich, OSF
Religious studies
25 000 vols; 120 curr per; 60 microforms; 400 av-mat; 2 000 sound-rec 35331

Clear Creek Baptist Bible College, Brooks Memorial Library, 300 Clear Creek Rd, ***Pineville***, KY 40977
T: +1 606 3371456; Fax: +1 606 3372372; URL: www.ccbbc.edu
Marge Cummings
40 000 vols; 256 curr per; 17 000 e-books 35332

Byzantine Catholic Seminary, 3605 Perrysville Ave, ***Pittsburgh***, PA 15214-2297
T: +1 412 3218383; Fax: +1 412 3229936; E-mail: bcsliboff@sgi.net
1950; Rose Schneider
Church Slavonic Language (Byzantine Catholic Liturgical Books Coll); Hist of the Byzantine Catholic Church in Carpatho-Ruthenia & in the United States; Ruthenian Cultural Coll; Ruthenian Hist Coll
30 201 vols; 75 curr per; 50 diss/theses; 50 music scores; 100 microforms; 75 sound-rec; 6 digital data carriers 35333

Pittsburgh Theological Seminary, Clifford E. Barbour Library, 616 N Highland Ave, ***Pittsburgh***, PA 15206-2596
T: +1 412 4413304; Fax: +1 412 3622329; URL: www.pts.edu; www.barbourlibrary.org
1794; Sharon Taylor
Hymnology (Warrington Coll), Reformation Theology (John M. Mason Coll)
285 000 vols; 870 curr per; 83 000 microforms; 3 323 sound-rec; 28 digital data carriers; 546 ft of mss, 345 film and video, 139 graphic
libr loan; ATLA 35334

Reformed Presbyterian Theological Seminary, Library, 7418 Penn Ave, ***Pittsburgh***, PA 15208-2594
T: +1 412 7318690; Fax: +1 412 7314834; URL: www.rpts.edu
1810; Thomas Reid
16th, 17th & 18th century rare theological works, Reformed (Covenanter) materials
70 000 vols; 240 curr per; 668 microforms; 1 107 sound-rec; 84 digital data carriers
libr loan; ATLA, OCLC 35335

Multnomah Bible College, John & Mary Mitchell Library, 8435 NE Glisan St, ***Portland***, OR 97220-5898
T: +1 503 2515321; Fax: +1 503 2541268; E-mail: library@multnomah.edu; URL: www.multnomah.edu/college/pageslibrary/library.asp
1936; Dr. Philip M. Johnson
Bible Coll
77 000 vols; 402 curr per; 11 000 e-books; 600 microforms; 10 digital data carriers
libr loan; ATLA 35336

Western Seminary, Cline-Tunnell Library, 5511 SE Hawthorne Blvd, ***Portland***, OR 97215-3367
T: +1 503 5171840; Fax: +1 503 5171801; URL: www.westernseminary.edu
1927; Dr. Robert A. Krupp
Baptist Hist; Instructional Resource Center
81 000 vols; 249 curr per; 33 000 microforms; 800 av-mat; 3 670 sound-rec; 14 digital data carriers
libr loan; ACL, ALA, ATLA 35337

Princeton Theological Seminary, Speer Library, Mercer St & Library Pl, ***Princeton***, NJ 08542; P.O. Box 111, Princeton, NJ 08542-0111
T: +1 609 4977940; Fax: +1 609 4971826; URL: www.ptsem.edu/library/
1812; Stephen D. Crocco
Baptist Controversy, Early American Theological Pamphlets (Sprague Coll), Hymnology (Benson Coll), Puritan Lit
534 000 vols; 4 022 curr per; 33 000 microforms; 25 digital data carriers; 100 000 mss 35338

Temple Beth El Congregation Sons of Israel & David, William G. Braude Library, 70 Orchard Ave, ***Providence***, RI 02906
T: +1 401 3316070 ext 111; Fax: +1 401 3318068; E-mail: rsilverman@temple-beth-el.org
1892; Reini Silverman
Latin American Jewish Studies Coll
25 000 vols; 73 curr per; 432 pamphlets 35339

Temple Emanu-El, Library, 99 Taft Ave, ***Providence***, RI 02906
T: +1 401 3311616; Fax: +1 401 4219279; E-mail: info@teprov.org; URL: www.teprov.org/education/library.php
1953
Judaica
9 000 vols; 20 curr per 35340

Sisters of The Immaculate Conception Convent Library, 600 Liberty Hwy, ***Putnam***, CT 06260
T: +1 860 9287955; Fax: +1 860 9281930
1944; Ona Strimaitis
Linguistics; Lithuanian Art, paintings, ceramics, wood-carvings, amber & woven art
15 000 vols; 25 curr per 35341

Rosicrucian Fraternity Library, Beverly Hall Corp, 5966 Clymer Rd, P.O. Box 220, ***Quakertown***, PA 18951
T: +1 215 5367048; Fax: +1 215 5367058; E-mail: bevhall@comcat.com; URL: www.rosecross.org
1906; Gerald E. Poesnecker
Religion, philosophy
20 000 vols; 16 curr per 35342

Assumption Abbey Library, 418 Third Ave W, ***Richardton***, ND 58652-7100; P.O. Box A, Richardton, ND 58652-0901
T: +1 701 9743315; Fax: +1 701 9743317
1899; Aaron Jensen
92 000 vols; 68 curr per
libr loan 35343

First Baptist Church Library, 2709 Monument Ave, ***Richmond***, VA 23220
T: +1 804 3558637; Fax: +1 804 3594000; E-mail: library@fbcrichmond.org; URL: www.fbcrichmond.org
David Jackson
16 000 vols; 30 curr per; 50 sound-rec; 2 VF drawers of clippings, 600 slides 35344

International Mission Board, Southern Baptist Convention – Jenkins Research Library, 3806 Monument Ave, ***Richmond***, VA 23230-3932; P.O. Box 6767, Richmond, VA 23230-0767
T: +1 804 2191429
1960
Southern Baptist Missions Hist
22 000 vols; 600 curr per; 25 sound-rec; 75 digital data carriers
libr loan; OCLC, SOLINET, SLA, ALA 35345

Reveille United Methodist Church Library, 4200 Cary Street Rd, ***Richmond***, VA 23221
T: +1 804 3596041; Fax: +1 804 3596090
Mrs William B. Guthrie
10 000 vols; 7 curr per; 4 VF drawers 35346

Union Theological Seminary & Presbyterian School of Christian Education, William Smith Morton Library, 3406 Chamberlayne Ave, ***Richmond***, VA 23227; mail address: 3401 Brook Rd, Richmond, VA 23227
T: +1 804 2784325; Fax: +1 804 2784375; URL: library.union-psce.edu
1806; Dr. Milton J. Coalter
William Blake (Norfleet); Religion (Records of Synod of Virginia), micro; Social Science; Presbyterian Church Arch
346 000 vols; 977 curr per; 68 maps; 31 000 microforms; 31 000 av-mat; 32 155 sound-rec; 234 digital data carriers; 623 Kits & Games
libr loan 35347

Minnesota Bible College, Library, 920 Mayowood Rd SW, ***Rochester***, MN 55902
T: +1 507 5353331; Fax: +1 507 2889046; E-mail: library@crossroadscollege.edu; URL: www.crossroadscollege.edu
1924; James M. Godsey
G H Cachiarias Memorial Libr (Disciples of Christ authors)
38 000 vols; 363 curr per; 800 microforms; 432 sound-rec; 12 digital data carriers; 14 slides, 74 overhead transparencies
libr loan 35348

Roberts Wesleyan College & Northeastern Seminary, Ora A. Sprague Library, 2301 Westside Dr, *Rochester*, NY 14624-1997
T: +1 585 5946017; Fax: +1 585 5946543; URL: www.roberts.edu/library/ 1866; Alfred Krober
Benjamin Titus Roberts Coll, Free Methodist Church Hist
129 000 vols; 806 curr per
libr loan   35349

Temple Beth El Library, 139 Winton Rd S, *Rochester*, NY 14610
T: +1 716 4731770 ext 27; Fax: +1 716 4732689; E-mail: ehagelberg@ tberochester.org; URL: tberochester.org
1946
10 000 vols; 29 curr per; 7 file drawers of pamphlets & clippings   35350

Temple B'Rith Kodesh, Feinbloom Library, 2131 Elmwood Ave, *Rochester*, NY 14618
T: +1 716 2447060; Fax: +1 716 2440557; E-mail: welcome@tbk.org; URL: www.tbk.org/study/library
1962
13 500 vols; 25 curr per   35351

Temple Sinai Library, 425 Roslyn Rd, *Roslyn Heights*, NY 11577
T: +1 516 6216800; Fax: +1 516 6256020; E-mail: TempleOffice@mysinai.org; URL: templesinairoslyn.com
Judaica
10 000 vols   35352

Michigan Lutheran Seminary, Library, 2777 Hardin St, *Saginaw*, MI 48602
T: +1 989 7931041; Fax: +1 989 7934213; E-mail: info@misem.org; URL: www.mlsem.org/home/
1952
10 000 vols; 48 curr per; 3 digital data carriers   35353

Saint Joseph Seminary College, Pere Rouquette Library, 75376 River Rd, *Saint Benedict*, LA 70457-9900
T: +1 985 8672237; Fax: +1 985 8672270; E-mail: rouquette@sjasc.edu; URL: www.sjasc.edu
1891; Bonnie Bess Wood
Diocesan Coll, Hist Natchez
60 000 vols; 155 curr per; 4 000 microforms; 212 sound-rec; 8 digital data carriers   35354

Holy Trinity Monastery Library, Hwy 80, Milepost 302, *Saint David*, AZ 85630, P.O. Box 298, Saint David, AZ 85630-0298
T: +1 520 7204754; Fax: +1 520 7204202; E-mail: trinitylib@theriver.com; URL: www.holytrinitymonastery.org
1974; Sister Corinne Fair
Monastic Coll; Southwest/Native American Coll
49 000 vols; 40 curr per; 1 400 sound-rec
libr loan   35355

St. Francis Seminary, Salzmann Library, 3257 S Lake Dr, *Saint Francis*, WI 53235-0905
T: +1 414 7476479; Fax: +1 414 7476483; URL: www.sfs.edu/ salzmann.html
1845; Kathy Frymark
Wisconsin Catholic church hist, Hagiology, Liturgy; Catholic Americana, Rembert Weakland
84 000 vols; 508 curr per; 1 380 diss/theses; 350 microforms; 3 000 sound-rec; 4 digital data carriers; 10 VF drawers   35356

Catholic Central Verein (Union) of America, Central Bureau Library, 3835 Westminster Pl, *Saint Louis*, MO 63108
T: +1 314 3711653; Fax: +1 314 3710889; E-mail: centbur@sbcglobal.net; URL: www.socialjusticereview.org
1908; Father Edward Krause
German-Americana
250 000 vols; 20 curr per; 2 000 microforms; 4 000 Pamphlets   35357

Concordia Historical Institute, Lutheran Church-Missouri Synod, Library, 804 Seminary Pl, *Saint Louis*, MO 63105-3014
T: +1 314 5057900; Fax: +1 314 5057901; E-mail: reference@chi.lcms.org; URL: chi.lcms.org

1927; N.N.
Arch of The Lutheran Church-Missouri Synod; mss colls on the hist of Lutheranism in America; Museum
59 000 vols; 180 curr per; 2 500 000 mss; 100 diss/theses; 3 500 music scores; 500 maps; 2 000 microforms; 300 av-mat; 2 000 sound-rec; 300 digital data carriers
libr loan   35358

Concordia Seminary, Library, 801 Seminary Pl, *Saint Louis*, MO 63105-3199
T: +1 314 5057031; Fax: +1 314 5057046; E-mail: libraryref@csl.edu; URL: www.csl.edu/library
1839; David O. Berger
Lutherana Coll; Peasant's War
215 000 vols; 1 070 curr per; 75 e-journals; 15 mss; 75 incunabula; 1 200 music scores; 275 maps; 49 000 microforms; 2 600 av-mat; 10 500 sound-rec; 45 digital data carriers; 34 694 Journals, 200 Bks on Deafness & Sign Lang
libr loan; ATLA   35359

Covenant Theological Seminary, J. Oliver Buswell Jr Library, 12330 Conway Rd, *Saint Louis*, MO 63141
T: +1 314 4344044; Fax: +1 314 4344819; E-mail: library@covenantseminary.edu; URL: library.covenantseminary.edu
1956; James Cotton Pakala
Religion, Counseling; Bible & Theology, Church Hist (Reformation & Puritan), Practical Theology
70 000 vols; 356 curr per; 13 592 e-journals; 272 diss/theses; 1 000 microforms; 3 424 sound-rec; 438 kits
libr loan; ATLA   35360

Kenrick-Glennon Seminary, Charles L. Souvay Memorial Library, 5200 Glennon Dr, *Saint Louis*, MO 63119
T: +1 314 7926100; Fax: +1 314 7926503; E-mail: souvaylibrary@kenrick.edu; URL: www.kenrick.edu/souvay-library
1893; Dr. Andrew J. Sopko
Roman Catholic theology, Scripture, Patristics, Liturgics, Canon law; Cuneiform Tablets, Thomas Merton Coll, Official Catholic Directory, Pre-Vatican II Catechism Coll, Rare Bk Coll, Code Coll of Catholic Americana
78 000 vols; 300 curr per; 893 microforms; 300 av-mat; 2 084 sound-rec; 1 354 VF mat   35361

Lutheran Church Missouri Synod, Central Library, 1333 S Kirkwood Rd, *Saint Louis*, MO 63122-7295
T: +1 314 9659000 ext 1298; Fax: +1 314 8228307; URL: www.lcms.org
1983
Theological Ref Coll
38 000 vols; 300 curr per; 600 av-mat; 400 sound-rec
libr loan   35362

Saul Brodsky Jewish Community Library, 12 Millstone Campus Dr, *Saint Louis*, MO 63146
T: +1 314 4323720; Fax: +1 314 4321277; E-mail: brodsky-library@jfedstl.org; URL: www.brodskylibrary.org
1983; Barbara Raznick
Israel; Jewish Children's Bks; Holocaust; Hebrew Lit; Nat & Internat Newspapers, Hebrew & Russian; Russian Lit
22 000 vols; 100 curr per; 150 av-mat; 300 sound-rec   35363

Shaare Emeth Temple, Fleischer Library, 11645 Ladue Rd, *Saint Louis*, MO 63141
T: +1 314 5690010; Fax: +1 314 5690271; E-mail: Library@shaare-emeth.org; URL: www.shaare-emeth.org
1960
15 000 vols; 11 curr per   35364

Webster University, Emerson Library, 101 Edgar Rd, *Saint Louis*, MO 63119; mail address: 470 E Lockwood Ave, Saint Louis, MO 63119-3194
T: +1 314 9686952; Fax: +1 314 9687113; URL: library.webster.edu
1969; Laura Rein
Harry James Cargas Coll; Children's Lit (Hochschild Coll); Reformed Church Hist (James I Good Coll)

242 000 vols; 1 500 curr per; 2 000 av-mat; 13 digital data carriers
libr loan   35365

Bethel Theological Seminary Library, 3949 Bethel Dr, *Saint Paul*, MN 55112
T: +1 651 6386184; Fax: +1 651 6386006; E-mail: sem-reference-desk@bethel.edu; URL: seminary.bethel.edu/library
1871; Sandra Oslund
Arch (Baptist General Conference); 19th C Pietism (Skarstedt); Klingberg Puritan Coll; Nelson-Lundquist Coll; Devotional Lit
220 000 vols; 630 curr per; 6 501 sound-rec   35366

Luther Seminary, Library, Gullixson Hall, 2375 Como Ave, *Saint Paul*, MN 55108
T: +1 651 6413447; Fax: +1 651 6413280; URL: www.luthersem.edu/library
1869; David R. Stewart
Reformation; Doving Hymnal Coll, Lutheran Brotherhood Foundation Reformation Film Libr, Preus Rare Bk Coll, Tanner Catechism Coll
249 000 vols; 621 curr per; 3 000 microforms; 7 digital data carriers; 2 238 slides, 175 art reproductions
libr loan; ATLA   35367

Mt. Zion Temple Library, 1300 Summit Ave, *Saint Paul*, MN 55105
T: +1 651 6983881; Fax: +1 651 6981263
1929; Robert A. Epstein
Children's Coll; Margolis Coll on Jewish Feminism
9 850 vols; 18 curr per; 230 av-mat; 150 sound-rec; 4 digital data carriers   35368

The Church of Jesus Christ of Latter-Day Saints, Church History Library & Archives, 50 E North Temple, *Salt Lake City*, UT 84150
T: +1 801 2402272; Fax: +1 801 2401845; URL: www.lds.org/churchhistory/ askalibrarian
1830; Christine Cox
300 000 vols; 799 curr per; 1 071 maps; 220 000 microforms; 3 000 av-mat; 290 000 minute bks & other handwritten vols, 12 000 transcriptions & tapes, 1 mio mss   35369

Church of Jesus Christ of Latter-Day Saints – Family History Library, 35 N West Temple St, RM 344, *Salt Lake City*, UT 84150-3400
T: +1 801 2402584; Fax: +1 801 2403718; E-mail: help@productsupport.familysearch.org; URL: www.familysearch.org
1894; Raymond Wright
Family Group Records coll (8 mil sheets); oral genealogy tapes; int coll of microfilm; two large databases on computer
888 980 vols; 4 500 curr per; 9 613 maps; 500 digital data carriers   35370

Alamo Heights United Methodist Church Library, 825 E Basse Rd, *San Antonio*, TX 78209-1832
T: +1 210 8263215; Fax: +1 210 8263897; URL: ahumc.org
Duane Johnson
Church history, christianity
10 000 vols   35371

Assumption Seminary, Library, 2600 W Woodlawn, *San Antonio*, TX 78228-5122; P.O. Box 28240, San Antonio, TX 78228-0201
T: +1 210 7345137; Fax: +1 210 7342324; E-mail: seminary78228@yahoo.com; URL: www.assumptionseminary.org
1952; Louise Cates
20 000 vols; 140 curr per; 1 000 pamphlets   35372

Hispanic Baptist Theological School Library, 8019 S Panam Expressway, *San Antonio*, TX 78224-1397
T: +1 210 9244338; Fax: +1 210 9242701; E-mail: reference.desk@bua.edu; URL: www.bua.edu
1955; Teresa Martinez
Theology in Spanish; Neal Coll
25 000 vols; 214 curr per; 155 microforms; 793 sound-rec   35373

San Antonio First Baptist Church, Wallace Library, 515 McCullough St, *San Antonio*, TX 78215
T: +1 210 2260363 ext 219; Fax: +1 210 2992633; E-mail: library@fbcsanan.org; URL: www.fbcsanan.org
1939
18 400 vols; 10 curr per; 500 microforms; 400 av-mat; 1 050 sound-rec; 80 VF drawers of archival mats   35374

Life Bible College, Alumni Library, 1100 W Covina Blvd, *San Dimas*, CA 91773-3203
T: +1 909 7063009; Fax: +1 909 5996690; E-mail: library@lifepacific.edu; URL: www.lifepacific.edu/library.html
1925; Keith Roberts Dawson
Arch of Int Church of the Foursquare Gospel
40 000 vols; 250 curr per; 1 139 sound-rec
ALA, CLA, ACL   35375

Rosicrucian Order, AMORC, Research Library, Rosicrucian Park, 1342 Naglee Ave, *San Jose*, CA 95191
T: +1 408 9473600; Fax: +1 408 9473677; E-mail: librarian@rosicrucian.org; URL: www.rosicrucian.org
1939; Julie Scott
Egypt
16 000 vols; 21 curr per   35376

Temple De Hirsch Sinai Library, 1511 E Pike, *Seattle*, WA 98122
T: +1 206 3157398; Fax: +1 206 3246772; E-mail: library@tdhs-nw.org; URL: www.tdhs-nw.org
Toby Harris
Judaism, Holocaust, history
10 000 vols; 20 curr per; 35 av-mat   35377

Central Baptist Theological Seminary, Pratt-Journeycake Library, 22074 W 66th St, mail address: 741 N 31st St, *Shawnee*, KS 66226
T: +1 913 4225789; Fax: +1 913 3718110; URL: www.cbts.edu
1901; Linda Kiesling
Kansas Qumram Project (Fred E. Young Qumran Coll); Anabaptist Foundation Coll
105 000 vols; 300 curr per; 10 000 microforms; 5 500 av-mat; 2 129 sound-rec; 12 digital data carriers
ALA, ATLA   35378

First Methodist Church, Bliss Memorial Library, 500 Common St, *Shreveport*, LA 71101; P.O. Box 1567, Shreveport, LA 71165-1567
T: +1 318 4247771; Fax: +1 318 4296888; E-mail: fumc@fumcShreveport.org; URL: www.fumcshreveport.org
1946
Religion, Children's lit, Fiction; Articles for Worship Ctrs, Church Arch
12 500 vols; 150 sound-rec   35379

Seventh Day Adventists General Conference Library, 12501 Old Columbia Pike, *Silver Spring*, MD 20904
T: +1 301 6806000; Fax: +1 301 6806090
Bert Haloviak
27 000 vols; 500 curr per   35380

North American Baptist Seminary, Kaiser-Ramaker Library, 1525 S Grange Ave, *Sioux Falls*, SD 57105-1526
T: +1 605 3359080; Fax: +1 605 3359090; E-mail: library@nabs.edu; URL: www.nabs.edu
1858; Paul A. Roberts
Theology, Religion, Marriage & Family studies; Harris Homiletics Coll, North American Baptist Arch
72 000 vols; 304 curr per; 739 microforms; 2 565 sound-rec; 4 digital data carriers; 2 VF drawers of pamphlets, 173 videos, 7 111 slides, 134 charts, 261 boxes of archival mats
libr loan   35381

Saint Paul Lutheran Church & School Library, 5201 Galitz, *Skokie*, IL 60077
T: +1 847 6735030; Fax: +1 847 6739828; URL: www.stpaulskokie.org
11 000 vols; 20 curr per   35382

Patriarch Saint Tikhon Library, Saint Tikhon's Orthodox Theological Seminary, St Tikhon's Rd, *South Canaan*, PA 18459; P.O. Box 130, South Canaan, PA 18459-0130
T: +1 570 9374411; Fax: +1 570 9373100; E-mail: library@stots.edu; URL: www.stots.edu/library.htm
1938; Sergei Arhipov
Russian & Church Slavic Theological & Lit Coll
36 000 vols; 230 curr per   35383

Gordon-Conwell Theological Seminary, Burton L. Goddard Library, 130 Essex St, *South Hamilton*, MA 01982-2317
T: +1 978 6464079; Fax: +1 978 6464567; E-mail: glibrary@gcts.edu; URL: www.youseemore.com/gordonconwell
1970; Dr. Freeman Barton
Assyro-Babylonian Mat (Mercer Coll), Rare Bibles (Babson Coll), Washburn Baptist Coll, John Bunyan Coll
174 000 vols; 700 curr per; 45 000 microforms; 1 700 av-mat; 400 sound-rec; 24 digital data carriers; 250 electronic media & resources, 23 021 journals
libr loan; ATLA   35384

Immaculate Conception Seminary, Library, Seton Hall University, 400 S Orange Ave, *South Orange*, NJ 07079
T: +1 973 7619336; Fax: +1 973 2752074; URL: theology.shu.edu/semlib.htm
1858; Father Lawrence Porter
Rare Sacral Bks, Cardinal Newman and Sources Chretiennes
63 000 vols; 450 curr per; 1 570 diss/theses; 2 000 microforms; 1 660 sound-rec; 25 digital data carriers
libr loan; ATLA, ALA, Catholic Theological Library Consortium   35385

New England Bible College, Library, 879 Sawyer St, *South Portland*, ME 04106; P.O. Box 2886, South Portland, ME 04116-2886
T: +1 207 7995979; Fax: +1 207 7996586; URL: www.nebc.edu
15 000 vols; 30 curr per   35386

Congregation Shaarey Zedek Library & Audio Visual Center, 27375 Bell Rd, *Southfield*, MI 48034
T: +1 248 3575544; Fax: +1 248 3570227; E-mail: csz.info@shaareyzedek.org; URL: www.shaareyzedek.org
1932; Janet Pont
Judaica, Holocaust, Juvenile delinquency, Lit; Holtzman Coll, Modern Hebrew Lit
40 000 vols; 12 curr per; 200 sound-rec; 4 digital data carriers; 20 VF drawers of pamphlets, clippings, pictures   35387

Saint Joseph's Abbey, Monastic Library, 167 N Spencer Rd, *Spencer*, MA 01562-1233
T: +1 508 8858700 ext 524; E-mail: monks@spencerabbey.org; URL: www.spencerabbey.org
1950
35 000 vols; 25 curr per   35388

Assemblies of God Theological Seminary, Cordas C. Burnett Library, 1435 N Glenstone Ave, *Springfield*, MO 65802-2131
T: +1 417 2681000; Fax: +1 417 2681001; E-mail: info@agseminary.edu; URL: www.agts.edu
1973; Joseph F. Marics Jr
91 000 vols; 405 curr per; 52 e-journals; 750 maps; 7 000 microforms; 4 355 sound-rec; 56 digital data carriers
libr loan   35389

Baptist Bible College, G.B.Vick Memorial Library, 730 E Kearney St, *Springfield*, MO 65803; mail address: 628 E Kearney St, Springfield, MO 65803-3426
T: +1 417 2686074; Fax: +1 417 2686690; E-mail: info@baptist.edu; URL: library.baptist.edu
1956; Gregory A. Smith
Baptist Hist
56 000 vols; 332 curr per; 24 mss; 2 592 govt docs; 234 maps; 20 000 microforms; 1 612 sound-rec; 5 digital data carriers
libr loan   35390

Central Bible College, Pearlman Memorial Library, 3000 N Grant Ave, *Springfield*, MO 65803
T: +1 417 8332551; Fax: +1 417 8335478; URL: www.cbcag.edu/library
1922; Lynn Anderson
Assemblies of God Coll
108 000 vols; 310 curr per; 2 330 e-journals; 8 000 e-books; 30 000 microforms; 900 av-mat; 80 sound-rec; 110 digital data carriers; 77 VF drawers of clippings & pamphlets, 285 linear feet of archival mat, 215 Bks on Deafness & Sign Language
libr loan   35391

Christian Life College, Library, 9023 N West Lane, *Stockton*, CA 95210
T: +1 209 4767882; Fax: +1 209 4767868; E-mail: info@clc.edu; URL: www.clc.edu
1955; Nancy Hunt
20 000 vols; 50 curr per   35392

Salvation Army School for Officer Training, Brengle Library, 201 Lafayette Ave, *Suffern*, NY 10901
T: +1 845 3687228; Fax: +1 845 3576644; URL: www.sfotusa.org
1936; Robin Rader
Salvation Army Publs
27 000 vols; 225 curr per   35393

The Master's Seminary Library, 13248 Roscoe Blvd, *Sun Valley*, CA 91352
T: +1 818 9095634; Fax: +1 818 9095680; URL: www.tms.edu
1986; Dennis M. Swanson
Archaeology, Biblical studies, Church hist, Theology
175 000 vols; 650 curr per; 26 e-journals; 500 av-mat; 500 DVDs   35394

Faith Evangelical Lutheran Seminary, Library, 3504 N Pearl St, *Tacoma*, WA 98407-2607; P.O. Box 7186, Tacoma, WA 98406-0186
T: +1 253 7522020; Fax: +1 253 7591790; E-mail: fsinfo@faithseminary.edu; URL: www.faithseminary.edu
1969; Bebhinn Horrigan
20 000 vols; 90 curr per   35395

The Abbey of Gethsemani Library, 3642 Monks Rd, *Trappist*, KY 40051
T: +1 502 5493117; Fax: +1 502 5494124; E-mail: trappists@monks.org; URL: www.monks.org
1848; Father Gaetan Blanchette
Religion, Theology, Philosophy; Thomas Merton Coll; Cistercian Monastic Hist & Liturgy, rare bks; Saint Bernard & DeRance, rare bks
40 000 vols   35396

United Theological Seminary, Library, 4501 Denlinger Rd, *Trotwood*, OH 45426
T: +1 937 5292201; Fax: +1 937 5292292; E-mail: library@united.edu; URL: library.united.edu
1871; Sarah D. Brooks Blair
Evangelical Church, Evangelical United Brethren Church, Edmund S Lorenz Hymnal Coll, J Allan Ranck Coll of Friendship Press, United Brethren in Christ Church, United Methodist Church, Waldensian-Methodist Coll
144 000 vols; 500 curr per; 1 327 sound-rec; 5 digital data carriers; 3 625 slides, 121 overhead transparencies
libr loan   35397

Temple Emanu-El, Library, 225 N Country Club Rd, *Tucson*, AZ 85716
T: +1 520 3274501; Fax: +1 520 3274504; URL: www.templeemanueltucson.org
1910
Biogr, Fiction, Holidays, Judaic hist, Relig practices, Youth
9 500 vols; 10 curr per; 50 av-mat; 170 sound-rec   35398

First United Methodist Church, Broadhurst Library, 1115 S Boulder, *Tulsa*, OK 74119-2492
T: +1 918 5879481; Fax: +1 918 4962117; E-mail: drkinard@cs.com
1964; Nancy J. Kinard
Methodist Hist; John Wesley Coll
18 000 vols; 10 curr per; 250 sound-rec; 15 digital data carriers; 25 other items   35399

Phillips Theological Seminary, Library, 901 N Mingo Rd, *Tulsa*, OK 74116
T: +1 918 2706437; Fax: +1 918 2706490; E-mail: ptslibrary@ptstulsa.edu; URL: www.ptstulsa.edu
1950; Sandy Shadley
Discipliana
94 000 vols; 450 curr per; 22 VF drawers of pamphlets   35400

Unity School of Christianity Library (Unity School Library), 1901 NW Blue Pkwy, *Unity Village*, MO 64065-0001
T: +1 816 5243550 ext 2370; Fax: +1 816 2513555; E-mail: library@unityworldhq.org; URL: www.unityworldhq.org/library
1965; Priscilla Richards
Unity School of Christianity Arch from 1889, bks, rec, flm
40 000 vols; 150 curr per; 250 microforms; 4 700 sound-rec; 1 075 video recordings
libr loan; ALA, SLA   35401

Old Cathedral Library, 205 Church St, *Vincennes*, IN 47591-1133
T: +1 812 8825638; Fax: +1 812 8824042
John Schipp
12 000 vols   35402

Beth Israel Synagogue, Beth Israel Community Library, 1015 E Park Ave, *Vineland*, NJ 08360
T: +1 856 6910852; Fax: +1 856 6921957
1926; Ruth Greenblatt
Hitler Period (Holocaust Lit)
9 000 vols; 25 curr per; 350 av-mat; 215 sound-rec
libr loan   35403

First Baptist Church, I.C. Anderson Library, 500 Webster St, *Waco*, TX 76706
T: +1 254 7523000; Fax: +1 254 7562237; URL: www.fbcwaco.org
1945
Church Hist
20 000 vols; 25 curr per; 35 av-mat; 540 sound-rec   35404

Southeastern Baptist Theological Seminary, Library, 114 N Wingate St, *Wake Forest*, NC 27587
T: +1 919 8638258; Fax: +1 919 8638150; E-mail: library@sebts.edu; URL: library.sebts.edu
1951; Shawn C. Madden
Baptists (Baptist Documents Coll), Education (Curriculum Lab, Lifeway Curriculum)
177 000 vols; 781 curr per   35405

Dominican College, Library, 487 Michigan Ave NE, *Washington*, DC 20017-1585
T: +1 202 4953821; Fax: +1 202 4953873; URL: www.dhs.edu
1905; Bernard Mulcahy
Dominican Hist, Liturgy & Authors; St Thomas Aquinas; Diss by Dominican Authors; Rare Bks
75 000 vols; 450 curr per; 1 000 microforms; 54 av-mat; 440 sound-rec; 2 digital data carriers
libr loan   35406

Franciscan Monastery Library, Commissariat of the Holy Land USA, 1400 Quincy St NE, *Washington*, DC 20017
T: +1 202 5266800; Fax: +1 202 5299889; URL: www.myfranciscan.com
1900; Heather K. Calloway
35 000 vols; 25 curr per   35407

United States Conference of Catholic Bishops Library, 3211 Fourth St NE, *Washington*, DC 20017-1194
T: +1 202 5413193; Fax: +1 202 5413322; E-mail: library@usccb.org; URL: www.usccb.org
1989; Anne Le Veque
Church hist, Theology, Human rights; Latin American Bishops, bks, pamphlets, Conference Publs
12 000 vols; 100 curr per   35408

Washington Theological Union, O'Toole Library, 6896 Laurel St NW, *Washington*, DC 20012-2016
T: +1 202 5415208; Fax: +1 202 7261716; URL: www.wtu.edu/resources/library.htm
1968; Alexander M. Moyer
Franciscana, Carmelitana, Augustiniana
100 000 vols; 450 curr per; 8 digital data carriers   35409

Wesley Theological Seminary, Library, 4500 Massachusetts Ave NW, *Washington*, DC 20016-5690
T: +1 202 8858695; Fax: +1 202 8858691; E-mail: library@wesleyseminary.edu; URL: library.wesleyseminary.edu
1882; William D. Faupel
Theology, Religion; Early American Methodism, Methodist Protestant Church, Wesleyana
142 000 vols; 575 curr per; 337 diss/theses; 11 000 microforms; 6 200 av-mat; 1 354 sound-rec; 5 digital data carriers
libr loan   35410

Whitefriars Hall, Order of Carmelites Library, 600 Webster St NE, *Washington*, DC 20017
T: +1 202 5261221; Fax: +1 202 5269217; E-mail: library@whitefriarshall.org; URL: www.loc.gov/rr/main/religion.wfh.html
1948; Father Patrick McMahon
50 000 vols; 60 curr per; 271 diss/theses; 100 microforms
libr loan
Research libr for scholars of the Carmelite Tradition in Roman Catholic Spirituality   35411

Woodstock Theological Center Library, Georgetown University, Lauinger Library, P.O. Box 571170, *Washington*, DC 20057-1170
T: +1 202 6877513; Fax: +1 202 6877473; E-mail: jlh3@georgetown.edu; URL: woodstock.georgetown.edu/library/index.htm
1869; J. Leon Hooper, S. J.
Theology & Jesuitica (Jogues, Shrub Oak, & Parsons Coll), bks; Palestinian Antiquities (Halpern Coll), engravings; Woodstock College Archives
190 000 vols; 1 932 curr per; 18 e-books; 2 000 mss; 27 incunabula; 8 diss/theses; 22 maps; 3 000 microforms; 3 av-mat; 15 digital data carriers; 400 pamphlets
libr loan   35412

Maranatha Baptist Bible College, Cedarholm Library & Resource Center, 745 W Main St, *Watertown*, WI 53094
T: +1 920 2062375; Fax: +1 920 2619109; URL: library.mbbc.edu
1967; Lois Oetken
136 000 vols; 356 curr per; 396 e-books; 223 av-mat   35413

Jewish Community Center of Metropolitan Detroit, Henry & Delia Meyers Memorial Library, 6600 W Maple, *West Bloomfield*, MI 48322
T: +1 248 4325546; Fax: +1 248 4325540; E-mail: fmenken@jccdet.org; URL: www.jccdet.org
1959; Francine Menken
Judaica & Non-Judaica for Children
11 000 vols; 35 curr per; 4 VF drawers   35414

Temple Israel Libraries & Media Center, 5725 Walnut Lake Rd, *West Bloomfield*, MI 48323
T: +1 248 6615700; Fax: +1 248 6611302; E-mail: TILibrary@aol.com; URL: www.temple-israel.org
1961; Rachel Erlich
Jewish Studies, Hebraica, Judaica, Judaism, Holocaust, Children's Lit
15 000 vols; 35 curr per; 20 av-mat; 175 sound-rec; 45 digital data carriers
ALA, MLA (Michigan Library Association), AJL   35415

Congregation Beth Israel, Deborah Library, 701 Farmington Ave, *West Hartford*, CT
T: +1 860 2338215; Fax: +1 860 5230223; URL: www.cbict.org/lifeLongLearning/library.html
1934
Israel, Holocaust, American Jewish hist

14 000 vols; 40 curr per; 12 diss/theses; 50 music scores; 12 maps; 400 av-mat; 278 sound-rec
AJL 35416

First Church of Christ Congregational, John P. Webster Library, 12 S Main St, ***West Hartford***, CT 06107
T: +1 860 2323893; Fax: +1 860 2328183; E-mail: jpwebsterdir@snet.net; URL: www.jpwlibrary.org/
1976; Patricia Malahan
Religion, Social issues, Psychology; The Practice of Parish Ministry
10 000 vols; 50 curr per; 500 av-mat
libr loan; CSLA 35417

Temple Israel Library of Judaica, 1901 N Flagler Dr, ***West Palm Beach***, FL 33407
T: +1 561 8338421; Fax: +1 561 8330571; E-mail: linda@temple-israel.com; URL: www.temple-israel.com
1958
Judaica, Judaism; Children's Coll
8 000 vols; 20 curr per; 1 000 pamphlets & clippings, 8 VF drawers 35418

Pope John XXIII National Seminary, Library, 558 South Ave, ***Weston***, MA 02493
T: +1 781 8995500; Fax: +1 781 8999057; E-mail: seminary@blessedjohnxxiii.edu; URL: www.blessedjohnxxiii.edu
1963; Sister Jacqueline Miller
Comprehensive English Language Theology, 1958 to date
66 000 vols; 7 600 sound-rec; 60 digital data carriers
libr loan; ATLA, ALA, NELA, MLA, CLA 35419

Saint Mary Seminary, Bruening-Marotta Library, 28700 Euclid Ave, ***Wickliffe***, OH 44092-2585
T: +1 440 9437665; Fax: +1 440 5853528
1848; Alan K. Rome
Theology (Horstmann Coll)
69 000 vols; 326 curr per; 281 diss/theses; 620 microforms; 759 sound-rec; 250 drafts & rpts, 5 boxes of pamphlets 35420

Payne Theological Seminary, Reverdy C. Ransom Memorial Library, 1230 Wilberforce-Clifton Rd, ***Wilberforce***, OH 45384; P.O. Box 474, Wilberforce, OH 45384-0474
T: +1 937 3762947; Fax: +1 937 3762888; E-mail: circulation@payne.edu; URL: www.payne.edu
1890; J. Dale Balsbaugh
African American Coll
25 000 vols; 67 curr per 35421

Asbury Theological Seminary, B.L. Fisher Library, 204 N Lexington Ave, ***Wilmore***, KY 40390-1199
T: +1 859 8582233; Fax: +1 859 8582330; E-mail: info_commons@asburyseminary.edu; URL: www.asburyseminary.edu
1939; Dr. Ken Boyd
Healing, Wesleyan/Holiness, World Council of Churches
260 000 vols; 1 054 curr per; 5 000 microforms; 9 051 sound-rec; 77 digital data carriers; 3 945 slides, 2 000 linear feet of arch mat 35422

Grace College & Grace Theological Seminary, Morgan Library, 200 Seminary Dr, ***Winona Lake***, IN 46590
T: +1 574 3725100; Fax: +1 574 3725176; URL: www.grace.edu/library
1937; William Darr
Biblical studies; Billy Sunday Papers, Grace Brethren Denominational Arch, Winona Railroad Special Coll
163 000 vols; 402 curr per; 497 maps; 850 av-mat; 3 267 sound-rec
libr loan 35423

Piedmont Bible College, George M. Manuel Memorial Library, 420 S Broad St, ***Winston-Salem***, NC 27101
T: +1 336 7258344; Fax: +1 336 7255522; URL: www.pbc.edu/ps/ps_ac_library.aspx
1947; Delores G. Fulton
58 000 vols; 225 curr per; 977 e-books; 1 936 maps 35424

Palmer Theological Seminary, Austen K. deBlois Library, Six Lancaster Ave, ***Wynnewood***, PA 19096
T: +1 610 6459318; E-mail: semlibr@eastern.edu; URL: www.palmerseminary.edu/library
1925; Melody Mazuk
Barbour Coll in Black Church Studies; MacBride Coll of Philosophy, Religion & Classical Lit; Soto-Fontanez Hispanic Studies
122 000 vols; 394 curr per; 500 av-mat; 1 063 sound-rec; 60 digital data carriers
libr loan; ATLA 35425

Philadelphia Archdiocesan Historical Research Center, Archives & Historial Collections, 100 E Wynnewood Rd, ***Wynnewood***, PA 19096-3001
T: +1 610 6672125; Fax: +1 610 6672730; E-mail: pahrc89@gmail.com; URL: www.pahrc.net
Joseph J. Casino
Immigration; American Catholic Newspapers (19th C) Religious American Coll, Popular Piety Coll, Catechism Coll
53 000 vols; 8 000 curr per 35426

Saint Charles Borromeo Seminary, Ryan Memorial Library, 100 E Wynnewood Rd, ***Wynnewood***, PA 19096
T: +1 610 7856277; Fax: +1 610 6647913; E-mail: libraryscs@adphila.org; URL: www.scs.edu/library
1832; Cait Kokolus
Catholic theology, Philosphy, Scripture studies; First & Second Vatican Councils, 19th c Devotional Lit, Liturgical Bks, Holy Cards, Rare Theology Bks
127 000 vols; 565 curr per; 2 000 e-journals; 7 253 sound-rec; 8 digital data carriers; 500 Paintings & Prints, 34 Electronic Media & Resources
libr loan 35427

Saint Joseph's Seminary, Corrigan Memorial Library, 201 Seminary Ave, ***Yonkers***, NY 10704
T: +1 914 9686200; Fax: +1 914 9688787
1896; Sister Monica Wood
85 000 vols; 275 curr per; 1 000 microforms; 6 digital data carriers
libr loan; ALA, ATLA 35428

Saint Vladimir's Orthodox Theological Seminary, Father Georges Florovsky Library, 575 Scarsdale Rd, ***Yonkers***, NY 10707-1699
T: +1 914 9618313; Fax: +1 914 9614507; E-mail: librarian@svots.edu; URL: www.svots.edu/library
1938; Eleana Silk
Byzantine Hist & Art; Russian Hist & Culture (incl 19th Century Russian theological periodicals); Theology, Hist & Culture of the Orthodox Church
138 000 vols; 355 curr per; 620 diss/theses; 2 000 microforms; 620 sound-rec; 7 digital data carriers
libr loan; ALA, OCLC 35429

Corporate, Business Libraries

Washington Savannah River Co, Savannah River Site Technical Library, Bldg 773A, ***Aiken***, SC 29808
T: +1 803 7255555; Fax: +1 803 7251169
Susan Isaacs-Bright
Nuclear, Safety, Technology
34 000 vols; 215 curr per 35430

Bridgestone/Firestone Research LLC, Bridgestone Americas Center for Research & Technology, 1200 Firestone Pkwy, ***Akron***, OH 44317
T: +1 330 3797630; Fax: +1 330 3797530
1945; David Koo
15 000 vols; 100 curr per 35431

FirstEnergy Corp, Business Information Center, 76 S Main St, A-GO-17, ***Akron***, OH 44308
T: +1 330 3844934; Fax: +1 330 2551099
1981; Susan R. Lloyd
Electric power, Energy, Industry utilities; Edison Electric Institute Rpts, Electric Power Res Institute Rpt Coll, Industry Standards
17 000 vols; 10 curr per 35432

OMNOVA Solutions Inc, Technical Information Center, 2990 Gilchrist Rd, ***Akron***, OH 44305
T: +1 330 7946382; Fax: +1 330 7946375; URL: www.omnova.com
1946; Barbara Hubal
Science & technology, Polymer chemistry, Rubber, Plastics, Adhesives
9 000 vols; 50 curr per; 6 digital data carriers 35433

McNamee, Lochner, Titus & Williams, PC, Law Library, 677 Broadway, ***Albany***, NY 12207; P.O. Box 459, Albany, NY 12201-0459
T: +1 518 4473200; Fax: +1 518 4264260; E-mail: mltw@mltw.com; URL: www.mltw.com
Dana Wantuch
12 000 vols 35434

Modrall, Sperling, Roehl, Harris & Sisk, Law Library, PO Box 2168, ***Albuquerque***, NM 87103
T: +1 505 8481800; Fax: +1 505 8481891; URL: www.modrall.com
Miriam Greenwood
10 000 vols; 40 curr per 35435

Sutin, Thayer & Browne, Law Library, Two Park Square Bldg, 10th Flr, 6565 Americas Pkwy NE, ***Albuquerque***, NM 87110
T: +1 505 8832500; Fax: +1 505 8886565; URL: www.sutinfirm.com
Richard P. McGoey
10 000 vols; 100 curr per 35436

Astre Corporate Group Library, 3801 Mount Vernon Ave, ***Alexandria***, VA 22313; P.O. Box 25766, Alexandria, VA 22213-5766
T: +1 703 5481343; Fax: +1 703 7381340; E-mail: info@astrecg.com; URL: www.astrecg.com
Roy A. Ackerman
76 000 vols; 1 500 curr per 35437

John McMullen Associates, Inc, JJMA Office Wide Library, 4300 King St, Ste 400, ***Alexandria***, VA 22302-1503
T: +1 703 9336645; Fax: +1 703 9336791; E-mail: mholland@jjma.com; URL: www.jjma.com
Margaret Holland
Marine engineering, Naval architecture; Society of Naval Architects & Marine Engineering, journals
20 000 vols 35438

VSE Corporation Library-BAV Division, 2550 Huntington Ave, ***Alexandria***, VA 22303-1499
T: +1 703 3175259; Fax: +1 703 9606599; URL: www.vsecorp.com
Mary Wantrobski
Mathematics, engineering, physics; military, industry and federal specifications and standards
37 000 vols; 10 curr per 35439

Air Products & Chemicals, Inc, Information & Library Services, 7201 Hamilton Blvd, ***Allentown***, PA 18195-1501
T: +1 610 4817288; Fax: +1 610 4816495; E-mail: info@airproducts.com; URL: www.airproducts.com
1953; Valerie Ryder
Family Safety & On-the-Job Safety Audio-Visual Coll
34 000 vols; 500 curr per; 648 microforms; 13 digital data carriers; 306 linear feet of vertical files
libr loan 35440

McDermott Technology, Inc, Library, 1562 Beeson St, ***Alliance***, OH 44601
T: +1 330 8297313; Fax: +1 330 8297777; E-mail: library@mcdermott.com
1947; James W. Carter
Chemical engineering, Heat transfer, Metallurgy, Steam generation
10 000 vols; 170 curr per 35441

PPG Industries, Inc, Technical Information Center, 4325 Rosanna Dr, ***Allison Park***, PA 15101
T: +1 412 4925443; Fax: +1 412 4925509; URL: www.ppg.com
Denise Callihan
Coatings, Polymers, Paints, Resins
10 000 vols; 200 curr per 35442

Gibson, Ochsner & Adkins LLP, Law Library, 701 S Taylor, Ste 500, ***Amarillo***, TX 79101-2400
T: +1 806 3789739; Fax: +1 806 3789797; E-mail: mlockman@goa-law.com; URL: www.goa-law.com
Melissa Lockman
10 000 vols; 20 curr per
AALL, SWALL 35443

Siemens Water Technology, Lawlor Library, 600 Arrasmith Trail, ***Ames***, IA 50010
T: +1 515 2688400; Fax: +1 515 2688500; E-mail: marlys.cooper@usfilter.com; URL: www.usfilter.com
Marlys A. Cooper
115 000 vols 35444

The Boeing Company, Library Services, 3370 Miraloma Ave AA14, ***Anaheim***, CA 92803
T: +1 714 7621081; Fax: +1 714 7620834; E-mail: teresa.g.powell@boeing.com
1955; Teresa G. Powell
Electronics, mathematics, solid state physics, computers, marine systems engineering, guidance, navigation and control, telecommunications
15 300 vols; 100 curr per; 100 000 microforms; 15 digital data carriers 35445

Philips Medical Systems Library, 3000 Minuteman Rd, MS 0115, ***Andover***, MA 01810-1099
T: +1 978 6592228; Fax: +1 978 6877941; E-mail: susan_saraidaridis@hsgmed.com
1917; Susan Saraidaridis
Cardiology, Computer science, Engineering, Business
9 000 vols; 250 curr per
libr loan 35446

Pfizer, Inc, Pfizer Global R&D Information Management, 2800 Plymouth Rd, ***Ann Arbor***, MI 48105
T: +1 734 6224493; Fax: +1 734 6227008; E-mail: libreq@aa.wl.com
1885; Sharon Lehman
Chemistry, Medicine, Microbiology, Pathology, Toxicology
21 600 vols; 780 curr per 35447

Exxon Research & Engineering Co, Clinton Information Center, Clinton Township, Rte 22 E, ***Annandale***, NJ 08801
T: +1 908 7302755; Fax: +1 908 7303021
Mary Latino
15 000 vols 35448

Oblon, Spivak, McClelland, Maier & Neustadt, Law Library, 1755 Jefferson Davis Hwy, Ste 400, ***Arlington***, VA 22202-3509
T: +1 703 4126391; Fax: +1 703 4132220; E-mail: jburke@oblon.com
1993; Jo Burke
10 000 vols; 145 curr per 35449

System Planning Corp, Charles S. Lerch Information Center, 1000 Wilson Blvd, ***Arlington***, VA 22209
T: +1 703 3518700; Fax: +1 703 3518663; URL: www.sysplan.com
1973; Nick Mercury
Arms control, Engineering for defense systems
15 000 vols; 100 curr per 35450

Verizon, Law Library, 1515 N Courthouse Rd, Ste 500, ***Arlington***, VA 22201
T: +1 703 3513173; Fax: +1 703 3513656
Natalie Davis
10 000 vols; 80 curr per 35451

Southern Highland Craft Guild, Robert W. Gray Library Collection-Folk Art Center, 382 Blue Ridge Pkwy, ***Asheville***, NC 28805; P.O. Box 9545, Asheville, NC 28815-0545
T: +1 828 2987928; Fax: +1 704 2987962; E-mail: library@craftguild.org; URL: www.southernhighlandguild.org
Deborah Schillo
Manuscript collections of mountain craft persons
9 000 vols; 45 curr per 35452

L3 Communications Library, 2116 Arlington Downs Rd, **Athens**, TX 76011-6382
T: +1 817 6193579; Fax: +1 817 6193590
1951; Kimberley Crator
Aeronautics, Aircraft simulators, Computers, Electrical & mechanical engineering
15 000 vols; 300 curr per   35453

Arnall, Golden & Gregory LLP, Law Library, 1201 W Peachtree St, Ste 2800, **Atlanta**, GA 30309-3450
T: +1 404 8738500; Fax: +1 404 8738501
Harriet Day
25 000 vols   35454

AT&T, Law Library, 675 W Peachtree St NE, Ste 4300, **Atlanta**, GA 30375
T: +1 404 3350746; Fax: +1 404 6883988
Julie Schein
27 000 vols; 50 curr per   35455

Bell South Corporation, Law Library, 1155 Peachtree St, Ste 1800, **Atlanta**, GA 30309-3610
T: +1 404 2492616; Fax: +1 404 2492895
Judith Krone
Securities
20 000 vols   35456

The Coca Cola Company, Strategic Information Management Department, One Coca-Cola Plaza, NW, **Atlanta**, GA 30313; P.O. Box 1734, Atlanta, GA 30301-1734
T: +1 404 5154636; Fax: +1 404 5152572; E-mail: koinfo@na.ko.com
1967; Peter Pearson
Food science & tech, Chemistry, Packaging
30 000 vols; 600 curr per; 20 diss/theses; 520 govt docs; 24 microforms; 25 digital data carriers   35457

Federal Reserve Bank of Atlanta, Research Library, 1000 Peachtree St NE, **Atlanta**, GA 30309-4470
T: +1 404 5218867; Fax: +1 404 5218572; E-mail: Barbara.frolik@att.frb.org; URL: www.frbatlanta.org
1938; Barbara F. Frolik
Federal Reserve Bank & Fderal Reserve Board Publs, Southwestern Regional Economics
8 000 vols; 1 500 curr per
libr loan   35458

Georgia Power Co-Southern Co, Business Information Center-Research Library, 241 Ralph McGill Blvd NE, Bin 10044, **Atlanta**, GA 30308
T: +1 404 5066633; Fax: +1 404 5066652; E-mail: bic@southernco.com
1957; Margo Surovik-Bohnert
Energy, Business & management; Annual Rpts, Company Hist, Energy Mats
8 000 vols; 160 curr per   35459

Jones Day, Law Library, 1420 Peachtree St NE, **Atlanta**, GA 30309
T: +1 404 5818118; Fax: +1 404 5818330
Jane Crawford
15 000 vols   35460

Kilpatrick Stockton, Law Library, 1100 Peachtree St, Ste 2800, **Atlanta**, GA 30309
T: +1 404 8156261; Fax: +1 404 8156555
1904; Louise Adams
30 000 vols   35461

King & Spalding, Law Library, 191 Peachtree St, **Atlanta**, GA 30303
T: +1 404 5724600; E-mail: lib.helpdesk@kslaw.com
MaryAnne C. Fry
50 000 vols   35462

Kutak & Rock, Law Library, 225 Peachtree St, Ste 2100, **Atlanta**, GA 30303
T: +1 404 2224600; Fax: +1 404 2224654
Mark Burgreen
10 000 vols; 20 curr per   35463

Long, Aldridge & Norman, Law Library, 303 Peachtree St, Ste 5300, One Peachtree Ctr, **Atlanta**, GA 30308
T: +1 404 5274057; Fax: +1 404 5278474
Cindy Adams
10 000 vols   35464

Paul, Hastings, Janofsky & Walker, Law Library, 600 Peachtree St NE, Ste 2400, **Atlanta**, GA 30308
T: +1 404 8152143; Fax: +1 404 8152424
10 000 vols; 50 curr per   35465

Powell, Goldstein LLP, Law Library & Information Center, 1201 W Peachtree St, 14th Flr, **Atlanta**, GA 30309
T: +1 404 5726696; Fax: +1 404 5726999; E-mail: rdesk@pogolaw.com; URL: www.pogolaw.com
Rita Treadwell
20 000 vols   35466

Schreeder, Wheeler & Flint LLP, Law Library, The Candler Bldg, 16th Fl, 127 Peachtree St NE, **Atlanta**, GA 30303-1845
T: +1 404 6813450; Fax: +1 404 6811046
Dionne Lyne
Real estate
10 000 vols; 10 curr per   35467

Smith, Currie & Hancock, Law Library, 2700 Marquis One Tower, 245 Peachtree Center Ave, NE, **Atlanta**, GA 30303-1227
T: +1 404 5828098; Fax: +1 404 6880671; URL: www.smithcurrie.com
Philleatra Gaylor
Construction law, government contracts
10 000 vols; 20 curr per   35468

Smith, Gambrell & Russell, Law Library, Prominade II, 1230 Peachtree St NE, Ste 3100, **Atlanta**, GA 30309
T: +1 404 8153618; Fax: +1 404 6856838; E-mail: jsims@sgrlaw.com
1893; Adrienne McElroy-Boone
50 000 vols; 100 curr per   35469

DaimlerChrysler Corp, Information Resources Center, 800 Chrysler Dr, CIMS 483-08-10, **Auburn Hills**, MI 48326-2757
T: +1 248 5768300; Fax: +1 248 5762349
1933; Gerald H. Rivers
Automotive engineering
10 000 vols; 350 curr per   35470

Illinois Mathematics & Science Academy, Leto M. Furnas Information Resources Center, Library, 1500 W Sullivan Rd, **Aurora**, IL 60506-1000
T: +1 630 9075920; Fax: +1 630 9075004; E-mail: irc@imsa.edu; URL: www.imsa.edu/team/irc
1986; Paula Garrett
IMSA Arch, Western Electric Libr Coll
42 000 vols; 120 curr per   35471

Brown McCarroll, LLP, Law Library, 111 Congress Ave, Ste 1400, **Austin**, TX 78701-4043
T: +1 512 4725456; Fax: +1 512 4791101; E-mail: jjames@mailbmc.com; URL: www.brownmccarroll.com
1938; Ms Jency J. James
10 000 vols; 75 curr per; 5 maps; 3 microforms; 50 av-mat; 300 sound-rec; 15 digital data carriers
libr loan; AALL, SLA, SWALL, ALL   35472

Clark, Thomas & Winters, Legal Research Center, 300 W Sixth St, Ste 1500, **Austin**, TX 78701; P.O. Box 1148, Austin, TX 78767-1148
T: +1 512 4728800; Fax: +1 512 4741129; URL: www.ctw.com
Priscilla Streightoff
10 000 vols; 45 curr per   35473

IBM Corp, Library Information Resource Center, 11400 Burnet Rd, B/908, Z/9081E020, **Austin**, TX 78758
T: +1 512 8381067
1967; Amanda Stites
Engineering, Electronics, Computer science, Business
11 000 vols; 300 curr per   35474

Jackson Walker LLP, Law Library, 100 Congress Ave, Ste 1100, **Austin**, TX 78701-4099
T: +1 512 2362306; Fax: +1 512 2362002; URL: www.jw.com
1980; Judith B. Hamner
Real estate
50 000 vols; 200 curr per   35475

Jenkens & Gilchrist, Law Library, 600 Congress Ave, Ste 2200, **Austin**, TX 78701-3278
T: +1 512 4993876; Fax: +1 512 4043520
Wendy Lyon
10 000 vols   35476

McGinnis, Lochridge & Kilgore, LLP, Law Library, 919 Congress Ave, Ste 1300, **Austin**, TX 78701
T: +1 512 4956000; Fax: +1 512 4956093; URL: www.mcginnislaw.com
Joan O'Mara
13 000 vols; 150 curr per   35477

DLA Piper US LLP, Law Library, 6225 Smith Ave, **Baltimore**, MD 21209-3600
T: +1 410 5803010; Fax: +1 410 5803261; URL: www.dlapiper.com
Ronelle Manger
28 000 vols; 70 curr per   35478

Fidelity & Deposit Company of Maryland, Law Library, 300 St Paul Pl, **Baltimore**, MD 21202; P.O. Box 1227, Baltimore, MD 21203-1227
T: +1 410 5282584, 5390080; Fax: +1 410 6593226; URL: www.fidelityanddeposit.com
1922; T. J. Ormseth
Law of Maryland dating back to 1800
42 000 vols   35479

Gordon, Feinblatt, Rothman, Hoffberger & Hollander, Law Library, 233 E Redwood St, **Baltimore**, MD 21202
T: +1 410 5764255; Fax: +1 410 5764246; URL: www.gfrlaw.com
Andy Zimmerman
10 000 vols; 100 curr per   35480

Loyola-Notre Dame Library, Inc, 200 Winston Ave, **Baltimore**, MD 21212
T: +1 410 6176801; Fax: +1 410 6176896; E-mail: askemail@loyola.edu; URL: www.loyola.edu/library
1973; John W. McGinty
English Lit (Gerard Manley Hopkins Coll), Book Decoration (Henry A. Knott Fore-Edge), Painting Coll
463 000 vols; 989 curr per; 24 000 e-journals; 256 000 e-books; 360 000 microforms; 10 080 av-mat; 2 200 digital data carriers   35481

Miles & Stockbridge PC Library, Ten Light St, **Baltimore**, MD 21202
T: +1 410 3853671; Fax: +1 410 3853700
1932; Anna B. Cole
15 000 vols; 200 curr per; 20 maps; 2 000 microforms; 30 av-mat; 10 digital data carriers
libr loan; AALL, SLA   35482

Niles, Barton & Wilmer Law Library, 111 S Calvert St, Ste 1400, **Baltimore**, MD 21202
T: +1 410 7836300; Fax: +1 410 7836363; URL: www.niles-law.com
Thea Warner
10 000 vols   35483

Northrop Grumman Corp, Electronic Sensors & Systems Division Library, P.O. Box 17319 MS A-255, **Baltimore**, MD 21297-1319
T: +1 410 9932463, 7752; Fax: +1 410 9937675, 1484; E-mail: es_communications@mail.northgrum.com; URL: www.es.northropgrumman.com
Susan Baker
Electronics, Marketing
34 000 vols; 400 curr per   35484

Ober, Kaler, Grimes & Shriver Law Library, 120 E Baltimore St, Ste 800, **Baltimore**, MD 21202-1643
T: +1 410 2307181; Fax: +1 410 5470699; URL: www.ober.com
Ginger Gerton
8 000 vols; 200 curr per   35485

Saint Paul Fire & Marine, Law Library, 5801 Smith Ave, MC 32, **Baltimore**, MD 21209
T: +1 410 2055990; Fax: +1 410 2056390; E-mail: wgrady@capcon.net
Lori Myers
46 000 vols; 114 curr per   35486

Saul Ewing LLP, 500 E Pratt St, 9th Flr, **Baltimore**, MD 21202
T: +1 410 3328832; Fax: +1 410 3328862; URL: www.saul.com
Stacey Digan
Commercial law, Corporate law, Banking, Real estate, Securities
10 000 vols; 40 curr per   35487

Semmes, Bowen & Semmes Library, 250 W Pratt St, 16th Fl, **Baltimore**, MD 21201
T: +1 410 3853936; Fax: +1 410 5395223; E-mail: ksweeney@mail.semmes.com
Kathleen Sweeney
20 000 vols; 400 curr per; 8 digital data carriers   35488

Tydings & Rosenberg LLP, Law Library, 100 E Pratt St, 26th Flr, **Baltimore**, MD 21202
T: +1 410 7529804; Fax: +1 410 7275460
Jean Hessenauer
10 000 vols; 50 curr per   35489

Venable LLP Library, 750 E Pratt St, 9th Flr, **Baltimore**, MD 21202
T: +1 410 2447502; Fax: +1 410 2447742; E-mail: lib01@venable.com; URL: www.venable.com
1900; John S. Nixdorff
30 000 vols; 50 curr per; 2 000 microforms; 25 digital data carriers
ALA, AALL, SLA   35490

Whiteford, Taylor & Preston, Law Library, Seven St Paul St, Ste 1500, **Baltimore**, MD 21202
T: +1 410 3478700; Fax: +1 410 7527092; URL: www.wtplaw.com
Anne B. Garrett
8 000 vols; 125 curr per   35491

ConocoPhillips Library Network, 122 PLB, **Bartlesville**, OK 74004
T: +1 918 6613433; Fax: +1 918 6622171
1947; Annabeth Robin
40 000 vols; 240 curr per   35492

Albemarle Corporation, Information Services, 451 Florida St, Rm 1425, **Baton Rouge**, LA 70801; P.O. Box 14799, Baton Rouge, LA 70898-4799
T: +1 225 3887822; Fax: +1 225 3887686; E-mail: kaye_french@albemarle.com; URL: www.albemarle.com
Kaye French
Chemical engineering, Chemistry
18 000 vols; 420 curr per   35493

Huey P. Long Memorial Law Library, State Capitol, 900 N Third St, 14th Flr, **Baton Rouge**, LA 70802; P.O. Box 94183, Baton Rouge, LA 70804
T: +1 225 3422414; Fax: +1 225 3422725
Arthur E. McEnany
10 000 vols; 50 curr per   35494

Tektronix, Inc Library, 14150 SW Karl Brawn Dr, **Beaverton**, OR 97005; P.O. Box 500, MS 50-510, Beaverton, OR 97077-0001
T: +1 503 6275385; Fax: +1 503 6274853; E-mail: jean.fryer@tektronix.com
1959; Jean Fryer
Electronics, Optics, Computer science
27 500 vols; 50 curr per   35495

Mitre Corp Bedford Library, 202 Burlington Rd, MS K206, **Bedford**, MA 01730
T: +1 781 2717834 ext 8447; Fax: +1 781 2713593; URL: www.mitre.org
1959; Betsy F. Cogliano
Computer science, Electronics, Engineering, Local hist, Mathematics
58 000 vols; 605 curr per
libr loan   35496

Bayer Healthcare, Berkeley Library & Information Services, 800 Dwight Way, *Berkeley*, CA 94701; P.O. Box 1986, Berkeley, CA 94701-1986
T: +1 510 7057817; Fax: +1 510 7057819
1948; Beatrice Yuan
Pharmaceutical science
29 000 vols   35497

Foster Associates, Inc Library, 4550 Montgomery Ave, Ste 350N, *Bethesda*, MD 20814
T: +1 301 6647800; Fax: +1 301 6647810; E-mail: fainfo@foster-fa.com; URL: www.foster-fa.com
1956; Peter McArthur
Energy, Telecommunications
18 000 vols; 150 curr per   35498

Jack Faucett Associates Library, 4550 Montgomery Ave, Ste 300N, *Bethesda*, MD 20814
T: +1 301 9618800; Fax: +1 301 4693001
1963; Donald Hutson
Economics, Energy, Environmental studies, Investing, Transportation
15 000 vols; 20 curr per   35499

Bethlehem Steel Corp – Bernard D. Broeker Law Library, 1170 Eighth Ave, Martin Tower, Rm 2027, *Bethlehem*, PA 18016-7699
T: +1 610 6945002; Fax: +1 610 6941447; URL: www.bethsteel.com
1954; Ethel H Vary
19 000 vols; 100 curr per   35500

Bethlehem Steel Corp – Schwab Information Center, 1170 Eighth Ave, Rm 100, *Bethlehem*, PA 18016-7699
T: +1 610 6943325; Fax: +1 610 6943290; E-mail: sterlein@bethsteel.com; URL: www.bethsteel.com
1949; Marie F. Sterlein
15 100 vols; 200 curr per; 10 digital data carriers; Engineering, Steel industry, Business & management
libr loan   35501

Thomson & Thomson, DeForest Research, 8383 Wilshire Blvd, Ste 450, *Beverly Hills*, CA 90211
T: +1 310 2732900; Fax: +1 310 8883270; URL: www.thomson-thomson.com
1913
Film, Television
10 000 vols; 12 curr per   35502

Crowley, Haughey, Hanson, Toole & Dietrich Library, 490 N 31st St, Ste 500, *Billings*, MT 59101-1288; P.O. Box 2529, Billings, MT 59103-2529
T: +1 406 2523441; URL: www.crowleylaw.com
Margaret Webster
Law
32 000 vols   35503

HKM Engineering Inc Library, 222 N 32nd, Ste 700, *Billings*, MT 59101; P.O. Box 31318, Billings, MT 59107-1318
T: +1 406 6566399; Fax: +1 406 6566398
Irene Nelson
13 000 vols   35504

Alabama Power Co, Research Services, 600 N 18th St, *Birmingham*, AL 35203-2206; P.O. Box 2641, Birmingham, AL 35291-0277
T: +1 205 2574466; Fax: +1 205 2572075; E-mail: w2xaplib@southernco.com
1936; Sherie Mattox
Computer science, Business & Management, Engineering
10 000 vols; 50 curr per
libr loan   35505

Balch & Bingham Attorneys Library, 1901 Sixth Ave N, Ste 1500, *Birmingham*, AL 35203
T: +1 205 2518100; Fax: +1 205 2268798
Christina Tabereaux
State Law
35 000 vols; 200 curr per   35506

Bradley, Arant, Rose & White, Law Library, One Federal Pl, 1819 Fifth Ave N, *Birmingham*, AL 35203
T: +1 205 5218000; Fax: +1 205 5218800; URL: www.bradleyarant.com
Lori D. Martin
27 000 vols   35507

Burr & Forman Library, Southtrust Tower, Ste 3100, 420 20th St N, *Birmingham*, AL 35203
T: +1 205 2513000; Fax: +1 205 4585100
Tim Lawson
Labor, Corporate law
18 000 vols; 25 curr per   35508

Sirote & Permutt, PC, Law Library, 2311 Highland Ave S, *Birmingham*, AL 35205-2792; P.O. Box 55727, Birmingham, AL 35255-5727
T: +1 205 9305233; Fax: +1 205 9305101
1946; William Preston Peyton
10 000 vols; 65 curr per
AALL   35509

Lummus Technology Library, 1515 Broad St, *Bloomfield*, NJ 07003
T: +1 973 8932257; Fax: +1 973 8932119; E-mail: cltb.library@cbi.com
1930; Elizabeth A. Schefler
Chemical engineering, petroleum refining, petrochemicals
11 000 vols; 75 curr per; 36 000 microforms; 20 digital data carriers   35510

Dickinson Wright PLLC Library, 38525 Woodward Ave, Ste 2000, *Bloomfield Hills*, MI 48304-2970
T: +1 248 4337200; Fax: +1 248 4337274
Jan Bissett
Law, US law
20 000 vols; 644 curr per   35511

Plunkett & Cooney, Law Library, 38585 Woodward Ave, *Bloomfield Hills*, MI 48304
T: +1 248 9014080; Fax: +1 248 9014040
Barbara R. Morrow
10 000 vols   35512

Larkin, Hoffman, Daly & Lindgren, Law Library, 7900 Xerxes Ave S, Ste 1500, mail address: 1500 Wells Fargo, *Bloomington*, MN 55431
T: +1 952 8353800; Fax: +1 952 8963333; URL: www.larkinhoffman.com
Marilynn Hallen
15 000 vols; 125 curr per   35513

Elam & Burke PA, Law Library, P.O. Box 1539, *Boise*, ID 83701-1539
T: +1 208 3435454; Fax: +1 208 3845844; URL: www.elamburke.com
Rochelle Tramell
Aviation
10 000 vols; 70 curr per   35514

Hawley Troxell Ennis & Hawley, Law Library, 877 Main St, Ste 1000, *Boise*, ID 83702-1617
T: +1 208 3446000; Fax: +1 208 3423829; URL: www.hteh.com
Allison Terry
12 000 vols; 50 curr per   35515

Maxim Technologies Inc Library, 3380 Americana Terrace, Ste 201, *Boise*, ID 83706-2519; P.O. Box 7777, Boise, ID 83707-1777
T: +1 208 3891030; Fax: +1 208 3891183
Paul Spillers
Chemistry
20 000 vols; 10 curr per   35516

Bingham McCutchen LLP, Law Library, 150 Federal St, *Boston*, MA 02110
T: +1 617 9514911; Fax: +1 617 9518543; URL: www.bingham.com
Gina Lynch
30 000 vols; 150 curr per   35517

Brown, Rudnick, Berlack, Israels LLP, Research & Information Center, One Financial Pl, *Boston*, MA 02111
T: +1 617 8568213; Fax: +1 617 8568201; E-mail: research@brownrudnick.com; URL: www.brownrudnick.com
10 000 vols; 400 curr per
AALL, SLA   35518

Choate, Hall & Stewart Library, Exchange Pl, 53 State St, *Boston*, MA 02109
T: +1 617 2485202; Fax: +1 617 2484000
Mary E. Rogalski
15 000 vols   35519

Edwards Angell Palmer & Dodge LLP, Law Library, 111 Huntington Ave, *Boston*, MA 02199
T: +1 617 2390254; Fax: +1 617 2274420; URL: www.eapdlaw.com
Nuchine Nobari
35 000 vols; 425 curr per
libr loan   35520

Federal Reserve Bank of Boston, Research Library, 600 Atlantic Ave, *Boston*, MA 02210-2204; P.O. Box 55882, Boston, MA 02205-5882
T: +1 617 9733397; Fax: +1 617 9734221; E-mail: boston.library@bos.frb.org; URL: www.bos.frb.org
1921; Joyce Hannan
Economics, Finance; Federal Reserve System Mat
70 000 vols   35521

Foley & Hoag LLP Library, 155 Seaport Blvd, *Boston*, MA 02210
T: +1 617 8327070; Fax: +1 617 8327000
Jeannette Tracy
Corporate law, Labor, Environmental law
50 000 vols; 250 curr per   35522

Goodwin Procter, Law Library, Exchange Pl, 53 State St, *Boston*, MA 02109
T: +1 617 5706868; Fax: +1 617 5231231; URL: www.goodwinprocter.com
40 000 vols; 2 500 curr per   35523

Goulston & Storrs, PC, Library Services, 400 Atlantic Ave, *Boston*, MA 02110
T: +1 617 4821776; URL: www.goulstonstorrs.com
Robert DeFabrizio
10 000 vols; 200 curr per
ALA, SLA   35524

Hale & Dorr Library, 60 State St, *Boston*, MA 02109
T: +1 617 5265900; Fax: +1 617 5265000; E-mail: boston.library@haledorr.com; URL: www.haledorr.com
Lynn K. Oser
Law, Govt
21 000 vols   35525

Hemenway & Barnes, Law Library, 60 State St, *Boston*, MA 02109
T: +1 617 2277940; Fax: +1 617 2270781; E-mail: claut@hembar.com
1863; Chris Laut
Massachusetts Statutes back to 1694
10 000 vols; 100 curr per   35526

Hill & Barlow Library, One International Pl, *Boston*, MA 02110
T: +1 617 4283000; Fax: +1 617 4283500
Julie Snyder
Real estate
13 000 vols; 25 curr per   35527

Holland & Knight LLP, Law Library, 10 Saint James Ave, *Boston*, MA 02116-3803
T: +1 617 5232700; Fax: +1 617 5236850; E-mail: mkearney@hklaw.com
Maureen C. Kearney
69 100 vols; 55 curr per   35528

John Hancock Mutual Life Insurance, Law Library, P.O. Box 111, *Boston*, MA 02117
T: +1 617 5726000; Fax: +1 617 5721565
Nancy McHugh
35 400 vols; 125 curr per   35529

John Snow, Inc, JSI Research & Training Institute Library, 44 Farnsworth St, *Boston*, MA 02210-1211
T: +1 617 4829485; Fax: +1 617 4820617; E-mail: jsinfo@jsi.com; URL: www.jsi.com
John Carper
Public health
10 000 vols; 200 curr per   35530

K&L Gates LLP, Law Library, State Street Financial Center, One Lincoln St, *Boston*, MA 02111-2950
T: +1 617 9519160; Fax: +1 617 2613175; URL: www.klgates.com
Elizabeth Labedz
8 000 vols; 100 curr per   35531

Nen-Life Science Products Inc, Library, 549 Albany St, *Boston*, MA 02118
T: +1 617 4829595 ext 9605; Fax: +1 617 3509658
1963; Judge Johnson
Cancer, Nuclear medicine, Chemistry
15 500 vols; 350 curr per   35532

Nixon Peabody LLP, Law Library, 100 Summer St, *Boston*, MA 02110-1832
T: +1 617 3451000; Fax: +1 617 3451300; URL: www.nixonpeabody.com
Joanne Santino
8 000 vols; 150 curr per   35533

Nutter, McClennen & Fish, Law Library, One International Pl, *Boston*, MA 02110-2699
T: +1 617 4392492, 2483; Fax: +1 617 9739748; URL: www.nutter.com
Susan Cleary
Corporate law, Labor, Planning
15 100 vols; 200 curr per   35534

Peabody & Arnold LLP, Law Library, Federal Reserve Plaza, 600 Atlantic Ave, *Boston*, MA 02110
T: +1 617 2615051; Fax: +1 617 9512125; URL: www.peabodyarnold.com
Brian Treanor
12 000 vols   35535

Rackemann, Sawyer & Brewster Library, 160 Federal St, *Boston*, MA 02110-1700
T: +1 617 8972287; Fax: +1 617 5427437; URL: www.rackemann.com
Tonika Farrell
Real estate; Zoning
12 000 vols; 100 curr per   35536

Ropes & Gray LLP Library, One International Pl, *Boston*, MA 02110-2624
T: +1 617 9517855; Fax: +1 617 9517050
Cornelia Trubey
Law
40 000 vols; 300 curr per   35537

Sullivan & Worcester, LLP, Law Library, One Post Office Sq, *Boston*, MA 02109
T: +1 617 3382888; Fax: +1 617 3382880; URL: www.sandw.com
W. Leslie Peat
Corporate law
25 000 vols; 100 curr per
libr loan   35538

Testa, Hurwitz & Thibeault, Law Library, High Street Tower, 125 High St, *Boston*, MA 02110
T: +1 617 2487000; Fax: +1 617 2487100; E-mail: bitman@tht.com; URL: www.tht.com
Leslie Bitman
10 000 vols; 425 curr per   35539

B. F. Goodrich Co, Charles Cross Goodrich Knowledge Center, 9921 Brecksville Rd, *Brecksville*, OH 44141-3289
T: +1 216 4475154; Fax: +1 216 4476247
1914; Linda Lamoun Hash
Aerospace, Plastics, Polymer tech, Rubber
8 600 vols; 450 curr per   35540

ATK Launch Systems, Library M/S LIB, PO Box 707, mail address: *Brigham City*, UT 84302-0707
T: +1 435 8636819; Fax: +1 435 8636023; URL: www.atk.com
1968; Diane Nielson
Solid propellant rocket technology
100 000 vols; 50 curr per
libr loan; ULA, ALA   35541

Hodgson Russ LLP, Law Library, One M&T Plaza, Ste 2000, *Buffalo*, NY 14203
T: +1 716 8481282; Fax: +1 716 8490349; URL: www.hodgsonruss.com
Joan T. White
18 000 vols   35542

Phillips, Lytle, Hitchcock, Blaine & Huber Library, 3400 Marine Midland Ctr, *Buffalo*, NY 14203
T: +1 716 8475470, 5471; Fax: +1 716 8526100
Angela Patti
Law, Legislation
25 000 vols; 120 curr per   35543

Roswell Park Cancer Institute Corp, Dr Edwin A. Mirand Library, Elm & Carlton Sts, **Buffalo**, NY 14263
T: +1 716 8455966; Fax: +1 716 8458699; E-mail: ill.library@roswellpark.org; URL: roswellpark.org
1898; Dr. Nancy A. Cunningham
92 000 vols; 1 200 curr per; 1 140 diss/theses; 1 sound-rec; 1 digital data carriers
libr loan; MLA   35544

Saperston & Day PC, Law Library, 1100 M&T Ctr, Three Fountain Plaza, **Buffalo**, NY 14203
T: +1 716 8565400; Fax: +1 716 8560139; E-mail: ag296@freenet.buffalo.edu; URL: www.sapterstonday.com
J. Jay Deveau
15 000 vols; 75 curr per   35545

Veridian Technical Information Center, 4455 Genesee St, **Buffalo**, NY 14225-0400; P.O. Box 400, Buffalo, NY 14225
T: +1 716 6316833; Fax: +1 716 6314119
1946; Susan Doughtie
Aerodynamics, Aeronautics, Applied physics, Automobiles, Defense, Electronics, Competitive intelligence, Space science, Transportation; Govt Res (NTIS Technical Reports), microfiche, all except DOE
200 000 vols; 50 curr per   35546

Phillips State Prison, Law & General Library, 2989 W Rock Quarry Rd, **Buford**, GA 30519-4198
T: +1 770 9324500; Fax: +1 770 9324676
Diane Secrist
8 000 vols; 45 curr per   35547

MSE, Inc, Technical Applications Library, Mike Mansfield Advanced Technology Center, **Butte**, MT 59702; P.O. Box 4078, Butte, MT 59702-4078
T: +1 406 4947417; Fax: +1 406 4947230
1977; Karen Henningson
Magnetohydrodynamics, Environmental management, Waste cleanup, Remediation
10 000 vols; 90 curr per; 500 standards   35548

Camp, Dresser & Mckee, Herman G. Dresser Information Center, 50 Hampshire St, **Cambridge**, MA 02139
T: +1 617 4526778; Fax: +1 617 4528000; E-mail: infocenter@cdm.com
1963; Cohen B. Stacie
Environmental engineering, Wastewater treatment
28 000 vols; 52 curr per   35549

Charles Stark Draper Laboratory, Inc, Technical Information Center, 555 Technology Sq, Mail Sta 74, **Cambridge**, MA 02139
T: +1 617 2583555; Fax: +1 617 2581919; E-mail: library@draper.com; URL: www.draper.com
1973; Laurie Rotman
Electronics, Engineering, Mathematics, Physics; C S Draper Laboratory Arch
16 000 vols; 100 curr per; 1 500 diss/theses; 5 digital data carriers; 700 tech rpts, 20 000 archival mats   35550

Pratt & Whitney Rocketdyne, Inc, Library Services, 6633 Canoga Ave, **Canoga Park**, CA 91309; P.O. Box 7922, 055-NA29, Canoga Park, CA 91309-7922
T: +1 818 5862575; Fax: +1 818 5869150; E-mail: grpcpcti@west.boeing.com
1955; Haroldeane E. Snell
Fluid mechanics, Materials, Nuclear engineering, Physics, Thermodynamics; Historic Rocketdyne photos
18 000 vols; 27 curr per; 150 000 microforms; 5 digital data carriers   35551

St Lawrence Supreme Court, Law Library, 48 Court St, **Canton**, NY 13617
T: +1 315 3792279; Fax: +1 315 3792424
Tammy Lomaki
14 000 vols   35552

Timken Co, Research Library-Res-06, 1835 Dueber Ave SW, **Canton**, OH 44706-0930
T: +1 330 4712049; Fax: +1 330 4712282; E-mail: cromip@timken.com
1966; Patricia A Cromi
Engineering, Ferrous metallurgical res
30 000 vols; 280 curr per; 400 av-mat   35553

Solutia Inc, Library & Information Services, 3000 Old Chemstrand Rd, **Cantonment**, FL 32533; P.O. Box 97, Cantonment, FL 32560-0097
T: +1 850 9688249; Fax: +1 850 9688248
1954; M. Janice LaMotte
Textiles, Polymer chemistry
12 275 vols; 250 curr per   35554

Westinghouse TRV Solutions Co, WIPP Technical Library, 4021 National Parks Hwy, **Carlsbad**, NM 88220; P.O. Box 2078, Carlsbad, NM 88221-5608
T: +1 505 2347618; Fax: +1 505 2347076; E-mail: desail@wipp.carlsbad.nm.us
Lata Desai
Waste management
11 000 vols; 275 curr per   35555

Rockwell Collins, Information Center, 400 Collins Rd NE, Mail Sta 105-160, **Cedar Rapids**, IA 52498-1000
T: +1 319 2953070; Fax: +1 319 2958417; E-mail: jaleavit@rockwellcollins.com
1942; Judith A Leavitt
Electronics, Telecommunication, Business & management
8 000 vols; 300 curr per   35556

Equistar, Chemical Library, 8280 Sheldon Rd, **Channelview**, TX 77530
T: +1 281 4528148; Fax: +1 281 4528153; URL: www.lyondel.com
1956; Linda L. Lozano
Design Manuals Coll
15 000 vols; 36 curr per   35557

TASC, Inc, Technical Library, 4801 Stonecraft Blvd, **Chantilly**, VA 20151-3822
T: +1 703 6338300 ext 4654; Fax: +1 703 4497648; E-mail: medionne@tasc.com; URL: www.tasc.com
1966; Martha Dionne
10 000 vols; 200 curr per   35558

Jackson Kelly, Law Library, 1600 Laidley Tower, **Charleston**, WV 25322; P.O. Box 553, Charleston, WV 25322-0553
T: +1 304 3401260; Fax: +1 304 3401261
Kimberly Adkins
8 000 vols   35559

Robinson & McElwee PLLC, Law Library, 400 Fifth Third Ctr, 700 Virginia St E, Ste 400, **Charleston**, WV 25301; P.O. Box 1791, Charleston, WV 25326-1791
T: +1 304 3478325; Fax: +1 304 3449566; E-mail: mda@ramlaw.com; URL: www.ramlaw.com
Mary Aldridge
8 000 vols   35560

Westvaco Corp, Charleston Research Information Services Center, 5600 Virginia Ave, **Charleston**, SC 29406-3612; P.O. Box 118005, Charleston, SC 29423-8005
T: +1 843 7453735; Fax: +1 843 7453718; E-mail: agstout@westraco.com
1955
Chemistry, Engineering, Paper, Pulp
12 000 vols; 100 curr per; 90 000 company archs   35561

Duke Energy Corp – David Nabow Library, 526 S Church St, MC EC06H, **Charlotte**, NC 28202; P.O. Box 1006, Charlotte, NC 28201-1006
T: +1 704 3824095; Fax: +1 704 3827826; E-mail: corporate_library_services@duke-energy.com
1967; Audrey W Caldwell
Automation, Business, Engineering, Technology; Standards (ANSI, NEMA, IEEE, EPRI, INPO)
50 000 vols; 200 curr per; 22 000 microforms; 2 digital data carriers
libr loan   35562

Moore & Van Allen PLLC, Law Library, Bank of America Corporate Ctr, 100 N Tryon, Ste 4700, **Charlotte**, NC 28202
T: +1 704 3311000; Fax: +1 704 3395946; E-mail: librarians@mvalaw.com
Tamara Reno
11 000 vols; 200 curr per   35563

Miller & Martin PLLC, Law Library, Volunteer Bldg, Ste 1000, 832 Georgia Ave, **Chattanooga**, TN 37402-2289
T: +1 423 7566600; Fax: +1 423 7858480; URL: www.millermartin.com
Gail Sisson
Real estate
30 000 vols; 20 curr per   35564

Altheimer & Gray, Library Information Services, 10 S Wacker, Ste 4000, **Chicago**, IL 60606
T: +1 312 7154000; Fax: +1 312 7154800; E-mail: library@altheimer.com; URL: www.altheimer.com
Susan L. Cochard
Law
20 000 vols; 600 curr per   35565

American Hospital Association Resource Center, One N Franklin, **Chicago**, IL 60606-3421
T: +1 312 4223000; Fax: +1 312 4224700; E-mail: rc@aha.org; URL: www.aha.org
1920; Jeanette Harlowe
Ray E. Brown Management Coll, Ctr for Hospital & Healthcare Administration Hist
64 000 vols; 250 curr per; 38 000 microforms; 8 digital data carriers   35566

Arnstein & Lehr LLP Library, 120 S Riverside Plaza, Ste 1200, **Chicago**, IL 60606-3910
T: +1 312 8767170; Fax: +1 312 8760288; URL: www.arnstein.com
Frank Drake
Law
12 000 vols; 400 curr per   35567

Bank One of Chicago – Corporate Information Center, One Bank One Plaza, Mailcode IL1-0477, **Chicago**, IL 60670
T: +1 312 7323532; Fax: +1 312 7327895
1931; Barbara Allamian
16 175 vols; 1 000 curr per   35568

Chicago Tribune, Research Center, 435 N Michigan Ave, Rm 529, **Chicago**, IL 60611
T: +1 312 2224265; URL: www.chicagotribune.com
1920; Debra K. Bade
Illinois, Great Lakes
10 000 vols; 100 curr per; 4 000 microforms   35569

Clausen Miller, PC Library, Ten S LaSalle St, 16th Flr, **Chicago**, IL 60603-1098
T: +1 312 6067887; Fax: +1 312 6067777; URL: www.clausen.com
1936; Nancy L. Tuohy
30 000 vols; 300 curr per; 2 000 microforms; 200 av-mat; 20 sound-rec; 3 digital data carriers   35570

DLA Piper US LLP, Law Library, 203 N LaSalle St, Ste 1800, **Chicago**, IL 60601
T: +1 312 9845763; Fax: +1 312 2515727; E-mail: librarychicago-1@dlapiper.com
John M. Klasey
Real estate law; Illinois Municipalities Zoning Ordinances Coll
15 000 vols; 350 curr per   35571

Drinker, Biddle & Reath, Library & Information Services, 191 N Wacker Dr, Ste 3700, **Chicago**, IL 60606-1698
T: +1 312 5691861; Fax: +1 312 5693000; URL: www.drinkerbiddle.com
Susan Yesnick
8 000 vols   35572

Dykema Gossett PLLC, Information Center, Ten S Wacker Dr, Ste 2300, **Chicago**, IL 60606
T: +1 312 8761700; Fax: +1 312 8761155; URL: www.rookspitts.com
Sam Wertime
16 000 vols   35573

Encyclopaedia Britannica Inc, Editorial Library, 310 S Michigan Ave, **Chicago**, IL 60604
T: +1 312 3477403; Fax: +1 312 2942162; E-mail: suddin@eb.com
1933; Shantha Uddin
Geography, Statistics
65 000 vols; 3 500 curr per   35574

Ernst & Young Company, Center for Business Knowledge, 233 S Wacker Dr, **Chicago**, IL 60606
T: +1 312 8792000; Fax: +1 312 8794035
1930; Kathy Fabianich
Consulting, Taxation, Accounting
11 000 vols; 150 curr per   35575

Federal Reserve Bank of Chicago Library, 230 S LaSalle St, P.O. Box 834, **Chicago**, IL 60690-0834
T: +1 312 3225826; Fax: +1 312 3225091; E-mail: bernarda.flores@chi.frb.org; URL: www.chicagofed.org
1920; Susan Chenoweth
18 500 vols; 500 curr per
libr loan   35576

Ferguson Publishing Co, Editorial Library, 200 W Jackson Blvd, 7th Fl, **Chicago**, IL 60606
T: +1 312 6921000 ext 230; Fax: +1 312 6920109; E-mail: jmloft@www.com
1958; Joyce M. Lofton
10 000 vols; 135 curr per   35577

Foley & Lardner, Law Library, 321 N Clark St, **Chicago**, IL 60610
T: +1 312 8324500; Fax: +1 312 8324700
1988; Christina Wagner
Real estate, Commodities, Securities
10 000 vols   35578

Hinshaw & Culbertson Library, 222 N LaSalle, Ste 300, **Chicago**, IL 60601-1081
T: +1 312 7043000; Fax: +1 312 7043951; URL: www.hinshawculbertson.com
Jennifer Kiszka
15 000 vols; 75 curr per   35579

Holland & Knight LLP, Law Library, 131 S Dearborn, 30th Flr, **Chicago**, IL 60603
T: +1 312 5786616; Fax: +1 312 5786666; URL: www.hklaw.com
Carolyn Hosticka
10 000 vols; 200 curr per   35580

Jenner & Block Library, 330 N Wabash Ave, **Chicago**, IL 60611
T: +1 312 2229350; Fax: +1 312 5270484
1914; Mitchell Klaich
Law
35 000 vols
libr loan   35581

Johnson Publishing Co Library, 820 S Michigan Ave, **Chicago**, IL 60605-2191; P.O. Box 690, Chicago
T: +1 312 3229200; Fax: +1 312 3220951; URL: www.ebonymag.com
1949; Pamela J. Cash
Hist, African-American, Music
17 000 vols; 100 curr per   35582

Katten, Muchin, Rosenman LLP Library, 525 W Monroe, **Chicago**, IL 60661-3693
T: +1 312 5778170; Fax: +1 312 6124013
1974; Susan P. Siebers
40 000 vols; 950 curr per   35583

K&L Gates LLP, 70 W Madison, Ste 3100, **Chicago**, IL 60602-4207
T: +1 312 3721121; Fax: +1 312 8278000; URL: www.klgates.com
John Fox
30 000 vols   35584

Lasalle Bank Building, Law Library, 135 S LaSalle St, Ste 1411, **Chicago**, IL 60603
T: +1 312 5583135; Fax: +1 312 5581929
Sherman Lewis
16 000 vols; 12 curr per   35585

Latham & Watkins, Law Library, Sears Tower, 233 S Wacker Dr, Ste 5800, *Chicago*, IL 60606
T: +1 312 9932620; Fax: +1 312 9939767; URL: www.lw.com
Deborah Rusin
12 000 vols; 142 curr per   35586

Lexecon Inc Library, 332 S Michigan Ave, Ste 1300, *Chicago*, IL 60604-4306
T: +1 312 3220622; Fax: +1 312 3220218
Debbie Zimmermann
15 260 vols; 82 curr per   35587

Locke Lord Bissell & Liddell LLP, Law Library, 111 S Wacker Dr, *Chicago*, IL 60606
T: +1 312 4430646; Fax: +1 312 4430336
Sandra Gold
10 000 vols   35588

Mayer, Brown & Platt, Law Library, 190 S LaSalle St, *Chicago*, IL 60603
T: +1 312 7017922; Fax: +1 312 7017711; E-mail: gmunden@mayerbrown.com
Gail Munden
40 000 vols; 60 curr per   35589

McDermott, Will & Emery Law Library, 227 W Monroe St, 46th Fl, *Chicago*, IL 60606-5096
T: +1 312 3722000; Fax: +1 312 9847700; E-mail: fzeiger@mwe.com; URL: www.mwe.com
Jerry Trenholm
50 000 vols; 500 curr per
libr loan; AALL, OCLC   35590

Montgomery Watson Harza Library, 175 Jackson Blvd, *Chicago*, IL 60604
T: +1 312 8313397; Fax: +1 312 8313999
Lori Potrykus
11 500 vols; 300 curr per   35591

Murphy-Jahn Library, 35 E Wacker Dr, *Chicago*, IL 60601
T: +1 312 4277300; Fax: +1 312 3320274
Joseph A. Stypka
Architecture, Construction, Planning
20 000 vols; 420 curr per   35592

Neal, Gerber & Eisenberg LLP, Law Library, Two N La Salle St, Ste 2200, *Chicago*, IL 60602
T: +1 312 2695220; Fax: +1 312 5781793; E-mail: library@ngelaw.com
1986; Monice M. Kaczorowski
Securities
40 000 vols; 300 curr per   35593

Pedersen & Houpt Library, 161 N Clark St, Ste 3100, *Chicago*, IL 60601
T: +1 312 6416888; Fax: +1 312 6416895
Joanne Brady
Law
10 000 vols   35594

Pepsico Beverages & Foods, Information Center, 555 W Monroe St, *Chicago*, IL 60661; P.O. Box 049001, Chicago, IL 60604-9001
T: +1 312 8211000; Fax: +1 312 8211987
Duncan McKenzie
150 000 vols; 362 curr per   35595

Peterson & Ross Library, 200 E Randolph Dr, Ste 7300, *Chicago*, IL 60601-6969
T: +1 312 8611400 ext 4686; Fax: +1 312 8611053; E-mail: peteross@class.org; URL: www.petersonross.com
Xiannian Ye
American law
10 000 vols   35596

Playboy Enterprises, Inc, Editorial Research Library, 680 N Lake Shore Dr, *Chicago*, IL 60611
T: +1 312 7518000 ext 2260; Fax: +1 312 7512818
Mark Duran
Civil rights, films, music, sexuality, sports
10 000 vols; 75 curr per   35597

Schiff, Hardin LLP Library, 233 S Wacker Dr, Ste 6600, *Chicago*, IL 60606
T: +1 312 2585500; Fax: +1 312 2585600; URL: www.schiffhardin.com
Ruth Bridges
40 000 vols   35598

Seyfarth Shaw, 55 E Monroe St, Ste 4200, *Chicago*, IL 60603-5803
T: +1 312 3468000; Fax: +1 312 2698869
1945; Gabrielle Lewis
Labor law, Environmental law, Securities; Arbitration Awards & Legal Memoranda (Seyfarth Shaw Coll)
38 000 vols; 484 curr per
AALL, SLA   35599

Sidley Austin LLP Library, One S Dearborn St, *Chicago*, IL 60603
T: +1 312 8537475; Fax: +1 312 8537036
Allyson D. Withers
Law
35 000 vols   35600

Skadden, Arps, Slate, Meagher & Flom (Illinois) Library, 333 W Wacker Dr, *Chicago*, IL 60606
T: +1 312 4070941; Fax: +1 312 4070924; E-mail: amorris@skadden.com; URL: www.skadd-lib.com
Ann Morris
20 000 vols   35601

Sonnenschein, Nath & Rosenthal, Law Library, 8000 Sears Tower, 233 S Wacker Dr, Ste 7800, *Chicago*, IL 60606-6404
T: +1 312 8768001; Fax: +1 312 8767934; URL: www.sonnenschein.com
Nancy Henry
US law
15 000 vols; 200 curr per
libr loan; CALL, SLA, AALL, ALA   35602

Vedder, Price, Kaufman & Kammholz, Law Library, 222 N LaSalle, *Chicago*, IL 60601
T: +1 312 6097500; Fax: +1 312 6095005
Kenneth Halicki
Corporate law
22 000 vols   35603

Wildman, Harrold, Allen & Dixon LLP, 225 W Wacker St, Ste 3000, *Chicago*, IL 60606
T: +1 312 2012000; Fax: +1 312 2012555; E-mail: library@wildmanharrold.com; URL: www.wildmanharrold.com
Cristina Eckl
30 000 vols   35604

Winston & Strawn LLP Library, 35 W Wacker Dr, *Chicago*, IL 60601
T: +1 312 5585740; Fax: +1 312 5585700; URL: www.winston.com
Kathy Lefco
Law
30 000 vols; 750 curr per; 20 digital data carriers
libr loan   35605

World Book Publishing, Research Library, 233 N Michigan Ave, 20th Flr, *Chicago*, IL 60601
T: +1 312 7295581; Fax: +1 312 7295600
1920; Stephanie Kitchen
Company Arch
16 000 vols; 100 curr per   35606

Dinsmore & Shohl Library, 255 E Fifth St, 1900 Chemed Ctr, *Cincinnati*, OH 45202-3172
T: +1 513 9778486; Fax: +1 513 9778141; E-mail: library@dinslaw.com
Mary Jo Merkowitz
Law
24 000 vols   35607

Frost Brown Todd LLC, Law Library, 2200 PNC Ctr, 201 E Fifth St, *Cincinnati*, OH 45202
T: +1 513 6516800; Fax: +1 513 6516981; URL: www.frostbrowntodd.com
Tracie Tiegs
27 000 vols; 225 curr per; 20 000 microforms; 10 digital data carriers   35608

General Electric Aircraft Engine, AEG Technical Information Center, One Neumann Way, *Cincinnati*, OH 45215-6301
T: +1 513 2434582; Fax: +1 513 2437426
1950; John Tebo
36 000 vols; 130 curr per; 805 000 microforms; tech doc, tech papers & patents (250 000)   35609

Graydon, Head & Ritchey LLP, Law Library, 1900 Fifth Third Center, 511 Walnut St, *Cincinnati*, OH 45202; P.O. Box 6464, Cincinnati, OH 45202
T: +1 513 6216464; Fax: +1 513 6513836; URL: www.graydon.com
Katherine Michniuk Steen
17 000 vols; 35 curr per   35610

Taft, Stettinius & Hollister Library, 425 Walnut St, Ste 1800, *Cincinnati*, OH 45202-3957
T: +1 513 3812838; Fax: +1 513 3810205
Barbara J. Davis
30 000 vols   35611

Thompson, Hine LLP, Law Library, 312 Walnut St, 14th Flr, *Cincinnati*, OH 45202
T: +1 513 3526528; Fax: +1 513 2414771
Barbara W. Silbersack
8 000 vols; 100 curr per   35612

Baker & Hostetler LLP Library, 3200 National City Ctr, 1900 E Ninth St, *Cleveland*, OH 44114-3485
T: +1 216 8617101; Fax: +1 216 6960740; URL: www.bakerlaw.com
1916; Alvin M. Podboy
54 000 vols   35613

Calfee, Halter & Griswold LLP, Resource Library, 1400 McDonald Investment Ctr, 800 Superior Ave, Ste 1800, *Cleveland*, OH 44114-2688
T: +1 216 6228208; Fax: +1 216 2410816; URL: www.caffee.com
Corine R. Corpora
18 000 vols; 100 curr per   35614

Federal Reserve Bank of Cleveland, Research Library, 1455 E Sixth St, *Cleveland*, OH 44101; P.O. Box 5620, Cleveland, OH 44101-0620
T: +1 216 5792052; Fax: +1 216 5793172; E-mail: 4D.Library@clev.frb.org
1918; Lee D. Faulhaber
22 000 vols; 600 curr per
libr loan   35615

Jones Day, Law Library, 901 Lakeside Ave, *Cleveland*, OH 44114
T: +1 216 5863939; Fax: +1 216 5790212
Suzanne Young
Corp law, Taxation
40 000 vols; 230 curr per   35616

McDonald Hopkins, LPA, Law Library, 600 Superior Ave E, Ste 2100, *Cleveland*, OH 44114
T: +1 216 3485400; Fax: +1 216 3485474
Mike Melillo
10 000 vols; 100 curr per   35617

Thompson, Hine LLP, Law Library, 3900 Key Ctr, 127 Public Sq, *Cleveland*, OH 44114-1291
T: +1 216 5665651; Fax: +1 216 5668565
30 000 vols; 150 curr per   35618

TRW Inc, Law Library, 1900 Richmond Rd, *Cleveland*, OH 44124
T: +1 216 2917391; Fax: +1 216 2917070
1974; Brenda Cherney
21 000 vols   35619

Tucker Ellis & West LLP, Law Library, 1150 Huntington Bldg, 925 Euclid Ave, *Cleveland*, OH 44115-1475
T: +1 216 5925000; Fax: +1 216 5925009
Ellen Smith
20 000 vols   35620

Walter & Haverfield, Law Library, 1300 Terminal Tower, *Cleveland*, OH 44113
T: +1 216 7811212; Fax: +1 216 5750911; E-mail: lstevens@com.aol; URL: www.walterhav.com
1932; Leon Stevens
10 000 vols; 100 curr per   35621

Weston, Hurd, Fallon, Paisley & Howley, LLP, Law Library, 2500 Terminal Tower, 50 Public Sq, *Cleveland*, OH 44113-2241
T: +1 216 2416602 ext 3333; Fax: +1 216 6218369; URL: www.westonhurd.com
Sandra Suchan
Insurance Law, Products Liability Law
15 000 vols   35622

ITT Industries, AV ACD Technical Library, 100 Kingsland Rd, MS 8545, *Clifton*, NJ 07014
T: +1 973 2843810; Fax: +1 973 2844141; E-mail: rita.reisman@itt.com
Rita Reisman
Electronics
10 000 vols; 296 curr per
SLA   35623

LexisNexis Law Library, 555 Middle Creek Pkwy, *Colorado Springs*, CO 80921-3630
T: +1 719 4817548; Fax: +1 719 4817387
1873; Sue McClendon
West Reporter System, official state court rpts, official codes & laws, legal per
160 000 vols; 20 curr per   35624

W. R. Grace & Co, Grace Davison Information Center Library, 7500 Grace Dr, *Columbia*, MD 21044
T: +1 410 5314146; Fax: +1 410 5314757
1953; Theo Jones-Quartey
12 000 vols; 100 curr per; 100 sound-rec   35625

McNair Law Firm, PA, Law Library, Wilbur Smith Bldg, 1301 Gervais St, *Columbia*, SC 29201; P.O. Box 11390, Columbia, SC 29211
T: +1 803 7999800; Fax: +1 803 7999804; URL: www.mcnair.net
1983; David Morgan
25 000 vols; 200 curr per   35626

Nelson, Mullins, Riley & Scarborough, Law Library, 1320 Main St, Ste 1700, *Columbia*, SC 29201; P.O. Box 11070, Columbia, SC 29211-1070
T: +1 803 2559367; Fax: +1 803 2557500
1982; Monica Wilson
16 000 vols; 124 curr per   35627

Wilbur Smith Associates Corporate Library, 1301 Gervais St, *Columbia*, SC 29201-3326; P.O. Box 92, Columbia, SC 29202-0092
T: +1 803 2512055; Fax: +1 803 2512064; E-mail: library@wilbursmith.com
Betty Phillips
17 000 vols; 70 curr per   35628

American Electric Power Service Corp, Library & Reference Service, One Riverside Plaza, *Columbus*, OH 43215
T: +1 614 7161000; Fax: +1 614 7161828; E-mail: aeplibrary@aep.com
1950
Sporn Coll (Hist Company File)
40 000 vols; 800 curr per; 142 e-books   35629

Bailey Cavalieri LLC, Law Library, Ten W Broad St, Ste 2100, *Columbus*, OH 43215-3422
T: +1 614 2213155; Fax: +1 614 2210479; URL: www.baileycavalieri.com
Pamela Muller
15 000 vols; 45 curr per   35630

Baker & Hostetler Library, 65 E State St, Ste 2100, *Columbus*, OH 43215-4260
T: +1 614 2281541; Fax: +1 614 4622616
Judith P. Rodgers
Legal mat
15 000 vols; 95 curr per   35631

FTI – SEA, Inc Library, 7349 Worthington-Galena Rd, *Columbus*, OH 43085
T: +1 614 8884160; Fax: +1 614 8858014; E-mail: msens@seaohio.com
Michael J. Sens
Industrial safety, Standardization; Stapp Car Crash Conference Proceedings, 1963-2000
12 000 vols; 25 curr per   35632

Kegler, Brown, Hill & Ritter, Law Library, 65 E State St, Ste 1800, *Columbus*, OH 43215
T: +1 614 4625400; Fax: +1 614 4642634
Keith S. Knopf
15 000 vols; 25 curr per; 100 av-mat; 30 digital data carriers   35633

Porter, Wright, Morris & Arthur, LLP, Law Library, Huntington Ctr – 41 S High St, **Columbus**, OH 43215-6194
T: +1 614 2271927; Fax: +1 614 2272100; URL: www.porterwright.com
Susan M. Schaefgen
28 000 vols; 1 000 curr per
libr loan   35634

Squire, Sanders & Dempsey, Law Library, 1300 Huntington Ctr, 41 S High St, Ste 1300, **Columbus**, OH 43215
T: +1 614 3652700; Fax: +1 614 3652499
Patricia Christian
13 000 vols; 50 curr per   35635

Bank of America, Technology & Operations Library, 1755 Grant, CA4-703-01-23, **Concord**, CA 94520
T: +1 925 6751361; Fax: +1 925 6753411
1978; Larry R. White
Data processing, Programming, Telecommunications
20 000 vols
libr loan   35636

Corning Inc, Technical Information Center, Sullivan Park, **Corning**, NY 14831
T: +1 607 9749000; Fax: +1 607 9742406
1936
Ceramics, Chemistry, Glass, Physics
22 000 vols; 320 curr per   35637

Celanese Dennis F. Ripple Technical Information Center, 1901 Clarkwood Rd, **Corpus Christi**, TX 78409; P.O. Box 9077, Corpus Christi, TX 78469-9077
T: +1 361 2424223; Fax: +1 361 2424251; E-mail: raumfleet@celanese.com
1947; Ruth Umfleet
Chemical engineering, Chemistry
13 000 vols; 78 curr per
SLA, ARMA   35638

CH2M Hill, Engineering Information Center Library, 2300 NW Walnut Blvd, **Corvallis**, OR 97330-3596; P.O. Box 428, Corvallis, OR 97339-0428
T: +1 541 7524271; Fax: +1 541 7520276
1946; Shirley Drake
12 000 vols; 250 curr per   35639

Hewlett-Packard, Research Center, 1000 NE Circle Blvd, MS 422A, **Corvallis**, OR 97330
T: +1 541 7150206 ext 2535; Fax: +1 541 7154015; URL: www.lib.utc.edu
Sharon Williams
Computer tech, Electronics, Applications software, Info management
13 000 vols; 460 curr per
libr loan   35640

Rutan & Tucker Library, 611 Anton, Ste 1400, **Costa Mesa**, CA 92626
T: +1 714 6415100; Fax: +1 714 5469035; URL: www.rutan.com/libinfo
Arlen Bristol
30 000 vols; 150 curr per
libr loan   35641

Akin, Gump, Strauss, Hauer & Feld Library, 1700 Pacific Ave, Ste 4100, **Dallas**, TX 75201-4618
T: +1 214 9694628; Fax: +1 214 9694343; URL: www.akingump.com
Joan Hass
40 000 vols; 100 curr per   35642

Carrington, Coleman, Solman & Blumenthal, LLP, Law Library, 200 Crescent Ct, Ste 1500, **Dallas**, TX 75201
T: +1 214 8553530; Fax: +1 214 8551333; URL: www.ccsb.com
Sue H. Johnson
19 000 vols; 50 curr per   35643

Degolyer & MacNaughton Library, One Energy Sq, Ste 400, **Dallas**, TX 75206
T: +1 214 3686391; Fax: +1 214 3694061
1939; Deborah Buchel
Engineering, Geology, Energy minerals, Petroleum, Economy
18 000 vols; 100 curr per   35644

Federal Reserve Bank of Dallas Library, 2200 N Pearl, **Dallas**, TX 75201
T: +1 214 9226000; Fax: +1 214 9225222
1921; Barbara Riley
19 000 vols; 750 curr per   35645

Gardere & Wynne, Law Library, 1601 Elm St, Ste 3000, **Dallas**, TX 75201
T: +1 214 9994738; Fax: +1 214 9993738; E-mail: whike@gardere.com; URL: www.gardere.com
Kellie Whitaker
8 000 vols; 200 curr per   35646

Haynes & Boone LLP, Law Library, 901 Main St, Ste 3100, **Dallas**, TX 75202-3789
T: +1 214 6515711; URL: www.haynesboone.com
1970; David Bader
60 000 vols; 130 curr per   35647

Hughes & Luce, LLP Library, 1717 Main St, Ste 2800, **Dallas**, TX 75201
T: +1 214 9395510; Fax: +1 214 9395849; URL: www.hughesluce.com
Thomas R. Austin
Law
30 000 vols; 400 curr per   35648

Jackson Walker LLP, Law Library, 901 Main St, Ste 6000, **Dallas**, TX 75202
T: +1 214 9536038; Fax: +1 214 9535822; URL: www.jw.com
Ann H. Jeter
35 000 vols; 200 curr per   35649

Jenkens & Gilchrist, Law Library, 1445 Ross Ave, Ste 3700, **Dallas**, TX 75202
T: +1 214 8554500; Fax: +1 214 8554300; URL: www.jenkens.com
Jane Reynolds
64 000 vols; 2 600 curr per   35650

Jones Day, Law Library, 2727 N Harwood St, **Dallas**, TX 75201-1515
T: +1 214 9694823; Fax: +1 214 9695100
Terri L. DiCenzo
50 000 vols; 10 digital data carriers   35651

Locke Liddell & Sapp, Law Library, 2200 Ross Ave, Ste 2200, **Dallas**, TX 75201-6776
T: +1 214 7408344; Fax: +1 214 7408800; URL: www.lockeliddell.com
1891; Barbara Fullerton
35 000 vols; 600 curr per   35652

Mobil Exploration & Producing Technical Center, Technical Library, 13777 Midway Rd, P.O. Box 819047, **Dallas**, TX 75244
T: +1 972 8518142; Fax: +1 972 8517116
1943; Asima Ali
Geology (Rocky Mountain) Coll, micro
60 000 vols; 150 curr per
libr loan   35653

Raytheon Systems Co, North Building Library, 13510 N Central Expressway, **Dallas**, TX 75243; P.O. Box 660246, MS211, Dallas, TX 75266
T: +1 972 3445036; Fax: +1 972 3445042; E-mail: k-nordhaus@raytheon.com
1950; Kathy Nordhaus
Computer science, Defense, Electrical engineering, Mathematics; IEEE Per from their beginning date
10 280 vols; 187 curr per   35654

Strasburger & Price LLP Library, 901 Main St, Ste 4300, **Dallas**, TX 75202
T: +1 214 6514300; Fax: +1 214 6514330; URL: www.strasburger.com
1939; Donna Bostic
Law, Texas Law
40 000 vols; 1 500 curr per
libr loan   35655

Texas Instruments Inc – SC Group Library, 7800 Banner Dr, **Dallas**, TX 75251; P.O. Box 655303, MS-8240, Dallas, TX 75265-5303
T: +1 972 9972138; Fax: +1 972 9972139; E-mail: Kenyon@ti.com; URL: www.ti.com
1958; J. Michael Kenyon
Electronics, Semi-conductor tech
9 000 vols; 100 curr per; 11 digital data carriers
libr loan   35656

Thompson & Knight, Law Library, 1722 Routh St, Ste 1500, **Dallas**, TX 75201
T: +1 214 9691427; Fax: +1 214 9691751; E-mail: Libraryr@tklaw.com; URL: www.tklaw.com
1914; Mariann Sears
CLE Coll, Federal Register 1971-Current
40 000 vols   35657

Grolier, Inc, Grolier Library, 6 Parklawn Dr, **Danbury**, CT 06801
T: +1 203 7973848; Fax: +1 203 7973428; E-mail: cchang@grolier.com
1936; Chun C. Chang
14 000 vols; 150 curr per   35658

Union Carbide Corp, Business Intelligence Center, 39 Old Ridgebury Rd, **Danbury**, CT 06817-0001
T: +1 203 7945314; Fax: +1 203 7945055; E-mail: houghtma@dow.com
1935; Margaret Houghton-Capozzi, Roger Miller
25 000 vols; 150 curr per
libr loan   35659

Lane & Waterman, Law Library, 220 N Main St, Ste 600, **Davenport**, IA 52801
T: +1 563 3243246; Fax: +1 563 3241616; E-mail: jthurow@l-wlaw.com; URL: www.l-wlaw.com
James Thurow
10 000 vols; 30 curr per   35660

Thompson, Hine LLP, Law Library, 2000 Courthouse Plaza NE, Ten W Second St, **Dayton**, OH 45402; P.O. Box 8801, Dayton, OH 45401-8801
T: +1 937 3316023; Fax: +1 937 4436635; URL: www.thompsonhine.com
Janie Hack
35 000 vols; 216 curr per   35661

A. E. Staley Manufacturing Co, Technical Information Center, 2200 E Eldorado St, **Decatur**, IL 62525-1801
T: +1 217 4212543; Fax: +1 217 4212519; E-mail: rewallace@tlna.com
1958; Richard E. Wallace
Nutrition
8 500 vols; 50 curr per; 100 maps; 250 microforms; 200 av-mat; 5 digital data carriers; 75 000 US & foreign patents
libr loan   35662

DuPont Company, Jackson Laboratory Library, Chambers Works, **Deepwater**, NJ 08023
T: +1 856 5402851; Fax: +1 856 5402344; E-mail: klaircm@al.jlcl01.umc.dupont.com
1918; Carolyn Klair
Chemical engineering, Petroleum
10 000 vols; 120 curr per
libr loan   35663

Davis, Graham & Stubbs LLP, Law Library, 1550 17th St, Ste 500, **Denver**, CO 80202
T: +1 303 8927306; Fax: +1 303 8931379; URL: www.dgslaw.com
Beth Mescall
24 000 vols; 300 curr per   35664

Faegre & Benson, LLP, Law Library, 3200 Wells Fargo Ctr" **Denver**, CO 80203
T: +1 303 6073500; Fax: +1 303 6073600; URL: www.faegre.com
Kaye Waelde
10 000 vols   35665

Holland & Hart, Law Library, 555 17th St, Ste 3200, **Denver**, CO 80201-3950; P.O. Box 8749, Denver, CO 80201-8749
T: +1 303 2967872, 2958091; Fax: +1 303 2958261; E-mail: hkulikowski@hollandhart.com
1948; Holly Kulikowski
23 000 vols   35666

Holme Roberts & Owen LLC, Law Library, 1700 Lincoln, Ste 4100, **Denver**, CO 80203
T: +1 303 8617000 ext 1380; Fax: +1 303 8660200; E-mail: mestes@csn.net
Mark E. Estes
35 000 vols   35667

Sherman & Howard, Law Library, 633 17th St, Ste 2400, **Denver**, CO 80202
T: +1 303 2998041; Fax: +1 303 2980940; E-mail: lrose@sah.com
1892; Nicole Fairless, Tom Seward
20 000 vols; 350 curr per
AALL, SWALL   35668

Bradshaw, Fowler, Proctor & Fairgrave, Law Library, 801 Grand Ave, Ste 3700, **Des Moines**, IA 50309
T: +1 515 2434191; Fax: +1 515 2465808
Pam Cumins
10 000 vols; 50 curr per   35669

Davis, Brown, Koehn, Shors & Roberts PC Library, 666 Walnut, Ste 2500, **Des Moines**, IA 50309
T: +1 515 2882500; Fax: +1 515 2430654; E-mail: info@lawiowa; URL: www.lawiowa
Sharon Kern
11 000 vols   35670

Bodman, LLP, Law Library, Ford Field, 6th Flr, 1901 Saint Antoine St, **Detroit**, MI 48226
T: +1 313 2597777; Fax: +1 313 3937579; URL: www.bodmanllp.com
Jeanne Stuart
10 000 vols; 15 curr per   35671

Clark Hill PLC, Law Library, 500 Woodward Ave, Ste 3500, **Detroit**, MI 48226-3435
T: +1 313 9658300; Fax: +1 313 9658252
Kathleen A. Gamache
10 000 vols; 200 curr per   35672

The Detroit News, Inc, George B. Catlin Memorial Library, 615 W Lafayette Blvd, **Detroit**, MI 48226
T: +1 313 2226029; Fax: +1 313 2222059; URL: www.detnews.com
1918; Linda Culpepper
Local hist; Detroit News 1873-Present, micro, clips & photos
15 000 vols; 100 curr per   35673

Dickinson Wright PLLC Library, 500 Woodward Ave, Ste 4000, **Detroit**, MI 48226-3425
T: +1 313 2233500; Fax: +1 313 2233598; URL: www.dickinson-wright.com
1878; Mark A. Heinrich
25 000 vols; 4 000 curr per   35674

Dykema Gossett PLLC, Law Library, 400 Renaissance Ctr, 38th Flr, **Detroit**, MI 48243
T: +1 313 5686714; Fax: +1 313 5686735; URL: www.dykema.com
1929; Patricia L. Orr
Tax Legislative Mat
15 000 vols; 510 curr per   35675

Honigman Miller Schwartz & Cohn LLP, Law Library, 2290 First National Bldg, 660 Woodward Ave, **Detroit**, MI 48226-3583
T: +1 313 4657169; Fax: +1 313 4658000
Trish Webster
15 000 vols; 350 curr per   35676

Miller, Canfield, Paddock & Stone Library, 150 W Jefferson, Ste 2500, **Detroit**, MI 48226
T: +1 313 9636420; Fax: +1 313 4968452
Penelope Damore
Education, Labor, Real estate, Securities
50 000 vols; 500 curr per   35677

Pepper, Hamilton LLP, Law Library, 100 Renaissance Ctr, Ste 3600, **Detroit**, MI 48243
T: +1 313 2597110; Fax: +1 313 2597926
Elizabeth Stajniak
12 000 vols; 20 curr per   35678

Philo, Atkinson, Stephens, Wright, Whitaker, Philo & Kayrouz, Law Library, 2920 E Jefferson Ave, **Detroit**, MI 48207
T: +1 313 2597092; Fax: +1 313 2597092; E-mail: lawyers@philoatconcern.com
Harry Maurice Philo Jr
10 000 vols; 70 curr per   35679

Amerind Foundation, Inc, Fulton-Hayden Memorial Library, 2100 North Amerind Rd, **Dragoon**, AZ 85609; P.O. Box 400, Dragoon, AZ 85609-0400
T: +1 520 5863666; Fax: +1 520 5864679; E-mail: amerind@amerind.org; URL: amerind.org
1937; Celia Skeeles
Archeology, anthropology, ethnology, hist; El Archivo de Hidalgo del Parral, 1631-1821, micro; Facsimile Editions of Major Mesoamerican Codices; Records of the Colonial Period of New Spain (northern Mexico); Southwest Americana
28 000 vols; 150 curr per; 250 maps; 5 965 microforms; 525 unpublished mss, 12 000 slides and photos, 2 260 pamphlets and reprints   35680

Ashland, Inc, Library & Information Services, 5200 Blazer Pkwy, **Dublin**, OH 43017; P.O. Box 2219, Columbus, OH 43216-2219
T: +1 614 7903281; Fax: +1 614 7904269; E-mail: asolis@ashland.com; URL: www.ashchem.com
1970; Joyce Klayman
Chemistry, chemical and related industries, marketing; US chemical patents & microfilm (1960 to present)
12 000 vols; 200 curr per; 4 digital data carriers
libr loan; OCLC   35681

Thomson Reuters West, Library Services, 610 Opperman Dr, **Eagan**, MN 55123
T: +1 651 8482760; Fax: +1 651 8482627; E-mail: eagan.library@thomsonreuters.com
Cynthia Schriber
Law
300 000 vols; 500 curr per   35682

ArcelorMittal Gloval Research & Development – East Chicago, Research Library, 3001 E Columbus Dr, **East Chicago**, IN 46312
T: +1 219 3996120; Fax: +1 219 3996562
1954; Barbara Minne Banek
12 000 vols; 35 curr per   35683

Levinson/Axelrod Library, Levinson Plaza, Two Lincoln Hwy, **Edison**, NJ 08818; P.O. Box 2095, Edison, NJ 08818-2095
T: +1 732 4942727; Fax: +1 732 4942712; E-mail: levinson@njlawyers.com; URL: www.njlawyers.com
Marilyn Campanella
10 000 vols; 23 curr per   35684

Revlon Research Center Library, 2121 Rte 27, **Edison**, NJ 08818
T: +1 723 2871400; Fax: +1 723 2487650
1955
Cosmetics, Dermatology, Pharmacology, Toxicology
11 000 vols; 75 curr per   35685

Kemp, Smith PC, Law Library, 221 N Kansas, Ste 1700, **El Paso**, TX 79901
T: +1 915 5465328; Fax: +1 915 5465360; E-mail: jrey@kempsmith.com; URL: www.kempsmith.com
1866; Becky McKenzie
16 000 vols; 200 curr per   35686

Mounce & Green, Meyers, Safi & Galatzan, Law Library, 100 N Stanton, Ste 1700, **El Paso**, TX 79901
T: +1 915 5322000; Fax: +1 915 5411597; URL: mgmsg.com
Sylvia T. Contreras
15 000 vols; 15 curr per   35687

Scott, Hulse, Marshall, Feuille, Finger & Thurmond, Law Library, 201 E Main, 11th Flr, **El Paso**, TX 79901; P.O. Box 99123, El Paso, TX 79999-9123
T: +1 915 5332493; Fax: +1 915 5468333; URL: scotthulse.com
16 000 vols; 57 curr per   35688

Aerospace Corp, Charles C. Lauritsen Library, 2350 E El Segundo Blvd, **El Segundo**, CA 90245-4691; P.O. Box 80966, Mail Station M1/199, Los Angeles, CA 90009-0966
T: +1 310 3366093; Fax: +1 310 3361467
1960; Patricia W. Green
Aerospace Corp Authors Coll
100 000 vols; 700 curr per; 500 maps; 27 000 microforms; 150 sound-rec; 10 digital data carriers; 166 000 tech rpts bd
libr loan   35689

Boeing Satellite Systems, El Segundo Library, Bldg S12-W348, 2020 E Imperial Hwy, **El Segundo**, CA 90245; P.O. Box 92919, Los Angeles, CA 90009-2919
T: +1 310 3646002; Fax: +1 310 3646424; URL: www.boeing.com/satellite
1963; Blair Hinz
Electronics, Radar, Physics, Mathematics, Computer science
30 000 vols; 200 curr per   35690

Raytheon Co, Space & Airborne Systems Technical Library, 2000 E El Segundo Blvd, **El Segundo**, CA 90245; Bldg E1/E117, P.O. Box 902, El Segundo, CA 90245-0902
T: +1 310 6472000; Fax: +1 310 6471312
1950; Kevin Liu
20 000 vols; 125 curr per   35691

Rodale Inc, Library & Information Services, 33 E Minor St, **Emmaus**, PA 18098
T: +1 610 9678729; Fax: +1 610 9677708
1976; Lynn Donches
Agriculture, Fitness, Gardening, Health, Nutrition, Sports
42 000 vols; 1 500 curr per   35692

IBM Corp, Endicott Site Technology Library, 1701 North St, **Endicott**, NY 13760
T: +1 607 7571487; Fax: +1 607 7571489; E-mail: emdeemie@us.ibm.com; URL: isiwww.endicott.ibm
1933; Eileen Deemie
Computers, Electronics, Engineering, Business management
12 500 vols; 275 curr per   35693

Unilever Bestfoods Information Center, 800 Sylvan Ave, **Englewood Cliffs**, NJ 07632-3201
T: +1 201 8947568; Fax: +1 201 8718265
1942; Karla Cicciari
Food industry, Food science, Nutrition; Tea
10 000 vols; 250 curr per   35694

Lord Corporation, Corporate Library, 2000 W Grandview Blvd, **Erie**, PA 16514; P.O. Box 10040, Erie, PA 16514-0040
T: +1 814 8680924; Fax: +1 814 8666323
1943; Dianne Howard
Adhesives, Chemistry, Coatings, Ploymeric mat, Rubber
21 000 vols; 562 curr per   35695

IBM Corp, Technical Library, Bldg 967B, 1000 River St, **Essex Junction**, VT 05452
T: +1 802 7696500; Fax: +1 802 7696501; E-mail: btvlib@us.ibm.com
1965; Karen Lynch
Computer science, Programming, Semiconductors, Solid state electronics, Solid state physics
9 000 vols; 30 curr per   35696

Exxonmobil Corp, Office of General Counsel-Legal Information Center, 3225 Gallows Rd, Rm 4C110, **Fairfax**, VA 22037
T: +1 703 8462821; Fax: +1 703 8461476
Sandra L. Dokachev
50 000 vols; 200 curr per
AALL   35697

Odin, Feldman & Pittleman Library, 9302 Lee Hwy, Ste 1100, **Fairfax**, VA 22031
T: +1 703 2182100; Fax: +1 703 2182160
Andrew Martin
11 000 vols; 30 curr per   35698

Eltech Systems Corp Library, 625 East St, **Fairport Harbor**, OH 44077
T: +1 440 3574000, 4002; Fax: +1 440 3574077; E-mail: daf@eltechsystems.com
Debbie Fultz
15 000 vols; 45 curr per   35699

NACCO Materials Handling Group, Inc, CBDC Technical Support Resource Center, 4000 NE Blue Lake Rd, **Fairview**, OR 97025
T: +1 503 7216234; Fax: +1 503 7211364; URL: www.nmhg.com
1961; Melissa Hardenbergh
Eng design for mat handling equipment, Construction safety, Domestic & foreign societies, Industrial safety
20 000 vols; 60 curr per   35700

Reed, Smith, Hazel & Thomas, Law Library, 3110 Fairview Park Dr, Ste 1400, P.O. Box 12001, **Falls Church**, VA 22042
T: +1 703 6414332; Fax: +1 703 6414340; URL: www.reedsmith.com
Anne Salzberg
10 000 vols; 250 curr per   35701

Levy & Droney, Law Library, 74 Batterson Park Rd, **Farmington**, CT 06032
T: +1 860 6763000; Fax: +1 860 6763200; URL: www.ldlaw.com
Patty Mackiewicz
15 000 vols   35702

Gale Group, Corporate Research Library, 27500 Drake Rd, **Farmington Hills**, MI 48331-3535
T: +1 248 6994253 ext 1223; Fax: +1 248 6998054; E-mail: tory.cariappa@galegroup.com
1956; Mrs Tory Cariappa
Biography, Business, Lit; Biographical Dictionaries & Encyclopedias
15 000 vols; 500 curr per   35703

Weyerhaeuser Library & Information Resources, 32901 Weyerhaeuser Way S, **Federal Way**, WA 98001; Library & Info Resources – WTC 1LIB, P.O. Box 9777, Federal Way, WA 98063-9777
T: +1 253 9243030; Fax: +1 253 9243612
Susan Smith
Forestry, Wood, Paper, Pulp, Printing; Annual Rpts, Trade Publs
8 000 vols; 300 curr per   35704

Maritz Inc Resource & Media Center, 1400 S Highway Dr, **Fenton**, MO 63099
T: +1 636 8272353; Fax: +1 636 8274248; E-mail: janet.finelli@maritz.com
1969; J. Finelli
Business, Graphic art, Incentives, Marketing, Organizational behavior, Quality, Training, Travel; Arch Coll
10 000 vols; 250 curr per   35705

Marathon Oil Co, Law Library, 539 S Main St, Rm 811-M, **Findlay**, OH 45840-3295
T: +1 419 4213392; Fax: +1 419 4213578; URL: mweb.fdy.mol.com; www.marathon.com
1930; Kim Haley
Energy law
8 000 vols   35706

Delphi Automotive System, Delphi Engineering Library, 1601 N Averill Ave, **Flint**, MI 48556
T: +1 810 2578183; Fax: +1 810 2572001
1925; Gloria Browning
Automotive engineering, Electronic engineering, Electronics, Mat Culture
8 000 vols; 10 curr per   35707

AT&T, Shannon Laboratory Library-Information Center, 180 Park Ave, Bldg 103, **Florham Park**, NJ 07932-0971
T: +1 973 3608160; E-mail: library@research.att.com
1997; Laurinda Jean Alcorn
Computer science, Mathematics, Networks, Signal processing, Speech, Statistics, Telephony
10 000 vols; 250 curr per   35708

Day Pitney LLP, Information Resource Center, 200 Campus Dr, **Florham Park**, NJ 07932; P.O. Box 1945, Morristown, NJ 07962-1945
T: +1 973 9666300; Fax: +1 973 9661015; URL: www.daypitney.com
1902; Julie L. Von Schrader
New Jersey law
35 000 vols; 200 curr per   35709

Drinker Biddle & Reath, Law Library, 500 Campus Dr, **Florham Park**, NJ 07932
T: +1 973 3601100; Fax: +1 973 3609831; URL: www.dbr.com
Kate Zuhusky
20 000 vols; 40 curr per   35710

Ruden, McClosky, Smith, Schuster & Russell, Law Library, 200 E Broward Blvd, P.O. Box 1900, **Fort Lauderdale**, FL 33302
T: +1 954 7646660; Fax: +1 954 3334053; URL: www.ruden.com
Sheryll I. Rappaport
10 000 vols; 25 curr per   35711

Kelly, Hart & Hallman, Law Library, Chase Bldg, 201 Main St, Ste 2500, **Fort Worth**, TX 76102
T: +1 817 3322500; Fax: +1 817 8789280; URL: www.khh.com
Amy E. Yawn
18 000 vols; 45 curr per   35712

Becton, Dickinson & Co, Information Resource Center, One Becton Dr, **Franklin Lakes**, NJ 07417-1884
T: +1 201 8477230; Fax: +1 201 8475377
Faina Menzul
Medicine, biomedicine, law
8 000 vols; 300 curr per   35713

Dow Chemical Library, Business Intelligence Center, Business Intelligence Ctr, B-1210, 2301 Brazosport Blvd, **Freeport**, TX 77541
T: +1 979 2384854
1944; Jeff Hart
12 000 vols; 47 curr per
libr loan   35714

Baker, Manock & Jensen Library, 5260 N Palm Ave, Ste 421, **Fresno**, CA 93704
T: +1 559 4325400
Lori Cain
20 000 vols   35715

McCormick, Barstow, Sheppard, Wayte & Carruth, Law Library, Five River Park Pl E, **Fresno**, CA 93720
T: +1 559 4332190; Fax: +1 559 4332485; URL: www.mccormickbarstow.com/
Lee Cabibi
12 000 vols   35716

Malcolm Pirnie Inc, Virtual Library, 20250 Century Blvd, **Germantown**, MD 20874
Fax: +1 301 5400903; E-mail: lgrossinger@pirnie.com
Lenore Grossinger
Engineering; Environmental Protection Agency
10 000 vols; 150 curr per   35717

The Futures Group Library, 80 Glastonbury Blvd, **Glastonbury**, CT 06033-4409
T: +1 860 6333501; Fax: +1 860 6573918; E-mail: k.willson@tfgi.com
1971; Katherine H. Willson
10 000 vols; 150 curr per; 25 digital data carriers
APLIC-I, SLA   35718

Knapp, Peterson & Clarke, Law Library, 500 N Brand Blvd, **Glendale**, CA 91203
T: +1 818 5475050; Fax: +1 818 5475329
Frances Brenland
14 000 vols; 28 curr per   35719

Walt Disney Imagineering, Information Research Center, 1401 Flower St, **Glendale**, CA 91201; P.O. Box 25020, Glendale, CA 91221-5020
T: +1 818 5446594; Fax: +1 818 5447845
1962; Aileen Kutaka
50 000 vols; 500 curr per   35720

Kraft Foods, Inc, Technical Information Group, 801 Waukegan Rd, **Glenview**, IL 60025-4312
T: +1 847 6463753; Fax: +1 847 6465150
1938; David Jourdan
Food science, Nutrition
13 000 vols; 300 curr per
libr loan   35721

Occidental Chemical Corporation, Technical Information Center, 2801 Long Rd, **Grand Island**, NY 14072
T: +1 716 7738531; Fax: +1 716 7738487
1916; A. Ben Wagner
Beilstein; Chemical Abstracts
25 000 vols; 550 curr per; 50 digital data carriers   35722

Mika Meyers Beckett & Jones, PLC, Law Library, 900 Monroe Ave NW, **Grand Rapids**, MI 49503
T: +1 616 6328000; Fax: +1 616 6328002; URL: www.mmbjlaw.com
Lana K. Ahumada
Energy, Oil, Gas & Water Law Coll
10 000 vols; 95 curr per; 10 e-books   35723

Varnum, Riddering, Schmidt & Howlett, Law Library, P.O. Box 352, **Grand Rapids**, MI 49501-0352
T: +1 616 3366000; Fax: +1 616 3367000
Marla Major
30 000 vols; 66 curr per   35724

Warner, Norcross & Judd, LLP Library, 900 Fifth Third Ctr, 111 Lyon St NW, **Grand Rapids**, MI 49503-2487
T: +1 616 7522236; Fax: +1 616 7522236; URL: www.wnj.com
Mary Lou Wilker
15 000 vols   35725

Owens-Corning Corp, Knowledge Resource Services, 2790 Columbus Rd, Rte 16, **Granville**, OH 43023-1200
T: +1 740 3215000; Fax: +1 740 3217255
Nancy Lemon
10 000 vols   35726

Smith Moore, LLP, Law Library, 300 N Greene St, **Greensboro**, NC 27401; P.O. Box 21927, Greensboro, NC 27420-1927
T: +1 336 3785272; Fax: +1 336 4337566
Anne Washburn
15 000 vols; 280 curr per   35727

Nelson, Mullins, Riley & Scarborough, Law Library, 104 S Main St, Ste 900, **Greenville**, SC 29601; P.O. Box 10084, Greenville, SC 29603-0084
T: +1 864 2502300; Fax: +1 864 2322925; URL: www.nmrs.com
Linda Gray
14 000 vols; 20 curr per   35728

Science Applications International Corporation, Foreign Systems Research Center, 6021 S Syracuse Way, Ste 300, **Greenwood Village**, CO 80111-4732
T: +1 303 7736900; Fax: +1 303 7703297
1979; Tom Banks
Int relations, Technology
14 000 vols; 175 curr per   35729

General Dynamics Corp, Electric Boat Company Library, Eastern Point Rd, Dept 455, **Groton**, CT 06340
T: +1 860 4333481; Fax: +1 860 4330647
1955; Theresa Morales
Oceanography
12 000 vols; 450 curr per; 25 000 tech rpts   35730

Archer & Greiner Library, One Centennial Sq, **Haddonfield**, NJ 08033-0968
T: +1 856 7952121; Fax: +1 856 7950574
Elizabeth Olson
12 000 vols   35731

Review & Herald Publishing Association, Editorial Library, 55 W Oak Ridge Dr, **Hagerstown**, MD 21740
T: +1 301 3934141; Fax: +1 301 3934055; E-mail: library@rhpa.org
1903; James Cavil
Theology, Church hist; Early Seventh-day Adventist Publs, Millerite Coll, Rare Bk Coll
40 000 vols; 115 curr per; 400 music scores; 4 000 microforms; 50 sound-rec; 2 000 pamphlets, 150 video-recordings
libr loan   35732

Public Service Electric & Gas Company, Nuclear Information Resource Center, Nuclear Admin Bldg, End of Alloway Creek Neck Rd, P.O. Box 236, PSE&G Nuclear IRC N02, **Hancocks Bridge**, NJ 08038
T: +1 856 3391135; Fax: +1 856 3391136; E-mail: virginia.swichel@pseg.com
1983; Virginia L Swichel
20 000 vols; 80 curr per   35733

McNess, Wallace & Nurick LLC, Information Center, 100 Pine St, **Harrisburg**, PA 17108; P.O. Box 1166, Harrisburg, PA 17108-1166
T: +1 717 2328000; Fax: +1 717 2375300; URL: www.mwn.com
Margaret J. Ross
15 000 vols; 200 curr per   35734

Tyco Electronics, Technology Information Center, PO Box 3608, **Harrisburg**, PA 17105-3608
T: +1 717 5640100; URL: www.tycoelectronics.com
1956
12 000 vols; 10 000 curr per   35735

Aetna Inc, Law Library, 151 Farmington Ave-RC4A, **Hartford**, CT 06156
T: +1 860 2738183; Fax: +1 860 2738340; URL: www.aetna.com
1975; Frances Bartelli, Susan Simonds
Corporate law, Health law
15 000 vols; 50 curr per   35736

Day, Berry & Howard Library, City Pl I, 185 Asylum St, **Hartford**, CT 06103-3499
T: +1 860 2750100; Fax: +1 860 2750343; URL: www.dbh.com
Sara Zagorski
Law
25 000 vols; 100 curr per   35737

Hartford Steam Boiler Inspection & Insurance Co., Virtual Information Center, One State St, 9th Fl, **Hartford**, CT 06102-5024
T: +1 860 7225486; Fax: +1 860 7225530; E-mail: peter_moon@hsb.com; URL: www.hsb.com
1988; Peter Moon
Insurance, property/casualty, engineering; ASME Boiler & Pressure Vessel Code
11 000 vols; 40 curr per; 5 diss/theses; 55 govt docs; 300 av-mat; 8 digital data carriers
libr loan; SLA   35738

Murtha, Cullina, Richter & Pinney Library, City Pl, **Hartford**, CT 06103
T: +1 860 2406092; Fax: +1 860 2406150
Judith Vanotta
20 000 vols; 300 curr per   35739

Robinson & Cole LLP Library, 280 Trumbull St, **Hartford**, CT 06103-3597
T: +1 860 2758200; Fax: +1 860 2758299
Nancy Marcove
31 000 vols; 222 curr per   35740

Titanium Metals Corporation of America (Laboratory), Henderson Technical Library, P.O. Box 2128, **Henderson**, NV 89009
T: +1 702 5642544 ext 396; Fax: +1 702 5649038; E-mail: htl@timet.com
Lynn Mooso
20 000 vols; 75 curr per   35741

Bristol-Myers Products, Strategic Information & Analysis Center, 1350 Liberty Ave, **Hillside**, NJ 07205
T: +1 908 8516053; Fax: +1 908 8516073
1946; Ann C. Swist
Pharmaceutical science, Pharmacology, Medicine, Microbiology
9 000 vols; 225 curr per   35742

Cades, Schutte, Fleming & Wright, Law Library, 1000 Bishop St, 14th Flr, **Honolulu**, HI 96813-4212; P.O. Box 939, Honolulu, HI 96808-0939
T: +1 808 5219200; Fax: +1 808 5405078; E-mail: cades@cades.com; URL: www.cades.com
1922; Debra Anne Oandasan
Corporate law, Securities, Real estate, Medicine; Ecology (Hawaiian Water Rights); Hawaii Legislative Rpts
15 000 vols; 215 curr per   35743

Carlsmith Ball LLP Library, ASB Tower, Ste 2200, 1001 Bishop St, **Honolulu**, HI 96813
T: +1 808 5232500; Fax: +1 808 5230842; URL: www.carlsmith.com
Grace Yamada
20 000 vols; 150 curr per   35744

IBM Corp, East Fishkill Site Library, 2070 Rte 52, **Hopewell Junction**, NY 12533
T: +1 845 8943198; Fax: +1 845 8926399
1964; Maris Kristapsons
Computer science, Electronics, Programming, Semiconductors, Solid state physics, Polymer chemistry, Business
20 000 vols; 300 curr per   35745

Baker & Botts LLP, Law Library, One Shell Plaza, 910 Louisiana St, **Houston**, TX 77002
T: +1 713 2291643; Fax: +1 713 2291522; URL: www.bakerbotts.com
1872; Suzanne Estep
Corporate Law Coll, Securities Coll, Tax Coll, Utilities Coll
80 000 vols; 20 digital data carriers
35746

BP, Library Information Center, 501 W Lake Park Blvd, 15-171, **Houston**, TX 77079; P.O. Box 3092, Houston, TX 77253
T: +1 281 3663387; Fax: +1 281 3663117
1972; Marva Coward
Petroleum, Engineering, Geology; Society Publs, Special Company Rpts, State Geological Survey Papers, USGS Publs
40 000 vols; 100 curr per
libr loan   35747

Chamberlain, Hrdlicka, White, Williams & Martin, Law Library, 1200 Smith St, Ste 1400, **Houston**, TX 77002
T: +1 713 6582547; Fax: +1 713 6582553; E-mail: firm@chamberlainlaw.com
Susan Earley
20 000 vols; 90 curr per   35748

Chevron Global Library Houston, 3901 Briarpark Dr, **Houston**, TX 77042
T: +1 713 9546007; Fax: +1 713 9546907; E-mail: libhou@chevron.com
1974; Nan Dubbelde
Petrochemical business, Petroleum engineering; Society of Petroleum Engineers, preprints
50 000 vols; 150 curr per   35749

Chevron Law Library, 1301 McKinney St, Rm 2810, **Houston**, TX 77010; P.O. Box 4553, Houston, TX 77210
T: +1 713 7543330; Fax: +1 713 7542288; E-mail: fari@chevron.com
1933; Frederick A. Riemann
55 000 vols   35750

Exxon Exploration Co – Exploration Library, Rm 174, P.O. Box 4778, **Houston**, TX 77210-4778
T: +1 281 6547188; Fax: +1 281 4237915
1971; Nancy Ramirez
Petroleum tech, Geology, Geophysics
8 312 vols; 64 curr per   35751

ExxonMobil Corp, Law Library, 800 Bell, Rm 1786, CORP-EMB-1786, **Houston**, TX 77002
T: +1 713 6564383; Fax: +1 713 6566770
Paula E. Howe
50 000 vols   35752

Fugro, Inc, Corporate Library, 6100 Hillcroft, **Houston**, TX 77081; P.O. Box 740010, Houston, TX 77274-0010
T: +1 713 3695500; Fax: +1 713 3695570
1966; Pat Farnell
Geotechnics (Company Rpts & Proceedings of Geotech Conferences); Related Publs; Special Tech Files
12 000 vols; 75 curr per; 1 000 microforms; 30 000 tech rpts, 550 slides
35753

Fulbright & Jaworski LLP, Law Library, 1301 McKinney St, **Houston**, TX 77010
T: +1 713 6515219; Fax: +1 713 6515246
1919; Jane D. Holland
55 000 vols   35754

Gardere, Wynne & Sewell Library, 1000 Louisiana, Ste 3400, **Houston**, TX 77002
T: +1 713 2765500; Fax: +1 713 2766736; E-mail: fabtr@gardere.com; URL: www.gardere.com
Trisha Fabugais
Law
10 000 vols; 50 curr per   35755

Halliburton Energy Services, Houston Technical Library, 3000 N Sam Houston Pkwy E, **Houston**, TX 77032; P.O. Box 60087, Houston, TX 77205
T: +1 281 8714544; Fax: +1 281 8714575
1979; Connie S. Bihon
Engineering, Electronics, Chemistry; Standards, Military Specifications, Society of Petroleum Engineers Coll, US Patents Coll
11 000 vols; 105 curr per   35756

Jackson Walker Law Library, 1401 McKinney, Ste 1900, **Houston**, TX 77010
T: +1 713 7524479; Fax: +1 713 7524221
Caren Zentner Luckie
15 000 vols; 70 curr per   35757

Occidental Oil & Gas Corp Library, 5 Greenway Pl, Ste 1600, **Houston**, TX 77046
T: +1 713 2157667; Fax: +1 713 2157528; E-mail: mitra_kia@oxy.com
1953; Mitra Kia
Exploration & production of petroleum
30 000 vols; 700 curr per   35758

Pennzoil Exploration & Production Co, Pepco Library, 700 Milam, P.O. Box 2967, **Houston**, TX 77002-2967
T: +1 713 5464000; Fax: +1 713 5468930; E-mail: peggy.peterson@pzlqs.com
1981; Peggy Peterson
10 000 vols; 65 curr per
SLA
Libr is not open to the public   35759

Shell Oil Company – Law Library, P.O. Box 2463, **Houston**, TX 77252-2463
T: +1 713 2413514; Fax: +1 713 2414722
1975; C. T. Cote
30 000 vols
AALL   35760

Shell Oil Company – Petro-Chemical Knowledge Center, 3333 Hwy 6 S, **Houston**, TX 77082-3199; P.O. Box 1378, Houston, TX 77251-1380
T: +1 281 5447510; Fax: +1 281 5448121; E-mail: kwok@shellus.com
1975; Elsie T. Kwok
Engineering
90 000 vols   35761

Shell Oil Company – Services Integration Group-EP Library Houston, 3737 Bellaire Blvd, **Houston**, TX 77025
T: +1 713 2457288; Fax: +1 713 2457108; E-mail: libreq@shell.com
1947; Francis K. Brown
30 000 vols; 250 curr per   35762

Shell Oil Company – Tax Library, P.O. Box 2463, **Houston**, TX 77252-2463
T: +1 713 2412155; Fax: +1 713 2417029; URL: www.shell.com
Lito Llamas
10 000 vols   35763

TCB Library, 5757 Woodway, Ste 101 W, **Houston**, TX 77057-1599
T: +1 713 2672777; Fax: +1 713 2672924; E-mail: library@tcb.aecom.com; URL: www.tcb.aecom.com
1968; Renee Eller
Civil engineering; Environmental Protection Agency Rpts Coll, micro
20 000 vols; 200 curr per   35764

Texaco, Upstream Technology Library, 3901 Briarpark, **Houston**, TX 77042
T: +1 713 9546007; Fax: +1 713 9546907; E-mail: libr@texaco.com
1961; Meena Baichan
Petroleum engineering, Geology, Geophysics; In-House Rpts, Central Files, Correspondence
40 000 vols; 300 curr per   35765

Vinson & Elkins, Law Library, 3055 First City Tower, 1001 Fannin, **Houston**, TX 77002-6760
T: +1 713 7582678; Fax: +1 713 6155211; URL: www.velaw.com
1917; Susan Yancey
145 000 vols   35766

Weil, Gotshal & Manges LLP, Law Library, 700 Louisiana St, Ste 1600, **Houston**, TX 77002
T: +1 713 5465131; Fax: +1 713 2249511; URL: www.weil.com
1985; Elizabeth Black Berry
8 000 vols   35767

Huddleston Bolen, LLP, Law Library, 611 Third Ave, **Huntington**, WV 25721-1308; P.O. Box 2185, Huntington, WV 25722-2185
T: +1 304 5296181; Fax: +1 304 5224312; URL: www.huddlestonbolen.com
Lauren Aldridge
35 000 vols   35768

Special Metals Corp, Technology Information Services, 3200 Riverside Dr, **Huntington**, WV 25705-1771
T: +1 304 5265433; Fax: +1 304 5265973; E-mail: cback@smcwv.com; URL: www.SpecialMetals.com
1964; Connie Back
8 000 vols; 100 curr per
libr loan; SLA   35769

Teledyne Brown Engineering Co, Technical Library, Cummings Research Park, 300 Sparkman Dr, *Huntsville*, AL 35807
T: +1 256 7261000; Fax: +1 256 7262747
1962; Judy Corvelle
10 000 vols; 300 curr per   35770

Battelle Energy Alliance, LLC, INL Technical Library, 1776 Science Center Dr, MS 2300, *Idaho Falls*, ID 83415-2300; mail address: 1765 N Yellowstone Hwy, University Pl, MS 2300, Idaho Falls, ID 83415-2300
T: +1 208 5261185; Fax: +1 208 5260211; E-mail: lib@inl.gov; URL: www.inl.gov/library
1951; Catherine Plowman
Chemistry, Electronics, Engineering, Geosciences, Mathematics, Nuclear science, Physics; AEC, ERDA & DOE Rpts
48 000 vols; 250 curr per; 2 000 e-books
libr loan   35771

Ferro Corp Library, Tech Center Library, 7500 E Pleasant Valley Rd, *Independence*, OH 44131-5592
T: +1 216 6416607, 6925; Fax: +1 216 7506953; E-mail: library@ferro.com
Deborah K Oberlander
Ceramics, Coatings, Plastics, Polymer science, Porcelain enamels, Refractories, Specialty chemistry
18 000 vols; 225 curr per
libr loan   35772

Baker & Daniels Library, Research Services, 300 N Meridian St, Ste 2700, *Indianapolis*, IN 46204
T: +1 317 2370300; Fax: +1 317 2371000
1952; Constance Matts
50 000 vols; 324 curr per; 30 digital data carriers
libr loan; AALL, SLA   35773

Bingham McHale LLP, Law Library, 2700 Market Tower, Ten W Market St, *Indianapolis*, IN 46204-4900
T: +1 317 9685354; Fax: +1 317 2369907
1917; Nikki Lynne Schofield
12 000 vols; 50 curr per   35774

Bose McKinney & Evans LLP, Law Library, 135 N Pennsylvania St, Ste 2700, *Indianapolis*, IN 46204
T: +1 317 6845166; Fax: +1 317 2230166; URL: www.boselaw.com
Cheryl Lynn Niemeier
Civil rights, Labor, Securities
27 000 vols; 150 curr per   35775

Eli Lilly & Co, KM Library & Information Services, Lilly Corp Ctr, Drop Code 0737, *Indianapolis*, IN 46285
T: +1 317 4330936, 2774225; Fax: +1 317 2764418; E-mail: pos@lilly.com
1890; Paula O. Schmidt
Biology, Chemistry, Medicine, Toxicology, Economics; Drug Encyclopedias Coll, Foreign Pharmacopeias Coll, Drug Product Info
10 000 vols; 1 650 curr per   35776

Ice Miller LLP, Law Library, One American Sq, Ste 2900, *Indianapolis*, IN 46282-0200
T: +1 317 2362414; Fax: +1 317 5924207; URL: www.icemiller.com
1923; Melanie A. Kelley
30 000 vols; 71 curr per; 3 000 microforms; 35 sound-rec
AALL, SLA   35777

Krieg DeVault LLP Library, One Indiana Sq, Ste 2800, *Indianapolis*, IN 46204-2017
T: +1 317 2386396; Fax: +1 317 6361507; URL: www.kriegdevault.com
Ann Levy
23 000 vols; 36 curr per
libr loan   35778

Locke, Reynolds, Boyd & Weisell Library, 201 N Illinois St, Ste 1000, 1000 Capital Center S, *Indianapolis*, IN 46204-4210
T: +1 317 2373945; Fax: +1 317 2373900; E-mail: lockereynolds@iquest.net; URL: www.locke.com
1913; Nancy Thoms
14 500 vols; 180 curr per
libr loan   35779

Rolls-Royce, Library Information Services, Mail Code S5, 2001 S Tibbs Ave, *Indianapolis*, IN 46241; Mail Code S5, P.O. Box 420, Indianapolis, IN 46206-0420
T: +1 317 2304751; Fax: +1 317 2308901
1941; Gabriel Hysong
Aerospace, Gas turbines, Metallurgy; Aeronautical Research Council Rpts, NACA Rpts, Allison Archs
12 000 vols; 350 curr per   35780

ExxonMobil Corp, Information Center, 5959 Las Colinas Blvd, *Irving*, TX 75039-2298
T: +1 972 4441321; Fax: +1 972 4441337; E-mail: libraries.corphq@exxonmobil.com
Kristin Sandefur
Petroleum industry, Economy
9 000 vols; 350 curr per   35781

Service Cost Information Center, 600 Hidden Ridge, *Irving*, TX 75038
T: +1 972 7185549, 7185263; Fax: +1 972 7182399; E-mail: charlotte.wixxclark@verizon.net
1981; Charlotte Wixx-Clark
Marketing, Business, Social sciences, Telecommunications
10 000 vols; 100 curr per   35782

Engelhard Corp, Technical Information Center, 25 Middlesex-Essex Turnpike, mail address: 101 Wood Ave S, *Iselin*, NJ 08830
T: +1 732 2055271; Fax: +1 732 2056900; E-mail: maurica.fedors@engelhard.com
Maurica Fedors
Nonmetallic Minerals, Raolin, & Clays; Precious Metals
40 000 vols   35783

Consumers Energy, Legal Library, One Energy Plaza, *Jackson*, MI 49201
T: +1 517 7881088; Fax: +1 517 7881682
1955; Betsy S. Domschot
32 000 vols; 25 curr per   35784

Phelps Dunbar, LLP, Law Library, 111 E Capitol St, Ste 600, *Jackson*, MS 39201-2122; P.O. Box 23066, Jackson, MS 39225-3060
T: +1 601 3522300; Fax: +1 601 3609777; URL: phelpsdunbar.com
Amanda Watson
Labor
10 000 vols; 45 curr per   35785

Watkins, Ludlam, Winter & Stennis, Law Library, 633 N State St, *Jackson*, MS 39202; P.O. Box 427, Jackson, MS 39205-0427
T: +1 601 9494792; Fax: +1 601 9494804; URL: www.watkinsludlam.com
Joe Xu
14 000 vols; 52 curr per   35786

CSX Transportation, Inc, Law Library, 500 Water St, J-150, *Jacksonville*, FL 32202
T: +1 904 3591258; Fax: +1 904 3591248
Ron Allen
10 000 vols; 23 curr per   35787

GRA Inc Library, 115 West Ave, Ste 201, *Jenkintown*, PA 19046
T: +1 215 8847500; Fax: +1 215 8841385; E-mail: dotf@gra-inc.com; URL: www.gra-inc.com
1975; Dorothy K. Finn
Economic & regulatory analysis, Transportation; US Airline Rpts
10 000 vols; 125 curr per   35788

Upjohn Company – Pharmacia at Upjohn Research Library, 301 Henrietta St, *Kalamazoo*, MI 49001-0199
T: +1 616 8330914; Fax: +1 616 8338603
1941; Fred Einspahr
24 000 vols; 700 curr per
libr loan   35789

Blackwell, Sanders, Peper, Martin LLP, Law Library, 2300 Main St, Ste 1100, *Kansas City*, MO 64108
T: +1 816 9838791; Fax: +1 816 9838080; E-mail: philkemeyer@blackwellsanders.com
Paula G. Hilkemeyer
10 000 vols; 250 curr per   35790

Burns & McDonnell Engineering Co, Central Library, 9400 Ward Pkwy, *Kansas City*, MO 64114
T: +1 816 8223550; Fax: +1 816 8223409; E-mail: library@burnsmcd.com; URL: www.burnsmcd.com
Gail Kammer
Mechanical engineering; Standards, Manufacturers' Cats
20 000 vols; 200 curr per   35791

Federal Reserve Bank of Kansas City, Research Library, One Memorial Dr, *Kansas City*, MO 64198
T: +1 816 8812970; Fax: +1 816 8812807; E-mail: research.library@kc.frb.org
Deng M. Pan
Economics, Finance, Statistics, Agriculture
16 000 vols; 20 curr per   35792

Hallmark Cards, Inc, Creative Resource Library, 2501 McGee, No 146, *Kansas City*, MO 64108
T: +1 816 2747470; Fax: +1 816 5452239; URL: www.hallmark.com
1930; Mark Spencer
Design, Fine arts
22 000 vols; 180 curr per   35793

Kansas City Star Library, 1729 Grand Blvd, *Kansas City*, MO 64108
T: +1 816 2344406; Fax: +1 816 2344817; E-mail: ddonovan@kcstar.com; URL: www.kcstar.com
Derek Donovan
Star/Times Newspaper Reference Coll, micro
6 000 vols; 40 curr per; 30 000 000 newspaper clips, 800 000 photos
SLA   35794

Lathrop & Gage LC Library, 2345 Grand Blvd, Ste 2800, *Kansas City*, MO 64108
T: +1 816 4605668; Fax: +1 816 2922001; E-mail: frankjc@lathropgage.com; URL: www.lathropgage.com
Jamie Frank
30 000 vols; 150 curr per   35795

Polsinelli Shughart, Law Library, 120 W 12th St, Ste 1800, *Kansas City*, MO 64105
T: +1 816 4213355; Fax: +1 816 3740509; URL: www.polsinelli.com
Avis C. Bates
12 000 vols; 37 curr per   35796

Polsinelli Shughart PC, Law Library, 700 W 47th St, Ste 1000, *Kansas City*, MO 64112
T: +1 816 7531000; Fax: +1 816 7531536; E-mail: informationcenter@pswslaw.com; URL: www.polsinelli.com
Michelle Rogers
15 000 vols; 150 curr per   35797

Shook, Hardy & Bacon, Law Library, 2555 Grand Blvd, 3rd Flr, *Kansas City*, MO 64108-2613
T: +1 816 4746550; Fax: +1 816 4215547
Lori Weiss
40 000 vols; 900 curr per   35798

Sprint Corporation, Corporate Research Center, 2330 Shawnee Mission Pkwy, *Kansas City*, KS 66205
T: +1 913 6248500; Fax: +1 913 6242286
Kathy Mobley
Engineering, Telecommunications, Advertising
35 000 vols; 500 curr per   35799

Stinson, Morrison, Hecker Library, 1201 Walnut St, No 2500, *Kansas City*, MO 64106-2149
T: +1 816 6912600; Fax: +1 816 6913495; URL: www.stinson.com
Dale Magariel
Law
21 000 vols; 50 curr per   35800

Schering-Plough, Library Information Center, 2015 Galloping Hill Rd, *Kenilworth*, NJ 07033-1300
T: +1 908 7407390; Fax: +1 908 7407015
1940; Allison Warzala
Medicine, Chemistry, Patents, Biology; Scholar-Proprietary Company Product Database
10 000 vols; 1 200 curr per   35801

Longwood Gardens Library, 409 Conservatory Rd, *Kennett Square*, PA 19348-1805; P.O. Box 501, Kennett Square, PA 19348-0501
T: +1 610 3881000; Fax: +1 610 3882078; E-mail: library@longwoodgardens.org
1960; Venice Bayrd
Horticulture and Botany (Curtis' Botanical Magazine, Volume One, 1787 – present)
25 000 vols; 350 curr per
libr loan; ALA, SLA, CBHL   35802

Arkema Inc, Information Resources, 900 First Ave, *King of Prussia*, PA 19406
T: +1 610 8786777; Fax: +1 610 8786270; E-mail: arkema.kopr-ir-help@arkemagroup.com; URL: www.arkema-inc.com
Susan Hunsicker
Ceramics, polymers; Organotin Index
25 000 vols; 300 curr per   35803

Atofina Chemicals Inc, Information Resources, 900 First Ave, *King of Prussia*, PA 19406
T: +1 610 8786760; Fax: +1 610 8786270; E-mail: 6072133@mcimail.com; URL: www.atofinachemicals.com
1916; Susan Hunsicker
Organotin Index
25 000 vols; 300 curr per   35804

GlaxoSmithKline Pharmaceuticals, Research & Development Library, UW2322, 709 Swedeland Rd, *King of Prussia*, PA 19406-2799; P.O. Box 1539, King of Prussia, PA 19406-0939
T: +1 610 2706400; Fax: +1 610 2704127
1947; Robert Guerrero
18 000 vols; 800 curr per   35805

Eastman Chemical Co, Library & Information Services, P.O. Box 1972, *Kingsport*, TN 37662-5150
T: +1 423 2293518; Fax: +1 423 2296114; URL: www.eastman.com
Gail Preslar
10 000 vols   35806

Hourigan, Kluger & Quinn, Law Library, 600 Third Ave, *Kingston*, PA 18704-5815
T: +1 570 2873000; Fax: +1 570 2878005; E-mail: hkq@hkqpc.com; URL: www.hkqpc.com
Michael J. Reilly
9 000 vols; 50 curr per   35807

ITT Industries – Advanced Engineering & Sciences Division, DTRIAC, 1680 Texas St SE, *Kirtland AFB*, NM 87117-5669
T: +1 505 8469448; Fax: +1 505 8469454
Connie Salus
Tech Rpts & Test Data Files on the Subject of Nuclear Weapons Effects
400 000 vols   35808

Tennessee Valley Authority – Legal Research Center, 400 W Summit Hill Dr, *Knoxville*, TN 37922
T: +1 865 6326613; Fax: +1 865 6326718
1935; Deborah Cherry
26 000 vols; 13 curr per   35809

Tennessee Valley Authority – Research Library-Knoxville, WT CC – K, 400 W Summit Hill Dr, *Knoxville*, TN 37902
T: +1 865 6323464; Fax: +1 865 6324475; E-mail: corplibknox@tva.gov
1933; Nancy Proctor
Electric Power, Engineering, Water resources; TVA Hist
17 000 vols; 12 curr per   35810

Copley Press Inc, James S. Copley Library, 1134 Kline St, *La Jolla*, CA 92038; P.O. Box 1530, La Jolla, CA 92038
T: +1 858 7298040; Fax: +1 858 7298051
1966; Carol Beales
American revolution, California; Samuel Langhorne Clemens, American Lit, Robinson Jeffers, John and Jessie Benton Fremont, Fine Press Bks
16 000 vols; 10 000 mss
ALA   35811

Onebane Law Firm APC, Law Library, 1200 Camellia Blvd, Ste 300, **Lafayette**, LA 70508
T: +1 337 2372660; Fax: +1 337 2661232; URL: www.onebane.com
Shelly LeBlanc
13 000 vols; 60 curr per   35812

Greenebaum, Doll & McDonald, Law Library, 300 W Vine St, Ste 1100, **Lexington**, KY 40507-1622
T: +1 859 2884717; Fax: +1 859 2552742; E-mail: dlf@gdm.com; URL: www.greenebaum.com; www.gdm.com
Lynn Fogle
10 000 vols; 150 curr per   35813

Stoll, Keenon & Park, Law Library, 300 W Vine St, Ste 2100, **Lexington**, KY 40507-1801
T: +1 859 2313000; Fax: +1 859 2531093
Jeffrey L. Frey
10 000 vols; 150 curr per   35814

University of Kentucky Libraries, Lexmark Library, Dept 990, Bldg 005-1, 740 New Circle Rd NW, **Lexington**, KY 40551
T: +1 859 2323783; Fax: +1 859 2325728; E-mail: ilibrary@lexmark.com; URL: lexmark.kyvl.org
1959; Alex Grigg
Business & management, Chemistry, Electronic engineering, Mechanical engineering, Metallurgy, Physics; Programming Trade Lit, Patents, Products Manuals
11 000 vols; 50 curr per   35815

Hewitt Associates Library, 100 Half Day Rd, **Lincolnshire**, IL 60069
T: +1 847 2955000; Fax: +1 847 7715805
1946; Patricia Ludwig Kuhl
Human resources, Benefits (labor), Compensation (labor)
17 000 vols; 1 550 curr per; 8 000 Proxy statements
libr loan   35816

Infineum USALP, Infineum Information Center, 1900 E Linden Ave, **Linden**, NJ 07036-1111; P.O. Box 536, Linden, NJ 07036-0536
T: +1 908 4742351; Fax: +1 908 4742020; URL: www.infineum.com
1920; Julie Kale
Patents
8 000 vols; 70 curr per   35817

Mitchell, Williams, Selig, Gates & Woodyard, Law Library, 320 W Capitol Ave, Ste 1800, **Little Rock**, AR 72201
T: +1 501 6888800; Fax: +1 501 6888807; E-mail: cbruhn@mwsgw.com
Catherine A. Bruhn
15 000 vols; 450 curr per   35818

Mackay & Co Library, One Imperial Pl, **Lombard**, IL 60148
T: +1 630 9166110; Fax: +1 708 9164661; URL: www.mackayco.com
Gloria Manata
100 000 vols; 30 curr per   35819

Keesal, Young & Logan, Law Library, 400 Oceangate, **Long Beach**, CA 90801
T: +1 562 4362000; Fax: +1 562 4367416; URL: www.kyl.com
Marilyn R. Wills
Employment, Securities
15 000 vols
SCALL, AALL   35820

SCS Engineers Library, 3900 Kilroy Airport Way, Ste 100, **Long Beach**, CA 90806-6816
T: +1 562 4269544; Fax: +1 562 4270805; URL: www.scsengineers.com
1970; Loran Bures
55 000 vols; 120 curr per; 10 000 microforms
SLA   35821

Consolidated Edison Company of New York, Corporate Library, 43-82 Vernon Blvd, **Long Island City**, NY 11101-6912
T: +1 718 4726054; Fax: +1 718 3491076; E-mail: tlclibrary@coned.com
1906; Peter Dietrich
10 000 vols; 25 curr per; 2 300 microforms; 500 av-mat; 150 sound-rec; 40 digital data carriers   35822

Metropolitan Life Insurance Co, Knowledge Center, 27-01 Queens Plaza N, **Long Island City**, NY 11101
T: +1 212 5783112; Fax: +1 212 7437614; URL: www.metlife.com
1910; Gregory Vailey
15 000 vols; 45 curr per   35823

Alschuler, Grossman, Stein & Kahan, Law Library, 2049 Century Park E, 39th Fl, **Los Angeles**, CA 90067
T: +1 310 2771226; Fax: +1 310 5526077; E-mail: dgrigst@agsk.com
Denise Julie Grigst
10 000 vols   35824

Arter & Hadden, Library, 725 S Figueroa St, Ste 3400, **Los Angeles**, CA 90017
T: +1 213 4303240; Fax: +1 213 6179255; E-mail: adelgado@arterhadden.com
Anna Delgado
10 000 vols; 40 curr per   35825

Baker & Hostetler, Law Library, 333 S Grand Ave, **Los Angeles**, CA 90071
T: +1 213 6242400; Fax: +1 213 9751740; E-mail: sannand@bakerlaw.com; URL: www.bakerlaw.com
Stewart Annand
30 000 vols   35826

Bank of America – Southern California Headquarters, Law Library, Legal Dept 24017, 555 S Flower St, **Los Angeles**, CA 90071
T: +1 213 2283148; Fax: +1 213 2284968; E-mail: elizabeth.s.corbarrubias@bankamerica.com
Elizabeth Cobarrubias
15 000 vols; 15 curr per   35827

Boeing Satellite System S24 Library, Bldg S24, M/S D538, 2020 E Imperial Hwy, P.O. Box 92919, **Los Angeles**, CA 90009-2919
T: +1 310 3646002; Fax: +1 310 3646424; E-mail: bsslibrary@boeing.com
1963; Patricia Adorno
Electronics, Physics, Mathematics, Chemistry, Business & management
20 000 vols; 300 curr per   35828

Brobeck, Phleger & Harrison Library, 550 S Hope St, Ste 2100, **Los Angeles**, CA 90071
T: +1 213 7453406; Fax: +1 213 7453345; E-mail: jmcmahon@brobeck.com
Jane McMahon
Corporate law; Bankruptcy
15 000 vols; 60 curr per   35829

Buchalter Nemer, Law Library, 1000 Wilshire Blvd, Ste 1500, **Los Angeles**, CA 90017
T: +1 213 8910700; Fax: +1 213 8960400; URL: www.buchalter.com
Michelle Kuczma
20 000 vols   35830

Cox, Castle & Nicholson LLP Library, 2049 Century Park E, 28th Flr, **Los Angeles**, CA 90067
T: +1 310 2774222; Fax: +1 310 2777889; URL: www.coxcastle.com
Janet Kasabian
17 000 vols; 500 curr per   35831

Economics Research Associates Library, 10990 Wilshire Blvd, Ste 1500, **Los Angeles**, CA 90024
T: +1 310 4779585; Fax: +1 310 4781950; URL: www.econres.com
1968; Ruth Behling
Recreation
17 000 vols; 150 curr per   35832

Foley & Lardner LLP, Los Angeles Library, 555 S Flower St, Ste 3500, **Los Angeles**, CA 90071-2411
T: +1 213 9724657; Fax: +1 213 4860065
Stefanie Frame
Intellectual property
8 000 vols; 142 curr per   35833

Fulbright & Jaworski LLP, Law Library, 555 S Flower St, 41st Flr, **Los Angeles**, CA 90017
T: +1 213 8929262; Fax: +1 213 8929494; URL: www.fulbright.com
Nina A. Clark
25 000 vols   35834

Gibson, Dunn & Crutcher, Law Library, 333 S Grand Ave, **Los Angeles**, CA 90071-3197
T: +1 213 2297000; Fax: +1 213 2297520
Dena Hollingsworth
50 000 vols   35835

Greenberg Glusker Fields Claman & Machtinger LLP Library, 1900 Avenue of the Stars, Ste 2100, **Los Angeles**, CA 90067
T: +1 310 5533610; Fax: +1 310 5530687
Marjorie Jay
21 000 vols; 700 curr per   35836

Hill, Farrer & Burrill, Law Library, 300 S Grand Ave, 37th Flr, **Los Angeles**, CA 90071-3147
T: +1 213 6200460; Fax: +1 213 6244840; URL: www.hillfarrer.com/
Robert S. Ryan
Labour, environmental law
14 000 vols; 67 curr per   35837

Howrey Simon Arnold & White LLP Library, 550 S Hope, Ste 1100, **Los Angeles**, CA 90071
T: +1 213 8921800; Fax: +1 213 8922300; URL: www.howrey.com
Sherri Cupplo
Corporate law
16 000 vols; 72 curr per   35838

Hurrell & Cantrall, Law Library, 660 S Figueroa St, 21st Flr, **Los Angeles**, CA 90017
T: +1 213 4262000; Fax: +1 213 42-2020; URL: www.forgeyhurrell.com
10 000 vols   35839

Irell & Manella LLP Library, 1800 Avenue of the Stars, Ste 500, **Los Angeles**, CA 90067
T: +1 310 2037010; Fax: +1 310 2037199; E-mail: llieb@irell.com; URL: www.irell.com
1968; Louise Lieb
70 000 vols   35840

Jones Day, Law Library, 555 W Fifth St, Ste 4600, **Los Angeles**, CA 90013-1025
T: +1 213 4893939; Fax: +1 213 2432539; URL: www.jonesday.com
Elizabeth Elliot
Law
10 000 vols   35841

Kaye Scholer LLP, Law Library, 1999 Avenue of the Stars, Ste 1700, **Los Angeles**, CA 90067
T: +1 310 7881000; Fax: +1 310 7881200
Elinor Martin
12 000 vols   35842

Los Angeles Times, Editorial Library, 202 W First St, **Los Angeles**, CA 90012-0267
T: +1 213 2377181; Fax: +1 213 2374641, 2374502; URL: www.latimes.com
1905; Dorothy Ingebretsen
Current events; Core Reference Coll, Los Angeles Times (clippings, film, photogs)
10 000 vols
libr loan   35843

Mayer, Brown & Platt, Law Library, 350 S Grand Ave, 25th Fl, **Los Angeles**, CA 90071
T: +1 213 2299565; Fax: +1 213 6250248
John Turner
13 000 vols   35844

Meserve, Mumper & Hughes, Law Library, 300 S Grand Ave, 24th Flr, **Los Angeles**, CA 90071-3185
T: +1 213 6200300; Fax: +1 213 6251930; URL: www.mmhllp.com
James Rollins
Insurance law, Employment law
10 000 vols
AALL, SCALL   35845

Milbank, Tweed, Hadley & McCloy, Law Library, 601 S Figueroa St, 30th Flr, **Los Angeles**, CA 90017
T: +1 213 8924000; Fax: +1 213 8924798; URL: www.milbank.com
10 000 vols; 218 curr per   35846

Mitchell Silberberg & Knupp LLP, Law Library, 11377 W Olympic Blvd, **Los Angeles**, CA 90064-1683
T: +1 310 3122000; Fax: +1 310 3123100; E-mail: library@msk.com; URL: www.msk.com
Carolyn A. Pratt
35 000 vols   35847

Morrison & Foerster LLP Library, 555 W Fifth St, Ste 3500, **Los Angeles**, CA 90013-1024
T: +1 213 8925359; Fax: +1 213 8925454; E-mail: lnemchek@mofo.com; URL: www.mofo.com
Lee R. Nemchek
Law, Labor, Real estate; California City Charters, California Municipal Planning & Zoning Ordinances
20 000 vols; 500 curr per   35848

Munger, Tolles & Olson LLP, Law Library, 355 S Grand Ave, 35th Flr, **Los Angeles**, CA 90071-1560
T: +1 213 6839100; Fax: +1 213 6835173
1963; Joan Schipper
American law
15 000 vols   35849

Musick, Peeler & Garrett Library, One Wilshire Bldg, 624 S Grand Ave, Ste 2000, **Los Angeles**, CA 90017
T: +1 213 6297600; Fax: +1 213 6241376; URL: www.mpgweb.com
Lisa L. Baker
Law, Labor, Real estate
20 000 vols; 400 curr per   35850

Nossaman, Guthner, Knox & Elliott Library, Union Bank Bldg, 31st Flr, 445 S Figueroa St, **Los Angeles**, CA 90071-1602
T: +1 213 6127822; Fax: +1 213 6127801; URL: www.nossaman.com
12 000 vols; 50 curr per   35851

O'Melveny & Myers LLP, Law Library, 400 S Hope St, **Los Angeles**, CA 90071-2899
T: +1 213 4306000; Fax: +1 213 4306407
1885; Cheryl Smith
45 000 vols; 500 curr per   35852

Paul, Hastings, Janofsky & Walker LLP, Law Library, 515 S Flower, 25th Flr, **Los Angeles**, CA 90071
T: +1 213 6835074; Fax: +1 213 6270705; URL: www.paulhastings.com
Susan Streiker
Labor law
30 000 vols; 400 curr per   35853

Proskauer Rose LLP, Law Library, 2049 Century Park E, Ste 3200, **Los Angeles**, CA 90067
T: +1 310 2845683; Fax: +1 310 5572193; URL: www.proskauer.com
Lisa Winslow
15 000 vols; 100 curr per   35854

Richards, Watson & Gershon Library, 333 S Hope St, 38th Fl, **Los Angeles**, CA 90071-1469
T: +1 213 6268484 ext 369; Fax: +1 213 6260078; E-mail: la@rwglaw.com; URL: www.rwglaw.com
Francine Biscardi
Labor, Pub law, Insurance, Environment, Real estate
13 000 vols; 100 curr per
AALL   35855

Riordan & McKinzie, Law Library, 300 S Grand Ave, Ste 2900, **Los Angeles**, CA 90071
T: +1 213 6294824; Fax: +1 213 2298550; E-mail: mckinzie@netcom.com; URL: www.riordan.com
1975; Mary Dryden
21 000 vols; 400 curr per   35856

Seyfarth Shaw Library, 2029 Century Park E, Ste 3300, **Los Angeles**, CA 90067
T: +1 310 2777200; Fax: +1 310 2015219
Beth Bernstein
10 000 vols   35857

Sheppard, Mullin, Richter & Hampton Library, 333 S Hope, 42nd flr, *Los Angeles*, CA 90071
T: +1 213 6174127; Fax: +1 213 6201398
Martin Korn
Banking, Antitrust law, Labor, Securities, Intellectual property
50 000 vols; 225 curr per; 250 e-books
35858

Sidley Austin LLP Library, 555 W Fifth St, Ste 4000, *Los Angeles*, CA 90013
T: +1 213 8966193; Fax: +1 213 8966600
1980; Elisabeth A. Lamartine
Bankruptcy, Labor, Litigation, Tax, Corporate, Intellectual Property Law, Securities
25 000 vols; 100 sound-rec
AALL, SLA, SCALL
35859

Stroock & Lavan, Law Library, 2029 Century Park E, Ste 1800, *Los Angeles*, CA 90067
T: +1 310 5565800; Fax: +1 310 5565959
12 000 vols
35860

Transamerica Occidental Life Insurance, Law Library, 1150 S Olive St, Ste T-2100, *Los Angeles*, CA 90015-2211
T: +1 213 7423123; Fax: +1 213 7416623
Pauline Leary
9 000 vols; 11 curr per
35861

Twentieth Century Fox Film Corp, Frances C. Richardson Research Center, 10201 W Pico Blvd, No 89/105, *Los Angeles*, CA 90035; P.O. Box 900, Beverly Hills, CA 90213-0900
T: +1 310 3692782; Fax: +1 310 3693645
1924; Lisa Fredsti
Art & architecture, Costume design, Hist; US & German Armies; WW II Combat photos
35 000 vols; 90 curr per
35862

Frost, Brown & Todd LLC, Law Library, 400 W Market St, 32nd Flr, *Louisville*, KY 40202-3363
T: +1 502 5895400; Fax: +1 502 5811087; E-mail: library-lou@fbtlaw.com; URL: www.fbtlaw.com
Gloria Jean Glasbrener
40 000 vols
35863

Greenebaum, Doll & McDonald, Law Library, 3500 National City Tower, *Louisville*, KY 40202-3140
T: +1 502 5894200; Fax: +1 502 5873695; URL: www.gdm.com
Will Mattingley
30 000 vols; 172 curr per
35864

Stites & Harbison, Law Library, 400 W Market St, Ste 1800, *Louisville*, KY 40202
T: +1 502 5873400; Fax: +1 502 5876391; URL: www.stites.com
Lynn H. Fritsch
Hist Kentucky Statutes (back to 1800s)
25 000 vols
35865

Süd-Chemie Inc., Technical Library, 1600 W. Hill St, *Louisville*, KY 40210
T: +1 502 6347400; Fax: +1 502 6347724; E-mail: dshobe@sud-chemieinc.com
1943; David S. Shobe
8 000 vols; 50 curr per
35866

Framatome ANP Inc, Technical Library, 3315 Old Forest Rd, P.O. Box 10935, *Lynchburg*, VA 24506
T: +1 434 8322476; Fax: +1 434 8322475
1955; Ella Carr-Payne
Chemistry, Computer applications, Mathematics, Nuclear science & technology
23 000 vols; 60 curr per
35867

General Electric Co, Aircraft Engines Library, 1000 Western Ave, Tech Info Ctr, 24001, *Lynn*, MA 01910
T: +1 781 5945363; Fax: +1 781 5941689; E-mail: sandra.moltz@ae.ge.com
1954; Sandra S. Moltz
Mechanical engineering, Mathematics, Metallurgy; Gas-turbine Lit (C.W. Smith Coll)
12 000 vols; 100 curr per
libr loan
35868

Brown & Williamson Tobacco Corp, Research Department Library, 2600 Weaver Rd, *Macon*, GA 31201
T: +1 478 4643565; Fax: +1 478 4644016
1958; Carol Lincoln
9 000 vols; 225 curr per
35869

Dewitt, Ross & Stevens SC, Law Library, Two E Mifflin St, Ste 600, *Madison*, WI 53703
T: +1 608 2835504; Fax: +1 608 2529243; URL: www.dewittross.com
Richard D. Hendricks
17 000 vols; 50 curr per
35870

RMT, Inc Library, 744 Heartland Trail, *Madison*, WI 53717
T: +1 608 6625385; Fax: +1 608 8313334
Mary Janeck
Environmental engineering
10 000 vols; 200 curr per
35871

HRL Laboratories, LLC Research Library, 3011 Malibu Canyon Rd, *Malibu*, CA 90265
T: +1 310 3175373; Fax: +1 310 3175624; E-mail: HRL_library@HRL.com
1959; Stephanie Tiffany
13 000 vols; 200 curr per; 5 900 microforms
35872

Lockheed Martin, Manassas Library, 9500 Godwin Dr, *Manassas*, VA 20110
T: +1 703 3676508; Fax: +1 703 3674698
1985; Jennifer Hatfield
Electrical & electronic engineering, Computer science, Naval science
10 000 vols; 80 curr per
35873

McLane, Graf, Raulerson & Middleton PA, Law Library, 900 Elm St, *Manchester*, NH 03101; P.O. Box 326, Manchester, NH 03105-0326
T: +1 603 6281428; Fax: +1 603 6255650; E-mail: library@mclane.com
Jennifer M. Finch
New Hampshire Legal, Legislative, Regulatory and Hist Information Coll
9 000 vols; 100 curr per
35874

Sheehan Phinney Bass + Green PA Library, 1000 Elm St, *Manchester*, NH 03105; P.O. Box 3701, Manchester, NH 03105-3701
T: +1 603 6680300; Fax: +1 603 6278121; URL: www.sheehan.com
Debra Piotrowski
10 000 vols
35875

Lockheed Martin Aeronautical Systems, Tech Information Center, D/48-11, MZ 0381, 86 South Cobb Drive, *Marietta*, GA 30063-0381
T: +1 770 4943947; Fax: +1 770 4940732; E-mail: daveledbetter@lmco.com; URL: www.lmco.com
1955; D. Ellis
Aeronautics, Electronics, Engineering, Economics; Foreign Country Data, Recreation Coll
50 000 vols; 450 curr per
35876

Watt, Tieder, Hoffar & Fitzgerald, 8405 Greensboro Dr, Ste 100, *McLean*, VA 22102
T: +1 703 7491019; Fax: +1 703 8938029; URL: www.wthf.com
Barbara Cumming
8 000 vols
35877

Holzmacher, McLendon & Murrell Library, H2M Group Library, 575 Broad Hollow Rd, *Melville*, NY 11747
T: +1 631 7568000 ext 1740; Fax: +1 631 6944122; E-mail: h2m_lib@h2m.com; URL: www.h2m.com
1960; Joan S Newlin
Environmental, Engineering, Architecture
14 300 vols; 240 curr per; 20 incunabula; 500 govt docs; 2 200 maps; 20 digital data carriers; 1 300 cat, 2 900 USEPA rpts, 2 600 Eng drawings, 100 slides
libr loan; ALA, SCLA, NCLA, LILRC, NYLA
35878

Buckman Laboratories International, Inc, Knowledge Resource Center, 1256 N McLean Blvd, *Memphis*, TN 38108
T: +1 901 2728585; Fax: +1 901 2728583; E-mail: krc@buckman.com; URL: www.buckman.com
1948; Cheryl Lamb

Microbiology, Pulp, Paper, Water treatment, Corrosion, Leather
20 000 vols; 600 curr per
libr loan; SLA, MALC
35879

Glankler Brown, Law Library, One Commerce Sq, Ste 1700, *Memphis*, TN 38103
T: +1 901 5251022; Fax: +1 901 5252389
Donna Windham
10 000 vols; 250 curr per
35880

Raychem Corp, Corporate Library, 300 Constitution Dr, *Menlo Park*, CA 94025
T: +1 415 3613282; Fax: +1 415 3612655; URL: www.raychem.com
Klaus Dahl
Radiation
10 000 vols; 450 curr per
libr loan
35881

Greenberg Traurig LLP, Research Center, 1221 Brickell Ave, *Miami*, FL 33131
T: +1 305 5790667; Fax: +1 305 7550904
1975; Margo M. Gustely
20 000 vols; 450 curr per
35882

Gunster, Yoakley, Valdes-Fauli & Stewart, PA, 2 S Biscayne Blvd, One Biscayne Tower, Ste 3400, *Miami*, FL 33131
T: +1 305 3766000; Fax: +1 305 3766010; E-mail: library@gunster.com
Law; Int Banking, Cuban Task Force Info-Bank
8 000 vols; 100 curr per
35883

Holland & Knight LLP, Law Library, 701 Brickell Ave, Ste 3000, *Miami*, FL 33131
T: +1 305 7897420; Fax: +1 305 7897799
Elizabeth Chifari
15 000 vols; 75 curr per
35884

Squire Sanders & Dempsey LLC Library, 200 S Biscayne Blvd, Ste 4000, *Miami*, FL 33131
T: +1 305 5772932; Fax: +1 305 5777001; URL: www.steelhector.com/library_banner.htm
Sid Kaskey
30 000 vols; 700 curr per
35885

Stearns, Weaver, Miller, Weissler, Alhadeff & Sitterson, Law Library, 2200 Museum Tower, 150 W Flagler St, *Miami*, FL 33130
T: +1 305 7893251; Fax: +1 305 7893395; URL: www.swmwas.com
Jeanne S. Korman
Corporate law, Banking, Securities
15 000 vols; 186 curr per; 10 000 microforms; 20 digital data carriers
35886

Walton, Lantaff, Schroeder & Carson, Law Library, 9350 S Dixie Hwy, 10th Fl, *Miami*, FL 33156
T: +1 305 6711300 ext 390; Fax: +1 305 6707065; E-mail: wlscmia@bellsouth.net
Daniel Linehan
30 000 vols
35887

Chemtura Corp, Information Services-Library, Benson Rd, *Middlebury*, CT 06749; P.O. Box 117, Waterbury, CT 06720-0117
T: +1 203 5734508; Fax: +1 203 5732890
1914; Patricia Ann Harmon
Chemistry, Rubber chemistry, Agricultural chemistry, Engineering
10 000 vols; 219 curr per; 10 microforms; 200 digital data carriers
libr loan; SLA
35888

AT&T Labs Library, 200 Laurel Ave, *Middletown*, NJ 07748-4801
T: +1 732 4205220; Fax: +1 732 4205208; E-mail: jshaw@att.com
John Shaw
30 000 vols; 400 curr per
35889

The Dow Chemical Co, Business Intelligence Center, 566 Bldg, *Midland*, MI 48667
T: +1 989 6362912; Fax: +1 989 6362524
1920; Nancy Cundiff
14 000 vols; 350 curr per
35890

Ecosystems International Inc Library, 1107 Dicus Mill Rd, *Millersville*, MD 21108; P.O. Box 225, Gambrills, MD 21054-0225
T: +1 410 9874976; Fax: +1 410 7291960; E-mail: wdc101@aol.com
Darlene Barckley
38 000 vols; 88 curr per
35891

Foley & Lardner, Law Library, 777 E Wisconsin Ave, *Milwaukee*, WI 53202-5306
T: +1 414 2712400; Fax: +1 414 2974900; URL: www.foley.com
1842; Susan O'Toole
32 000 vols; 450 curr per
35892

Johnson Controls, Inc, Corporate Library Services, 507 E Michigan St, *Milwaukee*, WI 53202
T: +1 414 5244687; Fax: +1 414 5244596
1973; Mary F Kaczmarek
American Society of Heating, Refrigeration & Air-Conditioning Engineers Material, dating back to 1900s
10 000 vols; 550 curr per
libr loan; OCLC
35893

Michael Best & Friedrich LLP, Information Services Department, 100 E Wisconsin Ave, *Milwaukee*, WI 53202-4108
T: +1 414 2716560; Fax: +1 414 2770656
1848; Jane B. Moberg
25 000 vols
35894

Northwestern Mutual Life Insurance Co, Northwestern Mutual Law Library, 720 E Wisconsin Ave, Rm 611, *Milwaukee*, WI 53202-4797
T: +1 414 6652422; Fax: +1 414 6252422
1892; Olivia Bradford Jaskolski
20 000 vols; 40 curr per
35895

Northwestern Mutual Life Insurance Co, Corporate Information Center, 720 E Wisconsin Ave, Rm 345, *Milwaukee*, WI 53202-4797
T: +1 414 6652492; Fax: +1 414 6657022
1951; Robert Duchek
10 000 vols; 350 curr per
35896

Quarles & Brady, Law Library, 411 E Wisconsin Ave, *Milwaukee*, WI 53202-4491
T: +1 414 2775000; Fax: +1 414 2713552; E-mail: aeb@quarles.com; URL: www.quarles.com
1910; Amy Bingenheimer
Code of Federal Regulations 1939 to date, Federal Register 1970 to date
10 000 vols; 100 curr per; 90 av-mat; 130 sound-rec; 12 digital data carriers
35897

Reinhart Boerner Van Deuren SC, Information Resource Center, 1000 N Water St, Ste 2100, *Milwaukee*, WI 53203-3400
T: +1 414 2988253; Fax: +1 414 2988097; URL: www.reinhartlaw.com
1975; Carol Bannen
Employee benefits
20 000 vols; 460 curr per
35898

Rockwell Automation Library, 1201 S Second St, *Milwaukee*, WI 53204
T: +1 414 3822342; Fax: +1 414 3822462; E-mail: rtbell@ra.rockwell.com
1942; Randy Bell
Indust automation, Engineering; Military standards & specifications
10 000 vols; 425 curr per
35899

Von Briesen & Roper Sc, Law Library, 411 E Wisconsin Ave, No 700, *Milwaukee*, WI 53202
T: +1 414 2761122; Fax: +1 414 2766281; URL: www.vonbriesen.com
JoAnne Meyer
10 000 vols
35900

Wisconsin Electric – Wisconsin Gas, Library & Research Services, 231 W Michigan St, Rm P160, *Milwaukee*, WI 53290-0001; P.O. Box 2046, Milwaukee, WI 53201
T: +1 414 2212580; Fax: +1 414 2212282
Mary Ann Barragry
Engineering
8 000 vols; 500 curr per
35901

Meyer, Suozzi, English & Klein, Law Library, 1505 Kellum Pl, *Mineola*, NY 11501-4824
T: +1 516 7416565; Fax: +1 516 7416706
Beth Murphy
10 000 vols   35902

Barr Engineering Co Library, 4700 W 77th St, *Minneapolis*, MN 55435
T: +1 952 8322600; Fax: +1 952 8322601; URL: www.barr.com
Karen Oakes
10 000 vols; 300 curr per   35903

Dorsey & Whitney, Information Resource Center, 50 S Sixth St, *Minneapolis*, MN 55402
T: +1 612 4925522; Fax: +1 612 4922868
Linda Will
60 000 vols; 400 curr per   35904

Faegre & Benson, LLP, Information Resources, 2200 Wells Fargo Ctr, 90 South Seventh St, *Minneapolis*, MN 55402-3901
T: +1 612 7667000; Fax: +1 612 7661600; E-mail: library@faegre.com
Kaye Waelde
32 000 vols; 515 curr per   35905

Federal Reserve Bank of Minneapolis, Research Library, 90 Hennepin Ave, *Minneapolis*, MN 55401-2171; P.O. Box 291, Minneapolis, MN 55480-0291
T: +1 612 2045509
1940
Economy, Finance, Monetary policy; Federal Reserve System Pubs Coll
15 000 vols; 2 000 curr per   35906

Felhaber, Larson, Fenlon & Vogt, Law Library, 220 S Sixth St, Ste 2200, *Minneapolis*, MN 55402-4302
T: +1 612 3396321; Fax: +1 612 3380535; URL: www.felhaber.com
Annette C. Borer
10 000 vols; 107 curr per   35907

Fredrikson & Bryon, Law Library, 200 S Sixth St, Ste 4000, *Minneapolis*, MN 55402
T: +1 612 4927842; Fax: +1 612 4927077; URL: www.fredlaw.com
Rebekah Anderson
20 000 vols; 300 curr per   35908

General Mills, Inc, James Ford Bell Library & Information Services, 9000 Plymouth Ave N, *Minneapolis*, MN 55427
T: +1 763 7646460; Fax: +1 763 7643166
1961; Fred Hulting
Food science
20 000 vols; 750 curr per   35909

Leonard, Street & Deinard, Law Library, 150 S Fifth St, Ste 2300, *Minneapolis*, MN 55402
T: +1 612 3351616; Fax: +1 612 3351657; E-mail: research.services@leonard.com
Patricia K. Cummings
15 000 vols; 150 curr per   35910

Mackall, Crounse & Moore, Law Library, 1400 AT&T Tower, 901 Marquette Ave, *Minneapolis*, MN 55402-2859
T: +1 612 3051687; Fax: +1 612 3051414; URL: www.mcmlaw.com
Noreen Shepard
11 000 vols; 92 curr per   35911

Oppenheimer Wolff & Donnelly Library, 45 S Seventh St, *Minneapolis*, MN 55402
T: +1 612 6077416; Fax: +1 612 6077100; URL: www.oppenheimer.com
Trudy Bush
Antitrust law, Corporate law, Int business law, Taxes
35 000 vols; 400 curr per   35912

Robins, Kaplan, Miller & Ciresi LLP, Law Library, 2800 LaSalle Plaza, mail address: 800 LaSalle Ave, *Minneapolis*, MN 55402
T: +1 612 3498529; Fax: +1 612 3394181; E-mail: cregcj@rkmc.com; URL: www.rkmc.com
Carol Crego
12 000 vols; 72 curr per   35913

Zelle, Hofmann, Voelbel, Mason & Gette, Law Library, 500 Washington Ave S, Ste 4000, *Minneapolis*, MN 55415
T: +1 612 3369129; Fax: +1 612 3369100; URL: www.zelle.com
Janet L. Rongitsch
10 000 vols; 200 curr per   35914

Deere & Co Library, One John Deere Pl, *Moline*, IL 61265
T: +1 309 7654733; Fax: +1 309 7654088
1958; Sean O'Hanlon
Engineering, Construction, Agriculture, Forestry, Finance, Marketing; Deere & Company Hist
12 000 vols; 800 curr per
libr loan   35915

PPG Industries, Inc, Chemicals Technical Information Center, 440 College Park Dr, *Monroeville*, PA 15146
T: +1 724 3255221; Fax: +1 724 3255289; URL: www.ppg.com
Denise Callihan
Organic chemistry
14 000 vols; 120 curr per   35916

USS Division of USX Corp, Knowledge Resource Center, One Tech Center Dr, MS-88, *Monroeville*, PA 15146
T: +1 412 8252345; Fax: +1 412 8252050; E-mail: mhdennis@uss.com
1928; Megan Dennis
Chemical technology, Metallurgy, Steelmaking; Hist Bks on Steel
30 000 vols; 200 curr per   35917

Fenwick & West LLP, Library, Silicon Valley Ctr, 801 California St, *Montain View*, CA 94041
T: +1 650 3357575; Fax: +1 650 9385200; E-mail: library@fenwick.com; URL: www.fenwick.com
1975
10 000 vols; 300 curr per   35918

CTB McGraw-Hill Library, 20 Ryan Ranch Rd, *Monterey*, CA 93940-5703
T: +1 831 3937008, 3930700; Fax: +1 831 3937825; URL: www.ctb.com
1965; Laurel Wilson
Education, Psychology, Statistics; Test Archs
10 500 vols; 150 curr per   35919

National Life Insurance Co Library, National Life Dr, *Montpelier*, VT 05604
T: +1 802 2293276; Fax: +1 802 2293743
JoAnn Morissette
Company Hist (Arch) bks, pictures, clippings, hists, etc; Looseleaf Services in Law, Pension, Employment, Taxes, Insurance
8 000 vols; 30 curr per   35920

Pfizer Inc, Information Center, 182 Tabor Rd, *Morris Plains*, NJ 07950
T: +1 973 3852875; Fax: +1 973 3854756
Catherine Christensen
15 000 vols; 100 curr per   35921

McElroy, Deutsch, Mulvaney & Carpenter, LLP, Law Library, 1300 Mt Kemble Ave, *Morristown*, NJ 07962
T: +1 973 4258810; Fax: +1 973 4250161; URL: www.mdmlaw.com; www.mdmc-law.com
Mary Ellen Kaas
10 000 vols   35922

Porzio, Bromberg & Newman Library, 100 Southgate Pkwy, *Morristown*, NJ 07962-1997
T: +1 973 5384006; Fax: +1 973 5385146; URL: www.pbnlaw.com
Janice Schouten
10 000 vols; 100 curr per   35923

Riker, Danzig, Scherer, Hyland & Perretti, Law Library, Headquarters Plaza, One Speedwell Ave, *Morristown*, NJ 07962
T: +1 973 5380800; Fax: +1 973 5381984
Karen Brunner
10 000 vols   35924

Caterpillar Inc, Technical Information Center, 14009 Old Galena Rd, *Mossville*, IL 61552; P.O. Box 225, Mossville, IL 61552-0225
T: +1 309 5786118; Fax: +1 309 5786732; URL: www.cat.com
1939; Dan Evans
Mechanical engineering
9 000 vols; 532 curr per
libr loan   35925

Capehart & Scatchard, PA Library, 8000 Midlantic Dr, Ste 300, *Mount Laurel*, NJ 08054
T: +1 856 2346800; Fax: +1 856 2352786; URL: www.capehart.com
Francine Viden
9 000 vols; 32 curr per   35926

Varian Medical, Ginzton Research Center, 2599 Garcia Ave, *Mountain View*, CA 94043-1111
T: +1 650 2138000; Fax: +1 650 4246988
1961; Sonya Shin
Electronics, Computer science, Chemistry
20 000 vols   35927

Boc Group, Inc, Information Center, 100 Mountain Ave, *Murray Hill*, NJ 07974
T: +1 908 7716498; Fax: +1 908 7716182; URL: www.boc.com
1918; Robert Yeager
Anesthesiology, Cryogenics, Health care, Indust gases, Metallurgy
10 000 vols; 61 curr per
libr loan   35928

Lucent Technologies, Murray Hill Library, 600 Mountain Ave, Rm 6A-200, *Murray Hill*, NJ 07974-2070
T: +1 908 5822601; Fax: +1 908 5827591
1925; M. E. Brennan
60 000 vols; 1 000 curr per
libr loan   35929

Nelson, Mullins, Riley & Scarborough, Law Library, 2411 N Oak St, Ste 301, *Myrtle Beach*, SC 29577-3165; P.O. Box 3939, Myrtle Beach, SC 29578-3939
T: +1 803 4483500; Fax: +1 803 4483437
Claire Engel
10 000 vols   35930

BP Information Services, BP 602 Library, MC F-1, 150 W Warrenville Rd, *Naperville*, IL 60563
T: +1 630 9617634; Fax: +1 630 4203697; URL: www.bp.com
1972; Joyce Fedeczko
Engineering, Chemistry, Petroleum, Environmental studies, Law, Business
10 000 vols; 150 curr per   35931

ONDEO Nalco Company Library, Information Services, One ONDEO Nalco Ctr, *Naperville*, IL 60563-1198
T: +1 630 3051000; Fax: +1 630 3052876; E-mail: nlibrary@ondeo-nalco.com
1928; Nancy Maloney
Chemistry, Water treatment
25 000 vols; 400 curr per   35932

BAE Systems, E & IS Technical Library, 95 Canal St, *Nashua*, NH 03061-2878; P.O. Box 868, Nashua, NH 03061-0868
T: +1 603 8852645; Fax: +1 603 8853214
1955; Mary-Ellen Reed
Engineering, Electronics, Computer science, Mathematics, Microwave, Physics, Telecommunications
25 000 vols; 75 curr per   35933

Boult, Cummings, Conners & Berry, Plc, Law Library, 414 Union St, Ste 1600, P.O. Box 198062, *Nashville*, TN 37219
T: +1 615 2523577; Fax: +1 615 2483031; E-mail: jjulian@boultcummings.com; URL: www.boultcummings.com
Julie L. Julian
10 000 vols; 81 curr per   35934

United Methodist Publishing House Library, 201 Eighth Ave S, *Nashville*, TN 37203
T: +1 615 7496335; Fax: +1 615 7496128
1945; Duane Diehl
Wesleyana
40 000 vols; 300 curr per; 1 000 microforms; 5 digital data carriers; 60 VF drawers
libr loan   35935

Gillette Co, Technical Information Center, 37 A St, *Needham*, MA 02494-2806
T: +1 781 2928407; URL: www.gillette.com
Rose Cunha
Chemistry, Cosmetics industry, Material culture
15 000 vols; 200 curr per   35936

Kimberly-Clark Corp Library, 2100 Winchester Rd, *Neenah*, WI 54957; P.O. Box 999, Neenah, WI 54957-0999
T: +1 920 7215262; Fax: +1 920 7216394
Cheryl R. Nelson
Chemistry, Paper, Pulp science, Technology, Textiles
20 000 vols; 450 curr per   35937

Utica International Insurance Group, Reference & Law Library, 180 Genesee St, *New Hartford*, NY 13413
T: +1 315 7342000; Fax: +1 315 7342680
Steve Barry
9 200 vols   35938

Tyler Cooper & Alcorn, LLP, Law Library, 555 Long Wharf Dr, 8th Flr, *New Haven*, CT 06509
T: +1 203 7848200; Fax: +1 203 7771181
Stephanie Steinmetz
15 000 vols
libr loan   35939

Wiggin & Dana Information Center, 265 Church St, *New Haven*, CT 06510
T: +1 203 4984413, 4400; Fax: +1 203 7822889; E-mail: amo@wiggin.com; URL: www.wiggin.com
Ana Oman
Law
17 000 vols   35940

Wiggin & Dana LLP, Wiggin & Dana Information Center, 265 Church St, *New Haven*, CT 06510; P.O. Box 1832, New Haven, CT 06508-1832
T: +1 203 4984400; Fax: +1 203 7822889; URL: www.wiggin.com
Ana Oman
17 000 vols   35941

Nestle PTC Information Center Library, 201 Housatonic Ave, *New Milford*, CT 06776
T: +1 860 3556213; Fax: +1 860 3556367; E-mail: martha.osborne@rdct.nestle.com
1981; Martha Osborn
Food science, Analytical chemistry; Nestle Nutrition Workshop Series
8 000 vols; 300 curr per   35942

Adams & Reese Law Library, One Shell Sq, 701 Poydras, Ste 4500, *New Orleans*, LA 70139
T: +1 504 5813234; Fax: +1 504 5660210; URL: www.adamsandreese.com
Brent Hightower
16 000 vols; 200 curr per   35943

Chaffe, McCall LLP, Law Library, 2300 Energy Centre, 1100 Poydras St, *New Orleans*, LA 70163-2300
T: +1 504 5857000; Fax: +1 504 5857075; URL: www.chaffe.com
15 000 vols; 10 curr per   35944

Deutsch, Kerrigan & Stiles, Law Library, 755 Magazine St, *New Orleans*, LA 70130-3672
T: +1 504 5815141; Fax: +1 504 5661201; URL: www.dkslaw.com
1926; Jean Barney
Great Britain
30 000 vols; 25 curr per   35945

Gordon, Arata, McCollam, Duplantis & Egan LLP, Law Library, 201 Saint Charles Ave, Ste 4000, *New Orleans*, LA 70170
T: +1 504 5821111; Fax: +1 504 5821121; URL: www.gamde.com
Edward Benefiel
10 000 vols; 50 curr per   35946

Jones, Walker, Waechter, Poitevent, Carrere & Denegre, Law Library, 201 Saint Charles Ave, **New Orleans**, LA 70170
T: +1 504 5828589; Fax: +1 504 5828567; URL: www.joneswalker.com
Tina Gambrell
30 000 vols; 300 curr per   35947

Lemle & Kelleher, Law Library, Pan Am Life Ctr, 21st Flr, 601 Poydras St, **New Orleans**, LA 70130
T: +1 504 5861241; Fax: +1 504 5849142; E-mail: info@lemle.com; URL: www.lemle.com
Elizabeth Cortez
15 000 vols; 125 curr per   35948

Phelps Dunbar LLP, Law Library, 365 Canal St, Ste 2000, **New Orleans**, LA 70130-6534
T: +1 504 5661311; Fax: +1 504 5689130
Cynthia Jones
50 000 vols; 75 curr per   35949

Sessions, Fishman, Nathan & Israel LLP Library, 201 St Charles Ave, Ste 3500, **New Orleans**, LA 70170-3500
T: +1 504 5821500; Fax: +1 504 5821555; URL: www.sessions-law.com
1958; Jeanne Claudel-Simoneaux
15 000 vols; 75 curr per
libr loan   35950

Simon, Peragine, Smith & Redfearn LLP, Law Library, Energy Ctr, 1100 Poydras St, 30th Flr, **New Orleans**, LA 70163
T: +1 504 5692030; Fax: +1 504 5692999; URL: www.spsr-law.com
Frank Schiavo
9 000 vols; 36 curr per   35951

Reed Elsevier New Providence Library, 121 Chanlon Rd, **New Providence**, NJ 07974
T: +1 908 6656761; Fax: +1 908 7718739
Nancy Upham
Publishing, Law
10 000 vols; 200 curr per   35952

Alston & Bird, LLP Library, 90 Park Ave, 12th Flr, **New York**, NY 10016
T: +1 212 2109526; Fax: +1 212 2109444
John H. Davey
German Law Mat
15 000 vols   35953

American Broadcasting Co, Inc, ABC News Research Center, 47 W 66th St, **New York**, NY 10023
T: +1 212 4563796; Fax: +1 212 4562450; E-mail: nrc@abc.com
1982; C. B. Hayden
Current events, Int relations, Media
35 000 vols; 600 curr per   35954

American Standards Testing Bureau, Inc, Sam Tour Memorial Library, 40 Water St, **New York**, NY 10004-2672
T: +1 212 9433160; Fax: +1 212 8252250; E-mail: worldteck@aol.com
1916; C. Chavis
Nat & Internat Standards
40 000 vols; 300 curr per   35955

Anderson, Kill & Olick, Law Library, 1251 Avenue of the Americas, **New York**, NY 10020-1182
T: +1 212 2781790; Fax: +1 212 2781733; E-mail: akony@andersonkill.com
Timothy Baran
20 000 vols; 150 curr per   35956

Baker & McKenzie Library, 805 Third Ave, 29th Flr, **New York**, NY 10022
T: +1 212 8913990; Fax: +1 212 7599133; URL: www.bakernet.com
1971; Leslee I. Budlong
Arbitration, Banking, Captive insurance, Finance, Intellectual property, Int law, Securities, Tax
25 000 vols; 70 curr per; 15 microforms; 8 digital data carriers
libr loan; AALL   35957

Bank of New York, Information Research Department, One Wall St, **New York**, NY 10286
T: +1 212 6351599; Fax: +1 212 6351568; E-mail: reicherter@bony.com
1920; Joan M. Reicherter
8 900 vols; 160 curr per   35958

Blank Rome Comisky & McCauley LLP Library, Chrysler Bldg, 405 Lexington Ave, **New York**, NY 10174
T: +1 212 8855000; Fax: +1 212 8855001; URL: www.blankrome.com
Anthony Rivitti
20 000 vols; 800 curr per   35959

Bryan Cave LLP, Law Library, 1290 Avenue of the Americas, **New York**, NY 10104
T: +1 212 5412166; Fax: +1 212 5411465; URL: www.bryancave.com
1950; Christine M. Wierzba
Law
10 000 vols; 300 curr per   35960

Cadwalader, Wickersham & Taft Library, 100 Maiden Lane, **New York**, NY 10038
T: +1 212 5046000; Fax: +1 212 4127041; URL: www.cadwalader.com
Rissa Peckar
Acquisitions, Banking, Commodities, Healthcare, Insolvency, Litigation, Mergers, Project finance, Real estate, Securities, Tax
40 000 vols   35961

Cahill, Gordon & Reindel LLP Library, 80 Pine St, **New York**, NY 10005
T: +1 212 7013542; Fax: +1 212 2695420; URL: www.cahill.com
Chan-Shen Lung
Antitrust, Corporate, Securities, Tax law; Legislative Hist & Law Reports Coll
45 000 vols   35962

Cahners Publishing Co, Frederic G. Melcher Library, 245 W 17th St, **New York**, NY 10011
T: +1 212 4636850
1962; Jean Peters
Book trade, Libr science; Books About Books
8 000 vols; 300 curr per   35963

Carter, Ledyard & Milburn Library, Two Wall St, **New York**, NY 10005
T: +1 212 2388851; Fax: +1 212 7323232
Emily Moog
20 000 vols; 30 curr per   35964

CBS News Reference Library, 524 W 57th St, Ste 510/1, **New York**, NY 10019
T: +1 212 9752877; Fax: +1 212 9757766; E-mail: st4@cbsnews.com
1940; Jean Stevenson
Biogr, Broadcasting, Current events, Govt, Hist, Politics; CBS News Broadcasts Coll (transcripts)
31 000 vols; 250 curr per   35965

Chadbourne & Parke Library, 30 Rockefeller Plaza, 33rd Fl, **New York**, NY 10112
T: +1 212 4081035; Fax: +1 212 7656923; E-mail: champton@chadbourne.com; URL: www.chadbourne.com
Lillian Arcuri
Insurance, Labor, Real estate, Tax
70 000 vols   35966

Chase Manhattan Bank, Legal Research & Information Services Unit, One Chase Manhattan Plaza, 25th Flr, **New York**, NY 10081
T: +1 212 5522499; Fax: +1 212 3830252
Tamar Raum
Banking law; Legislative Histories, microforms
35 000 vols; 250 curr per   35967

Cleary, Gottlieb, Steen & Hamilton Library, One Liberty Plaza, **New York**, NY 10006
T: +1 212 2253444; Fax: +1 212 2253449
Karol M. Sokol
25 000 vols   35968

Clifford Chance US LLP Library, 31 W 52nd St, **New York**, NY 10019
T: +1 212 8788095; Fax: +1 212 8783474; URL: www.cliffordchance.com
1871; Rosalinda Rupel
25 000 vols; 300 curr per   35969

Condon & Forsyth Library, Seven Times Sq, 18th Flr, **New York**, NY 10036
T: +1 212 4909100; Fax: +1 212 3704453; URL: www.condonlaw.com
Antonietta Tatta
Aviation
13 000 vols; 76 curr per   35970

Cooley Godward Kronish LLP, Law Library, 1114 Avenue of the Americas, **New York**, NY 10036
T: +1 212 4796025; Fax: +1 212 4796275
Gary Jaskula
15 000 vols; 100 curr per   35971

Coudert Brothers LLP Library, 1114 Avenue of the Americas, **New York**, NY 10036-7703
T: +1 212 6264796; Fax: +1 212 6264120; E-mail: library@coudert.com
Jane C. Rubens
Law
30 000 vols; 260 curr per   35972

Cravath, Swaine & Moore LLP, Law Library, 825 Eighth Ave, **New York**, NY 10019
T: +1 212 4743500; Fax: +1 212 4743556; URL: www.cravath.com
1819; Elsbeth Moller
Antitrust, Corporate, Litigation, Tax
50 000 vols; 400 curr per; 3 digital data carriers
libr loan; AALL, SLA   35973

Curtis, Mallet-Prevost, Colt & Mosle Library, 101 Park Ave, **New York**, NY 10178-0061
T: +1 212 6966138; Fax: +1 212 6971559
1900; John Kostecky
Central & South American law
16 000 vols   35974

Davis Polk & Wardwell Library, 450 Lexington Ave, **New York**, NY 10017
T: +1 212 4504266; Fax: +1 212 4505522
1891; Daniel J. Hanson
Antitrust, Banking, Securities, Tax; Int Law, Legislature
70 000 vols; 1 400 curr per   35975

Debevoise & Plimpton, Law Library, 919 Third Ave, **New York**, NY 10022
T: +1 212 9096275; Fax: +1 212 9091025; URL: www.debevoise.com
Stephen A. Lastres
Aviation, Corp law, Litigation, Real estate, Securities
30 000 vols; 175 curr per; 3 000 pamphlets   35976

Dechert Law Library, Law Library, 30 Rockefeller Plaza, 22nd Fl, **New York**, NY 10112
T: +1 212 6983500; Fax: +1 212 6983599
John Davey
Corp securities
20 000 vols; 300 curr per   35977

Dewey Ballantine LLP, Law Library, 1301 Avenue of the Americas, **New York**, NY 10019
T: +1 212 2596000; Fax: +1 212 2596333; E-mail: gseer@class.org; URL: www.deweyballantine.com
1915; Gitelle Seer
Antitrust, Corp, Environment, Securities, Tax, Technology; Legislative Histories
55 000 vols; 300 curr per   35978

Dewey Ballantine LLP Library, 1301 Avenue of the Americas, **New York**, NY 10019
T: +1 212 2596000; Fax: +1 212 2596679; E-mail: requests@deweyballantine.com; URL: www.deweyballantine.com
Gitelle Seer
Corporate law, securities; legislative histories
55 000 vols; 300 curr per   35979

Federal Reserve Bank of New York – Research Library, 33 Liberty St, Federal Reserve PO Sta, **New York**, NY 10045
T: +1 212 7205670, 8289; Fax: +1 212 7201372; E-mail: research.library@ny.frb.org
1916; Kathleen McKiernan
71 000 vols; 2 000 curr per   35980

FIND-SVP Inc Library, 625 Ave of the Americas, 2nd Fl, **New York**, NY 10011
T: +1 212 6454500; Fax: +1 212 6457681
1970; Melissa Hover
8 000 vols; 4 000 curr per; 200 500 microforms   35981

Fried, Frank, Harris, Shriver & Jacobson Library, One New York Plaza, **New York**, NY 10004
T: +1 212 8598302; Fax: +1 212 8598586; E-mail: nylibrary@ffhsj.com; URL: www.ffhsj.com
1960; Warren Gordon
30 000 vols; 110 curr per   35982

Greenberg Traurig LLP, Research Center Law Library, 200 Park Ave, **New York**, NY 10166
T: +1 212 8019200; URL: www.gtlaw.com
Sheila Sterling
10 000 vols   35983

Hawkins, Delafield & Wood, Law Library, One Chase Manhattan Plaza, **New York**, NY 10005
T: +1 212 8209444; Fax: +1 212 3446258
Kathryn McRae
10 000 vols; 50 curr per   35984

Holland & Knight LLP, Law Library, 195 Broadway, **New York**, NY 10007
T: +1 212 5133580; Fax: +1 212 3859010; E-mail: library@hklaw.com; URL: www.hklaw.com
Cristina K. Alvy
Maritime & Aviation Coll
25 000 vols; 60 curr per
libr loan; AALL, SLA   35985

Hughes, Hubbard & Reed Library, Law Library, One Battery Park Plaza, 16th Flr, **New York**, NY 10004
T: +1 212 8376666; Fax: +1 212 4224726; E-mail: library@hugheshubbard.com
1942; Patricia E. Barbone
30 000 vols; 1 200 curr per   35986

Hunton & Williams, Law Library, 200 Park Ave, 44th Fl, **New York**, NY 10166
T: +1 212 3091078; Fax: +1 212 3091100
Alina Alvarez-Lenda
25 000 vols; 5 000 curr per   35987

Jackson Lewis LLP, Law Library, 59 Maiden Lane, **New York**, NY 10038-4502
T: +1 212 5454033
Catherine M. Dillon
10 000 vols   35988

Katten Muchin Rosenman LLP, Law Library, 575 Madison Ave, **New York**, NY 10022
T: +1 212 9406709; Fax: +1 212 8945598; URL: www.kattenlaw.com
1946; Anthony J. Burgalassi
50 000 vols; 700 curr per; 1 000 microforms; 20 sound-rec; 20 digital data carriers   35989

Kaye Scholer LLP, Law Library, 425 Park Ave, **New York**, NY 10022
T: +1 212 8368141; Fax: +1 212 8366613
Shabeer Khan
48 000 vols; 1 168 curr per   35990

Kelley Drye & Warren, Law Library, 101 Park Ave, **New York**, NY 10178
T: +1 212 8087800; Fax: +1 212 8087897; URL: www.kelleydrye.com
Patricia Renze
25 000 vols; 100 curr per   35991

Kenyon & Kenyon LLP, Law Library, Library 10th Flr, One Broadway, **New York**, NY 10004-1007
T: +1 212 9086122; Fax: +1 212 9086113; URL: www.kenyon.com/flib.htm
Lucy Curci-Gonzalez
25 000 vols; 200 curr per   35992

Kramer, Levin, Naftalis & Frankel LLP, Law Library, 1177 Avenue of the Americas, **New York**, NY 10036
T: +1 212 7159321; Fax: +1 212 7158000; URL: www.kramerlevin.com
Daniel J. Pelletier
15 000 vols; 300 curr per   35993

Latham & Watkins, Law Library, 885 Third Ave, Ste 1000, **New York**, NY 10022
T: +1 212 9061200; Fax: +1 212 7514864
Anne Lewis
20 000 vols; 150 curr per   35994

Leboeuf, Lamb, Greene & Macrae, Law Library, 125 W 55th St, **New York**, NY 10019-5389
T: +1 212 4248000; Fax: +1 212 4248500
1929; Douglas Cinque
Federal Energy Regulatory Commission Coll, Insurance Laws, Securities & Exchange Commission Coll, Tax Coll, Water Coll
40 000 vols; 10 curr per   35995

Lighthouse International, Ruth M. Shellens Memorial Library, 111 E 59th St, **New York**, NY 10022
T: +1 212 8219681; Fax: +1 212 8219687; E-mail: gaks@lighthouse.org; URL: www.lighthouse.org
1967; Gloria Aks
Braille & Audio Bks Coll – popular fiction, non-fiction, computer skills Regular print – Vision, Orientation and Mobility related to vision impairment; Braille Transcription Dept, Recording Dept
10 000 vols; 50 curr per; 2 000 bks on tape, braille & large print bks (1 000)
ALA, SLA
Libr is limited to individuals who are legally blind and receive services from the Lighthouses. Research libr is limited to staff   35996

Linklaters, Law Library, Library, 19th Flr, 1345 Sixth Ave, **New York**, NY 10105
T: +1 212 4249000; Fax: +1 212 4249100; URL: www.linklaters.com
Francesco Gagliarai
Law
10 000 vols; 100 curr per   35997

Loeb & Loeb LLP, Law Library, 345 Park Ave, 18th flr, **New York**, NY 10154-0037
T: +1 212 4074961; Fax: +1 212 4074990; URL: www.loeb.com
Christine Weber
10 000 vols; 500 curr per   35998

Mendes & Mount, LLP, Law Library, 750 Seventh Ave, **New York**, NY 10019-6829
T: +1 212 2618338; Fax: +1 212 2618750; URL: www.mendes.com
Ray Jassin
10 000 vols   35999

Metropolitan Transportation Authority, Law Library & Corporate Library, 347 Madison Ave, 9th Flr, **New York**, NY 10017
T: +1 212 8787192; Fax: +1 212 8780147
Dennis Maffettone
8 000 vols; 80 curr per   36000

New York Life Insurance Co, Law Library, 51 Madison Ave, Rm 10-SB, **New York**, NY 10010
T: +1 212 5766458; Fax: +1 212 5766886
1946; Diana Louros
12 000 vols; 30 curr per   36001

New York Times, Reference Library, 620 8th Ave, 5th Flr, **New York**, NY 10018
T: +1 212 5567428; Fax: +1 212 5564448
Barbara Gray
Biogr, Journalism, Politics
40 000 vols; 200 curr per   36002

Newsweek, Inc, Research Center, 251 W 57th St, **New York**, NY 10019-1894
T: +1 212 4454680; Fax: +1 212 4454056
1933; Madeline Cohen
Current Affairs, Politics
20 000 vols; 800 curr per
libr loan; SLA, ALA, ASIS   36003

O'Melveny & Myers LLP, Law Library, Times Square Tower, Seven Times Sq, **New York**, NY 10036
T: +1 212 3262008; Fax: +1 212 3262061; E-mail: nylibrary@omm.com
Heide-Marie Bliss
Law
15 000 vols   36004

Orrick, Herrington & Sutcliffe, Law Library, 666 Fifth Ave, **New York**, NY 10103
T: +1 212 5065340; Fax: +1 212 5065151
Katharine Wolpe
15 000 vols   36005

O'Sullivan, LLP, Law Library, 30 Rockefeller Plaza, **New York**, NY 10112
T: +1 212 4082435; Fax: +1 212 4082467
Joanne Fazzari
Corp law
17 000 vols; 40 curr per   36006

Oxford University Press, Inc Library, 198 Madison Ave, **New York**, NY 10016
T: +1 212 7266000, 6013; Fax: +1 212 7266457
Jelani Wilson
13 000 vols; 30 curr per
The book coll consists of only titles published by Oxford University Press, Inc   36007

Patterson, Belknap, Webb & Tyler LLP Library, 1133 Avenue of the Americas, **New York**, NY 10036
T: +1 212 3362103; Fax: +1 212 3362222
Christina Senezak
25 000 vols; 115 curr per   36008

Paul, Weiss, Rifkind, Wharton & Garrison Library, 1285 Avenue of the Americas, **New York**, NY 10019-6064
T: +1 212 3732401; Fax: +1 212 3732268
Armando Gonzalez
Law
80 000 vols   36009

Pillsbury & Winthrop LLP, Law Library, 1540 Broadway, **New York**, NY 10036-4039
T: +1 212 8581000; Fax: +1 212 8581500; E-mail: info@pillsburywinthrop.com; URL: www.pillsburywinthrop.com
Linda Becker
40 000 vols   36010

Pryor, Cashman, Sherman & Flynn, Law Library, 410 Park Ave, **New York**, NY 10022
T: +1 212 4214100; Fax: +1 212 3260806
Michelle Victor
10 000 vols; 78 curr per   36011

Putney, Twombly, Hall & Hirson, Law Library, 521 Fifth Ave, 10th Flr, **New York**, NY 10175
T: +1 212 6820020; Fax: +1 212 6829380; URL: www.putneylaw.com
Jessica Lassa
40 000 vols; 30 curr per   36012

Reed Smith, Law Library, 599 Lexington Ave, **New York**, NY 10022
T: +1 212 5215200; Fax: +1 212 5215450
1920; David Adler
80 000 vols; 24 curr per   36013

Saint Matthew's & Saint Timothy's Neighborhood Center, Inc Library, Star Learning Center, 26 W 84th St, **New York**, NY 10024
T: +1 212 3626750; Fax: +1 212 7876196
1971; Rita Spano
Children's lit, Remedial reading
11 000 vols   36014

Satterlee, Stephens, Burke & Burke, Law Library, 230 Park Ave, **New York**, NY 10169
T: +1 212 8189200; Fax: +1 212 8189606; URL: www.ssbb.com
Dolores Fusik
13 000 vols; 25 curr per   36015

Scholastic Inc Library, 557 Broadway, **New York**, NY 10012
T: +1 212 3436175; Fax: +1 212 3893317
1929; Kerry Prendergast
Children's Magazines, bks, per
15 000 vols; 120 curr per; 10 digital data carriers; 1 000 pamphlets   36016

Schulte, Roth & Zabel LLP, Law Library, 919 Third Ave, **New York**, NY 10022
T: +1 212 7562302; Fax: +1 212 5935955; E-mail: carol.sergis@srz.com; URL: www.srz.com
1969; Carol K. Sergis
70 000 vols; 450 curr per   36017

Seward & Kissel LLP, Law Library, One Battery Park Plaza, **New York**, NY 10004
T: +1 212 5741478; Fax: +1 212 4808421; E-mail: sknyc@sewkis.com; URL: www.sewkis.com
Judith Koziara
Law
18 000 vols; 150 curr per   36018

Seyfarth & Shaw Law Library, 620 Eight Ave Flr33, **New York**, NY 10018-1405
T: +1 212 2185500; Fax: +1 212 2185526; URL: www.seyfarth.com
John P. Napoli
Labor law, taxation law
10 000 vols; 100 curr per   36019

Shearman & Sterling LLP Library, 599 Lexington Ave, **New York**, NY 10022-6069
T: +1 212 8484627; Fax: +1 646 8484627
1873; John Lai
Banking law, Labor law, Legislative hist, Securities
35 000 vols; 700 curr per   36020

Sidley, Austin, Brown & Wood LLP, Law Library, 787 Seventh Ave, 24th Flr, **New York**, NY 10019
T: +1 212 8395445; Fax: +1 212 8395599
Christine Lang
50 000 vols; 250 curr per   36021

Simpson, Thacher & Bartlett, Law Library, 425 Lexington Ave, **New York**, NY 10017-3954
T: +1 212 4552800; Fax: +1 212 4552502
1884; Peggy Martin
50 000 vols   36022

Skadden, Arps, Slate, Meagher & Flom Library, Four Times Sq, **New York**, NY 10036
T: +1 212 7353000; Fax: +1 212 7353244
1948; Janet Accardo
Law
75 000 vols   36023

Sotheby's Library, 1334 York Ave, **New York**, NY 10021
T: +1 212 6067265; Fax: +1 212 6067499
Jackie Chin
Decorative art, Painting; Auction Sales (American Art Assn, Anderson Galleries, Parke Bernet & Sotheby Parke Bernet since 1891)
20 000 vols; 200 curr per   36024

Standard & Poor's Library, Central Inquiry Center-Equity Research, 55 Water St, 43rd Fl, **New York**, NY 10041
T: +1 212 4387760; Fax: +1 212 4383429; E-mail: central_inquiry-unit@sandp.com; URL: www.spglobal.com
1917; Milva Luckenbach
Corp File, pamphlets; Corporations; Daily Stock Price Record, Finance Securities Investments; Financial Chronicles; Hist Pricing & Res
15 500 vols; 2 000 curr per   36025

Stroock & Stroock & Lavan, Library, 180 Maiden Lane, **New York**, NY 10038
T: +1 212 8065700; Fax: +1 212 8066006; URL: www.stroock.com
1876; June Berger
20 000 vols; 45 curr per; 300 govt docs
libr loan; AALL, SLA   36026

Sullivan & Cromwell LLP, Information Resources Center, 125 Broad St, **New York**, NY 10004
T: +1 212 5583780; Fax: +1 212 5583346; E-mail: library@sullcrom.com; URL: www.sullcrom.com
1879; Jennifer G. Rish
96 000 vols   36027

Swidler, Berlin, Shereff & Friedman, Law Library, 405 Lexington Ave, **New York**, NY 10174; mail address: 3000 K St, Ste 300, Washington, DC 20007
T: +1 212 9730111; Fax: +1 212 8919598
10 000 vols; 50 curr per   36028

Thelen, Reid & Priest Library, 40 W 57th St, **New York**, NY 10019
T: +1 212 6032265; Fax: +1 212 6032001; E-mail: rulferts@thelenreid.com; URL: www.thelenreid.com
Ruth Ulferts
Fed securities, Foreign, Latin America, Pub utility, Taxation
20 000 vols; 900 curr per   36029

Torys Law Library, 237 Park Ave, 20th Flr, **New York**, NY 10017
T: +1 212 8806177; Fax: +1 212 6820200; E-mail: info@torys.com; URL: www.torys.com
Michael B. Hoffman
10 000 vols; 50 curr per   36030

J. Walter Thompson Co, Knowledge Center, 466 Lexington Ave, **New York**, NY 10017
T: +1 212 2107267; Fax: +1 212 2107817; URL: www.jwt.com
1918; Carol Simas
Advertising, Communications, Marketing; Picture Ref Coll, Consumer Print Advertisements
8 000 vols; 400 curr per   36031

Warshaw, Burstein, Cohen, Schlesinger & Kuh, Law Library, 555 Fifth Ave, 11th Flr, **New York**, NY 10017
T: +1 212 9847730; Fax: +1 212 9729150; URL: www.wbcsk.com
Helen Akulich
15 000 vols   36032

Weil, Gotshal & Manges Library, 767 Fifth Ave, **New York**, NY 10153
T: +1 212 3108626; Fax: +1 212 3108786; URL: www.weil.com
Deborah G. Cinque
Business, Law
45 000 vols   36033

White & Case, Law Library, 1155 Ave of the Americas, **New York**, NY 10036
T: +1 212 8198200, 7567; Fax: +1 212 3548113
1901; John J. Banta
40 000 vols   36034

Willkie, Farr & Gallagher, Law Library, 787 Seventh Ave, **New York**, NY 10019
T: +1 212 7288700; Fax: +1 212 7283303; E-mail: dglessner@willkie.com
Debra Glessner
50 000 vols; 200 curr per   36035

Wilson, Elser, Moskowitz, Edelman & Dicker, Law Library, 150 E 42nd St, **New York**, NY 10017-5639
T: +1 212 4903000; Fax: +1 212 4903038
Jim Quinn
20 000 vols; 30 curr per   36036

Windels, Marx, Lane & Mittendorf, LLP Library, 156 W 56th St, **New York**, NY 10019
T: +1 212 2371000; Fax: +1 212 2621215; E-mail: jlsolomo@sprynet.com; URL: www.windelsmarx.com
Joel Solomon
Banking law, Corp law, Litigation
27 000 vols   36037

Winston & Strawn Library, 200 Park Ave, **New York**, NY 10166
T: +1 212 2944648; Fax: +1 212 2944700; URL: www.winston.com
1993; Winston McKenzie
Law
100 000 vols; 150 curr per
libr loan; AALL, SLA   36038

E. I. DuPont de Nemours & Co, Inc, Haskell Laboratory for Toxicology & Industrial Medicine Library, Elkton Rd, P.O. Box 50, Newark, DE 19714
T: +1 302 3665232; Fax: +1 302 3665732
1935; Scott Johnson
20 000 vols; 241 curr per
libr loan   36039

975

Gibbons, Del Deo, Dolan, Griffinger & Vecchione, Law Library, One Riverfront Plaza, **Newark**, NJ 07102
T: +1 973 5964500; Fax: +1 973 6396368; URL: www.gibbonslaw.com
Patrick Del Deo
35 000 vols   36040

W. L. Gore & Associates, Inc, Information Services Library, 551 Paper Mill Rd, **Newark**, DE 19714
T: +1 302 2924028; Fax: +1 302 2924159
Marilyn McAlack
Engineering, Chemistry, Electronics, Textiles
10 000 vols   36041

Matthew Bender & Company Inc, Library, 744 Broad St, **Newark**, NJ 07012
T: +1 973 8202627; Fax: +1 973 8202459; E-mail: nyo.library@bender.com; URL: www.bender.com
Marilyn Kahn
35 000 vols; 300 curr per   36042

McCarter & English, Law Library, Four Gateway Ctr, 100 Mulberry St, **Newark**, NJ 07102-4056
T: +1 973 6224444; Fax: +1 973 6247070; URL: www.mccarter.com
Scott L. Fisher
29 000 vols; 200 curr per   36043

McElroy, Deutsch, Mulvaney & Carpenter, LLP, Law Library, Three Gateway Ctr, 100 Mulberry St, **Newark**, NJ 07102
T: +1 973 6227711; Fax: +1 973 6225314
Lucy Faris
13 000 vols   36044

Public Service Electric & Gas Company, Corporate Library, 80 Park Plaza, Mailcode P3C, **Newark**, NJ 07101
T: +1 973 4307332; Fax: +1 973 6241551; E-mail: library@pseg.com
1911; Lisa Holland
27 140 vols; 775 curr per
libr loan   36045

Sills, Cummis, Et Al Law Library, One Riverfront Plaza, **Newark**, NJ 07102-5400
T: +1 973 6437000; Fax: +1 973 6436500; E-mail: lcruz@sillscommis.com
Lynne Cruz
25 500 vols; 500 curr per; 12 microforms; 15 sound-rec; 6 digital data carriers; video tapes 25   36046

St John & Wayne, Law Library, Two Penn Plaza E, **Newark**, NJ 07105
T: +1 973 4913300; Fax: +1 973 4913555; URL: www.stjohnlaw.com
1971; Pim Savetmalanond
8 000 vols   36047

Stryker, Tams & Dill, Law Library, Two Penn Plaza E, **Newark**, NJ 07105
T: +1 973 4919500; Fax: +1 973 4919692; URL: www.stryker.com
Judy Carberry
14 000 vols; 35 curr per   36048

Fort Belknap Archives, Inc Library, Rte 1, Box 27, **Newcastle**, TX 76372
T: +1 940 8463222
35 000 vols   36049

Newport Aeronautical Sales Corp Library, 1542 Monrovia Ave, **Newport Beach**, CA 92663-2807
T: +1 949 5744100; Fax: +1 949 5744106; E-mail: info@newportaero.com; URL: www.newportaero.com
1970; George M. Posey
Aerospace Technical Libr, NAVAIRS, Technical Orders, TM
300 000 vols; 1 500 sound-rec; 50 digital data carriers   36050

Stradling, Yocca, Carlson & Rauth, Law Library, 660 Newport Ctr, Ste 1600, **Newport Beach**, CA 92660
T: +1 949 7254000; Fax: +1 949 7254100; URL: www.sycr.com
Lynn Connor Merring
12 000 vols; 250 curr per   36051

Northrop Grumman Newport News, Library Services Department, 4101 Washington Ave, **Newport News**, VA 23607
T: +1 757 3802000; Fax: +1 757 3807794; URL: www.ngc.com
1947; Rosetta Lamb
54 000 vols; 800 curr per   36052

Lyondell Chemical Co, Technical Information Center, 3801 W Chester Pike, **Newtown Square**, PA 19073
T: +1 610 3592762; Fax: +1 610 3596431
1920; Marge Giorgianni
28 000 vols; 216 curr per   36053

General Electric Corporate Research & Development, Whitney Information Services, One Research Circle, **Niskayuana**, NY 12301-0008
T: +1 518 3877571; Fax: +1 518 3877593; E-mail: library@crd.ge.com
1900; James Lommel
Biology, Chemistry, Electronics, Engineering, Finance, Info science, Mathematics, Metallurgy, Physics
26 000 vols; 1 000 curr per
libr loan   36054

Northrop Grumman, ESL Research Library, 5441 Luce Ave, **North Highlands**, CA 95652-2417; P.O. Box 800, North Highlands, CA 95660-0800
T: +1 916 5704020
1966; Jane Talbot
Electronics, Telecommunications, Engineering; IEEE Conference Proceedings
11 000 vols; 150 curr per
libr loan   36055

Western Costume Co, Research Library, 11041 Vanowen St, **North Hollywood**, CA 91605
T: +1 818 5082148; Fax: +1 818 5082182; URL: www.westerncostume.com
1915; Bobi Garland
Godey, Peterson & Vogue, mags; Sears & Montgomery Ward Cats (1895 to present); Twentieth C Fox Costume Still Coll
14 000 vols; 60 curr per   36056

Triodyne Inc, Beth Hamilton Safety Library, 450 Skokie Blvd, **Northbrook**, IL 60062
T: +1 847 6774730; Fax: +1 847 6472047; E-mail: infoserv@triodyne.com; URL: www.triodyne.com/library
1979; Shelly Hamilton
Accident prevention, Ergonomics, Forensic engineering, Human factors, Automotive engineering; Expert Transcript Ctr
10 000 vols; 200 curr per
libr loan   36057

Wiss, Janney, Elstner Associates, Inc, 330 Pfingsten Rd, **Northbrook**, IL 60062
T: +1 847 7537202; Fax: +1 847 4980358; URL: www.wje.com
Penny Sympson
Engineering, cement
10 000 vols; 100 curr per; 5 diss/theses; 10 sound-rec; 5 digital data carriers
libr loan; NSLS, SLA   36058

R. T. Vanderbilt Co, Inc, Corporate Library, 30 Winfield St, **Norwalk**, CT 06855; P.O. Box 5150, Norwalk, CT 06856-5150
T: +1 203 8531400 ext 566; Fax: +1 203 8310648; URL: www.vanderbilt.com
1956; Anne Marie Dostilio
Plastics, Ceramics, Mineralogy
10 600 vols; 130 curr per   36059

Procter & Gamble Pharmaceuticals, Research Library, Rt 320, Woods Corners, **Norwich**, NY 13815; P.O. Box 191, Norwich, NY 13815
T: +1 607 3352947; Fax: +1 607 3352098
1961; Linda Slentz
Biology, Chemistry, Medicine, Pharm; Pharmacy (Drugs)
10 000 vols; 500 curr per; 150 microforms; 3 digital data carriers   36060

Factory Mutual Research Corp, Technical Information Center, 1151 Boston-Providence-Tpk, **Norwood**, MA 02062
T: +1 781 2554764; Fax: +1 781 7629375; E-mail: janet.green@fmglobal.com
1968; Janet B. Green
Technology; Company Hist
12 000 vols; 150 curr per   36061

Clayton Group Services, Inc, Library & Information Center, 22345 Roethel Dr, **Novi**, MI 48375-4710
T: +1 248 3441770; Fax: +1 248 3442654
1954; Bob Lieckfield
Pollution, Chemistry, Environmental engineering, Occupational safety, Public health, Toxicology; OSHA & EPA Govt Doc
10 000 vols; 150 curr per   36062

Hoffmann-La Roche, Inc Library, 340 Kingsland St, **Nutley**, NJ 07110-1199
T: +1 973 2353092; Fax: +1 973 2355477; E-mail: peter.borgula@roche.com
1930; Peter Borgulya
Pharmaceutical industry, Pharmacology, Marketing
16 000 vols; 200 curr per   36063

Oak Ridge National Laboratory, ORNL Research Library, Bldg 4500N, MS-6191, Bethel Valley Rd, **Oak Ridge**, TN 37830; P.O. Box 2008, Oak Ridge, TN 37831-6191
T: +1 865 5746744; Fax: +1 865 5746915; E-mail: library@ornl.gov; URL: www.ornl.gov/info/library
1946; Conrad Bob
Biology, Ceramics, Chemistry, Climate, Computing, Energy, Environmental sciences, Nuclear engineering, Mat science, Neutron science, Physics, Robotics; DOE Scientific & Tech Rpts
77 000 vols; 535 curr per; 9 000 e-books
libr loan   36064

Fitzgerald, Abbott & Beardsley, Law Library, 1221 Broadway, 21st Fl, **Oakland**, CA 94612
T: +1 510 4513300; Fax: +1 510 4511527; URL: www.fablaw.com
1906; Virginia Meadowcroft
12 000 vols   36065

Reed Smith LLP, Law Library, 1999 Harrison St, **Oakland**, CA 94612
T: +1 510 4666195; Fax: +1 510 2738832; URL: www.reedsmith.com
1967; Nora Levine
12 000 vols; 300 curr per; 20 digital data carriers   36066

Andrews, Davis, Legg, Bixler, Milsten & Price, Law Library, 100 N Broadway, Ste 3300, **Oklahoma City**, OK 73102
T: +1 405 2729241; Fax: +1 405 2358786; URL: www.andrewsdavis.com
Susan Fibus
12 000 vols; 50 curr per   36067

Macfee & Taft Law Offices, Law Library, Two Leadership Sq, 10th Flr, 211 N Robinson, **Oklahoma City**, OK 73102
T: +1 405 2359621; Fax: +1 405 2350439
1952; Patsy Trotter
14 000 vols; 50 curr per   36068

Tronox LLC, Technical Center Library, 3301 NW 150th St, **Oklahoma City**, OK 73134
T: +1 405 7755797; Fax: +1 405 7755632
1964; Vicki Vann
Patent-Chemicals Coll, Rare Earth Coll
12 000 vols; 100 curr per   36069

Kutak Rock LLP, Law Library, 1650 Farnam St, **Omaha**, NE 68102-2186
T: +1 402 3466000; Fax: +1 402 3461148; URL: www.kutakrock.com
Lynn Koperski
10 000 vols; 100 curr per   36070

Lamson, Dugan & Murray LLP, Law Library, 10306 Regency Pkwy Dr, **Omaha**, NE 68114
T: +1 402 3977300; Fax: +1 402 3977824; URL: www.ldmlaw.com
Evelyn Owens
20 000 vols; 50 curr per   36071

Akerman, Senterfitt & Eidson PA, Law Library, 420 S Orange Ave, Ste 1200, **Orlando**, FL 32801; P.O. Box 231, Orlando, FL 32802-0231
T: +1 407 4234000; Fax: +1 407 8436610; E-mail: info.library@akerman.com; URL: www.akerman.com
1922; Linda Fowlie
20 000 vols
AALL, SLA, SOLINET   36072

Holland & Knight LLP Orlando Library, 200 S Orange Ave, Ste 2600, **Orlando**, FL 32801; P.O. Box 1526, Orlando, FL 32802-1526
T: +1 407 4258500; Fax: +1 407 2445288; URL: www.hklaw.com
Margie Hawkins
Law
25 000 vols; 72 curr per   36073

Lockheed Martin Corp, Missiles & Fire Control Technical Information Center, 5600 Sand Lake Rd, MP 30, **Orlando**, FL 32819-8907
T: +1 407 3562051; Fax: +1 407 3563665
Patricia Puglisi
10 000 vols; 100 curr per   36074

Rumberger, Kirk & Caldwell, PA, Law Library, 300 S Orange Ave, Ste 1400, **Orlando**, FL 32801; P.O. Box 1873, Orlando, FL 32802-1873
T: +1 407 8727300; Fax: +1 407 8412133; URL: www.rumberger.com
1978; Dennis Herald
Expert Witness Deposition Bank, Florida Legislative Hist, Continuing Legal Education Audiovisuals, Engineering & Automotive Tech Publs, Skeletons & Organ Models
15 000 vols   36075

Oswego County Supreme Court, Law Library, 25 E Oneida St, **Oswego**, NY 13126
T: +1 315 3493297; Fax: +1 315 3493273
Catherine Upwood
10 000 vols   36076

Black & Veatch, Central Library, 11401 Lamar, **Overland Park**, KS 66211
T: +1 913 4587884; Fax: +1 913 4582934
1915; Kevin Nelson
Engineering
25 000 vols; 50 curr per   36077

Intersil Corp, Intersil/Harris-GCSD Library, 1650 Robert J Conlan Blvd NE, **Palm Bay**, FL 32905-3378
T: +1 321 7247733; Fax: +1 321 7291019; E-mail: library2@intersil.com
1952; Mary B. Briand
Engineering, Electronics
15 000 vols; 150 curr per   36078

Hewlett-Packard Laboratories, Research Library, 1501 Page Mill Rd, **Palo Alto**, CA 94303-0969; P.O. Box 10490, Palo Alto, CA 94303-0490
T: +1 650 8576620; Fax: +1 650 8528187; E-mail: research.library@hp.com; URL: lib.hpl.hp.com
1952; Eugenie Prime
Electronics, Physics, Marketing, Management
55 000 vols; 550 curr per
libr loan   36079

Wilson, Sonsini, Goodrich & Rosati, Library, 650 Page Mill Rd, **Palo Alto**, CA 94304
T: +1 650 4939300; Fax: +1 650 4936811
Leiza D. MacMorris
40 000 vols; 300 curr per   36080

Xerox Corp, Palo Alto Research Center Information Center, 3333 Coyote Hill Rd, **Palo Alto**, CA 94304-1314
T: +1 650 8124040; Fax: +1 650 8124028; E-mail: library.parc@xerox.com
1971; Katherine Jarvis
Electronics, Physics
15 000 vols; 800 curr per
libr loan   36081

Pearson Learning Group, 299 Jefferson Rd, *Parsippany*, NJ 07054
T: +1 973 7398000; Fax: +1 973 7398595
1885; Wendy Johnson
Computer education; Silver Burdett Publs from 1885
10 000 vols; 100 curr per   36082

Parsons Corp, Technical Library, 100 W Walnut, *Pasadena*, CA 91124
T: +1 626 4403998; Fax: +1 626 4402630; URL: www.parsons.com
Claire Hammond
Aviation, transportation
10 000 vols   36083

Parsons Corporation, Technical Library, 100 W Walnut St, *Pasadena*, CA 91124
T: +1 626 4403998; Fax: +1 626 4402630; E-mail: claire_hammond@parsons.com
Claire Hammond
Infrastructure, Transportation, Petrochemicals
10 000 vols; 38 000 curr per   36084

Lawler, Matusky & Skelly Engineers LLP Library, One Blue Hill Plaza, *Pearl River*, NY 10965-3104
T: +1 845 7358300 ext 334; Fax: +1 845 7357466; E-mail: vgalperin@lmseng.com
1965; Victoria Galperin
Water pollution, Water quality, Civil engineering, Ecology, Hazardous waste, Geology
10 500 vols; 150 curr per; 16 000 mss; 160 diss/theses; 170 000 microforms; 15 digital data carriers
libr loan   36085

Wyeth-Ayerst Research, Subba Row Memorial Library, 401 N Middletown Rd, *Pearl River*, NY 10965-1299
T: +1 914 7323401; Fax: +1 914 7325525; E-mail: prllib@war.wyeth.com
1915; Dr. Anne T. O'Brien
Biomedical sciences, Chemistry, Medicine, Pharmacology, Veterinary medicine
15 000 vols; 900 curr per   36086

Cleveland Electric Illuminating Company, Perry Power Plant Information & Resource Center, TEC 219 Bldg, 10 Center Rd, *Perry*, OH 44081
T: +1 440 2593737 ext 5864; Fax: +1 440 2808027
1984; Annette Sintracio
12 090 vols; 50 curr per
libr loan   36087

Ballard, Spahr, Andrews & Ingersoll, LLP Library, 1735 Market St, 51st Flr, *Philadelphia*, PA 19103-7599
T: +1 215 8648150; Fax: +1 215 8648999
David Proctor
25 000 vols; 400 curr per   36088

Beasley, Casey & Erbstein Library, 1125 Walnut St, *Philadelphia*, PA 19107-4997
T: +1 215 5921000; Fax: +1 215 5928360; URL: www.tortlaw.com
Joel Tuckman
Medicine
10 000 vols; 150 curr per   36089

Bio-Rad Laboratories, Inc, Informatics Division Library, 3316 Spring Garden St, *Philadelphia*, PA 19104
T: +1 215 3827800; Fax: +1 215 6620585
1966; Bernadette Steiner
Spectra for over 250,000 compounds
10 000 vols; 5 curr per   36090

Blank Rome LLP, Law Library, One Logan Sq, 18th & Cherry Sts, *Philadelphia*, PA 19103-6998
T: +1 215 5695500; Fax: +1 215 5695546; E-mail: librarygroup@blankrome.com; URL: www.blankrome.com
Mary Sheridan Newman
20 000 vols; 400 curr per   36091

Buchanan Ingersoll & Rooney PC, Philadelphia Branch, 1835 Market St, 14th Flr, *Philadelphia*, PA 19103
T: +1 215 6655311; Fax: +1 215 6658760; URL: www.bipc.com
Jeffrey Kreiling
18 000 vols; 312 curr per   36092

Dechert Library, 4000 Bell Atlantic Tower, 1717 Arch St, *Philadelphia*, PA 19103-2793
T: +1 215 9944000; Fax: +1 215 9942222; URL: www.dechert.com
Susan G. Alford
25 000 vols; 200 curr per   36093

Dilworth, Paxson LLP, Law Library, 1735 Market St, Mellon Banks Ctr, Ste 3200, *Philadelphia*, PA 19103-7595
T: +1 215 5757112; Fax: +1 215 5757200; E-mail: callahank@dilworthlaw.com; URL: www.dilworthlaw.com
Karen Callahan
15 000 vols; 50 curr per   36094

Drinker Biddle & Reath LLP, Law Library, One Logan Sq, 18th & Cherry St, *Philadelphia*, PA 19103
T: +1 215 9882952; Fax: +1 215 5641329
1970; Linda-Jean Schneider
20 000 vols; 350 curr per   36095

El Du Pont De Nemours & Co, Inc, Marshall Laboratory Library, 3401 Grays Ferry Ave, *Philadelphia*, PA 19146
T: +1 215 5396213; Fax: +1 215 5396305
1951; Dupont Marshell
Chemistry, Paint & coatings
12 000 vols; 225 curr per   36096

Duane Morris LLP Library, 30 S 17th St, *Philadelphia*, PA 19103-4196
T: +1 215 9791720; Fax: +1 215 9791020; URL: www.duanemorris.com
1904; Christine Scherzinger
18 000 vols   36097

Federal Reserve Bank of Philadelphia, Library & Research Center, 100 N Sixth St, Fourth Flr, *Philadelphia*, PA 19106
T: +1 215 5746540; Fax: +1 215 5743847; URL: www.philadelphiafed.org
1922; Christine Le
Federal Reserve System Publs
30 000 vols; 400 curr per   36098

Fox Rothschild LLP, Law Library, 2000 Market St, *Philadelphia*, PA 19103-3291
T: +1 215 2992732; Fax: +1 215 2992150; E-mail: library@foxrothschild.com
Catherine M. Monte
20 000 vols   36099

Harvey, Pennington, Cabot, Griffith & Renneisen, Ltd, Law Library, 11 Penn Ctr Plaza, 1835 Market St, 29th Fl, *Philadelphia*, PA 19103
T: +1 215 5634470 ext 236; Fax: +1 215 5681044; URL: www.harvpenn.com
Charlotte Braunstein
10 000 vols; 20 curr per   36100

Hoyle, Fickler, Herschel & Mathes LLP, Law Library, One S Broad St, Ste 1500, *Philadelphia*, PA 19107
T: +1 215 9815852; Fax: +1 215 9815959
1985; Gwendolyn R. Yohannan
8 000 vols; 25 curr per   36101

Klehr, Harrison, Harvey, Branzburg & Ellers, Law Library, 260 S Broad St, *Philadelphia*, PA 19102-5003
T: +1 215 5686060; Fax: +1 215 5686603; URL: www.klehr.com
Margaret S. Fallon
15 000 vols; 32 curr per   36102

Mesirov, Gelman, Jaffe, Cramer & Jamieson Library, 1600 Market St, *Philadelphia*, PA 19103-7598
T: +1 215 7512000, 9941000; Fax: +1 215 7512416
Linda Hauk
Law
10 000 vols; 40 curr per   36103

Montgomery, Mccracken, Walker & Rhoads LLP Library, 123 S Broad St, *Philadelphia*, PA 19109
T: +1 215 7727611; Fax: +1 215 7727620
Georgeanne H. Brown
Law
15 000 vols; 162 curr per   36104

Morgan, Lewis & Bockius LLP, Law Library, 1701 Market St, 13th Flr, *Philadelphia*, PA 19103-2921
T: +1 215 9635000; Fax: +1 215 9635001; URL: www.morganlewis.com
1873; Connie Smith
50 000 vols; 212 curr per   36105

Pepper, Hamilton LLP, Law Library, 3000 Two Logan Sq – 18th & Arch Sts, *Philadelphia*, PA 19103-2799
T: +1 215 9814105; Fax: +1 215 9814750; E-mail: library@pepperlaw.com
Robyn L. Beyer
35 000 vols; 500 curr per   36106

Philadelphia Newspapers, Inc, Inquirer & Daily News Library, 400 N Broad St, *Philadelphia*, PA 19130-4099; P.O. Box 8263, Philadelphia, PA 19101-8263
T: +1 215 8544660; Fax: +1 215 8545697
1926; Michael Panzer
Newspaper Clippings, photos, VF
8 000 vols; 30 curr per
libr loan   36107

Rawle & Henderson, Law Library, The Widener Bldg, One S Penn Sq, 17th Fl, *Philadelphia*, PA 19107
T: +1 215 5754480; Fax: +1 215 5632583
1783; Christine Harvan
14 000 vols   36108

Reed Smith LLP, Law Library, 2500 One Liberty Pl, 1650 Market St, *Philadelphia*, PA 19103
T: +1 215 8511413; Fax: +1 215 8511420; URL: www.reedsmith.com
Scott Demaris
8 000 vols; 100 curr per   36109

Saul Ewing LLP, Law Library, Centre Square W, 1500 Market St, 38th Flr, *Philadelphia*, PA 19102
T: +1 215 9727873; Fax: +1 215 9721945; URL: www.saul.com
Stacey Digan
24 000 vols; 100 curr per   36110

Schnader, Harrison, Segal & Lewis Library, 1600 Market St, Ste 3600, *Philadelphia*, PA 19103
T: +1 215 7512111; Fax: +1 215 7512205; URL: www.schnader.com
Bobbi Cross
Law
35 000 vols; 1 000 curr per   36111

Stradley, Ronon, Stevens & Young LLP Library, 2600 One Commerce Sq, *Philadelphia*, PA 19103-7098
T: +1 215 5648190; Fax: +1 215 5648120; E-mail: srsyljs@class.org
1972; Kathleen Caron
PA Pamphlet Laws Since 1700
10 000 vols; 250 curr per   36112

White & Williams, LLP, Law Library, 1800 One Liberty Pl, *Philadelphia*, PA 19103-7395
T: +1 215 8647000; Fax: +1 215 8647123; URL: www.whiteandwilliams.com
Evelyn Quillen
15 000 vols   36113

Mallinckrodt Baker Inc, The Library Center, 600 N Broad St, *Phillipsburg*, NJ 08865
T: +1 908 8592151 ext 9615; Fax: +1 908 8599454; E-mail: patrice.otani@mkg.com
1945; Patti Otani
Analytical chemistry, Biochemistry
12 500 vols; 100 curr per   36114

AMEC Earth & Environmental Technical Library, 3232 W Virginia Ave, *Phoenix*, AZ 85009
T: +1 602 2726848; Fax: +1 602 2727239; E-mail: erukkila@agraus.com
Elyse Rukkila
Engineering, Environment, Geology
11 300 vols; 120 curr per   36115

Burch & Cracchiolo, Law Library, 702 E Osborn Rd, Ste 200, *Phoenix*, AZ 85014; P.O. Box 16882, Phoenix, AZ 85014
T: +1 602 2348704; Fax: +1 602 2340341; E-mail: library@bcattorneys.com
Diane Abazarnia
12 000 vols   36116

Jennings, Strouss & Salmon, Law Library, The Collier Center, 11th Flr, 201 E Washington St, *Phoenix*, AZ 85004-2385
T: +1 602 2629511; Fax: +1 602 2533255; URL: www.jsslaw.com
Marcy McCahan
20 000 vols   36117

Lewis & Roca Library, Renaisonne Tower, No 2, 40 N Central Ave, Ste 1900, *Phoenix*, AZ 85004-4429
T: +1 602 2625303; Fax: +1 602 7343739; URL: lrlaw.com
Michael Reddy
Corporate law, Real estate
40 000 vols; 100 curr per
libr loan   36118

Maynard, Murray, Cronin, Erickson & Curran, Law Library, 3200 N Central, Ste 1800, *Phoenix*, AZ 85012
T: +1 602 2798500; Fax: +1 602 2638185
Heidi Burnett
13 000 vols   36119

Osborn Maledon, Law Library, 2929 W Central 21st Fl, *Phoenix*, AZ 85012-2794
T: +1 602 6409000; Fax: +1 602 6409050
Markita Martinez
10 000 vols; 300 curr per   36120

Perkins Coie Brown & Bain Library, 2901 N Central Ave, *Phoenix*, AZ 85012
T: +1 602 3518039; Fax: +1 602 6487039; URL: www.perkinscoie.com
Ellen Hepner
20 000 vols; 150 curr per   36121

Snell & Wilmer, Law Library, One Arizona Ctr, *Phoenix*, AZ 85004-0001
T: +1 602 3826056; Fax: +1 602 3826070; E-mail: bristoa@swlaw.com; URL: www.swlaw.com
1938; Arlen A. Bristol
25 000 vols
libr loan   36122

United States Enrichment Corp, X710 Technical Resource Center, 3930 US Rte 23S MS-1101 X710, P.O. Box 628, *Piketon*, OH 45661-0628
T: +1 740 8972925; Fax: +1 740 8972507
1953; Janet Lynch
Analytical & inorganic chemistry, Engineering, Science
40 000 vols; 300 curr per   36123

Colgate-Palmolive Co, Technology Information Center Library, 909 River Rd, *Piscataway*, NJ 08855
T: +1 732 8787574; Fax: +1 732 8787128; E-mail: miranda_scott@colpal.com
1898; Miranda Scott
Chemistry, Cosmetics, Dental, Medicine
25 000 vols; 400 curr per   36124

Telcordia Technologies Inc – Raritan River Software Systems Center Library, 444 Hoes Lane, Rm 1D124, *Piscataway*, NJ 08854-4182
T: +1 732 6992290; Fax: +1 732 3362940; E-mail: akneller@telcordia.com
1984; Anne N. Kneller
Telecommunications
14 000 vols; 300 curr per   36125

RFS Ecusta Inc, Technical Library, P.O. Box 200, *Pisgah Forest*, NC 28768-0200
T: +1 828 8772211, 2310; Fax: +1 828 8772385
1939; S.W. McCarty
Cellulose, Paper manufacture, Chemistry, Pulp, Air pollution, Water
9 946 vols; 65 curr per   36126

Bayer Corp Library, 100 Bayer Rd, *Pittsburgh*, PA 15205-9741
T: +1 412 7772782; Fax: +1 412 7772758; URL: www.bayer.com
1954; Nancy A. Alstadt
Science & technology, Coatings, Plastics, Market res; Chemical Economics Handbk; Manufacturing (State Industrial Directories); German Language (Varied, Language Courses), bk, flm, cassettes
8 000 vols; 25 curr per   36127

Buchanan Ingersoll & Rooney PC, One Oxford Ctr, 20th Flr, *Pittsburgh*, PA 15219-6498
T: +1 412 5628800; Fax: +1 412 5621041; URL: www.bipc.com
Lori Zilla
Law; Commerce Clearinghouse House Standard Federal Tax Reporter, Pennsylvania Bar Assn Seminar Coursebks
18 000 vols; 312 curr per   36128

977

Crucible Materials Corporation, Crucible Research Division Library, 6003 Campbells Run Rd, *Pittsburgh*, PA 15205
T: +1 412 9232955 ext 255; Fax: +1 412 7884665; E-mail: paducci@crucibleresearch.com
1954; Patricia J Aducci
12 000 vols 36129

Jones Day, Law Library, 500 Grant St, 31st Flr, *Pittsburgh*, PA 15219
T: +1 412 3947226; Fax: +1 412 3947959
Heather R. Love
15 000 vols 36130

PPG Industries Inc – Glass Technology Center, Library, Guys Run Rd, Harmar Township, *Pittsburgh*, PA 15238; P.O. Box 11472, Pittsburgh, PA 15238-0472
T: +1 412 8204936; Fax: +1 412 8208696; E-mail: dunckhorst@ppg.com; URL: www.ppg.com
1975; Martha Dunckhorst
Fiber glass; Patents Coll
25 000 vols; 150 curr per
libr loan 36131

Reed Smith LLP, Law Library, 435 Sixth Ave, *Pittsburgh*, PA 15219
T: +1 412 2883377; Fax: +1 412 2883063; URL: www.reedsmith.com
1970; Ronda W. Fisch
35 000 vols; 2 000 curr per 36132

Westinghouse Electric Corp – Energy Systems Business Unit Technical Library, P.O. Box 355, *Pittsburgh*, PA 15230
T: +1 412 3744816; Fax: +1 412 3744042
1955; Nancy Flury Carlson
25 000 vols; 210 curr per; 300 000 microforms; 3 digital data carriers; 50 000 rpts (hard copy), 5 000 standards
libr loan 36133

Harris, Beach PLLC, Law Library, 99 Garnsey Rd, *Pittsford*, NY 14534
T: +1 585 4198800; Fax: +1 585 4198814; URL: www.harrisbeach.com
Joan Pedzich
21 000 vols; 450 curr per 36134

Alcatel USA, Inc, Information Center, 1000 Coit Rd, *Plano*, TX 75075
T: +1 972 9965000; Fax: +1 972 5193115; E-mail: wanda.fox@alcatel.com; URL: www.alcatelusa.com
1953; Wanda Fox
Electronics, Computers, Telecommunications
55 000 vols; 300 curr per 36135

J. C. Penney Company Inc, Law Library, 6501 Legacy Dr, MS 1117, *Plano*, TX 75024
T: +1 972 4311254; Fax: +1 972 4311133; E-mail: dmmetivi@jcpenney.com
Donna M. Metivier
25 000 vols; 90 curr per 36136

Reader's Digest Association Inc Library, Editorial & Research Libraries, Reader's Digest Rd, *Pleasantville*, NY 10570-7000
T: +1 914 2445289; Fax: +1 914 2442444
Ann DiCesare
Reader's Digest Publications Coll
14 000 vols; 375 curr per; 3 digital data carriers 36137

Lanterman Developmental Center, Staff Library, 3530 W Pomona Blvd, P.O. Box 100, *Pomona*, CA 91769-0100
T: +1 909 4447264; Fax: +1 909 4447524; E-mail: inlib1@ix.netcom.com
1954; Treva Fredericks
Syndromes
15 000 vols; 131 curr per; 1 000 microforms; 600 sound-rec; 22 digital data carriers; 15 VF drawers of pamphlets, 20 slides
libr loan; MLGSCA 36138

Conoco Incorporated, Technology, Technical Information Center, 1000 S Pine, *Ponca City*, OK 74601-7509; P.O. Box 1267, Ponca City, OK 74602-1267
T: +1 580 7672334; Fax: +1 580 7674217
1950; Debra McConaghy
Chemistry, Coal, Engineering, Mineralogy, Petroleum
10 000 vols; 225 curr per
libr loan 36139

Ater & Wynne, LLP, Law Library, KOIN Ctr, Ste 1800, 222 SW Columbia, *Portland*, OR 97201
T: +1 503 2261191; Fax: +1 503 2260079; URL: www.aterwynne.com
Doreen Smith
20 000 vols; 50 curr per 36140

Bernstein, Shur, Sawyer & Nelson, Law Library, 100 Middle St, *Portland*, ME 04104; P.O. Box 9729, Portland, ME 04104-5029
T: +1 207 7741200; Fax: +1 207 7741127; URL: www.bernsteinshur.com
Christine Bertsch
8 000 vols; 200 curr per 36141

Bullivant, Houser & Bailey, Law Library, 300 Pioneer Tower, 888 SW Fifth Ave, *Portland*, OR 97204-2089
T: +1 503 2286351; Fax: +1 503 2950915; URL: www.bullivant.com
Laurie Daley
10 000 vols; 80 curr per 36142

Davis Wright Tremaine, Law Library, 1300 SW Fifth Ave, Ste 2300, *Portland*, OR 97201
T: +1 503 2412300; Fax: +1 503 7785299; E-mail: carolkreger@dwt.com; URL: www.dwt.com
Carol L. Kreger
10 000 vols; 40 curr per 36143

Lane Powell PC, Law Library, 601 SW Second Ave, No 2100, *Portland*, OR 97204
T: +1 503 7782100; Fax: +1 503 7782200; URL: www.lanepowell.com
Linda Tobiska
10 000 vols; 100 curr per 36144

Miller Nash LLP Library, 111 SW Fifth Ave, Ste 3400, *Portland*, OR 97204-3699
T: +1 503 2245858; Fax: +1 503 2240155; URL: www.millernash.com
Leslie Meserve
20 000 vols; 400 curr per
AALL 36145

Perkins Coie, Law Library, 1120 NW Couch St, 10th Flr, *Portland*, OR 97209-4128
T: +1 503 7272051; Fax: +1 503 7272222; URL: www.perkinscoie.com
Barbara Holt
10 000 vols; 50 curr per 36146

Pierce Atwood LLP, Law Library, One Monument Sq, *Portland*, ME 04101
T: +1 207 7911142; Fax: +1 207 7911350; URL: www.pierceatwood.com
Kami Bedard
20 000 vols; 120 curr per 36147

Schwabe, Williamson & Wyatt Library, Pacwest Center, 1211 SW Fifth Ave, Ste 1500, *Portland*, OR 97204-3795
T: +1 503 7962854; Fax: +1 503 7962900; URL: www.schwabe.com
LaJean Humphries
Oregon & Federal Law
20 000 vols; 218 curr per; 51 maps; 30 av-mat; 64 sound-rec; 3 digital data carriers 36148

Tonkon Torp LLP, Law Library, 888 SW Fifth Ave, Ste 1600, *Portland*, OR 97204-2099
T: +1 503 2211440; Fax: +1 503 9723789; URL: www.tonkon.com
Richard LaSasso
10 000 vols; 50 curr per 36149

Verrill & Dana Library, 2 Portland Sq, P.O. Box 586, *Portland*, ME 04112-0586
T: +1 207 7744000 ext 4856; Fax: +1 207 7747499; E-mail: areiman@verrilldana.com; URL: www.verrilldana.com
Anne M. Reiman
13 000 vols 36150

Raytheon Co, Research Library, 1847 W Main Rd, *Portsmouth*, RI 02871-1087
T: +1 401 8424372; Fax: +1 401 8425206
1960; Mark Baldwin
Electronics, Oceanography, Antisubmarine warfare, Sonar, Surface mount
25 000 vols; 110 curr per 36151

Strawbery Banke, Inc, Thayer Cumings Library & Archives, 454 Court St, *Portsmouth*, NH 03802-4603; P.O. Box 300, Portsmouth, NH 03802-0300
T: +1 603 4227524; Fax: +1 603 4331115; URL: www.strawberybanke.org
1970; Kimberly Alexander
Archaeology, Architecture, Art, Horticulture, Local hist; Mss Coll, Portsmouth Photo Coll (1870-1970)
8 000 vols; 20 curr per; 149 microforms 36152

IBM Corp, Poughkeepsie Library, 2455 South Rd, *Poughkeepsie*, NY 12601-5400
T: +1 845 4356029; Fax: +1 845 4329445
Maris Kristapsons
Computer science
12 000 vols; 200 curr per; 750 av-mat 36153

Drinker, Biddle & Reath LLP Library, 105 College Rd E, *Princeton*, NJ 08542
T: +1 609 7166500; Fax: +1 609 7997000; URL: www.dbr.com
Ann McCarron
55 000 vols 36154

Mathematica Policy Research Inc Library, 600 Alexander Park, *Princeton*, NJ 08543
T: +1 609 2752334; Fax: +1 609 7991654; URL: www.mathematica-mpr.com
Jan Watterworth
45 000 vols; 350 curr per 36155

Sarnoff Corporation Library, 201 Washington Rd, *Princeton*, NJ 08540; CN5300, Princeton, NJ 08543-5300
T: +1 609 7342850, 7342848; Fax: +1 609 7342339; E-mail: leubank@sarnoff.com
1942; Larry Eubank
Electronics
30 000 vols; 150 curr per 36156

Siemens Corporate Research, Inc, Research Library, 755 College Rd E, *Princeton*, NJ 08540
T: +1 609 7346506; Fax: +1 609 7346565; E-mail: rhw@scr.siemens.com; URL: www.scr.siemens.com
Ruth Weitzenfeld
Computer science
9 000 vols; 120 curr per 36157

Edwards & Angell, LLP Library, 2800 Financial Plaza, *Providence*, RI 02903
T: +1 401 2766521; Fax: +1 401 2766611
Mary M. Ames
10 000 vols 36158

Merck & Company, Inc, Library Rahway, LR86, R86-230, 126 E Lincoln Ave, *Rahway*, NJ 07065; P.O. Box 2000, Rahway, NJ 07065-2000
T: +1 732 5943747; Fax: +1 732 5942876
1953; Diana Komanesky
Chemistry, Computer science, Genetics, Medicine, Pharmacology
60 000 vols; 5 000 curr per 36159

Hunton & Williams, Law Library, One Bank of America Plaza, Ste 1400, 421 Fayetteville St, *Raleigh*, NC 27601
T: +1 919 8993057; Fax: +1 919 8336352; URL: www.hunton.com
1980; Susan Kaydos
10 000 vols; 70 curr per 36160

Parker, Poe, Adams & Bernstein, LLP, Law Library, Wachovia Capitol Ctr, 150 Fayetteville Street Mall, Ste 1400, *Raleigh*, NC 27601
T: +1 919 8280564; Fax: +1 919 8344564; E-mail: info@parkerpoe.com; URL: www.parkerpoe.com
Lisa W. Williams
9 000 vols; 39 curr per 36161

Poyner & Spruill, Law Library, 3600 Glenwood Ave, *Raleigh*, NC 27612-4945; P.O. Box 10096, Raleigh, NC 27605-0096
T: +1 919 7836400; Fax: +1 919 7831075; E-mail: aollam@poynerspruill.com; URL: www.poynerspruill.com
Lou Lamm
20 000 vols; 100 curr per 36162

Williams Mullen Library, 3200 Beechleaf Ct, Ste 500, *Raleigh*, NC 27604-1670; P.O. Drawer 19764, Raleigh, NC 27619-9764
T: +1 919 9814038; Fax: +1 919 9814300
Catherine V. Lambe
12 000 vols; 200 curr per; 20 av-mat; 20 sound-rec; 10 digital data carriers; 20 videos 36163

Carpenter Technology Corp, Research & Development Library, Center Ave & Union St, P.O. Box 14662, *Reading*, PA 19612-4662
T: +1 610 2082583; Fax: +1 610 2083256
1950; Wendy M Schmehl
14 100 vols; 160 curr per; 10 000 microforms; 8 500 doc unbound
libr loan 36164

Northrop Grumman Information Technology, TASC Library, 55 Walkers Brook Dr, *Reading*, MA 01867
T: +1 781 9422000; Fax: +1 781 9427100
Martha Dionne
Economics, Engineering, Finance, Mathematics, Meteorology, Software engineering
30 000 vols; 450 curr per 36165

TRW, Technical Information Center, One Space Park, S/1316, *Redondo Beach*, CA 90278-1099
T: +1 310 8124185; Fax: +1 310 8143203; E-mail: techinfo.center@trw.com
1954; Pat Artiago
Aerospace science, Electronics
25 000 vols; 300 curr per; 400 000 docs micro, 154 400 docs HC 36166

Ropers, Majeski, Kohn & Bentley, Law Library, 1001 Marshall St, *Redwood City*, CA 94063
T: +1 650 3648200; Fax: +1 650 7801701; URL: www.ropers.com
Carmen Callahan
10 000 vols; 12 curr per 36167

IBM Rtp Library, 3039 Cornwall Rd, *Research Triangle Park*, NC 27709
T: +1 919 5431299; Fax: +1 919 5431701
1965; Dorothy Huey
12 000 vols; 100 curr per 36168

Latham & Watkins LLP, Law Library, 11955 Freedom Dr, Ste 500, *Reston*, VA 20190
T: +1 703 4565249; Fax: +1 703 4561001; URL: www.lw.com
J. O. Wallace Jr
18 000 vols; 900 curr per 36169

Pacific Northwest National Laboratory – Legal Library, Federal Bldg, Rm 431, P.O. Box 800, *Richland*, WA 99352
T: +1 509 3766807; Fax: +1 509 3769039
Terri Pettibow
Environmental law & regulations
9 000 vols 36170

Christian & Barton, LLP Attorneys At Law, Law Library, 909 E Main St, *Richmond*, VA 23219
T: +1 804 6974100; Fax: +1 804 6974112; URL: www.cblaw.com/
Jane O. Bowe
10 000 vols; 200 curr per 36171

Division of Legislative Services Reference Center, General Assembly Bldg, 2nd Flr, 910 Capitol St, *Richmond*, VA 23219
T: +1 804 7863591; Fax: +1 804 3718705; URL: dls.state.va.us/lrc.htm
Cheryl Jackson
8 000 vols; 150 curr per 36172

Federal Reserve Bank of Richmond, Research Library, 701 E Byrd St, *Richmond*, VA 23219; P.O. Box 27622, Richmond, VA 23261-7622
T: +1 804 6978125; Fax: +1 804 6978134; E-mail: rich.reslib@rich.frb.org
1920; Anne R. Hallerman
Federal Reserve System Publs
29 000 vols; 325 curr per; 983 microforms; 7 digital data carriers; 144 VF drawers
libr loan 36173

Hunton & Williams, Law Library, Riverfront Plaza, E Tower, 951 E Byrd St, **Richmond**, VA 23219-4074
T: +1 804 7888272; Fax: +1 804 7888218
1901; Frosty Owen
Records & Briefs, Law Memoranda, Speeches of Henry W Anderson
65 000 vols; 150 curr per   36174

Philip Morris USA, Research Center Library, 4201 Commerce Rd, P.O. Box 26583, **Richmond**, VA 23261
T: +1 804 2742877; Fax: +1 804 2742160
1959; Carla L. Gregory
Tobacco
40 000 vols; 300 curr per   36175

Troutman, Sanders, Mays & Valentine LLP, Law Library, 1111 E Main St, **Richmond**, VA 23219; P.O. Box 1122, Richmond, VA 23218-1122
T: +1 804 6971200; Fax: +1 804 6971339; URL: www.troutmansanders.com
Bonnie K. Gates, Carol Goodman
10 000 vols   36176

Virginia Power Research, Research-Records Services, 701 E Cary St, P.O. Box 26666, **Richmond**, VA 23261
T: +1 804 7713659; Fax: +1 804 7713168; URL: www.vom.com
1937; Linda G. Royal
10 000 vols; 350 curr per   36177

Boehringer Ingelheim Pharmaceuticals, Inc, Scientific & Corporate Information Services, 900 Ridgebury Rd, **Ridgefield**, CT 06877
T: +1 203 7916172; Fax: +1 203 7916172; E-mail: webmaster@rdg.boehringer-ingelheim.com; URL: us.boehringer-ingelheim.com
1971; Len Sierra
10 000 vols; 800 curr per   36178

Schlumberger-Doll, Research Library, Old Quarry Rd, **Ridgefield**, CT 06877-4108
T: +1 203 4315600; Fax: +1 203 4315625; E-mail: library@ridgefield.sdr.slb.com
1947; Cherie Voris
15 000 vols; 170 curr per
libr loan   36179

Hayes, Seay, Mattern & Mattern, Inc Library, 1315 Franklin Rd, P.O. Box 13446, **Roanoke**, VA 24034
T: +1 540 8573244; Fax: +1 540 8573180; E-mail: chumphries@hsmm.com
1958; Cynthia Humphries
Engineering
13 810 vols; 284 curr per   36180

Woods Rogers PLC, Law Library, Wachovia Tower, Ste 1400, Ten S Jefferson St, **Roanoke**, VA 24011; P.O. Box 14125, Roanoke, VA 24038-4125
T: +1 540 9837531; Fax: +1 540 9837711
Jane Roth Baugh
10 000 vols; 150 curr per   36181

Chamberlain, D'Amanda, Oppenheimer & Greenfield, Law Library, 1600 Crossroads Bldg, Two State St, **Rochester**, NY 14614
T: +1 585 2323730; Fax: +1 585 2323882; E-mail: cab@cdog.com; URL: www.cdog.com
Christine Becker
8 000 vols; 15 curr per   36182

Harter, Secrest & Emery LLP, Law Library, 1600 Bausch & Lomb Pl, **Rochester**, NY 14604
T: +1 585 2311230; Fax: +1 585 2322152
Robert Salerno
15 000 vols; 25 curr per; 25 av-mat; 20 sound-rec; 20 digital data carriers   36183

IBM Corp, Rochester Information Resource Center Library, 3605 Highway 52 NW, Dept 205 005-1, **Rochester**, MN 55901-7829
T: +1 507 2534512; Fax: +1 507 2532593
Melanie Huntington
Computers, Engineering, Management
20 000 vols; 525 curr per   36184

Nixon Peabody LLP, Information Services Department, Clinton Sq, P.O. Box 31051, **Rochester**, NY 14603-1051
T: +1 716 2631000; Fax: +1 716 2631600
Kathy Tschetter
15 000 vols   36185

Underberg & Kessler Law Library, 300 Bausch & Lomb Pl, **Rochester**, NY 14604
T: +1 585 2582800; Fax: +1 585 2582821
Jane Snyder
8 000 vols   36186

Connell Foley Law Library, 85 Livingston Ave, **Roseland**, NJ 07068-3702
T: +1 973 5350500; Fax: +1 973 5359217; URL: www.connellfoley.com
Tae J. Yoo
Commercial law; tax, estates, trusts
22 000 vols   36187

Lowenstein Sandler PC Library, 65 Livingston Ave, **Roseland**, NJ 07068
T: +1 973 5972500; Fax: +1 973 5972400; E-mail: lawlibrary@lowenstein.com
Beth Petruska
25 000 vols   36188

Lum, Danzis, Drasco & Positan, Law Library, 103 Eisenhower Pkwy, **Roseland**, NJ 07068-1049
T: +1 973 4039000; Fax: +1 973 4039021
1869; Annette Davis
20 000 vols   36189

Unisys Corporation, TwinCities InfoCenter (MS 4613), 2470 Highcrest Rd, **Roseville**, MN 55113; P.O. Box 64942, Saint Paul, MN 55164
T: +1 651 6357211; Fax: +1 651 6357523; E-mail: theresa.mercado@unisys.com; URL: www.unisys.com
1965; Terri Mercado
15 000 vols; 450 curr per; 200 av-mat; 300 sound-rec; 20 digital data carriers; internal documentation (4 500)
libr loan; SLA   36190

ICF Jones & Stokes Library, 630 K St, Ste 400, **Sacramento**, CA 95814
T: +1 916 7373000; Fax: +1 916 7373030; E-mail: corporate_library@jsanet.com; URL: www.jonesandstokes.com
Sharon Hoepker
Water resources; JSA Rpts
18 000 vols; 500 curr per   36191

Weintraub, Genshlea, Chediak & Sproul, Law Library, 400 Capitol Mall, Ste 1100, **Sacramento**, CA 95814
T: +1 916 5586000; Fax: +1 916 4461611; URL: www.weintraub.com
Liana Chen-Knapp
10 000 vols; 52 curr per   36192

Downs Rachlin Martin PLLC, Law Library, 90 Prospect St, **Saint Johnsbury**, VT 05819; P.O. Box 99, Saint Johnsbury, VT 05819-0099
T: +1 802 7488324; Fax: +1 802 7484394; URL: www.drm.com
Wynne Browne
8 000 vols; 100 curr per   36193

Anheuser-Busch Co, Inc, Corporate Library, One Busch Pl, **Saint Louis**, MO 63118-1849; Bechtold Sta, P.O. Box 1828, Saint Louis, MO 63118-0828
T: +1 314 5772669; Fax: +1 314 5772006
1932; Mary Butler
40 000 vols; 1 000 curr per
SLA (Special Libraries Association)   36194

Blackwell, Sanders, Peper & Martin, Law Library, 720 Olive St, 24th Fl, **Saint Louis**, MO 63101
T: +1 314 3456871; Fax: +1 314 3456060; E-mail: lgenovese@blackwellsanders.com
Lee Ann Genovese
20 000 vols; 20 curr per   36195

Boeing Co, Technical Library, S034-1040, **Saint Louis**, MO 63166; P.O. Box 516, Saint Louis, MO 63166-0516
T: +1 314 3878023; Fax: +1 314 7772424
1948; Teresa Powell
34 000 vols; 500 curr per   36196

Covidien, Pharmaceutical & Chemical Library, 3600 N Second St, **Saint Louis**, MO 63147-3457
T: +1 314 6541511; Fax: +1 314 6541513; URL: www.covidien.com
Larry Walton
Pharmaceutical Chemistry; Pre-1900 German Chemistry & Pharmacy, Pre-1900 Pharmacopeias
12 000 vols; 150 curr per   36197

Federal Reserve Bank of Saint Louis, Research Library, One Federal Reserve Bank Plaza, **Saint Louis**, MO 63102-2005; mail address: 1421 Dr Martin Luther King Dr, Saint Louis, MO 63106-3716
T: +1 314 4448552; Fax: +1 314 4448694; E-mail: ref@stls.frb.org; URL: fraser.stlouisfed.org; liber8.stlouisfed.org/
1922; Katrina L. Stierholz
20 000 vols; 350 curr per
libr loan   36198

Gallop, Johnson & Neuman LC, Law Library, 101 S Hanley Rd, Ste 1700, **Saint Louis**, MO 63105
T: +1 314 6156000; Fax: +1 314 6156001; URL: www.gjn.com
Sue Dees
10 000 vols   36199

Greensfelder, Hemker & Gale, PC Library, Ten S Broadway, Ste 2000, **Saint Louis**, MO 63102
T: +1 314 2419090; Fax: +1 314 2418624; URL: www.greensfelder.com
Sally Crowley
13 000 vols; 400 curr per   36200

Hellmuth, Obata & Kassabaum, Inc, HOK, Inc Research Library, 211 N Broadway, Ste 700, **Saint Louis**, MO 63102
T: +1 314 7544217; Fax: +1 314 4216073
Architecture; HOK Hist
8 000 vols; 300 curr per; 130 000 slides   36201

Husch Blackwell Sanders LLP, Law Library, 190 Carondelet Plaza, Ste 600, **Saint Louis**, MO 63105
T: +1 314 4801500; Fax: +1 314 4801505; URL: www.huschblackwell.com
Karla A. Morris-Holmes
10 000 vols; 95 curr per   36202

Lashly & Baer PC, Law Library, 714 Locust St, **Saint Louis**, MO 63101
T: +1 314 6212939; Fax: +1 314 6216844; URL: www.lashlybaer.com
1928; Mimi Hubert
12 000 vols; 10 curr per
ALA, AALL, SWALL   36203

Lewis, Rice & Fingersh Law Library, 500 N Broadway, Ste 2000, **Saint Louis**, MO 63102-2147
T: +1 314 4447600; Fax: +1 314 6127681; URL: www.lewisrice.com
Helen Capdevielle
22 000 vols; 112 curr per; 30 digital data carriers   36204

May Department Stores Co, Law Library, 611 Olive St, Ste 1750, **Saint Louis**, MO 63101
T: +1 314 3426697; Fax: +1 314 3423066; E-mail: diane_burnett@may-co.com
Diane Burnett
10 000 vols; 55 curr per   36205

Thompson Coburn LLP, Law Library, One US Bank Plaza, **Saint Louis**, MO 63101
T: +1 314 5526275; Fax: +1 314 5527275; URL: www.thompsoncoburn.com
Mary Kay Jung
26 000 vols; 3 600 curr per
libr loan   36206

Vulcan Sports Media Inc, Sporting News Research Center, 10176 Corp Sq, Ste 200, **Saint Louis**, MO 63132
T: +1 314 9937787; Fax: +1 314 9970765; E-mail: tsnspg@aol.com; URL: www.sportingnews.com/archives
1987; James R. Meier
Arch of the Sporting News
10 000 vols   36207

Briggs & Morgan, Law Library, 332 Minnesota St, Ste 2200, **Saint Paul**, MN 55101
T: +1 651 2236600; Fax: +1 651 2236450; E-mail: morgan@email.briggs.com
Susan J. Redalen
53 000 vols; 214 curr per   36208

Minnesota Mining & Manufacturing Co, Law Library, Bldg 220-12E-02, 3M General Offices, **Saint Paul**, MN 55144-1000
T: +1 651 7331460; Fax: +1 651 7369469
Barbara Peterson
25 000 vols; 225 curr per   36209

Winthrop & Weinstine, Law Library, 3200 Minnesota World Trade Ctr, 30 E Seventh St, **Saint Paul**, MN 55101
T: +1 651 2908450; Fax: +1 651 2929347; URL: www.winthrop.com
Ann Turner
15 400 vols; 217 curr per   36210

Fabian & Clendenin, Law Library, 215 S State St, Ste 1200, **Salt Lake City**, UT 84111-2323; P.O. Box 510210, Salt Lake City, UT 84151-0210
T: +1 801 5318900; Fax: +1 801 5962814; URL: www.fabclen.com
1918; Allysson Rice Watkins
Historic Utah Coll
40 000 vols; 100 curr per
libr loan; AALL   36211

Leboeuf, Lamb, Greene & Macrae, Law Library, 1000 Kearns Bldg, 136 S Main St, **Salt Lake City**, UT 84101-1601
T: +1 801 3206700; Fax: +1 801 3598256; E-mail: mzthomas@llgm.com
Miki Thomas
10 000 vols   36212

Ray Quinney & Nebeker PC, Law Library, 36 S State St, Ste 1400, **Salt Lake City**, UT 84111; P.O. Box 45385, Salt Lake City, UT 84145-0385
T: +1 801 5321500; Fax: +1 801 5327543; URL: www.rqn.com
Gwendolyn Mulks
15 000 vols   36213

Akin, Gump, Strauss, Hauer & Feld LLP, 300 Convent St, Ste 1500, **San Antonio**, TX 78205
T: +1 210 2817130; Fax: +1 210 2242035; URL: www.akingump.com
20 000 vols   36214

Cox, Smith, Matthews Inc, Law Library, 112 E Pecan, Ste 1800, **San Antonio**, TX 78205-1521
T: +1 210 5545500; Fax: +1 210 2268395; URL: www.coxsmith.com
Elizabeth Prike
51 000 vols; 500 curr per   36215

Fulbright & Jaworski Library, 300 Convent St, Ste 2200, **San Antonio**, TX 78205-3792
T: +1 210 2245575; URL: www.fulbright.com
Kathy Darrow
30 000 vols; 60 curr per   36216

DLA Piper, Law Library, 401 B St, Ste 1700, **San Diego**, CA 92101-4297
T: +1 619 6992770; Fax: +1 619 6992701; E-mail: info@dlapiper.com; URL: www.dlapiper.com
N.N.
60 000 vols; 300 curr per   36217

Higgs, Fletcher & Mack LLP, Law Library, 401 West A St, Ste 2600, **San Diego**, CA 92101-7913
T: +1 619 2361551; Fax: +1 619 6961410; URL: www.higgslaw.com
1939; Glee Hotchkin
12 000 vols   36218

Luce, Forward, Hamilton & Scripps, Law Library, 600 W Broadway, Ste 2600, **San Diego**, CA 92101
T: +1 619 2361414; Fax: +1 619 2328311; URL: www.luce.com
Carmen Valero
Pacific Rim, Int Business
50 000 vols; 260 curr per   36219

NCR Corporation (Teradata Division) Library, Dept 8714, 17095 Via Del Campo, **San Diego**, CA 92127
T: +1 858 4853291; Fax: +1 858 4853567
Mary Hill
Electronics
16 000 vols; 70 curr per
libr loan
36220

Sempra Energy, Law Library, 101 Ash St, **San Diego**, CA 92101-3017
T: +1 619 6962034
18 000 vols; 600 curr per
36221

Bank of America, Law Library, 555 California St, Dept 13220, 8th Fl, **San Francisco**, CA 94104
T: +1 415 6222854; Fax: +1 415 6226601; E-mail: debra.martin@bankofamerica.com
Debra J. Martin
Commercial law
15 000 vols
36222

Bechtel Corporate Library, 50 Beale St, P.O. Box 193965, 50/202/C13, **San Francisco**, CA 94119-3965
T: +1 415 7685306; Fax: +1 415 7680837; E-mail: jxmah@bechtel.com
1951; Jeffery Mah
Engineering, Energy, Construction
40 000 vols; 250 curr per; 10 000 tech rpts
36223

Bingham McCutchen, Law Library, Three Embarcadero Ctr, **San Francisco**, CA 94111
T: +1 415 3932560; Fax: +1 415 3932286; URL: www.bingham.com
Jane Metz
50 000 vols
36224

Brobeck, Phleger & Harrison Library, One Market Plaza, Spear St Tower, 5th Fl, **San Francisco**, CA 94105-1100
T: +1 415 4420900, 9792619; Fax: +1 415 4421010; URL: www.brobeck.com
Alan R. MacDougall
39 100 vols
36225

Carroll, Burdick & McDonough, Law Library, 44 Montgomery St, Ste 400, **San Francisco**, CA 94104
T: +1 415 9895900; Fax: +1 415 9890932
Caren Doyle
10 000 vols
36226

Cooley Godward Kronish LLP Library, 101 California St, 5th Flr, **San Francisco**, CA 94111
T: +1 415 6932000; Fax: +1 415 6932222; URL: www.cooley.com
Margaret Baer
20 000 vols
36227

Cooper, White & Cooper, Law Library, 201 California St, 17th Flr, **San Francisco**, CA 94111-5002
T: +1 415 4331900; Fax: +1 415 4335530; URL: cwclaw.com
1896; Cindy Beck Weller
20 000 vols; 200 curr per
36228

Farella, Braun & Martel, Law Library, 235 Montgomery St, 19th Flr, **San Francisco**, CA 94104
T: +1 415 9544451; Fax: +1 415 9544480; URL: www.fbm.com
Mary Staats
20 000 vols
36229

Federal Reserve Bank of San Francisco, Research Library, 101 Market St, **San Francisco**, CA 94105-1579; P.O. Box 7702, Francisco, CA 94120-7702
T: +1 415 9743216; Fax: +1 415 9743429
Miriam Ciochon
35 000 vols; 350 curr per
36230

Gordon & Rees LLP, Law Library, 275 Battery St, 20th Flr, **San Francisco**, CA 94111
T: +1 415 9865900; Fax: +1 415 9868054; URL: www.gordonrees.com
Daniel Isaacs
Medicine
20 000 vols
36231

Hancock, Rothert & Bunshoft, Law Library, Four Embarcadero Ctr, 3rd Flr, **San Francisco**, CA 94111
T: +1 415 9815550; Fax: +1 415 9552599; URL: www.hrblaw.com
Leslie Hesdorfer
12 000 vols
36232

Hanson, Bridgett, Marcus, Vlahos & Rudy, Law Library, 333 Market St, 23rd Flr, Ste 2300, **San Francisco**, CA 94105
T: +1 415 9955150; Fax: +1 415 5419366; URL: www.hansonbridgett.com
Meredith Casteel
24 000 vols
36233

Hassard Bonnington, Law Library, Two Embarcadero Ctr, Ste 1800, **San Francisco**, CA 94111
T: +1 415 2889800; Fax: +1 415 2889801; URL: www.hassard.com
Diane Rodriguez
Medicine
14 000 vols; 200 curr per
36234

Heller, Ehrman, White & McCauliffe Library, 333 Bush St, **San Francisco**, CA 94104-2878
T: +1 415 7726812; Fax: +1 415 7726268; E-mail: lmak@hewm.com; URL: hewm.com
1921; Loretta Mak
30 004 vols; 396 curr per; 1 digital data carriers
36235

Howard, Rice, Nemerovski, Canady, Falk & Rabkin Library, Three Embarcadero Ctr, Ste 600, **San Francisco**, CA 94111
T: +1 415 3993043; Fax: +1 415 2175910; URL: www.hrice.com; www.howardrice.com
Marlowe Crawford
Law
13 000 vols
36236

Latham & Watkins LLP Library, 505 Montgomery St, Ste 1900, **San Francisco**, CA 94111
T: +1 415 3910600; Fax: +1 415 3958095; URL: www.lw.com
Ann Hardham
20 000 vols
36237

Leboeuf, Lamb, Greene & Macrae, LLP, Law Library, One Embarcadero Ctr, Ste 400, **San Francisco**, CA 94111
T: +1 415 9511306; Fax: +1 415 9511180; URL: www.llgm.com
1983; Nancy J. Carlin
15 000 vols; 50 curr per
36238

McKenna, Long & Aldridge, Law Library, 101 California St, Ste 4100, **San Francisco**, CA 94111-5886
T: +1 415 2674000; Fax: +1 415 2674198; URL: www.mckennalong.com
Eric Carlson
15 000 vols; 50 curr per
36239

Mills Law Library, 220 Montgomery St, 1st Flr, **San Francisco**, CA 94104
T: +1 415 7812665; Fax: +1 415 7811116; URL: www.millslibrary.org
1987; Jacob Koff
35 000 vols; 170 curr per
36240

Murphy, Sheneman, Julian & Rogers, Law Library, 101 California St, 39th Flr, **San Francisco**, CA 94111
T: +1 415 3984700; Fax: +1 415 4217879; E-mail: lafsmwb@well.com; URL: www.msjr.com
Leslie Ann Forrester
11 000 vols; 200 curr per
36241

Orrick, Herrington & Sutcliffe LLP, Law Library, The Orrick Bldg, 405 Howard St, **San Francisco**, CA 94105-2669
T: +1 415 7735700; Fax: +1 415 7735759; URL: www.orrick.com
1884; Peg LaFrance
California City Charters
30 000 vols; 350 curr per
36242

Pillsbury Winthrop LLP, Law Library, 50 Fremont St, **San Francisco**, CA 94105
T: +1 415 9831130; Fax: +1 415 9831200; URL: www.pillsburywinthrop.com
Debra Schwarz
Law Coll
30 000 vols; 500 curr per
libr loan
36243

Sedgwick, Detert, Moran & Arnold Library, Steuart Street Tower, One Market Plaza, 6th Flr, **San Francisco**, CA 94105
T: +1 415 7817900; Fax: +1 415 7812635
Mark Newman
15 000 vols
36244

Shearman & Sterling LLP, Library, 525 Market St, Ste 1500, **San Francisco**, CA 94105
T: +1 415 6161100; Fax: +1 415 6161199
Suzanne Glab
10 000 vols; 50 curr per
36245

Shearman & Sterling LLP, 525 Market St, Ste 1500, **San Francisco**, CA 94105
T: +1 415 6161100; Fax: +1 415 6161199
Suzanne Glab
10 000 vols; 50 curr per
36246

Thelen Reid Brown Raysman & Steiner LLP, Law Library & Information Resource Center, 101 Second St, Ste 1800, **San Francisco**, CA 94105-3672
T: +1 415 3697063; Fax: +1 415 6446512
Todd Bennett
30 000 vols; 200 curr per
36247

GE Nuclear Energy Library, 175 Curtner Ave, M/C 728, **San Jose**, CA 95125
T: +1 408 9253523; Fax: +1 408 9253536; E-mail: phyllis.reyburn@gene.ge.com
1955; Phyllis Reyburn
9 700 vols; 300 000 microforms; 75 000 documents
libr loan
36248

International Business Maschines Corp, Research Library, 650 Harry Rd, Dept K 74, Bldg 80, **San Jose**, CA 95120-6099
T: +1 408 9271580, 1060; Fax: +1 408 9273105
1952; Robert Scott
Engineering, Computer science
27 000 vols; 300 curr per
36249

Pratt & Whitney, CSD Library-Space Propulsion Operations, 600 Metcalf Rd, **San Jose**, CA 95138-9602
T: +1 408 7764957; Fax: +1 408 7765995; E-mail: schaffer@csd.com
1960; Karen Schaffer
Organic chemistry
9 500 vols; 100 curr per
36250

Lucasfilm Research Library, PO Box 2009, **San Rafael**, CA 94912
T: +1 415 6621912; E-mail: rlibs@lucasfilm.com
Paramount Studios Libr
20 000 vols; 100 curr per; 4 000 av-mat
36251

Kvaerner Metalls, Library Center, 12657 Alcosta Blvd, Ste 200, **San Ramon**, CA 94583
T: +1 925 8661166; Fax: +1 925 8666520
15 000 vols
36252

URS Corp Library, Library-Information Center, 2020 E First St, Ste 400, **Santa Ana**, CA 92705
T: +1 714 8356886 ext 500; Fax: +1 714 6677147; E-mail: ubherteo@wcc.com
1978; Karien Hudock
Engineering, Geology; Company Project Rpts
9 000 vols; 50 curr per
36253

Mission Research Corp, Technical Library, 735 State St, **Santa Barbara**, CA 93101
T: +1 805 9638761; Fax: +1 805 9628530; URL: www.mrcsb.com
Mary Sonronany
Technology
8 000 vols; 75 curr per
36254

Santa Barbara News Press Library, 715 Anacapa St, **Santa Barbara**, CA 93101-2203; P.O. Box 1359, Santa Barbara, CA 93102-1359
T: +1 805 5645200; Fax: +1 805 9666258; URL: www.newspress.com
Cass Cara
15 000 vols
36255

National Semiconductor Corp, Technical Library, 2900 Semiconductor Dr, MS-DT-05, **Santa Clara**, CA 95052-8090
T: +1 408 7213810; Fax: +1 408 7217060; E-mail: library@library.nsc.com
Mary Holland
Electronics, semiconductors
9 000 vols; 450 curr per
36256

Montgomery & Andrews, Law Library, 325 Paseo de Peralta, **Santa Fe**, NM 87501
T: +1 505 9823873; Fax: +1 505 9824289
Bertha Trujillo
Commercial law
15 000 vols
36257

Bryan Cave LLP, Law Library, 120 Broadway, Ste 300, **Santa Monica**, CA 90401
T: +1 310 5762100; Fax: +1 310 5762200
Karen Lasnick
60 000 vols
36258

Motorola, Inc, Research & Technical Library, 1301 E Algonquin Rd, Rm 1914, **Schaumburg**, IL 60196-1078
T: +1 847 5768580; Fax: +1 847 5764716
1978; Nancy Snyder
Electronics; Communications Coll
10 000 vols; 100 curr per
36259

KAPL Inc Libraries, 2401 River Rd, **Schenectady**, NY 12309; P.O. Box 7400, Schenectady, NY 12301-7400
T: +1 518 3954918; Fax: +1 518 3957761
Tony Oliveira
19 000 vols; 250 curr per
36260

Boeing Co, Technical Libraries, M/S 62-LC (P.O. Box 3707), **Seattle**, WA 98124-2207
T: +1 425 2378311; Fax: +1 425 2374582
1938; Barbara Whorton
Int Data Bank
120 000 vols; 5 500 curr per
36261

Carney, Badley & Spellman Library, Law Library, 701 Fifth Ave, Ste 3600, **Seattle**, WA 98104-7010
T: +1 206 6074149; Fax: +1 206 6228983; URL: www.carneylaw.com
Melissa Miller
10 000 vols; 30 curr per
36262

Davis Wright Tremaine LLP, Law Library, 1201 Third Ave, Ste 2200, **Seattle**, WA 98101-3045
T: +1 206 6223150; Fax: +1 206 6287699; E-mail: info@dwt.com; URL: www.dwt.com
Christy Leith
40 000 vols
36263

Foster Pepper PLLC, Research Center, 1111 Third Ave, Ste 3400, **Seattle**, WA 98101
T: +1 206 4476474; Fax: +1 206 4479700; E-mail: researchcenter@foster.com
Barbara Rothwell
13 000 vols
36264

Garvey, Schubert & Barer, Law Library, 1191 Second Ave, Ste 1800, **Seattle**, WA 98101-2939
T: +1 206 4643939; Fax: +1 206 4640125; URL: www.gsblaw.com
Jill Allyn
10 000 vols
36265

Graham & Dunn, Law Library, Pier 70, 2801 Alaskan Way, Ste 300, **Seattle**, WA 98121-1128
T: +1 206 9034801; Fax: +1 206 3409599; E-mail: library@grahamdunn.com
Katie Drake
10 000 vols
36266

Lane Powell PC Library, 1420 Fifth Ave, Ste 4100, **Seattle**, WA 98101-2388
T: +1 206 2236245; Fax: +1 206 2237107; URL: www.lanepowell.com
Karen Helde
14 000 vols; 500 curr per
36267

Perkins Coie Library, 1201 Third Ave, Ste 4800, **Seattle**, WA 98101
T: +1 206 3598444; Fax: +1 206 3599444; E-mail: library@perkinscoie.com
1912; Amy J. Eaton
50 000 vols; 1 000 curr per
libr loan
36268

Preston Gates & Ellis, Library & Research Services, 5000 Bank of America Tower, 701 Fifth Ave, *Seattle*, WA 98104-7078
T: +1 206 6237580; Fax: +1 206 6237022; E-mail: info@prestongates.com; URL: www.prestongates.com
Bridget Dacres
Law
28 000 vols   36269

Band, Lavis & Associates, Inc, Technical Library, 900 Ritchie Hwy, *Severna Park*, MD 21146
T: +1 410 5442800; Fax: +1 410 6473411
Shirley Wilson
15 500 vols; 25 curr per   36270

Worcester Foundation for Experimental Biology, George F. Fuller Research Library, 222 Maple Ave, *Shrewsbury*, MA 01545
T: +1 508 8428921 ext 282; Fax: +1 508 8420342
1945
Gregory Pincus Collected Papers
33 000 vols; 250 curr per; 4 VF drawers of patents, 4 VF drawers of subject bibliographies   36271

Amos Press, Inc Library, 911 Vandemark Rd, P.O. Box 4129, *Sidney*, OH 45365
T: +1 937 4982111 ext 276; Fax: +1 937 4980806; E-mail: khesselbein@amospress.com
1960; Krista Hesselbein
Numismatic mat, Philatelic mat
30 000 vols; 800 curr per; 24 maps; 700 microforms; 36 av-mat; 21 sound-rec; 2 digital data carriers
SLA   36272

Ethicon, Inc, Scientific Information Services, US Rte 22, P.O. Box 151, *Somerville*, NJ 08876
T: +1 908 2183259; Fax: +1 908 2183558
1956; Norma Bregen
Polymer chemistry, Surgery, Toxicology
9 000 vols; 375 curr per
libr loan   36273

Norris, McLaughlin & Marcus, Law Library, 721 Rte 202-206 N, *Somerville*, NJ 08876; P.O. Box 1018, Somerville, NJ 08876-1018
T: +1 908 7220700; Fax: +1 908 7220755
Janice S. Lustiger
10 000 vols   36274

Consol Energy, Inc, Technical Resource Center, 4000 Brownsville Rd, *South Park*, PA 15129
T: +1 412 8546599; Fax: +1 412 8546613
1947; Jacqueline Kniedler
Coal research
10 000 vols; 150 curr per; 200 govt docs
SLA   36275

Testing Engineers International, Inc, TEi-Library Services, 3455 S 500 West, *South Salt Lake*, UT 84115-4234; P.O. Box 57025, Murray, UT 84157-0025
T: +1 801 2622332; Fax: +1 801 2622363; E-mail: info@tei-libraryservices.com; URL: www.tei-libraryservices.com
1963; Matthew MacGregor
Science & technology; Cooking; Old Engineering & Science; Standards (25 000)
45 000 vols; 25 curr per   36276

Sommers, Schwartz, Silver & Schwartz, Law Library, 2000 Town Ctr, Ste 900, *Southfield*, MI 48075
T: +1 248 3550300; Fax: +1 248 7464001
14 000 vols; 250 curr per   36277

Milliken & Co, Library M-470, 920 Milliken Rd, *Spartanburg*, SC 29304
T: +1 864 5031589; Fax: +1 864 5032769
1960; Ginny Sikes
Chemistry, Textiles, Business
10 000 vols; 300 curr per   36278

Hanson Professional Services Inc, Technical Library, 1525 S Sixth St, *Springfield*, IL 62703-2886
T: +1 217 7882450; Fax: +1 217 7479416; E-mail: bhicks@hanson-inc.com; URL: www.hanson-inc.com
1975; Betty Lou Hicks
Geology (Illinois Coll); Illinois Topo, maps
19 582 vols; 225 curr per; 12 000 maps; 20 digital data carriers
libr loan; SLA   36279

Massachusetts Mutual Life Insurance Co – Law Library, 1295 State St, *Springfield*, MA 01111
T: +1 413 7442188; Fax: +1 413 7446279
Elenor Owczarski
23 000 vols   36280

Massachusetts Mutual Life Insurance Co Library, 1295 State St, *Springfield*, MA 01111
T: +1 413 7443361; Fax: +1 413 7448440; E-mail: sodonnell@massmutual.com; URL: www.massmutual.com
1929; Suzanne O'Donnell
Business & management, Investing
14 000 vols; 100 curr per   36281

Clairol Research Library, 2 Blachley Rd, *Stamford*, CT 06922
T: +1 203 3575001; Fax: +1 203 9692577
1968; Linda Massoni
14 175 vols; 315 curr per   36282

Cummings & Lockwood, Law Library, Six Landmark Sq, *Stamford*, CT 06901
T: +1 203 3514466; Fax: +1 203 3514534; URL: www.cl-law.com
Barbara Bentley
20 000 vols   36283

Xerox Corp, Law Library, 800 Long Ridge Rd, *Stamford*, CT 06904
T: +1 203 9684005; Fax: +1 203 9683446; E-mail: randall.wilcox@usa.xerox.com
1969; Randall Wilcox
15 000 vols   36284

General Dynamics Land Systems, Technical & Administrative Information Services (TAIS), 38500 Mound Rd, *Sterling Heights*, MI 48310
T: +1 586 8254402; Fax: +1 586 8254013
1985; Michaelene Iwanyckyj
Army Technical Manuals & DMWR's
16 000 vols
libr loan   36285

Austen Riggs Center, Inc, Austen Fox Riggs Library, 25 Main St, *Stockbridge*, MA 01262; P.O. Box 962, Stockbridge, MA 01262-0962
T: +1 413 2985511; Fax: +1 413 2984020; E-mail: info@austenriggs.org; URL: www.austenriggs.org
1919; Robert Difazio
Psychology, psychoanalysis psychiatry; David Rapaport Memorial Libr
13 000 vols; 85 curr per; 8 av-mat; 112 sound-rec; 2 digital data carriers
libr loan; NNLM, ALA   36286

Unocal Corp, Exploration & Production Library, 14141 Southwest Freeway, *Sugar Land*, P.O. Box 4570, Houston, TX 77210
T: +1 281 2875523; Fax: +1 281 2875141; E-mail: pvicars@unocal.com
Phyllis R. Vicars
Engineering
20 000 vols   36287

Hoechst Celanese Research Division, Technical Information Center, 86 Morris Ave, *Summit*, NJ 07901
T: +1 908 5227500; Fax: +1 908 5223902
1966; Bob Daly
Ceramics, Chemical engineering, Chemistry, Optics, Plastics, Polymer chemistry
12 000 vols; 300 curr per   36288

Novartis Pharmaceuticals, Netter Library, 556 Morris Ave, *Summit*, NJ 07901
T: +1 908 2775544; Fax: +1 908 2777999
1984; Deborah Juterbock
14 500 vols; 987 curr per
libr loan   36289

Fujitsu, Corporate Library, 1250 E Arques Ave, *Sunnyvale*, CA 94085
T: +1 408 7466000; Fax: +1 408 9922674; URL: www.fujitsu.com
70 000 vols; 250 curr per   36290

Bond, Schoeneck & King, PLLC, Law Library, One Lincoln Ctr, *Syracuse*, NY 13202-1355
T: +1 315 2188000; Fax: +1 315 2188100; URL: www.bsk.com
Maureen T. Kays
25 000 vols; 100 curr per   36291

Mackenzie, Hughes LLP, Law Library, 101 S Salina St, Ste 600, *Syracuse*, NY 13202-1399; P.O. Box 4967, Syracuse, NY 13221-4967
T: +1 315 4747571; Fax: +1 315 4746409; URL: www.mackenziehughes.com
Mike Campbell
Extensive Tax Coll
16 000 vols; 110 curr per   36292

Niagara Mohawk Power Corp, Technology Resource Center, 300 Erie Blvd W D-M, *Syracuse*, NY 13202-4250
T: +1 315 4285003; Fax: +1 315 4285004; E-mail: raymondb@niagaramohawk.com; URL: www.niagaramohawk.com
1992; Barbara J. Raymond
Lighting Res Ctr Mat, Energy Related Mat
15 000 vols; 75 curr per   36293

Weyerhaeuser Library, Technical Information Center, Box 2999, *Tacoma*, WA 98477-2999
T: +1 253 9246267; Fax: +1 253 9246870; E-mail: linda.martinez@weyerhaeuser.com
Linda Martinez
Forestry, Paper, Pulp, Wood
70 000 vols; 475 curr per   36294

Hopping, Green & Sams, Law Library, PO Box 6526, *Tallahassee*, FL 32314-6526
T: +1 850 2227500; Fax: +1 850 2248551; URL: www.hgslaw.com
Marisol Roberts
Environmental law, administrative law
8 000 vols; 10 curr per   36295

Carlton Fields, Law Library, 4221 W Boy Scout Blvd, Ste 1000, *Tampa*, FL 33607; P.O. Box 3239, Tampa, FL 33601-3239
T: +1 813 2237000; Fax: +1 813 2294133; URL: www.carltonfields.com
1915; Terry Psarras
30 000 vols; 200 curr per   36296

Fowler, White, Boggs & Banker, Law Library, 501 E Kennedy Blvd, Ste 1700, *Tampa*, FL 33602
T: +1 813 2287411; Fax: +1 813 2298313; URL: www.fowlerwhite.com
Elenita Lopez
Real estate, Securities, Immigration
25 000 vols; 10 curr per   36297

Gray-Robinson, PA, Law Library, 201 N Franklin St, Ste 2200, *Tampa*, FL 33602-5822; P.O. Box 3324, Tampa, FL 33601-3324
T: +1 813 2735294; Fax: +1 813 2735145; URL: www.gray-robinson.com
25 000 vols; 75 curr per   36298

Holland & Knight LLP, Library & Research Services Dept, 100 N Tampa St, Ste 4100, *Tampa*, FL 33602; P.O. Box 1288, Tampa, FL 33601-1288
T: +1 813 2278500; Fax: +1 813 2290134; URL: www.hklaw.com
Glenn Ross
30 000 vols   36299

Bayer Corporation, Information Resources & Services, 511 Benedict Ave, *Tarrytown*, NY 10591
T: +1 914 5242339; Fax: +1 914 5243075
1962; Gitta Benglas
Business, Management, Medicine, Science & technology
15 000 vols; 350 curr per
libr loan   36300

OSI Specialties/Crompton Corp, Library & Technical Information Service, 777 Old Saw Mill River Rd, *Tarrytown*, NY 10591-6799
T: +1 914 7844800; Fax: +1 914 7844935
1971; Barbara Lambert
Chemistry, Cryogenics, Engineering, Physics; Chemical Patents Coll, micro
40 000 vols; 350 curr per   36301

Honeywell Tempe Information Resource Center, 1300 W Warner Rd, 1207-3M, *Tempe*, AZ 85284-2896
T: +1 480 5927162; Fax: +1 480 5924663; E-mail: janet.lodge@honeywell.com
1986; Janet Lodge
Aerospace; Saftey Videos
10 000 vols; 260 curr per   36302

Raytheon Co, IDS Headquarters-Research Library – T3MA13, 50 Apple Hill Dr, *Tewksbury*, MA 01876
T: +1 978 8584700; Fax: +1 978 8584516; URL: www.raytheon.com
1952; Sandy Nawrocki
15 000 vols; 250 curr per; 100 000 microforms; 2 digital data carriers; 50 000 res rpts   36303

Rockwell Scientific Co, Technical Information Center, 1049 Camino Dos Rios, *Thousand Oaks*, CA 91360
T: +1 805 3734722; Fax: +1 805 3734296; E-mail: communications@teledyne-si.com
1962; Yolanda O. Fackler
Chemistry, Physics, Mathematics, Metallurgy
12 000 vols; 300 curr per   36304

Fuller & Henry Law Library, One Seagate, 17th Fl, *Toledo*, OH 43604-2606; P.O. Box 2088, Toledo, OH 43603
T: +1 419 2472891; Fax: +1 419 2472864
1892; Gail McCain
23 000 vols   36305

Libbey Owens Ford Co, Pilkington North America Corporate Library, 811 Madison Ave, *Toledo*, OH 43624
T: +1 904 3960660; Fax: +1 904 3966020; E-mail: jennifer.flint@us.pilkington.com
1950; Jennifer Anne Flint
Automotive industry, Glass
20 500 vols   36306

Marshall & Melhorn, Law Library, Four SeaGate, 8th Flr, *Toledo*, OH 43604
T: +1 419 2497100; Fax: +1 419 2497151
1895; Barbara Avery
14 000 vols; 26 curr per   36307

Dynamics Technology, Inc Library, 21311 Hawthorne Blvd, Ste 300, *Torrance*, CA 90503
T: +1 310 5435433; Fax: +1 310 5432117; E-mail: sthompson@dynatec.com; URL: www.dynatec.com
Seruia Thompson
16 000 vols; 74 curr per   36308

Honeywell International Inc, Engines & Systems, Technical Library, 2525 W 190th St, *Torrance*, CA 90504-6099
T: +1 310 5123666; Fax: +1 310 5121604; E-mail: momo.rhee@honeywell.com
1941; YangHoon Rhee
Heat Transfer, Aeronautics
13 000 vols; 125 curr per; 5 digital data carriers
SLA   36309

Osram Sylvania Library, Hawes St, *Towanda*, PA 18848
T: +1 570 2685322; Fax: +1 570 2685350
1956; Kathy Hammerly
Chemistry, Ceramics, Metallurgy
12 000 vols; 351 curr per   36310

D'Arcy, Masius, Benton & Bowles, Information Center, 3310 W Big Beaver, P.O. Box 5012, *Troy*, MI 48007-5012
T: +1 248 4588533; Fax: +1 248 4588520; E-mail: beth.callahan@darcyww.com
Beth Callahan
Advertising, Marketing
8 000 vols; 500 curr per   36311

Harness, Dickey & Pierce, PLC, Law Library, 5445 Corporate Dr, Ste 400, **Troy**, MI 48098
T: +1 248 6411600; Fax: +1 248 6410270; E-mail: hdpinfo@hdp.com; URL: www.hdp.com
10 000 vols; 37 curr per   36312

Raytheon Technical Library, 1151 E Hermanns Rd, Bldg 811/T, **Tucson**, AZ 85706; P.O. Box 11337, Tucson, AZ 85734
T: +1 520 7948807; E-mail: slmack@west.raytheon.com
Shannon Mack
Aeronautics, Electronics, Mathematics, Physics
18 070 vols; 300 curr per   36313

Gable & Gotwals, Inc, Law Library, 1100 Oneok Plaza, 100 W Fifth St, **Tulsa**, OK 74103-4217
T: +1 918 5954800; Fax: +1 918 5954990; E-mail: info@gablelaw.com; URL: gablelaw.com
1944; Terry Fisher
10 000 vols; 100 curr per   36314

The Williams Companies, Law Library, Legal Dept-Law Libr 41-3, One Williams Ctr, **Tulsa**, OK 74102
T: +1 918 5734738, 2000; Fax: +1 918 5733005
Vicki Ford
Corporate law
10 000 vols; 50 curr per   36315

Ramey & Flock, PC, Law Firm Library, 100 E Ferguson St, Ste 500, **Tyler**, TX 75702; P.O. Box 629, Tyler, TX 75710-0629
T: +1 903 5973301; Fax: +1 903 5972413; E-mail: ramey@ramey-flock.com; URL: www.ramey-flock.com
1922; Penny Brown
10 000 vols; 10 curr per   36316

International Flavors & Fragrances, Inc, IFF Research Information Services, 1515 Hwy 36, **Union Beach**, NJ 07735
T: +1 732 3352435; Fax: +1 732 3352657
1967; Susan Joseph
Complete Sets of American Chemical Society Journal, Biological Abstracts, Chemical Abstracts & US Chemical Patents (1966 to present), micro
11 000 vols; 200 curr per   36317

Rivkin Radler LLP, Law Library, 926 RexCorp Plaza, **Uniondale**, NY 11556-0926
T: +1 516 3573455; Fax: +1 516 3573333
Kathy Greco
10 000 vols; 50 curr per   36318

Ruskin, Moscou & Faltischek Pc, Law Library, E Tower, 15th Flr, 190 EAB Plaza, **Uniondale**, NY 11556-0190
T: +1 516 6636525; Fax: +1 516 6636725
Naomi Heftler
10 000 vols   36319

ITT Industries, Library, 5130 Doc Repository, P.O. Box 5728, **Vandenberg AFB**, CA 93437
T: +1 805 6063288; Fax: +1 805 7343140
Theresa Wilson
28 000 vols   36320

The RJS Group, Inc, Library, 549 W Randolph St, **Vernon Hills**, IL 60061
T: +1 312 8318200; Fax: +1 312 8318201
Joanne Maxwell
City & State Building Codes, NFPA Fire Codes
15 000 vols; 120 curr per   36321

Metcalf & Eddy Inc, Harry L. Kinsel Library, P.O. Box 4071, **Wakefield**, MA 01880-5371
T: +1 781 2465200; Fax: +1 781 2456293; E-mail: richard_mansfield@metcalfeddy.com
1908; Richard Mansfield
Environmental engineering
8 000 vols; 60 curr per   36322

Bristol-Myers Squibb Co, Knowledge Integration Resources, 5 Research Pkwy, P.O. Box 5103, **Wallingford**, CT 06492
T: +1 203 2846229; Fax: +1 203 6776006
Gina Addona
30 000 vols; 650 curr per   36323

GTE Laboratories, Inc, Library, 40 Sylvan Rd, **Waltham**, MA 02254
T: +1 781 4662952; Fax: +1 781 8905790; E-mail: swolfman@gte.com
1961; Sue Wolfman
Bellcore Docs; Artifical intelligence, computer science, telecommunications
20 000 vols; 450 curr per; 6 digital data carriers   36324

Simpson, Gumpertz & Heger, Inc Library, 41 Seyon St, Bldg No 1, Ste 500, **Waltham**, MA 02453
T: +1 781 9079000; Fax: +1 781 9079009; URL: www.sgh.com
Evelyn Neuburger
Structural engineering
20 000 vols; 45 curr per   36325

General Motors Corp, Information Research, GM Technical Ctr, MC 480-106-314, 30500 Mound Rd, **Warren**, MI 48090-9055
T: +1 586 9862000; Fax: +1 586 9862009
1917; Barbara Kunkel
30 000 vols; 200 curr per   36326

Akin, Gump, Strauss, Hauer & Feld LLP, Law Library, 1333 New Hampshire Ave NW, **Washington**, DC 20036-1564
T: +1 202 8874000; Fax: +1 202 8874288; E-mail: washdcinfo@akingump.com; URL: www.akingump.com
Annette Erbrecht
15 000 vols   36327

Arent Fox PLLC Library, 1050 Connecticut Ave NW, **Washington**, DC 20036-5339
T: +1 202 8576297; Fax: +1 202 8576395; URL: www.arentfox.com
Robert Dickey
60 000 vols; 400 curr per   36328

Arnold & Porter Library, 555 12th St NW, **Washington**, DC 20004-1206
T: +1 202 9425000; Fax: +1 202 9425999; URL: www.arnoldporter.com
James W. Shelar
Law, Legislation
70 000 vols; 350 curr per   36329

Baker Botts LLP, Law Library, 1299 Pennsylvania Ave NW, **Washington**, DC 20004-2400
T: +1 202 6397967; Fax: +1 202 6397890; URL: www.bakerbotts.com
Edward O'Rourke
25 000 vols; 250 curr per   36330

Baker & Hostetler, Law Library, Washington Square, Ste 1100, 1050 Connecticut Ave NW, **Washington**, DC 20036
T: +1 202 8611500; Fax: +1 202 8611783; URL: www.bakerlaw.com
Esther Koblenz
25 000 vols; 50 curr per   36331

Baker & McKenzie LLP Library, 815 Connecticut Ave NW, Ste 900, **Washington**, DC 20006-4078
T: +1 202 4527070; Fax: +1 202 4527074; URL: www.bakernet.com
Leslee Budlong
Int law & trade
20 000 vols; 70 curr per   36332

Ballard, Spahr, Andrews & Ingersoll LLP, Law Library, 601 13th St NW, Ste 1000, **Washington**, DC 20005-3807
T: +1 202 6612200; Fax: +1 202 6612299; URL: www.ballardspahr.com
Commercial law
10 000 vols   36333

Beveridge & Diamond, PC Library, 1350 I St NW, Ste 700, **Washington**, DC 20005-3311
T: +1 202 7896000; Fax: +1 202 7896190
Scott Larson
Environmental law
10 000 vols; 300 curr per   36334

Bracewell & Giuliani LLP, Law Library, 2000 K St NW, Ste 500, **Washington**, DC 20006-1872
T: +1 202 8285800; Fax: +1 202 2231225; URL: www.bracewellgiuliani.com
Ruth Mendelson
Corporate law, Environmental law
10 000 vols; 25 curr per   36335

Bryan Cave Library, 700 13th St NW, Ste 700, **Washington**, DC 20005-3960
T: +1 202 5086115; Fax: +1 202 5086200
Laura Green
10 000 vols; 200 curr per   36336

Cadwalader, Wickersham & Taft, Law Library, 1201 F St NW, Ste 1100, **Washington**, DC 20004
T: +1 202 8622217; Fax: +1 202 8622400; E-mail: cwtinfo@cwt.com; URL: www.cwt.com
Jane Platt-Brown
9 000 vols; 45 curr per   36337

Caplin & Drysdale Library, One Thomas Circle, NW, Ste 1100, **Washington**, DC 20005
T: +1 202 8625073; Fax: +1 202 4293301; E-mail: library@capdale.com; URL: www.caplindrysdale.com
1969; Nalini Rajguru
15 000 vols; 50 curr per   36338

Chadbourne & Parke LLP, Law Library, 1200 New Hampshire Ave NW, Ste 300, **Washington**, DC 20036
T: +1 202 9745695; Fax: +1 202 9745602; E-mail: washington@chadbourne.com; URL: www.chadbourne.com
Amy Ratchford
10 000 vols; 250 curr per   36339

Covington & Burling LLP, Law Library, 1201 Pennsylvania Ave NW, **Washington**, DC 20004-2401
T: +1 202 6626000; Fax: +1 202 7786658; URL: www.cov.com
1919; John Harbison
Trade
135 000 vols; 8 000 curr per   36340

Crowell & Moring, Law Library, 1001 Pennsylvania Ave NW, 10th Fl, **Washington**, DC 20004
T: +1 202 6242828; Fax: +1 202 6285116; URL: www.crowell.com
Annette Erbrecht
Govt Contracts, Legislative Hist
25 000 vols; 8 digital data carriers
                                     36341

Debevoise & Plimpton, Law Library, 555 13th St NW, Ste 1100 E, **Washington**, DC 20004
T: +1 202 3838075; Fax: +1 202 3838118; URL: www.debevoise.com
Vicki Bayer
15 000 vols; 65 curr per   36342

Dechert, Price & Rhoads, Law Library, 1775 Eye Ste, **Washington**, DC 20006
T: +1 202 2617764; Fax: +1 202 2613333; URL: www.dechert.com
Beth Matthews
Securities
10 000 vols   36343

Dewey & Ballantine Library, 1775 Pennsylvania Ave, Ste 300, **Washington**, DC 20006
T: +1 202 8621055; Fax: +1 202 8621093; URL: www.deweyballantine.com
Daria A. Proud
30 000 vols; 400 curr per   36344

Dewey & Lebouf Library, 1101 New York Ave NW, Ste 1100, **Washington**, DC 20005
T: +1 202 3468000; Fax: +1 202 3468102; URL: www.dl.com
Marie O'Brien
10 000 vols; 300 curr per   36345

Dickstein Shapiro LLP, Research Services, 1825 Eye St NW, **Washington**, DC 20006
T: +1 202 4204999; Fax: +1 202 4202201; URL: www.dsmo.com
Joe Meringolo
30 000 vols; 75 curr per   36346

DLA Piper US LLP, Law Library, 1200 19th St NW, **Washington**, DC 20036-2412
T: +1 202 6897010; Fax: +1 202 2232085; URL: www.dlapiper.com
Patricia Gudas Mitchell
Environmental law
8 000 vols; 150 curr per   36347

Dow, Lohnes & Albertson, Law Library, 1200 New Hampshire NW, Ste 600, **Washington**, DC 20036
T: +1 202 7762650; Fax: +1 202 7762222; URL: www.dlalaw.com
Elinor Russell
15 000 vols   36348

Export-Import Bank of the United States, Research Library, 811 Vermont Ave NW, Rm 966, **Washington**, DC 20571-0001
T: +1 202 5653980; Fax: +1 202 5653985; E-mail: research@exim.gov; URL: www.exim.gov
1946; Karen Krugman
Finance, Economics, Int trade, Law; Ex-Im Bank Arch, Congressional Mats, Export Credit Hist, Merchant Banking
20 000 vols; 100 curr per   36349

Federal Deposit Insurance Corp Library, 550 17th St NW, **Washington**, DC 20429
T: +1 202 8983631; Fax: +1 202 8983984; E-mail: dismith@fdic.gov
1934; Diana Smith
Economics, Banking, Real Estate, Law; Bank Rpts (State Bank Commissions Annual Rpts); FDIC Arch Mat
80 000 vols; 800 curr per; 19 500 microforms   36350

Finnegan, Henderson, Farabow, Garrett & Dunner, Law Library, 901 New York Ave NW, **Washington**, DC 20001-4413
T: +1 202 4084290; Fax: +1 202 4084400
Virginia A. McNitt
Patents, intellectual property law
12 000 vols; 40 curr per   36351

Foley & Lardner, Private Law Library, 3000 K St NW, Ste 500, **Washington**, DC 20007
T: +1 202 6725315; Fax: +1 202 6725399; E-mail: cknuth@foleylaw.com; URL: www.foleylardner.com
Charles M. Knuth
30 000 vols   36352

Fried, Frank, Harris, Shriver & Jacobson, Law Library, 1001 Pennsylvania Ave NW, Ste 900, **Washington**, DC 20004
T: +1 202 6397102; Fax: +1 202 6397008; E-mail: sandfdi@ffhsj.com
G. Diane Sandford
25 000 vols; 300 curr per   36353

Fried, Frank, Harris, Shriver & Jacobson LLP, Law Library, 1001 Pennsylvania Ave NW, Ste 800, **Washington**, DC 20004
T: +1 202 6397102; Fax: +1 202 6397008
Diane Sandford
25 000 vols; 300 curr per   36354

Hogan & Hartson LLP, Information Resource Center, 555 13th St NW, Ste 10W-100, **Washington**, DC 20004-1109
T: +1 202 6378701; Fax: +1 202 6375910; URL: www.hhlaw.com
1968; R. Austin Doherty
Commercial law, Corporate law, Health law
110 000 vols; 2 000 curr per; 100 Electronic Media & Resources   36355

Howrey LLP, Law Library, 1299 Pennsylvania Ave NW, 6th Flr, **Washington**, DC 20004
T: +1 202 3837149; Fax: +1 202 3836610; URL: www.howrey.com
Joan Marshman
Int trade, Patents
60 000 vols; 1 500 curr per; 25 digital data carriers   36356

Inter-American Development Bank, Felipe Herrera Library, 1300 New York Ave NW, Stop W-0102, **Washington**, DC 20577
T: +1 202 6233211; Fax: +1 202 6233183; E-mail: library@iadb.org; URL: www.iadb.org/lib
1960; Irene Münster
Internat Docs Coll (UN, OECD, World Bank, FAO, etc); Latin American & Caribbean Hist, Lit
100 000 vols; 1 500 curr per; 300 maps; 2 000 microforms
libr loan; OCLC, SLA, SALALM   36357

Jones Day, Law Library, 51 Louisiana Ave NW, *Washington*, DC 20001-2113
T: +1 202 8793939; Fax: +1 202 6261700; URL: www.jonesday.com
Harva L. Sheeler
20 000 vols					36358

Katten, Muchin & Rosenman, Law Library, 2900 K St NW, Ste 200, *Washington*, DC 20007
T: +1 202 6253500; Fax: +1 202 2987570; URL: www.kattenlaw.com
Lourie Russell
Energy, Aviation
10 000 vols; 100 curr per			36359

Kelley Drye Collier Shannon PLLC Library, 3050 K St NW, No 400, *Washington*, DC 20007
T: +1 202 3428675; Fax: +1 202 3428452; E-mail: jharbison@colliershannon.com; URL: www.colliershannon.com
John H. Harbison
Antitrust Docs Coll
20 000 vols; 320 curr per			36360

King & Spalding, Law Library, 1700 Pennsylvania Ave NW, Ste 200, *Washington*, DC 20006-4706
T: +1 202 7370500; Fax: +1 202 6263737; E-mail: kingspalding@kslaw.com; URL: www.kslaw.com
Sara Uehlein
10 000 vols					36361

Kirkland & Ellis Library, 655 15th St NW, Ste 1200, *Washington*, DC 20005-5793
T: +1 202 8795009; Fax: +1 202 8795200; E-mail: library.research.desk.washington@dc.kirkland.com
1951; Ansley Calhoun
Law, Legislation
10 000 vols; 100 curr per			36362

Kirkpatrick & Lockhart, Preston, Gates, Ellis, Law Library, 1601 K NW, *Washington*, DC 20006-1600
T: +1 202 7789000; Fax: +1 202 7789100; URL: www.klgates.com
Walker Chaffin
11 000 vols; 350 curr per			36363

K&L Gates LLP, Law Library, 1601 K Street NW, L-3, *Washington*, DC 20006
T: +1 202 6613715; Fax: +1 202 7789100; URL: www.klgates.com
Walker Chaffin
Maritime law, taxation law
10 000 vols; 250 curr per			36364

McKenna, Long & Aldridge, LLP, Law Library, 1900 K St NW, *Washington*, DC 20006
T: +1 202 4967579; Fax: +1 202 4967756
Kate Martin
Int law, Environmental law
30 000 vols; 210 curr per			36365

Milbank, Tweed, Hadley & McCloy LLP Library, 1850 K St, NW, Suite 1100, *Washington*, DC 20006
T: +1 202 8357578; Fax: +1 202 8357586; E-mail: gzsebi@milbank.com; URL: www.milbank.com/
Gabriele C. Zsebi
20 000 vols; 28 maps; 4 microforms; 2 av-mat; 2 sound-rec; 8 digital data carriers
libr loan; AALL				36366

Morgan, Lewis & Bockius LLP, Law Library, 1111 Pennsylvania Ave NW, *Washington*, DC 20004-2541
T: +1 202 7396424; Fax: +1 202 7393001; URL: www.morganlewis.com
Barbara Folensbee-Moore
Environmental law, Employment, Finance, Intellectual property; Legislative Hist, Federal Agency Releases
45 000 vols; 750 curr per			36367

O'Melveny & Myers, Law Library, 555 13th St NW, Ste 500 W, *Washington*, DC 20004
T: +1 202 3835311; Fax: +1 202 3835414; E-mail: dfisher@omm.com; URL: www.omm.com
Debra Fisher
Administrative law, Int trade, Ligitation; Legislative Hist Compilations
25 000 vols					36368

Patton & Boggs LLP, Law Library, 2550 M St NW, 8th Flr, *Washington*, DC 20037
T: +1 202 4576000; Fax: +1 202 4576315; URL: www.pattonboggs.com
Kevin McCall
20 000 vols; 200 curr per			36369

Paul, Weiss, Rifkind, Wharton & Garrison, Law Library, 1615 L St NW, Ste 1300, *Washington*, DC 20036-5694
T: +1 202 2237400; Fax: +1 202 2237420; E-mail: jeckel@paulweiss.com
Jennifer G. Eckel
Int trade
10 000 vols; 75 curr per			36370

Pension Benefit Guaranty Corporation, Corporate Library, 1200 K St NW, Ste 360, *Washington*, DC 20005-4026
T: +1 202 3264000; Fax: +1 202 3264011; E-mail: librarystaff2@pbgc.gov
1976; Judith Weiss
12 000 vols; 65 curr per			36371

Piper, Marbury, Rudnick & Wolfe LLP, Law Library, 1200 14th St NW, Ste 700, *Washington*, DC 20036
T: +1 202 8614171; Fax: +1 202 2232085
Pat G. Mitchell
20 000 vols					36372

Reed Smith LLP, Law Library, 1301 K St NW, Ste 1100, E Tower, *Washington*, DC 20005-3317
T: +1 202 4149415; Fax: +1 202 4149299; URL: www.reedsmith.com
Lorraine DeSouza
21 000 vols; 250 curr per			36373

Ross, Dixon & Bell, LLP, Law Library, 2001 K St NW, *Washington*, DC 20006
T: +1 202 6622142; Fax: +1 202 6622190; E-mail: mmaguire@rdblaw.com
1984; Mary M. Maguire
8 000 vols					36374

Seyfarth Shaw, Washington Branch Office Library, 975 F St NW, *Washington*, DC 20004
T: +1 202 8283559; Fax: +1 202 8285393
1978; Susan Ryan
Law, Labor law, Construction
11 000 vols; 150 curr per			36375

ShawPittman, LLP, Library & Information Center, 2300 N St NW, *Washington*, DC 20037
T: +1 202 4547499; Fax: +1 202 6638007; URL: www.shawpittman.com
Susan Mills
23 000 vols; 100 curr per			36376

Shearman & Sterling Library, 801 Pennsylvania Ave NW, Ste 900, *Washington*, DC 20004-2634
T: +1 202 5088055; Fax: +1 202 5088100; URL: www.shearman.com
Jill Sidford
Internatinal trade, securities
10 000 vols; 70 curr per			36377

Sidley Austin LLP, Law Library, 1501 K St NW, *Washington*, DC 20005
T: +1 202 7368525; Fax: +1 202 7368711
Jeffrey V. Bosh
Legislative Hist Coll
25 000 vols; 300 curr per
libr loan					36378

Skadden, Arps, Slate, Meagher & Flom LLP, Law Library, 1440 New York Ave NW, *Washington*, DC 20005
T: +1 202 3717760; Fax: +1 202 3935760; URL: www.skadden.com
Margaret M. Heath
Int trade, Mergers, Energy, Tax, Securities
30 000 vols; 400 curr per			36379

Sonnenschein, Nath & Rosenthal Library, 1301 K St NW, Ste 600E, *Washington*, DC 20005
T: +1 202 4086452; Fax: +1 202 4086399; URL: www.sonnenschein.com
Ann Green
15 000 vols; 25 curr per			36380

Spiegel & McDiarmid LLP, Law Library, 1333 New Hampshire Ave NW, 2nd Flr, *Washington*, DC 20036
T: +1 202 3932866; E-mail: library@spiegelmcd.com; URL: www.spiegelmcd.com
Jeffrey J. Berns
Environmental law
10 000 vols; 200 curr per
libr loan; AALL, SLA				36381

Squire, Sanders & Dempsey, Library, 1201 Pennsylvania Ave NW, Ste 500, *Washington*, DC 20044; P.O. Box 407, Washington, DC 20044-0407
T: +1 202 6266704; Fax: +1 202 6266780; URL: www.ssd.com
Scott Bailey
15 000 vols; 60 curr per			36382

Steptoe & Johnson Library, 1330 Connecticut Ave NW, *Washington*, DC 20036
T: +1 202 8283620; Fax: +1 202 4293902
1972; Ellen Brondfield
Law, Legislation
60 000 vols					36383

Sterne, Kessler, Goldstein & Fox Library, 1100 New York Ave NW, Ste 600, *Washington*, DC 20005-3934
T: +1 202 3712600 ext 557; Fax: +1 202 3712540; E-mail: kmartin@skgf.com
Kelly Hayes Martin
10 000 vols					36384

Sullivan & Cromwell LLP, Law Library, 1701 Pennsylvania Ave NW, *Washington*, DC 20006-5805
T: +1 202 9567538; Fax: +1 202 2936330; URL: www.sullcrom.com
Denise Noller
8 000 vols; 25 curr per			36385

Sutherland, Asbill & Brennan LLP Library, 1275 Pennsylvania Ave NW, 6th Flr, *Washington*, DC 20004-2415
T: +1 202 3830100; Fax: +1 202 6373593; URL: www.sablaw.com
Sara T. Stephens
Energy, Insurance, Patents, Tax & Trademarks Coll, Tax Legislative Hist Coll
30 000 vols					36386

Swidler, Berlin, Shereff & Friedman Library, 3000 K St NW, Ste 300, *Washington*, DC 20007-5116
T: +1 202 4247544; Fax: +1 202 4247643; URL: www.swidlaw.com
Laura Speer
10 000 vols; 50 curr per			36387

Van Ness Feldman Library, 1050 Thomas Jefferson St NW, *Washington*, DC 20007
T: +1 202 2981800; Fax: +1 202 3382416; URL: www.vnf.com
1976; George Bernard Kirlin
35 000 vols; 250 curr per			36388

Venable LLP Library, Washington, DC Office, 575 Seventh St, NW, *Washington*, DC 20004-1601
T: +1 202 3444612; Fax: +1 202 3448300
1981; Mohammad Nassim
28 000 vols; 50 curr per			36389

Verner, Lipfert, Bernard, McPherson & Hand, Law Library, 901 15th St NW, Ste 700, *Washington*, DC 20005-2327
T: +1 202 3716068; Fax: +1 202 3716279; URL: www.bernard.com
Cecelia Weinheimer
22 000 vols; 2 000 curr per			36390

Weil, Gotshal & Manges LLP, Law Library, 1300 Eye St NW, Ste 900, *Washington*, DC 20005
T: +1 202 6827117; Fax: +1 202 6827297; URL: www.weil.com
Ann Sloane
Int trade, Environmental law
30 000 vols					36391

White & Case LLP, Law Library, 701 13th St NW, *Washington*, DC 20005-3807
T: +1 202 6266475; Fax: +1 202 6399355; URL: www.whitecase.com
Richard Cousins
11 000 vols; 40 curr per			36392

Wiley Rein LLC Library, 1776 K St NW, *Washington*, DC 20006
T: +1 202 7193103; Fax: +1 202 7197049; URL: www.wrf.com; www.wileyrein.com
Carolyn Ahearn
25 000 vols; 200 curr per
AALL, SLA					36393

Wilkes, Artis, Hedrick & Lane, Chartered Library, 1150 18th St NW, *Washington*, DC 20036
T: +1 202 4577800; Fax: +1 202 4577814
Annette Erbrecht
Law, Real estate
20 000 vols					36394

Wilkinson Barker Knauer LLP, 2300 N St NW, Ste 700, *Washington*, DC 20037
T: +1 202 3833420; Fax: +1 202 7835851; URL: www.wbklaw.com
Louis C. Abramovitz
8 000 vols; 25 curr per			36395

Williams & Connolly Library, 725 12th St NW, *Washington*, DC 20005
T: +1 202 4345303; Fax: +1 202 4345029
1970; Ellen Feldman
Administrative law, Criminal law
75 000 vols; 300 curr per
libr loan					36396

Wilmer, Cutler, Hale & Dorr Library, 1875 Pennsylvania Ave NW, *Washington*, DC 20006
T: +1 202 6636771; Fax: +1 202 6636363; URL: www.wilmerhale.com
1963; Jane Huston
Antitrust law, Banking, Legislation, Securities
56 000 vols					36397

Carmody & Torrance, Law Library, 50 Leavenworth St, *Waterbury*, CT 06702
T: +1 203 5731200; Fax: +1 203 5752600; E-mail: ahodges@carmodylaw.com
Ann C. Hodges
12 000 vols; 100 curr per			36398

Beihoff Music Corporation, Sheet Music Departmant Library, 12040 W Feerick St, Unit H, *Wauwatosa*, WI 53222
T: +1 414 4383920; Fax: +1 414 4383939; URL: www.beihoffmusic.com
Kenneth Kunz
20 000 vols; 15 curr per			36399

ISP Management Co, Inc, Technical Information Services Library, 1361 Alps Rd, *Wayne*, NJ 07470
T: +1 973 6283234; Fax: +1 973 6283404; E-mail: inaznitsky@ispcorp.com; URL: www.ispcorp.com
1972; Ira Naznitsky
Acetylene chemistry, Building mat, Organic chemistry, Specialty chemicals
11 000 vols; 165 curr per			36400

Kearfott Guidance & Navigation Corp, Technical Information Center Library, 150 Totowa Rd, MS HQB33, *Wayne*, NJ 07474-0946
T: +1 973 7856481; Fax: +1 973 7856121
1954; Ed Hellmuth
Aerospace engineering, Computer science, Electrical & mechanical engineering; Patents, Symposia, Tech Rpts
27 000 vols; 110 curr per			36401

Marconi Aerospace Systems, CNI Division Technical Library, 164 Totowa Rd, *Wayne*, NJ 07474-0975
T: +1 973 6333438; Fax: +1 973 6334473; E-mail: aughey@systems.gec.com
Kathleen Aughey
10 000 vols; 300 curr per			36402

URS Greiner Woodward-Clyde Consultants Library, 201 Willowbrook Blvd, *Wayne*, NJ 07470-0290
T: +1 973 7850700; Fax: +1 973 7850023
Mary Tanner
Architectural engineering tech, Civil engineering, Environmental studies, Geology, Metallurgy
14 000 vols; 120 curr per			36403

DuPont De Nemours & Co, Inc, Benger Laboratory Library, 400 Du Pont Blvd, **Waynesboro**, VA 22980
T: +1 540 9492485; Fax: +1 540 9492949
1947; Becky Moomau
10 000 vols; 90 curr per 36404

Cargill, Inc – Law Library, 15407 McGinty Rd W, **Wayzata**, MN 55391-2399; P.O. Box 5624, Minneapolis, MN 55440-5624
T: +1 952 7426334; Fax: +1 952 7426349
Richard Wiebelhaus
8 000 vols; 120 curr per
libr loan 36405

Xerox Corp, TIC Resource Center, 800 Phillips Rd, Bldg 105-66C, **Webster**, NY 14580
T: +1 585 4222091; Fax: +1 585 4228299; E-mail: tic@crt.xerox.com
1958; Laura Tucker
Electronics, Engineering, Data processing, Photography, Chemistry, Business; Corp Tech Arch
28 000 vols; 200 curr per
libr loan 36406

Sun Life Assurance Company of Canada, Reference Library, One Sun Life Executive Park, **Wellesley Hills**, MA 02481
T: +1 781 4314926; Fax: +1 781 2371398; E-mail: PamelaMahaney@Sunlife.com
1973; Pamela Ann Mahaney
Accounting, Economics, Law, Real estate; Life Insurance Marketing & Research Assn Coll, Insurance Codes for all States
15 000 vols; 850 curr per; 20 digital data carriers
libr loan; SLA 36407

Optikon Research Laboratories Library, 62 River Rd, P.O. Box 259, **West Cornwall**, CT 06796-0259
T: +1 203 6726614; Fax: +1 203 6726615
William Covington
10 700 vols; 6 curr per 36408

J. C. Allen & Son, Inc, Photo Library, P.O. Box 1950, **West Lafayette**, IN 47996-1950
T: +1 765 4639614
John Allen
77 000 vols 36409

Lampf, Lipkind, Prupis & Petigrow, Law Library, 80 Main St, **West Orange**, NJ 07052-5482
T: +1 973 3252100; Fax: +1 973 3252839; E-mail: administrator@llpplaw.com; URL: www.llpplaw.com
Lucille Field
10 000 vols; 50 curr per 36410

Wolff & Samson, Law Library, One Boland Dr, **West Orange**, NJ 07052
T: +1 973 5302146; Fax: +1 973 5302346
Rosemary Walton
12 000 vols 36411

Merck & Co, Inc, Published Information Resources, Sumneytown Pike, **West Point**, PA 19486
T: +1 215 6526026; Fax: +1 215 6520721
1921; Sarah C. Williams
Biochemistry, Medicine, Microbiology, Pharmacology
31 000 vols; 4 000 e-books
libr loan 36412

Glencoe Publishing Company Library, 8787 O'Rion Pl, **Westerville**, OH 42240-4027
T: +1 800 8481567; Fax: +1 614 4304379; E-mail: darlene_yeager@mcgraw-hill.com
Darlene Yeager
Education; Textbks
10 000 vols 36413

Eveready Battery Company, Inc, Technical Information Center, 25225 Detroit Rd, P.O. Box 450777, **Westlake**, OH 44145-2510
T: +1 440 8357634; Fax: +1 440 8358479
1956; Elaine M. Balfe
Electrochemical energy sources
12 000 vols; 350 curr per 36414

Merck & Co, Inc, Law Library, One Merck Dr, **Whitehouse Station**, NJ 08889
T: +1 908 4235805; Fax: +1 908 7351147; E-mail: elizabeth_arnold@merck.com
1959; Elizabeth Arnold
13 000 vols; 272 curr per 36415

Cessna Aircraft Co, Engineering Library, One Cessna Blvd, **Wichita**, KS 67215; mail address: 2617 S Hoover Rd, Wichita, KS 67215-1200
T: +1 316 5178061; Fax: +1 316 5177437; E-mail: rinman@cessna.textron.com; URL: www.cessna.textron.com
Rhonda Inman
10 000 vols; 81 curr per 36416

Spirit AeroSystems, Inc, Technical Library, 3801 S Oliver, K78-38, **Wichita**, KS 67210; P.O. Box 780008, K78-38, Wichita, KS 67278-0008
T: +1 316 5263801; Fax: +1 316 5231169
1941; Tony Chavez
Aviation, Aerospace science, Engineering
10 000 vols; 68 curr per; 12 digital data carriers; 10 000 military specifications & standards, 6 000 tech rpts 36417

Lubrizol Corp Chemical Library, Information Center, 29400 Lakeland Blvd, **Wickliffe**, OH 44092
T: +1 440 9434200 ext 2509; Fax: +1 440 9435337
1946
10 200 vols; 210 curr per 36418

Rosenn, Jenkins & Greenwald Library LLP, 15 S Franklin St, **Wilkes-Barre**, PA 18711-0075
T: +1 570 8214727; Fax: +1 570 8317218; E-mail: library@rjglaw.com; URL: www.rjglaw.com
Beverly Ashmore
8 000 vols; 50 curr per 36419

AstraZeneca Pharmaceuticals, Information Science Libraries & Archives, 1800 Concord Pike, **Wilmington**, DE 19850-2902; P.O. Box 15437, Wilmington, DE 19850-5437
T: +1 302 8868232; Fax: +1 302 8865369
1918; June O'Brien
Pharmacology, Chemistry, Medicine
30 000 vols; 1 200 curr per 36420

DuPont Co, Lavoisier Library, Rte 141, **Wilmington**, DE 19803; P.O. Box 80301, Wilmington, DE 19880-0301
T: +1 302 6953200; Fax: +1 302 6951390
Chemistry, Chemical engineering, Petroleum, Toxicology, Biology, Medicine, Electronics, Marketing
40 000 vols; 1 400 curr per 36421

Hercules Incorporated – Law Department Library, Hercules Plaza 8330 SE, 1313 N Market St, **Wilmington**, DE 19894
T: +1 302 5945678; Fax: +1 302 5947038
21 000 vols; 30 curr per 36422

Hercules Incorporated – Research Center Library, 500 Hercules Rd, **Wilmington**, DE 19808-1599
T: +1 302 9953483; Fax: +1 302 9954101; E-mail: jhenderson@herc.com
1913; Susan Danko
Chemical technology
47 000 vols; 400 curr per 36423

Morris, Nichols, Arsht & Tunnell, LLP, Law Library, 1201 N Market St, **Wilmington**, DE 19801; P.O. Box 1347, Wilmington, DE 19899
T: +1 302 6589200; Fax: +1 302 6583989; URL: www.mnat.com
Elizabeth Stack
Corporate law, taxation law
10 000 vols; 50 curr per 36424

Prickett, Jones, Elliott, Law Library, 1310 King St, **Wilmington**, DE 19801; P.O. Box 1328, Wilmington, DE 19899-1328
T: +1 302 8886500; Fax: +1 302 6588111
Pamela L. Reed
10 000 vols 36425

Richards, Layton & Finger Library, 920 King St, **Wilmington**, DE 19801; P.O. Box 551, Wilmington, DE 19899-0551
T: +1 302 6517700; Fax: +1 302 4987700; URL: www.rlf.com
Robert Guerrero
Law
14 000 vols; 27 curr per 36426

Skadden, Arps, Slate, Meagher & Flom LLP Library, One Rodney Sq, 7th flr, **Wilmington**, DE 19801
T: +1 302 6513224; Fax: +1 302 6513001; URL: www.skadden.com
Leslie Corey Leach
Law
9 000 vols; 50 curr per 36427

Textron Defense Systems, Research Library, 201 Lowell St, **Wilmington**, MA 01887
T: +1 978 6572868; Fax: +1 978 6574975
1956; Joanne M. Campbell
Electronics
18 000 vols; 87 curr per 36428

United Technologies Corp, Information Network, One Hamilton Rd, Hamilton Sundstrand, MS 1-3BC52, **Windsor Locks**, CT 06096-1010
T: +1 860 6544352; Fax: +1 860 6543689
1957; Suzanne Cristina
Aerospace science, Electronics
12 054 vols; 264 curr per 36429

Womble, Carlyle, Sandridge & Rice, Law Library, One W Fourth St, **Winston-Salem**, NC 27101
T: +1 336 7474757; Fax: +1 336 7213660; URL: www.wcsr.com
Susan Garrison
20 000 vols; 50 curr per 36430

Stora Enso North America, Research Library, 300 N Biron Dr, **Wisconsin Rapids**, WI 54494
T: +1 715 4222368; Fax: +1 715 4222227
Cynthia Van Ert
10 000 vols; 700 curr per 36431

Greenbaum, Rowe, Smith, Ravin, Davis & Himmel LLP, Law Library, 99 Wood Ave S, **Woodbridge**, NJ 07095
T: +1 732 5495600; Fax: +1 732 5491881; URL: www.greenbaumlaw.com
Leigh DeProspo
Real estate
9 000 vols 36432

Wilentz, Goldman & Spitzer, Law Library, 90 Woodbridge Center Dr, **Woodbridge**, NJ 07095
T: +1 732 8556160; Fax: +1 732 7266525; URL: www.newjerseylaw.com
1919; Scott L. Fisher
20 000 vols; 110 curr per 36433

Litton Industries, Guidance & Control Systems, 5500 Canoga Ave, **Woodland Hills**, CA 91367
T: +1 818 7127355; Fax: +1 818 7127151
1954; Manuela Wood
Electronics, Engineering
40 500 vols
libr loan 36434

Bowditch & Dewey, Law Library, 311 Main St, **Worcester**, MA 01608; P.O. Box 15156, Worcester, MA 01615-0156
T: +1 508 9263331; Fax: +1 508 9293140; URL: www.bowditch.com
Byron C. Hill
10 000 vols; 400 curr per 36435

IBM Corp, Thomas J. Watson Research Center Library, 1101 Kitchawan Rd, **Yorktown Heights**, NY 10598
T: +1 914 9451415; Fax: +1 914 9454144; E-mail: watlib@us.ibm.com
Selena Thomas
45 000 vols; 700 curr per 36436

Special Libraries Maintained by Other Institutions

Abbott, Library Information Resources, 100 Abbott Park Rd, AP 6B, **Abbott Park**, IL 60064-6107
T: +1 847 9374600; Fax: +1 847

9376333
1888; Dee Mater
US, EPO & PCT Patents
8 000 vols; 1 500 curr per; 500 e-books
libr loan 36437

Chemical, Biological, Radiological & Nuclear Defense Information Analysis Center, CBRNIAC (a DTIC Information Analysis Center), 1204 Technology Drive, **Aberdeen**, MD 21001
T: +1 410 6769030; Fax: +1 410 6769703; E-mail: cbrniac@battelle.org; URL: www.cbrniac.apgea.army.mil
1986; James McNeely
Toxicology, demilitarization, Chemical Weapons Treaty Reference Collection. Biological Weapons Treaty Reference Collection; Analysis of Manufacturing Processes for Nuclear, Biological and Chemical (NBC) Defense Systems; Chemical and Physical Properties of CB Defense Materials; Chemical Identification; Combat Effectiveness; Counterproliferation; Counterterrorism; Decontamination; Defense Conversion and Dual-Use Technology Transfer; Demilitarization; Domestic Preparedness / Homeland Security; Environmental Fate and Effects; Force Protection; Individual and Collective Protection; International Technology Proliferation and Arms Control; Medical Effects and Treatment; Nuclear, Biological and Chemical Survivability; Radiological and Nuclear Defense; Smoke and Obscurants; Toxic Industrial Chemicals and Toxic Industrial Materials; Toxicology; Treaty Verification and Compliance; Warning and Identification
180 000 vols
libr loan 36438

US Army – Medical Research Institute of Chemical Defense, Wood Technical Library, 3100 Ricketts Point Rd, **Aberdeen Proving Ground**, MD 21010-5400
T: +1 410 4364135; Fax: +1 410 4363176; E-mail: mricd-wood.techlibrary@amedd.army.mil
1979; Barbara Schultz
Toxicology
18 160 vols; 152 curr per; 15 diss/theses; 229 govt docs; 20 maps; 26 755 microforms; 158 av-mat; 15 sound-rec; 4 digital data carriers
libr loan; FEDLINK, OCLC 36439

Ghost Ranch Conference Center Library, HC 77, Box 11, **Abiquiu**, NM 87510-9601
T: +1 505 6854333 ext 109; Fax: +1 505 6854519; E-mail: edd@ghostranch.org; URL: www.ghostranch.org
1955; Edgar Davy
Religion, Children's lit, Archeology, Paleontology, Fine arts, Ecology; Southwest Coll
20 000 vols; 35 curr per; 50 maps; 60 av-mat; 10 Bks on Deafness & Sign Lang 36440

Hawaii Agriculture Research Center Library, 99-193 Aiea Heights Dr, Ste 300, **Aiea**, HI 96701-3911
T: +1 808 4865370; Fax: +1 808 4865020; E-mail: amarsteller@harc-hspa.com
1895; Ann Marsteller
Plant physiology & pathology, Sugarcane, Tropical agriculture, Entomology, Chemistry; Project Files of Experiment Station
80 000 vols; 25 curr per 36441

Akron Art Museum, Martha Stecher Reed Art Reference Library, One South High, **Akron**, OH 44308
T: +1 330 3769186 ext 221; Fax: +1 330 3761180; E-mail: mail@AkronArtMuseum.org; URL: www.akronartmuseum.org
1972; Ellie Ward
Edwin C. Shaw Correspondence; Artist's Biographical Files
11 000 vols; 38 curr per; 100 av-mat; 5 000 exhibition cat, 11 350 slides
ARLIS/NA 36442

Akron Law Library Association, 209 S High St, 4th Flr, **Akron**, OH 44308-1675
T: +1 330 6432804; Fax: +1 330 5350077; E-mail:

lhostetler@akronlawlib.org; URL: www.akronlawlib.org
1888; Linda D. Hostetler
82 000 vols; 254 curr per
libr loan; AALL, ORALL, SLA   36443

Albany Institute of History & Art, 125 Washington Ave, *Albany*, NY 12210-2296
T: +1 518 4634478; Fax: +1 518 4635506; E-mail: library@albanyinstitute.org; URL: albanyinstitute.org
1791; Rebecca D. Rich-Wulfmeyer
Albany Imprints, Directories, Almanacs; Albany Social, Political & Business Hist, Dutch in the Upper Hudson Valley, 17th & 18th Centuries
14 000 vols; 50 curr per; 1 000 linear feet of mss   36444

New York State Library, Talking Book & Braille Library, Empire State Plaza, Cultural Education Ctr Basement, *Albany*, NY 12230
T: +1 518 4745935; Fax: +1 518 4861957
1896; Jane Somers
780 000 vols   36445

The Albuquerque Museum of Art & History, Reference Library, 2000 Mountain Rd NW, *Albuquerque*, NM 87104
T: +1 505 2437255; Fax: +1 505 7646546; URL: www.cabq.gov/museum
Connor O'Laughlin
Southwest art, Middle Rio Grande valley
12 000 vols; 23 curr per   36446

Lovelace Respiratory Research Institute, Sam White Library, 2425 Ridgecrest Dr SE, *Albuquerque*, NM 87108-5127
T: +1 505 3489414; Fax: +1 505 3484978; E-mail: libreq@lrri.org; URL: www.lrri.org
1974; Heather O'Daniel
25 000 vols; 75 curr per   36447

Menaul Historical Library of the Southwest, 301 Menaul Blvd NE, *Albuquerque*, NM 87107
T: +1 505 3437480; E-mail: mhlsw2@juno.com
1974; Nona Browne
Presbyterian churches of the Southern US
21 000 vols; 12 curr per; 20 Special Interest Per Sub   36448

Sandia National Laboratories, Technical Library, P.O. Box 5800, MS 0899, *Albuquerque*, NM 87185-0899
T: +1 505 8458287; Fax: +1 505 8443143; URL: www.sandia.gov/news-center/resources/tech-library/index.html
1948; Susan Stinchcomb
Aerodynamics, Electronics, Energy res, Explosives, Nuclear safety, Nuclear waste management, Solid-state physics; Sandia Tech Rpts
70 000 vols; 1 100 curr per   36449

Southwest Research and Information Center Library, 105 Stanford SE, P.O. Box 4524, *Albuquerque*, NM 87106
T: +1 505 2621862; Fax: +1 505 2621864; E-mail: info@sric.org; URL: www.sric.org
1971; Don Hancock
10 000 vols; 450 curr per; 7 cabinets of clippings, 100 sourcebks   36450

University of New Mexico – Bureau of Business & Economic Research Data Bank, 1919 Las Lomas NE, *Albuquerque*, NM 87106; MSC06 3510, One University of New Mexico, Albuquerque, NM 87131-0001
T: +1 505 2776626; Fax: +1 505 2772773; E-mail: dbinfo@unm.edu; URL: www.unm.edu/~bber/
1945; Kevin Kargacin
Economic development, New Mexico demographics; New Mexico's Economy Coll, New Mexico Social Statistics (1970, 1980 & 1990 Census Summary Tapes Coll), New Mexico Statistics
14 000 vols; 184 curr per; 60 digital data carriers   36451

Alcoa Technical Center Library, 100 Technical Dr, *Alcoa Center*, PA 15069-0001
T: +1 724 3375300; Fax: +1 724 3372394; URL: www.alcoa.com
1919; Christine Hennrich
Aluminium, Auto tech, Mat science, Surface science, Packaging; Alcoa Laboratories Hist, Alcoa Publs, Tech Translations
23 000 vols; 116 curr per; 16 Journals
PALINET   36452

Center for Naval Analyses Library, 4825 Mark Center Dr, *Alexandria*, VA 22311-1846
T: +1 703 8242117; Fax: +1 703 8242200; URL: www.cna.org
1962; Greg Kaminski
Military science, Political science, Operations res, Mathematics, Statistics, Systems analysis
15 000 vols; 300 curr per; 5 000 Congressional docs   36453

Educational Research Service, Information Resource Center, 1001 N Fairfax St, Ste 500, *Alexandria*, VA 22314
T: +1 703 2432100; Fax: +1 703 2431985; E-mail: ers@ers.org; URL: www.ers.org
1973; Brian Galvin
20 000 vols; 150 curr per   36454

Institute for Defense Analyses, Technical Information Services, 4850 Mark Center Dr, *Alexandria*, VA 22311
T: +1 703 8452087; Fax: +1 703 8207194; E-mail: refdesk@ida.org; URL: www.ida.org
1960; Russell I Fries
12 000 vols   36455

National Association of Chain Drug Stores, Resource Center, 413 N Lee St, *Alexandria*, VA 22314; P.O. Box 1417-D49, Alexandria, VA 22313-1480
T: +1 703 8374129; Fax: +1 703 7396079; E-mail: library@nacds.org
1933; Betsy Hageman
10 000 vols; 200 curr per   36456

Allentown Art Museum, Reference Library, 31 N. Fifth St, *Allentown*, PA 18101
T: +1 610 4324333; Fax: +1 610 4347409; E-mail: askus@allentownartmuseum.org; URL: www.allentownartmuseum.org
1959; Sofia Bakis
Art hist
15 460 vols; 78 curr per; 48 VF drawers   36457

Lehigh County Historical Society, Scott Andrew Trexler II Memorial Library, Old Court House, Fifth & Hamilton Sts, *Allentown*, PA 18101; P.0O. Box 1548, Allentown, PA 18105
T: +1 610 435-1072; Fax: +1 610 435-9812; E-mail: lchs@voicenet.com; URL: www.lchs.museum/
1906; Jan S. Ballard
Lehigh County; Allentown Imprints, City Directories, Early German Newspapers, Local Church Records
10 000 vols; 20 curr per; 500 maps; 225 microforms; 2 000 pamphlets, 65 000 photos & negatives, 300 linear feet of mss   36458

Texas Tech University Health Sciences Center at Amarillo, Harrington Library, 1400 Wallace Blvd, *Amarillo*, TX 79106
T: +1 806 3545448; Fax: +1 806 3545430; E-mail: dana.neeley@ttuhsc.edu; URL: www.ttuhsc.edu/libraries
1977; Dana M. Neeley
20 600 vols; 418 curr per; 2 150 e-journals; 3 800 e-books; 2 940 av-mat   36459

Amelia County Historical Society Library, Jackson Bldg, 16501 Church St, *Amelia*, VA 23002-0113; P.O. Box 113, Amelia, VA 23002-0113
T: +1 804 5613180
1957; Gary Austin
Gemstones Hist Coll, Indian Artifacts
13 000 vols; 30 curr per   36460

Iowa Department of Transportation Library, 800 Lincoln Way, *Ames*, IA 50010-6915
T: +1 515 2391200; Fax: +1 515 2337840
1970; Hank Zaletel
20 000 vols; 340 curr per
libr loan   36461

National Animal Disease Center Library, US Department of Agriculture, Agriculture Research Service, 2300 Dayton Ave, *Ames*, IA 50010; P.O. Box 70, Ames, IA 50010
T: +1 515 6637271; Fax: +1 515 6637458; E-mail: jeifling@nadc.ars.usda.gov
1961; Janice K Eifling
Veterinary Medicine
8 000 vols; 170 curr per
libr loan   36462

National Yiddish Book Center, Harry & Jeanette Weinberg Bldg, 1021 West St, *Amherst*, MA 01002-3375
T: +1 413 2564900; Fax: +1 413 2564700; URL: www.yiddishbookcenter.org
1980; Aaron Rubinstein
Yiddish language
1 500 000 vols; 800 music scores   36463

Braille Institute Library Services – Orange County Center, 527 N Dale Ave, *Anaheim*, CA 92801
T: +1 714 8215000; Fax: +1 714 5277621
Nancy Stanton
26 000 vols   36464

Anchorage Museum of History & Art Library & Archives, 121 W Seventh Ave, *Anchorage*, AK 99501
T: +1 907 3436191;
Fax: +1 907 3436149; URL: www.anchoragemuseum.org
1968; Tracy Leithauser
Alaskan anthropology, art, ethnology, hist; Alaska Railroad Coll, photos; Alexander Creek (Fred Winters Coll), diaries; Barrow & Diomede Islands (Eide Coll), photos; Ward Wells Coll, photos; Federal Aviation Administration Coll, photos; Reindeer Herding (Ickes Coll), photos; Steve McCutcheon, photos; Valdez Hist (Hinchley-Alagco Coll), photos
12 000 vols; 25 curr per; 500 maps; 20 000 sound-rec; 1 digital data carriers; 200 000 photos, 40 000 slides, 12 VF drawers
libr loan   36465

Special Education Service Agency (SESA), Library, 3501 Denati St, *Anchorage*, AK 99503
T: +1 907 5627372; Fax: +1 907 5620545; E-mail: afreitag@sesa.org; URL: www.sesa.org/
Anne K. Freitag
State Infant Learning Programm, Assistive Technology (ATA Coll), Alaska School Psychologist Assn
9 720 vols; 110 curr per   36466

World University Library, 222 W Seventh Ave, Rm 181, *Anchorage*, AK 99513-7586
T: +1 907 2715655; Fax: +1 907 2715640
1947; Catherine A. Davidson
Ancient history, Anthropology, Civilisation, Folklore, Metaphysics, Occult, Parapsychology, Peace studies, Theosophical hist, UFO Phenomena, Poetry; Rare bks in Religious Philosophy
26 000 vols; 12 curr per; 150 diss/theses; 60 av-mat; 175 sound-rec   36467

National Center for Manufacturing Sciences, Manufacturing Information Resource Center, 3025 Boardwalk, *Ann Arbor*, MI 48108
T: +1 734 9950300; Fax: +1 734 9953236; E-mail: info@ncms.org; URL: www.ncms.org
1989
12 000 vols; 500 curr per; 1 600 av-mat; 5 000 reports, 18 000 articles   36468

Fairfax County Public Schools, Education Library, 4414 Holborn Ave, *Annandale*, VA 22003
T: +1 703 5037420; Fax: +1 703 5037418; E-mail: jessica.foster@fcps.edu; URL: www.fcps.edu/specialed.htm
1968; Jessica S. Foster
17 000 vols; 250 curr per   36469

Alion Science & Technology, Technical Information Services, Joint Spectrum Ctr, 185 Admiral Cochrane Dr, *Annapolis*, MD 21401-7388
T: +1 410 5737075; Fax: +1 410 5737403
1962; Dee Zannowski
Electronics, Software engineering; Spectrum Signatures, Dept of Defense Frequency Allocation Applications
150 000 vols; 100 curr per; 40 000 microforms; 5 digital data carriers; 45 000 techl rpts, 12 000 equipment manuals, 10 000 standards   36470

Maryland State Archives Library, 350 Rowe Blvd, *Annapolis*, MD 21401
T: +1 410 2606400; Fax: +1 410 9743895; E-mail: archives@mdsa.net; URL: www.msa.md.gov
1935; Christine D. Alvey
Maryland, State hist, Natural resources, Genealogy, Hist; Archs, Laws of Maryland, Maryland State Publs
16 000 vols; 200 curr per   36471

Public Library of Anniston & Calhoun County – Library for the Blind & Physically Handicapped, 108 E Tenth St, *Anniston*, AL 36201; P.O. Box 308, Anniston, AL 36202
T: +1 256 2378501; E-mail: bandph@anniston.lib.al.us
Deenie M. Culver
12 000 vols   36472

California Thoroughbred Breeders Association, Carleton F. Burke Memorial Library, 201 Colorado Pl, *Arcadia*, CA 91007; P.O. Box 60018, Arcadia, CA 91066-6018
T: +1 626 4457800; Fax: +1 626 5740852; E-mail: ctbainfo@ctba.com; URL: www.ctba.com
1964; Vivian Montoya
American Breeding (C.C. Moseley Coll), Foreign Racing & Breeding (Edward Lasker Coll), Kent Cochran Coll
10 000 vols; 30 curr per; 600 microforms   36473

Los Angeles County Arboretum & Botanic Garden, Arboretum Library, 301 N Baldwin Ave, *Arcadia*, CA 91007-2697
T: +1 626 8213213; Fax: +1 626 4451217; URL: www.arboretum.org
1948; Susan C. Eubank
28 000 vols; 300 curr per   36474

Samuel Roberts Noble Foundation, Inc, Noble Foundation Library, 2510 Sam Noble Pkwy, *Ardmore*, OK 73401
T: +1 580 2246260; Fax: +1 580 2246265; URL: www.noble.org
1951; Pat Weaver-Meyers
Plant biology, Plant cell biology, Molecular plant pathology, Virology, Plant transformation, Agriculture, Agricultural economics, Agronomy, Horticulture, Livestock, Wildlife management, Forage, Plant genomics; Plant Specimen Identity Coll & Arch
34 000 vols; 430 curr per; 220 e-journals   36475

Argonne National Laboratory, Argonne Research Library, 9700 S Cass Ave, Bldg 203-D144, *Argonne*, IL 60439-4801
T: +1 630 2524275; Fax: +1 630 2525886; URL: www.library.anl.gov/
1946; Yvette N. Woell
Chemical engineering, Inorganic chemistry, Nuclear science, Physics, Mathematics; DOE, ERDA & AEC Tech Rpts
65 000 vols; 1 600 curr per; 600 000 govt docs; 1 000 000 microforms
libr loan   36476

American Psychiatric Association, Melvin Sabshin Library & Archives, 1000 Wilson Blvd, Ste 1825, *Arlington*, VA 22209-3901
T: +1 703 9078648; Fax: +1 703 9071084; E-mail: library@psych.org; URL: www.psych.org/library
1961; Gary McMillan
Hist of Psychiatry, rare bks, mss
8 000 vols; 300 curr per; 20 e-books; 92 av-mat; 188 sound-rec   36477

Anser Technical Library, Resource Center, 2900 S Quincy St, **Arlington**, VA 22206
T: +1 703 4162000; Fax: +1 703 4163386; URL: www.anser.org
1958
Computer science, Operations res, Military
14 000 vols; 222 curr per   36478

Armenian Cultural Foundation Library, 441 Mystic St, **Arlington**, MA 02474-1108
T: +1 781 6463090; Fax: +1 781 6463090
1945; Ara Ghazarian
33 000 vols; 10 curr per   36479

Insurance Institute for Highway Safety Library, 1005 N Glebe Rd, Ste 800, **Arlington**, VA 22201
T: +1 703 2471500; Fax: +1 703 2471678; E-mail: iihs-lib@cais.org; URL: www.iihs.org
1959; Ellen Sanders
Transportation; US Dept of Transportation Regulatory Docket Mat
20 000 vols; 150 curr per; 1 300 microforms; 6 000 research reports, 1 000 institute pubs   36480

Newspaper Association of America, Information Resource Center, 4401 Wilson Blvd, Suite 900, **Arlington**, VA 22203-1667
T: +1 571 3661000; Fax: +1 571 3661195; E-mail: jeff.sigmund@naa.org; URL: www.naa.org
1952; Paul Yachnes
Hist of Newspapers & Printing
8 000 vols; 400 curr per   36481

Rand Library, 1200 S Hayes St, **Arlington**, VA 22202-5050
T: +1 703 4131100 ext 5330; Fax: +1 703 4138111; URL: www.rand.org
1950
Education, Political science, Nat defense, Energy
10 000 vols; 250 curr per   36482

United States Department of Justice, Drug Enforcement Administration Library, 700 Army Navy Dr, **Arlington**, VA 22202; mail address: 8701 Morrissette Dr, Springfield, VA 22152
T: +1 202 3078932; Fax: +1 202 3078939; E-mail: dea.library@usdoj.gov
1960; RoseMary Russo
15 000 vols; 147 curr per
libr loan   36483

Walter T. McCarthy Law Library, Court House, 1425 N Courthouse Rd, Ste 1700, **Arlington**, VA 22201
T: +1 703 2287005; Fax: +1 703 2287360
1977; Patricia Petroccione
Continuing education; All Regional & Federal Reporters; Form Bks; Old Codes & Acts of Assembly, Virginia; Special Series; Various Digests; Virginia Treatises
22 000 vols; 12 curr per; 25 av-mat; 3 digital data carriers   36484

The Center: Resources for Teaching & Learning, The Center Library, 2626 S Clearbrook Dr, **Arlington Heights**, IL 6005-4626
T: +1 224 3668500; Fax: +1 224 3668514; URL: www.thecenterweb.org
1974; Kim Scannell
Early childhood education, Bilingual education, Literacy, Refugee assistance, Special education, Vocational education
24 000 vols; 60 curr per   36485

Atascadero State Hospital, Logan Professional Library, 10333 El Camino Real, **Atascadero**, CA 93422; P.O. Box 7003, Atascadero, CA 93423-7003
T: +1 805 4682491; Fax: +1 805 4683111
1957; Linda Fredericks
Sex Psychopathy & Forensic Psychiatry
9 000 vols; 106 curr per   36486

Atascadero State Hospital, Logan Patient's Library, 10333 El Camino Real, **Atascadero**, CA 93423-7003; P.O. Box 7003, Atascadero, CA 93423-7003
T: +1 805 4682520; Fax: +1 805 4683111
Linda Pacheco
Law
15 000 vols; 29 curr per   36487

Atlanta History Center, James G. Kenan Research Center, 3101 Andrews Dr NW, **Atlanta**, GA 30305; mail address: 130 W Paces Ferry Rd, Atlanta, GA 30305
T: +1 404 8144040; Fax: +1 404 8144175; E-mail: reference@atlantahistorycenter.com; URL: www.atlantahistorycenter.com
1926; Helen Matthews
Cookbks, Decorative Art Bks, Joel Chandler Harris Coll, Shutze Architecture Coll, Revolutionary War in Georgia
25 000 vols; 90 curr per
Society of American Archivists, RLG, ALA   36488

Fernbank Science Center Library, 156 Heaton Park Dr NE, **Atlanta**, GA 30307-1398
T: +1 678 8747116; Fax: +1 678 8747110; E-mail: fernbank@fernbank.edu; URL: fsc.fernbank.edu
1967; Mary Larsen
Science & tech, Environmental studies, Horticulture, Botany, Astronomy
22 000 vols; 330 curr per
libr loan   36489

Georgia Library for Accessible Services, 1150 Murphy Ave SW, **Atlanta**, GA 30310-3399
T: +1 404 7564476; Fax: +1 404 7564618; E-mail: glass@georgialibraries.org; URL: www.georgialibraries.org/public/glass.html
1931; Beverly Williams
Georgia Coll, tapes
300 000 vols; 57 000 av-mat; 57 000 Talking Bks   36490

Goethe-Institut, Library, 1197 Peachtree St NE, Colony Sq, Plaza Level, **Atlanta**, GA 30361-2401
T: +1 404 8922388; Fax: +1 404 8923832; E-mail: info@german-institute.orgt; URL: www.goethe.de/atlanta
1977
German language material
10 000 vols; 30 curr per; 1 000 av-mat; 650 sound-rec
libr loan; SOLINET   36491

High Museum of Art Library, 1280 Peachtree St, **Atlanta**, GA 30309
T: +1 404 7334480; URL: www.high.org
1985; Frances Francis
Pendley Coll, Auctions Cats
14 000 vols; 35 curr per; 8 000 artists' files, 10 000 slides
ARLIS/NA   36492

John Marshall Law School, Law Library, 1422 W Peachtree St NW, **Atlanta**, GA 30309
T: +1 404 8723593; Fax: +1 404 8733802; URL: www.johnmarshall.edu
1935; Michael J. Lynch
205 000 vols; 1 566 curr per; 91 000 microforms; 500 sound-rec; 100 digital data carriers   36493

Southern Regional Education Board Library, 592 Tenth St NW, **Atlanta**, GA 30318-5776
T: +1 404 8759211; Fax: +1 404 8721477; URL: www.sreb.org
1949; Debbie Curtis
Arch of SREB (publ items)
19 000 vols; 175 curr per; 25 av-mat; 50 sound-rec; 15 digital data carriers; 4 VF drawers of pamphlets
libr loan; SLA   36494

Federal Aviation Administration, Wm J. Hughes Technical Center, Technical Reference and Research Library, Atlantic City International Airport, **Atlantic City**, NJ 08405
T: +1 609 4855124; Fax: +1 609 4856088; E-mail: robert.mast@faa.gov; URL: www.faa.gov/about/office_org/headquarters_offices/ato/tc/library
1958; Robert Mast
Air traffic control, Aviation safety
12 000 vols; 400 curr per   36495

Princeton Antiques Bookservice, Art Marketing Reference Library, 2915-17 Atlantic Ave, **Atlantic City**, NJ 08401-6395
T: +1 609 3441943; Fax: +1 609 3441944; E-mail: princetn@earthlink.net; URL: www.princetonantiques.com
1974; Robert E. Ruffolo
Antiques, Gambling, Price info hist from 1900; Post Card Photo Libr Info Bank

Coll (1900-1950)
225 000 vols; 15 curr per   36496

Foundation Historical Association, Inc Library, Seward House, 33 South St, **Auburn**, NY 13021-3929
T: +1 315 2521283; Fax: +1 315 2533351; URL: www.sewardhouse.org
1955; Peter Wisbey
Civil War, Alaska, Local hist
10 000 vols   36497

Augusta Technical College, Jack B. Patrick Info Tech Center, 3200 Augusta Tech Dr, **Augusta**, GA 30906
T: +1 706 7714165; Fax: +1 706 7714110; E-mail: pbrucker@augustatech.edu; URL: www.augustatech.edu
1971; Patricia Brucker
Electronics
100 000 vols; 500 curr per; 2 256 sound-rec   36498

Maine Department of Transportation Library, 16 State House Sta, **Augusta**, ME 04333-0016
T: +1 207 6243230; Fax: +1 207 6243221
Pamela Dow-Shofner
12 000 vols   36499

Maine Office of Substance Abuse, Information & Resource Center, AMHI Complex Marquardt Bldg, 3rd Fl, 11 State House Sta, **Augusta**, ME 04333-0011
T: +1 207 2878900; Fax: +1 207 2878910; E-mail: osa.ircosa@maine.gov; URL: maineosa.org/irc
Jo McCaslin
8 000 vols   36500

Maine Regional Library for the Blind & Physically Impaired, c/o Maine State Library, Maine State Library, 64 State House Sta, **Augusta**, ME 04333-0064
T: +1 207 2875650; Fax: +1 207 2875654; URL: www.state.me.us/msl/outreach
1972; Melora Ranney Norman
French Language, cassette
135 000 vols; 85 123 sound-rec; 19 224 large print bks   36501

Riverview Psychiatric Center, Colonel Black Library, Arsenal St, P.O. Box 724, **Augusta**, ME 04332-0724
T: +1 207 2877266; Fax: +1 207 2877127; E-mail: jrclark@saturn.caps.maine.edu; URL: www.maine.gov/dhhs/riverview/about-us.html
John R Clark
16 750 vols; 57 curr per   36502

Austin Graduate School of Theology Library, 1909 University Ave, **Austin**, TX 78705
T: +1 512 4762772; Fax: +1 512 4763919; URL: www.austingrad.edu
1975; Jon Aho
Showalter Coll
25 000 vols; 120 curr per; 700 sound-rec; 1 digital data carriers   36503

Lyndon Baines Johnson Library and Museum, National Archives and Records Administration, 2313 Red River St, **Austin**, TX 78705-5702
T: +1 512 9165137; Fax: +1 512 916-5171 (main), -5170 (AV), -5142 (Archives); E-mail: johnson.library@nara.gov; URL: www.lbjlib.utexas.edu
1971; Dr. Betty Sue Flowers, Director
Oral hist (tapes, transcripts); Federal agency & commission records; personal papers of famous individuals (e.g., Drew Pearson, John Connally, Wm. Westmoreland); Audio-Visual arch; Museum coll; Recorded Telephone Conversations (tapes, transcripts)
17 615 vols; 6 curr per; 443 diss/theses; 29 microforms; 8 759 av-mat; 13 815 sound-rec; 1 digital data carriers; 45 million mss, 622 201 still photos
Interlibrary loan only for oral hist transcripts, finding aids, duplicates of specific archival mat, NOT for books   36504

Texas State Library & Archives Commission – Talking Book Program, 1201 Brazos, **Austin**, TX 78711-1938; P.O. Box 12927, Austin, TX 78711-2927
T: +1 512 4635458; Fax: +1 512 9360685; E-mail: tbp.services@tsl.state.tx.us; URL: www.TexasTalkingBooks.org
1931; Ava M. Smith
Spanish Coll, cassettes; Texas Coll, cassettes
795 000 vols; 710 660 av-mat; 22 650 Large Print bks, 710 660 Talking Bks   36505

Baltimore Metropolitan Council, Regional Information Center, 2700 Lighthouse Point E, Ste 310, **Baltimore**, MD 21224-4774
T: +1 410 7329570; Fax: +1 410 7329488; E-mail: ric@baltometro.org; URL: www.baltometro.org/content/view/43/86/
1963; Mary A. Logan
Economics, Environmental studies, Finance, Govt, Int relations, Transportation, Urban planning
10 000 vols; 200 curr per   36506

Baltimore Museum of Art, E. Kirkbride Miller Art Research Library, Ten Art Museum Dr, **Baltimore**, MD 21218-3898
T: +1 443 5731778; Fax: +1 443 5731781; URL: www.artbma.org
1929; Linda Tompkins-Baldwin
Papers and Libr of the Cone Sisters (early 20th c French art); White Libr (American decorative arts); Lucas Libr (19th c French art); Dunton Papers (American quilts)
50 000 vols; 300 curr per; 24 av-mat; 24 digital data carriers; 168 drawers of vertical files, 20 000 auction catalogs
ALA, ARLIS/NA, SAA   36507

Johns Hopkins Hospital, Wilmer Ophthalmological Institute, Jonas S. Friedenwald Library, Johns Hopkins Hospital, Woods Res Bldg, 600 N Wolfe St, Rm 3B-50, **Baltimore**, MD 21287-9105
T: +1 410 9553127; Fax: +1 410 9550046; URL: www.wilmer.jhu.edu
1925; Michael Piorunski
Ophthalmologic Drawings (Annette Burgess Art Coll), original art; Hist of Medicine (Wilmer Coll), rare bks
20 000 vols; 115 curr per; 300 av-mat; 10 digital data carriers; 8 000 slides   36508

Library Company of the Baltimore Bar, Baltimore Bar Library, 100 N Calvert St, Rm 618, **Baltimore**, MD 21202-1723
T: +1 410 7270280; Fax: +1 410 6854791; URL: www.barlib.org
1840; Joseph W. Bennett
200 000 vols; 314 curr per   36509

Maryland Historical Society, H. Furlong Baldwin Library, 201 W Monument St, **Baltimore**, MD 21201
T: +1 410 6853750; Fax: +1 410 3850487; E-mail: library@mdhs.org; URL: www.mdhs.org/explore/library.html
1844; Beatriz Hardy
Maryland; Original Mss of Star-Spangled Banner & Related Coll, Sheet Music (10 000 Maryland Pieces), Robert Merrick Print Coll, Unique Genealogical Finding Aids, Maryland Hist Society Arch
60 000 vols; 250 curr per; 3 000 maps; 1 700 microforms; 4 mio mss, 10 000 pamphlets, 8 000 prints, 28 000 rare bks, 500 000 photos, 26 VF drawers of clippings, 1 mio ephemera   36510

Maryland State Library for the Blind & Physically Handicapped, 415 Park Ave, **Baltimore**, MD 21201-3603
T: +1 410 2302424; Fax: +1 410 3332095; URL: www.lbph.lib.md.us
Jill Lewis
Maryland, Braille & Audio cassettes
272 000 vols; 328 000 Vols in braille, 10 134 Large Print Bks
libr loan   36511

Metropolitan Transition Center Library, 954 Forrest St, **Baltimore**, MD 21202
T: +1 410 2301472; Fax: +1 410 2301472
Ruth Mewborn

Correctional education
8 000 vols; 22 curr per 36512

National Institute on Aging, Gerontology Research Center Library, 5600 Nathan Shock Dr, ***Baltimore***, MD 21224-6825; mail address: 4940 Eastern Ave, Baltimore, MD 21224
T: +1 410 5588125; Fax: +1 410 5588224; E-mail: rrb-liss@grc.nia.nih.gov; URL: www.kisok.gov
1977; Carmen Harris
Coll of Basic, Clinical, Psychological Biomedicine
25 000 vols; 20 curr per; 200 microforms; 100 digital data carriers
libr loan; NLM, OCLC 36513

National Institute on Drug Abuse, Addiction Research Center Library, P.O. Box 5180, ***Baltimore***, MD 21224
T: +1 410 5501488; Fax: +1 410 5501438
1935
Drug chemistry, Biochemistry, Pharmacology, Psychiatry
8 700 vols; 250 curr per; 4 000 reprints 36514

Natural History Society of Maryland, Inc Library, 2643 N Charles St, ***Baltimore***, MD 21218
T: +1 410 2356116; URL: www.marylandnature.org
1929; Joe McSharry
20 200 vols; 20 curr per 36515

Space Telescope Science Institute Library, 3700 San Martin Dr, ***Baltimore***, MD 21218
T: +1 410 3384961; Fax: +1 410 3384767; E-mail: library@stsci.edu; URL: www.sesame.stsci.edu/library.html
1983; Jill Lagerstrom
Astrophysics; Palomar & European Southern Observatory/Science Research Council Photogr Sky Surveys
9 000 vols; 207 curr per; 210 linear feet of observatory publs 36516

Walters Art Museum Library, 600 N Charles St, ***Baltimore***, MD 21201-5185
T: +1 410 5479000; Fax: +1 410 7524797; E-mail: info@thewalters.org; URL: www.thewalters.org
1934; Christianne Henry
Hist; Art Auction Cat Coll
80 000 vols; 100 curr per 36517

Workforce & Technology Center, R.C. Thompson Library, 2301 Argonne Dr, ***Baltimore***, MD 21218-1696
T: +1 410 5549100; Fax: +1 410 5549112; E-mail: wtc@dors.state.md.us; URL: www.dors.state.md.us/DORS/ProgramServices/Business2
1973
Disability Coll, film/v-tapes
10 000 vols; 20 curr per; 500 av-mat
libr loan 36518

Barnstable Law Library, First District Court House, ***Barnstable***, MA 02630; P.O. Box 427, Barnstable, MA 02630-0427
T: +1 508 3628539; Fax: +1 508 3621374; E-mail: barnlaw45@hotmail.com; URL: www.lawlib.state.ma.us
1889; Margaret J. Hill
Massachusetts Continuing Legal Education Mat
15 000 vols; 30 curr per; 2 000 microforms; 7 digital data carriers
libr loan 36519

Sturgis Library, 3090 Main St, P.O. Box 606, ***Barnstable***, MA 02630-0606
T: +1 508 3626636; Fax: +1 508 3625467; E-mail: sturgislib@comcast.net; URL: www.sturgislibrary.org
1863; Lucy E. Loomis
Geography, Hist; Early Cape Cod Land Deeds, Some Indian (Stanley W. Smith Coll); Genealogy & Local Hist (Hooper Room); Local Authors (Cape Cod Coll); Maritime Hist (Kittredge Room); 19th C Lit
68 000 vols; 99 curr per; 45 maps; 165 microforms; 1 500 av-mat; 29 digital data carriers; 1 500 Deeds, 750 Audiotapes (books-on-tape)
libr loan; ALA 36520

Vermont Historical Society Library, Vermont History Ctr, 60 Washington St, ***Barre***, VT 05641-4209
T: +1 802 4798509; Fax: +1 802 4798510; E-mail: vhs-library@state.vt.us; URL: catalog.vermonthistory.org; www.vermonthistory.org
1838; Paul A. Carnahan
Broadsides, photos, Sheet Music
53 000 vols; 220 curr per; 1 000 maps; 300 microforms; 4 digital data carriers; 30 000 photos, 8 700 broadsides, 1 200 linear feet of mss 36521

Polk County Historical & Genealogical Library, Old Courthouse, 100 E Main St, ***Bartow***, FL 33830
T: +1 863 5344380; Fax: +1 863 5344382; URL: www.polk-county.net/subpage.aspx?menu_id=52&nav=res&id=4358
1940
Genealogy & Hist of the Southeastern US
40 000 media items; 40 curr per 36522

Fermi National Accelerator Laboratory Library, Information Resources Department, Kirk & Wilson Sts, ***Batavia***, IL 60510; P.O. Box 500, MS109, Batavia, IL 60510-5011
T: +1 630 8403401; Fax: +1 630 8404636; E-mail: library@fnal.gov; URL: library.fnal.gov
1967; Heath O'Connell
Energy, Engineering, Computer science; High-Energy Physics Preprints
20 000 vols; 160 curr per; 3 000 microforms; 3 digital data carriers
libr loan 36523

Maine Maritime Museum, Library Archives, 243 Washington St, ***Bath***, ME 04530
T: +1 207 4431316; Fax: +1 207 4431665; URL: www.mainemaritimemuseum.org
1964; Nathan Lipfert
Maritime hist; Ship Papers, original doc; Bath, Maine Built Vessels, photos
13 000 vols; 25 curr per; 500 000 mss docs, 17 000 photos, 30 000 ship plans 36524

State Library of Louisiana – Services for the Blind & Physically Handicapped, 701 N Fourth St, ***Baton Rouge***, LA 70802; P.O. Box 131, Baton Rouge, LA 70821-0131
T: +1 225 3424944; Fax: +1 225 3426817; E-mail: sbph@state.lib.la.us; URL: www.state.lib.la.us
1933; Margaret C. Harrison
Braille, Louisiana Cassettes, Descriptive Videos
100 000 vols; 10 000 large print bks 36525

Bedford Historical Society Library, 30 S Park St, P.O. Box 46282, ***Bedford***, OH 44146-0282
T: +1 440 2320796; URL: www.bedfordohiohistory.org/
1955; Richard J. Squire
Jacka 1876 Centennial Coll; Leonard Seigel Railroad Coll; Squire Coll of Lincolniana; Barnum Civil War Coll; Early Aviation Hist Coll; Arch for Local Govt, 78 rpm coll
9 150 vols; 11 curr per; 60 maps; 190 ledgers, record books, scrapbooks, albums, 160 file boxes and drawers of mss, archival mat, maps 36526

American Philatelic Research Library, 100 Match Factory Pl, ***Bellefonte***, PA 16875
T: +1 814 2373803; Fax: +1 814 2376128; E-mail: gini@stamps.org; URL: www.stamps.org
1968; Virginia Horn
AFDCS Arch; Postal Hist Society Libr; Ellis File; Piper File
20 000 vols; 2 000 curr per; 30 000 handbks & cat 36527

Kaiser-Permanente Medical Center, Health Sciences Library & Media Center, 9400 E Rosecrans Ave, ***Bellflower***, CA 90706
T: +1 562 4614247; Fax: +1 562 4614948
1965; Pam Lee
9 000 vols; 500 curr per 36528

McLean Hospital, Mental Health Sciences Library, 115 Mill St, ***Belmont***, MA 02478
T: +1 617 8552460; Fax: +1 617 8552414; E-mail: library@mclean.harvard.edu; URL: www.mclean.harvard.edu/library
1936; Marilyn Dietrich
11 600 vols; 145 curr per; 60 e-journals; 21 e-books; 650 av-mat; 2 digital data carriers
libr loan 36529

Thayer County Museum, Historical & Genealogical Library, 110 9th Street, ***Belvidere***, NE 68315
T: +1 402 7682147; E-mail: petechili@hotmail.com; URL: thayercountymuseum.com
1970; Jacqueline J. Williamson
Local hist, Genealogy
14 000 vols 36530

Institute of Buddhist Studies Library, 2140 Durant Ave, Suite 30, ***Berkeley***, CA 94704
T: +1 510 8091444; Fax: +1 510 8091443; URL: www.shin-ibs.edu/academics/libraries.php
1966; Irwin Feldman
Shin Buddhism; Pure Land Buddhism; Japanese Contemporary Writings in Jodoshinshu
14 838 vols; 130 curr per; 102 diss/theses 36531

International Child Resource Institute, 1581 Leroy Ave, ***Berkeley***, CA 94708
T: +1 510 6441000; Fax: +1 510 5254106; E-mail: icrichild@aol.com; URL: www.icrichild.org
Ken Jaffe
13 000 vols 36532

Judah L. Magnes Museum, Blumenthal Library, 2911 Russell St, ***Berkeley***, CA 94705
T: +1 510 5496950; Fax: +1 510 8493673; URL: www.magnes.org
1967; Aaron T. Kornblum
Jewish art, hist, lit, Jewish music; S. Belkin Papers, Jews in India, Early Hebrew Printed Bks; Heinrich Heine Coll
25 000 vols; 30 maps; 3 000 docs, 300 mss, 1 000 rare bks, 500 Music scores 36533

Lawrence Berkeley National Laboratory Library, Bldg 50B, Rm 4034, One Cyclotron Rd, ***Berkeley***, CA 94720-0001
T: +1 510 4865621; Fax: +1 510 4866406; E-mail: library@lbl.gov; URL: www-library.lbl-gov/teid/tmLib/aboutus/LibDefault.htm
1946; Jose Olivares
Bioinformatics, Chemistry, Earth science, Electronic engineering, Energy, Genomics, Nuclear medicine science, Physics; High Energy Physics Preprints Coll
25 000 vols; 600 curr per 36534

Wright Institute Library, 2728 Durant Ave, ***Berkeley***, CA 94704
T: +1 510 8419230 ext 140; Fax: +1 510 8410167; E-mail: library@wi.edu; URL: www.wi.edu/library/index.html
1968; Jason Strauss
Psychology, Research Methodology
10 000 vols; 125 curr per; 485 diss/theses; 278 av-mat; 5 digital data carriers; 142 audiotapes
libr loan; ALA 36535

Vermont Regional Library for the Blind & Physically Handicapped, Vermont Dept of Libraries Special Services Unit, 578 Paine Turnpike N, ***Berlin***, VT 05602
T: +1 802 8283273; Fax: +1 802 8282199; E-mail: ssu@mail.dol.state.vt.us; URL: dol.state.vt.us
1976; Teresa Faust
Handicaps Ref Mat
9 000 vols; 52 534 av-mat; 50 200 sound-rec; 123 Braille vols, 8 023 Large Print Bks 36536

National Library of Medicine, 8600 Rockville Pike, ***Bethesda***, MD 20894
T: +1 301 4966308; Fax: +1 301 4964450; E-mail: custserv@nlm.nih.gov; URL: www.nlm.nih.gov
1836; Dr. Donald A. B. Lindberg
Mss coll
2 500 000 vols; 22 500 curr per; 591 incunabula; 282 000 diss/theses; 592 000 microforms; 10 digital data carriers; 432 000 microforms of per, 2 451 mio mss, 56 000 art reproductions
libr loan 36537

The National Naval Medical Center, Edward Rhodes Stitt Library, 8901 Wisconsin Ave, ***Bethesda***, MD 20889-5600
T: +1 301 2951185; Fax: +1 301 2956001; URL: www.bethesda.med.navy.mil/professional/stitt_library/
1902; Jerry Meyer
Hist of Medicine Coll
90 000 vols; 550 curr per 36538

Moravian Archives, 41 W Locust St, ***Bethlehem***, PA 18018-2757
T: +1 610 8663255; Fax: +1 610 8669210; E-mail: info@moravianchurcharchives.org; URL: www.moravianchurcharchives.org
1751; Dr. Paul M. Peucker
Moravian church hist
20 000 vols; 12 curr per 36539

St Luke's Hospital & Health Network, W. L. Estes Jr Memorial Library, 801 Ostrum St, ***Bethlehem***, PA 18015
T: +1 610 9544650; Fax: +1 610 9544651; E-mail: estes.library@slhn.org; URL: www.mystlukesonline.org/locations/library-services/index.aspx
1947; Maria D. Collette
Nursing (Hist Coll)
10 000 vols; 300 curr per; 25 e-books; 186 av-mat; 375 sound-rec; 30 digital data carriers; 10 VF drawers, 66 slides 36540

Parents, Let's Unite for Kids, Training, Resource & Information Center, 516 N 32nd St, ***Billings***, MT 59101-6003
T: +1 406 2550540; Fax: +1 406 2550523; E-mail: info@pluk.org; URL: www.pluk.org
1984; Janice Sand
Disability-related issues
10 000 vols; 30 curr per 36541

Broome Developmental Disabilities Services Office Library, 249 Glenwood Rd, Broome Developmental Ctr, ***Binghamton***, NY 13905-1695
T: +1 607 7700410; Fax: +1 607 7700392; E-mail: mary.jeanne.perlmutter@omr.ny.state.us
1975; Mary Jeanne Perlmutter
12 480 vols; 60 curr per
libr loan 36542

Birmingham Museum of Art, Clarence B. Hanson Jr Library, 2000 Eighth Ave N, ***Birmingham***, AL 35203-2278
T: +1 205 2542982; Fax: +1 205 2542710; E-mail: library@artsbma.org; URL: www.artsbma.org
1966; Tatum Preston
Chellis Wedgwood Coll of bks and mss; Elizabeth Stewart Beeson Rare Book Room
26 000 vols; 75 curr per; 120 mss; 30 diss/theses; 50 digital data carriers; artists files, letters, ephemera, arch
ARLIS/NA, SAA, ALA 36543

Southern Research Institute, Thomas Martin Memorial Library, 2000 Ninth Ave S, ***Birmingham***, AL 35205; P.O. Box 55305, Birmingham, AL 35255-5305
T: +1 205 5812000; Fax: +1 205 5812008
1945; Richard J. Remy
Chemistry, Biology, Cancer, Engineering, Genetics
14 000 vols; 300 curr per; 2 000 microforms 36544

Medcenter One, Health Library, 622 Avenue A East, ***Bismarck***, ND 58501
T: +1 701 3235390; Fax: +1 701 3236967; URL: www.medcenterone.com
1927; Joan Bares
Int Pediatric Peptic Ulcer Registry
24 000 vols; 250 curr per; 4 VF drawers of pamphlets & reprints 36545

State Historical Society of North Dakota, State Archives & Historical Research Library, Heritage Ctr, Capitol Grounds, 612 E Boulevard Ave, ***Bismarck***, ND 58505
T: +1 701 3282091; Fax: +1 701 3282650; E-mail: archives@nd.gov; URL:

www.nd.gov/hist
1895; Jim Davis
North Dakota, Northern Great Plains, Native American Studies, hist
105 000 vols; 570 curr per; 1 700 mss; 42 000 govt docs; 10 545 maps; 15 000 microforms; 1 500 sound-rec; 150 000 photogr images, 2 800 archives research series, 14 300 rolls of 35 mm microfilm, 4 mio feet of motion picture film
libr loan 36546

Cranbrook Academy of Art Library, 39221 Woodward Ave, ***Bloomfield Hills***, MI 48304; P.O. Box 801, Bloomfield Hills, MI 48303-0801
T: +1 248 6453355; Fax: +1 248 6453464; E-mail: jdyki@cranbrook.edu; URL: www.cranbrookart.edu/library
1928; Judy Dyki
Artist's Bks, Booth Coll of Fine Arts Folios, Cranbrook Press Bks, Exhibition Cat, Fine Bindings, Folios, Theses
26 000 vols; 190 curr per; 1 400 av-mat; 700 digital data carriers; 50 000 Slides
libr loan 36547

Illinois Agricultural Association, Information Research Center, 1701 Towanda Ave, P.O. Box 2901, ***Bloomington***, IL 61701-2901
T: +1 309 5572552; Fax: +1 309 5573185; E-mail: sampson@ilfb.org; URL: www.ilfb.org
1960; Vince L. Sampson
Agricultural economics
15 000 vols; 500 curr per; 100 microforms; 500 av-mat; 35 digital data carriers
libr loan 36548

The Kinsey Institute for Research in Sex, Gender & Reproduction, Inc, Morrison Hall Rm 416, User Services, 1165 E Third St, ***Bloomington***, IN 47405
T: +1 812 8553058; Fax: +1 812 8558277; E-mail: libknsy@indiana.edu; URL: www.kinseyinstitute.org/library/index.html
1947
85 000 vols; 85 curr per; 335 microforms; 650 av-mat; 250 art reproductions 36549

McLean County Museum of History, Stevenson-Ives Library, 200 N Main, ***Bloomington***, IL 61701
T: +1 309 8270428; Fax: +1 309 8270100; E-mail: marketing@mchistory.org; URL: www.mchistory.org
1892; Greg Koos
McLean County Genealogigal Society Coll, Letitia Green Stevenson Chapter of N.S.D.A.R.
8 000 vols; 85 curr per; 40 digital data carriers
libr loan; ALA, SAA 36550

Bagaduce Music Lending Library, 3 Music Libr Lane, ***Blue Hill***, ME 04614; P.O. Box 829, Blue Hill, ME 04614-0829
T: +1 207 3745454; Fax: +1 207 3742733; E-mail: library@bagaducemusic.org; URL: www.bagaducemusic.org
1984; David Gautschi
State of Maine Music Coll
100 000 vols 36551

Idaho State Historical Society, Public Archives & Research Library, 2205 E Old Penitentiary Rd, Idaho History Ctr, ***Boise***, ID 83712
T: +1 208 3343356; Fax: +1 208 3343198; E-mail: lmorton-keithley@ishs.idaho.gov; URL: www.idahohistroy.net
1907; Linda Morton-Keithley
Idaho & Pacific Northwest Hist, Idaho Mss, Idaho Newspapers, Idaho photos
25 800 vols; 260 curr per; 18 800 mss; 20 000 maps; 25 000 microforms; 2 500 sound-rec; 8 000 cubic feet of mss, 300 000 photos 36552

Micron Technology, Inc Library, 8000 S Federal Way, Mail Stop 1-552, ***Boise***, ID 83707-0006
T: +1 208 3684258; Fax: +1 208 3683393
Nancy Elg
Information science
15 000 vols; 200 curr per 36553

Sam Rayburn Library & Museum, 800 W Sam Rayburn Dr, ***Bonham***, TX 75418-4103; P.O. Box 309, Bonham, TX 75418-0309
T: +1 903 5832455; Fax: +1 903 5837394; URL: www.cah.utexas.edu/divisions/Rayburn.html
1957; Dr. Patrick Cox
Life & Career of Speaker Sam Rayburn, hist memorabilia, official papers, interviews, micro
20 000 vols; 30 VF Drawers
libr loan 36554

Wisconsin Secure Program Facility Library, 1101 Morrison Dr, ***Boscobel***, WI 53805; P.O. Box 1000, Boscobel, WI 53805-1000
T: +1 608 3755656; Fax: +1 608 3755434
Linda Oatman
Corrections
43 000 vols; 20 curr per 36555

Appalachian Mountain Club Library, 5 Joy St, ***Boston***, MA 02108-1490
T: +1 617 5230636 ext 329; Fax: +1 617 5230722; URL: www.amcboston.org
1876; John Gerber
Sella Coll, White Mountains (Kilbourne Coll), Journals of mountain clubs around the world
8 000 vols; 16 curr per; 800 maps; 10 000 slides 36556

Benjamin Franklin Institute of Technology, Lufkin Memorial, 41 Berkeley St, Franklin Union Bldg, Rm 114, ***Boston***, MA 02116
T: +1 617 4234630 ext 123; E-mail: library@bfit.edu; URL: www.bfit.edu/pages/academics/LMLibrary.html
1908; Sharon B. Bonk
Architecture, Drafting, Engineering, Technology; Benjamin Franklin Coll, Photogr Science (Dr Leonard E. Ravich Coll)
23 000 vols; 120 curr per; 15 000 e-books; 23 digital data carriers 36557

Boston Architectural Center Library, Shaw & Stone Library, 320 Newbury St, ***Boston***, MA 02115
T: +1 617 5850155; Fax: +1 617 5850151; E-mail: library@the-bac.edu; URL: www.the-bac.edu/library
1966; Susan Lewis
Architectural Hist (Memorial Libr Coll)
35 000 vols; 150 curr per; 1 280 diss/theses; 40 digital data carriers; 40 000 slides
ARLIS/NA 36558

Boston Athenaeum, 10 1/2 Beacon St, ***Boston***, MA 02108-3777
T: +1 617 2270270; Fax: +1 617 2275266; URL: www.bostonathenaeum.org
1807; George Marshall Moriarty
Archs, Fine arts, Hist, Mss; American 19th C photos, American Prints & Drawings, Bks from the Libr of George Washington, Confederate Imprints, T.S. Eliot, Gypsy Lit
603 000 vols; 450 curr per 36559

Boston Children's Museum, Resource Center, 308 Congress St, ***Boston***, MA 02210
T: +1 617 4266500 ext 284; Fax: +1 617 4261944; E-mail: info@bostonchildrensmuseum.org; URL: www.bostonkids.org
1914; Jenny Zanger
Education, Children's lit, Ethnic studies, Natural science, Americana
15 000 vols; 50 curr per; 100 sound-rec; 300 multimedia kits 36560

Boston Medical Center, Health Sciences Library, 818 Harrison Ave, ***Boston***, MA 02118
T: +1 617 5345000 ext 4198
1864; Donna Rose
Hist of Medicine (Anthony Michelidakis Memorial Libr); Hist of Boston City Hospital Coll
16 500 vols; 140 curr per 36561

CRA International Library, 200 Clarendon St T-33, ***Boston***, MA 02116
T: +1 617 4253000; Fax: +1 617 4253132; URL: www.crai.com
1975; Doug Southard
Economics, Energy, Environmental studies, Metal working, Transportation
20 000 vols; 300 curr per 36562

The French Library & Alliance Francaise of Boston, 53 Marlborough St, ***Boston***, MA 02116-2099
T: +1 617 9120418; Fax: +1 617 9120450; E-mail: librarian@frenchlib.org; URL: www.frenchlib.org
1945; Parivashe Niamir
Lit & Hist of Francophone countries, Marcel Carne arch, Books published in NYC during WWII by French publishers
24 000 vols; 65 curr per; 645 av-mat; 1 100 sound-rec; 70 digital data carriers
libr loan 36563

Grand Lodge of Masons in Massachusetts, Samuel Crocker Lawrence Library, 186 Tremont St, ***Boston***, MA 02111
T: +1 617 4266040; Fax: +1 617 4266115; E-mail: library@massfreemasonry.org; URL: www.massfreemasonry.org
1814; Cynthia Alcorn
Freemassonry (John Paul Jones Coll), New England Towns & Cities, Hist, Anti-Masonry Coll
50 000 vols; 60 curr per; 60 microforms 36564

Harvard Musical Association Library, 57A Chestnut St, ***Boston***, MA 02108
T: +1 617 5232897; Fax: +1 617 5232897; E-mail: info@hmaboston.org; URL: www.hmaboston.org
1843; Natalie Palme
Chamber Music Parts, Two-piano Music
15 000 vols; 7 512 musical items 36565

Historic New England, Library & Archives, 141 Cambridge St, ***Boston***, MA 02114
T: +1 617 2273957; Fax: +1 617 9739050; URL: www.historicnewengland.org
1910; Emily R. Novak Gustainis
Hist of architecture & transportation in New England, daily life, occupations & recreation, regional photography, institutional arch
15 000 vols; 800 maps; 14 microforms; 600 ft of mss, 450 000 photos, 20 000 architectural drawings, 2 500 prints, 4 500 trade cards & cat, 5 000 pamphlets, 10 000 newspaper clippings
Open by appointment only 36566

Insurance Library Association of Boston, 156 State St, ***Boston***, MA 02109
T: +1 617 2272087; Fax: +1 617 7238524; URL: www.insurancelibrary.org
1887; Jean Lucey
Sanborn Fire Maps, Henry Belknap Coll
30 000 vols; 200 curr per; 30 digital data carriers
libr loan 36567

Lesley University, Art Institute of Boston Library, 700 Beacon St, ***Boston***, MA 02215-2598
T: +1 617 5856670; Fax: +1 617 5856655; URL: www.lesley.edu/library/
1969; Debbie Verhoff
Art hist, photography, humanities
9 000 vols; 75 curr per; 400 av-mat; 50 sound-rec; 1 digital data carriers; 45 000 slides, 18 000 picture files
libr loan 36568

Liberty Mutual Group, Law Library, 175 Berkeley St, 7th Flr, ***Boston***, MA 02117
T: +1 617 3579500; Fax: +1 617 5745830
1918; Rodney Koliha
Environmental law
130 000 vols; 120 curr per 36569

Lyman Library, Educator Resource Center, Museum of Science, One Science Park, ***Boston***, MA 02114-1099
T: +1 617 5890170; Fax: +1 617 5890494; E-mail: library@mos.org; URL: www.mos.org/library
1831; Jeff Mehigan
Education; Arch of the Boston Society of Natural Hist, 19th C Natural Hist (journal & mss)
17 000 vols; 75 curr per 36570

The Mary Baker Eddy Library for the Betterment of Humanity, Reference Dept, 200 Massachusetts Ave L03-10, ***Boston***, MA 02115-3017
T: +1 617 4507218; E-mail: reference@mbelibrary.org; URL: www.marybakereddylibrary.org
2000; Dr. Elaine R. Follis
Christian science, Church of Christ, Mary Baker Eddy, Christian healing, American religious hist; The Papers of Mary Baker Eddy; Bible Coll
18 000 vols; 60 curr per; 15 Special Interest Per Sub 36571

Massachusetts General Hospital – Treadwell Library, Bartlett Hall Ext – I, 55 Fruit St, ***Boston***, MA 02114-2696
T: +1 617 7268600; Fax: +1 617 7266784; E-mail: treadwellq&a@mgh.harvard.edu; URL: www.massgeneral.org/library
1858; Elizabeth Schneider
50 000 vols; 992 curr per 36572

Massachusetts General Hospital – Warren Library, 55 Fruit St, ***Boston***, MA 02114-2622
T: +1 617 7262253
Nancy Marshall
10 000 vols; 25 curr per 36573

Massachusetts Historical Society Library, 1154 Boylston St, ***Boston***, MA 02215-3695
T: +1 617 5361608; Fax: +1 617 8590074; E-mail: library@masshist.org; URL: www.masshist.org
1791; Peter Drummey
New England; hist mss, maps, photos, prints & broadsides, rare bks
200 000 vols; 160 curr per; mss colls (3 500) 36574

Museum of Fine Arts, Boston – W. Van Alan Clark Jr Library, 230 The Fenway, ***Boston***, MA 02115
T: +1 617 3693650; E-mail: library@smfa.edu
Aubrey Baer
Modern and contemporary art
21 000 vols 36575

Museum of Fine Arts, Boston – William Morris Hunt Memorial Library, 300 Massachusetts Ave, ***Boston***, MA 02115; mail address: 465 Huntington Ave, Boston, MA 02115
T: +1 617 3693385; Fax: +1 617 3694257; URL: www.mfa.org
1879; Maureen Melton
320 000 vols; 1 000 curr per; 154 000 pamphlets & auction cat 36576

New England Baptist Hospital, Paul E. Woodard Health Sciences Library, 125 Parker Hill Ave, ***Boston***, MA 02120-2847
T: +1 617 7545155; Fax: +1 617 7546414; E-mail: nebhweb@caregroup.harvard.edu; URL: www.nebh.org/health_library.asp
1963; Leonard L. Levin
Orthopedics; Nursing Coll, Orthopaedics & Hist of Medicine (Morton Smith-Petersen Coll), Otto Aufranc Coll
11 000 vols; 120 curr per
libr loan; MLA, NAHSL 36577

New England Historic Genealogical Society Library, 99-101 Newbury St, ***Boston***, MA 02116-3007
T: +1 617 2261234; Fax: +1 617 5367307; E-mail: nehgs@nehgs.org; URL: www.newenglandancestors.org
1845; Marie Daly
Coll of Canadian Genealogical Resource Mat, Irish Genealogical Resource Mat, English Genealogical Resource Mat; Manuscripts, Circulating Lib
200 000 vols; 700 curr per; 500 maps; 153 000 microforms; 140 digital data carriers; 3 500 linear feet of mss, 500 Electronic Media & Resources
libr loan; ALA 36578

Social Law Library, John Adams Courthouse, One Pemberton Sq, Ste 4100, ***Boston***, MA 02108
T: +1 617 5230018; Fax: +1 617 5232458; E-mail: rbrink@socialaw.com; URL: www.socialaw.com
1803; Robert J. Brink
Anglo-American Law Coll
450 000 vols; 967 curr per; 3 250 av-mat; 1 110 sound-rec 36579

State Transportation Library, Ten Park Plaza, **Boston**, MA 02116
T: +1 617 9738000; Fax: +1 617 9737153; E-mail: library@mbta.com; URL: www.stlibrary.org
1983; Lynn Matis
Boston Transportation Planning Review
Arch, MBTA Hist, Massachusetts Highway Hist
20 000 vols; 200 curr per   36580

Information Systems Management Office, Information Resources Division Library, 325 Broadway MC5, Radio Bldg, Rm 1202, **Boulder**, CO 80305-3328
T: +1 303 4973271; Fax: +1 303 4973890; E-mail: boulderlabs.main.library@noaa.gov; URL: library.bldrdoc.gov
1954; John Welsh
Meteorology, Electronics, Telecommunications, Cryogenics; Radio Science & Physics (Tech Rpts, 1940-1970), bks & micro
42 910 vols; 575 curr per   36581

National Indian Law Library, Native Americans Rights Fund, 1522 Broadway, **Boulder**, CO 80302-6217
T: +1 303 4478760; Fax: +1 303 4437776; E-mail: dselden@NARF.ORG; URL: www.narf.org
1972; David Selden
National Indian Law Libr (cases, briefs, pleadings, orders, legal opinions, rulings, memoranda, treatises, studies, articles, reports, and legislative histories pertinent to Federal Indian Law), Tribal Codes; Tribal Court Cases; Tribal codes and constitutions
1 000 vols; 68 curr per; 1 000 govt docs; 35 maps; 2 500 case files
libr loan; AALL, SLA   36582

Worcester County Horticultural Society, Tower Hill Botanic Garden Library, 11 French Dr, **Boylston**, MA 01505; P.O. Box 598, Boylston, MA 01505-0598
T: +1 508 8696111; Fax: +1 508 8690314; E-mail: librarian@towerhillbg.org; URL: www.towerhillbg.org
1842; Susannah Haney
Fruit Culture; Garden Hist; Landscape Design
9 000 vols; 26 curr per
CBHL   36583

Manatee County, Law Library, Manatee County Judicial Ctr, Rm 1101, 1051 Manatee Ave W, **Bradenton**, FL 34205; P.O. Box 3000, Bradenton, FL 34206-5400
T: +1 941 7414090; Fax: +1 941 7414085
1950; Judy Brand
28 000 vols; 68 sound-rec   36584

Brandeis-Bardin Institute, The Molle Library, 1101 Peppertree Lane, **Brandeis**, CA 93064
T: +1 805 5824450; Fax: +1 805 5261398; E-mail: info@thebbi.org; URL: www.thebbi.org
1973; Hannah R. Kuhn
Bks in Hebrew (Dr Shlomo Bardin)
10 000 vols; 2 000 sound-rec   36585

Cape Cod Museum of Natural History, Clarence L. Hay Library, 869 Main St, **Brewster**, MA 02631; P.O. Box 1710, Brewster, MA 02631-1710
T: +1 508 8963867; Fax: +1 508 8968844; URL: www.ccmnh.org
1958
Archeology Coll, Cape Cod Coll, Teachers' Coll, Scientific Specimens
10 000 vols; 35 curr per; 42 av-mat; 75 sound-rec; 4 VF drawers, 2 drawers of picture files   36586

Brockton Hospital Library, 680 Centre St, **Brockton**, MA 02302
T: +1 508 9417207; Fax: +1 508 9416412; E-mail: library@brocktonhospital.org; URL: www.brocktonhospital.com
Mary Conners
Medicine, nursing
9 000 vols; 268 curr per   36587

Brockton Law Library, 72 Belmont St, **Brockton**, MA 02301
T: +1 508 5867110; Fax: +1 508 5888483; E-mail: brocklaw72@hotmail.com; URL: www.lawlib.state.ma.us
Jean Smith
25 000 vols   36588

Enterprise Library, 60 Main St, **Brockton**, MA 02303; P.O. Box 1450, Brockton, MA 02303-1450
T: +1 508 5866200; Fax: +1 508 5866506; E-mail: newsroom@enterprisenews.com; URL: www.southofboston.com
1880; Beth Gould
Hist; News Clippings Coll
95 000 vols   36589

Bronx County Historical Society, Research Library, 3309 Bainbridge Ave, **Bronx**, NY 10467
T: +1 718 8818900; Fax: +1 718 8814827; E-mail: librarian@bronxhistoricalsociety.org; URL: www.bronxhistoricalsociety.org
1955; Dr. Gary Hermalyn
Edgar Allan Poe Coll, Local Newspaper Coll, Real Estate atlases for the Bronx 1860's-1940's; The Bronx County Archives; papers of business, associations and political leaders (10 000 linear ft)
7 000 vols; 200 curr per; 200 diss/theses; 200 maps; 400 microforms; 300 av-mat; 380 sound-rec; 100 digital data carriers; 50 000 photos, 1 242 postcards, 4 688 slides, 235 000 items such as clippings, magazine articles, etc
SLA   36590

Lincoln Medical Center, Health Sciences Library, 234 E 149th St, **Bronx**, NY 10451
T: +1 718 5795745; Fax: +1 718 5795170
1971; Inna Lipnitskaya
13 000 vols; 200 curr per   36591

The LuEsther T. Mertz Library, New York Botanical Garden, 2900 Southern Blvd, **Bronx**, NY 10458-5126
T: +1 718 8178980; Fax: +1 718 8178956; E-mail: libref@nybg.org; URL: www.nybg.org
1899; Susan Fraser
Botanical Art; Darwiniana (Charles F Cox Coll), mss & arch, Rosaceae (Jean Gordon Coll), artifacts, Lord & Burnham Co, drawings & correspondence
230 000 vols; 2 019 curr per; 66 incunabula; 31 000 microforms; 20 digital data carriers; 40 000 vertical files
CBHL   36592

Montefiore Medical Center, Tishman Learning Center Health Sciences Library, 111 E 210th St, **Bronx**, NY 10467
T: +1 718 9204666; Fax: +1 718 9204658; E-mail: medlib@montefiore.org
1926; Heather Barnabas
15 000 vols; 500 curr per; 250 e-books
libr loan   36593

Wildlife Conservation Society, Bronx Zoo Library, 2300 Southern Blvd, **Bronx**, NY 10460
T: +1 718 2206874; Fax: +1 718 2207114; E-mail: library@wcs.org
1908; Steven P. Johnson
Arch Mat
10 000 vols; 170 curr per; 10 digital data carriers   36594

Chicago Zoological Society, Brookfield Zoo Library, 3300 Golf Rd, **Brookfield**, IL 60513
T: +1 708 6888583; Fax: +1 708 6887583
1964; Carla Owens
8 000 vols; 300 curr per
libr loan   36595

International Foundation of Employee Benefit Plans, Information Center, 18700 W Bluemound Rd, **Brookfield**, WI 53045-2936; P.O. Box 69, Brookfield, WI 53008-0069
T: +1 262 7866710; Fax: +1 262 7868780; E-mail: infocenter@ifebp.org; URL: www.ifebp.org
1970; Dee B. Birschel
Employee benefits, compensation, human resources
13 000 vols; 500 curr per; 10 digital data carriers; 100 VF drawers of docs, clippings, 45 VF drawers of archival mats
SLA   36596

Brooklyn Bar Association Foundation Inc Library, 123 Remsen St, 2nd flr, **Brooklyn**, NY 11201-4212
T: +1 718 6240868; Fax: +1 718 7971713
1872
Govt publs
15 000 vols   36597

Brooklyn Botanic Garden Library, 1000 Washington Ave, **Brooklyn**, NY 11225
T: +1 718 6237302; Fax: +1 718 8572430; E-mail: library@bbg.org; URL: www.bbg.org/library
1911; Patricia Jonas
Lantern Slides, Botanical Art, Glass Plate Negatives
40 000 vols; 553 curr per   36598

Brooklyn Historical Society Library, 128 Pierrepont St, **Brooklyn**, NY 11201-2711
T: +1 718 2224111; Fax: +1 718 2223794; URL: www.brooklynhistory.org
1863; Jeffrey Burton
Brooklyn & Long Island hist, decorative arts, family hist, genealogy; newspaper clipping index; 19th c furniture and artifacts; 19th c paintings & prints coll
150 000 vols; 750 maps; 300 microforms; 1 700 linear feet of mss, 90 000 photos, 700 periodical titles, 350 newspapers   36599

Brooklyn Museum, Libraries & Archives, 200 Eastern Pkwy, **Brooklyn**, NY 11238
T: +1 718 5016307; Fax: +1 718 5016125; E-mail: library@brooklynmuseum.org; URL: www.brooklynmuseum.org
1823; Deirdre E. Lawrence
Original Costume & Fashion Sketches (1900-1950), photos of Mayan archaeological sites, artists' bks, museum Archives
250 000 vols; 2 000 curr per; 300 microforms; 12 digital data carriers; 100 VF drawers   36600

Brooklyn Museum of Art – Wilbour Library of Egyptology, 200 Eastern Pkwy, **Brooklyn**, NY 11238
T: +1 718 5016219; Fax: +1 718 5016125; E-mail: library@brooklynmuseum.org
1934; Mary Gow
50 000 vols; 200 curr per; 100 microforms; 7 000 pamphlets   36601

Coney Island Hospital, Harold Fink Memorial Library, 2601 Ocean Pkwy, **Brooklyn**, NY 11235
T: +1 718 6163000; Fax: +1 718 6164178
Munir U. Din
8 000 vols; 325 curr per   36602

World Jewish Genealogy Organization Library, Mechon Yochsin, PO Box 190420, **Brooklyn**, NY 11219-0009
T: +1 718 4354400; Fax: +1 718 6337050
25 000 vols; 1 000 digital data carriers   36603

Akiba Hebrew Academy, Joseph M. First Library, 270 S Bryn Mawr Ave, **Bryn Mawr**, PA 19066-1798
T: +1 610 6674070; Fax: +1 610 6679870; URL: www.akibaweb.org
1946; Wendie Gabay
20 000 vols; 39 curr per; 150 digital data carriers   36604

Speech & Language Development Center, Library, 8699 Holder St, **Buena Park**, CA 90620
T: +1 714 8213620; Fax: +1 714 8215683; E-mail: info@sldc.net; URL: www.sldc.net/Home.asp
1955; Carol Grundy
12 000 vols; 10 curr per   36605

Albright-Knox Art Gallery, G. Robert Strauss Jr Memorial Library, 1285 Elmwood Ave, **Buffalo**, NY 14222-1096
T: +1 716 2708240; Fax: +1 716 8826213; E-mail: artref@albrightknox.org; URL: akat.albrightknox.org; www.albrightknox.org
1905; Susana Tejada
Artist's bks, Illustrated bks, Rare book coll
45 000 vols; 100 curr per; 400 microforms; 100 av-mat; 1 digital data carriers; 20 000 other cataloged items, 10 000 vertical files   36606

Buffalo & Erie County Historical Society Research Library, 25 Nottingham Ct, **Buffalo**, NY 14216-3199
T: +1 716 8739644; Fax: +1 716 8738754; E-mail: library@bechs.org; URL: www.bechs.org
1862; Cynthia M. Van Ness
Local hist; Iconographic Coll, Mss Coll, Western New York Newspaper Coll
20 000 vols; 200 curr per   36607

Buffalo General Health System, A. H. Aaron Health Sciences Library, 100 High St, **Buffalo**, NY 14203
T: +1 716 8592878; Fax: +1 716 8591527
1920; Diane Schwartz
22 000 vols; 360 curr per   36608

Buffalo Museum of Science, Research Library, 1020 Humboldt Pkwy, **Buffalo**, NY 14211
T: +1 716 8965200; Fax: +1 716 8976723; E-mail: library@sciencebuff.org; URL: www.sciencebuff.org
1861; Linn Jean
Anthropology, Archaeology, Astronomy, Entomology, Geology, Botany, Zoology; Oriental Art & Archaeloy (Elizabeth W Hamlin Coll); Tifft Farm Oral Hist Coll
45 000 vols; 500 curr per; 7 500 maps; 2 000 microforms; 10 av-mat; 35 sound-rec
libr loan   36609

Erie County Medical Center Healthcare Network, W. Yerby Jones Memorial Library, 462 Grider St, **Buffalo**, NY 14215
T: +1 716 8983939; Fax: +1 716 8983291; E-mail: library@ecmc.edu; URL: lchib.buffalo.edu
1921; Edward J. Leisner
Trauma, Emergency medicine, Critical care, Surgery, Anesthesia, Orthopedics, Medicine
18 000 vols; 329 curr per; 600 av-mat; 35mm slide sets (350 titles)
libr loan; MLA   36610

Karpeles Manuscript Library Museum, 453 Porter Ave, **Buffalo**, NY 14201
T: +1 716 8854139; Fax: +1 716 8854139; E-mail: kmuseumbuf@aol.com; URL: www.karpeles.com
1998; Christopher Kelly
The William McKinley Room
15 000 vols; 1 200 000 mss   36611

Multidisciplinary Center for Earthquake Engineering Research, MCEER Information Service, 304 Capen Hall, State University of New York at Buffalo, **Buffalo**, NY 14260-2200
T: +1 716 6453377; Fax: +1 716 6453379; E-mail: mceeris@buffalo.edu; URL: mceer.buffalo.edu/infoservice
1986; Sofia A. Tangalos
Earthquake Engineering & Natural Hazards Mitigation, various formats
44 000 vols; 80 curr per
libr loan   36612

Research Institute on Addictions, Research Library, 1021 Main St, **Buffalo**, NY 14203-1016
T: +1 716 8872511; Fax: +1 716 8872490
1974; Ann Mina Sawusch
8 000 vols; 130 curr per   36613

The Women & Children's Hospital of Buffalo, Emily Foster Health Sciences Library, 118 Hodge Ave, **Buffalo**, NY 14222
T: +1 716 8787304; Fax: +1 716 8787547; E-mail: choblib@kaleidahealth.org; URL: library.kaleidahealth.org
1912; Elaine C. Mosher
10 000 vols; 80 curr per   36614

Providence Saint Joseph Medical Center, Health Science Library, 501 S Buena Vista St, **Burbank**, CA 91505-4866
T: +1 818 8473822; Fax: +1 818 8473823
1943; Lisa A. Marks
Hist of Medicine; Rare Book Coll

12 103 vols; 641 curr per; 12 digital data carriers
MLA, MLGSCA 36615

Southern California Genealogical Society, Family Research Library, 417 Irving Dr, **Burbank**, CA 91504-2408
T: +1 818 8437247; Fax: +1 818 8437262; E-mail: scgs@scgsgenealogy.com; URL: www.scgsgenealogy.com
1964; Fred Haughton
French-Canadian Coll, Alabama County Records, Massachusetts Town Records, Brossman Genealogical Columns, Pennsylvania Area Keys, Texas Robertson Colony Records, North Carolina Moravian Records, New England Hist & Genealogy Register, Joseph Brown Turner Coll, Ross Coll (Cornwall)
30 000 vols; 1 200 curr per; 40 drawers of mss, 15 file drawers of maps 36616

Warner Bros Studios, Research Library, 2777 N Ontario St, **Burbank**, CA 91504
T: +1 818 9775050; Fax: +1 818 5674366
1928; Phill Williams
Art & architecture, Costume design, Transportation, World War II; Historical Location Picture & Photogr Coll
100 000 vols; 92 curr per 36617

Lahey Clinic, Cattell Memorial Library, 41 Mall Rd, **Burlington**, MA 01805
T: +1 781 7448253; Fax: +1 781 7443615; URL: www.lahey.org/library
1965; Carol Spencer
8 000 vols; 160 curr per; 1 digital data carriers 36618

Harvard-Smithsonian Center for Astrophysics Library, John G. Wolbach Library, 60 Garden St, MS-56, **Cambridge**, MA 02138
T: +1 617 4965796; Fax: +1 617 4957199; E-mail: library@cfa.harvard.edu; URL: cfa-www.harvard.edu/library
1959; Amy L. Cohen
Publs of Astronomical Institutes, Early Observatory Publs
66 000 vols; 850 curr per; 40 000 microforms; 500 000 photogr plates 36619

Middlesex Law Library at Cambridge, Superior Courthouse, 40 Thorndike St, **Cambridge**, MA 02141
T: +1 617 4944148; Fax: +1 617 2250026; E-mail: midlawlib@yahoo.com; URL: www.lawlib.state.ma.us
1815; Linda Hom
Massachusetts, US & Federal Law
90 000 vols; 500 curr per; 23 digital data carriers 36620

US National Park Service, Longfellow National Historic Site Library, 105 Brattle St, **Cambridge**, MA 02138
T: +1 617 8764491; Fax: +1 617 4978718; URL: www.nps.gov/long
Naita Israel
14 000 vols 36621

Camden County Historical Society, Richard H. Hineline Research Center, Park Blvd & Euclid Ave, **Camden**, NJ 08103; P.O. Box 378, Collingswood, NJ 08108-0378
T: +1 856 9643333; Fax: +1 856 9640378; E-mail: cchsnj@cchsnj.org; URL: www.cchsnj.org
1899; Cope Linda
Hist (Boyer Coll & Morgan Coll), Dorwart Papers
20 000 vols; 15 curr per; 1 000 maps; 587 microforms; 900 slides, 20 VF drawers 36622

Dietrich Collection, 317 W Farms Rd, **Canaan**, NH 03741-7512
T: +1 603 6327156
1962; Dr. R. Krystyna Dietrich
Paintings, drawings & graphics of Alexander Orlowski, 1777 – 1832; American bks & periodicals with Information about A. Orlowski; Etchings of Daniel Chodowiecki, 1726-1801; bks about Chodowiecki; Etchings of Jean Pierre Norblin de la Gourdaine, 1745-1830; Polish translations of American Lit, 1790-1960; Polish works about American Lit; Art Coll
16 000 vols; 100 mss; 1 diss/theses; 6 000 uncatalogued items, 1 500 works of art, 200 35mm slides 36623

Stark County Law Library Association, 110 Central Plaza S, Ste 401, **Canton**, OH 44702
T: +1 330 4517380; Fax: +1 330 4517381; E-mail: info@starklawlibrary.org; URL: www.starklawlibrary.org
1890; Kendel Croston
65 000 vols; 40 curr per; 147 000 microforms; 160 av-mat; 696 sound-rec; 300 digital data carriers 36624

William McKinley Presidential Library & Museum, Ramsayer Research Library, 800 McKinley Monument Dr NW, **Canton**, OH 44708
T: +1 330 4557043; Fax: +1 330 4551137; E-mail: library@mckinleymuseum.org; URL: www.mckinleymuseum.org
Karl W. Ash
State history, city history, Don Mellett Murder & Trial Records, papers & photos
10 000 vols 36625

Panhandle-Plains Historical Museum, Research Center, 2503 Fourth Ave, **Canyon**, TX 79015; WT Box 60967, Canyon, TX 79016
T: +1 806 6512274; Fax: +1 806 6512250; URL: www.panhandleplains.org
1932; Cliff Vanderpool
Interviews with Early Settlers, Bob Wills Memorial Arch of Popular Music, Southwest Regional Architectural Drawings
15 000 vols; 800 maps; 2 000 microforms; 13 000 cubic feet of mss, 20 VF drawers of pamphlets, 300 000 photos 36626

Cumberland County Historical Society, Hamilton Library, 21 N Pitt, **Carlisle**, PA 17013-2945; P.O. Box 626, Carlisle, PA 17013-0626
T: +1 717 2497610; Fax: +1 717 2589332; E-mail: info@historicalsociety.com; URL: www.historicalsociety.com
1874; Linda Franklin Witmer
Cumberland County; Newspapers, 1749-present & Carlisle Indian School Publs, 1880-1917; A A Line, J N, Choate & Carlisle Indian School, photos; Cumberland County Firms Business Records; Carlisle; Imprints; Cartography; Judge James Hamilton; Official Cumberland County; Genealogy, VF
20 000 vols; 20 curr per; 400 maps; 1 500 microforms; 9 digital data carriers; 700 linear feet of Cumberland County doc, 200 linear feet of mss, 40 000 photos 36627

Richard T. Liddicoat Gemological Library & Information Center, GIA Library, 5345 Armada Dr, **Carlsbad**, CA 92008
T: +1 760 6034068; Fax: +1 760 6034256; E-mail: library@gia.edu; URL: www.gia.edu/library/1669/section_main_page.cfm
1931; Dona Mary Dirlam
Geology, Mineralogy; John & Marjorie Sinkankas Gemology & Mineralogy Libr, Auction Cat Coll, Cartier Rare Book Repository and Archives
38 000 vols; 300 curr per; 1 000 av-mat; 170 sound-rec; 6 000 reprints, 32 000 photos 36628

Institute for Advanced Studies of World Religions Library, 2020 Route 301, **Carmel**, NY 10512
T: +1 845 2251445; Fax: +1 845 2251485; E-mail: iaswr@aol.com; URL: www.iaswr.org
1972; Lena Lee Yang
Buddhism & Related Studies; Chinese Buddhist Texts; Asian Philosophy & Comparative Lit; Buddhist Philosophy & Chinese Art; Tibetan Bks Printed in India, Nepal & Bhutan; Buddhist Mss from Nepal; Chinese Mss from Tun-huang Coll
72 440 vols; 478 curr per; 481 maps; 68 015 microforms; 606 sound-rec
CEAL, OCLC, SENYLRC 36629

Legislative Counsel Bureau Research Library, 401 S Carson St, **Carson City**, NV 89701-4747
T: +1 775 6846827; Fax: +1 775 6846400; E-mail: library@lcb.state.nv.us; URL: www.leg.state.nev.us/lcb/research/library

1971; Nan Bowers
10 500 vols; 250 curr per 36630

Nevada Legislative Counsel Bureau, Research Library, 401 S Carson St, **Carson City**, NV 89701-4747
T: +1 775 6846827; Fax: +1 775 6846420; E-mail: library@lcb.state.nv.us; URL: www.leg.state.nv.us/lcb/research/library
Nan Bowers
Nevada Legislative History Coll
14 000 vols; 250 curr per 36631

Nevada State Library & Archives – Regional Library for the Blind & Physically Handicapped, 100 N Stewart St, **Carson City**, NV 89701-4285
T: +1 775 6843354; Fax: +1 775 6843355; URL: www.nevadaculture.org
1968; Keri Putnam
Nevada Authors
77 000 vols; 2 digital data carriers
libr loan 36632

Shawnee Library System – Southern Illinois Talking Book Center, 607 S Greenbriar Rd, **Carterville**, IL 62918-1602
T: +1 618 9858375; Fax: +1 618 9854211; E-mail: bphdept@shawls.lib.il.us; URL: www.shawls.lib.il.us/talkingbooks
Diana Brawley Sussman
28 000 vols; 100 curr per; 600 e-books 36633

Cordelia A. Greene Library, 11 S Main St, **Castile**, NY 14427; P.O. Box 208, Castile, NY 14427-0208
T: +1 585 4935466; Fax: +1 585 4935782; E-mail: caslib@pls-net.org; URL: www.castile.pls-net.org
Jacqueline Evadnie Hoyt
Local history
17 000 vols; 65 curr per; 300 av-mat 36634

VA Hudson Valley Health Care System, Library Service, Rte 90, **Castle Point**, NY 12511
T: +1 914 8312000 ext 5142; Fax: +1 914 8385246
Jeffrey Nicholas
14 000 vols; 124 curr per 36635

Genealogical Society of Linn County, Iowa, Linn County Genealogical Research Center, 813 1st Ave, SE, P.O. Box 175, **Cedar Rapids**, IA 52406
T: +1 319 3690022
1965; Marilyn Walsh
Linn County Cemetery & Court House Records
15 000 vols; 20 curr per; 200 digital data carriers 36636

Grand Lodge of Iowa, AF & AM, Iowa Masonic Library, 813 First Ave SE, **Cedar Rapids**, IA 52406; P.O. Box 279, Cedar Rapids, IA 52406-0279
T: +1 319 3651438; Fax: +1 319 3651439; E-mail: librarian@gl-iowa.org; URL: www.gl-iowa.org
1845; William R. Crawford
Abraham Lincoln Coll; Arthur Edward Waite Coll; Cedar Rapids Gazette, Early Cedar Rapids, Medical Coll; Prince Hall Masonic Coll; Robert Burns Coll
100 000 vols; 250 curr per 36637

National Czech & Slovak Museum & Library, 30 16th Ave SW, **Cedar Rapids**, IA 52404-5904
T: +1 319 3628500; Fax: +1 319 3632209; E-mail: dmuhlena@ncsml.org; URL: www.ncsml.org
1978; David Muhlena
Czech hist, Slovak hist, Czech-American hist, Slovak-American hist, Czech culture, Slovak culture, Czech-American culture, Slovak-American culture
30 000 vols; 100 curr per; 2 500 sound-rec 36638

Mercer County Law Library Association, Court House, Rm 206, N Main St, **Celina**, OH 45822
T: +1 419 5865669; Fax: +1 419 5864000
Bridget Weller
25 000 vols; 10 curr per 36639

Cherry Creek District Library, 14188 E Briarwood Ave, **Centennial**, CO 80112-4258
T: +1 720 8867080; Fax: +1 720 8867085; E-mail: distlib@cherrycreekschools.org; URL: www.ccsd.k12.co.us/admin/library/dist_lib.html
1968; Mark Ferguson
Textbook Preview, children & Young Adult Preview, Microcomputer Software Preview, Art Resource Center, Student resource Center
17 000 vols; 265 curr per; 40 000 microforms; 100 sound-rec; 50 digital data carriers
CLA, ALA 36640

Hazelden Foundation, Library & Information Resources, 15245 Pleasant Valley Rd, P.O. Box 11, CO-4, **Center City**, MN 55012
T: +1 651 2134093; Fax: +1 651 2134411; E-mail: info@hazelden.org; URL: www.hazelden.org/library
1966; Barbara S. Weiner
Chemical dependency, Treatment, Chronic illness, Spirituality, Twelve Steps, Recovery, Self-Help, Addictions; Pittman Arch at Hazelden, (Hist Alcohol Coll)
15 000 vols; 85 curr per; 500 av-mat; 600 sound-rec 36641

Museum of the Fur Trade Library, 6321 Hwy 20, **Chadron**, NE 69337
T: +1 308 4323843; Fax: +1 308 4325963; E-mail: museum@furtrade.org; URL: www.furtrade.org
Gail DeBuse-Potter
10 000 vols 36642

Kittochtinny Historical Society Library, 175 E King St, **Chambersburg**, PA 17201
T: +1 717 2641667
1898; Lillian Colletta
Fendrick Coll of Genealogical Records, Gabler Genealogy Coll, Church Hist
28 000 vols; 21 curr per 36643

Illinois State Water Survey, Research Center Library, 2204 Griffith Dr, **Champaign**, IL 61820
T: +1 217 3334956; Fax: +1 217 3336540; E-mail: library@sws.uiuc.edu; URL: www.sws.uiuc.edu/chief/library
1956; Patricia G. Morse
Water resources, Hydrology, Climatology; Global Climate Change
29 000 vols; 165 curr per 36644

Martin & Osa Johnson Safari Museum, Stott Explorers Library, 111 N Lincoln Ave, **Chanute**, KS 66720
T: +1 620 4312730; Fax: +1 620 4312730; E-mail: osajohns@safarimuseum.com; URL: www.safarimuseum.com
1980; Conrad G. Froehlich
Natural hist, Exploration photos, Africa, Pacific
10 000 vols 36645

Carolina Population Center, CPC Library & Research Translation Services, University of North Carolina, University Sq East, Room 302, 123 W Franklin St, **Chapel Hill**, NC 27516-2524
T: +1 919 9623081; Fax: +1 919 9627217; E-mail: cpclib@unc.edu; URL: www.cpc.unc.edu/aboutcpc/services/research/library
1967; Lori Delaney
Abortion File, rpts, papers; Area Files, rpts, papers; Bibliography File; Reprint File, rpts & papers
9 300 vols; 375 curr per; 860 microforms; 10 digital data carriers; 40 000 analytics, 15 000 docs, rpts, mss 36646

Law Library Association of Geauga County, 100 Short Court St, Ste BA, **Chardon**, OH 44024
T: +1 440 2852222; Fax: +1 440 2853603; E-mail: gcll@nls.net; URL: www.co.geauga.oh.us
Susan Proboski
20 000 vols; 100 curr per 36647

Charleston Library Society, 164 King St, **Charleston**, SC 29401
T: +1 843 7239912; Fax: +1 843 7233500; E-mail: librarysociety@bellsouth.net
1748; D. Carol Jones
Genealogy, hist, southern US lit; military

hist; Hinson Coll, Architecture (Staats Coll), 18th & 19th C Mss, Horticulture (Alken Garden Club Coll), Newspapers from 1732 to present, Timrod Scrapbks (Courtenay Coll), South Caroliniana, Jewish Hist
130 000 vols; 90 curr per; 300 mss; 200 diss/theses; 300 maps; 1 600 microforms; 2 000 av-mat; 1 840 sound-rec; 90 digital data carriers; 750 Large Print Bks
libr loan 36648

Charleston Museum Library, 360 Meeting St, ***Charleston***, SC 29403
T: +1 843 7222996;
Fax: +1 843 7221784; URL: www.charlestonmuseum.org
1773; Jennifer Scheetz
Hist & natural hist of South Carolina, Decorative arts, Historic archaeology; Heyward-Washington House Coll, Manigault House Coll, Gov Wm Aiken House Coll, Sheet Music Coll, Print Coll
10 000 vols; 600 maps; 60 000 photos, 300 docs, 200 pamphlets, 100 linear feet of mss 36649

South Carolina Historical Society Library, Fireproof Bldg, 100 Meeting St, ***Charleston***, SC 29401-2299
T: +1 843 7233225; Fax: +1 843 7238584; E-mail: info@schistory.org; URL: www.southcarolinahistoricalsociety.org
1855; Faye Jensen
Civil War (R. Lockwood Tower Coll)
30 000 vols; 2 000 mss; 100 diss/theses; 200 govt docs; 100 music scores; 7 500 maps; 1 000 microforms; 25 sound-rec; 25 digital data carriers; 15 000 Pamphlets, 40 000 photos, 2 200 Architectural Drawings
ALA, SAA 36650

West Virginia Archives & History Library, Cultural Ctr, 1900 Kanawha Blvd E, ***Charleston***, WV 25305-0300
T: +1 304 5580230; Fax: +1 304 5584193; URL: www.wvculture.org/history
1905; Joseph N. Geiger Jr
State Archs
65 000 vols; 326 curr per; 35 290 State Printed Doc 36651

Charlestown Boys' & Girls' Club, Charles Hayden Memorial Library, 15 Green St, ***Charlestown***, MA 02129
T: +1 617 2421775; Fax: +1 617 2413847; URL: www.bgcb.org
1893; Jenny Atkinson
8 000 vols; 15 curr per 36652

Mint Museum Library, 2730 Randolph Rd, ***Charlotte***, NC 28207
T: +1 704 3372000; Fax: +1 704 3372101; E-mail: library@mintmuseum.org; URL: www.mintmuseum.org/libraries.html
1958; Joyce Weaver
Fine arts, Craft, Design; Delhom-Gambrell Reference Libr (Decorative Arts, Emphasis on Ceramics)
15 000 vols; 75 curr per; 108 shelves of cat 36653

National Radio Astronomy Observatory Library, 520 Edgemont Rd, ***Charlottesville***, VA 22903-2475
T: +1 434 2960254; Fax: +1 434 2960278; E-mail: library@nrao.edu; URL: www.nrao.edu/library/
1959; Marsha Bishop
ESO/SEC Atlas of the Southern Sky Coll; Palomar Sky Survey; Observatory, Astronomical Institute & Govt Agency Coll
22 000 vols; 172 curr per; 3 000 e-books; 7 000 microforms; 29 digital data carriers 36654

Thomas Jefferson Foundation Inc, Jefferson Library, 1329 Kenwood Farm Lane, ***Charlottesville***, VA 22902; P.O. Box 316, Charlottesville, VA 22902
T: +1 434 9847540; Fax: +1 434 9847546; E-mail: library@monticello.org; URL: www.monticello.org/library/index.html
Jack Robertson
18th c American life
12 000 vols; 43 curr per 36655

Virginia Transportation Research Council Library, 530 Edgemont Rd, ***Charlottesville***, VA 22903
T: +1 434 2931902; Fax: +1 434 2934196; E-mail: Library.Circulation@VDOT.Virginia.gov; URL: vtrc.virginiadot.org
Ken Winter
Arch of Early Federal & Virginia Docs
32 000 vols; 87 curr per
libr loan 36656

Florida State Hospital, Library Services, Main Library Bldg 1049, ***Chattahoochee***, FL 32324; P.O. Box 1000, Chattahoochee, FL 32324-1000
T: +1 850 6637671; Fax: +1 850 6637303
Loretta Branwett
Religion, Music, Psychology, Law
35 000 vols; 18 curr per 36657

Erlanger Health System Library, 975 E Third St, ***Chattanooga***, TN 37403
T: +1 423 7787246; Fax: +1 423 7787247; E-mail: library@erlanger.org; URL: www.erlanger.org
1940; Langhorne Waterhouse
Hist of Medicine Coll
9 000 vols; 320 curr per 36658

Delaware County Historical Society, 408 Avenue of the States, ***Chester***, PA 19013
T: +1 610 8720502; Fax: +1 610 8720503; URL: www.delcohistory.org/dchs
1895; Margaret F. Johnson
1790-1920 Census of Delaware County; Baker Coll of Local Historic Data; Atlas Books of Early Land Holdings; Pennsylvania & New Jersey Arch
8 000 vols; 19 curr per; 4 500 files of clippings & pictures 36659

Wyoming State Archives, Historical Research Library, 2301 Central Ave, ***Cheyenne***, WY 82002
T: +1 307 7777826; Fax: +1 307 7777044; E-mail: wyarchive@state.wy.us; URL: wyoarchives.state.wy.us
1895; N.N.
Wyoming history
15 000 vols 36660

Adler School of Professional Psychology, Sol & Elaine Mosak Library, 17 North Dearborn, 15th Floor, ***Chicago***, IL 60602
T: +1 312 6626230; E-mail: library@adler.edu; URL: www.adler.edu/page/home/global/library
1952; Karen Drescher
Adlerian Athenaeum, Adlerian Psychology
10 000 vols; 165 curr per 36661

American Dental Association Department of Library Services, ADA Library, 211 E Chicago Ave, 6th Flr, ***Chicago***, IL 60611-2678
T: +1 312 4402653; Fax: +1 312 4402774; E-mail: library@ada.org; URL: www.ada.org
1927; Mary Kreinbring
Assn Arch; rare books
40 000 vols; 605 curr per
libr loan; MLA 36662

American Library Association Library, ALA Library, 50 E Huron St, ***Chicago***, IL 60611-2729
T: +1 312 2802153; Fax: +1 312 4409374; E-mail: library@ala.org; URL: www.ala.org/library
1924; Karen Muller
Libr & info science, Art & architecture; Library Buildings, Manuals, Policies, Public Relations; Dana Scrapbks
10 000 vols; 300 curr per; 10 000 microforms; 100 av-mat; 5 digital data carriers 36663

Art Institute of Chicago, Ryerson & Burnham Libraries, 111 S Michigan Ave, ***Chicago***, IL 60603
T: +1 312 4433671; Fax: +1 312 4430849; E-mail: ryerson@artic.edu; URL: www.artic.edu/aic/libraries/
1879; Jack Perry Brown
Architecture, 18th & 19th Century (Percier & Fontaine Coll); Catalan Art & Architectural (George R Collins Coll); Chicago & Midwestern Architecture, Chicago Art & Artists Scrapbook, Russian Art (Ernest Hamill Coll); Surrealism (Mary Reynolds Coll); Whistler (Walter Brewster Coll)
420 000 vols; 1 850 curr per; 22 400 microforms; 35 000 pamphlets & exhibition catalogs, 500 000 slides, 10 000 photogs
RLG, OCLC 36664

Balzekas Museum of Lithuanian Culture, Reference & Research Library, 6500 S Pulaski Rd, ***Chicago***, IL 60629
T: +1 773 5826500; Fax: +1 773 5825133; E-mail: balzekas@spekeasy.net
1966; Robert Balzekas
Lithuanian History, Lithuanian Immigration, Heraldry, Immigration Hist, Eastern European Genealogy & Heraldry, Rare Bks, Rare Maps
65 000 vols; 3 800 curr per; 100 mss; 25 diss/theses; 300 music scores; 645 maps; 2 100 sound-rec; 2 000 mss, 1 000 rare bks, 5 500 files of pamphlets & records, 500 slides, 1 000 art reproductions 36665

Center for Research Libraries, 6050 S Kenwood Ave, ***Chicago***, IL 60637-2804
T: +1 773 9554545; Fax: +1 773 9554339; E-mail: asd@crl.edu; URL: www.crl.edu
1949; Bernard Reilly
Africana, Latin America, South Asia, South East Asia, Middle East, Slavic; Rarely Held Current Serials; College Cats, Current & Retrospective; Foreign, Domestic & Ethnic Newspapers; Foreign Doctoral Diss; Russian Academy of Sciences Publs
2 126 000 vols; 10 548 curr per; 1 347 000 microforms
libr loan 36666

Chicago Academy of Sciences, Memorial Library, 2430 N Cannon Dr, ***Chicago***, IL 60614
T: +1 773 7555100; Fax: +1 773 7555199; URL: www.naturemuseum.org
1857; Steve Sullivan
Natural hist, Environmental studies, Geology
20 000 vols 36667

Chicago History Museum, Research Center, 1601 N Clark St, ***Chicago***, IL 60614-6099
T: +1 312 6424600; Fax: +1 312 2662076; E-mail: research@chicagohistory.org; URL: www.chicagohistory.org/research
1856; Debbie Vaughan
Chicago Directories, Trade Cats, Advertising Cards, Theater Programs; Sheet Music, Personal Papers & Records of Chicago Individuals & Organizations, J. Norman Jensen Coll of Lake & River Disasters
150 000 vols; 175 curr per; 9 860 maps; 11 000 microforms; 16 000 broadsides & posters, 725 atlases, 37 VF drawers of clippings, 300 000 archit drawings, 1,5 mio photogs, 53 044 prints 36668

Congregation Rodfei Zedek, Joseph & Dora Abbell Library, 5200 Hyde Park Blvd, ***Chicago***, IL 60615
T: +1 773 7522770; Fax: +1 773 7520330; URL: www.rodfei.org
1950; Thea Crook
Judaica, Americana, Lincolniana
8 000 vols; 30 curr per 36669

Cook County Hospital Libraries, 1900 W Polk St, 2nd fl, ***Chicago***, IL 60612
T: +1 312 8640506; E-mail: aclib2002@hotmail.com; URL: www.cchaclib.org
1953; Neera Kukreja
Rare Medical Bk Coll
8 000 vols; 500 curr per 36670

Dawson Technical Institute, Learning Resource Center, 3901 S State St, ***Chicago***, IL 60609
T: +1 773 4512087; Fax: +1 773 4512090; URL: www.ccc.edu
1972; Jacqueline Crosby
Nursing, Industrial arts, Food services business; Adult Education
16 000 vols; 90 curr per; 2 digital data carriers 36671

Edward Neisser Library, Erikson Institute Library, 451 N LaSalle St, Ste 210, ***Chicago***, IL 60654
T: +1 312 8937210; Fax: +1 312 8937213; E-mail: library@erikson.edu; URL: library.erikson.edu
1966; Dr. Janet Lynch Forde
Curriculum mat, Children's lit
15 000 vols; 85 curr per 36672

Field Museum of Natural History Library, 1400 S Lake Shore Dr, ***Chicago***, IL 60605-2498
T: +1 312 6657894; Fax: +1 312 6657893; E-mail: harlow@fieldmuseum.org; URL: www.fieldmuseum.org
1893; Christine Giannoni
Ayer Ornithology Libr, Laufer Coll of Far Eastern Studies, Schmidt Herpetology Libr
275 000 vols; 1 400 curr per; 100 e-journals
libr loan 36673

Gerber/Hart, Library and Archives, 1127 W Granville Ave, ***Chicago***, IL 60660
T: +1 773 3818030; Fax: +1 773 3818033; E-mail: info@gerberhart.org; URL: www.gerber.hart.org
1981; Susan Burnell
Homosexuality, Gay & Lesbian Life, AIDS, Homophobia; Records & Papers of Gay & Lesbian Life in Chicago & the Midwest
12 000 vols; 500 curr per 36674

Harrington College of Design, Design Library, 200 W Madison St, ***Chicago***, IL 60606
T: +1 312 6978021; Fax: +1 312 6978115; E-mail: library@interiordesign.edu; URL: www.interiordesign.edu
1975; Marc Gartler
Furniture Manufacturers Current Cats (Product Libr Coll), originals; Paint Color Cats; Plastic Laminate Samples; Ceramic Tile Samples; Samples of Building Mat
22 000 vols; 100 curr per; 10 000 e-journals; 7 500 e-books; 324 av-mat; 6 digital data carriers; 100 Journals
libr loan; ARLIS/NA 36675

Institute for Psychoanalysis, McLean Library, 122 S Michigan Ave 1300, Suite 1300, ***Chicago***, IL 60603
T: +1 312 9227474; Fax: +1 312 9225656; E-mail: admin@chicagoanalysis.org; URL: chicagoanalysis.org/mclean.php
1932; Nancy E. Harvey
Gitelson Film Libr Coll; Kohut Arch; Institute Arch
10 000 vols; 100 curr per; 4 500 microforms; 400 av-mat; 1 200 sound-rec 36676

James S. Todd Memorial Library, 515 N State St, 9th Flr, ***Chicago***, IL 60610
T: +1 312 4644855; Fax: +1 312 4645226
1911; Sandra R. Schefris
Hist of Medicine, AMA Publs
19 000 vols; 850 curr per; 75 digital data carriers
libr loan 36677

John G. Shedd Aquarium, Information Services, 1200 S Lake Shore Dr, ***Chicago***, IL 60605
T: +1 312 9392438; Fax: +1 312 9398069; E-mail: contactus@sheddaquarium.org; URL: www.sheddaquarium.org
1975
Great Lakes, Animals
10 000 vols; 200 curr per; 300 file folders of clippings, reprints, pamphlets, species files 36678

Legal Assistance Foundation of Chicago Library, 111 W Jackson, 3rd Flr, ***Chicago***, IL 60604
T: +1 312 3411070; Fax: +1 312 3411041
1974; Susan Sibert
Welfare law
18 000 vols; 160 curr per 36679

Lithuanian Research & Studies Center, Inc, 5600 S Claremont Ave, ***Chicago***, IL 60636-1039
T: +1 773 4344545; Fax: +1 773 4349363; E-mail: info@lithuanianresearch.org; URL: www.lithuanianresearch.org
1982; Dr. John A. Rackauskas
Dainauskas Hist Libr; Krupavicius Coll; Lithuanian Hist Society; Lithuanian Institute of Education; Marian Fathers Coll; Pakstas Coll; World Lithuanian Arch; Zilevicius-Kreivenas Lithuanian Musicology Arch
182 000 vols; 1 600 curr per; 800 mss;

82 diss/theses; 3200 govt docs; 8400 music scores; 586 maps; 200 av-mat; 325 sound-rec; 350 slides, 30 overhead transparencies, 100 art reproductions
libr loan   36680

Mercy Hospital & Medical Center, Medical Library, 2525 S Michigan Ave, **Chicago**, IL 60616-2477
T: +1 312 5672363; Fax: +1 312 5677086
1950; Timothy Oh
11 000 vols; 300 curr per   36681

Museum of Contemporary Art Library, 220 E Chicago Ave, **Chicago**, IL 60611-2604
T: +1 312 2802660; Fax: +1 312 3974099; URL: www.mcachicago.org
1981; Janice Dillard
Artists' Bks, Artist & Gallery Files
14 000 vols; 40 curr per; 300 av-mat; 300 sound-rec; 125 VF drawers of catalogs, 45 000 slides   36682

National Association of Realtors, Information Central, 430 N Michigan Ave, **Chicago**, IL 60611-4087
T: +1 312 3298201; Fax: +1 312 3295960; E-mail: infocentral@realtors.org; URL: www.realtor.org
1923; John Krukoff
15 000 vols; 500 curr per; 75 000 pamphlets   36683

National Opinion Research Center, Paul B. Sheatsley Library, c/o Univ of Chicago, 1155 E 60th St, Rm 281, **Chicago**, IL 60637-2667
T: +1 773 2566205; Fax: +1 773 7537886; URL: www.norc.uchicago.edu
1941; Ernest Tani
Population studies
8 700 vols   36684

Newberry Library, 60 W Walton St, **Chicago**, IL 60610-3305
T: +1 312 9439090; URL: www.newberry.org
1887; David Spadafora
Hist & Lit, Local Hist, Church Hist, Italian Renaissance, Philology, Bibliography, Music, Cartography; Hist of Printing; Western Americana; American Indian; Midwest Mss; Sherwood Anderson; Melville Coll; Railroad Arch
1 500 000 vols; 1 021 curr per; 5 000 000 mss; 65 000 maps; 225 000 microforms; 5 mio mss, 150 000 sheet music
IRLA, RLG   36685

Northeastern Illinois Planning Commission Library, 222 S Riverside Plaza, Ste 1800, **Chicago**, IL 60606-2642
T: +1 312 4540400 ext 610; Fax: +1 312 4540411; E-mail: tomasso@nipc.org; URL: www.nipc.cog.il.us
1958; Marc Thomas
Environmental studies; Northeastern Illinois Planning Commission Publs
9 000 vols; 50 curr per   36686

Polish Museum of America Library, 984 N Milwaukee Ave, **Chicago**, IL 60622-4101
T: +1 773 3843352; Fax: +1 773 3843799; E-mail: pma@polishmuseumofamerica.org; URL: www.polishmuseumofamerica.org
1913; Jan Lorys
Art, Poland; 16th-18th-c Royal Polish Mss, Polish American Newspapers & Publishers
50 000 vols; 35 maps; 70 sound-rec; 100 digital data carriers   36687

Prevention First Inc, Lura Lynn Ryan Preventive Research Library, 720 N Franklin St, Ste 500, **Chicago**, IL 60610
T: +1 312 9884646; Fax: +1 312 9887096; E-mail: mikolyt@prevention.org; URL: www.prevention.org
1980; Mary O'Brian
Preventive health, Substance abuse, Gambling, Violence against women
17 930 vols; 110 curr per; 2 486 av-mat   36688

Ukrainian Medical Association of North America, Ukrainske Likarske Tovaristvo Pivnichnoi Ameriki, Medical Archives & Library, 2247 W Chicago Ave, **Chicago**, IL 60622
T: +1 773 2786262; Fax: +1 773 2786962; E-mail: bbuniak@twcny.rr.com;

URL: www.umana.org
1950
18 000 vols   36689

Union League Club Library, 65 W Jackson Blvd, **Chicago**, IL 60604
T: +1 312 4277800; Fax: +1 312 6922322; E-mail: librarian@ulcc.org; URL: www.ulcc.org
Kimberly Kaufmann
American hist, Art & architecture, Business & management, Biology
8 000 vols   36690

Ironworld Discovery Center, Iron Range Research Center, 801 SW Hwy 169, Ste 1, **Chisholm**, MN 55719
T: +1 218 2547959; Fax: +1 218 2547971; E-mail: yourroots@ironworld.com; URL: www.ironrangeresearchcenter.org; www.ironworld.com
Jessica Oftelie
Labor, Immigration; US Steel Photo Coll, Superior National Forest Records
8 000 vols   36691

Cincinnati Art Museum, Mary R. Schiff Library, 953 Eden Park Dr, **Cincinnati**, OH 45202-1557
T: +1 513 6392978; Fax: +1 513 7210129; E-mail: library@cincyart.org; URL: www.cincinnatiartmuseum.org
1881; Mona L. Chapin
Art in Cincinnati, Cincinnati Artists' Coll
70 000 vols; 150 curr per; 500 av-mat; 250 000 Pamphlets & Clippings, Exhibition Catalogs, 16 000 Mounted Pictures   36692

Cincinnati Children's Hospital Medical Center Division of Developmental & Behavioral Pediatrics, Jack H. Rubinstein Library, D 2-60 MLC 3000, 3333 Burnet Ave, **Cincinnati**, OH 45229
T: +1 513 6364626; Fax: +1 513 6360107; URL: www.cincinnatichildrens.org
1957; Barbara Ann Johnson
Parents' Libr, Toy Libr for Children with Special Needs, Bibliotherapy Coll
8 000 vols; 120 curr per; 200 av-mat; 2 digital data carriers; 10 000 slides   36693

Cincinnati Law Library Association, Hamilton County Court House, 1000 Main St, Rm 601, **Cincinnati**, OH 45202
T: +1 513 9465300; Fax: +1 513 9465252; E-mail: reference@cms.hamilton-co.org; URL: www.hamilton-co.org/cinlawlib
1834; Mary Jenkins
Ohio Supreme Court Briefs & Records, bk, micro; Rare Legal Treatises; Regional Reporters Coll, CD-ROM network, ultrafiche; United States Session Laws; US Supreme Court Briefs & Records, micro
184 000 vols; 400 curr per; 200 digital data carriers   36694

Cincinnati Museum Center at Union Terminal, Cincinnati Historical Society Library, 1301 Western Ave, **Cincinnati**, OH 45203
T: +1 513 2877089; Fax: +1 513 2877095; E-mail: library@cincymuseum.org; URL: www.cincymuseum.org
1831; Ruby Rogers
James Albert Green Coll of William Henry Harrison; Peter G Thomson Ohio Coll; Cornelius J. Hauck Coll of arboreta & rare books
90 000 vols; 300 curr per; 2 500 maps; 5 000 microforms; 721 sound-rec; 20 000 cubic feet of mss, 14 000 slides, 1 mio photogs, 3 mio feet of film, 1 300 broadsides, 350 linear feet of clippings   36695

Cincinnati Regional Library for the Blind & Physically Handicapped, 800 Vine St, **Cincinnati**, OH 45202-2071
T: +1 513 3696999; Fax: +1 513 3693111; E-mail: lb@cincinnatilibrary.org; URL: www.cincinnatilibrary.org
Donna Foust
410 000 vols; 40 curr per; 388 303 av-mat   36696

Department of Veterans Affairs, Medical Center Library-142D, 3200 Vine St, **Cincinnati**, OH 45220-2213
T: +1 513 4756315; Fax: +1 513 4756545
11 170 vols
libr loan   36697

Lloyd Library & Museum, 917 Plum St, **Cincinnati**, OH 45202
T: +1 513 7213707; Fax: +1 513 7216575; E-mail: mheran@lloydlibrary.org; URL: www.lloydlibrary.org
1885; Maggie Heran
Horticulture, Pharmacy, Botany; Plant Chemistry & Floras, Linnean Lit Original Editions, Pharmacy, Mycology, Pharmacopoeias, Dispensatories, Formularies, Eclectic Medicine, Herbals
200 000 vols; 300 curr per; 50 mss; 1 incunabula; 300 diss/theses; 42 maps; 110 microforms; 30 digital data carriers; 50 000 Pamphlets
libr loan; SLA, CBHL   36698

Mercantile Library Association, 414 Walnut St, **Cincinnati**, OH 45202
T: +1 513 6210717; Fax: +1 513 6212023; URL: www.mercantilelibrary.com
1835; Albert Pyle
150 000 vols; 80 curr per   36699

TriHealth, Inc, Good Samaritan Hospital Library, 375 Dixmyth Ave, **Cincinnati**, OH 45220-2489
T: +1 513 8722433; Fax: +1 513 8724984
1915; Mike Douglas
Hist of Nursing Coll
10 000 vols   36700

University of Cincinnati Medical Center, Health Sciences Library, PO Box 670574, **Cincinnati**, OH 45267-0574
T: +1 513 5585628; Fax: +1 513 5582682; URL: www.aitl.uc.edu
1974; Leslie Schick
Rare Hist of Medicine Books; Albert B Sabin & Robert Kehoe Arch; Cincinnati Medical Hist of Nursing
284 000 vols; 4 075 curr per; 9 microforms; 2 444 av-mat; 1 764 sound-rec; 428 digital data carriers; 637 Slides, 86 Charts
libr loan; AAHSL, NNLM   36701

Rancho Santa Ana Botanic Garden Library, 1500 N College Ave, **Claremont**, CA 91711
T: +1 909 6258767; Fax: +1 909 6267670; URL: www.rsabg.org
1927; Patty Lindberg
Biology, Botany, Horticulture, Evolution; Californiana, Marcus E Jones Archival Mat
47 000 vols; 700 curr per; 75 diss/theses
CBHL   36702

Colombiere Center, Dinan Library, 9075 Big Lake Rd, **Clarkston**, MI 48346-1015
T: +1 248 6255611
1959; Stephen A. Meder
Greek & Latin Classics
10 000 vols; 50 curr per   36703

Anshe Chesed Fairmount Temple, Arthur J. Lelyveld Center for Jewish Learning, 23737 Fairmount Blvd, **Cleveland**, OH 44122
T: +1 216 4641330 ext 123; Fax: +1 216 4643628; E-mail: mail@fairmounttemple.org; URL: www.fairmounttemple.org
1927
Celia Smith Rogovin Childrens Library
19 000 vols; 30 curr per; 80 av-mat; 65 sound-rec; 5 digital data carriers
libr loan; AJL   36704

Cleveland Botanical Garden, Eleanor Squire Library, 11030 East Blvd, **Cleveland**, OH 44106
T: +1 216 7072812; Fax: +1 216 7212056; E-mail: gesmonde@cbgarden.org; URL: www.cbgarden.org/Learn/Library.html
1930; Gary Esmonde
Botany (Warren H Corning Coll); Flowering Plant Index of Ill & Info
16 000 vols; 175 curr per; 150 av-mat; 12 digital data carriers; 1 000 cat, 6 000 slides   36705

Cleveland Clinic Alumni Library, 9500 Euclid Ave, NA30, **Cleveland**, OH 44195-5243
T: +1 216 4445697; Fax: +1 216 4440271; E-mail: library@ccf.org
1921; Gretchen Hallerberg
36 500 vols; 850 curr per   36706

Cleveland Institute of Music, Robinson Music Library, 11021 East Blvd, **Cleveland**, OH 44106-1776
T: +1 216 7953114; Fax: +1 216 7913063; E-mail: jst4@po.cwru.edu; URL: www.cim.edu
1922; Jean Toombs
50 000 vols; 110 curr per; 15 897 sound-rec; 7 digital data carriers   36707

Cleveland Law Library Association, One W Lakeside Ave, 4th Flr, **Cleveland**, OH 44113-1023
T: +1 216 8615070; Fax: +1 216 8611606; E-mail: lawlib@clelaw.lib.oh.us; URL: www.clevelandlawlibrary.org
1869; Kathleen M. Sasala
Ohio State Supreme Court Records & Briefs, State Session Laws
185 000 vols; 676 curr per; 185 000 microforms; 676 sound-rec; 7 digital data carriers   36708

Cleveland Museum of Art, Ingalls Library, 11150 East Blvd, **Cleveland**, OH 44106-1797
T: +1 216 7072530; Fax: +1 216 4210921; E-mail: library@clevelandart.org; URL: www.clevelandart.org
1916; Elizabeth A. Lantz
Gemsheim photogs; Berenson Arch; Conway Libr; Witt Libr; Cicognara Libr; Archives
450 000 vols; 1 245 curr per; 50 000 microforms; 700 av-mat; 460 000 Slides, 400 000 Photogs, 80 000 Art Auction Catalogs, 5 000 Dealers Catalogs, 22 000 Clipping Files
libr loan; OCLC   36709

Cleveland Museum of Natural History, Harold T. Clark Library, University Circle, One Wade Oval Dr, **Cleveland**, OH 44106-1767
T: +1 216 2314600; Fax: +1 216 2315919; E-mail: library@cmnh.org; URL: www.cmnh.org/site/researchandcollections_library.aspx
1922; Wendy Wasman
Rare Bk Coll
60 000 vols; 650 curr per; 18 VF drawers of pamphlets   36710

Cleveland Public Library – Library for the Blind & Physically Handicapped, 17121 Lake Shore Blvd, **Cleveland**, OH 44110-4006
T: +1 216 6232911; Fax: +1 216 6237036; URL: www.cpl.org
1931; Barbara Mates
Ohio Coll, braille & cassettes; VF on Visual & Physical Disabilities
300 000 vols   36711

GrafTech International Ltd, Information Services, 12900 Snow Rd, **Cleveland**, OH 44130
T: +1 216 6762223
1945; Linda Riffle
Graphite tech, Carbon, High performance non-metallic mat
20 000 vols; 300 curr per   36712

Metrohealth Medical Center, Harold H. Brittingham Memorial Library, 2500 MetroHealth Dr, **Cleveland**, OH 44101-1998
T: +1 216 7785623; Fax: +1 216 7788242; E-mail: library@metrohealth.org; URL: www.metrohealth.org
1937; Christine A. Dziedzina
Arthritis & Rheumatism (Robert M. Stecher Coll), Highland View Hospital Libr Coll
9 000 vols; 450 curr per; 30 e-books; 4 VF drawers of pamphlets   36713

Ukrainian Museum-Archives Inc, 1202 Kenilworth Ave, **Cleveland**, OH 44113
T: +1 216 7814329; E-mail: staff@umacleveland.org; URL: www.umacleveland.org
1952; Andrew Fedynsky
Taras Shevchenko Coll; Ukrainian Revolution, Ukrainian Religion Per Coll outside of Ukraine, 1900-present
25 000 vols; 40 curr per   36714

US Book Exchange Library, 2969 W 25th St, **Cleveland**, OH 44113-5393
T: +1 216 2416960; Fax: +1 216 2416966; E-mail: usbe@usbe.com; URL: usbe.com
1948; John T. Zubal
Bibliogr
20 000 vols; 25 curr per   36715

Western Reserve Historical Society Library, 10825 East Blvd, *Cleveland*, OH 44106-1777
T: +1 216 7215722; Fax: +1 216 7215702; URL: www.wrhs.org
1867; Dr. John Grabowski
Charles Candee Baldwin Coll, Wallace H. Cathcart Coll of Shaker Lit, David Z. Norton Napoleon Coll, William P Palmer Civil War Coll, Society for American Baseball Res Libr, Jewish Arch, Irish Arch
238 000 vols; 325 curr per; 40 250 microforms; 50 500 Pamphlets, 20 000 000 Mss   36716

Department of Veterans Affairs, Coatesville VA Medical Center Library, 1400 Black Horse Hill Rd, *Coatesville*, PA 19320-2097
T: +1 610 3847711 ext 3902; Fax: +1 610 3830245; E-mail: Coatesville.Query@med.va.gov; URL: www.coatesville.va.gov
1931; Andrew Henry
12 039 vols; 350 curr per   36717

Florida Solar Energy Center, Research Library, 1679 Clearlake Rd, *Cocoa*, FL 32922-5703
T: +1 321 6381462; Fax: +1 321 6381463; URL: www.fsec.ucf.edu
1975; Iraida B. Rickling
13 000 vols; 95 curr per; 12 420 govt docs; 57 000 microforms; 211 sound-rec; 60 digital data carriers
libr loan; SOLINET   36718

Buffalo Bill Historical Center, McCracken Research Library, 720 Sheridan Ave, *Cody*, WY 82414
T: +1 307 5784059; Fax: +1 307 5276042; E-mail: hmrl@bbhc.org; URL: www.bbhc.org/hmrl
1927; Mary M. Robinson
20 000 vols; 96 curr per; 3 060 mss; 7 000 microforms; 1 532 sound-rec; 3 digital data carriers
libr loan; OCLC, CARL, ALA, SAA   36719

Park County Bar Association, Law Library, Court House, 1002 Sheridan Ave, *Cody*, WY 82414
T: +1 307 7542254; Fax: +1 307 5278687
Joseph Darrah
13 000 vols; 41 curr per   36720

Cold Spring Harbor Laboratory, Library & Archives, One Bungtown Rd, *Cold Spring Harbor*, NY 11724-2203
T: +1 516 3676872; Fax: +1 516 3676843; E-mail: libraryhelp@cshl.edu; URL: nucleus.cshl.org/CSHLlib
1890; Ludmila Pollack
Biochemistry, Cancer res, Cell biology, Genetics, Molecular biology, Neurobiology, Plant genetics, Virology; Hist Genetics-Eugenics Coll, Hist of Cold Spring Harbor Science
21 000 vols; 500 curr per   36721

American Institute of Physics, Niels Bohr Library, One Physics Ellipse, *College Park*, MD 20740-3843
T: +1 301 2093177; Fax: +1 301 2093144; URL: www.aip.org/history
1962; R. Joseph Anderson
Hist & Philosophy of Physics & Allied Sciences, Physics, Hist & Philosophy of Science
16 400 vols; 75 curr per   36722

Asian Studies Newsletter Archives, 9225 Limestone Pl, *College Park*, MD 20740-3943
T: +1 301 9355614
Frank Joseph Shulman
14 000 vols   36723

National Archives & Records Administration, National Archives Library, 8601 Adelphi Rd, Rm 2380, *College Park*, MD 20740
T: +1 301 8373415; Fax: +1 301 8370459; E-mail: alic@nara.gov; URL: www.archives.gov
1934; Jeffrey Hartley
Arch & Records Management Lit, Federal Govt Publs, US Hist
78 000 vols; 480 curr per   36724

American Numismatic Association Library, 818 N Cascade Ave, *Colorado Springs*, CO 80903-3279
T: +1 719 4829867; Fax: +1 719 6344085; E-mail: library@money.org; URL: www.money.org
1891; Amber Thompson
Arthur Braddan Coole Libr on Oriental Numismatics
50 000 vols; 100 curr per; 100 slides, 4 VF drawers of pamphlets & articles, 900-vol rare bk room, 20 000 auction catalogs   36725

Colorado Springs Fine Arts Center Library, 30 W Dale St, *Colorado Springs*, CO 80903
T: +1 719 4774341; Fax: +1 719 6340570
1936
Anthropology of the Southwest (Taylor Museum Coll), 20th C American Art, Arch
31 000 vols; 20 curr per   36726

United States Olympic Committee, Information Resource Center, 1750 E Boulder, *Colorado Springs*, CO 80909
T: +1 719 8664622; Fax: +1 719 6325352; E-mail: usoc.library@usoc.org; URL: www.usolympicteam.com
1981; Cindy Slater
Exercise, Athletic training, Coaches education, Biomechanics, Nutrition, Sports medicine
9 000 vols; 525 curr per
libr loan   36727

Western Museum of Mining & Industry Library, 225 N Gate Blvd, I-25 Exit 156-A, *Colorado Springs*, CO 80921
T: +1 719 4952182; Fax: +1 719 4889261; URL: www.wmmi.org
1970; Terry Girouard
14 000 vols; 11 curr per; 8 maps; 150 slides   36728

World Life Research Institute Library, 23000 Grand Terrace Rd, *Colton*, CA 92324
T: +1 909 8254773; Fax: +1 909 7833477
1959; Bruce Halstead
Nutrition, Marine biology, Pollution
30 000 vols; 50 mss; 350 000 reprints & pamphlets, 5 000 sci ills, 21 000 slides   36729

Columbia Museum of Art, Lorick Library, Main & Hampton, *Columbia*, SC 29201; P.O. Box 2068, Columbia, SC 29202-2068
T: +1 803 7992810; Fax: +1 803 3432150; URL: www.columbiamuseum.org
1980; Karen Brosius
Art Coll
14 000 vols; 30 curr per
ALSNA   36730

National Association of Watch & Clock Collections, Inc., Library & Research Center, 514 Poplar St, *Columbia*, PA 17512
T: +1 717 6848261 ext 224; Fax: +1 717 6840878; E-mail: sgordon@nawcc.org; URL: www.nawcc.org/index.php/library
1965; Sharon Gordon
Hamilton Watch Co Records & Publs
30 000 vols; 50 curr per; 21 000 American patents   36731

South Carolina State Library, Talking Book Services, 1430 Senate St, *Columbia*, SC 29201-3710; P.O. Box 821, Columbia, SC 29202-0821
T: +1 803 7344611; Fax: +1 803 7344610; E-mail: tbsbooks@statelibrary.sc.gov; URL: www.state.sc.us/scsl/bph
1973; Guynell Williams
South Caroliniana, cassettes, descriptive videotapes
349 000 vols; 45 160 av-mat; 16 800 Large Print Bks, 45 160 Talking Bks   36732

State Historical Society of Missouri Library, 1020 Lowry St, *Columbia*, MO 65201-7298
T: +1 573 8827083; Fax: +1 573 8844950; E-mail: shsofmo@umsystem.edu; URL: www.umsystem.edu/shs
1898; Dr. Gary R. Kremer

Church Hist; Joint Mss Coll with Univ of Missouri Western Hist Mss Coll; Mid-Western Hist; Missouri's Literary Heritage for Children & Youth; Missouri Newspapers (1808-present); Map Coll; Photogr Coll; United States Census Coll, J. Christian Bay Coll of Midwestern Americana
461 000 vols; 1 225 curr per; 1 394 music scores; 2 814 maps; 59 829 microforms; 2 390 av-mat; 1 451 sound-rec; 10 digital data carriers; 1 654 linear feet of mss, 302 newspaper subscriptions, 100 000 photos
libr loan   36733

William S. Hall Psychiatric Institute, Professional Library, 1800 Colonial Dr, *Columbia*, SC 29203-6827; P.O. Box 202, Columbia, SC 29202-0202
T: +1 803 8981735; Fax: +1 803 8981712; E-mail: nn570@wsphi.dmh.state.sc.us
1965; Neeta N. Shah
Annual Rpt, SC Dept Mental Health (1850-present); Hist of World Asylums Coll
13 200 vols; 350 curr per; 800 av-mat; 700 sound-rec; 12 VF drawers of pamphlets & reprints   36734

Bartholomew County Public Library – Subregional Library for the Blind & Physically Handicapped, 536 Fifth St, *Columbus*, IN 47201
T: +1 812 3791277; Fax: +1 812 3791275; E-mail: talkingbooks@barth.lib.in.us
Sharon Thompson
22 000 vols; 20 curr per   36735

Battelle Memorial Institute, Battelle Library, 505 King Ave, *Columbus*, OH 43201
T: +1 614 4246304; Fax: +1 614 4243607; E-mail: library@battelle.org; URL: www.battelle.org
1929; Kemberly Lang
Engineering, Chemistry, Physics, Biosciences
64 000 vols; 750 curr per; 500 e-journals; 750 e-books; 25 digital data carriers
libr loan; ASIS, SLA   36736

Byrd Polar Research Center, Goldthwait Polar Library, 176 Scott Hall, 1090 Carmack Rd, *Columbus*, OH 43210-1002
T: +1 614 2926715; Fax: +1 614 2924697; URL: www.library.osu.edu/sites/libinfo/PLR.html
Lynn B. Lay
Meteorology, geology, polar regions
12 000 vols; 215 curr per   36737

Center on Education & Training for Employment, 1900 Kenny Rd, *Columbus*, OH 43210-1090
T: +1 614 2926991; Fax: +1 614 2921260; E-mail: chambers.2@osu.edu; URL: www.cete.org
1965; Steve Chambers
68 000 vols; 171 curr per   36738

Columbus Law Library Association, 369 S High St, 10th Flr, *Columbus*, OH 43215-4518
T: +1 614 2214181; Fax: +1 614 2212115; E-mail: info@columbuslawlib.org; URL: columbuslawlib.org
1887; Keith Blough
116 000 vols; 1 100 curr per; 14 000 microforms; 20 av-mat; 269 sound-rec; 27 digital data carriers   36739

Grant Medical Center, Medical Library, 285 E State St, Ste 210, *Columbus*, OH 43215; mail address: 111 S Grant Ave, Columbus, OH 43215
T: +1 614 5669468; Fax: +1 614 5668451
1960; Stacy Gall
10 000 vols; 210 curr per
libr loan   36740

Mount Carmel, Health Sciences Library, 793 W State St, *Columbus*, OH 43222-1560
T: +1 614 2345214; Fax: +1 614 2341527; E-mail: library@mchs.com; URL: www.mccn.edu/library
1964; Stevo Roksandic
10 000 vols; 500 curr per   36741

Nationwide Library, One Nationwide Plaza 1-01-05, *Columbus*, OH 43215
T: +1 614 2496414; Fax: +1 614 2492218
1935; John W. Holtzclaw
Insurance, Business & management
10 000 vols; 150 curr per   36742

Ohio Health-Riverside Methodist Hospital, D.J. Vincent Medical Library, 3535 Olentangy River Rd, *Columbus*, OH 43214-3998
T: +1 614 5665230; Fax: +1 614 5666949; E-mail: medlib@ohiohealth.com
1960; Susi F. Miller
Health Education Coll
15 000 vols; 420 curr per   36743

Ohio Historical Society, Archives-Library, 1982 Velma Ave, *Columbus*, OH 43211-2497
T: +1 614 2972555; Fax: +1 614 2972546; E-mail: reference@ohiohistory.org; URL: www.ohiohistory.org
1885; Louise T. Jones
State Arch of Ohio
144 000 vols; 180 curr per; 99 000 microforms; 20 000 Newspaper Vols
libr loan   36744

Ohioana Library, 274 E First Ave, Ste 300, *Columbus*, OH 43201
T: +1 614 4663831; Fax: +1 614 7286974; E-mail: ohioana@ohioana.org; URL: www.ohioana.org
1929; Linda R. Hengst
Sherwood Anderson Coll, Mildred Wirt Benson Coll, Louis Bromfield Coll, Rollo W Brown Coll, George Randolph Chester Coll, Martha Finley Coll, Zane Gray Coll, W D Howells Coll, Dawn Powell Coll, R L Stine Coll, James Thurber Coll, Women's Hist in Ohio
46 000 vols; 10 000 musical compositions, 55 VF drawers   36745

Public Utilities Commission of Ohio, Library, 180 E Broad St, 12th Fl, *Columbus*, OH 43215
T: +1 614 4668054; Fax: +1 614 7288373; URL: www.puco.ohio.gov
1974
Staff Rpts, Public Utilities Commission Annual Rpts 1867 to present
16 000 vols; 220 curr per   36746

Rocky Mountain Arsenal, Technical Infromation Center, 5650 Havana St, *Commerce City*, CO 80022
T: +1 303 2890342; Fax: +1 303 2890205; E-mail: pao@rma.army.mil; URL: www.rma.army.mil/
1983
Environmental engineering; RMA Arch
15 000 vols; 75 curr per   36747

Franklin Pierce Law Center Library, Two White St, *Concord*, NH 03301
T: +1 603 2281541; Fax: +1 603 2280388; URL: www.piercelaw.edu/libsplash.htm
1973; Judith A. Gire
Intellectual Property
181 000 vols; 1 237 curr per; 491 000 microforms; 174 sound-rec; 49 digital data carriers   36748

New Hampshire Historical Society Library, Tuck Library, 30 Park St, *Concord*, NH 03301-6384
T: +1 603 8560643; Fax: +1 603 2240463; E-mail: library@nhhistory.org; URL: www.nhhistory.org
1823; Peter Wallner
Architecture, Decorative arts, Genealogy, Hist of New Hampshire; New Hampshire Church Records, 1700-1900, ms vols; New Hampshire Maps 1700-1900; New Hampshire Newspapers, 1790-1900, microfilm; New Hampshire photos, 1850-1990; New Hampshire Provincial Deeds, 1640-1770, micro; 19th & 20-c Account Bks & Diaries
50 000 vols; 150 curr per; 1 000 maps; 500 microforms; 1 500 000 Pages of Mss, 10 000 Broadsides, 5 000 Ephemera, 250 000 photos, 200 Music Scores   36749

993

**New Hampshire State Library –
Talking Book Services**, 117 Pleasant
St, Gallen State Office Park, Dolloff Bldg,
Concord, NH 03301-3852
T: +1 603 2713429; Fax: +1
603 2718370; E-mail:
marilyn.stevenson@dcr.nh.gov; URL:
www.state.nh.us/nhsl/talkbks
1970; Marilyn Stevenson
80 000 vols; 24 curr per; 460 av-mat;
96 520 sound-rec; 6 180 Large Print Bks
36750

New Hampshire Technical Institute,
Library, 31 College Drive, **Concord**,
NH 03301-7425
T: +1 603 2717186; Fax: +1 603
2717189; E-mail: nhtilibrary@nhctc.edu;
URL: www.nhti.edu/library
1965; Steven P. Ambra
Alcoholism counseling, Architecture,
Autism, Bus, Computer info systs, Dental
hygiene, Diagnostic ultrasound, Early
childhood educ, Electronic, Emergency
med care, Landscape architecture,
mechanical, Mental health, Nursing,
Radiologic tech
30 000 vols; 200 curr per; 400 maps;
32 000 microforms; 1 500 sound-rec; 23
digital data carriers; 2 000 slides, women
in literature clippings
libr loan
36751

**National Baseball Hall of Fame &
Museum, Inc**, Library & Archives, 25
Main St, **Cooperstown**, NY 13326-0590
T: +1 607 5470330; Fax: +1 607
5474094; URL: baseballhalloffame.org
1939; Jim Gates
Box Scores (1876-present); Schedules;
American League & National League
Performance Statistics; Arch
35 000 vols; 150 curr per; 100 mss;
50 000 incunabula; 1 000 diss/theses; 200
govt docs; 800 music scores; 10 maps;
700 microforms; 1 400 av-mat; 4 100
sound-rec; 100 digital data carriers; 2 000
scrapbooks, 1 500 pamphlets, 100 000
player data cards, 500 000 photos, 30 000
research files,
ALA, SLA
36752

New York State Historical Association,
Research Library, 5798 State Hwy 80,
Cooperstown, NY 13326; P.O. Box
800, Cooperstown, NY 13326-0800
T: +1 607 5471470; Fax: +1 607
5471405; E-mail: library@nysha.org;
URL: www.nysha.org/library
1899; Wayne Wright
New York State & Local Hist Coll, mss
89 000 vols; 320 curr per
libr loan
36753

Corning Museum of Glass, Juilette K.
& Leonard S. Rakow Research Library,
Five Museum Way, **Corning**, NY 14830
T: +1 607 9748649; Fax: +1 607
9748677; E-mail: rakow@cmog.org; URL:
www.cmog.org/library
1950; Diane Dolbashian
Antiquarian & Rare Bks, bks &
microfiche; Trade Cats
55 000 vols; 1 000 curr per; 14 000
microforms; 800 av-mat; 290 sound-rec;
800 digital data carriers; 900 docs,
160 000 slides, 33 VF drawers of
ephemera, 700 prints, 93 boxes of
company records
36754

Sherman Research Library, 614 Dahlia
Ave, **Corona del Mar**, CA 92625-2101
T: +1 949 6731880; Fax: +1 949
6755458; URL: slgardens.org
1966; W. O. Hendricks
Regional hist
25 000 vols
36755

CHRISTUS Spohn Health System,
Health Sciences Library, 2606 Hospital
Blvd, **Corpus Christi**, TX 78405
T: +1 361 9024197; Fax: +1 361
9024198
1972; Leta J. Dannelley
20 000 vols; 225 curr per
36756

**Corpus Christi Museum of Science
& History**, 1900 N Chaparral, **Corpus
Christi**, TX 78401
T: +1 361 8264650; Fax: +1 361
8847392; URL: www.ccmuseum.com
1957; Jesenia Guerra
Children's Fiction (Horatio Alger, Tom
Swift & others); Museological Coll;
Law Coll-19th & 20th Centuries; 1930's

Pictorial Hist of Corpus Christi; Natural
Hist (Netting Per Coll)
20 000 vols; 35 curr per; 325 maps; 180
sound-rec; 48 VF Drawers
36757

**Charles E. Stevens American
Atheist Library and Archives
Inc**, 225 Christiani St, **Cranford**,
NJ 07016-3214; P.O. Box 5733,
Parsippany, NJ 07054-6733
T: +1 908 2767300; Fax: +1 908
2767402; E-mail: info@atheists.org; URL:
www.atheists.org
1968; Ellen Johnson
Atheist & Freethought Magazines, pre-
Civil War to present, Haldeman-Julius
publs, Ingerson Coll, McCabe Coll,
Robertson Coll
20 000 vols; 60 curr per; 750 av-mat;
630 sound-rec; 450 000 pamphlets,
booklets, throw-aways, mss, docs,
clippings, leaflets
36758

Crazy Horse Memorial Library,
Ave of the Chiefs, **Crazy Horse**,
SD 57730-9506
T: +1 605 6734681; Fax: +1
605 6732185; E-mail:
memorial@crazyhorse.org; URL:
www.crazyhorse.org
1973; Anne Ziolkowski
22 800 vols; 14 curr per
36759

**Houston County Historical
Commission Archives**, Courthouse
Annex, 401 E Goliad, **Crockett**,
TX 75835-4035
T: +1 936 5443255; Fax: +1 936
5448053
Roy C. Smith
Indians, Civil War, Texana; Obituaries
1998-2004
27 000 vols; 49 curr per
36760

**Naval Surface Warfare Center
Dahlgren Div**, Technical Library, 17320
Dahlgren Rd Code B60, **Dahlgren**,
VA 22448-5100
T: +1 540 6538351; Fax: +1 540
6532292
Patricia Pulliam
43 000 vols; 218 curr per
36761

Dallas Historical Society, G. B.
Dealey Library, 3939 Grand Ave,
Dallas, TX 75210; P.O. Box 150038,
Hall of State in Fair Park, Dallas,
TX 75315-0038
T: +1 214 4214500; Fax: +1 214
4217500; URL: www.dallashistory.org
1922; Rachel Roberts
14 000 vols
36762

Dallas Museum of Art, Mildred R. &
Frederick M. Mayer Library, 1717 N
Harwood, **Dallas**, TX 75201
T: +1 214 9221277; Fax: +1
214 9540174; E-mail:
library@dallasmuseumofart.org; URL:
www.dallasmuseumofart.org
1938; Jacqueline Allen
80 000 vols; 110 curr per; 15 microforms;
60 VF Drawers af Artist Files, 19 800
Auction/Dealer Catalogs
libr loan
36763

Environmental Protection Agency,
Region 6 Library, 1445 Ross Ave, 6MD-
II, **Dallas**, TX 75202-2733
T: +1 214 6656424, 6656427; Fax: +1
214 6652714; E-mail: library-reg6@
epamail.epa.gov; URL: www.epa.gov/
libraries/region6.html
1971; Karen Denavit
ATSDR Toxicological Profiles; EPA Docs;
Hazardous Waste; OSWER Directives;
Risk Assessment; USDA Soil Surveys
for Arkansas, Louisiana, Oklahoma, New
Mexico & Texas
12 935 vols; 10 curr per
36764

**Graduate Institute of Applied
Linguistics**, Library, 7500 W Camp
Wisdom Rd, **Dallas**, TX 75236-5699
T: +1 972 7087416; Fax: +1 972
7087292; E-mail: library_dallas@gial.edu;
URL: www.gial.edu/library
1942; Ferne Weimer
Kenneth L Pike Coll, James Redden
African Languages & Linguistics Coll
42 000 vols; 121 curr per; 100 e-journals;
100 maps; 11 digital data carriers; 5 500
vertical files
36765

Parkland Health & Hospital System,
Fred Bonte Library, 5201 Harry Hines
Blvd, **Dallas**, TX 75235
T: +1 214 5900066; Fax: +1 214
5902720
Terry Napper
Radiology
8 000 vols; 50 curr per
36766

**IGFA Fishing Hall of Fame &
Museum**, E. K. Harry Library of Fishes,
300 Gulf Stream Way, **Dania Beach**,
FL 33004
T: +1 954 9272628; Fax: +1 954
9244299; E-mail: hq@igfa.org; URL:
www.igfa.org
1973; G. Morchower
Michael Lerner Coll, Joe Brooks Coll,
A.J. McClane Coll
15 000 vols; 150 curr per; 1 700 av-mat
36767

**Darlington County Historical
Commission**, 204 Hewitt St,
Darlington, SC 29532
T: +1 843 3984710; Fax: +1 843
3984742; E-mail: dchc1968@aol.com
1968; Doris G. Gandy
Family Name Files on Darlington County
Names
10 000 vols
36768

Figge Art Museum, Art Reference
Library, 225 West 2nd Street,
Davenport, IA 52801
T: +1 563 3267804; Fax: +1 563
3267876; URL: www.figgeartmuseum.org
1934; Kathy Phelan (interim)
Art history; American art; Haitian Art,
Mexican Colonial Art
8 500 vols; 26 curr per; 5 000 pamphlets
36769

**Putnam Museum of History & Natural
Science**, 1717 W 12th St, **Davenport**,
IA 52804
T: +1 563 3241933 ext 216;
Fax: +1 563 3246638; E-mail:
museum@putnam.org; URL:
www.putnam.org
1867; Christopher J. Reich
Steamboats (files, photos), Mss (A.
LeClaire, I. Hall, L. Summers, I.
Wetherby, R. Cram, Black, Store, Putnam
Family, James Grant), Local Hist
30 000 vols
36770

Scott County Bar Association,
Grant Law Library, 111 East Third St,
Davenport, IA 52801
T: +1 319 3264491; Fax: +1 319
3264498; URL: www.scottcountybar.org
1955; Ginger F. Wolfe
Iowa Supreme Court Abstracts
10 000 vols
36771

Calgene, LLC, Library Information Center,
1920 Fifth St, **Davis**, CA 95616
T: +1 530 7536313; Fax: +1
530 7922453; E-mail:
deanna.johnson@monsanto.com; URL:
www.calgene.com
1982; Deanna Johnson
9 425 vols; 174 curr per; 2 500
microforms; 3 digital data carriers
libr loan
36772

Dayton Art Institute, Louis Lott Memorial
Art Reference Library, 456 Belmonte
Park N, **Dayton**, OH 45405-4700
T: +1 937 2235277; Fax: +1
937 2233140; E-mail:
library@daytonartinstitute.org; URL:
www.daytonartinstitute.org
1922; Ellen Rohmiller
Lott-Schaeffer Memorial Architecture Libr,
Guy Elbert Allott and Gwen Jones Allott
Coll (Oriental Rugs)
20 000 vols; 50 curr per; 820 microforms;
2 digital data carriers; 5 000 auction
catalogs; 3 000 museum publs, 105
drawers of artist files
libr loan; ARLIS/NA, SLA
36773

Dayton Law Library, 41 N Perry St, Rm
505, **Dayton**, OH 45402; P.O. Box 972,
Dayton, OH 45422-2490
T: +1 937 2254496; Fax: +1 937
2255056
Joanne Beal
141 000 vols; 4 214 curr per
36774

Miami Valley Hospital, Craig Memorial
Library, One Wyoming St, **Dayton**,
OH 45409
T: +1 937 2082612; Fax: +1 937
2082569
1926; Shirley Sebald-Kinder
42 000 vols; 500 curr per; 1 200
sound-rec; 20 digital data carriers
36775

**Bureau of Braille & Talking Book
Library Services**, 420 Platt St,
Daytona Beach, FL 32114-2804
T: +1 386 2396000; Fax: +1 386
2396069; URL: dbs.myflorida.com/library/
index.shtml
1950; Michael G. Gunde
Womyn's Braille Press
2 400 000 vols
libr loan
36776

The Henry Ford, Benson Ford Research
Center, 20900 Oakwood Blvd, **Dearborn**,
MI 48124-4088; P.O. Box 1970,
Dearborn, MI 48121-1970
T: +1 313 9826070; Fax: +1
313 9826244; E-mail:
rescntr@thehenryford.org; URL:
www.thehenryford.org
1929; Judith E. Endelman
John Burroughs Papers; Henry Austin
Clark Coll; Detroit Publishing Company,
arch, photos; Edison Recording Artists;
Ephemera Coll, trade lit; Henry & Clara
Ford Papers; Ford Motor Company
Records; Fire Insurance Maps; Gebelein
Silversmiths; H.J. Heinz Co Records;
Images Ford Motor Company; Industrial
Design Coll; Stickley Furniture Co
Records
45 000 vols; 200 curr per; 380 av-mat;
20 000 trade cat, 25 000 linear feet of
archival mat, Ford Motor Co Photogr
Archives (350 000 images), graphics coll
(50 000 items), 2 200 journal titles
libr loan
36777

The Henry Ford, Benson Ford Research
Center, 20900 Oakwood Blvd, **Dearborn**,
MI 48124-5029; P.O. Box 1970,
Dearborn, MI 48121-1970
T: +1 313 9826020; Fax: +1
313 9826244; E-mail:
rescntr@thehenryford.org; URL:
www.thehenryford.org
Judith E. Endelman
History
45 000 vols; 200 curr per
36778

Oakwood Hospital Medical Library,
Ernest & Kellie Sorini Medical Library,
18101 Oakwood Blvd, **Dearborn**,
MI 48124-2500
T: +1 313 5937692; Fax: +1
313 4362699; E-mail:
medlibrary@oakwood.org; URL:
www.oakwood.org/medicallibrary
1953; Marilyn Kostrzewski
Obstetrics & Gynecology Coll
12 000 vols; 400 curr per; 275 e-books
libr loan; MLA
36779

Society of Manufacturing Engineers,
SME Library, 1 SME Dr, P.O. Box 930,
Dearborn, MI 48121
T: +1 313 2711500; Fax: +1 313
2408251; E-mail: library@sme.org; URL:
www.sme.org
1932; Carol Smith Tower
8 000 vols; 150 curr per; 18 000 technical
rpts
36780

National Park Service, Death Valley
National Park Research Libraries, P.O.
Box 579, **Death Valley**, CA 92328
T: +1 760 7862331 ext 287;
Fax: +1 760 7863283; E-mail:
blair_davenport@nps.gov; URL:
www.nps.gov/deva/index.htm
1958; Blair Davenport
Death Valley National Park, Natural
Sciences, Chiefly Geology
10 000 vols
36781

Atlanta VA Medical Center Library,
1670 Clairmont Rd, **Decatur**, GA 30033
T: +1 404 3216111 ext 7672; Fax: +1
404 7287781
1945; Shirley Avin
8 010 vols; 482 curr per
36782

Decatur Genealogical Society Library,
1255 W South Side Dr, **Decatur**,
IL 62521-4024; P.O. Box 1548, Decatur,
IL 62525-1548
T: +1 217 4290135; E-mail:
decaturgenealogicalsociety@msn.com;

URL: www.rootsweb.com/~ildecgs/library.htm
Mary Wilking
30 000 vols; 16 curr per 36783

Vesterheim Norwegian-American Museum, Special Library, 523 W Water St, ***Decorah***, IA 52101; P.O. Box 379, Decorah, IA 52101
T: +1 563 3829681; Fax: +1 563 3828828; E-mail: info@vesterheim.org; URL: vesterheim.org
1970; Laurann Gilbertson
Norwegian and Norwegian-American culture and emigration/immigration; Special colls: Norwegian Rosemaling; Norwegian-American Imprints
11 000 vols; 25 curr per 36784

Dedham Historical Society Library, 612 High St, ***Dedham***, MA 02027-0215
T: +1 781 3261385; Fax: +1 781 3265762; E-mail: society@dedhamhistorical.org; URL: www.dedhamhistorical.org
1859; Ronald F Frazier
Ames Family Coll; Church Records 1638-1890; Mann Family Coll; Original Dedham Grant Area Residents & Artifacts (including Nathaniel Ames-father & son); Records of Firms & Assns
10 000 vols; 10 curr per; 23 700 mss; 5 diss/theses; 800 govt docs; 200 music scores; 3 160 maps; 6 000 microforms; 15 av-mat; 10 digital data carriers 36785

Massachusetts Trial Court, Norfolk Law Library, 57 Providence Hwy, ***Dedham***, MA 02026
T: +1 781 7697483; Fax: +1 781 7697836; E-mail: agnesml@juno.com
1898; Carol Ewing, Agnes Leathe
State law; Legal videos for laypeople
15 000 vols; 24 curr per 36786

Historic Deerfield Inc & Pocumtuck Valley Memorial Association Libraries, 6 Memorial St, ***Deerfield***, MA 01342-9736; P.O. Box 53, Deerfield, MA 01342-0053
T: +1 413 7757125; Fax: +1 413 7757223; E-mail: library@historic-deerfield.org; URL: www.historic-deerfield.org/libraries.html
1870; David C. Bosse
Local hist, Decorative arts, Genealogy
50 350 vols; 40 curr per; 175 maps; 1 148 microforms
libr loan 36787

Colorado Agency for Jewish Education, Library, 300 S Dahlia St, Ste 101, ***Denver***, CO 80246
T: +1 303 3213191; Fax: +1 303 3215436; E-mail: library@caje-co.org; URL: www.caje-co.org
Susan Miller Rheins
Education, Judaica, Holocaust, Special education; Children's Judaica Libr, Judaic Teacher Mat, Video Coll
10 000 vols; 20 curr per 36788

Colorado Historical Society, Stephen H. Hart Library, 1300 Broadway, ***Denver***, CO 80203-2137
T: +1 303 8664600; Fax: +1 303 8664204; E-mail: research@chs.state.co.us; URL: www.coloradohistory.org
1879; Rebecca Lintz
Mining, Railroads, Social history; William Henry Jackson photogs; Denver and Rio Grande railroad photogs and papers; Colorado newspapers from 1859
22 000 vols; 491 curr per; 2 500 mss; 250 diss/theses; 5 000 maps; 29 000 microforms; 70 av-mat; 1 250 sound-rec; 2 000 mss, 600 000 photogs 36789

Colorado Mental Health Institute at Fort Logan – Children's Division Library, 3520 Oxford Ave, ***Denver***, CO 80236
T: +1 303 8667876; Fax: +1 303 8667696; E-mail: Sheridan.Garcia@state.co.us
Eileen Rice
Consumer/Parent Coll, Teen Issues
10 000 vols; 18 curr per 36790

Denver Art Museum Library, 414 14th St, ***Denver***, CO 80204; mail address: 100 W 14th Ave, Denver, CO 80204
T: +1 720 9130100; Fax: +1 720 9130001; E-mail: sferrer-vinent@denverartmuseum.org; URL: library.denverartmuseum.org
1935; Susan T. Ferrer-Vinent
Visual art, Architecture, design & graphics, Asian art, Film, Western American art, Modern and contemporary art, Native arts, Pre-colombian art, Spanish colonial art, Painting, Sculpture, Textile art, Art education
28 500 vols; 150 curr per
ARLIS/NA 36791

Denver Botanic Gardens, Helen Fowler Library, 909 York St, ***Denver***, CO 80206-3799
T: +1 720 8653570; Fax: +1 720 8653685; E-mail: library@botanicgardens.org; URL: www.botanicgardens.org
Deb Golanty
Botany, horticulture; Waring House Book Room
30 000 vols; 400 curr per 36792

Denver Museum of Nature & Science, Alfred M. Bailey Library & Archives, 2001 Colorado Blvd, ***Denver***, CO 80205-5798
T: +1 303 3706362; Fax: +1 303 3316492; URL: www.dmns.org/librarycatalog
1900; Katherine B. Gully
Institutional Arch for Museum; Photo Arch; Rare bks
46 000 vols; 190 curr per; 1 300 cubic feet of mss, 400 000 photogs, 300 Spec Interest Per Sub 36793

Denver Public School District, Professional Library, 1330 Fox St S, 3rd Flr, ***Denver***, CO 80204
T: +1 720 4238108; Fax: +1 720 4238100; URL: www.dpsk12.org
Jody Gehrig
74 000 vols; 10 curr per 36794

Health One Presbyterian-Saint Luke's Medical Center, Denver Medical Library, 1719 E 19th Ave, ***Denver***, CO 80218-1281
T: +1 303 8396670; Fax: +1 303 8691643; E-mail: info@denvermedlib.org; URL: www.denvermedlib.org
1893; Sharon Martin
Allied health, Cardiology, Geriatrics and gerontology, Neonatology, Obstetrics and gynecology, Pediatrics
56 000 vols; 425 curr per; 1 digital data carriers
libr loan 36795

National Theatre Conservatory, The Jones Library, 1101 13th St, ***Denver***, CO 80204
T: +1 303 4464869; Fax: +1 303 8252117; E-mail: eller@dcpa.org; URL: www.dcpa.org
1991; Linda M. Eller
Musical theatre, Performance, Theater; Denver Center Theater Company Production Docs, Playbills, Vinyl Records
35 000 vols; 15 curr per; 2 000 av-mat 36796

University of Colorado at Denver & Health Sciences Center, Denison Memorial Library, 4200 E Ninth Ave, ***Denver***, CO 80262
T: +1 303 3155127; Fax: +1 303 3156255
1924; Perry Gerald
Waring Hist of Medicine Coll, Indigenous Medicine Coll
107 000 vols; 5 581 curr per 36797

Des Moines Art Center Library, 4700 Grand Ave, ***Des Moines***, IA 50312-2099
T: +1 515 2774405; Fax: +1 515 2710357; E-mail: cdoolittle@desmoinesartcenter.org; URL: www.desmoinesartcenter.org
1950; Mary Morman-Graham
19th & 20th c art; Iowa Artists Coll
16 200 vols; 32 curr per; 15 digital data carriers; 80 file drawers of clippings/archives
ARLIS/NA 36798

Iowa Genealogical Society Library, 628 E Grand Ave, ***Des Moines***, IA 50309-1924
T: +1 515 2760287; Fax: +1 515 7271824; E-mail: igs@iowagenealogy.org; URL: www.iowagenealogy.org
Billie Murano
Genealogy, regional history
12 000 vols; 70 curr per 36799

Iowa Regional Library for the Blind & Physically Handicapped, 524 Fourth St, ***Des Moines***, IA 50309-2364
T: +1 515 2811333; Fax: +1 515 2811378; E-mail: library@blind.state.ia.us; URL: www.blind.state.ia.us
1960; Karen Keninger
Print Coll of Books about Blindness
311 000 vols; 75 curr per; 49 897 av-mat; 49 490 sound-rec; 8 digital data carriers; 49 900 Talking Bks, 6 450 Large Print Bks 36800

State Historical Society of Iowa – Des Moines Library, 600 E Locust, ***Des Moines***, IA 50319-0290
T: +1 515 2816200; Fax: +1 515 2820502; URL: www.iowahistory.org
1892; Carol Kirsch
Mss, State Arch
76 000 vols; 200 curr per; 3 000 maps 36801

Gas Technology Institute, Technical Information Center, 1700 S Mount Prospect Rd, ***Des Plaines***, IL 60018-1804
T: +1 847 7680664; Fax: +1 847 7680669; E-mail: library@gastechnology.org; URL: www.gastechnology.org
1941; Carol Worster
Natural gas; American Chemical Society Division of Fuel Chemistry, Preprints 1957 to present; Pipeline Simulation Interest Group, Proc; Energy Rpts (DOE, EPRI, GRI)
36 000 vols; 100 curr per; 1 000 diss/theses; 100 000 govt docs; 100 maps; 200 000 microforms; 5 050 patents
libr loan 36802

UOP Library & Information Services, 50 E Algonquin Rd, ***Des Plaines***, IL 60016-6102; P.O. Box 5016, Des Plaines, IL 60017-5016
T: +1 847 3912265; Fax: +1 847 3913330; E-mail: rdlibrary@uop.com
1926; Laura Claggett
Energy, Chemical engineering; Tech Oil Mission Rpts, Official Patent Publs
20 000 vols; 700 curr per; 100 microforms; 100 sound-rec; 7 digital data carriers; 450 000 patents 36803

Detroit Institute of Arts, Research Library & Archives, 5200 Woodward Ave, ***Detroit***, MI 48202
T: +1 313 8333460; Fax: +1 313 8336405; URL: www.dia.org/research/research_library/index.asp
1905; Maria Ketcham
Albert Kahn Architecture Libr; Puppetry (Paul McPharlin Coll); Grace Whitney Hoff Coll, fine bindings
180 000 vols; 200 curr per; 10 digital data carriers
libr loan 36804

International Union of United Automobile, Aerospace & Agricultural Implement Workers of America, UAW Research Library, 8731 E Jefferson Ave, ***Detroit***, MI 48214; mail address: 8000 E Jefferson Ave, Detroit, MI 48214
T: +1 313 9265000; Fax: +1 313 9265871; URL: www.uaw.org
1947; Maria Catalfio
Automobile industry; Coll of UAW-Hist
20 000 vols; 198 curr per; 1 000 govt docs
SLA, CIRL 36805

Southeast Michigan Council of Governments Library, 535 Griswold, Ste 300, ***Detroit***, MI 48226
T: +1 313 9614266; Fax: +1 313 9614869; E-mail: infocenter@semcog.org; URL: www.semcog.org
1972; Amanda G. Polanco
Detroit Regional Transportation & Land Use Study (TALUS), Detroit Metropolitan Area Regional Planning Commission; Master Plans for Southeast Michigan Communities
30 000 vols; 350 curr per
libr loan 36806

Third Judicial Circuit Court, Wayne County, Law Library, Two Woodward Ave, Ste 780, ***Detroit***, MI 48226-3461
T: +1 313 2245265; Fax: +1 313 9673562
Lynn Reeves
20 000 vols 36807

Kansas Heritage Center Library, 1000 N Second Ave, ***Dodge City***, KS 67801-4415; P.O. Box 1207, Dodge City, KS 67801-1207
T: +1 620 2271616; Fax: +1 620 2271701; E-mail: library@ksheritage.org; URL: www.ksheritage.org
1966; Jim Sherer
Kansas, American West; Historical Society Journals from the States of Kansas, Missouri, Colorado, New Mexico, Montana, & Oklahoma
9 000 vols; 45 curr per 36808

Delaware Division of Libraries-State Library – Delaware Library for the Blind & Physically Handicapped, 43 S Dupont Hwy, ***Dover***, DE 19901-7430
T: +1 302 7394748; Fax: +1 302 7396787; E-mail: debph@lib.de.us
Beth Landon
63 000 vols; 10 curr per 36809

Bucks County Historical Society, Spruance Library, 84 S Pine St, ***Doylestown***, PA 18901-4999
T: +1 215 3450210; Fax: +1 215 2300823; E-mail: mmlib@mercermuseum.org; URL: www.mercermuseum.org
1880; Beth Lander
American folk art, Antiques; Mss Colls, Durham Furnace Account Bks, Craftsmen's Account Bks, Turnpike Records, Handwritten School bks
18 000 vols; 2 000 microforms 36810

City of Hope National Medical Center, Lee Graff Medical & Scientific Library, 1500 E Duarte Rd, ***Duarte***, CA 91010-0269
T: +1 626 3018497; Fax: +1 626 3571929; E-mail: library@coh.org; URL: library.coh.org
1954; Janet Crum
Medicine, Biochemistry, Biology, Imunology, Genetics, Cancer
100 000 vols; 50 curr per; 22 259 e-journals; 5 134 e-books; 100 diss/theses
libr loan 36811

Department of Veterans Affairs, Carl Vinson VA Medical Center, 1826 Veterans Blvd, Vet Admin Med Ctr, ***Dublin***, GA 31021
T: +1 478 2772759; Fax: +1 478 2772771
1948; Jean Cannon
15 000 vols; 379 curr per; 750 microforms; 452 av-mat 36812

OCLC Library, 6565 Kilgour Pl, ***Dublin***, OH 43017; P.O. Box 7777, Dublin, OH 43017
T: +1 614 7938707; Fax: +1 614 7187336; E-mail: oclclibrary@oclc.org
1977; Lawrence Olszewski
OCLC archives
15 463 vols; 400 curr per; 225 e-journals; 17 maps; 782 microforms; 301 av-mat; 178 sound-rec; 10 digital data carriers
libr loan; ALA, SLA, ASIST, IFLA 36813

Mercy Medical Center – Dubuque, Anthony C. Pfohl Health Science Library, 250 Mercy Dr, ***Dubuque***, IA 52001-7398
T: +1 563 5899620; Fax: +1 563 5898185; URL: www.mercydubuque.com/educat/index.htm/library
1973; James H. Lander
10 000 vols; 300 curr per
libr loan 36814

Northeast Minnesota Historical Center, University of Minnesota-Duluth Library, 416 Library Dr, Annex 202, ***Duluth***, MN 55812
T: +1 218 7268526; Fax: +1 218 7266205; URL: www.d.umn.edu/lib
Patricia Maus
Mining, transportation
8 000 vols; 670 curr per 36815

Forest History Society Library, 701 William Vickers Ave, ***Durham***, NC 27701-3162
T: +1 919 6829319; Fax: +1 919 6822349; URL: www.foresthistory.org/research/library.html
1947; Cheryl Oakes
Forestry, environmental hist; Oral Hist Interviews, Historical Photo Coll; Archives of forest and conservation hist
9 000 vols; 200 curr per; 500 mss; 100 diss/theses; 700 maps; 693 sound-rec; 10 boxes of newsclippings
libr loan 36816

Rhine Research Center, Institute for Parapsychology, 2741 Campus Walk Ave, Bldg 500, **Durham**, NC 27705-3707
T: +1 919 3094600; Fax: +1 919 3094700; URL: www.rhine.org
Lauren McGlynn
Parapsychology; foreign language books on parapsychology, unpublished mss and theses
10 000 vols; 70 curr per   36817

New England Wireless & Steam Museum Inc Library, 1300 Frenchtown Rd, **East Greenwich**, RI 02818
T: +1 401 8850545; Fax: +1 401 8840683; E-mail: newsm@ids.net; URL: www.users.ids.net/~newsm
1964; Robert W. Merriam
Lloyd Espenshield, Thorn Mayes, Edward Raser, & A C Goodnow Textbk Colls
20 000 vols; 200 slides   36818

Nassau University Medical Center, Health Sciences Library, 2201 Hempstead Tpk, **East Meadow**, NY 11554
T: +1 516 5728742; Fax: +1 516 5725788; E-mail: rperelma@numc.edu; URL: www.nuhealth.net/education/library/index.asp
Rimma Perelman
8 500 vols; 950 curr per   36819

Department of Veterans Affairs Medical Center Library, 385 Tremont Ave, **East Orange**, NJ 07018-1095
T: +1 973 6761000 ext 1388; Fax: +1 973 3957234
1955; Sophie Winston
12 300 vols; 425 curr per   36820

Northampton County Historical & Genealogical Society, Mary Illick Memorial Library, 107 S Fourth St, **Easton**, PA 18042
T: +1 610 2531222; Fax: +1 610 2534701; URL: www.northamptonctymuseum.org
Jane S. Moyer
Local history, genealogy
10 000 vols   36821

Tompkins Memorial Library, 104 Courthouse Sq, **Edgefield**, SC 29824
T: +1 803 6374010; Fax: +1 803 6372116; E-mail: oedgs@aikenelectric.net; URL: oedgs.org
1929; Tonya Browder
Antiquities of England, Ireland, Wales, & Normandy; American, Colonial, South Carolina Hist, Genealogy Libr for Edgefield County; Antebellum Home Libr of James Madison Abney
17 000 vols   36822

Smithsonian Institution Libraries, Smithsonian Environmental Research Center Library, 647 Contees Wharf Rd, **Edgewater**, MD 21037; P.O. Box 28, Edgewater, MD 21037-0028
T: +1 443 4822273; Fax: +1 443 4822286; URL: www.sil.si.edu/libraries/serc
1972; Angela Haggins
Chesapeake Bay area ecology, Ecosysts, Environ management, Estuaries & landscape ecol, Evolution, Land use hist, Population & commun ecology, Solar UV light
12 000 vols   36823

El Paso Museum of Art, Algur H. Meadows Art Library, One Art Festival Plaza, **El Paso**, TX 79901
T: +1 915 5321707; Fax: +1 915 5321010; URL: www.elpasoartmuseum.org
1960; Amy Paoli
Renaissance & Baroque Artists Coll
11 000 vols; 24 curr per   36824

Ensanian Physicochemical Institute, Institute Library, Barden Brook Rd, P.O. Box 98, **Eldred**, PA 16731
T: +1 814 2253296
1964; Elisabeth Anahid Ensanian
Cosmology; Geotropism (Information Center for Gravitation Chemistry); Non-Destructive Testing of Materials; Physiochemical Robotic Sensors; Quantum Physics; Robotics & Robot Sensors (Tactile); Stored Energy in Metals (Electrotopography Information Center); Quantum Computation, Foundational Physics Group, National Metals Classification and Ranking Center
14 300 vols; 125 curr per; 2 015 microforms   36825

Church of the Brethren, Brethren Historical Library & Archives, 1451 Dundee Ave, **Elgin**, IL 60120-1694
T: +1 847 7425100; Fax: +1 847 7426103; URL: www.brethren.org/genbd/bhla/
1936; Kenneth M. Shaffer Jr
Arch & Mss Coll
10 000 vols; 847 microforms; 317 av-mat; 2 814 sound-rec; 5 921 slides   36826

Elgin Mental Health Center Library, FTP Library, 750 S State St, **Elgin**, IL 60123-7692
T: +1 847 7421040; Fax: +1 847 4294923
1872; David Hagerman
10 000 vols   36827

Essex County Historical Society, Brewster Library, 7590 Court St, **Elizabethtown**, NY 12932; P.O. Box 428, Elizabethtown, NY 12932-0428
T: +1 518 8738586; E-mail: echs@adkhistorycenter.org; URL: www.adkhistorycenter.org
1956; Margaret Gibbs
8 000 vols; 75 curr per; 400 maps; 32 microforms; 600 pamphlets, 24 VF drawer of ephemera, 350 mss, 34 drawers of cemetery records   36828

Keneseth Israel Reform Congregation, Meyers Library, 8339 Old York Rd, **Elkins Park**, PA 19027
T: +1 215 8878704; Fax: +1 215 8771070; E-mail: library@kenesethisrael.org; URL: www.kenesethisrael.org/index.htm
1870; Ellen Tilman
14 000 vols; 30 curr per   36829

Elmhurst Hospital Center, Medical Library, 79-01 Broadway, Rm D3-52A, **Elmhurst**, NY 11373
T: +1 718 3342040; Fax: +1 718 3345690; E-mail: ehelib3@medgate.metro.org
1965
19 500 vols; 200 curr per   36830

Elmhurst Memorial Hospital, Marquardt Memorial Library, 200 Berteau Ave, **Elmhurst**, IL 60126
T: +1 630 8331400 ext 42712; Fax: +1 630 7827834; E-mail: png@emhc.org
Pauline Ng
8 000 vols; 240 curr per; 850 av-mat; 18 series audiotapes, 800 files of pamphlets   36831

Department of Veterans Affairs Medical Center, Medical Library, 135 E 38th St, **Erie**, PA 16504-1559
T: +1 814 8688661; Fax: +1 814 8602469
Mary E. Nourse
Nursing, health care
35 000 vols; 125 curr per   36832

Young Men's Christian Association of the Rockies, Estes Park Center, Maude Jellison Library, P.O. Box 20550, **Estes Park**, CO 80511
T: +1 970 5863341; Fax: +1 970 5866078
1962; Cami Sebern
Colorado Hist; Natural Hist; Enos Mills; Mark Twain
18 000 vols; 5 curr per; 90 archival mat   36833

Society for Promoting & Encouraging Arts & Knowledge of the Church, Howard Lane Foland Library, 805 County Rd 102, **Eureka Springs**, AR 72632-9705
T: +1 501 2539701; Fax: +1 501 2531277; E-mail: speak@speakinc.org; URL: www.speakinc.org
1980
Bible, Eschatology, Theology, Pastoral relations, Liturgies
11 720 vols   36834

The Shakespeare Data Bank, Inc Library, 1217 Ashland Ave, **Evanston**, IL 60202-1103
T: +1 847 4757550
1981; Louis Marder
10 000 vols   36835

Evansville-Vanderburgh Public Library – Talking Books Service, 200 SE Martin Luther King Jr Blvd, **Evansville**, IN 47713-1604
T: +1 812 4288235; E-mail: tbs@evpl.org
Barbara Shanks
20 000 vols   36836

William H. Miller Law Library, 825 Sycamore, 207 City-County Courts Bldg, **Evansville**, IN 47708-1849
T: +1 812 4355175; Fax: +1 812 4261091; E-mail: evvlaw@evansville.net; URL: www.vanderburghgov.org/lawlibrary
1900; Helen S Reed
19 000 vols; 4 700 microforms; 1 digital data carriers
AALL   36837

Geophysical Institute, Keith B. Mather Library, Int Arctic Research Ctr, 930 Koyukuk, **Fairbanks**, AK 99775; P.O. Box 757355, Fairbanks, AK 99775-7355
T: +1 907 4747512; Fax: +1 907 4747290; E-mail: gilibrary@gi.alaska.edu; URL: www.gi.alaska.edu/services/library
1945; Julia Triplehorn
Geophysics, seismology, volcanology, polar regions, snow, glaciers, remote sensing, space physics; International Assn of Volcanology Chemistry of the Earth's Inerior Coll, Alaska Department of Transportation Libr
68 000 vols; 325 curr per; 60 digital data carriers
libr loan   36838

National Clearinghouse on Child Abuse & Neglect Information, 10530 Rosehaven St, Ste 400, **Fairfax**, VA 22030; mail address: 330 C St SW, Washington, DC 20447
T: +1 703 3857565; Fax: +1 703 3853206; E-mail: nccanch@caliber.com; URL: www.nccanch.acf.hhs.com
John Vogel
40 000 vols; 60 curr per   36839

National Clearinghouse on Child Abuse & Neglect Information, 10530 Rosehaven St, Ste 400, **Fairfax**, VA 22030; mail address: 330 C St SW, Washington, DC 20447
T: +1 703 3857565; Fax: +1 703 3853206; E-mail: nccanch@caliber.com; URL: nccanch.acf.hhs.gov
John Vogel
40 000 vols; 60 curr per   36840

Fairfield Historical Society Library, 636 Old Post Rd, **Fairfield**, CT 06430
T: +1 203 2591598; Fax: +1 203 2552716; URL: www.fairfieldhs.org/fairfield-museum-library.php
1903
Landscape architecture drawings, Local hist, Genealogy, Architectural drawings (local architects); Mss Coll
10 000 vols; 20 curr per; 100 mss; 150 maps; 10 microforms; 100 sound-rec; photos (40 unreel feet)   36841

Trial Court of Massachusetts, Fall River Law Library, Superior Courthouse, 441 N Main St, **Fall River**, MA 02720
T: +1 508 6768971; Fax: +1 508 6772966; E-mail: fallriver.lawlib@verizon.net; URL: www.lawlib.state.ma.us/locations.html
Madlyn Correa
31 000 vols; 37 curr per   36842

Inova Fairfax Hospital, Jacob D. Zylman Health Sciences Library, 3300 Gallows Rd, **Falls Church**, VA 22042-3300
T: +1 703 7763234; Fax: +1 703 7763353; E-mail: library@inova.com
1966; Lois H. Culler
9 000 vols; 450 curr per   36843

United States Golf Association Museum & Archives, Golf House, 77 Liberty Corner Rd, **Far Hills**, NJ 07931-2570; P.O. Box 708, Far Hills, NJ 07931
T: +1 908 2342300; Fax: +1 908 4705013; E-mail: library@usga.org; URL: www.usga.org
1936; Nancy Stulack
Rare Bks
30 000 vols; 100 curr per; 52 scrapbks of newspaper clippings   36844

Minnesota Braille & Talking Book Library, 388 SE Sixth Ave, **Faribault**, MN 55021-6340
T: +1 507 3334828; Fax: +1 507 3334832; E-mail: mn.btbl@state.mn.us; URL: education.state.mn.us/mde/index.html
Catherine Durivage
362 000 vols; 326 377 av-mat   36845

Hartford Medical Society, Steiner Library, 236 Farmington Ave, P.O. Box 4003, **Farmington**, CT 06034-4003
E-mail: huntmemorial@aol.com; URL: library.uchc.edu/hms
1889; Ira Spar
Gershom Bulkeley Mss; Hartford Imprints, Hist of Medicine
34 000 vols; 10 curr per; 75 mss
libr loan   36846

University of Connecticut Health Center, Lyman Maynard Stowe Library, 263 Farmington Ave, **Farmington**, CT 06034; P.O. Box 4003, Farmington, CT 06034-4003
T: +1 860 6794053; Fax: +1 860 6794046; E-mail: library@nso.uchc.edu; URL: library.uchc.edu
1965; Evelyn B. Morgen
Hist of Medicine, Learning Resources Ctr Coll in Audiovisuals
151 000 vols; 1 497 curr per; 204 digital data carriers; 5 drawers of pamphlets, 509 computer software programs
libr loan   36847

Holocaust Memorial Center, Morris and Emma Schaver Library-Archive, 28123 Orchard Lake Rd, **Farmington Hills**, MI 48334
T: +1 248 5532400; Fax: +1 248 5532433; E-mail: info@holocaustcenter.org; URL: holocaustcenter.org
1985; Feiga Weiss
Memorial Book Coll; Oral Hist Coll; Audio-Visual Documentaries Coll; Photogr Coll
12 400 vols; 35 curr per; 900 microforms; 60 linear feet of archival mat   36848

Arizona Historical Society Library, Northern Arizona Division, 2340 N Fort Valley Rd, **Flagstaff**, AZ 86001
T: +1 928 7746272; Fax: +1 928 7741596; E-mail: AHSFlagstaff@azhs.gov; URL: www.infomagic.net/~ahsnad
Joseph M. Meehan
50 000 vols   36849

Cross-Cultural Dance Resources Library, 518 S Agassiz St, **Flagstaff**, AZ 86001-5711
T: +1 928 7748108; Fax: +1 928 7748108; E-mail: jwk3@jan.ucc.nau.edu; URL: www.ccdr.org
1981
Eleanor King Coll, Gertrude Prokush Kurath Coll
12 000 vols; 129 curr per; 25 file drawers archives, 265 hours videos   36850

Lowell Observatory Library, 1400 W Mars Hill Rd, **Flagstaff**, AZ 86001
T: +1 928 2333216; E-mail: asb@lowell.edu; URL: www.lowell.edu/Research/library
1894; Antoinette Sansone-Beiser
Arch of Percival Lowell
10 000 vols; 50 curr per   36851

Museum of Northern Arizona-Harold S. Colton Memorial Library, Katharine Bartlett Learning Center, 3101 N Fort Valley Rd, **Flagstaff**, AZ 86001
T: +1 928 7745211; Fax: +1 928 7791527; E-mail: library@mna.mus.az.us; URL: www.musnaz.org
1928; Dr. Edward Evans
Archaeology, Ethnology, Natural hist; Southwestern Archaeology, Navajo & Hopi Indians, Geology of the Colorado Plateau
100 000 vols; 60 curr per; 26 500 pamphlets   36852

US Geological Survey Library, 2255 N Gemini Dr, **Flagstaff**, AZ 86001
T: +1 928 5567272; Fax: +1 928 5567237; E-mail: flag_lib@usgs.gov
1964; Jenny Prennace
Astro-Geology Coll
40 000 vols; 210 curr per   36853

Genesee County Circuit Court, Law Library, County Court House, Ste 204, 900 S Saginaw St, ***Flint***, MI 48502
T: +1 810 2573253; Fax: +1 810 2399280; URL: co.genesee.mi.us
Trea Poe
10 000 vols; 10 curr per 36854

Fond du Lac County Historical Society-Family Heritage Center, Adams House Resource Center, 336 Old Pioneer Rd, ***Fond du Lac***, WI 54935-6126; P.O. Box 1284, Fond du Lac, WI 54936-1284
T: +1 920 9221166; Fax: +1 920 9229099
Sally Albertz
Local hist
10 000 vols 36855

Kaiser-Permanente Medical Center, Health Sciences Library, 9961 Sierra Ave, ***Fontana***, CA 92335
T: +1 909 4275086; Fax: +1 909 4276288
Grace Johnston
Internal mediciine, dermatology
17 000 vols; 96 curr per; 164 e-books 36856

Evans Army Community Hospital, Lane Medical Library, 1650 Cochrane Circle, ***Fort Carson***, CO 80913-4604
T: +1 719 5267286; Fax: +1 719 5267113; E-mail: kathy.parker@amedd.army.mil; URL: evans.amedd.army.mil/lib
1952; Kathy Parker
Consumer Coll
13 815 vols; 220 curr per; 1 800 av-mat; 400 sound-rec; 4 digital data carriers
libr loan; MLA, FEDLINK, OCLC, BCR, NNLM, Southwest Voyager User's Group, Plains and Peaks Libr System, Pikes Peak Libr System 36857

Colorado Division of Wildlife, Research Center Library, 317 W Prospect, ***Fort Collins***, CO 80526-2097
T: +1 970 4724353; Fax: +1 970 4724457; E-mail: jackie.boss@state.co.us; URL: wildlife.state.co.us/Research/Library/
1967; Kay Horton Knudsen
Zoology, Environmental studies; Wildlife (Federal Aid in Fish & Wildlife Restoration, Colorado)
27 500 vols; 200 curr per; 450 diss/theses; 3 400 govt docs; 350 av-mat 36858

National Cancer Institute at Fredrick Scientific Library, Bldg 549, Sultan St, ***Fort Detrick***, MD 21702-8255; P.O. Box B, Frederick, MD 21702-1201
T: +1 301 8465843; Fax: +1 301 8466332; URL: www-library.ncifcrf.gov
1972; Susan W. Wilson
16 000 vols; 700 curr per; 3 000 microforms; 50 sound-rec; 10 digital data carriers 36859

International Swimming Hall of Fame, Henning Library, One Hall of Fame Dr, ***Fort Lauderdale***, FL 33316
T: +1 954 4626536; Fax: +1 954 5224521; E-mail: library@ishof.org; URL: www.ishof.org/library
1965; Preston Levi
Olympic Games Bk Coll, Image Coll, Photo Coll, Swimming & Related Aquatic Sports Coll
9 820 vols; 51 curr per; 42 mss; 70 diss/theses; 710 av-mat; 300 sound-rec
libr loan 36860

US Army Claims Service, JACS-ZXA, 4411 Llewellyn Ave, ***Fort Meade***, MD 20755-5360
T: +1 301 6777009 ext 335
George R Westerbeke
12 000 vols; 15 curr per
Libr is not open to the public 36861

Casemate Museum Library, 20 Bernard Rd, ***Fort Monroe***, VA 23651-1004; P.O. Box 51341, Fort Monroe, VA 23651-0341
T: +1 757 7883935; Fax: +1 757 7883886; URL: www-tradoc.monroe.army.mil/museum/
1976; Paul Morando
Civil War, military hist, peninsula Virginia; Coast Artillery School
12 000 vols; 12 curr per; 3 mss; 2 diss/theses; 1 000 govt docs; 11 music scores; 37 maps; 1 000 microforms; 91 sound-rec; 4 digital data carriers; 55 VCR, 4 reels 8 mm film 36862

Harbor Branch Oceanographic Institution, Inc, Library, 5600 US 1 N, ***Fort Pierce***, FL 34946
T: +1 772 4652400; Fax: +1 772 4652446; URL: www.hboi.edu
1975; Carla Robinson
30 000 vols; 1 000 microforms; 4 digital data carriers 36863

Allen County Law Library Association, Inc, Courthouse, Rm 105, 715 S Calhoun St, ***Fort Wayne***, IN 46802
T: +1 260 4497638; Fax: +1 260 4220791; E-mail: allencountylawlibrary@yahoo.com
1910; Cynthia Ripley
Supreme Ct Reporter, Regional Reporters
20 000 vols; 23 curr per; 15 digital data carriers 36864

Lincoln National Foundation, Lincoln Museum Library, 200 E Berry, P.O. Box 7838, ***Fort Wayne***, IN 46801
T: +1 260 4553031; Fax: +1 260 4556922; URL: TheLincolnMuseum.org
1928; Carolyn Texley
Abraham Lincoln & Contemporaries, broadsides, engravings, lithographs, mss, photogs, sheet music; Civil War, paintings & pamphlets; Richard W Thompson & Others, mss
18 000 vols; 50 curr per; 150 mss; 350 music scores; 10 microforms
libr loan; ALA, OCLC 36865

Parkview Hospital, Ridderheim Health Science Library, 2200 Randallia Dr, ***Fort Wayne***, IN 46805
T: +1 260 4846636 ext 22400; Fax: +1 260 3733692; E-mail: library@parkview.com; URL: www.parkview.com
Shannon Clever
Patient Info Coll
8 000 vols; 499 curr per 36866

Amon Carter Museum Library, 3501 Camp Bowie Blvd, ***Fort Worth***, TX 76107-2695
T: +1 817 9895040; Fax: +1 817 9895079; E-mail: library@cartermuseum.org; URL: www.cartermuseum.org/library
1961; Samuel Duncan
Western Americana, American art, Hist of photography; Eliot Porter Libr & Arch, Laura Gilpin Photogr Libr, M Knoedler Libr, 19th C Newspapers on Microfilm, 19th & Early 20th C American Art, American Illust Bks
50 000 vols; 120 curr per; 100 000 microforms; 7 000 microforms of per 36867

Botanical Research Institute of Texas Library, 500 E Fourth St, ***Fort Worth***, TX 76102
T: +1 817 3324441; Fax: +1 817 3324112; E-mail: info@brit.org
1987; Gary L. Jennings
Botany, Taxonomy (Lloyd Shinners Coll in Systematic Botany)
75 000 vols; 1 100 curr per
libr loan 36868

Harris Methodist Information Resources, 1301 Pennsylvania Ave, ***Fort Worth***, TX 76104
T: +1 817 8822917; Fax: +1 817 8823054; E-mail: coybowen@texashealth.org
Scarlett Burchfield
Surgery, internal medicine
11 000 vols; 397 curr per 36869

John Peter Smith Hospital, Medical Library, 1500 S Main St, ***Fort Worth***, TX 76104
T: +1 817 9213431; Fax: +1 817 9230718; E-mail: medicallibrary@jpshealth.org; URL: www.jpshealth.org
1960; Leslie Herman
20 000 vols; 300 curr per 36870

Kimbell Art Museum Library, 3333 Camp Bowie Blvd, ***Fort Worth***, TX 76107
T: +1 817 3328451; Fax: +1 817 8771264; URL: www.kimbellart.org
1967; Chia-Chun Shih
42 000 vols; 155 curr per; 23 000 microforms 36871

University of North Texas Health Science Center at Fort Worth, Lewis Health Science Library, 3500 Camp Bowie Blvd, ***Fort Worth***, TX 76107-2699
T: +1 817 7352491; Fax: +1 817 7630325; URL: www.library.hsc.unt.edu
1970; Bobby R. Carter
Hist of Osteopathic Medicine, William G Sutherland Coll, Texas Osteopathic Med Assn Arch
74 000 vols; 1 000 e-books; 48 diss/theses; 26 microforms; 3 106 av-mat; 511 sound-rec; 98 digital data carriers; 124 Models, 242 Computer Files
MLA 36872

Danforth Museum of Art, Marks Fine Arts Library, 123 Union Ave, ***Framingham***, MA 01702
T: +1 508 6200050; Fax: +1 508 8725542; URL: www.danforthmuseum.org
Katherine French
10 000 vols 36873

Metrowest Medical Center, Tedeschi Library & Information Center, 115 Lincoln St, ***Framingham***, MA 01702
T: +1 508 3831591; Fax: +1 508 8790471; URL: www.mwmc.com
1960
Framingham Union Hospital Arch
11 250 vols; 228 curr per; 220 av-mat; 250 sound-rec; 4 digital data carriers 36874

Kentucky Department of Public Advocacy Library, 100 Fair Oaks Lane, 3rd Flr, Ste 302, ***Frankfort***, KY 40601
T: +1 502 5648006; Fax: +1 502 5647890; E-mail: dpalibrary@ky.gov; URL: www.dpa.ky.gov
1974; Will Coy-Geeslin
Law; DPA Brief Bank & Motion Files, Kentucky Supreme Court & Court of Appeals Slip Opinions, Training Video Coll
35 000 vols 36875

Kentucky Historical Society Library, 100 W Broadway, ***Frankfort***, KY 40601
T: +1 502 5641792; Fax: +1 502 5644701; URL: www.history.ky.gov
James Kastner
101 000 vols; 197 curr per; 10 000 microforms; 1 000 cubic feet of mss, 125 cubic feet of maps, 75 000 photos 36876

Kentucky Regional Library for the Blind & Physically Handicapped, Kentucky Talking Book Library, 300 Coffee Tree Rd, ***Frankfort***, KY 40601; P.O. Box 537, Frankfort, KY 40602-0537
T: +1 502 5648300; Fax: +1 502 5645773; E-mail: ktbl.mail@ky.gov; URL: www.kdla.ky.gov/libsupport/ktbl/history.htm
1969; Barbara Penegor
Kentucky Coll
141 000 vols; 84 curr per; 132 316 av-mat; 150 000 sound-rec; 1 digital data carriers; 6 000 braille bks
libr loan 36877

The Franklin Mint, Information Research Center, US Rte 1, ***Franklin Center***, PA 19091
T: +1 610 4596294; Fax: +1 610 4597526; E-mail: irc@franklinmint.com; URL: www.franklinmint.com
1969; Cheryl Towne
45 000 vols; 100 curr per 36878

James Monroe Museum & Memorial Library, 908 Charles St, ***Fredericksburg***, VA 22401-5810
T: +1 540 6542113; Fax: +1 540 6541106; URL: www.umw.edu/jamesmonroemuseum/default.php
1927; John N. Pearce
Private Correspondence of James Monroe
10 000 vols; 10 curr per; 10 VF drawers, 27 000 arch items, 50 arch boxes, 15 pieces of artwork 36879

Monmouth County Historical Association Library and Archives, 70 Court St, ***Freehold***, NJ 07728
T: +1 732 4621466; Fax: +1 732 4628346; E-mail: mchalib@excite.com; URL: www.monmouthhistory.org
1898; Carla Z. Tobias
Allaire Papers (Howell works), Battle of Monmouth, Mott Family Papers, North American Phalanx, Philip Freneau, Steamship Coll, Monmouth County & New Jersey hist and genealogy, Pack Brothers Glass Plate Negative Coll
8 870 vols; 2 curr per; 25 000 mss; 100 maps 36880

Rutherford B. Hayes Presidential Center Library, Spiegel Grove, ***Fremont***, OH 43420-2796
T: +1 419 3322081; Fax: +1 419 3324952; E-mail: hayeslib@rbhayes.org; URL: www.rbhayes.org
1911; Rebecca B. Hill
Hist of the United States, 19th Century ; Manuscript Colls ; Ohio & Local Hist; Sandusky, River Valley & the Great Lakes, 19th centry Cookery, 19th c mss coll; Photographs, manuscripts coll
75 000 vols; 221 curr per; 1 000 000 mss; 6 000 microforms; 120 digital data carriers; 75 000 Photogs
libr loan 36881

Department of Veterans Affairs, Medical Center Library, 2615 E Clinton Ave, ***Fresno***, CA 93703-2286
T: +1 209 2285341; Fax: +1 209 2286924
1975; Cynthia Meyer
Reprint file on Coccidioidomycosis, 1880s-1972
13 250 vols; 370 curr per; 6 digital data carriers
libr loan 36882

FOI Services Inc Library, 704 Quince Orchard Rd, Ste 275, ***Gaithersburg***, MD 20878-1751
T: +1 301 9759400; Fax: +1 301 9750702; E-mail: infofoi@foiservices.com; URL: www.foiservices.com
John E. Carey
Pharmaceuticals (regulation, approval)
16 000 vols 36883

Nassau Academy of Medicine, John N. Shell Library, 1200 Stewart Ave, ***Garden City***, NY 11530
T: +1 516 8322300; Fax: +1 516 8328183
1964; Teresa Milone
Consumer Health Coll
10 000 vols; 250 curr per; 1 digital data carriers; 3 VF drawers 36884

Hastings Center, Robert S. Morison Memorial Library, 21 Malcolm Gordon Dr, ***Garrison***, NY 10524-5555
T: +1 845 4244040; Fax: +1 845 4244545; E-mail: mail@thehastingscenter.org; URL: www.thehastingscenter.org
1969
Bioethics, Med ethics, Environmental medicine, Medicine
8 000 vols; 220 curr per 36885

Gateway to the Panhandle, Gateway Library, Box 27, ***Gate***, OK 73844
T: +1 580 9342004
1976; L. E. Maphet
Children's lit, hist; Reader's Digest Coll
8 000 vols 36886

Gateway to the Panhandle Museum Library, Main St, ***Gate***, OK 73844; P.O. Box 27, Gate, OK 73844-0027
T: +1 580 9342004
Ernestine Maphet
Local history
8 000 vols; 1 curr per 36887

Bureau of Jewish Education, Milton Plesur Memorial Library, 2640 N Forest Rd, ***Getzville***, NY 14068
T: +1 716 6898844; Fax: +1 716 6898862; E-mail: Evie@bjebuffalo.org; URL: www.bjebuffalo.org
1928; Mark Cantor Horowitz
Jewish Heritage Video Coll
15 000 vols; 30 curr per 36888

Anthroposophical Society in America, Rudolf Steiner Library, 65 Fern Hill Rd, ***Ghent***, NY 12075
T: +1 518 6727690; Fax: +1 518 6725827; E-mail: rsteinerlibrary@taconic.net; URL: rslibrary.anthroposophy.org
1928; Judith Soleil
27 000 vols; 30 diss/theses; 500 mss, 30 Special Interest Per Sub 36889

Chicago Botanic Garden Library, 1000 Lake-Cook Rd, *Glencoe*, IL 60022
T: +1 847 8358201; Fax: +1 847 8356885; E-mail: library@chicagobotanic.org; URL: www.chicagobotanic.org/library
1950; Leora O. Siegel
25 000 vols; 400 curr per; 10 600 av-mat libr loan   36890

American Heritage Library & Museum, 600 S Central Ave, *Glendale*, CA 91204-2009
T: +1 818 2401775; E-mail: library@srcalifornia.com; URL: www.srcalifornia.com
1893; Richard H. Breithaupt Jr
D.A.R. Lineage Bks, GEC's Complete Peerage
25 000 vols; 2 500 family genealogies, 1 000 Electronic Media & Resources   36891

Glens Falls-Queensbury Historical Association, Chapman Historical Museum Library, 348 Glen St, *Glens Falls*, NY 12801
T: +1 518 7932826; Fax: +1 518 7932831; E-mail: contactus@chapmanmuseum.org; URL: www.chapmanmuseum.org
1967; Timothy Weidner
Seneca Ray Stoddard Coll (1864-1917), photos
30 000 vols; 20 000 photos, 30 000 Archs   36892

Scott Foresman Library, 1900 E Lake Ave, *Glenview*, IL 60025-2086
T: +1 847 4861616; Fax: +1 847 4863948
1942; Judith L. McNulty
Children's lit; Scott, Foresman Archs; Addison-Wesley Archs; Silver Burdett Ginn Archs
50 000 vols; 250 curr per   36893

American Alpine Club Library, 710 Tenth St, Ste 15, *Golden*, CO 80401
T: +1 303 3840110; Fax: +1 303 3840113; E-mail: library@americanalpineclub.org; URL: www.americanalpineclub.org
1902; Bridget Burke
Mountaineering, Glaciology; Himalayan Libr
17 000 vols; 150 curr per; 200 av-mat; 3 digital data carriers; slides, photos
libr loan; SLA   36894

Colorado Railroad Historical Foundation/Colorado Railroad Museum, Robert W. Richardson Railroad Library, 17155 W 44th Ave, P.O. Box 10, *Golden*, CO 80402-0010
T: +1 303 2794591; Fax: +1 303 2794229; E-mail: mail@crrm.org
1958; Kenton Forrest
Files, Office & Operating Records of Rio Grande Southern Railroad; Denver & Rio Grande Western, Colorado & Southern; Denver Tramway; Denver Union Termial
10 750 vols; 300 curr per; 10 000 maps; 20 sound-rec; 900 boxes, 15 file cabinets of mss   36895

Department of Human Services-Youth Corrections, Lookout Mountain Youth Services Center Library, MSCD Lab School Library, 2901 Ford St, *Golden*, CO 80401-1117
T: +1 303 2732767; Fax: +1 303 2732638
Milly Travis
8 000 vols; 15 curr per   36896

National Renewable Energy Laboratory Library, 1617 Cole Blvd, *Golden*, CO 80401-3393
T: +1 303 2754215; Fax: +1 303 2754222; E-mail: library@nrel.gov
1977; Mary Donahue
Alternative fuels, Biomass energy, Energy efficency, Energy policy, Photovoltaic cells, Renewable energy, Solar energy, Wind energy
80 000 vols; 700 curr per; 500 000 microforms
libr loan   36897

Goshen College – Mennonite Historical Library, 1700 S Main, *Goshen*, IN 46526
T: +1 574 5357418; Fax: +1 574 5357438; E-mail: mhl@goshen.edu; URL: www.goshen.edu/mhl

1906; Dr. John D. Roth
JD Hartzler Hymnal Coll; John Horsch Res Coll
65 000 vols; 400 curr per; 100 mss; 700 diss/theses; 3 000 music scores; 200 maps; 12 000 microforms; 250 sound-rec
libr loan; OCLC, INCOLSA   36898

Grand Canyon National Park Library, One Village Loop, *Grand Canyon*, AZ 86023; P.O. Box 129, Grand Canyon, AZ 86023-0129
T: +1 928 6387868; Fax: +1 928 6387797
1930; Betty Upchurch
Museum coll houses archives, photogr and other museum objects
12 000 vols; 25 curr per; 100 diss/theses; 1 000 govt docs; 10 maps; 1 000 microforms; 20 digital data carriers
libr loan; SLA   36899

Stuhr Museum of the Prairie Pioneer, Reynolds Research Center, 3133 W Hwy 34, *Grand Island*, NE 68801-7280
T: +1 308 3855316; Fax: +1 308 3855028; URL: www.stuhrmuseum.org/research/index.htm
1967; Russ Czaplewski
Arthur F. Bentley Coll, Judge Bayard H. Paine Coll
13 000 vols; 12 curr per; 2 100 music scores; 400 maps; 330 microforms; 260 sound-rec; 200 cubic feet & 250 boxes of docs, 22 000 photogs, 28 000 glass plates   36900

American Institute for Economic Research, E.C. Harwood Library, AIER Division St, P.O. Box 1000, *Great Barrington*, MA 01230-1000
T: +1 413 5281216; Fax: +1 413 5280103; E-mail: info@aier.org; URL: www.aier.org
1975; Eric Andelson
AIER's Economic Education Bulletins 1961-present; AIER'S Res Rpts 1934-present; The Annalist, 1923-1940; File of Harwood Papers; Commercial & Financial Chronicle, 1923-1974
15 000 vols; 250 curr per; 5 digital data carriers   36901

Montana School for the Deaf & Blind Library, 3911 Central Ave, *Great Falls*, MT 59405-1697
T: +1 406 7716051; Fax: +1 406 7716164
Staci Bechard
10 000 vols; 3 500 av-mat; 1 200 Braille Titles   36902

Massachusetts Trial Court, Franklin Law Library, Court House, 425 Main St, *Greenfield*, MA 01301
T: +1 413 7726580; Fax: +1 413 7720743; E-mail: franklinlawlib@hotmail.com; URL: www.lawlib.state.ma.us
1812; Howard Polonsky
30 000 vols; 200 curr per   36903

Center for Creative Leadership Library, One Leadership Pl, *Greensboro*, NC 27410; P.O. Box 26300, Greensboro, NC 27438-6300
T: +1 336 2864083; Fax: +1 336 2864087; E-mail: library@leaders.ccl.org; URL: www.ccl.org
Felecia Corbett
Psychology, organizational behavior, management science
8 000 vols; 150 curr per   36904

Westmoreland Museum of American Art, Art Reference Library, 221 N Main St, *Greensburg*, PA 15601-1898
T: +1 412 8371500; Fax: +1 412 8372921; E-mail: info@wmuseumaa.org; URL: www.wmuseumaa.org
1958
10 350 vols; 20 curr per; 25 000 exhibit cat, 10 000 museum exchange bulletins & rpts, 30 000 exhibit brochures   36905

Greenville Hospital System, Health Sciences Library, 701 Grove Rd, *Greenville*, SC 29605
T: +1 864 4557176; Fax: +1 864 4555696
1912; Fay Towell
1 branch libr
25 000 vols; 400 curr per; 2 000 e-journals; 30 e-books; 200 sound-rec   36906

George Junior Republic Library, 200 George Junior Rd, *Grove City*, PA 16127; P.O. Box 1058, Grove City, PA 16127-5058
Mary McKinley
Childrens and Juvenile Literature
20 000 vols; 20 curr per   36907

Motts Military Museum, Library, 5075 S Hamilton, *Groveport*, OH 43125
T: +1 614 8361500; Fax: +1 614 8365110; E-mail: info@mottsmilitarymuseum.org; URL: www.mottsmilitarymuseum.org
1988; Mel Gerhold
Civil War Time Life Series; World War II Time Life Series; Vietnam Time Life Series; Civil War Times; Blue & Gray; Nuremberg Trials
10 000 vols; 10 000 archival mat   36908

Butler County Law Library Association, Ten Journal Sq, Ste 200, *Hamilton*, OH 45011
T: +1 513 8873455; Fax: +1 513 8873696; URL: www.bclawlib.org
1889; Linda D. Hostetler
40 000 vols; 150 curr per; 35 digital data carriers   36909

Calhoun County Library, 109 Second St, *Hampton*, AR 71744; P.O. Box 1162, Hampton, AR 71744-1162
T: +1 870 7984492; Fax: +1 870 7984492; E-mail: calcolib@yahoo.com
Brenda Barfell
30 000 vols   36910

Department of Veterans Affairs Medical Center, Medical Library 142D, *Hampton*, VA 23667
T: +1 757 7229961 ext 3550; Fax: +1 757 7222988; E-mail: bird.j@hampton.va.gov
1870; Jacqueline Bird, Estelle Haskett
12 500 vols; 536 curr per   36911

Mennonite Historians of Eastern Pennsylvania, Mennonite Historical Library & Archives, 565 Yoder Rd, *Harleysville*, PA 19438; P.O. Box 82, Harleysville, PA 19438-0082
T: +1 215 2563020; Fax: +1 215 2563023; E-mail: info@mhep.org; URL: www.mhep.org
1967; Joel D. Alderfer
Franconia Mennonite Mission Board Coll, 1917-1971; Local Hist Coll, Mennonite Church Hist Colls
8 000 vols; 125 curr per   36912

US National Park Service, Effigy Mounds National Monument Library, 151 Hwy 76, *Harpers Ferry*, IA 52146-7519
T: +1 563 8733491; Fax: +1 563 8733743; E-mail: efmo_administration@nps.gov
Florenceia M. Wiles
Natural hist; Ellison Orr, mss
10 000 vols   36913

Commonwealth Court Library, 603 Irvis Office Bldg, Commonwealth & Walnut Aves, *Harrisburg*, PA 17120
T: +1 717 2551615; Fax: +1 717 2551784; URL: www.aopc.org/index/cwealth/indexcwealth.asp
1970; Mary Mills
20 000 vols   36914

Pennsylvania Department of Transportation, Library & Research Center, Commonwealth Keystone Bldg, 6th Flr, 400 North St, *Harrisburg*, PA 17120-0041; P.O. Box 3054, Harrisburg, PA 17105-3054
T: +1 717 7051546; Fax: +1 717 7839152; E-mail: penndot_library@state.pa.us; URL: www.dot.state.pa.us
1979
Transportation Research Board Series, Management and Supervisory, Personal Computer Skills, Self Improvement
24 000 vols; 240 curr per; 1 500 av-mat; 40 digital data carriers
SLA   36915

Pennsylvania Historical & Museum Commission, Reference Library, 400 North St, Plaza Level, *Harrisburg*, PA 17120-0053
T: +1 717 7839898; Fax: +1 717 2142989
1947; Sally Biel
22 000 vols; 100 curr per   36916

Connecticut Historical Society Library, One Elizabeth St, *Hartford*, CT 06105
T: +1 860 2365621; Fax: +1 860 2362664; E-mail: libchs@chs.org; URL: www.chs.org
1825; Nancy Milnor
Local hist, New England, Genealogy, Children's lit, Religion; Connecticut Imprints, Hist Mss, Juvenile (Bates & Caroline Hewins Coll); Sermons
100 000 vols; 200 curr per; 500 maps; 10 digital data carriers; 3 mio mss, 10 000 mss cards   36917

Harriet Beecher Stowe Center Library, 77 Forest St, *Hartford*, CT 06105-3296
T: +1 860 5229258; Fax: +1 860 5229259; E-mail: info@stowecenter.org; URL: www.harrietbeecherstowe.org
1965; Katherine Kane
African-American hist, 19th C American decorative arts, Women's hist; Lyman Beecher Family Coll, William H. Gillette Coll, Harriet Beecher Stowe Coll, Uncle Tom's Cabin Coll, Warner Family Coll, Hartford Architecture Conservancy Coll
15 000 vols; 40 curr per; 150 microfilms; 180 000 mss, 3 500 slides   36918

Hartford Hospital, Health Science Libraries, 80 Seymour St Conklin Bldg-3, P.O. Box 5037, *Hartford*, CT 06102-5037
T: +1 860 5452971; Fax: +1 860 5452572; E-mail: library@harthosp.org; URL: www.harthosp.org/library
1855; Shirley Gronholm
Hist of Nursing (Foley Coll), Consumer Health, Integrative Principles, Flight Nursing, Psychiatry; 1 branch libr
30 875 vols; 582 curr per; 728 av-mat   36919

Hartford Hospital – Institute of Living, Medical Library, 200 Retreat Ave, *Hartford*, CT 06106
T: +1 860 5457282; Fax: +1 860 5457275; URL: www.harthosp.org
1932
Hist of Psychiatry (Zilboorg & Norman Colls), Psychoanalysis (Jelliffe Coll)
15 000 vols; 150 curr per; 16 av-mat   36920

Saint Francis Hospital & Medical Center, Health Sciences Library, 114 Woodland St, *Hartford*, CT 06105
T: +1 860 7144773; Fax: +1 860 7148022; E-mail: library@stfranciscare.org
1976; Joseph M. Pallis
Hospital Arch
12 000 vols; 600 curr per; 12 digital data carriers   36921

Wadsworth Atheneum, Auerbach Art Library, 600 Main St, *Hartford*, CT 06103
T: +1 860 2782670; Fax: +1 860 5270803; URL: www.wadsworthatheneum.org
1934; John Teahan
Fine arts, Decorative arts, Costume design, Photography, Museology; Art (Watkinson Coll), Bookplates (Baker Coll)
44 000 vols; 90 curr per   36922

Fruitlands Museums Library, 102 Prospect Hill Rd, *Harvard*, MA 01451
T: +1 978 4563924; Fax: +1 978 4568078; URL: www.fruitlands.org
1914
American Indians Coll, 19th c American Paintings, Transcendentalism, Shaker Hist; Shaker & Transcendentalist Mss, American Hist & Lit Coll
11 000 vols; 15 curr per; 200 reports   36923

South Suburban Genealogical & Historical Society, Research Library, 3000 W 170th Pl, *Hazel Crest*, IL 60429-1174
T: +1 708 3353340; E-mail: ssghs@usa.net; URL: www.ssghs.org/sslib.htm
1972
Coll of Pullman car works employment records, Bishop family records, Calumet city naturalizations, Cone research records, Eddy family resarch records
8 000 vols; 500 curr per; 300 maps; 11 av-mat; 147 digital data carriers; 9 drawers fiche, 455 microfilms, pamphlet file, wills and bibles, cemetery readings etc   36924

Montana Historical Society, Research Center, 225 N Roberts St, ***Helena***, MT 59601-4514; P.O. Box 201201, Helena, MT 59620-1201
T: +1 406 4444739; Fax: +1 406 4445297; E-mail: mhslibrary@mt.gov; URL: montanahistoricalsociety.org
1865; Molly Kruckenberg
Cattle Industry (Huffmann Coll), photos; Genealogy (Daughters of American Revolution Coll); George Armstrong Custer (Edgar I Stewart Coll); Montana Newspapers; Montana State Arch; Range Cattle Industry, 1860-1945 (Teakle Coll); Yellowstone Park, Pacific Northwest & North Plains photos (Haynes Coll); 20th C Homesteading Photogr (Cameron Coll)
56 000 vols; 385 curr per; 55 000 govt docs; 4 000 music scores; 18 000 maps; 18 000 microforms; 3 500 sound-rec; 1 digital data carriers; 13 000 cubic feet of mss & rec, 1 000 slides, 120 art reproductions
libr loan 36925

Helen Keller Services for the Blind, Braille Library, One Helen Keller Way, ***Hempstead***, NY 11550
T: +1 516 4851234 ext 243; Fax: +1 516 5386785; E-mail: info@helenkeller.org; URL: www.helenkeller.org
1956
10 000 vols 36926

Airline Pilots Association International, Engineering & Air Safety Resource Center, 535 Herndon Pkwy, ***Herndon***, VA 20172-5226; P.O. Box 1169, Herndon, VA 20172-1169
T: +1 703 6894204; Fax: +1 703 4642104; E-mail: easlibrary@alpa.org; URL: www.alpa.org
Marvin Ramirez
Federal Aviation Regulations
60 000 vols 36927

Akwesasne Library & Culture Center, 321 State Rte 37, ***Hogansburg***, NY 13655
T: +1 518 3582240; Fax: +1 518 3582649; E-mail: akwlibr@northnet.org; URL: www.nc3r.org/akwlibr
1971; Valerie Garrow
American Indian Coll
29 000 vols; 51 curr per; 7 digital data carriers 36928

Himalayan International Institute of Yoga Science and Philosophy of the USA, Library, 952 Bethany Turnpik, ***Honesdale***, PA 18431
T: +1 570 2535551; Fax: +1 570 2539078; E-mail: info@himalayaninstitute.org; URL: www.himalayaninstitute.org
10 000 vols 36929

Bernice P. Bishop Museum Library & Archives, 1525 Bernice St, ***Honolulu***, HI 96817
T: +1 808 8484148; Fax: +1 808 8478241; E-mail: library@bishopmuseum.org; URL: www.bishopmuseum.org/research/library/libarch.html
1889; Duane Wenzel
Anthropology, Archaeology, Entomology, Botany, Malacology, Marine biology, Zoology, Hawaiiana, Exploration, Hist, Linguistics, Geology; Carter Coll of Hawaiana; Hawaiian Language Newspapers; Japanese Hawaii Imprints; Fuller Coll of Pacificana; Pacific Island Languages
115 000 vols; 1 100 curr per; 3 000 mss; 3 000 microforms; 90 av-mat; 15 000 sound-rec; 17 000 Pamphlets
libr loan; OCLC, CARL 36930

City & County of Honolulu, Department of Customer Services, Municipal Reference Library, City Hall Annex, 558 S King St, ***Honolulu***, HI 96813-3006
T: +1 808 7683759; Fax: +1 808 5234985; E-mail: library@honolulu.gov; URL: www.honolulu.gov/csd/lrmb/index.htm
1929; N.N.
Local govt, Transportation, Recreation; Publs of the City & County of Honolulu
32 000 vols; 115 curr per 36931

East-West Center, Research Information Services, 1601 East-West Rd, ***Honolulu***, HI 96848-1601
T: +1 808 9447345; Fax: +1 808 9447600; E-mail: ris@eastwestcenter.org; URL: www.eastwestcenter.org
1971; Phyllis Tabusa
Cultural change, Int politics & economics, Population studies, Energy economics, Environmental policy; Asian & Pacific Census, US Bureau of the Census, World Fertility Survey
66 000 vols; 40 curr per; 24 000 unpubl docs & reprints 36932

Hawaii State Archives, Iolani Palace Grounds, 364 S King St, ***Honolulu***, HI 96813
T: +1 808 5860329; Fax: +1 808 5860330; E-mail: archives@hawaii.gov; URL: www.hawaii.gov/dags/archives
Gina S. Vergara-Bautista
25 000 vols 36933

Hawaiian Historical Society Library, 560 Kawaiahao St, ***Honolulu***, HI 96813
T: +1 808 5376271; Fax: +1 808 5376271; E-mail: hhsbarb@lava.net; URL: www.hawaiianhistory.org/lib/libmain.html
1892; Barbara E. Dunn
Hist of Hawaii & Pacific Coll (late 18th & 19th c)
14 000 vols; 28 curr per 36934

Hawaiian Mission Children's Society Library, Mission Houses Museum Library, 553 S King St, ***Honolulu***, HI 96813
T: +1 808 5310481; Fax: +1 808 5452280; E-mail: info@missionhouses.org; URL: www.missionhouses.org
1908; Judith A. Kearney
19th C Hawaiiana, Hawaiian Language Publs 19th C
13 000 vols; 10 curr per; 100 microforms; 245 linear feet of mss 36935

Honolulu Academy of Arts, Robert C. Allerton Library, 900 S Beretania St, ***Honolulu***, HI 96814-1495
T: +1 808 5328755; Fax: +1 808 5323683; URL: www.honoluluacademy.org
1927; Dr. Ronald F. Chapman
Art Hist, esp Chinese & Japanese Art
45 000 vols; 45 curr per; 15 microforms; 4 000 pamphlets, 40 VF drawers of clippings 36936

National Marine Fisheries Service, Honolulu Laboratory Library, 2570 Dole St, ***Honolulu***, HI 96822-2396
T: +1 808 9835307; Fax: +1 808 9832902; E-mail: ani.au@noaa.gov
Ani D. Au
10 000 vols; 60 curr per; 3 digital data carriers
libr loan 36937

New Hampshire Antiquarian Society Library, 300 Main St, ***Hopkinton***, NH 03229
T: +1 603 7463825; E-mail: nhas@tds.net; URL: nhantiquarian.org
1859; Heather Mitchell
Music Bks, Hopkinton Town Rpts, Primitive Portraits, Costume Coll
14 000 vols 36938

Art Institute of Houston, Library, 1900 Yorktown, ***Houston***, TX 77056
T: +1 713 6232040; Fax: +1 713 9662700; URL: www.aih.aii.edu
Mary Sommerfeld
Applied art; Culinary Coll
30 000 vols; 200 curr per 36939

Hirsch Library, Museum of Fine Arts, Houston, 1001 Bissonnet, ***Houston***, TX 77005-1803; P.O. Box 6826, Houston, TX 77265-6826
T: +1 713 6397325; Fax: +1 713 6397707; E-mail: hirsch@mfah.org; URL: www.mfah.org/library
Jon Evans
Bayou Bend Libr; Museum Arch; Texana
110 000 vols; 300 curr per; 700 microforms; 12 digital data carriers; 165 VF drawers of exhibition cat, 25 000 auction cat, 100 000 slides 36940

Houston Academy of Medicine-Texas Medical Center Library, 1133 John Freeman Blvd, ***Houston***, TX 77030
T: +1 713 7954200; Fax: +1 713 7907052; E-mail: webmaster@exch.library.tmc.edu; URL: resource.library.tmc.edu
1949; Dr. Elizabeth K. Eaton
Hist of Medicine (McGovern Coll)
358 000 vols; 9 201 curr per; 6 510 e-journals; 5 000 e-books; 715 sound-rec; 200 digital data carriers
libr loan 36941

Houston Museum of Natural Science Library, One Hermann Circle Dr, ***Houston***, TX 77030-1799
T: +1 713 6394670; Fax: +1 713 6394767; URL: www.hmns.org
1969; Lisa Rebori
Malacology Coll
8 000 vols; 80 curr per 36942

Howrey LLP Library, 1111 Louisiana St, 25th Fl, ***Houston***, TX 77002
T: +1 713 7871543; Fax: +1 713 7871440; URL: www.howrey.com
Caralinn Cole
Patents, Trademarks, Copyrights
9 000 vols; 100 curr per 36943

Kellogg, Brown & Root Library, 601 Jefferson Ave, ***Houston***, TX 77002
T: +1 713 7538466; Fax: +1 713 7536226
John Galloway
Petroleum engineering, Petrochemistry, Chemistry, Technology, Business
20 000 vols; 200 curr per 36944

Lunar & Planetary Institute, Center for Information & Research Services, 3600 Bay Area Blvd, ***Houston***, TX 77058-1113
T: +1 281 4862136; Fax: +1 281 4862186; E-mail: library@lpi.usra.edu; URL: www.lpi.usra.edu/library/library.html
1970; Mary Ann Hager
Spacecraft photos & Maps Coll of the Moon, Mercury, Venus, Earth, Mars, Jupiter, Saturn, Uranus and Neptune; Regional Planetary Image Facilities
55 000 vols; 140 curr per; 4 000 maps; 20 000 microforms; 1 000 digital data carriers; 7 000 Slides
libr loan; OCLC 36945

Menil Foundation, The Menil Collection Library, 1500 Branard St, ***Houston***, TX 77006; mail address: 1511 Branard St, Houston, TX 77006
T: +1 713 5259426; Fax: +1 713 5259444
Phillip Thor Heagy
Rare Book room
20 000 vols; 100 curr per; 10 000 Art auction sales cat
libr loan; ARLIS/NA 36946

Huntington Museum of Art Library, 2033 McCoy Rd, ***Huntington***, WV 25701
T: +1 304 5292701; Fax: +1 304 5297447; E-mail: hma1@hmoa.org; URL: www.hmoa.org
Christopher Hatten
Fine Arts Coll, Fire Arms Hist & Manufacture Coll, Fastoria Glass Cat & Price Lists 1897-1980, Glass Hist & Technology Coll
21 000 vols; 55 curr per; 4 000 slides 36947

Boeing Co, Library & Information Services, 5301 Bolsa Ave, H010-B001, ***Huntington Beach***, CA 92647-2099
T: +1 714 8962319; Fax: +1 714 8961737
Sue Brewsaugh
11 000 vols; 176 curr per 36948

International Association of Educators for World Peace, Research Center for Intercultural Information Library, 2013 Orba Dr NE, ***Huntsville***, AL 35811-2414
T: +1 256 5345501; Fax: +1 256 5361018; E-mail: mercieca@knology.net; URL: www.earthportals.com/Portal_Messenger/mercieca.html
1969; Dr. Charles Mercieca
Disarmament, Education, Environ protection, Government hist, Human rights, Philosophy, Psychology, Securities, Sociology, Technology, United Nations
8 000 vols; 15 curr per 36949

Culinary Institute of America, Conrad N. Hilton Library, 1946 Campus Dr, ***Hyde Park***, NY 12538-1499
T: +1 845 4511322; Fax: +1 845 4511092; URL: www.ciachef.edu
1973; Eileen De Vries
Menus
71 000 vols; 280 curr per 36950

National Archives & Records Administration, Franklin D. Roosevelt Presidential Library, 4079 Albany Post Rd, ***Hyde Park***, NY 12538
T: +1 845 4861142; Fax: +1 845 4861147; E-mail: roosevelt.library@nara.gov; URL: www.fdrlibrary.marist.edu
1941; Cynthia M. Koch
American Hist; Early Juveniles Coll, Hudson River Valley Hist Coll, US Naval Hist Coll
47 000 vols; 12 curr per; 800 maps; 3 000 microforms; 17 million mss pages 36951

Idaho National Laboratory, INL Technical Library, 1765 N Yellowstone Hwy, ***Idaho Falls***, ID 83415-2300; P.O. Box 1625, Idaho Falls, ID 83415-2300
T: +1 208 5261185; Fax: +1 208 5261697; URL: www.inl.gov
1960; Catherine Plowman
Nuclear sciences
47 000 vols; 400 curr per; 2 000 e-books; 350 digital data carriers; 113 000 rpts 36952

Historical & Genealogical Society of Indiana County, 621 Wayne Ave, ***Indiana***, PA 15701-3042; Silas M Clark House, 200 S 6th St, Indiana, PA 15701
T: +1 724 4639600; Fax: +1 724 4639899; E-mail: ichs@ptd.net; URL: www.rootsweb.com/~paicgs
1939; Coleen Chambers
Frances Strong Helman Coll, Cecil Smith Coll
8 000 vols; 150 vols of newspapers on microfilm 36953

American Legion National Headquarters Library, 700 N Pennsylvania St, 4th Flr, ***Indianapolis***, IN 46204-1172; P.O. Box 1055, Indianapolis, IN 46206-1055
T: +1 317 6301366; Fax: +1 317 6301241; E-mail: library@legion.org; URL: www.legion.org
1923; Howard Trace
National defense, patriotism, veterans' affairs, children and youth; World War I & II Posters; Founding Fathers Exhibit; Arch of the American Legion
10 000 vols; 250 curr per; 20 digital data carriers; 1 140 VF drawers of pamphlets, mss, reports 36954

Dow Agrosciences, Information Management Center, 9330 Zionsville Rd, ***Indianapolis***, IN 46268
T: +1 317 3373517; Fax: +1 317 3373245
Margaret B. Hentz
Agronomy, organic chemistry
8 000 vols; 150 curr per 36955

Indiana Academy of Science, John Shepard Wright Memorial Library, Indiana State Library, 140 N Senate Ave, ***Indianapolis***, IN 46204-2296
T: +1 317 2323686; Fax: +1 317 2323728; URL: www.indianaacademyofscience.org
1885; Doug Conrads
Natural hist
13 000 vols 36956

Indiana Historical Society, William Henry Smith Memorial Library, 450 W Ohio St, ***Indianapolis***, IN 46202-3269
T: +1 317 2320321; Fax: +1 317 2340168; E-mail: shahn@indianahistory.org; URL: www.indianahistory.org
1934; Suzanne Hahn
Civil War, Indiana; Agricultural Hist, Architectural history, black history, charitable organizations, 19th c Indiana politics, Midwestern railroads, documentary photography
76 920 vols; 114 curr per; 4 500 mss; 1 051 maps; 1 659 microforms; 618 av-mat; 25 digital data carriers
ALA, AASLH, SAA 36957

Indiana State Archives, 6440 E 30th St, ***Indianapolis***, IN 46219
T: +1 317 5915222; Fax: +1 317 5915324; E-mail: arc@icpr.in.gov; URL: www.in.gov/icpr
Alan January
State history; Official Indiana State Records (1790-present), Indian Public Land Records, Ku Klux Klan
23 000 vols; 87 curr per 36958

Indianapolis Museum of Art, Stout Reference Library, 4000 Michigan Rd, ***Indianapolis***, IN 46208-3326
T: +1 317 9202647; Fax: +1 317 9268931; E-mail: ukolmstetter@ima-art.org; URL: web.ima-art.org/
1908; Ursula Kolmstetter
Prehistoric-present, Asian art, Decorative arts, Ethnography; Coll on Rugs & Carpets, Turner Coll
47 000 vols; 191 curr per; 150 000 slides, 30 000 artists' files
libr loan; ARLIS/NA, ALA, SLA 36959

West Virginia Division of Rehabilitation Services Library, West Virginia Department of Education & the Arts, Staff Library, West Virginia Rehabilitation Ctr, 1004 Barron Dr, ***Institute***, WV 25112; P.O. Box 1004, Institute, WV 25112-1004
T: +1 304 7664644; Fax: +1 304 7664913; URL: library.wvdrs.org
1968; Carol R. Johnson-Cyrus
Behavioral sciences, Social sciences, Medicine
16 000 vols; 13 curr per 36960

Interlochen Center for the Arts, John W. & Charlene B. Seabury Academic Library, 4000 M-137, ***Interlochen***, MI 49643; P.O. Box 199, Interlochen, MI 49643
T: +1 231 2767420; Fax: +1 231 2765232; URL: www.interlochen.org
1962; Evelyn R. Weliver
Music Libr Coll
27 000 vols; 80 curr per 36961

Act Information Resource Center, 200 ACT Dr, ***Iowa City***, IA 52243; P.O. Box 168, Iowa City, IA 52243-0168
T: +1 319 3371166; Fax: +1 319 3371538; E-mail: snider@act.org
1968; Jacqueline Snider
Education; ERIC Coll
30 000 vols; 600 curr per; 300 000 microforms; 4 digital data carriers 36962

State Historical Society of Iowa, 402 Iowa Ave, ***Iowa City***, IA 52240-1806
T: +1 319 3353916; Fax: +1 319 3353935; URL: www.iowahistory.org
1857; Carol Kirsch
Iowa Hist; Map Coll
153 000 vols; 575 curr per; 3 000 maps; 24 700 microforms; 4 000 linear feet of manuscripts
libr loan; AASLH 36963

Lawrence County Bar & Law Library Association, 111 South 4th St, ***Ironton***, OH 45638-1586
T: +1 740 5330582; Fax: +1 740 5331084
1911
15 000 vols 36964

National Safety Council Library, 1121 Spring Lake Dr, ***Itasca***, IL 60143
T: +1 630 2851121 ext 2199; Fax: +1 630 2850765; E-mail: library@nsc.org; URL: www.nsc.org
1915; Robert J. Marecek
Accident prevention, Occupational safety, Traffic safety; National Safety Council Arch
165 000 vols; 100 curr per; 5 700 microforms; 2 digital data carriers
libr loan; SLA 36965

Center for Religion, Ethics & Social Policy, Alternatives Library (Anne Carry Durland Memorial), Cornell University, 127 Anabel Taylor Hall, Rm 127, ***Ithaca***, NY 14853-1001
T: +1 607 2556486; Fax: +1 607 2559985; E-mail: alt-lib@cornell.edu; URL: www.alternativeslibrary.org
Lynn Andersen
Ecology, human rights, psychology; African Cinema and Literature, Native American Archives
8 000 vols; 320 curr per 36966

Paleontological Research Institution Library, 1259 Trumansburg Rd, ***Ithaca***, NY 14850
T: +1 607 2736623; Fax: +1 607 2736620; URL: www.priweb.org
1932; Dr. Warren Allmon
48 VF cabinets, 6 000 reprints & papers
60 000 vols; 13 curr per; 550 maps; 61 microforms 36967

Mississippi Museum of Art, Howorth Library, 201 E Pascagoula, ***Jackson***, MS 39201
T: +1 601 9601515; Fax: +1 601 9601505; URL: www.msmuseumart.org
45 000 vols 36968

Mississippi Museum of Natural Science Library, 2148 Riverside Dr, ***Jackson***, MS 39202
T: +1 601 3547303; Fax: +1 601 3547227; E-mail: library@mmns.state.ms.us; URL: www.mdwfp.com/museum
1974; Mary P. Stevens
18 000 vols; 104 curr per 36969

Mississippi Supreme Court, State Law Library, Carroll Gartin Justice Bldg, 450 High St, ***Jackson***, MS 39201; P.O. Box 1040, Jackson, MS 39215-1040
T: +1 601 3593672; Fax: +1 601 3592912; URL: www.mssc.state.ms.us
1838; Charlie Pearce
225 000 vols; 267 curr per 36970

University of Mississippi Medical Center, Rowland Medical Library, 2500 N State St, ***Jackson***, MS 39216-4505
T: +1 601 9841290; Fax: +1 601 9841251; URL: www.library.umc.edu
1955; Ada Seltzer
Hist of Health Sciences; Mississippi Health Sciences Info Network
73 000 vols; 2 202 curr per; 2 000 microforms; 74 digital data carriers
libr loan; MLA, SLA, AMIA 36971

Cummer Museum of Art Library, 829 Riverside Ave, ***Jacksonville***, FL 32204
T: +1 904 3566857; Fax: +1 904 3534101; URL: www.cummer.org
1961
Art hist; European Porcelains, esp Meissen
15 000 vols; 43 curr per; 10 000 art slides 36972

Mount Sinai Services-Queens Hospital Center Affiliation, Health Sciences Library, 82-68 164th St, ***Jamaica***, NY 11432
T: +1 718 8834021; Fax: +1 718 8836125
1960; Ruth Hoffenberg
18 000 vols; 425 curr per
libr loan 36973

Anne Carlsen Learning Center, 701 Third St NW, ***Jamestown***, ND 58401-2971
T: +1 701 9525169; Fax: +1 701 9525154; URL: www.annecenter.org
Mark Coppin
10 000 vols; 23 curr per 36974

Ashtabula County Law Library Association, County Courthouse, 25 W Jefferson St, ***Jefferson***, OH 44047
T: +1 440 5763690; Fax: +1 440 5765106
Vickie Lee Brown
40 000 vols; 10 curr per 36975

Missouri State Library, Wolfner Library for the Blind & Physically Handicapped, 600 W Main St, ***Jefferson City***, MO 65101-1532; P.O. Box 387, Jefferson City, MO 65102-0387
T: +1 573 7518720; Fax: +1 573 5262985; E-mail: wolfner@sos.mo.gov; URL: www.sos.mo.gov/wolfner
1924; Dr. Richard J. Smith
360 000 vols; 100 curr per; 76 140 av-mat; 403 000 sound-rec; 10 digital data carriers; 3 020 Large Print Bks, 41 830 Braille Vols 36976

National Park Service, Dinosaur National Monument Library, PO Box 128, ***Jensen***, UT 84035-0128
T: +1 435 7817700; Fax: +1 435 7811739
1960
Archaeology, Geology, Local hist, Zoology; Paleontology (Theodore White Coll)
15 000 vols; 10 curr per 36977

Alaska State Library – Alaska Historical Collections, 333 Willoughby Ave, ***Juneau***, AK 99801; P.O. Box 110571, Juneau, AK 99811-0571
T: +1 907 4652925; Fax: +1 907 4652990; E-mail: asl.historical@alaska.gov; URL: www.eed.state.ak.us/lam
1900; Gladi Kulp
Alaskana (Wickersham Coll); Alaska-Arctic Research; Alaska Juneau Mining Company Records, Alaska Packers Association Records, Marine Hist (L H Bayers), doc; Russian American Coll; Russian Hist-General and Military, (Dolgopolov Coll); Salmon Canneries, (Alaska Packers Assn Records); Trans-Alaska Pipeline Impact; Winter & Pond Photogr Coll; Vinokouroff Coll
37 000 vols; 400 mss; 2 000 maps; 6 000 microforms 36978

Alaska State Library – Archives & Museums, 333 Willoughby Ave, State Office Bldg, 8th Flr, ***Juneau***, AK 99801; P.O. Box 110571, Juneau, AK 99811-0571
T: +1 907 4652988; Fax: +1 907 4652151; E-mail: aslanc@alaska.gov; URL: www.library.state.ak.us
1957; Linda Thibodeau
Education, Alaska Hist, Libr & info science, State government; Alaska Hist (Wickersham Coll of Alaskana, Alaska Marine Hist (L H Bayers Coll), Salmon Canneries, Trans-Alaska Pipeline Impact; 3 branch libs
110 000 vols; 350 curr per; 390 av-mat; 500 sound-rec 36979

National Marine Fisheries Service, Auke Bay Laboratory Library, 11305 Glacier Hwy, ***Juneau***, AK 99801-8626
T: +1 907 7896009; Fax: +1 907 7896094; E-mail: paula.johnson@nona.gov
1960
Ecology, Fisheries
20 000 vols; 84 curr per 36980

National Tropical Botanical Garden Library, 3530 Papalina Rd, ***Kalaheo***, HI 96741
T: +1 808 3327324; Fax: +1 808 3329765; URL: www.ntbg.org/resources/library.php
1971; David H. Lorence
Botany, Horticulture; Tropical Botany Coll
17 000 vols; 400 curr per; 200 maps; 10 digital data carriers; 100 charts, 2 000 art reproductions, 10 500 slides
libr loan 36981

Borgess Library, 1521 Gull Rd, ***Kalamazoo***, MI 49048-1666
T: +1 616 2267360; Fax: +1 616 2266881; E-mail: jbarlow@borgess.com
1946; Jennifer Barlow
Patient Resource Room
8 000 vols; 400 curr per; 400 av-mat; 30 sound-rec; 20 digital data carriers 36982

Kalamazoo Institute of Arts, Mary & Edwin Meader Fine Arts Library, 314 S Park St, ***Kalamazoo***, MI 49007-5102
T: +1 269 3497775; Fax: +1 269 3499313; URL: www.kiarts.org/library
1956; Dennis Kreps
Weavers Guild of Kalamazoo Coll, German Expressionist Prints
11 000 vols; 52 curr per; 100 av-mat; 48 VF drawers, 10 000 slides 36983

W. E. Upjohn Institute for Employment Research Library, 300 S Westnedge Ave, ***Kalamazoo***, MI 49007-4686
T: +1 269 3435541; Fax: +1 269 3433308; URL: www.upjohninstitute.org
1960; Linda S. Richer
Employment, Unemployment, Labor economy
15 000 vols; 250 curr per 36984

Kansas City Art Institute Library, Jannes Library, 4438 Warwick Blvd, ***Kansas City***, MO 64111; mail address: 4415 Warwick Blvd, Kansas City, MO 64111-1874
T: +1 816 8023390; Fax: +1 816 8023338; E-mail: library@kcai.edu; URL: library.kcai.edu
1885; M.J. Poehler
Fine arts; Artists' Bks, Downing Rare Bks
30 000 vols; 110 curr per; 6 300 e-books; 14 digital data carriers 36985

Kansas University Medical Center, Clendening History of Medicine Library, 1020-1030 Robinson Bldg, 3901 Rainbow Blvd, ***Kansas City***, KS 66160-7311; University of Kansas Medical Center, MS 1024, 3901 Rainbow Blvd, Kansas City, KS 66160
T: +1 913 5887244; Fax: +1 913 5887060; E-mail: clendening@kumc.edu; URL: clendening.kumc.edu
1945; Dr. Christopher Crenner
Florence Nightingale & Joseph Lister Letters, Rudolph Virchow Mss, Samuel Crumbine Papers
26 000 vols; 40 curr per 36986

Linda Hall Library, 5109 Cherry St, ***Kansas City***, MO 64110
T: +1 816 3634600; Fax: +1 816 9268785; E-mail: requests@lindahall.org; URL: www.lindahall.org
1946; Brig McCoy
Mathematics, Astronomy, Physics, Chemistry, Geology, Biology, Pharmacy, Pharmacology, Engineering, Agriculture, Botany, Zoology; Hist of Science; NASA & DOE Tech Rpts; Sci-Tech Conference Procs; Soviet & European Sci-Tech Publs; Standards & Specifications; US Patent & Trademark Specifications, maps, govt docs
1 264 000 vols; 10 426 curr per; 37 incunabula; 1 100 000 govt docs; 66 000 maps; 1 200 000 microforms; 6 612 av-mat; 448 digital data carriers; 44 700 serial titles, 1 450 000 microforms of per
libr loan; ALA, IRLA, IATUL, SLA, MLA, OCLC, IFLA 36987

NAIC Research Library, 2301 McGee, Ste 800, ***Kansas City***, MO 64108
T: +1 816 7838250; Fax: +1 816 7838269; E-mail: reslib@naic.org; URL: www.naic.org
1871; Deborah K. Scott
NAIC Proceedings, 1871 to present
13 000 vols; 275 curr per 36988

Nelson-Atkins Museum of Art, Spencer Art Reference Library, 4525 Oak St, ***Kansas City***, MO 64111-1873
T: +1 816 7511216; Fax: +1 816 7510498; URL: www.nelson-atkins.org
1933; Marilyn Carbonell
Art hist, Asian art, Decorative arts, European art; Decorative Arts (per), Oriental Art, Prints & Drawings (The John H. Bender Libr), Henry Moore Photogr Arch & Study Ctr
152 000 vols; 625 curr per; 13 digital data carriers; 27 000 auction cat, 60 VF drawers 36989

Saint Luke's Hospital, Health Sciences Library, 4400 Wornall Rd, ***Kansas City***, MO 64111
T: +1 816 9322333; Fax: +1 816 9325197; E-mail: library@saint-lukes.org; URL: www.saint-lukes.org
1948
17 000 vols; 600 curr per
libr loan 36990

University of Kansas Medical Center, Archie Dykes Library of the Health Sciences, 2100 W 39th Ave, ***Kansas City***, KS 66160-7180; mail address: 3901 Rainbow Blvd, Mail Stop 1050, Kansas City, KS 66160
T: +1 913 5887300; Fax: +1 913 5887304; E-mail: dykesref@kumc.edu; URL: www.library.kumc.edu
1906; Karen Cole
SCAN microfiche, Calkins Educational Resource Ctr
62 000 vols; 1 000 curr per; 12 000 e-journals; 3 000 microforms; 21 digital data carriers; 52 177 monograph vols
libr loan 36991

Historical Society of Cheshire County, Wright Room Research Library, 246 Main St, P.O. Box 803, ***Keene***, NH 03431
T: +1 603 3521895; Fax: +1 603 3529226; E-mail: hscc@hsccnh.org; URL: www.hsccnh.org/library/default.cfm
1927
Cheshire County Hist, bks, mss, photos; New England Genealogy & Local Hist Coll
15 curr per; 3 000 vols on New England towns, counties, and family histories; 300 000 items of regional historical interest; business records; 11 500 Cheshire County photos; Beveridge Coll

(20 000 page computer printout of source material on 15 000 Cheshire County people and businesses   36992

The New England Electric Railway Historical Society, Seashore Trolley Museum Library, 195 Log Cabin Rd, *Kennebunkport*, ME 04046; P.O. Box A, Kennebunkport, ME 04046
T: +1 207 9672712; Fax: +1 207 9670867; E-mail: seashorelibrary@ramsdell.com; URL: www.trolleymuseum.org
Edward L. Ramsdell
10 000 vols   36993

Kenosha Public Museums Library, 5500 First Ave, *Kenosha*, WI 53140
T: +1 262 6534426; Fax: +1 262 6534437; URL: www.kenoshapublicmuseum.org
Paula Touhey
Art, American Civil War
10 000 vols; 21 curr per   36994

Holden Arboretum, Warren H. Corning Library, 9500 Sperry Rd, *Kirtland*, OH 44094
T: +1 440 9464400; Fax: +1 440 2565836; E-mail: holden@holdenarb.org; URL: www.holdenarb.org
1963; Susan Swisher
Warren H Corning Horticulture Classics (1200 vols)
9 000 vols; 60 curr per   36995

Idaho Maximum Security Institute Library, 13400 S Pleasant Valley Rd, *Kuna*, ID 83634; P.O. Box 51, Boise, ID 83707-0051
T: +1 208 3890269; Fax: +1 208 3449826
8 000 vols   36996

Gundersen Lutheran Health System, Adolf Gundersen, MD Health Sciences Library, 1900 South Ave, H01-011, *La Crosse*, WI 54601-9980
T: +1 608 7755410; Fax: +1 608 7756343; E-mail: library@gundluth.org
Melinda Orebaugh
8 000 vols; 650 curr per; 150 e-books
                    36997

Library Association of La Jolla, Athenaeum Music & Arts Library, 1008 Wall St, *La Jolla*, CA 92037
T: +1 619 4545872; Fax: +1 619 4545835; E-mail: kpeterson@ljathenaeum.org; URL: www.ljathenaeum.org
1899; Erika Torri
Artists' Bks, Bach Gesellschaft Coll
15 000 vols; 98 curr per; 5 462 music scores; 2 100 av-mat; 17 000 sound-rec; 30 digital data carriers; 2 400 clippings
MLA, ARLIS/NA   36998

Price-Pottenger Nutrition Foundation, Reference Library, P.O. Box 2614, *La Mesa*, CA 91943-2614
T: +1 619 4627600; E-mail: info@ppnf.org; URL: www.ppnf.org/catalog/ ppnf/ReferenceLibrary.htm
Ecology, Gardening
10 000 vols   36999

San Jacinto Museum of History, Albert & Ethel Herzstein Library, One Monument Circle, *La Porte*, TX 77571-9585
T: +1 281 4792421; Fax: +1 281 4792866; URL: www.sanjacinto-museum.org/herzstein_library
1939; Lisa A. Struthers
Texas Hist, San Jacinto, Battle of Texas 1836, Mexico Hist
20 000 vols; 35 curr per; 250 000 mss; 3 diss/theses; 37 music scores; 500 maps; 75 sound-rec; 50 digital data carriers; 500 pamphlets, 1 000 clippings, 300 slides, 2 500 photos
ALA, SAA, TLA   37000

Wilford Hall Medical Center Library, 59MD/MSTL, *Lackland AFB*, TX 78236-5300
T: +1 210 2927204; Fax: +1 210 2927030; URL: www.whmc.af.mil/ resources/index.asp
1950; Rita Smith
11 000 vols; 640 curr per   37001

Lafayette Natural History Museum & Planetarium, Research Library, 433 Jefferson St, *Lafayette*, LA 70501-7013
T: +1 337 2915544; Fax: +1 337 2915464; URL: www.lnhmpmuseum.org
Dr. Deborah J. Clifton
Natural history, natural sciences, astronomy
8 000 vols   37002

Saint Elizabeth Medical Center, Bannon Health Science Library, 1501 Hartford St, *Lafayette*, IN 47904-2126; P.O. Box 7501, Lafayette, IN 47903-7501
T: +1 765 4236143; Fax: +1 765 7425764; URL: www.ste.org
1919; Patricia A. Lunsford
Bioethics Coll
8 000 vols; 350 curr per; 250 books 700 sound-rec   37003

Tippecanoe County Historical Association, Alameda McCollough Research Library, 1001 South St, *Lafayette*, IN 47901
T: +1 765 4768411; Fax: +1 765 4768414; E-mail: library@tcha.mus.in.us; URL: www.tcha.mus.in.us
1925; Patty Bruinsma
Archs Coll
8 000 vols; 53 curr per; 50 diss/theses; 300 music scores; 250 maps; 575 microforms; 28 av-mat; 15 digital data carriers; 14 Journals
libr loan; ALA, AIM   37004

Medical Society of the State of New York, Albion O. Bernstein Library, 420 Lakeville Rd, *Lake Success*, NY 11042
T: +1 516 4886100 ext 388; Fax: +1 516 4881267; E-mail: library@mssny.org; URL: www.mssny.org
1975; Ella Abney
Socioeconomic & Clinical Medicine Colls
17 000 vols; 85 curr per   37005

Lakewood Hospital, Medical Library, 14519 Detroit Ave, *Lakewood*, OH 44107
T: +1 216 5214200 ext 7846; Fax: +1 216 5297094; URL: www.lakewoodhospital.org
13 880 vols; 154 curr per   37006

National Park Service, Library, 12795 W Alameda Pkwy, *Lakewood*, CO 80228; P.O. Box 25287, Lakewood, CO 80225-0287
T: +1 303 9692534; Fax: +1 303 9692557; E-mail: dsc-library@nps.gov; URL: www.library.nps.gov; www.nps.gov
1971; Carol Simpson
Architecture, Construction, Ecology, Hist, Landscape architecture, Engineering, Planning
30 000 vols; 120 curr per
libr loan   37007

Lancaster County Historical Society Library, 230 N President Ave, *Lancaster*, PA 17603-3125
T: +1 717 3924633; Fax: +1 717 2932739; E-mail: lchs@ptd.net; URL: lanclio.org
1886; Mary Virginia Shelley
18th and 19th c Law Libr of Judge Jasper Yeates; Lancaster County hist and Biography; some county records starting with 1729
12 540 vols; 85 curr per; 500 maps; 3 200 microforms; 20 digital data carriers; 276 mss groups   37008

Lancaster Mennonite Historical Society Library, 2215 Millstream Rd, *Lancaster*, PA 17602-1499
T: +1 717 3939745; Fax: +1 717 3938751; E-mail: lmhs@lmhs.org; URL: www.lmhs.org
1958; Beth E. Graybill
29 000 vols; 256 curr per; 200 maps; 420 microforms; 300 arch boxes, 200 000 vital statistical cards   37009

Landis Valley Museum, Reference Library, 2451 Kissel Hill Rd, *Lancaster*, PA 17601
T: +1 717 5690401; Fax: +1 717 5602147; URL: www.landisvalleymuseum.org
1925; Stephen S. Miller
Arts & crafts, Folklore, Hist
12 000 vols   37010

National Outdoor Leadership School (NOLS), Wilderness Education Resource Library, 284 Lincoln St, *Lander*, WY 82520-2848
T: +1 307 3321264; Fax: +1 307 3328811; E-mail: library@nols.edu; URL: www.nols.edu
1965
Wilderness Education, Environmental Education, Conservation
10 000 vols; 40 curr per
libr loan   37011

National Rehabilitation Information Center (NARIC), 8201 Corporate Dr, Ste 600, *Landover*, MD 20785
T: +1 301 4595900; Fax: +1 301 4594263; E-mail: naricinfo@heitechservices.com; URL: www.naric.com
1977; Mark X. Odum
20 000 vols; 108 curr per; 5 120 Bks on Deafness & Sign Lang   37012

Ingham Regional Medical Center, John W. Chi Memorial Medical Library, 401 W Greenlawn Ave, *Lansing*, MI 48910-2819
T: +1 517 3342270; Fax: +1 517 3342939; URL: www.irmc.org
1960; Judy Barnes
9 000 vols; 700 curr per
libr loan   37013

Library of Michigan Service for the Blind & Physically Handicapped, Michigan Library & Historical Ctr, 702 W Kalamazoo St, *Lansing*, MI 48915-1703; P.O. Box 30007, Lansing, MI 48909-7507
T: +1 517 3735614; Fax: +1 517 3735865; E-mail: sbph@michigan.gov; URL: www.michigan.gov/sbph
1931; Sue Chinault
Finnish Language (cassettes), Michigan Hist & Authors (cassettes)
315 000 vols; 60 000 av-mat; 60 000 talking Bks   37014

Saint Lawrence Hospital & Healthcare Services, Medical Library, 1210 W Saginaw St, *Lansing*, MI 48915
T: +1 517 3770354; Fax: +1 517 3770315; E-mail: claytorj@mlc.lib.mi.us
Jane B Claytor
15 200 vols; 210 curr per; 4 digital data carriers
libr loan   37015

American Heritage Center, Toppan Rare Books Library, 1000 E University Ave, Centennial Complex, Dept 3924, *Laramie*, WY 82071; P.O. Box 3924, Laramie, WY 82071
T: +1 307 7662565; Fax: +1 307 7665511; E-mail: amlane@uwyo.edu; URL: ahc.uwyo.edu/about/departments/ toppan.htm
Ann Marie Lane
50 000 vols   37016

Institute of Historical Survey Foundation Library, 3035 S Main, *Las Cruces*, NM 88005-3756; P.O. Box 36, Mesilla Park, NM 88047-0036
T: +1 575 5253035; Fax: +1 575 5250106; E-mail: ihsf@zianet.com
1970; Dr. Evan Davies
Hist, Philosophy, Religion, Art, Theater, Geography, Military
45 000 vols   37017

National Security Technologies, Nuclear Testing Archive, 755 C East Flamingo, *Las Vegas*, NV 89119; P.O. Box 98521, M/S 400, North Las Vegas, NV 89193-8521
T: +1 702 7945117; Fax: +1 702 7945198; E-mail: cic@nv.doe.gov; URL: www.nv.doe.gov
1981; Jeff Gordon
420 000 vols   37018

New York State Nurses Association Library, 11 Cornell Rd, *Latham*, NY 12110
T: +1 518 7829400 ext 266; Fax: +1 518 7829532; E-mail: library@nysna.org; URL: www.nysna.org
1972; Warren G. Hawkes
Assn Arch Mats
9 900 vols; 250 curr per; 67 microforms; 165 av-mat; 255 sound-rec; 4 digital data carriers; 8 VF drawers of pamphlets
ICIRN, NNLM   37019

Lauren Rogers Museum of Art Library, 565 N Fifth Ave, *Laurel*, MS 39440-3410; P.O. Box 1108, Laurel, MS 39441-1108
T: +1 601 6496374; Fax: +1 601 6496379; URL: www.lrma.org/library.html
1923; Donnelle Scott Conklin
American art, Georgian silver, Local hist, Mississippiana, Native American basket; Artists Clipping Files; Bookplates; Local Hist, photos, mss; Museum Arch; Postcard Coll; Rare Bks
10 000 vols; 75 curr per; 50 maps
ALA, ARLIS/NA, SLA   37020

Lawrence Law Library, Two Appleton St, *Lawrence*, MA 01840-1525
T: +1 978 6877608; Fax: +1 978 6882346; E-mail: lawrencelawlibrary@yahoo.com; URL: www.lawlib.state.ma.us
1905; Brian J. Archambault
50 000 vols   37021

Museum of the Great Plains, Research Center, 601 NW Ferris Ave, *Lawton*, OK 73507
T: +1 580 5813460; Fax: +1 580 5813458; E-mail: mgp@museumgreatplains.org; URL: www.museumgreatplains.org
1960; Deborah Baroff
Photogr Coll (80 000 images), Waldo Wedel Anthropological Libr (6 000 books)
30 000 vols; 80 curr per; 20 000 maps; 10 000 Slides   37022

Warren County Historical Society, Museum & Library, 105 South Broadway, *Lebanon*, OH 45036
T: +1 513 9321817; Fax: +1 513 9328560; E-mail: wchs@wchsmuseum.org; URL: www.wchsmuseum.org
1947
Genealogy & hist of Warren County & Southwest Ohio in general; Marcus Mote; Shaker; Russel Wright
10 000 vols; 20 curr per   37023

Department of Veterans Affairs, Medical Center Library, 421 N Main St, *Leeds*, MA 01053-9714
T: +1 413 5844040 ext 2432; Fax: +1 413 5823039
Dorothy Young
Patient Health Coll
10 260 vols; 255 curr per   37024

George C. Marshall Research Foundation Library, PO Drawer 1600, VMI Parade, *Lexington*, VA 24450-1600
T: +1 540 4637103; Fax: +1 540 4645229; E-mail: marshallfoundation@marshallfoundation.org; URL: www.marshallfoundation.org
1964; Paul B. Barron
Military-Diplomatic Hist 1900-1950 (250 mss colls); Cryptography (William Friedman Coll), oral hist, posters, photos, Women's Army Corps; Marshall Plan
25 000 vols; 66 curr per; 3 000 maps; 700 microforms
libr loan   37025

Keeneland Association, Keeneland Library, Keeneland Race Course, 4201 Verailles Rd, *Lexington*, KY 40588; P.O. Box 1690, Lexington, KY 40588-1690
T: +1 859 2543412; Fax: +1 859 2884191; E-mail: library@keeneland.com; URL: www.keeneland.com
Cathy Schenck
Horses, American Racing Coll, photog negative
10 000 vols; 45 curr per   37026

Massachusetts Institute of Technology, Lincoln Laboratory Library & Information Management, 244 Wood St, *Lexington*, MA 02420-9176
T: +1 781 9815500; E-mail: library@ll.mit.edu
1952; Marian Bremer
Aerospace science, Electronics, Engineering, Optics, Solid state physics
130 000 vols; 963 curr per; 4 000 e-journals; 10 000 e-books; 5 200 maps; 10 digital data carriers   37027

National Heritage Museum, Van Gorden-Williams Library, 33 Marrett Rd, *Lexington*, MA 02421
T: +1 781 8616559; Fax: +1 781 8619846; E-mail: library@nationalheritagemuseum.org;

URL: www.nationalheritagemuseum.org; vgw.library.net
1972; Kathy Bell
American clocks and furniture, American decorative arts, American folk art, American Freemasonry and Fraternalism, American maps; Scottish Rite Northern Supreme Council Arch
60 000 vols; 250 curr per; 100 maps; 8 000 Postcards, 600 Posters, 200 Autographs, 5 000 Slides 37028

Stonewall Jackson House, Garland Gray Research Center & Library, Stonewall Jackson House, Eight E Washington St, ***Lexington***, VA 24450
T: +1 540 4632552; Fax: +1 540 4634088; E-mail: sjh1@rockbridge.net
Michael Anne Lynn
11 000 vols 37029

Sam Houston Regional Library and Research Center, 650 FM 1011, ***Liberty***, TX 77575
T: +1 936 3368821; URL: www.tsl.state.tx.us/shc/
1977
Papers of Price Daniel, arch; Papers of Martin Dies, arch; Jean Houston Baldwin Coll of Sam Houston Images
11 504 vols; 51 curr per; 10 552 maps; 1 790 microforms; 49 av-mat; 16 000 cubic feet of mss & local govnt records, 19 245 artifacts, 257 slides, 2 570 art reproductions
libr loan 37030

Massachusetts Audubon Society, Hatheway Environmental Resource Library, 208 S Great Rd, ***Lincoln***, MA 01773
T: +1 781 2599506 ext 7255; Fax: +1 781 2598899; E-mail: edresources@massaudubon.org; URL: www.massaudubon.org
1967; Kristin Eldridge
Environmental Science; Natural Hist
9 000 vols; 40 curr per 37031

National Park Service, Midwest Archeological Center Library, 100 Centennial Mall N, Federal Blgd, Rm 474, ***Lincoln***, NE 68508
T: +1 402 4375392 ext 110; Fax: +1 402 4375098; E-mail: anne_vawser@nps.gov; URL: www.cr.nps.gov/mwac
1969; Anne Vawser
Archaeology Coll, mss
29 470 vols; 40 curr per 37032

Nebraska Library Commission, Talking Book & Braille Service, 1200 N St, Ste 120, ***Lincoln***, NE 68508-2023
T: +1 402 4714038; Fax: +1 402 4712083; E-mail: tbbs@nlc.state.ne.us; URL: www.nlc.state.ne.us/tbbs/
1952; David L. Oertli
Children's Braille, Nebraska Coll
150 000 vols; 56 760 av-mat; 56 620 sound-rec; 2 000 Braille vols 37033

Nebraska State Historical Society Library, Division of Library-Archives, 1500 R St, ***Lincoln***, NE 68508; P.O. Box 82554, Lincoln, NE 68501
T: +1 402 4714751; Fax: +1 402 4718922; E-mail: nshs.la@nebraska.gov; URL: www.nebraskahistory.org
1878; Andrea Faling
Solomon D. Butcher Coll, Eli S. Ricker Coll
80 000 vols; 300 curr per; 3 000 maps; 35 000 microforms; 1 500 sound-rec; 250 000 photos, 12 000 cubic feet of mss, 25 000 cubic feet of State Archives, 1 100 oral hist interviews 37034

Morton Arboretum, Sterling Morton Library, 4100 Illinois Rte 53, ***Lisle***, IL 60532-1293
T: +1 630 7192429; Fax: +1 630 7197950; URL: www.mortonarb.org; www.sterlingmortonlibrary.org
1922; Michael Thomas Stieber
Botany, Horticulture, Natural hist, Botanical art, Landscape architecture; Colls of the Papers of May T. Watts; Drawings, photos, & Letters of Jens Jensen, Landscape Designer; Drawings by Marshall Johnson, Jensen's Successor
27 000 vols; 250 curr per; 28 microforms; 3 digital data carriers; Print coll of more than 8 500 works of art, 150 Special Interest Per Sub
libr loan; Council on Botanical and Horticultural Libraries 37035

Litchfield Historical Society, H. J. Ingraham Memorial Research Library, Seven South St, ***Litchfield***, CT 06759-0385; P.O. Box 385, Litchfield, CT 06759-0385
T: +1 860 5674501; Fax: +1 860 5673565; E-mail: lhsoc@snet.net
1856; Catherine Fields
Local hist; Litchfield Female Academy Coll, Litchfield Law Coll, Local Economic Hist, American Hist
8 000 vols; 11 curr per; 600 linear feet of mss 37036

Arkansas Arts Center, Elizabeth Prewitt Taylor Memorial Library, MacArthur Park, 9th & Commerce, P.O. Box 2137, ***Little Rock***, AR 72203
T: +1 501 3960341; Fax: +1 501 3758053; E-mail: library@arkarts.com; URL: www.arkarts.com/library/
1963
John D. Reid Coll of Early American Jazz, George Fisher Cartoons Coll, Dr. W. Martin Eisele Coll of Oriental Carpets and Tapestries
9 000 vols; 125 curr per; 16 VF drawers of pamphlets, 300 pamphlet boxes of exhibit cat 37037

Arkansas Geological Commission Library, 3815 W Roosevelt Rd, ***Little Rock***, AR 72204-6369
T: +1 501 2961877; Fax: +1 501 6637360; E-mail: agc@mac.state.ar.us; URL: www.state.ar.us/agc/agc.htm
1923; Oleta Sproul
10 000 vols; 11 curr per 37038

Arkansas History Commission Library, One Capitol Mall, 2nd Flr, ***Little Rock***, AR 72201
T: +1 501 6826900; Fax: +1 501 6826916; URL: www.ark-ives.com
Dr. Wendy Richter
20 000 vols; 1 000 curr per 37039

Sandia National Laboratories, Technical Library, 7011 East Ave, ***Livermore***, CA 94550; P.O. Box 969, Livermore, CA 94551-0969
T: +1 925 2941029; Fax: +1 925 2942355; E-mail: dpowers@sandia.gov; URL: www.ca.sandia.gov
1956; Saundra Lormand
Chemistry, Physics, Engineering, Nuclear science, Energy, Mathematics
12 000 vols; 575 curr per
libr loan 37040

Department of Veterans Affairs Medical Center, Health Care Sciences Library, 5901 E Seventh St, Bldg 2, Rm 345, ***Long Beach***, CA 90822-5201
T: +1 562 4945465; Fax: +1 562 4945447; E-mail: vogel.karen@forum.va.gov
1946; Karen Bovel
Management Coll
9 405 vols; 500 curr per 37041

Los Alamos National Laboratory, Research Library, MS-P362, ***Los Alamos***; P.O. Box 1663, Los Alamos, NM 87544-7113
T: +1 505 6674448; Fax: +1 505 6656452; E-mail: library@lanl.gov; URL: library.lanl.gov
1943; Miriam Blake
Science & Technology; AEC Period Publs; Electronic Rpts; LANL Tech Rpts
132 000 vols; 6 710 curr per; 1 400 000 tech rpts, 14 000 full-image electronic rpts 37042

Autry National Center Institute for the Study of the American West – Autry Library, 4700 Western Heritage Way, ***Los Angeles***, CA 90027-1462
T: +1 323 6672000; Fax: +1 323 6605721; E-mail: rroom@autrynationalcenter.org; URL: www.autrynationalcenter.org/autry_library.php
1988; Marva Felchlin
25 000 vols; 100 curr per 37043

Autry National Center Institute for the Study of the American West – Braun Research Library, 234 Museum Dr, ***Los Angeles***, CA 90065
T: +1 323 2212164; Fax: +1 322 2218223; E-mail: library@southwestmuseum.org; URL: www.autrynationalcenter.org/braun.php
Kim Walters
50 000 vols 37044

Braille Institute Library Services, 741 N Vermont Ave, ***Los Angeles***, CA 90029-3594
T: +1 323 6603880; Fax: +1 323 6622440; E-mail: bils@braillelibrary.org; URL: www.braillelibrary.org
1919; Dr. Henry C. Chang
Blindness & Other Handicap Reference Mat, Southern California Coll; 4 branch libs
1 248 000 vols; 12 curr per; 145 incunabula; 7 maps; 1 033 724 sound-rec; 1 digital data carriers; 30 087 braille bks, 874 412 cassette bks, 217 231 disk bks
libr loan; ALA, CLA 37045

Cedars-Sinai Medical Center, Medical Library, South Tower, Rm 2815, 8700 Beverly Blvd, ***Los Angeles***, CA 90048-1865; P.O. Box 48956, Los Angeles, CA 90048-0956
T: +1 310 4233751; Fax: +1 310 4230138; E-mail: library@cshs.org; URL: www.csmc.edu/mlic
Janet L. Hobbs
10 000 vols; 475 curr per 37046

Fashion Institute of Design & Merchandising, Resource & Research Center Library, 919 S Grand Ave, ***Los Angeles***, CA 90015-1421
T: +1 213 6241200; Fax: +1 213 6249365; E-mail: rdodge@fidm.com; URL: www.fidm.com
1969; Robin Dodge
International Fashion Designers File; Rare books coll, Arnaud slide coll; International Video Division, Interior Design Workroom, Textile Division
21 862 vols; 588 curr per; 7 541 av-mat; 30 digital data carriers; 260 retail cat, 14 120 newspaper clipping files, 800 sets of slides 37047

Getty Research Institute, Research Library, 1200 Getty Center Dr, Ste 1100, ***Los Angeles***, CA 90049-1688
T: +1 310 4407395; Fax: +1 310 4407780; E-mail: reference@getty.edu; URL: www.getty.edu/gri
1983; Dr. Susan M. Allen
Archaelogy, Architecture, Archives, Drawings, Mss, photos, Prints, Rare bks; Display & Visual Resources for the Study of Art Hist, Historiography of Art, Modern Period
986 000 vols; 4 256 curr per; 26 incunabula; 423 music scores; 42 maps; 4 000 microforms; 450 av-mat; 76 sound-rec; 241 digital data carriers; 120 572 auction sales cat, 31 010 rare bks, artist bks, small press bks, rare serials, 2 mio study photos, 4 735 arch (incl MSS & architectural drawings), 741 rare prints, photos, charts
libr loan; ARLIS/NA, ARLIS/UK & Ireland, IFLA, SALALM 37048

Goethe-Institut, Mediothek, 5750 Wilshire Blvd, Ste 100, ***Los Angeles***, CA 90036
T: +1 323 5253388; Fax: +1 323 9343597; E-mail: info@losangeles.goethe.org; URL: www.goethe.de/uk/los
1983
8 400 vols; 32 curr per 37049

Griffith Observatory Library, 2800 E Observatory Rd, ***Los Angeles***, CA 90027
T: +1 213 4730800; Fax: +1 213 4730818; URL: www.griffithobs.org
1935; Dr. E. C. Krupp
8 000 vols; 20 curr per; 241 sound-rec; 17 digital data carriers; 10 000 slides 37050

Jewish Community Library of Los Angeles, Peter M. Khan Memorial Library, 6505 Wilshire Blvd, ***Los Angeles***, CA 90048
T: +1 323 7618644; Fax: +1 323 7618647; E-mail: jclref@yahoo.com; URL: www.jclla.org
1947; Abigail Yasgur
Israel, Judaism; Hist of the Local Jewish Community Docs (Arch of the Jewish Community of Los Angeles and, Vicinity)
30 000 vols; 75 curr per; 69 microforms; 1 000 av-mat; 40 digital data carriers; 20 cases of pamphlets,
libr loan 37051

Korean Cultural Center Library, 5505 Wilshire Blvd, ***Los Angeles***, CA 90036
T: +1 323 9367141; Fax: +1 323 9365712; E-mail: librarian@kccla.org; URL: kccla.org
1981; Na-yeon Kim
Korean art & culture
25 000 vols; 52 curr per; 300 govt docs; 10 music scores; 10 maps; 1 500 av-mat; 300 sound-rec; 200 digital data carriers; 300 slides 37052

LA84 Foundation, Sports Library, 2141 W Adams Blvd, ***Los Angeles***, CA 90018
T: +1 323 7304646; Fax: +1 323 7300546; E-mail: library@la84foundation.org; URL: www.la84foundation.org
1936; Shirley S. Ito
Olympic Games Coll, National Treck + Field Res Coll, sport photos, sport films; 30 000 PDF docs on the hist of sport and the Olympic movement on website
35 000 vols; 350 curr per; 100 diss/theses; 3 000 microforms; 5 000 av-mat; 50 digital data carriers
libr loan; IASI 37053

Legal Aid Foundation of Los Angeles, Law Library, 1102 Crenshaw Blvd, ***Los Angeles***, CA 90019
T: +1 323 8017940; Fax: +1 323 8017921
Pamela L. Hall
Bankruptcy, Consumer, Employment, Govt benefits, Housing, Immigration; Govt Agency Regulation Manuals
22 000 vols; 4 digital data carriers 37054

Los Angeles County, Martin Luther King Jr, Drew Medical Center Health Sciences, 1731 E 120th St, ***Los Angeles***, CA 90059
T: +1 323 5634869; Fax: +1 323 5634861; URL: www.kdhsl.cdrewu.edu/
1972; John Carney
Clinical Medicine Coll
48 000 vols; 760 curr per 37055

Los Angeles County Museum of Art, Mr & Mrs Allan C. Balch Art Research Library, 5905 Wilshire Blvd, ***Los Angeles***, CA 90036-4597
T: +1 323 8576118; Fax: +1 323 8574790; E-mail: library@lacma.org; URL: www.lacma.org
1965; Susan Chamberlin Trauger
Art hist; Knoedler Libr (Art), Rare Bks on Costume & Textiles, Rifkind Ctr for German Expressionist Studies; 1 branch libr
176 000 vols; 315 curr per; 24 000 microforms; 15 digital data carriers; 30 000 auction cat, 150 VF drawers exhibition cat
libr loan 37056

Los Angeles County-University of Southern California, General Hospital Medical Library, LAC & USC Healthcare Network, 1200 N State St, Rm 2050, ***Los Angeles***, CA 90033
T: +1 323 2263234; Fax: +1 323 2263360
1914; Bella Kwong
9 000 vols; 700 curr per; 1 000 sound-rec; 55 slide sets, 697 transparencies & filmstrips 37057

Natural History Museum of Los Angeles County, Research Library, 900 Exposition Blvd, ***Los Angeles***, CA 90007
T: +1 213 7633388; Fax: +1 213 7462999; URL: www.nhm.org
1913; Donald W. McNamee
Bookplate Coll, Lepidopterists' Society Libr, Southwestern US Newspaper Coll, Southern California Academy of Science Libr, Los Angeles Theatre Program Coll, Natural Hist Art & Ill
91 000 vols; 500 curr per; 20 000 maps; 857 microforms; 5 000 pamphlets, 300 environmental impact rpts
libr loan 37058

ONE Institute & Archives, ONE National Gay & Lesbian Archives (ONE-IGLA), 909 W Adams Blvd, ***Los Angeles***, CA 90007
T: +1 213 7410094; E-mail: askone@oneinstitute.org; URL: www.oneinstitute.org
1952; Dr. Joseph Hawkins
Homosexual Movement records from 1948 (incl the movement's orgn, social & polit hist)
25 000 vols; 32 VF drawers of mss, clippings, pamphlets, docs, 86 legal briefs, 30 boxes 37059

Philosophical Research Society Library, 3910 Los Feliz Blvd, ***Los Angeles***, CA 90027
T: +1 323 6632167; Fax: +1 323 6639443; E-mail: library@prs.org; URL: www.prs.org
1936; Maja D'Aoust
Edwin Parker Coll; Manly P Hall Coll; Philosophy & Psychology (Oliver L Reiser Coll); Le Plongeon Coll
19 000 vols; 3 curr per; 100 mss; 100 diss/theses; 50 music scores; 300 microforms 37060

Poland's Millennium Library in Los Angeles, 3424 W Adams Blvd, ***Los Angeles***, CA 90018
T: +1 310 2340279
1966; Danuta Zawadzki
Polish hist, lit, culture; Genocide in Poland, 1939-1944
16 000 vols
libr loan 37061

Reiss-Davis Child Study Center, Research Library, 3200 Motor Ave, ***Los Angeles***, CA 90034-3710
T: +1 310 2041666; Fax: +1 310 8384637; URL: www.vistadelmar.org
1950; Dawn Saunders
Freud Coll
14 000 vols; 95 curr per; 90 av-mat; 700 sound-rec; 10 VF drawers 37062

Simon Wiesenthal Center Library & Archives, 1399 S Roxbury Dr, mail address: 9760 W Pico Blvd, ***Los Angeles***, CA 90035-4709
T: +1 310 7727605; Fax: +1 310 2776568; E-mail: library@wiesenthal.net; URL: www.weisenthal.com/library
1978; Adaire J. Klein
Holocaust, Judaica; Primary Anti-semitica & Holocaust Denial; Books By & About Simon Wiesenthal
50 000 vols; 200 curr per; 50 diss/theses; 500 maps; 250 microforms; 400 av-mat; 515 sound-rec; 75 digital data carriers
libr loan; ALA, SLA, AJL 37063

Southern California Library for Social Studies & Research, 6120 S Vermont Ave, ***Los Angeles***, CA 90044
T: +1 323 7596063; Fax: +1 323 7592252; E-mail: archives@socallib.org; URL: www.socallib.org
1963; Yusef Omowale
Labor, Marxism, Socialism, Black, Chicano, Women's movements, Southern California Grassroots organizations; Civil Rights Congress Reords, Californian Democratic Council Records, Folk Music Colls
25 000 vols; 100 curr per; 3 500 sound-rec; 30 000 pamphlets, 500 000 clippings 37064

Southwest Museum, Braun Research Library, 234 Museum Dr, ***Los Angeles***, CA 90065
T: +1 323 2212164 ext 255; Fax: +1 323 2248223; E-mail: library@southwestmuseum.org; URL: www.southwestmuseum.org
1907; Kim G. Walters
Arizona & the Southwest (Munk Libr of Arizoniana); Archaeology (Hector Alliot Memorial Libr of, Archaeology); Californiana (George Wharton James Libr); Spanish Language, California and the Southwest, folklore and music (Charles Fletcher Lummis Libr); Frederick Webb Hodge Papers; Frank Hamilton Cushing; John Charles Fremont; George Bird Grinnell Papers; Photo Coll of Am Indian, Missions, Adobes, California & Arizona hist & archaeology of the Southwest
60 000 vols; 125 curr per; 800 mss; 3 000 maps; 2 000 sound-rec; 750 mss, 55 000 negatives, 140 000 photos
OCLC 37065

White Memorial Medical Center, Courville-Abbott Memorial Library, 1720 Cesar E Chavez Ave, ***Los Angeles***, CA 90033-2462
T: +1 323 2605715; Fax: +1 323 2605748
1920; Myrna Y. Uyengco-Harooch
Hist Coll, Hist of Medicine (Margaret & H James Hara Memorial Coll), Hist & Religion (Percy T Magan Coll)
43 000 vols; 350 curr per 37066

Filson Historical Society Library, 1310 S Third St, ***Louisville***, KY 40208
T: +1 502 6355083; Fax: +1 502 6355086; E-mail: filson@filsonhistorical.org; URL: www.filsonhistorical.org
1884; Dr. Mark V. Wetherington
Kentucky hist & genealogy, Civil War hist; KY Portraits, Silver (Kentucky Silversmiths Coll), Confederate Coll, Highbaugh Rare Book Coll, Sheet Music Coll, Temple-Bodley Coll of Jonathan Clark Papers, William Clark & Henry Clay Papers
50 000 vols; 100 curr per; 1 500 maps; 1,5 mio mss & private papers, 50 000 prints & photos, 3 700 scores 37067

Jefferson County Public Law Library, Old Jail Bldg, Ste 240, 514 W Liberty St, ***Louisville***, KY 40202-2806
T: +1 502 5745943; Fax: +1 502 5743483; E-mail: jcpll@bluegrass.net; URL: www.jcpll.com
1839; Linda Miller Robbins
Kentucky Coll, Indiana Coll
90 000 vols 37068

Louisville Academy of Music Library, 2740 Frankfort Ave, ***Louisville***, KY 40206-2669
T: +1 502 8937885; URL: www.laofm.org/library.htm
1954; Robert B. French
Lit, Local hist, Photography; Antique Piano Rolls, 19 C First Editions, Biogr on Local Musicians
8 000 vols; 16 000 sound-rec 37069

National Society of the Sons of the American Revolution, Historical & Genealogical Research Library, 1000 S Fourth St, ***Louisville***, KY 40203
T: +1 502 5891776; Fax: +1 502 5891671; E-mail: library@sar.org; URL: www.sar.org/geneal/library.htm
1889; Michael A. Christian
American Revolutionary War Hist, genealogy of the signers of the Declaration of Independence; George Washington Coll; United States Census (1790-1920)
58 000 vols; 20 curr per; 15 000 microforms; 270 digital data carriers 37070

Speed Art Museum Library, 2035 S Third St, ***Louisville***, KY 40208
T: +1 502 6342700; Fax: +1 502 6362899; E-mail: info@speedmuseum.org; URL: www.speedmuseum.org/
1927; Mary Jane Benedict
J B Speed's Lincoln Coll; Weygold Indian Coll
17 640 vols; 50 curr per; 54 VF drawers 37071

Massachusetts Trial Court, Lowell Law Library, Superior Court House,360 Gorham St, ***Lowell***, MA 01852
T: +1 978 4529301; Fax: +1 978 9702000; E-mail: lowlaw@meganet.net; URL: www.lawlib.state.ma.us/lowell.html
1815; Catherine Mello Alves
Law books – Legal Per statutes, case reporters, Digests, Legal treatises
35 000 vols; 53 curr per 37072

Texas Tech University Health Sciences Center, Preston Smith Library of the Health Sciences, 3601 Fourth St, ***Lubbock***, TX 79430-7781
T: +1 806 7432200; Fax: +1 806 7432218; URL: www.ttuhsc.edu/libraries
1971; Richard C. Wood
290 000 vols; 2 144 curr per 37073

George M. Jones Library Association, Jones Memorial Library, 2311 Memorial Ave, ***Lynchburg***, VA 24501
T: +1 434 8460501; Fax: +1 434 8461572; E-mail: webmaster@jmlibrary.org; URL: www.jmlibrary.org
1908; Phillip W. Rhodes
Genealogy, Architectural drawings, Local hist, Family histories; Lynchburg Architectural Arch, Lynchburg Business Arch
16 750 vols; 60 curr per; 3 500 microforms; 500 boxes of archival items, 12 000 drawings 37074

Saint Francis Medical Center, Medical Library, 3630 Imperial Hwy, ***Lynwood***, CA 90262
T: +1 310 6036045; Fax: +1 310 6395936
1971; Beth Araya
14 700 vols; 325 curr per 37075

University of Wisconsin-Madison – Water Resources Library, 1975 Willow Dr, 2nd Flr, ***Madison***, WI 53706-1177
T: +1 608 2623069; Fax: +1 608 2620591; E-mail: askwater@aqua.wisc.edu; URL: aqua.wisc.edu/waterlibrary
1966
30 000 vols; 60 curr per 37076

Wisconsin Historical Society Library, 816 State St, ***Madison***, WI 53706
T: +1 608 2646534; Fax: +1 608 2646520; URL: www.wisconsinhistory.org/libraryarchives; www.wisconsinhistory.org
1847; Peter Gottlieb
Ethnic hist, Labor hist, Military hist, Minority hist, Newspapers, Numismatics, Philately, Radical, Reform movements, Religion, Women's hist
3 900 000 vols 37077

American Institute for Chartered Property Casualty Underwriters & Insurance Institute of America, Nancy W. Spellman Memorial Library, 720 Providence Rd, Ste 100, ***Malvern***, PA 19355-3433
T: +1 610 6442100; Fax: +1 610 7250613; E-mail: cserv@cpcuiia.org; URL: www.aicpcu.org
1977
Insurance, Risk management, Economy; O D Dickerson Memorial Libr (Life Insurance)
10 000 vols; 40 curr per; 100 av-mat; 6 digital data carriers
SLA 37078

American-Canadian Genealogical Society Library, Four Elm St, ***Manchester***, NH 03103-7242; P.O. Box 6478, Manchester, NH 03108-6478
T: +1 603 6221554; E-mail: acgs@acgs.org; URL: www.acgs.org
1973; Gerry Savard
Loiselle Index of Quebec Marriages; Diocese of Moncton, New Brunswick; Massachusetts Vital Statistics; Vermont Vital Statistics; Drouin Marriage Indexes for Province of Quebec
9 000 vols 37079

Currier Museum of Art, Art Reference Library, 201 Myrtle Way, ***Manchester***, NH 03104
T: +1 603 6696144; Fax: +1 603 6694166; E-mail: libraryinfo@currier.org; URL: www.currier.org
1929; Alison Dickey
American painting, Decorative arts, Photography
15 000 vols; 25 curr per 37080

Franco-American Centre, Bibliothèque, 52 Concord St, ***Manchester***, NH 03101; P.O. Box 994, Manchester, NH 03101-0994
T: +1 603 6694045; Fax: +1 603 6690644; E-mail: icifranam@aol.com; URL: www.francoamericancentrenh.com
1918; Richard Charpentier
Acadians, Franco-Americans, French, Louisiana, Quebecois; record coll; 250 museum pieces; rare bks; sculptures, paintings, photos, old newspapers publ in French in the US, arch, rare bks
40 000 vols; 15 curr per; 280 microforms; 30 sound-rec; 200 VF drawers, 55 000 index cards 37081

Manomet Center for Conservation Sciences Library, PO Box 1770, ***Manomet***, MA 02345
T: +1 508 2246521; Fax: +1 508 2249220; URL: www.manomet.org
1969; Jack Halloran
Birds, Ecology
12 000 vols 37082

Ohio Genealogical Society Library, 713 S Main St, ***Mansfield***, OH 44907-1644
T: +1 419 7567294; Fax: +1 419 7568681; E-mail: ogs@ogs.org; URL: www.ogs.org
1959; Thomas Stephen Neel
County & State Source Material; Ohio Bible Records File; First Families of Ohio; (pre-1820 Settler Lineage Society); Society of Civil War Families of Ohio
32 000 vols; 150 curr per; 300 mss; 350 maps; 6 000 microforms; 200 sound-rec; 200 digital data carriers 37083

Information Masters Library, 37980 Reed Rd, ***Manzanita***, OR 97130; P.O. Box 525, Manzanita, OR 97130
T: +1 503 3686990; Fax: +1 503 3687118
Signe E. Larson
Business law, Business
11 000 vols; 50 curr per 37084

Marquette County Historical Society, John M. Longyear Research Library, 213 N Front St, ***Marquette***, MI 49855
T: +1 906 2263571; Fax: +1 906 2260919; URL: www.marquettecohistory.org
1918; Rosemary Michelin
Ethnology, Great Lakes, Mining, Railroads; Local Newspapers on microfilm from 1870s, Family Records, J. M. Longyear Coll, Burt Papers, Breitung-Kaufman Papers, Charles Thompson Harvey Papers, Business Records, Military (Local Service Men), Municipal Records
15 000 vols; 45 curr per; 1 000 maps; 75 microforms; 2 000 pamphlets, 35 VF drawers 37085

American Museum of Magic, Lund Memorial Library, 107 East Michigan Ave, ***Marshall***, MI 49068
T: +1 616 7817570; URL: www.americanmuseumofmagic.org
1943; Elaine Lund
Conjuring, confidence games, superstition; Irving Desfor Coll of Photos of Magicians
12 000 vols; 12 curr per; Newspaper clippings, letters, progs, photos, bus cards, mss & flyers (200 000) 37086

Nodaway County Historical Society / Mary H. Jackson Research Center, Genealogy Society Library, 110 N Walnut St, ***Maryville***, MO 64468-2251; P.O. Box 324, Maryville, MO 64468-0324
T: +1 660 5828176; Fax: +1 660 5623377; URL: www.nodawayhistorical.org
Tom Carneal
Local history, genealogy
12 000 vols 37087

Mashantucket Pequot Museum & Research Center, Information Resources, 110 Pequot Trail, ***Mashantucket***, CT 06339; P.O. Box 3180, Mashantucket, CT 06339-3180
T: +1 860 3966897; Fax: +1 860 3966874; E-mail: reference@mptn.org; URL: www.mpmrc.org
1998; Betsy Bahr Peterson
Native American Res Libr for Children; Popular Culture & Ephemera with Native American Themes; Tribal Archs
40 000 vols; 810 curr per 37088

ASM International Library, 9369 Kinsman Rd, ***Materials Park***, OH 44073
T: +1 440 3385155, 5151; Fax: +1 440 3388091; E-mail: asmlibr@po.asm-intl.org; URL: www.asm-intl.org
1959; Eleanor Baldwin
Mat tech, Metals; William Hunt Eisenman Rare Book Coll, Metallurgy Coll
15 000 vols; 2 400 curr per 37089

LMI Library, 2000 Corporate Ridge, ***McLean***, VA 22102-7805
T: +1 703 9177214; Fax: +1 703 9177474; E-mail: library@lmi.org
1965; Nancy Eichelman Handy
10 000 vols; 450 curr per; 2 000 rpts, 100 manuals & newsletters in binders 37090

Auerbach Central Agency for Jewish Education, Seidman Educational Resources Center, 7607 Old York Rd, ***Melrose Park***, PA 19027
T: +1 215 6358940; Fax: +1 215 6358946; E-mail: info@acaje-jop.org; URL: www.acaje.org

1979; Nancy M. Messinger
Holocaust Resources, Kossman Children's Lit Lending Libr
15 000 vols; 15 curr per; 1 000 av-mat; 200 sound-rec; 25 digital data carriers; 13 000 docs, 22 VF drawers of instructional materials 37091

Navistar, Inc, Information Center, 10400 W North Ave, ***Melrose Park***, IL 60160-1065
T: +1 708 8654004
William Warren
Automotive engineering
10 000 vols; 91 curr per 37092

University of Tennessee – Memphis, Health Sciences Library & Biocommunications Center, 877 Madison Ave, ***Memphis***, TN 38163
T: +1 901 4485634; Fax: +1 901 4487235; E-mail: utlibrary@uthsc.edu; URL: library.uthsc.edu
1913; Dr. Thomas Singarella
Hist of Medicine, Tennessee Authors (Wallace Memorial Coll)
187 000 vols; 1 400 curr per; 2 000 e-journals; 130 microforms; 306 Slide Programs
libr loan 37093

Exponent Failure Analysis Associates, Information Resources, 149 Commonwealth Dr, ***Menlo Park***, CA 94025
T: +1 650 6887171; Fax: +1 650 3299526; URL: www.exponent.com
Lee Pharis
Electronic engineering
8 000 vols; 25 curr per 37094

US Geological Survey Library, 345 Middlefield Rd, MS 955, ***Menlo Park***, CA 94025-3591
T: +1 650 3295027; Fax: +1 650 3295132; E-mail: men_lib@usgs.gov; URL: library.usgs.gov
1953; Angelica Bravos
Aerial photos, Biological science, California, Earth science, Maps; California Ctr Coll
400 000 vols; 1 500 curr per 37095

Mercy Medical Center Merced, Community Campus, William E. Fountain Health Sciences Library, 301 E 13th St, ***Merced***, CA 95340
T: +1 209 3857000; Fax: +1 209 3857038; URL: www.mercymedcares.org
Mary Silva
64 000 vols 37096

Lake County Public Library – Talking Book Service, 1919 W 81st Ave, ***Merrillville***, IN 46410-5382
T: +1 219 7693541
Dawn Mogle
28 000 vols; 53 curr per; 78 av-mat 37097

Center for the Advancement of Jewish Education, Adler Shinensky Library, 4200 Biscayne Blvd, ***Miami***, FL 33137-3279
T: +1 305 5764030 ext 154; Fax: +1 305 5760307; E-mail: info@caje-miami.org; URL: www.caje-miami.org
1948; Marci Wiseman
Hebrew Reference; Holocaust (Educational Resource Center Coll); Lit; Biblical Woman; Art Text Bks; Text Book Display; Video loan Depit; Maimonides; Hebrew Bks
55 000 vols; 70 curr per; 82 mss; 150 music scores; 100 maps; 1 000 av-mat; 125 digital data carriers; 1 000 video-cassettes
libr loan; AJL 37098

Fairchild Tropical Botanic Garden, Montgomery Library, 11935 Old Cutler Rd, ***Miami***, FL 33156
T: +1 305 6671651; Fax: +1 305 6694074; URL: www.fairchildgarden.org
1941; Nancy Korber
Botany, Horticulture; David Fairchild Coll, papers & photos; Florida Botanists
16 000 vols; 60 curr per 37099

Florida Center for Theological Studies (FCTS) Library, 111 NE First St, 7th Flr, ***Miami***, FL 33132-2517
T: +1 305 3793777; Fax: +1 305 3790012; E-mail: fctslib@fcts.edu; URL: www.fcts.edu
Dr. Jack Budrew
34 000 vols; 252 curr per 37100

Mount Sinai Medical Center, Medical Library, 4300 Alton Rd, ***Miami Beach***, FL 33140
T: +1 305 6742840; Fax: +1 305 6742843; URL: www.msmc.com
1946; Andre Peres
15 000 vols; 300 curr per; 400 sound-rec; 30 digital data carriers
libr loan; MLA 37101

National Sporting Library, Inc, 102 The Plains Rd, P.O. Box 1335, ***Middleburg***, VA 20118-1335
T: +1 540 6876542; Fax: +1 540 6878540; E-mail: nsl@nsl.org; URL: www.nsl.org
1954; Kenneth Y. Tomlinson
Foxhunting Papers (Harry Worcester Smith Coll); 16th-19th C Bks on Horses (Huth-Lonsdale-Arundel & von Hunersdorf Colls); Sporting Bks (John H & Martha Daniel Coll)
12 000 vols; 150 microforms
libr loan 37102

Connecticut Valley Hospital, Hallock Health Sciences Library, Silver St, P.O. Box 351, ***Middletown***, CT 06457
T: +1 860 2625059; Fax: +1 860 2625049; E-mail: pauline.kruk@po.state.ct.us
1950; Pauline A. Kruk, Stephen Curtin
Mental health
10 000 vols; 150 curr per; 150 av-mat; 400 sound-rec
libr loan; MLA, SLA, AMHL, SALIS 37103

Connecticut Valley Hospital Patients' Library, Willis Royle Library, Silver St, ***Middletown***, CT 06457; P.O. Box 351, Middletown, CT 06457
T: +1 860 2625520; Fax: +1 860 2625049
Pauline Kruk
Mental health, fiction, psychology
8 000 vols; 40 curr per; 100 av-mat 37104

Godfrey Memorial Library, 134 Newfield St, ***Middletown***, CT 06457-2534
T: +1 860 3464375; Fax: +1 860 3479874; E-mail: library@godfrey.org; URL: www.godfrey.org
Richard E. Black
History, genealogy
35 000 vols 37105

Haley Memorial Library & History Center, Nita Stewart Haley Memorial Library, 1805 W Indiana, ***Midland***, TX 79701
T: +1 432 6825785; Fax: +1 432 6853512; E-mail: haley-mail@att.net; URL: www.haleylibrary.com
1976; Nancy Jordan
Cowboys, Discovery & exploration, Indians, Railroads, Ranch hist, Rodeos, Texana, Western Americana
22 000 vols; 300 curr per 37106

Michigan Molecular Institute, Raymond F. Boyer Resource Center, 1910 W St Andrews Rd, ***Midland***, MI 48640
T: +1 517 8325555; Fax: +1 517 8325560; E-mail: eastland@mmi.org; URL: www.mmi.org
1972; Judy Eastland
Polymer Chemistry – Polymer Engineering, Physics + Math
8 500 vols; 134 curr per; 450 mss; 25 diss/theses; 75 digital data carriers
libr loan; SLA 37107

Cary Institute of Ecosystem Studies, Library, P.O. Box AB, 65 Sharon Tpk, ***Millbrook***, NY 12545
T: +1 845 6777600 ext 164; Fax: +1 845 6775976; E-mail: schulera@ecostudies.org; URL: www.ecostudies.org
1985; Amy Schuler
10 000 vols; 175 curr per 37108

Central State Hospital, Medical, Staff and Patients' Libraries, 620 Broad St, ***Milledgeville***, GA 31062-9989
T: +1 478 4456889; Fax: +1 478 4456699; E-mail: info@centralstatehospital.org; URL: www.centralstatehospital.org
Kathy Warner
15 600 vols; 180 curr per
libr loan 37109

Aurora Health Care Libraries, 2900 W Oklahoma Ave, ***Milwaukee***, WI 53215-4330; P.O. Box 2901, Milwaukee, WI 53201-2901
T: +1 414 6497356; Fax: +1 414 6497037; E-mail: aurora.libraries@aurora.org; URL: www.aurora.org
1967; Kathleen Strube
14 000 vols; 255 curr per; 330 e-books 37110

Columbia Hospital, Media Library, 2025 E Newport Ave, ***Milwaukee***, WI 53211
T: +1 414 9613858; Fax: +1 414 9613813
1950; Ruth Holst
Consumer Health Coll
10 000 vols; 250 curr per; 50 av-mat; 600 sound-rec
libr loan 37111

Milwaukee Art Museum Library, Art Research Library, 700 N Art Museum Dr, ***Milwaukee***, WI 53202
T: +1 414 2243270; Fax: +1 414 2717588; E-mail: library@mam.org; URL: www.mam.org
1916; Heather Lynn Winter
20 000 vols; 60 curr per 37112

Milwaukee County Historical Society, Library & Archives, 910 N Old World Third St, ***Milwaukee***, WI 53203
T: +1 414 2738288; Fax: +1 414 2733268; E-mail: mchs@prodigy.net; URL: www.milwaukeecountyhistsoc.org
1935; Steven Daily
Milwaukee County records, Episcopal Church of Milwaukee County records, Landmarks/Buildings records, Photogr Colls, Naturalization records for Milwaukee County
12 000 vols; 2 800 mss; 586 diss/theses; 783 maps; 585 microforms; 200 sound-rec
SAA, AASLH 37113

Milwaukee Institute of Art & Design Library, 273 E Erie St, ***Milwaukee***, WI 53202-6003
T: +1 414 8473342; Fax: +1 414 2918077; URL: www.miad.edu
1977; Cynthia D. Lynch
26 000 vols; 125 curr per; 1 000 av-mat; 45 000 slides 37114

Milwaukee Public Museum, Library & Archives, 800 W Wells St, ***Milwaukee***, WI 53233
T: +1 414 2782736; Fax: +1 414 2786100; URL: www.mpm.edu/collections/dept/library.php
1883; Susan Otto
Milwaukee Public Museum Arch
130 000 vols; 1 200 curr per; 2 300 maps; 800 microforms; 500 arch mat, primary institutional records
libr loan; OCLC 37115

Wisconsin Regional Library for the Blind & Physically Handicapped, 813 W Wells St, ***Milwaukee***, WI 53233-1436
T: +1 414 2863045; Fax: +1 414 2863102; E-mail: lbph@milwaukee.gov; URL: wmbph.mpl.org/opac; www.regionallibrary.wi.us
1960; Meredith J. Wittmann
160 000 vols; 20 curr per; 9 000 e-books; 137 778 av-mat; 234 170 sound-rec; 1 760 Braille vols, 20 Special Interest Per Sub, 70 Large Print Bks 37116

Allina Health System Library Services, 1801 Nicollet Ave, Nicollet Ave Ctr, ***Minneapolis***, MN 55403
T: +1 612 7757900; Fax: +1 612 7757946; E-mail: library@allina.com; URL: www.allina.com/library
1897; Betsy Moore
9 500 vols; 860 curr per 37117

American Swedish Institute Archives & Library, 2600 Park Ave, ***Minneapolis***, MN 55407
T: +1 612 8714907; Fax: +1 612 8718682; E-mail: info@americanswedishinst.org; URL: www.americanswedishinst.org
1929; Bruce N. Karstadt
Swedish Hist & Lit (Swan J. Turnblad Libr); Turnblad Lending Libr; Swedish Immigration Hist Coll (Victor Lawson Coll; microfilm church records; immigrant docs)
14 000 vols; 10 curr per; 1 000 microforms; 250 sound-rec 37118

BAE Systems Armament Systems, AS Library, 4800 E River Rd, M344, ***Minneapolis***, MN 55421-1498
T: +1 763 5727900; Fax: +1 763 5726037
Crystal Clift
8 000 vols; 200 curr per 37119

The Bakken – A. Library & Museum of Electricity in Life, 3537 Zenith Ave S, ***Minneapolis***, MN 55416
T: +1 612 9263878; Fax: +1 612 9277265; URL: www.thebakken.org
1975; David Rhees
Hist of electricity, electrotherapeutics and electrophysiology
12 000 vols; 30 curr per; 50 mss; 6 incunabula; 50 mss 37120

Hennepin County Medical Center, Health Sciences Library, Mail Code R2, 701 Park Ave, ***Minneapolis***, MN 55415
T: +1 612 8732714; Fax: +1 612 9044248; E-mail: media@hcmed.org; URL: www.hcmc.org/healthprofs/prof_resources.htm
1901; Kathleen Warner
8 000 vols; 530 curr per; 165 e-books
libr loan; MLA, OCLC 37121

Minneapolis Institute of Arts, Art Research & Reference Library, 2400 Third Ave S, ***Minneapolis***, MN 55404
T: +1 612 8703117; Fax: +1 612 8703004; URL: www.artsmia.org
1915; Janice Lurie
Art hist, Chinese bronzes, Drawing, English silver, Furniture, Jades, Painting, Porcelains, Prints, Sculpture, Textiles; Botany & Fashion (Minnich Coll); Five Hundred Years of Sporting Bks (John Daniels Coll), drawings; Hist of Printing (Lesli Coll)
55 000 vols; 120 curr per
ARLIS/NA 37122

Resource Center of the Americas, Penny Lernoux Memorial Library, 3019 Minnehaha Ave, S, ***Minneapolis***, MN 55406
T: +1 612 2760788; Fax: +1 612 2760898; E-mail: info@Americas.org; URL: www.americas.org
1983; Mary Swenson
Liberation Theology & Global Economy; Popular Education; Paulo Freire
10 000 vols; 70 curr per; 400 av-mat 37123

Walker Art Center, Staff Reference Library & Archives, 1750 Hennepin Ave, ***Minneapolis***, MN 55403
T: +1 612 3757680; Fax: +1 612 3757590; URL: www.walkerart.org
1950; Rosemary Furtak
Artists Bks, Audio Arch, Artist Cat 1940-present, Edmond R. Ruben Film Study Coll
35 000 vols; 110 curr per; 2 000 sound-rec; 40 digital data carriers; 20 000 cat, 290 000 slides
ARLIS/NA 37124

Montclair Art Museum, Le Brun Library, 3 S Mountain Ave, ***Montclair***, NJ 07042
T: +1 973 7465555 ext 223; Fax: +1 973 7460920; E-mail: library@montclairartmuseum.org; URL: www.montclair-art.org
1916; Jeffrey Guerrier
American art, North American Indians (Art & Archaeology); Montclair Art Soc & Museum sprapbks 1913-1990s, New Jersey Survey of Outdoor Sculpture (SOS) repository
45 000 vols; 25 curr per; 50 sound-rec; 25 000 catalogs, 14 000 artist & subject files, 20 000 35mm slides, 7 000 bookplates
libr loan; ARLIS/NA 37125

Defense Language Institute Foreign Language Center, Aiso Library, 543 Lawton Rd, Ste 617A, ***Monterey***, CA 93944-3214
T: +1 831 2426889; Fax: +1 831 2425816; URL: dlilibrary.monterey.army.mil/aisolib.htm
1943; Margaret J. Groner
Foreign language
115 000 vols; 800 curr per
libr loan 37126

Alabama Department of Archives & History Research Room, 624 Washington Ave, **Montgomery**, AL 36130-0100
T: +1 334 2424435; Fax: +1 334 2403433; URL: www.archives.alabama.gov
1901; Debbie Pendleton
State government, State hist; Alabama Newspapers, Hist Records of State of Alabama, mss, maps, pamphlets & photog
40 000 vols
37127

Society for the Study of Male Psychology & Physiology Library, 321 Iuka, **Montpelier**, OH 43543
T: +1 419 4853602; E-mail: jdbrg@bright.net
1974; Dr. Jerry Bergman
Male Psychology & Physiology, Discrimination, Male, Homosexuality, Male Rights movement
18 000 vols; 120 curr per; 4 200 mss; 127 diss/theses; 12 govt docs; 80 microforms; 214 av-mat; 871 sound-rec; 47 digital data carriers
libr loan
37128

Institute of Management Accountants, Inc., 10 Paragon Dr, **Montvale**, NJ 07645-1760
T: +1 201 5739000; Fax: +1 201 5739795; E-mail: library@imanet.org; URL: www.imanet.org
1919; Kathleen Muldowney
Trade Assn Accounting Manuals
10 000 vols; 300 curr per; 10 av-mat; 30 VF drawers
37129

Honeywell International, Technical Information Services, 101 Columbia Rd, CRL-1 Bldg, **Morristown**, NJ 07962-1021
T: +1 973 4552000; Fax: +1 973 4555295; E-mail: michael.keane@honeywell.com
1965; Michael Keane
Business, Chemistry, Economy, Energy industry, Environmental health, Medicine, Physics, Polymers, Toxicology; Annual Rpts, Consultants Rpts
30 000 vols; 40 curr per
37130

National Park Service, Morristown National Historical Park Library, 30 Washington Pl, **Morristown**, NJ 07960-4299
T: +1 908 7662821; Fax: +1 908 7664589; E-mail: morr_library@nps.gov; URL: www.nps.gov/morr
1933; Jude Pfister
18th c American life, 18th c culture; Ford Family Papers, Lidgerwood Hessian Transcripts, William Van Vleck Lidgerwood Bk Coll, Lloyd W. Smith Mss Coll, Washington Assn of New Jersey Book Coll, Records of the Washington Assn of New Jersey, Park Arch
45 000 vols; 83 curr per; 45 000 mss; 1 incunabula; 576 maps; 800 microforms; 500 Pamphlets
37131

California State University, Moss Landing Marine Laboratories Library, 8272 Moss Landing Rd, **Moss Landing**, CA 95039
T: +1 831 7714400; Fax: +1 831 6324403; E-mail: library@mlml.calstate.edu; URL: www.mlml.calstate.edu
1966; Joan Parker
Marine biology, Oceanography, Mammals, California; Elkhorn Slough Coll
15 000 vols; 120 curr per
37132

Ozark Folk Center Library, Ozark Cultural Resource Center, 1032 Park Ave, **Mountain View**, AR 72560; P.O. Box 500, Mountain View, AR 72560-0500
T: +1 870 2693280; Fax: +1 870 2692909; URL: www.ozarkfolkcenter.com
1976; Tricia Hearn
Ozark folklore, Crafts, Music
12 000 vols; 50 curr per
37133

Educational Information & Resource Center, South Jersey Technology Park, 107 Gilbreth Parkway, Suite 200, **Mullica Hill**, NJ 08062
T: +1 856 5827000 ext 140; Fax: +1 856 5824206; E-mail: lrc@eirc.org; URL: www.eirc.org
1969; Patricia Bruder
ERIC Coll, Bks on Special Education
10 000 vols; 90 curr per
37134

Middle Tennessee State University Center for Popular Music Library, 140 John Bragg Mass Communications Bldg, **Murfreesboro**, TN 37132; P.O. Box 41, Murfreesboro, TN 31733-0041
T: +1 615 8985513; Fax: +1 615 8985829; E-mail: ctrpopmu@mtsu.edu; URL: popmusic.mtsu.edu
1985; Paul Wells
American Broadside & Songsters; Ray Avery (jazz, black); Black Harmony Singing; Denominational Hymnals & gospel songbooks; Kenneth Goldstein Coll of American Broadsides & songsters; Alfred Moffatt Coll of Music Books; Brad McCuen Coll
18 000 vols; 275 curr per; 200 mss; 425 diss/theses; 65 000 music scores; 3 000 microforms; 125 000 sound-rec
MLA, Music OCLC Users Group
37135

International Fertilizer Development Center, Travis P. Hignett Memorial Library, P.O. Box 2040, **Muscle Shoals**, AL 35662
T: +1 205 3816600; Fax: +1 205 3817408; E-mail: library@ifdc.org; URL: www.ifdc.org/About/Library
1976; Jean S. Riley
Current Awareness Database; Developing Country File & International Agricultural Organization File; Fertaware Database; Patent Files; Training Programs
20 000 vols; 200 curr per; 85 maps; 646 av-mat; 24 digital data carriers; 948 patents; 211 training programs
libr loan
37136

Tennessee Valley Authority, TVA Research Library, PO Box 1010, **Muscle Shoals**, AL 35662-1010
T: +1 256 3862417; Fax: +1 256 3862453; URL: www.tva.gov
1961; Drucilla Sharp Gambrell
Alternative energy, Biomass, Chemistry, Chemical engineering, Environ sciences, Marketing, Waste management
12 000 vols
libr loan
37137

Mystic Seaport Museum, G. W. Blunt White Library, 75 Greenmanville Ave, **Mystic**, CT 06355; P.O. Box 6000, Mystic, CT 06355-6000
T: +1 860 5725367; Fax: +1 860 5725394; E-mail: collections@mysticseaport.org; URL: www.mysticseaport.org/library/home.cfm
1929; Paul J. O'Pecko
American Maritime Studies, mss; Strips Plans Coll (Tel. 572-5360)
75 000 vols; 404 curr per; 1 300 mss; 8 000 maps; 2 000 microforms; 1 250 logbooks
libr loan
37138

Lone Star Legal Aid, Law Library, 414 E Pillar St, **Nacogdoches**, TX 75961
T: +1 936 5601455; Fax: +1 936 5604795; URL: www.lsc.gov; www.lonestarlegal.org
1977; Diane Baker
6 branch libs
21 000 vols; 30 sound-rec; 3 VF drawers of poverty files, 232 briefs, 1 filing drawer of forms
37139

Nantucket Maria Mitchell Association, Maria Mitchell Science Library, 2 Vestal St, **Nantucket**, MA 02554-2699
T: +1 508 2289219; Fax: +1 508 2281031; E-mail: info@mmo.org; URL: www.mmo.org
1902; Patricia Hanley
Astronomy, Biology, Botany, Chemistry, Geology, Oceanography, Physics, Zoology; Maria Mitchell Memorabilia (Original Notebks & Papers)
9 000 vols; 42 curr per; 2 digital data carriers
37140

National Sea Grant Library, Pell Libr Bldg, URI-Bay Campus, **Narragansett**, RI 02882-1197
T: +1 401 8746114; Fax: +1 401 8746160; E-mail: nsgl@gso.uri.edu; URL: nsgl.gso.uri.edu
1970; Cynthia Murray
39 000 vols; 62 curr per; 70 av-mat; 10 digital data carriers
libr loan; IAMSLIC
37141

Nashotah House Library, 2777 Mission Rd, **Nashotah**, WI 53058-9793
T: +1 262 6466535; Fax: +1 262 6466504; E-mail: librarian@nashotah.edu; URL: www.nashotah.edu/library
1842; David Sherwood
Anglicana; Arch of Nashotah House, Nat Altar Guild Coll, Prayer Bks (Underwood Coll)
100 000 vols; 300 curr per
libr loan; ATLA
37142

Disciples of Christ Historical Society Library & Archives, 1101 Nineteenth Ave S, **Nashville**, TN 37212-2196
T: +1 615 3271444; Fax: +1 615 3271445; E-mail: mail@disciplehistory.org; URL: www.disciplehistory.org
1941; Sara Harwell
37 000 vols; 270 curr per; 1 000 mss; 2 500 sound-rec; 6 000 Slides
libr loan; ATLA
37143

Jewish Federation Libraries – Jewish Community Center Library, 801 Percy Warner Blvd, **Nashville**, TN 37205
T: +1 615 3563242 ext 255; Fax: +1 615 3520056; E-mail: library@jewishnashville.org; URL: www.jewishnashville.org
1902
Hebrew Coll, Holocaust Coll, Israel Coll, Judaica Coll, Large Print Coll, Russian Coll, Yiddish Coll
8 300 vols; 23 curr per
libr loan; AJL
37144

Lifeway Christian Resources of the Southern Baptist Convention, E.C. Dargan Research Library, One Lifeway Plaza, **Nashville**, TN 37234-0142
T: +1 615 2512126; Fax: +1 615 2778433; URL: www.library.lifeway.com
1933; Pat Brown
Scofield Photogr Coll, Sunday School Board Press Coll
41 000 vols; 100 curr per; 450 av-mat; 1 670 sound-rec
libr loan
37145

USW International Union Library, 3340 Perimeter Hill Dr, **Nashville**, TN 37211
T: +1 615 8348590; Fax: +1 615 8316792; URL: www.usw.org
1981; Mary Dimoff
UPIU Oral Hist Series, audio, microfilm, video cassette; Int Union; Allied Industrial Workers of America (AIW) Arch; Oil Chemical & Atomic Workers Int Union (OCAW) Arch
11 000 vols; 70 curr per; 550 govt docs; 10 000 microforms; 295 av-mat; 70 sound-rec; 3 digital data carriers
libr loan
37146

Office of Navajo Nation Library, Navajo Community Library, PO Box 22, **Navajo**, NM 87328-0022
T: +1 505 7772598; Fax: +1 505 7772817
Jolene Poyer
65 000 vols
37147

Moravian Historical Society, Museum & Library, 214 E Center St, **Nazareth**, PA 18064
T: +1 610 7595070; Fax: +1 610 7592461; URL: www.moravianhistoricalsociety.org
1857; Susan M. Dreydoppel
10 000 vols; 10 curr per; 400 maps
37148

Nebraska Center for the Education of Children Who Are Blind or Visually Impaired, 824 Tenth Ave, **Nebraska City**, NE 68410; P.O. Box 129, Nebraska City, NE 68410
T: +1 402 8735513; Fax: +1 402 8733463; URL: www.ncecbvi.org
1875; Karen Duffy
15 000 vols; 25 curr per
37149

American Herb Association Library, P.O. Box 1673, **Nevada City**, CA 95959-1673
T: +1 530 2659552; Fax: +1 530 2743140; URL: www.jps.net/ahaherb
1981; Kathi Keville
Medical Botany
6 000 vols; 40 curr per
37150

New Bedford Law Library, Superior Courthouse, 441 County St, **New Bedford**, MA 02740
T: +1 508 9928077; Fax: +1 508 9917411; URL: www.lawlib.state.ma.us/newbedford.html
1894; Jane E. Callahan
25 000 vols; 25 curr per; 100 sound-rec
37151

New Bedford Whaling Museum Research Library, 719 Purchase St, **New Bedford**, MA 02740; mail address: 18 Johnny Cake Hill, New Bedford, MA 02740
T: +1 508 9970046; Fax: +1 508 7176924; URL: www.whalingmuseum.org
1956; Micheal Dyer
Fine arts, Navigation, Technology, Voyages; Paintings & Drawings, photos, Prints, Whaling Mss
20 000 vols; 50 curr per; 20 000 mss; 10 incunabula; 100 diss/theses; 1 000 govt docs; 100 music scores; 500 maps; 300 microforms; 300 av-mat; 100 sound-rec; 10 Special Interest Per Sub
37152

Old Dartmouth Historical Society, New Bedford Whaling Museum Research Library, 791 Purchase St, **New Bedford**, MA 02740-6398
T: +1 508 9970046; Fax: +1 508 9944350; E-mail: research@whalingmuseum.org; URL: www.whalingmuseum.org/library/index.html
1903; Michael Dyer
Int Marine Arch, micro; Whaling Museum Logbook Coll; Whaling (Charles F Batchelder Coll); Maritime Hist (Charles A Goodwin Coll)
120 000 vols; 25 curr per; 750 maps; 2 000 microforms; 1 100 linear feet of mss
37153

New Britain General Hospital, Health Sciences Library, 100 Grand St, **New Britain**, CT 06050
T: +1 860 2245581; Fax: +1 860 2245970; URL: thocc.org/physicians/library.aspx
1946
13 500 vols; 250 curr per
libr loan; MLA
37154

Saint Peter's University Hospital Library, 254 Easton Ave, **New Brunswick**, NJ 08903
T: +1 732 7458508; Fax: +1 732 9376091; E-mail: mlibrary@saintpetersuh.com; URL: www.saintpetersuh.com/medicallibrary/
1907; Jeannine Creazzo
25 000 vols; 450 curr per
libr loan; ALA, SLA, HSLANJ
37155

University of Medicine & Dentistry of New Jersey, Robert Wood Johnson Library of Health Sciences, PO Box 19, **New Brunswick**, NJ 08903
T: +1 732 2357610; Fax: +1 732 2357826; URL: www2.umdnj.edu/rwjlbweb
Kerry O'Rourke
30 000 vols; 500 e-books
37156

Workingmen's Institute Library, 407 W Tavern St, **New Harmony**, IN 47631; P.O. Box 368, New Harmony, IN 47631-0368
T: +1 812 6824806; Fax: +1 812 6824806; URL: www.workingmensinstitute.org
1838; Sherry Graves
Education, Theater; 19th c Women's Rights (Workingmen's Institute), Mss Coll
32 000 vols; 20 curr per; 50 av-mat
37157

Connecticut Agricultural Experiment Station, Thomas B. Osborne Library, 123 Huntington St, **New Haven**, CT 06511-2000; P.O. Box 1106, New Haven, CT 06504-1106
T: +1 203 9748447; Fax: +1 203 9748502; E-mail: Michael.Last@ct.gov; URL: www.ct.gov/caes/site/default.asp
1875; Vickie Bomba-Lewandoski
Analytical chemistry, Biochemistry, Climatology, Entomology, Environmental studies, Genetics, Tobacco
11 000 vols; 500 curr per
37158

Haskins Laboratories Library, 270 Crown St, **New Haven**, CT 06511-6695 T: +1 203 8656163; Fax: +1 203 8658963; E-mail: story@haskins.yale.edu; URL: www.haskins.org 1937; Linda Story Speech communication, linguistics, experimental psychology; Acoustics; Speech Perception; Speech Physiology 8 200 vols; 101 curr per; 150 microforms 37159

Hospital of Saint Raphael, Health Sciences Library, 1450 Chapel St, **New Haven**, CT 06511 T: +1 203 7893330; Fax: +1 203 7895176 1941; Patricia L Wales 11 000 vols; 450 curr per; 5 digital data carriers 37160

New Haven Museum & Historical Society, Whitney Library, 114 Whitney Ave, **New Haven**, CT 06510-1025 T: +1 203 5624183; Fax: +1 203 5622002; URL: www.newhavenmuseum.org 1863; James W. Campbell Genealogy, Local hist; Dana Scrapbooks of New Haven, Mss Colls 30 000 vols; 20 curr per; 505 maps; 2 000 microforms; 4 800 architectural drawings 37161

Yale Center for British Art, Reference Library, 1080 Chapel St, **New Haven**, CT 06520; P.O. Box 208280, New Haven, CT 06520-8280 T: +1 203 4322814; Fax: +1 203 4329613; E-mail: ycba.reference@yale.edu; URL: ycba.yale.edu 1978; Kraig Binkowski Architecture, Art, Hist, Lit, Performing arts 65 000 vols; 75 curr per; 9 000 microforms; sales catalogs, diss on British art, pamphlets, artist file, photo arch 37162

Young Men's Institute Library, 847 Chapel St, **New Haven**, CT 06510 T: +1 203 5624045 1826; Rebecca McGaffin 30 000 vols 37163

Catholics United for Life, Library, c/o Dennis Musk, 3050 Gap Knob Rd, **New Hope**, KY 40052; P.O. Box 10, New Hope, KY 40052 T: +1 270 3253061; Fax: +1 270 3253091; E-mail: sales@newhope-ky.org; URL: www.newhope-ky.org/index.html Theo Stearns Religion, Theology, Philosophy, Law, Lit 15 000 vols; 15 curr per 37164

Long Island Jewish Medical Center, Health Sciences Library, 270-05 76th Ave, **New Hyde Park**, NY 11040 T: +1 718 4707070; Fax: +1 718 4706150; E-mail: medlib@lij.edu; URL: www.lij.edu/library 1954; Debra C. Rand 9 000 vols; 525 curr per 37165

Amistad Research Center, Library/Archives, Tulane Univ, Tilton Hall, 6823 St Charles Ave, **New Orleans**, LA 70118 T: +1 504 8623221; Fax: +1 504 8628961; E-mail: reference@admistadresearchcenter.org; URL: www.admistadresearchcenter.org 1966; Dr. Lee Hampton Ethnic minorities of America, Afro-American history & culture, Civil rights, Africa, Abolitionism, United Church of Christ; African-American Art Colls 20 000 vols; 200 curr per; 3 200 microforms; 10 mio mss, 1,5 mio clippings, 25 000 pamphlets 37166

Historic New Orleans Collection, William Research Center, 410 Chartres St, **New Orleans**, LA 70130-2102 T: +1 504 5987171; Fax: +1 504 5987168; E-mail: wrc@hnoc.org; URL: www.hnoc.org 1966; Alfred E. Lemmon All aspects of life pertaining to New Orleans, Louisiana & the Lower Mississippi Valley; Manuscripts Dept, Curatorial Dept, Publications Dept 16 000 vols; 30 curr per; 1 000 mss; 300 diss/theses; 1 000 music scores; 500 maps; 1 000 microforms; 300 sound-rec;

12 digital data carriers OCLC 37167

Louisiana State Museum – Louisiana Historical Center Library, 400 Esplanade Ave, 751 Chartres, P.O. Box 2448, **New Orleans**, LA 70176-2448 T: +1 504 5688214; Fax: +1 504 5682678; E-mail: kpage@crt.state.la.us; URL: lsm.crt.state.la.us 1930; Kathryn Page Local hist, Louisiana; Colonial Judicial Docs 30 000 vols; 5 000 music scores; 3 500 maps; 500 linear feet of mss 37168

Louisiana State Museum – New Orleans Jazz Club Collection, Old US Mint, 400 Esplanade Ave, **New Orleans**, LA 70176; P.O. Box 2448, New Orleans, LA 70165-2448 T: +1 504 5686968; Fax: +1 504 5684995; URL: lsm.crt.state.la.us/collections/jazz.htm Greg Lambousy New Orleans Jazz Club Coll 28 000 vols 37169

New Orleans Museum of Art, Felix J. Dreyfous Library, One Collins Diboll Circle City Park, **New Orleans**, LA 70124; P.O. Box 19123, New Orleans, LA 70179-0123 T: +1 504 6584117; Fax: +1 504 6584199; URL: www.noma.org 1972; Sheila A. Cork Rare & Limited-Edition Illust Books 20 000 vols; 70 curr per; 300 file boxes of pamphlets, 19 000 slides, 23 VF drawers of artist files libr loan 37170

Ochsner Clinic Foundation, Library and Archives, 1415 Jefferson Hwy, **New Orleans**, LA 70121-2429 T: +1 504 8423760; Fax: +1 504 8425339; E-mail: infodesk@ochsner.org; URL: www.ochsner.org/medical-library 1944; Ethel U. Madden Rare Books 29 000 vols; 554 curr per; 2 digital data carriers 37171

US Department of the Interior, Minerals Management Service – Regional Technical Library, 1201 Elmwood Park Blvd, MS-5031, **New Orleans**, LA 70123-2394 T: +1 504 7362521; Fax: +1 504 7362525; E-mail: stephen.pomes@mms.gov Stephen V. Pomes Environmental law, Geology, Geopysics, Paleontology, Oceanography; Environmental Impact Statements For Offshore Oil & Gas Issues 8 500 vols; 85 curr per libr loan 37172

Aesthetic Realism Foundation, Eli Siegel Collection, 141 Greene St, **New York**, NY 10012-3201 T: +1 212 7774490; Fax: +1 212 7774426; URL: www.aestheticrealism.org 1982 Lessons & lectures by Eli Siegel, founder of Aesthetic Realism; Poetry & prose of Eli Siegel, original mss, holograph; French, German & Spanish Lit, British and American poetry; early American Hist 25 000 vols; 500 curr per; 1 000 sound-rec; 500 19th and early 20th c periodicals 37173

American Academy of Arts & Letters Library, 633 W 155th St, **New York**, NY 10032 T: +1 212 3686361; Fax: +1 212 4914615; E-mail: academy@artsandletters.org; URL: www.artsandletters.org 1898; Virginia Dajani 24 000 vols; 500 sound-rec 37174

American Arbitration Association, Library & Information Center on the Resolution of Disputes, 1633 Broadway, 10th Flr, **New York**, NY 10019-6708 T: +1 212 4844127; Fax: +1 212 2459572; E-mail: referencedesk@adr.org; URL: www.adr.org 1954; Laura Ferris Brown Arbitration Awards; Archival Coll, rpt, letters, photos, pamphlets; State, Federal & Foreign ADR Statutes; Trade Assn & ADR Institution Rules

23 500 vols; 225 curr per; 19 500 microforms 37175

American Bible Society Library, 1865 Broadway, **New York**, NY 10023-9980 T: +1 212 4081203; Fax: +1 212 4088724; E-mail: library@americanbible.org 1816; Jacquelyn Sapiie Chicago Bible Society Scripture Coll, Bk Vols in 2150 Languages (Hist Bibles Coll) 70 000 vols; 250 curr per; 10 000 microforms; 15 digital data carriers; Archives 18 000 items 37176

American Folk Art Museum, Shirley K. Schlafer Library, 45 West 53rd St, **New York**, NY 10019 T: +1 212 2651040; Fax: +1 212 2652350; E-mail: library@folkartmuseum.org; URL: www.folkartmuseum.org 1961; James Mitchell US and international folk art; outsider art; art brut; colonial and early American art and decorative arts; ca 50 l.f. of institutional archives; Archival colls from the Historical Society of Early American Decoration; Henry Darger Study Center 10 000 vols; 180 curr per; 2 microforms; 150 av-mat; 100 sound-rec; ca 100 l.f. vertical files 37177

American Foundation for the Blind, M. Migel C. Memorial Library & Information Center, 11 Penn Plaza, Ste 300, **New York**, NY 10001 T: +1 212 5027661; Fax: +1 212 5027771; E-mail: info@afb.net; URL: www.afb.org 1921; Dr. Jaclyn Packer Helen Keller Arch, Rare Bk Coll 125 000 vols 37178

American Irish Historical Society Library, 991 Fifth Ave, **New York**, NY 10028 T: +1 212 2882263; Fax: +1 212 6287927; E-mail: info@aihs.org; URL: www.aihs.org 1897; N.N. Daniel Cohalan Papers, Friends of Irish Freedom Papers 10 000 vols; 200 linear feet of archives & mss 37179

American Jewish Committee, Blaustein Library, 165 E 56th St, Jacob Blaustein Bldg, **New York**, NY 10022 T: +1 212 7514000 ext 297; Fax: +1 212 8911470; E-mail: library@ajc.org; URL: www.ajc.org 1939; Cyma M. Horowitz 35 000 vols; 180 curr per; 1 450 microforms 37180

American Kennel Club Inc Library, AKC Library, 260 Madison Ave, 4th Flr, **New York**, NY 10016 T: +1 212 6968245; Fax: +1 212 6968281; E-mail: library@akc.org; URL: www.akc.org/about/library/index.cfm 1934; Barbara Kolk Domestic & foreign stud books, Training of dogs, Breeding & Care; Colls of prominent dog-fanciers 18 000 vols; 300 curr per; 5 diss/theses; 3 music scores; 85 microforms 37181

American Museum of Natural History, Osborn Library, Central Park West at 79th St, **New York**, NY 10024 T: +1 212 7695803; E-mail: expeditions@amnh.org; URL: www.amnh.org 1908; Susan K. Bell Vertebrate Paleontology 10 000 vols; 15 000 reprints 37182

American Museum of Natural History Library, Research Library, 79th St & Central Park W, **New York**, NY 10024-5192 T: +1 212 7695400; Fax: +1 212 7695009; E-mail: libref@amnh.org; URL: library.amnh.org 1869; Tom Baione Photogr & Arch Coll, Rare Bks & Mss Coll, Art & Memorabilia, Natural Hist Film Arch; 2 branch libs 489 000 vols; 4 004 curr per; 1 sound-rec; 40 digital data carriers; 80 000 slides libr loan 37183

American Numismatic Society Library, The Harry W. Bass Jr, 96 Fulton St, **New York**, NY 10038 T: +1 212 5714470 ext 1501; Fax: +1 212 5714479; E-mail: campbell@numismatics.org; URL: www.numismatics.org 1858; Francis D. Campbell David M. Bullowa Coll, George C. Miles Coll, Virgil M. Brand Arch, Garrett Numismatic Arch, New Netherlands Coin Company Arch, Norweb Coll Arch, Archer M. Huntington Coll 100 000 vols; 270 curr per; 350 microforms ARLIS/NA, SLA 37184

American Society for Psychical Research Inc Library, Five W 73rd St, **New York**, NY 10023 T: +1 212 7995050; Fax: +1 212 4962497; E-mail: aspr@aspr.com; URL: www.aspr.com 1885; Patrice Keane Alternative medicine, Parapsychology, Philosophy, Psychology, Religious studies, Spiritualism; Shaker Coll 8 000 vols; 110 curr per; 30 microforms; 300 av-mat; 700 sound-rec; 7 000 unbound periodicals, 5 000 pamphlets, 20 VF drawers, 50 diss, 300 linear feet of archives & mss 37185

Anti-Defamation League, Rita & Leo Greenland Library & Research Center, 605 Third Ave, **New York**, NY 10158 T: +1 212 8855844; Fax: +1 212 8855882; E-mail: librarian@adl.org; URL: www.adl.org 1939; Marianne Benjamin Human relations, Discrimination, Civil rights, Intergroup relations, Anti-Semitism, Political extremism; Anti-Semitic Per 15 000 vols; 300 curr per; 2 000 microforms; 10 000 pamphlets 37186

Association of the Bar of the City of New York Library, 42 W 44th St, **New York**, NY 10036 T: +1 212 3826666; Fax: +1 212 6267394; E-mail: library@nycbar.org; URL: www.abcny.org 1870; Richard Tuske Major coll of legal materials incl appellate court records & briefs, domestic law, early Am session laws 600 000 vols; 2 500 curr per; 20 incunabula; 1 000 govt docs; 200 microforms; 150 digital data carriers AALL, SLA 37187

Austrian Cultural Forum Library, 11 E 52nd St, **New York**, NY 10022 T: +1 212 3195300; Fax: +1 212 6448660; E-mail: library@acfny.org; URL: www.acfny.org 1962; Edeltrud Desmond Art & architecture, Education, Hist, Lit, Music, Performing arts; Austriaca Coll 10 000 vols; 22 curr per 37188

Beth Israel Medical Center, Seymour J. Phillips Health Sciences Library, 317 E 17th, **New York**, NY 10003; First Ave at 16th St, New York, NY 10003 T: +1 212 4202855; Fax: +1 212 4204640; E-mail: library@chpnet.org 1946; Maria Astifidis 15 000 vols; 600 curr per; 300 sound-rec; 3 digital data carriers libr loan; MLA, SLA 37189

Buttenwieser Library, 92nd Street Young Men's & Young Women's Hebrew Association, 1395 Lexington Ave, **New York**, NY 10128-1612 T: +1 212 4155542; Fax: +1 212 4276119; E-mail: library@92y.org; URL: www.92y.org 1874; Steven W. Siegel Judaica, Arch of the 92nd Street YM-YWHA, Poetry Coll, F.W. Greenfield Young People's Libr 30 000 vols; 90 curr per; 30 digital data carriers AJL 37190

Center for Jewish History, Leo Baeck Institute Library, 15 W 16 St, **New York**, NY 10011-6301 T: +1 212 7446400; Fax: +1 212 9881305; E-mail: lbaeck@lbi.cjh.org; URL: www.lbi.org 1955; Dr. Frank Mecklenburg Art Coll; Hist & Arch of German-speaking

Jewry of Central Europe, 18th-20th c; Extensive Coll of Lit by Jews in German Language
70 000 vols; 150 curr per; archives (1 300 linear ft)
libr loan; RLG 37191

The Century Association Library, 7 W 43rd St, ***New York***, NY 10036
T: +1 212 9440090; Fax: +1 212 8403609; E-mail: library@thecentury.org
1847; W. Gregory Gallagher
Platt Libr (Architecture)
25 000 vols; 90 curr per; 20 sound-rec 37192

Chancellor Robert R. Livingston Masonic Library of Grand Lodge, 71 W 23rd St, 14th Flr, ***New York***, NY 10010-4171
T: +1 212 3376620; Fax: +1 212 6332639; E-mail: info@nymasoniclibrary.org; URL: www.nymasoniclibrary.org
1865; Thomas M. Savini
NY City, bibliogr, directories & mss; NY City, State & County Hist; Esoteric & Hermetic Tradition; Int Masonic Per
60 000 vols; 150 curr per; 500 000 mss, 120 VF drawers
SLA 37193

Children's Book Council Library, 54 W 39th St, 14th Floor, ***New York***, NY 10018
T: +1 212 9661990; Fax: +1 212 9662073; E-mail: cbc.info@cbcbooks.org; URL: www.cbcbooks.org/about/library
1945; Paula Quint
Nat Children's Bk Week Posters 1919-1994, Selected Prizewinning Children's Bks
8 000 vols; 35 curr per 37194

City of New York Department of Records & Information Services, City Hall Library, 31 Chambers St, Rm 112, ***New York***, NY 10007
T: +1 212 7888590; Fax: +1 212 7888589; E-mail: chlibrary@records.nyc.gov; URL: www.nyc.gov/html/records
1913; Christine Bruzzese
Street name origins (street name file); Depository for N.Y.C. govt publs; Extensive clipping files on NYC affairs; Neighborhood files; Biography files on NYC civic personalities
250 000 vols; 90 curr per; 200 000 govt docs; 2 000 maps; 1 000 microforms; 3 digital data carriers
ALA, SLA 37195

City University of New York (CUNY), Hunter College, Centro de Estudios Puertorriquenos Library & Archives, East Bldg (inside Wexler Library), 3rd Fl, 695 Park Ave, ***New York***, NY 10065
T: +1 212 7724197; Fax: +1 212 6503628; URL: www.centropr.org
Dr. Alberto Hernandez
Puerto Rican Diss Coll
25 000 vols 37196

Collectors Club Library, 22 E 35th St, ***New York***, NY 10016-3806
T: +1 212 6830559; Fax: +1 212 4811269; URL: www.collectorsclub.org
1896; Miklos Tinther
Philately, Postal hist
30 000 vols; 30 curr per 37197

Cooper Union for Advancement of Science & Art Library, Seven E Seventh St, ***New York***, NY 10003; mail address: 30 Cooper Sq, New York, NY 10003-8001
T: +1 212 3534189; Fax: +1 212 3534017; URL: www.cooper.edu/facilities/library/library.html
1859; Ulla Volk
Architecture, Art, Engineering; Cooperana
100 000 vols; 321 curr per; 600 maps; 9 000 microforms; 108 av-mat; 6 digital data carriers; 90 000 pictures, 52 000 slides
libr loan 37198

Council on Foreign Relations Library, 58 E 68th St, ***New York***, NY 10021
T: +1 212 4349400; Fax: +1 212 4349824; E-mail: clibrary@cfr.org; URL: www.cfr.org
1930; Dr. Lilita Gusts
20 000 vols; 324 curr per; 100 microforms; 100 av-mat; 1 000 sound-rec; 8 digital data carriers; 1 000 pamphlets 37199

Department of Veterans Affairs, New York Harbor Healthcare System, New York Campus Library, 423 E 23rd St, ***New York***, NY 10010
T: +1 212 6867500 ext 7682; Fax: +1 212 9513367; E-mail: wiseman.karin@forum.va.gov
1956; Karin Wiseman
8 000 vols; 430 curr per 37200

Equitable Life Assurance Society of the United States Library, Law Libr, 12th Fl, 1290 Ave of Americas, ***New York***, NY 10104
T: +1 212 3143820; Fax: +1 212 7077858
Katherine Camesas
Corp law, Insurance law
10 000 vols; 75 curr per 37201

Explorers Club, James B. Ford Library, 46 E 70th St, ***New York***, NY 10021
T: +1 212 6288383; Fax: +1 212 2884449; E-mail: collections@explorers.org; URL: www.explorers.org
1905; Clare Flemming
Arctic Exploration & Studies (Peary Coll), 18th – 20th C Travel
15 000 vols; 50 curr per 37202

Fish & Neave IP Group of Ropes Gray LLP Library, 1251 Avenue of the Americas, ***New York***, NY 10020
T: +1 212 5969200; Fax: +1 212 5969090
Louise E. Studer
16 000 vols; 78 curr per 37203

French Institute-Alliance Francaise Library, Haskell Library, 22 E 60th St, ***New York***, NY 10022-1077
T: +1 646 3886656; Fax: +1 212 9354119; E-mail: library@fiaf.org; URL: www.fiaf.org
1911; Katharine Branning
Art & architecture, French language, Hist, Lit, Philosophy; Paris Coll
35 000 vols; 102 curr per; 2 500 av-mat; 2 500 sound-rec; 100 digital data carriers; 1 000 Large Print Bks
libr loan; ALA, SLA, NYLA 37204

The Frick Collection, Frick Art Reference Library, 10 East 71st St, ***New York***, NY 10021
T: +1 212 5470641; Fax: +1 212 8792091; E-mail: library@frick.org; URL: www.frick.org/library/index.htm
1920; N.N.
American & European fine arts; Photoarch, Artists' colls (res & mat: correspondence, papers, sketchbooks, diaries), Gallery arch
375 000 vols; 615 curr per; 3 000 diss/theses; 88 microforms; 47 digital data carriers; 70 000 auction sales cat
libr loan; IFLA, ARLIS/NA, RLG, VRA, ASIS, SLA 37205

General Society of Mechanics & Tradesmen, General Society Library, 20 W 44th St, ***New York***, NY 10036
T: +1 212 9211767; Fax: +1 212 8402046; E-mail: library@generalsociety.org; URL: www.generalsociety.org/library
1820; Dr. Janet Wells
History, Biogr, Fiction; Gilbert & Sullivan (Alma Watson Coll); 1 branch libr
153 000 vols; 45 curr per 37206

Goethe-Institut, Bibliothek (German Cultural Center), 1014 Fifth Ave, ***New York***, NY 10028
T: +1 212 4398688; Fax: +1 212 4398705; E-mail: library@newyork.goethe.org; URL: www.goethe.de/newyork
1957; Brigitte Doellgast
9 000 vols; 100 curr per; 800 av-mat; 500 sound-rec; 37 digital data carriers
libr loan; ALA, WESS, CARL, OCLC, LITA 37207

Goldwater Memorial Hospital, Health Sciences Library, 900 Main, Roosevelt Island, ***New York***, NY 10001
T: +1 212 3184376; Fax: +1 212 3184628
1939
14 000 vols; 386 curr per; 4 VF drawers 37208

Grolier Club of New York Library, 47 E 60th St, ***New York***, NY 10022
T: +1 212 8386690; Fax: +1 212 8382445; URL: www.grolierclub.org
1884; Fernando Pena
Bibliogr; Book Trade & Auction Cat; Hist of Printing & Book Collecting; Inventories of Private Libs, Authors, Printers & Artists
100 000 vols; 200 curr per 37209

The Hampden-Booth Theatre Library, The Players, 16 Gramercy Park South, ***New York***, NY 10003
T: +1 212 2281861; Fax: +1 212 2536473; E-mail: Hampdenboo@aol.com; URL: www.hampden-booth.org
1957; Raymond Wemmlinger
Edwin Booth Coll, Walter Hampden Coll, Robert B. Mantell Coll, Lawrence Barrett Coll, Sothern and Marlowe Coll, Union Square Theatre Coll
10 000 vols; 25 curr per; 20 microforms; 250 sound-rec; promptbks,letters, journals, photos
ALA 37210

Harvard Library in New York, 27 W 44th St, ***New York***, NY 10036
T: +1 212 8271246; Fax: +1 212 8271251 ext 1246; URL: hcny.com
1978; Adrienne G. Fischier
Harvardiana
24 000 vols
SLA 37211

Hispanic Society of America Library, 613 W 155th St, ***New York***, NY 10032
T: +1 212 9262234; Fax: +1 212 6900743; E-mail: library@hispanicsociety.org; URL: www.hispanicsociety.org
1904; Dr. John O'Neill
Coll of books printed before 1701; Coll of books printed before 1500; Spanish medieval mss
300 000 vols; 142 curr per; 200 000 mss; 250 incunabula; 113 VF drawers of clippings, 200 mss maps, 170 000 photos, 15 000 art prints, maps, globes, stamps, posters 37212

International Center of Photography Library, Concourse, 1114 Avenue of the Americas, ***New York***, NY 10036-7703
T: +1 212 8570004; Fax: +1 212 8570091; E-mail: library@icp.org; URL: www.icp.org/library/library.html
1977; Deirdre Donohue
15 000 vols; 75 curr per 37213

International Ladies' Garment Workers Union, Research Department Library, 1710 Broadway, ***New York***, NY 10019
T: +1 212 2657000 ext 207; Fax: +1 212 4897238
1937; Walter Mankoff
14 000 vols; 155 curr per; 75 VF drawers 37214

Istituto Italiano di Cultura, Biblioteca, 686 Park Ave, ***New York***, NY 10021-5009
T: +1 212 8794242; Fax: +1 212 8614018; E-mail: iicnewyork@esteri.it; URL: www.iicnewyork.esteri.it
1959; Paolo Barlera
34 500 vols; 110 curr per; 24 sound-rec; 5 digital data carriers 37215

Japan Society, C.V. Starr Library, 333 E 47th St, ***New York***, NY 10017
T: +1 212 8321155 ext 256; Fax: +1 212 7151262; URL: www.japansociety.com
1971; Reiko Sassa
14 000 vols; 116 curr per 37216

Knoedler Art Library, 19 E 70th St, ***New York***, NY 10021
T: +1 212 7940567; Fax: +1 212 7726932; E-mail: library@knoedlergallery.com
1846; Edye Weissler
60 000 vols; 26 000 microforms
ARLIS/NA 37217

KPMG LLP, Tax Library, 345 Park Ave, Lexington Level, ***New York***, NY 10154
T: +1 212 8726983; Fax: +1 212 7589819
Andrea Ying
10 000 vols; 100 curr per 37218

Kristine Mann Library, C G Jung Center of New York, 28 E 39th St, ***New York***, NY 10016
T: +1 212 6977877; Fax: +1 212 9861743; E-mail: info@junglibrary.org; URL: junglibrary.org
1945; Michele McKee
Carl Gustav Jung (Jung Press Arch), photostats of press clippings
21 000 vols; 40 curr per; 250 diss/theses; 1 000 sound-rec; 2 000 pamphlets and clippings
SLA 37219

Laboratory Institute of Merchandising Library, 216 E 45th St, 2nd Flr, ***New York***, NY 10017
T: +1 212 7521530; Fax: +1 212 7503453; URL: www.limcollege.edu
1939; George Sanchez
Merchandising (B. Earl Puckett Fund for Retail Education)
12 000 vols; 110 curr per 37220

Legal Aid Society, Central Library & Information Center, 199 Water St, ***New York***, NY 10038
T: +1 212 2985258; Fax: +1 212 6931149; URL: www.legal-aid.org
1876
64 000 vols; 80 curr per 37221

Lesbian Herstory Archives, 484 14th St, ***New York***, 11215-5702; P.O. Box 1258, New York, NY 10116-1258
T: +1 718 7683953; Fax: +1 718 7684663; URL: lesbianherstoryarchives.org
1974
Individual and organizational special collections (250)
15 000 vols; 1 500 curr per; 1 000 mss; 250 diss/theses; 1 000 av-mat; 200 sound-rec; 20 Special Interest Per Sub 37222

Marlborough Gallery Library, 40 W 57th St, ***New York***, NY 10019
T: +1 212 5414900; Fax: +1 212 5414948; E-mail: mny@marlboroughgallery.com; URL: www.marlboroughgallery.com
Kate Gilmartin
10 000 vols 37223

Memorial Sloan-Kettering Cancer Center Medical Library, Nathan Cummings Ctr, 1275 York Ave, ***New York***, NY 10021
T: +1 212 6397439; E-mail: beardslc@mskcc.org; URL: library.mskcc.org/scripts/portal/index.pl
Ann Robbins
Memorial Sloan-Kettering Cancer Ctr Arch
10 000 vols; 600 curr per; 4 digital data carriers; 18 drawers of microforms 37224

Metropolitan Club Library, One E 60th St, ***New York***, NY 10022
T: +1 212 8387400; Fax: +1 212 7556849; URL: www.metropolitanclubnyc.org
1891
10 000 vols; 20 curr per 37225

Metropolitan Hospital Center, Frederick M. Dearborn Library, 1901 First Ave & 97th St, ***New York***, NY 10029
T: +1 212 4236270; Fax: +1 212 4237961; E-mail: mhclibrary@yahoo.com
1906; Antoinette Drago
8 000 vols; 90 curr per; 3 200 mss
libr loan; NNLM 37226

The Metropolitan Museum of Art – Cloisters Library, Fort Tryon Park, ***New York***, NY 10040
T: +1 212 3965319; Fax: +1 212 7953640; E-mail: cloisters.library@metmuseum.org
1938; Michael K. Carter
European medieval art, Medieval architecture, Middle ages; Arch of The Cloisters; George Gray Barnard Papers; Harry Bober Papers; Sumner McKnight Crosby Papers
13 000 vols; 66 curr per; 3 digital data carriers; 20 000 slides, 140 000 photos 37227

The Metropolitan Museum of Art – Robert Goldwater Library, 1000 Fifth Ave, ***New York***, NY 10028-0198
T: +1 212 5703707; Fax: +1 212 5703879; E-mail: goldwater.library@metmuseum.org; URL: library.metmuseum.org
1982; Ross Day
Archeology, Art – African, Latin American, Indians, Oceania
20000 vols; 224 curr per 37228

The Metropolitan Museum of Art – Robert Lehman Collection Library, 1000 Fifth Ave, ***New York***, NY 10028
T: +1 212 5703915; Fax: +1 212 6502542; E-mail: lehman.library@metmuseum.org
1975; Pia Palladino
Western European Arts, Italian majolica, Hist of frames, Studies on illuminated mss; Archs, Photogr Coll
18000 vols
ALA, ARLIS/NA 37229

The Metropolitan Museum of Art – The Irene Lewisohn Costume Reference Library, Costume Institute, 1000 Fifth Ave, ***New York***, NY 10028
T: +1 212 6502723; Fax: +1 212 5703970; E-mail: thecostumeinstitute@metmuseum.org; URL: www.metmuseum.org
1951; Tatyana Pakhladzhyan
Fashion, Hist of costume; Norman Norell Coll (scrap bks), Mainbocher Arch
40000 vols; 75 curr per; 100 av-mat; 100000 pieces of ephemera 37230

The Metropolitan Museum of Art – Thomas J. Watson Library, 1000 Fifth Ave, ***New York***, NY 10028-0198
T: +1 212 6502175; Fax: +1 212 5703847; E-mail: watson.library@metmuseum.org; URL: library.metmuseum.org
1880; Kenneth Soehner
Archaeology, Architecture, Decorative art, Fine art; Art Auction & Exhibition cats
600000 vols; 2500 curr per; 30000 e-books; 2000 microforms; 75 digital data carriers; 103 VF drawers,
libr loan; RLG 37231

Museum of Jewish Heritage, A Living Memorial to the Holocaust, Library, 36 Battery Place, ***New York***, NY 10280
T: +1 646 4374200; URL: www.mjhnyc.org/index.htm
1986; Julia Bock
Memorial books of European Jewish communities (183 items), 36 linear feet archival colls
8000 vols; 10 curr per; 9 mss; 20 maps; 2800 sound-rec; 25 digital data carriers; 25 titles of archivial footage
libr loan; ALA, SLA 37232

Museum of Modern Art, Library, 11 W 53rd St, ***New York***, NY 10019-5498
T: +1 212 7089433; Fax: +1 212 3331122; E-mail: library@moma.org; URL: library.moma.org
1929; Milan Hughston
Artists' Bks, Artist Files, Dada & Surrealism (Eluard-Dausse Coll), Franklin Furnace/Museum of Modern Art, Latin American Art, Museum of Modern Art Publs, Political Art Documentation & Distribution (PADD)
200000 vols; 300 curr per
libr loan; ARLIS/NA, IFLA 37233

National Association for the Advancement of Colored People, NAACP Legal Defense & Educational Fund Law Library, 99 Hudson St, 16th Flr, ***New York***, NY 10013
T: +1 212 9652200; Fax: +1 212 2267592
Donna Gloeckner
Civil rights
21000 vols; 15 digital data carriers
37234

National Association for Visually Handicapped, Large Print Loan Library, 22 W 21st St, 6th Fl, ***New York***, NY 10010
T: +1 212 8893141; Fax: +1 212 7272931; E-mail: navh@navh.org; URL: www.navh.org
1978; Ann Illuzzi
12000 vols; 250 sound-rec 37235

New York Academy of Medicine Library, 1216 Fifth Ave, ***New York***, NY 10029-5293
T: +1 212 8227327; Fax: +1 212 4230266; E-mail: library@nyam.org; URL: www.nyam.org/library/index.shtml
1847; Janice Kaplan
Rare Medical Works, Medical Americana By & About J & W Hunter, Francesco Redi & Contemporaries, Medical Theses-16th-18th C, Cardiology, Engravings of Medical Men, 16-19th C Medals, German Psychology & Psychiatry, Anatomy & Surgery, Medical Economics, Foods & Cookery; Gladys Brooks Conservation Laboratory
550000 vols; 1000 curr per; 141 incunabula; 5 digital data carriers; 182804 pamphlets, 276000 parts & illus
libr loan; ALA, AMLA, SLA, International Greynet Association 37236

New York County Lawyers' Association Library, 14 Vesey St, ***New York***, NY 10007
T: +1 212 2676646; Fax: +1 212 7916437; URL: www.nycla.org
1908; Nuchine Nobari
American Citizenship Coll
200000 vols; 225 curr per; 7 digital data carriers
AALL 37237

New York Genealogical and Biographical Society Library, 122 E 58th St, 4th Flr, ***New York***, NY 10022-1939
T: +1 212 7558532; Fax: +1 212 7544218; E-mail: library@nygbs.org; URL: www.newyorkfamilyhistory.org
1869; William Potter
New York Church and cemetery records; Genealogical charts; Maps; Personal diaries; Family Bible records; Mss coll from professional genealogists; Rare Books, Mss Coll
75000 vols; 1320 curr per; 50000 mss; 250 incunabula; 5 diss/theses; 300 maps; 23000 microforms; 150 sound-rec; 105 digital data carriers
ALA, SLA 37238

New York Historical Society Library, 170 Central Park W, ***New York***, NY 10024-5152
T: +1 212 8733400; Fax: +1 212 8751591; URL: www.nyhistory.org
1804; Jean W. Ashton
American Almanacs, American Genealogy, American Indian (Accounts of & Captivities), Early American, Imprints, Early Travels in America, Early American Trials; Mss dept, dept of prints, photogr & architecture
350000 vols; 400 curr per; 2000000 mss; 30000 maps; 107500 microforms; 150000 Pamphlets, 25000 Broadsides
libr loan 37239

New York Law Institute Library, 120 Broadway, Rm 932, ***New York***, NY 10271-0043
T: +1 212 7328720
1828; Ralph Monoco
Records & Briefs: US Supreme Court; US Court of Appeals, 2nd Circuit; NY Court of Appeals; NY Supreme Court Appellate Div
270000 vols
AALL 37240

New York Legislative Service, Inc Library, 15 Maiden Lane, Ste 1000, ***New York***, NY 10038
T: +1 212 9622826; Fax: +1 212 9621420; E-mail: nylegal@nyls.org; URL: www.nyls.org
Laird Ehlert
New York State Legislative Docs
10000 vols 37241

New York Psychoanalytic Institute, Abraham A. Brill Library, 247 E 82nd St, ***New York***, NY 10028
T: +1 212 8796900; Fax: +1 212 8790588; E-mail: library@nypsa.org; URL: www.nypsa.org
Mathew von Unwerth
Freud's Writings in all Editions & Languages (Sigmund Freud Coll); Art (Arieti Papers); Hist, Lit &, Sociology (Ernst Kris Coll); Languages & Linguistics Coll
30000 vols; 55 curr per 37242

The New York Public Library – Astor, Lenox & Tilden Foundations – Andrew Heiskell Braille & Talking Book Library, 40 W 20th St (between 5th & 6th Avenues), ***New York***, NY 10011-4211
T: +1 212 2065400; Fax: +1 212 2065418; E-mail: ahlbph@nypl.org; URL: www.talkingbooks.nypl.org/
Bonnie Farrier
17000 vols; 960000 av-mat 37243

The New York Society Library, 53 E 79th St, ***New York***, NY 10021
T: +1 212 2886900; Fax: +1 212 7445832; E-mail: web@nysoclib.org; URL: www.nysoclib.org
1754; Mark Bartlett
Art, Biogr, Fiction, New York City, Travel, WWII; Alchemy & Chemistry (John Winthrop Libr); 18th & Early 19th C Fiction (Hammond Coll); 18th C Statesman (Goodhue Papers)
275000 vols; 200 curr per; 1500 Large Print Bks 37244

North General Hospital, Medical Library, 1879 Madison Ave, ***New York***, NY 10035
T: +1 212 4234476
Bruce Delman
Medicine, surgery
8000 vols; 40 curr per 37245

Parapsychology Foundation, Eileen J. Garrett Library, 228 E 71st St, ***New York***, NY 10021-5136
T: +1 212 6281550; Fax: +1 212 6281559; E-mail: INFO@parapsychology.org; URL: www.parapsychology.org
1951; Elizabeth Vajda
Rare Books Dealing with Psychical Research & Spiritualism; AV Department
11130 vols; 100 curr per; 98 mss; 500 av-mat; 11 VF drawers of pamphlets, clippings, reprints, 100 audio-tapes
SLA 37246

Pat Parker-Vito Russo Center Library, Lesbian & Gay Community Serv Ctr, 208 W 13th St, ***New York***, NY 10011
T: +1 212 6207310; E-mail: library@gaycenter.org; URL: www.gaycenter.org/library
David Chase
Literature produced by and for lesbian, gay, bisexual and transgender (LGBT), fiction, non-fiction
11000 vols; 20 curr per 37247

Pierpont Morgan Library, 29 E 36th St, ***New York***, NY 10016
T: +1 212 6850610; Fax: +1 212 4813484; E-mail: media@morganlibrary.org; URL: www.morganlibrary.org
1924; Charles E. Pierce Jr
Autogr mss, bk bindings, early children's bks, incunabula, letters, master drawings, medieval & Renaissance mss, Mesopotamian seals, musical mss
150000 vols; 300 curr per 37248

Pilsudski Institute of America Library, 180 Second Ave, ***New York***, NY 10003-5778
T: +1 212 5059077; Fax: +1 212 5059052; E-mail: info@pilsudski.org
1943; Anna Dyba
Polish hist & politics, US hist; Diplomatic & Military Docs of Jozef Pilsudski's Military Chancellery
22000 vols 37249

Polish Institute of Arts & Sciences in America, Inc, Research Library, 208 E 30th St, ***New York***, NY 10016
T: +1 212 6864164; Fax: +1 212 5451130; E-mail: piasany@verizon.net; URL: www.piasa.org
1942; Krystyna Baron
Polish Embassy in the US Coll; Oral Hist Coll; Latin-American Coll
24000 vols; 400 curr per 37250

Population Council Library, One Dag Hammarskjold Plaza, ***New York***, NY 10017
T: +1 212 3390533; Fax: +1 212 7556052; URL: www.popcouncil.org
1953; H. Neil Zimmerman
28000 vols; 350 curr per
SLA, APLIC-I 37251

Presbyterian Hospital, John M. Wheeler Library, Edward S. Harkness Eye Institute, 635 W 165th St, ***New York***, NY 10032
T: +1 212 3052916; Fax: +1 212 3053173
1933; Lijun Tian
Ophthalmology Memorabilia, Rare Bk Coll
14120 vols; 90 curr per; 485 sound-rec; 4250 reprints
libr loan 37252

Princeton Library in New York, 15 W 43rd St, 5th Flr, ***New York***, NY 10036
T: +1 212 5961250; Fax: +1 212 5961398; E-mail: info@princetonclub.com
1962; Betty Dornheim
Histories of New York; Princetoniana, Woodrow Wilson
10000 vols; 70 curr per 37253

Racquet and Tennis Club Library, 370 Park Ave, ***New York***, NY 10022-5968
T: +1 212 7539700
1905; Gerard J. Belliveau Jr
Court Tennis (Jeu de Paume Coll); Lawn Tennis; Early American Sports
19750 vols; 30 curr per 37254

Rockefeller Foundation Records & Library Services, 420 Fifth Ave, ***New York***, NY 10018-1600
T: +1 212 8528428; Fax: +1 212 8528443; URL: www.rockefellerfoundation.org
1913
Philanthropy, Social sciences, Biogr; Annual Rpts of Philanthropic Institutions
10000 vols; 600 curr per; 120 microforms; 4 digital data carriers 37255

Romanian Cultural Institute, 200 E 38th St, ***New York***, NY 10016
T: +1 212 6870180; Fax: +1 212 6870181; E-mail: roculture@aol.com; URL: www.icrny.org
1969; Bogdan Stefanescu
Bibliogr (Romanian Topics Coll)
22000 vols; 60 curr per; 500 sound-rec; 1000 slides 37256

Roosevelt Hospital, Medical Library, 1000 Tenth Ave, ***New York***, NY 10019
T: +1 212 5236100; Fax: +1 212 5236108
1955; Paul Barth
15000 vols; 500 curr per; 2 digital data carriers 37257

Shevchenko Scientific Society Inc, Library and Archives, 63 Fourth Ave, ***New York***, NY 10003
T: +1 212 2545130; Fax: +1 212 2545239; E-mail: library@shevchenko.org; URL: www.shevchenko.org
1952; Svitlana Andrushkiw
Ukrainian hist & lit; The Immigration of Ukrainians to North & South America, rare bks, arch & docs
45000 vols; 400 av-mat; 150 catalogued archives 37258

Sons of the Revolution in the State of New York Library, Fraunces Tavern Museum, 54 Pearl St, ***New York***, NY 10004
T: +1 212 4251778; Fax: +1 212 5093467; URL: www.fraucestavernmuseum.org
Andrew Batten
9000 vols 37259

South Street Seaport Museum Library, 213 Water St, ***New York***, NY 10038-2105; mail address: 12 Fulton St, New York, NY 10038-2106
Fax: +1 212 7488672; E-mail: sssmcurat@aol.com
1967; Jeff Remling
Maritime hist, Shipping, New York City; General Shipping (Port of New York), photogr & negatives
20000 vols; 50 mss; 1000 maps; 140 microforms; 20000 ship plans 37260

St Luke's-Roosevelt Hospital Center, Richard Walker Bolling Memorial Medical Library, 1111 Amsterdam Ave, ***New York***, NY 10025
T: +1 212 5234315; Fax: +1 212 5234313
1876; Dr. Nancy Panella
Surgical & Medical Hist Instruments, Photogr & Other Memorabilia, Hist of 19th/20th C Medicine – NYC; Hospital Ctr hist archs

10 000 vols; 275 curr per libr loan; IFLA, SLA, MLA   37261

Tibetan Buddhist Resource Center, Inc, TBRC at the Rubin Museum of Art, 150 W 17th St, **New York**, NY 10011
T: +1 212 6205000; Fax: +1 212 7272997; E-mail: info@tbrc.org; URL: tbrc.org
Ellis Gene Smith
Tibetan Literature
12 000 vols   37262

Union League Club Library, 38 E 37th St, **New York**, NY 10016
T: +1 212 6853800; Fax: +1 212 5450130; E-mail: info@unionleagueclub.org
1863; Carrie Hayter
American hist & biogr, Civil War
20 000 vols; 50 curr per   37263

United Nations Childrens Fund Library, Three UN Plaza, H-12C UNICEF House, **New York**, NY 10017
T: +1 212 3267064; Fax: +1 212 3037989
Howard Dale
UNICEF Board Docs, microfiche
25 000 vols; 400 curr per; 22 digital data carriers   37264

United Nations Dag Hammarskjold Library, United Nations, **New York**, NY 10017
T: +1 212 9637412; Fax: +1 212 9632388; E-mail: dhl_www@un.org; URL: www.un.org/depts/dhl/
1946; Linda Stoddart
Disarmament, Economy, Law, Political science, Int relations; League of Nations Docs, United Nations Docs, Specialized Agencies Docs, Int Affairs 1918-1945 (Woodrow Wilson Memorial Libr), Activities & Hist of the United Nations, Maps Coll, Govt Docs of Member States, Official Gazettes of all Countries
400 000 vols; 8 500 curr per; 80 000 maps; 323 000 microforms; 102 digital data carriers   37265

United Nations Dag Hammarskjold Library – Legal Collection, United Nations Plaza, UN Hq, Rm S-3455, **New York**, NY 10017
T: +1 212 9635372; Fax: +1 212 9631770; E-mail: dhllegal@un.org
Rosemary Noona
Int Law with Treaty Coll, Legal Per & Yearbks, New York Legislation & Decisions, US Federal Legislation & Decisions
13 000 vols   37266

United Nations Dag Hammarskjold Library – Statistical Collection, Two United Nations Plaza, Rm DC2-1143, **New York**, NY 10017
T: +1 212 9638727; Fax: +1 212 9630479
Luz Maria Saavedra
65 000 vols   37267

University Club Library, One W 54th St, **New York**, NY 10019
T: +1 212 5723418; Fax: +1 212 5723452; E-mail: library@universityclubny.org
1865; Laurie Bolger
Literature, History, Collegiana, Biography; Tinker, Darrow, Rudge Colls (Fine Printing & Printing Hist); Bicklehaupt Coll of Sporting Bks; Southern Society Coll
90 000 vols; 125 curr per; 4 digital data carriers; 40 VF drawers   37268

Whitney Museum of American Art, Frances Mulhall Achilles Library, 945 Madison Ave, **New York**, NY 10021
T: +1 212 5703648; Fax: +1 212 5707729; E-mail: carol_rusk@whitney.org; URL: www.whitney.org
1931; Carol Rusk
Artist's bks, Edward Hopper Research Coll
50 000 vols; 100 curr per; 100 microforms, 400 av-mat; 800 sound-rec; 160 Vertical file drawers of clippings
ARLIS/NA   37269

Xavier Society for the Blind, National Catholic Press & Lending Library for the Visually Impaired, 154 E 23rd St, **New York**, NY 10010-4595
T: +1 212 4737800; URL: www.xaviersociety.com
1900; Robert Nealon
Braille, Catholicism, Inspirational, Large print, Religious, Spirituals
20 000 vols; 10 curr per; 14 736 av-mat   37270

Yale Club Library, 50 Vanderbilt Ave, **New York**, NY 10017
T: +1 212 7162129; Fax: +1 212 7162158; E-mail: librarian@yaleclubnyc.org; URL: www.yaleclubnyc.org
Louise Jones
Yale Memorabilia & Publs
46 000 vols   37271

YIVO Institute for Jewish Research, Library & Archives, 15 W 16th St, **New York**, NY 10011
T: +1 212 2466080; Fax: +1 212 2921892; E-mail: yivomail@yivo.cjh.org; URL: www.yivo.org
1925; Aviva E. Astrinsky
Jewish Immigration (Hebrew Immigrant Aid Society – HIAS Coll), doc; Jewish Music Coll; Manuscript Coll, sheet music; Nazi Coll; Rabbinics (Vilna Coll); Yiddish Linguistics (Weinreich Libr Coll) Milwitzky Ladino Coll
350 000 vols; 200 curr per; 1 500 maps; 6 000 microforms; 1 500 sound-rec; 2 000 Slides, 3 000 art Reproductions   37272

Smithsonian Institution Libraries, Cooper-Hewitt, National Design Museum Library, Two E 91st St, 3rd Flr, **New York City**, NY 10128
T: +1 212 8498330; Fax: +1 212 8498339; URL: www.sil.si.edu/libraries/chm
Elizabeth Broman
18th & 19th C Line Engravings (George W Kubler Coll), Auction Cats, Archs, Therese Bonney photos, World's Fair Coll (1844-1893)
71 000 vols; 4 000 rare bks, 16 VF drawers   37273

Delaware Geological Survey Library, University of Delaware, Delaware Geological Survey Bldg, **Newark**, DE 19716-7501
T: +1 302 8312833; Fax: +1 302 8313579; E-mail: delgeosurvey@udel.edu; URL: www.udel.edu/dgs
Laura Wisk
Water resources, cartography, geology
10 000 vols   37274

Licking County Genealogical Society Library, 101 W Main St, **Newark**, OH 43055-5054
T: +1 740 3495510; E-mail: lcgs@npls.org; URL: www.npls.org/pub_lcgs_web
1972; G.R. Rose
Obit Files; Family Data Files; Ancestor Files
14 000 vols; 100 curr per; 1 056 microforms; 10 digital data carriers   37275

Licking County Law Library Association, 65 E Main St, **Newark**, OH 43055
T: +1 740 3496561; Fax: +1 740 3496561
1896; James W. Pyle
10 000 vols; 91 curr per   37276

Margaret S. Sterck School for the Deaf Library, 620 E Chestnut Hill Rd, **Newark**, DE 19713
T: +1 302 4542301; Fax: +1 302 4543493; URL: www.dsdhawks.org
1968; Arden Lantz
Large Print Books for the Visually Impaired; Captioned Filmstrip Coll; Professional Libr of mats on deafness, deaf culture & visual impairment
10 000 vols; 25 microforms   37277

New Jersey Historical Society Library, 52 Park Pl, **Newark**, NJ 07102-4302
T: +1 973 5968500; Fax: +1 973 5966957; E-mail: library@jerseyhistory.org; URL: www.jerseyhistory.org
1845; Julia Telonidis
Early New Jersey Imprint Coll, Ephemera Coll, Vertical File Coll, NJ Photo Negative Coll, Specialty Libr Card Indices
65 000 vols; 100 curr per; 200 diss/theses; 2 000 maps; 200 microforms; 3 000 linear feet of mss, 1 600 mss groups, 12 000 pamphlets, 1 000 rare bks, 400 Broadsides
Publishes New Jersey Historical journal.
Also museum artifact coll. Exhibition galleries, education programs   37278

Newark Museum Library, 49 Washington St, **Newark**, NJ 07102-3176
T: +1 973 5966625; Fax: +1 973 6420459; E-mail: library@newarkmuseum.org
1909; William A. Peniston
African, American & Asian art, Decorative arts, Natural science, Numismatics; Tibet Coll, Dana Coll; 11 branch libs
50 000 vols; 300 curr per; 10 000 Pamphlets   37279

Cinema Arts, Inc, Motion Picture Archives, 207 Lincoln Green Lane, **Newfoundland**, PA 18445; P.O. Box 452, Newfoundland, PA 18445-0452
T: +1 570 6764145; Fax: +1 570 6769194; E-mail: jeainc@gmail.com
Beverley Allen
World Wars I & II, education, travel, transportation
15 000 vols; 52 curr per   37280

ConnDOT Library & Information Center, 2800 Berlin Tpk, **Newington**, CT 06111-4113
T: +1 860 5943035; Fax: +1 860 5943039
1984; Betty Ambler
20 000 vols   37281

International Tennis Hall of Fame & Tennis Museum Library, 194 Bellevue Ave, **Newport**, RI 02840-3515
T: +1 401 8493990; Fax: +1 401 8517920; E-mail: research@tennisfame.com; URL: www.tennisfame.com
1954; Mark S. Young II
10 000 vols   37282

Newport Historical Society, Library, 82 Touro St, **Newport**, RI 02840
T: +1 401 8460813; Fax: +1 401 8461853; URL: www.newporthistorical.org
1854
Merchant Account Bks, Newport Imprints
13 000 vols; 800 microforms; 200 boxes of mss, 100 scrapbks   37283

Redwood Library & Athenaeum, 50 Bellevue Ave, **Newport**, RI 02840-3292
T: +1 401 8470295; Fax: +1 401 8470192; E-mail: redwood@redwoodlibrary.org; URL: www.redwoodlibrary.org
1747; Cheryl V. Helms
Cynthia Cary Coll of European pattern books, David King Coll of early medical texts, Original Coll, Newport Coll
200 000 vols; 221 curr per; 150 e-books; 7 incunabula; 300 maps; 1 764 av-mat; 1 480 sound-rec; 40 digital data carriers; 2 500 Large Print Bks
libr loan; ALA   37284

Library at the Mariners' Museum, 100 Museum Dr, **Newport News**, VA 23606-3759
T: +1 757 5917782; Fax: +1 757 5917310; E-mail: library@mariner.org; URL: www.mariner.org
1930; Susan Berg
Chesapeake Bay Watercraft, Civil War, Exploration and Discovery, Geography, Immigrant Passenger Ships, Marine Engineering, Maritime Commerce, Merchant Marine, Naval Science, Navigation, Navies, Pirates, Ports and Harbors, Recreational Boats, Shipbuilding, Shipping, Shipwrecks, Slave Trade, Steamships, Voyages and Travel, Yachts and Yachting; Chris-Craft Arch
92 000 vols; 225 curr per; 1 000 000 mss; 5 000 maps; 1 000 microforms; 60 000 plans and drawings, 600 000 photo and vertical files   37285

Virginia War Museum, Major George B. Collings Memorial Library, 9285 Warwick Blvd, **Newport News**, VA 23607
T: +1 757 2478523; Fax: +1 757 2478627; E-mail: info@warmuseum.org; URL: www.warmuseum.org
1923; John V. Quarstein
Spanish-American War, Civil War, World War I – Vietnam Film Coll, World Wars I & II, German Language Propaganda Publs, Uniform Regulations, Military Manuals
25 000 vols; 35 curr per; 500 maps; 300 microforms; 125 sound-rec; 25 oral hist tapes, 1 000 slides, 75 charts, 40 art reproductions   37286

Newton-Wellesley Hospital, Paul Talbot Babson Memorial Library, 2014 Washington St, **Newton Lower Falls**, MA 02462-1699
T: +1 617 2436279; Fax: +1 617 2436595; URL: www.nwh.org
1945; Christine L. Bell
Newton-Wellesley Hospital Arch
11 000 vols; 220 curr per   37287

Chrysler Museum of Art, Jean Outland Chrysler Library, 245 W Olney Rd, **Norfolk**, VA 23510-1587
T: +1 757 9652035; Fax: +1 757 6646201; E-mail: museum@chrysler.org; URL: www.chrysler.org
1929; Laura Christiansen
18th-20th c auction & exhibition cats; Knoedler Libr, Archs, Moses Myers Family Papers and Libr, Rare bks
80 000 vols; 250 curr per; 100 000 microforms; 5 digital data carriers; Videos
libr loan   37288

Sentara Norfolk General Hospital, Health Sciences Library, 600 Gresham Dr, **Norfolk**, VA 23507
T: +1 757 3883000
1942
16 000 vols; 260 curr per   37289

Illinois Lodge of Research, Louis L. Williams Masonic Library, 614 E Lincoln Ave, **Normal**, IL 61761
T: +1 309 4523109; E-mail: library@ilorlibrary.org; URL: www.ilorlibrary.org
Jeff Fox
Freemasonry
10 000 vols   37290

Bridgton Academy Library, Chadbourne Hill Rd, **North Bridgton**, ME 04057; P.O. Box 292, North Bridgton, ME 04057-0292
T: +1 207 6472121; Fax: +1 207 6473146; URL: www.bridgtonacademy.org
Linda Kautz
8 000 vols   37291

Department of Veterans Affairs Medical Center, Learning Resource Center, 3001 Green Bay Rd, **North Chicago**, IL 60064
T: +1 847 6881900; Fax: +1 847 5783819
1926; William Nielson
20 000 vols; 150 curr per   37292

Warren State Hospital, Library Services Department, 33 Main Dr, **North Warren**, PA 16365
T: +1 814 7264223; Fax: +1 814 7264562
Karen Ervin
Self-help/patient recovery Coll
10 000 vols; 80 curr per   37293

Norwegian-American Historical Association Archives, 1510 St Olaf Ave, **Northfield**, MN 55057
T: +1 507 6463221; Fax: +1 507 6463734; E-mail: naha@stolaf.edu; URL: www.naha.stolaf.edu
Kim Holland
8 000 vols   37294

First Parish Church of Norwell, The James Library Center for the Arts, 24 West St, **Norwell**, MA 02061; P.O. Box 164, Norwell, MA 02061
T: +1 781 6597100; E-mail: jameslibrary@verizon.net; URL: jameslibrary.org
1874; Caroline D. Chapin
First Parish Church, Norwell; Local Hist; North River Ship Building
20 000 vols   37295

Czechoslovak Heritage Museum, Library & Archives, 122 W 22nd St, **Oak Brook**, IL 60523
T: +1 630 4720500; Fax: +1 630 4721100
1974; Cary Mentzer
Arch of the Czechoslovak Society of America; Fraternal Publs; Original Minutes of the Society, 1854 to present; Sheet

Music; Hist of Czech, Moravian, & Slovak Immigration, 1643 to present
8 000 vols; 25 unbound rpts, 1 500 arch vols 37296

Oak Forest Hospital, Professional Library, 15900 S Cicero Ave, ***Oak Forest***, IL 60452
T: +1 708 6333551; Fax: +1 708 6333557
HuiLi Liu
Internal medicine, rehabilitation
24 000 vols; 250 curr per 37297

Children's Hospital Oakland, Gordon Health Sciences Library, 747 52nd St, ***Oakland***, CA 94609
T: +1 510 4283448; Fax: +1 510 6013963; E-mail: cholibrary@mail.cho.org
1938; Jane A. Irving
11 500 vols; 320 curr per; 6 diss/theses; 300 av-mat
libr loan; MLA 37298

Metropolitan Transportation Commission, Association of Bay Area Governments (ABAG) Library, 101 Eighth St, ***Oakland***, CA 94607
T: +1 510 4647836; Fax: +1 510 4647852; E-mail: library@mtc.ca.gov; URL: www.mtc.ca.gov
1972; Joan Friedman
City & regional planning, Transportation, Census; San Francisco Bay Region Planning & Transportation Hist; Environmental Impact Rpts
25 000 vols; 1 000 curr per 37299

Marion County Teachers Professional Library, Community Technology and Adult Education Bldg, 1014 SE 7th Rd, Suite 1, ***Ocala***, FL 34474
T: +1 352 6717759; Fax: +1 352 6717757; URL: www.marion.k12.fl.us/dept/LPC/library.cfm
Education; ERIC Docs, Kraus Curriculum Development Libr
11 750 vols 37300

Gulf Coast Research Laboratory, Gunter Library, 703 E Beach Dr, ***Ocean Springs***, MS 39564
T: +1 228 8724253; Fax: +1 228 8724264; URL: www.usm.edu/gcrl
1956; Joyce M. Shaw
Marine Biology (Expedition Rpts), bks, micro; Marine Invertebrate Zoology (Gunter Coll)
30 000 vols; 160 curr per; 382 diss/theses; 45 000 microforms; 20 000 reprints
libr loan; IAMSLIC 37301

Krotona Institute of Theosophy, Library, Krotona Hill, No 2, ***Ojai***, CA 93023
T: +1 805 6462653; Fax: +1 805 6465381; E-mail: krotona@dock.net; URL: www.theosophical.org/library/1814
1926
Rare & Out-of-print Bks of Major Religions
10 000 vols; 30 curr per; 420 av-mat 37302

National Cowboy & Western Heritage Museum, Donald C. & Elizabeth M. Dickinson Research Center, 1700 NE 63rd St, ***Oklahoma City***, OK 73111
T: +1 405 4782250 ext 276; Fax: +1 405 4786421; E-mail: chuckrand@nationalcowboymuseum.org; URL: www.nationalcowboymuseum.org/research/r_index.html
1997; Charles E. Rand
West (U.S.)-Hist; West (U.S.)-Popular culture; West in Art; Cowboy culture; Ranching; Rodeo hist; Native American Fine arts; Special colls: Image arch of over 139 000 records, Glenn D. Shirley Western American coll, Moving images libr; Archival & personal papers colls: Contemporary Western Artists, Native American Fine Arts
29 000 vols; 75 curr per; 26 mss; 3 182 av-mat; 1 921 sound-rec; 1 278 digital data carriers; Hall of Fame Inductee Files, Vertical files
SAA, SSA, OCLC 37303

Oklahoma Historical Society, Research Division, 2401 N Laird Ave, ***Oklahoma City***, OK 73105-4997
T: +1 405 5225209; Fax: +1 405 5220644; E-mail: mrarchives@okhistory.org; URL: www.okhistory.org
1893; William D. Welge
Barde Coll, Grant Foreman Papers, Indian-Pioneer Hist, Oklahoma Photogr Coll, David L Payne Papers, Alice Robertson Papers, Frederic B Severs Papers, Emmett Starr's Mss, Whipple Coll
86 000 vols; 80 curr per; 33 000 microforms; 7 000 000 doc bd, 1 800 000 photos, 6 000 oral hist tapes 37304

Oklahoma Historical Society – Research Center, 2100 N Lincoln Blvd, Wiley Post Historical Bldg, ***Oklahoma City***, OK 73105-4997
T: +1 405 5212491, 5225221; Fax: +1 405 5212492; E-mail: libohs@ok-history.mus.ok.us; URL: www.ok-history.mus.ok.us
1893; William Welge
Genealogy, Native American studies; Early Oklahoma College Yearbks, Sandborn Fire Insurance maps for Oklahoma
83 000 vols; 218 curr per; 100 diss/theses; 1 000 govt docs; 700 maps; 35 600 microforms; 7 000 av-mat
ALA, OLA 37305

University of Oklahoma Health Sciences Center, Robert M. Bird Health Sciences Library, 1000 Stanton L Young Blvd, ***Oklahoma City***, OK 73117-1213; P.O. Box 26901, Oklahoma City, OK 73126-0901
T: +1 405 2712285; Fax: +1 405 2713297; URL: library.ouhsc.edu
1928; Clinton M. Thompson
American Indian Health Coll, Medical Hist Coll, Rare Bks
317 000 vols; 2 825 curr per
libr loan; MLA 37306

New York Institute of Technology, New York College of Osteopathic Medicine Medical Library, Northern Blvd, ***Old Westbury***, NY 11568-8000
T: +1 516 6863743; Fax: +1 516 6863709
Jeanne Strausman
10 000 vols; 415 curr per 37307

Joslyn Art Museum, Milton R. & Pauline S. Abrahams Library, 2200 Dodge St, ***Omaha***, NE 68102-1296
T: +1 402 6613300; Fax: +1 402 3422376; E-mail: library@joslyn.org; URL: www.joslyn.org
1931; Peter Konin
Vertical files for over 5 000 artists
30 000 vols; 50 curr per; 618 microforms; 25 000 Slides, 45 Special Interest Per Sub
libr loan 37308

University of Nebraska Medical Center, McGoogan Library of Medicine, 600 S 42nd St, ***Omaha***, NE 68198-6705; mail address: 986705 Nebraska Medical Ctr, Omaha, NE 68198-6705
T: +1 402 5594006; Fax: +1 402 5595498; E-mail: askus@unmc.edu; URL: www.unmc.edu/library/
1902; Nancy N. Woelfl
Hist of Medicine, Nebraska Arch of Medicine, Consumer Health, Surgery Colls, Obstetrics & Gynecology Coll
234 000 vols; 1 427 curr per; 3 000 microforms; 25 VF drawers of archives
libr loan 37309

St. Joseph Hospital, Burlew Medical Library, 1100 W Stewart Dr, ***Orange***, CA 92868; P.O. Box 5600, Orange, CA 92863-5600
T: +1 714 7718291; Fax: +1 714 7448533; E-mail: jsmith@sjo.stjoe.org; URL: www.burlewmedicallibrary.org
1955; Julie L Smith
Nursing
14 000 vols; 600 curr per; 300 microforms; 1 500 av-mat; 1 650 sound-rec
libr loan; MLA, MLGSCA 37310

Nathan S. Kline Institute for Psychiatric Research, Health Sciences Library, 140 Old Orangeburg Rd, Bldg 35, ***Orangeburg***, NY 10962
T: +1 845 3986575; Fax: +1 845 3985551; URL: www.rfmh.org/nki
1952; Stuart Moss
Neuroscience, psychiatry; Family Resource Ctr
25 000 vols; 10 curr per
libr loan; MLA, AMHL 37311

Orange County Regional History Center, Joseph L. Brechner Research Center, 65 E Central Blvd, ***Orlando***, FL 32801
T: +1 407 8368541; Fax: +1 407 8368550; URL: www.thehistorycenter.org
Dr. Tana M. Porter
Joseph L Brechner Coll
8 000 vols; 11 curr per 37312

EAA Aviation Foundation, Boeing Aeronautical Library, P.O. Box 3065, ***Oshkosh***, WI 54903-3065
T: +1 920 4264848; Fax: +1 920 4264828; E-mail: library@eaa.org; URL: www.eaa.org/museum
1972; Susan A. Lurvey
Dwiggins Coll
10 000 vols; 30 curr per; 3 000 tech manuals, 3 000 tech rpts, 400 000 slides
libr loan 37313

Cleveland Chiropractic College, Ruth R. Cleveland Memorial Library, 10850 Lowell Ave, ***Overland Park***, KS 66210
T: +1 913 2340814; Fax: +1 913 2340901; URL: cleveland.edu/academicprograms/libraries/kc_library.aspx
1976; Marcia M. Thomas
Chiropractic Texts, Journals, Cleveland Arch
15 000 vols; 301 curr per; 18 000 microforms; 320 av-mat; 80 sound-rec; 14 000 slides
libr loan 37314

National Marine Fisheries Service, Southeast Fisheries Science Ctr, Cooperative Oxford Laboratory – Oxford Marine Library, 904 S Morris St, ***Oxford***, MD 21654-9724
T: +1 410 2265193; Fax: +1 410 2265925; E-mail: susie.hines@noaa.gov; URL: www.chbr.noaa.gov/default.aspx
1961; Susie K. Hines
Marine biology
11 530 vols; 75 curr per 37315

The Garden Library at Planting Fields, Planting Fields Arboretum, 1395 Planting Fields Rd, ***Oyster Bay***, NY 11771-1302
T: +1 516 9228691; URL: www.plantingfields.org
1975; Rosemarie Papayanopulos
Rare Books; L.I. Nursery catalogs; L.H. Bailey Special Coll; E.H. Wilson Special Coll; Reference Section, Rare Book Section
8 000 vols; 100 curr per; 2 diss/theses; 50 govt docs; 6 digital data carriers; 2 000 pamphlets + clippings, 5 videorecordings
CBHL 37316

George C. Wallace Community College Aviation Campus, Division of Wallace Community College of Dothan, Hwy 231 N, ***Ozark***, AL 36361; P.O. Box 1209, Ozark, AL 36361-1209
T: +1 334 7745113; Fax: +1 334 7746399; E-mail: aatc@snowhill.com; URL: www.wallace.edu
1980; A. P. Hoffman
Civil, Commercial & General Aviation
10 500 vols; 100 curr per; 26 000 microforms; 1 500 av-mat; 90 sound-rec; 3 digital data carriers
libr loan; ALA 37317

The Society of the Four Arts, King Library, Three Four Arts Plaza, ***Palm Beach***, FL 33480
T: +1 561 6552766; Fax: +1 561 6557233; E-mail: kinglibrary@fourarts.org; URL: www.fourarts.org
1936; Molly Charland
Addison Mizner Coll, John C. Jessup Coll, Henry P. McIntosh IV Coll
60 000 vols; 651 curr per 37318

Palm Springs Art Museum Library, 101 Museum Dr, ***Palm Springs***, CA 92262; P.O. Box 2310, Palm Springs, CA 92263-2310
T: +1 760 3257186; Fax: +1 760 3275069; URL: www.psmuseum.org
1938; Frank Lopez
Art, Comtemporary California, Western American Art, Native American, Mexican, Contemporary Glass, Prints, American photos, American Mid-20th C Architecture; Artist Files, Museum Hist Arch
8 000 vols; 20 curr per; 11 433 slides 37319

Department of Veterans Affairs Medical Center, Medical Library (142 D), 3801 Miranda Ave, ***Palo Alto***, CA 94304
T: +1 650 4935000 ext 65703; Fax: +1 650 8523258
1935; Susan Shyshka
12 000 vols; 450 curr per 37320

EPRI Library, 3420 Hillview Ave, ***Palo Alto***, CA 94304-1395
T: +1 650 8552354; Fax: +1 650 8552295; E-mail: library@epri.com; URL: epri.com
Judith Mills
Electric power, EPRI Reports
20 000 vols; 100 curr per 37321

Institute of Transpersonal Psychology, Library, 1069 East Meadow Circle, ***Palo Alto***, CA 94303
T: +1 650 4934430 ext 221; Fax: +1 650 8529780; E-mail: itpinfo@itp.edu; URL: www.itp.edu/library/index.php
1981
Complete Psychological Works of Sigmund Freud, Collected Works of C.G. Jung, Spirituality (Classics of Western Spirituality Coll)
15 000 vols; 170 curr per; 310 diss/theses; 700 microforms; 1 000 sound-rec; 4 digital data carriers; 110 ITP diss 37322

Palo Alto Medical Foundation, Barnett-Hall Library, Ames Bldg, 795 El Camino Real, ***Palo Alto***, CA 94301-2302
T: +1 650 3268120; Fax: +1 650 8532909; URL: www.pamf.org
1950; Debbie Schide
Clinical medicine
12 000 vols; 229 curr per; 13 VF drawers of pamphlets 37323

Armenian Missionary Association of America Library, 31 W Century Rd, ***Paramus***, NJ 07652
T: +1 201 2652607; Fax: +1 201 2656015; E-mail: amaa@amaa.org; URL: www.amaa.org
1918; Andrew Torigan
20 000 vols; 16 curr per 37324

American Farm Bureau Federation Library, 225 Touhy Ave, ***Park Ridge***, IL 60068
T: +1 847 6858781; Fax: +1 847 6858969; E-mail: sue@fb.com; URL: www.fb.com
1980; Susan J. Schultz
Agricultural economics, Agribusiness, Environmental economics
8 000 vols; 150 curr per 37325

American Society of Anesthesiologists, Wood Library-Museum of Anesthesiology, 520 N Northwest Hwy, ***Park Ridge***, IL 60068-2573
T: +1 847 8255586; Fax: +1 847 8251692; E-mail: communications@asahq.org; URL: www.asahq.org
1929; Patrick Sim
Hist of Anesthesiology Coll, Mesmerism Coll, Curare Coll
8 500 vols; 100 curr per; 20 shelf feet of mss, 40 VF drawers of pamphlets, photos, clippings 37326

Lutheran General Hospital, Advocate Health Sciences Library Network, 1775 Dempster St, ***Park Ridge***, IL 60068
T: +1 847 7235494; Fax: +1 847 6929576
1966; Marie T. Burns
16 000 vols; 250 curr per; 7 digital data carriers
libr loan 37327

California Institute of Technology – Jet Propulsion Laboratory Library, Archives & Records Section, 4800 Oak Grove Dr, MS 111-113, ***Pasadena***, CA 91109-8099
T: +1 818 3544200; Fax: +1 818 3936752; URL: beacon.jpl.nasa.gov
1948; Margo Young
Aerospace, Astronomy, Astrophysics, Physical Science, Business, Engineering, Communications
85 000 vols; 1 030 curr per
libr loan 37328

Observatories of the Carnegie Institution of Washington, Hale Library, 813 Santa Barbara St, ***Pasadena***, CA 91101
T: +1 626 3040228; Fax: +1 626 7958136; URL: www.ociw.edu
1904; John Grula
Astronomy, Astrophysic
31 000 vols; 28 curr per 37329

Pacific Asia Museum Library, 46 N Los Robles Ave, ***Pasadena***, CA 91101-2009
T: +1 626 4492742; Fax: +1 626 4492754; E-mail: library@pacificasiamuseum.org; URL: www.pacificasiamuseum.org
Sarah McKay
Asian and Pacific art and culture; India (Paul Sherbert Coll)
9 000 vols; 6 curr per 37330

Passaic County Historical Society, Edward B. Haines Local History Library, Lambert Castle, Valley Rd, ***Paterson***, NJ 07503-2932
T: +1 973 247-0085; Fax: +1 973 881-9434; E-mail: lambertcastle@verizon.net; URL: www.lambertcastle.org/
1926; Andrew Shick
Society for the Establishment of Useful Manufacturers Records; Abraham Hewitt Records; Family Group Sheets Coll; Hobart Coll
10 000 vols 37331

Patton State Hospital, Patients' Library, 3102 E Highland Ave, ***Patton***, CA 92369
T: +1 909 4256039; Fax: +1 909 4256162
Frederick Brenion
13 000 vols; 50 curr per 37332

Baptist Hospital, Medical Library, 1000 W Moreno St, ***Pensacola***, FL 32501
T: +1 904 4344877
1951; Elizabeth Richbourg
8 500 vols; 180 curr per 37333

US Navy – Naval Operational Medical Institute Library, 220 Hovey Rd, Code 31, ***Pensacola***, FL 32508-1047
T: +1 850 4522256; Fax: +1 850 4522304; E-mail: NOMI-Library@med.navy.mil; URL: www.med.navy.mil/sites/navmedmpte/nomi/Pages/NOMILibrary.aspx
Valerie S. McCann
Naval Aerospace Medical Res Laboratory Rpts Coll
10 000 vols; 98 curr per 37334

Department of Veterans Affairs, Medical Center Library, ***Perry Point***, MD 21902
T: +1 410 6422411 ext 5716; Fax: +1 410 6421103
1924; Kathy Vaughan
12 700 vols; 310 curr per; 200 av-mat; 150 sound-rec; 15 digital data carriers 37335

Central State Hospital, Medical-Patients Library, W Washington St, P.O. Box 4030, ***Petersburg***, VA 23803
T: +1 804 5247758; Fax: +1 804 5247308, 5247049; URL: www.csh.dbhds.virginia.gov
Marche Hale
8 000 vols; 100 curr per 37336

Academy of Natural Sciences of Philadelphia, Ewell Sale Stewart Library, 1900 Benjamin Franklin Pkwy, ***Philadelphia***, PA 19103-1195
T: +1 215 2991040; Fax: +1 215 2991144; E-mail: library@ansp.org; URL: www.ansp.org/library
1812; Robert M. Peck
Entomology (Libr of the American Entomological Society); Mss (Academy's Arch & Mss Coll); Photogr Coll; Portrait & Drawing Coll; Pre-Linnean Coll
200 000 vols; 2 500 curr per; 250 000 mss; 5 000 maps; 52 microforms; 20 digital data carriers
PALINET, OCLC 37337

American Philosophical Society Library, 105 South Fifth St, ***Philadelphia***, PA 19106-3386
T: +1 215 4403400; Fax: +1 215 4403423; URL: www.amphilsoc.org
1743; Dr. Matrin L. Levitt
Papers of Benjamin Franklin, Charles Darwin, Charles Willson Peale; Lewis & Clark Journals; American Indian Linguistics; Thomas Paine Coll; Simon Flexner Coll; GeneticsColl; Stephan Girard Papers; Franz Boas Coll; Hist of Quantum Physics; Arch
300 000 vols; 850 curr per; 134 240 microforms; 8 500 000 mss
IRLA, RLG, ALA 37338

Athenaeum of Philadelphia, 219 S Sixth St, East Washington Square, ***Philadelphia***, PA 19106-3794
T: +1 215 9252688; Fax: +1 215 9253755; E-mail: athena@philaathenaeum.org; URL: www.philaathenaeum.org; www.philadelphiabuildings.org
1814; Dr. Sandra L. Tatman
Architecture & design coll
98 000 vols; 200 microforms; 1 mio archit mss, 200 000 archit drawings, 300 000 archit photos
ARLIS/NA, ICAM 37339

Community Legal Services, Inc Library, 1424 Chestnut St, ***Philadelphia***, PA 19102
T: +1 215 9813771; Fax: +1 215 9810434; URL: www.clsphila.org
1968; Carl E. Mitchell
Civil law, Poverty law
15 000 vols; 120 curr per; 8 VF drawers 37340

Congregation Mikveh Israel Archives, Independence Mall E, 44 N Fourth St, ***Philadelphia***, PA 19106
T: +1 215 9225446; Fax: +1 215 9221550; E-mail: info@mikvehisrael.org; URL: www.mikvehisrael.org
Ruth Hoffman
40 000 vols 37341

Curtis Institute of Music, Library, 1726 Locust St, ***Philadelphia***, PA 19103
T: +1 215 8935252; Fax: +1 215 8939056; E-mail: elizabeth.walker@curtis.edu; URL: www.curtis.edu
1926; Elizabeth Walker
13 774 vols; 72 curr per; 500 mss; 2 incunabula; 25 diss/theses; 59 995 music scores; 33 microforms; 19 089 sound-rec; 5 digital data carriers; 572 VHS tapes, 326 laser disks, 336 piano rolls, 37 organ rolls
libr loan; MLA, IAML 37342

Department of Veterans Affairs, Library Services, University & Woodland Aves, ***Philadelphia***, PA 19104
T: +1 215 8235860; Fax: +1 215 8235108; E-mail: marchino@shrsys.hslc.org; URL: www.va.stars-and-stripes.com
Mark Marchino
9 890 vols; 453 curr per 37343

Donald F. & Mildred Topp Othmer Library of Chemical History, 315 Chestnut St, ***Philadelphia***, PA 19106
T: +1 215 8738269; Fax: +1 215 6295205; E-mail: reference@chemheritage.org; URL: chemheritage.org/library/library.html; othmerlib.chemheritage.org
1988; Elsa B. Atson
Hist of Chemistry
146 000 vols; 100 curr per; 1 000 e-books; 9 000 microforms; 400 av-mat; 1 800 sound-rec; 15 digital data carriers; 59 000 Journals
libr loan; SLA, ALA, IUG, OCLC 37344

Educational Management Corporation, Art Institute of Philadelphia Library, 1622 Chestnut St, ***Philadelphia***, PA 19103
T: +1 215 4056348
1974; Ruth Schachter
Visual communications, Interior design, Industrial design, Animation, Fashion marketing, Fashion design, Visual merchandising, Photography, Website design, Multimedia
28 000 vols; 190 curr per 37345

Franklin Institute, Science Museum Library, 222 N 20th St, ***Philadelphia***, PA 19103-1194
T: +1 215 4481239; Fax: +1 215 4481364; E-mail: icoffey@fi.edu; URL: www.fi.edu
1824; Irene D.Coffey
Mathematics, Physical science, Sience education; Underwater Man Coll, Ware Sugar Coll, Wright Brothers Aeronautical Eng Coll
19 500 vols; 150 curr per; 700 microforms; 4 000 av-mat 37346

Gay Lesbian Bisexual & Transgender Library/Archives of Philadelphia, William Way Community Ctr, 1315 Spruce St, ***Philadelphia***, PA 19107
T: +1 215 7322220; Fax: +1 215 7320770; URL: www.waygay.org
1976
Rare Bk Coll, Personal & Organizational Coll Related to GLBT People in the Greater Delaware Valley
11 000 vols; 10 curr per 37347

German Society of Pennsylvania, Joseph Horner Memorial Library, 611 Spring Garden St, ***Philadelphia***, PA 19123
T: +1 215 6272332; Fax: +1 215 6275297; E-mail: contact@germansociety.org; URL: www.germansociety.org
1817; Laurie Wolfe
German-American Coll, German-American Newspapers, Arch of German-American Singing Societies
70 000 vols; 12 curr per; 20 microforms 37348

Historical Society of Pennsylvania, Library Services, 1300 Locust St, ***Philadelphia***, PA 19107-5699
T: +1 215 7326200; Fax: +1 215 7322680; E-mail: library@hsp.org; URL: www.hsp.org
1824; Lee Arnold
The William Penn Family & Other important Colonial Families of Pennsylvania; Baker Coll of Washingtoniana; Frankliniana; Quakeriana; Genealogies; Eighteenth & Nineteenth Century Paintings; Prints & Drawings of Pennsylvania; Museum Objects incl Pieces owned by William Penn; manuscript dept, graphics dept
560 000 vols; 2 000 curr per; 16 000 000 mss; 10 incunabula; 130 music scores; 3 000 maps; 39 000 microforms; 164 av-mat; 60 digital data carriers; 300 000 Graphics
IRLA 37349

Independence Seaport Museum Library, 211 S Columbus Blvd, ***Philadelphia***, PA 19106
T: +1 215 9255439; Fax: +1 215 9256713; E-mail: library@phillyseaport.org; URL: phillyseaport.org
1974; Megan Hahn Fraser
Mss coll, Maritime related personal family & business papers, Map & Chart coll, Photogr coll, Rare book coll, Philadelphia shipbuilding
15 000 vols; 35 curr per; 750 maps; 1 000 microforms; 2 digital data carriers; 9 000 boat plans 37350

Library Company of Philadelphia, 1314 Locust St, ***Philadelphia***, PA 19107-5698
T: +1 215 5463181, 8229; Fax: +1 215 5465167; E-mail: refdept@librarycompany.org; URL: www.librarycompany.org
1731; John Van Horne
18th c & early 19th c science & medicine, Women's hist, American architecture, American political hist; Afro-American Coll, American Imprints to 1880, Judaica, American Technology & Business to 1860, Bks of Early American Bk Collectors, English & American Lit, German Americana to 1830, Philadelphiana
500 000 vols 37351

The Masonic Library & Museum of Pennsylvania, Masonic Temple, One N Broad St, ***Philadelphia***, PA 19107-2520
T: +1 215 9881933; Fax: +1 215 9881972; URL: www.pagrandlodge.org
1817; Andrew Zellers-Frederick
Rare Masonica coll
75 000 vols; 25 000 mss; 1 incunabula
ALA, ACRL 37352

National Park Service Independence National Historical Park, Library & Archives, Merchants Exchange Bldg, Third Flr, 143 S Third St, mail address: 313 Walnut St, ***Philadelphia***, PA 19106
T: +1 215 5978047; Fax: +1 215 5973969; URL: www.nps.gov/inde/library-and-archives.htm
1951; Karen Stevens
18th c Philadelphia, American Revolution, Constitution politics, Hist, Decorative arts; Horace Wells Sellers Coll, 1730-1930; Edwin Owen Lewis Papers, 1927-1974; The Morris Family Papers; National Museum Board of Managers Records, 1873-1918; Isidor Ostroff Papers, 1941-1968; Philadelphia Bureau of City Property
13 000 vols; 10 curr per 37353

National Railway Historical Society Library, 100 N 17th St, Ste 1203, ***Philadelphia***, PA 19103-2783; P.O. Box 58547, Philadelphia, PA 19102-8547
T: +1 215 5576606; Fax: +1 215 5576740; E-mail: info@nrhs.com; URL: www.nrhs.com
1977; Lynn Burshtin
National Railway Bulletin
10 000 vols 37354

Pennsylvania Horticultural Society Library, McLean Library, 100 N 20th St, 5th Fl, ***Philadelphia***, PA 19103-1495
T: +1 215 9888772; Fax: +1 215 9888783; E-mail: mcleanlibrary@pennhort.org; URL: www.pennsylvaniahorticulturalsociety.org/garden/libraryhome.html
1827; Janet Evans
Pennsylvania Horticulture, Herbals, Medical Botany, rare bks; 15th-20th C Horticultural Mat
14 000 vols; 200 curr per; 150 av-mat; 100 slides
libr loan 37355

Pennsylvania Hospital – Historic Library, Three Pine Ctr – 800 Spruce St, ***Philadelphia***, PA 19107-6192
T: +1 215 8295434; Fax: +1 215 8297155; URL: www.uphs.upenn.edu/paharc/collections/library.html
1762; Stacey C. Peeples
Benjamin Smith Barton, William Byrd of Westover, Lloyd Zachary, The Meig Family, Thomas Story Kirkbride
15 000 vols; 16 curr per 37356

Pennsylvania Hospital – Medical Library, Three Pine Ctr – 800 Spruce St, ***Philadelphia***, PA 19107-6192
T: +1 215 8293370; Fax: +1 215 8297155; E-mail: library.services@pahosp.com; URL: www.uphs.upenn.edu/pahedu/library
1940; Mary McCann
23 000 vols; 200 curr per 37357

Pew Charitable Trusts Library, One Commerce Sq – 2005 Market St, Ste 1700, ***Philadelphia***, PA 19103-7017
T: +1 215 5754922; Fax: +1 215 5754939; E-mail: pctlibrary@pewtrusts.org
Bruce C. Compton
Job listings, career opportunities
10 000 vols; 160 curr per 37358

Philadelphia Museum of Art Library, Ruth & Raymond G Perelman Bldg – 2525 Pennsylvania Ave, ***Philadelphia***, PA 19130; P.O. Box 7646, Philadelphia, PA 19101-7646
T: +1 215 6847650; Fax: +1 215 2360534; E-mail: library@philamuseum.org; URL: www.philamuseum.org/library/
1876; C. Danial Elliott
Johnson Coll (European Painting), Kienbusch Arms & Armor Coll
200 000 vols; 400 curr per; 50 microforms; 10 digital data carriers; 150 000 slides 37359

Philadelphia Orchestra Library, 300 S Broad St, ***Philadelphia***, PA 19102-4297
T: +1 215 6702343; Fax: +1 215 9850746
1900; Robert Grossman
Sheet music
10 000 vols 37360

Rosenbach Museum & Library, 2010 DeLancey Pl, ***Philadelphia***, PA 19103
T: +1 215 7321600; Fax: +1 215 5457529; E-mail: info@rosenbach.org; URL: www.rosenbach.org
1954; Derick Dreher
Americana, American & English lit; Rush-Williams-Biddle Family Papers; Marianne Moore Papers; Maurice Sendak Original Drawings; Rosenbach Company Arch

30 000 vols; 96 maps; 300 000 mss, 20 000 prints & drawings 37361

Ryerss Museum & Library, Burholme Park, 7370 Central Ave, **Philadelphia**, PA 19111-3055
T: +1 215 6850544; E-mail: ryerssmuseum@hotmail.com; URL: ryerssmuseum.org
1910; Peter Lurowist
Victoriana Coll
20 000 vols; 10 curr per 37362

Spiritual Frontiers Fellowship International, Lending Library, 3310 Baring St, **Philadelphia**, PA 19101-2332
T: +1 215 2220619; URL: www.spiritualfrontiers.org
Elizabeth W. Fenske
Psychology, Religion
15 000 vols 37363

Temple University Hospital, Episcopal Campus, Medical Library, Front St & Lehigh Ave, **Philadelphia**, PA 19125
T: +1 215 7070286; Fax: +1 215 7070291
Marita J. Krivda
Internal medicine
10 000 vols; 110 curr per 37364

Theodore F. Jenkins Memorial Law Library, 833 Chestnut St, Ste 1220, **Philadelphia**, PA 19107
T: +1 215 5747900; Fax: +1 215 5747920; E-mail: research@jenkinslaw.org; URL: www.jenkinslaw.org
1802; Regina A. Smith
Roman & Canon Law (John Marshall Gest Coll)
275 000 vols; 400 curr per
libr loan 37365

Union League of Philadelphia Library, 140 S Broad St, **Philadelphia**, PA 19102
T: +1 215 5875594; Fax: +1 215 5875598; E-mail: library@unionleague.org; URL: www.unionleague.org
1862; Jim Mundy
American Civil War, Lincoln, Philadelphia & Pennsylvania Hist Coll
26 000 vols; 61 curr per; 2 000 av-mat; 200 sound-rec; 500 digital data carriers 37366

Wagner Free Institute of Science Library, 1700 W Montgomery Ave, **Philadelphia**, PA 19121
T: +1 215 7636529; Fax: +1 215 7631299; E-mail: library@wagnerfreeinstitute.org; URL: www.wagnerfreeinstitute.org
1855; Lynn Dorwaldt
Hist of Science Coll; Natural Science Coll; Science Education Coll; William Wagner Coll; 19th Century US & State Geological Surveys
45 000 vols; 10 curr per; 1 000 maps; 500 linear ft of mss, 3 000 graphic items
libr loan; ALA, SLA, OCLC 37367

Banner Good Samaritan Medical Center, Merril W. Brown Health Sciences Library, 1111 E McDowell Rd, **Phoenix**, AZ 85006
T: +1 602 2394353; Fax: +1 602 2393493; URL: www.samaritan.edu
Sally Harvey
Consumer health, clinical medicine
10 000 vols; 750 curr per 37368

Foundation for Blind Children Library & Media Center, Arizona Instructional Resource Center, 1235 E Harmont Dr, **Phoenix**, AZ 85020-3864
T: +1 602 6785810; Fax: +1 602 6785811; URL: www.seeitourway.org
1958; Inge Durre
60 000 vols; 25 curr per; 100 e-books; 4 maps; 40 av-mat; 45 000 Braille vols, 100 High Interest/Low Vocabulary Bks, 8 000 Large Print Bks
All titles in collection are braille or large print or print and braille 37369

Heard Museum, Billie Jane Baguley Library and Archives, 2301 North Central Ave, **Phoenix**, AZ 85004-1323
T: +1 602 2528840; Fax: +1 602 2529757; E-mail: mario@heard.org; URL: www.heard.org
1929; Mario Nick Klimiades
Anthropology, Archaelogy, Native American art; Atlatl Coll; Fred Harvey Company Papers; R. Brownell McCrew

Coll; Native American Artists Resource Coll; North American Indian (Curtis Coll)
30 500 vols; 300 curr per; 750 av-mat; 650 sound-rec; 5 digital data carriers; 300 000 photos, slide libr 37370

Phoenix Art Museum Library, 1625 N Central Ave, **Phoenix**, AZ 85004-1685
T: +1 602 2572136; Fax: +1 602 2538662; E-mail: info@phxart.org; URL: www.phxart.org
1959; Sandra Wiles
Architecture, Contemporary Am paintings, Decorative arts, Graphics, Latin Am art, Sculpture; Int Auction Records, Museum Bulletins, Orme Lewis Coll of Rembrandt Etching Cat, Art Libraries Society Arch of Arizona Artists, One-Person Exhibition Coll
40 000 vols; 107 curr per; 7 000 sales & auction cat, 10 000 one-man show cat, 63 000 slides 37371

US Army, RDECOM-ARDEC, Technical Research Center, Bldg 59, Phipps Rd, AMSRD-AAR-EMK, **Picatinny Arsenal**, NJ 07806-5000
T: +1 973 7243757; Fax: +1 973 7243044
1929; Suseela Chandrasekar
Armament, Pyrotechnics, Propellants, Engineering; Archives for Picatinny Arsenal (NJ) and Frankford Arsenal (PA)
37 000 vols; 50 curr per; 600 e-books; 300 000 microforms
libr loan; SLA 37372

South Dakota Braille & Talking Book Library, McKay Bldg, 800 Governors Dr, **Pierre**, SD 57501-2294
T: +1 605 7733131; Fax: +1 605 7736962; E-mail: library@state.sd.us; URL: www.sdstatelibrary.com
1968; Dorothy Liegl
Dakota Language Coll, North Dakota & South Dakota Coll
179 000 vols 37373

South Dakota State Historical Society, State Archives, 900 Governors Dr, **Pierre**, SD 57501-2217
T: +1 605 7733804; Fax: +1 605 7736041; E-mail: archref@state.sd.us; URL: history.sd.gov/archives/Data/library/default.aspx
1901; LaVera Rose
Dakota & Western Indians Coll, photos, newspapers
13 770 vols; 188 curr per; 9 000 maps; 13 500 microforms; 990 doc bd
libr loan 37374

Central Louisiana State Hospital – Forest Glen Patient's Library, 242 W Shamrock St, **Pineville**, LA 71361; P.O. Box 5031, Pineville, LA 71360-5031
T: +1 318 4846364; Fax: +1 318 4846284
Deborah Boerdoom
10 000 vols; 20 curr per 37375

Seafarer's Harry Lundeberg School of Seamanship, Paul Hall Library & Maritime Museum, PO Box 75, **Piney Point**, MD 20674-0075
T: +1 301 9940010
1970; Janice M. Smolek
Maritime hist, Hist; Mss of Union Meetings, 1891-1907
19 000 vols; 100 curr per; 236 microforms; 420 av-mat; 10 digital data carriers; 8 VF drawers, 1 500 slides 37376

Allegheny General Hospital, Health Sciences Library, 320 E North Ave, **Pittsburgh**, PA 15212-4772
T: +1 412 3593040; Fax: +1 412 3594420; URL: www.wpahs.org; www.wphs.org/education/library/agh
1935; Susan Hoehl
13 000 vols; 750 curr per
libr loan 37377

Carnegie Museum of Natural History Library, 4400 Forbes Ave, **Pittsburgh**, PA 15213-4080
T: +1 412 6223284; Fax: +1 412 6228837; E-mail: cmnhlib@carnegiemnh.org; URL: www.carnegiemnh.org/library
1898; Xianghua Sun
Botany, Entomology, Paleontology, Ornithology, Mammology, Invertebrate Zoology, Molacology, Herpetology,

Anthropology; Institutional Arch
132 000 vols; 1 500 curr per
libr loan; OCLC, PALINET 37378

Children's Institute Library, 1405 Shady Ave, **Pittsburgh**, PA 15217-1350
T: +1 412 4202247; Fax: +1 412 4202510; URL: www.amazingkids.org
1972; Nancy J. Sakino-Spears
Pediatrics, Asthma, Burn injury, Cerebral palsy, Head injury, Prader-Willi, Spina bifida, Unusual syndromes, Children's lit
10 000 vols; 60 curr per 37379

Historical Society of Western Pennsylvania, Library & Archives, 1212 Smallman St, **Pittsburgh**, PA 15222
T: +1 412 4546364; Fax: +1 412 4546028; E-mail: library@hswp.org; URL: www.heinzhistorycenter.org
1879; Art Louderback
African-American Arch, Business Coll, Italian Arch, Jewish Arch, Polish Arch, Slovak Arch, Women's Coll
35 000 vols; 300 curr per; 500 maps; 265 microforms; 700 slides, 400 000 photos, 90 boxes of unbound newspapers, 28 VF drawers of pamphlets & clippings, 10 000 linear feet & 888 bound vols of mss 37380

Hunt Institute for Botanical Documentation, Hunt Botanical Library, Carnegie Mellon University, **Pittsburgh**, PA 15213-3890; mail address: 5000 Forbes Ave, Pittsburgh, PA 15213-3890
T: +1 412 2687301; Fax: +1 412 2685677; E-mail: huntinst@andrew.cmu.edu; URL: huntbot.andrew.cmu.edu
1961; Charlotte A. Tancin
Michel Adanson coll (bks, mss); Strandell coll of Linnaeana (bks, clippings)
29 000 vols; 250 curr per; 8 incunabula
libr loan; CBHL, EBHL
Limited interlibrary loan 37381

National Center for Juvenile Justice, Technical Assistance Resource Center, 710 5th Ave, **Pittsburgh**, PA 15219-3000
T: +1 412 2276950; Fax: +1 412 2276955; E-mail: ncjj@ncjj.org; URL: www.ncjj.org
1974
16 000 vols; 50 curr per 37382

National Institute for Occupational Safety & Health, Pittsburgh Library, Cochrans Mill Rd, **Pittsburgh**, PA 15236; P.O. Box 18070, Pittsburgh, PA 15236-8070
T: +1 412 3864431; Fax: +1 412 3864592; E-mail: kis2@cdc.gov
Kathleen Stabryla
165 000 vols; 100 curr per
libr loan 37383

Pittsburgh History & Landmarks Foundation, James D. Van Trump Library, 100 West Station Square Dr, Ste 450, **Pittsburgh**, PA 15219
T: +1 412 4715808; Fax: +1 412 4711633; E-mail: info@phlf.org; URL: www.phlf.org/james-d-van-trump-library/
1964
Historic preservation, Hist of architecture, Pittsburgh hist; Articles of James D. Van Trump & Walter C. Kidney
9 000 vols; 10 curr per; 50 sound-rec; 20 VF drawers of clippings & brochures; 40 VF drawers of slides, 20 drawers of architectural drawings & maps; 32 shelves of plat books, 50 mss 37384

Pittsburgh Zoo & Aquarium Library, One Wild Pl, **Pittsburgh**, PA 15206
T: +1 412 6653640; Fax: +1 412 6653661; E-mail: tgray@pittsburghzoo.org; URL: www.pittsburghzoo.org
1983
Ciguatera Poisoning, res papers; Gambierdiscus toxicus; Platanistidae Dolphins, res papers
14 000 vols; 15 curr per 37385

University of Pittsburgh Medical Center Shadyside, James Frazer Hillman Health Science Library, 5230 Centre Ave, **Pittsburgh**, PA 15232
T: +1 412 6232441; Fax: +1 412 6234155; URL: www.hsls.pitt.edu/about/libraries/shadyside/hillman/
Michele Klein-Fedyshin
8 000 vols; 225 curr per 37386

Massachusetts Trial Court, Berkshire Law Library, 76 East St, Court House, **Pittsfield**, MA 01201
T: +1 413 4425059; Fax: +1 413 4482474; E-mail: berkshirelawlib@hotmail.com; URL: www.lawlib.state.ma.us/libraries/locations/berkshire.html
1842; Janice B. Shotwell
Mass legal mat, Law for layman
10 000 vols; 40 curr per 37387

The Islamic Society of North America, 6555 S County Rd, **Plainfield**, IN 46168; P.O. Box 38, Plainfield, IN 46168
T: +1 317 8398157; Fax: +1 317 8391840; E-mail: isna@sorf.ici.com; URL: www.isna.net
1981; Dr. Sayyid M. Syeed
50 000 vols; 5 curr per 37388

Institute of Business Appraisers, Library, 7420 NW Fifth St, Ste 103, **Plantation**, FL 33317; P.O. Box 17410, Plantation, FL 33318
T: +1 954 5841144; Fax: +1 954 5841184; E-mail: hqiba@go-iba.org; URL: www.go-iba.org
1978
10 000 vols; 10 curr per 37389

Far Eastern Research Library, Nine First Ave NE, **Plato**, MN 55370-0181; P.O. Box 181, Plato, MN 55370-0181
T: +1 320 2382591; E-mail: laogan@fareasternlibrary.org; URL: www.fareasternlibrary.org
1969; Dr. Jerome Cavanaugh
Chinese Dialect Mat Coll, City of Tianjin (China) Coll (400 items), Chinese slang & maxims Coll (1 000 items)
56 000 vols; 248 curr per; 312 microforms; 110 sound-rec; 25 digital data carriers; 50 000 entry cards for a Taiwanese-Japanese dictionary, 4 000 Chinese Cultural Revolution Mat
libr loan; CEAL 37390

ECRI Library, 5200 Butler Pike, **Plymouth Meeting**, PA 19462
T: +1 610 8256000; Fax: +1 610 8347366; E-mail: ekuserk@ecri.org; URL: www.ecri.org
1969; Evelyn Kuserk
Medical Devices, Biomedical Engineering, Hospital Safety, Health Services Research, Health Technology Assessment; Health Devices Evaluation Services
10 000 vols; 2 000 curr per; 16 000 mss; 100 digital data carriers; 470 VF drawers of technical rpts & evaluation data
libr loan; NNLM 37391

Oakland County Library for the Visually & Physically Impaired, 1200 N Telegraph Rd, **Pontiac**, MI 48341-0482
T: +1 248 8585050; Fax: +1 248 8589313; E-mail: lvpi@oakgov.com; URL: www.oakgov.com/lvpi/index.html
1974; Stacey Boucher-Tabor
Braille
70 000 vols; 10 Special Interest Per Sub, 400 Large Print Bks 37392

Christus St. Mary Hospital, Health Science Library, 3600 Gates Blvd, **Port Arthur**, TX 77642-3850; P.O. Box 3696, Port Arthur, TX 77643-3696
T: +1 409 9895804; Fax: +1 409 9895137
Carol Bourland
22 000 vols; 132 curr per 37393

Huron City Museum Library, 7995 Pioneer Dr, **Port Austin**, MI 48467-9400
T: +1 517 4284123; Fax: +1 517 4284123; E-mail: huroncity@centturytel.net; URL: www.huroncitymuseums.org
1947; Patricia A Finan
Lit; Libr of William Lyon Phelps
10 000 vols 37394

Minisink Valley Historical Society Library, 138 Pike St, **Port Jervis**, NY 12771-1808; P.O. Box 659, Port Jervis, NY 12771-0659
T: +1 845 8562375; Fax: +1 845 8561049; E-mail: mvhs1889@magiccarpet.com; URL: www.minisink.org
1889; Peter Osborne III
Deerpark & Port Jervis arch mats, US

Census records
31 000 vols; 1 000 maps; 10 000 mss,
10 000 photogs   37395

Nassau County Museum Research Library, Jesse Merritt Memorial Library, Sands Point Preserve, 127 Middleneck Rd, **Port Washington**, NY 11050
T: +1 516 5717901; Fax: +1 516 5717909
1962; Gary Hammond
Long Island Photos; Queens County Deeds; Queens, Nassau & Suffolk Censuses; Queens County Estate Inventories
9 000 vols; 175 curr per   37396

Porterville Developmental Center, Residents' Resource Center, 26501 Avenue 140, **Porterville**, CA 93258-9109; P.O. Box 2000, Porterville, CA 93258-2000
T: +1 559 7822021
Vicki Boosalis
Developmental disabilities
10 000 vols   37397

Columbia River Inter-Tribal Fish Commission, StreamNet Library, 729 NE Oregon, Ste 190, **Portland**, OR 97232
T: +1 503 7311304; Fax: +1 503 7311260; E-mail: fishlib@critfc.org; URL: www.streamnet.org/library.html
1992; Lenora Oftedahl
15 000 vols; 25 curr per   37398

Genealogical Forum of Oregon, Inc Library, 1505 SE Gideon St, P.O. Box 42567, **Portland**, OR 97242
T: +1 503 9631932; E-mail: info@gfo.org; URL: www.gfo.org
1957
Records of Civil War Veterans, Willis Corbitt's Res Mat, Nellie Hiday's Res Mat
14 000 vols; 235 curr per; 5 000 vertical files   37399

Legacy Good Samaritan Hospital & Medical Center, Health Sciences Library, 1015 NW 22nd Ave, **Portland**, OR 97210
T: +1 503 4137335; Fax: +1 503 4138016
1941; Carolyn Adams
10 000 vols; 531 curr per; 6 digital data carriers
libr loan   37400

Maine Charitable Mechanic Association Library, 519 Congress St, **Portland**, ME 04101
T: +1 207 7738396
1820; Pat Larrabee
Fiction, Hist, Travel
32 000 vols; 2 curr per   37401

Maine Historical Society, Research Library, 485 Congress St, **Portland**, ME 04101-3401; mail address: 489 Congress St, Portland, ME 04101-3498
T: +1 207 7741822; Fax: +1 207 7754301; E-mail: rdesk@mainehistory.org; URL: www.mainehistory.org
1822; Richard D'Abate
Genealogy, New England; Records of Kennebec Proprietors, Fogg Autogr Coll, Northeast Boundary Coll, Papers of Governor William King, Portland Company Records
65 000 vols; 97 curr per; 3 000 maps; 300 microforms; 24 digital data carriers; 1 mio mss   37402

Multnomah Law Library, County Courthouse, 4th Flr, 1021 SW Fourth Ave, **Portland**, OR 97204
T: +1 503 9883394; Fax: +1 503 9883395
1890; Jacquelyn Jurkins
216 000 vols   37403

Nathan & Henry B. Cleaves Law Library, 142 Federal St, **Portland**, ME 04101
T: +1 207 7739712; Fax: +1 207 7732155; E-mail: info@cleaves.org; URL: www.cleaves.org
1811; Nancy Rabasca
Records & Briefs of the Supreme Judicial Court of the State of Maine
30 000 vols; 6 digital data carriers
                                     37404

Oregon Historical Society Research Library, 1200 SW Park Ave, **Portland**, OR 97205
T: +1 503 3065243; Fax: +1 503 2192040; E-mail: libreference@ohs.org; URL: www.ohs.org
1898; MaryAnn Campbell
Oregon Provisional Gov Papers, State & Regional Architecture, Ship & Vehicle Plans, Oregon Trail Diaries, Oregon Imprints by Belknap, Russian-American Studies
33 000 vols; 420 curr per; 30 000 maps; 20 000 microforms; 1 500 sound-rec; 12 000 linear feet of mss, 2 500 000 photos, 6 000 oral hist, 3 000 serials
                                     37405

Portland Art Museum, Crumpacker Family Library, 1219 SW Park Ave, **Portland**, OR 97205-2486
T: +1 503 2764215; E-mail: library@pam.org; URL: www.pam.org
1892; Debra Royer
Fine art, Art hist; Arts of the Pacific Northwest Coast Indians, Contemporary Art, English Silver Bks, Japanese Prints, Northwest Artists Arch
33 000 vols; 60 curr per; 135 av-mat; 540 Pamphlet Cases, 71 500 slides
ACRL, ALA, ARLIS/NA   37406

Portland VA Medical Center, Medical Library, 3710 SW US Veterans Hospital Rd, P.O. Box 1034, **Portland**, OR 97207
T: +1 503 2208262 ext 55955; Fax: +1 503 7217816; E-mail: portland.library@va.gov; URL: www.portland.va.gov/PORTLAND/medlibrary.asp
12 660 vols; 779 curr per   37407

Stoel Rives LLP, Law Library, 900 SW Fifth Ave, Ste 2600, **Portland**, OR 97204
T: +1 503 2949576; Fax: +1 503 2202480; URL: www.stoel.com
1906; Tony Haas
30 000 vols   37408

Portsmouth Athenaeum, Library – Museum – Gallery, 9 Market Sq, **Portsmouth**, NH 03801; P.O. Box 848, Portsmouth, NH 03821-0848
T: +1 603 4312538; Fax: +1 603 4317180; E-mail: athenaeum@juno.com
1817; Thomas Hardman Jr
John Langdon Papers, NH Gazette Coll, Pierce Papers, NH Fire & Marine Insurance Company Papers, Wendell Family Papers, Isles of Shoals Coll; Wentworth-by-the-Sea Hotel Papers
40 000 vols; 18 curr per; 75 mss; 246 maps; 168 microforms; photos, Postcards, Stereocards, Ephemera, CDV
libr loan; SAA   37409

Scioto County Bar & Law Library, Scioto County Court House, 3rd Flr, 602 Seventh St, **Portsmouth**, OH 45662
T: +1 740 3558259; E-mail: stjord@sciotowireless.net
Kevin Blume
25 000 vols   37410

Prescott Historical Society, Sharlot Hall Museum Library and Archives, 115 S McCormick St, **Prescott**, AZ 86301
T: +1 928 4453122; E-mail: archives@sharlot.org; URL: www.sharlot.org
1928; Michael Wurtz
Arch Coll, Arizona Hist Coll, newsp, mss; Early Arizona & Indian Coll, photog
10 000 vols; 12 curr per; 4 000 maps; 500 linear feet of docs, 100 000 photogs, 300 oral hist   37411

Bristol-Myers Squibb Pharmaceutical Research Institute Library, Rte 206 & Provence Line Rd, **Princeton**, NJ 08543; P.O. Box 4000, Princeton, NJ 08543-4000
T: +1 609 2524800; Fax: +1 609 2526280; URL: www.bms.com
1925; Mary Lasko
Biology, Chemistry, Medicine, Pharmacology
112 000 vols; 1 100 curr per
libr loan   37412

Educational Testing Service, Carl Campbell Brigham Library & Test Collection, Rosedale Rd, **Princeton**, NJ 08541
T: +1 609 7345667; Fax: +1 609 6837186; E-mail: internet_brigham@ets.org
1961; Karen McQuillen
Artificial intelligence, Behav science, Cognitive science, Education, Psychology, Social sciences, Statistics
18 000 vols; 300 curr per; 9 digital data carriers; 19 000 tests & related mat
                                     37413

Institute for Defense Analysis Library, 805 Bunn Dr, **Princeton**, NJ 08540
T: +1 609 9244600; Fax: +1 609 9243061
1959; Barbara Hamilton
Mathematics, Computer science, Speech
14 000 vols; 350 curr per; 2 digital data carriers   37414

Recording for the Blind Dyslexic (RFB&D), Library & Member Services Departments, 20 Roszel Rd, **Princeton**, NJ 08540
T: +1 609 4520606; Fax: +1 609 9878116; E-mail: reference@rfbd.org; URL: www.rfbd.org
1948; Pamela Johnson
98 775 vols; 80 000 sound-rec; 63 000 recorded educational textbks   37415

City of Providence, Parks Department, Roger Williams Park Museum of Natural History Library & Archives, Roger Williams Park, Elmwood Ave, **Providence**, RI 02907
T: +1 401 7859457; Fax: +1 401 4615146; URL: www.providenceri.com/museum
1896; Marilyn R. Massaro
Archs
10 000 vols   37416

Rhode Island Historical Society Library, 121 Hope St, **Providence**, RI 02906
T: +1 401 2738107; Fax: +1 401 7517930; URL: www.rihs.org/libraryhome.htm
1822; Kirsten Hammerstrom
Graphics Coll, Mss Coll, Film Coll
150 000 vols; 1 000 mss; 13 000 microforms   37417

Rhode Island Hospital, Peters Health Sciences Library, 593 Eddy St, **Providence**, RI 02902
T: +1 401 4448074; URL: www.lifespan.org/rih
1932; Marianne Slocomb
25 000 vols; 650 curr per   37418

Rhode Island School for the Deaf Library, One Corliss Park, **Providence**, RI 02908
T: +1 401 2227441; Fax: +1 401 2226998
Mary Cummings
10 000 vols   37419

State of Rhode Island Office of Library & Information Services – Talking Books Plus, One Capitol Hill, **Providence**, RI 02908
T: +1 401 2225800; Fax: +1 401 2224195; E-mail: tbplus@lori.ri.gov; URL: www.olis.ri.gov/tbp
Andrew I. Egan
828 000 vols; 24 curr per   37420

Vermont Institute of Natural Science Library, 6565 Woodstock Rd, Rte 4, POB 1281, **Quechee**, VT 05059
T: +1 802 3595000; URL: www.vinsweb.org
1972
Natural hist, Environmental studies, Ornithology; Henry Potter Papers, Olin Sewall Pettingill Ornithological Libr Billings – Kitteridge Herbarium
8 500 vols; 90 curr per   37421

National Fire Protection Association, Charles S. Morgan Technical Library, One Batterymarch Park, **Quincy**, MA 02169-7471
T: +1 617 9847445; Fax: +1 617 9847060; E-mail: library@nfpa.org; URL: www.nfpa.org/library
1945; Sue Marsh
Safety; Nat Fire Codes, Nat Fire Protection Assn Published Arch Coll
15 000 vols; 230 curr per; 11 000 microforms; 325 sound-rec; 9 000 tech rpts, 20 000 standards & patents   37422

United States Department of the Interior, National Park Service, Adams National Historical Park, Stone Library, 135 Adams St, **Quincy**, MA 02169-1749
T: +1 617 7731177; Fax: +1 617 4727562; URL: www.nps.gov/adam
1870; Kelly Cobble
Bks Owned & Used by President John Adams & President John Quincy Adams, Charles Frances Adams – Civil War Minister to England & his two sons, Henry & Brook Adams
14 000 vols   37423

Dorothea Dix Hospital, Walter A. Sikes Learning Resource Center, 820 S Boylan Ave, **Raleigh**, NC 27603-2176; mail address: MSC Center, Raleigh, NC 27699-3601
T: +1 919 7335111; Fax: +1 919 7339781; e-mail: james.osberg@dhhs.nc.gov; URL: www.dhhs.nc.gov
1957; Ella M. Williams
11 200 vols; 1 000 sound-rec; 1 digital data carriers
libr loan   37424

North Carolina Museum of Art, Art Reference Library, 2110 Blue Ridge Rd, **Raleigh**, NC 27607-6494; mail address: 4630 Mail Services Center, Raleigh, NC 27699-4630
T: +1 919 6646769; Fax: +1 919 7338034; URL: www.ncartmuseum.org/education/library.shtml
1956; Natalia J. Lonchyna
Decorative & fine arts
40 000 vols; 90 curr per; 20 000 auction cat, 10 500 vertikal files, 3 050 slides
ARLIS/NA   37425

North Carolina Regional Library for the Blind & Physically Handicapped, 1811 Capital Blvd, **Raleigh**, NC 27635-0001
T: +1 919 7334376; Fax: +1 919 7336910
1959; Francine Martin
311 000 vols   37426

North Carolina State Museum of Natural Sciences, H. H Brimley Memorial Library, 11 W Jones St, **Raleigh**, NC 27601-1029
T: +1 919 7337450; Fax: +1 919 7152356
1941; Janet G. Edgerton
12 000 vols; 80 curr per
ALA, SLA   37427

Wake Area Health Education Center, Medical Library, 3261 Atlantic Ave, Ste 212, **Raleigh**, NC 27604-8547
T: +1 919 3508529; Fax: +1 919 3508836; E-mail: medlibrary@wakemed.org; URL: www.wakeahec.org/library/
1965; Susan C. Corbett
8 000 vols; 170 curr per; 5 000 mss; 7 digital data carriers
libr loan; MLA, AMIA   37428

Braille Institute Library Services – Desert Center, 70-251 Ramon Rd, **Rancho Mirage**, CA 92270
T: +1 760 3211111; Fax: +1 760 3219715; URL: www.brailleinstitute.org
Gayle Wormell
12 000 vols   37429

Rapid City Regional Hospital, Medical Library, 353 Fairmont Blvd, **Rapid City**, SD 57701; P.O. Box 6000, Rapid City, SD 57709-6000
T: +1 605 7197101; Fax: +1 605 7191578; E-mail: library@rcrh.org; URL: www.rcrh.org/library
Patricia Hamilton
8 000 vols; 25 curr per; 30 e-books
                                     37430

Johnson & Johnson Pharmaceutical Research & Development, Hartman Library, 1000 Rte 202, **Raritan**, NJ 08869-0602; Box 300, Raritan, NJ 08869-0602
T: +1 908 7044919; Fax: +1 908 7079860
1944; Donna Wahl
Biology, Chemistry, Medicine
8 000 vols; 480 curr per   37431

1013

Historical Society of Berks County, Museum and Library, 940 Centre Ave, *Reading*, PA 19601
T: +1 610 3754375; Fax: +1 610 3754376; E-mail: society.library@verizon.net; URL: www.berksweb.com/histsoc
1869; Barbara A. Brophy
Becks County Family Histories; German & English newspapers of Reading & Berks County 1797-1907; Iron Hist of Berks County; Original Mss
11 000 vols; 20 curr per; 400 music scores; 600 maps; 600 microforms; 250 av-mat; 2 000 sound-rec; 50 digital data carriers
ALA   37432

San Mateo County Office of Education, The SMERC Library, 101 Twin Dolphin Dr, *Redwood City*, CA 94065-1064
T: +1 650 8025655; Fax: +1 650 8025665; URL: www.smcoe.k12.ca.us/smerc
Karol Thomas
30 000 vols; 400 curr per   37433

Nevada Historical Society, Museum-Research Library, 1650 N Virginia St, *Reno*, NV 89503-1799
T: +1 775 6881191; Fax: +1 775 6882917; E-mail: www@clan.lib.nv.us; URL: www.nevadaculture.org
1904; Peter L. Bandurraga
30 000 vols; 260 curr per; 55 000 maps; 8 000 microforms; 3 300 mss colls, 410 000 photogs
NLA   37434

CIIT Centers for Health Research, Golberg Library & eResource Center, 6 Davis Dr, P.O. Box 12137, *Research Triangle Park*, NC 27709-2137
T: +1 919 5581215; Fax: +1 919 5581300; E-mail: knight@ciit.org; URL: www.ciit.org
1979; Erin Knight
Environmental Health, Toxicology, Industrial Chemicals, Biochemistry, Epidemiology, Cell Biology, Analytical Chemistry
8 000 vols; 180 curr per   37435

National Institute of Environmental Health Sciences Library, Module A Basement, 111 T & W Alexander Dr, Bldg 101, *Research Triangle Park*, NC 27709; P.O. Box 12233, Mail Drop A0-01, Research Triangle Park, NC 27709
T: +1 919 5413426; Fax: +1 919 5410669; URL: library.niehs.nih.gov/home.htm
1967; W. Davenport Robertson
28 000 vols; 100 curr per; 4 500 e-journals; 300 e-books; 8 000 microforms; 350 av-mat; 40 sound-rec; 80 digital data carriers; 25 000 per vols unbound
libr loan; ALA, MLA, SLA   37436

RTI International, Information Services, 3040 Cornwallis Rd, *Research Triangle Park*, NC 27709; P.O. Box 12194, Research Triangle Park, NC 27709-2194
T: +1 919 5418787; Fax: +1 919 5411221; E-mail: mchristian@rti.org; URL: www.rti.org
1960; Mariel Christian
Chemistry, Education, Engineering, Environmental studies & engineering, Int studies, Medicine, Social sciences
50 000 vols; 1 125 curr per; 17 000 av-mat; 20 digital data carriers   37437

Pacific Northwest National Laboratory, Hanford Technical Library, 2770 University Dr, *Richland*, WA 99354; P.O. Box 999, MSIN P8-55, Richland, WA 99352-0999
T: +1 509 3727450; Fax: +1 509 3727426; E-mail: pnl.techlib@pnl.gov; URL: libraryweb.pnl.gov
1948; Annanaomi Sams
Dept of Energy Contractor Rpts
41 000 vols; 9 000 curr per
libr loan   37438

Chevron Information Technology Company, Division of Chevron USA, Inc, Global Library Richmond, 100 Chevron Way, Bldg 50, Rm 1212, *Richmond*, CA 94802
T: +1 510 2424755; Fax: +1 510 2425621; E-mail: wito@chevron.com

1920; Tora Williamsen-Berry
Patents, Chemical engineering, Fuels, Petroleum
30 000 vols; microforms: pages 2 500 000, patents 3 000 000
libr loan   37439

Grand Lodge of Virginia, A.F. & A.M., Allen E. Roberts Masonic Library and Museum, Inc., 4115 Nine Mile Rd, *Richmond*, VA 23223-4926
T: +1 804 2223110; Fax: +1 804 2224253; E-mail: grandlodge@rcn.com
1778; Marie M. Barnett
Original Proceedings of Grand Lodge of Virginia; Early Proceedings of Grand Lodges in the US; Local Lodge Histories; Arch; Transactions of AQC, ALR, Norcalore & Miscellanea; rare books
8 932 vols; 45 curr per; 10 music scores; 2 maps; 50 microforms; 20 av-mat; 60 boxes of reports, letters, mss, by-laws
                                               37440

Grand Lodge of Virginia AF&AM Library & Museum Historical Foundation, Allen E. Roberts Masonic Library & Museum, 4115 Nine Mile Rd, *Richmond*, VA 23223-4926
T: +1 804 2223110; Fax: +1 804 2224253; E-mail: library@grandlodgeofvirginia.org; URL: www.grandlodgeofvirginia.org
Marie M. Barnett
Masonic archival materials
9 000 vols; 65 curr per   37441

Museum of the Confederacy, Eleanor S. Brockenbrough Library, 1201 E Clay St, *Richmond*, VA 23219
T: +1 804 6491861; Fax: +1 804 6447150; E-mail: library@moc.org; URL: www.moc.org
1890; John Coski
Museum Archs; Confederate Bonds, Currency, Imprints, Miltary Unit Records; Jefferson Davis Coll
11 000 vols; 15 curr per   37442

Public Law Library, 101 E Franklin St, *Richmond*, VA 23219-2193
T: +1 804 7806500; E-mail: lawlibrary@richmondgov.com; URL: www.richmondpubliclibrary.org/content.asp?contentID=28
12 000 vols; 12 curr per   37443

Virginia Baptist Historical Society & the Center for Baptist Heritage & Studies Library, University of Richmond, P.O. Box 34, *Richmond*, VA 23173
T: +1 804 2898434; Fax: +1 804 2898953; E-mail: theheritagecenter@vbmb.org; URL: www.baptistheritage.org
1876; Fred Anderson
25 000 vols; 250 microforms; 7 000 mss
                                               37444

Virginia Historical Society Library, 428 North Blvd, *Richmond*, VA 23220; P.O. Box 7311, Richmond, VA 23221-0311
T: +1 804 3429688; Fax: +1 804 3552399; URL: www.vahistorical.org
1831; Charles F. Bryan Jr
Confederate Imprints, 17th & 18th-c English Architecture, Confederate Weaponry & Military
150 000 vols; 350 curr per; 5 000 maps; 7 000 000 mss   37445

Virginia Museum of Fine Arts Library, 200 N Boulevard, *Richmond*, VA 23220-4007
T: +1 804 3401495; Fax: +1 804 3401548; URL: www.vmfa.museum/library.html
1935; Dr. Suzanne Hill Freeman
American Arts (McGlothlin Coll), Oriental Arts (Weedon Coll & Coopersmith Coll), General Art Hist (Strause Coll), Faberge Coll, Paul Mellon Pre-1850 British Art Coll
144 000 vols; 248 curr per; 28 digital data carriers   37446

Suffolk County Historical Society Library, 300 W Main St, *Riverhead*, NY 11901-2894
T: +1 631 7272881; Fax: +1 631 7273467
1886
Churches, Deeds, Military, Shipwrecks, Towns & Wills Coll, VF, Daughters of the Revolution of 1776, Fullerton, Long

Island Hist Coll, The Talmage Weaving Coll
15 000 vols; 12 curr per; 300 microforms
                                               37447

Carilion Clinic, Health Sciences Library, Belleview at Jefferson St, *Roanoke*, VA 24033; P.O. Box 13367, Roanoke, VA 24033-3367
T: +1 540 9818039; Fax: +1 540 9818666
Karen Dillon
29 000 vols; 130 curr per; 21 000 e-books   37448

Fulton County Historical Society, Inc, Museum & Library, 37 E 375 N (just off N US 31), *Rochester*, IN 46975
T: +1 219 2234436; URL: www.fultoncountyhistory.org
1963
10 000 vols; 100 curr per   37449

George Eastman House, Richard & Ronay Menschel Library, 900 East Ave, *Rochester*, NY 14607
T: +1 585 2713361; Fax: +1 585 2713970; URL: www.eastmanhouse.org
1947; Rachel Stuhlman
Motion pictures, Still photos; 19th c Bks; Per Colls, Southward & Hawes Coll, Sipley/3M Coll, Alvin Langdon Coburn Coll, Lewis W. Hine Coll
36 000 vols; 320 curr per; 200 mss; 1 000 microforms; 940 sound-rec
libr loan   37450

Jewish Community Center of Greater Rochester, Philip Feinbloom Library, 1200 Edgewood Ave, *Rochester*, NY 14618-5408
T: +1 716 4612000 ext 607; Fax: +1 716 4610805; URL: www.jccrochester.org
1973; Ellen Steinberg
Judaica; Holocaust Coll, Jewish Children Coll, Dr Saul Moress Peace Coll
10 000 vols; 20 curr per   37451

Mayo Foundation, Medical School Library LRC, 200 First St SW, *Rochester*, MN 55905
T: +1 507 2843893; Fax: +1 507 2664065; URL: www.mayo.edu/medlib/medlib.html
Melissa Rethlefsen
Medicine
50 000 vols   37452

Monroe County Historian's Department Library, 115 South Ave, *Rochester*, NY 14604
T: +1 585 4288352; Fax: +1 585 4288353
Dr. Carolyn S. Vacca
10 000 vols   37453

National Braille Association, Inc, Braille Book Bank, 95 Allens Creek Rd, *Rochester*, NY 14618
T: +1 716 4278260; Fax: +1 716 4270263; URL: nationalbraille.org
Angela Coffaro
Braille College Level Textbks, Braille Music, Braille Tech Tables
14 100 vols   37454

Rochester Academy of Medicine Library, 1441 East Ave, *Rochester*, NY 14610-1665
T: +1 585 2711313; Fax: +1 585 2714172; E-mail: raom@choiceonemail.org; URL: www.raom.org
1900; Hechmat Tabechian
Historic Medical Libr
31 000 vols   37455

Rochester Historical Society Library, 485 East Ave, *Rochester*, NY 14607
T: +1 716 2712705; Fax: +1 716 2719089; URL: www.rochesterhistory.org
1861; Ann Salter
Rochester Hist Society Publs, Rochester City Directories, Rochester & Genesee Valley Hist Coll
15 000 vols   37456

Rochester Museum & Science Center, Schuyler C. Townson Library, 657 East Ave, *Rochester*, NY 14607
T: +1 585 2714320 ext 315; Fax: +1 716 2712119; E-mail: lea_kemp@rmsc.org; URL: www.rmsc.org
1917; Leatrice Kemp
Local hist, Anthropology, Antiques, Archaeology, Costume, Native Americans,

Natural hist; 19th C Per, Almanacs, Posters, Greeting Cards, Stone Coll – Glass Plate Negatives, Flour Milling Industry (Moseley & Motley Coll)
30 000 vols; 60 curr per; 100 maps; 50 av-mat; 45 000 slides, 40 000 photos
libr loan; Society of American Archivists
                                               37457

The Strong Museum Library, One Manhattan Sq, *Rochester*, NY 14607
T: +1 585 2632700; Fax: +1 585 2632493; URL: www.strongmuseum.org
1972; Carol Sandler
US social & cultural hist; Fore-Edge Paintings Coll, 19th & early 20th C Children's, 19th & 20th C Trade Cats, Patent Papers for Dolls & Toys, Winslow Homer Coll
40 000 vols; 225 curr per; 8 000 cat
                                               37458

University of Rochester Medical Center, Edward G. Miner Library, 601 Elmwood Ave, *Rochester*, NY 14642
T: +1 585 2753361; Fax: +1 585 7567762; URL: www.urmc.edu/hslt/miner
Julia Sollenberger
Medical history
240 000 vols; 1 200 curr per   37459

Visual Studies Workshop, Research Center, 31 Prince St, *Rochester*, NY 14607
T: +1 716 4428676 ext 211; Fax: +1 716 4421992; E-mail: researchcenter@vsw.org; URL: www.vsw.org/research_center/research_center.php
1970; William Johnson
19th and 20th c photogr mat, contemporary photographers, hist of art & social hist
17 000 vols; 214 curr per; 300 sound-rec; 3 800 artist's bks, 3 000 independent artists' videos, 30 000 slides, 16 000 original photos
libr loan   37460

Swenson Swedish Immigration Research Center, Augustana College, 3520 Seventh Ave, *Rock Island*, IL 61201; mail address: Augustana College, 639 38th St, Rock Island, IL 61201-2296
T: +1 309 7947204; Fax: +1 309 7947443; E-mail: sag@augustana.edu; URL: www.augustana.edu/swenson
1981; Dag Blanck
Swedish-American Newspapers, City Directories (Chicago, Minneapolis, St Paul), Name Indexes to Swedish Embarkation Ports of Gothenburg & Malmo
17 000 vols; 30 curr per
libr loan   37461

US Department of Health & Human Services – Food & Drug Administration, Medical Library, 5600 Fishers Lane, Rm 11B-40/HFD-230, *Rockville*, MD 20857
T: +1 301 8275701; Fax: +1 301 4436385
1948; Carol S. Cavanaugh
Pharmacology, Toxicology; Adverse Drug Effects, FDA Publs
30 000 vols; 1 500 curr per   37462

Westat, Inc Library, 1650 Research Blvd, *Rockville*, MD 20850
T: +1 301 2511500; Fax: +1 301 2942034
Maureen Stawick
Marketing, labor, economics
10 000 vols; 250 curr per   37463

Connecticut State Library – Library for the Blind & Physically Handicapped, 198 West St, *Rocky Hill*, CT 06067
T: +1 860 7212020; Fax: +1 860 7212056; E-mail: lbph@cslib.org; URL: www.cslib.org/lbph.htm
1968; Carol Taylor
Connecticut & New England Cassette Bks; 1 branch libr
232 000 vols; 219 902 av-mat; 12 600 braille vols, 220 000 talking bks   37464

Connecticut State Library – State Records Center, 198 West St, *Rocky Hill*, CT 06067
T: +1 860 7212041; Fax: +1 860 7212055; URL: www.cslib.org/records.htm
Donald P. Ballinger
60 000 vols   37465

Alion Science & Technology, AMPTIAC Library, 201 Mill St, **Rome**, NY 13440
T: +1 315 3397095; Fax: +1 315 3397107; E-mail: amptiac_librarian@alionscience.com; URL: amptiac.iitri.org/inforesources/docsearch.html
Perry Onderdonk
Ceramics, Composites, Metals, Plastics, Quality
30 000 vols; 20 curr per   37466

Chaves County District Court Library, P.O. Box 1776, **Roswell**, NM 88202
T: +1 505 6222565; Fax: +1 505 6249506
10 000 vols   37467

Roswell Museum & Art Center, Research Library, 100 W 11th St, **Roswell**, NM 88201
T: +1 505 6246744 ext 25; Fax: +1 505 6246765; E-mail: jordan@roswellmuseum.org; URL: www.roswellmuseum.org
1937; Candace Jordan Russell
Native American Art, Southwestern Archaeology
11 000 vols; 47 curr per; 200 mss; 8 diss/theses; 2 govt docs; 1 music scores; 44 maps; 2 microforms; 7 000 av-mat; 190 sound-rec; 4 digital data carriers   37468

Metropolitan Council for Educational Opportunity Library, 40 Dimock St, **Roxbury**, MA 02119
T: +1 617 4271545; Fax: +1 617 5410550; URL: www.metcoinc.org
1966; Jean McGuire
10 000 vols; 50 curr per   37469

William Beaumont Hospital, Medical Library, 3601 W 13 Mile Rd, **Royal Oak**, MI 48073-6769
T: +1 248 8981744; Fax: +1 248 8981060; E-mail: rodocdelivery@beaumont.edu
1956; Janet Zimmerman
Biomedical engineering
37 000 vols; 350 curr per; 150 e-books; 500 av-mat; 500 sound-rec; 5 digital data carriers   37470

Rusk State Hospital, Texas Department of Mental Health & Mental Retardation, Medical Library, Jacksonville Hwy N, P.O. Box 318, **Rusk**, TX 75785-0318
T: +1 903 6833421 ext 3240; Fax: +1 903 6832994
1976; Mildred Hataway
16 000 vols; 31 curr per   37471

Saco Museum, Dyer Library, 371 Main St, **Saco**, ME 04072
T: +1 207 2833861; Fax: +1 207 2830754; E-mail: astrassner@sacomuseum.org; URL: www.dyerlibrarysaco.org
1867; Andrea Strassner
Hist of York County & Maine, Lowell Innes Glass Libr, US Treasury Rpts
60 000 vols; 130 curr per; 500 av-mat; 500 sound-rec   37472

California Energy Commission Library, 1516 Ninth St, MS10, **Sacramento**, CA 95814-5512
T: +1 916 6544292; Fax: +1 916 6544046; E-mail: library@energy.state.ca.us; URL: www.energy.ca.gov/library/index.html
1975; Karen L. Kasuba
California Energy Commission Publs
18 000 vols; 150 curr per   37473

California Environmental Protection Agency, Library, 1001 I St, 2nd Floor, P.O. Box 2815, **Sacramento**, CA 95814
T: +1 916 327-0635; Fax: +1 916 322-7060; E-mail: CalEPALibrary@arb.ca.gov; URL: www.calepa.ca.gov/library
1976
Air Pollution Techn Info Ctr
13 500 vols; 600 curr per   37474

California State Railroad Museum, Library, 111 I St, **Sacramento**, CA 95814
T: +1 916 3238073; Fax: +1 916 3238073; E-mail: rrmuseumlibrary@parks.ca.gov; URL: www.csrmf.org/library-and-collections
1981; Ellen Halteman
10 000 vols; 90 curr per; 8 000 maps; 300 microforms; 225 sound-rec; 300 000 drawings, 25 videos, 10 000 slides   37475

California State Railroad Museum Library, 113 I St, **Sacramento**, CA 95814; mail address: 111 I St, Sacramento, CA 95814
T: +1 916 3238073; Fax: +1 916 3275655; E-mail: rrmuseumlibrary@parks.ca.gov; URL: www.californiastaterailroadmuseum.org
Ellen Louise Halteman
10 000 vols; 80 curr per   37476

Sierra Research Library, 1801 J St, **Sacramento**, CA 95814
T: +1 916 4444666; Fax: +1 916 4448373; E-mail: srinfo@sierraresearch.com; URL: www.sierraresearch.com
14 990 vols; 12 curr per   37477

State Water Resources Control Board, Law Library, 1001 I St, P.O. Box 100, **Sacramento**, CA 95814
T: +1 916 3415170; Fax: +1 916 6530428; E-mail: info@waterboards.ca.gov; URL: www.waterboards.ca.gov
Environmental law; California State Water Quality Orders, State Board Resolutions, State Water Rights Orders & Decisions
19 500 vols   37478

University of California, Davis – F. William Blaisdell Medical Library, 4610 X St, **Sacramento**, CA 95817
T: +1 916 7343533; Fax: +1 916 7347418; E-mail: mclref@ucdavis.edu; URL: www.lib.ucdavis.edu/hsl
1929; Rebecca Davis
Bioethics Coll, Birth Defects Coll
44 000 vols; 559 curr per; 4 000 microforms; 110 av-mat; 3 028 sound-rec
libr loan; NNLM   37479

Saint Augustine Historical Society, Research Library, 6 Artillery Lane, 2nd Fl, **Saint Augustine**, FL 32084; mail address: 271 Charlotte St, Saint Augustine, FL 32084-5099
T: +1 904 8242872; Fax: +1 904 8242569; E-mail: sahsdirector@bellsouth.net; URL: www.staugustinehistoricalsociety.org/library.html
1883; Charles A. Tingley
Card Calendar of Spanish Docs, 1512-1764; Cathedral Parish Records, Cuban Archs, Census Records for Northeast Florida, Colonial Office Records, Stetson Coll
13 000 vols; 75 curr per; 81 mss; 300 diss/theses; 18 000 govt docs; 100 music scores; 5 300 maps; 12 000 microforms; 110 av-mat; 250 sound-rec; 5 digital data carriers; 15 000 photos   37480

Illinois Youth Center – Saint Charles Library, 4450 Lincoln Hwy, **Saint Charles**, IL 60175
T: +1 630 5840506 ext 284; Fax: +1 630 5841126
Richard Fryer
12 000 vols; 14 curr per   37481

Department of Veterans Affairs Medical Center, Medical Library, 4801 Veterans Dr, **Saint Cloud**, MN 56303
T: +1 320 2556342; Fax: +1 320 2556493
1925; Jeanne Skaj
8 230 vols; 207 curr per; 650 av-mat; 60 digital data carriers
MLA   37482

Bryan Cave LLP, Law Library, One Metropolitan Sq, 211 N Broadway, Ste 3600, **Saint Louis**, MO 63102-2750
T: +1 314 2592298; Fax: +1 314 2592020; URL: www.bryancave.com
Judy Harris
22 000 vols; 100 curr per   37483

Central Institute for the Deaf, Speech, Hearing and Education Library, 909 South Taylor, **Saint Louis**, MO 63110; mail address: 4560 Clayton Ave, Saint Louis, MO 63110
T: +1 314 9770268; Fax: +1 314 9770046; E-mail: csarli@cid.wustl.edu; URL: www.cid.wustl.edu
1933; Cathy Sarli
Speech & Hearing (Max A Goldstein Coll), CID-Goldstein Historic Hearing Devices Coll
1 200 vols; 75 curr per; 4 500 mss; 500 diss/theses; 50 govt docs; 30 av-mat; 10 digital data carriers
SLA   37484

International Library, Archives & Museum of Optometry, 243 N Lindbergh Blvd, **Saint Louis**, MO 63141-7881
T: +1 314 9914100; Fax: +1 314 9914101; E-mail: ilamo@aoa.org; URL: www.aoa.org
1902; Ellen Dickman
Rare Books & Per, Topical Res Conducted at Libr
11 000 vols; 450 curr per   37485

Law Library Association of Saint Louis, 1300 Civil Courts Bldg, Ten N Tucker Blvd, **Saint Louis**, MO 63101
T: +1 314 6224386; Fax: +1 314 2410911; E-mail: rabtech@swbell.net; URL: tlc.library.net/lla
1838; Jean Moorleghen
70 000 vols; 120 curr per; 4 000 microforms; 22 digital data carriers   37486

Missouri Botanical Garden Library, 4500 Shaw Blvd, **Saint Louis**, MO 63110
T: +1 314 5775159; Fax: +1 314 5770840; E-mail: library@mobot.org; URL: www.mobot.org/mobot.molib/
1859; Douglas L. Holland
Pre-Linnean Botany (Sturtevant Coll), Bryology (Steere Coll), Ewan Coll, Folio Coll, Rare Bk Coll
178 000 vols; 800 curr per; 5 incunabula; 7 000 maps; 25 000 microforms; 1 digital data carriers; 100 000 Pamphlets, 500 Slides, 220 000 Archival Items, 3 059 Mss, 6 000 Art Works
libr loan; CBHL, SLA   37487

Missouri History Museum, Library & Research Center, 225 S Skinker Blvd, **Saint Louis**, MO 63105; P.O. Box 11940, Saint Louis, MO 63112-0040
T: +1 314 7464500; Fax: +1 314 4543162; E-mail: library@mohistory.org; URL: www.mohistory.org
1866; Emily Jaycox
Trade Catalogs; Western Americana; Theater Coll; Scrapbook Coll
81 000 vols; 250 curr per; 250 diss/theses; 8 500 music scores; 2 500 maps; 5 000 microforms; 42 Journals   37488

Monsanto Company, Library Services, 800 N Lindbergh Blvd, **Saint Louis**, MO 63167
T: +1 314 6944747; Fax: +1 314 6948748; E-mail: c.c.library@monsanto.com; URL: www.monsanto.com
1961; Gail Hoef
Agriculture, Biology, Biotech, Chemistry, Economy, Medicine, Pharmaceuticals
49 000 vols; 500 curr per; 500 av-mat; 150 sound-rec; 60 Large Print Bks   37489

Museum of Transportation, Reference Library, 3015 Barrett Station Rd, **Saint Louis**, MO 63122
T: +1 314 9658214; Fax: +1 314 9650242; URL: www.museumoftransport.org
1944; Nick Ohlman
Rail transportation; Transportation & Communication Coll
8 000 vols; 10 curr per; 30 000 engineering drawings and blueprints   37490

Saint John's Mercy Medical Center, Thomas F. Frawley Medical Center Library, Tower B, 621 S New Ballas Rd, Ste 1000, **Saint Louis**, MO 63141
T: +1 314 2516340; Fax: +1 314 2514299; E-mail: medlib@mercy.net; URL: www.stjohnsmercy.org/library
Jennifer P. Plaat
10 000 vols; 475 curr per; 110 e-books
libr loan; MLA   37491

Saint Louis Art Museum, Richardson Memorial Library, One Fine Arts Dr, Forest Park, **Saint Louis**, MO 63110-1380
T: +1 314 6555252; Fax: +1 314 7216172; E-mail: library@slam.org; URL: www.slam.org
1915; Marianne L. Cavanaugh
Art hist; Contemporary Art Ephemera, Louisiana Purchase Exposition 1904 rec, Museum Hist Coll, Rare Art Bks Coll
112 000 vols; 200 curr per; 3 digital data carriers; 41 000 pamphlets, 20 000 art auction cat, 55 000 slides
libr loan   37492

Saint Louis Mercantile Library at the University of Missouri-St Louis, Thomas Jefferson Library Bldg, One University Blvd, **Saint Louis**, MO 63121-4400
T: +1 314 5167247; Fax: +1 314 5167241; URL: www.umsl.edu/mercantile
1846; John Neal Hoover
Hist, Biography, Social sciences, Science, Fine arts, Fiction, Lit; American Railroads (John W Barriger III Coll), Early Western Americana; Nat Inland Waterways Coll, St Louis Hist Coll
250 000 vols; 2 500 000 Mss, 175 000 photos, 25 000 Newspapers   37493

Saint Louis Metropolitan Police Department, Saint Louis Police Library, 315 S Tucker Blvd, **Saint Louis**, MO 63102
T: +1 314 4445581; Fax: +1 314 4445689
1947; Barbara Miksicek
Annual Reports for Police Department dating from 1861 to present
30 000 vols; 120 curr per; 150 microforms; 10 digital data carriers
libr loan; SLA   37494

Saint Mary's Health Center, Health Sciences Library, 6420 Clayton Rd, **Saint Louis**, MO 63117
T: +1 314 7688112; Fax: +1 314 7688974
1933; Kathy Mullen
16 000 vols; 200 curr per
libr loan   37495

University of Missouri-Columbia – Missouri Institute of Mental Health Library, 5400 Arsenal St, **Saint Louis**, MO 63139-1403
T: +1 314 8776514; Fax: +1 314 8776521; URL: www.mimh.edu/library
1962; Christina Sullivan
Nat Assn of State Mental Health Program Directory Arch; SLSH Arch; MIMH Faculty Publs
10 000 vols; 260 curr per; 200 av-mat; 721 sound-rec
libr loan; MLA, AMHL, SALIS   37496

Chesapeake Bay Maritime Museum, Howard I. Chapelle Memorial Library, P.O. Box 636, **Saint Michaels**, MD 21663
T: +1 410 7452916; Fax: +1 410 7456088; E-mail: library@cbmm.org; URL: www.cbmm.org
1968; Joan Chlan
Ship/Boat Plans
9 800 vols; 18 curr per; 30 linear ft of vertical files, 25 000 photos, 150 linear feet of mss, 200 oral history tapes   37497

Chesapeake Bay Maritime Museum Library, PO Box 636, **Saint Michaels**, MD 21663-0636
T: +1 410 7452916; Fax: +1 410 7456088; E-mail: library@cbmm.org; URL: www.cbmm.org
Pete Lesher
Maritime history, Maryland, Virginia
10 000 vols; 17 curr per   37498

3M Information Research & Solutions, 3M Center Bldg, 201-1S 12 A' St", **Saint Paul**, MN 55144-1000
T: +1 651 7331110; Fax: +1 651 7366495; URL: www.mmm.com
Martha Ellison
100 000 vols; 1 550 curr per   37499

James J. Hill Reference Library, 80 W Fourth St, **Saint Paul**, MN 55102-1669
T: +1 651 2655453; Fax: +1 651 2655515; E-mail: info@jjhill.org; URL: www.jjhill.org
1921; Nicole Marchand
Business, Economics; James Jerome Hill Papers, Louis Warren Hill Papers
205 000 vols; 750 curr per; 900 Special Interest Per Sub   37500

Minnesota Department of Revenue Library, 600 N Robert St, ***Saint Paul***, MN 55101; Mail Sta 2230, Saint Paul, MN 55146
T: +1 651 2963529; Fax: +1 651 5563103
1986; Donna Davis
Taxation; CCH, RIA Tax Looseleaf Services
10 000 vols; 100 curr per; 2 e-journals; 2 maps; 50 sound-rec; 20 digital data carriers
libr loan; SLA 37501

Minnesota Historical Society Library, 345 Kellogg Blvd W, ***Saint Paul***, MN 55102-1906
T: +1 651 2962143; Fax: +1 651 2977436; E-mail: reference@mnhs.org; URL: www.mnhs.org
1849; Michael Fox
Great Northern & Northern Pacific Railroad Papers; Hubert H Humphrey Papers; Minnesota newspapers (1849-present); Minnesota State Arch, records of agencies of Minnesota state & local govt
411 890 vols; 1 300 curr per; 35 000 mss; 37 000 maps; 100 000 microforms; 2 500 av-mat; 6 digital data carriers; 6 500 newspaper titles
libr loan; RLG, MALC, SAA, ALA, SLA, Oral History Association 37502

Orphan Voyage & Concerned United Birth Parents, Kamman Dale Libraries, 57 N Dale St, ***Saint Paul***, MN 55102-2228
T: +1 651 2245160
1975; Jeanette G. Kamman
Genealogy, Hist, religion; Genealogies, bks, directories, hist mat, refs
50 000 vols 37503

Quatrefoil Library, 1619 Dayton Ave, Ste 105, ***Saint Paul***, MN 55104-6206
T: +1 651 6410969; E-mail: quatrefoillibrary@yahoo.com; URL: www.quatrefoillibrary.org
1983; Dan Hanson
Gay, Lesbian, and other Sexual Minority Materials
9 030 vols; 40 curr per; 1 000 av-mat; 300 sound-rec 37504

Quatrefoil Library, 1619 Dayton Ave, Ste 105, ***Saint Paul***, MN 55104-6206
T: +1 651 6410969; E-mail: quatrefoillibrary@yahoo.com; URL: www.qlibrary.org
Kathy Robbins
Lesbian, gay, bisexuality
11 000 vols; 40 curr per 37505

Science Museum of Minnesota, Louis S. Headley Memorial Library, 120 W Kellogg Blvd, ***Saint Paul***, MN 55102
T: +1 651 2219424; Fax: +1 651 2214750; URL: www.smm.org
1930; Kirsten Ellenbogen
Anthropology, Earth science, Natural hist; Egypt (Ames Coll), Geology (US Geological Survey), General Science (Museum Publs), Birds (Kate Farnham Memorial Coll)
25 000 vols; 50 curr per; 9 VF drawers of pamphlets, 208 file boxes of museum publs 37506

Saint Peter Regional Treatment Center Libraries, 100 Freeman Dr, ***Saint Peter***, MN 56082
T: +1 507 9317880; Fax: +1 507 9317177
1878; Thomas Fattichi
Alcoholic & drug dependents, Bks & media for retarded readers, Mentally ill patients
13 000 vols; 25 curr per 37507

Museum of Fine Arts, Reference Library, 255 Beach Dr NE, ***Saint Petersburg***, FL 33701-3498
T: +1 727 8962667; Fax: +1 727 8944638; URL: www.fine-arts.org
1962; Jordana Weiss
21 000 vols; 25 curr per 37508

Poynter Institute for Media Studies, Eugene Patterson Library, 801 Third St S, ***Saint Petersburg***, FL 33701
T: +1 727 8219494; Fax: +1 727 8989201; URL: www.poynter.org
1985; David Shedden
Mass communication, Journalism; Oral Hist of News Professionals, videotapes; Organization of News Ombudsmen Arch, vertical files; Don Murray Papers
11 000 vols; 100 curr per; 400 av-mat; 75 sound-rec 37509

Seneca Nation Library, 830 Broad St, ***Salamanca***, NY 14081
T: +1 716 9453157; Fax: +1 716 9459770; E-mail: smilibal@sni.org; URL: www.cclslib.org/snia/snia
1979; Pam Bowen
Native American Mat
18 000 vols; 374 curr per; 26 diss/theses; 9 govt docs; 23 maps; 110 microforms; 228 sound-rec; 11 digital data carriers
libr loan; ALA 37510

Essex Law Library, Superior Court House, 34 Federal St, ***Salem***, MA 01970
T: +1 978 7410674; Fax: +1 978 7457224; URL: www.lawlib.state.ma.us
Richard Adamo
30 000 vols; 200 curr per 37511

Oregon State Library Talking Book & Braille Services, 250 Winter St NE, ***Salem***, OR 97301-3950
T: +1 503 378-5389; Fax: +1 503 5858059; E-mail: susan.b.westin@state.or.us; URL: www.oregon.gov/osl/tbabs
1932; Susan Westin
Descriptive Video Coll
158 000 vols
libr loan 37512

Peabody Essex Museum, Phillips Library, East India Sq, 161 Essex St, ***Salem***, MA 01970-3783
T: +1 978 7459500; Fax: +1 978 7419012; URL: www.pem.org/museum/library.php
1799; Sidney E. Berger
Chinese Hist (Frederick T Ward Coll), East Asian books and manuscripts; ephemera; almanacs; postcards; clipper ship cards; 19th c. greeting cards; stereographs; cookery broadsides; genealogy Hawthorne military; early Essex Co. newspapers
400 000 vols; 200 curr per; 5 000 mss; 3 000 maps; 1 000 microforms; 1 mio drawing, mss, photogs & prints 37513

Salem Athenaeum, 337 Essex St, ***Salem***, MA 01970
T: +1 978 7442540; Fax: +1 978 7447536; E-mail: info@salemathenaeum.net; URL: www.salemathenaeum.net
1760; John M. Procious
Religion, Early sciences, Early children's lit, Early natural hist; Personal Libr of Dr Edward Holyoke, 18th to early 19th century; Philosophical Libr 1781; Social Libr of 1760
54 202 vols; 31 curr per 37514

Upper Colorado River Commission Library, 355 South 400 E, ***Salt Lake City***, UT 84111-2904
T: +1 801 5311150; Fax: +1 801 5319705
Don A. Ostler
9 000 vols 37515

Utah Department of Natural Resources Library, 1594 W North Temple, ***Salt Lake City***, UT 84114; P.O. Box 146100, Salt Lake City, UT 84114-6100
T: +1 801 5373333; Fax: +1 801 5373400; URL: dnrlibrary.utah.gov
Mage Yonetani
Geology
23 000 vols; 41 curr per 37516

Utah State Historical Society, Utah History Research Center Library, 300 Rio Grande, ***Salt Lake City***, UT 84101-1182
T: +1 801 5333535; Fax: +1 801 5333504; URL: www.history.utah.gov; www.historyresearch.utah.gov
1939; Doug Misner
Coll of Utah, Mormon and Western hist
50 000 vols; 205 curr per; 6 000 mss; 200 diss/theses; 25 000 music scores; 30 000 maps; 12 000 microforms; 10 000 sound-rec; 22 000 pamphlets 37517

Utah State Library Division – Program for the Blind & Disabled, 250 N 1950 West, Ste A, ***Salt Lake City***, UT 84116-7901
T: +1 801 7156789; Fax: +1 801 7156767; E-mail: blind@utah.gov; URL: blindlibrary.utah.gov
1957; Bessie Oakes
Mormon Lit Coll, Western Books Coll
470 000 vols; 140 curr per; 1 830 High Interest/Low Vocabulary Bks, 30 Bks on Deafness & Sign Lang 37518

Daughters of the Republic of Texas Library at the Alamo, 300 Alamo Plaza, ***San Antonio***, TX 78205; P.O. Box 1401, San Antonio, TX 78295-1401
T: +1 210 2251071; Fax: +1 210 2128514; E-mail: drtl@drtl.org; URL: drtl.org
1945; Elaine B. Davis
Battle of Flowers Assn Records (1895-); Beckmann family papers (1825-1973); Bustillo family papers (1772-1936); Cassiano-Perez family papers (1741-1954); Cumings family papers (1824-1926); James T. DeShields Coll (1863-1940); Leo M.J. Dielmann papers (1881-1969); Gentilz-Fretelliere family papers (1793-1962); Conrad A. Goeth papers (1869-1953); John Herndon James papers (1812-1938); Ernst Schuchard papers (1893-1972); John W. Smith Papers (1822-1934); Yanaguana Society records (1931-1960): records of a San Antonio historical society and hist coll of paintings by Theodore Gentilz and Carl G. von Iwonski
18 000 vols; 65 curr per; 700 maps; 121 microforms; 15 000 vertical files, 220 mss colls, 1 200 docs, 14 800 clipping files, 36 000 photogs
ALA, SAA, TLA 37519

McNay Art Museum Library, 6000 N New Braunfels Ave, ***San Antonio***, TX 78209; P.O. Box 6069, San Antonio, TX 78209-0069
T: +1 210 8051727; Fax: +1 210 8240218
1954; Ann Jones
Tobin Theater Arts Coll
26 000 vols; 41 curr per 37520

Southwest Research Institute, Thomas Baker Slick Memorial Library, 6220 Culebra Rd, MS 84, ***San Antonio***, TX 78238-5166
T: +1 210 5222125; Fax: +1 210 5225479; E-mail: library@swri.org; URL: www.swri.org
1948; Anita E. Lang
Bio-engineering, Environmental engineering, Science & technology
61 000 vols; 860 curr per; 260 e-journals; 3 900 Electronic Media & Resources
libr loan 37521

University of Texas Health Science Center at San Antonio, Briscoe Library, 7703 Floyd Curl Dr, MSC 7940, ***San Antonio***, TX 78229-3900
T: +1 210 5672410; Fax: +1 210 5672463; E-mail: askalibrarian@uthscsa.edu; URL: www.library.uthscsa.edu
1966; Mary Moore
Hist of Medicine Coll
222 000 vols; 3 614 curr per; 8 000 e-books; 1 digital data carriers
libr loan 37522

Ragusan Press, Croation Immigration Library, 2527 San Carlos Ave, ***San Carlos***, CA 94070
T: +1 650 5921190; Fax: +1 650 5921526; E-mail: croatians@aol.com; URL: www.croations.com
Adam S. Eterovich
Immigration; 70,000 Pioneers Coll (prior to 1900)
10 000 vols 37523

Braille Institute Library Services – San Diego Center, 4555 Executive Dr, ***San Diego***, CA 92121-3021
T: +1 858 4521111; Fax: +1 858 4521688
Louise Zuckerman
15 000 vols 37524

Salk Institute for Biological Studies, Salk Institute Library, 10010 N Torrey Pines Rd, ***San Diego***, CA 92037-1099; P.O. Box 85800, San Diego, CA 92186-5800
T: +1 858 4534100 ext 1235; Fax: +1 858 4527472; E-mail: library@salk.edu; URL: www.salk.edu
1962; Carol Bodas
AIDS, Biochemistry, Cancer, Genetics, Immunology, Molecular biology, Neuroscience, Virology
15 000 vols; 160 curr per; 1 digital data carriers 37525

San Diego Aero-Space Museum, Inc, N. Paul Whittier Historical Aviation Library, 2001 Pan American Plaza, Balboa Park, ***San Diego***, CA 92101-1636
T: +1 619 2348291; Fax: +1 619 2334526; E-mail: arenga@sdasm.org; URL: www.aerospacemuseum.org
1978; Pamela Gay
Pilot Logbooks, E. Cooper Air Mail Pioneers, The C. Ryan Libr, E.H. Heinemann Coll, Consolidated Company papers
24 000 vols; 18 curr per; 75 mss; 40 music scores; 500 maps; 100 000 microforms; 100 sound-rec; 1 200 airline insignia, 6 000 aircraft drawings, 4 000 manuals, 2 mio images, 150 scrapbooks, 300 brochures
libr loan; ALA, SAA 37526

San Diego Family History Center, Family History Center, 4195 Camino Del Rio S, ***San Diego***, CA 92108
T: +1 619 5847668; Fax: +1 619 5841225
1966; M. Lopez
13 000 vols; 42 000 microforms; 400 digital data carriers 37527

San Diego Historical Society, Research Archives, Balboa Park, 1649 El Prado, ***San Diego***, CA 92101; P.O. Box 81825, San Diego, CA 92138
T: +1 619 2326203; Fax: +1 619 2321059; E-mail: panter@sandiegohistory.org; URL: www.sandiegohistory.org
1929; John Panter
Architectural Drawings; Public Records (city, county, superior court); Oral Hist (1 100 titles); Photograph Dept (2 500 000 images)
14 000 vols; 32 curr per; 2 192 mss; 2 500 maps; 115 microforms; 151 av-mat; 1 000 linear feet of mss, 5 500 linear feet of govt rec 37528

San Diego Museum of Art Library, 1450 El Prado, ***San Diego***, CA 92101; P.O. Box 122107, San Diego, CA 92112-2107
T: +1 619 6961958; Fax: +1 619 2329367; E-mail: library@sdmart.org; URL: www.sdmart.org
1926; Dr. James Grebl
Binney Coll, bks & cats (Indian miniature painting)
30 000 vols; 85 curr per; 30 000 auction cat, 15 000 slides, 150 linear feet of vertical files
libr loan 37529

San Diego Museum of Man, Scientific Library, Balboa Park, 1350 El Prado, ***San Diego***, CA 92101
T: +1 619 2392001; Fax: +1 619 2392749; E-mail: library@museumofman.org; URL: www.museumofman.org
1915; Jane Bentley
North American Indians
33 000 vols; 300 curr per; 51 archival mss 37530

San Diego Natural History Museum, Research Library, Balboa Park, 1788 El Prado, ***San Diego***, CA 92101; P.O. Box 121390, San Diego, CA 92112-1390
T: +1 619 2550225; Fax: +1 619 2320248; E-mail: library@sdnhm.org; URL: www.sdnhm.org/research/library/index.html
1874; Margaret Dykens
Herpetology, Mammals, Palaeontology, Botany, Entomology; Bird Paintings (Sutton & Brooks Coll); Geology & Paleontology (Anthony W Vodges Coll); Herpetology (Lawrence Klauber Coll); Photo Arch; Wild Flower Paintings

(Valentien Coll)
56 000 vols; 5 000 maps; 150 microforms; 160 cubic feet of archival mat 37531

Scripps Mercy Hospital, Jean Farb Memorial Medical Library, 4077 Fifth Ave, ***San Diego***, CA 92103-2180
T: +1 619 2607024; Fax: +1 619 2607262; URL: www.scripps.org
1937; Penny T. Ward
Statewide Nursing Program Coll
12 000 vols; 300 curr per; 200 av-mat; 525 sound-rec; 4 digital data carriers
libr loan 37532

Zoological Society of San Diego Library, Arnold & Mable Beckman Ctr for Conservation & Research for Endangered Species, ***San Diego***, CA 92112; mail address: 15600 Psaqual Valley Rd, Escondido, CA 92027
Fax: +1 760 2915478; URL: library.sandiegozoo.org
1916; Linda L. Coates
Veterinary medicine, Horticulture; Ernst Schwarz Reprint Coll; Herpetology (Charles E. Shaw Coll); Zoo Publs
12 000 vols; 680 curr per; 5 digital data carriers; 15 000 reprints, 54 feet of archives, 76 oral hist tapes 37533

Art Institutes of California-San Francisco, 1170 Market St, ***San Francisco***, CA 94102-4908
T: +1 415 8650198; Fax: +1 415 8631121; URL: www.aisf.artinstitutes.edu
Jamie MacInnis
8 000 vols; 130 curr per 37534

Bohemian Club Library, 624 Taylor St, ***San Francisco***, CA 94102
T: +1 415 8852440; Fax: +1 415 5672332; E-mail: library@bc-owl.org
1872; Matthew W. Buff
San Francisco fine presses: Grabhorn, Taylor and Taylor, Nash, et al; Archs of the Bohemian Club
23 500 vols 37535

Bureau of Jewish Education, Jewish Community Library, 1835 Ellis St, mail address: 639 14th Ave, ***San Francisco***, CA 94115
T: +1 415 5673327; Fax: +1 415 5674542; E-mail: library@bjesf.org; URL: www.bjesf.org/library.htm
1974; Jonathan Schwartz
Holocaust Coll, Jewish Art Coll, Jewish Music Coll, Pedagogic Coll, Hebrew, Russian & Yiddish, Havas Children's Library
30 000 vols; 36 curr per; 200 music scores; 500 av-mat; 600 sound-rec; 25 digital data carriers
AJL 37536

California Academy of Sciences Library, Golden Gate Park. 55 Concourse Dr, ***San Francisco***, CA 94118
T: +1 415 3795493; Fax: +1 415 3795729; E-mail: library@calacademy.org; URL: www.calacademy.org/research/library
1853; Lawrence W. Currie
Academy Arch; Biodiversity Resource Center; Diatoms; Natural Hist of Baja California (Belvedere); Picture/Photo/Slide Coll
230 000 vols; 1 400 curr per; 250 e-journals; 75 000 maps; 5 000 microforms; 50 sound-rec; picture coll 1 mio items (40 000 slides), 600 linear feet of archives, 150 videos
libr loan 37537

California Historical Society, North Baker Research Library, 678 Mission St, ***San Francisco***, CA 94105
T: +1 415 3571848; Fax: +1 415 3571850; E-mail: reference@calhist.org; URL: www.californiahistoricalsociety.org
1922; Mary Morganti
Mss & Arch Coll, Photogr Coll, Western Printing and Publishing (Edward C. Kemble Coll)
57 000 vols; 400 mss; 100 incunabula; 400 music scores; 1 500 maps; 358 microforms; 80 Spec Interest Per Sub
libr loan 37538

Exploratorium Learning Commons, 3601 Lyon St, ***San Francisco***, CA 94123
T: +1 415 5610343; Fax: +1 415 5610370; E-mail: commons@exploratorium.edu; URL: www.exploratorium.edu/lc/
Deb Hunt
15 000 vols; 120 curr per 37539

Exploratorium Learning Studio, Exploratorium Museum, 3601 Lyon Sto, ***San Francisco***, CA 94123-1099
T: +1 415 5610343; Fax: +1 415 5610370; E-mail: studio@exploratorium.edu; URL: www.exploratorium.edu/ls
1985; Gilles Poitras
Science, Psychology, Art, Perception, Cognition, Informal Learning
12 000 vols; 150 curr per
libr loan 37540

Grand Lodge Free & Accepted Masons of California, Henry Wilson Coil Masonic Library & Museum, 1111 California St, ***San Francisco***, CA 94108
T: +1 415 2929114; Fax: +1 415 9290690; URL: www.freemason.org
1949; Adam Kendall
Annual Proceedings of Masonic Grand Lodge, California 1850-present
12 000 vols; 30 curr per 37541

Helen Brown Lombardi Library of International Affairs, World Affairs Council of Northern California, 312 Sutter St, Ste 200, mail address: 310 Sutter St, ***San Francisco***, CA 94108
T: +1 415 2934646; Fax: +1 415 9825028; E-mail: library@wacsf.org; URL: www.itsyourworld.org/library
1947; Mary Anne McGill
Int relations; Newsletters & Press Releases, US Dept of State Docs
8 000 vols; 100 curr per; 80 maps; 1 000 sound-rec; 4 digital data carriers; 200 docs, 3 000 pamphlets 37542

Holocaust Center of Northern California, The Laszlo N. Tauber Library and Research Center, 121 Steuart St, mail address: 639 14th Ave, ***San Francisco***, CA 94105
T: +1 415 7779060; Fax: +1 415 7779062; E-mail: info@hcnc.org; URL: www.hcnc.org
1979; Judy Janec
Memorial Bks on Destroyed Jewish Communities in Europe
14 000 vols; 30 maps; 1 000 av-mat; 50 art reproductions, 700 oral hist tapes, 2 000 photos 37543

Institute for Advanced Study of Human Sexuality, Research Library, 1523 Franklin St, ***San Francisco***, CA 94109
T: +1 415 9281133; Fax: +1 415 9288284; URL: www.iashs.edu
1976; Jerry Zientara
Lyle Stuart Libr of Sexual Science; Harry Mohne Coll
250 000 vols; 15 curr per; 200 diss/theses; 500 000 slides, 40 000 videos, 100 000 periodicals, 100 000 magazines, 10 unbound vols of journals 37544

Japanese American National Library, 1619 Sutter St, ***San Francisco***, CA 94109; P.O. Box 590598, San Francisco, CA 94159-0598
T: +1 415 5675006
1969; Karl K. Matsushita
National Repository of Japanese American Redress; Japanese American Vernacular Newspapers
20 000 vols; 55 curr per; 700 pamphlets, 4 000 clippings 37545

C. G. Jung Institute of San Francisco, Virginia Allen Detloff Library, 2040 Gough St, ***San Francisco***, CA 94109
T: +1 415 7718055 ext 207; Fax: +1 415 7718926; E-mail: library@sfjung.org; URL: www.sfjung.org
1960; Marianne Morgan
Jung Archive
16 000 vols; 25 curr per; 250 av-mat; 1 200 sound-rec; 1 000 mss
libr loan 37546

Mechanics' Institute Library, 57 Post St, ***San Francisco***, CA 94104-5003
T: +1 415 3930103; Fax: +1 415 4214192; E-mail: reference@milibrary.org; URL: www.milibrary.org
1854; Inez Shor Cohen
Art hist, Business, Chess, Economics, Finance, American & local hist, Social sciences, Languages & lit
165 000 vols; 500 curr per; 11 000 microforms; 4 000 av-mat; 4 000 sound-rec; 1 digital data carriers; 300 annual rpts 37547

Museum of Russian Culture, Inc Library, 2450 Sutter St, ***San Francisco***, CA 94115
T: +1 415 9214082; Fax: +1 415 9214082
Nicholas Koretsky
20 000 vols 37548

Pesticide Action Network North America Regional Center, International Information Program, 49 Powell St, Ste 500, ***San Francisco***, CA 94102
T: +1 415 9811771; Fax: +1 415 9811991; E-mail: panna@panna.org; URL: www.panna.org
1988
Pesticides, Agriculture & Int Development, bks
12 000 vols; 350 curr per 37549

San Francisco African-American Historical & Cultural Society, Library of San Francisco, 762 Fulton St, 2nd Flr, ***San Francisco***, CA 94102
T: +1 415 2926172; Fax: +1 415 4404231; URL: www.sfblackhistory.org
1955; Al Williams
Mary Ellen Pleasant Coll (African American Hist in San Francisco – Black Panthers)
36 000 vols; 10 curr per; 1 000 microforms; 50 000 photos 37550

San Francisco Art Institute, Anne Bremer Memorial Library, 800 Chestnut St, ***San Francisco***, CA 94133
T: +1 415 7494562; E-mail: library@sfai.edu; URL: www.sfai.edu
1871
Art, Photography, Film; Artists' Bk Coll; California Art
30 000 vols; 220 curr per 37551

San Francisco Botanical Garden Society at Strybing Arboretum, Helen Crocker Russell Library of Horticulture, 1199 Ninth Ave (Ninth Ave at Lincoln Way), ***San Francisco***, CA 94122-2384
T: +1 415 6611316; Fax: +1 415 6613539; E-mail: library@sfbotanicalgarden.org; URL: www.sfbotanicalgarden.org
1972; Barbara M. Pitschel
Nursery and Seed Cats, Children's Coll
25 000 vols; 500 curr per; 50 microforms; 50 av-mat; 100 sound-rec; 20 digital data carriers
CBHL, SLA 37552

San Francisco Law School Library, 20 Haight St, ***San Francisco***, CA 94102
T: +1 415 6265550; Fax: +1 415 6265584; URL: www.sfls.edu
1909; Jane Gamp
2 branch libs
21 000 vols 37553

San Francisco Maritime National Historical Park, J. Porter Shaw Library & Historic Documents Department, Fort Mason Ctr, Bldg E, 3rd Flr, ***San Francisco***, CA 94123
T: +1 415 5617030; Fax: +1 415 5561624; E-mail: safr_maritime_library@nps.gov; URL: www.nps.gov/safr/
1959; David Hull
Oral Hist, John Lyman Maritime Coll, San Francisco Marine Exchange Vessel Movement Records, Photogr Coll, Ship Registers, Ship Plans, Barbara Johnson Whaling Coll, Dean Mawdsley Coll of Worls War II Naval Hist
35 000 vols; 120 curr per; 1 000 maps; 2 000 microforms; 694 sound-rec; 250 000 photos, 120 000 Vessel plans, 79 videos, 4 694 pamphlets, 1 500 linear feet of microfiches, 2 900 logbooks, 260 scrapbooks
libr loan 37554

San Francisco Museum of Modern Art, Research Library, 151 Third St, ***San Francisco***, CA 94103-3107
T: +1 415 3574120; Fax: +1 415 3574038; E-mail: brominski@sfmoma.org; URL: www.sfmoma.org
1935; Barbara C. Rominski
Lit on Photography (Margery Mann & Sidney Tillim Colls); Modern Art, exhibition cats
60 000 vols; 425 curr per; 20 microforms; 40 digital data carriers; 50 000 cat, 448 VF drawers of clippings
ARLIS/NA 37555

San Francisco Performing Arts Library & Museum, 401 VanNess Ave, 4th Flr, ***San Francisco***, CA 94102
T: +1 415 2554800; Fax: +1 415 2551913; E-mail: info@sfpalm.org; URL: www.sfpalm.org
David Humpherey
Performing arts, dance;San Francisco Bay Area Performing Arts History
15 000 vols; 29 curr per 37556

Sierra Club, William E. Colby Memorial Library, 85 Second St, 2nd Flr, ***San Francisco***, CA 94105-3441
T: +1 415 9775506; Fax: +1 415 9775799; E-mail: colby.library@sierraclub.org; URL: www.sierraclub.org/library; www.sierraclub.org
1892; Ellen Byrne
Early International & North American Mountaineering (Sierra Club Mountaineering Coll)
11 000 vols; 350 curr per; 500 maps; 3 500 slides 37557

Society of California Pioneers, Alice Phelan Sullivan Library, 300 4th St, ***San Francisco***, CA 94107
T: +1 415 9571849; Fax: +1 415 9579858; E-mail: pkeats@californiapioneers.org; URL: www.californiapioneers.org/library.html
1850; Patricia L. Keats
Correspondence of Thomas Starr King; Letters of Jessie Benton Fremont, writer; Jacob Rink Snider Coll; Handwritten Diaries; Scrapbooks; Patterson Rapch papers, Cooper-Motera Papers
10 000 vols; 20 curr per; 1 000 mss; 25 diss/theses; 1 200 music scores; 600 maps; 30 sound-rec; 60 000 historic photos
libr loan; OCLC 37558

UCSF Medical Center at Mount Zion, Harris M. Fishbon Memorial Library, 1600 Divisadero St, Rm A-116, P.O. Box 7921, ***San Francisco***, CA 94115
T: +1 415 8857378; Fax: +1 415 7760689; E-mail: fishbon@ucsfmedctr.org; URL: mountzion.ucsfmedicalcenter.org/library/index.asp
Gail Sorrough
Hist of Medicine Coll, Psychiatry Coll
11 000 vols; 210 curr per; 1 300 av-mat; 3 digital data carriers 37559

Wells Fargo Bank Library – Historical Research Library, 420 Montgomery St, MAC-A0101-106, ***San Francisco***, CA 94163
T: +1 415 3964157; Fax: +1 415 3918644
Grace Evans
California Gold Rush and Mining, Californiana, San Francisco History, Staging and Western Transportation
8 000 vols; 15 curr per 37560

Santa Clara County Office of Education, Library Services, 1290 Ridder Park Dr, ***San Jose***, CA 95131
T: +1 408 4536800; Fax: +1 408 4536815; E-mail: professional_library@sccoe.org; URL: www.sccoe.org/depts/library
Donna Wheelehan
15 000 vols; 152 curr per 37561

Huntington Library, 1151 Oxford Rd, ***San Marino***, CA 91108
T: +1 626 4052191; Fax: +1 626 4495720; URL: www.huntington.org/LibraryDiv/LibraryHome.html
1919; David S. Zeidberg
British & American hist & lit
686 160 vols; 600 curr per; 244 040 microforms; 2,6 mio mss 37562

Marin Institute for the Prevention of Alcohol & Other Drug Problems, Resource Center, 24 Belvedere St, ***San Rafael***, CA 94901
T: +1 415 4565692; Fax: +1 415 4560491; URL: www.marininstitute.org
1989
10 000 vols; 60 curr per 37563

Braille Institute Library Services, Santa Barbara Center, 2031 De La Vina St, **Santa Barbara**, CA 93105
T: +1 805 6826222; Fax: +1 805 6876141
Nate Streeper
14 000 vols
37564

Cottage Hospital, David L. Reeves Medical Library at Bath St, P.O. Box 689, **Santa Barbara**, CA 93102
T: +1 805 5697240; Fax: +1 805 5697588; E-mail: reeves@sbch.org; URL: www.cottagehospital.org/
1943; Lucy B. Thomas
Hist of Medicine, Surgery, Internal Medicine
16 000 vols; 325 curr per; 300 av-mat; 500 sound-rec; 12 digital data carriers; 400 pamphlets
libr loan; ALA, MLA, MLGSCA
37565

Santa Barbara Botanic Garden Library, 1212 Mission Canyon Rd, **Santa Barbara**, CA 93105-2199
T: +1 805 6824726; Fax: +1 805 5630352; URL: www.sbbg.org
1942; Laurie Hannah
15 000 vols; 200 curr per; 19 mss; 700 maps; 60 sound-rec; 99 linear feet of reprints, 10 linear feet of nursery cat
CBHL
37566

Santa Barbara Mission, Archive-Library, 2201 Laguna St, **Santa Barbara**, CA 93105
T: +1 805 6824713; Fax: +1 805 6829323; E-mail: research@sbmal.org; URL: www.sbmal.org
1786; Lynn Bremer
California missions; Early California, Original Spanish & Mexican Missionary Coll, Spanish & Hispanic American Coll
23 000 vols
37567

Santa Barbara Museum of Art, Fearing Library, 1130 State St, **Santa Barbara**, CA 93101
T: +1 805 9634364; Fax: +1 805 9666840
1974; Heather Brodhead
Group Exhibition Cat Coll, Institutional Art Coll, Individual Artist File Coll, Sale Cat Coll
50 000 vols; 50 curr per; 292 microforms; 20 000 cat, 500 linear feet of arch
37568

Santa Barbara Museum of Natural History Library, 2559 Puesta del Sol Rd, **Santa Barbara**, CA 93105
T: +1 805 6824711; Fax: +1 805 5693170; URL: www.sbnature.org; www.centralcoastmuseums.org
1920; Terri Sheridan
Chumash Indians, Stillman Berry Malacology Coll, Pacific Voyages Coll, Natural Hist Art Coll, Channel Islands Arch, John Peabody Harrington California Indian Arch
40 000 vols; 420 curr per; 1 600 maps; 8 digital data carriers; 200 feet of reprints
37569

Institute of American Indian & Alaska Native Culture & Arts Development Library, 83 Avan Nu Po Rd, **Santa Fe**, NM 87508
T: +1 505 4245715; Fax: +1 505 4243131; URL: www.iaia.edu/college/library
1962; Sarah Kostelecky
Art Coll, Indian Coll, Indian Music Coll, Indian photogr Coll; Smithsonian Indian photogr Coll
33 000 vols; 150 curr per; 100 digital data carriers; 74 Electronic Media & Resources, 250 CDs, 1 000 Videos, 50 Special Interest Per Sub
37570

Museum of International Folk Art, The Bartlett Library, 706 Camino Lejo, P.O. Box 2087, **Santa Fe**, NM 87504-2087
T: +1 505 4761210; Fax: +1 505 4761300; E-mail: library@moifa.org; URL: www.moifa.org
1953
Ceramics, textiles, conservation, folk art, anthropology, religious iconography; Coll of folk music of Spanish New Mexico
13 000 vols; 85 curr per; 700 sound-rec; vertical files, postcards, photos, architectural drawings, ephemera, sound recordings, Exhibition records
libr loan; OCLC, IFLA
37571

Museum of New Mexico – Museum of Indian Arts & Culture-Laboratory of Anthropology, 708 Camino Lejo, **Santa Fe**, NM 87505; P.O. Box 2087, Santa Fe, NM 87504-2087
T: +1 505 4761264; Fax: +1 505 4761330; E-mail: miac.info@state.nm.us; URL: www.indianartsandculture.org
1929; Allison Colborne
Native Americans, Southwestern anthropology; Meso American Archaeology & Ethnohist Coll (Sylvanus G Morley Libr)
25 000 vols; 210 curr per
libr loan
37572

Museum of New Mexico – Palace of the Governors-Fray Angelico Chavez History Library, 120 Washington Ave, **Santa Fe**, NM 87501; P.O. Box 2087, Santa Fe, NM 87504
T: +1 505 4765090; Fax: +1 505 4765053; E-mail: histlib@mnm.state.nm.us; URL: www.palaceofthegovernors.org
1885; Tomas Jaehn
Manuscript Coll; New Mexico, maps & mss, newsp, rare bks; Southwest (Photo Arch); prints; Rare Books dept; Photo Archives
50 000 vols; 45 curr per; 3 200 maps; 2 000 microforms; 500 sound-rec; mss (375 linear feet), clipping files
37573

New Mexico State Library – Library for the Blind & Physically Handicapped, 1209 Camino Carlos Rey, **Santa Fe**, NM 87507-5166
T: +1 505 4769770; Fax: +1 505 4769776; E-mail: lbph@state.nm.us; URL: www.nmstatelibrary.org
John Mugford
292 000 vols
37574

Santa Fe Institute Library, 1399 Hyde Park Rd, **Santa Fe**, NM 87501
T: +1 505 9462708; Fax: +1 505 9820565; E-mail: mba@santafe.edu; URL: www.santafe.edu
1984; Margaret Alexander
Mathematics; Garrett Birkhoff Coll, Stanislav Ulum Coll
10 000 vols; 30 curr per
37575

School of Advanced Research, Catherine McElvain Library, 660 Garcia St, P.O. Box 2188, **Santa Fe**, NM 87504-2188
T: +1 505 9547234; Fax: +1 505 9547214; E-mail: library@sarsf.org; URL: www.sarweb.org
1907
Anthropology, Archeology, Ethnology, Southwest Indian Arts
8 703 vols; 65 curr per; 5 digital data carriers
37576

Rand Corporation Library, 1700 Main St, **Santa Monica**, CA 90407-2138; P.O. Box 2138, Santa Monica, CA 90407-2138
T: +1 310 3930411 ext 7788; Fax: +1 310 4516920; E-mail: library@rand.org; URL: www.rand.org
1948; Lucy S. Wegner
Science & technology, Education, Health science, Civil justice; Data Coll, Foreign Coll
51 000 vols; 2 300 curr per
libr loan
37577

Institute for Creation Research Library, 10946 Woodside Ave N, **Santee**, CA 92071
T: +1 619 4480090; Fax: +1 619 4483469; URL: www.icr.org
Richard LaHaye
Religion (Harald F J Ellingson Coll)
15 000 vols; 102 curr per
37578

Trudeau Institute Library, 154 Algonquin Ave, **Saranac Lake**, NY 12983
T: +1 518 8913080; Fax: +1 518 8915126; URL: www.trudeauinstitute.org
1964; Kelly Stanyon
Microbiology, Immunology
15 000 vols; 110 curr per
37579

John & Mable Ringling Museum of Art Library, 5401 Bayshore Rd, **Sarasota**, FL 34243
T: +1 941 3595700; Fax: +1 941 3607370; E-mail: library@ringling.org; URL: www.ringling.org
1946; Linda R. McKee
John Ringling Bk Coll, Rare Bks
72 000 vols; 125 curr per; 150 av-mat;

100 digital data carriers; 5 000 Ephemera vertical files, 10 000 Journals
libr loan; TBLC, SOLINET, OCLC, ARLIS/NA, ALA, ARLIS/Southeast, FLA
37580

Mote Marine Laboratory Library, Arthur Vining Davis Library, 1600 Ken Thompson Pkwy, **Sarasota**, FL 34236-1096
T: +1 941 3884441; E-mail: library@mote.org; URL: www.mote.org
1978; Susan M. Stover
Marine biology, Oceanography, Fisheries, Aquaculture; Mote Tech Rpts
10 000 vols; 400 curr per; 3 000 reprints
37581

Sarasota Memorial Hospital, Medical Library, 1700 S Tamiami Trail, **Sarasota**, FL 34239-3555
T: +1 941 9171730; Fax: +1 941 9171646
1956; Barbara Hartman
10 000 vols; 200 curr per; 2 digital data carriers
37582

Georgia Historical Society Library, 501 Whitaker St, **Savannah**, GA 31401
T: +1 912 6512125; Fax: +1 912 6512831; E-mail: ghslib@georgiahistory.com; URL: www.georgiahistory.com
1839; Nora E. Galler
Central of Georgia Railway Coll; Cordray-Foltz Photogr Coll; Walter C. Hartridge, Jr Genealogical Records; Revolutionary War & Civil War Colls; Savannah Jewish Archives
20 000 vols; 50 curr per
ALA, SAA
37583

Live Oak Public Libraries- Library for the Blind & Physically Handicapped, 2708 Mechanics Ave, **Savannah**, GA 31404
T: +1 912 3545864; Fax: +1 912 3545534
Linda Stokes
17 000 vols
37584

Skidaway Institute of Oceanography Library, John F. McGowan Library, Ten Ocean Science Circle, **Savannah**, GA 31411-1011
T: +1 912 5982474; Fax: +1 912 5982391; E-mail: library@skio.usg.edu; URL: www.skio.usg.edu/resources/library
1970; John Cruickshank
8 000 vols; 110 curr per; 6 digital data carriers; Access to 307 databases and 19 000 full-text/full-image journals
37585

Dudley Observatory Library, 107 Nott Terrace, Ste 201, **Schenectady**, NY 12308
T: +1 518 3827583; Fax: +1 518 3827584; E-mail: info@dudleyobservatory.org; URL: www.dudleyobservatory.org
1852; Nancy Langford
Astronomy & Mathematics Coll, Rare Bks, Benjamin A. Gould Coll
10 000 vols; 10 curr per; 2 incunabula; 50 maps
libr loan
37586

Marian & Ralph Feffer Library, 10460 N 56th St, **Scottsdale**, AZ 85253
T: +1 480 9510323; Fax: +1 480 9517150; E-mail: library@templebethisrael.org; URL: www.templebethisrael.com
1958; Carol Reynolds
Judaica Music Libr, Holocaust Coll, Children's Libr
12 000 vols; 15 curr per; 470 sound-rec; 26 VF drawers of pamphlets, clippings, maps
37587

Lackawanna Bar Association, Law Library, Courthouse, Ground Flr, 200 N Washington Ave, **Scranton**, PA 18503
T: +1 570 9636712; Fax: +1 570 3442944
1879; Marita E. Paparelli
25 000 vols; 127 curr per
37588

Penobscot Marine Museum, Stephen Phillips Memorial Library, 9 Church St, **Searsport**, ME 04974; P.O. Box 498, Searsport, ME 04974-0498
T: +1 207 5482529; Fax: +1 207 5482520; E-mail: cgood@pmm-maine.org; URL: www.penobscotmarinemuseum.org/library.html

1936; Cipperly Good
Genealogy; Maritime (Logbooks, Journals, Maritime Navigation & Law, Ship Registers); Maritime artifacts in museum collections
10 000 vols; 20 curr per; 1 000 mss; 10 diss/theses; 100 govt docs; 20 music scores; 100 maps; 120 microforms; 200 av-mat; 10 sound-rec; 15 digital data carriers; 15 000 photos, 3 000 Nautical Charts, mss (1 000 linear ft)
ICMM
37589

Art Institute of Seattle Library, North Campus, 5th Flr, 2323 Elliott Ave, **Seattle**, WA 98121
T: +1 206 2392359; Fax: +1 206 4413475; E-mail: ais.library@aii.edu; URL: ais.aiiresources.com
Andrew Harbison
Applied art, Design
24 000 vols; 290 curr per
37590

Museum of Flight Library, 9404 E Marginal Way S, **Seattle**, WA 98108-4097
T: +1 206 7645700; Fax: +1 206 7645707; E-mail: library@museumofflight.org; URL: www.museumofflight.org
1985; Dennis Parks
Aviation-Aerospace (G S Williams Photogr Coll), D D Hatfield Aviation Hist Coll, Elrey Jeppesen Aviation Hist Coll
30 000 vols; 85 curr per; 500 govt docs; 100 microforms; 7 000 av-mat; 110 sound-rec; 2 000 Pamphlets
37591

National Marine Fisheries Service, Northwest Fisheries Science Center Library, 2725 Montlake Blvd E, **Seattle**, WA 98112
T: +1 206 8603210; Fax: +1 206 8603442; URL: lib.nwfsc.noaa.gov
1931; Craig Wilson
16 000 vols; 250 curr per
37592

National Oceanic & Atmospheric Administration, NOAA Seattle Library, E/OC43, Bldg 3, 7600 Sand Point Way NE, **Seattle**, WA 98115
T: +1 206 5266241; Fax: +1 206 5264535; E-mail: Seattle.Library@noaa.gov; URL: www.wrclib.noaa.gov/lib/index.html
Chemistry, Climatology, Mathematics, Meteorology, Oceanography, Physics, Waste disposal; Puget Sound
10 000 vols
37593

Nordic Heritage Museum, Walter Johnson Memorial Library, 3014 NW 67th St, **Seattle**, WA 98117
T: +1 206 7895707; Fax: +1 206 7893271; E-mail: nordic@nordicmuseum.org; URL: www.nordicmuseum.org
Eric Nelson
Scandinavian languages, Finnish language
16 000 vols
37594

Seattle Art Museum, Dorothy Stimson Bullitt Library, 1300 First Ave, **Seattle**, WA 98101
T: +1 206 6543220; Fax: +1 206 6543135; URL: www.seattleartmuseum.org/learn/library/default.asp
1933; Traci Timmons
Archaeology, Art hist; Northwest Artists' Files, clippings
20 000 vols; 100 curr per; 6 000 exhibition cat, 70 000 slides
37595

Seattle Art Museum, McCaw Foundation Library of Asian Art, Seattle Asian Art Museum, 1400 E Prospect St, **Seattle**, WA 98112
T: +1 206 6543202; Fax: +1 206 6543191
1983; Jie Pan
15 000 vols; 100 curr per
ARLIS/NA
37596

Seattle Genealogical Society Library, 6200 Sand Point Way NE, **Seattle**, WA 98115; P.O. Box 15329, Seattle, WA 98115-0329
T: +1 206 5228658; E-mail: SeattleGenealogicalSociety@gmail.com; URL: www.rootsweb.com/~waseags/
Christine Schomaker
New Jersey (George C Kent Coll)
8 000 vols; 200 curr per
37597

Seattle Metaphysical Library, 1000 E Madison, Ste B, *Seattle*, WA 98122
T: +1 206 329-1794; E-mail: metaphysical@mindspring.com; URL: www.seattlemetaphysicallibrary.org/
1961; Margaret Bartley
Rare Parapsychology Coll
11 000 vols; 10 curr per; 50 av-mat; 450 sound-rec   37598

Seattle's Museum of History & Industry (MOHAI), Sophie Frye Bass Library, 2700 24th Ave E, *Seattle*, WA 98112
T: +1 206 3241126; Fax: +1 206 3241346; E-mail: library@seattlehistory.org; URL: www.seattlehistory.org
1952; Carolyn Marr, Mary Montgomery
PEMCO Webster & Stevens Photogr Coll, PSMHS Williamson Maritime Photogr Coll, Seattle Post-Intelligence Photogr Coll
10 000 vols; 150 curr per; 1 200 maps; 250 linear feet of mss, 250 linear feet of ephemera, 5 000 slides, 1 200 charts
37599

University of Washington Botanic Gardens, Elisabeth C. Miller Library, 3501 NE 41st St, *Seattle*, WA 98105; P.O. Box 354115, Seattle, WA 98195-4115
T: +1 206 5430415; Fax: +1 206 8971435; E-mail: hortlib@u.washington.edu; URL: www.millerlibrary.org
1985; Brian Thompson
Seed Cat Coll
15 000 vols; 400 curr per   37600

Washington Talking Book & Braille Library, 2021 Ninth Ave, *Seattle*, WA 98121-2783
T: +1 206 3861255; Fax: +1 206 6150441; E-mail: wtbbl@wtbbl.org; URL: www.wtbbl.org
1931; Gloria Leonard
Northwest Coll
311 000 vols   37601

World Research Foundation Library, 41 Bell Rock Plaza, *Sedona*, AZ 86351
T: +1 928 2843300; Fax: +1 928 2843530; E-mail: info@wrf.org; URL: www.wrf.org
1984; Steve Ross
Health & the Environment, Health Tools & Technologies available outside the US
20 000 vols   37602

Shawnee Mission Medical Center Library, 9100 W 74th St, *Shawnee Mission*, KS 66204
T: +1 913 6762101; Fax: +1 913 6762106
1963; Clifford L. Nestell
10 000 vols; 522 curr per
libr loan   37603

R. W. Norton Art Gallery, Reference-Research Library, 4747 Creswell Ave, *Shreveport*, LA 71106
T: +1 318 8654201; Fax: +1 318 8690435; E-mail: gallery@rwnaf.org; URL: www.rwnaf.org
1966; Ginger Specian
Fine arts, American hist, Louisiana & Virginia hist, lit, ornithology, bibliography, world hist; James M. Owens Memorial Coll of Virginiana; rare books
12 000 vols; 35 curr per   37604

General Conference of Seventh-Day Adventists, Rebok Memorial Library, 12501 Old Columbia Pike, *Silver Spring*, MD 20904
T: +1 301 6806495; Fax: +1 301 6806090
1983; Alan S. Hecht
Social Problems, Religion
14 000 vols   37605

George Meany Memorial Archives, Library, National Labor College, 10000 New Hampshire Ave, *Silver Spring*, MD 20903
T: +1 301 4315441; Fax: +1 301 4310385; E-mail: ldeloach@nlc.edu; URL: www.nlc.edu/archives/home.html
1916
Labor, Labor economics, Trade unions, Industrial relations; Nat & Int Trade-Union Proceedings Coll
14 000 vols; 400 curr per   37606

National Oceanic & Atmospheric Administration, Library & Information Services Division, 1315 East West Hwy, 2nd Flr, *Silver Spring*, MD 20910
T: +1 301 7132607; Fax: +1 301 7134598; E-mail: library.reference@noaa.gov; URL: www.lib.noaa.gov
1809; N.N.
Oceanography, Ocean engineering, Marine biology; Climatology Coll (Daily Weather Maps), bk, flm; Meteorology Coll (Data Publications), bk, flm, micro; Marine Fisheries Data (Tech Rpts), micro; Rare Bks; Miami Regional Library
1 500 000 vols; 350 e-journals; 258 mss; 2 incunabula; 141 diss/theses; 8 747 govt docs; 1 455 maps; 191 microforms; 623 digital data carriers
libr loan; ASLI   37607

National Archives & Records Administration, Ronald Reagan Presidential Library, 40 Presidential Dr, *Simi Valley*, CA 93065
T: +1 805 5228444; Fax: +1 805 5229621; E-mail: library@reagan.nara.gov
1991; Denise LeBeck
49 000 000 mss; 21 000 av-mat; 25 250 sound-rec   37608

Rob & Bessie Welder Wildlife Foundation Library, 12858 Hwy 77 N, *Sinton*, TX 78387; P.O. Box 1400, Sinton, TX 78387-1400
T: +1 361 3642643; Fax: +1 361 3642650; E-mail: welderwf@aol.com; URL: hometown.aol.com/welderwf/welderweb.html
1954; Lynn Drawe
Ornithology (Alexander Wetmore Coll), Rare Bk Coll
12 000 vols; 80 curr per; 4 500 mss; 325 diss/theses
libr loan   37609

Siouxland Heritage Museums, Pettigrew Museum Library, 200 W Sixth St, *Sioux Falls*, SD 57104
T: +1 605 3674210; Fax: +1 605 3676004
1926; William J. Hoskins
R.F. Pettigrew Papers & Private Libr, Northern League Baseball Records, Arthur C. Phillips Coll
10 000 vols; 10 curr per; 100 maps; 200 microforms; 10 sound-rec; 150 linear feet of mss, 500 slides, 10 000 photos
37610

Wegner Health Science Information Center, 1400 W 22nd St, Ste 100, *Sioux Falls*, SD 57105
T: +1 605 3571400; Fax: +1 605 3571490; E-mail: wegner@usd.edu; URL: www.usd.edu/wegner
1998; Carolyn Warmann
Clinical medicine, Pharmacology, Psychiatry
16 000 vols; 545 curr per; 2 291 other items   37611

Portland Cement Association, Library Services, 5420 Old Orchard Rd, *Skokie*, IL 60077-1083
T: +1 847 9729178; Fax: +1 847 9666221; URL: www.cement.org
1950; Connie N. Field
Cement, Concrete, Construction; Assn Publs, Foreign Lit Studies, Occupational Safety & Health Coll
105 000 vols; 200 curr per; 6 digital data carriers; 40 000 tech rpts, 100 000 abstracts   37612

Rand McNally, Map Library, 8255 N Central Park Ave, *Skokie*, IL 60076-2970
T: +1 847 3298100; Fax: +1 847 3296460
1949; Karen Cuiskelly
Cartography, Geography, Railroads
13 000 vols; 65 curr per; 300 000 maps; 150 sound-rec; 25 digital data carriers; 450 pamphlets, clippings, 3 500 Atlases, 2 500 Foreign Census
libr loan; ALA, SLA   37613

New Mexico Institute of Mining and Technology, Skeen Library, 801 Leroy Pl, *Socorro*, NM 87801
T: +1 505 8355614; Fax: +1 505 8355754; E-mail: PMartinez@admin.nmt.edu; URL: infohost.nmt.edu/~nmtlib/

1895; Dal Symes
US Geological Survey Publs, US Bureau of Mines Publs
350 000 vols; 600 curr per; 18 370 maps; 80 digital data carriers; 311 000 govt docs
libr loan   37614

Calvert Marine Museum Library, 14150 Solomons Island Rd, *Solomons*, MD 20688; P.O. Box 97, Solomons, MD 20688-0097
T: +1 410 3262042; Fax: +1 410 3266691; E-mail: information@calvertmarinemuseum.com; URL: www.calvertmarinemuseum.com
Paul L. Berry
Maritime history, ecology, local history; Seafood Processing, History of Solomons Island (Patuxent River Seafood Industries), Boat Building (M M Davis & Son Coll), blueprints, correspondence, clippings;B B Wills steamboat research files;Chesapeake Bay History (M V Brewington Research Coll);Tobacco Culture in Calvert County
8 000 vols; 35 curr per   37615

National Football Foundation's College, College Football Hall of Fame Library, 111 S St Joseph St, *South Bend*, IN 46601
T: +1 574 2355711; Fax: +1 574 2355720; URL: www.collegefootball.org
1978; Kent Stephens
12 000 vols; 68 curr per   37616

Northern Indiana Center for History, Vincent Bendix Reading Room, 808 W Washington St, *South Bend*, IN 46601
T: +1 574 2359664; Fax: +1 574 2359059; E-mail: archives@centerforhistory.org; URL: www.centerforhistory.org
1994; Randy Ray
All-American Girls Professional Baseball League photos, Docss, Memorabilia; Schuyler Colfax – U.S. Vicepresident 1864-1868 personal papers; Early Furtrade Ledeers 1800-1830s; local Industry Docs
9 000 vols; 33 curr per; 25 mss; 40 diss/theses; 500 govt docs; 150 music scores; 200 maps; 50 microforms; 200 av-mat; 250 sound-rec; 10 000 local history pamphlets/doc, 10 000 photos
37617

American Society of Military History Library, 1918 N Rosemeade Blvd, *South El Monte*, CA 91733
T: +1 626 4421776; Fax: +1 626 4431776; E-mail: tankland@aol.com
1962; Don Michelson
50 000 vols; 10 curr per   37618

American Society of Military History Library, 1918 N Rosemeade Blvd, *South El Monte*, CA 91733
T: +1 626 4421776; Fax: +1 626 4431776
Don Michelson
50 000 vols; 10 curr per   37619

Incorporated Long Island Chapter of the New York State Archaeological Association, Stanton Mott Memorial Library, 1080 Main Bayview Rd, Southold Indian Museum, P.O. Box 268, *Southold*, NY 11971
T: +1 631 7655577; Fax: +1 631 7655577; E-mail: indianmuseum@aol.com; URL: southoldindianmuseum.org/library.htm
1962; Ellen Barcel
Libr of the New York State Archaeologic Assn
10 000 vols; 50 curr per   37620

Intercollegiate Center for Nursing Education, Betty M. Anderson Library, 2917 W Fort George Wright Dr, *Spokane*, WA 99224-5290
T: +1 509 3247344; Fax: +1 509 3247349; E-mail: nursinglib@wsu.edu; URL: nursing.wsu.edu/library
1969; Robert M. Pringle Jr
Hist of Nursing
11 000 vols; 190 curr per; 4 000 microforms; 40 digital data carriers; 25 VF drawers
libr loan   37621

Northwest Museum of Arts and Culture / Eastern Washington State Historical Society, Research Library and Archives, 2316 W First Ave, *Spokane*, WA 99204
T: +1 509 3635313; Fax: +1 509 3635303; E-mail: archives@northwestmuseum.org; URL: www.northwestmuseum.org
1916; Rayette Wilder
Spokane Flour Mill; EXPO 74; Spokane Chamber of Commerce; Inland Empire Mining; Cutter Architectural Coll; Mae Awkright Hutton mss; Dr. Robert Ruby mss coll; American Indians colls
10 000 vols; 18 curr per; 600 mss; 750 sound-rec; 400 VHS tapes, 900 Oral Hist audo cassettes   37622

Johnson & Johnson Pharmaceutical Research & Development LLC, Welsh & McKean Rd, *Spring House*, PA 19477; P.O. Box 776, Spring House, PA 19477-0776
T: +1 215 6285623; Fax: +1 215 6285984
June Bente
Medicine, biology, chemistry
10 000 vols; 308 curr per   37623

Andrew McFarland Mental Health Center, Staff Library, 901 Southwind Rd, *Springfield*, IL 62703
T: +1 217 7860226; Fax: +1 217 7867725; E-mail: dhs741e@dhs.state.il.us
1968; Melanie Bock
12 250 vols; 49 curr per   37624

Baystate Medical Center, Health Sciences Library, 759 Chestnut St, *Springfield*, MA 01199
T: +1 413 7941865; Fax: +1 413 7941978; E-mail: library@bhs.org; URL: libraryinfo.bhs.org
Laurie Fornes
10 700 vols; 500 curr per   37625

Connecticut Valley Historical Museum, Genealogy & Local History Library, The Quadrangle, Edwards St, *Springfield*, MA 01103; mail address: 220 State St, Springfield, MA 01103
T: +1 413 2636800; Fax: +1 413 2636898; E-mail: cvhmgen@springfieldmuseums.org; URL: www.springfieldmuseums.org/cvhm.htm
1876; Margaret Humberston
Business & Personal Records of Connecticut Valley (1650-1950)
30 000 vols; 50 curr per; 1 500 000 mss; 70 maps; 5 000 microforms; 40 000 photos, 6 000 linear feet of archival records, 300 feet of vertical files, atlases & maps   37626

CoxHealth Libraries – David Miller Memorial Library, Cox Medical Ctr S, 3801 S National Ave, *Springfield*, MO 65807
T: +1 417 2693460; Fax: +1 417 2696140; E-mail: library@coxhealth.com
Pat Leembruggen
Medicine,nursing
9 000 vols; 292 curr per; 69 e-books
37627

CoxHealth Libraries – North Library, Cox Medical Ctr N, 1423 N Jefferson Ave, J-200, *Springfield*, MO 65802
T: +1 417 2693460; Fax: +1 417 2693492; E-mail: library@coxhealth.com; URL: www.coxhealth.com/body.cfm?id=2389
Wilma Bunch
Medicine, nursing
9 000 vols; 514 curr per; 69 e-books
37628

Illinois Historic Preservation Agency, Abraham Lincoln Presidential Library, 112 N Sixth St, *Springfield*, IL 62701
T: +1 217 5588841; Fax: +1 217 7856250; URL: www.alplm.org
1889; Kathryn M. Harris
Illinois, Civil War, Mormons; Illinois Newspapers, Lincolniana
350 000 vols; 1 200 curr per; 7 800 maps; 78 000 microforms; 10 000 000 mss
37629

Illinois State Library – Talking Book & Braille Service, 401 E Washington, *Springfield*, IL 62701-1207
T: +1 217 7829260; Fax: +1 888 2612709; URL: www.ilbph.org
Sharon Ruda
112 000 vols; 34 e-books   37630

Illinois State Museum Library, 502 S
Spring St, **Springfield**, IL 62706-5000
T: +1 217 7826623; Fax: +1
217 7821254; E-mail:
pburg@museum.state.il.us; URL:
www.museum.state.il.us
1877; Patricia Burg
Anthropology, Art, Ornithology,
Paleontology, Zoology
23 000 vols; 283 curr per   37631

**Sacred Heart Medical Center
at RiverBend**, Library Services,
3333 RiverBend Dr, **Springfield**,
OR 97477; P.O. Box 10905, Eugene,
OR 97440-2905
T: +1 541 2222280; E-mail:
libraryshmc@peacehealth.org
Kim Tyler
8 000 vols; 450 curr per   37632

St. John's Health Systems, Inc,
Medical Library, 1235 E Cherokee St,
Springfield, MO 65804-2263
T: +1 417 8202795; Fax: +1 417
8855399; E-mail: libstaff@mercy.net;
URL: www.stjohns.com/libraries/medlib/
1904; Holly Henderson
15 000 vols; 528 curr per; 28 000
microforms; 250 av-mat; 200 sound-rec;
10 digital data carriers
libr loan   37633

Stamford Historical Society Library,
1508 High Ridge Rd, **Stamford**,
CT 06903-4107
T: +1 203 3291183; Fax: +1 203
3221607; URL: www.stamfordhistory.org
1901; Ronald Marcus
Catherine Aiken School Coll, 1855-1913;
Eaton, Yale and Towne Coll on Yale
& Towne Manufacturing Company of
Stamford, 1868-1949; Anson Dickinson
Coll, 1779-1852
10 000 vols; 25 maps; 266 mss, 300
newspapers, 2 000 pictures, 1 000 slides,
16 VF drawers of docs & clippings
          37634

**Carnegie Foundation for the
Advancement of Teaching**, Information
Center, 51 Vista Lane, **Stanford**,
CA 94305
T: +1 650 5665100; Fax: +1
650 3260278; E-mail:
infocenter@carnegiefoundation.org; URL:
www.carnegiefoundation.org
1967; Megan Gutelius
Education, Public policy
11 000 vols; 100 curr per; 1 500
pamphlets & booklets
CFL   37635

**Center for Advanced Study in the
Behavioral Sciences Library**, 75 Alta
Rd, **Stanford**, CA 94305-8090
T: +1 650 3212052; Fax: +1 650
3211192; E-mail: info@casbs.stanford.edu;
URL: www.casbs.stanford.edu
1954; Tricia Soto
Psychiatry, Sociology, Anthropology,
Philosophy, Hist; Ralph Tyler Coll
10 000 vols; 290 curr per
ALA, SLA   37636

**International Institute for Sport &
Olympic History Library**, IISOH
Library, PO Box 732, **State College**,
PA 16804-0732; P.O. Box 175, State
College, PA 16804-0175
E-mail: olympicbks@aol.com;
URL: www.iisoh.org;
www.harveyabramsbooks.com/501c3.html
Harvey Lee Abrams
Sports, dance
10 000 vols   37637

Center for Migration Studies Library,
209 Flagg Pl, **Staten Island**, NY 10304
T: +1 718 3518800; Fax: +1 718
6674598; E-mail: library@cmsny.org;
URL: cmsny.library.net
1964; Diana Zimmerman
International migration, refugees, ethnic
groups; Ethnic Press Coll, Italian
American Arch; CMS Archives
32 000 vols; 100 curr per; 555
diss/theses; 700 microforms; 223
newsletters, 88 recordings
ALA, SLA   37638

**Staten Island Institute of Arts &
Sciences**, Archives & Library, 75
Stuyvesant Pl, **Staten Island**, NY 10301
T: +1 718 7271135; Fax: +1 718
2735683; URL: www.siiamuseum.org
1881; Patricia Salmon
Black Community in Staten Island,
Conservation Hist Coll, Daguerreotypes,
Environmental Coll, Mss Colls
30 000 vols; 12 000 maps; 99 microforms;
1 000 av-mat; 50 sound-rec; 45 000
photos, 1 200 prints   37639

Staten Island University Hospital,
Medical Library, 475 Seaview Ave,
Staten Island, NY 10305
T: +1 718 2269545; Fax: +1 718
2268582; E-mail: siuhlibrary@siuh.com;
URL: www.siuh.edu
1938; Yelena Friedman
22 000 vols; 560 curr per; 480 e-journals;
150 e-books; 1 736 sound-rec   37640

**Ezra Pound Institute of Civilization –
National Commission for Judicial
Reform**, Library, 126 Madison Pl,
Staunton, VA 24401
T: +1 540 8865580; Fax: +1 540
8865580
1989; Eustace Mullins
30 000 vols; 10 curr per   37641

**Ezra Pound Institute of Civilization
– National Council for Medical
Research Library**, 126 Madison Pl,
Staunton, VA 24401-4129
T: +1 540 8865580; URL:
www.eustacemullins.com
Eustace Mullins
40 000 vols   37642

**Molesworth Institute Library &
Archives**, 143 Hanks Hill Rd, **Storrs**,
CT 06268
T: +1 860 4297051; Fax: +1
860 4864521; E-mail:
norman.d.stevens@uconn.edu
1959; Norman D. Stevens
Bibliosmiles Reading Coll, postcards; Libr
Humor (Molesworth-Carberry Coll) bks
and ephemera
13 333 vols; 666 curr per; 10 sound-rec;
23 digital data carriers; 400 slides
          37643

Jessie Ball Dupont Memorial Library,
Stratford Hall, 483 Great House Rd,
Stratford, VA 22558
T: +1 804 4938038; Fax: +1 804
4938006; URL: www.stratfordhall.org
1980; Judith S. Hynson
Lee Family Hist, Colonial Virginia Hist;
Robert E Lee Memorial Assn Libr; Lee
Family Mss; Thomas Lee Shippen 1790
Inventory & Coll; Ditchley Coll of 16th-,
17th-, & 18th-c Books
10 000 vols; 25 curr per; 125 microforms;
332 cubic feet of archives   37644

Monroe County Historical Association,
Elizabeth D. Walters Library, 900 Main
St, **Stroudsburg**, PA 18360
T: +1 570 4217703; Fax: +1 570
4219199; E-mail: mcha@ptd.net; URL:
mcha-pa.org
1921; Amy Leiser
Monroe County Hist & Genealogy
10 000 vols; 17 curr per; 60 maps; 500
microforms; 10 sound-rec; 300 slides
          37645

Old Sturbridge Village, Research
Library, One Old Sturbridge Village Rd,
Sturbridge, MA 01566
T: +1 508 3473362; Fax: +1 508
3470375; URL: www.osv.org
1946; Jack Larkin
Arts & Crafts of Rural New England
1790-1860; Gravestone Rubbings; Hist
Agriculture; 19th & Early 20th Century
Massachusetts Townscapes; Water Power
Technology
35 000 vols; 80 curr per; 181 maps;
2 000 microforms; 400 shelf feet of
mss, 165 000 images, photogs, prints,
transparencies & negatives   37646

Smithsonian Institution Libraries,
Museum Support Center Library,
Smithsonian Museum Support Center,
Rm C-2000, MRC 534, 4210 Silver Hill
Rd, **Suitland**, MD 20746-2863
T: +1 301 2381030; Fax: +1 301
2383661; URL: www.sil.si.edu/libraries/
msc
1964; Gil Taylor

Conservation science
26 000 vols; 1 000 microforms; 30 000
reprints   37647

Smithsonian Institution Libraries, Vine
DeLoria, Jr Library, National Museum of
the American Indian, Cultural Resources
Ctr, MRC 538, 4220 Silver Hill Rd,
Suitland, MD 20746-0537
T: +1 301 2381376; Fax: +1 301
2383038; URL: www.sil.si.edu/libraries/
nmai
Lynne Altstatt
12 000 vols   37648

**Oklahoma School for the Deaf
Library**, 1100 E Oklahoma St, **Sulphur**,
OK 73086
T: +1 580 6224900; Fax: +1 580
6224959; URL: www.osd.k12.ok.us
Sue Galloway
9 000 vols; 53 curr per   37649

National Solar Observatory, Technical
Library, P.O. Box 62, **Sunspot**,
NM 88349
T: +1 505 4347024; Fax: +1 505
4347029; E-mail: library@sunspot.edu;
URL: www.nso.edu
1953; John Cornett
Publs of National Solar Observatory;
other Observatory publs (foreign & US);
Solar spectral atlases
10 500 vols; 60 curr per   37650

Everson Museum of Art Library,
Richard V. Smith Art Reference Library,
401 Harrison St, **Syracuse**, NY 13202
T: +1 315 4746064; Fax: +1
315 4746943; E-mail:
eversonadmin@everson.org; URL:
www.everson.org
1896; Mary Iversen
American Ceramics, mss; Arch of
Ceramic National Exhibitions, 1932-
present
16 000 vols; 20 curr per; 6 000 slides
          37651

**Onondaga Historical Association
Museum & Research Center**,
321 Montgomery St, **Syracuse**,
NY 13202-2098
T: +1 315 4281864; Fax: +1 315
4712133; URL: www.cnyhistory.org
1863; Thomas Hunter
8 020 vols; 3 curr per; 2 000 maps;
100 000 photos, 20 000 cubic feet of arch
mat   37652

**Washington State Historical Society
Research Center**, Special Collections
Division, 315 N Stadium Way, **Tacoma**,
WA 98403
T: +1 253 7985914; Fax: +1 253
5974186
1941; Edward W. Nolan
Asahel Curtis Negative Coll
14 000 vols; 2 000 maps; 1 000
microforms; 5 000 000 mss, 30 000
pamphlets, 30 000 ephemera   37653

**Alabama Institute for the Deaf &
Blind**, Library & Resource Center for
the Blind & Physically Handicapped, 705
South St, **Talladega**, AL 35160; P.O.
Box 698, Talladega, AL 35161
T: +1 256 7613237; Fax: +1 256
7613561; URL: www.aidb.org
1965; Teresa Lacy
Alabama Hist, cassettes; Reference Coll
on Blindness
60 000 vols; 62 curr per; 3 400 braille
titles, 25 000 cassette & disc titles, 170
large print titles   37654

Tall Timbers Research Station Library,
13093 Henry Beadel Dr, **Tallahassee**,
FL 32312-0918
T: +1 850 8934153; Fax: +1 850
6687781; URL: www.talltimbers.org
Carol Armstrong
Conservation, forestry, fire ecology
10 000 vols; 120 curr per   37655

**James A. Haley Veterans Hospital
Library**, 13000 Bruce B. Downs Blvd,
Tampa, FL 33612
T: +1 813 9722000 ext 6570; Fax: +1
813 9785917; URL: www.tampa.va.gov
1972; Nancy Bernal
8 120 vols; 360 curr per   37656

Museum of Science & Industry,
Science Library, 4801 E Fowler Ave,
Ste L, **Tampa**, FL 33617
T: +1 813 2733652; Fax: +1 813
9876381; URL: www.hcplc.org/hcplc/
liblocales/msi/
Cindy Nichols
11 000 vols; 13 curr per   37657

Providence Historical Society, 3980
Tampa Rd, Ste 207, **Tampa**, FL 34677
T: +1 813 8554635
Nancy Stewart
15 000 vols   37658

**Tampa-Hillsborough County Public
Library System – Talking Books
Library**, 3910 S Manhattan Ave,
Tampa, FL 33611
T: +1 813 2726024; Fax: +1 813
2726072
Ann Bush
35 000 vols   37659

Historic Hudson Valley's Library, 150
White Plains Rd, **Tarrytown**, NY 10591
T: +1 914 6318609; Fax: +1
914 6313591; E-mail:
librarian@hudsonvalley.org; URL:
www.hudsonvalley.org
1951; Catalina Hannan
Hudson River Valley hist, Architecture,
Decorative arts, Slavery; Washington
Irving Editions
25 000 vols; 90 curr per; 225 microforms;
75 VF drawers of maps, plans, graphics
          37660

Bristol Law Library, Superior Court
House, Nine Court St, **Taunton**,
MA 02780
T: +1 508 8247632; Fax: +1
508 8244723; E-mail:
bristollawlibrary@yahoo.com; URL:
www.lawlib.state.ma.us
1858; Cynthia Campbell
Complete Massachusetts Laws
28 000 vols; 24 curr per   37661

Branchville Training Center Library,
P.O. Box 500, **Tell City**, IN 47586-0500
T: +1 812 8435921; Fax: +1 812
8434262
C. Poehlein
15 000 vols; 70 curr per   37662

Arizona Historical Foundation, Hayden
Library, Hayden Library, ASU, **Tempe**,
AZ 85287; P.O. Box 871006, Tempe,
AZ 85287-1006
T: +1 480 9653283; Fax: +1 480
9661077
Susan Irwin
History of Arizona and Southwest,
Arizona Cattle & Cotton Growers
Association
12 000 vols; 10 curr per   37663

Salt River Project Library, PAB-ISB
Library, 1600 N Priest Dr, **Tempe**,
AZ 85281-1213
T: +1 602 2365676; Fax: +1 602
6298585; E-mail: isblib@srpnet.com
1963; Elizabeth Gouwens
Business, Management, Utilities industry,
Engineering, Water; Career Resources,
Electric Utility Company Annual Reports,
EPRI
30 000 vols; 200 curr per   37664

Scott & White Memorial Hospital,
Richard D. Haines Medical Library, 2401
S 31st, **Temple**, TX 76508
T: +1 254 7242387; Fax: +1 254
7244222; URL: www.sw.org/library/
library.htm
1923; Penny Worley
Nursing (Laura Cole Coll)
9 000 vols; 986 curr per   37665

SPJST, Library & Archives Museum, 520
N Main St, **Temple**, TX 76501
T: +1 254 7731575; Fax: +1 254
7747447
1968; Dorothy Pechal
Fiction, Genealogy, Geography, Hist;
Czech Language Coll
23 000 vols; 4 curr per   37666

**Terrell State Hospital – Medical
Library**, 1200 E Brin Ave, P.O. Box
70, **Terrell**, TX 75160
T: +1 972 5636452 ext 8620; Fax: +1
972 5518711
1964; Mary A. Griffith
17 300 vols; 50 curr per   37667

Terrell State Hospital – Patient Library, 1200 E Brin Ave, ***Terrell***, TX 75160-2938; P.O. Box 70, Terrell, TX 75160-9000
T: +1 972 5636452; Fax: +1 972 5518371
Judy Rasbury
12000 vols; 25 curr per 37668

Joe Buley Memorial Library, 35240 N Grant St, ***Third Lake***, IL 60030; P.O. Box 371, Third Lake, IL 60030
T: +1 847 2234300
Dr. Nicholas T. Groves
Serbian history and culture, Orthodox theology, Serbian archival mat
15000 vols; 50 curr per 37669

Fort Ticonderoga Museum, Thompson-Pell Research Center, Fort Rd, ***Ticonderoga***, NY 12883; P.O. Box 390, Ticonderoga, NY 12883-0390
T: +1 518 5852821; Fax: +1 518 5852210; E-mail: fort@fort-ticonderoga.org; URL: www.fort-ticonderoga.org
1908; Nicholas Westbrook
Colonial & Revolutionary hist
13000 vols; 20 curr per; 2500 mss; 1 digital data carriers 37670

Ticonderoga Historical Society, Library, Hancock House, Six Moses Circle, ***Ticonderoga***, NY 12883
T: +1 518 5857868; Fax: +1 518 5856367; E-mail: tihistory@verizon.net; URL: www.thehancockhouse.org/research_library.htm
1898; Robin Trudeau
Local Hist Bks, Account Bks, Diaries, Mss, photos, Newspapers
10000 vols 37671

Toledo Law Association Library, Lucas County Family Court Center, 905 Jackson St, ***Toledo***, OH 43604-5512
T: +1 419 2134747; Fax: +1 419 2134287; E-mail: librarian@toledolawlibrary.org; URL: www.toledolawlibrary.org
1870; Cathy Thomas
76000 vols; 300 curr per; 16000 microforms; 300 sound-rec; 8 digital data carriers 37672

Toledo Museum of Art, Art Reference Library, 2445 Monroe St, ***Toledo***, OH 43620; P.O. Box 1013, Toledo, OH 43697-1013
T: +1 419 2558000; Fax: +1 419 2555638; E-mail: information@toledomuseum.org; URL: www.toledomuseum.org/library/
1901; Anne O. Morris
Art hist, Music; George W. Stevens Coll (Hist of Writing)
70000 vols; 316 curr per; 409 microforms; 20000 vertical files on individual artists
libr loan 37673

Ocean County Historical Society, Richard Lee Strickler Research Center, 26 Hadley Ave, ***Toms River***, NJ 08753; P.O. Box 2191, Toms River, NJ 08754-2191
T: +1 732 3411880; Fax: +1 732 3414372; E-mail: oceancounty.history@verizon.net; URL: www.oceancountyhistory.org
1950; Linda Kay
Genealogy, Hist; Family Histories, Ocean County Cemeteries
8000 vols; 45 curr per; 1 digital data carriers; 100 docs, 100 mss, 100 nonbook items 37674

Grand Lodge of Kansas Library, 320 S W Eight Ave, ***Topeka***, KS 66601; P.O. Box 1217, Topeka, KS 66601-1217
T: +1 785 2345518; Fax: +1 785 3574036; E-mail: glksafam@alltel.net; URL: www.gl-ks.org
Robert Tomlinson
Freemasonry
13000 vols 37675

Kansas Department of Transportation Library, 2300 SW Van Buren, ***Topeka***, KS 66611
T: +1 785 2913854; Fax: +1 785 2962526; E-mail: library@ksdot.org; URL: www.ksdot.org
1962
Engineering
10200 vols; 10 curr per 37676

Kansas State Historical Society, Library & Archives Division, 6425 SW Sixth Ave, ***Topeka***, KS 66615-1099
T: +1 785 2728681; Fax: +1 785 2728682; E-mail: reference@kshs.org; URL: www.kshs.org
1875; Margaret Knecht
Kansas, American Indians, Civil War, Genealogy; Kansas State Arch
181000 vols; 475 curr per; 5900 mss; 25200 maps; 75000 microforms; 180 digital data carriers; 242360 pamphlets 37677

Kansas Supreme Court, Law Library, Kansas Judicial Ctr, 301 SW Tenth St, ***Topeka***, KS 66612-1502
T: +1 785 2963257; Fax: +1 785 2961863; E-mail: lawlibrary@kscourts.org; URL: www.kscourts.org/ctlib/
1855; Fred W. Knecht
Judicial Administration
201000 vols; 400 curr per 37678

Stormont – Vail HealthCare, Stauffer Health Sciences Library, 1500 SW 10th St, ***Topeka***, KS 66604-1353
T: +1 785 3545800; Fax: +1 785 3545059; E-mail: library2@stormontvail.org; URL: www.stormontvail.org
1889
8000 vols; 400 curr per 37679

Los Angeles County Harbor UCLA Medical Center, A.F. Parlow Library of Health Sciences, 1000 W Carson St, ***Torrance***, CA 90502-2059; P.O. Box 18, Torrance, CA 90507-0018
T: +1 310 2222372; Fax: +1 310 5335146; E-mail: libref@labiomed.org
1964; Mary Ann Berliner
Hospital administration, Medicine, Nursing, Social service
37200 vols; 515 curr per 37680

Ute Mountain Tribal Library, Education Ctr, 450 Sunset, ***Towaoc***, CO 81334; P.O. Box CC, Towaoc, CO 81334-0048
T: +1 970 5645348; Fax: +1 970 5645342
Martina King
Ute History
10000 vols; 15 curr per; 200 av-mat 37681

New Jersey State Library, New Jersey Library for the Blind & Physically Handicapped, 2300 Stuyvesant Ave, ***Trenton***, NJ 08618; P.O. Box 501, Trenton, NJ 08625-0501
T: +1 609 5304000; Fax: +1 609 5306384; E-mail: njlbh@njstatelib.org; URL: www.njlbh.org
1968; Adam Szczepaniak Jr
763000 vols; 90 curr per; 48000 av-mat; 32430 Braille vols, 12 Special Interest Per Sub, 150 Bks on Deafness & Sign Lang, 18920 Large Print Bks, 45400 Talking Bks 37682

Saint Francis Medical Center, Health Sciences Library, 601 Hamilton Ave, ***Trenton***, NJ 08629-1986
T: +1 609 5995068; Fax: +1 609 5995773; URL: www.stfrancismedical.com
1930; Donna Barlow
8000 vols; 300 curr per 37683

Arizona Historical Society, Library Archives-Southern Arizona Division, 949 E Second St, ***Tucson***, AZ 85719
T: +1 520 6285774; Fax: +1 520 6298966; E-mail: ahsref@vms.arizona.edu; URL: www.arizonahistoricalsociety.org
1884; Deborah Shelton
Arizona, Mexico; Charles B. Gatewood Military Coll, Byron Cummings Ethnological & Archaeological Coll, Aguiar Coll of Early 19th c Mexican Docs, Carl Hayen Biogr Files of Arizonans, Will C. Barnes Ranching & Forestry Papers
50000 vols; 40 curr per; 5000 maps; 10000 pamphlets, 750000 photos, 1000 mss 1200 oral histories 37684

Arizona State Museum Library, University of Arizona, 1013 E University Blvd, ***Tucson***, AZ 85721-0026; P.O. Box 210026, Tucson, AZ 85721-0026
T: +1 520 6214695;
Fax: +1 520 6212976; URL: www.statemuseum.arizona.edu
1957; Mary Graham
Anthropology, Archaelogy, Museology, Southwest
50000 vols; 30 curr per; 400 diss/theses 37685

Department of Veterans Affairs, Medical Center Library Service, 142-D, 3601 S Sixth Ave, ***Tucson***, AZ 85723
T: +1 520 6291836; Fax: +1 520 6294638
William E Azevedo
10827 vols; 290 curr per 37686

Edward F. Barrins Memorial Library, 2023 E Adams St, ***Tucson***, AZ 85719
T: +1 520 3277956
1979; Christine L. Taylor-Parsil
Hypnosis, Alternate Methods of Healing, Religion, Reincarnation, 20th C Fiction, Psychology, Braille; Catherine M. Willy Commemorative Coll; Phyllis C. Barrins Coll; Alexandria O'Sullivan Mystery Coll; Children's Books
9000 vols; 34 curr per; 24 linear feet of pamphlets, 24 linear feet of clippings, 15 VF drawers of research mss & transcripts, 15 cases of audiotapes 37687

National Optical Astronomy Observatories, Library, 950 N Cherry, ***Tucson***, AZ 85719; P.O. Box 26732, Tucson, AZ 85726-6732
T: +1 520 3188295; Fax: +1 520 3188360; E-mail: library@noao.edu; URL: www.noao.edu
1958; Mary Guerrieri
Astronomy, Astrophysics
15000 vols; 75 curr per 37688

Pima Council on Aging Library, 8467 E Broadway, ***Tucson***, AZ 85710
T: +1 520 7907262; Fax: +1 520 7907577; E-mail: help@pcoa.org; URL: www.pcoa.org
Melissa S. Morgan
10000 vols 37689

Tucson Family History Center, Regional Genealogical Library, 500 S Langley, ***Tucson***, AZ 85710
T: +1 520 2980905; Fax: +1 520 2982339
1968; Leonard Ingermanson
10000 vols; 20 curr per 37690

Tucson Museum of Art, Research Library, 140 N Main Ave, ***Tucson***, AZ 85701-8290
T: +1 520 6242333; Fax: +1 520 6247202; E-mail: library@tucsonmuseumofart.org; URL: www.tucsonmuseumofart.org
1974; Jill Ellen Provan
Pre-Columbian, African & Oceanic Arts (Fredrick Pleasant's Coll), bks, pamphlets, photos, slides; Spanish Colonial, Western & 20th C European & American Art
14000 vols; 18 curr per; 24000 slides 37691

Philbrook Museum of Art, H. A. & Mary K. Chapman Library, 2727 S Rockford Rd, ***Tulsa***, OK 74114-4104; P.O. Box 52510, Tulsa, OK 74152-0510
T: +1 918 7485306; Fax: +1 918 7485303; URL: www.philbrook.org
1940; Thomas Elton Young
American Indian (Roberta Campbell Lawson Coll)
18000 vols; 130 curr per; pamphlets, clippings
ARLIS/NA, OCLC 37692

Saint John Medical Center, Health Sciences Library, 1923 S Utica, ***Tulsa***, OK 74104
T: +1 918 7442970; Fax: +1 918 7443209; E-mail: library@sjmc.org
1946; James M. Donovan
Catholic bioethics
12000 vols; 135 curr per 37693

Thomas Gilcrease Institute of American History & Art, Library, 1400 Gilcrease Museum Rd, ***Tulsa***, OK 74127-2100
T: +1 918 5962700; Fax: +1 918 5962770; E-mail: amiller@ci.tulsa.ok.us
1942; April Miller
Native American hist, Western hist; Mss
40000 vols; 15 curr per 37694

Geological Survey of Alabama Library, Walter Bryan Jones Hall, 420 Hackberry Lane, P.O. Box 869999, ***Tuscaloosa***, AL 35486-6999
T: +1 205 2473634; Fax: +1 205 3492861; E-mail: library@gsa.state.al.us; URL: www.gsa.state.al.us
1873
Aerial Photography, Satellite Imagery for Alabama
150915 vols 37695

International Paper, Corporate Research Center – Knowledge Resource Center, 1422 Long Meadow Rd, ***Tuxedo Park***, NY 10987
T: +1 845 5777262; Fax: +1 845 5777559
1969; Bernadette Marasco
12000 vols; 250 curr per
SLA 37696

Brookhaven National Laboratory, Information Services Division, Research Library, Bldg 477, ***Upton***, NY 11973-5000
T: +1 631 3443483; Fax: +1 631 3442090; URL: www.bnl.gov/bnl.html
1947; Mary Petersen
Biology, Chemistry, Energy, Engineering, Medicine, Nuclear science, Physics
95000 vols; 800 curr per 37697

Champaign County Law Library Association, Champaign County Court House, 200 N Main St, ***Urbana***, OH 43078
T: +1 937 6532709; Fax: +1 937 6533538; E-mail: ccllccll@ctcn.net
Melinda Worthen
9000 vols; 163 curr per 37698

National Council of Teachers of English Library, 1111 W Kenyon Rd, ***Urbana***, IL 61801
T: +1 217 2783639; Fax: +1 217 2783761; URL: www.ncte.org
1959; Cheri Cameron
NCTE Monographs – English Language Arts (NCTE Arch Coll)
8385 vols; 90 curr per; 350 sound-rec; 1 digital data carriers 37699

Masonic Medical Research Laboratory Library, Max L. Kamiel Library, 2150 Bleecker St, ***Utica***, NY 13501-1787
T: +1 315 7352217; Fax: +1 315 7240963; E-mail: lib@mmrl.edu; URL: www.mmrl.edu
1959; Rebecca Warren
Early Medical References & Folklore (Van Gordon Coll)
13000 vols; 60 curr per; 250 microforms; 300 microcards
libr loan 37700

Munson-Williams-Proctor Arts Institute Library, Art Reference Library, 310 Genesee St, ***Utica***, NY 13502
T: +1 315 7970000; Fax: +1 315 7975608; E-mail: library@mwpi.edu; URL: www.mwpai.org/museum/library/#
1960; Kathryn L. Corcoran
Contemporary Artists' Coll; Fountain Elms Coll; autographs; bk plates
26000 vols; 80 curr per; 10 microforms; 2400 sound-rec; 330 digital data carriers; 700 videos, 25000 slides 37701

Westchester County Medical Center, Health Sciences Library, Eastview Hall, ***Valhalla***, NY 10595
T: +1 914 2857033
1925; Charlene Sikorski
13500 vols 37702

Tolstoy Foundation, Inc, Alexandra Tolstoy Memorial, 104 Lake Rd, ***Valley Cottage***, NY 10989-2339; P.O. Box 578, Valley Cottage, NY 10989-0578
T: +1 845 2686722; Fax: +1 845 2686937; E-mail: tfhq@aol.com; URL: www.tolstoyfoundation.org
Robert Whittaker
45000 vols 37703

Freedoms Foundation Library, 1601 Valley Forge Rd, ***Valley Forge***, PA 19482
T: +1 610 9338825; Fax: +1 610 9350522
1965; Hal Badger
US hist, 20th c totalitarianism, Modern economic & political systems; US Radical Movements Coll
21000 vols; 80 curr per; 1000 linear feet of unbound serials 37704

Archbold Biological Station Library, 123 Main Drive, ***Venus***, FL 33960-2039; P.O. Box 2057, Lake Placid, FL 33862
T: +1 863 4652571; Fax: +1 863 6991927; E-mail: library@archbold-station.org; URL: www.archbold-station.org
1941; Fred E. Lohrer
12 000 vols; 250 curr per; 250 diss/theses; 2 500 govt docs; 2 250 maps; 25 sound-rec; 15 000 slides, 15 000 reprints
libr loan; IAMSLIC, FLA 37705

A. Max Brewer Memorial Law Library, Brevard County Law Library, Moore Justice Ctr, 2825 Judge Fran Jamieson Way, ***Viera***, FL 32940
T: +1 321 617295; Fax: +1 321 6177303; URL: www.brev.org/locations/brevard_law/index.htm
1955; Annette Melnicove
32 000 vols; 20 curr per; 16 000 microforms; 400 sound-rec; 49 digital data carriers
libr loan 37706

Vineland Historical & Antiquarian Society Library, 108 S Seventh St, ***Vineland***, NJ 08360-4607; P.O. Box 35, Vineland, NJ 08362-0035
T: +1 856 6911111; Fax: +1 856 6916650; E-mail: vhas108@aol.com; URL: www.vinelandhistory.com
Charles J. Girard
Autographs 1750-1900
8 000 vols 37707

Association for Research & Enlightenment, Edgar Cayce Foundation Library, 215 67th St, ***Virginia Beach***, VA 23451
T: +1 757 4283588; Fax: +1 757 4224631; E-mail: library@edgarcayce.org; URL: www.edgarcayce.org
1940; Laura Hoff
Readings (Edgar Cayce Coll); Metaphysics (Andrew Jackson Davis Coll); Atlantis (Egerton Sykes Coll); San Francisco Metaphysical Libr Coll
66 000 vols; 75 curr per; 100 diss/theses; 3 500 sound-rec 37708

Virginia Beach Department of Public Libraries – Subregional Library for the Blind & Handicapped, Bayside Special Library Services, 936 Independence Blvd, ***Virginia Beach***, VA 23455
T: +1 757 3852680; E-mail: libssbh@vbgov.com
Carolyn Caywood
14 000 vols 37709

Masonic Grand Lodge Library & Museum of Texas, 715 Columbus, ***Waco***, TX 76701-1349; P.O. Box 446, Waco, TX 76703-0446
T: +1 254 7537395; Fax: +1 254 7532944; E-mail: library@grandlodgeoftexeas.org; URL: www.grandlodgeoftexas.org
1873; Barbara Mechell
Mss Coll of Masonic & Texas Mats
36 000 vols; 10 curr per; 50 av-mat; 10 420 Pamphlets, 1 425 Pictures 37710

The Alyce L. Haines Biomedical Library, Maui Memorial Medical Center, 221 Mahalani St, ***Wailuku***, HI 96793-2526
T: +1 808 2422337; Fax: +1 808 2427472
1967; Marilynn M L Wong
10 000 vols; 270 curr per; 3 000 govt docs; 2 000 microforms; 300 digital data carriers
MLA 37711

Kosciusko County Historical Society, Research Library & Archives, 121 N Indiana St, ***Warsaw***, IN 46581; P.O. Box 1071, Warsaw, IN 45681-1071
T: +1 574 2691078; E-mail: ksgenweb@embargmail.com
N.N.
County Records, 1830-present, County Newspapers (except Warsaw)
34 000 vols; 120 curr per 37712

Air Transport Association of America Library, 1301 Pennsylvania Ave NW, Ste 1100, ***Washington***, DC 20004-1707
T: +1 202 6264184; Fax: +1 202 6264181; E-mail: ata@airlines.org; URL: www.airlines.org
1944; Marion Mistrik
Aviation, Law, Statistics; Official Airline Guides, 1929-present
12 000 vols; 120 curr per; 350 microforms; 12 sound-rec; 10 digital data carriers; 12 VF drawers of pamphlets & Annual rpts, 500 report files 37713

Alexander Graham Bell Association for the Deaf & Hard of Hearing, Volta Bureau Library, 3417 Volta Pl NW, ***Washington***, DC 20007
T: +1 202 3375220; Fax: +1 202 3378314; E-mail: info@agbell.org; URL: www.agbell.org
1887
Deaf, Speech, Hearing; Alexander Graham Bell Coll (Hearing), mss
12 000 vols; 20 curr per; 300 microforms; 640 shelf feet of pamphlets, 30 VF drawers of clippings & reprints 37714

American Association of Retired Persons, Research Information Center, 601 E St, NW, Bldg B, 3rd Flr, ***Washington***, DC 20049
T: +1 202 4346220, 6233; Fax: +1 202 4346408; E-mail: info@aarp.com; URL: www.aarp.org
1964; Hugh O'Connor
Aging; White House Conference on Aging, 1961, 1971, 1981, 1995; Congressional Committees on Aging publs
30 000 vols; 350 curr per 37715

American Chemical Society, Library & Information Center, 1155 16th St NW, ***Washington***, DC 20036
T: +1 202 8724513; Fax: +1 202 8726257; E-mail: library@acs.org; URL: www.acs.org/
1876; Svetla Baykoucheva
16 000 vols; 550 curr per; 1 000 microforms; 7 digital data carriers; 300 000 photos 37716

American Health Care Association, Information Resource Center, 1201 L St, NW, ***Washington***, DC 20005
T: +1 202 8424444; Fax: +1 202 8423860; URL: www.ahca.org
Ann W. Williams
8 000 vols; 100 curr per 37717

American Institute of Architects Library & Archives, 1735 New York Ave NW, ***Washington***, DC 20006
T: +1 202 62674898; Fax: +1 202 6267587; E-mail: library@aia.org; URL: www.aia.org/library
1857; Nancy Hadley
Art & architecture, Urban planning; AIA Arch, Richard Morris Hunt Coll, Rare Bk Coll
40 000 vols; 200 curr per; 100 microforms; 5 digital data carriers; 130 000 slides, 3 500 ft of arch mat, 130 000 photos, 20 VF drawers of pamphlets & clippings 37718

American Insurance Association, Law Library, 1130 Connecticut Ave NW, Ste 1000, ***Washington***, DC 20036
T: +1 202 8287183; Fax: +1 202 2931219; E-mail: info@aiadc.org; URL: www.aiadc.org
1964; Allen K. Haddox
Health care, Property-casualty, Environment, Auto liability, Workmen's comp insurance
30 000 vols; 150 curr per; 110 digital data carriers; 215 VF drawers 37719

American Psychological Association, Arthur W. Melton Library, 750 First St NE, Rm 3012, ***Washington***, DC 20002-4242
T: +1 202 3365640; Fax: +1 202 3365643; URL: www.apa.org
1970; Wade E. Pickren
American Psychological Assn Arch, American Psychological Assn Central Office, Division & State Assn Publs, Classic bks in psychology
17 500 vols; 70 curr per; 35 av-mat; 17 sound-rec; 14 digital data carriers
libr loan 37720

American Public Transportation Association, Information Center, 1666 K St NW, Ste 1100, ***Washington***, DC 20005
T: +1 202 4964889; Fax: +1 202 4964326; E-mail: info@apta.com; URL: www.apta.com/research/info
APTA Publs, Fed Transit Agency
10 000 vols; 85 curr per 37721

American Society of International Law Library, 2223 Massachusetts Ave NW, ***Washington***, DC 20008-2864
T: +1 202 9396005; Fax: +1 202 3191670; URL: www.asil.org
1960; Kelly Vinopal
20 000 vols; 125 curr per; 2 digital data carriers 37722

Army & Navy Club Library, 901 17th St NW, ***Washington***, DC 20006-2503
T: +1 202 6288400 ext 386; Fax: +1 202 2968787; E-mail: anclibr@pop.dn.net; URL: www.armynavyclub.org
1885; Aleksandra M. Zajackowski
Reginald W Oakie Coll of Civil War Stereographs; Writings of Club Members
20 000 vols; 40 curr per 37723

Brookings Institution Library, 1775 Massachusetts Ave NW, ***Washington***, DC 20036
T: +1 202 7976240; Fax: +1 202 7972970; E-mail: circdesk@brookings.edu; URL: www.brookings.edu/lib/lib_hp.htm
1927; Cy Behroozi
Economics, Political science, Govt studies, Int relations
68 000 vols; 350 curr per; 14 digital data carriers 37724

Carnegie Endowment for International Peace Library, 1779 Massachusetts Ave NW, ***Washington***, DC 20036
T: +1 202 9392256; Fax: +1 202 4834462; E-mail: library@carnegieendowment.org; URL: www.carnegieendowment.org/about/library
Terezia Matus
Foreign policy
8 000 vols; 200 curr per 37725

Carnegie Institution of Washington, DTM-Geophysical Laboratory Library and Archives, 5241 Broad Branch Rd NW, ***Washington***, DC 20015
T: +1 202 4787960; Fax: +1 202 4787971; E-mail: library@dtm.ciw.edu; URL: www.library.gl.ciw.edu
1904; Shaun J. Hardy
Hist of Terrestrial Magnetism; Hist of Volcanology, Petrology and Physical Chemistry; Early 20th Century Exploration and Travel
10 000 vols; 275 curr per; 150 digital data carriers; 110 vertical file drawers of pamphlets, off prints, clippings, dissertations
libr loan; OCLC 37726

Child Welfare League of America, Dorothy L. Bernhard Library, 440 First St NW, Ste 310, ***Washington***, DC 20001-2085
T: +1 202 6382952; Fax: +1 202 6384004; URL: www.cwla.org
1920; Ming Wong
8 300 vols; 100 curr per; 2 digital data carriers 37727

Commodity Futures Trading Commission Library, 1155 21st St NW, ***Washington***, DC 20581
T: +1 202 4185255; Fax: +1 202 4185537
1976; Daniel May
Economics, Law; Commodity Exchange Act Coll, Trading Acts Coll, Legislative Hist
15 000 vols; 300 curr per 37728

Corcoran Gallery of Art/College of Art & Design Library, Corcoran Library, 500 17th St NW, ***Washington***, DC 20006
T: +1 202 4781544; Fax: +1 202 6287908; E-mail: library@corcoran.org; URL: www.corcoran.edu/library/index.asp
1869; Mario A. Ascencio
Fine arts, Graphic arts, Photography; Artists Books Coll
32 000 vols; 250 curr per; 167 videotapes, 20 000 slides
libr loan 37729

Department of Mental Health, St Elizabeth's Hospital, Frances Waldrop Health Sciences Library, CT6A, 2700 Martin Luther King Jr Ave SE, ***Washington***, DC 20032
T: +1 202 6457388; Fax: +1 202 6458353; E-mail: seh.library@dc.gov
Velora Jernigan-Pedrick
12 000 vols; 250 curr per 37730

Department of Veterans Affairs – Headquarters Library, 810 Vermont Ave NW, ***Washington***, DC 20420
T: +1 202 2738522; Fax: +1 202 2739125; E-mail: ginny.dupont@mail.va.gov
1923; Ginny DuPont
15 000 vols; 500 curr per; 30 000 microforms; 13 000 monogr
libr loan 37731

Department of Veterans Affairs – Office of the General Council Law Library, 810 Vermont Ave NW, ***Washington***, DC 20420
T: +1 202 2736558; Fax: +1 202 2736645; URL: www.va.gov/ogc
Susan Sokoll
25 000 vols; 21 curr per 37732

Edison Electric Institute, Information Resources Center, 701 Pennsylvania Ave NW, 3rd Flr, ***Washington***, DC 20004-2696
T: +1 202 5085623
1917; Susan Farkas
Electric power, Energy, Environ studies, Legislation; Edison Electric Institute Publications, Electrical World Directory of Electric Utilities (1912-present), Moody's Public Utility Manual (1928-present)
11 000 vols; 30 curr per; 10 000 microforms; 10 digital data carriers; 143 VF drawers
libr loan 37733

Embassy of Australia Library, 1601 Massachusetts Ave NW, ***Washington***, DC 20036
T: +1 202 7973377; Fax: +1 202 7973155; E-mail: library.@austemb.org; URL: www.austemb.org
1969; Melissa Elliott
8 000 vols; 300 curr per
libr loan 37734

Environmental Law Institute Library, 2000 L Street, NW, Suite 620, ***Washington***, DC 20036
T: +1 202 9393800; E-mail: law@eli.org; URL: www.eli.org
1971; Larry Ross
Wetlands
9 000 vols; 350 curr per; 21 000 microforms; 2 digital data carriers 37735

Environmental Protection Agency, Info Resources Center, 401 M St SW, Rm 2904, 3404, ***Washington***, DC 20460
T: +1 202 2609152, 6046; Fax: +1 202 2605153; E-mail: library-hq@epa.epa.gov; URL: www.epa.gov/libraries/hqirc/index.html
1971; Susan Westen-Barger
Hazardous Waste Coll
16 000 vols; 100 curr per
libr loan 37736

European Union, Delegation of the European Commission, Library, 2300 M St NW, ***Washington***, DC 20037
T: +1 202 8629500; Fax: +1 202 4291766; E-mail: relex-delusw-help@cec.eu.int; URL: www.eurunion.org
1963; Barbara Sloan
European Union Law
60 000 vols; 99 curr per; 20 034 microforms; 25 digital data carriers 37737

Fannie Mae, Research & Information Center, 3900 Wisconsin Ave NW, ***Washington***, DC 20016
T: +1 202 7527750; Fax: +1 202 7526134; URL: www.fanniemae.com
Otto Schultz
10 000 vols; 400 curr per
ALA, SLA 37738

Folger Shakespeare Library, 201 E Capitol St SE, ***Washington***, DC 20003-1094
T: +1 202 5444600; Fax: +1 202 5444623; E-mail: gpaster@folger.edu; URL: www.folger.edu
1932; Dr. Gail Paster
Shakespeare, Renaissance Drama, theatre hist in England, 16th-19th centuries, hist of western civilization, 16th-17th c; STC & Wing Colls; Exhibitions; Performing Arts; Publications
260 000 vols; 200 curr per; 55 000 mss; 500 incunabula; 115 000 microforms; 250 000 playbills, 20 000 prints & engravings, 4 000 works of art on paper
ALA, IRLA 37739

German Historical Institute Library, 1607 New Hampshire Ave NW, *Washington*, DC 20009-2562
T: +1 202 4833430; E-mail: library@ghi-dc.org; URL: www.ghi-dc.org
1987; Katharina Kloock
German Hist, German-American Relations
40 000 vols; 250 curr per; 500 microforms; 18 digital data carriers   37740

The Gunnery, Tisch Family Library, 99 Green Hill Rd, *Washington*, CT 06793
T: +1 860 8687334; Fax: +1 860 8680859
William Chase
Alumni Publs
15 000 vols; 90 curr per
libr loan   37741

Historical Society of Washington, DC, Kiplinger Research Library, 801 K Street, NW of Mt Vernon Sq, *Washington*, DC 20001
T: +1 202 3831850; Fax: +1 202 3831872; E-mail: library@historydc.org; URL: www.historydc.org/library/library.aspx
1894; Gail Redmann
Thomas G. Machen Coll of 19th c Prints & Engravings, J Harry Shannon Photogr Coll, Capital Traction Company Records
14 000 vols; 300 maps; 500 mss, 70 000 prints & slides   37742

Intelsat Library, 3400 International Dr, NW, *Washington*, DC 20008
T: +1 202 9446820; Fax: +1 202 9447319; E-mail: library@intelsat.int; URL: www.intelsat.com
1979; Rosa Liu
Satellite Communication, Communication Eng, Telecommunication Policy, Electrical Eng; Int Telecommunications Union Publs, 1976 to present
12 000 vols; 350 curr per   37743

Internal Revenue Service Library, 1111 Constitution Ave NW, Rm 4324, *Washington*, DC 20224
T: +1 202 6228050; Fax: +1 202 6225844
1917; Gail Henderson-Green
Econimics, Tax Law
90 000 vols; 300 curr per; 74 000 microforms; 15 digital data carriers
libr loan   37744

International Center for Research on Women, Library, 1717 Massachusetts Ave NW, Ste 302, *Washington*, DC 20036
T: +1 202 7970007; Fax: +1 202 7970020; E-mail: icrw@igc.apc.org; URL: www.icrw.org
1977
10 000 vols; 12 curr per; 50 av-mat; 5 digital data carriers   37745

International Food Policy Research Institute Library, 2033 K St NW, *Washington*, DC 20006-1002
T: +1 202 8625616; Fax: +1 202 4674439; E-mail: ifprilibrary@cgiar.org; URL: www.ifpri.org
1975; Luz Alvare
Developing countries, Environment, Food administration, Int trade, Nutrition
8 000 vols; 129 curr per   37746

Joint World Bank – International Monetary Fund Library, Joint Library, 700 19th St NW, *Washington*, DC 20431
T: +1 202 6237054; Fax: +1 202 6236417; E-mail: jointlib@imf.org; URL: jolis.worldbankimflib.org
1946; Pamela Tripp-Melby
Banking, Economic development, Int finance & economy; UN Documents Coll; Bretton Woods Coll
200 000 vols; 1 600 curr per; 3 000 e-journals; 500 e-books; 12 digital data carriers; 365 res paper series   37747

Library of Congress – African & Middle Eastern Division, Jefferson Bldg, Rm 220, 101 Independence Ave SE, *Washington*, DC 20540-4820
T: +1 202 7077937; Fax: +1 202 2523180; E-mail: amed@loc.gov; URL: www.loc.gov/rr/amed
Mary Jane Deeb
600 000 vols   37748

Library of Congress – National Library Service for the Blind & Physically Handicapped, 1291 Taylor St NW, *Washington*, DC 20542
T: +1 202 7075100; Fax: +1 202 7070712; E-mail: nls@loc.gov; URL: www.loc.gov/nls
Frank Kurt Cylke   37749

Library of Congress – Science, Technology & Business Division, Sci Reading Rm, John Adams Bldg, Rm 508, *Washington*, DC 20540-4750
T: +1 202 7071205; Fax: +1 202 7071925; URL: www.loc.gov/rr/scitech
Ronald Bluestone
American Nat Standards; British, Chinese, French, German & Japanese Colls; OSRD Reports on World War II Res & Development (Dept of Energy, Dept of Defense, Nat Aeronautics & Space Administration & Nat Tech Info Service)
3 750 000 vols   37750

Mathematica Policy Research, Inc Library, 600 Maryland Ave SW, Ste 550, *Washington*, DC 20024
T: +1 202 4844692; Fax: +1 202 8631763; URL: www.mathematica-mpr.com
Sally Henderson
Health sciences
15 000 vols; 300 curr per   37751

Metropolitan Club of the City of Washington Library, 1700 H St, NW, *Washington*, DC 20006
T: +1 202 8352556; Fax: +1 202 8352582; E-mail: library@metroclub.org; URL: www.metroclub.org
Michael J. Higgins
Local history
17 000 vols; 75 curr per   37752

Metropolitan Washington Council of Governments, Information Center, 777 N Capitol St NE, Ste 300, *Washington*, DC 20002-4239
T: +1 202 9623256; Fax: +1 202 9623308; E-mail: MWAQCPublicComment@mwcog.org; URL: www.mwcog.org
1957; Denise Pinchback
Local Docs of Counties & Cities Within Washington Region, 1990 Census Co State Data Ctr
10 000 vols; 124 curr per; 1 500 arch doc   37753

Middle East Institute, George Camp Keiser Library, 1761 N St NW, *Washington*, DC 20036
T: +1 202 7850183; Fax: +1 202 3318861; E-mail: library@mideasti.org; URL: www.mideasti.org/library
1946; Ruth Van Laningham
Arabic, Persian & Turkish Coll; 18th & 19th C travel accounts of the Middle East
25 000 vols; 300 curr per; 1 000 govt docs; 200 maps
ISKO, MELA   37754

National Academies, National Academy of Sciences, George E. Brown Jr Library, 500 5th St NW, 3rd Flr, *Washington*, DC 20001-2721
T: +1 202 3342125; Fax: +1 202 3341651; E-mail: nrclibrary@nas.edu; URL: www7.nationalacademies.org/nrclibrary/index.html
1945; Victoria Harriston
Nat Academy of Sciences, Nat Academy of Engineering, Nat Res Council & Institute of Medicine, rpts; Transportation Research Board Library
20 000 vols; 240 curr per; 13 000 e-journals; 50 Electronic Media & Resources, 75 Special Interest Per Sub   37755

National Academies, Transportation Research Board Library, 2001 Wisconsin Ave, GR314B, *Washington*, DC 20007; mail address: 2101 Constitution Ave NW, Washington, DC 20418
T: +1 202 3342990; Fax: +1 202 3342527; E-mail: bpost@nas.edu; URL: www.trb.org
1946; Barbara Post
Highway Res Board & Transportation Res Board Publs; Strategic Highway Res Board Publs
17 000 vols; 380 curr per; 20 digital data carriers
libr loan   37756

National Association of Broadcasters, Information & Resource Center, 1771 N St NW, *Washington*, DC 20036
T: +1 202 4295490; Fax: +1 202 4294199; E-mail: irc@nab.org; URL: www.nab.org
1946; Vivian A. Pollard
Assn Publs, Arch
8 000 vols; 150 curr per   37757

National Association of Home Builders, National Housing Resource Center, 1201 15th St NW, *Washington*, DC 20005
T: +1 202 2668296; Fax: +1 202 2668400; E-mail: nhl@nahb.com; URL: www.nahb.com
1955; Nancy Hunn
Construction, Business & management, Economics; NAHB Arch
40 000 vols; 250 curr per; 3 digital data carriers   37758

National Endowment for Democracy Library, Democracy Resource Center, 1025 F St NW, Ste 800, *Washington*, DC 20004
T: +1 202 3789700; Fax: +1 202 3789407; E-mail: drc@ned.org; URL: www.ned.org
1994; Allen Overland
Int affairs
25 000 vols; 300 curr per
SLA   37759

National Endowment for the Arts Library, 1100 Pennsylvania Ave NW, Rm 213, *Washington*, DC 20506
T: +1 202 6825485; Fax: +1 202 6825651; E-mail: webmgr@arts.gov; URL: www.arts.endow.gov
Joy Kiser
10 000 vols; 160 curr per   37760

National Gallery of Art Library, Fourth St & Constitution Ave, *Washington*, DC 20565; mail address: 2000B S Club Dr, Door 7, Landover, MD 20785-3230
T: +1 202 8426516; Fax: +1 202 7893068; URL: www.nga.gov/resources/dldesc.htm
1941; Neal Turtell
Art Exhibition Cat, Art Sales Records, Artists' Bk Coll, Museum & Private Art Coll, Photogr Arch of European & American Art, Leonardo da Vinci Coll
370 000 vols; 958 curr per; 25 000 microforms; Photo and slide archs
libr loan; ARLIS/NA, IFLA   37761

National Geographic Society Library, 1146 16th St NW, *Washington*, DC 20036; mail address: 1145 17th St NW, Washington, DC 20036
T: +1 202 8577783; Fax: +1 202 4295731; E-mail: library@ngs.org; URL: www.nationalgeographic.com
1920; Barbara Penfold Ferry
Hakluyt Society Publs, General A.W. Greely's Polar Libr, scrapbks
48 000 vols; 129 curr per   37762

National Labor Relations Board Library, 1099 14th St NW, Ste 8000, *Washington*, DC 20570-0001
T: +1 202 2733720; Fax: +1 202 2732906; URL: www.nlrb.gov
Kenneth Nero
37 000 vols; 30 curr per   37763

National League of Cities, Municipal Reference Service, 1301 Pennsylvania Ave NW, Ste 550, *Washington*, DC 20004
T: +1 202 6263130; Fax: +1 202 6263043; E-mail: mrs@nlc.org; URL: www.nlc.org
1963; Bruce Calvin
Serials & Rpts of State Leagues of Cities
20 000 vols; 300 curr per   37764

National Museum of American Jewish Military History Library, 1811 R St NW, *Washington*, DC 20009
T: +1 202 2656280; Fax: +1 202 2345662; E-mail: nmajmh@nmajmh.org; URL: www.nmajmh.org
Larry Richardson
22 000 vols   37765

National Museum of Women in the Arts, Library and Research Center, 1250 New York Ave, NW, *Washington*, DC 20005-3920
T: +1 202 7837365; Fax: +1 202 3933234; URL: www.nmwa.org/library
1982
Women Artists Files; Artists' Bks; Bookplate Coll; Libr of Irene Rice Pereira; Arch of the Int Conference of Women Artists in Copenhagen in 1985, Nairobi in 1990, & Beijing in 1995
18 500 vols; 70 curr per; 2 000 institution files, 450 subject files   37766

National Public Radio Broadcasting Library, 635 Massachusetts Ave NW, *Washington*, DC 20001
T: +1 202 5132060; Fax: +1 202 5133056; E-mail: broadcastlibrary@npr.org; URL: www.npr.org
1971; Laura Soto-Barra
Current Events, audiotape; Drama & Music Performances, audiotape
100 000 vols; 100 curr per; 800 000 av-mat   37767

National Reference Center for Bioethics Literature, Kennedy Institute of Ethics, Georgetown University, 37th & O St NW, *Washington*, DC 20057; Georgetown University, Box 571212, Washington, DC 20057-1212
T: +1 202 6873885; Fax: +1 202 6876770; E-mail: bioethics@georgetown.edu; URL: bioethics.georgetown.edu/
1973; Doris Goldstein
Kampelman Coll of Jewish Ethics; Shriver Coll of Christian Ethics; Arch Colls of Federal Bioethics & Human Experimentation Commissions; Curriculum Development Clearinghouse for Bioethics (Syllabus Exchange Coll)
28 000 vols; 500 curr per; 39 sound-rec; 160 000 articles file
MLA   37768

National Society of the Daughters of the American Revolution, DAR Library, 1776 D St NW, *Washington*, DC 20006-5303
T: +1 202 7772366; Fax: +1 202 8793227; E-mail: library@dar.org; URL: www.dar.org
1896; Eric G. Grundset
Local hist, Genealogy; American Indians
205 000 vols; 1 100 curr per; 50 000 mss; 60 000 microforms; 400 digital data carriers; 300 000 files of mss, 700 CDs   37769

Naval Research Laboratory, Ruth H. Hooker Research Library, 4555 Overlook Ave SW, Code 5596, *Washington*, DC 20375-5334
T: +1 202 7672357; Fax: +1 202 7673352; E-mail: ref@library.nrl.navy.mil; URL: infoweb.nrl.navy.mil
Suzanne Ryder
Chemistry, Physics, Biotechnology, Electronics
46 000 vols; 2 506 curr per
libr loan   37770

Nuclear Energy Institute Library, 1776 I St NW, Ste 300, *Washington*, DC 20006
T: +1 202 7398000; Fax: +1 202 7854019; E-mail: media@nei.org; URL: www.nei.org
1954
NUREG Coll, IAEA Publs
12 000 vols; 250 curr per   37771

Organization of American States, Columbus Memorial Library, 19th & Constitution Ave NW, *Washington*, DC 20006-4499
T: +1 202 4586041; Fax: +1 202 4583914; URL: www.oas.org
1890; Beverly Wharton-Lake
Inter-American System & OAS (1889 to the present), Rare Bks Coll, Arch of the General Secretariat; Archs, Docs Ctr
500 000 vols; 2 679 curr per; 800 music scores; 1 500 maps; 6 000 microforms; 100 sound-rec; 1 coll of mss, diss, 241 401 OAS docs, 45 000 photos
libr loan   37772

Pan American Health Organization, Headquarters Library, 525 23rd St NW, ***Washington***, DC 20037
T: +1 202 9743305; Fax: +1 202 9743623; E-mail: library@paho.org; URL: www.paho.org/english/DD/IKM/LI/library.htm
1926; Marcelo D'Agostino
World Health Organization & Pan American Health Organization Docs
50 000 vols; 30 curr per; 27 340 microforms
libr loan 37773

Peace Corps, ICE Resource Center, 1111 20th St NW, 5th Fl, ***Washington***, DC 20526
T: +1 202 6922640; Fax: +1 202 6922641; URL: www.peacecorps.gov
1966
Int development; Peace Corps Hist, Foreign Languages Mat
10 000 vols; 200 curr per; 50 diss/theses; 500 govt docs; 50 maps
libr loan; ALA, SLA 37774

The Phillips Collection Library, 1600 21st St NW, ***Washington***, DC 20009-1090
T: +1 202 3872151; Fax: +1 202 3872436; URL: www.phillipscollection.org
1976; Karen Schneider
19th & 20th C European & American Artists, monogr; Duncan Phillips Coll, mss; Hist of American Art Museums & Collecting; Exhibitions Cat
8 000 vols; 10 curr per 37775

Population Reference Bureau, Library, 1875 Connecticut Ave, NW – Ste 520, ***Washington***, DC 20009
T: +1 202 4831100; Fax: +1 202 3283937; E-mail: zuali@prb.org; URL: www.prb.org
1960; Zuali Malsawma
Demography, population studies, family planning, environment, health, development; Special colls: Statistical publs of countries; United Nations publs; United States Census materials; United States vital statistics
13 000 vols; 300 curr per; 100 digital data carriers; 2 500 reprints & papers, 15 VF drawers of pamphlets, clippings
libr loan; APLIC-I, SLA 37776

Roy A. Childs Jr Library, The Cato Institute, 1000 Massachusetts Ave NW, ***Washington***, DC 20001-5403
T: +1 202 8420200; Fax: +1 202 8423490; E-mail: cato@cato.org; URL: www.cato.org
Heidi Rasmussen
Public policy
30 000 vols; 200 curr per 37777

Scottish Rite Library, 1733 16th St NW, ***Washington***, DC 20009-3103
T: +1 202 2323579; Fax: +1 202 4640487; E-mail: info@culturaltourismdc.org; URL: www.scottishrite.org
1881; Joan K. Sansbury
Abraham Lincoln Coll, Masonic Coll, Panama Canal (Thatcher Coll), Goethe Coll, Burnsiana, J Edgar Hoover Coll
193 000 vols; 258 Special Interest Per Sub 37778

Small Business Administration, Reference Library, 409 Third St SW, ***Washington***, DC 20416
T: +1 202 2058885; E-mail: answerdesk@sba.gov; URL: www.sba.gov
1958
8 000 vols; 75 curr per 37779

Smithsonian Institution Libraries, Nat Museum of Natural Hist, Rm 22, MRC154, Tenth St & Constitution Ave NW, ***Washington***, DC 20560; P.O. Box 37012, Washington, DC 20013-7012
T: +1 202 6332240; Fax: +1 202 7862866; E-mail: libmail@si.edu; URL: www.sil.si.edu
1846; Nancy E. Gwinn
Natural history, tropical biology, ecology and environmental management, wildlife conservation, American ethnology and culture, American history, the history of science and technology, aviation history and space flight, postal history, design and decorative arts, African art, American art, modern and contemporary art, Asian art, horticulture, conservation, and museum administration, African American and Latino history and culture
1 588 000 vols; 6 000 curr per; 2 261 e-journals; 70 digital data carriers; 50 000 historically important rare books
libr loan; ARL, CIRLA, FLICC 37780

Smithsonian Institution Libraries – Botany & Horticulture Library, Nat Museum of Natural Hist, Rm W422, MRC 154, Tenth St & Constitution Ave NW, ***Washington***, DC 20560; P.O. Box 37012, Washington, DC 20013-7012
T: +1 202 6331245; Fax: +1 202 7862443; URL: www.sil.si.edu/libraries/bothort
Robin Everly
Agrostology (Hitchcock-Chase Coll); Algology (Dawson Coll); General Botany (John Donnell Smith Coll)
51 000 vols; 14 shelves of notebooks, 325 boxes of reprints, 21 herbaria on microfiche 37781

Smithsonian Institution Libraries – The Dibner Library of the History of Science & Technology, Nat Museum of American Hist, Rm 1041, MRC 672, 12th St & Constitution Ave NW, ***Washington***, DC 20560-0672; P.O. Box 37012, Washington, DC 20013-7012
T: +1 202 6333872; Fax: +1 202 6339102; E-mail: dibneylibrary@si.edu; URL: www.sil.si.edu/libraries/dibner
1976; Lilla VeKerdy
Physical Sciences, Natural Hist, Technology, Applied Arts; Smithson Coll; Wetmore Bequest (Ornithogy); Burndy Libr Donation (Science & Technology); Comegys Libr (19th-c Philadelphia Family Libr)
30 000 vols; 2 000 mss; 320 incunabula; 10 microforms 37782

Smithsonian Institution Libraries – Freer Gallery of Art & Arthur M. Sackler Gallery Library, Arthur M Sackler Gallery, Rm 2057, MRC 707, 12th St & Jefferson Dr SW, ***Washington***, DC 20560; P.O. Box 37012, Washington, DC 20013-7012
T: +1 202 6330477; Fax: +1 202 7862936; URL: www.asia.si.edu/visitor/library.htm
1923
Arts, Asia; Charles Lang Freer Coll, arch; Herzfeld Coll, arch; James M. Whistler & His Contemporaries
87 000 vols 37783

Smithsonian Institution Libraries – Hirshhorn Museum & Sculpture Garden Library, Seventh St & Independence Ave SW, Rm 427, MRC 361, ***Washington***, DC 20560; P.O. Box 37012, Washington, DC 20013-7012
T: +1 202 6332775; Fax: +1 202 7862682; E-mail: hmsglibmail@si.edu; URL: www.hmsg.si.edu
1969; Anna Brooke
Fine arts, European & American 20th-c painting & sculpture, American 19th-c painting
67 000 vols; 100 microforms; 367 sound-rec; 2 digital data carriers; 88 VF drawers 37784

Smithsonian Institution Libraries – John Wesley Powell Library of Anthropology, Nat Museum of Natural Hist, Rm 331, MRC 112, Tenth St & Constitution Ave NW, ***Washington***, DC 20560-0112; P.O. Box 37012, Washington, DC 20013-7012
T: +1 202 6331640; Fax: +1 202 7862443; URL: www.sil.si.edu/libraries/anth/
Margaret R. Dittemore
Asian Cultural Hist (Echols Coll), Mesoamerican Codices, Native American Languages/Linguistics, Physical Anthropology (Hrdlicka Coll), Bureau of American Ethnology Libr Coll
84 000 vols; 4 digital data carriers 37785

Smithsonian Institution Libraries – Museum Studies & Reference Library, Nat Museum of Natural Hist, Tenth St & Constitution Ave NW, Rm 27, MRC 154, ***Washington***, DC 20560; P.O. Box 37012, Washington, DC 20013-7012
T: +1 202 6331245; Fax: +1 202 7862443; URL: www.sil.si.edu/libraries/msrl
1982; Amy Levin
General reference, Libr & information science, Management/Administration, Social sciences; Smithsoniana; Nat Bibliogr
26 000 vols; 12 digital data carriers
libr loan 37786

Smithsonian Institution Libraries – National Air & Space Museum Library, 6th St & Independence Ave SW, National Air & Space Museum, Rm 3100, ***Washington***, DC 20013; P.O. Box 37012, Washington, DC 20013-7012
T: +1 202 6332320; Fax: +1 202 7862835; E-mail: baxterw@si.edu; URL: www.sil.si.edu/libraries/nasm-hp.htm
1972; William Baxter
Sherman Fairchild Photogr Coll, William A.M. Burden Coll (early ballooning works and aeronautica), Bella Landauer Aeronautical Sheet Music Coll
44 000 vols; 410 curr per; 125 microforms 37787

Smithsonian Institution Libraries – National Museum of American History Library, NMAH Rm 5016, MRC 630, 14th & Constitution Ave NW, ***Washington***, DC 20560; P.O. Box 37012, Washington, DC 20013-7012
T: +1 202 6333865; Fax: +1 202 3574256; URL: www.sil.si.edu/libraries/nmah
1968; Lucien R. Rossignol
Exhibitions & Expositions, Hist of Science & Technology, Radioana, Trade Lit
469 000 vols; 28 000 microforms; 500 000 trade cat
libr loan; IFLA, CIRLA 37788

Smithsonian Institution Libraries – National Museum of Natural History Library, Nat Museum of Natural Hist, Rm 51, MRC 154, Tenth St & Constitution Ave NW, ***Washington***, DC 20013-0712; P.O. Box 37012, Washington, DC 20013-7012
T: +1 202 6331245; Fax: +1 202 3571896; URL: www.sil.si.edu/libraries/nmnh/
1881
Over 40 000 historically important rare books found in the Dibner Library of the History of Science and Technology, the Joseph F. Cullman 3rd Library of Natural History, the National Air and space Museum Library's Ramsey Room and the Cooper-Hewitt, National Design Museum Library's Bradley Room; over 300 000 manufacturer's commercial trade catalogs representing 30 000 companies from the 19th and 20th centuries; and 2 000 mss groups
130 000 vols; 1 200 e-journals; 190 000 microforms
libr loan; RLG, ARL, CIRLA 37789

Smithsonian Institution Libraries – National Postal Museum Library, Two Massachusetts Ave NE, MRC 570, ***Washington***, DC 20560-0570; P.O. Box 37012, Washington, DC 20013-7012
T: +1 202 6335544; Fax: +1 202 6339371; URL: www.sil.si.edu/libraries/npm/
Paul McCutcheon
Philately; Postal Hist, US Post Office Dept Files
22 000 vols 37790

Smithsonian Institution Libraries – National Zoological Park Library, Nat Zoological Park, Education Bldg-Visitor Ctr, MRC 551, 3000 Block of Connecticut Ave NW, ***Washington***, DC 20008-0551; P.O. Box 37012, Washington, DC 20013-7012
T: +1 202 6731030; Fax: +1 202 6734900; E-mail: nzplibrary@si.edu; URL: www.sil.si.edu/libraries/nzp
Polly Lasker
14 000 vols 37791

Smithsonian Institution Libraries – Smithsonian American Art Museum/National Portrait Gallery Library, Victor Bldg, Rm 2100, MRC 975, 750 Ninth St NW, ***Washington***, DC 20560; P.O. Box 37012, Washington, DC 20013-7012
T: +1 202 6338240; Fax: +1 202 6338232; URL: www.sil.si.edu/libraries/aapg
1964; Cecilia H. Chin
California Art & Artists (Ferdinand Perret Art Reference Libr), scrapbks; Mallet Libr of Art Reproductions
136 000 vols; 13 000 microforms; 400 VF drawers 37792

Smithsonian Institution Libraries – Warren M. Robbins Library, National Museum of African Art, Nat Museum of African Art, Rm 2138, MRC 708, 950 Independence Ave SW, ***Washington***, DC 20560; P.O. Box 37012, Washington, DC 20013-7012
T: +1 202 6334680; Fax: +1 202 3574879; URL: www.sil.si.edu/libraries/nmafa
1971; Janet L. Stanley
30 000 vols; 500 vertical files 37793

Society of the Cincinnati Library, 2118 Massachusetts Ave NW, ***Washington***, DC 20008
T: +1 202 7852040 ext 426; Fax: +1 202 7850729; E-mail: emclark@societyofthecincinnati.org
1783; Ellen McCallister Clark
Arch, Coll of 18th c Art & War
45 000 vols; 100 curr per; 1 000 mss; 350 maps 37794

State Services Organization Library, Hall of the States, Ste 337, 444 North Capital St NW, ***Washington***, DC 20001
T: +1 202 6245485; URL: www.sso.org
1977; Marianne Reiff
State govt; State Bluebooks & Manuals Coll, microfiche
14 000 vols; 265 curr per 37795

Textile Museum, Arthur D. Jenkins Library, 2320 S St NW, ***Washington***, DC 20008
T: +1 202 6670441 ext 31; Fax: +1 202 4830994; E-mail: mmallia@textilemuseum.org; URL: www.textilemuseum.org
1926; Mary Mallia
Cultural Hist of the Americas, Asia, Africa, the Middle East & the Pacific Rim; Hist of Rugs, Textiles, Costume
20 000 vols; 144 curr per; 7 100 av-mat
OCLC 37796

Transafrica Forum, Arthur R. Ashe Jr Foreign Policy Library, 1629 K Street, NW, Ste 1100, ***Washington***, DC 20006
T: +1 202 2231960; Fax: +1 202 2231966; E-mail: info@transafrica.org; URL: www.transafrica.org
10 000 vols; 100 curr per
ALA, SLA 37797

United Nations Information Center, 1775 K St NW, Ste 400, ***Washington***, DC 20006
T: +1 202 3318670; Fax: +1 202 3319191; URL: www.unicwash.org/Library.aspx
1946; Jeanne Dixon
Int law, Social sciences, Economics, Human rights; Film Libr, UN Chronicles & Publs
10 000 vols 37798

United States Sentencing Commission Library, One Columbus Circle NE, Ste 2-500 S Lobby, ***Washington***, DC 20002-8002
T: +1 202 5024500; Fax: +1 202 5024699; URL: www.ussc.gov
Linda Baltrusch
10 000 vols 37799

Urban Institute Library, 2100 M St NW, ***Washington***, DC 20037-1207
T: +1 202 2615688; Fax: +1 202 2233043; E-mail: uilibrary@ui.urban.org; URL: www.urban.org
1968; Nancy L. Minter
Economics, Demography, Local & state govt, Sociology; Urban Institute Arch Mat, Census Coll
25 000 vols; 350 curr per; 6 000 microforms; 300 digital data carriers 37800

Urban Land Institute, Information Center, 1025 Thomas Jefferson St NW, Ste 500W, ***Washington***, DC 20007
T: +1 202 6247137; Fax: +1 202 6247140; URL: www.uli.org
1936; Joan Campbell
Real estate
10 000 vols; 275 curr per 37801

US Commission on Civil Rights, National Clearinghouse Library, 624 Ninth St NW, Ste 600, **Washington**, DC 20425
T: +1 202 3768110; Fax: +1 202 3767597; E-mail: pubs@usccr.gov; URL: www.usccr.gov
1957; Barbara Fontana
Black Law School Reviews; Census Mats; Civil Rights (US Commission on Civil Rights Coll); Federal Register, micro; Native American Law Reviews; Native American Per, micro; Spanish Speaking Background Law Reviews; Women's Law Reviews
70 000 vols; 95 curr per; 2 200 microforms; 60 000 other items
libr loan   37802

US Holocaust Memorial Museum Library, 100 Raoul Wallenberg Pl SW, **Washington**, DC 20024
T: +1 202 4799717; Fax: +1 202 4799726; E-mail: library@ushmm.org; URL: www.ushmm.org/research/library
1993; N.N.
80 000 vols; 100 curr per; 1 000 diss/theses; 200 microforms; 400 av-mat; 200 sound-rec; 25 digital data carriers
ALA, SLA   37803

US Institute of Peace, Jeannette Rankin Library Program, 1200 17th St, Ste 200, **Washington**, DC 20036-3011
T: +1 202 4571700; Fax: +1 202 4296063; E-mail: library@usip.org; URL: www.usip.org/library/
1988; Margarita Studemeister
Diplomacy, Meditation
11 000 vols; 150 curr per   37804

Walter Reed Army Medical Center – Post & Patient's Library, Bldg 1, Rm D 110, 6900 Georgia Ave NW, **Washington**, DC 20307-5001
T: +1 202 7826314; Fax: +1 202 7825094; URL: www.wramc.amedd.army.mil; wrpost.bytopia.amedd.army.mil
1920; Elizabeth Deal
Recreation, Education; Health Topics Coll
38 000 vols; 80 curr per
libr loan   37805

Washington Hospital Center, William B. Glew MD Health Sciences Library, 110 Irving St NW, Rm 2A-21, **Washington**, DC 20010-2975
T: +1 202 8776221; Fax: +1 202 8776757; E-mail: libraryreferenceservices@medstar.net; URL: www.whcenter.com
1958; Lynne Siemers
30 000 vols; 700 curr per; 15 digital data carriers
libr loan   37806

Washington National Cathedral, Cathedral Rare Book Library, Massachusetts & Wisconsin Aves NW, **Washington**, DC 20016-5098
T: +1 202 5376200; URL: www.nationalcathedral.org
1965; Anna Alston Donnelly
Carson Coll of American Bishops
9 000 vols; 21 curr per   37807

Woodrow Wilson International Center for Scholars, 1300 Pennsylvania Ave NW, **Washington**, DC 20004-3027
T: +1 202 6914150; Fax: +1 202 6914001; URL: www.wilsoncenter.org
1970; Janet Spikes
Ref, Russia
12 000 vols; 250 curr per   37808

Woodrow Wilson International Center for Scholars – Kennan Institute for Advanced Russian Studies, Library, 1300 Pennsylvania Ave NW, **Washington**, DC 20004-3027
T: +1 202 6914150
1975; Janet Spikes
Russia, Literary criticism, Economics
10 000 vols; 50 curr per   37809

The World Bank Group Library, 1818 H St NW, MSN MC-C3-220, **Washington**, DC 20433
T: +1 202 4732000; Fax: +1 202 5221160; E-mail: wbglibrary@worldbank.org; URL: www.jolis.worldbankimflib.org/external.htm
Marion Richards
85 000 vols; 525 curr per   37810

World Resources Institute, WRI Library and Information Center, 10 G St NE, Ste 800, **Washington**, DC 20002
T: +1 202 7297600; Fax: +1 202 7297610; E-mail: mmaguire@wri.org; URL: www.wri.org/library
1982; Mary Maguire
Environment, Natural Resources, Agriculture, Forestry, Climate, Energy, Biodiversity, Sustainable Development
10 000 vols; 200 curr per
SLA, ASIS   37811

David Library of the American Revolution, 1201 River Rd, P.O. Box 748, **Washington Crossing**, PA 18977-0748
T: +1 215 4936776; Fax: +1 215 4939276; E-mail: librarian@dlar.org; URL: www.dlar.org
1958; Meg McSweeney
American Revolution (Sol Feinstone Coll)
8 000 vols; 25 curr per; 2 500 mss; 250 diss/theses; 10 000 microforms; 20 digital data carriers   37812

Waterbury Hospital, Medical Library, 64 Robbins St, **Waterbury**, CT 06721
T: +1 203 5736136; Fax: +1 203 5736706; E-mail: library@wtbyhosp.org; URL: www.waterburyhospital.com/services/library/library.htm
8 800 vols; 293 curr per; 800 av-mat; 6 digital data carriers; 5 drawers of audiotapes   37813

Armenian Library & Museum of America, Inc, 65 Main St, **Watertown**, MA 02472
T: +1 617 9262562; Fax: +1 617 9260175; E-mail: info@armenianlibraryandmuseum.org; URL: www.almainc.org; www.armenianlibraryandmuseum.org
1971; Berj Chekijian
Early Armenian Printings (1514-1700 AD), Armenian Rug Society Data Bank, ALMA Oral Hist Colls (Genocide survivors)
22 000 vols   37814

Waukesha County Historical Society and Museum, Research Center – Library, 101 W Main St, **Waukesha**, WI 53186-4811
T: +1 262 5212859; Fax: +1 262 5212865; E-mail: info@wchsm.org; URL: www.waukeshacountymuseum.org
1914; Eric D. Vanden Heuvel
Pioneer Notebooks, County Naturalization Records, County High School Yearbooks
8 055 vols; 10 curr per; 178 maps; 84 VF drawers, 14 000 photos, 10 000 negatives, 1 800 slides   37815

Cargill, Inc – Information Center, 15407 McGinty Rd W, **Wayzata**, MN 55391; P.O. Box 5670, Minneapolis, MN 55440-5670
T: +1 952 7426498; Fax: +1 952 7426062
Peter Sidney
International trade, transportation
15 000 vols; 700 curr per   37816

Massachusetts Horticultural Society Library, 900 Washington St, Rte 16, **Wellesley**, MA 02482
T: +1 617 9334900; Fax: +1 617 9334901; URL: www.masshort.org/mhs-library
1829; Maureen Horn
Art, Hist; Print Coll, Rare Bk Coll
20 000 vols; 10 curr per; 1 000 microforms; 150 av-mat; 350 Historic Pamphlets, 43 000 Seed Cat, 4 000 Docs   37817

National Archives & Records Administration, Herbert Hoover Presidential Library, 210 Parkside Dr, **West Branch**, IA 52358-9685; P.O. Box 488, West Branch, IA 52358-04888
T: +1 319 6435301; Fax: +1 319 6436045; E-mail: library@hoover.archives.gov; URL: www.hoover.archives.gov
1962; Timothy Walch
Political science, Hist, Economics; Personal papers of Herbert Hoover
21 141 vols; 23 curr per; 8 400 000 mss; 3 250 microforms; 160 sound-rec; 343 oral hist transcripts
libr loan   37818

Pilgrim Psychiatric Center, Pilgrim Reading Room & Patients Library, Bldg 102, 998 Crooked Hill Rd, **West Brentwood**, NY 11717-1087
T: +1 631 7613813; Fax: +1 631 7613103; E-mail: pgetjxm@omh.state.ny.us
Jeanne Murphy
8 000 vols; 22 curr per; 15 av-mat   37819

Chester County Historical Society Library, 225 N High St, **West Chester**, PA 19380
T: +1 610 6924800; Fax: +1 610 6924357; URL: www.chestercohistorical.org
1893; Diane P. Rofini
Chester County Newspapers, Postal Hist (R F Brinton Coll), William Penn (A C Myers Coll), Art Hist (C Brinton Coll), Almanachs, Chester County Diaries, Paper Dolls & Paper Toys
25 000 vols; 50 curr per; 250 microforms; 3 digital data carriers; 152 VF drawers of clippings, 70 000 photos   37820

Waterton-Glacier International Peace Park, George C. Ruhle Library, Going to the Sun Hwy, Bldg 217, **West Glacier**, MT 59936; P.O. Box 128, West Glacier, MT 59936-0128
T: +1 406 8887932; Fax: +1 406 8885824; URL: www.library.nps.gov; www.nps.gov/glac/research/library.htm
1975
Plains Indians (James Willard Schultz Coll)
15 000 vols; 20 curr per; 500 maps; 12 000 slides   37821

Center for Early Education Library, 563 N Alfred St, **West Hollywood**, CA 90048
T: +1 323 6510707; Fax: +1 323 6510860
1965; Lucy Rafael
Child development, Children's lit, Education, Parenting
22 000 vols; 92 curr per; 100 digital data carriers   37822

Shakespeare Society of America, New Place Rare Book Library, 1107 N Kings Rd, **West Hollywood**, CA 90069
T: +1 323 6545623
1967; R. Thad Taylor
Renaissance lit; Early Science Coll
7 500 vols; 50 curr per; 100 sound-rec; 1 000 cat, 500 magazines & pamphlets, 450 clippings, 2 000 photos & slides   37823

Indiana Veteran's Home, Lawrie Library, 3851 North River Rd, **West Lafayette**, IN 47906
T: +1 765 4631502
1898; Sonya Hill
Lawrie, Indiana Hist
12 000 vols; 47 curr per; 4 047 sound-rec   37824

US Department of the Interior – National Park Service, Edison National Historic Site Library, Main St & Lakeside Ave, **West Orange**, NJ 07052-5515
T: +1 973 7360550 ext 13; Fax: +1 973 7368496; E-mail: edis_archives@nps.gov; URL: www.nps.gov/edis
Roger Durham
15 000 vols   37825

Genealogical Society of Palm Beach County, 3650 Summit Blvd, **West Palm Beach**, FL 33405; P.O. Box 17617, West Palm Beach, FL 33416-7617
T: +1 561 6163455; E-mail: ancestry@bellsouth.net; URL: www.pbcgs.org
1964; Mrs Alvin L. Lentsch
14 000 vols; 20 curr per   37826

The American Ceramic Society, James I. Mueller Memorial Library – Ceramic Information, 600 N Cleveland Ave, **Westerville**, OH 43082
T: +1 240 6467054; E-mail: customerservice@ceramics.org; URL: ceramics.org
1954; Greg Geiger
Ross Coffin Purdy Coll
10 500 vols; 900 curr per
libr loan   37827

Hitchcock Memorial Museum & Library, 1252 Rte 100, **Westfield**, VT 05874; P.O. Box 87, Westfield, VT 05874-0087
T: +1 802 7448258
Pati Austin-Kirk
Cinema
8 000 vols   37828

Wayne County Regional Library for the Blind & Physically Handicapped, 30555 Michigan Ave, **Westland**, MI 48186-5310
T: +1 734 7277300; Fax: +1 734 7277333; E-mail: wcrlbph@wayneregional.lib.mi.us; URL: wayneregional.lib.mi.us
1931; Reginald B. Williams
125 000 vols   37829

Exempla Healthcare Lutheran Medical Center, Medical Library, 8300 W 38th Ave, **Wheat Ridge**, CO 80033-8270
T: +1 303 4258662; Fax: +1 303 4678794; E-mail: wellsk@exempla.org
1961; Karen Wells
8 000 vols; 250 curr per; 8 digital data carriers; Consumer health coll
libr loan; MLA   37830

Colonel Robert R. McCormick Research Center, First Division Museum at Cantigny, One S 151 Winfield Rd, **Wheaton**, IL 60189-6097
T: +1 630 2608211; Fax: +1 630 2609298; E-mail: info@firstdivisionmuseum.org; URL: www.firstdivisionmuseum.org
1960; Eric Gillespie
US military hist, hist of the First Division of the US Army
14 000 vols; 75 curr per; 30 000 docs, 4 000 photos   37831

Theosophical Society in America, Henry S. Olcott Memorial Library, 1926 N Main St, **Wheaton**, IL 60187; P.O. Box 270, Wheaton, IL 60187-0270
T: +1 630 6681571; Fax: +1 630 6684976; E-mail: library@theosophical.org; URL: www.theosophical.org
1926; Marina Maestas
Rare Theosophical Journals Coll; Boris de Zirkoff Coll; Mary K Neff Coll
20 000 vols; 85 curr per; 60 microforms; 1 200 sound-rec; 700 videos   37832

Department of Veterans Affairs, Hospital Library, VA Domiciliary, 142-D, 8495 Crater Lake Hwy, **White City**, OR 97503
T: +1 541 8262111 ext 3690; Fax: +1 541 8303503
Sarah L. Fitzpatrick
World War II Coll
14 000 vols   37833

Mississippi State Hospital, Patient Library, Whitfield Rd, **Whitfield**, MS 39193
T: +1 601 3518000
Jane Hull
13 000 vols   37834

Midwest Historical and Genealogical Society, Inc, Library, 1203 N Main, P.O. Box 1121, **Wichita**, KS 67201-1121
T: +1 316 2643611
1966; Donna Woods
20 000 vols; 45 curr per; 800 maps; 600 microforms; 100 scrapbks, 24 VF drawers   37835

Via Christi Libraries, Saint Francis Campus, 929 N Saint Francis, **Wichita**, KS 67214-1315
T: +1 316 2685979; Fax: +1 316 2688694; URL: www.via-christi.org/rmclibrary
1938; Kristin Sen
10 000 vols; 500 curr per
libr loan   37836

Wichita Art Museum, The Emprise Bank Research Library, 1400 W Museum Blvd, **Wichita**, KS 67203-3296
T: +1 316 2684918; Fax: +1 316 2684980; E-mail: library@wichitaartmuseum.org; URL: www.wichitaartmuseum.org
1963; Lois F. Crane
American Art, Art Hist; Elizabeth S. Navas Papers
13 000 vols; 32 curr per; 1 digital data carriers; 5 600 slides, 21 VF drawers
ARLIS/NA   37837

Wichita Public Library – Subregional Library for the Blind & Physically Handicapped, 223 S Main St, ***Wichita***, KS 67202
T: +1 316 2618574; Fax: +1 316 2624540
Brad Reha
27 000 vols 37838

Wilkes-Barre Law & Library Association, Max Rosen Memorial Law Library, Luzerne County, Court House, Rm 23, 200 N River St, ***Wilkes-Barre***, PA 18711-1001
T: +1 570 8226712; Fax: +1 570 8228210; E-mail: law.library@luzernecounty.org; URL: www.wblawlibrary.org
1866; Joseph Burke III
40 000 vols; 40 curr per 37839

Colonial Williamsburg Foundation – John D. Rockefeller Jr Library, 313 First St, ***Williamsburg***, VA 23185-4306; P.O. Box 1776, Williamsburg, VA 23187-1776
T: +1 757 5658500; Fax: +1 757 5658508; E-mail: libref@cwf.org; URL: www.history.org
1933; James Horn
Colonial Williamsburg House Histories and Research Reports; Research Query Files, 1927-present; 18th-c history of the Chesapeake Region, 18th-c hist of western industrial arts, culture; Rockefeller Jr Library-Special Collections, Rockefeller Jr Library-Visual Resources
83 000 vols; 400 curr per; 50 000 mss; 2 incunabula; 6 000 microforms; 100 sound-rec; 20 digital data carriers; 52 000 photos, slides, negatives, digital images, 50 000 architectural drawings, 800 postcards, 12 000 rare books
libr loan; ARLIS/NA, ALA 37840

Colonial Williamsburg Foundation – John D. Rockefeller Jr Library-Special Collections, 313 First St, ***Williamsburg***, VA 23185-4306; P.O. Box 1776, Williamsburg, VA 23185-1776
T: +1 757 5658520; Fax: +1 757 5658528; E-mail: speccoll@cwf.org
Douglas Mayo
History of Williamsburg (Virginia) and Chesapeake region in colonial, eaarly American periods
12 000 vols 37841

Eastern State Hospital, Library Services, 4601 Ironbound Rd, ***Williamsburg***, VA 23188; P.O. Box 8791, Williamsburg, VA 23187-8791
T: +1 757 2535387; Fax: +1 757 2535192; E-mail: libraryservices@esh; URL: www.esh.dmhmrsas.virginia.gov
1843; Judy Harrell
Galt papers, mss; Staff/Medical Library; Patients' Library
18 000 vols; 343 curr per; 4 500 av-mat; 2 800 sound-rec; 300 digital data carriers
ALA, VLA 37842

National Center for State Courts Library, 300 Newport Ave, ***Williamsburg***, VA 23185-4147
T: +1 757 2597590; Fax: +1 757 2591530; E-mail: library@ncsc.dni.us; URL: www.ncsconline.org
1973; Joan Cochet
National Center for State Courts Reports, ICM Course Notebooks, Court Admininstration video coll, State of the Judiciary messages; Arligton Branch, Denver Branch
40 000 vols; 2 000 microforms; 5 digital data carriers
libr loan; AALL 37843

Omohundro Institute of Early American History & Culture, Kellock Library, P.O. Box 8781, ***Williamsburg***, VA 23187-8781
T: +1 757 2211126; Fax: +1 757 2211047; E-mail: pvhigg@facstaff.wm.edu
Patricia Higgs
8 650 vols; 27 curr per; 2 000 microforms; 2 000 reels of microfilm 37844

Sterling & Francine Clark Art Institute Library, 225 South St, ***Williamstown***, MA 01267; P.O. Box 8, Williamstown, MA 01267-0008
T: +1 413 4589545; Fax: +1 413 4589542; E-mail: library@clarkart.edu; URL: www.clarkart.edu/library
1962; Susan Roeper
European & American art; Auction Sales catalogs; Mary Ann Beinecke Decorative Art Coll; Robert Sterling Clark Rare Bk Coll; Artist Books Coll; David A. Hanson Collection of the History of Photomechanical Reproduction
200 000 vols; 700 curr per; 3 000 microforms; 110 digital data carriers; 145 000 slides, 1 mio art reproductions
libr loan; ARLIS/NA, IFLA 37845

Congregation Beth Emeth, William, Vitellia & Topkis Library, 300 W Lea Blvd, ***Wilmington***, DE 19802
T: +1 302 7642393; Fax: +1 302 7642395
Barry Wexler
8 000 vols 37846

Delaware Academy of Medicine, Inc, Lewis B. Flinn Library, 1925 Lovering Ave, ***Wilmington***, DE 19806
T: +1 302 6566398, 1629; Fax: +1 302 6560470; E-mail: plg@delamed.org; URL: www.delamed.org
1930; P.J. Grier
Medical Journals; 1 branch libr
20 000 vols; 265 curr per
libr loan; OCLC 37847

Delaware Art Museum, Helen Farr Sloan Library, 2301 Kentmere Pkwy, ***Wilmington***, DE 19806
T: +1 302 5719590; Fax: +1 302 5710220; URL: www.delart.org
1912; Sarena Fletcher
Extensive Coll of Monogr & Clippings on Contemporary Artists, Howard Pyle Arch, Bancroft Pre-Raphaelite Coll, John Sloan Arch
40 000 vols; 50 curr per; 10 000 mss; 700 microforms; 150 av-mat; 19th & 20th C ill bks & per
libr loan; ARLIS/NA, MARAC 37848

Delaware Historical Society Research Library, 505 Market St, ***Wilmington***, DE 19801
T: +1 302 6557161; Fax: +1 302 6557844; E-mail: deinfo@dehistory.org; URL: www.dehistory.org
1864; Dr. Constance Cooper
Delaware, Genealogy, American hist; Civil War Coll, Wilmington Businesses Coll
32 000 vols; 73 curr per; 1 000 maps; 1 000 microforms; 2 mio mss, 500 newspapers, 10 000 pamphlets, 500 000 photos 37849

Delaware Museum of Natural History Library, 4840 Kennett Pike, ***Wilmington***, DE 19807; P.O. Box 3937, Wilmington, DE 19807-0937
T: +1 302 6589111; Fax: +1 302 6582610; URL: www.delmnh.org
1972; Dr. Jean Woods
Mollusk dept, Birds dept
10 000 vols; 130 curr per; 300 maps; 10 microforms; 5 digital data carriers; 14 000 reprints
Access by appointment only 37850

Hagley Museum & Library, 298 Buck Rd E, ***Wilmington***, DE 19807; P.O. Box 3630, Wilmington, DE 19807-0630
T: +1 302 6582400; Fax: +1 302 6580568; URL: www.hagley.org
1955; Terry Snyder
Hist – business, industrial, technological, naval, computer, French Revolution, Explosives, Aeronautics; Business & Industrial Records
250 000 vols; 259 curr per; 1 200 maps; 2 000 microforms; 7 300 sound-rec; 20 000 mss arch (linear feet), 1 mio photos, 10 000 slides
libr loan 37851

Oregon State Penitentiary Library, Coffee Creek Correctional Facility, 24499 SW Grahams Ferry Rd, ***Wilsonville***, OR 97070
T: +1 503 5706783; Fax: +1 503 5706786
Angela Wheeler
8 000 vols 37852

Flag Research Center Library, Three Edgehill Rd, ***Winchester***, MA 01890-3915; P.O. Box 580, Winchester, MA 01890-0880
T: +1 781 7299410; Fax: +1 781 7214817; E-mail: vexor@comcast.net; URL: flagsmith.com
1962; Whitney Smith
Heraldry, Symbolism
13 000 vols; 60 curr per; 2 000 flags 37853

Office of Navajo Nation Library, Hwy 264, Post Office Loop Rd, P.O. Box 9040, ***Window Rock***, AZ 86515-9040
T: +1 928 8716376, 8716526; Fax: +1 928 8717304; E-mail: inelso@citlink.net; URL: nnlib.org
1941; Irving Nelson
Archaeology; Navajo Hist & Culture
60 000 vols; 135 curr per 37854

Connecticut Aeronautical Historical Association, Inc, John W. Ramsay Research Library, Bradley Int Airport, ***Windsor Locks***, CT 06096
T: +1 860 6233305; Fax: +1 860 6272820
1960; Robert Foster
Burnelli Aircraft Coll; Igor Sikorsky Coll; Early New England Aviation Hist
18 000 vols; 200 maps; 400 microforms; 3 500 tech manuals, 66 000 bound & unbound periodicals, 30 000 photogs & slides 37855

Old Salem Museums & Gardens, Museum of Early Southern Decorative Arts, 924 S Main St, ***Winston-Salem***, NC 27101
T: +1 336 7217372; Fax: +1 336 7217367; E-mail: library@oldsalem.org
1965; Kathryn Schlee
20 000 vols; 125 curr per; 3 000 microforms; data file of 330 000 cards, 110 VF drawers 37856

The Winterthur Library, Rte 52, ***Winterthur***, DE 19735
T: +1 302 8884681; Fax: +1 302 8884870; E-mail: reference@winterthur.org; URL: www.winterthur.org
1951; Bert Denker
Edward Deming Andrews Memorial Shaker Coll, Waldron Phoenix Belknap, Jr, .Research Libr of American Painting, Decorative Arts Photogr Coll, Joseph Downs Coll of Mss and Printed Ephemera; Winterthur Archives
82 000 vols; 300 curr per; 13 000 microforms; 172 000 slides, mss 2 500 rec groups, 165 000 photos
libr loan; ALA, IRLA 37857

Ohio Agricultural Research & Development Center Library, 1680 Madison Ave, ***Wooster***, OH 44691-4096
T: +1 330 2633690; Fax: +1 330 2633689; E-mail: library_oardc@osu.edu; URL: www.oardc.ohio-state.edu/library
1892; Constance Britton
70 000 vols; 250 curr per; 2 digital data carriers
libr loan
Univ of Toronto Libraries System includes central, branch, and college libs 37858

American Antiquarian Society Library, 185 Salisbury St, ***Worcester***, MA 01609-1634
T: +1 508 7555221; Fax: +1 508 7533311; E-mail: library@mwa.org; URL: www.americanantiquarian.org
1812; Ellen S. Dunlap
American hist, Lit; Americana (Early American Imprints), Children's Lit, Genealogies, Graphics Arts, Local & State Histories, Music
690 000 vols; 1 200 curr per; 271 000 microforms; 11 digital data carriers; 2 000 linear feet of mss 37859

University of Massachusetts Medical School, Lamar Soutter Medical Library, 55 Lake Ave N, ***Worcester***, MA 01655-2397
T: +1 508 8566099; Fax: +1 508 8565039; URL: library.umassmed.edu
1973; Elaine Martin
Massachusetts Medical Hist (Worcester Medical Libr)
280 000 vols; 200 e-books; 27 000 microforms; 4 000 digital data carriers
libr loan; MLA 37860

Worcester Art Museum Library, 55 Salisbury St, ***Worcester***, MA 01609-3196
T: +1 508 7994406; Fax: +1 508 7985646; E-mail: library@worcesterart.org; URL: www.worcesterart.org
1909; Debby Aframe
American, Asian & European art, Prints
45 000 vols; 396 linear feet of cat, 30 000 slides 37861

Worcester Historical Museum, Research Library, 30 Elm St, ***Worcester***, MA 01605
T: +1 508 7538278; Fax: +1 508 7539070; URL: www.worcesterhistory.org/library.html
1875; William D. Wallace
Local hist; Anti-Slavery (Kelley-Foster Coll), mss; Architectural Drawings Coll; City of Worcester, mss; Diner Industry; Howland Valentines; Local Info (Worcester pamphlet files); Out-of-Print Worcester Newspaper & Per
10 000 vols; 20 curr per; 4 000 linear feet of photos 37862

National Park Service-Yellowstone Association, Yellowstone Research Library & Archives, Albright Visitor Ctr, Mommoth Hot Springs, P.O. Box 168, ***Yellowstone National Park***, WY 82190
T: +1 307 3442264; Fax: +1 307 3442323; E-mail: yell_research_library@nps.gov; URL: wyldweb.state.wy.us/yrl
1933; Alissa Cherry, Barbara Zafft
Montana Hist Society Coll (1876-present); Yellowstone Area (Rare Bks Coll)
10 700 vols; 55 curr per; 100 mss; 150 diss/theses; 5 000 govt docs; 25 music scores; 500 maps; 100 microforms; 250 av-mat; 300 sound-rec; 5 digital data carriers
libr loan 37863

Wellspan Health at York Hospital, Philip A. Hoover MD Library, 1001 S George St, ***York***, PA 17405
T: +1 717 8513323; Fax: +1 717 8512487; E-mail: library@wellspan.org
1931; Suzanne M. Shultz
31 000 vols; 375 curr per; 2 000 e-journals 37864

York County Heritage Trust, Historical Society of York County Library & Archives, 250 E Market St, ***York***, PA 17403
T: +1 717 8481587; Fax: +1 717 8121204; E-mail: library@yorkheritage.org; URL: www.yorkheritage.org
1895; Lila Fourhman-Shaull
Circus & Theater (James Shettle Circus & Theater in America Coll); Art (Lewis Miller Folk Drawing Coll); Governmental Arch; General Jacob Devers Coll
25 000 vols; 50 curr per 37865

BOCES – Putnam-Northern Westchester, BOCES Professional Library, 200 BOCES Dr, ***Yorktown Heights***, NY 10598
T: +1 914 2482392; Fax: +1 914 2482419; URL: www.pnwboces.org/library/default.htm
John P. Monahan
Education
14 000 vols; 190 curr per 37866

George Kurian Reference Books, Editorial Library, 3689 Campbell Ct, ***Yorktown Heights***, NY 10598-1808; P.O. Box 519, Baldwin Place, NY 10505-0519
T: +1 914 9623287; Fax: +1 914 9623287; E-mail: gtkurian@aol.com
Sarah Claudine
Third World, Politics, Religion
28 270 vols; 40 curr per; 47 digital data carriers 37867

Yosemite National Park Service, Research Library, Museum Bldg, P.O. Box 577, ***Yosemite National Park***, CA 95389-0577
T: +1 209 3720280; Fax: +1 209 3720255
1923
10 000 vols; 80 curr per 37868

Mahoning Law Library Association, Courthouse 4th Flr, 120 Market St, ***Youngstown***, OH 44503-1752
T: +1 330 7402295; Fax: +1 330 7441406; E-mail: mlladir@mahoninglawlibrary.org; URL: www.mahoninglawlibrary.org
1906; Anna E. Paczelt
Ohio Legal Journals
25 000 vols; 140 curr per; 64 000 microforms; 10 digital data carriers 37869

Lincoln Memorial Library, 240 California Dr, *Yountville*, CA 94599-1445; P.O. Box 1200, Yountville, CA 94599-1297
T: +1 707 9444915; Fax: +1 707 9444915; E-mail: carole.debell@cdva.ca.gov
1886; Carole DeBell
World War I & II; Spanish American War
40 000 vols; 62 curr per; 970 Audio Bks, 10 000 Large Print Bks   37870

Public Libraries

Vermilion Parish Library, 405 E Saint Victor, *Abbeville*, LA 70510-5101; P.O. Drawer 640, Abbeville, LA 70511-0640
T: +1 337 8932655; Fax: +1 337 8980526; E-mail: www.vermilion.lib.la.us; URL: www.vermilion.lib.la.us
1941; Charlotte Trosclair
Louisiana Coll; 8 branch libs
174 000 vols; 97 curr per; 4 957 av-mat; 4 960 Audio Bks, 81 Bks on Deafness & Sign Lang, 7 755 Large Print Bks 37871

Alexander Mitchell Public Library, Aberdeen Public Library, 519 S Kline St, *Aberdeen*, SD 57401-4495
T: +1 605 6267097; Fax: +1 605 6263506; E-mail: library@aberdeen.sd.us; URL: ampl.sdln.net
1884; Pamla J. Lingor
AFRA (American Family Record Assn), Dakota Art, Dakota Maps, Genealogy, Germans from Russia, Railroad Hist, L Frank Baum
108 000 vols; 364 curr per
libr loan   37872

Abilene Public Library, 202 Cedar St, *Abilene*, TX 79601-5793
T: +1 325 6766021; Fax: +1 325 6766028; URL: www.abilenetx.com/apl
1899; Ricki V. Brown
Genealogy, Business, Science; 1 branch libr
288 000 vols; 224 curr per
libr loan   37873

Abilene Public Library, 209 NW Fourth, *Abilene*, KS 67410-2690
T: +1 785 2633082; Fax: +1 785 2632274; E-mail: abplib@sbcglobal.net; URL: abilene.mykansaslibrary.org/
1908; Judy Leyerzapf
Mamie Eisenhower Doll Coll, WPA Doll Coll
61 000 vols; 106 curr per
libr loan   37874

Washington County Public Library, 205 Oak Hill St, *Abingdon*, VA 24210
T: +1 276 6766383; Fax: +1 276 6766235; URL: www.wcpl.net
1954; Charlotte L. Parsons
4 branch libs
108 000 vols; 240 curr per   37875

Abington Public Library, 600 Gliniewicz Way, *Abington*, MA 02351
T: +1 781 9822139; Fax: +1 781 8787361; E-mail: ablib@ocln.org; URL: abingtonpl.org
1878; Deborah Grimmett
Arnold Civil War Coll, Large Print Coll
61 000 vols; 88 curr per   37876

Abington Township Public Library, Abington Free Library, 1030 Old York Rd, *Abington*, PA 19001-4594
T: +1 215 8855180; Fax: +1 215 8859242; URL: abingtonfreelibrary.org
1966; Nancy Hammeke Marshall
1 branch libr
154 000 vols; 300 curr per   37877

Eastern Shore Public Library, 23610 Front St, *Accomac*, VA 23301; P.O. Box 360, Accomac, VA 23301-0360
T: +1 757 8245151; Fax: +1 757 7872241; URL: www.espl.org
1957; Carol Vincent
Boat building, Sailing; Local Hist Coll
128 000 vols; 108 curr per   37878

Acton Memorial Library, 486 Main St, *Acton*, MA 01720
T: +1 978 2649641; Fax: +1 978 6350073; E-mail: acton@minlib.net; URL: www.actonmemoriallibrary.org
1890; Marcia Rich
Arthur Davis Paintings
128 000 vols; 215 curr per   37879

Ada Public Library, Hugh Warren Memorial Library, 124 S Rennie, *Ada*, OK 74820
T: +1 580 4368123; Fax: +1 580 4360534; URL: www.ada.lib.ok.us
1939; Jennifer Greenstreet
Large print, Oklahoma, Music
60 000 vols; 282 curr per   37880

Addison Public Library, Four Friendship Plaza, *Addison*, IL 60101
T: +1 630 4583178; Fax: +1 630 5436645; E-mail: librarydirector@addisonlibrary.org; URL: www.addisonlibrary.org
1962; Mary A. Medjo-Me-Zengue
Italian, Spanish & Polish language, Careers, Genealogy
122 000 vols; 165 curr per; 72 microforms; 1 246 av-mat; 5 147 sound-rec; 197 digital data carriers
libr loan; ALA, ABA   37881

Adrian Public Library, 143 E Maumee St, *Adrian*, MI 49221-2773
T: +1 517 2652265; Fax: +1 517 2658847; URL: www.adrian.lib.mi.us
1868; Carol A. Souchock
80 000 vols; 145 curr per
libr loan   37882

Lenawee County Library, 4459 W US Rte 223, *Adrian*, MI 49221-1294; P.O. Box 609, Ridgeway
T: +1 517 2631011; Fax: +1 517 2637109; URL: www.lenawee.lib.mi.us
1935; Teresa Calderone
7 branch libs
130 000 vols; 503 curr per   37883

Agawam Public Library, 750 Cooper St, *Agawam*, MA 01001
T: +1 413 7891550; Fax: +1 413 7891552; URL: www.agawamlibrary.org
1979; Judith M. Clini
124 000 vols; 243 curr per   37884

County of Los Angeles Public Library – Agoura Hills Library, 29901 Ladyface Ct, *Agoura Hills*, CA 91301
T: +1 818 8892278; Fax: +1 818 9915019; URL: www.colapublib.org/libs/lasvirgenes
Raya Sagi
115 000 vols; 152 curr per   37885

Hawaii State Public Library System – Aiea Public Library, 99-143 Moanalua Rd, *Aiea*, HI 96701-4009
T: +1 808 4837333; Fax: +1 808 4837336
Arlene Ching
77 000 vols; 64 curr per   37886

Aiken-Bamberg-Barnwell-Edgefield Regional Library System, 314 Chesterfield St SW, *Aiken*, SC 29801-7171
T: +1 803 6427575; Fax: +1 803 6427597
1958; Mary Jo Dawson
South Carolina Coll; 14 branch libs
233 000 vols; 367 curr per   37887

Akron-Summit County Public Library, 60 S High St, *Akron*, OH 44326
T: +1 330 6439010; Fax: +1 330 6439160; URL: www.akronlibrary.org
1874; David Jennings
17 branch libs
1 641 000 vols; 1 838 curr per
libr loan   37888

Akron-Summit County Public Library – Ellet, 2470 E Market St, *Akron*, OH 44312
T: +1 330 7842019; Fax: +1 330 7846692
Anita Marky
54 000 vols   37889

Akron-Summit County Public Library – Fairlawn-Bath, 3101 Smith Rd, *Akron*, OH 44333
T: +1 330 6664888; Fax: +1 330 6668741
Jill Stroud
56 000 vols   37890

Akron-Summit County Public Library – Goodyear, 60 Goodyear Blvd, *Akron*, OH 44305-4487
T: +1 330 7847522; Fax: +1 330 7846599
Deborah Catrone
50 000 vols   37891

Akron-Summit County Public Library – Kenmore, 2200 14th St SW, *Akron*, OH 44314-2302
T: +1 330 7456126; Fax: +1 330 7459947
Kathy Forsthoffer
53 000 vols   37892

Akron-Summit County Public Library – Northwest Akron, 1720 Shatto Ave, *Akron*, OH 44313
T: +1 330 8361081; Fax: +1 330 8361574
Gladys Rossi
59 000 vols   37893

Albert L. Scott Library, 100 Ninth St NW, *Alabaster*, AL 35007-9172
T: +1 205 6646822; Fax: +1 205 6646839; URL: www.shelbycounty-al.org
Nan Abbott
American Libs, Libr Journal, Booklist, Kirkus Review, School Libr Journal
51 000 vols; 128 curr per; 1 658 av-mat
                                   37894

Alameda Free Library, 1550 Oak St, *Alameda*, CA 94501-2932
T: +1 510 7477777; Fax: +1 510 8651230; URL: www.alamedafree.org
1877; Jane Chisaki
Asian Languages, Chinese & Japanese Language Coll, Bks on Tape, Alameda Hist; 2 branch libs
179 000 vols; 422 curr per; 5 000 maps; 4 000 microforms; 3 500 sound-rec; 10 digital data carriers; 90 High Interest/Low Vocabulary Bks, 60 Bks on Deafness & Sign Lang, 1 720 Large Print Bks
libr loan; ALA, CLA   37895

Alameda County Library – Albany Branch, 1247 Marin Ave, *Albany*, CA 94706-1796
T: +1 510 5263720; Fax: +1 510 5268754
Ronnie Davis
80 000 vols; 158 curr per   37896

Albany Public Library, 1390 Waverly Dr SE, *Albany*, OR 97322
T: +1 541 9177582; Fax: +1 541 9177586; URL: library.cityofalbany.net
1907; Ed Gallagher
1 branch libr
133 000 vols; 270 curr per; 1 219 sound-rec; 30 digital data carriers; 3 202 Talking Bks
libr loan   37897

Albany Public Library, 161 Washington Ave, *Albany*, NY 12210
T: +1 518 4274300;
Fax: +1 518 4493386; URL: www.albanypubliclibrary.org
1833; Jeffrey Cannell
Business, Management; Hist Coll; 5 branch libs
300 000 vols; 3 472 curr per; 29 000 microforms; 35 623 sound-rec; 100 digital data carriers; 2 020 Large Print Bks
libr loan   37898

Dougherty County Public Library, 300 Pine Ave, *Albany*, GA 31701-2533
T: +1 229 4203200; Fax: +1 229 4203215; URL: www.docolib.org
1905; Teresa Cole
5 branch libs
318 000 vols; 118 curr per
libr loan   37899

Albert Lea Public Library, 211 E Clark St, *Albert Lea*, MN 56007
T: +1 507 3774350; Fax: +1 507 3774339; URL: www.city.albertlea.org
1897; Peggy Havener
Obituary Index to Local Newspaper
71 000 vols; 300 curr per   37900

Shelter Rock Public Library, 165 Searingtown Rd, *Albertson*, NY 11507
T: +1 516 2487343; Fax: +1 516 2484897; E-mail: shelterrock@srpl.org; URL: www.nassaulibrary.org/shelter
1962; Andrea Meluskey
149 000 vols; 474 curr per; 16 800 av-mat; 12 870 sound-rec; 1 020 digital data carriers; 1 690 Large Print Bks
                                   37901

Albertville Public Library, 200 Jackson St, P.O. Box 430, *Albertville*, AL 35950
T: +1 256 8918290; Fax: +1 256 8918295; E-mail: lrowell@albertvillelibrary.org; URL: www.albertvillelibrary.org
1963; Lisa Rowell
Civil War Coll, Rare Books Room
85 000 vols; 100 curr per   37902

Albion District Library, 501 S Superior St, *Albion*, MI 49224
T: +1 517 6293993; Fax: +1 517 6295354; E-mail: info@albionlibrary.org; URL: www.albionlibrary.org
1919; Colleen Verge
Spanish Coll
50 000 vols; 149 curr per   37903

Noble County Public Library, Central Library, 813 E Main St, *Albion*, IN 46701
T: +1 219 6367197; Fax: +1 219 6363321; URL: www.nobleco.lib.in.us/central_library.html
1914; Sandy Petrie
2 branch libs
92 000 vols; 210 curr per   37904

Alexandria Library, 5005 Duke St, *Alexandria*, VA 22304
T: +1 703 5195900; Fax: +1 703 5195916; URL: www.alexandria.lib.va.us
1794; Rose T. Dawson
7 branch libs
266 000 vols; 704 curr per; 4 000 e-books; 14 910 av-mat; 14 500 Talking Bks
libr loan   37905

Douglas County Library, 720 Fillmore St, *Alexandria*, MN 56308-1763
T: +1 320 7623014; Fax: +1 320 7623036; E-mail: library@douglascounty.lib.mn.us; URL: www.douglascountylibrary.org
1878; Patricia Conroy
Kensington Runestone
60 000 vols; 176 curr per   37906

Rapides Parish Library, 411 Washington St, *Alexandria*, LA 71301-8338
T: +1 318 4456436; Fax: +1 318 4456478; URL: www.rpl.org
1942; Stephen L. Rogge
9 branch libs
315 000 vols; 9 067 av-mat; 11 600 sound-rec; 366 High Interest/Low Vocabulary Bks, 5 782 Large Print Bks, 145 Bks on Deafness & Sign Lang
                                   37907

Algonquin Area Public Library District, 2600 Harnish Dr, *Algonquin*, IL 60102-5900
T: +1 847 6594343; Fax: +1 847 4589359; URL: www.aapld.org
1921; Randall Vlcek
182 000 vols; 404 curr per; 13 000 e-books   37908

Alhambra Public Library, 410 W Main St, *Alhambra*, CA 91801-3432
T: +1 626 5705028; Fax: +1 626 4571104; E-mail: refdesk@alhambralibrary.org; URL: alhambralibrary.org
Carmen M. Hernandez
158 000 vols; 213 curr per   37909

Alice Public Library, 401 E Third St, *Alice*, TX 78332
T: +1 361 6649506; Fax: +1 361 6683248; URL: www.youseemore.com/alice/default.asp
1932; Alicia Salinas
Spanish language, Texas hist
99 000 vols; 175 curr per   37910

B. F. Jones Memorial Library, Aliquippa District Library Center, 663 Franklin Ave, *Aliquippa*, PA 15001-3736
T: +1 724 3752900; Fax: +1 724 3753274; E-mail: bfjones@aliquippa.lib.pa.us; URL: www.aliquippa.lib.pa.us
1926; Mary Elizabeth Colombo
LPDR for Nuclear Reg Com for Beaver Valley I & II Power Stations; PA Airhelp Resource Ctr
66 000 vols; 181 curr per
libr loan   37911

Orange County Public Library – Aliso Viejo Branch, One Journey, *Aliso Viejo*, CA 92656-3333
T: +1 949 3601730; Fax: +1 949 3601728
Hilary Keith
61 000 vols; 57 curr per 37912

Allen Public Library, 300 N Allen Dr, *Allen*, TX 75013
T: +1 214 5094900; Fax: +1 214 5094950; URL: www.allenlibrary.org
1967; Jeff Timbs
124 000 vols; 150 curr per 37913

Allen Park Public Library, 8100 Allen Rd, *Allen Park*, MI 48101
T: +1 313 3812425; Fax: +1 313 3812124; URL: www.allen-park.lib.mi.us
1927; Sandi Blakney
80 000 vols; 220 curr per 37914

Allendale-Hampton-Jasper Regional Library, Allendale County Library, 158 McNair St, *Allendale*, SC 29810-0280; P.O. Box 280, Allendale, SC 29810-0280
T: +1 803 5843513; Fax: +1 803 5848134; URL: www.ahjlibrary.org
1905; Beth McNeer
61 000 vols 37915

Allentown Public Library, 1210 Hamilton St, *Allentown*, PA 18102
T: +1 610 8202400; Fax: +1 610 8200640; URL: www.allentownpl.org
1912; Kathryn Stephanoff
1 branch libr
221 000 vols; 580 curr per
libr loan; ALA 37916

Parkland Community Library, 4422 Walbert Ave, *Allentown*, PA 18104-1619
T: +1 610 3981333; Fax: +1 610 3983538; E-mail: info@parklandlibrary.org; URL: parklandlibrary.org
1973; Karen Gartner
Jobs, Ocupations, Large print; Local Hist containing most of the Publs of the Penna-German Society (formerly Penna-German Folklore Society), Birdsbiro, PA
73 000 vols; 118 curr per 37917

Rodman Public Library, 215 E Broadway St, *Alliance*, OH 44601-2694
T: +1 330 8212665; Fax: +1 330 8215053; URL: www.rodmanlibrary.com
1900; Madge Engle
1 branch libr
169 000 vols; 589 curr per; 59 000 microforms; 14 479 av-mat; 28 000 sound-rec; 1 361 digital data carriers; 6 300 Large Print Bks
libr loan 37918

Boston Public Library – Honan-Allston Branch, 300 N Harvard St, *Allston*, MA 02134
T: +1 617 7876313; Fax: +1 617 8592185
Sarah Markell
50 000 vols; 65 curr per; 300 av-mat 37919

Alma Public Library, 351 N Court, *Alma*, MI 48801-1999
T: +1 989 4633966; Fax: +1 989 4665901; E-mail: ill@alma.lib.mi.us; URL: www.alma.lib.mi.us
1909; Bryan E. Dinwoody
Republic Truck Photogr Coll
69 000 vols; 158 curr per; 2 061 av-mat 37920

Alpena County George N. Fletcher Library, 211 N First St, *Alpena*, MI 49707
T: +1 989 3566188; Fax: +1 989 3562765; URL: www.alpenalibrary.org
1967
Adult Literacy, Cooperating Coll for the Foundation of New York, Education Info Ctr (Job Launch) Educational Media, Genealogy Ctr, Michigan Coll
82 000 vols; 240 curr per; 1 400 av-mat; 899 sound-rec; 86 digital data carriers; 123 art repros
libr loan; ALA 37921

Atlanta-Fulton Public Library System – Alpharetta Library, 238 Canton St, *Alpharetta*, GA 30004
T: +1 770 7402425; Fax: +1 770 7402427
Leona Bolch
79 000 vols 37922

Atlanta-Fulton Public Library System – Dr. Robert E. Fulton Regional at Ocee, 5090 Abbotts Bridge Rd, *Alpharetta*, GA 30005-4601
T: +1 770 3608897; Fax: +1 770 3608892
Gayle Holloman
129 000 vols 37923

Atlanta-Fulton Public Library System – Northeast-Spruill Oaks Regional Library, 9560 Spruill Rd, *Alpharetta*, GA 30022
T: +1 770 3608820; Fax: +1 770 3608823
Carla Burton
192 000 vols 37924

Alpine Public Library, 203 N Seventh St, *Alpine*, TX 79830
T: +1 432 8372621; Fax: +1 432 8372501; E-mail: alpinepl@sbcglobal.net; URL: alpinepubliclibrary.org
1947; Anitra J. Clausen
SouthWest coll (3 106 vols)
50 000 vols; 263 mss; 28 govt docs; 47 music scores; 87 av-mat; 5 270 sound-rec; 16 digital data carriers; 732 audio bks, 87 videos 37925

Alsip-Merrionette Park Public Library District, 11960 S Pulaski Rd, *Alsip*, IL 60803-1197
T: +1 708 3715666; Fax: +1 708 3715672; E-mail: ampl@sslic.net; URL: www.alsip-mp.lib.il.us
1973; Ruthann Swanson
118 000 vols; 276 curr per 37926

Altadena Library District, 600 E Mariposa St, *Altadena*, CA 91001
T: +1 626 7980833; Fax: +1 626 7985351; E-mail: ref@altadenalibrary.org; URL: altadenalibrary.org
1908; Barbara Pearson
Altadena Hist Coll; 1 branch libr
142 000 vols; 230 curr per; 350 microforms; 4 705 av-mat; 2 500 sound-rec; 1 230 Large Print Bks
libr loan 37927

Hayner Public Library District, 326 Belle St, *Alton*, IL 62002; mail address: 401 State St, Alton, IL 62002-6137
T: +1 618 4620677; Fax: +1 618 4624919; E-mail: main.library@haynerlibrary.org; URL: www.haynerlibrary.org
1891; Jeffrey A. Owen
Hist; Illinois Hist; 2 branch libs
208 000 vols; 374 curr per 37928

Altoona Area Public Library, 1600 Fifth Ave, *Altoona*, PA 16602-3693
T: +1 814 9460417; Fax: +1 814 9463230; URL: www.altoonalibrary.org
1927; Deborah Weakland
Adult literacy, railroad hist; railroad photos
135 000 vols; 220 curr per; 6 978 av-mat; 7 307 sound-rec
libr loan; OCLC 37929

Modoc County Library, 212 W Third St, *Alturas*, CA 96101
T: +1 530 2336340; Fax: +1 530 2333375; URL: www.infopeople.org/modoc
1906; Cheryl Baker
California Indian Libr, Modoc County Hist Coll; 4 branch libs
63 000 vols; 65 curr per 37930

Altus Public Library, 421 N Hudson, *Altus*, OK 73521
T: +1 580 4772890; Fax: +1 580 4773626; E-mail: spls@spls.lib.ok.us; URL: www.spls.lib.ok.us
1936; Donna Smith
60 000 vols; 75 curr per 37931

Southern Prairie Library System, 421 N Hudson, *Altus*, OK 73521; P.O. Drawer U, Altus
T: +1 580 4772890; Fax: +1 580 4773626
1973; Katherine Hale
English-As-A-Second Language, Literacy
52 000 vols; 76 curr per
libr loan 37932

Alva Public Library, 504 Seventh St, *Alva*, OK 73717
T: +1 580 3271833; Fax: +1 580 3275329
Larry Thorne
Daughters of the American Revolution (DAR) Coll
56 000 vols; 147 curr per; 200 av-mat 37933

Brazoria County Library System – Alvin Branch, 105 S Gordon, *Alvin*, TX 77511
T: +1 281 3884300; Fax: +1 281 3884305; E-mail: alvin@bcls.lib.tx.us
Danna Kay Wilson
66 000 vols 37934

Amarillo Public Library, 413 E Fourth Ave, *Amarillo*, TX 79101; P.O. Box 2171, Amarillo, TX 79105-2171
T: +1 806 3783053; Fax: +1 806 3789327; URL: www.amarillolibrary.org
1902; Donna Littlejohn
Southwestern Hist (William H Bush & Laurence J Fitzsimon Coll), Genealogy; 4 branch libs
722 000 vols; 51 curr per 37935

Amarillo Public Library – East Branch, 2232 E 27th St, *Amarillo*, TX 79103
T: +1 806 3421589; Fax: +1 806 3421591
Jan Tortoriello
121 000 vols 37936

Amarillo Public Library – North Branch, 1504 NE 24th St, *Amarillo*, TX 79107
T: +1 806 3817931; Fax: +1 806 3817929
Zetta Riles
150 000 vols 37937

Amarillo Public Library – Northwest Branch, 6100 W 9th, *Amarillo*, TX 79106-0700
T: +1 806 3592035; Fax: +1 806 3592037
Cindi Dockery
212 000 vols 37938

Amarillo Public Library – Southwest Branch, 6801 W 45th St, *Amarillo*, TX 79109
T: +1 806 3592094; Fax: +1 806 3592096
Valisa McHugh
548 000 vols 37939

American Fork City Library, 64 S 100 East, *American Fork*, UT 84003
T: +1 801 7633070; Fax: +1 801 7633073; URL: www.afcity.org
Sheena Parker
90 000 vols; 106 curr per 37940

Amery Public Library, 801 Keller Ave S, *Amery*, WI 54001-1096
T: +1 715 2689340; Fax: +1 715 2688659; E-mail: amerypl@spacestar.net
Barbara Sorenson
62 000 vols; 65 curr per 37941

Ames Public Library, 515 Douglas Ave, *Ames*, IA 50010
T: +1 515 2395630; Fax: +1 515 2324571; URL: www.amespubliclibrary.org
1903; Art Weeks
Photography
200 000 vols; 150 curr per; 50 High Interest/Low Vocabulary Bks, 20 Bks on Deafness & Sign Lang
libr loan 37942

Amesbury Public Library, 149 Main St, *Amesbury*, MA 01913
T: +1 978 3888148; Fax: +1 978 3882662; E-mail: mam@mvic.org; URL: www.amesburylibrary.org
1856; Patty DiTullo
Amesbury Carriage Hist Mat, Charles H Davis Painting Coll, John Greenleaf Whittier Mat
72 000 vols; 120 curr per 37943

Amherst Public Library, 350 John James Audubon Pkwy, *Amherst*, NY 14228
T: +1 716 6894922; Fax: +1 716 6896116
1842; Mary F. Bobinski
4 branch libs
255 000 vols 37944

Amherst Public Library, 221 Spring St, *Amherst*, OH 44001
T: +1 440 9884230; Fax: +1 440 9884115; E-mail: amherstlibrary@hotmail.com; URL: www.amherst.lib.oh.us
1906; Robin Wood
57 000 vols; 142 curr per 37945

Amherst Town Library, 14 Main St, *Amherst*, NH 03031-2930
T: +1 603 6732288; Fax: +1 603 6726063; E-mail: library@amherst.lib.nh.us; URL: www.amherst.lib.nh.us
1891; Amy LaPointe
73 000 vols; 136 curr per 37946

Jones Library, Inc, 43 Amity St, *Amherst*, MA 01002-2285
T: +1 413 2593090; Fax: +1 413 2564096; E-mail: info@joneslibrary.org; URL: www.joneslibrary.org
1919; Bonnie J. Isman
English language; Early Textbks & Children's Bks, Emily Dickinson Coll, Genealogy, Harlan Fiske Stone Coll, Julius Lester Coll, Ray Stannard Baker Coll, Robert Frost Coll, Sidney Waugh Coll; 2 branch libs
209 000 vols; 316 curr per 37947

Tangipahoa Parish Library, 200 E Mulberry St, *Amite*, LA 70422
T: +1 985 7487559; Fax: +1 985 7482812; URL: www.tangipahoa.lib.la.us
1944; Mary Battles
6 branch libs
195 000 vols; 622 curr per 37948

Tangipahoa Parish Library – Amite Branch, 739 W Oak St, *Amite*, LA 70422
T: +1 985 7487151; Fax: +1 985 7485476
Sherri Alford
52 000 vols; 86 curr per 37949

Amityville Public Library, Oak & John Sts, *Amityville*, NY 11701
T: +1 631 2640567; Fax: +1 631 2642000; URL: www.suffolk.lib.ny.us/libraries/amty
1906; Nora Schual
112 000 vols; 255 curr per 37950

Anacortes Public Library, 1220 Tenth St, *Anacortes*, WA 98221-1922
T: +1 360 2931910; Fax: +1 360 2931929; URL: library.cityofanacortes.org
1911; Cynthia Harrison
71 000 vols; 300 curr per 37951

Anaheim Public Library, 500 W Broadway, *Anaheim*, CA 92805-3699
T: +1 714 7651880; Fax: +1 714 7651730; URL: www.anaheim.net/library.html
1901; Carol Stone
Anaheim Hist Coll; 3 branch libs
461 000 vols 37952

Anaheim Public Library – Canyon Hills, 400 Scout Trail, *Anaheim*, CA 92807-4763
T: +1 714 9747630; Fax: +1 714 9981468
Karen Gerloff
110 000 vols 37953

Anaheim Public Library – Elva L. Haskett Branch, 2650 W Broadway, *Anaheim*, CA 92804
T: +1 714 7655075
Marianne Hugo
57 000 vols 37954

Anaheim Public Library – Sunkist, 901 S Sunkist St, *Anaheim*, CA 92806-4739
T: +1 714 7653576; Fax: +1 714 7653574
Maria Luz Cayabyab
61 000 vols 37955

Anderson City, Anderson, Stony Creek & Union Townships Public Library, 111 E 12th St, *Anderson*, IN 46016-2701
T: +1 765 6412456; Fax: +1 765 6412197; E-mail: cr@and.lib.in.us; URL: www.and.lib.in.us
1891; Sarah Later
Govt Docs, Wendell Hall Music Coll, Literacy; 1 branch libr
437 000 vols; 673 curr per
libr loan 37956

Anderson County Library,
300 N McDuffie St, **Anderson,**
SC 29621-5643; P.O. Box 4047,
Anderson, SC 29622-4047
T: +1 864 2604500; Fax: +1 864
2604510; URL: www.andersonlibrary.org
1958; Faith A. Line
Foundation Ctr Cooperating Coll; 8
branch libs
336 000 vols; 1 079 curr per; 100 e-
books; 6 450 Large Print Bks   37957

Memorial Hall Library, Elm Sq,
Andover, MA 01810
T: +1 978 6238401; Fax: +1 978
6238407; E-mail: rdesk@mhl.org; URL:
www.mhl.org
1873; James E. Sutton
Law, Art, Business, Maps, Travel; Civil
War Coll
186 000 vols; 526 curr per; 9 590 av-mat
37958

Andrews County Library, 109 NW First
St, **Andrews,** TX 79714
T: +1 432 5239819; Fax: +1 432
5234570; URL: www.andrews.lib.tx.us
1950; Liz Stottlemyre
67 000 vols; 99 curr per   37959

Brazoria County Library System,
451 N Velasco, Ste 250, **Angleton,**
TX 77515; Bldg 29-A, Ste 250, 111 E
Locust St, Angleton, TX 77515-4642
T: +1 979 8641505; Fax: +1 979
8641298; URL: bcls.lib.tx.us
1941; Catherine H. Threadgill
11 branch libs
595 000 vols; 1 044 curr per   37960

**Brazoria County Library System –
Angleton Branch,** 401 E Cedar St,
Angleton, TX 77515-4652
T: +1 979 8641520; Fax: +1 979
8641518; E-mail: angleton@bcls.lib.tx.us
Layna L. Lewis
85 000 vols   37961

**Carnegie Public Library of Steuben
County,** 322 S Wayne St, **Angola,**
IN 46703
T: +1 260 6653362; Fax: +1 260
6658958; E-mail: info@steuben.lib.in.us;
URL: www.steuben.lib.in.us
1915; Sonya Dintaman
64 000 vols; 145 curr per; 250
microforms; 2 680 av-mat; 1 701
sound-rec; 561 digital data carriers;
100 slides, 245 art reproductions
libr loan; ALA   37962

Kirkendall Public Library, 1210 NW
Prairie Ridge Dr, **Ankeny,** IA 50021
T: +1 515 9656463; Fax: +1 515
9656474; URL: www.ci.ankeny.ia.us/
library
1960; Myrna L. Anderson
Large type print
65 000 vols; 275 curr per; 1 849 av-mat
37963

Ann Arbor District Library, 343 S Fifth
Ave, **Ann Arbor,** MI 48104
T: +1 734 3274295; URL: aadl.org
1856; Josie Parker
Art, Business, African-American studies,
Ethnic studies, Basic education, Music;
Foreign Language Coll, Art Prints; 4
branch libs
355 000 vols; 1 475 curr per; 16 000 e-
books; 7 240 av-mat; 51 030 sound-rec;
34 350 digital data carriers; 3 540 Large
Print Bks   37964

**Public Library Association of
Annapolis & Anne Arundel County,
Inc,** Headquarters, Five Harry S Truman
Pkwy, **Annapolis,** MD 21401
T: +1 410 2227371; Fax: +1 410
2227188; URL: www.aacpl.net
1936; Nancy J. Choice
15 branch libs
1 225 000 vols   37965

**Public Library Association of
Annapolis & Anne Arundel County,
Inc – Annapolis Area,** 1410 West St,
Annapolis, MD 21401
T: +1 410 2221750; Fax: +1 410
2221116; E-mail: ann@aacpl.net
Gloria Davis
163 000 vols   37966

**Public Library of Anniston & Calhoun
County,** 108 E Tenth St, **Anniston,**
AL 36201; P.O. Box 308, Anniston
T: +1 256 2378501; Fax: +1 256
2380474; URL: www.anniston.lib.al.us
1965; Bonnie G. Seymour
Alabama Hist, Genealogy; Alabama
Room, Anniston Room Coll; 2 branch
libs
115 000 vols; 225 curr per   37967

Ansonia Library, 53 S Cliff St, **Ansonia,**
CT 06401-1909
T: +1 203 7346275; Fax: +1 203
7324551; URL: www.biblio.org/ansonia
1896; Joyce Ceccarelli
Daughters of American Revolution
84 000 vols; 144 curr per; 115 e-books;
592 av-mat   37968

Antigo Public Library, 617 Clermont St,
Antigo, WI 54409-1894
T: +1 715 6233724; Fax: +1 715
6272317; E-mail: antigopl@wvls.lib.wi.us;
URL: wvls.lib.wi.us/AntigoPL
1900; Cynthia Taylor
72 000 vols; 180 curr per; 522 av-mat;
605 sound-rec   37969

Antioch District Library, 757 Main St,
Antioch, IL 60002
T: +1 847 3950874; Fax: +1 847
3955399
1921; Kathy LaBuda
Parenting, Cookung, Crafts
138 000 vols; 138 curr per   37970

**Contra Costa County Library –
Antioch Community Library,** 501
W 18th St, **Antioch,** CA 94509-2292
T: +1 925 7579224; Fax: +1 925
4278540
Kathy Middleton
80 000 vols   37971

Nashville Public Library – Southeast,
2325 Hickory Highlands Dr, **Antioch,**
TN 37013-2101
T: +1 615 8625871; Fax: +1 615
8625756
Edward Tood
78 000 vols   37972

**Wake County Public Library System
– Eva H. Perry Regional,** 2100
Shepherd's Vineyard Dr, **Apex,**
NC 27502
T: +1 919 3872100; Fax: +1 919
3874320
Christina H. Piscitello
204 000 vols   37973

**Orange County Library District –
North Orange,** 1211 E Semoran Blvd,
Apopka, FL 32703
Carolyn Rosenblum
109 000 vols   37974

Appleton Public Library, 225 N Oneida
St, **Appleton,** WI 54911-4780
T: +1 920 8326170; Fax: +1 920
8326182; URL: www.apl.org
1897; Terry P. Dawson
403 000 vols; 600 curr per
ALA   37975

**Santa Cruz City-County Library
System Headquarters – Aptos
Branch,** 7695 Soquel Dr, **Aptos,**
CA 95003-3899
T: +1 831 4205309; Fax: +1 831
6614828
Julie Richardson
54 000 vols   37976

Arab Public Library, 325 Second St
NW, **Arab,** AL 35016-1999
T: +1 256 5863366; Fax: +1 256
5865638; E-mail: library@arabcity.org
Kathy Handle
52 000 vols; 40 curr per   37977

Arcadia Public Library, 20 W Duarte
Rd, **Arcadia,** CA 91006
T: +1 626 8214326; Fax: +1 626
4478050; E-mail: ref247@ci.arcadia.ca.us;
URL: library.ci.arcadia.ca.us
1920; Janet Sporleder
166 000 vols; 253 curr per; 2 e-books; 2
av-mat
libr loan   37978

Bienville Parish Library, 2768 Maple St,
Arcadia, LA 71001-3699
T: +1 318 2637410; Fax: +1
318 2637428; E-mail:
admin.g1bv@pelican.state.lib.la.us; URL:
www.bienville.lib.la.us
1964; Peggy Walls
Large Print Coll, Louisiana Mat; 1 branch
libr
88 000 vols; 226 curr per   37979

Arcanum Public Library, 101 W North
St, **Arcanum,** OH 45304-1126
T: +1 937 6928484; Fax: +1
937 6928916; E-mail:
library@arcanumpubliclibrary.org; URL:
www.arcanumpubliclibrary.org
1911; Marilyn Walden
54 000 vols; 204 curr per   37980

Ardmore Public Library, 320 E St NW,
Ardmore, OK 73401
T: +1 580 2239524; Fax: +1 580
2232033; URL: www.ardmorelibrary.org
1906; Daniel Gibbs
Eliza Cruce Hall Doll Museum Coll,
McGalliard Local Hist Coll, Tomlinson
Geological Coll
79 000 vols   37981

Chickasaw Regional Library System,
601 Railway Express, **Ardmore,**
OK 73401
T: +1 580 2233164; Fax: +1 580
2233280; URL: www.regional-sys.lib.ok.us
1960; Lynn McIntosh
7 branch libs
209 000 vols; 307 curr per   37982

Lower Merion Library System,
75 E Lancaster Ave, **Ardmore,**
PA 19003-2388
T: +1 610 6456110; Fax: +1 610
6498835; E-mail: lmls@lmls.org; URL:
www.lmls.org
1935; Christine Steckel
Art & architecture, Music, Horticulture
425 000 vols; 350 curr per   37983

Arkadelphia Public Library, 609 Caddo
St, **Arkadelphia,** AR 71923
T: +1 870 2462271; Fax: +1 870
2464189; URL: www.clark-library.com/
arkadelphiaLOC.php
1897
1 branch libr
90 000 vols; 80 curr per   37984

**Arlington County Department of
Libraries,** Arlington Public Library, 2100
Clarendon Blvd, Ste 406, **Arlington,**
VA 22201
T: +1 703 2283346; Fax: +1 703
2283354; E-mail: libraries@arlingtonva.us;
URL: www.arlingtonva.us/departments/
libraries/librariesmain.aspx
1937; Diane Kresh
Children's Illustrators, Virginiana Coll
580 000 vols; 24 000 e-books; 23 086
Audio Bks, 52 Electronic Media &
Resources
libr loan   37985

**Arlington County Department of
Libraries Central Library,** 1015 N
Quincy St, **Arlington,** VA 22201-4661
T: +1 703 2285940; Fax: +1 703
2285962; E-mail: libraries@arlingtonva.us;
URL: www.arlingtonva.us
Chang Lui
Vietnamese;College (Career
Coll);Children's Illustrators
300 000 vols   37986

**Arlington County Department of
Libraries – Columbia Pike,** 816 S
Walter Reed Dr, **Arlington,** VA 22204
T: +1 703 2285710; Fax: +1 703
2285559
Jason Rodgers
Vocational Coll
69 000 vols   37987

Arlington Public Library System,
George W. Hawkes Central Library, 101
E Abram St, MS 10-0100, **Arlington,**
TX 76010-1183
T: +1 817 4596900; Fax: +1
817 4596936; URL: www.pub-
lib.ci.arlington.tx.us
1923; Cary Siegfried
5 branch libs
570 000 vols; 842 curr per; 14 154
av-mat; 14 160 Audio Bks, 8 160 Large
Print Bks   37988

**Arlington Public Library System –
East Arlington,** 1624 New York Ave,
Arlington, TX 76010-4795
T: +1 817 2753321; Fax: +1 817
7950726
Marc Marchand
Children's Learning Center
Coll;GED/ESL;Multicultural Coll
62 000 vols; 86 curr per; 2 522 av-mat
37989

**Arlington Public Library System –
Lake Arlington,** 4000 W Green Oaks
Blvd, **Arlington,** TX 76016-4442
T: +1 817 4783762; Fax: +1 817
5619823
Debbie Potts
74 000 vols; 88 curr per; 2 360 av-mat
37990

**Arlington Public Library System
– Northeast,** 1905 Brown Blvd,
Arlington, TX 76006-4605
T: +1 817 2775573; Fax: +1 817
2768649
Marc Marchand
70 000 vols; 106 curr per; 2 226 av-mat
37991

**Arlington Public Library System –
Woodland West,** 2837 W Park Row
Dr, **Arlington,** TX 76013-2261
T: +1 817 2775265; Fax: +1 817
7954741
Debbie Potts
51 000 vols; 68 curr per; 1 988 av-mat
37992

Robbins Library, 700 Massachusetts Ave,
Arlington, MA 02476
T: +1 781 3163200; Fax: +1 781
3163209; E-mail: arlington@minlib.net;
URL: www.robbinslibrary.org
1835; Maryellen Remmert-Loud
Robbins Print Coll, etchings, lithogr,
prints; 1 branch libr
205 000 vols; 296 curr per   37993

Sno-Isle Libraries Arlington Branch,
135 N Washington Ave, **Arlington,**
WA 98223
T: +1 360 4353033; Fax: +1 360
4353854
Kathy Bullene
52 000 vols   37994

Arlington Heights Memorial Library,
500 N Dunton Ave, **Arlington Heights,**
IL 60004-5966
T: +1 847 3920100; Fax: +1 847
3920136; URL: www.ahml.info
1926; Paula M. Moore
Business, Local Hist/Genealogy, Large
Type, Foreign Language; 1 bookmobile
331 000 vols; 1 114 curr per; 19 794
av-mat; 28 166 sound-rec; 13 783 digital
data carriers
libr loan; ALA, ILA, PLA, ULC   37995

North Castle Public Library, 19
Whippoorwill Rd E, **Armonk,** NY 10504
T: +1 914 2733887; Fax: +1 914
2735572; E-mail: armref@wlsmail.org;
URL: www.northcastlelibrary.org
1938; M. Cristina Ansnes
63 670 vols; 150 curr per   37996

Artesia Public Library, 306
W Richardson Ave, **Artesia,**
NM 88210-2499
T: +1 505 7464252, 4692;
Fax: +1 505 7463075; E-mail:
apublib@pvtnetwworks.net
1902
60 000 vols; 166 curr per
libr loan   37997

Asbury Park Free Public Library, 500
First Ave, **Asbury Park,** NJ 07712
T: +1 732 7744221; Fax: +1
732 9886101; E-mail: apl-
info@asburyparklibrary.org; URL:
www.asburyparklibrary.org
1878; Robert W. Stewart
114 000 vols; 406 curr per   37998

Randolph Public Library, Headquarters,
201 Worth St, **Asheboro,** NC 27203
T: +1 336 3186800; Fax: +1 336
3186823; URL: www.randolphlibrary.org
1940; Suzanne Tate
Micro Film Coll; 7 branch libs
245 000 vols; 672 curr per; 4 820 av-mat;
320 sound-rec; 700 digital data carriers;
5 700 Large Print Bks   37999

Buncombe County Public Libraries, 67 Haywood St, **Asheville**, NC 28801
T: +1 828 2504711; Fax: +1 828 2504746; E-mail: library@buncombecounty.org; URL: www.buncombecounty.org
1879; Edward J. Sheary
Thomas Wolfe Coll; 11 branch libs, 1 bookmobile
495 000 vols; 830 curr per   38000

Ashland Public Library, 224 Claremont Ave, **Ashland**, OH 44805
T: +1 419 2898188; Fax: +1 419 2818552; URL: www.ashland.lib.oh.us
1893; Pamela A. Jordan
130 000 vols; 250 curr per   38001

Boyd County Public Library, 1740 Central Ave, **Ashland**, KY 41101
T: +1 606 3290518; Fax: +1 606 3290578; URL: www.thebookplace.org
1935; Debbie Cosper
Arnold Hanners Photo Coll, Art Gallery, Laguage Tape Coll; 2 branch libs
175 000 vols; 500 curr per; 300 e-books   38002

Ashtabula County District Library, 335 W 44th St, **Ashtabula**, OH 44004-6897
T: +1 440 9979341; Fax: +1 440 9981198; E-mail: ashref@oplin.org; URL: www.acdl.info
1813; Donna M. Wall
1 branch libr
180 000 vols; 326 e-books; 9 250 av-mat; 4 480 sound-rec; 1 900 digital data carriers   38003

Pitkin County Library, 120 N Mill St, **Aspen**, CO 81611
T: +1 970 4291900; Fax: +1 970 9253935; E-mail: libraryinfo@co.pitkin.co.us; URL: www.pitcolib.org
1940; Kathy Chandler
Music Records & Scores, Aspen Newspapers, 1888-present
82 000 vols; 202 curr per   38004

Astoria Public Library, 450 Tenth St, **Astoria**, OR 97103
T: +1 503 3257323; URL: astorialibrary.org
1892; Jane Tucker
53 000 vols; 30 curr per   38005

Atchison Public Library, 401 Kansas Ave, **Atchison**, KS 66002-2495
T: +1 913 3671902; Fax: +1 913 3672717; URL: www.atchisonlibrary.org
P. J. Capps
Hist, Music
63 000 vols; 80 curr per   38006

Camden County Library System – South County Regional Branch, 35 Coopers Folly Rd, **Atco**, NJ 08004
T: +1 856 7532537; Fax: +1 856 7537289; E-mail: southcounty@camden.lib.nj.us
Nancy Bennett
77 000 vols; 150 curr per   38007

Athens Regional Library System – Athens-Clarke County Library, 2025 Baxter St, **Athens**, GA 30606-6331
T: +1 706 6133650; Fax: +1 706 6133660; URL: www.clarke.public.lib.ga.us
1936; Kathryn S. Ames
11 branch libs
203 000 vols; 161 curr per; 10 940 av-mat; 5 680 sound-rec; 291 digital data carriers; 7 090 Large Print Bks   38008

Edward Gauche Fisher Public Library, 1289 Ingleside Ave, **Athens**, TN 37303
T: +1 423 7457782; Fax: +1 423 7451763; URL: www.fisherlibrary.org
1969; Beth Allen Mercer
60 000 vols; 70 curr per   38009

Fort Loudoun Regional Library Center, 718 George St NW, **Athens**, TN 37303-2214
T: +1 423 7455194; Fax: +1 423 7458086; URL: www.state.tn.us/sos/statelib/p&d/fortloudoun/
1939; Lynette Sloan
1 bookmobile
265 000 vols   38010

Henderson County, Clint W. Murchison Memorial Library, 121 S Prairieville, **Athens**, TX 75751
T: +1 903 6777295; Fax: +1 903 6777275; E-mail: librarian@co.henderson.tx.us; URL: www.hendersoncountylibrary.org
1972; Terry Warren
53 000 vols; 57 curr per   38011

Nelsonville Public Library – Athens Public, 30 Home St, **Athens**, OH 45701
T: +1 740 5924272; Fax: +1 740 5944204
Marilyn Zwayer
84 000 vols   38012

Athol Public Library, 568 Main St, **Athol**, MA 01331
T: +1 978 2499515; Fax: +1 978 2497636; E-mail: info@athollibrary.org; URL: athollibrary.org
1882; Debra Blanchard
Coll of Local Art Originals
53 000 vols; 58 curr per; 1843 av-mat   38013

Atlanta-Fulton Public Library System, One Margaret Mitchell Sq NW, **Atlanta**, GA 30303-1089
T: +1 404 7304636; Fax: +1 404 7301990; E-mail: comments@co.fulton.ga.us; URL: www.af.public.lib.ga.us
1901; John F. Szabo
Margaret Mitchell Coll, African American Hist & Culture, Children's Lit; 33 branch libs
2 220 000 vols; 6 000 curr per
libr loan   38014

Atlanta-Fulton Public Library System – Auburn Avenue Research Library on African-American Culture & History, 101 Auburn Ave NE, **Atlanta**, GA 30303
T: +1 404 7304001; Fax: +1 404 7305879
Akilah Nosakhere
54 000 vols   38015

Atlanta-Fulton Public Library System – Buckhead Library, 269 Buckhead Ave NE, **Atlanta**, GA 30305
T: +1 404 8143500; Fax: +1 404 8143503
Nancy Powers
114 000 vols   38016

Atlanta-Fulton Public Library System – Cleveland Avenue, 47 Cleveland Ave, **Atlanta**, GA 30315
T: +1 404 7624116; Fax: +1 404 7624118
Gloria Dennis
56 000 vols   38017

Atlanta-Fulton Public Library System – Northside Library, 3295 Northside Pkwy NW, **Atlanta**, GA 30327
T: +1 404 8143508; Fax: +1 404 8143511
N.N.
65 000 vols   38018

Atlanta-Fulton Public Library System – Peachtree Library, 1315 Peachtree St NE, **Atlanta**, GA 30309
T: +1 404 8857830; Fax: +1 404 8557833
Mary Silver
55 000 vols   38019

Atlanta-Fulton Public Library System – Ponce de Leon Library, 980 Ponce de Leon Ave NE, **Atlanta**, GA 30306
T: +1 404 8857820; Fax: +1 404 8857822
Bill Munro
86 000 vols   38020

Atlanta-Fulton Public Library System – Southwest Regional Library, 3665 Cascade Rd SW, **Atlanta**, GA 30331
T: +1 404 6996363; Fax: +1 404 6996381
Eugene Haston
133 000 vols   38021

Georgia Library for Accessible Services, 1150 Murphy Ave SW, **Atlanta**, GA 30310-3399
T: +1 404 7564476; Fax: +1 404 7564618; E-mail: glass@georgialibraries.org; URL: www.georgialibraries.org/public/glass.html
1931; Beverly Williams
Georgia Coll, tapes
300 000 vols; 57 000 av-mat; 57 000 Talking Bks   38022

Atlantic City Free Public Library, One N Tennessee Ave, **Atlantic City**, NJ 08401
T: +1 609 3452269; Fax: +1 609 3455570; URL: www.acfpl.org
1902; Maureen Sherr Frank
Casino Gambling; 1 branch libr
135 000 vols; 350 curr per
libr loan   38023

Atmore Public Library, 700 E Church St, **Atmore**, AL 36502
T: +1 251 3685234; Fax: +1 251 3684130; URL: www.atmorelibrary.com
1923; Cathy J. McKinley
Forestry (Atmores Industries Coll), Cancer & Heart Coll, Scout Books Coll
50 000 vols; 54 curr per   38024

Attleboro Public Library, Joseph L. Sweet Memorial, 74 N Main St, **Attleboro**, MA 02703
T: +1 508 2220157; Fax: +1 508 2263326; URL: www.sailsinc.org/attleboro/apl.asp
1885; Walter Stitt
90 000 vols; 120 curr per   38025

Auburn Public Library, 49 Spring St, **Auburn**, ME 04210
T: +1 207 3336640; Fax: +1 207 3336644; E-mail: email@auburnpubliclibrary.org; URL: www.auburnpubliclibrary.org
1890; Rosemary Waltos
62 000 vols; 150 curr per   38026

Auburn Public Library, 369 Southbridge St, **Auburn**, MA 01501
T: +1 508 8327790; Fax: +1 508 8327792; URL: www.auburnlibrary.org
1872; Joan Noonan
79 000 vols; 132 curr per; 4 082 av-mat   38027

Auburn Public Library, 749 E Thach Ave, **Auburn**, AL 36830
T: +1 334 5013190; URL: www.auburnalabama.org/library
Margie Huffman
58 000 vols; 135 curr per; 850 av-mat   38028

Auburn-Placer County Library, 350 Nevada St, **Auburn**, CA 95603-3789
T: +1 530 8864510; Fax: +1 530 8864555; URL: www.placer.ca.gov/library
1937; Elaine Paez Reed
11 branch libs
115 000 vols; 6 microforms; 5 000 av-mat; 6 900 sound-rec; 17 digital data carriers
libr loan   38029

Eckhart Public Library, 603 S Jackson St, **Auburn**, IN 46706-2298
T: +1 260 9252414; Fax: +1 260 9259376; E-mail: reference@epl.lib.in.us; URL: www.epl.lib.in.us
1910; Janelle Graber
De Kalb County Hist Coll, bks & genealogy
68 000 vols; 240 curr per   38030

Auburn Hills Public Library, 3400 E Seyburn Dr, **Auburn Hills**, MI 48326-2759
T: +1 248 3709466; Fax: +1 248 3709364; URL: www.auburn-hills.lib.mi.us
1986; Karrie Waarala
60 000 vols   38031

East Central Georgia Regional Library – Augusta-Richmond County Public Library, 902 Greene St, **Augusta**, GA 30901
T: +1 706 8212637; Fax: +1 706 8212629; E-mail: main@ecgrl.org; URL: www.ecgrl.org
1848; Gary Swint
15 branch libs
618 000 vols; 188 curr per; 2 000 e-books; 33 836 av-mat; 202 sound-rec; 823 Audio Bks   38032

East Central Georgia Regional Library – Jeff Maxwell Branch, 1927 Lumpkin Rd, **Augusta**, GA 30906
T: +1 706 7932020; Fax: +1 706 7901023; E-mail: maxwell@ecgrl.org
Linda Beck
60 000 vols; 37 curr per   38033

Lithgow Public Library, 45 Winthrop St, **Augusta**, ME 04330-5599
T: +1 207 6262415; Fax: +1 207 6262419; URL: www.lithgow.lib.me.us
1896; Elizabeth L. Pohl
58 000 vols; 105 curr per   38034

Maine Regional Library for the Blind & Physically Impaired, c/o Maine State Library, Maine State Library, 64 State House Sta, **Augusta**, ME 04333-0064
T: +1 207 2875650; Fax: +1 207 2875654; URL: www.state.me.us/msl/outreach
1972; Melora Ranney Norman
French Language, cassette
135 000 vols; 85 123 sound-rec; 19 224 large print bks   38035

Aurora Public Library, One E Benton St, **Aurora**, IL 60505-4299
T: +1 630 2644100; Fax: +1 630 8963209; URL: www.aurora.lib.il.us
1881; Eva Luckinbill
2 branch libs
397 000 vols; 8 891 curr per   38036

Aurora Public Library, Administration – Department of Library, Recreation & Cultural Services, 14949 E Alameda Pkwy, **Aurora**, CO 80012
T: +1 303 7396628; Fax: +1 303 7396586; URL: www.auroragov.org; auroralibrary.org
1929; Patti Bateman
Spanish language; 6 branch libs
506 000 vols   38037

Aurora Public Library, Eola, 555 S Eola Rd, **Aurora**, IL 60504-8992
T: +1 630 2643400; Fax: +1 630 8985220
Elizabeth Bumgarner
88 000 vols; 120 curr per   38038

Aurora Public Library District, 414 Second St, **Aurora**, IN 47001-1384
T: +1 812 9260646; Fax: +1 812 9260665; E-mail: weblib@eapld.org; URL: www.eapld.org
1901; Mary Alice Horton
54 000 vols; 181 curr per; 510 av-mat   38039

Portage County District Library – Aurora Memorial, 115 E Pioneer Trail, **Aurora**, OH 44202-9349
T: +1 330 5626502; Fax: +1 330 5622084
Cheryl Chlysta
76 000 vols   38040

Austin Public Library, 323 Fourth Ave NE, **Austin**, MN 55912-3773
T: +1 507 433-2391; Fax: +1 507 433-8787; E-mail: ahokanson@selco.info;
URL: www.austinpubliclibrary.org/
1904; Ann Hokanson
3 branch libs
90 000 vols; 324 maps
libr loan   38041

Austin Public Library, Faulk Central, 800 Guadalupe St, **Austin**, TX 78701; P.O. Box 2287, Austin, TX 78768-2287
T: +1 512 9747400; Fax: +1 512 9747403; E-mail: aplmail@ci.austin.tx.us; URL: www.ci.austin.tx.us/library
1926; Brenda Branch
Austin Hist Center; 21 branch libs
1 444 000 vols; 1 571 curr per
libr loan   38042

Austin Public Library – Little Walnut Creek, 835 W Rundberg Lane, **Austin**, TX 78758
T: +1 512 8368975; Fax: +1 512 8350361
Michael Abramov
54 000 vols   38043

Austin Public Library – Manchaca Road, 5500 Manchaca Rd, **Austin**, TX 78745
T: +1 512 4476651; Fax: +1 512 4445132
Anita Fudell
70 000 vols; 155 curr per   38044

Austin Public Library – Spicewood Springs, 8637 Spicewood Springs Rd, **Austin**, TX 78759
T: +1 512 2589070; Fax: +1 512 3314435
Nancy Toombs
Chinese (Traditional & Simplified)
55 000 vols; 110 curr per   38045

Austin Public Library – Windsor Park, 5833 Westminster Dr, **Austin**, TX 78723
T: +1 512 9280333; Fax: +1 512 9290654
D. J. Harris
54 000 vols   38046

Westbank Community Library District, 1309 Westbank Dr, **Austin**, TX 78746
T: +1 512 3143582; Fax: +1 512 3273074; E-mail: askus@westbank.lib.tx.us;
URL: www.lauraslibrary.org; www.westbank.lib.tx.us
1984; Beth Wheeler Fox
67 000 vols; 160 curr per; 15 e-books; 100 maps; 5 622 av-mat; 2 368 sound-rec; 140 digital data carriers
ALA   38047

Avon Free Public Library, 281 Country Club Rd, **Avon**, CT 06001
T: +1 860 6739712; Fax: +1 860 6756364; URL: www.avonctlibrary.info
1798; Virginia Vocelli
79 000 vols; 274 curr per; 2 420 av-mat
libr loan   38048

Avon-Washington Township Public Library, 498 N State Rd 267, **Avon**, IN 46123
T: +1 317 2724818; Fax: +1 317 2727302; E-mail: reference@avon.lib.in.us;
URL: www.avon.lib.in.ua
Laurel Setser
91 000 vols; 208 curr per   38049

Eagle Valley Library District – Avon Public Library, 200 Benchmark Rd, **Avon**, CO 81620; P.O. Box 977, Avon, CO 81620-0977
T: +1 970 9496797; Fax: +1 970 9490233
N.N.
60 000 vols; 265 curr per   38050

Avon Lake Public Library, 32649 Electric Blvd, **Avon Lake**, OH 44012-1669
T: +1 440 9337857; Fax: +1 440 9335659; E-mail: refdesk@avonlake.lib.oh.us; URL: www.alpl.org; www.avonlake.lib.oh.us
1930; Mary Crehore
Hands-on Learning Ctr
91 000 vols; 251 curr per; 4 137 av-mat; 3 541 Large Print Bks, 3 674 Audio Bks
libr loan   38051

Ayer Library, 26 E Main St, **Ayer**, MA 01432
T: +1 978 7728250; Fax: +1 978 7728251; URL: www.ayerlibrary.org
Mary Anne Lucht
50 000 vols; 70 curr per   38052

Azusa City Library, 729 N Dalton Ave, **Azusa**, CA 91702-2586
T: +1 626 8125232; Fax: +1 626 3344368; E-mail: library_staff@ci.azusa.ca.us; URL: ci.azusa.ca.us
1898; Albert Tovar
Indians of North America Coll, Spanish Language Coll
118 000 vols; 163 curr per
libr loan   38053

Babylon Public Library, 24 S Carll Ave, **Babylon**, NY 11702
T: +1 631 6691624; Fax: +1 631 6697826; E-mail: babllib@suffolk.lib.ny.us; URL: www.suffolk.lib.ny.us/libraries/babl
1895; N.N.
66 000 vols; 4 157 curr per   38054

Southwest Georgia Regional Library, Gilbert H. Gragg Library, 301 S Monroe St, **Bainbridge**, GA 31717
T: +1 229 2482665; Fax: +1 229 2482670; E-mail: librarian@swgrl.org; URL: www.swgrl.org/decatur.htm
1902; Shelley Sudderth
Georgia Author Coll, Jack Wingate Hunters & Anglers Coll; 3 branch libs
99 670 vols; 99 curr per   38055

Kitsap Regional Library – Bainbridge Island Branch, 1270 Madison Ave N, **Bainbridge Island**, WA 98110-2747
T: +1 206 8424162; Fax: +1 206 7805310
Rebecca Judd
89 000 vols   38056

East Baton Rouge Parish Library – Baker Branch, 3501 Groom Rd, **Baker**, LA 70714
T: +1 225 7785980; Fax: +1 225 7785949
Becky Andrews
93 000 vols   38057

Baker County Public Library, 2400 Resort St, **Baker City**, OR 97814-2798
T: +1 541 5236419; Fax: +1 541 5239088; E-mail: ask@bakerlib.org; URL: www.bakerlib.org
1906; Perry N. Stokes
Baker County Coll, Oregon Hist Coll; 4 branch libs, 1 bookmobile
117 000 vols; 349 curr per; 3 000 e-books; 8 621 av-mat; 3 400 sound-rec; 3 285 Large Print Bks   38058

Kern County Library, 701 Truxtun Ave, **Bakersfield**, CA 93301-4816
T: +1 661 8680700; Fax: +1 661 8680799; E-mail: kclweb@kerncountylibrary.org; URL: www.kerncountylibrary.org
1900; Diane R. Duquette
Technology, Natural Science, Hist; Kern County Hist Coll, California Geology-Mining-Petroleum Coll; 26 branch libs
992 000 vols; 575 curr per; 276 e-books; 8 127 maps; 23 000 microforms; 10 703 av-mat; 28 600 sound-rec; 1 018 digital data carriers
libr loan   38059

Kern County Library – Beale Memorial, 701 Truxtun Ave, **Bakersfield**, CA 93301-4816
T: +1 661 8680760; Fax: +1 661 8680831; E-mail: kclweb@kerncountylibrary.org; URL: www.kerncountylibrary.org
Nila Stearns
Kern County History Coll;Genealogy Coll;Geology Mining & Petroleum Coll
292 000 vols   38060

Kern County Library – Northeast Bakersfield, 3725 Columbus St, **Bakersfield**, CA 93306-2719
T: +1 661 8719017
Sarah Bleyl
56 000 vols   38061

Kern County Library – Southwest Branch, 8301 Ming Ave, **Bakersfield**, CA 93311-2020
T: +1 661 6647716; Fax: +1 661 6647717
Heather Eddy
80 000 vols   38062

Bala-Cynwyd Library, 131 Old Lancaster Rd, **Bala Cynwyd**, PA 19004-3037
T: +1 610 6641196; Fax: +1 610 6645534; E-mail: balacynwydlibrary@lmls.org; URL: lmls.org
1915; Jean Knapp
Music Coll, bks, cassettes, scores, compact discs; Judaica Coll; Children's Hist Book Coll
122 000 vols; 188 curr per
libr loan   38063

Newport Beach Public Library – Balboa Branch, 100 E Balboa Blvd, **Balboa**, CA 92661
T: +1 949 9173800; Fax: +1 949 6758524
Phyllis Scheffler
Nautical Coll
50 000 vols   38064

Baldwin Public Library, 2385 Grand Ave, **Baldwin**, NY 11510-3289
T: +1 516 2236228; Fax: +1 516 6237991; E-mail: director_ba@nassaulibrary.org; URL: www.nassaulibrary.org/baldwin
1922; H. Maria Sysak
Special education; Long Island Coll
155 000 vols; 2 700 curr per; 4 000 e-books; 15 000 Large Print Bks
libr loan   38065

County of Los Angeles Public Library – Baldwin Park Library, 4181 Baldwin Park Blvd, **Baldwin Park**, CA 91706-3203
T: +1 626 9626947; Fax: +1 626 3376631; URL: www.colapublib.org/libs/baldwinpark
Rafael Gonzalez
105 000 vols; 135 curr per   38066

Baldwinsville Public Library, 33 E Genesee St, **Baldwinsville**, NY 13027-2575
T: +1 315 6355631; Fax: +1 315 6356760; E-mail: info@bville.lib.ny.us; URL: www.bville.lib.ny.us
1948; Marilyn R. Laubacher
Newspaper Coll (1846-present), micro
92 000 vols; 274 curr per
libr loan   38067

Sprague Public Library, One Main St, **Baltic**, CT 06330-1320; P.O. Box 162, Baltic, CT 06330-0162
T: +1 860 8223012; Fax: +1 860 8223013; E-mail: librarian@ctsprague.org; URL: www.ctsprague.org
Barbaranne Warner
90 000 vols; 213 av-mat   38068

Baltimore County Public Library – Arbutus, 1581 Sulphur Spring Rd, Ste 105, **Baltimore**, MD 21227-2598
T: +1 410 8871451; Fax: +1 410 5360328
Gail Ross
82 000 vols   38069

Baltimore County Public Library – Catonsville, 1100 Frederick Rd, **Baltimore**, MD 21228-5092
T: +1 410 8870951; Fax: +1 410 7888166; E-mail: catonsville@bcpl.net
Robert Maranto
157 000 vols   38070

Baltimore County Public Library – Essex, 1110 Eastern Blvd, **Baltimore**, MD 21221-3497
T: +1 410 8870295; Fax: +1 410 6870075; E-mail: essex@bcpl.net
Mollie Fein
85 000 vols   38071

Baltimore County Public Library – North Point, 1716 Merritt Blvd, **Baltimore**, MD 21222-3295
T: +1 410 8877255; Fax: +1 410 2823272
Beth McGraw-Wagner
95 000 vols   38072

Baltimore County Public Library – Parkville-Carney, 9509 Harford Rd, **Baltimore**, MD 21234-3192
T: +1 410 8875353; Fax: +1 410 6683678
Edward Woznicki
91 000 vols   38073

Baltimore County Public Library – Perry Hall, 9440 Belair Rd, **Baltimore**, MD 21236-1504
T: +1 410 8875195; Fax: +1 410 5299430; E-mail: perryhall@bcpl.net
Sarah Stanhope
53 000 vols   38074

Baltimore County Public Library Pikesville, 1301 Reisterstown Rd, **Baltimore**, MD 21208-4195
T: +1 410 8871234; Fax: +1 410 4862782; E-mail: pikesville@bcpl.net
Alan McWilliams
105 000 vols   38075

Baltimore County Public Library – Rosedale, 6105 Kenwood Ave, **Baltimore**, MD 21237-2097
T: +1 410 8870512; Fax: +1 410 8664299; E-mail: rosedale@bcpl.net
Judith Kaplan
63 000 vols   38076

Baltimore County Public Library – White Marsh, 8133 Sandpiper Circle, **Baltimore**, MD 21236-4973
T: +1 410 8875097; Fax: +1 410 9319229; E-mail: whitemarsh@bcpl.net
David LaPenotiere
101 000 vols   38077

Baltimore County Public Library – Woodlawn, 1811 Woodlawn Dr, **Baltimore**, MD 21207-4074
T: +1 410 8871336; Fax: +1 410 2819584; E-mail: woodlawn@bcpl.net
Michelle Hamiel
70 000 vols   38078

Enoch Pratt Free Library, 400 Cathedral St, **Baltimore**, MD 21201-4484
T: +1 410 3965395; Fax: +1 410 3961441; E-mail: GenInfo@prattlibrary.org; URL: www.prattlibrary.org
1886; Dr. Carla D. Hayden
Afro-American; Baltimore Views; Printed Book, Illustrations to 1900; Greeting Cards; Insurance (Baltimore Life Underwriters); Stamps; War Posters; Cookery, Gastronomy & Wines; Job & Career Information; 22 branch libs
2 240 000 vols; 14 293 curr per; 300 e-journals; 23 000 e-books; 15 438 av-mat
libr loan   38079

Maryland State Library for the Blind & Physically Handicapped, 415 Park Ave, **Baltimore**, MD 21201-3603
T: +1 410 2302424; Fax: +1 410 3332095; URL: www.lbph.lib.md.us
Jill Lewis
Maryland, Braille & Audio cassettes
272 000 vols; 328 000 Vols in braille, 10 134 Large Print Bks
libr loan   38080

Bandon Public Library, 1204 11th St SW, **Bandon**, OR 97411; P.O. Box 128, Bandon, OR 97411-0128
T: +1 541 3473221; Fax: +1 541 3479363; URL: info.ccllsd.org/ban
Deirdre S. Krumper
Oral History on the Bandon Fire of 1936
63 000 vols; 93 curr per   38081

Bangor Public Library, 145 Harlow St, **Bangor**, ME 04401-1802
T: +1 207 9478336; Fax: +1 207 9456694; E-mail: bplill@bpl.lib.me.us; URL: www.bpl.lib.me.us
1883; Barbara Rice McDade
Hist; Mountaineering, Ornithology, World War II
466 000 vols; 597 curr per
libr loan   38082

Banning Public Library, 21 W Nicolet St, **Banning**, CA 92220
T: +1 909 8493192; Fax: +1 909 8496355; E-mail: bld@banninglibrarydistrict.org; URL: www.banninglibrarydistrict.org/banning/
1916; Nancy Kerr
88 005 vols; 75 maps; 28 sound-rec   38083

Baraboo Public Library, 230 Fourth Ave, **Baraboo**, WI 53913
T: +1 608 3566166; Fax: +1 608 3552779; URL: www.scls.lib.wi.us/bar
1903; Richard MacDonald
61 000 vols; 198 curr per; 2 896 av-mat   38084

Barberton Public Library, 602 W Park Ave, **Barberton**, OH 44203-2458
T: +1 330 7451194; Fax: +1 330 7458261; URL: www.barbertonlibrary.org
1903; Julianne Bedel
1 branch libr
105 000 vols; 231 curr per; 34 000 microforms; 7 830 av-mat; 11 319 sound-rec; 1 157 digital data carriers; 3 870 Large Print Bks
libr loan; ALA   38085

Knox County Public Library, 206 Knox St, **Barbourville**, KY 40906
T: +1 606 5465339; Fax: +1 606 5463602; URL: www.knoxpubliclibrary.com
Lana Hale
Kentucky Coll
52 000 vols; 48 curr per   38086

Nelson County Public Library, 201 Cathedral Manor, **Bardstown**, KY 40004-1515
T: +1 502 3483714; Fax: +1 502 3485578; E-mail: nelsoncopublib@hotmail.com; URL: www.nelsoncopublib.org
1967; Sharon Shanks
2 branch libs
54 000 vols; 58 curr per; 1 291 av-mat   38087

Barnesville Hutton Memorial Library, 308 E Main St, **Barnesville**, OH 43713-1410
T: +1 614 4251651; Fax: +1 614 4253504; E-mail: brownbr@oplin.org; URL: www.barnesvillehutton.lib.oh.us
1924; Brenda Brown
Religious studies, Ohio hist
53 900 vols; 148 curr per   38088

Sturgis Library, 3090 Main St, **Barnstable**, MA 02630; P.O. Box 606, Barnstable, MA 02630-0606
T: +1 508 3626636; Fax: +1 508 3625467; E-mail: sturgislib@comcast.net; URL: www.sturgislibrary.org
Lucy E. Loomis
Nineteenth Century Literature;Local Authors (Cape Cod Coll);Genealogy & Local History (Hooper Room);Maritime History (Kittredge Room), bks, micro;Early Cape Cod Land Deeds, Some Indian (Stanley W Smith Coll)
68 000 vols; 55 curr per   38089

Aldrich Public Library, 6 Washington, **Barre**, VT 05641-4227
T: +1 802 4767550; Fax: +1 802 4790450; E-mail: AldrichLibrary@charter.net; URL: www.aldrich.lib.vt.us
1907
52 170 vols; 135 curr per   38090

Barrington Public Library, 281 County Rd, **Barrington**, RI 02806
T: +1 401 2471920; Fax: +1 401 2473763; URL: www.barringtonlibrary.org
1880; Debbie Barchie
125 000 vols; 182 curr per
libr loan   38091

Barrington Public Library District, 505 N Northwest Hwy, **Barrington**, IL 60010
T: +1 847 3821300;
Fax: +1 847 3821261; URL: www.barringtonarealibrary.org
1913; Barbara L. Sugden
269 000 vols; 455 curr per; 6 000 e-books; 11 500 av-mat; 11 600 sound-rec; 6 420 digital data carriers   38092

Bartlesville Public Library, 600 S Johnstone, **Bartlesville**, OK 74003
T: +1 918 3384161; Fax: +1 918 3375338; URL: www.bartlesville.lib.ok.us
1913; Joan Singleton
Local Hist Museum
116 000 vols; 161 curr per; 2 000 e-books; 127 maps; 4 000 microforms; 3 771 av-mat; 2 557 sound-rec; 303 digital data carriers
libr loan; ALA, Mountain Plains Library Association, OLA   38093

Bartlett Public Library District, 800 S Bartlett Rd, **Bartlett**, IL 60103
T: +1 630 8372855; Fax: +1 630 8372669; E-mail: bpldref@bartlett.lib.il.us; URL: www.bartlett.lib.il.us
1972; Todd Morning
107 000 vols   38094

Memphis Public Library & Information Center – Bartlett Branch, 5884 Stage Rd, **Bartlett**, TN 38134
T: +1 901 3868968; Fax: +1 901 3862358
Gay Cain
782 000 vols   38095

Alpha Park Public Library District, 3527 S Airport Rd, **Bartonville**, IL 61607-1799
T: +1 309 6973822; Fax: +1 309 6979681; E-mail: alpha@alphapark.org; URL: www.alphapark.org
1972; John D. Richmond
Peoria State Hospital Coll
73 000 vols; 378 curr per   38096

Big Horn County Library, 430 West C St, **Basin**, WY 82410; P.O. Box 231, Basin
T: +1 307 5682388; Fax: +1 307 5682011; URL: will.state.wy.us/bighorn
1903; Becky L. Hawkins
99 000 vols; 147 curr per; 65 microforms; 1200 av-mat; 200 sound-rec
libr loan   38097

Bernards Township Library, 32 S Maple Ave, **Basking Ridge**, NJ 07920-1216
T: +1 908 2043031; Fax: +1 908 7661580; URL: www.bernards.org/library
1898; Anne Meany
Performing arts coll
110 000 vols; 290 curr per; 192 digital data carriers
libr loan; ALA   38098

Morehouse Parish Library, 524 E Madison Ave, **Bastrop**, LA 71220
T: +1 318 2813696; Fax: +1 318 2813683; E-mail: tlmh@state.lib.la.us; URL: www.youseemore.com/morehouse
1940; Ellen M. Highsmith
6 branch libs
53 000 vols; 101 curr per   38099

Batavia Public Library District, Ten S Batavia Ave, **Batavia**, IL 60510-2793
T: +1 630 8791393; Fax: +1 630 8799118; E-mail: askus@bataviapubliclibrary.org; URL: www.bataviapubliclibrary.org
1873; George H. Scheetz
139 000 vols; 265 curr per   38100

Clermont County Public Library, 326 Broadway St, **Batavia**, OH 45103
T: +1 513 7322736; Fax: +1 513 7323177; URL: www.clermont.lib.oh.us
1955; Sue Riggs
10 branch libs
536 000 vols; 467 curr per; 56 388 av-mat; 30 970 sound-rec; 52 digital data carriers; 280 Bks on Deafness & Sign Lang, 15 520 Large Print Bks   38101

Richmond Memorial Library, 19 Ross St, **Batavia**, NY 14020
T: +1 585 3439550; Fax: +1 585 3444651; E-mail: btvdir@nioga.org; URL: www.batavialibrary.org
1889; Diana Wyrwa
98 000 vols; 177 curr per   38102

White River Regional Library, 368 E Main St, **Batesville**, AR 72501
T: +1 870 7938814; Fax: +1 870 7938896; E-mail: indcolib@hotmail.com
1978; Debra Sutterfield
5 branch libs
152 000 vols; 462 curr per   38103

Patten Free Library, 33 Summer St, **Bath**, ME 04530
T: +1 207 4435141; Fax: +1 207 4433514; E-mail: pff@patten.lib.me.us; URL: www.patten.lib.me.us
1847; Anne Phillips
Hist Preservation, Maine Hist, Maritime Hist, Native Americans
56 000 vols; 243 curr per   38104

East Baton Rouge Parish Library, Main Library, 7711 Goodwood Blvd, **Baton Rouge**, LA 70806-7625
T: +1 225 2317520; Fax: +1 225 2313788; E-mail: admin.c1eb@pelican.state.lib.la.us; URL: www.ebr.lib.la.us
1939; David Farrar
12 branch libs
1 618 000 vols; 3 100 curr per; 13 000 e-books; 38 799 av-mat; 31 750 sound-rec; 7 960 digital data carriers; 500 Braille vols, 2 940 High Interest/Low Vocabulary Bks, 54 480 Large Print Bks
libr loan   38105

East Baton Rouge Parish Library – Bluebonnet Regional, 9200 Bluebonnet Blvd, **Baton Rouge**, LA 70810
T: +1 225 7632250; Fax: +1 225 7632253
Melinda Newman
Genealogy Coll
241 000 vols   38106

East Baton Rouge Parish Library – Central, 11260 Joor Rd, **Baton Rouge**, LA 70818
T: +1 225 2622650; Fax: +1 225 2622649
Heather Harrison
106 000 vols   38107

East Baton Rouge Parish Library – Delmont Gardens, 3351 Lorraine St, **Baton Rouge**, LA 70805
T: +1 225 3547050; Fax: +1 225 3547049
Charlotte Pringle
104 000 vols   38108

East Baton Rouge Parish Library – Greenwell Springs Road Regional, 11300 Greenwell Springs Rd, **Baton Rouge**, LA 70814
T: +1 225 2744450; Fax: +1 225 2744454
Geralyn Davis
164 000 vols   38109

East Baton Rouge Parish Library – Jones Creek Regional, 6222 Jones Creek Rd, **Baton Rouge**, LA 70817
T: +1 225 7561150; Fax: +1 225 7561153
Yvonne Hull
237 000 vols   38110

East Baton Rouge Parish Library – River Center, 120 St Louis St, **Baton Rouge**, LA 70802; P.O. Box 1471, Baton Rouge, LA 70821-1471
T: +1 225 3894967; Fax: +1 225 3898910
Lewis Diane
Baton Rouge History Coll;Career Center;Foundation Center Cooperating Coll
128 000 vols   38111

East Baton Rouge Parish Library – Scotlandville, 7373 Scenic Hwy, **Baton Rouge**, LA 70807
T: +1 225 3547550; Fax: +1 225 3547551
Chaundra Carroccio
92 000 vols   38112

Willard Library, Seven W Van Buren St, **Battle Creek**, MI 49017-3009
T: +1 269 9688166; Fax: +1 269 9683284; E-mail: info@willard.lib.mi.us; URL: www.willard.lib.mi.us
1870; Rick Hulsey
180 000 vols   38113

Bay City Public Library, 1100 Seventh St, **Bay City**, TX 77414
T: +1 979 2456931; Fax: +1 979 2452614
1913; Jana Prock
70 000 vols; 65 curr per; 57 digital data carriers
libr loan   38114

Bay County Library System, 500 Center Ave, **Bay City**, MI 48708
T: +1 989 8942837; Fax: +1 989 8942021; URL: www.baycountylibrary.org
1974; Linda Heemstra
5 branch libs
305 000 vols; 781 curr per; 25 485 Audio Bks, 43 Electronic Media & Resources
libr loan   38115

Bay County Library System – Alice & Jack Wirt Public Library, 500 Center Ave, **Bay City**, MI 48708-5989
T: +1 989 8939566; Fax: +1 989 8939799
Jane Anderson
94 000 vols   38116

Bay County Library System – Sage Branch, 100 E Midland St, **Bay City**, MI 48706
T: +1 989 8928555; Fax: +1 989 8921516
Sarah Wohlschlag
77 000 vols   38117

Hancock County Library System, 312 Hwy 90, **Bay Saint Louis**, MS 39520-3595
T: +1 228 4676836; Fax: +1 228 4675503; E-mail: hcls@hancock.lib.ms.us; URL: www.hancock.lib.ms.us
1934; Paul Eddy
Mississippi-Louisianna (Mississippiana Coll); 3 branch libs
109 000 vols; 170 curr per   38118

Cuyahoga County Public Library – Bay Village Branch, 502 Cahoon Rd, **Bay Village**, OH 44140-2179
T: +1 440 8716392; Fax: +1 440 8715320
Patricia King
66 000 vols; 4 251 av-mat   38119

Free Public Library of Bayonne, 697 Avenue C, **Bayonne**, NJ 07002
T: +1 201 8586160; Fax: +1 201 8586678; URL: www.bayonnenj.org/library
1893; Sneh P. Bains
Local Hist; 2 branch libr
179 000 vols; 312 curr per; 2 990 Large Print Bks   38120

Queens Borough Public Library – Bay Terrace, 18-36 Bell Blvd, **Bayside**, NY 11360
T: +1 718 4237004
Eve Hammer
57 000 vols   38121

Queens Borough Public Library – Bayside Branch, 214-20 Northern Blvd, **Bayside**, NY 11361
T: +1 718 2291834; Fax: +1 718 2258547
Janice Chan
95 000 vols   38122

Queens Borough Public Library – Windsor Park, 79-50 Bell Blvd, **Bayside**, NY 11364
T: +1 718 4688300; Fax: +1 718 4682274
Michelle Chan
64 000 vols   38123

Sterling Municipal Library, Wilbanks Ave, **Baytown**, TX 77520
T: +1 281 4277331; Fax: +1 281 4205347; E-mail: smlib@hpl.lib.tx.us; URL: www.sml.lib.tx.us
1961; Katherine Skinner Brown
2 bookmobiles
180 000 vols; 150 curr per   38124

Bayville Free Library, 34 School St, **Bayville**, NY 11709
T: +1 516 6282765; Fax: +1 516 6282738; URL: www.nassaulibrary.org/bayville/
Richard Rapecis
56 000 vols; 70 curr per   38125

Ocean County Library – Berkeley, 30 Station Rd, **Bayville**, NJ 08721-2198
T: +1 732 2692144; Fax: +1 732 2372955
Heather Andolsen
65 000 vols   38126

Cuyahoga County Public Library – Beachwood Branch, 25501 Shaker Blvd, **Beachwood**, OH 44122-2306
T: +1 216 8316868; Fax: +1 216 8310412
Caroline Vicchiarelli
Holocaust Coll
62 000 vols; 5 691 av-mat   38127

Bear Library, 101 Governors Pl, **Bear**, DE 19701
T: +1 302 8383300; Fax: +1 302 8383307; URL: www.nccdelib.org
Susan Menson
Delawareana
99 000 vols; 273 curr per; 8 392 av-mat   38128

Bear Public Library, 101 Governors Pl, **Bear**, DE 19701
T: +1 302 8383300; Fax: +1 302 8383307; URL: www.lib.de.us, www2.nccde.org/libraries/Bear/default.aspx
1998
77 490 vols; 214 curr per   38129

Beatrice Public Library, 100 N 16th St, **Beatrice**, NE 68310-4100
T: +1 402 2233584; Fax: +1 402 2233913
1893; Laureen Riedesel
Nebraska State Genealogical Society Coll
94 000 vols; 165 curr per   38130

Beaufort County Library, 311 Scott St, **Beaufort**, SC 29902-5591
T: +1 843 4706504; Fax: +1 843 4706542; URL: www.co.beaufort.sc.us/bftlib/desault
1963; Hillary Barnwell
4 branch libs
181 000 vols; 353 curr per   38131

Beaumont Library District, 125 E Eighth St, **Beaumont**, CA 92223-2194
T: +1 951 8451357; Fax: +1 951 8456217; E-mail: beaumontlib@telis.org; URL: www.bld.lib.ca.us
1911; Clara DiFelice
56 000 vols; 75 curr per; 820 av-mat   38132

Beaumont Public Library System, 801 Pearl St, **Beaumont**, TX 77701; P.O. Box 3827, Beaumont
T: +1 409 8386606; Fax: +1 409 8386838
1926; Geri Roberts
Genealogy & Texana (Tyrell Hist Libr); 6 branch libs
345 000 vols; 126 curr per   38133

Beaver Area Memorial Library, 100 College Ave, **Beaver**, PA 15009-2794
T: +1 724 7751132; Fax: +1 724 7756982
1948; Diane Wakefield
52 000 vols; 67 curr per   38134

Beaver Dam Community Library,
311 N Spring St, *Beaver Dam*,
WI 53916-2043
T: +1 920 8874631; Fax: +1 920
8874633; URL: www.beaverdam.lib.wi.us
1884; Susan Mary Mevis
105 000 vols; 195 curr per   38135

Carnegie Free Library, 1301 Seventh
Ave, *Beaver Falls*, PA 15010-4219
T: +1 724 8464340; Fax: +1 724
8460370; URL: www.co.beaver.pa.us/
library/main.html
1902; Jean Ann Barsotti
Resource & Res Ctr for Local Hist
50 000 vols; 204 curr per   38136

**Greene County Public Library –
Beavercreek Community Library**,
3618 Dayton-Xenia Rd, *Beavercreek*,
OH 45432-2884
T: +1 937 3524001; Fax: +1 937
4260481
Toni White
136 000 vols   38137

Bedford Free Public Library, 7 Mudge
Way, *Bedford*, MA 01730
T: +1 781 2759440; Fax: +1 781
2753590; E-mail: bedford@minlib.net;
URL: www.bedfordlibrary.net
1876; Richard Callaghan
Bedford Coll, Parent's Coll
98 000 vols; 308 curr per   38138

Bedford Public Library, 1323 K St,
Bedford, IN 47421
T: +1 812 2754471; Fax: +1 812
2785244; E-mail: bpl@bedlib.com; URL:
www.bedlib.org
1898; Susan A. Miller
Genealogy Coll, Hist of the Civil War
Coll
90 000 vols; 190 curr per   38139

Bedford Public Library, Three
Meetinghouse Rd, *Bedford*, NH 03110
T: +1 603 4723023; Fax: +1 603
4722978; URL: www.bedford.lib.nh.us
1789; Mary Ann Senatro
Sheet Music Coll
69 000 vols; 124 curr per   38140

Bedford Public Library, 1805 L Don
Dodson Dr, *Bedford*, TX 76021
T: +1 817 9522330; Fax: +1 817
9522396; URL: www.bedfordlibrary.org
1964; Maria Redburn
Bedford Hist Coll
95 000 vols; 239 curr per; 26 000 e-
books; 2 379 av-mat; 2 379 Audio Bks,
10 Special Interest Per Sub, 34 Bks on
Deafness & Sign Lang, 3 208 Large Print
Bks
libr loan; ALA   38141

Bedford Public Library System, 321 N
Bridge St, *Bedford*, VA 24523-1924
T: +1 540 5868911; Fax: +1 540
5868875; URL: www.library.bedford.va.us
1900; Peggy Bias
World War II; 5 branch libs
199 000 vols; 157 curr per   38142

Bedford Park Public Library District,
7816 W 65th Pl, *Bedford Park*,
IL 60501
T: +1 708 4586826; Fax: +1 708
4589827; URL: www.bplib.net
1963; Anne Murphy
88 000 vols; 223 curr per   38143

Clarence Dillon Public Library, 2336
Lamington Rd, *Bedminster*, NJ 07921
T: +1 908 2342325; Fax: +1 908
7819402; E-mail: ref@dillonlibrary.org;
URL: www.clarencedillonpl.org
Sumner Putnam
Civil War
93 000 vols; 7 156 av-mat   38144

Beech Grove Public Library, 1102 Main
St, *Beech Grove*, IN 46107
T: +1 317 7884203; Fax: +1
317 7880489; E-mail:
bgplreference@bgpl.lib.in.us; URL:
www.bgpl.lib.in.us
1949; Diane Burns
78 000 vols; 263 curr per; 391 av-mat
          38145

Harford County Public Library,
1221-A Brass Mill Rd, *Belcamp*,
MD 21017-1209
T: +1 410 5756761; Fax: +1 410
2735606; URL: www.hcplonline.info
1946; Audra L. Caplan
Maryland; Juvenile Hist Coll, Ripken
Literacy Coll, Rolling Reader Coll; 11
branch libs
740 000 vols; 2 023 curr per; 2 000 e-
books; 76 282 av-mat; 48 120 sound-rec;
24 610 digital data carriers; 250 Bks on
Deafness & Sign Lang, 9 700 Large Print
Bks   38146

Bellaire City Library, 5111 Jessamine,
Bellaire, TX 77401-4498
T: +1 713 6628160; Fax: +1 713
6628169; URL: www.ci.bellaire.tx.us
1951; Mary Alford
62 000 vols; 136 curr per   38147

Bellaire Public Library, 330 32nd St,
Bellaire, OH 43906
T: +1 740 6769421; Fax: +1 740
6767940; E-mail: bellaire@oplin.org;
URL: www.bellaire.lib.oh.us
1927; John T. Kniesner
79 000 vols; 135 curr per; 33 000
microforms; 900 sound-rec; 100 digital
data carriers
libr loan; ALA, PLA   38148

Logan County District Library, 220 N
Main St, *Bellefontaine*, OH 43311-2288
T: +1 937 5994189; Fax: +1 937
5995503; E-mail: lcdlref@oplin.org; URL:
www.loganco.lib.oh.us
1901; Judith A. Goodrich
6 branch libs
169 000 vols; 287 curr per   38149

**Centre County Library & Historical
Museum**, 203 N Allegheny St,
Bellefonte, PA 16823-1601
T: +1 814 3551516; Fax: +1
814 3552700; E-mail:
refdesk@centrecountylibrary.org; URL:
www.centrecountylibrary.org
1938; Lisa Erickson
Hist; County Docs, Genealogy (Spangler
Coll); 3 branch libs, 1 bookmobile
200 000 vols; 256 curr per
libr loan   38150

**Queens Borough Public Library –
Bellerose Branch**, 250-06 Hillside Ave,
Bellerose, NY 11426
T: +1 718 8318644
I-mei Lee
72 000 vols   38151

Belleville Public Library, 121 E
Washington St, *Belleville*, IL 62220
T: +1 618 2340441; Fax: +1
618 2349474; E-mail:
mainlibrary@bellevillepubliclibrary.org;
URL: www.bellevillepubliclibrary.org
1836; Kenda Burrack
1 branch libr
135 000 vols; 443 curr per; 7 000
microforms; 2 742 av-mat; 2 650
sound-rec; 60 digital data carriers
libr loan; ALA, ILA   38152

**Belleville Public Library & Information
Center**, 221 Washington Ave, *Belleville*,
NJ 07109-3189
T: +1 973 4503434; Fax: +1 973
4509518; URL: www.bellepl.org
1902; Joan Barbara Taub
96 000 vols; 156 curr per   38153

Fred C. Fischer Library, 167 Fourth St,
Belleville, MI 48111
T: +1 734 6993291; Fax: +1
734 6996352; E-mail:
fredcfisherlibrary@gmail.com; URL:
www.belleville.lib.mi.us
1920; Debra L. Green
61 000 vols; 171 curr per   38154

Bellevue Public Library, 224 E Main St,
Bellevue, OH 44811-1467
T: +1 419 4834769; Fax: +1 419
4830158; URL: www.bellevue.lib.oh.us
1904; Molly Carver
61 000 vols   38155

**County of Los Angeles Public
Library – Clifton M. Brakensiek
Library**, 9945 E Flower St, *Bellflower*,
CA 90706-5486
T: +1 562 9255543; Fax: +1 562
9209249; URL: www.colapublib.org/libs/
brakensiek
Sarah Comfort
119 000 vols; 127 curr per   38156

Bellingham Public Library, 210 Central
Ave, *Bellingham*, WA 98225; mail
address: 210 Central Ave, CS-9710,
Bellingham, WA 98227-9710
T: +1 360 7787220;
Fax: +1 360 7787295; URL:
www.bellinghampubliclibrary.org
1904; Pamela Nyberg Kiesner
1 branch libr
259 000 vols; 520 curr per; 16 156
av-mat; 9 160 sound-rec; 4 570 digital
data carriers; 8 220 Large Print Bks,
6 920 Talking Bks   38157

Whatcom County Library System,
5205 Northwest Dr, *Bellingham*,
WA 98226-9050
T: +1 360 3544883; Fax: +1 360
3844947; URL: www.wcls.org
1945; Joan Airoldi
9 branch libs
193 000 vols; 1 151 curr per; 12 153
av-mat   38158

Bellmore Memorial Library, 2288
Bedford Ave, *Bellmore*, NY 11710
T: +1 516 7852990; Fax: +1 516
7838550; URL: www.nassaulibrary.org/
bellmore
1948; Steven Bregman
102 000 vols; 295 curr per   38159

South Country Library, 22 Station Rd,
Bellport, NY 11713
T: +1 631 2860818; Fax: +1 631
2864873; E-mail: sctyref@sctylib.org;
URL: sctylib.org/libraryinfo.htm
1921; Mary Haines
83 000 vols; 6 357 curr per; 7 717 av-mat;
5 000 sound-rec; 350 digital data carriers
libr loan; ALA, NYLA   38160

Suffolk Cooperative Library System,
627 N Sunrise Service Rd, *Bellport*,
NY 11713; P.O. Box 9000, Bellport
T: +1 631 2861600; Fax: +1 631
2861647; URL: www.suffolk.lib.ny.us
1961; Gerald D. Nichols
Adult New Readers, Auto & Home
Appliance Repair Manuals, Disability
Ref Coll, Talking Bks, Multi-Language
Coll; 1 branch libr
143 000 vols; 47 201 curr per; 8 000 e-
books; 60 Bks on Deafness & Sign Lang
          38161

Bellwood Public Library, 600 Bohland
Ave, *Bellwood*, IL 60104-1896
T: +1 708 5477393; Fax: +1 708
5479352; E-mail: bws@bellwoodlibrary.org;
URL: www.bellwoodlibrary.org
1932; Mrs Jimmi Wooten
93 420 vols; 136 curr per   38162

Bellwood-Antis Public Library, 526
Main St, *Bellwood*, PA 16617
T: +1 814 742-8234; Fax: +1 814 742-
8235; E-mail: hab@tome.bapl.lib.pa.us;
URL: tome.bapl.lib.pa.us/
1965; Hazel A. Bilka
50 270 vols; 75 curr per   38163

Belmont Public Library, 336 Concord
Ave, *Belmont*, MA 02478-0904; P.O.
Box 125, Belmont, MA 02478-0125
T: +1 617 9932850; Fax: +1 617
9932894; URL: www.belmont.lib.ma.us
1868; Maureen Conners
1 branch libr
145 000 vols; 236 curr per   38164

**San Mateo County Library – Belmont
Branch**, 1110 Alameda de las Pulgas,
Belmont, CA 94002
T: +1 650 5918286; Fax: +1 650
5912763; URL: www.belmontlibrary.org
Julie Finklang
100 000 vols   38165

Beloit Public Library, 409 Pleasant St,
Beloit, WI 53511
T: +1 608 3642905; Fax: +1
608 3642907; E-mail:
reference@beloitlibrary.info; URL:
www.als.lib.wi.us/BPL
1902; Daniel Zack
132 000 vols; 246 curr per
libr loan   38166

Ida Public Library, 320 N State St,
Belvidere, IL 61008-3299
T: +1 815 5443838; Fax: +1 815
5448909; URL: www.idapubliclibrary.org
1885; Connie Harrison
77 000 vols; 250 curr per   38167

Warren County Library, Court House
Annex, 199 Hardwick St, *Belvidere*,
NJ 07823
T: +1 908 4756321; Fax: +1 908
4751706; URL: www.warrenlib.org
1931; Richard Moore
3 branch libs
226 000 vols; 571 curr per   38168

Deschutes Public Library District, 507
NW Wall St, *Bend*, OR 97701-2698
T: +1 541 3121021; Fax: +1 541
3892982; URL: www.dpls.us
1920; Michael Gaston
5 branch libs
396 000 vols; 849 curr per
libr loan   38169

**Deschutes Public Library District –
Bend Branch**, 601 NW Wall St, *Bend*,
OR 97701
T: +1 541 6177040; Fax: +1 541
6177044
Heather McNeil
216 000 vols; 274 curr per   38170

Benicia Public Library, 150 East L St,
Benicia, CA 94510-3281
T: +1 707 7464343; Fax: +1 707
7478122; URL: www.BeniciaLibrary.org
1911; Diane Smikahl
Art & architecture; Antiques Coll,
California Coll, North Point Press Coll
106 000 vols; 277 curr per; 291 govt
docs; 30 maps; 2 000 microforms; 1 838
sound-rec; 1 digital data carriers; 2 625
videorecordings
libr loan; CLA, PLA, ALA   38171

Marlboro County Library, 200 John
Corry Rd, *Bennettsville*, SC 29512
T: +1 843 4795630; Fax: +1
843 4795645; E-mail:
marlbcolibrary@yahoo.com; URL:
marlborocountylibrary.org
1901; Sharon Clontz Rowe
South Caroliniana, Large print
54 000 vols; 88 curr per   38172

Bennington Free Library, 101 Silver St,
Bennington, VT 05201
T: +1 802 4429051; URL:
benningtonfreelibrary.org
1865
Benninton Banner, 1903-present, flm
56 110 vols; 187 curr per   38173

**Bucks County Free Library –
Bensalem Branch**, 3700 Hulmeville
Rd, *Bensalem*, PA 19020-4449
T: +1 215 6382030; Fax: +1 215
6382192
Lisa Kern
110 000 vols; 89 curr per   38174

Bensenville Community Public Library,
200 S Church Rd, *Bensenville*,
IL 60106
T: +1 630 7664642; Fax: +1 630
7660788; E-mail: bcplweb@gmail.com;
URL: www.bensenville.lib.il.us
1960; Jill Rodriguez
Large Print Bks
87 000 vols; 370 curr per   38175

Saline County Public Library, 224 W
South St, *Benton*, AR 72015
T: +1 501 7784766; Fax: +1 501
7780536; URL: www.saline.lib.ar.us
1928; Erin Waller
Arkansas; 1 branch libr
60 000 vols; 105 curr per   38176

Benton Harbor Public Library,
213 E Wall St, *Benton Harbor*,
MI 49022-2499
T: +1 269 9266139; Fax: +1
269 9261674; E-mail:
staff@bentonharborlibrary.com; URL:
www.bentonharborlibrary.com
1899; Frederick J. Kirby
Ethnic studies; Biological Sciences (Don
Farnum Coll), Black Studies (Martin
Luther King Jr Coll), Civil War (Randall
Perry Coll), Judaica (Lillian Faber Coll),
Theater (Helen Polly Klock Coll)
101 000 vols; 80 curr per   38177

Bentonville Public Library, 405 S Main
St, *Bentonville*, AR 72712
T: +1 479 2713192; Fax: +1
479 2716775; E-mail:
library@bentonvillear.com
1947; Hadi Dudley
Arkansas Hist, Benton County Democrat
62 000 vols; 205 curr per; 3 500 av-mat
          38178

Cuyahoga County Public Library – Berea Branch, Seven Berea Commons, *Berea*, OH 44017-2524
T: +1 440 2345475; Fax: +1 440 2342932
Donna Meyers
63 000 vols; 4 840 av-mat   38179

Bergenfield Public Library, 50 W Clinton Ave, *Bergenfield*, NJ 07621-2799
T: +1 201 3874040; Fax: +1 201 3879004; E-mail: bfldcirc@bccls.org; URL: www.bccls.org/bergenfield
1920; Mary Riskind
Education; English as a Second Language, Spanish Language Bks & Magazines
137 000 vols; 229 curr per
libr loan; ALA   38180

Berkeley Public Library, 2090 Kittredge St, *Berkeley*, CA 94704
T: +1 510 9816100; Fax: +1 510 9816176; E-mail: director@berkeleypubliclibrary.org; URL: www.berkeleypubliclibrary.org
1893; Donna Corbeil
Ethnic studies, Art, Architecture, Civil rights, Music, Feminism; Berkeley Hist (Swingle Coll); 4 branch libs
447 000 vols; 1 524 curr per
libr loan   38181

Berkeley Heights Free Public Library, 290 Plainfield Ave, *Berkeley Heights*, NJ 07922
T: +1 908 4649333; Fax: +1 908 4647098; E-mail: reference@bhplnj; URL: www.bhplnj.org
1953; Stephanie Bakos
Art & architecture
82 000 vols; 220 curr per; 1 696 av-mat   38182

Berkley Public Library, 3155 Coolidge Hwy, *Berkley*, MI 48072
T: +1 248 6583440; Fax: +1 248 6583441; E-mail: library@berkley.lib.mi.us; URL: www.berkley.lib.mi.us
1927; Celia B. Morse
Art & architecture
71 000 vols; 170 curr per; 4 000 av-mat   38183

Berlin Public Library, 121 W Park Ave, *Berlin*, WI 54923
T: +1 920 3615420; Fax: +1 920 3615424; E-mail: director@berlinlibrary.org; URL: www.berlinlibrary.org
1903; Diane Disterhaft
Spanish Language Mat Coll, Large Print Coll, Literacy Coll
50 000 vols; 101 curr per   38184

Bernardsville Public Library, One Anderson Hill Rd, *Bernardsville*, NJ 07924
T: +1 908 7660118; Fax: +1 908 7662464; E-mail: library@bernardsvillelibrary.org; URL: www.bernardsvillelibrary.org
1902; Karen Brodsky
71 990 vols; 143 curr per   38185

Berne Public Library, 166 N Sprunger St, *Berne*, IN 46711-1595
T: +1 260 5892809; Fax: +1 260 5892940; E-mail: bpl@bernepl.lib.in.us; URL: www.bernepl.lib.in.us
1935; Marvel Zuercher
Mennonite Hist Coll
70 000 vols; 128 curr per   38186

Berrien Springs Community Library, 215 W Union St, *Berrien Springs*, MI 49103-1077
T: +1 269 4717074; Fax: +1 269 4714433; E-mail: bsclibrary@comcast.net; URL: bsclibrary.org
1906; Judy Berry
60 000 vols; 120 curr per; 2 051 av-mat   38187

Berwyn Public Library, 2701 Harlem Ave, *Berwyn*, IL 60402
T: +1 708 7958000; Fax: +1 708 7958101; URL: www.berwynlibrary.net
Bill Hensley
Czechoslovakian Language Coll
158 000 vols; 736 curr per   38188

Bessemer Public Library, 701 Ninth Ave N, *Bessemer*, AL 35020
T: +1 205 4287882; Fax: +1 205 4287885; URL: bessemerlibrary.org
1908
90 000 vols; 123 curr per   38189

Metropolitan Library System in Oklahoma County – Bethany Library, 3510 N Mueller, *Bethany*, OK 73008-3952
T: +1 405 7898363; Fax: +1 405 6063239; E-mail: bethany@metrolibrary.org
Katrina Prince
50 000 vols; 103 curr per   38190

Bethel Public Library, 189 Greenwood Ave, *Bethel*, CT 06801-2598
T: +1 203 7948756; Fax: +1 203 7948761; E-mail: adult@bethellibrary.org; URL: www.bethellibrary.org
1909; Lynn M. Rosato
93 000 vols; 209 curr per; 2 155 av-mat   38191

Bethel Park Public Library, 5100 W Library Ave, *Bethel Park*, PA 15102
T: +1 412 8316800; Fax: +1 412 8359360; E-mail: bethpark@einetwork.net; URL: www.bethelparklibrary.org
1955; Cheryl Napsha
101 000 vols; 123 curr per   38192

Bethlehem Area Public Library, 11 W Church St, *Bethlehem*, PA 18018
T: +1 610 8673761; Fax: +1 610 8672767; E-mail: info@bapl.org; URL: www.bapl.org
1901; Janet S. Fricker
Literacy Coll, Spanish Coll; 1 branch libr, 1 bookmobile
197 000 vols; 357 curr per; 14 140 av-mat; 13 161 sound-rec; 2 230 digital data carriers; 3 750 Large Print Bks
libr loan   38193

Bethpage Public Library, 47 Powell Ave, *Bethpage*, NY 11714-3197
T: +1 516 9313907; Fax: +1 516 9313926; E-mail: bethpage@nassaulibrary.org
1927; Lois Lovisolo
186 000 vols; 309 curr per   38194

Bettendorf Public Library Information Center, 2950 Learning Campus Dr, *Bettendorf*, IA 52722; P.O. Box 1330, Bettendorf, IA 52722-1330
T: +1 563 3444183; Fax: +1 563 3444185; E-mail: info@bettendorflibrary.com; URL: www.bettendorflibrary.com
1957; Steven Nielsen
Iowa Hist Coll
132 000 vols; 348 curr per; 7 000 e-books   38195

Beverly Public Library, 32 Essex St, *Beverly*, MA 01915-4561
T: +1 978 9216062; Fax: +1 978 9228329; E-mail: beverly.library@noblenet.org; URL: www.noblenet.org
1855; Patricia Cirone
Will Barnet Coll of Original Lithogr; 1 branch libr
180 000 vols; 125 curr per   38196

Beverly Hills Public Library, 444 N Rexford Dr, *Beverly Hills*, CA 90210-4877
T: +1 310 2882222; Fax: +1 310 2783387; E-mail: library@beverlyhills.org; URL: www.bhpl.org
1929; Nancy Hunt-Coffey
Art, Dance; 19th & 20th C Art & Artists; Beverly Hills Coll
305 000 vols; 597 curr per   38197

Citrus County Library System, 425 W Roosevelt Blvd, *Beverly Hills*, FL 34465-4281
T: +1 352 7469077; Fax: +1 352 7469493; E-mail: suggestions@citruslibraries.org
Flossie Benton Rogers
160 000 vols; 161 curr per; 30 000 e-books; 111 av-mat   38198

Bexley Public Library, 2411 E Main St, *Bexley*, OH 43209
T: +1 614 2318795; URL: www.bexlib.org
1924; Robert M. Stafford
206 000 vols; 481 curr per   38199

Bicknell-Vigo Township Public Library, 201 W Second St, *Bicknell*, IN 47512-2109
T: +1 812 7352317; Fax: +1 812 7352018; URL: bicknell-vigo.lib.in.us
1926; Ginger Rogers
65 000 vols; 65 curr per; 450 av-mat   38200

McArthur Public Library, 270 Main St, *Biddeford*, ME 04005; P.O. Box 346, Biddeford, ME 04005-0346
T: +1 207 2844181; Fax: +1 207 2846761; E-mail: reference@mcarthur.lib.me.us; URL: www.mcarthur.lib.me.us
1863; Dora St. Martin
62 000 vols; 174 curr per; 1 259 av-mat   38201

Big Rapids Community Library, 426 S Michigan Ave, *Big Rapids*, MI 49307
T: +1 231 7965234; Fax: +1 231 7961078; E-mail: librarian@bigrapids.lib.mi.us; URL: www.bigrapids.lib.mi.us
1903; Gaylynn Rorabaugh
67 000 vols; 125 curr per   38202

Howard County Library – Dora Roberts Library, 500 Main St, *Big Spring*, TX 79720-2532
T: +1 432 2642260; Fax: +1 432 2642263; URL: www.howard-county.lib.tx.us
1907; Hollis McCright
Texana
70 000 vols; 64 curr per   38203

Lonesome Pine Regional Library – C. Bascom Slemp Memorial, 11 Proctor St N, *Big Stone Gap*, VA 24219
T: +1 276 5231334; Fax: +1 276 5235306; E-mail: cbslib@lprlibrary.org
Christine Smith
54 000 vols; 65 curr per; 1 284 av-mat   38204

Billerica Public Library, 15 Concord Rd, *Billerica*, MA 01821
T: +1 978 6710948; Fax: +1 978 6674242; URL: www.billericalibrary.org
Barbara A. Flaherty
120 000 vols; 230 curr per   38205

Parmly Billings Library, 510 N Broadway, *Billings*, MT 59101-1196
T: +1 406 6578258; Fax: +1 406 6578293; E-mail: refdesk@billings.lib.mt.us; URL: www.billings.lib.mt.us
1901; William M. Cochran
Genealogy; Montana Room
260 000 vols; 332 curr per; 22 810 av-mat; 18 760 sound-rec; 5 370 digital data carriers; 90 Bks on Deafness & Sign Lang, 14 100 Large Print Bks
libr loan   38206

Broome County Public Library, 185 Court St, *Binghamton*, NY 13901-3503
T: +1 607 7786410; Fax: +1 607 7786429; URL: www.bclibrary.info
1902; Donna Riegel
196 000 vols
libr loan   38207

Birmingham Public Library, 2100 Park Pl, *Birmingham*, AL 35203
T: +1 205 2263600; Fax: +1 205 2263743; URL: www.bplonline.org
1902; Barbara C. Sirmans
Genealogy; Dance Coll, Cartography, Drama Coll, Rare Books, Affiliate Agency-Alabama Data Center; 20 branch libs
899 000 vols; 2 500 curr per; 217 e-books; 57 648 av-mat; 35 304 sound-rec; 1 200 Journals, 19 540 Large Print Bks
libr loan   38208

Birmingham Public Library – Springville Road, 1224 Springville Rd, *Birmingham*, AL 35215
T: +1 205 2264081; Fax: +1 205 8560825; URL: www.bplonline.org
Rochelle Sides-Renda
59 000 vols; 2 880 av-mat   38209

North Shelby County Library, 5521 Cahaba Valley Rd, *Birmingham*, AL 35242
T: +1 205 4395508; Fax: +1 205 4395503; URL: www.northshelbylibrary.org
Carol Farr
58 000 vols; 95 curr per   38210

Cochise County Library District, Old High School, 2nd Flr, 100 Clawson, *Bisbee*, AZ 85603; P.O. Drawer AK, Bisbee
T: +1 520 4328930; Fax: +1 520 4327339; URL: cochise.lib.az.us
1970; Lise Gilliland
5 branch libs
120 000 vols   38211

Bismarck Veterans Memorial Public Library, 515 N Fifth St, *Bismarck*, ND 58503-4081
T: +1 701 2226410; Fax: +1 701 2216854; URL: www.bismarcklibrary.org
1917; Thomas T. Jones
193 000 vols; 553 curr per; 10 000 e-books; 1 520 av-mat; 1 520 Audio Bks, 2 800 CDs
libr loan   38212

Camden County Library System – Gloucester Township, 15 S Black Horse Pike, *Blackwood*, NJ 08012
T: +1 856 2280022; Fax: +1 856 2289085; E-mail: glouce@camden.lib.nj.us
Anne Ackroyd
55 000 vols; 50 curr per   38213

Anoka County Library, 707 County Rd 10 NE, *Blaine*, MN 55434-2398
T: +1 763 7853695; Fax: +1 763 7173262; E-mail: anoka@anoka.lib.mn.us; URL: www.anoka.lib.mn.us
1958; Marlene Moulton Janssen
8 branch libs
614 000 vols; 1 377 curr per   38214

Blanchester Public Library, 110 N Broadway, *Blanchester*, OH 45107-1250
T: +1 937 7833585; Fax: +1 937 7832910; URL: www.blanchester.lib.oh.us
1935; Chris Owens
51 000 vols; 130 curr per   38215

Blauvelt Free Library, 541 Western Hwy, *Blauvelt*, NY 10913
T: +1 845 3592811; Fax: +1 845 3980017; E-mail: blv@rcls.org; URL: www.rcls.org/blv
1909; Mary E. Behringer
Budke Coll; Blauvelt Family Hist
63 000 vols; 417 curr per; 260 microforms   38216

Bloomfield Public Library, 90 Broad St, *Bloomfield*, NJ 07003
T: +1 973 5666200; Fax: +1 973 5666206; URL: www.bplnj.org
1924; Hasija C. Gian
Business & Labor Coll, Music Coll
160 000 vols
libr loan; NJLA   38217

Prosser Public Library, One Tunxis Ave, *Bloomfield*, CT 06002-2476
T: +1 860 2439721; Fax: +1 860 2421629; URL: www.prosserlibrary.info
1901; Beverly Lambert
1 branch libr
96 000 vols; 183 curr per   38218

Bloomingdale Public Library, 101 Fairfield Way, *Bloomingdale*, IL 60108-1579
T: +1 630 5293120; Fax: +1 630 5293243; E-mail: bdref@mybpl.org; URL: www.mybpl.org
1974; Timothy Jarzemsky
108 000 vols; 202 curr per; 1 000 e-books; 5 308 av-mat   38219

Monroe County Public Library, 303 E Kirkwood Ave, *Bloomington*, IN 47408
T: +1 812 3493090; Fax: +1 812 3493051; URL: www.mcpl.info
Sara Laughlin
VITAL – Volunteers in Tutoring Adult Learners; Indiana & Monroe County History (Indiana Coll), bks, mag, microfilm, newsp, pamphlets, hist tapes, maps, video & audio cassettes; CATS – Community Access Television
380 000 vols; 1 078 curr per   38220

Sullivan County Public Library, 1655 Blountville Blvd, *Blountville*, TN 37617; P.O. Box 510, Blountville, TN 37617-0510
T: +1 423 2792714; Fax: +1 423 2792836; URL: www.wrlibrary.org/sullivan
Theresa McMahan
Sullivan County History
123 000 vols; 161 curr per   38221

Public Library of Cincinnati & Hamilton County – Blue Ash Branch, 4911 Cooper Rd, *Blue Ash*, OH 45242
T: +1 513 3696051; Fax: +1 513 3694464
Robert Burdick
76 000 vols   38222

Wissahickon Valley Public Library, 650
Skippack Pike, **Blue Bell**, PA 19422
T: +1 215 6431320; Fax: +1 215
6436611; E-mail: library@wvpl.org; URL:
www.wvpl.org
1934; David J. Roberts
1 branch libr
118 000 vols; 220 curr per; 8 603 av-mat;
6 260 sound-rec; 2 370 digital data
carriers; 2 102 Large Print Bks   38223

City of Blue Island Public Library,
2433 York St, **Blue Island**,
IL 60406-2011
T: +1 708 3881078; Fax: +1
708 3881143; E-mail:
bipl@blueislandlibrary.org; URL:
www.blueislandlibrary.org
1897; N.N.
82 000 vols; 315 curr per
libr loan   38224

Bayport-Blue Point Public Library,
203 Blue Point Ave, **Blue Point**,
NY 11715-1217
T: +1 631 3636133; Fax: +1 631
3636133; E-mail: bprtlib@suffolk.lib.ny.us;
URL: bprt.suffolk.lib.ny.us/
1938; John O'Hare
110 000 vols; 378 curr per
libr loan   38225

Craft Memorial Library, 600 Commerce
St, **Bluefield**, WV 24701
T: +1 304 3253943; Fax: +1 304
3253702; E-mail: cml@mail.mln.lib.wv.us;
URL: craftmemorial.lib.wv.us
1972; Eva McGuire
Eastern Regional Coal Arch, West
Virginia & Virginia Hist Coll
103 000 vols; 115 curr per; 2 480 Audio
Bks   38226

Wells County Public Library, 200 W
Washington St, **Bluffton**, IN 46714-1999
T: +1 260 8241612; Fax: +1 260
8243129; E-mail: wcpl@wellscolibrary.org;
URL: www.wellscolibrary.org
1902; Stephanie Davis
Large Print Coll, Literacy, Young Teen
Fiction; 2 branch libs
102 000 vols; 321 curr per; 3 000 e-
books; 2 000 microforms; 5 481 av-mat;
1 719 sound-rec; 379 digital data carriers;
140 art prints 1 442 recorded bks
libr loan; ALA   38227

Palo Verde Valley District Library, 125
W Chanslorway, **Blythe**, CA 92225-1293
T: +1 760 9225371; Fax: +1 760
9225371; E-mail: pvvdl@global101.com;
URL: www.paloverdevalleylibrary.com
1959; Brenda S. Lugo
52 000 vols; 79 curr per
libr loan   38228

Mississippi County Library System,
200 N Fifth St, **Blytheville**,
AR 72315-2791
T: +1 870 7622431; Fax: +1
870 7622442; E-mail:
misscolibrary@yahoo.com
1921; Joseph F. Ziolko
7 branch libs
200 000 vols; 250 curr per; 700 av-mat;
500 sound-rec   38229

**Mississippi County Library System
– Blytheville Public**, 200 N Fifth St,
Blytheville, AR 72315
T: +1 870 7622431; Fax: +1
870 7622442; E-mail:
misscolibrary@yahoo.com
Jay Ziolko
150 000 vols; 100 curr per; 250 av-mat   38230

Boaz Public Library, 404 Thomas Ave,
Boaz, AL 35957
T: +1 256 5938056; Fax: +1 256
5938153; E-mail: library@cityofboaz.org;
URL: www.cityofboaz.org
1971; Lynn Burgess
Paperback Coll
60 000 vols; 100 curr per   38231

Boca Raton Public Library, 200
NW Boca Raton Blvd, **Boca Raton**,
FL 33432-3706
T: +1 561 3937906; Fax: +1 561
3937823; URL: www.bocalibrary.org
1938; Catherine A. O'Connell
151 000 vols; 188 curr per; 3 143 av-mat;
625 High Interest/Low Vocabulary Bks,
100 Bks on Deafness & Sign Lang,
4 921 Large Print Bks, 3 143 Talking Bks
libr loan   38232

Connetquot Public Library, 760 Ocean
Ave, **Bohemia**, NY 11716
T: +1 631 5675079; Fax: +1 631
5675137; URL: www.connetquot.lib.ny.us
1974; Kimberly DeCristofaro
230 000 vols; 490 curr per   38233

Ada Community Library, 10664 W
Victory Rd, **Boise**, ID 83709
T: +1 208 3620181; Fax: +1 208
3620303; URL: www.adalib.org
1984; Mary DeWalt
Horses; 2 branch libs
138 000 vols; 200 curr per; 7 000 e-
books; 9 000 av-mat
libr loan   38234

Boise Public Library, 715 S Capitol
Blvd, **Boise**, ID 83702-7195
T: +1 208 3844449; Fax: +1
208 3844021; E-mail:
askalibrarian@cityofboise.org; URL:
www.boisepubliclibrary.org
1895; Kevin Wayne Booe
1 branch libr, 1 bookmobile
357 000 vols; 299 curr per; 5 043 Large
Print Bks   38235

Fountaindale Public Library District,
300 W Briarcliff Rd, **Bolingbrook**,
IL 60440-2844
T: +1 630 7592102; Fax: +1 630
7599519; URL: www.fountaindale.lib.il.us
1970; Karen T. Anderson
1 branch libr
286 000 vols   38236

Polk County Library, 1690 W Broadway
St, **Bolivar**, MO 65613
T: +1 417 3264531; Fax: +1 417
3264366
1947; Terri York
55 000 vols; 57 curr per   38237

**Lee County Library System – Bonita
Springs Public**, 26876 Pine Ave,
Bonita Springs, FL 34135-5009
T: +1 239 9920101; Fax: +1 239
9926680
Maureen Pollock
69 000 vols   38238

Ericson Public Library, 702 Greene St,
Boone, IA 50036
T: +1 515 4323727; Fax: +1 515
4321103; E-mail: ericson@boone.lib.ia.us;
URL: www.boone.lib.ia.us
1901; Barbara A. Rardin
70 000 vols; 182 curr per; 3 940
sound-rec; 12 digital data carriers; 135
art repros
libr loan   38239

Watauga County Public Library, 140
Queen St, **Boone**, NC 28607
T: +1 828 2648784; Fax: +1 828
2641794; URL: www.arlibrary.org
1932; John Blake
86 000 vols; 110 curr per
libr loan   38240

**Boonville-Warrick County Public
Library**, 611 W Main St, **Boonville**,
IN 47601-1544
T: +1 812 8971500; Fax: +1 812
8971508
1911; Lois A. Aigner
Lincoln Coll; 3 branch libs
123 000 vols; 221 curr per; 2 143 av-mat;
500 sound-rec   38241

Hutchinson County Library, 625 N
Weatherly, **Borger**, TX 79007-3621
T: +1 806 2730126; Fax: +1 806
2730128
1938; Carolyn Wilkinson
Genealogy, Large Print, Texas
59 000 vols; 175 curr per   38242

Bossier Parish Central Library, 2206
Beckett St, **Bossier City**, LA 71111
T: +1 318 7461693; Fax: +1
318 7467768; E-mail:
admin.g1bs@pelican.state.lib.la.us; URL:
www.bossierlibrary.org
1940; Delbert Terry
7 branch libs
206 000 vols; 916 curr per   38243

Boston Public Library, 700 Boylston St,
Boston, MA 02117-0286
T: +1 617 5365400; Fax: +1 617
2364306; E-mail: info@bpl.org; URL:
www.bpl.org
1852; Amy Ryan
American & English Hist & Lit,
Rare Editions, American Accounting,
Astronomy, Mathematics, Navigation,
Defoe, Engravings, German Poetry,
Rare Editions, Joan of Arc Coll, Military
Science, Baseball Players, Prints,
Theology & Religion; 26 branch libs
8 895 000 vols; 4 673 curr per; 4 100 e-
books
libr loan   38244

**Boston Public Library – South
End**, 685 Tremont St, **Boston**,
MA 02118-3198
T: +1 617 5368241; Fax: +1 617
2668993; E-mail: south_end@bpl.org
Anne Smart
50 000 vols; 20 curr per; 300 av-mat   38245

Boulder Public Library, 1000 Canyon
Blvd, PO Drawer H, **Boulder**, CO 80306
T: +1 303 4413100; Fax: +1 303
4421808; URL: www.boulder.lib.co.us
Tony Tallent
Boulder Arts Resource Coll, Children's Lit
Ref; 3 branch libs
378 000 vols; 1 040 curr per; 20 000 e-
books   38246

Boulder City Library, 701 Adams Blvd,
Boulder City, NV 89005-2697
T: +1 702 2931281; Fax: +1 702
2930239; URL: www.bclibrary.org
1933; Lynn Schofield-Dahl
100 000 vols; 250 curr per; 3 000 av-mat   38247

Bound Brook Memorial Library, 402 E
High St, **Bound Brook**, NJ 08805
T: +1 732 3560043; Fax: +1
732 3561379; E-mail:
hkerwin@bblibrary.blogspot.com; URL:
www.bblibrary.blogspot.com
1897; Hannah Kerwin
75 000 vols; 150 curr per; 7 digital data
carriers   38248

Davis County Library – South, 725 S
Main St, **Bountiful**, UT 84010
T: +1 801 2958732
Bradley Maurer
154 000 vols   38249

Jonathan Bourne Public Library, 19
Sandwich Rd, **Bourne**, MA 02532-3699
T: +1 508 7590647; Fax: +1 508
7590647; URL: www.bournelibrary.org
1896; Patrick W. Marshall
Cape Cod Coll; Army Nat Guard Base
(Otis) Hazardous Waste Cleanup;
Genealogy Coll; Large Print Bk Coll;
Young Adult Coll, A-V mat equip
53 000 vols; 107 curr per   38250

Warren County Public Library, 1225
State St, **Bowling Green**, KY 42101
T: +1 270 7814882; Fax: +1 270
7817323; URL: www.warrenpl.org
1940; Lisa R. Rice
Local authors, Rare bks; 3 branch libs
136 000 vols; 225 curr per   38251

Wood County District Public Library,
251 N Main St, **Bowling Green**,
OH 43402-2477
T: +1 419 3525104; Fax: +1 419
3538013; URL: wcdpl.lib.oh.us
1875; Elaine Paulette
1 branch libr, 1 bookmobile
156 000 vols; 251 curr per; 6 000 e-
books   38252

Boxford Town Library, Ten Elm St,
Boxford, MA 01921
T: +1 978 8877323; Fax: +1 978
8876352; E-mail: boxford@mvlc.org;
URL: www.boxfordtownlib.org
1966; Diane H. C. Giarrusso
55 000 vols; 311 curr per; 2 375 av-mat;
1 862 sound-rec; 10 digital data carriers
libr loan; MLA-ALA   38253

Southside Regional Library, 316
Washington St, **Boydton**, VA 23917;
P.O. Box 10, Boydton, VA 23917-0010
T: +1 434 7386580; Fax: +1 434
7386070; URL: www.srlib.org
1944; John Walter
Mecklenburg County Newspaper, late
1800s, micro
130 000 vols; 365 curr per; 3 245 av-mat   38254

Boynton Beach City Library, 208
S Seacrest Blvd, **Boynton Beach**,
FL 33435
T: +1 561 7426380; Fax: +1 561
7426381; URL: www.boyntonlibrary.org
1961; Craig B. Clark
Large print bks, Gardening, Investing;
Telephone Bks, Song Bks, New Reader's
Series
129 000 vols; 200 curr per
libr loan   38255

Bozeman Public Library, 626 E Main
St, **Bozeman**, MT 59715
T: +1 406 5822408; Fax: +1 406
5822424; E-mail: mtb@mtlib.org; URL:
www.bozemanlibrary.org
1891; Alice Meister
Montana Room Coll
101 000 vols; 235 curr per; 9 000 e-
books; 3 089 av-mat
libr loan   38256

**Manatee County Public Library
System**, 1301 Barcarrota Blvd W,
Bradenton, FL 34205
T: +1 941 7485555; Fax: +1 941
7497191; URL: www.co.manatee.fl.us/
library/master.html
1971; John C. Van Berkel
Spanish languages mat; Talking Bks,
Oral Hist Coll; 5 branch libs
436 000 vols; 8 170 curr per; 48
Electronic Media & Resources
libr loan   38257

Bradford Area Public Library,
67 W Washington St, **Bradford**,
PA 16701-1234
T: +1 814 3626527; Fax: +1 814
3624168; E-mail: bapublib@atlanticbb.net;
URL: www.bradfordlibrary.org
1900
Dr T E Hanley Coll (Art & Lit)
55 700 vols; 100 curr per   38258

Fossil Ridge Public Library, 386
Kennedy Rd, **Braidwood**, IL 60408
T: +1 815 4582187; Fax: +1 815
4582042; E-mail: ayackle@fossilridge.org;
URL: www.fossilridge.org
1970; Anna Yackle
50 000 vols; 196 curr per; 600 av-mat
ALA, ILA   38259

Brainerd Public Library, 416 S Fifth St,
Brainerd, MN 56401
T: +1 218 8295574; Fax: +1 218
8290055; E-mail: brainerd@krls.org; URL:
www.krls.org
1882
80 000 vols   38260

**Central Mississippi Regional Library
System**, 104 Office Park Dr, **Brandon**,
MS 39042-2404; P.O. Box 1749, Brandon
T: +1 601 8250100; Fax: +1 601
8250199; URL: www.cmrls.lib.ms.us
1986; Kaileen Thieling
295 000 vols; 530 curr per; 10 e-books;
26 131 av-mat; 10 130 sound-rec   38261

**Tampa-Hillsborough County Public
Library System – Brandon Regional
Branch**, 619 Vonderburg Dr, **Brandon**,
FL 33511-5972
Fax: +1 813 7445632
Lorri Robinson
161 000 vols   38262

James Blackstone Memorial Library,
Blackstone Library, 758 Main St,
Branford, CT 06405-3697
T: +1 203 4881441; Fax: +1
203 4881260; E-mail:
library@blackstone.lioninc.org; URL:
www.blackstonelibrary.org
1893; Kathy Rieger
75 000 vols; 175 curr per   38263

Brooks Memorial Library, 224 Main St,
Brattleboro, VT 05301
T: +1 802 2545290; Fax: +1 802
2572309; E-mail: brattlib@brooks.lib.vt.us;
URL: www.brooks.lib.vt.us
1882; Jerry Carbone
Fine Arts Coll, Lawson Genealogy Coll,
Nuclear Energy, Windam World Affairs
Council Archs
68 140 vols; 350 curr per   38264

Brawley Public Library, 400 Main St, *Brawley*, CA 92227-2491
T: +1 760 3441891; Fax: +1 760 3440212
1927; Marjo Mello
51 000 vols; 102 curr per; 950 av-mat; 9 digital data carriers
libr loan   38265

Orange County Public Library – Brea Branch, One Civic Center Circle, *Brea*, CA 92821-5784
T: +1 714 6711722; Fax: +1 714 9900581
Cheryl Nakaji
75 000 vols; 78 curr per   38266

Kitsap Regional Library, 1301 Sylvan Way, *Bremerton*, WA 98310-3498
T: +1 360 4059158; Fax: +1 360 4059128; URL: www.krl.org
1955; Jill Jean
9 branch libs, 1 bookmobile
442 000 vols; 435 curr per; 484 e-books
libr loan   38267

Kitsap Regional Library – Sylvan Way Branch, 1301 Sylvan Way, *Bremerton*, WA 98310-3466
T: +1 360 4059100; Fax: +1 360 4059128; URL: www.krl.org
Ruth Bond
118 000 vols   38268

Brentwood Library & Center for Fine Arts, 8109 Concord Rd, *Brentwood*, TN 37027
T: +1 615 3710090; Fax: +1 615 3712238; E-mail: reference@brentwood-tn.org; URL: www.brentwood-tn.org/library
1978; Chuck Sherrill
Civil War, bks, maps, charts, graphs; Tennessee Hist (James A Crutchfield), bks, maps, charts, graphs
114 000 vols; 225 curr per
libr loan; ALA   38269

Brentwood Public Library, 8765 Eulalie Ave, *Brentwood*, MO 63144
T: +1 314 9638631; Fax: +1 314 9628675; E-mail: circulation@bplmo.org; URL: www.brentwood.lib.mo.us
1939; John Furlong
54 000 vols; 107 curr per; 90 av-mat
libr loan   38270

Transylvania County Library, 212 S Gaston, *Brevard*, NC 28712
T: +1 828 8843151; Fax: +1 828 8774230; URL: www.transylvania.lib.nc.us
1912; Anna Yount
North Carolina Coll
95 000 vols; 208 curr per
libr loan   38271

Queens Borough Public Library – Briarwood Branch, 85-12 Main St, *Briarwood*, NY 11435
T: +1 718 6581680
Barry Ernst
63 000 vols   38272

Ocean County Library – Brick Branch, 301 Chambers Bridge Rd, *Brick*, NJ 08723-2803
T: +1 732 4774513; Fax: +1 732 9209314
147 000 vols   38273

Bridgeport Public Library, 925 Broad St, *Bridgeport*, CT 06604
T: +1 203 5767403; Fax: +1 203 5768255; URL: www.bridgeportpubliclibrary.org
1881; Ann Osbon
Lit, Technology, Business, Art; PT Barnum Circus Coll; 4 branch libs
515 000 vols
libr loan   38274

Bridgeport Public Library, 1200 Johnson Ave, *Bridgeport*, WV 26330
T: +1 304 8428248; Fax: +1 304 8424018; E-mail: library@bridgeportwv.com; URL: www.bridgeportwv.com
1956; Sharon R. Saye
Michael Benedum Coll, bks, clippings, memorabilia, per, scrapbks; West Virginia Coll, bk, pamphlets
98 000 vols; 175 curr per
libr loan   38275

Bridgeport Public Library – North, 3455 Madison Ave, *Bridgeport*, CT 06606
T: +1 203 5767423; Fax: +1 203 5767752
Paula Keegan
50 000 vols   38276

Bridgeton Free Public Library, 150 E Commerce St, *Bridgeton*, NJ 08302-2684
T: +1 856 4512620; URL: www.clueslibs.org
1811; Gail S. Robinson
American Indians (Lenni-Lenapes), South Jersey Hist
65 000 vols; 75 curr per   38277

Cumberland County Library, 800 E Commerce St, *Bridgeton*, NJ 08302-2295
T: +1 856 4532210; Fax: +1 856 4511940; E-mail: ref@clueslibs.org; URL: www.clueslibs.org
1963; Nancy Forester
Adult Basic Education Mat; 1 bookmobile
117 000 vols; 135 curr per   38278

Saint Louis County Library – Bridgeton Trails Branch, 3455 McKelvey Rd, *Bridgeton*, MO 63044
T: +1 314 2917570; Fax: +1 314 2917593
Trudy Williams
93 000 vols   38279

Bridgeview Public Library, 7840 W 79th St, *Bridgeview*, IL 60455-1496
T: +1 708 4582880; Fax: +1 708 4583553; E-mail: bvs@mls.lib.il.us; URL: www.bridgeviewlibrary.org
1966; Rose Taylor
72 000 vols; 172 curr per; 3934 sound-rec
libr loan; ALA   38280

Bridgewater Public Library, 15 South St, *Bridgewater*, MA 02324-2593
T: +1 508 6973331; Fax: +1 508 2791467; E-mail: bwpl@sailsinc.org; URL: www.bridgewaterpubliclibrary.org
1881; Elizabeth L. Gregg
Adaptive Equipment & Mat for Disabled
139 000 vols; 110 curr per
libr loan   38281

Somerset County Library System, One Vogt Dr, *Bridgewater*, NJ 08807-2136; P.O. Box 6700, Bridgewater
T: +1 908 5264016; Fax: +1 908 5265221; URL: www.somerset.lib.nj.us
1930; James M. Hecht
New Jersey Room; 8 branch libs
813 000 vols; 680 curr per
libr loan   38282

Brigham City Library, 26 E Forest, *Brigham City*, UT 84302-2198
T: +1 435 7235850; Fax: +1 435 7232813; URL: bcpl.lib.ut.us
1915; Susan H. Hill
Mormon Hist; Box Elder Hist, bks & pamphlets; Mormon Religion
57 000 vols; 176 curr per   38283

Boston Public Library – Brighton Branch, 40 Academy Hill Rd, *Brighton*, MA 02135-3316
T: +1 617 7826032; Fax: +1 617 8592405
Paula Posnick
75 000 vols; 50 curr per   38284

Brighton District Library, 100 Library Dr, *Brighton*, MI 48116
T: +1 810 2296571; Fax: +1 810 2298924; URL: www.brightonlibrary.info
1992; Nancy B. Johnson
Career Coll
78 000 vols; 300 curr per   38285

Rangeview Library District – Brighton Branch, 575 S Eighth Ave, *Brighton*, CO 80601-3122
T: +1 303 6592572; Fax: +1 303 6540793
Alex Villagran
57 000 vols; 107 curr per; 1400 av-mat
38286

Bay Shore-Brightwaters Public Library, One S Country Rd, *Brightwaters*, NY 11718-1517
T: +1 631 6654350; Fax: +1 631 6654958; E-mail: bsbwlib@suffolk.lib.ny.us; URL: bayshore.suffolk.lib.ny.us
1901; Eileen J. Kavanagh
Long Island, ESL Coll
139 000 vols; 620 curr per   38287

Bristol Public Library, 701 Goode St, *Bristol*, VA 24201-4199
T: +1 276 6458781; Fax: +1 276 6695593; URL: www.bristol-library.org
1909; Jud B. Barry
1 branch libr
140 000 vols; 295 curr per; 3000 e-books; 9873 av-mat; 4570 sound-rec; 1700 digital data carriers
38288

Bristol Public Library, Five High St, *Bristol*, CT 06010
T: +1 860 5847787; Fax: +1 860 5847696; E-mail: bristollibraryrefdept@ci.bristol.ct.us; URL: www.ci.bristol.ct.us
1892; Francine Petosa
Agriculture, Art, Music, Education, Environmental studies; 1 branch libr
142 000 vols; 328 curr per   38289

Margaret R. Grundy Memorial Library, 680 Radcliffe St, *Bristol*, PA 19007-5199
T: +1 215 7887891; Fax: +1 215 7884976; URL: www.buckslib.org/libraries/bristol
1966; Mary Jane Mannherz
Bucks County Census 1790-1910, microfilm; Bucks County Courier Time 1911-Present, microfilm
75 000 vols; 121 curr per; 91 av-mat
38290

Bristol Public Library, 1855 Greenville Rd, *Bristolville*, OH 44402-9700; P.O. Box 220, Bristolville, OH 44402-0220
T: +1 330 8893651; Fax: +1 330 8899794; E-mail: bristol@oplin.org; URL: www.youseemore.com/bristolpl/
1910; Mona Saltzmann
60 000 vols; 243 curr per   38291

Broadview Public Library District, 2226 S 16th Ave, *Broadview*, IL 60155-4000
T: +1 708 3451325; Fax: +1 708 3455024; E-mail: brs@sls.lib.il.us; URL: www.sls.lib.il.us/brs
1955; Carl J. Caruso
51 000 vols; 125 curr per   38292

Brockport Seymour Library, 161 East Ave, *Brockport*, NY 14420-1987
T: +1 585 6371050; Fax: +1 585 6371051; URL: www.seymourlibraryweb.org
Kathleen Phillips
Bks on the Erie Canal; Census Coll, micro
52 000 vols; 90 curr per   38293

Brockton Public Library System, 155 W Elm St, *Brockton*, MA 02301-5390
T: +1 508 5807890; Fax: +1 508 5807898; URL: www.brocktonpubliclibrary.org
1867; Keith Choquette
2 branch libs
140 000 vols; 120 curr per   38294

Western Pocono Community Library, 2000 Pilgrim Way, *Brodheadsville*, PA 18322; P.O. Box 318, Brodheadsville, PA 18322-0318
T: +1 570 9927934; Fax: +1 570 9927915; E-mail: wpcl@ptd.net; URL: www.wpcl.lib.pa.us
1974; Carol Kern
53 000 vols; 102 curr per   38295

Tulsa City-County Library System – Broken Arrow Branch, 300 W Broadway, *Broken Arrow*, OK 74012
T: +1 918 2515359; Fax: +1 918 2580324
Ann Gaebe
61 000 vols   38296

Levy County Public Library System, 612 Hathaway Ave, *Bronson*, FL 32621; P.O. Box 1210, Bronson, FL 32621-1210
T: +1 352 4865552; Fax: +1 352 4865553
Bonnie Tollefson
76 000 vols   38297

The New York Public Library – Astor, Lenox & Tilden Foundations – Baychester Branch, 2049 Asch Loop N (north of Bartow Ave), *Bronx*, NY 10475
T: +1 718 3796700; Fax: +1 718 6712836; E-mail: baychester@nypl.org; URL: www.nypl.org/branch/local/bx/bar.cfm
Irina A. Kuharets
60 000 vols   38298

The New York Public Library – Astor, Lenox & Tilden Foundations – Mosholu Branch, 285 E 205th St (near Perry Ave), *Bronx*, NY 10467
T: +1 718 8828239; Fax: +1 718 5470434; E-mail: mosholu@nypl.org; URL: www.nypl.org/branch/local/bx/mo.cfm
Jimmie L. Pate
62 000 vols   38299

The New York Public Library – Astor, Lenox & Tilden Foundations – Pelham Bay Branch, 3060 Middletown Rd (north of Crosby Ave), *Bronx*, NY 10461
T: +1 718 7926744; Fax: +1 718 7926744; E-mail: pelham_bay@nypl.org; URL: www.nypl.org/branch/local/bx/pm.cfm
Angela Calderella
57 000 vols   38300

The New York Public Library – Astor, Lenox & Tilden Foundations – Spuyten Duyvil Branch, 650 W 235th St (@ Independence Ave), *Bronx*, NY 10463
T: +1 718 7961202; Fax: +1 718 7962351; E-mail: spuyten_duyvil@nypl.org; URL: www.nypl.org/branch/local/bx/dy.cfm
Tim P. Tureski
57 000 vols   38301

Bronxville Public Library, 201 Pondfield Rd, *Bronxville*, NY 10708
T: +1 914 3377680; Fax: +1 914 3370332; E-mail: leckley@wlsmail.org; URL: www.bxvlibrary.org
1906; Laura Eckley
Fine Arts & Japanese Coll, paintings; Burtnett Coll
65 000 vols; 135 curr per   38302

Cuyahoga County Public Library – Brook Park Branch, 6155 Engle Rd, *Brook Park*, OH 44142-2105
T: +1 216 2675250; Fax: +1 216 2673776
Kevin Payne
85 000 vols; 4450 av-mat   38303

Brookfield Free Public Library, 3609 Grand Blvd, *Brookfield*, IL 60513
T: +1 708 4856917; Fax: +1 708 4855172; E-mail: director@brookfieldpubliclibrary.info; URL: www.brookfieldpubliclibrary.info
1913; Kimberly Litland
76 000 vols; 191 curr per   38304

The Brookfield Library, 182 Whisconier Rd, *Brookfield*, CT 06804
T: +1 203 7756241; Fax: +1 203 7407723; URL: www.brookfieldlibrary.org
1951; Anita Barney
55 000 vols; 60 curr per   38305

Brookfield Public Library, 1900 N Calhoun Rd, *Brookfield*, WI 53005
T: +1 262 7824140; Fax: +1 262 7966670; E-mail: brookfieldpubliclibrary@ci.brookfield.wi.us; URL: www.brookfieldlibrary.com
1960; Edell Schaefer
Adult Literacy, Large Print Bks, Bks on Tape, Interactive Multimedia
171 000 vols; 415 curr per   38306

Brookings Public Library, 515 3rd St, *Brookings*, SD 57006
T: +1 605 6929407; Fax: +1 605 6929386; E-mail: elandau@sdln.net; URL: www.brookingslibrary.org
1913; Elvita Landau
South Dakota Coll
83 000 vols; 215 curr per
libr loan   38307

Chetco Community Public Library, 405 Alder St, *Brookings*, OR 97415
T: +1 541 4697738; Fax: +1 541 4696746; URL: www.chetcolibrary.org
1947; Susana Fernandez
Large Print Coll
51 000 vols; 118 curr per; 2644 av-mat
38308

Public Library of Brookline, 361 Washington St, *Brookline*, MA 02445
T: +1 617 7302370; Fax: +1 617 7302160; URL: www.brooklinelibrary.com
1857; James C. Flaherty
2 branch libs
320 000 vols; 820 curr per   38309

**Public Library of Brookline –
Coolidge Corner**, 31 Pleasant St,
Brookline, MA 02446
T: +1 617 7302380; Fax: +1 617
7344565
Catherine Dooley
Coll of Chinese language books,
magazines, audio; Coll of Russina
language books, magazines, audio
94 000 vols
libr loan   38310

Brooklyn Public Library, Grand Army
Plaza, **Brooklyn**, NY 11238-5698
T: +1 718 2302137;
Fax: +1 718 3983947; URL:
www.brooklynpubliclibrary.org
1897; Dionne Mack-Harvin
58 branch libs
6 971 000 vols   38311

Brooklyn Public Library – Arlington,
203 Arlington Ave, **Brooklyn**, NY 11207
T: +1 718 2776105; Fax: +1 718
2776177
Cynthia Woronowicz
77 000 vols   38312

Brooklyn Public Library – Bay Ridge,
7223 Ridge Blvd, **Brooklyn**, NY 11209
T: +1 718 7485709; Fax: +1 718
7487095
Maureen McCoy
102 000 vols   38313

Brooklyn Public Library – Bedford,
496 Franklin Ave, **Brooklyn**, NY 11238
T: +1 718 6230012; Fax: +1 718
6229919
62 000 vols   38314

**Brooklyn Public Library – Borough
Park**, 1265 43rd St, **Brooklyn**,
NY 11219
T: +1 718 4374085; Fax: +1 718
4373021
Ellen Fecher
119 000 vols   38315

**Brooklyn Public Library – Brighton
Beach**, 16 Brighton First Rd, **Brooklyn**,
NY 11235
T: +1 718 9462917; Fax: +1 718
9466176
Eileen Kassab
88 000 vols   38316

**Brooklyn Public Library – Brooklyn
Heights**, 280 Cadman Plaza W,
Brooklyn, NY 11201
T: +1 718 6237100; Fax: +1 718
2225681
Susan Phillis
101 000 vols   38317

**Brooklyn Public Library – Brower
Park**, 725 Saint Marks Ave, **Brooklyn**,
NY 11216
T: +1 718 7737208; Fax: +1 718
7737838
53 000 vols   38318

Brooklyn Public Library – Bushwick,
340 Bushwick Ave, **Brooklyn**, NY 11206
T: +1 718 6021348; Fax: +1 718
6021352
82 000 vols   38319

Brooklyn Public Library – Business,
280 Cadman Plaza W, **Brooklyn**,
NY 11201
T: +1 718 6237100; Fax: +1 718
2225681
Susan Phillis
139 000 vols   38320

Brooklyn Public Library – Canarsie,
1580 Rockaway Pkwy, **Brooklyn**,
NY 11236
T: +1 718 2576547; Fax: +1 718
2576557
Vicki Hill
72 000 vols   38321

**Brooklyn Public Library – Carroll
Gardens**, 396 Clinton St, **Brooklyn**,
NY 11231
T: +1 718 5966972; Fax: +1 718
5960370
Susan Asis
80 000 vols   38322

Brooklyn Public Library Clarendon,
2035 Nostrand Ave, **Brooklyn**,
NY 11210
T: +1 718 4211159; Fax: +1 718
4211244
Nanella Warren
59 000 vols   38323

**Brooklyn Public Library – Clinton
Hill**, 380 Washington Ave, **Brooklyn**,
NY 11238
T: +1 718 3988713; Fax: +1 718
3988715
Sharon Tidwell
64 000 vols   38324

**Brooklyn Public Library – Coney
Island**, 1901 Mermaid Ave, **Brooklyn**,
NY 11224
T: +1 718 2653220; Fax: +1 718
2655026
57 000 vols   38325

Brooklyn Public Library – Cortelyou,
1305 Cortelyou Rd, **Brooklyn**,
NY 11226
T: +1 718 6937763; Fax: +1 718
6937874
Paula Menzics
95 000 vols   38326

**Brooklyn Public Library – Crown
Heights**, 560 New York Ave, **Brooklyn**,
NY 11225
T: +1 718 7731180; Fax: +1 718
7730144
Anthony Robertson
85 000 vols   38327

**Brooklyn Public Library – Cypress
Hills**, 1197 Sutter Ave, **Brooklyn**,
NY 11208
T: +1 718 2776004; Fax: +1 718
2776009
Hardeep Sareen
77 000 vols   38328

**Brooklyn Public Library – DeKalb
Branch**, 790 Bushwick Ave, **Brooklyn**,
NY 11221
T: +1 718 4553898; Fax: +1 718
4554071
Paul Van Linden Tol
85 000 vols   38329

Brooklyn Public Library – Dyker, 8202
13th Ave, **Brooklyn**, NY 11228
T: +1 718 7486261; Fax: +1 718
7486370
Yvonne Zhow
54 000 vols   38330

**Brooklyn Public Library – East
Flatbush**, 9612 Church Ave, **Brooklyn**,
NY 11212
T: +1 718 9220927; Fax: +1 718
9222394
Tracey Mantrone
67 000 vols   38331

**Brooklyn Public Library – Eastern
Parkway**, 1044 Eastern Pkwy,
Brooklyn, NY 11213
T: +1 718 9534225; Fax: +1 718
9533970
Sandra Sutton
77 000 vols   38332

Brooklyn Public Library – Flatbush,
22 Linden Blvd, **Brooklyn**, NY 11226
T: +1 718 8560813; Fax: +1 718
8560899
Negla Ross-Parris
70 000 vols   38333

Brooklyn Public Library – Flatlands,
2065 Flatbush Ave, **Brooklyn**,
NY 11234
T: +1 718 2534409; Fax: +1 718
2535018
Karen Keith
84 000 vols   38334

**Brooklyn Public Library – Fort
Hamilton**, 9424 Fourth Ave, **Brooklyn**,
NY 11209
T: +1 718 7486919; Fax: +1 718
7487335
Phyllis Bornstein
59 000 vols   38335

**Brooklyn Public Library – Gerritsen
Beach**, 2808 Gerritsen Ave, **Brooklyn**,
NY 11229
T: +1 718 3681435; Fax: +1 718
3681506
Ed Flanagan
66 000 vols   38336

Brooklyn Public Library – Gravesend,
303 Avenue X, **Brooklyn**, NY 11223
T: +1 718 3825792; Fax: +1 718
3825926
Boris Ioselev
74 000 vols   38337

Brooklyn Public Library – Greenpoint,
107 Norman Ave, **Brooklyn**, NY 11222
T: +1 718 3498504; Fax: +1 718
3498790
Mel Gooch
78 000 vols   38338

Brooklyn Public Library – Highlawn,
1664 W 13th St, **Brooklyn**, NY 11223
T: +1 718 2347208; Fax: +1 718
2347238
Shapiro Danielle
72 000 vols   38339

Brooklyn Public Library – Homecrest,
2525 Coney Island Ave, **Brooklyn**,
NY 11223
T: +1 718 3825924; Fax: +1 718
3825955
Lyubov Klavansky
76 000 vols   38340

**Brooklyn Public Library – Jamaica
Bay**, 9727 Seaview Ave, **Brooklyn**,
NY 11236
T: +1 718 2413571; Fax: +1 718
2411981
Genya Konny
62 000 vols   38341

Brooklyn Public Library – Kensington,
410 Ditmas Ave, **Brooklyn**, NY 11218
T: +1 718 4359431; Fax: +1 718
4359491
Elizabeth Ridler
72 000 vols   38342

**Brooklyn Public Library – Kings
Bay**, 3650 Nostrand Ave, **Brooklyn**,
NY 11229
T: +1 718 3681709; Fax: +1 718
3681410
Mirian Rivera-Shapiro
120 000 vols   38343

Brooklyn Public Library – Leonard, 81
Devoe St, **Brooklyn**, NY 11211
T: +1 718 4863365; Fax: +1 718
4863370
Morris Denmark
65 000 vols   38344

Brooklyn Public Library – Marcy, 617
DeKalb Ave, **Brooklyn**, NY 11216
T: +1 718 9350032; Fax: +1 718
9350045
Phyllis Lu
64 000 vols   38345

**Brooklyn Public Library – McKinley
Park**, 6802 Fort Hamilton Pkwy,
Brooklyn, NY 11219
T: +1 718 7488001; Fax: +1 718
7487746
Scott Bowman
88 000 vols   38346

Brooklyn Public Library – Midwood,
975 E 16th St, **Brooklyn**, NY 11230
T: +1 718 2520967; Fax: +1 718
2521263
Rosita McCleavey
100 000 vols   38347

Brooklyn Public Library – Mill Basin,
2385 Ralph Ave, **Brooklyn**, NY 11234
T: +1 718 2413973; Fax: +1 718
2411957
Christina Armieri
68 000 vols   38348

**Brooklyn Public Library – New
Lots**, 665 New Lots Ave, **Brooklyn**,
NY 11207
T: +1 718 6490311; Fax: +1 718
6490719
Karen Lorde
83 000 vols   38349

**Brooklyn Public Library – New
Utrecht**, 1743 86th St, **Brooklyn**,
NY 11214
T: +1 718 2364086; Fax: +1 718
2347702
Edward Jelen
88 000 vols   38350

Brooklyn Public Library – Pacific, 25
Fourth Ave, **Brooklyn**, NY 11217
T: +1 718 6381531; Fax: +1 718
6381580
Gerard Costa
62 000 vols   38351

Brooklyn Public Library – Paerdegat,
850 E 59th St, **Brooklyn**, NY 11234
T: +1 718 2413994; Fax: +1 718
2411335
Ianthee Williams
66 000 vols   38352

Brooklyn Public Library – Park Slope,
431 Sixth Ave, **Brooklyn**, NY 11215
T: +1 718 8321853; Fax: +1 718
8329024
Uldis Skrodelis
80 000 vols   38353

Brooklyn Public Library – Red Hook,
Seven Wolcott St, **Brooklyn**, NY 11231
T: +1 718 9350203; Fax: +1 718
9350160
Salvador Salame
61 000 vols   38354

Brooklyn Public Library – Rugby,
1000 Utica Ave, **Brooklyn**, NY 11203
T: +1 718 5660054; Fax: +1 718
5660059
Sheila Eastmond
69 000 vols   38355

Brooklyn Public Library – Ryder, 5902
23rd Ave, **Brooklyn**, NY 11204
T: +1 718 3312962; Fax: +1 718
3313445
Michael McKegney
63 000 vols   38356

Brooklyn Public Library – Saratoga,
Eight Thomas S Boyland St, **Brooklyn**,
NY 11223
T: +1 718 5735224; Fax: +1 718
5735402
Monica Williams
87 000 vols   38357

**Brooklyn Public Library – Sheepshead
Bay**, 2636 E 14th St, **Brooklyn**,
NY 11235
T: +1 718 3681815; Fax: +1 718
3681872
Yelena Litinskaya
84 000 vols   38358

**Brooklyn Public Library – Spring
Creek**, 12143 Flatlands Ave, **Brooklyn**,
NY 11207
T: +1 718 2576571; Fax: +1 718
2576588
Kerwin Pilgrim
67 000 vols   38359

**Brooklyn Public Library – Stone
Avenue**, 581 Mother Gaston Blvd,
Brooklyn, NY 11212
T: +1 718 4858347
Patricia Howard
100 000 vols   38360

**Brooklyn Public Library – Sunset
Park**, 5108 Fourth Ave, **Brooklyn**,
NY 11220
T: +1 718 5672806; Fax: +1 718
5672810
Roxana Benavides
98 000 vols   38361

Brooklyn Public Library – Ulmer Park,
2602 Bath Ave, **Brooklyn**, NY 11214
T: +1 718 2653443; Fax: +1 718
2655115
Marina Arane
72 000 vols   38362

**Brooklyn Public Library – Walt
Whitman Branch**, 93 Saint Edwards
St, **Brooklyn**, NY 11205
T: +1 718 9350244; Fax: +1 718
9350284
52 000 vols   38363

**Brooklyn Public Library – Washington
Irving Branch**, 360 Irving Ave,
Brooklyn, NY 11237
T: +1 718 6288378; Fax: +1 718
6288439
David Camara
61 000 vols   38364

**Brooklyn Public Library –
Williamsburgh**, 240 Division Ave,
Brooklyn, NY 11211
T: +1 718 3023485; Fax: +1 718
3023499
Jennifer Gellmann
119 000 vols   38365

**Brooklyn Public Library – Windsor
Terrace**, 160 E Fifth St, **Brooklyn**,
NY 11218
T: +1 718 6869707; Fax: +1 718
6860162
Leslie Ogan
58 000 vols   38366

Hernando County Public Library System – Lykes Memorial Library, 238 Howell Ave, *Brooksville*, FL 34601
T: +1 352 7544042; Fax: +1 352 7544044; URL: www.hcpl.lib.fl.us/home.htm
1926; Barbara Shiflett
5 branch libs
197 000 vols; 402 curr per; 29 000 e-books   38367

Marple Public Library, 2599 Sproul Rd, *Broomall*, PA 19008-2399
T: +1 610 3561510; Fax: +1 610 3563589; E-mail: marple@delco.lib.pa.us; URL: www.marplepubliclibrary.org
1951; Deborah Parsons
Delaware County Literacy Council Coll
92 000 vols; 90 curr per   38368

Mamie Doud Eisenhower Public Library, Broomfield Public, Three Community Park Rd, *Broomfield*, CO 80020-3781
T: +1 720 8872300; Fax: +1 720 8871384; E-mail: info@broomfield.org; URL: www.broomfieldlibrary.org; www.ci.broomfield.co.us/library
1960; Depp Roberta
Mamie Doud Eisenhower Coll
144 000 vols; 248 curr per; 308 maps; 2 000 microforms; 998 av-mat; 2 543 sound-rec; 41 digital data carriers; 6 161 pamphlets
libr loan   38369

Brown Deer Public Library, 5600 W Bradley Rd, *Brown Deer*, WI 53223-3510
T: +1 414 3570106; Fax: +1 414 3570156; URL: www.browndeerwi.org/departments/library/library.htm
1969; Arnold Gutkowski
78 000 vols; 200 curr per   38370

Burlington County Library – Pemberton Community Library, 16 Broadway, *Browns Mills*, NJ 08015
T: +1 609 8938262; Fax: +1 609 8937547; E-mail: p@bcls.lib.nj.us
Nancy Breece
61 000 vols   38371

Brownsburg Public Library, 450 S Jefferson St, *Brownsburg*, IN 46112-1310
T: +1 317 8523167; Fax: +1 317 8526039; E-mail: reference@brownsburg.lib.in.us; URL: www.brownsburg.lib.in.us
1917; Wanda L. Pearson
Audio Visual Coll, Large Print Coll
102 000 vols; 307 curr per; 2 000 e-books   38372

Brownsville Public Library, 2600 Central Blvd, *Brownsville*, TX 78520-8824
T: +1 956 5481055; Fax: +1 956 5480684; URL: www.brownsville.lib.tx.us
Jesus Campos
144 000 vols; 250 curr per   38373

Brownwood Public Library, 600 Carnegie Blvd, *Brownwood*, TX 76801-7038
T: +1 325 6460155; Fax: +1 325 6466503; E-mail: genealogy@brownwoodpubliclibrary.com; URL: www.bwdpublib.org
1904; Mathew P. McConnell
78 000 vols; 120 curr per   38374

Brunswick Public Library Association, Captain John Curtis Memorial Library, 23 Pleasant St, *Brunswick*, ME 04011-2295
T: +1 207 7255242; Fax: +1 207 7256313; E-mail: info@curtislibrary.com; URL: www.curtislibrary.com
1883; Elisabeth Doucett
134 000 vols   38375

Bryan+College Station Public Library System, 201 E 26th St, *Bryan*, TX 77803-5356
T: +1 979 2095614; Fax: +1 979 2095610; URL: www.bcslibrary.org
1903; Clara B. Mounce
Ana Ludmilla Ballet Coll; 2 branch libs
235 000 vols; 524 curr per; 2 855 av-mat; 3 062 sound-rec   38376

Williams County Public Library, 107 E High St, *Bryan*, OH 43506-1702
T: +1 419 6366734; Fax: +1 419 6300408; URL: www.williamsco.lib.oh.us
1882; Jeffrey A. Yahraus
5 branch libs
120 000 vols; 610 curr per   38377

Ludington Public Library, Five S Bryn Mawr Ave, *Bryn Mawr*, PA 19010-3471
T: +1 610 5251776; Fax: +1 610 5251783; URL: www.lmls.org
1916; Margery Hall
Art & architecture, Horticulture
150 000 vols; 300 curr per   38378

Fontana Regional Library, 33 Fryemont St, *Bryson City*, NC 28713
T: +1 828 4882382; Fax: +1 828 4882638; E-mail: info@fontanalib.org; URL: www.fontanalib.org
1944; Karen Wallace
1 bookmobile
167 000 vols; 459 curr per; 13 825 av-mat; 7 810 sound-rec; 1 480 digital data carriers; 490 High Interest/Low Vocabulary Bks, 100 Bks on Deafness & Sign Lang, 4 630 Large Print Bks   38379

Buchanan District Library, 128 E Front St, *Buchanan*, MI 49107
T: +1 269 6953681; Fax: +1 269 6950004; E-mail: pasatlib@qtm.net; URL: www.buchananlibrary.org
Pam Salo
52 000 vols; 100 curr per   38380

Upshur County Public Library, Rte 6, *Buckhannon*, WV 26201; Box 480, Buckhannon
T: +1 304 4734219; Fax: +1 304 4734221; E-mail: upshur@clark.lib.wv.us; URL: upshurcounty.lib.wv.us
1956; Sandra Bumgardner
69 000 vols; 81 curr per
libr loan; ALA, PLA   38381

Buena Park Library District, 7150 La Palma Ave, *Buena Park*, CA 90620-2547
T: +1 714 8264100; Fax: +1 714 8265052; E-mail: library@buenapark.lib.ca.us; URL: www.buenaparklibrary.org
1919; Louise S. Mazerov
118 000 vols; 200 curr per; 6 323 av-mat; 3 630 sound-rec; 1 200 Large Print Bks
libr loan   38382

Buffalo & Erie County Public Library System, One Lafayette Sq, *Buffalo*, NY 14203-1887
T: +1 716 8588900; Fax: +1 716 8586211; URL: www.buffalolib.org
1836; Bridget Quinn-Carey
Mark Twain Coll, Niagara Falls Prints, World War I & II Posters; 19 branch libs
1 534 000 vols; 4 101 curr per   38383

Johnson County Library, 171 N Adams Ave, *Buffalo*, WY 82834
T: +1 307 6845546; Fax: +1 307 6847888; URL: wyld.state.wy.us/john
1909; Cynthia R. Twing
Western Hist Coll
57 000 vols; 158 curr per   38384

Burbank Public Library, 110 N Glenoaks Blvd, *Burbank*, CA 91502-1203
T: +1 818 2385600; Fax: +1 818 2385553; URL: www.burbank.lib.ca.us
1938; Sharon Cohen
5 000 Large Print Bks; 2 branch libs
370 000 vols; 600 curr per; 174 000 av-mat
libr loan   38385

Burbank Public Library – Buena Vista, 300 N Buena Vista St, *Burbank*, CA 91505-3208
T: +1 818 2385620; Fax: +1 818 2385623; URL: www.burbanklibrary.com
Christine Rodriguez
117 000 vols   38386

Burbank Public Library – Northwest, 3323 W Victory Blvd, *Burbank*, CA 91505-1543
T: +1 818 2385640; Fax: +1 818 2385642
Patrice Samko
61 000 vols   38387

Prairie Trails Public Library District, 8449 S Moody, *Burbank*, IL 60459
T: +1 708 4303688; Fax: +1 708 4305596; E-mail: pts@mls.lib.il.us; URL: www.prairietrailslibrary.org
1969
57 670 vols; 145 curr per   38388

Pender County Public Library, 103 S Cowan St, *Burgaw*, NC 28425; P.O. Box 879, Burgaw, NC 28425-0879
T: +1 910 2591234; Fax: +1 910 2590656; URL: www.youseemore.com/penderpl
1942; Michael Taylor
American Indian (Arnold Coll)
95 000 vols; 200 curr per   38389

Burley Public Library, 1300 Miller Ave, *Burley*, ID 83318-1729
T: +1 208 8787708; Fax: +1 208 8787018; E-mail: library@bplibrary.org; URL: www.bplibrary.org
1922
Idaho Coll
54 230 vols   38390

Burlingame Public Library, 480 Primrose Rd, *Burlingame*, CA 94010-4083
T: +1 650 5587417; Fax: +1 650 3426295; URL: www.burlingame.org/library
1909; Alfred H. Escoffier
1 branch libr
217 000 vols; 340 curr per
libr loan   38391

Alamance County Public Libraries, 342 S Spring St, *Burlington*, NC 27215
T: +1 336 2293588; Fax: +1 336 2293592; URL: www.alamancelibraries.org
1962; Judy Clayton Cobb
4 branch libs
227 000 vols; 309 curr per   38392

Boone County Public Library District, 1786 Burlington Pike, *Burlington*, KY 41005
T: +1 859 3422665; Fax: +1 859 6890435; E-mail: info@bcpl.org; URL: www.bcpl.org
1973; Lucinda A. Brown
346 000 vols; 690 curr per   38393

Burlington Public Library, 22 Sears St, *Burlington*, MA 01803
T: +1 781 2701690; Fax: +1 781 2290406; URL: www.burlingtonpubliclibrary.org
1857; Laura Hodgson
Business & management, Law
91 000 vols; 168 curr per   38394

Burlington Public Library, 210 Court St, *Burlington*, IA 52601
T: +1 319 7531647; Fax: +1 319 7530789; URL: www.burlington.lib.ia.us
1868; Rhonda J. Frevert
111 000 vols; 457 curr per; 3 740 av-mat   38395

Burlington Public Library, 166 E Jefferson St, *Burlington*, WI 53105
T: +1 262 7637623; Fax: +1 262 7631938; E-mail: director@burlington.lib.wi.us; URL: www.burlington.lib.wi.us
1908; Gayle A. Falk
Strangite Mormon Newspapers & Chronicles
61 000 vols; 184 curr per   38396

Coffey County Library, 410 Juniatta St, *Burlington*, KS 66839
T: +1 620 3642010; Fax: +1 620 3642603; URL: www.cclibraryks.org
1987; Valerie Williams
6 branch libs
87 000 vols; 288 curr per   38397

Fletcher Free Library, 235 College St, *Burlington*, VT 05401
T: +1 802 8657217; Fax: +1 802 8657227; URL: www.fletcherfree.org
1873; Robert Resnik
119 000 vols; 275 curr per; 3 152 sound-rec
libr loan   38398

Burnet County Library System, Herman Brown Free Library, 100 E Washington, *Burnet*, TX 78611
T: +1 512 7562328; Fax: +1 512 7562610; E-mail: bcls@burnetcountylibrary.org
1948; Ann Brock
80 000 vols; 250 curr per   38399

Avery-Mitchell-Yancey Regional Library System, 289 Burnsville School Rd, *Burnsville*, NC 28714; P.O. Drawer 310, Burnsville
T: +1 828 6824476; Fax: +1 828 6826277; URL: www.avery.lib.nc.us
1961; Dr. Daniel Barron
Video Colls; Census Records on microfilm; Minerals, incl Geology, Precious Stones, Gem Cutting, Jewelry, Pottery & Handicraft Ceramics; 1 bookmobile
161 000 vols; 198 curr per; 1 897 av-mat; 10 000 Large Print Bks, 2 010 Talking Bks   38400

Town of Ballston Community Library, Burnt Hills-Ballston Lake, Two Lawmar Lane, *Burnt Hills*, NY 12027
T: +1 518 3998174; Fax: +1 518 3991687; E-mail: bur-director@sals.edu; URL: burnthills.sals.org
1952; Karen DeAngelo
59 000 vols; 155 curr per   38401

Burton Public Library, 14588 W Park St, *Burton*, OH 44021; P.O. Box 427, Burton, OH 44021-0427
T: +1 440 8344466; Fax: +1 440 8340128; E-mail: email@burton.lib.oh.us; URL: www.burton.lib.oh.us
1910; Holly M. Lynn
Amish Coll, Circulating Software, Puppets
77 000 vols; 161 curr per
libr loan   38402

Choctaw County Public Library, 124 N Academy Ave, *Butler*, AL 36904
T: +1 205 4592542; E-mail: ccpl1@hotmail.com; URL: choctawbookworm.wetpaint.com; www.pinebelt.net/~ccpl
1954; Ashley Kay Taylor
Alabama; 2 branch libs
75 000 vols; 30 curr per
libr loan   38403

Butte-Silver Bow Public Library, 226 W Broadway St, *Butte*, MT 59701-9297
T: +1 406 7232138; Fax: +1 406 7821825
1890; Ann Drew
97 000 vols; 226 curr per; 2 578 av-mat
libr loan   38404

Byron Public Library District, 109 N Franklin St, *Byron*, IL 61010; P.O. Box 434, Byron, IL 61010-0434
T: +1 815 2345107; Fax: +1 815 2345582; E-mail: library@byron.lib.il.us; URL: www.byron.lib.il.us
1916; Penny O'Rourke
51 000 vols; 108 curr per; 345 e-books; 1 700 av-mat   38405

Kent District Library – Byron Township Branch, 8191 Byron Center Ave SW, *Byron Center*, MI 49315
T: +1 616 6473830; Fax: +1 616 8783933
Marie Van Fleet
52 000 vols   38406

Cadillac-Wexford Public Library, 411 S Lake St, *Cadillac*, MI 49601
T: +1 231 7756541; Fax: +1 231 7756778; E-mail: liedekea@cadillaclibrary.org; URL: www.cadillaclibrary.org
1906; Kathy Kirch
4 branch libs
147 000 vols; 275 curr per   38407

Puskarich Public Library, 200 E Market St, *Cadiz*, OH 43907-1185
T: +1 740 9422623; Fax: +1 740 9428047; URL: www.harrison.lib.oh.us
1880; Sandi Thompson
51 000 vols; 185 curr per   38408

Cairo Public Library, 1609 Washington Ave, *Cairo*, IL 62914; P.O. Box 151, Cairo, IL 62914-0151
T: +1 618 7341840; Fax: +1 618 7349346
Monica L. Smith
Army & Navy Records, Civil War Coll, Jesuit Relations
50 000 vols; 60 curr per
libr loan   38409

Roddenbery Memorial Library, 320 N Broad St, *Cairo*, GA 39828-2109
T: +1 229 3773632; Fax: +1 229 3777204; E-mail: rml@rmlibrary.org; URL: www.rmlibrary.org
1939; Alan L. Kaye
64 000 vols; 127 curr per; 47 av-mat
38410

Caldwell Public Library, 1010 Dearborn, *Caldwell*, ID 83605-4195
T: +1 208 4593242; Fax: +1 208 4597344; E-mail: caldwellweb@fiberpipe.net; URL: www.caldwell.lili.org
1887; Elaine Leppert
95 000 vols; 160 curr per
38411

State Library of Ohio – Southeastern Ohio Library Center, 40780 Marietta Rd, *Caldwell*, OH 43724
T: +1 740 7835705; URL: seoweb.seo.lib.oh.us
Dianna Clark
200 000 vols
38412

Camarena Memorial Library, 850 Encinas Ave, *Calexico*, CA 92231
T: +1 760 7682170; Fax: +1 760 3570404
1919; Sandra Tauler
Hist of Imperial Valley; 1 branch libr
75 000 vols; 104 curr per; 357 microforms; 100 av-mat
38413

Bunnvale Public Library, Seven Bunnvale Rd, Rte 513, *Califon*, NJ 07830
T: +1 908 6388884
Marie Taluba
58 000 vols; 50 curr per; 250 av-mat
38414

Calumet City Public Library, 660 Manistee Ave, *Calumet City*, IL 60409
T: +1 708 5626220; Fax: +1 708 8620872; URL: www.calumetcitypl.org
1927; William Pixley
127 000 vols; 327 curr per; 1 371 av-mat; 2 020 Large Print Bks, 1 371 Talking Bks
38415

Ventura County Library – Camarillo Library, 4101 Las Posas Rd, *Camarillo*, CA 93010
T: +1 805 3885222; Fax: +1 805 3885822
Sandi Banks
183 000 vols
38416

Camas Public Library, 625 NE Fourth Ave, *Camas*, WA 98607
T: +1 360 8344692; Fax: +1 360 8340199; E-mail: library@ci.camas.wa.us; URL: www.ci.camas.wa.us/library
1929; David Zavortink
62 000 vols; 200 curr per
38417

Cambridge Public Library, 359 Broadway, *Cambridge*, MA 02139
T: +1 617 3494044; Fax: +1 617 3494028; URL: www.cambridgema.gov/~cpl
Susan Flannery
6 branch libs
259 000 vols; 537 curr per
38418

Cambridge Public Library – Central Square Branch, 45 Pearl St, *Cambridge*, MA 02139
T: +1 617 3494010; Fax: +1 617 3494418
Esme Green
55 000 vols
38419

Dorchester County Public Library, 303 Gay St, *Cambridge*, MD 21613
T: +1 410 2287331; Fax: +1 410 2286313; E-mail: infodesk@dorchesterlibrary.org; URL: www.dorchesterlibrary.org
1922; Jean S. Del Sordo
Drama Coll, Education Coll, Maryland Coll; 1 branch libr
97 000 vols; 153 curr per
38420

East Central Regional Library, 244 S Birch, *Cambridge*, MN 55008-1588
T: +1 763 6897390; Fax: +1 763 6897389; E-mail: ecregion@ecrl.lib.mn.us; URL: ecrl.lib.mn.us
1959; Robert Boese
Minnesota
320 000 vols; 160 curr per
38421

Guernsey County District Public Library, 800 Steubenville Ave, *Cambridge*, OH 43725-2385
T: +1 740 4325946; Fax: +1 740 4327142; URL: www.gcdpl.lib.oh.us
1832; Richard E. Goodwin
2 branch libr, 1 bookmobile
130 000 vols; 320 curr per
38422

Camden Free Public Library, 418 Federal St, *Camden*, NJ 08103
T: +1 856 7577640; Fax: +1 856 7577631; URL: www.geocities.com/cfpl418/index.htm
1904; Theresa Gorman
Black Hist & Culture Coll
124 000 vols; 155 curr per; 250 av-mat
38423

Camden Public Library, 55 Main St, *Camden*, ME 04843-1703
T: +1 207 2363440; Fax: +1 207 2366673; URL: www.camden.lib.me.us
1896; Nikki Maounis
Edna St Vincent Millay Coll
60 000 vols; 114 curr per; 18 e-books
38424

Kershaw County Library, Camden Branch, 1304 Broad St, *Camden*, SC 29020-3595; mail address: 632 DeKalb St, Ste 109, Camden, SC 29020-4254
T: +1 803 4242352; Fax: +1 803 4242046; URL: www.kershawcountylibrary.org
1936; Amy Schofield
South Carolina; 2 branch libs, 1 bookmobile
88 000 vols; 190 curr per; 5 594 av-mat; 330 sound-rec; 300 digital data carriers; 4 280 Large Print Bks
38425

Camden County Library District, Camdenton Library, 89 Rodeo Rd, *Camdenton*, MO 65020; P.O. Box 1320, Camdenton, MO 65020-1320
T: +1 573 3465954; Fax: +1 573 3461263; URL: www.ccld.us
Carolyn F. Chittenden
99 000 vols; 227 curr per; 1 842 av-mat; ALA
38426

Cameron Parish Library, 498 Marshal St, P.O. Box 1130, *Cameron*, LA 70631-1130
T: +1 337 7755421; Fax: +1 337 7755346; E-mail: bmorgan@cameron.lib.la.us; URL: www.cameron.lib.la.us
1958; Bobbie Morgan
Louisiana; Cookbks
50 000 vols; 100 curr per
38427

El Dorado County Library – Cameron Park Branch, 2500 Country Club Dr, *Cameron Park*, CA 95682
T: +1 530 6215500; Fax: +1 530 6721346
Suzanne Plessinger-Skiar
78 000 vols
38428

De Soto Trail Regional Library, 145 E Broad St, *Camilla*, GA 31730-1842
T: +1 229 3368372; Fax: +1 229 3369353; E-mail: camillalibrary@yahoo.com; URL: www.georgialibraries.org/~desoto/
Lisa Rigsby
5 branch libs
50 000 vols; 76 curr per
38429

Cleve J. Fredricksen Library, 100 N 19th St, *Camp Hill*, PA 17011-3900
T: +1 717 7613900; Fax: +1 717 7615493; E-mail: fredricksen@ccpa.net; URL: www.ccpa.net/fredricksenpl/
1957; Bonnie Goble
120 000 vols
38430

Santa Clara County Library – Campbell Public, 77 Harrison Ave, *Campbell*, CA 95008-1433
T: +1 408 8661991; Fax: +1 408 8661433
Theresa Lehan
190 000 vols
38431

Taylor County Public Library, 205 N Columbia, *Campbellsville*, KY 42718
T: +1 270 4652527; Fax: +1 270 4658026; E-mail: taybooks@kyol.net; URL: www.taylorcountypubliclibrary.org
1974; Elaine J. Munday
58 000 vols; 21 curr per; 1 693 slides
38432

Canal Fulton Public Library, 154 Market St NE, *Canal Fulton*, OH 44614-1196
T: +1 330 8544148; Fax: +1 330 8549520; E-mail: info@canalfultonlibrary.org; URL: www.canalfultonlibrary.org
1937; Sandra E. Murphy
54 000 vols; 200 curr per
38433

Wood Library, 134 N Main St, *Canandaigua*, NY 14424
T: +1 716 3941381; Fax: +1 716 3942954; E-mail: woodlibrary@owwl.org; URL: woodlibrary.org
1857; Carol R. Shama
68 000 vols; 153 curr per
38434

Canastota Public Library, 102 W Center St, *Canastota*, NY 13032
T: +1 315 6977030; Fax: +1 315 6978653; URL: www.canastotalibrary.org
1896; Elizabeth Metzger
Children's Bks for Parents (Dorothy Canfield Fisher Award Coll)
58 050 vols; 150 curr per
38435

Canby Public Library, 292 N Holly St, *Canby*, OR 97013-3732
T: +1 503 2663394; Fax: +1 503 2661709; E-mail: caref@lincc.lib.or.us; URL: www.ci.canby.or.us/canbylibrary/home.htm
Beth Saul
Emma Wakefield Coll (mat on herbs)
60 000 vols; 70 curr per
38436

Los Angeles Public Library System – Canoga Park Branch, 20939 Sherman Way, *Canoga Park*, CA 91303
T: +1 818 8870320; Fax: +1 818 3461074; E-mail: cngopk@lapl.org
Renee Ardon
59 000 vols
38437

Canon City Public Library, 516 Macon Ave, *Canon City*, CO 81212-3380
T: +1 719 2699020; Fax: +1 719 2699031; E-mail: ccplrequests@hotmail.com; URL: ccpl.lib.co.us/
1886; Susan Ooton
69 000 vols; 130 curr per
38438

Canton Free Library, Eight Park St, *Canton*, NY 13617; P.O. Box 150, Canton, NY 13617-0150
T: +1 315 3863712; Fax: +1 315 3864131; E-mail: canton@northnet.org; URL: www.nc3r.org/canlib
1896; Carolyn J. Swafford
54 000 vols; 175 curr per
38439

Canton Public Library, 786 Washington St, *Canton*, MA 02021-3029
T: +1 781 8215027; Fax: +1 781 8215029; E-mail: caill@ocln.org; URL: www.town.canton.ma.us/library
1902; Mark Lague
Art & architecture, Business & management, Law
99 000 vols; 147 curr per
38440

Canton Public Library, 40 Dyer Ave, *Canton*, CT 06019
T: +1 860 6935800; Fax: +1 860 6935804; URL: www.cantonpubliclibrary.org
Kathleen Cockcroft
60 000 vols
38441

Canton Public Library, 1200 S Canton Center Rd, *Canton*, MI 48188-1600
T: +1 734 3970999; Fax: +1 734 3971130; URL: www.cantonpl.org
1980; Eva Davis
296 000 vols; 750 curr per; 19 000 e-books
38442

Madison County Library System, 102 Priestley St, *Canton*, MS 39046-4599
T: +1 601 8597733; Fax: +1 601 8590041; E-mail: feedback@mcls.ms; URL: www.mad.lib.ms.us
Sandra Sanders
African-American hist, Mississippi writers; Local Picture Coll, Madison County Hist Mat; 5 branch libs
177 000 vols; 224 curr per; 12 642 av-mat; 5 040 sound-rec
libr loan
38443

Madison County Library System – Canton Public Library, 102 Priestley St, *Canton*, MS 39046
T: +1 601 8593202; Fax: +1 601 8592728
Christine Greenwood
59 000 vols; 82 curr per
38444

Parlin Ingersoll Public Library, 205 W Chestnut St, *Canton*, IL 61520-2499
T: +1 309 6470328; Fax: +1 309 6478117; E-mail: parlin@parliningersoll.org; URL: www.parliningersoll.org
Fulton County Hist Coll
76 000 vols; 125 curr per; 5 300 av-mat; 9 500 sound-rec
libr loan
38445

Sequoyah Regional Library System, R. T. Jones Memorial Library, Headquarters, 116 Brown Industrial Pkwy, *Canton*, GA 30114-2899
T: +1 770 4793090; Fax: +1 770 4793069; URL: www.sequoyahregionallibrary.org
1956; Nicholas Fogarty
Career Ctr, Bks on Tape, Homeschooling Ctr, Large Print Bks, Spanish Ctr; 7 branch libs
301 000 vols
38446

Stark County District Library, 715 Market Ave N, *Canton*, OH 44702-1018
T: +1 330 4583152; Fax: +1 330 4520403; E-mail: contactus@starklibrary.org; URL: www.starklibrary.org
1884; Kent Oliver
10 branch libs
714 000 vols; 2 445 curr per; 81 770 av-mat; 47 260 sound-rec; 14 550 digital data carriers; 18 160 Large Print Bks
libr loan
38447

Cape Canaveral Public Library, 201 Polk Ave, *Cape Canaveral*, FL 32920-3067
T: +1 321 8681101; Fax: +1 321 8681103; URL: www.brev.org
1966; Isabel M. Matos-Escapa
Theater (Play Coll)
56 000 vols; 96 curr per; 2 170 av-mat; 868 sound-rec
libr loan; FLA, ALA
38448

Lee County Library System – Cape Coral-Lee County Public, 921 SW 39th Terrace, *Cape Coral*, FL 33914-5721
T: +1 239 5423953; Fax: +1 239 5422711
Sharon Myers
234 000 vols
38449

Cape Girardeau Public Library, 711 N Clark St, *Cape Girardeau*, MO 63701
T: +1 573 3345279; URL: www.capelibrary.org
1922; Elizabeth Martin
DAC & DAR Coll, Mississippi River Valley, Groves Genealogy, Hirsch Foreign Language for Children
99 000 vols; 154 curr per; 6 000 e-books
38450

Cape May County Library, 30 Mechanic St, mail address: 4 Moore Rd, DN 2030, *Cape May Court House*, NJ 08210
T: +1 609 4636350; Fax: +1 609 4653895; URL: www.cape-may.county.lib.nj.us
1925; Andrew Martin
Environmental studies, Marine sciences; 6 branch libs, 1 bookmobile
359 000 vols; 700 curr per
38451

Carbondale Public Library, 405 W Main St, *Carbondale*, IL 62901-2995
T: +1 618 4570354; Fax: +1 618 4570353; E-mail: cpllib@shawls.lib.il.us; URL: www.carbondale.lib.il.us
1923; Connie Steudel
95 000 vols; 202 curr per; 801 e-books
38452

Bosler Free Library, 158 W High St, *Carlisle*, PA 17013-2988
T: +1 717 2434642; Fax: +1 717 2438281
1900; Linda K. Rice
98 000 vols; 172 curr per; 130 av-mat
libr loan
38453

Carlsbad City Library, 1775 Dove Lane, *Carlsbad*, CA 92011-4048
T: +1 760 6022011; Fax: +1 760 6027942; E-mail: librarian@ci.carlsbad.ca.us; URL: www.ci.carlsbad.ca.us
1956; Bill Richmond
Genealogy, Local hist; 2 branch libs
287 000 vols; 1 400 curr per
libr loan
38454

Carlsbad Public Library, 101 S Halagueno St, *Carlsbad*, NM 88220
T: +1 505 8856776; Fax: +1 505 8858809; E-mail: carnold@elinlib.org; URL: carlsbadpubliclibrary.org
1897; Cassandra Arnold
Family Hist, New Mexico Hist, Waste Isolation Pilot Project (WIPP)
50 000 vols; 225 curr per   38455

Carmel Clay Public Library, 55 Fourth Ave SE, *Carmel*, IN 46032-2278
T: +1 317 8443362; Fax: +1 317 5714285; URL: www.carmel.lib.in.us
1904; Wendy A. Phillips
Indiana Coll
281 000 vols; 399 curr per   38456

Harrison Memorial Library, Carmel Public Library, Ocean Ave & Lincoln St, *Carmel*, CA 93921; P.O. Box 800, Carmel, CA 93921-0800
T: +1 831 6244629; Fax: +1 831 6240407; URL: www.hm-lib.org
1906; Janet Cubbage
Carmel Hist Coll (photogr), Edward Weston photos, Robinson Jeffers
80 000 vols; 217 curr per
libr loan   38457

Caro Area District Library, 840 W Frank St, *Caro*, MI 48723
T: +1 989 6734329; Fax: +1 989 6734777; E-mail: info@carolibrary.org; URL: www.carolibrary.org
1904; Marcia Dievendorf
63 000 vols; 162 curr per; 1 527 av-mat   38458

Carol Stream Public Library, 616 Hiawatha Dr, *Carol Stream*, IL 60188
T: +1 630 6530755; Fax: +1 630 6536809; URL: www.cslibrary.org
1962; Ann L. Kennedy
Arts & Crafts Coll, Vocational Advancement Coll, French Coll
167 000 vols; 390 curr per
libr loan   38459

Carroll Public Library, 118 E Fifth St, *Carroll*, IA 51401
T: +1 712 7923432; Fax: +1 712 7920141; E-mail: carrollpublic@yahoo.com; URL: www.carroll-library.org
1893; Linda Reida
Science fiction, Large type print
79 000 vols; 181 curr per   38460

Carroll County District Library, 70 Second St NE, *Carrollton*, OH 44615
T: +1 330 6272613; Fax: +1 330 6272523; E-mail: carroll@oplin.org; URL: www.carroll.lib.oh.us
1935; Robert Antill
Sports, Large print bks
80 000 vols; 200 curr per
libr loan   38461

Carrollton Public Library, 1700 Keller Springs Rd, *Carrollton*, TX 75006
T: +1 972 4664800; Fax: +1 972 4664265; URL: www.cityofcarrollton.com/library
1963; Jan Sapp
247 000 vols; 565 curr per   38462

Pickens County Cooperative Library, Hwy 17 S, *Carrollton*, AL 35447; P.O. Box 489, Carrollton, AL 35447-0489
T: +1 205 3678407; E-mail: pccl@nctv.com
Lori Ward Smith
Pickens County Historical Coll
50 000 vols   38463

County of Los Angeles Public Library – Carson Library, 151 E Carson St, *Carson*, CA 90745-2797
T: +1 310 8300901; Fax: +1 310 8306181; URL: colapublib.org/libs/carson
Leticia Tan
185 000 vols; 297 curr per   38464

Carson City Library, 900 N Roop St, *Carson City*, NV 89701
T: +1 775 8872244; Fax: +1 775 8872273; E-mail: cclb@ci.carson-city.nv.us; URL: www.carson-city.nv.us/library
1966; Sara Jones
Large print, Nevada
125 000 vols; 327 curr per   38465

Carteret Public Library, 100 Cooke Ave, *Carteret*, NJ 07008
T: +1 732 5413830; Fax: +1 732 5416948; E-mail: cfpl@lmxac.org; URL: www.carteretfreepubliclibrary.org
1931; Samuel Latini
71 000 vols; 140 curr per   38466

Bartow County Public Library System, 429 W Main St, *Cartersville*, GA 30120
T: +1 770 3824203; Fax: +1 770 3863056; URL: www.bartowlibraryonline.org/
1981; Carmen Melinda Sims
3 branch libs
81 000 vols; 121 curr per   38467

Bartow County Public Library System – Cartersville Main Street, 429 W Main St, *Cartersville*, GA 30120
T: +1 770 3824203; Fax: +1 770 3863056
Joe Byrne
56 000 vols   38468

Carthage Public Library, 612 S Garrison Ave, *Carthage*, MO 64836
T: +1 417 2377040; Fax: +1 417 2377041; E-mail: carthage@carthagelibrary.org; URL: carthage.lib.mo.us
1902; Jennifer Seaton
53 000 vols; 139 curr per; 7 000 e-books; 1 135 av-mat   38469

Moore County Library, 101 Saunders St, *Carthage*, NC 28327; P.O. Box 400, Carthage, NC 28327-0400
T: +1 910 9475335; Fax: +1 910 9473660; URL: www.srls.info/moore/mooreindex.html
Nancy Garner
86 000 vols; 88 curr per; 1 973 av-mat   38470

Sammy Brown Library, 522 West College, *Carthage*, TX 75633
T: +1 903 6936741; Fax: +1 903 6934503; URL: www.carthagetexas.com/Library/index.htm
1962
50 000 vols; 52 curr per   38471

Carver Public Library, Two Meadowbrook Way, *Carver*, MA 02330-1278
T: +1 508 8663415; Fax: +1 508 8663416; URL: www.carverpl.org
1895; Carole A. Julius
Local Genealogy Coll
54 000 vols; 100 curr per; 300 av-mat   38472

Cary Area Public Library District, 1606 Three Oaks Rd, *Cary*, IL 60013-1637
T: +1 847 6394210; Fax: +1 847 6398890; E-mail: librarybd@cary.lib.il.us; URL: www.caryarealibrary.info; www.cary.lib.il.us
1951; Diane R. McNulty
86 000 vols; 203 curr per; 9 000 e-books   38473

Wake County Public Library System – Cary Branch, 310 S Academy St, *Cary*, NC 27511
T: +1 919 4603350; Fax: +1 919 4603362
Liz Bartlett
115 000 vols; 114 curr per   38474

Wake County Public Library System – West Regional, 4000 Louis Stephens Dr, *Cary*, NC 27519
T: +1 919 4638500
Terri Luke
153 000 vols   38475

Natrona County Public Library, 307 E Second St, *Casper*, WY 82601
T: +1 307 2374935; Fax: +1 307 2663734; URL: www.natronacountylibrary.org
1910; Bill Nelson
Selective Govt Docs; 2 branch libs, 1 bookmobile
130 000 vols; 200 curr per   38476

Cass District Library, 319 Michigan Rd 62 N, *Cassopolis*, MI 49031-1099
T: +1 269 4453400; Fax: +1 269 4458795; E-mail: jray@cass.lib.mi.us; URL: www.cass.lib.mi.us
1940; Jennifer Ray
5 branch libs
56 649 vols
libr loan   38477

Douglas County Libraries, 100 S Wilcox, *Castle Rock*, CO 80104
T: +1 303 6887620; Fax: +1 303 6887655; URL: www.douglascountylibraries.org
1966; Jamie LaRue
6 branch libs
540 000 vols; 1 203 curr per; 550 e-books; 41 457 av-mat; 25 digital data carriers; 41 460 Audio Bks
libr loan; ALA, PLA   38478

Alameda County Library – Castro Valley Branch, 20055 Redwood Rd, *Castro Valley*, CA 94546-4382
T: +1 510 6706280; Fax: +1 510 5375991
Carolyn Moskovitz
90 000 vols; 233 curr per   38479

Cazenovia Public Library Society, Inc, 100 Albany St, *Cazenovia*, NY 13035
T: +1 315 6559322; Fax: +1 315 6555935; E-mail: cazenovia@midyork.org; URL: www.midyork.org/cazenovia
1886; Elizabeth Kennedy
53 000 vols; 75 curr per
libr loan   38480

Knox County Public Library System – Cedar Bluff Branch, 9045 Cross Park Dr, *Cedar Bluff*, TN 37923
T: +1 865 4707033; Fax: +1 865 4700927
Jackie Hill
86 000 vols   38481

Cedar City Public Library in the Park, 303 N 100 East, *Cedar City*, UT 84720
T: +1 435 5866661; Fax: +1 435 8657280; URL: cedarcitylibrary.org
1914; Steven D. Decker
Rare Book Coll; Local Newspaper Coll, micro
66 000 vols; 68 curr per   38482

Cedar Falls Public Library, 524 Main St, *Cedar Falls*, IA 50613-2830
T: +1 319 2738643; Fax: +1 319 2738648; E-mail: cfplref@gmail.com; URL: www.cedarfallspubliclibrary.org
1865; Sheryl Groskurth
116 000 vols; 253 curr per; 7 700 av-mat; 2 180 Audio Bks   38483

Zula-Bryant-Wylie Library, 225 Cedar St, *Cedar Hill*, TX 75104-2655
T: +1 972 72917323; Fax: +1 972 2915361; E-mail: library@cedarhilltx.com; URL: www.zulabwylielib.org
1948; Pat Bonds
55 000 vols; 110 curr per; 28 000 e-books   38484

Lake County Public Library – Cedar Lake Branch, 10010 W 133rd Ave, *Cedar Lake*, IN 46303
T: +1 219 3747121; Fax: +1 219 3746333
Linda Johnsen
56 000 vols; 49 curr per; 81 av-mat   38485

Cedar Park Public Library, 550 Discovery Blvd, *Cedar Park*, TX 78613
T: +1 512 4015608; Fax: +1 512 2595236; URL: www.cedarparktx.us
1981; Pauline P. Lam
90 000 vols; 217 curr per; 2 781 av-mat   38486

Cedar Rapids Public Library, 500 First St SE, *Cedar Rapids*, IA 52401-2095
T: +1 319 3985123; Fax: +1 319 3980476; E-mail: info@crlibrary.org; URL: www.crlibrary.org
1896; Tamara Glise
US Census Coll; 1 branch lib
272 000 vols; 412 curr per
libr loan   38487

Cedarburg Public Library, W63 N583 Hanover Ave, *Cedarburg*, WI 53012
T: +1 262 3757640; Fax: +1 262 3757618
1911; Mary Marquardt
76 000 vols; 205 curr per; 1 852 av-mat   38488

Mercer County District Library, 303 N Main St, *Celina*, OH 45822
T: +1 419 5864442; Fax: +1 419 5863222; E-mail: mercer@oplin.org; URL: www.mercer.lib.oh.us
1899; Austin R. Schneider
Children's lit
98 000 vols; 125 curr per   38489

Arapahoe Library District – Castlewood Xpress Library, 6739 S Uinta St, *Centennial*, CO 80112
Fax: +1 303 7713264
Laurie Christensen
71 000 vols   38490

Center Moriches Free Public Library, 235 Main St, *Center Moriches*, NY 11934
T: +1 631 8780940; Fax: +1 631 8780067; URL: centermoricheslibrary.org
1920; Nan Peel
80 000 vols; 115 curr per   38491

Middle Country Public Library, 101 Eastwood Blvd, *Centereach*, NY 11720
T: +1 631 5859393; Fax: +1 631 5855035; URL: www.mcpl.lib.ny.us
1960; Sandra Feinberg
Business & Finance Coll, Career Info Ctr, Children's Braille, Children's Foreign Language Coll, Sign Language, Large Print Bks, Medical Coll, Law Coll, Parents Ctr; 1 branch libr
348 000 vols; 108 307 curr per   38492

Washington-Centerville Public Library, 111 W Spring Valley Rd, *Centerville*, OH 45458
T: +1 937 4338091; Fax: +1 937 4331366; E-mail: cvref@wcpl.lib.oh.us; URL: www.wcpl.lib.oh.us
1930; N.N.
Chinese and Japanese language coll; 1 branch libr
383 000 vols; 437 curr per; 2 000 e-books
libr loan; ALA   38493

Central Islip Public Library, 33 Hawthorne Ave, *Central Islip*, NY 11722
T: +1 631 2349333; Fax: +1 631 2349386; E-mail: cisplib@suffolk.lib.ny.us; URL: www.suffolk.lib.ny.us/libraries/cisp/
1952; Anne Pavlak
121 000 vols; 232 curr per   38494

Centralia Regional Library District, 515 E Broadway, *Centralia*, IL 62801
T: +1 618 5325222; Fax: +1 618 5328578; URL: www.centralialibrary.org
1874; Joyce Jackson
Legislative Docs, State Daughters of the American Revolution Genealogy Coll; 4 branch libs
105 000 vols; 203 curr per   38495

Queen Anne's County Free Library, 121 S Commerce St, *Centreville*, MD 21617
T: +1 410 7580980; Fax: +1 410 7580614; URL: www.quan.lib.md.us
1909; Charles V. Powers
Decoy Carving & Boat Building Coll, Waterfowl; 1 branch libr
95 000 vols; 50 curr per   38496

Cerritos Library, 18025 Bloomfield Ave, *Cerritos*, CA 90703
T: +1 562 9161342; Fax: +1 562 9161375; E-mail: library@ci.cerritos.ca.us; URL: www.ci.cerritos.ca.us/library
1973; Don Buckley
Artists' Books, First Ladies' Coll, Performing Arts
220 000 vols; 387 curr per; 506 e-books; 21 350 av-mat; 6 300 sound-rec; 660 Large Print Bks
libr loan   38497

Cuyahoga County Public Library – Chagrin Falls Branch, 100 E Orange St, *Chagrin Falls*, OH 44022-2735
T: +1 440 2473556; Fax: +1 440 2470179
Kimberly Dressel
53 000 vols; 3 336 av-mat   38498

Coyle Free Library, 102 N Main St, *Chambersburg*, PA 17201
T: +1 717 2631054; Fax: +1 717 2632248; E-mail: dbigham@fclspa.org
1924
85 000 vols; 205 curr per
libr loan   38499

Grove Family Library, 101 Ragged Edge Rd S, *Chambersburg*, PA 17201
T: +1 717 2649663; Fax: +1 717 2646055; E-mail: realexlib@yahoo.com; URL: www.rel/rel.htm; www.fclspa.org/rel/rel.htm
1948; Louisa Cowles
51 000 vols
libr loan   38500

Champaign Public Library, 200 W Green St, *Champaign*, IL 61820-5193
T: +1 217 4032070; Fax: +1 217 4032073; URL: www.champaign.org
1876; Marsha Grove
1 branch libr
187 000 vols; 564 curr per
libr loan   38501

Chandler Public Library, 22 S Delaware, *Chandler*, AZ 85225; MS601, P.O. Box 4008, Chandler, AZ 85244-4008
T: +1 480 7822803; Fax: +1 480 7822823; URL: www.chandlerlibrary.org
1954; Brenda Brown
Arizona Indian Coll, Large Print Coll, New Reader Coll, Spanish Language Coll; 3 branch libs
471 000 vols; 270 curr per   38502

Chandler Public Library – Basha, 5990 S Val Vista Dr, *Chandler*, AZ 85249; MS 920, P.O. Box 4008, Chandler, AZ 85244-4008
T: +1 480 7822850; Fax: +1 480 7822855
Mary Sagar
55 000 vols; 91 curr per   38503

Chandler Public Library – Hamilton, 3700 S Arizona Ave, *Chandler*, AZ 85248-4500; MS917, P.O. Box 4008, Chandler, AZ 85244-4008
T: +1 480 7822828; Fax: +1 480 7822833
George Delalis
66 000 vols; 90 curr per   38504

Chandler Public Library – Sunset, 4930 W Ray Rd, *Chandler*, AZ 85226-6219; MS918, P.O. Box 4008, Chandler, AZ 85244-4008
T: +1 480 7822846; Fax: +1 480 7822848
Susan Hoffman
99 000 vols; 104 curr per   38505

Three Rivers Public Library District, 25207 W Channon Dr, *Channahon*, IL 60410-5028; P.O. Box 300, Channahon, IL 60410-0300
T: +1 815 4674010; Fax: +1 815 4674012; URL: www.three-rivers-library.org
1976; Erik R. Blomstedt
1 branch libr
80 000 vols; 203 curr per; 744 av-mat
                                     38506

Fairfax County Public Library – Chantilly Regional, 4000 Stringfellow Rd, *Chantilly*, VA 20151-2628
T: +1 703 5023883
Bonnie Worcester
168 000 vols   38507

Chanute Public Library, 111 N Lincoln, *Chanute*, KS 66720-1819
T: +1 620 4313820; Fax: +1 620 4313848; E-mail: chanulib@chanuteks.com
1905; Susan Willis
Esther Clark Hill Coll; Nora B Cunningham Coll, letters
60 000 vols; 130 curr per   38508

Chapel Hill Public Library, 100 Library Dr, *Chapel Hill*, NC 27514
T: +1 919 9682780; Fax: +1 919 9682838; E-mail: library@townofchapelhill.org; URL: www.chapelhillpubliclibrary.org
1958; Kathleen L. Thompson
110 000 vols; 150 curr per   38509

Chappaqua Central School District Public Library, 195 S Greeley Ave, *Chappaqua*, NY 10514
T: +1 914 2384779; Fax: +1 914 2383597; E-mail: chappaweb@westchesterlibraries.org; URL: www.chappaqualibrary.org
1922; Pamela Thornton
Special education; Horace Greeley
117 000 vols; 316 curr per   38510

Charles City Public Library, 106 Milwaukee Mall, *Charles City*, IA 50616
T: +1 641 2576319; Fax: +1 641 2576325; E-mail: director@charles-city.lib.ia.is; URL: www.charles-city.lib.ia.us
1904
Mooney Art Coll, Iowa Hist
52 000 vols; 156 curr per   38511

Charles Town Library, 200 E Washington St, *Charles Town*, WV 25414
T: +1 304 7252208; Fax: +1 304 7256618; E-mail: octldirector@frontier.net; URL: www.ctlibrary.org/octl
1927; P. Douglas Perks
Civil War Hist, Genealogy (Perry Room); Large Type Coll, West Virginia Hist
95 000 vols; 150 curr per   38512

Charleston Carnegie Public Library, 712 Sixth St, *Charleston*, IL 61920
T: +1 217 3454913; Fax: +1 217 3485616; E-mail: information@charlestonlibrary.org; URL: www.charlestonlibrary.org
1896
Education, Medicine, Religion
52 000 vols; 206 curr per   38513

Charleston County Public Library, 68 Calhoun St, *Charleston*, SC 29401
T: +1 843 8056930; Fax: +1 843 7273741; URL: www.ccpl.org
1930; Thomas Raines
Ethnic studies; 15 branch libs, 2 bookmobiles
1 112 000 vols; 2 361 curr per   38514

Charleston County Public Library – West Ashley, 45 Windermere Blvd, *Charleston*, SC 29407
T: +1 843 7666635
Beth Bell
56 000 vols   38515

Kanawha County Public Library, 123 Capitol St, *Charleston*, WV 25301
T: +1 304 3434646; Fax: +1 304 3486530; URL: kanawha.lib.wv.us
1909; Alan Engelbert
11 branch libs
624 000 vols   38516

Charlestown-Clark County Public Library, 51 Clark Rd, *Charlestown*, IN 47111-1997
T: +1 812 2563337; Fax: +1 812 2563890; URL: www.clarkco.lib.in.us/branch/charlestown.htm
1966; Tamsie Meurer
Lexicography Coll
51 000 vols; 125 curr per   38517

Charlotte Public Library, 226 S Bostwick, *Charlotte*, MI 48813
T: +1 517 5438859; Fax: +1 517 5438868; E-mail: charlib@ameritech.net; URL: www.charlottelibrary.org
1895; William Siarny
65 000 vols; 112 curr per   38518

Public Library of Charlotte & Mecklenburg County, 310 N Tryon St, *Charlotte*, NC 28202-2176
T: +1 704 4160105; Fax: +1 704 4160677; E-mail: webmaster@plcmc.org; URL: www.plcmc.org
1903; Charles Brown
Business Management Coll, Mecklenburg Res Room; 28 branch libs
1 322 000 vols; 5 000 Large Print Bks
libr loan   38519

Public Library of Charlotte & Mecklenburg County – Independence Regional, 6000 Conference Dr, *Charlotte*, NC 28212
T: +1 704 5683151; URL: www.plcmc.org/locations/branches.asp?id=10
Chris Bates
65 000 vols   38520

Public Library of Charlotte & Mecklenburg County – Morrison Regional Library, 7015 Morrison Blvd, *Charlotte*, NC 28211
T: +1 704 3362109; E-mail: morrison@plcmc.org; URL: www.plcmc.org/Locations/branches.asp?id=14
Elly Tomlinson
104 000 vols   38521

Public Library of Charlotte & Mecklenburg County – South County Regional, 5801 Rea Rd, *Charlotte*, NC 28277
T: +1 704 4166600; URL: www.plcmc.org/Locations/branches.asp?id=20
Tammy Baggett
118 000 vols   38522

Public Library of Charlotte & Mecklenburg County – Steele Creek, 13620 Steele Creek Rd, *Charlotte*, NC 28273
T: +1 704 5884345; E-mail: st@plcmc.org; URL: www.plcmc.org/Locations/branches.asp?id=21
Susan McDonald
53 000 vols   38523

Public Library of Charlotte & Mecklenburg County – University City Regional, 301 East W T Harris Blvd, *Charlotte*, NC 28262
T: +1 704 5959828; E-mail: univcity@plcmc.org; URL: www.plcmc.org/Locations/branches.asp?id=22
Julia Smith
85 000 vols   38524

Jefferson-Madison Regional Library, Central Library, 201 E Market St, *Charlottesville*, VA 22902-5287
T: +1 434 9797151; Fax: +1 434 9717035; E-mail: central@jmrl.org; URL: www.jmrl.org
1921; John Halliday
Charlottesville/Albemarle Hist Coll; 7 branch libs
522 000 vols   38525

Carver County Library, Four City Hall Plaza, *Chaska*, MN 55318-1963
T: +1 952 4489395; Fax: +1 952 4489392; URL: www.carverlib.org
1975; Melissa Brechon
5 branch libs
193 000 vols; 680 curr per   38526

Chatham Area Public Library District, 600 E Spruce St, *Chatham*, IL 62629
T: +1 217 4832713; Fax: +1 217 4832361; URL: ghs.bcsd.k12.il.us/library
1987; Linda Meyer
62 000 vols; 124 curr per
libr loan; ALA   38527

Chatham Public Library, 11 Woodbridge Ave, *Chatham*, NY 12037-1399
T: +1 518 3923666; Fax: +1 518 3921546; E-mail: chathampubliclibrary@chatham.k12.ny.us; URL: chatham.lib.ny.us
1884; Wendy Fuller
64 000 vols; 80 curr per   38528

Eldredge Public Library, 564 Main St, *Chatham*, MA 02633-2296
T: +1 508 9455170; Fax: +1 508 9455173; URL: www.eldredgelibrary.org
1895; Irene B. Gillies
Genealogy (Edgar Francis Waterman Coll), Life Saving Service Rpts
55 000 vols; 120 curr per; 16 e-books
                                     38529

The Library of The Chathams, 214 Main St, *Chatham*, NJ 07928
T: +1 973 6350603; Fax: +1 973 6357827; E-mail: Diane.OBrien@mainlib.org; URL: www.chatham-library.org
1907; Diane O'Brien
Psychology, Earth science, Health
90 000 vols; 220 curr per; 5 000 av-mat
libr loan   38530

Pittsylvania County Public Library, 24 Military Dr, *Chatham*, VA 24531
T: +1 434 4323271; Fax: +1 434 4321405; E-mail: pittslibrary@hotmail.com; URL: www.pcplib.org
1939; Diane S. Adkins
Pittsylvania County local hist and genealogy
113 000 vols; 156 curr per; 8 microforms; 2 114 av-mat; 3 100 sound-rec; 52 digital data carriers; 3 000 other items
libr loan   38531

Washington County Public Library, 14102 Saint Stephens Ave, *Chatom*, AL 36518; P.O. Box 1057, Chatom, AL 36518-1057
T: +1 251 8472097; Fax: +1 251 8472098
Jessica Ross
55 000 vols; 116 curr per   38532

Chattanooga-Hamilton County Bicentennial Library, 1001 Broad St, *Chattanooga*, TN 37402-2652
T: +1 423 7575315; Fax: +1 423 7574343; E-mail: library@lib.chattanooga.gov; URL: www.lib.chattanooga.gov
1905; David F. Clapp
Interviews Chattanooga & Hamilton County Hist, Genealogy, Tennesseana (Tennessee Rm Coll)
479 000 vols; 952 curr per   38533

Chattanooga-Hamilton County Bicentennial Library – Eastgate, 5900 Bldg, 5705 Marlin Rd, Ste 1500, *Chattanooga*, TN 37411
T: +1 423 8552689; Fax: +1 423 8552696
Margaret Curtis
53 000 vols   38534

Cheektowaga Public Library, Julia Boyer-Reinstein Library, 1030 Losson Rd, *Cheektowaga*, NY 14227
T: +1 716 6684991; Fax: +1 716 6684806
1938; Salvatore Bordonaro
3 branch libs
159 000 vols; 517 curr per   38535

Cheektowaga Public Library – Reinstein Memorial, 2580 Harlem Rd, *Cheektowaga*, NY 14225
T: +1 716 8928089; Fax: +1 716 8923370
Christine Bazan
58 000 vols   38536

Chelmsford Public Library, 25 Boston Rd, *Chelmsford*, MA 01824-3088
T: +1 978 2565521; Fax: +1 978 2568511; E-mail: askus@mvlc.org; URL: www.chelmsfordlibrary.org
1894; Becky Herrmann
1 branch libr
118 000 vols; 238 curr per; 39 microforms; 2 555 av-mat; 8 342 sound-rec; 20 digital data carriers; 147 art reproductions   38537

Chelsea Public Library, 569 Broadway, *Chelsea*, MA 02150-2991
T: +1 617 4664355; Fax: +1 617 4664359; E-mail: coclibrary@chelseama.gov; URL: www.chelseama.gov
1870; Robert E. Collins
Large Print Coll, Massachusetts Annotated Laws, Massachusetts Law, Spanish Coll
75 000 vols; 75 curr per
libr loan   38538

Cherry Hill Public Library, 1100 Kings Hwy N, *Cherry Hill*, NJ 08034-1911
T: +1 856 9031218; Fax: +1 856 6679503; URL: www.chplnj.org
1957; Manuel A. Paredes
Business & investing, Foreign trade; New Jersey, Telephone Directories Coll
145 000 vols; 471 curr per
libr loan   38539

Chesapeake Public Library, 298 Cedar Rd, *Chesapeake*, VA 23322-5512
T: +1 757 4107101; Fax: +1 757 4107150; URL: www.chesapeake.lib.va.us
1961; Elizabeth Fowler
Family, Hist, Law; Wallace Memorial Libr Coll (Local Hist), City Records Management; 5 branch libs
609 000 vols; 1 038 curr per; 16 000 e-books; 30 180 av-mat; 32 300 sound-rec
                                     38540

Cheshire Public Library, 104 Main St, *Cheshire*, CT 06410-2499
T: +1 203 2722245; Fax: +1 203 2727714; E-mail: cheshire@cheshirelibrary.org; URL: www.cheshirelibrary.org
1892; Ramona Harten
97 000 vols; 266 curr per; 9 720 av-mat; 6 500 sound-rec   38541

Chester County Library, 100 Center St, *Chester*, SC 29706
T: +1 803 3778145; Fax: +1 803 3778146; URL: www.chesterlibsc.org
1900; Marguerite Dube
97 000 vols; 90 curr per   38542

Chester Public Library, 1784 Kings Hwy, *Chester*, NY 10918
T: +1 845 4694252; Fax: +1 845 4697583; URL: www.rcls.org/chs
Lynn Coppers
55 000 vols; 120 curr per   38543

J. Lewis Crozer Library, 620 Engle St, *Chester*, PA 19013-2199
T: +1 610 4943459; Fax: +1 610 4948954; E-mail: crozerlibrary@delco.lib.pa.us
1894; Katie C. Newell
Black hist, Delaware County hist
52 000 vols; 90 curr per; 250 av-mat
38544

Chesterfield County Library, 119 W Main St, *Chesterfield*, SC 29709
T: +1 843 6237489; Fax: +1 843 6233295; E-mail: library@chesterfield.lib.sc.us
1969
69 140 vols; 229 curr per
38545

Chesterfield County Public Library, 9501 Lori Rd, *Chesterfield*, VA 23832; P.O. Box 297, Chesterfield, VA 23832-0297
T: +1 804 7481603; Fax: +1 804 7514679; URL: www.library.chesterfield.gov
1965; Michael R. Mabe
Law Libr Coll; 9 branch libs
739 000 vols; 908 curr per
38546

Westchester Public Library, Thomas Library, 200 W Indiana, *Chesterton*, IN 46304-3122
T: +1 219 9267696; Fax: +1 219 9266424; URL: www.wpl.lib.in.us
1972; Phil Baugher
1 branch libr
154 000 vols; 125 curr per; 11 850 av-mat; 8 750 sound-rec; 4 200 digital data carriers
38547

Kent County Public Library, 408 High St, *Chestertown*, MD 21620-1312
T: +1 410 7783636; Fax: +1 410 7786756; E-mail: referencedesk@kent.lib.md.us; URL: www.kentcountylibrary.org/index.php
1961
2 branch libs
60 710 vols; 110 curr per
38548

Laramie County Library System, 2200 Pioneer Ave, *Cheyenne*, WY 82001-3610
T: +1 307 6343561; Fax: +1 307 6342082; URL: www.lclsonline.org
1886; Lucie P. Osborn
Western Hist Coll, Elk of North America; 2 branch libs, 1 bookmobile
318 000 vols; 436 curr per; 20 910 av-mat; 9 390 sound-rec; 410 Large Print Bks
38549

Chicago Public Library, 400 S State St, *Chicago*, IL 60605
T: +1 312 7474396; URL: www.chicagopubliclibrary.org
1872; Craig Davis
87 branch libs
5 527 000 vols
libr loan
38550

Chicago Ridge Public Library, 10400 Oxford Ave, *Chicago Ridge*, IL 60415-1507
T: +1 708 4237753; Fax: +1 708 4232758; E-mail: kmcswain@chicagoridge.lib.il.us; URL: www.chicagoridge.lib.il.us
1966; Kathleen McSwain
53 170 vols; 456 curr per
38551

Chickasha Public Library, 527 Iowa Ave, *Chickasha*, OK 73018
T: +1 405 2226075; Fax: +1 405 2226072; E-mail: chicklib@chickasha.lib.ok.us
Catharine Cook
52 000 vols; 93 curr per
38552

Chicopee Public Library, 449 Front St, *Chicopee*, MA 01013
T: +1 413 5941800; Fax: +1 413 5941819; E-mail: cpl@chicopeepubliclibrary.org; URL: www.chicopeepubliclibrary.org
1853; Nancy M. Contois
Polish language; 3 branch libs
107 000 vols; 300 curr per
38553

Chillicothe & Ross County Public Library, 140-146 S Paint St, *Chillicothe*, OH 45601; P.O. Box 185, Chillicothe, OH 45601-0185
T: +1 740 7024160; Fax: +1 740 7024153; E-mail: crcpl@oplin.org; URL: www.crcpl.org

1859; Jennifer Thompson
Ross County Census Records, 1820 through 1900, micro; Burton E Stevenson Coll; 6 branch libs
66 000 vols; 1 080 av-mat; 3 063 sound-rec; 2 190 digital data carriers; 2 690 Large Print Bks
38554

Livingston County Library, 450 Locust St, *Chillicothe*, MO 64601-2597
T: +1 660 6460547; Fax: +1 660 6465504; E-mail: librarian@livingstoncountylibrary.org; URL: www.livingstoncountylibrary.org
1921; Robin S. Westphal
Missouri Hist (George Somerville Coll)
54 000 vols; 100 curr per; 6 000 e-books
38555

Washington County Library, 1444 Jackson Ave, *Chipley*, FL 32428
T: +1 850 6381314; Fax: +1 850 6389499; URL: www.pplcs.org
Linda Norton
150 000 vols
38556

Chippewa Falls Public Library, 105 W Central, *Chippewa Falls*, WI 54729-2397
T: +1 715 7231146; Fax: +1 715 7206922; E-mail: cflib@ifls.lib.wi.us; URL: www.chippewafallslibrary.org
1893; Rosemary Kilbridge
Wisconsin (Hist Coll), bk, microfilm, fs
95 000 vols; 215 curr per; 13 000 e-books
38557

Chula Vista Public Library, Civic Center, 365 F St, *Chula Vista*, CA 91910-2697
T: +1 619 6915069; Fax: +1 619 4274246; URL: www.chulavistalibrary.com
1891; David Palmer
2 branch libs
426 000 vols; 16 841 av-mat; 16 810 Audio Books, 600 Electronic Media & Resources
libr loan
38558

Chula Vista Public Library – Eastlake, 1120 Eastlake Pkwy, *Chula Vista*, CA 91915-2102
T: +1 619 6560314
Scott Love
58 000 vols
38559

Chula Vista Public Library – South Chula Vista, 389 Orange Ave, *Chula Vista*, CA 91911-4116
T: +1 619 5855750; Fax: +1 619 4201591
Ramiro Gonzalez
188 000 vols
38560

Cicero Public Library, 5225 W Cermak Rd, *Cicero*, IL 60804
T: +1 708 6528084; Fax: +1 708 6528095; E-mail: ciceropublic@yahoo.com; URL: cicerolibrary.org
1921; Jane Schoen
Polish language, Spanish language; 1 branch libr
112 000 vols; 465 curr per; 12 Spec Interest Per Sub
38561

Hamilton North Public Library, 209 W Brinton, *Cicero*, IN 46034
T: +1 317 9845623; Fax: +1 317 9847505; URL: www.hnpl.lib.in.us
Samuel Mitchell
88 000 vols; 200 curr per; 2 000 av-mat
38562

Cincinnati Regional Library for the Blind & Physically Handicapped, 800 Vine St, *Cincinnati*, OH 45202-2071
T: +1 513 3696999; Fax: +1 513 3693111; E-mail: lb@cincinnatilibrary.org; URL: www.cincinnatilibrary.org
Donna Foust
410 000 vols; 40 curr per; 388 303 av-mat
38563

Public Library of Cincinnati & Hamilton County, Cincinnati & Hamilton County Public Library, 800 Vine St, *Cincinnati*, OH 45202-2009
T: +1 513 3696900; Fax: +1 513 3693123; E-mail: info@cincinnatilibrary.org; URL: www.cincinnatilibrary.org
1853; Kimber L. Fender
55 branch libs
4 191 000 vols; 13 912 curr per; 112 997 av-mat; 272 320 sound-rec; 89 570 Large Print Bks
libr loan
38564

Public Library of Cincinnati & Hamilton County – Anderson, 7450 State Rd, *Cincinnati*, OH 45230
T: +1 513 3696030; Fax: +1 513 3694444
Patricia Peterson
94 000 vols
38565

Public Library of Cincinnati & Hamilton County – Bond Hill, 1740 Langdon Farm Rd at Jordan Crossing, *Cincinnati*, OH 45237
T: +1 513 3694445; Fax: +1 513 3694532
Holbrook Sample
52 000 vols
38566

Public Library of Cincinnati & Hamilton County – Covedale, 4980 Glenway Ave, *Cincinnati*, OH 45238
T: +1 513 3694460; Fax: +1 513 3694461
Eileen Mallory
60 000 vols
38567

Public Library of Cincinnati & Hamilton County – Delhi Township, 5095 Foley Rd, *Cincinnati*, OH 45238
T: +1 513 3696019; Fax: +1 513 3694453
Susan Hamrick
78 000 vols
38568

Public Library of Cincinnati & Hamilton County – Green Township, 6525 Bridgetown Rd, *Cincinnati*, OH 45248
T: +1 513 3696095; Fax: +1 513 3694482
Sarah Connatser
77 000 vols
38569

Public Library of Cincinnati & Hamilton County – Groesbeck, 2994 W Galbraith Rd, *Cincinnati*, OH 45239
T: +1 513 3694454; Fax: +1 513 3694455
Ned Heeger-Brehm
76 000 vols
38570

Public Library of Cincinnati & Hamilton County – North Central, 11109 Hamilton Ave, *Cincinnati*, OH 45231
T: +1 513 3696068; Fax: +1 513 3694459
Dale Snair
66 000 vols
38571

Public Library of Cincinnati & Hamilton County – Sharonville, 10980 Thornview Dr, *Cincinnati*, OH 45241
T: +1 513 3696049; Fax: +1 513 3694504
Chris Holt
68 000 vols
38572

Burlington County Library – Cinnaminson Branch, 1619 Riverton Rd, *Cinnaminson*, NJ 08077
T: +1 856 8299340; Fax: +1 856 8292243; E-mail: cb@bcls.lib.nj.us
Eileen Rauth
83 000 vols
38573

Chilton Clanton Library, 100 First Ave, *Clanton*, AL 35045
T: +1 205 7551768; Fax: +1 205 7551374; URL: ccpl.lib.al.us
1963; Mary Jo Abernathy
Genealogy; 3 branch libs
69 000 vols; 25 curr per
38574

County of Los Angeles Public Library – Claremont Library, 208 N Harvard Ave, *Claremont*, CA 91711
T: +1 909 6214902; Fax: +1 909 6212366; URL: www.colapublib.org/libs/claremont
Donald Slaven
137 000 vols; 213 curr per
38575

Will Rogers Library, 1515 N Florence Ave, *Claremore*, OK 74017
T: +1 918 3411564; Fax: +1 918 3420362; URL: www.claremorecity.com/index.aspx?nid=168
1936; Sherry Beach
Indians, Will Rogers
75 000 vols; 92 curr per
38576

Clarence Public Library, Three Town Pl, *Clarence*, NY 14031
T: +1 716 7412650; Fax: +1 716 7411243; URL: www.buffalolib.org
1933; Roseanne Butler-Smith
61 000 vols
38577

Clarendon Hills Public Library, Seven N Prospect Ave, *Clarendon Hills*, IL 60514
T: +1 630 3238188; Fax: +1 630 3238189; E-mail: cns@clarendonhillslibrary.org
1963; Margaret Brandon
Travel
51 000 vols; 146 curr per
38578

Clark Public Library, 303 Westfield Ave, *Clark*, NJ 07066
T: +1 732 3885999; Fax: +1 732 3887866; E-mail: director@clarklibrary.org; URL: www.clarklibrary.org
1961; Maureen Baker Wilkinson
60 000 vols; 190 curr per
libr loan
38579

Northeast Georgia Regional Library, Clarksville-Habersham County Library, 178 E Green St, *Clarksville*, GA 30523; P.O. Box 2020, Clarksville
T: +1 706 7540416; Fax: +1 706 7543479; E-mail: webadmin@negeorgialibraries.org; URL: www.negeorgialibraries.org
1938; Emerson G. Murphy
5 branch libs
200 000 vols; 426 curr per; 7 442 av-mat; 7 450 Talking Bks
38580

Abington Community Library, 1200 W Grove St, *Clarks Summit*, PA 18411-9501
T: +1 570 5873440; URL: www.lackawannacountylibrarysystem.org
1961; Leah Rudolph
Audio-Visual Coll
70 000 vols; 100 curr per
38581

Clarksburg-Harrison Public Library, 404 W Pike St, *Clarksburg*, WV 26301
T: +1 304 6272236; Fax: +1 304 6272239; URL: www.clarksburglibrary.info
1907; Beth Nicholson
UFO (Gray Barker Coll)
72 000 vols
38582

Carnegie Public Library of Clarksdale & Coahoma County, 114 Delta Ave, *Clarksdale*, MS 38614
T: +1 662 6244461; Fax: +1 662 6274344
1914
Archaeology, artifacts; Delta Blues Music Coll
65 000 vols; 189 curr per; 350 microforms; 1 500 av-mat
libr loan
38583

Independence Township Public Library, 6495 Clarkston Rd, *Clarkston*, MI 48346-1501
T: +1 248 6252212; Fax: +1 248 6258852; URL: www.indelib.org
1955; Julia Meredith
100 000 vols; 250 curr per
38584

Clarksville-Montgomery County Public Library, 350 Pageant Lane, Ste 501, *Clarksville*, TN 37040
T: +1 931 6488826; Fax: +1 931 6488831; URL: www.clarksville.org
1894; Stephen Lesnak
Brown Harvey Genealogy Room; Family & Tennessee Hist, bks & microfilm
270 000 vols; 128 curr per; 2 266 sound-rec; 11 digital data carriers
38585

Jeffersonville Township Public Library – Clarksville Branch, 1312 Eastern Blvd, *Clarksville*, IN 47129-1704
T: +1 812 2855640; Fax: +1 812 2855642; URL: jefferson.lib.in.us
Kathleen Rosga
50 000 vols; 82 curr per
38586

Blair Memorial Library, 416 N Main St, *Clawson*, MI 48017-1599
T: +1 248 5885500; Fax: +1 248 5883114; E-mail: clawill@tln.lib.mi.us; URL: www.clawson.lib.mi.us
1929; Elizabeth Kelman
Auto Repair Manuals
65 000 vols; 130 curr per; 2 060 sound-rec; 11 digital data carriers; 3 011 videocassettes + 31 DVDs
38587

Contra Costa County Library – Clayton Library, 6125 Clayton Rd, *Clayton*, CA 94517
T: +1 925 6730659; Fax: +1 925 6730359
Karen Hansen-Smith
69 000 vols
38588

Public Library of Johnston County & Smithfield – Hocutt-Ellington Memorial, 100 S Church St, *Clayton*, NC 27520
T: +1 919 5535542; Fax: +1 919 5531529
Betty Coats
50 000 vols; 57 curr per; 1212 av-mat
38589

Davis County Library – North, 562 S 1000 East, *Clearfield*, UT 84015
T: +1 801 8256662
Patricia York
94 000 vols
38590

Clearwater Public Library System, 100 N Osceola Ave, *Clearwater*, FL 33755
T: +1 727 5624970; Fax: +1 727 5624977; URL: www.myclearwater.com/cpl
1915; Barbara Pickell
4 branch libs
230 000 vols; 427 curr per
38591

Clearwater Public Library System – Countryside, 2741 State Rd 580, *Clearwater*, FL 33761
T: +1 727 6691290; Fax: +1 727 6691289
Georgiana Ata
116 000 vols; 129 curr per
38592

Clearwater Public Library System – East, 2251 Drew St, *Clearwater*, FL 33765
T: +1 727 6691280; Fax: +1 727 6691281
Ann Scheffer
102 000 vols; 140 curr per
38593

Cooper Memorial Library, 2525 Oakley Seaver Dr, *Clermont*, FL 32711
T: +1 352 5362275; URL: www.mylakelibrary.org/libraries/cooper_memorial_library.aspx
1914; Bruce Boyd
Large type print
52 000 vols; 109 curr per; 50 maps
38594

Bolivar County Library System – Robinson-Carpenter Memorial Library, 104 S Leflore Ave, *Cleveland*, MS 38732
T: +1 662 8432774; Fax: +1 662 8434701; URL: www.bolivar.lib.ms.us
1958; Linda Kern
6 branch libs
102 000 vols; 188 curr per
38595

Cleveland Public Library, 795 Church St NE, *Cleveland*, TN 37311-5295
T: +1 423 4722163; Fax: +1 423 3399791; E-mail: info@clevelandlibrary.org; URL: www.clevelandlibrary.org
1923; Andrew Hunt
Corn Cherokee coll, Tennessee genealogy coll; History Branch
113 000 vols; 154 curr per; 24 000 e-books; 54 mss; 6 000 microforms; 9 772 av-mat; 4 719 sound-rec; 126 digital data carriers; 3 017 vertical files, 22 art reproductions, 64 toys
libr loan; TLA
38596

Cleveland Public Library, 325 Superior Ave, *Cleveland*, OH 44114-1271
T: +1 216 6232827; Fax: +1 216 6237015; E-mail: info@cpl.org; URL: www.cpl.org
1869; N.N.
Chess, Folklore & Orientalia (John G White Coll), OAS, Photogr Coll, US Patent Coll, Architecture Coll, Cookbks, Visual Arts Coll; 29 branch libs
4 500 000 vols; 5 502 curr per; 16 000 e-books; 78 290 av-mat; 166 210 sound-rec; 39 290 digital data carriers
libr loan
38597

Cleveland Public Library – Library for the Blind & Physically Handicapped, 17121 Lake Shore Blvd, *Cleveland*, OH 44110-4006
T: +1 216 6232911; Fax: +1 216 6237036; URL: www.cpl.org
1931; Barbara Mates
Ohio Coll, braille & cassettes; VF on Visual & Physical Disabilities
300 000 vols
38598

Cleveland Heights-University Heights Public Library, 2345 Lee Rd, *Cleveland Heights*, OH 44118-3493
T: +1 216 9320932; Fax: +1 216 9320932; URL: www.heightslibrary.org
1916; Nancy S. Levin
Deafness; Parenting Coll; 3 branch libs
284 000 vols; 1 033 curr per; 16 000 e-books; 5 174 av-mat; 7 060 Audio Bks, 60 Electronic Media & Resources, 500 High Interest/Low Vocabulary Bks, 1 000 Bks on Deafness & Sign Lang
38599

Hendry County Library System – Clewiston Public Library (Headquarters), 120 W Osceola Ave, *Clewiston*, FL 33440
T: +1 863 9831493; Fax: +1 863 9839194
1962; Barbara Oeffner
71 000 vols; 152 curr per
38600

Cliffside Park Free Public Library, 505 Palisade Ave, *Cliffside Park*, NJ 07010
T: +1 201 9452867; Fax: +1 201 9451016; E-mail: clpcirc@bccls.org; URL: cliffsidepark.bccls.org/kids.htm
1913; Ana Chelariu
65 000 vols; 161 curr per
38601

Clifton Park-Halfmoon Public Library, 475 Moe Rd, *Clifton Park*, NY 12065-3808
T: +1 518 3718622; Fax: +1 518 3713799; URL: www.cphlibrary.org
1969; Josephine L. Piracci
Local Hist (Howard I Becker Memorial Coll); Job, Business & Finance Info
124 000 vols; 264 curr per
38602

Audubon Regional Library, 12220 Woodville St, *Clinton*, LA 70722; P.O. Box 8389, Clinton, LA 70722-8389
T: +1 225 6838753; Fax: +1 225 6834634; E-mail: admin.c1ar@state.lib.la.us; URL: www.youseemore.com/audubon
1963; Mary Bennett Lindsey
4 branch libs
89 000 vols; 212 curr per
38603

Bigelow Free Public Library, 54 Walnut St, *Clinton*, MA 01510
T: +1 978 3654160; Fax: +1 978 3654161; E-mail: bigelowlibrary@yahoo.com; URL: www.bigelowlibrary.org
Christine Flaherty
138 000 vols; 60 curr per
libr loan
38604

Clinton Public Library, 118 S Hicks St, *Clinton*, TN 37716-2826
T: +1 865 4570519; Fax: +1 865 4574233; E-mail: clinton_library@comcast.net
Jane Giles
Genealogy Res Coll
50 000 vols; 52 curr per
38605

Henry Carter Hull Library, Inc, Ten Killingworth Tpk, *Clinton*, CT 06413
T: +1 860 6692342; Fax: +1 860 6698318; E-mail: askus@hchlibrary.org; URL: www.hchlibrary.org
1925; Maribeth Breen
84 000 vols; 130 curr per
38606

Henry County Library, 123 E Green St, *Clinton*, MO 64735-1462
T: +1 660 8852612; Fax: +1 660 8858953; URL: tacnet.missouri.org/hcl
1946; Elizabeth A. Cashell
80 000 vols; 96 curr per; 2 115 av-mat
38607

Hunterdon County Library – North County, 65 Halstead St, *Clinton*, NJ 08809
T: +1 908 7306135; Fax: +1 908 7306467
Barbara Riesenfeld
99 000 vols
38608

Vespasian Warner Public Library District, 310 N Quincy, *Clinton*, IL 61727
T: +1 217 9355174; Fax: +1 217 9354425; URL: www.warner.lib.il.us
1901; Tom Rudasill
C H Moore Coll (Early Illinois Hist & Geography)
80 000 vols; 200 curr per
38609

Western Plains Library System – Clinton Public Library, 721 Frisco, *Clinton*, OK 73601-3320
T: +1 580 3232165; Fax: +1 580 3237884; E-mail: clinton.public@wplibs.com
Kathy Atchley
75 000 vols; 50 curr per
38610

Clinton-Macomb Public Library, 40900 Romeo Plank Rd, *Clinton Township*, MI 48038-2955
T: +1 586 2265020; Fax: +1 586 2265008; E-mail: info@cmpl.org; URL: www.cmpl.org
1992; Larry Neal
Low vision
278 000 vols
38611

Macomb County Library, 16480 Hall Rd, *Clinton Township*, MI 48038-1132
T: +1 586 4125987; Fax: +1 586 4125958; E-mail: mclweb@libcoop.net; URL: www.libcoop.net/mcl
1946; Darlene LaBelle
Business, Investment, Career info, Consumer health, Foundations, Fund raising; 1 branch libr
151 000 vols; 393 curr per
libr loan
38612

Clintonville Public Library, 75 Hemlock St, *Clintonville*, WI 54929-1461
T: +1 715 8234563; Fax: +1 715 8237134; E-mail: cpl@mail.owls.lib.wi.us; URL: www.clintonvillelibrary.org
1905; Kathleen Mitchell
61 000 vols; 173 curr per; 14 000 e-books
ALA
38613

Lonesome Pine Regional Library – Jonnie B. Deel Memorial, 198 Chase St, *Clintwood*, VA 24228; P.O. Box 650, Clintwood, VA 24228-0650
T: +1 276 9266617; Fax: +1 276 9266795; E-mail: jbdlib@lprlibrary.org
Sheila Phipps
64 000 vols; 65 curr per; 1 358 av-mat
38614

Cloquet Public Library, 320 14th St, *Cloquet*, MN 55720-2100
T: +1 218 8791531; Fax: +1 218 8796531; URL: www.cloquet.lib.mn.us
1895; Mary Lukkarila
54 000 vols; 155 curr per
38615

Closter Public Library, 280 High St, *Closter*, NJ 07624-1898
T: +1 201 7684197; Fax: +1 201 7684220; E-mail: cltrcirc@bccls.org; URL: www.bccls.org/closter
1956; Laura DeAngelis
57 000 vols; 127 curr per; 2 130 av-mat; 50 digital data carriers
libr loan; ALA, NSLA
38616

Clovis-Carver Public Library, 701 N Main, *Clovis*, NM 88101
T: +1 505 7697840; Fax: +1 505 7697842; URL: www.cityofclovis.org/library
1949; Marilyn Belcher
New Mexico Docs
115 000 vols; 139 curr per; 1 761 av-mat
38617

Fresno County Public Library – Clovis Regional, 1155 Fifth St, *Clovis*, CA 93612-1391
T: +1 559 2999531
Joseph Augustino
65 000 vols
38618

Coal City Public Library District, 85 N Garfield St, *Coal City*, IL 60416
T: +1 815 6344552; Fax: +1 815 6342950; E-mail: ccpld@coalcity.lib.il.us; URL: www.ccpld.org
1886; Jolene Franciskovich
51 000 vols; 332 curr per; 1 500 sound-rec
38619

Coalinga-Huron USD Library District, Coalinga District Library, 305 N Fourth St, *Coalinga*, CA 93210
T: +1 559 9351676; Fax: +1 559 9351058
1912; Carol Kreamer
1 branch libr
85 000 vols; 123 curr per; 1 833 av-mat
38620

Coatesville Area Public Library, Dr. Michael Margolies Library, 501 E Lincoln Hwy, *Coatesville*, PA 19320-3413
T: +1 610 3844115; Fax: +1 610 3847551; E-mail: coastaff@ccls.org; URL: www.ccls.org/othlibs/coats.htm
1936; Mike Geary
65 000 vols; 85 curr per; 1 960 av-mat; 500 sound-rec; 12 digital data carriers
libr loan
38621

Baltimore County Public Library – Cockeysville Area Branch, 9833 Greenside Dr, *Cockeysville*, MD 21030-2188
T: +1 410 8877750; Fax: +1 410 6660325; E-mail: cockeysville@bcpl.net
Margaret Prescott
157 000 vols
38622

Brevard County Library System, 308 Forrest Ave, 2nd Fl, ADMIN, *Cocoa*, FL 32922-7781
T: +1 321 6331801; Fax: +1 321 6331798; URL: www.brev.org
1972; Catherine Schweinsberg
1 243 000 vols; 2 505 curr per; 108 000 av-mat; 28 640 sound-rec
38623

Central Brevard Library & Reference Center, 308 Forrest Ave, *Cocoa*, FL 32922
T: +1 321 6331793; Fax: +1 321 6331964; URL: www.brev.org
1895; Catherine Schweinsberg
Large print
162 000 vols; 364 curr per; 60 718 av-mat; 5 992 digital data carriers; 4 400 Audio Bks, 60 720 Talking Bks
38624

Cocoa Beach Public Library, 550 N Brevard Ave, *Cocoa Beach*, FL 32931
T: +1 321 8681104; Fax: +1 321 8681107; URL: www.cocoabeachpubliclibrary.org
1955; Ray Dickinson
Art (Cocoa Beach Artists), oil, watercolor, batik & bronze
104 000 vols; 296 curr per
38625

Park County Library, 1500 Heart Mountain St, *Cody*, WY 82414
T: +1 307 5271880; Fax: +1 307 5271888; URL: parkcountylibrary.org/cody/
1906; Frances Clymer
68 000 vols; 130 curr per
38626

Coeur D'Alene Public Library, 702 E Front Ave, *Coeur d'Alene*, ID 83814
T: +1 208 7692315; Fax: +1 208 7692381; E-mail: info@cdalibrary.org; URL: www.cdalibrary.org
1904
Human Rights Coll, Idaho Coll
67 770 vols; 143 curr per
38627

Coffeyville Public Library, 311 W 10th St, *Coffeyville*, KS 67337
T: +1 620 2511370; Fax: +1 620 2511512; E-mail: Cvillepl@coffeyvillepl.org; URL: www.cvillepublib.org
1911; Jennifer Dalton
Coll of Genealogy
72 000 vols; 124 curr per; 15 maps; 573 microforms; 1 321 av-mat; 997 sound-rec; 147 digital data carriers; 34 art repros
libr loan; ALA
38628

Paul Pratt Memorial Library, 35 Ripley Rd, *Cohasset*, MA 02025-2097
T: +1 781 3831348; Fax: +1 781 3831698; E-mail: library@cohassetlibrary.org; URL: www.cohassetlibrary.org
1879; Jackie Rafferty
54 000 vols; 164 curr per; 52 av-mat; 675 sound-rec; 1 digital data carriers
38629

Burnham Memorial Library, 898 Main St, *Colchester*, VT 05446
T: +1 802 8797576; Fax: +1 802 8795079; E-mail: info@burnham.lib.vt.us; URL: www.burnham.lib.vt.us
1902; Martha Reid
Large print bks
50 000 vols; 95 curr per
38630

Cragin Memorial Library, Eight Linwood Ave, *Colchester*, CT 06415; P.O. Box 508, Colchester, CT 06415-0508
T: +1 860 5375752; Fax: +1 860 5374559; E-mail: craginml@yahoo.com; URL: colchesterct.net
Siobhan M. Grogan
55 000 vols; 160 curr per
38631

Campbell County Public Library District, 3920 Alexandria Pike, *Cold Spring*, KY 41076
T: +1 859 7816166; Fax: +1 859 5725032; URL: www.cc-pl.org
1978; J. C. Morgan
2 branch libs
196 000 vols; 394 curr per; 105 e-books; 19 630 av-mat; 14 740 music data carriers; 40 Bks on Deafness & Sign Lang
libr loan   38632

Cold Spring Harbor Library, 75 Goose Hill Rd, *Cold Spring Harbor*, NY 11724
T: +1 631 6926820; Fax: +1 631 6926827; URL: www.cshlibrary.org
1886
Large Print Books
50 000 vols; 139 curr per   38633

Coldwater Branch Library, 10 E Chicago St, *Coldwater*, MI 49036-1615
T: +1 517 2782341;
Fax: +1 517 2797134; URL: www.branchdistrictlibrary.org/coldwater
1881
Genealogical Res Mat; 5 branch libs
71 470 vols; 95 curr per   38634

Grant Parish Library, 300 Main St, *Colfax*, LA 71417-1830
T: +1 318 6279920; Fax: +1 318 6279900; E-mail: admin.h1gr@pelican.state.lib.la.us; URL: www.grant.lib.la.us
1959; Doris Lively
4 branch libs
62 000 vols; 131 curr per   38635

Whitman County Library, Colfax (Main) Branch, 102 S Main St, *Colfax*, WA 99111-1863
T: +1 509 3974366; Fax: +1 509 3976156; E-mail: info@whitco.lib.wa.us; URL: www.whitco.lib.wa.us
1945; Kristie Kirkpatrick
78 000 vols; 3 000 sound-rec; 50 digital data carriers
libr loan   38636

Bryan+College Station Public Library System – Larry J. Ringer Library, 1818 Harvey Mitchell Pkwy S, *College Station*, TX 77845
T: +1 979 7643416; Fax: +1 979 7646379
Amy Beck
80 000 vols   38637

Collingswood Free Public Library, 771 Haddon Ave, *Collingswood*, NJ 08108-3714
T: +1 856 8580649; Fax: +1 856 8585016; URL: www.collingswood.lib.nj.us
1910; Bradley A. Green
Southern New Jersey Coll, bks, blue prints, engravings, maps, trade cat, photos
65 000 vols; 120 curr per   38638

Collinsville Memorial Public Library District, 408 W Main St, *Collinsville*, IL 62234
T: +1 618 3441112; Fax: +1 618 3456401; E-mail: cve@lcls.org; URL: www.collinsvillelibrary.org
1915; Barbara Rhodes
Programming
65 000 vols; 250 curr per; 7 000 e-books   38639

Coloma Public Library, 151 W Center St, *Coloma*, MI 49038-0430
T: +1 269 4683431; Fax: +1 269 4688077
1963; Charles Dickinson
55 000 vols; 204 curr per   38640

Pikes Peak Library District, 20 N Cascade Ave, *Colorado Springs*, CO 80903; P.O. Box 1579, Colorado Springs, CO 80901-1579
T: +1 719 5316333; Fax: +1 719 3898989; URL: www.ppld.org
1905; Paula J. Miller
11 branch libs
919 000 vols; 2 101 curr per; 2 000 e-books
libr loan   38641

Pikes Peak Library District – Ruth Holley Branch, 685 N Murray Blvd, *Colorado Springs*, CO 80915
T: +1 719 5975377; Fax: +1 719 5971214
Janet Cox
60 000 vols   38642

Colton Public Library, 656 N Ninth St, *Colton*, CA 92324
T: +1 909 3705083; Fax: +1 909 4220873; URL: www.ci.colton.ca.us/pages/librarywelcome.htm
1906; Diana Lynn Fraser
1 branch libr
100 000 vols; 300 curr per
libr loan   38643

Blue Grass Regional Library Center, 104 E Sixth St, *Columbia*, TN 38401-3359
T: +1 931 3889282; Fax: +1 931 3881762
1954; Marion K. Bryant
Large print bks; 1 branch libr
140 000 vols   38644

Howard County Library, 6600 Cradlerock Way, *Columbia*, MD 21045-4912
T: +1 410 3137750; Fax: +1 410 3137742; URL: www.hclibrary.org
1940; Valerie Gross
Health sciences; Literacy Mat, County Detention Libr, Mat for the Disabled; 7 branch libs
953 000 vols; 1 324 curr per
libr loan   38645

Howard County Library – Central, 10375 Little Patuxent Pkwy, *Columbia*, MD 21044-3499
T: +1 410 3137800; Fax: +1 410 3137864
Nina Krzysko
301 000 vols   38646

Howard County Library – East Columbia, 6600 Cradlerock Way, *Columbia*, MD 21045-4912
T: +1 410 3137700; Fax: +1 410 3137741
N.N.
167 000 vols   38647

Maury County Public Library, 211 W Eighth St, *Columbia*, TN 38401
T: +1 931 3756501;
Fax: +1 931 3756519; URL: www.maurycountylibrary.org
Elizabeth Potts
79 000 vols; 214 curr per   38648

Richland County Public Library, 1431 Assembly St, *Columbia*, SC 29201-3101
T: +1 803 9293400; Fax: +1 803 9293439; URL: www.myrcpl.com
1934; C. David Warren
Columbia; South Carolina and Richland County (South Carolina) USA hist colls; 9 branch libs
1 259 000 vols; 2 763 curr per; 2 000 e-books; 1 235 maps; 67 000 microforms; 27 130 av-mat; 67 900 sound-rec; 8 170 digital data carriers
libr loan   38649

South Mississippi Regional Library – Columbia Marion County Library, 900 Broad St, *Columbia*, MS 39429
T: +1 601 7365516; Fax: +1 601 7361379
1972; Linda Gail Bracey
2 branch libs
52 000 vols; 228 curr per   38650

Peabody Public Library, 1160 E Hwy Rd 205, *Columbia City*, IN 46725; P.O. Box 406, Columbia City, IN 46725-0406
T: +1 260 2445541; Fax: +1 260 2445653; URL: ppl.lib.in.us
1901; Janet M. Scank
Large print bks
84 000 vols; 195 curr per   38651

Columbia Heights Public Library, 820 4th Ave NE, *Columbia Heights*, MN 55421
T: +1 763 7063690; Fax: +1 763 7063691; URL: www.anoka.lib.mn.us
1928; Rebecca Loader
71 000 vols; 110 curr per   38652

Columbiana Public Library, 332 N Middle St, *Columbiana*, OH 44408
T: +1 330 4825509; Fax: +1 330 4829669; E-mail: columlib@oplin.org; URL: www.columbiana.lib.oh.us
1934; Carol Cobbs
58 000 vols; 187 curr per   38653

Bartholomew County Public Library, 536 Fifth St, *Columbus*, IN 47201-6225
T: +1 812 3791266; Fax: +1 812 3791275; E-mail: library@barth.lib.in.us; URL: www.barth.lib.in.us
1899; Elizabeth Booth-Poor
American hist, Hist, Architecture; Talking Bks Coll; 2 branch libs
345 000 vols; 400 curr per; 25 000 av-mat
libr loan   38654

Chattahoochee Valley Regional Library System, Columbus Public Library, Headquarters, 3000 Macon Rd, *Columbus*, GA 31906-2201
T: +1 706 2432609; Fax: +1 706 2432710; E-mail: reference@cvrls.net; URL: www.thecolumbuslibrary.org
1908; Claudya Muller
8 branch libs
328 000 vols; 1 100 curr per   38655

Columbus Metropolitan Library, 96 S Grant Ave, *Columbus*, OH 43215-4781
T: +1 614 8491209; Fax: +1 614 8491309; E-mail: webmaster@columbuslibrary.org; URL: www.columbuslibrary.org
1872; Patrick A. Losinski
Black Heritage Coll; 23 branch libs
2 176 000 vols; 4 628 curr per
libr loan   38656

Columbus Metropolitan Library – Hilltop, 511 S Hague Ave, *Columbus*, OH 43204
T: +1 614 4793431; Fax: +1 614 4794439
Joe Yersavich
120 000 vols   38657

Columbus Metropolitan Library – Karl Road, 5590 Karl Rd, *Columbus*, OH 43229
T: +1 614 6452275; Fax: +1 614 4794259
Gay Banks
170 000 vols   38658

Columbus Metropolitan Library – Livingston Branch, 3434 E Livingston Ave, *Columbus*, OH 43227
T: +1 614 6452275; Fax: +1 614 4794339
Shirley Freeman
90 000 vols   38659

Columbus Metropolitan Library – Main Library, 96 S Grant Ave, *Columbus*, OH 43215-4781
T: +1 614 6452275; Fax: +1 614 8491110
Susan Studebaker
Black Heritage Coll;Columbus & Ohio (Local History Coll), bks, micro, VF
781 000 vols   38660

Columbus Metropolitan Library – Northern Lights, 4093 Cleveland Ave, *Columbus*, OH 43224
T: +1 614 6452275; Fax: +1 614 4794249 .
Gay Banks
73 000 vols   38661

Columbus Metropolitan Library – Northwest, 2280 Hard Rd, *Columbus*, OH 43235
T: +1 614 8072650;
Fax: +1 614 8072659; URL: www.worthingtonlibraries.org
Cathy Allen
170 000 vols   38662

Columbus Metropolitan Library – South High, 3540 S High St, *Columbus*, OH 43207
T: +1 614 4793361; Fax: +1 614 6452275
Sandee Wagle
84 000 vols   38663

Columbus Metropolitan Library – Southeast, 3980 S Hamilton Rd, *Columbus*, OH 43125
T: +1 614 6452275; Fax: +1 614 4794359
Nancy Sullivan
112 000 vols   38664

Columbus Metropolitan Library – Whetstone Branch, 3909 N High St, *Columbus*, OH 43214
T: +1 614 4793151; Fax: +1 614 6452275
Greg Denby
170 000 vols   38665

Columbus Metropolitan Library – Whitehall Branch, 4371 E Broad St, *Columbus*, OH 43213
T: +1 614 4793321; Fax: +1 614 6452275
Deborah Replogle
63 000 vols   38666

Columbus Public Library, 2504 14th St, *Columbus*, NE 68601-4988
T: +1 402 5647116; Fax: +1 402 5633378; URL: www.megavision.net/library
1900; R. C. Trautwein
Play & Theatre Coll
89 000 vols; 170 curr per; 3 400 av-mat   38667

Columbus-Lowndes Public Library, 314 N Seventh St, *Columbus*, MS 39701
T: +1 662 3295300; Fax: +1 662 3295156; E-mail: pres@lowndes.lib.ms.us; URL: www.lowndes.lib.ms.us
1940; Benjamin E. Petersen
Eudora Welty Special Coll, Margaret Latimer Buckley Room, Annie Laurie Leech Coll (Genealogy), Lowndes County Arch, Tennessee Williams; 3 branch libs
95 000 vols; 151 curr per   38668

Grandview Heights Public Library, 1685 W First Ave, *Columbus*, OH 43212
T: +1 614 4862951; Fax: +1 614 4817021; URL: www.ghpl.org
1924; Mary Ludlum
136 000 vols; 400 curr per; 7 000 e-books; 3 600 Audio Bks   38669

Upper Arlington Public Library, 2800 Tremont Rd, *Columbus*, OH 43221
T: +1 614 4860900; Fax: +1 614 4864530; URL: www.ualibrary.org
1967; Ann R. Moore
Foreign language; 2 branch libs
465 000 vols; 506 curr per; 4 000 e-books
libr loan   38670

Colusa County Free Library, 738 Market St, *Colusa*, CA 95932
T: +1 530 4587671; Fax: +1 530 4587358; E-mail: sevenlibraries@countyofcolusa.org
Ellen H. Brow
6 branch libs
88 000 vols; 77 curr per; 300 av-mat   38671

City of Commerce Public Library, 5655 Jillson St, *Commerce*, CA 90040-1485
T: +1 323 7226660; Fax: +1 323 7241978; E-mail: referencel@ci.commerce.ca.us; URL: www.ci.commerce.ca.us/library.asp
1961; Vilko Domic
Coatings (Varnish & Paint)
132 000 vols; 399 curr per; 3 127 av-mat; 1 246 sound-rec; 20 digital data carriers
libr loan; ALA, SLA, PLA   38672

County of Los Angeles Public Library – Compton Library, 240 W Compton Blvd, *Compton*, CA 90220-3109
T: +1 310 6370202; Fax: +1 310 5371141; URL: www.colapublib.org/libs/compton
Sharon Johnson
106 000 vols; 125 curr per   38673

Comstock Township Library, 6130 King Hwy, *Comstock*, MI 49041; P.O. Box 25, Comstock, MI 49041-0025
T: +1 269 3450136; Fax: +1 269 3450138; E-mail: adultlibrarian@yahoo.com; URL: www.comstocklibrary.org
1938; Margaret King-Sloan
79 000 vols; 125 curr per
libr loan; ALA, MLA   38674

Kent District Library, 814 W River Center Dr, *Comstock Park*, MI 49321
T: +1 616 7842016; Fax: +1 616 6479211; URL: www.kdl.org
1936; Martha Smart
19 branch libs
881 000 vols; 2 355 curr per; 2 000 e-books; 34 079 av-mat; 57 849 sound-rec; 64 880 digital data carriers; 111 Braille vols, 1 060 Bks on Deafness & Sign Lang, 12 362 Large Print Bks   38675

Cabarrus County Public Library, 27 Union St N, *Concord*, NC 28025-4793
T: +1 704 9202050; Fax: +1 704 7843822; URL: www.cabarruscounty.us/library
1911; Thomas W. Dillard
Holt Coll, original art; 3 branch libs
260 000 vols; 537 curr per; 7 130 av-mat; 3 000 Large Print Bks
libr loan   38676

Concord Free Public Library, 129 Main St, *Concord*, MA 01742-2494
T: +1 978 3183301; Fax: +1 978 3183344; URL: www.concordlibrary.org
1873; Barbara Powell
American lit, Hist; Henry David Thoreau Coll, Nathaniel Hawthorne Coll, Ralph Waldo Emerson Coll; 1 branch libr
232 000 vols; 570 curr per; 3 340 Large Print Bks   38677

Concord Public Library, 45 Green St, *Concord*, NH 03301-4294
T: +1 603 2258670; Fax: +1 603 2303693; E-mail: library@onconcord.com; URL: www.onconcord.com/library
1855; Patricia A. Immen
1 branch libr
124 000 vols; 356 curr per   38678

Contra Costa County Library – Concord Library, 2900 Salvio St, *Concord*, CA 94519-2597
T: +1 925 6465455; Fax: +1 925 6465453
Maureen Kilmurray
80 000 vols   38679

Conneaut Public Library, 304 Buffalo St, *Conneaut*, OH 44030-2658
T: +1 440 5931608; Fax: +1 440 5934470; E-mail: conneaut@oplin.org; URL: www.conneaut.lib.oh.us
1908; Deborah Zingaro
Large Print; Local newspapers, 1835-1982; New Grove Directory of Music & Musicians; Old Radio Shows Coll; Young Adult
53 000 vols; 127 curr per   38680

Carnegie Free Library, 299 S Pittsburgh St, *Connellsville*, PA 15425-3580
T: +1 724 6281380; Fax: +1 724 6285636; E-mail: carnegie@zoominternet.net; URL: www.carnegiefreelib.org
1903; Julia Allen
Local & Fayette County Hist (Pennsylvania Coll)
64 000 vols; 108 curr per; 921 av-mat   38681

Fayette County Public Library, 828 N Grand Ave, *Connersville*, IN 47331
T: +1 765 8270883; Fax: +1 765 8254592; E-mail: fcplconnersville@yahoo.com; URL: www.fcplibrary.lib.in.us
Marilyn Robinson
90 000 vols; 240 curr per; 1 623 av-mat   38682

Montgomery County Memorial Library System, 104 I-45 N, *Conroe*, TX 77301-2720
T: +1 936 7888377; Fax: +1 936 7888398; URL: www.countylibrary.org
1948; Jerilynn A. Williams
5 branch libs
478 000 vols   38683

Faulkner-Van Buren Regional Library System, 1900 Tyler St, *Conway*, AR 72032
T: +1 501 3277482; Fax: +1 501 3279098; E-mail: fcl@fcl.org; URL: www.fcl.org
1938; Ruth Ann Voss
Arkansas Coll, Faulkner County Law Libr Coll; 6 branch libs
183 000 vols; 309 curr per; 530 e-books   38684

Conyers-Rockdale Library System, 864 Green St, *Conyers*, GA 30012
T: +1 770 3885040; Fax: +1 770 3885043
Deborah S. Manget
100 000 vols; 129 curr per   38685

Upper Cumberland Regional Library, 208 E Minnear St, *Cookeville*, TN 38501-3949
T: +1 931 5264016; Fax: +1 931 5283311; URL: www.state.tn.us/sos/statelib/p&d/upper
1946; Jennifer Cowan-Henderson
119 000 vols   38686

Anoka County Library – Crooked Lake, 11440 Crooked Lake Blvd, *Coon Rapids*, MN 55433-3441
T: +1 763 5765972; Fax: +1 763 5765973
Kathy Petron
63 000 vols; 137 curr per   38687

Coos Bay Public Library, 525 Anderson St, *Coos Bay*, OR 97420-1678
T: +1 541 2691101; Fax: +1 541 2697567; URL: bay.cooslibaries.org; www.cooslibraries.org
1910; Carol Ventgen
Oregon Hist (Helene Stack Bower Oregon Coll)
119 000 vols; 253 curr per   38688

Copiague Memorial Public Library, 50 Deauville Blvd, *Copiague*, NY 11726-4100
T: +1 631 6911111; Fax: +1 631 6915098; E-mail: copglib@suffolk.lib.ny.us; URL: www.copiaguelibrary.org
1961; Alicja Feitzinger
Foreign Language
138 000 vols; 8 500 curr per; 162 av-mat; 1 520 Large Print Bks, 170 Talking Bks   38689

Coppell Public Library, William T. Cozby Library, 177 N Hertz Rd, *Coppell*, TX 75019
T: +1 972 3043065; Fax: +1 972 3043622; E-mail: library@ci.coppell.tx.us; URL: www.ci.coppell.tx.us
Kathleen Metz Edwards
85 000 vols; 189 curr per; 28 000 e-books   38690

Copperas Cove Public Library, 501 S Main St, *Copperas Cove*, TX 76522
T: +1 254 5473826; Fax: +1 254 5427279; URL: www.ci.copperas-cove.tx.us
1959; Cherri Shelnutt
Texana
56 000 vols
libr loan   38691

Coralville Public Library, 1401 5th St, *Coralville*, IA 52241
T: +1 319 2481850; E-mail: agalstad@coralville.lib.ia.us; URL: www.coralvillepubliclibrary.org
1965; Alison Ames Galstad
66 730 vols; 169 curr per   38692

Cordell Public Library, 208 S College, *Cordell*, OK 73632
T: +1 580 8323530; Fax: +1 580 8323530; E-mail: cordell.public@wplibs.com; URL: www.wplibs.com/branches/cordell
Rhonda Schmidt
50 000 vols; 12 curr per   38693

Memphis Public Library & Information Center – Cordova Branch, 8457 Trinity Rd, *Cordova*, TN 38018
T: +1 901 7548443; Fax: +1 901 7546874
Caroline Barnett
114 000 vols   38694

Corfu Free Library, Seven Maple Ave, *Corfu*, NY 14036; P.O. Box 419, Corfu, NY 14036-0419
T: +1 585 5993321; Fax: +1 585 5993321; E-mail: corfuref@nioga.org
Kelly A. March
75 000 vols; 25 curr per   38695

Northeast Regional Library, 1023 Fillmore St, *Corinth*, MS 38834-4199
T: +1 662 2877311; Fax: +1 662 2868010; URL: www.nereg.lib.ms.us
1951; William L. McMullin
Genealogy; 13 branch libs
200 000 vols; 211 curr per; 2 000 microforms; 6 967 av-mat; 4 799 sound-rec; 69 Art Reproductions
libr loan   38696

Northeast Regional Library – Corinth Public Library, 1023 Fillmore St, *Corinth*, MS 38834-4199
T: +1 662 2872441; Fax: +1 662 2868010
Ann F. Coker
64 000 vols   38697

Southeast Steuben County Library, 300 Civic Center Plaza, Ste 101, *Corning*, NY 14830
T: +1 607 9363713; Fax: +1 607 9361741; URL: www.stls.org/corning
2000; Lise Gilliland
Glass, Literacy; Caldecott & Newbery Winners & Honors, Feature Film Coll
118 000 vols; 125 curr per   38698

Corona Public Library, 650 S Main St, *Corona*, CA 92882
T: +1 951 7362382; Fax: +1 951 7362499; URL: www.coronapubliclibrary.org
1899; Julie Fredericksen
Corona Newspaper Arch, photos
160 000 vols; 400 curr per
libr loan   38699

Queens Borough Public Library – Lefrak City, 98-30 57th Ave, *Corona*, NY 11368
T: +1 718 5927677
Jiang Liu
58 000 vols   38700

Coronado Public Library, 640 Orange Ave, *Coronado*, CA 92118-1526
T: +1 619 5227390; Fax: +1 619 4354205; URL: coronado.lib.ca.us
1890; Christian R. Esquevin
Gardening, World War II; Coronado Govnt Docs
152 000 vols; 344 curr per
libr loan   38701

Corpus Christi Public Libraries, Central Library, 805 Comanche, *Corpus Christi*, TX 78401
T: +1 361 8807000; Fax: +1 361 8807005; E-mail: library@ccpl.ci.corpus-christi.tx.us; URL: www.library.ci.corpus-christi.tx.us
1909; Herbert G. Canales
4 branch libs
376 000 vols; 622 curr per
libr loan   38702

Corry Public Library, 117 W Washington St, *Corry*, PA 16407
T: +1 814 664-7611/-4404; Fax: +1 814 6630742; E-mail: cplib@tbscc.com; URL: www.corrylibrary.org
1900; Frances Church
50 000 vols; 95 curr per   38703

Corsicana Public Library, 100 N 12th St, *Corsicana*, TX 75110
T: +1 903 6544813; Fax: +1 903 6544814; E-mail: library@ci.corsicana.tx.us; URL: www.ci.corsicana.tx.us
1901; Rodney L. Bland
64 000 vols; 119 curr per   38704

Marin County Free Library – Corte Madera Branch, 707 Meadowsweet, *Corte Madera*, CA 94925-1717
T: +1 415 9243515
Nancy Davis
89 000 vols   38705

Cortland Free Library, 32 Church St, *Cortland*, NY 13045
T: +1 607 7531042; Fax: +1 607 7587329; URL: www.flls.org/cortlandlib
1886; Kay Zaharis
90 000 vols; 195 curr per; 60 digital data carriers; 1 107 Audio Bks, 60 CDs, 900 Large Print Bks   38706

Community District Library, Administration Office, 231 N Shiawassee St, *Corunna*, MI 48817
T: +1 989 7433287; Fax: +1 989 7435496; E-mail: cdlinformation@yahoo.com; URL: www.communitydistrictlibrary.org
1965; Betsy Hull
Large type print; 5 branch libs
53 000 vols   38707

Corvallis-Benton County Public Library, 645 NW Monroe Ave, *Corvallis*, OR 97330
T: +1 541 7666928; Fax: +1 541 7666726; URL: www.ci.corvallis.or.us/library/; www.thebestlibrary.net
1899; Carolyn Rawles-Heiser
3 branch libs, 1 bookmobile
370 000 vols; 1 072 curr per   38708

Coshocton Public Library, 655 Main St, *Coshocton*, OH 43812-1697
T: +1 740 6220956; Fax: +1 740 6224331; E-mail: coshpl@oplin.org; URL: www.coshoctonpl.org
1904; Ann Miller
1 branch libr
122 000 vols; 236 curr per; 307 e-books
libr loan   38709

Orange County Public Library – Costa Mesa Branch, 1855 Park Ave, *Costa Mesa*, CA 92627-2778
T: +1 949 6468845; Fax: +1 949 6313112
Dolores Madrigal
63 000 vols; 79 curr per   38710

Orange County Public Library – Mesa Verde, 2969 Mesa Verde Dr E, *Costa Mesa*, CA 92626-3699
T: +1 714 5465274; Fax: +1 714 5408413
Grace Barnes
59 000 vols; 92 curr per   38711

Washington County Library – Park Grove Branch, 7900 Hemingway Ave S, *Cottage Grove*, MN 55016-1833
T: +1 651 4592040; Fax: +1 651 4597051
Carol Warner
85 000 vols   38712

Cottonwood Public Library, 100 S Sixth St, *Cottonwood*, AZ 86326
T: +1 928 6347559; Fax: +1 928 6340253; E-mail: joneill@ci.cottonwoodaz.gov; URL: www.ctwpl.info/cottonwood
1960; John R. O'Neill
Large type print
63 000 vols; 104 curr per   38713

Council Bluffs Public Library, 400 Willow Ave, *Council Bluffs*, IA 51503-4269
T: +1 712 3237553; Fax: +1 712 3231269; URL: www.cbpl.lib.ia.us
1866; Barbara Peterson
Lewis Carroll Coll, Railways, Woman Suffrage
144 000 vols; 332 curr per; 4 218 av-mat; 4 218 Talking Bks
libr loan   38714

Blackwater Regional Library, Walter Cecil Rawls Library, 22511 Main St, *Courtland*, VA 23837
T: +1 757 6532821; Fax: +1 757 6539374; E-mail: courtland@blackwaterlib.org; URL: www.blackwaterlib.org
1958; Patricia Ward
Cary Close Memorial Coll (Civil War Ref Mat); 8 branch libs
246 000 vols; 326 curr per
libr loan   38715

Coventry Public Library, 1672 Flat River Rd, *Coventry*, RI 02816
T: +1 401 8229100; Fax: +1 401 8229133; URL: www.coventrylibrary.org
1972; Lynn H. Blanchette
Adult literacy; Civil War Coll, High-Low Mat, Literacy Mat for Tutors & Students Incl ESL
100 000 vols; 115 curr per   38716

County of Los Angeles Public Library – Charter Oak Library, 20540 E Arrow Hwy, Ste K, *Covina*, CA 91724-1238
T: +1 626 3392151; Fax: +1 626 3392799; URL: www.colapublib.org/libs/charteroak
Denise Dilley
53 000 vols; 92 curr per   38717

Covina Public Library, 234 N Second Ave, *Covina*, CA 91723-2198
T: +1 626 3324575; Fax: +1 626 9158915; URL: ci.covina.ca.us/library
1887; Roger Possner
94 000 vols; 175 curr per; 1 878 av-mat   38718

J. R. Clarke Public Library, 102 E Spring St, *Covington*, OH 45318
T: +1 937 4732226; Fax: +1 937 4738118
1917; Marjorie F. Mutzner
Civil War, Hist, Large print; J R Clarke Coll
53 000 vols; 138 curr per; 749 av-mat
38719

Kenton County Public Library, 502 Scott Blvd, *Covington*, KY 41011
T: +1 859 9624084; Fax: +1 859 9624096; URL: www.kentonlibrary.org
1967; David E. Schroeder
2 branch libs
458 000 vols; 1 069 curr per; 8 000 e-books; 15 675 av-mat
38720

Newton County Library System, 7116 Floyd St NE, *Covington*, GA 30014
T: +1 770 7873231; Fax: +1 770 7842092; E-mail: library@newtonlibrary.org; URL: www.newtonlibrary.org
1944; Greg Heid
Heritage Room, Porter Foundation Garden Coll
148 000 vols; 192 curr per; 3 955 av-mat; 270 sound-rec; 517 digital data carriers
libr loan
38721

Saint Tammany Parish Library, 310 W 21st Ave, *Covington*, LA 70433
T: +1 985 8936280; Fax: +1 985 8711224; E-mail: admin.c1tm@pelican.state.lib.la.us; URL: www.sttammany.lib.la.us
1950; Janice Butler
11 branch libs
497 000 vols; 1 000 curr per
38722

Saint Tammany Parish Library – Covington Branch, 310 W 21st Ave, *Covington*, LA 70433
T: +1 985 8936280; Fax: +1 985 8936283; E-mail: covington@mail.sttammany.lib.la.us
Brent Geiger
564 000 vols
38723

Tipton County Public Library, 300 W Church Ave, *Covington*, TN 38019-2729
T: +1 901 4768289; Fax: +1 901 4760008; E-mail: tiptonpl@covingtones.com
1938; Susan Cheairs
MacArthur Tapes, VHS Tapes of Local TV News 1989-93, WW II Tapes
50 000 vols; 30 curr per
38724

Cranberry Public Library, 2525 Rochester Rd, Ste 300, *Cranberry Township*, PA 16066-6423
T: +1 724 7769100; Fax: +1 724 7762490; E-mail: cranberry@bcfls.org; URL: www.bcfls.org/cranberry
1974; Carol B. Troese
86 000 vols; 133 curr per
38725

Cranston Public Library, 140 Sockanosset Cross Rd, *Cranston*, RI 02920-5539
T: +1 401 9439080; Fax: +1 401 9465079; URL: cranstonlibrary.org
1966; David Macksam
Boats & boating, Parents, Child, Careers; Civil Service Test Bks, Italian Language Bks; 5 branch libs
331 000 vols; 511 curr per
38726

Crawfordsville District Public Library, 205 S Washington St, *Crawfordsville*, IN 47933-2444
T: +1 765 3622242; Fax: +1 765 3627986; E-mail: web@cdpl.lib.in.us; URL: www.cdpl.lib.in.us
1897; Laurence Hathaway
114 000 vols; 311 curr per; 2 226 av-mat; 1 300 sound-rec; 400 Large Print Bks
38727

Cresskill Public Library, 53 Union Ave, *Cresskill*, NJ 07626
T: +1 201 5673521; Fax: +1 201 5675067; E-mail: chi@bccls.org; URL: www.bccls.org/cresskill
1930; Alice Chi
64 780 vols; 200 curr per
38728

Crestline Public Library, 324 N Thoman St, *Crestline*, OH 44827-1410
T: +1 419 6833909; Fax: +1 419 6833022; URL: crestlinepubliclibrary.org
1925; Cheryl Swihart
Railroad printed mat; County Census Records, County Court Records, Filmed Per, Hist Picture Coll, Large Print Bks
69 000 vols; 215 curr per
38729

Crestwood Public Library District, 4955 W 135th St, *Crestwood*, IL 60445
T: +1 708 3714090; Fax: +1 708 3714127; E-mail: cws@crestwoodlibrary.org
1973; Suzanne Bleskin
50 000 vols; 100 curr per
38730

Crete Public Library District, 1177 N Main St, *Crete*, IL 60417
T: +1 708 6728017; Fax: +1 708 6723529; E-mail: ctscirc@sslic.net; URL: www.cretelibrary.org
1985
Antiques, Collectibles
52 150 vols; 97 curr per
38731

Cromwell Belden Public Library, 39 West St, *Cromwell*, CT 06416
T: +1 860 6323460; Fax: +1 860 6323484; URL: cromwellct.com/library
1888; Eileen Branciforte
52 000 vols; 150 curr per; 2 375 av-mat
38732

Lake Agassiz Regional Library – Crookston Public Library, 110 N Ash St, *Crookston*, MN 56716-1702
T: +1 218 2814522; Fax: +1 218 2814523; E-mail: crookston@larl.org; URL: www.larl.org
1903
57 000 vols; 85 curr per
38733

Croton Free Library, 171 Cleveland Dr, *Croton-on-Hudson*, NY 10520
T: +1 914 2716612; Fax: +1 914 2710931; URL: www.crotonfreelibrary.org
1937; Mary C. Donnery
Railroad Coll
85 000 vols; 156 curr per; 89 maps; 788 microforms; 112 av-mat; 4 112 sound-rec; 4 digital data carriers
38734

Acadia Parish Library, 1125 N Parkerson, *Crowley*, LA 70526; P.O. Drawer 1509, Crowley, LA 70527-1509
T: +1 337 7881881; Fax: +1 337 7883759; E-mail: admin.b1ac@pelican.state.lib.la.us; URL: www.acadia.lib.la.us
1945; Lyle C. Johnson
Paul Freeland's Crowley Coll; 6 branch libs
127 000 vols; 310 curr per
38735

Crown Point Community Library, 214 S Court St, *Crown Point*, IN 46307-3975
T: +1 219 6630271; Fax: +1 219 6630403; URL: cat.crownpoint.lib.in.us
1906; Lynn M. Frank
1 branch libr
94 000 vols; 212 curr per; 3 262 av-mat; 2 819 sound-rec; 538 digital data carriers
38736

Crystal Lake Public Library, 126 Paddock St, *Crystal Lake*, IL 60014
T: +1 815 4591687; Fax: +1 815 4599581; URL: www.crystallakelibrary.org
1913; Kathryn I. Martens
177 000 vols; 360 curr per
38737

Cudahy Family Library, 3500 Library Dr, *Cudahy*, WI 53110; P.O. Box 100450, Cudahy, WI 53110-6107
T: +1 414 7692245; Fax: +1 414 7692252; URL: www.cudahyfamilylibrary.org
1906; Rebecca Roepke
100 000 vols; 200 curr per
38738

Cullman County Public Library System, 200 Clark St NE, *Cullman*, AL 35055
T: +1 256 7341068; Fax: +1 256 7346902
1928; Sharon Townson
Alabam, Civil War; Genealogy Colls, photos; 5 branch libs
88 000 vols; 167 curr per; 45 000 e-books
38739

Culpeper County Library, 271 Southgate Shopping Ctr, *Culpeper*, VA 22701
T: +1 540 8258691; Fax: +1 540 8257486; E-mail: cclva@cclva.org; URL: tlc.library.net/culpeper
1946; Susan J. Keller
Art Coll, Civil War Coll
66 000 vols; 150 curr per
libr loan
38740

County of Los Angeles Public Library – Culver City Julian Dixon Library, 4975 Overland Ave, *Culver City*, CA 90230-4299
T: +1 310 5591676; Fax: +1 310 5592994; URL: www.colapublib.org/libs/culvercity
Laura Frakes
Judaica Coll
198 000 vols; 321 curr per
38741

Allegany County Library System, 31 Washington St, *Cumberland*, MD 21502
T: +1 301 7771200; Fax: +1 301 7777299; E-mail: mainpl@allconet.org; URL: www.alleganycountylibrary.info
1924; John Taube
5 branch libs
56 000 vols; 144 curr per
38742

Cumberland Public Library, Edward J. Hayden Library, 1464 Diamond Hill Rd, *Cumberland*, RI 02864-5510
T: +1 401 3332552; Fax: +1 401 3340578; E-mail: reference@cumberlandlibrary.org; URL: www.cumberlandlibrary.org
1976; Janet A. Levesque
81 000 vols; 204 curr per
38743

Forsyth County Public Library, Cumming Library & FCPL Headquarters, 585 Dahlonega Rd, *Cumming*, GA 30040-2109
T: +1 770 7819840; Fax: +1 770 7818089; URL: www.forsythpl.org
1956; Jon McDaniel
1 branch libr
226 000 vols; 328 curr per; 9 614 av-mat; 9 614 Audio Bks, 22 Electronic Media & Resources, 2 240 Large Print Bks
38744

Forsyth County Public Library, Sharon Forks, 2820 Old Atlanta Rd, *Cumming*, GA 30041
T: +1 770 7819840
Brenda Johnson
98 000 vols; 158 curr per; 4 553 av-mat
38745

Santa Clara County Library – Cupertino Public, 10800 Torre Ave, *Cupertino*, CA 95014-3254
T: +1 408 4461677; Fax: +1 408 2528749
Jah-Lih Lee
Local History (California Western Americana Coll)
307 000 vols
38746

Cushing Public Library, 215 N Steele, *Cushing*, OK 74023-3319; P.O. Box 551, Cushing, OK 74023-0551
T: +1 918 2254188; Fax: +1 918 2256201
1939; Ruth Ann Johnson
Genealogy (Cushing Family), Payne County Law Bks
65 000 vols; 82 curr per
38747

Cuyahoga Falls Library, Taylor Memorial Association, 2015 Third St, *Cuyahoga Falls*, OH 44221-3294
T: +1 330 9282117; Fax: +1 330 9282535; E-mail: mail@cuyahogafallslibrary.org; URL: www.cuyahogafallslibrary.org
1911; Kevin M. Rosswurm
147 000 vols; 363 curr per
38748

Cynthiana-Harrison County Public Library, 104 N Main St, *Cynthiana*, KY 41031
T: +1 859 2344881; Fax: +1 859 2340059; E-mail: info@cynthianalibrary.org; URL: www.cynthianalibrary.org
1932; Pat Barnes
Economics, Health sciences, Hist, Home economics, Cookbk Coll, Civil War Coll
57 200 vols; 42 curr per
38749

Harris County Public Library – Cy-Fair College Branch, 9191 Barker Cypress Rd, *Cypress*, TX 77433
T: +1 281 2903210; Fax: +1 281 2905288
Mick Stafford
76 000 vols
38750

Harris County Public Library – Northwest, 11355 Regency Green Dr, *Cypress*, TX 77429
T: +1 281 8902665; Fax: +1 281 4694718
Deborah Sica
84 000 vols
38751

Orange County Public Library – Cypress Branch, 5331 Orange Ave, *Cypress*, CA 90630-2985
T: +1 714 8260350; Fax: +1 714 8281103
Helen Richardson
104 000 vols; 93 curr per
38752

Horseshoe Bend Regional Library, 207 N West St, *Dadeville*, AL 36853
T: +1 256 8259232; Fax: +1 256 8254314; E-mail: horseshoebend@bellsouth.net; URL: www.mindspring.com/hbrl/hbrl.html
1940; Susie Anderson
120 000 vols; 32 curr per
38753

Lincoln Heritage Public Library, 105 Wallace St, *Dale*, IN 47523-9267; P.O. Box 784, Dale, IN 47523-0784
T: +1 812 9377170; Fax: +1 812 9377102; URL: www.lincolnheritage.lib.in.us
Lynn Mehringer
55 000 vols; 115 curr per
38754

Back Mountain Memorial Library, 96 Huntsville Rd, *Dallas*, PA 18612
T: +1 570 6751182; Fax: +1 570 6745863; E-mail: backmtlb@epix.net; URL: backmountainlibrary.org
1945; Martha Butler
76 000 vols; 145 curr per; 982 av-mat
38755

Dallas Public Library, 950 Main St, *Dallas*, OR 97338
T: +1 503 6232633; Fax: +1 503 6237357; E-mail: Donna.Zehner@ci.dallas.or.us; URL: www.ci.dallas.or.us/index.aspx?nid=102
1908; Donna E. Zehner
68 000 vols; 113 curr per
38756

Dallas Public Library, 1515 Young St, *Dallas*, TX 75201-5499
T: +1 214 6707809; Fax: +1 214 6701738; E-mail: info@dallaslibrary.org; URL: dallaslibrary.org
1901; Laurie Evans
Business Histories, Children's Lit Coll, Classical Lit, Classical Recordings, Dallas Black Hist, Dance, Fashion, Genealogy, Grants, Hist of Printing, Standards & Specifications, Theater, US Marshal Clinton T. Peoples Coll, US Serial Set; 23 branch libs
2 365 000 vols; 4 874 curr per; 70 190 av-mat
libr loan
38757

Dallas Public Library, 1302 N Justin Ave, *Dallas*, TX 75211-1142
T: +1 214 6706446; Fax: +1 214 6707502; URL: dallaslibrary.org/arcadia.htm
Sergio Pineda
68 000 vols
38758

Dallas Public Library – Audelia Road, 10045 Audelia Rd, *Dallas*, TX 75238-1999
T: +1 214 6701350; Fax: +1 214 6700790; E-mail: audeliaroad@dallaslibrary.org; URL: dallaslibrary.org/audelia.htm
Becky Hubbard
95 000 vols
38759

Dallas Public Library – Casa View, 10355 Ferguson Rd, *Dallas*, TX 75228-3099
T: +1 214 6708403; Fax: +1 214 6708405; URL: dallaslibrary.org/casa.htm
Becky Hubbard
70 000 vols
38760

Dallas Public Library – Fretz Park, 6990 Belt Line Rd, *Dallas*, TX 75240-7963
T: +1 214 6706421; Fax: +1 214 6706621; E-mail: fretzpark@dallaslibrary.org; URL: dallaslibrary.org/fretz.htm
Kitty Stone
76 000 vols   38761

Dallas Public Library – Hampton-Illinois, 2210 W Illinois Ave, *Dallas*, TX 75224-1699
T: +1 214 6707646; Fax: +1 214 6707652; E-mail: hamptonillinois@dallaslibrary.org; URL: dallaslibrary.org/hampton.htm
Linda Holland
69 000 vols   38762

Dallas Public Library – Lakewood, 6121 Worth St, *Dallas*, TX 75214-4497
T: +1 214 6701376; Fax: +1 214 6705701; E-mail: lakewood@dallaslibrary.org; URL: dallaslibrary.org/lakewood.htm
Christina Worden
53 000 vols   38763

Dallas Public Library – Mountain Creek, 6102 Mountain Creek Pkwy, *Dallas*, TX 75249
T: +1 214 6706704; Fax: +1 214 6706780; E-mail: mountaincreek@dallaslibrary.org; URL: dallaslibrary.org/mountain.htm
Linda Holland
60 000 vols   38764

Dallas Public Library – North Oak Cliff, 302 W Tenth St, *Dallas*, TX 75208-4617
T: +1 214 6707555; Fax: +1 214 6707548; E-mail: northoakcliff@dallaslibrary.org; URL: dallaslibrary.org/north.htm
Leonardo Melo
74 000 vols   38765

Dallas Public Library – Park Forest, 3421 Forest Lane, *Dallas*, TX 75234-7776
T: +1 214 6706333; Fax: +1 214 6706623; URL: dallaslibrary.org/park.htm
Lynne Craddock
65 000 vols   38766

Dallas Public Library – Pleasant Grove, 1125 S Buckner Blvd, *Dallas*, TX 75217-4399
T: +1 214 6700965; Fax: +1 214 6700320; E-mail: pleasantgrove@dallaslibrary.org; URL: dallaslibrary.org/pleasant.htm
Linda Holland
53 000 vols   38767

Dallas Public Library – Polk-Wisdom, 7151 Library Lane, *Dallas*, TX 75232-3899
T: +1 214 6701947; Fax: +1 214 6700589; E-mail: polkwisdom@dallaslibrary.org; URL: dallaslibrary.org/polk.htm
Sharon Martin
64 000 vols   38768

Dallas Public Library – Preston Royal, 5626 Royal Lane, *Dallas*, TX 75229-5599
T: +1 214 6707128; Fax: +1 214 6707135; E-mail: prestonroyal@dallaslibrary.org; URL: dallaslibrary.org/preston.htm
Jo Guidice
72 000 vols   38769

Dallas Public Library – Renner Frankford Branch, 6400 Frankford Rd, *Dallas*, TX 75252-5747
T: +1 214 6706100; Fax: +1 214 6706090; E-mail: rennerfrankford@dallaslibrary.org; URL: dallaslibrary.org/renner.htm
Becky Hubbard
80 000 vols   38770

Dallas Public Library – Skillman Southwestern, 5707 Skillman St, *Dallas*, TX 75206
T: +1 214 6706078; Fax: +1 214 6706184; E-mail: skillmansouthwestern@dallaslibrary.org; URL: dallaslibrary.org/skillman.htm
Paula Husky
57 000 vols   38771

Dallas Public Library – Skyline, 6006 Everglade, *Dallas*, TX 75227-2799
T: +1 214 6700938; Fax: +1 214 6700321; E-mail: skyline@dallaslibrary.org; URL: dallaslibrary.org/skyline.htm
Becky Hubbard
61 000 vols   38772

Northwest Georgia Regional Library System, 310 Cappes St, *Dalton*, GA 30720
T: +1 706 8761360; Fax: +1 706 2722977; URL: www.ngrl.org
1924; Joe B. Forsee
3 branch libs
210 000 vols; 65 curr per   38773

Daly City Public Library, 40 Wembley Dr, *Daly City*, CA 94015-4399
T: +1 650 9918025; Fax: +1 650 9915726; E-mail: dcplref@plsinfo.org; URL: www.dalycitylibrary.org
1911; Carol E. Simmons
Daly City History Coll; 3 branch libs
176 000 vols; 288 curr per; 10 000 e-books; 21 542 av-mat; 6815 sound-rec
                                     38774

Daly City Public Library – Westlake, 275 Southgate Ave, *Daly City*, CA 94015-3471
T: +1 650 9918071; Fax: +1 650 9918180
Tom Goward
60 000 vols   38775

Orange County Public Library – Dana Point Branch, 33841 Niguel Rd, *Dana Point*, CA 92629-4010
T: +1 949 4965517; Fax: +1 949 2407650
John Dunham
82 000 vols; 92 curr per   38776

Danbury Public Library, 170 Main St, *Danbury*, CT 06810
T: +1 203 7974505; Fax: +1 203 7974501; URL: danburylibrary.org
1869; Mark Hasskarl
95 000 vols; 300 curr per   38777

Killingly Public Library, 25 Westcott Rd, *Danielson*, CT 06239
T: +1 860 7795383; Fax: +1 860 7791823; URL: www.killinglypubliclibrary.org
1854; Marie C. Chartier
68 000 vols; 170 curr per   38778

Peabody Institute Library of Danvers, 15 Sylvan St, *Danvers*, MA 01923-2735
T: +1 978 7740554; Fax: +1 978 7620251; E-mail: dan@noblenet.org; URL: www.danverslibrary.org
1866; Douglas W. Rendell
Anti Slavery (Parker Pillsbury Coll); Danvers (Town Records Coll), mss; Witchcraft (Ellerton J. Brehart Coll)
119 000 vols; 240 curr per; 1680 Large Print Bks   38779

Boyle County Public Library, Danville Library Inc, 1857 S Danville Bypass, *Danville*, KY 40422
T: +1 859 2368466; Fax: +1 859 2367692; URL: www.boylepublib.org
1893; Karl Benson
Kentucky, Shakers
105 000 vols; 145 curr per; 210 maps; 725 microforms; 5825 av-mat; 2625 sound-rec; 6 digital data carriers
libr loan   38780

Contra Costa County Library – Danville Library, 400 Front St, *Danville*, CA 94526-3465
T: +1 925 8374889; Fax: +1 925 8311299
Seng Lovan
71 000 vols   38781

Danville Public Library, 511 Patton St, *Danville*, VA 24541
T: +1 434 7995195; Fax: +1 434 7925172; E-mail: library5@ci.danville.va.us; URL: www.danville-va.gov; www.danvillelibrary.org
1923; Dr. O. Douglas Alexander
2 branch libs
131 000 vols   38782

Danville Public Library, 319 N Vermilion St, *Danville*, IL 61832
T: +1 217 4775228; Fax: +1 217 4775230; URL: www.danville.lib.il.us
1883; Barbara J. Nolan
Genealogy, Gardening
161 000 vols; 353 curr per; 6366 av-mat; 6370 Audio Bks   38783

Danville-Center Township Public Library, 101 S Indiana St, *Danville*, IN 46122-1809
T: +1 317 7452604; Fax: +1 317 7450756; URL: www.dpl.lib.in.us
1903; Loren Malloy
Genealogy
59 000 vols; 167 curr per; 1206 av-mat
                                     38784

Arkansas River Valley Regional Library System, Headquarters, 501 N Front St, *Dardanelle*, AR 72834-3507
T: +1 479 2294418; Fax: +1 479 2292595; E-mail: darlib@centurytel.net; URL: www.arvrls.com
1959; Donna McDonald
Arkansas Coll; 6 branch libs
267 000 vols; 250 curr per   38785

Darien Library, 1441 Post Rd, *Darien*, CT 06820-5419
T: +1 203 6551234; Fax: +1 203 6551547; E-mail: askus@darienlibrary.org; URL: www.darienlibrary.org
1894; Louise Berry
169 000 vols; 257 curr per; 7877 e-journals; 10 634 av-mat; 10 711 sound-rec; 8230 digital data carriers; 40 Electronic Media & Resources, 2500 Large Print Bks   38786

Indian Prairie Public Library District, 401 Plainfield Rd, *Darien*, IL 60561-4207
T: +1 630 8878760; Fax: +1 630 8871018; E-mail: ippl@indianprairielibrary.org; URL: www.indianprairielibrary.org
1988; Jamie Bukovac
154 000 vols; 451 curr per
libr loan; ILA   38787

Darlington County Library, 204 N Main St, *Darlington*, SC 29532
T: +1 843 3984940; Fax: +1 843 3984942; URL: www.darlington-lib.org
1893; Sue Rainey
NUREG; 3 branch libs
140 000 vols; 422 curr per   38788

Dartmouth Public Libraries – Southworth Library, 732 Dartmouth St, *Dartmouth*, MA 02748
T: +1 508 9990726; Fax: +1 508 9929914; URL: www.dartmouthpubliclibraries.org
Denise Medeiros
2 branch libs
118 000 vols; 200 curr per   38789

Davenport Public Library, 321 Main St, *Davenport*, IA 52801-1490
T: +1 563 3267832; Fax: +1 563 3267809; URL: www.davenportlibrary.com
1877; LaWanda Roudebush
Economics, Business & management; Iowa Authors Coll; 2 branch libs
305 000 vols; 681 curr per
libr loan   38790

Dayton Metro Library, 215 E Third St, *Dayton*, OH 45402-2103
T: +1 937 4632665; Fax: +1 937 4964300; URL: www.daytonmetrolibrary.org
1847; Timothy Kambitsch
21 branch libs
2 207 000 vols; 919 curr per
libr loan   38791

Bureau of Braille & Talking Book Library Services, 420 Platt St, *Daytona Beach*, FL 32114-2804
T: +1 386 2396000; Fax: +1 386 2396069; URL: dbs.myflorida.com/library/index.shtml
1950; Michael G. Gunde
Womyn's Braille Press
2 400 000 vols
libr loan   38792

Volusia County Public Library, 1290 Indian Lake Rd, *Daytona Beach*, FL 32124
T: +1 386 2576037; Fax: +1 386 2481746; URL: www.vcpl.lib.fl.us
1961; Lucinda Colee
Genealogy Coll; 15 branch libs
812 000 vols; 2111 curr per; 3000 e-books; 123 712 av-mat; 52 500 sound-rec
                                     38793

Beauregard Parish Public Library, 205 S Washington Ave, *De Ridder*, LA 70634
T: +1 337 4636217; Fax: +1 337 4625434; E-mail: w1bg@beau.org; URL: library.beau.org/lib/index.html
1947
6 branch libs
80 000 vols; 138 curr per   38794

Dearborn Public Library, Henry Ford Centennial Library, 16301 Michigan Ave, *Dearborn*, MI 48126
T: +1 313 9432330; Fax: +1 313 9433063; E-mail: library@ci.dearborn.mi.us; URL: www.dearbornlibrary.org
1919; Maryanne Bartles
3 branch libs
100 000 vols; 21 751 av-mat; 13 720 sound-rec; 9550 digital data carriers; 2488 Large Print Bks   38795

Dearborn Heights City Libraries – Caroline Kennedy Library, 24590 George St, *Dearborn Heights*, MI 48127
T: +1 313 7913804; Fax: +1 313 7913801; URL: www.ci.dearborn-heights.mi.us/library/caroline/caroline.htm
1961; Michael McCaffery
59 000 vols; 120 curr per   38796

Dearborn Heights City Libraries – John F. Kennedy Jr Library, 24602 Van Born Rd, *Dearborn Heights*, MI 48125
T: +1 313 7916050; Fax: +1 313 7916051; URL: www.ci.dearborn-heights.mi.us/library/john/john.htm
Michael Wrona
55 000 vols; 120 curr per   38797

Adams Public Library System, 128 S Third St, *Decatur*, IN 46733-1691
T: +1 260 7242605; Fax: +1 260 7242877; URL: www.apls.lib.in.us
1905; Kelly A. Ehinger
Adams County Genealogy, Gene Stratton Porter Coll (Indiana), Large Print Coll
74 000 vols; 253 curr per   38798

Decatur Public Library, 504 Cherry St NE, *Decatur*, AL 35601
T: +1 256 3532993; Fax: +1 256 3506736; E-mail: wheelerbasin@prodigy.net; URL: www.decatur.lib.al.us
1905; Sandra Sherman McCandless
Genealogy, Local history
98 000 vols; 200 curr per; 153 maps; 76 microforms; 2726 av-mat; 10 digital data carriers; 2730 Talking Bks   38799

Decatur Public Library, 130 N Franklin St, *Decatur*, IL 62523-1327
T: +1 217 4219732; Fax: +1 217 2334071; URL: www.decatur.lib.il.us
1876; Lee Ann Fisher
2 bookmobiles
262 000 vols; 791 curr per   38800

DeKalb County Public Library, 215 Sycamore St, 4th Flr, *Decatur*, GA 30030
T: +1 404 3708450; Fax: +1 404 3708469; URL: www.dekalblibrary.org
1925; Darro Willey
Lit, Genealogy; Global Res Ctr; 24 branch libs
813 000 vols; 1899 curr per; 77 074 av-mat; 40 369 sound-rec; 5789 digital data carriers; 19 700 Large Print Bks
                                     38801

Van Buren District Library, Webster Memorial Library, 200 N Phelps St, *Decatur*, MI 49045-1086
T: +1 269 4234771; Fax: +1 269 4238373; URL: www.vbdl.org
1941; David Tate
6 branch libs
154 000 vols; 303 curr per; 11 000 e-books; 4719 av-mat; 4404 Large Print Bks, 4719 Audio Bks
libr loan   38802

Decorah Public Library, 202 Winnebago St, **Decorah**, IA 52101
T: +1 563 3823717; Fax: +1 563 3824524; E-mail: dpllib@decorah.lib.ia.us; URL: www.decorah.lib.ia.us
1893; Lorraine Borowski
Vera Harris Large Print Book Coll; Genealogy Libr, Local Hist Arch, Toy-lending Libr, RSVP-Retired Senior Volunteer Program
74 000 vols; 200 curr per; 4 548 av-mat; 2 500 sound-rec
libr loan; ALA   38803

Dedham Public Library, 43 Church St, **Dedham**, MA 02026
T: +1 781 7519284; Fax: +1 781 7519080; E-mail: dedham@mln.lib.net; URL: www.dedhamlibrary.org
1872; Patricia Lambert
Art & architecture, Business, Hist, Social sciences; Dedham Hist Coll; 1 branch libr
88 000 vols; 325 curr per   38804

Deer Park Public Library, 44 Lake Ave, **Deer Park**, NY 11729-6047
T: +1 631 5863000; Fax: +1 631 5863006; URL: dprk.suffolk.lib.ny.us
1964; Dina M. Reilly
100 000 vols; 410 curr per   38805

Deer Park Public Library, 3009 Center St, **Deer Park**, TX 77536-5099
T: +1 281 4787208;
Fax: +1 281 4787212; URL: catalog.library.deerparktx.org/polaris/
1962; Rebecca Pool
CAER Coll
57 000 vols; 250 curr per   38806

Deerfield Public Library, 920 Waukegan Rd, **Deerfield**, IL 60015
T: +1 847 9453311; Fax: +1 847 9453402; E-mail: info@deerfieldlibrary.org; URL: www.deerfieldlibrary.org
1927; Mary Pergander
Art
140 000 vols; 347 curr per
libr loan   38807

Defiance Public Library, 320 Fort St, **Defiance**, OH 43512-2186
T: +1 419 7821456; Fax: +1 419 7826235; URL: www.defiancelibrary.org
Marilyn Hite
Slocum;Ohioana
141 000 vols; 283 curr per   38808

DeKalb Public Library, Haish Memorial Library, 309 Oak St, **DeKalb**, IL 60115-3369
T: +1 815 7569568; Fax: +1 815 7567837; E-mail: dkplref@dkpl.org; URL: www.dkpl.org
1893; Dee Coover
Donn V Hart Southeast Asian Coll, American & English Lit Colls
142 000 vols; 261 curr per; 2 118 sound-rec; 10 digital data carriers
libr loan   38809

Val Verde County Library, 300 Spring St, **Del Rio**, TX 78840
T: +1 830 7747595; Fax: +1 830 7747607; E-mail: library@vvcl.lib.tx.us; URL: vvcl.lib.tx.us
1940; Willie Braudaway
John R Brinkley Coll, Del Rio Coll, Texana, Bks on Tape
57 000 vols; 200 curr per   38810

Delafield Public Library, 500 Genessee St, **Delafield**, WI 53018-1895
T: +1 262 6466232; Fax: +1 262 6466232; URL: delafieldlibrary.org
1907; Terry Zignego
60 000 vols; 90 curr per   38811

Delaware County District Library, 84 E Winter St, **Delaware**, OH 43015
T: +1 740 3637277; Fax: +1 740 3690196; E-mail: askus@delawarelibrary.org; URL: www.delawarelibrary.org
1906; Mary Jane Santos
2 branch libs
218 000 vols; 284 curr per   38812

Delphi Public Library, 222 E Main St, **Delphi**, IN 46923
T: +1 765 5642929; Fax: +1 765 5644746; E-mail: dplibrar@carrollnet.com; URL: www.carrollnet.org/dpl
1904; Kelly D. Currie
61 000 vols; 215 curr per; 300

microforms; 3 428 av-mat; 1 658 sound-rec; 119 digital data carriers; 146 slides
libr loan; ALA, PLA   38813

Delphos Public Library, 309 W Second St, **Delphos**, OH 45833-1695
T: +1 419 6954015; Fax: +1 419 6954025; URL: www.delphos.lib.oh.us
1901; Nancy Mericle
Delphos Newspapers, 1872-1946
82 000 vols; 230 curr per   38814

Delray Beach Public Library, 100 W Atlantic Ave, **Delray Beach**, FL 33444
T: +1 561 2660196; Fax: +1 561 2669757; URL: www.delraylibrary.org
1939; Alan Kornblau
250 000 vols; 350 curr per   38815

Delta Public Library, 402 Main St, **Delta**, OH 43515-1304
T: +1 419 8223110; Fax: +1 419 8225310; URL: www.deltaohio.com/library
1911; Patricia Grover
53 000 vols; 180 curr per   38816

Marshall Memorial Library, 110 S Diamond St, **Deming**, NM 88030-3698
T: +1 505 5469202; Fax: +1 505 5469649; E-mail: demingpl@zianet.com; URL: www.zianet.com/demingpl/
1917; Donna George
Southwest Coll
68 000 vols; 100 curr per   38817

Jasper County Public Library – DeMotte Branch, 901 Birch St SW, **DeMotte**, IN 46310; P.O. Box 16, DeMotte, IN 46310-0016
T: +1 219 9872221; Fax: +1 219 9872220
Brenda Thompson
53 000 vols   38818

Livingston Parish Library – Denham Springs – Walker Branch, 8101 US Hwy 190, **Denham Springs**, LA 70726
T: +1 225 6658118; Fax: +1 225 7916325
Felicia West
55 000 vols; 75 curr per   38819

Livingston Parish Library – Watson Branch, 36581 Ontboek Rd, **Denham Springs**, LA 70786; P.O. Box 149, Watson, LA 70786-0149
T: +1 225 6643963; Fax: +1 225 6641949
Stacie Davis
50 000 vols; 60 curr per   38820

Denison Public Library, 300 W Gandy, **Denison**, TX 75020-3153
T: +1 903 4651797; Fax: +1 903 4651130; E-mail: arbailey@texoma.net; URL: www.barr.org/denison.htm
1936; Alvin R. Bailey
Area Hist & Bks by Area Authors (Texoma Coll), print, bks, pamphlet
80 710 vols; 160 curr per   38821

Norelius Community Library, 1403 1st Ave South, **Denison**, IA 51442
T: +1 712 2639355; Fax: +1 712 2638578; E-mail: norlib@frontiernet.net; URL: www.denison.lib.ia.us
50 000 vols; 176 curr per   38822

Caroline County Public Library, 100 Market St, **Denton**, MD 21629
T: +1 410 4792254; Fax: +1 410 4794935; E-mail: info@carolib.org; URL: www.carolib.org
1961; George A. Sands
2 branch libs
84 000 vols; 284 curr per
libr loan   38823

Denton Public Library – Emily Fowler Central Library, 502 Oakland St, **Denton**, TX 76201
T: +1 940 3498754; Fax: +1 940 3498123; E-mail: library@cityofdenton.com; URL: www.dentonlibrary.com
Eva D. Poole
Texas & Denton History Coll, archives, bks, clippings, maps, micro, pamphlets, photog;Genealogy Coll
197 000 vols; 464 curr per   38824

Denver Public Library, Ten W 14th Ave Pkwy, **Denver**, CO 80204-2731
T: +1 720 8651325; Fax: +1 720 8652087; URL: denverlibrary.org
1889; Shirley Amore
Aeronautics Coll, Fine Printing, Mountaineering, Napoleon, Original Western Art Coll, World War II 10th Mountain Division; 22 branch libs
2 042 000 vols; 5 560 curr per; 20 000 e-books
libr loan   38825

Denville Free Public Library, 121 Diamond Spring Rd, **Denville**, NJ 07834
T: +1 973 6276555; Fax: +1 973 6271913; URL: www.denvillelibrary.org
1921; Elizabeth L. Kanouse
70 000 vols; 162 curr per   38826

James H. Johnson Memorial Library, Deptford Public Library, 670 Ward Dr, **Deptford**, NJ 08096
T: +1 856 8489149; Fax: +1 856 8481813; E-mail: information@deptfordpubliclibrary.com; URL: www.deptfordpubliclibrary.org
1961; Arn Ellsworth Winter
Antiques, Hearing impaired
60 000 vols; 212 curr per   38827

Derby Neck Library, 307 Hawthorne Ave, **Derby**, CT 06418-1199
T: +1 203 7341492; Fax: +1 203 7322913; E-mail: derbynecklibrary@biblio.org; URL: www.derbynecklibrary.org
1897; Judith W. Augusta
Audiobooks; Large Print Books; Videotapes
50 000 vols; 130 curr per; 86 e-books; 75 maps; 4 000 microforms; 3 900 av-mat; 1 611 sound-rec; 26 digital data carriers; 200 pamphlets, 2 000 clippings
libr loan; ALA   38828

Derby Public Library, 611 N Mulberry Rd, Ste 200, **Derby**, KS 67037
T: +1 316 7880760; Fax: +1 316 7887313; URL: www.derbylibrary.com
1957; Judy K. Bennett
Kansas, Parenting
67 000 vols; 100 curr per   38829

Beauregard Parish Library, 205 S Washington Ave, **DeRidder**, LA 70634
T: +1 337 4636217; Fax: +1 337 4625434; E-mail: admin@beau.org; URL: www.beau.lib.la.us
Lilly F. Smith
83 000 vols; 210 curr per   38830

Derry Public Library, 64 E Broadway, **Derry**, NH 03038-2412
T: +1 603 4326140; Fax: +1 603 4326128; URL: www.derry.lib.nh.us
1905; Cheryl Lynch
Houses of Derry (Harriet Newell Coll), Tasha Tudor Drawings, Robert Frost Mat
96 000 vols; 184 curr per; 2 415 av-mat
                                    38831

Des Moines Public Library, 1000 Grand Ave, **Des Moines**, IA 50309
T: +1 515 2834152; Fax: +1 515 2371654; E-mail: reference@desmoineslibrary.com; URL: www.desmoineslibrary.com
1866; Saul Amdursky
Iowa, Sheet music; Foundation Ctr; 5 branch libs
557 000 vols; 999 curr per
libr loan   38832

Des Moines Public Library – East Side, 2559 Hubbell Ave, **Des Moines**, IA 50317
T: +1 515 2834152; Fax: +1 515 2486256; E-mail: eastlib@netins.net
Carolyn Greufe
58 000 vols   38833

Des Moines Public Library – Franklin Avenue, 5000 Franklin Ave, **Des Moines**, IA 50310
T: +1 515 2834152; Fax: +1 515 2718734
Kathleen Bognanni
116 000 vols   38834

Des Moines Public Library – North Side, 3516 Fifth Ave, **Des Moines**, IA 50313
T: +1 515 2834152; Fax: +1 515 2422684
Chris C. Shelton
50 000 vols   38835

Des Moines Public Library – South Side, 1111 Porter Ave, **Des Moines**, IA 50315
T: +1 515 2834152; Fax: +1 515 2562567; E-mail: southlib@netins.net
Nyla Wobig
71 000 vols   38836

Iowa Regional Library for the Blind & Physically Handicapped, 524 Fourth St, **Des Moines**, IA 50309-2364
T: +1 515 2811333; Fax: +1 515 2811378; E-mail: library@blind.state.ia.us; URL: www.blind.state.ia.us
1960; Karen Keninger
Print Coll of Books about Blindness
311 000 vols; 75 curr per; 49 897 av-mat; 49 490 sound-rec; 8 digital data carriers; 49 900 Talking Bks, 6 450 Large Print Bks   38837

Des Plaines Public Library, 1501 Ellinwood St, **Des Plaines**, IL 60016-4553
T: +1 847 8275551; Fax: +1 847 8277974; URL: www.dppl.org
1906; Sandra K. Norlin
1 bookmobile
229 000 vols; 567 curr per; 21 422 sound-rec; 181 digital data carriers; 231 art prints   38838

DeSoto Public Library, 211 E Pleasant Run Rd, Ste C, **DeSoto**, TX 75115
T: +1 972 2309656; Fax: +1 972 2305797; URL: www.ci.desoto.tx.us/index.aspx?NID=110
1943
70 000 vols; 125 curr per   38839

Detroit Public Library, 5201 Woodward Ave, **Detroit**, MI 48202
T: +1 313 8331415; Fax: +1 313 8332327; E-mail: is@detroit.lib.mi.us; URL: www.detroitpubliclibrary.org
1865; Nancy Skowronski
National Automotive Hist Coll, Michigan, Great Lakes, Northwest Territory (Burton Hist Coll), Black American in the Performing Arts (E Azalia Hackley Coll), Patents, Rare Books, Genealogy, National Bibliography; 25 branch libs
7 904 000 vols
libr loan   38840

Detroit Public Library – Wilder, 7140 E Seven Mile/Van Dyke, **Detroit**, MI 48234-3096
T: +1 313 8524285; Fax: +1 313 8524313
Patricia Petrone
64 000 vols   38841

Dexter District Library, 8040 Fourth St, **Dexter**, MI 48130
T: +1 734 4267731; Fax: +1 734 4261217; URL: www.dexter.lib.mi.us
1964; Paul McCann
Bone Science Fiction Coll
63 000 vols; 90 curr per   38842

County of Los Angeles Public Library – Diamond Bar Library, 1061 S Grand Ave, **Diamond Bar**, CA 91765-2299
T: +1 909 8614978; Fax: +1 909 8603054; URL: www.colapublib.org/libs/diamondbar
Irene Wang
87 000 vols; 184 curr per   38843

Dickinson Public Library, 139 Third St W, **Dickinson**, ND 58601
T: +1 701 4567700; Fax: +1 701 4567702; E-mail: Dickinson.library@sendit.nodak.edu; URL: www.dickinsonlibrary.com
1908
Dickinson Press, 1883-present; book mobile
55 000 vols; 150 curr per   38844

Dickson County Public Library, 206 Henslee Dr, **Dickson**, TN 37055-2020
T: +1 615 4468293; Fax: +1 615 4469130
1933; N.N.
77 000 vols; 110 curr per; 4 000 e-books; 247 maps; 2 000 microforms; 3 233 av-mat; 741 sound-rec; 2 digital data carriers; 1 665 slides
libr loan   38845

Dillon County Library, 600 E Main St, *Dillon*, SC 29536
T: +1 843 7740330; Fax: +1 843 7740733; E-mail: dilloncountylibrary@hotmail.com; URL: www.dillon.lib.sc.us
Yolanda Manning McCormick
93 000 vols; 50 curr per   38846

Half Hollow Hills Community Library, 55 Vanderbilt Pkwy, *Dix Hills*, NY 11746
T: +1 631 4214530; Fax: +1 631 4210730; E-mail: hhhllib@suffolk.lib.ny.us; URL: hhhlibrary.org
1959; Michele Lauer-Bader
Business & management, Economy, Religious studies; 1 branch libr
341 000 vols; 10 750 av-mat; 7 000 sound-rec; 5 670 digital data carriers
                    38847

Dixon Public Library, 221 S Hennepin Ave, *Dixon*, IL 61021-3093
T: +1 815 2847261; Fax: +1 815 2887323; E-mail: maillibrary@dixonpubliclibrary.org; URL: www.dixonpubliclibrary.org
1872; Lynn A. Roe
90 000 vols; 125 curr per   38848

Dodge City Public Library, 1001 N Second Ave, *Dodge City*, KS 67801-4484
T: +1 620 2250248; Fax: +1 620 2252761; E-mail: library@trails.net; URL: www.dcpl.info
1905; Cathy Reeves
Kansas, Hist, Spanish language; Cookbk Coll, Large Print Coll
127 000 vols; 202 curr per; 25 e-books; 6 400 digital data carriers; 5 879 Audio Bks
libr loan   38849

Dolton Public Library District, 14037 Lincoln, *Dolton*, IL 60419-1091
T: +1 708 8492385; Fax: +1 708 8412725; E-mail: doltonpublib@hotmail.com
1963; Allison Heard
African-American hist; Adult New Readers Coll
100 000 vols; 155 curr per; 1 282 av-mat; 2 199 Large Print Bks, 1 200 Talking Bks
                    38850

Ascension Parish Library,
Donaldsonville Branch, 500 Mississippi St, *Donaldsonville*, LA 70346-2535
T: +1 225 4738052; Fax: +1 225 4739522; E-mail: admin.c1ac@pelican.state.lib.la.us; URL: www.ascension.lib.la.us
1960; Angelle Deshautelles
Lower Mississippi Flood Control, docs; US Army Corps of Engineers; 2 branch libs
224 000 vols; 540 curr per   38851

Boston Public Library – Codman Square, 690 Washington St, *Dorchester*, MA 02124-3598
T: +1 617 4368214; Fax: +1 617 8592425
Janice Knight
84 000 vols; 70 curr per; 1 500 av-mat
                    38852

Houston Love Memorial Library, 212 W Burdeshaw St, *Dothan*, AL 36303; P.O. Box 1369, Dothan
T: +1 334 7939767; Fax: +1 334 7936645; URL: www.houstonlovelibrary.org
1900; Bettye Forbus
3 branch libs
192 000 vols; 116 curr per; 3 706 av-mat; 9 357 Large Print Bks   38853

Converse County Library, 300 E Walnut St, *Douglas*, WY 82633
T: +1 307 3583644;
Fax: +1 307 3586743; URL: www.conversecountylibrary.org; conversecountylibrary.com
1905; Paul Pidde
Large print, Vietnam conflict, Quilting; Western American Coll
54 000 vols; 131 curr per   38854

Delaware Division of Libraries – State Library, 43 S Dupont Hwy, *Dover*, DE 19901-7430
T: +1 302 7394748; Fax: +1 302 7396787; URL: www.state.lib.de.us
1901; Annie E. Norman
1 branch libr
100 000 vols; 77 curr per
libr loan   38855

Delaware Division of Libraries-State Library – Delaware Library for the Blind & Physically Handicapped, 43 S Dupont Hwy, *Dover*, DE 19901-7430
T: +1 302 7394748; Fax: +1 302 7396787; E-mail: debph@lib.de.us
Beth Landon
63 000 vols; 10 curr per   38856

Dover Public Library, 45 S State St, *Dover*, DE 19901
T: +1 302 7367033; Fax: +1 302 7365087; E-mail: answerline_kent@lib.de.us; URL: www.doverpubliclibrary.org
1885; N.N.
115 000 vols; 320 curr per; 4 000 microforms; 1 486 av-mat; 3 500 sound-rec; 9 digital data carriers
libr loan; ALA   38857

Dover Public Library, 73 Locust St, *Dover*, NH 03820-3785
T: +1 603 5166082; Fax: +1 603 5166053; URL: www.dover.lib.nh.us
1883; Cathleen C. Beaudoin
New Hampshire & New England Hist & Genealogical Mat
104 000 vols; 294 curr per; 4 207 av-mat; 5 260 sound-rec; 1 000 digital data carriers; 1 400 Large Print Bks   38858

Dover Public Library, 525 N Walnut St, *Dover*, OH 44622
T: +1 330 3436123; Fax: +1 330 3432087; URL: www.doverlibrary.org
1923; Daniel R. Cooley
93 000 vols; 320 curr per   38859

Dover Town Library, 56 Dedham St, *Dover*, MA 02030-2214; P.O. Box 669, Dover, MA 02030-0669
T: +1 508 7858113; Fax: +1 508 7850138; E-mail: dover@minlib.net; URL: library.doverma.org
1894; Kathy Killeen
Military & Aviation Coll
58 000 vols; 118 curr per; 1 234 av-mat; 947 sound-rec; 65 digital data carriers
libr loan; ALA   38860

Downers Grove Public Library, 1050 Curtiss St, *Downers Grove*, IL 60515
T: +1 630 9601200; Fax: +1 630 9609374; E-mail: referencedesk@downersgrovelibrary.org; URL: www.downersgrovelibrary.org
1891; Christopher Bowen
Humanities, Art, Education
282 000 vols; 576 curr per; 1 870 e-books; 15 850 av-mat; 17 242 sound-rec
                    38861

County of Los Angeles Public Library, 7400 E Imperial Hwy, *Downey*, CA 90242-3375; P.O. Box 7011, Downey, CA 90241-7011
T: +1 562 9408462; Fax: +1 562 8033032; E-mail: colapl@library.lacounty.gov; URL: www.colapublib.org
1912; Terri Maguire
American Indian Resource Ctr, Asian Pacific Resource Ctr, Black Resource Ctr, Chicano Resource Ctr, Granger Poetry Coll, HIV Information Ctr, Judaica; 88 branch libs
6 528 000 vols; 10 777 curr per
libr loan   38862

Bucks County Free Library, 150 S Pine St, *Doylestown*, PA 18901-4932
T: +1 215 3481866; Fax: +1 215 3484760; URL: www.bucks.lib.org
1956; Martina Kominiarek
Foreign fiction, Large print; The Woods Handicapped & Gifted Coll; 7 branch libs
925 000 vols; 1 013 curr per
libr loan   38863

Bucks County Free Library – Library Center at Doylestown, 150 S Pine St, *Doylestown*, PA 18901-4932
T: +1 215 3489081; Fax: +1 215 3489489
N.N.
192 000 vols; 130 curr per   38864

Moses Greeley Parker Memorial Library, 28 Arlington St, *Dracut*, MA 01826
T: +1 978 4545474; Fax: +1 978 4549120; E-mail: mdr@mvlc.org; URL: www.dracutlibrary.org
1922; Dana Mastroianni
80 000 vols; 105 curr per   38865

County of Los Angeles Public Library – Duarte Library, 1301 Buena Vista St, *Duarte*, CA 91010-2410
T: +1 626 3581865; Fax: +1 626 3034917; URL: www.colapublib.org/libs/duarte
Robert Johnston
76 000 vols; 82 curr per   38866

Alameda County Library – Dublin Branch, 200 Civic Plaza, *Dublin*, CA 94568
T: +1 925 8281315; Fax: +1 925 8289296
Eileen Jouthas
130 000 vols; 234 curr per   38867

Columbus Metropolitan Library – Dublin Branch, 75 N High St, *Dublin*, OH 43017
T: +1 614 6452275; Fax: +1 614 4794179
Michael Blackwell
140 000 vols   38868

Oconee Regional Library, Laurens County, 801 Bellevue Ave, *Dublin*, GA 31021; P.O. Box 100, Dublin, GA 31040-0100
T: +1 478 2725710; Fax: +1 478 2755381; URL: www.ocrl.org
1904; Leard Daughety
Georgia Coll; 4 branch libs
93 000 vols; 135 curr per   38869

Carnegie-Stout Public Library, 360 W 11th St, *Dubuque*, IA 52001
T: +1 563 5894225; Fax: +1 563 5894217; E-mail: cspl@stout.dubuque.lib.ia.us; URL: www.dubuque.lib.ia.us
1902; Susan Henricks
200 000 vols; 487 curr per   38870

Duluth Public Library, 520 W Superior St, *Duluth*, MN 55802
T: +1 218 7304200; Fax: +1 218 7233822; E-mail: webmail@duluth.lib.mn.us; URL: www.duluth.lib.mn.us
1890; Carla Powers
Great Lakes, Minnesota; Adaptive Toys, Minnesota Coll, Sign Language Videos
419 310 vols; 559 curr per
libr loan   38871

Duluth Public Library – Mount Royal, 105 Mount Royal Shopping Circle, *Duluth*, MN 55803
T: +1 218 73-290; Fax: +1 218 7233846
Dennis Sherman
57 000 vols; 68 curr per   38872

Killgore Memorial Library, Moore County Library, 124 S Bliss Ave, *Dumas*, TX 79029-3889
T: +1 806 9354941; Fax: +1 806 9353324; E-mail: killgore@moorecountytexas.com; URL: www.mocolib.net
1936; Angela Dacus
58 000 vols; 60 curr per   38873

Duncan Public Library, 2211 North Hwy 81, *Duncan*, OK 73533
T: +1 580 2550636; Fax: +1 580 2556136; E-mail: jcole@duncan.lib.ok.us; URL: www2.youseemore.com/duncan/Default.asp?
1921; Jan Cole
59 480 vols; 125 curr per   38874

Duncanville Public Library, 201 James Collins Blvd, *Duncanville*, TX 75116
T: +1 972 7805052; Fax: +1 972 7804958; URL: www.youseemore.com/duncanville
1955; Carla Wolf Bryan
Texas Heritage Resource Ctr
94 000 vols; 176 curr per   38875

Dundee Township Public Library District, 555 Barrington Ave, *Dundee*, IL 60118-1496
T: +1 847 4283661; Fax: +1 847 4280521; URL: www.dundeelibrary.info
1879; Elisa Topper
110 000 vols; 336 curr per   38876

Dunedin Public Library, 223 Douglas Ave, *Dunedin*, FL 34698
T: +1 727 2983080; Fax: +1 727 2983088; URL: www.ci.dunedin.fl.us/dunedin/library/htm
1895; Anne M. Shepherd
Health care; Multimedia Coll
112 000 vols; 334 curr per
libr loan   38877

Dunkirk Free Library, 536 Central Ave, *Dunkirk*, NY 14048
T: +1 716 366-2511; Fax: +1 716 366-2525; E-mail: dunkirklibrary@gmail.com; URL: www.cclslib.org/Dunkirk/
1904; Janice Dekoff
50 000 vols; 114 curr per   38878

Durango Public Library, 1188 E Second Ave, *Durango*, CO 81301
T: +1 970 3753381;
Fax: +1 970 3753398; URL: www.durangopubliclibrary.org
1889; Sherry Taber
Archaeology Coll, Southwestern Hist Coll
86 000 vols; 219 curr per; 6 000 e-books; 3 321 av-mat
libr loan   38879

Robert L. Williams Public Library, 323 W Beech, *Durant*, OK 74701
T: +1 580 9243486; Fax: +1 580 9248843; E-mail: dlibrary@redriverok.com
1925; Alice Moore
67 000 vols; 25 curr per   38880

Durham County Library, Headquarters, 300 N Roxboro St, *Durham*, NC 27701; P.O. Box 3809, Durham, NC 27702-3809
T: +1 919 5600189; Fax: +1 919 5600106; E-mail: librarywebmaster@durhamcountync.gov; URL: www.durhamcountylibrary.org
1897; Hampton M. Auld
Adult education, Vocational education, Early childhood; African-American Coll; 7 branch libs
555 000 vols; 1 219 curr per
libr loan   38881

Durham Public Library, Seven Maple Ave, *Durham*, CT 06422
T: +1 860 3499544; Fax: +1 860 3491897; URL: www.durhamlibrary.org
1894; Valerie Kilmartin
59 000 vols; 85 curr per   38882

Duxbury Free Library, 77 Alden St, *Duxbury*, MA 02332
T: +1 781 93427211; Fax: +1 781 9340663; URL: www.duxburyfreelibrary.org
1889; Elaine Winquist
Graton Coll on Book Collecting
106 000 vols; 250 curr per; 8 200 av-mat; 1 000 sound-rec   38883

Dakota County Library System, 1340 Wescott Rd, *Eagan*, MN 55123-1099
T: +1 651 4502925; Fax: +1 651 4502915; E-mail: askalibrarian@co.dakota.mn.us; URL: www.co.dakota.mn.us/library
1959; Ken Behringer
8 branch libs
922 000 vols; 1 730 curr per; 7 000 e-books   38884

Eagle Public Library, 100 N Stierman Way, *Eagle*, ID 83616-5162
T: +1 208 9396814; Fax: +1 208 9391359; E-mail: eaglelibrary@cityofeagle.org; URL: www.eaglepubliclibrary.org
1969; Ronald J. Baker
54 000 vols; 174 curr per   38885

Eagle Valley Library District – Eagle Public Library, 600 Broadway, P.O. Box 240, *Eagle*, CO 81631-0240
T: +1 970 3288800; Fax: +1 970 3286901; E-mail: ajohnson@marmot.org; URL: www.evld.org
1993; Robyn Bryant
52 600 vols; 425 curr per
libr loan   38886

Pickens County Library System, 304 Biltmore Rd, *Easley*, SC 29640; P.O. Box 8010, Easley
T: +1 864 8507077; Fax: +1 864 8507088; E-mail: reference@pickens.lib.sc.us; URL: www.pickens.lib.sc.us
1935; Marguerite D. Keenan
South Caroliniana
146 000 vols; 326 curr per   38887

East Alton Public Library District, 250 Washington, *East Alton*, IL 62024-1547
T: +1 618 2590787; Fax: +1 618 2590788; E-mail: eae@lcls.org; URL: www.eastaltonlibrary.org
1936; Richard Chartrand
62 000 vols; 186 curr per
libr loan   38888

East Bridgewater Public Library, 32 Union St, *East Bridgewater*, MA 02333-1598
T: +1 508 3781616; Fax: +1 508 3781617; E-mail: ebpl@sailsinc.org; URL: www.sailsinc.org/ebpl
Manny Leite
51 000 vols; 105 curr per; 361 e-books; 1 106 av-mat   38889

East Brunswick Public Library, Two Jean Walling Civic Center, *East Brunswick*, NJ 08816-3599
T: +1 732 3906761; Fax: +1 732 3906869; URL: www.ebpl.org
1967; Carol Nersinger
Chinese language, Holocaust; Indian Languages Fiction Coll, Mystery Classics
139 000 vols; 340 curr per
libr loan   38890

East Chicago Public Library, 2401 E Columbus Dr, *East Chicago*, IN 46312-2998
T: +1 219 3972453; Fax: +1 219 3976715; URL: www.ecpl.org
1909; Manuel Montalvo
Ethnic studies; 1 branch libr
295 000 vols; 583 curr per; 14 271 av-mat; 7 890 sound-rec; 390 Large Print Bks
libr loan   38891

East Chicago Public Library – Robert A. Pastrick Branch, 1008 W Chicago Ave, *East Chicago*, IN 46312
T: +1 219 3975505; Fax: +1 219 3982827
Dr. James M. Rajchel
73 000 vols; 135 curr per; 419 av-mat   38892

East Cleveland Public Library, 14101 Euclid Ave, *East Cleveland*, OH 44112-3891
T: +1 216 5414128; Fax: +1 216 5411790; URL: www.ecpl.lib.oh.us
1916; Gregory L. Reese
Black Heritage Coll, Illust Children's Bk Coll, Holograph Letters of the Presidents from George Washington to James E Carter; 2 branch libs
210 000 vols; 200 curr per
libr loan   38893

Queens Borough Public Library – East Elmhurst Branch, 95-06 Astoria Blvd, *East Elmhurst*, NY 11369
T: +1 718 4242619; Fax: +1 718 6517045
Johnnie Dent
52 000 vols   38894

Kent District Library – East Grand Rapids Branch, 746 Lakeside Dr SE, *East Grand Rapids*, MI 49506
T: +1 616 6473880; Fax: +1 616 9403680
Dawn Lewis
63 000 vols   38895

East Greenbush Community Library, Ten Community Way, *East Greenbush*, NY 12061
T: +1 518 4777476;
Fax: +1 518 4776692; URL: www.eastgreenbushlibrary.org
1948; Deborah Graves Shoup
71 000 vols; 196 curr per; 2 000 e-books; 10 maps; 6 910 av-mat; 2 000 sound-rec; 50 digital data carriers
libr loan   38896

East Greenwich Free Library, 82 Peirce St, *East Greenwich*, RI 02818
T: +1 901 8849511;
Fax: +1 401 8843790; URL: www.eastgreenwichlibrary.org
1869; Karen A. Taylor
55 000 vols; 110 curr per; 370 e-books   38897

East Hampton Library, 159 Main St, *East Hampton*, NY 11937
T: +1 631 3240222; Fax: +1 631 3295947; E-mail: info@easthamptonlibrary.org; URL: www.easthamptonlibrary.org
1897; Dennis Fabiszak
Thomas Moran Biogr Coll, Long Island Hist Coll
64 000 vols; 96 curr per   38898

East Hampton Public Library, 105 Main St, *East Hampton*, CT 06424
T: +1 860 2676621; Fax: +1 860 2674427; E-mail: ehplct@hotmail.com; URL: www.easthamptonct.org
1898; Susan Berescik
East Hampton Room
53 000 vols; 85 curr per   38899

East Hanover Township Free Public Library, 415 Ridgedale Ave, *East Hanover*, NJ 07936
T: +1 973 4283075;
Fax: +1 973 4287253; URL: www.easthanoverlibrary.com
1959; Gayle B. Carlson
68 000 vols; 112 curr per   38900

East Hartford Public Library, Raymond Memorial Library, 840 Main St, *East Hartford*, CT 06108
T: +1 860 2896429; Fax: +1 860 2919166; URL: www.easthartford.lib.ct.us
1879; Patrick Michael Jones
Aviation Coll, Tobacco Coll; 3 branch libs
290 000 vols; 400 curr per   38901

Hagaman Memorial Library, East Haven Public Library, 227 Main St, *East Haven*, CT 06512-3003
T: +1 203 4683890; Fax: +1 203 4683892; URL: www.leaplibraries.org/ehaven
1909; Ellen Gambini
67 000 vols; 133 curr per; 2 040 av-mat   38902

East Islip Public Library, 381 E Main St, *East Islip*, NY 11730-2896
T: +1 631 5819200; Fax: +1 631 5812245; E-mail: eipl@suffolk.lib.ny.us; URL: www.eipl.org
1960; Guy P. Edwards
Art Originals, Lighthouses Coll
178 000 vols; 494 curr per   38903

East Lansing Public Library, 950 Abbott Rd, *East Lansing*, MI 48823-3105
T: +1 517 3512420; Fax: +1 517 3519536; E-mail: elplcirc@cityofeastlansing.com; URL: www.elpl.org
1923; Sylvia Marabate
150 000 vols; 316 curr per   38904

Carnegie Public Library, 219 E Fourth St, *East Liverpool*, OH 43920-3143
T: +1 330 3852048; Fax: +1 330 3857600; E-mail: eastliv@oplin.org; URL: www.carnegie.lib.oh.us
1900; Melissa A. W. Percic
Pottery/Ceramics Coll
85 000 vols; 255 curr per   38905

East Longmeadow Public Library, 60 Center Sq, *East Longmeadow*, MA 01028-2459
T: +1 413 5255400; Fax: +1 413 5250344; URL: www.eastlongmeadow.org/library
1896; Susan M. Peterson
75 000 vols; 139 curr per
libr loan   38906

East Meadow Public Library, 1886 Front St, *East Meadow*, NY 11554-1700
T: +1 516 7942570; Fax: +1 516 7948536; E-mail: contactus@eastmeadow.info; URL: www.eastmeadow.info
1955; Carol Probeyahn
Behavioral sciences, civil service, religious studies, repair manuals; Long Island hist, literary criticism
272 000 vols; 18 013 sound-rec; 20 digital data carriers   38907

East Moline Public Library, 740 16th Ave, *East Moline*, IL 61244-2122
T: +1 309 7559614; Fax: +1 309 7553901; E-mail: emp@empl.lib.il.us; URL: www.empl.lib.il.us
1917; Cynthia K. Coe
65 000 vols; 130 curr per; 6 000 e-books; 6 236 av-mat   38908

East Orange Public Library, 21 S Arlington Ave, *East Orange*, NJ 07018-3892
T: +1 973 2665607; Fax: +1 973 6741991; E-mail: feedback@eopl.org; URL: www.eopl.org
1900; Carolyn Ryan Reed
Ethnic studies; New Jerseyana; 3 branch libs
405 000 vols; 634 curr per   38909

Fondulac Public Library District, 140 E Washington St, *East Peoria*, IL 61611-2598
T: +1 309 6993917; Fax: +1 309 6997851; E-mail: mail@fondulaclibrary.org; URL: www.fondulaclibrary.org
1935; Nancy Gillfillan
72 000 vols; 108 curr per; 325 e-books   38910

Atlanta-Fulton Public Library System – East Point Library, 2757 Main St, *East Point*, GA 30344
T: +1 404 7624842; Fax: +1 404 7624844
Michael Hickman
68 000 vols   38911

East Providence Public Library, 41 Grove Ave, *East Providence*, RI 02914
T: +1 401 4342719;
Fax: +1 401 4354997; URL: www.eastprovidencelibrary.org
Eileen Socha
East Bay Literacy; 3 branch libs
304 000 vols; 320 curr per   38912

East Providence Public Library – Fuller, 260 Dover Ave, *East Providence*, RI 02914
T: +1 401 4341136; Fax: +1 401 4343896
Sharon Branch
62 000 vols; 150 curr per   38913

East Providence Public Library – Riverside, 475 Bullocks Point Ave, *East Providence*, RI 02915
T: +1 401 4334877; Fax: +1 401 4334820
Sarah Capobianco
55 000 vols; 115 curr per   38914

East Providence Public Library – Rumford, 1392 Pawtucket Ave, *East Providence*, RI 02916
T: +1 401 4348559; Fax: +1 401 4341808
Denise Inman
63 000 vols; 110 curr per   38915

County of Los Angeles Public Library – East Rancho Dominguez Library, 4205 E Compton Blvd, *East Rancho Dominguez*, CA 90221-3664
T: +1 310 6326193; Fax: +1 310 6080294; URL: www.colapublib.org/libs/dominguez
Betty Marlow
54 000 vols; 66 curr per   38916

East Rockaway Public Library, 477 Atlantic Ave, *East Rockaway*, NY 11518
T: +1 516 5991664; Fax: +1 516 5960154; E-mail: eastrockpl@yahoo.com; URL: www.nassaulibrary.org/eastrock
Ellen Rockmuller
Special Needs Parent Coll
64 000 vols; 95 curr per   38917

East Saint Louis Public Library, 5300 State St, *East Saint Louis*, IL 62203
T: +1 618 3970991; Fax: +1 618 3971260; E-mail: esa@lcls.org
1872; Jean L. Brezger
Metro-East Journal since 1889, micro
51 000 vols; 40 curr per   38918

Mercer County Library – Twin Rivers Branch, 276 Abbington Dr, *East Windsor*, NJ 08520
T: +1 609 4431880; Fax: +1 609 4900186
Rebecca Sloan
50 000 vols   38919

Eastchester Public Library, 11 Oak Ridge Pl, *Eastchester*, NY 10709
T: +1 914 7935055; Fax: +1 914 7937862; URL: www.eastchesterlibrary.org
1947; Catherine L. McDowell
94 000 vols; 1940 curr per   38920

Ocmulgee Regional Library System – Dodge County Library (System Headquarters), 535 Second Ave, *Eastman*, GA 31023
T: +1 478 3744711; Fax: +1 478 3745646
Stephen Whigham
210 000 vols; 262 curr per   38921

Easton Area Public Library & District Center, 515 Church St, *Easton*, PA 18042-3587
T: +1 610 2582917; Fax: +1 610 2532231; E-mail: director@eastonpl.org; URL: www.eastonpl.org
1811; Jennifer Stocker
2 branch libs
188 000 vols; 416 curr per
libr loan   38922

Easton Public Library, 691 Morehouse Rd, *Easton*, CT 06612; P.O. Box 2, Easton, CT 06612-0002
T: +1 203 2610134; Fax: +1 203 2610708; E-mail: leastonp@optonline.net; URL: www.eastonlibrary.org
1934; Bernadette Baldino
53 000 vols   38923

Talbot County Free Library, 100 W Dover St, *Easton*, MD 21601-2620
T: +1 410 8221626; Fax: +1 410 8208217; E-mail: askus@talb.lib.md.us; URL: www.talb.lib.md.us
1925; Robert T. Horvath
1 branch libr
107 000 vols; 150 curr per   38924

Eastpointe Memorial Library, 15875 Oak St, *Eastpointe*, MI 48021-2390
T: +1 586 4455096; Fax: +1 586 7750150; E-mail: eplweb@libcoop.net; URL: www.ci.eastpointe.mi.us/library
1939; Carol Sterling
Automobile Manuals Coll
62 000 vols; 124 curr per   38925

Preble County District Library, 450 S Barron St, *Eaton*, OH 45320-2402
T: +1 937 4564250; Fax: +1 937 4566092; E-mail: pcdllibrary@oplin.org; URL: www.pcdl.lib.oh.us
1959; Abigail Noland
9 branch libs
174 000 vols; 980 curr per   38926

L. E. Phillips Memorial Public Library, 400 Eau Claire St, *Eau Claire*, WI 54701
T: +1 715 8587096; Fax: +1 715 8393822; E-mail: librarian@eauclaire.lib.wi.us; URL: www.eauclaire.lib.wi.us
1875; Michael Golrick
294 000 vols; 15 450 av-mat; 5 985 sound-rec; 3 974 digital data carriers; 10 995 Large Print Bks
libr loan   38927

Rockingham County Public Library, 527 Boone Rd, *Eden*, NC 27288
T: +1 336 6271106; Fax: +1 336 6231258; E-mail: reference@library.rcpl.org; URL: www.rcpl.org
1934; Jay Stephens
African-American, Business, State hist, Large print, Literacy; 5 branch libs
294 000 vols; 624 curr per   38928

Edinburg Public Library, 401 E Cano St, *Edinburg*, TX 78539
T: +1 956 3836247; Fax: +1 956 2922026; URL: www.edinburg.lib.tx.us
1967; Leticia Salas Leija
Bilingual-Bicultural (Spanish) Coll, Literacy & Adult Basis Education, Texana
104 000 vols; 228 curr per   38929

Shenandoah County Library, 514 Stoney Creek Blvd, *Edinburg*, VA 22824
T: +1 540 9848200; Fax: +1 540 9848207; E-mail: scl@shentel.net; URL: www.shenandoah.co.lib.va.us
1984; Robert L. Pasco
107 000 vols; 115 curr per; 300 e-books   38930

Edison Township Free Public Library, 340 Plainfield Ave, *Edison*, NJ 08817
T: +1 732 2872298; Fax: +1 732 8199134; URL: www.lmxac.org/edisonlib
1928; Judith Mansbach
2 branch libs
281 000 vols; 146 curr per; 7 067 av-mat; 2 492 sound-rec   38931

Edison Township Free Public Library – Clara Barton Branch, 141 Hoover Ave, *Edison*, NJ 08837
T: +1 732 7380096; Fax: +1 732 7388325
Margaret Vellucci
57 000 vols; 99 curr per; 428 av-mat
38932

Edison Township Free Public Library – North Edison Branch, 777 Grove Ave, *Edison*, NJ 08820
T: +1 732 5483045; Fax: +1 732 5495171
Evan T. Davis
101 000 vols; 146 curr per; 2 194 av-mat
38933

Metropolitan Library System in Oklahoma County -, Ten South Blvd, *Edmond*, OK 73034-3798
T: +1 405 3419282; Fax: +1 405 6063411; E-mail: edmond@metrolibrary.org
Karen Bays
124 000 vols; 143 curr per
38934

Sno-Isle Libraries – Edmonds Branch, 650 Main St, *Edmonds*, WA 98020
T: +1 425 7711933; Fax: +1 425 7711977
Lesly Kaplan
112 000 vols
38935

Edwardsville Public Library, 112 S Kansas St, *Edwardsville*, IL 62025
T: +1 618 6927556; Fax: +1 618 6929566; E-mail: ede@lcls.org; URL: edwardsvillelibrary.org
1818; Deanne W. Holshouser
Madison County Genealogical Society Coll
90 000 vols
38936

Helen Matthes Library, 100 E Market Ave, *Effingham*, IL 62401
T: +1 217 3422464; Fax: +1 217 3422413; E-mail: hmlib@effinghamlibrary.org; URL: www.effinghamlibrary.org
1942; Amanda E. Standerfer
Genealogy, World War II
69 000 vols; 80 curr per
38937

San Diego County Library – El Cajon Branch, 201 E Douglas, *El Cajon*, CA 92020
T: +1 619 5883718; Fax: +1 619 5883701
Cheryl Doty
148 000 vols
38938

San Diego County Library – Rancho San Diego, 11555 Via Rancho San Diego, *El Cajon*, CA 92019
T: +1 619 6605370; Fax: +1 619 6606327
Brenna Ring
91 000 vols
38939

Imperial County Free Library, 1331 S Clark, Bldg 24, *El Centro*, CA 92243; mail address: 1125 Main St, El Centro, CA 92243-2748
T: +1 760 3396460; Fax: +1 760 3396465; URL: www.co.imperial.ca.us/library
1912; Connie Barrington
53 000 vols; 45 curr per; 300 sound-rec; 20 digital data carriers
libr loan
38940

Barton Library, 200 E Fifth St, *El Dorado*, AR 71730-3897
T: +1 870 8635447; Fax: +1 870 8623944; E-mail: inquiries@bartonlibrary.org; URL: www.bartonlibrary.org
1958; Nancy Arn
Arkansas Coll, Genealogy Coll, Large Print Coll; 5 branch libs
73 000 vols; 163 curr per
38941

Bradford Memorial Library, 611 S Washington St, *El Dorado*, KS 67042
T: +1 316 3213363; Fax: +1 316 3215546; E-mail: bmlibadm@eldoks.com; URL: skyways.lib.ks.us/library/bradford
1897; Hollis Helmeci
Connell Ornithology Coll
66 000 vols; 108 curr per
38942

County of Los Angeles Public Library – El Monte Library, 3224 Tyler Ave, *El Monte*, CA 91731-3356
T: +1 626 4449506; Fax: +1 626 4435864; URL: www.colapublib.org/libs/elmonte
Tony Ramirez
85 000 vols; 100 curr per
38943

County of Los Angeles Public Library – Norwood Library, 4550 N Peck Rd, *El Monte*, CA 91732-1998
T: +1 626 4433147; Fax: +1 626 3506099; URL: www.colapublib.org/libs/norwood
Martin Delgado
69 000 vols; 81 curr per
38944

El Paso Public Library, 501 N Oregon St, *El Paso*, TX 79901
T: +1 915 5435410; Fax: +1 915 5435410; URL: www.elpasolibrary.org
1894; Carol A. Brey-Casiano
Raza Coll; 10 branch libs
950 000 vols; 425 curr per
38945

El Reno Carnegie Library, 215 E Wade, *El Reno*, OK 73036-2753
T: +1 405 2622409; Fax: +1 405 4222136; E-mail: library@elrenolibrary.org; URL: www.elreno.lib.ok.us
Kate Shaklee
Edna May Armold Arch
55 000 vols; 149 curr per
38946

El Segundo Public Library, 111 W Mariposa Ave, *El Segundo*, CA 90245-2299
T: +1 310 5242722; Fax: +1 310 6487560; URL: www.elsegundo.org
1930; Debra F. Brighton
153 000 vols; 340 curr per; 1 000 e-books; 79 000 microforms; 4 700 sound-rec; 110 digital data carriers; 500 High Interest/Low Vocabulary Bks, 20 Special Interest Per Sub, 20 Bks on Deafness & Sign Lang, 1 500 Talking Bks, 5 000 Large Print Bks
libr loan; ALA
38947

Elbert County Public Library, 345 Heard St, *Elberton*, GA 30635
T: +1 706 2835375; Fax: +1 706 2835456
1925; Peggy Jane Johnson
72 000 vols; 118 curr per
38948

Scott County Library System, 200 N Sixth Ave, *Eldridge*, IA 52748
T: +1 563 2854794; Fax: +1 563 2854743; URL: www.scottcountrylibrary.org
1950; Paul H. Seelau
Scott County Hist; 6 branch libs
136 000 vols; 645 curr per
38949

Gail Borden Public Library District, 270 N Grove Ave, *Elgin*, IL 60120-5596
T: +1 847 4294690; Fax: +1 847 7420485; URL: www.gailborden.info
1873; Carole Medal
Genealogy Coll, Spanish Language Mats
404 000 vols; 682 curr per; 7 000 e-books; 29 812 av-mat; 27 070 sound-rec; 1 170 High Interest/Low Vocabulary Bks, 190 Bks on Deafness & Sign Lang
38950

Elizabeth Public Library, 11 S Broad St, *Elizabeth*, NJ 07202
T: +1 908 3546060; Fax: +1 908 3545845; URL: www.njpublib.org
1908; Dorothy Key
Elizabethtown Rm Coll; 3 branch libs
300 000 vols; 587 curr per
38951

East Albemarle Regional Library, 100 E Colonial Ave, *Elizabeth City*, NC 27909-0303
T: +1 252 3352511; Fax: +1 252 3352386; URL: earlibrary.org
1964; Jackie King
North Carolina Coll; 1 bookmobile
170 000 vols; 350 curr per; 7 945 maps
38952

Pasquotank-Camden Library, 100 E Colonial Ave, *Elizabeth City*, NC 27909
T: +1 252 3357536; Fax: +1 252 3317449; URL: www.earlibrary.org/pasquotank-camden/index.html
1930; Jackie King
North Carolina hist
62 000 vols; 88 curr per; 1 795 av-mat
38953

Elizabethton-Carter County Public Library, 201 N Sycamore St, *Elizabethton*, TN 37643
T: +1 423 5476360; Fax: +1 423 5421510; URL: www.eccpl.org
1929; Joyce H. White
Tennessee Hist Coll
59 000 vols; 75 curr per
38954

Bladen County Public Library, 111 N Cyprus St, *Elizabethtown*, NC 28337; P.O. Box 1419, Elizabethtown, NC 28337-1419
T: +1 910 8626990; Fax: +1 910 8628777; E-mail: bcpl@bladenco.org; URL: www.youseemore.com/bladen
1939; Anta Maria Hebert
76 000 vols; 85 curr per; 539 av-mat
38955

Hardin County Public Library, 100 Jim Owen Dr, *Elizabethtown*, KY 42701
T: +1 270 7696337; Fax: +1 270 7690437; E-mail: hcplm@kvnet.org; URL: www.hcpl.info
1958; Brenda G. Macy
1 branch libr
98 000 vols; 211 curr per; 319 e-books
38956

Elk Grove Village Public Library, 1001 Wellington Ave, *Elk Grove Village*, IL 60007
T: +1 847 4390447; Fax: +1 847 4390475; E-mail: contact@egvpl.org; URL: www.egvpl.org
1959; Lee James Maternowski
European languages, Civil War, World War II
270 000 vols; 522 curr per; 25 963 av-mat; 27 050 sound-rec; 12 550 digital data carriers; 21 Braille vols, 271 High Interest/Low Vocabulary Bks, 50 Bks on Deafness & Sign Lang, 7 431 Large Print Bks
38957

Elkhart Public Library, 300 S Second St, *Elkhart*, IN 46516-3184
T: +1 574 5222665; Fax: +1 574 2939213; E-mail: admin@elkhart.lib.in.us; URL: www.elkhart.lib.in.us
1903; Connie Jo Ozinga
4 branch libs
385 000 vols; 1 980 curr per; 23 125 av-mat; 110 sound-rec; 590 Large Print Bks
libr loan
38958

Northwestern Regional Library, 111 N Front St, *Elkin*, NC 28621
T: +1 336 8354894; Fax: +1 336 8351356; E-mail: nwrl@nwrl.org; URL: www.nwrl.org/mta.asp
1959; John Hedrick
319 000 vols; 225 curr per; 48 000 av-mat; 17 030 sound-rec; 445 digital data carriers; 18 860 Large Print Bks
38959

Elkins Park Free Library, 563 E Church Rd, *Elkins Park*, PA 19027-2499
T: +1 215 6355000; Fax: +1 215 6355844; E-mail: elkinspark@mclinc.org; URL: www.cheltenhamlibraries.org
1958; Cathleen McCoy
Art, Multicultural mat, Large print bks
52 000 vols; 166 curr per
38960

Howard County Library – Elkridge Branch, 6540 Washington Blvd, *Elkridge*, MD 21075
T: +1 410 3135077; Fax: +1 410 3135090
Phil Lord
78 000 vols
38961

Cecil County Public Library, Elkton Central Library, 301 Newark Ave, *Elkton*, MD 21921-5441
T: +1 410 9961055; Fax: +1 410 9965604; E-mail: cecilref@ccplnet.org; URL: www.cecil.ebranch.info
1947; Denise Davis
6 branch libs
204 000 vols; 436 curr per
38962

Ellenville Public Library & Museum, 40 Center St, *Ellenville*, NY 12428
T: +1 845 6475530; Fax: +1 845 6473554; E-mail: epl@rcls.org; URL: eplm.org/default.aspx
1893; Pamela Stocking
50 900 vols; 99 curr per
38963

Howard County Library – Miller Branch, 9421 Frederick Rd, *Ellicott City*, MD 21042-2119
T: +1 410 3131950; Fax: +1 410 3131999
Susan Stonesifer
196 000 vols
38964

Elmhurst Public Library, 125 S Prospect, *Elmhurst*, IL 60126-3298
T: +1 630 2798696; Fax: +1 630 5161364; E-mail: reference@elmhurst.org; URL: www.elmhurstpubliclibrary.org
1916; Marilyn Boria
Art & architecture, Business & management
283 000 vols; 547 curr per; 10 170 av-mat; 3 413 digital data carriers; 11 585 CDs
libr loan
38965

Queens Borough Public Library – Elmhurst Branch, 86-01 Broadway, *Elmhurst*, NY 11373
T: +1 718 2711020; Fax: +1 718 6998069
Yasha Hu
135 000 vols
38966

Chemung County Library District, Steele Memorial Library, 101 E Church St, *Elmira*, NY 14901-2799
T: +1 607 7338607; Fax: +1 607 7339176; URL: www.ccld.lib.ny.us
1893; James G. Sleeth
Art Coll, Genealogy Coll, Large Print Coll, Census Coll; 4 branch libs
267 000 vols; 224 curr per
38967

Elmont Public Library, 700 Hempstead Turnpike, *Elmont*, NY 11003-1896
T: +1 516 3545280; Fax: +1 516 3543276; URL: www.nassaulibrary.org/elmont
1939; Janis A. Schoen
2 branch libs
191 000 vols; 299 curr per
38968

Elmwood Park Public Library, One Conti Pkwy, *Elmwood Park*, IL 60707
T: +1 708 4537645; Fax: +1 708 4534671; E-mail: eps@elmwoodparklibrary.org; URL: www.elmwoodparklibrary.org
1936; Shawn Strecker
Italian, Spanish & Polish language, Lit
77 000 vols; 503 curr per
38969

Elmwood Park Public Library, 210 Lee St, *Elmwood Park*, NJ 07407
T: +1 201 7962497; Fax: +1 201 7031425; E-mail: elpkcirc@bccls.org; URL: www.bccls.org/elmwoodpark
1953; Ethan Galvin
66 000 vols; 125 curr per; 1 950 av-mat
38970

North Madison County Public Library System – Elwood Public Library, 1600 Main St, *Elwood*, IN 46036
T: +1 765 5525001; Fax: +1 765 5520955
1898; Jamie Scott
Wendell L. Wilkie Coll
86 000 vols; 109 curr per
38971

Elyria Public Library System, 320 Washington Ave, *Elyria*, OH 44035-5199
T: +1 440 3220461; Fax: +1 440 3231078; E-mail: epl@elyria.lib.oh.us; URL: www.elyria.lib.oh.us
1870; Janet Stoffer
Ely Papers, Nonprofit Info Ctr; 3 branch libs
384 000 vols; 715 curr per
libr loan
38972

Elyria Public Library System – West River, 1194 West River Rd N, *Elyria*, OH 44035
T: +1 440 3242270; Fax: +1 440 3244766
Jennifer Jung Gallant
147 000 vols
38973

Emmaus Public Library, 11 E Main St, *Emmaus*, PA 18049
T: +1 610 9659284; Fax: +1 610 9656446; E-mail: emmauspl@cliu.org; URL: www.emmauspl.org
1966; Frances A. Larash
Art & Reference Bks (Roeder Coll), Shelter House Coll (Local Hist)
90 000 vols; 120 curr per
38974

Emmett Public Library, 275 S Hayes, *Emmett*, ID 83617-2972
T: +1 208 3656057; Fax: +1 208 3656060; URL: www.emmettlibrary.com
1924; Alyce Kelley
61 000 vols; 39 curr per   38975

Emporia Public Library, 110 E Sixth Ave, *Emporia*, KS 66801-3960
T: +1 620 3406462; Fax: +1 620 3406444; URL: www.emporialibrary.org
1869; Sue Blechl
96 000 vols; 203 curr per   38976

San Diego County Library – Encinitas Branch, 540 Cornish Dr, *Encinitas*, CA 92024-4599
T: +1 760 7537376; Fax: +1 760 7530582
Sandy Housley
63 000 vols   38977

George F. Johnson Memorial Library, 1001 Park St, *Endicott*, NY 13760
T: +1 607 7575350; Fax: +1 607 7572491; E-mail: en.web@4cls.org; URL: www.gfjlibrary.org
1915; Edward Andrew Dunscombe
George F. Johnson Memorabilia
90 000 vols; 177 curr per; 1945 av-mat   38978

Arapahoe Library District, 12855 E Adam Aircraft Circle, *Englewood*, CO 80112
T: +1 303 5427279; Fax: +1 303 7982485; URL: www.arapahoelibraries.org
1966; Eloise May
8 branch libs
477 000 vols; 1 691 curr per; 37 047 av-mat; 25 050 sound-rec; 6 802 Large Print Bks   38979

Englewood Public Library, 31 Engle St, *Englewood*, NJ 07631
T: +1 201 5682215; Fax: +1 201 5686895; URL: www.englewoodlibrary.org
1901; Donald Jacobsen
African-American studies, Careers, Jazz, Judaica
124 000 vols; 255 curr per   38980

Englewood Public Library, 1000 Englewood Pkwy, *Englewood*, CO 80110
T: +1 303 7622550; Fax: +1 303 7836890; URL: www.englewoodgov.org
1920; Hank Long
105 000 vols; 310 curr per   38981

Enumclaw Public Library, 1700 First St, *Enumclaw*, WA 98022
T: +1 360 8252938; Fax: +1 360 8250825; E-mail: library@ci.enumclaw.wa.us; URL: www.enumclaw.lib.wa.us
1922
53 760 vols; 145 curr per   38982

Ephrata Public Library, 550 S Reading Rd, *Ephrata*, PA 17522
T: +1 717 7389291; Fax: +1 717 7213003; URL: www.ephratapubliclibrary.org
1962; Joseph Zappacosta
72 000 vols; 35 curr per   38983

Erie County Public Library, 160 E Front St, *Erie*, PA 16507
T: +1 814 4516952; Fax: +1 814 4516969; E-mail: reference@erielibrary.org; URL: www.erielibrary.org
1895; Margaret Z. Stewart
Genealogy Coll (Western Pennsylvania); 6 branch libs
486 000 vols; 427 curr per
libr loan   38984

High Plains Library District – Erie Community Library, 400 Powers St, *Erie*, CO 80516
T: +1 720 6855200; Fax: +1 720 6855201
Tony Brewer
54 000 vols   38985

Kenton County Public Library – Erlanger Branch, 401 Kenton Lands Rd, *Erlanger*, KY 41018
T: +1 859 9624000; Fax: +1 859 9624010
Susan Banks
150 000 vols; 200 curr per   38986

Escanaba Public Library, 400 Ludington St, *Escanaba*, MI 49829
T: +1 906 7860942; E-mail: epl@uproc.lib.mi.us; URL: www.uproc.lib.mi.us/epl
1903; Carolyn Stacey
73 000 vols; 136 curr per; 800 av-mat   38987

Escondido Public Library, East Valley, 2245 E Valley Pkwy, *Escondido*, CA 92027
T: +1 760 8394394; Fax: +1 760 7462052; URL: www.library.escondido.org
Jeff Wyner
53 000 vols; 73 curr per   38988

Espanola Public Library, 313 North Paseo de Oñate, *Espanola*, NM 87532
T: +1 505 7476087; Fax: +1 505 7535543; E-mail: espanolapublic@yahoo.com; URL: www.espanolaonline.com/library
1969
Southwest Coll
56 780 vols; 105 curr per   38989

Brownell Library, Six Lincoln St, *Essex Junction*, VT 05452-3154
T: +1 802 8786954; Fax: +1 802 8786946; E-mail: brownell_library@yahoo.com; URL: www.brownelllibrary.org
1897; Penelope D. Pillsbury
67 000 vols; 184 curr per; 486 e-books; 3 554 av-mat   38990

Estacada Public Library, 825 NW Wade St, *Estacada*, OR 97023
T: +1 503 6308273; Fax: +1 503 6308282; URL: www.estacada.lib.or.us
Katinka Bryk
Estacada History
59 000 vols; 223 curr per; 2 538 av-mat   38991

Lee County Library System – South County Regional, 21100 Three Oaks Pkwy, *Estero*, FL 33928-3020
T: +1 239 3903200; Fax: +1 239 4986424
Elizabeth Nitch
160 000 vols; 116 curr per; 4 154 av-mat   38992

Estes Park Public Library, 335 E Elkhorn Ave, *Estes Park*, CO 80517; P.O. Box 1687, Estes Park, CO 80517-1687
T: +1 970 5868116; Fax: +1 970 5860189; E-mail: reference@esteslibrary.org; URL: www.esteslibrary.org
1922; Claudine Perrault
61 000 vols; 200 curr per   38993

Estherville Public Library, 613 Central, *Estherville*, IA 51334
T: +1 712 3627731; Fax: +1 712 3623509; E-mail: main@esthervillepubliclibrary.com; URL: www.esthervillepubliclibrary.com
1903; Carolyn L. Walz
Iowa; Expand Your Horizons, (Books for Literacy Groups)
56 870 vols; 85 curr per   38994

Euclid Public Library, 631 E 222nd St, *Euclid*, OH 44123-2091
T: +1 216 2615300; Fax: +1 216 2610575; URL: www.euclidlibrary.org
1935; Donna L. Perdzock
279 000 vols; 472 curr per
libr loan   38995

Euless Public Library, 201 N Ector Dr, *Euless*, TX 76039-3595
T: +1 817 6851679; Fax: +1 817 2671979; URL: www.ci.euless.tx.us/dept/library/library.htm
1961; Kate Lyon
Genealogy
100 000 vols; 290 curr per   38996

Humboldt County Library, 1313 Third St, *Eureka*, CA 95501-0553
T: +1 707 2691900; URL: www.humlib.org
1878; Victor Zazueta
American Indians Coll, Humboldt County Hist Coll, NRC Nuclear Power Plants LDR; 10 branch libs
240 000 vols; 332 curr per
libr loan   38997

Humboldt County Library – Eureka Main Library, 1313 Third St, *Eureka*, CA 95501
T: +1 707 2691900
Carolyn Stacey
162 000 vols   38998

Eustis Memorial Library, 120 N Center St, *Eustis*, FL 32726-3598
T: +1 352 3575686; Fax: +1 352 3575450; E-mail: contactus@eustismemoriallibrary.org; URL: www.eustismemoriallibrary.org
1902; E. Steven Benetz
Circulating Art Prints, Florida Coll
96 000 vols; 177 curr per; 1 929 av-mat   38999

East Central Georgia Regional Library – Columbia County Public Library, 7022 Evans Town Center Blvd, *Evans*, GA 30809
T: +1 706 863194; Fax: +1 706 8683351
Christina Rice
92 000 vols; 119 curr per; 250 av-mat   39000

Evanston Public Library, 1703 Orrington, *Evanston*, IL 60201
T: +1 847 8660300; Fax: +1 847 8660313; URL: www.epl.org
1873; N.N.
Antique Silver (Berg Coll), Music (Sadie Coe Coll); 2 branch libs
465 000 vols; 918 curr per   39001

Uinta County Library, 701 Main St, *Evanston*, WY 82930
T: +1 307 7892770; Fax: +1 307 7890148; URL: www.uintalibrary.org
1904; Dale E. Collum
90 000 vols; 249 curr per   39002

Evansville-Vanderburgh Public Library, 200 SE Martin Luther King Jr Blvd, *Evansville*, IN 47713-1604
T: +1 812 4288204; Fax: +1 812 4288397; E-mail: comments@evpl.org; URL: www.evpl.org
1911; Marcia Learned-Au
Agriculture, Economics, Education, Religion; Judaica Coll, Libr Science; 9 branch libs
653 000 vols; 2 231 curr per; 2 000 e-books; 20 124 av-mat; 13 600 Large Print Bks, 20 150 Talking Bks
libr loan   39003

Evansville-Vanderburgh Public Library – McCollough Branch, 5115 Washington Ave, *Evansville*, IN 47715
T: +1 812 4288236; Fax: +1 812 4730877; E-mail: mc@evpl.org
Glynis Rosendall
75 000 vols; 218 curr per   39004

Evansville-Vanderburgh Public Library – North Park, 960 Koehler Dr, *Evansville*, IN 47710
T: +1 812 4288237; Fax: +1 812 4288243; E-mail: np@evpl.org
Michael Campese
70 000 vols; 189 curr per   39005

Evansville-Vanderburgh Public Library – Oaklyn, 3001 Oaklyn Dr, *Evansville*, IN 47711
T: +1 812 4288234; Fax: +1 812 4288245; E-mail: oa@evpl.org
Pamela Locker
67 000 vols; 203 curr per   39006

Evansville-Vanderburgh Public Library – Red Bank, 120 S Red Bank Rd, *Evansville*, IN 47712
T: +1 812 4288205; Fax: +1 812 4288240; E-mail: rb@evpl.org
Nancy Higgs
61 000 vols; 280 curr per   39007

Willard Library of Evansville, 21 First Ave, *Evansville*, IN 47710-1294
T: +1 812 4254309; Fax: +1 812 4219742; E-mail: willard@willard.lib.in.us; URL: www.willard.lib.in.us
1885; Gregory M. Hager
Thrall Art Bk Coll (Architecture), 19th C Per Lit
135 000 vols; 270 curr per
libr loan   39008

Everett Public Libraries, Frederick E. Parlin Memorial, 410 Broadway, *Everett*, MA 02149
T: +1 617 3942300; Fax: +1 617 3891230; E-mail: eve@noblenet.org; URL: www.noblenet.org/everett
1879; Deborah V. Abraham
1 branch libr
85 000 vols; 120 curr per   39009

Everett Public Library, 2702 Hoyt Ave, *Everett*, WA 98201-3556
T: +1 425 2577615; Fax: +1 425 2578016; URL: www.epls.org
1894; Eileen D. Simmons
Fore Edge Bks (Baker Coll), Northwest Coll; 1 branch libr
254 000 vols; 856 curr per; 14 274 av-mat; 11 020 sound-rec; 8 120 digital data carriers; 140 Bks on Deafness & Sign Lang, 10 870 Large Print Bks   39010

Evergreen Public Library, 201 Park St, *Evergreen*, AL 36401
T: +1 251 5782670; Fax: +1 251 5782316; E-mail: evergreen-conecuhlib@barbe-sassy.com
Vern Steenwyk
Heritage Section
50 000 vols; 10 curr per   39011

Evergreen Park Public Library, 9400 S Troy Ave, *Evergreen Park*, IL 60805-2383
T: +1 708 4228522; Fax: +1 708 4228665; URL: www.evergreenparklibrary.org
1944; Nicolette Seidl
70 000 vols   39012

Hawaii State Public Library System – Ewa Beach Public & School Library, 91-950 North Rd, *Ewa Beach*, HI 96706
T: +1 808 6891204; Fax: +1 808 6891349
Gerald Goff
86 000 vols; 74 curr per   39013

Mercer County Library – Ewing Branch, 61 Scotch Rd, *Ewing*, NJ 08628
T: +1 609 8823130; Fax: +1 609 5380212
Jackie Huff
103 000 vols   39014

Exeter Public Library, One Founders Park, *Exeter*, NH 03833
T: +1 603 7723101; Fax: +1 603 7727548; E-mail: epl@exeterpl.org; URL: www.exeterpl.org
1853; Hope Godino
New Hampshire, NE Antiquities Res Assn, Rockingham County Genealogical Coll
80 000 vols; 194 curr per   39015

Chester County Library System, 450 Exton Square Pkwy, *Exton*, PA 19341-2496
T: +1 610 2802600; Fax: +1 610 2802694; URL: www.ccls.org
1928; John Venditta
Large Type Colls, Computer Software Coll, Rare Bks, Career Info Coll; 1 bookmobile
283 000 vols; 1 117 curr per
libr loan   39016

Maurice M. Pine Free Public Library, 10-01 Fair Lawn Ave, *Fair Lawn*, NJ 07410
T: +1 201 7963400; Fax: +1 201 7946344; E-mail: fair1@bccls.org; URL: www.bccls.org/fairlawn
1933; Timothy H. Murphy
Art
179 000 vols; 204 curr per   39017

Fairbanks North Star Borough Public Library & Regional Center, Noel Wien Library, 1215 Cowles St, *Fairbanks*, AK 99701
T: +1 907 4591020; Fax: +1 907 4591024; E-mail: fairbanks@fnsblibrary.us; URL: library.fnsb.lib.ak.us
1909; Greg Hill
Alaska; 1 branch lib
294 000 vols; 999 curr per
libr loan; OCLC   39018

Greene County Public Library – Fairborn Community Library, One E Main St, *Fairborn*, OH 45324-4798
T: +1 937 8789383; Fax: +1 937 8780374
Robin Weinstein
103 000 vols   39019

Fairfax County Public Library – Administrative Offices, 12000 Government Center Pkwy, Ste 324, *Fairfax*, VA 22035-0012
T: +1 703 3243100; Fax: +1 703 3248365; E-mail: wwwlibr@fairfaxcounty.gov; URL: www.fairfaxcounty.gov/library
1939; Edwin S. Clay
Virginia Hist Coll; Access services for people with disabilities, 20 branch libs
2 615 000 vols
libr loan   39020

Marin County Free Library – Fairfax Branch, 2097 Sir Francis Drake Blvd, *Fairfax*, CA 94930-1198
T: +1 415 4538151
Gail Wiemann
82 000 vols   39021

Fairfield Public Library, 1080 Old Post Rd, *Fairfield*, CT 06824
T: +1 203 2563160;
Fax: +1 203 2563198; URL: www.fairfieldpubliclibrary.org
1877; Maura Ritz
1 branch libr
299 000 vols; 672 curr per
libr loan   39022

Fairfield Public Library, 104 W Adams, *Fairfield*, IA 52556
T: +1 641 4726551; Fax: +1 641 4723249; URL: www2.youseemore.com/fairfield/default.asp?
1853; Rebecca Higgins
69 000 vols; 233 curr per; 300 maps; 2 500 av-mat; 3 000 sound-rec; 100 digital data carriers
libr loan   39023

Fairfield Public Library – Anthony Pio Costa Memorial Library, 261 Hollywood Ave, *Fairfield*, NJ 07004
T: +1 973 2273575; Fax: +1 973 2277305; E-mail: ffpl@ffpl.org; URL: www.ffpl.org
1968; John J. Helle
61 000 vols; 130 curr per; 650 av-mat   39024

Fairfield Public Library – Fairfield Woods, 1147 Fairfield Woods Rd, *Fairfield*, CT 06825
T: +1 203 2557307; Fax: +1 203 2557311
Anne Marie Carey
67 000 vols; 154 curr per   39025

Solano County Library, 1150 Kentucky St, *Fairfield*, CA 94533-5799
T: +1 707 7841500; Fax: +1 707 4217474; URL: www.solanolibrary.com
1914; Ann Cousineau
Donovan McCune Coll (Printing); 8 branch libs
572 000 vols; 1 261 curr per; 6 000 e-books; 12 412 Audio Bks
libr loan   39026

Solano County Library – Fairfield Civic Center, 1150 Kentucky St, *Fairfield*, CA 94533
T: +1 707 4216500; Fax: +1 707 4217207
Cara Swartz
105 000 vols; 2 212 av-mat   39027

Millicent Library, 45 Centre St, *Fairhaven*, MA 02719; P.O. Box 30, Fairhaven, MA 02719-0030
T: +1 508 9925342; Fax: +1 508 9937288; URL: www.millicentlibrary.org
1893; Carolyn Longworth
Art & architecture; Mark Twain (Letters to the Rogers Family), Manjiro Nakahama (Journal of His Voyages 1840)
65 000 vols; 175 curr per   39028

Fairhope Public Library, 501 Fairhope Ave, *Fairhope*, AL 36532
T: +1 251 9287483; Fax: +1 251 9289717; E-mail: reference@fairhopelibrary.org; URL: www.fairhopelibrary.org
1894; Ilse Guilds Krick
Theosophy; Alabama Poetry (Frances Ruffin Durham), Area Hist Coll, Local Authors Coll
52 000 vols; 128 curr per; 3 030 av-mat
libr loan   39029

Fairport Public Library, One Village Landing, *Fairport*, NY 14450
T: +1 585 2239091; URL: www.fairportlibrary.org
1906; Betsy Gilbert
102 000 vols; 150 curr per; 18 digital data carriers   39030

Fairview Heights Public Library, 10017 Bunkum Rd, *Fairview Heights*, IL 62208-1703
T: +1 618 4892070; Fax: +1 618 4892079; E-mail: fhpl@fairviewheightslibrary.org; URL: www.fairviewheightslibrary.org
1972
50 000 vols; 156 curr per   39031

Cuyahoga County Public Library – Fairview Park Branch, 21255 Lorain Rd, *Fairview Park*, OH 44126-2120
T: +1 440 3334700; Fax: +1 440 3330697
Rebecca Wills
166 000 vols; 8 423 av-mat   39032

Duplin County Library, Emily S. Hill Library, 106 Park Circle Dr, *Faison*, NC 28341; P.O. Box 129, Faison, NC 28341-0129
T: +1 910 2670601; Fax: +1 910 2670601
Glenda Hooks
50 000 vols   39033

Brooks County Library, Ed Rachal Memorial Library, 203 S Calixto Mora Ave, *Falfurrias*, TX 78355
T: +1 361 3252144; Fax: +1 361 3253743
1960; Enola Garza
53 000 vols; 28 curr per; 58 av-mat   39034

Fall River Public Library, 104 N Main St, *Fall River*, MA 02720
T: +1 508 3242700; Fax: +1 508 3242707; URL: sailsinc.org/fallriver/main.htm
1860; Keith Stavely
Portuguese Language Mat Coll, Dr David S Greer Coll for Peace, Science & Education, Lizzie Borden Coll; 2 branch libs
200 000 vols; 245 curr per; 361 e-books; 6 200 av-mat; 5 150 sound-rec; 1 500 digital data carriers; 6 000 Large Print Bks   39035

San Diego County Library – Fallbrook Branch, 124 S Mission Rd, *Fallbrook*, CA 92028-2896
T: +1 760 7282373; Fax: +1 760 7284731
Ann Jones
65 000 vols   39036

Churchill County Library, 553 S Maine St, *Fallon*, NV 89406-3387
T: +1 775 4237581; Fax: +1 775 4237766; E-mail: churchillinfo@clan.lib.nv.us; URL: www.clan.lib.nv.us
1932; Barbara Mathews
Nevada Hist
93 000 vols; 80 curr per   39037

Mary Riley Styles Public Library, 120 N Virginia Ave, *Falls Church*, VA 22046
T: +1 703 2485030; Fax: +1 703 2485144; URL: www.falls-church.lib.va.us
1898; Mary W. McMahon
Falls Church Hist Coll
157 000 vols; 225 curr per; 16 000 e-books   39038

The Falls City Library & Arts Center, 1400 Stone St, *Falls City*, NE 68355
T: +1 402 2452913; Fax: +1 402 2453031; E-mail: fclib@sentco.net; URL: www.fallscitylibrary.org
Hope Schawang
Cakepan Coll; Local Artists Coll, paintings
50 000 vols; 100 curr per   39039

Queens Borough Public Library – Far Rockaway Branch, 1637 Central Ave, *Far Rockaway*, NY 11691
T: +1 718 3272549; Fax: +1 718 3374184
Ken Schubert
72 000 vols   39040

Buckham Memorial Library, Faribault Public Library, 11 Division St E, *Faribault*, MN 55021-6000
T: +1 507 3342089; Fax: +1 507 3840503; URL: www.faribault.org/lib
1897; Renee Lowery
85 000 vols; 225 curr per   39041

Farmers Branch Manske Library, 13613 Webb Chapel, *Farmers Branch*, TX 75234-3756
T: +1 972 2472511; Fax: +1 972 2479600; URL: www.farmersbranch.info
1961; Danita Barber
108 000 vols; 250 curr per   39042

Union Parish Library, 202 W Jackson St, *Farmerville*, LA 71241-2799
T: +1 318 3689291; Fax: +1 318 3689224; E-mail: admin.t1un@pelican.state.lib.la.us; URL: www.youseemore.com/unionparish
1956; Stephanie Herrmann
55 000 vols; 61 curr per; 856 av-mat   39043

Farmingdale Public Library, 116 Merritts Rd, *Farmingdale*, NY 11735
T: +1 516 2499090; Fax: +1 516 6949697; E-mail: fdaleref@nassaulibrary.org; URL: www.nassaulibrary.org/farmingdle/index.html
1923; Debbie Podolski
226 000 vols; 381 curr per   39044

Davis County Library, 38 S 100 East, *Farmington*, UT 84025; P.O. Box 115, Farmington, UT 84025-0115
T: +1 801 4512322; Fax: +1 801 4519561; URL: www.co.davis.ut.us/library
1946; Chris Sanford
6 branch libs
618 000 vols   39045

Farmington Community Library – Farmington Branch, 23500 Liberty St, *Farmington*, MI 48335-3570
T: +1 248 5530300; Fax: +1 248 4746915
Nina M. Harris
70 000 vols; 200 curr per   39046

The Farmington Library, Six Monteith Dr, *Farmington*, CT 06032; P.O. Box 407, Farmington, CT 06034-0407
T: +1 860 6736791; Fax: +1 860 6757148; URL: www.farmingtonlibct.org
1890; Jay Johnston
Toy Lending Libr; 1 branch libr
190 000 vols; 375 curr per
libr loan; ALA, NELA   39047

Farmington Community Library, 32737 W 12 Mile Rd, *Farmington Hills*, MI 48334-3302
T: +1 248 5538678; Fax: +1 248 5536892; URL: www.farmlib.org
1955; Tina M. Theeke
Business, Law, Parenting, Teacher; 1 branch libr
260 000 vols; 395 curr per
libr loan   39048

Central Virginia Regional Library, Farmville-Prince Edward Library, 217 W Third St, *Farmville*, VA 23901
T: +1 434 3926924;
Fax: +1 434 3929784; URL: www.centralvirginiaregionallibrary.org
1993; Peggy Epperson
60 000 vols; 45 curr per
libr loan   39049

Farnhamville Public Library, 240 Hardin, *Farnhamville*, IA 50538; P.O. Box 216, Farnhamville, IA 50538-0216
T: +1 515 5443660; Fax: +1 515 5443703; E-mail: director@farnhamville.lib.ia.us; URL: www.farnhamville.lib.ia.us
Sharon Vogel
52 000 vols; 20 curr per   39050

Knox County Public Library System – Farragut Branch, 417 N Campbell Station Rd, *Farragut*, TN 37922
T: +1 865 7771750; Fax: +1 865 7771751
Marilyn Jones
69 000 vols   39051

Cumberland County Public Library & Information Center, Headquarters, 300 Maiden Lane, *Fayetteville*, NC 28301-5000
T: +1 910 4831580; Fax: +1 910 4866661; E-mail: library@cumberland.lib.nc.us; URL: www.cumberland.lib.nc.us
1932; Jody A. Risacher
Foreign Language Ctr; 7 branch libs
611 000 vols; 1 102 curr per; 1 175 maps; 29 microforms; 17 770 av-mat; 21 468 sound-rec; 22 digital data carriers
libr loan   39052

Cumberland County Public Library & Information Center – Bordeaux, 3711 Village Dr, *Fayetteville*, NC 28304-1530
T: +1 910 4244008; Fax: +1 910 4231456; URL: www.cumberland.lib.nc.us/locations/bor/bordeaux.htm
Robin Deffendall
51 000 vols   39053

Cumberland County Public Library & Information Center – Cliffdale, 6882 Cliffdale Rd, *Fayetteville*, NC 28314-1936
T: +1 910 8643800; Fax: +1 910 4879090; URL: www.cumberland.lib.nc.us/locations/clf/cliffdale.htm
Pamela Kource
79 000 vols   39054

Cumberland County Public Library & Information Center – East Regional, 4809 Clinton Rd, *Fayetteville*, NC 28301-8992
T: +1 910 4852955; Fax: +1 910 4855492; URL: www.cumberland.lib.nc.us/locations/erl/eastregional.htm
Judy Brown
71 000 vols   39055

Cumberland County Public Library & Information Center – North Regional, 855 McArthur Rd, *Fayetteville*, NC 28311-1961
T: +1 910 8221998; Fax: +1 910 4800030; E-mail: nrl@cumberland.lib.nc.us; URL: www.cumberland.lib.nc.us/Locations/nrl/northregional
Mary Campbell
92 000 vols   39056

Fayetteville Public Library, Blair Library, 401 W Mountain St, *Fayetteville*, AR 72701
T: +1 479 8567000; Fax: +1 479 5710222; E-mail: questions@faylib.org; URL: www.faylib.org
1916; Louise L. Schaper
Literacy, Spanish language
238 000 vols; 357 curr per
libr loan   39057

Flint River Regional Library – Fayette County Public Library, 1821 Heritage Park Way, *Fayetteville*, GA 30214
T: +1 770 4618841
Christeen Snell
115 000 vols   39058

Washington County Library System, 1080 W Clydesdale Dr, *Fayetteville*, AR 72701
T: +1 479 4426253; Fax: +1 479 4426812; E-mail: info@wcls.lib.ar.us
Glenda Audrain
350 000 vols   39059

Township Library of Lower Southampton, 1983 Bridgetown Pike, *Feasterville*, PA 19053-4493
T: +1 215 3551183; Fax: +1 215 3645735; URL: www.buckslib.org
1956; Sally Pollock
71 000 vols; 86 curr per   39060

Fergus Falls Public Library, 205 E Hampden, *Fergus Falls*, MN 56537-2930
T: +1 218 7399387; Fax: +1 218 7365131; E-mail: library@fergusfalls.lib.mn.us
1891; Walter J. Dunlap
94 000 vols; 195 curr per   39061

Ferguson Municipal Public Library, 35 N Florissant Rd, *Ferguson*, MO 63135
T: +1 314 5214820; Fax: +1 314 5211275; URL: www.ferguson.lib.mo.us
1933; Joan G. Henderson
Ferguson Hist coll
80 000 vols; 126 curr per; 1 579

av-mat; 6 digital data carriers; 3 352
art reproductions
libr loan; MLA   39062

**Nassau County Public Library System
– Fernandina Beach Branch**, 25
N Fourth St, *Fernandina Beach*,
FL 32034-4123
T: +1 904 4913622; Fax: +1 904
2777366
Dawn S. Bostwick
166 000 vols; 115 curr per   39063

Ferndale Public Library, 222 E Nine
Mile Rd, *Ferndale*, MI 48220
T: +1 248 5462504; Fax: +1
248 5455840; E-mail:
ferndalepubliclibrary@gmail.com; URL:
www.ferndale.lib.mi.us
1930; Mary K. Trenner
90 000 vols; 125 curr per; 1 000 av-mat
                                           39064

Concordia Parish Library, 1609 Third
St, *Ferriday*, LA 71334-2298
T: +1 318 7573550; Fax: +1
318 7571941; E-mail:
admin.t1cn@pelican.state.lib.la.us; URL:
www.concordia.lib.la.us
1928; Amanda Taylor
2 branch libs
85 000 vols; 95 curr per   39065

**Findlay-Hancock County District
Public Library**, 206 Broadway, *Findlay*,
OH 45840-3382
T: +1 419 4221712; Fax: +1 419
4220638; URL: www.findlaylibrary.org
1888; Carol Dunn
1 branch libr, 1 bookmobile
176 000 vols; 306 curr per   39066

Hamilton East Public Library, Fishers
Branch, Five Municipal Dr, *Fishers*,
IN 46038-1574
T: +1 317 5790300; Fax: +1 317
5790309
David L. Cooper
253 000 vols   39067

Augusta County Library, 1759 Jefferson
Hwy, *Fishersville*, VA 22939
T: +1 540 9496354;
Fax: +1 540 9435965; URL:
www.augustacountylibrary.org
1977; Diantha McCauley
1 branch libr, 1 bookmobile
170 000 vols; 310 curr per; 5 570 av-mat
                                           39068

Fitchburg Public Library, 610 Main St,
Fitchburg, MA 01420-3146
T: +1 978 3459635; Fax: +1 978
3459631; E-mail: fplref@cwmars.org;
URL: www.fitchburgpubliclibrary.org
1859; Ann Wirtanen
148 000 vols; 254 curr per; 531 e-books;
3 136 av-mat; 1140 Talking Bks   39069

**Flagstaff City-Coconino County
Public Library System**, 300 W Aspen,
Flagstaff, AZ 86001
T: +1 928 7797670;
Fax: +1 928 7749573; URL:
www.flagstaffpubliclibrary.org
1890; Kay Whitaker
Arizona; Southwest Coll, Arizona Coll
207 000 vols; 400 curr per
libr loan   39070

Mount Olive Public Library, 202
Flanders-Drakestown Rd, *Flanders*,
NJ 07836
T: +1 973 6918686; Fax: +1 973
6910226; URL: www.mopl.org
1976; Rita L. Hilbert
72 000 vols; 138 curr per   39071

Hunterdon County Library, 314 State
Rte 12, *Flemington*, NJ 08822
T: +1 908 7881437; Fax: +1 908
8065179; E-mail: reference@hclibrary.us;
URL: www.hunterdon.lib.nj.us
1928; Mark Titus
2 branch libs, 1 bookmobile
385 000 vols; 591 curr per   39072

Flint Public Library, 1026 E Kearsley St,
Flint, MI 48502-1994
T: +1 810 2327111; Fax: +1 810
2492635; E-mail: askus@fpl.info; URL:
www.flint.lib.mi.us
1851; Jo Anne G. Mondowney
3 branch libs
408 000 vols; 976 curr per; 300 e-books
libr loan   39073

Genesee District Library, G-4195 W
Pasadena Ave, *Flint*, MI 48504
T: +1 810 7320110; Fax: +1 810
7321161; URL: www.thegdl.org
1942; Carolyn Nash
Genealogy; American Indians, Civil War
Coll; 20 branch libs
671 000 vols; 1 305 curr per; 1 000 e-
books; 27 515 av-mat; 16 517 sound-rec;
53 130 digital data carriers; 13 240 Large
Print Bks
libr loan   39074

Floral Park Public Library, 17 Caroline
Place, *Floral Park*, NY 11002
T: +1 516 3266330; Fax: +1 516
4376959; URL: www.nassaulibrary.org/
fpark/floralpark_databases.php
1923; Tracey Simon
86 960 vols; 150 curr per   39075

Florence County Library System, 509
S Dargan St, *Florence*, SC 29506
T: +1 843 6628424; Fax: +1
843 6617544; E-mail:
reference@florencelibrary.org
1925; Ray McBride
Caroliniana; 5 branch libs
310 000 vols; 446 curr per
libr loan   39076

Florence-Lauderdale Public Library,
350 N Wood Ave, *Florence*, AL 35630
T: +1 256 7646564; Fax: +1 256
7646629; URL: flpl.lib.al.us
1945; Nancy Sanford
91 000 vols; 100 curr per; 25 000 e-
books   39077

Siuslaw Public Library District, 1460
Ninth St, *Florence*, OR 97439-0022
T: +1 541 9973132; Fax: +1 541
9974007; URL: www.siuslawlibrary.org
1985; Stephen C. Skidmore
Newspapers published in Florence, 1891-
1974, micro; Oregon Past & Present
Coll; Reference Libr of Frank Herbert
80 000 vols; 293 curr per; 4 000 e-books
                                           39078

Flossmoor Public Library, 1000 Sterling
Ave, *Flossmoor*, IL 60422-1295
T: +1 708 7983600; Fax: +1 708
7983603; E-mail: flossref@sslic.net; URL:
www.flossmoorlibrary.org
1953; Megan Heligas
56 000 vols; 270 curr per; 3 756
sound-rec; 14 digital data carriers   39079

Flower Mound Public Library, 3030
Broadmoor Lane, *Flower Mound*,
TX 75022
T: +1 972 8746200; Fax: +1 972
8746466; E-mail: fmpl@flower-mound.com;
URL: www.fmlibrary.net
1984
51 030 vols; 176 curr per   39080

**Queens Borough Public Library –
Auburndale**, 25-55 Francis Lewis Blvd,
Flushing, NY 11358
T: +1 718 3522027
Paul Tam
78 000 vols   39081

**Queens Borough Public Library –
Flushing Branch**, 41-17 Main St,
Flushing, NY 11355
T: +1 718 6611200; Fax: +1 718
6611290
Ramona Rendon
314 000 vols   39082

**Queens Borough Public Library
– Hillcrest**, 187-05 Union Turnpike,
Flushing, NY 11366
T: +1 718 4542786; Fax: +1 718
2647567
Alice Norris
79 000 vols   39083

**Queens Borough Public Library –
Kew Gardens Hills**, 72-33 Vleigh Pl,
Flushing, NY 11367
T: +1 718 2616654
Margalit Susser
85 000 vols   39084

**Queens Borough Public Library –
McGoldrick**, 155-06 Roosevelt Ave,
Flushing, NY 11354
T: +1 718 4611616
Susan Xie
77 000 vols   39085

**Queens Borough Public Library –
Mitchell-Linden**, 29-42 Union St,
Flushing, NY 11354
T: +1 718 5392330
Farzaneh Momeni
62 000 vols   39086

**Queens Borough Public Library –
Pomonok**, 158-21 Jewel Ave, *Flushing*,
NY 11365
T: +1 718 5914343
Elizabeth Eshun
54 000 vols   39087

**Queens Borough Public Library –
Queensboro Hill**, 60-05 Main St,
Flushing, NY 11355
T: +1 718 3598332
Ai-Hua Chen
61 000 vols   39088

Folsom Public Library, 411 Stafford St,
Folsom, CA 95630
T: +1 916 3557374; Fax: +1 916
3557332; E-mail: libstaff@folsom.ca.us;
URL: www.folsomlibrary.com
1993; Katy Curl
75 000 vols; 130 curr per
libr loan   39089

Ridley Township Public Library, 100 E
MacDade Blvd, *Folsom*, PA 19033-2592
T: +1 610 5830593; Fax: +1
610 5839505; E-mail:
ridleytownship@delco.lib.pa.us; URL:
www.delcolibraries.org/ridleytwp
1957; Hal Rosenberg
Law (Pennsylvania Annotated Code & US
Code), Irish Lit
74 000 vols; 199 curr per   39090

Fond Du Lac Public Library, 32
Sheboygon St, *Fond du Lac*, WI 54935
T: +1 920 9297080; Fax: +1 920
9297082; E-mail: reference@fdlpl.org;
URL: www.fdlpl.org
1876; Kenneth D. Hall
Art & architecture; Mertes Children's Folk
& Fairy Tale Coll; 1 bookmobile
176 000 vols; 372 curr per
libr loan   39091

**Free Public Library of Woodbridge –
Fords Branch**, 211 Ford Ave, *Fords*,
NJ 08863
T: +1 732 7267071; Fax: +1 732
7267081
Anne Caldwell Taylor
55 000 vols   39092

Forest Grove City Library, 2114 Pacific
Ave, *Forest Grove*, OR 97116-9019
T: +1 503 9923337; Fax: +1 503
9923333; URL: www.fglibrary.plinkit.org
1909; Colleen Winters
Large Print Coll, Spanish Language Coll
88 000 vols; 212 curr per   39093

**Queens Borough Public Library –
Forest Hills Branch**, 108-19 71st Ave,
Forest Hills, NY 11375
T: +1 718 2687934; Fax: +1 718
2681614
Hwai-Min Chen-Wood
100 000 vols   39094

**Queens Borough Public Library –
North Forest Park**, 98-27 Metropolitan
Ave, *Forest Hills*, NY 11375
T: +1 718 2615512
Frances Tobin
61 000 vols   39095

Forest Lake Library, 220 N Lake St,
Forest Lake, MN 55025
T: +1 651 4644088; Fax: +1 651
4644296; URL: www.co.washington.mn.us/
info_for_residents/library
1941; Amy Worwa
66 000 vols   39096

**Washington County Library –
Hardwood Creek Branch**, 19955
Forest Road North, *Forest Lake*,
MN 55025
T: +1 651 2757300
Amy Worwa
50 000 vols   39097

**Clayton County Library System –
Forest Park Branch**, 696 Main St,
Forest Park, GA 30297
T: +1 404 3660850; Fax: +1 404
3660884
Lydia Bigard
58 000 vols; 55 curr per   39098

Forest Park Public Library, 7555
Jackson Blvd, *Forest Park*, IL 60130
T: +1 708 3667171;
Fax: +1 708 3667293; URL:
www.forestparkpubliclibrary.org
1906; Rodger Brayden
80 000 vols; 180 curr per   39099

**Ocean County Library – Lacey
Township**, Ten Lacey Rd, *Forked
River*, NJ 08731-3626
T: +1 609 6938566; Fax: +1 609
9718973
Kathy Lanzim
68 000 vols   39100

Dwight Foster Public Library, 102
E Milwaukee Ave, *Fort Atkinson*,
WI 53538-2049
T: +1 920 5637790; Fax: +1 920
5637774; URL: www.fortlibrary.org
1890; Connie Meyer
Lorine Niedecker Coll
100 000 vols; 237 curr per   39101

Fort Collins Public Library,
201 Peterson St, *Fort Collins*,
CO 80524-2990
T: +1 970 2216526; Fax: +1 970
4162140; E-mail: refdesk@fcgov.com;
URL: www.fcgov.com/library
1900; Brenda Carns
4 308 e-books, 12 175 Audio Bks, 45
Electronic Media & Resources; 1 branch
libr
397 000 vols; 632 curr per; 4 000 e-
books; 12 175 av-mat   39102

Fort Dodge Public Library, 424 Central
Ave, *Fort Dodge*, IA 50501
T: +1 515 5738167; Fax: +1 515
5735422; E-mail: fdpl@fortdodge.lib.ia.us;
URL: www.fortdodgeiowa.org/library
1890; Larry Koeninger
81 000 vols; 200 curr per   39103

Broward County Division of Libraries,
Main Library, 100 S Andrews Ave, *Fort
Lauderdale*, FL 33301
T: +1 954 3577443; Fax: +1
954 3575733; E-mail:
answer@browardlibrary.org; URL:
www.broward.org/library
1974; Robert E. Cannon
Bienes Rare Book Coll, Florida
Diagnostic & Learning Resource System,
Judaica, Small Business Resource Ctr;
40 branch libs
2 414 000 vols; 6 727 curr per; 15 000
e-books; 253 036 av-mat; 290 900
sound-rec; 9 000 Bks by Mail, 410 Braille
vols, 86 300 Large Print Bks
libr loan   39104

Fort Madison Public Library, Cattermole
Memorial Library, 1920 Avenue E, *Fort
Madison*, IA 52627
T: +1 319 3725721; Fax: +1 319
3725726; E-mail: fmpl@ft-madison.lib.ia.us;
URL: www.fortmadisonlibrary.org
1894; Sarah Clendineng
Black Hist (Dr Harry D. Harper Sr Coll);
Railroad (Chester S Gross Memorial
Coll); 1 branch libr
85 000 vols; 150 curr per   39105

Lee County Library System, 2345
Union St, *Fort Myers*, FL 33901-3917
T: +1 239 4797380; Fax: +1 239
4612919; URL: www.lee-county.com/
library
1964; Sheldon Kaye
14 branch libs
1 406 000 vols; 2 932 curr per; 6 586 e-
journals; 64 000 e-books; 55 190 av-mat;
85 845 sound-rec; 37 357 digital data
carriers; 5 428 bks by mail, 272 Braille
vols, 8 730 high interest/low vocabulary
bks, 503 bks on deafness & sign
language, 48 200 large print bks
libr loan   39106

**Lee County Library System –
Fort Myers-Lee County Public**,
2050 Central Ave, *Fort Myers*,
FL 33901-3917
T: +1 239 4794636; Fax: +1 239
4794634
Madeleine Plummer
171 000 vols; 524 curr per   39107

Lee County Library System – Lakes Regional, 15290 Bass Rd, *Fort Myers*, FL 33919
T: +1 239 5334000; Fax: +1 239 5334040
Joanne Fischer
186 000 vols   39108

Fort Myers Beach Library, 2755 Estero Blvd, *Fort Myers Beach*, FL 33931
T: +1 941 7658162; Fax: +1 941 4638776; E-mail: library@fmb.lib.fl.us;
URL: www.fmb.lib.fl.us
1955; Dr. Leroy Hommerding
94 710 vols; 145 curr per   39109

DeKalb County Public Library, 504 Grand Ave NW, *Fort Payne*, AL 35967
T: +1 256 8452671; Fax: +1 256 8452671; URL: web2.lmn.lib.al.us
Elizabeth Tucker
Indian Coll
81 000 vols; 34 curr per   39110

Saint Lucie County Library System – Fort Pierce Branch, 101 Melody Lane, *Fort Pierce*, FL 34950-4402
T: +1 772 4621964; Fax: +1 772 4622750; URL: www.st-lucie.lib.fl.us
1953; Susan Kilmer
Local Coll, Florida Coll, Black Hist; 3 branch libs
306 000 vols; 737 curr per   39111

Saint Lucie County Library System – Lakewood Park, 7605 Santa Barbara Dr, *Fort Pierce*, FL 34951
T: +1 772 4626870; Fax: +1 772 4626874
Carol Shroyer
50 000 vols   39112

Fort Smith Public Library, 3201 Rogers Ave, *Fort Smith*, AR 72903
T: +1 479 7830229; Fax: +1 479 7828571; URL: www.fortsmithlibrary.org
1906; Jennifer Goodson
Spanish; Vietnamese Coll, Mathew C. Clark American Sign Language Coll; 3 branch libs
253 000 vols; 475 curr per; 3 000 microforms; 2 000 av-mat; 1 500 sound-rec; 252 digital data carriers   39113

Fort Stockton Public Library, 500 N Water St, *Fort Stockton*, TX 79735
T: +1 432 3363374; Fax: +1 432 3366648; E-mail: info@fort-stockton.lib.tx.us; URL: www.fort-stockton.lib.tx.us
1911; Elva Valadez
Texana & the West
52 000 vols; 225 curr per   39114

Peach Public Libraries, 315 Martin Luther King Dr, *Fort Valley*, GA 31030-4196
T: +1 478 8251640; Fax: +1 478 8252061; URL: www.peach.public.lib.ga.us
1915
2 branch libs
69 614 vols; 124 curr per   39115

Fort Walton Beach Library, 185 Miracle Strip Pkwy SE, *Fort Walton Beach*, FL 32548
T: +1 850 8339590; Fax: +1 850 8339659; E-mail: fwblibr@fwb.org; URL: www.fwb.org/library
1955; Patricia Gould
62 000 vols; 184 curr per
libr loan   39116

Upper Dublin Public Library, 805 Loch Alsh Ave, *Fort Washington*, PA 19034
T: +1 215 6288744; E-mail: upperdublinlibrary@mclinc.org; URL: www.upperdublinlibrary.org
1932; Cherilyn Fiory
American art, Local History
110 000 vols; 125 curr per; 5 400 av-mat; 6 000 sound-rec
libr loan   39117

Allen County Public Library, 900 Library Plaza, *Fort Wayne*, IN 46802; P.O. Box 2270, Fort Wayne, IN 46801-2270
T: +1 260 4211235; Fax: +1 260 4211386; URL: www.acpl.lib.in.us
1895; Jeffrey R. Krull
Fine Arts, bks & slides; 13 branch libs
2 745 000 vols; 8 218 curr per; 24 790 Electronic Media & Resources, 46 700 Large Print Bks
libr loan   39118

Allen County Public Library – Aboite, 5630 Coventry Lane, *Fort Wayne*, IN 46804
T: +1 260 4211310; Fax: +1 260 4322394; URL: www.acpl.lib.in.us/aboite
Susan Hunt
84 000 vols   39119

Allen County Public Library – Dupont, 536 E Dupont Rd, *Fort Wayne*, IN 46825
T: +1 260 4211315; Fax: +1 260 4897756; URL: www.acpl.lib.in.us/dupont
Rebecca Wolfe
102 000 vols   39120

Allen County Public Library – Georgetown, 6600 E State Blvd, *Fort Wayne*, IN 46815
T: +1 260 4211320; Fax: +1 260 7498513; URL: www.acpl.lib.in.us/georgetown
Lisa Upchurch
102 000 vols   39121

Allen County Public Library – Little Turtle, 2201 Sherman Blvd, *Fort Wayne*, IN 46808
T: +1 260 4211335; Fax: +1 260 4245170; URL: www.acpl.lib.in.us/littleturtle
Rosie Stier
53 000 vols   39122

Allen County Public Library – Shawnee, 5600 Noll Ave, *Fort Wayne*, IN 46806
T: +1 260 4211355; Fax: +1 260 4561871; URL: www.acpl.lib.in.us/shawnee
Pamela Martin-Diaz
70 000 vols   39123

Allen County Public Library – Waynedale, 2200 Lower Huntington Rd, *Fort Wayne*, IN 46819
T: +1 260 4211365; Fax: +1 260 7474123; URL: www.acpl.lib.in.us/waynedale
Don Fisher
61 000 vols   39124

Fort Worth Public Library, 500 W Third St, *Fort Worth*, TX 76102
T: +1 817 8717705; Fax: +1 817 8717734; E-mail: LibraryWebMail@FortWorthGov.org; URL: www.fortworthgov.org/Library/
1901; Dr. Gleneice A. Robinson
Bookplates, Early Children's Books, Earth Science, Popular Sheet Music; 14 branch libs
2 241 000 vols; 1 392 curr per
libr loan   39125

Fort Worth Public Library – East Berry, 4300 E Berry St, *Fort Worth*, TX 76105
T: +1 817 5361945; Fax: +1 817 5366253
Cindy Upchurch
52 000 vols   39126

Fort Worth Public Library – East Regional, 6301 Bridge St, *Fort Worth*, TX 76105
T: +1 817 8716436; Fax: +1 817 8716440
Barbara Henderson
57 000 vols   39127

Fort Worth Public Library – Meadowbrook, 5651 E Lancaster, *Fort Worth*, TX 76112
T: +1 817 4510916; Fax: +1 817 4968931
Joycelyn Claer
67 000 vols   39128

Fort Worth Public Library – Ridglea, 3628 Bernie Anderson Ave, *Fort Worth*, TX 76116
T: +1 817 7376619; Fax: +1 817 7638404
Rebecca Caldow
95 000 vols   39129

Fort Worth Public Library – Riverside, 2913 Yucca, *Fort Worth*, TX 76111
T: +1 817 8386931; Fax: +1 817 8385403
Barbara Grisell
55 000 vols   39130

Fort Worth Public Library – Seminary South, 501 E Bolt St, *Fort Worth*, TX 76110
T: +1 817 9260215; Fax: +1 817 9261703
Sally McCoy
66 000 vols   39131

Fort Worth Public Library – Southwest Regional, 4001 Library Lane, *Fort Worth*, TX 76109
T: +1 817 7829853; Fax: +1 817 7328714
Shelia Barnett
131 000 vols   39132

Fort Worth Public Library – Wedgwood, 3816 Kimberly Lane, *Fort Worth*, TX 76133
T: +1 817 2923368; Fax: +1 817 3461862
Marion Edwards
81 000 vols   39133

Fortville-Vernon Township Public Library, 625 E Broadway, *Fortville*, IN 46040-1549
T: +1 317 4856402; Fax: +1 317 4854084; URL: www.fortville.lib.in.us
1917; Richard Bell
62 000 vols; 131 curr per; 391 av-mat   39134

San Mateo County Library – Foster City Branch, 1000 E Hillsdale Blvd, *Foster City*, CA 94404
T: +1 650 5744842; Fax: +1 650 5721875; URL: www.fostercitylibrary.org
Barbara Escoffier
145 000 vols   39135

Kaubisch Memorial Public Library, 205 Perry St, *Fostoria*, OH 44830
T: +1 419 4352813; Fax: +1 419 4355350; URL: www.fostoria.lib.oh.us
1892; Dee Conine
75 000 vols; 400 curr per   39136

Orange County Public Library – Fountain Valley Branch, 17635 Los Alamos St, *Fountain Valley*, CA 92708-5299
T: +1 714 9621324; Fax: +1 714 9648164
Jane Deeley
84 000 vols; 88 curr per   39137

Fox Lake District Library, 255 E Grand Ave, *Fox Lake*, IL 60020-1697
T: +1 847 5870198; Fax: +1 847 5879493; URL: www2.youseemore.com/foxlake/default.asp?
1939; Harry J. Bork
70 000 vols; 179 curr per   39138

Boyden Library, Ten Bird St, *Foxborough*, MA 02035
T: +1 508 5431245; Fax: +1 508 5431193; URL: www.boydenlibrary.org
1870; Jerry Cirillo
Pace Genealogy Coll
91 000 vols; 194 curr per   39139

Frankfort Community Public Library, 208 W Clinton St, *Frankfort*, IN 46041
T: +1 765 6548746; Fax: +1 765 6548747; E-mail: fcpl@accs.net; URL: www.accs.net/fcpl
1880; Michelle Bradley
Agriculture, Gardening, Nuclear energy; Genealogy Coll; 3 branch libs
142 000 vols; 200 curr per; 3 473 av-mat; 2 410 Large Print Bks, 2 601 Talking Bks
libr loan   39140

Frankfort Public Library District, 21119 S Pfeiffer Rd, *Frankfort*, IL 60423-9302
T: +1 815 4692423; Fax: +1 815 4699307; URL: www.frankfortlibrary.org
1966
87 000 vols; 298 curr per   39141

Kentucky Regional Library for the Blind & Physically Handicapped, Kentucky Talking Book Library, 300 Coffee Tree Rd, *Frankfort*, KY 40601; P.O. Box 537, Frankfort, KY 40602-0537
T: +1 502 5648300; Fax: +1 502 5645773; E-mail: ktbl.mail@ky.gov; URL: www.kdla.ky.gov/libsupport/ktbl/history.htm
1969; Barbara Penegor
Kentucky Coll
141 000 vols; 84 curr per; 132 316 av-mat; 150 000 sound-rec; 1 digital data carriers; 6 000 braille bks
libr loan   39142

Paul Sawyier Public Library, 319 Wapping St, *Frankfort*, KY 40601
T: +1 502 3522665; Fax: +1 502 2272250; URL: www.pspl.org
1965; Donna Riis Gibson
109 000 vols; 235 curr per; 12 407 av-mat; 9 120 sound-rec; 4 170 Large Print Bks   39143

Franklin Public Library, 118 Main St, *Franklin*, MA 02038
T: +1 508 5204940; URL: www.franklin.ma.us/auto/town/library
1790; Betsy Ferry
Benjamin Franklin Special Coll, First Bks of the Franklin Libr
87 000 vols; 230 curr per   39144

Franklin-Springboro Public Library, 44 E Fourth St, *Franklin*, OH 45005
T: +1 937 7462665; Fax: +1 937 7462847; URL: www.franklin.lib.oh.us
1923; Anita G. Carroll
1 branch lib
151 000 vols; 381 curr per; 12 805 av-mat; 7 060 sound-rec; 2 500 digital data carriers; 3 380 Large Print Bks
libr loan   39145

Johnson County Public Library, 401 State St, *Franklin*, IN 46131-2545; mail address: 49 E Monroe St, Franklin, IN 46131-2312
T: +1 317 7389835; Fax: +1 317 7389635; E-mail: webmaster@jcplin.org; URL: www.jcplin.org
1911; Beverly A. Martin
Travel, Consumer, Careers; 3 branch libs
362 000 vols; 556 curr per; 16 e-books; 34 000 microforms; 3 963 av-mat; 14 012 sound-rec; 28 digital data carriers; 56 art reproductions, 8 200 Large Print Bks
libr loan   39146

Macon County Public Library, 819 Siler Rd, *Franklin*, NC 28734
T: +1 828 5243600; Fax: +1 828 5249550; URL: www.fontanalib.org
1890; Karen Wallace
60 000 vols   39147

Saint Mary Parish Library, 206 Iberia St, *Franklin*, LA 70538-4906
T: +1 337 8285364; Fax: +1 337 8282329; E-mail: admin.b1my@pelican.state.lib.la.us; URL: www.stmary.lib.la.us
1953; Julie Champagne
5 branch libs
178 000 vols; 239 curr per   39148

Williamson County Public Library, 1314 Columbia Ave, *Franklin*, TN 37064-3626
T: +1 615 5951243; Fax: +1 615 5951202; E-mail: ref@williamson-tn.org; URL: lib.williamson-tn.org
1927; Janice E. Keck
African-American Genealogy & Photogr Coll, Civil War Coll, Local Authors Coll; 4 branch libs
161 000 vols; 171 curr per; 7 000 e-books; 13 011 av-mat; 7 055 sound-rec; 1 930 digital data carriers; 1 120 High Interest/Low Vocabulary Bks, 140 Bks on Deafnes & Sign Lang, 2 280 Large Print Bks   39149

Franklin Lakes Free Public Library, 470 DeKorte Dr, *Franklin Lakes*, NJ 07417
T: +1 201 8912224; Fax: +1 201 8915102; URL: www.franklinlakeslibrary.org
1968; Gerry McMahon
80 000 vols; 255 curr per
libr loan   39150

Franklin Park Public Library District, 10311 Grand Ave, *Franklin Park*, IL 60131
T: +1 847 4556016; Fax: +1 847 4556299; URL: www.franklinparklibrary.org
1962; Marie Saeli
Science & technology
150 000 vols; 246 curr per   39151

Franklin Square Public Library, 19 Lincoln Rd, *Franklin Square*, NY 11010
T: +1 516 4883444; Fax: +1 516 3543368; E-mail: esplref@lilrc.org; URL: www.nassaulibrary.org
1938; Carol Ahrens
106 000 vols; 198 curr per   39152

Washington Parish Library System, 825 Free St, *Franklinton*, LA 70438
T: +1 985 8397806; Fax: +1 985 8397808; E-mail: admin.c1wa@pelican.state.lib.la.us; URL: www.washington.lib.la.us
1946; Joseph Sbisa
4 branch libs
109 000 vols   39153

Fraser Public Library, 16330 Fourteen Mile Rd, *Fraser*, MI 48026-2034
T: +1 586 2932055; Fax: +1 586 2945777; E-mail: fraserlibrary@hotmail.com; URL: www.ci.fraser.mi.us
Regina Slivka
63 000 vols; 120 curr per; 739 av-mat
39154

Frederick County Public Libraries, 110 E Patrick St, *Frederick*, MD 21701
T: +1 301 6001613; Fax: +1 301 6003789; URL: www.fcpl.org
1937; Darrell Batson
Maryland Hist (Maryland Coll), bk, micro; 7 branch libs
475 000 vols; 670 curr per
39155

Central Rappahannock Regional Library, Headquarters, 1201 Caroline St, *Fredericksburg*, VA 22401-3761
T: +1 540 3721144; Fax: +1 540 3739411; URL: www.librarypoint.org
1969; Donna Cote
7 branch libs
450 000 vols; 1 192 curr per
39156

Darwin R. Barker Library, Seven Day St, *Fredonia*, NY 14063
T: +1 716 6728051; Fax: +1 716 6793547; E-mail: barker@netsync.net
Joy Harper
55 000 vols; 180 curr per
39157

Freeport Community Library, Ten Library Dr, *Freeport*, ME 04032
T: +1 207 8653307; Fax: +1 207 8651395; URL: www.freeportlibrary.com
Beth Edmonds
Sportman's Coll
56 000 vols; 157 curr per; 1 965 av-mat
39158

Freeport Memorial Library, 144 W Merrick Rd & S Ocean Ave, *Freeport*, NY 11520
T: +1 516 3793274; Fax: +1 516 8689741; E-mail: reffr@nassau.library.org; URL: www.nassaulibrary.org/freeport
1884; Dave Opatow
Long Island Hist, Vocational & Careers
217 000 vols; 434 curr per; 297 maps; 2 489 av-mat; 9 950 sound-rec; 4 200 digital data carriers; 2 500 Audio Bks, 4 200 CDs, 5 630 High Interest/Low Vocabulary Bks, 2 370 Large Print Bks
39159

Freeport Public Library, 100 E Douglas St, *Freeport*, IL 61032
T: +1 815 2333000; Fax: +1 815 2331099; E-mail: information@freeportpubliclibrary.org; URL: www.freeportpubliclibrary.org
1874; Carol Dickerson
120 000 vols; 396 curr per
39160

Alameda County Library, 2450 Stevenson Blvd, *Fremont*, CA 94538-2326
T: +1 510 7451500; Fax: +1 510 7932987; URL: www.aclibrary.org
1910; Linda Carroll
Spanish Language, Business & Management; 11 branch libs
1 011 000 vols; 2 056 curr per; libr loan
39161

Birchard Public Library of Sandusky County, 423 Croghan St, *Fremont*, OH 43420
T: +1 419 3347101; Fax: +1 419 3344788; URL: www.birchard.lib.oh.us
1873; Mary Anne Culbertson
3 branch libs
175 000 vols; 390 000 microforms; 3 430 av-mat; 5 255 sound-rec
39162

Fremont Area District Library, 104 E Main, *Fremont*, MI 49412
T: +1 231 9243480; Fax: +1 231 9242355; E-mail: fmt@llcoop.org; URL: www.fremontlibrary.net
1996; Ray Arnett
Harry L. Spooner Local Hist Coll
81 000 vols; 200 curr per
39163

Keene Memorial Library, Fremont Public Library, 1030 N Broad St, *Fremont*, NE 68025-4199
T: +1 402 7272694; Fax: +1 402 7272693; URL: www.keene.lib.ne.us
1901; Ann E. Stephens
Classics (Taylor)
106 000 vols; 235 curr per
39164

Queens Borough Public Library – Fresh Meadows Branch, 193-20 Horace Harding Expressway, *Fresh Meadows*, NY 11365
T: +1 718 4547272; Fax: +1 718 4545820
Julia Hua
104 000 vols
39165

Fresno County Public Library, 2420 Mariposa St, *Fresno*, CA 93721-2285
T: +1 559 4883195; Fax: +1 559 4881971; URL: www.fresnolibrary.org
1910; Karen Bosch Cobb
Japanese language, California, Genealogy, Leo Politi Coll, Native American Coll; 37 branch libs
1 132 000 vols; 1 950 curr per; 280 e-books; 5 443 maps; 61 141 av-mat; 24 100 sound-rec; 15 700 Large Print Bks
libr loan
39166

Fresno County Public Library – Central Branch, 2420 Mariposa St, *Fresno*, CA 93721-2285
T: +1 559 4883195; Fax: +1 559 4881971
Karen Bosch Cobb
California History Room;Mother Goose Coll;William Saroyan Coll
303 000 vols
39167

Fresno County Public Library – Fig Garden Regional, 3071 W Bullard, *Fresno*, CA 93711
T: +1 559 4384071
Joy Sentman-Paz
51 000 vols
39168

Fresno County Public Library – Woodward Park Regional, 944 E Perrin Ave, *Fresno*, CA 93720
T: +1 559 4333136; Fax: +1 559 4331348
Martha Connor
Chinese Language Materials
76 000 vols
39169

San Juan Island Library District, 1010 Guard St, *Friday Harbor*, WA 98250-9612
T: +1 360 3782798; Fax: +1 360 3782706; E-mail: sjlib@sjlib.org; URL: www.sjlib.org
1922; Laura Tretter
60 000 vols; 170 curr per
39170

Anoka County Library – Mississippi, 410 Mississippi St NE, *Fridley*, MN 55432-4416; mail address: 707 County Rd 10 NE, Blaine, MN 55434-2398
T: +1 763 5711934; Fax: +1 763 5711935
Beth Werking
53 000 vols; 155 curr per
39171

Friendswood Public Library, 416 S Friendswood Dr, *Friendswood*, TX 77546-3897
T: +1 281 4827135; Fax: +1 281 4822685; E-mail: frpublib@friendswood.lib.tx.us; URL: www.friendswood.lib.tx.us
1968; Mary Booker Perroni
83 000 vols; 143 curr per; 2 000 microforms; 2 807 av-mat; 900 sound-rec; 200 digital data carriers
libr loan; ALA, PLA, TLA
39172

Summit County Library, 37 County Rd 1005, *Frisco*, CO 80443; P.O. Box 770, Frisco, CO 80443-0770
T: +1 970 6685555; Fax: +1 970 6685556; URL: www.co.summit.co.us/library
1962; Joyce Dierauer
Skiing, Colorado Hist
106 000 vols; 220 curr per; libr loan
39173

Samuels Public Library, 538 Villa Ave, *Front Royal*, VA 22630
T: +1 540 6353153; Fax: +1 540 6357229; URL: www.samuelslibrary.net
1952; Barbara Ecton
Virginia Coll
76 000 vols; 129 curr per; 3 microforms; 3 720 av-mat; 1 170 sound-rec; 10 digital data carriers; 1
libr loan
39174

Fulton Public Library, 312 Main St, *Fulton*, KY 42041
T: +1 270 4723439; Fax: +1 270 4726241; E-mail: fultonpl@bellsouth.net; URL: www.fultonlibrary.com
1965; Elaine Allen
Civil War Records; 1 branch libr
70 000 vols; 150 curr per
libr loan
39175

Gadsden-Etowah County Library – Gadsden Public Library, 254 College St, *Gadsden*, AL 35901
T: +1 256 5494699; E-mail: gploffice@library.gadsden.com; URL: www.library.gadsden.com; www.librarycat.gadsden.com
1906; Amanda Jackson
Alabama Hist
202 000 vols
39176

Cherokee County Public Library, 300 E Rutledge Ave, *Gaffney*, SC 29340
T: +1 864 4872711; Fax: +1 864 4872752; E-mail: cherokeelib@spiritcom.net; URL: www.cherokeecountylibrary.org
1902
South Carolina genealogy; Arthur Gettys Genealogy Coll, Gladys Coker Fort Fine Arts Coll, Heritage Rm, June Carr Photogr Coll, Raymond & Bright Parker Story Tape Coll, Ruby Cash Garvin South Carolina Coll
92 000 vols; 141 curr per
39177

Columbus Metropolitan Library – Gahanna Branch, 310 Granville St, *Gahanna*, OH 43230
T: +1 614 6452275
Mary Kelly
143 000 vols
39178

Alachua County Library District, Headquarters Library, 401 E University Ave, *Gainesville*, FL 32601-5453
T: +1 352 3344700; Fax: +1 352 3341252; E-mail: ysref@aclib.us; URL: www.aclib.us
1906; Sol M. Hirsch
10 branch libs
692 000 vols; 1 964 curr per; 29 000 e-books; 42 566 av-mat; 15 580 sound-rec; 25 300 digital data carriers; 230 Braille vols, 13 771 High Interest/Low Vocabulary Bks, 80 Spec Interes Per Sub, 790 Bks on Deafness & Sign Lang, 24 876 Large Print Bks
39179

Cooke County Library, 200 S Weaver St, *Gainesville*, TX 76240-4790
T: +1 940 6685530; Fax: +1 940 6685533; URL: cookecountylibrary.org
1893; Jennifer Johnson-Spence
Local Artists Coll
65 000 vols; 53 curr per
39180

Hall County Library System, 127 Main St NW, *Gainesville*, GA 30501-3699
T: +1 770 5323311; Fax: +1 770 5324305; E-mail: refdesk@hallcountylibrary.org; URL: www.hallcountylibrary.org
1997; Adrian Mixson
James Longstreet Coll, papers; 4 branch libs
297 000 vols; 408 curr per
39181

Hall County Library System – Blackshear Place, 2927 Atlanta Hwy, *Gainesville*, GA 30507
T: +1 770 5323311; Fax: +1 770 2873653
Barbara Perry
57 000 vols; 97 curr per
39182

Galax-Carroll Regional Library, Galax Public Library, 610 W Stuart Dr, *Galax*, VA 24333
T: +1 276 2362042; Fax: +1 276 2365153; URL: galaxcarroll.lib.va.us
1938; Laura A. Bryant
60 000 vols; 131 curr per
39183

Stone County Library, 106 E Fifth St, *Galena*, MO 65656; P.O. Box 225, Galena, MO 65656-0225
T: +1 417 3576410; Fax: +1 417 3576695; URL: www.stonecountylibrary.org
1948; David R. Doennig
1 branch libr
55 000 vols; 60 curr per
39184

Galesburg Public Library, 40 E Simmons St, *Galesburg*, IL 61401-4591
T: +1 309 3436118; Fax: +1 309 3434877; E-mail: director@galesburglibrary.org; URL: www.galesburglibrary.org
1874; Pamela Van Kirk
Humanities
183 000 vols; 333 curr per; 2 000 e-books; 3 350 av-mat; 4 084 sound-rec
libr loan; JLA
39185

Galion Public Library Association, 123 N Market St, *Galion*, OH 44833
T: +1 419 4683203; Fax: +1 419 4687298; URL: www.galion.lib.oh.us; www.galionlibrary.org
1901; Larry Ostrowski
100 000 vols; 150 curr per
39186

Gallia County District Library, Dr. Samuel L. Bossard Memorial Library, Seven Spruce St, *Gallipolis*, OH 45631
T: +1 740 4467323; Fax: +1 740 4461701; E-mail: bossard@oplin.org; URL: www.bossard.lib.oh.us
1899; Deborah Saunders
O O McIntyre Coll, Genealogy Coll
115 000 vols; 250 curr per; 4 000 av-mat; 4 000 Audio Bks, 50 Bks on Deafness & Sign Lang, 4 000 Large Print Bks
39187

Octavia Fellin Public Library, 115 W Hill Ave, *Gallup*, NM 87301
T: +1 505 8631291; Fax: +1 505 7225090; URL: ofpl.ci.galllup.nm.us
1928; Mary Browder
Alcohol Abuse Video Coll, Indians of the Southwest Coll, Memorial Bk Coll; Children's Libr
160 000 vols; 166 curr per; 12 000 av-mat; 1 670 digital data carriers; 1 460 Large Print Bks, 12 000 Talking Bks
39188

Rosenberg Library, 2310 Sealy Ave, *Galveston*, TX 77550
T: +1 409 7638854; Fax: +1 409 7630275; E-mail: admin@rosenberg-library.org; URL: www.rosenberg-library.org
1904; John Augelli
Maritime Hist; Rare Books; Galveston & Texas Hist; Museum Coll, art, artifacts
161 000 vols; 354 curr per
39189

Finney County Public Library, 605 E Walnut St, *Garden City*, KS 67846
T: +1 620 2723680; Fax: +1 620 2723682; URL: www.fcpl.homestead.com
1897; Lyn Eckland
Spanish & Vietnamese Lit Coll
100 000 vols; 300 curr per
libr loan
39190

Garden City Public Library, 60 Seventh St, *Garden City*, NY 11530-2891
T: +1 516 7428405; Fax: +1 516 7422675; E-mail: gcplref@lilrc.org; URL: www.gardencitypl.org
1952; N.N.
Garden City Arch, Long Island Hist
153 000 vols; 442 curr per
39191

Garden City Public Library, 31735 Maplewood Rd, *Garden City*, MI 48135
T: +1 734 7931830; Fax: +1 734 7931831; URL: garden-city.lib.mi.us
N.N.
50 000 vols; 80 curr per
39192

Orange County Public Library – Chapman, 9182 Chapman Ave, *Garden Grove*, CA 92841-2590
T: +1 714 5392115; Fax: +1 714 5309363
Wendy Crutcher
53 000 vols; 61 curr per
39193

Orange County Public Library – Garden Grove Regional Library, 11200 Stanford Ave, *Garden Grove*, CA 92840-5398
T: +1 714 5300711; Fax: +1 714 5300961
Su Chung Chay
162 000 vols; 133 curr per
39194

Orange County Public Library – West Garden Grove Branch, 11962 Bailey St, *Garden Grove*, CA 92845-1104
T: +1 714 8972594; Fax: +1 714 8952761
Dennis McGuire
53 000 vols; 63 curr per
39195

County of Los Angeles Public Library – Gardena Mayme Dear Library, 1731 W Gardena Blvd, *Gardena*, CA 90247-4726
T: +1 310 3236363; Fax: +1 310 3270992; URL: www.colapublib.org/libs/gardena
Ruth Morse
121 000 vols; 132 curr per   39196

County of Los Angeles Public Library- Masao W. Satow Library, 14433 S Crenshaw Blvd, *Gardena*, CA 90249-3142
T: +1 310 6790638; Fax: +1 310 9700275; URL: www.colapublib.org/libs/satow
Elaine Fukumoto
72 000 vols; 108 curr per   39197

Gardendale Martha Moore Public Library, 995 Mt Olive Rd, *Gardendale*, AL 35071
T: +1 205 6316639; Fax: +1 205 6310146; URL: www.gardendale.lib.al.us
1959; Connie Smith
51 000 vols; 120 curr per   39198

Levi Heywood Memorial Library, 55 W Lynde St, *Gardner*, MA 01440
T: +1 978 6325298; Fax: +1 978 6302864
1886; Gail P. Landy
Furniture
100 000 vols; 155 curr per   39199

Garfield Free Public Library, 500 Midland Ave, *Garfield*, NJ 07026
T: +1 973 4783800; Fax: +1 973 4787162; URL: www.bccls.org/garfield
1923; Kathleen Zalenski
Black Hist, Persian Art
85 000 vols; 147 curr per   39200

Cuyahoga County Public Library – Garfield Heights Branch, 5409 Turney Rd, *Garfield Heights*, OH 44125-3203
T: +1 216 4758178; Fax: +1 216 4751015
Melanie Rapp Weiss
57 000 vols; 3 109 av-mat   39201

Nicholson Memorial Library System – Central Library, 625 Austin St, *Garland*, TX 75040-6365
T: +1 972 2052543; Fax: +1 972 2052523; URL: www.nmls.lib.tx.us
1933; Claire Bausch
4 branch libs
384 000 vols; 987 curr per; 3 870 Large Print Bks
libr loan   39202

Wake County Public Library System – Southeast Regional, 908 Seventh Ave, *Garner*, NC 27529
T: +1 919 6622250; Fax: +1 919 6622270
E. Gail Harrell
133 000 vols   39203

Haverstraw Kings Daughters Public Library, Ten W Ramapo Rd, *Garnerville*, NY 10923
T: +1 845 7863800; Fax: +1 845 7863791; E-mail: information@hkdpl.org; URL: www.hkdpl.org
Joanne Sininsky
Local History Coll (focus on Haverstraw & the Hudson Valley)
116 000 vols; 325 curr per; 6 595 av-mat   39204

Portage County District Library, 10482 South St, *Garrettsville*, OH 44231
T: +1 330 5275082; Fax: +1 330 5274370; URL: www.portagecounty.lib.oh.us
1935; Cecilia Swanson
Large print; 6 branch libs
259 000 vols; 530 curr per   39205

Portage County District Library – Garrettsville Branch, 10482 South St, *Garrettsville*, OH 44231
T: +1 330 5274378; Fax: +1 330 5274370
Kathleen Kozup
65 000 vols   39206

Gary Public Library, 220 W Fifth Ave, *Gary*, IN 46402-1215
T: +1 219 8862484; Fax: +1 219 8829528; URL: www.gary.lib.in.us
1908; Roma Ivey
5 branch libs
777 000 vols; 582 curr per; 22 314 av-mat; 302 Large Print Bks
libr loan   39207

Gary Public Library – John F. Kennedy Branch, 3953 Broadway, *Gary*, IN 46408-1799
T: +1 219 8878112; Fax: +1 219 8875967
Dorothy Swain
76 000 vols; 56 curr per   39208

Gary Public Library – Ora L. Wildermuth Branch, 501 S Lake, *Gary*, IN 46403-2408
T: +1 219 9383941; Fax: +1 219 9388759
Patience Ojomo
60 000 vols; 52 curr per   39209

Gary Public Library – W. E. B. Du Bois Branch, 1835 Broadway, *Gary*, IN 46407-2298
T: +1 219 8869120; Fax: +1 219 8869391
Kokuleeba S. Lwanga
Afro-American Rare Book Coll, micro-fiche
74 000 vols; 68 curr per   39210

Gaston-Lincoln Regional Library, 1555 E Garrison Blvd, *Gastonia*, NC 28054
T: +1 704 8682164; Fax: +1 704 8536012; URL: www.glrl.lib.nc.us
1964; Lucinda W. Moose
573 000 vols; 657 curr per; 40 Electronic Media & Resources   39211

Lonesome Pine Regional Library – Scott County Public, 297 W Jackson St, *Gate City*, VA 24251
T: +1 276 3863302; Fax: +1 276 3862977; E-mail: scplib@lprlibrary.org
Rita Walters
79 000 vols; 77 curr per; 2 318 av-mat   39212

Otsego County Library, 700 S Otsego Ave, *Gaylord*, MI 49735-1723
T: +1 989 7325841; Fax: +1 989 7329401; E-mail: ocl@otsego.org; URL: otsego.lib.mi.us
Maureen Derenzy
2 branch libs
59 000 vols; 355 curr per   39213

Geneva Public Library, 244 Main St, *Geneva*, NY 14456-2370
T: +1 315 7895303; Fax: +1 315 7899835; URL: genevapubliclibrary.net
1905; Michael Nyerges
78 000 vols; 137 curr per; 450 av-mat; 1 722 sound-rec; 6 digital data carriers   39214

Geneva Public Library District, 127 James St, *Geneva*, IL 60134
T: +1 630 2320780; Fax: +1 630 2320881; URL: www.geneva.lib.il.us
1894; Matt Teske
142 000 vols; 270 curr per; 9 616 av-mat; 9 616 Audio Bks, 300 High Interest/Low Vocabulary Bks, 15 Bks on Deafness & Sign Lang, 2 800 Large Print Bks   39215

Genoa Public Library District, 232 W Main St, *Genoa*, IL 60135
T: +1 815 7842627; Fax: +1 815 7842627; E-mail: genoalbry2@hotmail.com; URL: www.genoalibrary.org
Susan Walker
55 000 vols; 63 curr per   39216

Live Oak County Library, 402 Houston St, P.O. Box 698, *George West*, TX 78022-0698
T: +1 361 4491124
1955
50 000 vols; 30 curr per   39217

Georgetown County Library, 405 Cleland St, *Georgetown*, SC 29440-3200
T: +1 843 5453300; Fax: +1 843 5453395; URL: georgetowncountylibrary.sc.gov
1799; Dwight McInvaill
2 branch libs, 1 bookmobile
142 000 vols; 201 curr per; 6 953 av-mat   39218

Georgetown Public Library, 808 Martin Luther King St, *Georgetown*, TX 78626-5527
T: +1 512 9303627; Fax: +1 512 9303764; E-mail: gpl@georgetowntx.org; URL: www.georgetowntx.org
1968; Eric P. Lashley
58 000 vols; 118 curr per   39219

Scott County Public Library, 104 S Bradford Lane, *Georgetown*, KY 40324-2335
T: +1 502 8633566; Fax: +1 502 8639621; URL: www.scottpublib.org
1928; Earlene Hawkins Arnett
92 000 vols; 167 curr per; 6 526 av-mat   39220

Germantown Community Library, 112 N 16957 W Mequon Rd, *Germantown*, WI 53022; P.O. Box 670, Germantown, WI 53022-0670
T: +1 262 2537760; Fax: +1 262 2537763; URL: www.germantown-library.org/
1963; Roberta M. Olson
130 000 vols; 250 curr per; 175 digital data carriers
libr loan   39221

Germantown Community Library, 1925 Exeter Rd, *Germantown*, TN 38138-2815
T: +1 901 7577323; Fax: +1 901 7569940; URL: www.germantown-library.org
Melody Pittman
156 000 vols; 194 curr per   39222

Germantown Public Library, 51 N Plum St, *Germantown*, OH 45327
T: +1 937 8554001; Fax: +1 937 8556098; URL: www.germantown.lib.oh.us
1888; Joe Knueven
87 490 vols; 88 curr per   39223

Adams County Library System, Central Library, 140 Baltimore St, *Gettysburg*, PA 17325-2311
T: +1 717 3345716; Fax: +1 717 3347992; E-mail: adams@adamslibrary.org; URL: www.adamslibrary.org
1945; Robin Lesher
Art; Eisenhower Rm Coll (Civil War & Local Hist)
141 000 vols; 354 curr per   39224

Northern Tier Library Association, Richland Center, 4015 Dickey Rd, *Gibsonia*, PA 15044-9713
T: +1 724 4492665; Fax: +1 724 4436755; URL: www.northerntierlibrary.org
1954; Sharon L. Dawe
61 000 vols; 65 curr per   39225

Campbell County Public Library System, 2101 S 4-J Rd, *Gillette*, WY 82718-5205
T: +1 307 6870009; Fax: +1 307 6864009; URL: www.ccpls.org
1928; Patricia Myers
Western Art, US Geological Survey Map Depot; 1 branch libr
129 000 vols; 358 curr per
libr loan   39226

Long Hill Township Free Public Library, 917 Valley Rd, *Gillette*, NJ 07933
T: +1 908 6472088; Fax: +1 908 6472098; URL: www.longhilllibrary.org
1957; Peggy Neubig
Cooking, Gardening, Large print, Parenting
50 000 vols; 140 curr per   39227

Upshur County Library, 702 W Tyler St, *Gilmer*, TX 75644
T: +1 903 8435001; Fax: +1 903 8433995
1929; Mark Warren
Libr of America
78 000 vols; 76 curr per   39228

Santa Clara County Library – Gilroy Library, 7387 Rosanna St, *Gilroy*, CA 95020-6193
T: +1 408 8428207; Fax: +1 408 8420489
Lani Yoshimura
142 000 vols   39229

Girard Free Library, 105 E Prospect St, *Girard*, OH 44420-1899
T: +1 330 5452508; Fax: +1 330 5458213; URL: www.girard.lib.oh.us
1919; Rose Ann Lubert
Music nostalgia, WW II
76 000 vols; 217 curr per   39230

Gladstone Public Library, 135 E Dartmouth St, *Gladstone*, OR 97027-2496
T: +1 503 6562411; Fax: +1 503 6552438; E-mail: glref@lincc.lib.or.us; URL: www.gladstone.lib.or.us
Mary Nixon
55 000 vols; 120 curr per   39231

Gladwin County District Library, 555 W Cedar Ave, *Gladwin*, MI 48624
T: +1 989 4268221; Fax: +1 989 4266958; E-mail: Director@gladwinlibrary.org; URL: www.gladwinlibrary.org
1934
Local newspapers on micro; 1 branch libr
50 100 vols; 126 curr per   39232

Mary Wood Weldon Memorial Library, 107 W College St, *Glasgow*, KY 42141
T: +1 270 6512824; Fax: +1 270 6512824; E-mail: public_library@glascow-ky.com; URL: weldonpubliclibrary.org
1925; Jim Hyatt
Genealogy (Kentucky Coll)
61 000 vols; 115 curr per   39233

Gloucester County Library System – Glassboro Public, Two Center St, *Glassboro*, NJ 08028-1995
T: +1 856 8810001; Fax: +1 856 8819338; URL: www.gcls.org
1956; Robert S. Wetherall
African-American Coll
57 000 vols; 149 curr per   39234

Welles-Turner Memorial Library, 2407 Main St, *Glastonbury*, CT 06033
T: +1 860 6527720; Fax: +1 860 6527721; URL: www.wtmlib.com
1895; Barbara J. Bailey
Parenting, Gardening; Connecticut Down Syndrome Congress
137 000 vols; 312 curr per   39235

County of Henrico Public Library – Glen Allen Branch, 10501 Staples Mill Rd, *Glen Allen*, VA 23060
T: +1 804 7567523; Fax: +1 804 7551702
Andrea Brown
50 000 vols; 80 curr per   39236

County of Henrico Public Library – Twin Hickory, 5001 Twin Hickory Rd, *Glen Allen*, VA 23059
T: +1 804 3641400
Ahmed Tabib
57 000 vols; 91 curr per   39237

Glen Cove Public Library, Four Glen Cove Ave, *Glen Cove*, NY 11542-2885
T: +1 516 6762130; Fax: +1 516 6762788; E-mail: glencove@lilrc.org; URL: www.nassaulibrary.org/glencove
1894; Maija Sperauskas
Long Island Hist (Glen Cove)
136 000 vols; 250 curr per; 3 500 av-mat; 1 658 sound-rec; 101 digital data carriers
libr loan; NYLA, NCLA   39238

Glen Ellyn Public Library, 400 Duane St, *Glen Ellyn*, IL 60137-4508
T: +1 630 4690879; Fax: +1 630 4691086; URL: www.gepl.org
1907; Dawn A. Bussey
201 000 vols; 800 curr per   39239

Queens Borough Public Library – Glen Oaks Branch, 256-04 Union Tpk, *Glen Oaks*, NY 11004
T: +1 718 8318636; Fax: +1 718 8318635
Jeffrey Berger
65 000 vols   39240

Glen Ridge Free Public Library, 240 Ridgewood Ave, *Glen Ridge*, NJ 07028
T: +1 973 7485482; Fax: +1 973 7489350; E-mail: jbreuergrlib@gmail.com; URL: www.glenridgelibrary.org
1912; Jennifer Breuer
Glen Ridge Hist (Barrows Coll & Glen Ridge Coll), pictures, slides, clippings, memorabilia
50 800 vols; 205 curr per   39241

Glen Rock Public Library, 315 Rock Rd, *Glen Rock*, NJ 07452-1795
T: +1 201 6703970; Fax: +1 201 4450872; E-mail: glrkcirc@bccls.org; URL: glenrock.bccls.org
1922; Roz Pelcyger
85 000 vols; 100 curr per; 90 maps; 2 000 microforms; 4 846 av-mat; 1 175 sound-rec; 284 digital data carriers
libr loan; ALA   39242

Glencoe Public Library, 320 Park Ave, **Glencoe**, IL 60022-1597
T: +1 847 8355056; Fax: +1 847 8355648; URL: www.glencoe.lib.il.us
1909; Margaret M. Hamil
83 000 vols; 270 curr per; 554 e-books
39243

Glendale Public Library, 222 E Harvard St, **Glendale**, CA 91205-1075
T: +1 818 5482027; Fax: +1 818 5487225; URL: library.ci.glendale.ca.us; www.glendalepubliclibrary.org
1906; N.N.
Art Coll, Sheet Music Coll, Piano Roll Coll; 6 branch libs
717 000 vols; 781 curr per; 616 412 sound-rec
libr loan
39244

Glendale Public Library – Brand Art & Music, 1601 W Mountain St, **Glendale**, CA 91201-1209
T: +1 818 5482051; Fax: +1 818 5485079
Alyssa Resnick
105 000 vols
39245

Glendale Public Library – Casa Verdugo, 1151 N Brand Blvd, **Glendale**, CA 91202-2503
T: +1 818 5482047; Fax: +1 818 5488052
Mary Alice Wollam
55 000 vols
39246

North Shore Library, 6800 N Port Washington Rd, **Glendale**, WI 53217
T: +1 414 3513461; Fax: +1 414 3513528; URL: www.mcfls.org
1980; Richard Nelson
125 000 vols; 237 curr per
39247

Queens Borough Public Library – Glendale Branch, 78-60 73rd Pl, **Glendale**, NY 11385
T: +1 718 8214980; Fax: +1 718 8217160
Ann-Marie R. Josephs
52 000 vols
39248

Glenside Public Library District, 25 E Fullerton Ave, **Glendale Heights**, IL 60139-2697
T: +1 630 2601550; Fax: +1 630 2601433; E-mail: ghdadmin@glensidepld.org; URL: glensidepld.org/gpld.htm
1974; Liz Fitzgerald
Deaf; Learning Games Coll, Signed English Children's Bks
124 000 vols; 243 curr per; 4 000 e-books
libr loan; ALA
39249

Glendora Public Library & Cultural Center, 140 S Glendora Ave, **Glendora**, CA 91741
T: +1 626 8524891; Fax: +1 626 8524899; E-mail: library@glendoralibrary.org; URL: www.glendoralibrary.org
1912; Robin Weed-Brown
114 000 vols; 244 curr per; 8 000 e-books; 4 000 microforms; 3 182 av-mat; 1 469 sound-rec; 75 digital data carriers
39250

Crandall Public Library, 251 Glen St, **Glens Falls**, NY 12801-3593
T: +1 518 7926508; Fax: +1 518 7925251; E-mail: info@crandalllibrary.org; URL: www.crandalllibrary.org
1892; Christine McDonald
Architecture, Art; Americana, Folklife & Local Hist of Northern New York Adirondacks & Upper Hudson Valley
173 000 vols; 374 curr per; 2 880 Large Print Bks
39251

Shaler North Hills Library, 1822 Mt Royal Blvd, **Glenshaw**, PA 15116
T: +1 412 4860211; Fax: +1 412 4868286; URL: www.shalerlibrary.org
1942
Puppets Coll
96 970 vols; 216 curr per; 268 microforms; 6 180 sound-rec; 381 digital data carriers; AV (31 kits), 69 art reproductions
libr loan
39252

Cheltenham Township Library System, 215 S Keswick Ave, **Glenside**, PA 19038-4420
T: +1 215 8850457; Fax: +1 215 8851239; URL: www.cheltenhamlibraries.org
1966; Dorothy L. Sutton
Multicultural Mat, Music CD's, Educational CD-ROM Software
124 000 vols; 466 curr per
libr loan
39253

Glenside Free Library, 215 S Keswick Ave, **Glenside**, PA 19038-4420
T: +1 215 8850455; Fax: +1 215 8851019; URL: www.cheltenhamlibraries.org/glenside/glenside.php
1928
Business Coll
55 030 vols; 182 curr per
39254

Glenview Public Library, 1930 Glenview Rd, **Glenview**, IL 60025-2899
T: +1 847 7297500; Fax: +1 847 7297558; E-mail: info@glenviewpl.org; URL: www.glenviewpl.org
1930; Vickie L. Novak
Medicine
266 000 vols
libr loan
39255

Gloucester County Library, 6920 Main St, **Gloucester**, VA 23061; P.O. Box 2380, Gloucester, VA 23061-2380
T: +1 804 6932998; Fax: +1 804 6931477; URL: www.gloucesterva.info/lib/home.html
Melissa Malcolm
82 000 vols; 150 curr per
39256

Gloucester, Lyceum & Sawyer Free Library, Two Dale Ave, **Gloucester**, MA 01930-5906
T: +1 978 2819763; Fax: +1 978 2819770; E-mail: sflib@sawyerfreelibrary.org; URL: www.sawyerfreelibrary.org
1830; David McArdle
Art & architecture; Charles Olson Coll, US Census, T.S. Eliot Coll
118 000 vols; 194 curr per
39257

Gloucester City Library, 50 N Railroad Ave, **Gloucester City**, NJ 08030
T: +1 856 4564181; Fax: +1 856 4566724; E-mail: gc@gcpl.us; URL: www.gloucestercitylibrary.org
1925; Elizabeth J. Egan
How-to-do-it & Mechanics Coll
61 000 vols; 113 curr per
39258

Gloversville Public Library, 58 E Fulton St, **Gloversville**, NY 12078-3219
T: +1 518 7252819; Fax: +1 518 7730292; E-mail: //gpl@sals.edu; URL: www.gloversvillelibrary.org
1880; Barbara Madonna
70 000 vols; 118 curr per
39259

Wayne County Public Library, Inc, 1001 E Ash St, **Goldsboro**, NC 27530
T: +1 919 7351824; Fax: +1 919 7312889; URL: www.wcpl.org
1907; Jane Rustin
4 branch libs
100 000 vols; 212 curr per
39260

Gorham Public Library, 35 Railroad St, **Gorham**, NH 03581
T: +1 603 4662525; E-mail: gpublib@ncia.net
1895; Suzanne Colburn
New Hampshire Bks Coll
95 000 vols; 28 curr per; 65 av-mat
39261

Grafton – Midview Public Library, 983 Main St, **Grafton**, OH 44044-1492
T: +1 440 9263317; Fax: +1 440 9263000; URL: www.graftonpl.lib.oh.us
1944; Terry Cook
Automotive Repair Manuals Coll
59 000 vols; 150 curr per
39262

USS Liberty Memorial Public Library, 1620 11th Ave, **Grafton**, WI 53024-2404
T: +1 262 3755315; Fax: +1 262 3755317; E-mail: grafton@esls.lib.wi.us; URL: www.grafton.lib.wi.us
1958; John Hanson
70 000 vols; 165 curr per; 8 000 e-books; 1 600 av-mat; 1 800 sound-rec; 55 digital data carriers
libr loan
39263

Los Angeles Public Library System – Granada Hills Branch, 10640 Petit Ave, **Granada Hills**, CA 91344-6305
T: +1 818 3685687; Fax: +1 818 7569286
Pamela Rhodes
61 000 vols
39264

Granby Public Library, 15 N Granby Rd, **Granby**, CT 06035
T: +1 860 8445275; Fax: +1 860 6530241; E-mail: staff@granbypl.libct.org
1869; Joan Fox
Large type print; 1 branch libr
55 000 vols; 251 curr per
39265

Grand County Library District, 225 E Jasper Ave, **Granby**, CO 80446; P.O. Box 1050, Granby, CO 80446-1050
T: +1 970 8879411; Fax: +1 970 8873227; E-mail: library@rkymtnhi.com; URL: www.gcld.org
1933; Mary Anne Hanson-Wilcox
5 branch libs
74 000 vols; 100 curr per
39266

Grand Forks Public City-County Library, 2110 Library Circle, **Grand Forks**, ND 58201-6324
T: +1 701 7728116; Fax: +1 701 7721379; URL: www.grandforksgov.com/library
1900; N.N.
Agriculture
370 000 vols; 349 curr per; 21 320 av-mat; 6 000 sound-rec; 10 000 digital data carriers; 3 000 Bks-by-Mail
39267

Loutit District Library, 1051 S Beacon Blvd, **Grand Haven**, MI 49417-2607
T: +1 616 8425560; Fax: +1 616 8470570; E-mail: gdh@llcoop.org; URL: www.loutitlibrary.org
1910; Sandra Knes
Local hist and genealogy
101 000 vols; 115 curr per; 15 maps; 75 microforms; 625 sound-rec; 50 digital data carriers
libr loan; ALA
39268

Grand Island Memorial Library, 1715 Bedell Rd, **Grand Island**, NY 14072
T: +1 716 7737124; Fax: +1 716 7741146; URL: www.buffalolib.org
Lynn Alan Konovitz
63 000 vols; 142 curr per; 5 303 av-mat
39269

Grand Island Public Library, Edith Abbott Memorial Library, 211 N Washington St, **Grand Island**, NE 68801-5855
T: +1 308 3855333; Fax: +1 308 3855339; E-mail: refdesk@gi.lib.ne.us; URL: www.gi.lib.ne.us
1884; Steve Fosselman
Genealogy (Lue R Spencer State DAR Coll & Ella Sprague Coll); Abbott Sisters Research Ctr
123 000 vols; 300 curr per; 2 063 av-mat; 2 748 sound-rec; 113 digital data carriers; 217 JUV kits
libr loan; ALA, PLA
39270

Mesa County Public Library District, 530 Grand Ave, **Grand Junction**, CO 81502; P.O. Box 20000-5019, Grand Junction, CO 81502-5019
T: +1 970 2415251; Fax: +1 970 2434744; URL: www.mcpld.org
1967; Terry Pickens
Railroads, Agriculture; Western Fiction; 7 branch libs
270 000 vols; 425 curr per
libr loan
39271

Grand Ledge Area District Library, 131 E Jefferson St, **Grand Ledge**, MI 48837-1534
T: +1 517 6277014; Fax: +1 517 6276276; URL: grandledge.lib.mi.us
1911; Suzanne E. Bowles
50 000 vols; 76 curr per
39272

Grand Prairie Public Library System, 901 Conover Dr, **Grand Prairie**, TX 75051
T: +1 972 2375702; Fax: +1 972 2375750; E-mail: infodesk@gptx.org; URL: www.gptx.org/library; golibrarygo.blogspot.com
1937; Kathy Ritterhouse
1 branch libr
166 000 vols; 425 curr per
39273

Grand Rapids Public Library, 111 Library St NE, **Grand Rapids**, MI 49503-3268
T: +1 616 9885400; Fax: +1 616 9885419; URL: www.grpl.org
1871; Marcia A. Warner
Art, Business, Education, Hist, Music, Social sciences; Children's Lit Coll, Furniture, Gardening, Landscape Architecture, Picture Bks; 7 branch libs
722 000 vols; 1 163 curr per
libr loan
39274

Kent District Library, 2650 Five Mile Rd NE, **Grand Rapids**, MI 49525
T: +1 616 6473930; Fax: +1 616 3611007
David Stracke
81 000 vols
39275

Kent District Library – Cascade Township Branch, 2870 Jacksmith Ave SE, **Grand Rapids**, MI 49546
T: +1 616 6473850; Fax: +1 616 9403075
Diane Cutler
92 000 vols
39276

Kent District Library – Grandville Branch, 4055 Maple St SW, **Grandville**, MI 49418
T: +1 616 6473890; Fax: +1 616 5304653
Patrice Vrona
72 000 vols
39277

Six Mile Regional Library District, 2001 Delmar St, **Granite City**, IL 62040-4590
T: +1 618 4526238; Fax: +1 618 8766317; URL: sixmilerld.org
1912; Lester McKiernan
1 branch libr
177 000 vols; 405 curr per
39278

Josephine Community Libraries, Inc, 200 NW C St, **Grants Pass**, OR 97526-2094; P.O. Box 1684, Grants Pass
T: +1 541 4760571; E-mail: info@josephinelibrary.org; URL: www.co.josephine.or.us; www.solis.lib.or.us
1913; Russell Long
3 branch libs
156 000 vols; 157 curr per; 9 252 av-mat; 4 798 sound-rec
libr loan
39279

Granville Public Library, 217 E Broadway, **Granville**, OH 43023-1398
T: +1 740 5870196; Fax: +1 740 5870197; E-mail: chansen@granvillelibrary.org; URL: www.granvillelibrary.org
1912; Charlie Hansen
94 000 media items; 196 curr per   39280

Grapevine Public Library, 1201 Municipal Way, **Grapevine**, TX 76051
T: +1 817 4103400; Fax: +1 817 4103084; URL: www.grapevine.lib.tx.us
1923; Janis Roberson
Large Print Bks, Texas Coll
130 000 vols; 185 curr per; 2 000 e-books; 3 695 av-mat
ALA, TLA
39281

Terrebonne Parish Library – North Terrebonne, 4130 W Park Ave, **Gray**, LA 70359
T: +1 985 8683050; Fax: +1 985 8689404; E-mail: north.b1tb@state.lib.la.us
N.N.
57 000 vols
39282

Crawford County Library System – Devereaux Memorial Library, 201 Plum St, **Grayling**, MI 49738
T: +1 989 3489214; Fax: +1 989 3489294; E-mail: ccls-director@crawfordco.lib.mi.us; URL: www.grayling_mi.com/library.htm
Bambi Mansfield-Sanderson
54 000 vols; 72 curr per; 14 000 e-books; 4 024 av-mat
39283

Grayslake Area Public Library District, 100 Library Lane, **Grayslake**, IL 60030-1684
T: +1 847 2235313; Fax: +1 847 2236482; URL: www.grayslake.info
1931; Roberta Thomas
128 000 vols; 374 curr per; 8 612 sound-rec
39284

Great Bend Public Library, 1409 Williams St, *Great Bend*, KS 67530-4090
T: +1 620 7922409; Fax: +1 620 7925495; URL: www.ckls.org/~gbpl
1908; James A. Swan
Kansas & Federal Law Coll, American Petroleum Institute Coll
100 000 vols; 300 curr per; 2 000 microforms
libr loan   39285

Great Falls Public Library, 301 Second Ave N, *Great Falls*, MT 59401-2593
T: +1 406 4530349; Fax: +1 406 4530181; E-mail: questions@greatfallslibrary.org; URL: www.greatfallslibrary.org
1890; Jim Heckel
Montana Hist Coll; 1 bookmobile
150 000 vols; 318 curr per   39286

Great Neck Library, 159 Bayview Ave, *Great Neck*, NY 11023-1938
T: +1 516 4668055; Fax: +1 516 8298297; E-mail: comments@greatnecklibrary.org; URL: www.greatnecklibrary.org
1889; Arlene Nevens
Art & architecture, Behav sciences, Social sciences; 3 branch libs
394 000 vols; 996 curr per   39287

Greece Public Library, Two Vince Tofany Blvd, *Greece*, NY 14612
T: +1 585 2258951; Fax: +1 585 2252777; E-mail: grwebmst@libraryweb.org; URL: www.rochester.lib.ny.us/greece
1958; Bernadette Foster
1 branch libr
97 000 vols   39288

High Plains Library District, 1939 61st Ave, *Greeley*, CO 80634-7940
T: +1 970 5068500; Fax: +1 970 5068551; URL: www.mibiblioteca.us; www.mylibrary.us
1931; Janine Reid
Small Business & Non-Profit; 4 branch libs
576 000 vols; 1 178 curr per; 4 000 e-books
libr loan   39289

High Plains Library District – Centennial Park, 2227 23rd Ave, *Greeley*, CO 80634-6632
T: +1 970 5068656; Fax: +1 970 5068601
Cindy Osborne
76 000 vols   39290

High Plains Library District – Farr Regional Library, 1939 61st Ave, *Greeley*, CO 80634
T: +1 970 5068500
Jody Hungenberg
114 000 vols   39291

Brown County Library, 515 Pine St, *Green Bay*, WI 54301
T: +1 920 4484400; Fax: +1 920 4484376; E-mail: bc_library@co.brown.wi.us; URL: www.browncountylibrary.org
1968; Lynn Stainbrook
Hmong Language Coll; 8 branch libs, 1 bookmobile
468 000 vols; 1 128 curr per
libr loan   39292

Pima County Public Library – Joyner-Green Valley, 601 N La Canada Dr, *Green Valley*, AZ 85614
T: +1 520 6258660; Fax: +1 520 7915247
Amber Mathewson
71 000 vols   39293

Greenburgh Public Library, 177 Hillside Ave, *Greenburgh*, NY 10607
T: +1 914 9931612; Fax: +1 914 9931613; URL: www.greenburghlibrary.org
1962; Eugenie Contrata
186 000 vols; 422 curr per; 9 652 av-mat
libr loan   39294

Putnam County Public Library, 103 E Poplar St, *Greencastle*, IN 46135-1655; P.O. Box 116, Greencastle, IN 46135-0116
T: +1 765 6532755; Fax: +1 765 6532756; E-mail: library@putnam.lib.in.us; URL: www.putnam.lib.in.us
1902; Alice Greenburg

66 000 vols; 269 curr per; 29 e-books; 4 252 av-mat
libr loan   39295

Greendale Public Library, 5647 Broad St, *Greendale*, WI 53129-1887
T: +1 414 4232136; Fax: +1 414 4232139; URL: www.greendale.org/library/l-admin.htm
1938; Gary Warren Niebuhr
56 000 vols; 80 curr per   39296

Greeneville Green County Public Library, 210 N Main St, *Greeneville*, TN 37745-3816
T: +1 423 6385034; Fax: +1 423 6383841; E-mail: grv@ggcpl.org; URL: www.ggcpl.org
1908; Madge Walker
50 000 vols; 125 curr per   39297

Greenfield Public Library, 402 Main St, *Greenfield*, MA 01301
T: +1 413 7721544; Fax: +1 413 7721589; E-mail: librarian@greenfieldpubliclibrary.org; URL: www.greenfieldpubliclibrary.org
1881; Sharon Sharry
Town Histories-Franklin County, Genealogy
53 000 vols; 134 curr per   39298

Greenfield Public Library, 7215 W Coldspring, *Greenfield*, WI 53220
T: +1 414 3219595; Fax: +1 414 3218595; E-mail: librarian@greenfieldpubliclibrary.org; URL: www.greenfieldlibrary.org
Sheila O'Brien
99 000 vols; 120 curr per   39299

Hancock County Public Library, 900 W McKenzie Rd, *Greenfield*, IN 46140-1741
T: +1 317 4625141; Fax: +1 317 4625711; E-mail: hcpl@hcplibrary.org; URL: www.hcplibrary.org
1898; Dianne Osborne
188 000 vols; 275 curr per   39300

Harborfields Public Library, 31 Broadway, *Greenlawn*, NY 11740-1382
T: +1 631 7574200; Fax: +1 631 7574266; E-mail: harblib@suffolk.lib.ny.us; URL: harb.suffolk.lib.ny.us
1970; Carol Albano
Business, Career, Consumer info, Health, Parenting, Teacher, Travel
102 000 vols; 1 138 curr per
libr loan; ALA   39301

Greensboro Public Library, 219 N Church St, *Greensboro*, NC 27402-3178
T: +1 336 3732471; Fax: +1 336 3336781; URL: www.greensborolibrary.org
1902; Sandra M. Neerman
6 branch libs
542 000 vols; 790 curr per
libr loan   39302

Greensboro Public Library – Blanche S. Benjamin Branch, 1530 Benjamin Pkwy, *Greensboro*, NC 27408
T: +1 336 3737540; Fax: +1 336 5455954
Velma Shoffner
55 000 vols   39303

Greensboro Public Library – Kathleen Clay Edwards Family Branch, 1420 Price Park Rd, *Greensboro*, NC 27410
T: +1 336 3732923; Fax: +1 336 8515047
Susan Vermeulen
53 000 vols; 70 curr per   39304

Greensburg Hempfield Area Library, 237 S Pennsylvania Ave, *Greensburg*, PA 15601-3086
T: +1 724 8375620; Fax: +1 724 8360160; URL: www.ghal.org
1936; Cesare J. Muccari
Pennsylvania Rm (Pa Hist & Geneology)
83 000 vols; 149 curr per; 1 511 av-mat
libr loan; ALA   39305

Greensburg-Decatur County Public Library, 1110 E Main St, *Greensburg*, IN 47240
T: +1 812 6632826; Fax: +1 812 6635617; E-mail: grefdesk@greensburg.lib.in.us; URL: www.greensburglibrary.org
1903; Andrea Ingmire
1 branch libr
87 000 vols; 233 curr per
libr loan; INCOLSA   39306

Greenup County Public Libraries, 614 Main St, *Greenup*, KY 41144-1036
T: +1 606 4736514; Fax: +1 606 4736514
1969; Dorothy K. Griffith
Jesse Stuart Coll, photos; 2 branch libs
96 000 vols; 290 curr per   39307

Flat River Community Library, 200 W Judd St, *Greenville*, MI 48838-2225
T: +1 616 7546359; Fax: +1 616 7541398; E-mail: gre@flatriverlibrary.org; URL: www.flatriverlibrary.org
1868; James P. Hibler
61 370 vols; 165 sound-rec   39308

Greenville County Library System, Hughes Main Library, 25 Heritage Green Pl, *Greenville*, SC 29601-2034
T: +1 864 2425000; Fax: +1 864 2358375; URL: www.greenvillelibrary.org
1921; Beverly A. James
11 branch libs
774 000 vols; 1 715 curr per; 8 e-books
libr loan; ALA   39309

Greenville Public Library, 520 Sycamore St, *Greenville*, OH 45331-1438
T: +1 937 5483915; Fax: +1 937 5483837; URL: www.greenville-publiclibrary.org
1883; John L. Vehre
Genealogy of Parke County Coll, Annie Oakley Coll, Saint Clair Coll, Sheet Music Coll, Signed Limited Eds
100 000 vols; 211 curr per   39310

Greenville Public Library, 573 Putnam Pike, *Greenville*, RI 02828-2195
T: +1 401 9493630; Fax: +1 401 9490530; E-mail: yourlibrary@lori.ri.gov; URL: www.yourlibrary.ws
1882; Christopher LaRoux
Arts, Crafts; High/Low Young Adult & Adult New Reader Mat
70 000 vols; 160 curr per; 112 e-books; 3 587 av-mat
libr loan; ALA   39311

Muhlenberg County Libraries, 117 S Main, *Greenville*, KY 42345
T: +1 270 3384760; Fax: +1 270 3384000; E-mail: hmlib@mcplib.org; URL: www.mcplib.org
1970; Anniesse Williams
2 branch libs
90 000 vols; 140 curr per; 2 200 av-mat   39312

Sheppard Memorial Library, 530 S Evans St, *Greenville*, NC 27858-2308
T: +1 252 3294063; Fax: +1 252 3294587; URL: www.sheppardlibrary.org
1929; Willie E. Nelms
4 branch libs
228 000 vols; 389 curr per; 17 973 av-mat; 10 300 sound-rec   39313

Washington County Library System – William Alexander Percy Memorial Library, 341 Main St, *Greenville*, MS 38701-4097
T: +1 662 3352331; Fax: +1 662 3904758; URL: www.washington.lib.ms.us
1964; Kay Clanton
Oral Hist Coll, State Docs; Genealogy dept, 5 branch libs
281 000 vols; 394 curr per; 1 201 sound-rec; 3 digital data carriers; 224 Art Reproductions
libr loan; MLA   39314

Greenwich Library, 101 W Putnam Ave, *Greenwich*, CT 06830-5387
T: +1 203 6227944; Fax: +1 203 6227959; URL: www.greenwichlibrary.org
1878; Carol Mahoney
2 branch libs
361 000 vols; 739 curr per; 8 000 e-books; 11 581 av-mat; 52 681 sound-rec; 14 394 digital data carriers   39315

Greenwood County Library, 106 N Main St, *Greenwood*, SC 29646
T: +1 864 9414650;
Fax: +1 864 9414651; URL: www.greenwoodcountylibrary.org
1901; Prudence A. Taylor
2 branch libs, 1 bookmobile
104 000 vols; 400 curr per; 722 e-books; 78 Electronic Media & Resources   39316

Greenwood Public Library, 310 S Meridian St, *Greenwood*, IN 46143-3135
T: +1 317 8834224; Fax: +1 317 8811963; E-mail: reflib@mail.greenwood.lib.in.us; URL: www.greenwood.lib.in.us
1917; Margaret L. Hamilton
100 000 vols; 360 curr per   39317

Greenwood-Leflore Public Library System, 405 W Washington, *Greenwood*, MS 38930-4297
T: +1 662 4533634; Fax: +1 662 4530683
1914; Susan Harris
Mae Wilson McBee Genealogy Coll; 2 branch libs
88 000 vols; 176 curr per   39318

Scott-Sebastian Regional Library, 18 N Adair, *Greenwood*, AR 72936; P.O. Box 400, Greenwood
T: +1 479 9962856; Fax: +1 479 9962236
1954; Judy Beth Clevenger
Large Print Coll, Arkansas Hist Coll; 5 branch libs
92 000 vols; 150 curr per   39319

Scott-Sebastian Regional Library – Sebastian County Library, 18 N Adair, *Greenwood*, AR 72936; P.O. Box 400, Greenwood, AR 72936
T: +1 479 9962856
Gayle Taylor
78 000 vols; 145 curr per   39320

Elizabeth Jones Library, 1050 Fairfield Ave, *Grenada*, MS 38901-3605; P.O. Box 130, Grenada, MS 38902-0130
T: +1 662 2262072; Fax: +1 662 2268747; E-mail: grenadapubliclibrary@elizabeth.lib.ms.us; URL: www.elizabeth.lib.ms.us
1933; Crystal M. Osborne
50 000 vols; 80 curr per   39321

Flint River Regional Library, 800 Memorial Dr, *Griffin*, GA 30223
T: +1 770 4124770; URL: www.frrls.net
1949; Walter H. Murphy
Georgia; 8 branch libs
455 000 vols
libr loan   39322

Flint River Regional Library – Griffin-Spalding County Library, 800 Memorial Dr, *Griffin*, GA 30223
T: +1 770 4124770
Walter H. Murphy
103 000 vols   39323

Lake County Public Library – Griffith Branch, 940 N Broad St, *Griffith*, IN 46319-1528
T: +1 219 8382825
Mark Furukawa
60 000 vols; 75 curr per   39324

Stewart Library, 926 Broad St, *Grinnell*, IA 50112-2046; P.O. Box 390, Grinnell, IA 50112-0390
T: +1 641 2362661; Fax: +1 641 2362667; E-mail: grinlib@iowatelecom.net; URL: www.grinnell.lib.ia.us
1901; Lorna Caulkins
62 000 vols; 181 curr per; 1 559 av-mat   39325

Grosse Pointe Public Library, Ten Kercheval at Fisher Rd, *Grosse Pointe Farms*, MI 48236-3693
T: +1 313 3432074; Fax: +1 313 3432437; URL: www.gp.lib.mi.us
1929; Vickey Bloom
Business, Music, Medicine; 2 branch libs
173 000 vols; 276 curr per   39326

Groton Public Library, 52 Newtown Rd, *Groton*, CT 06340
T: +1 860 4416750; Fax: +1 860 4480363; E-mail: reference@grotonpl.org; URL: www.seconnlib.org
1959; Alan G. Benkert
119 000 vols; 234 curr per   39327

Groton Public Library, 99 Main St, *Groton*, MA 01450
T: +1 978 4488000; Fax: +1 978 4481169; E-mail: info@gpl.org; URL: www.gpl.org
1854; Missus Owen Smith Shuman
57 000 vols; 147 curr per; 1 833 av-mat   39328

Southwest Public Libraries, SPL Admin, 3359 Park St, ***Grove City***, OH 43123
T: +1 614 8756716; Fax: +1 614 8752219; URL: www.spl.lib.oh.us
1891; Mark M. Shaw
1 branch libr
267 000 vols; 380 curr per
libr loan 39329

East Central Georgia Regional Library – Euchee Creek Library, 5907 Euchee Creek Dr, ***Grovetown***, GA 30813
T: +1 706 5560594; Fax: +1 706 5562585; E-mail: euchee@ecgrl.org
John Welch
55 000 vols; 61 curr per; 150 av-mat 39330

Buchanan County Public Library, Rte 2, Poetown Rd, ***Grundy***, VA 24614-9613; Box 3, Grundy, VA 24614-9613
T: +1 276 9356582; Fax: +1 276 9356292; E-mail: bcpl@bcplnet.org; URL: www.bcplnet.org
1961; Edgar F. Talbott
90 000 vols; 155 curr per 39331

Guilderland Public Library, 2228 Western Ave, ***Guilderland***, NY 12084-9701
T: +1 518 4562400; Fax: +1 518 4560923; E-mail: guilgen@uhls.lib.ny.us; URL: guilderlandpublic.info
1957; Barbara Nichols Randall
Altamont Enterprise microfilm, 1892 current
98 000 vols; 270 curr per 39332

Guilford Free Library, 67 Park St, ***Guilford***, CT 06437
T: +1 203 4538282; Fax: +1 203 4538288; URL: www.guilfordfreelibrary.org
1888; Sandra Ruoff
Poetry; Spanish Language Coll
117 000 vols; 209 curr per 39333

Gulfport Public Library, 5501 28th Ave S, ***Gulfport***, FL 33707
T: +1 727 8931073; Fax: +1 727 8931072; E-mail: gulfport@tblc.org; URL: tblc.org/gpl/
1935; Catherine Smith
Russian Bks Coll (popular)
70 000 vols; 100 curr per 39334

Harrison County Library System, 2600 24th Ave, No 6, ***Gulfport***, MS 39501-2081
T: +1 228 8681383; Fax: +1 228 8637433; URL: www.harrison.lib.ms.us
1916; Robert Lipscomb
Art, Careers, Maps, Genealogy; 8 branch libs
276 000 vols; 608 curr per; 150 maps; 79 digital data carriers; 350 slides, 25 art reproductions 39335

Harrison County Library System – Gulfport Temporary Library, 47 Maples Dr, Trailer No 1, ***Gulfport***, MS 39507
T: +1 228 8717171; Fax: +1 228 8717067
Frank Murphy
85 000 vols; 150 curr per 39336

Guntersville Public Library, 1240 O'Brig Ave, ***Guntersville***, AL 35976
T: +1 256 5717595; Fax: +1 256 5717596; URL: www.guntersvillelibrary.org
1947; Joanne Savoie
Genealogical Mat of Marshall County, Hist Coll of Guntersville's Newspapers
50 000 vols; 65 curr per; 51 maps; 1 800 av-mat; 1 120 sound-rec; 52 digital data carriers
libr loan; ALA 39337

Warren-Newport Public Library District, 224 N O'Plaine Rd, ***Gurnee***, IL 60031
T: +1 847 2445150; Fax: +1 847 2443499; URL: www.wnpl.info
1973; Stephen Bero
178 000 vols; 420 curr per; 13 553 av-mat; 90 High Interest/Low Vocabulary Bks, 140 Bks on Deafness & Sign Lang, 2 460 Large Print Bks, 13 560 Talking Bks 39338

County of Los Angeles Public Library – Hacienda Heights Library, 16010 La Monde St, ***Hacienda Heights***, CA 91745-4299
T: +1 626 9689356; Fax: +1 626 3363126; URL: www.colapublib.org/libs/haciendahts
Wen Wen Zhang
110 000 vols; 189 curr per 39339

Johnson Free Public Library, 274 Main St, ***Hackensack***, NJ 07601-5797
T: +1 201 3434169; Fax: +1 201 3431395; E-mail: hackref@bccls.org; URL: www.bccls.org/hackensack
1901; Sharon Castanteen
Art, Careers, Law, Music; Bergen County Hist Society Coll, New Jersey Hist Coll
153 000 vols; 335 curr per 39340

Haddon Heights Public Library, 608 Station Ave, ***Haddon Heights***, NJ 08035-1907; P.O. Box 240, Haddon Heights, NJ 08035-0240
T: +1 856 5477132;
Fax: +1 856 5472867; URL: www.haddonheights.lib.nj.us
1902; Robert J. Hunter
Drama
54 000 vols; 97 curr per 39341

Haddonfield Public Library, 60 Haddon Ave, ***Haddonfield***, NJ 08033-2422
T: +1 856 4291304; Fax: +1 856 4293760; URL: www.haddonfieldlibrary.org
1803; Douglas B. Rauschenberger
75 000 vols; 202 curr per 39342

Washington County Free Library, 100 S Potomac St, ***Hagerstown***, MD 21740
T: +1 301 7393250; Fax: +1 301 7397603; E-mail: webmaster@washcolibrary.org; URL: www.washcolibrary.org
1898; Mary Catherine Baykan
7 branch libs
329 000 vols; 580 curr per; 3 000 av-mat; 800 Audio Bks, 500 High Interest/Low Vocabulary Bks, 200 Bks on Deafness & Sign Lang, 5 000 Large Print Bks, 3 000 Talking Bks 39343

Western Maryland Public Libraries, 100 S Potomac St, ***Hagerstown***, MD 21740
T: +1 301 7393250; Fax: +1 301 7397603; URL: www.pilot.wash.lib.md.us
1968; Mary C. Baykan
62 000 vols 39344

San Mateo County Library – Half Moon Bay Branch, 620 Correas St, ***Half Moon Bay***, CA 94019
T: +1 650 7262316; Fax: +1 650 7269282
Anne Marie Malley
83 000 vols 39345

Halifax County Library, 33 S Granville St, ***Halifax***, NC 27839; P.O. Box 97, Halifax, NC 27839-0097
T: +1 252 5833631; Fax: +1 252 5838661; URL: www.halifaxnc.com
1941; Virginia Orvedahl
Large Print Coll, Audio Bks, Videocassettes
65 000 vols; 180 curr per 39346

Halifax County-South Boston Regional Library, 177 S Main St, ***Halifax***, VA 24558; P.O. Box 1729, Halifax, VA 24558-1729
T: +1 434 4763357; Fax: +1 434 4763359; URL: www.halifaxlibrary.org
1961; Kenneth Paul Johnson
Art, Business, Large print; 1 branch libr, 1 bookmobile
130 000 vols; 151 curr per; 6 945 av-mat; 5 760 sound-rec; 1 000 digital data carriers; 2 570 Large Print Bks 39347

Forked Deer Regional Library Center, 220 N Front St, ***Halls***, TN 38040-1559; P.O. Box 68, Halls
T: +1 731 8365812; Fax: +1 731 8367085; E-mail: fdeer@mail.state.tn.us
1965; Robert Toth
133 000 vols 39348

Haltom City Public Library, 4809 Haltom Rd, ***Haltom City***, TX 76117-3622; P.O. Box 14277, Haltom City, TX 76117-0277
T: +1 817 2227790; Fax: +1 817 8341446; URL: www.haltomcitytx.com/library
1961; Lesly M. Smith
105 000 vols 39349

Hamburg Public Library, 102 Buffalo St, ***Hamburg***, NY 14075-5097
T: +1 716 6494415; Fax: +1 716 6494160; E-mail: ham@buffalolib.org; URL: www.buffalolib.org/
1897; John Edson
Antiques, Art
54 000 vols; 193 curr per 39350

Hamburg Township Library, 10411 Merrill Rd, ***Hamburg***, MI 48139; P.O. Box 247, Hamburg, MI 48139-0247
T: +1 810 2311771; Fax: +1 810 2311520; E-mail: hamb@tln.lib.mi.us; URL: www.hamburglibrary.org
1966; Christine Weber
Arts, Civil War, Michigan; Depot for Speigleburg Rasmussen Sites, EPA
55 000 vols; 88 curr per 39351

Hamden Public Library, Miller Memorial Central Library, 2901 Dixwell Ave, ***Hamden***, CT 06518-3135
T: +1 203 2872680; Fax: +1 203 2872685; URL: www.hamdenlibrary.org
1944; Robert Gualtieri
Business Resource Ctr, Career Resource Ctr, Literacy; 2 branch libs
196 000 vols; 818 curr per; 200 High Interest/Low Vocabulary Bks 39352

Hamilton Township Public Library, One Municipal Dr, ***Hamilton***, NJ 08619
T: +1 609 5814060; Fax: +1 609 5814067
1923; George Conwell
270 000 vols; 440 curr per; 3 843 av-mat; 3 810 Large Print Bks, 2 850 Talking Bks 39353

Lane Public Library, 300 N Third St, ***Hamilton***, OH 45011-1629
T: +1 513 8947156; Fax: +1 513 8446535; E-mail: comments@lanepl.org; URL: www.lanepl.org
1866; Mary Pat Essman
Theatre & Drama (CC Fracker Memorial & Kathleen Stuckley Memorial Coll); 2 branch libs
506 000 vols; 643 curr per; 32 070 av-mat; 18 850 sound-rec; 31 560 digital data carriers
libr loan 39354

Hammond Public Library, 564 State St, ***Hammond***, IN 46320-1532
T: +1 219 9315100; Fax: +1 219 9313474; URL: www.hammond.lib.in.us
1902; Margaret Evans
2 branch libs
207 000 vols; 466 curr per
libr loan 39355

Pender County Public Library – Hampstead Branch, 75 Library Dr, ***Hampstead***, NC 28443
T: +1 910 2704603; Fax: +1 910 2705015
Marsha Dees
North Carolina Coll
90 000 vols; 65 curr per 39356

Lane Memorial Library, Two Academy Ave, ***Hampton***, NH 03842
T: +1 603 9263368; Fax: +1 603 9261348; E-mail: library@hampton.lib.nh.us; URL: www.hampton.lib.nh.us
1881; Catherine M. Redden
New Adult Readers
56 000 vols; 279 curr per; 4 556 av-mat 39357

Hampton Bays Public Library, 52 Ponquogue Ave, ***Hampton Bays***, NY 11946-0207
T: +1 631 7286241; Fax: +1 631 7280166; E-mail: hbaylib@suffolk.lib.ny.us; URL: hbay.suffolk.lib.ny.us
1960; Michael Firestone
Art & architecture; Long Island Hist
65 000 vols; 6 334 curr per; 2 545 sound-rec; 13 digital data carriers 39358

Hamtramck Public Library, Albert J. Zak Memorial Library, 2360 Caniff St, ***Hamtramck***, MI 48212
T: +1 313 3657050; Fax: +1 313 3650160; URL: hamtramck.lib.mi.us
1924; E. Tamara Sochacka
Polish, Ukrainian, Russian (Foreign Language Coll) & City of Hamtramck Hist File, bks, clippings, microfilm & newspapers
60 000 vols; 200 curr per; 830 av-mat 39359

Kings County Library, 401 N Douty St, ***Hanford***, CA 93230
T: +1 559 5820262; Fax: +1 559 5836163; URL: www.kingscountylibrary.org
1911; Louise E. Hodges
6 branch libs
192 000 vols; 249 curr per
libr loan 39360

Kings County Library – Hanford Branch, 401 N Douty St, ***Hanford***, CA 93230
T: +1 559 5820261; Fax: +1 559 5836163
Gail Lucas
160 000 vols 39361

Hannibal Free Public Library, 200 S Fifth St, ***Hannibal***, MO 63401
T: +1 573 2210222; Fax: +1 573 2210369; E-mail: webmaster@hannibal.lib.mo.us; URL: hannibal.lib.mo.us
1845; Hallie Yundt Silver
74 000 vols; 189 curr per; 1 000 sound-rec; 50 digital data carriers
libr loan; ALA 39362

Guthrie Memorial Library – Hanover's Public Library, Two Library Pl, ***Hanover***, PA 17331-2283
T: +1 717 6325183; Fax: +1 717 6327565; E-mail: gulibrary@yorklibraries.org; URL: www.guthrielibrary.org
1911; Roberta Greene
92 000 vols; 174 curr per 39363

Howe Library, 13 South St, ***Hanover***, NH 03755
T: +1 603 6434120; Fax: +1 603 6430725; E-mail: howe.library@thehowe.org; URL: www.howelibrary.org
1900; Mary H. White
67 000 vols; 219 curr per 39364

John Curtis Free Library, 534 Hanover St, ***Hanover***, MA 02339-2228
T: +1 781 8262972; Fax: +1 781 8263130; E-mail: lwelsh@ocln.org; URL: www.hanovermass.com/library
1907; Lorraine Welsh
Hist Coll
52 620 vols; 72 curr per 39365

Pamunkey Regional Library, 7527 Library Dr, ***Hanover***, VA 23069; P.O. Box 119, Hanover, VA 23069-0119
T: +1 804 5376211; Fax: +1 804 5376389; E-mail: ask@pamunkeylibrary.org; URL: www.pamunkeylibrary.org
1941; Fran Freimarck
Civil War, Virginiana; 10 branch libs
400 000 vols; 430 curr per; 12 494 av-mat; 100 Bks on Deafness & Sign Lang, 4 910 Large Print Bks, 12 500 Talking Bks 39366

Breckinridge County Public Library, 112 S Main St, P.O. Box 248, ***Hardinsburg***, KY 40143
T: +1 502 7562323; Fax: +1 502 7565634; URL: www.bcplibrary.org
1953; Holly Gregory
Children's lit, Young adult; 2 branch libs
51 710 vols; 86 curr per 39367

Harlan Community Library, 718 Court St, ***Harlan***, IA 51537
T: +1 712 7555934; Fax: +1 712 7553952; E-mail: harlanpl@harlannet.com; URL: www.harlan.lib.ia.us
Michael Burris
Iowa Hist
50 000 vols; 126 curr per; 1 650 av-mat; 2 000 sound-rec 39368

Harlan County Public Library, Bryan W. Whitfield Jr Public Library, 107 N Third St, ***Harlan***, KY 40831
T: +1 606 5735220; Fax: +1 606 5735220; URL: harlancountylibraries.org
Richard Haynes
Kentucky, Genealogy, Coal; 2 branch libs
77 000 vols; 94 curr per; 962 av-mat 39369

Harlingen Public Library, 410 76 Dr, ***Harlingen***, TX 78550
T: +1 956 2165800; Fax: +1 956 4306654; URL: www.myharlingen.us/library1.htm
1920; Ruben Rendon
171 000 vols; 265 curr per; 4 316 av-mat; 4 050 Talking Bks
libr loan; ALA 39370

Dauphin County Library System, 101 Walnut St, *Harrisburg*, PA 17101
T: +1 717 2344961; Fax: +1 717 2347479; URL: www.dcls.org
1889; Richard Bowra
Job & Carees Ctr, Grants Info; 8 branch libs
349 000 vols; 594 curr per; 34 920 av-mat; 25 120 sound-rec; 5 150 Large Print Bks
libr loan   39371

Dauphin County Library System – East Shore Area Library, 4501 Ethel St, *Harrisburg*, PA 17109
T: +1 717 6529380; Fax: +1 717 6525012
Darlene Ford
102 000 vols; 82 curr per; 88 av-mat
                                             39372

Boone County Library, 221 W Stephenson, *Harrison*, AR 72601-4225
T: +1 870 7415913; Fax: +1 870 7415946; E-mail: boonecolibrary@hotmail.com; URL: bcl.state.ar.us
1903; LaVoyce Ewing
81 000 vols   39373

Harrison Public Library, Bruce Ave, *Harrison*, NY 10528
T: +1 914 8350324; Fax: +1 914 8351564; URL: www.harrisonpl.org
1905; Virginia Weimer Vogl
Italian & Japanese Language Coll; 1 branch libr
115 000 vols
libr loan   39374

Public Library of Cincinnati & Hamilton County – Harrison Branch, 10398 New Haven Rd, *Harrison*, OH 45030
T: +1 513 3694442; Fax: +1 513 3694443
Maria Bach
62 000 vols   39375

Massanutten Regional Library, 174 S Main St, *Harrisonburg*, VA 22801
T: +1 540 4344475; Fax: +1 540 4344382; E-mail: info@mrlib.org; URL: www.mrlib.org
1928; Phillip T. Hearne
8 branch libs
160 000 vols; 305 curr per   39376

Cass County Public Library, Administration Office, 400 E Mechanic, *Harrisonville*, MO 64701
T: +1 816 3804600; Fax: +1 816 8842301; E-mail: asklib@casscolibrary.org; URL: www.casscolibrary.org
1947; Christie Kessler
Genealogy; Cass County & Missouri, hist doc; 7 branch libs
214 000 vols; 586 curr per   39377

Alcona County Library System, 312 W Main, *Harrisville*, MI 48740; P.O. Box 348, Harrisville
T: +1 989 7246796; Fax: +1 989 7246173; E-mail: alcona1@northland.lib.mi.us
1940; Carol Luck
3 branch libs
50 000 vols; 50 curr per   39378

Mercer County Public Library, 109 W Lexington St, *Harrodsburg*, KY 40330-1542
T: +1 859 7343680; Fax: +1 859 7347524; E-mail: webmistress@mcplib.info; URL: www.mcplib.info
1970; Robin Singer Ison
Kentucky, Genealogy; Draper Mss Coll
63 000 vols; 121 curr per   39379

Hartford Public Library, 115 N Main St, *Hartford*, WI 53027-1596
T: +1 262 6738240; Fax: +1 262 6738300; URL: www.hartfordlibrary.org/
1904; Michael J. Gelhausen
All US Census Records for Wisconsin-Washington County; Hist Room Coll; Local Papers (1864-2002), micro
102 000 vols; 200 curr per; 200 microforms; 3 381 av-mat; 3 150 sound-rec; 97 digital data carriers; 2 140 Large Print Bks, 3 380 Talking Bks
libr loan; ALA   39380

Hartford Public Library, 15 Franklin St, *Hartford*, MI 49057; P.O. Box 8, Hartford, MI 49057-0008
T: +1 616 6213408; Fax: +1 616 6213073; E-mail: hartfordlibrary2000@yahoo.com
Stephanie Daniels
Hartford Day Spring Newspaper 1881-1973, microfilm
50 000 vols; 48 curr per; 575 av-mat
                                             39381

Hartford City Public Library, 314 N High St, *Hartford City*, IN 47348-2143
T: +1 765 3481720; Fax: +1 765 3485090; E-mail: hartfordcitylibrary@yahoo.com; URL: www.hcpubliclibrary.org
1903; Vicki Cecil
Music (George Leonard Fulton Memorial Record Libr)
63 000 vols; 321 curr per; 71 av-mat
                                             39382

Cromaine District Library, 3688 N Hartland Rd, P.O. Box 308, *Hartland*, MI 48353-0308
T: +1 810 6325200; Fax: +1 810 6327351; E-mail: cromaine@cromaine.org; URL: www.cromaine.org
1927; Sharon W. Hupp
Hist Docs of Hartland & Livingston County (J. R. Crouse), autogr, letters, doc, bks, photos, art works
65 600 vols; 354 curr per
libr loan   39383

Hartland Public Library, 110 E Park Ave, *Hartland*, WI 53029
T: +1 262 3673350; Fax: +1 262 3692251; URL: www.hartlandlibrary.org
1897; Nancy Massnick
Parenting, bks, prints; Vintage Videos
65 000 vols; 103 curr per   39384

Darlington County Library – Hartsville Memorial, 147 W College St, *Hartsville*, SC 29550
T: +1 843 3325115; Fax: +1 843 3327071
Audrey Tripp
Nuclear Regulatory Deposit
90 000 vols; 110 curr per   39385

Wright County Library, Administrative Headquarters, 125 Court Sq, *Hartville*, MO 65667-9998; P.O. Box 70, Hartville, MO 65667-0070
T: +1 417 7417595; Fax: +1 417 7417927; E-mail: htvlib@getgoin.net; URL: www.wrightcountylibrary.org
Judy Epperly
73 000 vols; 71 curr per   39386

Hart County Library, 150 Benson St, *Hartwell*, GA 30643
T: +1 706 3764655; Fax: +1 706 3761157; E-mail: info@hartcountylibrary.com
1936; Richard Sanders
Hist of Hart County, A-tapes
52 000 vols; 32 curr per   39387

Harvard Public Library, Four Pond Rd, *Harvard*, MA 01451-1647
T: +1 978 4564114; Fax: +1 978 4564115; URL: www.harvardpubliclibrary.org
1886; Mary Wilson
Shakers; Sears Coll
56 000 vols; 158 curr per; 2 739 sound-rec; 68 digital data carriers
libr loan; ALA, NELA, MLA   39388

Harvey Public Library District, 15441 Turlington Ave, *Harvey*, IL 60426-3683
T: +1 708 3310757; Fax: +1 708 3312835; E-mail: has@harvey.lib.il.us; URL: www.harvey.lib.il.us
1903; Jay Kalman
Ethnic studies
91 000 vols; 375 curr per   39389

Brooks Free Library, 739 Main St, *Harwich*, MA 02645
T: +1 508 4307562; Fax: +1 508 4307564; E-mail: bfl_mail@clamsnet.org; URL: www.brooksfreelibrary.org
1855; Ginny Hewitt
Childrens Drama (Harwich Junior Theatre Coll), scripts; Local Hist Room
62 500 vols; 205 curr per   39390

Eisenhower Public Library District, 4652 N Olcott, *Harwood Heights*, IL 60706
T: +1 708 8677828; Fax: +1 708 8671535; E-mail: ess@mls.lib.il.us; URL: www.eisenhowerlibrary.org
1972; Ronald V. Stoch
114 000 vols; 762 curr per; 56 e-books; 3 920 av-mat; 7 490 sound-rec
libr loan; ALA   39391

Free Public Library of Hasbrouck Heights, 320 Boulevard, *Hasbrouck Heights*, NJ 07604
T: +1 201 2880484; Fax: +1 201 2886653; URL: hasbrouckheights.bccls.org/
1916; Mimi Hui
50 000 vols; 115 curr per
libr loan   39392

Hastings Public Library, 517 W Fourth St, *Hastings*, NE 68901-7560; P.O. Box 849, Hastings, NE 68902-0849
T: +1 402 4612346; Fax: +1 402 4612359; E-mail: staff@hastings.lib.ne.us; URL: www.hastings.lib.ne.us
1903; N.N.
Toys, Puppets
143 000 vols; 157 curr per
libr loan   39393

Pleasant Hill Library, 1490 S Frontage Rd, *Hastings*, MN 55033
T: +1 651 4380200; URL: www.co.dakota.mn.us
80 000 vols; 130 curr per   39394

Hastings-on-Hudson Public Library, 7 Maple Ave, *Hastings-on-Hudson*, NY 10706
T: +1 914 4783307; Fax: +1 914 4784813; E-mail: has@westchesterlibraries.org; URL: www.hastingslibrary.org
1913; Susan Feir
51 000 vols; 152 curr per   39395

The Library of Hattiesburg, Petal, Forrest County, 329 Hardy St, *Hattiesburg*, MS 39401-3496
T: +1 601 5824461; Fax: +1 601 5825338; E-mail: rooms@hpfc.lib.ms.us; URL: www.hpfc.lib.ms.us
1916; Pamela J. Pridgen
Adult New Reader's Coll; 1 branch libr
150 000 vols; 222 curr per   39396

Haverstraw Kings Daughters Public Library – Village Library, 85 Main St, *Haverstraw*, NY 10927
T: +1 845 4293445; Fax: +1 845 4297313
1895; Vivien Maisey
Careers, Spanish language; Haverstraw Bay Photo Arch, North Rockland Hist
98 000 vols; 295 curr per   39397

Havre Hill County Library, 402 Third St, *Havre*, MT 59501
T: +1 406 2652123; Fax: +1 406 2621091; URL: www.havrehilllibrary.org
1983
70 240 vols; 103 curr per   39398

County of Los Angeles Public Library – Hawthorne Library, 12700 S Grevillea Ave, *Hawthorne*, CA 90250-4396
T: +1 310 6798193; Fax: +1 310 6794846; URL: www.colapublib.org/libs/hawthorne
Donald Rowe
142 000 vols; 115 curr per   39399

Louis Bay 2nd Library & Community Center, Hawthorne Public Library, 345 Lafayette Ave, *Hawthorne*, NJ 07506-2599
T: +1 973 4275745; Fax: +1 973 4275269; E-mail: hthn1@bccls.org; URL: www.bccls.org/hawthorne
1913; Thomas Frawley
Cinema, Hist, Quilting; Hawthorne Hist Coll, Early Childhood & Parenting, Literacy, Seniors, Deafness & Sign Language
76 000 vols; 194 curr per   39400

Hays Public Library, 1205 Main, *Hays*, KS 67601-3693
T: +1 785 6259014; Fax: +1 785 6258683; URL: www.hayspublib.org
1899; Melanie Miller
146 000 vols; 200 curr per; 105 High Interest/Low Vocabulary Bks, 160 Bks on Deafness & Sign Lang, 5 700 Large Print Bks   39401

Hayward Public Library – Weekes Branch, 27300 Patrick Ave, *Hayward*, CA 94544
T: +1 510 7822155; Fax: +1 510 2590429
Johnny Melesha
52 000 vols   39402

Perry County Public Library, 289 Black Gold Blvd, *Hazard*, KY 41701
T: +1 606 4362475; Fax: +1 606 4360191; E-mail: library@perrylib.org; URL: www.perrycountylibrary.org
1970
58 210 vols; 47 curr per   39403

Grande Prairie Public Library District, 3479 W 183rd St, *Hazel Crest*, IL 60429
T: +1 708 7985563; Fax: +1 708 7985874; E-mail: gps@sslic.net; URL: grandeprairie.org
1960; Susan K. Roberts
Fiction Works by African American Writers
78 000 vols; 210 curr per; 821 e-books
                                             39404

Hazel Park Memorial Library, 123 E Nine Mile Rd, *Hazel Park*, MI 48030
T: +1 248 5420940; Fax: +1 248 5464083; URL: www.hazel-park.lib.mi.us
1936; Joan E. Ludlow
Large Print Coll
87 000 vols; 130 curr per   39405

Monmouth County Library – Hazlet Branch, 251 Middle Rd, *Hazlet*, NJ 07730
T: +1 732 2647164; Fax: +1 732 7391556
Beth Henderson
54 000 vols   39406

Hazleton Area Public Library, 55 N Church St, *Hazleton*, PA 18201-5893
T: +1 570 4542961; Fax: +1 570 4540630; URL: www.hazletonlibrary.org
1907; Jim Reinmiller
4 branch libs
158 000 vols; 471 curr per; 3 796 av-mat; 1 610 sound-rec; 352 Large Print Bks
                                             39407

Wasatch County Library, 465 East 1200 South, *Heber*, UT 84032
T: +1 435 6541511; Fax: +1 435 6546456; E-mail: kbowcutt@co.wasatch.ut.us; URL: www.wasatch.lib.ut.us/pages/contactUs.htm
1919; Kristin Bowcutt
50 000 vols; 55 curr per   39408

Phillips-Lee-Monroe Regional Library – Phillips County Library, 623 Pecan St, *Helena*, AR 72342
T: +1 870 3383537; Fax: +1 870 3388855; E-mail: reflib@hnb.com; URL: www.geocities.com/athens/ithaca/4022
1961; Linda Bennett
Arkansas, Genealogy, Local hist; 5 branch libs
220 000 vols; 121 curr per; 1 301 sound-rec
libr loan   39409

Hemet Public Library, 300 E Latham, *Hemet*, CA 92543
T: +1 951 7652440; Fax: +1 951 7652446; URL: www.cityofhemet.org
1907; Wayne Disher
Large Print Bks Coll, California Coll
103 000 vols; 192 curr per; 228 maps; 50 000 microforms; 3 720 av-mat; 1 319 sound-rec
libr loan   39410

Hempstead Public Library, 115 Nichols Ct, *Hempstead*, NY 11550-3199
T: +1 516 4816990; Fax: +1 516 4816719; E-mail: hempstead@nassaulibrary.org; URL: www.nassaulibrary.org/hempstd
1889; Irene A. Duszkiewicz
Ethnic studies, Hist; Adult Multi-Media, Black Studies Coll, Early American Textbk, Foreign Language Coll, Hispanic Studies Coll, Job & Education Info Ctr, Walt Whitman Coll
212 000 vols; 331 curr per
libr loan   39411

Henderson County Public Library, 101 S Main St, *Henderson*, KY 42420-3599 T: +1 270 8263712; Fax: +1 270 8274226; URL: www.hcpl.org 1904; Donald L. Wathen 97 000 vols; 124 curr per; 5 586 av-mat
39412

Henderson District Public Libraries, James I. Gibson Library, 280 Water St, *Henderson*, NV 89015 T: +1 702 5649261; Fax: +1 702 5658832 Mae Giaimo Small business administration; Nevada Hist; 3 branch libs 138 000 vols
39413

Henderson District Public Libraries, Paseo Verde Library, 280 S Green Valley Pkwy, *Henderson*, NV 89012 T: +1 702 4927252; Fax: +1 702 4921711; URL: www.hdpl.org Tom Fay 108 000 vols
39414

H. Leslie Perry Memorial Library, 205 Breckenridge St, *Henderson*, NC 27536 T: +1 252 4383316; Fax: +1 252 4383744; URL: www.perrylibrary.org 1924; Jeanne Walton Fox 104 000 vols; 200 curr per; 4 280 av-mat
39415

Rusk County Library, 106 E Main St, *Henderson*, TX 75652 T: +1 903 6578557; Fax: +1 903 6577637; URL: www.rclib.org/ 1937; Pamela Pipkin Texas Heritage Resource Ctr 70 000 vols; 136 curr per
39416

Henderson County Public Library, 301 N Washington St, *Hendersonville*, NC 28739 T: +1 828 6974725; Fax: +1 828 6974700; E-mail: reference@henderson.lib.nc.us; URL: www.henderson.lib.nc.us 1914; William E. Snyder 4 branch libs 250 000 vols; 250 curr per
39417

Deaf Smith County Library, 211 E Fourth St, *Hereford*, TX 79045 T: +1 806 3641206; Fax: +1 806 3637063; URL: www.hlc-lib.org/hereford 1910; Rebecca Walls Adult Basic Education, Bks on Tape, Spanish Coll, Texas Hist 72 000 vols; 141 curr per; 1 511 av-mat
39418

Nashville Public Library – Hermitage Branch, 3700 James Kay Lane, *Hermitage*, TN 37076-3429 T: +1 615 8803951; Fax: +1 615 8803955 Gloria Coleman 105 000 vols
39419

Hershey Public Library, 701 Cocoa Ave, *Hershey*, PA 17033 T: +1 717 5336555; Fax: +1 717 5341666; E-mail: library@derrytownship.org; URL: www.hersheylibrary.org 1913; Barbara S. Ellis Chocolate, Pennsylvania Coll 109 000 vols; 193 curr per
39420

Hewlett-Woodmere Public Library, 1125 Broadway, *Hewlett*, NY 11557-0903 T: +1 516 3741967; Fax: +1 516 5691229; URL: www.hwpl.org 1947; Susan O. DeSciora Art Coll, Music Coll 191 000 vols; 459 curr per
39421

Hialeah–John F. Kennedy Library, 190 W 49th St, *Hialeah*, FL 33012-3798 T: +1 305 8212700; Fax: +1 305 8189144; E-mail: jfklib@hialeahfl.gov; URL: www.hialeahfl.gov/library 1928; Diane Diaz Art & architecture, Business & management, Science & technology, Spanish, Music; 4 branch libr 163 000 vols; 200 curr per; 14 000 e-books; 1 000 av-mat; 5 000 sound-rec; 5 000 digital data carriers libr loan
39422

Hibbing Public Library, 2020 E Fifth Ave, *Hibbing*, MN 55746-1702 T: +1 218 3625959; Fax: +1 218 3129779; E-mail: hibbingpl@arrowhead.lib.mn.us; URL: www.hibbing.lib.mn.us 1908; Ginny Richmond Bob Dylan Coll 80 000 vols; 125 curr per
39423

Catawba County Library – Saint Stephens, 3225 Springs Rd, *Hickory*, NC 28601-9700 T: +1 828 2563030; Fax: +1 828 2566029 Debbie Hosford 78 000 vols; 46 curr per
39424

Hickory Public Library, Patrick Beaver Memorial Library, 375 Third St NE, *Hickory*, NC 28601-5126 T: +1 828 3040500; Fax: +1 828 3040023; URL: www.hickorygov.com/library 1922; Mary Montgomery Sizemore Career Enhancement Coll; 1 branch libr 112 000 vols; 354 curr per
39425

Hicksville Public Library, 169 Jerusalem Ave, *Hicksville*, NY 11801 T: +1 516 9311417; Fax: +1 516 8225672; E-mail: hilmail@nassaulibrary.org; URL: www.nassaulibrary.org/hicksv 1926; Carol Ahrens 240 000 vols; 584 curr per
39426

High Point Public Library, 901 N Main St, *High Point*, NC 27262; P.O. Box 2530, High Point, NC 27261-2530 T: +1 336 8839317; Fax: +1 336 8833636; URL: www.highpointpubliclibrary.com 1926; Kem B. Ellis Design, Furniture, Genealogy; Afro-American Hist, Family Literacy 268 000 vols; 650 curr per
39427

Jefferson County Library, 5678 State Rd PP, *High Ridge*, MO 63049-2216 T: +1 636 6778689; Fax: +1 636 6771769; E-mail: info@jeffersoncountylibrary.org; URL: www.jeffersoncountylibrary.org 1989; Pamela R. Klipsch Business, Parenting, Genealogy; 3 branch libs 154 000 vols; 323 curr per; 440 microforms; 2 780 sound-rec; 14 digital data carriers; 1 011 High Interest/Low Vocabulary Bks, 50 Bks on Deafness & Sign Lang
39428

Highland Township Public Library, 444 Beach Farm Circle, *Highland*, MI 48357; P.O. Box 277, Highland, MI 48357-0277 T: +1 248 8872218; Fax: +1 248 8875179; E-mail: htplreply@highland.lib.mi.us; URL: www.highlandlibrary.info 1927; Jude Halloran 66 000 vols; 165 curr per; 2 328 av-mat
39429

Lake County Public Library – Highland Branch, 2841 Jewett St, *Highland*, IN 46322-1617 T: +1 219 8382394; Fax: +1 219 9235886 Kathy Deal 84 000 vols; 53 curr per; 119 av-mat
39430

Highland Park Public Library, 31 N Fifth Ave, *Highland Park*, NJ 08904 T: +1 732 5722750; Fax: +1 732 8199046; URL: www.hpplnj.org 1922; Jane Stanley Drama 60 000 vols; 170 curr per
39431

Highland Park Public Library, 494 Laurel Ave, *Highland Park*, IL 60035-2690 T: +1 847 4320216; Fax: +1 847 4329139; E-mail: hppla@hplibrary.org; URL: www.hplibrary.org Jane Conway 194 000 vols; 350 curr per
39432

Douglas County Libraries – Highlands Ranch Library, 9292 Ridgeline Blvd, *Highlands Ranch*, CO 80129 T: +1 303 7917323; Fax: +1 720 3489510 N.N. 138 000 vols; 150 curr per
39433

Highwood Public Library, 102 Highwood Ave, *Highwood*, IL 60040-1597 T: +1 847 4325404; Fax: +1 847 4325806; E-mail: highwoodlibrary@yahoo.com 1887; Joan Retnauer Italian Lang, Mozart 51 000 vols; 93 curr per libr loan
39434

Columbus Metropolitan Library – Hilliard Branch, 4772 Cemetery Rd, *Hilliard*, OH 43026 T: +1 614 6452275; Fax: +1 614 4794149 Grace Kendall 170 000 vols
39435

Highland County District Library, Ten Willettsville Pike, *Hillsboro*, OH 45133 T: +1 937 3933114; Fax: +1 937 3932985; E-mail: highlandco@highlandco.org; URL: www.highlandco.org 1878; Jennifer West 3 branch libs 157 000 vols; 365 curr per
39436

Hillsboro Public Library, 2850 NE Brookwood Pkwy, *Hillsboro*, OR 97124 T: +1 503 6156500; Fax: +1 503 6156601; E-mail: library@ci.hillsboro.or.us; URL: www.ci.hillsboro.or.us/library 1914; N.N. Spanish & Japanese language mat; 2 branch libs 240 000 vols; 700 curr per; 16 731 av-mat; 16 730 Talking Bks, 19 650 Videos libr loan
39437

Hillsboro Public Library – Shute Park, 775 SE Tenth Ave, *Hillsboro*, OR 97123 T: +1 503 6156500; Fax: +1 503 6156501 N.N. Spanish Language Coll 130 000 vols
39438

Hyconeechee Regional Library, 300 W Tryon St, *Hillsborough*, NC 27278 T: +1 919 6443011; Fax: +1 919 6443003 1948; Brenda W. Stephens Large print, Antiques, Crafts, Geography 211 000 vols; 907 curr per
39439

Orange County Public Library, 300 W Tryon St, *Hillsborough*, NC 27278 T: +1 919 2452525; Fax: +1 919 6443003; URL: www.co.orange.nc.us/library 1912; Brenda W. Stephens Crafts, Large Print Bks Coll 86 000 vols; 420 curr per
39440

Hillsdale Free Public Library, 509 Hillsdale Ave, *Hillsdale*, NJ 07642 T: +1 201 3585072; Fax: +1 201 3585074; E-mail: hldlcirc@bccls.org; URL: www.bccls.org/hillsdale 1935; David J. Franz Fiction, Lit, Literary criticism 60 000 vols; 120 curr per
39441

Hillside Public Library, 405 N Hillside Ave, *Hillside*, IL 60162-1295 T: +1 708 4497510; Fax: +1 708 4496119; E-mail: jchesham@hillsidelibrary.org; URL: www.hillsidelibrary.org 1962; Douglas Losey 52 300 vols; 192 curr per
39442

Hillside Public Library, John F. Kennedy Plaza, Hillside & Liberty Aves, *Hillside*, NJ 07205-1893 T: +1 973 9234413; Fax: +1 973 9230506; URL: www.hillsidepl.org 1947; Miriam Bein 126 000 vols; 109 curr per
39443

Hawaii State Public Library System – Hilo Public Library, 300 Waianuenue Ave, *Hilo*, HI 96720-2447 T: +1 808 9338888; Fax: +1 808 9338895 Claudine Fujii 222 000 vols; 404 curr per
39444

Parma Public Library, 7 West Ave, *Hilton*, NY 14468 T: +1 585 3928350; Fax: +1 585 3929870; URL: www.rochester.lib.ny.us/parma 1885 70 000 vols; 170 curr per
39445

Live Oak Public Libraries – Liberty County Branch, 236 Memorial Dr, *Hinesville*, GA 31313 T: +1 912 3684003; Fax: +1 912 3697148 Robin Shader 54 000 vols
39446

Hingham Public Library, 66 Leavitt St, *Hingham*, MA 02043 T: +1 781 7411405; Fax: +1 781 7490956; E-mail: href@ocln.org; URL: www.hingham-ma.com/library 1869; Dennis Corcoran Typography (W.A. Dwiggins Coll) 175 000 vols; 348 curr per libr loan
39447

Hinsdale Public Library, 20 E Maple St, *Hinsdale*, IL 60521 T: +1 630 9861976; Fax: +1 630 9869720; E-mail: reference@hinsdalelibrary.info; URL: www.hinsdalelibrary.info 1892; Lynn Elam 126 000 vols; 295 curr per
39448

Lake County Public Library – Hobart Branch, 100 Main St, *Hobart*, IN 46342-4391 T: +1 219 9422243; Fax: +1 219 9471823 Carol Daumer-Gutjahr 84 000 vols; 107 curr per; 83 av-mat
39449

Hobbs Public Library, 509 N Shipp, *Hobbs*, NM 88240 T: +1 505 3979328; Fax: +1 505 3971508; URL: hobbspublib.leaco.net 1939; Cristine Adams Large Print Bk Coll, Petroleum, Southwest Coll 109 000 vols; 125 curr per; 2 502 av-mat libr loan
39450

Hoboken Public Library, 500 Park Ave, *Hoboken*, NJ 07030 T: +1 201 4202346; Fax: +1 201 4202299; E-mail: hobk1@bccls.org; URL: www.bccls.org/hoboken 1890; Lina Podles Hoboken Hist Coll 74 000 vols; 166 curr per
39451

Hockessin Public Library, 1023 Valley Rd, mail address: 87 Reads Way, *Hockessin*, DE 19707 T: +1 302 2395160; Fax: +1 302 2391519; URL: www.nccdelib.org 1977; Diann Colose 71 000 vols; 150 curr per; 65 531 mss; 3 429 sound-rec; 4 digital data carriers libr loan
39452

Sachem Public Library, 150 Holbrook Rd, *Holbrook*, NY 11741 T: +1 631 5885024; Fax: +1 631 5885064; E-mail: sachlib@suffolk.lib.ny.us; URL: sachemlibrary.org 1961; Judith M. Willner 251 000 vols; 550 curr per; 1 000 microforms; 17 055 av-mat; 10 113 sound-rec; 2 137 digital data carriers; 30 123 mat in electronic format libr loan
39453

Gale Free Library, 23 Highland St, *Holden*, MA 01520-2599 T: +1 508 8290228; Fax: +1 508 8290232; E-mail: sscott@townofholden.net; URL: www.townofholden.net/Pages/HoldenMA_library/index 1888; Susan Scott Local newspapers, micro 59 700 vols; 142 curr per
39454

Holdrege Public Library System, 604 E Ave, *Holdrege*, NE 68949 T: +1 308 9956556; Fax: +1 308 9955732; E-mail: info@holdregelibrary.org; URL: www.holdregelibrary.org 1895; Pam Soreide 68 050 vols; 123 curr per libr loan
39455

Herrick District Library, 300 S River Ave, *Holland*, MI 49423-3290 T: +1 616 3553726; Fax: +1 616 3553083; URL: www.herrickdl.org 1867; Thomas J. Genson Dutch & Spanish Per, Indo-Chinese Language Coll, Spanish Language; 2 branch libs 248 000 vols; 586 curr per
39456

Toledo-Lucas County Public Library – Holland Branch, 1032 S McCord Rd, *Holland*, OH 43528
T: +1 419 2595240; Fax: +1 419 8656706
Colleen Lehmann
126 000 vols   39457

San Benito County Free Library, 470 Fifth St, *Hollister*, CA 95023-3885
T: +1 831 6364107; Fax: +1 831 6364099; E-mail: sanbenlib@sbcglobal.net; URL: www.sanbenitofl.org
1918; Nora Conte
California, Spanish language, Japanese Language, Art
82 000 vols; 220 curr per   39458

Holliston Public Library, 752 Washington St, *Holliston*, MA 01746
T: +1 508 4290617; Fax: +1 508 4290625; E-mail: lmcdonnell@minlib.net; URL: www.hollistonlibrary.org
1879; Leslie McDonnell
55 000 vols; 110 curr per   39459

Wake County Public Library System – Holly Springs Branch, 300 W Ballentine St, *Holly Springs*, NC 27540
T: +1 919 5771660
Elena Owens
50 000 vols   39460

Holyoke Public Library, 335 Maple St, *Holyoke*, MA 01040-4999
T: +1 413 3225640; Fax: +1 413 5324230; E-mail: library@ci.holyoke.ma.us; URL: www.holyokelibrary.org
1870; Maria G. Pagan
US Volleyball Assn
64 000 vols; 110 curr per   39461

Homewood Public Library, 1721 Oxmoor Rd, *Homewood*, AL 35209-4085
T: +1 205 3326600; Fax: +1 205 8026424; E-mail: homewood@bham.lib.al.us; URL: www.homewood.lib.al.us
1942; Edith C. Harwell
88 000 vols; 245 curr per; 47 e-journals; 472 e-books; 59 Electronic Media & Resources   39462

Homewood Public Library District, 17917 Dixie Hwy, *Homewood*, IL 60430-1703
T: +1 708 7980121; Fax: +1 708 7980662; E-mail: hws@mls.lib.il.us; URL: homewoodlibrary.org
1927; Judi Wolinsky
161 000 vols; 260 curr per
libr loan   39463

Hawaii State Public Library System, Office of the State Librarian, 44 Merchant St, *Honolulu*, HI 96813
T: +1 808 5863621; Fax: +1 808 5863715; URL: www.librarieshawaii.org
1961; Richard Burns
Asian Language Mat, Federal Docs, Hawaii & Pacific Coll, Patent & Trademark Depot; 51 branch libs
3 399 000 vols; 5 282 curr per; 6 000 e-books; 47 500 av-mat; 71 600 sound-rec; 61 000 digital data carriers   39464

Hawaii State Public Library System -Aina Haina Public Library, 5246 Kalanianaole Hwy, *Honolulu*, HI 96821
T: +1 808 3772456; Fax: +1 808 3772455
Hueyduan Kwok
64 000 vols; 75 curr per   39465

Hawaii State Public Library System – Hawaii State Library, 478 S King St, *Honolulu*, HI 96813
T: +1 808 5863621; Fax: +1 808 5863943
Diane Eddy
553 000 vols; 1 114 curr per   39466

Hawaii State Public Library System – Hawaii-Kai Public Library, 249 Lunalilo Home Rd, *Honolulu*, HI 96825
T: +1 808 3975833; Fax: +1 808 3975832
Colleen Lashway
74 000 vols; 72 curr per   39467

Hawaii State Public Library System – Kaimuki Public Library, 1041 Koko Head Ave, *Honolulu*, HI 96816-3707
T: +1 808 7338422; Fax: +1 808 7338426
Daniel Roffman
114 000 vols; 113 curr per   39468

Hawaii State Public Library System – Kalihi-Palama Public Library, 1325 Kalihi St, *Honolulu*, HI 96819
T: +1 808 8323466; Fax: +1 808 8323469
Marcia Nakama
59 000 vols; 103 curr per   39469

Hawaii State Public Library System – Liliha Public Library, 1515 Liliha St, *Honolulu*, HI 96817-3526
T: +1 808 5877577; Fax: +1 808 5877579
Sylvia Mitchell
77 000 vols; 79 curr per   39470

Hawaii State Public Library System – McCully-Moiliili Public Library, 2211 S King St, *Honolulu*, HI 96826
T: +1 808 9731099; Fax: +1 808 9731095
Hillary Chang
99 000 vols; 81 curr per   39471

Hawaii State Public Library System – Salt Lake-Moanalua Public Library, 3225 Salt Lake Blvd, *Honolulu*, HI 96818
T: +1 808 8316831; Fax: +1 808 8316834
Duane Wenzel
69 000 vols; 52 curr per   39472

Hood River County Library, 502 State St, *Hood River*, OR 97031
T: +1 541 3862535; Fax: +1 541 3863835; E-mail: gorge.link@co.hood-river.or.us; URL: www.co.hood-river.or.us/library
1912; June M. Knudson
69 000 vols; 196 curr per   39473

Hoover Public Library, 200 Municipal Dr, *Hoover*, AL 35216
T: +1 205 4447800; Fax: +1 205 4447878; URL: www.hoover.lib.al.us
1983; Linda R. Andrews
172 000 vols; 347 curr per   39474

Southwest Arkansas Regional Library, 202 S Main St, *Hope*, AR 71801; P.O. Box 1388, Hope
T: +1 870 7772957; Fax: +1 870 7229956; URL: library.nevada.ar.us/regionallibraries.html
1947; N.N.
19 branch libs
232 000 vols; 340 curr per   39475

Southwest Arkansas Regional Library – Hempstead County Library, 500 S Elm St, *Hope*, AR 71801
T: +1 870 7774564; Fax: +1 870 7772915
Judy Sooter
227 000 vols; 334 curr per   39476

Cumberland County Public Library & Information Center – Hope Mills Branch, 3411 Golfview Rd, *Hope Mills*, NC 28348-2266
T: +1 910 4258455; Fax: +1 910 4230997; URL: www.cumberland.lib.nc.us/locations/hpm/hopemills.htm
Lisa Olsen
56 000 vols   39477

Appomattox Regional Library, Maude Nelson Langhorne Library, 209 E Cawson St, *Hopewell*, VA 23860
T: +1 804 4586329; Fax: +1 804 4584349; URL: www.arls.org
1974; Charles Koutnik
6 branch libs
171 000 vols; 371 curr per; 2 200 av-mat; 2 200 Talking Bks   39478

Hopkinsville-Christian County Public Library, 1101 Bethel St, *Hopkinsville*, KY 42240
T: +1 270 8874262; Fax: +1 270 8874264; URL: hccpl.org
1874
McCarroll Genealogy Coll
99 743 vols; 104 curr per   39479

Chemung County Library District, Horseheads Free Library, 405 S Main St, *Horseheads*, NY 14845
T: +1 607 7394581; Fax: +1 607 7394592; E-mail: horseheads@stls.org
1944; Maureen Ferrell
53 000 vols; 30 curr per   39480

Horsham Township Library, 435 Babylon Rd, *Horsham*, PA 19044-1224
T: +1 215 4432609; Fax: +1 215 4432697; URL: www.horshamlibrary.org
Laurie Tynan
80 000 vols; 133 curr per   39481

Garland County Library, 1427 Malvern Ave, *Hot Springs*, AR 71901
T: +1 501 6234161; Fax: +1 501 6235647; E-mail: gclhsar@hotmail.com; URL: www.garland.lib.ar.us
1948; John W. Wells
Arkansas Hist, Genealogy Coll
110 000 vols; 450 curr per   39482

Cary Library, 107 Main St, *Houlton*, ME 04730
T: +1 207 5321302; Fax: +1 207 5324350; URL: www.cary.lib.me.us
1903; Linda Faucher
Aroostook Oral Hist Project (128 cassette tapes)
50 000 vols; 120 curr per
libr loan   39483

Terrebonne Parish Library, 151 Library Dr, *Houma*, LA 70360
T: +1 985 8765158; Fax: +1 985 8765864; E-mail: main.b1tb@state.lib.la.us; URL: www.terrebonne.lib.la.us
1939; Mary Cosper LeBoeuf
Petroleum; 8 branch libs
302 000 vols; 1 008 curr per; 1 000 e-books; 5 840 av-mat; 7 420 sound-rec; 6 000 digital data carriers   39484

Harris County Public Library, 8080 El Rio, *Houston*, TX 77054
T: +1 713 7499000; Fax: +1 713 7499090; URL: www.hcpl.net
1921; Rhoda L. Goldberg
26 branch libs
1 799 000 vols; 3 307 curr per; 54 000 e-books
libr loan   39485

Harris County Public Library – Aldine Branch, 11331 Airline Dr, *Houston*, TX 77037
T: +1 281 4455560; Fax: +1 281 4458625
Clara Maynard
59 000 vols   39486

Harris County Public Library – Clear Lake City-County Freeman Branch, 16616 Diana Lane, *Houston*, TX 77062
T: +1 281 4881906; Fax: +1 281 2863931
Karen Akkerman
117 000 vols   39487

Harris County Public Library – Katherine Tyra Branch, 16719 Clay Rd, *Houston*, TX 77084
T: +1 281 5500885; Fax: +1 281 5503304
Sandra Silvey
104 000 vols   39488

Harris County Public Library – North Channel, 15741 Wallisvillle Rd, *Houston*, TX 77049
T: +1 281 4571631
Carolyn Dial
70 000 vols   39489

Harris County Public Library – Parker Williams Branch, 10851 Scarsdale Blvd, Ste 510, *Houston*, TX 77089
T: +1 281 4842036; Fax: +1 281 4810729
Mary Murray
70 000 vols   39490

Harris County Public Library – Spring Branch Memorial, 930 Corbindale, *Houston*, TX 77024
T: +1 713 4641633; Fax: +1 713 9732654
Karen Hayes
67 000 vols   39491

Houston Public Library, 7200 Keller, *Houston*, TX 77012
T: +1 832 3932480
LeRoy Robinson
53 000 vols   39492

Houston Public Library – Acres Homes, 8501 W Montgomery Rd, *Houston*, TX 77088
T: +1 832 3931700
Sanya Bunton
50 000 vols   39493

Houston Public Library – Bracewell Branch, 10115 Kleckley Dr, *Houston*, TX 77075
T: +1 832 3932580; Fax: +1 832 3932581; E-mail: Bracewell.Branch@cityofhouston.net
James Maynard
80 000 vols   39494

Houston Public Library – Carnegie Regional, 1050 Quitman, *Houston*, TX 77009
T: +1 832 3931720; Fax: +1 832 3931721
Jim Pearson
125 000 vols   39495

Houston Public Library – Clayton Library Center for Genealogical Research, 5300 Caroline, *Houston*, TX 77004-6896
T: +1 832 3932600; E-mail: Clayton.Branch@cityofhouston.net; URL: www.houstonlibrary.org/clayton
Susan D. Kaufman
85 000 vols; 2 648 curr per   39496

Houston Public Library – Collier Regional, 6200 Pinemont, *Houston*, TX 77092
T: +1 832 3931740; Fax: +1 832 3931741
John Tuggle
112 000 vols   39497

Houston Public Library – Fine Arts & Recreation Department, 500 McKinney Ave, *Houston*, TX 77002-2534
T: +1 832 3931313; URL: www.hpl.lib.tx.us
Blaine Davis
Artist information file, auction and exhibition catalogs
110 000 vols   39498

Houston Public Library – Flores Branch, 110 N Milby, *Houston*, TX 77003
T: +1 832 3931790
Elvia Pillado
62 000 vols   39499

Houston Public Library – Frank Branch, 6440 W Bellfort Ave, *Houston*, TX 77035
T: +1 832 3932410
Lon LeMaster
110 000 vols   39500

Houston Public Library – Freed-Montrose Branch, 4100 Montrose, *Houston*, TX 77006
T: +1 832 3931800; Fax: +1 832 3931801; E-mail: Montrose.Branch@cityofhouston.net
Mark Morrow
90 000 vols   39501

Houston Public Library – Heights, 1302 Heights Blvd, *Houston*, TX 77008
T: +1 832 3931810
Laurie Covington
54 000 vols   39502

Houston Public Library – Henington-Alief Regional, 7979 S Kirkwood, *Houston*, TX 77072
T: +1 832 3931820; Fax: +1 832 3931821
Rebecca Hubert
146 000 vols; 75 curr per   39503

Houston Public Library – Hillendahl Branch, 2436 Gessner Dr, *Houston*, TX 77080
T: +1 832 3931940
Alice M. Depot
78 000 vols   39504

Houston Public Library – Humanities Department, 500 McKinney Ave, *Houston*, TX 77002-2534
T: +1 832 3931313
Beatrice Temp
276 000 vols   39505

Houston Public Library – Jungman Branch, 5830 Westheimer Rd, *Houston*, TX 77057
T: +1 832 3931860
Susan Seidensticker
133 000 vols   39506

Houston Public Library – Kendall Branch, 14330 Memorial Dr, *Houston*, TX 77079
T: +1 832 3931880
Elizabeth Kelly
80 000 vols   39507

Houston Public Library – Meyer Branch, 5005 W Bellfort, **Houston**, TX 77035
T: +1 832 3931840; E-mail: Meyer.Branch@cityofhouston.net
Suzanne Rickles
82 000 vols   39508

Houston Public Library – Moody Branch, 9525 Irvington Blvd, **Houston**, TX 77076
T: +1 832 3931960
Sergio Pineda
58 000 vols   39509

Houston Public Library – Oak Forest, 1349 W 43rd St, **Houston**, TX 77018
T: +1 832 3931960; URL: www.hpl.lib.tx.us/branches/oak_home.html
Candace Sawyer
63 000 vols   39510

Houston Public Library – Park Place Regional, 8145 Park Place Blvd, **Houston**, TX 77017
T: +1 832 3931970
Regina Stemmer
103 000 vols   39511

Houston Public Library – Ring Branch, 8835 Long Point Rd, **Houston**, TX 77055
T: +1 832 3932000
Amelia Juresko
85 000 vols   39512

Houston Public Library – Robinson-Westchase Branch, 3223 Wilcrest, **Houston**, TX 77042
T: +1 832 3932011
Michael McNamara
101 000 vols   39513

Houston Public Library – Scenic Woods Regional, 10677 Homestead Rd, **Houston**, TX 77016
T: +1 832 3932030
77 000 vols   39514

Houston Public Library – Social Sciences Department, 500 McKinney Ave, **Houston**, TX 77002-2534
T: +1 832 3931313
Blaine Davis
153 000 vols   39515

Houston Public Library – Stanaker Branch, 611 S Sergeant Macario Garcia Dr, **Houston**, TX 77011
T: +1 832 3932080
Marian Warby
72 000 vols   39516

Houston Public Library – Stimley-Blue Ridge Branch, 7007 W Fuqua, **Houston**, TX 77489
T: +1 832 3932370
59 000 vols   39517

Houston Public Library – Tuttle Branch, 702 Kress, **Houston**, TX 77020
T: +1 832 3932100; E-mail: Tuttle.Branch@cityofhouston.net
Beatriz DeAngulo
59 000 vols   39518

Houston Public Library – Vinson Branch, 3100 W Fuqua, **Houston**, TX 77045
T: +1 832 3932120; E-mail: Vinson.Branch@cityofhouston.net
N.N.
74 000 vols   39519

Houston Public Library – Walter Branch, 7660 Clarewood, **Houston**, TX 77036
T: +1 713 2723661; E-mail: Walter.Branch@cityofhouston.net
LeeRoy Robinson
92 000 vols   39520

Houston Public Library – Young Neighborhood Library, 5260 Griggs Rd, **Houston**, TX 77021
T: +1 832 3932140; E-mail: Young.Branch@cityofhouston.net
N.N.
53 000 vols   39521

Texas County Library, 117 W Walnut, **Houston**, MO 65483
T: +1 417 9672258; Fax: +1 417 9672262; E-mail: hlibrary@train.missouri.org; URL: train.missouri.org
1946; Audrey Barnhart
3 branch libs
60 000 vols; 60 curr per; 729 av-mat
                                                        39522

Queens Borough Public Library – Howard Beach Branch, 92-06 156th Ave, **Howard Beach**, NY 11414
T: +1 718 6417086
McGann Jane
60 000 vols   39523

Howell Carnegie District Library, 314 W Grand River Ave, **Howell**, MI 48843
T: +1 517 5460720; Fax: +1 517 5461494; URL: www.howelllibrary.org
1875; Kathleen Zaenger
119 000 vols; 401 curr per   39524

Hubbard Public Library, 436 W Liberty St, **Hubbard**, OH 44425
T: +1 330 5343512; Fax: +1 330 5347836; URL: www.hubbard.lib.oh.us
1937
60 000 vols; 284 curr per
libr loan   39525

Hills Memorial Library, 18 Library St, **Hudson**, NH 03051-4244
T: +1 603 8866030; Fax: +1 603 5952850; URL: www.hillsml.lib.nh.us
1908; Mary P. Weller
Zylonis Lithuanian Heritage Coll
50 000 vols; 130 curr per   39526

Hudson Library & Historical Society, 96 Library St, **Hudson**, OH 44236-5122
T: +1 330 6536658; Fax: +1 330 6503373; URL: www.hudsonlibrary.org
1910; E. Leslie Polott
John Brown, Abolitionist Leader (Clarence S Gee) bks, holographs, pictures, clippings
99 000 vols; 681 curr per   39527

Hudson Public Library, Three Washington St at The Rotary, **Hudson**, MA 01749-2499
T: +1 978 5689644;
Fax: +1 978 5689646; URL: www.hudsonpubliclibrary.com
1868; Patricia Desmond
65 000 vols; 149 curr per; 2 000 av-mat
                                                        39528

Pasco County Library System, 8012 Library Rd, **Hudson**, FL 34667
T: +1 727 8343635; Fax: +1 727 8613025; URL: pascolibraries.org
1980; Linda Allen
7 branch libs
625 000 vols; 1 684 curr per; 18 000 e-books; 18 020 av-mat; 17 220 sound-rec
                                                        39529

Harris County Public Library – Atascocita, 19520 Pinehurst Trails Dr, **Humble**, TX 77346
T: +1 281 8122162; Fax: +1 281 8122135
Beth Krippel
71 000 vols   39530

Harris County Public Library – Baldwin Boettcher Branch, 22248 Aldine Westfield Rd, **Humble**, TX 77338
T: +1 281 8211320; Fax: +1 281 4438068
Elise Shell
50 000 vols   39531

Harris County Public Library- Octavia Fields Memorial, 1503 S Houston Ave, **Humble**, TX 77338
T: +1 281 4463377; Fax: +1 281 4464203
Fayth Brady
68 000 vols   39532

Public Library of Charlotte & Mecklenburg County – North County Regional, 16500 Holly Crest Lane, **Huntersville**, NC 28078
T: +1 704 8954020; E-mail: nocounty@plcmc.org; URL: www.plcmc.org/Locations/branches.asp?id=16
John Zika
94 000 vols   39533

Huntingdon County Library, 330 Penn St, **Huntingdon**, PA 16652-1487
T: +1 814 6430200; Fax: +1 814 6430132; E-mail: library@huntingdon.net; URL: www.huntingdon.net/library
1935; Nancy Holland
Pennsylvania Rm
74 000 vols; 95 curr per   39534

Huntingdon Valley Library, 625 Red Lion Rd, **Huntingdon Valley**, PA 19006-6297
T: +1 215 9475138; Fax: +1 215 9385894; URL: www.hvlibrary.org
1953; Jerry Szpila
90 000 vols; 115 curr per
PLA   39535

Cabell County Public Library, 455 Ninth Street Plaza, **Huntington**, WV 25701
T: +1 304 5285660; Fax: +1 304 5285739; E-mail: library@cabell.lib.wv.us; URL: www.cabell.lib.wv.us
1902; Judy K. Rule
8 branch libs
186 000 vols; 328 curr per   39536

Huntington City Township Public Library, 200 W Market St, **Huntington**, IN 46750-2655
T: +1 260 3560824; Fax: +1 260 3563073; URL: huntingtonpub.lib.in.us
1874; Kathryn Holst
Trains Coll
148 000 vols; 186 curr per; 3 000 av-mat
libr loan   39537

Huntington Beach Public Library System, Information & Cultural Resource Center, 7111 Talbert Ave, **Huntington Beach**, CA 92648
T: +1 714 8424481; Fax: +1 714 3755180; E-mail: library@hbpl.org; URL: www.hbpl.org
1909; Ronald Hayden
Careers, Business & management, Technology; Genealogy Coll; 4 branch libs
438 000 vols; 433 curr per; 4 558 av-mat
libr loan   39538

County of Los Angeles Public Library – Huntington Park Library, 6518 Miles Ave, **Huntington Park**, CA 90255-4388
T: +1 323 5831461; Fax: +1 323 5872061; URL: www.colapublib.org/libs/huntingtonpark
Norma Montero
American Indian Resource Center
126 000 vols; 137 curr per   39539

South Huntington Public Library, 145 Pidgeon Hill Rd, **Huntington Station**, NY 11746
T: +1 631 5494411; Fax: +1 631 5491266; URL: shpl.info
1961; Kenneth Weil
Education
253 000 vols; 300 curr per   39540

Huntley Area Public Library District, 11000 Ruth Rd, **Huntley**, IL 60142-7155
T: +1 847 6695386; Fax: +1 847 6695439; URL: www.huntleylibrary.org
1989; Virginia Maravilla
71 000 vols; 188 curr per   39541

Huntsville Public Library, 1216 14th St, **Huntsville**, TX 77340
T: +1 936 2915470; Fax: +1 936 2915418; URL: www.huntsville.lib.tx.us
1967; Linda Dobson
Adult Education, Genealogy, Ornithology
65 000 vols; 100 curr per   39542

Huntsville-Madison County Public Library, 915 Monroe St, **Huntsville**, AL 35801; P.O. Box 443, Huntsville, AL 35804-0443
T: +1 256 5325984; Fax: +1 256 5325997; URL: www.hpl.lib.al.us
1818; Laurel Best
Foreign language; Civil War & Southern Hist (Zeitler Room Coll), Heritage Room Coll; 12 branch libs
575 000 vols; 1 211 curr per; 11 000 e-books   39543

Margaret E. Heggan Free Public Library of the Township of Washington, 208 E Holly Ave, **Hurffville**, NJ 08080
T: +1 856 5893334; Fax: +1 856 5822042; URL: www.hegganlibrary.org
1965; Linda H. Snyder
72 000 vols; 145 curr per; 75 maps; 893 av-mat; 2 686 sound-rec; 12 digital data carriers
libr loan; ALA   39544

Huron Public Library, 521 Dakota Ave S, **Huron**, SD 57350
T: +1 605 3538530; Fax: +1 605 3538531; URL: hpllib.sdln.net
1907
86 700 vols; 159 curr per   39545

Putnam County Library, 4219 State Rte 34, **Hurricane**, WV 25526
T: +1 304 7577308; Fax: +1 304 7577384; E-mail: putnam@cabell.lib.wv.us; URL: putnam.lib.wv.us
1959; Jacquelin S. Chaney
93 000 vols; 101 curr per   39546

Hurst Public Library, 901 Precinct Line Rd, **Hurst**, TX 76053
T: +1 817 7887300; Fax: +1 817 5909515; URL: www.hurst.lib.tx.us
1959; Susan Andrews
128 000 vols; 392 curr per   39547

Hutchinson Public Library, 901 N Main, **Hutchinson**, KS 67501-4492
T: +1 620 6635441; Fax: +1 620 6631583; URL: www.hutchpl.org
Gregg Wamsley
289 000 vols; 396 curr per   39548

Hyannis Public Library Association, 401 Main St, **Hyannis**, MA 02601-3019
T: +1 508 7752280; Fax: +1 508 7900087; URL: www.hyannislibrary.org
1865; Ann-Louise Harries
60 000 vols; 175 curr per   39549

Prince George's County Memorial Library System, 6532 Adelphi Rd, **Hyattsville**, MD 20782-2098
T: +1 301 6993500; Fax: +1 301 9855494; URL: www.pgcmls.info
1946; N.N.
Horses & Horse Racing, American Blacks, Maryland Room, Planned Communities & Consumers' Cooperatives; 19 branch libs
233 000 vols; 559 curr per   39550

Boston Public Library – Hyde Park Branch, 35 Harvard Ave, **Hyde Park**, MA 02136-2862
T: +1 617 3612524; Fax: +1 617 8592489
Barbara S. Wicker
50 000 vols; 100 curr per   39551

Idaho Falls Public Library, 457 W Broadway, **Idaho Falls**, ID 83402
T: +1 208 6128334; Fax: +1 208 6128467; E-mail: rwright@ifpl.org; URL: www.ifpl.org
1909; Robert Wright
Vardis Fisher Coll
222 000 vols; 225 curr per; 300 e-books; 4 643 av-mat; 1 540 Large Print Bks, 4 650 Talking Bks   39552

Ilion Free Public Library, 78 West St, **Ilion**, NY 13357
T: +1 315 8945028; Fax: +1 315 8949980; URL: www.midyork.org/Ilion
1893; Thomasine Jennings
Ilion, New York (Seamans Coll), photos, slides
50 000 vols; 98 curr per   39553

Cuyahoga County Public Library – Independence Branch, 6361 Selig Dr, **Independence**, OH 44131-4926
T: +1 216 4470160; Fax: +1 216 4471371
Valerie Kocin
50 000 vols; 3 715 av-mat   39554

Inyo County Free Library, 168 N Edwards St, **Independence**, CA 93526; P.O. Drawer K, Independence, CA 93526-0610
T: +1 760 8780359; Fax: +1 760 8780360; E-mail: inyocolib@qnet.com
1913; Nancy Masters
Mining; Mary Hunter Austen Coll; 6 branch libs
102 000 vols; 129 curr per
libr loan   39555

The Kansas City Public Library – Trails West, 11401 E 23rd St, *Independence*, MO 64052
T: +1 816 7013483; Fax: +1 816 7013493
Ritchie Momon
78 000 vols   39556

Kenton County Public Library – Independence, 6477 Taylor Mill Rd, *Independence*, KY 41051
T: +1 859 9624030; Fax: +1 859 9624037
Anita Carroll
70 000 vols; 100 curr per   39557

Mid-Continent Public Library, 15616 E US Hwy 24, *Independence*, MO 64050
T: +1 816 8365200; Fax: +1 816 5217253; E-mail: info@mcpl.lib.mo.us;
URL: www.mcpl.lib.mo.us
1965; Richard J. Wilding
U.S. Census on microfilm 1790-1920 all states, Handicapped Coll; 30 branch libs
3 106 000 vols; 8 283 curr per; 1 328 maps; 2 806 000 microforms; 195 820 av-mat; 275 468 sound-rec; 36 330 digital data carriers; 7 000 High Interest/Low Vocabulary Bks, 129 800 Large Print Bks
libr loan; MLA, ALA   39558

Mid-Continent Public Library – Midwest Genealogy Center, 3440 S Lee Summit Rd, *Independence*, MO 64055-1923
T: +1 816 2527228; Fax: +1 816 2547114
Janice Schultz
62 000 vols; 600 curr per   39559

Indiana Free Library, Inc, 845 Philadelphia St, *Indiana*, PA 15701-3908
T: +1 724 4658841; Fax: +1 724 4659902; E-mail: indpub@arin.k12.pa.us;
URL: www.indianafreelibrary.org
Kate Geiger
72 000 vols; 123 curr per   39560

Indianapolis-Marion County Public Library, 2450 N Meridian St, *Indianapolis*, IN 46208; P.O. Box 211, Indianapolis, IN 46206-0211
T: +1 317 2754910; Fax: +1 317 2294510; URL: www.imcpl.org
1873; Laura Bramble
Foundation Coll, Old Cookbks, Early Children's Lit; 22 branch libs
158 000 vols; 3 231 curr per; 27 500 music scores; 107 980 av-mat; 180 666 sound-rec; 5 442 pamphlets
libr loan; ALA   39561

Indianapolis-Marion County Public Library – Decatur, 5301 Kentucky Ave, *Indianapolis*, IN 46221-6540
T: +1 317 2754330
Tia Jah Wynne Ayers
54 000 vols   39562

Indianapolis-Marion County Public Library – Eagle, 3325 Lowry Rd, *Indianapolis*, IN 46222-1240
T: +1 317 2754340
Mary Agnes Hylton
66 000 vols   39563

Indianapolis-Marion County Public Library – Franklin Road, 5500 S Franklin Rd, *Indianapolis*, IN 46239
T: +1 317 2754380
Jill Wetnight
80 000 vols   39564

Indianapolis-Marion County Public Library – Glendale, Glendale Mall, Upper Level South, 6101 N Keystone Ave, *Indianapolis*, IN 46220
T: +1 317 2754410
Joyce Karnes
126 000 vols   39565

Indianapolis-Marion County Public Library – Irvington, 5625 E Washington St, *Indianapolis*, IN 46219-6411
T: +1 317 2754450
Sue Kennedy
73 000 vols   39566

Indianapolis-Marion County Public Library – Lawrence, 7898 N Hague Rd, *Indianapolis*, IN 46256-1754
T: +1 317 2754460
Betsy Crawford
119 000 vols   39567

Indianapolis-Marion County Public Library – Nora, 8625 Guilford Ave, *Indianapolis*, IN 46240-1835
T: +1 317 2754470
Sharon Bernhardt
112 000 vols   39568

Indianapolis-Marion County Public Library – Pike, 6525 Zionsville Rd, *Indianapolis*, IN 46268-2352
T: +1 317 2754480
Carol Schlake
108 000 vols   39569

Indianapolis-Marion County Public Library – Southport, 2630 E Stop 11 Rd, *Indianapolis*, IN 46227-8899
T: +1 317 2754510
Mike Williams
116 000 vols   39570

Indianapolis-Marion County Public Library – Warren, 9701 E 21st St, *Indianapolis*, IN 46229-1707
T: +1 317 2754550
Ruth Hans
95 000 vols   39571

Indianapolis-Marion County Public Library – Wayne, 198 S Girls School Rd, *Indianapolis*, IN 46231-1120
T: +1 317 2754530
Sharon Smith
88 000 vols   39572

Sunflower County Library System, 201 Cypress Dr, *Indianola*, MS 38751-2499
T: +1 662 8872298; Fax: +1 662 8871618
1938; Alice Shands
6 branch libs
94 000 vols; 147 curr per
libr loan   39573

Inglewood Public Library, 101 W Manchester Blvd, *Inglewood*, CA 90301-1771
T: +1 310 4125620; Fax: +1 310 4128848; E-mail: publiclibrary@cityofinglewood.org; URL: www.inglewoodlibrary.org
1962; Richard Joseph Siminski
African American studies; Hispanic Services, Large Print Bks Coll, Spanish Bks; 2 branch libs
470 000 vols; 1 590 curr per; 200 govt docs; 100 maps; 41 000 microforms; 3 604 av-mat; 12 107 sound-rec; 10 digital data carriers; 146 art repros
libr loan   39574

Southeast Kansas Library System, 218 E Madison Ave, *Iola*, KS 66749
T: +1 620 3655136; Fax: +1 620 3655137; URL: www.sekls.lib.ks.us
1966; Roger L. Carswell
Genealogy; Art Prints Coll, Kansas Census
68 000 vols; 20 curr per   39575

Iowa City Public Library, 123 S Linn St, *Iowa City*, IA 52240
T: +1 319 3565200; Fax: +1 319 3565494; E-mail: icpl@icpl.org; URL: www.icpl.org
1896; Susan Craig
195 000 vols; 546 curr per; 53 Electronic Media & Resources   39576

Ipswich Public Library, 25 N Main St, *Ipswich*, MA 01938-2287
T: +1 978 3566648; Fax: +1 978 3566647; E-mail: ipswich@mvlc.org; URL: www.town.ipswich.ma.us/library
1868; Victor Dyer
Historical Coll
88 000 vols; 136 curr per; 237 microforms; 1 334 av-mat; 2 145 sound-rec; 33 digital data carriers
libr loan; ALA   39577

Dickinson County Library, 401 Iron Mountain St, *Iron Mountain*, MI 49801-3435
T: +1 906 7741218; Fax: +1 906 7744079; E-mail: dcl@dcl-lib.org; URL: www.dcl-lib.org
1902
Education & Health (Claire A Lilja Memorial), Wood Industry (Abbott Fox Memorial)
86 673 vols
libr loan   39578

West Iron District Library, 116 W Genesee St, *Iron River*, MI 49935-1437; P.O. Box 328, Iron River, MI 49935-0328
T: +1 906 2652831; Fax: +1 906 2652062; E-mail: lbbartel@uproc.lib.mi.us; URL: rpa.uproc.lib.mi.us/westiron.htm; www.uproc.lib.mi.us/widl
1967; Barbara Bartel
Large Print Bks
980 000 vols; 90 curr per   39579

Briggs Lawrence County Public Library, Ironton Library, 321 S Fourth St, *Ironton*, OH 45638
T: +1 740 5321124; Fax: +1 740 5324948; E-mail: irontonbranch@briggslibrary.org; URL: www.briggslibrary.com
1881; Joseph Jenkins
4 branch libs
141 000 vols; 293 curr per
libr loan   39580

Ozark Regional Library, 402 N Main St, *Ironton*, MO 63650
T: +1 573 5462615; Fax: +1 573 5467225
1947; John F. Mertens
Eastern US genealogy; 8 branch libs
145 000 vols; 250 curr per; 880 microforms; 1 340 av-mat; 7 578 sound-rec; 201 digital data carriers
libr loan; ALA   39581

Orange County Public Library – Heritage Park Regional Library, 14361 Yale Ave, *Irvine*, CA 92604-1901
T: +1 949 9364040; Fax: +1 949 5519283
Barbara Brook
133 000 vols; 350 curr per   39582

Orange County Public Library – University Park, 4512 Sandburg Way, *Irvine*, CA 92612-2794
T: +1 949 7864001; Fax: +1 949 8571029
Joy Johnson
91 000 vols; 115 curr per   39583

Irving Public Library, 801 W Irving Blvd, *Irving*, TX 75015; P.O. Box 152288, Irving
T: +1 972 7212440; Fax: +1 972 7214771; E-mail: reference@cityofirving.org; URL: www.irvinglibrary.org
1961; Patricia Landers
Business & management; 4 branch libs, 1 bookmobile
569 000 vols; 1 088 curr per; 50 661 av-mat; 31 860 sound-rec; 9 432 digital data carriers
libr loan   39584

Irving Public Library – Southwest, 2216 W Shady Grove, *Irving*, TX 75060; P.O. Box 152288, Irving, TX 75060-2288
T: +1 972 7212546; Fax: +1 972 7213638
Carol Danielson
58 000 vols   39585

Irving Public Library – Valley Ranch, 401 Cimmaron Trail, *Irving*, TX 75063-4680; P.O. Box 152288, Irving, TX 75015-2288
T: +1 972 7214669; Fax: +1 972 8310672
Patty Mount
72 000 vols   39586

Irvington Public Library, Civic Sq, *Irvington*, NJ 07111-2498
T: +1 973 3726400; Fax: +1 973 3726860; URL: www.irvingtonpubliclibrary.org
1914; Joan Whittaker
Adult Literacy
200 000 vols; 160 curr per; 500 av-mat; 500 Audio Bks, 500 Large Print Bks   39587

Irvington Public Library, Guiteau Foundation Library, 12 S Astor St, *Irvington*, NY 10533
T: +1 914 5917840; Fax: +1 914 5910347; URL: www.irvingtonlibrary.org
1866; Pamela Strachan
52 000 vols; 85 curr per   39588

Norwin Public Library Association Inc, 100 Caruthers Ln, *Irwin*, PA 15642
T: +1 724 8634700; Fax: +1 724 8636195; E-mail: norwinpl@nb.net; URL: norwinlibrary.nb.net; www.norwinpubliclibrary.org
1937; Falk Diana
61 000 vols; 55 curr per   39589

Ishpeming Carnegie Public Library, 317 N Main St, *Ishpeming*, MI 49849-1994
T: +1 906 4864381; Fax: +1 906 4866226; URL: www.uproc.lib.mi.us/ish
Cindy Kariniemi
72 000 vols; 78 curr per; 270 av-mat   39590

Island Park Public Library, 176 Long Beach Rd, *Island Park*, NY 11558
T: +1 516 4320122; Fax: +1 516 8893584; E-mail: ilandpk@lilrc.org; URL: www.nassaulibrary.org/islandp
1938; Ronnie Swett
53 000 vols; 126 curr per   39591

Island Trees Public Library, 38 Farmedge Rd, *Island Trees*, NY 11756-5200
T: +1 516 7312211; Fax: +1 516 7312398; E-mail: islandtreespubliclibrary@yahoo.com; URL: www.islandtreespubliclibrary.org
1967; Jessica Koenig
Child Psychology (Carol Cass Memorial Coll)
62 000 vols; 180 curr per   39592

Islip Public Library, 71 Monell Ave, *Islip*, NY 11751-3999
T: +1 631 5815933; Fax: +1 631 2778429; E-mail: islplib@suffolk.lib.ny.us; URL: www.isliplibrary.org
1924; Mary Schubart
170 000 vols; 1 389 curr per; 920 Large Print Bks
libr loan; ALA/PLA   39593

King County Rural Library District – King County Library System, 960 Newport Way NW, *Issaquah*, WA 98027
T: +1 425 3693200; Fax: +1 425 3693255; URL: www.kcls.org
1943; Bill Ptacek
45 branch libs
917 000 vols; 11 000 curr per; 207 011 av-mat; 200 300 sound-rec; 1 904 digital data carriers
libr loan   39594

Itasca Community Library, 500 W Irving Park Rd, *Itasca*, IL 60143
T: +1 630 7731699; Fax: +1 630 7731707; E-mail: itascal@linc.lib.il.us; URL: www.itasca.lib.il.us
1957; Elizabeth Adamowski
69 000 vols; 191 curr per   39595

Finger Lakes Library System, 119 E Green St, *Ithaca*, NY 14850
T: +1 607 2734074; Fax: +1 607 2733618; E-mail: kiraci@flls.org; URL: www.flls.org
1958; Kimberly A. Iraci
94 510 vols; 46 curr per
libr loan   39596

Tompkins County Public Library, 101 E Green St, *Ithaca*, NY 14850-5613
T: +1 607 2724557; Fax: +1 607 2728111; E-mail: reference@tcpl.org; URL: www.tcpl.org
1864; Janet E. Steiner
Central Bk Aid Coll (Finger Lakes Libr System)
258 000 vols; 334 curr per   39597

Amador County Library, 530 Sutter St, *Jackson*, CA 95642
T: +1 209 2236400; E-mail: library@co.amador.ca.us; URL: www.co.amador.ca.us/depts/library
Laura Einstadter
Amador County Hist, Bancrofts Works & Hist of Calif, Index of Amador County Newspapers on microfilm, Mines & Mineral Res Coll; 6 branch libs
88 000 vols; 71 curr per   39598

Breathitt County Public Library, 1024 College Ave, *Jackson*, KY 41339
T: +1 606 6665541; Fax: +1 606 6668166; E-mail: breathitt@bellsouth.net; URL: www.breathittcountylibrary.com
1967; Stephen Bowling
Genealogical Res Libr
52 000 vols; 50 curr per   39599

Jackson/Hinds Library System, Eudora Welty Library, 300 N State St, *Jackson*, MS 39201-1799
T: +1 601 9685825; Fax: +1 601 9685806; E-mail: welty@jhlibrary.com; URL: www.jhlibrary.com
1986; Carolyn McCallum
Mississippi Writers Coll; 14 branch libs
604 000 vols; 941 curr per   39600

Jackson-Madison County Library, 433 E Lafayette St, *Jackson*, TN 38301-6386
T: +1 731 4258600; Fax: +1 731 4258609; URL: www.jmcl.tn.us
1903; Richard Salmons
Genealogy, bks, micro; Jackson Area Business Hist Coll; Local & State Hist (Tennessee Rm Coll), bks, micro
116 000 vols; 211 curr per   39601

Mississippi Library Commission, 3881 Eastwood Dr, *Jackson*, MS 39211
T: +1 601 4324153; Fax: +1 601 4324478; E-mail: mlcref@mlc.lib.ms.us; URL: www.mlc.lib.ms.us
1926; Sharman Smith
Large print, Libr science, Mississippiana; Fed, US Patent & Trademark
70 000 vols; 107 curr per; 641 e-books; 80 101 av-mat
libr loan   39602

Ocean County Library – Jackson Township, Two Jackson Dr, *Jackson*, NJ 08527-3601
T: +1 732 9284400; Fax: +1 732 8330615
John Glace
120 000 vols   39603

Riverside Regional Library, 204 S Union St, *Jackson*, MO 63755-1949; P.O. Box 389, Jackson, MO 63755-0389
T: +1 573 2438141; Fax: +1 573 2438142; URL: www.riversideregionallibrary.org
1955; Nancy Howland
Large Print Coll; 5 branch libs
148 000 vols; 135 curr per; 11 450 av-mat; 2 080 sound-rec
libr loan   39604

Shiloh Regional Library, 573 Old Hickory Blvd, *Jackson*, TN 38305-2901
T: +1 731 6680710; Fax: +1 731 6686663
1956; Carla Jacobs
116 000 vols   39605

Teton County Library, 125 Virginian Lane, P.O. Box 1629, *Jackson*, WY 83001-1629
T: +1 307 7332164; Fax: +1 307 7334568; E-mail: tetnref@tclib.org; URL: tclib.org
1940; Deb Adams
61 930 vols; 252 curr per   39606

Queens Borough Public Library – Jackson Heights Branch, 35-51 81st St, *Jackson Heights*, NY 11372
T: +1 718 8992500; Fax: +1 718 8997003
Harriet Benjamin
150 000 vols   39607

Central Arkansas Library System – Esther Nixon Library, 308 W Main St, *Jacksonville*, AR 72076-4507
T: +1 501 9825533
Kathy Seymour
62 000 vols   39608

Jacksonville Public Library, 502 S Jackson, *Jacksonville*, TX 75766
T: +1 903 5867664; Fax: +1 903 5863397; E-mail: director@jacksonvillelibrary.com; URL: www.jacksonvillelibrary.com
1913; Barbara Crossman
58 000 vols; 105 curr per   39609

Jacksonville Public Library, 201 W College Ave, *Jacksonville*, IL 62650-2497
T: +1 217 2435435; Fax: +1 217 2432182; E-mail: japl@csj.net; URL: japl.lib.il.us/library
1889; Sharon R. Zuiderveld
91 000 vols; 115 curr per; 4 112 av-mat
libr loan   39610

Jacksonville Public Library, 200 Pelham Rd S, *Jacksonville*, AL 36265
T: +1 256 4356332; Fax: +1 256 4354459; E-mail: jplkids@hotmail.com; URL: www.jacksonvillepubliclibrary.org
1957; Barbara Rowell
Jacksonville Hist Museum Coll, John Francis Papers (Civil War Roster & Letters), Col John Pelham Papers
52 000 vols; 33 curr per; 611 av-mat   39611

Jacksonville Public Library – Bradham-Brooks Northwest Branch, 1755 Edgewood Ave W, *Jacksonville*, FL 32208-7206
T: +1 904 7655402; Fax: +1 904 7687609
Pat Doyle
105 000 vols   39612

Jacksonville Public Library – Highlands Regional, 1826 Dunn Ave, *Jacksonville*, FL 32218-4712
T: +1 904 7577702; Fax: +1 904 6964328
Donna Thomas
157 000 vols   39613

Jacksonville Public Library – Laura Street, 303 N Laura St, *Jacksonville*, FL 32202
T: +1 904 6302793; Fax: +1 904 6302734; E-mail: libdir@coj.net; URL: jaxpubliclibrary.org
1903; Barbara A. B. Gubbin
Frederick Delius Memorial Fund, African-American Coll, Holocaust Coll; 21 branch libs
2 629 000 vols; 855 curr per; 28 000 e-books; 238 534 av-mat; 339 534 sound-rec
libr loan   39614

Jacksonville Public Library – Mandarin, 3330 Kori Rd, *Jacksonville*, FL 32257-5454
T: +1 904 2625201; Fax: +1 904 2921029
Michael Sullivan
170 000 vols   39615

Jacksonville Public Library – Old Middleburg Road, 7973 Old Middleburg Rd S, *Jacksonville*, FL 32222
T: +1 904 5733164; Fax: +1 904 5733162
Lynne Baldwin
61 000 vols   39616

Jacksonville Public Library – Pablo Creek, 13295 Beach Blvd, *Jacksonville*, FL 32246
T: +1 904 9927101; Fax: +1 904 9923987
Carol Bailey
161 000 vols   39617

Jacksonville Public Library – Regency Square, 9900 Regency Square Blvd, *Jacksonville*, FL 32225-6539
T: +1 904 7265142; Fax: +1 904 7265153
Marshelle Berry
153 000 vols   39618

Jacksonville Public Library – San Marco, 1513 LaSalle St, *Jacksonville*, FL 32207-3107
T: +1 904 8582907; Fax: +1 904 3062182
Erica Brown
70 000 vols   39619

Jacksonville Public Library – South Mandarin, 12125 San Jose Blvd, *Jacksonville*, FL 32223
T: +1 904 2886385; Fax: +1 904 2886399
Ed Murray
103 000 vols   39620

Jacksonville Public Library – Southeast Regional, 10599 Deerwood Park Blvd, *Jacksonville*, FL 32256
T: +1 904 9960325; Fax: +1 904 9960340
Carole Schwartz
217 000 vols   39621

Jacksonville Public Library – University Park, 3435 University Blvd N, *Jacksonville*, FL 32277
T: +1 904 7442265; Fax: +1 904 7446892
Michael Rouse
79 000 vols   39622

Jacksonville Public Library – Webb Wesconnett Regional, 6887 103rd St, *Jacksonville*, FL 32210-6897
T: +1 904 7787305; Fax: +1 904 7772262
Theresa Barmer
147 000 vols   39623

Jacksonville Public Library – West Regional, 1425 Chaffee Rd S, *Jacksonville*, FL 32221-1119
T: +1 904 6931448
Jane Harris
185 000 vols   39624

Onslow County Public Library, 58 Doris Ave E, *Jacksonville*, NC 28540
T: +1 910 4557350; Fax: +1 910 4551661; URL: www.co.onslow.nc.us/library
1936; Philip Cherry
4 branch libs
75 000 vols; 131 curr per; 1 347 av-mat; 2 500 Large Print Bks, 1 350 Talking Bks
libr loan   39625

Queens Borough Public Library, Queens Library, 89-11 Merrick Blvd, *Jamaica*, NY 11432
T: +1 718 9900778; Fax: +1 718 2918936; URL: www.queenslibrary.org
1896; Thomas Galante
Long Island Hist Coll; 70 branch libs
5 677 000 vols; 8 032 curr per; 129 700 av-mat; 148 840 sound-rec
libr loan   39626

Queens Borough Public Library – Business, Science & Technology Division, 89-11 Merrick Blvd, *Jamaica*, NY 11432
T: +1 718 9900760; Fax: +1 718 6588342
Nelson Lu
Sam's PhotoFacts;Automobile Shop Manuals;Consumer Health Coll;Small Business Center
200 000 vols; 582 curr per   39627

Queens Borough Public Library – Fine Arts & Recreation Division, 89-11 Merrick Blvd, *Jamaica*, NY 11432
T: +1 718 9900819; Fax: +1 718 6588342
Esther Lee
150 000 vols; 380 curr per; 14 800 av-mat   39628

Queens Borough Public Library – Literature & Languages Division, 89-11 Merrick Blvd, *Jamaica*, NY 1432
T: +1 718 9900851; Fax: +1 718 6588342
Rachel Donner
Black Authors Fiction Coll;African-American Literature Coll
364 000 vols; 225 curr per   39629

Queens Borough Public Library – Social Sciences Division, 89-11 Merrick Blvd, *Jamaica*, NY 11432
T: +1 718 9900762
Monica Rhodd
New York City Commission publications (Includes Community District Needs of all boroughs);Standardized Tests (Civil Service Employment);New York City Civil Service Test;New York Times Microfilm, (1851 to present);Road Maps (US Cities & World Cities);New Y
329 000 vols; 485 curr per   39630

Queens Borough Public Library – Youth Services Division, 89-11 Merrick Blvd, *Jamaica*, NY 11432
T: +1 718 9900767
Lynn T. Gonen
African American Culture, History & Biography;Augusta Baker Reference Coll;Hispanic Heritage Coll
218 000 vols; 75 curr per   39631

Alfred Dickey Public Library, 105 SE Third St, *Jamestown*, ND 58401
T: +1 701 2522990; Fax: +1 701 2526030; E-mail: adpl@daktel.com; URL: www.adpl.org
Daphne Drewello
Louis L'Amour Memorial Coll
51 000 vols; 119 curr per   39632

Chautauqua-Cattaraugus Library System, 106 W Fifth St, *Jamestown*, NY 14702-0730
T: +1 716 4847135; Fax: +1 716 4836880; E-mail: cway@cclslib.org; URL: www.cclslib.org
1960; Catherine A. Way
148 000 vols; 16 curr per; 20 digital data carriers   39633

James Prendergast Library Association, 509 Cherry St, *Jamestown*, NY 14701
T: +1 716 4847135; Fax: +1 716 4871148; URL: www.prendergastlibrary.org
1880; Catherine A. Way
Art Gallery
380 000 vols; 293 curr per   39634

Hedberg Public Library, 316 S Main St, *Janesville*, WI 53545
T: +1 608 7586581; Fax: +1 608 7586583; E-mail: referencedesk@hedbergpubliclibrary.org; URL: www.hedbergpubliclibrary.org
1884; Bryan J. McCormick
Janesville Rm
228 000 vols; 579 curr per
libr loan   39635

Carl Elliott Regional Library System, 98 E 18th St, *Jasper*, AL 35501
T: +1 205 2212568; E-mail: ill_cerls@hotmail.com
1957; Sandra Underwood
Lit (Musgrove Coll)
98 000 vols; 132 curr per   39636

Jasper Public Library, 98 18th Street East, *Jasper*, AL 35501
T: +1 205 2218512; URL: www.jaspercity.com/boards/library/index.htm
Colleen Miller
60 000 vols; 124 curr per   39637

Jasper-Dubois County Contractual Public Library, 1116 Main St, *Jasper*, IN 47546-2899
T: +1 812 4822712; Fax: +1 812 4827123; URL: www.jdcpl.lib.in.us
1934; Rita Douthitt
Large type print, Hist, Indiana; 1 branch libr
95 000 vols; 220 curr per; 3 800 av-mat   39638

Henderson Memorial Public Library Association, 54 E Jefferson St, *Jefferson*, OH 44047-1198
T: +1 440 5763761; Fax: +1 440 5768402; URL: www.henderson.lib.oh.us
1883; Kathleen L. Jozwiak
Local Hist (Jefferson Gazette & Ashtabula Company Genealogical Coll), microfilm
55 000 vols   39639

Missouri River Regional Library, 214 Adams St, *Jefferson City*, MO 65101-3244; P.O. Box 89, Jefferson City, MO 65102-0089
T: +1 573 6342464; Fax: +1 573 6347028; URL: www.mrrl.org
1994; Bill Rodgers
1 branch libr, 1 bookmobile
191 000 vols; 500 curr per; 6 014 Audio Bks, 6 252 CDs   39640

Jeffersonville Township Public Library, 211 E Court Ave, *Jeffersonville*, IN 47130; P.O. Box 1548, Jeffersonville, IN 47131-1548
T: +1 812 2855631; Fax: +1 812 2821264; URL: www.jefferson.lib.in.us
1900; Libby Pollard
1 branch libr
178 000 vols; 287 curr per; 12 086 av-mat; 7 980 sound-rec; 1 590 digital data carriers
libr loan   39641

LaSalle Parish Library, 3108 N First St, *Jena*, LA 71342-3199
T: +1 318 9925675; Fax: +1 318 9927374; E-mail: admin.h1ls@state.lib.la.us; URL: www.lasalle.lib.la.us
1952; Andrea Book
1 branch libr
53 440 vols; 150 curr per
libr loan   39642

Georgetown Charter Township Library, 1525 Baldwin St, *Jenison*, MI 49428
T: +1 616 4579620; Fax: +1 616 4573666; E-mail: jen@lakeland.lib.mi.us; URL: www.georgetown-mi.gov/library
1965; Pamela A. Myers
98 000 vols; 313 curr per; 45 microforms; 5 472 av-mat; 4 981 sound-rec; 254 digital data carriers
libr loan 39643

Jefferson Davis Parish Library, 118 W Plaquemine St, *Jennings*, LA 70546-5856
T: +1 337 8241210; Fax: +1 337 8245444; E-mail: admin.b1jd@pelican.state.lib.la.us; URL: www.jefferson-davis.lib.la.us
1968; Linda LeBert
Louisiana, Indians; 4 branch libs
101 000 vols; 254 curr per; 2 317 av-mat 39644

Jennings Carnegie Public Library, 303 Cary Ave, *Jennings*, LA 70546-5223
T: +1 337 8215517; Fax: +1 337 8215527; E-mail: jcpl303@yahoo.com
1885; Harriet Shultz
55 000 vols; 86 curr per; 156 av-mat 39645

Jericho Public Library, One Merry Lane, *Jericho*, NY 11753
T: +1 516 9356790; Fax: +1 516 4339581; E-mail: info@jericholibrary.org; URL: www.jericholibrary.org
1964; John Bosco
173 000 vols; 363 curr per 39646

Jersey City Free Public Library, 472 Jersey Ave, *Jersey City*, NJ 07302-3499
T: +1 201 5474500; Fax: +1 201 5474584; URL: www.jclibrary.org
1889; Priscilla Gardner
New Jersey Coll; 12 branch libs, 1 bookmobile
463 000 vols; 831 curr per 39647

Jerseyville Public Library, 105 N Liberty St, *Jerseyville*, IL 62052-1512
T: +1 618 4989514; Fax: +1 618 4983036; E-mail: jee@lcls.org; URL: jerseyvillelibrary.org
1894; Anita Driver
60 000 vols; 144 curr per; 806 av-mat 39648

Grant County Library, 507 S Canyon Blvd, *John Day*, OR 97845-1050
T: +1 541 5751992; E-mail: grant047@centurytel.net
Melody Jackson
60 000 vols; 24 curr per 39649

Johnsburg Public Library District, 3000 N Johnsburg Rd, *Johnsburg*, IL 60050
T: +1 815 3440077; Fax: +1 815 3443524; URL: www.johnsburglibrary.org
1982; Maria Zawacki
Homeschool Resource Ctr
54 000 vols; 108 curr per 39650

Johnson City Public Library, 100 W Millard St, *Johnson City*, TN 37604
T: +1 423 4344450; Fax: +1 423 4344469; E-mail: info@jcpl.net; URL: www.jcpl.net
1895; Nelson Worley
134 000 vols; 395 curr per; 1 299 av-mat; 2 258 sound-rec; 7 digital data carriers; 136 art reproductions
libr loan 39651

Your Home Public Library, Johnson City Library, 107 Main St, *Johnson City*, NY 13790
T: +1 607 7974816; Fax: +1 607 7988895
1917; Steven J. Bachman
54 000 vols; 54 curr per 39652

Johnston Public Library, 6700 Merle Hay Rd, *Johnston*, IA 50131-0327; P.O. Box 327, Johnston, IA 50131-0327
T: +1 515 2784975; Fax: +1 515 2785233; E-mail: info@johnstonlibrary.com; URL: www.johnstonlibrary.com
1988; Willona Goers
56 000 vols; 188 curr per 39653

Marian J. Mohr Memorial Library, One Memorial Ave, *Johnston*, RI 02919-3221
T: +1 401 2314980; Fax: +1 401 2314984; E-mail: mohr@lori.state.ri.us; URL: web.provlib.org/johlib
1961; Jon R. Anderson
80 000 vols; 137 curr per 39654

Cambria County Library System & District Center, 248 Main St, *Johnstown*, PA 15901
T: +1 814 5365131; Fax: +1 814 5354140; E-mail: campub@cclib.lib.pa.us; URL: cclib.lib.pa.us
1870; Lyn Meek
Pennsylvania Room, vertical file for local hist, govt docs, Home Schoolers Resource Ctr; Childrens Preschool Libr
146 000 vols; 176 curr per; 37 490 govt docs; 500 maps; 18 000 microforms; 2 633 av-mat; 1 140 sound-rec; 90 digital data carriers; 2 633 audio bks
libr loan; ALA, PLA 39655

Clayton County Library System, 865 Battlecreek Rd, *Jonesboro*, GA 30236
T: +1 770 4733850; Fax: +1 770 4733858; E-mail: branchhq@claytonpl.org; URL: www.claytonpl.org
1941; Carol Johnson Stewart
4 branch libs
439 000 vols; 184 curr per; 7 200 Audio Bks 39656

Clayton County Library System – Jonesboro Branch, 124 Smith St, *Jonesboro*, GA 30236
T: +1 770 4787120; Fax: +1 770 4733846
Martha Caldwell
61 000 vols; 64 curr per 39657

Craighead County Jonesboro Public Library, 315 W Oak Ave, *Jonesboro*, AR 72401-3513
T: +1 870 9355133; Fax: +1 870 9357987; E-mail: reference@libraryinjonesboro.org; URL: www.libraryinjonesboro.org
1917; Phyllis G. Burkett
Genealogy; Arkansas Hist Coll; 3 branch libs
99 000 vols; 420 curr per; 3812 av-mat; 2 736 sound-rec; 643 digital data carriers; 3 709 Large Print Bks 39658

Crowley Ridge Regional Library, 315 W Oak Ave, *Jonesboro*, AR 72401
T: +1 870 9355133; Fax: +1 870 9357987; URL: www.libraryinjonesboro.org
1966; Phyllis Burkett
186 000 vols; 416 curr per; 59 Electronic Media & Resources
libr loan 39659

Jackson Parish Library, 614 S Polk Ave, *Jonesboro*, LA 71251-3442
T: +1 318 2595697; Fax: +1 318 2598984; E-mail: admin.t1ja@pelican.state.lib.la.us; URL: www.jacksonparishlibrary.org; www.jackson.lib.la.us
1960; Robin Toms
Jennifer Blake Coll
63 000 vols; 162 curr per 39660

Washington County – Jonesborough Library, 200 Sabin Dr, *Jonesborough*, TN 37659-1306
T: +1 423 7531800; Fax: +1 423 7531802; URL: www.wrlibrary.org/Libraries/washco.htm
1896; Patricia H. Beard
74 000 vols; 72 curr per; 773 e-books 39661

Joplin Public Library, 300 S Main, *Joplin*, MO 64801
T: +1 417 6245465; Fax: +1 417 6254728; URL: www.joplinpubliclibrary.org
1902; Susan Wray
Fine & Decorative Arts (Winfred L & Elizabeth C Post Memorial Art Reference Libr), Genealogy Coll
115 000 vols; 253 curr per; 320 sound-rec 39662

Dorothy Bramlage Public Library, 230 W Seventh, *Junction City*, KS 66441-3097
T: +1 785 2384311; Fax: +1 785 2387873; E-mail: jclibrary@jclib.org; URL: www.jclib.org
1907; Susan Moyer
65 000 vols; 170 curr per 39663

Juneau Public Libraries, 292 Marine Way, *Juneau*, AK 99801
T: +1 907 5865324; Fax: +1 907 5863419; URL: www.juneau.org/library
1913; Barbara Berg
Alaska; 2 branch libs
119 000 vols; 364 curr per 39664

Northeast Missouri Library Service, 207 W Chestnut, *Kahoka*, MO 63445
T: +1 660 7272327; Fax: +1 660 7272327
1961; Cathy James
Genealogy; Large Print Mat; 4 branch libs
119 000 vols; 49 curr per
libr loan 39665

Northeast Missouri Library Service – H. E. Sever Memorial, 207 W Chestnut, *Kahoka*, MO 63445
T: +1 660 7273262; Fax: +1 660 7271055
Brenda Spriggs
Large Print Materials;Four County (Knox, Lewis, Clark, Schuyler) Histories & Genealogy Coll, micro
59 000 vols; 112 curr per 39666

Hawaii State Public Library System – Kahului Public Library, 90 School St, *Kahului*, HI 96732-1627
T: +1 808 8733097; Fax: +1 808 8733094
Sana Daliva
99 000 vols; 216 curr per 39667

Hawaii State Public Library System – Kailua Public Library, 239 Kuulei Rd, *Kailua*, HI 96734
T: +1 808 2669911; Fax: +1 808 2669915
Patti Meerians
80 000 vols; 114 curr per 39668

Hawaii State Public Library System – Kailua-Kona Public Library, 75-138 Hualalai Rd, *Kailua-Kona*, HI 96740-1704
T: +1 808 3274327; Fax: +1 808 3274326
Irene Horvath
54 000 vols; 65 curr per 39669

Kalamazoo Public Library, 315 S Rose St, *Kalamazoo*, MI 49007-5264
T: +1 269 5537806; Fax: +1 269 5537999; URL: www.kpl.gov
1872; Ann Rohrbaugh
Hist, Martin Luther King Coll; 7 branch libs
410 000 vols; 875 curr per; 655 e-books
libr loan 39670

Kalamazoo Public Library – Oshtemo, 7265 W Main St, *Kalamazoo*, MI 49009
T: +1 269 5537980; Fax: +1 269 3756610
Martha Lohrstorfer
71 000 vols 39671

Flathead County Library, 247 First Ave E, *Kalispell*, MT 59901
T: +1 406 7585819; Fax: +1 406 7585868; URL: www.flatheadcountylibrary.org
1943; Kim Crowley
Montana Author & Subject Coll; 4 branch libs
207 000 vols; 231 curr per; 3 000 e-books 39672

Hawaii State Public Library System – Kaneohe Public Library, 45-829 Kamehameha Hwy, *Kaneohe*, HI 96744
T: +1 808 2335676; Fax: +1 808 2335672
Tom Churma
114 000 vols; 153 curr per 39673

Kankakee Public Library, 201 E Merchant St, *Kankakee*, IL 60901
T: +1 815 9394564; Fax: +1 815 9399057; URL: www.lions-online.org
1899; Cynthia Fuerst
90 000 vols; 250 curr per 39674

Kansas City, Kansas Public Library, 625 Minnesota Ave, *Kansas City*, KS 66101
T: +1 913 5513280; Fax: +1 913 5513243; URL: www.kckpl.lib.ks.us
1892; Teresa Garrison
Kansas; Fine Arts Coll, Spanish Language, Connelly Coll (Wyandot Indians); 3 branch libs
488 000 vols; 980 curr per; 2 200 Electronic Media & Resources 39675

Kansas City, Kansas Public Library – Argentine Branch, 2800 Metropolitan Ave, *Kansas City*, KS 66106
T: +1 913 7227400; Fax: +1 913 7227402
Helen Rigdon
Spanish Language Coll
67 000 vols; 87 curr per 39676

The Kansas City Public Library, 14 W 10th St, *Kansas City*, MO 64105
T: +1 816 7013433; Fax: +1 816 7013401; URL: www.kclibrary.org
1873; Lillie Brack
Music, Statistics, photos; African American Hist Coll, Civil War Coll, Kansas City Latino Heritage Coll; 9 branch libs
859 000 vols; 19 157 curr per
libr loan 39677

The Kansas City Public Library – Lucile H. Bluford Branch, 3050 Prospect Ave, *Kansas City*, MO 64128
T: +1 816 7013482; Fax: +1 816 7013492
Oliver Clark
68 000 vols 39678

The Kansas City Public Library – North-East, 6000 Wilson Rd, *Kansas City*, MO 64123
T: +1 816 7013485; Fax: +1 816 7013495
Claudia Visnich
66 000 vols 39679

The Kansas City Public Library – Plaza, 4801 Main St, *Kansas City*, MO 64112-2765
T: +1 816 7013481; Fax: +1 816 7013491; URL: www.kclibrary.org/plaza
Joel Jones
150 000 vols 39680

The Kansas City Public Library – Waldo Community, 201 E 75th St, *Kansas City*, MO 64114
T: +1 816 7013486; Fax: +1 816 7013496
Alicia Ahlvers
73 000 vols 39681

Hawaii State Public Library System – Kapolei Public Library, 1020 Manawai St, *Kapolei*, HI 96707
T: +1 808 6937050; Fax: +1 808 6937062
Stacie Kanno
111 000 vols; 203 curr per 39682

Katonah Village Library, 26 Bedford Rd, *Katonah*, NY 10536-2121
T: +1 914 2323508; Fax: +1 914 2320415; E-mail: katref@wlsmail.org; URL: www.katonahlibrary.org
1880; Van Kozelka
Art, Fishing, Lit, Poetry
69 000 vols; 196 curr per; 196 maps; 300 av-mat 39683

Fort Bend County Libraries – Cinco Ranch, 2620 Commercial Center Blvd, *Katy*, TX 77494
T: +1 281 3951311; Fax: +1 281 3956377
Cindy Ruggeri
89 000 vols; 188 curr per; 2 737 av-mat 39684

Harris County Public Library – Katy Branch, 5414 Franz Rd, *Katy*, TX 77493
T: +1 281 3913509; Fax: +1 281 3911927
Cecillia Shearron-Hawkins
55 000 vols 39685

Harris County Public Library – Maud Smith Marks Branch, 1815 Westgreen Blvd, *Katy*, TX 77450
T: +1 281 4928592; Fax: +1 281 4923420
Sylvia Powers
73 000 vols 39686

Kaukauna Public Library, 111 Main Ave, *Kaukauna*, WI 54130-2436
T: +1 920 7666340; Fax: +1 920 7666343; E-mail: kau@mail.owls.lib.wi.us; URL: www.kaukaunalibrary.org
1899; Margaret J. Waggoner
60 000 vols; 200 curr per
libr loan 39687

Kearney Public Library, 2020 First Ave, **Kearney**, NE 68847
T: +1 308 2333282; Fax: +1 308 2333291; URL: www.kearneylib.org
1890; Williams R. Williams
Kearney Coll, VF; 1 bookmobile
110 000 vols; 169 curr per   39688

Salt Lake County Library Services – Kearns Branch, 5350 S 4220 West, **Kearns**, UT 84118-4314
T: +1 801 9447612; Fax: +1 801 9678958
Kent Johnson
100 000 vols   39689

Kearny Public Library, 318 Kearny Ave, **Kearny**, NJ 07032
T: +1 201 9982666; Fax: +1 201 9981141; E-mail: admin.@kearnylibrary.org; URL: www.kearnylibrary.org
1906; Julie McCarthy
70 000 vols; 147 curr per   39690

Keene Public Library, 60 Winter St, **Keene**, NH 03431-3360
T: +1 603 3520157; Fax: +1 866 7430446; URL: www.keenepubliclibrary.org
1857; Nancy Vincent
117 000 vols; 233 curr per
libr loan   39691

Keller Public Library, 640 Johnson Rd, **Keller**, TX 76248
T: +1 817 4319011; Fax: +1 817 4313887; URL: www.kellerlib.org
1972; Lisa Harper Wood
82 000 vols; 160 curr per   39692

Lincoln County Library, 519 Emerald, **Kemmerer**, WY 83101
T: +1 307 8776961; Fax: +1 307 8774147; URL: will.state.wy.us/lincoln
1983; Brenda McGinnis
3 branch libs
111 000 vols; 243 curr per; 7 390 Audio Bks   39693

Kenai Community Library, 163 Main St Loop, **Kenai**, AK 99611-7723
T: +1 907 2834378; Fax: +1 907 2832266; E-mail: kenailibrary@ci.kenai.ak.us; URL: www.kenalibrary.org
1949; Mary Jo Joiner
Alaska, Fishing, Genealogy
78 000 vols; 275 curr per   39694

Duplin County Library, Dorothy Wightman Library, 107 Bowden Dr, **Kenansville**, NC 28349-0930; P.O. Box 930, Kenansville, NC 28349-0930
T: +1 910 2962117; Fax: +1 910 2962172; URL: www.duplincountync.com
1920; Linda Hadden
55 000 vols; 200 curr per   39695

Kendallville Public Library, 221 S Park Ave, **Kendallville**, IN 46755-2248
T: +1 260 3432010; Fax: +1 260 3432011; E-mail: info@kendallvillelibrary.org; URL: www.kendallvillelibrary.org
1913; Jenny Draper
Gene Stratton-Porter Coll, M. F. Owen Scrapbk; 1 branch libr
68 000 vols; 226 curr per
libr loan; ALA   39696

Town of Tonawanda Public Library, Kenmore Branch, 160 Delaware Rd, **Kenmore**, NY 14217
T: +1 716 8732842; Fax: +1 716 8738416; URL: www.buffalolib.org
1925; Kate Weeks
Large print, Newspapers, micro; 1 branch libr
182 000 vols   39697

Cobb County Public Library System – Kennesaw Branch, 2250 Lewis St, **Kennesaw**, GA 30144
T: +1 770 5282529; Fax: +1 770 5282593
Jill Tempest
54 000 vols   39698

Cobb County Public Library System – West Cobb Regional, 1750 Dennis Kemp Lane, **Kennesaw**, GA 30152
T: +1 770 5284699; Fax: +1 770 5284619
Steve Powell
96 000 vols   39699

Dunklin County Library, 209 N Main, **Kennett**, MO 63857
T: +1 573 8883561; Fax: +1 573 8886393; URL: dunklin-co.lib.mo.us
1947; JoNell Minton
8 branch libs
180 000 vols; 150 curr per; 1 000 av-mat; 1 000 sound-rec   39700

Bayard Taylor Memorial Library, 216 E State St, **Kennett Square**, PA 19348-3112; P.O. Box 730, Kennett Square, PA 19348-0730
T: +1 610 4442988; Fax: +1 610 4441752; URL: www.bayardtaylor.org
1895; Donna L. Murray
Antiques (Harlan R. Cole Coll), Art & Architecture (Atlantis Coll), Lit (Bayard Taylor Coll), Wildflowers (Botanica Coll)
51 000 vols; 161 curr per; 992 av-mat   39701

Mid-Columbia Libraries, 405 S Dayton, **Kennewick**, WA 99336
T: +1 509 5824745; Fax: +1 509 7376349; URL: www.midcolumbialibraries.org
1949; Danielle Krol
Business, Economy, Mexican-American studies; 11 branch libs, 1 bookmobile
424 000 vols; 852 curr per
libr loan   39702

Kenosha Public Library, 812 56th St, **Kenosha**, WI 53140-3735; P.O. Box 1414, Kenosha, WI 53141-1414
T: +1 262 5646101; Fax: +1 262 5646370; URL: www.kenosha.lib.wi.us
1895; Douglas Baker
Reading Readiness Resource Coll; 4 branch libs, 1 bookmobile
403 000 vols; 1 643 curr per; 4 000 e-books; 7 644 Large Print Bks, 3 550 Talking Bks
libr loan   39703

Kenosha Public Library – Northside Library, 1500 27th Ave, **Kenosha**, WI 53140-4679; P.O. Box 1414, Kenosha, WI 53141-1414
T: +1 252 5646100
Therese O'Halloran
124 000 vols   39704

Mary Lou Johnson Hardin County District Library, 325 E Columbus St, **Kenton**, OH 43326-1546
T: +1 419 6732278; Fax: +1 419 6744321; E-mail: mljref@oplin.org; URL: mljlibrary.oplin.org
1853; Heather O'Donnell
240 Circulating Hand Puppets; Ellison dia machine + dias
63 000 vols; 180 curr per; 519 microforms; 4 241 av-mat; 2 306 sound-rec; 240 digital data carriers; 2 809 vertical files, 66 baskets
libr loan; ALA   39705

Kent District Library – Kentwood Branch, 4700 Kalamazoo Ave SE, **Kentwood**, MI 49508
T: +1 616 6473910; Fax: +1 616 4552528
Cheryl Cammenga
93 000 vols   39706

Winkler County Library, 307 S Poplar St, **Kermit**, TX 79745-4300
T: +1 915 5863841; Fax: +1 915 5862462
1929; Laurie Shropshire
Genealogy Coll
62 000 vols; 50 curr per   39707

Forsyth County Public Library – Kernersville Branch, 130 E Mountain St, **Kernersville**, NC 27284
T: +1 336 7032930; Fax: +1 336 9935216
Lisa Elmore
56 000 vols; 102 curr per; 33 e-books   39708

Butt-Holdsworth Memorial Library, 505 Water St, **Kerrville**, TX 78028
T: +1 830 2578422; Fax: +1 830 7925552; URL: Library.Webmaster@kerrvilletx.gov; URL: www.bhmlibrary.org
1967; Antonio Martinez
Texana Coll
52 000 vols; 212 curr per   39709

Ketchikan Public Library, 629 Dock St, **Ketchikan**, AK 99901
T: +1 907 2253331; Fax: +1 907 2250153; E-mail: library@firstcitylibraries.org; URL: www.firstcitylibraries.org
1901; Judith L. Anglin
60 000 vols; 130 curr per   39710

Community Library Association, 415 Spruce Ave N, P.O. Box 2168, **Ketchum**, ID 83340-2168
T: +1 208 7263493; Fax: +1 208 7260756; E-mail: info@thecommunitylibrary.org; URL: www.thecommunitylibrary.org
1955
Sun Valley Ski Coll, Astrology & Occult Sciences (John Lister Coll); Children's Libr; Regional Hist Dept
82 000 vols; 175 curr per   39711

Kewanee Public Library District, 102 S Tremont St, **Kewanee**, IL 61443
T: +1 309 8520111; Fax: +1 309 8524466; URL: www.kewaneelibrary.org
1875; John E. Sayers
57 000 vols; 175 curr per; 2 218 av-mat   39712

Monroe County Public Library, 700 Fleming St, **Key West**, FL 33040
T: +1 305 2923595; Fax: +1 305 2953626; URL: www.keyslibraries.org
1892; Norma Kula
4 branch libs
184 000 vols; 120 curr per   39713

Keyport Free Public Library, 109 Broad St, **Keyport**, NJ 07735-1202
T: +1 732 2640543; Fax: +1 732 2640875; E-mail: keyport@shore.co.momouth.nj.us; URL: www.keyportlibrary.org
1914; Jackie LaPolla
War of the Rebellion: A Compilation of the Official Records of the Union & Confederate Armies
61 000 vols; 80 curr per   39714

Hawaii State Public Library System – Kihei Public Library, 35 Waimahaihai St, **Kihei**, HI 96753-8015
T: +1 808 8756833; Fax: +1 808 8756834
Janet Fehr
63 000 vols; 68 curr per   39715

Killeen City Library System, 205 E Church Ave, **Killeen**, TX 76541
T: +1 254 5018304; Fax: +1 254 5017704; E-mail: library@ci.killeen.tx.us; URL: portal.ci.killeen.tx.us; www.ci.killeen.tx.us/?section=50
1958; Deanna A. Frazee
114 000 vols; 112 curr per   39716

Kimberly Public Library – James J. Siebers Memorial Library, 515 W Kimberly Ave, **Kimberly**, WI 54136
T: +1 920 7887515; Fax: +1 920 7887516; E-mail: KIM@mail.owls.lib.wi.us; URL: www.kimlit.org
1907; Barbara Wentzel
Toys, Hand Puppets 300
105 000 vols; 150 curr per   39717

Lewis Egerton Smoot Memorial Library, 9533 Kings Hwy, **King George**, VA 22485
T: +1 540 7757951; Fax: +1 540 7755292; E-mail: smootlib@crosslink.net; URL: www.smoot.org
1969; Rita Schepmoes
52 000 vols; 90 curr per   39718

Upper Merion Township Library, 175 W Valley Forge Rd, **King of Prussia**, PA 19406-2399
T: +1 610 2651196; Fax: +1 610 2653398; E-mail: uppermerionlibrary@mclinc.org; URL: www.umtownship.org/library/index.html
1963; Karl Helicher
100 000 vols; 180 curr per   39719

Mohave County Library District, 3269 N Burbank Rd, **Kingman**, AZ 86401; P.O. Box 7000, Kingman
T: +1 928 6922665; Fax: +1 928 6925790; URL: www.co.mohave.az.us/library
1926; Robert Shupe
Arizona; 8 branch libs
237 000 vols; 640 curr per   39720

Mauney Memorial Library, 100 S Piedmont Ave, **Kings Mountain**, NC 28086
T: +1 704 7392371; Fax: +1 704 7344499
1936; Sharon Stack
Genealogy coll; local history, family histories
58 000 vols; 45 curr per; 2 808 maps; 100 microforms; 5 000 sound-rec; 100 digital data carriers
libr loan; ALA   39721

Kingsport Public Library & Archives, J. Fred Johnson Memorial Library, 400 Broad St, **Kingsport**, TN 37660-4292
T: +1 423 2242539; Fax: +1 423 2242558; E-mail: kptlib@wrlibrary.org; URL: www.kingsportlib.org
1921; Helen Whittaker
Kingsport City Arch, First Tennessee Bank Small Business Ctr, Palmer Regional Hist Coll; 1 branch libr
119 000 vols; 388 curr per; 2 582 av-mat; 2 582 Audio Bks   39722

Hoyt Library, 284 Wyoming Ave, **Kingston**, PA 18704-3597
T: +1 570 2872013; Fax: +1 570 2832081; E-mail: hoytlib@ptd.net; URL: www.hoytlibrary.org
1928; David Marks
Early Americana (William Brewster Coll), maps, bks; Holocaust (Reuben Levy Coll); Jewish Hist (Levison Coll)
105 000 vols; 80 curr per   39723

Kingston Library, 55 Franklin St, **Kingston**, NY 12401
T: +1 845 3394260; Fax: +1 845 3317981; E-mail: kingstonlibrary@hvc.rr.com; URL: www.kingstonlibrary.org
1899; Bruce George
95 000 vols; 100 curr per   39724

Kingston Public Library, Six Green St, **Kingston**, MA 02364
T: +1 781 5850517; Fax: +1 781 5850521; E-mail: kilib@kingstonpubliclibrary.org; URL: www.kingstonpubliclibrary.org
1898; Sia Stewart
Parenting, Travel; Hist of Kingston, Hist of Plymouth County
50 000 vols; 107 curr per
libr loan   39725

Williamsburg County Library, 215 N Jackson, **Kingstree**, SC 29556-3319
T: +1 843 3559486; Fax: +1 843 3559991; URL: www.wlbg.lib.sc.us
1967; Norris Wootton
55 000 vols; 99 curr per   39726

Robert J. Kleberg Public Library, 220 N Fourth St, **Kingsville**, TX 78363-4410
T: +1 361 5926381; Fax: +1 361 5927461; E-mail: kpldirector@kleberglibrary.com; URL: www.youseemore.com/rjkleberg/about.asp
1927; Robert R. Rodriguez
65 000 media items; 164 curr per   39727

Harris County Public Library – Kingwood Branch, 4102 Rustic Woods Dr, **Kingwood**, TX 77345
T: +1 281 3606804; Fax: +1 281 3602093
Christi Whittington
80 000 vols   39728

Kinnelon Public Library, 132 Kinnelon Rd, **Kinnelon**, NJ 07405-2393
T: +1 973 8381321; Fax: +1 973 8380741; URL: www.kinnelonlibrary.org
1962; Barbara Owens
68 000 vols; 115 curr per; 4 322 av-mat   39729

Kinsman Free Public Library, 6420 Church St, **Kinsman**, OH 44428-9702; P.O. Box 166, Kinsman, OH 44428-0166
T: +1 330 8762461; Fax: +1 330 8763335; E-mail: Reference@kinsmanlibrary.org; URL: www.kinsmanlibrary.org
1885; Cheryl Bugnone
Clarence Darrow Coll, Dr Ernest L Scott Coll
62 000 vols; 190 curr per   39730

Neuse Regional Library, Kinston-Lenoir County Public Library (Headquarters), 510 N Queen St, ***Kinston***, NC 28501
T: +1 252 5277066; Fax: +1 252 5279235; E-mail: nrl@neuselibrary.org; URL: www.neuselibrary.org
1962; Agnes W. Ho
224 000 vols; 448 curr per; 109 maps; 12 057 av-mat; 11 000 sound-rec; 523 digital data carriers; 184 art reproductions
libr loan 39731

Adair County Public Library, One Library Lane, ***Kirksville***, MO 63501
T: +1 660 6656038; Fax: +1 660 6270028; URL: youseemore.com/adaircpl
1986; Glenda Hunt
61 000 vols; 85 curr per 39732

Kirkwood Public Library, 140 E Jefferson Ave, ***Kirkwood***, MO 63122
T: +1 314 8215770; Fax: +1 314 8223755; E-mail: kirkwoodlibrary@yahoo.com; URL: kpl.lib.mo.us
1924; Wicky Sleight
Storytelling; Early Missouri, St Louis City, County & Kirkwood Hist; Louisiana Exposition; Nonprofit Resources Coll
81 000 vols; 342 curr per 39733

Rice Public Library, Eight Wentworth St, ***Kittery***, ME 03904
T: +1 207 4391553; Fax: +1 207 4391765; E-mail: arabella@rice.lib.me.us; URL: www.rice.lib.me.us
1875; Nancy Johnson
Kittery Hist & Genealogy, bks, photos
58 000 vols; 123 curr per; 45 microforms; 1 445 sound-rec; 5 digital data carriers; 901 video recordings
libr loan 39734

Klamath County Library Services District, 126 S Third St, ***Klamath Falls***, OR 97601-6394
T: +1 541 8828894; Fax: +1 541 8826166; URL: www.klamathlibrary.plinkit.org
1913; Andy Swanson
12 branch libs
177 000 vols; 275 curr per
libr loan 39735

Wake County Public Library System – East Regional, 946 Steeple Square Ct, ***Knightdale***, NC 27545
T: +1 919 2175301; Fax: +1 919 2175327
Ann Burlingame
117 000 vols 39736

Starke County Public Library System – Henry F. Schricker (Main Library), 152 W Culver Rd, ***Knox***, IN 46534-2220
T: +1 574 7727323; E-mail: webmaster@scpl.lib.in.us; URL: www.scpl.lib.in.us
1919; Ellen A. Dodge
3 branch libs
102 000 vols; 287 curr per 39737

Knox County Public Library System – Lawson McGhee Library-East Tennessee History Center, 500 W Church Ave, ***Knoxville***, TN 37902-2505
T: +1 865 2158750; Fax: +1 865 2158742; URL: www.knoxlib.org
1886; Larry Frank
18 branch libs
1 046 000 vols; 1 122 curr per
libr loan; ALA 39738

Knox County Public Library System – West Knoxville Branch, 100 Golf Club Rd, ***Knoxville***, TN 37919-4801
T: +1 865 5888813; Fax: +1 865 5887580
Rebecca Dames
62 000 vols 39739

A. Holmes Johnson Memorial Library, Kodiak Public Library, 319 Lower Mill Bay Rd, ***Kodiak***, AK 99615
T: +1 907 4868686; Fax: +1 907 4868681; URL: www.city.kodiak.ak.us
1946; Joseph D'Elia
Fisheries; Alaska Coll
68 000 vols; 275 curr per 39740

Kokomo-Howard County Public Library, 220 N Union St, ***Kokomo***, IN 46901-4614
T: +1 765 4573242; Fax: +1 765 4573683; E-mail: info@khcpl.org; URL: www.khcpl.org
1885; Charles N. Joray
Hoosier Art Coll; 2 branch libs
301 000 vols; 799 curr per; 7 000 e-books; 9 140 av-mat; 384 sound-rec; 201 Large Print Bks
libr loan 39741

Kountze Public Library, 800 S Redwood Ave, ***Kountze***, TX 77625
T: +1 409 2462826; Fax: +1 409 2464659; URL: www.kountzelibrary.org/www/index.htm
Crysel Laverne
52 000 vols; 25 curr per 39742

Kuna Community Library, 457 N Locust, ***Kuna***, ID 83634-1926; P.O. Box 129, Kuna, ID 83634-0129
T: +1 208 9221025; Fax: +1 208 9221026; URL: www.lili.org/kuna
1964; Anne Hankins
68 000 vols; 65 curr per; 1 350 av-mat 39743

County of Los Angeles Public Library – La Canada Flintridge Library, 4545 N Oakwood Ave, ***La Canada Flintridge***, CA 91011-3358
T: +1 818 7903330; Fax: +1 818 9521754; URL: www.colapublib.org/libs/lacanada
Kathleen Coakley
110 000 vols; 120 curr per 39744

County of Los Angeles Public Library – La Crescenta Library, 4521 La Crescenta Ave, ***La Crescenta***, CA 91214-2999
T: +1 818 2485313; Fax: +1 818 2481289; URL: www.colapublib.org/libs/lacrescenta
Victoria Guagliardo
62 000 vols; 21 curr per 39745

La Crosse Public Library, 800 Main St, ***La Crosse***, WI 54601
T: +1 608 7897167; Fax: +1 608 7897106; E-mail: refdesk@lacrosse.lib.wi.us; URL: www.lacrosselibrary.org
1888; Kelly Kreig-Sigman
2 branch libs
213 000 vols; 552 curr per; 11 180 av-mat; 8 590 sound-rec
libr loan 39746

La Grande Public Library, 2006 Fourth St, ***La Grande***, OR 97850-2496
T: +1 541 9621339; Fax: +1 541 9621338; URL: www.ci.la-grande.or.us/dept_library.cfm
1912; Jo E. Cowling
61 000 vols; 108 curr per; 3 337 av-mat 39747

La Grange Public Library, 10 W Cossitt Ave, ***La Grange***, IL 60525
T: +1 708 3520576; Fax: +1 708 3521620; E-mail: lgs@lagrangelibrary.org; URL: www.lagrangelibrary.org
1906; Jeannie Dilger-Hill
Business Coll
100 000 vols; 1 010 curr per
libr loan 39748

Oldham County Public Library, Duerson, 106 E Jefferson St, ***La Grange***, KY 40031
T: +1 502 2419899; Fax: +1 502 2416048; URL: www.oldhampl.org
1968; Susan Eubank
Kentucky Coll, Genealogy Coll, Census Coll, Learn Resource; 2 branch libs
66 000 vols; 166 curr per 39749

La Grange Park Public Library District, 555 N LaGrange Rd, ***La Grange Park***, IL 60526-5644
T: +1 708 3520100; Fax: +1 708 3521606; E-mail: info@lplibrary.org; URL: www.lplibrary.org
1975; Dixie M. Conkis
70 000 vols; 204 curr per 39750

Orange County Public Library – La Habra Branch, 221 E La Habra Blvd, ***La Habra***, CA 90631-5437
T: +1 714 5267728; Fax: +1 562 6918043
Jill Patterson
87 000 vols; 98 curr per 39751

La Marque Public Library, 1011 Bayou Rd, ***La Marque***, TX 77568-4195
T: +1 409 9389270; Fax: +1 409 9389277; URL: www.ci.la-marque.tx.us
1946; Marilee Neale
63 000 vols; 50 curr per 39752

San Diego County Library – La Mesa Branch, 8074 Allison Ave, ***La Mesa***, CA 91941-5001
T: +1 619 4692151; Fax: +1 619 6973751
Elizabeth Hildreth
75 000 vols 39753

County of Los Angeles Public Library – La Mirada Library, 13800 La Mirada Blvd, ***La Mirada***, CA 90638-3098
T: +1 562 9430277; Fax: +1 562 9433920; URL: www.colapublib.org/libs/lamirada
Jennifer McCarty
114 000 vols; 115 curr per 39754

Orange County Public Library – La Palma Branch, 7842 Walker St, ***La Palma***, CA 90623-1721
T: +1 714 5238585; Fax: +1 714 5215551
Susan Sassone
60 000 vols; 77 curr per 39755

Charles County Public Library, La Plata Branch, Two Garrett Ave, ***La Plata***, MD 20646-5959
T: +1 301 9349001; Fax: +1 301 9342297; URL: www.ccplonline.org
1923; Emily Ferren
Southern MD. Genealogy; 2 branch libs
125 000 vols; 264 curr per; 144 e-books; 16 743 av-mat; 16 452 sound-rec; 1 digital data carriers; 100 Bks on Deafness & Sign Lang
libr loan; ALA, MLA 39756

La Porte County Public Library, 904 Indiana Ave, ***La Porte***, IN 46350-3407
T: +1 219 3626156; Fax: +1 219 3626158; URL: www.lapcat.org
1897; Judy R. Hamilton
6 branch libs
300 000 vols; 255 curr per; 8 000 av-mat; 2 000 High Interest/Low Vocabulary Bks, 30 000 Large Print Bks, 8 000 Talking Bks
libr loan 39757

County of Los Angeles Public Library – La Puente Library, 15920 E Central Ave, ***La Puente***, CA 91744-5499
T: +1 626 9684613; Fax: +1 626 3690294; URL: www.colapublib.org/libs/lapuente
Jeanette Freels
81 000 vols; 82 curr per 39758

County of Los Angeles Public Library – Sunkist Library, 840 N Puente Ave, ***La Puente***, CA 91746-1316
T: +1 626 9602707; Fax: +1 626 3385141; URL: www.colapublib.org/libs/sunkist
Yaa Sefa-Boakye
79 000 vols; 88 curr per 39759

La Vergne Public Library, 5063 Murfreesboro Rd, ***La Vergne***, TN 37086-0177
T: +1 615 7937303; Fax: +1 615 7937307; E-mail: lavergnelibrary@hotmail.com; URL: www.lavergne.org/library/index.htm
1979; Faye Toombs
56 000 vols; 30 curr per; 1 071 av-mat 39760

County of Los Angeles Public Library – La Verne Library, 3640 D St, ***La Verne***, CA 91750-3572
T: +1 909 5961934; Fax: +1 909 5967303; URL: www.colapublib.org/libs/laverne
George May
66 000 vols; 75 curr per 39761

La Vista Public Library, 9110 Giles Rd, ***La Vista***, NE 68128
T: +1 402 5373900; Fax: +1 402 5373902; E-mail: lvlibrary@lavistamail.mccneb.edu; URL: www.ci.la-vista.ne.us
1972; Rose Schinker
52 000 vols; 85 curr per; 29 000 e-books; 900 av-mat 39762

Cherokee Regional Library System – La Fayette-Walker County Public Library, 305 S Duke St, ***LaFayette***, GA 30728-2936
T: +1 706 6382064; Fax: +1 706 6383979; URL: www.walker.public.lib.ga.us
1938; Valinda Oliver
4 branch libs
67 000 vols; 18 curr per; 1 244 av-mat 39763

Contra Costa County Library – Lafayette Library, 952 Moraga Rd, ***Lafayette***, CA 94549-4594
T: +1 925 2833872; Fax: +1 925 2838231
Susan Weaver
60 000 vols 39764

Lafayette Public Library, 301 W Congress, ***Lafayette***, LA 70501-6866; P.O. Box 3427, Lafayette
T: +1 337 2615757; Fax: +1 337 2615782; E-mail: admin.b1lf@pelican.state.lib.la.us; URL: www.lafayette.lib.la.us
1946; Sona J. Dombourian
Adult New Readers, Dolls, Large Print, Louisiana, Genealogy; 9 branch libs
338 000 vols; 501 curr per; 6 528 sound-rec; 18 digital data carriers
libr loan 39765

Tippecanoe County Public Library, 627 South St, ***Lafayette***, IN 47901-1470
T: +1 765 4290100; Fax: +1 765 4290150; URL: www.tcpl.lib.in.us
1882; Leanne York
Mental health; Indiana
302 000 vols; 1 947 curr per; 1 870 av-mat; 180 sound-rec; 500 Large Print Bks 39766

William Jeanes Memorial Library, 2391 Harts Lane, ***Lafayette Hill***, PA 19444
T: +1 610 8280441; Fax: +1 610 8284049; E-mail: jeanesinfo@mclinc.org; URL: jeaneslibrary.org
1933
Art, Cookbks, Quaker hist
51 000 vols; 210 curr per 39767

LaGrange County Public Library, 203 W Spring St, ***LaGrange***, IN 46761-1845
T: +1 260 4632842; Fax: +1 260 4632841; E-mail: info@lagrange.lib.in.us; URL: www.lagrange.lib.in.us
1919; Mary Hooley
Large type print; 3 branch libs
100 000 vols; 150 curr per 39768

Orange County Public Library – Laguna Beach Branch, 363 Glenneyre St, ***Laguna Beach***, CA 92651-2310
T: +1 949 4971733; Fax: +1 949 4972876
Marianna Hof
75 000 vols; 78 curr per 39769

Orange County Public Library – Laguna Niguel Branch, 30341 Crown Valley Pkwy, ***Laguna Niguel***, CA 92677-6326
T: +1 949 2495252; Fax: +1 949 2495258
Loretta Farley
77 000 vols; 82 curr per 39770

Lake Bluff Public Library, 123 Scranton Ave, ***Lake Bluff***, IL 60044
T: +1 847 2342540; Fax: +1 847 2342649; E-mail: mwomack@lakeblufflibrary.org; URL: www.lakeblufflibrary.org
1926; Matt Womack
51 310 vols; 144 curr per 39771

Calcasieu Parish Public Library System, 301 W Claude St, ***Lake Charles***, LA 70605-3457
T: +1 337 7217116; Fax: +1 337 4758806; E-mail: cen@calcasieu.lib.la.us; URL: www.calcasieu.lib.la.us
1944; Michael Sawyer
14 branch libs
406 000 vols; 2 425 curr per; 130 Bks on Deafness & Sign Lang, 7 440 Large Print Bks
libr loan 39772

Calcasieu Parish Public Library System – Central Library, 301 W Claude St, ***Lake Charles***, LA 70605
T: +1 337 4758792; Fax: +1 337 4758797
Geraldine Harris
124 000 vols 39773

Columbia County Public Library, 308 NW Columbia Ave, ***Lake City***, FL 32055
T: +1 386 7581018; Fax: +1 386 7582135; URL: www.ccpl.sirsi.net
Deborah J. Paulson
2 branch libr
132 000 vols; 220 curr per
libr loan 39774

Lake Forest Library, 360 E Deerpath Ave, *Lake Forest*, IL 60045-2252
T: +1 847 6154316; Fax: +1 847 2341453; E-mail: reference@lfl.alibrary.com; URL: www.lakeforestlibrary.org
1898; Kaye Grabbe
Art, Gardening
121 000 vols; 342 curr per; 3 000 e-books; 4 800 DVDs
libr loan   39775

Orange County Public Library – El Toro Branch, 24672 Raymond Way, *Lake Forest*, CA 92630-4489
T: +1 949 8558173; Fax: +1 949 5867412
Phyllis Brown
112 000 vols; 101 curr per   39776

Brazoria County Library System – Lake Jackson Branch, 250 Circle Way, *Lake Jackson*, TX 77566
T: +1 979 4152590; Fax: +1 979 4152993
Nancy Hackney
96 000 vols   39777

Lake Oswego Public Library, 706 Fourth St, *Lake Oswego*, OR 97034-2399
T: +1 503 6367628; Fax: +1 503 6752536; URL: www.lakeoswegolibrary.org
1930; N.N.
Northwest Coll, bks & files
153 000 vols; 282 curr per   39778

Lake Villa District Library, 1001 E Grand Ave, *Lake Villa*, IL 60046
T: +1 847 3567711; Fax: +1 847 2659595; URL: www.lvdl.org
1952; Robert Watson
145 000 vols; 463 curr per; 11 e-books
                                     39779

Newton County Public Library, Lake Village Memorial Township Library, 9444 N 315 W, *Lake Village*, IN 46349; P.O. Box 206, Lake Village, IN 46349-0206
T: +1 219 9923490; Fax: +1 219 9929198; E-mail: lakevil@netnitco.net
1947; Mary K. Emmrich
2 branch libs
71 000 vols; 176 curr per; 50 govt docs; 50 maps; 16 microforms; 1 452 av-mat; 250 sound-rec
libr loan; INCOLSA   39780

Lake Wales Public Library, 290 Cypress Garden Lane, *Lake Wales*, FL 33853
T: +1 863 6784004; Fax: +1 863 6784051; E-mail: library@cityoflakewales.com; URL: www.cityoflakewales.com/library
1919; Tina M. Peak
Business, Careers, Genealogy, Florida, Large print; Libr of Congress Talking Bks
61 000 vols; 110 curr per   39781

Lake Worth Public Library, 15 North M St, *Lake Worth*, FL 33460
T: +1 561 5337354; Fax: +1 561 5861651; E-mail: lwlibrary@lakeworth.org; URL: www.lakeworth.org
1912; Vickie Joslin
Large Print Books Coll
60 000 vols; 102 curr per; 14 000 e-books   39782

Ela Area Public Library District, 275 Mohawk Trail, *Lake Zurich*, IL 60047
T: +1 847 4383433; Fax: +1 847 4389290; E-mail: elaref1@eapl.org; URL: www.eapl.org
1972; Mary Elizabeth Campe
184 000 vols; 150 curr per; 5 355 Large Print Bks
libr loan   39783

Ocean County Library – Manchester Township, 21 Colonial Dr, *Lakehurst*, NJ 08733-3801
T: +1 732 6577600; Fax: +1 732 3239246
Suzanne Scro
82 000 vols   39784

Lakeland Public Library, 100 Lake Morton Dr, *Lakeland*, FL 33801-5375
T: +1 863 8344280; Fax: +1 863 8344293; E-mail: publiclibrary@lakelandgov.net; URL: www.lakelandgov.net/library
1927; Lisa Lilyquist
1 branch libr, 1 bookmobile
261 000 vols; 402 curr per
libr loan   39785

Lake County Library, Lakeport Library, 1425 N High St, *Lakeport*, CA 95453-3800
T: +1 707 2638816; Fax: +1 707 2636796; URL: www.co.lake.ca.us/library/library.html
1974; Susan Jean Clayton
Pomo Indians, Geothermal Resources; 3 branch libs
140 000 vols; 174 curr per   39786

North Arkansas Regional Library, 319 Hwy 14, *Lakeview*, AR 72642
T: +1 870 4495808; Fax: +1 870 4495808
N.N.
60 000 vols   39787

County of Los Angeles Public Library – Angelo M. Iacoboni Library, 4990 Clark Ave, *Lakewood*, CA 90712-2676
T: +1 562 8661777; Fax: +1 562 8661217; URL: www.colapublib.org/libs/iacoboni
Eileen Tokar
181 000 vols; 154 curr per   39788

County of Los Angeles Public Library – George Nye Jr Library, 6600 Del Amo Blvd, *Lakewood*, CA 90713-2206
T: +1 562 4218497; Fax: +1 562 4963943; URL: www.colapublib.org/nye
Carol Burke
61 000 vols; 100 curr per   39789

Jefferson County Public Library, 10200 W 20th Ave, *Lakewood*, CO 80215
T: +1 303 2329507; Fax: +1 303 2752202; URL: www.jefferson.lib.co.us
1952; Marcellus Turner
Colorado, Art, Consumer health, Folk music, Law, Railroads, Gov docs; 11 branch libs
1 200 000 vols; 4 357 curr per; 3 000 e-books   39790

Ocean County Library – Lakewood Branch, 301 Lexington Ave, *Lakewood*, NJ 08701
T: +1 732 3631435; Fax: +1 732 3631438
1872; Marvelene Beach
110 000 vols; 372 curr per; 2 000 av-mat; 3 183 sound-rec   39791

County of Los Angeles Public Library – Lancaster Library, 601 W Lancaster Blvd, *Lancaster*, CA 93534
T: +1 661 9485029; Fax: +1 661 9450480; URL: www.colapublib.org/libs/lancaster
Judy Hist
277 000 vols; 64 curr per   39792

Fairfield County District Library, 219 N Broad St, *Lancaster*, OH 43130-3098
T: +1 740 6532745; Fax: +1 740 6534199; URL: www.fcdlibrary.org
1878; Marilyn C. Steiner
3 branch libs
230 000 vols; 431 curr per; 27 e-books
libr loan   39793

Lancaster County Library, 313 S White St, *Lancaster*, SC 29720
T: +1 803 2851502; Fax: +1 803 2856004; E-mail: lanclib@comporium.net; URL: www.lanclib.org
1907; Judy Hunter
1 branch libr
143 000 vols; 282 curr per; 6 264 av-mat; 3 400 sound-rec; 820 digital data carriers; 42 Bks on Deafness & Sign Lang, 2 500 Large Print Bks   39794

Lancaster Public Library, 5466 Broadway, *Lancaster*, NY 14086
T: +1 716 6831120; Fax: +1 716 6860749
1895; James Stelzle
Parenting Resource Ctr
78 000 vols   39795

Lancaster Veterans Memorial Library, 1600 Veterans Memorial Pkwy, *Lancaster*, TX 75134
T: +1 972 2271080; Fax: +1 972 2275560; URL: www.lancastertxlib.org
1939/1951; Cami Loucks
Mildred & Welton Fail Genealogical Resource Center
63 000 vols; 99 curr per
libr loan; ALA, TLA, PLA   39796

Thayer Memorial Library, 717 Main St, *Lancaster*, MA 01523-2248; P.O. Box 5, Lancaster, MA 01523-0005
T: +1 978 3688928; Fax: +1 978 3688929
1862; Joseph J. Mule
Botany, Hist Reference Coll, Rare Bks
51 000 vols; 55 curr per; 7 000 e-books; 3 969 mss; 1 microforms; 775 av-mat; 789 sound-rec
libr loan   39797

Fremont County Library System, 451 N Second St, *Lander*, WY 82520-2316
T: +1 307 3321600; Fax: +1 307 3321504; URL: www.fremontcountylibraries.org
1907; Jill Rourke
Western Americana; 2 branch libs
188 000 vols; 228 curr per   39798

Bucks County Free Library – Pennwood, 301 S Pine St, *Langhorne*, PA 19047-2887
T: +1 215 7572510; Fax: +1 215 7579579
Jan Dickler
73 000 vols; 84 curr per   39799

Lansdowne Public Library, 55 S Lansdowne Ave, *Lansdowne*, PA 19050-2804
T: +1 610 6230239; Fax: +1 610 6236825; E-mail: lansdown@delco.lib.pa.us; URL: www.delco.lib.pa.us
1898; Amy Gillespie
54 000 vols; 87 curr per; 1 500 sound-rec; 300 digital data carriers
libr loan; ALA   39800

Capital Area District Library, 401 S Capitol Ave, *Lansing*, MI 48933; P.O. Box 40719, Lansing, MI 48909-7919
T: +1 517 3676328; Fax: +1 517 3741068; E-mail: comments@cadl.org; URL: www.cadl.org
1998; Susan J. Hill
13 branch libs
504 000 vols; 1 365 curr per; 542 e-books; 25 740 av-mat; 23 370 sound-rec; 18 950 digital data carriers; 12 540 Large Print Bks
libr loan   39801

Lapeer District Library, 201 Village West Dr S, *Lapeer*, MI 48446-1699
T: +1 810 6649521; Fax: +1 810 6648527; URL: www.library.lapeer.org
1939; Kate A. Pohjola
8 branch libs
102 000 vols; 246 curr per   39802

Lapeer District Library – Marguerite deAngeli Main Branch, 921 W Nepessing St, *Lapeer*, MI 48446
T: +1 810 6646971; Fax: +1 810 6645581
June Mendel
Career Resource Center;Collection & Exhibit on Marguerite deAngeli;Genealogy Coll
80 000 vols; 121 curr per   39803

Saint John the Baptist Parish Library, 2920 New Hwy 51, *LaPlace*, LA 70068
T: +1 985 6526857; Fax: +1 985 6528005; URL: www.stjohn.lib.la.us
1966; Randy A. De Soto
Parish Hist, photos; 1 branch libr
144 000 vols; 186 curr per; 3 670 av-mat; 444 sound-rec; 63 digital data carriers; 5 670 Large Print Bks   39804

Harris County Public Library – LaPorte Branch, 600 S Broadway, *LaPorte*, TX 77571
T: +1 281 4714022; Fax: +1 281 4700839
Myra Wilson
54 000 vols   39805

Prairie-River Library District, 103 N Main St, *Lapwai*, ID 83540; P.O. Box 1200, Lapwai, ID 83540-1200
T: +1 208 8437254
1959
7 branch libs
92 000 vols; 50 curr per; 1 287 av-mat
                                     39806

Albany County Public Library, 310 S Eighth St, *Laramie*, WY 82070-3969
T: +1 307 7212580; Fax: +1 307 7212584; E-mail: albyref@will.state.wy.us; URL: acpl.lib.wy.us
1887; Susan M. Simpson
76 000 vols; 357 curr per; 2 850 Large Print Bks
libr loan   39807

Larchmont Public Library, 121 Larchmont Ave, *Larchmont*, NY 10538
T: +1 914 8342281; Fax: +1 914 8340351; E-mail: larchmontlibrary@hotmail.com; URL: www.larchmontlibrary.org
1926
90 420 vols; 231 curr per; 10 microforms; 2 681 av-mat; 3 194 sound-rec; 40 digital data carriers
libr loan   39808

Laredo Public Library, 1120 E Calton Rd, *Laredo*, TX 78041
T: +1 956 7952400; Fax: +1 956 7952403; URL: www.laredolibrary.org
1951; N.N.
Foundation Ctr Cooperating Coll; 1 branch libr, 1 bookmobile
249 000 vols; 185 curr per; 363 e-books
                                     39809

Largo Public Library, 120 Central Park Dr, *Largo*, FL 33771
T: +1 727 5876715; Fax: +1 727 5867353; URL: www.asklargo.com
1916; Casey McPhee
Large print, Literacy, Parenting, Arts & crafts
231 000 vols; 198 curr per
libr loan   39810

Larkspur Public Library, 400 Magnolia Ave, *Larkspur*, CA 94939
T: +1 415 9275005; Fax: +1 415 9275136; E-mail: library@larkspurcityhall.org; URL: www.ci.larkspur.ca.us/209.html
55 000 vols; 100 curr per   39811

Las Cruces Public Library, Thomas Branigan Memorial Library, 200 E Picacho Ave, *Las Cruces*, NM 88001-3499
T: +1 575 5284000; Fax: +1 575 5284030; URL: library.las-cruces.org
1935; Kathleen Teaze
1 bookmobile
145 000 vols; 502 curr per; 1 000 e-books; 1 mss; 3 715 sound-rec; 94 digital data carriers; 50 Bks on Deafness & Sign Lang, 7501 Talking Bks
libr loan; ALA   39812

Carnegie Public Library, 500 National Ave, *Las Vegas*, NM 87701
T: +1 505 4263304; E-mail: rellis@ci.lasa-vegas.nm.us; URL: www.lasvegasnm.gov/library/home.htm
1904
Southwest Coll
62 000 vols
libr loan   39813

Las Vegas-Clark County Library District, 833 Las Vegas Blvd N, *Las Vegas*, NV 89101
T: +1 702 5073611; Fax: +1 702 5073609; URL: www.lvccld.org
1985; Daniel L. Walters
Nevada State Data Ctr, State Publication Distribution Ctr, Int Languages Coll; 26 branch libs
2 810 000 vols; 4 915 curr per; 32 000 e-books
libr loan   39814

Adams Memorial Library, 1112 Ligonier St, *Latrobe*, PA 15650
T: +1 724 5391972; Fax: +1 724 5370338; E-mail: library@adamslib.org; URL: www.adamslib.org
1927; Tracy Trotter
Large Print, Bk Tapes, Toy Libr, Videos; 2 branch libs, 1 bookmobile
110 000 vols; 50 curr per; 2 316 sound-rec; 50 digital data carriers
libr loan; ALA, PLA   39815

Howard County Library – Savage, 9125 Durness Lane, *Laurel*, MD 20723-5991
T: +1 410 8805975; Fax: +1 410 8805999
Diane Li
80 000 vols   39816

Laurel-Jones County Library, 530 Commerce St, *Laurel*, MS 39440 T: +1 601 4284313; Fax: +1 601 4280597; URL: www.laurel.lib.ms.us 1919; Mary-Louise D. Breland 2 branch libs 82 000 vols; 250 curr per　39817

Queens Borough Public Library – Laurelton Branch, 134-26 225th St, *Laurelton*, NY 11413 T: +1 718 5282822; Fax: +1 718 7236837 Daniel Nkansah 60 000 vols　39818

Laurens County Library, 1017 W Main St, *Laurens*, SC 29360 T: +1 864 6817323; Fax: +1 864 6810598; URL: www.lcpl.org 1929; Ann Szypulski 1 branch libr, 1 bookmobile 122 000 vols; 296 curr per; 2 920 av-mat; 5 194 sound-rec libr loan　39819

Scotland County Memorial Library, 312 W Church St, *Laurinburg*, NC 28352-3720; P.O. Box 369, Laurinburg, NC 28353-0369 T: +1 910 2760563; Fax: +1 910 2764032; URL: www.scotlandcolibrary.com 1941; Robert Busko Heritage Room (North Carolina Coll) 60 000 vols; 125 curr per; 2 047 av-mat　39820

Lawrence Public Library, 707 Vermont St, *Lawrence*, KS 66044-2371 T: +1 785 8433833; Fax: +1 785 8433368; E-mail: custserv@lawrence.lib.ks.us; URL: www.lawrencepubliclibrary.org 1904; Bruce Flanders 218 000 vols; 450 curr per; 4 000 Large Print Bks, 5 800 Talking Bks　39821

Lawrence Public Library, 51 Lawrence St, *Lawrence*, MA 01841 T: +1 978 6203621; Fax: +1 978 6883142; URL: www.lawrencefreelibrary.org 1872; Maureen L. Nimmo Career Opportunity Ctr, Adult Basic Education, Old Radio Shows; 1 branch libr 150 000 vols; 300 curr per; 2 000 Large Print Bks　39822

Peninsula Public Library, 280 Central Ave, *Lawrence*, NY 11559 T: +1 516 2393262; Fax: +1 516 2398425; E-mail: pplmail@peninsulapublic.org; URL: www.nassaulibrary.org/peninsula 1951; Arleen J. Reo Judaica Coll; 1 bookmobile 120 000 vols; 12 000 curr per libr loan　39823

Anderson County Public Library, 114 N Main St, *Lawrenceburg*, KY 40342 T: +1 502 8396420; Fax: +1 502 8397243; URL: www.andersonpubliclibrary.org 1908; Jeffrey Sauer 56 000 vols; 81 curr per　39824

Lawrenceburg Public Library District, 150 Mary St, *Lawrenceburg*, IN 47025-1995 T: +1 812 5372775; Fax: +1 812 5372810; E-mail: lawplib@lpld.lib.in.us; URL: www.lpld.lib.in.us 1910; Sally Stegner 126 000 vols; 310 curr per　39825

Gwinnett County Public Library, 1001 Lawrenceville Hwy NW, *Lawrenceville*, GA 30045 T: +1 770 8224522; Fax: +1 770 8225379; URL: www.gwinnettpl.org 1935; Nancy Stanbery-Kellam 12 branch libs 884 000 vols; 2 354 curr per　39826

Meherrin Regional Library, Brunswick County Library, 133 W Hicks St, *Lawrenceville*, VA 23868 T: +1 434 8482418; Fax: +1 434 8484786; E-mail: bcl@meherrinlib.org; URL: meherrinlib.org 1940; Marilyn S. Marston 80 000 vols; 200 curr per　39827

Mercer County Library – Lawrence Headquarters, 2751 Brunswick Pike, *Lawrenceville*, NJ 08648 T: +1 609 9896920; Fax: +1 609 5381208 Ellen Brown 145 000 vols　39828

Lawton Public Library, 110 SW Fourth St, *Lawton*, OK 73501-4034 T: +1 580 5813450; Fax: +1 580 2480243; URL: www.cityof.lawton.ok.us/library 1904; David Snider Oklahoma (Voices of Oklahoma); Southwest Oklahoma Genealogical Res Coll; 1 branch libr, 1 bookmobile 121 000 vols; 327 curr per libr loan　39829

Davis County Library – Central, 155 N Wasatch Dr, *Layton*, UT 84041 T: +1 801 5470729 Marilyn Getts 139 000 vols　39830

Helen Hall Library, 100 W Walker, *League City*, TX 77573-3899 T: +1 281 5541101; Fax: +1 281 5541117; URL: www.leaguecitylibrary.org 1972; Shelley Leader 175 000 vols; 230 curr per; 14 032 av-mat; 5 850 sound-rec; 4 370 Large Print Bks　39831

Crawford County Public Library, Breeden Memorial, 529 Old State Rd 62, *Leavenworth*, IN 47137; P.O. Box 100, Leavenworth, IN 47137-0100 T: +1 812 3382606; Fax: +1 812 9514152; E-mail: breedenlibrary@yahoo.com; URL: www.ccpl.lib.in.us/breeden.htm C. Ramsey 65 000 vols; 55 curr per　39832

Leavenworth Public Library, 417 Spruce St, *Leavenworth*, KS 66048 T: +1 913 6825666; Fax: +1 913 6821248; URL: www.leavenworthpubliclibrary.org 1895; Kimberly Baker 129 000 vols; 235 curr per　39833

Johnson County Library – Leawood Pioneer Branch, 4700 Town Center Dr, *Leawood*, KS 66211 T: +1 913 3440250; Fax: +1 913 3440253 Sandra Sutter 77 000 vols; 117 curr per　39834

Jonathan Trumbull Library, 580 Exeter Rd, P.O. Box 145, *Lebanon*, CT 06249 T: +1 860 6427761; Fax: +1 860 6424880; E-mail: librarian@lebanonctlibrary.org; URL: www.lebanonctlibrary.org 1896 50 000 vols; 60 curr per　39835

Lebanon Community Library, 125 N Seventh St, *Lebanon*, PA 17046-5000 T: +1 717 2737624; Fax: +1 717 2732719; URL: www.lclibs.org/lebanon 1925; Jayne Tremaine 102 000 vols; 120 curr per; 2 721 av-mat　39836

Lebanon Public Library, 104 E Washington St, *Lebanon*, IN 46052 T: +1 765 4823460; Fax: +1 765 8735059; E-mail: lpldir@bccn.boone.in.us; URL: www.bccn.boone.in.us/lpl 1905; Kay K. Martin Abraham Lincoln Coll 60 340 vols; 176 curr per　39837

Lebanon Public Library, 101 S Broadway, *Lebanon*, OH 45036 T: +1 513 9322665; Fax: +1 513 9327323; URL: www.lebanonlibrary.com 1902; Julie S. Florence 151 000 vols; 70 curr per　39838

Lebanon-Laclede County Library, 915 S Jefferson Ave, *Lebanon*, MO 65536-3667 T: +1 417 5324212; Fax: +1 417 5327424; E-mail: askref@lebanon-laclede.lib.mo.us; URL: www.lebanon-laclede.lib.mo.us Cathy Dame 1 branch libr, 2 bookmobiles 93 000 vols; 119 curr per; 490 av-mat; 338 sound-rec; 21 Journals, 215 Large Print Bks　39839

Lebanon-Wilson County Library, 108 S Hatton Ave, *Lebanon*, TN 37087-3590 T: +1 615 4440632; Fax: +1 615 4440535; E-mail: leblibrary@charter.net; URL: lebanonlibrary.net Alesia Burnley 55 000 vols; 80 curr per　39840

Ledyard Public Libraries, Bill Library, 718 Colonel Ledyard Hwy, *Ledyard*, CT 06339; P.O. Box 225, Ledyard, CT 06339-0225 T: +1 860 4649912; Fax: +1 860 4649927; E-mail: bill-lib@ledyard.lioninc.org; URL: www.lioninc.org/ledyard 1863; Gale F. Bradbury 1 branch libr 85 000 vols; 178 curr per　39841

Leesburg Public Library, 100 E Main St, *Leesburg*, FL 34748 T: +1 352 7289790; Fax: +1 352 7289794; URL: leesburgflorida.gov/library 1883; Barbara J. Morse 133 000 vols; 300 curr per; 3 008 av-mat; 2 000 sound-rec; 5 520 Large Print Bks, 8 950 Talking Bks　39842

Loudoun County Public Library, Admin Offices, 908A Trailview Blvd SE, *Leesburg*, VA 20175-4415 T: +1 703 7770368; Fax: +1 703 7715238; E-mail: libraries@loudoun.gov; URL: www.lcpl.lib.va.us 1973; Douglas A. Henderson American Sign Lang Coll, English as a Second Lang Coll; 8 branch libs 488 000 vols; 1114 curr per; 41 000 e-books; 310 Bks on Deafness & Sign Lang, 8 510 Large Print Bks libr loan; ALA　39843

Vernon Parish Library, 1401 Nolan Trace, *Leesville*, LA 71446 T: +1 337 2392027; Fax: +1 337 2380666; E-mail: admin.w1vr@pelican.state.lib.la.us; URL: www.vernon.lib.la.us 1956; Howard L. Coy Genealogy, Louisiana, Civil War, World War II; 2 branch libs 81 000 vols; 10 137 av-mat; 1 030 sound-rec　39844

Lehi City Library, 120 N Center St, *Lehi*, UT 84043 T: +1 801 7687150; Fax: +1 801 7668856; E-mail: library@lehi-ut.gov; URL: www.lehicity.com 1917; Kristi Seely 71 000 vols; 108 curr per　39845

San Diego County Library – Lemon Grove Branch, 8073 Broadway, *Lemon Grove*, CA 91945-2599 T: +1 619 4639819; Fax: +1 619 4638069 Amparo Madera 51 000 vols　39846

Lemont Public Library District, 50 E Wend St, *Lemont*, IL 60439-6439 T: +1 630 2576541; Fax: +1 630 2577737; E-mail: info@lemontlibrary.org; URL: www.lemontlibrary.org 1943; Sandra Pointon 91 000 vols; 213 curr per　39847

Johnson County Library – Lackman, 15345 W 87th St Pkwy, *Lenexa*, KS 66219 T: +1 913 4957540; Fax: +1 913 4957556 Leslie Nord 96 000 vols; 165 curr per　39848

County of Los Angeles Public Library – Lennox Library, 4359 Lennox Blvd, *Lennox*, CA 90304-2398 T: +1 310 6740385; Fax: +1 310 6736508; URL: www.colapublib.org/libs/lennox Peter Hsu 58 000 vols; 114 curr per　39849

Caldwell County Public Library, 120 Hospital Ave, *Lenoir*, NC 28645-4454 T: +1 828 7571270; Fax: +1 828 7571413; URL: www.ccpl.us 1930; Jimmy D. McKee 2 branch libs 85 000 vols; 399 curr per; 7 132 av-mat; 4 620 sound-rec; 500 digital data carriers　39850

Lenox Library Association, 18 Main St, *Lenox*, MA 01240 T: +1 413 6370197; Fax: +1 413 6372115; URL: lenoxlib.org 1856; Denis J. Lesieur Art Exhibits, Music Study Scores 67 000 vols; 137 curr per　39851

Leominster Public Library, 30 West St, *Leominster*, MA 01453 T: +1 978 5347522; Fax: +1 978 8403357; E-mail: leomref@cwmars.org; URL: www.leominsterlibrary.org 1856; Susan Theriault Shelton Career Info Ctr, Parent Resource Ctr 105 000 vols; 203 curr per　39852

Saint Mary's County Memorial Library, 23250 Hollywood Rd, *Leonardtown*, MD 20650 T: +1 301 4752846; Fax: +1 301 8844415; URL: www.stmalib.org 1950; Kathleen Reif Genealogy Society Coll; 2 branch libs 177 000 vols; 444 curr per　39853

Leonia Public Library, 227 Ft Lee Rd, *Leonia*, NJ 07605 T: +1 201 5925770; Fax: +1 201 5925775; E-mail: leoncirc@bccls.org; URL: www.leonianj.gov/content/Library.aspx 1923 60 000 vols; 135 curr per　39854

Woodward Memorial Library, Seven Wolcott St, *LeRoy*, NY 14482 T: +1 585 7688300; Fax: +1 585 7684768; E-mail: wmlib@nioga.org; URL: www.woodwardmemoriallibrary.org Roberta Bialasik Lit (Woodward Coll) 50 000 vols; 97 curr per libr loan　39855

Bucks County Free Library – Levittown Branch, 7311 New Falls Rd, *Levittown*, PA 19055-1006 T: +1 215 9492324; Fax: +1 215 9490643 Mary Ann Bursk 169 000 vols; 130 curr per　39856

Levittown Public Library, One Bluegrass Lane, *Levittown*, NY 11756-1292 T: +1 516 7315728; Fax: +1 516 7353168; E-mail: levtown@nassaulibrary.org; URL: www.nassaulibrary.org/levtown/index.html 1950; Celeste Watman Local Hist, Natural science 261 000 vols; 672 curr per　39857

Lewes Public Library, 111 Adams Ave, *Lewes*, DE 19958 T: +1 302 6452733; Fax: +1 302 6456235; E-mail: info@leweslibrary.org; URL: www.leweslibrary.org Chrys Dudbridge 51 000 vols; 119 curr per　39858

Public Library for Union County, 255 Reitz Blvd, *Lewisburg*, PA 17837-9211 T: +1 570 5231172; Fax: +1 570 5247771; URL: www.publibuc.org 1910; Kathleen Vellam 84 000 vols; 90 curr per libr loan　39859

Union County Library System, 255 Reitz Blvd, *Lewisburg*, PA 17837 T: +1 570 5231172; Fax: +1 570 5247779; URL: publibuc.org 96 100 vols; 176 curr per　39860

Union County Library System, 255 Reitz Blvd, *Lewisburg*, PA 17837-9211 T: +1 570 5231172; Fax: +1 570 5247771 Kathleen O'Brien Vellam 96 000 vols; 176 curr per　39861

Lewiston City Library, 428 Thain Rd, *Lewiston*, ID 83501-5399 T: +1 208 7436519; Fax: +1 208 7984446; E-mail: library@cityoflewiston.org; URL: www.cityoflewiston.org/library 1901; Dawn Wittman 1 branch libr 75 000 vols; 128 curr per libr loan; ILA　39862

Lewiston Public Library, 200 Lisbon St, *Lewiston*, ME 04240
T: +1 207 7840135; Fax: +1 207 7843011; URL: www.lplonline.org
1902; Richard A. Speer
French Lit Coll
128 000 vols; 292 curr per; 747 High Interest/Low Vocabulary Bks   39863

Mifflin County Library, 123 N Wayne St, *Lewistown*, PA 17044-1794
T: +1 717 2422391; Fax: +1 717 2422825; E-mail: mifflincolib@mifflincountylibrary.org; URL: www.mifflincountylibrary.org
1842; Dr. Carol J. Veitch
91 000 vols; 159 curr per   39864

Lewisville Public Library System, 1197 W Main at Civic Circle, *Lewisville*, TX 75067; P.O. Box 299002, Lewisville
T: +1 972 2193570; Fax: +1 972 2195094; URL: library.cityoflewisville.com
1968; Ann Loggins
Children's Enrichment, bk, flm
169 000 vols; 128 curr per   39865

Cary Memorial Library, 1874 Massachusetts Ave, *Lexington*, MA 02420
T: +1 781 8626288; Fax: +1 781 8627355; URL: www.carylibrary.org
1868; Connie Rawson
Original Prints Coll, American Revolutionary War Coll; 1 branch libr
213 000 vols; 303 curr per; 4 320 Audio Bks, 300 Electronic Media & Resources, 2 160 Large Print Bks   39866

Davidson County Public Library System, 602 S Main St, *Lexington*, NC 27292
T: +1 336 2422063; Fax: +1 336 2498161; URL: www.co.davidson.nc.us/library
1928; Ruth Ann Copley
Lit (Richard Walser Coll)
295 000 vols; 3 629 curr per; 2 000 e-books   39867

Lexington County Public Library System, 5440 Augusta Rd, *Lexington*, SC 29072
T: +1 803 7852600; Fax: +1 803 7852601; URL: www.lex.lib.sc.us
1948; Daniel S. MacNeill
8 branch libs
517 000 vols; 2 630 curr per; 38 707 av-mat
libr loan   39868

Lexington Public Library, 140 E Main St, *Lexington*, KY 40507-1376
T: +1 859 2315504; Fax: +1 859 2315598; URL: www.lexpublib.org
1898; Kathleen Imhoff
African-American Coll, Early Kentucky Bks & Newspapers, Grants Coll, Large Print Coll, Lexington Urban County Doc Coll; 5 branch libs
648 000 vols; 1 366 curr per   39869

Rockbridge Regional Library, 138 S Main St, *Lexington*, VA 24450-2316
T: +1 540 4634324; Fax: +1 540 4644824; E-mail: rrl@rrlib.net; URL: www.rrlib.net
1934; Linda L. Krantz
4 branch libs
161 000 vols; 271 curr per; 9 351 av-mat; 5 780 sound-rec
libr loan   39870

Liberal Memorial Library, 519 N Kansas, *Liberal*, KS 67901
T: +1 620 6260180; Fax: +1 620 6260182; E-mail: director@lmlibrary.org; URL: lmlibrary.org
1904; Jill Pannkuk
Map Coll of Southwest Kansas & Oklahoma
70 000 vols; 170 curr per   39871

Cook Memorial Public Library District, 413 N Milwaukee Ave, *Libertyville*, IL 60048-2280
T: +1 847 3622330; Fax: +1 847 3622354; URL: www.cooklib.org
1921; Frederick H. Byergo
Genealogy Reference Coll; Lake County Hist Coll
266 000 vols; 709 curr per; 254 e-books; 843 music scores; 950 microforms; 15 383 av-mat; 11 045 sound-rec; 907 digital data carriers; 128 art reproductions, 18 928 pamphlets
libr loan; ALA, PLA, ILA   39872

Ligonier Valley Library Association, Inc, 120 W Main St, *Ligonier*, PA 15658-1243
T: +1 724 2386451; Fax: +1 724 2386989; E-mail: lvlibrary@wpa.net; URL: www.ligonierlibrary.org
1946; M. Janet Hudson
Alcohol education, Drug info, Hist; Large Print Coll, Pennsylvania Rm
68 000 vols; 123 curr per; 2 477 av-mat   39873

Hawaii State Public Library System – Lihue Public Library, 4344 Hardy St, *Lihue*, HI 96766
T: +1 808 2413222; Fax: +1 808 2413225
Carolyn Larson
82 000 vols; 100 curr per   39874

Harnett County Public Library, 601 S Main St, *Lillington*, NC 27546-6107; P.O. Box 1149, Lillington, NC 27546-1149
T: +1 910 8933446; Fax: +1 910 8933001; URL: www.harnett.org/library
1941; Melanie Collins
Cooking; Alcoholism Coll, Photo Coll; 5 branch libs, 1 bookmobile
175 000 vols; 412 curr per; 5 182 av-mat; 4 461 Audio Bks, 30 High Interest/Low Vocabulary Bks
libr loan; ALA, NCLA   39875

Lima Public Library, 650 W Market St, *Lima*, OH 45801
T: +1 419 2285113; Fax: +1 419 2242669; URL: www.limalibrary.com
1884; Scott L. Shafer
Judaica, Art & Architecture; 6 branch libs
364 000 vols; 808 curr per; 218 e-books; 13 344 av-mat; 23 866 sound-rec; 18 780 digital data carriers; 12 480 Large Print Bks
libr loan   39876

Lincoln City Libraries, Bennett Martin Public Library, 136 S 14th St, *Lincoln*, NE 68508-1899
T: +1 402 4418525; Fax: +1 402 4418586; E-mail: library@lincolnlibraries.org; URL: www.lincolnlibraries.org
1877; Pat Leach
Nebraska Authors; 7 branch libs
776 000 vols; 1 810 curr per; 12 000 e-books; 93 499 av-mat; 67 160 sound-rec; 9 990 digital data carriers
libr loan   39877

Lincoln Public Library, Three Bedford Rd, *Lincoln*, MA 01773
T: +1 781 2598465; Fax: +1 781 2591056; URL: www.lincolnpl.org
1883; Barbara Myles
87 000 vols; 216 curr per   39878

Lincoln Public Library, 145 Old River Rd, *Lincoln*, RI 02865
T: +1 401 3332422; Fax: +1 401 3334154; URL: www.lincolnlibrary.com
1875; Becky A. Boragine
Descriptive Videos for Blind/Visually Handicapped
111 000 vols; 161 curr per   39879

Nebraska Library Commission, Talking Book & Braille Service, 1200 N St, Ste 120, *Lincoln*, NE 68508-2023
T: +1 402 4714038; Fax: +1 402 4712083; E-mail: tbbs@nlc.state.ne.us; URL: www.nlc.state.ne.us/tbbs/
1952; David L. Oertli
Children's Braille, Nebraska Coll
150 000 vols; 56 760 av-mat; 56 620 sound-rec; 2 000 Braille vols   39880

Driftwood Public Library, 801 SW Hwy 101, Ste 201, *Lincoln City*, OR 97367-2720
T: +1 541 9962277; Fax: +1 541 9961262; E-mail: librarian@driftwoodlib.org; URL: www.driftwoodlib.org
1965
Pacific Northwest
72 000 vols; 150 curr per   39881

Lincoln Park Public Library, 1381 Southfield Rd, *Lincoln Park*, MI 48146
T: +1 313 3810374; Fax: +1 313 3812205; URL: www.lincoln-park.lib.mi.us
1925; Theresa Powers
56 000 vols; 120 curr per   39882

Lincoln Park Public Library, 12 Boonton Tpk, *Lincoln Park*, NJ 07035
T: +1 973 6948283; Fax: +1 973 6945515
1922; Francis Kaiser
New Jersey Coll
52 000 vols; 129 curr per   39883

Vernon Area Public Library District, 300 Olde Half Day Rd, *Lincolnshire*, IL 60069-2901
T: +1 847 6343650; Fax: +1 847 6348667; URL: www.vapld.info
1974; Allen Meyer
243 000 vols; 924 curr per; 21 000 e-books   39884

Lincoln County Public Library, Charles R. Jonas Library, 306 W Main St, *Lincolnton*, NC 28092
T: +1 704 7358044; Fax: +1 704 7329042; URL: www.glrl.lib.nc.us
1925; Gary Hoyle
Lincoln County Hist & Genealogical Coll
117 000 vols   39885

Lincolnwood Public Library District, 4000 W Pratt Ave, *Lincolnwood*, IL 60712
T: +1 847 6775277; Fax: +1 847 6771937; URL: www.lincolnwoodlibrary.org
1978
Literacy Coll, David Zemsky Low Vision Coll
60 000 vols; 121 curr per; 2 159 sound-rec; 7 digital data carriers   39886

Linden Free Public Library, 31 E Henry St, *Linden*, NJ 07036
T: +1 908 2983830; Fax: +1 908 4862636; URL: www.lindenpl.org
1925; Dennis Patrick Purves
89 000 vols; 185 curr per   39887

Lindenhurst Memorial Library, One Lee Ave, *Lindenhurst*, NY 11757-5399
T: +1 631 9577755; Fax: +1 631 9570993; E-mail: lindlib@suffolk.lib.ny.us; URL: lml.suffolk.lib.ny.us
1946; Christine Salita
224 000 vols; 520 curr per   39888

Lepper Public Library, 303 E Lincoln Way, *Lisbon*, OH 44432-1400
T: +1 330 4243117; Fax: +1 330 4247343; E-mail: lepper@oplin.org; URL: www.lepper.lib.oh.us
1897; Nancy J. Simpson
67 000 vols; 184 curr per   39889

Lisle Library District, 777 Front St, *Lisle*, IL 60532-3599
T: +1 630 9711675; Fax: +1 630 9711701; URL: www.lislelibrary.org
1967; B. Strecker
Oriental Art artifacts
160 000 vols; 800 curr per; 134 maps; 128 microforms; 11 271 sound-rec; 316 digital data carriers; 6 047 videotapes, 351 art reproductions
libr loan; ALA   39890

Oliver Wolcott Library, Litchfield Public Library, 160 South St, *Litchfield*, CT 06759-0187; P.O. Box 187, Litchfield, CT 06759-0187
T: +1 860 5678030; Fax: +1 860 5674784; E-mail: owlibrary@owlibrary.org; URL: www.owlibrary.org
1862; Ann Marie White
50 000 vols; 99 curr per   39891

The Wagnalls Memorial Library, 150 E Columbus St, *Lithopolis*, OH 43136; P.O. Box 217, Lithopolis, OH 43136-0217
T: +1 614 8374765; Fax: +1 614 8370781; URL: www.wagnallslibrary.org
1925; Erma Storts
Paintings (John Ward Dunsmore Coll), Poetry Hand Written & Framed (Edwin Markham Coll), Letters (O Henry to Mabel Wagnalls Jones Coll)
90 000 vols; 200 curr per   39892

Lititz Public Library, 651 Kissel Hill Rd, *Lititz*, PA 17543
T: +1 717 6262255; Fax: +1 717 6274191; URL: www.lititzlibrary.org
Susan Miller Tennant
55 000 vols; 50 curr per   39893

Kimberly Public Library – Gerard H. Van Hoof Library, 625 Grand Ave, *Little Chute*, WI 54140
T: +1 920 7887825; Fax: +1 920 7887827; E-mail: lit@mail.owls.lib.wi.us
Barbara Wentzel
50 000 vols; 150 curr per   39894

Little Falls Public Library, 8 Warren St, *Little Falls*, NJ 07424
T: +1 973 2562784; Fax: +1 973 2566312; E-mail: pelak@littlefallslibrary.org; URL: www.littlefallslibrary.org
1905; Patricia Pelak
New Jersey
50 000 vols; 130 curr per   39895

Queens Borough Public Library – Douglaston-Little Neck Branch, 249-01 Northern Blvd, *Little Neck*, NY 11363
T: +1 718 2258414; Fax: +1 718 6318829
Elizabeth Shinouda
51 000 vols   39896

Central Arkansas Library System, 100 Rock St, *Little Rock*, AR 72201-4698
T: +1 501 9183000; Fax: +1 501 3761830; E-mail: refdesk@cals.lib.ar.us; URL: www.cals.org
1910; Dr. Bobby Roberts
Butler Center for Arkansas Studies, Jay Miller Aviation Hist Coll
826 000 vols; 1 599 curr per; 71 000 microforms; 11 digital data carriers
ALA, ULC   39897

Central Arkansas Library System – Adolphine Fletcher Terry Branch, 2015 Napa Valley Dr, *Little Rock*, AR 72212
T: +1 501 2280129
Leslie Blanchard
104 000 vols   39898

Central Arkansas Library System – Dee Brown Branch, 6325 Baseline, *Little Rock*, AR 72209-4810
T: +1 501 5687494
Sarah McClure
63 000 vols   39899

Central Arkansas Library System – John Gould Fletcher Branch, 823 N Buchanan St, *Little Rock*, AR 72205-3211
T: +1 501 6635457
Kate Matthews
95 000 vols   39900

County of Los Angeles Public Library – Littlerock Branch, 35119 80th St E, *Littlerock*, CA 93543
T: +1 661 9444138; Fax: +1 661 9444150; URL: www.colapublib.org/libs/littlerock
Trisha Pritchard
51 000 vols; 79 curr per   39901

Edwin A. Bemis Public Library, Littleton Public Library, 6014 S Datura St, *Littleton*, CO 80120-2636
T: +1 303 7953961; Fax: +1 303 7953996; E-mail: bemislib@earthlink.net; URL: www.littletongov.org/bemis
1897; Margery Smith
177 000 vols; 305 curr per; 100 Braille vols   39902

Reuben Hoar Library, 41 Shattuck St, *Littleton*, MA 01460-4506
T: +1 978 5402600; Fax: +1 978 9522323; E-mail: mli@mvlc.org; URL: www.littletonlibrary.org
Laura Zalewski
75 000 vols; 200 curr per   39903

Suwannee River Regional Library, 1848 Ohio Ave S, Dr Martin Luther King Jr Ave S, *Live Oak*, FL 32064-4517
T: +1 386 3622317; Fax: +1 386 3646071; URL: www.neflin.org/srrl
1958; John D. Hales Jr
7 branch libs
161 000 vols; 507 curr per   39904

Livermore Public Library, 1188 S Livermore Ave, *Livermore*, CA 94550
T: +1 925 3735509; Fax: +1 925 3735502; E-mail: www.livermore.lib.ca.us; URL: www.livermore.lib.ca.us
1896; Susan R. Gallinger
Story & Folk Tale Coll; 2 branch libs
220 000 vols; 400 curr per; 5 000 e-books; 15 295 av-mat; 6 626 sound-rec; 20 digital data carriers; 3 030 Large Print Bks
libr loan   39905

Liverpool Public Library, 310 Tulip St, *Liverpool*, NY 13088-4997
T: +1 315 4351800; Fax: +1 315 4537867; E-mail: info@mailbox.lpl.org; URL: www.lpl.org
1893; N.N.
Local Hist Video Coll
103 000 vols; 340 curr per   39906

Livingston-Park County Public Library, 228 W Callender St, *Livingston*, MT 59047-2618
T: +1 406 2220862; Fax: +1 406 2226522; E-mail: lpcpublib@ycsi.net; URL: library.ycsi.net
1901; Milla Cummins
Montana
53 000 vols; 239 curr per   39907

Livonia Public Library, 32777 Five Mile Rd, *Livonia*, MI 48154-3045
T: +1 734 4662450; Fax: +1 734 4586011; URL: livonia.lib.mi.us
1958
4 branch libs
256 000 vols; 672 curr per   39908

Livonia Public Library – Alfred Noble Branch, 32901 Plymouth Rd, *Livonia*, MI 48150-1793
T: +1 734 4216600; Fax: +1 734 4216606; URL: livonia.lib.mi.us/noble.html
Rachel Charette
51 000 vols; 162 curr per   39909

Livonia Public Library – Civic Center, 32777 Five Mile Rd, *Livonia*, MI 48154-3045
T: +1 734 4662491; Fax: +1 734 4586011; URL: livonia.lib.mi.us/civic.html
Kathleen L. Monroe
163 000 vols; 125 curr per   39910

Annie Halenbake Ross Library, 232 W Main St, *Lock Haven*, PA 17745-1241
T: +1 570 7483321; Fax: +1 570 7481050; E-mail: ross1@rosslibrary.org; URL: www.rosslibrary.org
1910; Diane Whitaker
Local and Pennsylvania hist colls; John Sloan Museum; 2 branch libs
126 000 vols; 141 curr per; 10 maps; 846 microforms; 3 238 av-mat; 150 sound-rec; 380 digital data carriers; 5 000 Large Print Bks
libr loan; ALA   39911

Lockport Public Library, 23 East Ave, *Lockport*, NY 14094; P.O. Box 475, Lockport, NY 14095-0475
T: +1 716 4335935; Fax: +1 716 4390198; URL: www.lockportlibrary.org
1897; Marie E. Bindeman
Freemasonry & Anti-Masonic
124 000 vols; 253 curr per   39912

Nioga Library System, 6575 Wheeler Rd, *Lockport*, NY 14094
T: +1 716 4346167; Fax: +1 716 4348231; URL: www.nioga.org
1959; Thomas C. Bindeman
209 000 vols; 577 curr per   39913

Locust Valley Library, 170 Buckram Rd, *Locust Valley*, NY 11560-1999
T: +1 516 6711837; Fax: +1 516 6768164
1910; Janis A. Schoen
Parenting (Carol Tilliston Holmboe Coll)
65 000 vols; 126 curr per
libr loan   39914

Lodi Memorial Library, One Memorial Dr, *Lodi*, NJ 07644-1692
T: +1 973 3654044; Fax: +1 973 3650172; URL: www.bccls.org/lodi
1924; Anthony P. Taormina
Italy Coll
99 000 vols; 281 curr per; 800 av-mat   39915

Logan Library, 255 N Main, *Logan*, UT 84321-3914
T: +1 435 7169123; Fax: +1 435 7169145; E-mail: libstaff@loganutah.org; URL: library.loganutah.org
1916; Ronald K. Jenkins
179 000 vols; 155 curr per; 9 500 e-books; 6 989 av-mat; 2 820 digital data carriers; 1 720 Large Print Bks   39916

Logansport-Cass County Public Library, 616 E Broadway, *Logansport*, IN 46947-3187
T: +1 574 7536383; Fax: +1 574 7225889; E-mail: library@logan.lib.in.us; URL: www.logan.lib.in.us
1894; Dave Ivey
1 branch libr
213 000 vols; 711 curr per; 1 400 av-mat; 110 sound-rec; 37 Journals, 140 Large Print Bks   39917

Helen M. Plum Memorial Library, Lombard Public Library, 110 W Maple St, *Lombard*, IL 60148-2594
T: +1 630 6270316; Fax: +1 630 6270336; URL: www.plum.lib.il.us
1928; Robert A. Harris
Music, Art
212 000 vols; 432 curr per; 1 720 music scores; 9 585 av-mat; 10 520 sound-rec; 77 high Interest/Low Vocabulary Bks, 2 877 Large Print Bks   39918

Laurel County Public Library District, 120 College Park Dr, *London*, KY 40741
T: +1 606 8645759; Fax: +1 606 8649061; E-mail: mail@laurellibrary.org; URL: laurellibrary.org
1915; Lori Acton
Children's branch libr
120 000 vols; 100 curr per   39919

Leach Library, 276 Mammoth Rd, *Londonderry*, NH 03053-3097
T: +1 603 4321132; Fax: +1 603 4376610; URL: www.londonderry.nh.org
1889; Barbara Ostertag-Holtkamp
67 000 vols; 147 curr per
libr loan; ALA   39920

Douglas County Libraries – Lone Tree Library, 8827 Lone Tree Pkwy, *Lone Tree*, CO 80124-8961
T: +1 303 7917323; Fax: +1 303 7994275
Sharon Nemechek
60 000 vols   39921

Long Beach Public Library, 111 W Park Ave, *Long Beach*, NY 11561-3326
T: +1 516 4327258; Fax: +1 516 4321477; E-mail: lblibrary@yahoo.com; URL: www.longbeachlibrary.org
1928; George Trepp
Holocaust; Foreign Language, Congressman Allard K Lowenstein Memorabilia Coll; 2 branch libs
173 000 vols; 312 curr per; 6 100 av-mat; 8 490 sound-rec; 1 640 digital data carriers   39922

Long Beach Public Library & Information Center, 101 Pacific Ave, *Long Beach*, CA 90822-1097
T: +1 562 5706053; Fax: +1 562 5707408; URL: www.lbpl.org
1896; Glenda Williams
Petroleum, Video cassettes; Bertrand L. Smith Rare Books, Marilyn Horne Coll; 11 branch libs
870 000 vols; 8 192 av-mat; 25 421 sound-rec; 215 digital data carriers
libr loan   39923

Long Beach Public Library & Information Center – Bret Harte Branch, 1595 W Willow St, *Long Beach*, CA 90810
T: +1 562 5701044
Karol Seehaus
53 000 vols   39924

Long Beach Public Library & Information Center – Burnett, 560 E Hill St, *Long Beach*, CA 90806
T: +1 562 5701001
Mary Hopman
55 000 vols   39925

Long Beach Public Library & Information Center – Dana, 3680 Atlantic Ave, *Long Beach*, CA 90807
T: +1 562 5701042
Jennifer Songster
52 000 vols   39926

Long Beach Public Library & Information Center – El Dorado, 2900 Studebaker Rd, *Long Beach*, CA 90815
T: +1 562 5703136
Nancy Paradise
64 000 vols   39927

Long Beach Public Library & Information Center – Los Altos, 5614 Britton Dr, *Long Beach*, CA 90815
T: +1 562 5701045
Clifford Phillips
51 000 vols   39928

Long Branch Free Public Library, 328 Broadway, *Long Branch*, NJ 07740
T: +1 732 2223900; Fax: +1 732 2223799; E-mail: ibruck@lmxac.org; URL: www.lmxac.org/longbranch
1916; Ingrid Bruck
Spanish & Portuguese Coll, Bks on Tape
76 890 vols; 106 curr per   39929

Queens Borough Public Library – Broadway, 40-20 Broadway, *Long Island City*, NY 11103
T: +1 718 7212462
Tatyana Magazinnik
108 000 vols   39930

Queens Borough Public Library – Steinway, 21-45 31st St, *Long Island City*, NY 11105
T: +1 718 7281965; Fax: +1 718 9563575
Christine Gina
76 000 vols   39931

Queens Borough Public Library – Sunnyside, 43-06 Greenpoint Ave, *Long Island City*, NY 11104
T: +1 718 7843033
Anne Bagnall
81 000 vols   39932

Washington Township Free Public Library, 37 E Springtown Rd, *Long Valley*, NJ 07853
T: +1 908 8763596; Fax: +1 908 8763541; URL: www.wtpl.org
1968; Virginia Scarlatelli
66 000 vols; 143 curr per   39933

Richard Salter Storrs Library, 693 Longmeadow St, *Longmeadow*, MA 01106
T: +1 413 5654181; Fax: +1 413 5654183; URL: www.longmeadow.org/library/overview.htm
1908; Carl L. Sturgis
80 000 vols; 130 curr per   39934

Longmont Public Library, 409 Fourth Ave, *Longmont*, CO 80501-6006
T: +1 303 6518472; Fax: +1 303 7744365; URL: www.ci.longmont.co.us/library.htm
1871; Tony Brewer
Automotive, Business, Career, Consumer, Spanish
360 000 vols; 358 curr per   39935

Longview Public Library, 222 W Cotton St, *Longview*, TX 75601-6348
T: +1 903 2371354; Fax: +1 903 2371327; URL: www.longviewlibrary.com
1932; Kara Spitz
East Texas, Oil Field Production Records
121 000 vols; 345 curr per
libr loan   39936

Longview Public Library, 1600 Louisiana St, *Longview*, WA 98632-2993
T: +1 360 4425310; Fax: +1 360 4425954; URL: www.longviewlibrary.org
1926; Chris Skaugset
Rudolf Steiner Coll, Longview Coll
158 000 vols; 300 curr per; 4 650 sound-rec; 2 800 digital data carriers   39937

Lonoke Prairie County Regional Library Headquarters, 204 E Second St, *Lonoke*, AR 72086-2858
T: +1 501 6766635; Fax: +1 501 6767687; URL: www.lpregional.lib.ar.us
1937; Leroy Gattin
8 branch libs
131 000 vols; 240 curr per
libr loan   39938

Lorain Public Library System, 351 Sixth St, *Lorain*, OH 44052
T: +1 440 2441192; Fax: +1 440 2441733; URL: www.lorain.lib.oh.us
1901; Joanne Eldridge
Ohio Hist, Census Coll, Philosophy & Religion (Hageman Memorial Coll); 5 branch libs
518 000 vols; 1 301 curr per; 50 790 av-mat; 52 930 sound-rec; 17 130 digital data carriers; 10 Braille vols
libr loan   39939

Los Alamos County Library System, 2400 Central Ave, *Los Alamos*, NM 87544
T: +1 505 6628250; Fax: +1 505 6628246; URL: library.lac-nm.us
1943; Charlie Kalogeros-Chattan
Southwest Coll
152 000 vols; 514 curr per; 4 000 e-books   39940

Santa Clara County Library – Los Altos Main Library, 13 S San Antonio Rd, *Los Altos*, CA 94022-3049
T: +1 650 9487683; Fax: +1 650 9416308
Cheryl Houts
244 000 vols   39941

County of Los Angeles Public Library – A. C. Bilbrew Library, 150 E El Segundo Blvd, *Los Angeles*, CA 90061-2356
T: +1 310 5383350; Fax: +1 310 3270824; URL: www.colapublib.org/libs/bilbrew
Alice Tang
Black Resource Center
81 000 vols; 114 curr per   39942

County of Los Angeles Public Library – Anthony Quinn Library, 3965 Cesar E Chavez Ave, *Los Angeles*, CA 90063
T: +1 323 2647715; Fax: +1 323 2627121; URL: www.colapublib.org/libs/quinn
Joshua Cloner
63 000 vols; 94 curr per   39943

County of Los Angeles Public Library – City Terrace Library, 4025 E City Terrace Dr, *Los Angeles*, CA 90063-1297
T: +1 323 2610295; Fax: +1 323 2611790; URL: www.colapublib.org/libs/cityterrace
Jing Li
55 000 vols; 118 curr per   39944

County of Los Angeles Public Library – East Los Angeles Library, 4837 E Third St, *Los Angeles*, CA 90022-1601
T: +1 323 2640155; Fax: +1 323 2645465; URL: www.colapublib.org/libs/eastla
Alice Medina
Chicano Resource Center
123 000 vols; 221 curr per   39945

County of Los Angeles Public Library – Florence Library, 1610 E Florence Ave, *Los Angeles*, CA 90001-2522
T: +1 323 5818028; Fax: +1 323 5873240; URL: www.colapublib.org/libs/florence
Judy Weigel
52 000 vols; 105 curr per   39946

County of Los Angeles Public Library – Graham Library, 1900 E Firestone Blvd, *Los Angeles*, CA 90001-4126
T: +1 323 5822903; Fax: +1 323 5818478; URL: www.colapublib.org/libs/graham
Cindy Singer
52 000 vols; 112 curr per   39947

County of Los Angeles Public Library – View Park Library, 3854 W 54th St, *Los Angeles*, CA 90043-2297
T: +1 323 2935371; Fax: +1 323 2924330; URL: www.colapublib.org/libs/viewpark
Linda Dickerson
66 000 vols; 98 curr per   39948

County of Los Angeles Public Library – Woodcrest Library, 1340 W 106th St, *Los Angeles*, CA 90044-1626
T: +1 323 7579373; Fax: +1 323 7564907; URL: www.colapublib.org/libs/woodcrest
Donald Rowe
65 000 vols; 107 curr per   39949

Los Angeles Public Library System, 630 W Fifth St, *Los Angeles*, CA 90071-2097
T: +1 213 2287515; Fax: +1 213 2287429; URL: www.lapl.org
1872; Anne Connor
California Hist, Japanese Prints, Automotive Repair Manuals, Language Study, Film Study; 70 branch libs
6 219 000 vols; 91 000 maps; 19 000 sound-rec; 244 digital data carriers
libr loan   39950

Los Angeles Public Library System – Chinatown, 639 N Hill St, *Los Angeles*, CA 90012-2317
T: +1 213 6200925; Fax: +1 213 6209956
Shan Liang
85 000 vols   39951

Los Angeles Public Library System – El Sereno, 5226 Huntington Dr S, *Los Angeles*, CA 90032
T: +1 323 2259201; Fax: +1 323 4410112
Eugene Estrada
53 000 vols   39952

Los Angeles Public Library System – Frances Howard Goldwyn-Hollywood Regional Branch, 1623 N Ivar Ave, *Los Angeles*, CA 90028-6305
T: +1 323 4671821; Fax: +1 323 4675707
Judy Hermann
91 000 vols   39953

Los Angeles Public Library System – Harbor City-Harbor Gateway, 24000 S Western Ave, *Los Angeles*, CA 90710
T: +1 310 5349520; Fax: +1 310 9521932
Jene Brown
56 000 vols   39954

Los Angeles Public Library System – Lake View Terrace, 12002 Osborne St, *Los Angeles*, CA 91342
T: +1 818 8907404; Fax: +1 818 8972738
Constance Dosch
52 000 vols   39955

Los Angeles Public Library System – Little Tokyo, 203 1s Los Angeles St, *Los Angeles*, CA 90012
T: +1 213 6120525; Fax: +1 213 6120424
Hitoshi Ohta
66 000 vols   39956

Los Angeles Public Library System – Los Feliz, 1801 Hillhurst Ave, *Los Angeles*, CA 90027-2794
T: +1 323 9134710; Fax: +1 323 9134714
Pearl Yonezawa
54 000 vols   39957

Los Angeles Public Library System – Pio Pico-Koreatown, 694 S Oxford, *Los Angeles*, CA 90005-2872
T: +1 213 3687282; Fax: +1 213 6120433
Myungcha Miki Lim
83 000 vols   39958

Los Angeles Public Library System – West Los Angeles Regional, 11360 Santa Monica Blvd, *Los Angeles*, CA 90025-3152
T: +1 310 5758323; Fax: +1 310 3128309
N.N.
52 000 vols   39959

Los Angeles Public Library System – Will & Ariel Durant Branch, 7140 W Sunset Blvd, *Los Angeles*, CA 90046
T: +1 323 8762741; Fax: +1 323 8760485
Hannah Kramer
58 000 vols   39960

Los Angeles Public Library System – Wilmington, 1300 N Avalon Blvd, *Los Angeles*, CA 90744
T: +1 310 8341082; Fax: +1 310 5487418
John Pham
52 000 vols   39961

Santa Clara County Library, 14600 Winchester Blvd, *Los Gatos*, CA 95032
T: +1 408 2932326; Fax: +1 408 3640161; E-mail: melinda.cervantes@lib.sccgov.org; URL: www.santaclaracountylib.org
1912; Melinda Cervantes
California Western Americana; 8 branch libs
1 319 000 vols; 2 831 curr per; 30 498 av-mat; 3 871 High Interest/Low Vocabulary Bks, 34 520 Large Print Bks, 30 500 Talking Bks   39962

Loudonville Public Library, 122 E Main St, *Loudonville*, OH 44842
T: +1 419 9945531; Fax: +1 419 9944321; E-mail: hansench@loudonvillelibrary.org; URL: www.loudonvillelibrary.org
1905; Charlie Hansen
50 000 vols; 195 curr per   39963

Franklin County Library – Louisburg Main Library, 906 N Main St, *Louisburg*, NC 27549-2199
T: +1 919 4962111; Fax: +1 919 4975821; URL: www.fcnclibrary.org
1937; Holt Kornegay
3 branch libs
105 000 vols; 183 curr per   39964

Louisville Public Library, 306 E Broad St, *Louisville*, GA 30434
T: +1 478 6253751;
Fax: +1 478 6257683; URL: www.jefferson.public.lib.ga.us/librarybranches.php
1954; Charlotte Rogers
Children's lit; Georgia Hist Coll, Southern Lit Coll; 2 branch libs
81 000 vols; 143 curr per   39965

Louisville Public Library, 700 Lincoln Ave, *Louisville*, OH 44641-1474
T: +1 330 8751696; Fax: +1 330 8753530; URL: www.louisvillelibrary.org
1935; Mike Snyder
120 000 vols; 225 curr per   39966

Louisville Public Library, 951 Spruce St, *Louisville*, CO 80027
T: +1 303 3354849; Fax: +1 303 3354833; E-mail: ref_desk@ci.louisville.co.us; URL: www.ci.louisville.co.us/library
1925; Anne Mojo
70 000 vols; 186 curr per   39967

Loveland Public Library, 300 N Adams Ave, *Loveland*, CO 80537-5754
T: +1 970 9622402; Fax: +1 970 9622905; URL: www.ci.loveland.co.us/library/libmain.htm
1905; Ted Schmidt
Oral Hist Coll, Western Americana Coll
150 000 vols; 319 curr per; 4 477 av-mat; 1 500 High Interest/Low Vocabulary Bks, 4 500 Talking Bks   39968

Public Library of Cincinnati & Hamilton County – Symmes Township, 11850 E Enyart Rd, *Loveland*, OH 45140
T: +1 513 3696001; Fax: +1 513 3694481
Tara Kressler
71 000 vols   39969

North Suburban Library-Loves Park, 6340 N Second St, *Loves Park*, IL 61111
T: +1 815 6334247;
Fax: +1 815 6334249; URL: www.northsuburbanlibrary.org
1944; Ann Powell
1 branch libr
198 000 vols; 354 curr per; 6 500 sound-rec; 5 digital data carriers
libr loan   39970

Lovington Public Library, 115 Main St, *Lovington*, NM 88260
T: +1 505 3963144; Fax: +1 505 3967189; URL: lovingtonpublib.leaco.net
1931
55 600 vols; 45 curr per   39971

Lowell Public Library, 1505 E Commercial Ave, *Lowell*, IN 46356-1899
T: +1 219 6967704; Fax: +1 219 6965280; URL: www.lowellpl.lib.in.us
Sandra Morgan
Large type print, Indiana
87 000 vols; 169 curr per   39972

Pollard Memorial Library, 401 Merrimack St, *Lowell*, MA 01852
T: +1 978 9704120; Fax: +1 978 9704117; URL: www.pollardml.org
1844; N.N.
Hist of Lowell
166 000 vols; 580 curr per   39973

Lubbock Public Library – Godeke, 6601 Quaker Ave, *Lubbock*, TX 79413
T: +1 806 7926566; Fax: +1 806 7673762
Tammy Brawn
87 000 vols   39974

Lubbock Public Library – Mahon (Main), 1306 Ninth St, *Lubbock*, TX 79401
T: +1 806 7752834; Fax: +1 806 7752827; URL: www.lubbocklibrary.com
1966; Jane Clausen
3 branch libs
384 000 vols; 164 curr per   39975

Mason County District Library, Ludington Branch, 217 E Ludington Ave, *Ludington*, MI 49431-2118; P.O. Box 549, Ludington, MI 49431-0549
T: +1 231 8438465; Fax: +1 231 8431491; E-mail: librarian@masoncounty.lib.mi.us; URL: www.masoncounty.lib.mi.us
1905; Robert T. Dickson
Genealogical Coll of Mason County; 1 branch libr
120 000 vols; 200 curr per   39976

Hubbard Memorial Library, 24 Center St, *Ludlow*, MA 01056-2795
T: +1 413 5833408; Fax: +1 413 5835646; E-mail: info@hubbardlibrary.org; URL: www.hubbardlibrary.org
1891; Judy Kelly
Vietnamese Conflict, WW II
57 000 vols; 95 curr per   39977

Kurth Memorial Library, 706 S Raguet, *Lufkin*, TX 75904
T: +1 936 6300560;
Fax: +1 936 6392487; URL: www.kurthmemoriallibrary.com
1932; Sue Randleman
60 000 vols; 80 curr per   39978

Saint Charles Parish Library, West Regional Branch, 105 Lakewood Dr, *Luling*, LA 70070; P.O. Box 949, Luling, LA 70070-0949
T: +1 985 7858471; Fax: +1 985 7858499; URL: www.stcharles.lib.la.us
1955; Mary Des Bordes
4 branch libs
222 000 vols; 600 curr per   39979

Robeson County Public Library, 101 N Chestnut St, *Lumberton*, NC 28358-5639; P.O. Box 988, Lumberton
T: +1 910 7384859; Fax: +1 910 7398321
1967; Robert F. Fisher
Indian mat; 5 branch libs, 1 bookmobile
143 000 vols; 129 curr per   39980

St. James Parish Library, 1879 W Main St, *Lutcher*, LA 70071
T: +1 225 8693618; Fax: +1 225 8698435; URL: www.stjames.lib.la.us
1966
Louisiana Coll; 1 branch libr
66 670 vols; 155 curr per   39981

Tampa-Hillsborough County Public Library System – Lutz Branch, 101 Lutz Lake Fern Rd W, *Lutz*, FL 33548-7220
Fax: +1 813 2643907
Janet Spearel
63 000 vols   39982

Lynbrook Public Library, 56 Eldert St, *Lynbrook*, NY 11563
T: +1 516 5998630; Fax: +1 516 5961312; E-mail: lynbrook@nassaulibrary.org
1929; Natalie Lapp
84 000 vols; 152 curr per; 983 av-mat
libr loan   39983

Lynchburg Public Library, 2315 Memorial Ave, *Lynchburg*, VA 24501
T: +1 434 4556300; Fax: +1 434 8471578; URL: www.lynchburgva.gov/publiclibrary
1966; Lynn L. Dodge
1 branch libr
150 000 vols; 241 curr per; 700 microforms; 3 000 sound-rec
libr loan; ALA   39984

Lyndhurst Free Public Library, 355 Valley Brook Ave, *Lyndhurst*, NJ 07071
T: +1 201 8042478; Fax: +1 201 9397677; E-mail: lyndcirc@mail.bccls.org; URL: www.bccls.org/lyndhurst
1914; Donna M. Romeo
65 000 vols; 114 curr per; 5 maps; 232 microforms
libr loan   39985

Lynnfield Public Library, 18 Summer St, *Lynnfield*, MA 01940-1837
T: +1 781 3345411; Fax: +1 781 3342164; E-mail: lfd@noblenet.org; URL: www.noblenet.org/lynnfield
1892; Sue Koronowski
Learning Resources Coll
55 000 vols; 173 curr per   39986

Sno-Isle Libraries – Lynnwood Branch, 19200 44th Ave W, *Lynnwood*, WA 98036
T: +1 425 7782148; Fax: +1 425 7747764; E-mail: lynref@sno-isle.org
Michael Delury
183 000 vols   39987

County of Los Angeles Public Library – Lynwood Library, 11320 Bullis Rd, *Lynwood*, CA 90262-3661
T: +1 310 6357121; Fax: +1 310 6354967; URL: www.colapublib.org/libs/lynwood
Glorieta Navo
106 000 vols; 140 curr per   39988

Lyons Public Library, 4209 Joliet Ave, *Lyons*, IL 60534-1597
T: +1 708 4473577; Fax: +1 708 4473589; E-mail: lyons@lyonslibrary.org; URL: www.lyonslibrary.org
1938; Sarah C. Horn
17th, 18th, 19th C Passenger Lists
60 000 vols; 115 curr per   39989

Cobb County Public Library System – South Cobb, 805 Clay Rd, *Mableton*, GA 30126
T: +1 678 3985828; Fax: +1 678 3985833
Vicki Green
88 000 vols   39990

Macomb Public Library District, 235 S Lafayette St, *Macomb*, IL 61455-2231
T: +1 309 8332714; Fax: +1 309 8332714; E-mail: library@macomb.com; URL: www.macomb.lib.il.us
1881; Dennis Danowski
60 000 vols; 200 curr per   39991

Clinton-Macomb Public Library, North, 16800 24 Mile Rd, *Macomb Township*, MI 48042
T: +1 586 2265080; Fax: +1 586 2265088
Emily Kubash
53 000 vols   39992

Middle Georgia Regional Library System – Genealogical & Historical Room & Georgia Archives, 1180 Washington Ave, *Macon*, GA 31201-1790; P.O. Box 6334, Macon, GA 31208-6334
T: +1 478 7440841; Fax: +1 478 7440840
Gail Moon
132 000 vols   39993

Middle Georgia Regional Library System – Riverside Branch, Rivergate Shopping Ctr, 110 Holiday Dr N, *Macon*, GA 31210
T: +1 478 7578900; Fax: +1 478 7571094
Vivecca Jackson
53 000 vols   39994

Middle Georgia Regional Library System – Washington Memorial (Main Library), 1180 Washington Ave, *Macon*, GA 31201-1790
T: +1 478 7440841; Fax: +1 478 7423161; URL: www.co.bibb.ga.us/library/
1889; Thomas Jones
African-American; Genealogy Dept; 15 branch libs
483 000 vols
libr loan   39995

Middle Georgia Regional Library System – West Bibb Branch, Northwest Commons, 5580 Thomaston Rd, *Macon*, GA 31220-8118
T: +1 478 7440818; Fax: +1 478 7440819
Iona Foreman
72 000 vols   39996

Public Library of Cincinnati & Hamilton County – Madeira Branch, 7200 Miami Ave, *Madeira*, OH 45243
T: +1 513 3696028; Fax: +1 513 3694501
Kathy Kennedy-Brunner
77 000 vols   39997

Gulf Beaches Public Library, 200 Municipal Dr, *Madeira Beach*, FL 33708
T: +1 727 3912828; Fax: +1 727 3992840; URL: www.tblc.org/gulfbeaches
1952; Jan L. Horah
Florida Coll
70 000 vols; 127 curr per   39998

Madera County Library, 121 North G St, *Madera*, CA 93637-3592
T: +1 559 6757871; Fax: +1 559 6757998; URL: www.sjvls.org/madera
1910; John Taylor
103 000 vols; 240 curr per
libr loan   39999

Boone-Madison Public Library, 375 Main St, *Madison*, WV 25130-1295
T: +1 304 3697842; Fax: +1 304 3697842; E-mail: mischlar@mail.mln.lib.wv.us; URL: boone.lib.wv.us
1974; Susan Mischler
50 870 vols; 75 curr per; 50 maps; 330 av-mat; 626 sound-rec   40000

Dane County Library Service, 201 W Mifflin St, *Madison*, WI 53703-2597
T: +1 608 2666388; E-mail: dcljac@scls.lib.wi.us; URL: www.scls.lib.wi.us/dcl
1966; Julie Anne Chase
56 000 vols; 34 curr per   40001

Huntsville-Madison County Public Library – Madison Public Library, 130 Plaza Blvd, *Madison*, AL 35758
T: +1 256 4610046; Fax: +1 256 4610530
Sarah Sledge
80 000 vols   40002

Madison Public Library, 201 W Mifflin St, *Madison*, WI 53703
T: +1 608 2666302; Fax: +1 608 2664230; E-mail: madcirc@scls.lib.wi.us; URL: www.madisonpubliclibrary.org
1875; Barbara Dimick
8 branch libs
786 000 vols; 2 647 curr per; 7 000 e-books; 59 571 av-mat; 63 940 sound-rec; 21 340 digital data carriers; 25 Braille vols, 1 670 High Interest/Low Vocabulary Bks, 103 Bks on Deafness & Sign Lang, 21 220 Large Print Bks   40003

Madison Public Library, 39 Keep St, *Madison*, NJ 07940
T: +1 973 3770722; Fax: +1 973 3773142; URL: www.rosenet.org/library/
1900; Nancy S. Adamczyk
American lit, British lit; Golden Hind Press Pubs
127 000 vols; 249 curr per   40004

Madison Public Library, 6111 Middle Ridge Rd, *Madison*, OH 44057-2818
T: +1 440 4282189; Fax: +1 440 4287402; E-mail: info@madison-library.info; URL: www.madison-library.info
1915; Nancy Currie
106 000 vols; 310 curr per; 190 High Interest/Low Vocabulary Bks, 5 000 Large Print Bks   40005

Madison Public Library – Alicia Ashman Branch, 733 N High Point Rd, *Madison*, WI 53717
T: +1 608 8241780; Fax: +1 608 8241790
Margie Navarre-Saaf
72 000 vols; 199 curr per; 3 227 av-mat   40006

Madison Public Library – Pinney Branch, 204 Cottage Grove Rd, *Madison*, WI 53716
T: +1 608 2247100; Fax: +1 608 2247102
Norma Hanson
90 000 vols; 223 curr per; 2 625 av-mat   40007

Madison Public Library – Sequoya Branch, 513 S Midvale Blvd, *Madison*, WI 53711
T: +1 608 2666385; Fax: +1 608 2667353
Ann Michalski
109 000 vols; 246 curr per; 3 798 av-mat   40008

Nashville Public Library – Madison Branch, 610 Gallatin Pike S, *Madison*, TN 37115-4013
T: +1 615 8625868; Fax: +1 615 8625889
DeAnza Williams
110 000 vols   40009

E. C. Scranton Memorial Library, 801 Boston Post Rd, *Madison*, CT 06443
T: +1 203 2457365; Fax: +1 203 2457821; E-mail: scrantonlibrary@madisonct.org; URL: www.scrantonlibrary.org
1900; Marcia Sokolnicki
101 000 vols; 215 curr per   40010

Uncle Remus Regional Library System, 1131 East Ave, *Madison*, GA 30650
T: +1 706 3424974; Fax: +1 706 3424510; URL: www.uncleremus.org
1952; Steve W. Schaefer
Joel Chandler Harris Coll; 8 branch libs
215 000 vols; 126 curr per; 15 741 av-mat; 9 341 sound-rec; 1 081 digital data carriers   40011

Madison Heights Public Library, 240 W 13 Mile Rd, *Madison Heights*, MI 48071-1894
T: +1 248 8372852; Fax: +1 248 5882470; E-mail: library@madison-heights.org; URL: www.madison-heights.org/library
1954; Roslyn Yerman
1 branch libr
117 000 vols; 177 curr per   40012

Hopkins County-Madisonville Public Library, 31 S Main St, *Madisonville*, KY 42431
T: +1 270 8252680; Fax: +1 270 8252777; E-mail: library@publiclibrary.org; URL: www.publiclibrary.org
1974; Lisa Wigley
Kentucky Coll, Rare Bks; 1 branch libr
85 000 vols; 108 curr per; 800 av-mat   40013

Jefferson County Library District, 241 SE Seventh St, *Madras*, OR 97741-1611
T: +1 541 4753351; Fax: +1 541 4757434; E-mail: library@jcld.org; URL: www.jcld.org
1915; Sally Beesley
Jefferson County Hist
50 000 vols; 88 curr per
libr loan; ALA   40014

Salt Lake County Library Services – Magna Branch, 8339 W 3500 South, *Magna*, UT 84044-1870
T: +1 801 9447626; Fax: +1 801 2506927
Trish Hull
86 000 vols   40015

Columbia County Library, Asa C. Garrett Memorial Library, 220 E Main St, *Magnolia*, AR 71753
T: +1 870 2341991; Fax: +1 870 2345077; URL: colcnty.lib.ar.us
1947; Laura Cleveland
Arkansas, Genealogy; 6 branch libs
125 000 vols; 58 curr per   40016

Mahopac Public Library, 668 Rte 6, *Mahopac*, NY 10541
T: +1 845 6282009; Fax: +1 845 6280672; E-mail: library@mahopaclibrary.org; URL: www.mahopaclibrary.org
1952; Patricia Kaufman
Putnam County Ref Ctr, Foundation Ctr, Health Info Ctr, Job & Education Info Ctr, Parenting Coll
104 000 vols; 246 curr per   40017

Mahwah Public Library, 100 Ridge Rd, *Mahwah*, NJ 07430
T: +1 201 5297323; Fax: +1 201 5299027; E-mail: mahw@bccls.org; URL: www.bccls.org/mahwah
1912; Kenneth W. Giaimo
Poetry, Theater; Videocassette Coll, Business Directories
85 000 vols; 268 curr per   40018

Maitland Public Library, 501 S Maitland Ave, *Maitland*, FL 32751-5672
T: +1 407 6477700; URL: www.maitlandpubliclibrary.org
1896; Karen Potter
Natural Hist & Environment (Audubon Coll)
80 000 vols; 225 curr per   40019

County of Los Angeles Public Library – Malibu Library, 23519 W Civic Center Way, *Malibu*, CA 90265-4804
T: +1 310 4566438; Fax: +1 310 4568681; URL: www.colapublib.org/libs/malibu
Elaine Adler
Arkel Erb Memorial Mountaineering Coll
72 000 vols; 39 curr per   40020

Malvern-Hot Spring County Library, Mid-Arkansas Regional Library, 202 E Third St, *Malvern*, AR 72104
T: +1 501 3325441; Fax: +1 501 3326679; E-mail: hotspringcountylibrary@yahoo.com; URL: www.hsc.lib.ar.us
1928; Regina C. Cortez
74 000 vols; 70 curr per   40021

Malverne Public Library, 61 Saint Thomas Pl, *Malverne*, NY 11565
T: +1 516 5990750; Fax: +1 516 5993320; E-mail: malverne@nassaulibrary.org; URL: www.nassaulibrary.org/malverne
1928; Joan Kelleher
55 000 vols; 107 curr per   40022

Mamaroneck Public Library District, 136 Prospect Ave, *Mamaroneck*, NY 10543
T: +1 914 6981250; Fax: +1 914 3813088; URL: www.mamaronecklibrary.org
1922; Joan Grott
94 000 vols; 218 curr per   40023

Mono County Free Library, Mammoth Lakes, 960 Forest Trail, *Mammoth Lakes*, CA 93546; P.O. Box 1120, Mammoth Lakes, CA 93546-1120
T: +1 760 9344777; Fax: +1 760 9346268; URL: www.monocolibraries.org
Bill Michael
110 000 vols; 120 curr per   40024

Ocean County Library – Stafford, 129 N Main St, *Manahawkin*, NJ 08050-2933
T: +1 609 5973381; Fax: +1 609 9780770
Sharon Osborn
53 000 vols   40025

Manchester City Library, Carpenter Memorial Bldg, 405 Pine St, *Manchester*, NH 03104-6199
T: +1 603 6246550; Fax: +1 603 6246559; URL: www.manchester.lib.nh.us
1854; Denise van Zanten
New Hampshire Coll; 1 branch libr
235 000 vols; 401 curr per   40026

Manchester Public Library, Mary Cheney Library, 586 Main St, *Manchester*, CT 06040
T: +1 860 6450821; Fax: +1 860 6439453; URL: library.ci.manchester.ct.us
1871; Douglas McDonough
1 branch libr
191 000 vols; 255 curr per   40027

Pine Mountain Regional Library, 218 Perry St NW, *Manchester*, GA 31816-1317; P.O. Box 709, Manchester, GA 31816-0709
T: +1 706 8463851; Fax: +1 706 8469632
Charles B. Gee
119 000 vols   40028

Manhasset Public Library, 30 Onderdonk Ave, *Manhasset*, NY 11030
T: +1 516 6272300; Fax: +1 516 3653466; URL: www.nassaulibrary.org/manhass
1945; Marian P. Robertson
Puppet Coll, Career Ctr, Long Island Coll, New York Coll, Poetry for Children Coll
116 000 vols; 341 curr per; 3 970 av-mat; 2 310 sound-rec; 2 210 digital data carriers; 75 Special Interest Per Sub, 530 Bks on Deafness & Sign Lang
libr loan   40029

Manhattan Public Library, 629 Poyntz Ave, *Manhattan*, KS 66502-6086
T: +1 785 7764741; Fax: +1 785 7761545; URL: www.manhattan.lib.ks.us
1904; Fred Atchison
174 000 vols; 325 curr per; 5 065 av-mat; 5 070 Audio Bks
libr loan   40030

Manhattan Public Library District, 240 Whitson St, *Manhattan*, IL 60442; P.O. Box 53, Manhattan, IL 60442-0053
T: +1 815 4783987; Fax: +1 815 4783988; URL: www.mpld.org
1909; Judith Ann Pet
53 000 vols; 122 curr per   40031

North Central Kansas Libraries System, 629 Poyntz Ave, *Manhattan*, KS 66502-6086
T: +1 785 7764741; Fax: +1 785 7761545; URL: www.manhattan.lib.ks.us/nckl
1968; Fred D. Atchison Jr
1 branch libr
58 000 vols   40032

County of Los Angeles Public Library – Manhattan Beach Library, 1320 Highland Ave, *Manhattan Beach*, CA 90266-4789
T: +1 310 5458595; Fax: +1 310 5455394; URL: www.colapublib.org/libs/manhattan
Don Gould
107 000 vols; 182 curr per   40033

Manistee County Library System, Manistee County Library, 95 Maple St, *Manistee*, MI 49660
T: +1 231 7232519; Fax: +1 231 7238270; E-mail: admin@manisteelibrary.org; URL: www.manisteelibrary.org
1903; Charles Haemker
Victorian lit; 5 branch libs
110 000 vols; 220 curr per; 2 246 av-mat
libr loan   40034

Manitowoc Public Library, 707 Quay St, *Manitowoc*, WI 54220
T: +1 920 6834863; Fax: +1 920 6834873; E-mail: mplref@mcls.lib.wi.us; URL: www.manitowoc.lib.wi.us
1900; Patty Dwyer Wanninger
Submarines; Art & Gardening (Ruth West Libr of Beauty), Behnke Hist Photo Coll, World War II Personal Narratives
221 000 vols; 359 curr per
libr loan   40035

Blue Earth County Library Services, 100 E Main St, *Mankato*, MN 56001
T: +1 507 3044001; Fax: +1 507 3044009; URL: www.beclibrary.org
1902; Tim Hayes
Minnesota Print Mat; 2 branch libs
300 000 vols; 520 curr per
libr loan   40036

DeSoto Parish Library, 109 Crosby St, *Mansfield*, LA 71052
T: +1 318 8726100; Fax: +1 318 8726120; URL: www.desotoparishlibrary.com/Index.aspx
1941
Civil War, World War I & II; Louisiana Hist Coll; 3 branch libs
100 000 vols; 116 curr per   40037

Mansfield Public Library, 104 S Wisteria St, *Mansfield*, TX 76063
T: +1 817 4734391; Fax: +1 817 4534975; E-mail: library@ci.mansfield.tx.us; URL: www.mansfield-tx.gov/departments/library
Steven R. Standefer
55 000 vols; 120 curr per   40038

Mansfield Public Library, 255 Hope St, *Mansfield*, MA 02048-2353
T: +1 508 2617380; Fax: +1 508 2617422; E-mail: mansfieldlibrary2001@yahoo.com; URL: www.sailsinc.org/mansfield
1884; Janet Campbell
77 000 vols; 188 curr per; 361 e-books; 6 160 av-mat; 2 721 sound-rec; 224 digital data carriers; 244 puppets and kits
libr loan   40039

Mansfield-Richland County Public Library, 43 W Third St, *Mansfield*, OH 44902-1295
T: +1 419 5213140; Fax: +1 419 5254750; URL: www.mrcpl.org
1887; Joseph C. Palmer
Personal Libr of Senator John Sherman; 8 branch libs
516 000 vols; 233 000 av-mat; 48 000 sound-rec
libr loan   40040

Mansfield Public Library, 54 Warrenville Rd, *Mansfield Center*, CT 06250
T: +1 860 4232501; Fax: +1 860 4239856; E-mail: mansfield@biblio.org; URL: www.biblio.org/mansfield
1906; Louise Bailey
80 000 vols; 36 digital data carriers
40041

Stockton-San Joaquin County Public Library – Manteca Branch, 320 W Center St, *Manteca*, CA 95336
Fax: +1 209 8252394
Diane Bills
98 000 vols
40042

Dare County Library, Highway 64 and Burnside Rd, P.O. Box 1000, *Manteo*, NC 27954
T: +1 252 4732372; Fax: +1 252 4736034; URL: www.earlibrary.org/manteo/about-us.html
1935
85 000 vols; 180 curr per
40043

Sabine Parish Library, 705 Main St, *Many*, LA 71449-3199
T: +1 318 2564150; Fax: +1 318 2564154; E-mail: admin.g1sb@pelican.state.lib.la.us; URL: www.sabine.lib.la.us
1933; Rebecca Morris
7 branch libs
64 000 vols; 249 curr per
40044

Cuyahoga County Public Library – Maple Heights Branch, 5225 Library Lane, *Maple Heights*, OH 44137-1242
T: +1 216 4755000; Fax: +1 216 5877284
Vicki Adams-Cook
208 000 vols; 8 281 av-mat
40045

Burlington County Library – Maple Shade Branch, 200 Stiles Ave, *Maple Shade*, NJ 08052
T: +1 856 7799767; Fax: +1 856 7790033; E-mail: ma@bcls.lib.nj.us
Mike Bennett
53 000 vols; 85 curr per
40046

Maplewood Memorial Library, 51 Baker St, *Maplewood*, NJ 07040-2618
T: +1 973 7621671; Fax: +1 973 7620762; URL: www.maplewoodlibrary.org
1913; Jane Kennedy
Asher Brown Durand Coll, American artists; James Ricalton (teacher, world traveller, photographer); 1 branch libr
132 000 vols; 180 curr per
libr loan; The Eastern N.J. Regional Library Cooperative
40047

Bollinger County Library, 302 Conrad St, *Marble Hill*, MO 63764; P.O. Box 919, Marble Hill, MO 63764-0919
T: +1 573 2382713; Fax: +1 573 2382879; E-mail: bollinger@yahoo.com; URL: www.outreach.missouri.edu/bollinger/library
1947; Eva M. Dunn
90 000 vols; 50 curr per
40048

Abbot Public Library, 235 Pleasant St, *Marblehead*, MA 01945
T: +1 781 6311481; Fax: +1 781 6390558; E-mail: mar@noblenet.org; URL: www.abbotlibrary.org
1878; Bonnie J. Strong
Govt; Yachts & Yachting
116 000 vols; 258 curr per
40049

Marengo Public Library District, 200 S State St, *Marengo*, IL 60152
T: +1 815 5688236; Fax: +1 815 5685209; URL: www.marengopubliclibrary.org
1878; Mary Hanson
53 000 vols
40050

Jackson County Public Library System, 2929 Green St, *Marianna*, FL 32446
T: +1 850 4829631; Fax: +1 850 4829632
1977; Jo-Ann Rountree
78 000 vols; 192 curr per
40051

Cobb County Public Library System, 266 Roswell St, *Marietta*, GA 30060-2004
T: +1 770 5282320; Fax: +1 770 5282349; URL: www.cobbcat.org
1958; Tamara George
Georgia Room; 16 branch libs
301 000 vols
40052

Cobb County Public Library System – East Marietta, 2051 Lower Roswell Rd, *Marietta*, GA 30068
T: +1 770 5092711; Fax: +1 770 5092714
N.N.
58 000 vols
40053

Cobb County Public Library System – Kemp Memorial, 4029 Due West Rd NW, *Marietta*, GA 30064
T: +1 770 5282527; Fax: +1 770 5282592
Colleen Moses
52 000 vols
40054

Cobb County Public Library System – Merchant's Walk, 1315 Johnson Ferry Rd, *Marietta*, GA 30068
T: +1 770 5092730; Fax: +1 770 5092733
N.N.
71 000 vols
40055

Cobb County Public Library System – Mountain View, 3320 Sandy Plains Rd, *Marietta*, GA 30066
T: +1 770 5092725; Fax: +1 770 5092726
Ansie Krige
124 000 vols
40056

Washington County Public Library, 615 Fifth St, *Marietta*, OH 45750-1973
T: +1 740 3731057; Fax: +1 740 3762171; URL: www.wcplib.lib.oh.us
1901; Justin Mayo
4 branch libs, 1 bookmobile
218 000 vols; 245 curr per
40057

Monterey County Free Libraries, 188 Seaside Ctr, *Marina*, CA 93933-2500
T: +1 831 8837573; Fax: +1 831 8837574; URL: www.co.monterey.ca.us/library/
1912; Jayanti Addleman
401 e-bks; 17 branch libs
404 000 vols; 431 curr per; 382 e-books
40058

County of Los Angeles Public Library – Lloyd Taber-Marina del Rey Library, 4533 Admiralty Way, *Marina del Rey*, CA 90292-5416
T: +1 310 8213415; Fax: +1 310 3063372; URL: www.colapublib.org/libs/marina
Winona Phillabaum
Nautical Coll
57 000 vols; 117 curr per
40059

Marinette County Consolidated Public Library Service, Stephenson Public Library, 1700 Hall Ave, *Marinette*, WI 54143-1709
T: +1 715 7327570; Fax: +1 715 7327575; E-mail: mrt@mail.nfls.lib.wi.us; URL: www.nfls.lib.wi.us/mrt
Tim Dirks
6 branch libs
200 000 vols; 550 curr per
40060

Marion Carnegie Library, 206 S Market St, *Marion*, IL 62959-2519
T: +1 618 9935935; Fax: +1 618 9976485; URL: www.marioncarnegielibrary.org
1912; Linda J. Mathias
Small Businesses, Civil War
62 000 vols; 80 curr per; 1 200 av-mat
40061

Marion County Library, 101 E Court St, *Marion*, SC 29571-3699
T: +1 843 4238300; Fax: +1 843 4238302; E-mail: marionlibr@infoave.net; URL: www.marioncountylibrary.org
1898; Salley B. Davidson
Hist; South Carolina, bks & micro
82 000 vols; 202 curr per
40062

Marion Public Library, 1095 Sixth Ave, *Marion*, IA 52302
T: +1 319 3773412; Fax: +1 319 3770113; E-mail: meckerle@marion.lib.in.us; URL: www.marion.lib.in.us/
Mary Eckerle
156 000 vols; 303 curr per; 10 470 av-mat; 17 615 sound-rec
40063

Marion Public Library, 445 E Church St, *Marion*, OH 43302-4290
T: +1 740 3870992; Fax: +1 740 3879768; URL: www.marion.lib.oh.us
1886; Mark Rose
4 branch libs
245 000 vols; 321 curr per; 14 810 av-mat; 11 900 sound-rec; 4 780 digital data carriers; 4 740 Large Print Bks
40064

Marion Public Library, 600 S Washington St, *Marion*, IN 46953-1992
T: +1 765 6682900; Fax: +1 765 6682911; E-mail: mpl@marion.lib.in.us; URL: www.marion.lib.in.us
1884; Mary Theresa Eckerle
Genealogy Museum
137 000 vols; 436 curr per; 547 digital data carriers
libr loan
40065

McDowell County Public Library, 90 W Court St, *Marion*, NC 28752
T: +1 828 6523858; Fax: +1 828 6522098; E-mail: mcdowellcountypubliclibrary@yahoo.com; URL: www.main.nc.us/libraries/mcdowell
1960; Jean Krause
1 branch libr
110 000 vols; 252 curr per
40066

Smyth-Bland Regional Library, 118 S Sheffey St, *Marion*, VA 24354
T: +1 276 7832323; Fax: +1 276 7835279; URL: www.sbrl.org
1972; Patricia M. Hatfield
Sherwood Anderson Coll
117 000 vols; 252 curr per
40067

Avoyelles Parish Library, 104 N Washington St, *Marksville*, LA 71351-2496
T: +1 318 2537559; Fax: +1 318 2536361; E-mail: admin.h1av@pelican.state.lib.la.us; URL: www.avoyelles.lib.la.us
1949; Theresa Thevenote
Louisiana Room; 7 branch libs
80 000 vols; 200 curr per
40068

Monmouth County Library – Marlboro Branch, One Library Ct, *Marlboro*, NJ 07746-1102
T: +1 732 5369406; Fax: +1 732 5364708
Jennifer King
100 000 vols
40069

Marlborough Public Library, 35 W Main St, *Marlborough*, MA 01752-5510
T: +1 508 6246900; Fax: +1 508 4851494; URL: www.marlborough-ma.gov
1871; Salvatore Genovese
Horatio Alger Coll
110 000 vols; 275 curr per
40070

Pocahontas County Free Library, 500 Eighth St, *Marlinton*, WV 24954-1227
T: +1 304 7996000; Fax: +1 304 7993988; URL: www.pocahontaslibrary.org
Allen R. Johnson
65 000 vols; 23 curr per
40071

Burlington County Library – Evesham Branch, Evesham Municipal Complex, 984 Tuckerton Rd, *Marlton*, NJ 08053
T: +1 856 9831444; Fax: +1 856 9834939; E-mail: ev@bcls.lib.nj.us
Susan Szymanik
95 000 vols
40072

Peter White Public Library, 217 N Front St, *Marquette*, MI 49855
T: +1 906 2264311; Fax: +1 906 2261783; E-mail: pwpl@uproc.lib.mi.us; URL: www.pwpl.info
1891; Pamela R. Christensen
Children's Hist Bk coll, Guns, Railroads, Ships (Miller coll), Merrit coll, Nadeau coll, Shiras coll, Finnish Coll
206 000 vols; 350 curr per; 66 e-books; 7 250 av-mat; 16 815 sound-rec; 1 020 digital data carriers; 370 Large Print Bks
libr loan
40073

Madison County Public Library, 1335 N Main St, *Marshall*, NC 28753-6901
T: +1 828 6493741; Fax: +1 828 6493504; URL: www.madisoncountylibrary.org
Kathleen Phillips
52 000 vols; 75 curr per; 2 486 av-mat
40074

Marshall District Library, 124 W Green St, *Marshall*, MI 49068
T: +1 616 7817821; Fax: +1 616 7817090; E-mail: semiferoa@marshalldistrictlibrary.org; URL: www.marshalldistrictlibrary.org
1912
55 300 vols; 120 curr per
40075

Marshall Public Library, 300 S Alamo Blvd, *Marshall*, TX 75670
T: +1 903 9354465; Fax: +1 903 9354463; URL: www.marshallpubliclibrary.org
1970; Anna Lane
Music, tapes
64 000 vols; 53 curr per
40076

Marshall-Lyon County Library, 301 W Lyon St, *Marshall*, MN 56258
T: +1 507 5377003; E-mail: library@marshalllyonlibrary.org; URL: www.marshalllyonlibrary.org
1886; Wendy Wendt
Large print; Cake Pans; 2 branch libs
82 000 vols; 188 curr per
libr loan
40077

Marshalltown Public Library, 36 N Center St, *Marshalltown*, IA 50158-4911
T: +1 641 7545780; Fax: +1 641 7545708; E-mail: ask@marshalltown.lib.ia.us; URL: www.marshalltownlibrary.org
1898; Carole Booker Winkleblack
Genealogy (Holdings of the Central Iowa Genealogical Society & Holdings of the Marshalltown DAR), Marshall County Heritage Room
78 000 vols; 225 curr per; 3 676 av-mat
40078

Marshfield Public Library, 211 E Second St, *Marshfield*, WI 54449
T: +1 715 3878494; Fax: +1 715 3876909; E-mail: busoff@marshfieldlibrary.org; URL: www.marshfieldlibrary.org
1864; Lori Belongia
Spanish Coll
118 000 vols; 255 curr per; 7 000 e-books; 7 270 av-mat; 5 310 sound-rec; 2 070 digital data carriers; 350 High Interest/Low Vocabulary Bks, 3 440 Large Print Bks
40079

Ventress Memorial Library, Library Plaza, *Marshfield*, MA 02050
T: +1 781 8345535; Fax: +1 781 8378362; URL: www.ventresslibrary.org
1895
81 539 vols; 200 curr per
40080

Reelfoot Regional Library Center, 542 N Lindell St, *Martin*, TN 38237; P.O. Box 168, Martin
T: +1 731 5872347; Fax: +1 731 5870027
1942; Susan N. Blakely
120 000 vols
40081

Martins Ferry Public Library, 20 James Wright Pl, *Martins Ferry*, OH 43935
T: +1 740 6330314; Fax: +1 740 6330935; URL: www.mfpl.org
1927; Yvonne O. Myers
James Wright Poetry Coll; 5 branch libs, 1 bookmobile
193 000 vols; 280 curr per; 916 microforms; 4 819 sound-rec; 14 digital data carriers; 22 High Interest/Low Vocabulary Bks
libr loan; ALA
40082

Martinsburg-Berkeley County Public Library – Martinsburg Public Library, 101 W King St, *Martinsburg*, WV 25401
T: +1 304 2678933; Fax: +1 304 2679720; URL: www.youseemore.com/martinsburgberkeley
1926; Pamela Coyle
3 branch libs
150 000 vols; 310 curr per; 6 000 av-mat; 4 300 sound-rec; 500 digital data carriers; 500 Large Print Bks
40083

Blue Ridge Regional Library, 310 E Church St, *Martinsville*, VA 24112-2999; P.O. Box 5264, Martinsville, VA 24115-5264
T: +1 276 4035450; Fax: +1 276 6321660; E-mail: martinsville@brrl.lib.va.us; URL: www.brrl.lib.va.us

1923; Hal Hubener
5 branch libs
262 000 vols; 150 curr per; 9 954 av-mat;
9 960 Talking Bks
libr loan   40084

Morgan County Public Library, 110 S
Jefferson St, *Martinsville*, IN 46151
T: +1 765 3423451; Fax: +1 765
3429992; URL: www.morg.lib.in.us
1906; Krista Ledbetter
166 000 vols; 225 curr per   40085

Marysville Public Library, 231 S Plum
St, *Marysville*, OH 43040-1596
T: +1 937 6421876; Fax: +1 937
6423457; URL: www.marysvillelib.org
1910; R. McDonnell
90 000 vols; 118 curr per   40086

Sno-Isle Libraries, 7312 35th Ave NE,
Marysville, WA 98271-7417
T: +1 360 6517000; Fax: +1 360
6517151; URL: www.sno-isle.org
1945; Jonalyn Woolf-Ivory
20 branch libs
900 000 vols   40087

Sno-Isle Libraries – Marysville Branch,
6120 Grove St, *Marysville*, WA 98270
T: +1 360 6585000; Fax: +1 360
6595050
Eric Spencer
166 000 vols   40088

Yuba County Library, 303 Second St,
Marysville, CA 95901-6099
T: +1 530 7497380; Fax: +1 530
7413098; E-mail: library@co.yuba.ca.us;
URL: library.yuba.org
1858; Loren MccRory
Hist, Travel; California Hist
155 000 vols; 115 curr per   40089

Blount County Public Library, 508 N
Cusick St, *Maryville*, TN 37804-5714
T: +1 865 9820981; Fax: +1 865
9771142; URL: www.kornet.org/bcpl
1919; Kathryn Pagles
Genealogy Coll
158 000 vols; 411 curr per; 3 000 e-
books
libr loan   40090

Maryville Public Library, 509 N Main,
Maryville, MO 64468
T: +1 660 5825281; Fax: +1
660 5822411; E-mail:
admin@maryvillepubliclibrary.org; URL:
www.maryvillepubliclibrary.lib.mo.us
1904; Stephanie Patterson
Cookbks
55 000 vols; 45 curr per; 1 114 av-mat
                                  40091

Mason City Public Library, 225 Second
St SE, *Mason City*, IA 50401
T: +1 641 4213669; Fax: +1 641
4232615; E-mail: librarian@mcpl.org;
URL: www.mcpl.org
1876; Mary Markwalter
Lee P Loomis Arch of Mason City Hist,
Prairie School Architecture Coll
109 000 vols; 325 curr per   40092

**Queens Borough Public Library –
Maspeth Branch**, 69-70 Grand Ave,
Maspeth, NY 11378
T: +1 718 6395228
Usha Pinto
71 000 vols   40093

Massapequa Public Library, Bar Harbor
Branch, 40 Harbor Lane, *Massapequa
Park*, NY 11762
T: +1 516 7990770; Fax: +1 516
7957528; E-mail: mpl2@nassaulibrary.org;
URL: www.nassaulibrary.org/massapq
1952; Patricia Page
1 branch libr
189 000 vols; 590 curr per   40094

Massena Public Library, Warren
Memorial Library, 41 Glenn St,
Massena, NY 13662
T: +1 315 7699914; Fax: +1 315
7695978; URL: www.northnet.org/
massenalibrary/mpl.htm
1897; Paul L. Schaffer
68 000 vols; 156 curr per; 800 av-mat
libr loan   40095

Massillon Public Library, 208 Lincoln
Way E, *Massillon*, OH 44646-8416
T: +1 330 8329831; Fax: +1 330
8302182; E-mail: mpl.ref@gmail.com;
URL: www.massillonlibrary.org
1897; Camille J. Leslie
Cooking, Gardening, Early Ohio &
Quaker Hist (Rotch-Wales Coll, mss;
Lillian Gish Coll; 2 branch libs
170 000 vols; 199 curr per   40096

Matawan-Aberdeen Public Library, 165
Main St, *Matawan*, NJ 07747
T: +1 732 5839100; Fax: +1 732
5839360; URL: www.lmxac.org/mata/
index.html
1903; Susan Pike
90 000 vols; 153 curr per   40097

Matteson Public Library, 801 S School
St, *Matteson*, IL 60443-1897
T: +1 708 7484431; Fax: +1 708
7480510; E-mail: mtslib@sslic.net; URL:
www.mattesonpubliclibrary.org
1964; Kathy Berggren
80 000 vols; 210 curr per   40098

**Public Library of Charlotte &
Mecklenburg County – Matthews
Branch**, 230 Matthews Station St,
Matthews, NC 28105
T: +1 704 8476691; E-mail:
matthews@plcmc.org; URL:
www.plcmc.org/Locations/
branches.asp?id=12
Debbie McWreath
66 000 vols   40099

Mattituck-Laurel Library, 13900 Main
Rd, *Mattituck*, NY 11952; P.O. Box
1437, Mattituck, NY 11952-0991
T: +1 631 2984134; Fax: +1 631
2984764; URL: www.suffolk.lib.ny.us/
libraries/matt/
Kay Zegel
54 000 vols; 300 curr per   40100

Mattoon Public Library, 1600 Charleston
Ave, *Mattoon*, IL 61938-3635
T: +1 217 2342621; Fax: +1 217
2342660; URL: www.mattoonlibrary.org
1893; Jennie Cisna
55 000 vols; 170 curr per
libr loan   40101

**Toledo-Lucas County Public Library
– Maumee Branch**, 501 River Rd,
Maumee, OH 43537
T: +1 419 2595360; Fax: +1 419
2595203; URL: www.toledolibrary.org
Joseph Ludwig
118 000 vols; 200 curr per; 500 av-mat
                                  40102

Graves County Public Library, 601 N
17th St, *Mayfield*, KY 42066
T: +1 270 2472911; Fax: +1 270
2472990; URL: www.gcpl.org
1940; Diane Bennett
53 940 vols; 108 curr per   40103

**Cuyahoga County Public Library –
Mayfield Branch**, 6080 Wilson Mills Rd,
Mayfield Village, OH 44143-2103
T: +1 440 4730350; Fax: +1 440
4730774
William Rubin
194 000 vols; 6 844 av-mat   40104

Maynard Public Library, 77 Nason St,
Maynard, MA 01754
T: +1 978 8971010; Fax: +1 978
8979884; URL: web.maynard.ma.us/library
1881; Stephen Weiner
Maynard Hist, Parenting
50 000 vols; 126 curr per; 800 av-mat;
Town Newspaper (microfilm)
libr loan; ALA   40105

Atlantic County Library, 40 Farragut
Ave, *Mays Landing*, NJ 08330-1750
T: +1 609 6252776; Fax: +1 609
6258143; URL: www.atlanticlibrary.org
1926; Bill Paullin
Afro-American Coll, Music Coll,
Jerseyana; 9 branch libs, 1 bookmobile
338 000 vols
libr loan   40106

Mason County Public Library, 218 E
Third St, *Maysville*, KY 41056
T: +1 606 5643286; Fax: +1
606 5645408; E-mail:
masoncolibrary@bellsouth.net; URL:
www.masoncountylibrary.com
1876; Valerie Zempter
52 000 vols; 58 curr per; 2 377 av-mat
                                  40107

Mayville Public Library, 111 N Main St,
Mayville, WI 53050
T: +1 920 3877910; Fax: +1 920
3877917; E-mail: maylib@mwfls.org;
URL: www.mayville.lib.wi.us
1904; Alixe M. Bielot
57 000 vols; 120 curr per   40108

Maywood Public Library, 459 Maywood
Ave, *Maywood*, NJ 07607-1909
T: +1 201 8452915; Fax: +1 201
8457387; E-mail: maywcirc@bccls.org;
URL: www.maywoodreads.org
1951; Diane Rhodes
Automobile manuals
61 000 vols; 81 curr per   40109

Maywood Public Library District, 121
S Fifth Ave, *Maywood*, IL 60153
T: +1 708 3431847; Fax: +1 708
3432115; URL: www.maywood.org
1874; Stan Huntington
Hist, Lit, African-American
72 000 vols; 116 curr per; 1 115 av-mat
                                  40110

McAlester Public Library, 401 N
Second St, *McAlester*, OK 74501
T: +1 918 4260930; Fax: +1 918
4235731; URL: www.mcalester.lib.ok.us
Christine Sauro
Local Coll (newspapers back to 1890's)
70 000 vols; 115 curr per   40111

**Southeastern Public Library System
of Oklahoma**, 401 N Second St,
McAlester, OK 74501
T: +1 918 4260456; Fax: +1 918
4260543; URL: www.sepl.lib.ok.us
1967; June Doyle
All County Newspapers on Microfilm; 6
branch libs
383 000 vols; 648 curr per
libr loan   40112

McAllen Memorial Library, 601 N Main,
McAllen, TX 78501-4666
T: +1 956 6883300; Fax: +1
956 6883301; E-mail:
ReferenceLibrarian@mcallen.net; URL:
www.mcallenlibrary.net
1932; Jose A. Gamez
Mexican-American Coll, Spanish; 2
branch libs
359 000 vols; 1 366 curr per; 76 e-books
                                  40113

Herbert Wescoat Memorial Library,
120 N Market St, *McArthur*,
OH 45651-1218
T: +1 740 5965691; Fax: +1 740
5962477; E-mail: hwmlib@oplin.org; URL:
www.vintoncountypublic.lib.oh.us
Clint Walker
50 000 vols; 171 curr per   40114

McComb Public Library, 113 S Todd
St, *McComb*, OH 45858-0637
T: +1 419 2932425; Fax: +1 419
2932748; E-mail: mccomb@oplin.org;
URL: mccombpl.org
1935; Jane Schaffner
Art Coll, prints; Civil War (Andrews
Raiders), displays
54 720 vols; 113 curr per   40115

**Pike-Amite-Walthall Library System
– McComb Public Library
(Headquarters)**, 1022 Virginia Ave,
McComb, MS 39648
T: +1 601 6847034; Fax: +1 601
2501213
1964; Gabriel Morley
8 branch libs
129 000 vols; 277 curr per; 1 172 av-mat;
1 180 Audio Bks, 2 780 Large Print Bks
                                  40116

**Kate Love Simpson Morgan
County Library**, 358 E Main St,
McConnelsville, OH 43756-1130
T: +1 740 9622533; Fax: +1 740
9623316; E-mail: katelove@oplin.org;
URL: www.morgan.lib.oh.us
1920; Laura Walker
61 000 vols; 77 curr per
libr loan   40117

**Henry County Library System –
McDonough Public Library**, 1001
Florence McGarity Blvd, *McDonough*,
GA 30252-2981
T: +1 770 9542806; Fax: +1 770
9542808
Kaye West
56 000 vols; 82 curr per; 1 027 av-mat
                                  40118

McHenry Public Library District, 809 N
Front St, *McHenry*, IL 60050
T: +1 815 3850036; Fax: +1
815 3857085; E-mail:
webmaster@mchenrylibrary.org; URL:
www.mchenrylibrary.org
1943; James C. Scholtz
Large-Type Bks Coll
123 000 vols; 429 curr per; 3 000 e-
books; 5 661 av-mat   40119

Carnegie Library of McKeesport,
1507 Library Ave, *McKeesport*,
PA 15132-4796
T: +1 412 6720625; Fax: +1 412
6727860; URL: www.einetwork.net/ein/
mckeespt
1902; Jo Ellen Kenney
114 000 vols; 211 curr per; 3 742 av-mat
libr loan   40120

McKinney Memorial Public Library,
101 E Hunt St, *McKinney*, TX 75069
T: +1 972 5477323;
Fax: +1 972 5420868; URL:
www.mckinneypubliclibrary.org
1928; Beth A. Scudder
130 000 vols   40121

McMinnville Public Library, 225 N
Adams St, *McMinnville*, OR 97128
T: +1 503 4355555; Fax: +1
503 4355560; E-mail:
libref@ci.mcminnville.or.us; URL:
www.maclibrary.org
1912; Jill Poyer
82 500 vols; 168 curr per   40122

Peters Township Public Library,
616 E McMurray Rd, *McMurray*,
PA 15317-3495
T: +1 724 9419430; Fax: +1 724
9419438; URL: www.ptlibrary.org
1957; Pier M. Lee
121 000 vols; 200 curr per   40123

McPherson Public Library, 214 W
Marlin, *McPherson*, KS 67460-4299
T: +1 620 2452570; Fax: +1 620
2452567; E-mail: library@macpl.org;
URL: www.macpl.org
1902; Steven D. Read
Kansas Coll, Small Business &
Entrepreneurship
70 000 vols; 185 curr per   40124

**Crawford County Federated Library
System**, 848 N Main St, *Meadville*,
PA 16335-2689
T: +1 814 3361773; Fax: +1 814
3338173
John Brice
103 000 vols; 271 curr per; 3 171 av-mat
                                  40125

Meadville Public Library, 848 N Main
St, *Meadville*, PA 16335-2689
T: +1 814 3361773; Fax: +1 814
3338173; URL: www.meadvillelibrary.org
1879; John J. Brice
Hist; Crawford County Hist Coll, books &
mss
90 000 vols; 249 curr per; 3 591 av-mat
                                  40126

Joseph T. Simpson Public Library,
16 N Walnut St, *Mechanicsburg*,
PA 17055-3362
T: +1 717 7660171; Fax: +1
717 7660152; E-mail:
mechanicsburg@ccpa.net; URL:
www.ccpa.net/simpson
1961; Sue Erdman
Irving College, Mechanicsburg,
Pennsylvania
82 000 vols; 135 curr per
libr loan   40127

Pamunkey Regional Library – Atlee,
9161 Atlee Rd, *Mechanicsville*,
VA 23116
T: +1 804 5590654; Fax: +1 804
5590645
Toni M. Heer
50 000 vols; 100 curr per   40128

Medfield Memorial Public Library, 468
Main St, *Medfield*, MA 02052-2008
T: +1 508 3594544; Fax: +1 508
3598124; E-mail: info@medfieldlibrary.org;
URL: www.medfieldlibrary.org
1872; Dan Brassell
67 000 vols; 104 curr per   40129

Burlington County Library – Pinelands, 39 Allen Ave, **Medford**, NJ 08055
T: +1 609 6546113; Fax: +1 609 9532142; E-mail: mf@bcls.lib.nj.us
Judy Aley
72 000 vols; 75 curr per   40130

Jackson County Library Services, 205 S Central Ave, **Medford**, OR 97501-2730
T: +1 541 7746421; Fax: +1 541 7746748; E-mail: infolib@jcls.org; URL: www.jcls.org
1908; Denise Galarraga
Music Coll, Oregon Coll; 14 branch libs
567 000 vols; 1 560 curr per
libr loan   40131

Medford Public Library, 111 High St, **Medford**, MA 02155
T: +1 781 3957950; Fax: +1 781 3912261; URL: www.medfordlibrary.org
1825; Brian G. Boutilier
150 000 vols; 250 curr per; 500 Large Print Bks   40132

Lee-Whedon Memorial Library, 620 West Ave, **Medina**, NY 14103
T: +1 585 7983430; Fax: +1 585 7984398; URL: www.nioga.org/medina
1928; Mary G. Zangerle
60 000 vols; 134 curr per   40133

Medina County District Library, 210 S Broadway, mail address: 887 W Liberty St, **Medina**, OH 44256
T: +1 330 7226235; Fax: +1 330 7252053; URL: www.medina.lib.oh.us
1905; Carole Kowell
5 branch libs
576 000 vols; 1 289 curr per   40134

Medway Public Library, 26 High St, **Medway**, MA 02053
T: +1 508 5333217; Fax: +1 508 5333219; E-mail: mwydir@minlib.net; URL: medwaylib.org
1860; N.N.
56 000 vols; 131 curr per   40135

Eau Gallie Public Library, 1521 Pineapple Ave, **Melbourne**, FL 32935-6594
T: +1 321 2554304; Fax: +1 321 2554323; URL: www.brev.org
1939; Sharon K. Dwyer
73 000 vols; 213 curr per   40136

Melbourne Public Library, 540 E Fee Ave, **Melbourne**, FL 32901
T: +1 321 9524514; Fax: +1 321 9524518; URL: www.brev.org
1918; Geraldine Prieth
Investing Coll
108 000 vols; 298 curr per   40137

Melrose Public Library, 69 W Emerson St, **Melrose**, MA 02176
T: +1 781 6652313; Fax: +1 781 6624229; E-mail: mel@noblenet.org; URL: www.melrosepubliclibrary.org
1871; Dennis J. Kelley
Fine Arts (Felix A Gendrot Coll), Sadie & Alex Levine Coll
110 000 vols; 95 curr per
libr loan   40138

Melrose Park Public Library, 801 N Broadway, **Melrose Park**, IL 60160
T: +1 708 3433391; Fax: +1 708 5315327; E-mail: mps@sls.lib.il.us; URL: www.melroseparklibrary.org
1898; Patrick Italia
Cinema Coll
87 000 vols; 183 curr per; 1 834 sound-rec; 10 digital data carriers
libr loan   40139

Melvindale Public Library, 18650 Allen Rd, **Melvindale**, MI 48122
T: +1 313 4291090; Fax: +1 313 3880432; E-mail: kieltyka@melvindale.lib.mi.us; URL: www.melvindale.lib.mi.us
1928; Theresa Kieltyka
50 000 vols; 100 curr per   40140

Memphis Public Library & Information Center – Benjamin L. Hooks Central Library, 3030 Poplar Ave, **Memphis**, TN 38111-3527
T: +1 901 4152702; Fax: +1 901 3237637; URL: www.memphislibrary.org
1893; E. Keenon McCloy
Afro-American Memphis Colls, Commerce & Industry Colls, Folk Hist Colls, Politics & Govt, Public Health Colls, War Hist Colls; 25 branch libs
1 638 000 vols; 16 e-books   40141

Memphis Public Library & Information Center – Parkway Village, 4655 Knight Arnold Rd, **Memphis**, TN 38118-3234
T: +1 901 3638923; Fax: +1 901 7942344
Sara Ellen Reid
55 000 vols   40142

Memphis Public Library & Information Center – Poplar-White Station, 5094 Poplar, **Memphis**, TN 38117-7629
T: +1 901 6821616; Fax: +1 901 6828975
Cris Mitchell
53 000 vols   40143

Memphis Public Library & Information Center – Whitehaven, 4120 Mill Branch Rd, **Memphis**, TN 38116
T: +1 901 3969700; Fax: +1 901 3326150
Karen Hall
104 000 vols   40144

Menasha Public Library, Elisha D. Smith Public Library, 440 First St, **Menasha**, WI 54952-3191
T: +1 920 9673690; Fax: +1 920 9675159; E-mail: reference@menashalibrary.org; URL: www.menashalibrary.org
1896; Tasha Saecker
140 000 vols; 200 curr per; 50 High Interest/Low Vocabulary Bks, 35 Bks on Deafness & Sign Lang   40145

Menlo Park Public Library, 800 Alma St, Alma & Ravenswood, **Menlo Park**, CA 94025-3460
T: +1 650 3302500; Fax: +1 650 3277030; URL: www.menloparklibrary.org
1916; Susan Holmer
1 branch libr
150 000 vols
libr loan   40146

Spies Public Library, 940 First St, **Menominee**, MI 49858-3296
T: +1 906 8633911; Fax: +1 906 8635000; E-mail: spies@uproc.lib.mi.us; URL: www.uproc.lib.mi.us/spies
1903; Cheryl Hoffman
57 000 vols; 126 curr per; 35 e-books   40147

Menomonee Falls Public Library, W 156th N, 8436 Pilgrim Rd, **Menomonee Falls**, WI 53051
T: +1 262 5328900; Fax: +1 262 5328939; URL: www.mf.lib.wi.us
1906; Jane Schall
Hist Photogr Coll
140 000 vols; 362 curr per; 220 microforms; 5 000 sound-rec; 500 digital data carriers
libr loan; ALA   40148

Menomonie Public Library, 600 Wolske Bay Rd, **Menomonie**, WI 54751
T: +1 715 2322164; Fax: +1 715 2322324; E-mail: mpl@lfls.lib.wi.us; URL: www.menomonielibrary.org
1986; Ted Stark
Libr Bks by Mail Dept
86 000 vols; 192 curr per; 2 572 sound-rec; 8 digital data carriers
libr loan; ALA   40149

Mentor Public Library, 8215 Mentor Ave, **Mentor**, OH 44060
T: +1 440 2558811; Fax: +1 440 2550520; E-mail: askalibrarian@mentorpl.org; URL: www.mentorpl.org
1819; Lynn Hawkins
2 branch libs
247 000 vols; 346 curr per; 65 e-books 40150

Frank L. Weyenberg Library of Mequon-Thiensville, 11345 N Cedarburg Rd, **Mequon**, WI 53092-1998
T: +1 262 2422593; Fax: +1 262 4783200; E-mail: admin@flwlib.org; URL: www.flwlib.org
1954; Linda Bendix
140 000 vols; 225 curr per   40151

Merced County Library, 2100 O St, **Merced**, CA 95340-3637
T: +1 209 3857484; Fax: +1 209 7267912; URL: www.mercedcountylibrary.org
1910; Jacque Meriam
Cookery Coll; 15 branch libs
373 000 vols; 211 curr per   40152

The Fendrick Library, 20 N Main St, **Mercersburg**, PA 17236
T: +1 717 3289233; E-mail: fendricklibrary@comcast.net; URL: www.fendricklibrary.com/
50 000 vols; 18 curr per   40153

Fendrick Library, 20 N Main St, **Mercersburg**, PA 17236-1612
T: +1 717 3289233; URL: www.fendricklibrary.com
Cheryl Custer
50 000 vols; 18 curr per   40154

Meriden Public Library, 105 Miller St, **Meriden**, CT 06450
T: +1 203 6306352; Fax: +1 203 2386950; URL: www.meridenlibrary.org
1903; Karen Roesler
Silver Industry, US Census Bureau
235 000 vols; 125 curr per   40155

Meridian District Library, 1326 W Cherry Lane, **Meridian**, ID 83646
T: +1 208 8884451; Fax: +1 208 8840745; URL: www.mld.org
1924; Patricia Younger
Idaho Coll (bks by & about Idaho authors); 1 branch libr
140 000 vols; 175 curr per   40156

Meridian-Lauderdale County Public Library, 2517 Seventh St, **Meridian**, MS 39301
T: +1 601 6936771; Fax: +1 601 4862260; E-mail: library@meridian.lib.ms.us; URL: meridian.lib.ms.us
1913; D. Steven McCartney
176 000 vols; 124 curr per; 1 767 av-mat; 1 770 Talking Bks   40157

Johnson County Library – Antioch, 8700 Shawnee Mission Pkwy, **Merriam**, KS 66202; P.O. Box 2933, Shawnee Mission, KS 55201-1333
T: +1 913 2612300; Fax: +1 913 2612320
Ken Werne
111 000 vols; 131 curr per   40158

Merrick Library, 2279 Merrick Ave, **Merrick**, NY 11566-4398
T: +1 516 3776112; Fax: +1 516 3771108; E-mail: merricklibrary@merricklibrary.org; URL: www.merricklibrary.org
1891; Ellen Firer
Ctr for Intellectual Freedom & Censorship Mat
78 000 vols; 233 curr per   40159

T. B. Scott Library, Merrill Public Library, 106 W First St, **Merrill**, WI 54452-2398
T: +1 715 5367191; Fax: +1 715 5361705; URL: wvls.lib.wi.us/merrillpl
1891; Beatrice Lebal
96 000 vols; 177 curr per   40160

Lake County Public Library, 1919 W 81st Ave, **Merrillville**, IN 46410-5488
T: +1 219 7693541; Fax: +1 219 7569358; E-mail: webmastr@lakeco.lib.in.us; URL: www.lakeco.lib.in.us
1959; Lawrence Acheff
12 branch libs
1 255 000 vols; 1 013 curr per; 3 416 maps; 7 000 microforms; 42 985 av-mat; 58 916 sound-rec; 2 530 digital data carriers; 110 Journals, 460 Large Print Bks
libr loan   40161

Merrimack Public Library, 470 Daniel Webster Hwy, **Merrimack**, NH 03054
T: +1 603 4245021; Fax: +1 603 4247312; E-mail: mmkpl@merrimack.lib.nh.us; URL: www.merrimack.lib.nh.us
1892; Janet Angus
78 000 vols; 200 curr per   40162

City of Mesa Library, 64 E First St, **Mesa**, AZ 85201-6768
T: +1 480 6442723; Fax: +1 480 6442991; URL: www.mesalibrary.org
1926; Heather Wolf
Spanish, Large type print, Small business; 2 branch libs
585 000 vols; 521 curr per; 76 e-books; 32 691 av-mat; 39 412 sound-rec; 3 349 digital data carriers
libr loan   40163

Mesquite Public Library, 300 W Grubb Dr, **Mesquite**, TX 75149
T: +1 972 2166220; Fax: +1 972 2166740; E-mail: mainbr@library.mesquite.tx.us; URL: www.library.mesquite.tx.us
1963; John Williams
1 branch libr
177 000 vols; 272 curr per   40164

Jefferson Parish Library, East Bank Regional Administrative Headquarters, 4747 W Napoleon Ave, **Metairie**, LA 70001
T: +1 504 8381127; Fax: +1 504 8381121; E-mail: admin.b1jn@pelican.state.lib.la.us; URL: www.jefferson.lib.la.us
1949; Lon R. Dickerson
Louisiana; Foreign Language; 13 branch libs
850 000 vols   40165

Illinois Prairie District Public Library, 208 E Partridge, **Metamora**, IL 61548; P.O. Box 770, Metamora, IL 61548-0770
T: +1 309 3674594; Fax: +1 309 3672687; URL: www.ipdpl.org
1950; Grant A.. Fredericksen
Hist, Education, Agriculture, Antiques; 6 branch libs
127 000 vols; 381 curr per
libr loan   40166

Nevins Memorial Library, 305 Broadway, **Methuen**, MA 01844
T: +1 978 6864080; Fax: +1 978 6868669; E-mail: kmcleod@mvlc.org; URL: www.nevinslibrary.org
1883; Krista I. McLeod
Elise Nevins Morgan Meditation Series, mss & bks
61 831 vols; 128 curr per   40167

Metuchen Public Library, 480 Middlesex Ave, **Metuchen**, NJ 08840
T: +1 732 6328526; Fax: +1 732 6328535; URL: www.lmxac.org, www.metuchen.com/library
1870; Melody B. Kokola
Chinese Language Coll, Large Print
75 000 vols; 165 curr per; 300 av-mat   40168

Mexico-Audrain County Library District, 305 W Jackson St, **Mexico**, MO 65265
T: +1 573 5814939; Fax: +1 573 5817510; E-mail: mexicoaudrain@netscape.net; URL: mexico-audrain.lib.mo.us
1912; Ray Hall
Genealogy Coll; 4 branch libs
150 000 vols; 154 curr per
libr loan   40169

Miami Memorial-Gila County Library, 1052 Adonis Ave, **Miami**, AZ 85539-1298
T: +1 928 4732621; Fax: +1 928 4732567
Jeanne Michie
70 000 vols; 10 curr per   40170

Miami Public Library, 200 N Main, **Miami**, OK 74354
T: +1 918 5412292; Fax: +1 918 5429363; URL: www.miami.lib.ok.us
1920; Marcia Johnson
51 000 vols; 118 curr per   40171

Miami-Dade Public Library System, 101 W Flagler St, **Miami**, FL 33130-1523
T: +1 305 3752665; Fax: +1 305 3753048; URL: www.mdpls.org
1937; Raymond Santiago
40 branch libs
4 000 000 vols; 3 957 curr per
libr loan   40172

Brockway Memorial Library, 10021 NE Second Ave, *Miami Shores*, FL 33138 T: +1 305 7588107; Fax: +1 305 7547660; URL: www.brockwaylibrary.org 1949; Elizabeth Esper 63 000 vols; 129 curr per; 16 000 e-books; 3 000 av-mat; 1 800 sound-rec
40173

Michigan City Public Library, 100 E Fourth St, *Michigan City*, IN 46360-3393 T: +1 219 8733478; Fax: +1 219 8733475; E-mail: Refdesk@mclib.org; URL: www.mclib.org 1897; Don Glossinger 137 000 vols; 420 curr per
40174

Queens Borough Public Library – Middle Village Branch, 72-31 Metropolitan Ave, *Middle Village*, NY 11379 T: +1 718 3261390 Steven Nobel 54 000 vols
40175

Middleborough Public Library, 102 N Main St, *Middleborough*, MA 02346 T: +1 508 9462470; Fax: +1 508 9462473; E-mail: midlib@sailsinc.org; URL: www.midlib.org 1875; Danielle Bowker Cranberry Culture Coll 77 000 vols; 214 curr per
40176

Cuyahoga County Public Library – Middleburg Heights Branch, 15600 E Bagley Rd, *Middleburg Heights*, OH 44130-4830 T: +1 440 2343600; Fax: +1 440 2340849 Barbara Stiber 66 000 vols; 3 184 av-mat
40177

Ilsley Public Library, 75 Main St, *Middlebury*, VT 05753 T: +1 802 3884095; Fax: +1 802 3884367; URL: www.ilsleypubliclibrary.org 1866; David Clark Vermont Coll 54 000 vols; 108 curr per; 100 microforms libr loan; VLA, PLA, ALA
40178

Middlebury Public Library, 30 Crest Rd, *Middlebury*, CT 06762 T: +1 203 7582634; Fax: +1 203 5774164; URL: www.biblio.org/middlebury; www.middlebury-ct.org/library.shtml 1794; Jane O. Gallagher 65 000 vols; 90 curr per
40179

Middlesborough-Bell County Public Library, 126 S 20th St, *Middlesboro*, KY 40965-1212; P.O. Box 1677, Middlesboro, KY 40965-1677 T: +1 606 2484812; Fax: +1 606 2488766; E-mail: mborolib@tcnet.net; URL: www.bellcountypubliclibraries.org 1912; Beverly Greene 89 000 vols; 1 600 curr per
40180

Middlesex Public Library, 1300 Mountain Ave, *Middlesex*, NJ 08846 T: +1 732 3566602; Fax: +1 732 3568420; E-mail: info@middlesexlibrarynj.org; URL: www.middlesexlibrarynj.org 1963; Dr. May Lein Ho 58 000 vols; 147 curr per libr loan
40181

Middleton Public Library, 7425 Hubbard Ave, *Middleton*, WI 53562-3117 T: +1 608 8277425; Fax: +1 608 8365724; E-mail: mid@scls.lib.wi.us; URL: www.midlibrary.org 1926; Paul Nelson 83 000 vols; 293 curr per; 2 807 av-mat
40182

Middletown Public Library, 700 W Main Rd, *Middletown*, RI 02842-6391 T: +1 401 8461573; Fax: +1 401 8463031; E-mail: robertbt@lori.ri.gov; URL: middletownpubliclibrary.org 1848; Robert L. Balliot Rhode Island-Middletown Hist (Rhode Island Hist Coll) 58 000 vols; 100 curr per
40183

Middletown Public Library, 125 S Broad St, *Middletown*, OH 45044 T: +1 513 4241251; Fax: +1 513 4246585; URL: www.middletownlibrary.org 1911; Douglas J. Bean 2 branch libs 169 000 vols; 578 curr per libr loan
40184

Middletown Thrall Library, 11-19 Depot St, *Middletown*, NY 10940 T: +1 845 3415454; Fax: +1 845 3415480; URL: www.thrall.org 1901; Kevin Gallagher Orange County Hist Coll 175 000 vols; 654 curr per
40185

Middletown Township Public Library, 55 New Monmouth Rd, *Middletown*, NJ 07748 T: +1 732 6713700; Fax: +1 732 6715839; E-mail: reference@mtpl.org; URL: www.mtpl.org 1921; Susan O'Neal 3 branch libs 215 000 vols; 338 curr per; 5 000 e-books; 13 920 av-mat; 15 457 sound-rec; 1 408 digital data carriers; 1 980 Large Print Bks
40186

Russell Library, 123 Broad St, *Middletown*, CT 06457 T: +1 860 3472520; Fax: +1 860 3476690; URL: www.russelllibrary.org 1875; Arthur S. Meyers Adult Basic Education (ABE), Literacy Mats, NRC Haddam Neck Plant 162 000 vols; 342 curr per libr loan; ALA, CLA, NELA
40187

Grace A. Dow Memorial Library, 1710 W St Andrews Ave, *Midland*, MI 48640-2698 T: +1 989 8373449; Fax: +1 989 8373468; E-mail: gadml@midland-mi.org; URL: www.midland-mi.org/gracedowlibrary 1895; Melissa Barnard Genealogy; Fine Arts (Alden B Dow Coll), bks, rec, art prints 231 000 vols; 595 curr per
40188

Midland County Public Library, 301 W Missouri, *Midland*, TX 79701 T: +1 432 6884320; Fax: +1 432 6884939; E-mail: reference@co.midland.tx.us/; URL: www.co.midland.tx.us/library 1903; E. Denise Johnson 1 branch libr 266 000 vols; 445 curr per libr loan
40189

Midland Park Memorial Library, 250 Godwin Ave, *Midland Park*, NJ 07432 T: +1 201 4442390; Fax: +1 201 4442813; E-mail: mipk@mail.bccls.org; URL: www.bccls.org 1937; Jean Scott 55 000 vols; 155 curr per
40190

Midlothian Public Library, 14701 S Kenton Ave, *Midlothian*, IL 60445-4122 T: +1 708 5352027; Fax: +1 708 5352053; E-mail: mds@midlothianlibrary.org; URL: www.midlothianlibrary.org 1931; Mary Beth Sharples 59 000 vols; 261 curr per libr loan
40191

Salt Lake County Library Services – Ruth V. Tyler Branch, 8041 S Wood St, *Midvale*, UT 84047-7559 T: +1 801 9447641; Fax: +1 801 5658012 Lorraine Jeffrey 69 000 vols
40192

Metropolitan Library System in Oklahoma County – Midwest City Library, 8143 E Reno, *Midwest City*, OK 73110-3999 T: +1 405 7324828; Fax: +1 405 6063451; E-mail: midwestcity@metrolibrary.org Deborah Willis 89 000 vols; 186 curr per
40193

Juniata County Library, Inc, 498 Jefferson St, *Mifflintown*, PA 17059-1424 T: +1 717 4366378; Fax: +1 717 4369324; E-mail: juniata@acsworld.com; URL: www.juniatacountylibrary.org 1966; Thomas Hipple 55 000 vols; 98 curr per
40194

Milan-Berlin Township Public Library, 19 E Church St, *Milan*, OH 44846; P.O. Box 1550, Milan, OH 44846-1550 T: +1 419 4994117; Fax: +1 419 4994697; URL: www.milan-berlin.lib.oh.us 1877; Wendy Harper 96 000 vols; 199 curr per; 2 000 av-mat; 2 730 sound-rec; 5 digital data carriers libr loan
40195

Grant County Public Library, 207 E Park Ave, *Milbank*, SD 57252-2497 T: +1 605 4326543; Fax: +1 605 4324635; E-mail: gclibrary21@hotmail.com; URL: grantcountylibrary.com 1978; Robin M. Schrupp Maps; Old County Newpapers, microfilm 52 000 vols; 75 curr per; 1 883 libr loan
40196

Miles City Public Library, One S Tenth St, *Miles City*, MT 59301-3398 T: +1 406 2341496; Fax: +1 406 2342095; E-mail: mcpl@midrivers.com; URL: ntserver.mtmc.mt.lib.com 1902; N.N. Extensive coll of rare and out of print Montana hist titles as well as rare and out of print fiction by Montana authors 60 000 vols; 70 curr per; 88 maps; 908 av-mat; 5 000 sound-rec; 88 digital data carriers; 222 slides libr loan; ALA
40197

Milford Public Library, 57 New Haven Ave, *Milford*, CT 06460 T: +1 203 7833292; Fax: +1 203 8771072; URL: milfordlibrary.org 1895; Jean Tsang 102 000 vols; 228 curr per; 2 005 Large Print Bks
40198

Milford Town Library, 80 Spruce St, *Milford*, MA 01757 T: +1 508 4730651; Fax: +1 508 4738651; E-mail: milfref@cwmars.org; URL: www.milfordtownlibrary.org 1986; Jennifer Perry Milford Hist Coll 107 000 vols; 147 curr per; 7 000 e-books; 3 053 av-mat; 1 500 sound-rec; 40 digital data carriers
40199

Pike County Public Library, 201 Broad St, *Milford*, PA 18337-1398 T: +1 570 2968211; Fax: +1 570 2968987; E-mail: admin@pcpl.org; URL: www.pcpl.org 1902; Ellen Schaffner 67 000 vols
40200

Wadleigh Memorial Library, 49 Nashua St, *Milford*, NH 03055-3753 T: +1 603 6732408; Fax: +1 603 6726064; E-mail: wadleigh@wadleigh.lib.nh.us; URL: www.wadleigh.lib.nh.us 1868; Michelle Sampson Fairy tales; Hutchinson Family (singers), Rothovius (scholarly res) 67 000 vols; 180 curr per; 900 e-books; 1 977 av-mat
40201

Hawaii State Public Library System – Mililani Public Library, 95-450 Makaimoimo St, *Mililani*, HI 96789-3018 T: +1 808 6277470; Fax: +1 808 6277309 Wendi Woodstrup 86 000 vols; 64 curr per
40202

Sno-Isle Libraries – Mill Creek Branch, 15429 Bothell-Everett Hwy, *Mill Creek*, WA 98012 T: +1 425 3374822; Fax: +1 425 3373567 Darlene Weber 74 000 vols
40203

San Mateo County Library – Millbrae Branch, One Library Ave, *Millbrae*, CA 94030 T: +1 650 6977607; Fax: +1 650 6924747; URL: www.millbraelibrary.org Kathleen Beasley 111 000 vols
40204

Millburn Free Public Library, 200 Glen Ave, *Millburn*, NJ 07041 T: +1 973 3761006; Fax: +1 973 3760104; URL: www.millburn.lib.nj.us 1938; William Swinson 107 000 vols; 293 curr per; 4 561 av-mat; 4 570 Audio Bks libr loan
40205

Twin Lakes Library System – Mary Vinson Memorial Library – Headquarters, 151 S Jefferson St SE, *Milledgeville*, GA 31061-3419 T: +1 478 4520677; Fax: +1 478 4520680; URL: www.ccmi.1.com/mvml Barry Reese 75 000 vols
40206

Holmes County District Public Library, 3102 Glen Dr, *Millersburg*, OH 44654 T: +1 330 6745972; Fax: +1 330 6741938; URL: www.holmes.lib.oh.us 1928; Arlene Victoria Radden Amish & Mennonite Genealogies Coll; 4 branch libs, 2 bookmobiles 145 000 vols; 128 curr per
40207

Millville Public Library, 210 Buck St, *Millville*, NJ 08332 T: +1 609 8257087; Fax: +1 609 3278572; URL: www.millvillepubliclibrary.org 1864 Career & College Ctr, New Jersey Coll 70 386 vols; 180 curr per
40208

Santa Clara County Library – Milpitas Public, 160 N Main St, *Milpitas*, CA 95035-4323 T: +1 408 2621171; Fax: +1 408 2625806 Linda Arbaugh 232 000 vols
40209

Milton Public Library, 476 Canton Ave, *Milton*, MA 02186-3299 T: +1 617 6985757; Fax: +1 617 6980441; E-mail: miref@ocln.org; URL: www.miltonlibrary.org 1871; Philip McNulty 1 branch libr 104 000 vols; 168 curr per; 20 e-books
40210

Santa Rosa County Library System, 6568 Caroline St, Ste 101, *Milton*, FL 32570 T: +1 850 6232043; Fax: +1 850 6232138 Linda Hendrix 125 000 vols; 40 curr per
40211

Milwaukee County Federated Library System, 709 N Eighth St, *Milwaukee*, WI 53233-2414 T: +1 414 2863210; Fax: +1 414 2863209 James Gingery 4 750 000 vols
40212

Milwaukee Public Library, 814 W Wisconsin Ave, *Milwaukee*, WI 53233-2385 T: +1 414 2863082; Fax: +1 414 2862794; E-mail: mailbox@mpl.org; URL: www.mpl.org 1878; Paula Kiely Railroad Hist, Trans-Miss Am; American Maps, Cookery, Eastman Fairy Tale Coll, Rare Bird Prints; 12 branch libs 2 557 000 vols; 8 519 curr per libr loan
40213

Milwaukee Public Library – Atkinson, 1960 W Atkinson Ave, *Milwaukee*, WI 53209 T: +1 414 2863068; Fax: +1 414 2868469 David Allen Sikora 73 000 vols
40214

Milwaukee Public Library – Bay View, 2566 S Kinnickinnic Ave, *Milwaukee*, WI 53207 T: +1 414 2863019; Fax: +1 414 2868459 Christopher Gawronski 77 000 vols
40215

Milwaukee Public Library – Capitol, 3969 N 74th St, *Milwaukee*, WI 53216 T: +1 414 2863006; Fax: +1 414 2868432 Acklen J. Banks 116 000 vols
40216

Milwaukee Public Library – Center Street, 2727 W Fond du Lac Ave, *Milwaukee*, WI 53210 T: +1 414 2863090; Fax: +1 414 2868467 Kirsten Thompson 73 000 vols
40217

Milwaukee Public Library – East, 1910 E North Ave, *Milwaukee*, WI 53202 T: +1 414 2863058; Fax: +1 414 2868431 Nancy Torphy 100 000 vols
40218

Milwaukee Public Library – Forest Home, 1432 W Forest Home Ave, *Milwaukee*, WI 53204
T: +1 414 2863083; Fax: +1 414 2868461
N.N.
79 000 vols   40219

Milwaukee Public Library – Martin Luther King Branch, 310 W Locust St, *Milwaukee*, WI 53212
T: +1 414 2863098; Fax: +1 414 2868465
Rachel Collins
93 000 vols   40220

Milwaukee Public Library – Mill Road, 6431 N 76th St, *Milwaukee*, WI 53223
T: +1 414 2863088; Fax: +1 414 2868454
Joy Kilimann
90 000 vols   40221

Milwaukee Public Library – Tippecanoe, 3912 S Howell Ave, *Milwaukee*, WI 53207
T: +1 414 2863085; Fax: +1 414 2868405
Neal Kaluzny
93 000 vols   40222

Milwaukee Public Library – Villard Avenue, 3310 W Villard Ave, *Milwaukee*, WI 53209
T: +1 414 2863079; Fax: +1 414 2868473
Brian Williams-Vanklooster
77 000 vols   40223

Milwaukee Public Library – Washington Park, 2121 N Sherman Blvd, *Milwaukee*, WI 53208
T: +1 414 2863066; Fax: +1 414 2868471
Enid Gruszka
61 000 vols   40224

Milwaukee Public Library – Zablocki, 3501 W Oklahoma Ave, *Milwaukee*, WI 53215
T: +1 414 2863055; Fax: +1 414 2868430
Linda Vincent
78 000 vols   40225

Wisconsin Regional Library for the Blind & Physically Handicapped, 813 W Wells St, *Milwaukee*, WI 53233-1436
T: +1 414 2863045; Fax: +1 414 2863102; E-mail: lbph@milwaukee.gov; URL: wmbph.mlp.org/opac; www.regionallibrary.wi.us
1960; Meredith J. Wittmann
160 000 vols; 20 curr per; 9 000 e-books; 137 778 av-mat; 234 170 sound-rec; 1 760 Braille vols, 20 Special Interest Per Sub, 70 Large Print Bks   40226

Ledding Library of Milwaukie, 10660 SE 21st Ave, *Milwaukie*, OR 97222
T: +1 503 7867580; Fax: +1 503 6599497; URL: www.milwaukie.lib.or.us
1934; Josef Sandfort
89 000 vols; 210 curr per   40227

Douglas County Public Library, Minden Library, 1625 Library Lane, *Minden*, NV 89423-4420; P.O. Box 337, Minden
T: +1 775 7829841; Fax: +1 775 7826766; URL: www.douglas.lib.nv.us
1967; Linda L. Deacy
1 branch libr
114 000 vols; 140 curr per; 835 av-mat; 835 Talking Bks   40228

Webster Parish Library, 521 East & West St, *Minden*, LA 71055
T: +1 318 3713080; Fax: +1 318 3713081; E-mail: admin.g1wb@pelican.state.lib.la.us; URL: www.webster.lib.la.us
1929; Beverly Hammett
Art & architecture, Business & management, Religion, Technology; Louisiana Coll; 6 branch libs
85 000 vols; 203 curr per   40229

Mineola Memorial Library, 195 Marcellus Rd, *Mineola*, NY 11501
T: +1 516 7468488; Fax: +1 516 2946459; E-mail: mineola@nassaulibrary.org; URL: www.nassaulibrary.org/mineola
Charles Sleefe
85 000 vols; 150 curr per   40230

Minerva Public Library, 677 Lynnwood Dr, *Minerva*, OH 44657-1200
T: +1 330 8684101; Fax: +1 330 8684267; URL: www.minerva.lib.oh.us
1910; Tom Dillie
Gypsies Coll
94 000 vols; 193 curr per   40231

Hennepin County Library, 12601 Ridgedale Dr, *Minnetonka*, MN 55305-1909
T: +1 952 8478800; Fax: +1 952 8478653; URL: www.hclib.org
1922; Lois Langer Thompson
World Languages; 26 branch libs
4 540 000 vols; 8 018 curr per; 22 000 e-books; 22 950 Large Print Bks, 67 450 Talking Bks
libr loan   40232

Minot Public Library, 516 Second Ave SW, *Minot*, ND 58701-3792
T: +1 701 8521045; Fax: +1 701 8522595; URL: www.minotlibrary.org
1908; Jerry Kaup
Literacy, North Dakota, Genealogy
134 000 vols; 300 curr per; 14 000 e-books; 200 maps; 2 000 microforms; 5 356 sound-rec; 105 digital data carriers; 4 050 Large Print Bks
libr loan   40233

Ward County Public Library, 405 Third Ave SE, *Minot*, ND 58701-4020
T: +1 701 8525388; Fax: +1 701 8374960; E-mail: library@co.ward.nd.us; URL: www.co.ward.nd.us/library
1960; Jan Murphy
53 000 vols; 35 curr per; 500 av-mat   40234

Public Library of Charlotte & Mecklenburg County – Mint Hill Branch, 6840 Matthews-Mint Hill Rd, *Mint Hill*, NC 28227
T: +1 704 5734054; URL: www.plcmc.org/Locations/branches.asp?id=13
Neily Trump
51 000 vols   40235

Mishawaka-Penn-Harris Public Library, 209 Lincoln Way E, *Mishawaka*, IN 46544-2084
T: +1 574 2595277; Fax: +1 574 2558489; URL: www.mphpl.org
1907; David J. Eisen
Heritage Ctr Coll; 2 branch libr
262 000 vols; 842 curr per; 1 000 e-books; 9 650 av-mat; 4 714 digital data carriers
libr loan   40236

Mishawaka-Penn-Harris Public Library – Bittersweet, 602 Bittersweet Rd, *Mishawaka*, IN 46544-4155
T: +1 574 2590392; Fax: +1 574 2590399
Linda Sears
61 000 vols; 71 curr per   40237

Speer Memorial Library, Mission Public Library, 801 E 12th St, *Mission*, TX 78572
T: +1 956 5808750; Fax: +1 956 5808756; E-mail: library@mission.lib.tx.us; URL: www.mission.lib.tx.us
Harold R. Dove
Genealogy, Spanish, Texana, Large Print, Bilingual-English & Spanish
79 000 vols; 103 curr per   40238

Mission Viejo Library, 100 Civic Ctr, *Mission Viejo*, CA 92691
T: +1 949 8307100; Fax: +1 949 5868447; URL: www.cmvl.org
1997; Valerie Maginnis
155 000 vols; 211 curr per; 253 e-books; 800 Large Print Bks   40239

Missoula Public Library, 301 E Main, *Missoula*, MT 59802-4799
T: +1 406 7212665; Fax: +1 406 7285900; E-mail: mslaplib@missoula.lib.mt.us; URL: www.missoulapubliclibrary.org
1894; Honore Bray
Montana Hist Coll, Northwest Hist Coll; 2 branch libs
222 000 vols; 571 curr per; 7 244 av-mat; 7 250 Audio Bks, 1 000 Music Scores   40240

Fort Bend County Libraries – Missouri City Branch, 1530 Texas Pkwy, *Missouri City*, TX 77489-2170
T: +1 281 4994100; Fax: +1 281 2615829
Katryna Russell
83 000 vols; 151 curr per; 1 521 av-mat   40241

Mitchell Community Public Library, 804 Main St, *Mitchell*, IN 47446
T: +1 812 8492412; Fax: +1 812 8492665; E-mail: mitlib@mitlib.org; URL: www.mitlib.org
Patty Vahey
50 000 vols; 112 curr per   40242

Mitchell Public Library, 221 N Duff St, *Mitchell*, SD 57301-2596
T: +1 605 9958480; Fax: +1 605 9958482; URL: mitlib.sdln.net
1903; Jackie Hess
South Dakota Coll; Mitchell Area Arch, newspapers
72 000 vols; 159 curr per   40243

Little Dixie Regional Libraries, 111 N Fourth St, *Moberly*, MO 65270-1577
T: +1 660 2634426; Fax: +1 660 2634024; E-mail: sysadmin@little-dixie.lib.mo.us; URL: www.little-dixie.lib.mo.us
1966; Karen Hayden
Civil War Coll; 3 branch libs
170 000 vols; 214 curr per
libr loan   40244

Mobile Public Library, 700 Government St, *Mobile*, AL 36602
T: +1 251 2087106; Fax: +1 251 2085865; URL: www.mplonline.org
1928; Spencer Watts
Genealogy, Mobile Hist (1702-present), Mobile Mardi Gras Coll; 5 branch libs
472 000 vols; 800 curr per
libr loan   40245

Mobile Public Library – Monte L. Moorer-Spring Hill Branch, Four S McGregor Ave, *Mobile*, AL 36608-1827
T: +1 251 4707770; Fax: +1 251 4707774; E-mail: moorerbranch@mplonline.org
Janette Curry
76 000 vols   40246

Mobile Public Library – West Regional, 5555 Grelot Rd, *Mobile*, AL 36609
T: +1 251 3408555; Fax: +1 251 3042160; E-mail: westregionalbranch@mplonline.org
Margie Calhoun
190 000 vols   40247

Davie County Public Library, 371 N Main St, *Mocksville*, NC 27028-2115
T: +1 336 7512023; Fax: +1 336 7511370; E-mail: info@library.daviecounty.org; URL: www.library.daviecounty.org
1943; Ruth A. Hoyle
66 000 vols   40248

Stanislaus County Free Library, 1500 I St, *Modesto*, CA 95354-1166
T: +1 209 5587801; Fax: +1 209 5294779; E-mail: refquest@scfl.lib.ca.us; URL: www.stanislauslibrary.org
1912; Vanessa Czopek
Californiana, Spanish & Vietnamese, Song File, Stanislaus County Hist Coll; 13 branch libs
770 000 vols; 1 019 curr per   40249

Akron-Summit County Public Library – Mogadore Branch, 144 S Cleveland Ave, *Mogadore*, OH 44260
T: +1 330 6289228; Fax: +1 330 6283256
Karen Steiner
57 000 vols   40250

Mokena Community Public Library District, 11327 W 195th St, *Mokena*, IL 60448
T: +1 708 4799663; Fax: +1 708 4799684; URL: www.mokena.lib.il.us
1976; Phyllis A. Jacobek
Large print, Parenting
130 000 vols; 325 curr per   40251

Beaver County Library System, One Campus Dr, *Monaca*, PA 15061-2523
T: +1 724 7283737; Fax: +1 724 7288024; URL: www.beaverlibraries.org
Patricia Smith
Large Print Coll, Record Album Coll
309 000 vols; 50 curr per   40252

Ward County Library, 409 S Dwight, *Monahans*, TX 79756
T: +1 432 9433332; Fax: +1 432 9433332; E-mail: wardlibrary@hotmail.com; URL: www.wcl.lib.tx.us
Bonnie Moore
64 000 vols; 25 curr per   40253

Berkeley County Library System, 1003 Hwy 52, *Moncks Corner*, SC 29461
T: +1 843 7194223; URL: www.berkeley.lib.sc.us
1936; Donna Osborne
3 branch libs
229 000 vols; 240 curr per
libr loan; ALA   40254

Monessen Public Library & District Center, 326 Donner Ave, *Monessen*, PA 15062-1182
T: +1 724 6844750; Fax: +1 724 6847077; E-mail: reference@monpldc.org; URL: www.monpldc.org
1936; S. Fred Natale
109 000 vols; 345 curr per
libr loan   40255

Monessen Public Library & District Center, 326 Donner Ave, *Monessen*, PA 15062-1182
T: +1 724 6844750; Fax: +1 724 6847077; E-mail: reference@monpldc.org; URL: www.monpldc.org
1936; S. Fred Natale
109 000 vols; 345 curr per; 1 500 High Interest/Low Vocabulary Bks   40256

Barry-Lawrence Regional Library, 213 Sixth St, *Monett*, MO 65708-2147
T: +1 417 2356646; Fax: +1 417 2356799; E-mail: execdir@blrlibrary.org; URL: tlc.library.net/bll
1954; Jean H. Berg
10 branch libs
229 000 vols; 396 curr per; 9 210 av-mat; 6 619 sound-rec; 3 655 digital data carriers; 9 040 Large Print Bks   40257

Monmouth Public Library, 168 S Ecols St, *Monmouth*, OR 97361; P.O. Box 10, Monmouth, OR 97361-0010
T: +1 503 8381932; Fax: +1 503 8383899; E-mail: library@ci.monmouth.or.us; URL: www.ccrls.org/monmouth
1934; Phelps Shepard
67 000 vols; 174 curr per   40258

Warren County Public Library District, 62 Public Sq, *Monmouth*, IL 61462-1756
T: +1 309 7343166; Fax: +1 309 7345955; E-mail: wcpl@wcplibrary.org; URL: www.wcplibrary.org
1868; Larisa Good
Agriculture, Law; Lincoln Coll; 3 branch libs
96 000 vols; 161 curr per; 2 762 av-mat   40259

South Brunswick Public Library, 110 Kingston Lane, *Monmouth Junction*, NJ 08852
T: +1 732 3294000; Fax: +1 732 3290573; URL: www.sbpl.info
1967; Christopher Carbone
Adult basic reading, English as a second language, Arabic, Chinese & Spanish language; 1 bookmobile
137 000 vols; 263 curr per   40260

Monona Public Library, 1000 Nichols Rd, *Monona*, WI 53716-2531
T: +1 608 2226127; Fax: +1 608 2228590; E-mail: library@scls.lib.wi.us; URL: www.mononalibrary.info
1964; Demita Gerber
Living Hist of Historic Blooming Grove
61 000 vols; 80 curr per   40261

Edith Wheeler Memorial Library, 733 Monroe Tpk, *Monroe*, CT 06468
T: +1 203 4522850; E-mail: monroe-ref@biblio.org; URL: www.biblio.org/monroe
1954; Robert Simon
Environmental Resources Info Ctr
75 000 vols; 110 curr per   40262

Monroe Free Library, 44 Millpond Pkwy, *Monroe*, NY 10950
T: +1 845 7834411; Fax: +1 845 7824707; E-mail: mfl@rcls.org; URL: www.monroelibrary.org
1908; Marilyn McIntosh
50 000 vols; 135 curr per   40263

Monroe Public Library, 925 16th Ave, **Monroe**, WI 53566-1497
T: +1 608 3287010; Fax: +1 608 3294657; E-mail: ludemail@scls.lib.wi.us; URL: www.monroepubliclibrary.org
1904; Barbara Brewer
81 000 vols; 177 curr per   40264

Ouachita Parish Public Library, 1800 Stubbs Ave, **Monroe**, LA 71201
T: +1 318 3271490; Fax: +1 318 3271373; URL: www.oplib.org
1940; Cheryl Mouliere
6 branch libs
377 000 vols; 1 209 curr per; 5 000 e-books; 24 200 av-mat; 21 240 sound-rec; 8 970 digital data carriers; 260 Bks on Deafness & Sign Lang, 18 450 Large Print Bks   40265

Sno-Isle Libraries – Monroe Branch, 1070 Village Way, **Monroe**, WA 98272
T: +1 360 7947851; Fax: +1 360 7940292
Betsy Lewis
80 000 vols   40266

Union County Public Library, 316 E Windsor St, **Monroe**, NC 28112
T: +1 704 2838184; Fax: +1 704 2820657; URL: www.union.lib.nc.us
1930; Martie Smith
4 branch libs
192 000 vols; 330 curr per   40267

Monroe Township Public Library, Four Municipal Plaza, **Monroe Township**, NJ 08831-1900
T: +1 732 5215000; Fax: +1 732 5214766; URL: www.monroetwplibrary.org
1989; Irene Goldberg
Holocaust (Henry Ricklis Memorial Coll)
102 000 vols; 248 curr per; 600 microforms; 3 130 sound-rec; 6 digital data carriers
libr loan   40268

Monroeville Public Library, 4000 Gateway Campus Blvd, **Monroeville**, PA 15146-3381
T: +1 412 3720500; Fax: +1 412 3721168; URL: www.einetwork.net/ein/monroevl; www.monroevillelibrary.org
1960; Christy Fusco
New Reader Coll
104 000 vols; 221 curr per; 7 000 microforms; 5 184 av-mat; 4 228 sound-rec; 252 digital data carriers; 80 art repros
libr loan   40269

Monrovia Public Library, 321 S Myrtle Ave, **Monrovia**, CA 91016-2848
T: +1 626 2568274; Fax: +1 626 2568255; URL: www.ci.monrovia.ca.us
1895; Monica Greening
111 000 vols; 280 curr per
libr loan   40270

Montclair Free Public Library, 50 S Fullerton Ave, **Montclair**, NJ 07042
T: +1 973 7440500; Fax: +1 973 7445268; URL: www.montlib.com
1893; David Hinkley
Art & Music Coll, College & Career Coll, Folk Arts, Delahinty Irish; 1 branch libr
170 000 vols; 277 curr per   40271

County of Los Angeles Public Library – Montebello Library, 1550 W Beverly Blvd, **Montebello**, CA 90640-3993
T: +1 323 7226551; Fax: +1 323 7223018; URL: www.colapublib.org/libs/montebello
Lisa Castaneda
Asian-Pacific Resource Center
169 000 vols; 286 curr per   40272

Monterey Park Bruggemeyer Library, Monterey Park Public Library, 318 S Ramona Ave, **Monterey Park**, CA 91754-3399
T: +1 626 3071366; Fax: +1 626 2884251; URL: ci.monterey-park.ca.us/library
1929; Linda Wilson
Beatrix Potter Figurine Coll, Chinese, Vietnamese, Korean & Japanese Language Colls
164 000 vols; 156 curr per; 7 150 av-mat; 4 900 sound-rec; 1 118 Large Print Bks
libr loan   40273

Montevideo-Chippewa County Public Library, 224 S First St, **Montevideo**, MN 56265-1425
T: +1 320 2696501; Fax: +1 320 2698696; URL: www.montelibrary.org
David Lauritsen
16mm, Film; Spanish Language Uruguayan Mat
58 000 vols; 155 curr per; 175 av-mat   40274

Alabama Public Library Service, 6030 Monticello Dr, **Montgomery**, AL 36130
T: +1 334 2133900; Fax: +1 334 2133993; URL: www.apls.state.al.us
1959; Rebecca S. Mitchell
Alabama; Audiovisual Coll; 1 branch libr
147 000 vols; 1 249 curr per; 306 212 av-mat; 306 200 Talking Bks
libr loan   40275

Monticello Union Township Public Library, 321 W Broadway, **Monticello**, IN 47960-2047
T: +1 574 5835643; Fax: +1 574 5832782; E-mail: mutpl@monticello.lib.in.us; URL: www.monticello.lib.in.us
1903; Bill Caddell
63 000 vols; 131 curr per   40276

Southeast Arkansas Regional Library, 107 E Jackson St, **Monticello**, AR 71655
T: +1 870 3678584; Fax: +1 870 3675166
1947; Kim Patterson
14 branch libs
183 000 vols; 495 curr per   40277

Wilderness Coast Public Libraries, 280 W Washington St, **Monticello**, FL 32344; P.O. Box 551, Monticello
T: +1 850 9264571; Fax: +1 850 9265157
1992; Cheryl Turner
Florida Coll, Jefferson County Elder Black People Recollections
107 000 vols; 158 curr per; 24 000 e-books; 4 346 av-mat; 2 034 sound-rec; 10 digital data carriers
libr loan; ALA   40278

Bear Lake County Free Library, 138 N Sixth St, **Montpelier**, ID 83254-1556
T: +1 208 8471664; Fax: +1 208 8471664; E-mail: blkcolib@dcdi.net; URL: www.lili.org/bearlake; bearlake.lili.org
1959; Mary Nate
60 000 vols; 120 curr per   40279

Kellogg-Hubbard Library, 135 Main St, **Montpelier**, VT 05602
T: +1 802 2233338; Fax: +1 802 2233338; E-mail: info@kellogghubbard.org; URL: www.kellogghubbard.org
1894; Robin Sales
71 310 vols; 100 curr per   40280

Montpelier Public Library, 216 E Main St, **Montpelier**, OH 43543-1199
T: +1 419 4853287; Fax: +1 419 4855671; URL: montpelierpubliclibrary.oplin.org
Gloria Osburn
Coop learning, Sci fairs
60 000 vols; 70 curr per   40281

Glendale Public Library – Montrose-Crescenta Branch, 2465 Honolulu Ave, **Montrose**, CA 91020-1803
T: +1 818 5482048; Fax: +1 818 2486987
Patricia Zeider
66 000 vols   40282

Hendrick Hudson Free Library, 185 Kings Ferry Rd, **Montrose**, NY 10548
T: +1 914 7395654; Fax: +1 914 7395659; URL: www.henhudfreelibrary.org
1937; M. Jill Davis
54 000 vols; 99 curr per   40283

Montrose Library District, 320 S Second St, **Montrose**, CO 81401-3909
T: +1 970 2499656; Fax: +1 970 2401901; URL: www.montroselibrary.org
1969; Paul H. Paladino
2 branch libs
112 000 vols; 250 curr per   40284

Montville Township Public Library, 90 Horseneck Rd, **Montville**, NJ 07045-9626
T: +1 973 4020900; Fax: +1 973 4020592; URL: www.montvillelib.org
1921; Patricia K. Anderson
95 000 vols; 150 curr per
ALA   40285

Moon Township Public Library, 1700 Beaver Grade Rd, Ste 100, **Moon Township**, PA 15108-2984
T: +1 412 2690334; Fax: +1 412 2690136; E-mail: moontwp@einetwork.net; URL: www.moonlibrary.org
Leslie Pallotta
58 000 vols; 110 curr per; 596 av-mat   40286

Pioneer Library System – Moore Public, 225 S Howard, **Moore**, OK 73160
T: +1 405 7935100; Fax: +1 405 7938755
Lisa Wells
89 000 vols   40287

Hardy County Public Library, 102 N Main St, **Moorefield**, WV 26836
T: +1 304 5386560; Fax: +1 304 5382639; URL: hardycounty.martin.lib.wv.us
1939; Carol See
56 000 vols; 75 curr per   40288

Moorestown Public Library, 111 W Second St, **Moorestown**, NJ 08057-2481
T: +1 856 2340333; Fax: +1 856 7789536; E-mail: reference@moorestown.lib.nj.us; URL: www.moorestown.lib.nj.us
1853; N.N.
142 000 vols; 228 curr per
libr loan; ALA   40289

Mooresville Public Library, 220 W Harrison St, **Mooresville**, IN 46158-1633
T: +1 317 8317323; Fax: +1 317 8317383; E-mail: wecare@mooresville.lib.in.us; URL: www.mooresvillelib.org
1912; Diane Huerkamp
63 000 vols; 245 curr per   40290

Mooresville Public Library, 304 S Main St, **Mooresville**, NC 28115
T: +1 704 6642927; Fax: +1 704 6603292; URL: www.ci.mooresville.nc.us/library
1894; John Pritchard
73 000 vols; 198 curr per; 110 maps; 1 228 av-mat   40291

Lake Agassiz Regional Library, 118 S Fifth St, **Moorhead**, MN 56560-2756; P.O. Box 900, Moorhead
T: +1 218 2333757; Fax: +1 218 2337556; E-mail: lakeagassiz@larl.org; URL: www.larl.org
1961; Kathy Fredette
13 branch libr
332 000 vols; 299 curr per   40292

Lake Agassiz Regional Library – Moorhead Public Library, 118 S Fifth St, **Moorhead**, MN 56561-2756
T: +1 218 2337594; Fax: +1 218 2367405; E-mail: moorhead@larl.org; URL: www.larl.org/branch/moorhead.html
1906
123 000 vols; 299 curr per   40293

Contra Costa County Library – Moraga Library, 1500 Saint Mary's Rd, **Moraga**, CA 94556-2099
T: +1 925 3766852; Fax: +1 925 3763034
Linda Waldroup
67 000 vols   40294

Rowan County Public Library, 185 E First St, **Morehead**, KY 40351
T: +1 606 7847137; Fax: +1 606 7843917; URL: www.rowancountylibrary.org
1952; Helen E. Williams
52 000 vols; 125 curr per; 809 av-mat   40295

Moreno Valley Public Library, 25480 Alessandro Blvd, **Moreno Valley**, CA 92553
T: +1 951 4133880; Fax: +1 951 4133895; URL: www.moreno-valley.ca.us
Rebecca Guillan
98 000 vols; 118 curr per   40296

Santa Clara County Library – Morgan Hill Branch, 660 W Main Ave, **Morgan Hill**, CA 95037-4128
T: +1 408 7793196; Fax: +1 408 7790883
Rosanne Macek
161 000 vols   40297

Burke County Public Library, 204 S King St, **Morganton**, NC 28655-3535
T: +1 828 4375638; Fax: +1 828 4331914; E-mail: library@bcpls.org; URL: www.bcpls.org
1924; Jim Wilson
Art, Environmental studies; North Carolina Hist
98 000 vols; 210 curr per   40298

Morgantown Public Library – Morgantown Public Library Service Center, 373 Spruce St, **Morgantown**, WV 26505
T: +1 304 2917425; Fax: +1 304 2917427; URL: morgantown.lib.wv.us
1929; Sharon Turner
3 branch libs
103 000 vols; 85 curr per; 10 000 av-mat; 8 300 Audio Bks   40299

Morris Area Public Library District, 604 Liberty St, **Morris**, IL 60450
T: +1 815 9426880; Fax: +1 815 9426415; URL: www.morrislibrary.com
1913; Pamela J. Wilson
54 000 vols; 130 curr per; 1 830 av-mat   40300

Morris Public Library, 102 E Sixth St, **Morris**, MN 56267-1211
T: +1 320 5891634; Fax: +1 320 5898892; URL: www.viking.lib.mn.us
Rita Mulcahy
50 000 vols; 100 curr per   40301

The Morristown & Morris Township Library, One Miller Rd, **Morristown**, NJ 07960
T: +1 973 5386161; Fax: +1 973 2674064; URL: www.jfpl.org; www.morristownmorristwplibrary.info
1917; Susan H. Gulick
Printing Coll, American Hist, Mss Coll; 1 branch libr, 1 bookmobile
181 000 vols; 573 curr per   40302

The Morristown & Morris Township Library – North Jersey History & Genealogy Department, One Miller Rd, **Morristown**, NJ 07960
T: +1 973 5383473; Fax: +1 973 2674064
Susan Honeywell Gulick
Homer Davenport Coll;New Jersey Historical Material;Eastern US Genealogical Resources;John DePol Coll;Archival Colls, church, club, govt, local bus, orgn & sch recs, deeds, family & personal papers, hist presv res mat;Thomas Nast Coll;A B Frost Coll
59 000 vols; 350 curr per   40303

Morristown-Hamblen Library, 417 W Main St, **Morristown**, TN 37814-4686
T: +1 423 5866410; Fax: +1 423 5876226; E-mail: library@lcs.net; URL: www.discoveret.org/mhlib
1925; Pamela R. Mullins
Hist & Genealogy (Meta Turley goodson Rm); 1 branch libr
128 000 vols; 153 curr per; 28 e-books   40304

Nolichucky Regional Library, 315 McCrary Dr, **Morristown**, TN 37814
T: +1 423 5866251; Fax: +1 423 5867741; URL: www.tennessee.gov/tsla/regional/NRL/index.htm
1941; Donald B. Reynolds Jr
Business Coll, Head Start Resource Coll, Professional Libr Coll, Rural Coll
166 000 vols; 37 curr per; 4 925 av-mat; 5 000 Talking Bks   40305

Clayton County Library System – Morrow Branch, 6225 Maddox Rd, **Morrow**, GA 30260
T: +1 404 3667749; Fax: +1 404 3634569
June Shapiro
68 000 vols; 58 curr per   40306

Salem Township Public Library, 535 W Pike St, **Morrow**, OH 45152
T: +1 513 8992588; Fax: +1 513 8999420; E-mail: salemtwppl@salem-township.lib.oh.us; URL: www.salem-township.lib.oh.us
1884; Jerri A. Short
85 000 vols; 226 curr per; 14 601 av-mat   40307

Morton Public Library District, 315 W Pershing St, **Morton**, IL 61550
T: +1 309 2632200; Fax: +1 309 2669604; E-mail: mortonlibrary@hotmail.com; URL: www.mortonlibrary.org
1925; Janice Sherman
84 000 vols; 111 curr per
libr loan   40308

Morton Grove Public Library, 6140 Lincoln Ave, **Morton Grove**, IL 60053-2989
T: +1 847 9654220; Fax: +1 847 9657903; E-mail: info@webrary.org; URL: www.webrary.org
1938; Benjamin Hall Schapiro
118 000 vols; 7 740 av-mat; 5 500 sound-rec
libr loan   40309

Latah County Library District, 110 S Jefferson, **Moscow**, ID 83843-2833
T: +1 208 8823925; Fax: +1 208 8825098; E-mail: moscow@latahlibrary.org; URL: norby.latah.lib.id.us
1901; Anne Cheadle
Science fiction; 6 branch libs
99 000 vols; 237 curr per   40310

Moultrie-Colquitt County Library, 204 Fifth St SE, **Moultrie**, GA 31768; P.O. Box 2828, Moultrie
T: +1 229 9856540; Fax: +1 229 9850936; URL: colquitt.k12.ga.us/public_lib
1907; Melody S. Jenkins
Ellen Payne Odom Genealogy Libr (genealogical info on the US & arch & genealogical records of approximately 110 Scottish clans & family organizations; 1 branch libr, 1 bookmobile
120 000 vols; 66 curr per; 448 microforms; 800 av-mat; 2 845 sound-rec; 7 digital data carriers; 117 Slides, 86 Art reproductions
libr loan   40311

Moundsville-Marshall County Public Library, 700 Fifth St, **Moundsville**, WV 26041-1993
T: +1 304 8456911; Fax: +1 304 8456912
1917; Susan Reilly
2 branch libs
175 000 vols; 170 curr per   40312

Mount Airy Public Library, 145 Rockford St, **Mount Airy**, NC 27030-4759
T: +1 336 7895108; Fax: +1 336 7865838; E-mail: mta@nwrl.org; URL: www.nwrl.org
1930; Pat Gwyn
58 000 vols; 25 curr per   40313

Mount Carmel Public Library, 727 Mulberry St, **Mount Carmel**, IL 62863-2047
T: +1 618 2633531; Fax: +1 618 2624243; URL: www.sirin.lib.il.us
1911; Louise Taylor
50 000 vols; 128 curr per   40314

Mount Clemens Public Library, 150 Cass Ave, **Mount Clemens**, MI 48043
T: +1 586 4696200; Fax: +1 586 4696668; E-mail: mcpl@libcoop.net; URL: www.libcoop.net/mountclemens
1865; Donald E. Worrell
124 000 vols; 212 curr per   40315

Mount Kisco Public Library, 100 Main St, **Mount Kisco**, NY 10549
T: +1 914 6668041; Fax: +1 914 6663899; URL: www.mountkiscolibrary.org
1913; Susan Riley
Job Information; Large Print Coll
74 000 vols; 148 curr per; 86 maps; 128 microforms; 500 av-mat; 9 076 sound-rec; 25 digital data carriers
libr loan   40316

Mount Laurel Library, 100 Walt Whitman Ave, **Mount Laurel**, NJ 08054
T: +1 856 2347319; Fax: +1 856 2346916; E-mail: ref@mtlaurel.lib.nj.us; URL: www.mtlaurel.lib.nj.us
1970; Joan E. Bernstein
Adult New Reader, Alice Paul Coll
99 000 vols; 289 curr per; 2 000 digital data carriers   40317

Brown County Public Library, 720 N High St, **Mount Orab**, OH 45154; P.O. Box 527, Mount Orab, OH 45154-0527
T: +1 937 4440181; Fax: +1 937 4446052; E-mail: bookly@oplin.org
Lynn A. Harden
103 000 vols; 380 curr per   40318

Chippewa River District Library, Veterans Memorial Library, 301 S University Ave, **Mount Pleasant**, MI 48858-2597
T: +1 989 7733242; Fax: +1 989 7723280; URL: www.crdl.org
1909; Lise Mitchell
Native American Culture Coll; 5 branch libs
161 000 vols; 300 curr per; 60 microforms; 1 944 av-mat; 2 522 sound-rec; 72 digital data carriers
libr loan; ALA, MLA, OCLC, MLC   40319

Mount Pleasant Public Library, 307 E Monroe Suite 301, **Mount Pleasant**, IA 52641
T: +1 319 3851490; Fax: +1 319 3851491; E-mail: directormppl@iowatelecom.net; URL: www.mountpleasantiowalibrary.com
1901; Gayle Trede
53 000 vols; 100 curr per   40320

Mount Prospect Public Library, Ten S Emerson St, **Mount Prospect**, IL 60056
T: +1 847 2535675; Fax: +1 847 2530642; URL: www.mppl.org
1943; Marilyn Genther
344 000 vols; 548 curr per; 545 incunabula; 22 626 govt docs; 700 music scores; 420 maps; 60 000 microforms; 15 808 av-mat; 16 970 sound-rec; 480 digital data carriers; 262 art reproductions
libr loan; ALA   40321

Alexandrian Public Library, 115 W Fifth St, **Mount Vernon**, IN 47620-1869
T: +1 812 8383286; Fax: +1 812 8389639; E-mail: alexpl@evansville.net; URL: www.apl.lib.in.us
1895; Marissa Priddis
Curriculum Enrichment
85 000 vols; 114 curr per; 4 000 microforms; 3 500 av-mat; 8 950 sound-rec; 53 digital data carriers   40322

C. E. Brehm Memorial Public Library District, 101 S Seventh St, **Mount Vernon**, IL 62864
T: +1 618 2426322; Fax: +1 618 2420810; URL: www.mtvbrehm.lib.il.us
1899; Marilyn Konold
Genealogy Coll, Southern Illinois Hist
74 000 vols; 195 curr per; 1 000 e-books
libr loan   40323

Mount Vernon City Library, 315 Snoqualmie St, **Mount Vernon**, WA 98273
T: +1 360 3366209; Fax: +1 360 3366259; URL: www.ci.mount-vernon.wa.us
1908; Brian M. Soneda
Art & architecture, Aviation, Hist; Spanish Language Coll
69 000 vols; 327 curr per; 1 419 av-mat
                                             40324

Mount Vernon Public Library, 28 S First Ave, **Mount Vernon**, NY 10550
T: +1 914 6681840; Fax: +1 914 6681018; URL: mountvernonpubliclibrary.org
1854; Rodney J. Lee
Black Heritage (Haines Coll), Mills Law Coll
615 000 vols; 4 320 curr per
libr loan   40325

Public Library of Mount Vernon & Knox County, 201 N Mulberry St, **Mount Vernon**, OH 43050-2413
T: +1 740 3922665; Fax: +1 740 3973866; E-mail: library@knox.net; URL: www.knox.net
1888; John K. Chidester
3 branch libs
166 000 vols; 651 curr per; 5 121 sound-rec
                                             40326

Emmet O'Neal Library, 50 Oak St, **Mountain Brook**, AL 35213
T: +1 205 8790492; Fax: +1 205 8795388; E-mail: the@eolib.org; URL: www.eolib.org
1964; Susan DeBrecht Murrell
Gardening Coll
124 000 vols; 132 curr per   40327

Baxter County Library, 300 Library Hill, **Mountain Home**, AR 72653
T: +1 870 5809887; URL: www.baxtercountylibrary.org
75 000 vols; 250 curr per   40328

Baxter County Library, 424 W Seventh St, **Mountain Home**, AR 72653
T: +1 870 4253598; Fax: +1 870 4257226; E-mail: baxlib@baxtercountylibrary.org; URL: www.baxtercountylibrary.org
Gwen Khayat
73 000 vols; 190 curr per   40329

Mountain View Public Library, 585 Franklin St, **Mountain View**, CA 94041-1998
T: +1 650 9036335; Fax: +1 650 9620438; URL: www.library.ci.mtnview.ca.us
1905; Karen E. Burnett
Automobiles (Maintenance & Repair)
229 000 vols; 326 curr per; 2 000 e-books
libr loan   40330

Mountainside Public Library, Constitution Plaza, **Mountainside**, NJ 07092
T: +1 908 2330115; Fax: +1 908 2327311; E-mail: info@mountainsidelibrary.org; URL: www.mountainsidelibrary.org
1934; Michael Banick
53 000 vols; 120 curr per; 1 413 av-mat
                                             40331

Sno-Isle Libraries – Mountlake Terrace Branch, 23300 58th Ave W, **Mountlake Terrace**, WA 98043
T: +1 425 7768722; Fax: +1 425 7763411
Rosy Brewer
64 000 vols   40332

Sno-Isle Libraries – Mukilteo Branch, 4675 Harbour Pointe Blvd, **Mukilteo**, WA 98275
T: +1 425 4938202; Fax: +1 425 4931601
Jane Crawford
80 000 vols   40333

Mukwonago Community Library, 300 Washington Ave, **Mukwonago**, WI 53149-1909
T: +1 262 3636456; Fax: +1 262 3636457; URL: www.wcfls.lib.wi.us/mukcom/
Kathleen McBride
69 000 vols; 152 curr per   40334

Gloucester County Library System, 389 Wolfert Station Rd, **Mullica Hill**, NJ 08062
T: +1 856 2236000; Fax: +1 856 2236039; E-mail: gloucester@gcls.org; URL: www.gcls.org
1976; Robert S. Wetherall
7 branch libs
213 000 vols; 186 curr per; 420 e-books
13 080 av-mat; 16 390 sound-rec   40335

Gloucester County Library System – Mullica Hill (Headquarters), 389 Wolfert Station Rd, **Mullica Hill**, NJ 08062
T: +1 856 2236000
Robert Wetherall
214 000 vols; 275 curr per; 271 e-books
                                             40336

Carnegie Library, 301 E Jackson St, **Muncie**, IN 47305-1878
T: +1 765 7478208; Fax: +1 765 7475156; URL: www.munpl.org
1874
Goddard Memorial Coll (Indiana Authors)
89 000 vols; 805 curr per   40337

Muncie Center Township Public Library, 2005 S High St, **Muncie**, IN 47302
T: +1 765 7478204; Fax: +1 765 7478211; URL: www.munpl.org
Virginia Nilles
204 000 vols; 494 curr per; 76 e-books; 4 930 av-mat   40338

Fremont Public Library District, 1170 N Midlothian Rd, **Mundelein**, IL 60060
T: +1 847 5668702; Fax: +1 847 5660204; E-mail: ref@fremontlibrary.org; URL: www.fremontlibrary.org
1955; Scott Davis
Local govt, American Indians
170 000 vols; 200 curr per
libr loan; ALA   40339

Carnegie Library of Homestead, 510 E Tenth Ave, **Munhall**, PA 15120-1910
T: +1 412 4623444; Fax: +1 412 4624669
Tyrone Ward
50 000 vols; 132 curr per; 903 av-mat
                                             40340

Lake County Public Library – Munster Branch, 8701 Calumet Ave, **Munster**, IN 46321-2526
T: +1 219 8368450; Fax: +1 219 8365694
Linda Dunn
78 000 vols; 71 curr per   40341

Linebaugh Public Library System of Rutherford County, 105 W Vine St, **Murfreesboro**, TN 37130-3673
T: +1 615 8934131; Fax: +1 615 8485038; URL: www.linebaugh.org
1948; Laurel Smith Best
2 branch libr
141 000 vols; 100 curr per; 497 e-books
                                             40342

Sallie Logan Public Library, 1808 Walnut St, **Murphysboro**, IL 62966
T: +1 618 6843271; Fax: +1 618 6842392; E-mail: asksallielogan@shawls.lib.il.us; URL: www.murphysboro.lib.il.us
1936; Donella L. Odum
58 000 media items; 52 curr per   40343

Calloway County Public Library, 710 Main St, **Murray**, KY 42071
T: +1 270 7532288; Fax: +1 270 7538263
1967; Ben Graves
63 000 vols; 150 curr per   40344

Murray Public Library, 166 E 5300 South, **Murray**, UT 84107
T: +1 801 2642580; Fax: +1 801 2642586; E-mail: librarian@murray.utah.gov; URL: murraylibrary.org
1910
Utah State Historical Quarterly
84 110 vols; 263 curr per
libr loan   40345

Murrieta Public Library, Eight Town Sq, **Murrieta**, CA 92562
T: +1 951 4616127; Fax: +1 951 6960165; URL: www.murrieta.org/services/library
Loretta McKinney
75 000 vols; 200 curr per   40346

Murrysville Community Library, 4130 Sardis Rd, **Murrysville**, PA 15668-1120
T: +1 724 3271102; Fax: +1 724 3277142; E-mail: murrysvillelibrary@comcast.net; URL: www.murrysvillelibrary.org
1922; Denise S. Sticha
Large Print Coll
56 000 vols; 123 curr per
libr loan   40347

Westmoreland County Federated Library System, 4130 Sardis Rd, **Murrysville**, PA 15668
T: +1 724 3271677; Fax: +1 724 3271697
Denise Sticha
760 000 vols; 2 000 curr per   40348

Musser Public Library, 304 Iowa Ave, **Muscatine**, IA 52761-3875
T: +1 563 2633472; Fax: +1 563 2641033; E-mail: refmus@muscatinelibrary.us; URL: www.muscatinelibrary.us
1901; Pam Collins
Area Servicemen, World War II, Laura Ingalls Wilder Coll
136 000 vols; 225 curr per   40349

Muskego Public Library, S73 W16663 Janesville Rd, **Muskego**, WI 53150; P.O. Box 810, Muskego, WI 53150-0810
T: +1 262 9712112; Fax: +1 262 9712115; E-mail: refd@ci.muskego.wi.us; URL: www.ci.muskego.wi.us/library
1960; Holly Sanhuber
111 000 vols; 300 curr per
libr loan   40350

Hackley Public Library, 316 W Webster Ave, *Muskegon*, MI 49440
T: +1 231 7227276; Fax: +1 231 7264724; E-mail: askus@hackleylibrary.org; URL: www.hackleylibrary.org
1888; Martha Ferriby
Civil War, Genealogy, Lumbering
165 000 vols; 315 curr per   40351

Muskegon Area District Library, 4845 Airline Rd, Unit 5, *Muskegon*, MI 49444-4503
T: +1 231 7376248; Fax: +1 231 7376307; URL: www.madl.org
1938; Stephen Dix
10 branch libs
232 000 vols; 554 curr per; 4 463 Large Print Bks   40352

Muskegon Area District Library – Norton Shores Jacob O. Funkhouser Branch, 705 Seminole Rd, *Muskegon*, MI 49441-4797
T: +1 231 7808841; Fax: +1 231 7805436; E-mail: nor@llcoop.org; URL: www.madl.org/shores
Mark Ames
53 000 vols; 134 curr per   40353

Eastern Oklahoma District Library System, 814 W Okmulgee, *Muskogee*, OK 74401-6839
T: +1 918 6832846; Fax: +1 918 6830436; URL: www.eodls.lib.ok.us
1973; Mary J. Moroney
Grant Foreman Coll (incl bks on Indians & Early Oklahoma Hist), Local Hist (Essa Gladney Coll)
472 000 vols; 667 curr per; 19 420 av-mat; 14 140 sound-rec
ALA   40354

Muskogee Public Library, 801 W Okmulgee, *Muskogee*, OK 74401
T: +1 918 6826657; Fax: +1 918 6829466; URL: www.eok.lib.ok.us
1909; Jan Bryant
Local hist (Grant Foreman Room) & Genealogy
247 000 vols; 625 curr per; 2 794 av-mat; 1 196 sound-rec
libr loan   40355

Chapin Memorial Library, 400 14th Ave N, *Myrtle Beach*, SC 29577-3612
T: +1 843 9181275; Fax: +1 843 9181288; URL: www.chapinlibrary.org
1949; Catherine B. Wiggins
89 000 vols; 138 curr per   40356

Horry County Memorial Library – Socastee, 141 707-Connector Rd, *Myrtle Beach*, SC 29588
T: +1 843 2154700; Fax: +1 843 2152801; E-mail: socasteelibrary@horrycounty.org
Sharon Eels
59 000 vols   40357

Nacogdoches Public Library, 1112 North St, *Nacogdoches*, TX 75961-4482
T: +1 936 5592970; Fax: +1 936 5698282; URL: npl.sfasu.edu
1974; Anne Barker
88 000 vols; 84 curr per; 2 673 av-mat   40358

Nampa Public Library, 101 11th Ave S, *Nampa*, ID 83651
T: +1 208 4685800; Fax: +1 208 4652277; E-mail: info@nampalibrary.org; URL: www.nampalibrary.org
1904; Karen Ganske
110 000 vols; 260 curr per   40359

Mill Memorial Library, 495 E Main St, *Nanticoke*, PA 18634-1897
T: +1 570 7353030; Fax: +1 570 7350340; URL: www.gnasd.com/millmemorial.htm
1945; Cliff Farides
57 000 vols; 142 curr per; 92 av-mat   40360

Naperville Public Library – 95th Street, 3015 Cedar Glade Dr, *Naperville*, IL 60564
Fax: +1 630 6374870
John Spears
209 000 vols   40361

Naperville Public Library – Naper Boulevard, 2035 S Naper Blvd, *Naperville*, IL 60565-3353
T: +1 630 9614100
Olya Tymciurak
153 000 vols   40362

Naperville Public Library – Nichols Library, 200 W Jefferson Ave, *Naperville*, IL 60540-5374
T: +1 630 9614100; Fax: +1 630 6376149; URL: www.naperville-lib.org
1897; Donna Dziedzic
2 branch libr
703 000 vols
libr loan   40363

Collier County Public Library, 2385 Orange Blossom Dr, *Naples*, FL 34109
T: +1 239 5930334; Fax: +1 239 2548167; URL: www.gov.net/library
Kathleen Teaze
632 000 vols; 1 300 curr per   40364

Napoleon Public Library, 310 W Clinton St, *Napoleon*, OH 43545-1472
T: +1 419 5922531; Fax: +1 419 5991472
1906; Pamela J. Lieser
County Papers from 1852, microfilm; 2 branch libs
145 000 vols; 252 curr per; 1 160 av-mat   40365

Assumption Parish Library, 293 Napoleon Ave, *Napoleonville*, LA 70390
T: +1 504 3697070; Fax: +1 504 3696019; URL: www.assumption.lib.la.us/index.htm
1968; Dr. Teri Maggio
Assumption Pioneer (1850-), micro; French Language Mat Coll (childrens & adult); Southern Louisiana Genealogy; 2 branch libs
58 000 vols; 400 curr per   40366

Nappanee Public Library, 157 N Main St, *Nappanee*, IN 46550
T: +1 574 7737919; Fax: +1 574 7737910; E-mail: readmore@nappanee.lib.in.us; URL: www.nappanee.lib.in.us
1921; Linda Yoder
66 000 vols; 195 curr per; 298 av-mat; 1 117 sound-rec; 60 digital data carriers
libr loan   40367

Maury Loontjens Memorial Library, 35 Kingstown Rd, *Narragansett*, RI 02882
T: +1 401 7899507; Fax: +1 401 7820677; URL: web.provlib.org/narlib
1903; Patti Arkwright
90 000 vols; 125 curr per; 2 565 av-mat   40368

Nashua Public Library, Two Court St, *Nashua*, NH 03060
T: +1 603 5894635; Fax: +1 603 5894640; E-mail: administration@nashualibrary.org; URL: www.nashualibrary.org
1867; Joseph R. Dionne
Architecture, Art, Business, Music; Hunt Room Coll; 1 branch libr, 1 bookmobile
241 000 vols; 250 curr per; 4 820 sound-rec; 4 820 digital data carriers; 4 450 Large Print Bks   40369

Brown County Public Library, 205 Locust Lane, *Nashville*, IN 47448; P.O. Box 8, Nashville, IN 47448-0008
T: +1 812 9882850; Fax: +1 812 9888119; URL: www.browncounty.lib.in.us
1919; Yvonne C. Oliger
Brown County Artists & Authors
60 000 vols; 115 curr per   40370

Nashville Public Library, 615 Church St, *Nashville*, TN 37219-2314
T: +1 615 8625760; Fax: +1 615 8625771; URL: www.library.nashville.org
1904; Donna D. Nicely
Children's Int Coll, Deaf Services Coll, Govt Arch, Naff Drama Coll; 23 branch libr
1 570 000 vols
libr loan   40371

Nashville Public Library – Bellevue, 650 Colice Jeanne Rd, *Nashville*, TN 37221-2811
T: +1 615 8625854; Fax: +1 615 8625758
Deborah Hynes
55 000 vols   40372

Nashville Public Library – Bordeaux, 4000 Clarksville Pike, *Nashville*, TN 37218-1912
T: +1 615 8625856; Fax: +1 615 8625748
Verlon Malone
88 000 vols   40373

Nashville Public Library – Donelson, 2315 Lebanon Rd, *Nashville*, TN 37214-3410
T: +1 615 8625859; Fax: +1 615 8625799
Barbara Franklin
52 000 vols   40374

Nashville Public Library – Edmondson Pike, 5501 Edmondson Pike, *Nashville*, TN 37211-5808
T: +1 615 8803957; Fax: +1 615 8803961
Susan Perry
113 000 vols   40375

Nashville Public Library – Green Hills, 3701 Benham Ave, *Nashville*, TN 37215-2121
T: +1 615 8625863; Fax: +1 615 8625881
Claudia Schauman
108 000 vols   40376

Nashville Public Library – Richland Park, 4711 Charlotte Ave, *Nashville*, TN 37209-3404
T: +1 615 8625870; Fax: +1 615 8625897
Victoria J. Elliott
52 000 vols   40377

Homochitto Valley Library Service, 220 S Commerce St, *Natchez*, MS 39120
T: +1 601 4458862; Fax: +1 601 4467795
1883
Natchez Mississippi Coll; 3 branch libs
92 000 vols; 230 curr per; 500 sound-rec   40378

Natchitoches Parish Library, 450 Second St, *Natchitoches*, LA 71457-4649
T: +1 318 3573280; Fax: +1 318 3577073; E-mail: info@natchitoches.lib.la.us; URL: www.youseemore.com/natchitoches
1939; Robert E. Black
Natchitoches Authors, Louisiana Hist
90 000 vols; 120 curr per   40379

Morse Institute Library, 14 E Central St, *Natick*, MA 01760
T: +1 508 6476520; Fax: +1 508 6476527; URL: www.morseinstitute.org
1873; Paula M. Polk
163 000 vols; 300 curr per; 1 000 microforms; 3 948 av-mat; 3 042 sound-rec; 84 digital data carriers
libr loan   40380

Howard Whittemore Memorial Library, 243 Church St, *Naugatuck*, CT 06770-4198
T: +1 203 7294591; Fax: +1 203 7231820; URL: www.biblio.org/whittemore
1894; Joan Lamb
Large Print Bks
65 000 vols; 120 curr per   40381

Memorial Library of Nazareth & Vicinity, 295 E Center St, *Nazareth*, PA 18064-2298
T: +1 610 7594932; Fax: +1 610 7599513; URL: www.nazarethlibrary.org
1949; Lynn Snodgrass-Pilla
59 000 vols; 103 curr per; 2 000 e-books   40382

Marion & Ed Hughes Public Library, 2712 Nederland Ave, *Nederland*, TX 77627-7015
T: +1 409 7221255; Fax: +1 409 7215469
1930; Victoria L. Klehn
Large Print Coll
55 000 vols; 72 curr per; 93 microforms; 4 410 av-mat; 1 604 sound-rec; 169 digital data carriers
libr loan   40383

Needham Free Public Library, 1139 Highland Ave, *Needham*, MA 02494-3298
T: +1 781 4557559; Fax: +1 781 4557591; E-mail: neemail1@minlib.net; URL: www.needhamma.gov/library
1888; Ann C. MacFate
Business & management; Benjamin Franklin, N.C. Wyeth Art Coll
132 000 vols; 250 curr per; 4 200 Large Print Bks   40384

Neenah Public Library, 240 E Wisconsin Ave, *Neenah*, WI 54956-3010; P.O. Box 569, Neenah, WI 54957-0569
T: +1 920 8866300; Fax: +1 920 8866324; E-mail: library@neenahlibrary.org; URL: www.neenahlibrary.org
1884; Stephen Lewis Proces
200 000 vols; 371 curr per; 13 000 e-books; 15 000 Audio Bks, 800 Electronic Media & Resources, 300 High Interest/Low Vocabulary Bks, 3 000 Large Print Bks   40385

Nelsonville Public Library, Athens County Library Services, 95 W Washington, *Nelsonville*, OH 45764-1177
T: +1 740 7532118; Fax: +1 740 7533543; E-mail: nelpl@athenscounty.lib.oh.us; URL: www.myacpl.org
1936; Lauren Miller
60 000 vols; 189 curr per   40386

Neosho Newton County Library, 201 W Spring St, *Neosho*, MO 64850
T: +1 417 4514231; Fax: +1 417 4516438; URL: www.neosholibrary.org
1956; Ginny Ray
51 000 vols; 132 curr per; 618 av-mat   40387

Neptune Public Library, 25 Neptune Blvd, *Neptune*, NJ 07753-1125
T: +1 732 7758241; Fax: +1 732 7741132; E-mail: library@neptunetownship.org; URL: www.neptunepubliclibrary.org
1924; Marian R. Bauman
75 000 vols; 104 curr per; 15 microforms; 5 200 av-mat; 3 200 sound-rec; 100 digital data carriers   40388

Jacksonville Public Library – Beaches Regional, 600 Third St, *Neptune Beach*, FL 32266-5014
T: +1 904 2411141; Fax: +1 904 2414965
Sally Doherty
Joe Gill Business Coll
143 000 vols   40389

Nevada Public Library, 631 K Ave, *Nevada*, IA 50201
T: +1 515 3822628; Fax: +1 515 3823552; E-mail: npl@nevada.lib.ia.us; URL: nevada.lib.ia.us
1896; Beth Williams
60 000 vols; 130 curr per   40390

Columbus Metropolitan Library – New Albany Branch, 200 Market St, *New Albany*, OH 43054
T: +1 614 6452275; Fax: +1 614 4794945
Christopher Korenowsky
120 000 vols   40391

New Albany-Floyd County Public Library, 180 W Spring St, *New Albany*, IN 47150-3692
T: +1 812 9448464; Fax: +1 812 9493532; URL: www.nafcpl.lib.in.us
1884; Stephen T. Day
1 bookmobile
226 000 vols; 405 curr per
libr loan   40392

Union County Library, Jennie Stephens Smith Library, 219 King St, *New Albany*, MS 38652; P.O. Box 846, New Albany, MS 38652-0846
T: +1 662 5341991; Fax: +1 662 5341937
1933; Kay Sappington
Mississippi hist; Genealogy Coll
70 000 vols; 75 curr per; 90 microforms; 298 sound-rec
libr loan   40393

New Berlin Public Library, 15105 Library Lane, *New Berlin*, WI 53151
T: +1 262 7854980; Fax: +1 262 7854984; E-mail: nbinfo@wcfls.lib.wi.us; URL: www.wcfls.lib.wi.us/newberlin
1969; Katie Schulz
State Selected Depot Libr
150 000 vols; 253 curr per   40394

Craven-Pamlico-Carteret Regional Library System, 400 Johnson St, ***New Bern***, NC 28560
T: +1 252 6387800; Fax: +1 252 6387817; URL: newbern.cpclib.org/index.html
1962; Jackie Beach
Genealogy East North Carolina, North Carolina Coll
260 000 vols; 200 curr per 40395

New Bern-Craven County Public Library, 400 Johnson St, ***New Bern***, NC 28560-4098
T: +1 252 6387800; Fax: +1 252 6387817; URL: newbern.cpclib.org
1912; Joanne Straight
North Carolina Coll
100 000 vols; 223 curr per
libr loan 40396

New Braunfels Public Library, 700 E Common St, ***New Braunfels***, TX 78130-5689
T: +1 830 6082150; Fax: +1 830 6082151; URL: www.nbpl.lib.tx.us
Ben Pensiero
97 000 vols; 236 curr per; 3 413 av-mat 40397

New Britain Public Library, 20 High St, ***New Britain***, CT 06051-4226
T: +1 860 2243155; Fax: +1 860 8265191; URL: www.nbpl.info
1858; Candice Brown
Foreign Language Mats, Elihu Burritt Coll; 2 branch libs
229 000 vols; 297 curr per 40398

New Brunswick Free Public Library, 60 Livingston Ave, ***New Brunswick***, NJ 08901-2597
T: +1 732 7455108; Fax: +1 732 8460226; URL: www.nbfpl.org
1883; Robert Belvin
Hungarian Language Coll
80 000 vols; 254 curr per 40399

New Buffalo Township Public Library, 33 N Thompson St, ***New Buffalo***, MI 49117
T: +1 269 4692933; Fax: +1 269 4693521; URL: www.nbtpl.org
Julie Grynwich
50 000 vols; 112 curr per 40400

New Canaan Library, 151 Main St, ***New Canaan***, CT 06840
T: +1 203 5945010; Fax: +1 203 5945032; E-mail: onlineref@newcanaanlibrary.org; URL: newcanaanlibrary.org
1877; Alice Knapp
Broadcast Journalism, European Art, Gardening/Landscaping, Howard Schless Medieval Coll, Nature, World War II (Chester Hansen Coll)
185 000 vols; 345 curr per; 4 820 av-mat; 6 320 sound-rec; 437 digital data carriers
libr loan 40401

Garfield County Public Library System, 402 W Main, ***New Castle***, CO 81647; P.O. Box 320, New Castle
T: +1 970 9842347; Fax: +1 970 9842487; URL: www.garfieldlibraries.org
1938; Anne Moore
6 branch libs
157 000 vols; 292 curr per 40402

Garfield County Public Library System – New Castle Branch, 402 W Main, ***New Castle***, CO 81647; P.O. Box 320, New Castle, CO 81647-0320
T: +1 970 9842346; Fax: +1 970 9842081
Ann Honchell
53 000 vols 40403

Lawrence County Federated Library System, 207 E North St, ***New Castle***, PA 16101-3691
T: +1 724 6586659; Fax: +1 724 6587209
Susan E. Walls
171 000 vols; 125 curr per 40404

New Castle County Public Library System, 77 Reads Way, ***New Castle***, DE 19720
T: +1 302 3955600; Fax: +1 302 3955592; URL: www.nccdelib.org
1975; Diana Brown
1 000 000 vols; 2 568 curr per
libr loan 40405

New Castle Public Library, 207 E North St, ***New Castle***, PA 16101-3691
T: +1 724 6586659; Fax: +1 724 6589012; URL: www.ncdlc.org
1908; Susan Walls
Hist; Architecture (Jane Jackson Coll), Brotherhood (Joshua A. Kaplan Coll), Gardening & Landscape (Wylie McCaslin Coll), Judaism, Pharmaceutical Coll, Polish Falcons Coll, Women's World
121 000 vols; 125 curr per
libr loan 40406

New City Free Library, 220 N Main St, ***New City***, NY 10956
T: +1 845 6344997; Fax: +1 845 6344401; URL: www.newcitylibrary.org
1933; Charles Emery McMorran
171 000 vols; 634 curr per; 11 939 av-mat; 10 460 sound-rec; 53 690 digital data carriers; 3 480 Large Print Bks 40407

New Cumberland Public Library, One Benjamin Plaza, ***New Cumberland***, PA 17070-1597
T: +1 717 7747820; Fax: +1 717 7747824; E-mail: newcumberland@ccpa.net; URL: www.ccpa.net
1941; Judith Dillen
60 000 vols; 131 curr per; 491 av-mat 40408

New Hartford Public Library, 2 Library Lane, ***New Hartford***, NY 13413
T: +1 315 7331535; Fax: +1 315 7330795; URL: www.newhartfordpubliclibrary.org
1976
52 270 vols; 209 curr per 40409

New Hartford Public Library, Two Library Lane, ***New Hartford***, NY 13413-2815
T: +1 315 7331535; Fax: +1 315 7330795; E-mail: newhartford@midyork.org; URL: www.newhartfordpubliclibrary.org
Hans J. Plambeck
68 000 vols; 209 curr per 40410

New Haven Free Public Library, 133 Elm St, ***New Haven***, CT 06510
T: +1 203 9468130; Fax: +1 203 9468140; URL: www.cityofnewhaven.com/library
1887; James C. Welbourne
4 branch libs
640 000 vols; 350 curr per
libr loan 40411

New Haven Free Public Library – Mitchell, 37 Harrison St, ***New Haven***, CT 06515
T: +1 203 9468117
Diane Carvalho
51 000 vols 40412

Hillside Public Library, 155 Lakeville Rd, ***New Hyde Park***, NY 11040
T: +1 516 3557850; E-mail: hillsidelibrary@yahoo.com; URL: www.nassaulibrary.org/hillside/index.htm
1962; Charlene Noll
Cooking, Crafts
65 000 vols; 90 curr per 40413

Iberia Parish Library, 445 E Main St, ***New Iberia***, LA 70560-3710
T: +1 337 3647024; Fax: +1 337 3647042; URL: www.iberia.lib.la.us
1947; Kathleen Miles
Drob-Bunk Johnson Jazz Coll, I. A. & Carroll Martin Photo Coll; 6 branch libs
250 000 vols; 222 curr per; 3 161 av-mat; 4 952 Large Print Bks, 1 090 Audio Bks
libr loan 40414

Peoples Library, 880 Barnes St, ***New Kensington***, PA 15068-6235
T: +1 724 3391021; Fax: +1 724 3392027; URL: www.peopleslink.org
1928
59 000 vols; 36 curr per 40415

New Lenox Public Library, 120 Veterans Pkwy, ***New Lenox***, IL 60451
T: +1 815 4852605; Fax: +1 815 4852548; E-mail: director@newlenoxlibrary.org; URL: www.newlenoxlibrary.org
1936; Kate Hall
Quilting
94 600 vols; 220 curr per; 43 microforms; 1 221 av-mat; 1 915 sound-rec; 43 digital data carriers
libr loan 40416

Perry County District Library, 117 S Jackson St, ***New Lexington***, OH 43764-1330
T: +1 740 3424194; Fax: +1 740 3424204; E-mail: webmaster@pcdl.org; URL: www.pcdl.org
1935; Melissa A. Marolt
6 branch libs
88 053 vols; 246 curr per; 8 421 e-books; 7 685 av-mat; 3 836 sound-rec; 282 digital data carriers
ALA 40417

New London Public Library, 406 S Pearl St, ***New London***, WI 54961-1441
T: +1 920 9828519; Fax: +1 920 9828617; E-mail: nlp@mail.owls.lib.wi.us; URL: www.owls.lib.wi.us/nlp
1895; Ann Hunt
Hist Coll
55 000 vols; 119 curr per 40418

New London Public Library, 63 Huntington St, ***New London***, CT 06320
T: +1 860 4471411; Fax: +1 860 4432083; URL: lioninc.org/newlondon
1891; Peter F. Ciparelli
83 000 vols; 155 curr per; 2 000 microforms; 1 803 av-mat; 1 249 sound-rec; 12 digital data carriers
libr loan 40419

New Milford Public Library, 24 Main St, ***New Milford***, CT 06776
T: +1 860 3551191; Fax: +1 860 3509579; URL: www.biblio.org/newmilford/
1898; Carl DeMilia
Original Lithographs (Newton Coll), Pepper Coll
105 000 vols; 200 curr per 40420

New Milford Public Library, 200 Dahlia Ave, ***New Milford***, NJ 07646-1812
T: +1 201 2621221; Fax: +1 201 2625639; E-mail: nmilcirc@bccls.org
1936; Terri McColl
Consumer info
54 000 vols; 105 curr per; 1 042 av-mat 40421

New Orleans Public Library, 219 Loyola Ave, ***New Orleans***, LA 70112-2044
T: +1 504 5962560; Fax: +1 504 5962609; E-mail: AskNOPL@gno.lib.la.us; URL: www.nutrias.org
1896; Rica Trigs
City Archs, Louisiana Div; 14 branch libs
795 000 vols; 999 curr per 40422

New Orleans Public Library – Algiers Regional, 3014 Holiday Dr, ***New Orleans***, LA 70131
T: +1 504 5962641; Fax: +1 504 5962661
Seale Patterson
56 000 vols 40423

New Orleans Public Library – Milton H. Latter Memorial Branch, 5120 St Charles Ave, ***New Orleans***, LA 70115-4941
T: +1 504 5962625; Fax: +1 504 5962665
Missy Abbott
60 000 vols; 20 curr per 40424

Tuscarawas County Public Library, 121 Fair Ave NW, ***New Philadelphia***, OH 44663-2600
T: +1 330 3644474; Fax: +1 330 3648217; E-mail: tuscwref@oplin.org; URL: www.tusclibrary.org
1905; Michelle Ramsell
4 branch libs
117 000 vols; 372 curr per; 9 570 av-mat; 6 760 sound-rec; 1 985 digital data carriers 40425

New Port Richey Public Library, 5939 Main St, ***New Port Richey***, FL 34652
T: +1 727 8531279; Fax: +1 727 8531280; URL: www.tblc.org/newport
1919; Susan D. Dillinger
Avery Coll
87 000 vols; 202 curr per; 6 000 e-books 40426

New Providence Memorial Library, 377 Elkwood Ave, ***New Providence***, NJ 07974
T: +1 908 6650311; Fax: +1 908 6652319; E-mail: library@newprov.org; URL: www.newprovidencelibrary.org; www.newprovidencelibrary.org
1921; James Keehbler
80 000 vols; 120 curr per 40427

Pointe Coupee Parish Library, New Roads (Main Branch), 201 Claiborne St, ***New Roads***, LA 70760-3403
T: +1 225 6389841; Fax: +1 225 6389847; E-mail: c1pc@pelican.state.lib.la.us; URL: www.pointe-coupee.lib.la.us
Melissa Hymel
Art, Geology, Humanities; Louisiana Studies; 4 branch libs
80 000 vols; 216 curr per 40428

New Rochelle Public Library, One Library Plaza, ***New Rochelle***, NY 10801
T: +1 914 6327878; Fax: +1 914 6320262; URL: www.nrpl.org
1894; Tom Geoffino
Art Slides, Fine Art Bks (Retrospective), Libretti Scores, Local Newspapers from 1861, Opera, Picture Coll; 1 branch libr
271 000 vols; 3 055 curr per
libr loan 40429

New Ulm Public Library, 17 N Broadway, ***New Ulm***, MN 56073-1786
T: +1 507 3598332; Fax: +1 507 3543255; URL: www.newulmlibrary.org
Larry B. Hlavsa
81 000 vols; 173 curr per 40430

Mercantile Library of New York Center for Fiction, 17 E 47th St, ***New York***, NY 10017
T: +1 212 7556710; Fax: +1 212 8240831; E-mail: info@mercantilelibrary.org; URL: www.mercantilelibrary.org
1820; Brenda Wegener
Lit; 19th C Fiction & Nonfiction
150 000 vols; 66 curr per 40431

The New York Public Library – Astor, Lenox & Tilden Foundations – 58th Street Branch, 127 E 58th St (between Park & Lexington Aves), ***New York***, NY 10022-1211
T: +1 212 7597358; Fax: +1 212 7586858; URL: www.nypl.org/branch/man/fe.html
John D. Bhagwandin
60 000 vols 40432

The New York Public Library – Astor, Lenox & Tilden Foundations – 96th Street Branch, 112 E 96th St (near Lexington Ave), ***New York***, NY 10128-2597
T: +1 212 2890908; Fax: +1 212 4104564; E-mail: 96st_branch@nypl.org; URL: www.nypl.org/branch/local/man/nsr.cfm
William J. Seufert
59 000 vols 40433

The New York Public Library – Astor, Lenox & Tilden Foundations – Asian & Middle Eastern Division, Fifth Ave & 42nd St, Rm 219 & 220, ***New York***, NY 10018-2788
T: +1 212 9300616; Fax: +1 212 9300551; E-mail: asiaref@nypl.org; URL: www.nypl.org/research/chss/index.html
John M. Lundquist
Chinese Mohammedan Literature (Mason Coll);Chinese Rare Books of Ming & Ching Dynasties from the Personal Library of James Legge, incl ms;Materials from the Schiff Coll & Manuscripts Coll;Periodicals in all Literature Languages of Asia & the Middle East;A
464 000 vols; 1 400 curr per 40434

The New York Public Library – Astor, Lenox & Tilden Foundations – Dorot Jewish Division, Fifth Ave & 42nd St, Rm 111, ***New York***, NY 10018-2788
T: +1 212 9300601; Fax: +1 212 6420141; URL: www.nypl.org/research/chss/jws/jewish.html
Michael Terry
Early Hebrew Printing;Oral Histories of the American Jewish Experience;Jewish Newspapers on Microfilm;Yiddish Theater Manuscripts & Ephemera
308 000 vols 40435

The New York Public Library – Astor, Lenox & Tilden Foundations – Epiphany Branch, 228 E 23rd St (near Second Ave), ***New York***, NY 10010-4672
T: +1 212 6792645; Fax: +1 212 7794624; E-mail: epiphany@nypl.org; URL: www.nypl.org/branch/local/man/

ep.cfm
Karen A. Weis-Pullen
57 000 vols   40436

The New York Public Library – Astor, Lenox & Tilden Foundations – Fort Washington Branch, 535 W 179th St (between St Nicholas & Audubon Aves), *New York*, NY 10033-5799
T: +1 212 9273533; Fax: +1 212 7408601; E-mail: fort_washington@nypl.org; URL: www.nypl.org/branch/local/man/fw.cfm
Dorota A. Socha
70 000 vols   40437

The New York Public Library – Astor, Lenox & Tilden Foundations – General Research Division, Fifth Ave & 42nd St, Rm 315, *New York*, NY 10018-2788
T: +1 212 9300830; Fax: +1 212 9300572; E-mail: grdref@nypl.org; URL: www.nypl.org/research/chss/index.html
Ruth Carr
Spanish Civil War (David McKelvy White Coll);World Wars I & II;Small Press Poetry;Baseball (Spalding Coll);Chess Coll;Native Americans;Folklore of Americas;Literature (Bunyan, Cervantes, Dante, Milton & Shakespeare Colls)
4 500 000 vols; 11 000 curr per   40438

The New York Public Library – Astor, Lenox & Tilden Foundations – Irma & Paul Milstein Division of US History, Local History & Genealogy, 5th Ave & 42nd St, Rm 121, *New York*, NY 10018-2788
T: +1 212 9300828; E-mail: histref@nypl.org; URL: www.nypl.org/research/chss/lhg/genea.html
Ruth A. Carr
Postcards & Scrapbooks of United States Local Views;Political Campaign Ephemera, broadsides, candidate position papers, pamphlets;Local History Coll;Photographic Views of New York City
317 000 vols   40439

The New York Public Library – Astor, Lenox & Tilden Foundations – Mid-Manhattan Library, 455 Fifth Ave (at 40th St), *New York*, NY 10016-0122; mail address: 5 E 39th St, New York, NY 10016-0122
T: +1 212 3400830; Fax: +1 212 5760048; URL: www.nypl.org/branch/central/mml/
Anne Hofmann
Picture coll
1 555 000 vols   40440

The New York Public Library – Astor, Lenox & Tilden Foundations – Miriam & Ira D. Wallach Division of Art, Prints & Photographs, Fifth Ave & 42nd St, *New York*, NY 10018
T: +1 212 9300817; Fax: +1 212 9300530; E-mail: prints:prnref@nypl.org; URL: www.nypl.org/research/chss/spe/art/print/print.htmlpencer
Clayton Kirking
George Washington Portraits (McAlpin Coll);New York City Views (Eno Coll);Exhibition Catalogs;Individual Artists & Architects, clipping file;Sales Catalogs;American & European Political Cartoons & Caricatures;American Views (Phelps Stokes Coll);Milton & P
435 000 vols   40441

The New York Public Library – Astor, Lenox & Tilden Foundations – Music Division, 40 Lincoln Center Plaza, *New York*, NY 10023-7498
T: +1 212 8701649; Fax: +1 212 8701794; URL: www.nypl.org/research/lpa/mus/mus.html
George Boziwick
Individual & Corporate Archival Coll (Marcella Sembrich, Arturo Toscanini, Town Hall, the New Music Society & Composers Forum);Drexel Coll;Joseph Muller Coll, fine prints;European Composers 17th-20th Centuries (Toscanini Memorial Archives), autograph scor
644 000 vols; 850 curr per   40442

The New York Public Library – Astor, Lenox & Tilden Foundations – New Amsterdam Branch, 9 Murray St (between Broadway and Church St), *New York*, NY 10007-2223
T: +1 212 7328186; Fax: +1

212 7328815; E-mail: new_amsterdam@nypl.org; URL: www.nypl.org/branch/local/man/lm.cfm
N.N.
58 000 vols   40443

The New York Public Library – Astor, Lenox & Tilden Foundations – New York Public Library for the Performing Arts Circulating Collections, 40 Lincoln Center Plaza, *New York*, NY 10023-7498
T: +1 212 8701618; Fax: +1 212 8701704; E-mail: lpacirc@nypl.org; URL: www.nypl.org/research/lpa/circ/circ.html
Don Francis Baldini
220 000 vols   40444

The New York Public Library – Astor, Lenox & Tilden Foundations – Rare Books Division, Fifth Ave & 42nd St, Rm 328, *New York*, NY 10018-2788
T: +1 212 9300801; Fax: +1 212 3024815; E-mail: rbkref@nypl.org; URL: www.nypl.org/research/chss/spe/rbk/rbooks.html
Early English Cookbooks (Whitney Coll);German Literature (Axel Rosin Coll);18th Century American Newspapers & Periodicals;Korean History Collection, 1870-1948;Kelmscott Press, Ashendene, Doves, Golden Cockerel, Grabhorn, Nonesuch, Vale, Bruce Rogers
140 000 vols   40445

The New York Public Library – Astor, Lenox & Tilden Foundations – Riverside Branch, 127 Amsterdam Ave (@ W 65th St), *New York*, NY 10023-6447
T: +1 212 8701810; Fax: +1 212 8701819; URL: www.nypl.org/branch/local/man/rs.cfm
Avril Opinante
69 000 vols   40446

The New York Public Library – Astor, Lenox & Tilden Foundations – Schomburg Center for Research in Black Culture, 515 Malcolm X Blvd, *New York*, NY 10037-1801
T: +1 212 4912200; Fax: +1 212 4916760; E-mail: scgenref@nypl.org; URL: www.nypl.org/research/sc/sc.html
Howard Dodson
Papers of John E Bruce, Civil Rights Congress, International Labor Defense, Carnegie-Myrdal research memoranda, Alexander Crummell, Oakley Johnson, National Association of Colored Graduate Nurses, National Negro Congress, Phelps-Stokes Fund, Central Afric
150 000 vols   40447

The New York Public Library – Astor, Lenox & Tilden Foundations – Seward Park Branch, 192 E Broadway (@ Jefferson St), *New York*, NY 10002-5597
T: +1 212 4776770; Fax: +1 212 4776770; E-mail: seward_park@nypl.org; URL: nypl.org/branch/local/man/se.cfm
Mary M. Jones
59 000 vols   40448

The New York Public Library – Astor, Lenox & Tilden Foundations – Slavic & Baltic Division, Fifth Ave & 42nd St, Rm 216-217, *New York*, NY 10018-2788
T: +1 212 9300714; Fax: +1 212 9300940; E-mail: slavicref@nypl.org; URL: www.nypl.org/research/chss/slv/slav.balt.html
Edward Kasinec
Baltic Slavic Language Monographic & Serial Materials;Russian & Ukrainian Poster Coll;Rare Books;Art Publications;Glaser Microfilm Coll;Illustrated Books;Bates/Pantuhoff Coll of Russian Imprints & Art;Russian Imperial Association Copies;Early Russian Imp
480 000 vols; 1 200 curr per   40449

The New York Public Library – Astor, Lenox & Tilden Foundations – Yorkville Branch, 222 E 79th St (between Second & Third Aves), *New York*, NY 10021-1295
T: +1 212 7445824; Fax: +1 212 7445929; E-mail: yorkville@nypl.org; URL: www.nypl.org/branch/local/man/yv.cfm
Gladys D. Sanders-Valdes
58 000 vols   40450

Alameda County Library – Newark Branch, 6300 Civic Terrace Ave, *Newark*, CA 94560-3795
T: +1 510 7952627; Fax: +1 510 7973019
Kathleen Steel-Sabo
71 000 vols; 159 curr per   40451

Licking County Library, 101 W Main St, *Newark*, OH 43055-5054
T: +1 740 3495500; Fax: +1 740 3495535; URL: www.lickingcountylibrary.info
1908; Steven Hawk
4 branch libs, 1 bookmobile
428 000 vols; 361 curr per   40452

Newark Free Library, 750 Library Ave, *Newark*, DE 19711-7146
T: +1 302 7317550; Fax: +1 302 7314019; URL: www.nccdelib.org
1897; Martha Birchenall
Spanish Coll
107 000 vols; 326 curr per   40453

Newark Public Library, Five Washington St, *Newark*, NJ 07101; P.O. Box 630, Newark, NJ 07101-0630
T: +1 973 7337800; Fax: +1 973 7335919; URL: www.npl.org
1888; Wilma J. Grey
Art Colls, Children's Bks, Puerto Rico Reference Coll, Fine Printing, New Jersey Rare Bks, Medieval Mss; 11 branch libs
1 684 000 vols; 1 275 curr per; 3 000 000 govt docs; 15 789 music scores; 1 319 maps; 12 000 microforms; 410 av-mat; 13 561 sound-rec; 3 406 digital data carriers
libr loan   40454

Newberg Public Library, 503 E Hancock St, *Newberg*, OR 97132-2899
T: +1 503 5387323; Fax: +1 503 5389720; E-mail: nplibrary@ci.newberg.or.us; URL: www.ci.newberg.or.us/library
1907; Leah Griffith
70 000 vols; 268 curr per; 2 500 av-mat   40455

Newberry County Library, 1300 Friend St, *Newberry*, SC 29108-3400
T: +1 803 2760854; Fax: +1 803 2767478; URL: www.newberrylibrary.org
Tucker Neel Taylor
58 000 vols; 115 curr per
libr loan   40456

Ohio Township Public Library System, 4111 Lakeshore Dr, *Newburgh*, IN 47630-2274; P.O. Box 850, Newburgh
T: +1 812 8535468; Fax: +1 812 8530509; URL: www.ohio.lib.in.us
1897; Stephen Thomas
2 branch libs
105 000 vols; 308 curr per   40457

Ohio Township Public Library System – Newburgh Library, 30 W Water St, *Newburgh*, IN 47630
T: +1 812 8581437; Fax: +1 812 8589390
Caryl Hulgus
57 000 vols; 72 curr per   40458

Thousand Oaks Library Newbury Park Branch, 2331 Borchard Rd, *Newbury Park*, CA 91320-3206
T: +1 805 4982139; Fax: +1 805 4987034
66 000 vols   40459

Newburyport Public Library, 94 State St, *Newburyport*, MA 01950-6619
T: +1 978 4654428; Fax: +1 978 4630394; URL: www.newburyportpl.org
1854; Dorothy R. LaFrance
102 000 vols; 174 curr per   40460

County of Los Angeles Public Library – Newhall Library, 22704 W Ninth St, *Newhall*, CA 91321-2808
T: +1 661 2590750; Fax: +1 661 2545760; E-mail: ac@104.colapl.org; URL: www.colapublib.org/libs/newhall
Paula Hock
50 000 vols; 76 curr per   40461

Lucy Robbins Welles Library, 95 Cedar St, *Newington*, CT 06111-2645
T: +1 860 6658730; Fax: +1 860 6671255; URL: www.newington.lib.ct.us
1752; Marian Amodeo
Index of Local Newspaper
104 000 vols; 217 curr per; 25 microforms; 4 912 sound-rec; 50 digital data carriers
libr loan; CRLC (Capitol Region Library Council)   40462

Avery County Morrison Public Library, 150 Library Place, *Newland*, NC 28657
T: +1 828 7339393; Fax: +1 828 6826277; E-mail: acpl@amyregionallibrary.org; URL: www.amyregionallibrary.org/avery/index.html
Phyllis Burroughs
Robert Morrison Ref Coll
70 000 vols   40463

Newport Public Library, 35 NW Nye St, *Newport*, OR 97365-3714
T: +1 541 2652153; Fax: +1 541 5749496; E-mail: reference@newportlibrary.org; URL: www.newportlibrary.org
1945; Rebecca Cohen
Oregon
75 000 vols; 183 curr per; 1 719 av-mat
libr loan; ALA, OLA   40464

Newport Public Library, 300 Spring St, *Newport*, RI 02840
T: +1 401 8478720; Fax: +1 401 8420841; E-mail: nptref@lori.state.ri.us; URL: www.newportlibraryri.org
1868; Regina Slezak
Afro-American Coll, Chinese Rm Coll, Cookbks
126 000 vols; 284 curr per   40465

Vermillion County Public Library, 385 E Market St, *Newport*, IN 47966; P.O. Box 100, Newport, IN 47966-0100
T: +1 765 4923555; Fax: +1 765 4929558; E-mail: newport_library@hotmail.com; URL: www.newportlibrary.info
1929; Brigit Steinbrenner
62 000 vols; 50 curr per; 400 av-mat   40466

Newport Beach Public Library, 1000 Avocado Ave, *Newport Beach*, CA 92660-6301
T: +1 949 7173800; Fax: +1 949 6405681; E-mail: nbplref@city.newport-beach.ca.us; URL: www.newportbeachlibrary.org
1920; Linda Katsouleas
Nautical Coll; 4 branch libs
326 000 vols; 708 curr per
libr loan   40467

Newport Beach Public Library – Central Library, 1000 Avocado Ave, *Newport Beach*, CA 92660-6301
T: +1 949 7173800; Fax: +1 949 6405681
Linda Katsouleas
227 000 vols   40468

Newport Beach Public Library – Mariners, 2005 Dover Dr, *Newport Beach*, CA 92660
T: +1 949 7173800; Fax: +1 949 6424848
Mary Ellen Bowman
65 000 vols   40469

Newport News Public Library System, 700 Town Center Dr, Ste 300, *Newport News*, VA 23606
T: +1 757 9261350; Fax: +1 757 9261365; URL: www.nngov.com/library
1891; Izabela M. Cieszynski
Martha Woodroof Hiden Virginiana Coll, bks & micro; Old Dominion Land Company Coll; 7 branch libs
354 000 vols; 472 curr per; 8 347 av-mat; 16 309 sound-rec   40470

Catawba County Library, 115 West C St, *Newton*, NC 28658
T: +1 828 4658664; Fax: +1 828 4658983; URL: www.catawbacountync.gov/library/
1936; Karen Foss
6 branch libs
207 000 vols; 763 curr per; 11 893 av-mat   40471

Newton Public Library, 100 N Third Ave W, **Newton**, IA 50208; P.O. Box 746, Newton, IA 50208-0746
T: +1 641 7924108; Fax: +1 641 7910729; E-mail: newtonpl@newton.lib.ia.us; URL: www.newton.lib.ia.us
1896; Sue Padilla
80 000 vols; 141 curr per; 3 200 sound-rec
libr loan   40472

Newton Public Library, 720 N Oak, **Newton**, KS 67114
T: +1 316 2832890; Fax: +1 316 2832916; E-mail: library@newtonplks.org; URL: www.newtonplks.org
1886
Genealogy; Kansas Coll, Large Print, Spanish Coll
87 100 vols; 318 curr per   40473

Sussex County Library System, 125 Morris Turnpike, **Newton**, NJ 07860-0076
T: +1 973 9483660; Fax: +1 973 9482071; E-mail: sussexref@sussexcountylibrary.org
1942; Stan Pollakoff
County Hist, Delaware Water Gap Nat Recreation Area (Tocks Island Regional Advisory Council Libr); 5 branch libs
270 000 vols; 539 curr per; 10 000 microforms; 3 534 av-mat; 7 169 sound-rec; 15 digital data carriers; 326 Art reproductions   40474

Newton Free Library, 330 Homer St, **Newton Centre**, MA 02459-1429
T: +1 617 7961360; Fax: +1 617 9658457; URL: www.ci.newton.ma.us/library
1870; Nancy Perlow
4 branch libs
521 000 vols; 775 curr per; 990 Electronic Media & Resources   40475

Cyrenius H. Booth Library, 25 Main St, **Newtown**, CT 06470
T: +1 203 4268552; Fax: +1 203 4262196; E-mail: chbooth@biblio.org; URL: www.biblio.org/chbooth/chbooth.htm
1932; Janet Woycik
Brush Genealogy Coll, John Angel Art Coll, Swanbery Coll, Newton authors coll, Wasserman early lighting coll
130 000 vols; 507 curr per; 1 000 maps; 6 000 microforms; 50 digital data carriers
libr loan; ALA   40476

Niagara Falls Public Library, 1425 Main St, **Niagara Falls**, NY 14305
T: +1 716 2864880;
Fax: +1 716 2864912; URL: www.niagarafallspubliclib.org
1838; Betty Babanoury
1 branch libr
403 000 vols
libr loan   40477

East Lyme Public Library, Inc, 39 Society Rd, **Niantic**, CT 06357-1100
T: +1 860 7396926; Fax: +1 860 6910020; E-mail: elpl@ely.lioninc.org; URL: www.lioninc.org/eastlyme
1868; William Deakyne
Ridder Music Coll
103 000 vols; 218 curr per; 73 000 microforms; 1 224 av-mat; 332 sound-rec
libr loan   40478

Niceville Public Library, 206 N Partin Dr, **Niceville**, FL 32578
T: +1 850 7294090; Fax: +1 850 7294053; E-mail: ncvlibrary@okaloosa.lib.fl.us; URL: www.cityofniceville.org/library.html
1974; Sheila K. Bishop
541 000 vols; 149 curr per; 2 986 av-mat
                                       40479

Jessamine County Public Library, 600 S Main St, **Nicholasville**, KY 40356
T: +1 859 8853523; Fax: +1 859 8855164; URL: www.jesspublib.org
1968; Susan B. Lawrence
Kentucky Room
61 000 vols; 115 curr per; 4 216 av-mat
                                       40480

McKinley Memorial Library, 40 N Main St, **Niles**, OH 44446-5082
T: +1 330 6521704; Fax: +1 330 6525788; E-mail: mckinley@mcklib.org; URL: www.mckinley.lib.oh.us
1908; Patrick E. Finan
William McKinley, bk & music artifacts

72 000 vols; 120 curr per; 2 200 av-mat
ALA   40481

Niles District Library, 620 E Main St, **Niles**, MI 49120
T: +1 269 6838545; Fax: +1 269 6830075; URL: www.nileslibrary.com
1903; Nancy Studebaker
Ring Lardner (Complete Works); Niles Newspapers, 1834 to date, microflm
127 000 vols; 250 curr per   40482

Niles Public Library District, 6960 Oakton St, **Niles**, IL 60714
T: +1 847 6631234; Fax: +1 847 6631360; E-mail: books@nileslibrary.org; URL: www.nileslibrary.org
1958; Linda Weiss
206 000 vols; 421 curr per   40483

Hamilton East Public Library, Noblesville Library, One Library Plaza, Cumberland Rd, **Noblesville**, IN 46060-5639
T: +1 317 7731384; Fax: +1 317 7766936; URL: www.hepl.lib.in.us
1909; David L. Cooper
525 000 vols; 799 curr per
libr loan; ALA   40484

Nogales-Santa Cruz County Public Library, Nogales Public Library, 518 Grand Ave, **Nogales**, AZ 85621
T: +1 520 2873343; Fax: +1 520 2874823; URL: www.nogales-santacruz.lib.az.us
1923; Suzanne Haddock
Arizona & Southwest Hist Coll, Korean Books, Vocational Guidance & Career Coll; 3 branch libs
62 565 vols; 105 curr per
libr loan   40485

Norfolk Public Library, 308 Prospect Ave, **Norfolk**, NE 68701
T: +1 402 8442100; E-mail: jhilkema@ci.norfolk.ne.us; URL: www.ci.norfolk.ne.us/library
1906
Genealogy, Nebraska hist, Poetry
84 080 vols; 191 curr per   40486

Norfolk Public Library, Kirn Memorial Library, 301 E City Hall Ave, **Norfolk**, VA 23510-1776
T: +1 757 6647337; Fax: +1 757 6647320; URL: www.npl.lib.va.us
1870; Norman Maas
Virginiana (Sargeant Memorial Room), bks, film & micro; 11 branch libs
317 000 vols; 1 249 curr per   40487

Norfolk Public Library – Lafayette, 1610 Cromwell Rd, **Norfolk**, VA 23509
T: +1 757 4412842; Fax: +1 757 4411454
Jennifer Johnson
52 000 vols; 88 curr per   40488

Norfolk Public Library – Mary D. Pretlow Anchor Library, 111 W Ocean View Ave, **Norfolk**, VA 23503-1608
T: +1 757 4411750; Fax: +1 757 4411748
Cynthia Seay
53 000 vols; 101 curr per   40489

Normal Public Library, 206 W College Ave, **Normal**, IL 61761; P.O. Box 325, Normal, IL 61761-0325
T: +1 309 4521757; Fax: +1 309 4525312; URL: normal-library.org
1939; Robert L. Wegman
101 000 vols; 300 curr per   40490

Pioneer Library System, 225 N Webster Ave, **Norman**, OK 73069-7133
T: +1 405 7012600; Fax: +1 405 7012649; URL: www.pioneer.lib.ok.us
1957; Anne Masters
Oklahoma Coll; 9 branch libs
455 000 vols
libr loan   40491

Pioneer Library System – Norman Public, 225 N Webster Ave, **Norman**, OK 73069
T: +1 405 7012600; Fax: +1 405 7012608
Susan Gregory
184 000 vols   40492

Montgomery County-Norristown Public Library, 1001 Powell St, **Norristown**, PA 19401-3817
T: +1 610 2785100; Fax: +1 610 2785110; URL: www.mc-npl.org
1794; Kathleen Arnold-Yergel
Carolyn Wicker Field Coll (Autographed Children's Bks & Correspondence from Children's Autors & III), Old Fiction, Steinbright Local Hist Coll; 4 branch libs
479 000 vols; 831 curr per; 675 High Interest/Low Vocabulary Bks, 10 080 Large Print Bks
libr loan   40493

North Adams Public Library, 74 Church St, **North Adams**, MA 01247
T: +1 413 6623133; Fax: +1 413 6623039; E-mail: napl@bcn.net; URL: www.naplibrary.org
1884; Marcia Gross
57 000 vols; 193 curr per; 930 e-books; 3 024 av-mat
libr loan   40494

Stevens Memorial Library, 345 Main St, **North Andover**, MA 01845
T: +1 978 6889505; Fax: +1 978 6889507; E-mail: msm@mvlc.org; URL: www.stevensmemlib.org
1907; Mary Rose Quinn
Essex County Ma; Poetry (Anne Bradstreet Coll)
98 000 vols; 178 curr per   40495

North Arlington Free Public Library, 210 Ridge Rd, **North Arlington**, NJ 07031
T: +1 201 9555640; Fax: +1 201 9917850; E-mail: noarl@bccls.org; URL: www.bccls.org/northarlington
1939; Maria Puszkar
95 000 vols; 60 curr per; 700 av-mat
                                       40496

Messenger Public Library of North Aurora, 113 Oak St, **North Aurora**, IL 60542
T: +1 630 8960240; Fax: +1 630 8964654; URL: www.messengerpl.org
1937; G. Kevin Davis
54 000 vols; 210 curr per   40497

North Babylon Public Library, 815 Deer Park Ave, **North Babylon**, NY 11703-3812
T: +1 631 6694020; Fax: +1 631 6693432; E-mail: nbablib@suffolk.lib.ny.us; URL: www.suffolk.lib.ny.us/libraries/nbab/
1960; Marc David Horowitz
European hist & travel; Newsday on microfilm from 1944
114 000 vols; 205 curr per; 2 000 microforms; 13 488 sound-rec; 14 digital data carriers   40498

North Bend Public Library, 1800 Sherman Ave, **North Bend**, OR 97459
T: +1 541 7560400; Fax: +1 541 7561073; URL: info.cclsd.org/non
1914; Gary Sharp
Oregoniana (Oregon Coll), bks, clippings, pamphlets
112 000 vols; 214 curr per; 47 microforms; 8 digital data carriers
libr loan; ALA   40499

North Bergen Free Public Library, 8411 Bergenline Ave, **North Bergen**, NJ 07047-5097
T: +1 201 8694715; Fax: +1 201 8697626; URL: www.nbpl.org
1951; Sai Rao
Foreign language, Lit, Large Print, New Jerseyana
166 000 vols; 226 curr per   40500

North Branford Library Department, Atwater Memorial, 1720 Foxon Rd, **North Branford**, CT 06471; P.O. Box 258, North Branford, CT 06471-0258
T: +1 203 3156020; Fax: +1 203 3156021
Robert V. Hull
1 branch libr
60 000 vols; 210 curr per   40501

North Brunswick Free Public Library, 880 Hermann Rd, **North Brunswick**, NJ 08902
T: +1 732 2463545; Fax: +1 732 2461341; E-mail: refdesk@northbrunswicklibrary.org; URL: www.northbrunswicklibrary.org/
1966; Cheryl McBride
Antiques; Adult Basic Reading for

Literacy Program
120 000 vols; 240 curr per   40502

North Canton Public Library, 185 N Main St, **North Canton**, OH 44720-2595
T: +1 330 4994712; Fax: +1 330 4997356; E-mail: reference@northcantonlibrary.org; URL: www.ncantonlibrary.org
1926; Karen Sonderman
Art
106 000 vols; 415 curr per   40503

North Chicago Public Library, 2100 Argonne Dr, **North Chicago**, IL 60064
T: +1 847 6890125; Fax: +1 847 6899117; E-mail: info@ncplibrary.org; URL: www.ncplibrary.org
1916; Joan Battley
58 000 vols; 140 curr per   40504

Ames Free Library, Easton's Public Library, 53 Main St, **North Easton**, MA 02356
T: +1 508 2382000; Fax: +1 508 2382980; E-mail: library@easton.ma.org; URL: www.amesfreelibrary.org
1879; Madeline Miele Holt
Decorative arts; 19th C Per, H.H. Richardson Mat; 1 branch libr
53 000 vols; 167 curr per   40505

Lee County Library System – North Fort Myers Public, 2001 N Tamiami Trail NE, **North Fort Myers**, FL 33903-2802
T: +1 239 9970320; Fax: +1 239 6567949
Maryellen Woodside
69 000 vols   40506

North Haven Memorial Library, 17 Elm St, **North Haven**, CT 06473
T: +1 203 2395803; Fax: +1 203 2342130; URL: www.leaplibraries.org/nhaven
1894; Lois D. Baldini
95 000 vols; 225 curr per   40507

Los Angeles Public Library System – Mid-Valley Regional, 16244 Nordhoff St, **North Hills**, CA 91343
T: +1 818 8953650; Fax: +1 818 8953657
Yvonne Wong
133 000 vols   40508

Los Angeles Public Library System – North Hollywood Regional, 5211 Tujunga Ave, **North Hollywood**, CA 91601-3179
T: +1 818 7667185; Fax: +1 818 7569135
Arthur Pond
55 000 vols   40509

North Kansas City Public Library, 2251 Howell St, **North Kansas City**, MO 64116
T: +1 816 2213360; Fax: +1 816 2218298; URL: www.northkclibrary.org
1939; Steve Campbell
72 000 vols; 186 curr per; 1 144 av-mat
                                       40510

North Kingstown Free Library, 100 Boone St, **North Kingstown**, RI 02852-5176
T: +1 401 2943306; Fax: +1 401 2941690; URL: www.nklibrary.org
1898; Susan Aylward
103 000 vols   40511

North Las Vegas Library District, 2300 Civic Center Dr, **North Las Vegas**, NV 89030
T: +1 702 6331070; Fax: +1 702 6492576; URL: www.nlvld.org
1962; Anita Laruy
Auto repair manuals; Kiel Ranch Coll, Nevada Coll
115 000 vols; 102 curr per   40512

North Logan City Library, 475 E 2500 North, **North Logan**, UT 84341-1523
T: +1 435 7557169; Fax: +1 435 2270032; URL: www.northloganlibrary.org
Sue Randleman
54 000 vols; 78 curr per   40513

North Manchester Public Library, 405 N Market St, **North Manchester**, IN 46962
T: +1 260 9824773; Fax: +1 260 9826342; E-mail: nmpl@nman.lib.in.us; URL: www.nman.lib.in.us
1912; Theresa Tyner
59 000 vols; 165 curr per   40514

North Merrick Public Library, 1691 Meadowbrook Rd, *North Merrick*, NY 11566
T: +1 516 3787474;
Fax: +1 516 3782523; URL: www.northmerrickpubliclibrary.org
1965; Margaret Cincotta
Art, Cookery
98 000 vols; 218 curr per; 7 000 av-mat; 5 064 sound-rec 40515

North Miami Public Library, E. May Avil Library, 835 NE 132nd St, *North Miami*, FL 33161
T: +1 305 8915535; Fax: +1 305 8920843; E-mail: library@northmiamifl.gov; URL: www.northmiamifl.gov/community/library
1949; Joyce Pernicone
Paranting, Careers, Large print; Art Coll, Civil War Coll, Filipiana Coll, Literacy Coll, Stage & Studio Coll, Bicentennial of the Constitution Coll
90 000 vols; 186 curr per; 2 749 av-mat 40516

North Miami Beach Public Library, 1601 NE 164th St, *North Miami Beach*, FL 33162-4099
T: +1 305 9482970; Fax: +1 305 7876007; URL: www.citynmb.com
Florence Simkins Brown
59 000 vols; 200 curr per 40517

Cuyahoga County Public Library – North Olmsted Branch, 27403 Lorain Rd, *North Olmsted*, OH 44070-4037
T: +1 440 7776211; Fax: +1 440 7774312
Kacie Armstrong
58 000 vols; 4 257 av-mat 40518

Somerset County Library System – North Plainfield Memorial, Six Rockview Ave, *North Plainfield*, NJ 07060
T: +1 908 7557909; Fax: +1 908 7558177
Richard Stevens
85 000 vols; 175 curr per 40519

North Platte Public Library, 120 W Fourth St, *North Platte*, NE 69101-3993
T: +1 308 5358036; Fax: +1 308 5358296; E-mail: library@ci.north-platte.ne.us; URL: www.ci.north-platte.ne.us/library
1907; Cecelia C. Lawrence
Circulating Original Art Coll, Nebraska Hist Coll
121 000 vols; 215 curr per; 2 398 av-mat; 600 sound-rec; 250 digital data carriers
libr loan; ALA, NLA 40520

North Port Public Library, 13800 S Tamiami Trail, *North Port*, FL 34287-2030
T: +1 941 8611300; Fax: +1 941 4266564; URL: www.sclibs.net
1975; Janita Wisch
55 000 vols; 122 curr per 40521

North Providence Union Free Library, Mayor Salvatore Mancini Union Free Public Library & Cultural Center, 1810 Mineral Spring Ave, *North Providence*, RI 02904
T: +1 401 3535600; URL: www.nplib.org
1869; Mary Ellen Hardiman
Careers-Vocational Guidance, vocational mat; Computer Software (Rhode Island Coll), Elderly Concerns; Large Print
128 000 vols; 230 curr per 40522

Flint Memorial Library, 147 Park St, *North Reading*, MA 01864
T: +1 978 6644942; Fax: +1 978 6640812; E-mail: mnr@mvlc.lib.ma.us; URL: www.flintmemoriallibrary.org
1872; Helena Minton
Genealogy & Hist (Clara Burnham, North Reading & George Root Colls) mss, maps
70 000 vols; 200 curr per 40523

North Richland Hills Public Library, 9015 Grand Ave, *North Richland Hills*, TX 76180
T: +1 817 4276816; Fax: +1 817 4276808; URL: www.library.nrhtx.com
1971; Steven L. Brown
Digital Hist Photo Arch
179 000 vols; 188 curr per 40524

Lorain Public Library System – North Ridgeville Branch, 35700 Bainbridge Rd, *North Ridgeville*, OH 44039
T: +1 440 3278326
Ken Cromer
66 000 vols; 366 curr per 40525

Cuyahoga County Public Library – North Royalton Branch, 14600 State Rd, *North Royalton*, OH 44133-5120
T: +1 440 2373800; Fax: +1 440 2376149
Jeanne Cilenti
67 000 vols; 5 357 av-mat 40526

North Tonawanda Public Library, 505 Meadow Dr, *North Tonawanda*, NY 14120-2888
T: +1 716 6933009; Fax: +1 716 6930719; E-mail: ntwref@nioga.org; URL: www.ntlibrary.org
1893; Margaret A. Waite
Carousels coll
163 000 vols; 268 curr per; 7 060 av-mat; 22 141 sound-rec; 3 230 digital data carriers
libr loan; ALA 40527

Jennings County Public Library, 2375 N State Hwy 3, *North Vernon*, IN 47265-1596
T: +1 812 3462091; Fax: +1 812 3462127; E-mail: jlibrary@seidata.com; URL: www.jenningscounty.lib.in.us
1813; Mary Hougland
Genealogy
90 000 vols; 236 curr per; 25 maps; 4 000 microforms; 5 000 av-mat; 2 400 sound-rec; 10 digital data carriers; 1 200 Large Print Bks
libr loan; ALA 40528

Appalachian Regional Library, 215 Tenth St, *North Wilkesboro*, NC 28659
T: +1 336 8382818; Fax: +1 336 6672638
Louise Humphrey
250 000 vols; 410 curr per 40529

Wilkes County Public Library, 215 Tenth St, *North Wilkesboro*, NC 28659
T: +1 336 8382818; Fax: +1 336 6672638; URL: www.arlibrary.org
1909; Jennifer Lee Murphy
99 000 vols; 110 curr per; 3 061 av-mat 40530

Forbes Library, 20 West St, *Northampton*, MA 01060-3798
T: +1 413 5871011; Fax: +1 413 5871015; URL: www.forbeslibrary.org
1894; Janet Moulding
Art & architecture, Music; Calvin Coolidge, Connecticut Valley Hist, Genealogy Coll
236 000 vols; 250 curr per; 3 000 Large Print Bks 40531

Northampton Area Public Library, 1615 Laubach Ave, *Northampton*, PA 18067-1597
T: +1 610 2627537; Fax: +1 610 2624356; URL: www.northamptonapl.org
1965; Mary Beller
54 000 vols; 95 curr per; 1 020 av-mat
libr loan; ALA 40532

Northborough Free Library, 34 Main St, *Northborough*, MA 01532-1942
T: +1 508 3935025; Fax: +1 508 3935027
1868; Jean M. Langley
71 000 vols; 238 curr per; 2 572 av-mat 40533

Northbrook Public Library, 1201 Cedar Lane, *Northbrook*, IL 60062-4581
T: +1 847 2726224; Fax: +1 847 2725362; URL: www.northbrook.info
1952; Chadwick Raymond
Art, Landscape architecture, Science & technology
273 000 vols; 613 curr per 40534

Northeast Harbor Library, One Joy Rd, *Northeast Harbor*, ME 04662; P.O. Box 279, Northeast Harbor, ME 04662-0279
T: +1 207 2763333; Fax: +1 207 2763315; E-mail: info@nehlibrary.org; URL: www.nehlibrary.org
Robert R. Pyle
50 000 vols; 62 curr per 40535

Akron-Summit County Public Library – Nordonia Hills, 9458 Olde Eight Rd, *Northfield*, OH 44067-1952
T: +1 330 4678595; Fax: +1 330 4674332
Janet Stavole
60 000 vols 40536

Northfield Public Library, 210 Washington St, *Northfield*, MN 55057
T: +1 507 6456606; Fax: +1 507 6451820; URL: www.northfield.mn.info
1857; Lynne Young
77 000 vols; 260 curr per; 15 000 e-books; 100 sound-rec; 15 digital data carriers
libr loan 40537

Rangeview Library District – Northglenn Branch, 10530 N Huron, *Northglenn*, CO 80234-4011
T: +1 303 4527534; Fax: +1 303 4502578
N.N.
69 000 vols; 177 curr per 40538

Northport-East Northport Public Library, Northport Public Library, 151 Laurel Ave, *Northport*, NY 11768
T: +1 631 2616930; Fax: +1 631 2616718; E-mail: nenpl@suffolk.lib.ny.us; URL: www.nenpl.org
1914; Stephanie Heineman
Architecture, Art, Boating; Jack Kerouac Coll; 1 branch libr
249 000 vols; 820 curr per 40539

Los Angeles Public Library System – Northridge Branch, 9051 Darby Ave, *Northridge*, CA 91325-2708
T: +1 818 8863640; Fax: +1 818 8866850
Leslie Chudnoff
58 000 vols 40540

Los Angeles Public Library System – Porter Ranch, 11371 Tampa Ave, *Northridge*, CA 91326
T: +1 818 3605706; Fax: +1 818 3603106
Shayeri Tangri
58 000 vols 40541

Northville District Library, 212 W Cady St, *Northville*, MI 48167-1560
T: +1 248 3493020; Fax: +1 248 3498250; URL: www.northville.lib.mi.us
Large type print
65 000 vols; 158 curr per 40542

Akron-Summit County Public Library – Norton Branch, 3930 S Cleveland-Massillon Rd, *Norton*, OH 44203-5563
T: +1 330 8257800; Fax: +1 330 8255155
Cindy Howe
58 000 vols 40543

Northwest Kansas Library System, Two Washington Sq, *Norton*, KS 67654-1615
T: +1 785 8775148; Fax: +1 785 8775697; E-mail: nwkls@ruraltel.net; URL: skyways.lib.ks.us/nwkls/norwest.html
1966; Leslie Bell
Blind & Physically Handicapped Dept
67 000 vols; 11 curr per; 1 898 av-mat
libr loan 40544

Norton Public Library, One Washington Sq, *Norton*, KS 67654-1615; P.O. Box 446, Norton, KS 67654-0446
T: +1 785 8772481; E-mail: nortonpl@ruraltel.net
1909; Kay LeBeau
Masonic Coll
78 000 vols; 40 curr per 40545

Norton Public Library, L. G. & Mildred Balfour Memorial, 68 E Main St, *Norton*, MA 02766
T: +1 508 2862694; Fax: +1 508 2850266; E-mail: dbriody@sailsinc.org; URL: www.nortonlibrary.org
Elaine Jackson
54 000 vols; 137 curr per 40546

County of Los Angeles Public Library – Norwalk Library, 12350 Imperial Hwy, *Norwalk*, CA 90650-3199
T: +1 562 8680775; Fax: +1 562 9291130; URL: www.colapublib.org/libs/norwalk
Sue Kane
Edelman Public Policy Coll;Business Subject Specialty Center
234 000 vols; 231 curr per 40547

Norwalk Public Library, 46 W Main St, *Norwalk*, OH 44857
T: +1 419 6686063; Fax: +1 419 6632190; E-mail: norwalk@oplin.org; URL: www.norwalk.lib.oh.us
1861; Martin L. Haffey
Huron County Hist & Genealogy; Local Newspapers, micro
85 000 vols; 210 curr per 40548

Norwalk Public Library, One Belden Ave, *Norwalk*, CT 06850
T: +1 203 8992780; Fax: +1 203 8574410; URL: www.norwalkpubliclibrary.org
1895; Les Kozerowitz
1 branch libr
225 000 vols; 410 curr per 40549

Norwell Public Library, 64 South St, *Norwell*, MA 02061-2433
T: +1 781 6592015; Fax: +1 781 6596755; E-mail: noref@ocln.org; URL: www.norwellpubliclibrary.org
Rebecca Freer
62 000 vols; 115 curr per 40550

Guernsey Memorial Library, Three Court St, *Norwich*, NY 13815
T: +1 607 3344034; Fax: +1 607 3363901; URL: www.guernseylibrary.org
1908; Melanie Battoe
79 000 vols; 109 curr per 40551

Otis Library, Two Cliff St, *Norwich*, CT 06360
T: +1 860 8892365; Fax: +1 860 8864744; URL: www.otislibrarynorwich.org
1850; Bob Farwell
Careers, Business & management; Large Print Coll
99 000 vols; 158 curr per
libr loan 40552

Morrill Memorial Library, 33 Walpole St, *Norwood*, MA 02062-1206; P.O. Box 220, Norwood, MA 02062-0220
T: +1 781 7690200; Fax: +1 781 7696083; E-mail: norwood@minlib.net; URL: www.ci.norwood.ma.us/library/index.php
1873; Charlotte Canelli
Norwood Coll
93 000 vols; 187 curr per 40553

Marin County Free Library – Novato Branch, 1720 Novato Blvd, *Novato*, CA 94947
T: +1 415 8971141
Donna Mettier
106 000 vols 40554

Novi Public Library, 45245 W Ten Mile Rd, *Novi*, MI 48375
T: +1 248 3490720; Fax: +1 248 3496520; URL: www.novi.lib.mi.us
1960; Julie Farkas
Career, Finance, Law; Adult Basic Readers
124 000 vols; 153 curr per; 6 059 av-mat; 5 016 sound-rec 40555

Nutley Free Public Library, 93 Booth Dr, *Nutley*, NJ 07110-2782
T: +1 973 6670405; Fax: +1 973 6670408; E-mail: librarydirector@nutleynj.org; URL: nutley.bccls.org
1913; JoAnn A. Tropiano
86 000 vols; 215 curr per; 2 793 av-mat 40556

The Nyack Library, 59 S Broadway, *Nyack*, NY 10960
T: +1 845 3583370; Fax: +1 845 3586429; E-mail: libref@nyack.lib.ny.us; URL: nyack.lib.ny.us
1879; James J. Mahoney
Haitian interest
96 000 vols; 211 curr per; 7 550 sound-rec; 10 digital data carriers 40557

Oak Brook Public Library, 600 Oak Brook Rd, *Oak Brook*, IL 60523
T: +1 630 3687702; Fax: +1 630 9904509; URL: www.oak-brook.lib.il.us
1960; Margaret G. Klinkow Hartmann
94 000 vols; 10 080 curr per 40558

Oak Creek Public Library, 8620 S Howell Ave, *Oak Creek*, WI 53154
T: +1 414 7644400; Fax: +1 414 7686583; URL: www.mcfls.org/ocpl
1972; Ross Talis
65 000 vols; 195 curr per; 12 000 e-books; 2 000 microforms; 3 682 sound-rec; 33 digital data carriers
libr loan 40559

Acorn Public Library District, 15624 S Central Ave, *Oak Forest*, IL 60452-3299
T: +1 708 6873700; Fax: +1 708 6873712; E-mail: acorn@acornlibrary.org;
URL: www.acornlibrary.org
1966; Mary Tuytschaevers
64 000 vols; 180 curr per   40560

Clackamas County Library, Oak Lodge, 16201 SE McLoughlin Blvd, *Oak Grove*, OR 97267-4653
T: +1 503 6558543; URL: www.co.clackamas.us/lib/
1938; Doris Grolbert
142 000 vols; 276 curr per; 3 000 e-books   40561

Sno-Isle Libraries – Oak Harbor Branch, 1000 SE Regatta Dr, *Oak Harbor*, WA 98277
T: +1 360 6755115; Fax: +1 360 6793761
Mary Campbell
108 000 vols   40562

Fayette County Public Libraries, 531 Summit St, *Oak Hill*, WV 25901
T: +1 304 4650121; Fax: +1 304 4655306; URL: www.fayette.lib.wv.us
1959; Judy Gunsaulis
83 000 vols; 26 curr per   40563

Oak Lawn Public Library, 9427 S Raymond Ave, *Oak Lawn*, IL 60453-2434
T: +1 708 4224990; Fax: +1 708 4225061; URL: www.oaklawnlibrary.org
1943; Dr. James B. Casey
Law, Careers; US College Cats, microfiche
281 000 vols; 735 curr per;
libr loan   40564

Oak Park Public Library, 834 Lake St, *Oak Park*, IL 60301
T: +1 708 3838200; Fax: +1 708 6976917; URL: www.oppl.org
1903; Deirdre Brennan
Art & architecture; Frank Lloyd Wright & Ernest Hemingway, Hist; 2 branch libs
289 000 vols; 500 curr per; 35 000 govt docs; 1 000 maps; 3 000 microforms
libr loan   40565

Oak Park Public Library, 14200 Oak Park Blvd, *Oak Park*, MI 48237-2089
T: +1 248 6917480; Fax: +1 248 6917155; URL: www.oakpark-mi.com/library/library.htm
1957; John Martin
Judaica; Arabic & Russian language materials
98 000 vols; 185 curr per; 3 200 av-mat; 1 500 sound-rec
libr loan   40566

Jefferson Township Public Library, 1031 Weldon Rd, *Oak Ridge*, NJ 07438
T: +1 973 2086244; Fax: +1 973 6977051; E-mail: JTPL@jeffersonlibrary.net; URL: www.jeffersonlibrary.net/
1960; Seth Stephens
52 000 vols; 97 curr per;
libr loan   40567

Oak Ridge Public Library, Civic Center, 1401 Oak Ridge Tpk, *Oak Ridge*, TN 37830-6224; P.O. Box 1, Oak Ridge, TN 37831-0001
T: +1 865 4253455; Fax: +1 865 4253429; URL: www.orpl.org
1944; Kathy E. McNeilly
108 000 vols; 237 curr per   40568

Monmouth County Library – Township of Ocean, 701 Deal Rd, *Oakhurst*, NJ 07755
T: +1 732 5315092; Fax: +1 732 5315092
Beth Miller
64 000 vols   40569

Oakland Public Library, Two Municipal Plaza, *Oakland*, NJ 07436
T: +1 201 3373742; Fax: +1 201 3370261; E-mail: oakpl@bccls.org; URL: www.bccls.org/oakland/
1910; Carolyn R. Stefani
Large Print Bks
60 000 vols; 180 curr per; 980 av-mat; 2 294 sound-rec; 1 548 digital data carriers
libr loan; NJLA   40570

Oakland Public Library, 125 14th St, *Oakland*, CA 94612
T: +1 510 2386719; Fax: +1 510 2386722; URL: www.oaklandlibrary.org
1878; Carmen L. Martinez
Local history; Tool Lending Library; African American Museum and Library at Oakland, 2nd Start Adult Literacy, 15 branch libs
1 149 000 vols; 3 359 curr per; 7 000 e-books; 9 030 govt docs; 40 000 music scores; 140 000 microforms; 23 038 av-mat; 65 500 sound-rec; 34 digital data carriers; 4 881 Large Print Bks
libr loan; ALA, PLA, ULC   40571

Oakland Public Library – Asian, 388 Ninth St, Ste 190, *Oakland*, CA 94612
T: +1 510 2383400; Fax: +1 510 2384732
Madeleine Lee
Asian American Coll, english;Asian Studies Coll;Asian Materials in Asian Languages (Chinese, Japanese, Korean, Vietnamese, Thai, Cambodian, Tagalog & Laotian) gen subj titles, major ref titles
75 000 vols   40572

Oakland Public Library – Dimond, 3565 Fruitvale Ave, *Oakland*, CA 94602
T: +1 510 4827844; Fax: +1 510 4827824
Catherine Nichols
American Indian Coll
76 000 vols   40573

Oakland Public Library – Main Library, 125 14th St, *Oakland*, CA 94612
T: +1 510 2383134; Fax: +1 510 2382232
Douglas Smith
355 000 vols   40574

Oakland Public Library – Rockridge, 5366 College Ave, *Oakland*, CA 94618
T: +1 510 5975017; Fax: +1 510 5975067
Patricia Lichter
84 000 vols   40575

Oakland Public Library – West Oakland, 1801 Adeline St, *Oakland*, CA 94607
T: +1 510 2387352; Fax: +1 510 2387551
Christine Saed
52 000 vols   40576

Ruth Enlow Library of Garrett County, Six N Second St, *Oakland*, MD 21550-1393
T: +1 301 3343996; Fax: +1 301 3344152; E-mail: info@relib.net; URL: www.relib.net
1946; Cathy A. Ashby
Western Maryland, Garrett County; 4 branch libs
71 000 vols; 225 curr per   40577

Wright Memorial Public Library, 1776 Far Hills Ave, *Oakwood*, OH 45419-2598
T: +1 937 2948572; Fax: +1 937 2948578; URL: www.wrightlibrary.org
1913; Ann Snively
Literary criticism
168 000 vols; 520 curr per   40578

Allen Parish Libraries, 320 S Sixth St, *Oberlin*, LA 70655; P.O. Box 400, Oberlin, LA 70655-0400
T: +1 337 6394315; Fax: +1 337 6392654; URL: www.allen.lib.la.us
1957; Karen Teigen
2 branch libs
90 000 vols; 181 curr per; 1 886 av-mat   40579

Oberlin Public Library, 65 S Main St, *Oberlin*, OH 44074-1626
T: +1 440 7754790; Fax: +1 440 7742880; URL: www.oberlinpl.lib.oh.us
1947; Darren McDonough
Children's lit, Folklore, Ohio
151 000 vols; 235 curr per   40580

Marion County Public Library System, 2720 E Silver Springs Blvd, *Ocala*, FL 34470
T: +1 352 3684500; Fax: +1 352 3684518; URL: library.marioncountyfl.org
1961; Julia H. Sieg
Southeast States Genealogy; 9 branch libs
490 000 vols; 700 curr per   40581

Ocean City Free Public Library, 1735 Simpson Ave, *Ocean City*, NJ 08226
T: +1 609 3992434; Fax: +1 609 3988944; URL: www.oceancitylibrary.org
Christopher Maloney
90 000 vols; 250 curr per   40582

Oceanside Library, 30 Davison Ave, *Oceanside*, NY 11572-2299
T: +1 516 7662360; Fax: +1 516 7661895; URL: www.oceansidelibrary.com
1938; Evelyn Rothschild
Ethnic Coll, Holocaust Coll
174 000 vols
libr loan   40583

Oceanside Public Library, 330 N Coast Hwy, *Oceanside*, CA 92054-2824
T: +1 760 4355600; Fax: +1 760 4359614; E-mail: public.library@ci.oceanside.ca.us; URL: www.oceansidepubliclibrary.org
1904; Deborah Polich
Children's lit, Art & architecture, Employment, Parent; Caldecott-Newberry Memorial Coll, Martin Luther King Coll, Parents Resource Coll, Teachers Resource Ctr, Samoan Culture Coll, 280 000 vols; 426 curr per; 2 000 e-books; 10 980 av-mat
libr loan   40584

Oconomowoc Public Library, 200 South St, *Oconomowoc*, WI 53066-5213
T: +1 262 5692194; Fax: +1 262 5692176
1893; Ray McKenna
Art, Biogr
100 000 vols; 240 curr per   40585

Ector County Library, 321 W Fifth St, *Odessa*, TX 79761-5066
T: +1 432 3320633; Fax: +1 432 3324211; URL: www.ector.lib.tx.us
1938; Rebbecca Taylor
129 000 vols; 320 curr per   40586

Tampa-Hillsborough County Public Library System – Austin Davis Branch, 17808 Wayne Rd, *Odessa*, FL 33556
Fax: +1 813 2643903
Carol Hershman
61 000 vols   40587

O'Fallon Public Library, 120 Civic Plaza, *O'Fallon*, IL 62269-2692
T: +1 618 6323783; Fax: +1 618 6323759; E-mail: ofa@lcls.org; URL: www.ofallonlibrary.org
1943; Mary L. Smith
Family Reading Ctr, Learning Activities Resource Ctr, Rotary Grants Coll
57 000 vols; 165 curr per; 1 873 av-mat   40588

Saint Charles City County Library District – Deer Run Branch, 1300 N Main, *O'Fallon*, MO 63366-2013
T: +1 636 9783251; Fax: +1 636 9783209
Tim DeGhelder
86 000 vols   40589

Saint Charles City County Library District – Middendorf-Kredell Branch, 2750 Hwy K, *O'Fallon*, MO 63368-7859
T: +1 636 2724999; Fax: +1 636 9787998
Patricia Kern
179 000 vols   40590

Weber County Library System, 2464 Jefferson Ave, *Ogden*, UT 84401-2464
T: +1 801 3372617; Fax: +1 801 3372615; URL: www.weberpl.lib.ut.us
1903; Lynnda Wangsgard
Ava J Cooper Spec Coll (Utah & Western Hist); 3 branch libs
330 000 vols; 998 curr per; 8 591 av-mat; 660 Bks on Deafness & Sign Lang, 1 750 High Interest/Low Vocabulary Bks, 3 600 Large Print Bks, 7 750 Talking Bks
libr loan; ALA   40591

Ogdensburg Public Library, 312 Washington St, *Ogdensburg*, NY 13669-1518
T: +1 315 3934325; Fax: +1 315 3934344; E-mail: ogdlib@nnyln.net; URL: www.nc3r.org/ogdensburg
1893; David A. Franz
General Newton Martin Curtis Civil War Coll; Ogdensburg Hist (Ogdensburg Arch), bk, mss, flm
72 000 vols; 190 curr per; 5 000 microforms; 600 av-mat; 1 275 sound-rec; 12 digital data carriers   40592

Oil City Library, Two Central Ave, *Oil City*, PA 16301-2795
T: +1 814 6783072; Fax: +1 814 6768028; URL: www.oilcitylibrary.org
1904; Bruce George
Hist of oil; Genealogy (Selden Coll)
96 000 vols; 250 curr per; 1 910 av-mat   40593

Oil City Library, Two Central Ave, *Oil City*, PA 16301-2795
T: +1 814 6783072; Fax: +1 814 6768028; URL: www.oilcitylibrary.org
1995; Bruce George
Professional Coll, V-tape, bks on tape
96 000 vols; 250 curr per; 5 426 av-mat; 1 910 sound-rec; 481 digital data carriers; 2 250 Large Print Bks   40594

Okeechobee County Public Library, 206 SW 16th St, *Okeechobee*, FL 34974
T: +1 863 7633536; Fax: +1 863 7635368; URL: www.myhlc.org
1967; Margaret S. Taylor
52 000 vols; 73 curr per; 4 241 av-mat; 1 246 sound-rec; 20 digital data carriers; 500 High Interest/Low Vocabulary Bks, 21 Bks on Deafness & Sign Lang, 2 443 Large Print Bks   40595

Metropolitan Library System in Oklahoma County, 300 Park Ave, *Oklahoma City*, OK 73102
T: +1 405 6063726; Fax: +1 405 6063722; E-mail: director@metrolibrary.org; URL: www.metrolibrary.org
1965; Donna Morris
Local Black Hist (Black Chronicles Coll); 18 branch libs
1 013 000 vols; 2 569 curr per
libr loan   40596

Metropolitan Library System in Oklahoma County – Belle Isle Library, 5501 N Villa, *Oklahoma City*, OK 73112-7164
T: +1 405 8439601; Fax: +1 405 8434560; E-mail: belleisle@metrolibrary.org
Priscilla Doss
96 000 vols; 183 curr per   40597

Metropolitan Library System in Oklahoma County – Ronald J. Norick Downtown Library – Ronald J. Norick Downtown Library, 300 Park Ave, *Oklahoma City*, OK 73102
T: +1 405 2318650; Fax: +1 405 6063895; E-mail: downtown@metrolibrary.org
Mary Patton
168 000 vols; 984 curr per   40598

Metropolitan Library System in Oklahoma County – Southern Oaks Library, 6900 S Walker, *Oklahoma City*, OK 73139-7299
T: +1 405 6314468; Fax: +1 405 6063484; E-mail: southernoaks@metrolibrary.org
Wayland Randy
102 000 vols; 161 curr per   40599

Metropolitan Library System in Oklahoma County – The Village Library, 10307 N Pennsylvania Ave, *Oklahoma City*, OK 73120
T: +1 405 7550710; Fax: +1 405 6063502; E-mail: village@metrolibrary.org
Lavetta Dent
67 000 vols; 183 curr per   40600

Metropolitan Library System in Oklahoma County – Warr Acres Library- Warr Acres Library, 5901 NW 63, *Oklahoma City*, OK 73132-2401
T: +1 405 7212616; Fax: +1 405 6063534; E-mail: warracres@metrolibrary.org
Barbara Beasley
67 000 vols; 99 curr per   40601

Oklahoma Department of Libraries, 200 NE 18th St, *Oklahoma City*, OK 73105
T: +1 405 5212502; Fax: +1 405 5257804; URL: www.odl.state.ok.us
1890; Susan McVey
Law; Oklahoma Coll
270 000 vols; 1 604 curr per; 575 e-books; 2 536 347 govt docs; 300 000 microforms
libr loan   40602

Okmulgee Public Library, 218 S Okmulgee Ave, *Okmulgee*, OK 74447
T: +1 918 7561448; Fax: +1 918 7581148; E-mail: books@okmulgeelibrary.org; URL: www.okmulgeelibrary.org
1907; Kristin Cunningham
68 000 vols; 150 curr per   40603

Olathe Public Library, 201 E Park St, *Olathe*, KS 66061
T: +1 913 9716850; Fax: +1 913 9716809; URL: www.olathelibrary.org
1909; Emily F. Baker
Kansas Room, Grandparents Coll; 1 branch libr
181 000 vols   40604

Olathe Public Library – Indian Creek, 12990 S Black Bob Rd, *Olathe*, KS 66062
T: +1 913 9715235; Fax: +1 913 9715239
Kathleen O'Leary
92 000 vols   40605

Old Bridge Public Library, One Old Bridge Plaza, *Old Bridge*, NJ 08857-2498
T: +1 732 7215600; Fax: +1 732 6790556; URL: www.oldbridgelibrary.org
1970; Allan Kleiman
1 branch libr
170 000 vols; 538 curr per; 6 900 av-mat; 6 900 Talking Bks   40606

Perrot Memorial Library, 90 Sound Beach Ave, *Old Greenwich*, CT 06870
T: +1 203 6373870; Fax: +1 203 6982620; URL: www.perrotlibrary.org
1905; Kevin McCarthy
Gardening, Sailing, Cooking
70 000 vols; 2 500 av-mat; 1 800 sound-rec; 25 digital data carriers   40607

Acton Public Library, 60 Old Boston Post Rd, *Old Saybrook*, CT 06475-2200
T: +1 860 3953184; Fax: +1 860 3952462; URL: www.oldsaybrookct.org
1873; Janet M. Crozier
Large type print
71 000 vols; 133 curr per; 1 520 av-mat   40608

Old Town Public Library, 46 Middle St, *Old Town*, ME 04468
T: +1 207 8273972; Fax: +1 207 8273978; E-mail: cindy.jennings@old-town.org; URL: old-town.lib.me.us
1904; Cynthia Jennings
50 000 vols   40609

Olean Public Library, 134 N Second St, *Olean*, NY 14760-2583
T: +1 716 3720200; Fax: +1 716 3728651; E-mail: info@oleanlibrary.org; URL: www.oleanlibrary.org
1871; Lance Chaffee
99 000 vols; 373 curr per
libr loan   40610

Olney Public Library, 400 W Main St, *Olney*, IL 62450
T: +1 618 3923711; Fax: +1 618 3923139; E-mail: ruthchil@shawls.lib.il.us; URL: www.olney.lib.il.us
1872; Judy Whitaker
Civil War Coll, Collectibles & Antiques
50 000 vols; 90 curr per   40611

Omaha Public Library, W. Dale Clark Library, 215 S 15th St, *Omaha*, NE 68102-1629
T: +1 402 4444800; Fax: +1 402 4444504; E-mail: webdesk@omahapubliclibrary.org; URL: www.omahapubliclibrary.org
1872; Rivkah K. Sass
African-American hist, Foreign language, Genealogy; 9 branch libs
898 000 vols; 2 282 curr per; 49 618 av-mat; 47 500 sound-rec; 17 000 Large Print Bks
libr loan; ULC, ALA   40612

Omaha Public Library – Benson, 2918 N 60th, *Omaha*, NE 68104
T: +1 402 4444846; Fax: +1 402 4446595
Susan Thornton
58 000 vols   40613

Omaha Public Library – Millard, 13214 Westwood Lane, *Omaha*, NE 68144
T: +1 402 4444848; Fax: +1 402 4446623
Mary Griffin
129 000 vols   40614

Omaha Public Library – Milton R. Abrahams Branch, 5111 N 90th St, *Omaha*, NE 68134
T: +1 402 4446284; Fax: +1 402 4446590
Sarah Watson
95 000 vols   40615

Omaha Public Library – W. Clarke Swanson Branch, 9101 W Dodge Rd, *Omaha*, NE 68114
T: +1 402 4444852; Fax: +1 402 4446651
Pam Scott
Children's Historic Literature Coll
103 000 vols   40616

Oneida Public Library, 220 Broad St, *Oneida*, NY 13421
T: +1 315 3633050; Fax: +1 315 3634217; E-mail: oneida@mail.midyork.org; URL: www.midyork.org/oneida
1924; Carolyn Gerakopoulos
58 000 vols; 110 curr per   40617

Huntington Memorial Library, 62 Chestnut, *Oneonta*, NY 13820-2498
T: +1 607 4321980; URL: www.4cls.org
1893; Marie Bruni
DAR Lineage Coll; Railroads (Beach Coll), pictures
80 000 vols; 215 curr per   40618

Ontario City Library, 215 East C St, *Ontario*, CA 91764
T: +1 909 3952004; Fax: +1 909 3952043; URL: www.ci.ontario.ca.us/library
1885; Judy Evans
209 000 vols; 686 curr per; 575 maps; 13 300 av-mat; 7 300 sound-rec; 2 500 digital data carriers; 30 585 pamphlets
libr loan   40619

Ontario City Library – Colony High, 3850 E Riverside Dr, *Ontario*, CA 91761-1623
T: +1 909 3952014; Fax: +1 909 9300836
Holly Kurtz
56 000 vols; 57 curr per   40620

Chattanooga-Hamilton County Bicentennial Library – Ooltewah-CollegedaleOoltewah-Collegedale, 9318 Apison Pike, *Ooltewah*, TN 37363
T: +1 423 3969300; Fax: +1 423 3969334
Joanne Stanfield
58 000 vols   40621

Lewis Cooper Junior Memorial Library, 200 S Sixth St, *Opelika*, AL 36801
T: +1 334 7055380; Fax: +1 334 7055381; URL: www.opelika.org/depts/library/index.html
1941; Susan Delmas
Genealogy Coll
62 000 vols; 60 curr per   40622

Opelousas-Eunice Public Library, 212 E Grolee St, *Opelousas*, LA 70570; P.O. Box 249, Opelousas, LA 70571-0249
T: +1 337 9483693; Fax: +1 337 9485200
1967
Large Print Coll
77 000 vols; 139 curr per; 740 av-mat   40623

Oradell Free Public Library, 375 Kinderkamack Rd, *Oradell*, NJ 07649-2122
T: +1 201 2622613; Fax: +1 201 2629112; E-mail: oradcirc@bccls.org; URL: www.bccls.org/oradell
1913; Lori Ann Barnes
Bks Illust by Charles Livingston Bull
69 000 vols; 314 curr per   40624

Case Memorial Library, 176 Tyler City Rd, *Orange*, CT 06477-2498
T: +1 203 8912170; Fax: +1 203 8912190; URL: www.leaplibraries.org/orange
1956; Meryl P. Farber
Large type print
89 000 vols; 146 curr per   40625

Orange County Library, 146A Madison Rd, *Orange*, VA 22960
T: +1 540 6723811; Fax: +1 540 6725040; URL: orangecountyva.gov; tlc.library.net/orange
1903; Kathryn Hill
2 branch libs
117 000 vols; 150 curr per; 8 698 av-mat; 3 334 sound-rec; 1 507 digital data carriers   40626

Orange Public Library, 348 Main St, *Orange*, NJ 07050-2794
T: +1 973 6730153; Fax: +1 973 6731847; URL: www.orangepl.org
1884; Doris Theresa Walker
Black Lit & Hist
195 000 vols; 128 curr per; 403 microforms; 1 280 av-mat; 5 240 sound-rec; 6 digital data carriers   40627

Orange Public Library & History Center, 407 E Chapman Ave, mail address: 230 E Chapman Ave, *Orange*, CA 92866-1509
T: +1 714 2882400; Fax: +1 714 7716126; URL: www.cityoforange.org/library
1885; Nora Jacob
Children's lit, Biology; Spanish Coll; 2 branch libs
221 000 vols; 177 curr per; 1 114 maps; 14 000 microforms; 537 digital data carriers   40628

Orange Public Library & History Center – El Modena Branch, 380 S Hewes St, *Orange*, CA 92869-4060
T: +1 714 2882454; Fax: +1 714 9971041
Irma Morales
57 000 vols   40629

Wheeler Memorial Library, 49 E Main St, *Orange*, MA 01364-1267
T: +1 508 5442495; Fax: +1 508 5441116; E-mail: wowens@orangelib.org; URL: www.orangelib.org
1847; Walt Owens
55 100 vols   40630

Clay County Public Library System – Headquarters Library, 1895 Town Center Blvd, *Orange Park*, FL 32003; P.O. Box 10109, Fleming Island, FL 32006-0109
T: +1 904 2783720; Fax: +1 904 2784747
Arnold Weeks
268 000 vols   40631

Clay County Public Library System – Orange Park Public Library, 2054 Plainfield Ave, *Orange Park*, FL 32073-5498
T: +1 904 2784750; Fax: +1 904 2783618; URL: ccpl.lib.fl.us
Walter Brown
84 000 vols   40632

Orangeburg County Library, 510 Louis St, *Orangeburg*, SC 29115-5030; P.O. Box 1367, Orangeburg, SC 29116-1367
T: +1 803 5335857; Fax: +1 803 5335860; E-mail: OCLSNotify@orangeburgcounty.org; URL: www.orangeburgcounty.org/library
1937; Paula F. Paul
120 000 vols; 310 curr per
libr loan   40633

Orchard Park Public Library, 4570 South Buffalo St, *Orchard Park*, NY 14127
T: +1 716 6629851; Fax: +1 716 6773098; URL: www.buffalolib.org/libraries/orchardpark
1935
65 880 vols; 129 curr per   40634

Oregon Public Library, 256 Brook St, *Oregon*, WI 53575
T: +1 608 8353656; Fax: +1 608 8352856; E-mail: oregon@scls.lib.wi.us; URL: www.oregonpubliclibrary.org
52 420 vols; 147 curr per   40635

Toledo-Lucas County Public Library – Oregon Branch, 3340 Dustin Rd, *Oregon*, OH 43616
T: +1 419 2595250; Fax: +1 419 6913341
Nicole Naylor
110 000 vols   40636

Oregon City Public Library, 362 Warner Milne Rd, *Oregon City*, OR 97045
T: +1 503 6578269; Fax: +1 503 6573702; URL: www.oregoncity.lib.or.us
1904; Scott Archer
Genealogy; Oregon hist (The Oregon Coll), microforms, photos, newspaper clippings
100 000 vols; 208 curr per; 450 microforms; 5 000 av-mat; 6 200 sound-rec; 530 digital data carriers
libr loan; ALA, OLA   40637

Orem Public Library, 58 N State St, *Orem*, UT 84057-5596
T: +1 801 2297175; Fax: +1 801 2297130; URL: www.oremlibrary.org
1940; Louise Wallace
Film classics, Religious studies
280 000 vols; 255 curr per; 6 250 av-mat; 18 700 sound-rec; 5 digital data carriers   40638

Contra Costa County Library – Orinda Library, 26 Orinda Way, *Orinda*, CA 94563-2555
T: +1 925 2542184; Fax: +1 925 2538629
Caroline Gick
66 000 vols   40639

Orland Free Library, 333 Mill St, *Orland*, CA 95963
T: +1 530 8651640; E-mail: orlandfreelibrary@hotmail.com; URL: www.orlandfreelibrary.net
1909; Marilyn Cochran
Hist
62 000 vols; 80 curr per   40640

Orland Park Public Library, 14921 Ravinia Ave, *Orland Park*, IL 60462
T: +1 708 4285100; Fax: +1 708 4285183; E-mail: askoppl@orlandparklibrary.org; URL: www.orlandparklibrary.org
Sharon Wsol
124 000 vols; 402 curr per   40641

Orange County Library District, 4324 E Colonial Dr, *Orlando*, FL 32803
Edward Booker
72 000 vols   40642

Orange County Library District – Alafaya, 12000 E Colonial Dr, *Orlando*, FL 32826
Lisa Stewart
124 000 vols   40643

Orange County Library District – Edgewater, 5049 Edgewater Dr, *Orlando*, FL 32810-4743
Kelly Pepo
110 000 vols; 20 curr per   40644

Orange County Library District – Hiawassee, 2768 N Hiawassee Rd, *Orlando*, FL 32818
Fax: +1 407 5212461
Ken Gibert
114 000 vols   40645

Orange County Library District – Orlando Public Library, 101 E Central Blvd, *Orlando*, FL 32801
T: +1 407 8357426; URL: www.ocls.info
1923; Mary Anne Hodel
Walt Disney World Coll, Florida Coll; 15 branch libs
2 300 000 vols; 2 097 curr per; 21 000 e-books
libr loan   40646

Orange County Library District – South Creek, 1702 Deerfield Blvd, *Orlando*, FL 32837
Julie Ventura
104 000 vols   40647

Orange County Library District – South Trail, 4600 S Orange Blossom Trail, *Orlando*, FL 32839
Carolyn McClendon
94 000 vols   40648

Orange County Library District – Southeast/Semoran Blvd, 5575 S Semoran Blvd, *Orlando*, FL 32822
Paolo Melillo
90 000 vols   40649

Orange County Library District – Southwest/Della Drive, 7255 Della Dr, *Orlando*, FL 32819
Bethany Corbett
108 000 vols
40650

Orange County Library District – Washington Park, 5151 Raleigh St, Ste A, *Orlando*, FL 32811
Fax: +1 407 5212468
Patsy Williams
52 000 vols
40651

Snow Library, 67 Main St, *Orleans*, MA 02653-2413
T: +1 508 2403760; Fax: +1 508 2555701; URL: www.snowlibrary.org
1877; Mary Reuland
H.K. Cummings Coll of Hist Photography (1870-1890)
55 000 vols; 92 curr per; 2 401 av-mat
40652

Oro Valley Public Library, 1305 W Naranja Dr, *Oro Valley*, AZ 85737-9762
T: +1 520 2295300; Fax: +1 520 2295319; URL: www.orovalleylib.com
Shirley Hall Dornberg
77 000 vols; 227 curr per; 2 751 av-mat
40653

Butte County Library, 1820 Mitchell Ave, *Oroville*, CA 95966-5387
T: +1 530 5387326; Fax: +1 530 5387235; E-mail: lib@buttecounty.net; URL: www.buttecounty.net/bclibrary
1913; Derek Wolfgram
5 branch libs
277 000 vols; 526 curr per; 4 e-books
libr loan
40654

Orrville Public Library, 230 N Main St, *Orrville*, OH 44667
T: +1 330 6831065; Fax: +1 330 6831984; E-mail: askus@orrville.lib.oh.us; URL: www.orrville.lib.oh.us
1925; Leslie Picot
Railroads
64 000 vols; 181 curr per; 16 000 e-books; 2 985 av-mat
libr loan
40655

Brandon Township Public Library, 304 South St, *Ortonville*, MI 48462
T: +1 248 6271464; Fax: +1 248 6279880; E-mail: reference@brandonlibrary.org; URL: www.brandonlibrary.org
1926; Paula J. Gauthier
Unabridged audiobooks
68 000 vols; 182 curr per; 4 000 e-books; 2 419 av-mat; 956 sound-rec
libr loan; ALA
40656

Oshkosh Public Library, 106 Washington Ave, *Oshkosh*, WI 54901-4985
T: +1 920 2365203; Fax: +1 920 2365228; URL: www.oshkoshpubliclibrary.org
1895; Jeff Gilderson-Duwe
US Census Coll; 1 bookmobile
270 000 vols; 560 curr per
libr loan
40657

Oskaloosa Public Library, 301 S Market St, *Oskaloosa*, IA 52577
T: +1 641 6730441; Fax: +1 641 6736237; E-mail: opl@opl.oskaloosa.org; URL: www.opl.oskaloosa.org
1903; Susan Holland
County & Area Hist Coll, Genealogy (DAR Lineage Bks Coll)
55 000 vols; 198 curr per
libr loan
40658

Ossining Public Library, 53 Croton Ave, *Ossining*, NY 10562-4903
T: +1 914 9412416; Fax: +1 914 9417464; E-mail: opldirector@wlsmail.org; URL: ossininglibrary.org
1898; Elizabeth Bermel
Job Info Ctr, bks, per & pamphlets
112 000 vols; 310 curr per
40659

Oswego Public Library District, 32 W Jefferson St, *Oswego*, IL 60543
T: +1 630 5543150; Fax: +1 630 5548445
1964; Sarah Skilton
Collectibles & Antiques; Illinois Census, micro
145 000 vols
40660

Oswego School District Public Library, 120 E Second St, *Oswego*, NY 13126
T: +1 315 3415867; Fax: +1 315 2166492; E-mail: oswegopl@northnet.org; URL: www.oswegopubliclibrary.org
1854; Carol Ferlito
55 000 vols; 1 188 curr per; 1 000 microforms
40661

Otsego District Public Library, 219 S Farmer St, *Otsego*, MI 49078-1313
T: +1 269 6949690; Fax: +1 269 6949129; E-mail: otslib@otsegolibrary.org; URL: www.otsegolibrary.org
1844; Ryan S. Wieber
Hist of Otsego, 12 vol Coll, bks & micro; Michigan Pioneer Coll
53 000 vols; 110 curr per; 1 149 av-mat
40662

Ottawa Library, 105 S Hickory St, *Ottawa*, KS 66067-2306
T: +1 785 2423080; Fax: +1 785 2428789; E-mail: ottawalibraryreference@yahoo.com; URL: www.ottawa.lib.ks.us
1872; Robin Flory
58 000 vols; 124 curr per
40663

Putnam County District Library, The Educational Service Ctr, 124 Putnam Pkwy, *Ottawa*, OH 45875-1471; P.O. Box 230, Ottawa
T: +1 419 5233747; Fax: +1 419 5236477; URL: www.putnamco.lib.oh.us
Kelly Ward
164 000 vols; 329 curr per
40664

Reddick Library, 1010 Canal St, *Ottawa*, IL 61350
T: +1 815 4340509; Fax: +1 815 4342634; URL: www.reddicklibrary.org
1888; Victoria Trupiano
Illinois Coll
73 000 vols; 179 curr per
40665

Ottumwa Public Library, 102 W Fourth St, *Ottumwa*, IA 52501
T: +1 641 6827563; Fax: +1 641 6824970; URL: www.ottumwapubliclibrary.org
1872; Mary Ann Lemon
Ottumwa & Wapello County Hist (Iowa Coll), bks, clippings, pictures
70 000 vols; 120 curr per; 1 400 av-mat; 400 sound-rec
40666

Johnson County Library, 9875 W 87th St, *Overland Park*, KS 66212; P.O. Box 2933, Shawnee Mission, KS 66201-1333
T: +1 913 4952400; Fax: +1 913 4952460; URL: www.jocolibrary.org
1952; Donna Lauffer
13 branch libs
1 196 000 vols; 2 851 curr per; 1 000 e-books; 26 898 digital data carriers; 41 174 CDs, 79 864 Videos, 23 566 Large Print Bks
libr loan
40667

Johnson County Library – Blue Valley, 9000 W 151st, *Overland Park*, KS 66221
T: +1 913 4953850; Fax: +1 913 4953821
Cyndi Chappell
110 000 vols; 123 curr per
40668

Johnson County Library – Central Resource, 9875 W 87th St, *Overland Park*, KS 66212; P.O. Box 2933, Shawnee Mission, KS 66201-1333
Fax: +1 913 4952480
Carolyn Weeks
289 000 vols; 874 curr per
40669

Johnson County Library – Oak Park, 9500 Bluejacket, *Overland Park*, KS 66214
T: +1 913 7528700; Fax: +1 913 7528709
Magaly Vallazza
121 000 vols; 165 curr per
40670

Tulsa City-County Library System – Owasso Branch, 103 W Broadway, *Owasso*, OK 74055
T: +1 918 5914566; Fax: +1 918 5914568
Barbara Barnes
52 000 vols
40671

Owatonna Public Library, 105 N Elm Ave, *Owatonna*, MN 55060-2405; P.O. Box 387, Owatonna, MN 55060-0387
T: +1 507 4442460; Fax: +1 507 4442465; E-mail: info@owatonna.info; URL: www.owatonna.info
1900; Mary Kay Feltes
1 branch libr
169 000 vols; 6 040 av-mat; 4 100 sound-rec; 400 digital data carriers; 200 Toys
libr loan
40672

Daviess County Public Library, 2020 Frederica St, *Owensboro*, KY 42301
T: +1 270 6840211; Fax: +1 270 6840218; URL: www.dcplibrary.org
1909; Deborah Mesplay
186 000 vols; 327 curr per; 12 307 av-mat; 7 390 sound-rec
40673

Shiawassee District Library, Owosso Branch, 502 W Main St, *Owosso*, MI 48867-2607
T: +1 989 7255134; Fax: +1 989 7235444; URL: www.sdl.lib.mi.us
1910; Steven Flayer
James Oliver Curwood Coll, bks, mss, pictures; 1 branch libr
60 000 vols; 158 curr per; 7 000 microforms; 1 402 av-mat; 2 897 sound-rec; 56 art reproductions
libr loan
40674

Granville County Library System, Richard H. Thornton Library, 210 Main St, *Oxford*, NC 27565-3321; P.O. Box 339, Oxford
T: +1 919 6931121; Fax: +1 919 6932244; URL: www.granville.lib.nc.us
1935; Tresia Dodson
72 000 vols; 267 curr per; 1 992 av-mat
40675

Oxford Free Library, 339 Main St, *Oxford*, MA 01540
T: +1 508 9876003; Fax: +1 508 9873896
1903; Timothy A. Kelley
53 000 vols; 111 curr per; 1 084 av-mat
40676

Oxnard Public Library, 251 South A St, *Oxnard*, CA 93030
T: +1 805 3857500; Fax: +1 805 3857585; URL: www.oxnard.org
1907; Barbara J. Murray
Spanish language; Local govt; 2 branch libs
374 000 vols; 3 615 curr per
libr loan; ALA
40677

Oyster Bay-East Norwich Public Library, 89 E Main St, *Oyster Bay*, NY 11771
T: +1 516 9221212; Fax: +1 516 2648693; E-mail: oysterbay@nassaulibrary.org; URL: www.nassaulibrary.org/oysterbay
1901; Suzanne Koch
Career development; Presidential (Theodore Roosevelt Coll)
86 000 vols; 1 703 curr per
40678

Christian County Library, 1005 N Fourth Ave, *Ozark*, MO 65721
T: +1 417 5812432; Fax: +1 417 5818855; E-mail: info@christiancounty.lib.mo.us; URL: christiancounty.lib.mo.us
1956; Mabel Gaye Phillips
Antiques, Crafts; Southwest Missouri & the Ozarks
76 000 vols; 203 curr per; 8 000 e-books; 1 613 av-mat
libr loan; ALA
40679

Queens Borough Public Library – Ozone Park Branch, 92-24 Rockaway Blvd, *Ozone Park*, NY 11417
T: +1 718 8453127; Fax: +1 718 8481082
Maryellen Borello
72 000 vols
40680

Pacific Grove Public Library, 550 Central Ave, *Pacific Grove*, CA 93950-2789
T: +1 831 6485762; Fax: +1 831 3733268; URL: www.pacificgrove.lib.ca.us
1908; Barbara Morrison
98 000 vols; 206 curr per; 489 maps; 9 000 microforms; 1 461 av-mat; 3 416 sound-rec; 41 digital data carriers; 1 943 video tapes, 30 art reproductions; 6 984 pamphlets
libr loan
40681

San Mateo County Library – Pacifica Branch, 104 Hilton, *Pacifica*, CA 94044
T: +1 650 3555196; Fax: +1 650 3556658; URL: www.pacificalibrary.org
Thom Ball
53 000 vols
40682

Los Angeles Public Library System – Pacoima Branch, 13605 Van Nuys Blvd, *Pacoima*, CA 91331-3697
T: +1 818 8995203; Fax: +1 818 8995336
Erik Surber
53 000 vols
40683

McCracken County Public Library, 555 Washington St, *Paducah*, KY 42003-1735
T: +1 270 4422510; Fax: +1 270 4439322; URL: www.mclib.net
1901; Marie Liang
Kentucky Coll, Irvin S Cobb Lit Coll
128 000 vols; 280 curr per; 2 000 e-books
40684

Page Public Library, 479 S Lake Powell Blvd, *Page*, AZ 86040; P.O. Box 1776, Page, AZ 86040-1776
T: +1 928 6454270; Fax: +1 928 6455804; URL: www.youseemore.com/page
1959; Debbie Winlock
Arizona Hist Coll, Native American Coll
59 000 vols; 50 curr per
libr loan
40685

Johnson County Public Library, 444 Main St, *Paintsville*, KY 41240
T: +1 606 7894355; Fax: +1 606 7896758; E-mail: johnsonlibrary@bellsouth.net
Karen Daniel
55 000 vols; 63 curr per; 973 av-mat
40686

Palatine Public Library District, 700 N North Ct, *Palatine*, IL 60067-8159
T: +1 847 3585881; Fax: +1 847 3587192; URL: www.ppld.alibrary.com
1923; Daniel G. Armstrong
2 branch libs
298 000 vols; 631 curr per
40687

Palestine Public Library, 1101 N Cedar St, *Palestine*, TX 75801
T: +1 903 7298087; Fax: +1 903 7294062; URL: www.palestine.lib.tx.us
1882; Carol A. Herrington
Texana Coll, Anderson Co, Civil War Coll
77 000 vols; 99 curr per
40688

Palisades Park Free Public Library, 257 Second St, *Palisades Park*, NJ 07650
T: +1 201 5854150; Fax: +1 201 5852151; E-mail: palpcirc@bccls.org; URL: palisadespark.bccls.org
1922; Megan Doyle
Korean language; Cookery Coll
54 000 vols; 75 curr per
NJLA
40689

Franklin T. Degroodt Library, 6475 Minton Rd SW, *Palm Bay*, FL 32908
T: +1 321 9526317; Fax: +1 321 9526320; URL: www.brev.org
1980; Patricia A. Portnowitz
68 000 vols; 205 curr per; 2 144 av-mat
libr loan
40690

Flagler County Public Library, 2500 Palm Coast Pkwy NW, *Palm Coast*, FL 32137
T: +1 386 4466763; Fax: +1 386 4466773; E-mail: AskaLibrarian@flaglercounty.org; URL: www.flaglerlibrary.org
1980; Holly Albanese
78 000 vols; 150 curr per
libr loan; FLA
40691

Palm Harbor Library, 2330 Nebraska Ave, *Palm Harbor*, FL 34683
T: +1 727 7843332; Fax: +1 727 7856534; URL: www.palmharborlibrary.org
1978; Gene Coppola
Business, Handicrafts, Genealogy; Florida Coll, French, German, Literacy, Plays
215 000 vols; 45 curr per
40692

Palm Springs Public Library, 300 S Sunrise Way, *Palm Springs*, CA 92262-7699
T: +1 760 3238285; Fax: +1 760 3209834; E-mail: library.info@palmsprings-ca.gov; URL: www.palmspringslibrary.org
1940; Barbara L. Roberts
Hist, Art & architecture; Robinson Jeffers Coll, Southern Californiana
124 000 vols; 212 curr per; 16 Electronic Media & Resources
libr loan   40693

Palmdale City Library, 700 E Palmdale Blvd, *Palmdale*, CA 93550
T: +1 661 2675600; Fax: +1 661 2675606; URL: www.palmdalelibrary.org
1977; Nancy Quelland
Aerospace Coll, Genealogy Coll, Law; 1 branch libr
128 000 vols; 453 curr per
libr loan   40694

Palmer Public Library, 1455 N Main St, *Palmer*, MA 01069
T: +1 413 2833330; Fax: +1 413 2839970; URL: www.palmer.lib.ma.us
1878; Nancy E. Bauer
111 000 vols; 147 curr per   40695

Palmyra Public Library, Borough Bldg, 325 S Railroad St, *Palmyra*, PA 17078-2492
T: +1 717 8381347; Fax: +1 717 8381236; URL: www.lclibs.org/palmyra
1954; Karla J. Marsteller
54 000 vols; 121 curr per; 190 av-mat 40696

Palo Alto City Library, 1213 Newell Rd, *Palo Alto*, CA 94303-2907; P.O. Box 10250, Palo Alto, CA 94303-0250
T: +1 650 3292436; Fax: +1 650 3272033; E-mail: pa.library@cityofpaloalto.org; URL: www.cityofpaloalto/library
1904; Diane Jennings
4 branch libs
240 000 vols; 681 curr per
libr loan   40697

Palo Alto City Library – Mitchell Park, 3700 Middlefield Rd, *Palo Alto*, CA 94303
Fax: +1 650 8567925
Marilyn Gillespie
83 000 vols   40698

Palos Heights Public Library, 12501 S 71st Ave, *Palos Heights*, IL 60463
T: +1 708 4481473; Fax: +1 708 4488950; E-mail: phlibrary@palosheightslibrary.org; URL: www.palosheightslibrary.org
1944; Elaine Savage
70 000 vols; 223 curr per   40699

Green Hills Public Library District, 8611 W 103rd St, *Palos Hills*, IL 60465
T: +1 708 5988446; Fax: +1 708 5980856; E-mail: ghs@greenhills.lib.il.us; URL: www.greenhills.lib.il.us
1962; Annette T. Armstrong
Literary criticism, Polish language, Large print bks
60 000 vols; 138 curr per; 1 416 av-mat 40700

Northwest Regional Library System – Bay County Public Library, 898 W 11 St, *Panama City*, FL 32401; P.O. Box 59625, Panama City
T: +1 850 5222100; Fax: +1 850 5222138; URL: www.nwrls.com
1942; Joyce Dannecker
6 branch libs
187 000 vols; 483 curr per   40701

Carson County Public Library, 401 Main, *Panhandle*, TX 79068; P.O. Box 339, Panhandle, TX 79068-0339
T: +1 806 5373742; Fax: +1 806 5373780
1937; Terri Koetting
52 000 vols; 106 curr per   40702

Sump Memorial Library, 222 N Jefferson, *Papillion*, NE 68046
T: +1 402 5972040; Fax: +1 402 3398019; URL: www.papillion.ne.us/library.htm
1921; Sally Payne
55 000 vols; 169 curr per; 2 337 av-mat 40703

County of Los Angeles Public Library – Paramount Library, 16254 Colorado Ave, *Paramount*, CA 90723-5085
T: +1 562 6303171; Fax: +1 562 6303968; URL: www.colapublib.org/libs/paramount
Cherie Shih
59 000 vols; 70 curr per   40704

Paramus Public Library, E 116 Century Rd, *Paramus*, NJ 07652-4398
T: +1 201 5991305; Fax: +1 201 5990059; E-mail: para1@bccls.org; URL: www.bccls.org/paramus
1954; Leonard LoPinto
Fine & Performing Arts Coll; 1 branch libr
104 000 vols; 338 curr per; 7 maps; 154 000 microforms; 6 096 av-mat; 5 091 sound-rec; 40 digital data carriers
libr loan   40705

Parchment Community Library, 401 S Riverview Dr, *Parchment*, MI 49004-1200
T: +1 269 3437747; Fax: +1 269 3437749; URL: www.parchmentlibrary.org
1963; Teresa L. Stannard
53 000 vols; 134 curr per; 925 av-mat 40706

Paris Public Library, 326 S Main St, *Paris*, TX 75460
T: +1 903 7858531; Fax: +1 903 7846325; URL: www.paristexaslibrary.com
1926; Priscilla McAnally
Texana Coll
87 000 vols; 104 curr per   40707

Park City Library, 1255 Park Ave, *Park City*, UT 84060; P.O. Box 668, Park City, UT 84060-0668
T: +1 435 6155600; Fax: +1 435 6154903; URL: www.parkcitylibrary.org
1917; Linda L. Tillson
Skiing-Mountaineering-Outdoors Coll
60 000 vols; 196 curr per; 2 000 e-books
libr loan   40708

Park Forest Public Library, 400 Lakewood Blvd, *Park Forest*, IL 60466
T: +1 708 7483731; Fax: +1 708 7488829; E-mail: pfs@sslic.net; URL: www.pfpl.org
1955; Barbara Byrne Osuch
Civil War, Biogr, Drama, Cookbks
170 000 vols; 345 curr per; 425 High Interest/Low Vocabulary Bks, 3 570 Large Print Bks
libr loan   40709

Park Ridge Public Library, 20 S Prospect, *Park Ridge*, IL 60068-4188
T: +1 847 7203271; Fax: +1 847 8250001; E-mail: librarydirector@prpl.org; URL: www.parkridgelibrary.org
1910; Janet Van De Carr
213 000 vols; 420 curr per; 1 000 e-books; 7 000 microforms; 7 250 av-mat; 8 913 sound-rec; 2 570 digital data carriers   40710

Parkersburg & Wood County Public Library, Wood County Service Center, 3100 Emerson Ave, *Parkersburg*, WV 26104-2414
T: +1 304 4204587; Fax: +1 304 4204589; E-mail: info@park.lib.wv.us; URL: parkersburg.lib.wv.us
1905; Brian E. Raitz
4 branch libs, 1 bookmobile
170 000 vols; 224 curr per; 1 000 microforms; 12 419 av-mat; 26 811 sound-rec; 1 060 digital data carriers
libr loan   40711

Sayreville Public Library, 1050 Washington Rd, *Parlin*, NJ 08859
T: +1 732 7270212; Fax: +1 732 5530775; URL: www.lmxac.org/sayreville
1931; Joseph E. Lyons
97 000 vols; 196 curr per; 1 355 av-mat
                                             40712

Cuyahoga County Public Library, 2111 Snow Rd, *Parma*, OH 44134-2728
T: +1 216 7499512; Fax: +1 216 3981748; URL: www.cuyahogalibrary.org
1922; Sari Feldman
28 branch libs
2 333 000 vols; 8 502 curr per; 2 000 e-books; 126 110 av-mat
libr loan   40713

Cuyahoga County Public Library – Parma-Snow, 2121 Snow Rd, *Parma*, OH 44134-2728
T: +1 216 6614240; Fax: +1 216 6611019
Dianne Discenzo
58 000 vols; 3 545 av-mat   40714

Cuyahoga County Public Library – Parma-South Branch, 7335 Ridge Rd, *Parma*, OH 44129-6602
T: +1 440 8855362; Fax: +1 440 8842263
Dorothy Lettus
171 000 vols; 9 477 av-mat   40715

Cuyahoga County Public Library – Parma Heights Branch, 6206 Pearl Rd, *Parma Heights*, OH 44130-3045
T: +1 440 8842313; Fax: +1 440 8842713
Nicholas Cronin
52 000 vols; 4 192 av-mat   40716

Parsippany-Troy Hills Free Public Library, 449 Halsey Rd, *Parsippany*, NJ 07054
T: +1 973 8878907; Fax: +1 973 8870062; URL: www.parsippanylibrary.org
1968; Jayne Beline
2 branch libs
210 000 vols; 210 curr per
libr loan   40717

Parsons Public Library, 311 S 17th St, *Parsons*, KS 67357
T: +1 620 4215920; Fax: +1 620 4213951; E-mail: parsonslibrary@hotmail.com; URL: skyways.lib.ks.us/library/parsons/
1908; Dayna Williams-Capone
70 000 vols; 110 curr per   40718

Pasadena Public Library, 285 E Walnut St, *Pasadena*, CA 91101
T: +1 626 7444223; Fax: +1 626 5858396; URL: www.cityofpasadena.net
1882; Jan Sanders
9 branch libs
711 000 vols; 1 145 curr per; 156 000 microforms; 600 av-mat; 6 077 sound-rec
                                             40719

Pasadena Public Library, 1201 Jeff Ginn Memorial Dr, *Pasadena*, TX 77506-4895
T: +1 713 4770276; Fax: +1 713 4739640; URL: www.ci.pasadena.tx.us/library/home.html
1953; Sheila Ross Henderson
1 branch libr
324 000 vols; 1 103 curr per; 16 916 av-mat   40720

Pasadena Public Library – Hastings, 3325 E Orange Grove Blvd, *Pasadena*, CA 91107
T: +1 626 7447262; Fax: +1 626 4400222
Thelma Watson
64 000 vols   40721

Pasadena Public Library – La Pintoresca, 1355 N Raymond Ave, *Pasadena*, CA 91103
T: +1 626 7447268; URL: www.ci.pasadena.ca.us/library/lapintoresca.asp
Elizabeth Brooks
57 000 vols   40722

Pasadena Public Library – Lamanda Park, 140 S Altadena Dr, *Pasadena*, CA 91107
T: +1 626 7447266; URL: www.ci.pasadena.ca.us/library/lamanda.asp
Diane Walker
52 000 vols   40723

Jackson-George Regional Library System, 3214 Pascagoula St, *Pascagoula*, MS 39567
T: +1 228 7693060; Fax: +1 228 7693146; URL: www.jgrls.org
1940; Carol Hewlett
Genealogy Coll; 8 branch libs
289 000 vols; 936 curr per; 19 137 av-mat; 12 180 Large Print Bks, 10 080 Talking Bks   40724

Paso Robles Public Library, 1000 Spring St, *Paso Robles*, CA 93446-2207
T: +1 805 2373870; Fax: +1 805 2383665; URL: www.prcity.com
1903; Julie Dahlen
Paso Robles Newspapers from 1892, micro
56 000 vols; 120 curr per
libr loan   40725

Passaic Public Library, Julius Forstmann Library, 195 Gregory Ave, *Passaic*, NJ 07055
T: +1 973 7790474; Fax: +1 973 7790889; E-mail: circ@passaicpubliclibrary.org; URL: www.passaicpubliclibrary.org
1887; Kathleen Mollica
Passaic Hist, Jewish Studies, Cooking, Art; 1 branch libr, 1 bookmobile
110 000 vols; 365 curr per
libr loan   40726

Pataskala Public Library, 101 S Vine St, *Pataskala*, OH 43062
T: +1 740 9279986; Fax: +1 740 9646204
1937; Matthew J. Nojonen
Accords Coll
70 000 vols; 53 curr per   40727

Patchogue-Medford Library, 54-60 E Main St, *Patchogue*, NY 11772
T: +1 631 6544700; Fax: +1 631 2893999; E-mail: ptchlib@pmlib.org; URL: pmlib.org
1900; Geraldine Chrils
Law, Martial Arts (Maccarrone-Kresge Coll), Music, Opera (Sara Courant Coll)
300 000 vols; 36 450 curr per; 3 626 e-journals; 1 608 music scores; 40 000 microforms; 17 792 sound-rec; 1 015 digital data carriers; 531 High Interest/Low Vocabulary Bks
libr loan   40728

Paterson Free Public Library – Danforth Memorial Library, 250 Broadway, *Paterson*, NJ 07501
T: +1 973 3211223; Fax: +1 973 3211205; URL: www.palsplus.org/patersonpl
1885; Cynthia Czesak
Genealogy, Spanish; Arabic/Islamic Coll; 2 branch libs
183 000 vols; 340 curr per   40729

Paulding County Carnegie Library, 205 S Main St, *Paulding*, OH 45879-1492
T: +1 419 3992032; Fax: +1 419 3992114; URL: www.pauldingcountylibrary.com
1898; Susan Pieper
Paulding County Press (1859 to 1959)
80 000 vols; 140 curr per   40730

Paw Paw District Library, 609 W Michigan Ave, *Paw Paw*, MI 49079-1072
T: +1 269 6573800; Fax: +1 269 6572603; E-mail: ppdl49079@yahoo.com; URL: www.pawpaw.lib.mi.us
1920; John Mohney
Michigan Coll
67 000 vols; 142 curr per   40731

Pawtucket Public Library, 13 Summer St, *Pawtucket*, RI 02860
T: +1 401 7253714; Fax: +1 401 7282170; E-mail: pawlib@lori.ri.gov; URL: www.pawtucketlibrary.org
1852; Susan L. Reed
Polish Coll, bks, rec, per; Local & Rhode Island Hist; 1 bookmobile
176 000 vols; 210 curr per   40732

Payson Public Library, 328 N McLane Rd, *Payson*, AZ 85541
T: +1 928 4749260; Fax: +1 928 4742679
1923; Terry Morris
Southwest Coll, American Indians
85 000 vols; 90 curr per   40733

Peabody Institute Library, 82 Main St, *Peabody*, MA 01960-5592
T: +1 978 5310100; Fax: +1 978 5321797; URL: www.peabodylibrary.org
1852; Martha H. Holden
2 branch libs
163 000 vols; 331 curr per   40734

South Kingstown Public Library, Peace Dale Library, 1057 Kingstown Rd, *Peace Dale*, RI 02879-2434
T: +1 401 7834085; Fax: +1 401 7826370; E-mail: skiref@skpl.org; URL: skpl.org
1975; Shirley Long
Adult New Reader Coll, Bks on Tape Coll
68 000 vols; 274 curr per   40735

Flint River Regional Library – Peachtree City Library, 201 Willowbend Rd, *Peachtree City*, GA 30269
T: +1 770 6312520
Jill Prouty
83 000 vols   40736

Pearl Public Library, 2416 Old Brandon Rd, *Pearl*, MS 39208-4601
T: +1 601 9322562; Fax: +1 601 9323535; E-mail: pearl@cmrls.lib.ms.us
Cecelia Sandifer
53 000 vols; 120 curr per   40737

Hawaii State Public Library System – Pearl City Public Library, 1138 Waimano Home Rd, *Pearl City*, HI 96782
T: +1 808 4536566; Fax: +1 808 4536570
Floriana Cofman
162 000 vols; 107 curr per   40738

Pearl River Public Library, 80 Franklin Ave, *Pearl River*, NY 10965
T: +1 845 7354084; Fax: +1 845 7354041; E-mail: prlibrary@ucs.net; URL: www.rcls.org/pearlriverlibrary.org
1935; Carolyn E. Johnson
Art & architecture
88 000 vols; 296 curr per; 1624 av-mat   40739

Brazoria County Library System – Pearland Branch, 3522 Liberty Dr, *Pearland*, TX 77581
T: +1 281 4854876; Fax: +1 281 4855576; E-mail: pearland@bcls.lib.tx.us
Elisabeth Williams
114 000 vols   40740

Field Library of Peekskill, Four Nelson Ave, *Peekskill*, NY 10566-2138
T: +1 914 7371212; Fax: +1 914 7370714; E-mail: pek@westchesterlibraries.org; URL: www.peekskill.org
1887; Sibyl Canaan
Local Hist, Lincoln, Down Syndrome, Job Information
95 000 vols; 135 curr per; 155 maps; 3 000 av-mat; 6 000 sound-rec; 200 digital data carriers
libr loan   40741

Pekin Public Library, 301 S Fourth St, *Pekin*, IL 61554-4284
T: +1 309 3477111; Fax: +1 309 3476587; E-mail: library@pekin.net; URL: www.pekin.net/library
1896; Jeff Brooks
129 000 vols; 196 curr per
libr loan   40742

Pelham Public Library, 3160 Pelham Pkwy, *Pelham*, AL 35124; P.O. Box 1627, Pelham, AL 35124-5627
T: +1 205 6206418; Fax: +1 205 6206469; E-mail: library@pelhamonline.com; URL: www.pelhamlibrary.com
1975; Barbara Roberts
50 000 vols; 50 curr per   40743

Pell City Public Library, 1923 First Ave N, *Pell City*, AL 35125
T: +1 205 8841015; Fax: +1 205 8144798; URL: www.pc.lib.al.us
Danny Stewart
56 000 vols   40744

Pella Public Library, 603 Main St, *Pella*, IA 50219-1592
T: +1 641 6284268; Fax: +1 641 6281735; E-mail: pplill@cityofpella.com; URL: www.cityofpella.com/librarybody.htm
1903; Wendy Street
63 000 vols; 159 curr per; 2 980 av-mat   40745

Pemberville Public Library, 375 E Front St, *Pemberville*, OH 43450
T: +1 419 2874012; Fax: +1 419 2874620; URL: www.pembervillelibrary.org
1937
55 000 vols; 145 curr per   40746

Pembroke Public Library, 142 Center St, *Pembroke*, MA 02359-2613
T: +1 781 2936771; Fax: +1 781 2940742; URL: www.pembrokepubliclibrary.org
1878; Deborah Wall
69 000 vols; 117 curr per; 22 av-mat   40747

Pendleton Community Library, 595 E Water St, *Pendleton*, IN 46064-1070
T: +1 765 7787527; Fax: +1 765 7787529; URL: www.pendleton.lib.in.us
1912; Lynn Hobbs
53 000 vols; 175 curr per   40748

Pendleton Public Library, 502 SW Dorion, *Pendleton*, OR 97801
T: +1 541 9660380; Fax: +1 541 9660382; URL: www.pendleton.plinkit.org
1987
Northeast Oregon, Pacific Northwest, Rodeo & Western Lit
56 730 vols; 231 curr per   40749

Penfield Public Library, 1985 Baird Rd, *Penfield*, NY 14526
T: +1 585 3408720; Fax: +1 585 3408748; URL: www.penfieldlibrary.org
Mary E. Maley
114 000 vols; 1 216 curr per; 7 875 sound-rec
libr loan   40750

Penn Yan Public Library, 214 Main St, *Penn Yan*, NY 14527-1796
T: +1 315 5366114; Fax: +1 315 5360131; E-mail: lhovergaard@pypl.org; URL: www.pypl.org
1895; Lynn Overgaard
51 100 vols; 141 curr per   40751

Mercer County Library – Hopewell Township, 245 Pennington Titusville Rd, *Pennington*, NJ 08534
T: +1 609 7372610; Fax: +1 609 7377419
Andrea Merrick
59 000 vols   40752

Lonesome Pine Regional Library – Lee County Public, 406 Joslyn Ave, *Pennington Gap*, VA 24277
T: +1 276 5461141; Fax: +1 276 5465136; E-mail: lcplib@lprlibrary.org
Audrey Evans
65 000 vols; 66 curr per; 1 801 av-mat   40753

West Florida Regional Library, 200 W Gregory, *Pensacola*, FL 32502
T: +1 850 4365060; Fax: +1 850 4365039; URL: www.wfrl.lib.fl.us
1937; Eugene T. Fischer
6 branch libs
471 000 vols; 751 curr per; 29 000 e-books; 45 818 av-mat   40754

Peoria Public Library -, 107 NE Monroe St, *Peoria*, IL 61602-1070
T: +1 309 4972153; Fax: +1 309 6740116; URL: www.peoriapubliclibrary.org
1880; Edward Szynaka
5 branch libs
635 000 vols; 5 810 sound-rec; 1 026 digital data carriers   40755

Peoria Public Library – Lakeview, 1137 W Lake, *Peoria*, IL 61614-5935
T: +1 309 4972200; Fax: +1 309 4972211
Jennifer Sevier
80 000 vols; 1 343 av-mat   40756

Peotone Public Library District, 515 N First St, *Peotone*, IL 60468
T: +1 708 2583436; Fax: +1 708 2589796; E-mail: information@peotone.lib.il.us; URL: www.peotone.lib.il.us
Cynthia Cooper
65 000 vols; 100 curr per   40757

Lawrence Library, 15 Main St (Rte 113), mail address: *Pepperell*, MA 01463-1616
T: +1 978 4330330; Fax: +1 978 4330317; E-mail: dspratt@cwmars.org; URL: www.lawrencelibrary.org
1900; Debra Sprätt
Hist, Genealogy
54 940 vols; 130 curr per
libr loan   40758

Bucks County Free Library – Samuel Pierce Branch, 491 Arthur Ave, *Perkasie*, PA 18944-1033
T: +1 215 2579718; Fax: +1 215 2570759
Elizabeth Anderson
70 000 vols; 90 curr per   40759

Houston County Public Library System, 1201 Washington Ave, *Perry*, GA 31069
T: +1 478 9873050; Fax: +1 478 9874572; URL: www.houston.public.lib.ga.us
1974; Marsha Christy
3 branch libs
200 000 vols; 125 curr per   40760

Perry Public Library, 3753 Main St, *Perry*, OH 44081
T: +1 440 2593300; Fax: +1 440 2593977; E-mail: askus@perry.lib.oh.us; URL: www.perrypubliclibrary.org
1929; Virginia Sharp March
66 350 vols; 250 curr per   40761

Way Public Library, 101 E Indiana Ave, *Perrysburg*, OH 43551
T: +1 419 8743135; Fax: +1 419 8746129; E-mail: wayref@oplin.org; URL: www.way.lib.oh.us
1881; Nancy J. Kelley
101 000 vols; 222 curr per   40762

Riverside Regional Library – Perryville Branch, 800 City Park Dr, Ste A, *Perryville*, MO 63775
T: +1 573 5476508; Fax: +1 573 5473715
Julie Sauer
62 000 vols; 15 curr per; 103 av-mat   40763

Perth Amboy Free Public Library, 196 Jefferson St, *Perth Amboy*, NJ 08861
T: +1 732 8262600; Fax: +1 732 3248079; URL: www.lmxac.org; www.ci.perthamboy.nj.us
1903; Patricia Gandy
Large print bks, Spanish language; cassette coll, old microfilms
176 000 vols; 444 curr per
libr loan   40764

Peru Public Library, 102 E Main St, *Peru*, IN 46970-2338
T: +1 765 4733069; Fax: +1 765 4733060; E-mail: ppl@peru.lib.in.us; URL: www.peru.lib.in.us
1902; Charles A. Wagner
55 000 vols; 185 curr per; 1 500 av-mat   40765

Peru Public Library, 1409 11th St, *Peru*, IL 61354
T: +1 815 2230229; Fax: +1 815 2231559; URL: www.htls.lib.il.us/pub
1911; Shirley Sharpe
50 000 vols; 140 curr per   40766

Petersburg Public Library, 137 S Sycamore St, *Petersburg*, VA 23803
T: +1 804 7332387; Fax: +1 804 7337972; URL: www.ppls.org
1924; Wayne M. Crocker
Newspapers since 1800; 2 branch libs
153 000 vols; 229 curr per; 1 555 av-mat; 1 555 Talking Bks   40767

Pewaukee Public Library, 210 Main St, *Pewaukee*, WI 53072-3596
T: +1 262 6915670; Fax: +1 262 6915673; E-mail: pwlib@pewaukee.lib.wi.us; URL: www.pewaukeelibrary.org
1904; Jennie Stoltz
95 000 vols; 150 curr per   40768

Pharr Memorial Library, 121 E Cherokee St, *Pharr*, TX 78577-4826
T: +1 956 7873966; Fax: +1 956 7873345; E-mail: info@pharr.lib.tx.us; URL: www.pharr.lib.tx.us
1960; Jose A. Gamez
Large Print Coll, Spanish Coll (Spanish language mat), Texas Coll
85 000 vols; 70 curr per
libr loan   40769

Phenix City-Russell County Library, 1501 17th Ave, *Phenix City*, AL 36867
T: +1 334 2971139; Fax: +1 334 2988452; E-mail: phenixcitylibrary@gmail.com
1957; Martha Noyes
69 000 vols; 95 curr per; 2 000 av-mat; 300 sound-rec   40770

Free Library of Philadelphia, 1901 Vine St, *Philadelphia*, PA 19103-1189
T: +1 215 6865360; Fax: +1 215 5633628; URL: www.library.phila.gov
1891; Siobhan Reardon
74 branch libs
4 902 000 vols; 19 278 curr per; 1 256 398 av-mat; 1 108 930 sound-rec
libr loan   40771

Free Library of Philadelphia – Central Children's Department, 1901 Vine St, *Philadelphia*, PA 19103-1189
T: +1 215 6865369
Irene Wright
Bibliographies;Bibliographies for the Adult Researcher;Folklore Coll;Original Children's Book Illustrations & Manuscripts;Children's Books in 60 Different Languages;Kathrine McAlarney Coll of Illustrated Children's Books;Historical Bibliography: Books abo
52 000 vols; 57 curr per   40772

Free Library of Philadelphia – Interlibrary Loan, 1901 Vine St, *Philadelphia*, PA 19103-1189
T: +1 215 6865360; Fax: +1 215 5633628
Larry Richards
304 000 vols   40773

Free Library of Philadelphia – Lucien E. Blackwell West Philadelphia Regional, 125 S 52nd St, *Philadelphia*, PA 19139-3408
T: +1 215 6857424; Fax: +1 215 6857438
Roben Manker
77 000 vols; 181 curr per; 691 av-mat   40774

Free Library of Philadelphia – Northeast Regional, 2228 Cottman Ave, *Philadelphia*, PA 19149-1297
T: +1 215 6850500; Fax: +1 215 7423225
Kathryn Whitacre
103 000 vols; 210 curr per; 918 av-mat   40775

Free Library of Philadelphia – Social Science & History, 1901 Vine St, *Philadelphia*, PA 19103-1189
T: +1 215 6865396
Jim DeWalt
Confederate Imprints (Simon Gratz Coll);American Indian (Wilberforce Eames Coll);Rowing (Lewis H Kenney Coll);American Imprint Series, micro;Chess (Charles Willing Coll)
51 000 vols; 122 curr per   40776

Friends Free Library of Germantown, 5418 Germantown Ave, *Philadelphia*, PA 19144
T: +1 215 9512355; Fax: +1 215 9512697; URL: www.germantownfriends.org
1845; Katherine St Clair
Quaker hist, African-American studies, Natural science; Irvin C Poley Theatre Coll 1900-1975
64 000 vols; 198 curr per   40777

Phillipsburg Free Public Library, 200 Frost Ave, *Phillipsburg*, NJ 08865
T: +1 908 4543712; Fax: +1 908 8594667; URL: www.pburglib.org
1923; Patricia Lawson
Education
104 000 vols; 252 curr per
libr loan   40778

Maricopa County Library District, 2700 N Central Ave, Ste 700, *Phoenix*, AZ 85004
T: +1 602 6523000; Fax: +1 602 6523071; URL: www.mcldaz.org
1929; Harry R. Courtright
Arizona Hist (Southwest Coll); 13 branch libs
755 000 vols; 1 372 curr per; 9 e-books; 24 773 av-mat; 17 800 Large Print Bks   40779

Phoenix Public Library, Burton Barr Central Library, 1221 N Central Ave, *Phoenix*, AZ 85004-1820
T: +1 602 2624636; Fax: +1 602 2618836; URL: www.phxlib.org
1898; Toni Garvey
Arizona Hist Coll, Center for Children's Lit, Art of Bk Coll, Libr of American Civilization Coll, Rare Bk Room; 13 branch libs

1790 000 vols; 21 000 e-books; 3 100 Large Print Bks
libr loan   40780

Phoenixville Public Library, 183 Second Ave, *Phoenixville*, PA 19460-3420
T: +1 610 9333013; Fax: +1 610 9334338; URL: www.phoenixvillelibrary.org
1896; John Kelley
70 000 vols; 120 curr per   40781

Pearl River County Library System – Margaret Reed Crosby Memorial Library, 900 Goodyear Blvd, *Picayune*, MS 39466
T: +1 601 7985081; Fax: +1 601 7985082; URL: www.pearlriver.lib.ms.us
1926; Linda Tufaro
1 branch libr
90 000 vols; 70 curr per   40782

Pickerington Public Library, 201 Opportunity Way, *Pickerington*, OH 43147-1296
T: +1 614 8331004; Fax: +1 614 8378425; URL: www.pickerington.lib.oh.us
1909; Suellen Goldsberry
103 000 vols; 296 curr per   40783

County of Los Angeles Public Library – Pico Rivera Library, 9001 Mines Ave, *Pico Rivera*, CA 90660-3098
T: +1 562 9427394; Fax: +1 562 9427779; URL: www.colapublib.org/libs/picorivera
Rosemary Gurrola
63 000 vols; 76 curr per   40784

Rawlins Municipal Library, 1000 E Church St, *Pierre*, SD 57501
T: +1 605 7737421; Fax: +1 605 7737423; URL: rpllib.sdln.net
1905; Beverly Lewis
Hist (South Dakota Coll), bks, pamphlets, pictures
79 000 vols; 120 curr per   40785

South Dakota Braille & Talking Book Library, McKay Bldg, 800 Governors Dr, *Pierre*, SD 57501-2294
T: +1 605 7733131; Fax: +1 605 7736962; E-mail: library@state.sd.us; URL: www.sdstatelibrary.com
1968; Dorothy Liegl
Dakota Language Coll, North Dakota & South Dakota Coll
179 000 vols   40786

Pike County Public Library District, 119 College St, Ste 3, *Pikeville*, KY 41501-1787; P.O. Box 1197, Pikeville, KY 41502-1197
T: +1 606 4329977; Fax: +1 606 4329908; E-mail: pcpldao@pikelibrary.org; URL: www.informationplace.org
1970; Leean L. Allen
5 branch libs
168 000 vols; 296 curr per   40787

Charles H. Stone Memorial Library, 319 W Main St, *Pilot Mountain*, NC 27041; P.O. Box 10, Pilot Mountain, NC 27041-0010
T: +1 336 3682370; Fax: +1 336 3689587; E-mail: pilotmtlibrarian@yahoo.com; URL: www.nwrl.org
Heather Elliott
70 000 vols; 75 curr per   40788

Arkansas Department of Correction Library System, 6814 Princeton Pike, *Pine Bluff*, AR 71611; P.O. Box 8707, Pine Bluff
T: +1 870 2676277; Fax: +1 870 2676363
Dennice B. Alexander
250 000 vols   40789

Pine Bluff & Jefferson County Library System, Main Library, 200 E Eighth Ave, *Pine Bluff*, AR 71601
T: +1 870 5344818; Fax: +1 870 5348707; E-mail: pbjc-lib@pbjc-lib.state.ar.us; URL: pbjc-lib.state.ar.us; www.pbjclibrary.org
1913; Dave Burdick
Genealogy Coll, Arkansas Coll; 4 branch libs
97 000 vols; 145 curr per; 4 600 av-mat; 4 600 sound-rec; 3 500 Large Print Bks
                                      40790

Kitchigami Regional Library, 310 Second St N, *Pine River*, MN 56474; P.O. Box 84, Pine River
T: +1 218 5872171; Fax: +1 218 5874855; URL: krls.org
1969; Marian Ridge
American Indians
300 000 vols; 525 curr per; 7 307 av-mat; 10 700 Large Print Bks, 7 310 Talking Bks
                                      40791

Sublette County Library, 155 S Tyler Ave, *Pinedale*, WY 82941; P.O. Box 489, Pinedale, WY 82941-0489
T: +1 307 3674115;
Fax: +1 307 3676722; URL: www.sublettecountylibrary.org; pinedaleonline.com/library
1967; Daphne Platts
Film Bks, Rocky Mountain Fur Trade, photos
66 000 vols; 236 curr per; 2 200 av-mat
                                      40792

Pinellas Park Public Library, 7770 52nd St, *Pinellas Park*, FL 33781-3498
T: +1 727 5410714; Fax: +1 727 5410818; URL: pppl.tblc.lib.fl.us
1948; Barbara S. Ponce
111 000 vols; 315 curr per; 37 000 e-books
libr loan   40793

Pineville-Bell County Public Library, 214 Walnut St, P.O. Box 1490, *Pineville*, KY 40977-1490
T: +1 606 3373422; Fax: +1 606 3379862; E-mail: director@bellcolib.org; URL: www.bellcountypubliclibraries.org
1933; Beverly Greene
80 830 vols; 164 curr per   40794

Wyoming County Public Library, Castle Rock Ave, *Pineville*, WV 24874; P.O. Box 130, Pineville, WV 24874-0130
T: +1 304 7326228; URL: wyoming.lib.wv.us
1966; Carolyn M. Gaddis
65 000 vols; 95 curr per   40795

Flesh Public Library, 124 W Greene St, *Piqua*, OH 45356-2399
T: +1 937 7736753; Fax: +1 937 7735981; URL: www.piqua.lib.oh.us
1890; James C. Oda
136 000 vols; 165 curr per   40796

Piscataway Township Free Public Library, John F. Kennedy Memorial Library, 500 Hoes Lane, *Piscataway*, NJ 08854
T: +1 732 4633159; Fax: +1 732 4639022; URL: www.piscatawaylibrary.org
1961; Anne Roman
Chinese language; Art Bks (Silkotch Coll), Gujarti Coll, Hindi Coll; 1 branch libr, 1 bookmobile
156 000 vols; 308 curr per   40797

Pittsburg Public Library, 308 N Walnut, *Pittsburg*, KS 66762-4732
T: +1 620 2318110; Fax: +1 620 2322258; URL: pittsburgpubliclibrary.org
1902; Pat Clement
Medicine, Religion; Crawford County Genealogical Society Coll
62 000 vols; 214 curr per
libr loan   40798

Carnegie Library of Pittsburgh – Downtown & Business, 612 Smithfield St, *Pittsburgh*, PA 15222-2506
T: +1 412 2815945; URL: www.carnegielibrary.org/locations/downtown
Karen Rossi
Industrial & Trade Directories;Pittsburgh Company Index
61 000 vols; 145 curr per   40799

Carnegie Library of Pittsburgh – East Liberty, 130 S Whitfield St, *Pittsburgh*, PA 15206-3408
T: +1 412 3638232; URL: www.carnegielibrary.org/locations/eastliberty
Charmaine Mozlack
77 000 vols; 97 curr per   40800

Carnegie Library of Pittsburgh – Squirrel Hill, 5801 Forbes Ave, *Pittsburgh*, PA 15217-1601
T: +1 412 4229650; URL: www.carnegielibrary.org/locations/squirrelhill
Holly McCullough
Jewish History & Culture (Olender Foundation)
107 000 vols; 147 curr per   40801

Lauri Ann West Memorial Library, 1220 Powers Run Rd, *Pittsburgh*, PA 15238-2618
T: +1 412 8289520; Fax: +1 412 8284960; E-mail: lawest@einetwork.net; URL: www.lauriannwestlibrary.org
Susan Holmes
81 000 vols; 237 curr per   40802

Mt Lebanon Public Library, 16 Castle Shannon Blvd, *Pittsburgh*, PA 15228-2252
T: +1 412 5311912; Fax: +1 412 5311161; E-mail: mtlebanon@einetwork.net; URL: www.mtlebanonlibrary.org
1932; Cynthia K. Richey
Small business, Study guides; Pennsylvania Coll
158 000 vols; 303 curr per   40803

Northland Public Library, 300 Cumberland Rd, *Pittsburgh*, PA 15237-5455
T: +1 412 3668100; Fax: +1 412 3662064; E-mail: northland@einetwork.net; URL: www.northlandlibrary.org
1968; Sandra A. Collins
167 000 vols; 369 curr per; 15 maps; 6 700 av-mat; 22 951 sound-rec; 2 940 digital data carriers; 2 130 Large Print Bks
libr loan   40804

Upper Saint Clair Township Library, Upper Saint Clair, 1820 McLaughlin Run Rd, *Pittsburgh*, PA 15241-2397
T: +1 412 8355540; Fax: +1 412 8356763; URL: www.twpusc.org/library
1957
81 270 vols; 193 curr per
libr loan   40805

Whitehall Public Library, 100 Borough Park Dr, *Pittsburgh*, PA 15236-2098
T: +1 412 8826622; Fax: +1 412 8829556; E-mail: whitehall@einetwork.net; URL: www.whitehallpubliclibrary.org
1963; Robyn Hammer-Clarey
Large Print Coll, Toddler Board & Cloth Bks Coll, Bks on tapes, College Career Info, Parenting Coll, Business Coll
62 000 vols; 100 curr per   40806

Wilkinsburg Public Library, 605 Ross Ave, *Pittsburgh*, PA 15221-2195
T: +1 412 2442940; Fax: +1 412 2436943; URL: www.wilkinsburglibrary.org
1899
Mysteries, Biogr
82 000 vols; 140 curr per   40807

Berkshire Athenaeum, Pittsfield Public Library, One Wendell Ave, *Pittsfield*, MA 01201-6385
T: +1 413 4999480; Fax: +1 413 4999489; E-mail: pittsref@cwmars.org; URL: www.pittsfieldlibrary.org
1871; Ronald B. Latham
Hermann Melville Memorial Room, Berkshire Authors Room, Morgan Ballet Coll
220 000 vols; 1 233 curr per   40808

Pittsford Community Library, 24 State St, *Pittsford*, NY 14534
T: +1 585 2486275; Fax: +1 585 2486259; URL: www.rochester.lib.ny.us/pittsford
1920; Marjorie Shelly
91 000 vols   40809

Placentia Library District, 411 E Chapman Ave, *Placentia*, CA 92870
T: +1 714 5281906; Fax: +1 714 5288236; E-mail: reference@placentialibrary.org; URL: www.placentialibrary.org
1919; Jeanette Contreras
California Coll, Arabic Language Coll
112 000 vols; 141 curr per   40810

El Dorado County Library, 345 Fair Lane, *Placerville*, CA 95667
T: +1 530 6215540; Fax: +1 530 6223911; E-mail: lib-pl@eldoradolibrary.org; URL: www.eldoradolibrary.org
1948; Jeanne Amos
California; 5 branch libs
350 000 vols; 263 curr per; 100 art reproductions   40811

Plain City Public Library, 305 W Main St, *Plain City*, OH 43064-1148
T: +1 614 8738364; Fax: +1 614 8738364; E-mail: clong@plaincitylib.org; URL: www.plain.lib.oh.us
1946; Chris Long
54 900 vols; 114 curr per   40812

Plainfield Public Library, 800 Park Ave, *Plainfield*, NJ 07060-2594
T: +1 908 7572305; Fax: +1 908 7540063; URL: www.plainfieldlibrary.info
1881; Joseph Hugh DaRold
Hist Architectural Plans of Plainfield area, Photogr Coll
190 000 vols; 275 curr per   40813

Plainfield Public Library District, 15025 S Illinois St, *Plainfield*, IL 60544
T: +1 815 4366639; Fax: +1 815 4392878; E-mail: jmmilavec@plainfield.lib.il.us; URL: plainfield.lib.il.us
Julie Milavec
73 850 vols; 173 curr per   40814

Plainsboro Free Public Library, 641 Plainsboro Rd, *Plainsboro*, NJ 08536-0278
T: +1 609 2752897; Fax: +1 609 7995883; E-mail: plibrary@lmxac.org; URL: www.lmxac.org/plainsboro
1964; Virginia Baeckler
Chinese Cultural Exchange, German Coll, JFK Coll, Large Print Coll, Hindi Coll, Gujarati Coll
76 000 vols; 199 curr per   40815

Plainview-Old Bethpage Public Library, 999 Old Country Rd, *Plainview*, NY 11803-4995
T: +1 516 9380077; Fax: +1 516 4334645; E-mail: website@poblib.org; URL: www.poblib.org
1955; Gretchen Brown
Career, Career-Job Learning, Consumer Info, Law, LI Jewish Genealogical Society Coll, New & Used Automobiles, Plainview Old Bethpage Coll
194 000 vols; 3 235 curr per; 168 maps; 15 302 av-mat; 12 340 sound-rec; 1 440 digital data carriers; 160 High Interest/Low Vocabulary Bks, 60 Bks on Deafness & Sign Lang, 4 200 Large Print Bks   40816

Unger Memorial Library, 825 Austin St, *Plainview*, TX 79072-7235
T: +1 806 2961148; Fax: +1 806 2911245; E-mail: johnsigwald@texasonline.net; URL: unger.myplainview.com/index.htm
1960; John Sigwald
Plainview & Hale County Hist
50 000 vols; 110 curr per   40817

Plainville Public Library, 56 E Main St, *Plainville*, CT 06062
T: +1 860 7931446; Fax: +1 860 7932241; URL: www.plainvillelibrary.org
1894; Peter F. Chase
94 000 vols; 164 curr per   40818

Charles A. Ransom District Library, 180 S Sherwood Ave, *Plainwell*, MI 49080-1896
T: +1 269 6858024; Fax: +1 269 6852266; E-mail: info@ransomlibrary.org; URL: www.ransomlibrary.org
1868; Katie Bell Moore
Adult education
55 000 vols; 79 curr per   40819

Plano Public Library System, 2501 Coit Rd, *Plano*, TX 75075
T: +1 972 7694208; Fax: +1 972 7694269; URL: www.planolibrary.org
1965; Joyce Baumbach
6 branch libs
714 000 vols; 1 026 curr per; 6 349 digital data carriers   40820

Bruton Memorial Library, 302 McLendon St, *Plant City*, FL 33563
T: +1 813 7579215; Fax: +1 813 7579217; URL: www.hcplc.org/hcplc/bru/
1960; Anne T. Haywood
Civil War Coll; Florida Hist Coll; Florida Shelf Coll
118 000 vols; 175 curr per; 989 sound-rec; 140 digital data carriers
libr loan   40821

Helen B. Hoffman Plantation Library,
501 N Fig Tree Lane, *Plantation*,
FL 33317
T: +1 954 7972140; Fax: +1 954
7972767; E-mail: library@plantation.org;
URL: www.plantation.org
1963; Dee Anne Merritt
Art, Architecture, Sculpture; Florida Coll,
Large Print Coll
70 000 vols; 600 av-mat; 900 sound-rec;
125 digital data carriers
ALA   40822

Iberville Parish Library, 24605 J
Gerald Berret Blvd, *Plaquemine*,
LA 70764; P.O. Box 736, Plaquemine,
LA 70765-0736
T: +1 225 6874397; Fax: +1
225 6879719; E-mail:
admin.c1il@pelican.state.lib.la.us; URL:
www.iberville.lib.la.us
1951; Dannie J. Ball
7 branch libs
175 000 vols; 594 curr per; 7 685 av-mat;
1 630 sound-rec; 2880 Large Print Bks
                                       40823

**Clinton-Essex-Franklin Library
System**, 33 Oak St, *Plattsburgh*,
NY 12901-2810
T: +1 518 5635190; Fax: +1 518
5630421; URL: www.cefls.org
1954; Ewa Jankowska
68 000 vols; 71 curr per; 1 790 av-mat
libr loan   40824

Plattsburgh Public Library, 19 Oak St,
Plattsburgh, NY 12901-2810
T: +1 518 5630921; Fax: +1
518 5631681; E-mail:
sransom@cityofplattsburgh-ny.gov; URL:
www.plattsburghlib.org
1894; Stanley A. Ransom
Cook bks; Hist (Clinton County &
Plattsburgh)
99 500 vols
libr loan   40825

Pleasant Grove Public Library,
30 E Center St, *Pleasant Grove*,
UT 84062-2234
T: +1 801 7853950; Fax: +1 801
7859734; URL: www.plgrove.org/library/
index.html
April H. Harrison
85 000 vols; 50 curr per   40826

Contra Costa County Library,
1750 Oak Park Blvd, *Pleasant Hill*,
CA 94523-4497
T: +1 925 9273224; Fax: +1 925
6466461; E-mail: libadmin@ccclib.org;
URL: www.ccclib.org
1913; Anne Cain
Vincent Davi Coll (Food Technology),
Contra Costa Hist; 23 branch libs
1 293 000 vols; 2 234 curr per
libr loan   40827

**Contra Costa County Library –
Pleasant Hill Library**, 1750 Oak Park
Blvd, *Pleasant Hill*, CA 94523-4497
T: +1 925 6466434; Fax: +1 925
6466040
Debbie Tyler
196 000 vols   40828

Pleasanton Public Library, 400 Old
Bernal Ave, *Pleasanton*, CA 94566-7012
T: +1 925 9313400; Fax: +1 925
8468517; URL: www.ci.pleasanton.ca.us/
library.html
1999; Julie Farnsworth
168 000 vols; 329 curr per   40829

Mount Pleasant Public Library,
350 Bedford Rd, *Pleasantville*,
NY 10570-3099
T: +1 914 7690548; Fax: +1
914 7696149; E-mail:
info@mountpleasantlibrary.org; URL:
www.mountpleasantlibrary.org
1893; John Fearon
Art, Humanities; Lachenbruch Coll
(Children's Bks for Storytellers); 1 branch
libr
118 000 vols; 150 curr per   40830

Pettigrew Regional Library, 201 E Third
St, *Plymouth*, NC 27962; P.O. Box
906, Plymouth
T: +1 252 7932875; Fax: +1
252 7932818; E-mail:
headquarters@pettigrewlibraries.org; URL:
www.pettigrewlibraries.org
1955; Kay E. Davis

119 000 vols; 229 curr per; 4 652 av-mat
                                       40831

Plymouth District Library, 223 S Main
St, *Plymouth*, MI 48170-1687
T: +1 734 4530750; Fax: +1 734
4530733; E-mail: info@plymouthlibrary.org;
URL: www.plymouthlibrary.org
1923; Patricia Thomas
180 000 vols; 400 curr per   40832

Plymouth Public Library, 201 N Center
St, *Plymouth*, IN 46563
T: +1 574 9362324; Fax: +1 574
9367423; E-mail: info@plymouth.lib.in.us;
URL: www.plymouth.lib.in.us
1910; Susie Reinholt
113 000 vols; 251 curr per; 3 100 av-mat
                                       40833

Plymouth Public Library, 130 Division
St, *Plymouth*, WI 53073-1802
T: +1 920 8924416; Fax: +1 920
8926295
1909; Martha Suhfras
71 000 vols; 180 curr per; 7 000 e-books;
500 av-mat
libr loan; ALA, WLA   40834

Plymouth Public Library, 132 South St,
Plymouth, MA 02360-3309
T: +1 508 8304250;
Fax: +1 508 8304258; URL:
www.plymouthpubliclibrary.org
1811; Dinah L. O'Brien
Plymouth Coll, Irish Coll; 1 branch libr
136 000 vols; 294 curr per; 2 e-books
                                       40835

Marshall Public Library, 113 S Garfield,
Pocatello, ID 83204-5722
T: +1 208 2321263; Fax: +1 208
2329266; URL: www.marshallpl.org
1905; Mike Doellman
Idaho Coll
152 000 vols; 198 curr per   40836

Mason County Library System,
508 Viand St, *Point Pleasant*,
WV 25550-1199
T: +1 304 6750894; Fax: +1 304
6750895; URL: masoncounty.lib.wv.us
1930; Debbie Hopson
52 000 vols; 50 curr per; 129 microforms;
105 sound-rec   40837

**Public Library of Youngstown &
Mahoning County – Poland Branch**,
311 S Main St, *Poland*, OH 44514
T: +1 330 7571852; Fax: +1 330
7571570
Kathleen Austrino
70 000 vols   40838

Meigs County District Public Library,
216 W Main, *Pomeroy*, OH 45769-1032
T: +1 740 9925813; Fax: +1 740
9926140; E-mail: meigscty@oplin.org;
URL: www.meigs.lib.oh.us
1881; Kristi Eblin
3 branch libs
110 000 vols; 300 curr per   40839

Pomona Public Library, 625 S Garey
Ave, *Pomona*, CA 91766-3322; P.O.
Box 2271, Pomona, CA 91769-2271
T: +1 909 6202043; Fax: +1 909
6203713; E-mail: library@ci.pomona.ca.us;
URL: www.pomonalibrary.org;
www.youseemore.com/pomona
1887; Gregory Shapton
Orange Crate Labels, Postcard & Photo
coll, Laura Ingalls Wilder Coll
286 000 vols; 211 curr per; 95 000
microforms; 283 av-mat; 7 320 sound-rec;
76 digital data carriers; 3 819 Large Print
Bks
libr loan   40840

**Pequannock Township Public
Library**, 477 Newark Pompton Turnpike,
Pompton Plains, NJ 07444
T: +1 973 8357460; Fax: +1 973
8351928; URL: www.pequannocklibrary.org
1962; Rosemary Garwood
Landsberger Holocaust Coll
85 000 vols; 200 curr per; 3 542 av-mat
libr loan   40841

Ponca City Library, 515 E Grand,
Ponca City, OK 74601
T: +1 580 7670345; Fax: +1 580
7670377; URL: www.poncacity.lib.ok.us
1904; Holly La Bossiere
Oriental & 20th C Western Paintings
(Matzene Art Coll); Genealogy dept
75 000 vols; 264 curr per; 2 368 av-mat;
2 628 sound-rec; 40 digital data carriers
libr loan   40842

Pontiac Public Library, 60 E Pike St,
Pontiac, MI 48342
T: +1 248 7583942; Fax: +1 248
7583990; E-mail: pont@tln.lib.mi.us; URL:
www.pontiac.lib.mi.us
Stephanie McCoy
Hist
90 000 vols; 150 curr per   40843

Dixie Regional Library System, 111 N
Main St, *Pontotoc*, MS 38863-2103
T: +1 662 4893960; Fax: +1 662
4897777; E-mail: pclib@dixie.lib.ms.us;
URL: www.dixie.lib.ms.us
1961; Judy McNeece
8 branch libs
133 000 vols; 143 curr per   40844

Poquoson Public Library, 500 City Hall
Ave, *Poquoson*, VA 23662-1996
T: +1 757 8683060; Fax: +1 757
8683106; E-mail: library@poquoson-
va.gov; URL: www.poquoson-va.gov/library
1976; Elizabeth L. Tai
52 000 vols; 174 curr per; 1 409 av-mat
                                       40845

West Baton Rouge Parish Library,
830 N Alexander Ave, *Port Allen*,
LA 70767-2327
T: +1 225 3427920; Fax: +1
225 3427918; E-mail:
admin.c1we@pelican.state.lib.la.us; URL:
www.wbr.lib.la.us
1965; Beth Vandersteen
Louisiana Mat
84 000 vols; 180 curr per; 1 400 av-mat
                                       40846

North Olympic Library System,
2210 S Peabody St, *Port Angeles*,
WA 98362-6536
T: +1 360 4178501; Fax: +1 360
4572690; URL: www.nols.org
1919; Paula Simpson Barnes
Clallam County Hist, Pacific Northwest
Hist; 4 branch libs
225 000 vols; 987 curr per; 7 569
sound-rec; 152 Bks on Deafness &
Sign Lang
libr loan   40847

**North Olympic Library System – Port
Angeles Branch**, 2210 S Peabody St,
Port Angeles, WA 98362
T: +1 360 4178500; Fax: +1 360
4572581
Sandra J. Hill
143 000 vols   40848

Port Arthur Public Library, 4615 Ninth
Ave, *Port Arthur*, TX 77642
T: +1 409 9858838; Fax: +1 409
9855969; URL: www.pap.lib.tx.us
1918; Raymond D. Cline
Port Arthur Hist Archs
140 000 vols; 368 curr per; 2 000 av-mat;
1 000 sound-rec; 4 000 Large Print Bks
                                       40849

Charlotte County Library System, 2050
Forrest Nelson Blvd, *Port Charlotte*,
FL 33952
T: +1 941 6133200; Fax: +1 941
6133196; URL: www.charlottecountyfl.com/
library
1963; Angie Patteson
Large Print Bk Coll; 5 branch libs
205 000 vols; 615 curr per; 29 000 e-
books; 11 800 Audio Bks   40850

Ida Rupp Public Library, 310 Madison
St, *Port Clinton*, OH 43452
T: +1 419 7323212; E-mail: idarupp@
oplin.lib.oh.us; URL: www.idarupp.lib.oh.us
1913; James Crawford
Bataan Memorial Coll
90 000 vols; 254 curr per   40851

**Jefferson County Rural Library
District**, 620 Cedar Ave, *Port Hadlock*,
WA 98339-9514
T: +1 360 3856544; Fax: +1 360
3857921; URL: www.jclibrary.info
1980; Raymond Serebrin
54 000 vols; 120 curr per; 2 200 av-mat
                                       40852

**Ventura County Library – Ray D.
Prueter Library**, 510 Park Ave, *Port
Hueneme*, CA 93041
Cathy Thomason
74 000 vols   40853

Saint Clair County Library System,
210 McMorran Blvd, *Port Huron*,
MI 48060-4098
T: +1 810 9877323; Fax: +1 810
9877874; URL: www.sccl.lib.mi.us
1917; James F. Warwick
Can & Prov, Michigan & the Great
Lakes (W. L. Jenks Hist Coll); 13 branch
libs
433 000 vols; 560 curr per   40854

**Saint Clair County Library System –
Main Library**, 210 McMorran Blvd, *Port
Huron*, MI 48060
T: +1 810 9877323; Fax: +1 810
9877874
James F. Warwick
158 000 vols   40855

Port Jefferson Free Library, 100
Thompson St, *Port Jefferson*,
NY 11777-1897
T: +1 631 4730631; Fax: +1 631
4734765; URL: pjfl.suffolk.lib.ny.us
1908; Estherine Bonanno
146 000 vols; 251 curr per; 17 000
av-mat; 2 432 digital data carriers; 350
High Interest/Low Vocabulary Bks, 12
Special Interest Per Sub, 2 500 Large
Print Bks   40856

Comsewogue Public Library, 170
Terryville Rd, *Port Jefferson Station*,
NY 11776
T: +1 631 9281212; Fax: +1 631
9286307; URL: cpl.suffolk.lib.ny.us
1966; Brandon Pantorno
Video Cassette Titles
166 000 vols; 670 curr per; 10 000
av-mat; 7 500 sound-rec
libr loan; ALA   40857

Calhoun County Library, 200 W Mahan,
Port Lavaca, TX 77979
T: +1 361 5527323; Fax: +1 361
5524926; E-mail: staff@cclibrary.org;
URL: www.cclibrary.org
1962; Noemi Cruz
56 000 vols; 122 curr per
libr loan   40858

Effie & Wilton Hebert Public Library,
2025 Merriman, *Port Neches*, TX 77651
T: +1 409 7224554; Fax: +1 409
7194296; URL: www.ptn.lib.tx.us
1934; Coleen Molden
60 000 vols; 105 curr per   40859

**Kitsap Regional Library – Port
Orchard Branch**, 87 Sidney Ave, *Port
Orchard*, WA 98366-5249
T: +1 360 8762224; Fax: +1 360
8769588
Kathleen Wilson
61 000 vols   40860

W. J. Niederkorn Library, Port
Washington Public, 316 W Grand Ave,
Port Washington, WI 53074-2293
T: +1 262 2845031;
Fax: +1 262 2847680; URL:
www.portwashington.lib.wi.us; www.ci.port-
washington.wi.us/library/library.htm
David Nimmer
55 000 vols; 202 curr per; 8 000 e-books;
2 417 av-mat   40861

Port Washington Public Library,
One Library Dr, *Port Washington*,
NY 11050
T: +1 516 8834400; Fax: +1 516
8837927; E-mail: library@pwpl.org; URL:
www.pwpl.org
1892; Nancy Curtin
Sinclair Lewis, Long Island Coll, Robert
Hamilton Ball Theatre Coll, Nautical Ctr
Coll
153 000 vols; 915 curr per
libr loan; ALA, NYLA, NCLA   40862

Portage District Library, 300 Library
Lane, *Portage*, MI 49002
T: +1 269 3294544; Fax: +1 269
3249222; E-mail: info@portagelibrary.info;
URL: www.portagelibrary.info
1962; Christine A. Berro
John Todd Coll, aerial photos
136 000 vols; 211 curr per
libr loan   40863

Portage Public Library, 253 W Edgewater St, *Portage*, WI 53901
T: +1 608 7424959; Fax: +1 608 7423819; E-mail: porill@scls.lib.wi.us; URL: www.scls.lib.wi.us/portage
1902; Hans W. Jensen
Zona Gale Coll
63 000 vols; 149 curr per; 8 000 e-books
40864

Porter County Public Library System – Portage Public, 2665 Irving St, *Portage*, IN 46368-3504
T: +1 219 7631508; Fax: +1 219 7620101; E-mail: porref@pepls.lib.in.us
Nancy Clark
84 000 vols
40865

New Madrid County Library, 309 E Main St, *Portageville*, MO 63873
T: +1 573 3793583; Fax: +1 573 3799220
1948; Tom Sadler
3 branch libs
86 000 vols; 52 curr per
40866

Porterville Public Library, 41 W Thurman, *Porterville*, CA 93257-3652
T: +1 559 7840177; Fax: +1 559 7814396; URL: www.sjvls.org; www.ci.porterville.ca.us
1904; N.N.
70 000 vols; 151 curr per
40867

Cedar Mill Community Library, 12505 NW Cornell Rd, *Portland*, OR 97229
T: +1 503 6440043; Fax: +1 503 6443964; E-mail: askuscml@wccls.org; URL: www.cedarmill.org/library
1975; Peter Leonard
Oregon & Pacific Northwest, bks, per; Parent-Teacher Res Coll
184 000 vols; 350 curr per; 120 e-books
40868

Jay County Public Library, 315 N Ship St, *Portland*, IN 47371
T: +1 260 7267890; Fax: +1 260 7267317; URL: www.jaycpl.lib.in.us
1898; Rosalie Clamme
81 000 vols; 220 curr per; 287 microforms; 9 000 av-mat; 3 951 sound-rec; 4 digital data carriers; 64 art reproductions
libr loan
40869

Multnomah County Library, 205 NE Russell St, *Portland*, OR 97212-3796
T: +1 503 9885402; Fax: +1 503 9885441; URL: www.multcolib.org
1864; N.N.
Oregon Coll, Roses, McCormack Coll, John Wilson Room; 17 branch libs
1 697 000 vols
40870

Portland Library, 20 Freestone Ave, *Portland*, CT 06480
T: +1 860 3426770; Fax: +1 860 3426778; URL: www.portlandlibraryct.org
1895; Janet Nocek
60 000 vols; 256 curr per
40871

Portland Public Library, Five Monument Sq, *Portland*, ME 04101
T: +1 207 8711700; Fax: +1 207 8711703; E-mail: admin@portland.lib.me.us; URL: www.portlandlibrary.com
1867; Stephen J. Podgajny
Art, Consumer health, Ireland; Hugh Thomson Coll of Antique Children's Bks, Jacob Abbott, Jewish Oral Hist Program, Press Bks, State of Maine; 5 branch libs
311 000 vols; 1 700 curr per
libr loan
40872

Portsmouth Free Public Library, 2658 E Main Rd, *Portsmouth*, RI 02871
T: +1 401 6839457; Fax: +1 401 6835013; E-mail: porlib@yahoo.com; URL: www.portsmouthlibrary.org
1898; Carolyn B. Magnus
70 000 vols; 115 curr per
40873

Portsmouth Public Library, 1220 Gallia St, *Portsmouth*, OH 45662-4185
T: +1 740 3545688; Fax: +1 740 3531249; URL: www.portsmouth.lib.oh.us
1878; Beverly L. Cain
Northwest Territories Coll; 5 branch libs
214 000 vols; 390 curr per
40874

Portsmouth Public Library, 601 Court St, *Portsmouth*, VA 23704-3604
T: +1 757 3938973; URL: www.portsmouthpubliclibrary.org
1914; Susan H. Burton
Lighthouses & Lightships; 3 branch libs
339 000 vols; 460 curr per; 5 e-books; 13 000 av-mat; 6 500 sound-rec; 5 000 digital data carriers; 7 910 Large Print Bks
libr loan
40875

Portsmouth Public Library, 175 Parrott Ave, *Portsmouth*, NH 03801-4452
T: +1 603 7661720; Fax: +1 603 4330981; E-mail: info@lib.cityofportsmouth.com; URL: www.cityofportsmouth.com/library
Susan F. McCann
110 000 vols; 406 curr per
40876

Portsmouth Public Library – Churchland, 3215 Academy Ave, *Portsmouth*, VA 23703
T: +1 757 6862538
Clint Rudy
58 000 vols; 46 curr per; 1 707 av-mat
40877

Washington County Library, 235 E High St, *Potosi*, MO 63664
T: +1 573 4384691; Fax: +1 573 4386423; E-mail: washingtonlibrary@hotmail.com; URL: washington.mogenweb.org/wclib.html
1948; Dorothy A. Lore
Hist Coll, bks & micro
67 000 vols; 138 curr per
40878

Potsdam Public Library & Reading Center, Civic Center, Two Park St, *Potsdam*, NY 13676
T: +1 315 2657230; Fax: +1 315 2680306; URL: www.northnet.org/potsdamlib
1896; Patricia Musante
Employment Info & Micro-Enterprize Coll
60 000 vols; 141 curr per
40879

Pottstown Public Library, 500 High St, *Pottstown*, PA 19464-5656
T: +1 610 9706551; Fax: +1 610 9706553; E-mail: pottstownlibrary@mclinc.org; URL: www.ppl.mclinc.org
1921; Lynn Burkholder
Limerick Nuclear Power Plant
64 000 vols; 78 curr per
40880

Pottsville Free Public Library, 215 W Market St, *Pottsville*, PA 17901-4304
T: +1 570 6228880; Fax: +1 570 6222157; E-mail: potpublib@iu29.org; URL: www.pottsvillelibrary.org
1911; Nancy Smink
Anthracite Coll, Lincoln Coll, Molly Maguires Coll
110 000 vols; 200 curr per; 42 000 govt docs; 4 000 microforms; 4 200 av-mat; 7 500 sound-rec; 400 digital data carriers; 4 250 Large Print Bks
libr loan
40881

Mid-Hudson Library System, 103 Market St, *Poughkeepsie*, NY 12601-4098
T: +1 845 4716060; Fax: +1 845 4545940; E-mail: webaccount@midhudson.org; URL: www.midhudson.org
1859; Joshua Cohen
2 099 000 vols
libr loan
40882

Poughkeepsie Public Library District, Adriance Memorial Library, 18 Bancroft Rd, *Poughkeepsie*, NY 12601
T: +1 845 4853445; Fax: +1 845 4853789; E-mail: info@poklib.org; URL: www.poklib.org
1841; Thomas A. Lawrence
Foundation Ctr; 2 branch libs
180 000 vols; 460 curr per; 300 High Interest/Low Vocubulary Bks
40883

Kitsap Regional Library – Poulsbo Branch, 700 N E Lincoln St, *Poulsbo*, WA 98370-7688
T: +1 360 7792915; Fax: +1 360 7791051
Sharon Lee
Scandinavian Coll
68 000 vols
40884

Hiram Halle Memorial Library, 271 Westchester Ave, *Pound Ridge*, NY 10576-1714
T: +1 914 7645085; Fax: +1 914 7645319; URL: www.poundridgelibrary.org
1953; Marilyn Tinter
Art; Phonograph Records
57 000 vols; 98 curr per
40885

San Diego County Library – Poway Branch, 13137 Poway Rd, *Poway*, CA 92064-4687
T: +1 858 5132900; Fax: +1 858 5132922; URL: www.ci.poway.ca.us
Judy Chatterjee
124 000 vols
40886

Cobb County Public Library System – Powder Springs Branch, 4181 Atlanta St, Bldg 1, *Powder Springs*, GA 30127
T: +1 770 4393600; Fax: +1 770 4393620
Bruce Thompson
51 000 vols
40887

Powhatan County Public Library, 2270 Mann Rd, *Powhatan*, VA 23139-5748
T: +1 804 5985670; Fax: +1 804 5985671; URL: www.powhatanlibrary.org
Kim Armentrout
54 000 vols; 20 curr per
40888

Johnson County Library – Corinth, 8100 Mission Rd, *Prairie Village*, KS 66208
T: +1 913 9678650; Fax: +1 913 9678663
Stuart Hinds
161 000 vols; 170 curr per
40889

Autauga-Prattville Public Library, 254 Doster St, *Prattville*, AL 36067-3933
T: +1 334 3653396; Fax: +1 334 3653397; E-mail: autlib@excite.com; URL: www.appl.info
1956; Janice Yates Earnest
Alabama Hist Coll, Business Reference Ctr
85 000 vols; 145 curr per
40890

Prescott Public Library, 215 E Goodwin St, *Prescott*, AZ 86303
T: +1 928 7771500; Fax: +1 928 7715829; URL: www.prescottlib.info
1903; Toni Kaus
Southwest Coll
154 000 vols; 263 curr per; 8 000 e-books
40891

Prescott Valley Public Library, 7501 Civic Circle, *Prescott Valley*, AZ 86314
T: +1 928 7593040; Fax: +1 928 7593121; URL: www.pvlib.net
Stuart Mattson
80 000 vols; 81 curr per
FOL
40892

Mark & Emily Turner Memorial Library, 39 Second St, *Presque Isle*, ME 04769
T: +1 207 7642571; Fax: +1 207 7685756; URL: www.presqueisle.lib.me.us
1908; Sonja Morgan
Local Newspapers, microfilm
65 000 vols; 98 curr per; 1 400 av-mat
40893

Floyd County Public Library, 161 N Arnold Ave, *Prestonsburg*, KY 41653
T: +1 606 8862981; Fax: +1 606 8862284; URL: www.fclib.org
1957; Homer L. Hall
Kentucky Coll
71 000 vols; 50 curr per
40894

Price City Library, 159 E Main St, *Price*, UT 84501-3046
T: +1 435 6363188; Fax: +1 435 6372905; URL: www.priceutah.net
1914; Norma Rae Procarione
50 000 vols; 52 curr per
40895

Prichard Public Library, 300 W Love Joy Loop, *Prichard*, AL 36610
T: +1 251 4527847; Fax: +1 251 4527935; E-mail: prichardlib@bellsouth.net
Betty J. Hall
77 000 vols; 215 curr per
40896

Calvert County Public Library, Calvert Library, 850 Costley Way, *Prince Frederick*, MD 20678
T: +1 410 5350291; Fax: +1 410 5353022; URL: www.calvert.lib.md.us
1959; Patricia Hofmann
Maryland; 3 branch libs
153 000 vols; 221 curr per; 1 000 e-books
40897

Prince William Public Library System, 13083 Chinn Park Dr, *Prince William*, VA 22192-5073
T: +1 703 7926100; Fax: +1 703 7924875; URL: www.pwcgov.org/library
1952; Richard Murphy
Management & Govt Info Ctr; 10 branch libs
735 000 vols
40898

Somerset County Library System, 11767 Beechwood St, *Princess Anne*, MD 21853
T: +1 410 6510852; Fax: +1 410 6511388; URL: www.some.lib.md.us
1967; Renee Croft
Maryland, Hist; 2 branch libs
105 000 vols; 67 curr per
40899

Princeton Public Library, 124 S Hart, *Princeton*, IN 47670
T: +1 812 3854564; Fax: +1 812 3861662; E-mail: director@princetonpl.lib.in.us; URL: www.princetonpl.lib.in.us
1883; Brenda Williams
Genealogy & Local Hist Room, Children's Room
60 000 vols; 91 curr per; 450 microforms; 300 av-mat; 300 sound-rec
libr loan
40900

Princeton Public Library, Mercer Memorial Library, 205 Center St, *Princeton*, WV 24740-2932
T: +1 304 4875045; Fax: +1 304 4875046; URL: princetonlibrarywv.com/site
1922
Print Coll
83 000 vols; 149 curr per
40901

Mercer County Library – West Windsor, 333 N Post Rd, *Princeton Junction*, NJ 08550
T: +1 609 7990462; Fax: +1 609 9361206
Kaija Greenberg
82 000 vols
40902

Hawaii State Public Library System – Princeville Public Library, 4343 Emmalani Dr, *Princeville*, HI 96722
T: +1 808 8264310; Fax: +1 808 8264311
Jennifer Relacion
52 000 vols; 80 curr per
40903

Prospect Heights Public Library District, 12 N Elm St, *Prospect Heights*, IL 60070-1450
T: +1 847 2593500; Fax: +1 847 2594602; URL: www.phpl.info
1955; William McCully
Antiques, Glass
86 000 vols; 221 curr per; 317 e-books; 10 000 microforms; 3 500 av-mat; 9 200 sound-rec; 250 digital data carriers; 137 games
libr loan
40904

Providence Public Library, 150 Empire St, *Providence*, RI 02903-3283
T: +1 401 4558046; Fax: +1 401 4558065; E-mail: pplref@provlib.org; URL: www.provlib.org
1875; Dale Thompson
Rhode Island, Business, Art, Music; Architecture Coll, Band Music, Checkers Coll, Irish Culture, Italian Coll, Jewelry Coll, Textile Coll, Whaling; 9 branch libs
896 000 vols; 828 curr per; 256 e-books; 915 409 govt docs; 76 000 microforms; 20 744 sound-rec; 15 digital data carriers; 525 Art Repros, 28 020 CDs
libr loan
40905

Providence Public Library – Knight Memorial, 275 Elmwood Ave, *Providence*, RI 02907
T: +1 401 4558102; Fax: +1 401 7814712
Kathleen Vernon
RI History;Arnold Tombstone Records
60 000 vols; 40 curr per
40906

Providence Public Library – Smith Hill, 31 Candace St, *Providence*, RI 02908
T: +1 401 4558104; Fax: +1 401 3312913
Alan Gunther
250 000 vols
40907

1095

State of Rhode Island Office of Library & Information Services – Talking Books Plus, One Capitol Hill, **Providence**, RI 02908
T: +1 401 2225800; Fax: +1 401 2224195; E-mail: tbplus@lori.ri.gov; URL: www.olis.ri.gov/tbp
Andrew I. Egan
828 000 vols; 24 curr per   40908

Provo City Library, 550 N University Ave, **Provo**, UT 84601-1618
T: +1 801 8526661; Fax: +1 801 8526688; URL: www.provo.lib.ut.us; www.provolibrary.com; www.provocitylibrary.com
1904; Gene Nelson
240 000 vols   40909

Pueblo City-County Library District, Robert Hoag Rawlings Public Library, 100 E Abriendo Ave, **Pueblo**, CO 81004-4290
T: +1 719 5625601; Fax: +1 719 5625610; URL: www.pueblolibrary.org
1891; Jon Walker
Hispanic studies, Genealogy; Western Res Coll, Southeastern Colorado Coll (Photos), Frank I. Lamb Memorial Coll (Business); 3 branch libs
486 000 vols; 739 curr per; 402 e-books
libr loan   40910

Pueblo City-County Library District – Frank I. Lamb Branch, 2525 S Pueblo Blvd, **Pueblo**, CO 81005-2700
T: +1 719 5625670; Fax: +1 719 5625675
Diane Logie
53 000 vols   40911

Pueblo City-County Library District – Frank & Marie Barkman Branch, 1300 Jerry Murphy Rd, **Pueblo**, CO 81001-1858
T: +1 719 5625620; Fax: +1 719 5625685
Carol Rooney
60 000 vols   40912

Pulaski County Public Library System, 60 W Third St, **Pulaski**, VA 24301
T: +1 540 9942453; Fax: +1 540 9807775; URL: www.pclibs.org
1937; Dot M. Ogburn
82 000 vols; 165 curr per; 14 000 e-books; 2 412 av-mat   40913

Neill Public Library, Pullman Public Library, 201 N Grand Ave, **Pullman**, WA 99163-2693
T: +1 509 3344555, 3343595; Fax: +1 509 3346051; E-mail: library@neill-lib.org; URL: www.neill-lib.org
1921; Michael Pollastro
English as a Second Language Coll
57 700 vols; 163 curr per   40914

Puyallup Public Library, 324 S Meridian, **Puyallup**, WA 98371
T: +1 253 8415452; Fax: +1 253 8415483; E-mail: puylib@ci.puyallup.wa.us; URL: www.puyalluplibrary.org
1912; Mary Jo Torgeson
113 000 vols; 135 curr per; 6 211 av-mat   40915

Bucks County Free Library – James A. Michener Branch, 401 W Mill St, **Quakertown**, PA 18951-1248
T: +1 215 5363306; Fax: +1 215 5368397
Diane M. Davis
95 000 vols; 105 curr per   40916

Queens Borough Public Library – Queens Village Branch, 94-11 217th St, **Queens Village**, NY 11428
T: +1 718 7766800; Fax: +1 718 4794609
Sinisa Bibin
80 000 vols   40917

Gadsden County Public Library, 7325 Pat Thomas Pkwy, **Quincy**, FL 32351
T: +1 850 6277106; Fax: +1 850 6277775
1979; Jane Mock
85 000 vols; 76 curr per   40918

Plumas County Library, 445 Jackson St, **Quincy**, CA 95971-9410
T: +1 530 2836310; Fax: +1 530 2833242; E-mail: pclibq@psln.com; URL: www.psln.com/pclibq
1916; Margaret Miles
Botany; Local Mining Coll
73 000 vols; 220 curr per; 805 av-mat   40919

Thomas Crane Public Library, 40 Washington St, **Quincy**, MA 02269-9164
T: +1 617 3761319; Fax: +1 617 3761313; E-mail: quref@ocln.org; URL: thomascranelibrary.org
1871; Ann E. McLaughlin
Art hist, Music; 3 branch libs
255 000 vols; 543 curr per; 7 000 microforms; 3 600 av-mat; 4 601 sound-rec; 118 digital data carriers
libr loan   40920

Brooks County Public Library, 404 Barwick Rd, **Quitman**, GA 31643
T: +1 229 2634412; Fax: +1 229 2638002; URL: www.brooks.public.lib.ga.us
Laura Harrison
Genealogy, African-American
60 000 vols; 45 curr per   40921

East Mississippi Regional Library System, 116 Water St, **Quitman**, MS 39355-2336
T: +1 601 7763881; Fax: +1 601 7766599; E-mail: 4pea@emrl.lib.ms.us; URL: www.emrl.lib.ms.us
1966; Susan T. Byra
9 branch libs
63 000 vols; 56 curr per; 2 423 av-mat   40922

Racine Public Library, 75 Seventh St, **Racine**, WI 53403
T: +1 262 6369241; Fax: +1 262 6369260; URL: www.racinelibrary.info/
1897; Jessica MacPhail
Early Childhood Resource Coll; Bbookmobile
267 000 vols; 465 curr per; 9 000 e-books; 8 367 av-mat; 4 640 Large Print Bks, 8 370 Talking Bks   40923

Radford Public Library, 30 W Main St, **Radford**, VA 24141
T: +1 540 7313621; Fax: +1 540 7314857; E-mail: tonicox@radford.va.us; URL: www.radfordpl.org
1941; Toni Cox
Adult Low Reading Level Coll
76 570 vols; 109 curr per   40924

Rahway Public Library, Two City Hall Plaza, **Rahway**, NJ 07065
T: +1 732 3401551; Fax: +1 732 3400393; URL: www.rahwaylibrary.org
1858; Gail Miller
90 000 vols; 210 curr per   40925

Rainbow City Public Library, 3702 Rainbow Dr, **Rainbow City**, AL 35906
T: +1 256 4428477; Fax: +1 256 4424128; E-mail: rbclibrary@bellsouth.net; URL: www.rbclibrary.org
1981; Tina M. Brooks
51 000 vols; 80 curr per; 3 900 av-mat; 125 sound-rec; 938 books-on-tape   40926

North Carolina Regional Library for the Blind & Physically Handicapped, 1811 Capital Blvd, **Raleigh**, NC 27635-0001
T: +1 919 7334376; Fax: +1 919 7336910
1959; Francine Martin
311 000 vols   40927

Wake County Public Library System – Athens Drive Community, 1420 Athens Dr, **Raleigh**, NC 27606
T: +1 919 2334000; Fax: +1 919 2334082
Betty Williams
50 000 vols   40928

Wake County Public Library System – Cameron Village Regional, 1930 Clark Ave, **Raleigh**, NC 27605
T: +1 919 8566710; Fax: +1 919 8566722
Dale Cousins
North Carolina Coll
180 000 vols   40929

Wake County Public Library System – Duraleigh Road, 5800 Duraleigh Rd, **Raleigh**, NC 27612
T: +1 919 8811344; Fax: +1 919 8811317
Linda Cooper
66 000 vols   40930

Wake County Public Library System – Green Road, 4101 Green Rd, **Raleigh**, NC 27604
T: +1 919 7903200; Fax: +1 919 7903250
Travis Horton
70 000 vols   40931

Wake County Public Library System – North Regional, 7009 Harps Mill Rd, **Raleigh**, NC 27615
T: +1 919 8704000; Fax: +1 919 8704007
Michael Wasilick
175 000 vols   40932

Wake County Public Library System – Richard B. Harrison Branch, 1313 New Bern Ave, **Raleigh**, NC 27610
T: +1 919 8565720; Fax: +1 919 8566943
Wanda Cox-Bailey
Black Literature (Mollie H Lee Coll)
53 000 vols   40933

Ramsey Free Public Library, 30 Wyckoff Ave, **Ramsey**, NJ 07446
T: +1 201 3271445; Fax: +1 201 3273687; E-mail: ramsref@hotmail.com; URL: www.ramseylibrary.org; www.bccls.org/ramsey
1921; Wendy B. Bloom
Large print; Sidoroff Language Learning Coll
95 000 vols; 250 curr per   40934

Rancho Cucamonga Public Library, 7368 Archibald Ave, **Rancho Cucamonga**, CA 91730
T: +1 909 4772720; Fax: +1 909 4772721; E-mail: reference@cityofrc.com; URL: www.rcpl.lib.ca.us
1994; Robert Karatsu
137 000 vols; 215 curr per; 1 826 av-mat; 1 451 High Interest/Low Vocabulary Bks, 12 Bks on Deafness & Sign Lang, 2 110 Large Print Bks, 1 830 Talking Bks   40935

Rancho Mirage Public Library, 71-100 Hwy 111, **Rancho Mirage**, CA 92270
T: +1 760 3417323; Fax: +1 760 3415213; E-mail: librarian@ranchomiragelibrary.org; URL: www.ranchomiragelibrary.org
1994; David Bryant
85 000 vols; 453 curr per; 68 e-books   40936

Orange County Public Library – Rancho Santa Margarita Branch, 30902 La Promesa, **Rancho Santa Margarita**, CA 92688-2821
T: +1 949 4596094; Fax: +1 949 4598391
Trish Noa
103 000 vols; 130 curr per   40937

Baltimore County Public Library – Randallstown Area Branch, 8604 Liberty Rd, **Randallstown**, MD 21133-4797
T: +1 410 8870770; Fax: +1 410 5213614; E-mail: randallstown@bcpl.net
Darcy Cahill
114 000 vols   40938

Randolph Township Free Public Library, 28 Calais Rd, **Randolph**, NJ 07869
T: +1 973 8953556; Fax: +1 973 8954946; E-mail: anita.freeman@mainlib.org; URL: www.randolphnj.org/library
1964; Anita S. Freeman
95 000 vols; 335 curr per; 1 100 microforms; 1 500 av-mat; 2 000 sound-rec; 300 digital data carriers
libr loan   40939

Turner Free Library, 2 North Main St, **Randolph**, MA 02368
T: +1 617 9610932; Fax: +1 617 9610933; E-mail: sslymon@randolph-ma.gov
1874; Sara Slymon
60 000 vols; 126 curr per   40940

Rantoul Public Library, 106 W Flessner, **Rantoul**, IL 61866
T: +1 217 8933955; Fax: +1 217 8933961; URL: www.rantoul.lib.il.us
1934; Holly Thompson
Aero-Space Coll
50 000 vols; 175 curr per   40941

Rapid City Public Library, 610 Quincy St, **Rapid City**, SD 57701-3630
T: +1 605 3946139; Fax: +1 605 3944064; URL: www.rapidcitylibrary.org
1903; Greta Chapman
Large print; Rapid City Society for Genealogical Res Coll
158 000 vols; 392 curr per; 9 000 e-books; 10 000 av-mat
libr loan   40942

Reed Memorial Library, Ravenna Public Library, 167 E Main St, **Ravenna**, OH 44266-3197
T: +1 330 2962827; Fax: +1 330 2963780; E-mail: refreed@oplin.org; URL: www.reed.lib.oh.us
1915; Phyllis Cettomai
Genealogy/Local hist coll
94 000 vols; 206 curr per; 667 microforms; 3 700 av-mat; 5 025 sound-rec; 2 digital data carriers; 243 cake pans, 1 659 posters, 370 toys, 430 puppets
libr loan; OLA, ALA   40943

Carbon County Library System, 215 W Buffalo St, **Rawlins**, WY 82301
T: +1 307 3282618; Fax: +1 307 3282615; URL: www.carboncolibrary.state.wy.us
1925; Vicki Hitchcock
93 000 vols; 75 curr per   40944

Richland Parish Library, 1410 Louisa St, **Rayville**, LA 71269-3299
T: +1 318 7284806; Fax: +1 318 7286108; URL: www.richland.lib.la.us
1926; Brenda Doran
Ruth Hatch Lit Coll, 1927 Flood Coll, Innis Morris Ellis Coll; 2 branch libs
66 000 vols; 118 curr per   40945

Berks County Public Libraries, 1037F MacArthur Rd, **Reading**, PA 19605
T: +1 610 3785260; Fax: +1 610 3781525
Julie Rinehart
533 000 vols; 1 166 curr per   40946

Reading Public Library, 100 S Fifth St, **Reading**, PA 19602
T: +1 610 6556350; Fax: +1 610 6556609; E-mail: reference@reading.lib.pa.us; URL: www.reading.lib.pa.us
1763; Frank Kasprowicz
John Updike Coll, Berks Authors Coll, Local Imprints, Pennsylvania German Coll; 3 branch libs
312 000 vols; 305 curr per   40947

Reading Public Library, 64 Middlesex Ave, **Reading**, MA 01867-2550
T: +1 781 9440840; Fax: +1 781 9429106; E-mail: readingpl@noblenet.org; URL: www.readingpl.org
1867; Ruth Urell
110 000 vols; 515 curr per; 40 e-books, 15 000 av-mat; 6 500 sound-rec; 180 Electronic Media & Resources   40948

Los Angeles Public Library System – West Valley Regional, 19036 Vanowen St, **Reseda**, CA 91335
T: +1 818 3459806; Fax: +1 818 3454288
Paul Montgomerie
76 000 vols   40949

Red Bank Public Library, Eisner Memorial Library, 84 W Front St, **Red Bank**, NJ 07701
T: +1 732 8420690; Fax: +1 732 8424191; URL: www.lmxac.org/redbank
1878; Deborah Griffin-Sadel
53 000 vols; 184 curr per
libr loan   40950

Tehama County Library, 645 Madison St, **Red Bluff**, CA 96080-3383
T: +1 530 5270607; Fax: +1 530 5271562; URL: www.tehamacountylibrary.org
1916; Caryn Brown
122 000 vols; 45 curr per   40951

Red Wing Public Library, 225 East Ave, **Red Wing**, MN 55066
T: +1 651 3853673; E-mail: rwpl@selco.info; URL: www.redwing.lib.mn.us
1894; James Lund
69 600 vols; 340 curr per   40952

Shasta County Library, 1855 Shasta St, **Redding**, CA 96001
T: +1 530 2255754; Fax: +1 530 2417169; E-mail: contactus@shastacountylibrary.org; URL: www.shastacountylibrary.org
1949; Janice Erickson
2 branch libs
205 000 vols; 249 curr per
libr loan; ALA   40953

Redford Township District Library, 25320 W Six Mile, *Redford*, MI 48240
T: +1 313 5315960; Fax: +1 313 5311721; E-mail: rtdl@redfordlibrary.org; URL: redfordlibrary.org
1947; Kimberley Potter
130 000 vols; 181 curr per; 4 800 av-mat; 5 500 sound-rec; 1 130 digital data carriers; 380 High Interest/Low Vocabulary Bks, 50 Bks on Deafness & Sign Lang, 600 Large Print Bks
libr loan; ALA   40954

A. K. Smiley Public Library, 125 W Vine St, *Redlands*, CA 92373
T: +1 909 798565; Fax: +1 909 7987566; E-mail: admin@akspl.org; URL: www.akspl.org
1894; Larry E. Burgess
Art & architecture, Civil War; Californiana, Redlands Heritage; Lincoln (Watchorn Memorial Shrine); Smiley Family Letters
122 000 vols; 300 curr per; 1 529 av-mat
                                  40955

Deschutes Public Library District – Redmond Branch, 827 SW Deschutes Ave, *Redmond*, OR 97756
T: +1 541 3121050; Fax: +1 541 5486358
Todd Dunkelberg
86 000 vols; 133 curr per   40956

Redondo Beach Public Library, 303 N Pacific Coast Hwy, *Redondo Beach*, CA 90277
T: +1 310 3180675; Fax: +1 310 3183809; URL: www.redondo.org/library
1908; Jean M. Scully
1 branch libr
176 000 vols; 323 curr per
libr loan   40957

Reedsburg Public Library, 370 Vine St, *Reedsburg*, WI 53959
T: +1 608 7687323; URL: www.scls.lib.wi.us/ree
1898; Susan Steiner
Large Print Bks Coll
56 340 vols; 195 curr per; 5 microforms; 2 600 av-mat; 952 sound-rec; 143 digital data carriers
libr loan   40958

Queens Borough Public Library – Rego Park Branch, 91-41 63rd Dr, *Rego Park*, NY 11374
T: +1 718 4595140
Inna Yangarber
81 000 vols   40959

Rehoboth Beach Public Library, 226 Rehoboth Ave, *Rehoboth Beach*, DE 19971-2141
T: +1 302 2278044; Fax: +1 302 2270597; E-mail: rehoboth-beach-director@lib.de.us; URL: www.rehobothlibrary.org
Margaret LaFond
Bolton Bird Coll, Parks Wedgewood Coll, Sussex Gardener's Coll
50 000 vols; 85 curr per   40960

Baltimore County Public Library – Reisterstown Branch, 21 Cockeys Mill Rd, *Reisterstown*, MD 21136-1285
T: +1 410 8871165; Fax: +1 410 8338756; E-mail: reisterstown@bcpl.net
Barbara Salit-Michel
98 000 vols   40961

Washoe County Library System, 301 S Center St, *Reno*, NV 89501-2102; P.O. Box 2151, Reno, NV 89505-2151
T: +1 775 4241845; Fax: +1 775 4241840; E-mail: info@washoe.lib.nv.us; URL: www.washoe.lib.nv.us
1904; Arnold Maurins
Gambling, Nevada hist, Literacy; 15 branch libs
959 000 vols; 2 109 curr per; 15 600 digital data carriers; 35 000 Videos
libr loan   40962

Washoe County Library System – Downtown Reno Library, 301 S Center, *Reno*, NV 89501; P.O. Box 2151, Reno, NV 89505-2151
T: +1 775 3278312; Fax: +1 775 3278390
Scottie Wallace
178 000 vols   40963

Washoe County Library System – North Valleys Library, 1075 N Hills Blvd, No 340, *Reno*, NV 89506; P.O. Box 2151, Reno, NV 89505-2151
T: +1 775 9720281; Fax: +1 775 9726810
N.N.
61 000 vols   40964

Washoe County Library System – Sierra View Library, 4001 S Virginia St, Reno Town Mall, *Reno*, NV 89502; P.O. Box 2151, Reno, NV 89505-2151
T: +1 775 8273555; Fax: +1 775 8278792
Kristin Cannard
New Adult Readers
143 000 vols   40965

Washoe County Library System – South Valleys Library, 15650A Wedge Pkwy, *Reno*, NV 89511; P.O. Box 2151, Reno, NV 89505-2151
T: +1 775 8515190; Fax: +1 775 8515188
Tammy Cirrincione
66 000 vols   40966

Jasper County Public Library, Rensselaer Public, 208 W Susan St, *Rensselaer*, IN 47978
T: +1 219 8665881; Fax: +1 219 8667378; URL: www.jasperco.lib.in.us
1904; Patty Stringfellow
Humanities, Natural science; Civil War Coll; 2 branch libs
59 000 vols; 530 curr per; 583 Electronic Media & Resources   40967

Renton Public Library, 100 Mill Ave S, *Renton*, WA 98057
T: +1 425 4306610; Fax: +1 425 4306833; URL: rentonwa.gov/living
1914; Bette Anderson
1 branch libr
125 000 vols; 425 curr per; 12 500 av-mat; 1 500 sound-rec; 1 040 digital data carriers; 1 730 Large Print Bks
                                  40968

Columbus Metropolitan Library – Reynoldsburg Branch, 1402 Brice Rd, *Reynoldsburg*, OH 43068
T: +1 614 4793341; Fax: +1 614 6452275
Christopher Korenowsky
170 000 vols   40969

Rhinelander District Library, 106 N Stevens St, *Rhinelander*, WI 54501-3193
T: +1 715 3651070; Fax: +1 715 3651076; URL: wvls.lib.wi.us/rhinelanderdistrictlibrary
1898; Kris Adams Wendt
Art (Ruth Smith Bump Coll)
73 000 vols; 149 curr per   40970

Richardson Public Library, 900 Civic Center Dr, *Richardson*, TX 75080
T: +1 972 7444363;
Fax: +1 972 7445806; URL: www.richardsonpubliclibrary.com
1959; Steve Benson
Chinese & Korean Lang Mat
236 000 vols; 450 curr per; 84 music scores; 500 maps; 700 microforms; 6 500 av-mat; 8 452 sound-rec; 281 digital data carriers
libr loan   40971

Free Library of Northampton Township, 25 Upper Holland Rd, *Richboro*, PA 18954-1514
T: +1 215 3573050; URL: www.buckslib.org/northampton
1970; Virginia Volkman
98 000 vols; 128 curr per   40972

Akron-Summit County Public Library – Richfield Branch, 3761 S Grant St, *Richfield*, OH 44286-9603
T: +1 330 6594343; Fax: +1 330 6596205
Elizabeth Reed
50 000 vols   40973

Pulaski County Library District, 111 Camden St, *Richland*, MO 65556; P.O. Box 340, Richland, MO 65556-0340
T: +1 573 7653642;
Fax: +1 573 7655395; URL: www.pulaskicounty.lib.mo.us
1965; Marcia Crandall
50 000 vols; 27 curr per   40974

Richland Public Library, 1270 Lee Blvd, *Richland*, WA 99352-3539
T: +1 509 9427454; Fax: +1 509 9427447; URL: www.richland.lib.wa.us
1951; Ann L. Roseberry
Nuclear Regulatory Commission, Repository (WPPSS Docs)
148 000 vols; 286 curr per; 571 e-books
                                  40975

Brewer Public Library, Richland Center Public, 325 N Central Ave, *Richland Center*, WI 53581-1802
T: +1 608 6476444; Fax: +1 608 6476797; E-mail: richlandcenterpl@yahoo.com; URL: www.rc.swls.org
1900; Michele Nolen-Karras
Richland County Hist Coll
52 000 vols; 111 curr per   40976

County of Henrico Public Library, 1001 N Laburnum Ave, *Richmond*, VA 23223
T: +1 804 6523200; Fax: +1 804 2225566; URL: www.co.henrico.va.us/library
1966; Gerald M. McKenna
10 branch libs, 1 bookmobile
896 000 vols
libr loan   40977

County of Henrico Public Library – Dumbarton Area, 6800 Staples Mill Rd, *Richmond*, VA 23228
T: +1 804 2626507; Fax: +1 804 2668986; E-mail: library@henrico.lib.va.us; URL: www.co.henrico.lib.va.us
Deborah Lammers
106 000 vols; 98 curr per   40978

County of Henrico Public Library – Fairfield Area, 1001 N Laburnum Ave, *Richmond*, VA 23223
T: +1 804 6523251; Fax: +1 804 2221958
Sharon Crenshaw
73 000 vols; 102 curr per   40979

County of Henrico Public Library – Gayton, 10600 Gayton Rd, *Richmond*, VA 23238
T: +1 804 7402747; Fax: +1 804 7414495
Thomas Bruno
74 000 vols; 83 curr per   40980

County of Henrico Public Library – Tuckahoe Area, 1901 Starling Dr, *Richmond*, VA 23229
T: +1 804 2709578; Fax: +1 804 3460985
Janet Woody
155 000 vols; 130 curr per   40981

Fort Bend County Libraries – George Memorial Library, 1001 Golfview Dr, *Richmond*, TX 77469-5199
T: +1 281 3412606; Fax: +1 281 3412689; URL: www.fortbend.lib.tx.us
1947; N.N.
Civil War, Regional Hist Resource Dept of Texas, Restoration (George Carriage Coll); 8 branch libs
156 000 vols; 396 curr per; 23 000 e-books; 17 000 av-mat; 12 000 sound-rec; 2 020 digital data carriers   40982

Library & Resource Center, Department for the Blind & Vision Impaired, 395 Azalea Ave, *Richmond*, VA 23227-3633
T: +1 804 3713661; Fax: +1 804 3713508; URL: www.vdbvi.org
1958; Barbara McCarthy
Print Res Mat on Blindness
203 000 vols; 16 curr per; 7 229 braille bks, 500 bks in print, 58 420 cassettes, 2 140 large print bks, 64 000 talking bks
ALA   40983

Madison County Public Library, 507 W Main St, *Richmond*, KY 40475
T: +1 859 6236704; Fax: +1 859 6232023; E-mail: richmond@madisonlibrary.org; URL: madisonlibrary.org
1988; Sue Hays
1 branch libr
95 000 vols; 217 curr per; 4 502 av-mat
                                  40984

Ray County Library, 215 E Lexington, *Richmond*, MO 64085-1834
T: +1 816 7765104; Fax: +1 816 7765103; E-mail: raycopublibrary@yahoo.com; URL: www.raycountylibrary.homestead.com
1947; Steve Meyer
64 000 vols; 105 curr per; 1 139 av-mat; 500 sound-rec; 10 digital data carriers
                                  40985

Richmond Public Library, 101 E Franklin St, *Richmond*, VA 23219-2193
T: +1 804 6464256;
Fax: +1 804 6467685; URL: www.richmondpubliclibrary.org
1924; Harriet H. Coalter
Rare Children's Bks, Richmond Authors; 9 branch libs
419 000 vols
libr loan   40986

Richmond Public Library – West End, 5420 Patterson Ave, *Richmond*, VA 23226
T: +1 804 6462559; Fax: +1 804 6467685
E. Brooke Spieldenner
61 000 vols   40987

Wayne Township Library, Morrisson-Reeves Library, 80 N Sixth St, *Richmond*, IN 47374-3079
T: +1 765 9668291; Fax: +1 765 9621318; E-mail: library@mrlinfo.org; URL: www.mrlinfo.org
1864; Carol B. McKey
Cookbks, Pop Music; 1 branch libr
263 000 vols; 330 curr per; 760 av-mat; 300 sound-rec; 360 Large Print Bks
libr loan   40988

Richmond Heights Memorial Library, 8001 Dale Ave, *Richmond Heights*, MO 63117
T: +1 314 6456202; Fax: +1 314 7813434; E-mail: gfl000@mail.connect.more.net; URL: rhml.lib.mo.us
1935; Jeanette Moore Piquet
61 000 vols; 117 curr per; 448 av-mat; 500 sound-rec
libr loan   40989

Queens Borough Public Library – Lefferts, 103-34 Lefferts Blvd, *Richmond Hill*, NY 11419
T: +1 718 8435950
Weiqing Dai
81 000 vols   40990

Queens Borough Public Library – Richmond Hill Branch, 118-14 Hillside Ave, *Richmond Hill*, NY 11418
T: +1 718 8497150; Fax: +1 718 8494717
Susan Wetjen
76 000 vols   40991

Richwood North Union Public Library, Four E Ottawa St, *Richwood*, OH 43344-1296
T: +1 740 9433054; Fax: +1 740 9439211
1882; Judith Lawler
74 000 vols; 149 curr per   40992

Kern County Library – Ridgecrest Branch, 131 E Las Flores Ave, *Ridgecrest*, CA 93555-3648
T: +1 760 3845870; Fax: +1 760 3843211
Marsha Lloyd
Petroglyphs; Local History Coll; Flora & Fauna of Indian Wells Valley
54 000 vols   40993

Ridgefield Library Association Inc, 472 Main St, *Ridgefield*, CT 06877-4585
T: +1 203 4386960; Fax: +1 203 4384558; URL: www.ridgefieldlibrary.org
1901; Christina B. Nolan
130 000 vols; 240 curr per   40994

Ridgefield Public Library, 527 Morse Ave, *Ridgefield*, NJ 07657
T: +1 201 9410192; Fax: +1 201 9419354; E-mail: rsldcirc@mail.bccls.org; URL: www.bccls.org/ridgefield
1930; Jane Forte
Ridgefield Hist Coll
61 000 vols; 200 curr per   40995

Ridgefield Park Free Public Library, 107 Cedar St, *Ridgefield Park*, NJ 07660
T: +1 201 6410689; Fax: +1 201 4401058; E-mail: rfpk@bccls.org; URL: www.ridgefieldpark.org
1890; Eileen Karpoff
62 000 vols; 95 curr per   40996

Madison County Library System – Ridgeland Public Library, 397 Hwy 51 N, *Ridgeland*, MS 39157
T: +1 601 8564536; Fax: +1 601 8563748
Nan Crosby
53 000 vols; 80 curr per   40997

Queens Borough Public Library – Ridgewood Community, 20-12 Madison St, *Ridgewood*, NY 11385
T: +1 718 8214770; Fax: +1 718 6286263
David Roycroft
77 000 vols   40998

Ridgewood Public Library, 125 N Maple Ave, *Ridgewood*, NJ 07450-3288
T: +1 201 6705600; Fax: +1 201 6700293; URL: www.ridgewoodlibrary.org
1923; Nancy K. Greene
131 000 vols   40999

Ringwood Public Library, 30 Cannici Dr, *Ringwood*, NJ 07456
T: +1 973 9626256; Fax: +1 973 9627799; E-mail: ringwoodpl@hotmail.com; URL: www.ringwoodlibrary.org
1960; Andrea R. Cahoon
Ringwood Hist Coll, microfiche, clippings, bks & pamphlets; Local Minutes of Public Agencies Coll
54 000 vols; 107 curr per   41000

Rio Rancho Public Library, 755 Loma Colorado Dr NE, *Rio Rancho*, NM 87124; P.O. Box 15670, Rio Rancho, NM 87174-0670
T: +1 505 8968818; Fax: +1 502 8924782; URL: www.ci.rio-rancho.nm.us/library.htm
1974; Toni Beatty
Southwest Coll
150 000 vols; 111 curr per   41001

Jackson County Library, 208 N Church St, *Ripley*, WV 25271
T: +1 304 3725343; Fax: +1 304 3727935; URL: jackson.park.lib.wv.us
1949
52 890 vols; 88 curr per
libr loan   41002

Ripon Public Library, 120 Jefferson St, *Ripon*, WI 54971
T: +1 920 7486160; Fax: +1 920 7486298; URL: www.riponlibrary.org
1885; Desiree Bongers
57 000 vols; 170 curr per   41003

River Edge Free Public Library, 685 Elm Ave, *River Edge*, NJ 07661
T: +1 201 2611663; Fax: +1 201 9860214; E-mail: rive1@bccls.org; URL: www.bccls.org/riveredge
1953; Daragh O'Connor
Language Coll, Korean Coll
75 000 vols; 171 curr per; 685 av-mat   41004

River Falls Public Library, 140 Union St, P.O. Box 45, *River Falls*, WI 54022
T: +1 715 4250905; Fax: +1 715 4250914; E-mail: nancym@riverfallspubliclibrary.org; URL: www.riverfallspubliclibrary.org
Nancy Y. Miller
65 000 vols; 200 curr per   41005

River Forest Public Library, 735 Lathrop Ave, *River Forest*, IL 60305-1883
T: +1 708 3665205; Fax: +1 708 3668699; E-mail: reference@riverforestlibrary.org; URL: www.riverforestlibrary.org
1899; Sophia Anastos
67 000 vols; 166 curr per   41006

River Vale Free Public Library, 412 Rivervale Rd, *River Vale*, NJ 07675
T: +1 201 3912323; Fax: +1 201 3916599; E-mail: rivl1@bccls.org; URL: www.bccls.org/rivervale
1964; Holly Deni
52 000 vols; 149 curr per   41007

Clayton County Library System – Riverdale Branch, 420 Valley Hill Rd, *Riverdale*, GA 30274
T: +1 770 4728100; Fax: +1 770 4728106
Vivian Chandler
78 000 vols; 72 curr per   41008

Riverhead Free Library, 330 Court St, *Riverhead*, NY 11901-2885
T: +1 631 7273228; Fax: +1 631 7274762; URL: river.suffolk.lib.ny.us
1896; Lisa Jacobs
Tanger Wellness Coll
166 000 vols; 334 curr per   41009

Riverside County Library System, 4080 Lemon St, 4th Flr, *Riverside*, CA 92501-3679
T: +1 951 9551100; Fax: +1 951 9551105
Nancy Johnson
1 034 000 vols; 875 curr per   41010

Riverside Public Library, 3581 Mission Inn Ave, *Riverside*, CA 92501; P.O. Box 468, Riverside, CA 92502-0468
T: +1 951 8265201; Fax: +1 951 7881528; URL: www.riversideca.gov/library
1888; N.N.
Dorothy Daniels Memorial Coll (Historical Children's Bks), Sight Handicapped Coll, Black Hist Coll, Mexican American Coll; 5 branch libs
522 000 vols; 997 curr per   41011

Riverside Public Library, 1 Burling Rd, *Riverside*, IL 60546
T: +1 708 4426366; Fax: +1 708 4429462; E-mail: RSS@RiversideLibrary.org; URL: www.riversidelibrary.org
1930; Janice Fisher
Landscape architecture
67 000 vols; 125 curr per   41012

Riverside Public Library – Arlington, 9556 Magnolia, *Riverside*, CA 92503-3698
T: +1 951 6896612; Fax: +1 951 6896612
Charlene Swanson
51 000 vols   41013

Riverside Public Library – La Sierra, 4600 La Sierra, *Riverside*, CA 92505-2722
T: +1 951 6887740; Fax: +1 951 3527578
Linda Taylor
78 000 vols   41014

Fremont County Library System – Riverton Branch, 1330 W Park, *Riverton*, WY 82501
T: +1 307 8563556; Fax: +1 307 8573722; E-mail: riverton@will.state.wy.us
Gloria Brodle
75 000 vols; 150 curr per   41015

Salt Lake County Library Services – Riverton Branch, 12877 S 1830 West, *Riverton*, UT 84065-3204
T: +1 801 9447677; Fax: +1 801 4668601
Royce Nielsen
131 000 vols   41016

Riverview Public Library, 14300 Sibley Rd, *Riverview*, MI 48193
T: +1 734 2831250; Fax: +1 734 2836843; URL: riverviewpubliclibrary.com
1962; Kirk A. Borger
57 000 vols; 120 curr per   41017

Tampa-Hillsborough County Public Library System – Riverview Branch, 10509 Riverview Dr, *Riverview*, FL 33569-4367
Fax: +1 813 6717793
Janet Molen
70 000 vols   41018

Riviera Beach Public Library, 600 W Blue Heron Blvd, *Riviera Beach*, FL 33404-4398; P.O. Box 11329, Riviera Beach, FL 33419-1329
T: +1 561 8454195; Fax: +1 561 8817308; URL: www.rivierabch.com
1950; Anne Sutton
Black Studies Coll, Florida Coll
86 000 vols; 126 curr per   41019

Botetourt County Library, 28 Avery Row, *Roanoke*, VA 24012
T: +1 540 9773433; Fax: +1 540 9772407; URL: www.botetourt.org
1979; Stephen C. Vest
115 000 vols; 220 curr per; 4 400 av-mat   41020

Roanoke County Public Library, 3131 Electric Rd SW, *Roanoke*, VA 24018-6496
T: +1 540 7767327; Fax: +1 540 7722131; URL: www.yourlibrary.us
1945; Diana Rosapepe
Kirkwood Professional Coll; 6 branch libs
315 000 vols; 567 curr per; 47 000 e-books; 9 000 av-mat; 2 779 digital data carriers
libr loan   41021

Roanoke County Public Library – Hollins, 6624 Peters Creek Rd, *Roanoke*, VA 24019
T: +1 540 5618024; Fax: +1 540 5638902
Ann Tripp
105 000 vols; 101 curr per   41022

Roanoke Public Libraries, 706 S Jefferson St, *Roanoke*, VA 24016-5191
T: +1 540 8532473; Fax: +1 540 8531781; E-mail: main.library@roanokeva.gov; URL: www.roanokegov.com/library
1921; Sheila Umberger
6 branch libs, 1 bookmobile
340 000 vols; 905 curr per   41023

Robinson Public Library District, 606 N Jefferson St, *Robinson*, IL 62454-2665
T: +1 618 5442917; Fax: +1 618 5447172; E-mail: rpladmin@shawls.lib.il.us; URL: www.sirin.lib.il.us/docs/rob/docs/lib
1906; Marilyn Manning
3 branch libs
70 000 vols; 80 curr per; 710 av-mat   41024

Flagg-Rochelle Public Library District, 619 Fourth Ave, *Rochelle*, IL 61068
T: +1 815 5623431; Fax: +1 815 5623432; E-mail: library@rochelle.net
1889; Barbara A. Kopplin
53 000 vols; 147 curr per   41025

Brighton Memorial Library, 2300 Elmwood Ave, *Rochester*, NY 14618
T: +1 585 7845300; Fax: +1 585 7845333; E-mail: briref1@libraryweb.org; URL: www.brightonlibrary.org
1953; Angela Bonazinga
Judaica
111 000 vols   41026

Chili Public Library, 3333 Chili Ave, *Rochester*, NY 14624-5494
T: +1 585 8892200; Fax: +1 585 8895819; URL: www.chililibrary.org
1962; Jeff Baker
86 000 vols; 198 curr per
libr loan   41027

Fulton County Public Library, Rochester Library, 320 W Seventh St, *Rochester*, IN 46975-1332
T: +1 574 2232713; Fax: +1 574 2235102; URL: www.fulco.lib.in.us
1906; Linn Landis
Hist, Indiana, Business & management, Agriculture; 2 branch libs
142 000 vols; 239 curr per
libr loan   41028

Gates Public Library, 1605 Buffalo Rd, *Rochester*, NY 14624
T: +1 585 2476446; Fax: +1 585 4265733; URL: www.gateslibrary.org
1960; Judy MacKnight
Italian Coll, bks
127 000 vols; 236 curr per   41029

Irondequoit Public Library, Pauline Evans Branch, 45 Cooper Rd, *Rochester*, NY 14617
T: +1 585 3366062; Fax: +1 585 3366066; URL: www.irondequoit.org/library
1947; Terrence Buford
1 branch libr
157 000 vols; 250 curr per   41030

Rochester Hills Public Library, 500 Olde Towne Rd, *Rochester*, MI 48307-2043
T: +1 248 6507174; Fax: +1 248 6507121; URL: www.rhpl.org
1924; Christine Lind Hage
Technology; Photogr Arch
208 000 vols; 404 curr per; 24 464 Large Print Bks   41031

Rochester Public Library, 65 S Main St, *Rochester*, NH 03867-2707
T: +1 603 3321428; Fax: +1 603 3357582; URL: www.rpl.lib.nh.us
1894; John Fuchs
Rochester Courier Coll
89 000 vols; 220 curr per   41032

Rochester Public Library, 115 South Ave, *Rochester*, NY 14604-1896
T: +1 585 4288160; Fax: +1 585 4288353; URL: www.libraryweb.org
1911; Carol Nersinger
Black Hist Coll, Reynolds Audio Visual; 10 branch libs
938 000 vols; 213 607 curr per
libr loan   41033

York County Library, Rock Hill Public, 138 E Black St, *Rock Hill*, SC 29731; P.O. Box 10032, Rock Hill
T: +1 803 9815831; Fax: +1 803 9815866; URL: www.yclibrary.org
1884; Shasta Brewer
South Carolina Colls, Catawba Indians; 4 branch libs, 1 bookmobile
277 000 vols; 577 curr per; 18 Electronic Media & Resources   41034

Rock Island Public Library, 401 19th St, *Rock Island*, IL 61201
T: +1 309 7327341; Fax: +1 309 7327342; URL: rbls.lib.il.us
1872; Ava L. Ketter
Lit, Literary criticism; Large Print Bks; 2 branch libs
190 000 vols; 382 curr per   41035

Atchison County Library, 200 S Main St, *Rock Port*, MO 64482-1532
T: +1 660 7445404; Fax: +1 660 7442861
1946; Janice S. Rosenbohm
2 branch libs
61 000 vols; 95 curr per; 1 555 av-mat   41036

Rockaway Township Free Public Library, 61 Mount Hope Rd, *Rockaway*, NJ 07866
T: +1 973 6272344; Fax: +1 973 6277658; E-mail: rockawaytwplibrary@rtlibrary.org; URL: www.rtlibrary.org
1966; Joy Kauffman
Business; 1 branch libr
107 000 vols; 260 curr per; 4 900 av-mat; 2 400 sound-rec; 612 digital data carriers   41037

Queens Borough Public Library – Peninsula, 92-25 Rockaway Beach Blvd, *Rockaway Beach*, NY 11693
T: +1 718 6341110; Fax: +1 718 3185253
Karen Lowenstein
60 000 vols; 1 451 curr per   41038

Queens Borough Public Library – Seaside Community, 116-15 Rockaway Beach Blvd, *Rockaway Park*, NY 11694
T: +1 718 6341876; Fax: +1 718 6348711
Caroline Stark
61 000 vols   41039

Rockford Public Library, 215 N Wyman St, *Rockford*, IL 61101-1023
T: +1 815 9876600; Fax: +1 815 9650866; URL: www.rockfordpubliclibrary.org
1872; Frank R. Novak
Art, Architecture, Business & management, Economics; 5 branch libs
438 000 vols; 20 553 curr per   41040

Auburn-Placer County Library – Rocklin Branch, 5460 Fifth St, *Rocklin*, CA 95677-2547
T: +1 916 6243133; Fax: +1 916 6329152
Mary George
57 000 vols   41041

Aransas County Public Library, 701 E Mimosa, *Rockport*, TX 78382
T: +1 361 7900153; Fax: +1 361 7900150; URL: www.acplibrary.org
1956; Iris Sanchez
50 000 vols; 21 curr per   41042

Spencer County Public Library, 210 Walnut St, *Rockport*, IN 47635
T: +1 812 6494866; Fax: +1 812 6494018; E-mail: reference@rockport-spco.lib.in.us; URL: www.rockport-spco.lib.in.us
1917
Religion, Hist, Large type print; Abraham Lincoln Coll; Genealogy dept
81 460 vols; 150 curr per
libr loan   41043

Rockville Centre Public Library, 221 N Village Ave, *Rockville Centre*, NY 11570
T: +1 516 7666257; Fax: +1 516 7666090; E-mail: reference@rvcpl.org; URL: www.rvclibrary.org
1882; Maureen Chiofalo
Education, Feminism, Hist, Industry
179 000 vols; 300 curr per; 3504 Audio Bks, 4380 Large Print Bks   41044

Rockwall County Library, 1215 E Yellowjacket Lane, *Rockwall*, TX 75087
T: +1 972 2046900; Fax: +1 972 2046909; E-mail: rocklib@rocklib.com; URL: www.rocklib.com
1945; Marcine McCulley
50 000 vols; 106 curr per   41045

Connecticut State Library – State Records Center, 198 West St, *Rocky Hill*, CT 06067
T: +1 860 7212041; Fax: +1 860 7212055; URL: www.cslib.org/records.htm
Donald P. Ballinger
60 000 vols   41046

Cora J. Belden Library, 33 Church St, *Rocky Hill*, CT 06067
T: +1 860 2587621; Fax: +1 860 2587624; E-mail: mhogan@ci.rocky-hill.ct.us; URL: www.rockyhilllibrary.info
1794; Mary Hogan
71 000 vols; 102 curr per   41047

Somerset County Library System – Mary Jacobs Memorial, 64 Washington St, *Rocky Hill*, NJ 08553
T: +1 609 9247073; Fax: +1 609 9247668; E-mail: mjlstaff@hublib.lib.nj.us
Helen Morris
50 000 vols; 130 curr per   41048

Braswell Memorial Public Library, 727 N Grace St, *Rocky Mount*, NC 27804-4842
T: +1 252 4421951; Fax: +1 252 4427366; URL: www.braswell-library.org
1922; Steve Farlow
African-American, Civil War, Large print; Parent-Teacher Coll
123 000 vols; 150 curr per; 21 000 e-books; 4570 Talking Bks   41049

Franklin County Public Library, 355 Franklin St, *Rocky Mount*, VA 24151
T: +1 540 4833098; Fax: +1 540 4836652; URL: www.franklincountyva.org/library
1975; David E. Bass
90 000 vols; 220 curr per   41050

Rocky River Public Library, 1600 Hampton Rd, *Rocky River*, OH 44116-2699
T: +1 440 3337610; Fax: +1 440 3334184; URL: www.rrpl.org
1928; John A. Lonsak
Cowan Pottery Museum
113 000 vols; 830 curr per   41051

Johnson County Library – Cedar Roe, 5120 Cedar, *Roeland Park*, KS 66205
T: +1 913 3848590; Fax: +1 913 3848597
Meredith Roberson
77 000 vols   41052

Rogers Public Library, 711 S Dixieland Rd, *Rogers*, AR 72758
T: +1 479 6211152; Fax: +1 479 6211165; E-mail: library@rpl.lib.ar.us; URL: library.rpl.lib.ar.us
1904; Judy Faye Casey
96 000 vols; 228 curr per; 4000 e-books; 3400 av-mat; 3400 sound-rec
libr loan   41053

Presque Isle District Library, 181 E Erie St, *Rogers City*, MI 49779-1709
T: +1 989 7342477; Fax: +1 989 7344899; E-mail: pidldir@i2k.net; URL: www.pidl.org
Janis K. Stevenson
Michigan Coll;Great Lakes Nautical Coll
65 000 vols; 65 curr per   41054

Rohnert Park-Cotati Regional Library, 6250 Lynne Conde Way, *Rohnert Park*, CA 95428
T: +1 707 5849121; Fax: +1 707 5848561; URL: www.sonomalibrary.org/branches/Rohnert.html
55 000 vols; 50 curr per   41055

Rolla Free Public Library, 900 Pine St, *Rolla*, MO 65401
T: +1 573 3642604; Fax: +1 573 3415768; E-mail: rollalib@fidnet.com; URL: rollapubliclibrary.org
1938; Cheryl A. Goltz
Missouriana
53 000 vols; 121 curr per   41056

Palos Verdes Library District, Peninsula Center Library, 701 Silver Spur Rd, *Rolling Hills Estates*, CA 90274
T: +1 310 3779584; Fax: +1 310 5410095; URL: www.pvld.org
1928; Katherine R. Gould
2 branch libs
230 000 vols; 795 curr per; 2000 e-books; 2500 music scores; 1632 maps; 70 000 microforms; 39 000 av-mat; 18 000 sound-rec; 13 Electronic Media & Resources, 4500 Large Print Bks
libr loan; ALA   41057

Jervis Public Library Association, Inc, 613 N Washington St, *Rome*, NY 13440-4296
T: +1 315 3364570; URL: www.jervislibrary.org
1894; Lisa M. Matte
Engineering (John B. Jervis Papers), Revolutionary War (Bright-Huntington Coll)
120 000 vols; 220 curr per; 110 High Interest/Low Vocabulary Bks, 32 Special Interest Per Sub, 50 Bks on Deafness & Sign Lang, 1900 Large Print Bks, 7000 Talking Bks   41058

Hampshire County Public Library, 153 W Main St, *Romney*, WV 26757
T: +1 304 8223185;
Fax: +1 304 8223955; URL: www.hampshirecopubliclib.com
1942; Brenda Riffle
Census of the County 1810-1920, Local Newspaper Coll (1884-1994)
52 000 vols; 63 curr per   41059

Romulus Public Library, 11121 Wayne Rd, *Romulus*, MI 48174
T: +1 734 9427589; Fax: +1 734 9413575
Shelley DeLano
51 000 vols; 200 curr per   41060

Duchesne County Library, 70 W Lagoon 44-4, *Roosevelt*, UT 84066-2841
T: +1 435 7224441; Fax: +1 435 7223386; URL: www.duchesne.gov.net/library.html
Lori Womack
52 000 vols; 45 curr per   41061

Roosevelt Public Library, 27 W Fulton Ave, *Roosevelt*, NY 11575
T: +1 516 3780222; Fax: +1 516 3781011; URL: www.nassaulibrary.org/roosevelt
1934; Joyce F. Scott
African-American studies
60 000 vols   41062

Douglas County Library System, 1409 NE Diamond Lake Blvd, *Roseburg*, OR 97470
T: +1 541 4404311; Fax: +1 541 9577798; URL: www.co.douglas.or.us/library
1955; Max Leek
Community needs; 10 branch libs
257 000 vols
libr loan; ALA, OLA   41063

Roseland Free Public Library, 20 Roseland Ave, *Roseland*, NJ 07068-1235
T: +1 973 2268636; Fax: +1 973 2266429; URL: www.bccls.org; www.roselandnj.org/library/library.htm
1961; Judith Yankielun Lind
Railroadiana
65 000 vols; 80 curr per   41064

Roselle Free Public Library, 104 W Fourth Ave, *Roselle*, NJ 07203
T: +1 908 2455809; Fax: +1 908 2988881; E-mail: roselle@lmxac.org; URL: www.lmxac.org/roselle
1917; W. Keith McCoy
56 000 vols; 125 curr per   41065

Roselle Public Library District, 40 S Park St, *Roselle*, IL 60172-2020
T: +1 630 5291641; Fax: +1 630 5297579; URL: www.roselle.lib.il.us
1940; Kenneth L. Gross
97 000 vols; 248 curr per; 4753 av-mat
                                             41066

Roselle Park, Veterans Memorial Library, 404 Chestnut St, *Roselle Park*, NJ 07204-1506
T: +1 908 2452456; Fax: +1 908 2459204; URL: www.roselleparklibrary.org
1920; Susan Calantone
Jones Memorial Science Coll, Weissman Coll (decorating)
55 000 vols; 132 curr per   41067

County of Los Angeles Public Library – Rosemead, 8800 Valley Blvd, *Rosemead*, CA 91770-1788
T: +1 626 5735220; Fax: +1 626 2808523; URL: www.colapublib.org/libs/rosemead
Desiree Lee
Californiana Coll
149 000 vols; 133 curr per   41068

Roseville Public Library, 225 Taylor St, *Roseville*, CA 95678-2681
T: +1 916 7745221; Fax: +1 916 7735594; E-mail: library@roseville.ca.us; URL: www.roseville.ca.us/library
1912; Dianne Bish
Railroads; 1 branch libr
159 000 vols; 474 curr per; 300 maps; 14 000 microforms; 2000 av-mat; 2400 sound-rec
libr loan   41069

Bryant Library, Two Paper Mill Rd, *Roslyn*, NY 11576-2193
T: +1 516 6212240; Fax: +1 516 6212542; E-mail: info@bryantlibrary.org; URL: www.nassaulibrary.org/bryant
1878; Elizabeth McCloat
Christopher Morley Coll, Roslyn Architecture Coll, William Cullen Bryant Coll
164 000 vols; 10 000 av-mat; 1480 digital data carriers; 60 Electronic Media & Resources, 1670 Large Print Bks
libr loan   41070

Rossford Public Library, 720 Dixie Hwy, *Rossford*, OH 43460-1289
T: +1 419 6660924; Fax: +1 419 6661989; E-mail: rossford@oplin.org; URL: www.rossfordlibrary.org
1936; Jeannine Wilbarger
65 000 vols; 174 curr per; 52 e-books; 3000 av-mat; 1650 sound-rec; 10 digital data carriers
libr loan; ALA   41071

Atlanta-Fulton Public Library System – Roswell Regional Library, 115 Norcross St, *Roswell*, GA 30075
T: +1 770 6403075; Fax: +1 770 6403077
Louise Conti
142 000 vols   41072

Roswell Public Library, 301 N Pennsylvania Ave, *Roswell*, NM 88201
T: +1 575 6227101;
Fax: +1 575 6227107; URL: www.roswellpubliclibrary.org
1906; Betty Long
Southwest Coll
164 000 vols; 178 curr per; 2966 av-mat; 2970 Audio Bks, 72 Bks on Deafness & Sign Lang, 3640 Large Print Bks
libr loan   41073

Round Lake Area Public Library District, 906 Hart Rd, *Round Lake*, IL 60073
T: +1 847 5467060; Fax: +1 847 5467104; E-mail: webmaster@rlalibrary.org; URL: www.rlalibrary.org
1972; Elizabeth Novak
113 000 vols; 167 curr per; 10 500 av-mat   41074

Round Rock Public Library System, 216 E Main St, *Round Rock*, TX 78664
T: +1 512 2187003; Fax: +1 512 2183272; URL: www.ci.round-rock.tx.us/library
1963; Dale Ricklefs
Williamson County Genealogical Society Coll, Fire Dept Hist Rm
132 000 vols; 427 curr per   41075

County of Los Angeles Public Library – Rowland Heights Library, 1850 Nogales St, *Rowland Heights*, CA 91748-2945
T: +1 626 9125348; Fax: +1 626 8103538; URL: www.colapublib.org/libs/rowlandhts
Susana Rogers
137 000 vols; 186 curr per   41076

Rowlett Public Library, 3900 Main St, *Rowlett*, TX 75088-5075
T: +1 972 4126161; Fax: +1 972 4126153; URL: www.rowlett.lib.tx.us
Kathleen A. Cockcroft
93 000 vols
libr loan   41077

Boston Public Library – Dudley Literacy Center, 65 Warren St, *Roxbury*, MA 02119-3206
T: +1 617 8592446
JoAnn E. Butler
78 000 vols; 40 curr per   41078

Royal Oak Public Library, 222 E Eleven Mile Rd, *Royal Oak*, MI 48067-2633; P.O. Box 494, Royal Oak, MI 48068-0494
T: +1 248 2463700; Fax: +1 248 2463701; URL: www.ropl.org
1922; Metta T. Lansdale
Auto Repair Manuals, Royal Oak Hist Coll
121 000 vols; 250 curr per; 4440 sound-rec; 1800 digital data carriers; 1420 Large Print Bks   41079

Ruidoso Public Library, 107 Kansas City Rd, *Ruidoso*, NM 88345
T: +1 505 2583704; Fax: +1 505 2584619; E-mail: library@ruidoso-nm.gov; URL: www.youseemore.com/ruidosopl/
1950; Beverly McFarland
50 000 vols
FOL   41080

Tampa-Hillsborough County Public Library System – Ruskin Branch, One Dickman Dr SE, *Ruskin*, FL 33570-4314
Fax: +1 813 6717698
Charlanne Purdy
62 000 vols   41081

Tampa-Hillsborough County Public Library System – SouthShore Regional, 15816 Beth Shields Way, *Ruskin*, FL 33573-4093
Fax: +1 813 3721150
Eloise Hurst
118 000 vols   41082

Pope County Library System – Russellville Headquarters Branch, 116 E Third St, *Russellville*, AR 72801
T: +1 479 9684368; Fax: +1 479 9683222
1920; Judy Mays
Local hist; Arkansas Hist, Genealogy; 3 branch libs
115 000 vols; 142 curr per; 3000 microforms; 650 av-mat; 761 sound-rec; 6 digital data carriers
libr loan   41083

Campbell County Public Library, 684 Village Hwy, Lower Level, *Rustburg*, VA 24588; P.O. Box 310, Rustburg, VA 24588-0310
T: +1 434 3329560; Fax: +1 434 3329697; URL: tlc.library.net/campbell
1968; Nan Carmack
3 branch libs, 1 bookmobile
153 000 vols; 285 curr per; 6000 av-mat; 3600 sound-rec   41084

Lincoln Parish Library, 910 N Trenton St, *Ruston*, LA 71270-3328
T: +1 318 5136409; Fax: +1 318 5136446; E-mail: administrator@mylpl.org; URL: www.mylpl.org
1962; Vivian McCain
98 000 vols; 279 curr per; 5 004 av-mat
41085

Rutherford Free Public Library, 150 Park Ave, *Rutherford*, NJ 07070
T: +1 201 9398600; Fax: +1 201 9394108; E-mail: ruth1circ@bccls.org; URL: bccls.org/rutherford
1894; Jane Fisher
William Carlos Williams Coll
115 000 vols; 230 curr per
41086

Rutland Free Library, Ten Court St, *Rutland*, VT 05701-4058
T: +1 802 7731860; Fax: +1 802 7731825; E-mail: rutlandfree@rutlandfree.org; URL: rutlandfree.org
1886; Paula J. Baker
Vermont Hist
80 000 vols; 231 curr per
libr loan
41087

Rutland Free Public Library, 280 Main St, *Rutland*, MA 01543
T: +1 508 8864108; Fax: +1 508 8864141; URL: www.rutlandlibrary.org
Kerry J. Remington
50 000 vols; 100 curr per
41088

Rye Free Reading Room, 1061 Boston Post Rd, *Rye*, NY 10580
T: +1 914 2313161; Fax: +1 914 9675522; E-mail: rfrr@westchesterlibraries.org; URL: www.ryelibrary.org
1884; N.N.
80 000 vols; 177 curr per
libr loan
41089

Dyer Library, 371 Main St, *Saco*, ME 04072
T: +1 207 2833861; E-mail: cspaulding@dyer.lib.me.us; URL: www.dyerlibrarysacomuseum.org/lib_home.shtml
1881; Gerard Morin
Hist (Maine Coll), newspapers & photos (18th-19th c)
60 000 vols; 120 curr per
41090

California Department of Corrections Library System, c/o The Office of Correctional Education, The Office of Correctional Education, 1515 S St, Rm 221 N, *Sacramento*, CA 95814-7243
T: +1 916 3244615; Fax: +1 916 3241416
1989; Janice Cesolini-Stuter
32 branch libs
470 000 vols; 1 660 curr per
libr loan
41091

Sacramento Public Library, 828 I St, *Sacramento*, CA 95814
T: +1 916 2642920; Fax: +1 916 2642755; E-mail: askus@saclibrary.org; URL: www.saclibrary.org
1857; N.N.
Art, Californiana, Hist of Printing, Music, Business and management; 25 branch libs
2 100 000 vols
41092

Saddle Brook Free Public Library, 340 Mayhill St, *Saddle Brook*, NJ 07663
T: +1 201 8433287; Fax: +1 201 8435512; URL: www.bccls.org/saddlebrook
1945; Alma J. Henderson
63 000 vols; 125 curr per
41093

Safety Harbor Public Library, 101 Second St N, *Safety Harbor*, FL 34695
T: +1 727 7241525; Fax: +1 727 7241533; URL: www.tblc.org/shpl/
1938; Lana Bullian
Deafness & Sign Language Coll
65 000 vols; 113 curr per; 4 000 e-books; 3 456 av-mat
libr loan
41094

Safford City-Graham County Library, 808 Seventh Ave, *Safford*, AZ 85546
T: +1 520 4281531; Fax: +1 520 343209; URL: www.saffordlibrary.org
1962; Jan Elliott
Adult Basic Education Coll, Arizona Coll
93 000 vols; 68 curr per
41095

Public Libraries of Saginaw, Hoyt Main Library, 505 Janes Ave, *Saginaw*, MI 48607
T: +1 989 7550904; Fax: +1 989 7559829; E-mail: saginaw@saginawlibrary.org; URL: www.saginawlibrary.org
1890; Kimberly White
African Heritage Coll; 4 branch libs
412 000 vols; 762 curr per; 6 Electronic Media & Resources
41096

Public Libraries of Saginaw – Butman-Fish, 1716 Hancock, *Saginaw*, MI 48602
T: +1 989 7999160; Fax: +1 989 7998149
Paul Lutenske
Large Print Books;African Heritage Coll;Hispanic Heritage Coll
93 000 vols
41097

Public Libraries of Saginaw – Zauel Memorial Library, 3100 N Center Rd, *Saginaw*, MI 48603
T: +1 989 7992771; Fax: +1 989 7991771
Thomas H. Birch
African Heritage Coll;Hispanic Heritage Coll;Parenting Coll
117 000 vols
41098

Thomas Township Library, 8207 Shields Dr, *Saginaw*, MI 48609-4814
T: +1 989 7813770; Fax: +1 989 7813881; URL: www.thomastwp.org/library.htm
Tari L. Dusek
54 000 vols; 130 curr per
41099

Saint Johns County Public Library System – Southeast Branch Library & Administrative Headquarters, 6670 US 1 South, *Saint Augustine*, FL 32086
T: +1 904 8276925; Fax: +1 904 8276930; E-mail: libse@co.st-johns.fl.us; URL: www.sjcpls.org
1875; Mary Jane Little
3 branch libs
398 000 vols
41100

Saint Charles City County Library District – Kathryn Linnemann Branch, 2323 Elm St, *Saint Charles*, MO 63301
T: +1 636 9466294; Fax: +1 636 9470692
Ann King
193 000 vols
41101

Saint Charles City County Library District – Kisker Road Branch, 1000 Kisker Rd, *Saint Charles*, MO 63304-8726
T: +1 636 9267323; Fax: +1 636 9260869
Ken Cascio
100 000 vols
41102

Saint Charles City County Library District – McClay, 2760 McClay Rd, *Saint Charles*, MO 63303-5427
T: +1 636 4417577; Fax: +1 636 4415898
Martha Radginski
95 000 vols
41103

St Charles Public Library District, One S Sixth Ave, *Saint Charles*, IL 60174-2105
T: +1 630 5840076; Fax: +1 630 5843448; E-mail: adultref@stcharleslibrary.org; URL: www.stcharleslibrary.org
1906; Diana M. Brown
Adult New Reader 350 vols
265 000 vols; 1 197 curr per; 16 000 e-books; 75 maps; 41 microforms; 23 071 sound-rec; 5 012 digital data carriers
libr loan; ALA
41104

Saint Clair Shores Public Library, 22500 11 Mile Rd, *Saint Clair Shores*, MI 48081-1399
T: +1 586 7719020; Fax: +1 586 7718935; URL: www.libcoop.net/stclairshores
1935; Rosemary Orlando
Careers, Great Lakes, Michigan; Great Lakes Hist
114 000 vols; 385 curr per
41105

Saint Clairsville Public Library, 108 W Main St, *Saint Clairsville*, OH 43950-1225
T: +1 740 6952062; Fax: +1 740 6956420; URL: www.scpl.lib.oh.us
1941; Sheila D. Perkins
57 000 vols; 161 curr per
41106

Great River Regional Library, 1300 W St Germain St, *Saint Cloud*, MN 56301-3667
T: +1 320 6502500; Fax: +1 320 6502501; URL: www.griver.org
1969; Kirsty Smith
Readmobile-Children's Mat
747 000 vols; 1 554 curr per; 33 000 av-mat
libr loan
41107

Saint Cloud Public Library, 1300 W Saint Germain St, *Saint Cloud*, MN 56301-3667
T: +1 320 6502500; Fax: +1 320 6502501; URL: www.griver.org
Mark Troendle
194 000 vols; 421 curr per
41108

Dorchester County Library, 506 N Parler Ave, *Saint George*, SC 29477-2297
T: +1 843 5639189; Fax: +1 843 5637823; URL: www.dcl.lib.sc.us
1953; Angus M. Prim
South Carolina Hist; 1 branch libr, 1 bookmobile
126 000 vols; 217 curr per; 5 905 av-mat; 5 910 Audio Bks, 750 Electronic Media & Resources
41109

Washington County Library, 88 West 100 South, *Saint George*, UT 84770-3490
T: +1 435 6345737; Fax: +1 435 6345798; URL: www.washco.lib.ut.us
1912; Brenda Brown
WPA Pioneer Diary Coll; 5 branch libs
107 000 vols; 250 curr per
41110

Saint Helena Public Library, George & Elsie Wood Public Library, 1492 Library Lane, *Saint Helena*, CA 94574-1143
T: +1 707 9635244; Fax: +1 707 9635264; E-mail: admin@shpl.org; URL: www.shpl.org
1892; Larry Hlavsa
Napa Valley Wine Libr
92 000 vols; 132 curr per; 5 000 e-books; 35 diss/theses; 32 microforms; 14 000 av-mat; 3 800 sound-rec; 102 digital data carriers
libr loan
41111

Watonwan County Library, 125 Fifth St S, *Saint James*, MN 56081
T: +1 507 3751278; Fax: +1 507 3755415; E-mail: libtwa@tds.lib.mn.us; URL: www.tds.sirsi.net; www.tds.lib.mn.us
1943; Cheryl Bjoin
Agriculture, Tractors; 4 branch libs
83 000 vols; 169 curr per
41112

Apache County Library District, 245 W First S, *Saint Johns*, AZ 85936; P.O. Box 2760, Saint Johns
T: +1 928 3373067; Fax: +1 928 3373960
1976; Judith Pepple
51 000 vols; 93 curr per; 2 250 av-mat
41113

Northeast Regional Library, Vermont Dept of Libraries, 23 Tilton Rd, *Saint Johnsbury*, VT 05819
T: +1 802 7483420; E-mail: NERL@dol.state.vt.us
1936; Michael Roche
75 000 vols
41114

Maud Preston Palenske Memorial Library, Saint Joseph Public Library, 500 Market St, *Saint Joseph*, MI 49085
T: +1 269 9837167; Fax: +1 269 9835804; E-mail: sjlibrarymi@yahoo.com; URL: www.stjoseph.lib.mi.us
1903; Mary Kynast
Cook Nuclear Plant NCR Docs
113 000 vols; 180 curr per
41115

Rolling Hills Consolidated Library, 1904 N Belt Hwy, *Saint Joseph*, MO 64506-2201
T: +1 816 2325479; Fax: +1 816 2362133; URL: www.rollinghills.lib.mo.us
Barbara Read
77 000 vols
41116

St Joseph Public Library, 927 Felix St, *Saint Joseph*, MO 64501-2799
T: +1 816 2324038; Fax: +1 816 2327516; URL: sjpl.lib.mo.us
1891; Mary Beth Revels
Lit (Eugene Field Coll), Dr Wayne Toothaker Medical Libr; 3 branch libs
238 000 vols; 241 curr per; 2 000 e-books; 4 400 Large Print Bks
41117

St Joseph Public Library – East Hills Library, 502 N Woodbine Rd, *Saint Joseph*, MO 64506
T: +1 816 2362136; Fax: +1 816 2361429
Steven Kent Olson
74 000 vols
41118

Saint Louis County Library, 1640 S Lindbergh Blvd, *Saint Louis*, MO 63131-3598
T: +1 314 9943300; Fax: +1 314 9977602; URL: www.slcl.org
1946; Charles Pace
20 branch libs
2 112 000 vols; 6 550 curr per; 22 297 digital data carriers; 55 593 Audio Bks, 1 420 Electronic Media & Resources, 75 230 CDs, 66 492 Videos
libr loan
41119

Saint Louis Public Library, 1301 Olive St, *Saint Louis*, MO 63103-2325
T: +1 314 2412288; Fax: +1 314 5390393; E-mail: webmaster@slpl.org; URL: www.slpl.org
1865; Waller McGuire
Architecture (Steedman Coll), Genealogy Coll, Black Hist (Julia Davis Coll), NJ Werner Coll of Typography, St Louis Media Arch; 16 branch libs
4 680 000 vols; 7 637 curr per
41120

Saint Louis Public Library – Buder, 4401 Hampton Ave, *Saint Louis*, MO 63109-2237
T: +1 314 3522900; Fax: +1 314 3525387
James Moses
87 000 vols
41121

Saint Louis Public Library – Carondelet, 6800 Michigan Ave, *Saint Louis*, MO 63111
T: +1 314 7529224; Fax: +1 314 7527794
Jennifer Halla-Sindelar
54 000 vols
41122

Saint Louis Public Library – Carpenter, 3309 S Grand Ave, *Saint Louis*, MO 63118
T: +1 314 7726586; Fax: +1 314 7721871
Cynthia E. Jones
66 000 vols
41123

Saint Louis Public Library – Julia Davis Branch, 4415 Natural Bridge Rd, *Saint Louis*, MO 63115
T: +1 314 3833021; Fax: +1 314 3830251
Floyd Council
72 000 vols
41124

Saint Louis Public Library – Machacek, 6424 Scanlan Ave, *Saint Louis*, MO 63139
T: +1 314 7812948; Fax: +1 314 7818441
Nancy Duerhoff
58 000 vols
41125

Saint Louis Public Library – Schlafly, 225 N Euclid Ave, *Saint Louis*, MO 63108
T: +1 314 3674120; Fax: +1 314 3674814
Leandrea Lucas
50 000 vols
41126

Saint Martin Parish Library, 201 Porter St, *Saint Martinville*, LA 70582
T: +1 337 3942207; Fax: +1 337 3942248; URL: www.stmartin.lib.la.us
1955; Jeanne A. Essmeier
Louisiana, Hist, Genealogy; 6 branch libs
143 000 vols; 278 curr per
41127

Pottawatomie Wabaunsee Regional Library, 306 N Fifth St, *Saint Marys*, KS 66536-1404
T: +1 785 4372778; Fax: +1 785 4372778; E-mail: illpowab@oct.net; URL: skyways.lib.ks.us/library/pottwablib
1962; Judith Cremer
3 branch libs
115 000 vols; 141 curr per; 10 000 e-books; 3 000 av-mat; 2 800 sound-rec; 7 107 Large Print Bks
libr loan; ALA, KLA
41128

St Marys Community Public Library, 140 S Chestnut St, *Saint Marys*, OH 45885-2307
T: +1 419 3947471; Fax: +1 419 3947291; URL: www.stmarys.lib.oh.us
1922; Susan Heckler Pittman
Jim Tully Coll
65 000 vols; 135 curr per; 1 432 av-mat
41129

Saint Paul Public Library, 90 W Fourth St, mail address: 2109 Wilson Ave, *Saint Paul*, MN 55102-1668
T: +1 651 2667000; Fax: +1 651 2667060; URL: www.sppl.org
1882; Melanie Huggins
12 branch libs
999 000 vols; 1 634 curr per
libr loan
41130

Saint Paul Public Library – Hayden Heights, 1456 White Bear Ave, *Saint Paul*, MN 55106-2405
T: +1 651 7933934; Fax: +1 651 7933936; E-mail: branch.hayden-heights@ci.stpaul.mn.us
Patty Krezowski
63 000 vols; 98 curr per
41131

Saint Paul Public Library- Highland Park, 1974 Ford Pkwy, *Saint Paul*, MN 55116-1922
T: +1 651 6953700; Fax: +1 651 6953701; E-mail: branch.highland@ci.stpaul.mn.us
Pat Gerlach
127 000 vols; 129 curr per
41132

Saint Paul Public Library – Lexington Outreach, 1080 W University Ave, *Saint Paul*, MN 55104-4707
T: +1 651 6420359; Fax: +1 651 6420381
Alice Neve
64 000 vols; 98 curr per
41133

Saint Paul Public Library – Merriam Park, 1831 Marshall Ave, *Saint Paul*, MN 55104-6010
T: +1 651 6420385; Fax: +1 651 6420391; E-mail: branch.merriam@ci.stpaul.mn.us
Linda Valen
69 000 vols; 114 curr per
41134

Saint Paul Public Library – Sun Ray, 2105 Wilson Ave, *Saint Paul*, MN 55119-4033
T: +1 651 5016300; Fax: +1 651 5016303; E-mail: branch.sunray@ci.stpaul.mn.us
Karen Kolb Peterson
70 000 vols; 77 curr per
41135

St Pete Beach Public Library, 365 73rd Ave, *Saint Pete Beach*, FL 33706-1996
T: +1 727 3639238; Fax: +1 727 5521760; E-mail: libadmin@stpetebeach.org; URL: www.tblc.org/spb
Roberta L. Whipple
52 000 vols; 60 curr per; 350 av-mat
41136

Saint Charles City County Library District, 77 Boone Hills Dr, *Saint Peters*, MO 63376-0529; P.O. Box 529, Saint Peters, MO 63376-0529
T: +1 636 4412300; Fax: +1 636 4413132; E-mail: library@stchlibrary.org; URL: www.youranswerplace.org
1973; Carl Sandstedt
12 branch libs
657 000 vols; 8 507 curr per; 8 000 e-books
41137

Saint Charles City County Library District – Spencer Road Branch, 427 Spencer Rd, *Saint Peters*, MO 63376; P.O. Box 529, Saint Peters, MO 63376-0529
T: +1 636 4410522; Fax: +1 636 9263948

Jim Brown
Business Services
185 000 vols
41138

Saint Petersburg Public Library, 3745 Ninth Ave N, *Saint Petersburg*, FL 33713
T: +1 727 8937724; Fax: +1 727 8925432; URL: splibraries.org
1910; Mary Gaines
5 branch libs
436 000 vols; 886 curr per
41139

Roanoke County Public Library – Glenvar, 3917 Daugherty Rd, *Salem*, VA 24153
T: +1 540 3876163; Fax: +1 540 3803951
John Vest
51 000 vols; 91 curr per
41140

Salem Public Library, 821 E State St, *Salem*, OH 44460-2298
T: +1 330 3320042; Fax: +1 330 3324488; E-mail: library@salem.lib.oh.us; URL: www.salem.lib.oh.us
1895; Bradley K. Stephens
Columbiana County & Salem Hist, Quaker Hist & Biogr
75 000 vols; 225 curr per; 3 056 av-mat
libr loan; ALA
41141

Salem Public Library, 370 Essex St, *Salem*, MA 01970-3298
T: +1 978 7440860; Fax: +1 978 7458616; E-mail: sal@noblenet.org; URL: www.noblenet.org
1888; Lorraine Jackson
129 000 vols; 150 curr per
41142

Salem Public Library, 585 Liberty St SE, *Salem*, OR 97301; P.O. Box 14810, Salem
T: +1 503 5886060; Fax: +1 503 5886055; E-mail: library@cityofsalem.net; URL: www.salemlibrary.org
1904; Gail Warner
Oregon Hist Coll, Original Art, Salem Hist photogr Coll; 1 branch libr, 1 bookmobile
485 000 vols; 839 curr per
libr loan; ALA
41143

Salem Public Library, 28 E Main St, *Salem*, VA 24153
T: +1 540 3753089; Fax: +1 540 3897054; E-mail: library@salemva.gov; URL: www.salemlibrary.info
1969; Janis C. Augustine
Lit for Visually Handicapped (Listening Libr), cassettes, phonodiscs
113 000 vols; 138 curr per; 26 000 e-books; 4 800 av-mat; 6 500 sound-rec; 2 452 digital data carriers; 55 Electronic Media & Resources
41144

Stanislaus County Free Library – Salida Branch, 4835 Sisk Rd, *Salida*, CA 95368-9445
T: +1 209 5437353; Fax: +1 209 5437318
Stacey Chen
78 000 vols
41145

Salina Public Library, 301 W Elm St, *Salina*, KS 67401
T: +1 785 8254624; Fax: +1 785 8230706; E-mail: webmaster@salpublib.org; URL: www.salpublib.org
1897; Joe McKenzie
Hist, Education, Art; Kansas Hist
264 000 vols; 352 curr per
41146

Salinas Public Library – Cesar Chavez Library, 615 Williams Rd, *Salinas*, CA 93905
T: +1 831 7587345; Fax: +1 831 7589172; E-mail: cclib@salinas.lib.ca.us
Maria Roddy
Chicano Cultural Resource Center
55 000 vols; 86 curr per
41147

Salinas Public Library – El Gabilan, 1400 N Main St, *Salinas*, CA 93906
T: +1 831 7587302; Fax: +1 831 4420817; E-mail: eglib@salinas.lib.ca.us
Maria Roddy
54 000 vols; 55 curr per
41148

Salinas Public Library – John Steinbeck Library, 350 Lincoln Ave, *Salinas*, CA 93901
T: +1 831 7587314; Fax: +1 831 7587336; E-mail: library@salinas.lib.ca.us; URL: www.salinas.lib.ca.us
1909; Elizabeth Martinez
John Steinbeck Coll; 2 branch libs
205 000 vols; 271 curr per
41149

Saline District Library, 555 N Maple Rd, *Saline*, MI 48176
T: +1 734 9440600; E-mail: leslee@salinelibrary.org; URL: saline.lib.mi.us
1900; Leslee Niethammer
52 480 vols; 120 curr per
41150

Rowan Public Library, 201 W Fisher St, *Salisbury*, NC 28144-4935; P.O. Box 4039, Salisbury, NC 28145-4039
T: +1 704 2168256; Fax: +1 704 2168246; E-mail: info@co.rowan.nc.us; URL: www.rowanpubliclibrary.org
1911; Phillip K. Barton
2 branch libs, 1 bookmobile
227 000 vols; 386 curr per; 555 microforms; 3 855 sound-rec; 210 digital data carriers; 10 Electronic Media & Resources
libr loan; ALA
41151

Wicomico Public Library, 122 S Division St, *Salisbury*, MD 21801
T: +1 410 7493612; Fax: +1 410 5482968; E-mail: askus@wicomico.org; URL: www.wicomicolibrary.org
Tom Hehman
155 000 vols; 281 curr per
41152

Salt Lake City Public Library, 210 E 400 S, *Salt Lake City*, UT 84111-3280
T: +1 801 5248200; Fax: +1 801 3228196; URL: www.slcpl.org
1898; Beth Elder
Zine Libr, Hist of Salt Lake City; 5 branch libs
662 000 vols; 2 670 curr per; 50 microforms; 80 000 sound-rec; 2 134 digital data carriers
41153

Salt Lake County Library Services, 2197 E Fort Union Blvd, *Salt Lake City*, UT 84121-3139
T: +1 801 9434636; Fax: +1 801 9426323; URL: www.slco.lib.ut.us
1938; James D. Cooper
19 branch libs
2 000 000 vols; 7 000 curr per
41154

Salt Lake County Library Services- Calvin S. Smith Branch, 810 E 3300 South, *Salt Lake City*, UT 84106-1534
T: +1 801 9447630; Fax: +1 801 4853243
Danette Hantla
52 000 vols
41155

Salt Lake County Library Services – East Millcreek, 2266 E Evergreen Ave, *Salt Lake City*, UT 84109-2927
T: +1 801 9447623; Fax: +1 801 2789016
Suzanne Tronier
96 000 vols
41156

Salt Lake County Library Services – Holladay, 2150 E Murray-Holladay Rd, *Salt Lake City*, UT 84117-5241
T: +1 801 9447629; Fax: +1 801 2788947
Steve Pierson
114 000 vols
41157

Salt Lake County Library Services – Hunter Branch, 4740 W 4100 South, *Salt Lake City*, UT 84120-4948
T: +1 801 9447594; Fax: +1 801 9688350
Ruby Cheesman
180 000 vols
41158

Salt Lake County Library Services – Whitmore Branch, 2197 E Fort Union Blvd, *Salt Lake City*, UT 84121-3139
T: +1 801 9447666; Fax: +1 801 9447534
Kent Dean
207 000 vols
41159

Calaveras County Library, 891 Mountain Ranch Rd, *San Andreas*, CA 95249
T: +1 209 7546510; Fax: +1 209 7546512; URL: www.calaveraslibrary.com
1939; Maurie Hoekstra
Mining
85 000 vols; 105 curr per
41160

Tom Green County Library System, 113 W Beauregard, *San Angelo*, TX 76903
T: +1 915 6557321; Fax: +1 915 6594027; E-mail: infodesk@co.tom-green.tx.us; URL: www.co.tom-green.tx.us
1923; Larry D. Justiss
2 branch libs
304 000 vols; 368 curr per; 3 338 av-mat; 3 338 Audio Bks, 4 100 Large Print Bks, 3 340 Talking Bks
41161

San Antonio Public Library, 600 Soledad, *San Antonio*, TX 78205-2786
T: +1 210 2072617; Fax: +1 210 2072622; E-mail: librarywebadmin@sanantonio.gov; URL: www.sanantonio.gov/library
1903; Ramiro S. Salazar
Texana Coll, Hertzberg Circus Coll, Latino Coll; 30 branch libs
2 156 000 vols; 2 000 curr per
libr loan; ALA, PLA, TLA
41162

San Bernardino County Library, 104 W Fourth St, *San Bernardino*, CA 92415-0035
T: +1 909 3875531; Fax: +1 909 3875724; URL: www.sbcounty.gov/library
1914; Ed Kieczykowski
San Bernardino County Hist Coll, Mines & Mineral Resources Coll; 28 branch libs
1 041 000 vols
41163

San Bernardino Public Library, Norman F. Feldheym Public Library, 555 W Sixth St, *San Bernardino*, CA 92410-3001
T: +1 909 3818226; Fax: +1 909 3818229; URL: www.sbpl.org
1891; Ophelia Roop
3 branch libs
219 000 vols; 489 curr per
libr loan
41164

San Bruno Public Library, 701 Angus Ave W, *San Bruno*, CA 94066-3490
T: +1 650 6167078; Fax: +1 650 8760848; E-mail: sbpl@plsinfo.org; URL: www.sanbrunolibrary.org
1916; Terry Jackson
120 000 vols
41165

San Mateo County Library – San Carlos Branch, 610 Elm St, *San Carlos*, CA 94070
T: +1 650 5910341; Fax: +1 650 5911585; URL: www.sancarloslibrary.org
Chet Mulawka
116 000 vols
41166

Orange County Public Library – San Clemente Branch, 242 Avenida Del Mar, *San Clemente*, CA 92672-4005
T: +1 949 4923493; Fax: +1 949 4985749
Terry Pringle
73 000 vols; 112 curr per
41167

San Diego County Library, 5555 Overland Ave, Ste 1511, *San Diego*, CA 92123-1245
T: +1 619 5883715; Fax: +1 858 4955981; E-mail: info@sdcl.org; URL: www.sdcl.org
Jose A. Aponte
Tagalog, Laotian, Italian, Russian, Armenian, Mandarin, Hindi, Hmong, Hebrew, Cantonese, German, Miao, Mon-Khmer, Cambodian, Punjabi, Thai, Urdu, African Languages, Japanese, Pashto, Persian, Arabic, French, Greek, Korean, Spanish, Scandinavian Languages, Vietnamese, Portuguese
1 163 000 vols; 5 533 curr per; 1 000 e-books
41168

San Diego Public Library, Central Library, 820 E St, *San Diego*, CA 92101-6478
T: +1 619 2365800; Fax: +1 619 2386639; URL: www.sandiego.gov/public-library
1882; Deborah Barrow
California Colls, Rare Books & Hist of Printing (Wangenheim Coll); 34 branch libs
3 256 000 vols; 4 124 curr per; 7 000 e-books
41169

San Diego Public Library – Mountain View-Beckwourth, 721 San Pasqual St, *San Diego*, CA 92113
T: +1 619 5273404; Fax: +1 619 5273408; E-mail: bwstaff@sandiego.gov
Catherine Greene
60 000 vols
41170

County of Los Angeles Public Library – San Dimas Library, 145 N Walnut Ave, **San Dimas**, CA 91773-2603
T: +1 909 5996738; Fax: +1 909 5924490; URL: www.colapublib.org/libs/sandimas
Pui-Ching Ho
110 000 vols; 143 curr per   41171

County of Los Angeles Public Library – San Fernando Library, 217 N Maclay Ave, **San Fernando**, CA 91340-2433
T: +1 818 3656928; Fax: +1 818 3653820; URL: www.colapublib.org/libs/sanfernando
Sue Yamamoto
70 000 vols; 15 curr per   41172

San Francisco Public Library, 100 Larkin St, **San Francisco**, CA 94102-4733
T: +1 415 5574406; Fax: +1 415 5574424; E-mail: info@sfpl.org; URL: www.sfpl.org
1878; Luis Herrera
Calligraphy Coll, Chinese Language, Spanish Language, Gay & Lesbian Arch, Science Fiction; 27 branch libs
20 889 000 vols; 10 103 curr per; 21 000 e-books
libr loan   41173

County of Los Angeles Public Library – San Gabriel Library, 500 S Del Mar Ave, **San Gabriel**, CA 91776-2408
T: +1 626 2870761; Fax: +1 626 2852610; URL: www.colapublib.org/libs/sangabriel
Julie Sorensen
80 000 vols; 75 curr per   41174

San Jose Public Library – Almaden, 6445 Camden Ave, **San Jose**, CA 95120
T: +1 408 8083040; Fax: +1 408 9971212
Pamela Crider
101 000 vols   41175

San Jose Public Library – Berryessa, 3355 Noble Ave, **San Jose**, CA 95132-3198
T: +1 408 9233336; Fax: +1 408 9233222
Donna Ward
116 000 vols   41176

San Jose Public Library – Biblioteca Latino Americana, 921 S First St, **San Jose**, CA 95110-2939
T: +1 408 2941237; Fax: +1 408 2974278
Linda Mendez-Ortiz
77 000 vols   41177

San Jose Public Library – Calabazas, 1230 Blaney Ave, **San Jose**, CA 95129-3799
T: +1 408 8651873; Fax: +1 408 2557597
Debbie Erwin
92 000 vols   41178

San Jose Public Library – Cambrian, 1780 Hillsdale Ave, **San Jose**, CA 95124-3199
T: +1 408 8083080; Fax: +1 408 2641894
Hannah Slocum
133 000 vols   41179

San Jose Public Library – Dr Roberto Cruz – Alum Rock, 3090 Alum Rock Ave, **San Jose**, CA 95127
T: +1 408 8083090; Fax: +1 408 9285628
Judith Gregg
105 000 vols   41180

San Jose Public Library – Educational Park, 1770 Educational Park Dr, **San Jose**, CA 95133-1703
T: +1 408 2723663; Fax: +1 408 2516971
Meg Omainsky
88 000 vols   41181

San Jose Public Library – Evergreen, 2635 Aborn Rd, **San Jose**, CA 95121-1294
T: +1 408 2380384; Fax: +1 408 2380548
Angela McCarren
109 000 vols   41182

San Jose Public Library – Hillview, 1600 Hopkins Dr, **San Jose**, CA 95122-1199
T: +1 408 7299516; Fax: +1 408 7299518
Kim Nguyen
80 000 vols   41183

San Jose Public Library – Rose Garden, 1580 Naglee Ave, **San Jose**, CA 95126-2094
T: +1 408 8083070; Fax: +1 408 9990909
Gayleen Thomas
79 000 vols   41184

San Jose Public Library – Seventrees, 3597 Cas Dr, **San Jose**, CA 95111-2499
T: +1 408 2775317; Fax: +1 408 3651736
Keye Luke
68 000 vols   41185

San Jose Public Library – Tully Community, 880 Tully Rd, **San Jose**, CA 95111
T: +1 408 8083030; Fax: +1 408 9773113
Carol DaSilva
118 000 vols   41186

San Jose Public Library – Vineland, 1450 Blossom Hill Rd, **San Jose**, CA 95118
T: +1 408 8083000; Fax: +1 408 9781080
Mary Ellen Westmoreland
148 000 vols   41187

San Jose Public Library West Valley, 1243 San Tomas Aquino Rd, **San Jose**, CA 95117-3399
T: +1 408 2774905; Fax: +1 408 9843736
Daisy Porter
121 000 vols   41188

Orange County Public Library – San Juan Capistrano Regional Library, 31495 El Camino Real, **San Juan Capistrano**, CA 92675-2600
T: +1 949 4931752; Fax: +1 949 2407680
Teri Garza
82 000 vols; 103 curr per   41189

San Leandro Public Library, 300 Estudillo Ave, **San Leandro**, CA 94577
T: +1 510 5773970; Fax: +1 510 5773967; URL: www.sanleandrolibrary.org
1906; David R. Bohne
3 branch libs
255 000 vols; 463 curr per; 5 000 govt docs; 2 500 maps; 4 000 microforms; 10 000 av-mat; 4 221 sound-rec; 111 digital data carriers; 6 500 pamphlets; 2 500 local hist photos   41190

Alameda County Library – San Lorenzo Branch, 395 Paseo Grande, **San Lorenzo**, CA 94580-2491
T: +1 510 6706283; Fax: +1 510 3178497
Anthony Dos Santos
105 000 vols; 302 curr per   41191

San Luis Obispo City-County Library, 995 Palm St, **San Luis Obispo**, CA 93401; P.O. Box 8107, San Luis Obispo, CA 94303-8107
T: +1 805 7815774; Fax: +1 805 7811320; URL: www.slolibrary.org
1919; Brian A.. Reynolds
14 branch libs
439 000 vols; 766 curr per
libr loan   41192

San Diego County Library – San Marcos Branch, No 2 Civic Center Dr, **San Marcos**, CA 92069-2949
T: +1 760 8913000; Fax: +1 760 8913015
Ann Terrell
97 000 vols   41193

San Marcos Public Library, 625 E Hopkins, **San Marcos**, TX 78666
T: +1 512 3938200; E-mail: smpl@ci.san-marcos.tx.us; URL: ci.san-marcos.tx.us/library.htm
1966; Stephanie Langenkamp
140 000 vols; 230 curr per   41194

San Marino Public Library, 1890 Huntington Dr, **San Marino**, CA 91108-2595
T: +1 626 3000777; Fax: +1 626 2840766; URL: www.sanmarinopl.org
1953; Carolyn Crain
California, Travel, Art
90 000 vols; 200 curr per   41195

San Mateo County Library, 125 Lessingia Ct, **San Mateo**, CA 94402-4000
T: +1 650 3125258; Fax: +1 650 3125382; URL: www.smcl.org
1915; Martin Gomez
Libr & Info Science; 11 branch libs
619 000 vols; 1 729 curr per
libr loan   41196

San Mateo Public Library, 1100 Park Ave, **San Mateo**, CA 94403-7108
T: +1 650 5227818; Fax: +1 650 5227801; URL: www.smplibrary.org
1899; Benjamin Ocon
2 branch libs
280 000 vols; 794 curr per
libr loan   41197

Los Angeles Public Library System – San Pedro Regional, 931 S Gaffey St, **San Pedro**, CA 90731-1349
T: +1 310 5487779; Fax: +1 310 5487453
Brenda Hicks
80 000 vols   41198

Marin County Free Library, 3501 Civic Center Dr, Rm 414, **San Rafael**, CA 94903-4177
T: +1 415 4993220; Fax: +1 415 4993726; E-mail: library@co.marin.ca.us; URL: www.marinlibrary.org
1926; Gail Haar
Frank Lloyd Wright Coll, CA Indian Libr Coll, San Quentin Prison Coll; 11 branch libs
416 000 vols; 9 000 e-books
libr loan   41199

Marin County Free Library – Civic Center Branch, 3501 Civic Center Dr, Rm 427, **San Rafael**, CA 94903-4177
T: +1 415 4996056
Damon Hill
San Quentin Prison Coll;Frank Lloyd Wright Coll;CA Indian Library Coll
100 000 vols   41200

San Rafael Public Library, 1100 E St, **San Rafael**, CA 94901-1900
T: +1 415 4853319; Fax: +1 415 4853403
1887; David Dodd
Audio Bks, California Coll
128 000 vols; 302 curr per; 2 039 av-mat; 650 sound-rec; 2 650 Large Print Bks, 2 040 Talking Bks   41201

Contra Costa County Library – San Ramon Library, 100 Montgomery St, **San Ramon**, CA 94583-4707
T: +1 925 9732850; Fax: +1 925 8666720
Anna Koch
87 000 vols   41202

East Bonner County Free Library District, 1407 Cedar St, **Sandpoint**, ID 83864-2052
T: +1 208 2636930; Fax: +1 208 2638320; URL: www.ebcl.lib.id.us/ebcl/
1912; Wayne L. Gunter
Northwest Coll; 2 branch libs
77 000 vols; 215 curr per   41203

Sandusky Library, 114 W Adams St, **Sandusky**, OH 44870
T: +1 419 6253834; Fax: +1 419 6254574; E-mail: comments@sandusky.lib.oh.us; URL: www.sandusky.lib.oh.us
1895; Julie Brooks
2 branch libs
208 000 vols; 320 curr per   41204

Sandwich Public Library, 142 Main St, **Sandwich**, MA 02563
T: +1 508 8880625; Fax: +1 508 8331076; E-mail: spllib@comcast.net; URL: www.sandwichpubliclibrary.com/
1891; Joanne Lamothe
Glass technology; Glass Bks
58 000 vols; 288 curr per   41205

Salt Lake County Library Services – Sandy Branch, 10100 S Petunia Way, **Sandy**, UT 84092-4380
T: +1 801 9447574; Fax: +1 801 5728247
Cheryl Mansen
229 000 vols   41206

Atlanta-Fulton Public Library System – Sandy Springs Regional Library, 395 Mount Vernon Hwy, **Sandy Springs**, GA 30328
T: +1 404 3036130; Fax: +1 404 3036133
Dorothy Parker
123 000 vols   41207

Louis B. Goodall Memorial Library, Sanford Library Association, 952 Main St, **Sanford**, ME 04073
T: +1 207 3244714; Fax: +1 207 3245982; URL: www.lbgoodall.org
Jackie McDougal
70 000 vols; 127 curr per   41208

Seminole County Public Library System, 150 N Palmetto Ave, **Sanford**, FL 32771; mail address: 1101 E First St, Sanford, FL 32771
T: +1 407 6651550; Fax: +1 407 6651610; URL: www.seminolecountyfl.gov/lls/library
1987; J. Suzy Goldman
5 branch libs
531 000 vols; 631 curr per; 27 000 e-books   41209

Fresno County Public Library – Sanger Branch, 1812 Seventh St, **Sanger**, CA 93657-2805
T: +1 559 8752435
Barbara Light
57 000 vols   41210

Sanibel Public Library District, 770 Dunlop Rd, **Sanibel**, FL 33957
T: +1 239 4722483; Fax: +1 239 4729524; URL: www.sanlib.org
1962; Margaret Mohundro
62 000 vols; 226 curr per   41211

Orange County Public Library, 1501 E St Andrew Pl, **Santa Ana**, CA 92705-4048
T: +1 714 5663000; Fax: +1 714 5663042; URL: www.ocpl.org
1921; John M. Adams
Business, Popular music, Chinese, Japanese, Korean & Spanish language; 32 branch libs
2 500 000 vols; 5 070 curr per; 2 860 av-mat; 2 860 Talking Bks   41212

Santa Ana Public Library, 26 Civic Center Plaza, **Santa Ana**, CA 92701-4010
T: +1 714 6475250; Fax: +1 714 6475356; URL: www.santa-ana.org/library
1891; Rob Richard
California Hist Coll, Business, Spanish Language Coll, Hispanic & Asian American Heritage; 1 branch libr
305 000 vols; 275 curr per   41213

Santa Ana Public Library – Newhope Library Learning Center, 122 N Newhope, **Santa Ana**, CA 92703
T: +1 714 6476992; Fax: +1 714 5549633
69 000 vols; 50 curr per   41214

Santa Barbara Public Library, 40 E Anapamu St, **Santa Barbara**, CA 93101-2722; P.O. Box 1019, Santa Barbara, CA 93102-1019
T: +1 805 9627653; Fax: +1 805 5645626; URL: www.sbplibrary.org
Carol L. Keator
8 branch libs
183 000 vols; 405 curr per   41215

County of Los Angeles Public Library – Canyon Country Jo Anne Darcy Library, 18601 Soledad Canyon Rd, **Santa Clarita**, CA 91351
T: +1 661 2512720; Fax: +1 661 2987137; URL: www.colapublib.org/libs/canyoncountry
Marta Wiggins
100 000 vols; 321 curr per   41216

County of Los Angeles Public Library – Valencia Library, 23743 W Valencia Blvd, **Santa Clarita**, CA 91355-2191
T: +1 661 2598942; Fax: +1 616 2597187; URL: www.colapublib.org/libs/valencia/
Sue Yamamoto
215 000 vols; 119 curr per   41217

Santa Cruz City-County Library System Headquarters – Branciforte, 230 Gault St, *Santa Cruz*, CA 95062-2599
T: +1 831 4206330; Fax: +1 831 4206332
Linda Gault
53 000 vols   41218

Santa Cruz City-County Library System Headquarters – Central, 224 Church St, *Santa Cruz*, CA 95060-3873
T: +1 831 4205700; Fax: +1 831 4205701
Barbara Snider
213 000 vols   41219

Santa Cruz City-County Library System Headquarters – Live Oak, 2380 Portola Dr, *Santa Cruz*, CA 95062
T: +1 831 4205359; Fax: +1 831 4657227
Paula Contreras
54 000 vols   41220

New Mexico State Library – Library for the Blind & Physically Handicapped, 1209 Camino Carlos Rey, *Santa Fe*, NM 87507-5166
T: +1 505 4769770; Fax: +1 505 4769776; E-mail: lbph@state.nm.us; URL: www.nmstatelibrary.org
John Mugford
292 000 vols   41221

Santa Fe Public Library, 145 Washington Ave, *Santa Fe*, NM 87501
T: +1 505 9556789; Fax: +1 505 9556676; E-mail: library@ci.santa-fe.nm.us; URL: www.santafelibrary.org
1896; Patricia C. Hodapp
1 branch libr
336 000 vols; 568 curr per; 400 music scores; 5 200 av-mat; 44 000 sound-rec; 2 820 digital data carriers; 413 High Interest/Low Vocabulary Bks, 3 050 Large Print Bks   41222

Santa Fe Springs City Library, 11700 E Telegraph Rd, *Santa Fe Springs*, CA 90670-3600
T: +1 562 8687738; Fax: +1 562 9293680; E-mail: library@santafesprings.org; URL: www.santafesprings.org/depts/community_serv/library/default.asp
1961; Hilary Gordon Keith
Whittier Area Genealogical Society Coll
88 000 vols; 225 curr per   41223

Santa Maria Public Library, 421 S McClelland St, *Santa Maria*, CA 93454-5116
T: +1 805 9250994; Fax: +1 805 9222330; E-mail: libraryreference@ci.santa-maria.ca.us; URL: www.ci.santa-maria.ca.us/210.shtml
1909; Jack Buchanan
3 branch libs
167 314 vols; 436 curr per; 2 039 e-journals; 6 966 govt docs; 1 354 microforms; 19 000 av-mat; 7 000 sound-rec; 14 digital data carriers
libr loan; AMIGOS, OCLC   41224

Santa Monica Public Library, 601 Santa Monica Blvd, *Santa Monica*, CA 90401
T: +1 310 4588606; Fax: +1 310 3948951; URL: www.smpl.org
1890; Greg Mullen
3 branch libs
350 000 vols; 850 curr per
libr loan   41225

Blanchard-Santa Paula Public Library District, Blanchard Community Library, 119 N Eighth St, *Santa Paula*, CA 93060-2709
T: +1 805 5253615; Fax: +1 805 9332324; URL: www.rain.org/~stapaula
1909; Daniel O. Robles
Hispanic Bilingual Coll
80 000 vols; 79 curr per; 85 av-mat
libr loan; ALA   41226

Sonoma County Library, Third & E Sts, *Santa Rosa*, CA 95404-4400
T: +1 707 5450831; Fax: +1 707 5750437; URL: www.sonomalibrary.org
1965; Sandra M. Cooper
Sonoma County Wine Libr; 13 branch libs
722 000 vols; 1 758 curr per
libr loan   41227

San Diego County Library – Santee Branch, 9225 Carlton Hills Blvd, No 17, *Santee*, CA 92071-3192
T: +1 619 4481863; Fax: +1 619 4481497
Penny Taylor
79 000 vols   41228

Fruitville Public Library, 100 Coburn Rd, *Sarasota*, FL 34240
T: +1 941 8612500; Fax: +1 941 8612528; URL: www.sclibs.net
Ann Ivey
Spanish Language;Florida Coll
75 000 vols; 170 curr per; 3 000 av-mat
                                41229

Gulf Gate Public Library, 7112 Curtiss Ave, *Sarasota*, FL 34231
T: +1 941 8611230; Fax: +1 941 3161221; URL: www.sclibs.net
Shirley Birkett
90 000 vols; 175 curr per   41230

Selby Public Library, 1331 First St, *Sarasota*, FL 34236-4899
T: +1 941 8611120; Fax: +1 941 3161188; URL: suncat.co.sarasota.fl.us
1907; Liz Nolan
193 000 vols; 372 curr per
libr loan   41231

Santa Clara County Library Saratoga Community Library, 13650 Saratoga Ave, *Saratoga*, CA 95070-5099
T: +1 408 8676129; Fax: +1 408 8679806
Barbara Morrow Williams
205 000 vols   41232

Saratoga Springs Public Library, 49 Henry St, *Saratoga Springs*, NY 12866
T: +1 518 5847860; Fax: +1 518 5847866; URL: www.sspl.org
1950; Harry Dutcher
Balneology, Hydrotherapy, Saratogiana
207 000 vols; 400 curr per   41233

Southern Adirondack Library System, 22 Whitney Pl, *Saratoga Springs*, NY 12866-4596
T: +1 518 5847300; Fax: +1 518 5875589; URL: www.sals.edu
1958; Sara Dallas
94 000 vols; 25 curr per
libr loan   41234

Satellite Beach Public Library, 751 Jamaica Blvd, *Satellite Beach*, FL 32937
T: +1 321 7794004; Fax: +1 321 7794036
1966; Nancy Grout
Florida
76 860 vols; 216 curr per   41235

Saugus Public Library, 295 Central St, *Saugus*, MA 01906
T: +1 781 2314168; Fax: +1 781 2314169; E-mail: sau@noblenet.org; URL: www.saugus.ma.us/Library
1888
Ceramics & Glass (Dorothy E Lunt Coll), Civil War (Franklin P. Bennett Jr Coll), Music (Lt Col Harry J. Jenkins, ret, Coll)
70 000 vols; 150 curr per   41236

Bayliss Public Library, 541 Library Dr, *Sault Sainte Marie*, MI 49783
T: +1 906 6329331; Fax: +1 906 6350210; E-mail: kmiller@uproc.lib.mi.us; URL: www.baylisslib.org
1903; Ken Miller
Hist (Judge Joseph H Steere Coll), bks, mss; 5 branch libs
84 080 vols; 192 curr per   41237

Sausalito Public Library, 420 Litho St, *Sausalito*, CA 94965-1933
T: +1 415 2894121; Fax: +1 415 3317943; URL: www.ci.sausalito.ca.us/library
1907; Augie Webb
Boats
61 000 vols; 185 curr per   41238

Scott County Library System, 13090 Alabama Ave S, *Savage*, MN 55378-1479
T: +1 952 7071760; Fax: +1 952 7071775; URL: www.scott.lib.mn.us
1969; Vanessa J. Birdsey
7 branch libs
266 000 vols; 366 curr per; 5 361 av-mat; 402 toys
libr loan   41239

Live Oak Public Libraries, 2002 Bull St, *Savannah*, GA 31401
T: +1 912 6523606; Fax: +1 912 6523638; URL: www.liveoakpl.org
1903; Christian Kruse
19 branch libs
302 000 vols; 1 148 curr per; 18 404 av-mat; 18 410 Talking Bks
libr loan   41240

Live Oak Public Libraries – Oglethorpe Mall Branch, Seven Mall Annex, *Savannah*, GA 31406
T: +1 912 9255432; Fax: +1 912 9252031
Michael Hill
81 000 vols   41241

Sayville Library, 11 Collins Ave, *Sayville*, NY 11782-3199
T: +1 631 5894440; Fax: +1 631 5896128; E-mail: sayvlib@suffolk.lib.ny.us; URL: www.sayville.suffolk.lib.ny.us
1914; Alice Lepore
109 000 vols; 255 curr per; 3 369 av-mat
libr loan   41242

Scarborough Public Library, 48 Gorham Rd, *Scarborough*, ME 04074
T: +1 207 8834723; Fax: +1 207 8839728; E-mail: askspl@scarborough.lib.me.us; URL: www.library.scarborough.me.us
1899; Nancy E. Crowell
56 000 vols; 9 401 curr per   41243

Scarsdale Public Library, 54 Olmsted Rd, *Scarsdale*, NY 10583
T: +1 914 7221300; Fax: +1 914 7221305; URL: www.scarsdalelibrary.org
1928; Stephanie C. Sarnoff
ESL, Japanese language, Large print
168 000 vols; 4 418 curr per   41244

Schaumburg Township District Library, 130 S Roselle Rd, *Schaumburg*, IL 60193
T: +1 847 9854000; Fax: +1 847 9233207; URL: www.stdl.org
1963; Michael J. Madden
Business & management; 2 branch libs
518 000 vols; 1 361 curr per; 740 e-books   41245

Mohawk Valley Library System, 858 Duanesburg Rd, *Schenectady*, NY 12306-1095
T: +1 518 3552010; Fax: +1 518 3550674; URL: www.mvlas.info
1959; Carol Clingan
119 000 vols   41246

Schenectady County Public Library, 99 Clinton St, *Schenectady*, NY 12305-2083
T: +1 518 3884518; Fax: +1 518 3862241; URL: www.scpl.org/
1894; Andrew L. Kulmatiski
9 branch libs
550 000 vols; 1 652 curr per; 1 810 music scores; 16 000 av-mat; 9 500 sound-rec; 10 400 digital data carriers; 9 030 Large Print Bks   41247

Lake County Public Library – Dyer-Schererville Branch, 1001 W Lincoln Hwy, *Schererville*, IN 46375-1552
T: +1 219 3224731; Fax: +1 219 8655478
Pam Maud
103 000 vols; 88 curr per; 139 av-mat
                                41248

Schiller Park Public Library, 4200 Old River Rd, *Schiller Park*, IL 60176-1699
T: +1 847 6780433; Fax: +1 847 6780567; URL: www.schillerparklibrary.org
1962; Tina J. Setzer
80 000 vols; 150 curr per
libr loan   41249

Scituate Town Library, 85 Branch St, *Scituate*, MA 02066
T: +1 781 5458727; Fax: +1 781 5458728; E-mail: info@scituatetownlibrary.org; URL: www.scituatetownlibrary.org
1893; Kathleen Meeker
75 000 vols; 121 curr per   41250

Scotch Plains Public Library, 1927 Bartle Ave, *Scotch Plains*, NJ 07076-1212
T: +1 908 3225007; Fax: +1 908 3220490; E-mail: info@scotlib.org; URL: www.scotlib.org
1888; Meg Kolaya
Cookbks; Poetry & Prose (Robert Frost), Local & State Hist (New Jersey)
75 000 vols; 196 curr per   41251

Scottsbluff Public Library, 1809 Third Ave, *Scottsbluff*, NE 69361-2493
T: +1 308 6306250; Fax: +1 308 6306293; URL: www.scottsbluff.org/lib
1917
Western Americana (Western Hist Coll)
64 710 vols; 159 curr per; 3 600 microforms; 1 700 av-mat; 2 460 sound-rec; 12 digital data carriers
libr loan   41252

Scott County Public Library, 108 S Main St, *Scottsburg*, IN 47170
T: +1 812 7522751; Fax: +1 812 7522878; E-mail: info@scott.lib.in.us; URL: www.scottcounty.org/arts/scl.html
1921; Andrew Rowden
Hist, Civil War, Indiana; Carl R Bogardus Sr, MD Coll
79 000 vols; 209 curr per; 300 av-mat
                                41253

Scottsdale Public Library System, 3839 N Drinkwater Blvd, *Scottsdale*, AZ 85251-4467; P.O. Box 1000 LI 101, Scottsdale
T: +1 480 3127323; Fax: +1 480 3127993; URL: library.scottsdale.az.gov
1959; Rita Hamilton
Art & architecture, Business & management; Southwest Coll; 3 branch libs
531 000 vols; 815 curr per; 35 000 e-books
libr loan   41254

Scranton Public Library, Albright Memorial Library, Albright Memorial Bldg, 500 Vine St, *Scranton*, PA 18509-3298
T: +1 570 3483000; Fax: +1 570 3483020; URL: www.lclshome.org/albright
1893; Jack Finnerty
3 branch libs
180 000 vols; 300 curr per
libr loan   41255

Seaford District Library, 402 Porter St, *Seaford*, DE 19973
T: +1 302 6292524; Fax: +1 302 6299181; URL: www.seaford.lib.de.us
1902; Leigh Ann DePope
Classic Cars, Delaware Hist (Delawareana)
51 000 vols; 103 curr per; 2 700 av-mat; 4 318 sound-rec; 25 digital data carriers
libr loan   41256

Seaford Public Library, 2234 Jackson Ave, *Seaford*, NY 11783
T: +1 516 2211334; Fax: +1 516 8268133; E-mail: splreference@nassaulibrary.org; URL: www.nassaulibrary.org/seaford
1956; Marilyn J. Griffin
74 000 vols   41257

Orange County Public Library – Los Alamitos-Rossmoor, 12700 Montecito Rd, *Seal Beach*, CA 90740-2745
T: +1 562 4301048; Fax: +1 562 4312931
Jane Deeley
79 000 vols; 91 curr per   41258

Orange County Public Library – Seal Beach-Mary Wilson Branch, 707 Electric Ave, *Seal Beach*, CA 90740-6196
T: +1 562 4313584; Fax: +1 562 4313374
Diane Gayton
55 000 vols; 80 curr per   41259

White County Regional Library System, 113 E Pleasure Ave, *Searcy*, AR 72143-7798
T: +1 501 2792870; Fax: +1 501 2685682
1896; Susie Boyett
6 branch libs
114 000 vols; 720 curr per; 1 200 av-mat
                                41260

White County Regional Library System – Searcy Public, 113 E Pleasure Ave, ***Searcy***, AR 72143
Fax: +1 501 2685682
Joyce Turley
59 000 vols 41261

Monterey County Free Libraries – Seaside Branch, 550 Harcourt Ave, ***Seaside***, CA 93955
T: +1 831 8992055; Fax: +1 831 8992735
Sharon Freed
72 000 vols 41262

The Seattle Public Library, 1000 Fourth Ave, ***Seattle***, WA 98104-1109
T: +1 206 3864636; Fax: +1 206 3864604; E-mail: amy.lawson@spl.org; URL: www.spl.org
1891; Susan Hildreth
27 branch libs
926 000 vols
libr loan 41263

The Seattle Public Library – Lake City, 12501 28th Ave NE, ***Seattle***, WA 98125
T: +1 206 6847518
Andy Bates
55 000 vols 41264

The Seattle Public Library – North East, 6801 35th Ave NE, ***Seattle***, WA 98115
T: +1 206 6847539
Marion Scichilone
63 000 vols 41265

Washington Talking Book & Braille Library, 2021 Ninth Ave, ***Seattle***, WA 98121-2783
T: +1 206 3861255; Fax: +1 206 6150441; E-mail: wtbbl@wtbbl.org; URL: www.wtbbl.org
1931; Gloria Leonard
Northwest Coll
311 000 vols 41266

North Indian River County Library, 1001 Sebastian Blvd, CR 512, ***Sebastian***, FL 32958
T: +1 772 5891355; Fax: +1 772 3883697; URL: www.sebastianlibrary.com
1983; Lynn Walsh
Florida Coll
109 000 vols; 233 curr per; 244 microforms; 4 514 sound-rec; 44 digital data carriers 41267

Sebastopol Regional Library, 7140 Bodega Ave, ***Sebastopol***, CA 95472-3712
T: +1 707 8237691; Fax: +1 707 8237172
50 000 vols; 130 curr per 41268

Highlands County Library System – Sebring Public Library, Sebring Public Library Bldg, 319 W Center Ave, ***Sebring***, FL 33870-3109
T: +1 863 4026716; Fax: +1 863 3852883; URL: www.myhlc.org
1926; Mary Myers
Florida Coll
102 000 vols; 145 curr per 41269

Secaucus Free Public Library, 1379 Patterson Plank Rd, ***Secaucus***, NJ 07094
T: +1 201 3302083; Fax: +1 201 6171695; E-mail: seca1@bccls.org; URL: www.bccls.org/secaucus
1957; Katherine Steffens
58 000 vols; 82 curr per 41270

Security Public Library, 715 Aspen Dr, ***Security***, CO 80911-1807
T: +1 719 3913191; Fax: +1 719 3927641; URL: www.wsd3.k12.co.us/spl/home.html
1961; Barbara Hudson
Colorado hist
60 000 vols; 135 curr per; 2 103 av-mat 41271

Boonslick Regional Library, 219 W Third St, ***Sedalia***, MO 65301-4347
T: +1 660 8277111; Fax: +1 660 8274668; URL: brl.lib.mo.us
1953; Linda Allcorn
4 branch libs
200 000 vols; 175 curr per 41272

Sedalia Public Library, 311 W Third St, ***Sedalia***, MO 65301-4399
T: +1 660 8261314; Fax: +1 660 8260396; URL: www.sedalialibrary.com
1895; Pam Hunter
82 000 vols; 155 curr per; 7 000 e-books 41273

Sedona Public Library, 3250 White Bear Rd, ***Sedona***, AZ 86336
T: +1 928 2827714; Fax: +1 928 2825789; E-mail: library@sedonalibrary.org; URL: www.sedonalibrary.org
1958; David W. Keeber
Arizona Coll
80 000 vols; 230 curr per 41274

Sedro-Woolley Public Library, 802 Ball Ave, ***Sedro-Woolley***, WA 98284-2008
T: +1 360 8551166; URL: www.yousee.more.com/sedro-woolley
Debra D. Peterson
50 000 vols; 140 curr per 41275

Seekonk Public Library, 410 Newman Ave, ***Seekonk***, MA 02771
T: +1 508 3368230; Fax: +1 508 3367062; E-mail: library@seekonkpl.org; URL: www.seekonkpl.org
1899; Sharon E. Saint-Hilaire
75 000 vols; 245 curr per; 361 e-books 41276

Seguin-Guadalupe County Public Library, 707 E College St, ***Seguin***, TX 78155-3217
T: +1 830 4012422; Fax: +1 830 4012477; E-mail: seguinlibrary@seguin.lib.tx.us; URL: www.seguin.lib.tx.us
1930; Jacki Gross
Guadalupe County land records, micro; hist photography (Kubala Coll); Old Seguin Newspapers, micro-filmed
69 000 vols; 104 curr per 41277

Snyder County Libraries, Community Bldg, One N High St, ***Selinsgrove***, PA 17870-1599
T: +1 570 3747163; Fax: +1 570 3742120; URL: www.snydercountylibraries.org
1976; Pam Ross
75 000 vols; 172 curr per 41278

Public Library of Selma & Dallas County, 1103 Selma Ave, ***Selma***, AL 36703-4445
T: +1 334 8741725; Fax: +1 334 8741729; E-mail: info@selmalibrary.org; URL: www.selmalibrary.org
1903; Becky Nichols
73 000 vols; 61 curr per 41279

Seminole Community Library, 9200 113th St N, ***Seminole***, FL 33772
T: +1 727 3946905; Fax: +1 727 3983113; URL: www.spcollege.edu/scl
1959; Michael Bryan
Parent-Teacher Coll
76 000 vols; 128 curr per; 735 microforms; 1 800 av-mat; 2 325 sound-rec; 10 digital data carriers
libr loan 41280

Seneca Public Library District, 210 N Main St, ***Seneca***, IL 61360
T: +1 815 3576566; Fax: +1 815 3576568; URL: www.senecalibrary.org
1938; Margie Nolan
World War II, Senaca LST Shipyard
51 000 vols; 120 curr per; 1 081 av-mat 41281

North Olympic Library System – Sequim Branch, 630 N Sequim Ave, ***Sequim***, WA 98382
T: +1 360 6831161; Fax: +1 360 6817811
Jola Nicola
62 000 vols 41282

Emma S. Clark Memorial Library, 120 Main St, ***Setauket***, NY 11733-2868
T: +1 631 9414080; E-mail: reference@emmaclark.org; URL: www.emmaclark.org
1892; Edward Elenausky
185 000 vols; 378 curr per 41283

Sevier County Public Library System, 321 Court Ave, ***Sevierville***, TN 37862
T: +1 865 7746033; Fax: +1 865 7746024; URL: www.sevierlibrary.org
1920; Kaurri C. Williams
70 000 vols; 177 curr per 41284

Sewickley Public Library, Inc, 500 Thorn St, ***Sewickley***, PA 15143-1333
T: +1 412 7416920; Fax: +1 412 7416099; E-mail: sewickley@einetwork.net; URL: sewickleylibrary.org
1873; Carolyn Toth
89 000 vols; 220 curr per 41285

Jackson County Public Library, Seymour Library, 303 W Second St, ***Seymour***, IN 47274-2147
T: +1 812 5223412; Fax: +1 812 5225456; E-mail: admin@myjclibrary.org; URL: www.myjclibrary.org
1904; Julia Aker
2 branch libs
91 000 vols; 326 curr per 41286

Seymour Public Library, 46 Church St, ***Seymour***, CT 06483
T: +1 203 8883903; Fax: +1 203 8884099; E-mail: seymourpubliclibrary@biblio.org; URL: www.seymourpubliclibrary.org
1892; Carol Ralston
Hist Reference Coll
62 000 vols; 111 curr per; 200 e-books; 1 194 av-mat 41287

Shaker Heights Public Library, 16500 Van Aken Blvd, ***Shaker Heights***, OH 44120-5318
T: +1 216 9912030; Fax: +1 216 9915951; URL: www.shpl.lib.oh.us; www.shakerlibrary.org
1936; Luren E. Dickinson
1 branch libr
145 000 vols; 693 curr per; 12 000 e-books; 25 645 Audio Bks 41288

Shaker Heights Public Library – Bertram Woods Branch, 20600 Fayette Rd, ***Shaker Heights***, OH 44122-2979
T: +1 216 9912421; Fax: +1 216 9913124
Lynne Miller
69 000 vols 41289

Sharon Public Library, 11 N Main St, ***Sharon***, MA 02067-1299
T: +1 781 7841578; Fax: +1 781 7844728; E-mail: info@sharonpubliclibrary.org; URL: www.sharonpubliclibrary.org
1877; Barbra Nadler
Deborah Sampson Coll
90 000 vols; 125 curr per
libr loan; ALA 41290

Shenango Valley Community Library, 11 N Sharpsville Ave, ***Sharon***, PA 16146
T: +1 724 9814360; Fax: +1 724 9815208; E-mail: svalleylib@yahoo.com
1923; Karen L. Spak
60 000 vols; 125 curr per 41291

Johnson County Library – Shawnee Branch, 13811 Johnson Dr, ***Shawnee***, KS 66216
T: +1 913 9623800; Fax: +1 913 9623809
Roxanne Belcher
69 000 vols; 122 curr per 41292

Pioneer Library System – Shawnee Public, 101 N Philadelphia, ***Shawnee***, OK 74801
T: +1 405 2756353; Fax: +1 405 2730590
Julia Harmon
59 000 vols 41293

Mead Public Library, 710 N Eighth St, ***Sheboygan***, WI 53081-4563
T: +1 920 4593400; Fax: +1 920 4590204; URL: www.sheboygan.lib.wi.us
1897; Sharon L. Winkle
US & State Census Coll
298 000 vols; 569 curr per; 14 000 e-books
libr loan 41294

Sheffield Public Library, 316 N Montgomery Ave, ***Sheffield***, AL 35660
T: +1 256 3865633; Fax: +1 256 3865608; URL: www.lmn.lib.al.us
Beth Ridgeway
54 000 vols; 80 curr per; 27 000 e-books 41295

Lorain Public Library System – Domonkas Branch, 4125 E Lake Rd, ***Sheffield Lake***, OH 44054
T: +1 440 9497410; Fax: +1 440 9497741
Elizabeth Fahnert
50 000 vols; 179 curr per 41296

Cleveland County Library System, 104 Howie Dr, ***Shelby***, NC 28150; P.O. Box 1120, Shelby
T: +1 704 4879069; Fax: +1 704 4874856; URL: www.ccml.org
1909; Carol H. Wilson
140 000 vols; 96 curr per
libr loan 41297

Marvin Memorial Library, 29 W Whitney Ave, ***Shelby***, OH 44875-1252
T: +1 419 3475576; Fax: +1 419 3477285; URL: www.shelbymm.lib.oh.us
1897; Ann Bavin
59 000 vols; 142 curr per 41298

Shelby Township Library, 51680 Van Dyke, ***Shelby Township***, MI 48316-4448
T: +1 586 7397414; Fax: +1 586 7260535; URL: www.libcoop.net/shelby
1972; Judith Chambers
Women's studies
116 000 vols; 262 curr per; 3 755 av-mat 41299

Argie Cooper Public Library, 100 S Main St, ***Shelbyville***, TN 37160-3984
T: +1 931 6847323; Fax: +1 931 6854848; URL: www.acolibrary.com
1966; Pat Hastings
Hist (Early Editions of Newspapers in Bedford County), micro
50 000 vols; 72 curr per 41300

Shelbyville-Shelby County Public Library, 57 W Broadway, ***Shelbyville***, IN 46176
T: +1 317 8352653; Fax: +1 317 3984430; URL: www.sscpl.lib.in.us/library/
1898; Janet Wallace
93 000 vols; 267 curr per; 1 402 av-mat
libr loan 41301

North Bingham County District Library, 197 W Locust St, ***Shelley***, ID 83274-1139
T: +1 208 3577801; Fax: +1 208 3572272; URL: www.ida.net/org/nbcdl
Heidi Riddoch
50 000 vols; 600 av-mat 41302

Plumb Memorial Library, 65 Wooster St, ***Shelton***, CT 06484
T: +1 203 9241580; Fax: +1 203 9248422; URL: www.plumblibrary.org
1896; C. Elspeth Lydon
1 branch libr
134 000 vols; 223 curr per; 11 e-books; 2 090 av-mat; 2 090 Audio Bks, 1 400 Large Print Bks 41303

Plumb Memorial Library – Huntington Branch, 41 Church St, ***Shelton***, CT 06484-5804
T: +1 203 9260111; Fax: +1 203 9260181
N.N.
51 000 vols; 106 curr per; 500 av-mat 41304

Shenandoah Public Library, 201 S Elm St, ***Shenandoah***, IA 51601
T: +1 712 2462315; Fax: +1 712 2465847; E-mail: shenlib@heartland.net; URL: www.shenandoah.lib.ia.us
1904; Jan Frank-de Ois
55 000 vols 41305

Bullitt County Public Library, Ridgway Memorial Library, 127 N Walnut St, ***Shepherdsville***, KY 40165-6083; P.O. Box 99, Shepherdsville, KY 40165-0099
T: +1 502 5437675; Fax: +1 502 5435487; E-mail: bcpl@iglou.com; URL: www.bcplib.org
1954; Randy Matlow
3 branch libs
129 000 vols; 131 curr per 41306

Sherborn Library, 4 Sanger St, ***Sherborn***, MA 01770-1499
T: +1 508 6530770; Fax: +1 508 6509243; E-mail: elizabeth.johnston2@comcast.net; URL: library.sherbornma.org
1860; Elizabeth Johnston
52 730 vols; 169 curr per 41307

Sherman Public Library, 421 N Travis, ***Sherman***, TX 75090-5975
T: +1 903 8927240; Fax: +1 903 8927101; URL: www.barr.org/sherman.htm
1911; Jacqueline Banfield
Hilmer H Flemming Mss, Mattie Davis Lucas Mss, Grayson County Hist Resources for Texas State Libr
108 000 vols; 220 curr per 41308

Los Angeles Public Library System – Sherman Oaks Branch, 12511 Moorpark St, ***Sherman Oaks***, CA 91423
T: +1 818 2059716; Fax: +1 818 2059866
Mary Beaumont
58 000 vols 41309

Shippensburg Public Library, 73 W King St, ***Shippensburg***, PA 17257-1299
T: +1 717 5324508; Fax: +1 717 5322454; E-mail: shippensburg@ccpa.net; URL: www.ccpa.net
1933; Susan Sanders
60 000 vols; 107 curr per 41310

Mastics-Moriches-Shirley Community Library, 407 William Floyd Pkwy, ***Shirley***, NY 11967
T: +1 631 3991511; Fax: +1 631 2814442; E-mail: mmshlib@suffolk.lib.ny.us; URL: www.communitylibrary.org
1974; Kerri Rosalia
Large print, Italian language, Spanish
272 000 vols; 27 141 curr per 41311

Shoreham-Wading River Public Library, 250 Rte 25A, ***Shoreham***, NY 11786-9677
T: +1 631 9294488; Fax: +1 631 9294551; E-mail: nspl@suffolk.lib.ny.us; URL: www.nspl.suffolk.lib.ny.us
1975; Laura Hawrey
Shoreham Nuclear Power Plant Documents
97 510 vols; 337 curr per 41312

Shorewood Public Library, 3920 N Murray Ave, ***Shorewood***, WI 53211-2385
T: +1 414 8472670; URL: www.shorewoodlibrary.org
1932; Elizabeth Carey
51 000 vols; 110 curr per 41313

Shorewood-Troy Public Library District, 650 Deerwood Dr, ***Shorewood***, IL 60404
T: +1 815 7251715; Fax: +1 815 7251722; E-mail: reference@shorewoodtroylibrary.org; URL: www.shorewood.lib.il.us
1975; Jennie Cisna Mills
Township Records
54 000 vols; 129 curr per 41314

Shreve Memorial Library, 424 Texas St, ***Shreveport***, LA 71101; P.O. Box 21523, Shreveport, LA 71120-1523
T: +1 318 2265871; Fax: +1 318 2264780; E-mail: admin.g1sh@pelican.state.lib.la.us; URL: www.shreve-lib.org
1923; James R. Pelton
Geology, Petroleum; Louisiana Coll; 20 branch libs
725 000 vols; 660 curr per; 34 000 av-mat; 19 000 sound-rec; 18 950 digital data carriers; 15 180 Large Print Bks 41315

Shreve Memorial Library – Broadmoor Branch, 1212 Captain Shreve Dr, ***Shreveport***, LA 71105
T: +1 318 8690120; Fax: +1 318 8689464
Betty Cannon
79 000 vols 41316

Shreve Memorial Library – Hamilton/South Caddo Branch, 2111 Bert Kouns Industrial Loop, ***Shreveport***, LA 71118
T: +1 318 6876824; Fax: +1 318 6860971
Bob Gullion
75 000 vols 41317

Monmouth County Library – Eastern, 1001 Rte 35, ***Shrewsbury***, NJ 07702
T: +1 732 8425995; Fax: +1 732 2190140
Janet Kranis
145 000 vols 41318

Shrewsbury Public Library, 609 Main St, ***Shrewsbury***, MA 01545
T: +1 508 8420081; Fax: +1 508 8418540; URL: www.shrewsbury-ma.gov/library
1872; Ellen M. Dolan
Early New England Hist & Biogr (Artemas Ward Coll)
134 000 vols; 180 curr per 41319

John C. Hart Memorial Library, 1130 Main St, ***Shrub Oak***, NY 10588
T: +1 914 2455262; Fax: +1 914 2452216; E-mail: pbarresi@wlsmail.org; URL: www.yorktownlibrary.org
1920; Patricia Barresi
Special Education (Jerome Thaler Wheather Coll)
100 000 vols; 248 curr per 41320

Shelby County Libraries, 230 E North St, ***Sidney***, OH 45365-2785
T: +1 937 4928354; Fax: +1 937 4929229; E-mail: libraram@oplin.org; URL: www.shelbyco.lib.oh.us
1869; Suzanne Cline
Lois Lenski Coll, Bessie Schiff Coll
169 000 vols; 234 curr per; 3 573 av-mat 41321

Sidney Memorial Public Library, Eight River St, ***Sidney***, NY 13838
T: +1 607 5631200; Fax: +1 607 5637675; URL: www.lib.4cty.org/sidney/smpl.htm
1887; Mary Grace Flaherty
51 000 vols; 154 curr per 41322

Sierra Madre Public Library, 440 W Sierra Madre Blvd, ***Sierra Madre***, CA 91024-2399
T: +1 626 3557186; Fax: +1 626 3556218; E-mail: ref@sierramadre.lib.ca.us; URL: www.sierramadre.lib.ca.us
1887; Toni Buckner
60 000 vols; 99 curr per; 800 sound-rec; 3 digital data carriers
libr loan 41323

Sierra Vista Public Library, 2600 E Tacoma, ***Sierra Vista***, AZ 85635-1399
T: +1 520 4584239; Fax: +1 520 4585377; URL: www.ci.sierra-vista.az.us/svlibrary
1958; David L. Gunckel
Arizona (Margaret Carmichael Coll)
100 000 vols; 180 curr per 41324

Silver City Public Library, 515 W College Ave, ***Silver City***, NM 88061
T: +1 505 5383672; Fax: +1 505 3883757; E-mail: info@townofsilvercity.org; URL: www.townofsilvercity.org
1952; Cheryl Ward
54 000 vols; 67 curr per 41325

Silver Falls Library District, 410 S Water St, ***Silverton***, OR 97381-2137
T: +1 503 8738796; Fax: +1 503 8736227; E-mail: silvfals@ccrls.org; URL: www.ccrls.org/sil
1911; Marlys Swalboski
Homer Davenport Coll (turn of the c political cartoonist)
62 000 vols; 134 curr per; 100 maps; 150 microforms; 2 589 av-mat; 1 200 sound-rec; 220 digital data carriers
libr loan 41326

Ventura County Library – Simi Valley Library, 2969 Tapo Canyon Rd, ***Simi Valley***, CA 93063
T: +1 805 5261735; Fax: +1 805 5261738
Gabriel Lundeen
154 000 vols 41327

Simsbury Public Library, 725 Hopmeadow St, ***Simsbury***, CT 06070
T: +1 860 6587663; Fax: +1 860 6586732; URL: www.simsburylibrary.info
1890; Susan Bullock
140 000 vols 41328

Sioux City Public Library, Wilbur Aalfs Main Library, 529 Pierce St, ***Sioux City***, IA 51101-1203
T: +1 712 2552933; Fax: +1 712 2796432; E-mail: questions@siouxcitylibrary.org; URL: www.siouxcitylibrary.org
1877; Betsy J. Thompson
2 branch libs
127 000 vols; 144 curr per; 200 maps; 7 500 sound-rec; 200 digital data carriers; 2 487 Large Print Bks
libr loan; OCLC 41329

Siouxland Libraries, 200 N Dakota Ave, ***Sioux Falls***, SD 57104; P.O. Box 7403, Sioux Falls, SD 57117-7403
T: +1 605 3678701; Fax: +1 605 3674312; URL: www.siouxlandlib.org
1886; Sally Felix
Foreign language, Lit; 11 branch libs, 2 bookmobiles
334 000 vols; 785 curr per; 14 000 e-books; 24 Electronic Media & Resources, 13 Special Interest Per Sub 41330

Kettleson Memorial Library, 320 Harbor Dr, ***Sitka***, AK 99835-7553
T: +1 907 7478708; Fax: +1 907 7478755; E-mail: library@cityofsitka.com; URL: www.cityofsitka.com/dept/library/library.html
1923; Cheryl Pearson
Alaskana Local Hist Coll; Louise Brightman Room
60 000 vols; 283 curr per; 111 govt docs; 57 maps; 273 microforms; 4 435 sound-rec; 9 digital data carriers; 4 347 other items
libr loan; ALA 41331

Skokie Public Library, 5215 Oakton St, ***Skokie***, IL 60077-3680
T: +1 847 6733733; Fax: +1 847 6737797; E-mail: tellus@skokielibrary.info; URL: www.skokielibrary.info
1941; Carolyn A. Anthony
Lit, Art, Business, Foreign language, Holocaust
441 000 vols; 887 curr per; 5 000 e-books 41332

North Smithfield Public Library, 20 Main St, ***Slatersville***, RI 02876; P.O. Box 950, Slatersville, RI 02876-0898
T: +1 401 7672780; Fax: +1 401 7672782; E-mail: nsmlibrary@yahoo.com; URL: www.nsmlibrary.org
1928; Carol H. Brouwer
Child care, Parenting
50 000 vols; 124 curr per
libr loan 41333

East Smithfield Public Library, 50 Esmond St, ***Smithfield***, RI 02917-3016
T: +1 401 2315150; Fax: +1 401 2312940; E-mail: esm@lori.ri.gov; URL: web.prov.lib.org/esmlib
1916; Elodie E. Blackmore
Cookery, Medicine for Layman
52 000 vols; 119 curr per 41334

Public Library of Johnston County & Smithfield, 305 E Market St, ***Smithfield***, NC 27577-3919
T: +1 919 9348146; Fax: +1 919 9348084; E-mail: webmaster@pljcs.org; URL: www.pljcs.org
1966; Margaret Marshall
97 000 vols; 75 curr per; 5 000 av-mat 41335

Smithtown Library, One N Country Rd, ***Smithtown***, NY 11787
T: +1 631 2652072; Fax: +1 631 2652044; URL: www.smithlib.org
1905; Robert Lusak
Behav sciences, Social sciences, Business & management; Long Island Coll; 3 branch libs
506 000 vols; 2 309 curr per
libr loan 41336

Linebaugh Public Library System of Rutherford County – Smyrna Public, 400 Enon Springs Rd W, ***Smyrna***, TN 37167
T: +1 615 4594884; Fax: +1 615 4592370
Carol Kersey
67 000 vols; 160 curr per 41337

Smyrna Public Library, 100 Village Green Circle, ***Smyrna***, GA 30080-3478
T: +1 770 4312860; Fax: +1 770 4312862; E-mail: slibrary@ci.smyrna.ga.us; URL: www.smymalibrary.net; www.smyrnalibrary.com
1936; Michael E. Seigler
Antique bks; Fed depository; genealogy; Georgia hist
100 000 vols; 125 curr per; 3 610 govt docs; 17 000 microforms; 3 800 av-mat; 2 630 sound-rec; 161 digital data carriers
libr loan 41338

Sno-Isle Libraries – Snohomish Branch, 311 Maple Ave, ***Snohomish***, WA 98290
T: +1 360 5682898; Fax: +1 360 5681922
Eileen McDonnell
124 000 vols 41339

Worcester County Library, 307 N Washington St, ***Snow Hill***, MD 21863
T: +1 410 6322600; Fax: +1 410 6321159; E-mail: worc@worc.lib.md.us; URL: www.worcesterlibrary.org
1959; Mark Thomas
Real estate; 5 branch libs
151 000 vols; 464 curr per; 5 500 av-mat 41340

Scurry County Library, 1916 23rd St, ***Snyder***, TX 79549-1910
T: +1 325 5735572; Fax: +1 325 5731060; E-mail: scurrycl@snydertex.com
1958; L. Jane Romine
66 000 vols; 62 curr per 41341

San Diego County Library – Solana Beach Branch, 157 Stevens Ave, ***Solana Beach***, CA 92075-1873
T: +1 858 7551404; Fax: +1 858 7559327
Rebecca Lynn
71 000 vols 41342

Somers Library, Reis Park, Rte 139, P.O. Box 443, ***Somers***, NY 10589
T: +1 914 2325717; Fax: +1 914 2321035; E-mail: miller@wlsmail.org; URL: www.somerslibrary.org
1876; Pat Miller
Coll of Circus Bks
56 570 vols; 236 curr per; 25 maps; 106 microforms; 5 141 av-mat; 79 digital data carriers
libr loan 41343

Franklin Township Free Public Library, 485 DeMott Lane, ***Somerset***, NJ 08873
T: +1 732 8738700; Fax: +1 732 8730746; URL: www.franklintwp.org
1957; January Adams
African American Coll, Coll of Local Hist & Regional Genealogy, Career & Business, Foreign Language, Law
137 000 vols; 200 maps; 500 microforms; 4 500 av-mat; 3 000 sound-rec
libr loan 41344

Pulaski County Public Library, 107 N Main St, ***Somerset***, KY 42501-1402; P.O. Box 36, Somerset, KY 42502-0036
T: +1 606 6798401; Fax: +1 606 6791779; URL: www.pcpl.lib.ky.us
1905; Judith Burdine
Genealogy; Kentucky Coll; 4 branch libs
93 000 vols; 270 curr per 41345

Somerset County Federated Library System, 6022 Glades Pike, Ste 120, ***Somerset***, PA 15501-0043
T: +1 814 4455907; Fax: +1 814 4430650; E-mail: somerset@somersetcountypalibraries.org
Eve Kline
144 000 vols; 297 curr per 41346

Somerset County Library, 6022 Glades Pike, Ste 120, ***Somerset***, PA 15501-4300
T: +1 814 4455907; Fax: +1 814 4430650; E-mail: somerset@somersetcountypalibraries.org; URL: somersetcountypalibraries.org
1947; Eve Kline
Large Print, Videos
100 000 vols; 176 curr per; 2 309 av-mat 41347

Somerset Public Library, 1464 County St, ***Somerset***, MA 02726
T: +1 508 6462829; Fax: +1 508 6462831; E-mail: somersetpl@sailsinc.org; URL: www.sailsinc.org/somerset/
1897; Bonnie Davis Mendes
74 000 vols; 210 curr per; 50 govt docs; 25 microforms
libr loan; MLA, ALA 41348

Somerville Public Library, 79 Highland Ave, ***Somerville***, MA 02143
T: +1 617 6235000; Fax: +1 617 6284052; E-mail: somerville@minlib.net; URL: www.somervillepubliclibrary.org
1873; Nancy Milnor
Art, Genealogy, Travel, Women's studies; New England Hist Coll; 2 branch libs
217 000 vols; 819 curr per; 240 Electronic Media & Resources 41349

Somerville Public Library, 35 West End Ave, **Somerville**, NJ 08876
T: +1 908 7251336; Fax: +1 908 2310608; URL: www.lmxac.org/somervillelib/
1871; Melissa A. Banks
59 000 vols; 120 curr per; 1 010 av-mat
41350

Sonoma Valley Regional Library, 755 W Napa St, **Sonoma**, CA 95476
T: +1 707 9965217; Fax: +1 707 9965918; URL: www.sonoma.lib.ca.us/branches/Sonoma.html
53 000 vols; 20 curr per
41351

Tuolumne County Free Library, 480 Greenley Rd, **Sonora**, CA 95370-5956
T: +1 209 5335507; Fax: +1 209 5330936; E-mail: libref@co.tuolumne.ca.us; URL: www.tuolcolib.org; www.library.co.tuolumne.ca.us
1917; Constance J. Corcoran
Mining; Tuolumne County Hist, tapes
96 000 vols; 135 curr per
libr loan
41352

Sadie Pope Dowdell Library of South Amboy, 100 Hoffman Plz, **South Amboy**, NJ 08879
T: +1 732 7216060; Fax: +1 732 7211054; E-mail: info@dowdell.org; URL: www.dowdell.org
1917; Elaine R. Gaber
Railroads, New Jersey hist
68 000 vols; 107 curr per
41353

Saint Joseph County Public Library, 304 S Main, **South Bend**, IN 46601-2125
T: +1 574 2824617; Fax: +1 574 2802763; URL: www.libraryforlife.org
1888; Donald J. Napoli
Large print bks; Adult Reading Ctr; 8 branch libs
524 000 vols; 3 726 curr per
libr loan
41354

Halifax County-South Boston Regional Library – South Boston Public Library, 509 Broad St, **South Boston**, VA 24592
T: +1 434 5754228; Fax: +1 434 5754229; E-mail: sobolib@halifax.com
Woodson Hughes
50 000 vols
41355

South Burlington Community Library, 550 Dorset St, **South Burlington**, VT 05403
T: +1 802 6527080; Fax: +1 802 6527013; URL: sbcl.sbschools.net
Louise Murphy
52 000 vols; 123 curr per
41356

South Charleston Public Library, 312 Fourth Ave, **South Charleston**, WV 25303-1297
T: +1 304 7446561; Fax: +1 304 7448808; E-mail: reference@scpl.wvnet.edu; URL: www.infospot.org
1966; Jennifer Soule
72 000 vols; 70 curr per; 4 680 av-mat
41357

Cuyahoga County Public Library – South Euclid-Lyndhurst Branch, 4645 Mayfield Rd, **South Euclid**, OH 44121-4018
T: +1 216 3824880; Fax: +1 216 3824584
Kris Gill
54 000 vols; 3 070 av-mat
41358

County of Los Angeles Public Library – Hollydale Library, 12000 S Garfield Ave, **South Gate**, CA 90280-7894
T: +1 562 6340156; Fax: +1 562 5319530; URL: www.colapublib.org/libs/hollydale
Jenny Wrenn
54 000 vols; 86 curr per
41359

County of Los Angeles Public Library – Leland R. Weaver Library, 4035 Tweedy Blvd, **South Gate**, CA 90280-6199
T: +1 323 5678853; Fax: +1 323 5631046; URL: www.colapublib.org/libs/weaver
Glorieta Navo
108 000 vols; 162 curr per
41360

Hamilton-Wenham Public Library, 14 Union St, **South Hamilton**, MA 01982
T: +1 508 4685577; Fax: +1 508 4685535; E-mail: jdempsey@mvlc.org; URL: www.hwlibrary.org
1891; Jan Dempsey
85 000 vols; 161 curr per
41361

South Haven Memorial Library, 314 Broadway, **South Haven**, MI 49090
T: +1 616 6372403; Fax: +1 616 6391685; E-mail: shml@shmlibrary.org; URL: www.shmlibrary.org
1910; Deborah Jones
53 000 vols; 105 curr per
41362

South Holland Public Library, 16250 Wausau Ave, **South Holland**, IL 60473
T: +1 708 3315262; Fax: +1 708 3316557; E-mail: library@southhollandlibrary.org; URL: www.southhollandlibrary.org
1961; Alma J. DeYoung
98 000 vols; 259 curr per; 270 digital data carriers
41363

South Central Kansas Library System, 321 N Main, **South Hutchinson**, KS 67505-1146
T: +1 620 6633211; Fax: +1 620 6639797; URL: wwwsckls.info
1969; Paul Hawkins
1 branch libr
50 000 vols; 20 curr per; 200 Audio Bks
libr loan
41364

Salt Lake County Library Services – South Jordan Branch, 10673 S Redwood Rd, **South Jordan**, UT 84095-8697
T: +1 801 9447650; Fax: +1 801 2549047
Dina Wyatt
90 000 vols
41365

El Dorado County Library – South Lake Tahoe Branch, 1000 Rufus Allen Blvd, **South Lake Tahoe**, CA 96150
T: +1 530 5733185; Fax: +1 530 5448954
Sally Neitling
60 000 vols
41366

Lyon Township Public Library, 27005 S Milford Rd, **South Lyon**, MI 48178
T: +1 248 4378800; Fax: +1 248 4374621; E-mail: lyon@lyon.lib.mi.us; URL: www.lyon.lib.mi.us
Holly Teasdle
50 000 vols; 120 curr per
41367

Salem-South Lyon District Library, 9800 Pontiac Trail, **South Lyon**, MI 48178-1307
T: +1 248 4376431; Fax: +1 248 4376593; URL: www.salemsouthlyonlibrary.info
1939; Doreen Hannon
58 000 vols; 103 curr per; 2 342 av-mat
41368

South Milwaukee Public Library, 1907 Tenth Ave, **South Milwaukee**, WI 53172
T: +1 414 7688195; Fax: +1 414 7688072; URL: www.southmilwaukee.org
1899; Bob Pfeiffer
128 000 vols; 202 curr per; 403 e-books; 7 670 Audio Bks, 2 510 Large Print Bks
41369

Norwalk Public Library – South Norwalk Branch, Ten Washington St, **South Norwalk**, CT 06854
T: +1 203 8992790; Fax: +1 203 8992788
Reginald St Fort
72 000 vols
41370

South Orange Public Library, 65 Scotland Rd, **South Orange**, NJ 07079
T: +1 973 7620230; Fax: +1 973 7621469; E-mail: librarian@sopl.org; URL: www.sopl.org
1864; Melissa Kopecky
92 000 vols; 223 curr per
41371

South Park Township Library, 2575 Brownsville Rd, **South Park**, PA 15129-8527
T: +1 412 8335585; Fax: +1 412 8337368; URL: www.einpgh.org/ein/southprk
1970; Sharon Jean Bruni
50 000 vols; 105 curr per; 1 436 sound-rec
libr loan
41372

South Pasadena Public Library, 1100 Oxley St, **South Pasadena**, CA 91030-3198
T: +1 626 4037340; Fax: +1 626 4037331; URL: library.ci.south-pasadena.ca.us
1895; Steven Warren Fjeldsted
Drama, Art; Plays
143 000 vols; 400 curr per; 1 000 av-mat; 2 000 sound-rec; 1 000 digital data carriers; 300 high interest/low vocabulary bks, 1 000 large print bks
libr loan; CLA, ALA
41373

South Plainfield Free Public Library, 2484 Plainfield Ave, **South Plainfield**, NJ 07080
T: +1 908 7547885; Fax: +1 908 7533846; URL: www.southplainfield.lib.nj.us
1935; Sundra L. Randolph
55 000 vols; 215 curr per
41374

South Portland Public Library, 482 Broadway, **South Portland**, ME 04106
T: +1 207 7677660; Fax: +1 207 7677626; E-mail: splib@southportland.org; URL: www.southportlandlibrary.com
1965; Kevin Davis
Early American Children's Bks, Cape Elizabeth Hist Records; 1 branch libr
80 000 vols; 50 curr per
41375

South Saint Paul Public Library, 106 Third Ave N, **South Saint Paul**, MN 55075-2098
T: +1 651 5543240; Fax: +1 651 5543241; URL: www.southstpaul.org/departments/library
1922; Jane Kroschel
Original art, State artists
90 000 vols; 185 curr per
libr loan
41376

Salt Lake County Library Services – Columbus Branch, 2530 S 500 East, **South Salt Lake City**, UT 84106-1316
T: +1 801 9447625; Fax: +1 801 4120944
Darlene Dineen
56 000 vols
41377

South San Francisco Public Library, 840 W Orange Ave, **South San Francisco**, CA 94080-3125
T: +1 650 8293860; Fax: +1 650 8293865; E-mail: ssfpladm@plsinfo.org; URL: www.ssf.net/library
1916; Valerie Sommer
1 branch lib
166 000 vols; 315 curr per; 6 431 e-journals; 8 200 av-mat; 24 Electronic Media & Resources
41378

South Windsor Public Library, 1550 Sullivan Ave, **South Windsor**, CT 06074
T: +1 860 6441541; Fax: +1 860 6447645; URL: www.southwindsorlibrary.org
1898; Mary J. Etter
124 000 vols; 350 curr per
41379

Yarmouth Town Libraries, 312 Old Main St, **South Yarmouth**, MA 02664
T: +1 508 7604822; Fax: +1 508 7602699; URL: www.yarmouthlibraries.org
1866; Jacqueline Adams
86 000 vols; 399 curr per; 19 e-books
41380

Rogers Memorial Library, 91 Coopers Farm Rd, **Southampton**, NY 11968
T: +1 631 2830774; Fax: +1 631 2876537; URL: www.myrml.org
1893; Debra Engelhardt
Long Island Coll
100 000 vols; 275 curr per; 3 000 av-mat; 1 300 sound-rec; 13 digital data carriers
libr loan
41381

Southampton Free Library, 947 Street Rd, **Southampton**, PA 18966
T: +1 215 3221415; Fax: +1 215 3969375; URL: www.southamptonpa.com/library
1921; Thais Gardy
Large Print Bks
70 000 vols; 132 curr per; 780 CD's
libr loan; ALA, PLA
41382

Southborough Public Library, 25 Main St, **Southborough**, MA 01772
T: +1 508 4855031; Fax: +1 508 2294451; URL: www.southboroughlibrary.org
1852; Kimberley Ivers
63 000 vols; 100 curr per
41383

Jacob Edwards Library, 236 Main St, **Southbridge**, MA 01550-2598
T: +1 508 7645426; Fax: +1 508 7645428; URL: www.jacobedwardslibrary.org
1914; Margaret Morrissey
61 580 vols; 145 curr per; 50 maps; 5 931 microforms; 2 197 sound-rec; 59 digital data carriers; 208 children's dept kits, 96 art reproductions
libr loan; ALA, NELA, MLA
41384

Southbury Public Library, 100 Poverty Rd, **Southbury**, CT 06488
T: +1 203 2620626; Fax: +1 203 2626734; URL: www.southburylibrary.org
1969; Shirley Thorson
Large type print
89 000 vols; 127 curr per
41385

Southern Pines Public Library, 170 W Connecticut Ave, **Southern Pines**, NC 28387-4819
T: +1 910 6928235; Fax: +1 910 6951037; E-mail: lrb@southernpines.net; URL: www.sppl.net
1922; D. Lynn Thompson
North Carolina, Large Print
62 000 vols; 132 curr per
41386

Southfield Public Library, David Stewart Memorial Library, 26300 Evergreen Rd, **Southfield**, MI 48076; P.O. Box 2055, Southfield, MI 48037-2055
T: +1 248 7964208; Fax: +1 248 7964305; URL: www.sfldlib.org
1960; Dave Ewick
Business Ref Coll, Folklore & Fairy Tales; 1 branch libr
255 000 vols; 445 curr per; 2 000 microforms; 4 700 av-mat; 2 700 sound-rec; 11 digital data carriers
libr loan
41387

Southgate Veterans Memorial Library, 14680 Dix-Toledo Rd, **Southgate**, MI 48195
T: +1 734 2583002; Fax: +1 734 2849477; URL: www.southgate.lib.mi.us
1951; Joyce Farkas
77 000 vols; 122 curr per
libr loan
41388

Southington Public Library & Museum, 255 Main St, **Southington**, CT 06489
T: +1 860 6280947; Fax: +1 860 6280488; URL: www.southingtonlibrary.org
1902; Susan Smayda
120 000 vols; 240 curr per
41389

Brunswick County Library, Margaret & James Harper Library, 109 W Moore St, **Southport**, NC 28461
T: +1 910 4576237; URL: library.brunsco.net
1912; Maurice T. Tate
100 000 vols
41390

Spanish Fork Public Library, 49 S Main St, **Spanish Fork**, UT 84660-2030
T: +1 801 7985010; Fax: +1 801 7985014; URL: www.spanishfork.org
Pam Jackson
60 000 vols; 91 curr per; 3 300 av-mat
41391

Washoe County Library System – Spanish Springs Library, 7100A Pyramid Lake Hwy, **Sparks**, NV 89436-6669; P.O. Box 2151, Reno, NV 89505-2151
T: +1 775 4241800; Fax: +1 775 4241840
Corinne Dickman
89 000 vols
41392

Washoe County Library System – Sparks Library, 1125 12th St, **Sparks**, NV 89431; P.O. Box 2151, Reno, NV 89505-2151
T: +1 775 3523205; Fax: +1 775 3523207
Julie Machado
134 000 vols
41393

Caney Fork Regional Library, 25 Rhea St, **Sparta**, TN 38583
T: +1 931 8362209; Fax: +1 931 8363469; URL: www.state.tn.us/sos/statelib/p&d/caneyfork
1957; Faith A. Holdredge
108 000 vols; 10 curr per
41394

Sparta Free Library, 124 W Main St, *Sparta*, WI 54656; P.O. Box 347, Sparta, WI 54656-0347
T: +1 608 2692010; Fax: +1 608 2691542; E-mail: spartalibrary@wrlsweb.org; URL: www.spartalibrary.org
1861; Pamela Westby
51 000 vols; 150 curr per   41395

Sparta Public Library, 22 Woodport Rd, *Sparta*, NJ 07871
T: +1 973 7293101; Fax: +1 973 7291755; URL: www.sparta.library.com
1841; Carol Boutilier
68 000 vols; 200 curr per   41396

Sparta Township Library, 80 N Union St, *Sparta*, MI 49345
T: +1 616 8879937; Fax: +1 616 8870179; URL: www.sparta.llcoop.org
1917; Lois Lovell
60 000 vols; 48 curr per   41397

Spartanburg County Public Libraries, 151 S Church St, *Spartanburg*, SC 29306-3241
T: +1 864 5963507; Fax: +1 864 5963518; URL: www.infodepot.org
1885; Todd Stephens
9 branch libs
848 000 vols
libr loan   41398

Grace Balloch Memorial Library, 625 N Fifth St, *Spearfish*, SD 57783
T: +1 605 6421330; URL: www.cityofspearfish.com/library
1945; Amber Wilde
South Dakota Hist Coll
51 700 vols; 146 curr per
libr loan   41399

Speedway Public Library, 5633 W 25th St, *Speedway*, IN 46224-3899
T: +1 317 2438959; Fax: +1 317 2439373; URL: www.speedway.lib.in.us
1965; Darsi Bohr
78 000 vols; 126 curr per; 6 386 av-mat   41400

Owen County Public Library, 10 S Montgomery St, *Spencer*, IN 47460-1713
T: +1 812 8293392; Fax: +1 812 8296165; E-mail: vfreeland@owenlib.org; URL: www.owenlib.org
1912; Vickey Freeland
52 914 vols; 174 curr per; 10 maps; 392 microforms; 2 300 av-mat; 1 503 sound-rec; 110 digital data carriers
libr loan   41401

Richard Sugden Library, Eight Pleasant St, *Spencer*, MA 01562
T: +1 508 8857513; Fax: +1 508 8857523
Mary Baker-Wood
Hist Mat of Spencer & Massachusetts
55 000 vols; 100 curr per   41402

Ogden Farmer's Library, Farmers' Library Company of Ogden, 269 Ogden Center Rd, *Spencerport*, NY 14559
T: +1 585 3522141; Fax: +1 585 3523406; URL: www.ogdenny.com
1817; Patricia Uttaro
64 000 vols; 136 curr per   41403

Rutherford County Library, 255 Callahan Koon Rd, *Spindale*, NC 28160
T: +1 828 2876115; Fax: +1 828 2876119; URL: www.rutherfordcountync.gov/dept/library/main.php
1939; Martha Schatz
North Carolina Hist & Genealogy Coll; 2 branch libs
74 000 vols; 58 curr per; 23 000 e-books; 2 812 av-mat; 2 820 Talking Bks   41404

Spokane County Library District, Administrative Offices, 4322 N Argonne Rd, *Spokane*, WA 99212-0791
T: +1 509 8938230; Fax: +1 509 8938472; E-mail: admin@scld.org; URL: www.scld.org
1942; Michael J. Wirt
10 branch libs
362 000 vols; 1 421 curr per; 10 740 sound-rec; 7 530 digital data carriers; 2 350 Electronic Media & Resources, 11 320 Large Print Bks, 11 350 Talking Bks, 8 000 Videos   41405

Spokane Public Library, 906 W Main Ave, *Spokane*, WA 99201-0976
T: +1 509 4445333; Fax: +1 509 4445365; E-mail: info@spokanelibrary.org; URL: www.spokanelibrary.org
1891; Pat Partovi
Book Hist, African-American Coll, Adult Literacy; 5 branch libs
527 000 vols; 300 e-books; 200 govt docs; 6 200 maps; 190 000 microforms; 36 850 sound-rec; 2 065 digital data carriers; 18 000 Videos, 400 Vertical File, 35 Electronic Media & Resources
libr loan; ALA, Washington Library Association, ULC   41406

Harris County Public Library – Barbara Bush Branch, 6817 Cypresswood Dr, *Spring*, TX 77379
T: +1 281 3764610; Fax: +1 281 3760820; E-mail: cc@hcpl.net; URL: www.hcpl.net/branchinfo/cc/ccinfo.htm
Nancy Agafitei
152 000 vols   41407

Spring Lake District Library, 123 E Exchange St, *Spring Lake*, MI 49456-2018
T: +1 616 8465770; Fax: +1 616 8442129; URL: www.sllib.org
Claire Sheridan
69 000 vols; 175 curr per   41408

Spring Lake Public Library, 1501 Third Ave, *Spring Lake*, NJ 07762
T: +1 732 4496654; E-mail: spklib1@ve; URL: www.springlake.org
1920; Kateri Quinn
Large Print Coll
56 000 vols; 50 curr per   41409

Finkelstein Memorial Library, 24 Chestnut St, *Spring Valley*, NY 10977-5594
T: +1 845 3525700; Fax: +1 845 3522319; URL: finkelsteinlibrary.org
1917; Robert S. Devino
Foreign language, Holocaust, Education
225 000 vols; 518 curr per; 3 000 av-mat; 3 000 Audio Bks   41410

San Diego County Library – Spring Valley Branch, 836 Kempton St, *Spring Valley*, CA 91977
T: +1 619 4633006; Fax: +1 619 4638917
Angelica Guerrero
54 000 vols   41411

Springdale Public Library, 405 S Pleasant St, *Springdale*, AR 72764
T: +1 479 7508180; Fax: +1 479 7508182; URL: www.springdalelibrary.org
1923; Marcia Ransom
141 000 vols; 277 curr per
libr loan; ALA, PLA   41412

Clark County Public Library, 201 S Fountain Ave, *Springfield*, OH 45506; P.O. Box 1080, Springfield, OH 45501-1080
T: +1 937 3286901; Fax: +1 937 3286908; URL: www.ccpl.lib.oh.us
1872; John McConagha
Children's Lit (Lois Lenski Coll); 5 branch libs
437 000 vols; 429 curr per; 250 e-books; 75 Bks on Deafness & Sign Lang, 200 High Interest/Low Vocabulary Bks   41413

Illinois State Library – Talking Book & Braille Service, 401 E Washington, *Springfield*, IL 62701-1207
T: +1 217 7829260; Fax: +1 888 2612709; URL: www.ilbph.org
Sharon Ruda
112 000 vols; 34 e-books   41414

Lincoln Library – The Public Library of Springfield, Illinois, 326 S Seventh St, *Springfield*, IL 62701
T: +1 217 7534900; Fax: +1 217 7535329; URL: www.lincolnlibrary.info
1886; Nancy Huntley
Hist, Political science, Music; 2 branch libs
303 000 vols; 682 curr per
libr loan   41415

Lincoln Library – West, 1251 W Washington, *Springfield*, IL 62702
T: +1 217 7534985; Fax: +1 217 7534984
Lois Morse
67 000 vols   41416

Springfield City Library, Central Branch, 220 State St, *Springfield*, MA 01103
T: +1 413 2636828; Fax: +1 413 2636817; URL: www.springfieldlibrary.org
1857; Emily B. Bader
Aston Coll of American Wood Engravings, David A. Wells Economics Coll, Genealogy Coll, WW I & II Propaganda Coll; 9 branch libs
692 000 vols; 28 568 curr per; 33 818 av-mat; 33 818 Audio Bks, 4 178 Electronic Media & Resources, 22 825 Videos, 1 330 Special Interest Per Sub
libr loan   41417

Springfield City Library – Forest Park Branch, 380 Belmont Ave, *Springfield*, MA 01108
T: +1 413 2636843; Fax: +1 413 2636845; URL: www.springfieldlibrary.org/branches/fp.html
Reginald Wilson
68 000 vols   41418

Springfield City Library – Pine Point Branch, 204 Boston Rd, *Springfield*, MA 01109
T: +1 413 2636855; Fax: +1 413 2636857; URL: www.springfieldlibrary.org/branches/pp.html
Norma Couture
62 000 vols   41419

Springfield City Library – Sixteen Acres Branch, 1187 Parker St, *Springfield*, MA 01129
T: +1 413 2636858; Fax: +1 413 2636860; URL: www.springfieldlibrary.org/branches/sa.html
Norma Couture
71 000 vols   41420

Springfield Free Public Library, 66 Mountain Ave, *Springfield*, NJ 07081-1786
T: +1 973 3764930; Fax: +1 973 3761334; E-mail: questions@springfieldpubliclibrary.com; URL: www.springfieldpubliclibrary.com
1931; Susan Permahos
Local Hist (Sarah Bailey), Local Hist Artifacts (Donald B. Palmer Museum)
71 000 vols; 219 curr per   41421

Springfield Public Library, 225 Fifth St, *Springfield*, OR 97477-4697
T: +1 541 7263766; Fax: +1 541 7263747; E-mail: library@ci.springfield.or.us; URL: wheremindsgrow.org
1908; Robert Everett
Mystery & Detective Fiction Coll (1920 to present)
142 000 vols; 122 curr per; 265 govt docs; 380 maps; 5 630 sound-rec; 20 digital data carriers; 3 660 Pamphlets, Picture Files, Art Prints
libr loan   41422

Springfield Township Library, 70 Powell Rd, *Springfield*, PA 19064-2495
T: +1 610 5432113; Fax: +1 610 5431356; E-mail: springfield@delco.lib.pa.us; URL: www.delcolibraries.org
1937; Audrey Blossic
110 000 vols; 143 curr per   41423

Springfield-Greene County Library District, 4653 S Campbell, *Springfield*, MO 65810-1723; P.O. Box 760, Springfield, MO 65801-0760
T: +1 417 8835341; Fax: +1 417 8892547; URL: thelibrary.org
1903; Regina Cooper
Genealogy & Missouri Hist, Max Hunter Folk Songs Coll, Turnbo Papers, Music Coll; 9 branch libs
531 000 vols; 1 190 curr per; 6 500 e-books; 22 610 digital data carriers; 44 540 Audio Bks, 24 760 CDs, 17 000 Videos
libr loan   41424

Springville Public Library, 50 S Main St, *Springville*, UT 84663
T: +1 801 4892722; Fax: +1 801 4892709; URL: www.springvillelibrary.org
1916; Pamela Vaughn
73 000 vols; 150 curr per; 100 maps; 1 microforms; 8 000 sound-rec; 800 digital data carriers; 1 000 other items
libr loan   41425

Gentry County Library, 304 N Park, *Stanberry*, MO 64489
T: +1 660 7832335; Fax: +1 660 7832335; URL: www.gentrycountylibrary.org
1955; Judy Garrett
Large print, Missouri Titles, Genealogy
54 000 vols; 70 curr per; 200 microforms; 431 av-mat; 265 sound-rec; 116 digital data carriers
libr loan; MLA, ALA   41426

Orange County Public Library – Stanton Branch, 7850 Katella Ave, *Stanton*, CA 90680-3195
T: +1 714 8983302; Fax: +1 714 8980040
Tom Fitch
62 000 vols; 79 curr per   41427

Starkville-Oktibbeha County Public Library System, 326 University Dr, *Starkville*, MS 39759
T: +1 662 3232783; Fax: +1 662 3239140; E-mail: starkvillelibrary@yahoo.com
Virginia Holtcamp
Genealogy (Katie-Prince Eskar Coll); 2 branch libs
59 000 vols; 115 curr per   41428

Schlow Centre Region Library, 211 S Allen St, *State College*, PA 16801-4806
T: +1 814 2376236; Fax: +1 814 2388508; E-mail: refdesk@schlowlibrary.org; URL: www.schlowlibrary.org
1957; Elizabeth Allen
110 000 vols; 202 curr per   41429

The New York Public Library – Astor, Lenox & Tilden Foundations – Dongan Hills Branch, 1617 Richmond Rd (between Seaview & Liberty Aves), *Staten Island*, NY 10304
T: +1 718 3511444; Fax: +1 718 9876883; E-mail: dongan_hills@nypl.org; URL: www.nypl.org/branch/local/si/dh.cfm
Mary L. Pyrak
55 000 vols   41430

The New York Public Library – Astor, Lenox & Tilden Foundations – New Dorp Branch, 309 New Dorp Lane, *Staten Island*, NY 10306
T: +1 718 3512977; Fax: +1 718 3514993; E-mail: new_dorp@nypl.org; URL: www.nypl.org/branch/local/si/ndr.cfm
Yolanda Gleason
87 000 vols   41431

The New York Public Library – Astor, Lenox & Tilden Foundations – Richmondtown Branch, 200 Clarke Ave (@ Amber St), *Staten Island*, NY 10306
T: +1 718 6680413; Fax: +1 718 6681889; E-mail: richmondtown@nypl.org; URL: www.nypl.org/branch/local/si/rt.cfm
Nancy Avrin
66 000 vols   41432

The New York Public Library – Astor, Lenox & Tilden Foundations – St George Library Center, Five Central Ave (near Borough Hall), *Staten Island*, NY 10301
T: +1 718 4428560; Fax: +1 718 4472703; E-mail: st_george@nypl.org; URL: www.nypl.org/branch/local/si/sgc.cfm
Scott Lambdin
86 000 vols   41433

Statesboro Regional Library System, 124 S Main St, *Statesboro*, GA 30458
T: +1 912 7641341; Fax: +1 912 7641348; URL: www.srls.public.lib.ga.us
1937; Lois Roberts
196 000 vols; 303 curr per   41434

Iredell County Public Library, 201 N Tradd St, *Statesville*, NC 28677; P.O. Box 1810, Statesville, NC 28687-1810
T: +1 704 8785140; Fax: +1 704 8785449; URL: www.iredell.lib.nc.us
1967; Steve L. Messick
1 branch libr, 1 bookmobile
186 000 vols; 290 curr per; 70 Bks on Deafness & Sign Lang   41435

Staunton Public Library, One Churchville Ave, **Staunton**, VA 24401
T: +1 540 3323902; Fax: +1 540 3323906; E-mail: library@ci.staunton.va.us; URL: www.stauntonlibrary.org
1930; Ruth S. Arnold
148 000 vols; 212 curr per; 73 maps; 2 000 microforms; 28 000 av-mat; 40 000 sound-rec; 2 760 digital data carriers; 124 Art Reproductions, 930 High Interest/Low Vocabulary Bks, 200 Bks on Deafness & Sign Lang, 3 430 Large Print Bks
libr loan; ALA, PLA, VLA   41436

Bud Werner Memorial Library, 1289 Lincoln Ave, **Steamboat Springs**, CO 80487
T: +1 970 8790240; Fax: +1 970 8793476; URL: www.steamboatlibrary.org
1967; Christine Painter
Western Coll
73 000 vols; 250 curr per; 4 321 av-mat   41437

Menominee County Library, S319 Railroad St, **Stephenson**, MI 49887; P.O. Box 128, Stephenson, MI 49887-0128
T: +1 906 7536923; Fax: +1 906 7534678; E-mail: mcl@uproc.lib.mi.us; URL: www.uproc.lib.mi.us/menominee
1944; Patricia F. Cheski
1 branch libr
59 000 vols; 99 curr per; 1 454 av-mat   41438

Sterling Public Library, 102 W Third St, **Sterling**, IL 61081-3504
T: +1 815 6251370; Fax: +1 815 6257037; E-mail: director@sterlingpubliclibrary.org; URL: www.sterlingpubliclibrary.org
Jennifer Slaney
56 000 vols; 132 curr per; 3 202 av-mat   41439

Sterling Heights Public Library, 40255 Dodge Park Rd, **Sterling Heights**, MI 48313-4140
T: +1 586 4462668; Fax: +1 586 2764067; URL: www.shpl.net
1971; Tammy Turgeon
Careers, Children's lit, Large print; Int Language Coll, Polish Lit Coll
190 000 vols; 359 curr per; 15 560 Audio Bks, 52 Electronic Media & Resources   41440

Public Library of Steubenville & Jefferson County, 407 S Fourth St, **Steubenville**, OH 43952-2942
T: +1 740 2829782; Fax: +1 740 2822919; E-mail: steubnvl@oplin.org; URL: www.steubenville.lib.oh.us
1899; Alan Craig Hall
6 branch libs, 1 bookmobile
179 000 vols; 199 curr per; 5 000 e-books; 12 000 Electronic Media & Resources, 5 200 Large Print Bks
libr loan   41441

Portage County Public Library, Charles M White Library Bldg, 1001 Main St, **Stevens Point**, WI 54481-2860
T: +1 715 3461545; Fax: +1 715 3461239; URL: library.uwsp.edu/pcl/
1895; Robert J. Stack
3 branch libs
158 000 vols; 391 curr per   41442

Lincoln Township Public Library, 2099 W John Beers Rd, **Stevensville**, MI 49127
T: +1 269 4299575; Fax: +1 269 4293500; URL: www.lincolntownshiplibrary.org
1959; Dina M. Reilly
Gardening; Bartz Poetry Coll
81 000 vols; 214 curr per   41443

Stickney-Forest View Public Library District, 6800 W 43rd St, **Stickney**, IL 60402
T: +1 708 7491050; Fax: +1 708 7491054; URL: www.sfvpld.org/
1953; Tina Williams
Czechoslovakian Language Coll
56 000 vols; 189 curr per   41444

Stillwater Public Library, 224 N Third St, **Stillwater**, MN 55082
T: +1 651 2754338; Fax: +1 651 2754342; E-mail: splinfo@ci.stillwater.mn.us; URL: www.stillwaterlibrary.org
Lynne Bertalmio
83 000 vols; 271 curr per   41445

Stillwater Public Library, 1107 S Duck St, **Stillwater**, OK 74074
T: +1 405 3723633; Fax: +1 405 6240552; URL: library.stillwater.org
1923; Lynda Reynolds
Genealogy, Stillwater Coll
89 000 vols; 243 curr per   41446

Stockton-San Joaquin County Public Library, 605 N El Dorado St, **Stockton**, CA 95202
T: +1 209 9378683; Fax: +1 209 9378683; E-mail: library@ci.stockton.ca.us; URL: www.stockton.lib.ca.us
1880; Natalie Rencher
Southeast Asian Language Coll, Spanish Language Coll, Online Community Info Database; 10 branch libs
915 000 vols; 1 590 curr per; 678 av-mat; 680 Talking Bks   41447

Stockton-San Joaquin County Public Library – Margaret K. Troke Branch, 502 W Benjamin Holt Dr, **Stockton**, CA 95207
Fax: +1 209 9377721
David Gouker
136 000 vols   41448

Stoneham Public Library, 431 Main St, **Stoneham**, MA 02180
T: +1 781 4381324; Fax: +1 781 2793836; E-mail: todd@noblenet.org; URL: www.stonehamlibrary.org
1859; Mary Todd
18th-20th c (Stoneham Coll), docs on micro
80 592 vols; 226 curr per   41449

Stoughton Public Library, 304 S Fourth St, **Stoughton**, WI 53589
T: +1 608 8736281; Fax: +1 608 8730108; URL: www.scls.lib.wi.us/sto
1901
82 260 vols; 196 curr per   41450

Stoughton Public Library, 84 Park St, **Stoughton**, MA 02072-2974
T: +1 781 3442711; Fax: +1 781 3447340; E-mail: stlib@ocln.org; URL: www.stoughtonlibrary.org
1874; Patricia Basler
102 000 vols; 196 curr per; 1 000 av-mat; 1 586 sound-rec; 3 digital data carriers; 5 colls of microforms
libr loan   41451

Tredyffrin Public Library, 582 Upper Gulph Rd, **Strafford**, PA 19087-2096
T: +1 610 6887092; Fax: +1 610 6882014; E-mail: tredyffrinpubliclibr@ccls.org; URL: www.tredyffrinlibraries.org; www.ccls.org
1965; Joseph L. Sherwood
91 000 vols; 412 curr per; 38 e-books   41452

Stratford Library Association, 2203 Main St, **Stratford**, CT 06615
T: +1 203 3814166; Fax: +1 203 3812079; URL: www.stratford.lib.ct.us
1886; Susan Morong
Careers; ESL Coll
148 000 vols; 305 curr per; 2 744 av-mat; 4 485 sound-rec; 1 711 digital data carriers; 2 744 Audio Bks, 22 Electronic Media & Resources, 5 533 Videos, 43 Braille vols, 795 High Interest/Low Vocabulary Bks, 154 Bks on Deafness & Sign Lang, 1 768 Large Print Bks
libr loan   41453

Wiggin Memorial Library, Ten Bunker Hill Ave, **Stratham**, NH 03885
T: +1 603 7724346; Fax: +1 603 7724071; E-mail: wigginml@comcast.net; URL: www.wigginml.org
1891; Lesley Kimball
50 000 vols   41454

Cuyahoga County Public Library – Strongsville Branch, 18700 Westwood Dr, **Strongsville**, OH 44136-3431
T: +1 440 2385530; Fax: +1 440 5728685
Lucinda Bereznay
78 000 vols; 6 472 av-mat   41455

Eastern Monroe Public Library, 1002 N Ninth St, **Stroudsburg**, PA 18360
T: +1 570 4210800; Fax: +1 570 4210212; E-mail: reference@monroepl.org; URL: www.monroepl.org
1913; Barbara J. Keiser
Monroe County Hist Coll; 2 branch libs, 1 bookmobile
162 000 vols; 341 curr per
libr loan   41456

Martin County Library System -, 2351 SE Monterey Rd, **Stuart**, FL 34996
T: +1 772 2211402; Fax: +1 772 2211388; URL: www.library.martin.fl.us
1957; Donna M. Tunsoy
Maritime Coll, Travel Essay Coll; 6 branch libs
267 000 vols; 7 941 curr per; 24 324 av-mat; 24 330 Audio Bks   41457

Los Angeles Public Library System – Studio City Branch, 4400 Babcock Ave, **Studio City**, CA 91604-1399
T: +1 818 7557873; Fax: +1 818 7557873
Fagheh Mofidi
60 000 vols   41458

Door County Library, 107 S Fourth Ave, **Sturgeon Bay**, WI 54235
T: +1 920 7436578; Fax: +1 920 7436697; URL: www.dcl.lib.wi.us
1950; Rebecca N. Berger
8 branch libs
154 000 vols; 466 curr per; 2 500 av-mat; 2 062 sound-rec; 150 digital data carriers   41459

Sturgis District Library, 255 North St, **Sturgis**, MI 49091
T: +1 269 6597224; Fax: +1 269 6514534; URL: www.sturgis-library.org
1846; Todd Reed
Business & management
57 000 vols; 160 curr per   41460

Stuttgart Public Library – Arkansas County Library Headquarters, 2002 S Buerkle St, **Stuttgart**, AR 72160-6508
T: +1 870 6731966; Fax: +1 870 6734295; E-mail: legalbookie@lycos.com
1922; Ted T. Campbell
Agriculture, American Indians, Antiques, Genealogy; Arkansas Coll, Queeny Coll (Rare Books)
84 000 vols; 135 curr per; 27 av-mat   41461

Roxbury Township Public Library, 103 Main St, **Succasunna**, NJ 07876
T: +1 973 5842400; Fax: +1 973 5845484; URL: www.roxburylibrary.org
1960; N.N.
Children's Hist Fiction (Mary Wolfe Thompson Coll), Hist Ref (New Jersey)
70 000 vols; 217 curr per; 7 000 microforms; 1 200 av-mat; 1 550 sound-rec; 43 digital data carriers
libr loan   41462

Goodnow Library, 21 Concord Rd, **Sudbury**, MA 01776-2383
T: +1 978 4431035; Fax: +1 978 4431047; E-mail: goodnow@sudbury.ma.us; URL: library.sudbury.ma.us
1862; William Talentino
93 000 vols; 130 curr per   41463

Suffern Free Library, 210 Lafayette Ave, **Suffern**, NY 10901
T: +1 845 3571237; Fax: +1 845 3573156; URL: www.suffernfreelibrary.org
1926; Ruth Bolin
135 000 vols; 280 curr per   41464

Kent Memorial Library, 50 N Main St (Junction of Rtes 75 & 168), **Suffield**, CT 06078-2117
T: +1 860 6683896; Fax: +1 860 6683895; URL: www.suffield-library.org
1884; James McShane
75 000 vols; 106 curr per; 634 e-books   41465

Suffolk Public Library System, Morgan Memorial Library, 443 W Washington St, **Suffolk**, VA 23434
T: +1 757 5147323; Fax: +1 757 5397155; URL: www.suffolk.lib.va.us
1959; Elliot A. Drew
Black Arts & Lit (Reid Coll); 3 branch libs
175 000 vols; 166 curr per   41466

Fort Bend County Libraries – First Colony, 2121 Austin Pkwy, **Sugar Land**, TX 77479-1219
T: +1 281 2654444; Fax: +1 281 2654440
David Lukose
107 000 vols; 155 curr per; 2 744 av-mat   41467

Fort Bend County Libraries – Sugar Land Branch, 550 Eldridge Rd, **Sugar Land**, TX 77478
T: +1 281 2778934; Fax: +1 281 2778945
Virginia Harrell
97 000 vols; 149 curr per; 2 300 av-mat   41468

Sullivan County Public Library, 100 S Crowder St, **Sullivan**, IN 47882
T: +1 812 2684957; Fax: +1 812 2685370; URL: sullivan.lib.in.us
1904; Rebecca C. Cole
Indiana Coll
99 000 vols; 346 curr per; 4 500 av-mat; 500 sound-rec; 50 High Interst/Low Vocabulary Bks, 2 600 Large Print Bks, 1 500 Talking Bks   41469

Calcasieu Parish Public Library System – Sulphur Regional Library, 1160 Cypress St, **Sulphur**, LA 70663
T: +1 337 5277200; Fax: +1 337 5277200; E-mail: sul@calcasieu.lib.la.us
Esther Pennington
66 000 vols   41470

Sulphur Springs Public Library, 611 N Davis St, **Sulphur Springs**, TX 75482
T: +1 903 8854926; Fax: +1 903 4391052; URL: www.sslibrary.org
Kathryn St Claire
56 000 vols; 125 curr per   41471

Chattooga County Library, 360 Farrar Dr, **Summerville**, GA 30747
T: +1 706 8572553; Fax: +1 706 8577841; E-mail: sstephens@chattoogacountylibrary.org; URL: www.chattoogacountylibrary.org
1941; Susan Stephans
52 010 vols; 91 curr per   41472

Summit Free Public Library, 75 Maple St, **Summit**, NJ 07901-9984
T: +1 908 2730350; Fax: +1 908 2730031; URL: www.summitlibrary.org
1874; Glenn E. Devitt
Business, Management
127 000 vols; 469 curr per; 5 384 av-mat
libr loan   41473

Sumter County Library, 111 N Harvin St, **Sumter**, SC 29150
T: +1 803 7737273; Fax: +1 803 7734875; E-mail: sumtercolib@spiritcom.net; URL: www.midnet.sc.edu/sumtercls
1917; Robert Harden
2 branch libs, 1 bookmobile
175 000 vols; 250 curr per; 325 microforms; 1 600 av-mat; 4 291 sound-rec; 332 digital data carriers
libr loan   41474

Sumter County Public Library System, Clark Maxwell Jr. Library, 1405 County Rd, 526A, **Sumterville**, FL 33585
T: +1 352 5683074; Fax: +1 352 5683376
50 000 vols   41475

Sun City Library, Bell Library, 16828 N 99th Ave, **Sun City**, AZ 85351-1299
T: +1 623 9742569; Fax: +1 623 8760283; URL: www.sclib.com
1961; Grace Melody
Investing; Large Print Coll, Arizona Coll; 1 branch libr
90 000 vols; 220 curr per   41476

Sun Prairie Public Library, 1350 Linnerud Dr, **Sun Prairie**, WI 53590-2631
T: +1 608 8257323; Fax: +1 608 8253936; E-mail: sun@scls.lib.wi.us; URL: www.sunprairiepubliclibrary.org
Tracy Herold
92 000 vols; 278 curr per   41477

Degenstein Community Library, 40 S Fifth St, **Sunbury**, PA 17801
T: +1 570 2862461; Fax: +1 570 2864203; E-mail: kauflib@ptd.net; URL: www.degensteinlibrary.org
1937; Gail E. Broome
Pennsylvania Hist Rm
64 000 vols; 98 curr per   41478

Crook County Library, 414 Main St, *Sundance*, WY 82729; P.O. Box 910, Sundance, WY 82729-0910
T: +1 307 2831006; Fax: +1 307 2831006; E-mail: crookcountylib@rangeweb.net; URL: will.state.wy.us/crook
1937; Jill A. Mackey
77 000 vols; 278 curr per; 1 090 av-mat
libr loan; ALA   41479

Sunnyvale Public Library, 665 W Olive Ave, *Sunnyvale*, CA 94086-7622; P.O. Box 3714, Sunnyvale, CA 94088-3714
T: +1 408 7307314; Fax: +1 408 7358767; E-mail: adultref@ci.sunnyvale.ca.us; URL: www.sunnyvalelibrary.org
1960; N.N.
Electronics, Business & management, Art & architecture
252 000 vols; 445 curr per
libr loan   41480

Superior Public Library, 1530 Tower Ave, *Superior*, WI 54880-2532
T: +1 715 3948860; Fax: +1 715 3948870; URL: www.ci.superior.wi.us
1888; Janet I. Jennings
Anna Butler Art Coll, photos; Learning Disabilities (Burton Ansell Memorial Coll); Rare Bks (Henry S Butler); 1 branch libr
132 000 vols; 234 curr per
libr loan, ALA, WLA   41481

Pauline Haass Public Library, N64 W23820 Main St, *Sussex*, WI 53089-3120
T: +1 262 2465180; Fax: +1 262 2465236; URL: www.wcfls.lib.wi.us/phpl
1988; Kathy B. Klager
76 000 vols; 161 curr per   41482

Swampscott Public Library, 61 Burrill St, *Swampscott*, MA 01907
T: +1 781 5968867; Fax: +1 781 5968826; E-mail: swa@noblenet.org; URL: www.noblenet.org/swampscott
1853; Alyce Deveau
Town Hist (Henry Sill Baldwin Coll), Railroads & Model Railroads (Albert W. Lalime Coll)
92 000 vols; 100 curr per   41483

Onslow County Public Library – Swansboro Branch, 1460 W Corbett Ave, *Swansboro*, NC 28584
T: +1 910 3264888; Fax: +1 910 3266682
Michelle King
223 000 vols; 29 curr per; 641 av-mat   41484

Swansea Public Library, 69 Main St, *Swansea*, MA 02777
T: +1 508 6749609; Fax: +1 508 6755444; E-mail: Library@swansealibrary.org; URL: www.swansealibrary.org/swansea
1896; Kevin Lawton
1 branch libr
52 000 vols; 117 curr per   41485

B. B. Comer Memorial Library, 314 N Broadway, *Sylacauga*, AL 35150-2528
T: +1 256 2490961; URL: www.sylacauga.net/library
1939; Dr. Shirley K. Spears
Alabama Hist Coll
100 000 vols; 170 curr per   41486

Screven-Jenkins Regional Library, 106 S Community Dr, *Sylvania*, GA 30467
T: +1 912 5647526; Fax: +1 912 5647580; URL: www.sjrls.org
1951; Wendy Weinberger
1 branch libr
144 000 vols; 86 curr per   41487

Toledo-Lucas County Public Library – Sylvania Branch, 6749 Monroe St, *Sylvania*, OH 43560
T: +1 419 8822089; Fax: +1 419 8828993
John Cleveland
147 000 vols   41488

Syosset Public Library, 225 S Oyster Bay Rd, *Syosset*, NY 11791-5897
T: +1 516 9217161; Fax: +1 516 9218771; E-mail: administration-syosset@nassaulibrary.org; URL: www.syossetlibrary.org
1961; Judith Lockman
Business
222 000 vols; 349 curr per; 339 maps   41489

Davis County Library – Syracuse Northwest Branch, 1875 S 2000 West, *Syracuse*, UT 84075-9359
T: +1 801 8257080; Fax: +1 801 8257083
Carrie Murphy
61 000 vols   41490

Onondaga County Public Library – Robert P. Kinchen Central Library, The Galleries of Syracuse, 447 S Salina St, *Syracuse*, NY 13202-2494
T: +1 315 4351800; Fax: +1 315 4358533; E-mail: reference@onlib.org; URL: www.ocpl.lib.ny.us
1852; Elizabeth J. Dailey
Foundation Ctr; 10 branch libs
1 700 000 vols   41491

Syracuse Turkey Creek Township Public Library, 115 E Main St, *Syracuse*, IN 46567
T: +1 574 4573022; Fax: +1 574 4578971; URL: www.syracuse.lib.in.us
1909; John Castleman
57 000 vols; 83 curr per; 323 av-mat   41492

Pierce County Library System, 3005 112th St E, *Tacoma*, WA 98446-2215
T: +1 253 5826040; Fax: +1 253 5374600; URL: www.piercecountylibrary.org
1946; Neel Parikh
Korean & Spanish Coll; 17 branch libs, 1 bookmobile
1 108 000 vols; 2 950 curr per
libr loan   41493

Tacoma Public Library, 1102 Tacoma Ave S, *Tacoma*, WA 98402-2098
T: +1 253 5915606; Fax: +1 253 5915470; URL: www.tacomapubliclibrary.org
1886; Jerome Myers
Lincoln Coll, Northwest Hist, Photography Coll, John B Kaiser World War I Coll; 9 branch libs
2 123 000 vols; 1 638 curr per   41494

Tacoma Public Library – Fern Hill, 765 S 84th St, *Tacoma*, WA 98444
Janet Myers
91 000 vols   41495

Tacoma Public Library – Kobetich, 212 Brown's Point Blvd NE, *Tacoma*, WA 98422
Ruie Miller
51 000 vols   41496

Tacoma Public Library – Martin Luther King Jr Branch, 1902 S Cedar St, *Tacoma*, WA 98405
Barbara Scott
52 000 vols   41497

Tacoma Public Library – Moore Branch, 215 S 56th St, *Tacoma*, WA 98408
Vicki Armstrong
104 000 vols   41498

Tacoma Public Library – Mottet Branch, 3523 East G St, *Tacoma*, WA 98404
Vicki Armstrong
55 000 vols   41499

Tacoma Public Library – South Tacoma, 3411 S 56th St, *Tacoma*, WA 98409
Janet Myers
64 000 vols   41500

Tacoma Public Library – Swasey Branch, 7001 Sixth Ave, *Tacoma*, WA 98406
Barbara Scott
97 000 vols   41501

Tacoma Public Library – Wheelock Branch, 3722 N 26th St, *Tacoma*, WA 98407
Cheryl Towne
133 000 vols   41502

Tahlequah Public Library, 120 S College Ave, *Tahlequah*, OK 74464
T: +1 918 4562581; Fax: +1 918 4580590; URL: www.tahlequah.lib.ok.us
Robin Mooney
50 000 vols; 20 curr per   41503

Takoma Park Maryland Library, 101 Philadelphia Ave, *Takoma Park*, MD 20912
T: +1 301 8917259; Fax: +1 301 2708794; URL: www.takomapark.info/library/
1935; Ellen Arnold-Robbins
Children's lit, English language, Horticulture, Parenting, Restoration; Takoma Park Hist (Takoma Journal 1923-1955), unbd issues, microfilm
63 000 vols; 201 curr per   41504

Talladega Public Library, 202 South St E, *Talladega*, AL 35160
T: +1 256 3624211; Fax: +1 256 3620653; E-mail: talladeg@hiwaay.net
1906; N.N.
Alabama Coll
80 000 vols; 110 curr per; 507 microforms; 1 500 sound-rec; 25 digital data carriers; 44 art reproductions
libr loan   41505

LeRoy Collins Leon County Public Library System, 200 W Park Ave, *Tallahassee*, FL 32301-7720
T: +1 850 6062665; Fax: +1 850 6062601; E-mail: answersquad@leoncountyfl.gov; URL: www.leoncountylibrary.org
1955; Cay Hohmeister
4 branch libs
624 620 vols; 330 curr per; 31 037 e-books; 108 192 sound-rec
libr loan   41506

Akron-Summit County Public Library – Tallmadge Branch, 90 Community Rd, *Tallmadge*, OH 44278
T: +1 330 6334345; Fax: +1 330 6336324
Karen Wiper
55 000 vols   41507

Tampa-Hillsborough County Public Library System, 900 N Ashley Dr, *Tampa*, FL 33602-3704
T: +1 813 2733652; Fax: +1 813 2733707; E-mail: board@thpl.org; URL: thpl.org
1915; Joe Stines
25 branch libs
2 242 000 vols; 4 276 curr per   41508

Tampa-Hillsborough County Public Library System – Charles J. Fendig Library, 3909 W Neptune St, *Tampa*, FL 33629
Fax: +1 813 2768561
Keith Allen
64 000 vols   41509

Tampa-Hillsborough County Public Library System – Jan Kaminis Platt Regional, 3910 S Manhattan Ave, *Tampa*, FL 33611-1214
Fax: +1 813 2726071
Clarice Ruder
137 000 vols   41510

Tampa-Hillsborough County Public Library System – Jimmie B. Keel Regional, 2902 W Bearss Ave, *Tampa*, FL 33618-1828
Fax: +1 813 2643834
Lauren Levy
152 000 vols   41511

Tampa-Hillsborough County Public Library System – New Tampa Regional, 10001 Cross Creek Blvd, *Tampa*, FL 33647-2581
Fax: +1 813 9032289
Virginia Zurflieh
150 000 vols   41512

Tampa-Hillsborough County Public Library System – North Tampa, 8916 North Blvd, *Tampa*, FL 33604-1299
Fax: +1 813 9752057
Peggy Callahan
76 000 vols   41513

Tampa-Hillsborough County Public Library System – Seminole Heights, 4711 Central Ave, *Tampa*, FL 33603-3934
Fax: +1 813 2733670
Joseph O'Sullivan
56 000 vols   41514

Tampa-Hillsborough County Public Library System – Town 'N Country Regional, 7606 Paula Dr, *Tampa*, FL 33615-4116
Fax: +1 813 5545121
Keith Allen
133 000 vols   41515

Tampa-Hillsborough County Public Library System – Upper Tampa Bay Regional, 1121 Countryway Blvd, *Tampa*, FL 33626-2624
Fax: +1 813 9642967
Jodi Cohen
71 000 vols   41516

Taos Public Library, 402 Camino de La Placita, *Taos*, NM 87571
T: +1 505 7583063; Fax: +1 505 7372586; URL: www.taoslibrary.org
1923; Dorothy Kethler
Fine arts, Southwest; D. H. Lawrence Coll, bks
65 000 vols; 140 curr per   41517

Edgecombe County Memorial Library, 909 Main St, *Tarboro*, NC 27886
T: +1 252 8231141; Fax: +1 252 8237699; URL: www.edgecombelibrary.org
1920; Daniel Swartout
North Carolina Hist Coll
102 000 vols; 300 curr per; 4 000 av-mat
libr loan   41518

Community Library of Allegheny Valley, 400 Lock St, *Tarentum*, PA 15084
T: +1 724 2260770; Fax: +1 724 2263526; URL: www.alleghenyvalleylibrary.org
1923; Kathy Firestone
55 000 vols; 77 curr per   41519

Tarpon Springs Public Library, 138 E Lemon St, *Tarpon Springs*, FL 34689
T: +1 727 9434922; Fax: +1 727 9434926; E-mail: tslref@tblc.org; URL: www.tblc.org/tarpon/
1916; N.N.
Florida, Garden, Greek language
96 000 vols; 270 curr per; 4 000 e-books; 2 009 av-mat
libr loan   41520

Warner Library, 121 N Broadway, *Tarrytown*, NY 10591
T: +1 914 6317734; Fax: +1 914 6312324; URL: westchesterlibraries.org
1929; Kristin Weltzheimer
Out of State Telephone Bks, Washington Irving Coll, Large Print Bks
88 000 vols; 194 curr per
libr loan   41521

Los Angeles Public Library System – Encino-Tarzana Branch, 18231 Ventura Blvd, *Tarzana*, CA 91356-3620
T: +1 818 3431983; Fax: +1 818 3437867
Melissa Potter
60 000 vols   41522

Taunton Public Library, 12 Pleasant St, *Taunton*, MA 02780
T: +1 508 8211412; Fax: +1 508 8211414; E-mail: cotlib01@tmlp.net; URL: www.tauntonlibrary.org
1866; Susanne Costa Duquette
American lit, Art & architecture, Hist, World War II; American-Portuguese Genealogical Coll, Literacy Ctr, Portuguese Coll, Young Adult Coll
201 000 vols; 167 curr per; 6 500 av-mat; 350 sound-rec; 1 660 digital data carriers; 3 830 Audio Bks, 92 Electronic Media & Resources   41523

Lake County Library System, 2401 Woodlea Rd, *Tavares*, FL 32778; P.O. Box 7800, Tavares, FL
T: +1 352 2536180; Fax: +1 352 2536184; URL: www.lakeline.lib.fl.us
1982; Wendy R. Breeden
Genealogy; Florida Environment Coll; 6 branch libs
479 000 vols; 1 439 curr per
libr loan; ALA, FLA   41524

Alexander County Library, 77 First Ave SW, *Taylorsville*, NC 28681
T: +1 828 6324058; Fax: +1 828 6321094; URL: www.alexanderlibrary.org
1967; Gary Hoyle
57 000 vols; 149 curr per   41525

Salt Lake County Library Services – Park, 4870 S 2700 West, *Taylorsville*, UT 84118-2138
T: +1 801 9447638; Fax: +1 801 9653907
Jan Elkins
106 000 vols   41526

Tazewell County Public Library, 310 E Main St, *Tazewell*, VA 24651; P.O. Box 929, Tazewell, VA 24651-0929
T: +1 276 9882541; Fax: +1 276 9885980; URL: www.tcplweb.org
1964; Laurie S. Roberts
2 branch libs
83 000 vols; 260 curr per; 1768 av-mat
41527

Teaneck Public Library, 840 Teaneck Rd, *Teaneck*, NJ 07666
T: +1 201 8374171; Fax: +1 201 8370410; E-mail: tean1@bccls.org; URL: www.teaneck.org
1922; Michael McCue
African-American studies, Judaica
124 000 vols; 375 curr per; 2 300 av-mat; 975 Large Print Bks, 2 300 Talking Bks
41528

Tecumseh District Library, 215 N Ottawa St, *Tecumseh*, MI 49286-1564
T: +1 517 4232238; Fax: +1 517 4235519; E-mail: questions@tecumseh.lib.mi.us; URL: www.tecumseh.lib.mi.us
1883; S. Gayle Hazelbaker
Civil War, Indians
53 000 vols; 157 curr per
41529

Indian Valley Public Library, 100 E Church Ave, *Telford*, PA 18969
T: +1 215 7239109; Fax: +1 215 7230583; URL: www.ivpl.org
1963; Linda Beck
Pennsylvania Arch, Charles Price Genealogy Coll, Local Newspaper 1881-date, Chinese Culture Coll
130 000 vols; 130 curr per; 925 Large Print Bks
41530

Tell City-Perry County Public Library, 2328 Tell St, *Tell City*, IN 47586
T: +1 812 5472661; Fax: +1 812 5473038; E-mail: library@tcpclibrary.org; URL: www.tcpclibrary.org
1905; Larry Oathout
72 000 vols; 215 curr per; 250 microforms; 964 sound-rec; 55 digital data carriers; pamphlets
libr loan; INCOLSA
41531

San Miguel County Public Library District 1, Wilkinson Public Library, 100 W Pacific Ave, *Telluride*, CO 81435-2189; P.O. Box 2189, Telluride
T: +1 970 7284519; Fax: +1 970 7283340; E-mail: askus@telluridelibrary.org; URL: www.telluridelibrary.org
Barbara Brattin
Books about Film; Classic Movie Video Coll
64 000 vols; 199 curr per
41532

Tempe Public Library, 3500 S Rural Rd, *Tempe*, AZ 85282
T: +1 480 3505511; Fax: +1 480 3505544; URL: www.tempe.gov/library
1935; Teri Metros
Foreign Language Coll
468 000 vols; 1 450 curr per
41533

Temple Public Library, 100 W Adams Ave, *Temple*, TX 76501-7641
T: +1 254 2985702; Fax: +1 254 2985328; URL: www.templelibrary.us
1900; Judy Duer
Adult Education, Career & Job Info, Genealogy, Large-Print, Local Authors; 1 bookmobile
144 000 vols; 204 curr per; 8 100 av-mat; 8 000 sound-rec; 10 180 Large Print Bks
41534

County of Los Angeles Public Library – Temple City Library, 5939 Golden West Ave, *Temple City*, CA 91780-2292
T: +1 626 2852136; Fax: +1 626 2852314; URL: www.colapublib.org/libs/templecity
Marie Schreb-Clift
99 000 vols; 103 curr per
41535

Temple Terrace Public Library, 202 Bullard Pkwy, *Temple Terrace*, FL 33617-5512
T: +1 813 9897164; Fax: +1 813 9897069; URL: www.hcplc.org
1960; Mary H. Satterwhite
75 000 vols; 123 curr per
libr loan
41536

Tenafly Public Library, 100 River Edge Rd, *Tenafly*, NJ 07670-2087
T: +1 201 5688680; Fax: +1 201 5685475; E-mail: tenfl@bccls.org; URL: tenaflynj.org
1920; Stephen R. Wechtler
Gardening, East Asian Culture, 2 Hist Local Newspapers on microfilm, Mitchell Young Adult Fiction Coll, Music Coll
84 000 vols; 212 curr per; 3 050 av-mat
41537

Terra Alta Public Library, 701-B E State Ave, *Terra Alta*, WV 26764-1204
T: +1 304 7892724; Fax: +1 304 7892724
1972; Ima Thomas
63 000 vols; 73 curr per
41538

Vigo County Public Library, Seventh & Poplar St, One Library Square, *Terre Haute*, IN 47807-3609
T: +1 812 2321113; Fax: +1 812 2323208; URL: www.vigo.lib.in.us
1882; Nancy E. Dowell
Community Archs Coll, Eugene V Debbs Coll, Lifelong Learning Ctr Literacy Mats, Max Ehrmann Coll; 4 branch libs
172 000 vols; 1 607 curr per; 4 000 e-books; 5 591 av-mat; 6 600 Audio Bks, 942 Electronic Media & Resources, 9 730 Large Print Bks
libr loan
41539

Riter C. Hulsey Public Library, 301 N Rockwall, *Terrell*, TX 75160-2618
T: +1 972 5516663; Fax: +1 972 5516662; E-mail: library@cityofterrell.org; URL: www.terrellpl.org
1904; Rebecca W. Sullivan
71 000 vols; 81 curr per; 1774 av-mat
41540

Terryville Public Library, 238 Main St, *Terryville*, CT 06786
T: +1 860 5823121; Fax: +1 860 5854068; E-mail: tplstaff@biblio.org; URL: www.terryvillepl.info
1895; Lynne White
Literacy, Large type print; Career Corner, Terryville-Plymouth Room
51 000 vols; 101 curr per
41541

Tewksbury Public Library, 300 Chandler St, *Tewksbury*, MA 01876
T: +1 978 6404490; E-mail: mte@mvlc.org; URL: www.tewksburypl.org
1877; Elisabeth Desmarais
90 000 vols; 175 curr per
41542

Texarkana Public Library, 600 W Third St, *Texarkana*, TX 75501-5054
T: +1 903 7942149; Fax: +1 903 7942139; URL: www.txar-publib.org
1925; Alice Coleman
90 000 vols; 139 curr per
41543

Moore Memorial Public Library, 1701 Ninth Ave N, *Texas City*, TX 77590
T: +1 409 6435979; Fax: +1 409 9481106; URL: www.texascity.library.org
1928; Beth Ryker Steiner
120 000 vols; 180 curr per
41544

The Colony Public Library, 6800 Main St, *The Colony*, TX 75056-1133
T: +1 972 6251900; Fax: +1 972 6242245; E-mail: library@thecolony.lib.tx.us; URL: www.thecolony.lib.tx.us
Joan Sveinsson
70 000 vols; 146 curr per
41545

The Dalles-Wasco County Library, 722 Court St, *The Dalles*, OR 97058-2270
T: +1 541 2962815; Fax: +1 541 2964179; E-mail: cityinfo@ci.the-dalles.or.us; URL: www.ci.the-dalles.or.us
1909; Sheila Dooley
Oregon hist
65 000 vols; 114 curr per; 2 326 av-mat
41546

Montgomery County Memorial Library System – South Branch, 2101 Lake Robbins Dr, *The Woodlands*, TX 77380
T: +1 281 2989110; Fax: +1 936 7888372
Catherine Pell
151 000 vols
41547

Hot Springs County Library, 344 Arapahoe, *Thermopolis*, WY 82443-0951; P.O. Box 951, Thermopolis, WY 82443-0951
T: +1 307 8643104; Fax: +1 307 8645416; URL: will.state.wy.us/hotsprings
1919; Chrissy Bendlin
Local Newspapers from 1905, micro; Wyoming Hist Coll
55 000 vols; 80 curr per; 2 mss; 1 diss/theses; 600 govt docs; 150 maps; 58 microforms; 1 399 av-mat; 800 sound-rec; 10 digital data carriers; uncataloged vertical file
libr loan
41548

Lafourche Parish Public Library, 303 W Fifth St, *Thibodaux*, LA 70301-3123
T: +1 985 4461163; Fax: +1 985 4463848; E-mail: admin.b1a@pelican.state.lib.la.us; URL: www.lafourche.org
1947; Susanna LeBouef
6 branch libs
234 000 vols; 400 curr per
41549

Northwest Regional Library, 210 LaBree Ave N, *Thief River Falls*, MN 56701; P.O. Box 593, Thief River Falls
T: +1 218 6811066; Fax: +1 218 6811095; URL: www.nwrlib.org
Barbara Jauquet-Kalinoski
High interest/low vocabulary, Large print, Toddler's, Literacy
143 000 vols; 400 curr per; 843 av-mat
41550

Thomaston Public Library, 248 Main St, *Thomaston*, CT 06787
T: +1 860 2834339; Fax: +1 860 2834330; URL: www.biblio.org/thomaston
1898; Jane T. Kendrick
Career Info, Art Techniques (Bradshaw Coll), Conklin Coll of the Arts
50 000 vols; 115 curr per
41551

Davidson County Public Library System – Thomasville Public, 14 Randolph St, *Thomasville*, NC 27360-4638
T: +1 336 4742690; Fax: +1 336 4724690
Janet Malliett
97 000 vols; 132 curr per
41552

Thomas County Public Library System, 201 N Madison St, *Thomasville*, GA 31792-5414
T: +1 229 2255252; Fax: +1 229 2255258; URL: www.tcpls.net
1988; Nancy Tillinghast
Art; Lt Henry O. Flipper Black Hist Coll; 5 branch libs
59 000 vols; 81 curr per; 4 500 av-mat; 1 012 sound-rec; 63 digital data carriers; 1 820 Audio Bks
libr loan
41553

Rangeview Library District, 8992 Washington St, *Thornton*, CO 80229-4537
T: +1 303 2882001; Fax: +1 303 2875971; URL: www.rangeviewld.org
1953; Pam Sandlian Smith
7 841 Audio Bks, 6 447 CDs. 6 907 Videos; 6 branch libs
287 000 vols; 678 curr per; 116 microforms; 7 841 av-mat; 8 700 sound-rec; 1 424 digital data carriers; 2 084 Videos, 593 Puzzles, 846 High Interest/Low Vocabulary Bks, 200 Bks on Deafness & Sign Lang, 6 026 Large Print Bks
libr loan
41554

Thousand Oaks Library, 1401 E Janss Rd, *Thousand Oaks*, CA 91362-2199
T: +1 805 4492660; Fax: +1 805 4958485; URL: www.tol.lib.ca.us
1982; Nancy Sevier
Radio & TV Broadcasting; 1 branch lib
345 000 vols; 589 curr per
41555

Belvedere-Tiburon Library, 1501 Tiburon Blvd, *Tiburon*, CA 94920
T: +1 415 7892665; Fax: +1 415 7892650; E-mail: dmazzolini@bel-tib-lib.org; URL: www.thelibrary.info
1997; Deborah Mazzolini
59 800 vols; 413 curr per
41556

Coastal Plain Regional Library – Headquarters, 2014 Chestnut Ave, *Tifton*, GA 31794
T: +1 229 3863400; Fax: +1 229 3867007; URL: www.cprl.org
1956; Carrie C. Zeiger
5 branch libs
221 000 vols; 186 curr per
41557

Tigard Public Library, 13500 SW Hall Blvd, *Tigard*, OR 97223-8111
T: +1 503 7182517; Fax: +1 503 5987515; URL: www.tigard-or.gov/library/default.asp
1964; Margaret Barnes
118 000 vols; 242 curr per; 3 165 av-mat
41558

Tillamook County Library, 1716 Third St, *Tillamook*, OR 97141
T: +1 503 8424792; Fax: +1 503 8158194; URL: www.tillamook.plinkit.org
1947; Sara Charlton
5 branch libs, 1 bookmobile
150 000 vols; 300 curr per; 600 govt docs; 750 music scores; 600 maps; 13 microforms; 5 000 av-mat; 3 000 sound-rec; 2 digital data carriers; 9 newspapers
libr loan
41559

Tinley Park Public Library, 7851 Timber Dr, *Tinley Park*, IL 60477-3398
T: +1 708 5320160; Fax: +1 708 5322981; E-mail: tp_library@tplibrary.org; URL: www.tplibrary.org
1959; Rich Wolff
German Language Bks Coll
142 000 vols; 813 curr per
41560

Tipp City Public Library, 11 E Main St, *Tipp City*, OH 45371
T: +1 937 6673826; Fax: +1 937 6677968; URL: www.tippcitylibrary.org
1922; Marcus A. Mabelitini
95 000 vols; 126 curr per; 48 digital data carriers
libr loan; ALA
41561

Tipton County Public Library, 127 E Madison St, *Tipton*, IN 46072
T: +1 765 6758761; Fax: +1 765 6754475; E-mail: tipton@tiptonpl.lib.in.us; URL: www.tiptonpl.lib.in.us
1902
Indians; 1 branch libr
89 000 vols; 318 curr per
41562

Benson Memorial Library, 213 N Franklin St, *Titusville*, PA 16354-1788
T: +1 814 8272913; Fax: +1 814 8279836
1904; Gail K. Myer
55 000 vols; 238 curr per; 1 255 av-mat
41563

Titusville Public Library, 2121 S Hopkins Ave, *Titusville*, FL 32780
T: +1 321 2645026; Fax: +1 321 2645030; URL: www.brev.org
1906; Pamela D. Boddy
93 000 vols; 292 curr per
41564

Tiverton Library Services, 238 Highland Rd, *Tiverton*, RI 02878
T: +1 401 6256796; Fax: +1 401 6255499; E-mail: arust@tivertonlibrary.org; URL: www.tivertonlibrary.org
Ann Grealish-Rust
55 000 vols; 45 curr per
41565

Toledo-Lucas County Public Library, 325 N Michigan St, *Toledo*, OH 43604-6614
T: +1 419 2595272; Fax: +1 419 2551332; URL: www.toledolibrary.org
1970; Clyde Scoles
18 branch libs
2 880 000 vols; 4 731 curr per; 34 647 av-mat; 34 647 Audio Bks, 35 813 Large Print Bks
libr loan
41566

Toledo-Lucas County Public Library – Heatherdowns, 3265 Glanzman, *Toledo*, OH 43614
T: +1 419 2595270; Fax: +1 419 3823231
Kathleen Lundberg
147 000 vols
41567

Toledo-Lucas County Public Library – Kent, 3101 Collingwood Blvd, *Toledo*, OH 43610
T: +1 419 2595340; Fax: +1 419 2436536
Faith Hairston
68 000 vols
41568

Toledo-Lucas County Public Library – Locke, 703 Miami St, **Toledo**, OH 43605
T: +1 419 2595310; Fax: +1 419 6913237
Mary Beth Gratop
59 000 vols   41569

Toledo-Lucas County Public Library – Mott, 1085 Dorr St, **Toledo**, OH 43607
T: +1 419 2595230; Fax: +1 419 2554237
Judith Jones
64 000 vols   41570

Toledo-Lucas County Public Library – Point Place, 2727 117th St, **Toledo**, OH 43611
T: +1 419 2595390; Fax: +1 419 7295363
Hannah Lammie
84 000 vols   41571

Toledo-Lucas County Public Library – Reynolds Corners, 4833 Dorr St, **Toledo**, OH 43615
T: +1 419 2595320; Fax: +1 419 5314076
Marilee McSweeny
110 000 vols   41572

Toledo-Lucas County Public Library – Sanger, 3030 W Central Ave, **Toledo**, OH 43606
T: +1 419 2595370; Fax: +1 419 5369573
Erin Connolly
131 000 vols   41573

Toledo-Lucas County Public Library – Toledo Heights, 423 Shasta Dr, **Toledo**, OH 43609
T: +1 419 2595220; Fax: +1 419 3859297
Mary Chwialkowski
52 000 vols   41574

Toledo-Lucas County Public Library – Washington, 5560 Harvest Lane, **Toledo**, OH 43623
T: +1 419 2595330; Fax: +1 419 4724991
Barbara Lough
125 000 vols   41575

Toledo-Lucas County Public Library – West Toledo, 1320 Sylvania Ave, **Toledo**, OH 43612
T: +1 419 2595290; Fax: +1 419 4760892
Susan Schafer
114 000 vols   41576

Tolland Public Library, 21 Tolland Green, **Tolland**, CT 06084
T: +1 860 8713620; Fax: +1 860 8713626; E-mail: bbutler@tolland.org; URL: www.tolland.org/library
1899; Barbara Butler
52 000 vols; 80 curr per   41577

Tomah Public Library, 716 Superior Ave, **Tomah**, WI 54660-2098
T: +1 608 3747470; Fax: +1 608 3747471; E-mail: tomah_public_library@yahoo.com; URL: www.tomah.com/tomahpl
1876; Cathy W. Peterson
51 000 vols
libr loan   41578

Tomahawk Public Library, 300 W Lincoln Ave, **Tomahawk**, WI 54487
T: +1 715 4532455; Fax: +1 715 4531630
Mary E. Dunn
56 000 vols; 22 curr per   41579

Harris County Public Library – Tomball Branch, 30555 Tomball Pkwy, **Tomball**, TX 77375
T: +1 832 5594200; Fax: +1 832 5594248
Wendy Schneider
83 000 vols   41580

Ocean County Library, 101 Washington St, **Toms River**, NJ 08753
T: +1 732 3496200;
Fax: +1 732 3490478; URL: www.oceancountylibrary.org
1925; Elaine McConnell
Nathanial Holmes Bishop papers, local Ocean County hist, Genealogy (New Jersey Coll); 19 branch libs
972 000 vols; 3 274 curr per; 125 maps; 8 000 microforms; 1 226 digital data

carriers; 804 DVDs, 3 932 feature films, 13 226 videos, net libr (1 600)
libr loan   41581

Tooele City Public Library, 128 W Vine St, **Tooele**, UT 84074-2059
T: +1 435 8822182; Fax: +1 435 8432159; URL: www.tooelecity.org
1910; Karen Emery
57 000 vols   41582

Topeka & Shawnee County Public Library, 1515 SW Tenth Ave, **Topeka**, KS 66604-1374
T: +1 785 5804400; Fax: +1 785 5804496; URL: www.tscpl.org
1870; Gina Millsap
Art, Illust bks, Music; Miniature Bks Coll; Subregional Libr for the Blind & Physically Handicapped
386 000 vols; 1 301 curr per; 28 340 av-mat; 27 070 Large Print Bks, 28 340 Talking Bks
libr loan   41583

Torrance Public Library, Katy Geissert Civic Center, 3301 Torrance Blvd, **Torrance**, CA 90503
T: +1 310 6185959; Fax: +1 310 6185952; URL: www.library.torrnet.com
1967; Paula Weiner
Radio, Hist, Art; 5 branch libs
540 000 vols; 997 curr per
libr loan   41584

Torrington Library, 12 Daycoeton Pl, **Torrington**, CT 06790-6399
T: +1 860 4896684; Fax: +1 860 4824664; E-mail: info@torringtonlibrary.org; URL: www.torringtonlibrary.org
1864; Karen Worrall
Large Print Coll
55 000 vols; 74 curr per; 800 av-mat
                                          41585

Baltimore County Public Library, 320 York Rd, **Towson**, MD 21204-5179
T: +1 410 8876100; Fax: +1 410 8876103; E-mail: bcpl@bcpl.info; URL: www.bcpl.info
1948; James H. Fish
Spanish Colls, African American Colls, Korean Colls, Russian Colls; 17 branch libs
1 559 000 vols; 3 895 curr per   41586

Baltimore County Public Library – Towson Area Branch, 320 York Rd, **Towson**, MD 21204-5179
T: +1 410 8876166; Fax: +1 410 8873170; E-mail: towson@bcpl.net
Jennifer Haire
188 000 vols   41587

Stockton-San Joaquin County Public Library – Tracy Branch, 20 E Eaton Ave, **Tracy**, CA 95376
Fax: +1 209 8314252
Kathleen Buffleben
Kiersh Memorial Music;Tugel Memorial Natural History
119 000 vols   41588

Traverse Area District Library, 610 Woodmere Ave, **Traverse City**, MI 49686
T: +1 231 9328502; Fax: +1 231 9328578; E-mail: libadmin@tadl.tcnet.org; URL: www.tadl.org
1897; Michael McGuire
Deaf & Hearing impaired, Sheet music; 3 branch libs
410 000 vols; 808 curr per
libr loan   41589

Trenton Free Public Library, 120 Academy St, **Trenton**, NJ 08608-1302
T: +1 609 3927188; Fax: +1 609 3967655; E-mail: ref@trentonlib.org; URL: www.trenton.lib.nj.us
1750; Kimberly Bray
4 branch libs
375 000 vols; 770 curr per   41590

Trenton Veterans Memorial Library, 2790 Westfield Rd, **Trenton**, MI 48183-2482
T: +1 734 6769777; Fax: +1 734 6769895; E-mail: tren@tln.lib.mi.us; URL: www.trenton.lib.mi.us
1928; Francene Sanak
89 000 vols; 210 curr per; 1 610 av-mat
                                          41591

Bradford County Library System, RR 3, Box 320, **Troy**, PA 16947-9440
T: +1 570 2972436; Fax: +1 570 2974197
1941; Diane Sadler
194 000 vols   41592

Montgomery County Public Library, 215 W Main, **Troy**, NC 27371
T: +1 910 5721311; Fax: +1 910 5765565; URL: www.srls.info/montgomery/montindex.html
David R. Atkins
62 000 vols; 200 curr per   41593

Troy Public Library, 300 N Three Notch Rd, **Troy**, AL 36081
T: +1 334 5661314; Fax: +1 334 5664392; URL: www.troycitylibrary.org
50 000 vols; 250 curr per   41594

Troy Public Library, 510 W Big Beaver Rd, **Troy**, MI 48084-5289
T: +1 248 5243549; Fax: +1 248 5240112; URL: www.libcoop.net/troy
1962; Brian Stoutenburg
Frances Teasdale Civil War Coll, Oakland Co Genealogical Society Coll, Morgan – West White House Memorabilia
206 000 vols; 656 curr per; 4 000 e-books; 26 890 av-mat; 21 660 digital data carriers; 26 890 Audio Bks, 85 Electronic Media & Resources   41595

Troy Public Library, 100 Second St, **Troy**, NY 12180-4005
T: +1 518 2747071; Fax: +1 518 2719154; E-mail: troyref@uhls.lib.ny.us; URL: www.uhls.org/troy
1835; Paul Hicok
2 branch libs
131 000 vols; 412 curr per; 1 457 maps; 2 000 microforms; 1 344 sound-rec; 1 digital data carriers   41596

Trumbull Library, 33 Quality St, **Trumbull**, CT 06611
T: +1 203 4525197; Fax: +1 203 4525125; URL: www.trumbullct-library.org
1975; Susan Horton
Framed Art & photos, Sculpture, Records; 1 branch libr
172 000 vols; 200 curr per; 1 700 av-mat; 11 279 Videos, 1 700 Talking Bks   41597

Trussville Public Library, 201 Parkway Dr, **Trussville**, AL 35173
T: +1 205 6552022; Fax: +1 205 6611645; URL: www.trussvillelibrary.com
Brenda Brasher
Alabama Hist Coll
60 000 vols; 95 curr per   41598

Truth or Consequences Public Library, 325 Library Lane, **Truth Or Consequences**, NM 87901-2375
T: +1 505 8943027; Fax: +1 505 8942068; E-mail: torclibrary@torcnm.org; URL: www.youseemore.com/torcnm/
1933; Pat O'Hanlon
Southwest Coll
50 000 vols; 180 curr per   41599

Tualatin Public Library, 8380 SW Nyberg St, **Tualatin**, OR 97062
T: +1 503 6913071; Fax: +1 502 6929720; URL: www.tualatinlibrary.org; www.ci.tualatin.or.us
1977; Darrel Condra
68 000 vols; 180 curr per   41600

Jackson County Library, Tuckerman Branch, 200 W Main St, **Tuckerman**, AR 72473; P.O. Box 1117, Tuckerman, AR 72473-1117
T: +1 870 3495336; Fax: +1 870 3495336; E-mail: tuckerman.librarian@jacksoncolibrary.net
Shirley Manuel
60 000 vols   41601

Pima County Public Library, 101 N Stone Ave, **Tucson**, AZ 85701
T: +1 520 7914391; Fax: +1 520 7913213; URL: www.library.pima.gov
1883; Nancy Ledeboer
Business & management, Arizona, govt; Southwestern Lit for Children; 25 branch libs
1 428 000 vols; 3 403 curr per; 35 641 av-mat
libr loan   41602

Pima County Public Library – Columbus, 4350 E 22nd St, **Tucson**, AZ 85711
T: +1 520 7914081; Fax: +1 520 7913213
Terry Rill
76 000 vols   41603

Pima County Public Library – Dusenberry River Center, 5605 E River Rd, No 105, **Tucson**, AZ 85750
T: +1 520 7914979; Fax: +1 520 7914982
Kathleen Dannreuther
85 000 vols   41604

Pima County Public Library – George Miller-Golf Links, 9640 E Golf Links Rd, **Tucson**, AZ 85730
T: +1 520 7915524; Fax: +1 520 7915770
Sharla Darby
82 000 vols   41605

Pima County Public Library – Joel D. Valdez Main Library, 101 N Stone Ave, **Tucson**, AZ 85701
T: +1 520 7914393; Fax: +1 520 7915248
Karyn Prechtel
186 000 vols   41606

Pima County Public Library – Kirk Bear Canyon, 8959 E Tanque Verde Rd, **Tucson**, AZ 85749
T: +1 520 7915021; Fax: +1 520 7915024
Margaret Wilke
77 000 vols   41607

Pima County Public Library – Mission, 3770 S Mission Rd, **Tucson**, AZ 85713
T: +1 520 7914811; Fax: +1 520 7915330
Martin Rivera
66 000 vols   41608

Pima County Public Library – Nanini, 7300 N Shannon Rd, **Tucson**, AZ 85741
T: +1 520 7914626; Fax: +1 520 7913213
Kristi Bradford
118 000 vols   41609

Pima County Public Library – Valencia, 202 W Valencia Rd, **Tucson**, AZ 85706
T: +1 520 7914531; Fax: +1 520 7915342
Elaine Valenzuela
84 000 vols   41610

Pima County Public Library – Wilmot-Murphy, 530 N Wilmot Rd, **Tucson**, AZ 85711
T: +1 520 7914627; Fax: +1 520 7913213
Daphne Daly
123 000 vols   41611

Pima County Public Library – Woods Memorial Branch, 3455 N First Ave, **Tucson**, AZ 85719
T: +1 520 7914548; Fax: +1 520 7913213
Joan Biggar
99 000 vols   41612

Tulare Public Library, 113 North F St, **Tulare**, CA 93274-3857
T: +1 559 6852341; Fax: +1 559 6852345; URL: www.sjvls.lib.ca.us/tularepub/
1878; Michael C. Stowell
Art, California, Religion, Music, Hist
110 000 vols; 138 curr per; 239 e-books
                                          41613

Coffee County Lannom Memorial Public Library, 312 N Collins St, **Tullahoma**, TN 37388-3229
T: +1 931 4552460; Fax: +1 931 4542300; URL: www.lannom.org
1947; Susan Stovall
Genealogy Coll
70 000 vols; 106 curr per   41614

Tulsa City-County Library System Central Library, 400 Civic Ctr, **Tulsa**, OK 74103
T: +1 918 5967897
Suanne Wymer
508 000 vols   41615

United States of America: Public Libraries 41616 – 41659

Tulsa City-County Library System – Hardesty Regional Library, 8316 E 93rd St, *Tulsa*, OK 74133
T: +1 918 2507307; Fax: +1 918 2507843
Louix Escobar-Matute
164 000 vols 41616

Tulsa City-County Library System – Martin Regional Library, 2601 S Garnett, *Tulsa*, OK 74129
T: +1 918 6696340; Fax: +1 918 6696344
Theresa Fowler
110 000 vols 41617

Tulsa City-County Library System – Peggy V. Helmerich Library, 5131 E 91st, *Tulsa*, OK 74114
T: +1 918 5962466; Fax: +1 918 5962468
Marilyn Neal
67 000 vols 41618

Tulsa City-County Library System – Rudisill Regional Library, 1520 N Hartford, *Tulsa*, OK 74106
T: +1 918 5967280; Fax: +1 918 5967283
Keith Jemison
71 000 vols 41619

Tulsa City-County Library System – Zarrow Regional Library, 2224 W 51st, *Tulsa*, OK 74107
T: +1 918 5914366
Barry Hensley
58 000 vols 41620

Timberland Regional Library, 415 Tumwater Blvd SW, *Tumwater*, WA 98501-5799
T: +1 360 9435001; Fax: +1 360 5866838; URL: www.trlib.org
1968; Jodi Reng
27 branch libs
1 092 000 vols; 3 601 curr per; 41 085 av-mat; 41 085 Talking Bks 41621

Lee County Library, 219 N Madison St, *Tupelo*, MS 38804-3899
T: +1 662 8419029; Fax: +1 662 8407615; E-mail: circulation@li.lib.ms.us; URL: www.li.lib.ms.us
1942; Jan Willis
Genealogy; 1 branch libr
142 000 vols; 279 curr per; 2 794 av-mat; 2 800 Audio Bks 41622

Goff-Nelson Memorial Library, Tupper Lake Public Library, 41 Lake St, *Tupper Lake*, NY 12986
T: +1 518 3599421; Fax: +1 518 3599421; E-mail: goffnelson@adelphia.net
1932; Linda Auclair
Adirondack Hist, micro
55 000 vols; 150 curr per 41623

Stanislaus County Free Library – Turlock Branch, 550 Minaret Ave, *Turlock*, CA 95380-4198
T: +1 209 6648100; Fax: +1 209 6648101
Diane Bartlett
89 000 vols 41624

Tuscaloosa Public Library, 1801 Jack Warner Pkwy, *Tuscaloosa*, AL 35401-1027
T: +1 205 3455820; Fax: +1 205 7581735; URL: www.tuscaloosa-library.org
1921; Nancy C. Pack
3 branch libs
184 000 vols; 709 curr per; 63 Bks on Deafness & Sign Lang, 4 000 Talking Bks 41625

Orange County Public Library – Tustin Branch, 345 E Main St, *Tustin*, CA 92780-4491
T: +1 714 5447725; Fax: +1 714 8324279
Emily Moore
117 000 vols 41626

Twin Falls Public Library, 201 Fourth Ave E, *Twin Falls*, ID 83301-6397
T: +1 208 7332964; Fax: +1 208 7332965; E-mail: tfpl@lib.tfid.org; URL: www.twinfallspubliclibrary.org
1908; Susan Ash
Agriculture, Business, Travel, Music, Hist; Large Print Coll, Musical Scores, Pacific Northwest Coll
148 000 vols; 348 curr per 41627

Twinsburg Public Library, 10050 Ravenna Rd, *Twinsburg*, OH 44087-1796
T: +1 330 4254268; Fax: +1 330 4253622; URL: www.twinsburglibrary.org
1910; Karen D. Tschudy
129 000 vols; 419 curr per; 16 000 e-books; 6 128 sound-rec; 453 digital data carriers; 241 DVDs, 6 454 Videos, 6 664 Bks on Tape, 46 Puzzles 41628

Lester Public Library, 1001 Adams St, *Two Rivers*, WI 54241
T: +1 920 7938888; Fax: +1 920 7937150; E-mail: lesref@esls.lib.wi.us; URL: www.tworivers.lib.wi.us
1891; Jeff Dawson
55 000 vols; 215 curr per; 525 microforms; 2 700 av-mat; 2 171 sound-rec; 40 digital data carriers
libr loan 41629

Tyler Public Library, 201 S College Ave, *Tyler*, TX 75702-7381
T: +1 903 5937323; Fax: +1 903 5311329; URL: www.tylerlibrary.com
1899; Chris Albertson
1 bookmobile
188 000 vols; 447 curr per 41630

Mendocino County Library, 105 N Main St, *Ukiah*, CA 95482
T: +1 707 4634493; Fax: +1 707 4635472; URL: www.mendolibrary.org
1964; Melanie Webber Lightbody
Genealogy, Indians, Forestry, Fishing, Wines, Art; Native American of Mendocino County; 4 branch libs
160 000 vols; 110 curr per 41631

Grant County Library, 215 E Grant Ave, *Ulysses*, KS 67880-2958
T: +1 620 3561433; Fax: +1 620 3561344; E-mail: frances@pld.com; URL: users.pld.com/frances/
1915; Frances Roberts
Religion, Art, Social sciences, Pioneer life; Spanish Coll, Large Print Bks, Local Newspapers (to 1989)
50 000 vols; 140 curr per
libr loan 41632

Scenic Regional Library of Franklin, Gasconade & Warren Counties, Union Service Center, 308 Hawthorne Dr, *Union*, MO 63084
T: +1 636 5833224; Fax: +1 636 5836519; URL: www.scenicregional.org
1959; Kenneth J. Rohrbach
7 branch libs
238 000 vols; 464 curr per; 2 586 av-mat; 1 660 sound-rec; 750 digital data carriers; 7 920 Audio Bks, 4 680 Videos 4 760 Large Print Bks 41633

Union County Carnegie Library, 300 E South St, *Union*, SC 29379-2392
T: +1 864 4277140; URL: www.unionlibrary.org
1904; Nancy Rosenwald
South Caroliniana Coll, bks & micro
59 000 vols; 123 curr per 41634

Union Township Public Library, 1980 Morris Ave, *Union*, NJ 07083-3578
T: +1 908 8515452; Fax: +1 908 8514671; URL: www.youseemore.com/unionpl
1927; Laurie D. Sansone
1 branch libr
169 000 vols; 238 curr per 41635

Alameda County Library – Union City Branch, 34007 Alvarado-Niles Rd, *Union City*, CA 94587-4498
T: +1 510 7451464; Fax: +1 510 4877241
Mira Geroy
95 000 vols; 174 curr per 41636

Atlanta-Fulton Public Library System – South Fulton Regional Library, 4055 Flatshoals Rd SW, *Union City*, GA 30291
T: +1 770 3063092; Fax: +1 770 3063127
Clay Payne
101 000 vols 41637

Uniondale Public Library, 400 Uniondale Ave, *Uniondale*, NY 11553-1995
T: +1 516 4892220; Fax: +1 516 4894005; E-mail: upl@lilrc.org; URL: www.nassaulibrary.org/uniondale
1954; Susan K. Kern
115 000 vols 41638

Akron-Summit County Public Library – Green, 4046 Massillon Rd, *Uniontown*, OH 44685-4046
T: +1 330 8969074; Fax: +1 330 8969412
Mary Miller
60 000 vols 41639

Uniontown Public Library, 24 Jefferson St, *Uniontown*, PA 15401
T: +1 724 4371165; Fax: +1 724 4395689; URL: www.uniontownlib.org
1928; Lynne E. Tharan
Pennsylvania Room Coll, local hist and Fayette County genealogical research mat
101 000 vols; 100 curr per; 1 000 microforms; 500 sound-rec; 40 digital data carriers
libr loan 41640

Dorothy W. Quimby Library, 90 Quaker Hill Rd, *Unity*, ME 04988; P.O. Box 167, Unity, ME 04988-0167
T: +1 207 9483131; Fax: +1 207 9482795; E-mail: library@unity.edu; URL: www.unity.edu/library/
1966; Robert J. Doan
Environmental sciences, North American Indians
61 000 vols; 325 curr per 41641

Upland Public Library, 450 N Euclid Ave, *Upland*, CA 91786-4732
T: +1 909 9314200; Fax: +1 909 9314209; URL: www.ci.upland.ca.us/asp/site/library/about/general/index.asp
1913; Kathy Bloomberg-Rissman
2 330 Audio Bks, 3 303 Videos, 1 600 Large Print Bks
148 000 vols; 195 curr per; 2 454 av-mat; 1 540 sound-rec; 531 digital data carriers 41642

Upper Darby Township & Sellers Memorial Free Public Library, Upper Darby Sellers, 76 S State Rd, *Upper Darby*, PA 19082
T: +1 610 7894440; Fax: +1 610 7895319; E-mail: upperdarby@delco.lib.pa.us; URL: www.udlibraries.org
1930; Nancy L. Hallowell
2 branch libs
122 000 vols; 251 curr per 41643

Upper Saddle River Public Library, 245 Lake St, *Upper Saddle River*, NJ 07458
T: +1 201 3272583; Fax: +1 201 3273966; E-mail: usdrc..irc@bccls.org; URL: www.uppersaddleriverlibrary.org
1960
60 000 vols; 175 curr per 41644

Champaign County Library, 1060 Scioto St, *Urbana*, OH 43078
T: +1 937 6533811; Fax: +1 937 6535679; E-mail: champref@oplin.org; URL: www.champaign.lib.oh.us
1890; Zara Liskowiak
98 000 vols; 249 curr per 41645

The Urbana Free Library, 210 W Green St, *Urbana*, IL 61801-5326
T: +1 217 3674405; Fax: +1 217 5317089; URL: www.urbanafreelibrary.org
1874; Debra Lissak
House construction, restoration & repair; mystery series; Folklore; City of Urbana Docs (600 000 pages) & Records
236 000 vols; 4 355 av-mat; 3 800 Audio Bks, 12 Electronic Media & Resources, 13 640 CDs, 4 790 Large Print Bks
libr loan; CRL 41646

Urbandale Public Library, 3520 86th St, *Urbandale*, IA 50322-4056
T: +1 515 3314488; Fax: +1 515 3316737; URL: www.urbandalelibrary.org
1961; Sara L. Pearson
131 000 vols; 362 curr per; 2 000 av-mat; 3 664 sound-rec; 165 art reproductions 41647

Mid-York Library System, 1600 Lincoln Ave, *Utica*, NY 13502
T: +1 315 7358328; Fax: +1 315 7350943; URL: www.midyork.org
1960; Mary Lou Caskey
85 000 vols; 20 curr per 41648

Utica Public Library, 303 Genesee St, *Utica*, NY 13501
T: +1 315 7352279; Fax: +1 315 7341034; URL: www.uticapubliclibrary.org
1893; Darby O'Brien
151 000 vols; 98 curr per; 126 249 govt docs; 18 000 av-mat; 26 000 sound-rec; 380 digital data carriers
libr loan 41649

El Progreso Memorial Library, 301 W Main St, *Uvalde*, TX 78801
T: +1 830 2782017; Fax: +1 830 2784940; URL: www.elprogreso.org
1903; Susan Anderson
Uvalde (Uvalde Hist Coll), bks & VF mat
55 000 vols; 90 curr per; 1 440 av-mat 41650

Solano County Library – Vacaville Public Library-Cultural Center, 1020 Ulatis Dr, *Vacaville*, CA 95687
Fax: +1 707 4510987
Jeanette Stevens
115 000 vols; 3 031 av-mat 41651

Vail Public Library, 292 W Meadow Dr, *Vail*, CO 81657
T: +1 970 4792185; Fax: +1 970 4792192; E-mail: info@vaillibrary.com; URL: www.vaillibrary.com
1972
Mountain Environment (Alpine Coll), Skiing Coll
52 000 vols; 300 curr per 41652

South Georgia Regional Library System – Valdosta-Lowndes County Public, 300 Woodrow Wilson Dr, *Valdosta*, GA 31602-2592
T: +1 229 3330086; Fax: +1 229 3337669; URL: www.sgrl.org
1876; David C. Gibson
Birds Coll; 6 branch libs
230 000 vols; 280 curr per 41653

Solano County Library – John F. Kennedy Branch, 505 Santa Clara St, *Vallejo*, CA 94590
T: +1 707 5535092; Fax: +1 707 5535667
Linda Matchette
118 000 vols; 2 569 av-mat 41654

H. Grady Bradshaw Chambers County Library, 3419 20th Ave, *Valley*, AL 36854
T: +1 334 7682161; Fax: +1 334 7682272; E-mail: chamberscountylibrary@yahoo.com
1976; Mary H. Hamilton
Cobb Memorial Arch; 1 branch libr
63 000 vols; 153 curr per 41655

San Diego County Library – Valley Center Branch, 29200 Cole Grade Rd, *Valley Center*, CA 92082-5880
T: +1 760 7491305; Fax: +1 760 7491764
Sandy Puccio
51 000 vols 41656

Valley City-Barnes County Public Library, 410 N Central Ave, *Valley City*, ND 58072-2949
T: +1 701 8453821; Fax: +1 701 8454884; E-mail: vcbcpl@csicable.net; URL: www.kleinonline.com/library.htm
1903; Mary E. Fischer
Large Print Coll, verticle file; North Dakota Coll; Childrens dept
65 000 vols; 50 maps; 2 035 sound-rec; 12 digital data carriers; 1 300 video tapes
NDLA 41657

Valley Cottage Free Library, 110 Rte 303, *Valley Cottage*, NY 10989
T: +1 845 2687700; Fax: +1 845 2687760; E-mail: vclref@rcls.org; URL: www.vclib.org
1959; Amelia Kalin
70 000 vols 41658

Henry Waldinger Memorial Library, Valley Stream Public Library, 60 Verona Pl, *Valley Stream*, NY 11582-3011
T: +1 516 8256422; Fax: +1 516 8256551; E-mail: hwmlcontact@hotmail.com; URL: www.nassaulibrary.org/valleyst
1932; Mamie Eng
123 000 vols; 181 curr per; 2 500 av-mat; 2 464 sound-rec; 145 digital data carriers
libr loan 41659

Porter County Public Library System, 103 Jefferson St, *Valparaiso*, IN 46383-4820
T: +1 219 4620524; Fax: +1 219 4774866; URL: www.pcpls.lib.in.us
1905; James D. Cline
5 branch libs
434 000 vols; 1 199 curr per; 1 500 High Interst/Low Vocabulary Bks, 300 Bks on Deafness & Sign Lang   41660

Porter County Public Library System – Valparaiso Public (Central), 103 Jefferson St, *Valparaiso*, IN 46383-4820
T: +1 219 4620524; Fax: +1 219 4774867
Connie Sullivan
178 000 vols   41661

Tampa-Hillsborough County Public Library System – Bloomingdale Regional Pulbic – Bloomingdale Regional Pulbic, 1906 Bloomingdale Ave, *Valrico*, FL 33594-6206
Fax: +1 813 6351646
Julie Beamguard
82 000 vols   41662

Los Angeles Public Library System – Van Nuys Branch, 6250 Sylmar Ave Mall, *Van Nuys*, CA 91401-2787
T: +1 818 7568453; Fax: +1 818 7569291; E-mail: vnnuys@lapl.org
Janet Metzler
63 000 vols   41663

Brumback Library, 215 W Main St, *Van Wert*, OH 45891-1695
T: +1 419 2382168; Fax: +1 419 2383180; E-mail: brumback@brumbacklib.com; URL: www.brumbacklib.com
1901; John Carr
Rare Bks; Van Wert Newspapers, 1855 to present on microfilm; 5 branch libs
189 000 vols; 310 curr per; 30 e-books; 13 000 av-mat; 902 digital data carriers; 5 980 Audio Bks   41664

Fort Vancouver Regional Library District, 1007 E Mill Plain Blvd, *Vancouver*, WA 98663
T: +1 360 6951566; Fax: +1 360 6932681; E-mail: contact@fvrl.org; URL: www.fvrl.org
1950; Bruce Ziegman
13 branch libs
785 000 vols   41665

Fort Vancouver Regional Library District – Three Creeks Community Library, 800 C NE Tenney Rd, *Vancouver*, WA 98685
T: +1 360 5179696; Fax: +1 360 5746429
Gwen Scott-Miller
82 000 vols; 238 curr per   41666

Fort Vancouver Regional Library District – Vancouver Community Library (Main Library), 1007 E Mill Plain Blvd, *Vancouver*, WA 98663
T: +1 360 6951564; Fax: +1 360 6998823
Karin Ford
279 000 vols; 611 curr per   41667

Venice Public Library, 300 S Nokomis Ave, *Venice*, FL 34285-2416
T: +1 941 8611330; Fax: +1 941 4862345; URL: www.sclibs.net
1964; Ann Hall
110 000 vols; 230 curr per   41668

Ventura County Library, 646 County Square Dr, Ste 150, *Ventura*, CA 93003
T: +1 805 4777331; Fax: +1 805 4777340; URL: www.vencolibrary.org
1915; Jackie Griffin
15 branch libs
874 000 vols; 963 curr per; 5 000 e-books; 25 150 Audio Bks, 16 800 Videos
libr loan   41669

Ventura County Library – E. P. Foster Library, 651 E Main St, *Ventura*, CA 93001
T: +1 805 6482715; Fax: +1 805 6483696
Mary Stewart
140 000 vols   41670

Ventura County Library – H. P. Wright Library, 57 Day Rd, *Ventura*, CA 93003
T: +1 805 6420336; Fax: +1 805 6448725
Mary Lynch
95 000 vols   41671

Ritter Public Library, 5680 Liberty Ave, *Vermilion*, OH 44089-1196
T: +1 440 9673798; Fax: +1 440 9675482; E-mail: info@ritter.lib.oh.us; URL: www.ritter.lib.oh.us
1912; Janet L. Ford
61 000 vols; 278 curr per; 12 000 e-books   41672

Uintah County Library, 155 E Main St, *Vernal*, UT 84078-2695
T: +1 435 7890091; Fax: +1 435 7896822; URL: www.uintah.lib.ut.us; www.catalog.uintah.lib.ut.us
1908; Dr. Frank Alan Bruno
Regional hist center & coll
98 000 vols; 159 curr per; 5 000 e-books; 379 microforms; 4 455 sound-rec; 253 digital data carriers; 12 264 videos   41673

Rockville Public Library, George Maxwell Memorial Library, 52 Union St, *Vernon*, CT 06066
T: +1 860 8755892; Fax: +1 860 8759795; E-mail: rockvillewebmaster@biblio.org; URL: www.rockvillepubliclibrary.org
1896; Donna Enman
51 600 vols; 116 curr per   41674

Indian River County Library System, 1600 21st St, *Vero Beach*, FL 32960
T: +1 772 7705060; Fax: +1 772 7705066; E-mail: refdesk@irclibrary.org; URL: Indian-river.lib.fl.us
1915; Mary Lou Rethman
Hist, Lit, Business; Florida Authors; 1 branch libr
443 000 vols; 1 912 curr per; 25 000 e-books; 21 820 Audio Bks, 70 Electronic Media & Resources, 35 120 Large Print Bks, 6 750 Talking Bks   41675

Verona Public Library, 17 Gould St, *Verona*, NJ 07044-1928
T: +1 973 8574848; Fax: +1 973 8574851; URL: www.veronalibrary.org
1912; James A. Thomas
60 000 vols; 175 curr per   41676

Verona Public Library, 500 Silent St, *Verona*, WI 53593
T: +1 608 8457180; Fax: +1 608 8458917; URL: www.veronapubliclibrary.org
1959; Brian Simons
499 000 vols; 139 curr per; 8 000 e-books   41677

Four County Library System, 304 Clubhouse Rd, *Vestal*, NY 13850-3713
T: +1 607 7238236; Fax: +1 607 7231722; URL: www.4cls.org
1960; David J. Karre
170 000 vols
libr loan   41678

Vestal Public Library, 320 Vestal Pkwy E, *Vestal*, NY 13850-1632
T: +1 607 7544244; Fax: +1 607 7547936; E-mail: ve.ref@4cls.org; URL: lib.4cty.org/vestal-r.html
1970; Carol Boyce
David Ross Locke Bk Coll
116 000 vols; 256 curr per   41679

Vestavia Hills-Richard M. Scrushy Public Library, 1112 Montgomery Hwy, *Vestavia Hills*, AL 35216
T: +1 205 9780155; Fax: +1 205 9780156; URL: www.vestavia.lib.al.us
1969; Jeff Hammack
American Heritage, National Geographic, 1955-82
82 000 vols; 135 curr per   41680

Warren County-Vicksburg Public Library, 700 Veto St, *Vicksburg*, MS 39180-3595
T: +1 601 6366411; Fax: +1 601 6344809; URL: www.warren.lib.ms.us
1915; Deborah Mitchell
Civil War, State of MS, MS River, Mystery novels
128 000 vols; 169 curr per; 2 000 microforms; 3 500 av-mat; 2 358 sound-rec; 43 digital data carriers
libr loan   41681

Victoria Public Library, 302 N Main, *Victoria*, TX 77901-6592
T: +1 361 4853302; Fax: +1 361 4853295; URL: www.victoriapubliclibrary.org
1932; James B. Stewart
Great Plains-Texas, New Mexico & Arizona (Claude K McCan Coll of the Great Plains)
149 000 vols; 748 curr per; 2 000 e-books
libr loan   41682

Ohoopee Regional Library System – Vidalia-Toombs County Library Headquarters, 610 Jackson St, *Vidalia*, GA 30474-2835
T: +1 912 5379283; Fax: +1 912 5373735; E-mail: vidalialib@ohoopeelibrary.org; URL: www.ohoopeelibrary.org
1938; Dusty Gres
5 branch libs, 1 bookmobile
130 000 vols; 110 curr per   41683

Vienna Public Library, 2300 River Rd, *Vienna*, WV 26105
T: +1 304 2957771; Fax: +1 304 2957776; E-mail: viennapl@park.lib.wv.us; URL: vienna.park.lib.wv.us
1959; Alice C. Thomas
54 000 vols; 72 curr per   41684

Villa Park Public Library, 305 S Ardmore Ave, *Villa Park*, IL 60181-2698
T: +1 630 8341164; Fax: +1 630 8340489; E-mail: vppladmin@linc.lib.il.us; URL: www.villapark.lib.il.us
1928; Sandra Hill
110 000 vols; 286 curr per   41685

Evangeline Parish Library, 242 W Main St, *Ville Platte*, LA 70586
T: +1 337 3631369; Fax: +1 337 3632353; E-mail: admin.b1ev@pelican.state.lib.la.us; URL: www.evangeline.lib.la.us
1948; Mary Foster-Galasso
1 branch libr
65 000 vols; 655 curr per   41686

Knox County Public Library, 502 N Seventh St, *Vincennes*, IN 47591-2119
T: +1 812 8864380; Fax: +1 812 8860342; E-mail: publib@kcpl.lib.in.us; URL: www.kcpl.lib.in.us
1889; Emily Cooper Bunyan
124 000 vols; 253 curr per; 4 800 av-mat; 350 sound-rec; 350 High Interest/Low Vocabulary Bks, 110 Large Print Bks   41687

Vineland Free Public Library, 1058 E Landis Ave, *Vineland*, NJ 08360
T: +1 856 7944244; Fax: +1 856 6910366; URL: www.vineland.lib.nj.us
1901; Gloria Urban
106 000 vols; 110 curr per   41688

Roanoke County Public Library – Vinton Branch, 800 E Washington Ave, *Vinton*, VA 24179
T: +1 540 8575043; Fax: +1 540 3443285
Jamie Rowles-Channell
81 000 vols; 88 curr per   41689

Arrowhead Library System, 701 11th St North, *Virginia*, MN 55792-2298
T: +1 218 7413840; Fax: +1 218 7413519; E-mail: als@arrowhead.lib.mn.us; URL: www.arrowhead.lib.mn.us
1966; Jim Weikum
Coll of 'Described' videos
92 550 vols; 33 curr per; 7 064 sound-rec; 16 digital data carriers
libr loan; ALA, PLA   41690

Virginia Public Library, 215 Fifth Ave S, *Virginia*, MN 55792-2642
T: +1 218 7487525; Fax: +1 218 7487527; URL: www.virginia.lib.mn.us
1905; Nancy Maxwell
60 000 vols; 170 curr per; 1 600 av-mat; 1 605 sound-rec; 69 art reproductions   41691

Virginia Beach Department of Public Libraries, Bldg 19, Municipal Ctr, 2nd Flr, *Virginia Beach*, VA 23456
T: +1 757 3850167; Fax: +1 757 3854220; URL: www.vbgov.com
1959; Marcy Sims
Princess Anne Hist Coll; 15 branch libs
718 000 vols
libr loan   41692

Tulare County Library – Visalia Headquarters Branch, 200 W Oak Ave, *Visalia*, CA 93291
T: +1 559 7336954; Fax: +1 559 7302524
Mike Drake
318 000 vols; 448 curr per; 239 e-books; 3 256 av-mat   41693

San Diego County Library – Vista Branch, 700 Eucalyptus Ave, *Vista*, CA 92084-6245
T: +1 760 6435100; Fax: +1 760 6435127
Alice Rigg
157 000 vols   41694

Camden County Library System, 203 Laurel Rd, *Voorhees*, NJ 08043
T: +1 856 7721636; Fax: +1 856 7726105; E-mail: ref@camden.lib.nj.us; URL: www.camdencountylibrary.org
1922; William Brahms
5 branch libs
430 000 vols; 895 curr per; 93 Electronic Media & Resources
libr loan   41695

Wabash Carnegie Public Library, 188 W Hill St, *Wabash*, IN 46992-3048
T: +1 260 5632972; Fax: +1 260 5630222; E-mail: general@wabash.lib.in.us; URL: www.wabash.lib.in.us
1903; Ware Wimberly
Gene Stratton-Porter Coll
67 000 vols; 168 curr per   41696

Waco-McLennan County Library System, 1717 Austin Ave, *Waco*, TX 76701-1794
T: +1 254 7505943; Fax: +1 254 7505940; E-mail: referencewaco@ci.waco.tx.us; URL: www.waco-texas.com/city_depts/libraryservices/libraryservices.htm
1898; James A. Karney
Grants Resource Ctr; 3 branch libs
295 000 vols; 767 curr per; 31 000 e-books   41697

Waco-McLennan County Library System – R. B. Hoover Library, 1428 Wooded Acres, Ste 104, *Waco*, TX 76710
T: +1 254 7456018; Fax: +1 254 7456019
Mary Ellen Wright
50 000 vols   41698

Hampton B. Allen Library, 120 S Greene St, *Wadesboro*, NC 28170
T: +1 704 6945177; Fax: +1 704 6945178; URL: www.srls.info/anson/ansonindex.html
1923; Phoebe Midlan
52 000 vols; 120 curr per   41699

Ella M. Everhard Public Library, Wadsworth Public Library, 132 Broad St, *Wadsworth*, OH 44281-1897
T: +1 330 3351294; Fax: +1 330 3346605; E-mail: director@wadsworthlibrary.com; URL: www.wadsworthlibrary.com
C. Allen Nichols
157 000 vols; 339 curr per   41700

Hawaii State Public Library System – Wahiawa Public Library, 820 California Ave, *Wahiawa*, HI 96786
T: +1 808 6226345; Fax: +1 808 6226348
Anthony Hooper
51 000 vols; 97 curr per   41701

Hawaii State Public Library System – Waianae Public Library, 85-625 Farrington Hwy, *Waianae*, HI 96792
T: +1 808 6977868; Fax: +1 808 6977870
Faith Arakawa
57 000 vols; 29 curr per   41702

Hawaii State Public Library System – Wailuku Public Library, 251 High St, *Wailuku*, HI 96793
T: +1 808 2435766; Fax: +1 808 2435768
Susan Werner
69 000 vols; 99 curr per   41703

Hawaii State Public Library System – Waipahu Public Library, 94-275 Mokuola St, *Waipahu*, HI 96797
T: +1 808 6750358; Fax: +1 808 6750360
Lorna Miyasaki
67 000 vols; 48 curr per   41704

Wake County Public Library System – Wake Forest Branch, 400 E Holding Ave, ***Wake Forest***, NC 27587
T: +1 919 5548498; Fax: +1 919 5548499
Yvonne T. Allen
61 000 vols 41705

Lucius Beebe Memorial Library, 345 Main St, ***Wakefield***, MA 01880-5093
T: +1 781 2466334; Fax: +1 781 2466385; E-mail: wakefieldlibrary@noblenet.org; URL: www.wakefieldlibrary.org
1856; Sharon A. Gilley
Wakefield Authors, Rifles, Riflery & Target Shooting (Keough Coll)
110 000 vols 41706

Waldwick Public Library, 19-21 E Prospect St, ***Waldwick***, NJ 07463-2099
T: +1 201 6525104; Fax: +1 201 6526233; E-mail: wald1@bccls.org; URL: www.bccls.org/waldwick/
1954; Patricia D. Boyd
Italian-American Coll
56 000 vols; 97 curr per 41707

Oconee County Library, 501 W South Broad St, ***Walhalla***, SC 29691
T: +1 864 6384133; Fax: +1 864 6384132; E-mail: ocpldirector@InfoAve.Net; URL: ocplibrary.org
1948; Martha B. Baily
3 branch libs, 1 bookmobile
202 000 vols; 368 curr per 41708

Kent District Library – Walker Branch, 4293 Remembrance Rd NW, ***Walker***, MI 49544
T: +1 616 6473970; Fax: +1 616 4534622
Jane Seitz
54 000 vols 41709

Monmouth County Library – Wall, 2700 Allaire Rd, ***Wall***, NJ 07719
T: +1 732 4498877; Fax: +1 732 4491732
Pamela Sawall
104 000 vols 41710

Walla Walla Public Library, 238 E Alder, ***Walla Walla***, WA 99362
T: +1 509 5274550; Fax: +1 509 5273748; URL: www.ci.walla-walla.wa.us
1897; Martha A. Van Pelt
111 000 vols; 202 curr per 41711

Walled Lake City Library, 1499 E West Maple Rd, ***Walled Lake***, MI 48390
T: +1 248 6243772; Fax: +1 248 6240041; E-mail: admin@walledlakelibrary.org; URL: www.walledlakelibrary.org
1963; Donna Rickabaugh
50 000 vols; 70 curr per 41712

Wallingford Public Library, 200 N Main St, ***Wallingford***, CT 06492-3791
T: +1 203 2656754; Fax: +1 203 2695698
1881; Leslie Scherer
Holocaust, Oneida Community; 1 branch libr
196 000 vols; 320 curr per; 4 699 av-mat; 7 600 sound-rec; 3 160 digital data carriers; 4 300 Audio Bks 41713

County of Los Angeles Public Library – Walnut Library, 21155 La Puente Rd, ***Walnut***, CA 91789-2017
T: +1 909 5950757; Fax: +1 909 5957553; URL: www.colapublib.org/libs/walnut
Jenny Cheng
87 000 vols; 82 curr per 41714

Contra Costa County Library – Walnut Creek – Park Place Library, 1395 Civic Dr, ***Walnut Creek***, CA 94596-4297
T: +1 925 6466773; Fax: +1 925 6466048
Cindy Brittain
71 000 vols 41715

Contra Costa County Library – Ygnacio Valley (Thurman G. Casey Memorial), 2661 Oak Grove Rd, ***Walnut Creek***, CA 94598-3627
T: +1 925 9381481; Fax: +1 925 6466026
Cindy Brittain
74 000 vols 41716

Lawrence County Library, 1315 W Main St, ***Walnut Ridge***, AR 72476
T: +1 870 8863222; Fax: +1 870 8869520; E-mail: lawcolib@yahoo.com
1942; Ashley Burris
2 branch libs
62 000 vols; 92 curr per 41717

Walpole Public Library, 65 Common St, ***Walpole***, MA 02081
T: +1 508 6607341; Fax: +1 508 6602714; URL: www.walpole.ma.us/library.htm
1876; Jerry Romelczyk
102 000 vols; 150 curr per 41718

Colleton County Memorial Library, 600 Hampton St, ***Walterboro***, SC 29488-4098
T: +1 843 5495621; Fax: +1 843 5495122; URL: www.colletonlibrary.org
1820; Sylvia N. Rowland
110 000 vols; 177 curr per 41719

Waltham Public Library, 735 Main St, ***Waltham***, MA 02451
T: +1 781 3143425; Fax: +1 781 3143426; URL: www.waltham.library.ma.us
1865; Thomas N. Jewell
196 000 vols; 417 curr per 41720

Auglaize County Public District Library, 203 S Perry St, ***Wapakoneta***, OH 45895-1999
T: +1 419 7382921; Fax: +1 419 7385168; E-mail: acpdl@oplin.org; URL: auglaize.oplin.org
1925; M. Jo Derryberry
5 branch libs, 1 bookmobile
156 000 vols; 321 curr per; 3 000 e-books; 4 300 av-mat; 1 030 sound-rec; 2 335 digital data carriers 41721

Wareham Free Library, 59 Marion Rd, ***Wareham***, MA 02571
T: +1 508 2952343;
Fax: +1 508 2952678; URL: www.warehamfreelibrary.org
1891; Mary Jane Pillsbury
Wareham Coll
101 000 vols; 110 curr per 41722

Warminster Township Free Library, 1076 Emma Lane, ***Warminster***, PA 18974
T: +1 215 6724362; Fax: +1 215 6723604; URL: www.buckslib.org/libraries/warminster/index.htm
1960; Caroline C. Gallis
101 000 vols; 91 curr per; 1 600 av-mat; 18 digital data carriers
libr loan 41723

Somerset County Library System Warren Township Branch, 42 Mountain Blvd, ***Warren***, NJ 07059
T: +1 908 7545554; Fax: +1 908 7542899
Elaine Whiting
100 000 vols; 250 curr per 41724

Warren Library Association, Warren Public Library, 205 Market St, ***Warren***, PA 16365
T: +1 814 7234650; Fax: +1 814 7234521; URL: www.warrenlibrary.org
1873; Patricia Sherbondy
Petroleum Hist Coll, Sheet Music, Popular Show Tunes 1834-1955 (Robertson Music Coll
167 000 vols; 242 curr per 41725

Warren Public Library, Civic Center Library, One City Sq, Ste 100, ***Warren***, MI 48093-2396
T: +1 586 5744564; URL: www.libcoop.net/warren
1958; Amy Henderstein
Adaptive Devices (visually impaired), Large Print Bks; 4 branch libs
259 000 vols; 525 curr per; 1 366 govt docs; 2 548 music scores; 85 microforms; 41 494 sound-rec; 2 917 digital data carriers; 12 087 videos, 10 412 pamphlets, 1 346 puppets and toys, 120 art repros
libr loan; ALA, PLA 41726

Warren-Trumbull County Public Library, 444 Mahoning Ave NW, ***Warren***, OH 44483
T: +1 330 3998807; Fax: +1 330 3953988; URL: www.wtcpl.lib.oh.us
1890; N.N.
5 branch libs, 1 bookmobile
409 000 vols; 602 curr per; 25 807 sound-rec; 664 digital data carriers
libr loan 41727

Trails Regional Library, 432 N Holden St, ***Warrensburg***, MO 64093
T: +1 660 7471699; Fax: +1 660 7475774; URL: www.trl.lib.mo.us
1957; Karen Hicklin
8 branch libs
195 000 vols; 432 curr per; 10 000 av-mat; 4 900 Audio Bks 41728

Fauquier County Public Library, 11 Winchester St, ***Warrenton***, VA 20186-2825
T: +1 540 3478750;
Fax: +1 540 3493278; URL: www.library.fauquiercounty.gov
1922; Maria Del Rosso
2 branch libs
107 000 vols; 291 curr per; 32 000 e-books; 4 991 av-mat
libr loan 41729

Warrenville Public Library District, 28 W 751 Stafford Pl, ***Warrenville***, IL 60555
T: +1 630 3931171; Fax: +1 630 3931688; E-mail: director@warrenville.com; URL: www.warrenville.com
1979; Sandy Whitmer
Visual art; Local Artists (Albright Coll), Original Fine Arts Coll
71 090 vols; 200 curr per 41730

Warsaw Community Public Library, 310 E Main St, ***Warsaw***, IN 46580-2882
T: +1 574 2676011; Fax: +1 574 2697739; E-mail: info@warsawlibrary.org; URL: www.warsawlibrary.org
1885; Ann M. Zydek
157 000 vols; 334 curr per; 5 000 e-books; 4 000 microforms; 7 732 sound-rec; 453 digital data carriers
libr loan; ALA, IALA 41731

Albert Wisner Public Library, Two Colonial Ave, ***Warwick***, NY 10990-1191
T: +1 845 9861047; Fax: +1 845 9871228; E-mail: warref@rcls.org; URL: www.albertwisnerlibrary.org
1974; Rosemary Cooper
Job Info Ctr
50 000 vols; 84 curr per 41732

Warwick Public Library, 600 Sandy Lane, ***Warwick***, RI 02889-8298
T: +1 401 7395440; Fax: +1 401 7322055; URL: www.warwickpl.org
1965; Douglas A. Pearce
3 branch libs
202 000 vols; 444 curr per
libr loan 41733

Waseca-Le Sueur Regional Library, 408 N State St, ***Waseca***, MN 56093
T: +1 507 8352910; Fax: +1 507 8353700; URL: www.tds.lib.mn.us
1965; Theresa Meadows
8 branch libs
150 000 vols; 460 curr per 41734

Bartram Trail Regional Library – Mary Willis Library Headquarters, 204 E Liberty St, ***Washington***, GA 30673
T: +1 706 6787736; Fax: +1 706 6781615; URL: www.wilkes.public.lib.ga.us
1888; Lillie Crowe
2 branch libs
101 000 vols; 75 curr per 41735

Beaufort, Hyde & Martin County Regional Library, Old Court House, 158 N Market St, ***Washington***, NC 27889
T: +1 252 9466401; Fax: +1 252 9460352; URL: www.bhmlib.org
1941; Maryjane Carbo
7 branch libs
117 000 vols; 275 curr per 41736

Citizens Library, 55 S College St, ***Washington***, PA 15301
T: +1 724 2222400; Fax: +1 724 2257303; E-mail: citlib@citlib.org; URL: www.citlib.org
1870; Kathy Kennedy
Iams Coll (Genealogy & Local Hist)
133 000 vols; 236 curr per
libr loan 41737

District of Columbia Public Library, 901 G St NW, ***Washington***, DC 20001-4599
T: +1 202 7271101; Fax: +1 202 7271129; URL: www.dclibrary.org
1896; Bridget Bradley
African-american Studies, Washingtoniana; 24 branch libs
2 800 000 vols 41738

District of Columbia Public Library – Capitol View, 5001 Central Ave SE, ***Washington***, DC 20019
T: +1 202 6450755
Winnell Morris Montague
68 000 vols 41739

District of Columbia Public Library – Chevy Chase, 5625 Connecticut Ave NW, ***Washington***, DC 20015
T: +1 202 2820021
Burrows Martin
107 000 vols 41740

District of Columbia Public Library – Cleveland Park, 3310 Connecticut Ave NW, ***Washington***, DC 20008
T: +1 202 2823080
Brian Brown
99 000 vols 41741

District of Columbia Public Library – Francis A. Gregory Branch, 3660 Alabama Ave SE, ***Washington***, DC 20020
T: +1 202 6454297
Lessie Owens Mtewa
50 000 vols 41742

District of Columbia Public Library – Georgetown, 3260 R St NW, ***Washington***, DC 20007
T: +1 202 2820220
N.N.
90 000 vols 41743

District of Columbia Public Library – Juanita E. Thornton/Shepherd Park Neighborhood Library, 7420 Georgia Ave NW, ***Washington***, DC 20012
T: +1 202 5416100; E-mail: jet-spk.dcpl@dc.gov; URL: dclibrary.org
Jameely Dahma
67 000 vols 41744

District of Columbia Public Library – Lamond Riggs, 5401 S Dakota Ave NE, ***Washington***, DC 20011
T: +1 202 5416255
Norberta Winborne
50 000 vols 41745

District of Columbia Public Library – Martin Luther King Jr Memorial, 901 G St NW, ***Washington***, DC 20001-4599
T: +1 202 7270321; Fax: +1 202 7273856
Pamela Stovall
The Children's Illustrator Coll;The District of Columbia Community Archives;The Black Studies Center;Washingtoniana
876 000 vols 41746

District of Columbia Public Library – Mount Pleasant, 3160 16th St NW, ***Washington***, DC 20010
T: +1 202 6710200; Fax: +1 202 6732184
Roman A. Santillan
60 000 vols 41747

District of Columbia Public Library – Northeast/7th Street, 330 Seventh St NE, ***Washington***, DC 20002
T: +1 202 6983320
Karen Butler
58 000 vols 41748

District of Columbia Public Library – Palisades, 4901 V St NW, ***Washington***, DC 20007
T: +1 202 2823139
Lucy Thrasher
82 000 vols 41749

District of Columbia Public Library – Petworth, 4200 Kansas Ave NW, ***Washington***, DC 20011
T: +1 202 5416300
Anthony Porter
73 000 vols 41750

District of Columbia Public Library – Southeast/7th Street, 403 Seventh St SE, ***Washington***, DC 20003
T: +1 202 6983377; E-mail: soe.dcpl@dc.gov
April King
Hist of Eastern Mkt & Capital Hill Communities
78 000 vols 41751

District of Columbia Public Library – Southwest/Wesley Place, 900 Wesley Pl SW, ***Washington***, DC 20024
T: +1 202 7244752
Diane Henry
70 000 vols 41752

**District of Columbia Public Library –
Washington Highlands**, 115 Atlantic St
SW, *Washington*, DC 20032
T: +1 202 6455880
Laura Gonzales
67 000 vols   41753

**District of Columbia Public Library
– West End**, 1101 24th St NW,
Washington, DC 20037
T: +1 202 7248707
Edward Robinson-El
72 000 vols   41754

**District of Columbia Public Library
– Woodridge**, 1801 Hamlin St NE,
Washington, DC 20018
T: +1 202 5416226
Mary Cooper
88 000 vols   41755

George H. & Laura E. Brown Library,
122 Van Norden St, *Washington*,
NC 27889
T: +1 252 9464300; Fax: +1 252
9752015; URL: washington-nc.com/
library.aspx
1911; Gloria Moore
55 000 vols; 75 curr per   41756

Romeo District Library, Graubner
Library, 65821 Van Dyke, *Washington*,
MI 48095
T: +1 586 7520603; Fax: +1 586
7528416; E-mail: romeo@libcoop.net;
URL: www.libcoop.net/romeo
1909; Mary Elizabeth Harper
1 branch libr
84 000 vols; 311 curr per; 4 679 av-mat
                                              41757

Washington District Library, 301
Walnut St, *Washington*, IL 61571
T: +1 309 4442241; Fax: +1 309
4444711; E-mail: washlib@mtco.com;
URL: washington.lib.il.us
1937; Pamela Tomka
65 000 vols; 150 curr per   41758

Washington-Carnegie Public Library,
300 W Main St, *Washington*,
IN 47501-2698
T: +1 812 2544586;
Fax: +1 812 2544585; URL:
www.washingtonpubliclibrary.org
Elizabeth Dowling
53 000 vols; 159 curr per; 88 av-mat
                                              41759

Carnegie Public Library, 127 S North
St, *Washington Court House*,
OH 43160
T: +1 740 3352540; Fax: +1 740
3358409; URL: www.cplwcho.org
1891; Susan J. McDaniel
Genealogy, bks, flm
69 000 vols; 98 curr per
libr loan   41760

Wasilla Public Library, 391 N Main St,
Wasilla, AK 99654-7085
T: +1 907 3765913; Fax: +1 907
3762347; E-mail: library@ci.wasilla.ak.us;
URL: www.cityofwasilla.com/library
1938; K. J. Martin-Albright
Alaskana
56 000 vols; 350 microforms; 2 072
av-mat; 940 sound-rec
libr loan   41761

Silas Bronson Library, 267 Grand St,
Waterbury, CT 06702-1981
T: +1 203 5748222; Fax: +1 203
5748055; URL: www.bronsonlibrary.org
1869; J. Emmett McSweeney
1 branch libr
256 000 vols; 195 curr per; 20 Electronic
Media & Resources   41762

Waterford Public Library, 49 Rope
Ferry Rd, *Waterford*, CT 06385
T: +1 860 4445805; Fax: +1
860 4371685; E-mail:
waterfordlibrary@juno.com; URL:
www.waterfordpubliclibrary.org
1923; Roslyn Rubinstein
Travel Coll
85 000 vols; 125 curr per; 3 300 av-mat;
1 000 sound-rec; 750 digital data carriers;
2 500 Large Print Bks   41763

Waterloo Public Library, 415
Commercial St, *Waterloo*,
IA 50701-1385
T: +1 319 2914497; Fax: +1
319 2919013; E-mail:
infowiz@wplwloo.lib.ia.us; URL:
www.waterloopubliclibrary.org
1896; Sheryl Groskurth
Aids Lending Libr
206 000 vols; 376 curr per
libr loan   41764

North Country Library System,
22072 Country Rte 190, *Watertown*,
NY 13601-1066; P.O. Box 99, Watertown
T: +1 315 7825540; Fax: +1 315
7826883; URL: www.nclsweb.org
1947; Stephen B. Bolton
Large print
239 000 vols; 40 curr per; 2 403 av-mat;
2 400 Talking Bks
libr loan   41765

Roswell P. Flower Memorial Library,
229 Washington St, *Watertown*,
NY 13601-3388
T: +1 315 7857705;
Fax: +1 315 7882584; URL:
www.flowermemoriallibrary.org
1904; Barbara J. Wheeler
194 000 vols; 140 curr per; 30 e-journals;
4 500 av-mat; 1 400 Audio Bks   41766

Watertown Library Association, 470
Main St, *Watertown*, CT 06795
T: +1 860 9455360; Fax: +1
860 9455367; E-mail:
wtnlib@watertownlibrary.org; URL:
www.watertownlibrary.org
1865; Joan K. Rintelman
1 branch libr
107 000 vols; 180 curr per   41767

Watertown Public Library, 100 S Water
St, *Watertown*, WI 53094-4320
T: +1 920 2624090; Fax: +1 920
2618943; E-mail: askrefwt@mwfls.org;
URL: www.watertown.lib.wi.us
1907; Peg Checkai
99 000 vols   41768

Watertown Regional Library,
160 Sixth St NE, *Watertown*,
SD 57201-2778; P.O. Box 250,
Watertown, SD 57201-0250
T: +1 605 8826220; Fax: +1 605
8826221; E-mail: adminwat@sdln.net;
URL: watweb.sdln.net
1899; Michael C. Mullin
Dakota Coll
67 000 vols; 226 curr per   41769

**Toledo-Lucas County Public Library –
Waterville Branch**, 800 Michigan Ave,
Waterville, OH 43566
T: +1 419 8783055; Fax: +1 419
8784688
Campbell Brady
87 000 vols   41770

Waterville Public Library, 73 Elm St,
Waterville, ME 04901-6078
T: +1 207 8725433; Fax: +1
207 8734779; E-mail:
helpdesk1@waterville.lib.me.us; URL:
www.waterville.lib.me.us
1896; Sarah Sugden
92 000 vols; 76 curr per   41771

Wauconda Area Library, 801 N Main
St, *Wauconda*, IL 60084
T: +1 847 5266225; Fax: +1 847
5266244; E-mail: library@wauclib.org;
URL: www.wauclib.org
1939; Tom Kern
87 000 vols; 500 curr per   41772

Waukegan Public Library, 128 N
County St, *Waukegan*, IL 60085
T: +1 847 6232041; Fax: +1 847
6232094; URL: www.waukeganpl.org
1898; Richard Lee
Social Problems, US Census, Local
Authors
276 000 vols; 395 curr per   41773

Waupaca Area Public Library, 107 S
Main St, *Waupaca*, WI 54981-1521
T: +1 715 2584416; E-mail:
wau@mail.owls.lib.wi.us; URL:
www.waupacalibrary.org
1900; Peg Burington
Large Print Coll, Susan Christenson
Memorial Women's Studies Coll,
Wisconsin Coll
64 000 vols; 120 curr per   41774

Waupun Public Library, 123 S Forest
St, *Waupun*, WI 53963; P.O. Box 391,
Waupun, WI 53963-0391
T: +1 920 3247925; E-mail:
wpl@mwfls.org; URL:
www.waupunpubliclibrary.org
1857; Bret Jaeger
60 000 vols; 153 curr per
libr loan   41775

Marathon County Public Library, 300
N First St, *Wausau*, WI 54403-5405
T: +1 715 2617230; Fax: +1 715
2617204; URL: www.mcpl.us
1905; Mary Bethke
Old Popular Sheet Music; 8 branch libs
338 000 vols; 726 curr per   41776

Wautaga Public Library, 7109 Whitley
Rd, *Wautaga*, TX 76148-2024
T: +1 817 5145855; Fax: +1 817
5813910; E-mail: lewell@cowtx.org; URL:
www.ci.watauga.tx.us/library/libindex.htm
1983; Lana Ewell
71 500 vols; 160 curr per   41777

Wauwatosa Public Library, 7635 W
North Ave, *Wauwatosa*, WI 53213-1718
T: +1 414 4718484; Fax: +1 414
4798984; E-mail: tosa.mail@mcfls.org;
URL: tpublib.fp.execpc.com
1886; Mary Murphy
227 000 vols; 302 curr per   41778

**Garnet A. Wilson Public Library
of Pike County**, 207 N Market St,
Waverly, OH 45690-1176
T: +1 740 9474921; Fax: +1 740
9472918; URL: www.pike.lib.oh.us
1939; Thomas S. Adkins
89 000 vols; 248 curr per
libr loan; ALA   41779

Waverly Public Library, 1500 W Bremer
Ave, *Waverly*, IA 50677-2836
T: +1 319 3521223; Fax: +1
319 3520872; E-mail:
waverly@waverly.lib.ia.us; URL:
www.waverlyia.com/library.asp
1859; Sarah Meyer-Ryerson
Farm Mats
68 000 vols; 162 curr per; 800 av-mat;
198 sound-rec; 38 digital data carriers;
7 000 pamphlets   41780

Nicholas P. Sims Library, 515 W Main,
Waxahachie, TX 75165-3235
T: +1 972 9372671; Fax: +1 972
9374409; E-mail: info@simslib.org; URL:
www.simslib.org
1904; Susan A. Maxwell
130 000 vols; 198 curr per; 2 800 av-mat
                                              41781

Okefenokee Regional Library, 401 Lee
Ave, *Waycross*, GA 31501
T: +1 912 2874978; Fax: +1 912
2842533
1955; Midge Galentine-Steis
4 branch libs
230 000 vols; 123 curr per   41782

Wayland Free Public Library, 5
Concord Rd, *Wayland*, MA 01778
T: +1 508 3582311; Fax: +1
508 3585249; E-mail:
wayland@waylandlibrary.org; URL:
www.waylandlibrary.org
1848; Ann F. Knight
Gardening
75 070 vols; 235 curr per
libr loan; ALA, PLA, MLA, NELA   41783

**Memorial Library of Radnor
Township**, 114 W Wayne Ave, *Wayne*,
PA 19087-4098
T: +1 610 6871124; Fax: +1 610
6871454; URL: www.radnorlibrary.org
Barbara Casini
Pennsylvania History;Reader Development
Coll
120 000 vols; 231 curr per; 1 251 av-mat
                                              41784

Wayne Public Library, 3737 S Wayne
Rd, *Wayne*, MI 48184
T: +1 888 6728983; Fax: +1 734
7210341; URL: wayne.lib.mi.us
1927; Paulette Medvecky
95 000 vols; 138 curr per
libr loan   41785

Wayne Public Library, 461 Valley Rd,
Wayne, NJ 07470
T: +1 973 6944272; Fax: +1
973 6920637; E-mail:
wplcomments@waynepubliclibrary.org;
URL: www.waynepubliclibrary.org
1922; Jody C. Treadway
Business Ref, New Jersey Hist Coll; 1
branch libr
221 000 vols; 294 curr per
libr loan   41786

Waynesboro Public Library, 600 S
Wayne Ave, *Waynesboro*, VA 22980
T: +1 540 9426746; Fax: +1 540
9426753; URL: www.waynesboro.va.us/
library.html
1912; Zahir M. Mahmoud
Charles Smith Art Coll, prints & papers;
George Speck Art Coll, prints
178 000 vols; 141 curr per; 1 500 music
scores; 10 000 av-mat; 710 sound-rec;
1 330 digital data carriers; 6 340 Large
Print Bks   41787

Eva K. Bowlby Public Library, 311 N
West St, *Waynesburg*, PA 15370-1238
T: +1 724 6279776; Fax: +1 724
8521900; E-mail: bowlby@alltel.net; URL:
www.alltel.net/bowlby
1943; Barbara Ferguson
55 000 vols; 100 curr per
libr loan   41788

Mary L. Cook Public Library, 381 Old
Stage Rd, *Waynesville*, OH 45068
T: +1 513 8974826; Fax: +1 513
8979215; URL: mlcook.lib.oh.us
1917; Linda Swartzel
Early Quaker Theology
70 000 vols; 7 000 av-mat   41789

Weatherford Public Library,
1014 Charles St, *Weatherford*,
TX 76086-5098
T: +1 817 5984159; Fax: +1 817
5984161; URL: www.weatherford.lib.tx.us
1959; Dale Fleeger
Mary Martin Coll; Parker County Hist &
Genealogical Coll, multi-media
89 000 vols; 350 curr per; 3 700 av-mat
libr loan   41790

**Western Plains Library System –
Weatherford Public Library**, 219 E
Frankin, *Weatherford*, OK 73096-5134
T: +1 580 7723591; Fax: +1
580 7723591; E-mail:
weatherford.public@wplibs.com
Jamie Hudson
63 000 vols; 40 curr per   41791

Chester C. Corbin Public Library,
Webster Public Library, 2 Lake St,
Webster, MA 01570
T: +1 508 9493880; Fax: +1 508
9490537; E-mail: dgallagher@cwmars.org;
URL: www.corbinlibrary.org
1889; Dan Gallagher
81 070 vols; 69 curr per   41792

Webster Public Library, Webster Plaza,
980 Ridge Rd, *Webster*, NY 14580
T: +1 585 8727075; URL:
www.websterlibrary.org
1929; Theresa Bennett
138 000 vols; 243 curr per   41793

Kendall Young Library, 1201 Willson
Ave, *Webster City*, IA 50595-2294
T: +1 515 8329100; Fax: +1 515
8329102; E-mail: info@kendall-
young.lib.ia.us; URL: www.kendall-
young.lib.ia.us
1898; Cynthia A. Weiss
MacKinlay Kantor Coll; Clark R
Mollenhoff Coll
55 000 vols; 140 curr per; 1 433 av-mat
                                              41794

Webster Groves Public Library, 301
E Lockwood Ave, *Webster Groves*,
MO 63119-3102
T: +1 314 9613784; Fax: +1 314
9614233; E-mail: info@wgpl.org; URL:
www.wgpl.org
1928; Tom Cooper
69 000 vols; 152 curr per; 6 529 av-mat
                                              41795

Mary H. Weir Public Library, 3442 Main St, **Weirton**, WV 26062
T: +1 304 7978510; Fax: +1 304 7978526; URL: www.weirton.lib.wv.us
1958; Richard G. Rekowski
113 000 vols; 110 curr per; 1 diss/theses; 25 000 govt docs; 200 maps; 10 000 microforms; 2 300 av-mat; 7 500 sound-rec; 600 digital data carriers
libr loan; ALA   41796

Herrick Memorial Library, 101 Willard Memorial Sq, **Wellington**, OH 44090-1342
T: +1 440 6472120; Fax: +1 440 6472103; URL: www.youseemore.com/herrick
1873; Janet L. Hollingsworth
Wellington Hist Photo Coll
57 000 vols; 168 curr per   41797

Wellington Public Library, 121 W Seventh St, **Wellington**, KS 67152-3898
T: +1 620 3262011; Fax: +1 620 3268193; E-mail: wlibrary@idir.net; URL: skyways.lib.ks.us/towns/wellington/library.html
1916; Donna McNeil
53 000 vols; 105 curr per
libr loan   41798

David A. Howe Public Library, 155 N Main St, **Wellsville**, NY 14895
T: +1 585 5933410; Fax: +1 585 5934176; E-mail: wellsville@stls.org; URL: www.davidahowelibrary.org
1894; Brian M. Hildreth
Lit; Bird's Egg Coll (Charles Munson Coll) Indian Artifacts (Avery Mosher Coll), Children's Ref Bks
96 000 vols; 91 curr per   41799

North Central Regional Library, 16 N Columbia St, **Wenatchee**, WA 98801-8103
T: +1 509 6625021; Fax: +1 509 6628554; E-mail: ncrl@ncrl.org; URL: www.ncrl.org
1961; Dean Marney
28 branch libs
527 000 vols; 898 curr per; 55 digital data carriers; 50 753 Videos, 23050 Talking Bks
libr loan   41800

North Central Regional Library – Wenatchee Public (Headquarters), 310 Douglas St, **Wenatchee**, WA 98801-2864
T: +1 509 6625021; Fax: +1 509 6639731; E-mail: wenatchee@ncrl.org
1961; Katy Sessions
83 000 vols; 803 av-mat   41801

Saint Charles City County Library District – Corporate Parkway Branch, 1200 Corporate Pkwy, **Wentzville**, MO 63385-4828
T: +1 636 3274010; Fax: +1 636 3270548
Diana Tucker
162 000 vols   41802

Weslaco Public Library, 525 S Kansas Ave, **Weslaco**, TX 78596-6215
T: +1 956 9733144; Fax: +1 956 9694069; E-mail: askthelibrarian@weslaco.lib.tx.us; URL: www.weslaco.lib.tx.us
1949; Michael Fisher
Texana Coll; Adult learning ctr
70 000 vols; 133 curr per; 1 400 av-mat; 2 913 sound-rec; 40 digital data carriers
libr loan; ALA   41803

West Allis Public Library, 7421 W National Ave, **West Allis**, WI 53214-4699
T: +1 414 3028500; Fax: +1 414 3028545; URL: www.ci.west-allis.wi.us/library
1898; Michael Koszalka
216 000 vols; 372 curr per
libr loan   41804

West Babylon Public Library, 211 Rte 109, **West Babylon**, NY 11704
T: +1 631 6695445; Fax: +1 631 6696539; URL: wbab.suffolk.lib.ny.us
1982
86 330 vols; 248 curr per   41805

West Bend Community Memorial Library, 630 Poplar St, **West Bend**, WI 53095-3380
T: +1 262 3355151; Fax: +1 262 3355150; E-mail: libref@west-bendlibrary.org; URL: www.west-bendlibrary.org
1901; John Reid
125 000 vols; 360 curr per; 470 microforms; 2 000 av-mat; 1 000 sound-rec; 50 digital data carriers
libr loan   41806

Berlin Township Library, John J. McPeak Library, 201 Veteran's Ave, **West Berlin**, NJ 08091
T: +1 856 7670439; Fax: +1 856 7536729; URL: www.berlintwp.com
Mary Holt
75 000 vols; 26 curr per; 227 av-mat   41807

West Bloomfield Township Public Library, 4600 Walnut Lake Rd, **West Bloomfield**, MI 48323
T: +1 248 2322315; Fax: +1 248 2322291; E-mail: webmaster@wblib.org; URL: www.wblib.org
1938; Clara Nalli Bohrer
1 branch libr
226 000 vols; 525 curr per   41808

Beaman Memorial Public Library, Eight Newton St, **West Boylston**, MA 01583
T: +1 508 8353711; Fax: +1 508 8354770; E-mail: beaman@cwmars.org; URL: beamanlibrary.org
1912; Louise Howland
57 000 vols; 125 curr per   41809

West Bridgewater Public Library, 80 Howard St, **West Bridgewater**, MA 02379-1710
T: +1 508 8941255; Fax: +1 508 8941258; URL: www.sailsinc.org/westbridgewater
1879; Beth Roll Smith
West Bridgewater Hist Coll, World War II autobiographies
70 000 vols; 161 curr per; 50 maps; 1 100 av-mat; 600 sound-rec; 30 digital data carriers
libr loan; MLA, NELA, ALA   41810

Middletown Public Library – West Chester Branch, 7900 Cox Rd, **West Chester**, OH 45069
T: +1 513 7773131;
Fax: +1 513 7778452; URL: www.westchesterlibrary.org/wc.htm
Steven Mayhugh
101 000 vols; 404 curr per   41811

West Chicago Public Library District, 118 W Washington St, **West Chicago**, IL 60185-2803
T: +1 630 2311552; Fax: +1 630 2311709; URL: www.westchicago.lib.il.us
1927; Coleman Melody
Railroads; Book-plates Coll
83 000 vols; 281 curr per
libr loan; ALA   41812

County of Los Angeles Public Library – West Covina Library, 1601 W Covina Pkwy, **West Covina**, CA 91790-2786
T: +1 626 9623541; Fax: +1 626 9621507; URL: www.colapublib.org/libs/wcovina
Beth Wilson
230 000 vols; 203 curr per   41813

West Deptford Public Library, 420 Crown Point Rd, **West Deptford**, NJ 08086-9598
T: +1 856 8455593; Fax: +1 856 8483689; E-mail: admin@westdeptford.lib.nj.us; URL: www.westdeptford.lib.nj.us
1965; Marie Downes
Environ, Health, Nutrition
87 000 vols; 227 curr per; 1 000 av-mat; 2 000 sound-rec; 50 digital data carriers
libr loan   41814

West Des Moines Public Library, 4000 Mills Civic Pkwy, **West Des Moines**, IA 50265-2049
T: +1 515 2223404; Fax: +1 515 2223401; URL: www.wdmlibrary.org
1940; Ray Vignovich
142 000 vols; 317 curr per   41815

West Fargo Public Library, 109 Third St E, **West Fargo**, ND 58078
T: +1 701 4335460; Fax: +1 701 4335479; URL: www.westfargolibrary.org
1971; Sandra Hannahs
60 000 vols; 150 curr per; 1 025 sound-rec   41816

Lawrenceburg Public Library District – North Dearborn Branch, 25969 Dole Rd, **West Harrison**, IN 47060
T: +1 812 6370777; Fax: +1 812 6370797
Criss Green
149 000 vols; 482 curr per; 208 e-books
323 av-mat   41817

West Hartford Public Library, Noah Webster Memorial Library, 20 S Main St, **West Hartford**, CT 06107-2432
T: +1 860 5616950;
Fax: +1 860 5616976; URL: www.westhartfordlibrary.org
1897; Patricia Holloway
Connecticut Reference Coll, Noah Webster Coll; 2 branch libs
203 000 vols; 701 curr per
libr loan   41818

West Haven Public Library, 300 Elm St, **West Haven**, CT 06516-4692
T: +1 203 9374233; Fax: +1 203 9317827; URL: www.westhavenpl.org
1906; Donna Lolos
2 branch libs
160 000 vols; 303 curr per   41819

West Hempstead Public Library, 500 Hempstead Ave, **West Hempstead**, NY 11552
T: +1 516 4816591; Fax: +1 516 4812608; E-mail: westhempsteadpl@yahoo.com; URL: www.whplibrary.org
1965; Regina Mascia
Large Type Bks, Computer Bks
86 000 vols; 156 curr per
libr loan   41820

County of Los Angeles Public Library – West Hollywood Library, 715 N San Vicente Blvd, **West Hollywood**, CA 90069-5020
T: +1 310 6525340; Fax: +1 310 6522580; URL: www.colapublib.org/libs/whollywood
Susan Anderson
Ron Shipton HIV Information Center
75 000 vols; 44 curr per   41821

West Islip Public Library, Three Higbie Lane, **West Islip**, NY 11795-3999
T: +1 631 6617080; Fax: +1 631 6616573; E-mail: wisplib@suffolk.lib.ny.us; URL: www.wipublib.org
1957; Andrew Hamm
Job Info Ctr
181 000 vols; 415 curr per; 13 500 av-mat; 7 800 sound-rec; 2 440 digital data carriers; 4 430 Audio Bks, 92 High Interest/Low Vocabulary Bks, 175 Bks on Deafness & Sign Lang, 3 350 Large Print Bks
libr loan   41822

Ashe County Public Library, 148 Library Dr, **West Jefferson**, NC 28694
T: +1 336 2462041; Fax: +1 336 2467503; URL: www.ashelibrary.com
1932; Jim McQueen
52 000 vols; 72 curr per   41823

Hurt-Battelle Memorial Library of West Jefferson, 270 Lily Chapel Rd, **West Jefferson**, OH 43162-1202
T: +1 614 8798448; Fax: +1 614 8798668; URL: www.hbmlibrary.org
1914; Cathy Allen
53 000 vols; 50 curr per   41824

Salt Lake County Library Services – Bingham Creek, 4834 W 9000 South, **West Jordan**, UT 84088-2213
T: +1 801 9447684; Fax: +1 801 2820943; E-mail: binghamc@slco.lib.ut.us
Darin Butler
155 000 vols   41825

Salt Lake County Library Services – West Jordan Branch, 1970 W 7800 South, **West Jordan**, UT 84088-4025
T: +1 801 9447646; Fax: +1 801 5628761
Nanette Alderman
103 000 vols   41826

West Lafayette Public Library, 208 W Columbia St, **West Lafayette**, IN 47906
T: +1 765 7432261; Fax: +1 765 7430540; E-mail: refdesk@wlaf.lib.in.us; URL: www.westlafayettepubliclibrary.org
1922; Nick Schenkel
Dickey Coll (Children's Lit Award Winners), Reisner Cookbks Coll, Large Print Coll, Lit Coll, Crafts Coll, Play Coll
118 000 vols; 484 curr per; 600 av-mat
                                   41827

West Linn Public Library, 1595 Burns St, **West Linn**, OR 97068-3231
T: +1 503 6567853; Fax: +1 503 6562746; E-mail: wlrefpage@lincc.org; URL: www.westlinn.lib.or.us
Sarah McIntyre
82 000 vols; 285 curr per   41828

West Melbourne Public Library, 2755 Wingate Blvd, **West Melbourne**, FL 32904
T: +1 321 9524508; Fax: +1 321 9524510; URL: www.brev.org
1970; Marian Hallett Griffin
68 000 vols; 98 curr per   41829

West Memphis Public Library, 213 N Avalon, **West Memphis**, AR 72301
T: +1 870 7327590; Fax: +1 870 7327676
Caroline Redfearn
50 000 vols; 47 curr per   41830

Ouachita Parish Public Library – Ouachita Valley, 581 McMillian Rd, **West Monroe**, LA 71294
T: +1 318 3271471; Fax: +1 318 3271473
Julie Crump
79 000 vols; 125 curr per; 4 197 av-mat
                                   41831

Ouachita Parish Public Library – West Monroe Branch, 315 Cypress, **West Monroe**, LA 71291
T: +1 318 3271365; Fax: +1 318 3294062
Glenda Patterson
53 000 vols; 89 curr per; 653 av-mat
                                   41832

West New York Public Library, 425 60th St, **West New York**, NJ 07093-2211
T: +1 201 2955137; Fax: +1 201 6621473; E-mail: wnyploffice@hplhub.org
1916; Weiliang Lai
Family literacy, Jerseyana, Large print, Parenting, Spanish
62 000 vols; 125 curr per; 200 av-mat
libr loan   41833

West Orange Free Public Library, 46 Mount Pleasant Ave, **West Orange**, NJ 07052-4903
T: +1 973 7360198; Fax: +1 973 7361655; E-mail: admin@westorangelibrary.org; URL: www.wopl.org
1948; Mary Romance
153 000 vols; 278 curr per; 1 650 Large Print Bks   41834

Palm Beach County Library System, 3650 Summit Blvd, **West Palm Beach**, FL 33406-4198
T: +1 561 2332600; Fax: +1 561 2332692; E-mail: pbclref@pbclibrary.org; URL: www.pbclibrary.org
1967; John Callahan
Audubon Coll, Large Print Coll, Florida Coll; 14 branch libs
1 249 000 vols; 4 163 curr per; 2 000 e-books
libr loan   41835

West Palm Beach Public Library, 100 Clematis St, **West Palm Beach**, FL 33401
T: +1 561 8687700; Fax: +1 561 8687706; URL: www.mycitylibrary.org
1921; Christopher Murray
Floridiana
112 000 vols; 406 curr per   41836

West Plains Public Library, 750 W Broadway, **West Plains**, MO 65775-2369
T: +1 417 2564775; Fax: +1 417 2568316; URL: www.westplains.net
1948; Sherry Russell
57 000 vols; 121 curr per; 901 av-mat
FOL   41837

**Tombigbee Regional Library System –
Bryan Public Library, Headquarters**,
338 Commerce, *West Point*, MS 39773;
P.O. Box 675, West Point
T: +1 662 4944872; Fax: +1 662
4940300; URL: www.tombigbee.lib.ms.us
1916; Mary Helen Waggoner
9 branch libs
122 000 vols; 276 curr per   41838

**Boston Public Library – West
Roxbury Branch**, 1961 Centre St,
West Roxbury, MA 02132-2595
T: +1 617 3253147; Fax: +1 617
8592465
Sheila G. Scott
105 000 vols; 50 curr per   41839

West Seneca Public Library, 1300
Union Rd, *West Seneca*, NY 14224
T: +1 716 6742928; Fax: +1 716
6749206
1935; Mary Moore
65 000 vols; 314 curr per   41840

West Springfield Public Library, 200
Park St, *West Springfield*, MA 01089
T: +1 413 7364561; Fax: +1 413
7366469; URL: www.wspl.org
1916; Antonia Golinski-Foisy
102 000 vols; 196 curr per; 2 937
sound-rec; 72 digital data carriers; 50
Art Reproductions, 12 Vertical Files,
1 032 Puppets, Kits
libr loan; ALA   41841

**Salt Lake County Library Services
– West Valley Branch**, 2880 W
3650 South, *West Valley City*,
UT 84119-3743
T: +1 801 9447631; Fax: +1 801
9691782
Maggie Mills
83 000 vols   41842

West Warwick Public Library, Robert
H. Champlin Memorial Library, 1043 Main
St, *West Warwick*, RI 02893
T: +1 401 8283750; Fax: +1 401
8288493; E-mail: ref@wwlibrary.org;
URL: www.wwlibrary.org
1967; Frances Farrell-Bergeron
Science Fiction, Rhode Island
94 000 vols; 142 curr per   41843

Burlington County Library, Five Pioneer
Blvd, *Westampton*, NJ 08060
T: +1 609 2679660; Fax: +1 609
2674091; E-mail: hq@bcls.lib.nj.us; URL:
www.bcls.lib.nj.us
1921; Gail W. Sweet
Local Newspapers from 1835; 7 branch
libs
956 000 vols; 991 curr per; 3 000 e-
books; 22 000 av-mat; 15 000 sound-rec;
17 320 digital data carriers; 18 320 Large
Print Bks, 120 Braille Vols   41844

Westborough Public Library, 55 W
Main St, *Westborough*, MA 01581
T: +1 508 3663050; Fax: +1 508
3663049; E-mail: westboro@cwmars.org;
URL: www.westboroughlib.org
1859; Maureen Ambrosino
Art, Genealogy; Local Hist (Reed Coll)
81 000 vols; 223 curr per   41845

Walker Memorial Library, 800 Main St,
Westbrook, ME 04092
T: +1 207 8540630; Fax: +1
207 8540629; E-mail:
walkerlibrary@westbrook.me.us; URL:
www.westbrookmaine.com
1894; Karen Valley
Large type print
54 000 vols; 93 curr per   41846

Westbury Memorial Public Library, 445
Jefferson St, *Westbury*, NY 11590
T: +1 516 3330176; Fax: +1
516 3331752; E-mail:
contactus@westburylibrary.org; URL:
www.westburylibrary.org
1924; Cathleen Towey
Art Coll, Old English & American
Children's Bks
100 000 vols; 100 curr per
libr loan   41847

Westchester Public Library, 10700
Canterbury St, *Westchester*, IL 60154
T: +1 708 5623573; Fax: +1 708
5621298; E-mail: wcs@westchesterpl.org;
URL: www.westchesterpl.org
1956; Ruth McCrank
81 000 vols; 267 curr per; 621 e-books
          41848

Westerly Public Library, 44 Broad St,
Westerly, RI 02891
T: +1 401 5962877; Fax: +1 401
5965600; URL: www.westerlylibrary.org
1894; Kathryn T. Taylor
Children's Lit, Granite Carving &
Quarrying Gallery, Museum Coll
160 000 vols; 376 curr per
libr loan   41849

Thomas Ford Memorial Library,
Western Springs Library, 800 Chestnut
Ave, *Western Springs*, IL 60558
T: +1 708 2460520; Fax: +1 708
2460403; E-mail: info@fordlibrary.org;
URL: www.fordlibrary.org
1932; Anne Kozak
Play & Theatre Coll
75 000 vols; 300 curr per   41850

Westerville Public Library, 126 S State
St, *Westerville*, OH 43081-2095
T: +1 614 8827277; Fax: +1 614
8824160; URL: www.westervillelibrary.org
1930; Don W. Barlow
Temperance Coll; 1 branch libr
283 000 vols; 526 curr per
libr loan   41851

Westfield Athenaeum, Six Elm St,
Westfield, MA 01085-2997
T: +1 413 5680638; Fax: +1 413
5680988; URL: www.westath.org
1864; Christopher J. Lindquist
Edward Taylor Colonial Poetry Coll
119 000 vols; 132 curr per
libr loan   41852

Westfield Memorial Library, 550 E
Broad St, *Westfield*, NJ 07090
T: +1 908 7894090; Fax: +1 908
7890921; URL: www.wmlnj.org
1872; Philip Israel
196 000 vols; 241 curr per; 1 200
sound-rec; 200 digital data carriers
          41853

**Westfield Washington Public
Library**, 333 W Hoover St, *Westfield*,
IN 46074-9283
T: +1 317 8969391; Fax: +1 317
8963702; URL: www.wwpl.lib.in.us
1901; Sheryl A. Sollars
American Quaker Genealogy Coll
128 000 vols; 330 curr per   41854

J. V. Fletcher Library, 50 Main St,
Westford, MA 01886-2599
T: +1 978 6925557; Fax: +1 978
6924418; E-mail: mwf@mvlc.org; URL:
www.westfordlibrary.org
1797; Ellen Downey Rainville
Genealogical Data Coll, Hist Doc Coll
(pertaining to hist of Merrimack Valley),
Textile Mill Hist Coll
101 000 vols; 369 curr per   41855

Westlake Porter Public Library,
27333 Center Ridge Rd, *Westlake*,
OH 44145-3925
T: +1 440 2505470; Fax: +1
440 8716969; E-mail:
porter@westlakelibrary.org; URL:
www.westlakelibrary.org
1884; Mary Worthington
177 000 vols; 520 curr per; 10 522
av-mat; 19 000 sound-rec; 14 500 digital
data carriers; 10 070 Audio Bks, 18 450
CDs, 18 450 Videos, 7 330 Large Print
Bks
libr loan   41856

**County of Los Angeles Public Library
– Westlake Village Library**, 31220
W Oak Crest Dr, *Westlake Village*,
CA 91361
T: +1 818 8659230; Fax: +1 818
8650724; URL: www.colapublib.org/libs/
westlake/
Mark Totten
66 000 vols; 108 curr per   41857

**Wayne County Regional Library for
the Blind & Physically Handicapped**,
30555 Michigan Ave, *Westland*,
MI 48186-5310
T: +1 734 7277300; Fax: +1
734 7277333; E-mail:
wcrlbph@wayneregional.lib.mi.us; URL:
wayneregional.lib.mi.us
1931; Reginald B. Williams
125 000 vols   41858

Carroll County Public Library, 115
Airport Dr, *Westminster*, MD 21157
T: +1 410 3864488; Fax: +1 410
3864489; URL: library.carr.org
1958; K. Lynn Wheeler
5 branch libs
584 000 vols; 1 461 curr per   41859

**Orange County Public Library –
Westminster Branch**, 8180 13th St,
Westminster, CA 92683-8118
T: +1 714 8935057; Fax: +1 714
8980229
Mary Ann Hutton
Vietnamese Language Coll
141 000 vols; 127 curr per   41860

Westminster Public Library, College
Hill Library, 3705 W 112th Ave,
Westminster, CO 80031
T: +1 303 4045128; Fax: +1 303
4045135; URL: www.westminster.lib.co.us
1951; Mary Grace Barrick
1 branch libr
221 000 vols; 300 curr per   41861

**Camden County Library System –
William G. Rohrer Memorial Library
– Haddon Township**, 15 MacArthur
Blvd, *Westmont*, NJ 08108
T: +1 856 8542752; Fax: +1
856 8548825; E-mail:
haddon@camden.lib.nj.us
Nan Rosenthal
57 000 vols; 100 curr per   41862

Westmont Public Library, 428 N Cass,
Westmont, IL 60559-1502
T: +1 630 9695625; Fax: +1
630 9696490; E-mail:
wms@westmontlibrary.org; URL:
westmontlibrary.org
1943; Christine Kuhn
90 000 vols; 175 curr per; 3 800
sound-rec; 15 digital data carriers
libr loan; ALA   41863

Weston Public Library, 87 School St,
Weston, MA 02493
T: +1 781 8933312; Fax: +1 781
5290174; E-mail: weston@minlib.net;
URL: www.westonlibrary.org
1857; Susan Brennan
92 000 vols; 221 curr per   41864

Weston Public Library, 13153 Main
St, *Weston*, OH 43569; P.O. Box 345,
Weston
T: +1 419 6693415; Fax: +1 419
6693216; URL: library.norweld.lib.oh.us/
weston
1942; Shelen A. Stevens
72 000 vols; 180 curr per   41865

Westport Library Association, Westport
Public Library, Arnold Bernhard Plaza, 20
Jesup Rd, *Westport*, CT 06880
T: +1 203 2914840; Fax: +1 203
2914820; URL: www.westportlibrary.org
1908; Maxine Bleiweis
Picture file of over 500 000 photos and
clippings
191 000 vols; 307 curr per; 5 865 av-mat;
6 640 sound-rec; 177 digital data carriers;
5 900 Audio Bks, 16 784 Videos
libr loan   41866

Westwood Free Public Library, 49 Park
Ave, *Westwood*, NJ 07675
T: +1 201 6640583; Fax: +1 201
6646088; E-mail: westwood@bccls.org;
URL: www.bccls.org/westwood
1919; Martha Urbiel
Health & Popular Medicine Coll, Literacy
Program Coll
53 000 vols; 90 curr per; 1 000
microforms   41867

Westwood Public Library, 668 High St,
Westwood, MA 02090
T: +1 781 3267562; Fax: +1 781
3265383; URL: www.westwoodlibrary.org/
home
1898; Thomas P. Viti
Sauter Art Coll; 1 branch libr
97 000 vols; 196 curr per   41868

Wethersfield Public Library, 515 Silas
Deane Hwy, *Wethersfield*, CT 06109
T: +1 860 7212985; Fax: +1
860 7212991; E-mail:
library@wethersfieldlibrary.org; URL:
www.wethersfieldlibrary.org
1893; Laurel Goodgion
Art Coll
98 000 vols; 270 curr per   41869

Tufts Library, Public Libraries of
Weymouth, 46 Broad St, *Weymouth*,
MA 02188
T: +1 781 3371402; Fax: +1 781
6826123
1879; Joanne Lamothe
Teachers' Professional Libr; 3 branch libs
155 000 vols; 400 curr per   41870

Wharton County Library, 1920 N Fulton,
Wharton, TX 77488
T: +1 979 5328080; URL:
www.whartonco.lib.tx.us
1938; Barbara J. Goodell
86 000 vols; 114 curr per
FOL   41871

Platte County Public Library, 904 Ninth
St, *Wheatland*, WY 82201-2699
T: +1 307 3222689; Fax: +1
307 3223540; E-mail:
platcircmgr@will.state.wy.us; URL: www-
wsl.state.wy.us/platte/
1894; Julie Henion
63 000 vols; 94 curr per
libr loan   41872

Wheaton Public Library, 225 N Cross
St, *Wheaton*, IL 60187-5376
T: +1 630 6683097; Fax: +1
630 6681465; E-mail:
askref@wheatonlibrary.org; URL:
www.wheatonlibrary.org
1891; Sarah Meisels
DuPage County Hist Coll
389 000 vols; 475 curr per; 55 e-books;
12 090 Audio Bks, 1 732 Electronic Media
& Resoruces, 4 560 Large Print Bks
          41873

Indian Trails Public Library District,
355 S Schoenbeck Rd, *Wheeling*,
IL 60090
T: +1 847 4594100; Fax: +1 847
4594760; URL: www.itpld.lib.il.us;
www.indiantrailslibrary.org
1959; Tamiye Meehan
213 000 vols; 410 curr per; 500 e-books
          41874

Ohio County Public Library, 52 16th
St, *Wheeling*, WV 26003-3696
T: +1 304 2320244; Fax: +1
304 2326848; E-mail:
ocplweb@weirton.lib.wv.us; URL:
wheeling.weirton.lib.wv.us
1882; Dottie Thomas
128 000 vols; 304 curr per; 2 870 Audio
Bks, 6 980 Large Print Bks, 28 100
Talking Bks   41875

Morris County Library, 30 E Hanover
Ave, *Whippany*, NJ 07981
T: +1 973 2856951; Fax: +1 973
2856962; URL: www.mclib.info
1922; Joanne Kares
Mysteries, New Adult Readers, New
Jersey Hist, Original art, Sheet music,
Bks on Tape, New Jersey Coll
203 000 vols; 947 curr per; 55 000
microforms; 16 000 av-mat; 42 000
sound-rec; 17 000 digital data carriers;
1 644 Music Scores, 400 High
Interest/Low Vocabulary Bks, 150 Bks
on Deafness & Sign Lang, 6 182 Large
Print Bks
libr loan   41876

White Lake Township Library,
7527 E Highland Rd, *White Lake*,
MI 48383-2938
T: +1 248 6984942; Fax: +1
248 6982550; E-mail:
reference@whitelakelibrary.org; URL:
www.whitelakelibrary.org
1980; Lawrence Ostrowski
50 000 vols; 85 curr per   41877

White Plains Public Library,
100 Martine Ave, *White Plains*,
NY 10601-2599
T: +1 914 4221400; Fax: +1 914
4221462; E-mail: cybrary@wppl.lib.ny.us;
URL: www.whiteplainslibrary.org
1899; Sandra Miranda
White Plains Room (Westchester County
hist), Percy Grainger Music Coll, Folklore
Coll, Children's Lit Res Coll
309 000 vols; 2 300 curr per; 1 600 music
scores; 500 maps; 16 000 microforms;
12 000 av-mat; 16 000 sound-rec; 250
digital data carriers; 28 000 Pamphlets
libr loan   41878

White Settlement Public Library, 8215 White Settlement Rd, *White Settlement*, TX 76108-1604
T: +1 817 3670166; Fax: +1 817 2468184; E-mail: wsplib@wstx.us; URL: www.whitesettlement.lib.tx.us
Teresa McBrayer
70 000 vols; 80 curr per   41879

Whitefish Bay Public Library, 5420 N Marlborough Dr, *Whitefish Bay*, WI 53217
T: +1 414 9644380; Fax: +1 414 9645733; URL: www.wfblibrary.org
1936; Kristen Hewitt
72 000 vols; 144 curr per   41880

Whitehall Township Public Library, 3700 Mechanicsville Rd, *Whitehall*, PA 18052-3399
T: +1 610 4324339; Fax: +1 610 4329387; E-mail: whitehallpl@cliu.org; URL: whitehall.lib.pa.us
1964; Nancy J. Adams
Braille Bible, Juvenile Braille Picture Bks, Framed Art Prints
106 000 vols; 168 curr per   41881

Dunham Public Library, 76 Main St, *Whitesboro*, NY 13492
T: +1 315 7369734; Fax: +1 315 7363265; E-mail: whref@mail.midyork.org; URL: www.whitesborolibrary.info
1926; Francis R. McBride
Helen Dunham Coll
60 000 vols; 138 curr per   41882

Letcher County Public Libraries, Harry Caudill Memorial Library, 220 Main St, *Whitesburg*, KY 41858
T: +1 606 6337547; Fax: +1 606 6333407; E-mail: letcolib@bellsouth.net; URL: www.lcld.org
1952; Angelina Tidal
Genealogy mat, family hist; 3 branch libs
88 000 vols; 191 curr per; 4 300 av-mat
libr loan; ALA   41883

Queens Borough Public Library – Whitestone Branch, 151-10 14th Rd, *Whitestone*, NY 11357
T: +1 718 7678010
Nonyem Illobachie
59 000 vols   41884

Columbus County Public Library, Carolyn T. High Memorial Library, 407 N Powell Blvd, *Whiteville*, NC 28472
T: +1 910 6413976; Fax: +1 910 6423839; URL: www.columbusco.org
1921; Morris D. Pridgen
5 branch libs
120 000 vols; 321 curr per   41885

Irvin L. Young Memorial Library, 431 W Center St, *Whitewater*, WI 53190-1915
T: +1 262 4730530; Fax: +1 262 4730539; URL: www.whitewater.lib.wi.us
1899; Stacey L. Lunsford
Achen Photosw
66 000 vols; 287 curr per   41886

Whiting Public Library, 1735 Oliver St, *Whiting*, IN 46394-1794
T: +1 219 4734700; Fax: +1 219 6595833; URL: www.whiting.lib.in.us
1906; Christina Young
91 000 vols; 245 curr per; 1 120 av-mat   41887

County of Los Angeles Public Library – South Whittier Library, 14433 Leffingwell Rd, *Whittier*, CA 90604-2966
T: +1 562 9464415; Fax: +1 562 9416138; URL: www.colapublib.org/libs/swhittier/
Verna Dunning
71 000 vols; 104 curr per   41888

Whittier Public Library, 7344 S Washington Ave, *Whittier*, CA 90602
T: +1 562 4643450; Fax: +1 562 4643569; E-mail: lib@whittierpl.org; URL: www.whittierpl.org
1900; Paymaneh Maghsoudi
Margaret Fulmer Peace Coll, Whittier Hills Archs; 1 branch lib
252 000 vols
libr loan   41889

Wichita Public Library, 223 S Main St, *Wichita*, KS 67202
T: +1 316 2618508; Fax: +1 316 2624540; E-mail: admin@wichita.lib.ks.us; URL: www.wichita.lib.ks.us
1876; Cynthia Berner Harris
Music, Genealogy, Auto repair; American Indian Coll, Motor Manuals, Philatelic Coll; 10 branch libs
666 000 vols; 47 000 av-mat; 30 000 sound-rec; 7 900 digital data carriers; 2 075 Electronic Media & Resources, 24 590 Talking Bks
libr loan   41890

Wichita Public Library – Evergreen, 2601 N Arkansas, *Wichita*, KS 67204
T: +1 316 3038181
Dawn Williams
52 000 vols   41891

Wichita Falls Public Library, 600 11th St, *Wichita Falls*, TX 76301-4604
T: +1 940 7670868; Fax: +1 940 7206659; E-mail: ill@wfpl.net; URL: www.wfpl.net
1918; Lesley Daly
Texana & Southwest (Texas Coll)
135 000 vols; 300 curr per   41892

Wickliffe Public Library, 1713 Lincoln Rd, *Wickliffe*, OH 44092
T: +1 440 9446010; Fax: +1 440 9447264; URL: www.wickliffe.lib.oh.us
1934; Nancy D. Fisher
104 000 vols; 318 curr per; 12 000 e-books; 12 691 av-mat; 667 High Interest/Low Vocabulary Bks, 2 792 Large Print Bks, 12 691 Talking Bks
libr loan   41893

Wilbraham Public Library, 25 Crane Park Dr, *Wilbraham*, MA 01095-1799
T: +1 413 5966141; Fax: +1 413 5965090; E-mail: reference@wilbrahamlibrary.org; URL: www.wilbrahamlibrary.org
1892; Christine Bergquist
65 000 vols; 131 curr per; 540 microforms; 3 400 av-mat; 3 500 sound-rec; 90 digital data carriers
libr loan; ALA   41894

Osterhout Free Library, 71 S Franklin St, *Wilkes-Barre*, PA 18701-1287
T: +1 570 8230156; Fax: +1 570 8235477; URL: www.luzerneco.lib.pa.us
1889; Hansen Hansen
3 branch libs
200 000 vols; 300 curr per   41895

Willard Memorial Library, Six W Emerald St, *Willard*, OH 44890-1417
T: +1 419 9338564; Fax: +1 419 9334783; E-mail: willard@oplin.org; URL: willardlibrary.oplin.org
1921; Cinda S. Bretz Wallace
Railroad
71 000 vols; 126 curr per; 22 000 microforms; 2 242 sound-rec; 1 177 historical photogs
libr loan; NORWELD   41896

Williamsburg Regional Library, 7770 Croaker Rd, *Williamsburg*, VA 23188
T: +1 757 2597720; Fax: +1 757 2594079; URL: www.wrl.org
1908; Dr. John A. Moorman
2 branch libs
299 000 vols
libr loan   41897

Williamsburg Regional Library – Williamsburg Library, 515 Scotland St, *Williamsburg*, VA 23185
T: +1 757 2594050
Genevieve S. Owens
150 000 vols   41898

James V. Brown Library of Williamsport & Lycoming County, 19 E Fourth St, *Williamsport*, PA 17701-6390
T: +1 570 3260536; Fax: +1 570 3261671; URL: www.jvbrown.edu
1905; Janice Trapp
Adoption, Parenting; American Sign Language Video Coll
165 000 vols; 240 curr per
libr loan   41899

David & Joyce Milne Public Library, 1095 Main St, *Williamstown*, MA 01267-2627
T: +1 413 4585369; Fax: +1 413 4583085; E-mail: pmcleod@williamstown.net; URL: www.milnelibrary.org
1876; Patricia McLeod
53 710 vols; 1 395 curr per   41900

Free Public Library of Monroe Township, 306 S Main St, *Williamstown*, NJ 08094
T: +1 856 6291212; Fax: +1 856 8750191; E-mail: mtlibrary@buyrite.com; URL: www.monroetownshiplibrary.org
1969; Elizabeth L. Lillie
Motion pictures
74 000 vols; 147 curr per; 2 500 av-mat   41901

Willimantic Public Library, 905 Main St, *Willimantic*, CT 06226; P.O. Box 218, Willimantic
T: +1 860 4653079; Fax: +1 860 4653083; E-mail: wpldir@biblio.org; URL: www.biblio.org/willimantic/
1854; Ted Perch
60 000 vols; 137 curr per   41902

Willingboro Public Library, 220 Willingboro Pkwy, At The Town Center, *Willingboro*, NJ 08046
T: +1 609 8776668; Fax: +1 609 8351699; E-mail: wipl@willingboro.org; URL: www.willingboro.org
1960; Christine H. King
African American Hist Coll, Hist (Sanford Soren Spanish Civil War Hist Coll)
82 000 vols; 200 curr per   41903

Williston Community Library, 1302 Davidson Dr, *Williston*, ND 58801
T: +1 701 7748805; Fax: +1 701 5721186; URL: www.willistonlibrary.org
1983; Debbie Slais
American Indian Coll, North Dakota
55 000 vols; 100 curr per; 1 000 av-mat   41904

Pioneerland Library System, 410 Fifth St SW, *Willmar*, MN 56201; P.O. Box 327, Willmar
T: +1 320 2356106; Fax: +1 320 2140187; URL: www.pioneerland.lib.mn.us
1983; John Houlahan
675 000 vols; 415 curr per; 11 500 av-mat; 6 040 sound-rec   41905

Willmar Public Library, 410 Fifth St SW, *Willmar*, MN 56201-3298
T: +1 320 2353162; Fax: +1 320 2353169; E-mail: willmar@willmar.lib.mn.us; URL: www.willmarpubliclibrary.org
1904; Christine Beyerl
73 000 vols; 260 curr per   41906

Upper Moreland Free Public Library, 109 Park Ave, *Willow Grove*, PA 19090
T: +1 215 6590741; Fax: +1 215 8301223; E-mail: uppermoreland@mclinc.org; URL: www.uppermorelandlibrary.org
1959
57 164 vols; 120 curr per   41907

Willoughby-Eastlake Public Library, 263 E 305th St, *Willowick*, OH 44095
T: +1 440 9432203; Fax: +1 440 9432383; URL: www.wepl.lib.oh.us
1827; Kathy Dugan
4 branch libs
215 000 vols; 477 curr per   41908

Willows Public Library, 201 N Lassen St, *Willows*, CA 95988-3010
T: +1 530 9345156; Fax: +1 530 9342225
1906; Don Hampton
62 000 vols; 100 curr per; 1 387 av-mat
libr loan   41909

Wilmette Public Library District, 1242 Wilmette Ave, *Wilmette*, IL 60091-2558
T: +1 847 2565025; Fax: +1 847 2566911; E-mail: wilref@wilmettelibrary.info; URL: www.wilmettelibrary.info
1901; Ellen Boates Clark
Art, Travel; Corporations Rpts
239 000 vols; 668 curr per   41910

Brandywine Hundred Branch, 1300 Foulk Rd, *Wilmington*, DE 19803
T: +1 302 4773150; Fax: +1 302 4774545; URL: www.nccdelib.org
1959; Thomas M. Weaver
Holocaust, Coin Coll, Delawareana
152 000 vols; 330 curr per   41911

Kirkwood Highway Library, 6000 Kirkwood Hwy, *Wilmington*, DE 19808-4817
T: +1 302 9957663; Fax: +1 302 9957687; URL: www.librarytechnology.org
1967; Sally Brown
84 000 vols; 206 curr per   41912

New Hanover County Public Library, 201 Chestnut St, *Wilmington*, NC 28401
T: +1 910 7986300; Fax: +1 910 7986312; URL: www.nhclibrary.org
1906; David M. Paynter
Civil War Mat, Fales Coll, North Carolina; 4 branch libs
389 000 vols; 569 curr per; 9 000 av-mat; 6 080 sound-rec; 3 500 digital data carriers
libr loan   41913

Wilmington Institute Library, Tenth & Market St, *Wilmington*, DE 19801; P.O. Box 2303, Wilmington, DE 19899-2303
T: +1 302 5717421; Fax: +1 302 6549132; URL: www.wilmlib.org
1788; David H. Burdash
3 branch libs
335 000 vols; 700 curr per; 18 000 av-mat; 7 800 sound-rec; 5 000 digital data carriers
libr loan   41914

Wilmington Memorial Library, 175 Middlesex Ave, *Wilmington*, MA 01887-2779
T: +1 978 6942099; Fax: +1 978 6589699; E-mail: mwlinfo@mclc.org; URL: www.wilmlibrary.org
1871; Christina A. Stewart
64 000 vols; 175 curr per   41915

Wilmington Public Library of Clinton County, 268 N South St, *Wilmington*, OH 45177-1696
T: +1 937 3822417; Fax: +1 937 3821692; URL: www.wilmington.lib.oh.us
1899; Nancy Ehas
Ohio Coll, Wilmington News-Journal (newsp)
69 000 vols; 204 curr per; 943 av-mat
libr loan   41916

Wilson County Public Library, 249 Nash St W, *Wilson*, NC 27893-3801
T: +1 252 2375355; Fax: +1 252 2655569; URL: www.wilson-co.com/library.html
1937; Greg Needham
Drama; Genealogy Coll, North Carolina; 5 branch libs
174 000 vols; 250 curr per   41917

Wilsonville Public Library, 8200 SW Wilsonville Rd, *Wilsonville*, OR 97070
T: +1 503 6822744; Fax: +1 503 6828685; E-mail: wvref@lincc.org; URL: www.wilsonville.lib.or.us
1982; Patrick Duke
Japanese Language Children's Bks
83 000 vols; 181 curr per; 100 maps; 15 microforms; 3 000 av-mat; 3 800 sound-rec; 126 digital data carriers
libr loan   41918

Wilton Library Association, 137 Old Ridgefield Rd, *Wilton*, CT 06897-3019
T: +1 203 7627196; Fax: +1 203 8341166; E-mail: library@wiltonlibrary.org; URL: www.wiltonlibrary.org
1895; Kathy Leeds
Richard Gilmore Knott Memorial Record Coll, Brubeck Jazz Coll
123 000 vols; 227 curr per; 4 510 Audio Bks   41919

Pulaski County Public Library, 121 S Riverside Dr, *Winamac*, IN 46996-1596
T: +1 574 9463432; Fax: +1 574 9466598; URL: www.pulaski-libraries.lib.in.us
1905; Katherine Scott
53 000 vols; 162 curr per   41920

Clark County Public Library, 370 S Burns Ave, **Winchester**, KY 40391 T: +1 859 7445661; Fax: +1 859 7445993; E-mail: clarkbooks@gmail.com; URL: www.clarkpublib.org
1950; Julie Maruskin
89 000 vols; 300 curr per   41921

Handley Regional Library, 100 W Piccadilly St, **Winchester**, VA 22601; P.O. Box 58, Winchester, VA 22601-0058 T: +1 540 6629041; Fax: +1 540 7224769; E-mail: hlref@hrl.lib.state.va.us; URL: www.hrl.lib.state.va.us
1913; Trish Ridgeway
Northern Shenandoah Valley hist, Civil War; 2 branch libs, 1 bookmobile
240 000 vols; 315 curr per; 312 microforms; 4 969 av-mat; 5 287 sound-rec; 27 digital data carriers
libr loan   41922

Winchester Community Library, 125 N East St, **Winchester**, IN 47394-1698 T: +1 765 5844824; Fax: +1 765 5843624; E-mail: wincomlib@yahoo.com
1912; Jana Barnes
Abraham Lincoln Coll, Large Print Bks
55 000 vols; 69 curr per   41923

Winchester Public Library, 80 Washington St, **Winchester**, MA 01890 T: +1 781 7217177; Fax: +1 781 7217101; URL: winpublib.org
1859; Lynda J. Wills
Civil War Hist (Lincoln & Lee Coll)
107 000 vols; 382 curr per   41924

Piedmont Regional Library, 189 Bell View St, **Winder**, GA 30680-1706 T: +1 770 8672762; Fax: +1 770 8677483; URL: library.barrow.public.lib.ga.us
1954; Alan Harkness
10 branch libs
182 000 vols; 262 curr per   41925

Windsor Public Library, 323 Broad St, **Windsor**, CT 06095 T: +1 860 2851918; Fax: +1 860 2851889; E-mail: library@townofwindsorct.com; URL: www.windsorlibrary.com
1888; Laura Kahkonen
Career Ctr, Health Info Ctr, Travel Ctr, Parenting Ctr; 1 branch libr
93 000 vols; 212 curr per   41926

Windsor-Severance Library District, 720 Third St, **Windsor**, CO 80550-5109 T: +1 970 6869955; Fax: +1 970 6862502; URL: www.wsld.info
1922; Carol A. Engel
Colorado
65 000 vols; 123 curr per
libr loan; ALA   41927

Windsor Locks Public Library, 28 Main St, **Windsor Locks**, CT 06096 T: +1 860 6271495; Fax: +1 860 6271496; URL: www.windsorlockslibrary.org
1907; Gloria Malec
Arts, Crafts
65 000 vols; 70 curr per   41928

Northwest Regional Library, 185 Ashwood Dr, **Winfield**, AL 35594-5436; P.O. Box 1527, Winfield, AL 35594-1527 T: +1 205 4872330; Fax: +1 205 4874815; E-mail: nwrl@dlis.net
1961; Ann Lynn
9 branch libs
80 000 vols; 10 curr per   41929

Winfield Public Library, 605 College St, **Winfield**, KS 67156-3199 T: +1 620 2214470; Fax: +1 620 2216135; E-mail: library@wpl.org; URL: www.wpl.org
1912; Joan Cales
61 000 vols; 168 curr per   41930

Humboldt County Library, 85 E Fifth St, **Winnemucca**, NV 89445 T: +1 775 6236388; Fax: +1 775 6236438
1923; Sharon Allen
55 000 vols; 171 curr per   41931

Winnetka-Northfield Public Library District, 768 Oak St, **Winnetka**, IL 60093-2583 T: +1 847 4467220; Fax: +1 847 4465085; URL: www.wpld.alibrary.com
1884; Barbara J. Aron
1 branch libr
127 000 vols; 146 curr per; 3 000 e-books; 6 181 av-mat; 6 180 Audio Bks, 422 Electronic Media & Resources, 2 220 Videos   41932

Winn Parish Library, 205 W Main St, **Winnfield**, LA 71483-2718 T: +1 318 6284478; Fax: +1 318 6289820; E-mail: admin.h1wn@pelican.state.lib.la.us; URL: www.winn.lib.la.us
1937; Mary Doherty
Genealogy Coll, Louisiana Coll; 4 branch libs
98 000 vols; 250 curr per   41933

Fairfield County Library, 300 Washington St, **Winnsboro**, SC 29180 T: +1 803 6354971; Fax: +1 803 6357715; URL: www.fairfield.lib.sc.us
1938; Sarah D. McMaster
Hist
82 000 vols; 174 curr per   41934

Franklin Parish Library, 705 Prairie St, **Winnsboro**, LA 71295-2629 T: +1 318 4354336; Fax: +1 318 4351990; E-mail: admin.t1fr@pelican.state.lib.la.us; URL: www.franklinparishlibrary.org
1950; Carolyn Slint
Louisiana Coll; 1 branch libr
80 000 vols; 87 curr per   41935

Winona Public Library, 151 W Fifth St, **Winona**, MN 55987-3170; P.O. Box 1247, Winona, MN 55987-7247 T: +1 507 4524860; Fax: +1 507 4525842; URL: www.selco.lib.mn.us/winona/default.htm
1899; James P. Stetina
154 000 vols; 320 curr per; 1 868 av-mat; 2 330 Large Print Bks, 1 870 Talking Bks   41936

Beardsley & Memorial Library, 40 Munro Pl, **Winsted**, CT 06098 T: +1 860 3796043; Fax: +1 860 3793621; E-mail: director@beardsleyandmemorial.org; URL: www.beardsleyandmemorial.org
1898; Mary Lee Bulat
Genealogy, Hist, Humanities, Economics, Science fiction, Antiques, Large type print
58 000 vols; 105 curr per; 3 000 microforms; 1 700 sound-rec; 6 digital data carriers
libr loan   41937

Forsyth County Public Library, 660 W Fifth St, **Winston-Salem**, NC 27101 T: +1 336 7032665; Fax: +1 336 7272549; URL: www.forsythlibrary.org
1903; Sylvia Sprinkle-Hamlin
North Carolina, Frank Jones Photogr Print Coll, H Kapp Ogborn Philatelic Coll; 11 branch libs
675 000 vols; 700 curr per; 6 000 e-books; 7 830 Large Print Bks
libr loan   41938

Forsyth County Public Library – Reynolda Manor, 2839 Fairlawn Dr, **Winston-Salem**, NC 27106 T: +1 336 7032960; Fax: +1 336 7483318
Laura Weigand
57 000 vols; 100 curr per; 40 e-books   41939

Orange County Library District – Winter Garden Branch, 805 E Plant St, **Winter Garden**, FL 34787
Glenda Houck
66 000 vols   41940

Winter Haven Public Library, 325 Avenue A, NW, **Winter Haven**, FL 33881 T: +1 863 2915880; Fax: +1 863 2915889; E-mail: jkovac@mywinterhaven.com; URL: whpl.mywinterhaven.com
1910; Jennifer Kovac
60 000 vols; 182 curr per   41941

Winter Park Public Library, 460 E New England Ave, **Winter Park**, FL 32789-4493 T: +1 407 6233300; Fax: +1 407 6233489; URL: www.wppl.org
1885; Robert G. Melanson
175 000 vols; 326 curr per   41942

Winthrop Public Library & Museum, Two Metcalf Lane, **Winthrop**, MA 02152-3157 T: +1 617 8461703; Fax: +1 617 8467083
1885; John Cronin
Local Hist (Museum Coll), Lincoln Memorabilia
84 000 vols; 35 curr per   41943

Albemarle Regional Library, 303 W Tryon St, **Winton**, NC 27986; P.O. Box 68, Winton, NC 27986-0068 T: +1 252 3587832; Fax: +1 252 3587868; URL: www.arlnc.org
1948; Sue D. Williams
7 branch libs
195 000 vols; 319 curr per; 300 av-mat; 11 000 sound-rec; 9 670 digital data carriers; 50 High Interest/Low Vocabulary Bks, 1 250 Large Print Bks
libr loan   41944

McMillan Memorial Library, 490 E Grand Ave, **Wisconsin Rapids**, WI 54494-4898 T: +1 715 4231040; Fax: +1 715 4232665; E-mail: askmcm@scls.lib.wi.us; URL: www.mcmillanlibrary.org
1889; Ronald McCabe
103 000 vols; 288 curr per   41945

Lonesome Pine Regional Library, 124 Library Rd SW, **Wise**, VA 24293-5907 T: +1 276 3288325; Fax: +1 276 3281739; E-mail: reglib@lprlibrary.org; URL: www.lprlibrary.org
1958; Amy Bond
9 branch libs
486 000 vols; 576 curr per; 11 983 av-mat; 11 100 Talking Bks
libr loan   41946

Lonesome Pine Regional Library – Wise County Public, 124 Library Rd SW, **Wise**, VA 24293 T: +1 276 3288061; Fax: +1 276 3281739; E-mail: wcplib@lprlibrary.org
Linda Scarborough
135 000 vols; 158 curr per; 2 193 av-mat   41947

Woburn Public Library, 45 Pleasant St, **Woburn**, MA 01801; P.O. Box 298, Woburn, MA 01801-0298 T: +1 781 9330148; Fax: +1 781 9387860; URL: www.woburnpubliclibrary.org
1855; Kathleen O'Doherty
Genealogy
80 000 vols; 152 curr per   41948

Wolcott Public Library, 469 Bound Line Rd, **Wolcott**, CT 06716 T: +1 203 8798110; Fax: +1 203 8798109; URL: www.wolcottlibrary.org
1916; Kathy Giotsas
Job Resource Ctr, Foreign Language Cassettes
54 000 vols; 72 curr per; 819 av-mat   41949

Roosevelt County Library, 220 Second Ave S, **Wolf Point**, MT 59201 T: +1 406 6532411; Fax: +1 406 6531365; E-mail: ask@rooseveltcountylibrary.org; URL: www.rooseveltcountylibrary.org
1950
3 branch libs
58 150 vols; 108 curr per
libr loan   41950

Wood Dale Public Library District, 520 N Wood Dale Rd, **Wood Dale**, IL 60191 T: +1 630 7666762; Fax: +1 630 7665715; URL: www.wooddalelibrary.org
1958; Yvonne Bergendorf
Large print
86 000 vols; 193 curr per   41951

Wood River Public Library, 326 E Ferguson Ave, **Wood River**, IL 62095-2098 T: +1 618 2544832; Fax: +1 618 2544836; URL: woodriverlibrary.org/
1920; Diane Steele
65 000 vols   41952

Free Public Library of Woodbridge, George Frederick Plaza, **Woodbridge**, NJ 07095 T: +1 732 6344450; Fax: +1 732 6341569; URL: www.woodbridgelibrary.org
1964; John Hurley
3 branch libs
463 000 vols; 510 curr per; 4 000 e-books; 2 313 av-mat; 390 High Interest/Low Vocabulary Bks, 7 160 Large Print Bks, 2 320 Talking Bks   41953

Woodbridge Town Library, 10 Newton Rd, **Woodbridge**, CT 06525 T: +1 203 3893433; Fax: +1 203 3893457; E-mail: tfabian@ci.woodbridge.ct.us; URL: www.woodbridge.lioninc.org
1940; Todd Fabian
Travel
79 800 vols   41954

Woodburn Public Library, 280 Garfield St, **Woodburn**, OR 97071-4698 T: +1 503 9825263; Fax: +1 503 9822808; E-mail: woodburn@ccrls.org; URL: www.woodburnlibrary.org
1914; Anna Stavinoha
Large print; Language (Russian & Spanish Coll)
69 000 vols; 172 curr per
libr loan; PNLA, ALA   41955

Washington County Library, 8595 Central Park Pl, **Woodbury**, MN 55125-9453 T: +1 651 2758500; Fax: +1 651 2758509; URL: www.co.washington.mn.us/library
1967; Patricia M. Conley
8 branch libs
391 000 vols; 1 263 curr per   41956

Washington County Library – R. H. Stafford Branch, 8595 Central Park Pl, **Woodbury**, MN 55125-9613 T: +1 651 7311320; Fax: +1 651 2758562
Chad Lubbers
127 000 vols   41957

Woodbury Public Library, 33 Delaware St, **Woodbury**, NJ 08096 T: +1 856 8452611; Fax: +1 856 8455280; E-mail: wod@jersey.net; URL: www.woodburylibrary.org
1790; Jean Wipf
59 000 vols; 90 curr per   41958

Woodbury Public Library, 269 Main St S, **Woodbury**, CT 06798 T: +1 203 2633502; Fax: +1 203 2630571; URL: www.biblio.org/woodbury
1902; Patricia Lunn
90 000 vols; 145 curr per
libr loan; ALA, NELA, CLA   41959

Queens Borough Public Library – Woodhaven Branch, 85-41 Forest Pkwy, **Woodhaven**, NY 11421 T: +1 718 8491010
Rebecca Babirye-Alibat
54 000 vols   41960

Woodland Public Library, 250 First St, **Woodland**, CA 95695-3411 T: +1 530 6615981; Fax: +1 530 6665408; URL: www.cityofwoodland.org/library
1904; Paul Miller
82 000 vols; 189 curr per; 2 508 av-mat
libr loan   41961

Los Angeles Public Library System – Platt, 23600 Victory Blvd, **Woodland Hills**, CA 91367 T: +1 818 3409386; Fax: +1 818 3409645
Lynn Light
61 000 vols   41962

Rampart Public Library District, Woodland Park Public Library, 218 E Midland St, **Woodland Park**, CO 80863; P.O. Box 336, Woodland Park, CO 80866-0336 T: +1 719 6879281; Fax: +1 719 6876631; URL: rampart.colibraries.org
Sharon Quay
56 000 vols; 140 curr per   41963

Woodridge Public Library, Three Plaza Dr, **Woodridge**, IL 60517-5014
T: +1 630 9647899; Fax: +1
630 9640175; E-mail:
wrs@woodridgelibrary.org; URL:
www.woodridgelibrary.org
1967; Mary Sue Brown
150 000 vols; 424 curr per   41964

Monroe County District Library, 96 Home Ave, **Woodsfield**, OH 43793
T: +1 740 4721954; Fax: +1 740
4721110; E-mail: mcdl@oplin.org; URL:
www.monroecounty.lib.oh.us
1939; Kathy South
63 000 vols; 140 curr per; 1 100 av-mat
41965

Queens Borough Public Library – Woodside Branch, 54-22 Skillman Ave, **Woodside**, NY 11377
T: +1 718 4294700
Jingru Pei
100 000 vols; 120 curr per   41966

Woodstock Library, Five Library Lane, **Woodstock**, NY 12498-1299
T: +1 845 6792213; Fax: +1 845
6797149; E-mail: info@woodstock.org;
URL: www.woodstock.org
1913; Diana B. Stern
Art, Architecture, Music, Hist, Belles lettres
60 000 vols   41967

Woodstock Public Library, 414 W Judd, **Woodstock**, IL 60098-3195
T: +1 815 3380542; Fax: +1 815
3342296; E-mail: library@woodstockil.gov;
URL: www.woodstockil.gov/library
1890; Margaret E. Crane
106 000 vols; 242 curr per; 1 000
microforms; 6 000 sound-rec; 12 digital
data carriers
libr loan   41968

Woonsocket Harris Public Library, 303 Clinton St, **Woonsocket**, RI 02895
T: +1 401 7674124; Fax: +1 401
7674120; URL: woonsocketlibrary.org
1863; Leslie Page
127 000 vols; 142 curr per   41969

Wayne County Public Library, 220 W Liberty St, **Wooster**, OH 44691-3593;
P.O. Box 1349, Wooster, OH 44691-7086
T: +1 330 8044659; Fax: +1 330
2621352; URL: wcpl.info
1897; Greg Lubelski
City Law; 6 branch libs
306 000 vols; 932 curr per; 12 000 e-books; 21 000 av-mat; 17 000 sound-rec;
11 280 digital data carriers; 1 100
Electronic Media & Resources, 6 170
Large Print Bks
libr loan   41970

Worcester Public Library, Three Salem Sq, **Worcester**, MA 01608
T: +1 508 7991655; Fax: +1 508
7991652; URL: www.worcpublib.org
1859; Penelope B. Johnson
US Hist (Libr of American Civilization),
micro; 3 branch libs
589 000 vols; 1 148 curr per; 796 e-books
libr loan   41971

Washakie County Library System, 1019 Coburn Ave, **Worland**, WY 82401
T: +1 307 3472231; Fax: +1 307
3472248; URL: www.wsl.state.wy.us/
washakie
1914; Dolores Koch
61 000 vols; 44 curr per   41972

Nobles County Library, 407 12th St, **Worthington**, MN 56187; P.O. Box
1049, Worthington, MN 56187-5049
T: +1 507 3722981; Fax: +1 507
3722982; URL: www.plumcreeklibrary.org/
Adrian/page11.html
1947; Roger E. Spillers
1 branch libr
72 000 vols; 100 curr per; 2 054 av-mat
41973

Plum Creek Library System, 290 S Lake St, **Worthington**, MN 56187; P.O.
Box 697, Worthington, MN 56187-0697
T: +1 507 3765803; Fax: +1 507
3769244
Mark Ranum
655 000 vols   41974

Worthington Libraries, Old Worthington Library, 820 High St, **Worthington**,
OH 43085
T: +1 614 8072620;
Fax: +1 614 8072642; URL:
www.worthingtonlibraries.org
1925; Meribah Mansfield
1 branch libr
409 000 vols
libr loan   41975

Fiske Public Library, 110 Randall Rd, **Wrentham**, MA 02093; P.O. Box 340,
Wrentham, MA 02093-0340
T: +1 508 3845440; Fax: +1 508
3845443; URL: www.fiskelib.org
1892; Mary Tobichuk
55 000 vols; 110 curr per; 2 446 av-mat
41976

Wyandanch Public Library, 14 S 20th St, **Wyandanch**, NY 11798
T: +1 631 6434848; Fax: +1 631
6439664; URL: www.suffolk.lib.ny.us/
libraries/wyan
1974; Corey Fleming
60 000 vols; 96 curr per   41977

Bacon Memorial District Library, 45 Vinewood, **Wyandotte**, MI 48192-5221
T: +1 734 2468357; Fax: +1 734
2821540; URL: www.baconlibrary.org
1869; Janet Cashin
Military Hist
56 000 vols; 198 curr per; 1 406 av-mat
41978

Wyckoff Public Library, 200 Woodland Ave, **Wyckoff**, NJ 07481
T: +1 201 8914866; Fax: +1 201
8913892; URL: www.wyckoff-nj.com
1921; Judy Schmitt
76 000 vols; 90 curr per   41979

Free Library of Springfield Township, 1600 Paper Mill Rd, **Wyndmoor**,
PA 19038
T: +1 215 8365300; Fax: +1
215 8362404; E-mail:
springfieldlibrary@mclinc.org; URL:
fls.mclinc.org
1965; Marycatherine McGarvey
Art (Malta Fund), Business, Outdoor
Sports (Brodsky Fund), Performing Arts,
Popular Medicine (Ramsey Coll)
71 000 vols; 133 curr per; 1 300 av-mat;
1 580 sound-rec; 3 digital data carriers
libr loan   41980

Kent District Library – Wyoming Branch, 3350 Michael Ave SW,
Wyoming, MI 49509
T: +1 616 6473980; Fax: +1 616
5344822
Mary Hollinrake
111 000 vols   41981

Greene County Public Library, 76 E Market St, **Xenia**, OH 45385-3100; P.O.
Box 520, Xenia, OH 45385-0520
T: +1 937 3524000; Fax: +1 937
3724673; URL: www.greenelibrary.info
1878; Karl Colon
7 branch libs
548 000 vols; 635 curr per; 24 000
av-mat; 31 047 sound-rec; 1 611 digital
data carriers; 7 700 Large Print Bks
libr loan   41982

Greene County Public Library – Xenia Community Library, 76 E Market St,
Xenia, OH 45385-0520
T: +1 937 3524000; Fax: +1 937
3765523
Brenda Charny
139 000 vols   41983

Yadkin County Public Library, 233 E Main St, **Yadkinville**, NC 27055; P.O.
Box 607, Yadkinville, NC 27055-0607
T: +1 336 6798792; Fax: +1 336
6794625; E-mail: ydk@nwrl.org; URL:
www.nwrl.org
1942; Malinda S. Sells
58 000 vols; 10 curr per   41984

Yakima Valley Libraries, 102 N Third St, **Yakima**, WA 98901-2759
T: +1 509 4528541; Fax: +1 509
5752093; E-mail: reference@yvrl.org;
URL: www.yvrls.org
1951; Monica Weyhe
Northwest Americans & Indians of the
Pacific Northwest (Relander Coll); 18
branch libs
398 000 vols; 581 curr per   41985

Yankton Community Library, 515 Walnut, **Yankton**, SD 57078-4042
T: +1 605 6685275; Fax: +1 605
6685277; URL: ycllib.sdln.net
1902; James C. Scholtz
75 000 vols; 160 curr per   41986

Bucks County Free Library – Yardley-Makefield Branch, 1080 Edgewood Rd,
Yardley, PA 19067-1648
T: +1 215 4939020; Fax: +1 215
4930279
Jeannie Alexander
89 000 vols; 117 curr per   41987

Delphi Public Library, Northwest Carroll Branch, 164 W Forest St, **Yeoman**,
IN 47997
T: +1 574 9652382
Jane Cruz
100 000 vols   41988

Lyon County Library System, 20 Nevin Way, **Yerington**, NV 89447
T: +1 775 4636645; Fax: +1 775
4636646; URL: www.lyon-county.org
Diane Brigham
108 000 vols; 65 curr per   41989

Yorba Linda Public Library, 18181 Imperial Hwy, **Yorba Linda**,
CA 92886-3437
T: +1 714 7772873; Fax: +1 714
7770640; E-mail: ylpl@ylpl.lib.ca.us; URL:
www.ylpl.lib.ca.us
1913; Melinda Steep
131 000 vols; 277 curr per; 6 927 av-mat;
6930 Talking Bks   41990

Martin Memorial Library, 159 E Market St, **York**, PA 17401-1269
T: +1 717 8465300; Fax: +1 717
8482330; URL: www.yorklibraries.org
1935; William Schell
150 000 vols; 200 curr per
libr loan   41991

York County Library System, 159 E Market St, 3rd Flr, **York**, PA 17402
T: +1 717 8465300; Fax: +1 717
8496999; URL: www.yorklibraries.org
Patricia Calvani
5 branch libs
454 000 vols; 661 curr per   41992

York Public Library – Kilgore Memorial Library, 520 Nebraska Ave,
York, NE 68467-3095
T: +1 402 3632620; Fax: +1 402
3632627; E-mail: co51537@alltel.net
1885; Stan Schulz
55 000 vols; 110 curr per
libr loan   41993

York County Public Library – Tabb Library, 100 Long Green Blvd,
Yorktown, VA 23693
T: +1 757 8905120; Fax: +1 757
8905127
Linda Blanchard
156 000 vols; 260 curr per   41994

York County Public Library – Yorktown Library, 8500 George
Washington Memorial Hwy, **Yorktown**,
VA 23692
T: +1 757 8905207; Fax: +1 757
8902956; URL: www.yorkcounty.gov/
library
1968; Norma Colton
1 branch libr
60 000 vols; 260 curr per   41995

Public Library of Youngstown & Mahoning County, 305 Wick Ave,
Youngstown, OH 44503-1079
T: +1 330 7448636; Fax: +1 330
7443355; URL: www.libraryvisit.org
1880; Carlton A. Sears
16 branch libs
686 000 vols; 1 122 curr per; 2 000 e-books; 25 209 av-mat; 12 200 sound-rec;
17 850 digital data carriers; 39 970
Videos, 25 210 Talking Bks   41996

Public Library of Youngstown & Mahoning County – Austintown,
600 S Raccoon Rd, **Youngstown**,
OH 44515
T: +1 330 7926982
Ann Martini
64 000 vols   41997

Public Library of Youngstown & Mahoning County – Boardman, 7680
Glenwood Ave, **Youngstown**, OH 44512
T: +1 330 7581414; Fax: +1 330
7587918
Mary Frum
79 000 vols   41998

Ypsilanti District Library, Whittaker Road Library, 5577 Whittaker Rd,
Ypsilanti, MI 48197
T: +1 734 4824110; Fax: +1 734
4820047; URL: www.ypsilibrary.org
Jill Morey
Ypsilanti & Michigan History Coll
237 000 vols; 359 curr per   41999

Siskiyou County Public Library, 719 Fourth St, **Yreka**, CA 96097
T: +1 530 8414175; Fax: +1
530 8427001; E-mail:
siskiyoulibrary@snowcrest.net; URL:
www.siskiyoulibrary.info
1915; Betsy Emry
Gold mining; 11 branch libs
140 000 vols; 260 curr per
libr loan; OCLC   42000

Sutter County Free Library, 750 Forbes Ave, **Yuba City**, CA 95991-3891
T: +1 530 8227137; Fax: +1 530
6716539; E-mail: suttlibr@yahoo.com;
URL: www.saclibrary.org
1917; Roxanna Parker
98 000 vols; 159 curr per
libr loan   42001

Yuma County Library District, 350 Third Ave, **Yuma**, AZ 85364
T: +1 928 7821871; Fax: +1
928 3736470; E-mail:
librarian@yumalibrary.org; URL:
www.yumalibrary.org
1921; Susan M. Evans
Arizona Govt Docs Coll; 6 branch libs
160 000 vols; 289 curr per; 2 000 e-books
libr loan   42002

East Baton Rouge Parish Library – Zachary Branch, 1900 Church St,
Zachary, LA 70791
T: +1 225 6581850; Fax: +1 225
6581844
Lula Pride
98 000 vols   42003

Muskingum County Library System, 220 N Fifth St, **Zanesville**,
OH 43701-3587
T: +1 740 4530391; Fax: +1 740
4556357; URL: www.muskingumlibrary.org
1903; Sandi Plymire
Zanesville & Muskingum County Hist,
Hist (Ohio Coll), Business & Industry; 5
branch libs
304 000 vols; 185 curr per   42004

Howard Miller Public Library, 14 S Church St, **Zeeland**, MI 49464-1728
T: +1 616 7720874; Fax: +1 616
7723253; E-mail: zee@llcoop.org; URL:
www.hmpl.org
1969; Dennis M. Martin
67 000 vols; 184 curr per
libr loan   42005

Zion-Benton Public Library District, 2400 Gabriel Ave, **Zion**, IL 60099
T: +1 847 8724680; Fax: +1 847
8724942; E-mail: library@zblibrary.org;
URL: www.zblibrary.org
1937; Nann Blaine Hilyard
128 000 vols; 88 curr per; 2 000 e-books
libr loan   42006

Hussey-Mayfield Memorial Public Library, 250 N Fifth St, **Zionsville**,
IN 46077-1324; P.O. Box 840, Zionsville,
IN 46077-0840
T: +1 317 8733149; Fax: +1
317 8738339; E-mail:
askalib@zionsville.lib.in.us; URL:
www.zionsville.lib.in.us
Martha E. Catt
101 000 vols; 260 curr per; 11 234
sound-rec; 174 digital data carriers
42007

Uruguay

National Libraries

Biblioteca Nacional del Uruguay
(Uruguay National Library), Av 18 de
Julio 1790, CC 452, 11200 *Montevideo*
T: +598 24005385; Fax: +598 24096902;
E-mail: bibliotecanacional@bibna.gub.uy;
URL: www.bibna.gub.uy
1816; Tomás de Mattos
Attached: Centro Nacional de
Documentación Científica, Técnica y
Económica
900 000 vols
libr loan   42008

University Libraries, College Libraries

Escuela Naval, Armada Nacional,
Biblioteca, Rambla Tomás Berreta s/n,
Carrasco, CP 11500 *Montevideo*
T: +598 26006021; Fax: +598 26005225;
E-mail: esnal_biblioteca@armada.mil.uy;
URL: www.armada.mil.uy
1907; Clary Rymer
20 630 vols; 20 curr per; 84 diss/theses;
75 maps; 149 av-mat; 38 digital data
carriers
libr loan   42009

Escuela Universitaria de Música,
Biblioteca, Guayabo 1773, 11200
Montevideo
T: +598 29002375; Fax: +598 29007204;
E-mail: eme @ eumus.edu.uy; URL:
www.eumus.edu.uy/eme/
1974; Elena Parentini Amarillo
Instruments dept
3 500 vols; 25 curr per; 18 000 music
scores; 70 av-mat; 3 800 sound-rec; 10
digital data carriers
libr loan   42010

Instituto de Profesores Artigas,
Biblioteca, Av Libertador Brig. Gral.
Lavalleja 2025, *Montevideo*
T: +598 29247074; Fax: +598 29247074;
E-mail: perelac@adinet.com.uy
1977; Elena Isabel Laclau Milán
Education
75 000 vols; 10 curr per; 30 maps; 50
sound-rec; 20 digital data carriers
libr loan   42011

**Universidad Católica del Uruguay
'Damaso Antonio Larrañaga'**,
Biblioteca, Av 8 de Octubre 2738, 11600
Montevideo
T: +598 2472717; Fax: +598 2802124;
E-mail: aterra@ucu.edu.uy; URL:
www.ucu.edu.uy
1979; Fernando Sorondo
Humanities, religion, information science
45 000 vols; 300 curr per; 150
diss/theses; 3 500 av-mat; 250 sound-rec
                                        42012

Universidad de la República,
Montevideo
– Escuela Universitaria de Bibliotecología y
Ciencias Afines, Biblioteca, Emilio Frugoni
1427, 11200 Montevideo
1943; Teresa Fittipaldi
12 000 vols; 400 curr per; 229
diss/theses
libr loan   42013

– Facultad de Agronimía, Departamento de
Documentacion y Biblioteca, Av Garzón
780, Montevideo
1906; Raquel Schneider
65 000 vols
libr loan   42014

– Facultad de Arquitectura, Biblioteca, Bd
Artigas 1031, Montevideo
1915; Ana María Chiacchio
34 500 vols   42015

– Facultad de Derecho y Ciencias Sociales,
18 de Julio 1824, Piso 1, Montevideo
Raquel Ortiz de Balbis
450 000 vols   42016

– Facultad de Ingeniería, Departamento de
Documentación y Biblioteca, Julio Herrera
y Reissig 565, Casilla de correo 30,
11300 Montevideo
T: +598 27110595, 27110383; Fax: +598
27108647; E-mail: bibliofi @ fing.edu.uy;
URL: www.fing.edu.uy/biblioteca/
1888; Susana Gil Gelós
Classic Coll
60 000 vols; 700 curr per
libr loan; ISTEC   42017

– Facultad de Química, Biblioteca, Av
General Flores 2124, Montevideo
1929
22 000 vols   42018

– Facultad de Veterinaria, Departamento
de Documentación y Biblioteca, Alberto
Lasplaces 1550, Distrito 6, Casilla 16062,
11600 Montevideo
T: +598 26226409; Fax: +598 26226409;
E-mail: bibliovet@fvet.edu.uy; URL:
www.fvet.edu.uy/fvbiblio
1907; Beatriz Sarachaga
27 200 vols; 120 curr per; 10 e-journals;
46 diss/theses; 80 av-mat; 200 digital
data carriers; 1 400 pamphlets
libr loan   42019

Universidad del Trabajo del Uruguay,
Departamento de Biblioteca Of. 27, San
Salvador 1674, *Montevideo*
T: +598 2481526
1971; Maria Gloria de Béjar
7 000 vols; 292 curr per
libr loan   42020

– Instituto Superior de Electrotécnica,
Electrónica y Computación, Biblioteca, C/
Joaquín Peguena 1931, Montevideo
T: +598 2492520
Marcial Peon
10 000 vols   42021

Government Libraries

Embajada Argentina en el Uruguay,
Biblioteca, 19 de Abril 3309,
Montevideo
T: +598 2382725
Blanca Cosenza de Caffera
Hist of Argentine
8 200 vols; 400 curr per; 250 av-mat;
210 sound-rec
libr loan   42022

Junta Departamental de Montevideo,
Biblioteca 'José Artigas', 25 de Mayo
609, CP 11000, *Montevideo*
T: +598 29152126; Fax: +598 29163614;
E-mail: ilago@juntamvd.gub.uy
1962; Irene Lago
Public admini, law, hist, architecture
17 000 vols; 70 curr per; 110 av-mat;
press clippings   42023

Ministerio de Ganadería y Agricultura,
Departamento de Prensa y Biblioteca,
Cerrito 315, *Montevideo*
1926
108 000 vols   42024

Ministerio del Interior, Departamento
de Biblioteca, Mercedes 993, 11100
Montevideo
T: +598 2931004; Fax: +598 2931004
1946; Beatriz Almeida de Loro
Law
16 000 vols; 140 curr per; 3 diss/theses
libr loan   42025

Palacio Legislativo, Biblioteca del Poder
Legislativo, Av de las Leyes s/n, Palacio
Legislativo, piso 2, *Montevideo*
T: +598 2142-2425/-2378; E-mail:
biblioteca@parlamento.gub.uy; URL:
www.parlamento.gub.uy/palacio3/
index.asp?e=0&w=1152
1929
Jurisprudence
250 000 vols
Legal deposit libr in conjunction with
National Libr   42026

**Servicio de Extensión Cultural y
Educativo**, Biblioteca, Reconquista 535,
Montevideo
T: +598 2950103
1976; Vilma Igorra de Arruabarrena
10 200 vols   42027

Servicio de Información Comercial,
Centro de Información, Cuareim 1384,
Casilla de Correo 10771, *Montevideo*
T: +598 2920319
1974; María Teresa Castilla
12 500 vols; 420 curr per
libr loan   42028

Ecclesiastical Libraries

**Facultad de Teología del Uruguay
'Mons. Mariano Soler'**, Biblioteca, San
Fructuoso 1019, 11800 *Montevideo*;
Casilla 1234, 11800 Montevideo
T: +598 22000289; E-mail: bibteol@
ucu.edu.uy
1967; Oscar Adolfo Chappar
Theol, philos, hist, lit
30 000 vols; 200 curr per; 25 maps
                                        42029

Corporate, Business Libraries

Banco de Seguros del Estado,
Biblioteca, Mercedes y Av Lobertador
Brig. Gral. Lavalleja, Casilla de Correo
473, *Montevideo*
T: +598 2904933
1933; Marta Fernández de Aguzzi
30 400 vols; 97 curr per
libr loan   42030

Banco Hipotecario del Uruguay,
Biblioteca, Av Daniel Fernández Crespo
1508, *Montevideo*
T: +598 24090000; Fax: +598
24093482; E-mail: info@bhu.net; URL:
www.bhu.com.uy
1940; Lucia Olariaga
40 000 vols
libr loan   42031

Special Libraries Maintained by Other Institutions

INIA La Estanzuela, Biblioteca, Ruta
50, Km. 11, La Estanzuela, Casilla de
Correo 39173, 70000 *Colonia*
T: +598 5224060 ext 136;
Fax: +598 5224061; E-mail:
achiacchio@tb.inia.org.uy; URL:
www.inia.org.uy
1914; Ana María Chiacchio Di Paula
Agriculture, animal husbandry, crop
production, soil science, dairy science,
biological control, meteorology
11 000 vols; 483 curr per
libr loan   42032

**Administración de Ferrocarriles del
Estado, AFE**, Biblioteca, Rondeau 1921
esq. Lima, Edif. BAALBEK, *Montevideo*
T: +598 2916132; E-mail: paginaweb@
afe.com.uy; URL: www.afe.com.uy
1958; Carlos Izquierdo
15 500 vols; 203 curr per
libr loan   42033

**Administracion Nacional de
Combustibles, Alcohol y Portland,
ANCAP**, Biblioteca, Centro de
Información, Av Libertador Brig. Gral.
Lavalleja y Paysandú, Casilla de Correo
1090, *Montevideo*
T: +598 2989355
1934; María Haydée Prunell
11 000 vols; 300 curr per   42034

Alliance Française, Biblioteca (Alianza
Francesa), Br. Artigas 1229, 11200
Montevideo
T: +598 24000505 ext 224;
Fax: +598 24000505 ext 220; E-mail:
bibliotheque@alliancefrancaise.edu.uy;
URL: www.alliancefrancaise.edu.uy
1923; Rosario Barba Duthilleu
French civilisation, lit, art, hist, humanities
20 000 vols; 55 curr per; 255 govt docs;
1 500 av-mat; 1 500 sound-rec; 185 digital
data carriers   42035

Archivo General de la Nación,
Biblioteca, Calle Convención 1474, 11100
Montevideo
T: +598 229007232;
Fax: +598 229081330; E-mail:
consultas@agn.gub.uy; URL:
www.agn.gub.uy
14 000 vols   42036

**Asociación de Empleados Bancarios
del Uruguay (AEBU)**, Biblioteca,
Camacuá 575, *Montevideo*
T: +598 29161060; E-mail:
biblioteca@aebu.org.uy; URL:
www.aebu.org.uy/biblioteca
1959
18 000 vols; 30 curr per
libr loan   42037

**Asociación de Escribanos del
Uruguay**, Biblioteca, Av 18 de Julio
1730, Piso 11, CP 11200, *Montevideo*
T: +598 24006400; Fax: +598 24021943;
E-mail: biblioteca@aeu.org.uy; URL:
www.aeu.org.uy
1882
Law-economics, sociology; Depósito de
Documentos de Propiedad Horizontal,
Depósito de Títulos Provenientes de
Bancos Liquidados e Intervenidos
18 600 vols; 110 curr per; 1 450 av-mat;
121 sound-rec; 4 digital data carriers;
2 410 consultas jurídicas
libr loan   42038

**Asociación Latinoamericana de
Integración (ALADI)**, Biblioteca (Latin
American Integration Association (LAIA)),
Cebollatí 1461, 11200 *Montevideo*
T: +598 24001121; Fax: +598
24090649; E-mail: bibliot@aladi.org;
URL: www.aladi.org
1960; Elena Ravina de Gazzolo
Economics, politics, law, integration,
international integration
40 000 vols; 200 curr per
libr loan   42039

Asociación Odontológica Uruguaya,
Biblioteca, Av Durazno 937-39,
Montevideo
T: +598 29001572; URL: www.aou.org.uy
1971
6 000 vols; 350 curr per
libr loan   42040

Asociación Rural del Uruguay,
Biblioteca, Av Uruguay 864, *Montevideo*
T: +598 29020484; Fax: +598 29020489;
URL: www.aru.org.uy
1871; Isabel Triay
3 000 vols
libr loan   42041

Banco Central del Uruguay, Biblioteca,
Diagonal Fabini 777, Piso 5° – Oficina
508, Casilla de Correo 1467, 11100
Montevideo
T: +598 21967ext 1515 or 1523;
Fax: +598 21967ext 1525; E-mail:
info@bcu.gub.uy; URL: www.bcu.gub.uy/
a5542.html
1967
4 700 vols; 120 curr per
libr loan   42042

Biblioteca 'José Batlle y Ordóñez',
Ministerio de Educación y Cultura, José
Belloni 4370, *Montevideo*
T: +598 2224412; Fax: +598 2224412
1975; Prof. José Cozzo
12 000 vols; 50 music scores   42043

**Biblioteca Nacional de Medicina
(BINAME)**, Centro Nacional de
Documentación e Información en
Medicina y Ciencias de la Salud,
CENDIM, Falcultad de Medicina,
Universidad de la República, Av Gral.
Flores 2125, *Montevideo*
T: +598 29243414 ext 3454; Fax: +598
29243416; E-mail: biname@fmed.edu.uy;
URL: www.biname.edu.uy
1875; Stella M. Launy
Hist of medicine coll
130 000 vols; 3 300 curr per; 6 773
diss/theses; 51 microforms; 16 digital
data carriers
libr loan   42044

Biblioteca Pedagógica Central, Consejo
de Educación Primaria, Plaza Cagancha
1175, *Montevideo*
T: +598 29020915; Fax: +598 29084131;
URL: www.cep.edu.uy
1889
Anales de Instrucció Primaria,
Enciclopedia de la Educación; Libr for
the blind, libs in 2 schools
119 000 vols; 885 curr per
libr loan   42045

Centro Artigas-Washington, Alianza Cultural Uruguay Estados Unidos, Paraguay 1217, 11100 *Montevideo*
T: +598 29017423; Fax: +598 29021621; E-mail: centroaw@alianza.edu.uy
1943; Fernanda del Castillo
13 600 vols; 110 curr per; 1 437 music scores; 258 maps; 12 microforms; 600 av-mat; 1 085 sound-rec; 28 digital data carriers; 230 pamphlets   42046

Centro de Altos Estudios Nacionales, Biblioteca, Bvar. Artigas 1488, *Montevideo*
T: +598 2794988; Fax: +598 2775314
1978; Luis Leivas
9 000 vols; 52 curr per; 300 diss/theses; 51 maps; 110 sound-rec
libr loan   42047

Centro Interamericano de Investigación y Documentación sobre Formación Profesional (CINTERFOR), Organización Internacional del Trabajo, Servicio de Información y Documentación, Av Uruguay 1238, Casilla de Correo 1761, *Montevideo*
T: +598 29086023; Fax: +598 29021305; E-mail: dirmvd@cinterfor.org.uy; URL: www.cinterfor.org.uy
1964; Diana Fernandez, Andrès Tellagorry
Vocational and rehabilitation training
23 000 vols; 1 200 curr per; 120 av-mat; 52 digital data carriers
libr loan   42048

Centro Latinoamericano de Economía Humana, CLAEH, Biblioteca, Zelmar Michelini 1220, 11100 *Montevideo*; Casilla de Correo 5021, Montevideo
T: +598 29007194; Fax: +598 2921127; E-mail: info@claeh.org.uy; URL: www.claeh.org.uy
1958; Rita Grisolia
Soc sci, internat polit, Latin America, Uruguay
10 500 vols; 265 curr per; Deposit Library of IDB (Inter – American Development Bank)
libr loan   42049

Comisión Económica para América Latina y el Caribe (CEPAL), Economic Commission for Latin America and the Caribbean (ECLAC), Biblioteca, Juncal 1305, Piso 10, Casilla de Correo 1207, *Montevideo*
T: +598 2961576
1970; Graciela Schusselin
Social & econo development in Latin America
10 000 vols; 9 curr per
libr loan   42050

Consejo de Educación Secundaria, Biblioteca Central y Publicaciones, Eduardo Acevedo 1427, Piso 1°, *Montevideo*
T: +598 2483051; Fax: +598 2481252
1885; David Yudchak
Hist of French Revolution, lit
110 000 vols; 60 curr per; 200 av-mat; 50 sound-rec; 800 rare bks   42051

Cooperativa de Consumo CUTE-ANTEL, Biblioteca, Av Agraciada 2336, Piso 2, Casilla de Correo 6549, *Montevideo*
T: +598 29243472; Fax: +598 29243518
1983; Janet Fernández
Recreation, cooperatives, electricity, textbks
11 300 vols; 27 curr per   42052

Departamento de Documentación Nuclear, Biblioteca, Mercedes 1041, 11100 *Montevideo*
T: +598 2906919; Fax: +598 2921619
1955; Ana E. Rebellato Rodriguez
4 000 vols; 130 curr per; 3 000 microforms; 6 digital data carriers   42053

Departamento de Estudios Históricos, Comando General del Ejército, Ministerio de Defensa Nacional, Biblioteca, Garibaldi 2313, *Montevideo*
T: +598 2281543 ext 201; Fax: +598 2236600
1957; Manuel L. Saavedra
Military docs
25 000 vols; 30 curr per   42054

Dirección Nacional de Ciencia, Tecnología e Innovación (DINACYT), Unidad de Biblioteca y Documentación (National Direction of Science, Technology and Innovation), Paraguay 1470, piso 2, 11100 *Montevideo*; Casilla de Correos 1869 – Correo Central, 11100 Montevideo
T: +598 29014285 ext 124; Fax: +598 29024870; E-mail: biblioteca@dinacyt.gub.uy
1992; Susana Maggioli
Science and technology policies, Innovation policies, Metric Studies of Information; Coll of final rpts of consultancy activities related to science, technology and innovation programs
8 000 vols; 250 curr per; 600 diss/theses; 100 govt docs; 30 microforms; 30 av-mat; 100 sound-rec; 50 digital data carriers
libr loan   42055

Instituto Artigas del Servicio Exterior, Ministerio de Relaciones Exteriores, Biblioteca, Colonia 1206.Piso 1°, 11100 *Montevideo*
T: +598 2903742; Fax: +598 2921327; E-mail: iasel15@mrree.gub.uy; URL: www.mrree.gub.uy
Mercedes Rodríguez, Stella Nogueira
Recueil des cours: collected courses of The Hague Academy of International Law
18 000 vols; 200 curr per; atlases, folletos, recortes de prensa etc.   42056

Instituto de Cultura Uruguayo – Brasileño, Embajada de Brasil, Biblioteca, 18 de Julio 994, Piso 6, *Montevideo*
T: +598 29082622; Fax: +598 29082621
1941; Patricia Daniela Petroccelli
14 100 vols; 98 curr per; 420 music scores; 581 microforms
libr loan   42057

Instituto de Investigaciones Biológicas 'Clemente Estable', Biblioteca, Av Italia 3318, *Montevideo*
1927; Blanca Ferreyra
12 000 vols   42058

Instituto Interamericano del Niño, Biblioteca (Inter-American Children's Institute), Av 8 de Octubre 2904, *Montevideo*
T: +598 2472150; Fax: +598 2473242; E-mail: iin@chasque.apc.org
1927; Julio Rosenblatt
Education, pediatrics, child care and child development, law of minorities, family affairs
50 000 vols; 600 curr per   42059

Instituto Uruguayo de Normas Técnicas, Biblioteca, Pza Independencia 812, fiso 2 C.P. 11000, *Montevideo*
T: +598 29012048; Fax: +598 29021681; E-mail: unit@adinet.com.uy; URL: www.unit.org.uy
1939; Pablo Benia
350 000 vols; 50 curr per   42060

Istituto Italiano di Cultura, Biblioteca, C/ Paraguay 1177, *Montevideo*
1950; Dr. Renata Gerone
11 400 vols   42061

Museo Histórico Nacional, Biblioteca, Casa Rivera, C Rincón 437, 11000 *Montevideo*
T: +598 2951051
1940; Elisa Silva Cazet
American hist, hist of art; Libr an Uruguayan colls of Dr. Pablo Blanco Acevedo
150 000 vols; 314 curr per; 4 000 mss   42062

Museo Nacional de Historia Natural, Biblioteca, Buenos Aires 652, Casilla de Correo 399, 11000 *Montevideo*
T: +598 29160908; Fax: +598 29170215; E-mail: mnhn@internet.com.uy
1837; Alvaro Mones, museum director
250 000 vols; 6 300 curr per   42063

Museo y Archivo Histórico Municipal, Biblioteca, Palacio del Cabildo, c/ Juan Carlos Gómez 1362, *Montevideo*
T: +598 2982826
1915; Dr. Armando D. Pirotto
9 000 vols   42064

Oficina Regional de Ciencia y Tecnología de la UNESCO para América Latina y el Caribe, Biblioteca, Av Brasil 2697, Casilla 859, 11000 *Montevideo*
T: +598 2772023; Fax: +598 2772140
1949; Susana Gianelli de Blasco
16 000 vols; 450 curr per   42065

Sociedad Uruguaya de Pediatría, Biblioteca, Av Libertador Brigadier General Lavalleja 1464, piso 13°, Casilla 10906, *Montevideo*
10 000 vols   42066

Instituto de Formación Docente de Trinidad, Consejo Directivo Central, Administración Nacional de Educación Pública, Biblioteca Pedagógica 'Juan Pedro Lageard', Alfredo J. Puig 520, *Trinidad*, Flores
1970; María Amali Hiriart de Irazábal
10 424 vols; 12 curr per; 30 sound-rec   42067

Public Libraries

Centro Protección de Choferes de Montevideo, Biblioteca, Soriano 1227, Piso 3, *Montevideo*
1912; Raúl Diego Muiños
60 000 vols; 14 curr per   42068

Uzbekistan

National Libraries

Alisher Navoi State Public Library of Uzbekistan, Mustakillik maidoni, 5, 700078 *Toshkent*
T: +998 71 1391858, 1394341; Fax: +998 371 1330908; E-mail: navoi@phusic.uzsci.net; URL: www.ula.uzsci.net/
1870; Zuhriddin Isomiddinov
4 908 924 vols; 400 curr per; 148 mss; 10 incunabula; 60 100 music scores; 3 362 maps; 12 936 sound-rec; 100 digital data carriers
IFLA   42069

General Research Libraries

Republican Library of Kara-Kalpak, Doslyk guzari, 92, 742000 *Nukus*
T: +998 36122 22074
1937; Karshiga Zh. Nauryzbaeva
303 585 vols; 92 curr per; 29 diss/theses; 1 478 govt docs; 3 155 music scores; 10 sound-rec
libr loan   42070

Republican Library for Science and Technology, Buyuk Turonkuch 17, 700000 *Toshkent*
T: +998 371 1391167
1957; Munira A. Maksumova
2 000 000 vols   42071

Republican Scientific and Technical Library, Sharaf Rachidov ul 16, 700000 *Toshkent*
T: +998 3712 330922; Fax: +998 3712 1394040, 1391167; E-mail: stlib@physic.uzsci.net
1957; Munira A. Maksumova
2 000 000 vols; 400 curr per; 4 756 diss/theses; 163 324 microforms
libr loan   42072

Uzbek Academy of Sciences, Central Library, 2, Mavlyanova St, 700084 *Toshkent*
T: +998 71 1376055; Fax: +998 71 1376054; E-mail: info@uzsci.net; URL: www.uzsci.net/
1934; Nabi G. Umarov
1 500 000 vols; 15 394 curr per; 92 400 music scores; 4 380 maps; 300 microforms   42073

University Libraries, College Libraries

Andizhan Institute of Medicine, Library, prosp. Navoi 136, 710000 *Andizhan*
T: +998 37422 23682
1955
105 000 vols; 281 curr per   42074

Pedagogical Institute of Languages, Library, pr. Zhdanova, 5, 710011 *Andizhan*
T: +998 37422 5694
1966
46 000 vols; 185 curr per   42075

Pedagogical Institute, Library, ul. Petrovskogo, 24, *Bukhara* 1
1930
515 000 vols   42076

Fergana Polytechnic Institute, Library, Ferganskaya st 86, block 8, 712022 *Fergana*
T: +998 732 221260; Fax: +998 732 222781; E-mail: monitoring@vodiy.uz; URL: ferpi.dem.ru/eng/biblio.htm
1968; Muhtaraly Turdalievich Turdaliev
214 000 vols   42077

Fergana State University, Ulugbek Library, *Fergana*
T: +998 3732 242623; Fax: +998 3732 243532
1930; Boborokhimov
Golden Coll (500 books in arabic and cyrillic); Music dept, Mathematic dept, Foreign lit dept, Russian lit depts
734 308 vols; 114 curr per; 1 902 diss/theses; 1 200 govt docs; 2 240 music scores; 13 800 pamphlets   42078

Agricultural Institute, Library, *Institutskoe*, Tashkentskaya obl.
1932
538 000 vols; 1 000 mss; 300 slides   42079

Pedagogical Institute, Library, Uychinskaya, 333, *Namangan*
T: +998 3662 5251
1946
127 000 vols   42080

Pedagogical Institute, Library, ul. Kalinina, 106, *Nukus*
1934
187 000 vols; 245 curr per; 200 maps; 200 sheet music   42081

Pedagogical Institute, Library, Krasnoarmeyskoe shosse, 166, *Samarkand* 46
T: +998 3662 40848
1967
129 000 vols   42082

Samarkand Agricultural Institute, Library, Khamza 77, 703003 *Samarkand*
T: +998 3662 43320; Fax: +998 3662 310786; E-mail: rector@agri.samuni.silk.org
1929
533 800 vols; 3 500 curr per   42083

Samarkand I.P. Pavlov State Medical Institute, Library, ul. Amir Temur 18, 703000 *Samarkand*
T: +998 3662 330766
1930
340 000 vols; 410 curr per   42084

Samarkand State Architectural and Civil Engineering Institute, Library, Lyalyazar 70, *Samarkand* 47
T: +998 3662 3715983; Fax: +998 3662 310452; E-mail: achilov@uni.uzsci.net
1966
400 000 vols   42085

Samarkand State Institute of Foreign Languages, Library, Akhunbabaev 93, 703004 *Samarkand*
T: +998 3662 2337843; Fax: +998 3662 2356619; E-mail: rector@bisr.silk.org
9 000 vols   42086

Samarkand State University, Central Library, bul. University, 15, 703004 *Samarkand*
T: +998 3662 32406
1933; G. Valieva
1 632 000 vols; 787 curr per; 9 000 mss   42087

Pedagogical Institute, Library, Frunze, 43, *Termez*
1965
76 000 vols; 136 curr per   42088

Electrotechnical University of Communications, Library, Amir Temur 108, 700084 *Toshkent*
T: +998 3712 350934; Fax: +998 3712 357771
1955
400 000 vols   42089

First Toshkent State Medical Institute, Library, Khamza 103, 700048 **Toshkent**
T: +998 3712 676307
1920
600 000 vols
42090

Institute for Continuing Education in Medicine, Library, ul. im. generala Petrova, 51, 700007 **Toshkent**
T: +998 3712 677287; Fax: +998 3712 681744
1938; I.N. Bogoyavlenskaya
Hist of medicine
116 000 vols; 55 curr per; 404 diss/theses
42091

Institute of Mathematics, Library of V.I. Romanovsky, ul. F. Khodzhaeva, 29, 700143 **Toshkent**
T: +998 371 1627544; E-mail: root@im.tashkent.su
1947; Lola Sharipova
bks and letters signed by famous mathematicians, ancient textbks of mathematics, geophysics, mechanics
17 200 vols; 11 curr per; 10 mss; 349 diss/theses; 20 maps; 50 microforms
42092

Institute of Postgraduate Teacher Training, Library, Chimbay, 3, **Toshkent** 95
1947
58 000 vols; 104 curr per
42093

Institute of Russian Language and Literature, Library, Literaturnaya, 4, **Toshkent** 110
1964
160 000 vols; 121 curr per
42094

Institute of Theatrical Art and Art Institute, Library, ul. Germana Lopatina, 77, 700031 **Toshkent**
1945
79 000 vols; 110 curr per; 3 700 music scores
42095

Pedagogical Institute, Library, Pedagogicheskaya, 103, **Toshkent** 64
T: +998 3712 550500
1935
808 000 vols; 500 curr per
42096

Pedagogical Institute of Foreign Languages, Library, ul. Mukimi, 104, **Toshkent**
T: +998 3712 772541
1948
180 000 vols; 138 curr per
42097

Polytechnic Institute, Library, ul. Navoi, 13, **Toshkent**
T: +998 3712 391160
1929
756 000 vols; 1 000 curr per
42098

Toshkent Electrotechnical University of Communications, Library, ul. Amir Temor 108, 70084 **Toshkent** 87
T: +998 3712 350934; Fax: +998 3712 351040; E-mail: teic@uzpak.uz
1955
500 000 vols
42099

Toshkent Institute of Agricultural Engineering and Irrigation, Library, ul. Kary Niyazova 39, 700000 **Toshkent**
T: +998 3712 334685; Fax: +998 3712 442603
1934
864 000 vols; 437 curr per
42100

Toshkent Institute of Railway Engineers, Library, Oboronnaya 1, 700045 **Toshkent**
T: +998 3712 911440
1931
500 000 vols; 416 curr per
42101

Toshkent Institute of Textile and Light Industry, Library, Gorbunova 5, 700100 **Toshkent**
T: +998 3712 530606; Fax: +998 3712 533617
1932
644 000 vols
42102

Toshkent Pharmaceutical Institute, Library, Kafanova 35, 700015 **Toshkent**
T: +998 3712 563839
1940
100 000 vols; 223 curr per
42103

Toshkent State Agrarian University, Library, Universitetskaya 5, 700183 **Toshkent**, Kibray
T: +998 3712 637600; E-mail: gulumov@atabah.silk.org
196 000 vols
42104

Toshkent State Conservatoire, Library, Pushkinskaya 31, 700000 **Toshkent**
T: +998 371 1335274; Fax: +998 371 1331035
1936
243 000 vols
42105

Toshkent State Economics University, Library, Almazar 183, 700063 **Toshkent**
T: +998 3712 452626
1931; Shachida Ganieva
300 000 vols
42106

Toshkent V.I. Lenin State University, Central Library, Vuzgorodok, 700095 **Toshkent**
T: +998 3712 29145
1918; L.S. Jugai
2 460 000 vols; 13 500 curr per; 900 mss; 200 microforms
42107

Uzbek Institute of Physical Culture, Library, Vystavochnaya, 28, **Toshkent** 95
1955
87 000 vols
42108

Government Libraries

Ministry of Meat and Dairy Products, Library, Timiryazeva, 23, **Toshkent**
1959
19 000 vols
42109

Ministry of Power and Electrification, Library, Chorezmskaya, 6, **Toshkent**
T: +998 3712 391350
1950
53 000 vols; 26 curr per
42110

State Committee for Broadcasting and Television, Library, Chorezmskaya, 49, **Toshkent**
T: +998 3712 3995505
1930
60 000 vols; 25 000 music scores
42111

Supreme Council of the Republic of Uzbekistan, Library, Oliy Kengashining Kotibiyati, Mustalik Madoni, 5, 7000008 **Toshkent**
T: +998 3712 398749; Fax: +998 3712 394151
1940; Irina Internovna Tyuryabayeva
24 000 vols; 100 curr per
42112

Special Libraries Maintained by Other Institutions

Center for Education in Sanitation, Library, Stroitelnaya, 1, **Bukhara** 25
T: +998 36522 43082
1925
19 000 vols
42113

Scientific Research Institute of Crop Husbandry, Library, **Galljaaral**, Samarkand obl.
1913
35 000 vols; 155 curr per; 200 mss
42114

Institute of Animal Husbandry, Library, **Krasny Vodopad**, Tashkentskaya obl.
1940
52 500 vols; 457 curr per; 2 000 mss
42115

Cotton Institute, Library, **Kuigan-Yar**, Andizhan obl.
1964
31 000 vols; 37 curr per
42116

Institute for Research in the Silk Industry, Library, K. Marksa, 202, **Margilan**, Ferganskaya obl.
1937
10 000 vols; 92 curr per
42117

Karakalpak Art Museum, Library, pr. Doslyka 127, 742000 **Nukus**, Karakalpakstan
T: +998 36122 222256; Fax: +998 36122 222256; E-mail: rashkhal@miras.nukus.silk.org
1966; Nataliya Mescherinova
8 745 vols
42118

Institute for Research in Gardening, Viticulture and Oenology, Library, **Samarkand**
T: +998 3662 31091
1932
11 000 vols; 36 curr per
42119

L. M. Isaev Research Institute of Medical Parasitology, Library, Isaeva 38, 703005 **Samarkand**
T: +998 3662 374242
1924
Epidemiology, prophylactic medicine
45 000 vols; 212 curr per
42120

Museum of History, Culture and Arts of Uzbekistan, Library, Sovetskaya, 51, **Samarkand**
T: +998 3662 32872
1911
35 000 vols; 315 curr per
42121

Research Institute of Karakul Sheep Breeding, Library, ul. Mirzo Ulugbek 47, **Samarkand**
T: +998 3662 333279; Fax: +998 3662 333481; E-mail: mukimov@samuni.silk.org
1930
56 000 vols; 1 470 mss
42122

Samarkand Co-operative Institute, Library, Kommunisticheskaya,41, 703000 **Samarkand**
T: +998 3662 31285
1924
206 000 vols
42123

Museum of Regional Studies, Library, Festivalnaya pl, 1, **Termez**
1933
20 000 vols; 151 curr per
42124

Kh. M. Abdullaev Institute of Geology and Geophysics, Uzbek Academy of Sciences, Library, Akad. Ch. Suleiman St, 33, 700017 **Toshkent**
T: +998 3712 331280
1945
26 000 vols; 101 curr per
42125

Agricultural Planning Institute, Library, Volgogradskaya, **Toshkent**
1952
30 000 vols; 92 curr per
42126

Astronomical Institute, Uzbek Academy of Sciences, Library, Astronomicheskaya, 33, 700059 **Toshkent**
T: +998 3712 350638
1873
44 000 vols; 152 curr per; 500 mss
42127

Botanical Garden, Uzbek Academy of Sciences, Library, D. Abidovoy, 232, 700053 **Toshkent**
T: +998 3712 350801
1958; R.S. Sharafutdinova
19 800 vols; 30 curr per; 880 mss; 40 diss/theses
libr loan
42128

Bureau for the Design of Cotton Growing Equipment, Library, 4-i Tyulpanovsky pr., 5, **Toshkent**
T: +998 3712 662916
1954
338 000 vols; 176 curr per
42129

Central Asian Research Hydrometeorological Institute (SANIGMI), Library, 72, K. Makhsumov str, 700052 **Toshkent**
T: +998 3712 358611; Fax: +998 3712 1332025; E-mail: sanigmi@meteo.uz
1943; Mitina S.A.
Hydrometeorology, hydrology, glaciology, ecology, weather modification, agrometeorology, environmental sciences
30 000 vols; 72 curr per; 98 diss/theses; 280 govt docs; 100 maps; 1 020 microforms
42130

Central State Archives of Uzbekistan, Library, Chilanzar, 2, **Toshkent** 43
1919
17 000 vols; 57 curr per
42131

Historical Museum of the People of Uzbekistan, Library, Kuibysheva, 15, 700047 **Toshkent**
T: +998 3712 335733
1876
15 000 vols
42132

Hydrotechnology Planning and Research Institute, Library, B. Khmelnitskogo, 20, **Toshkent** 64
1939
54 000 vols; 63 curr per
42133

Institut for Research in Geology and Mineral Resources, Library, Shevchenko, 15, **Toshkent**
T: +998 3712 3311549
1957
23 000 vols; 16 curr per
42134

Institute for Dermatology and Venereology, Library, Farabi 3, 700109 **Toshkent**
T: +998 3712 460807
1932
Physiology, biochemistry
14 000 vols; 161 curr per
42135

Institute for Planning in the Food Industry, Library, ul. Navoi, 9, **Toshkent**
T: +998 3712 442133
1951
45 000 vols; 26 curr per
42136

Institute for Planning in the Processing of Cotton and Bast, Library, ul. Shahrisabskaya, 85, 700047 **Toshkent**
T: +998 71 1331604; Fax: +998 71 1336846
1937; Tatjana Turina
68 000 vols; 90 curr per; 851 govt docs
42137

Institute for Research and Planning in Nonferrous Metals, Library, Sh. Rustaveli, 45, **Toshkent** 100
T: +998 3712 551206
1951
75 000 vols
42138

Institute for Research and Planning of Housing and Public Buildings, Library, ul. A. Khidoyatova, **Toshkent**
T: +998 3712 440272
1959
20 000 vols; 152 curr per
42139

Institute for Research in Balneology and Physical Therapy, Library, ul. Khurshida, 13, **Toshkent** 84
T: +998 3712 350013
1919
24 000 vols; 39 curr per
42140

Institute for Research in Building Materials, Library, Pushkina, 88, **Toshkent**
T: +998 3712 336718
1935
77 000 vols
42141

Institute for Research in Hygiene, Sanitation and Occupational Illness, Library, Tsiolkovskogo, 325, **Toshkent**
1946
23 000 vols; 63 curr per
42142

Institute for Research in Natural Gas, Library, ul. Mukimi, 98, **Toshkent** 115
T: +998 3712 771112
1960
17 000 vols; 129 curr per
42143

Institute for Research in Plant Protection, Library, 3-i tupik Nasyrova, 20, **Toshkent**
T: +998 3712 352218
1958
15 000 vols; 65 curr per; 600 mss
42144

Institute for Research in Roentgenology, Radiology and Oncology, Library, Chigatay, 383, **Toshkent** 169
T: +998 3712 42765
1960
13 000 vols; 78 curr per
42145

Institute for Research in the Cotton Industry, Library, Sh. Rustaveli, 8, **Toshkent**
T: +998 3712 550800
1932
52 000 vols; 130 curr per
42146

Institute for Research in Vaccines and Sera, Library, Timiryazeva 37, 700084 **Toshkent**
T: +998 3712 437953
1918
39 000 vols; 126 curr per
42147

Institute for the Design of Power Networks, Library, Sh. Rustaveli, 41, **Toshkent** 70
T: +998 3712 559064
1962
45 000 vols; 105 curr per   42148

Institute of Civil Engineering, Library, ul. Navoi, 18, **Toshkent**
1951
10 000 vols   42149

Institute of Fat Technology for the Food Industry, Library, ul. Navoi, 9, **Toshkent**
T: +998 3712 442133
1952
43 000 vols; 17 curr per   42150

Institute of Fine Arts, Library, ul. Abdully Tukaeva, 1, **Toshkent** 29
T: +998 3712 391191
1943
85 000 vols; 32 curr per; 850 microforms   42151

Institute of Geology and Petroleum Exploration, Library, U. Nasyr str 114, **Toshkent** F-59
T: +998 3712 509239; Fax: +998 3712 509213
1960; Batrova Nailya Karimovna
18 689 vols; 18 689 mss; 12 427 periodicals   42152

Institute of Hydrology and Engineering Geology, Library, Hodzhibaeva 64, 700041 **Toshkent**
T: +998 3712 626215; Fax: +998 3712 624553
1960
18 000 vols; 51 curr per   42153

Institute of Industrial Planning – Heavy Industry, Library, Pushkin St, 88, **Toshkent**
T: +998 3712 336718
1941
85 000 vols; 200 curr per   42154

Institute of Irrigation and Land Improvement, Library, ul. Navoi, 44, **Toshkent**
1929
59 000 vols   42155

Institute of Power Manufacture, Library, Sh. Rustaveli, 41, **Toshkent** 70
T: +998 3712 559885
1934
45 000 vols; 62 curr per   42156

Institute of Road Engineering, Library, ul. 40 let Komsomola, 22, **Toshkent** 170
1959
11 000 vols; 70 curr per   42157

Institute of Seismology, Uzbek Academy of Sciences, Khurshida 3, 700128 **Toshkent**
T: +998 3712 415170; Fax: +998 3712 415314
1966
43 000 vols   42158

Republican Library for Science and Technology, Buyuk Turonkuch 17, 700000 **Toshkent**
T: +998 371 1391167
1957; Munira A. Maksumova
2 000 000 vols   42159

Scientific-Technical Library on Irrigation, Galaba, 1, **Toshkent**
T: +998 3712 481051; Fax: +998 3712 481003
1958; Ludmila Fatykhova
Agricultural machines construction, Irrigational Equipment
71 920 vols; 324 curr per; 180 govt docs; 20 maps; 5 000 microforms
libr loan   42160

State Scientific Medical Library (SSML), Kuibysheva, 22, **Toshkent**
T: +998 71 133-3822/-1183/-3554;
Fax: +998 71 1338367; E-mail: lrctas@bcc.com.uz; URL: medlib.freenet.uz
1935
WHO library-depository in Uzbekistan; 12 branch libs
1 000 000 vols
libr loan   42161

Transport Planning Research Institute, Library, ul. Poltoratskogo, 23a, **Toshkent**
1953
15 000 vols   42162

Tuberculosis Institute, Library, ul. Medzhlisi, 1, **Toshkent**
T: +998 3712 772556
1938
Microbiology
14 000 vols; 78 curr per   42163

Uzbek Institute of Rehabilitation and Physiotherapy, Semashko Institute, Library, Khurshida 4, **Toshkent**
T: +998 3712 345500
1919
16 210 vols   42164

Uzbek Orthopaedics and Traumatology Research Institute, Library, Khamza 78, **Toshkent**
T: +998 3712 331030
1946
Physiology
28 000 vols   42165

Uzbek State Museum of Art, Library, Proletarskaya, 16, 700060 **Toshkent**
T: +998 3712 323444
1918
23 000 vols; 147 curr per; 500 mss   42166

Institute of Atomic Energy, Library, **Ulugbek**, Tashkentskaya obl.
1957
86 000 vols; 944 curr per   42167

Institute for Research in Rural Electrification and Mechanization, Library, **Yangiyul**, Tashkentskaya obl.
1927
54 000 vols; 298 curr per; 2 150 mss   42168

Agricultural Institute, Library, **Zhanbas-Zhan**, Chimbay
1958
Crop husbandry, agricultural economics
11 000 vols; 210 curr per   42169

Public Libraries

Z. M. Babura District Library, Fitrat, 232, 710002 **Andizhan** P/O 2
T: +998 74 255010
1906; A.B. Kabulov
Dept of foreign language lit, periodicals, tech and agricultural lit, art
538 100 vols; 113 curr per; 3 870 sheet music, 1 850 records   42170

District Library Ibn-Sin, Bibliotechnaya, 4, **Bukhara**
T: +998 36522 42223
1921
204 500 vols; 250 curr per   42171

District Library, ul. Lenina, 29, **Fergana**
1899
191 500 vols   42172

Gorky District Library, ul. Lenina, 297, **Karshi**
1943
145 000 vols; 150 curr per   42173

Regional Library, Krasnoarmeyskaya, 10, **Namangan**
T: +998 3662 4832
1918
143 000 vols   42174

A. S. Pushkin District Library, Engels St, 62, **Samarkand**
T: +998 3662 30574
1911
226 000 vols   42175

District Library, Pervomaisky per., 3, **Termez**
1936
209 000 vols; 360 curr per   42176

Toschkent District Research Library 'Turon', ul. Maksi-Duzlik, 25, **Toshkent** 2
T: +998 3712 404997; E-mail: Turon@physic.uz.sci.net
1918; Vasira Sabirov
376 200 vols; 238 curr per
libr loan   42177

Regional Library, Kirova, 16, **Urgench**
1939
198 000 vols; 474 curr per   42178

Vanuatu

National Libraries

National Library of Vanuatu, P.O. Box 184, **Port-Vila**
T: +678 22129; Fax: +678 26590; E-mail: nasonal.laebri@vanuatuculture.org; URL: www.vanuatuculture.org
1962; Anne Naupa
Vanuata (formerly New Hebrides), Pacific Islands
8 000 vols; 30 curr per; 100 diss/theses; clippings coll
Vanuatu Library Association   42179

University Libraries, College Libraries

University of the South Pacific – Emalus Campus Library, PMB 072, **Port Vila**
T: +678 22748; Fax: +678 22633; E-mail: Murgatroyd_p@Vanuatu.usp.ac.fj; URL: www.vanuatu.usp.ac.fj/library
Peter Murgatroyd
libr loan; VLA   42180

Government Libraries

Palemen, Library (Parliament), Parliament House, PMB 052, **Port-Vila**
T: +678 22229 ext 267; Fax: +678 24530
1985
500 vols   42181

Vatican City

General Research Libraries

Biblioteca Apostolica Vaticana, Cortile del Belvedere, 00120 **Città del Vaticano**
T: +39 0669879402; Fax: +39 0669884795; E-mail: bav@vatlib.it; URL: www.vaticanlibrary.vatlib.it
1451; Cardinal Archivist and Librarian Prof. Don Raffaele Farina
Depts: Mss, Printed Books, Numismatics, Drawings and Engravings
1 600 000 vols; 75 000 curr per; 72 000 mss; 8 300 incunabula; 20 000 maps; 75 000 archival vols, 100 000 engravings, 330 000 coins and medals
IFLA, LIBER, AIB   42182

University Libraries, College Libraries

Pontificia Facoltà Teologica di San Bonaventura, Biblioteca, Via del Serafico 1, 00142 **Città del Vaticano**
T: +39 0651503206; Fax: +39 065192067
Bonaventura Danza, OFM
160 000 vols; 452 curr per; 954 mss; 970 rare bks   42183

Pontificia Facoltà Teologica Marianum, Biblioteca, Viale Trenta Aprile 6, 00153 **Città del Vaticano**
T: +39 0665814441; Fax: +39 0665880292
1930; Silvano Danieli
Acta Beatificationis et Canonizationis S. Congregationis Rituum; Hist of Servite Order, Mariology
95 000 vols; 450 curr per; 1 320 mss; 23 incunabula; 1 350 diss/theses; 2 300 microforms; 50 av-mat; 100 sound-rec; 1 500 engravings, 90 brass plates
libr loan; ABEI   42184

Pontificia Facoltà Teologica Teresianum, Biblioteca, Piazza San Pancrazio 5A, 00152 **Città del Vaticano**
T: +39 06585401; Fax: +39 065809050
1956; Eugenio Duque
280 000 vols; 1 650 curr per; 125 mss; 12 incunabula; 10 000 diss/theses; 135 microfilms, 140 rare bks   42185

Pontificia Universitá Lateranese, Biblioteca Beato Pio IX, Piazza San Giovanni in Laterano 4, 00100 **Città del Vaticano**
T: +39 0669886107; Fax: +39 0669886107; E-mail: info@pul.it; URL: www.pul.it
1937
Fondo del Pontificio Istituto Pastorale
550 000 vols; 710 curr per; 392 mss; 40 incunabula   42186

– Biblioteca Claretianum, Istituto di Teologia della vita Consacrata, Largo L. Mossa 4, 00165 Città del Vaticano
T: +39 0666102504; Fax: +39 0666102503; E-mail: istvitcons@libero.it; URL: www.claretianum.org
1959; P. Josep Rovira
58 000 vols; 342 curr per; 5 incunabula   42187

– Istituto Patristico 'Augustinianum', Biblioteca, Via Paolo VI 25, 00193 Città del Vaticano
T: +39 06680069; Fax: +39 0668006298
José M. Guirau
43 000 vols; 230 curr per; 5 incunabula   42188

Pontificia Università San Tommaso d'Aquino, Biblioteca (St. Thomas Aquinas Pontifical University), Largo Angelicum 1, 00184 **Città del Vaticano**
T: +39 066702348; Fax: +39 06790407, 066702270; E-mail: biblio@pust.urbe.it; URL: www.angelicum.org
1580; Miguel Itza
Religion, theology
220 000 vols; 490 curr per; 8 mss; 18 incunabula
URBE, ABEI   42189

Pontificia Università Urbaniana, Biblioteca (Pontifical Urbanian University Library), Via Urbano, VIII, 16, 00186 **Città del Vaticano**
T: +39 0669889676; Fax: +39 0669889663; E-mail: biblioteca@urbaniana.edu; URL: www.urbaniana.edu/biblio/it/index.htm
1627; P. Roberto Streit
Coll in 530 non-European languages; Pontifical Missionary Libr
350 000 vols; 832 curr per; 1 150 mss; 21 incunabula; 5 604 diss/theses; 300 maps; 51 000 microforms; 230 digital data carriers
AIB   42190

– Pontificia Biblioteca Missionaria della Sacra Congregazione per L'Evangelizzazione dei Popoli, Via Urbano VIII,16, 00186 Città del Vaticano
T: +39 0668308361
1925; Willi Henkel
Coll in 530 non-European languages
300 000 vols; 850 curr per; 1 150 mss; 18 incunabula; 5 104 diss/theses; 263 maps; 1 000 microforms   42191

Pontificio Istituto di Archeologia Cristiana, Biblioteca, Via Napoleone III, 1, 00185 **Città del Vaticano**
T: +39 064465574; Fax: +39 064469197; E-mail: piac.biblio@piac.it; URL: www.piac.it/biblioteca/biblioteca.htm
1924; Giorgio Nestori
60 000 vols; 150 curr per   42192

Università Pontificia Salesiana, Biblioteca, Piazza dell' Ateneo Salesiano, 1, 00139 **Roma**
T: +39 0687290402; Fax: +39 0687290318; E-mail: rettore@ups.urbe.it
1904; Don Juan Picca
500 000 vols; 4 500 curr per; 500 mss; 11 incunabula; 12 000 diss/theses; 1 000 govt docs; 2 000 music scores   42193

Ecclesiastical Libraries

Basilica di S. Paolo, Biblioteca, Via Ostiense 186, 00146 **Città del Vaticano**
T: +39 065410341
P. Cesario D'Amato, OSB
100 000 vols; 50 curr per; 400 mss; 45 incunabula   42194

Collegio Americano del Nord, Biblioteca, Via Gianicolo 14, 00120 **Città del Vaticano**
T: +39 066794435
30 000 vols; 250 curr per   42195

Pontificio Collegio Armeno, Biblioteca, Via S.N. da Tolentino 17, 00187 *Città del Vaticano*
T: +39 064824883
1883; Don Giorgio Zabarian
30 000 vols; 50 mss   42196

Pontificio Collegio Belga, Biblioteca, Via Giambattista Pagano 35, 00167 *Città del Vaticano*
T: +39 066230476
1844; Werner Quintens
20 000 vols; 20 curr per   42197

Pontificio Collegio Croato di S. Girolamo, Biblioteca, Piazza Augusto Imperatore, 4, 00186 *Città del Vaticano*
T: +39 066878284
1901; Dr. Ratko Perić
28 000 vols; 50 curr per   42198

Pontificio Collegio Greco di S. Atanasio, Biblioteca, Via del Babuino 149, 00187 *Città del Vaticano*
T: +39 066795355; Fax: +39 066796120
1557; P. Oliviero Raquez, OSB
40 000 vols; 45 curr per; 320 mss; 10 incunabula; Numerous rare bks   42199

Pontificio Collegio Pio Latino Americano, Biblioteca, Via Aurelia Antica 408, 00165 *Città del Vaticano*
T: +39 0666693160; E-mail: bibliopiolatino@libero.it
1858; Marta Pavón Ramírez
CELAM
60 000 vols; 110 curr per; 5 incunabula; 300 diss/theses   42200

Pontificio Seminario Francese, Biblioteca, Via S. Chiara 42, 00186 *Città del Vaticano*
T: +39 0668801526
1856; Marcel Martin
30 000 vols; 120 curr per   42201

Pontificium Collegium Germanicum et Hungaricum, Biblioteca, Via S. Nicola da Tolentino 13, 00187 *Roma*
T: +39 0642119-403; Fax: +39 0642119-125; E-mail: biblioteca@cgu.it; URL: www.cgu.it
1818; Br. Markus Pillat SJ
Bibliotheca Landgrafiana (Theologie der Frühscholastik)
95 000 vols; 85 curr per; 150 rare bks
AKThB   42202

Venezuela

National Libraries

Biblioteca Nacional de Venezuela, Servicios de Bibliotecas (National Library of Venezuela), Parroquia Altagracia, Final Av Pantéon, Esquina Fe a Remedios, *Caracas* 1010
T: +58 212 505-9125; Fax: +58 212 505-9124; URL: www.bnv.gob.ve/
1883; Fernando Báez
7 131 700 vols; 2 500 000 records on database
libr loan   42203

University Libraries, College Libraries

Instituto Universitario Politécnico, Biblioteca, Av Copra Guaico, Apdo 539, *Barquisimeto*, Lara
1962
13 000 vols   42204

Universidad Nacional Experimental Politécnica 'Antonio José de Sucre', Biblioteca, Apdo Postal 539, *Barquisimeto*, Edo Lara
1962
19 000 vols; 9 000 periodicals   42205

Universidad Católica 'Andrés Bello', Biblioteca Central, Montalbán, La Vega, Apdo 29068, *Caracas*
1956; Emilio Píriz Pérez
111 600 vols   42206

Universidad Central de Venezuela, Biblioteca Central, Ciudad Universitaria, Los Chaguaramos, *Caracas* 1040
T: +58 212 6054175; Fax: +58 212 6054212; E-mail: bibcentral@sicht.ucv.ve; URL: www.sicht.ucv.ve:8080/bc/
1850; Neisa Guevara
280 000 vols; 3 500 curr per   42207

– Centro de Estudios del Desarrollo (CENDES), Biblioteca (Development Studies Centre), Apdo 47604, Caracas 1041A
T: +58 212 7523475; Fax: +58 212 7512691; E-mail: cendes@reaccium.ve
1961; Georgina Licea
35 000 vols   42208

– Facultad de Agronomía, Biblioteca, Apdo 4579, Maracay
1951; Carmela Di Stefano
Botany and Zoology colls
150 000 vols; 600 curr per; 1 051 diss/theses; 24 maps; 980 sound-rec
libr loan; AGRINTER, AIBDA   42209

– Facultad de Arquitectura y Urbanismo, Biblioteca, Caracas
T: +58 212 6624616
1955; Lilian Martinez
Incunabula section
11 000 vols; 400 curr per; 1 000 diss/theses   42210

– Facultad de Ciencias Económicas y Sociales, Bibliotecas 'Salvador de la Plaza', Antigua Residencia Vargas Ciudad Universitaria, Caracas
1958
30 500 vols   42211

– Facultad de Ciencias Jurídicas y Políticas, Biblioteca, Chacao, Edo Miranda, Apdo 61591, Caracas
1948; Elke N. de Stockhausen
51 600 vols; 560 curr per; 2 000 incunabula; 709 diss/theses
libr loan   42212

– Facultad de Economía, Biblioteca, Caracas 105
1946
16 000 vols   42213

– Facultad de Farmacia, Biblioteca, Ciudad Universitaria, Apdo 40109, Caracas 1040A
1941; Zoraida Zamora de Baena
10 000 vols; 386 curr per; 263 diss/theses
libr loan   42214

– Facultad de Ingeniería, Centro de Información y Documentación Técnica, Biblioteca, Apdo 50656, Caracas 1051-A
E-mail: bibliofi@dino.conicit.ve
1953; Connie J. Zurita
Industrial standards, electronics devices and vendor catalogs
22 200 vols
libr loan; ACURIL   42215

– Instituto de Medicina Experimental, Biblioteca 'Humberto García Arocha', Ciudad Universitaria Sabana Grande, Apdo 50587, Caracas 1051
T: +58 212 6931862; Fax: +58 212 6931260; E-mail: aacosta@conicit.ve
1940; Trina Yanes de Ramírez
History of Medicine
30 200 vols; 4 100 curr per; 700 diss/theses; 78 microforms; 20 digital data carriers
libr loan   42216

Universidad Simón Bolívar, Biblioteca Central, Apdo 89000, *Caracas* 1080
T: +58 212 9063125; Fax: +58 212 9063131; E-mail: bib@usb.ve; URL: www.bib.usb.ve
1970; Dra Rosario Gassol de Horowitz
Mathematics, chemistry, physics, computer science and electronics
130 000 vols; 1 200 curr per; 500 mss; 30 incunabula; 6 056 diss/theses; 2 376 govt docs; 246 maps; 2 790 microforms; 2 800 av-mat; 50 sound-rec
ACURIL, IFLA   42217

Universidad de Oriente, Biblioteca Central, Av Gran Mariscal, Edif.del Rectorado, Apdo de Correos 094, *Cumaná* 1010-A
T: +58 93 301158; Fax: +58 93 301358; E-mail: bashiru@re.udo.edu.ve; URL: www.udo.edu.ve
1960; Rosa González de López
154 600 vols   42218

Universidad de los Trabajadores de América Latina, Centro de Información y Documentación, Vía el Limón Edif. UTAL, San Antonio de los Altos, *Los Teques*
28 000 vols   42219

Universidad del Zulia, Biblioteca Central, Edif. del Rectorado, Av 16 (Guajira) con Calle 67B, *Maracaibo*, Estado Zulia
T: +58 261 7598441; Fax: +58 261 7598515; URL: www.luz.edu.ve
1946; Egla Ortega
19 000 vols   42220

– Facultad de Agronomía, Biblioteca Ing. Agr. Hugo Gonzalez Rincón, Apdo 526, Maracaibo, Zulia
E-mail: serbiluz@dino.conicit.ve
1959
3 depts
20 100 vols
ANABISAI   42221

– Facultad de Arquitectura, Biblioteca 'Rafael Puig Gomez', Av, 16 Goajira (antes Ziruma), Maracaibo, Zulia
1962; Daysi R. Aguado Bracho
14 220 vols
libr loan   42222

– Facultad de Ciencias Económicas y Sociales, Biblioteca-Hemeroteca 'Salvador de la Plaza', Ciudad Universitaria, Apdo 526, Maracaibo, Zulia
1958
26 255 vols
libr loan   42223

– Facultad de Derecho, Biblioteca 'Dr. Jesus Enrique Losada', Ciudad Universitaria Ziruma, Maracaibo, Zulia
1949
Law, Social Work
25 000 vols
libr loan   42224

– Facultad de Humanidades y Educación, Biblioteca 'Dr. Raul Osorio Lazo', Ctra de Ziruma, Apdo 526, Maracaibo, Zulia
1964
37 000 vols   42225

– Facultad de Ingeniería, Sub Sistema de Biblioteca, Av 16 Ziruma, Apdo 4011-A526, Maracaibo, Zulia
1973
21 600 vols   42226

Universidad de Los Andes, SERBIULA, Edif. Administrativo, 5° piso, *Mérida* 5101; mail address: Nucleo Pedro Rincón Gutierrez, Edif. C, PB, La Hechicera, Merida 5101
T: +58 274 2401227; Fax: +58 274 2401224; E-mail: adquis@ula.ve; URL: www.ula.ve
1889; María Elena Colls
Coll of 3 000 16th-17th c books
500 000 vols; 9 637 curr per; 22 000 diss/theses; 5 200 govt docs; 604 maps; 927 microforms; 3 360 av-mat; 94 sound-rec   42227

– Instituto de Geografía y Conservación de Recursos Naturales, Biblioteca, V. Chorros de Milla, Mérida 5101
T: +58 74 401645; Fax: +58 74 401503
1964; Maria Elena Márquez
19 900 vols; 575 curr per; 245 diss/theses   42228

– Servicios Bibliotecarios de Ciencias Jurídicas y Políticas, Edif. del Rectorado, Escuela del Derecho, Mérida
T: +58 74 520011
1946; Florencia M. de Krupij
8 500 vols   42229

– Servicios Bibliotecarios de Farmacia, Campo de Oro, Facultad de Farmacia, Mérida
T: +58 74 403454; Fax: +58 74 403454; E-mail: cfuentes@ing.ula.ve
1942; Crisálida Fuentes Guzmán
13 600 vols; 240 curr per; 96 diss/theses
libr loan   42230

– Servicios Bibliotecarios de Humanidades y Educación, Av Universidad, Facultad de Humanidades y Educación, Mérida
1958; Cecilia P. de Chersi
45 000 vols; 300 curr per
libr loan; ACURIL   42231

Universidad de Carabobo, Fundación Centro de Información y Documentación, Biblioteca Central, Av Andrés Eloy Blanco c/c 137, Urb. Prebo., *Valencia* 2001, Edo Carobobo
T: +58 241 8222613, 8222608, 8240871; Fax: +58 2418 212121; E-mail: fundacid@uc.edu.ve; URL: www.cid.uc.edu.ve
1987; Ing Eva Monagas
350 000 vols; 1 200 curr per; 15 e-journals; 7 200 diss/theses; 3 microforms
libr loan; ISTEC   42232

Government Libraries

Congreso, Biblioteca, Plaza del Capitolio, *Caracas* 1010
T: +58 212 4832344; Fax: +58 212 4832904
1915; Lourdes García
56 000 vols
libr loan   42233

Dirección General de Desarrollo Urbanistico, Ministerio de Desarrollo Urbano, Biblioteca, Edif. Banco de Venezuela, piso 5, *Caracas*
1946
20 000 vols   42234

Ministerio de Agricultura y Cría, Biblioteca, Av Lecuna, Parque Central, Torre Este, 6° piso, *Caracas*
1936; Tusnelda Crespo Pietri
70 000 vols   42235

Ministerio de Energía y Minas, Biblioteca Central 'Juan Pablo Pérez Alfonzo', Av Lecuna, Torre Oeste, 2° piso, Parque Central, Apdo 1061, *Caracas*
T: +58 212 5075-400/-369; Fax: +58 212 5075400
1951; Silvia Pernia C.
28 000 vols; 200 curr per; 150 diss/theses; 4 500 govt docs; 4 maps; 11 digital data carriers
libr loan; RRIAN   42236

Ministerio de Sanidad y Asistencia Social, Biblioteca, c/o Instituto Nacional de Higiene, Universidad Central de Venezuela, Ciudad Universitaria, *Caracas*
1936
9 500 vols   42237

Tribunales del Distrito Federal 'Fundación Rojas Astudillo', Biblioteca, Edif. Gradillas, piso 3°, Letra 'B' Gradillas a San Jacinto, Apdo 344, *Caracas* 1010
1950; Aura C. López Rivas
52 000 vols
libr loan   42238

Corporate, Business Libraries

Banco Central de Venezuela, Biblioteca 'Ernesto Peltzer', Av Urdaneta, esq. de Carmelitas, Torre Financiera, piso 16, Apdo 2017, *Caracas* 1010
T: +58 212 8015692; Fax: +58 212 8610048; E-mail: info@bcv.org.ve; URL: www.bcv.org.ve
1941; Javier Bringas
Coll of ancient bks, coll of official gazettes, World Bank Depository Libr
100 000 vols; 1 000 curr per
libr loan; ANABISAI   42239

Electridad de Caracas, Biblioteca, Av Vollmer, Apdo 2299, *Caracas*
1935
19 000 vols   42240

PDVSA Servicios, Biblioteca, Edif. PDVSA Servicios, Los Chaguaramos, Apdo 889, *Caracas* 1010-A
T: +58 212 6064770; Fax: +58 212 6063637; E-mail: dalessandroe@pdu.com; URL: www.pdvsa.com
1946; Maria Elene d'Alessandro
Administration, Management, Oil, Environmental, Hist, Industrial Security, Hygiene
16 500 vols; 350 curr per; 10 digital data carriers   42241

Special Libraries Maintained by Other Institutions

Biblioteca Técnica Científica Centralizado para el Desarrollo de la Región, Centro Occidental, Av Corpalwayco, Apdo 254, *Barquisimeto*, Lara
E-mail: fudeco@dino.conicit.ve
1966; Cecilia Vega Febres
Regional development in Venezuela (Region Centro Occidental)
50 000 vols; 800 curr per; 5 000 diss/theses; 8 500 maps; 2 000 microforms
libr loan   42242

Academia de Ciencias Físicas, Matemáticas y Naturales, Biblioteca Científica y Tecnológica 'Jesús Muñoz Tebar', El Silencio Torre Sur, Local 5, Nivel Av, Apdo 1421, *Caracas* 1010
T: +58 212 4834133
1979; Alberto E. Olivares
20 000 vols   42243

Academia Nacional de la Historia, Biblioteca, Av Universidad, Palacio de las Academias, Bolsa a San Francisco, *Caracas* 1010
T: +58 212 4823849; E-mail: j0027186-1@cantv.net
1888; Dra Ermila de Veracoechea
120 000 vols   42244

Academia Nacional de Medicina, Biblioteca Médica Venezolana 'Dr. Ricardo Archila', Palacio de las Academias, Bolsa a San Francisco, Apdo 804, *Caracas* 1010
T: +58 212 4818939
1904; Dr. Tulio Briceño Maaz
8 000 vols   42245

Academia Venezolana, Biblioteca, Palacio de las Academias, Bolsa a San Francisco, *Caracas* 1010
T: +58 212 4818716
1883; Mario Torrealba Lossi
Ayacucho coll, 'El Coyo Ilustrado', Venezuelan classics, dictionaries
25 000 vols   42246

Alianza Francesa en Venezuela, Chacao, Apdo 61000, Los Caobos, *Caracas* 1060-A
T: +58 212 7823013
1976
17 000 vols   42247

Banco del Libro, Centro de Documentación e Información, Biblioteca, Final Av Luis Roche, Altamira Sur, Apdo Postal 5893, *Caracas* 1010-A
T: +58 212 2663621; Fax: +58 212 2663621; E-mail: blibro@reaccion.ve; URL: www.bancodellibro.org.ve
1975; Marta Martinez
Poster coll (521)
24 161 vols; 124 curr per; 8 489 mss; 27 diss/theses; 80 digital data carriers
libr loan; IBBY   42248

Centro Regional para la Educación Superior en América Latina y El Caribe, Unesco/Cresalc, Servicio de Información y Documentación, Apdo 68394, *Caracas* 1062-A
T: +58 212 2845075; Fax: +58 212 2831411
1978; José Antonio Quinteiro Goris
20 000 vols; 300 curr per
libr loan   42249

Colección Ornitológica Phelps (Phelps Ornithological Collection), Av Abraham Lincoln, entre 1a y 2a Calle de Bello Monte, Ed Gran Sabana, Piso 3, *Caracas*; Apdo 2009, Carmelitas, Caracas 1010A
T: +58 212 7615631;
Fax: +58 212 7633695; E-mail: fundacionphelps@cantv.net
1938; Margarita Martínez
Originals of sketch maps of explorations
10 000 vols; 233 curr per; 6 000 mss; 900 maps; 30 posters, 20 calendars
                                   42250

Fundación La Salle de Ciencias Naturales, Dirección Nacional de Bibliotecas, Documentación y Archivos, Edif. Fundación la Salle 5 Piso, Apdo 1930, *Caracas* 1010-A
T: +58 212 7828711, 7934255;
Fax: +58 212 7937493; E-mail: mireya.viloria@fundacionlasalle.org.ve
1942; Mireya Viloria
Natural sciences, sociology, ethnology, anthropology; Archivo de Indias de Sevilla, documentos referentes a Venezuela
30 800 vols; 1 404 mss; 20 diss/theses; 323 microforms
libr loan   42251

Goethe-Institut, Asociación Cultural Humboldt, Biblioteca 'Andrés Bello', Av Jorge Washington cruce con Av Juan Germán Roscio San Bernardino, Apdo 60501, *Caracas* 1060-A
T: +58 212 5526445; Fax: +58 212 5525621; E-mail: info@caracas.goethe.org; URL: www.goethe.de/ins/ve/car/inz/deindex.htm
1949; Hannelore Minkert
AV Dept
9 555 vols; 23 curr per; 10 diss/theses; 50 maps; 135 av-mat; 1 206 sound-rec; 32 digital data carriers
libr loan   42252

Instituto Nacional de Nutrición, Biblioteca, Esq El Carmen, Apdo 2049, *Caracas*
T: +58 212 4833099
1949
10 000 vols   42253

Instituto Venezolano de Investigaciones Científicas (IVIC), Biblioteca Marcel Roche, Altos de Pipe, Ctra Panamericana, km 11, Apdo 21827, *Caracas* 1020-A
T: +58 212 50415-15/-16; Fax: +58 212 5041681; E-mail: bibliotk@ivic.ve; URL: biblioteca.ivic.ve
1959; Xiomara Jayaro
Rare bks coll
500 000 vols; 5 817 curr per; 720 diss/theses; 800 govt docs; 45 digital data carriers
ACURIL, FID, ANABISAI   42254

Instituto Venezolano Tecnológico del Petróleo (Intevep), Centro de Investigación y Desarrollo de Petróleos de Venezuela, SA, Apdo 76343, *Caracas* 1070A
T: +58 212 9086111
1979
80 000 vols; 1 400 curr per   42255

Sociedad de Obstetricia y Ginecología de Venezuela, Biblioteca "Dr. M.A. Sánchez Carvajal", Maternidad Concepción Palacios, Av San Martín, Apdo 20081, *Caracas* 1020A
T: +58 212 4515955; Fax: +58 212 4510895; E-mail: sogvzla@cantv.net; URL: www.sogvzla.org
1940; Dra Leonor Zapata
8 500 vols; 123 curr per; 20 incunabula; 100 diss/theses
libr loan   42256

Sociedad Venezolana de Ciencias Naturales, Biblioteca, C/ Arichuna y Cumaco, El Marqués, Apdo 1521, *Caracas* 1010A
1931; Dr. Ricardo Muñoz Tébar
12 000 vols; 4 000 curr per   42257

Sociedad Venezolana de Ingenieros Forestales, Biblioteca, c/o Colegio de Ingenieros de Venezuela, Apdo 2006, *Caracas*
T: +58 212 5713122 ext 167
1960
8 000 vols   42258

Salón de Lectura 'Armando Zuloaga Blanco', Biblioteca, C Sucre, *Cumaná*
1940
10 000 vols   42259

Biblioteca Central, *Maracay*
T: +58 43 452491; Fax: +58 43 454320; E-mail: biblioteca_ceniap@impsat.net; URL: www.ceniap.gov.ve
1937
300 000 vols; 2 441 curr per; 466 diss/theses; 300 maps; 50 microforms
                                   42260

Centro Nacional de Investigaciones Agropecuarias, Biblioteca Central, Apdo A-4653, *Maracay* 2101
1937; Nancy Garcés de Hernández
200 000 vols; 374 curr per; 164 diss/theses; 300 maps; 50 microforms
libr loan   42261

Estación de Investigaciones Marinas (EDIMAR), Biblioteca, Apdo 144, *Porlamar*, Punta de Piedras, Isla de Margaritas, Edo Nueva Esparta
1960
18 000 vols   42262

Vietnam

National Libraries

National Library of Vietnam (NLV), 31 Trang Thi St, 10000 *Hanoi*
T: +84 4 38254938, 39349023; Fax: +84 4 38253357; E-mail: info@nlv.gov.vn; URL: www.nlv.gov.vn
1919; Mr Pham The Khang
1 000 000 vols; 10 000 titles of IndoChina books before 1954 (microfiches), 1 500 titles of books and newspapers (microfilm)   42263

General Research Libraries

Mang Thông Thin Khoa Hoc & Công Nghê Viêt Nam, Library (Vietnam Information for Science and Technology Advance), 24-26 Ly Thuong Kiet, *Hanoi*
T: +84 4 9349111; Fax: +84 8246325; E-mail: webAdmin@vista.gov.vn; URL: www.vista.gov.vn
1990; Dr. Ta Ba Hung
350 000 vols; 4 900 curr per   42264

General Sciences Library, 69 Lý Tu Trong, District 1, BP341, *Ho Chi Minh City*
T: +84 8 225055; Fax: +84 8 299318; E-mail: gsl.hcmc@hcm.vnn.vn; URL: www.gslhcm.org.vn
1868; Nguyen Thi Bac
1 369 216 vols; 63 527 curr per; 11 581 diss/theses; 955 maps; 13 544 microforms; 644 sound-rec; 480 digital data carriers   42265

University Libraries, College Libraries

Tay Nguyen University, Library, Dai hoc Tay, *Buon Ma Thuot*, Daklak Province
T: +84 505 52290
1977; Ta thi Minh Phuong
Animal sciences, medicine, forestry, agriculture
32 000 vols   42266

Cantho University, Learning Resource Center, Campus II, 3/2 St, *Cantho*
T: +84 71 831530-8525; Fax: +84 71 831565; E-mail: httrang@ctu.edu.vn; URL: www.lrc.ctu.edu.vn/IntroductionE.aspx
1968; Huynh Thi Trang
FAO, WHO, IRRI coll
200 000 vols; 885 curr per; 10 pamphlets, 6 newsletters (for agricultural extension work)   42267

Hanoi Agricultural University, Library, Gia Lam District, *Hanoi*
T: +84 4 8276906; Fax: +84 4 8276554; E-mail: qhqt.hau@fpt.vn
1956
25 000 vols   42268

Hanoi College of Construction, Library, Tu Liem District, *Hanoi*
1966
150 000 vols   42269

Hanoi College of Pharmacy, Library, 13 Le Thanh Tong St, *Hanoi*
1961
20 000 vols   42270

Hanoi Medical School, Library, Ton That Tung St, Dong Da, *Hanoi*
T: +84 4 8524752; Fax: +84 4 525115
1902
50 000 vols   42271

Hanoi University of Civil Engineering, Library, 5 Giaiphong Rd, *Hanoi*
T: +84 4 8691684; Fax: +84 4 8691684; E-mail: dngoaidhxd@hn.vnn.vn
1966
380 000 vols   42272

Hanoi University of Finance and Accountancy, Library, Tú Liêm, *Hanoi*
T: +84 4 8362293; Fax: +84 4 8362171
1963; Nguyen Van Thuc
20 000 vols; 130 curr per; 150 diss/theses; 1 500 govt docs
libr loan   42273

Hanoi University of Technology, Truong Dai Hoc Bach Khoa Ha Noi, Library, 1 Dai Co Viet Rd, *Hanoi*
T: +84 4 8692243; Fax: +84 4 8681643; E-mail: linc-office@mail.hut.edu.vn; URL: library.hut.edu.vn
1956; Nguyen Kim Khanh
700 000 vols   42274

Institute of Literature, Library, 20 Ly Thai To St, *Hanoi*
T: +84 4 8253548; Fax: +84 4 8250385
1959; Do Duc Hai
Literary magazines
60 000 vols; 50 curr per; 1 500 mss; 200 diss/theses   42275

Institute of Social Science Information, Central Social Sciences Library, 26 Lý Thuòng Kiêt, *Hanoi*
1975; Dao Van Tan
United Nations depository libr
300 000 vols; 1 200 curr per   42276

Vietnam National University, Library and Information Center, 19 Le Thanh Tong, *Hanoi*
T: +84 4 7680545; Fax: +84 4 7680600; E-mail: LIC@vnu.edu.vn; www.lic.vnu.edu.vn
1956; Nguyen Huy Chuong
1 400 000 vols
libr loan   42277

College of Technical Teacher Training, Library, 1 Vo van Ngan, Thu Duc, *Ho Chi Minh City*
T: +84 8 8968641; Fax: +84 8 8964922
1962; Ngo Thi Hoa
210 000 vols   42278

University of Agriculture and Forestry, Library, Thu Duc District, *Ho Chi Minh City*
T: +84 8 8960711; Fax: +84 8 8960173; E-mail: bctuyen@hcm.vnn.vn; URL: www.hcmuaf.edu.vn
1959
66 000 vols   42279

University of Thu viên Dai Hoc Khoa hoc Huê, Library, 20 Lê Loi, *Hué*
T: +84 54 832447, 822440; Fax: +84 54 824901; E-mail: Dtqthu@yahoo.com
1957; Ng-ô Thi Hiên
Dept of preservation
140 000 vols; 30 mss; 78 diss/theses; 1 av-mat; 3 sound-rec; 1 digital data carriers; 1 video
libr loan   42280

University of Fisheries, Library, 2 Nguyen Dinh Chieu, *Nha-Trang*, Khanh Hoa
T: +84 58 831145; Fax: +84 58 831147
1959
18 000 vols; 500 curr per   42281

Bac Thai Medical College, Library, *Thai Nguyen City*, Bac Thai
T: +84 28 52671; Fax: +84 28 55710
1968
25 000 vols   42282

Government Libraries

National Assembly, Library, Thú viên Quôc hôi, Só 35 Ngo Quyên, *Hanoi*
T: +84 4 582613658; Fax: +84 4 253763
1979; Dào Van Thach
12 000 vols; 120 curr per   42283

Special Libraries Maintained by Other Institutions

Centre for Nuclear Research, Library, P.O. Box 60, *Dalat*
1962
10 000 vols; 25 000 microforms   42284

Research Institute of Marine Products, Library, 170 Le Lai St, 35000 *Haiphong*
T: +84 31 836664; Fax: +84 31 836812
1961
Fisheries oceanography
12 000 vols
42285

Central Institute for Medical Science Information, Central Medical Library, 13-15 Lê Thánh Tông, *Hanoi*
T: +84 4 8264040; Fax: +84 4 8242668; E-mail: vttyh@hn.vnn.vn; URL: www.cimsi.org.vn
1979; Nguyen Tuan Khoa
60 000 vols; 730 curr per; 478 diss/theses; 10 av-mat; 500 digital data carriers
42286

Hydraulic Engineering Consultants Corporation No 1, Library, 299 Tay Son St, Dong Da, *Hanoi*
T: +84 4 8530209; Fax: +84 4 8522374
1956
9 000 vols
42287

Institiute of World Economy, Information and Library Section, 176 Thai Ha St, Dong Da, *Hanoi*
T: +84 4 8572295; Fax: +84 4 8574316; E-mail: iwevn@netnam.org.vn
1983; Ta Kim Ngoc
14 500 vols
42288

Institute for Building Science and Technology, Library, Nghia Tan, Tu Liem, *Hanoi*
T: +84 4 8344196; Fax: +84 4 8361197
1974
20 000 vols
42289

Institute for the Protection of the Mother and Newborn Child, Library, 43 Trang Thi St, *Hanoi*
T: +84 4 8259281; Fax: +84 4 8254638; E-mail: ipmn@hn.vnn.vn
1966; Nguyen Ann Thu
Family planning, obstetrics, gynecology, perinatalogy, birth control, in vitro fertilization, sexually transmitted diseases, gynaecological endoscopy, IVF Center, pediatrics
5 150 vols; 25 curr per; 40 mss; 100 diss/theses; 1 digital data carriers; 1 Micro computer
42290

Institute of Economic Management, Library, 27 Tran Xuan Soan St, *Hanoi*
T: +84 4 8243951; Fax: +84 4 8261632; E-mail: hue@ie-ncss.ac.vn
1960; Nguyen Thi Hue
10 000 vols; 6 000 docs
42291

Institute of Ethnology, Department of Information, Documentation and Library, 27 Tran Xuan Soan St, *Hanoi*
T: +84 4 9711251; Fax: +84 4 9711435; E-mail: nhi@FPT.vn
1968; Nguyen Thi Hong Nhi
Review of Ethnology: 120 copies (4 numbers/year)
10 000 vols; 250 curr per; 60 diss/theses; 200 govt docs; 10 av-mat; 5 000 photos
42292

Institute of International Relations, Library, Lang Thuong, Dong Da, *Hanoi*
T: +84 4 8343543; Fax: +84 4 8343543
1960
25 000 vols
42293

Institute of Machiniery and Industrial Instruments, Library, 34 Lang Ha St, Dong Da, *Hanoi*
T: +84 4 344565; Fax: +84 4 344975
1973
10 000 vols
42294

Institute of Mathematics, Library, Bo Ho, P.O. Box 631, 10000 *Hanoi*
T: +84 4 8361317; Fax: +84 4 8343303; E-mail: tdvan@ioit.nest.ac.vn
1969
11 000 vols; 350 curr per
42295

Institute of Mechanics, Library, 224 Doi Can St, Ba Dinh, *Hanoi*
T: +84 4 8263641
1979
Basic and applied mechanics
14 200 vols
42296

Institute of Mining and Metallurgy, Library, 30B Doan Thi Diem St, *Hanoi*
T: +84 4 8256994; Fax: +84 4 8256983
1967
10 000 vols
42297

Institute of Philosophy, Library, 25 Lang Ha St, Ba Dinh, *Hanoi*
T: +84 4 5140530; Fax: +84 4 5140530
1964; Le Ngoc Anh
50 000 vols
42298

Institute of Research on Mining Technology, Ministry of Energy, Library, Phuong Liet, Thanh Xuan, *Hanoi*
T: +84 4 8642024; Fax: +84 4 8641564; E-mail: tttthan@hn.vnn.vn
1979; Thïeu Thi Hoan
9 000 vols
42299

National Institute for Educational Science, Library, 101 Tran Hung Dao St, *Hanoi*
T: +84 4 8256978; Fax: +84 4 8221521; E-mail: dinhphuong@bdvn.vnmail.vnd.net
40 000 vols
42300

National Institute of Hygiene and Epidemiology, Library, 1 Yersin St, 10000 *Hanoi*
T: +84 4 8213241; Fax: +84 4 8210853; E-mail: nihe@netnam.org.vn
1924
12 000 vols
42301

National Institute of Tuberculosis and Respiratory Diseases, Library, 120 Hoang Hoa Tham St, *Hanoi*
T: +84 4 8326249; Fax: +84 4 8326162
1957
10 000 vols
42302

Vietnam History Museum, Library, 1 Trang Tien, *Hanoi*
T: +84 4 8252853
1958; Hoang Thi Vinh
22 000 vols
libr loan
42303

Vietnam Institute of Traditional Medicine, Library, 29 Nguyen Binh Khiem, 600 *Hanoi*
T: +84 4 8263668; Fax: +84 4 8229353; E-mail: yhcotruyen@hn.vnn.vn
1957
Vietnamese and chinese medicine in Vietnam
20 000 vols; 5 curr per; 2 maps; 2 transparencies, pamphlets
42304

Vietnam Revolution Museum, Library, 25 Tong Dan St, *Hanoi*
T: +84 4 8254323
1959
21 000 vols; 18 000 photos
42305

Institute of Agricultural Science of South Vietnam, Library, 121 Nguyen Binh Khiem St, *Ho Chi Minh City*
T: +84 8 8291746; Fax: +84 8 8297650; E-mail: ias@netnam2.org.vn
1952; Quach Bich Ngoc
12 000 vols; 27 curr per; 76 diss/theses; 32 govt docs; 22 maps; 56 microforms; 3 digital data carriers
libr loan
42306

Social Sciences Library, 34 Ly-Tú-Trong, *Ho Chi Minh City*
T: +84 8 8296744; Fax: +84 8 8223735
1975; Tran Minh Duc
Colls of the former Aechaeological Research Institute
145 000 vols
42307

Institute of Oceanography, Viên Nghiên Cuu Biên, Library, 01 Cau Da, *Nha-Trang*, Khanh Hoa
T: +84 58 881151, 881153;
Fax: +84 58 881152; E-mail: haiduong@nhatrang.teltic.com.vn
1922; Do Minh Thu
60 000 vols
42308

Virgin Islands (U.S.)

University Libraries, College Libraries

University of The Virgin Islands, Saint Croix Campus Library, RR 2, Box 10000, *Kingshill*, VI 00850-9781
T: +1 340 6924130; Fax: +1 340 6924135; URL: library.uvi.edu
1969; Jennifer Jackson
Caribbean mat, VI docs
56 000 vols; 171 curr per
42309

University of The Virgin Islands, Ralph M. Paiewonsky Library, Two John Brewers Bay, *Saint Thomas* VI 00802-9990
T: +1 340 6931367; Fax: +1 340 6931365; URL: library.uvi.edu
1963; Linda Robinson Barr
Caribbean Coll; Eastern Caribbean Center, Melchior Center for Local VI Hist
97 000 vols; 717 curr per; 201 diss/theses; 15 786 govt docs; 1 000 maps; 347 000 microforms; 100 sound-rec; 10 500 pamphlets
ACURIL, ALA
42310

Government Libraries

Legislative Council of the Virgin Islands, Library, Road Town, *Tortola*
T: +1 284 4944757; Fax: +1 284 4944544; E-mail: bvilegislatire@bvigovernment.org
42311

Special Libraries Maintained by Other Institutions

Virgin Islands Division of Libraries, Archives & Museums – Regional Library for the Blind & Physically Handicapped, 3012 Golden Rock, Christiansted, *Saint Croix*, VI 00820
T: +1 340 7722250; Fax: +1 340 7723545; E-mail: reglib@viaccess.net
Letitia G. Gittens
25 000 vols; 50 curr per; 46 821 av-mat
42312

Island Resources Foundation, Library, 6292 Estate Nazareth, No 100, *St. Thomas*, VI 00802-1104
T: +1 340 7756225; Fax: +1 340 7792022; E-mail: etowle@irf.org; URL: www.irf.org
1972; Bruce Potter
Environment, fisheries, maritime hist, tourism, West Indian hist
10 000 vols; 30 digital data carriers; 3 500 slides, 3 500 photos, 400 linear feet of doc & ref mat
42313

Public Libraries

Athalie McFarlane Petersen Public Library, 604 Strand St, Frederiksted, *Saint Croix*, VI 00840
T: +1 340 7720315; Fax: +1 340 7194380
1920; Sylvie Renaud
16 000 vols
42314

Florence A.S. Williams Public Library, 1122 King St, Christiansted, *Saint Croix*, VI 00820-4951
T: +1 340 7735715; Fax: +1 340 7735327; E-mail: wallacewilliams@msn.com; URL: www.virginislandspace.com/fwplweb
1920; Wallace Williams
20 000 vols; 15 curr per
42315

Virgin Islands Division of Libraries, Archives & Museums – Regional Library for the Blind & Physically Handicapped, 3012 Golden Rock, Christiansted, *Saint Croix*, VI 00820
T: +1 340 7722250; Fax: +1 340 7723545; E-mail: reglib@viaccess.net
Letitia G. Gittens
25 000 vols; 50 curr per; 46 821 av-mat
42316

Division of Libraries, Archives and Museums, 23 Dronningens Gade, *Saint Thomas*, VI 00802
T: +1 340 7743407; Fax: +1 340 7751887; URL: libraries.gov.vi
1920; Ingrid Bough
Von Scholten Coll
169 000 vols; 180 curr per
42317

Enid M. Baa Library & Archives, 5424 Store Tvaer Gade, *Saint Thomas*, VI 00802-6947
T: +1 340 7740630; Fax: +1 340 7751887; E-mail: dlamdir@vipowernet.net; URL: www.library.gov.vi
Ingrid Bough
Caribbeana;UN & VI Documents Coll;Building Houses Von Scholten
51 000 vols; 37 curr per
42318

Elaine Ione Sprauve Library and Museum, Enighed Estate, Cruz Bay, P.O. Box 30, *St. John*, VI 00831
T: +1 340 7766359; Fax: +1 340 6937375; E-mail: friendsprlib@att.net; URL: www.library.gov.vi/sprauve
1957; Carol McGuinness
12 000 vols; 37 curr per
libr loan
42319

Enid M. Baa Library and Archives, 20 Dronningens Gade, *St. Thomas*, VI 80802
T: +1 340 7740630; Fax: +1 340 7751887
1924; N.N.
Caribbean Special Coll
40 000 vols; 20 curr per
libr loan; ALA, ACURIL
42320

Yemen

Special Libraries Maintained by Other Institutions

British Council, Library, Sana'a Trade Centre, Algiers St, Administrative Tower, 3rd Fl, P.O. Box 2157, *San'a*
T: +967 1 448-356/-357; Fax: +967 1 448-360; E-mail: information@ye.britishcouncil.org; URL: www.britishcouncil.org/yemen
1974; Adrian Chadwick
Arab and Middle East coll
10 000 vols
42321

Zambia

University Libraries, College Libraries

Chipata Teacher Training College, Library, P.O. Box 510189, *Chipata*
T: +260 62 221214, 221710; Fax: +260 62 221717, 221214; E-mail: cttc@zamtel.zm
1966; Zacchaeus Phiri
16 000 vols; 120 curr per; 40 mss; 4 diss/theses; 105 govt docs; 30 music scores; 35 maps; 15 sound-rec; 10 digital data carriers; 70 mat on HIV/AIDS
libr loan; ZLA
42322

David Livingstone Teachers' College, Resource Centre, PB 1, *Livingstone*
T: +260 3 320024; Fax: +260 3 321987; E-mail: dalitco@zamtel.zm
1978; E.M. Katungula
25 000 vols; 10 curr per; 150 govt docs; 100 maps; 300 av-mat; 308 sound-rec; 8 digital data carriers; 2 200 pictures, 200 pamphlets
libr loan
42323

Evelyn Hone College of Applied Arts and Commerce, Library, Church Rd, P.O. Box 30029, *Lusaka*
T: +260 1 211557
1963; J. Masempel N'Gandu
20 000 vols
42324

National Institute of Public Administration, Library, P.O. Box 31990, 10101 *Lusaka*
T: +260 1 228802
1963; A.G. Kasonso
28 000 vols
42325

Natural Resources Development College, Library, P.O. Box 310099, *Lusaka*
T: +260 1 281328; Fax: +260 1 224639
1965; M.M. Misengo
Agriculture, animal science, human nutrition, water engineering, agricultural education
32 000 vols; 600 curr per
libr loan; SCOHLZA
42326

University of Zambia, Library, Great East Rd, P.O. Box 32379, *Lusaka*
T: +260 1 295220; E-mail: library@unza.zm; URL: www.unza.zm/
1965; Dr. Hudwell C. Mwacalimba
Zambiana, UN publ, Univ arch, oral lit
2 500 000 vols; 100 curr per; 1 000 diss/theses; 26 400 govt docs; 20 music scores; 754 maps; 10 000 microforms; 5 500 sound-rec; 28 digital data carriers
libr loan; IFLA   42327

– School of Medicine, Medical Library, University Teaching Hospital, P.O. Box 50110, Lusaka
T: +260 1 250801; Fax: +260 1 250753; E-mail: medlib@unza.zm; URL: www.medguide.org.zm
1968; Norah M. Mumba
2 218 WHD docs
45 000 vols; 46 curr per; 120 govt docs; 128 av-mat; 12 digital data carriers
libr loan; AHILA   42328

– Veterinary Medicine Library, P.O. Box 32379, Lusaka
Fax: +260 1 250753
Mr Wina Simonda
7 000 vols   42329

Mufulira Teachers Training College, Library, P.O. Box 40400, *Mufulira*
T: +260 412340; Fax: +260 412157
1963; James B. Chikonde
9 000 vols; 7 curr per; 5 maps   42330

Northern Technical College, Library, P.O. Box 250093, *Ndola*
T: +260 2 680423; Fax: +260 2 680288
1966; M.C. Banda
Mechanical, electrical, automotive and heavy-duty engineering
20 000 vols; 70 curr per
ZLA   42331

Government Libraries

Attorney General's Library, Ministry of Legal Affairs, 15101 Ridgeway, P.O. Box 50106, *Lusaka*
T: +260 1 251588; Fax: +260 1 253695
Teddy R. Malunga
18 000 vols; 50 curr per
libr loan; ZLA   42332

Office of the Auditor-General, Library, 271 Ueale St, *Muckleneuk Pretoria*; P.O. Box 446, Pretoria 0001
T: +260 12 4268176; Fax: +260 12 4268333
1991; E.A. Dutoit
2 500 vols; 170 curr per; 500 govt docs; 100 av-mat; 70 sound-rec; 10 digital data carriers
libr loan   42333

Ecclesiastical Libraries

Jesuit Theological Library, P.O. Box 310085, *Lusaka*
T: +260 1 281906
1972; Roy Thaden
24 000 vols   42334

Special Libraries Maintained by Other Institutions

Central Fisheries Research Institute, Library, P.O. Box 350100, *Chilanga*
T: +260 278680
1965
4 500 vols   42335

Mount Makulu Agricultural Research Station, Library, PB 7, *Chilanga*
T: +260 1 278655
1953; Jeneviver Namangala
Fao depository section
30 000 vols; 200 curr per; 8 000 mss; 100 diss/theses; 20 000 govt docs; 60 maps; 200 microforms; 20 digital data carriers; 15 000 reprints, 18 000 rpts
libr loan; ZLA, AGRIS, IAALD   42336

Pan African Institute for Development, East and Southern Africa, Library, P.O. Box 80448, *Kabwe*
T: +260 5 223651; Fax: +260 5 223451
1980
7 500 vols; 120 curr per; 80 mss; 60 govt docs; 6 maps; 8 av-mat; 20 sound-rec
libr loan; ZLA   42337

Dag Hammarskjold Memorial Library, Mindolo Ecumenical Foundation, Box 2149, 260 *Kitwe*
T: +260 2 19012, 11488; Fax: +260 2 11001; E-mail: daglib@zamnet.zm
1961; Dunstan Chikonka
Arch of the Ecumenical Movement in Africa, MEF records; British Council Coll (1 200 vols)
19 vols; 21 curr per; 6 diss/theses; 37 govt docs; 50 maps; 30 real to real tapes, 4 audio tapes, 150 video cassettes, 300 slides
libr loan; ZLA   42338

Division of Forest Research, Library, P.O. Box 22099, *Kitwe*
T: +260 2 220456; Fax: +260 2 224110
1956
7 800 vols; 250 curr per   42339

Zambia Institute of Technology, Library, Jambo Drive, Riverside, P.O. Box 21993, *Kitwe*
T: +260 2 212066
1972; E.N. Chitwamali
Electrical, mining and construction engineering, British, Indian and Zambian Standards, business studies
ZLA   42340

Livingstone Museum, Library, Mosi-oa-Tunya Rd, P.O. Box 60498, *Livingstone*
T: +260 3 324427; Fax: +260 3 320991; E-mail: livmus@zamnet.zm
1934
Ethnography, hist, natural hist, letters and relics of David Livingstone
20 000 vols; 200 curr per   42341

National Heritage Conservation Commission, Library, P.O. Box 60124, *Livingstone*
T: +260 3 320481; Fax: +260 3 324509
1948; Jane Munyenyembe
4 000 vols; 1 500 curr per; 5 diss/theses; 500 govt docs; 3 000 av-mat; 2 500 negatives, 1 000 slides, 1 000 pictures
libr loan; IFLA, ZLA   42342

Geological Survey of Zambia, Ministry of Mines and Minerals Development, Library, CNR Government/Nationalist RDS – Ridgeway, 10101 *Lusaka*; P.O. Box 50135, Ridgeway, 10101 Lusaka
T: +260 1 254485, 250174; Fax: +260 1 250174; E-mail: gsd@zamnet.zm
1952; Davies Muunga
44 000 vols; 60 curr per; 40 mss; 2 incunabla; 45 diss/theses; 1 070 govt docs; 3 000 maps; 15 digital data carriers; geophysical data – maps, mosaics
libr loan; IFLA, ZLA   42343

National Archives of Zambia, Library, Governmemt Rd, 10101 *Lusaka*; P.O. Box 50010, 10101 Lusaka
T: +260 1 254081; Fax: +260 1 254080; E-mail: naz@zamnet.zm
1947; N. Mutiti
Public Arch
17 000 vols; 11 000 curr per; 10 000 maps; 11 000 periodicals
libr loan; IFLA
Reference and legal deposit libr for all printed publ in Zambia   42344

National Council for Scientific Research, Documentation and Scientific Information Centre, Chelston, P.O. Box 310158, *Lusaka* 15302
T: +260 1 281081; Fax: +260 1 283502
1967
9 200 vols; 100 curr per
libr loan; FID, Zambia Library Association
                42345

National Institute for Scientific and Industrial Research, Library, P.O. Box 310158, 'Chelston, 15302 *Lusaka*
T: +260 1 281081; Fax: +260 1 283502; E-mail: nisiris@zamnet.zm
9 200 vols; 100 curr per   42346

Parliamentary Information and Research Library, P.O. Box 31299, *Lusaka*
T: +260 1 29242531; Fax: +260 1 292252; E-mail: tcmtine@zamnet.zm
1967; Mrs Tembi C.C. Mtine
Parliamentary papers, social sciences
28 000 vols
libr loan; COMLA, ZLA   42347

Zambia National Library & Cultural Centre for the Blind, P.O. Box 35583, *Lusaka*
T: +260 1 260516; Fax: +260 1 260516; E-mail: liblind@zamtel.zm   42348

International Red Locust Control Organization for Central and Southern Africa, Library, P.O. Box 240252, *Ndola*
T: +260 2 612433; Fax: +260 2 614285
1970
3 000 vols; 35 curr per   42349

Council for Scientific and Industrial Research (CSIR), Division of Information Services, P.O. Box 395, *Pretoria* 0001
T: +260 12 8412000; Fax: +260 12 862869
1945
110 000 vols; 22 000 pamphlets   42350

Public Libraries

KMC Town Centre Public Library, Freedom Way, P.O. Box 80424, *Kabwe*
T: +260 5 222381, 222389; Fax: +260 5 224467
1957; Christine Kunda
13 064 vols; 50 curr per; 60 mss; 20 diss/theses; 13 govt docs; 8 maps; braille bks, cassettes
ZLA   42351

Kasama Provincial Library, Zambia Library Service, P.O. Box 98, *Kasama*
T: +260 5 221399
1972; M.E. Sakanya
libr loan   42352

Kitwe Public Library, Kaunda Sq, P.O. Box 20070, *Kitwe*
T: +260 2 213685
1954; F.B.A. Sichali
33 000 vols
libr loan; ZLA   42353

Lusaka City Libraries, Katondo Rd, P.O. Box 31304, *Lusaka*
T: +260 1 227282
1943; J.C. Nkole
Coll of mat on Zambia; 3 branch libs, 1 mobile libr
145 000 vols; 200 curr per; 320 maps
libr loan; ZLA   42354

Zambia Library Service (ZLS) HQ, Haile Selassie Av, Long Acres, P.O. Box 30802, *Lusaka*
T: +260 1 254993; Fax: +260 1 254993
1962; E.M. Moadabwe
Community inf; 6 regional libs, 18 branch libs
500 000 vols; 6 curr per
libr loan   42355

Zambia National Library & Cultural Centre for the Blind, P.O. Box 35583, *Lusaka*
T: +260 1 260516; Fax: +260 1 260516; E-mail: liblind@zamtel.zm   42356

Bophuthatswana Library Service, PB X2044, *Mmabatho*, 8681
T: +260 140 292374; Fax: +260 140 22063
1978; S. Khutsoane
180 000 vols; 111 curr per
libr loan   42357

Ndola Public Library, 218 Independence Way, P.O. Box 70388, *Ndola*
T: +260 2 620599
1934; Mr Micky Saili
Braille bks
95 000 vols; 145 curr per
libr loan; ZLA   42358

Zambesi District Library, Zambia Library Service, P.O. Box 150022, *Zambesi*
T: +260 8 371055
1965; Musa Nyirongo
120 000 vols
libr loan   42359

Zimbabwe

National Libraries

National Archives of Zimbabwe – Library and Reference Services, 12th Av South Park, P.O. Box 1773, *Bulawayo*
T: +263 9 62359; Fax: +263 9 77662
1944; H.R. Ncube
Institute of Bankers in Zimbabwe (Bulawayo Branch), Matabeleland Turf Club, Coll of Portuguese books on the history of the country from 1514-1830, Journals of Robert Moffat
100 000 vols
libr loan; ZLA   42360

University Libraries, College Libraries

Bulawaya Polytechnic, Park Rd, 12th Av, P.O. Box 1392, *Bulawayo*
T: +263 9 63181; Fax: +263 9 71165
1927
44 200 vols; 55 curr per; AV materials
libr loan; RLA, ZLA   42361

Midlands State University, Library, Senga Township, *Gweru*; PB 9055, Gweru
T: +263 54 60445; Fax: +263 54 60311, 60753; E-mail: sheilanzw@yahoo.com
2000; Sheila Ndlovu
Economics, History & Soicial Studies, Zimbabweana, Govt Publs, Conference papers, Past examination papers; Home economics, Tourism, Commerce
52 000 vols; 15 curr per; 20 digital data carriers
libr loan; ZULC   42362

Domboshawa National Training Centre, Library, PB 7746, Causeway, *Harare*
T: +263 4 882245; Fax: +263 4 882245
1957; K.K. Muwonde
Agriculture, economics, finance, management, public administration, rural development
13 123 vols; 6 970 curr per; 23 diss/theses; 325 govt docs; 54 maps; 6 microforms; 8 av-mat; 14 sound-rec
libr loan   42363

Harare Polytechnic, Library, Herbert Chitepo Av West, P.O. Box CY 407, Causeway, *Harare*
T: +263 4 752311, 752319; Fax: +263 4 720155
1927; F. Mutungira
Printing
60 000 vols; 145 curr per; 56 av-mat; 41 sound-rec
libr loan   42364

School of Social Work, Cnr Grant St / Chinhoyi St, PB 66022, Kopje, *Harare*
T: +263 4 75-2965 to 2967; Fax: +263 4 751903; E-mail: ssw@esanet.zw
1964; Mr Enoch Chipunza
9 200 vols; 44 curr per; 350 mss; 115 diss/theses; 50 govt docs; 4 maps; 70 av-mat; 5 digital data carriers; 5 info kits/teaching packages
libr loan; ZLA   42365

University of Zimbabwe, Library, P.O. Box M45, Mt Pleasant, *Harare*
T: +263 4 303211; Fax: +263 4 335383; E-mail: infocentre@uzlib.uz.ac.zw; URL: www.uz.ac.zw/
1956; Dr. B. Mbambo
Godlonton Coll (Zimbabwe), Doke Coll (African Langs), Astor Coll (American Civil War)
430 000 vols; 110 637 periodical titles
libr loan; IFLA, ZLA   42366

– Education Library, P.O. Box MP 45, Harare
T: +263 4 303211 ext 1254
9 000 vols   42367

– Institute of Development Studies, Campus Mount Pleasant, Box 880, Harare
T: +263 4 333341; Fax: +263 4 333345; E-mail: ids_library@ids1.uz.zw
1983; Audrey Mhlanga
Development in social science
12 000 vols; 50 curr per; 800 microforms; 5 digital data carriers
libr loan   42368

– Medical Library, Mount Pleasant, P.O. Box MP45, Harare
T: +263 4 791631; Fax: +263 4 795019; E-mail: patrikios@healthnnet.zw
1963; Mrs H. Patrikios
Medical Africana coll, Utano coll (local unpublished health lit), Medical hist coll; Malaria Information Resource Centre, Sub-Regional
54 000 vols; 77 curr per; 378 diss/theses; 137 av-mat; 10 digital data carriers
libr loan; AHILA   42369

Chibero College of Agriculture, Library, PB 901, *Norton*
T: +263 162 2230/-2238; Fax: +263 162 2233
1961; L. Jonhera
animal husbandry, crop production, agric. engineering and farm management & economics
21 000 vols; 500 curr per; 120 mss; 100 diss/theses; 1 000 govt docs   42370

School Libraries

Mkoba Teachers College, P.O. Box MK 20, *Gweru*
T: +263 54 50-041/-044; Fax: +263 54 50020; E-mail: htkwe@hotmail.com
1976; Mr Hosea Tokwe
46 300 vols; 300 curr per; 8 diss/theses; 30 govt docs; 4 maps; 1 av-mat; 6 sound-rec; 4 digital data carriers
libr loan; ZLA   42371

Government Libraries

Education Services Centre, Library, Head Office, Ambassador House, Union Av Causeway, P.O. Box MP 133, *Harare*
T: +263 4 333818
1967; Fabian Charkiadza
15 000 vols; 150 curr per
libr loan   42372

Ministry of Lands and Agriculture, Central Library, 1 Borrowdale Rd, PB 7701, Causeway, *Harare*
T: +263 4 70-6081/-6089; Fax: +263 4 734646
1906; Mr A. Mangau
SADC Coll, Agriculture, Training and

Extension Services, FAO publs
8 700 vols; 240 curr per; 73 diss/theses; 10 maps; 4 100 microforms; 26 510 reports, pamphlets, photocopies, clippings
libr loan; ZLA, IAALD, AGRIS   42373

Ministry of Transport and Energy, Library, P.O. Box CY595 Causeway, *Harare*
T: +263 4 70-0991/-0999; Fax: +263 4 708225
1953; Richard Mahlahla
18 033 vols; 19 curr per; 3 mss; 9 diss/theses; 12 govt docs; 7 maps; 2 digital data carriers
libr loan; ZLA   42374

Parliament of Zimbabwe, Library, Cnr Baker Av and 3rd St, P.O. Box Cy 298, Causeway, *Harare*
T: +263 4 700181 ext 2131; Fax: +263 4 795548
1923; Nelson Masawi
Zimbabweana Coll, Parliamentary and Govt Docs Coll, political science, administration, education, hist, science and technology, English lit, pamphlets
115 000 vols; 100 curr per; 70 000 govt docs; 400 maps
libr loan; ZLA, IFLA   42375

Ecclesiastical Libraries

Regional Seminary, Library, P.O. Box 1139, *Harare*
T: +263 4 491097; Fax: +263 4 491003
1960; A.F. Parish
Philosophy, religion
21 000 vols; 50 curr per; 50 diss/theses; 100 govt docs   42376

Special Libraries Maintained by Other Institutions

National Archives of Zimbabwe, Borrowdale Rd, PB 7729 Causeway, *Harare*
T: +263 4 792741; Fax: +263 4 792398; E-mail: archives@gta.gov.zw; URL: www.gta.gov.zw
1936; D. Sibanda
48 142 vols; 6 663 curr per; 3 469

mss; 400 diss/theses; 6 000 maps; 1 650 microforms; 1 204 av-mat; 11 816 sound-rec; 6 digital data carriers; 39 875 photos, 550 posters
libr loan; SALA, CILIP, Library Association/Zimbabwe, IFLA   42377

National Gallery of Zimbabwe, Thomas Meikle Library, Julius Nyerere Way, Box CY 848, *Harare*
T: +263 4 704666; Fax: +263 4 704668; E-mail: ngallery@ecoweb.co.zw
1957; Luness Mpunwa
12 000 vols; 4 diss/theses; 48 govt docs; 2 maps; 1 000 av-mat; 30 sound-rec; 20 digital data carriers; 50 000 periodicals, 40 000 exhibition cat, special colls of photos, slides and CDs
ZLA   42378

Standards Association of Zimbabwe, Standards Information, Northend Close, Northridge Park, Borrowdale, P.O. Box 2259, *Harare*
T: +263 4 88-5511/-5512; Fax: +263 4 882020; E-mail: sazinfo@mweb.co.zw
1957; R. Marunda
70 000 vols; 10 curr per; 500 govt docs; 30 av-mat; 15 digital data carriers
ISO   42379

Tobacco Research Board, Library, Kutsaga Research Station, P.O. Box 1909, *Harare*
T: +263 4 575289; Fax: +263 4 575288; E-mail: tobres@kutsaga.co.zw
1954; K.B. Giesswein
12 000 vols; 250 curr per
libr loan; IAALD   42380

Public Libraries

Bulawayo Public Library, Fort St/ 8th Av, P.O. Box 586, *Bulawayo*
T: +263 9 60965; Fax: +263 9 60965; E-mail: bpl@netconnect.co.zw; URL: www.angelfire.com/ky/bpl
1896; R.W. Doust
Historic reference coll; Braille Libr
100 000 vols; 200 curr per; 40 diss/theses; 2 500 govt docs; 30 music scores; 600 maps; 50 microforms; 800 av-mat; 550 sound-rec; 10 digital data carriers
libr loan   42381

Gweru Memorial Library, 8th St, P.O. Box 41, *Gweru*
T: +263 54 22628
1898; Filen Bayison
Ref div, Children lending libr
50 000 vols; 15 curr per; 120 000 maps; 40 sound-rec
libr loan; ZLA   42382

Harare City Library, P.O. Box 1087, *Harare*
T: +263 4 75-1834/-1835
1902; Mr Try Simango
Children's Libr; Free Reference Libr, Textbook Libr
200 000 vols; 30 curr per
libr loan; ZLA   42383

Zimbabwe National Library and Documentation Service, P.O. Box 758, *Harare*
T: +263 4 252122; Fax: +263 4 252121; E-mail: nat.free@telconet.co.zw
1972; S.R. Dube
World Bank Depository; Zimbabwe Coll; Everyman's Libr Classics
98 000 vols; 336 curr per; World Bank Publ (1 408 vols), World Bank Pamphlets colls (2 490)
libr loan; IFLA   42384

City of Mutare Public Libraries, P.O. Box 48, *Mutare*
T: +263 20 60823; Fax: +263 20 61002
1970; D. Mandowo
Sakubua Public Libr (27 000 vols), Dangamvura Public Libr (18 000 vols)
90 000 vols; 24 curr per; 12 govt docs; 16 maps; 4 digital data carriers
libr loan; UNAL, ZLA   42385

Turner Memorial Library, Queensway, Civic Complex, P.O. Box 48, Kingsway, *Mutare*
T: +263 20 63412; Fax: +263 20 67335; E-mail: csd@mutare.intersol.co.zw
1904; Mr D. Mandowo
Zimbabweana Coll; Textbook Reference Libr, Children's Libr
50 000 vols; 24 curr per; 4 mss; 1 diss/theses; 65 govt docs; 10 music scores; 30 maps; 14 av-mat
libr loan; ZLA, UNAL   42386

Alphabetical Index of Libraries

1-y gosudarstvenny institut po proektirovaniyu predpriyati pishchevoi promyshlennosti (Gipropishcheprom-1), Moskva 23161
II. Gimnazija Maribor, Maribor 24884
II. Grammar School Maribor, Maribor 24884
II. Rákóczi Ferenc Megyei Könyvtár, Miskolc 13540
15th May's Public Library, Cairo 07196
3M Information Research & Solutions, Saint Paul 37499
A. and M. Miskiniai Public LIbrary of Utena, Utena 18372
A. D. Gordon Agriculture, Nature and Kinnereth Valley Study Institute, Emek Ha-Yarden 14716
A. D. Xenopol Institute of History, Iaşi 22042
A Dunamelléki Református Egyházkerület, Budapest 13326
A. E. Staley Manufacturing Co, Decatur 35662
A. Holmes Johnson Memorial Library, Kodiak 39740
A. I. Karaev Institute of Physiology, Baku 01715
A. K. Smiley Public Library, Redlands 40955
A. M. Gorky Memorial Museum, Moskva 23196
A. M. Razmadze Mathematical Institute, Tbilisi 09273
A. Max Brewer Memorial Law Library, Viera 37706
A. N. Sinha Institute of Social Studies, Patna 14180
A. T. Still University of the Health Sciences, Kirksville 31513
A. V. Shchusev State Research Museum of Architecture, Moskva 23374
Aabenraa Bibliotekerne, Aabenraa 06838
Aalborg Bibliotekerne, Aalborg 06840
Aalborg Hospital, Aalborg 06766
Aalborg Katedralskole, Aalborg 06726
Aalborg Public Libraries, Aalborg 06840
Aalborg Sygehus, Aalborg 06766
Aalborg Universitet, Aalborg 06633
– Aalborg Universitetsbibliotek Esbjerg, Esbjerg 06634
Aalborg University, Aalborg 06633
– Aalborg University Library Esbjerg, Esbjerg 06634
Aalestrup Bibliotek, Ålestrup 06844
Aalto City Central Library, Vyborg 24357
Aalto University Library, Espoo 07331
Aalto-yliopiston kirjasto, Espoo 07331
Äänekosken kaupunginkirjasto, Äänekoski 07558
Aargauer Kantonsbibliothek, Aarau 26962
Aarhus School of Architecture, Århus 06649
Aarhus School of Business, Århus 06650
Aarhus School of Social Work, Århus 06654
Aarhus Technical Library, Århus 06768
Aarhus Universitet, Århus
– Æstetikbiblioteket, Århus 06637
– Bibliotek for Matematiske Fag, Århus 06638
– Biblioteket for Sprog, Litteratur og Kultur, Århus 06639
– Biologisk Institut, Århus 06640
– Fysisk og Astronomisk Bibliotek, Århus 06641
– Geologisk Institut, Århus 06642
– Institut for Statskundskab, Århus 06643
– Kemisk Instituts Bibliotek, Århus 06644
– Økonomisk Bibliotek, Århus 06645
– Psykologisk Institut, Århus 06646
– Ringgadebiblioteket, Århus 06647
– Det Teologiske Fakultetsbibliotek, Århus 06648
Aarhus University, Århus
– Department of Biological Sciences, Århus 06640
– Department of Chemistry Library, Århus 06644
– Department of Earth Sciences, Århus 06642
– Department of Political Science, Århus 06643
– Department of Psychology, Århus 06646
– Library for Language, Literature and Culture, Århus 06639
– Library of Aesthetics, Århus 06637
– Library of Mathematical Sciences, Århus 06638
– Physics and Astronomy Library, Århus 06641
– School of Economics and Management, Århus 06645
– Theological Faculty Library, Århus 06648
AarhusKarlshamn Sweden AB, Karlshamn 26537
Aaron Copeland School of Music, Flushing 31148
Aars Folkebibliotek, Aars 06841
Aba Teachers' College, Wenchuan 05646
Abadan Institute of Chemical and Petroleum Technology, Ahvaz 14423
Abadia de Montserrat, Montserrat 25799
Abadía de Santa Cruz, San Lorenzo del Escorial 25816
Abadía de Santo Domingo de Silos, Santo Domingo de Silos 25824
AB-AESON Insurance Company, Budapest 13351
Abakan State University, Abakan 22201
Abakanski gosudarstvenny universitet, Abakan 22201
ABB Lummus Global BV, Den Haag 19330
Abbaye Bénédictine Sainte-Marie, Paris 08210
Abbaye Cistercienne de Timadeuc, Rohan 08230
Abbaye de Bellefontaine, Bégrolles-en-Mauge 08177
Abbaye de Clervaux, Clervaux 18400
Abbaye de Fleury, Saint-Bénoît-sur-Loire 08232
Abbaye de la Coudre, Laval 08193
Abbaye de la Pierre qui Vire, Saint-Léger-Vauban 08234
Abbaye de Maredsous, Denée 02458
Abbaye de Melleray, La Melleraye-de-Bretagne 08190
Abbaye de Saint-Guenole, Landevennec 08191
Abbaye de Saint-Maurice, St.-Maurice 27273
Abbaye d'Orval, Villers-devant-Orval 02492
Abbaye du Bec-Hellouin, Le Bec-Hellouin 08195
Abbaye la Joie Notre-Dame, Campénéac 08180
Abbaye Notre Dame de Fontgombault, Fontgombault 08187
Abbaye Notre-Dame, Ganagobie 08188
Abbaye Notre-Dame, Tournay 08245
Abbaye Notre-Dame de Belloc, Urt 08246
Abbaye Saint Louis du Temple, Vauhallan 08249
Abbaye Sainte-Anne-de-Kergonan, Plouharnel 08226
Abbaye Saint-Martin de Mondaye, Juaye-Mondaye 08189
Abbaye Saint-Paul, Wisques 08251
Abbazia Benedettina di Monte Maria, Malles Venosta 15886
Abbazia del Santuario di Montenero, Livorno 15877
Abbazia di Grottaferrata, Grottaferrata 15856
Abbazia di S. Benedetto, Seregno 16066
Abbazia di S. Miniato al Monte, Firenze 15824
Abbazia di S. Trinità, Firenze 15825
Abbazia di Vallombrosa, Vallombrosa 16117
The Abbey of Gethsemani Library, Trappist 35396
Abbey of Regina Laudis Library, Bethlehem 35135
Abbot Public Library, Marblehead 40049
Abbott, Abbott Park 36437
ABC- und Selbstschutzschule, Sonthofen 12496
ABCER – Association Bibliothèque Culture et Religion (CTM), Meylan 08205
ABD – Archives départementales des Bouches-du-Rhône, Marseille 08403
Abdij Affligem, Hekelgem 02468
Abdij der Norbertijnen, Averbode 02450
Abdij Keizersberg, Leuven 02470
Abdij Mariënkroon, Nieuwkuyk 19312
Abdij O. L. Vrouw van Koningsoord, Berkel-Enschot 19298
Abdij Onze-Lieve-Vrouw van het Heilig Hart, Westmalle 02494
Abdij Sint Benedictusberg, Lemiers 19311
Abdij van Berne, Heeswijk 19308
Abdij van Middelburg, Middelburg 19282
Abdij van 't Park, Leuven 02471
Abdul Hameed Shoman Public Library, Amman 17886
Kh. M. Abdullaev Institute of Geology and Geophysics, Toshkent 42125
Abdus Salam International Centre for Theoretical Physics, Trieste 16617
Abegg-Stiftung, Riggisberg 27426
Aberdare Library, Aberdare 29921
Aberdeen City Council, Aberdeen 29922
Aberdeen College, Aberdeen 28923
Aberdeenshire Libraries, Oldmeldrum 30056
Aberystwyth University / Prifysgol Aberystwyth, Aberystwyth 28927
– Institute of Biological, Environmental and Rural Sciences (IBERS), Aberystwyth 28928
– Old College Library, Aberystwyth 28929
– Thomas Parry Library, Aberystwyth 28930
ABF, a.s., Praha 06474
Abgeordnetenhaus von Berlin, Berlin 10784
Abia State University, Uturu 19982
Abilene Christian University, Abilene 30163
Abilene Public Library, Abilene 37873
Abilene Public Library, Abilene 37874
Abington Community Library, Clarks Summit 38581
Abington Public Library, Abington 37876
Abington Township Public Library, Abington 37877
Abkhazian D.I. Gulia Institute of Language, Literature and History, Sukhumi 09218
Åbo Akademi, Åbo 07328
– BioCity Library, Turku 07329
– Ekonomiska biblioteket, Åbo 07330
Åbo Akademi University, Åbo 07328
– Library of Economics and Business Administration, Åbo 07330
Aboriginal Affairs Department, Perth 00709
Aboriginal and Torres Strait Islander Commission (ATSIC), Woden 01051
Abraham Baldwin Agricultural College, Tifton 33799
Abraka College of Education, Abraka 19919
Abramtsevo Museum, Abramtsevo 22982
Abtei der Missionsbenediktiner St. Georgenberg-Fiecht, Fiecht 01424
Abtei Ettal, Ettal 11211
Abtei Himmerod, Großlittgen 11229
Abtei Königsmünster, Meschede 11273
Abtei Maria Frieden, Dahlem 11186
Abtei Mariawald, Heimbach 11243
Abtei Münsterschwarzach, Münsterschwarzach 11290
Abtei Neuburg, Heidelberg 11240
Abtei Scheyern, Scheyern 11323
Abtei Scheyern – Byzantinisches Institut, Scheyern 11324
Abtei Speinshart, Speinshart 11331
Abtei St. Hildegard, Rüdesheim 11319
Abtei St. Maria, Fulda 11219
Abtei St. Matthias, Trier 11345
Abtei St. Mauritius, Tholey 11343
Abu Qir Company for Fertilizers and Chemical Industries Co. S.A.E., Alexandria 07132
Abukabar Tafawa Balewa University of Technology, Bauchi 19926
Abul Kalam Azad Oriental Research Institute, Hyderabad 13994
Acad. G. Tsereteli Institute of Oriental Studies, Tbilisi 09223
Academia Argentina de Letras, Buenos Aires 00229
Academia Belgica, Roma 16437
Academia Brasileira de Letras, Rio de Janeiro 03327
Academia Cearense de Letras, Fortaleza 03297
Academia Colombiana de Historia, Santafé de Bogotá 06018
Academia Colombiana de la Lengua, Santafé de Bogotá 06019
Academia das Ciências de Lisboa, Lisboa 21763
Academia de artă teatrală, Târgu Mureş 21946
Academia de arte vizuale 'Ion Andreescu', Cluj-Napoca 21927
Academia de Ciencias Físicas, Matemáticas y Naturales, Caracas 42243
Acadèmia de Ciències Mèdiques de Catalunya i Balears, Barcelona 25877
Academia de Geografía e Historia de Guatemala, Guatemala City 13179
Academia de Guerra Aérea, Santiago 04904
Academia de Historia del Norte de Santander, Cúcuta 06011
Academia de la Historia de Cartagena de Indias, Cartagena 06010
Academia de Letras da Bahia, Salvador 03375
Academia de Letras e Artes do Planalto, Luziânia 03305
Academia de Muzică 'George Dima', Cluj-Napoca 21928
Academia de ştiinţe agricole şi silvice, Bucureşti 21965
Academia de studii economice, Bucureşti 21920
Academia de Studii Social-Politice, Bucureşti 21966
Academia de Teatru şi Film, Bucureşti 21921
Academia Diplomática de Chile, Santiago 04857
Academia Dominicana de la Lengua, Santo Domingo 06964
Academia General del Aire, San Javier 25742
Academia General Militar, Zaragoza 25756
Academia Historica, Taipei 27594
Academia Mexicana de Ciencias, México 18830
Academia Mexicana de la Lengua, A.C., México 18831
Academia Militar das Agulhas Negras, Resende 03228
Academia Nacional de Belas-Artes, Lisboa 21764
Academia Nacional de Ciencias de Bolivia, La Paz 02934
Academia Nacional de Ciencias Económicas, Buenos Aires 00230
Academia Nacional de Ciencias en Córdoba, Córdoba 00296
Academia Nacional de Ciencias Morales y Políticas, Buenos Aires 00231
Academia Nacional de la Historia, Buenos Aires 00232
Academia Nacional de la Historia, Caracas 42244
Academia Nacional de Medicina, Caracas 42245
Academia Nacional de Medicina, Rio de Janeiro 03328
Academia Nacional de Medicina, Santafé de Bogotá 06020
Academia Nacional de Medicina de Buenos Aires, Buenos Aires 00233
Academia Nacional de Medicina de México, México 18832
Academia Naţională de Educaţie Fizică şi Sport, Bucureşti 21952
Academia Paulista de Letras, São Paulo 03392
Academia Pernambucana de Letras, Recife 03320
Academia Portuguesa da Historia, Lisboa 21765
Academia Română, Bucureşti 21915
Academia Română – Filiala din Cluj, Cluj-Napoca 21917
Academia Salvadoreña de la Historia, San Salvador 07208
Academia Sinica, Beijing 04952
Academia Sinica, Taipei 27522
Academia Sinica – Institute of Chemistry, Taipei 27595
Academia Sinica – Institute of Economics, Taipei 27596
Academia Sinica – Institute of Ethnology, Taipei 27597
Academia Sinica – Institute of History and Philology, Nankang 27591
Academia Sinica – Institute of Mathematics, Taipei 27598
Academia Sinica – Institute of Modern History, Taipei 27599
Academia Sinica – Institute of Physics, Taipei 27600
Academia Sinica – Life Science Library, Taipei 27601
Academia Venezolana, Caracas 42246
Academia Wychowania Fizycznego im. Jędrzeja Śniadeckiego, Gdańsk 21035
Academic Choreographical School, Moskva 22680
Academic Information Service, Center of Doshisha University, Kyoto 16998
– Faculty of Law, Kyoto 16999
– Institute of the Study of Humanities and Social Sciences, Kyoto 17000
Académie d'Agriculture de France, Paris 08459
Académie des Sciences, Belles-Lettres et Arts de Lyon, Lyon 08390
Académie des Sciences d'Outre-Mer, Paris 08460
Académie Florimontane, Annecy 08278
Académie Nationale de Médecine, Paris 08461
Académie Nationale Malgache, Antananarivo 18450
Académie royale des Sciences, des Lettres et des Beaux-Arts de Belgique, Bruxelles 02258
Académie Saint-Anselme, Aosta 15705
Academie van Bouwkunst, Amsterdam 19052
Academie van Bouwkunst, Rotterdam 19193
Academy for Social and Political Studies, Bucureşti 21966
Academy Library, UNSW@ADFA, Canberra 00411
Academy of Agricultural and Forestry Sciences, Bucureşti 21965
Academy of Agricultural Business and Management of the Nechernozyom Zone of Russia, Sankt-Peterburg 22511
Academy of Agricultural Sciences of the Republic of Belarus, Minsk 01994
Academy of Agriculture, Kraków 20836
Academy of American Franciscan History Library, Berkeley 35132
Academy of Art College, San Francisco 33728
Academy of Arts, Tiranë 00008
Academy of Arts, Architecture and Design in Prague, Praha 06406
Academy of Athens, Athinai 13069
Academy of Athens – "Ioannis Sykoutris" Library, Athinai 13070
Academy of Business Administration, Yogyakarta 14306
Academy of Civil Aviation, Sankt-Peterburg 22509

Academy of Culture, Sankt-Peterburg 22510
Academy of Drama, Film and Television Arts, Budapest 13265
Academy of Dramatic Art, Târgu Mureş 21946
Academy of Economic Sciences, Bucureşti 21920
Academy of Economic Studies of Moldova Republic, Chişinău 18881
Academy of Education, Maribor 24866
Academy of Fine Arts, Gdańsk 20804
Academy of Fine Arts, Kraków 20837
Academy of Fine Arts, Praha 06382
Academy of Fine Arts, Warszawa 20950
Academy of Fine Arts, Wrocław 21025
Academy of Fine Arts and Design, Bratislava 24663
Academy of Fine Arts Vienna, Wien 01291
Academy of Finland, Helsinki 07442
Academy of Forestry and Timber Technology, Sankt-Peterburg 22527
Academy of Korean Studies, Songnam-si 18121
Academy of Labour and Social Relations, Moskva 22327
Academy of Medical Sciences, London 29709
Academy of Medical Sciences of Ukraine Inst. of Gerontology, Kyiv 28327
Academy of Motion Picture Arts & Sciences, Beverly Hills 30423
Academy of Music, Beograd 24456
Academy of Music, Katowice 20815
Academy of Music, Ljubljana 24809
Academy of Music, Łódź 20873
Academy of Music, Poznań 20901
Academy of Music, Wrocław 21357
Academy of Music and Dramatic Arts, Bratislava 24662
Academy of Music Library, Aalborg 06727
Academy of Music, Theatre and Fine Arts, Chişinău 18882
Academy of Music, Theatre and Fine Arts, Chişinău 18883
Academy of Natural Sciences of Philadelphia, Philadelphia 37337
Academy of Our Lady, Agana 13169
Academy of Performing Arts, Praha 06381
Academy of Physical Education, Wrocław 21001
Academy of Physical Education and Sport, Tiranë 00009
Academy of Romania, Cluj-Napoca Branch, Cluj-Napoca 21917
Academy of Sanskrit Research, Melkote 14076
Academy of Sciences, Tiranë 00015
Academy of Sciences of the Czech Republic, Oriental Institute, Praha 06512
Academy of Sciences of the Republic of Kazakhstan, Almaty 17889
Academy of Sciences of Turkmenistan, Ashkhabad 27904
Academy of Scientific Research and Technology, Cairo 07150
Academy of the Arabic Language, Cairo 07151
Academy of the Chinese People's Armed Police Force, Langfang 05372
Academy of the General Staff, Warszawa 21238
Academy of the Kingdom of Morocco, Rabat 18953
Academy of Theatrical Arts of Russia, Moskva 22326
Academy of Veterinary Medicine, Lviv 28032
Academy of Veterinary Medicine and Biotechnology, Saratov 22570
Academy of Visual Arts 'Ion Adreescu', Cluj-Napoca 21927
Academy of Zoology, Agra 13916
Acadia Parish Library, Crowley 38735
Acadia University, Wolfville 03976
Accademia Albertina delle Belle Arti di Torino, Torino 15482
Accademia Alfonsiana e dei Padri Redentoristi, Roma 16438
Accademia Britannica – International House, Roma 16439
Accademia dei Concordi, Rovigo 14813
Accademia dei Fisiocritici onlus, Siena 16576
Accademia dei Georgofili, Firenze 16248
Accademia della Crusca, Firenze 16249
Accademia delle Scienze dell'Istituto di Bologna, Bologna 16184
Accademia delle Scienze di Ferrara, Ferrara 16244
Accademia delle Scienze di Torino, Torino 16588
Accademia di Agricoltura di Torino, Torino 16589
Accademia di Agricoltura, Scienze e Lettere di Verona, Verona 16664
Accademia di Belle Arti di Bologna, Bologna 15614
Accademia di Belle Arti di Brera, Milano 15624
Accademia di Belle Arti di Carrara, Carrara 16220
Accademia di Belle Arti di Firenze, Firenze 15620
Accademia di Belle Arti di Venezia, Venezia 16641
Accademia di Belle Arti Pietro Vannucci, Perugia 16413
Accademia di Danimarca, Roma 16440
Accademia di Francia, Roma 16441
Accademia di Medicina di Torino, Torino 16590
Accademia di Romania in Roma, Roma 16442
Accademia di Scienze, Lettere ed Arti, Palermo 16394
Accademia di Ungheria, Roma 16443
Accademia Filarmonica di Bologna, Bologna 15615
Accademia Filarmonica di Verona, Verona 16665
Accademia Filarmonica Romana, Roma 15365
Accademia Galileiana di Scienze, Lettere ed Arti in Padova, Padova 14802
Accademia Georgica, Treia 16610
Accademia Gioenia di Catania, Catania 16224
Accademia Ligure di Scienze e Lettere, Genova 14794
Accademia Marchigiana di Scienze, Lettere ed Arti, Ancona 16164
Accademia Medica di Roma, Roma 16444
Accademia Militare di Modena, Modena 15651
Accademia Musicale Chigiana, Siena 15633
Accademia Navale, Livorno 15646
Accademia Nazionale dei Lincei e Corsiniana, Roma 14808
Accademia Nazionale di Agricoltura, Bologna 16185
Accademia Nazionale di Scienze, Lettere e Arti, Modena 14799
Accademia Nazionale Virgiliana di Scienze Lettere e Arti, Mantova 14796
Accademia Peloritana dei Pericolanti, Messina 16315
Accademia Petrarca di Lettere, Arti e Scienze, Arezzo 16169
Accademia Polacca delle Scienze, Roma 14812
Accademia Pontaniana, Napoli 16364
Accademia 'Raffaello', Urbino 16638
Accademia Senese degli Intronati, Siena 16577
Accademia Spoletina, Spoleto 16582
Accademia Tadini, Lovere 16307
Accademia Tiberina, Roma 16445
Accademia Toscana di Scienze e Lettere 'La Colombaria', Firenze 16250
Accra Central Library, Accra 13063
Accra Polytechnic, Accra 12993
Accra Technical Training Centre, Accra 12994
Accrington and Rossendale College, Accrington 28931
ACERALIA, Aviles 25851
Acharya Nagarjuna University, Nagarjunanagar 13716
Acharya Narendra Dev Pustakalaya, Lucknow 14227
Achimota School Library, Accra 13022
Acorn Public Library District, Oak Forest 40560
Acos Especiais Itabira, Timoteo 03266
ACT Electricity and Water Library, Canberra 00887
ACT Government and Assembly Library, Canberra City 00652
ACT Government Library, Canberra 00635
ACT Health Library, Woden 01052
Act Information Resource Center, Iowa City 36962
ACT Library and Information Service, Canberra 01083
Acton College, Acton 29446
Acton Memorial Library, Acton 37879
Acton Public Library, Old Saybrook 40608
ACTS Library, Langley 04208
ACTU Library, Carlton South 00901
Ada Community Library, Boise 38234
Ada Public Library, Ada 37880
Adair County Public Library, Kirksville 39732
Adalbert-Stifter-Institut des Landes Oberösterreich, Linz 01557
Adam Mickiewicz Museum of Literature, Warszawa 21319
Adam Smith College, Kirkcaldy 29148
Adams County Law Library, Gettysburg 34298
Adams County Library System, Gettysburg 39224
Adams Memorial Library, Latrobe 39815
Adams Public Library System, Decatur 38798
Adams & Reese Law Library, New Orleans 35943
Adams State College, Alamosa 30177
Adamson University, Manila 20708
Adams-Pratt Oakland County Law Library, Pontiac 34688
Adat Shalom Synagogue, Farmington Hills 35203
Addis Ababa University, Addis Ababa 07261
– Central Medical Library, Addis Ababa 07262
– Faculty of Technology, Addis Ababa 07263
– Faculty of Technology – South Library, Addis Ababa 07264
– Institute of Ethiopian Studies, Addis Ababa 07265
– Institute of Public Administration, Addis Ababa 07266
– John F. Kennedy Memorial Library, Addis Ababa 07267
Addison Public Library, Addison 37881
Adelaide City Council, Adelaide 01056
Adelaide College of Divinity, Brooklyn Park 00756
Adelaide College of Ministries, Klemzig 00449
Adelaide Institute of TAFE, Adelaide 00559
Adelphi University, Garden City 31205
Adirondack Community College, Queensbury 33666
ADISSEO, Commentry 08341
Adler Museum of the History of Medicine, Johannesburg 25131
Adler School of Professional Psychology, Chicago 36661
Administração do Porto de Lisboa, Lisboa 21718
Administración de Ferrocarriles del Estado, AFE, Montevideo 42033
Administracion Nacional de Combustibles, Alcohol y Portland, ANCAP, Montevideo 42034
Administration of Lenenergo, Sankt-Peterburg 22947
Det Administrative Bibliotek, København 06758
Administrative Bibliothek des Bundes (AB), Wien 01392
Administrative Court of Appeal, Jönköping 26492
Administrative Court of Appeal in Stockholm, Stockholm 26506
The Administrative Library, København 06758
Administrative Office of the United States Courts Library, Washington 34977
Administrative Staff College of India, Hyderabad 13806
Administrative Staff College of Nigeria, Topo-Badagry 19995
Admiral G.I. Nevelsky Far Eastern Marine State Unioversity, Vladivostok 22637
Adolf Fredriks musikklasser, Stockholm 26480
Adolf Reichwein-Bibliothek, Berlin 12654
Adolfo Ibáñez University, Peñalolen 04856
Adomas Mickevicius Vilnius County Public LIbrary, Vilnius 18375
Adrian College, Adrian 30170
Adrian Public Library, Adrian 37882
Adult Migrant Education Service, Adelaide 00845
Adult Migrant English Service, Hobart 00931
Adult Migrant English Service, Sydney 01018
Advanced Centre for Treatment, Research & Education in Cancer, Navi Mumbai 14109
Advanced Military Course for Naval Officers, Sankt-Peterburg 22561
Advanced Teacher Training Institute, Kyiv 28023
Adveniat-Bibliothek, Essen 11208
ADVN – Archief en Documentatiecentrum voor het Vlaams-Nationalisme, Antwerpen 02506
Advocates Library, Edinburgh 29651
Ady Endre Public Library, Baja 13527
Ady Endre Városi Könyvtár, Baja 13527
Adyar Lending Library, Sydney 01019
Adyar Library and Research Centre, Chennai 13955
Adygeiskaya respublikanskaya detskaya biblioteka, Maikop 24163
Adygeiskaya respublikanskaya yunosheskaya biblioteka, Maikop 24164
Adygeiski nauchno-issledovatelski institut yazyka, literatury i istorii, Maikop 23150
Adygey State University, Maikop 22318
Adygeyski gosudarstvenny universitet, Maikop 22318
AECL Library Services, Chalk River 04251
Aegean University Central Library, Bornova 27851
Ærø Folkebibliotek, Ærøskøbing 06842
Aerogeodesy Research Company, Sankt-Peterburg 23780
Aeronautical Research and Test Institute, Praha 06526
Aerospace Corp, El Segundo 35689
Aerospace Museum Association of Calgary Library, Calgary 04316
Aesthetic Realism Foundation, New York 37173
Aetna Inc, Hartford 35736
Affaires étrangères, Commerce extérieur et Coopération au Développement, Bruxelles 02440
Afghanistan Institut und Afghanistan-Museum, Bubendorf 27338
Afghanistan-Museum, Bubendorf 27338
AFIC National Facilities, Sydney 01020
AFPA – Association nationale pour la formation professionnelle des adultes, Montreuil 08425
Africa Institute of South Africa, Pretoria 25153
African Development Bank, Abidjan 06133
African Regional Centre for Technology / Centre Régional Africain de Technologie, Dakar 24430
African Studies Centre, Leiden 19473
African Union Commission, Addis Ababa 07275
Afrika Centrum, Cadier en Keer 19406
Afrikabibliothek, Trier 11344
Afrikamissionare, Trier 11344
Afrika-Studiecentrum (ASC), Leiden 19473
Afro Health Sciences Library and Documentation Centre, Brazzaville 06098
Aftonbladet, Stockholm 26553
AFTRS (Australian Film, Television and Radio School), Stawberry Hills 00516
Ag Research Wallaceuille Animal Research Centre Library, Upper Hutt 19844
Aga Khan University, Karachi 20437
– Faculty of Health Sciences – Medical College, Karachi 20438
Agawam Public Library, Agawam 37884
Age Concern England (ACE), London 29710
Agence de l'Eau Seine-Normandie, Nanterre 08439
Agencia Española de Cooperación Internacional – Biblioteca Hispánica, Madrid 25972
Agencia Española de Cooperación Internacional para el Desarrollo (AECID) – Biblioteca Islámica Félix María Pareja, Madrid 25973
Agencia Nacional de Meteorología, Madrid 25974
Agency for Educational and Cultural Research and Development, Jakarta 14357
Agency for Forestry Research and Development, Jakarta 14356
Agency of Industrial Science and Technology, Tokyo 17311
AGH University of Science and Technology, Kraków 20824
– Faculty of Applied Mathematics, Kraków 20832
– Faculty of Geology, Geophysics and Environmental Protection, Kraków 20827
– Faculty of Management, Krakow 20835
– Faculty of Mechanical Engineering and Robotics, Kraków 20830
– Faculty of Metallurgy and Materials Science, Kraków 20831
– Faculty of Mining and Geoengineering, Kraków 20828
– Faculty of Mining, Surveying and Environmental Engineering, Kraków 20826
Aginskaya okruzhnaya natsionalnaya biblioteka im. Ts. Zhamtsarano, Aginskoe 24100
Agnes Scott College, Decatur 30977
Agraren universitet, Plovdiv 03456
Agrarny universitet, Sankt-Peterburg 22507
AgResearch, Mosgiel 19839
AgResearch Ltd, Hamilton 19833
Agricultura Ganaderia y Desarrollo Rural, México 18833
Agricultural Academy Maize Research Institute, Knezha 03504
Agricultural and Food Library, Praha 06523
Agricultural College and Research Institute Library, Coimbatore 13632
Agricultural College Library, Bapatla 13614
Agricultural Economics Research Institute, Moskva 23385
Agricultural Electrification and Mechanization Research Institute, Yakimivka 28482
Agricultural Engineering and Land Reclamation College, Nova Kakhovka 28125

Agricultural Engineering Institute, Kirovograd 28291
Agricultural Information Centre, Brasília 03220
Agricultural Information Division, Moka 18694
Agricultural Institute, Blagoveshchensk 22220
Agricultural Institute, Institutskoe 42079
Agricultural Institute, Irkutsk 22259
Agricultural Institute, Ivanovo 22265
Agricultural Institute, Kamyanets-Podilsk 27983
Agricultural Institute, Kherson 28001
Agricultural Institute, Kirov 22292
Agricultural Institute, Krasnoyarsk 22309
Agricultural Institute, Kursk 22314
Agricultural Institute, Lugansk 28030
Agricultural Institute, Novosibirsk 23656
Agricultural Institute, Odesa 28060
Agricultural Institute, Penza 22467
Agricultural Institute, Saratov 22578
Agricultural Institute, Solyanka 23978
Agricultural Institute, Stavropol 22586
Agricultural Institute, Ulyanovsk 22617
Agricultural Institute, Uralsk 17924
Agricultural Institute, Zhanbas-Zhan 42169
Agricultural Institute of Slovenia, Ljubljana 24922
Agricultural Irrigation Research Institute, Kherson 28288
Agricultural Museum, Cairo 07152
Agricultural Planning Institute, Toshkent 42126
Agricultural Research Centre, Tripoli 18256
Agricultural Research Corporation, Wad Medani 26359
Agricultural Research Council of Nigeria, Ibadan 20014
Agricultural Research Council – Research Centre for the Soil-Plant System (CRA-RPS), Roma 16480
Agricultural Research Institute, Hovd 18908
Agricultural Research Institute, Kroměříž 06459
Agricultural Research Institute, Reykjavík 13573
Agricultural Science Information Center, Taipei 27602
Agricultural Services Library, East Melbourne 00670
Agricultural State Academy, Nizhni Novgorod 22421
Agricultural University, Plovdiv 03456
Agricultural University of Athens, Athinai 13072
Agricultural University of Hebei, Baoding 04986
Agricultural University of Iceland, Borgarnes 13560
Agricultural University of Tirana, Tiranë 00011
Agricultural University Peshawar, Peshawar 20451
Agriculture & Agri-Food Canada, Lethbridge 04092
Agriculture & Agri-Food Canada, Ottawa 04428
Agriculture & Agri-Food Canada, Sainte-Foy 04149
Agriculture & Agri-Food Canada – Plant Research Library, Ottawa 04429
Agriculture Canada, Saint-Jean-sur-Richelieu 04154
Agriculture, Forestry and Fisheries Finance Corporation, Tokyo 17474
Agriculture, Forestry and Fisheries Research Information Center, Tsukuba 17805
Agrifood Research and Technology Centre of Aragon, Zaragoza 26190
Agritec, Research, Breeding and Services, Ltd., Sumperk 06539
Agrocampus Rennes, Rennes 08019
Agrochemical Research Institute, Kyiv 28313
Agrofizicheski nauchno-issledovatelski institut, Sankt-Peterburg 23745
Agro-inzhenerny universitet V.P. Goryachkina, Moskva 22324
Agromechanics, Gyanzha 01724
Agronomical Research Institute, Bari 16175
AgroParisTech – Ecole Nationale du Génie Rural, des Eaux et des Forêts, Paris 07836
AgroParisTech – Ecole Nationale du Génie Rural, des Eaux et des Forêts (ENGREF), Nancy 07783
AgroParisTech – Institut des sciences et industries du vivant et de l'environnement, Paris 08462
Agroprojekt, Praha 06423
Ägyptisches Museum, Berlin 11563
Ahfad University for Women, Omdurman 26350
Ahmadu Bello University, Zaria
– Advanced Teachers' College, Zaria 19984
– Center of Islamic Legal Studies, Zaria 19985
– College of Agriculture, Zaria 19986
– Institue for Agricultural Research Library, Zaria 19987
– Institute of Administration, Zaria 19988
– Kashim Ibrahim Library, Zaria 19989
– LABU Teaching Hospital, Zaria 19990
– Lee T. Railsback Library, Zaria 19991
– National Animal Production Research Institute (NAPRI), Zaria 19992
Ahmedabad Textile Industry's Research Association, Ahmedabad 13917
AIB – Vinçotte Belgium vzw, Brussel 02500
Aichi Arts Center / Aichi Prefectural Library, Nagoya 17850
Aichi Cancer Center, Nagoya 17612
Aichi Daigaku, Toyohashi 17218
Aichi Gakuin Daigaku, Aichi 16898
– Keiei Kenkyujo, Aichi 16899
– Shigakubu Bunkan, Nagoya 16900
Aichi Gakuin University, Aichi 16898
– Research Institute of Business Administration, Aichi 16899
– School of Dentistry, Nagoya 16900
Aichi Geijutsu Bunka Senta, Nagoya 17850
Aichi Prefectural Assembly, Aichi 17282
Aichi Prefectural Education Center, Aichi 17509
Aichi Prefectural Labour Center, Nagoya 17613
Aichi University, Toyohashi 17218
Aichiken Gan Senta, Nagoya 17612
Aichiken Gikai, Nagoya 17282
Aichiken Kinrokaikan Rodo Toshoshiryoshitsu, Nagoya 17613
Aichiken Kyoiku Senta, Aichi 17509
AIDS Fonds, Amsterdam 19380
Aiken Technical College, Graniteville 33382
Aiken-Bamberg-Barnwell-Edgefield Regional Library System, Aiken 37887
Aims Community College, Greeley 33386
Ain Shams University, Cairo 07034
– Faculty of Agriculture, Cairo 07035
– Faculty of Arts, Cairo 07036
– Faculty of Commerce, Cairo 07037
– Faculty of Education, Cairo 07038
– Faculty of Engineering, Cairo 07039
– Faculty of Law, Cairo 07040
– Faculty of Medicine, Cairo 07041
– Faculty of Sciences, Cairo 07042
– The Women's College, Cairo 07043
Air Force Research Laboratory, Eglin AFB 34169
Air Headquarters Technical Reference Library, New Delhi 13865
Air Products & Chemicals, Inc, Allentown 35440
Air Traffic and Airport Administration, Airport Library and Documentation Bureau, Budapest 13320
Air Transport Association of America Library, Washington 37713
Air-Ix Oy, Tampere 07431
Airlangga University Library, Surabaya 14298
– Faculty of Law, Surabaya 14299
– Faculty of Medicine, Surabaya 14300
Airline Pilots Association International, Herndon 36927
Airservices Australia Aviation Business Intelligence Services, Canberra 00636
Aisberg Central Design Bureau, Sankt-Peterburg 23922
Ajax Public Library, Ajax 04565
Ajia Daigaku, Musashino 17048
Ajia-Afurika Toshokan, Mitaka 17598
Ajtte, Svenskt Fjäll- och Samemuseum, Jokkmokk 26594
Ajtte, Swedish Mountain- and Sámi Museum, Jokkmokk 26594
Aju University, Suwon 18100
Ajuntament de Barcelona, Barcelona 25680
Ajuntament de Barcelona – Centro de Documentación Estadística, Barcelona 25681
AK Bibliothek Wien für Sozialwissenschaften, Wien 01586
AK Niederösterreich, Wien 01615
Akaa Public Library, Toijala 07613
Akaan kaupunginkirjasto, Toijala 07613
Akademi Akutansi, Yogyakarta 14304
Akademi Militer, Magelang 14337
Akademi Pembangunan Masjarakat Desa, Yogyakarta 14305
Akademi Pimpinan Perusahaan, Yogyakarta 14306
Akademia Athenon, Athinai 13069
Akademia Athenon – Vivliothiki "Ioannis Sykoutris", Athinai 13070
Akademia e Arteve, Tiranë 00008
Akademia e Edukimit Fizik dhe e Sportit 'Vojo Kushi', Tiranë 00009
Akademia e Shkencave, Tiranë 00015
Akademia Ekonomiczna, Kraków 20822
Akademia Ekonomiczna, Poznań 20899
Akademia Ekonomiczna im. Karola Adamieckiego, Katowice 20814
Akademia Górniczo-Hutnicza, Kraków 20823
Akademia Górniczo-Hutnicza im. Stanislawa Staszica, Kraków 20824
– Wydział Fizyki i Informatyki Stosowanej, Kraków 20825
– Wydział Geodezji Górniczej i Inżynierii Środowiska, Kraków 20826
– Wydział Geologii, Geofizyki i Ochrony Srodowiska, Kraków 20827
– Wydział Górnictwa i Geoinżynierii, Kraków 20828
– Wydział Humanistyczny, Kraków 20829
– Wydział Inżynierii Mechanicznej i Robotyki, Kraków 20830
– Wydział Inżynierii Metali i Informatyki Przemyslowej, Kraków 20831
– Wydział Matematyki Stosowanej, Kraków 20832
– Wydział Odlewnictwa, Kraków 20833
– Wydział Wiertnictwa, Nafty i Gazu, Kraków 20834
– Wydział Zarządzania, Krakow 20835
Akademia Marynarki Wojennej im. Bohaterów Westerplatte, Gdynia 20809
Akademia Medyczna, Białystok 20793
Akademia Medyczna, Wrocław 21000
Akademia Medyczna im. Karola Marcinkowskiego w Poznaniu, Poznań 20900
Akademia Medyczna w Gdańsku, Gdańsk 20803
Akademia Morska, Gdynia 20810
Akademia Muzyczna, Łódź 20873
Akademia Muzyczna im. Ignacego Jana Paderewskiego, Poznań 20901
Akademia Muzyczna im. Karola Lipińskiego, Wrocław 21357
Akademia Muzyczna im. Karola Szymanowskiego, Katowice 20815
Akademia Pedagogiki Specjalnej im. Marii Grzegorzewskiej, Warszawa 20949
Akademia Rolnicza im. Hugona Kołłątaja, Kraków 20836
Akademia Sztabu Generalnego im. Gen. Broni Karola Świerczewskiego, Warszawa 21238
Akademia Sztuk Pięknych, Gdańsk 20804
Akademia Sztuk Pięknych, Kraków 20837
Akademia Sztuk Pięknych, Warszawa 20950
Akademia Wychowania Fizycznego, Wrocław 21001
Akademia Wychowania Fizycznego im. Bronisława Czecha w Krakowie, Kraków 20838
Akademia Wychowania Fizycznego im. Eugeniusza Piaseckiego, Poznań 20902
Akademia Wychowania Fizycznego Józefa Piłsudskiego, Warszawa 20951
Akademicheski Maly Teatr, Moskva 23162
Akademie Auswärtiger Dienst, Berlin 10788
Akademie der Bildenden Künste, München 10358
Akademie der bildenden Künste Wien, Wien 01291
Akademie der Bundeswehr für Information und Kommunikation, Strausberg 12509
Akademie der Wissenschaften und der Literatur Mainz, Mainz 12229
Akademie für Bildende Künste, Mainz 10265
Akademie für Gesundheitsberufe Heidelberg gGmbH, Heidelberg 10030
Akademie für Öffentliches Gesundheitswesen in Düsseldorf, Düsseldorf 10721
Akademie für Politische Bildung, Tutzing 12566
Akademie Klausenhof, Hamminkeln 10737
Akademie múzických umění, Praha 06381
Akademie věd České Republiky, Praha 06351
Akademie výtvarných umění, Praha 06382
Akademija za Glasbo, Ljubljana 24809
Akademiska sjukhuset, Uppsala 26705
Akademiya fizicheski kultury im. P.F. Lesgafta, Sankt-Peterburg 22508
Akademiya grazhdanskoi aviatsii, Sankt-Peterburg 22509
Akademiya kommunalnogo khozyaistva im. K.D. Pamfilova, Moskva 22325
Akademiya kultury, Sankt-Peterburg 22510
Akademiya menedzhmenta i agrobisnesa Nechernozemnoi zony Rossii, Sankt-Peterburg 22511
Akademiya nauk Belarusi – Institut mekhaniki metallopolimernykh sistem im. V.A. Belogo, Gomel 01951
Akademiya nauk Belarusi – Otdelenie instituta matematiki, Gomel 01952
Akademiya teatralnogo iskusstva Rossii, Moskva 22326
Akademiya truda i sotsialnykh otnosheni, Moskva 22327
Akademiya veterinarnoi meditsiny i biotekhnologii, Saratov 22570
Akaki Tsereteli State University Scientific Library, Kutaisi 09191
Akanu Ibiam Federal Polytechnic, Unwana, Afikpo 19921
Akdeniz Üniversitesi, Antalya 27850
Akerman, Senterfitt & Eidson PA, Orlando 36072
Åkersberga bibliotek, Åkersberga 26723
Akershus County Library, Kjeller 20354
Akershus University College, Kjeller 20097
– Faculty of Social Education, Lillestrøm 20099
Akershus University Hospital, Nordbyhagen 20238
Akiba Hebrew Academy, Bryn Mawr 36604
Akin, Gump, Strauss, Hauer & Feld Library, Dallas 35642
Akin, Gump, Strauss, Hauer & Feld LLP, San Antonio 36214
Akin, Gump, Strauss, Hauer & Feld LLP, Washington 36327
Akita Daigaku, Akita 16901
– Igakubu Bunkan, Akita 16902
Akita Kenritsu Hokubutsukan, Akita 17510
Akita Kenritsu Nogyo Tanki Daigaku, Akita 17241
Akita Prefectural Administrative Record Office, Akita 17251
Akita Prefectural Junior College of Agriculture, Akita 17241
Akita Prefectural Library, Akita 17834
Akita Prefectural Museum, Akita 17510
Akita University, Akita 16901
– Medical Branch, Akita 16902
Akitaken Somubu Bunshokohoka, Akita 17251
Akmenės rajono savivaldybės viešoji biblioteka, N. Akmenė 18343
Akmola Civil Engineering Institute, Akmola 17891
Akron Art Museum, Akron 36442
Akron Law Library Association, Akron 36443
Akron-Summit County Public Library, Akron 37888
Akron-Summit County Public Library – Ellet, Akron 37889
Akron-Summit County Public Library – Fairlawn-Bath, Akron 37890
Akron-Summit County Public Library – Goodyear, Akron 37891
Akron-Summit County Public Library – Green, Uniontown 41639
Akron-Summit County Public Library – Kenmore, Akron 37892
Akron-Summit County Public Library – Mogadore Branch, Mogadore 40250
Akron-Summit County Public Library – Nordonia Hills, Northfield 40536
Akron-Summit County Public Library – Northwest Akron, Akron 37893
Akron-Summit County Public Library – Norton Branch, Norton 40543
Akron-Summit County Public Library – Richfield Branch, Richfield 40973
Akron-Summit County Public Library – Tallmadge Branch, Tallmadge 41507
Aktsionerno tovarishchestvo 'Naukovo-doslidny instytut po teplu ta gazopostachannyu, kompleksnomu blagoustriyu mist i sil Ukrainy', Kyiv 28155
Aktsionernoe obshchestvo Bashkirgeologia, Ufa 22966
Aktsionernoe tovarishchestvo instytut reabilitatsiyi Sechenova, Yalta 28172
Akureyri Hospital, Medical Library, Akureyri 13566
Akusticheski institut im. N.N. Andreeva RAN, Moskva 23163
Akwesasne Library & Culture Center, Hogansburg 36928
Al al-Bayt University, Mafraq 17877
Ål bibliotek, Ål 20303
Al Zahiriah, Damascus 27520
Alabama Agricultural & Mechanical University, Huntsville 31375
Alabama Department of Archives & History Research Room, Montgomery 37127
Alabama Institute for the Deaf & Blind, Talladega 37654
Alabama Power Co, Birmingham 35505
Alabama Public Library Service, Montgomery 40275
Alabama Southern Community College, Monroeville 33551
Alabama Southern Community College, Thomasville 32814
Alabama State University, Montgomery 31884
Alabama Supreme Court & State Law Library, Montgomery 34553
Alabin Samara Museum of Regional Studies, Samara 23744

Alacheevsk Metallurgical Plant Joint-Stock Company, Alchevsk 28135
Alachua County Library District, Gainesville 39179
Al-Ahliyya Amman University, Amman 17871
Al-Ahram Institution, Cairo 07153
Al-Akhbar Institution, Cairo 07154
Alamance County Public Libraries, Burlington 38392
Alameda County Library, Fremont 39161
Alameda County Library – Albany Branch, Albany 37896
Alameda County Library – Castro Valley Branch, Castro Valley 38479
Alameda County Library – Dublin Branch, Dublin 38867
Alameda County Library – Newark Branch, Newark 40451
Alameda County Library – San Lorenzo Branch, San Lorenzo 41191
Alameda County Library – Union City Branch, Union City 41636
Alameda Free Library, Alameda 37895
Alamo Heights United Methodist Church Library, San Antonio 35371
The Åland Parliament, Mariehamn 07404
Ålands lagting, Mariehamn 07404
Ålands Museum Bibliotek, Mariehamn 07528
Alanus Hochschule für Kunst und Gesellschaft, Alfter 09391
Alanus University of Arts and Social Sciences, Alfter 09391
Al-Aqsa Mosque Library, Jerusalem 14696
Al-Arab Medical University, Benghazi 18236
Alaska Resources Library & Information Services (ARLIS), Anchorage 30219
Alaska State Court Law Library, Anchorage 33919
Alaska State Department of Corrections, Eagle River 34162
Alaska State Library – Alaska Historical Collections, Juneau 30137
Alaska State Library – Archives & Museums, Juneau 30138
AL-Assad National Library, Damascus 27510
Al-Awqaf Central Library, Baghdad 14451
Al-Azhar University, Cairo 07044
– Faculty of Arabic Studies, Cairo 07045
– Faculty of Engineering, Madeenet Nasr 07046
– Faculty of Islamic Law, Cairo 07047
– Faculty of Linguistics and Translation, Cairo 07048
– Islamic College for Women, Madeenet Nasr 07049
Al-Baath University, Homs 27515
Albanian Assembly, Juridical Department, Tiranë 00013
Albany County Public Library, Laramie 39807
Albany Institute of History & Art, Albany 36444
Albany Law School, Albany 30178
Albany Medical College, Albany 30179
Albany Museum, Grahamstown 25124
Albany Public Library, Albany 37897
Albany Public Library, Albany 37898
Albany State University, Albany 30180
Albemarle Corporation, Baton Rouge 35493
Albemarle Regional Library, Winton 41944
Albert C. Wagner Youth Correctional Facility Library, Bordentown 33996
Albert Einstein College of Medicine, Bronx 30554
Albert L. Scott Library, Alabaster 37894
Albert Lea Public Library, Albert Lea 37900
Albert Schweitzer Ziekenhuis, Dordrecht 19418
Albert Wisner Public Library, Warwick 41732
Alberta Bible College, Calgary 04199
Alberta College of Art & Design, Calgary 03980
Alberta Department of Energy – Alberta Geological Survey, Edmonton 04060
Alberta Department of Environment Library, Edmonton 04061
Alberta Department of Justice – Corporated Reference Center, Edmonton 04062
Alberta Distance Learning Centre Library, Barrhead 04311
Alberta Economic Development Library, Edmonton 04063
Alberta Education, Edmonton 04332
Alberta Energy & Utilities Board Library, Calgary 04317
Alberta Government Library, Edmonton 04333
Alberta Government Library, Edmonton 04064
Alberta Government Library, Edmonton 04334
Alberta Government Library – Food Processing Development Centre, Leduc 04091

Alberta Government Library – Neil Crawford Provincial Centre Library, Edmonton 04065
Alberta Historical Resources Foundation Library, Edmonton 04335
Alberta Justice, Grande Prairie 04080
Alberta Justice – Law Library, Edmonton 04066
Alberta Justice Law Society Library, Lethbridge, Lethbridge 04373
Alberta Law Libraries, Edmonton 04067
Alberta Law Society Libraries, Calgary 04053
Alberta Law Society Libraries, Edmonton 04336
Alberta Law Society Libraries, Peace River 04455
Alberta Law Society Libraries, Red Deer 04465
Alberta Legislature Library, Edmonton 04068
Alberta Research Council, Edmonton 04337
Alberta Research Council Library, Vegreville 04177
Alberta School for the Deaf Library, Edmonton 03986
Alberta Teachers' Association Library, Edmonton 04338
Alberta Union of Provincial Employees Library, Edmonton 04069
Alberta Wilderness Association Library, Calgary 04318
Albertina, Wien 01587
Albert-Ludwigs-Universität Freiburg, Freiburg 09758
– Bibliothek der Technischen Fakultät, Freiburg 09759
– Bibliothek Forstwissenschaft, Freiburg 09760
– Bibliothek für Chemie und Pharmazie, Freiburg 09761
– Bibliothek für Rechtswissenschaft, Freiburg 09762
– Bibliothek Geographie, Hydrologie und Völkerkunde, Freiburg 09763
– Deutsches Seminar I, Arbeitsbereich Badisches Wörterbuch, Freiburg 09764
– Deutsches Seminar I, Institut für Deutsche Sprache und Ältere Literatur, Freiburg 09765
– Deutsches Seminar II, Neuere Deutsche Literatur, Freiburg 09766
– Englisches Seminar I und II, Freiburg 09767
– Fachbereichsbibliothek Philosophie und Erziehungswissenschaft, Freiburg 09768
– Fakultätsbibliothek Biologie I (Zoologie), Freiburg 09769
– Fakultätsbibliothek Biologie II + III (Botanik), Freiburg 09770
– Forschungsstelle für Kirchenrecht und Staatskirchenrecht, Freiburg 09771
– Historisches Seminar, Abt. Landesgeschichte, Freiburg 09772
– IGPP-Bibliothek, Freiburg 09773
– Institut für Archäologische Wissenschaften, Freiburg 09774
– Institut für Geowissenschaften, Freiburg 09775
– Institut für Geschichte der Medizin, Freiburg 09776
– Institut für Kriminologie und Wirtschaftsstrafrecht, Freiburg 09777
– Institut für Physik, Freiburg 09778
– Institut für Psychologie, Freiburg 09779
– Institut für Rechtsgeschichte und Geschichtliche Rechtsvergleichung, Freiburg 09780
– Institut für Sport und Sportwissenschaft, Freiburg 09781
– Institut für Staatswissenschaft und Rechtsphilosophie, Freiburg 09782
– Institut für Systematische Theologie, Arbeitsbereich Moraltheologie, Freiburg 09783
– Institut für Ur- und Frühgeschichte und Archäologie des Mittelalters, Freiburg 09784
– Institut für Volkskunde, Freiburg 09785
– Kunstgeschichtliches Institut, Freiburg 09786
– Mathematisches Institut, Freiburg 09787
– Musikwissenschaftliches Seminar, Freiburg 09788
– Orientalisches Seminar, Freiburg 09789
– Romanisches Seminar, Freiburg 09790
– Seminar für Alte Geschichte, Freiburg 09791
– Seminar für Klassische Philologie, Freiburg 09792
– Seminar für Lateinische Philologie des Mittelalters, Freiburg 09793
– Slavisches Seminar, Freiburg 09794
– Sprachwissenschaftliches Seminar, Freiburg 09795
– Theologische Fakultät, Freiburg 09796

– Universitätsklinikum Freiburg – Neurozentrum, Freiburg 09797
– Verbundbibliothek im KG IV, Freiburg 09798
– Volkswirtschaftliches Seminar, Freiburg 09799
– Zentrum für Kinderheilkunde und Jugendmedizin, Freiburg 09800
Alberton Public Library, Alberton 25182
Albertslund Bibliotek, Albertslund 06843
Albertson College of Idaho, Caldwell 30607
Albertus Magnus College, New Haven 31970
Albertus-Magnus-Gymnasium, Rottweil 10765
Albertus-Magnus-Institut, Bonn 11620
Albertville Public Library, Albertville 37902
Albert-Westmorland-Kent Regional Library, Moncton 04688
Albion College, Albion 30185
Albion District Library, Albion 37903
Albright College, Reading 32418
W. F. Albright Institute of Archaeological Research, Jerusalem 14722
Albright-Knox Art Gallery, Buffalo 36606
The Albuquerque Museum of Art & History, Albuquerque 36446
Alcan International Ltd, Kingston 04260
Alcatel Lucent, Antwerpen 02496
Alcatel SEL AG, Stuttgart 11427
Alcatel USA, Inc, Plano 36135
Alcoa of Australia Ltd, Booragoon 00811
Alcoa Technical Center Library, Alcoa Center 36452
ALCOA World Alumina Australia Information Services, Applecross 00808
Alcohol and other Drugs Council of Australia, Deakin 00907
Alcohol + Drug Service, Brisbane 00875
Alcona County Library System, Harrisville 39378
Alcorn State University, Alcorn State 30192
Alderney Library, Alderney 29924
Alderson-Broaddus College, Philippi 32286
Aldrich Public Library, Barre 38090
Alemannisches Institut Freiburg e.V., Freiburg 11883
Alemaya University of Agriculture, Dire Dawa 07270
– Agricultural Research Center, Debre Zeit 07271
Aleš South Bohemian Gallery, Hluboká nad Vltavou 06449
Ålesund bibliotek, Ålesund 20304
Alexander County Library, Taylorsville 41525
Alexander Graham Bell Association for the Deaf & Hard of Hearing, Washington 37714
Alexander Mitchell Public Library, Aberdeen 37872
Alexander Turnbull Library, Wellington 19760
'Alexandr & Aristia Aman' Country Library, Craiova 22087
Alexandria Library, Alexandria 37905
Alexandria Municipal Library, Alexandria 07195
Alexandria University, Alexandria 07002
– Faculty of Agriculture, Alexandria 07003
– Faculty of Arts, Alexandria 07004
– Faculty of Commerce, Alexandria 07005
– Faculty of Dentistry, Alexandria 07006
– Faculty of Education, Alexandria 07007
– Faculty of Engineering, Alexandria 07008
– Faculty of Law, Alexandria 07009
– Faculty of Medicine, Alexandria 07010
– Faculty of Physical Education (for Girls), Alexandria 07011
– Faculty of Sciences, Alexandria 07012
– Higher Agricultural Institute for Cotton Technology, Alexandria 07013
– Higher Institute of Nursing, Alexandria 07014
– Higher Institute of Public Health, Alexandria 07015
Alexandrian Public Library, Mount Vernon 40322
N.N. Alexandrov National Cancer Center of Belarus, p. Lesnoy-2 01967
Al-Farabi Kazakh State National University, Almaty 17904
Al-Fateh University, Tripoli 18244
– Faculty of Agriculture, Tripoli 18245
– Faculty of Education, Tripoli 18246
– Faculty of Engineering, Tripoli 18247
– Faculty of Science, Tripoli 18248
Alfred Dickey Public Library, Jamestown 39632
Alfred Hospital, Prahran 00988
Alfred University, Alfred 30194
– Scholes Library of Ceramics, Alfred 30195
Alfred Wegener Institute for Polar and Marine Research, Bremerhaven 11693
Alfred-Wegener-Institut für Polar- und Meeresforschung in der Helmholtz-Gemeinschaft, Bremerhaven 11693
Algemeen Archief der Vlaamse Minderbroeders (APB), Sint-Truiden 02488

Algemeen Belgisch Vakverbond (ABVV), Brussel 02554
Algemeen Burgerlijk Pensioenfonds (ABP), Heerlen 19458
Algemeen Christelijk Vakverbond (ACV), Brussel 02555
Algemeen Rijksarchief, Bruxelles 02560
Algemeen Rijksarchief (ARA), Brussel 02556
Algemene Centrale der Liberale Vakbonden van België (ACLVB) / Centrale Générale des Syndicats Liberaux de Belgique (CGSLB), Gent 02642
Algemene Rekenkamer, Den Haag 19250
Algoma University College, Sault Sainte Marie 03858
Algonquin Area Public Library District, Algonquin 37908
Algonquin College of Applied Arts & Technology, Ottawa 04015
Alhambra Public Library, Alhambra 37909
Al-Husn Polytechnic, Al-Husn 17870
Aliança Francesa, Rio de Janeiro 03329
Alianza Cultural Colombo-Francesa, Santafé de Bogotá 06021
Alianza Francesa, Buenos Aires 00234
Alianza Francesa, Montevideo 42035
Alianza Francesa de A Coruña, A Coruña 25946
Alianza Francesa en Venezuela, Caracas 42247
Alice Lloyd College, Pippa Passes 33645
Alice Public Library, Alice 37910
Alice Salomon Hochschule Berlin, Berlin 09401
Aligarh Muslim University, Aligarh 13598
Alingsås bibliotek, Alingsås 26724
Alion Science & Technology, Annapolis 36470
Alion Science & Technology, Rome 37466
Alisher Navoi State Public Library of Uzbekistan, Toshkent 42069
All Hallows College, Dublin 14464
All India Congress Committee Library, New Delhi 13866
All India Institute of Hygiene and Public Health, Kolkata 13681
All India Institute of Medical Sciences, New Delhi 13721
All India Radio, New Delhi 14110
All Pakistan Educational Conference, Karachi 20439
All Russian Research Institute of Sugar Beet and Sugar, Ramon 23714
All Saints Catholic Church, Dallas 35174
All Souls College, Oxford 29299
Allahabad Public Library, Allahabad 14223
Allama Iqbal Open University, Islamabad 20430
Allameh Tabatabai University, Tehran 14399
– Faculty of Social Science, Tehran 14400
Allami Biztosító, Budapest 13351
Allan Hancock College, Santa Maria 33746
A'llatorvostudományi Egyetem, Hódmezővásárhely 13277
Allatorvos-tudományi Könyvtár, Levéltár és Múzeum, Budapest 13217
– Kosáry Domokos Könyvtár és Levéltár, Gödöllo 13218
Allatorvos-tudományi Kutatóintézete Könyvtár, Budapest 13432
Állattenyésztési és Takarmányozási Kutatóintézet, Herceghalom 13479
Allegany College of Maryland, Cumberland 33267
Allegany County Circuit Court Law Library, Cumberland 34120
Allegany County Library System, Cumberland 38742
Allegheny College, Meadville 31781
Allegheny County Court of Common Pleas Library, Pittsburgh 34677
Allegheny County Health Department Library, Pittsburgh 34678
Allegheny County Law Library, Pittsburgh 34679
Allegheny General Hospital, Pittsburgh 37377
Allegheny Wesleyan College Library, Salem 32546
Allen Community College, Iola 33429
Allen Correctional Institution Library, Lima 34455
Allen County Law Library, Lima 34456
Allen County Law Library Association, Inc, Fort Wayne 36864
Allen County Public Library, Fort Wayne 39118
Allen County Public Library – Aboite, Fort Wayne 39119
Allen County Public Library – Dupont, Fort Wayne 39120
Allen County Public Library – Georgetown, Fort Wayne 39121
Allen County Public Library – Little Turtle, Fort Wayne 39122

Allen County Public Library – Shawnee, Fort Wayne 39123
Allen County Public Library – Waynedale, Fort Wayne 39124
Allen Library, West Hartford 33020
Allen Parish Libraries, Oberlin 40579
Allen Park Public Library, Allen Park 37914
Allen Public Library, Allen 37913
J. C. Allen & Son, Inc, West Lafayette 36409
Allendale-Hampton-Jasper Regional Library, Allendale 37915
Allens Arthur Robinson Library Brisbane, Brisbane 00812
Allens Arthur Robinson Library Melbourne, Melbourne 00819
Allens Arthur Robinson Library Sydney, Sydney 00831
Allentown Art Museum, Allentown 36457
Allentown Public Library, Allentown 37916
Allerød Bibliotek, Allerød 06845
Allgemeine Lesegesellschaft Basel, Basel 27472
Allgemeines Krankenhaus Barmbek, Hamburg 11967
Alliance Française, Accra 13039
Alliance Française, Antananarivo 18455
Alliance Française, Bangkok 27773
Alliance Française, Buenos Aires 00234
Alliance Française, Bujumbura 03626
Alliance Française, Colombo 26302
Alliance Française, Dakar 24431
Alliance Française, Dublin 14520
Alliance Française, Glasgow 29671
Alliance Française, Kinshasa 06067
Alliance Française, Montevideo 42035
Alliance Française, Mumbai 14077
Alliance Française, New Delhi 14111
Alliance Française, New York 37204
Alliance Française, Paris 08128
Alliance Française, Port-au-Prince 13203
Alliance Française, Quito 06989
Alliance Israélite Universelle (AIU), Paris 08463
Alliance Sud, Bern 27317
Alliant International University, Alhambra 30197
Alliant International University, Fresno 31177
Alliant International University, San Diego 32575
Alliant International University, San Francisco 32583
Allianz Deutschland AG, München 11418
Allina Health System Library Services, Minneapolis 37117
Alloa Library, Alloa 29925
Alloy Steels Plant, Durgapur 13903
All-Pakistan Educational Conference Library, Karachi 20490
All-Russia Research Institute of Horse Breeding, Ryazan 23731
All-Russia Scientific Research Institute of Agroforestry Reclamation, Volgograd 24063
All-Russia Scientific Research Institute of Fats, Sankt-Peterburg 23933
All-Russian Chamber of Commerce, Moskva 23544
All-Russian Construction and Technological Institute of Mechanical Engineering, Moskva 23595
All-Russian D.I. Mendeleev Institute of Metrology, Ekaterinburg 23038
All-Russian Drilling Technology Research Institute, Moskva 23557
All-Russian Extra-Mural Institute of Agriculture, Balashikha 22209
All-Russian Extra-Mural Institute of Communication Engineering, Moskva 22408
All-Russian Geological Oil Prospecting Research Institute, Moskva 23549
All-Russian Institute of Heat Engineering, Moskva 23593
All-Russian Institute of Poultry-Farming, Sergyev Posad 23966
All-Russian Institute of Research of New Antibiotics, Moskva 23547
All-Russian Institute of the Economics of Mineral Resources and Geological Prospecting, Moskva 23558
All-Russian Oil and Land-Prospecting Institute, Perm 23693
All-Russian Research and Construction Institute of Food Technology and Engineering, Moskva 23550
All-Russian Research Construction and Technological Institute of Hydroengineering, Moskva 23590
All-Russian Research Inst. for Butter-and Cheesemaking, Uglich 22969
All-Russian Research Institute for Hydraulic Engineering and Land Reclamation, Moskva 23563
All-Russian Research Institute for the Ball Bearing Industry, Moskva 23551
All-Russian Research Institute of Animal Husbandry, Dubrovitsy 23026
All-Russian Research Institute of Building Materials and Construction P.P. Budnikov, Kraskovo 23109
All-Russian Research Institute of Cinematography, Moskva 23589
All-Russian Research Institute of Electromechanics, Moskva 23164
All-Russian Research Institute of Equipments, Moskva 23591
All-Russian Research Institute of Fertilizers and Soil Science, Moskva 23586
All-Russian Research Institute of Fish Farming, Rybnoe 23737
All-Russian Research Institute of Fodder Crops, Lugovaya 23144
All-Russian Research Institute of Forestry and Mechanization of Forestry, Pushchino 23710
All-Russian Research Institute of Fuel and Energy Problems (VNIIKTEP), Moskva 23165
All-Russian Research Institute of Geophysical Prospecting, Sankt-Peterburg 23944
All-Russian Research Institute of Heat Engineering, Moskva 23555
All-Russian Research Institute of Medical Plants, Vilp 24044
All-Russian Research Institute of Mineral Synthesis, Aleksandrov 22984
All-Russian Research Institute of Natural Gas & Gas Technologies, Vidnoe 24043
All-Russian Research Institute of Phytopathology, Bolshie Vyazemy 23006
All-Russian Research Institute of Pulp & Paper Industry (VNIIB) Scientific-Technical Library, Sankt-Peterburg 23934
All-Russian Research Institute of Vitamins, Moskva 23594
All-Russian Research Textile Institute, Serpukhov 23967
All-Russian Scientific Research Institute for Electrification of Agriculture (VIESH), Moskva 23559
All-Russian Theatrical Society, Kabardino-Balkarian Department, Nalchik 23610
All-Russian Theatrical Society, Krasnodar Department, Krasnodar 23120
All-Russian Theatrical Society, Kursk Department, Kursk 23138
All-Russian Theatrical Society, Primorsk Department, Vladivostok 24061
All-Union Institute for Scientific and Technical Information, Moskva 23546
Alma College, Alma 30202
Alma Public Library, Alma 37920
Almaty Institute of Power Engineering and Telecommunication, Almaty 17896
Almaty Kurman-gazy State Conservatoire, Almaty 17897
Almedalsbiblioteket, Visby 26372
Älmhults bibliotek, Älmhult 26725
Almo Collegio Capranica, Roma 16015
Al-Mustansiriyah University, Baghdad 14445
Alnwick Castle Library, Alnwick 29594
Alpena Community College, Alpena 33125
Alpena County George N. Fletcher Library, Alpena 37921
Alpen-Adria-Universität Klagenfurt, Klagenfurt am Wörthersee 01243
Alpha Park Public Library District, Bartonville 38096
Alpine Club, London 29711
Alpine Public Library, Alpine 37925
Al-Quds University, Jerusalem 14615
Alschuler, Grossman, Stein & Kahan, Los Angeles 35824
Alsip-Merrionette Park Public Library District, Alsip 37926
Alšova jihočeská galerie, Hluboká nad Vltavou 06449
Alston & Bird, LLP Library, New York 35953
Alta bibliotek, Alta 20305
Altadena Library District, Altadena 37927
Altai Agricultural University, Barnaul 22213
Altai Experimental Station, Barnaul 22996
Altai Museum of Regional Studies, Barnaul 22997
Altai N.K. Krupskaya District Children Library, Barnaul 24109
Altai Polzunov Polytechnic Institute, Barnaul 22215
Altai Regional Universal Scientific Library named after Shishkov, Barnaul 22119
Altai Scientific Research Institute of Agriculture, Barnaul 22999
Altai State Medical University, Barnaul 22211
Altai State Museum of Applied Arts, Barnaul 22995
Altai State Pedagogical Academy, Barnaul 22212
Altai State University, Barnaul 22214
Altai Tractor Joint-Stock Company, Rubtsovsk 22900
Altaiskaya gosudarstvenny pedagogicheskaya akademiya, Barnaul 22212
Altaiskaya kraevaya detskaya biblioteka im. N.K. Krupskoi, Barnaul 24109
Altaiskaya kraevaya universalnaya nauchnaya biblioteka im. V.Ya. Shishkova, Barnaul 22119
Altaiskaya plodovo-yagodnaya opytnaya stantsiya, Barnaul 22996
Altaiski agrarni universitet, Barnaul 22213
Altaiski gosudarstvenny universitet, Barnaul 22214
Altaiski kraevedcheski muzei, Barnaul 22997
Altaiski mezhotraslevoi territorialny tsentr nauchno-tekhnicheskoi informatsii i propagandy, Barnaul 22998
Altaiski nauchno-issledovatelski institut selskogo khozyaistva, Barnaul 22999
Altaiski politekhnicheski institut im. I.I. Polzunova, Barnaul 22215
Altheimer & Gray, Chicago 35565
Althingi Bókasafn, Reykjavík 13561
Altmärkisches Museum Stendal, Hansestadt Stendal 12502
Altonaer Museum, Hamburg 12015
Altoona Area Public Library, Altoona 37929
Altus Public Library, Altus 37931
Aluminium Plant, Trade Union Library, Zaporizhzhya 28884
Alunnato Benedittino, Assisi 15711
Alupka State Palace and Park Preserve, Alupka 28178
Alures S.C.p.A. (ISML), Novara 16386
Alushta City Centralized Library System, Sergeev-Tsenky Main Library, Alushta 28489
Alushtinska miska TsBS, Tsentralna biblioteka im. S.M. Sergeeva-Tsenkoho, Alushta 28489
Alva Public Library, Alva 37933
Alvan Ikoku College of Education, Owerri 19974
Alvernia College, Reading 32419
Alverno College, Milwaukee 31833
Alvesta bibliotek, Alvesta 26726
Älvsbyns kommunbibliotek, Älvsbyn 26727
The Alyce L. Haines Biomedical Library, Wailuku 37711
Alytaus Jurgio Kunčino viešoji biblioteka, Alytus 18315
Alytaus rajono savivaldybės viešoji biblioteka, Alytus 18316
Alyuminevi kombinat, Zaporizhzhya 28884
Alyuminievy zavod im. S.M. Kirova, Volkhov 22973
Amador County Library, Jackson 39598
Amagasaki Shiritsu Chiiki Kenkyu Shiryokan, Amagasaki 17511
Åmåls bibliotek, Åmål 26728
Amani Medical Research Centre, Amani 27671
Amanzimtoti Public Library, Amanzimtoti 25183
Amarillo College, Amarillo 33126
Amarillo Public Library, Amarillo 37935
Amarillo Public Library – East Branch, Amarillo 37936
Amarillo Public Library – North Branch, Amarillo 37937
Amarillo Public Library – Northwest Branch, Amarillo 37938
Amarillo Public Library – Southwest Branch, Amarillo 37939
Amathole Museum, King William's Town 25143
Amauta Bibliothek, Wien 01588
Ambasciata del Brasile, Roma 16446
Ambassade der Verenigde Staten, Brussel 02557
Ambedkar Memorial Library, Nagarjunanagar 13716
Ambrose Alli University, Ekpoma 19935
AMEC Earth & Environmental Technical Library, Phoenix 36115
AMEC Energy & Mining, Calgary 04235
Amecea Pastoral Institute, Eldoret 18028
Amelia County Historical Society Library, Amelia 36460
Americal School of Classical Studies, Athinai 13110
American Academy in Rome, Roma 16447
American Academy of Arts & Letters Library, New York 37174
American Alpine Club Library, Golden 36894
American Antiquarian Society Library, Worcester 37859
American Arbitration Association, New York 37175
American Association of Retired Persons, Washington 37715
American Baptist Churches-USA, Atlanta 35112
American Baptist College of the American Baptist Theological Seminary, Nashville 35295
American Baptist Historical Society, Atlanta 35113
American Bible Society Library, New York 37176
American Broadcasting Co, Inc, New York 35954
American Center Library, Alexandria 07146
American Center Library, Colombo 26303
American Center Library, Nairobi 18032
American Center Library, New Delhi 14112
The American Ceramic Society, Westerville 37827
American Chemical Society, Washington 37716
American College of Obstetricians & Gynecologists, Washington 32953
American Congregational Association, Boston 35143
American Cultural Center, Dakar 24432
American Cultural Center, Freetown 24579
American Dental Association Department of Library Services, Chicago 36662
American Electric Power Service Corp, Columbus 35629
American Farm Bureau Federation Library, Park Ridge 37325
American Folk Art Museum, New York 37177
American Fork City Library, American Fork 37940
American Foundation for the Blind, New York 37178
American Health Care Association, Washington 37717
American Herb Association Library, Nevada City 37150
American Heritage Center, Laramie 37016
American Heritage Library & Museum, Glendale 36891
American Hospital Association Resource Center, Chicago 35566
American Indian College, Phoenix 35330
American Information Resource Center, Mumbai 14078
American Information Resource Center, New Delhi 14113
American Institute for Chartered Property Casualty Underwriters & Insurance Institute of America, Malvern 37078
American Institute for Economic Research, Great Barrington 36901
American Institute of Architects Library & Archives, Washington 37718
American Institute of Physics, College Park 36722
American Insurance Association, Washington 37719
American International College, Springfield 32702
American Irish Historical Society Library, New York 37179
American Islamic College, Chicago 30756
American Jewish Committee, New York 37180
American Jewish Historical Society, Newton Centre 35312
American Kennel Club Inc Library, New York 37181
American Legion National Headquarters Library, Indianapolis 36954
American Library, Brussel 02557
American Library Association Library, Chicago 36663
American Library in Paris / Bibliothèque américaine à Paris, Paris 08464
American Library of Montpellier, Montpellier 08421
American Library Resource Centre, Singapore 24608
American Museum of Magic, Marshall 37086
American Museum of Natural History, New York 37182
American Museum of Natural History Library, New York 37183
American Numismatic Association Library, Colorado Springs 36725
American Numismatic Society Library, New York 37184
American Philatelic Research Library, Bellefonte 36527
American Philosophical Society Library, Philadelphia 37338
American Psychiatric Association, Arlington 36477
American Psychological Association, Washington 37720
American Public Transportation Association, Washington 37721

American Research Center in Egypt, Cairo 07155
American Resource Center, Helsinki 07443
American Resource Center, Taipei 27603
American River College, Sacramento 33707
American Samoa-Office of Library Services, Pago Pago 00095
American School Foundation A.C., México 18811
American School of Classical Studies at Athens, Athinai 13096
American Society for Psychical Research Inc Library, New York 37185
American Society of Anesthesiologists, Park Ridge 37326
American Society of International Law Library, Washington 37722
American Society of Military History Library, South El Monte 37618
American Society of Military History Library, South El Monte 37619
American Standards Testing Bureau, Inc, New York 35955
American Studies Research Centre, Hyderabad 13995
American Swedish Institute Archives & Library, Minneapolis 37118
The American University, Washington 32954
American University Alumni, Bangkok 27683
American University in Bulgaria, Blagoevgrad 03453
American University in Cairo (AUC), Cairo 07050
– Institute of Solid State and Materials Research Center, Cairo 07051
American University of Beirut, Beirut 18201
– Saab Medical Library, Beirut 18202
American University of Paris, Paris 07837
American University – Washington College of Law, Washington 32955
American-Canadian Genealogical Society Library, Manchester 37079
Amerika-Haus, München 12276
Amerind Foundation, Inc, Dragoon 35680
Amery Public Library, Amery 37941
Ames Flagstaff Centre, Melbourne 00575
Ames Free Library, North Easton 40505
Ames Public Library, Ames 37942
Amesbury Public Library, Amesbury 37943
AMF Centre, Glenside 00923
AMG International Library, Chattanooga 35154
Amherst College, Amherst 30211
– Keefe Science Library, Amherst 30212
– Vincent Morgan Music Library, Amherst 30213
Amherst Public Library, Amherst 37944
Amherst Public Library, Amherst 37945
Amherst Town Library, Amherst 37946
Amir-ud-Daula Public Library, Lucknow 14228
Amistad Research Center, New Orleans 37166
Amityville Public Library, Amityville 37950
Amman University College for Applied Engineering, Amman 17872
Amon Carter Museum Library, Fort Worth 36867
Amos Anderson Art Museum, Helsinki 07444
Amos Andersonin taidemuseo, Helsinki 07444
Amos Press, Inc Library, Sidney 36272
Amravati University, Amravati 13602
Amsab – Institute of Social History, Gent 02643
Amsab – Instituut voor Sociale Geschiedenis, Gent 02643
Amstelland Bibliotheken, Amstelveen 19549
Amstelland Bibliotheken – Bibliotheek Amstelveen Stadsplein, Amstelveen 19550
Amstelveen Library, Amstelveen 19550
Amsterdams Historisch Museum, Amsterdam 19346
Amsterdams NME-Centrum, Amsterdam 19359
Amsterdamse Hogeschool voor de Kunsten – Academie van Bouwkunst, Amsterdam 19052
Amsterdamse Hogeschool voor de Kunsten – Reinwardt Academie, Amsterdam 19057
Amt der Niederösterreichischen Landesregierung, St. Pölten 01577
Amt der Oberösterreichischen Landesregierung in Linz – Statistischer Dienst, Linz 01386
Amt der Österreichischen Landesregierung in Linz – Amtsbibliothek, Linz 01387
Amt der Salzburger Landesregierung, Salzburg 01390
Amt der VELKD, Hannover 11235
Amt der Vorarlberger Landesregierung, Bregenz 01384
Amt für Geoinformationswesen, Traben-Trarbach 11112
Amt für Lehrerbildung (AfL), Fuldatal 11907
Amt für Stadtentwicklung und Statistik, Köln 10982
Amt für Statistik Berlin-Brandenburg, Berlin 10785
Amt für Statistik Berlin-Brandenburg, Potsdam 11080
Amt für Strategische Steuerung, Stadtforschung und Statistik, Wiesbaden 11117
AMT Management Services Library, Winnipeg 04306
Amtsbibliothek, Ziesar 11138
Amtsgericht München, München 11022
Amtsgericht Nürnberg, Nürnberg 11066
Amtsgericht Tiergarten, Berlin 10786
Amur Regional Children Library, Blagoveshchensk 24113
Amur Regional Research Library, Blagoveshchensk 22122
Amur Research Centre, Blagoveshchensk 23002
Amur State University, Blagoveshchensk 22218
Amurskaya oblastnaya detskaya biblioteka, Blagoveshchensk 24113
Amurskaya oblastnaya nauchnaya biblioteka im. N.N. Muraveva-Amurskogo, Blagoveshchensk 22122
Amurski Nauchni Tsentr, Blagoveshchensk 23002
Amurski oblastnoi muzei kraevedenia im. 'Novikova-Daurskogo', Blagoveshchensk 23003
AMVC-Letterenhuis, Antwerpen 02507
AMVC-Literary Centre, Antwerpen 02507
AMWU – Australian Manufacturing Workers Union, Granville 00925
Anacortes Public Library, Anacortes 37951
Anaheim Public Library, Anaheim 37952
Anaheim Public Library – Canyon Hills, Anaheim 37953
Anaheim Public Library – Elva L. Haskett Branch, Anaheim 37954
Anaheim Public Library – Sunkist, Anaheim 37955
Anambra State Library Board, Enugu 20049
Ananda Mohan College, Kolkata 13810
Anarchistische Bücherei im Haus der Demokratie Berlin, Berlin 11494
Anatolia College, Thessaloniki 13092
Anchorage Museum of History & Art Library & Archives, Anchorage 36465
Ancient Orient Museum, Tokyo 17669
Anderson City, Anderson, Stony Creek & Union Townships Public Library, Anderson 37956
Anderson College, Anderson 30222
Anderson County Library, Anderson 37957
Anderson County Public Library, Lawrenceburg 39824
Anderson, Kill & Olick, New York 35956
Anderson University, Anderson 30223
Andhra Medical College, Visakhapatnam 13847
Andhra Pradesh Agricultural University, Hyderabad 13654
Andhra Sahitya Parishat Library, Kakinada 14019
Andhra University, Waltair 13787
Andizhan Institute of Medicine, Andizhan 42074
Andorra National Library, Andorra la Vella 00097
Andover Newton Theological School, Newton Centre 35313
András Keve Library for Ornithology and Nature Conservation, Budapest 13369
Andrássy Gyula Budapesti Német Nyelvü Egyetem / Andrássy Gyula Deutschsprachige Universität Budapest, Budapest 13219
Andrássy Gyula Deutschsprachige Universität Budapest, Budapest 13219
Andrássy Gyula German-speaking University Budapest, Budapest 13219
Andreas-Möller-Bibliothek, Freiberg 10727
Andreev Institute of Acoustics of the Russian Academy of Sciences, Moskva 23163
Andrew College, Cuthbert 33270
Andrew Inglis Clark Law Library, Hobart 00818
Andrew McFarland Mental Health Center, Springfield 37624
Andrews County Library, Andrews 37959
Andrews, Davis, Legg, Bixler, Milsten & Price, Oklahoma City 36067
Andrews University, Berrien Springs 30415
– Architectural Resource Center, Berrien Springs 30416
– Music Materials Center, Berrien Springs 30417
Ang Pambansang Aklatan, Manila 20681
Ånge folkbibliotek, Ånge 26729
Angeles University Foundation, Angeles City 20683
Ängelholms stadsbibliotek, Ängelholm 26730
Angelina College, Lufkin 31694
Angelo State University, San Angelo 32564
Angestelltenkammer Bremen, Bremen 11669
Anglia Ruskin University, Cambridge 28983
Anglia Ruskin University, Chelmsford 29063
Anglo American Corporation of South Africa Ltd, Marshalltown 25094
Angolan Directorate of Geological and Mining Services, Luanda 00106
Anguilla Public Library, The Valley 00110
Angus Council, Forfar 29982
Anhaltische Landesbücherei Dessau, Dessau-Roßlau 12702
Anheuser-Busch Co, Inc, Saint Louis 36194
Anhui Agricultural University, Hefei 05259
Anhui Agrotechnical Teachers College, Fengyang 05171
Anhui College of Finance and Trade, Bengbu 05057
Anhui College of Traditional Chinese Medicine, Hefei 05260
Anhui Commercial College, Maanshan 05413
Anhui Economy and Management Cadres Institute, Hefei 05261
Anhui Institute of Education, Hefei 05262
Anhui Institute of Mechanical and Electrical Technology, Wuhu 05666
Anhui Institute of Technology, Hefei 05263
Anhui Medical University, Hefei 05264
Anhui Normal University, Wuhu 05667
Anhui Provincial Library, Hefei 04961
Anhui University, Hefei 05265
Anil Products Limited, Ahmedabad 13918
Animal and Poultry Farms Equipment Construction and Repairs Research Institute, Minsk 02044
Animal Health and Industry Training Institute, Kabete 18011
Animal Health and Veterinary Laboratories Agency, Addlestone 29592
Animal Health Research Centre, Entebbe 27922
Animal Husbandry and Veterinary Institute, Almaty 17937
Animal Improvement Institute, Irene 25129
Animal Research Institute, Accra 13040
Anjuman Taraqqi-e-Urdu Pakistan, Karachi 20491
Anjuman-i-Islam Urdu Research Institute, Mumbai 14079
Ankara Üniversitesi, Ankara 27846
Ann Arbor District Library, Ann Arbor 37964
Anna Lindh-biblioteket, Stockholm 26424
Anna Maria College, Paxton 32239
An-Najah National University, Nablus 14643
Annamalai University, Annamalainagar 13606
Annapolis Valley Regional Library, Bridgetown 04587
Anna-Seghers-Bibliothek, Berlin 12648
Anne Arundel Community College, Arnold 33132
Anne Arundel County Circuit Court, Annapolis 33927
Anne Carlsen Learning Center, Jamestown 36974
Annie Halenbake Ross Library, Lock Haven 39911
Anoka County Law Library, Anoka 33930
Anoka County Library, Blaine 38214
Anoka County Library – Crooked Lake, Coon Rapids 38687
Anoka County Library – Mississippi, Fridley 39171
Anoka-Ramsey Community College, Coon Rapids 33257
Anotati Scholi Kalon Texnon, Athinai 13073
Anqing Teachers' College, Anqing 04980
ANS Institute of Social Studies, Patna 14176
Anser Technical Library, Arlington 36478
Anshan Normal University, Anshan 04981
Anshan Vocational College, Anshan 04982
Anshe Chesed Fairmount Temple, Cleveland 36704
Ansonia Library, Ansonia 37968
Antal Reguly Memorial Library, Zirc 13525
Antelope Valley College, Lancaster 33471
Anthropological Survey of India, Kolkata 14027
Anthropos Institut e.V., Sankt Augustin 12459
Anthroposophical Society in America, Ghent 36889
Anthroposophical Society in the Netherlands, Den Haag 19430
Antibiotics Factory, Razgrad 03496
Anti-Defamation League, New York 37186
Antigo Public Library, Antigo 37969
Antikenmuseum Basel und Sammlung Ludwig, Basel 27299
Antikensammlung, Berlin 11564
Antioch College, Yellow Springs 33103
Antioch District Library, Antioch 37970
Antioch New England Graduate School Library, Keene 31490
Antiplague Research Institute, Rostov-na-Donu 23725
Antiplague Research Institute 'Microbe', Saratov 23961
Anti-Slavery International, London 29712
Anton de Kom Universiteit van Suriname, Paramaribo 26360
– Medische Bibliotheek, Paramaribo 26361
Anton Melik Geographical Institute of the Scientific Research Centre of the Slovenian Academy of Sciences and Arts (GI SRC SASA), Ljubljana 24916
Anton-Bruckner-Privatuniversität, Linz 01250
Antoni van Leeuwenhoek Ziekenhuis, Amsterdam 19371
Antoniana Library, Padova 15952
Antonina Marek Town Library in Turnov, Turnov 06615
Antonov Military Advanced School of Topography, Sankt-Peterburg 22566
Anton-und-Katharina-Kippenberg-Stiftung, Düsseldorf 11794
Antopol Village Library, Antopol 02092
Antopolskaya gorposelkovaya bibliyateka, Antopol 02092
Antroposofische Vereniging in Nederland, Den Haag 19430
Antwerp Municipal Port Authority, Antwerpen 02508
Anvil Mountain Correctional Center Library, Nome 34611
Anyang Teachers' College, Anyang 04983
Anyang University, Anyang 04984
Anykščių rajono savivaldybės L. ir S. Didžiulių viešoji biblioteka, Anykščiai 18317
AO Alacheevsky metalurgiyny zavod, Alchevsk 28135
AO Avtokran, Ivanovo 22794
AO Avtonormal, Belebei 22779
AO Baltiskaya bumaga, Sankt-Peterburg 22901
AO Barnaultransmash, Barnaul 22777
AO Bolshaya Kostromskaya Inyanaya manufaktura, Kostroma 22812
AO Buguruslanski zavod Radiator, Buguruslan 22781
AO Dalenergomash, Khabarovsk 22804
AO Elektromekhanicheski zavod im. Vladimira Ilicha, Moskva 22819
AO Glukhovski tekstil, Noginsk 22875
AO Gukovugol, Gukovo 22793
AO instituta Gidroproekt, Moskva 22820
AO KamAZ, ONTI, Naberezhnye Chelny 22870
AO Khimprom, Kemerovo 22801
AO Kizelugol, Kizel 22808
AO Kondopoga, Kondopoga 22810
AO Konus, Mozhga 22869
AO Krangormash, Novomoskovsk 22879
AO Krasny Vyborzhets, Sankt-Peterburg 22902
AO LESINVEST, Sankt-Peterburg 22903
AO Molodaya Gvardiya, Informatsionno-bibliograficheski tsenter, Moskva 24167
AO Morshanskhimmash, Morshansk 22818
AO Mosgiprobum, Moskva 22821
AO Nauchno-issledovatelski institut rezinotekhnicheskogo mashinostroeniya, Tambov 22960
AO NIITEKhIM, Moskva 22822
AO Nizhnetagilskogo metallurgicheski kombinat, Nizhni Tagil 22873
AO OKB Sredstv avtomatizatsii, Moskva 22823
AO OKB Stankostroenie, Sankt-Peterburg 22904
AO Otkrytogo Tipa Moskvich, Moskva 22824
AO Pavlovski Avtobus, Pavlovo 22888
AO po vypusku tyazhyolykh ekskavatorov, Voronezh 22975
AO Polimermash, Tambov 22961
AO PROMSTROIPROEKT, Moskva 22825
AO Promstroiproekt, Sankt-Peterburg 22905
AO Promyshlennykh Traktorov, Cheboksary 22782
AO Rudgormash, Voronezh 22976
AO Saranski zavod Avtosamosvalov, Saransk 22952
AO Shcherbakov, Moskva 22826
AO Sokolski tsellyulozno-bumazhny kombinat, Sokol 22956
AO Solombalski tsellyulozno-bumazhny kombinat, Arkhangelsk 22773
AO Tryokhgornaya manufaktura, Moskva 22827
AO Tsellyulozny zavod Pitkyaranta, Pitkyaranta 22892
AO Tsentralny nauchno-issledovatelski institut mekhanicheskoi obrabotki drevesiny, Arkhangelsk 22774
AO Tsvetmetavtomatika, Moskva 22828
AO Tulazheldormash, Tula 22963
AO Tutaevski motorny zavod, Tutaev 22964

AO Uralski avtomotorny zavod, Novouralsk 22882
AO Uraltrak, Biblioteka profkoma, Chelyabinsk 24121
AO Vagonmash-zavod im. Egorova, Sankt-Peterburg 22906
AO Vagonostroitelny zavod, Tver 22965
AO VakuumMash, Kazan 22799
AO VNIIinstrument, Moskva 22829
AO Volgogradski traktorny zavod, Volgograd 22972
AO Vserossiski nauchno-issledovatelski institut gidrotekhniki im. B.E. Vedeneeva, Sankt-Peterburg 22907
AO Zavod Tyazhpromarmatura, Aleksin 22772
AOK-Bundesverband, Berlin 11486
Aomori Prefectural Animal Husbandry Experiment Station, Kamikita 17564
Aomori Prefectural Assembly, Aomori 17252
Aomori Prefectural Fisheries Experiment Station, Nishitsugaru 17625
Aomori Prefecture Agricultral Experiment Station, Kuroishi 17584
Aomori Prefecture Planning Department, Aomori 17253
Aomoriken Chikusan Chikenjo Kenkyu Konrishitsu, Kamikita 17564
Aomoriken Gikai, Aomori 17252
Aomoriken Kikakubu, Aomori 17253
Aomoriken Nogyo Shikenjo, Kuroishi 17584
Aomoriken Suisan Shikenjo, Nishitsugaru 17625
AOOT Avtoagregat, Kineshma 22805
AOOT Azovski zavod kuznechno-pressovykh avtomatov, Azov 22775
AOOT Dinamo, Moskva 22830
AOOT INEUM, Moskva 23166
AOOT Informenergo, Moskva 22831
AOOT Kran, Uzlovaya 22970
AOOT Kuznetski nauchno-issledovatelski ugolny institut, Prokopevsk 22894
AOOT MPO Plastik, Moskva 22832
AOOT NPT i EN Orgstankinprom, Novosibirsk 22881
AOOT Pishchepromproekt, Ekaterinburg 22788
AOOT PTF IKAR, Voronezh 22977
AOOT Radiator, Orenburg 22884
AOOT Rostovgiproshakht, Rostov-na-Donu 22897
AOOT Tsellyulozno-bumazhny zavod, Sovetsk 22957
AOOT Voronezhski Stanko-Zavod, Voronezh 22978
AOOT Zavod avtobetonovozov, Tuimazy 22962
Aoyama Gakuin University, Tokyo 17135
AOZT Neivo-Rudyanski lesokhimicheski zavod, Kirovograd 22806
AOZT YUZHURALMASH, Orsk 22886
AP Kompozit, Sankt-Peterburg 22908
Apache County Library District, Saint Johns 41113
Apiculture Research Institute, Rybnoe 23736
Apostolivska raionna TsBS, Tsentralna biblioteka, Apostolove 28490
Apostolove Regional Centralized Library System, Main Library, Apostolove 28490
Appalachian Mountain Club Library, Boston 36556
Appalachian Regional Library, North Wilkesboro 40529
Appalachian School of Law Library, Grundy 31270
Appalachian State University, Boone 30476
Appleton Public Library, Appleton 37975
Applied Arts and Crafts Research Institute, Moskva 23405
Applied Economics Research Centre, Karachi 20492
Applied Mathematics and Mechanics Institute, Donetsk 28222
Applied Physics Institute, Minsk 02024
Appomattox Regional Library, Hopewell 39478
Aquaproiect, Bucureşti 21967
Aquinas College, Grand Rapids 31235
Aquinas College of Higher Studies, Colombo 26268
Aquinas Library, Masaka 27921
Aquinas University, Legazpi City 20705
Arab Bureau for Education in the Gulf States, Riyadh 24399
Arab Bureau for Engineering Design and Councils, Cairo 07135
Arab Contractors Company, Giza 07144
Arab Gulf States Information and Documentation Center, Baghdad 14452
Arab Institute of Aviation Technology, Cairo 07156
Arab Language Academy of Damascus, Damascus 27512
Arab League, Cairo 07157
Arab League Documentation and Information Centre (ALDOC), Tunis 27827
Arab League Educational, Cultural and Scientific Organization (ALECSO), Tunis 27828
Arab Library, Chinguetti 18684
Arab Medical University, Benghazi 18237
Arab Planning Institute, Kuwait, Safat 18129
Arab Public Library, Arab 37977
ARAL Forschung, Bochum 11376
Aransas County Public Library, Rockport 41042
Arany János Múzeum, Nagykörös 13492
Aranzadi Society of Sciences, Donostia 25953
Aranzadi Zientzia Elkartea, Donostia 25953
Arapahoe Community College, Littleton 33490
Arapahoe Library District, Englewood 38979
Arapahoe Library District – Castlewood Xpress Library, Centennial 38490
Arapai Agricultural College, Soroti 27917
Arbeiderbevegelsens arkiv og bibliotek, Oslo 20240
Arbeidsforskningsinstituttet, Oslo 20241
Arbeidshof, Antwerpen 02424
Arbeiterkammer Niederösterreich, Wien 01615
Arbeiterkammer Steiermark, Graz 01521
Arbeiterkammer Tirol, Innsbruck 01671
Arbejderbevægelsens Bibliotek og Arkiv, København 06783
Arbetarrörelsens arkiv och bibliothek, Stockholm 26639
Arbetets museum, Norrköping 26621
Arbogas stadsbibliotek, Arboga 26731
Arbuzov Institute of Organic and Physical Chemistry of the Russian Academy of Sciences, Kazan 23081
ARC Irene Campus Information Centre, Irene 25130
Arcada, Helsinki 07337
Arcada University of Applied Sciences, Helsinki 07337
Arcadia Public Library, Arcadia 37978
Arcadia University, Glenside 31220
Arcadis Euroconsult/BMB, Arnhem 19391
ARCADIS Heidemij Advies BV, Arnhem 19392
Arcanum Public Library, Arcanum 37980
ArcelorMittal Eisenhüttenstadt GmbH, Eisenhüttenstadt 11390
ArcelorMittal Gloval Research & Development – East Chicago, East Chicago 35683
Archaelogical Survey Department of Sri Lanka, Colombo 26289
Archaeological and Ethnographic Museum, Łódź 21163
Archaeological Institute and Museum, Sofiya 03520
Archaeological Institute of Aegean Studies, Rhodos 13130
Archaeological Institute of the Serbian Academy of Sciences and Arts, Beograd 24498
Archaeological Museum, Poznań 21203
Archaeological Museum, Split 06221
Archaeological Museum, Stavanger 20293
Archaeological Museum, Varna 03584
Archaeological Museum, Zadar 06228
Archaeological Museum, Zagreb 06231
Archaeological Museum of Bologna, Bologna 16195
Archaeological Museum of Catalonia, Barcelona 25920
Archaeological Survey of India, Nagpur 14106
Archaeological Survey of India, New Delhi 14114
Archaiologiki Hetairia, Athinai 13097
Archäologie für Westfalen, Münster 12349
Archäologie Schweiz, Basel 27300
Archäologie und Museum, Liestal 27238
Archäologische Staatssammlung, München 12277
Archäologischer Park Carnuntum, Bad Deutsch-Altenburg 01510
Archäologisches Landesmuseum, Schleswig 12465
Archäologisches Museum, Frankfurt am Main 11835
Archäologisches Museum, Münster 10484
Archbishop Marsh's Library, Dublin 14521
Archbishop O'Brien Library, Hartford 35224
Archbishop Simor Library, Esztergom 13336
Archbishop Vehr Theological Library, Denver 35184
Archbishopric Library, Nicosia 06333
Archbold Biological Station Library, Venus 37705
The Archdiocesan Center at St. Thomas Seminary – Archbishop O'Brien Library, Hartford 35224
Archdiocesan Library, Zagreb 06209
Archenhold-Sternwarte, Berlin 11487
Archeological Museum in Gdańsk, Gdańsk 21099
Archeologický ústav AV ČR, Brno 06425
Archeologický ústav AV ČR, Praha, v.v.i., Praha 06475
Archeologisch-Antropologisch Instituut Nederlandse Antillen (AAINA), Willemstad 06319
Archer & Greiner Library, Haddonfield 35731
Archidiocesan Library of Eger, Eger 13334
Archief en Cultuurcentrum van Arenberg, Enghien 02640
Archief en Documentatiecentrum voor het Vlaams-Nationalisme, Antwerpen 02506
Archief en Museum van het Vlaams Leven te Brussel, Brussel 02558
Archiepiscopal Library, Veszprém 13350
Archiepiscopal Theological College, Eger 13333
Architects' Union of Romania, Bucureşti 22016
Architects' Union of Russia – St. Petersburg Branch, Sankt-Peterburg 23901
Architectural Association (AA), London 29170
Architectural Institute of Japan, Tokyo 17726
Architectural Society of China, Beijing 05817
Architecture and Building Foundation, Praha 06474
Archiv der Akademie der Künste, Berlin 11488
Archiv der Arbeiterjugendbewegung, Oer-Erkenschwick 12381
Archiv der deutschen Frauenbewegung, Kassel 12105
Archiv der deutschen Jugendbewegung, Witzenhausen 12604
Archiv der Diözese Gurk, Klagenfurt 01442
Archiv der Evangelischen Brüder-Unität, Herrnhut 11246
Archiv der Evangelischen Kirche im Rheinland, Düsseldorf 11199
Archiv der Hansestadt Lübeck, Lübeck 12210
Archiv der Hansestadt Rostock, Rostock 12440
Archiv der Hansestadt Wismar, Wismar 12598
Archiv der Max-Planck-Gesellschaft, Berlin 11489
Archiv der Parteien und Massenorganisationen der DDR im Bundesarchiv, Berlin 11586
Archiv der Stadt Linz, Linz 01558
Archiv der Stadt Salzburg, Salzburg 01562
Archiv der Universität Wien, Wien 01317
Archiv des Heimatvereins Vilsbiburg, Vilsbiburg 12574
Archiv Frau und Musik, Frankfurt am Main 11836
Archiv für deutschsprachige Literatur seit 1945, Sulzbach-Rosenberg 12547
Archiv für Kulturpolitik / Europäische Kulturdokumentation beim Zentrum für Kulturforschung (ZfKf), Bonn 11621
Archiv Grünes Gedächtnis, Berlin 11490
Archív hlavného mesta SR Bratislavy, Bratislava 24764
Archiv hlavního města Prahy, Praha 06476
Archív města Brna, Brno 06426
Archiv und Kulturzentrum von Arenberg, Enghien 02640
Archiv zur Geschichte der schweizerischen Frauenbewegung, Worblaufen 27444
Archival Museum, Peshawar 20521
Archivamt für Westfalen, Münster 11059
The Archive General of Indias Library, Sevilla 26144
Archive of Serbia, Beograd 24500
Archive of the German Women's Movement, Kassel 12105
Archive-Library-Museum of Freemasonry, Den Haag 19432
Archives cantonales vaudoises, Chavannes-près-Renens 27341
Archives de la Critique d'Art, Châteaugiron 08329
Archives de la Marne et de la Province de Champagne, Châlons-en-Champagne 08320
Archives de la Ville de Bruxelles, Bruxelles 02559
Archives de la Ville de Genève, Genève 27354
Archives de la Ville de Lausanne, Lausanne 27390
Archives de l'Etat, Neuchâtel 27241
Archives de l'Etat à Arlon, Arlon 02539
Archives de l'Etat à Liège, Liège 02676
Archives de l'Etat à Mons, Mons 02688
Archives de l'Etat à Namur, Namur 02690
Archives de l'Etat de Berne, Bern 27332
Archives de Maurice, Petite Rivière 18698
Archives de Paris, Paris 08465
Archives départementales de la Charente-Maritime, La Rochelle 08369
Archives départementales de la Corrèze, Tulle 08703
Archives départementales de la Côte-d'Or, Dijon 08346
Archives départementales de la Creuse, Guéret 08361
Archives départementales de la Dordogne, Périgueux 08638
Archives départementales de la Drôme, Valence 08705
Archives départementales de la Gironde, Bordeaux 08301
Archives départementales de la Guadeloupe, Basse-Terre 13161
Archives départementales de la Haute-Marne, Chaumont 08332
Archives départementales de la Haute-Saône, Vesoul 08712
Archives départementales de la Haute-Savoie, Annecy 08279
Archives départementales de la Haute-Vienne, Limoges 08386
Archives départementales de la Loire, Saint-Etienne 08666
Archives départementales de la Loire-Atlantique, Nantes 08442
Archives départementales de la Martinique, Fort-de-France 18678
Archives départementales de la Mayenne, Laval 08371
Archives départementales de la Meuse, Bar-le-Duc 08291
Archives départementales de la Sarthe, Le Mans 08378
Archives départementales de la Savoie, Chambéry 08322
Les Archives départementales de la Seine-et-Marne, Melun 08413
Archives départementales de la Seine-Saint-Denis, Bobigny 08299
Archives départementales de la Vendée, La Roche-sur-Yon 08370
Archives départementales de la Vienne, Poitiers 08646
Archives départementales de l'Allier, Yzeure 08719
Archives départementales de l'Aube, Troyes 08699
Archives départementales de l'Aveyron, Rodez 08657
Archives départementales de l'Essonne, Chamarande 08321
Archives départementales de l'Eure, Evreux 08349
Archives départementales de l'Hérault, Montpellier 08420
Archives départementales de Loir-et-Cher, Blois 08298
Archives départementales de l'Oise, Beauvais 08293
Archives départementales de l'Orne, Alençon 08273
Archives départementales de Maine-et-Loire, Angers 08274
Archives départementales de Meurthe-et-Moselle, Nancy 08434
Archives départementales de Tarn-et-Garonne, Montauban 08417
Archives départementales des Alpes de Haute-Provence, Digne-les-Bains 08345
Archives départementales des Alpes-Maritimes, Nice 08446
Archives départementales des Ardennes, Charleville-Mézières 08325
Archives départementales des Bouches-du-Rhône, Marseille 08403
Archives départementales des Côtes-d' Armor, Saint-Brieuc 08662
Archives départementales des Pyrénées-Atlantiques, Pau 08636
Archives départementales des Pyrénées-Orientales, Perpignan 08641
Archives départementales des Yvelines et de l'ancienne Seine-et-Oise, Montigny-le-Bretonneux 08419
Archives départementales d'Eure-et-Loir, Chartres 08326
Archives départementales d'Ille-et-Vilaine, Rennes 08651
Archives départementales d'Indre-et-Loire, Tours 08698
Archives départementales du Calvados, Caen 08313
Archives départementales du Cantal, Aurillac 08286
Archives départementales du Doubs, Besançon 08294
Archives départementales du Finistère, Quimper 08648
Archives départementales du Gard, Nîmes 08451
Archives départementales du Gers, Auch 08285

Archives départementales du Haut-Rhin, Colmar 08339
Archives départementales du Lot-et-Garonne, Agen 08265
Archives départementales du Morbihan, Vannes 08708
Archives départementales du Nord, Lille 08382
Archives départementales du Pas-de-Calais, Dainville 08342
Archives départementales du Puy-de-Dôme, Clermont-Ferrand 08336
Archives départementales du Rhône, Lyon 08391
Archives d'Etat, Genève 27355
Archives du Loiret, Orléans 08454
Archives du Wilayat de Tizi-Ouzou, Tizi-Ouzou 00085
Archives du Wilayate d'Algers, Alger 00057
Archives du Wilayate d'Annaba, Annaba 00072
Archives du Wilayate de Batna, Batna 00074
Archives du Wilayate de Saida, Saïda 00083
Archives du Wilayate de Tiaret, Tiaret 00084
Archives du Wilayate d'Oran, Oran 00081
Archives et Centre Culturel d'Arenberg / Archief en Cultuurcentrum van Arenberg / Archiv und Kulturzentrum von Arenberg, Enghien 02640
Archives Générales du Royaume / Algemeen Rijksarchief, Bruxelles 02560
Archives municipales, Marseille 08404
Archives municipales, Metz 08414
Archives National du Sénégal, Dakar 24409
Archives Nationales, Antananarivo 18456
Archives nationales, Brazzaville 06097
Archives nationales, Conakry 13188
Archives Nationales, Kinshasa 06068
Archives Nationales, Nouakchott 18685
Archives Nationales, Paris 08466
Archives Nationales, Yaoundé 03652
Archives Nationales de la République du Bénin, Porto-Novo 02906
Archives nationales de Luxembourg, Luxembourg 18406
Archives nationales du Québec, Chicoutimi 04326
Archives Nationales du Québec, Sainte-Foy 04483
Archives of Bosnia and Herzegovina, Sarajevo 02960
Archives of Literature and Fine Arts, Moskva 23167
Archives of Novgorod District, Veliki Novgorod 24039
Archives of Ontario Library, Toronto 04503
Archives of the Astrakhan Region, Astrakhan 22992
Archives of the Capital of the Slovak Republic Bratislava, Bratislava 24764
Archives of the City of Brno, Brno 06426
Archives of the Kostroma Region, Kostroma 23107
Archives of the Republic of Slovenia, Ljubljana 24912
Archives of the Ulyanovsk Region, Ulyanovsk 24035
Archives Office of Azerbaijan, Baku 01693
Archives Office of Kazakhstan, Almaty 17938
Archivio, Biblioteca, Museo Civico, Altamura 16163
Archivio Centrale dello Stato, Roma 16448
Archivio di Stato di Alessandria, Alessandria 16162
Archivio di Stato di Bellinzona, Bellinzona 27315
Archivio di Stato di Bologna, Bologna 16186
Archivio di Stato di Bolzano, Bolzano 16201
Archivio di Stato di Brescia, Brescia 16206
Archivio di Stato di Cagliari, Cagliari 16212
Archivio di Stato di Campobasso, Campobasso 16217
Archivio di Stato di Catania, Catania 16225
Archivio di Stato di Firenze, Firenze 16251
Archivio di Stato di Genova, Genova 16281
Archivio di Stato di Gorizia, Gorizia 16293
Archivio di Stato di Lucca, Lucca 16308
Archivio di Stato di Macerata, Macerata 16310
Archivio di Stato di Massa, Massa 16313
Archivio di Stato di Messina, Messina 16316
Archivio di Stato di Milano, Milano 16319
Archivio di Stato di Modena, Modena 16358
Archivio di Stato di Napoli, Napoli 16365
Archivio di Stato di Novara, Novara 16387
Archivio di Stato di Palermo, Palermo 16395
Archivio di Stato di Pallanza, Pallanza 16407
Archivio di Stato di Parma, Parma 16408
Archivio di Stato di Pisa, Pisa 16418
Archivio di Stato di Reggio Emilia, Reggio Emilia 16432
Archivio di Stato di Roma, Roma 16449
Archivio di Stato di Savona, Savona 16573
Archivio di Stato di Siena, Siena 16578
Archivio di Stato di Teramo, Teramo 16585
Archivio di Stato di Torino, Torino 16591
Archivio di Stato di Trieste, Trieste 16618
Archivio di Stato di Udine, Udine 16634
Archivio di Stato di Venezia, Venezia 16642
Archivio di Stato di Vercelli, Vercelli 16661
Archivio Generale Padri Maristi, Roma 16016
Archivio Storico Arcivescovile, Capua 15775
Archivio Storico Arcivescovile, Ravenna 16004
Archivio Storico dei PP. Somaschi, Genova 15840
Archivio Storico del Comune, Messina 16317
Archivio Storico delle Arti Contemporanee della Biennale di Venezia, Venezia 16643
Archivio Storico Diocesano, Milano 16320
Archivo de la Corona de Aragón, Barcelona 25878
Archivo del Reino de Valencia, Valencia 26167
Archivo do Reino de Galicia, A Coruña 25947
Archivo General de Centro América, Guatemala City 13180
Archivo General de Indias, Sevilla 26144
Archivo General de la Nación, México 18834
Archivo General de la Nación, Montevideo 42036
Archivo General de Simancas, Simancas 26158
Archivo Histórico Municipal, Girona 25956
Archivo Histórico Municipal de Murcia, Murcia 26107
Archivo Histórico Nacional, Madrid 25975
Archivo Histórico Provincial de Valladolid, Valladolid 26178
Archivo Iberoamericano, Madrid 25976
Archivo Municipal, Sevilla 26145
Archivo Municipal, Vitoria-Gasteiz 26184
Archivo Municipal de Burgos, Burgos 25936
Archivo Nacional de Colombia, Santafé de Bogotá 06022
Archivo Nacional de Cuba, La Habana 06288
Archivo Nacional de Nicaragua, Managua 19894
Archivo Real y General de Navarra, Pamplona 26121
Archivo y Biblioteca Nacionales de Bolivia, Sucre 02917
Archivo-Biblioteca de la Función Legislativa, Quito 06984
Archivos Nacionales, Santiago 04922
Archivschule Marburg, Marburg 12252
Archivum Helveto-Polonicum, Bourguillon 27336
Archiwum Archidiecezjalne, Gniezno 21052
Archiwum Główne Akt Dawnych, Warszawa 21239
Archiwum Pánstwowe, Kraków 21126
Archiwum Państwowe Miasta Stołecznego Warszawy i Województwa Warszawskiego, Warszawa 21241
Archiwum Państwowe w Bydgoszczy, Bydgoszcz 21078
Archiwum Państwowe w Gdańsku, Gdańsk 21093
Archiwum Państwowe w Łodzi, Łódź 21157
Archiwum Państwowe w Poznaniu, Poznań 21192
Archiwum Państwowe w Radomiu, Radom 21215
Archiwum Państwowe w Szczecinie, Szczecin 21227
Arcidiocesi di Pescara-Penne, Penne 15971
ARC-Infruitec Centre for Fruit Technology, Stellenbosch 25181
ARC-Institute for Soil, Climate and Water, Pretoria 25154
ARC-Plant Protection Research Institute, Pretoria 25155
ARC-Roodeplaat Vegetable and Ornamental Plant Institute Library, Pretoria 25156
Arctic and Antarctic Research Institute, Sankt-Peterburg 23746
Arctic Centre, Information Service, Rovaniemi 07538
Arcybiskupie Wyzsze Seminarium Duchowne, Szczecin 20926
Årdal bibliotek, Årdalstangen 20307
Ardhi Institute, Dar es Salaam 27650
Ardmore Public Library, Ardmore 37981
ARD-Werbung Sales & Services GmbH Media Perspektiven, Frankfurt am Main 11837
Årebiblioteken, Järpen 26802
Arecibo Observatory, Arecibo 21901
Arellano University, Manila 20709
Arendal bibliotek, Arendal 20308
Arent Fox PLLC Library, Washington 36328
Argentine Agricultural Society, Buenos Aires 00292
Argentine Chemical Association, Buenos Aires 00238
Argentine Industrial Union, Buenos Aires 00294
Argentine Library for the Blind, Buenos Aires 00241
Argentine Neurological, Neurosurgical and Psychiatric Society, Buenos Aires 00286
Argentine Scientific Society, Buenos Aires 00289
Argentine Society of Authors, Buenos Aires 00290
Argentine Standardization Certification Institute, Buenos Aires 00260
Argentinian Society of Geographical Studies, Buenos Aires 00287
Argeş District Museum, Piteşti 22052
Argie Cooper Public Library, Shelbyville 41300
Argonne National Laboratory, Argonne 36476
Argyll & Bute Council Library Information Service, Sandbank 30068
Arheološki Institut, Beograd 24498
Arheološki Muzej, Split 06221
Arheološki Muzej, Zadar 06228
Arheološki Muzej, Zagreb 06231
Arheološki Muzej Istre, Pula 06216
Arhiv Bosne i Hercegovine, Sarajevo 02960
Arhiv Jugoslavije, Beograd 24499
Arhiv Republike Slovenije, Ljubljana 24912
Arhiv Srbije, Beograd 24500
Arhiv Vojvodine, Novi Sad 24529
Århus Kommunes Biblioteker, Århus 06846
Århus Tekniske Bibliotek, Århus 06768
ARIAS State Industrial Corporation, Sankt-Peterburg 22912
Arid Zone Research Institute, Alice Springs 00604
Aristotelio Panepistemio Thessalonikes, Thessaloniki 13085
Aristotle University of Thessaloniki, Thessaloniki 13085
Arizona Department of Corrections – Adult, Tucson 34941
Arizona Department of Corrections – Adult Institutions, Buckeye 34016
Arizona Department of Corrections – Adult Institutions, Douglas 34152
Arizona Department of Corrections – Adult Institutions, Florence 34206
Arizona Department of Corrections – Adult Institutions, Florence 34207
Arizona Department of Corrections – Adult Institutions, Goodyear 34304
Arizona Department of Corrections – Adult Institutions, Marana 34503
Arizona Department of Corrections – Adult Institutions, Phoenix 34668
Arizona Department of Corrections – Adult Institutions, Safford 34769
Arizona Department of Corrections – Adult Institutions, Tucson 34942
Arizona Department of Corrections – Adult Institutions, Winslow 35077
Arizona Department of Corrections, Arizona State Prison Complex-Safford, Fort Grant 34233
Arizona Department of Economic Security, Phoenix 34669
Arizona Department of Education, Phoenix 34670
Arizona Department of Environmental Quality, Phoenix 34671
Arizona Department of Transportation Library, Phoenix 34672
Arizona Historical Foundation, Tempe 37663
Arizona Historical Society, Tucson 37684
Arizona Historical Society Library, Flagstaff 36849
Arizona State Library, Phoenix 30145
Arizona State Museum Library, Tucson 37685
Arizona State Schools for the Deaf & the Blind Library, Tucson 33808
Arizona State University, Tempe 32803
– Architecture & Environmental Design Library, Tempe 32804
– Daniel E. Noble Science & Engineering Library, Tempe 32805
– Music Library, Tempe 32806
– John J. Ross – William C. Blakley Law Library, Tempe 32807
Arizona State University – West, Glendale 31216
Arizona Western College & Nau-Yuma, Yuma 33890
Årjängs folkbibliotek, Årjäng 26732
Arjo Wiggins Ltd, Beaconsfield 29566
Arkadelphia Public Library, Arkadelphia 37984
Arkansas Arts Center, Little Rock 37037
Arkansas Baptist College, Little Rock 31623
Arkansas Department of Correction Library System, Pine Bluff 40789
Arkansas Geological Commission Library, Little Rock 37038
Arkansas History Commission Library, Little Rock 37039
Arkansas River Valley Regional Library System, Dardanelle 38785
Arkansas School for the Deaf Library, Little Rock 33489
Arkansas State Library, Little Rock 30141
Arkansas State University, Jonesboro 31469
Arkansas State University – Beebe, Beebe 30371
Arkansas Supreme Court Library, Little Rock 34466
Arkansas Tech University, Russellville 32484
Arkchangelsk Regional Children's Library, Arkhangelsk 24104
Arkema Inc, King of Prussia 35803
Arkeologisk museum i Stavanger, Stavanger 20293
Arkhangelsk Forestry Engineering College, Arkhangelsk 22664
Arkhangelsk Regional Museum, Arkhangelsk 22989
Arkhangelsk Regional Scientific Library, Arkhangelsk 22117
Arkhangelsk State Museum of Fine Arts, Arkhangelsk 22988
Arkhangelsk State Technical University, Arkhangelsk 22202
Arkhangelskaya oblastnaya detskaya biblioteka im. A.P. Gaidara, Arkhangelsk 24104
Arkhangelskaya oblastnaya nauchnaya biblioteka im. N.A. Dobrolyubova, Arkhangelsk 22117
Arkhangelski lesotekhnicheski kolledzh, Arkhangelsk 22664
Arkhangelski oblastnoi kraevedcheski muzei, Arkhangelsk 22989
Arkheologicheski institut s muzei, Sofiya 03520
Arkhitekturno-stroitelny universitet, Tomsk 22593
Arkhiv Astrakhanskoi oblasti, Astrakhan 22992
Arkhiv Kostromskoi oblasti, Kostroma 23107
Arkhiv literatury i iskusstva, Moskva 23167
Arkhiv Mogilevskoi oblasti, Mogilev 02059
Arkhiv Penzenskoi oblasti, Penza 23682
Arkhiv Ulyanovskoi oblasti, Ulyanovsk 24035
Arkhiv Vladimirskoi oblasti, Vladimir 24047
Arkitektskolen Aarhus, Århus 06649
Arkitekturmuseet, Stockholm 26640
Arkticheski i Antarkticheski nauchno-issledovatelski institut, Sankt-Peterburg 23746
Arktinen keskus, Rovaniemi 07538
Arlington Baptist College, Arlington 35110
Arlington County Department of Libraries, Arlington 37985
Arlington County Department of Libraries Central Library, Arlington 37986
Arlington County Department of Libraries – Columbia Pike, Arlington 37987
Arlington Heights Memorial Library, Arlington Heights 37995
Arlington Public Library System, Arlington 37988
Arlington Public Library System – East Arlington, Arlington 37989
Arlington Public Library System – Lake Arlington, Arlington 37990
Arlington Public Library System – Northeast, Arlington 37991
Arlington Public Library System – Woodland West, Arlington 37992
Armadale Public Library & Information Service, Armadale 01057
Armagh Observatory, Armagh 29597
Armament Research and Development ESTT, Pune 14184
Armaturny zavod, Penza 22889
Armed Forces Academy, Magelang 14337
Armed Forces Institute of Pathology, Washington 34978
Armed Forces Library, Kuala Lumpur 18566
Armed Forces Library Services, Accra 13027
Armed Forces Museum, Oslo 20189
Armémuseum, Stockholm 26641
Armenian Agricultural Institute, Yerevan 00328
Armenian Apostolic Church of America, New York 35304
Armenian Architects' Union, Yerevan 00341
Armenian Composers' Union, Yerevan 00342
Armenian Cultural Foundation Library, Arlington 36479
Armenian House of Artists, Yerevan 00343
Armenian Institute of Sport, Yerevan 00329
Armenian Library & Museum of America, Inc, Watertown 37814
Armenian Missionary Association of America Library, Paramus 37324
Armenian Museum of All Saviour's Cathedral, Isfahan 14420
Armenian Patriarchate, Jerusalem 14697

Armenian Pedagogical Institute, Yerevan 00330
Armenian Pedagogical Institute, Yerevan 00331
Armenian Research Centre for Maternal & Child Health Care, Yerevan 00344
Armenian Research Institute for Scientific and Technical Information, Yerevan 00327
Armenian Scientific Research Institute of Agriculture, Echmiadzin 00340
Armenian State Picture Gallery, Yerevan 00345
Armenian Writers' Union, Yerevan 00346
Armidale Dumaresq Council, Armidale 00606
The Armitt Collection, Ambleside 29595
Armstrong Atlantic State University, Savannah 32621
Army & Navy Club Library, Washington 37723
Arnall, Golden & Gregory LLP, Atlanta 35454
Arnamagnean Institute in Iceland, Reykjavík 13575
Arnold Bergstraesser Institute of Socio-Cultural Research at the University of Freiburg, Freiburg 11884
Arnold & Porter Library, Washington 36329
Arnold-Bergstraesser-Institut für Kulturwissenschaftliche Forschung e.V. an der Albert-Ludwigs Universität Freiburg im Breisgau, Freiburg 11884
Arnstein & Lehr LLP Library, Chicago 35567
ARO Volcani Center, Bet Dagan 14714
Arquivo Distrital de Leiria, Leiria 21836
Arquivo do Estado de São Paulo, São Paulo 03393
Arquivo Histórico de Moçambique, Maputo 18980
Arquivo Histórico de São Tomé e Principe, São Tomé 24386
Arquivo Histórico Militar, Lisboa 21766
Arquivo Histórico Nacional, Luanda 00104
Arquivo Histórico Ultramarino, Lisboa 21767
Arquivo Nacional, Rio de Janeiro 03330
Arquivo Regional da Madeira, Funchal 21760
Arquivos Nacionalis Torre do Tombo, Lisboa 21768
ARRB Group Ltd, Vermont South 01041
Arrondissementsrechtbank Breda, Breda 19245
Arrondissementsrechtbank Utrecht, Utrecht 19289
Arrondissementsrechtbank Zutphen, Zutphen 19293
Arrowhead Library System, Virginia 41690
Arsip Nasional Republik Indonesia, Jakarta 14355
Art Center College of Design, Pasadena 32231
Art, Folklore and Ethnography Institute, Minsk 02018
Art Gallery, Split 06222
Art Gallery of Bosnia and Herzegovina, Sarajevo 02962
Art Gallery of New South Wales, Sydney 01021
Art Gallery of Ontario, Toronto 04504
Art Gallery of South Australia, Adelaide 00846
Art Gallery of Western Australia, Perth 00981
Art Historians Society, Warszawa 21346
Art Institute of Atlanta, Atlanta 30284
Art Institute of Chicago, Chicago 36664
The Art Institute of Dallas, Dallas 30939
Art Institute of Fort Lauderdale, Fort Lauderdale 31157
Art Institute of Houston, Houston 36939
Art Institute of Pittsburgh, Pittsburgh 32295
Art Institute of Portland Library, Portland 32334
Art Institute of Seattle Library, Seattle 37590
Art Institutes of California-San Francisco, San Francisco 37534
Art Library, Stockholm 26658
Art Library Milan City Council, Milano 16324
Art Library of the State Russian Museum, Sankt-Peterburg 23767
Art Literature Library, Riga 18186
Art Museum, Göteborg 26576
Art Museum of Estonia, Tallinn 07231
Art Museum of Norrköping, Norrköping 26623
Artem Industrial Corporation, Kyiv 28161
Artem Industrial Corporation Trade Union Committee, Kyiv 28685
Arter & Hadden, Los Angeles 35825
Artes Escénicas de Andalucía, Sevilla 26147
Artesia Public Library, Artesia 37997
Artesis Hogeschool Antwerpen, Antwerpen
– Departement Bedrijfskunde – Lerarenopleiding – Sociaal Werk, Antwerpen 02260
– Departement Gezondheidszorg, Merksem 02261
– Departement Industriele Wetenschappen en Technologie (2 cycli), Antwerpen 02262
– Department Ontwerpwetenschappen & Koninklijke Academie voor Schone Kunsten, Antwerpen 02263
– Hoger Instituut voor Vertalers en Tolken, Antwerpen 02264
– Koninklijk Conservatorium Antwerpen, Antwerpen 02265
Arteveldehogeschool, Gent
– Mediatheek Grafische Bedrijven, Mariakerke 02314
– Mediatheek Handelswetenschappen en Bedrijfskunde, Gent 02315
– Mediatheek Kantienberg, Gent 02316
– Mediatheek Opleiding Sociaal Werk, Gent 02317
ArtEZ Akademie voor Art & Design, Enschede 19118
ArtEZ Akademie voor Art & Design, Zwolle 19232
Artificial Fibre Industrial Corporation, Main Trade Union Library, Mogilev 02205
Artillery Officers Military School, Sankt-Peterburg 22559
Artis Bibliotheek / Bijzondere Collecties Universiteitsbibliotheek, Amsterdam 19347
Artis Library – Special Collections University Library, University of Amsterdam, Amsterdam 19347
Artistic Library of the City of Brussels, Bruxelles 02570
Artium, Vitoria-Gasteiz 26185
Arts Com & Sci College, Cambay 13795
Les Arts décoratifs, Paris 08509
Arts et Métiers ParisTech, Paris 07838
Arts Library, Melbourne 00946
Artsen Zonder Grenzen (AZG) / Médecins Sans Frontières (MSF), Brussel 02561
Aruba National Library, Oranjestad 00363
Arup Information and Library Services, London 29572
ARUP Melbourne Library, Melbourne 00947
Arus Public Library, Barcelona 25886
Arusha International Conference Centre (AICC), Arusha 27672
Arutunoff Library, Siloam Springs 32671
Arvika bibliotek, Arvika 26733
Arxiu del Regne de Mallorca, Palma de Mallorca 26110
Arxiu Històric de la Ciutat de Barcelona, Barcelona 25879
Arxiu Històric de Protocols de Barcelona, Barcelona 25880
Arxiu Històric de Sabadell, Sabadell 26127
Arxiu Provincial de l'Escola Pia de Catalunya, Barcelona 25764
Arzemju mākslas muzejs, Riga 18180
Arzobispado de Granada, Granada 25779
Ärztebücherei der Bezirksärztekammer Nordwürttemberg, Stuttgart 12510
Ärztekammer Hamburg – Bibliothek des Ärztlichen Vereins, Hamburg 11968
Ärztliche Zentralbibliothek, Hamburg 09955
Asahi Broadcasting Corp, Enterprise Division, Osaka 17403
Asahi Garasu K. K. Kenkyukaihatsubu, Yokohama 17501
Asahi Glass Co, Ltd, Research and Development Division, Yokohama 17501
Asahi Hoso K. K., Jigyokyoku Hoso Shiryobu, Osaka 17403
Asahi Mutual Life Insurance Co, Tokyo 17433
Asahi Seimei Hoken Sogogaisha, Tokyo 17433
Asahi Shimbun Co, Ltd, Osaka Main Office, Osaka 17404
Asahi Shimbun Osaka Honsha, Osaka 17404
Asahikawa Medical College, Asahikawa 16903
Asamblea de Madrid, Madrid 25696
Asamblea Legislativa, Panamá 20536
Asamblea Legislativa, San José 06114
Asamblea Legislativa, San Salvador 07205
Asamblea Nacional, Managua 19892
Asamblea Nacional de Rectores, Lima 20655
Asamblea Nacional del Poder Popular, La Habana 06280
M. M. Asatiani Institute of Psychiatry, Tbilisi 09224
Asbury College, Wilmore 33076
Asbury Park Free Public Library, Asbury Park 37998
Asbury Theological Seminary, Wilmore 35422
Asbury Village Jaycee Library, Hong Kong 05868
Ascension Lutheran Church Library, Milwaukee 35284
Ascension Parish Library, Donaldsonville 38851
N.N. Aseev Kursk Regional Research Library, Kursk 22147
Åsele bibliotek, Åsele 26735
Ashanti Regional Library, Kumasi 13064
Ashburton Public Library, Ashburton 19858
Ashe County Public Library, West Jefferson 41823
Asheville-Buncombe Technical Community College, Asheville 33134
Ashfield Municipal Library, Ashfield 01058
Ashikaga Gakko Iseki Toshokan, Ashikaga 17514
Ashikaga School Remains Library, Ashikaga 17514
Ashland Community & Technical College, Ashland 33135
Ashland, Inc, Dublin 35681
Ashland Public Library, Ashland 38001
Ashland Theological Seminary, Ashland 35111
Ashland University, Ashland 30264
Ashtabula County District Library, Ashtabula 38003
Ashtabula County Law Library Association, Jefferson 36975
Asia University, Musashino 17048
Asia-Africa Library, Mitaka 17598
Asian and Pacific Development Centre (APDC), Kuala Lumpur 18579
Asian Development Bank, Manila 20765
Asian Institute of Management, Makati City 20770
Asian Institute of Technology, Klong Luang 27794
Asian Social Institute, Manila 20772
Asian Studies Newsletter Archives, College Park 36723
Asian Vegetable Research and Development Center (AVRDC), Shanhua 27592
Asia-Pacific Cultural Center for UNESCO (ACCU), Tokyo 17670
Asiatic Society, Kolkata 14028
Asiatic Society Mumbai, Mumbai 14080
Asker bibliotek, Asker 20311
Askersunds kommunbibliotek, Askersund 26736
Askham Bryan College, York 29442
ASKLEPIOS Fachklinikum Brandenburg, Brandenburg an der Havel 11654
Asklepios Fachklinikum Stadtroda GmbH, Stadtroda 12498
Asklepios Klinik Harburg, Hamburg 11969
Asklepios Medical School, Hamburg 09912
Askov Folk High School Library, Vejen 06750
Askov Højskoles Bibliotek, Vejen 06750
Askøy folkebibliotek, Kleppestø 20356
ASM International Library, Materials Park 37089
Asmara Public Library, Asmara 07216
Åsnes folkebibliotek, Flisa 20329
Asnuntuck Community College, Enfield 33321
Asociación Argentina de la Cultura Inglesa, Buenos Aires 00235
Asociación 'Bernardino Rivadavia', Bahía Blanca 00318
Asociación Dante Alighieri, Buenos Aires 00236
Asociación de Empleados Bancarios del Uruguay (AEBU), Montevideo 42037
Asociación de Escribanos del Uruguay, Montevideo 42038
Asociación de Prensa Profesional, Barcelona 25881
Asociación Española de Normalización y Certificación (AENOR), Madrid 25977
Asociación Gran Peña, Madrid 25978
Asociación Latinoamericana de Instituciones Financieras de Desarrollo (ALIDE), Lima 20656
Asociación Latinoamericana de Integración (ALADI), Montevideo 42039
Asociación Médica Argentina, Buenos Aires 00237
Asociación Nacional de Industriales (ANDI), Medellín 06013
Asociación Nacional de Universidades e Institutos de Educacion Superior, México 18722
Asociación Odontológica Uruguaya, Montevideo 42040
Asociación para la Prevención de Accidentes, San Sebastián 26134
Asociación Química Argentina, Buenos Aires 00238
Asociación Rural del Uruguay, Montevideo 42041
Asociaţia de drept internaţional şi relaţii internaţionale din România, Bucureşti 21968
Asociaţia generală a inginerilor din România (AGIR), Bucureşti 21969
Assaf Ha-Rofeh Medical Centre, Zerifin 14773
Assam Agricultural University, Jorhat 13671
– Khanpara Campus, Gauhati 13672
Assemblea Regionale Siciliana, Palermo 15653
Assemblée Nationale, Paris 08142
Assemblée Nationale du Burundi, Bujumbura 03625
Assemblée Nationale du Cameroun, Yaoundé 03642
Assemblée Nationale du Gabon, Libreville 09170
Assemblée Nationale du Mali, Bamako 18643
Assemblée Nationale du Québec, Québec 04127
Assemblée Nationale du Sénégal, Dakar 24424
Assemblée Permanente des Chambres d'Agriculture (APCA), Paris 08467
Assemblée Permanente des Chambres de Métiers (APCM), Paris 08468
Assembléia da República, Lisboa 21719
Assembléia Legislativa do Estado de Minas Gerais, Belo Horizonte 03213
Assembleia Nacional, São Tomé 24385
Assemblies of God Theological Seminary, Springfield 35389
Assembly of Moscow People's Deputies, Moskva 22737
Assembly of the Republic of Macedonia, Skopje 18440
Assens Bibliotek, Assens 06848
Assicurazioni Generali, Trieste 16157
AssiDomän Skärblacka AB, Skärblacka 26548
Assiniboine Community College, Brandon 03978
Assistance Publique – Hôpitaux de Paris (APHP), Paris 08469
Assiut University, Assiut 07016
– Civil Department, Assiut 07017
– Electricity Department, Assiut 07018
– Faculty of Agriculture, Assiut 07019
– Faculty of Commerce, Assiut 07020
– Faculty of Education, Assiut 07021
– Faculty of Engineering, Assiut 07022
– Faculty of Law, Assiut 07023
– Faculty of Medicine, Assiut 07024
– Faculty of Pharmacology, Assiut 07025
– Faculty of Physical Education, Assiut 07026
– Faculty of Science, Assiut 07027
– Faculty of Veterinary Medicine, Assiut 07028
– Higher Institute of Nursing, Assiut 07029
– Mechanics Department, Assiut 07030
– Teachers' College, Assiut 07031
Associação Brasileira de Educação e Cultura (ABEC), Brasília 03197
Associação Brasileira de Imprensa, Rio de Janeiro 03331
Associação Brasileira de Metallurgia, Materiais e Mineração (ABM), São Paulo 03394
Associação Comercial do Rio de Janeiro, Rio de Janeiro 03332
Associação de Cultura Franco-Brasileira, Rio de Janeiro 03333
Associação de Ensino, Ribeirão Prêto 03325
Associação de Ensino Unificado do Distrito Federal, Brasília 03198
Associação dos Advogados, São Paulo 03395
Associação dos Arqueólogos Portugueses (AAP), Lisboa 21769
Associação Escola de Agrimensura, Araraquara 03268
Associação Industrial Portuguesa (AIP), Lisboa 21770
Associação Portuguesa de Bibliotecários, Arquivistas e Documentalistas, Lisboa 21771
Associação Protectora dos Diabéticos de Portugal (APDP), Lisboa 21772
Associació d'Enginyers Industrials de Catalunya, Barcelona 25882
Associated Mennonite Biblical Seminary, Elkhart 35196
Association canadienne pour l'intégration communautaire, North York 04425
Association Culturelle Franco-Japonaise de Tenri, Paris 08470
Association des transports du Canada, Ottawa 04453
Association Egyptologique Reine Elisabeth, Bruxelles 02562
Association for Farmers' Education, Koibuchi Gakuen Library, Ibaraki 17553
Association for Nature Conservation in South-West Tavastland, Forssa 07438

Association for Research & Enlightenment, Virginia Beach 37708
Association for the Blind of Western Australia, Victoria Park 01042
Association Française de Normalisation (AFNOR), Saint-Denis La Plaine 08664
Association Internationale des Universités, Paris 08576
Association Nationale pour la Formation Professionnelle des Adultes, Montreuil 08425
Association of African Universities (AAU), Accra 13041
Association of Angolan Writers, Luanda 00108
Association of Commonwealth Universities (ACU), London 29713
Association of Dissemination of Sciences, Budapest 13462
Association of Finnish Local Authorities, Helsinki 07490
Association of Indian Universities, New Delhi 14115
Association of International Law and International Relations of Romania, Bucureşti 21968
Association of Polish Journalists, Warszawa 21345
Association of Surgeons of India, Chennai 13956
Association of the Bar of the City of New York Library, New York 37187
Association of the Tractor Industry, Ursus 21236
Association pour la Conservation du château d'Oron, Oron-le-Châtel 27421
Association pour la Promotion Technique-Culturelle Belgo-Latino-Américaine, Bruxelles 02563
Association pour le Bien des Aveugles et malvoyants, Genève 27356
Association Valentin Haüy – Bibliothèque Braille et Le Livre Parlé, Paris 08471
Associazione Archeologica Romana, Roma 16450
Associazione Bancaria Italiana, Roma 16151
Associazione Culturale Famiglia Meneghina, Milano 16321
Associazione Culturale Italo-Francese, Trieste 16619
Associazione fra le Società Italiane per Azioni, Roma 16451
Associazione Italiana Biblioteche, Roma 16452
Associazione per lo Sviluppo dell'Industria nel Mezzogiorno (Svimez), Roma 16453
Assumption Abbey Library, Richardton 35343
Assumption College, Makati City 20747
Assumption College, Worcester 33093
Assumption Parish Library, Napoleonville 40366
Assumption Seminary, Plumpton 00791
Assumption Seminary, San Antonio 35372
Assumption University, Bangkok 27684
Astan Quds Razavi, Mashhad 14386
Aston University, Birmingham 28947
Astoria Public Library, Astoria 38005
Åstorps bibliotek, Åstorp 26737
Astra Draco AB, Lund 26540
Astra Hässle AB, Mölndal 26543
ASTRA National Museum Complex, Sibiu 22056
Astra Zeneca RoD Södertälje, Södertälje 26551
Astrakhan State Conservatoire, Astrakhan 22208
Astrakhanskaya oblastnaya detskaya biblioteka, Astrakhan 24106
Astrakhanskaya oblastnaya nauchnaya biblioteka im. N.K. Krupskoi, Astrakhan 22118
Astrakhanskaya oblastnaya yunosheskaya biblioteka, Astrakhan 24107
Astrakhanski gosudarstvenny tekhnicheski universitet, Astrakhan 22206
AstraZeneca Pharmaceuticals, Wilmington 36420
Astre Corporate Group Library, Alexandria 35437
Astrium GmbH, Bremen 11380
Astrobibliothek, Garching 11909
Astrological Research Library of Canada, Toronto 04505
Astronomical Institute, Toshkent 42127
Astronomical Observatory of Belgrade, Beograd 24501
Astronomicheskaya observatoriya im. Engelgardta, Zelenodolsk 24095
Astronomicheskaya observatoriya im. V.P. Engelgardta, Stanitsa 23979
Astronomska Opservatorija u Beogradu, Beograd 24501
Astrophysical Institute, Almaty 17939

Astrophysical Observatory, Abastumani 09209
Astrophysikalisches Institut Potsdam, Potsdam 12401
AT Svietlahorski CKK, Svetlogorsk 01929
Atascadero State Hospital, Atascadero 36486
Atascadero State Hospital, Atascadero 36487
Atatürk Kitaphgı, Istanbul 27901
Atchison County Library, Rock Port 41036
Atchison Public Library, Atchison 38006
Atef Danial Library, Qalhat 18213
Atelier Parisien d'Urbanisme (APUR), Paris 08472
Ateneo Barcelonès, Barcelona 25883
Ateneo Científico, Literario y Artístico, Madrid 25979
Ateneo Científico, Literario y Artístico, Maó 26103
Ateneo de Davao University, Davao City 20696
Ateneo de Macorís, San Pedro de Macorís 06962
Ateneo de Manila University, Manila 20710
Ateneo di Brescia, Brescia 16207
Ateneo di Salò, Salò 16571
Ateneo di Scienze, Lettere ed Arti di Bergamo, Bergamo 16179
Ateneo Mercantil de Valencia, Valencia 26258
Ateneo Puertorriqueño, San Juan 21903
Ateneo Veneto, Venezia 16644
Ateneu Comercial, Porto 21822
Ater & Wynne, LLP, Portland 36140
Athabasca University, Athabasca 03673
Athalie McFarlane Petersen Public Library, Saint Croix 42314
The Athenaeum, London 29714
Athenaeum, Stade 10768
Athenaeum Liverpool Library, Liverpool 29701
Athenaeum of Ohio, Cincinnati 30798
Athenaeum of Philadelphia, Philadelphia 37339
Athens Archaeological Society, Athinai 13097
Athens College – Psychico College, Athinai 13090
Athens County Law Library, Athens 33939
Athens Regional Library System – Athens-Clarke County Library, Athens 38008
Athens School of Fine Arts (ASFA), Athinai 13073
Athens State University, Athens 30271
Athens Technical College Library, Athens 33137
Athens University of Economics and Business, Athinai 13077
Athol Public Library, Athol 38013
ATK Launch Systems, Brigham City 35541
Atlanta Christian College, East Point 31066
Atlanta College of Art, Atlanta 30285
Atlanta History Center, Atlanta 36488
Atlanta Metropolitan College, Atlanta 33138
Atlanta University Center Inc, Atlanta 30286
Atlanta VA Medical Center Library, Decatur 36782
Atlanta-Fulton Public Library System, Atlanta 38014
Atlanta-Fulton Public Library System – Alpharetta Library, Alpharetta 37922
Atlanta-Fulton Public Library System – Auburn Avenue Research Library on African-American Culture & History, Atlanta 38015
Atlanta-Fulton Public Library System – Buckhead Library, Atlanta 38016
Atlanta-Fulton Public Library System – Cleveland Avenue, Atlanta 38017
Atlanta-Fulton Public Library System – Dr. Robert E. Fulton Regional at Ocee, Alpharetta 37923
Atlanta-Fulton Public Library System – East Point Library, East Point 38911
Atlanta-Fulton Public Library System – Northeast-Spruill Oaks Regional Library, Alpharetta 37924
Atlanta-Fulton Public Library System – Northside Library, Atlanta 38018
Atlanta-Fulton Public Library System – Peachtree Library, Atlanta 38019
Atlanta-Fulton Public Library System – Ponce de Leon Library, Atlanta 38020
Atlanta-Fulton Public Library System – Roswell Regional Library, Roswell 41072
Atlanta-Fulton Public Library System – Sandy Springs Regional Library, Sandy Springs 41207
Atlanta-Fulton Public Library System – South Fulton Regional Library, Union City 41637
Atlanta-Fulton Public Library System – Southwest Regional Library, Atlanta 38021
Atlantic Cape Community College, Mays Landing 33522
Atlantic City Free Public Library, Atlantic City 38023

Atlantic County Library, Mays Landing 40106
Atlantic Food and Horticulture Research Centre Library, Kentville 04366
Atlantic Provinces Special Education Authority Library, Halifax 04081
Atlantic Research Institute of Fisheries and Oceanography (AtlantNIRO), Kaliningrad 23066
Atlantic School of Theology Library, Halifax 03715
Atlantic Union College, South Lancaster 32688
Atlanticheski nauchno-issledovatelski institut rybnogo khozyaistva i okeanografii, Kaliningrad 23066
Atma Jaya Catholic University of Indonesia, Jakarta 14257
Atmore Public Library, Atmore 38024
Atofina Chemicals Inc, King of Prussia 35804
ATOL vzw, Kessel-Lo 02667
Atomic Energy Authority, Cairo 07105
Atomic Energy Centre, Dhaka 01755
Atomic Energy Commission, Buenos Aires 00251
Atomic Energy Corporation of South Africa Ltd, Pretoria 25071
Atomic Energy Council, Yonghe 27625
Atomic Energy of Canada Ltd, Mississauga 04093
Atomic Weapons Establishment (AWE), Reading 29883
AT&T, Atlanta 35455
AT&T, Florham Park 35708
AT&T Labs Library, Middletown 35889
Attica Correctional Facility School Library, Attica 33945
Attleboro Public Library, Attleboro 38025
Attorney General and Ministry of Legal Affairs, Port-of-Spain 27808
Attorney General's Chamber, Nairobi 18019
Attorney Generals Chambers, Kuala Lumpur 18556
Attorney General's Chambers, Singapore 24595
Attorney General's Department, Barton 00607
Attorney General's Department, Colombo 26290
Attorney General's Department, Hobart 00683
Attorney General's Department, Melbourne 00694
Attorney General's Library, Accra 13028
Attorney General's Library, Lilongwe 18480
Attorney General's Library, Lusaka 42332
Attorney General's Office, Baltimore 33962
Attorney General's Office, Olympia 34633
Attorney-General's Department, Sydney 00724
Åtvidabergs kommunbibliotek, Åtvidaberg 26738
Atwater Library & Computer Centre, Westmount 04551
Aua Language Center, Bangkok 27774
Auburn Hills Public Library, Auburn Hills 38031
Auburn Library Service, Auburn 01059
Auburn Public Library, Auburn 38026
Auburn Public Library, Auburn 38027
Auburn Public Library, Auburn 38028
Auburn University, Auburn
– The Library of Architecture, Design & Construction, Auburn 30298
– Ralph Brown Draughon Library, Auburn 30299
– Veterinary Medical Library, Auburn 30300
Auburn University, Montgomery 31885
Auburn-Placer County Library, Auburn 38029
Auburn-Placer County Library – Rocklin Branch, Rocklin 41041
Auckland Art Gallery Toi o Tāmaki, Auckland 19819
Auckland City Libraries, Auckland 19859
Auckland District Law Society, Auckland 19820
Auckland Regional Council, Auckland 19821
Auckland University of Technology, Auckland 19761
Auckland War Memorial Museum, Auckland 19822
Audencia Nantes, Nantes 07803
Audiencia Provincial de Santa Cruz de Tenerife, Santa Cruz de Tenerife 25743
Audit Commission, Bristol 29620
Audren Cohen College, New York 32024
Audubon Regional Library, Clinton 38603
Auerbach Central Agency for Jewish Education, Melrose Park 37091
Augenklinik, München 10432
Auglaize County Law Library, Wapakoneta 34974
Auglaize County Public District Library, Wapakoneta 41721
Augsburg College, Minneapolis 31843
Augusta County Library, Fishersville 39068
Augusta State University, Augusta 30301

Augusta Technical College, Augusta 36498
Augustana College, Rock Island 32471
Augustana College, Sioux Falls 32677
– Center for Western Studies Library, Sioux Falls 32678
Augustana-Hochschule, Neuendettelsau 10514
August-Bebel-Institut (ABI), Berlin 11491
Augustijnenklooster Sint-Stefanus, Gent 02461
Augustijns Historisch Instituut, Leuven 02472
Augustiner-Chorherrenstift, St. Florian 01471
Augustiner-Chorherrenstift, Vorau 01478
Augustiner-Chorherrenstift Herzogenburg, Herzogenburg 01435
Augustinerkloster, Würzburg 11363
Augustinerkloster St. Michael, Münnerstadt 11287
Augustinermuseum, Freiburg 11885
Augustinian Central Library, Brookvale 00757
Augustinian Historical Institute, Leuven 02472
Augustinian Library, Valladolid 25837
Auraria Library, Denver 30992
Aurora Health Care Libraries, Milwaukee 37110
Aurora Public Library, Aurora 38036
Aurora Public Library, Aurora 38037
Aurora Public Library, Aurora 04571
Aurora Public Library, Aurora 38038
Aurora Public Library District, Aurora 38039
Aurora Research Institute, Aurora College, Inuvik 04365
Aurora University, Aurora 30304
Außenpolitische Bibliothek, Wien 01393
Aust-Agder bibliotek og kultur-formidling, Arendal 20309
Austen Riggs Center, Inc, Stockbridge 36286
Austin Clarke Library, Dublin 14522
Austin College, Sherman 32665
Austin Community College, Austin 33143
Austin Graduate School of Theology Library, Austin 36503
Austin Health Library, Heidelberg 00929
Austin Peay State University, Clarksville 30822
Austin Presbyterian Theological Seminary, Austin 35118
Austin Public Library, Austin 38041
Austin Public Library, Austin 38042
Austin Public Library – Little Walnut Creek, Austin 38043
Austin Public Library – Manchaca Road, Austin 38044
Austin Public Library – Spicewood Springs, Austin 38045
Austin Public Library – Windsor Park, Austin 38046
Australia Council, Red Fern 00993
Australia Council for the Arts Research Centre, Surry Hills 01015
Australia-Japan Foundation Australian Resource Centre, Tokyo 17300
Australian Agency for International Development, Canberra 00637
Australian Antarctic Division Library, Kingston 00935
Australian Broadcasting Corporation, Southbank 01010
Australian Broadcasting Corporation, Ultimo 01038
Australian Bureau of Statistics, Belconnen 00612
Australian Capital Territory Health Authority, Acton 00594
Australian Catholic University, Brisbane
– Aquinas Campus Callinan Library, Ballarat 00392
– Banyo Library, Banyo 00393
– Mackillop Campus Victor Couch Library, North Sydney 00394
– Mount St. Mary Campus, Strathfield 00395
– Saint Patrick's Campus, East Melbourne 00396
– Signadou Campus, Watson 00397
Australian Centre for the Moving Image, Flinders Lane 00918
Australian College of Applied Psycholgy, Strawberry Hills 00517
Australian College of Ministries, Kenmore 00769
Australian College of Physical Education Ltd., Sydney Olympic Pard 00544
Australian Communications and Media Authority, Melbourne 00695
Australian Competition and Consumer Commission Library, Canberra 00888
Australian Conservation Foundation, Carlton 00897
Australian Council for Educational Research, Camberwell 00884
Australian Council of Trade Unions, Carlton South 00901

Australian Education Union – Queensland Branch Grulke Library, Milton BC 00961
Australian Education Union – Victorian Branch Library, Abbotsford 00843
Australian Federal Police, Barton 00869
Australian Film, Television and Radio School, Stawberry Hills 00516
Australian Geological Survey Organisation, Symonston 01033
Australian Goverment Solicitor National Library Service, Barton 00608
Australian Government Solicitor, Brisbane 00614
Australian Graduate School of Management, Sydney 00519
Australian High Commission Library, Port Moresby 20593
Australian Industrial Registry Principal Library, Melbourne 00696
Australian Institute of Aboriginal and Torres Strait Islander Studies (AIATSIS), Acton 00844
Australian Institute of Criminology, Griffith 00926
Australian Institute of Family Studies, Melbourne 00948
Australian Institute of Management, Saint Kilda 00997
Australian Institute of Management, Spring Hill 01012
Australian Institute of Marine Science, Townsville 01036
Australian Institute of Music Resource Centre, Surry Hills 01016
Australian Institute of Police Management, Manly 00693
Australian Lutheran College, North Adelaide 00493
Australian Maritime College, Newnham 00492
Australian Mineral Foundation Centre, Glenside 00923
Australian Museum, Sydney 01022
Australian National Botanic Gardens, Canberra 00889
Australian National Maritime Museum, Pyrmont 00991
Australian National University, Canberra 00412
– Art Library, Canberra 00413
Australian Nuclear Science and Technology Organisation (ANSTO), Lucas Heights 00943
Australian Public Service Commission, Phillip 00985
Australian Radiation Protection and Nuclear Safety Agency, Yallambie 00746
Australian Red Cross, Carlton 00898
Australian Security Intelligence Organisation, Canberra 00890
Australian Society for Indigenous Languages, Palmerston 00973
Australian Sports Commission, Bruce 00883
Australian Taxation Office, Canberra 00638
Australian War Memorial, Campbell 00885
Austrian Academy of Sciences (AAS), Wien 01187
Austrian Archives for Adult Education, Wien 01646
Austrian Cultural Forum, London 29715
Austrian Cultural Forum Library, New York 37188
Austrian Horticultural Society, Wien 01631
Austrian Institute of Construction Engineering, Wien 01640
Austrian Institute of Economic Research, Wien 01641
Austrian Leather Library, Hofkirchen 01539
Austrian National Library, Wien 01180
Austrian Parliamentary Library, Wien 01410
Austrian Patent Office, Wien 01409
Austrian Society of Contemporary History, Wien 01633
Austrian Study Center for Peace and Conflict Resolution, Stadtschlaining 01580
Austrian Zoological-Botanical Society, Wien 01665
Auswärtiges Amt, Berlin 10787
Auswärtiges Amt – Akademie Auswärtiger Dienst, Berlin 10788
Auswärtiges Amt, Sprachendienst, Berlin 10789
ausZeiten – Bildung, Information, Forschung und Kommunikation für Frauen e.V., Bochum 11616
Autauga-Prattville Public Library, Prattville 40890
Authority for Books and National Documents, Cairo 07106
Authority for Geological Surveying and Metallurgical Projects, Cairo 07107
Authority for Marine Education and Training, Akko 14677
Authority for Meteorology, Cairo 07108
Authority for Research & Conservation of Cultural Heritage (ARCCH), Addis Ababa 07279
Autoglass Plant, Kostyantynivka 28150
Autókut – Autóipari Kutató és Fejlesztø Vállalat, Budapest 13370
Autókut Research and Development Company for the Automotive Industry, Budapest 13370
Automatic Control Systems Scientific and Industrial Corporation, Severodonetsk 28465
Automatic Hydraulic Machine Plant, Borisov 01838
Automation Institute Scientific and Industrial Corporation, Kyiv 28379
Automatisation Methods Design Bureau, Sankt-Peterburg 23805
Automobile and Railway Transport Industrial Corporation, Lviv 28423
Automobile Factory, Lviv 28163
Automobile Plant, Serpukhov 22954
Automobile Plant, Main Trade Union Library, Mogilev 02192
Automobile Testing and Adjustment Research Centre, Dmitrov 23024
Automobile Transport Institute, Moskva 23382
Automobile Transport Professional School, Kyiv 28115
Automobile Units Industrial Corporation, Kineshma 22805
Automotive Production Lines Plant, Gomel 01865
Automotive Research Association of India, Pune 14185
Automóvil Club Argentino, Buenos Aires 00239
Autry National Center Institute for the Study of the American West – Autry Library, Los Angeles 37043
Autry National Center Institute for the Study of the American West – Braun Research Library, Los Angeles 37044
Av Geral de Energia, Lisboa 21773
Avangard Industrial and Research Corporation, Sankt-Peterburg 23862
Avans Hogeschool, Breda 19099
– Hogeschool West-Brabant – Sector Pedagogisch Onderwijs, Breda 19100
Avans Hogeschool, Faculteit Techniek & Natuur, Tilburg 19201
Ave Maria School of Law Library, Ann Arbor 30228
Ave Maria University, Ave Maria 30327
Avele College, Apia 24379
Averett University, Danville 30956
Avery County Morrison Public Library, Newland 40463
Avery-Mitchell-Yancey Regional Library System, Burnsville 38400
Avesta bibliotek, Avesta 26739
Aviation and Cosmonautics Centre, Moskva 23507
Aviatsionno-tekhnologicheski universitet im. K.E. Tsiolkovskogo, Moskva 22328
Avila College, Kansas City 31480
AVL List GmbH, Graz 01497
Avon Free Public Library, Avon 38048
Avon Lake Public Library, Avon Lake 38051
Avon Park Correctional Facility, Avon Park 33960
Avondale College, Cooranbong 00427
Avon-Washington Township Public Library, Avon 38049
Avoyelles Correctional Center Library, Cottonport 34115
Avoyelles Parish Library, Marksville 40068
Avtobusny zavod, Lviv 28163
'Avtokran' Industrial Corporation, Ivanovo 22794
Avtomobilny zavod im. I.A. Likhacheva – Dvorets kultury, Moskva 24168
Avtonormal Automobile Joint Stock Company, Belebei 22779
Avtozavod, Mogilev 02192
Avtozavod, Serpukhov 22954
Avvocatura dello Stato, Roma 15656
Awadhesh Pratap Singh University, Rewa 13753
Awapatent AB, Malmö 26541
Ayasofya (Saint Sophia) Museum, Istanbul 27883
Ayer Library, Ayer 38052
Ayuntamiento de Madrid, Madrid 25697
Azabu Daigaku, Sagamihara 17092
Azabu University, Sagamihara 17092
Azerbaijan Agricultural Institute, Gandja 01692
Azerbaijan Institute of Sports, Baku 01694
Azerbaijan Medical Institute, Baku 01682
Azerbaijan National Library named after M. F. Akhundov, Baku 01679
Azerbaijan Pedagogical Institute, Baku 01683
Azerbaijan Pedagogical Institute for Languages, Baku 01684
Azerbaijan Research Institute of Animal Husbandry, Gyanzha 01725
Azerbaijan Research Institute of Cotton Crops, Gyanzha 01726
Azerbaijan Research Institute of Hydrotechnics and Improvement, Baku 01695
Azerbaijan Scientific and Technical Library, Baku 01680
Azerbaijan State Petroleum Academy, Baku 01685
Azerbaijan State Theatrical Institute, Baku 01686
Azerbaijan State University of Art, Baku 01687
Azerbaijan Technical University, Baku 01688
Azerbaijan Veterinary Institute, Baku 01697
Azienda Ospedaliera, Roma 16454
Azot Industrial Corporation, Dniprodzerzhinsk 28137
Azov and Black Sea Region Agricultural Mechanization Institute, Zernograd 22662
Azov Automatic Forging Equipment Plant, Azov 22775
Azovo-Chernomorski institut mekhanizatsii selskogo khozyaistva, Zernograd 22662
Azusa City Library, Azusa 38053
B. B. Comer Memorial Library, Sylacauga 41486
B. F. Goodrich Co, Brecksville 35540
B. F. Jones Memorial Library, Aliquippa 37911
B. Khmelnytsky Cherkasy State University, Cherkasy 27955
B. P. Library of Motoring, Beaulieu 29605
Dr. Babasaheb Ambedkar Marathwada University, Aurangabad 13607
Babasaheb Bhimrao Ambedkar Bihar University, Muzaffarpur 13713
The Babraham Institute, Cambridge 29623
Babson College, Babson Park 30328
Z. M. Babura District Library, Andizhan 42170
Babylon Public Library, Babylon 38054
Bac Thai Medical College, Thai Nguyen City 42282
Bach-Archiv Leipzig, Leipzig 12178
Back Mountain Memorial Library, Dallas 38755
Bacon Memorial District Library, Wyandotte 41978
Bacone College, Muskogee 33567
Badan Penelitian dan Pengembangan Kehutanan, Jakarta 14356
Badan Penelitian dan Pengembangan Pendidikan dan Kebudayaan, Jakarta 14357
Badia di Santa Maria del Monte, Cesena 15792
Badische Landesbibliothek, Karlsruhe 09321
Badischer Landesverein für Naturkunde und Naturschutz e.V., Freiburg 11886
Badisches Landesmuseum Karlsruhe, Karlsruhe 12084
Badisches Landesmuseum Karlsruhe, Außenstelle Südbaden, Staufen 12499
BAE Systems, Nashua 35933
BAE Systems Advanced Technology Centres, Chelmsford 29569
BAE Systems Armament Systems, Minneapolis 37119
Bæjar- og héra sbókasafnið Selfossi, Selfoss 13584
Bæjar- og héraðsbókasafnið Ísafirði, Ísafjörður 13580
Baendasamtök Islands, Reykjavík 13568
Bærum bibliotek, Bekkestua 20312
Bagaduce Music Lending Library, Blue Hill 36551
Baháí World Centre Library, Haifa 14695
Bahauddin Zakariya University, Multan 20449
AB Bahco Ventilation, Enköping 26528
Bahirdar University, Bahirdar 07268
– Education Faculty Library, Bahirdar 07269
Bahrain Centre for Studies and Research, Manama 01740
Bai Jerbai Wadia Library, Pune 13835
Baikov Institute of Metallurgy, Moskva 23284
Bailey Cavalieri LLC, Columbus 35630
Baillie's Library, Glasgow 29672
Bainbridge College, Bainbridge 33146
Baker & Botts LLP, Houston 35746
Baker Botts LLP, Washington 36330
Baker College of Muskegon, Muskegon 31924
Baker College of Owosso, Owosso 32220
Baker College System Libraries, Flint 33339
Baker County Public Library, Baker City 38058
Baker & Daniels Library, Indianapolis 35773
Baker & Hostetler, Los Angeles 35826
Baker & Hostetler, Washington 36331
Baker & Hostetler Library, Columbus 35631
Baker & Hostetler LLP Library, Cleveland 35613
Baker, Manock & Jensen Library, Fresno 35715
Baker & McKenzie Library, Bangkok 27769
Baker & McKenzie Library, New York 35957
Baker & McKenzie LLP Library, Washington 36332
Baker Medical Research Institute, Prahran 00989
Baker University, Baldwin City 30331
Bakersfield College, Bakersfield 33147
Bakhchisarai Historical and Cultural State Preserve, Bakhchisarai 28182
The Bakken – A. Library & Museum of Electricity in Life, Minneapolis 37120
Bakony Mountains Natural History Museum, Zirc 13524
Bakonyi Természettudomány Múzeum, Zirc 13524
Baku State University, Baku 01689
Bakul Institute for Superhard Materials, Kyiv 28341
Bakulev Institute of Heart and Vascular Surgery, Moskva 23322
Bala-Cynwyd Library, Bala Cynwyd 38063
Balai Besar Penelitian dan Pengembangan Industri Hasil Pertanian, Bogor 14345
Balai Penelitian Perikanan Laut, Jakarta 14358
Balai Penelitian Perkebunan (RISPA), Medan 14376
Balai Penelitian Veteriner, Bogor 14346
Balai Penjelidikan Pertanian C.P.V., Bogor 14347
Balakliska raionna TsB S, Tsentralna biblioteka, Balakliya 28491
Balakliya Regional Centralized Library System, Main Library, Balakliya 28491
Balashov State Pedagogical Institute, Balashov 22210
Balassa Bálint Múzeum, Esztergom 13471
Balassi Bálint Intézet Könyvtára, Budapest 13220
Balassi Bálint Könyvtar, Salgótarján 13545
Balch & Bingham Attorneys Library, Birmingham 35506
Baldwin Public Library, Baldwin 38065
Baldwinsville Public Library, Baldwinsville 38067
Baldwin-Wallace College, Berea 30384
– Riemenschneider Bach Institute Library, Berea 30385
Bălgarska akademiya na naukite, Sofiya 03452
Bălgarska akademiya na naukite – Etnografski institut s muzej, Sofiya 03521
Bălgarska akademiya na naukite – Institute of Balkan Studies, Sofiya 03522
Bălgarska dărzhavna konservatoriya, Sofiya 03461
Bălgarska tărgovsko-promischlena palata, Sofiya 03523
Bălgarska telegrafna agentsiya, Sofiya 03524
Bălgarsko geologichesko druzhestvo, Sofiya 03525
Ball State University, Muncie 31918
– Architecture Library, Muncie 31919
– Science-Health Science Library, Muncie 31920
Ballarat Health Services Base Hospital, Ballarat 00867
Ballarat Mechanics' Institute, Ballarat 00868
Ballard, Spahr, Andrews & Ingersoll LLP, Washington 36333
Ballard, Spahr, Andrews & Ingersoll, LLP Library, Philadelphia 36088
Ballarmine University, Louisville 31681
Ballerup Bibliotek, Ballerup 06849
Ballygunje Institute, Kolkata 14029
Balob Teachers' College, Lae 20556
Balochi Academy and Research Centre, Quetta 20524
Baltic Institute of Tourism, Sankt-Peterburg 23747
Baltic Paper Joint-Stock Company, Sankt-Peterburg 22901
Baltic Plant, Sankt-Peterburg 22513
Baltic State University of Technology, Sankt-Peterburg 22512
Baltic Waterways Trade Union Administration – Sailors Recreation Centre, Sankt-Peterburg 24245
Baltimore City Community College, Baltimore 33148
Baltimore County Circuit Court, Towson 34931
Baltimore County Public Library, Towson 41586
Baltimore County Public Library – Arbutus, Baltimore 38069
Baltimore County Public Library – Catonsville, Baltimore 38070

Baltimore County Public Library
– Cockeysville Area Branch,
Cockeysville 38622
Baltimore County Public Library –
Essex, Baltimore 38071
Baltimore County Public Library –
North Point, Baltimore 38072
Baltimore County Public Library –
Parkville-Carney, Baltimore 38073
Baltimore County Public Library –
Perry Hall, Baltimore 38074
Baltimore County Public Library
Pikesville, Baltimore 38075
Baltimore County Public Library
– Randallstown Area Branch,
Randallstown 40938
Baltimore County Public Library –
Reisterstown Branch, Reisterstown 40961
Baltimore County Public Library –
Rosedale, Baltimore 38076
Baltimore County Public Library –
Towson Area Branch, Towson 41587
Baltimore County Public Library –
White Marsh, Baltimore 38077
Baltimore County Public Library –
Woodlawn, Baltimore 38078
Baltimore Hebrew Congregation,
Baltimore 35121
Baltimore International College,
Baltimore 33149
Baltimore Metropolitan Council,
Baltimore 36506
Baltimore Museum of Art, Baltimore 36507
Baltiski federalny universitet im.
Immanuila Kanta, Kaliningrad 22270
Baltiski gosudarstvenny tekhnicheski
universitet, Sankt-Peterburg 22512
Baltiski institut turizma, Sankt-
Peterburg 23747
Baltiski zavod, Sankt-Peterburg 22513
Balzekas Museum of Lithuanian
Culture, Chicago 36665
BAM Bundesanstalt für Material-
forschung und -prüfung, Berlin 11492
Bamberg State Library, Bamberg 09300
Bamble bibliotek, Langesund 20363
Banana Shire Library & Information
Service, Biloela 01070
Banaras Hindu University, Varanasi 13783
Banaras Hindu University, Varanasi 13781
– Institute of Medical Science,
Varanasi 13782
Banasthali Vidyapith, Banasthali 13608
Banca d'Italia, Roma 16152
Banca Nazionale del Lavoro (BNL),
Roma 16455
Banco Bilbao Vizcaya Argentaria,
Madrid 25860
Banco Central de Bolivia, La Paz 02933
Banco Central de Chile, Santiago 04913
Banco Central de Costa Rica, San
José 06118
Banco Central de la República
Argentina, Buenos Aires 00240
Banco Central de Nicaragua,
Managua 19893
Banco Central de Reserva del Perú
(BCRP), Lima 20648
Banco Central de Venezuela, Caracas 42239
Banco Central del Ecuador, Quito 06990
Banco Central del Uruguay,
Montevideo 42042
Banco Central do Brasil, Brasília 03254
Banco de Desenvolvimento de
Minas Gerais – BDMG S.A., Belo
Horizonte 03251
Banco de Desenvolvimento do Paraná
S.A., Curitiba 03292
Banco de España, Madrid 25858
Banco de Guatemala, Guatemala City 13178
Banco de la Provincia de Buenos
Aires, Buenos Aires 00224
Banco de la República, Santafé de
Bogotá 06006
Banco de México, México 18822
Banco de Portugal, Lisboa 21744
Banco de Seguros del Estado,
Montevideo 42030
Banco del Libro, Caracas 42248
Banco di Sardegna, Sassari 16156
Banco di Sicilia, Palermo 16150
Banco do Brasil SA, Rio de Janeiro 03259
Banco Exterior de España, Barcelona 25852
Banco Hipotecario del Uruguay,
Montevideo 42031
Banco Nacional de Cuba, La Habana 06286
Banco Totta & Açores, Lisboa 21745
Banco Urquijo, Madrid 25859
Banco Wiese Ltdo, Lima 20649
Band, Lavis & Associates, Inc,
Severna Park 36270
Bandaranayake Centre for
International Studies, Colombo 26304
Bandon Public Library, Bandon 38081
Bandung Institute of Technology,
Bandung 14239

Banff Centre, Banff 04310
Bang Phra Agricultural College,
Chonburi 27735
Bangabandu Sheikh Mujib Medical
University, Dhaka 01746
Bangalore State Central Library,
Bangalore 13587
Bangko Sentral ng Pilipinas, Manila 20766
Bangkok Metropolis Special Library,
Bangkok 27775
Bangkok Metropolitan Bank, Bangkok 27770
Bangkok Patana School Library,
Bangkok 27754
Bangkok University, Bangkok 27685
Bangla Academy, Dhaka 01756
Bangladesh Agricultural University,
Mymensingh 01751
Bangladesh Atomic Energy
Commission, Dhaka 01757
Bangladesh Central Public Library,
Dhaka 01772
Bangladesh College of Textile
Technology, Dhaka 01747
Bangladesh Institute of Development
Studies (BIDS), Dhaka 01758
Bangladesh Medical Association,
Dhaka 01759
Bangladesh National Museum, Dhaka 01760
Bangladesh National Scientific and
Technical Documentation Centre
(BANSDOC), Dhaka 01761
Bangladesh Parliament, Dhaka 01753
Bangladesh Public Administration
Training Centre, Dhaka 01762
Bangladesh University of Engineering
and Technology, Dhaka 01748
Bangor Public Library, Bangor 38082
Bangor Theological Seminary, Bangor 35126
Bangor Theological Seminary, Bangor 35127
Bangor University, Bangor 28934
Bangunan Sultan Abdul Samad, Kuala
Lumpur 18580
Bank Bumiputra Centre, Kuala
Lumpur 18570
Bank Bumiputra Malaysia Bhd, Kuala
Lumpur 18570
Bank Indonesia, Jakarta 14341
Bank Markazi Jomhoori, Tehran 14421
Bank Markazi Jomhowri Eslami Iran,
Tehran 14422
Bank Negara Malaysia, Kuala Lumpur 18571
Bank of Alexandria, Cairo 07136
Bank of America, Concord 35636
Bank of America, San Francisco 36222
Bank of America – Southern California
Headquarters, Los Angeles 35827
Bank Of Canada Library, Ottawa 04104
Bank of Ceylon, Colombo 26299
Bank of England Information Centre,
London 29573
Bank of Finland Library, Helsinki 07492
Bank of Ghana, Accra 13037
Bank of Greece, Athinai 13125
Bank of Guyana, Georgetown 13195
Bank of Israel, Jerusalem 14705
Bank of Japan, Institute for Monetary
and Economic Studies, Tokyo 17464
Bank of New York, New York 35958
Bank of Nova Scotia Business
Information Services, Toronto 04277
Bank of Sierra Leone, Freetown 24578
Bank of Thailand, Bangkok 27771
Bank of Tokyo, Economic Research
Dept, Tokyo 17486
Bank of Yokohama, Yokohama 17826
Bank One of Chicago – Corporate
Information Center, Chicago 35568
Bank Street College of Education,
New York 32025
Bank Widya Gama, Malang 14271
Bankers' Library, Tokyo 17675
Bánki Donát Műszaki Főiskola,
Budapest 13221
Bánki Donát Technical College,
Budapest 13221
Bankovski kolledzh, Sankt-Peterburg 22693
Bankstown City Library and
Information Service, Bankstown 01062
Banner Good Samaritan Medical
Center, Phoenix 37368
Banning Public Library, Banning 38083
Banque Centrale de Tunisie, Tunis 27824
Banque Centrale du Congo, Kinshasa 06064
Banque Centrale du Luxembourg,
Luxembourg 18407
Banque de France, Paris 08260
Banque d'Etat du Maroc, Rabat 18946
Banque Européenne d'Investissement
(BEI), Luxembourg 18403
Banque Nationale de Belgique (BNB),
Bruxelles 02564
Banque Nationale de Paris, Paris 08261
Banque Nationale du Rwanda, Kigali 24373
Banverket Biblioteket, Borlänge 26563
Baoding Financial College, Baoding 04987
Baoding Teachers' College, Baoding 04988

Baoji College of Arts and Science,
Baoji 04992
Baoshan Teachers' College, Baoshan 04993
Baotou Medical College, Baotou 04994
Baptist Bible College, Springfield 35390
Baptist Bible College and Seminary,
Clarks Summit 35162
Baptist Hospital, Pensacola 37333
Baptist Missionary Association
Theological Seminary, Jacksonville 35245
Baptist Theological College of
Southern Africa, Randburg 25045
Baptist Theological College of
Western Australia, Bentley 00752
Dr. Baquir's Library, Lahore 20508
Bar Ilan University, Ramat-Gan 14645
– Chemistry Library, Ramat-Gan 14646
– Department of Mathematics and
Computer Sciences, Ramat-Gan 14647
– English Department, Ramat-Gan 14648
– Faculty of Jewish Studies, Ramat-
Gan 14649
– Faculty of Law, Ramat-Gan 14650
– Faculty of Social Sciences, Ramat-
Gan 14651
– French Department, Ramat-Gan 14652
– Hebrew Literature Department,
Ramat-Gan 14653
– History Department, Ramat-Gan 14654
– Library of Education, Ramat-Gan 14655
– Life Sciences Library, Ramat-Gan 14656
– Philosophy Department, Ramat-
Gan 14657
– Psychology Department, Ramat-
Gan 14658
The Bar Library, Belfast 29607
Bar Regional Centralized Library
System, Main Library, Bar 28492
Baraboo Public Library, Baraboo 38084
Baranovichi Central City Children's
Library, Baranovichi 02095
Baranovichi Central City Library,
Baranovichi 02094
Baranovichi Central Railway Station,
Baranovichi 01943
Baranovichi City Hospital, Baranovichi 01941
Baranovichi City Library no 1,
Baranovichi 02093
Baranovichi-Polesskie Railway
Station, Baranovichi 01942
Baranovichiskaya gorodskaya
bibliyateka no 1, Baranovichi 02093
Baranovichiskaya gorodskaya
bolnitsa, Baranovichi 01941
Baranovichiskaya tsentralnaya
gorodskaya bibliyateka im. Tavlaya,
Baranovichi 02094
Baranovichiskaya tsentralnaya
gorodskaya detskaya bibliyateka,
Baranovichi 02095
Baranya Megyei Könyvtár, Pécs 13544
Barbados Community College, St.
Michael 01777
Barbados Department of Archives, St.
James 01780
Barbados Museum and Historical
Society, St. Michael 01781
Barber Scotia College, Concord 30917
Barberton Public Library, Barberton 38085
Barbican Centre, London 30016
Barcelona Athenaeum Library,
Barcelona 25883
Barcelona Bar association library,
Barcelona 25894
Barclay College, Haviland 31316
Bard College, Annandale-on-Hudson 30250
– Center for Curatorial Studies,
Annandale-on-Hudson 30251
– Levy Economics Institute,
Annandale-on-Hudson 30252
Bard College at Simon's Rock, Great
Barrington 31244
The Bard Graduate Center for Studies
in the Decorative Arts, Design, and
Culture, New York 32026
Bare Hill Correctional Facility Library,
Malone 34494
Bareilly College Library, Bareilly 13792
Barkatullah University Bhopal, Bhopal 13618
Barking Central Library, Barking 29933
Barksdale Air Force Base Library,
Barksdale AFB 33968
Barnard College, New York 32027
Barnaul Transport Engineering Joint-
Stock Company, Barnaul 22777
Barnesville Hutton Memorial Library,
Barnesville 38088
Barnet Libraries, London 30017
Barnsley Central Library, Barnsley 29934
Barnsley College, Barnsley 28935
Barnsley District General Hospital,
Barnsley 29602
Barnstable Law Library, Barnstable 36519
Barossa Council, Nuriootpa 01136
Barr Engineering Co Library,
Minneapolis 35903

Barrie Public Library, Barrie 04575
Barrier Reef Institute of TAFE,
Townsville 00548
Barrington Public Library, Barrington 38091
Barrington Public Library District,
Barrington 38092
Barry University, Miami 31805
Barry-Lawrence Regional Library,
Monett 40257
Barska raionna TsBS, Tsentralna
biblioteka, Bar 28492
Barstow College, Barstow 33154
Barsumian and Derian Libraries,
Beirut 18204
Bartholomew County Public Library,
Columbus 38654
Bartlesville Public Library, Bartlesville 38093
Bartlett Public Library District, Bartlett 38094
Barton College, Wilson 33077
Barton County Community College,
Great Bend 33385
Barton Institute of TAFE, Moorabbin 00579
Barton Library, El Dorado 38941
Bartow County Public Library System,
Cartersville 38467
Bartow County Public Library
System – Cartersville Main Street,
Cartersville 38468
Bartram Trail Regional Library –
Mary Willis Library Headquarters,
Washington 41735
Baruch College-CUNY, New York 32028
Barvinkove Regional Centralized
Library System, Main Library,
Barvinkove 28493
Barvinkovska raionna TsBS,
Tsentralna biblioteka, Barvinkove 28493
BASF SE, Ludwigshafen 11416
Bashkir Agricultural Institute, Ufa 22610
Bashkir Institute of Industrial
Construction, Ufa 24023
Bashkir Medical Institute, Ufa 22608
Bashkir Pedagogical Institut, Ufa 22609
Bashkir Research Institute of Machine
Building for Oil Industry, Ufa 24025
Bashkir Scientific Research Institute of
Agriculture, Ufa 24024
Bashkir State Archives, Ufa 24019
Bashkir State Museum of Arts, Ufa 24020
Bashkir State University, Ufa 22607
Bashkirski gosudarstvenny
khudozhestvenny muzei im. M.V.
Nesterova, Ufa 24020
Bashkirski gosudarstvenny universitet,
Ufa 22607
Bashkirski meditsinski institut, Ufa 22608
Bashkirski mezhotraslevoi territorialny
tsentr nauchno-tekhnicheskoi
informatsii i propagandy, Ufa 24021
Bashkirski nauchno-issledovatelski i
proektny institut nefti (Bashnipineft),
Ufa 24022
Bashkirski nauchno-issledovatelski
institut Promstroi, Ufa 24023
Bashkirski nauchno-issledovatelski
institut selskogo khozyaistva, Ufa 24024
Bashkirski pedagogicheski institut, Ufa 22609
Bashkirski selskokhozyaistvenny
institut, Ufa 22610
Basilica di S. Nicola, Bari 15723
Basilica di S. Paolo, Città del Vaticano 42194
Basilica di Superga, Torino 16091
Basilica Santuario Santa Maria di
Campagna, Piacenza 15978
Basingstoke College of Technology,
Basingstoke 29450
Basler Afrika Bibliographien, Basel 27301
Basler Papiermühle, Basel 27302
Basler Zeitung (BaZ), Basel 27282
BasNII Neftemash, Ufa 24025
Bast Fibres Initial Treatment Research
Institute, Minsk 02043
Batavia Public Library District, Batavia 38100
Batchelor Institute of Indigenous
Tertiary Education, Batchelor 00385
Bates College, Lewiston 31589
Bath and North East Somerset
Council, Bath 29936
Bath Spa University, Bath 28936
Bathurst Library, Bathurst 01064
Battelle Energy Alliance, LLC, Idaho
Falls 35771
Battelle Memorial Institute, Columbus 36736
Batu Lintang Teachers' Training
College, Kuching 18538
Bauchi College of Arts and Science,
Bauchi 19927
Bauchi State Library Board, Bauchi 20047
Bauhaus Dessau, Dessau-Roßlau 11736
Bauhaus-Archiv Berlin, Berlin 11493
Bauhaus-Universität Weimar, Weimar 10682
Bauman Moscow State Engineering
University, Moskva 22370
Bauman Veterinary Institute, Kazan 22281
Bavarian Environment Agency,
Augsburg 10778

Bavarian Environment Agency, Hof 10951
Bavarian Natural History Collections, München 12288
Bavarian State Library, München 09294
Bavovnyany kombinat, Kherson 28147
Bavovnyany kombinat – Biblioteka profkomu, Kherson 28604
Baxter County Library, Mountain Home 40328
Baxter County Library, Mountain Home 40329
Bay City Public Library, Bay City 38114
Bay County Library System, Bay City 38115
Bay County Library System – Alice & Jack Wirt Public Library, Bay City 38116
Bay County Library System – Sage Branch, Bay City 38117
Bay De Noc Community College, Escanaba 33324
Bay Path College, Longmeadow 31640
Bay Shore-Brightwaters Public Library, Brightwaters 38287
Bayard Taylor Memorial Library, Kennett Square 39701
Bayard-Presse, Paris 08262
Bayer Corp Library, Pittsburgh 36127
Bayer Corporation, Tarrytown 36300
Bayer Healthcare, Berkeley 35497
Bayerische Akademie der Schönen Künste, München 12278
Bayerische Akademie der Wissenschaften, München 12279
Bayerische Akademie für Naturschutz und Landschaftspflege, Laufen 12175
Bayerische Armeebibliothek, Ingolstadt 12073
Bayerische Blindenhörbücherei e.V., München 12280
Bayerische Forstschule / Bayerische Technikerschule für Waldwirtschaft, Lohr a. Main 10754
Bayerische Landesanstalt für Weinbau und Gartenbau (LWG), Veitshöchheim 10772
Bayerische Motoren Werke AG, München 11419
Bayerische Staatsbibliothek, München 09294
Bayerische Staatsgemäldesammlungen, München 12281
Bayerische Staatskanzlei, München 11023
Bayerische Staatssammlung für Paläontologie und Geologie, München 12282
Bayerische Verwaltung der Staatlichen Schlösser, Gärten und Seen, München 11024
Bayerischer Landesverein für Heimatpflege e.V., München 12283
Bayerischer Landtag, München 11025
Bayerischer Rundfunk, München 12284
Bayerischer Verwaltungsgerichtshof, München 11026
Bayerisches Armeemuseum, Ingolstadt 12073
Bayerisches Geologisches Landesamt, München 11027
Bayerisches Hauptstaatsarchiv, München 12285
Bayerisches Landesamt für Denkmalpflege, München 11028
Bayerisches Landesamt für Statistik und Datenverarbeitung, München 11029
Bayerisches Landesamt für Steuern – Dienststelle München, München 11030
Bayerisches Landesamt für Umwelt (LFU), Augsburg 10778
Bayerisches Landesamt für Umwelt (LFU), Hof 10951
Bayerisches Landessozialgericht, München 11031
Bayerisches Nationalmuseum, München 12286
Bayerisches Staatsinstitut für Hochschulforschung und Hochschulplanung, München 12287
Bayerisches Staatsministerium der Finanzen, München 11032
Bayerisches Staatsministerium der Justiz und für Verbraucherschutz, München 11033
Bayerisches Staatsministerium des Innern, München 11034
Bayerisches Staatsministerium für Arbeit und Sozialordnung, Familie und Frauen, München 11035
Bayerisches Staatsministerium für Ernährung, Landwirtschaft und Forsten, München 11036
Bayerisches Staatsministerium für Umwelt und Gesundheit, München 11037
Bayerisches Staatsministerium für Unterricht und Kultus, München 11038
Bayerisches Staatsministerium für Wirtschaft, Infrastruktur, Verkehr und Technologie, München 11039
Bayerisches Verwaltungsgericht Ansbach, Ansbach 10777
Bayero University, Kano 19956
Bayliss Public Library, Sault Sainte Marie 41237
Baylor University, Waco 32938
– Armstrong Browning Library, Waco 32939
– Baylor Collections of Political Materials, Waco 32940
– Crouch Fine Arts Library, Waco 32941
– J. M. Dawson Church-State Research Center Library, Waco 32942
– Jesse H. Jones Library, Waco 32943
– Sheridan & John Eddie Williams Legal Research & Technology Center, Waco 32944
– Texas Collection, Waco 32945
Bayport-Blue Point Public Library, Blue Point 38225
Bayside City Council, Brighton 01071
Bayside State Prison Library, Leesburg 34452
Baystate Medical Center, Springfield 37625
Bayswater – Public Library & Information Service, Morley 01126
Bayville Free Library, Bayville 38125
BBVA – Banco Bilbao Vizcaya Argentaria, Madrid 25860
BCC-UCF Joint Use Library, Cocoa 33246
BCM Biblioteca Civica di Mestre, Venezia 16853
BCT – Biblioteca Comunale Terni, Terni 16835
Beacon Bay Public Library, Beacon Bay 25184
Beacon University Library, Columbus 30891
Beaconsfield Public Library, Beaconsfield 04576
Beaman Memorial Public Library, West Boylston 41809
Bear Lake County Free Library, Montpelier 40279
Bear Library, Bear 38128
Bear Public Library, Bear 38129
Beardsley & Memorial Library, Winsted 41937
Beasley, Casey & Erbstein Library, Philadelphia 36089
Beatrice Public Library, Beatrice 38130
Beaufort County Library, Beaufort 38131
Beaufort, Hyde & Martin County Regional Library, Washington 41736
Beaumont Library District, Beaumont 38132
Beaumont Public Library System, Beaumont 38133
Beauregard Parish Library, DeRidder 38830
Beauregard Parish Public Library, De Ridder 38794
Beaver Area Memorial Library, Beaver 38134
Beaver Commonwealth College, Monaca 31875
Beaver County Law Library, Beaver 33977
Beaver County Library System, Monaca 40252
Beaver Dam Community Library, Beaver Dam 38135
Bechtel Corporate Library, San Francisco 36223
Becker College, Leicester 33481
Becker College, Worcester 33883
Becton, Dickinson & Co, Franklin Lakes 35713
Bedales Memorial Library, Petersfield 29486
Bedford Free Public Library, Bedford 38138
Bedford Historical Society Library, Bedford 36526
Bedford Institute of Oceanography Library, Dartmouth 04330
Bedford Park Public Library District, Bedford Park 38143
Bedford Public Library, Bedford 38139
Bedford Public Library, Bedford 38140
Bedford Public Library, Bedford 38141
Bedford Public Library System, Bedford 38142
Bedfordshire Libraries, Bedford 29937
Bedřich Smetana Library, Praha 06495
Bedřich Smetana Museum, Praha 06498
Beech Grove Public Library, Beech Grove 38145
Beeld en Geluid, Hilversum 19463
Beethoven-Haus, Bonn 11622
Bega Vallex Shire Library, Bega 01066
Behörde für Justiz und Gleichstellung, Hamburg 10912
Behramjee Jijeebhai Medical College, Pune 13741
Behrend College, Erie 31095
Beihoff Music Corporation, Wauwatosa 36399
Beijing Agricultural College, Beijing 04995
Beijing Broadcasting Institute, Beijing 04996
Beijing Capital Library, Beijing 04951
Beijing Clothing Arts and Crafts Institute, Beijing 04997
Beijing Coal Mining Management Institute, Beijing 04998
Beijing College of Acupuncture and Orthopedics, Beijing 04999
Beijing College of Economics, Beijing 05000
Beijing College of Planning and Labour Management, Beijing 05001
Beijing Film Academy, Beijing 05002
Beijing Foreign Studies University, Beijing 05003
Beijing Forestry University, Beijing 05004
Beijing Institute of Agricultural Mechanization, Beijing 05005
Beijing Institute of Chemical Technology, Beijing 05006
Beijing Institute of Civil Engineering and Architecture, Beijing 05007
Beijing Institute of Commerce, Beijing 05008
Beijing Institute of Finance and Trade, Beijing 05009
Beijing Institute of Light Industry, Beijing 05010
Beijing Institute of Mechanical Industry, Beijing 05011
Beijing Institute of Meteorology, Beijing 05012
Beijing Institute of Physical Culture, Beijing 05013
Beijing Institute of Printing Technology, Beijing 05014
Beijing Institute of Technology, Beijing 05015
Beijing International Studies University, Beijing 05016
Beijing Language and Culture University, Beijing 05017
Beijing Materials Institute, Beijing 05018
Beijing Medical University, Beijing 05019
Beijing Metallurgical Management Cadres Institute, Beijing 05020
Beijing Natural History Museum, Beijing 05818
Beijing Normal University (BNU), Beijing 05021
Beijing Power Engineering and Economics Institute, Beijing 05022
Beijing Teachers' College of Physical Culture, Beijing 05023
Beijing University of Aeronautics and Astronautics, Beijing 05024
Beijing University of Chemical Technology, Beijing 05025
Beijing University of Posts and Telecommunications, Beijing 05026
Beijing University of Technology, Beijing 05027
Beijing Youth and Politics Institute, Beijing 05028
Beilstein-Institut zur Förderung der Chemischen Wissenschaften, Frankfurt am Main 11838
Beirut Arab University, Beirut 18203
Beit Berl, Beit Berl 14602
Békés County Library, Békéscsaba 13528
Békés Megyei Könyvtár, Békéscsaba 13528
Belarus Academy of Sciences – Institute of Biochemistry, Grodno 01962
Belarus Academy of Sciences – Institute of Mathematics, Gomel 01952
Belarus Academy of Sciences – Metal-Polymer Research Institute, Gomel 01951
Belarus Agricultural Academy, Gorki 01958
Belarus Agricultural Mechanization Scientific and Industrial Corporation, Minsk 01893
Belarus Agricultural Research Institute, Zhodino 02089
Belarus Automobile Plant, Zhodino 01940
Belarus Cardiology Research Institute, Minsk 01996
Belarus Central State Archive – Grodno Archive Department, Grodno 01959
Belarus Central State Art and Literature Archive and Museum, Minsk 01993
Belarus Commercial Construction for Trade and Catering Planning Institute, Minsk 01982
Belarus Conservatory, Minsk 01801
Belarus Construction and Mechanization Planning and Design Institute, Minsk 01983
Belarus Fish Industry Scientific and Industrial Corporation, Minsk 01988
Belarus Fuel Industry Research Institute, Minsk 01986
Belarus Gallurgy Research Institute, Minsk 01984
Belarus Haematology and Blood Tansfusion Research Institute, Minsk 01987
Belarus Land Utilization Planning Institute, Minsk 01990
Belarus Local Industries Planning Institute, Minsk 02010
Belarus Ministry of Education – Republican Centre for Young Technicians, Minsk 01824
Belarus Ministry of Higher Education and Sprecial Training, Minsk 01822
Belarus Ministry of Light Industry – Republican Work Coordination and Production Management Centre, Minsk 01825
Belarus Municipal Services Construction Planning Institute, Minsk 01981
Belarus Music Instruments Industrial Corporation, Borisov 01835
Belarus National Institute for Continuing Education in Medicine, Minsk 01799
Belarus Oil-extraction Industrial Corporation, Prospecting Station, Gomel 01847
Belarus Pedagogy Research Institute, Minsk 02042
Belarus Plant Protection Research Institute, Priluki 02072
Belarus Potash Extraction Industrial Corporation, Main Trade Union Library, Soligorsk 02238
Belarus Potash Extraction Industrial Corporation, Scientific and Technical Library, Soligorsk 01928
Belarus Potato, Fruit and Vegetable-Cultivation Research Institute, Samokhvalovichi 02076
Belarus Power Engineering and Electrical Transmission Research and Planning Institute, Minsk 01985
Belarus Power Engineering and Electricity Central Board, Minsk 02002
Belarus Railways, Railway Scientific and Technical Library, Minsk 01972
Belarus Regional Beetroot-cultivation Research and Experimental Station, Nesvizh 02068
Belarus Road Building Research Institute, Minsk 01973
Belarus Rubber Technology Industrial Corporation, Bobruysk 01832
Belarus Sanitary and Hygiene Research Institute, Minsk 01970
Belarus State Agrarian Technical University, Minsk 01802
Belarus State Agricultural Construction Research and Planning Institute, Minsk 01977
Belarus State Art and Drama Institute, Minsk 01980
Belarus State Building Industry Administration – Construction and Architecture Institute, Minsk 02005
Belarus State Building Industry Planning Institute, Minsk 01989
Belarus State Economic University, Minsk 01991
Belarus State Food Production Industry Planning Institute, Minsk 01979
Belarus State Industrial Construction Planning Institute, Minsk 01976
Belarus State Light Industry Planning Institute, Minsk 01974
Belarus State Medical University, Minsk 01803
Belarus State Motorways and Bridges Planning and Research Institute, Minsk 01978
Belarus State Museum of Great Patriotic War, Minsk 01992
Belarus State Plannng Committee – Automatic Planning Systems and Management Research Institute, Minsk 02009
Belarus State Water Supply and Land Reclamation Engineering Planning Institute, Minsk 01975
Belarus Stock-breeding Research Institute, Zhodino 02090
Belarus Television and Radio Head Quarters, Minsk 02047
Belarus Trade Union Organizations, Republican Main Trade Union Library, Minsk 02157
Belarus Venerology and Skin Disorders Research Institute, Minsk 01997
Belarusian State University, Minsk 01804
Belarusian State University of Informatics and Radioelectronics, Minsk 01805
Belarusian University of Culture, Minsk 01971
Belaruskaya gosudarstvennaya politekhnicheskaya akademiya, Minsk 01798
Belaruskaya selskogospodarchaya akademiya, Gorki 01958
Belaruskaya zheleznaya doroga, Minsk 01972

Belaruskaya zonalnaya opytnaya stantsiya po sakharnoi svekle, Nesvizh 02068
Belaruski dorozhny nauchno-issledovatelski institut, Minsk 01973
Belaruski gosudarstvenny institut po proektirovaniyu predpriyati legkoi promyshlennosti, Minsk 01974
Belaruski gosudarstvenny institut po proektirovaniyu vodokhozyaistvennogo i meliorativnogo stroitelstva, Minsk 01975
Belaruski gosudarstvenny institut promyshlennogo proektirovaniya, Minsk 01976
Belaruski gosudarstvenny institut usovershenstvovaniya vrachei, Minsk 01799
Belaruski gosudarstvenny nauchno-issledovatelski i proektny institut po stroitelstvy na sele, Minsk 01977
Belaruski gosudarstvenny nauchno-issledovatelski institut po izyskaniyu i proektirovaniyu avtomobilnykh dorog i mostov, Minsk 01978
Belaruski gosudarstvenny proektny institut pishchevoi promyshlennosti, Minsk 01979
Belaruski gosudarstvenny teatralno-khudozhestvenny institut, Minsk 01980
Belaruski institut inzhenerov zheleznodorozhnogo transporta, Gomel 01790
Belaruski institut proektirovaniya obektov kommunalnogo khozyaistva (Belkommunproekt), Minsk 01981
Belaruski institut proektirovaniya predpriyati torgovli i obshchestvennogo pitaniya (Belgiprotorg), Minsk 01982
Belaruski konstruktorsko-tekhnologicheski institut stroimekhanizatsii, Minsk 01983
Belaruski nauchno-issledovatelski i proektny institut galurgii, Minsk 01984
Belaruski nauchno-issledovatelski institut energosetproekt, Minsk 01985
Belaruski nauchno-issledovatelski institut toplivnoi promyshlennosty, Minsk 01986
Belaruski NII gematologii i perelivaniya krovi, Minsk 01987
Belaruski NPO rybnogo khozyaistva, Minsk 01988
Belaruski PO Muzykalnykh instrumentov (Belmuzprom), Borisov 01835
Belaruski proektny institut Belgosproekt, Minsk 01989
Belaruski proektny institut po zemleustroistvu, Minsk 01990
Belarusneft – Normativno-issledovatelskaya stantsiya, Gomel 01847
Belarussian Agricultural Academy, Mogilev 01810
Belarussian State Polytechnical Academy, Minsk 01798
Belarussian State Technological University, Minsk 01800
Belarusski gosudarstvenny institut narodnogo khozyaistva, Minsk 01991
Belarusski gosudarstvenny muzei Velikoi Otechestvennoi voiny, Minsk 01992
Belarusski tsentralny gosudarstvenny arkhiv-muzei literatury i iskusstva, Minsk 01993
Belarusski tsentralny gosudarstvenny istoricheski arkhiv – Otdelenie v Grodnenskoi oblasti, Grodno 01959
Belaryski avtomobilny zavod, Zhodino 01940
Belasting en Douane Museum, Rotterdam 19506
Belastingdienst/CFD, Utrecht 19290
Belau National Museum, Koror 20531
BelavtoMAZ Industrial Corporation – Minsk Automobile Plant, Minsk 01896
Belfast Institute of Further and Higher Education, Belfast 28938
Belfast Public Library, Belfast 29938
Belgian Institute for Higher Chinese Studies, Brussel 02583
Belgian Road Research Centre (BRRC), Sterrebeek 02700
Belgisch Instituut voor Normalisatie (BIN), Brussel 02565
Belgisch Olympisch en Interfederaal Comité (BOIC), Brussel 02566
Belgisch Stripcentrum / Centre Belge de la Bande Dessinée, Brussel 02567
Belgorod State Universal Research Library, Belgorod 22120
Belgorod State University, Belgorod 22217
Belgorodskaya gosudarstvennaya universalnaya nauchnaya biblioteka (BGUNB), Belgorod 22120
Belgorodskaya oblastnaya detskaya biblioteka im. A.P. Gaidara, Belgorod 24110
Belgorodskaya oblastnaya yunosheskaya biblioteka, Belgorod 24111
Belgorodski gosudarstvenny pedagogicheski universitet im. Olminskogo, Belgorod 22216
Belgorodski gosudarstvenny universitet, Belgorod 22217
Belgrad City Museum, Beograd 24516
Belgrade City Library, Beograd 24535
Belhaven College, Jackson 31437
Belinsky Regional Universal Scientific Library of Sverdlovsk, Ekaterinburg 22128
Belinsky Teacher Training College in Penza, Penza 22466
Bell Flavors & Fragrances Duft und Aroma GmbH, Leipzig 11415
A. K. Bell Library, Perth 30058
Bell South Corporation, Atlanta 35456
belladonna, Bremen 11670
Bellaire City Library, Bellaire 38147
Bellaire Public Library, Bellaire 38148
Belle-Idée, Chêne-Bourg 27342
Belleville Public Library, Belleville 38152
Belleville Public Library, Belleville 04579
Belleville Public Library & Information Center, Belleville 38153
Bellevue Community College, Bellevue 33164
Bellevue Public Library, Bellevue 38155
Bellevue University, Bellevue 30373
Bellingham Public Library, Bellingham 38157
Bellmore Memorial Library, Bellmore 38159
Bellville Public Library, Bellville 25185
Bellwood Public Library, Bellwood 38162
Bellwood-Antis Public Library, Bellwood 38163
Belmont Abbey College, Belmont 30376
Belmont County Law Library, Saint Clairsville 34772
Belmont Public Library, Belmont 38164
Belmont Technical College, Saint Clairsville 32495
Belmont University, Nashville 31932
BelNII lesnogo khozyaistva, Gomel 01953
Belogirsk Regional Centralized Library System, Main Library, Belogirsk 28494
Belogirska raionna TsBS, Tsentralna biblioteka, Belogirsk 28494
Belogorka North-Western Agricultural Research Institute, Sankt-Peterburg 23900
Beloit College, Beloit 30378
Beloit Public Library, Beloit 38166
Beloruskaya konservatoriya, Minsk 01801
Belorussian State University of Transport (BelGUT), scientific-technical library, Gomel 01790
Belorusskaja selskokhozjaistvennaja biblioteka, Minsk 01994
Belorusski gosudarstvenny agrarny tekhnicheski universitet, Minsk 01802
Belorusski gosudarstvenny meditsinski universitet, Minsk 01803
Belorusski gosudarstvenny universitet, Minsk 01804
Belorusski gosudarstvenny universitet informatiki i radioelektroniki, Minsk 01805
Belorusski nauchno-issledovatelski institut gigieny i profpatologii, Minsk 01995
Belorusski nauchno-issledovatelski institut kardiologii, Minsk 01996
Belorusski nauchno-issledovatelski institut kartofelevodstva i plodoovoshchevodstva, Samokhvalovichi 02076
Belorusski nauchno-issledovatelski institut kozhno-venerologicheskikh zabolevani, Minsk 01997
Belorusski nauchno-issledovatelski institut perelivaniya krovi, Minsk 01998
Belorusski nauchno-issledovatelski institut zashchity rasteni, Priluki 02072
Belorusski nauchno-issledovatelski institut zemledeliya, Zhodino 02089
Belorusski nauchno-issledovatelski institut zhivotnovodstva, Zhodino 02090
Belsovprof, Minsk 02157
Belvedere Research Center, Wien 01630
Belvedere-Tiburon Library, Tiburon 41556
Bemidji State University, Bemidji 30380
Ben Gurion University of the Negev, Beer-Sheva 14599
– Jacob Blaustein Institute for Desert Research, Be'er-Sheva 14600
– Soroka Medical Center, Be'er-Sheva 14601
Benaki Museum, Athinai 13098
Benaki Phytopathological Institute, Athinai 13099
Benakio Phytopathologiko Instituto, Athinai 13099
Bendel State Library, Benin City 19915
Bendigo TAFE, Bendigo 00388
Bendra Lietuvių literatūros ir tautsakos bei Lietuvių kalbos institutų mokslinė biblioteka, Vilnius 18299
Benedict College, Columbia 30869
Benedictine Archabbay Library at Pannonhalma, Pannonhalma 13341
Benedictine College, Atchison 30270
Benedictine Community of New Norcia, New Norcia 00784
Benedictine University, Lisle 31622
Bénédictines du Saint Sacrement, Rouen 08231
Benediktinenstift Nonnberg, Salzburg 01458
Benediktinerabtei Braunau, Rohr i. NB 11317
Benediktinerabtei Ettal, Ettal 11211
Benediktinerabtei Gerleve, Billerbeck 11173
Benediktinerabtei Kornelimünster, Aachen 11142
Benediktinerabtei Maria Laach, Glees 11225
Benediktinerabtei Metten, Metten 11274
Benediktinerabtei Michaelbeuern, Michaelbeuern 01455
Benediktinerabtei Neresheim, Neresheim 11291
Benediktinerabtei Neuburg, Heidelberg 11240
Benediktinerabtei Niederaltaich, Niederaltaich 11294
Benediktinerabtei Ottobeuren, Alte Bibliothek (Bibliotheca Ottenburana), Ottobeuren 11302
Benediktinerabtei Ottobeuren, Neue Bibliothek, Ottobeuren 11303
Benediktinerabtei Plankstetten, Berching 11162
Benediktiner-Abtei Schweiklberg, Vilshofen 11356
Benediktinerabtei St. Matthias, Trier 11345
Benediktinerabtei St. Otmarsberg, Uznach 27275
Benediktiner-Abtei Unserer Lieben Frau zu den Schotten, Wien 01479
Benediktiner-Abtei Weingarten, Weingarten 11358
Benediktiner-Erzabtei St. Peter, Salzburg 01459
Benediktinergemeinschaft Fischingen, Fischingen 27258
Benediktinerinnen-Abtei Frauenwörth im Chiemsee, Frauenchiemsee 11214
Benediktinerinnen-Abtei Kloster Engelthal, Altenstadt 11148
Benediktinerinnen-Abtei St. Gertrud, Tettenweis 11342
Benediktinerinnen-Abtei Varensell, Rietberg 11316
Benediktinerinnenabtei vom Hl. Kreuz Herstelle, Beverungen 11170
Benediktinerinnenkloster Fahr (AG), Unterengstringen 27274
Benediktinerkloster Engelberg, Engelberg 27255
Benediktinerkloster Mariastein, Mariastein 27265
Benediktinerkloster St. Stephan, Augsburg 11150
Benediktiner-Kollegium, Sarnen 27270
Benediktinerstift Admont, Admont 01414
Benediktinerstift Göttweig, Göttweig 01426
Benediktinerstift Lambach, Lambach 01446
Benediktinerstift Melk, Melk 01454
Benediktinerstift St. Paul, St. Paul 01473
Bengal Engineering and Science University, Shibpur, Howrah 13653
Bengbu College, Bengbu 05058
Bengbu Medical College, Bengbu 05059
Benghazi Public Library, Benghazi 18265
Bengtsfors bibliotek, Bengtsfors 26742
Ben-Gurion House, Tel-Aviv 14758
Benicia Public Library, Benicia 38171
Benjamin Franklin Institute of Technology, Boston 36557
Benjamin Franklin Library, Guadalajara 18827
Bennett College, Greensboro 31251
Bennett Jones LLP Library, Calgary 04319
Bennington College, Bennington 30382
Bennington Free Library, Bennington 38173
Beno Župančič Public Library, Postojna 24977
Benoni Public Library, Benoni 25186
Bensenville Community Public Library, Bensenville 38175
Benson Memorial Library, Titusville 41563
Bentley College, Waltham 32948
Benton Harbor Public Library, Benton Harbor 38177
Bentonville Public Library, Bentonville 38178
Benxi Metallurgical College, Benxi 05060
Benxi University, Benxi 05061
Berdichiv City Centralized Library System, Main Library, Berdichiv 28495
Berdichivska miska TsBS, Tsentralna biblioteka, Berdichiv 28495
Berdyansk City Centralized Library System, Main Library, Berdyansk 28496
Berdyansk City Technical Library, Berdyansk 28184
Berdyanska miska TsBS, Tsentralna biblioteka, Berdyansk 28496
Berdyanskaya gorodskaya tekhnicheskaya biblioteka, Berdyansk 28184
N. A. Berdzenishvili Kutaisi State Museum of History and Ethnography, Kutaisi 09213
Berea College, Berea 30386
Beregove Regional Centralized Library System, Main Library, Beregove 28497
Beregovska raionna TsBS, Tsentralna biblioteka, Beregove 28497
Berenson Library, Firenze 14990
Bereza Regional Central Library, Bereza 02097
Berezhani Regional Centralized Library System, Main Library, Berezhani 28498
Berezhanska raionna TsBS, Tsentralna biblioteka, Berezhani 28498
Berezino Regional Central Library, Berezino 02098
Berezinsкaya tsentralnaya raionnaya bibliyateka, Berezino 02098
Berezovskaya tsentralnaya raionnaya bibliyateka, Bereza 02097
Bergen College of Higher Education, Bergen
Bergen Community College, Paramus 33618
Bergen County Law Library, Hackensack 34327
Bergen Maritime Museum, Bergen 20210
Bergen Offentlige Bibliotek, Bergen 20313
Bergen Public Library, Bergen 20313
Bergenfield Public Library, Bergenfield 38180
Bergens sjøfartsmuseum, Bergen 20210
Bergische Universität Wuppertal, Wuppertal 10694
Bergs Bibliotek, Svenstavik 26920
Berhampur University, Berhampur 13615
Beritashvili Institute of Physiology, Tbilisi 09225
Berkeley College, West Paterson 33860
Berkeley County Library System, Moncks Corner 40254
Berkeley Heights Free Public Library, Berkeley Heights 38182
Berkeley Public Library, Berkeley 38181
Berklee College of Music, Boston 30478
Berkley Public Library, Berkley 38183
Berks County Law Library, Reading 34730
Berks County Public Libraries, Reading 40946
Berks Lehigh Valley College, Fogelsville 31150
Berkshire Athenaeum, Pittsfield 40808
Berkshire Community College, Pittsfield 33648
Berks-Lehigh Valley College, Reading 32420
Berlin Mission, Berlin 11163
Berlin Public Library, Berlin 38184
Berlin School of Economics and Law, Campus Lichtenberg, Berlin 09448
Berlin School of Economics and Law, Campus Schöneberg, Berlin 09449
Berlin Township Library, West Berlin 41807
Berlin-Brandenburg Academy of Sciences and Humanities, Berlin 09301
Berlin-Brandenburgische Akademie der Wissenschaften, Berlin 09301
Berliner Missionswerk, Berlin 11163
Bermuda College Library, Paget 02910
Bermuda National Library, Hamilton 02909
Bern University of Applied Sciences – Bern University of the Arts – Music Library, Bern 27016
The Bernadotte Library, Stockholm 26642
Bernadottebiblioteket, Stockholm 26642
Bernard E. Witkin Alameda Country Law Library, Oakland 34623
Bernard E. Witkin Alameda County Law Library – South County Branch, Hayward 34347
Bernards Township Library, Basking Ridge 38098
Bernardsville Public Library, Bernardsville 38185
Berne Public Library, Berne 38186
Berner Fachhochschule – Architektur, Holz und Bau/Technik und Informatik, Burgdorf 27061
Berner Fachhochschule – Hochschule der Künste Bern – Musikbibliothek, Bern 27016
Berner Fachhochschule – Soziale Arbeit, Bern 27017
Berner Fachhochschule – Technik und Informatik, Biel 27054
Berner Fachhochschule – Wirtschaft und Verwaltung, Bern 27018
Bernhard Nocht Institute for Tropical Medicine, Hamburg 11970
Bernhard-Nocht-Institut für Tropenmedizin, Hamburg 11970
Bernice P. Bishop Museum Library & Archives, Honolulu 36930

Bernstein, Shur, Sawyer & Nelson, Portland 36141
Berrien Springs Community Library, Berrien Springs 38187
Berry College, Mount Berry 31912
Bershad Regional Centralized Library System, Main Library, Bershad 28499
Bershadska raionna TsBS, Tsentralna biblioteka, Bershad 28499
Berufliches Schulzentrum Großenhain, Großenhain 10732
Berufsakademie Heidenheim, Heidenheim 10038
Berufsakademie Lörrach, Lörrach 10248
Berufsakademie Mannheim, Mannheim 10292
Berufsakademie Mosbach, Mosbach 10356
Berufsakademie Sachsen – Staatliche Studienakademie Bautzen, Bautzen 10707
Berufsakademie Sachsen – Staatliche Studienakademie Glauchau, Glauchau 09834
Berufsakademie Sachsen – Staatliche Studienakademie Leipzig, Leipzig 10241
Berufsakademie Sachsen – Staatliche Studienakademie Riesa, Riesa 10764
Berufsakademie Villingen-Schwenningen, Villingen-Schwenningen 10681
Berufsgenossenschaft der Bauwirtschaft, Frankfurt am Main 11395
Berufsgenossenschaft der Chemischen Industrie, Heidelberg 12038
Berwyn Public Library, Berwyn 38188
Berzeliusskolan, Linköping 26474
Berzsenyi Daniel Könyvtár, Szombathely 13551
Berzsenyi Dániel Library, Szombathely 13551
Berzsenyi Daniel Tanárképző Főiskola, Szombathely 13309
Bessemer Public Library, Bessemer 38189
Bessenyei György College of Nyíregyháza, Nyíregyháza 13286
Bessenyei György Tanárképző Főiskola, Nyíregyháza 13287
Bestuursdienst Rotterdam, Rotterdam 19287
Beta-Kutató Kft., Sopronhorpács 13500
Beta-Research Ltd., Sopronhorpács 13500
Beth Emet Synagogue, Evanston 35201
Beth Israel Medical Center, New York 37189
Beth Israel Synagogue, Vineland 35403
Beth Tzedec Congregation, Toronto 04506
Bethany Bible College, Sussex 04231
Bethany College, Bethany 30418
Bethany College, Lindsborg 31621
Bethany College, Scotts Valley 32626
Bethany Lutheran College, Mankato 31753
Bethel College, McKenzie 31776
Bethel College, Mishawaka 31866
Bethel College, North Newton 32146
– Mennonite Library & Archives, North Newton 32147
Bethel Park Public Library, Bethel Park 38192
Bethel Public Library, Bethel 38191
Bethel Theological Seminary Library, Saint Paul 35366
Bethel University, Saint Paul 32528
Bethlehem Area Public Library, Bethlehem 38193
Bethlehem Steel Corp – Bernard D. Broeker Law Library, Bethlehem 35500
Bethlehem Steel Corp – Schwab Information Center, Bethlehem 35501
Bethlehem University, Jerusalem 14616
Bethpage Public Library, Bethpage 38194
Bethune-Cookman College, Daytona Beach 30972
Bettendorf Public Library Information Center, Bettendorf 38195
Beulah Heights Bible College, Atlanta 35114
Beuth Hochschule für Technik Berlin – University of Applied Sciences, Berlin 09402
Beveridge & Diamond, PC Library, Washington 36334
Beverly Hills Public Library, Beverly Hills 38197
Beverly Public Library, Beverly 38196
Bevill State Community College, Fayette 31128
Bexar County Law Library, San Antonio 34800
Bexley Library Service, Bexleyheath 29939
Bexley Public Library, Bexley 38199
Beyazit Devlet Kütüphanesi, Istanbul 27844
Beyazit State Library, Istanbul 27844
Beylorussian State Center for Testing of Agricultural Machinery, Privolny 01925
Bezalel Academy of Arts and Design, Jerusalem 14723
Bežigrad Public Library, Ljubljana 24966
Bezirk Oberbayern, München 11040
Bezirksamt Charlottenburg-Wilmersdorf von Berlin, Verwaltungsinformationszentrum (VIZ), Berlin 10790
Bezirksärztekammer Nordwürttemberg, Stuttgart 12510
Bezirksregierung Detmold, Detmold 10855
Bezirksregierung Düsseldorf, Düsseldorf 10867
Bezirksregierung Köln, Köln 10983
Bezirksregierung Münster, Münster 11052
Bezirkszentralbibliothek Philipp-Schaeffer, Berlin 12652
Bhabha Atomic Research Centre, Mumbai 14081
Bhagalpur University, Bhagalpur 13616
Bhairahawa Multiple Campus Library, Rupandehi 19026
Bhandarkar Oriental Research Institute, Pune 14186
Bharat Heavy Electricals Ltd, Bhopal 13902
Bharat Heavy Electricals Ltd, Hyderabad 13904
Bharat Heavy Electricals Ltd (BHEL), Bangalore 13927
Bharat Kala Bhavan, Varanasi 13783
Bharata Ganita Parisad, Lucknow 14063
Bharata Itihasa Samshodhaka Mandala, Pune 14187
Bharathiar University, Coimbatore 13633
Bharatiya Vidya Bhavan, Mumbai 13701
Bhavnagar University, Bhavnagar 13617
Bhilai Steel Plant, Bhilai 13901
Dr. Bhim Rao Ambedkar University, Agra 13592
Bhogilal Leherchand Institute of Indology, Delhi 13980
BHP Research, Port Kembla 00986
BHP Research, Wollongong 00842
BI Handelshøyskolen, Oslo 20111
BI Pharma GmbH & Co KG, Ingelheim 11410
Bialystok Technical University, Białystok 20794
Bibiana, International House of Art for Children, Bratislava 24683
Bibiana medzinárodný dom umenia pre deti, Bratislava 24683
Bible College of New Zealand, Auckland 19810
Bible College of Queensland, Toowong 00801
Bible College of South Australia, Malvern 00777
Bible College of Victoria, Lilydale 00776
Bible College of Western Australia Inc., Wattle Grove 00804
Biblical Theological Seminary, Hatfield 35226
Bibliobus de l'Université populaire jurassienne, Delémont 27483
Bibliobus neuchâtelois, La Chaux-de-Fonds 27489
Bibliografijos ir knygotyros centras, Vilnius 18300
Bibliographical Institute, Ramat-Efal 14752
Bibliothèque municipale, Saint-Lô 09069
Bibliomedia, Lausanne 27490
Bibliomedia Schweiz, Solothurn 27498
BiblioNova – Openbare Bibliotheek Westelijke Mijnstreek, Vestiging Born, Born 19570
BiblioNova – Openbare Bibliotheek Westelijke Mijnstreek, Vestiging Echt, Echt 19589
BiblioNova – Openbare Bibliotheek Westelijke Mijnstreek, Vestiging Geleen, Geleen 19604
BiblioNova – Openbare Bibliotheek Westelijke Mijnstreek, Vestiging Sittard, Sittard 19698
BiblioNova – Openbare Bibliotheek Westelijke Mijnstreek, Vestiging Stein, Stein 19706
BiblioNova – Openbare Bibliotheek Westelijke Mijnstreek, Vestiging Susteren, Susteren 19707
Bibliorura – Basisbibliotheek Roermond, Roermond 19690
Bibliorura – Locatie Montfort, Montfort 19663
Biblioservice Gelderland, Arnhem 19554
Biblioteca Abbaziale, Nonantola 15941
Biblioteca Agnesiana e Diocesana, Vercelli 16130
Biblioteca Agrícola Nacional, Lima 20627
Biblioteca Agropecuaria de Colombia, Santafé de Bogotá 06023
Biblioteca 'Alfredo M. Aguayo', La Habana 06289
Biblioteca Aloisianum, Gallarate 15839
Biblioteca Amador-Washington, Panamá 20539
Biblioteca Ambrosiana, Milano 16322
Biblioteca Ambulantes, San Salvador 07212
Biblioteca Americana J.F. Kennedy, Catania 16226
Biblioteca Ana Maria Poppovic, São Paulo 03403
Biblioteca Andrighetti Marcello, Venezia 15577
Biblioteca Angelica, Roma 14809
Biblioteca Antoniana, Padova 15952
Biblioteca Apostolado de la Oración, Escuela Batalla de Tucumán 00222
Biblioteca Apostolica Vaticana, Città del Vaticano 42182
Biblioteca Archeologica e Numismatica, Milano 16323
Biblioteca, Archivo y Colección Arqueológica Municipal, Jerez de la Frontera 25963
Biblioteca Arcivescovile, Bari 15724
Biblioteca Arcivescovile, Bologna 15740
Biblioteca Arcivescovile, Cosenza 15804
Biblioteca Arcivescovile, Matera 15893
Biblioteca Arcivescovile, Otranto 15951
Biblioteca Arcivescovile, Taranto 16082
Biblioteca Arcivescovile, Trento 16105
Biblioteca Arcivescovile "Antonio Lombardi", Catanzaro 15787
Biblioteca Arcivescovile Cardinale Pietro Maffi, Pisa 15985
Biblioteca Arcivescovile di Udine, Udine 16115
Biblioteca Arcivescovile "Luigi Sodo", Cerreto Sannita 15791
Biblioteca Arcivescovile "Mons. A. Lanza", Reggio Calabria 16008
Biblioteca Argentina 'Dr. Juan Alvarez', Rosario 00324
Biblioteca Argentina para Ciegos, Buenos Aires 00241
Biblioteca "Aula A. De Felice" degli avvocati e procuratori, Salerno 16569
Biblioteca Azcona, Tafalla 26160
Biblioteca Balmes, Barcelona 25765
Biblioteca "Basso", Roma 16456
Biblioteca Benedetto Croce, Napoli 16369
Biblioteca Berenson, Firenze 14990
Biblioteca "Bertini Frassoni", Roma 16457
Biblioteca Borja, Sant Cugat del Vallès 25820
Biblioteca Braille, Cagliari 16213
Biblioteca Bucovinei 'I.G. Sbiera', Suceava 22106
Biblioteca Caixa de Sabadell, Sabadell 26246
Biblioteca Can Pedrals, Granollers 26217
Biblioteca cantonale, Bellinzona 27474
Biblioteca cantonale di Locarno, Locarno 26973
Biblioteca cantonale di Lugano, Lugano 26974
Biblioteca cantonale e del Liceo di Mendriso, Mendrisio 26976
Biblioteca Capitolare, Biancavilla 15735
Biblioteca Capitolare, Gualdo Tadino 15858
Biblioteca Capitolare, Ortona a Mare 15946
Biblioteca Capitolare, Padova 15953
Biblioteca Capitolare, Reggio Emilia 16011
Biblioteca Capitolare, San Giovanni in Persiceto 16052
Biblioteca Capitolare, Treviso 16110
Biblioteca Capitolare, Verona 16131
Biblioteca Capitolare Fabroniana, Pistoia 15988
Biblioteca Capitolare "Finia", Gravina di Puglia 15855
Biblioteca Capitolare Parrocchiale, San Chirico Raparo 16049
Biblioteca Cardinale Giulio Alberoni, Piacenza 15979
La Biblioteca "Carlo Manzia", SJ, Anagni 15700
Biblioteca Casanatense, Roma 14810
Biblioteca Castilla y León, Burgos 25937
Biblioteca Central, Maracay 42260
Biblioteca Central da Marinha, Lisboa 21774
Biblioteca Central de Cantabria, Santander 25256
Biblioteca Central de Educação, Rio de Janeiro 03334
Biblioteca Central de la Armada, Buenos Aires 00204
Biblioteca Central de Macau, Macao 04966
Biblioteca Central de Vigo, Vigo 26262
Biblioteca Central d'Igualada, Igualada 26221
Biblioteca Central Militar, Madrid 25698
Biblioteca Centrală Universitară, București 21916
Biblioteca Centrală Universitară 'Lucian Blaga', Cluj-Napoca 21929
Biblioteca Centrala Universitara 'Mihai Eminescu', Iași 21935
Biblioteca Centrale Cappuccini, Roma 16017
Biblioteca Centrale della Marina Militare, Roma 16458
Biblioteca Centrale della Regione Siciliana, Palermo 14803
Biblioteca Centrale dell'Aeronautica Militare, Roma 16459
Biblioteca Centrale delle Ferrovie dello Stato, Roma 15657
Biblioteca centrale di Medicina, Trieste 15548
Biblioteca Centrale Diocesana, Acireale 15695
Biblioteca Centrale Pietro Lincoln Cadioli, Sesto San Giovanni 16828
Biblioteca Centro Cultural 'Ramon Alonso Luzzy', Cartagena 26207
Biblioteca Centro Elis, Roma 16460
Biblioteca Charitas, Paola-Santuario 15964
Biblioteca Circolante Don Luigi Geromin, Venezia 16854
Biblioteca Circolante Don Natale Perini, Udine 16846
Biblioteca Circolante Mons. Enrico Da Ronco, Udine 16847
Biblioteca Circulant Braille, Barcelona 25884
Biblioteca Cisneros de la Conferencia de Franziscanos, Madrid 25786
Biblioteca Città di Arezzo, Arezzo 16688
Biblioteca Civica, Bassano del Grappa 16696
Biblioteca Civica, Busto Arsizio 16707
Biblioteca Civica, Parma 16789
Biblioteca Civica, Tortona 16838
Biblioteca Civica, Vercelli 16856
Biblioteca Civica, Verona 14818
Biblioteca Civica Angelo Mai, Bergamo 14787
Biblioteca Civica "Antonio Delfini", Modena 16776
Biblioteca Civica Attilio Hortis, Trieste 14816
Biblioteca Civica 'Barrili', Savona 16822
Biblioteca Civica Berio, Genova 16749
Biblioteca Civica Bertoliana, Vicenza 16858
Biblioteca Civica Bonetta, Pavia 16790
Biblioteca Civica Bruno Emmert, Arco 16687
Biblioteca Civica 'C. Alliaudi', Pinerolo 16795
Biblioteca Civica 'C. Nigra', Ivrea 16759
Biblioteca Civica "C. Sabbadino", Chioggia 16725
Biblioteca Civica Carlo Negroni, Novara 16780
Biblioteca Civica Cesare Battisti, Bolzano 16703
Biblioteca Civica del Comune di Pordenone, Pordenone 16797
Biblioteca civica della città di Novi Ligure, Novi Ligure 16781
Biblioteca Civica di Alessandria, Alessandria 16683
Biblioteca Civica di Belluni, Belluno 16697
Biblioteca Civica di Biella, Biella 16699
Biblioteca Civica di Cologno Monzese, Cologno Monzese 16727
Biblioteca Civica di Cosenza, Cosenza 16731
Biblioteca Civica di Cuneo, Cuneo 16733
Biblioteca Civica di Fossano, Fossano 16746
Biblioteca Civica di Massa, Massa 16772
Biblioteca Civica di Mestre, Venezia 16853
Biblioteca Civica di Mondovi, Mondovì 16778
Biblioteca Civica di Monza, Monza 16779
Biblioteca Civica di Padova, Padova 16786
Biblioteca Civica di Riva del Garda, Riva del Garda 16809
Biblioteca Civica di Saluzzo, Saluzzo 16813
Biblioteca Civica di Storia dell'Arte "Luigi Poletti", Modena 16359
Biblioteca Civica di Treviglio, Treviglio 16842
Biblioteca Civica di Vimercate, Vimercate 16859
Biblioteca Civica 'Dott. Francesco Corradi', Sanremo 16817
Biblioteca Civica e Archvi storici di Rovereto, Rovereto 16810
Biblioteca Civica 'Farinone-Centa', Varallo Sesia 16850
Biblioteca Civica "Francesco Gallino", Genova 16750
Biblioteca Civica 'G. Ferrero', Alba 16681
Biblioteca Civica "G. Gambirasio", Seriate 16826
Biblioteca Civica Gambalunga, Rimini 16808
Biblioteca Civica "Giovanni Canna", Casale Monferrato 16714
Biblioteca Civica Giovanni Verga, Ragusa 16803
Biblioteca Civica Internazionale, Bordighera 16704
Biblioteca Civica "L. Lagorio", Imperia 16758
Biblioteca civica "Luigi Majno", Gallarate 16747
Biblioteca Civica 'M. A. Martini', Scandicci 16823
Biblioteca Civica Musicale Andrea della Corte, Torino 16592
Biblioteca Civica "N. Francone" di Chieri, Chieri 16723
Biblioteca Civica "Pietro Acclavio", Taranto 16833
Biblioteca Civica Pietro Ceretti, Verbania-Pallanza 16855
Biblioteca Civica 'Pio Rajna', Sondrio 16831
Biblioteca Civica Queriniana, Brescia 14790
Biblioteca Civica "Renato Bortoli" di Schio, Schio 16824
Biblioteca Civica Ricottiana, Voghera 16862
Biblioteca Civica Romolo Spezioli, Fermo 16738

Biblioteca Civica "Ubaldo Mazzini", La Spezia 16761
Biblioteca Civica Uberto Pozzoli, Lecco 16764
Biblioteca Civica 'V. Joppi', Udine 16848
Biblioteca Comunal, Canillo 00099
Biblioteca Comunal, Encamp 00100
Biblioteca Comunal, La Massana 00101
Biblioteca Comunale, Agrigento 16678
Biblioteca Comunale, San Gimignano 16814
Biblioteca Comunale, Sassari 16819
Biblioteca Comunale "A. Manzoni", Trezzo sull'Adda 16844
Biblioteca Comunale "A. Ruggiero", Caserta 16715
Biblioteca Comunale "A. Saffi" – Sezione Moderna "A. Schiavi", Forli 16744
Biblioteca Comunale "Alessandro Lazzerini", Prato 16801
Biblioteca Comunale Antonelliana, Senigallia 16825
Biblioteca Comunale Antonio Baldini, Santarcangelo di Romagna 16818
Biblioteca Comunale Ariostea, Ferrara 14792
Biblioteca Comunale Augusta, Perugia 16791
Biblioteca Comunale Aurelio Saffi, Forli 16745
Biblioteca Comunale Centrale, Milano 16775
Biblioteca Comunale Centrale, Milano 14797
Biblioteca Comunale Chelliana, Grosseto 16753
Biblioteca Comunale "D. Topa", Palmi 16788
Biblioteca Comunale degli Ardenti, Viterbo 16860
Biblioteca Comunale degli Intronati, Siena 16829
Biblioteca Comunale dell'Archiginnasio, Bologna 14788
Biblioteca Comunale di Ala, Ala 16680
Biblioteca Comunale di Alzano Lombardo, Alzano Lombardo 16684
Biblioteca Comunale di Bagno a Ripoli, Bagno a Ripoli 16693
Biblioteca Comunale di Castelfranco Veneto (BCCV), Castelfranco Veneto 16717
Biblioteca Comunale di Cattolica, Cattolica 16721
Biblioteca Comunale di Como, Como 16728
Biblioteca Comunale di Crema, Crema 16732
Biblioteca Comunale di Enna, Enna 16734
Biblioteca Comunale di Fabriano, Fabriano 16735
Biblioteca Comunale di Foligno, Foligno 16742
Biblioteca Comunale di Iesi, Iesi 16756
Biblioteca Comunale di Imola, Imola 16757
Biblioteca Comunale di Mazara del Vallo, Mazara del Vallo 16774
Biblioteca Comunale di Palermo, Palermo 14804
Biblioteca Comunale di Siracusa, Siracusa 16830
Biblioteca Comunale di Studi Sardi, Cagliari 16214
Biblioteca Comunale di Trento, Trento 16841
Biblioteca Comunale di Treviso, Treviso 16843
Biblioteca Comunale di Urbania, Urbania 16849
Biblioteca Comunale di Varese, Varese 16851
Biblioteca Comunale di Velletri "Augusto Tersenghi", Velletri 16852
Biblioteca Comunale e Archivio Storico "A. Minuziano", San Severo 16816
Biblioteca Comunale e dell'Accademia Etrusca, Cortona 16234
Biblioteca Comunale "E. Rogadeo", Bitonto 16700
Biblioteca Comunale "Emanuele Taranto", Caltagirone 16708
Biblioteca Comunale "F. Angelini De Miccolis", Putignano 16802
Biblioteca Comunale "F. Trinchera Seniore", Ostuni 16785
Biblioteca Comunale Fabrizio Trisi, Lugo 16769
Biblioteca Comunale Federiciana, Fano 16737
Biblioteca Comunale "Filippo De Nobili", Catanzaro 16720
Biblioteca Comunale Foresiana, Portoferraio 16798
Biblioteca Comunale Forteguerriana, Pistoia 16796
Biblioteca Comunale "Francesco Cini", Osimo 16784
Biblioteca Comunale "G. Marconi", Viareggio 16857
Biblioteca Comunale "G. Panunzio", Molfetta 16777
Biblioteca Comunale "Giosuè Carducci", Spoleto 16832
Biblioteca Comunale "Giovanni Bovio", Trani 16839
Biblioteca Comunale "Giulio Cesare Croce", San Giovanni in Persiceto 16815
Biblioteca Comunale "Giulio Einaudi", Correggio 16729
Biblioteca Comunale "Giulio Gabrielli", Ascoli Piceno 16689
Biblioteca Comunale Giuseppe Taroni, Bagnacavallo 16692
Biblioteca Comunale "L. Benincasa", Ancona 16685
Biblioteca Comunale Labronica "F. D. Guerrazzi", Livorno 16765
Biblioteca Comunale Laudense, Lodi 16766
Biblioteca Comunale Leonardiana, Vinci 16675
Biblioteca Comunale Liciniana, Termini Imerese 16834
Biblioteca Comunale Luciano Scarabelli, Caltanissetta 16709
Biblioteca Comunale Luigi Fumi, Orvieto 16783
Biblioteca Comunale M. Leoni, Fidenza 16739
Biblioteca Comunale Malatestiana, Cesena 16722
Biblioteca Comunale Manfrediana di Faenza, Faenza 16736
Biblioteca Comunale "Mozzi-Borgetti", Macerata 16770
Biblioteca Comunale "Natale Cionini" di Sassuolo, Sassuolo 16820
Biblioteca Comunale Paroniana, Rieti 16807
Biblioteca Comunale Passerini-Landi, Piacenza 16794
Biblioteca Comunale "Pietro De Nava", Reggio Calabria 16805
Biblioteca Comunale "Ruggero Bonghi" di Lucera, Lucera 16768
Biblioteca Comunale "Sabino Loffredo", Barletta 16695
Biblioteca Comunale Simone Corleo, Salemi 16811
Biblioteca Comunale Sperelliana, Gubbio 16755
Biblioteca Comunale Terni, Terni 16835
Biblioteca Comunale "Ugo Bernasconi", Cantù 16712
Biblioteca Comunale Vincenzo Balzano, Castel di Sangro 16716
Biblioteca Comunale Vincenzo Bellini, Catania 16718
Biblioteca Comunidad, San Lorenzo del Escorial 25817
Biblioteca Consorziale Astense, Asti 16690
Biblioteca Cultural Altino Arantes, Ribeirão Prêto 03326
Biblioteca da Casa do Douro, Peso da Régua 21821
Biblioteca Dante Alighieri, Venezia 16645
Biblioteca D'Arco, Mantova 16311
Biblioteca d'Arte Beato Angelico, Roma 16461
Biblioteca d'Arte, CaSVA, Milano 16324
Biblioteca d'Artiglieria e Genio del Ministero della Difesa, Roma 15658
Biblioteca de Andalucía, Granada 25248
Biblioteca de Asturias Ramón Pérez de Ayala, Oviedo 25254
Biblioteca de Autores Nacionales 'Carlos A. Rolando', Guayaquil 06987
Biblioteca de Autores Nacionales 'Fray Vicente Solano', Cuenca 06986
Biblioteca de Castilla y León, Valladolid 25259
Biblioteca de Castilla-La Mancha, Toledo 25257
Biblioteca de Catalunya, Barcelona 25247
Biblioteca de Comerç i Turisme, Barcelona 25682
Biblioteca de Cultura Artesana, Palma de Mallorca 26111
Biblioteca de Extremadura, Badajoz 25246
Biblioteca de Investigadores de la Provincia de Guadalajara, Guadalajara 25695
Biblioteca de la Civiltà Cattolica, Roma 16018
Biblioteca de la Municipalidad de Lima, Lima 20642
Biblioteca de La Rioja, Logroño 25249
Biblioteca de México, México 18871
Biblioteca de Pedagogía, Madrid 25981
Biblioteca de Temas Gaditanos, Cádiz 25856
Biblioteca degli Avvocati e Procuratori, Roma 15659
Biblioteca degli Uffizi, Firenze 16252
Biblioteca dei Benedittini di Muri-Gries, Bolzano 15750
Biblioteca dei Cappuccini, Ancona 15702
Biblioteca dei Cappuccini, Cagliari 15761
Biblioteca dei Cappuccini, Castelbuono 15783
Biblioteca dei Cappuccini, Chiaravalle 15793
Biblioteca dei Cappuccini, Corinaldo 15802
Biblioteca dei Cappuccini, Fermo 15817
Biblioteca dei Cappuccini, Firenze 15826
Biblioteca dei Cappuccini, Iesi 15859
Biblioteca dei Cappuccini, Jesi 15864
Biblioteca dei Cappuccini, Lecco 15875
Biblioteca dei Cappuccini, Loreto 15881
Biblioteca dei Cappuccini, Padova 15954
Biblioteca dei Cappuccini, Piacenza 15980
Biblioteca dei Cappuccini, Salemi 16044
Biblioteca dei Cappuccini, Scandiano 16063
Biblioteca dei Cappuccini, Spoleto 16075
Biblioteca dei Carmelitani Scalzi, Milano 15902
Biblioteca dei Civici Musei, Reggio Emilia 16433
Biblioteca dei Civici Musei Scientifici, Trieste 16620
Biblioteca dei Domenicani, Pistoia 15989
Biblioteca dei Domenicani, Reggio Calabria 16009
Biblioteca dei Domenicani di S. Maria di Castello, Genova 15841
Biblioteca dei Francescani, Cles 15799
Biblioteca dei Frati Francescani, Massafra 15892
Biblioteca dei Frati Minori Conventuali, Quartu S. Elena 16001
Biblioteca dei Frati Minori Francescani, Busto Arsizio 15760
Biblioteca dei Frati Minori Francescani di San Bernadino, Verona 16132
Biblioteca dei Gesuiti, Napoli 15922
Biblioteca dei Magistrati, Roma 15660
Biblioteca dei Minori Francescani, Caldaro 15765
Biblioteca dei Minori Francescani, Dongo 15809
Biblioteca dei Minori Francescani, Mezzolombardo 15901
Biblioteca dei Minori Francescani, Pergine Valsugana 15973
Biblioteca dei Minori Francescani del Deserto, Venezia 16121
Biblioteca dei Passionisti, Soriano Nel Cimino 16074
Biblioteca dei PP. Barnabiti, San Felice a Cancello 16051
Biblioteca dei PP. Barnabiti, Trani 16103
Biblioteca dei PP. Cappuccini, Cosenza 15805
Biblioteca dei PP. Cappuccini, Mazzarino 15896
Biblioteca dei PP. Cappuccini, Troina 16114
Biblioteca dei PP. Domenicani, Napoli 15923
Biblioteca dei PP. Passionisti, Manduria 15887
Biblioteca dei PP. Redentoristi, Mercato San Severino 15897
Biblioteca dei PP. Redentoristi, Venezia 16122
Biblioteca dei Servi, Milano 15903
Biblioteca del Arzobispado de Sevilla, Sevilla 25826
Biblioteca del Capitolo della Cattedrale, Atri 15719
Biblioteca del Centro dantesco, Ravenna 16427
Biblioteca del Clero della Chiesa del Gesù, Castellammare di Stabia 15784
Biblioteca del Congreso, Lima 20643
Biblioteca del Cottolengo, Torino 16092
Biblioteca del Poder Legislativo, Montevideo 42026
Biblioteca del Sacro Eremo Tuscolano, Monteporzio Catone 15921
Biblioteca del Santuario, Biella 15736
Biblioteca del Seminario, Casale Monferrato 15777
Biblioteca della Casa Santi Martiri (Padri Gesuiti), Torino 16093
Biblioteca della Certosa, Serra San Bruno 16067
Biblioteca della Ghisa, Follonica 16743
Biblioteca della Misericordia, Fermo 15818
Biblioteca della Provincia di Torino, Torino 16836
Biblioteca della Regione Piemonte, Torino 16593
Biblioteca della Santa Casa, Loreto 15882
Biblioteca dell'Amministrazione Provinciale, Pisa 16419
Biblioteca dell'Arcispedale Santa Maria Nuova, Reggio Emilia 16434
Biblioteca dello Studio Filosofico Francescano, Casale Monferrato 15778
Biblioteca dello Studio Teologico per Laici, Firenze 15827
Biblioteca Demonstrativa de Brasília FBN/minc, Brasília 03426
Biblioteca Departamental de Calí, Calí 06040
Biblioteca di Americanistica, Iberistica e Slavistica (AMERIBE), Venezia 15580
Biblioteca di Casa Pascoli, Barga 16172
Biblioteca di Castelcapuano, Napoli 16366
Biblioteca di Geografia, Firenze 16274
Biblioteca di S. Lorenzo Maggiore, Napoli 15924
Biblioteca di San Nicola dei Padri Domenicani, Bari 15725
Biblioteca di Scienze del Linguaggio (SC-LING), Venezia 15581
Biblioteca di Scienze dell'Educazione, Modena 16360
Biblioteca di Stato, San Marino 24384
Biblioteca di Storia Contemporanea A. Oriani, Ravenna 16428
Biblioteca di Storia dell'Arte, Genova 16282
Biblioteca di storia moderna e vontemporanea, Roma 16462
Biblioteca di Studi Eurasiatici (EURASIA), Venezia 15582
Biblioteca di Studi Europei e Postcolonali (SLLEP), Venezia 15583
Biblioteca di Studi Meridionali Giustino Fortunato, Roma 16463
Biblioteca di Studi sull'Asia Orientale (ASIA-OR), Venezia 15584
Biblioteca Diocesana, Ancona 15703
Biblioteca Diocesana, Foggia 15834
Biblioteca Diocesana, Ivrea 15863
Biblioteca Diocesana, Livorno 15878
Biblioteca Diocesana, Pozzuoli 15997
Biblioteca Diocesana, Recanati 16006
Biblioteca Diocesana, Salerno 16045
Biblioteca Diocesana, Sulmona 16078
Biblioteca Diocesana, Trani 16104
Biblioteca Diocesana "A. Sanfelice", Nardò 15936
Biblioteca Diocesana di Ogliastra, Lanusei 15866
Biblioteca Diocesana di Susa, Susa 16081
Biblioteca Diocesana "Mons. E. Biancheri", Rimini 16014
Biblioteca Diocesana "Mons. Luigi Roba", Genova 15842
Biblioteca Diocesana Tridentina "Antonio Rosmini", Trento 16106
Biblioteca Dipartimentale della Marina Militare, La Spezia 15644
Biblioteca documentară a Arhivelor Naționale, București 21970
Biblioteca documentară de istorie a medicinei, București 21971
Biblioteca documentară Năsăud, Năsăud 22047
Biblioteca Documentară Teleki-Bolyai, Târgu Mureş 22060
Biblioteca documentară 'Timotei Cipariu', Blaj 21960
Biblioteca Domenicana, Firenze 15828
Biblioteca Dominicana, Santo Domingo 06965
Biblioteca Dominicini, Perugia 15974
Biblioteca Dominicos San Pedro Mártir, Madrid 25982
Biblioteca e Archivio del Capitolo Metropolitano, Milano 15904
Biblioteca e Casa Carducci, Bologna 16701
Biblioteca Ecuatoriana 'Aurelio Espinosa Pólit', Quito 06999
Biblioteca ed Archivio del Risorgimento, Firenze 16253
Biblioteca ed Archivo del Seminario, Rossano 16042
Biblioteca Egidiana San Nicola, Tolentino 16090
Biblioteca 'El Ateneo', Masaya 19901
Biblioteca Emidiana, Agnone 16160
Biblioteca "Emilio Sereni" – Archivio Storico Nazionale dei Movimenti Contadini – Biblioteca del Museo Cervi, Reggio Emilia 16435
Biblioteca Engiadinaisa, Sils Maria 27434
Biblioteca Episcopal de Vic, Vic 25839
Biblioteca "Eredi Gargallo di Castel Lentini", Siracusa 16580
Biblioteca Española, Paris 08575
Biblioteca Española de Música y Teatro Contemporáneos, Madrid 26024
Biblioteca Estadual de Agricultura, Curitiba 03293
Biblioteca Estadual de Niterói, Niterói 02982
Biblioteca Estense Universitaria, Modena 14800
Biblioteca Etnografica "Pitrè", Palermo 16396
Biblioteca Euclides da Cunha, Rio de Janeiro 03438
Biblioteca Fardelliana, Trapani 16840
Biblioteca Filippo Monaco, Marcianise 16312
Biblioteca Forense, Catania 16227
Biblioteca Forense, Cosenza 16235
Biblioteca Fornasini, Castenedolo 16223
Biblioteca Francesca Serrado, Martorelles 26104
Biblioteca Francescana, Artena 15709
Biblioteca Francescana, Assisi 15712
Biblioteca Francescana, Falconara 15813
Biblioteca Francescana dei Frati Minori Conventuali, Palermo 15960
Biblioteca Francescana dei SS. XII Apostoli, Roma 16019
Biblioteca Francescana di S. Maria di Loreto, Paduli 15958

Biblioteca Francescana Provinciale "Fra Landolfo Caracciolo" dei Frati Minori Conventuali, Napoli 15925
Biblioteca Francescana "S. Pietro In Silki", Sassari 16060
Biblioteca Francescano-Cappuccina Provinciale, Milano 15905
Biblioteca Francisco de Zabálburu, Madrid 25983
Biblioteca Franzoniana, Genova 15843
Biblioteca Fundaziun Tscharner Zernez, Zernez 27446
Biblioteca "G. L. Lercari", Genova 16751
Biblioteca Gaetano Ricchetti, Bari 16694
Biblioteca 'Gard. Cicognani', Faenza 15812
Biblioteca General de Navarra, Pamplona 25255
Biblioteca General Medinaceli, Madrid 25984
Biblioteca 'George Alexander', São Paulo 03443
Biblioteca "Giulia", Trivento 16113
Biblioteca Giuseppe Dossetti della Fondazione per le Scienze Religiose Giovanni XXIII, Bologna 16187
Biblioteca Gonzalo Rojas, Bremen 11680
Biblioteca Gregoriana, Belluno 15728
Biblioteca Guarnacci, Volterra 16676
Biblioteca Hispánica – Agencia Española de Cooperación Internacional, Madrid 25972
Biblioteca Histórica Municipal, Madrid 25250
Biblioteca Infantil Santa Creu, Barcelona 25885
Biblioteca Instituto 'Don Bosco', Viedma 00223
Biblioteca Insular, Las Palmas 26240
Biblioteca Insular Especializada en Discapacidad, Santa Cruz de Tenerife 26137
Biblioteca Isimbardi, Milano 16325
Biblioteca Islámica Félix María Pareja – Agencia Española de Cooperación Internacional, Madrid 25973
Biblioteca Istituto Regina Elena Centro di Conoscenza "R. Maceratini", Roma 16464
Biblioteca Italiana per i Ciechi 'Regina Margherita' – ONLUS, Monza 16363
Biblioteca Italo Calvino, Paris 08577
Biblioteca "Italo Grassi", Varallo Sesia 16640
Biblioteca Jaura, Valchava 27443
Biblioteca Jenny Klabin Segall, São Paulo 03397
Biblioteca João Paulo II, Lisboa 21665
Biblioteca 'José A. Echeverría', La Habana 06290
Biblioteca 'José Batlle y Ordóñez', Montevideo 42043
Biblioteca José Saramago, Leiria 21652
Biblioteca Judeţeană, Bacău 22075
Biblioteca Judeţeană, Bistriţa 22077
Biblioteca Judeţeană, Constanţa 22085
Biblioteca Judeţeană, Giurgiu 22092
Biblioteca Judeţeană, Satu Mare 22101
Biblioteca Judeţeană, Vaslui 22112
Biblioteca Judeţeană 'A.D. Xenopol', Arad 22074
Biblioteca Judeţeană Alba, Alba Iulia 22072
Biblioteca Judeţeană 'Alexandru Odobescu', Călăraşi 22083
Biblioteca Judeţeană AMAN-DOLJ, Craiova 22087
Biblioteca Judeţeană 'Antim Ivireanul' Râmnicu Vâlcea, Râmnicu Vâlcea 22099
Biblioteca Judeţeană Arges, Piteşti 22097
Biblioteca Judeţeană 'Astra', Sibiu 22103
Biblioteca Judeţeană 'Christian Tell' Gorj, Tîrgu Jiu 22110
Biblioteca Judeţeană Covasna, Sfântu Gheorghe 22102
Biblioteca Judeţeană Dîmbovita, Târgovişte 22107
Biblioteca Judeţeană 'Duiliu Zamfirescu', Focşani 22090
Biblioteca Judeţeană 'G. T. Kirileanu' Neamţ, Piatra-Neamţ 22096
Biblioteca Judeţeană 'George Baritiu' Brasov, Braşov 22080
Biblioteca Judeţeană 'Gh. Asachi', Iaşi 22093
Biblioteca Judeţeană 'Gheorghe Sincai' Bihor, Oradea 22095
Biblioteca Judeţeană Harghita, Miercurea Ciuc 22094
Biblioteca Judeţeană Ialomiţa, Slobozia 22105
Biblioteca Judeţeană 'Ioan N. Roman' Constanţa, Constanţa 22086
Biblioteca Judeteana 'Ion Minulescu', Slatina 22104
Biblioteca Judeţeană 'Ioniţă Scipione Bădescu', Zalău 22113
Biblioteca Judeţeană Mehedinţi, Drobeta-Turnu Severin 22089
Biblioteca Judeţeană 'Mihai Eminescu', Botoşani 22078
Biblioteca Judeţeană Mureş, Târgu Mureş 22108
Biblioteca Judeţeană 'N. Iorga Prahova', Ploieşti 22098
Biblioteca Judeţeană 'Octavian Goga' Cluj, Cluj-Napoca 22084
Biblioteca Judeţeană "Ovid Densusianu", Deva 22088
Biblioteca Judeţeană 'Panait Istrati', Brăila 22079
Biblioteca Judeţeană "Pandit Cerna" Tulcea, Tulcea 22111
Biblioteca Judeţeană 'Paul Iorgovici', Reşiţa 22100
Biblioteca Judeţeană 'Petre Dulfu' Baia Mare-Maramures, Baia Mare 22076
Biblioteca Judeţeană Teleorman, Alexandria 22073
Biblioteca Judeţeană Timiş, Timişoara 22109
Biblioteca Judeţeană 'V. Voiculescu', Buzău 22082
Biblioteca Karl A. Boedecker, São Paulo 03404
Biblioteca "L. Cozza", Viterbo 16138
Biblioteca "La Pace" dei Frati Minori, Sassoferrato 16061
Biblioteca Lambert Mata, Ripoll 26245
Biblioteca Lancisiana, Roma 16465
Biblioteca Lucchesiana, Agrigento 16679
Biblioteca Ludovico Jacobilli, Foligno 15835
Biblioteca Magistrale del Provveditora agli Studi, Bologna 16188
Biblioteca 'Manuel Sanguily', La Habana 06281
Biblioteca 'Maria SS dello Splendore', Giulianova 15852
Biblioteca Mario Sturzo, Piazza Armerina 15983
Biblioteca Marucelliana, Firenze 14793
Biblioteca Mediateca Gino Baratta, Mantova 16771
Biblioteca Medica, Cremona 16236
Biblioteca Medica Giuseppe Pasta, Bergamo 16180
Biblioteca Medica 'Mario Segale', Genova 16283
Biblioteca Medica Statale, Roma 16466
Biblioteca Medicea Laurenziana, Firenze 16254
Biblioteca Menéndez Pelayo, Santander 26252
Biblioteca Metropolitană Bucureşti, Bucureşti 22081
Biblioteca Militare Centrale dello Stato Maggiore dell'Esercito, Roma 15661
Biblioteca Militare del Presidio, Milano 15648
Biblioteca Militare del Presidio, Napoli 15652
Biblioteca Militare del Presidio, Palermo 15654
Biblioteca Militare di Presidio, Cagliari 15640
Biblioteca "Misciattelli", Roma 16467
Biblioteca Monastica, Borutta 15753
Biblioteca Monastica, Matera 15894
Biblioteca Monastica dei PP. Benedettini, Parma 15965
Biblioteca Monastica della Badia dei Benedittini, Finalpia 15823
Biblioteca Monastica 'Madonna della Scala', Noci 15939
Biblioteca Moncayo, Zaragoza 26189
Biblioteca Mons. G. M. Radini Tedeschi, Bergamo 15731
Biblioteca Monumentale, Perugia 15975
Biblioteca Morcelliana, Chiari 16229
Biblioteca Multimediale Arturo Loria, Carpi 16713
Biblioteca Municipal, Bluefields 19899
Biblioteca Municipal, Elvas 21833
Biblioteca Municipal, León 19900
Biblioteca Municipal, Luanda 00109
Biblioteca Municipal, La Orotava 26236
Biblioteca Municipal, Petrópolis 03436
Biblioteca Municipal, Santander 26253
Biblioteca Municipal, Santo Domingo 06974
Biblioteca Municipal Camões, Santarém 21842
Biblioteca Municipal Central de Lisboa, Lisboa 21837
Biblioteca Municipal de Beja – José Saramago, Beja 21831
Biblioteca Municipal de Donostia San Sebastián, San Sebastián 26248
Biblioteca Municipal de Estudos Locais, A Coruña 25948
Biblioteca Municipal de Ponferrada, Ponferrada 26242
Biblioteca Municipal Florbela Espanca, Matosinhos 21839
Biblioteca Municipal 'Gabriel Morillo', Moca 06973
Biblioteca Municipal 'Orígenes Les sa', Lençois Paulista 03433
Biblioteca Municipal 'Pedro Carbo', Guayaquil 06998
Biblioteca Municipale A. Panizzi, Reggio Emilia 16806
Biblioteca Musical, Madrid 25986
Biblioteca Musical de Compositores Valencianos, Valencia 26168
Biblioteca Nacional, Rio de Janeiro 02981
Biblioteca Nacional Agropecuaria y Forestal (BINAF), La Habana 06291
Biblioteca Nacional Aruba, Oranjestad 00363
Biblioteca Nacional d'Andorra, Andorra la Vella 00097
Biblioteca Nacional de Aeronáutica, Buenos Aires 00242
Biblioteca Nacional de Angola, Luanda 00102
Biblioteca Nacional de Antropología e Historia 'Dr. Eusebio Dávalos Hurtado', México 18835
Biblioteca Nacional de Antropología e Historia 'Dr. Eusebio Dávalos Hurtado', México 18836
Biblioteca Nacional de Belice, Belize City 02896
Biblioteca Nacional de Cabo Verde, Praia 04830
Biblioteca Nacional de Chile, Santiago 04842
Biblioteca Nacional de Ciencia y Técnologiá, La Habana 06292
Biblioteca Nacional de Colombia, Santafé de Bogotá 05929
Biblioteca Nacional de El Salvador, San Salvador 07200
Biblioteca Nacional de España, Madrid 25245
Biblioteca Nacional de Guatemala "Luis Cardoza y Aragon", Guatemala City 13172
Biblioteca Nacional de Honduras "Juan Ramón Molina", Tegucigalpa 13207
Biblioteca Nacional de la República Argentina, Buenos Aires 00112
Biblioteca Nacional de Maestros, Buenos Aires 00243
Biblioteca Nacional de Medicina (BINAME), Montevideo 42044
Biblioteca Nacional de México (BNM), México 18707
Biblioteca Nacional de Moçambique, Maputo 18975
Biblioteca Nacional de Portugal (BNP), Lisboa 21619
Biblioteca Nacional de Salud (BNS), Managua 19895
Biblioteca Nacional de Venezuela, Caracas 42203
Biblioteca Nacional del Perú, Lima 20612
Biblioteca Nacional del Uruguay, Montevideo 42008
Biblioteca Nacional Ernesto J. Castillero, Panamá 20534
Biblioteca Nacional "Eugenio Espejo", Quito 06975
Biblioteca Nacional Forestal 'Ing. Roberto Villaseñor Angeles', México 18837
Biblioteca Nacional "José Martí", La Habana 06266
Biblioteca Nacional Miguel Obregón Lizano, San José 06105
Biblioteca Nacional Militar, Buenos Aires 00205
Biblioteca Nacional Pedro Henríquez Ureña, Santo Domingo 06953
Biblioteca Nacional "Rubén Darío" de Nicaragua, Managua 19887
Biblioteca Naţională a României, Bucureşti 21914
Biblioteca Naţională a României – Filiala Batthyaneum, Alba Iulia 21913
Biblioteca Naturalistica Calabrese, Tarsia 16584
Biblioteca Nazionale Braidense, Milano 14798
Biblioteca Nazionale Centrale della Meteorologia Italiana, Roma 16468
Biblioteca Nazionale Centrale di Firenze, Firenze 14784
Biblioteca Nazionale Centrale di Roma, Roma 14785
Biblioteca Nazionale di Potenza, Potenza 16799
Biblioteca Nazionale Marciana, Venezia 14817
Biblioteca Nazionale "Sagarriga Visconti Volpi", Bari 14786
Biblioteca nazionale svizzera, Bern 26961
Biblioteca Nazionale Vittorio Emanuele III, Napoli 14801
Biblioteca "Nicolò V", Sarzana 16059
Biblioteca Oasis, Perugia 15976
Biblioteca Oliveriana, Pesaro 16792
Biblioteca "P. Sandicchi", Reggio Calabria 16430
Biblioteca 'Padre Billini', Baní 06972
Biblioteca Padre Clemente Benedettucci, Recanati 16007
Biblioteca Padre Lebret, São Paulo 03247
Biblioteca Painiana, Messina 15900
Biblioteca Palatina, Parma 14805
Biblioteca Palatina, Sezione Musicale, Parma 14806
Biblioteca Panamericana, Cuenca 06996
Biblioteca Parrocchiale Villadicanese, Castiglione di Sicilia 15786
Biblioteca Pedagógica Central, Montevideo 42045
Biblioteca Pedagogică Naţională 'I. C. Petrescu', Bucureşti 21972
Biblioteca pedagogica nazionale dell' Agenzia Nazionale per lo Sviluppo dell' Autonomia Scolastica, Firenze 16255
Biblioteca Pirro Marconi, Agrigento 16161
Biblioteca Planettiana, Jesi 16760
Biblioteca Popolare Comunale, Albino 16682
Biblioteca Popular de la Caixa d'Estalvis Laietana, Mataró 26235
Biblioteca Popular de Lisboa, Lisboa 21838
Biblioteca Popular 'El Porvenir', Concepción del Uruguay 00321
Biblioteca Popular 'P. Fidel Fita', Arenys de Mar 26201
Biblioteca Porziuncola, Assisi 15713
Biblioteca Privata Ferraironi, Triora 16632
Biblioteca Privata Leopardi, Recanati 16429
Biblioteca Provincial de Córdoba, Córdoba 26210
Biblioteca Provincial de Granada, Granada 26216
Biblioteca Provincial Elvira Cape, Santiago de Cuba 06315
Biblioteca Provincial 'Rubén Martínez Villena', La Habana 06314
Biblioteca Provinciale, Benevento 16698
Biblioteca Provinciale, Potenza 16800
Biblioteca Provinciale "Angelo Camillo De Meis", Chieti 16724
Biblioteca Provinciale Anselmo Anselmi, Viterbo 16861
Biblioteca Provinciale Cappuccini, Bologna 15741
Biblioteca Provinciale Cappuccini, Reggio Emilia 16012
Biblioteca Provinciale Cappuccini, Venezia 16123
Biblioteca Provinciale Cappucini, Torino 16094
Biblioteca Provinciale dei Cappuccini di Puglia, Bari 15726
Biblioteca Provinciale dei Cappuccini Liguri, Genova 15844
Biblioteca Provinciale dei Frati Cappucini, Trento 16107
Biblioteca Provinciale dei Minori Francescani, Firenze 15829
Biblioteca Provinciale di Brindisi, Brindisi 16706
Biblioteca Provinciale di Foggia, Foggia 16741
Biblioteca Provinciale di Roma, Roma 14811
Biblioteca Provinciale di Salerno, Salerno 16812
Biblioteca Provinciale Difilosofia s. Tommaso d'Aquino, Torino 16095
Biblioteca provinciale francescana, Napoli 15926
Biblioteca Provinciale "Gabriele D'Annunzio", Pescara 16793
Biblioteca Provinciale Laurenziana, Napoli 15927
Biblioteca Provinciale "Melchiorre Delfico", Teramo 14814
Biblioteca Provinciale "N. Bernardini", Lecce 16763
Biblioteca Provinciale "P. Albino", Campobasso 16711
Biblioteca Provinciale "Salvatore Tommasi", L'Aquila 16762
Biblioteca Provinciale Scipione e Giulio Capone, Avellino 16691
Biblioteca Provinciale Tommaso Stigliani, Matera 16773
Biblioteca Pubblica Arcivescovile "Annibale de Leo", Brindisi 15759
Biblioteca Pubblica e Casa della Cultura – Fondazione "A. Marazza", Borgomanero 16705
Biblioteca pubblica "Ernesto Ragionieri", Sesto Fiorentino 16827
Biblioteca Pública, Castellón 26208
Biblioteca Pública, Lleida 26225
Biblioteca Pública, Tarragona 26257
Biblioteca Pública, Valladolid 26261
Biblioteca Pública Arús, Barcelona 25886
Biblioteca Pública de Braga, Braga 21832
Biblioteca Pública de Burgos, Burgos 26205
Biblioteca Pública de Ciudad Real, Ciudad Real 26209
Biblioteca Pública de Évora, Évora 21834
Biblioteca Pública de Girona, Girona 26215
Biblioteca Pública de l'Estat a Palma – Can Sales, Palma de Mallorca 26239
Biblioteca Pública de Maó, Maó 26234
Biblioteca Pública de Valencia, Valencia 26259

Biblioteca Pública del Estado, Albacete 26197
Biblioteca Pública del Estado, Huelva 26219
Biblioteca Pública del Estado, Huesca 26220
Biblioteca Pública del Estado, Málaga 26232
Biblioteca Publica del Estado, Pontevedra 26243
Biblioteca Pública del Estado, Segovia 26254
Biblioteca Pública del Estado, Zamora 26265
Biblioteca Pública del Estado 'A. Rodríguez Moñino/M. Brey', Cáceres 26206
Biblioteca Pública del Estado en Alicante "José Martínez Ruíz, 'Azorín", Alicante 26198
Biblioteca Pública del Estado en Guadalajara, Guadalajara 26218
Biblioteca Pública del Estado en Las Palmas, Las Palmas 26241
Biblioteca Pública del Estado en León, León 26224
Biblioteca Pública del Estado en Lugo, Lugo 26226
Biblioteca Pública del Estado en Orense, Ourense 26237
Biblioteca Pública del Estado en Palencia, Palencia 26238
Biblioteca Pública del Estado en Salamanca, Salamanca 26247
Biblioteca Pública del Estado en Santa Cruz de Tenerife, Santa Cruz de Tenerife 26250
Biblioteca Pública del Estado en Soria, Soria 26256
Biblioteca Pública del Estado en Vitoria-Gasteiz, Vitoria-Gasteiz 26264
Biblioteca Pública del Estado 'Fermín Caballero', Cuenca 26213
Biblioteca Pública do Estado da Bahia, Salvador 03441
Biblioteca Pública do Estado de Santa Catarina, Florianópolis 03431
Biblioteca Pública do Estado de Sergipe, Aracaju 03424
Biblioteca Pública do Estado do Rio de Janeiro, Rio de Janeiro 03439
Biblioteca Pública do Paraná, Curitiba 03428
Biblioteca Pública e Arquivo Regional de Angra do Heroísmo, Angra do Heroismo 21830
Biblioteca Pública e Arquivo Regional de Ponta Delgada, Ponta Delgada 21840
Biblioteca Pública Episcopal del Seminari de Barcelona, Barcelona 25766
Biblioteca Pública Estadual de Pernambuco, Recife 03437
Biblioteca Pública Estadual Luiz de Bessa, Belo Horizonte 03425
Biblioteca Publica 'Francesca Bonnemaison', Barcelona 26202
Biblioteca Pública General San Martín, Mendoza 00322
Biblioteca Pública João Bosco Evangelista, Manaus 03434
Biblioteca Pública Miguel González Garcés, A Coruña 26211
Biblioteca Pública Morales, Panamá 20543
Biblioteca Pública Moratalaz, Madrid 26227
Biblioteca Pública Municipal, Cuenca 06997
Biblioteca Pública Municipal, Gijón 26214
Biblioteca Pública Municipal, Setúbal 21843
Biblioteca Pública Municipal, Vila Nova de Gaia 21844
Biblioteca Pública Municipal Alvenir Peixoto de Exu, Exu 03430
Biblioteca Pública Municipal Central, Jerez de la Frontera 26223
Biblioteca Pública Municipal Central, Valencia 26260
Biblioteca Pública Municipal de Alzira, Alzira 26200
Biblioteca Pública Municipal de Santa Cruz de Tenerife, Santa Cruz de Tenerife 26251
Biblioteca Pública Municipal do Porto, Porto 21841
Biblioteca Pública Municipal Doutor Getulio Vargas de Tomazina, Tomazina 03445
Biblioteca Pública Municipal Dr. Gladstone Osorio Marsico, Erechim 03429
Biblioteca Pública Municipal 'Dr. Joaquín Menéndez', Pergamino 00323
Biblioteca Pública Municipal 'Mário de Andrade', São Paulo 03444
Biblioteca Pública Municipal Miguel Hernández, Villena 26263
Biblioteca Pública Municipal 'Monteiro Lobato', São Bernardo do Campo 03442
Biblioteca Pública Municipal Monteiro Lobato de Horizontina, Horizontina 03432
Biblioteca Pública Municipal 'Pedro Fernandes Tomás', Figueira da Foz 21835
Biblioteca Pública Municipal Rui Barbosa de Colombo, Colombo 03427
Biblioteca Pública No 1 'Santiago Severín', Valparaíso 04949
Biblioteca Pública Pelotense, Pelotas 03435
Biblioteca Pública Piloto de Medellín para América Latina, Medellín 06041
Biblioteca Pública Provincial de Almería "Francisco Villaespesa", Almería 26199
Biblioteca Pública Provincial de Jaén, Jaén 26222
Biblioteca Publica Provincial 'Infanta Elena', Sevilla 26255
Biblioteca Pública Regional da Madeira, Santo António 21828
Biblioteca Pública Retiro, Madrid 26228
Biblioteca Pública 'Roberto Noble', Buenos Aires 00319
Biblioteca Pública Usera José Hierro, Madrid 26229
Biblioteca "R. Caracciolo", Lecce 15873
Biblioteca Reale, Torino 14815
Biblioteca Regional de Madrid, Madrid 25251
Biblioteca Regional de Murcia, Murcia 25253
Biblioteca Repiblicană de Informaţie Tehnico-Ştiinţifică, Chişinău 18879
Biblioteca Republicana Stiintifica Agricola, Chişinău 18884
Biblioteca Riccardiana, Firenze 16740
Biblioteca Riccardo e Fernanda Pivano, Milano 16326
Biblioteca Rio-Grandense, Rio Grande 03440
Biblioteca Romana, Roma 16469
Biblioteca Romana Antonio Sarti, Roma 16470
Biblioteca Roncioniana, Prato 14807
Biblioteca Rosminiana, Rovereto 16565
Biblioteca Rosminiana, Stresa 16076
Biblioteca 'Rui Barbosa', Bauru 02987
Biblioteca S. Antonio Dottore, Nocera Inferiore 15937
Biblioteca S. Chiara, L'Aquila 15867
Biblioteca S. Chiara, Napoli 15928
Biblioteca S. Domenico, Bologna 15742
Biblioteca S. Giacomo, Pontida 15991
Biblioteca S. Maria della Catena, Laurignano 15872
Biblioteca sabatini di San Venditto, Pescocostanzo 16416
Biblioteca 'Sacro Cuore' dei Padri Cappuccini, Campobasso 15773
Biblioteca Salaborsa, Bologna 16702
Biblioteca Salesiana, L'Aquila 15868
Biblioteca Salesiana, Torino 16096
Biblioteca Salita dei frati, Lugano 27262
Biblioteca San Bernardino, L'Aquila 15869
Biblioteca San Bernardo, Trento 16108
Biblioteca San Francesco, Bologna 15743
Biblioteca "San Francesco", Ravello 16003
Biblioteca San Giuseppe, Viterbo 16139
Biblioteca San Ignacio, Santiago 04911
Biblioteca Sant'Alfonso, Frosinone 15838
Biblioteca Sant'Anselmo Badia Primaziale, Roma 16020
Biblioteca Scientifica "Carlo Livi", Reggio Emilia 16436
Biblioteca Scolopica di S. Pantaleo, Roma 16021
Biblioteca Seminari de Girona, Girona 25778
Biblioteca Sezione W.W.F. L'Aquila, L'Aquila 16300
Biblioteca Sir Robert Ho Tung, Macao 05880
Biblioteca Sportiva Nazionale, Roma 16471
Biblioteca Statale, Lucca 16767
Biblioteca statale annessa, Padova 15955
Biblioteca Statale di Cremona, Cremona 14791
Biblioteca Statale di Trieste, Trieste 16845
Biblioteca Statale Isontina e Civica, Gorizia 14795
Biblioteca Statale Oratoriana del Monumento Nazionale dei Girolamini, Napoli 15929
Biblioteca Teatrale del Burcardo, Roma 16472
Biblioteca Técnica Científica Centralizado para el Desarrollo de la Región, Barquisimeto 42242
Biblioteca Teologica Francescana, Levanto 15876
Biblioteca Tornquist, Buenos Aires 00240
Biblioteca Trivulziana e Archivio Storico Civico, Milano 16327
Biblioteca Universitaria 'Alfonso Reyes', San Nicolas de los Garza 18802
Biblioteca Universitaria y Técnica Antón Ramírez, Madrid 25418
Biblioteca Universitatii Petrol – Gaze din Ploieşti, Ploieşti 21942
Biblioteca "'V. A. Urechia", Galaţi 22091
Biblioteca Valenciana, Valencia 25258
Biblioteca Valentiniana, Camerino 16710
Biblioteca Vallicelliana, Roma 16473
Biblioteca Vescovile, Castellaneta 15785
Biblioteca Vescovile, Nuoro 15944
Biblioteca Vescovile, San Sepolcro 16056
Biblioteca Vescovile, Venosa 16129
Biblioteca Vescovile 'A. Marena', Bitonto 15739
Biblioteca Vescovile 'San Gaudenzio', Novara 15943
Biblioteca y Archivo Nacionales del Paraguay, Asunción 20601
Biblioteca Zelantea, Acireale 16158
Biblioteca Zenoni, Venezia 16646
Biblioteca-Hemeroteca de Castilla y León y las Autonomías Españolas, Burgos 25938
Bibliotecas Públicas de Madrid, Madrid 26230
Bibliotecas Públicas Municipales, Zaragoza 26266
Biblioteche Civiche e Raccolte Storiche, Torino 16837
Biblioteche e Archivi del Vittoriale, Gardone Riviera 16279
Biblioteche riunite Civica e Ursino Recupero, Catania 16719
Bibliotechniy Tsentr dlya detey i yunoshestva 'Chitai-Gorod', Veliki Novgorod 24341
Bibliotechny tekhnikum, Sankt-Peterburg 22694
Biblioteek voor Hedendaagse Dokumentatie, Sint-Niklaas 02697
Bibliotek & Kultur, Karlshamn 26808
Biblioteka Administratsii Prezidenta Rossikoi Federatsii, Moskva 22732
Biblioteka Aprilov – Palauzov, Gabrovo 03593
Biblioteka 'Braka Miladinovci', Skopje 18448
Biblioteka Czartoryskich, Kraków 21127
Biblioteka Dimitrije Tucovič, Lazarevac 24543
Biblioteka dlya ditei im. A. Gaidara, Kyiv 28643
Biblioteka dlya ditei im. G. Kotovskoho, Kyiv 28644
Biblioteka dlya ditei im. K. Chukovskoho, Kyiv 28645
Biblioteka dlya ditei im. M. Kotsyubinskoho, Kyiv 28646
Biblioteka dlya ditei im. N. Zabili, Kyiv 28647
Biblioteka dlya ditei im. O. Pirogovskoho, Kyiv 28648
Biblioteka dlya ditei im. P. Usenka, Kyiv 28649
Biblioteka dlya ditei im. P. Verdhigori, Kyiv 28650
Biblioteka dlya ditei im. Yu. Gagarina, Kyiv 28651
Biblioteka dlya ditei no 115, Kyiv 28652
Biblioteka dlya yunatstva no 11, Kyiv 28653
Biblioteka Đorde Jovanović, Stari Grad 06260
Biblioteka Državnog Muzeja Crne Gore, Cetinje 18914
Biblioteka e Shtetit, Durrës 00019
Biblioteka e Shtetit, Elbasan 00020
Biblioteka e Shtetit, Gjirokastër 00021
Biblioteka e Shtetit, Korçë 00022
Biblioteka Gdańska PAN, Gdańsk 20788
Biblioteka Gligorije Vozarević, Sremska Mitrovica 24553
Biblioteka Grada Beograda, Beograd 24535
Biblioteka I. Kudri, Kyiv 28654
Biblioteka im. B. Kuchera, Kyiv 28655
Biblioteka im. Gertsena, Kyiv 28656
Biblioteka im. I.I. Lense, Sankt-Peterburg 24215
Biblioteka im. K. Simonova, Kyiv 28657
Biblioteka im. Makhtumkuli, Kyiv 28658
Biblioteka im. M.E. Saltykova-Shchedrina, Sankt-Peterburg 24216
Biblioteka im. N. Gogolya, Kyiv 28659
Biblioteka im. N.G. Chernyshevskogo, Sankt-Peterburg 24217
Biblioteka im. O. Bloka, Kyiv 28660
Biblioteka im. O. Novikova-Priboya, Kyiv 28661
Biblioteka im. P. Mirnoho, Kyiv 28662
Biblioteka im. V. Yana, Kyiv 28663
Biblioteka im. Zielinskich, Płock 21189
Biblioteka Instituta Ekologii Rasteniy i Zhivotnykh, Ekaterinburg 22235
Biblioteka Instytutu Sztuki PAN, Warszawa 21242
Biblioteka Kapitulna Wrocław, Wrocław 21064
Biblioteka khimicheskoi literatury RAN, Moskva 23168
Biblioteka Kirovskikh ostrovov, Sankt-Peterburg 24218
Biblioteka Kombëtare e Shoipërisë, Tiranë 00006
Biblioteka Kórnicka PAN, Kórnik 20790
Biblioteka M. Rilskoho, Kyiv 28664
Biblioteka Matice srpske, Novi Sad 24454
Biblioteka Milutin Bojić, Beograd 24536
Biblioteka mizhnarodnoho tsentru kultury ta mystetstv profspilnok Ukrainy, Kyiv 28665
Biblioteka Narodowa, Warszawa 20787
Biblioteka Nationala a Republicii Moldova, Chişinău 18878
Biblioteka natsionalnoi spilky pysmennykhiv Ukrainy, Kyiv 28302
Biblioteka Naukowa PAU i PAN w Krakowie, Kraków 20791
Biblioteka no 5, Kyiv 28666
Biblioteka no 13, Kyiv 28667
Biblioteka Pedagogiczna, Biała Podlaska 21072
Biblioteka Pedagogiczna, Gorzów Wielkopolski 21108
Biblioteka Pedagogiczna, Legnica 21155
Biblioteka Pedagogiczna, Tarnobrzeg 21230
Biblioteka Pedagogiczna, Tarnów 21231
Biblioteka Pedagogiczna im. Zenona Klemensiewicza, Koszalin 21124
Biblioteka Pedagogiczna w Legnicy – Filia w Jeleniej Górze, Jelenia Góra 21111
Biblioteka Pedagogiczna w Łomży, Łomża 21167
Biblioteka Pedagogiczna w Radomiu, Radom 21216
Biblioteka Pedagogiczna w Toruniu, Toruń 21232
Biblioteka Pedagogiczna w Wałbrzychu, Wałbrzych 21237
Biblioteka Pisarzy Towarzystwa Jezusowego, Warszawa 21061
Biblioteka po estestvennym naukam Rossiskoi Akademii Nauk, Moskva 23169
Biblioteka Polska w Paryżu, Paris 08496
Biblioteka Publiczna, Będzin 21378
Biblioteka Publiczna, Czerwieńsk 21405
Biblioteka Publiczna, Częstochowa 21406
Biblioteka Publiczna, Głubczyce 21417
Biblioteka Publiczna, Radzymin 21546
Biblioteka Publiczna, Tczew 21587
Biblioteka Publiczna, Trzcianka 21590
Biblioteka Publiczna, Tychy 21594
Biblioteka Publiczna, Zgorzelec 21616
Biblioteka Publiczna Gminy Kłobuck im. Jana Długosza, Kłobuck 21446
Biblioteka Publiczna im. Jana Daniela Janockiego, Międzychód 21492
Biblioteka Publiczna im. Marii Konopnickiej, Suwałki 21576
Biblioteka Publiczna im. Stefana Rowińskiego, Ostrów Wielkopolski 21521
Biblioteka Publiczna Miasta i Gminy, Wschowa 21611
Biblioteka Publiczna Miasta i Gminy w Słubicach, Słubice 21559
Biblioteka Publiczna m.st. Warszawy – Biblioteka Główna Województwa Mazowieckiego, Warszawa 21599
Biblioteka Publiczna w Pasłęku, Pasłęk 21525
Biblioteka Publike "Marin Barleti, Shkodër 00023
Biblioteka Raczyńskich, Poznań 21532
Biblioteka Rossiskoi akademii nauk (BAN), Sankt-Peterburg 22171
Biblioteka Sejmowa, Warszawa 21041
Biblioteka Seminarium Duchownego, Włocławek 21063
Biblioteka Śląska, Katowice 20789
Biblioteka 'Sv. Kliment Ohridski', Bitola 18429
Biblioteka Sveti Sava, Zemun 24559
Biblioteka Uniwersytecka im. Jerzego Giedroycia w Bialymstoku, Białystok 20795
Biblioteka Uniwersytecka we Wrocławiu, Wrocław 21002
– Biblioteka Studium Praktycznej Nauki Języków Obcych, Wrocław 21003
– Instytut Chemii, Wrocław 21004
– Instytut Filologii Angielskiej, Wrocław 21005
– Instytut Filologii Germańskiej, Wrocław 21006
– Instytut Filologii Klasycznej i Kultury Antycznej, Wrocław 21007
– Instytut Filologii Polskiej, Wrocław 21008
– Instytut Filologii Romańskiej, Wrocław 21009
– Instytut Filologii Słowiańskiej, Wrocław 21010
– Instytut Fizyki, Wrocław 21011
– Instytut Geografii, Wrocław 21012
– Instytut Historii Sztuki, Wrocław 21013
– Instytut Historyczny, Wrocław 21014
– Instytut Informacji Naukonej i Bibliotekoznawstwa, Wrocław 21015
– Instytut Informatyki, Wrocław 21016
– Instytut Matematyczny, Wrocław 21017
– Instytut Nauk Geologicznych, Wrocław 21018
– Instytut Pedagogiki, Wrocław 21019
– Instytut Psychologii, Wrocław 21020
– Instytut Zoologiczny, Wrocław 21021
– Katedra Etnologii i Antropologii Kulturowej, Wrocław 21022
– Katedra Kulturoznawstwa, Wrocław 21023
– Wydział Prawa, Administracji i Ekonomii, Wrocław 21024

Biblioteka Uniwersytetu Śląskiego, Katowice 20816
Biblioteka-filial no 1, Krivi Rig 28633
Biblioteka-filial no 14, Krivi Rig 28634
Biblioteka-filial no 16, Krivi Rig 28635
Biblioteka-klubs 'Imanta', Riga 18197
Biblioteka-Knizhkova svitlitsya dlya ditei, Kyiv 28668
Biblioteket, Oslo 20111
Biblioteket for det Folkelige Arbejde, Vester Skerninge 06836
Biblioteket i Dragør, Dragør 06854
Biblioteket i Kopparberg, Kopparberg 26821
Biblioteket Nordborg, Nordborg 06906
Bibliotheca Alexandrina, Alexandria 07001
Bibliotheca Augustiniana – Forschungsbibliothek der Deutschen Augustiner (BADA), Würzburg 11364
Bibliotheca Bipontina, Zweibrücken 09345
Bibliotheca Cappuccinorum, Salzburg 01460
Bibliotheca Carmelitana Provinciae Germaniae OCD, Würzburg 11365
Bibliotheca Carmelitarum Moguntiacum, Mainz 11268
Bibliotheca Carmelitarum Moguntiacum, Mainz 11269
Bibliotheca Collegii Georgiani Monacensis, München 11277
Bibliotheca Hertziana, Roma 16474
Bibliotheca Minima, Novoli 16389
Bibliotheca Mozartiana, Salzburg 01563
Bibliotheca Philosophica Hermetica, Amsterdam 19348
Bibliotheca Rosenthaliana, Amsterdam 19349
Bibliotheca Tessinensis, Cieszyn 21087
Bibliotheca Wasiana vzw, Sint-Niklaas 02698
Bibliotheca Wittockiana, Brussel 02568
Bibliotheek, Eindhoven 19111
Bibliotheek aan den IJssel – Vestiging Capelle Centrum, Capelle aan den IJssel 19576
Bibliotheek aan den IJssel – Vestiging Krimpen, Krimpen a/d IJssel 19644
Bibliotheek Almelo, Almelo 19545
Bibliotheek Almere, Almere 19546
Bibliotheek Amstelveen Stadsplein, Amstelveen 19550
Bibliotheek Arnhem, Arnhem 19555
Bibliotheek Assen, Assen 19556
Bibliotheek Augustijns Instituut, Eindhoven 19307
Bibliotheek Baarn, Baarn 19559
Bibliotheek Balie Antwerpen, Antwerpen 02509
Bibliotheek Barendrecht, Barendrecht 19560
Bibliotheek Beek-Ubbergen, Beek-Ubbergen 19563
Bibliotheek Bilthoven, Bilthoven 19568
Bibliotheek Bloemendaal, Bloemendaal 19569
Bibliotheek Brakkenstein, Nijmegen 19496
Bibliotheek Breda, Breda 19571
Bibliotheek Buitenpost, Buitenpost 19573
Bibliotheek Burgum, Burgum 19574
Bibliotheek Capelle Centrum, Capelle aan den IJssel 19576
Bibliotheek Culemborg, Culemborg 19579
Bibliotheek De Groene Venen – Vestiging Waddinxveen, Waddinxveen 19734
Bibliotheek De Kempen – Vestiging Valkenswaard, Valkenswaard 19719
Bibliotheek Den Haag, Den Haag 19611
Bibliotheek Deurne, Deurne 19582
Bibliotheek Dinkelland – Vestiging Weerselo, Weerselo 19737
Bibliotheek Doetinchem, Doetinchem 19584
Bibliotheek Dommeldal – Vestiging Geldrop, Geldrop 19603
Bibliotheek Duinrand – Vestiging Bloemendaal, Bloemendaal 19569
Bibliotheek Duiven, Duiven 19588
Bibliotheek Eijsden, Eijsden 19591
Bibliotheek Emmen-centrum, Emmen 19595
Bibliotheek Ermelo, Ermelo 19598
Bibliotheek Ets Haim – Livraria Montezinos, Amsterdam 19350
Bibliotheek Etten-Leur, Etten-Leur 19600
Bibliotheek Gelderland Zuid – Bibliotheek Malden, Malden 19661
Bibliotheek Gelderland Zuid – Centrale Bibliotheek De Mariënburg, Nijmegen 19667
Bibliotheek Gelderland Zuid – Vestiging Beek-Ubbergen, Beek-Ubbergen 19563
Bibliotheek Geldermalsen, Geldermalsen 19602
Bibliotheek Geldrop, Geldrop 19603
Bibliotheek Genk, Genk 02764
Bibliotheek Goes, Goes 19605
Bibliotheek Gorredijk, Gorredijk 19607
Bibliotheek Harderwijk, Harderwijk 19614
Bibliotheek Harelbeke, Harelbeke 02773
Bibliotheek Heemstede, Heemstede 19615
Bibliotheek Heerenveen, Heerenveen 19617
Bibliotheek Hellevoetsluis, Hellevoetsluis 19622
Bibliotheek Hemond-Peel – Vestiging Deurne, Deurne 19582
Bibliotheek Hengelo, Hengelo 19625
Bibliotheek Het Markiezaat – Vestiging Markiezenhof, Bergen op Zoom 19566
Bibliotheek Hilversum, Hilversum 19629
Bibliotheek Hoogeveen, Hoogeveen 19631
Bibliotheek Hoorn – Centrale, Hoorn 19633
Bibliotheek Horst, Horst 19634
Bibliotheek Hulst, Hulst 19638
Bibliotheek IJmond Noord – Centrale Bibliotheek Beverwijk, Beverwijk 19567
Bibliotheek infopunt, Oostkamp 02838
Bibliotheek Katwijk, Katwijk 19642
Bibliotheek Kennemerwaard, Alkmaar 19544
Bibliotheek Kerkrade, Kerkrade 19643
Bibliotheek Krimpen, Krimpen a/d IJssel 19644
Bibliotheek Kris Lambert, Oostende 02837
Bibliotheek Landgraaf – Centrale Bibliotheek, Landgraaf 19645
Bibliotheek Langedijk, Zuid-Scharwoude 19749
Bibliotheek Le Sage ten Broek, Nijmegen 19497
Bibliotheek Leeuwarden Centrum, Leeuwarden 19649
Bibliotheek Leidschendam, Leidschendam 19652
Bibliotheek Leidschendam-Voorburg, Locatie Leidschendam, Leidschendam 19652
Bibliotheek Leidschendam-Voorburg, Locatie Voorburg, Voorburg 19730
Bibliotheek Lisse, Lisse 19655
Bibliotheek Maarssen-Dorp, Maarssen-Dorp 19657
Bibliotheek Maasbracht, Maasbracht 19658
Bibliotheek Malden, Malden 19661
Bibliotheek Markiezenhof, Bergen op Zoom 19566
Bibliotheek Meppel, Meppel 19662
Bibliotheek Montfort, Montfort 19663
Bibliotheek Mu.ZEE, Oostende 02692
Bibliotheek Naarden-Bussum, Centrale Bibliotheek, Bussum 19575
Bibliotheek Nederweert, Nederweert 19664
Bibliotheek Nijkerk, Nijkerk 19666
Bibliotheek Nijverdal, Nijverdal 19669
Bibliotheek Noord-Veluwe, Novetheek Elburg, Elburg 19593
Bibliotheek Noord-Veluwe, Novetheek Heerde, Heerde 19616
Bibliotheek Noord-Veluwe, Novetheek Oldebroek, Oldebroek 19673
Bibliotheek Noordwijk, Noordwijk 19671
Bibliotheek Oost-Achterhoek, Bibliotheek Winterswijk, Winterswijk 19743
Bibliotheek Oostburg, Oostburg 19676
Bibliotheek Oosterbeek, Oosterbeek 19677
Bibliotheek Oosterwolde, Oosterwolde 19679
Bibliotheek plus – Centrum voor kunst en kultuur, Centrale Bibliotheek, Leiden 19650
Bibliotheek plus – Centrum voor kunst en kultuur, Locatie Leiderdorp, Leiderdorp 19651
Bibliotheek Purmerend, Purmerend 19682
Bibliotheek Putten, Putten 19683
Bibliotheek Raalte, Raalte 19684
Bibliotheek Ridderkerk, Ridderkerk 19686
Bibliotheek Rijen, Rijen 19687
Bibliotheek Rijn en Venen, Alphen aan den Rijn 19547
Bibliotheek Rijssen, Rijssen 19688
Bibliotheek Rijssen-Holten, Vestiging Rijssen, Rijssen 19688
Bibliotheek Rijswijk, Rijswijk 19689
Bibliotheek Rivierenland – Vestiging Tiel, Tiel 19710
Bibliotheek Roermond, Roermond 19690
Bibliotheek Rotterdam, Rotterdam 19692
Bibliotheek Sas van Gent, Sas van Gent 19694
Bibliotheek ServiceCentrum, Houten 19635
Bibliotheek Sint Maartensdijk, Sint Maartensdijk 19697
Bibliotheek Soest, Soest 19702
Bibliotheek Staphorst, Staphorst 19705
Bibliotheek Tessenderlo, Tessenderlo 02865
Bibliotheek Tiel, Tiel 19710
Bibliotheek Tilburg Centrum, Tilburg 19711
Bibliotheek Tubbergen, Tubbergen 19713
Bibliotheek Uithoorn, Amstelveen 19549
Bibliotheek Utrecht, Utrecht 19717
Bibliotheek Valkenburg, Valkenburg 19718
Bibliotheek Valkenswaard, Valkenswaard 19719
Bibliotheek van de Provincie Overijssel, Zwolle 19294
Bibliotheek van het Cultureel Centrum van Anderlecht (CCA), Brussel 02735
Bibliotheek van het Parlament, Bruxelles 02425
Bibliotheek Velsen – Centrale Bibliotheek IJmuiden, IJmuiden 19639
Bibliotheek Venlo, Venlo 19724
Bibliotheek Venray, Venray 19725
Bibliotheek Vleuten, Vleuten 19727
Bibliotheek Vlissingen, Vlissingen 19728
Bibliotheek Voorburg, Voorburg 19730
Bibliotheek Voorschoten-Wassenaar, Vestiging Wassenaaar, Wassenaar 19736
Bibliotheek Vroomshoop, Vroomshoop 19732
Bibliotheek Waddinxveen, Waddinxveen 19734
Bibliotheek Wassenaar, Wassenaar 19736
Bibliotheek Waterweg – Openbare Bibliotheek Schiedam, Schiedam 19696
Bibliotheek Weerselo, Weerselo 19737
Bibliotheek Weert, Weert 19738
Bibliotheek Winterswijk, Winterswijk 19743
Bibliotheek Zeist, Zeist 19744
Bibliotheek Zevenaar, Zevenaar 19745
Bibliotheek Zoetermeer – Hoofdbibliotheek, Zoetermeer 19746
Bibliotheek Zuid-Hollandse Delta – Vestiging Hellevoetsluis, Hellevoetsluis 19622
Bibliotheek Zutphen, Zutphen 19750
Bibliotheek Zwijndrecht – Vestiging Anjerstraat, Zwijndrecht 19751
Bibliotheek Zwolle, Zwolle 19752
Bibliothek am Guisanplatz / Bibliothèque Am Guisanplatz, Bern 27318
Bibliothek am Luisenbad, Berlin 12653
Bibliothek Amauta, Wien 01588
Bibliothek Bartkowiak, Hamburg 11971
Bibliothek Baudenkmalpflege, Halle (Saale) 10907
Bibliothek Bauwesen, Soziale Arbeit und Gesundheit, Hildesheim 10041
Bibliothek der Abtei Ettal, Ettal 11211
Bibliothek der Abteilung für Rechtsvergleichung, München 10405
Bibliothek der Barmherzigen Brüder, Trier 11346
Bibliothek der Freien, Berlin 11494
Bibliothek der Hansestadt Lübeck, Lübeck 12833
Bibliothek der Kapuziner, Münster 11288
Bibliothek der Modesammlung, Wien 01589
Bibliothek der Redemptoristen, Gars a. Inn 11222
Bibliothek der Schönstattpatres, Vallendar 11354
Bibliothek der Stadt Ilmenau, Ilmenau 12794
Bibliothek des Ärztlichen Vereins, Hamburg 11968
Bibliothek des Börsenvereins des Deutschen Buchhandels, Frankfurt am Main 09292
Bibliothek des Diakonischen Werkes der EKD (Bestand Berlin), Berlin 11164
Bibliothek des Diözesan-Zentrums (BDZ), Villars-sur-Glâne 27276
Bibliothek des Evangelischen Ministeriums, Erfurt 11207
Bibliothek des Jugendhauses Düsseldorf e.V., Düsseldorf 11200
Bibliothek des Ministeriums für Innovation, Wissenschaft und Forschung des Landes Nordrhein-Westfalen und des Ministeriums für Schule und Weiterbildung des Landes Nordrhein-Westfalen, Düsseldorf 10868
Bibliothek des Oberlandesgerichtes und Bibliothek der Grupenschen Stiftung, Celle 10846
Bibliothek des Ruhrgebiets, Bochum 11617
Bibliothek des Wissenschaftsparks Albert Einstein, Potsdam 12402
Bibliothek für Bildungsgeschichtliche Forschung, Berlin 11495
Bibliothek für Gestaltung Basel, Basel 27219
Bibliothek für Internationale Entwicklungs- und Umweltforschung und Osteuropäische Geschichte, Gießen 09831
Bibliothek für Österreichische Literatur und Kultur (BIBLIOfÖLK), Saarbrücken 10568
Bibliothek für Psychoanalyse und Sexualforschung, Salzburg 01564
Bibliothek für Zeitgeschichte in der Württembergischen Landesbibliothek, Stuttgart 12511
Bibliothek Gestaltung, Erhaltung von Kulturgut, Hildesheim 10042
Bibliothek HALLE 14, Leipzig 12179
Bibliothek Haus Villigst, Schwerte 11327
Bibliothek Hör- und Sprachgeschädigtenwesen, Leipzig 12180
Bibliothek im Basler Missionshaus, Basel 27244
Bibliothek im Klinikum Bremen-Ost, Bremen 11671
Bibliothek im Wissenschaftszentrum Ost- und Südosteuropa, Regensburg 12425
Bibliothek Landsberg-Velen, Balve 11471
Bibliothek Management, Soziale Arbeit, Bauen, Holzminden 10046
Bibliothek MeteoSchweiz Zürich, Zürich 27449
Bibliothek Museumsinsel, Berlin 11570
Bibliothek Naturwissenschaften und Technik, Göttingen 09903
Bibliothek Oberschlema, Bad Schlema 12641
Bibliothek Otto Schäfer, Schweinfurt 12479
Bibliothek Ressourcenmanagement, Göttingen 09902
Bibliothek Schloss Arenenberg, Salenstein 27428
Bibliothek St. Albertus Magnus, Köln 12123
Bibliothek Teterow, Teterow 12958
Bibliothek Theologicum, Katholisch-Theologisches Seminar, Tübingen 10631
Bibliothek und Informationszentrum für Statistik / Bibliothèque et centre d'information pour la statistique, Neuchâtel 27415
Bibliothek Werner Oechslin, Einsiedeln 27351
Bibliothek "Wissenschaft und Weisheit", Mönchengladbach 11275
Bibliothek zur Geschichte der DDR, Bonn 11649
Bibliotheken der Stadt Dortmund – Stadt- und Landesbibliothek, Dortmund 09310
Bibliotheken der Stadt Mainz, Wissenschaftliche Stadtbibliothek, Mainz 09327
Bibliotheken Eemland – Centrale Bibliotheek, Amersfoort 19548
Bibliotheken Eemland – Vestiging Baarn, Baarn 19559
Bibliotheks- und Medienzentrum Nordelbien, Hamburg 11231
Bibliothekszentrum Klosterbau, Sammlung Friedberg und Wetterau, Friedberg (Hessen) 11903
Bibliothèque administrative de la Ville de Paris (BAVP), Paris 08143
Bibliothèque Aequatoria, Kinshasa 06069
Bibliothèque Africaine, Bruxelles 02569
Bibliothèque Africaine et Malgache, Paris 08473
Bibliothèque Albert le Grand, Montreal 04210
Bibliothèque Albert-le-Grand, Montreal 04211
Bibliothèque Alsatique du Crédit Mutuel, Strasbourg 08679
Bibliothèque am Guisanplatz, Bern 27318
Bibliothèque américaine à Paris, Paris 08464
Bibliothèque Américaine d'Angers, Angers 08275
Bibliothèque Américaine de Montpellier, Montpellier 08421
Bibliothèque André Malraux, Paris 08996
Bibliothèque Armand Salacrou, Le Havre 08909
Bibliothèque Artistique de la Ville de Bruxelles, Bruxelles 02570
Bibliothèque Beaugrenelle, Paris 08997
Bibliothèque Bernheim, Nouméa 19759
Bibliothèque Biblique BOSEB, Paris 08211
Bibliothèque Boris Vian, Tremblay-en-France 09118
Bibliothèque Bozidar Kantuser, Paris 08474
Bibliothèque Braille Romande et livre parlé, Genève 27356
La Bibliothèque Broca-Droit, Paris 07899
Bibliothèque Buffon, Paris 08998
Bibliothèque cantonale et universitaire Fribourg (BCU) / Kantons- und Universitätsbibliothek Freiburg (KUB), Fribourg 26968
Bibliothèque cantonale jurassienne, Porrentruy 26978
Bibliothèque centrale de la Province de Namur, Namur 02831
Bibliothèque centrale de la Province du Hainaut, La Louvière 02810
Bibliothèque centrale de prêt de Martinique, Fort-de-France 18679
Bibliothèque centrale de prêt de Mayotte, Mamoudzou 18706
Bibliothèque Centrale du Service de Santé des Armées, Paris 08475
Bibliothèque centrale Elsa Triolet, Bobigny 08775
Bibliothèque centrale Mériadeck, Bordeaux 08778
Bibliothèque Centrale Municipale, Alger 00086
Bibliothèque centrale Robert-Desnos, Montreuil 08971
Bibliothèque Charles H. Blais, Sillery 04758
Bibliothèque Charles-Edouard-Mailhot, Victoriaville 04812
Bibliothèque Chiroux, Liège 02801
Bibliothèque communale, Bernissart 02722

Bibliothèque communale, Boussu 02729
Bibliothèque communale, Malmédy 02816
Bibliothèque communale, Namur 02832
Bibliothèque communale, Oupeye 02840
Bibliothèque communale, Seraing 02855
Bibliothèque communale "Buxin-Simon", Florennes 02761
Bibliothèque communale du Centre, Uccle 02871
Bibliothèque communale d'Uccle-Montjoie, Uccle 02872
Bibliothèque communale et Centre de Lecture publique, Hannut 02772
Bibliothèque communale G. Spailier, Spa 02861
Bibliothèque communale locale de la Posterie, Courcelles 02745
Bibliothèque communale locale pivot "Henri Matelart", Châtelet 02744
Bibliothèque communale locale-pivot et principale, Morlanwelz 02829
Bibliothèque Communale Maurice Carême, Wavre 02880
Bibliothèque communale Ulysse Cordier, Quaregnon 02846
Bibliothèque Couronnes, Paris 08999
Bibliothèque d'Agglomération, Saint-Omer 09076
Bibliothèque d'Agglomération, Saint-Quentin-en-Yvelines 09079
Bibliothèque d'Agglomeration de Marne-la-Vallée – Val-Maubuée, Torcy 09110
Bibliothèque d'Amiens Métropole, Amiens 08730
Bibliothèque d'art et d'archéologie, Genève 27357
Bibliothèque de Brossard, Brossard 04589
Bibliothèque de Charlesbourg, Charlesbourg 04606
Bibliothèque de Culture Religieuse, Pau 08225
Bibliothèque de Documentation Internationale Contemporaine (BDIC), Nanterre 07794
Bibliothèque de Dollard-des-Ormeaux, Dollard-des-Ormeaux 04626
Bibliothèque de Dorval, Dorval 04627
La Bibliothèque de Douai, Douai 08853
Bibliothèque de Farnham, Inc., Farnham 04637
Bibliothèque de Genève, Genève 26969
Bibliothèque de la Bourgeoisie de Berne, Bern 27319
Bibliothèque de la C.A.P.I., Villefontaine 09144
Bibliothèque de la Cité, Genève 27485
Bibliothèque de la Commission de la Fonction Publique, Ottawa 04450
Bibliothèque de la Communauté d'Agglomération d'Annecy, Annecy 08735
Bibliothèque de la Communauté Urbaine de Casablanca, Casablanca 18972
Bibliothèque de la Maison de l'Orient et de la Méditerranée, Lyon 08392
Bibliothèque de la Nièvre, Varennes-Vauzelles 09130
Bibliothèque de la Sorbonne, Paris 07906
Bibliothèque de la Source, Rabat 18954
Bibliothèque de la Ville, Biel 27478
Bibliothèque de la Ville de Colmar, Colmar 08832
Bibliothèque de la Ville de Tournai, Tournai 02869
Bibliothèque de la Ville du Locle, Le Locle 27493
Bibliothèque de l'Abbaye Saint-Benoit, Saint Benoît-du-Lac 04222
Bibliothèque de Lachine, Lachine 04675
Bibliothèque de l'Alcazar, Marseille 08944
Bibliothèque de l'Arrondissement du Vieux-Longueuil, Longueuil 04682
Bibliothèque de l'Arsenal, Paris 07625
Bibliothèque de l'Atrium, Chaville 08824
Bibliothèque de Montciel, Lons-le-Saunier 08388
Bibliothèque de Montréal, Montreal 04690
Bibliothèque de Montréal – Jean-Corbeil, Anjou 04570
Bibliothèque de philosophie des sciences, Paris 07839
Bibliothèque de Riom Communauté, Riom 09041
Bibliothèque de Sciences religieuses, Annecy 08174
Bibliothèque de Valpré, Ecully 08186
Bibliothèque de Vendôme, Vendôme 09131
Bibliothèque départementale de la Charente, Confolens 08839
Bibliothèque départementale de la Charente-Maritime, Saintes 09059
Bibliothèque départementale de la Haute-Marne, Chamarandes-Choignes 08805

Bibliothèque départementale de la Mayenne, Saint-Berthevin 09053
Bibliothèque départementale de la Meuse, Bar-le-Duc 08757
Bibliothèque départementale de la Somme, Amiens 08731
Bibliothèque départementale de la Vienne, Poitiers 09025
Bibliothèque départementale de l'Ain, Bourg-en-Bresse 08781
Bibliothèque départementale de l'Ariège, Foix 08870
Bibliothèque départementale de l'Aube, Saint-André-les-Vergers 09051
Bibliothèque Départementale de l'Aude, Carcassonne 08796
Bibliothèque départementale de l'Aveyron, Rodez 09044
Bibliothèque départementale de l'Indre, Châteauroux 08818
Bibliothèque départementale de l'Isère, Saint-Martin-d'Hères 09072
Bibliothèque départementale de Lot-et-Garonne, Villeneuve-sur-Lot 09148
Bibliothèque départementale de l'Yonne, Saint-Georges-sur-Baulche 09064
Bibliothèque départementale de prêt de Corse-du-Sud, Ajaccio 08723
Bibliothèque départementale de prêt de Haute-Corse, Corte 08841
Bibliothèque départementale de prêt de la Corrèze, Tulle 09121
Bibliothèque départementale de prêt de la Côte-d'Or, Saint-Apollinaire 09052
Bibliothèque Départementale de Prêt de la Creuse, Guéret 08885
Bibliothèque Départementale de Prêt de la Dordogne, Périgueux 09019
Bibliothèque départementale de prêt de la Gironde, Saint-Médard-en-Jalles 09074
Bibliothèque départementale de prêt de la Guadeloupe, Basse-Terre 13162
Bibliothèque départementale de prêt de la Guyane, Cayenne 09163
Bibliothèque départementale de prêt de la Haute-Loire, Le Puy-en-Velay 08918
Bibliothèque départementale de prêt de la Haute-Vienne, Limoges 08929
Bibliothèque départementale de prêt de la Loire-Atlantique, Carquefou 08799
Bibliothèque départementale de prêt de la Lozère, Mende 08951
Bibliothèque départementale de prêt de la Manche, Saint-Lô 09070
Bibliothèque départementale de prêt de la Marne, Châlons-en-Champagne 08803
Bibliothèque départementale de prêt de la Moselle, Metz 08954
Bibliothèque départementale de prêt de la Réunion, Saint-Denis 21911
Bibliothèque départementale de prêt de la Sarthe, Le Mans 08910
Bibliothèque Départementale de Prêt de l'Aisne, Soissons 09097
Bibliothèque départementale de prêt de l'Ardèche 'André Malraux', Veyras 09138
Bibliothèque départementale de prêt de l'Essonne, Evry 08865
Bibliothèque départementale de prêt de l'Eure, Evreux 08863
Bibliothèque départementale de prêt de Loir-et-Cher, Blois 08773
Bibliothèque départementale de prêt de Seine-Maritime, Mont-Saint-Aignan 08973
Bibliothèque départementale de prêt des Alpes-de-Haute-Provence, Digne-les-Bains 08849
Bibliothèque départementale de prêt des Ardennes, Charleville-Mézières 08812
Bibliothèque départementale de prêt des Vosges, Epinal 08861
Bibliothèque départementale de prêt du Calvados, Ranville 09035
Bibliothèque départementale de prêt du Gard, Nîmes 08986
Bibliothèque départementale de prêt du Gers, Auch 08743
Bibliothèque départementale de prêt du Jura, Lons-le-Saunier 08933
Bibliothèque départementale de prêt du Loiret, Orléans 08990
Bibliothèque départementale de prêt du Lot, Cahors 08791
Bibliothèque départementale de prêt du Maine-et-Loire, Avrille 08753
Bibliothèque départementale de prêt du Puy-de-Dôme, Clermont-Ferrand 08828
Bibliothèque départementale de prêt du Vaucluse, Sorgues 09099
Bibliothèque départementale de Saône-et-Loire, Charnay-les-Mâcon 08814

Bibliothèque départementale de Touraine, Tours 09116
Bibliothèque départementale de Vendée, La Roche-sur-Yon 08898
Bibliothèque départementale des Bouches-du-Rhône, Marseille 08945
Bibliothèque départementale des Deux-Sèvres, Niort 08987
Bibliothèque départementale des Hautes-Alpes, Gap 08878
Bibliothèque départementale des Pyrénées-Atlantiques, Pau 09018
Bibliothèque départementale des Yvelines, Le Mesnil-Saint-Denis 08914
Bibliothèque départementale d'Eure-et-Loire, Mainvilliers 08940
Bibliothèque départementale d'Ille-et-Vilaine, Rennes 09038
Bibliothèque Départementale du Bas-Rhin, Truchtersheim 09120
Bibliothèque départementale du Doubs, Besançon 08768
Bibliothèque départementale du Tarn, Albi 08725
Bibliothèque départementale du Val d'Oise, Pontoise 09028
Bibliothèque départementale Franconie, Cayenne 09164
Bibliothèque des Arts Décoratifs (BAD), Paris 08476
Bibliothèque des Arts et Métiers / Gewerbebibliothek, Fribourg 27221
Bibliothèque des Bollandistes, Bruxelles 02571
Bibliothèque des Cèdres, Lausanne 27391
Bibliothèque des Côtes d'Armor, Plerin 09023
Bibliothèque des Dominicains, L'Arbresle 08192
Bibliothèque des Dominicains, Toulouse 08243
Bibliothèque des Dominicains de l'Annonciation, Paris 08212
Bibliothèque des Littératures Policières (BILIPO), Paris 08477
Bibliothèque des Musées de Strasbourg, Strasbourg 08680
Bibliothèque des Pasteurs, Neuchâtel 27266
Bibliothèque des Quatres Pilliers, Bourges 08783
Bibliothèque des Riches Claires Bruxelles I, Bruxelles 02736
Bibliothèque des Sciences de la Santé et Centre de Documentation / AFRO Health Sciences Library and Documentation Centre, Brazzaville 06098
Bibliothèque Deschatelets, Ottawa 04430
Bibliothèque d'Etude et d'Information, Cergy-Pontoise 08802
Bibliothèque d'Etude et du Patrimoine, Toulouse 09113
Bibliothèque d'Études Augustiennes, Paris 07954
Bibliothèque d'études et de conservation, Besançon 08769
Bibliothèque d'Histoire Africaine, Yaoundé 03653
Bibliothèque diocésaine, Aix-en-Provence 08171
Bibliothèque diocésaine, Arras 08175
Bibliothèque diocésaine, Bayonne 08176
Bibliothèque diocésaine, Besançon 08178
Bibliothèque diocésaine, Chamalières 08181
Bibliothèque diocésaine, Chartres 08182
Bibliothèque diocésaine, Coutances 08184
Bibliothèque Diocésaine, Laval 08194
Bibliothèque diocésaine, Marseille 08202
Bibliothèque Diocésaine, Metz 08204
Bibliothèque diocésaine, Poitiers 08227
Bibliothèque Diocésaine, Quimper 08228
Bibliothèque diocésaine, Reims 08229
Bibliothèque diocésaine, Saintes 08233
Bibliothèque diocésaine, Soissons 08235
Bibliothèque diocésaine, Vannes 08248
Bibliothèque Diocésaine d'Amiens, Amiens 08173
Bibliothèque diocésaine de Nancy, Villers-les-Nancy 08250
Bibliothèque Diocésaine de Sciences Religieuses, Valence 08247
Bibliothèque diocésaine Guillaume Briçonnet, Meaux 08203
Bibliothèque Diocésaine Gustave Bardy, Dijon 08185
Bibliothèque du Centre diocésain (BCD) / Bibliothek des Diözesan-Zentrums (BDZ), Villars-sur-Glâne 27276
Bibliothèque du Centre Ville, Saint-Germain-en-Laye 09065
Bibliothèque du Centre-Ville, Dunkerque 08857
Bibliothèque du Château, Châtellerault 08820
Bibliothèque du Cher, Bourges 08784
Bibliothèque du Chevalier de Cessole, Nice 08447

Bibliothèque du cinéma François Truffaut, Paris 08478
Bibliothèque du Conservatoire de Fribourg (COF), Granges-Paccot 27093
Bibliothèque du Film (BIFI), Paris 08479
Bibliothèque du Finistère Levraoueg Penn-Ar-Bed, Quimper 08649
Bibliothèque du Gouvernement, Luxembourg 18393
Bibliothèque du Grand Séminaire, Aosta 15706
Bibliothèque du Ministère de la Culture et des Communications du Québec, Québec 04128
Bibliothèque du Parlement / Bibliotheek van het Parlement, Bruxelles 02425
Bibliothèque du Personnel de l'Education Nationale, Paris 08480
Bibliothèque du Saulchoir, Paris 08213
Bibliothèque Dumont, Aulnay-sous-Bois 08745
Bibliothèque Edmond Rostand, Paris 09000
Bibliothèque Eschduerf, Eschdorf 18405
Bibliothèque et Archives Canada, Ottawa 03666
Bibliothèque et Archives nationale du Québec (BAnQ), Montreal 03665
Bibliothèque et centre d'information pour la statistique, Neuchâtel 27415
Bibliothèque et Cercles Saint-Nizier, Lyon 08937
Bibliothèque et Ludothèque communales de Braine-l'Alleud, Braine-l'Alleud 02730
Bibliothèque Fabien-LaRochelle, Shawinigan 04754
Bibliothèque Faidherbe, Paris 09001
Bibliothèque Forney, Paris 08481
Bibliothèque Franciscaine Provinciale (BFP), Paris 08214
Bibliothèque franciscaine provinciale des Capucins, Montreal 04212
Bibliothèque Franco-allemande, Bordeaux 08305
Bibliothèque François Truffaut, Paris 08478
Bibliothèque François Villon, Paris 09002
Bibliothèque francophone multimédia, Limoges 08930
Bibliothèque Gabrielle Roy, Québec 04730
Bibliothèque Gaston-Miron, Paris 08482
Bibliothèque Gatien-Lapointe, Trois-Rivières 04804
Bibliothèque Générale et Archives, Tétouan 18974
Bibliothèque générale et Archives du Maroc, Rabat 18916
Bibliothèque Georges Brassens, Paris 09003
Bibliothèque Georges Pompidou, Châlons-en-Champagne 08804
Bibliothèque Georges-Brassens, Chelles 08825
Bibliothèque Goutte d'Or, Paris 09004
Bibliothèque Guillaume Apollinaire, Pontoise 09029
Bibliothèque Haitienne des F.I.C., Port-au-Prince 13204
Bibliothèque Henri Feulard, Paris 08549
Bibliothèque historique centrale de la Marine et service historique de la Défense, Paris 08144
Bibliothèque historique de la ville de Paris (BHVP), Paris 08484
Bibliothèque Humaniste et Municipale, Sélestat 08674
Bibliothèque Inguimbertine, Carpentras 08798
Bibliothèque Issa-Berr, Mopti 18644
Bibliothèque Italie, Paris 08485
Bibliotheque Jacques Duquesne, Liévin 08927
Bibliothèque Jean-Pierre Melville, Paris 09006
Bibliothèque Juive de Genève "Gérard Nordmann", Genève 27358
Bibliothèque Jules Verne, Houilles 08889
Bibliothèque Kandinsky, Paris 08486
Bibliothèque Landowski, Boulogne-Billancourt 08779
Bibliothèque Lesettle Morim, Municipale de Rimouski, Rimouski 04737
Bibliothèque l'Etoile des Alpes, Digne-les-Bains 08850
Bibliothèque libre de Froidmont, Rixensart 02484
Bibliothèque locale, Verviers 02873
Bibliothèque locale de la Province du Hainaut, La Louvière 02811
Bibliothèque Louis Aragon, Chartres 08815
Bibliothèque Louis Nucéra, Nice 08984
Bibliothèque Louis-Notari, Monaco 18898
Bibliothèque Marguerite Durand (BMD), Paris 08489
Bibliothèque Marxiste de Paris, Paris 08490
Bibliothèque Mazarine, Paris 08491
Bibliothèque Medem, Paris 08492

Bibliothèque Médiathèque de Bayonne, Bayonne 08761
La Bibliothèque Médiathèque Hermeland, Saint-Herblain 09066
Bibliothèque Médiathèque municipale, Saint-Junien 09068
Bibliothèque Médiathèque Municipale, Vevey 27502
Bibliothèque médiathèque municipale G. Perec, Gagny 08877
Bibliothèque Meteorologique, Kinshasa 06070
Bibliothèque Montarville-Boucher-de-la-Bruère, Boucherville 04581
Bibliothèque Mouffetard-Contrescarpe, Paris 09007
Bibliothèque multimédia intercommunale Epinal – Golbey, Epinal 08862
Bibliothèque municipale, Abbeville 08720
Bibliothèque municipale, Abidjan 06139
Bibliothèque municipale, Ajaccio 08724
Bibliothèque municipale, Angers 08732
Bibliothèque municipale, Anglet 08733
Bibliothèque municipale, Angoulême 08734
Bibliothèque municipale, Annaba 00087
Bibliothèque municipale, Annonay 08736
Bibliothèque municipale, Antananarivo 18466
Bibliothèque municipale, Antibes 08737
Bibliothèque municipale, Auch 08744
Bibliothèque municipale, Autun 08749
Bibliothèque Municipale, Auxerre 08750
Bibliothèque municipale, Bagnères-de-Bigorre 08754
Bibliothèque municipale, Bastia 08759
Bibliothèque municipale, Beauvais 08762
Bibliothèque municipale, Bègles 08764
Bibliothèque municipale, Béjaia 00088
Bibliothèque municipale, Belfort 08765
Bibliothèque municipale, Bergerac 08767
Bibliothèque municipale, Bois-Colombes 08776
Bibliothèque municipale, Bonneuil-sur-Marne 08777
Bibliothèque municipale, Boulogne-sur-Mer 08780
Bibliothèque municipale, Bourg-en-Bresse 08782
Bibliothèque municipale, Bourg-les-Valence 08785
Bibliothèque municipale, Brest 08786
Bibliothèque municipale, Cahors 08792
Bibliothèque municipale, Calais 08793
Bibliothèque municipale, Carcassonne 08797
Bibliothèque municipale, Castres 08800
Bibliothèque municipale, Chantilly 08810
Bibliothèque municipale, Charenton-le-Pont 08811
Bibliothèque municipale, Châtillon-sur-Seine 08822
Bibliothèque municipale, Clichy 08830
Bibliothèque municipale, Cognac 08831
Bibliothèque municipale, Concarneau 08838
Bibliothèque municipale, Constantine 00089
Bibliothèque municipale, Cosne-sur-Loire 08842
Bibliothèque municipale, Créteil 08845
Bibliothèque municipale, Dammare-les-Lys 08847
Bibliothèque municipale, Dieppe 08848
Bibliothèque municipale, Dreux 08856
Bibliothèque municipale, Fleury-les-Aubrais 08868
Bibliothèque municipale, Floirac 08869
Bibliothèque municipale, Fontainebleau 08871
Bibliothèque municipale, Fougères 08873
Bibliothèque municipale, Franconville-la-Garenne 08874
Bibliothèque municipale, Fresnes 08876
Bibliothèque municipale, Gap 08879
Bibliothèque municipale, Gérardmer 08882
Bibliothèque municipale, Hérouville-Saint-Clair 08888
Bibliothèque municipale, Joigny 08894
Bibliothèque municipale, La Seyne-sur-Mer 08900
Bibliothèque municipale, Laon 08902
Bibliothèque municipale, Laval 08903
Bibliothèque municipale, Le Cateau-Cambrésis 08907
Bibliothèque municipale, Le Puy-en-Velay 08919
Bibliothèque municipale, Les Pavillons-sous-Bois 08921
Bibliothèque municipale, Les Sables-d'Olonne 08922
Bibliothèque municipale, Levallois-Perret 08925
Bibliothèque municipale, Lille 08928
Bibliothèque municipale, Longwy-Haut 08932
Bibliothèque municipale, Lons-le-Saunier 08934
Bibliothèque municipale, Lyon 08938
Bibliothèque municipale, Mâcon 08939
Bibliothèque municipale, Maubeuge 08948
Bibliothèque municipale, Meknès 18973
Bibliothèque municipale, Melun 08950
Bibliothèque municipale, Menton 08953
Bibliothèque municipale, Millau 08957
Bibliothèque municipale, Montargis 08959
Bibliothèque municipale, Montbrison 08963
Bibliothèque municipale, Montélimar 08966
Bibliothèque municipale, Montrouge 08972
Bibliothèque municipale, Morlaix 08974
Bibliothèque municipale, Morlaix 08975
Bibliothèque municipale, Mostaganem 00091
Bibliothèque municipale, Nancy 08978
Bibliothèque municipale, Nantes 08981
Bibliothèque municipale, Narbonne 08982
Bibliothèque municipale, Olivet 08989
Bibliothèque municipale, Oran 00092
Bibliothèque municipale, Oullins 08993
Bibliothèque municipale, Oyonnax 08994
Bibliothèque municipale, Périgueux 09020
Bibliothèque municipale, Pierrefitte-sur-Seine 09022
Bibliothèque municipale, Puteaux 09032
Bibliothèque municipale, Quimper 09033
Bibliothèque municipale, Rillieux-la-Pape 09040
Bibliothèque municipale, Roanne 09042
Bibliothèque municipale, Roubaix 09048
Bibliothèque municipale, Rouen 09049
Bibliothèque municipale, Saintes 09060
Bibliothèque municipale, Saint-Etienne 09061
Bibliothèque municipale, Saint-Jean-de-Luz 09067
Bibliothèque municipale, Saint-Malo 09071
Bibliothèque municipale, Saint-Ouen 09077
Bibliothèque municipale, Saint-Quentin 09078
Bibliothèque municipale, Salins-les-Bains 09082
Bibliothèque municipale, Salon-de-Provence 09083
Bibliothèque municipale, Sarreguemines 09085
Bibliothèque municipale, Sartrouville 09087
Bibliothèque municipale, Sceaux 09089
Bibliothèque municipale, Senlis 09092
Bibliothèque municipale, Sens 09093
Bibliothèque municipale, Soissons 09098
Bibliothèque municipale, Strasbourg 09100
Bibliothèque municipale, Tergnier 09105
Bibliothèque municipale, Terrebonne 04777
Bibliothèque municipale, Thiers 09106
Bibliothèque municipale, Thionville 09107
Bibliothèque municipale, Thonon-les-Bains 09108
Bibliothèque municipale, Tlemcen 00093
Bibliothèque municipale, Toulon 09112
Bibliothèque municipale, Tours 09117
Bibliothèque Municipale, Val d'Or 04807
Bibliothèque municipale, Valenciennes 09125
Bibliothèque municipale, Valognes 09126
Bibliothèque municipale, Vannes 09128
Bibliothèque Municipale, Vaudreuil-Dorion 04809
Bibliothèque municipale, Vesoul 09136
Bibliothèque municipale, Vienne 09140
Bibliothèque municipale, Ville-d'Avray 09143
Bibliothèque municipale, Villeparisis 09150
Bibliothèque municipale, Vire 09154
Bibliothèque municipale, Vitry-le-François 09157
Bibliothèque municipale, Wattrelos 09159
Bibliothèque municipale, Yerres 09160
Bibliothèque municipale Abbé-Grégoire, Blois 08774
Bibliothèque municipale Adélard-Berger, Saint-Jean-Sur-Richelieu 04748
Bibliothèque municipale Anatole France, Villeneuve-le-Roi 09147
Bibliothèque municipale André Malraux, Les Lilas 08920
Bibliothèque municipale André Malraux, Maisons-Alfort 08941
Bibliothèque municipale Andrè Malraux, Sevran 09095
Bibliothèque municipale Antonin Perbosc, Montauban 08960
Bibliothèque municipale Armand Salacrou, Le Havre 08909
Bibliothèque municipale centrale – Saint-Charles, Marseille 08946
Bibliothèque municipale Chamonix Mont-Blanc, Chamonix 08808
Bibliothèque municipale Come-Saint-Germain, Drummondville 04628
Bibliothèque municipale Condorcet, Montivilliers 08967
Bibliothèque Municipale d'Alma, Alma 04567
Bibliothèque municipale d'Amos, Amos 04569
Bibliothèque municipale Danielle et François Mitterrand, Bailleul 08756
Bibliothèque municipale d'Aylmer, Aylmer 04572
Bibliothèque Municipale de Baie Comeau, Baie Comeau 04574
Bibliothèque Municipale de Beauport, Beauport 04578
Bibliothèque municipale de Carouge, Carouge 27481
Bibliothèque municipale de Colomiers, Colomiers 08835
Bibliothèque municipale de Dijon, Dijon 08851
Bibliothèque municipale de Gatineau, Gatineau 04646
Bibliothèque municipale de Gatineau, Hull 04662
Bibliothèque municipale de Jonquière, Jonquiere 04665
Bibliothèque municipale de Lausanne, Lausanne 27491
Bibliothèque municipale de l'Horloge, Cergy 08801
Bibliothèque Municipale de Repentigny, Repentigny 04734
Bibliothèque Municipale de Rouyn-Noranda, Rouyn-Noranda 04739
Bibliothèque Municipale de Saint-Bruno-de-Montarville, Saint Bruno-de-Montarville 04742
Bibliothèque Municipale de Sainte-Foy, Sainte-Foy 04746
Bibliothèque Municipale de Sainte-Thérèse, Sainte Therese 04745
Bibliothèque Municipale de Sept-Iles, Sept Iles 04752
Bibliothèque municipale de Versailles, Versailles 09135
Bibliothèque municipale Denis Diderot, Sèvres 09096
Bibliothèque municipale d'Esch-sur-Alzette, Esch-sur-Alzette 18424
Bibliothèque municipale d'étude et d'information, Grenoble 08884
Bibliothèque municipale Edouard Le Héricher, Avranches 08752
Bibliothèque Municipale Elsa Trioter, Saint-Etienne-du-Rouvray 09062
Bibliothèque municipale Emile Bernard, Asnières-sur-Seine 08741
Bibliothèque Municipale Eva-Senecal, Sherbrooke 04755
Bibliothèque municipale Florian, Rambouillet 09034
Bibliothèque municipale François Mitterand, Le Pré-Saint-Gervais 08917
Bibliothèque municipale François Mitterand, Pontault-Combault 09027
Bibliothèque municipale François Rabelais, Gennevilliers 08881
Bibliothèque municipale Guy-Belisle Saint-Eustache, Saint Eustache 04743
Bibliothèque Municipale Jacques Prévert, Cherbourg 08826
Bibliothèque municipale Jacques Termeau, La Flèche 08896
Bibliothèque municipale Jean d'Ormesson, Marignane 08943
Bibliothèque municipale Lacépède, Agen 08721
Bibliothèque municipale Lamartine, Mende 08952
Bibliothèque municipale Léon Gabriel Gros, Marseille 08947
Bibliothèque municipale Louise Labé, Saint-Chamond 09055
Bibliothèque municipale Luxembourg, Luxembourg 18425
Bibliothèque municipale Manoir de Ferron, Dinan 08852
Bibliothèque municipale Marcel Arland, Langres 08901
Bibliothèque municipale Marcel Pagnol, Aubagne 08742
Bibliothèque municipale Maurice Genevoix, Eaubonne 08858
Bibliothèque municipale Méjanes, Aix-en-Provence 08722
Bibliothèque municipale Nelson Mandela, Vitry-sur-Seine 09158
Bibliothèque municipale Paul Guth, Villeneuve-sur-Lot 09149
Bibliothèque municipale Paul Langevin, Saint-Martin-d'Hères 09073
Bibliothèque municipale Pierre et Marie Curie, Nanterre 08980
Bibliothèque Musicale de la Ville de Genève, Genève 27359
Bibliothèque Nationale, Abidjan 06127
Bibliothèque Nationale, N'Djamena 04835
Bibliothèque Nationale d'Algérie, Alger 00024
Bibliothèque nationale de France, Paris 07623
Bibliothèque nationale de France – Centre National de la litterature pour la jeunesse – La Joie par les Livres, Paris 08493
Bibliothèque Nationale de Guinée, Conakry 13184
Bibliothèque Nationale de la République Démocratique du Congo (BNC), Kinshasa 06044
Bibliothèque nationale de Luxembourg, Luxembourg 18379
Bibliothèque Nationale de Madagascar, Antananarivo 18449
Bibliothèque Nationale de Tunisie, Tunis 27813
Bibliothèque Nationale d'Haïti, Port-au-Prince 13198
Bibliothèque Nationale du Bénin, Porto-Novo 02901
Bibliothèque Nationale du Burkina, Ouagadougou 03611
Bibliothèque Nationale du Burundi, Bujumbura 03622
Bibliothèque Nationale du Cameroun, Yaoundé 03631
Bibliothèque Nationale du Laos, Vientiane 18164
Bibliothèque Nationale du Liban, Beirut 18200
Bibliothèque Nationale du Mali, Bamako 18641
Bibliothèque Nationale du Mauritania, Nouakchott 18681
Bibliothèque Nationale du Rwanda, Kigali 24366
Bibliothèque Nationale du Togo, Lomé 27795
Bibliothèque Nationale et Documentation Gabonaise, Libreville 09167
Bibliothèque Nationale et Universitaire de Strasbourg, Strasbourg 07624
Bibliothèque Nationale Populaire, Brazzaville 06093
La Bibliothèque nationale suisse, Bern 26961
Bibliothèque Nubarian, Paris 08494
Bibliothèque Pablo Neruda, Echirolles 08859
Bibliothèque Parmentier, Paris 09009
Bibliothèque Paul-O.-Trepanier, Granby 04650
Bibliothèque Picpus, Paris 08495
Bibliothèque Pierre-Georges-Roy, Lévis 04678
Bibliothèque Pierre-Monbeig, Paris 07985
Bibliothèque Polonaise de Paris / Biblioteka Polska w Paryžu, Paris 08496
Bibliothèque principale communale "Arthur Rimbaud", Charleroi 02743
Bibliothèque principale communale 'Jean de La Fontaine', Ath 02717
Bibliothèque principale de Bruxelles II, Bruxelles 02737
Bibliothèque Publique, Beaconsfield 04576
Bibliothèque publique, Bukavu 06087
Bibliothèque publique, Kabinda 06088
Bibliothèque publique, Kikwit 06089
Bibliothèque publique, Kinshasa 06090
Bibliothèque publique, Kisangani 06091
Bibliothèque publique, Lulualbourg 06092
Bibliothèque publique centrale de la Communauté française (Brabant wallon), Nivelles 02836
Bibliothèque publique centrale de la Province de Luxembourg, Marché-en-Famenne 02817
Bibliothèque Publique de Chicoutimi, Chicoutimi 04611
Bibliothèque Publique de Mouscon Locale et Principale, Mouscron 02830
Bibliothèque Publique de Pointe-Claire, Pointe-Claire 04721
Bibliothèque publique d'Ottawa, Ottawa 04713
Bibliothèque Publique du Grand Sudbury, Sudbury 04772
Bibliothèque Publique d'Yverdon-les-Bains, Yverdon-les-Bains 27506
Bibliothèque publique et universitaire, Neuchâtel 26977
Bibliothèque Publique Juive / Jewish Public Library, Montreal 04385
Bibliothèque publique libre de Hesbaye, Waremme 02878
Bibliothèque publique libre Saint-Hadelin, Visé 02876
Bibliothèque publique locale, Woluwe-Saint-Lambert 02886
Bibliothèque publique locale de la Ville de Huy, Huy 02785
Bibliothèque Queteletfonds, Bruxelles 02572
Bibliothèque Régionale, Aosta 16686
Bibliothèque Rennes Métropole, Rennes 09039
Bibliothèque Robert-Bourassa, Outremont 04714
Bibliothèque royale Albert Ier, Bruxelles 02257
Bibliothèque royale de Belgique / Koninklijke Bibliotheek van België, Bruxelles 02257
Bibliothèque Russe Tourguénev, Paris 08497
Bibliothèque Saint-Blaise, Paris 09010
Bibliothèque Saint-Corneille, Compiègne 08837
Bibliothèque Saint-Eloi, Paris 09011

Bibliothèque Sanitaire et Sociale de la Ville de Paris, Paris 08498
Bibliothèque Schoelcher, Fort-de-France 18680
Bibliothèque Sèvres, Paris 08215
Bibliothèque Sonore Romande (BSR), Lausanne 27392
Bibliothèque St. Clément, Kananga 06063
Bibliothèque Thiers, Paris 08499
Bibliothèque Tony Bourg, Troisvierges 18422
Bibliothèque Trocadéro, Paris 09012
Bibliothèque Ukrainienne Symon Petlura (BUSP), Paris 08500
Bibliothèque universitaire de Caen, Caen 07678
Bibliothèque universitaire de la Défense, Brussel 02297
Bibliothèque Universitaire de Santé, Clermont-Ferrand 07684
Bibliothèque Valeyre, Paris 09013
Bibliothèque Vandamme, Paris 09014
Bibliothèque Vaugirard, Paris 09015
Bibliothèque Ville de Mont-Royal, Mont Royal 04689
Bibliothèque-discothèque Elsa-Triolet et Louis-Aragon, Argenteuil 08739
Bibliothèque-Discothèque Georges-Duhamel, Mantes-la-Jolie 08942
Bibliothèque-Discothèque municipale, Le Creusot 08908
Bibliothèque-Discothèque municipale, Le Plessis-Robinson 08916
Bibliothèque-Discothèque municipale Jacques Prévert, Colombes 08834
Bibliothèque-Ludothèque Pestalozzi, Neuchâtel 27416
Bibliothèque-Médiathèque, Evreux 08864
Bibliothèque-Médiathèque, Montbéliard 08962
Bibliothèque/médiathèque d'Ivry-sur-Seine, Ivry-sur-Seine 08893
Bibliothèque-médiathèque "Hergé", Etterbeek 02758
Bibliothèque-musée de l'Opéra, Paris 08501
Bibliothèques de Caen, Caen 08790
Bibliothèques de la Ville de Paris – Bureau des bibliothèques, de la lecture publique et du multimedia, Paris 09016
Bibliothèques – Médiathèques de Metz, Metz 08955
Bibliothèques publiques, Tunis 27842
Bibliothèques publiques de Laval, Laval 04676
Bibliotheques-discotheques de la cotecom, Verdun 09132
B.I.C.C. – College Library, Bradford 29453
Bicherbus, Diekirch 18423
Bicknell-Vigo Township Public Library, Bicknell 38200
Bicol Christian College of Medicine, Legazpi City 20746
Bicol University, Legazpi City 20706
Bidhan Chandra Krishi Viswa Vidyalaya, Mohanpur 13699
BIDOC De Bron, Harelbeke 02662
Bidyapati Janahit Library, Mahottari 19045
Bienville Parish Library, Arcadia 37979
Bifröst University, Borgarnes 13555
Big Bend Community College, Moses Lake 33560
Big Horn County Library, Basin 38097
Big Rapids Community Library, Big Rapids 38202
Bigelow Free Public Library, Clinton 38604
Bihar Research Society, Patna 14177
Bihar Secretariat Library, Patna 13896
Bihar Veterinary College, Patna 13834
Bihor District Library 'Gheorghe Sincai', Oradea 22095
Bijie Teachers' College, Bijie 05062
Bila Tserkva Agricultural University, Bila Tserkva 27953
Bila Tserkva City Centralized Library System, Main Library, Bila Tserkva 28500
Bildagentur für Kunst, Kultur und Geschicht, Berlin 11497
Bilderbuchmuseum der Stadt Troisdorf, Troisdorf 12554
Bildungs- und Wissenschaftszentrum der Bundesfinanzverwaltung, Münster 11053
Bildungs- und Wissenschaftszentrum der Bundesfinanzverwaltung, Sigmaringen 11099
Bildungs- und Wissenschaftszentrum der Bundesfinanzverwaltung, Werder (Havel) 11116
Bildungswerk der Evangelischen Kirche in Berlin-Brandenburg, Berlin 11496
Bildungszentrum der Thüringer Steuerverwaltung, Gotha 10730
Bildungszentrum und Archiv zur Frauengeschichte Baden-Württembergs e.V. (BAF), Tübingen 12556
Bildungszentrum Zofingen, Zofingen 27228
bildwechsel, Hamburg 11972
Bilkent Universitesi Kütüphanesi, Ankara 27847
Bilkent University, Ankara 27847
Bill Robertson Library, Dunedin 19776
Billerica Public Library, Billerica 38205
Bilotserkivska miska TsBS, Tsentralna biblioteka, Bila Tserkva 28500
Bilotserkivsky derzhavny agrarny universytet, Bila Tserkva 27953
Bilozerka Regional Centralized Library System, Main Library, Bilozerka 28501
Bilozerkska raionna TsBS, Tsentralna biblioteka, Bilozerka 28501
Bingham McCutchen, San Francisco 36224
Bingham McCutchen LLP, Boston 35517
Bingham McHale LLP, Indianapolis 35774
Binzhou Medical College, Binzhou 05063
Binzhou Teachers' College, Binzhou 05064
Bioforsk, Plant Health and Plant Protection Division, Ås 20208
Bioforsk Plantehelse, Ås 20208
Biogal Gyógyszergyár, Debrecen 13360
BIOGAL Pharmaceutical Works Co, Inc, Debrecen 13360
Biola University, La Mirada 31536
Biological Institute of Inland Waters, Tolyatti 23993
Biological Research Institute, Cluj-Napoca 22022
Bioorganic Chemistry Institute, Minsk 02011
Bio-Rad Laboratories, Inc, Philadelphia 36090
BIOS – Bermuda Institute of Ocean Sciences, Saint Georges 02912
Birchard Public Library of Sandusky County, Fremont 39162
BIREME – Centro Latino-Americano de Informação em Ciências da Saúde, São Paulo 03398
Birkbeck College, London 29233
Birkenhead Public Library, Birkenhead 19860
Birla Industrial and Technological Museum, Kolkata 14030
Birla Institute of Technology and Science, Pilani 13739
Birlinghovener Informationsdienste Fraunhofer, Sankt Augustin 12460
Birmingham and Midland Institute, Birmingham 29610
Birmingham Central Library, Birmingham 29942
Birmingham City University, Birmingham 28948
– Birmingham Conservatoire Library, Birmingham 28949
Birmingham Conservatoire Library, Birmingham 28949
Birmingham Heartlands Hospital, Birmingham 29611
Birmingham Law Society, Birmingham 29612
Birmingham Museum of Art, Birmingham 36543
Birmingham Public Library, Birmingham 38208
Birmingham Public Library – Springville Road, Birmingham 38209
Birmingham-Southern College, Birmingham 30430
Biro Data Kependudukan, Jakarta 14359
Biro Organisasi Setwilda, Yogyakarta 14338
Biro Pusat Statistik, Jakarta 14360
Birobidzhan Regional Science Library named after Sholom-Aleikhem, Birobidzhan 22121
Birobidzhanskaya oblastnaya universalnaya nauchnaya biblioteka im. Sholom-Aleikhema, Birobidzhan 22121
Birsa Agricultural University, Ranchi 13751
Birštono viešoji biblioteka, Birštonas 18318
Birzeit University, Birzeit 14603
Biržų rajono savivaldybės viešoji biblioteka, Biržai 18319
Bischhöfliches Priesterseminar, Regensburg 11309
Bischöfliche Akademie des Bistums Aachen, Aachen 11143
Bischöfliche Bibliothek, Passau 11307
Bischöfliche Bibliothek Chur, Chur 27252
Bischöfliche Zentralbibliothek, Regensburg 11310
Bischöfliches Diözesanarchiv, Aachen 11144
Bischöfliches Generalvikariat, Osnabrück 11300
Bischöfliches Gymnasium, Graz 01372
Bischöfliches Hilfswerk Misereor e.V. – Midoc, Aachen 11439
Bischöfliches Priesterseminar, Fulda 09810
Bischöfliches Priesterseminar St. German, Speyer 11332
Bischöfliches Priesterseminar Trier, Trier 11347
Bischöfliches Seminar Graz, Graz 01427
Bischöfliches Seminar 'Vinzentinum', Brixen 15756
Bisdom van Breda, Breda 19302
Bishop Burton College, Beverley 28945
Bishop Grosseteste University College, Lincoln 29165
Bishop State Community College, Mobile 33545
Bishop's Library, Székesfehérvár 13346
Bishopsgate Library, London 29716
Bismarck State College, Bismarck 33174
Bismarck Veterans Memorial Public Library, Bismarck 38212
Bisschoppelijk Grootseminarie Brugge, Brugge 02452
Bisschoppelijk Seminarie, Gent 02462
Bistumsgeschichtliche Bibliothek der Diözese Augsburg, Augsburg 11151
Bitterfelder Umweltbibliothek im Hause der Wolfener Analytik GmbH, Bitterfeld-Wolfen 11613
BIUM – Bibliothèque interuniversitaire de médecine, Paris 07840
B.J. Medical College, Ahmedabad 13789
Björkängsgymnasiet, Borås 26468
Bjugn folkebibliotek, Bjugn 20315
Bjugn Public Library, Bjugn 20315
Bjuvs kommunbibliotek, Bjuv 26744
Black Hawk College, Moline 33548
Black Hills State University, Spearfish 32696
Black & Veatch, Overland Park 36077
Blackburn College, Blackburn 28959
Blackburn College, Carlinville 30675
Blackburn Royal Infirmary, Blackburn 29615
Blackpool and The Fylde College, Blackpool 28960
Blackpool and The Fylde College, Fleetwood 29111
Blacktown City Libraries, Bankstown 01063
Blackwater Regional Library, Courtland 38715
Blackwell, Sanders, Peper & Martin, Saint Louis 36195
Blackwell, Sanders, Peper, Martin LLP, Kansas City 35790
Bladen County Public Library, Elizabethtown 38955
Blaenau Gwent Country Borough Council, Brynmar 29951
Blagonravova Institute of Machine-Building of the Russian Academy of Sciences, Moskva 23283
Blair County Courthouse, Hollidaysburg 34357
Blair Memorial Library, Clawson 38587
Blake Dawson Library, Sydney 00832
Blanchard-Santa Paula Public Library District, Santa Paula 41226
Blanche Knopf Central Library, Recife 03322
Blanchester Public Library, Blanchester 38215
Bland Correctional Center, Bland 33990
Blank Rome Comisky & McCauley LLP Library, New York 35959
Blank Rome LLP, Philadelphia 36091
Blantyre Teachers' College, Limbe 18472
Blauvelt Free Library, Blauvelt 38216
Blekinge hospital library, Karlskrona 26599
Blekingesjukhuset, Karlshamn 26597
Blekingesjukhusets bibliotek, Karlskrona 26599
BLIK – Græsted Bibliotek, Græsted 06868
Blindiana Library, Pretoria 25157
Blinn College, Brenham 33186
Bloemfontein City Libraries, Bloemfontein 25187
Bloemfontein Regional Library / Bloemfonteinse Streekbiblioteek, Bloemfontein 25188
Bloemfonteinse Streekbiblioteek, Bloemfontein 25188
O. Blok Library, Kyiv 28660
Bloomfield College, Bloomfield 30443
Bloomfield Public Library, Bloomfield 38217
Bloomingdale Public Library, Bloomingdale 38219
Bloomsburg University of Pennsylvania, Bloomsburg 30464
Bloomsbury Healthcare Library, London 29717
Blount County Public Library, Maryville 40090
Blue Cloud Abbey, Marvin 35276
Blue Earth County Library Services, Mankato 40036
Blue Grass Regional Library Center, Columbia 38644
Blue Mountain College, Blue Mountain 30465
Blue Mountain Community College, Pendleton 33628
Blue Mountains City, Springwood 01160
Blue Ridge Community College, Flat Rock 33338
Blue Ridge Community College, Weyers Cave 33864
Blue Ridge Regional Library, Martinsville 40084
Bluefield College, Bluefield 30466
Bluefield State College, Bluefield 30467
Bluescope Steel Research, Port Kembla 00987
Bluffton University, Bluffton 30468
BMC Hungarian Music Information Center, Budapest 13371
bmi, bibliothèque multimédia intercommunale Epina-Golbey, Epinal 08862
BMT Group Limited, Teddington 29584
BNA, Inc, Arlington 33932
Bo Teachers' Training College, Bo 24564
Board of Audit, Tokyo 17306
Board of Governors of the Federal Reserve System – Law Library, Washington 34979
Board of Governors of the Federal Reserve System – Research Library, Washington 34980
Boaz Public Library, Boaz 38231
Bob Jones University, Greenville 31258
Bobrinets Regional Centralized Library System, Main Library, Bobrinets 28502
Bobrinetska raionna TsBS, Tsentralna biblioteka, Bobrinets 28502
Bobruysk Central City Children's Library 'Gaidar', Bobruysk 02106
Bobruysk Central City Library 'Gorky', Bobruysk 02105
Bobruysk City Children's Library 'Krupskaya', Bobruysk 02102
Bobruysk City Children's Library no 5, Bobruysk 02103
Bobruysk City Children's Library no 6 'Ostrovski', Bobruysk 02104
Bobruysk City Library no 3, Bobruysk 02100
Bobruysk Machine Building Plant, Bobruysk 01831
Bobruysk Timber Industrial Corporation – FANDOKA, Bobruysk 01830
Bobruysk Timber Plant, Main Trade Union Library, Bobruysk 02107
Bobruysk Tyre Production Industrial Corporation, Bobruysk 01833
Bobruyskaya gorodskaya bibliyateka im. Gogolya, Bobruysk 02099
Bobruyskaya gorodskaya bibliyateka no 3, Bobruysk 02100
Bobruyskaya gorodskaya bibliyateka no 7 im. Shogentsukova, Bobruysk 02101
Bobruyskaya gorodskaya detskaya bibliyateka im. Krupskoi, Bobruysk 02102
Bobruyskaya gorodskaya detskaya bibliyateka no 5, Bobruysk 02103
Bobruyskaya gorodskaya detskaya bibliyateka no 6 im. Ostrovskogo, Bobruysk 02104
Bobruyskaya tsentralnaya gorodskaya bibliyateka im. Gorkogo, Bobruysk 02105
Bobruyskaya tsentralnaya gorodskaya detskaya bibliyateka im. Gaidara, Bobruysk 02106
Bobruyskdrev – Golovnoe predpriyatie FANDOKA, Bobruysk 01830
Bobruyski mashinostroitelny zavod, Bobruysk 01831
Boc Group, Inc, Murray Hill 35928
Boca Raton Public Library, Boca Raton 38232
Bocconi University, Milano 15103
BOCES – Putnam-Northern Westchester, Yorktown Heights 37866
Bochum University of Applied Sciences, Bochum 09463
Bochumer Zentrum für Stadtgeschichte, Bochum 11619
Bodens stadsbibliotek, Boden 26745
Bodenseebibliothek, Friedrichshafen 11904
Bodleian Library, Oxford 29305
Bodman, LLP, Detroit 35671
Bodø Bibliotek, Bodø 20316
Bodø University College, Bodø 20084
Boehringer Ingelheim Pharma KG, Biberach an der Riß 11374
Boehringer Ingelheim Pharmaceuticals, Inc, Ridgefield 36178
Boeing Co, Huntington Beach 36948
Boeing Co, Saint Louis 36196
Boeing Co, Seattle 36261
The Boeing Company, Anaheim 35445
Boeing North American Inc, Canoga Park 35551
Boeing Satellite System S24 Library, Los Angeles 35828
Boeing Satellite Systems, El Segundo 35690
Boekman Foundation Library, Amsterdam 19351
Boekmanstichting-Bibliotheek, Amsterdam 19351
Boerenbond, Leuven 02670
Bogasky Institute of Physical Chemistry, Odesa 28436
Boğaziçi Üniversitesi, Istanbul 27854
'Bogdan Ogrizović' Knjižnica i Čitaonica, Zagreb 06262
Bogdan Ogrizovič Library and Reading Room, Zagreb 06262

Bogdanovich Minsk City Library no 8, Minsk 02166
Bogdanovich Molodechno Central City Library, Molodechno 02208
Bogodukhiv Regional Centralized Library System, Main Library, Bogodukhiv 28503
Bogodukhivska raionna TsBS, Tsentralna biblioteka, Bogodukhiv 28503
Bogolyubov Institue for Theoretical Physics, Kyiv 28350
Bogomoletz Institute of Physiology, Kyiv 28303
Bogor Agricultural University, Bogor 14247
Bogoslovsk Aluminium Plant, Krasnoturinsk 22814
Bogoslovski alyuminievoi zavod, Krasnoturinsk 22814
Boguslav Regional Centralized Library System, Main Library, Boguslav 28504
Boguslavska raionna TsBS, Tsentralna biblioteka, Boguslav 28504
Bohemian Club Library, San Francisco 37535
Böhler Edelstahl GmbH, Kapfenberg 01500
Bohuslav Martinů Foundation, Praha 06501
Boise Bible College, Boise 35142
Boise Public Library, Boise 38235
Boise State University, Boise 30474
Bókasafn Hafnarfjarðar, Hafnarfjörður 13579
Bókasafn Kópavogs, Kópavogur 13581
Bókasafn Norræna hússins, Reykjavík 13569
Bókasafn Reykjanesbæjar, Reykjanesbær 13582
Bókasafn Vestmannaeyja, Vestmannaeyjar 13585
Boksburg Public Library, Boksburg 25191
Boliden Mineral AB, Skelleftehamn 26549
Bolivar County Library System – Robinson-Carpenter Memorial Library, Cleveland 38595
Bolivian National Academy of Sciences, La Paz 02934
Bollinger County Library, Marble Hill 40048
Bollnäs bibliotek, Bollnäs 26746
Bolnitsa no 15, Sankt-Peterburg 23748
Bolnitsa no 30 im. S.P. Botkina, Sankt-Peterburg 23749
Bolnitsa no 31, Sankt-Peterburg 23750
Bolsa de Cereales, Buenos Aires 00244
Bolsa Oficial de Comercio de Madrid, Madrid 25861
Bolshevik Scientific and Industrial Corporation, Main Library, Kyiv 28678
Bolton Central Library, Bolton 29943
The Bolton Library, Cashel 14518
Bomana Police Training College, Boroko 20571
Bomann-Museum, Celle 11706
Bombardier Aerospace, Dorval 04252
Bombardier Transportation GmbH, Hennigsdorf 11407
Bombay Natural History Society, Mumbai 14082
Bombay Textile Research Association, Mumbai 14083
Bombay Veterinary College, Mumbai 13819
Bømlo folkebibliotek, Bremnes 20320
Bonch-Bruevich Institute of Electrical Engineering, Sankt-Peterburg 22514
Bond Beter Leefmilieu, Brussel 02573
Bond, Schoeneck & King, PLLC, Syracuse 36291
Bond University, Robina 00504
– Law Library, Robina 00505
Bonn-Rhein-Sieg University of Applied Sciences, Sankt Augustin 10579
Book Development Centre, Bangkok 27776
Book Fund Department of Jasnaya Polyana Country Estate Museum of L.N. Tolstoy, Yasnaya Polyana 24091
Book Palace Publishing House, Moskva 23481
Boone County Library, Harrison 39373
Boone County Public Library District, Burlington 38393
Boone-Madison Public Library, Madison 40000
Boonslick Regional Library, Sedalia 41272
Boonville-Warrick County Public Library, Boonville 38241
Booth Museum of Natural History, Brighton 29619
Bophuthatswana Library Service, Mmabatho 42357
Borås hospital, Borås 26561
Borås stadsbibliotek, Borås 26747
Bord na Mona, Droichead Nua 14519
Börde-Museum Burg Ummendorf, Ummendorf 12570
Borden Ladner Gervais LLP Library, Toronto 04278
Borden Ladner Gervais LLP Library, Vancouver 04299
Borden Ladner Gervais LLP-Montreal, Montreal 04266
Boreham Library, Fort Smith 33348
Boreskov Institute of Catalysis, Novosibirsk 23642
Borgå domkapitels bibliotek, Borgå 07408
Borgå gymnasiums bibliotek, Borgå 07383
Borgå stadsbibliotek, Porvoo 07601
Borgarbókasafn Reykjavíkur, Reykjavík 13583
Borgarfjördur Municipal Library, Borgarnes 13578
Borge bibliotek, Sellebakk 20396
Borgess Library, Kalamazoo 36982
Boricua College, Brooklyn 30566
Boricua College, New York 32029
Borislav Regional Centralized Library System, Main Library, Borislav 28505
Borislavska raionna TsBS, Tsentralna biblioteka, Borislav 28505
Borisoglebsk State Pedagogical Institute, Borisoglebsk 22221
Borisoglebski gosudarstvenny pedagogicheski institut, Borisoglebsk 22221
Borisov Central Children's City Library, Borisov 02110
Borisov Central City Library, Borisov 02109
Borisov City Library no 1, Borisov 02108
Borisov Power Plant, Borisov 01836
Borisov Timber Industrial Corporation, Borisov 01837
Borisovskaya gorodskaya bibliyateka no 1, Borisov 02108
Borisovskaya tsentralnaya gorodskaya bibliyateka, Borisov 02109
Borisovskaya tsentralnaya gorodskaya detskaya bibliyateka, Borisov 02110
Borisovski zavod agregatov, Borisov 01836
Borivska raionna TsBS, Tsentralna biblioteka, Borova 28507
Borlänge bibliotek, Borlänge 26749
Bornholms Centralbibliotek, Rønne 06920
Borno State Library Board, Maiduguri 20060
Borochov Library, Haifa 14774
Borodino State Museum-Preserve of Military History, Borodino 23007
Borodyanka Regional Centralized Library System, Main Library, Borodyanka 28506
Borodyanska raionna TsBS, Tsentralna biblioteka, Borodyanka 28506
City of Boroondara, Camberwell 01078
Borough of Manhattan Community College, New York 33577
Borova Regional Centralized Library System, Main Library, Borova 28507
Borregaard Industries Ltd, Sarpsborg 20207
BORSODCHEM RT, Kazincbarcika 13361
Bose Institute, Kolkata 14031
Bose McKinney & Evans LLP, Indianapolis 35775
Bosler Free Library, Carlisle 38453
Bossier Parish Central Library, Bossier City 38243
Bossier Parish Community College, Bossier City 33180
Boston Architectural Center Library, Boston 36558
Boston Athenaeum, Boston 36559
Boston Baptist College Library, Boston 30479
Boston Children's Museum, Boston 36560
Boston College Libraries, Chestnut Hill 30746
– Bapst Art Library, Chestnut Hill 30747
– Catherine B. O'Connor Library, Weston 30748
– Educational Resource Center, Chestnut Hill 30749
– Graduate School of Social Work Library, Chestnut Hill 30750
– John J. Burns Library of Rare Books & Special Collections, Chestnut Hill 30751
– Law Library, Newton Centre 30752
– Thomas P. O'Neill Jr Library (Central Library), Chestnut Hill 30753
The Boston Conservatory, Boston 30480
Boston Medical Center, Boston 36561
Boston Public Library, Boston 38244
Boston Public Library – Brighton Branch, Brighton 38284
Boston Public Library – Codman Square, Dorchester 38852
Boston Public Library – Dudley Literacy Center, Roxbury 41078
Boston Public Library – Honan-Allston Branch, Allston 37919
Boston Public Library – Hyde Park Branch, Hyde Park 39551
Boston Public Library – South End, Boston 38245
Boston Public Library – West Roxbury Branch, West Roxbury 41839
Boston University, Boston
– African Studies Library, Boston 30481
– Alumni Medical Library, Boston 30482
– Frederick S. Pardee Management Library, Boston 30483
– Mugar Memorial Library, Boston 30484
– Music Library, Boston 30485
– Pappas Law Library, Boston 30486
– Pickering Educational Resources Library, Boston 30487
– School of Theology Library, Boston 30488
– Science & Engineering Library, Boston 30489
Boston University – Stone Science Library, Boston 30490
Botanic Gardens of Adelaide and State Herbarium, Adelaide 00847
Botanical Garden, Bishkek 18144
Botanical Garden, Donetsk 28211
Botanical Garden, Machindzhauri 09216
Botanical Garden, Sukhumi 09219
Botanical Garden, Tbilisi 09226
Botanical Garden, Toshkent 42128
Botanical Garden, Warszawa 21328
Botanical Garden and Museum, Oslo 20242
Botanical Garden of Babeş-Bolyai University, Cluj-Napoca 22020
Botanical Gardens, Faculty of Science, University of Tokyo, Tokyo 17765
Botanical Institute and Garden of the University of Belgrade, Beograd 24502
Botanical Research Institute of Texas Library, Fort Worth 36868
Botanical Society of Japan, Tokyo 17733
Botanical Survey of India, Kolkata 14032
Botanicheski institut im. V.L. Komarova RAN, Sankt-Peterburg 23751
Botanichny sad, Donetsk 28211
Botanički Zavod i Bašte, Beograd 24502
Botanický ústav – Slovenská Akadémia Vied, Bratislava 24684
Botanikos institutas, Vilnius 18301
Botanische Staatssammlung, München 12288
Botanischer Garten, Hamburg 09929
Botanischer Garten, Münster 10481
Botanischer Garten Berlin-Dahlem, Berlin 09412
Botanisches Museum Berlin-Dahlem, Berlin 09412
Botanisk hage og museum, Oslo 20242
City of Botany Bay, Maroubra 01120
Botetourt County Library, Roanoke 41020
Botkin Hospital no 30, Sankt-Peterburg 23749
Botoşani District Museum, Botoşani 21961
Botswana College of Agriculture, Gaborone 02965
Botswana Institute of Administration and Commerce, Gaborone 02975
Botswana National Archives and Records Services, Gaborone 02976
Botswana National Assembly, Gaborone 02969
Botswana National Library Service, Gaborone 02964
Botswana Technology Centre, Gaborone 02973
Boulder City Library, Boulder City 38247
Boulder Public Library, Boulder 38246
Boult, Cummings, Conners & Berry, Plc, Nashville 35934
Bound Brook Memorial Library, Bound Brook 38248
Bournemouth and Poole College, Poole 29487
Bournemouth University, Poole 29376
Bouwfonds Nederlandse Gemeenten, Hoevelaken 19276
Boverket, Karlskrona 26494
Bowditch & Dewey, Worcester 36435
Bowdoin College, Brunswick 30581
Bowes Museum, Barnard Castle 29601
Bowie State University, Bowie 30527
Bowling Green State University, Bowling Green 30528
– Center for Archival Collections, Bowling Green 30529
– Curriculum Resource Center Library, Bowling Green 30530
– Frank Ogg Science & Health Library, Bowling Green 30531
– Government Documents, Bowling Green 30532
– Historical Collections of the Great Lakes, Bowling Green 30533
– Music Library & Sound Recordings Archives, Bowling Green 30534
– Ray & Pat Browne Popular Culture Library, Bowling Green 30535
Bowling Green State University, Huron 31379
Bowman Gilfillan Hayman Godfrey Inc, Johannesburg 25132
Box Hill Institute, Box Hill 00874
Box Hill Institute of TAFE, Box Hill 00391
Boxford Town Library, Boxford 38253
Boyd County Public Library, Ashland 38002
Boyden Library, Foxborough 39139
Boyle County Public Library, Danville 38780
Boynton Beach City Library, Boynton Beach 38255
Bozeman Public Library, Bozeman 38256
Božena Němcová Library Domažlice, Domažlice 06565
BP, Houston 35747
BP Information Services, Naperville 35931
bpk – Bildagentur für Kunst, Kultur und Geschichte, Berlin 11497
BPO Belarusrezinotekhnika, Bobruysk 01832
Brabant Pers BV, Best 19326
Brabants Conservatorium, Tilburg 19202
Brabants Historisch Informatie Centrum, 's-Hertogenbosch 19460
Bracewell & Giuliani LLP, Washington 36335
Bradford Area Public Library, Bradford 38258
Bradford County Library System, Troy 41592
Bradford Memorial Library, El Dorado 38942
Bradford Metropolitan Council, Bradford 29945
Bradley, Arant, Rose & White, Birmingham 35507
Bradley University, Peoria 32243
Bradshaw, Fowler, Proctor & Fairgrave, Des Moines 35669
Brahms-Institut an der Musikhochschule Lübeck, Lübeck 12211
Braille Institute Library Services, Los Angeles 37045
Braille Institute Library Services, Santa Barbara 37564
Braille Institute Library Services – Desert Center, Rancho Mirage 37429
Braille Institute Library Services – Orange County Center, Anaheim 36464
Braille Institute Library Services – San Diego Center, San Diego 37524
Brain Research Institute of the Medical Academy, Moskva 23414
Brainerd Public Library, Brainerd 38260
Brakpan Transitional Local Council, Brakpan 25192
Brampton Library, Brampton 04583
Bramson Ort College, Forest Hills 31153
Branch Centre for Research and Development of Electric Machines, Katowice 21114
Branch Institute of Solid State Physics and Semiconductors, Vitebsk 02081
Branchville Training Center Library, Tell City 37662
Brandeis University, Waltham 32949
Brandeis-Bardin Institute, Brandeis 36585
Brandenburg University of Applied Sciences, Brandenburg an der Havel 09590
Brandenburgische Technische Universität Cottbus, Cottbus 09621
Brandenburgisches Landesamt für Denkmalpflege und Archäologisches Landesmuseum, Teilbibliothek Archäologie, Zossen 11139
Brandenburgisches Landesamt für Denkmalpflege und Archäologisches Landesmuseum, Zentralabteilung, Teilbibliothek Denkmalpflege, Zossen 11140
Brandenburgisches Landeshauptarchiv, Potsdam 12403
Brandenburgisches Oberlandesgericht, Brandenburg an der Havel 10835
Brandon Township Public Library, Ortonville 40656
Brandon University, Brandon 03677
Brandywine Hundred Branch, Wilmington 41911
Brantford Public Library, Brantford 04586
Branżowy Ośrodek Badawczo-Rozwojowy Maszyn Elektrycznych, Katowice 21114
Brasenose College, Oxford 29300
Brasilian Metallurgy, Materials and Mining Association, São Paulo 03394
Braslav Regional Central Library, Braslav 02111
Braslavskaya tsentralnaya raionnaya bibliyateka, Braslav 02111
Braşov History Museum, Braşov 21964
Braswell Memorial Public Library, Rocky Mount 41049
Bratislava Institute of Geodesy and Cartography, Bratislava 24690
Braunschweig University, Braunschweig 09592
– Chemistry Library, Braunschweig 09593
Braunschweigisches Landesmuseum, Braunschweig 11658
Brawijaya University, Malang 14274
– Faculty of Engineering, Malang 14275
Brawijaya University, Malang 14272
Brawley Public Library, Brawley 38265
Bray Library, Bray 14564
Brazilian Centre for Physics Research, Rio de Janeiro 03335
Brazilian Historical and Geographical Institute, Rio de Janeiro 03358
Brazilian Institute for the Environment and Renewable Natural Resources, Brasília 03281

Brazilian Institute of Geography and Statistics (IBGE), Rio de Janeiro 03350
Brazilian Press Association, Rio de Janeiro 03331
Brazoria County Law Library, Angleton 33924
Brazoria County Library System, Angleton 37960
Brazoria County Library System – Alvin Branch, Alvin 37934
Brazoria County Library System – Angleton Branch, Angleton 37961
Brazoria County Library System – Lake Jackson Branch, Lake Jackson 39777
Brazoria County Library System – Pearland Branch, Pearland 40740
Brazosport College, Lake Jackson 33467
Breathitt County Public Library, Jackson 39599
Breckinridge County Public Library, Hardinsburg 39367
Břeclav Public Library, Břeclav 06555
Breda's Museum, Breda 19402
C. E. Brehm Memorial Public Library District, Mount Vernon 40323
Bremanger folkebibliotek, Svelgen 20406
Bremer Institute of TAFE Bundamba, Booval 00560
Bremer Landesmuseum für Kunst- und Kulturgeschichte, Bremen 11674
Bremer Theater, Bremen 11672
Bremische Bürgerschaft, Bremen 10839
Brenau University, Gainesville 31189
Brent Library Service, Wembley 30094
Brenthurst Library, Houghton 25127
Brentwood Library & Center for Fine Arts, Brentwood 38269
Brentwood Public Library, Brentwood 38270
Brescia College Library, London 03742
Brescia University, Owensboro 32218
Brest Agricultural Machine Building Plant, Brest 01844
Brest Central City Children's Library, Brest 02117
Brest Central City Library 'Pushkin', Brest 02116
Brest Central Railway Station, Main Railway Technical Library, Brest 01948
Brest City Library 'Ya. Kupala', Brest 02112
Brest District State Archive, Brest 01945
Brest East Railway Station, Railway Technical Library, Brest 01949
Brest Hosiery Factory, Brest 01840
Brest Polytechnic Institute, Brest 01786
Brest Region Agricultural Research and Experimental Station, Pruzhany 02073
Brest Region Children's Library, Brest 02115
Brest Regional Advanced Teacher Training College, Brest 01788
Brest Regional Library, Brest 02114
Brest Regional Medical Library, Brest 01944
Brest State University, Brest 01787
Brestskaya gorodskaya bibliyateka im. Ya. Kupaly, Brest 02112
Brestskaya gorodskaya bibliyateka Yunost, Brest 02113
Brestskaya oblastnaya bibliyateka im. Gorkogo, Brest 02114
Brestskaya oblastnaya detskaya bibliyateka, Brest 02115
Brestskaya oblastnaya meditsinskaya bibliyateka, Brest 01944
Brestskaya oblastnaya sel-skokhozyaistvennaya opytnaya stantsiya, Pruzhany 02073
Brestskaya tsentralnaya gorodskaya bibliyateka im. Pushkina, Brest 02116
Brestskaya tsentralnaya gorodskaya detskaya bibliyateka im. Pushkina, Brest 02117
Brestski chulochny kombinat, Brest 01840
Brestski gosudarstvenny universitet, Brest 01787
Brestski oblastnoi institut usovershenstvovaniya uchitelei, Brest 01788
Brevard College, Brevard 30544
Brevard Community College, Melbourne 33526
Brevard County Library System, Cocoa 38623
Brewer Public Library, Richland Center 40976
Brewton-Parker College, Mount Vernon 33564
Brežice Public Library, Brežice 24954
Brian O'Malley Central Library and Arts Centre, Rotherham 30067
Briar Cliff University, Sioux City 32675
Bridgend Library and Information Service, Bridgend 29946
Bridgeport Public Library, Bridgeport 38274
Bridgeport Public Library, Bridgeport 38275
Bridgeport Public Library – North, Bridgeport 38276
Bridgestone Co, Ltd, Technical Center, Kodaira 17377
Bridgestone/Firestone Research LLC, Akron 35431
Bridgeton Free Public Library, Bridgeton 38277
Bridgeview Public Library, Bridgeview 38280
Bridgewater College, Bridgewater 30546
Bridgewater Public Library, Bridgewater 38281
Bridgewater State College, Bridgewater 30547
Bridgton Academy Library, North Bridgton 37291
Bridgwater College Library, Bridgwater 29454
Briercrest Family of Schools, Caronport 04203
Briggs Lawrence County Public Library, Ironton 39580
Briggs & Morgan, Saint Paul 36208
Brigham City Library, Brigham City 38283
Brigham Young University, Provo
– Harold B. Lee Library, Provo 32391
– Howard W. Hunter Law Library, Provo 32392
Brigham Young University – Hawaii, Laie 31542
Brigham Young University-Idaho, Rexburg 32430
Brighton Central Library, Brighton 29948
Brighton District Library, Brighton 38285
Brighton Memorial Library, Rochester 41026
Brigitte Reimann Stadtbibliothek, Hoyerswerda 12792
Brimbank Libraries, Sunshine 01163
Brisbane Catholic Education Centre, Dutton Park 00910
Brisbane City Council, Brisbane 01072
Brisbane North Institute of TAFE, Bracken Ridge 00561
Brisbane North Institute of TAFE Learning HUB, Eagle Farm BC 00568
Bristol City Council, Bristol 29949
Bristol Community College, Fall River 33330
Bristol Law Library, Taunton 37661
Bristol Public Library, Bristol 38288
Bristol Public Library, Bristol 38289
Bristol Public Library, Bristolville 38291
Bristol-Myers Products, Hillside 35742
Bristol-Myers Squibb Co, Wallingford 36323
Bristol-Myers Squibb Pharmaceutical Research Institute Library, Princeton 37412
Britannia Royal Naval College, Dartmouth 29079
British Antarctic Survey, Cambridge 29624
British Architectural Library, London 29718
British Broadcasting Corporation, London 29719
British Columbia Archives, Victoria 04548
British Columbia Courthouse Library Society, Prince George 04457
British Columbia Department of the Attorney General Library, Vancouver 04173
British Columbia Genealogical Society, Richmond 04473
British Columbia Institute of Technology Library, Burnaby 03979
British Columbia Legislative Library, Victoria 04178
British Columbia Ministry of Employment & Investment, Victoria 04179
British Columbia Ministry of Energy & Mines, Victoria 04180
British Columbia Ministry of Health & Audiovisual Library, Victoria 04181
British Columbia Ministry of Sustainable Resource Management, Victoria 04182
British Columbia Research Inc Library, Vancouver 04536
British Columbia Teachers' Federation, Vancouver 04537
British Council, Abuja 20007
British Council, Accra 13042
British Council, Addis Ababa 07280
British Council, Ahmedabad 13919
British Council, Amman 17883
British Council, Ankara 27872
British Council, Asmara 07215
British Council, Athinai 13100
British Council, Bangalore 13928
British Council, Beograd 24503
British Council, Brussel 02574
British Council, București 21973
British Council, Cairo 07158
British Council, Chandigarh 13950
British Council, Chennai 13957
British Council, Chittagong 01754
British Council, Colombo 26305
British Council, Dar es Salaam 27673
British Council, Dhaka 01763
British Council, Dubai 28909
British Council, Freetown 24580
British Council, Gabarone 02974
British Council, Hong Kong 05869
British Council, Hyderabad 13996
British Council, Islamabad 20477
British Council, Istanbul 27884
British Council, Jakarta 14361
British Council, Kandy 26323
British Council, Karachi 20493
British Council, Kathmandu 19032
British Council, Khartoum 26355
British Council, Kolkata 14033
British Council, Kota Kinabalu 18576
British Council, Kuala Lumpur 18581
British Council, Kumasi 13059
British Council, Lahore 20509
British Council, Lefkosia (Nicosia) 06334
British Council, Lilongwe 18487
British Council, Lisboa 21775
British Council, Madrid 25987
British Council, Manama 01741
British Council, México 18838
British Council, Mombasa 18031
British Council, Moskva 23170
British Council, Mumbai 14084
British Council, Muscat 20423
British Council, Nairobi 18033
British Council, New Delhi 14116
British Council, Penang 18612
British Council, Peshawar 20522
British Council, Porto 21823
British Council, Providencia 04921
British Council, Pune 14188
British Council, Rabat 18955
British Council, Ramat Gan 14750
British Council, Riyadh 24400
British Council, Rose Hill 18701
British Council, Safat 18130
British Council, San'a 42321
British Council, Santafé de Bogotá 06024
British Council, Sofiya 03526
British Council, Tunis 27829
British Council, Vilnius 18302
British Dental Association, London 29720
British Film Institute, London 29721
British Gas, Reading 29884
British Geological Survey, Edinburgh 29652
British Geological Survey, London 29722
British Geological Survey, Nottingham 29862
British Institute Library, Sevilla 26146
British Institute of Archaeology at Ankara, Ankara 27873
British Institute of Florence, Firenze 16256
British Institute of Persian Studies, Tehran 14425
The British Library, London 28913
The British Library – Asia, Pacific and Africa Collections, London 28914
The British Library – Humanities Reading Rooms, London 28915
The British Library – Librarianship & Information Sciences Service (LIS), London 28916
The British Library – Manuscripts Collections, London 28917
The British Library – Map Collections, London 28918
The British Library – Music Collections, London 28919
The British Library – Newspaper Library, London 28920
British Library of Political and Economic Science, London 29202
The British Library – Science Technology & Business Collections Development Policy, London 28921
The British Library – Sound Archive Information Service, London 28922
British Medical Association, London 29732
The British Museum, London 29733
The British Museum – Centre of Anthropology, London 29734
British Orthodox Church, London 29550
British School at Athens, Athinai 13101
British School of Archaeology in Jerusalem, Jerusalem 14617
British Senegalese Institute, Dakar 24433
British Standards Institution, London 29735
Britų tarybos informacijos centras, Vilnius 18302
Brno Municipal Museum, Brno 06433
Brno University of Technology, Brno 06365
Broadcasting Dept, Film Library, Kuala Lumpur 18588
Broadgreen Hospital, Liverpool 29702
Broadview Public Library District, Broadview 38292
Brobeck, Phleger & Harrison Library, Los Angeles 35829
Brobeck, Phleger & Harrison Library, San Francisco 36225
Brock University, St. Catharines 03864
Brockport Seymour Library, Brockport 38293
Brockton Hospital Library, Brockton 36587
Brockton Law Library, Brockton 36588
Brockton Public Library System, Brockton 38294
Brockville Public Library, Brockville 04588
Brockway Memorial Library, Miami Shores 40173
Brodi Regional Centralized Library System, Main Library, Brody 28508
Brodivska raionna TsBS, Tsentralna biblioteka, Brody 28508
Brodsky Library, Saint Louis 35363
Bródy Sándor County and City Library, Eger 13535
Bródy Sándor Megyei és Városi Könyvtár, Eger 13535
Broken Hill City Library, Broken Hill 01073
Bromley Central Library, Bromley 29950
Bromley College of Further and Higher Education, Bromley 28977
Bromölla folkbibliotek, Bromölla 26751
Brøndbyøster Bibliotek, Brøndby 06851
Brønderslev Bibliotek, Brønderslev 06852
Brönnkyrka gymnasium, Hägersten 26470
Brønnøy folkebibliotek, Brønnøysund 20321
Bronx Community College, Bronx 33191
Bronx County Historical Society, Bronx 36590
Bronxville Public Library, Bronxville 38302
Brookdale Community College, Lincroft 33488
Brookdale JDC Library, Jerusalem 14724
Brookfield Free Public Library, Brookfield 38304
The Brookfield Library, Brookfield 38305
Brookfield Public Library, Brookfield 38306
Brookhaven College, Farmers Branch 33331
Brookhaven National Laboratory, Upton 37697
Brookings Institution Library, Washington 37724
Brookings Public Library, Brookings 38307
Brooklyn Bar Association Foundation Inc Library, Brooklyn 36597
Brooklyn Botanic Garden Library, Brooklyn 36598
Brooklyn College, Brooklyn 30567
Brooklyn Correctional Institution Library, Brooklyn 34008
Brooklyn Historical Society Library, Brooklyn 36599
Brooklyn Law School Library, Brooklyn 30568
Brooklyn Museum, Brooklyn 36600
Brooklyn Museum of Art – Wilbour Library of Egyptology, Brooklyn 36601
Brooklyn Public Library, Brooklyn 38311
Brooklyn Public Library – Arlington, Brooklyn 38312
Brooklyn Public Library – Bay Ridge, Brooklyn 38313
Brooklyn Public Library – Bedford, Brooklyn 38314
Brooklyn Public Library – Borough Park, Brooklyn 38315
Brooklyn Public Library – Brighton Beach, Brooklyn 38316
Brooklyn Public Library – Brooklyn Heights, Brooklyn 38317
Brooklyn Public Library – Brower Park, Brooklyn 38318
Brooklyn Public Library – Bushwick, Brooklyn 38319
Brooklyn Public Library – Business, Brooklyn 38320
Brooklyn Public Library – Canarsie, Brooklyn 38321
Brooklyn Public Library – Carroll Gardens, Brooklyn 38322
Brooklyn Public Library Clarendon, Brooklyn 38323
Brooklyn Public Library – Clinton Hill, Brooklyn 38324
Brooklyn Public Library – Coney Island, Brooklyn 38325
Brooklyn Public Library – Cortelyou, Brooklyn 38326
Brooklyn Public Library – Crown Heights, Brooklyn 38327
Brooklyn Public Library – Cypress Hills, Brooklyn 38328
Brooklyn Public Library – DeKalb Branch, Brooklyn 38329
Brooklyn Public Library – Dyker, Brooklyn 38330
Brooklyn Public Library – East Flatbush, Brooklyn 38331
Brooklyn Public Library – Eastern Parkway, Brooklyn 38332
Brooklyn Public Library – Flatbush, Brooklyn 38333
Brooklyn Public Library – Flatlands, Brooklyn 38334
Brooklyn Public Library – Fort Hamilton, Brooklyn 38335
Brooklyn Public Library – Gerritsen Beach, Brooklyn 38336
Brooklyn Public Library – Gravesend, Brooklyn 38337
Brooklyn Public Library – Greenpoint, Brooklyn 38338
Brooklyn Public Library – Highlawn, Brooklyn 38339

Brooklyn Public Library – Homecrest, Brooklyn 38340
Brooklyn Public Library – Jamaica Bay, Brooklyn 38341
Brooklyn Public Library – Kensington, Brooklyn 38342
Brooklyn Public Library – Kings Bay, Brooklyn 38343
Brooklyn Public Library – Leonard, Brooklyn 38344
Brooklyn Public Library – Marcy, Brooklyn 38345
Brooklyn Public Library – McKinley Park, Brooklyn 38346
Brooklyn Public Library – Midwood, Brooklyn 38347
Brooklyn Public Library – Mill Basin, Brooklyn 38348
Brooklyn Public Library – New Lots, Brooklyn 38349
Brooklyn Public Library – New Utrecht, Brooklyn 38350
Brooklyn Public Library – Pacific, Brooklyn 38351
Brooklyn Public Library – Paerdegat, Brooklyn 38352
Brooklyn Public Library – Park Slope, Brooklyn 38353
Brooklyn Public Library – Red Hook, Brooklyn 38354
Brooklyn Public Library – Rugby, Brooklyn 38355
Brooklyn Public Library – Ryder, Brooklyn 38356
Brooklyn Public Library – Saratoga, Brooklyn 38357
Brooklyn Public Library – Sheepshead Bay, Brooklyn 38358
Brooklyn Public Library – Spring Creek, Brooklyn 38359
Brooklyn Public Library – Stone Avenue, Brooklyn 38360
Brooklyn Public Library – Sunset Park, Brooklyn 38361
Brooklyn Public Library – Ulmer Park, Brooklyn 38362
Brooklyn Public Library – Walt Whitman Branch, Brooklyn 38363
Brooklyn Public Library – Washington Irving Branch, Brooklyn 38364
Brooklyn Public Library – Williamsburgh, Brooklyn 38365
Brooklyn Public Library – Windsor Terrace, Brooklyn 38366
Brooklyn Technical College, Birmingham 29452
Brooks County Library, Falfurrias 39034
Brooks County Public Library, Quitman 40921
Brooks Free Library, Harwich 39390
Brooks Institute, Santa Barbara 32604
Brooks Memorial Library, Brattleboro 38264
Broome Community College, Binghamton 33172
Broome County Public Library, Binghamton 38207
Broome Developmental Disabilities Services Office Library, Binghamton 36542
Broomfield Public Library, Broomfield 38369
Brotherhood of Saint Laurence, Fitzroy 00916
The Brothers Waga Łomza Scientific Society, Łomża 21168
Brovari Regional Centralized Library System, Main Library, Brovari 28509
Brovarska raionna TsBS, Tsentralna biblioteka, Brovari 28509
Brovst Bibliotek, Brovst 06853
Broward Community College, Davie 30960
Broward County Division of Libraries, Fort Lauderdale 39104
Broward County Law Library, Fort Lauderdale 34243
Brown County Library, Green Bay 39292
Brown County Public Library, Mount Orab 40318
Brown County Public Library, Nashville 40370
Brown Deer Public Library, Brown Deer 38370
Brown McCarroll, LLP, Austin 35472
Brown, Rudnick, Berlack, Israels LLP, Boston 35518
Brown University, Providence 32382
– John Carter Brown Library, Providence 32383
– Orwig Music Library, Providence 32384
– Sciences Library, Providence 32385
Brown & Williamson Tobacco Corp, Macon 35869
Brownell Library, Essex Junction 38990
Brownsburg Public Library, Brownsburg 38372
Brownsville Public Library, Brownsville 38373
Brownwood Public Library, Brownwood 38374
Bruce County Public Library, Port Elgin 04724
Brüder-Grimm-Museum Kassel, Kassel 12097
Brumback Library, Van Wert 41664
Brunei Museum, Bandar Seri Begawan 03449
Brunel University, Uxbridge 29429
Brunnsviks Bokstuga, Ludvika 26610
Brunswick Correctional Center, Lawrenceville 34445
Brunswick County Library, Southport 41390
Brunswick Public Library Association, Brunswick 38375
Bruton Memorial Library, Plant City 40821
Bryan Cave Library, Washington 36336
Bryan Cave LLP, New York 35960
Bryan Cave LLP, Saint Louis 37483
Bryan Cave LLP, Santa Monica 36258
Bryan College, Dayton 30966
Bryan+College Station Public Library System, Bryan 38376
Bryan+College Station Public Library System – Larry J. Ringer Library, College Station 38637
Bryansk Institute of Technology, Bryansk 22222
Bryanskaya oblastnaya detskaya biblioteka, Bryansk 24117
Bryanskaya oblastnaya nauchnaya biblioteka im. F.I. Tyutcheva, Bryansk 22123
Bryanskaya oblastnaya yunosheskaya biblioteka, Bryansk 24118
Bryanski tekhnologicheski institut, Bryansk 22222
Bryanski tsentr nauchno-tekhnicheskoi informatsii, Bryansk 23010
Bryant Library, Roslyn 41070
Bryant Universtiy, Smithfield 32683
Bryn Athya College, Bryn Athyn 30582
Bryn Mawr College, Bryn Mawr 30583
– Lois & Reginald Collier Science Library, Bryn Mawr 30584
– Rhys Carpenter Library for Art, Archaeology, and Cities, Bryn Mawr 30585
BSI Knowledge Centre, London 29735
BTJ Företagsbibliotek, Lund 26612
Buchach Regional Centralized Library System, Main Library, Buchach 28510
Buchalter Nemer, Los Angeles 35830
Buchanan County Public Library, Grundy 39331
Buchanan District Library, Buchanan 38380
Buchanan Ingersoll & Rooney PC, Philadelphia 36092
Buchanan Ingersoll & Rooney PC, Pittsburgh 36128
Buchatska raionna TsBS, Tsentralna biblioteka, Buchach 28510
Bücherei der Barmherzigen Schwestern, Innsbruck 01436
Bücherei Schatzkiste im Kath. Forum für Erwachsenen- und Familienbildung, Krefeld 12161
Büchereien Wien, Wien 01677
Büchereien Wien – Hauptbücherei am Gürtel, Wien 01678
Bücherhallen Hamburg, Hamburg 12772
Buchkunst, Stiftung, Frankfurt am Main 11874
Buckham Memorial Library, Faribault 39041
Buckingham Correctional Center, Dillwyn 34149
Buckinghamshire County Council, Aylesbury 29929
Buckinghamshire New Unversity, High Wycombe 29134
Buckman Laboratories International, Inc, Memphis 35879
Bucknell University, Lewisburg 31587
Bucks County Community College, Newtown 33583
Bucks County Free Library, Doylestown 38863
Bucks County Free Library – Bensalem Branch, Bensalem 38174
Bucks County Free Library – James A. Michener Branch, Quakertown 40916
Bucks County Free Library – Levittown Branch, Levittown 39856
Bucks County Free Library – Library Center at Doylestown, Doylestown 38864
Bucks County Free Library – Pennwood, Langhorne 39799
Bucks County Free Library – Samuel Pierce Branch, Perkasie 40759
Bucks County Free Library – Yardley-Makefield Branch, Yardley 41987
Bucks County Historical Society, Doylestown 36810
Bucks County Law Library, Doylestown 34156
Bucovinian State Medical University, Chernivtsi 27958
Bud Werner Memorial Library, Steamboat Springs 41437
Budai Görög Keleti Szerb Püspökség, Szentendre 13347
Budapest Business School, Budapest 13224
Budapest Business School – College of Finance and Accountancy, Budapest 13226
Budapest Business School – Faculty of Commerce, Catering and Tourism, Budapest 13225
Budapest History Museum, Budapest 13372
Budapest Slowak-Speaking Lutheran Parish, Budapest 13327
Budapest Teacher Training College, Budapest 13230
Budapest Tech – Politechnical Institution, Budapest 13229
Budapest University of Technology and Economics, National Technical Information Centre and Library, Budapest 13227
– Faculty of Water Management, Baja 13228
Budapesti Corvinus Egyetem, Budapest 13222
Budapesti Corvinus Egyetem, Közgazgazás-tudományi Kar Könyvtár, Budapest 13223
Budapesti Gazdasági Főiskola, Budapest 13224
Budapesti Gazdasági Főiskola – Kereskedelmi és Vendéglátóipari és Idegenforfalmi Főiskola Kar, Budapest 13225
Budapesti Gazdasági Főiskola – Pénzügyi és Számviteli Főiskola, Budapest 13226
Budapesti Mûszaki és Gazdaságtudományi Egyetem, Budapest 13227
– Pollack Mihály Műszaki Főiskola, Baja 13228
Budapesti Müszaki Föiskola, Budapest 13229
Budapesti Szlovákajku Evangélikus Gyülekezet, Budapest 13327
Budapesti Tanítóképző Főiskola, Budapest 13230
Budapesti Történeti Múzeum, Budapest 13372
Buddhist and Pali University of Sri Lanka, Homagama 26275
Buddhist Institute, Phnom Penh 03630
Buddhist Library of China, Hong Kong 05870
Budivelny tekhnikum, Kyiv 28156
Budyonny Military Academy of Telecommunications, Sankt-Peterburg 22552
Buena Park Library District, Buena Park 38382
Buena Vista Correctional Complex Library, Buena Vista 34017
Buena Vista University, Storm Lake 32754
Buenos Aires City Bar Association, Buenos Aires 00249
Buenos Aires Province Ministry of the Interior, La Plata 00219
Buenos Aires Provincial Bank, Buenos Aires 00224
Buffalo Bill Historical Center, Cody 36719
Buffalo City Library, East London 25198
Buffalo & Erie County Historical Society Research Library, Buffalo 36607
Buffalo & Erie County Public Library System, Buffalo 38383
Buffalo General Health System, Buffalo 36608
Buffalo Museum of Science, Buffalo 36609
Buguruslan Radiator Plant Joint-Stock Company, Buguruslan 22781
Building Center of Japan, Tokyo 17727
Building Research Association of New Zealand, Judgeford 19834
Building Research Institute, Tsukuba 17790
Building Research Institute, Warszawa 21307
Building Structures Research Institute, Kyiv 28363
Bartlett Built Environment Library, London 29225
Bukit Merah Community Library, Singapore 24624
Bukkyo University, Kyoto 16997
Bukovinski derzhavni medichny universitet, Chernivtsi 27958
Bukowina-Institut an der Universität Augsburg, Augsburg 11458
Bulawaya Polytechnic, Bulawayo 42361
Bulawayo Public Library, Bulawayo 42381
Bulgarian Academy of Sciences, Sofiya 03452
Bulgarian Academy of Sciences – Ethnographic Institute with Museum, Sofiya 03521
Bulgarian Chamber of Commerce & Industry, Sofiya 03523
Bulgarian Geological Society, Sofiya 03525
Bulgarian National Bank, Sofiya 03497
Bulgarian Telecommunications Company, Sofiya 03578
Bulgarian Telegraph Agency, Sofiya 03524
Bulgarsko Natsionalno Radio (BNR), Sofiya 03527
Bull, Housser & Tupper Library, Vancouver 04300
Bullitt County Public Library, Shepherdsville 41306
Bullivant, Houser & Bailey, Portland 36142
Bulolo University College, Bulolo 20549
Bumazhnaya fabrika, Dobrush 01846
Bumazhnaya fabrika no 2, Sankt-Peterburg 22909
Bumazhno-lesokhimicheski zavod, Gomel 01848
Buncombe County Public Libraries, Asheville 38000
Bunda College of Agriculture, Lilongwe 18470
Bundaberg Regional Library Service, Bundaberg 01074
Bundesagentur für Arbeit – Regionaldirektion Niedersachsen-Bremen, Hannover 10930
Bundesakademie für musikalische Jugendbildung, Trossingen 12555
Bundesakademie für Wehrverwaltung und Wehrtechnik, Mannheim 10291
Bundesamt für Bauwesen und Raumordnung (BBR), Bonn 10822
Bundesamt für Bevölkerungsschutz und Katastrophenhilfe, Bonn 10819
Bundesamt für den Zivildienst, Köln 10984
Bundesamt für Eich- und Vermessungswesen, Wien 01394
Bundesamt für Gesundheit, Liebefeld 27237
Bundesamt für Güterverkehr, Köln 10985
Bundesamt für Kartographie und Geodäsie – Außenstelle Leipzig, Leipzig 10998
Bundesamt für Kartographie und Geodäsie (BKG), Frankfurt am Main 10897
Bundesamt für Migration und Flüchtlinge, Nürnberg 11067
Bundesamt für Naturschutz (BfN), Bonn 10820
Bundesamt für Seeschifffahrt und Hydrographie, Hamburg 10913
Bundesamt für Seeschifffahrt und Hydrographie (BSH), Rostock 11084
Bundesamt für Sozialversicherungen, Bern 27230
Bundesamt für Sport, Magglingen 27115
Bundesamt für Strahlenschutz, Oberschleißheim 11073
Bundesamt für Strahlenschutz, Salzgitter 11090
Bundesamt für Strahlenschutz – Dienststelle Berlin, Berlin 10791
Bundesamt für Strassen (ASTRA), Bern 27231
Bundesamt für Verbraucherschutz und Lebensmittelsicherheit (BVL), Berlin 10792
Bundesamt für Wehrtechnik und Beschaffung (BWB), Koblenz 10973
Bundesanstalt für Agrarwirtschaft, Wien 01395
Bundesanstalt für Finanzdienstleistungsaufsicht (BaFin), Bonn 10821
Bundesanstalt für Geowissenschaften und Rohstoffe (BGR), Hannover 12023
Bundesanstalt für Gewässerkunde, Koblenz 10974
Bundesanstalt für Materialforschung und -prüfung, Berlin 11492
Bundesanstalt für Straßenwesen (BASt), Bergisch Gladbach 11485
Bundesanstalt für Wasserbau (BAW), Karlsruhe 12085
Bundesarbeitsgericht, Erfurt 10886
Bundesarchiv, Koblenz 12119
Bundesarchiv – Filmarchiv, Berlin 11498
Bundesarchiv – Lastenausgleichsarchiv, Bayreuth 11478
Bundesarchiv – Militärarchiv, Freiburg 11887
Bundes-Blindenerziehungsinstitut, Wien 01590
Bundesbriefmuseum, Schwyz 27432
Bundesdenkmalamt, Wien 01396
Bundesfinanzdirektion Südost, Nürnberg 11068
Bundesfinanzdirektion West, Köln 10986
Bundesfinanzhof, München 11041
Bundesforschungs- und Ausbildungszentrum für Wald, Naturgefahren und Landschaft (BFW), Wien 01591
Bundesforschungsinstitut für Ernährung und Lebensmittel – Standort Detmold, Detmold 11739
Bundesforschungsinstitut für Ernährung und Lebensmittel – Standort Karlsruhe (Hauptsitz), Karlsruhe 12092

Bundesforschungsinstitut für Ernährung und Lebensmittel – Standort Kiel, Kiel 12114
Bundesforschungsinstitut für Ernährung und Lebensmittel – Standort Kulmbach, Kulmbach 12167
Bundesforschungsinstitut für Kulturpflanzen (JKI), Quedlinburg 12416
Bundesforschungsinstitut für Ländliche Räume, Wald und Fischerei – Fachinformationszentrum Fischerei, Hamburg 11997
Bundesforschungsinstitut für Ländliche Räume, Wald und Fischerei – Fachinformationszentrum Fischerei, Rostock 12441
Bundesforschungsinstitut für Ländliche Räume, Wald und Fischerei – Fachinformationszentrum Ländliche Räume, Braunschweig 11664
Bundesforschungsinstitut für Ländliche Räume, Wald und Fischerei – Fachinformationszentrum Wald, Hamburg 11998
Bundesforschungszentrum für Kulturpflanzen, Kleinmachnow 12118
Bundesgericht, Luzern 27239
Bundesgerichtshof, Karlsruhe 10955
Bundesinstitut für Arzneimittel und Medizinprodukte (BfArM), Bonn 11623
Bundesinstitut für Bau-, Stadt- und Raumforschung (RBSR) im Bundesamt für Bauwesen und Raumordnung (BBR), Bonn 10822
Bundesinstitut für Berufsbildung, Bonn 11624
Bundesinstitut für Erwachsenenbildung St. Wolfgang, Strobl 01583
Bundesinstitut für Infektionskrankheiten und nicht übertragbare Krankheiten, Berlin 11558
Bundesinstitut für Risikobewertung (BfR), Berlin 11499
Bundesinstitut für Sozialpädagogik, Baden 01368
Bundeskanzleramt, Berlin 10793
Bundeskartellamt, Bonn 10823
Bundeskriminalamt, Wiesbaden 11118
Bundesministerium der Finanzen, Berlin 10794
Bundesministerium der Justiz, Berlin 10795
Bundesministerium der Verteidigung, Bonn 10824
Bundesministerium des Inneren, Berlin 10796
Bundesministerium für Arbeit, Soziales und Konsumentenschutz, Wien 01397
Bundesministerium für Arbeit und Soziales, Berlin 10797
Bundesministerium für auswärtige Angelegenheiten, Wien 01393
Bundesministerium für Bildung und Forschung, Bonn 10825
Bundesministerium für Ernährung, Landwirtschaft und Verbraucherschutz, Bonn 10826
Bundesministerium für Familie, Senioren, Frauen und Jugend, Berlin 10798
Bundesministerium für Finanzen, Wien 01398
Bundesministerium für Gesundheit, Bonn 10827
Bundesministerium für Justiz, Wien 01399
Bundesministerium für Land- und Forstwirtschaft, Umwelt und Wasserwirtschaft (BMLFUW), Wien 01400
Bundesministerium für Landesverteidigung und Sport, Wien 01401
Bundesministerium für Umwelt, Naturschutz und Reaktorsicherheit, Bonn 10828
Bundesministerium für Unterricht, Kunst und Kultur, Wien 01402
Bundesministerium für Verkehr, Bau und Stadtentwicklung, Berlin 10799
Bundesministerium für Verkehr, Bau- und Stadtentwicklung, Bonn 10829
Bundesministerium für Wirtschaft und Technologie, Berlin 10800
Bundesministerium für wirtschaftliche Zusammenarbeit und Entwicklung, Bonn 10830
Bundesnetzagentur, Bonn 10831
Bundespolizeidirektion Wien, Wien 01403
Bundespräsident-Theodor-Heuss-Haus, Stuttgart 12543
Bundespräsidialamt, Berlin 10801
Bundesrat, Berlin 10802
Bundessortenamt, Hannover 10931
Bundessozialgericht, Kassel 10959
Bundessprachenamt, Hürth 10953
Bundesstaatliche Pädagogische Bibliothek beim Landesschulrat für Niederösterreich, St. Pölten 01391
Bundesverband der Betriebskrankenkassen, Essen 11821
Bundesverband Güterkraftverkehr, Logistik und Entsorgung (BGL) e.V., Frankfurt am Main 11839
Bundesverfassungsgericht, Karlsruhe 10956
Bundesversicherungsamt, Bonn 10832
Bundesverwaltungsgericht, Leipzig 10999
Bundeswehr, Wilhelmshaven 11134
Bundeswehr Scientific and Technical Information Center, Bonn 10834
Bundeswehr, Wehrbereichskommando III, Leipzig 11000
Bundeswehrkrankenhaus Berlin, Berlin 11500
Bundeswehrkrankenhaus Hamburg, Hamburg 11974
Bundeszentrale für Gesundheitliche Aufklärung, Köln 12124
Bündner Volksbibliothek, Chur 27482
Bunker Hill Community College, Boston 33181
Bunnvale Public Library, Califon 38414
Burapha University, Chonburi 27736
Burbank Public Library, Burbank 38385
Burbank Public Library – Buena Vista, Burbank 38386
Burbank Public Library – Northwest, Burbank 38387
Burch & Cracchiolo, Phoenix 36116
Burdekin Shire Council, Ayr 01060
Burdenko Institute of Neurosurgery, Moskva 23295
Bureau de Recherches Géologiques et Minières (BRGM), Orléans 08455
Bureau des Longitudes, Paris 08502
Bureau for the Design of Cotton Growing Equipment, Toshkent 42129
Bureau Goudappel Coffeng BV (BGC), Deventer 19327
Bureau International d'Education, Le Grand-Saconnex 27402
Bureau international du travail (BIT), Genève 27367
Bureau of Braille & Talking Book Library Services, Daytona Beach 36776
Bureau of Indian Standards (BIS), New Delhi 14117
Bureau of Jewish Education, Getzville 36888
Bureau of Jewish Education, San Francisco 37536
Bureau of Land Management, Springfield 34879
Bureau of Land Management Library, Denver 34134
Bureau of Meteorology, Melbourne 00949
Bureau of Plant Industry, Manila 20767
Bureau of Prisons, El Reno 34178
Bureau of Public Enterprises, New Delhi 13867
Bureau of Statistics, Suva 07312
Bureau Régional de l'UNESCO pour l'Education en Afrique (BREDA) / UNESCO Regional Office for Education in Africa, Dakar 24434
Burevestnik Instrument-Making Research and Industrial Corporation, Sankt-Peterburg 22919
Burg Giebichenstein – Hochschule für Kunst und Design, Halle (Saale) 09910
Burgas University of Technology, Burgas 03454
Burgenländische Landesbibliothek, Eisenstadt 01182
Burgenländisches Landesmuseum, Eisenstadt 01516
Burgerbibliothek Bern / Bibliothèque de la Bourgeoisie de Berne, Bern 27319
Bürgerschaft der Freien und Hansestadt Hamburg, Hamburg 10914
Burijisuton K. K., Kodaira 17377
Burke County Public Library, Morganton 40298
Burleigh College, Unley 00802
Burley Public Library, Burley 38390
Burlingame Public Library, Burlingame 38391
Burlington County College, Pemberton 33627
Burlington County Library, Westampton 41844
Burlington County Library – Cinnaminson Branch, Cinnaminson 38573
Burlington County Library – Evesham Branch, Marlton 40072
Burlington County Library – Maple Shade Branch, Maple Shade 40046
Burlington County Library – Pemberton Community Library, Browns Mills 38371
Burlington County Library – Pinelands, Medford 40130
Burlington Public Library, Burlington 04590
Burlington Public Library, Burlington 38394
Burlington Public Library, Burlington 38395
Burlington Public Library, Burlington 38396
Burnaby Public Library, Burnaby 04591
Burnet County Library System, Burnet 38399
Burnet, Duckworth & Palmer, LLP, Calgary 04236
Burnham Memorial Library, Colchester 38630
Burnley College, Richmond 00469
Burns & McDonnell Engineering Co, Kansas City 35791
Burnside Library, Glenside 01097
Burr & Forman Library, Birmingham 35508
Burton Public Library, Burton 38402
Burundi National Library, Bujumbura 03622
Burundi Parliamentary Library, Bujumbura 03625
Burwood Library, Burwood 01075
Bury Central Library, Bury 29952
Buryat Agricultural Institute, Ulan-Ude 22614
Buryat Complex Research Institute of the Russian Academy of Sciences, Ulan-Ude 24032
Buryat Geological Administration, Ulan-Ude 22768
Buryat National Library, Ulan-Ude 22185
Buryat State Archives, Ulan-Ude 24031
Buryatski kompleksny nauchno-issledovatelski institut RAN, Ulan-Ude 24032
Buryatski mezhotraslevoi territorialny tsentr nauchno-tekhnicheskoi informatsii i propagandy, Ulan-Ude 24033
Buryatski selskokhozyaistvenny institut, Ulan-Ude 22614
Buryatskoe geologicheskoe upravlenie, Ulan-Ude 22768
Business and Consume Affairs Library, Sydney 00725
Busk Regional Centralized Library System, Main Library, Busk 28511
Buska raionna TsBS, Tsentralna biblioteka, Busk 28511
Buskerod College, Hønefoss 20096
Buskerud Central Hospital, Drammen 20220
Buskerud College, Kongsberg, Kongsberg 20100
Buskerud sentralsykehus, Drammen 20220
Butler County Community College, Butler 33200
Butler County Community College, El Dorado 33313
Butler County Law Library, Butler 34023
Butler County Law Library Association, Hamilton 36909
Butler University, Indianapolis 31385
– Ruth Lilly Science Library, Indianapolis 31386
Butte Community College, Oroville 33608
Butte County Law Library, Oroville 34641
Butte County Library, Oroville 40654
Buttenwieser Library, New York 37190
Butte-Silver Bow Public Library, Butte 38404
Butt-Holdsworth Memorial Library, Kerrville 39709
Butwal Multiple Campus Library, Rupandehi 19027
Bvumbwe Research Library, Limbe 18489
Býarbókasavnið / Tórshavns Folkebibliotek, Tórshavn 06942
Bydgoskie Towarzystwo Naukowe, Bydgoszcz 21079
Bydgoszcz Scientific Society, Bydgoszcz 21079
Byrd Polar Research Center, Columbus 36737
Byrnes-Quanbeck Library at Mayville State University, Mayville 31774
Byron Public Library District, Byron 38405
Byuro morskogo mashinostroeniya Malakhit, Sankt-Peterburg 23752
Byzantine Catholic Seminary, Pittsburgh 35333
C. D. Howe Institute Library, Toronto 04515
C. E. Brehm Memorial Public Library District, Mount Vernon 40323
C. E. des Cheminots Région de Montpellier, Béziers 08297
C. E. des Cheminots Région Metz-Nancy, Chalindrey 08253
C. G. Jung Institute of San Francisco, San Francisco 37546
C. G. Jung-Institut Zürich, Küsnacht 27386
C. J. Langenhoven Memorial Library, Oudtshoorn 25218
Cabarrus County Public Library, Concord 38676
Cabell County Public Library, Huntington 39536
CABI Europe – UK, Egham 29667
Cabinet Legislation Bureau, Tokyo 17318
Cabinet Office Library, Kampala 27918
Cabrillo College, Aptos 33131
Cabrini College, Radnor 32407
Cades, Schutte, Fleming & Wright, Honolulu 35743
Cadillac-Wexford Public Library, Cadillac 38407
Cadwalader, Wickersham & Taft, Washington 36337
Cadwalader, Wickersham & Taft Library, New York 35961
Cahill, Gordon & Reindel LLP Library, New York 35962
Cahners Publishing Co, New York 35963
Cairns Diocesan Education Services, Cairns 00759
Cairns Libraries, Cairns 01077
Cairo Demographic Centre, Cairo 07159
Cairo Public Library, Cairo 38409
Cairo University, Giza 07055
– African Studies and Research Institute, Giza 07056
– Department of Library Science, Giza 07057
– Faculty of Agriculture, Giza 07058
– Faculty of Arts, Giza 07059
– Faculty of Commerce, Giza 07060
– Faculty of Dar al-Ulum, Cairo 07061
– Faculty of Economics and Political Science, Giza 07062
– Faculty of Engineering, Giza 07063
– Faculty of Law, Giza 07064
– Faculty of Medicine, Cairo 07065
– Faculty of Pharmacology, Cairo 07066
– Faculty of Science, Giza 07067
– Faculty of Veterinary Medicine, Giza 07068
– Statistics Studies and Research Institute, Giza 07069
Caixa Económica Federal, Brasília 03278
Caixa Geral do Depósitos SA (CGD), Lisboa 21746
Caja de Ahorros del Mediterráneo – Biblioteca Gabriel Miró, Alicante 25870
Caja de Compensación Familiar (COMFENALCO), Medellín 06042
Caja de Compensación Familiar de Antioquia (COMFAMA), Medellín 06005
Caja de Compensación Familiar del Atlántico, Barranquilla 06004
Caja España de Inversiones, León 25967
Caja Nacional de Ahorro y Seguro, Buenos Aires 00225
Calaveras County Law Library, San Andreas 34799
Calaveras County Library, San Andreas 41160
Calcasieu Parish Public Library System, Lake Charles 39772
Calcasieu Parish Public Library System – Central Library, Lake Charles 39773
Calcasieu Parish Public Library System – Sulphur Regional Library, Sulphur 41470
Calcutta Mathematical Society, Kolkata 14034
Calderdale College, Halifax 29467
Calderdale Metropolitan Borough Council, Halifax 29991
Caldwell College, Caldwell 30608
Caldwell Community College & Technical Institute, Hudson 33421
Caldwell County Public Library, Lenoir 39850
Caldwell Public Library, Caldwell 38411
Caledon Public Library, Bolton 04580
Calfee, Halter & Griswold LLP, Cleveland 35614
Calgary Academy Library, Calgary 03679
Calgary Board of Education, Calgary 04320
Calgary Public Library, Calgary 04592
Calgary Public Library – Country Hills Library, Calgary 04593
Calgary Public Library – Crowfoot Library, Calgary 04594
Calgary Public Library – Louise Riley Branch, Calgary 04595
Calgary Public Library – Nose Hill, Calgary 04596
Calgary Public Library – Shawnessy Library, Calgary 04597
Calgary Public Library – Signal Hill, Calgary 04598
Calgary Public Library – Village Square, Calgary 04599
Calgary Public Library – W. R. Castell Central Library, Calgary 04600
Calgene, LLC, Davis 36772
Calhoun Community College, Decatur 33285
Calhoun Correction Institution Library, Blountstown 33991
Calhoun County Library, Hampton 36910
Calhoun County Library, Port Lavaca 40858
California Academy of Sciences Library, San Francisco 37537
California Baptist University, Riverside 32453
California Christian College, Fresno 31178
California College of Arts & Crafts, Oakland 32171
California College of the Arts Libraries, San Francisco 32584
California Court of Appeal, San Diego 34805
California Court of Appeal Fifth Appellate District Library, Fresno 34287
California Department of Corrections – Deuel Vocational Institution Library, Tracy 34932
California Department of Corrections Library System, Avenal 33958

California Department of Corrections Library System, Blythe 33992
California Department of Corrections Library System, Blythe 33993
California Department of Corrections Library System, Calipatria 34026
California Department of Corrections Library System, Chino 34076
California Department of Corrections Library System, Chowchilla 34078
California Department of Corrections Library System, Chowchilla 34079
California Department of Corrections Library System, Coalinga 34091
California Department of Corrections Library System, Corcoran 34112
California Department of Corrections Library System, Crescent City 34116
California Department of Corrections Library System, Delano 34132
California Department of Corrections Library System, Imperial 34376
California Department of Corrections Library System, Jamestown 34396
California Department of Corrections Library System, Lancaster 34432
California Department of Corrections Library System, Norco 34612
California Department of Corrections Library System, Represa 34737
California Department of Corrections Library System, Represa 34738
California Department of Corrections Library System, Sacramento 41091
California Department of Corrections Library System, San Luis Obispo 34828
California Department of Corrections Library System, San Luis Obispo 34829
California Department of Corrections Library System, San Quentin 34831
California Department of Corrections Library System, Susanville 34905
California Department of Corrections Library System, Wasco 34976
California Department of Justice, Sacramento 34759
California Department of Justice Library, Los Angeles 34472
California Department of Justice Library, San Diego 34806
California Department of Pesticide Regulation Library, Sacramento 34760
California Energy Commission Library, Sacramento 37473
California Environmental Protection Agency, Sacramento 37474
California Historical Society, San Francisco 37538
California Institute of Integral Studies Library, San Francisco 32585
California Institute of Technology – Caltech Library System 1-32, M/C 1-32, Pasadena 32232
– Astrophysics Library, Pasadena 32233
– Earthquake Engineering Research Library, Pasadena 32234
– Sherman Fairchild Library of Engineering & Applied Science, Pasadena 32235
California Institute of Technology – Jet Propulsion Laboratory Library, Archives & Records Section, Pasadena 37328
California Institute of the Arts, Valencia 32921
California Lutheran University, Thousand Oaks 32816
California Maritime Academy Library, Vallejo 32923
California Pacific Medical Center-University of the Pacific School of Dentistry, San Francisco 32586
California Polytechnic State University, San Luis Obispo 32599
California Province of the Society of Jesus, Los Gatos 35269
California School for the Deaf Library, Riverside 33687
California Second District Court of Appeals, Los Angeles 34473
California State Court of Appeal, Riverside 34744
California State Court of Appeal, Sacramento 34761
California State Department of Corporations Libary, Los Angeles 34474
California State Department of Food & Agriculture, Sacramento 34762
California State Department of Transportation, Sacramento 34763
California State Department of Water Resources, Sacramento 34764
California State Legislative Counsel, Sacramento 34765
California State Library, Sacramento 30150

California State Library – Sutro Library, San Francisco 30155
California State Polytechnic University Library, Pomona 32327
California State Railroad Museum, Sacramento 37475
California State Railroad Museum Library, Sacramento 37476
California State University, Bakersfield 30330
California State University, Fullerton 31183
California State University, Hayward 31320
California State University, Moss Landing 37132
California State University, Sacramento 32487
– Tsakopoulos Hellenic Collection, Sacramento 32488
California State University, San Bernardino 32574
California State University, San Marcos 32600
California State University, Seaside 32630
California State University, Turlock 32855
California State University, Chico, Chico 30796
California State University Dominguez Hills, Carson 30680
California State University, Fresno, Fresno 31179
California State University, Long Beach, Long Beach 31637
California State University, Los Angeles, Los Angeles 31644
– Roybal Institute for Applied Gerontology Library, Los Angeles 31645
California State University, Northridge, Northridge 32157
California Thoroughbred Breeders Association, Arcadia 36473
California University of Pennsylvania, California 30610
California Western School of Law Library, San Diego 32576
California Youth Authority – N. A. Chaderjian Youth Correctional Facility Library, Stockton 34897
California Youth Authority – O. H. Close Youth Correctional Facility Library, Stockton 34898
Callahan Library, Saint Joseph's College, Patchogue 32238
Calloway County Public Library, Murray 40344
Calumet City Public Library, Calumet City 38415
Calumet College of Saint Joseph, Whiting 33052
Calvary Baptist Theological Seminary, Lansdale 35258
Calvary Bible College Theological Seminary, Kansas City 35248
Calvert County Public Library, Prince Frederick 40897
Calvert Marine Museum Library, Solomons 37615
Calvin College & Theological Seminary, Grand Rapids 35218
K. R. Cama Oriental Institute, Mumbai 14085
Camara de Comércio e Indústria do Porto, Porto 21824
Cámara de Comercio y Producción del Distrito Nacional, Santo Domingo 06966
H. Cámara de Senadores, México 18813
Cámara di Diputados, México 18814
Câmara dos Deputados, Brasília 03214
Câmara Municipal de São Paulo, São Paulo 03238
Cámara Oficial de Comercio e Industria de Madrid, Madrid 25988
Cámara Oficial de Comercio, Industria y Navegación, A Coruña 25949
Cámara Oficial de Comercio, Industria y Navegación, Santander 26138
Camarena Memorial Library, Calexico 38413
Camas Public Library, Camas 38417
Camberwell College of Arts, London 29171
Cambra de Comerç de Barcelona, Barcelona 25887
Cambra Oficial de Comerç, Indústria i Navegació de Barcelona, Barcelona 25887
Cambria County Free Law Library, Ebensburg 34165
Cambria County Library System & District Center, Johnstown 39655
Cambrian College of Applied Arts & Technology, Sudbury 03871
Cambridge Libraries and Galleries, Cambridge 04601
Cambridge Public Library, Cambridge 38418
Cambridge Public Library – Central Square Branch, Cambridge 38419
Cambridge Refrigeration Technology, Cambridge 29625
Camden Council, Camden 01079

Camden County College Library, Blackwood 33175
Camden County Historical Society, Camden 36622
Camden County Library District, Camdenton 38426
Camden County Library System, Voorhees 41695
Camden County Library System – Gloucester Township, Blackwood 38213
Camden County Library System – South County Regional Branch, Atco 38007
Camden County Library System – William G. Rohrer Memorial Library – Haddon Township, Westmont 41862
Camden Free Public Library, Camden 38423
Camden Public Library, Camden 38424
Camden Theological Library, North Parramatta 00786
Camera dei Deputati, Roma 15662
Camera del Lavoro, Milano 15649
Camera di Commercio, Industria, Artigianato e Agricoltura, Bergamo 16181
Camera di Commercio, Industria, Artigianato e Agricoltura, Bologna 16189
Camera di Commercio, Industria, Artigianato e Agricoltura, Bolzano 16202
Camera di Commercio, Industria, Artigianato e Agricoltura, Brescia 16208
Camera di Commercio, Industria, Artigianato e Agricoltura, Cagliari 16215
Camera di Commercio, Industria, Artigianato e Agricoltura, Carrara 16221
Camera di Commercio, Industria, Artigianato e Agricoltura, Como 16232
Camera di Commercio, Industria, Artigianato e Agricoltura, Cremona 16237
Camera di Commercio, Industria, Artigianato e Agricoltura, Cuneo 16238
Camera di Commercio, Industria, Artigianato e Agricoltura, Ferrara 16245
Camera di Commercio, Industria, Artigianato e Agricoltura, Forli 16275
Camera di Commercio, Industria, Artigianato e Agricoltura, Gorizia 16294
Camera di Commercio, Industria, Artigianato e Agricoltura, Grosseto 16297
Camera di Commercio, Industria, Artigianato e Agricoltura, La Spezia 16299
Camera di Commercio, Industria, Artigianato e Agricoltura, Milano 16328
Camera di Commercio, Industria, Artigianato e Agricoltura, Napoli 16367
Camera di Commercio, Industria, Artigianato e Agricoltura, Pisa 16420
Camera di Commercio, Industria, Artigianato e Agricoltura, Roma 16475
Camera di Commercio, Industria, Artigianato e Agricoltura, Salerno 16570
Camera di Commercio, Industria, Artigianato e Agricoltura, Sassari 16572
Camera di Commercio, Industria, Artigianato e Agricoltura, Teramo 16586
Camera di Commercio, Industria, Artigianato e Agricoltura, Torino 16594
Camera di Commercio, Industria, Artigianato e Agricoltura, Trieste 16621
Camera di Commercio, Industria, Artigianato e Agricoltura, Vercelli 16662
Camera di Commercio, Industria, Artigianato e Agricoltura, Verona 16666
Camera di Commercio, Industria, Artigianato e Agricoltura, Bari 16173
Camera Naţionala a Carţii din Republica Moldova, Chişinău 18885
Cameron County Law Library, Emporium 34190
Cameron Parish Library, Cameron 38427
Cameron University, Lawton 31582
Camosun College, Victoria 03951
Camp, Dresser & Mckee, Cambridge 35549
Campaspe Regional Library, Echuca 01093
Campbell County Public Library, Rustburg 41084
Campbell County Public Library District, Cold Spring 38632
Campbell County Public Library System, Gillette 39226
Campbell University, Buies Creek 30600
– School of Law Library, Buies Creek 30601
Campbellsville University, Campbellsville 30666
Campbelltown City Library Services, Campbelltown 01080
Campbelltown Public Library Services, Campbelltown 01081
Campion College Library, Regina 03829
Campus Universitario Duques de Soria, Soria 25629
Campus-Bibliothek für Informatik und Mathematik, Saarbrücken 12450
CAN Bibliotek, Stockholm 26643
Canada Agriculture & Agri-Food Canada, Swift Current 04158
Canada Aviation Museum, Ottawa 04431

Canada College, Redwood City 33678
Canada Department of Finance & Treasury Board Library, Ottawa 04105
Canada Department of Fisheries and Oceans, Moncton 04382
Canada Department of Fisheries & Oceans, Sydney 04159
Canada Department of Indian Affairs & Northern Development, Ottawa 04106
Canada Department of Justice, Edmonton 04070
Canada Department of Justice, Montreal 04094
Canada Department of Justice, Vancouver 04174
Canada Department of Justice Library, Ottawa 04107
Canada Department of National Defence, Dartmouth 04057
Canada Department of National Defence, Halifax 04355
Canada Department of National Defence – Base Borden Public & Military Library, Borden 04051
Canada Department of National Defence – National Defense Library Services, Ottawa 04108
Canada Fisheries & Oceans Library, Ottawa 04432
Canada Public Service Labour Relations Board Library, Ottawa 04433
Canada Science & Technology Museum, Ottawa 04434
Canada-Manitoba Business Service Centre Library, Winnipeg 04188
Canada-Newfoundland & Labrador Business Service Centre Library, St. John's 04499
Canadian Association for Community Living / Association canadienne pour l'intégration communautaire, North York 04425
Canadian Bible College – Canadian Theological Seminary, Calgary 04200
Canadian Broadcasting Corporation – Reference Library, Toronto 04507
Canadian Centre for Architecture, Montreal 04388
Canadian Centre for Ecumenism Library, Montreal 04213
Canadian Centre for Occupational Health & Safety, Hamilton 04361
The Canadian Children's Book Centre, Toronto 04508
Canadian Conservation Institute, Ottawa 04435
Canadian Cultural Centre, Paris 08505
Canadian Department of Fisheries and Oceans, Saint John's 04148
Canadian Department of National Defence – Canadian Forces Medical Service School, Borden 04052
Canadian Forces College, Toronto 03879
Canadian Forest Service – Great Lakes Forestry Centre, Sault Sainte Marie 04155
Canadian Forest Service – Laurentian Forestry Centre, Sainte-Foy 04485
Canadian Forest Service – Northern Forestry Centre, Edmonton 04071
Canadian Forest Service – Pacific Forestry Centre, Victoria 04183
Canadian Heritage-Parks Canada, Cornwall 04329
Canadian Housing Information Centre, Ottawa 04436
Canadian Imperial Bank of Commerce, Toronto 04279
Canadian International Development Agency (CIDA), Hull 04364
Canadian International Trade Tribunal Library, Ottawa 04437
Canadian Lesbian & Gay Archives, Toronto 04509
Canadian Lutheran Bible Institute Library, Camrose 04202
Canadian Memorial Chiropractic College, Toronto 03880
Canadian Mennonite University, Winnipeg 03959
Canadian Mental Health Association, Calgary 04054
Canadian Museum of Civilization Library, Gatineau 04350
Canadian Museum of Nature, Ottawa 04438
Canadian Music Centre / Centre de Musique Canadienne, Toronto 04510
Canadian National Institute for the Blind, Toronto 04511
Canadian Nuclear Safety Commission, Ottawa 04109
Canadian Pacific Railway, Calgary 04237
Canadian Police College, Ottawa 04016
Canadian Psychoanalytic Society Library, Montreal 04386

Canadian Radio-Television & Telecommunications Commission Library, Ottawa 04439
Canadian Tax Foundation, Toronto 04512
Canadian Transportation Agency Library, Gatineau 04351
Canadian University College, Lacombe 03734
Canadian War Museum, Ottawa 04440
Canadore College, North Bay 03807
Canal Fulton Public Library, Canal Fulton 38433
Canastota Public Library, Canastota 38435
Canberra Institute of Technology (CIT), Canberra 00414
– Bruce Library, Canberra 00415
– Weston Library, Canberra 00416
Canby Public Library, Canby 38436
Cancer Center & Institute of Oncology, Warszawa 21253
Cancer Research UK, London 29736
Caney Fork Regional Library, Sparta 41394
Canisius College, Buffalo 30589
Canisiushaus der Gesellschaft Jesu und Kroatisches Historisches Institut, Wien 01480
Canisius-Kolleg, Berlin 10708
City of Canning, Riverton 01148
Canon City Public Library, Canon City 38438
Cánovas del Castillo, Málaga 26233
Canspec Group, Inc, Edmonton 04253
Canterbury Cathedral Library, Canterbury 29538
Canterbury Christ Church University College, Canterbury 29054
Canterbury City Council, Campsie 01082
Canterbury College of Technology, Canterbury 29455
Canterbury District Law Society, Christchurch 19829
Canterbury Museum, Christchurch 19830
Canterbury Public Library, Christchurch 19862
Cantho University, Cantho 42267
Canton Free Library, Canton 38439
Canton Public Library, Canton 38440
Canton Public Library, Canton 38441
Canton Public Library, Canton 38442
CapcMusée d'Art Contemporain, Bordeaux 08302
Cape Breton Regional Library, Sydney 04775
Cape Canaveral Public Library, Cape Canaveral 38448
Cape Cod Community College, West Barnstable 33856
Cape Cod Museum of Natural History, Brewster 36586
Cape Fear Community College, Wilmington 33874
Cape Girardeau Public Library, Cape Girardeau 38450
Cape May County Library, Cape May Court House 38451
Cape Peninsula University of Technology, Wellington 25057
Cape Provincial Library Service, Cape Town 24997
Cape Technikon, Cape Town 25005
Cape Town City Libraries, Cape Town 25193
Capehart & Scatchard, PA Library, Mount Laurel 35926
Capilano College, North Vancouver 04012
Capital Area District Library, Lansing 39801
Capital Institute of Medicine, Beijing 05029
Capital Normal University, Beijing 05030
Capital University, Columbus 30892
– Law School Library, Columbus 30893
Capitol College, Laurel 31568
Capitolo della Cattedrale, Bergamo 15732
Caplin & Drysdale Library, Washington 36338
Capricornia Electricity Board, Rockhampton 00827
Capuchin College, Washington 32956
Capuchin Monastery, Škofja Loka 24901
Capuchin Provincial Library, Floriana 18662
Capuchins of the Province Navarra, Cantabria, Aragon, Pamplona 25808
Carbon County Law Library, Jim Thorpe 34401
Carbon County Library System, Rawlins 40944
Carbondale Public Library, Carbondale 38452
Carburetors Factory, Moskva 22838
Carburettor Fittings Plant Industrial Corporation, Sankt-Peterburg 22914
Cardif University, Cardiff 29056
Cardiff Council, Cardiff 29954
Cardiff Naturalists' Society, Cardiff 29629
Cardiff University, Cardiff 29057
Cardiff University, Cardiff 29058
Cardinal Stefan Wyszyński University, Warszawa 20973
Cardinal Stritch University, Milwaukee 31834

Cardiology Research Complex, Moskva 23171
Cargill, Inc – Information Center, Wayzata 37816
Cargill, Inc – Law Library, Wayzata 36405
Carhop, Bruxelles 02578
Caribbean Community Secretariat, Greater Georgetown 13197
Caribbean Graduate School of Theology, Kingston 16880
Carilion Clinic, Roanoke 37448
Carinthia University of Applied Sciences – Campus Library Feldkirchen, Feldkirchen 01191
Carinthia University of Applied Sciences – Campus Library Klagenfurt, Klagenfurt 01240
Carinthia University of Applied Sciences – Campus Library Spittal, Spittal 01286
Carinthia University of Applied Sciences – Campus Library Villach, Villach 01290
Carl Elliott Regional Library System, Jasper 39636
Firma Carl Freudenberg, Weinheim 11433
Carl Robinson Correctional Institution Library, Enfield 34191
Carl Sandburg College, Galesburg 33363
Carl von Ossietzky State and University Library of Hamburg, Hamburg 09317
– Departments of Language, Literature and Media, Library, Hamburg 09941
– International Tax Institute, Hamburg 09948
Carl von Ossietzky Universität Oldenburg, Oldenburg 10525
Carleton College, Northfield 32153
Carleton University, Ottawa 03809
Carletonville Public Library, Carletonville 25195
Carlisle Cathedral Library, Carlisle 29539
Carlisle College, Carlisle 29061
Carlos Albizu University, San Juan 21904
Carlow College, Pittsburgh 32296
Carlow County Library, Carlow 14565
Carlsbad City Library, Carlsbad 38454
Carlsbad Public Library, Carlsbad 38455
Carlsberg Forskningscenter, Valby 06835
Carlsberg Research Center, Valby 06835
Carl-Schurz-Haus, Freiburg 11888
Carlsmith Ball LLP Library, Honolulu 35744
Carl-Thiem-Klinikum Cottbus gemeinnützige GmbH, Cottbus 11716
Carlton Fields, Tampa 36296
Carl-von-Ossietzky-Oberschule, Berlin 10709
CARMABI Foundation Library, Piscadera Baai 06318
Carmarthenshire County Libraries, Carmarthen 29956
Carmel Clay Public Library, Carmel 38456
Carmelite Fathers, Brugge 02455
Carmelite Library, Middle Park 00779
Carmelite Monastery, Baltimore 35122
Carmelite Monastery Library, Beacon 35130
Carmody & Torrance, Waterbury 36398
Carnegie College, Dunfermline 29086
Carnegie Endowment for International Peace Library, Washington 37725
Carnegie Foundation for the Advancement of Teaching, Stanford 37635
Carnegie Free Library, Beaver Falls 38136
Carnegie Free Library, Connellsville 38681
Carnegie Institution of Washington, Washington 37726
Carnegie Library, Ayr 29930
Carnegie Library, Curepipe 18702
Carnegie Library, Muncie 40337
Carnegie Library, Suva 07324
Carnegie Library of Homestead, Munhall 40340
Carnegie Library of McKeesport, McKeesport 40120
Carnegie Library of Pittsburgh – Downtown & Business, Pittsburgh 40799
Carnegie Library of Pittsburgh – East Liberty, Pittsburgh 40800
Carnegie Library of Pittsburgh – Squirrel Hill, Pittsburgh 40801
Carnegie Mellon University, Pittsburgh 32297
– Software Engineering Institute Library, Pittsburgh 32298
Carnegie Museum of Natural History Library, Pittsburgh 37378
Carnegie Public Library, East Liverpool 38905
Carnegie Public Library, Las Vegas 39813
Carnegie Public Library, Washington Court House 41760
Carnegie Public Library of Clarksdale & Coahoma County, Clarksdale 38583
Carnegie Public Library of Steuben County, Angola 37962
Carnegie-Stout Public Library, Dubuque 38870

Carney, Badley & Spellman Library, Seattle 36262
Caro Area District Library, Caro 38458
Carol Stream Public Library, Carol Stream 38459
Carolina Population Center, Chapel Hill 36646
Caroline County Public Library, Denton 38823
Carpenter Technology Corp, Reading 36164
Carpet Production Industrial Corporation, Brest 01843
Carpet Production Industrial Corporation, Vitebsk 01933
Carré d'Art – Musée d'art contemporain, Nîmes 08452
Carrington, Coleman, Solman & Blumenthal, LLP, Dallas 35643
Carroll, Burdick & McDonough, San Francisco 36226
Carroll College, Helena 31322
Carroll College, Waukesha 33001
Carroll Community College, Westminster 33862
Carroll County Court Library, Westminster 35060
Carroll County District Library, Carrollton 38461
Carroll County Public Library, Westminster 41859
Carroll Public Library, Carroll 38460
Carrollton Public Library, Carrollton 38462
Carson City Library, Carson City 38465
Carson County Public Library, Panhandle 40702
Carson-Newman College, Jefferson City 31460
Cartagena Academy of History, Cartagena 06010
Carter, Ledyard & Milburn Library, New York 35964
Carter Observatory, Wellington 19845
Carteret Public Library, Carteret 38466
Carthage College, Kenosha 31494
Carthage Public Library, Carthage 38469
Cartuja de Aula Dei, Zaragoza 25848
Carver Bible College, Atlanta 35115
Carver County Library, Chaska 38526
Carver Public Library, Carver 38472
Cary Area Public Library District, Cary 38473
Cary Institute of Ecosystem Studies, Millbrook 37108
Cary Library, Houlton 39483
Cary Memorial Library, Lexington 39866
Casa Central de los Paules, Madrid 25787
Casa de Cervantes, Bologna 16190
Casa de Colón, Las Palmas 26117
Casa de Estudios de los Mercedarios, Madrid 25788
Casa de Rui Barbosa, Rio de Janeiro 03345
Casa de Velázquez, Madrid 25989
Casa del Obispado de Urgell, La Seu d'Urgell 25825
Casa della Missione, Genova 15845
Casa di Dante in Roma, Roma 16476
Casa di Goldoni, Venezia 16647
Casa do Brasil, Madrid 25419
Casa Giorgio Cini, Ferrara 16246
Casa Museo Pérez Galdos, Las Palmas 26118
Casa Museo Unamuno, Salamanca 26129
Casa Provincial de los Paules, Santa Marta de Tormes 25822
Casa-Museo Azorín, Monóvar 26106
Case Memorial Library, Orange 40625
Case Western Reserve University, Cleveland 30827
– Lillian & Milford Harris Library, Cleveland 30828
– School of Law Library, Cleveland 30829
Casemate Museum Library, Fort Monroe 36862
Casey College of TAFE, Dandenong 00567
Casey-Cardinia Library Corporation, Cranbourne 01089
Casper College, Casper 30683
Cass County Public Library, Harrisonville 39377
Cass District Library, Cassopolis 38477
Cassa di Risparmio delle Provincie Lombarde, Milano 16145
Cassa di Risparmio di Parma e Monte di Credito su Pegno di Busseto, Busseto 16143
Cassa di Risparmio in Bologna, Bologna 16142
Cassels, Brock & Blackwell Library, Toronto 04280
Cast Iron Plant, Moskva 22833
Castello del Buonconsiglio, Trento 16611
Casting Equipment Industrial Corporation, Pinsk 01920
Casting Research Institute and Industrial Corporation, Sankt-Peterburg 23879
Castle Library, Lázně Kynžvart 06460

Castle Matrix Library, Rathkeale 14561
Castlegar & District Public Library, Castlegar 04605
Castleton State College, Castleton 30685
Catasto e Servizi Tecnici Erariali, Roma 15663
Catawba College, Salisbury 32556
Catawba County Library, Newton 40471
Catawba County Library – Saint Stephens, Hickory 39424
Catawba Valley Community College, Hickory 33408
Catedral de Pamplona, Pamplona 25807
Catering Technology Professional School, Sankt-Peterburg 22712
Caterpillar Inc, Mossville 35925
Cathedral Library, Rochester 29562
The Cathedral Llibrary in Gniezno, Gniezno 21052
Cathedral School of Saint John the Divine Library, New York 35305
Catholic Central Library, Farnborough 29545
Catholic Central Verein (Union) of America, Saint Louis 35357
Catholic Diocesan Archives of Toowoomba, Toowoomba 01035
Catholic Education Office, Kirwan 00773
Catholic Education Office, Leederville 00774
Catholic Education Office, Leichhardt 00775
Catholic Education Office, Manuka 00778
Catholic Education Office, Thebarton 00798
Catholic Education Office, Warragul 00803
Catholic Institute of Sydney, Strathfield 00796
Catholic Medical College, Seoul 18075
Catholic Newman Center, Lexington 35261
Catholic Resource + Information Service, Thebarton 00799
Catholic Theological College, East Melbourne 00764
Catholic Theological Institute, Boroko 20581
Catholic Theological Union, Chicago 35155
Catholic University, Milano 15102
Catholic University – Brescia, Brescia 14912
Catholic University of America, Washington 32957
– Engineering-Architecture & Mathematics Library, Washington 32958
– Judge Kathryn J. DuFour Law Library, Washington 32959
– Music Library, Washington 32960
– Nursing-Biology Library, Washington 32961
– Oliveira Lima Library, Washington 32962
– Rare Books / Special Collections, Washington 32963
– Reference & Instructional Services Division, Washington 32964
– Semitics/ICOR Library, Washington 32965
Catholic University of Applied Science, Berlin 09451
Catholic University of Cordoba, Córdoba 00147
Catholic University of Lille, Lille 07729
Catholic University of Lublin, Lublin 20882
– Department of Polish Language, Lublin 20886
– Faculty of Christian Philosophy, Lublin 20884
– Faculty of Humanities, Lublin 20885
Catholic University of Portugal in Lisbon, Lisboa 21665
Catholic University of Puerto Rico, Guayama 21854
Catholic University of Rio de Janeiro, Rio de Janeiro 03099
Catholic University of Taegu-Hyosung, Kyongsan-si 18069
Catholic University of the Diocese Linz, Linz 01262
Catholics United for Life, New Hope 37164
Cavan County Library, Cavan 14567
Cayey University College Library, Cayey 21852
Cayuga Community College, Auburn 33139
Cazenovia College, Cazenovia 33205
Cazenovia Public Library Society, Inc, Cazenovia 38480
CBS News Reference Library, New York 35965
CCBC Essex, Baltimore 33151
CCFE Fusion Library, Abingdon 29591
CDVEC Curriculum Development Unit, Dublin 14523
CE, solutions for environment, economy and technology, Delft 19407
Ceará Institute, Fortaleza 03298
Cebu CFI Law Library, Cebu City 20769
Cecil Community College, North East 33585
Cecil County Public Library, Elkton 38962
Cedar City Public Library in the Park, Cedar City 38482
Cedar Crest College, Allentown 30199
Cedar Falls Public Library, Cedar Falls 38483

Cedar Mill Community Library, Portland 40868
Cedar Park Public Library, Cedar Park 38486
Cedar Rapids Public Library, Cedar Rapids 38487
Cedar Valley College, Lancaster 33472
Cedarburg Public Library, Cedarburg 38488
Cedars-Sinai Medical Center, Los Angeles 37046
Cedarville University, Cedarville 30691
CEDOFOR – Centre de documentation et de formation religieuses, Carouge 27250
Cégep Beauce-Appalaches, Saint-Georges de Beauce 03845
Cégep de Baie-Comeau, Baie Comeau 03675
Cégep de Chicoutimi, Chicoutimi 03984
Cégep de Jonquière, Jonquière 03996
Cégep de La Pocatière, La Pocatière 04000
Cégep de l'Abitibi – Temiscamingue, Rouyn-Noranda 03836
Cégep de Sainte-Foy, Sainte-Foy 03843
Cégep de Saint-Félicien, Saint-Félicien 03844
Cégep de Saint-Hyacinthe, Saint-Hyacinthe 04223
Cégep de Saint-Laurent, Saint-Laurent 04027
Cégep de Valleyfield, Valleyfield 04042
Cégep de Victoriaville, Victoriaville 04043
Cegep du Vieux Montreal, Montreal 04387
Cégep Marie Victorin, Montreal 04004
Cégep Regional de Lanaudière à Joliette, Joliette 03995
Cégep Rivière-du-Loup-Bibliothèque, Rivière-du-Loup 03835
Cégep St.-Jean-sur-Richelieu, Saint-Jean-Sur-Richelieu 04025
CEGES Bibliothèque, Bruxelles 02575
Celanese Dennis F. Ripple Technical Information Center, Corpus Christi 35638
Celia – Library for the Visually Impaired, Helsinki 07445
Celia – Näkövammaisten kirjasto, Helsinki 07445
Celje Public Library, Celje 24955
Cell Biology and Genetic Engineering Institute, Kyiv 28331
Cellule Information et Documentation (CID), Ouagadougou 03616
Cemagref, Cestas 08319
Cement Concrete and Aggregates Australia, Saint Leonards 00829
Cement och Betong Institutet, Stockholm 26644
CENEAM, Valsain 26181
Census Operations Library, Chennai 13958
Centar za istraživanje mora, Rovinj 06217
Centenary College, Hackettstown 31275
Centenary College of Louisiana, Shreveport 32667
Centennial College of Applied Arts & Technology, Scarborough 04031
Center for Advanced Study in the Behavioral Sciences Library, Stanford 37636
Center for Agricultural Library and Technology Dissemination, Bogor 14348
Center for American Studies, Roma 16483
Center for American Studies (CAS), Bruxelles 02576
Center for Creative Leadership Library, Greensboro 36904
Center for Creative Studies Library, Detroit 31004
Center for Early Education Library, West Hollywood 37822
Center for Economic Research and Graduate Education – Economics Institute, Praha 06477
Center for Education in Sanitation, Bukhara 42113
Center for information om kvinde- og kønsforskning, København 06808
Center for International Cooperation and Developing Special Library, Ljubljana 24913
Center for Jewish History, New York 37191
Center for Judaic Studies Library, Philadelphia 32250
Center for Marine Research, Rovinj 06217
Center for Migration Studies Library, Staten Island 37638
Center for Naval Analyses Library, Alexandria 36453
Center for Puerto Rican Studies, New York 37196
Center for Religion, Ethics & Social Policy, Ithaca 36966
Center for Research Libraries, Chicago 36666
Center for Sanitation Education, Sankt-Peterburg 23758
Center for Socio-Economic Documentation and Publications, Tehran 14426
Center for the Advancement of Jewish Education, Miami 37098
Center for Undervisningsmidler i Sydslesvig, Flensburg 11832
Center Moriches Free Public Library, Center Moriches 38491
Center of Folk Art, Astrakhan 22993
Center of Theater Studies and Cinematography at the Academy of Theater, Radio, Film and Television, Ljubljana 24914
Center on Education & Training for Employment, Columbus 36738
The Center: Resources for Teaching & Learning, Arlington Heights 36485
Center za mednarodno sodelovanje in razvoj knjižnica, Ljubljana 24913
Center za teatrologijo in filmologijo pri AGRFT, Ljubljana 24914
Centers for Disease Control & Prevention Information Center, Atlanta 33940
Centraal Bureau voor Genealogie (CBG), Den Haag 19431
Centraal Historisch Archief, Willemstad 06320
Centraal Museum, Utrecht 19524
Centraal Planbureau (CPB), Den Haag 19251
Central Academy of Arts and Design, Beijing 05031
Central Academy of Drama, Beijing 05032
Central Aerological Observatory, Dolgoprudny 23025
Central Agency for Organization and Management, Cairo 07160
Central Agricultural Library, Bydgoszcz 21080
Central Agricultural Library, Puławy 21212
Central Agricultural Library, Sofiya 03581
Central Agricultural Library, Warszawa 21315
Central Agricultural Research Institute, Monrovia 18231
Central Alabama Community College, Alexander City 33124
Central American Institute of Business Administration, Alajuela 06119
Central American Technical Institute, Santa Tecla 07204
Central Archaeological Library, New Delhi 14118
Central Architects Club, Moskva 23506
Central Archive of Tatarstan Republic, Kazan 23088
Central Archives for Finnish Business Records, Mikkeli 07532
The Central Archives for the History of the Jewish People, Jerusalem 14725
Central Archives of Historical Records, Warszawa 21239
Central Arid Zone Research Institute (ICAR), Jodhpur 14017
Central Arizona College, Coolidge 30923
Central Arkansas Library System, Little Rock 39897
Central Arkansas Library System – Adolphine Fletcher Terry Branch, Little Rock 39898
Central Arkansas Library System – Dee Brown Branch, Little Rock 39899
Central Arkansas Library System – Esther Nixon Branch, Jacksonville 39608
Central Arkansas Library System – John Gould Fletcher Branch, Little Rock 39900
Central Artists Club, Moskva 23509
Central Asian Plague Prevention Research Institute, Almaty 17940
Central Asian Research Hydrometeorological Institute (SANIGMI), Toshkent 42130
Central Astronomical Observatory of the Russian Academy of Sciences, Sankt-Peterburg 23764
Central Automation Research Institute, Moskva 23526
Central Bank of Barbados, Bridgetown 01779
Central Bank of Brazil, Brasília 03254
Central Bank of Egypt, Cairo 07137
Central Bank of Hungary – Research Library, Budapest 13357
Central Bank of Iceland, Reykjavík 13564
Central Bank of Malaysia, Kuala Lumpur 18571
Central Bank of Malta, Valletta 18668
Central Bank of Nigeria, Lagos 20003
Central Bank of Russia, Moskva 22733
Central Bank of Sri Lanka, Colombo 26300
Central Bank of the Netherlands, Library, Amsterdam 19322
Central Bank of the Philippines, Manila 20766
Central Baptist College, Conway 30918
Central Baptist Theological Seminary, Shawnee 35378
Central Bible College, Springfield 35391
Central Biomedical Library, Yangon 19000
Central Bohemian Research Library in Kladno, Kladno 06346
Central Botanical Garden, Minsk 02057
Central Botanical Garden, Snigiri 23975
Central Botanical Garden of the Russian Academy of Sciences, Moskva 23192
Central Brevard Library & Reference Center, Cocoa 38624
Central Building Research Institute, Roorkee 14206
Central Building Research Institute, Rourkela 14208
Central Bureau of Statistics, Jakarta 14360
Central Bureau of Statistics, Jerusalem 14679
Central Bureau of Statistics, Nairobi 18020
Central Bureau of Statistics, Zagreb 06207
Central Bureau of Text-book, Delhi 13981
Central Capuhin Library, Roma 16017
Central Carolina Community College, Sanford 33739
Central Casting Plant, Gomel 01866
Central Catholic Library, Dublin 14508
Central Children's City Library, Bishkek 18153
Central China Institute of Technology, Wuhan 05902
Central Christian College, Moberly 31871
Central Christian College of Kansas, McPherson 31779
Central City Children's Library, Moskva 24176
Central City Children's library named by A.S. Pushkin, Sankt-Peterburg 24312
Central City Library, Nikolaevsk-na-Amure 24183
Central City Library, Novokuznetsk 24188
Central City Library, Novorossisk 24189
Central City Library, Novosibirsk 24192
Central City Library, Petropavlovsk-Kamchatski 24200
Central City Library, Rubtsovsk 24208
Central City Library, Rybinsk 24212
Central City Library, Shadrinsk 24317
Central City Library, Velikie Luki 24343
Central City Library of Dzhalal-Abad, Dzhalal-Abad 18155
Central City Library of Osh, Osh 18157
Central City Library of Rybache, Rybache 18160
Central City Library of Talas, Talas 18161
Central College, Pella 32240
Central College of Commerce, Glasgow 29113
Central Connecticut State University, New Britain 31953
Central Conservatory of Music, Beijing 05033
Central Construction Bureau 'Proektmashdetal', Moskva 22859
Central Cotton Research Institute, Multan 20520
Central Council for Research in Homoeopathy (CCRH), New Delhi 14119
Central Council of Scientific and Technical Unions, Sofiya 03580
Central Design Bureau, Minsk 02056
Central Design Office, Minsk 01999
Central Directorate for the Nile's Co, Giza 07145
Central Drug Research Institute, Lucknow 14064
Central Electricity Authority, New Delhi 13868
Central Electrochemical Research Institute, Karaikudi 14023
Central Epidemiology Institute, Moskva 23524
Central European University, Budapest 13231
Central Finland Health Care District, Medical Library, Jyväskylä 07516
Central Finland Museum, Jyväskylä 07515
Central Fish Health Laboratory, Nir-David 14748
Central Fisheries Research Institute, Chilanga 42335
Central Florida Community College, Ocala 33595
Central Food Research Institute, Budapest 13391
Central Food Technological Research Institute, Mysore 14103
Central Forest Library, Kathmandu 19033
Central Fuel Research Institute, Dhanbad 13987
Central Geological Survey, MOEA, Taipei 27604
Central Glass and Ceramic Research Institute, Kolkata 14035
Central Highlands Regional Library Cooperation, North Ballarat 01132
Central Hindi Directorate, New Delhi 14120
Central Historical Library, Tbilisi 09227
Central Hospital, Karlstad 26602
Central Hospital in Mikkeli, Medical Library, Mikkeli 07530
Central Hospital in Seinäjoki, Scientific Library Ostrobothnia, Seinäjoki 07540
Central Hospital Library, Västerås 26716
Central Hospital of Satakunta, Medical Library, Pori 07535
Central House of Journalists, Moskva 23510
Central Inland Capture Fisheries Research Institute, Barrackpore 13944
Central Institut for Meteorology and Geophysics, Wien 01664
Central Institute for Continuing Education in Medicine, Moskva 22394
Central Institute for Cultural Relations Israel, Ibero-America, Spain and Portugal, Jerusalem 14726
Central Institute for Labour Protection, Warszawa 21247
Central Institute for Medical Science Information, Hanoi 42286
Central Institute for Research in Farm Mechanization and Electrification, Minsk 02000
Central Institute for Research on Cotton Technology, Mumbai 14086
Central Institute for the Deaf, Saint Louis 37484
Central Institute of Economics and Mathematics, Moskva 23511
Central Institute of Education Library, Delhi 13982
Central Institute of English and Foreign Languages, Hyderabad 13655
Central Institute of Fine Arts, Beijing 05034
Central Institute of Fisheries Education, Mumbai 13702
Central Institute of Fisheries Technology, Kochi 14026
Central Institute of Indian Languages, Mysore 13828
Central Institute of Scientific and Technical Information and Research on Chemical and Petro-Engineering, Moskva 23513
Central Institute of Tool Design, Hyderabad 13997
Central Intelligence Agency, Washington 34981
Central Islip Public Library, Central Islip 38494
Central Laboratory for Batteries and Cells, Poznań 21193
Central Lakes College, Brainerd 30541
Central Land Council, Alice Springs 00605
Central Law Courts Library, Perth 00710
Central Leather Research Institute, Chennai 13959
Central Leprosy Teaching and Research Institute, Chengalpattu 13954
Central Library, Bahawalpur 20526
Central Library, Sovetski 24321
Central Library, Vadodara 14234
Central Library, Yoshkar-Ola 24364
Central Library Cambridge, Cambridge 29953
Central Library for Social Work, Ramat Gan 14751
Central Library for the Blind, Kyiv 28387
Central Library for the Blind, Visually and Physical Handicapped, Netanya 14747
Central Library Kranj, Kranj 24964
Central Library of Agricultural Science, Rehovot 14754
Central Library of Communications, Budapest 13380
Central Library of Corvinus University of Budapest, Budapest 13222
Central Library of Geography and Environmental Protection, Warszawa 21244
Central LIbrary of Ignalina District Municipality, Ignalina 18324
Central library of Novomoskovsk, Novomoskovsk 28756
Central Library of Oil Industry, Moskva 23497
Central Library of Physical Training, Praha 06524
Central Library of Sport Science, Köln 10123
Central Library of St. Lucia, Castries 26337
Central Library of the Food Industry, Moskva 23498
Central Library of the Islamic Republic of Iran, Tehran 14421
Central Library of the Navy, Buenos Aires 00204
Central Library of the Slovenian Academy of Sciences and Arts, Karst Research Institute, Postojna 24952
Central Library of the University of Medicine and Pharmacy, Iaşi 22045
Central Library of Vilnius Municipality, Vilnius 18376
Central Library Services, Pretoria 25158
Central Library Vuk Karadžić, Prijepolje 24549

Central Louisiana State Hospital
– Forest Glen Patient's Library, Pineville 37375
Central Luzon Polytechnic College, Cabanatuan City 20744
Central Luzon State University, Nueva Ecija 20729
Central Marine Research and Design Institute, Ltd (CNIIMF), Sankt-Peterburg 23918
Central Mechanical Engineering Research Institute (CMERI), Durgapur 13991
Central Medical Library, Lagos 20029
Central Medical Library, Seeb 20425
Central Medical Library, Sofiya 03476
Central Medical Library, Warszawa 21258
Central Methodist College, Fayette 31129
Central Michigan University, Mount Pleasant 31913
– Clarke Historical Library, Mount Pleasant 31914
Central Military Archives, Warszawa 21243
Central Military Library, Bandung 14330
Central Military Library, Warszawa 21246
Central Military Museum, Bucureşti 22006
Central Mindanao University, Bukidnon 20688
Central Mine Planning and Design Institute, Ranchi 14203
Central Mining and Ore Rectification Industrial Corporation, Krivi Rig 28154
Central Mining Institute, Katowice 20817
Central Mining Research Institute, Dhanbad 13988
Central Mississippi Regional Library System, Brandon 38261
Central Missouri State University, Warrensburg 32951
Central Museum of Armed Forces, Moskva 23515
Central National Library of the Republic of Montenegro, Cetinje 18912
Central Naval Library, Sankt-Peterburg 23915
Central Naval Museum, Sankt-Peterburg 23921
Central Northern Regional Library, Tamworth 01168
Central Ohio Technical College, Newark 32099
Central Oregon Community College, Bend 33165
Central Organisation for Traffic Safety in Finland, Helsinki 07446
Central Paper Research Institute, Pravdinsk 23705
Central Philippine University, Iloilo City 20699
Central Piedmont Community College, Charlotte 33213
Central Planning and Research Institute for Grain Industry, Moskva 23529
Central Planning Office, Suva 07300
Central Potato Research Institute, Simla 14210
Central Production and Research Station for Sericulture, Bucureşti 22014
Central Public Library, Floriana 18670
Central Public Library, Isa Town 01743
Central Public Library, Tehran 14441
Central Public Library of Komsomolsk, Komsomolsk-na-Amure 24147
Central Public Library 'Shaar Zion', Tel-Aviv 14782
Central Public Works Department Library, New Delhi 13869
Central Queensland TAFE, Gladstone 00442
Central Queensland TAFE, Mackay 00451
Central Queensland TAFE, Rockhampton 00506
Central Queensland University, North Rockhampton 00494
Central Rappahannock Regional Library, Fredericksburg 39156
Central Remedial Clinic, Dublin 14524
Central Repair-Mechanical Factory, Moskva 22860
Central Research and Design Institute of Fuel Equipment for Automobile and Tractor Engines, Sankt-Peterburg 23925
Central Research Centre of Haematology, Moskva 23180
Central Research Institute, Kasauli 14025
Central Research Institute for Animal Sciences, Bogor 14353
Central Research Institute for Cotton Industry, Moskva 23539
Central Research Institute for Customer Services, Moskva 23520
Central Research Institute for Jute and Allied Fibres, Kolkata 14036
Central Research Institute for Textile Machinery Components, Moskva 23528
Central Research Institute of Automobilea and Automobile Engines, Moskva 23516
Central Research Institute of Electric Power Industry, Abiko 17508
Central Research Institute of Electric Power Industry, Head Office, Tokyo 17672
Central Research Institute of Ferrous Metallurgy, Moskva 23521
Central Research Institute of Ferrous Metallurgy, Moskva 23522
Central Research Institute of Geodesy, Aerial Photography and Cartography, Moskva 23525
Central research institute of geological prospecting for base and precious metals, Moskva 23517
Central Research Institute of Mechanical Engineering Technology, Moskva 23535
Central Research Institute of Prosthetics and Prosthesis Design, Moskva 23532
Central Research Institute of the Hosiery Industry, Moskva 23537
Central Research Institute of the Wool Industry, Moskva 23533
Central Research Institute of Tuberculosis, Moskva 23538
Central Research Institute of Work Ability Testing and Invalid Rehabilitation, Moskva 23523
Central Research Library of Communications, Moskva 23502
Central Research Library of Inland Water Transport, Moskva 23500
Central Research Organization, Yangon 19001
Central Research Scientific Library for Civil Engineering and Architecture, Moskva 23499
Central Research Station for Agricultural Plants on Sandy Soils, Dăbuleni 22031
Central Rice Research Institute, Cuttack 13974
Central Road Research Institute, New Delhi 14121
Central Saint Martins College of Art & Design, London 29172
Central Salt & Marine Chemicals Research Institute, Bhavnagar 13946
Central Sanskrit Research Institute, Allahabad 13926
Central School of Speech and Drama, London 29173
Central Science Laboratory, York 29918
Central Scientific Agricultural Library, Krasnoobsk 23121
Central Scientific Agricultural Library of the Russian Academy of Agricultural Sciences, Moskva 23493
Central Scientific and Technical Library for Light Industry, Moskva 23496
Central Scientific and Technical Library of Coal Industry, Moskva 23504
Central Scientific Instruments Organization, Chandigarh 13951
Central Scientific Library of the Far Eastern Branch of the Russian Academy of Sciences, Vladivostok 22192
Central Scientific Library of the Uralian Department of the Russian Academy of Sciences, Ekaterinburg 22129
Central Scientific Medical Library, Moskva 23492
Central Scientific Research and Project Design Institute for Mechanization and Power Supply of Forst Industry (TsNIIME), Khimki 23100
Central Scientific Technical Library for Machine Tool Manufacture, Moskva 23540
Central Scientific Technical Library for Nonferrous Metallurgy, Moskva 23454
Central Secretariat Library, New Delhi 13870
Central Shanxi Teachers' College, Yuci 05763
Central Social Sciences Library, Hanoi 42276
Central South University of Finance and Economics, Wuhan 05650
Central St Martins College of Art and Design, London 29174
Central State Archives of Ancient Acts and Documents, Moskva 23477
Central State Archives of Georgia, Tbilisi 09228
Central State Archives of Kyrgyzia, Bishkek 18145
Central State Archives of Military History, Moskva 23512
Central State Archives of Republic Udmurtiya, Izhevsk 23063
Central State Archives of the Abkhazian Region, Sukhumi 09220
Central State Archives of the Tajik Republic, Dushanbe 27641
Central State Archives of Uzbekistan, Toshkent 42131
Central State Historical Archives, Yerevan 00347
Central State Hospital, Milledgeville 37109
Central State Hospital, Petersburg 37336
Central State Library, Patiala 14231
Central State University, Wilberforce 33059
Central Statistical Organisation, New Delhi 13871
Central Statistics Office, Cork 14498
Central Tafe, Northbridge 00496
Central TAFE, Perth 00582
Central Temperance Library, Helsinki 07485
Central Texas College, Killeen 31506
Central Theological College, Budapest 13252
Central Tobacco Research Institute, Rajahmundry 14201
Central Union for Child Welfare, Helsinki 07471
Central Union of Agricultural Cooperatives, Tokyo 17780
Central Union of Agricultural Cooperatives, Central Cooperative College, Machida 17593
Central University Library, Bucureşti 21916
Central University Library 'Lucian Blaga', Cluj-Napoca 21929
Central University Library 'Mihai Eminescu', Iaşi 21935
Central University of Nationalities, Beijing 05035
Central University of Technology, Free State, Bloemfontein 25001
Central Virginia Community College, Lynchburg 33507
Central Virginia Regional Library, Farmville 39049
Central Washington University, Ellensburg 31084
Central Water and Power Research Station, Pune 14189
Central Water Comission, New Delhi 14122
Central West Libraries, Orange 01137
Central Western Co-Operative Public Library, Orange 01138
Central Writers Club, Moskva 23508
Central Wyoming College, Riverton 33689
Central Zionist Archives, Jerusalem 14727
Centrale Administratieve Bibliotheek (CAB), Hasselt 02444
Centrale Bibliotheek Beverwijk, Beverwijk 19567
Centrale Bibliotheek De Mariënburg, Nijmegen 19667
Centrale Bibliotheek Hoofddorp, Hoofddorp 19630
Centrale Bibliotheek IJmuiden, IJmuiden 19639
Centrale bibliotheek Oosterhout, Oosterhout 19678
Centrale de l'Enseignement du Québec, Québec 04458
Centrale Générale des Syndicats Liberaux de Belgique, Gent 02642
Centrale Openbare Bibliotheek, Leuven 02800
Centrale organisatie werk en inkomen, Amsterdam 19353
Centrale Raad voor het Bedrijfsleven (CRB), Brussel 02577
Centralia College, Centralia 33207
Centralia Correctional Center Library, Centralia 34053
Centralia Regional Library District, Centralia 38495
Centralized Library System of Yelizovo District, Yelizovo 24361
Centrallasarettet, Västerås 26716
Centralna Biblioteka Geografii i Ochrony Środowiska, Warszawa 21244
Centralna Biblioteka Rolnicza, Bydgoszcz 21080
Centralna Biblioteka Rolnicza, Puławy 21212
Centralna Biblioteka Statystyczna im. Stefana Szulca, Warszawa 21245
Centralna Biblioteka Wojskowa im. Marszałka Józefa Piłsudskiego, Warszawa 21246
Centralna Narodna Biblioteka Crne Gore "Đurđe Crnojevic", Cetinje 18912
Centralne geografska biblioteka, Zagreb 06232
Centralne Laboratorium Akumulatorów i Ogniw, Poznań 21193
Centralne Muzeum Morskie, Gdańsk 21094
Centralne Muzeum Włókiennictwa, Łódź 21158
Centralny Instytut Ochrony Pracy, Warszawa 21247
Centralny Ośrodek Informacji Budownictwa, Warszawa 21248
Centralny Ośrodek Informatyki Górnictwa SA, Katowice 21115
Centralsjukhuset, Karlstad 26602
Central-South China Institute of Mining and Metallurgy, Changsha 05090
Centre africain de formation et de recherche administratives pour le développement (CAFRAD), Tanger 18966
Centre Belge de la Bande Dessinée, Brussel 02567
Centre Canadien d'Architecture / Canadian Centre for Architecture, Montreal 04388
Centre Canadien d'Etudes et de Cooperation Internationale (CECI), Montreal 04389
Centre Charles Péguy, Orléans 08456
Centre College of Kentucky, Danville 30957
Centre County Law Library, Bellefonte 33981
Centre County Library & Historical Museum, Bellefonte 38150
Centre Cultural del Comú d'Escaldes-Engordany, Escaldes-Engordany 00098
Centre Culturel Albert Camus, Antananarivo 18457
Centre Culturel Algérien, Paris 08503
Centre Culturel Américain, Douala 03648
Centre culturel André Malraux, Verrières-le-Buisson 09134
Centre Culturel Calouste Gulbenkian, Paris 08504
Centre culturel canadien, Paris 08505
Centre Culturel Egyptien, Rabat 18956
Centre Culturel Espagnol, Tétouan 18971
Centre Culturel et de Cooperation Linguistique (CCCL), Bremen 11673
Centre Culturel Français, Abidjan 06134
Centre Culturel Français, Annaba 00073
Centre Culturel Français, Bamako 18645
Centre Culturel Français, Bangui 04832
Centre Culturel Français, Brazzaville 06099
Centre Culturel Français, Bujumbura 03627
Centre Culturel Français, Constantine 00076
Centre Culturel Français, Cotonou 02904
Centre Culturel Français, Dakar 24435
Centre Culturel Français, Douala 03649
Centre Culturel Français, Haifa 14719
Centre Culturel Français, Helsinki 07447
Centre Culturel Français, Jakarta 14362
Centre Culturel Français, Khartoum 26356
Centre Culturel Français, Kinshasa 06071
Centre Culturel Français, Libreville 09173
Centre Culturel Français, Lomé 27799
Centre Culturel Français, Luxembourg 18408
Centre Culturel Français, Nairobi 18034
Centre Culturel Français, Niamey 19909
Centre Culturel Français, Nicosia 06335
Centre Culturel Français, Palermo 16397
Centre Culturel Français, Roma 16477
Centre Culturel Français, Saint-Louis 24449
Centre Culturel Français, Torino 16595
Centre Culturel Français, Tripoli 18257
Centre Culturel Français, Vilnius 18303
Centre Culturel Français, Yaoundé 03654
Centre Culturel Français Antoine de Saint-Exupéry, Nouakchott 18686
Centre Culturel Français d'Alexandrie, Alexandria 07147
Centre Culturel Français de Groningue, Groningen 19427
Centre Culturel Français Freiburg, Freiburg 11889
Centre Culturel Français G. Méliès, Ouagadougou 03617
Centre Culturel Irlandais, Paris 08506
Centre Culturel Syrien, Paris 08507
Centre d'accès à l'Information Juridique-Bibliothèque de Québec, Québec 04129
Centre d'Animation et de Recherche en Histoire Ouvrière et Populaire (Carhop), Bruxelles 02578
Centre de Coopération Internationale en Recherche Agronomique pour le Développement (CIRAD), Montpellier 08422
Centre de Diffusion Culturelle et de Documentation Pédagogique de Tunis, Tunis 27830
Centre de documentation A.-G. Haudricourt, Villejuif 08714
Centre de Documentation Agricole (CDA), Kinshasa 06072
Centre de Documentation Collégiale, La Salle 03732
Centre de Documentation des Experts Comptables et des Commissaires aux Comptes, Paris 08508
Centre de documentation des musées Les Arts décoratifs, Paris 08509
Centre de documentation du Soir, Bruxelles 02501
Centre de Documentation et d'Animation à la Culture Catalane (CDACC), Perpignan 08642
Centre de Documentation et d'Animation Interculturelles (CDAIC), Luxembourg 18409

Centre de Documentation et de Recherche Religieuses, Namur 02482
Centre de Documentation et de Recherche sur la Resistance (CDRR), Luxembourg 18410
Centre de Documentation et d'Information Pédagogiques, Porto-Novo 02907
Centre de Documentation Médico-Pharmaceutique, Paris 08510
Centre de Documentation Nationale, Tunis 27831
Centre de Documentation Negocia-EAP, Paris 08511
Centre de Documentation sur l'Education des Adultes et la Condition Feminine, Montreal 04390
Centre de Documentation Toxicologique Fernand Widal, Paris 08512
Centre de Documentation UNESCO-ICOMOS, Paris 08513
Centre de foresterie des Laurentides, Sainte-Foy 04485
Centre de Formation Administrative, Constantine 00051
Centre de Formation Administrative, Oran 00052
Centre de Formation des Journalistes (CFJ), Paris 08129
Centre de Formation Pédagogique, Saint-Louis 24450
Centre de Hautes Etudes d'Art Dramatique, Hammamet 27820
Centre de Lectura de Reus, Reus 26244
Centre de Lecture publique, Gembloux 02763
Centre de Lecture publique, Lobbes 02805
Centre de Musique Canadienne, Toronto 04510
Centre de Recherche Agro-Alimentaire (CRAA), Lubumbashi 06083
Centre de Recherche Agronomique de Bébédjia, Moundou 04837
Centre de Recherche en Sciences Humaines (CRSH), Kinshasa 06073
Centre de Recherche Géographique et de Production Cartographique, Brazzaville 06100
Centre de Recherche Lionel-Groulx, Outremont 04454
Centre de Recherche pour l'Etude et l'Observation des Conditions de Vie (CREDOC), Paris 08514
Centre de Recherche Public Gabriel Lippmann, Belvaux 18404
Centre de Recherches Anthropologiques, Préhistoriques et Ethnographiques (CRAPE), Alger 00058
Centre de Recherches et de Documentation du Sénégal (CRDS), Saint-Louis 24451
Centre de Recherches Géologiques et Minières, Kinshasa 06074
Centre de Recherches Océanologiques (CRO), Abidjan 06135
Centre de Recherches Routières (CRR), Sterrebeek 02700
Centre de Recherches Sociologiques sur le Droit et les Institutions Pénales (CESDIP), Guyancourt 08363
Centre de ressources et de documentation pédagogiques (CRDP), Genève 27360
Centre de Santé et de Services Sociaux du Nord de Lanaudière Bibliothèque, Saint Charles Borromee 04477
Centre d'Echanges Culturels Franco-Rwandais, Kigali 24376
Centre d'Enseignement Supérieur de Brazzaville, Brazzaville 06095
Centre d'Enseignement Supérieur de la Marine (CESM), Paris 07841
Centre Départemental de Documentation Pédagogique (CDDP Isère), Grenoble 08353
Centre Départemental de Documentation Pédagogique de la Haute-Marne, Chaumont 08333
Centre Départemental de Documentation Pédagogique de la Haute-Saône, Vesoul 08133
Centre départemental de documentation pédagogique de l'Aube, Troyes 08700
Centre départemental de documentation pédagogique de Meurthe-et-Moselle, Nancy 08435
Centre Départemental de Documentation Pédagogique des Hautes-Pyrénées, Tarbes 08690
Centre Départemental de Documentation Pédagogique des Hauts-de-Seine, Boulogne-Billancourt 08310

Centre Départemental de Documentation Pédagogique du Finistère, Quimper 08650
Centre Départemental de Documentation Pédagogique du Tarn, Albi 08272
Centre Départemental d'Histoire des Familles (CDHF), Guebwiller 08360
Centre des littératures, des paralittératures, de la bande dessinée et du cinéma, Embourg 02639
Centre d'Estudis Jurídics i Formació Especialitzada, Barcelona 25888
Centre d'Etude d'Afrique Noire, Pessac 08644
Centre d'Etude de la Peinture du XVE siecle dans les Pays bas Meridionaux et la Principaute de Liège, Brussel 02579
Centre d'étude de l'Energie Nucléaire, Mol 02687
Centre d'Etude et de Documentation, Bruxelles 02426
Centre d'Etude et de Recherche en Histoire de l'Education, Saint-Brieuc 08663
Centre d'Etude et de Recherches Vétérinaires et Agrochimiques (CERVA), Brussel 02582
Centre d'Etude et de Réflexion Chrétienne, Orléans 08208
Centre d'Etudes Africaines et Arabes de Toulouse, Toulouse 08693
Centre d'Etudes de Carthage, Tunis 27815
Centre d'Etudes de Saclay (CENS), Gif-sur-Yvette 08350
Centre d'Etudes Economiques et Sociales de l'Afrique de l'Ouest, Bobo-Dioulasso 03614
Centre d'Etudes et de Documentation Economique, Juridiques et Sociale, Cairo 07161
Centre d'Etudes et de Formation Pédagogiques CFP, Paris 07842
Centre d'Etudes et de Recherches Economiques et Sociales (CERES), Tunis 27832
Centre d'Etudes et de Recherches Européennes (CERE) Robert Schuman, Luxembourg 18411
Centre d'Etudes et de Recherches sur le Moyen-Orient Contemporain (CERMOC), Beirut 18217
Centre d'Etudes et de Recherches sur les Mouvements Trotskistes et Révolutionnaires internationaux (CERMTRI), Paris 08515
Centre d'Etudes et de Recherches sur les Qualifications (CEREQ), Marseille 08405
Centre d'Etudes Islamiques, Taroudant 18970
Centre d'Etudes ISTINA, Paris 08516
Centre d'Etudes Linguistiques et Historiques par Tradition Orale, Niamey 19910
Centre d'Etudes Œcuméniques (CEO), Strasbourg 08236
Centre d'Etudes Préhistoire Antiquité Moyen-Age, Valbanne 08704
Centre d'Etudes Prospectives et d'Informations Internationales (CEPII), Paris 08517
Centre d'Etudes Techniques de l'Equipment, Le Grand-Quevilly 08376
Centre d'Exécution de Programmes Sociaux et Economiques (CEPSE), Lubumbashi 06084
Centre d'Histoire de la Résistance et de la Déportation (CHRD), Lyon 08393
Centre d'Information des Nations Unies, Brazzaville 06101
Centre d'Information et de Documentation des Femmes "Thers Bode", Luxembourg 18412
Centre d'Information et de Documentation Economique et Sociale (CIDES), Niamey 19911
Centre d'Information et d'Etudes sur les Migrations Internationales (CIEMI), Paris 08518
Centre DJOLIBA recherche-formation-documentation pour le développement, Bamako 18646
Centre Dominicain de Recherches Socio-Culturelles, Yaoundé 03655
Centre Européen de Documentation Musicale, Strassen 18421
Centre europén pour l'enseignement supérieur (CEPES), Bucureşti 22015
Centre Excursionista de Catalunya, Barcelona 25889
Centre for Addiction & Mental Health Library, Toronto 04513
Centre for Advanced Aacademic Studies (CAAS), Dubrovnik 06214

Centre for Advanced Studies in Applied Mathematics, Kolkata 14037
Centre for Arab Unity Studies, Beirut 18218
Centre for Asia Minor Studies, Athinai 13102
Centre for Automotive Safety, Adelaide 00848
Centre for Black and African Arts and Civilisation, Lagos 20030
Centre for Chemistry of Drugs, Moskva 23172
Centre for Culture Ethnicity and Health, Carlton 00899
Centre for Development Information, Colombo 26306
Centre for Development Studies, Thiruvananthapuram 14215
Centre for Disability Services and Research, Calgary 04324
Centre for Ecology and Hydrology, Penicuik 29874
The Centre for Environment, Fisheries and Aquaculture Science, Lowestoft 29531
Centre for Environmental Planning and Technology, Ahmedabad 13594
Centre for Environmental Science, Madrid 25990
Centre for Historical Research and Documentation on War and Contemporary Society, Bruxelles 02575
Centre for Hygiene Education, Grodno 01961
Centre for Hygiene Education, Vitebsk 02080
Centre for Independent Studies, Saint Leonards 00998
Centre for Industrial Technology Information Services, Katubedda 26297
Centre for Industrial Technology Information Services, Moratuwa 26325
Centre for Iranian Anthropology, Tehran 14427
Centre for Kentish Studies, Maidstone 29848
Centre for Library and Information Science, Budapest 13388
Centre for Local Economic Strategies, Manchester 29850
Centre for Management Development, Ikeja 20024
Centre for Molecular and Macromolecular Studies, Łódź 21159
Centre for Nuclear Research, Dalat 42284
Centre for Policy on Ageing, London 29737
Centre for Propagation of Science and Technology, Sankt-Peterburg 23757
Centre for Reformation and Renaissance Studies, Toronto 03881
Centre for Regional Studies, Békéscsaba 13367
Centre for Research in Chemistry, Sofiya 03579
Centre for Research in Physics, Sofiya 03529
Centre for South Asian Studies, Lahore 20510
Centre for Strategic and International Studies, Jakarta 14363
The Centre for Swedish Folk Music and Jazz Research, Stockholm 26690
Centre for the study of Fifteenth-Century Painting in the Southern Netherlands and the Principality of Liege, Brussel 02579
Centre Français de Culture et de Coopération (CFCC), Cairo 07162
Centre Français de Droit Comparé (CFDC), Paris 08519
Centre Guy Mollet, Paris 08520
Centre Historique Minier, Lewarde 08381
Centre Hôspitalier du Mans, Le Mans 08379
Centre Hospitalier Robert-Giffard Bibliothèque, Beauport 04312
Centre Hôspitalier Sainte-Anne, Paris 08521
Centre Hôspitalier Spécialisé de Saint-Egrève, Saint-Egrève 08665
Centre International de Crimologie Comparée, Montreal 04391
Centre International de Recherche sur le Cancer (CIRC), Lyon 08398
Centre International de Recherches Sur l'Anarchisme (CIRA), Lausanne 27393
Centre International des Civilisations Bantu, Libreville 09174
Centre International des Recherches Médicales de Franceville, Franceville 09172
Centre International d'Etudes Pédagogiques (CIEP), Sèvres 08677
Centre International d'Information de Musique Contemporaine, Paris 08474
Centre International du Vitrail, Chartres 08327
Centre international francophone de documentation et du formation (CIFDI), Bordeaux 08303
Centre interrégional de perfection-nement (CIP) / Interregionales Fortbildungszentrum, Tramelan 27441
Centre Jeanne-d'Arc, Orléans 08457

Centre La Baume-les-Aix, Aix-en-Provence 08172
Centre Lebret, Dakar 24426
Centre Michel Foucault d'études françaises, Warszawa 20989
Centre Militaire, Diekirch 18392
Centre Multimédia Don Bosco, Liège 02802
Centre National d'Appui a la Lutte contre la Maladie (CNAM), Bamako 18647
Centre National de Documentation et d'Information Scientifique et Technique, Brazzaville 06102
Centre National de Documentation Pédagogique (CNDP), Paris 08522
Centre National de Formation Météorologique, Kinshasa 06075
Centre National de la Bande Dessinée et de l'Image (CNBDI), Angoulême 08276
Centre National de la litterature pour la jeunesse, Paris 08493
Centre National de la Recherche Scientifique et Technologique, Ouagadougou 03618
Centre national de littérature (Maison Servais), Mersch 18417
Centre National de Recherches Agronomiques (CNRA), Bambey 24428
Centre National de Recherches Océanographiques, Nosy-Bé 18465
Centre National de Recherches Préhistoriques, Anthropologiques et Historiques, Alger 00059
Centre National de Recherches Zootechniques, Bamako 18648
Centre National d'Etudes et d'Analyses pour la Population et le Developpement (CENEAPED), Alger 00060
Centre National d'Etudes Historiques, Kouba 00080
Centre National d'Etudes Supérieures de Sécurité Sociale (CNESSS), Saint-Etienne 08667
Centre National du Machinisme Agricole, du Genie Rural, des Eaux et des Forêts, Cestas 08319
Centre National du Machinisme Agricole, du Génie Rural, des Eaux et des Forêts (CEMAGREF), Antony 08280
Centre national supérieur de formation et de recherche pour l'éducation des jeunes handicapés et les enseignements adaptés (INS HEA), Suresnes 08069
Centre Nationale de Documentation et de Recherches Scientifique, Moroni 06043
Centre of Bibliography and Book Science within the Martynas Mažvydas National Library of Lithuania, Vilnius 18300
Centre of International and European Economic Law, Thessaloniki 13131
Centre of Librarian's and Cultural Services, Chomutov 06561
Centre of Planning and Economic Research (KEPE), Athinai 13103
Centre of Scientific and Technical Information, Bryansk 23010
Centre of Scientific and Technical Information and Popularization, Barnaul 22998
Centre of Scientific and Technical Information and Popularization, Cheboksary 23015
Centre of Scientific and Technical Information and Popularization, Chita 23020
Centre of Scientific and Technical Information and Popularization, Irkutsk 23054
Centre of Scientific and Technical Information and Popularization, Ivanovo 23060
Centre of Scientific and Technical Information and Popularization, Kaliningrad 23068
Centre of Scientific and Technical Information and Popularization, Kaluga 23076
Centre of Scientific and Technical Information and Popularization, Kemerovo 23091
Centre of Scientific and Technical Information and Popularization, Khabarovsk 23098
Centre of Scientific and Technical Information and Popularization, Kirov 23102
Centre of Scientific and Technical Information and Popularization, Kostroma 23108
Centre of Scientific and Technical Information and Popularization, Krasnodar 23113

Centre of Scientific and Technical Information and Popularization, Krasnoyarsk 23127
Centre of Scientific and Technical Information and Popularization, Kurgan 23130
Centre of Scientific and Technical Information and Popularization, Lipetsk 23141
Centre of Scientific and Technical Information and Popularization, Makhachkala 23151
Centre of Scientific and Technical Information and Popularization, Nizhni Novgorod 23614
Centre of Scientific and Technical Information and Popularization, Novosibirsk 23653
Centre of Scientific and Technical Information and Popularization, Omsk 23667
Centre of Scientific and Technical Information and Popularization, Orenburg 23674
Centre of Scientific and Technical Information and Popularization, Oryol 23679
Centre of Scientific and Technical Information and Popularization, Penza 23685
Centre of Scientific and Technical Information and Popularization, Rostov-na-Donu 23728
Centre of Scientific and Technical Information and Popularization, Ryazan 23735
Centre of Scientific and Technical Information and Popularization, Samara 23741
Centre of Scientific and Technical Information and Popularization, Smolensk 23974
Centre of Scientific and Technical Information and Popularization, Tambov 23991
Centre of Scientific and Technical Information and Popularization, Tomsk 23999
Centre of Scientific and Technical Information and Popularization, Tula 24005
Centre of Scientific and Technical Information and Popularization, Tver 24009
Centre of Scientific and Technical Information and Popularization, Tyumen 24016
Centre of Scientific and Technical Information and Popularization, Ufa 24021
Centre of Scientific and Technical Information and Popularization, Ulan-Ude 24033
Centre of Scientific and Technical Information and Popularization, Ulyanovsk 24037
Centre of Scientific and Technical Information and Popularization, Vladikavkaz 24045
Centre of Scientific and Technical Information and Popularization, Vladimir 24050
Centre of Scientific and Technical Information and Popularization, Volgograd 24064
Centre of Scientific and Technical Information and Popularization, Voronezh 24078
Centre of Scientific and Technical Information and Popularization, Yaroslavl 24090
Centre of Scientific and Technical Information of the Civil Aviation, Moskva 23489
Centre of Traumatology and Hygiene, Donetsk 28218
Centre orthodoxe du patriarcat œcuménique, Chambésy 27251
Centre Penitentiaire de Luxembourg, Schrassig 18420
Centre Pierre Mendès-France, Paris 07909
Centre Pompidou, Paris 09017
Centre pour la Recherche Interdisciplinaire sur le Développement, Bruxelles 02580
Centre Psychothérapique, Laxou 08372
Centre Régional Africain de Technologie, Dakar 24436
Centre Régional Africain de Technologie, Dakar 24430
Centre régional de documentation en art contemporain, Nîmes 08453
Centre Régional de Documentation Pédagogique (CRDP), Toulouse 08694
Centre Régional de Documentation Pédagogique de Clermont-Ferrand (CRDP), Clermont-Ferrand 08337
Centre Régional de Documentation Pédagogique de Haute-Normandie, Mont-Saint-Aignan 08427
Centre Régional de Documentation Pédagogique de l'Académie de Lyon, Lyon 08394
Centre Régional de Documentation Pédagogique du Nord – Pas de Calais, Arras 08283
Centre Régional de Documentation Pédagogique du Nord – Pas de Calais, Calais 08316
Centre Régional de Documentation Pédagogique du Nord – Pas de Calais, Dunkerque 08348
Centre Régional de Documentation Pédagogique du Nord – Pas de Calais, Lille 08385
Centre Régional de Documentation Pédagogique du Nord – Pas de Calais, Valenciennes 08706
Centre Régional de Documentation Pédagogique Poitou-Charentes, Poitiers 08647
Centre Régional de Recherche Sud-Bénin, Attogon 02903
Centre Regional de Service aux Bibliothèque Publique de Prêt Gaspesie Isle de la Madelene, Cap-Chat 04604
Centre Regional de Services aux Bibliothèques Publiques, Trois-Rivières 04805
Centre Regional de Services aux Bibliothèques Publiques (CRSBP) de la Côte-Nord, Inc, Sept Iles 04753
Centre Régional de Services aux Bibliothèques Publiques (CRSBP) du Bas-Saint-Laurent, Rivière-du-Loup 04738
Centre Regional de Services aux Bibliothèques Publiques de la Capitale-Nationale et de la Chaudiere-Appalaches Inc, Charny 04608
Centre Regional de Services aux Bibliothèques Publiques de la Monteregie Inc, La Prairie 04674
Centre Regional de Services aux Bibliothèques Publiques de l'Abitibi-Temiscamingue-Nord-du-Québec, Rouyn-Noranda 04740
Centre Scientifique et Technique de la Construction, Limelette 02683
Centre Scientifique et Technique du Bâtiment (CSTB), Paris 08523
Centre St Irénée, Lyon 08199
Centre Suisse de Coordination pour la Recherche en Education (CSRE), Aarau 27297
Centre Technique des Industries Mécaniques (CETIM), Senlis 08675
Centre Technique National d'Etudes et de Recherches sur les Handicaps et les Inadaptions (CTNERHI), Paris 08524
Centre Territorial de Recherche et de Documentation Pédagogiques de Nouvelle-Calédonie, Nouméa 19755
Centre Virtuel de la Connaissance sur l'Europe (CVCE), Sanem 18419
Centres Jeunesse de Montréal, Montreal 04392
Centro Agronomico Tropical de Investigación y Eseñanza (CATIE), Turrialba 06126
Centro Applicazione Militare Energie Nucleari (CAMEN), San Piero a Grado 15689
Centro Argentino de Ingenieros, Buenos Aires 00245
Centro Artigas-Washington, Montevideo 42046
Centro Asociado a la UNED de Correos, Madrid 25669
Centro Brasileiro de Pesquisas Físicas, Rio de Janeiro 03335
Centro Camuno di Studi Preistorici, Capo di Ponte Valcamonica 16218
Centro Cientifico e Cultural de Macau, Lisboa 21776
Centro Colombo Americano, Medellín 06014
Centro Comunicazioni Tecniche e Scientifiche (della 3M Italia Spa), Ferrania 16144
Centro Cultural Brasil-Estados Unidos, Santos 03384
Centro Cultural Costarricense-Norteamericano, San José 06120
Centro Cultural Hispánico, Damascus 27519
Centro Cultural Paraguayo-Americano, Asunción 20608
Centro Culturale Tedesco, Torino 16604
Centro de Altos Estudios Nacionales, Montevideo 42047
Centro de Arte Moderna José de Azeredo Perdigão, Lisboa 21777
Centro de Ciências, Letras e Artes, Campinas 03288
Centro de Ciencias Medioambientales, Madrid 25990
Centro de Cooperación Regional para la Educación de Adultos en América Latina y el Caribe (CREFAL), Pátzcuaro 18868
Centro de Documentação e Informação – Estação Zootécnica Nacional, Santarém 21827
Centro de Documentação Técnica e Científica, São Tomé 24387
Centro de Documentación de las Artes Escénicas de Andalucía, Sevilla 26147
Centro de Documentación del Transporte, Madrid 25699
Centro de Documentación Internacional, Buenos Aires 00246
Centro de Documentación Médica de Cataluña, Barcelona 25890
Centro de Energia Nuclear na Agricultura (CENA), Piracicaba 03314
Centro de Enseñanza Técnica y Superior (CETYS), Mexicali 18720
Centro de Enseñanzas Integradas, Cáceres 25663
Centro de Enseñanzas Integradas, Cheste 25664
Centro de Enseñanzas Integradas, Eibar 25665
Centro de Enseñanzas Integradas, Haciadama-Elburgo 25667
Centro de Enseñanzas Integradas, Huesca 25668
Centro de Enseñanzas Integradas, Toledo 25675
Centro de Ensino Superior de Juiz de Fora, Juiz de Fora 03200
Centro de Estudios Aplicados al Desarrollo Nuclear (CEADEN), La Habana 06293
Centro de Estudios Avanzados de Puerto Rico y el Caribe, San Juan 21867
Centro de Estudios Borjanos, Borja 25935
Centro de Estudios de Historia y Organización de la Ciencia 'Carlos J. Finlay', La Habana 06294
Centro de Estudios Educativos, México 18839
Centro de Estudios Extremeños, Badajoz 25873
Centro de Estudios Histórico-Militares del Perú, Lima 20644
Centro de Estudios Históricos, Madrid 25991
Centro de Estudios Jurídicos, Madrid 25992
Centro de Estudios Jurídicos y Formación Especializada, Barcelona 25888
Centro de Estudios Norteamericano, Valencia 26169
Centro de Estudios Políticos y Constitucionales, Madrid 25993
Centro de Estudios Regionales Andinos Bartolomé de las Casas, Cuzco 20653
Centro de Estudios Sociales de la Abadía de Santa Cruz del Valle de los Caídos, San Lorenzo del Escorial 26132
Centro de Estudios Teológicos de la Amazonia Peruana (CETA), Loreto 20679
Centro de Estudios y Experimentación de Obras Públicas – Centro de Estudios de Técnicas Aplicadas, Madrid 25700
Centro de Estudo e Apoio à Criança e à Familia, Lisboa 21778
Centro de Estudos Africanos, Maputo 18977
Centro de Estudos da Guiné-Bissau, Bissau 13190
Centro de Estudos e Formaçaõ des Portug, Lisboa 21779
Centro de Estudos Egas Moniz, Lisboa 21780
Centro de Física Miguel A. Catalán, Madrid 25994
Centro de Formación de la Policía, Ávila 25679
Centro de Información de las Naciones Unidas para Colombia, Ecuador y Venezuela, Santafé de Bogotá 06025
Centro de Informacion de Recursos Naturales (CIREN), Santiago 04923
Centro de Información Documental de Archivos, Madrid 25995
Centro de Información Geo-Biologica del Noroeste Argentino, San Miguel de Tucumán 00315
Centro de Información MICONS, La Habana 06282
Centro de Información Técnica de la Construcción, La Habana 06295
Centro de Información y Documentación Agraria, Sevilla 26148
Centro de Información y Documentación Científica (CINDOC), Madrid 25996
Centro de Informações e Biblioteca em Educação (CIBEC), Brasília 03279
Centro de Intervenção para o Desenvolvimento Amilcar Cabral, Lisboa 21781
Centro de Investigación de la Caña de Azucar de Colombia (CENICAÑA), Florida 06012
Centro de Investigación de Relaciones Internacionales y Desarrollo, Barcelona 25892
Centro de Investigación y de Estudios Avanzados – Depto de Investigaciones Educativas, México 18723
Centro de Investigación y Desarrollo de la Educación (CIDE), Santiago 04924
Centro de Investigación y Desarrollo de Petróleos de Venezuela, SA, Caracas 42255
Centro de Investigación y Docencia Económicas, México 18724
Centro de Investigación y Educación Popular (CINEP), Santafé de Bogotá 06026
Centro de Investigación y Promoción del Campesinado (CIPCA), Piura 20680
Centro de Investigación y Tecnología Agroalimentaria de Aragón, Zaragoza 26190
Centro de Investigaciones Biológicas, Madrid 25997
Centro de Investigaciónes de Recursos Naturales, Castelar 00295
Centro de Investigaciones Económicas, Buenos Aires 00247
Centro de Investigaciones Energéticas, Medioambientales, Tecnológicas (CIEMAT), Madrid 25998
Centro de Investigaciones Sociológicas (CIS), Madrid 25999
Centro de Investigaciones y Capacitación Forestales del INDAF, La Habana 06296
Centro de Letras e Artes da UNI-RIO, Rio de Janeiro 03201
Centro de Pesquisa Agroflorestal da Amazonia Oriental (CPATU), Belém 03269
Centro de Pesquisa Agropecuaria do Meio-Norte, Teresina 03423
Centro de Pesquisa Agropecuária do Trópico Umido (CPATU), Belém 03270
Centro de Pesquisa e Gestão de Recursos Pesqueiros Continentais, Pirassununga 03315
Centro de Pesquisas e Desenvolvimento (CEPED), Camaçari 03287
Centro de Pesquisas e Desenvolvimento Leopoldo A. Miguez de Mello, Rio de Janeiro 03260
Centro de Pesquisas Gonçalo Moniz, Salvador 03376
Centro de Recursos Educativos, Madrid 25701
Centro de Tecnologías Físicas "Leonardo Torres Quevedo", Madrid 26000
Centro Deportivo Israelita, A.C., México 18840
Centro di Cultura Ebraica, Roma 16478
Centro di Documentazione della Fondazione G. Agnelli, Torino 16596
Centro di Documentazione e Ricerca Visiva, Livorno 16303
Centro di Lingua e Cultura tedesca, Palermo 16399
Centro di ricerca e documentazione Arti Visive, Roma 16479
Centro di Ricerca 'G. Dorso' per lo Studio del Pensiero Meridionalistico, Avellino 16171
Centro di Ricerca per la Selvicoltura, Arezzo 16170
Centro di Ricerca per lo studio delle Relazioni tra Pianta e Suolo (CRA-RPS), Roma 16480
Centro di Ricerca per lo Studio delle Relazioni tra Planta e Suolo, Roma 16560
Centro di Servizi Culturali, Pescara 15655
Centro di Studi sull'Ordine dei Servi di Maria, Bologna 15744
Centro di Studio per la Polarografia, Padova 16391
Centro didattico cantonale, Bellinzona 27316
Centro didattico cantonale, Massagno 27410
Centro d'Investigació i Desenvolupament, Barcelona 25891
Centro Espacial San Miguel, San Miguel 00314
Centro Estatal de Documentación e Información de Servicios Sociales, Madrid 25702
Centro Evangelizzazione e Catechesi Don Bosco, Cascine Vica 15780

Centro Federal de Educação Tecnológica do Paraná, Curitiba 03294
Centro Fonseca, A Coruña 25774
Centro Fusione e Applicazioni Laser, Frascati 16276
Centro Interamericano de Administración del Trabajo (CIAT), Lima 20657
Centro Interamericano de Investigación y Documentación sobre Formación Profesional (CINTERFOR), Montevideo 42048
Centro Internacional de Agricultura Tropical (CIAT), Calí 06008
Centro Internacional de Estudios Superiores de Comunicación para América Latina (CIESPAL), Quito 06980
Centro Internacional de la Papa (CIP), Lima 20658
Centro Internacional del Libro Infantil y Juvenil, Salamanca 26130
Centro Internazionale di Studi di Architettura "Andrea Palladio", Vicenza 16671
Centro Latinoamericano de Economía Humana, CLAEH, Montevideo 42049
Centro Librario e Bibliotecario 'Cristo Re', Portici 16424
Centro Mazziano, Verona 16667
Centro Missionario Pime, Milano 15906
Centro Museo Vasco de Arte Contemoporáneo, Vitoria-Gasteiz 26185
Centro Nacional de Alimentación, Majadahonda 26099
Centro Nacional de Biotecnología (CSIC), Madrid 26001
Centro Nacional de Documentação e Informação de Moçambique, Maputo 18981
Centro Nacional de Documentação e Investigação Histórico, Luanda 00105
Centro Nacional de Documentación Científica y Tecnológica, La Paz 02935
Centro Nacional de Documentación e Información Educativa, Buenos Aires 00248
Centro Nacional de Educación Ambiental (CENEAM), Valsain 26181
Centro Nacional de Investigación Teatral 'Rodolfo Usigli' – CITRU, México 18841
Centro Nacional de Investigaciones Agropecuarias, Maracay 42261
Centro Nacional de Investigaciones Científicas, La Habana 06297
Centro Nacional de Investigaciones Metalúrgicas (CENIM), Madrid 26002
Centro Nacional de Pesquisa de Arroz e Feijão (EMBRAPA), Santo Antônio de Goiás 03383
Centro Nacional de Protecção da Produção Agrícola, Lisboa 21720
Centro Nacional de Tecnología Agropecuaria (CENTA), San Salvador 07209
Centro Nazionale di Studi Manzoniani, Milano 16329
Centro Panamericano de Ingeniería Sanitaria y Ciencias del Ambiente (CEPIS), Lima 20659
Centro Pedagogico Meridionale, Bari 16174
Centro Pignatelli, Zaragoza 26191
Centro Protección de Choferes de Montevideo, Montevideo 42068
Centro Química Orgánica "Manuel Lora Tamayo", Madrid 26003
Centro Regional de Investigaciones Científicas y Tecnológicas, Mendoza 00304
Centro Regional para la Educación Superior en América Latina y El Caribe, Caracas 42249
Centro Regional Universitário de Espírito Santo do Pinhal, Espírito Santo do Pinhal 03199
Centro Ricerche Energia Casaccia, Roma 16481
Centro Roberto Bellarmino, Santiago 04925
Centro Simulazione Validazione dell'Esercito, Civitavecchia 15619
Centro Sperimentale di Cinematografia, Roma 16482
Centro Sperimentale Zootecnico Veterinario Farmitalia, Nerviano 16385
Centro Studi Americani, Roma 16483
Centro Studi del Teatro Stabile di Torino, Torino 16597
Centro Studi Educativi, Milano 16330
Centro Studi Francescani per la Liguria, Genova 15846
Centro Studi Nazionale CISL, Firenze 16257
Centro Studi Piero Gobetti, Torino 16598
Centro Studi Saint-Louis-de-France, Roma 16484
Centro Studi Salentini, Lecce 16302
Centro Studi Sull'Arte Licia e Carlo Ludovico Ragghianti, Lucca 16309
Centro Superior de Estudios de la Defensa Nacional (CESEDEN), Madrid 25703
Centro Sviluppo Materiali S.p.A., Roma 16485
Centro Sviluppo Materiali S.p.A., Roma 16486
Centro Técnico Aero-Espacial, São José dos Campos 03390
Centro Teológico, Las Palmas 25806
Centro Teologico, Torino 16097
Centro Universitário Salesiano, São Paulo 03248
Centrul de documentare medicală 'Dr Dimitrie Nanu', București 21974
Centrum Astronomiczne im. Mikołaja Kopernika PAN, Warszawa 21249
Centrum Badán Kosmicznych PAN, Warszawa 21250
Centrum Badan Molekularnych i Makromolekularnych PAN, Łódź 21159
Centrum Doskonalenia Nauczycieli Szkół Rolniczych, Brwinów-Pszczelin 21077
Centrum Educatieve Dienstverlening / Pedologisch Instituut, Rotterdam 19507
Centrum Edukacji Nauczycieli w Białymstoku, Białystok 21073
Centrum Medyczne Kształcenia Podyplomowego, Warszawa 21251
Centrum Naukowo-Techniczne Kolejnictwa, Warszawa 21252
Centrum Onkologii, Warszawa 21253
Centrum Techniki Okrętowej, Gdańsk 21095
Centrum voor Innovatie van Opleidingen, 's-Hertogenbosch 19461
Centrum voor Kandidaat Vreemdelingen, Brussel 02581
Centrum voor Onderzoek in Diergeneeskunde en Agrochemie (CODA) / Centre d'Etude et de Recherches Vétérinaires et Agrochimiques (CERVA), Brussel 02582
Centrum voor Sportcultuur, Hofstade 02664
Centrum voor Wiskunde en Informatica (CWI), Amsterdam 19352
The Century Association Library, New York 37192
Century Community & Technical College, White Bear Lake 33867
Century Correctional Institution Library, Century 34054
CERAM, Sophia-Antipolis 08041
Cercle litteraire, Lausanne 27492
Cereal Research Non-Profit Ltd, Szeged 13504
Ceredigion County Library, Aberystwyth 29923
CERGE-EI, Praha 06477
CERN – European Organization for Nuclear Research, Genève 27361
Cerrados Agricultural Research Center, Planaltina 03316
Cerritos College, Norwalk 33587
Cerritos Library, Cerritos 38497
Certosa di Farneta, Maggiano 15885
Cēsis Association of Museums, Cēsis 18179
Česká společnost entomologická, Praha 06478
Česká společnost zoologická, Praha 06479
Česká zemědělská univerzita, Praha 06383
České Vysoké Učení Technické v Praze, Praha 06384
Český hydrometeorologický ústav, Praha 06480
Český rozhlas, Praha 06481
Český statistický úřad, Praha 06482
Český svaz včelařů, Praha 06483
Cessna Aircraft Co, Wichita 36416
Cessnock City Library, Cessnock 01085
Ceylon Chamber of Commerce, Colombo 26307
CFP – Technique, Petit-Lancy 27226
Ch. Valikhanov Pedagogical Institute, Kokchetav 17911
CH2M Hill, Corvallis 35639
Chabot College, Hayward 33402
Chacra Experimental de Barrow, Tres Arroyos 00317
Chadbourne & Parke Library, New York 35966
Chadbourne & Parke LLP, Washington 36339
Chadron State College, Chadron 30696
Chaffe, McCall LLP, New Orleans 35944
Chaffey College, Rancho Cucamonga 33671
Chaikovsky School of Music, Ekaterinburg 22667
P.I. Chaikovsky State Conservatoire, Kyiv 28008
Chaim Sheba Medical Center, Tel Hashomer 14757
Chaleur Library Region, Campbellton 04603
Challenger Tafe, Beaconsfield 00387
Chalmers tekniska högskola, Göteborg 26375
– School of Architecture, Göteborg 26376
Chalmers University of Technology, Göteborg 26375
Chalyabinsk Regional Scientific Medical Library, Chelyabinsk 23017
Chamber of Commerce and Production, Santo Domingo 06966
Chamber of Commerce for Vienna, Wien 01663
Chamber of Deputies, México 18814
Chamber of Geological Engineers of Turkey, Ankara 27876
Chamber of Mines of South Africa, Marshalltown 25145
Chamberlain, D'Amanda, Oppenheimer & Greenfield, Rochester 36182
Chamberlain, Hrdlicka, White, Williams & Martin, Houston 35748
Chambre d'Agriculture de Perpignan, Perpignan 08643
Chambre de commerce, d'industrie, d'artisanat et d'agriculture d'antananarivo (CCIAA), Antananarivo 18458
Chambre de Commerce d'Industrie de la Region du Cap Vert, Dakar 24437
Chambre de Commerce d'Industrie et d'Agriculture de Dakar, Dakar 24438
Chambre de Commerce, d'Industrie et des Mines du Cameroun, Douala 03650
Chambre de Commerce et d'Industrie, Marseille 08406
Chambre de Commerce et d'Industrie de Bordeaux, Bordeaux 08304
Chambre de Commerce et d'Industrie de Paris (CCIP), Paris 08525
Chambre de Commerce et d'Industrie de Rouen, Rouen 08659
Chambre des Députés, Luxembourg 18394
Chambre des Députés, Tunis 27822
Chambre des Représentants, Rabat 18938
Chambre Régionale de Commerce et d'Industrie de Basse-Normandie, Caen 08314
Chaminade University of Honolulu, Honolulu 31345
Champaign County Law Library Association, Urbana 37698
Champaign County Library, Urbana 41645
Champaign Public Library, Champaign 38501
Champlain College, Burlington 33199
Champlain Regional College, Saint-Lambert 04026
Chancellor Robert R. Livingston Masonic Library of Grand Lodge, New York 37193
Chandler Public Library, Chandler 38502
Chandler Public Library – Basha, Chandler 38503
Chandler Public Library – Hamilton, Chandler 38504
Chandler Public Library – Sunset, Chandler 38505
Chang Gung Memorial Hospital, Taipei 27605
Changchun Architecture College, Changchun 05066
Changchun Auto Industry College, Changchun 05067
Changchun City Library, Changchun 05908
Changchun College of Geology, Changchun 05068
Changchun College of Traditional Chinese Medicine, Changchun 05069
Changchun Finance College, Changchun 05070
Changchun Institute of Applied Chemistry, Changchun 05855
Changchun Institute of Optics and Precision Instruments, Changchun 05071
Changchun Institute of Physics, Changchun 05856
Changchun Tax Institute, Changchun 05072
Changchun Teachers' College, Changchun 05073
Changchun University, Changchun 05074
Changchun University of Science and Technology, Changchun 05075
Changchun Water Conservancy and Power College, Changchun 05076
Changde College, Changde 05805
Changde Teachers' College, Changde 05089
Changhua County Library, Changhua 27626
Changsha Communications Institute, Changsha 05091
Changsha Railway University, Changsha 05092
Changsha University, Changsha 05093
Changsha Vocational Teachers' College, Changsha 05094
Changshu College, Changshu 05107
Changwei Teachers' College, Weifang 05643
Changzhi Medical College, Changzhi 05108
Changzhou City Library, Changzhou 05909
Changzhou Institute of Industrial Technology, Changzhou 05110
Chanute Public Library, Chanute 38508
Chaohu Teachers' College, Chaohu 05111
Chapel Hill Public Library, Chapel Hill 38509
Chapelwood United Methodist Church, Houston 35233
Chapin Memorial Library, Myrtle Beach 40356
Chapman University, Orange 32201
Chapman University School of Law, Orange 32202
Chappaqua Central School District Public Library, Chappaqua 38510
Chapter Library of Borgå Diocese, Borgå 07408
Charité – Universitätsmedizin Berlin, Berlin 09403
Charles A. Ransom District Library, Plainwell 40819
Charles B. Swartwood Supreme Court Library, Elmira 34185
Charles City Public Library, Charles City 38511
Charles County Circuit Court, La Plata 34426
Charles County Public Library, La Plata 39756
Charles Darwin University, Darwin 00432
Charles E. Stevens American Atheist Library and Archives Inc, Cranford 36758
Charles Evans Inniss Memorial Library, Brooklyn 30569
Charles H. Stone Memorial Library, Pilot Mountain 40788
Charles Stark Draper Laboratory, Inc, Cambridge 35550
Charles Sturt Library Service, Woodville 01179
Charles Sturt University, Bathurst 00386
Charles Town Library, Charles Town 38512
Charles University, Praha 06386
– 1st Faculty of Medicine, Praha 06396
– Astronomical Institute, Praha 06387
– Faculty of Education, Praha 06400
– Faculty of Law, Praha 06401
– Faculty of Mathematics and Physics, Praha 06399
– Faculty of Pharmacy, Hradec Králové 06392
– Faculty of Physical Training and Sport, Praha 06391
– Faculty of Science, Praha 06402
– History Institute and Archive of Charles University, Praha 06403
– Protestant Theological Faculty, Praha 06388
– Second Faculty of Medicine, Praha 06397
– Third Faculty of Medicine, Praha 06398
Charles-Edouard-Mailhot Library, Victoriaville 04812
Charleston Carnegie Public Library, Charleston 38513
Charleston County Public Library, Charleston 38514
Charleston County Public Library – West Ashley, Charleston 38515
Charleston Library Society, Charleston 36648
Charleston Museum Library, Charleston 36649
Charleston Southern University, Charleston 30713
Charlestown Boys' & Girls' Club, Charlestown 36652
Charlestown-Clark County Public Library, Charlestown 38517
Charlotte County Library System, Port Charlotte 40850
Charlotte Public Library, Charlotte 38518
Chartered Institute of Building, Ascot 29598
The Chartered Institute of Logistics and Transport (UK), Corby 29637
Chartered Institute of Management Accountants, London 29738
Chartered Institute of Personnel and Development, London 29739
Chartered Insurance Institute, London 29741
Chartered Management Institute, Corby 29638
Chase Manhattan Bank, New York 35967
Château de Chantilly – Musée Condé, Chantilly 08323
Château d'Oron, Oron-le-Châtel 27422
Chateauguay Municipal Library, Chateauguay 04609
Chatham Area Public Library District, Chatham 38527
Chatham College, Pittsburgh 32299
Chatham Public Library, Chatham 38528
Chatham-Kent Public Library, Chatham 04610
Chattahoochee Valley Regional Library System, Columbus 38655
Chattanooga State Community College, Chattanooga 33215

Chattanooga-Hamilton County Bicentennial Library, Chattanooga 38533
Chattanooga-Hamilton County Bicentennial Library – Eastgate, Chattanooga 38534
Chattanooga-Hamilton County Bicentennial Library – Ooltewah-CollegedaleOoltewah-Collegedale, Ooltewah 40621
Chattooga County Library, Summerville 41472
Chaudhary Charan Singh University, Meerut 13697
Chautauqua County Law Library, Mayville 34514
Chautauqua-Cattaraugus Library System, Jamestown 39633
Chavain National Library, Yoshkar-Ola 22199
Chaves County District Court Library, Roswell 37467
CHE – Christelijke Hogeschool Ede, Ede 19108
Cheb Regional Museum, Cheb 06443
Cheboksarski tekstilny tekhnikum, Cheboksary 22665
Cheboksary Textile Professional School, Cheboksary 22665
Chechelnik Regional Centralized Library System, Main Library, Chechelnik 28513
Chechelnikska raionna TsBS, Tsentralna biblioteka, Chechelnik 28513
Checheno-Ingush A.P. Chekhov Library, Grozny 22132
Checheno-Ingush State University, Grozny 22252
Checheno-Ingushskaya respublikanskaya biblioteka im. A.P. Chekhov, Grozny 22132
Checheno-Ingushski universitet, Grozny 22252
Cheektowaga Public Library, Cheektowaga 38535
Cheektowaga Public Library – Reinstein Memorial, Cheektowaga 38536
Cheikh Anta Diop University, Dakar 24410
Chelan County Law Library, Wenatchee 35052
Chelmsford Public Library, Chelmsford 38537
Chełmska Biblioteka Publiczna, Chełm 21396
Chelsea College of Art & Design, London 29175
Chelsea Public Library, Chelsea 38538
Cheltenham Township Library System, Glenside 39253
Chelyabinsk Regional Library for Youth, Chelyabinsk 24123
Chelyabinsk State University, Chelyabinsk 22226
Chelyabinskaya oblastnaya detskaya biblioteka im. V. Mayakovskogo, Chelyabinsk 24122
Chelyabinskaya oblastnaya nauchnaya meditsinskaya biblioteka, Chelyabinsk 23017
Chelyabinskaya oblastnaya universalnaya nauchnaya biblioteka, Chelyabinsk 22125
Chelyabinskaya oblastnaya yunosheskaya biblioteka, Chelyabinsk 24123
Chelyabinski gosudarstvenny universitet, Chelyabinsk 22226
Chemeketa Community College, Salem 33716
Chemerivtsi Regional Centralized Library System, Main Library, Chemerivtsi 28514
Chemerovetska raionna TsBS, Tsentralna biblioteka, Chemerivtsi 28514
Chemical and Biological Defence Establishment, Salisbury 29889
Chemical, Biological, Radiological & Nuclear Defense Information Analysis Center, Aberdeen 36438
Chemical Engineering College of Beijing Union University, Beijing 05036
Chemical Engineering Plant, Ekaterinburg 22789
Chemical Fibers Industrial Corporation, Grodno 01869
Chemical Fibre Industrial Corporation, Scientific and Technical Library, Svetlogorsk 01930
Chemical Fibres Industrial Corporation, Scientific and Technical Library, Mogilev 01903
Chemical Institute, Almaty 17941
Chemical Plant, Dimitrovgrad 03495
Chemical Plant, Gomel 02130
Chemical Plant, Novomoskovsk 22880
Chemical Plant, Ufa 22967
Chemical Plants 'POLICE', Police 21067
Chemical Process Engineering Library, Toulouse 08080
Chemically Pure Reagents for Electronics Industry Research Institute, Donetsk 28229
Chemistry Administration Library, Cairo 07109
Chemistry Museum of the Hungarian Museum for Science and Technology, Várpalota 13518
Chemtura Canada Co/Cie, Guelph 04257
Chemtura Corp, Middlebury 35888
Chemung County Library District, Elmira 38967
Chemung County Library District, Horseheads 39480
Chengde Medical College, Chengde 05112
Chengde Petroleum College, Chengde 05113
Chengdu City Library, Chengdu 05910
Chengdu Coal-Mining Management College, Chengdu 05114
Chengdu Institute of Textile Industry, Chengdu 05115
Chengdu Teachers' College, Chengdu 05116
Chengdu University, Chengdu 05117
Chengdu University of Traditional Chinese Medicine, Chengdu 05118
Cherkasi City Centralized Library System, Main Library, Cherkasy 28515
Cherkaska miska TsBS, Tsentralna biblioteka, Cherkasy 28515
Chernigiv City Centralized Library, Main Library, Chernigiv 28518
Chernigiv State Technological University, Chernigiv 27956
Chernigivska miska TsBS, Tsentralna biblioteka, Chernigiv 28518
Chernigivsky basovy medychny koledzh, Chernigiv 28095
Chernigivsky derzhavny tekhnologichny universytet, Chernigiv 27956
Chernihiv Medical College, Chernigiv 28095
Chernivetska miska TsBS, Tsentralna biblioteka, Chernivtsi 28522
Chernivetsky kraeznavchy muzei, Chernivtsi 28194
Chernivtsi City Centralized Library System, Main Library, Chernivtsi 28522
Chernivtsi Local Museum, Chernivtsi 28194
Chernivtsi State University, Chernivtsi 27959
Chernomorski Ship Building Industrial Corporation, Trade Union Library, Mykolaiv 28741
Chernorechenski khimicheski zavod im. M.I. Kalinina, Derzhinsk 22786
Chernyakhiv Regional Centralized Library System, Main Library, Chernyakhiv 28525
Chernyakhivska raionna TsBS, Tsentralna biblioteka, Chernyakhiv 28525
Chernyshevskogo Pedagogical Chernyshevsky Institute, Chita 22233
Chernyshevsky Library, Sankt-Peterburg 24217
Chernyshevsky Memorial Museum, Saratov 23955
Cherokee County Public Library, Gaffney 39177
Cherokee Regional Library System – La Fayette-Walker County Public Library, LaFayette 39763
Cherry Creek District Library, Centennial 36640
Cherry Hill Public Library, Cherry Hill 38539
Chervona zirka Industrial Corporation, Kirovograd 28149
Chervonograd Regional Centralized Library System, Main Library, Chervonograd 28526
Chervonogradska raionna TsBS, Tsentralna biblioteka, Chervonograd 28526
Chervonozavod Regional Centralized Library System of Kharkiv City, Main Library, Kharkiv 28598
Chervonozavodska raionna TsBS mista Kharkiva, Tsentralna biblioteka, Kharkiv 28598
Chesapeake Bay Maritime Museum, Saint Michaels 37497
Chesapeake Bay Maritime Museum Library, Saint Michaels 37498
Chesapeake College, Wye Mills 33100
Chesapeake Public Library, Chesapeake 38540
Cheshire County Council, Chester 29959
Cheshire Public Library, Cheshire 38541
Chester Beatty Library, Dublin 14525
Chester C. Corbin Public Library, Webster 41792
Chester College of Higher Education, Chester 29066
Chester County Historical Society Library, West Chester 37820
Chester County Law Library, West Chester 35055
Chester County Library, Chester 38542
Chester County Library System, Exton 39016
Chester Public Library, Chester 38543
Chesterfield College, Chesterfield 29458
Chesterfield County Library, Chesterfield 38545
Chesterfield County Public Library, Chesterfield 38546
Chestnut Hill College, Philadelphia 32251
Chetco Community Public Library, Brookings 38308
Chetham's Library, Manchester 29851
Chevalier Resource Centre, Kensington 00770
Chevron Canada Resources Library, Calgary 04238
Chevron Global Library Houston, Houston 35749
Chevron Information Technology Company, Division of Chevron USA, Inc, Richmond 37439
Chevron Law Library, Houston 35750
Cheyney University, Cheyney 30755
Chhatrapati Shahu Ji Maharaj University, Kanpur 13675
Chiang Mai Teacher College, Chiang Mai 27723
Chiang Mai University, Chiang Mai 27724
– Faculty of Agriculture Library, Chiang Mai 27725
– Faculty of Dentistry, Chiang Mai 27726
– Faculty of Engineering Library, Chiang Mai 27727
– Faculty of Humanities, Chiang Mai 27728
– Faculty of Medicine, Chiang Mai 27729
– Faculty of Social Sciences, Chiang Mai 27730
– Tribal Research Library, Chiang Mai 27731
Chiba Economic Center, Chiba 17518
Chiba Keizai Senta, Chiba 17518
Chiba Prefectural Assembly, Chiba 17254
Chiba Prefectural Central Library, Chiba 17836
Chiba Prefectural General Education Center, Chiba 17521
Chiba Prefectural Institute of Public Health, Chiba 17519
Chiba Prefecture Administrative Data Room, Chiba 17255
Chiba Prefecture Agricultural Experiment Station, Chiba 17520
Chiba University, Chiba 16904
– Faculty of Horticulture, Matsudo 16905
– Institute of Chemobiodynamics, Chiba 16907
– Life Science Branch, Chiba 16906
Chiba University Library, Chiba 16904
– Engeigakubu Bunkan, Matsudo 16905
– Inohana Bunkan, Chiba 16906
– Seibutsu Kassei Kenkyujo, Chiba 16907
Chibaken Eisei Kenkyujo, Chiba 17519
Chibaken Gikan, Chiba 17254
Chibaken Gyosei Shiryasihitsu, Chiba 17255
Chibaken Nogyo Shikenjo, Chiba 17520
Chibaken Sogo Kyoiku Senta, Chiba 17521
Chibero College of Agriculture, Norton 42370
Chicago Academy of Sciences, Chicago 36667
Chicago Botanic Garden Library, Glencoe 36890
Chicago History Museum, Chicago 36668
Chicago Public Library, Chicago 38550
Chicago Ridge Public Library, Chicago Ridge 38551
Chicago School of Professional Psychology Library, Chicago 30757
Chicago State University, Chicago 30758
Chicago Theological Seminary, Chicago 35156
Chicago Transit Authority-Law Library, Chicago 34067
Chicago Tribune, Chicago 35569
Chicago Zoological Society, Brookfield 36595
Chickasaw Regional Library System, Ardmore 37982
Chickasha Public Library, Chickasha 38552
Chicopee Public Library, Chicopee 38553
A. S. Chikobava Institute of Linguistics, Tbilisi 09229
Child Accident Prevention Trust (CAPT), London 29740
Child Welfare League of America, Washington 37727
Children and Youth Library Center "Reading-Town", Veliki Novgorod 24341
Children Library, Khabarovsk 24143
Children, Youth and Womens Health Service Library, North Adelaide 00968
Children's Book Council Library, New York 37194
Children's Hospital no 1, Sankt-Peterburg 23754
Children's Hospital Oakland, Oakland 37298
Children's Institute Library, Pittsburgh 37379
Children's Library, Ekaterinburg 24126
Children's Library, Rostov-na-Donu 24206
Children's Library, Stavropol 24323
Children's Library, Tambov 24329
Children's Library, Tula 24334
Children's Library, Ulyanovsk 24340
Children's Library no 115, Kyiv 28652
Children's Playground Association, Hong Kong 05871
Chili Public Library, Rochester 41027
Chillicothe Correctional Institution Library, Chillicothe 34073
Chillicothe & Ross County Public Library, Chillicothe 38554
Chilton Clanton Library, Clanton 38574
Chimatech, Sofiya 03528
China Academy, Hwa Kang 27588
China Academy of Fine Arts, Hangzhou 05233
China Academy of Traditional Chinese Medicine, Beijing 05037
China Art Gallery, Beijing 05819
China Association for Science and Technology (CAST), Beijing 05820
China Central Radio and Television University, Beijing 05038
China Civil Aviation Flight College, Guanghan 05199
China Evangelical Seminary, Taipei 27582
China Institute of Civil Aviation, Tianjin 05606
China Institute of Finance, Beijing 05039
China Insurance Management Cadre Institute, Changsha 05095
China Law Society, Beijing 05821
China Medical College, Taichung 27536
China Medical University, Shenyang 05542
China Pharmaceutical University, Nanjing 05433
China Society of Fisheries, Beijing 05822
China University of Geosciences, Wuhan 05651
China University of Mining and Technology, Beijing 05040
China University of Mining and Technology, Xuzhou 05730
China University of Political Science and Law, Beijing 05041
Chinese Academy od Sciences, Taipei 27522
Chinese Academy of Agricultural Sciences, Beijing 05823
Chinese Academy of Forestry, Beijing 05824
Chinese Academy of Sciences – Documentation and Information Centre (DICCAS), Beijing 04952
Chinese Academy of Surveying and Mapping, Beijing 05825
Chinese Coal Economics College, Yantai 05746
Chinese Culture University, Taipei 27542
Chinese Library, Brussel 02583
Chinese Medical Association, Beijing 05826
Chinese People's Liberation Army, Beijing 05042
Chinese People's University of Public Security, Beijing 05043
Chinese Petroleum Corporation, Refining & Manufacturing, Chia-yi 27584
Chinese Petroleum Society, Beijing 05827
Chinese Postal Museum, Taipei 27606
Chinese Sericulture Society, Zhenjiang 05907
Chinese Society of Agricultural Machinery, Beijing 05828
Chinese Society of Metals, Beijing 05829
Chinese Society of Mineralogy, Petrology and Geochemistry, Giuyang 05862
Chinese Society of Tropical Crops, Danzhou 05860
Chinese Traditional Opera College, Beijing 05044
Chinoin Pharmaceutical and Chemical Works Co, Budapest 13352
Chinook Regional Library, Swift Current 04774
Chios Dimotiki Vivliothiki, Chios 13142
Chipata Teacher Training College, Chipata 42322
Chipola Junior College, Marianna 33515
Chippewa Falls Public Library, Chippewa Falls 38557
Chippewa River District Library, Mount Pleasant 40319
Chippewa Valley Technical College, Eau Claire 33309
Chirurgische Klinik und Poliklinik Innenstadt, München 10433
Chirurgische Klinik und Poliklinik Innenstadt, Röntgenabteilung, München 10434
Chirurgisches Zentrum / Klinik und Poliklinik für Allgemein-, Viszeral-, Thorax- und Gefäßchirurgie, Klinik für Unfallchirurgie, Urologische Klinik, Bonn 09510
Chisholm Institute, Dandenong 00430
Chita Polytechnic Institute, Chita 22231
Chita Region Library named after A.S. Pushkin, Chita 22127

Chitedze Agricultural Research Station, Lilongwe 18488
Chitinskaya oblastnaya biblioteka im. A.S. Pushkina, Chita 22127
Chitinskaya oblastnaya detskaya biblioteka, Chita 24125
Chitinski mezhotraslevoi territorialny tsentr nauchno-tekhnicheskoi informatsii i propagandy, Chita 23020
Chittagong Divisional Government Public Library, Chittagong 01771
Chiyoda Corporation, Yokohama 17502
Chiyoda Kako Kensetsu K. K. Kikaku Hyojunbu Raiburari, Yokohama 17502
Chiyoda Mutual Life Insurance Co, Tokyo 17434
Chiyoda Seimei Hoken Sogo Kaisha Sogo Kikakushitsu, Tokyo 17434
D.I. Chizhevsky Regional Universal Research Library, Kirovograd 27935
Chizhou Teachers' College, Guichi 05212
Choate, Hall & Stewart Library, Boston 35519
Choctaw County Public Library, Butler 38403
An Chomhairle Leabharlanna, Dublin 14526
Chonbuk National University, Chonju 18059
Chongqing Architecture University, Chongqing 05129
Chongqing College of Iron and Steel Technology, Chongqing 05130
Chongqing Commercial College, Chongqing 05131
Chongqing Institute of Civil Engineering Architecture, Chongqing 05132
Chongqing Institute of Industrial Management, Chongqing 05133
Chongqing Institute of Post and Telecommunications, Chongqing 05134
Chongqing Library, Chongqing 04956
Chongqing Teachers' College, Chongqing 05135
Chongqing University, Chongqing 05136
Chongqing University of Medical Sciences, Chongqing 05137
Chonnam National University, Kwangju 18065
Chornukhi Regional Centralized Library System, Main Library, Chornukhi 28527
Chornukhinska raionna TsBS, Tsentralna biblioteka, Chornukhi 28527
Chosun University, Kwangju 18066
Chowan University, Murfreesboro 31921
Chr. Michelsen Institute, Bergen 20211
Christ College, Sandy Bay 00794
Christ the King Seminary, East Aurora 35193
Christchurch College of Education, Christchurch 19772
Christchurch Polytechnic Institute of Technology (CPIT), Christchurch 19795
Christelijk Pedagogisch Studiecentrum (CPS), Amersfoort 19343
Christelijke Agrarische Hogeschool, Dronten 19107
Christelijke Bibliotheek voor Blinden en Slechtzienden (CBB), Ermelo 19424
Christelijke Hogeschool Ede, Ede 19108
Christelijke Hogeschool Windesheim, Zwolle 19233
Christendom College, Front Royal 31181
Christian Agricultural College, Dronten 19107
Christian & Barton, LLP Attorneys At Law, Richmond 36171
Christian Brothers University, Memphis 31789
Christian County Library, Ozark 40679
Christian Heritage College, El Cajon 31077
Christian Heritage College, Mansfield 00453
Christian Leaders' Training College, Mount Hagen 20561
Christian Life College, Stockton 35392
Christian Medical College and Hospital, Vellore 13786
Christian Museum Library, Esztergom 13472
Christian Theological Academy, Warszawa 20952
Christian Theological Seminary, Indianapolis 35238
Christian Union Bible College Library, Greenfield 35221
Christian-Albrechts-Universität zu Kiel, Kiel 10063
– Englisches Seminar, Kiel 10064
– Fachbibliothek Anatomie/Biochemie, Kiel 10065
– Fachbibliothek für Wirtschaftswissenschaften, Kiel 10066
– Fachbibliothek Germanistik, Kiel 10067
– Fachbibliothek Geschichte, Kiel 10068
– Fachbibliothek Mathematik/Informatik/Rechenzentrum, Kiel 10069
– Fachbibliothek Ökologie und Wasserwirtschaft, Kiel 10070
– Fachbibliothek Physikzentrum, Kiel 10071
– Fachbibliothek Theologie, Kiel 10072
– Geographisches Institut, Kiel 10073
– Ingenieurwissenschaftliche Abteilung der Zentralbibliothek / Fachbibliothek Ingenieurwissenschaften an der Technischen Fakultät, Kiel 10074
– Institut für Europäisches und Internationales Privat- und Verfahrensrecht, Kiel 10075
– Institut für Geowissenschaften, Kiel 10076
– Institut für Humanernährung und Lebensmittelkunde, Kiel 10077
– Institut für Klassische Altertumskunde, Kiel 10078
– Institut für Landwirtschaftliche Verfahrenstechnik, Kiel 10079
– Institut für Osteuropäisches Recht, Kiel 10080
– Institut für Pädagogik, Kiel 10081
– Institut für Pflanzenbau und Pflanzenzüchtung, Kiel 10082
– Institut für Pflanzenernährung und Bodenkunde, Kiel 10083
– Institut für Phonetik und digitale Sprachverarbeitung, Kiel 10084
– Institut für Physikalische Chemie, Kiel 10085
– Institut für Phytopathologie, Kiel 10086
– Institut für Polarökologie, Kiel 10087
– Institut für Psychologie, Kiel 10088
– Institut für Sanktionenrecht und Kriminologie, Kiel 10089
– Institut für Slavistik, Kiel 10090
– Institut für Sozialwissenschaften, Kiel 10091
– Institut für Sport- und Sportwissenschaften, Kiel 10092
– Institut für Tierernährung und Stoffwechselphysiologie, Kiel 10093
– Institut für Tierzucht und Tierhaltung, Kiel 10094
– Institut für Ur- und Frühgeschichte, Kiel 10095
– Institut für Wirtschafts- und Steuerrecht einschl. Wirtschaftsstrafrecht, Kiel 10096
– Institute für Anorganische und Organische Chemie, Kiel 10097
– Juristisches Seminar, Kiel 10098
– Kunsthalle zu Kiel, Kiel 10099
– Kunsthistorisches Institut, Kiel 10100
– Lektorat Deutsch als Fremdsprache, Kiel 10101
– Medizin- und Pharmaziehistorische Sammlung Kiel, Kiel 10102
– Musikwissenschaftliches Institut, Kiel 10103
– Nordisches Institut, Kiel 10104
– Pharmazeutisches Institut, Abt. Pharmazeutische Biologie, Kiel 10105
– Pharmazeutisches Institut, Abt. Pharmazeutische Chemie und Abt. Pharmazeutische Technologie, Kiel 10106
– Philosophisches Seminar, Kiel 10107
– Romanisches Seminar, Kiel 10108
– Seminar für Allgemeine und Vergleichende Sprachwissenschaft, Kiel 10109
– Seminar für Europäische Ethnologie/Volkskunde, Kiel 10110
– Seminar für Orientalistik (Bereich: Indologie), Kiel 10111
– Seminar für Orientalistik (Bereich: Islamwissenschaft), Kiel 10112
– Seminar für Orientalistik (Bereich: Sinologie), Kiel 10113
– Walther-Schücking-Institut für Internationales Recht, Kiel 10114
Christianeum, Hamburg 10733
Christian-Weise-Bibliothek Zittau, Zittau 12989
Christie Hospital NHS Trust, Manchester 29852
Christlich-Islamische Begegnungs- und Dokumentationsstelle (CIBEDO), Frankfurt am Main 11840
Christopher Newport University, Newport News 32115
Christophorus-Diakoniewerk gGmbH, Ueckermünde 12569
Christ's College, Cambridge 28993
CHRISTUS Spohn Health System, Corpus Christi 36756
Christus St. Mary Hospital, Port Arthur 37393
Chrysler Museum of Art, Norfolk 37288
Chrześcijańska Akademia Teologiczna, Warszawa 20952
CHU Sainte-Justine, Montreal 04393
G. N. Chubinashvili Institute of History of Georgian Art, Tbilisi 09230
Chubu Daigaku, Kasugai 16968
Chubu Denryoku K. K., Kikakushitsu Toshoshitsu, Nagoya 17390
Chubu Electric Power Co, Inc, Nagoya 17390
Chubu Nippon Broadcasting Co, Ltd, Nagoya 17391
Chubu Nippon Hosa K. K., Nagoya 17391
Chubu University, Miura Memorial Library, Kasugai 16968
Chugoku Denryoku K. K., Keizai Kenkyujo, Hiroshima 17358
Chugoku Electric Power Co, Inc, Economic Research Institute, Hiroshima 17358
Chugoku Kogyo Gijutsu Shikenjo, Kure 17583
Chugoku National Agricultural Experiment Station, Fukuyama 17539
Chugoku Nogyo Shikenjo, Fukuyama 17539
Chugoku Post Office Department, Hiroshima 17262
Chugoku Yusei Kyoku, Hiroshima 17262
Chugunoliteyny zavod 'Stankolit', Moskva 22833
Chukotskaya okruzhnaya detskaya biblioteka, Anadyr 24101
Chukovski Tuva Republican Children's Library, Kyzyl 24156
K. Chukovsky Children's Library, Kyiv 28645
Chukyo Daigaku, Nagoya 17054
– Toyota Library, Toyota 17055
Chukyo University, Nagoya 17054
Chula Vista Public Library, Chula Vista 38558
Chula Vista Public Library – Eastlake, Chula Vista 38559
Chula Vista Public Library – South Chula Vista, Chula Vista 38560
Chulalongkorn University, Bangkok 27686
– Faculty of Arts, Bangkok 27687
– Faculty of Commerce and Accountancy, Bangkok 27688
– Faculty of Communication Arts, Bangkok 27689
– Faculty of Education, Bangkok 27690
– Faculty of Engineering, Bangkok 27691
– Faculty of Medicine, Bangkok 27692
– Faculty of Political Science, Bangkok 27693
– Faculty of Science, Bangkok 27694
– Faculty of Veterinary Science, Bangkok 27695
– Thailand Information Center, Bangkok 27696
Chulochno-trikotazhnaya fabrika, Gomel 01849
CHUM, Hôpital Notre-Dame, Montreal 04394
Chung Cheng College of Technology, T'aoyüan 27577
Chung Hwa Free Clinic, Singapore 24609
Chung Yuan Christian University, Chungli 27526
Chung-ang University, Seoul 18076
Chungbuk National University, Cheongju 18058
Chungcheng Military Academy, Kaohsiung 27567
Chungnam National University, Taejon 18101
Chunichi Shimbunsha Editorial Bureau, Nagoya 17392
Chunichi Shimbunsha Henshykuyoku Shiryobu, Nagoya 17392
Chunpuk College, Cheongju 18103
Chuo Daigaku, Hachioji 16934
– Keizai Kenkyujo, Hachioji 16935
Chuo University, Hachioji 16934
– Institute of Economic Research, Hachioji 16935
Chuokoronsha Library, Tokyo 17435
Chuokoronsha Shiryoshitsu, Tokyo 17435
Church of Denmark Center for Theology and Religious Education, Løgumkloster 06763
Church of God Theological Seminary, Cleveland 35163
Church of Ireland (Anglican), Dublin 14513
Church of Ireland College of Education, Dublin 14465
Church of Jesus Christ of Latter-Day Saints, Mesa 35281
The Church of Jesus Christ of Latter-Day Saints, Salt Lake City 35369
Church of Jesus Christ of Latter-Day Saints – Family History Library, Salt Lake City 35370
Church of the Brethren, Elgin 36826
Church of the Incarnation, Dallas 35175
Church of the Redeemer, Andalusia 35108
Churches of Christ Theological College, Mulgrave 00782
Churchill County Library, Fallon 39037
Chusho Kigyo Jigyodan, Tokyo 17436
Chusho Kigyo Kinyu-Koko Chosabu Shiryo Tokeika, Tokyo 17437
Chusho Kigyo Shin-yo Hoken Koko, Tokyo 17438
Chuvash Agricultural Institute, Cheboksary 22225
Chuvash Art Museum, Cheboksary 23012
Chuvash Central State Archives, Cheboksary 23013
Chuvash Medical Library, Cheboksary 23014
Chuvash National Library, Cheboksary 22124
Chuvash State University, Cheboksary 22223
Chuvash Teacher-Training Institute, Cheboksary 22224
Chuvashkaya respublikanskaya meditsinskaya biblioteka, Cheboksary 23014
Chuvashski gosudarstvenny universitet, Cheboksary 22223
Chuvashski mezhotraslevoi territorialny tsentr nauchno-tekhnicheskoi informatsii i propagandy, Cheboksary 23015
Chuvashski pedagogicheski institut, Cheboksary 22224
Chuvashski selskokhozyaistvenny institut, Cheboksary 22225
Chuxiong Teachers' College, Chuxiong 05142
O. O. Chuyko Institute of Surface Chemistry, Kyiv 28312
Chuzhou University, Chuzhou 05143
Chuzhou University Library, Chuzhou 05143
CIBEDO, Frankfurt am Main 11840
Cicero Public Library, Cicero 38561
CID-Femmes – Centre d'Information et de Documentation des Femmes "Thers Bode", Luxembourg 18412
CIDOB – Centro de Investigación de Relaciones Internacionales y Desarrollo, Barcelona 25892
Cidoc / Semarnap, Jardines en la Montaña 18829
Ciencias Sociales 'Mirta Aguirre', La Habana 06298
Cieszyn Historical Library, Cieszyn 21087
CII Knowledge Services, London 29741
CIIT Centers for Health Research, Research Triangle Park 37435
CILAG AG, Schaffhausen 27290
Cimentos de Portugal S.A. (CIMPOR), Lisboa 21747
CIMMYT Library, Texcoco 18870
Cincinnati Art Museum, Cincinnati 36692
Cincinnati Children's Hospital Medical Center Division of Developmental & Behavioral Pediatrics, Cincinnati 36693
Cincinnati Christian University, Cincinnati 35159
Cincinnati Law Library Association, Cincinnati 36694
Cincinnati Museum Center at Union Terminal, Cincinnati 36695
Cincinnati Regional Library for the Blind & Physically Handicapped, Cincinnati 36696
CINDOC (Centro de Información y Documentación Científica), Madrid 25996
Cinema Arts, Inc, Newfoundland 37280
Cinema 'Delmiro de Caralt' de la Filmoteca, Barcelona 25893
Cinéma François Truffaut, Paris 08478
Cinemateca Portuguesa – Museu do Cinema, Lisboa 21782
Cinémathèque Québécoise, Montreal 04395
Cinémathèque Royale de Belgique, Bruxelles 02584
Cinémathèque suisse / Dokumentationsstelle Zürich, Zürich 27450
Cinémathèque Suisse (SCL) / Schweizer Filmarchiv, Lausanne 27394
Cineteca del Friuli, Gemona 16280
Circolo Filologico Milanese, Milano 16331
Círculo de Bellas Artes, Madrid 26005
Círculo Oficiales de Mar Profesional y Mutual, Buenos Aires 00206
Circus and Variety Arts Professional School, Moskva 22683
Ciril Kosmač Public Library Tolmin, Tolmin 24986
Cisterciënzerabdij Lilbosch, Echt 19305
Cisterciënzerabdij Maria Toevlucht, Zundert 19319
Cisterciënzerabdij O.-L. Vrouw van Sion, Diepenveen 19303
Cisterciënzer-Abdij Onze Lieve Vrouw Onbevlekt Ontvangen, Tegelen 19316
Cisterciënzerabdij Onze Lieve Vrouw van Koningshoeven, Berkel-Enschot 19299
Cisterciënzerabdij Sint-Bernardus, Bornem 02451
The Citadel, Charleston 30714
Cité de la musique, Paris 07843
Cité de l'Architecture et du Patrimoine, Paris 08526
Cité des Sciences et de l'Industrie, Paris 08527
Cité-Bibliothèque Luxembourg, Luxembourg 18426
Citizens Library, Washington 41737
CITO National Institute for Educational Measurement, Arnhem 19393
Citog, Arnhem 19393

Citrus College, Glendora 33375
Citrus County Library System, Beverly Hills 38198
Citrus Research & Education Center, Lake Alfred 31543
City and County of Swansea, Swansea 30086
City and District Public Library named after Maria Konopnicka, Lubań 21485
City and Islington College, London 29475
City and Regional Municipal Library, Nagykanizsa 13542
City and University Library Osijek, Osijek 06145
– Agricultural Institute Library, Osijek 06148
– Faculty of Economy, Osijek 06146
– Faculty of Law, Osijek 06149
– Faculty of Philosophy, Osijek 06147
City Archives and Athenaeum Library, Deventer 19047
City Archives of Amsterdam, Amsterdam 19378
City Archives of Geel – Historical Society of Geel, Geel 02641
City Business Library, London 30018
City Central Library, Bisk 24112
City Central Library, Magnitogorsk 24162
City College of San Francisco, San Francisco 33729
City College of the City University of New York, New York 32030
– Dominican Studies Institute Research Library & Archives, New York 32031
City Council, Sankt-Peterburg 22766
City & County of Honolulu, Department of Customer Services, Honolulu 36931
City & County of San Francisco – Law Library, San Francisco City Attorney, San Francisco 34818
City Hall Library, København 06800
City Hall Reference Library, Hong Kong 05916
The City Historical Archive Barcelona, Barcelona 25879
City Library, Angarsk 24102
City Library, Balashov 24108
City Library, Bełchatów 21379
City Library, Białogard 21381
City Library, Biłgoraj 21386
City Library, Bochnia 21387
City Library, Bolesławiec 21388
City Library, Brzesko 21390
City Library, Brzozów 21391
City Library, Bucureşti 22081
City Library, Busko-Zdrój 21392
City Library, Bystrzyca Kłodzka 21394
City Library, Bytów 21395
City Library, Chełmno 21397
City Library, Chmielnik 21398
City Library, Chodzież 21399
City Library, Chojnice 21400
City Library, Choszczno 21401
City Library, Chrzanów 21402
City Library, Czarnków 21404
City Library, Człuchów 21408
City Library, Dąbrowa Tarnowska 21409
City Library, Drawsko Pomorskie 21410
City Library, Działdowo 21411
City Library, Ełk 21413
City Library, Engelsk 24128
City Library, Eskilstuna 26760
City Library, Garwolin 21414
City Library, Głogów 21416
City Library, Goleniów 21418
City Library, Golub-Dobrzyń 21419
City Library, Gorlice 21420
City Library, Grajewo 21421
City Library, Grodków 21422
City Library, Grójec 21423
City Library, Grudziądz 21424
City Library, Gryfice 21425
City Library, Gryfino 21426
City Library, Hafnarfjörður 13579
City Library, Hajnówka 21427
City Library, Janów Lubelski 21431
City Library, Jarosław 21432
City Library, Jasło 21433
City Library, Jędrzejów 21434
City Library, Kalisz 21436
City Library, Kamień Pomorski 21438
City Library, Kamienna Góra 21439
City Library, Kartuzy 21440
City Library, Kazimierza Wielka 21441
City Library, Kępno 21442
City Library, Kętrzyn 21443
City Library, Kłobuck 21446
City Library, Kluczbork 21448
City Library, Kolbuszowa 21449
City Library, Kolno 21450
City Library, Koło 21451
City Library, Kołobrzeg 21452
City Library, Końskie 21454
City Library, Kościan 21455
City Library, Kościerzyna 21456
City Library, Kozienice 21458
City Library, Koźle 21459
City Library, Kraśnik 21461
City Library, Krasnystaw 21462
City Library, Krotoszyn 21465
City Library, Kutno 21466
City Library, Łańcut 21467
City Library, Łapy 21468
City Library, Łask 21469
City Library, Lębork 21470
City Library, Łęczyca 21471
City Library, Leszno 21474
City Library, Lidzbark Warmiński 21476
City Library, Limanowa 21477
City Library, Lipno 21478
City Library, Łobez 21479
City Library, Łosice 21482
City Library, Łowicz 21483
City Library, Lubaczów 21484
City Library, Lubartów 21486
City Library, Lubliniec 21488
City Library, Łuków 21489
City Library, Lwówek Śląski 21490
City Library, Miastko 21491
City Library, Międzychód 21492
City Library, Międzyrzecz 21493
City Library, Mielec 21494
City Library, Milicz 21495
City Library, Mińsk Mazowiecki 21496
City Library, Mława 21497
City Library, Mogilno 21498
City Library, Mónki 21499
City Library, Morąg 21500
City Library, Mrągowo 21501
City Library, Myślibórz 21502
City Library, Newcastle upon Tyne 30048
City Library, Nidzica 21503
City Library, Niemodlin 21504
City Library, Nowa Ruda 21506
City Library, Nowa Sòl 21507
City Library, Nowe Miasto Lubawskie 21508
City Library, Nowy Dwór Mazowiecki 21509
City Library, Oława 21512
City Library, Oleśnica 21514
City Library, Oleszno 21515
City Library, Olkusz 21516
City Library, Opoczno 21518
City Library, Ostrów Mazowiecka 21520
City Library, Ostrzeszów 21522
City Library, Oświęcim 21523
City Library, Parczew 21524
City Library, Pińczów 21527
City Library, Pisz 21528
City Library, Pleszew 21529
City Library, Poddębice 21531
City Library, Port Louis 18703
City Library, Pretoria 25228
City Library, Proszowice 21533
City Library, Prudnik 21534
City Library, Pruszcz Gdański 21535
City Library, Przasnysz 21536
City Library, Przeworsk 21538
City Library, Pszczyna 21539
City Library, Puławy 21540
City Library, Pułtusk 21541
City Library, Pyrzyce 21542
City Library, Radomsko 21545
City Library, Rawicz 21548
City Library, Rypin 21550
City Library, Sandomierz 21552
City Library, Sanok 21553
City Library, Siemiatycze 21554
City Library, Skierniewice 21557
City Library, Sławno 21558
City Library, Sochaczew 21562
City Library, Sokółka 21563
City Library, Sokołów Podląski 21564
City Library, Śrem 21565
City Library, Środa Śląska 21566
City Library, Środa Wielkopolska 21567
City Library, Stargard Szczeciński 21569
City Library, Staszów 21570
City Library, Strzelce Krajeńskie 21571
City Library, Strzelce Opolskie 21572
City Library, Strzelin 21573
City Library, Suącin 21574
City Library, Sulechów 21575
City Library, Świebodzin 21578
City Library, Szubin 21582
City Library, Taganrog 24327
City Library, Tarnów 21584
City Library, Tarnowskie Góry 21586
City Library, Tash-Kumyr 18162
City Library, Tomaszów Lubelski 21588
City Library, Trzebnica 21591
City Library, Tuchola 21592
City Library, Turek 21593
City Library, Västerås 26951
City Library, Wabrzeźno 21595
City Library, Wadowice 21596
City Library, Wągrowiec 21597
City Library, Węgrów 21600
City Library, Wieluń 21601
City Library, Wieruszów 21602
City Library, Włocławek 21603
City Library, Włodawa 21605
City Library, Włoszczówa 21606
City Library, Wodzisław 21607
City Library, Wołów 21608
City Library, Września 21610
City Library, Wschowa 21611
City Library, Wyrzysk 21612
City Library, Ząbkowice 21613
City Library, Zakopane 21614
City Library, Złotów 21618
City Library in Mariánské Lázně, Mariánske Lázně 06587
City library Ivan Goran Kovacic, Karlovac 06257
City Library Krnov, Krnov 06581
City Library no 2, Moskva 24170
City Library of Gothenburg, Göteborg 26775
City Library of Ostrava, Ostrava 06596
City Library of Plzeň, Plzeň 06598
City Library of Reykjavik, Reykjavík 13583
City Library of Słubice, Słubice 21559
City Library of Umeå, Umeå 26937
City Library of Umeå, ILL (interlibrary loan) Department, Umeå 26938
City Library of Zagreb, Zagreb 06265
City Library Prostějov, Prostějov 06602
City Library Trutnov, Trutnov 06614
City Museum, Jičin 06453
City Museum, Trenčín 24757
City Museum, Ustí nad Labem 06546
City Museum of Gothenburg, Göteborg 26578
City Museum of Stockholm, Stockholm 26683
City of Blue Island Public Library, Blue Island 38224
City of Chicago, Chicago 34068
City of Commerce Public Library, Commerce 38672
City of Detroit, Detroit 34146
City of Edmonton, Edmonton 04072
City of Gothenburg, The Government Library, Göteborg 26487
City of Helsinki Urban Facts, Helsinki 07395
City of Hope National Medical Center, Duarte 36811
City of Houston – Legal Department Library, Houston 34366
City of Mesa Library, Mesa 40163
City of Mutare Public Libraries, Mutare 42385
City of New York Department of Records & Information Services, New York 37195
City of Prague Museum, Praha 06499
City of Providence, Parks Department, Providence 37416
City of Savannah, Savannah 34843
City of Sunderland Metropolitan District Council, Sunderland 30084
City of Winnipeg Libraries, Winnipeg 04823
City of Winnipeg Water & Waste Department, Winnipeg 04189
City of York Libraries, York 30104
City Public Library, Brezno 24770
City Public Library, Dubnica nad Váhom 24773
City Public Library, Jelenia Góra 21435
City Public Library, Kielce 21444
City Public Library, Łomża 21481
City Public Library, Púchov 24791
City Public Library, Słupsk 21560
City Public Library, Żary 21615
City Public Library, Želiezovce 24803
City Scientific Medical Library, Sevastopol 28461
City Trade, Catering and Municipal Services Committee – Professional School, Sankt-Peterburg 22698
City Trade Union Committee, Main Library, Kyiv 28357
City Trade Union Committee of the Roads and Auto Transport Workers, Moskva 24171
City University, Bellevue 30374
City University of Hong Kong, Hong Kong 05279
– Chung Chi College, Hong Kong 05280
– New Asia College Library, Hong Kong 05281
– United College Library, Hong Kong 05282
City University of New York, New York 32032
City University of New York (CUNY), Bronx 30555
City University of New York (CUNY), Hunter College, New York 37196
City University of New York Hunter College Libraries, New York 32033
– Health Professions Library, New York 32034
– School of Social Work Library, New York 32035
City-archive Antwerp, Antwerpen 02513
Civic Museums of History and Art, Trieste 16622
Civica Biblioteca Glemonense "Don Valentino Baldissera", Gemona 16748
Civica Biblioteca Popolare 'L. Ricca', Codogno 16726
Civica Raccolta delle Stampe Achille Bertarelli, Milano 16332
Civici Musei di Storia ed Arte, Trieste 16622
Civici Musei e Gallerie di Storia ed Arte, Udine 16635
Civico Museo Revoltella, Trieste 16623
Civico Museo Teatrale 'Carlo Schmidl', Trieste 16624
Civil Aviation Air Worthiness Section, Boroko 20585
Civil Aviation Authority, Adelaide 00595
Civil Aviation Authority (CAA), Gatwick Airport South 29515
Civil Aviation Authority of Fiji, Nadi Airport 07311
Civil Aviation Department, Kuala Lumpur 18552
Civil Aviation Department Headquarters, New Delhi 13872
Civil Aviation Training School, Lumbadzi 18477
Civil Court of the City of New York Library, New York 34590
Civil Engineering Institute, Minsk 02046
Civil Engineering Institute, Moskva 22342
Civil Engineering Institute, Penza 22464
Civil Service College, Ascot 29448
Civil Service Training Centre, Dar es Salaam 27674
Clackamas Community College, Oregon City 33606
Clackamas County Library, Oak Grove 40561
Clackmannanshire Libraries, Alloa 29925
Claflin University, Orangeburg 32205
Clairol Research Library, Stamford 36282
Clare College, Cambridge 28984
Clare County Library, Ennis 14575
Claremont Colleges Libraries, Claremont 30814
– Ella Strong Denison Library, Claremont 30815
– Norman F. Sprague Memorial Library, Claremont 30816
– Seeley G. Mudd Science Library, Claremont 30817
Claremont Graduate University, Claremont 30818
Claremont School of Theology Library, Claremont 30819
Clarence Dillon Public Library, Bedminster 38144
Clarence Public Library, Clarence 38577
Clarence Regional Library, South Grafton 01157
Clarendon Hills Public Library, Clarendon Hills 38578
Clarendon Parish Library, May Pen 16890
Clarington Public Library, Bowmanville 04582
Clarion University of Pennsylvania, Clarion 30821
Clarion University of Pennsylvania, Oil City 32182
Clark College, Vancouver 33821
Clark County Law Library, Las Vegas 34440
Clark County Law Library, Vancouver 34958
Clark County Public Library, Springfield 41413
Clark County Public Library, Winchester 41921
Clark Hill PLC, Detroit 35672
Clark Public Library, Clark 38579
Clark State Community College, Springfield 33775
Clark, Thomas & Winters, Austin 35473
Clark University, Worcester 33094
– Special Collections-Archives, Worcester 33095
Clarke College, Dubuque 31028
J. R. Clarke Public Library, Covington 38719
Clarksburg-Harrison Public Library, Clarksburg 38582
Clarkson University, Potsdam 32350
Clarksville-Montgomery County Public Library, Clarksville 38585
Clatsop Community College, Astoria 33136
Clausen Miller, PC Library, Chicago 35570
Clausthal University of Technology, Clausthal-Zellerfeld 09615
– Institute of Management and Economics, Clausthal-Zellerfeld 09619
Clay County Public Library System – Headquarters Library, Orange Park 40631
Clay County Public Library System – Orange Park Public Library, Orange Park 40632
Clayton County Library System, Jonesboro 39656
Clayton County Library System – Forest Park Branch, Forest Park 39098
Clayton County Library System – Jonesboro Branch, Jonesboro 39657
Clayton County Library System – Morrow Branch, Morrow 40306

Clayton County Library System – Riverdale Branch, Riverdale 41008
Clayton Group Services, Inc, Novi 36062
Clayton State College & State University, Morrow 31909
Clear Creek Baptist Bible College, Pineville 35332
Clearfield County Law Library, Clearfield 34084
Clearwater Christian College, Clearwater 30824
Clearwater Public Library System, Clearwater 38591
Clearwater Public Library System – Countryside, Clearwater 38592
Clearwater Public Library System – East, Clearwater 38593
Cleary, Gottlieb, Steen & Hamilton Library, New York 35968
Clemson University, Clemson 30825
– Gunnin Architecture Library, Clemson 30826
Clermont County Law Library Association, Batavia 33970
Clermont County Public Library, Batavia 38101
Cleve J. Fredricksen Library, Camp Hill 38430
Cleveland Botanical Garden, Cleveland 36705
Cleveland Chiropractic College, Los Angeles 31646
Cleveland Chiropractic College, Overland Park 37314
Cleveland Clinic Alumni Library, Cleveland 36706
Cleveland College of Art and Design, Middlesbrough 29273
Cleveland County Library System, Shelby 41297
Cleveland Electric Illuminating Company, Perry 36087
Cleveland Health Sciences Library, Cleveland 30830
– Dittrick Medical History Center, Cleveland 30831
Cleveland Heights-University Heights Public Library, Cleveland Heights 38599
Cleveland Institute of Art, Cleveland 30832
Cleveland Institute of Music, Cleveland 36707
Cleveland Law Library Association, Cleveland 36708
Cleveland Museum of Art, Cleveland 36709
Cleveland Museum of Natural History, Cleveland 36710
Cleveland Public Library, Cleveland 38596
Cleveland Public Library, Cleveland 38597
Cleveland Public Library – Library for the Blind & Physically Handicapped, Cleveland 36711
Cleveland Scientific Institution, Guisborough 29680
Cleveland State Community College, Cleveland 33235
Cleveland State University, Cleveland
– Cleveland-Marshall Law Library, Cleveland 30833
– University Library, Cleveland 30834
Clifford Chance US LLP Library, New York 35969
Cliffside Park Free Public Library, Cliffside Park 38601
Clifton Park-Halfmoon Public Library, Clifton Park 38602
Climate and Pollution Agency, Oslo 20192
Clínica Nuestra Señora de la Concepción, Madrid 26006
Clinical and Experimental Surgery Research Institute, Kyiv 28371
Clinical Hospital 'Losenez', Sofiya 03571
Clinique d'ophtalmologie, Genève 27087
Clinton Community College, Plattsburgh 33650
Clinton County Law Library, Wilmington 35072
Clinton Public Library, Clinton 38605
Clinton-Essex-Franklin Library System, Plattsburgh 40824
Clinton-Macomb Public Library, Clinton Township 38611
Clinton-Macomb Public Library, Macomb Township 39992
Clintonville Public Library, Clintonville 38613
Clock Factory, Moskva 22861
Cloisters Library, New York 37227
Cloquet Public Library, Cloquet 38615
Closter Public Library, Closter 38616
Clovis Community College, Clovis 30841
Clovis-Carver Public Library, Clovis 38617
Club Alpino Italiano, Torino 16599
Club Alpino Italiano – Sezione di Milano, Milano 16333
Club of Composers, Sankt-Peterburg 23756
Clube de Engenharia, Rio de Janeiro 03336
Clube Naval, Rio de Janeiro 03337

Clusius Library, Leiden 19155
CMI (Chr. Michelsen Institute), Bergen 20211
CNRA Centre National de Recherche Agronomique, Bouaké 06137
CNRS – Centre de Documentation en Sciences Sociales, Marseille 08407
Coahoma Community College, Clarksdale 33233
Coal City Public Library District, Coal City 38619
Coal Research Institute, Donetsk 28231
Coalinga-Huron USD Library District, Coalinga 38620
Coastal Carolina Community College, Jacksonville 33436
Coastal Georgia Community College, Brunswick 33196
Coastal Plain Regional Library – Headquarters, Tifton 41557
Coatbridge College, Coatbridge 29460
Coatbridge Library, Coatbridge 29962
Coatesville Area Public Library, Coatesville 38621
Coatesville VA Medical Center Library, Coatesville 36717
Cobb County Public Library System, Marietta 40052
Cobb County Public Library System – East Marietta, Marietta 40053
Cobb County Public Library System – Kemp Memorial, Marietta 40054
Cobb County Public Library System – Kennesaw Branch, Kennesaw 39698
Cobb County Public Library System – Merchant's Walk, Marietta 40055
Cobb County Public Library System – Mountain View, Marietta 40056
Cobb County Public Library System – Powder Springs Branch, Powder Springs 40887
Cobb County Public Library System – South Cobb, Mableton 39990
Cobb County Public Library System – West Cobb Regional, Kennesaw 39699
Cobourg Public Library, Cobourg 04613
The Coca Cola Company, Atlanta 35457
Cochin University of Science and Technology, Kochi 13679
Cochise College, Douglas 33299
Cochise County Law Library, Bisbee 33987
Cochise County Library District, Bisbee 38211
Cockburn City, Bibra Lake DC 01069
Cocoa Beach Public Library, Cocoa Beach 38625
Cocoa Research Institute of Ghana, Tafo 13062
Cocoa Research Institute of Nigeria, Ibadan 20015
Coconino County Superior Court, Flagstaff 34204
Coconut Research Institute, Lunuwila 26324
CODA Centrale Bibliotheek, Apeldoorn 19552
Codarts, Hogeschool voor de Kunsten, Rotterdam 19187
Codarts, University for the Arts, Rotterdam 19187
Coe College, Cedar Rapids 30689
Coeur D'Alene Public Library, Coeur d'Alene 38627
Coffee County Lannom Memorial Public Library, Tullahoma 41614
Coffee Research and Experimental Station, Moshi 27678
Coffey County Library, Burlington 38397
Coffeyville Public Library, Coffeyville 38628
Coffs Harbour City, Coffs Harbour 01087
Cogswell Polytechnical College, Sunnyvale 32763
Coke Chemical Technology Plant, Trade Union Library, Mariupol 28721
Coker College, Hartsville 31307
Il lustre Col legi d'Advocats de Barcelona, Barcelona 25894
Colby College Libraries, Waterville 32997
– Bixler Art & Music Library, Waterville 32998
– Science Library, Waterville 32999
Colby-Sawyer College, New London 31996
Colchester-East Hants Regional Library, Truro 04806
Cold Spring Harbor Laboratory, Cold Spring Harbor 36721
Cold Spring Harbor Library, Cold Spring Harbor 38633
Coldwater Branch Library, Coldwater 38634
Colección de Prehistoria, Madrid 26007
Colección Ornitológica Phelps, Caracas 42250
Coleg Llandrillo Cymru, Colwyn Bay 29461
Colégio Cristo Rei, São Leopoldo 03133
Colegio de Abogados de la Ciudad Buenos Aires, Buenos Aires 00249
Colegio de Abogados de La Plata, La Plata 00159

Colegio de Escribanos, Buenos Aires 00250
Colegio de Escribanos de la Provincia de Buenos Aires, La Plata 00160
Colegio de España, Paris 08528
Colegio de Graduados en Ciencias Económicas, Buenos Aires 00117
El Colegio de Jalisco, Zapopan 18809
El Colegio de la Frontera Sur-Eldsur, San Cristóbal de las Casas 18812
El Colegio de México, México 18725
Colegio Interamericano de Defensa, Washington 32986
Colegio Máximo de la Compañía de Jesús, Santafé de Bogotá 05990
Colegio Mayor de Nuestra Señora del Rosario, Santafé de Bogotá 05961
Colégio Militar, Lisboa 21721
Colegio Nacional, México 18726
Colegio Notarial de Granada, Granada 25960
Colegio Notarial de Madrid, Madrid 26031
Colegio Oficial de Arquitectos de Madrid, Madrid 26008
Colegio Oficial de Arquitectos Vasco-Navarro – Delegación Navarra, Pamplona 26122
Colegio Rio Branco, São Paulo 03205
Colegio Sagrado Corazón de Jesús, Barcelona 25661
Colegio San Estanislao de Kostka, Salamanca 25810
Colegio Seminario del Corpus Christi el Patriarca, Valencia 25834
Colegio Territorial de Arquitectos de Valencia, Valencia 26170
Colegio Universitario Cardenal Gil de Albornoz, Cuenca 25374
Colegio Universitario de Estudios Financieros – CUNEF, Madrid 25420
Colegio Universitario del Este, Carolina 21851
Colgate University, Hamilton 31279
Colgate-Palmolive Co, Piscataway 36124
Collectors Club Library, New York 37197
Collège Ahuntsic, Montreal 03762
College at New Paltz, New Paltz 32017
College at Old Westbury, Old Westbury 32189
Collège Bourget, Rigaud 04022
Collège Camerounais des Arts, des Sciences et de la Technologie, Bambili 03632
College Church in Wheaton, Wheaton 33042
Collège d'Alfred de l'Université de Guelph, Alfred 03669
Collège de Bois-de-Boulogne, Montreal 03763
Collège de France, Paris 07844
– Bibliothèque byzantine, Paris 07845
– Bibliothèque d'Anthropologie sociale, Paris 07846
– Bibliothèque d'Assyriologie, Paris 07847
– Bibliothèque de'Études arabes, turques et islamiques, Paris 07848
– Bibliothèque d'Égyptologie, Paris 07849
– Bibliothèque d'Études sémitiques, Paris 07850
– Bibliothèque d'Histoire des christianismes orientaux, Paris 07851
– Bibliothèques d'Extrême-Orient, Paris 07852
Collège de Granby (Cégep), Granby 03712
Collège de l'Assomption, L'Assomption 03736
Collège de Lévis, Lévis 03740
Collège de Limoilou – Campus de Charlesbourg, Charlesbourg 03688
Collège de Maisonneuve, Montreal 03764
Collège de Montréal, Montreal 03765
Collège de Rosemont (Cégep), Montreal 04005
Collège de Sainte-Anne-de-la-Pocatière, La Pocatière 04001
Collège de Sherbrooke, Sherbrooke 03860
Collège d'Enseignement General & Professionnel de Limoilou, Québec 04018
Collège des Médecins du Québec, Montreal 04006
Collège Dominicain de Philosophie et de Théologie, Ottawa 03810
Collège Edouard Montpetit, Saint Hubert 03840
Collège Edouard-Montpetit, Longueuil 03753
– Ecole Nationale d'Aerotechnique, Saint Hubert 03754
College for Arabic and Islamic Studies, Omdurman 26351
College for Greek-Orthodox Priests, Athinai 13091
College for Roman Catholic Priests, Esztergom 13335
College House Institute of Theology, Christchurch 19813
Collège Jean-de-Brebeuf, Montreal 03766
Collège Jésus-Marie de Sillery, Sillery 04229
Collège Lasalle, Montreal 03767
Collège Mérici, Québec 03823

Collège Militaire Royal de Saint-Jean, Saint-Jean-Sur-Richelieu 03846
College Misericordia, Dallas 30940
Collège Montmorency, Laval 03737
Collège Notre Dame, Montreal 03768
College of Aeronautics, East Elmhurst 33306
College of African Wildlife Management, Moshi 27657
College of Agricultural Engineering, Kirovograd 28105
College of Agricultural Studies, Khartoum 26342
College of Agriculture, Indore 13807
College of Agriculture, Pune 13836
College of Agriculture and Education, Siedlce 20924
College of Agriculture, Science and Education (CASE), Port Antonio 16873
College of Alameda, Alameda 33116
College of Arms, London 29476
P. S. G. College of Arts and Science, Coimbatore 13634
College of Arts and Science of Beijing United University, Beijing 05045
College of Arts, Science and Technology, Kingston 16864
College of Banking, Sankt-Peterburg 22693
College of Business and Commerce, Ulaanbaatar 18902
College of Business Education, Dar es Salaam 27660
College of Calamity Prevention Technology, Sanhe 05495
College of Catholic Philosophy and Theology, Maumere 14339
College of Charleston, Charleston 30715
College of Commerce, Kherson 28102
College of Commerce of Yangzhou University, Yangzhou 05737
College of Dairying, Ebetsu 16914
College of Dupage, Glen Ellyn 31215
College of Eastern Utah, Price 32361
College of Economics, Ulaanbaatar 18903
College of Education, Banja Luka 02948
College of Education, Słupsk 20925
College of Education, Split 06169
College of Engineering, Bangalore 13790
College of Engineering, Koszalin 20821
College of Engineering, Zielona Góra 21034
College of Europe, Brugge 02281
College of External Studies, Konedobu 20555
College of Finance and Accountancy, Budapest 13226
College of Fine Arts, Paddington 00524
College of Food Industry, Szeged 13308
College of Further Education Plymouth, Plymouth 29369
College of Health Sciences, Manama 01736
College of Industrial Education, Rijeka 06159
College of Lake County, Grayslake 33384
College of Literature and Human Sciences, Shiraz 14396
College of Marin, Kentfield 33454
College of Marin, Indian Valley Campus, Novato 32168
College of Marine Military Science, Gdynia 20809
College of Medicine, Chichiri 18468
College of Medicine, Sevastopol 28129
College of Medicine and Allied Health Sciences, Freetown 24566
College of Micronesia-FSM, Kolonia 18873
College of Mount Saint Joseph, Cincinnati 30799
College of Mount Saint Vincent, Bronx 30556
College of New Caledonia, Prince George 04017
The College of New Jersey, Ewing 31114
College of New Rochelle, New Rochelle 32019
College of North East London, London 29176
College of North West London, London 29177
College of Notaries, Buenos Aires 00250
College of Nursing, Burwood 00407
College of Osteopathic Medicine, Tulsa 32848
College of Our Lady of the Elms, Chicopee 30797
College of Physical Culture, Gdańsk 21035
College of Physicians of Philadelphia, Philadelphia 32252
College of Physicians & Surgeons of British Columbia, Vancouver 03930
College of Renewable Natural Resources, Sunyani 13026
College of Saint Benedict, Saint Joseph 32499
College of Saint Catherine, Minneapolis 31844
College of Saint Catherine, Saint Paul 32529
College of Saint Elizabeth, Morristown 31906
College of Saint John the Evangelist and Trinity Methodist College, Auckland 19811

College of Saint Joseph, Rutland 32486
College of Saint Mary, Omaha 32193
College of Saint Rose, Albany 30181
College of Saint Scholastica, Duluth 31032
College of San Mateo, San Mateo 33736
College of Santa Fe, Santa Fe 32613
College of Social Work, Mumbai 13820
College of Southern Idaho, Twin Falls 32866
College of Southern Maryland, La Plata 33465
College of Staten Island, Staten Island 32735
College of Stord/Haugesund, Rommetveit 20156
College of Technical Teacher Training, Ho Chi Minh City 42278
College of Technology, Budapest 13266
College of Technology of Shanghai University, Shanghai 05497
College of Textiles and Clothing, Qingdao 05482
College of the Albemarle, Elizabeth City 33316
College of the Atlantic, Bar Harbor 30351
College of The Bahamas, Nassau 01732
College of the Canyons, Santa Clarita 33744
College of the Desert, Palm Desert 33614
College of the Holy Cross – Dinand Library, Worcester 33096
College of the Holy Cross – O'Callahan Science Library, Worcester 33097
College of the Holy Spirit, Manila 20763
College of the Mainland, Texas City 32812
College of the Marshall Islands, Majuro 18672
College of the North Atlantic, St. John's 03865
College of the Ozarks, Point Lookout 32326
College of the Redwoods, Eureka 33327
College of the Rockies, Cranbrook 03985
College of the Sequoias, Visalia 33828
College of the Siskiyous, Weed 33850
College of the Southwest, Hobbs 31341
College of Traditional Indonesian Arts, Surakarta 14302
College of Veterinary Science and Animal Husbandry, Mhow 13698
College of Veterinary Science and Animal Husbandry C.S.A., Mathura 13696
College of West Anglia, King's Lynn 29144
College of William and Mary in Virginia – Earl Gregg Swem Library, Williamsburg 33065
College of William & Mary, Gloucester Point 31222
College of William & Mary in Virginia – Marshall-Wythe Law Library, Williamsburg 33066
The College of Wooster Libraries, Wooster 33091
Le Collège Presbytérien à l'Université McGill, Montreal 04217
Collège Saint-Alphonse, Sainte-Anne-de-Beaupré 04024
Collège Stanislas, Outremont 03818
Collège Universitaire de Saint-Boniface, Winnipeg 03960
College voor Zorgverzekeringen (CVZ), Diemen 19415
Col.legi d'Aparelladors, Arquitectes Tècnics i Enginyers d'Edificació de Barcelona, Barcelona 25895
Collegi d'Arquitectes de Catalunya, Girona 25957
Collegi d'Arquitectes de Catalunya, Demarcació de Barcelona, Barcelona 25896
Collegi dels Jesuites de Casp, Barcelona 25662
Collegio Americano del Nord, Città del Vaticano 42195
Collegio degli Ingegneri e Architetti, Milano 15625
Collegio delle Missioni Estere dei Frati Minori Conventuali, Palermo 15961
Collegio di S. Clemente dei Dominicani Irlandesi, Roma 16022
Collegio di S. Isidoro, Roma 16023
Collegio di Spagna, Bologna 15745
Collegio Emiliani, Genova 15022
Collegio Ghislieri, Pavia 15310
Collegio Missionario Teologico dei Frati Minori Conventuali, Assisi 15714
Collegio Nazareno, Roma 16024
Collegio San Bonaventura, Grottaferrata 15857
Collegio San Francesco d'Assisi dei Cappuccini, Milano 15907
Collegio San Francesco dei Padri Barnabiti, Lodi 15879
Collegio San Giuseppe, Torino 15634
Collegio San Luigi dei Barnabiti, Bologna 15746
Collegio Teologico, Parma 15966
Collegio Teologico dei Carmelitani Scalzi, Firenze 15830
Collegio Teologico dei PP. Barnabiti, Roma 16025
Collegio Teutonico di S. Maria dell'Anima, Roma 16026
Collegium Albertinum, Bonn 11175
Collegium Borromaeum, Freiburg 11216
Collegium Budapest Library, Budapest 13232
Collegium Canisianum, Innsbruck 01437
Collegium Carolinum, München 12289
Collegium Hungaricum, Wien 01592
Collegium Medicum im. Ludwika Rydygiera, Bydgoszcz 20797
Collegium Oecumenicum, München 11278
Collegium pro America Latina (COPAL), Leuven 02473
Colleton County Memorial Library, Walterboro 41719
Collier County Public Library, Naples 40364
Collin County Community College District, Plano 33649
Collin County Law Library, McKinney 34519
Collingswood Free Public Library, Collingswood 38638
Collingwood Public Library, Collingwood 04616
Collins Correctional Facility Library, Collins 34093
Collinsville Memorial Public Library District, Collinsville 38639
Cologne University of Applied Sciences, Köln 10124
Coloma Public Library, Coloma 38640
Colombiere Center, Clarkston 36703
Colombo National Museum, Colombo 26308
Colombo Observatory, Colombo 26309
Colombo Plan Staff College for Technician Education, Manila 20711
Colonel Robert R. McCormick Research Center, Wheaton 37831
Colonial Williamsburg Foundation – John D. Rockefeller Jr Library, Williamsburg 37840
Colonial Williamsburg Foundation – John D. Rockefeller Jr Library-Special Collections, Williamsburg 37841
The Colony Public Library, The Colony 41545
Colorado Agency for Jewish Education, Denver 36788
Colorado Christian University, Lakewood 31549
Colorado College, Colorado Springs 30867
Colorado Department of Corrections, Crowley 34119
Colorado Department of Corrections, Denver 34135
Colorado Department of Corrections, Limon 34458
Colorado Department of Corrections, Pueblo 34712
Colorado Department of Corrections – Centennial Correctional Facility Library, Canon City 34039
Colorado Department of Corrections – Colorado State Penitentiary Library, Canon City 34040
Colorado Department of Corrections – Four Mile Correctional Center Library, Canon City 34041
Colorado Department of Corrections – Fremont Correctional Facility Library, Canon City 34042
Colorado Department of Corrections – Sterling Correctional Facility Library-East Side, Sterling 34893
Colorado Department of Corrections – Sterling Correctional Facility Library-West Side, Sterling 34894
Colorado Department of Transportation Library, Denver 34136
Colorado Division of Wildlife, Fort Collins 36858
Colorado Historical Society, Denver 36789
Colorado Mental Health Institute at Fort Logan – Children's Division Library, Denver 36790
Colorado Railroad Historical Foundation/Colorado Railroad Museum, Golden 36895
Colorado School for the Deaf & the Blind, Colorado Springs 33247
Colorado School of Mines, Golden 31223
Colorado Springs Fine Arts Center Library, Colorado Springs 36726
Colorado State University, Fort Collins 31154
– Veterinary Teaching Hospital, Fort Collins 31155
Colorado State University, Pueblo 32393
Colorado Supreme Court Library, Denver 34137
Colorado Territorial Correctional Facility Library, Canon City 34043
Colton Public Library, Colton 38643
Columban Mission Institute, North Turramurra 00787
Columbia Basin College, Pasco 33623
Columbia Bible College, Abbotsford 04198
Columbia College, Chicago 30759
Columbia College, Columbia 30870
Columbia College, Columbia 30871
Columbia College, Sonora 33767
Columbia College Library, Columbia 30872
Columbia Correctional Institution Library, Portage 34697
Columbia County Library, Magnolia 40016
Columbia County Public Library, Lake City 39774
Columbia Environmental Research Center Library, Columbia 34095
Columbia Heights Public Library, Columbia Heights 38652
Columbia Hospital, Milwaukee 37111
Columbia International University, Columbia 30873
Columbia Museum of Art, Columbia 36730
Columbia River Inter-Tribal Fish Commission, Portland 37398
Columbia State Community College, Columbia 33249
Columbia Theological Seminary, Decatur 35183
Columbia Union College, Takoma Park 32788
Columbia University, New York 32036
– African Studies Library, New York 32037
– Augustus C. Long Health Sciences Library, New York 32038
– Avery Architectural & Fine Arts Library, New York 32039
– Biological Sciences Library, New York 32040
– The Burke Library at Union Theological Seminary, New York 32041
– Butler Library Reference Department, New York 32042
– Chemistry Library, New York 32043
– Diamond Law Library, New York 32044
– Engineering Library, New York 32045
– Gabe M. Wiener Music and Arts Library, New York 32046
– Geology Library, New York 32047
– Lamont-Doherty Geoscience Library, Palisades 32048
– Lehman Social Science Library, New York 32049
– Mathematics-Science Library, New York 32050
– Middle East and Jewish Studies Library, New York 32051
– Philip L. Milstein Family College Library, New York 32052
– Physics-Astronomy Library, New York 32053
– Psychology Library, New York 32054
– Rare Book & Manuscript Library, New York 32055
– Russian, Eurasian & European Studies Library, New York 32056
– Social Work Library, New York 32057
– South & Southeast Asia Library, New York 32058
– C. V. Starr East Asian Library, New York 32059
– Watson Library of Business & Economics, New York 32060
Columbia University – Teachers College, New York 32061
Columbia-Greene Community College, Hudson 33422
Columbiana County Law Library, Lisbon 34465
Columbiana Public Library, Columbiana 38653
Columbus College of Art & Design, Columbus 30894
Columbus County Public Library, Whiteville 41885
Columbus Law Library Association, Columbus 36739
Columbus Metropolitan Library, Columbus 38656
Columbus Metropolitan Library – Dublin Branch, Dublin 38868
Columbus Metropolitan Library – Gahanna Branch, Gahanna 39178
Columbus Metropolitan Library – Hilliard Branch, Hilliard 39435
Columbus Metropolitan Library – Hilltop, Columbus 38657
Columbus Metropolitan Library – Karl Road, Columbus 38658
Columbus Metropolitan Library – Livingston Branch, Columbus 38659
Columbus Metropolitan Library – Main Library, Columbus 38660
Columbus Metropolitan Library – New Albany Branch, New Albany 40391
Columbus Metropolitan Library – Northern Lights, Columbus 38661
Columbus Metropolitan Library – Northwest, Columbus 38662
Columbus Metropolitan Library – Reynoldsburg Branch, Reynoldsburg 40969
Columbus Metropolitan Library – South High, Columbus 38663
Columbus Metropolitan Library – Southeast, Columbus 38664
Columbus Metropolitan Library – Whetstone Branch, Columbus 38665
Columbus Metropolitan Library – Whitehall Branch, Columbus 38666
Columbus Public Library, Columbus 38667
Columbus State Community College, Columbus 33252
Columbus State University, Columbus 30895
Columbus-Lowndes Public Library, Columbus 38668
Colusa County Free Library, Colusa 38671
Combine Building Industrial Corporation, Kherson 28148
Comédie-Française, Paris 08529
Comenius Higher School of Teachers' Training, Sárospatak 13292
Comenius Museum, Uhersky Brod 06545
Comenius State Library of Education, Praha 06508
Comenius University in Bratislava, Bratislava 24637
– Faculty of Education, Bratislava 24660
– Faculty of Law Library, Bratislava 24658
– Faculty of Mathematics, Physics and Informatics, Bratislava 24653
– Faculty of Medicine, Bratislava 24659
– Faculty of Pharmacy, Bratislava 24655
– Faculty of Philosophy, Bratislava 24656
– Faculty of Physical Education and Sport, Bratislava 24654
– Roman Catholic Faculty of Theology of Cyrill and Methodius, Bratislava 24661
Comenius-Institut, Münster 12342
B. B. Comer Memorial Library, Sylacauga 41486
Comfamiliares Unidas del Valle (Comfaunion Palmira), Palmira 06016
Comintern Machine Tools Plant, Vitebsk 01936
Comisao Nacional de Energia Nuclear, Rio de Janeiro 03338
Comisión Chilena de Energía Nuclear, Santiago 04926
Comisión de las Comunidades Europeas, Madrid 25704
Comisión Económica para América Latina y el Caribe (CEPAL), Montevideo 42050
Comisión Nacional de Energía Atómica, Bariloche 00228
Comisión Nacional de Energía Atómica (CNEA), Buenos Aires 00251
Comisión Nacional Española de Cooperación con la UNESCO, Madrid 26009
Comissão de Coordenação da Região do Norte, Porto 21741
Comissão de Coordenação e Desenvolvimento da Regional do Centro, Coimbra 21757
Comissão Executiva do Plano da Lavoura Cacaueira, Itabuna 03301
Comissão Nacional da FAO, Lisboa 21722
Comissão para a Igualdade e Direitos das Mulheres, Lisboa 21783
Comité d'Etablissement SNCF de Périgueux, Périgueux 08639
Comité Intergouvernemental de Recherches Urbaines et Régionales, Toronto 04517
Comité international de la Croix-Rouge / International Committee of the Red Cross / Internationales Kommitee vom Roten Kreuz, Genève 27362
Commercial Library & Reading Room, Kolkata 13859
Commercial Machine Building Plant, Baranovichi 01829
Commercial School Slagelse, Slagelse 06725
Commercial Technical Institute at Mataria, Cairo 07052
Commerzbibliothek der Handelskammer Hamburg, Hamburg 11975
Commissariat à l'Energie Atomique, Bagnols-sur-Cèze 08289
Commissariat à l'Energie atomique (CEA), Gif-sur-Yvette 08351
Commission de la Construction du Québec, Montreal 04095
Commission de la Santé et de la Sécurité du Travail, Montreal 04096
Commission de Toponymie du Québec, Québec 04459
Commission des Valeurs Mobilières du Québec, Montreal 04097
Commission du Bassin du Lac Tchad, N'Djamena 04838

Commission Européenne, Luxembourg 18395
Commission for Integrated Survey of Natural Resources, Beijing 05830
Commission of the European Communities, Dublin 14527
Commission Scolaire de la Capitale, Québec 04460
Commissioner of Official Languages Library, Ottawa 04441
Committee of Forests, Sofiya 03488
Committee on Meteorites of the Russian Academy of Sciences, Moskva 23342
Commodity Futures Trading Commission Library, Washington 37728
Common Services Agency, Edinburgh 29653
Commonwealth Court Library, Harrisburg 36914
Commonwealth Director of Pulic Prosecutions, Canberra 00639
Commonwealth of Massachusetts, Boston 33997
Commonwealth of Massachusetts Trial Court, Northampton 34618
Commonwealth of Puerto Rico – Legislative Reference Library, San Juan 21894
Commonwealth of Puerto Rico – Office of the Attorney General, San Juan 21895
Commonwealth of Puerto Rico – Supreme Court Library, San Juan 21896
Commonwealth Secretariat, London 29742
Communauté d'Agglomération Béziers Méditerranée, Béziers 08771
Communauté des Dominicains, Paris 08216
Communauté du Saint-Sacrement, Paris 08217
Communauté Fanciscaine de Strasbourg, Strasbourg 08237
Communications Research Laboratory, Koganei 17272
Community College of Allegheny County, Monroeville 33552
Community College of Allegheny County, Pittsburgh 33647
Community College of Allegheny County, West Mifflin 33859
Community College of American Samoa, Pago Pago 00094
Community College of Baltimore County, Baltimore 33150
Community College of Baltimore County Catonsville, Catonsville 33204
Community College of Baltimore County – Essex Campus, Baltimore 33151
Community College of Beaver County, Monaca 33549
Community College of Philadelphia, Philadelphia 33634
Community College of Rhode Island, Lincoln 33487
Community College of Rhode Island, Warwick 33839
Community College of Southern Nevada, Las Vegas 33478
Community Development Library, Dhaka 01764
Community District Library, Corunna 38707
Community Legal Services, Inc Library, Philadelphia 37340
Community Library Association, Ketchum 39711
Community Library of Allegheny Valley, Tarentum 41519
Community of Christ, Independence 35237
Compagnie de Jésus, Montreal 04214
Compagnie de Saint-Sulpcie, Paris 08218
Compagnie Générale de Géophysique (CGG), Massy 08258
Companhia de Tecnologia de Saneamento Ambiental Endereço (CETESB), São Paulo 03399
Companhia Energética de São Paulo (CESP), São Paulo 03400
Companhia Paranense de Energia (COPEL), Curitiba 03258
Companhia Siderúrgica Paulista (COSIPA), Cubatão 03257
Compañia de Jesús, La Paz 02931
Compania de Stat Teleradio-Moldova, Chişinău 18892
Competition Commission, London 29743
Complejo Museográfico 'Enrique Udaondo', Luján 00302
Complexul muzeal Goleşti, Goleşti-Ştefăneşti 22035
Complexul Naţional Muzeal ASTRA, Sibiu 22056
Complutense University of Madrid, Madrid 25439
– Faculty of Chemistry, Madrid 25456
– School of Social Work, Pozuelo de Alarcón (Madrid) 25448

Composers' and Musicologists' Union of Romania, Bucureşti 22018
Compton Community College, Compton 33255
Comsewogue Public Library, Port Jefferson Station 40857
Comstock Township Library, Comstock 38674
Comunidad de Agustinos de San Manuel y San Benito, Madrid 25789
Comunidad de Canónigos Regulares Lateranenses, Oñati 25801
Comunidad de Franciscanos, Santiago de Compostela 25823
Comunidad de Franciscanos, Zarautz 25849
Comunidad de Jesuítas, Alicante 25758
Comunidad de Jesuítas, Villafranca de los Barros 25842
Comunidad de Madrid, Madrid 25705
Comunità Ebraica, Firenze 15831
Comunità Ebraica, Torino 16098
Comunità Ebraica, Trieste 16625
Comunità Ebraica, Venezia 16124
Comunità Missionaria 'Paradiso', Bergamo 15733
Conception Abbey & Seminary, Conception 35169
Concord Free Public Library, Concord 38677
Concord Public Library, Concord 38678
Concord Repatriation General Hospital, Concord 00906
Concord University, Athens 30272
Concordia College, Bronxville 30561
Concordia College, Moorhead 31896
Concordia College, Selma 32656
Concordia Historical Institute, Saint Louis 35358
Concordia Parish Library, Ferriday 39065
Concordia Seminary, Saint Louis 35359
Concordia Theological Seminary, Fort Wayne 35207
Concordia University, Ann Arbor 30229
Concordia University, Irvine 31411
Concordia University, Portland 32335
Concordia University, River Forest 32450
Concordia University, Saint Paul 32530
Concordia University, Seward 32659
Concordia University College, Edmonton 03694
Concordia University Libraries, Montreal 03769
Concordia University Wisconsin, Mequon 31803
Concregation Beth Shalom, Overland Park 35320
Concremat Engenharia e Tecnologia S.A., Rio de Janeiro 03261
Concrete Information Ltd, Camberley 29622
Concrete-Mixer Lorry Plant Joint-Stock Company, Tuimazy 22962
Condon & Forsyth Library, New York 35970
Conestoga College of Applied Arts & Technology, Kitchener 03999
Coney Island Hospital, Brooklyn 36602
Confederação Nacional da Indústria, Rio de Janeiro 03339
Confederación Española de Cajas de Ahorros, Madrid 26021
Confederation Centre Public Library, Charlottetown 04607
Confederation College of Applied Arts & Technology, Thunder Bay 04035
Confédération des Caisses Populaires et d'Economie des Jardins du Québec, Lévis 04374
Confederation of Norwegian Business and Industry (NHO), Oslo 20202
Confederazione Generale del Commercio e del Turismo, Roma 16487
Confederazione Generale dell'Agricoltura Italiana ' Zappi Recordati', Roma 16488
Confederazione Generale dell'Industria Italiana, Roma 16489
Confederazione Generale Italiana del Lavoro, Roma 16490
Conférence des Recteurs et des Principaux des Universités du Québec, Montreal 04396
Conferencia Episcopal Española, Madrid 25790
Confindustria, Roma 16491
Congregatie Broeders Hiëronymieten, Sint-Niklaas 02486
Congregatie der Assumptionisten, Boxtel 19301
Congregation Beth Emeth, Wilmington 37846
Congregation Beth Israel, West Hartford 35416
Congregation Beth Shalom, Oak Park 35315
Congregation Beth Yeshurun, Houston 35234
Congregation Emanu-El, New York 35306
Congregation Emanu-el B'ne Jeshurun Library, Milwaukee 35285
Congregation Emanuel Library, Denver 35185

Congregation Mikveh Israel Archives, Philadelphia 37341
Congregation Rodeph Shalom, Philadelphia 35326
Congregation Rodfei Zedek, Chicago 36669
Congregation Shaarey Zedek Library & Audio Visual Center, Southfield 35387
Congregation Solel Library, Highland Park 35229
Congregational Library, London 29551
Congregazione Mechitaristi, Venezia 16125
Congregazione Religiosa dei PP. Redentoristi, Pagani 15959
Congreso, Caracas 42233
Congreso de la Nación Argentina, Buenos Aires 00207
Congreso de los Diputados, Madrid 25706
Congreso Nacional, Santafé de Bogotá 05991
Congreso Nacional, Santo Domingo 06961
Congreso Nacional, Tegucigalpa 13210
Congreso Nacional de Bolivia, La Paz 02926
Congreso Nacional de Chile, Santiago 04905
Congreso Nacional del Ecuador, Quito 06985
Congress of the Federated States of Micronesia, Pohnpei 18875
Congressional Budget Office Library, Washington 34982
ConnDOT Library & Information Center, Newington 37281
Conneaut Public Library, Conneaut 38680
Connecticut Aeronautical Historical Association, Inc, Windsor Locks 37855
Connecticut Agricultural Experiment Station, New Haven 37158
Connecticut College, New London 31997
– Greer Music Library, New London 31998
Connecticut Historical Society Library, Hartford 36917
Connecticut Judicial Branch, Bridgeport 34004
Connecticut Judicial Branch, Danbury 34125
Connecticut Judicial Branch, New Haven 34581
Connecticut Judicial Branch, Norwich 34620
Connecticut Judicial Branch, Putnam 34714
Connecticut Judicial Branch, Rockville 34750
Connecticut Judicial Branch, Stamford 34888
Connecticut Judicial Department, Hartford 34344
Connecticut Legislative Library, Hartford 34345
Connecticut State Judicial Department, Waterbury 35046
Connecticut State Library, Hartford 30132
Connecticut State Library – Library for the Blind & Physically Handicapped, Rocky Hill 37464
Connecticut State Library – State Records Center, Rocky Hill 37465
Connecticut Valley Historical Museum, Springfield 37626
Connecticut Valley Hospital, Middletown 37103
Connecticut Valley Hospital Patients' Library, Middletown 37104
Connell Foley Law Library, Roseland 36187
Connemara (State Central) Public Library, Chennai 14224
Connetquot Public Library, Bohemia 38233
Connors State College, Warner 33836
Conoco Incorporated, Technology, Ponca City 36139
ConocoPhillips Library Network, Bartlesville 35492
Conrad Grebel University College, Waterloo 03954
Conseil de l'Europe, Strasbourg 08163
Conseil d'Etat, Paris 08145
Conseil d'Etat du Grand-Duché de Luxembourg, Luxembourg 18396
Conseil Economique et Social (CES), Paris 08146
Conseil Général de la Moselle, Saint-Julien-Lès-Metz 08672
Conseil Général de Paris, Paris 08147
Conseil Général des Alpes-Maritimes, Nice 08140
Conseil Général du Loiret, Orléans 08141
Conseil Général du Val-de-Marne, Créteil 08136
Conseil oecuménique des Églises (COE), Genève 27363
Conseil pour le Développement de la Recherche Economique et Social en Afrique (CODESRIA), Dakar 24439
Conseil Régional de Provence-Alpes-Côte d'Azur, Marseille 08139
Conseil Superieur de l'Education, Sainte-Foy 04150
Consejería de Gobernación, Sevilla 25746
Consejería de Obras Públicas, Urbanismo y Transportes, Madrid 25707
Consejería de Salud, Sevilla 25747
Consejo de Bogotá, Santafé de Bogotá 05992

Consejo de Educación Secundaria, Montevideo 42051
Consejo de Estado, Madrid 25708
Consejo de Recursos Minerales (CEDOREM), México 18842
Consejo Federal de Inversiones, Buenos Aires 00252
Conselho de Reitores das Universidades Brasileiras, Brasília 03003
Conservation and Environment Library, Winnipeg 04554
Conservation Council of South Australia, Adelaide 00849
Conservatoire, Žilina 24670
Conservatoire de Fribourg (COF), Granges-Paccot 27093
Conservatoire de Lausanne HEM, Lausanne 27095
Conservatoire de Lille, Lille 07725
Conservatoire de Lyon, Lyon 07737
Conservatoire de musique, Genève 27085
Conservatoire de Musique de la Ville de Luxembourg, Luxembourg 18380
Conservatoire de Musique de Québec, Québec 03824
Conservatoire de Musique et de Déclamation, Alger 00025
Conservatoire de Strasbourg, Strasbourg 08043
Conservatoire et Jardin botaniques de la Ville de Genève, Chambésy-Genève 27340
Conservatoire National de Musique de Dijon, Dijon 07698
Conservatoire National de Région, Toulouse 08072
Conservatoire National des Arts et Métiers, Paris 07853
Conservatoire National Supérieur d'Art Dramatique, Paris 07854
Conservatoire National Supérieur de Musique et de Danse de Paris (CNSMDP), Paris 07855
Conservatoire National Supérieur Musique et Danse de Lyon, Lyon 07738
Conservatoire of Music, Praha 06385
Conservatoire Royal de Musique de Liège, Liège 02372
Conservatoire Royal de Musique de Mons, Mons 02409
Conservatori Municipal de Música de Barcelona, Barcelona 25292
Conservatorio 'C. Pollini', Padova 15172
Conservatorio di Musica Antonio Vivaldi, Alessandria 15612
Conservatorio di Musica 'G. Pierluigi da Palestrina', Cagliari 14917
Conservatorio di Musica G. Verdi, Milano 15626
Conservatorio di Musica 'Gioachino Rossini', Pesaro 15630
Conservatorio di Musica 'Giovan Battista Martini', Bologna 15616
Conservatorio di Musica 'Niccoló Piccinni', Bari 15613
Conservatorio di Musica S. Pietro a Maiella, Napoli 15628
Conservatorio di Musica Santa Cecilia, Roma 15631
Conservatorio di Musica 'Vincenzo Bellini', Palermo 15629
Conservatorio G. Tartini di Trieste, Trieste 15637
Conservatorio Nacional de Música, Lima 20618
Conservatorio Nacional de Música, México 18727
Conservatorio Nazionale di Musica 'Benedetto Marcello', Venezia 15578
Conservatorio Nicolò Paganini, Genova 16284
Conservatorio Profesional de Música de Córdoba, Córdoba 25354
Conservatorio Statale di Musica 'Claudio Monteverdi', Bolzano 15617
Conservatorio Statale di Musica 'E.F. Dall'Abaco', Verona 15593
Conservatorio Statale di Musica 'Giuseppe Verdi', Torino 15483
Conservatorio statale di Musica Luigi Cherubini, Firenze 15621
Conservatorio Superior de Música de Murcia "Manuel Massotti Littel", Murcia 25511
Conservatorio Superior de Música de Valencia, Valencia 25616
Conservatorio Superior de Música "Manuel Castillo" de Sevilla, Sevilla 25585
Conservatorium, Gent 02322
Conservatorium, Groningen 19128
Conservatory in Brno, Brno 06354
Consiglio di Stato, Roma 15664
Consiglio Nazionale delle Ricerche (CNR), Roma 16492
Consiglio Nazionale dell'Economia e del Lavoro, Roma 15665

Consiglio Regionale, Firenze 15643
Consiglio Regionale del Piemonte, Torino 15691
Consiglio Regionale del Veneto, Venezia 15694
Consiglio Regionale della Sardegna, Cagliari 15641
Consistorium der Französischen Kirche, Berlin 11165
Consol Energy, Inc, South Park 36275
Consolidated Edison Company of New York, Long Island City 35822
Consorzio per la Pubblica Lettura 'S. Satta', Nuoro 16782
Constanta County Library, Constanța 22086
Constantine Archaeological Society, Constantine 00079
Constantine the Philosopher University, Nitra 24668
Construction and Assembly Industrial Corporation, Gomel 01858
Construction Equipment Research Institute, Kobylka 21119
Construction Professional School, Kyiv 28156
Consultative Council, Riyadh 24397
Stichting Consument en Veiligheid (SCV), Amsterdam 19381
Consumer Safety Institute, Amsterdam 19381
Consumers Energy, Jackson 35784
Continental Reifen Deutschland GmbH, Hannover 11404
Continental Theological Seminary, Sint-Pieters-Leeuw 02487
Contra Costa College, San Pablo 33737
Contra Costa County Law Library, Martinez 34507
Contra Costa County Library, Pleasant Hill 40827
Contra Costa County Library – Antioch Community Library, Antioch 37971
Contra Costa County Library – Clayton Library, Clayton 38588
Contra Costa County Library – Concord Library, Concord 38679
Contra Costa County Library – Danville Library, Danville 38781
Contra Costa County Library – Lafayette Library, Lafayette 39764
Contra Costa County Library – Moraga Library, Moraga 40294
Contra Costa County Library – Orinda Library, Orinda 40639
Contra Costa County Library – Pleasant Hill Library, Pleasant Hill 40828
Contra Costa County Library – San Ramon Library, San Ramon 41202
Contra Costa County Library – Walnut Creek – Park Place Library, Walnut Creek 41715
Contra Costa County Library – Ygnacio Valley (Thurman G. Casey Memorial), Walnut Creek 41716
Convent of the Sacred Heart, Saint Julian's 18660
Convento Cappuccini, Bigorio 27249
Convento de Capuchinos, Barcelona 25767
Convento de Carmelitas, Bilbao 25769
Convento de Carmelitas de la Orden de María del Monte Carmelo, Vitoria-Gasteiz 25844
Convento de Carmelitas Descalzos, Madrid 25791
Convento de Carmelitas Descalzos, Salamanca 25811
Convento de la Recoleta, Sucre 02932
Convento de Santo Domingo del Real, Madrid 25792
Convento dei Cappuccini, Faido 27257
Convento dei Cappuccini, Modena 15912
Convento dei Cappuccini, Vasto 16119
Convento dei Cappuccini di Terzolas, Terzolas 16088
Convento dei Domenicani, San Domenico di Fiesole 16050
Convento dei Francescani, San Candido 16048
Convento dei Frati Minori, Assisi 15715
Convento dei Frati Minori, Vicenza 16135
Convento dei Minori Francescani di S. Vigilio, Cavalese 15790
Convento dei Padri Carmelitani Scalzi, Brescia 15755
Convento dei PP. Cappuccini 'S. Francesco da Paola', Sulmona 16079
Convento dei PP. Cappuccini 'San Felice', Cava dei Tirreni 15789
Convento dei PP. Carmelitani di Sant'Anna, Genova 15847
Convento dei PP. Francescani, Borgo Valsugana 15752
Convento dei Santuari Antoniani, Camposampiero 15774
Convento della SS. Annunziata, Parma 15967
Convento dell'Osservanza, Siena 16069
Convento di Nostra Signora di Bonaria, Cagliari 15762
Convento di S. Giuliano, L'Aquila 15870
Convento di S. Mauro, Cagliari 15763
Convento di S. Paolo in Monte-Osservanza, Bologna 15747
Convento di San Francesco, Fiesole 15822
Convento di San Marco, Firenze 15832
Convento di San Matteo, San Marco in Lamis 16054
Convento di San Pasquale, Taranto 16083
Convento di San Torpè, Pisa 15986
Convento di Santa Maria, Potenza 15995
Convento di Santa Maria in Colleromano dei Frati Minori, Penne 15972
Convento di Sant'Angelo della Pace, Lanciano 15865
Convento Francescano di S. Lucia al Monte, Napoli 15930
Convento Frati Monori, Santa Maria degli Angeli 16058
Convento pp minimi 'Santa Maria della Stella', Napoli 15931
Convento S. Francesco della Vigna, Venezia 16126
Convento S. Lorenzo Maggiore, Napoli 15932
Convento 'S. Mario del Paradiso', Tocco da Casauria 16089
Convento 'San Domenico', Soriano Calabro 16073
Convento 'San Francesco d'Albaro', Genova 15848
Convento San Giuseppe da Copertino, Osimo 15948
Convento San Pedro Mártir, Madrid 25982
Convento Santa Maria degli Angeli, Nocera Superiore 15938
Convento 'Sant'Antonio', Afragola 15696
Convento 'Sant'Antonio' dei Frati Minori, Sulmona 16080
Convento 'Sant'Antonio' dei frati minori conventuali, Portici 15994
Converse College, Spartanburg 32693
Converse County Library, Douglas 38854
Convitto Nazionale, Palermo 15962
Convitto Nazionale 'C. Colombo', Genova 15849
Convitto Nazionale 'Cicognini', Prato 15999
Conyers-Rockdale Library System, Conyers 38685
Coober Pedy School, Coober Pedy 00565
Cook County Hospital Libraries, Chicago 36670
Cook County Law Library, Chicago 34069
Cook Memorial Public Library District, Libertyville 39872
Cook School for Christian Leadership, Tempe 33795
Cooke County Library, Gainesville 39180
Cooley Godward Kronish LLP, New York 35971
Cooley Godward Kronish LLP Library, San Francisco 36227
Cooloola Sunshine Institute of TAFE, Nambour 00580
Cooper Memorial Library, Clermont 38594
Cooper Union for Advancement of Science & Art Library, New York 37198
Cooper, White & Cooper, San Francisco 36228
Cooperativa de Consumo CUTE-ANTEL, Montevideo 42052
Cooperative Academic Library Services (CALS), Spokane 32697
Co-operative College, Moshi 27658
Cooperative College of Malaysia, Petaling Jaya 18544
Cooperative Institute, Karaganda 17907
Cooperative Research Institute, Budapest 13459
Cooper-Hewitt, National Design Museum Library, New York City 37273
Coordinación Nacional de Restauración del Patrimonio Cultural, México 18853
Coos Bay Public Library, Coos Bay 38688
COPENE Petroquímica do Nordeste, Camaçari 03256
Copenhagen Business School, Frederiksberg 06657
Copenhagen College of Social Work, Frederiksberg 06659
Copenhagen Public Libraries, København 06894
N. Copernicus Astronomical Centre of the Polish Academy of Sciences, Warszawa 21249
Copernicus Provincial and Municipal Library, Toruń 21589
Copiague Memorial Public Library, Copiague 38689
Copiah-Lincoln Community College, Wesson 33855
Copley Press Inc, La Jolla 35811
Coppell Public Library, Coppell 38690
Copper Smelting Factory, Moskva 22844
Copperas Cove Public Library, Copperas Cove 38691
Coppin State College, Baltimore 30332
Coptic Museum, Cairo 07163
Coquitlam Public Library, Coquitlam 04617
Cora J. Belden Library, Rocky Hill 41047
Coralville Public Library, Coralville 38692
Corangamite Regional Library Corporation, Colac 01088
Corban College, Salem 32547
Corcoran Gallery of Art/College of Art & Design Library, Washington 37729
Cordelia A. Greene Library, Castile 36634
Cordell Public Library, Cordell 38693
Corfu Free Library, Corfu 38695
Corfu Reading Society, Kerkira 13127
Corgialenios Library, Argostoli 13138
Corgialenios Vivliothiki Argostoliou, Argostoli 13138
Cork City Library, Cork 14569
Cork County Library & Arts Service, Cork 14570
Cork Institute of Technology, Cork 14461
Cork University Hospital, Cork 14463
Cornell College, Mount Vernon 31916
Cornell University, Ithaca 31415
– Adelson Library, Laboratory of Ornithology, Ithaca 31416
– Albert R. Mann Library, Ithaca 31417
– L. H. Bailey Hortorium Library, Ithaca 31418
– Division of Rare & Manuscript Collections, Ithaca 31419
– Edna McConnell Clark Physical Sciences Library, Ithaca 31420
– Engineering Library, Ithaca 31421
– Fine Arts, Ithaca 31422
– Flower-Sprecher Veterinary Library, Ithaca 31423
– John Henrik Clarke Africana Studies Library, Ithaca 31424
– Johnson Graduate School of Management Library, Ithaca 31425
– Law School Library, Ithaca 31426
– Mathematics Library, Ithaca 31427
– Nestle Hotel School Library, Ithaca 31428
– New York State Agricultural Experiment Station, Geneva 31429
– Olin-Kroch-Uris Library, Ithaca 31430
– Population & Development Program Research & Reference Library, Ithaca 31431
– School of Industrial & Labor Relations, Ithaca 31432
– Sidney Cox Library of Music & Dance, Ithaca 31433
– Weill Cornell Medical Library, New York 31434
Cornerstone University, Grand Rapids 31236
Corning Community College, Corning 33259
Corning Inc, Corning 35637
Corning Museum of Glass, Corning 36754
Cornish College of the Arts, Seattle 32631
Cornwall Public Library, Cornwall 04618
Corona Public Library, Corona 38699
Coronado Public Library, Coronado 38701
Corporación Autónoma Regional del Cauca (CVC), Calí 06009
Corporación Educativa Mayor del Desarrollo 'Simón Bolívar', Barranquilla 06007
Corporación Nacional del Cobre (CODELCO), Los Andes 04920
Corporación Universitaria de la Costa, Barranquilla 05931
Corpus Christi College, Cambridge 29626
Corpus Christi College, Oxford 29312
Corpus Christi Museum of Science & History, Corpus Christi 36757
Corpus Christi Public Libraries, Corpus Christi 38702
Correctional Institution for Women, Clinton 34088
Correctional Service of Canada, Bath 04050
Correctional Service of Canada, Laval 04090
Correctional Service of Canada, Saskatoon 04486
Correctional Service of Canada – Pacific Region, Agassiz 04049
Corrective Services Academy, Eastwood 00914
Corrs Chambers Westgarth, Brisbane 00813
Corrs Chambers Westgarth, Melbourne 00820
Corry Public Library, Corry 38703
Corsicana Public Library, Corsicana 38704
Corte Costituzionale, Roma 15666
Corte d'Appello, Milano 15650
Corte d'Appello, Torino 15692
Corte dei Conti, Roma 15667
Corte Suprema de Justicia, Santiago 04906
Corte Suprema de Justicia de la Nación, Buenos Aires 00208
Cortes de Aragón, Zaragoza 25757
Cortes de Castilla y León, Fuensaldaña 25694
Cortes Valencianas, Valencia 25751
Cortland Free Library, Cortland 38706
Corus Technology BV, Velsen 19340
Corus UK Limited, Middlesbrough 29855
Corus UK Limited, Rotherham 29580
Corvallis-Benton County Public Library, Corvallis 38708
Coshocton Public Library, Coshocton 38709
Cosmo Research Institute, Kikakatsushika 17372
Cossowary Coast Libraries, innisfail 01102
Costal Carolina University, Conway 30919
Cosumnes River College, Sacramento 33708
Côte-Saint-Luc Public Library, Côte Saint-Luc 04620
Cottage Hospital, Santa Barbara 37565
Cottey College, Nevada 31951
Cotton and Durum Wheat Research Institute, Chirpan 03501
Cotton and Paper Factory, Orekhovo-Zuevo 22883
Cotton and Paper Plant, Cheboksary 22783
Cotton Industrial Corporation, Kherson 28147
Cotton Industrial Corporation – Trade Union Library, Kherson 28604
Cotton Institute, Kuigan-Yar 42116
Cotton Production Corporation, Main Trade Union Library, Baranovichi 02096
Cotton Production Corporation, Scientific and Technical Library, Baranovichi 01826
Cottonwood Public Library, Cottonwood 38713
Coudert Brothers LLP Library, New York 35972
Council Bluffs Public Library, Council Bluffs 38714
Council for Economic Planning and Development, Taipei 27580
Council for Geoscience, Pretoria 25159
Council for Labor Efficiency Improvement, Kyiv 28382
Council for Scientific and Industrial Research (CSIR), Pretoria 42350
Council for Scientific and Industrial Research (CSIR), Pretoria 25160
Council for the Protection of Rural England (CPRE), London 29744
Council of Adult Education, Melbourne 00950
Council of Agiculture (COA), Taipei 27607
Council of Brazilian University Rectors, Brasília 03003
Council of Scientific and Industrial Research, New Delhi 14123
Council of State, Den Haag 19268
Council of State Governments, Lexington 34454
Council on Foreign Relations Library, New York 37199
Count Károly Esterházy Castle and Regional Museum, Pápa 13493
Country Estate Museum 'Arkhangelskoe', Arkhangelskoe 22990
Country Library 'Ovid Densusianu', Deva 22088
Country Library 'Panait Istrati', Brăila 22079
Countryside Agency, Cheltenham 29634
Countryside Council for Wales / Cyngor Cefn Gwlad Cymru, Bangor 29600
County Administrative Board, Stockholm 26508
County and City Library, Kaposvár 13538
County College of Morris, Randolph 33672
County Hospital, Trollhättan 26701
County Library "Antim Ivireanul" Valcea, Râmnicu Vâlcea 22099
County Library of the Nógrád County, Salgótarján 13545
County Library of Västerbotten, Umeå 26936
County Mueum of Satu Mare, Satu Mare 22054
The County Museum of Gotland, Visby 26721
County of Carleton Law Library, Ottawa 04110
County of Henrico Public Library, Richmond 40977
County of Henrico Public Library – Dumbarton Area, Richmond 40978
County of Henrico Public Library – Fairfield Area, Richmond 40979
County of Henrico Public Library – Gayton, Richmond 40980
County of Henrico Public Library – Glen Allen Branch, Glen Allen 39236
County of Henrico Public Library – Tuckahoe Area, Richmond 40981
County of Henrico Public Library – Twin Hickory, Glen Allen 39237
County of Los Angeles Public Library, Downey 38862
County of Los Angeles Public Library – A. C. Bilbrew Library, Los Angeles 39942

County of Los Angeles Public Library – Agoura Hills Library, Agoura Hills 37885
County of Los Angeles Public Library – Angelo M. Iacoboni Library, Lakewood 39788
County of Los Angeles Public Library – Anthony Quinn Library, Los Angeles 39943
County of Los Angeles Public Library – Baldwin Park Library, Baldwin Park 38066
County of Los Angeles Public Library – Canyon Country Jo Anne Darcy Library, Santa Clarita 41216
County of Los Angeles Public Library – Carson Library, Carson 38464
County of Los Angeles Public Library – Charter Oak Library, Covina 38717
County of Los Angeles Public Library – City Terrace Library, Los Angeles 39944
County of Los Angeles Public Library – Claremont Library, Claremont 38575
County of Los Angeles Public Library – Clifton M. Brakensiek Library, Bellflower 38156
County of Los Angeles Public Library – Compton Library, Compton 38673
County of Los Angeles Public Library – Culver City Julian Dixon Library, Culver City 38741
County of Los Angeles Public Library – Diamond Bar Library, Diamond Bar 38843
County of Los Angeles Public Library – Duarte Library, Duarte 38866
County of Los Angeles Public Library – East Los Angeles Library, Los Angeles 39945
County of Los Angeles Public Library – East Rancho Dominguez Library, East Rancho Dominguez 38916
County of Los Angeles Public Library – El Monte Library, El Monte 38943
County of Los Angeles Public Library – Florence Library, Los Angeles 39946
County of Los Angeles Public Library – Gardena Mayme Dear Library, Gardena 39196
County of Los Angeles Public Library – George Nye Jr Library, Lakewood 39789
County of Los Angeles Public Library – Graham Library, Los Angeles 39947
County of Los Angeles Public Library – Hacienda Heights Library, Hacienda Heights 39339
County of Los Angeles Public Library – Hawthorne Library, Hawthorne 39399
County of Los Angeles Public Library – Hollydale Library, South Gate 41359
County of Los Angeles Public Library – Huntington Park Library, Huntington Park 39539
County of Los Angeles Public Library – La Canada Flintridge Library, La Canada Flintridge 39744
County of Los Angeles Public Library – La Crescenta Library, La Crescenta 39745
County of Los Angeles Public Library – La Mirada Library, La Mirada 39754
County of Los Angeles Public Library – La Puente Library, La Puente 39758
County of Los Angeles Public Library – La Verne Library, La Verne 39761
County of Los Angeles Public Library – Lancaster Library, Lancaster 39792
County of Los Angeles Public Library – Leland R. Weaver Library, South Gate 41360
County of Los Angeles Public Library – Lennox Library, Lennox 39849
County of Los Angeles Public Library – Littlerock Library, Littlerock 39901
County of Los Angeles Public Library – Lloyd Taber-Marina del Rey Library, Marina del Rey 40059
County of Los Angeles Public Library – Lynwood Library, Lynwood 39988
County of Los Angeles Public Library – Malibu Library, Malibu 40020
County of Los Angeles Public Library – Manhattan Beach Library, Manhattan Beach 40033
County of Los Angeles Public Library-Masao W. Satow Library, Gardena 39197
County of Los Angeles Public Library – Montebello Library, Montebello 40272
County of Los Angeles Public Library – Newhall Library, Newhall 40461
County of Los Angeles Public Library – Norwalk Library, Norwalk 40547
County of Los Angeles Public Library – Norwood Library, El Monte 38944
County of Los Angeles Public Library – Paramount Library, Paramount 40704
County of Los Angeles Public Library – Pico Rivera Library, Pico Rivera 40784
County of Los Angeles Public Library – Rosemead Library, Rosemead 41068
County of Los Angeles Public Library – Rowland Heights Library, Rowland Heights 41076
County of Los Angeles Public Library – San Dimas Library, San Dimas 41171
County of Los Angeles Public Library – San Fernando Library, San Fernando 41172
County of Los Angeles Public Library – San Gabriel Library, San Gabriel 41174
County of Los Angeles Public Library – South Whittier Library, Whittier 41888
County of Los Angeles Public Library – Sunkist Library, La Puente 39759
County of Los Angeles Public Library – Temple City Library, Temple City 41535
County of Los Angeles Public Library – Valencia Library, Santa Clarita 41217
County of Los Angeles Public Library – View Park Library, Los Angeles 39948
County of Los Angeles Public Library – Walnut Library, Walnut 41714
County of Los Angeles Public Library – West Covina Library, West Covina 41813
County of Los Angeles Public Library – West Hollywood Library, West Hollywood 41821
County of Los Angeles Public Library – Westlake Village Library, Westlake Village 41857
County of Los Angeles Public Library – Woodcrest Library, Los Angeles 39949
County of Prince Edward, Picton 04720
County of Simcoe Library Co-Operative, Midhurst 04686
County Reference & Information Library, Truro 30089
Cour d'Appel de Paris, Paris 08148
Cour de Cassation, Paris 08149
Cour de justice des Communautés européennes, Luxembourg 18397
Cour des Comptes Européenne, Luxembourg 18398
Court International de Justice, Den Haag 19255
Court of Appeal for Southern Norrland, Sundsvall 26524
Court of Appeal for Western Sweden, Göteborg 26488
Court of Appeals Eleventh Circuit Library, Atlanta 33941
Court of Appeals for Sixth District of Texas, Texarkana 34922
Court of the Justice of the European Communities, Luxembourg 18397
The Courtney Library, Truro 29911
Courtneypark Branch Library, Mississauga 04687
Courts Administration Authority, Adelaide 00596
Couvent des Capucins, Fribourg 27259
Couvent des Capucins, Saint-Maurice 27269
Couvent des Cordeliers / Franziskanerkloster, Fribourg 27260
Couvent des Franciscains, Strasbourg 08238
Couvent des Pères Dominicains, Strasbourg 08239
Couvent Pères Capucins, Strasbourg 08240
Covasna District Public Library, Sfântu Gheorghe 22102
Covenant College, Lookout Mountain 31642
Covenant Theological Seminary, Saint Louis 35360
Coventry City Council, Coventry 29963
Coventry Public Library, Coventry 38716
Coventry University, Coventry 29074
Covidien, Saint Louis 36197
Covina Public Library, Covina 38718
Covington & Burling LLP, Washington 36340
Cowes Library and Maritime Museum, Cowes 29640
Cox, Castle & Nicholson LLP Library, Los Angeles 35831
Cox, Smith, Matthews Inc, San Antonio 36215
CoxHealth Libraries – David Miller Memorial Library, Springfield 37627
CoxHealth Libraries – North Library, Springfield 37628
Coyle Free Library, Chambersburg 38499
CPA Australia, Melbourne 00951
CQUniversity, North Rockhampton 00494
C.R.A. – Experimental Institute for Cereal Crop Production – Rice Research Section, Vercelli 16663
CRA International Library, Boston 36562
CRA – SEL Centro di Ricerca per la Selvicoltura, Arezzo 16170
Cracow Salt-Works Museum Wieliczka, Wielczka 21354
Cracow University of Economics, Kraków 20822
Cracow University of Technology, Kraków 20839
Craft Memorial Library, Bluefield 38226
Crafton Hills College, Yucaipa 33889
Crafts Museum, New Delhi 14124
Cragin Memorial Library, Colchester 38631
Craighead County Jonesboro Public Library, Jonesboro 39658
CRAL – Observatoire Astronomique de Lyon, Saint-Genis-Laval 08668
Cranberry Public Library, Cranberry Township 38725
Cranbrook Academy of Art Library, Bloomfield Hills 36547
Cranbrook Public Library, Cranbrook 04622
Crandall Public Library, Glens Falls 39251
Crane Construction Industrial Corporation, Uzlovaya 22970
Cranfield University, Cranfield 29076
Cranston Public Library, Cranston 38726
Cravath, Swaine & Moore LLP, New York 35973
Craven-Pamlico-Carteret Regional Library System, New Bern 40395
Crawford County Federated Library System, Meadville 40125
Crawford County Law Library, Meadville 34521
Crawford County Library System – Devereaux Memorial Library, Grayling 39283
Crawford County Public Library, Leavenworth 39832
Crawfordsville District Public Library, Crawfordsville 38727
Crazy Horse Memorial Library, Crazy Horse 36759
Credito Italiano, Milano 16146
Creighton University, Omaha 32194
– Health Sciences Library, Omaha 32195
– Klutznick Law Library, Omaha 32196
Crescent Road library, Reading 29382
Cresskill Public Library, Cresskill 38728
Crestline Public Library, Crestline 38729
Crestmont College Salvation Army, Rancho Palos Verdes 34726
Crestwood Public Library District, Crestwood 38730
Crete Public Library District, Crete 38731
Crichton College, Memphis 31790
Crime and Misconduct Commission, Brisbane 00615
Crimea State Agricultural Experimental Station, Klepinino 28293
Crimean Astrophysical Observatory, Nauchne 28434
Crimean Centre of Scientific and Technical Information, Simferopol 28472
Crimean Medical Institute, Simferopol 28074
Crimean Republican Universal Scientific Library, Simferopol 27946
Crimean Republican Youth Library, Simferopol 28816
Crimean State University of Agriculture, Simferopol 28072
Criminal Court of the City of New York Library, Bronx 34007
Criş County Museum, Oradea 22049
CRIST – Central Research Institute of Shipbuilding Technology, Sankt-Peterburg 23753
Criswell College, Dallas 30941
Crnogorska Akademija Nauka i Umjetnosti, Podgorica 18913
Croatian Academy of Sciences and Arts, Zagreb 06247
Croatian Academy of Sciences and Arts, Library, Zagreb 06144
Croatian Historical Museum, Zagreb 06240
Croatian Library for the Blind, Zagreb 06238
Croatian Natural History Museum, Zagreb 06241
Croatian Parliament, Zagreb 06208
Croatian School Museum, Zagreb 06242
Croatian State Archives, Zagreb 06239
Croix-Rouge Française (CRF), Paris 08530
Cromaine District Library, Hartland 39383
Cromwell Belden Public Library, Cromwell 38732
Cronulla Fisheries Research Centre, Hunter Region MC 00689
Crook County Library, Sundance 41479
Crosna Scientific and Industrial Corporation, Moskva 22851
Cross City Correctional Institution Library, Cross City 34118
Cross-Cultural Dance Resources Library, Flagstaff 36850
Crossroads Bible College, Indianapolis 35239
Croton Free Library, Croton-on-Hudson 38734
Crowder College, Neosho 33573
Crowell & Moring, Washington 36341
Crowley, Haughey, Hanson, Toole & Dietrich Library, Billings 35503
Crowley Ridge Regional Library, Jonesboro 39659
Crown College, Saint Bonifacius 32493
Crown Point Community Library, Crown Point 38736
Crowther Mission Education Centre Library, Oxford 29561
Croydon College, Croydon 29077
Croydon Libraries, Croydon 29964
J. Lewis Crozer Library, Chester 38544
CRS Australia, Brisbane 00876
CRS Australia, Brisbane 00877
CRSBP Outaouais Inc, Gatineau 04647
Crucible Materials Corporation, Pittsburgh 36129
Cruz Roja Española, Madrid 26010
Crystal Lake Public Library, Crystal Lake 38737
Csepel Metal Works, Budapest 13353
Csepeli Fémmü Rt, Budapest 13353
Csiki Székely Museum, Ciuc Szekler Museum, Miercurea Ciuc 22046
CSIR, Pretoria 25097
CSIR – Mining Technology, Auckland Park 25104
CSIRO, Clayton 00903
CSIRO Animal Production, Prospect 00990
CSIRO Atmospheric Research, Aspendale 00866
CSIRO Black Mountain Library, Canberra 00891
CSiro Centre for Environment and Life Sciences, Wembley 01045
CSIRO Cunningham Library, Saint Lucia 01001
CSIRO Division of Forestry, Yarralumla 01055
CSIRO Division of Soils, Urrbrae 01040
CSIRO Energy Technology, Newcastle 00967
CSIRO Ian Wark Library, Clayton 00904
CSIRO Institute of Minerals, Energy and Construction, North Ryde 00970
CSIRO Manufacturing and Materials, Highett 00930
CSIRO Marine and Atmospheric, Hobart 00932
CSIRO Pastoral Research Laboratory, Armidale 00865
CSIRO Sustainable Ecosystems, Canberra 00892
CSIRO Telecommunication and Industrial Physics, Marsfield 00944
CSIRO Telecommunications and Industrial Physics, Lindfield 00939
CSIRO Textile and Fibre Technology, Belmont 00872
CSIRO Urrbrae, Glen Osmond 00922
CSIRO Western Australian Laboratories, Floreat Park 00919
CSL Behring AG, Bern 27284
Csorba Győző County Library, Pécs 13544
CSX Transportation, Inc, Jacksonville 35787
CTB McGraw-Hill Library, Monterey 35919
CTT Correios de Portugal, Lisboa 21784
Cuartel General de la Armada, Madrid 25709
Cuartel General del Aire, Madrid 25710
Cuartel General del Ejército, Madrid 25711
Cuba National Assembly, La Habana 06280
Cudahy Family Library, Cudahy 38738
Cuesta College, San Luis Obispo 33734
Culinary Institute of America, Hyde Park 36950
Cullman County Public Library System, Cullman 38739
Culpeper County Library, Culpeper 38740
Cultural Center of the Philippines, Pasay City 20777
Cultural Heritage Office of the Republic of Slovenia, Ljubljana 24940
Cultural Office of the Egyptian Embassy in Madrid, Madrid 26048
Cultural Palace of Kuznetsk Metal Workers, Novokuznetsk 23633
Cultural Research Institute, Kolkata 14038
Culture History Museum, Lund 26613
Culture House of the I.A. Likhachev Automobile Plant, Moskva 24168
Cultureel Centrum Suriname, Paramaribo 26365
Cultureel Maçonniek Centrum 'Prins Frederik', Den Haag 19432
Cultuurbibliotheek, Brugge 02545
Cultuurnetwerk Nederland, Utrecht 19525
Culver-Stockton College, Canton 30667
Cumberland County College, Vineland 33826
Cumberland County Historical Society, Carlisle 36627
Cumberland County Law Library, Carlisle 34046
Cumberland County Library, Bridgeton 38278
Cumberland County Public Library & Information Center, Fayetteville 39052
Cumberland County Public Library & Information Center – Bordeaux, Fayetteville 39053

Cumberland County Public Library & Information Center – Cliffdale, Fayetteville 39054
Cumberland County Public Library & Information Center – East Regional, Fayetteville 39055
Cumberland County Public Library & Information Center – Hope Mills Branch, Hope Mills 39477
Cumberland County Public Library & Information Center – North Regional, Fayetteville 39056
Cumberland Public Library, Cumberland 38743
Cumberland Regional Library Board, Amherst 04568
Cumberland University, Lebanon 31584
Cumbernauld & Kilsyth District Council, Cumbernauld 29965
Cumbria County Council, Carlisle 29955
Cumhuriyet Üniversitesi, Sivas 27868
Cummer Museum of Art Library, Jacksonville 36972
Cummings & Lockwood, Stamford 36283
Cumnock and Doon Valley District Library, Cumnock 29966
Curaçao Public Library Foundation / Stichting Openbare Bibliotheek Curaçao, Willemstad 06321
Curia Generalizia della compagnia di Gesù, Roma 16027
Curia Provincial de Hermanos Capuchinos de la Provincia de Navarra-Cantabria-Aragón, Pamplona 25808
M. Curie-Skłodowska University, Lublin 20888
Curriculum Development Centre, Kuala Lumpur 18603
Currie Library, Greensboro 35222
Currier Museum of Art, Manchester 37080
Curry College, Milton 31832
Curtin University of Technology, Bentley 00389
Curtin University of Technology, Kalgoorlie 00447
Curtis Institute of Music, Philadelphia 37342
Curtis, Mallet-Prevost, Colt & Mosle Library, New York 35974
Cushing Public Library, Cushing 38747
Customs national Museum Library, Bordeaux 08309
Cuttington University College, Monrovia 18229
Cuyahoga Community College – Eastern Campus Library, Cleveland 33236
Cuyahoga Community College – Metropolitan Campus, Cleveland 33237
Cuyahoga Community College – Western Campus Library, Cleveland 33238
Cuyahoga County Public Library, Parma 40713
Cuyahoga County Public Library – Bay Village Branch, Bay Village 38119
Cuyahoga County Public Library – Beachwood Branch, Beachwood 38127
Cuyahoga County Public Library – Berea Branch, Berea 38179
Cuyahoga County Public Library – Brook Park Branch, Brook Park 38303
Cuyahoga County Public Library – Chagrin Falls Branch, Chagrin Falls 38498
Cuyahoga County Public Library – Fairview Park Branch, Fairview Park 39032
Cuyahoga County Public Library – Garfield Heights Branch, Garfield Heights 39201
Cuyahoga County Public Library – Independence Branch, Independence 39554
Cuyahoga County Public Library – Maple Heights Branch, Maple Heights 40045
Cuyahoga County Public Library – Mayfield Branch, Mayfield Village 40104
Cuyahoga County Public Library – Middleburg Heights Branch, Middleburg Heights 40177
Cuyahoga County Public Library – North Olmsted Branch, North Olmsted 40518
Cuyahoga County Public Library – North Royalton Branch, North Royalton 40526
Cuyahoga County Public Library – Parma Heights Branch, Parma Heights 40716
Cuyahoga County Public Library – Parma-Snow, Parma 40714
Cuyahoga County Public Library – Parma-South Branch, Parma 40715
Cuyahoga County Public Library – South Euclid-Lyndhurst Branch, South Euclid 41358
Cuyahoga County Public Library – Strongsville Branch, Strongsville 41455
Cuyahoga Falls Library, Cuyahoga Falls 38748
Cuyamaca College, El Cajon 33311
CVU Alpha, Jydsk Pædagog-Seminarium, Risskov, Risskov 06745
CWI – Centrale organisatie werk en inkomen, Amsterdam 19353
Cyngor Cefn Gwlad Cymru, Bangor 29600
Cynthiana-Harrison County Public Library, Cynthiana 38749
Cypress College, Cypress 33271
Cyprus American Archaeological Research Institute, Nicosia 06336
Cyprus Library, Nicosia 06322
Cyprus Museum, Nicosia 06337
Cyprus Turkish National Library, Lefke 06340
Cyrenius H. Booth Library, Newtown 40476
Cyril Taylor Library, Richmond 29385
Cytec Austria GmbH, Graz 01498
Czech Academy of Sciences, Main Library, Praha 06351
Czech Botanical Society, Praha 06484
Czech Entomological Society, Praha 06478
Czech Hydrometeorological Institute, Praha 06480
Czech radio, Praha 06481
Czech Statistical Office – Central Statistical Library, Praha 06482
Czech Union of Bee-Keepers, Praha 06483
Czech Univerzita of Agriculture, Praha 06383
Czech Zoological Society, Praha 06479
Czechoslovak Heritage Museum, Oak Brook 37296
Czenstochowa Museum, Częstochowa 21089
Czestochowa University of Technology, Częstochowa 20800
D. Dr. Otto-Beuttenmüller-Bibliothek, Bretten 11698
D. S. Senanayake Memorial Public Library, Kandy 26332
D. Tsenov Economic University, Svishtov 03483
Dabney S. Lancaster Community College, Clifton Forge 33240
Dachau Concentration Camp Memorial Site, Dachau 11718
Dacian and Roman Civilization Museum, Deva 22032
Daemen College, Amherst 30214
Dag Hammarskjold Memorial Library, Kitwe 42338
Dagestan Agricultural Institute, Makhachkala 22322
Dagestan Institute of Microbiological Substrate Production, Makhachkala 23152
Dagestan Republic Scientific Library, Makhachkala 22151
Dagestan Scientific Research Institute of Agriculture, Makhachkala 23153
Dagestan State Pedagogic University, Makhachkala 22319
Dagestanskaya respublikanskaya detskaya biblioteka, Makhachkala 24166
Dagestanskaya respublikanskaya nauchnaya biblioteka im. A.S. Pushkina, Makhachkala 22151
Dagestanski gosudarstvenny pedagogicheski universitet, Makhachkala 22319
Dagestanski gosudarstvenny universitet, Makhachkala 22320
Dagestanski meditsinski institut, Makhachkala 22321
Dagestanski mezhotraslevoi territorialny tsentr nauchno-tekhnicheskoi informatsii i propagandy, Makhachkala 23151
Dagestanski nauchno-issledovatelski institut po proizvodstvu pitatelnykh sred, Makhachkala 23152
Dagestanski nauchno-issledovatelski institut selskogo khozyaistva, Makhachkala 23153
Dagestanski selskokhozyaistvenny institut, Makhachkala 22322
Daghestan State University, Makhachkala 22320
Daheshite Museum and Library, Beirut 18219
Daigoji Reihokan, Kyoto 17337
Daiichi College of Pharmaceutical Sciences, Fukuoka 16917
Dai-ichi Mutual Life Insurance Co, Research Dept, Tokyo 17439
Daiichi Seimei Hoken Sogogaisha Chosabu, Tokyo 17439
Daiichi Seiyaku Co, Ltd, Research Institute, Tokyo 17440
Daiichi Seiyaku K. K. Chuo Kenkyujo, Tokyo 17440
Daiichi Yakka Daigaku, Fukuoka 16917
Daimler-Benz Archiv, Stuttgart 11428
DaimlerChrysler Corp, Auburn Hills 35470
Dai-Nippon Ink and Chemicals Inc, Takaishi 17427
Dainippon Inki Kangaku Kogyo K. K., Takaishi 17427
Daini-Tokyo Bar Assn, Tokyo 17671
Daini-Tokyo Bengoshikai, Tokyo 17671
Dairy Institute, Vologda 24072
Dairy Research Institute, Vidin 03589
Daito Bunka University, Tokyo 17136
Daiwa Bank Ltd, Osaka 17405
Daiwa Ginko K. K. Chosabu, Osaka 17405
Dakota County Library System, Eagan 38884
Dakota State University, Madison 31707
Dakota Wesleyan University, Mitchell 31870
Dakshina Bharat Hindi Prachar Sabha, Chennai 13628
Dalens sjukhus, Enskededalen 26567
Dalhousie University, Halifax 03716
– Faculty of Law, Halifax 03717
– W. K. Kellogg Health Sciences Library, Halifax 03718
Dalhousie-University of Nova Scotia, Halifax 03719
Dali Medical College, Dali 05144
Dali Teachers' College, Dali 05145
Dalian City Library, Dalian 05911
Dalian Fisheries College, Dalian 05146
Dalian Institute of Chemical Physics, Dalian 05859
Dalian Institute of Foreign Languages, Dalian 05147
Dalian Institute of Light Industry, Dalian 05148
Dalian Maritime University, Dalian 05149
Dalian Medical University, Dalian 05150
Dalian Railway University, Dalian 05151
Dalian University of Technology, Dalian 05152
Dalkeith Library, Dalkeith 29968
Dallas Baptist University, Dallas 30942
Dallas Christian College, Dallas 30943
Dallas County Law Library – Civil Law Collection, Dallas 34123
Dallas County Law Library – Criminal Law Collection, Dallas 34124
Dallas Historical Society, Dallas 36762
Dallas Museum of Art, Dallas 36763
Dallas Public Library, Dallas 38756
Dallas Public Library, Dallas 38757
Dallas Public Library, Dallas 38758
Dallas Public Library – Audelia Road, Dallas 38759
Dallas Public Library – Casa View, Dallas 38760
Dallas Public Library – Fretz Park, Dallas 38761
Dallas Public Library – Hampton-Illinois, Dallas 38762
Dallas Public Library – Lakewood, Dallas 38763
Dallas Public Library – Mountain Creek, Dallas 38764
Dallas Public Library – North Oak Cliff, Dallas 38765
Dallas Public Library – Park Forest, Dallas 38766
Dallas Public Library – Pleasant Grove, Dallas 38767
Dallas Public Library – Polk-Wisdom, Dallas 38768
Dallas Public Library – Preston Royal, Dallas 38769
Dallas Public Library – Renner Frankford Branch, Dallas 38770
Dallas Public Library – Skillman Southwestern, Dallas 38771
Dallas Public Library – Skyline, Dallas 38772
Dallas Theological Seminary, Dallas 35176
The Dalles-Wasco County Library, The Dalles 41546
Dalnaya Svyaz Telecommunications Research and Industrial Corporation, Sankt-Peterburg 23863
Dalnevostochnaya gosudarstvennaya nauchnaya biblioteka, Khabarovsk 22140
Dalnevostochnaya zheleznaya doroga, Khabarovsk 23093
Dalnevostochnoe territorialnoe geologicheskoe upravlenie, Khabarovsk 22731
Dalnevostochny gosudarstvenny tekhnicheski universitet, Vladivostok 22632
Dalnevostochny gosudarstvenny universitet, Vladivostok 22633
Dalnevostochny institut torgovli, Vladivostok 22634
Dalnevostochny khudozhestvenny muzei, Khabarovsk 23094
Dalnevostochny nauchno-issledovatelski gidromet meteorologicheski institut, Vladivostok 24051
Dalnevostochny nauchno-issledovatelski institut lesnogo khozyaistva, Khabarovsk 23095
Dalnevostochny nauchno-issledovatelski institut selskogo khozyaistva, Khabarovsk 23096
Dalnevostochny nauchno-issledovatelski veterinarny institut, Blagoveshchensk 23004
Dalnevostochny pedagogicheski institut iskusstv, Vladivostok 22635
Dalton McCaughey Library, Parkville 00976
Dalton State College, Dalton 33276
Daly City Public Library, Daly City 38774
Daly City Public Library – Westlake, Daly City 38775
Damietta Spinning and Weaving Company 'Damiatex', Damietta 07142
Damjanich János Múzeum, Szolnok 13511
Dana College, Blair 30441
Danbury Public Library, Danbury 38777
Danderyd Hospital Professional Library, Stockholm 26645
Danderyds bibliotek, Danderyd 26753
Danderyds sjukhus Fackbibliotek, Stockholm 26645
Dandong Teachers' College, Dandong 05157
Dandong Textile Engineering College, Dandong 05158
Dane County Library Service, Madison 40001
Danforth Museum of Art, Framingham 36873
Daniel Webster College, Nashua 31930
D. Danielopolu Institute of Normal and Pathological Physiology, București 21990
Danilovskaya manufaktura, Moskva 22834
Danilovskaya Textile Mill, Moskva 22834
Danish Building Research Institute Library, Hørsholm 06782
Danish Center for International Studies and Human Rights, København 06791
Danish Central Library of South Schleswig, Flensburg 11833
Danish Centre for Information on Women and Gender, København 06808
The Danish Design School Library, København 06665
Danish Emigration Archives, Aalborg 06767
Danish Film Institute, København 06788
Danish Folklore Archives, København 06787
Danish Institute for Fisheries Research, Department of Seafood Research, Lyngby 06827
Danish Institute for Food and Veterinary Research, Søborg 06833
Danish Institute for Health Sevices Research, København 06794
Danish Maritime Library, København 06823
Danish Maritime Museum, Helsingør 06778
Danish Meteorological Institute Library, København 06785
Danish Museum of Books and Printing, København 06625
Danish Museum of Science and Technology, Helsingør 06777
The Danish Music Museum, Musikhistorisk Museum & The Carl Claudius Collection, København 06813
Danish National Archives Library, København 06819
Danish National Art Library, København 06784
Danish National Business Archives, Århus 06769
Danish National Institute of Social Research Library, København 06822
Danish National Institute of Social Research, København 06820
Danish National Library for Persons with Print Disabilities, København 06817
Danish Patent and Trademark Office Library, Taastrup 06761
The Danish Respository Library for Public Libraries, Århus 06847
Danish Road Directorate, Hedehusene 06776
Danish School of Journalism, Århus 06651
Danish Standards Association, Charlottenlund 06774
Danish State Railways Library and Photo Archives, København 06793
Dankook University, Seoul 18077
Danmarks Biblioteksskole, København 06664
Danmarks Bogmuseum, København 06625
Danmarks Designskoles Bibliotek, København 06665
Danmarks Fiskeriundersøgelser, Afdeling for Fiskeindustriel Forskning, Lyngby 06827
Danmarks Kunstbibliotek, København 06784
Danmarks Meteorologiske Instituts Bibliotek, København 06785
Danmarks Miljøundersøgelser (DMU), Kalø, Rønde 06830
Danmarks Nationalbibliotek, København 06623
Danmarks Pædagogiske Bibliotek, København 06666
Danmarks Statistiks Bibliotek og Information, København 06786
Danmarks Tekniske Museum, Helsingør 06777

Danmarks Tekniske Universitet, Lyngby 06713
– Center for Trafik og Transport, Lyngby 06714
– Institut for Informatik og Matematisk Modellering, Lyngby 06715
Danmarks Tekniske Universitet, Institut for Akvatiske Ressourcer, Charlottenlund 06656
Dansacademie, Tilburg 19202
Dansk Bibliotek, Husum 12069
Dansk Center for Internationale Studier og Menneskerettigheder, København 06791
Dansk Centralbibliotek for Sydslesvig, Flensburg 11833
Dansk Folkemindesamling, København 06787
Dansk Standard, Charlottenlund 06774
Det Danske Filminstituts Bibliotek, København 06788
Den Danske Frimurerorden, København 06789
Det Danske Kunstindustrimuseums Bibliotek, København 06790
Det danske Udvandrerarkiv, Aalborg 06767
Danube Delta-Institute, Tulcea 22068
Danville Area Community College, Danville 33279
Danville Community College, Danville 33280
Danville Public Library, Danville 38782
Danville Public Library, Danville 38783
Danville-Center Township Public Library, Danville 38784
Daqing Petroleum Institute, Anda 04979
Daqing Teachers' College, Daqing 05161
Dar al-Kuttub, Cairo 07000
Dar es Salaam Institute of Technology, Dar es Salaam 27651
Dar es Salaam School of Accountancy, Dar es Salaam 27652
D'Arcy, Masius, Benton & Bowles, Troy 36311
Dare County Library, Manteo 40043
Darebin Libraries, Preston 01145
Darien Library, Darien 38786
Darlington College, Darlington 29078
Darlington County Historical Commission, Darlington 36768
Darlington County Library, Darlington 38788
Darlington County Library – Hartsville Memorial, Hartsville 39385
Darnitska raionna TsBS m. Kyiva, Tsentralna biblioteka im. V. Mayakovskoho, Kyiv 28669
Darnitski Region Central Library System of the City of Kiev – Mayakovsky Main Library, Kyiv 28669
Dartington College of Arts, Totnes 29427
Dartmouth College, Hanover 31288
– Baker-Berry Library, Hanover 31289
– Biomedical Libraries, Hanover 31290
– Feldberg Business Administration & Engineering Library, Hanover 31291
– Kresge Physical Sciences Library & Cook Mathematics Collection, Hanover 31292
– Paddock Music Library, Hanover 31293
– Rauner Special Collections Library, Hanover 31294
– Sanborn English Library, Hanover 31295
– Sherman Art Library, Hanover 31296
Dartmouth Public Libraries – Southworth Library, Dartmouth 38789
Darton College, Albany 33118
Darwin City Council Libraries, Darwin 01090
Darwin R. Barker Library, Fredonia 39157
Data Processing Centre, Tbilisi 09231
Datong Medical College, Datong 05162
Daugavpils Universitates, Daugavpils 18169
Daugavpils University, Daugavpils 18169
Daughters of the Republic of Texas Library at the Alamo, San Antonio 37519
Dauli Teachers' College, Tari 20569
Dauphin County Law Library, Harrisburg 34339
Dauphin County Library System, Harrisburg 39371
Dauphin County Library System – East Shore Area Library, Harrisburg 39372
Davenport Public Library, Davenport 38790
Davenport University, Dearborn 30975
Davenport University, Grand Rapids 31237
David A. Howe Public Library, Wellsville 41799
David & Joyce Milne Public Library, Williamstown 41900
David Library of the American Revolution, Washington Crossing 37812
David Livingstone Teachers' College, Livingstone 42323
David Lubin Memorial Library, Roma 16503
David M. Stewart Museum Library, Montreal 04397
David N. Myers College, Cleveland 30835
David Wade Correctional Center, Homer 34360
David Yellin College of Education, Jerusalem 14618
Davidson College, Davidson 30959
Davidson County Community College, Lexington 33483
Davidson County Public Library System, Lexington 39867
Davidson County Public Library System – Thomasville Public, Thomasville 41552
Davie County Public Library, Mocksville 40248
Davies, Ward, Phillips & Vineberg, Toronto 04281
Daviess County Public Library, Owensboro 40673
Davis, Brown, Koehn, Shors & Roberts PC Library, Des Moines 35670
Davis & Co, Vancouver 04301
Davis County Library, Farmington 39045
Davis County Library – Central, Layton 39830
Davis County Library – North, Clearfield 38590
Davis County Library – South, Bountiful 38249
Davis County Library – Syracuse Northwest Branch, Syracuse 41490
Davis & Elkins College, Elkins 31082
Davis, Graham & Stubbs LLP, Denver 35664
Davis Memorial Library, Fayetteville 31131
Davis Polk & Wardwell Library, New York 35975
Davis Wright Tremaine, Portland 36143
Davis Wright Tremaine LLP, Seattle 36263
Dawood College of Engineering and Technology, Karachi 20440
Dawson College, Westmount 04045
Dawson Creek Municipal Public Library, Dawson Creek 04625
Dawson Technical Institute, Chicago 36671
Daxian Teachers' College, Daxian 05165
Day, Berry & Howard Library, Hartford 35737
Day Pitney LLP, Florham Park 35709
Dayalbagh Educational Institute, Agra 13593
Dayanand College of Arts and Science, Sholapur 13842
Dayton Art Institute, Dayton 36773
Dayton Law Library, Dayton 36774
Dayton Metro Library, Dayton 38791
Daytona Beach Community College, Daytona Beach 33283
DB Mobility Logistics AG, DB-Museum Nürnberg, Nürnberg 12369
DB-Museum Nürnberg, Nürnberg 12369
DC Court of Appeals Library, Washington 34983
DCISM – Dansk Center for Internationale Studier og Menneskerettigheder, København 06791
De Anza College, Cupertino 33269
De Grandpré Chait Library, Montreal 04398
De Havilland College, Welwyn Garden City 29501
De La Salle University, Manila 20712
De Montfort University, Leicester 29160
– Polhill Campus Library, Bedford 29161
De Soto Trail Regional Library, Camilla 38429
Deaf Smith County Library, Hereford 39418
Deakin University, Geelong 00439
– Burwood Campus Library, Burwood 00440
Dean College, Franklin 31171
Dearborn Heights City Libraries – Caroline Kennedy Library, Dearborn Heights 38796
Dearborn Heights City Libraries – John F. Kennedy Jr Library, Dearborn Heights 38797
Dearborn Public Library, Dearborn 38795
Debaltseve Regional Centralized Library System, Main Library, Debaltseve 28528
Debaltsevska raionna TsBS, Tsentralna biblioteka, Debaltseve 28528
Debevoise & Plimpton, New York 35976
Debevoise & Plimpton, Washington 36342
Debrecen University of Agriculture, Debrecen 13268
– College of Agriculture, Hódmezővásárhely 13269
– Gödöllői College Faculty for Mechanical Production Engineering in Agriculture, Mezőtur 13270
Debrecen University of Medicine, Debrecen 13272
Debreceni Agrártudományi Egyetem, Debrecen 13268
– Mezőgazdasági Főiskolai Kar, Hódmezővásárhely 13269
– Mezőgazdasági Gépészeti Főiskolai Kar, Mezőtur 13270
Debreceni Egyetem, Debrecen 13271
Debreceni Orvostudományi Egyetem, Debrecen 13272
Debreceni Református Teológiai Akadémia, Debrecen 13273
Debreceni Városi Könyvtár, Debrecen 13533
Decatur Genealogical Society Library, Decatur 36783
Decatur Public Library, Decatur 38799
Decatur Public Library, Decatur 38800
Deccan College, Pune 13742
DECHEMA, Gesellschaft für Chemische Technik und Biotechnologie e.V., Frankfurt am Main 11396
DECHEMA Society for Chemical Engineering and Biotechnology, Frankfurt am Main 11396
Dechert Law Library, New York 35977
Dechert Library, Philadelphia 36093
Dechert, Price & Rhoads, Washington 36343
Decorah Public Library, Decorah 38803
Dedham Historical Society Library, Dedham 36785
Dedham Public Library, Dedham 38804
Deendayal Upadhyaya Gorakhpur University, Gorakhpur 13649
Deer Park Public Library, Deer Park 38805
Deer Park Public Library, Deer Park 38806
Deere & Co Library, Moline 35915
Deerfield Correctional Center, Capron 34045
Deerfield Public Library, Deerfield 38807
Defence College of Management and Technology, Swindon 29423
Defence Electronics Research Laboratory, Hyderabad 13998
Defence Forces Technical Research Centre, Lakiala 07526
Defence Library Service, Hohola 20575
Defence R&D Canada-Valcartier Library, Val-Belair 04172
Defence Research and Development Laboratory, Hyderabad 13999
Defence Science and Technology Organisation, Edinburgh 00915
Defence Science Library, Delhi 13854
Defence Technology Agency Library, Auckland 19823
Defence Technology Tower Library, Singapore 24596
Defense Intelligence Agency, Washington 34984
Defense Intelligence Agency – Joint Military Intelligence College, Washington 34985
Defense Language Institute Foreign Language Center, Monterey 37126
Defiance College, Defiance 30981
Defiance County Law Library, Defiance 34130
Defiance Public Library, Defiance 38808
DEG – Deutsche Investitions- und Entwicklungsgesellschaft mbH, Köln 11413
Degenstein Community Library, Sunbury 41478
Degolyer & MacNaughton Library, Dallas 35644
Degussa AG, Hanau 11402
Deichman Library, Oslo Public Library, Oslo 20384
Deichmanske bibliotek, Oslo 20384
Deineka Kursk Art Gallery, Kursk 23134
DeKalb County Public Library, Decatur 38801
DeKalb County Public Library, Fort Payne 39110
DeKalb Public Library, DeKalb 38809
Del Mar College, Corpus Christi 33260
Delafield Public Library, Delafield 38811
DeLaSalle Schule Strebersdorf, Wien 01481
Delaware Academy of Medicine, Inc, Wilmington 37847
Delaware Art Museum, Wilmington 37848
Delaware County Commonwealth College, Media 31785
Delaware County Community College, Media 33525
Delaware County District Library, Delaware 38812
Delaware County Historical Society, Chester 36659
Delaware County Supreme Court Library, Delhi 34133
Delaware Department of Transportation Library, Dover 34153
Delaware Division of Libraries – State Library, Dover 38855
Delaware Division of Libraries-State Library – Delaware Library for the Blind & Physically Handicapped, Dover 36809
Delaware Geological Survey Library, Newark 37274
Delaware Historical Society Research Library, Wilmington 37849
Delaware Museum of Natural History Library, Wilmington 37850
Delaware State Law Library in Kent County, Dover 34154
Delaware State University, Dover 31022
Delaware Technical & Community College – Stanton Campus Library, Newark 33580
Delaware Technical & Community College – Stephen J. Betze Library, Georgetown 33370
Delaware Technical & Community College, Wilmington Campus, Wilmington 33875
Delaware Valley College, Doylestown 31026
Délégation aux Arts Plastiques (DAP), Paris 08531
Delft University of Technology, Delft 19104
Delgado Community College, New Orleans 33576
Delhi Public Library, New Delhi 14229
Delhi School of Economics, Delhi 13800
Delo d.d., Ljubljana 24915
'Delo' Magazine Concern, Ljubljana 24915
Delphi Automotive System, Flint 35707
Delphi Public Library, Delphi 38813
Delphi Public Library, Yeoman 41988
Delphos Public Library, Delphos 38814
Delray Beach Public Library, Delray Beach 38815
Delta College, University Center 33816
Delta Public Library, Delta 38816
Delta State University, Cleveland 30836
Delta Steel Co Ltd, Warri 20006
Deltares, Delft 19408
Democritus University of Thrace, Komotini 13081
Demokritos National Centre for Scientific Research, Athinai 13104
Demokritos University of Thrace, Xanthi 13089
Demosia Vivliothiki Rethimnou, Rethimnon 13151
Dengen Kaihatsu K. K., Tokyo 17441
Denison Public Library, Denison 38821
Denison University Libraries, Granville 31242
Deniz Bilimleri ve İşletmeci, Istanbul 27885
Denkmalschutzamt der Freien und Hansestadt Hamburg, Hamburg 10915
Denmark's International Study Program (DIS), København 06792
Denryoku Chuo Kenkyujo, Keizai-Kenkyujo, Tokyo 17672
Denryoku Chuo Kenkyusho Jimubu, Abiko 17508
Densi Gijyutu Sogo Kenkyujo, Tsukuba 17785
Denton Public Library – Emily Fowler Central Library, Denton 38824
Dentsu Advertising Ltd, Marketing Division, Osaka 17406
Dentsu Inc Corporate Communications Division, Tokyo 17442
Dentsu K. K. Osakashisha Maketingu-kyoku Shiryobu, Osaka 17406
Denver Art Museum Library, Denver 36791
Denver Botanic Gardens, Denver 36792
Denver Museum of Nature & Science, Denver 36793
Denver Public Library, Denver 38825
Denver Public School District, Denver 36794
Denver Seminary, Littleton 35265
Denville Free Public Library, Denville 38826
Departament morskogo transporta, Moskva 23173
Departamento Administrativo de la Presidencia de la República, Santafé de Bogotá 05993
Departamento Administrativo Nacional de Estadística (DANE), Santafé de Bogotá 05994
Departamento de Arqueologia y Museologia, Arica 04916
Departamento de Avaliação, Prospectiva e Planeamento, Lisboa 21723
Departamento de Documentación Nuclear, Montevideo 42053
Departamento de Educación, Universidades e Investigación, Vitoria-Gasteiz 25753
Departamento de Estudios Históricos, Montevideo 42054
Departamento de Presidencia. Asesoria Jurídica, Pamplona 25740
Departamento de Prospectiva e Planeamento e Relações Internacionais, Lisboa 21724
Departamento do Ensino Secundário, Lisboa 21725
Departamento Forestal, Santiago 04927
Departamento Médico do Serviço Civil do Estado, São Paulo 03239
Departamento Nacional de Produção Mineral, Brasília 03215
Departement de Recherches Agronomiques de la République Malgache, Antananarivo 18459
Departement Kesehatan, Jakarta 14331

Departement Leefmilieu, Natuur en Energie, Brussel 02427
Departemento de Planejamento Ambiental (FEEMA), Rio de Janeiro 03229
Department for Communities and Local Government, London 29516
Department for Community Development Library, East Perth 00673
Department for Environment, Food and Rural Affairs, London 29517
Department for the Distribution of Scientific and Technical Information at the French Cultural Centre in Algiers, Alger 00071
Department for Transport, Energy and Infrastructure, Adelaide 00597
Department for Work and Pensions, London 29518
Department of Agricultural and Scientific Research and Extension, Amman 17880
Department of Agricultural Research, Gaborone 02970
Department of Agriculture, Bangalore 13849
Department of Agriculture, Lami 07298
Department of Agriculture, Njala 24577
Department of Agriculture, South Perth 01007
Department of Agriculture and Environmental Affairs – Free State Province, Glen 25062
Department of Agriculture and Livestock, Konedobu 20577
Department of Agriculture and Rural Development, Belfast 29508
Department of Agriculture, Fisheries and Forestry, Barton 00870
Department of Agriculture, Food and Forestry, Dublin 14499
Department of Agriculture, Forestry and Fisheries, Pretoria 25072
Department of Alternative Energy Development and Efficiency (DEDE), Bangkok 27755
Department of Antiquities Library, Tripoli 18258
Department of Arts, Culture, Science and Technology, Pretoria 25073
Department of Attorney General, Waigani 20578
Department of Biological Publications of the Russian Academy of Sciences, Moskva 23455
Department of Canadian Heritage, Hull 04087
Department of Census and Statistics, Colombo 26291
Department of Chemical Technology, Mumbai 13862
Department of Community Services, Government of Yukon, Whitehorse 04186
Department of Conservation, Wellington 19797
Department of Conservation and Land Management, Como 00658
Department of Conservation and Natural Resources, East Melbourne 00671
Department of Correctional Education, State Farm 34890
Department of Cross-Cultural and Regional Studies – Section of Asian Studies, København 06686
Department of Defence, Cerberus 00655
Department of Defence, East Fremantle 00668
Department of Defence, Enoggera 00678
Department of Defence, Forest Hill 00679
Department of Defence, Laverton 00691
Department of Defence, Moorebank 00703
Department of Defence, Nowra 00705
Department of Defence, Puckapunyal 00719
Department of Defence, Southbank 00722
Department of Defence, Sydney 00726
Department of Defence, Townsville 00736
Department of Defence, Waverton 00739
Department of Defence, Defence Library Service, East Sale 00677
Department of Defence, Defence Library Service Canberra – Russell E., Canberra 00640
Department of Defence, Defence Library Service – Canberra Vane Green Library, Canberra 00417
Department of Defence, Defence Library Service – Canungra, Canungra 00653
Department of Dialectology, Onomastics and Folklore Research in Gothenburg, Göteborg 26584
Department of Education, Canberra 00641
Department of Education, Dublin 14500
Department of Education and Children's Services, Hectorville 00682
Department of Education, Culture & Sports, Pasig City 20760
Department of Employment and Industrial Relations, Brisbane 00616
Department of Employment and Workplace Relations, Canberra 00642
Department of Employment, Economic Development and Innovation Library, Queensland, Brisbane 00617
Department of Enterprise Trade and Investment, Belfast 29509
Department of Environment and Conservation, Waigani 20579
Department of Environmental Affairs and Tourism, Pretoria 25074
Department of Environmental Planning, Rio de Janeiro 03229
Department of Environmental Protection, Hamilton 02911
Department of Environmental Protection, Perth 00711
Department of Extension, Promotion and Scientific Information, Skierniewice 21222
Department of Families, Housing, Community Services and Indigenous Affairs, Canberra 00643
Department of Finance, Pretoria 25075
Department of Finance and Deregulation, Terrace Parces 00735
Department of Fisheries and Oceans Canada, Mont-Joli 04384
Department of Fisheries and Oceans Canada, Winnipeg 04555
Department of Foreign Affairs, Dublin 14501
Department of Foreign Affairs, Pasay City 20758
Department of Foreign Affairs, Pretoria 25076
Department of Foreign Affairs and Trade, Barton 00609
Department of Foreign Languages, Moskva 23341
Department of Geography, Zagreb 06232
Department of Geological Literature of the Russian Academy of Sciences, Moskva 23456
Department of Geology and Mineral Survey, Kabul 00002
Department of Health, Jakarta 14331
Department of Health, London 29519
Department of Health, Pretoria 25077
Department of Health and Aging, Woden 00743
Department of Health and Children, Dublin 14502
Department of Health and Human Services, Hobart 00684
Department of Housing and Urban Development, Adelaide 00598
Department of Human Resources Development Canada-Library, Ottawa 04111
Department of Human Services (Housing Group), Adelaide 00850
Department of Human Services-Youth Corrections, Golden 36896
Department of Indian Affairs & Northern Development (DIAND), Whitehorse 04552
Department of Industrial Relations, Boroko 20572
Department of Infrastructure, Transport, Regional Development and Local Government, Canberra 00644
Department of Innovation, Industry, Science and Research, Canberra 00645
Department of Internal Affairs, Wellington 19798
Department of Justice, Hong Kong 05811
Department of Justice, Manila 20754
Department of Justice, Melbourne 00697
Department of Justice, Pretoria 25078
Department of Justice, St. John's 04156
Department of Labour, Colombo 26292
Department of Labour, New Delhi 13873
Department of Main Roads, Brisbane 00618
Department of Maritime Transport, Moskva 23173
Department of Mental Health, St Elizabeth's Hospital, Washington 37730
Department of Musical History of the AS CR, library, Praha 06494
Department of National Archives, Colombo 26310
Department of Natural Resources, Parramatta 00979
Department of Oriental Studies, Yerevan 00348
Department of Physical Engineering and Mathematical Sciences of the Russian Academy of Sciences, Moskva 23484
Department of Planning, Sydney 00727
Department of Plant Protection, Karachi 20494
Department of Premier and Cabinet, Perth 00712
Department of Primary Industries, Brisbane 00619
Department of Primary Industries, Water and Environment, Hobart 00685
Department of Public Works, Boroko 20573
Department of Public Works and Services, Sydney 00728
Department of Religious Affairs, Yangon 19002
Department of Revenue Library, New Delhi 13874
Department of Science Service, Bangkok 27756
Department of Sinology, Institute of Social Sciences, Moskva 23485
Department of Social Sciences, Yerevan 00349
Department of Social Security, Tuggeranong 00738
Department of Social Welfare, Dublin 14503
Department of State Library of Florida, Tallahassee 34910
Department of Statistics, Kuala Lumpur 18554
Department of STI, Žilina 24760
Department of Studies and Researches on Africa and Arab Countries, Napoli 16384
Department of the Environment, Dublin 14504
Department of the Environment, Water, Heritage and the Arts, Canberra 00646
Department of the Parliamentary Library, Canberra 00647
Department of the Prime Minister and Cabinet, Barton 00610
Department of Trade and Industry – Legal Library & Information Centre, London 29520
Department of Veterans Affairs, Cincinnati 36697
Department of Veterans Affairs, Coatesville 36717
Department of Veterans Affairs, Dublin 36812
Department of Veterans Affairs, Fresno 36882
Department of Veterans Affairs, Leeds 37024
Department of Veterans Affairs, Perry Point 37335
Department of Veterans Affairs, Philadelphia 37343
Department of Veterans Affairs, San Juan 21905
Department of Veterans Affairs, Tucson 37686
Department of Veterans Affairs, White City 37833
Department of Veterans' Affairs, Woden 00744
Department of Veterans Affairs – Headquarters Library, Washington 37731
Department of Veterans Affairs Medical Center, Erie 36832
Department of Veterans Affairs Medical Center, Hampton 36911
Department of Veterans Affairs Medical Center, Long Beach 37041
Department of Veterans Affairs Medical Center, North Chicago 37292
Department of Veterans Affairs Medical Center, Palo Alto 37320
Department of Veterans Affairs Medical Center, Saint Cloud 37482
Department of Veterans Affairs Medical Center Library, East Orange 36820
Department of Veterans Affairs, New York Harbor Healthcare System, New York 37200
Department of Veterans Affairs – Office of the General Council Law Library, Washington 37732
Department of Water Affairs and Forestry, Pretoria 25079
Department Perhubungan Darat & Ptt, Jakarta 14332
Departmental Archives of Creuse, Guéret 08361
DePaul University, Chicago 30760
– Law Library, Chicago 30761
– Loop, Chicago 30762
DePauw University, Greencastle 31249
Depotbiblioteksfunktionen, Århus 06847
Dept of Business, Industry + Resource Development, Darwin 00662
Deputazione di Storia Patria, L'Aquila 16301
Deputazione di Storia Patria per la Toscana, Firenze 16258
Deputazione di Storia Patria per le Antiche Provincie Modenesi, Modena 16361
Deputazione di Storia Patria per le Venezie, Venezia 16648
Deputazione di Storia Patria per l'Umbria, Perugia 16414
Deputazione Subalpina di Storia Patria, Torino 16600
DERA Malvern, Malvern 29849
Derby Neck Library, Derby 38828
Derby Public Library, Derby 38829
Derbyshire County Council, Matlock 30041
Deree College Library, Aghia Paraskevi 13071
Dérer Hospital and Polyclinic, Bratislava 24685
Dérerova Nemocnica s Poliklinikou, Bratislava 24685
Derevoobrabatyvayushchee obedinenie Pinskdrev, Pinsk 01917
Déri Múzeum, Debrecen 13467
Dermatovenerological Research Institute, Almaty 17942
Derrick Learning Support Centre, Suva 07290
Derry Public Library, Derry 38831
Derzhavna akademiya budivnitstva ta arkhitekturi, Odesa 28049
Derzhavna akademiya kholoda, Odesa 28050
Derzhavna akademiya miskoho gospodarstva, Kharkiv 27984
Derzhavna akademiya tekhnologiyi ta organizatsiyi kharchuvaniya, Kharkiv 28249
Derzhavna biblioteka Ukrainy dlya ditei, Kyiv 28670
Derzhavna biblioteka Ukrainy dlya yunatstva, Kyiv 28671
Derzhavna istorichna biblioteka Ukrainy (DIBU), Kyiv 28304
Derzhavna konservatoriya im. Nezhdanovoi, Odesa 28051
Derzhavna konservatoriya im. P.I. Chaikovskoho, Kyiv 28008
Derzhavna konservatoriya im. Prokofeva, Donetsk 27970
Derzhavna medychna akademiya, Dnipropetrovsk 27961
Derzhavna metallurgina akademiya Ukrainy, Dnipropetrovsk 27962
Derzhavna naukova arkhitekturno-budivelna biblioteka, Kyiv 28305
Derzhavna naukova medychna biblioteka, Kharkiv 28250
Derzhavna naukova silskohospo-darska biblioteka UAAN, Kyiv 28306
Derzhavna naukovo-pedahohichna biblioteka Ukrainy im. V. O. Sukhomlynskoho, Kyiv 27936
Derzhavna naukovo-tekhnichna biblioteka Ukrainy, Kyiv 27937
Derzhavna oblasna biblioteka, Rivne 28796
Derzhavna oblasna biblioteka, Ternopil 28841
Derzhavna oblasna biblioteka dlya ditei, Dnipropetrovsk 28533
Derzhavna oblasna biblioteka dlya ditei, Donetsk 28543
Derzhavna oblasna biblioteka dlya ditei, Lutsk 28705
Derzhavna oblasna biblioteka dlya ditei, Lviv 28709
Derzhavna oblasna biblioteka dlya ditei, Odesa 28762
Derzhavna oblasna biblioteka dlya ditei, Zaporizhzhya 28885
Derzhavna oblasna biblioteka dlya ditei, Zhytomyr 28897
Derzhavna oblasna biblioteka dlya ditei im. D. U. Dobroi, Kyiv 28672
Derzhavna oblasna biblioteka dlya ditei im. Franka, Vinnytsya 28868
Derzhavna oblasna biblioteka dlya ditei im. Vakarova, Uzhgorod 28858
Derzhavna oblasna biblioteka dlya yunatstva, Donetsk 28544
Derzhavna oblasna biblioteka dlya yunatstva, Kharkiv 28599
Derzhavna oblasna biblioteka dlya yunatstva, Lutsk 28706
Derzhavna oblasna biblioteka dlya yunatstva, Lviv 28710
Derzhavna oblasna biblioteka dlya yunatstva, Vinnytsya 28869
Derzhavna oblasna biblioteka dlya yunatstva, Zaporizhzhya 28886
Derzhavna oblasna biblioteka dlya yunatstva, Zhytomyr 28898
Derzhavna oblasna biblioteka dlya yunatstva im. M. Svetlova, Dnipropetrovsk 28534
Derzhavna oblasna biblioteka dlya yunatstva im. Vaidi, Uzhgorod 28859
Derzhavna oblasna bilbioteka dlya yunatstva, Odesa 28763
Derzhavna oblasna universalna biblioteka, Cherkasy 27932
Derzhavna oblasna universalna biblioteka, Chernigiv 28519
Derzhavna oblasna universalna biblioteka, Chernivtsi 28523

Derzhavna oblasna universalna naukova biblioteka, Kharkiv 27933
Derzhavna oblasna universalna naukova biblioteka, Khmelnitski 28610
Derzhavna oblasna universalna naukova biblioteka, Lviv 27940
Derzhavna oblasna universalna naukova biblioteka, Odesa 27943
Derzhavna oblasna universalna naukova biblioteka, Uzhgorod 27948
Derzhavna oblasna universalna naukova biblioteka, Zhytomyr 28899
Derzhavna oblasna universalna naukova biblioteka im. Gorkoho, Zaporizhzhya 27950
Derzhavna oblasna universalna naukova biblioteka im. N. K. Krupskoi, Donetsk 28545
Derzhavna oblasna universalna naukova biblioteka im. O. Pchilki, Lutsk 27939
Derzhavna oblasnaya biblioteka dlya yunatstva, Ivano-Frankivsk 28583
Derzhavne khoreografichne uchilishche, Kyiv 28009
Derzhavne muzichne uchilishche, Kyiv 28010
Derzhavny agrarny universytet, Dnipropetrovsk 27963
Derzhavny arkhiv Chernivetskoi oblasti, Chernivtsi 28195
Derzhavny arkhiv Ivano-Frankivskoyi oblasti, Ivano-Frankivsk 28240
Derzhavny arkhiv Volynskoyi oblasti, Lutsk 28401
Derzhavny ekonomichny universytet, Kharkiv 27985
Derzhavny gidrometeorologichny instytut, Odesa 28052
Derzhavny instytut fizychnoyi kulturi, Lviv 28033
Derzhavny instytut fizychnoyi kulturi ta sportu, Dnipropetrovsk 28198
Derzhavny instytut kulturi, Kyiv 28673
Derzhavny instytut kulturi, Rivne 28797
Derzhavny instytut proektuvannya zavodov silskogospodarskogo mashinobuduvannya, Kyiv 28308
Derzhavny instytut udoskonalennya likariv, Kyiv 28011
Derzhavny instytut vdoskonalennya likariv im. Gorkoho, Zaporizhzhya 28086
Derzhavny istoriko-kulturny zapovednik, Kerch 28247
Derzhavny komitet Ukrainy po vugilnoyi promislovosti, Donetsk 28212
Derzhavny komunalny proektny kompleks Kyivproekt, Kyiv 28157
Derzhavny medichny instytut, Lviv 28405
Derzhavny medychny universytet im. N.I. Pirogova, Odesa 28053
Derzhavny Nikitsky botanichny sad, Yalta 28483
Derzhavny pedahohichny instytut, Berdyansk 27952
Derzhavny pedahohichny instytut, Glukhiv 27977
Derzhavny pedahohichny instytut, Krivi Rig 28006
Derzhavny pedahohichny instytut, Melitopol 28043
Derzhavny pedahohichny instytut, Mykolaiv 28045
Derzhavny pedahohichny instytut, Rivne 28068
Derzhavny pedahohichny instytut, Slovyansk 28075
Derzhavny pedahohichny instytut im. Makarenka, Sumy 28076
Derzhavny pedahohichny instytut im. Pushkina, Kirovograd 28005
Derzhavny pedahohichny universytet, Chernigiv 27957
Derzhavny pedahohichny universytet, Uman 28080
Derzhavny prirodoznavchi muzei, Lviv 28406
Derzhavny proektno-konstruktorsky ta eksperementalny instytut kompleksnoyi mekhanizatsiyi shakht, Donetsk 28213
Derzhavny proektny instytut Dongiproshakht, Donetsk 28214
Derzhavny proektny instytut po tsivilnomu i promislovomu budivnitstvu, Kyiv 28309
Derzhavny tekhnichny universytet, Dniprodzerzhinsk 27960
Derzhavny universytet, Chernivtsi 27959
Derzhavny universytet im. Frunze, Simferopol 28073
Derzhavny universytet im. I.I. Mechnikova, Odesa 28054
Derzhavny universytet Lvivska politekhnika, Lviv 28034
Derzhkharchoprom Ukrainy, Kyiv 28133
Des Moines Area Community College, Ankeny 33127
Des Moines Art Center Library, Des Moines 36798
Des Moines Public Library, Des Moines 38832
Des Moines Public Library – East Side, Des Moines 38833
Des Moines Public Library – Franklin Avenue, Des Moines 38834
Des Moines Public Library – North Side, Des Moines 38835
Des Moines Public Library – South Side, Des Moines 38836
Des Plaines Public Library, Des Plaines 38838
Desales University, Center Valley 30692
Deschutes Public Library District, Bend 38169
Deschutes Public Library District – Bend Branch, Bend 38170
Deschutes Public Library District – Redmond Branch, Redmond 40956
Desenvolvimento Econômico e Social – BNDES, Rio de Janeiro 03262
Desert Research Institute, Reno 32423
Design Academy Eindhoven, Eindhoven 19109
Design Bureau for Construction in Textile Industry, Sankt-Peterburg 23781
Design Bureau for North-Western Area Fish Industry, Sankt-Peterburg 23906
Design Center Stuttgart, Stuttgart 12512
Design Institute for Industrial Construction Joint-Stock Company, Moskva 22825
Design Institute for Water Resources Engineering, Bucureşti 21967
Design museum Gent, Gent 02644
Design Office for Urban Planning and Architecture, Sankt-Peterburg 23838
Design Office of Automatic Systems Ltd, Moskva 22823
Design-Bibliothek, Stuttgart 12512
Designmuseo, Helsinki 07448
Desjardins, Ducharme, Stein & Morast, Montreal 04267
DeSoto Parish Library, Mansfield 40037
DeSoto Public Library, DeSoto 38839
DESY Zeuthen, Zeuthen 12623
Detroit Baptist Theological Seminary, Allen Park 35104
Detroit Institute of Arts, Detroit 36804
The Detroit News, Inc, Detroit 35673
Detroit Public Library, Detroit 38840
Detroit Public Library – Wilder, Detroit 38841
Detskaya biblioteka im. K.I. Chukovskogo Respubliki Tyva, Kyzyl 24156
Detskaya bolnitsa no 1, Sankt-Peterburg 23754
Detsko-yunosheskaya biblioteka Respubliki Kareliya, Petrozavodsk 24202
Deutsch, Kerrigan & Stiles, New Orleans 35945
Deutsch-Amerikanisches Institut e.V., Freiburg 11888
Deutsch-Amerikanisches Institut, Nürnberg 12370
Deutsch-Amerikanisches Institut, Tübingen 12557
Deutsch-Amerikanisches Institut (DAI), Heidelberg 12039
Deutsche Akademie der Naturforscher Leopoldina e.V., Halle (Saale) 11955
Deutsche Bank AG, Hannover 12024
Deutsche Bibliothek für Kurzschrift, Textverarbeitung und Maschinenschreiben, Bayreuth 11479
Deutsche Bibliothek / Saksalainen kirjasto, Helsinki 07449
Deutsche Blinden-Bibliothek, Emil-Krückmann-Bücherei (EKB), Marburg 12253
Deutsche Blindenstudienanstalt e.V., Marburg 12253
Deutsche Büchereizentrale und Zentralbücherei Apenrade, Aabenraa 06765
Deutsche Bundesbank, Frankfurt am Main 11397
Deutsche Elektronen-Synchrotron, Hamburg 11976
Deutsche Esperanto-Bibliothek, Aalen 11446
Deutsche Flugsicherung GmbH, Langen (Hessen) 12172
Deutsche Gesellschaft für Auswärtige Politik e.V., Berlin 11501
Deutsche Gesellschaft für Internationale Zusammenarbeit (GIZ) GmbH, Bonn 11625
Deutsche Glastechnische Gesellschaft e.V., Offenbach 12382
Deutsche Hochschule für Verwaltungswissenschaften Speyer, Speyer 10589
Deutsche Investitions- und Entwicklungsgesellschaft mbH, Köln 11413
Deutsche Katholische Blindenbücherei GmbH, Bonn 11626
Deutsche Kinemathek, Berlin 11502
Deutsche Kinemathek – Museum für Film und Fernsehen, Berlin 11502
Deutsche Messebibliothek, Berlin 11503
Deutsche Meteorologische Bibliothek, Offenbach 12383
Deutsche Nationalbibliothek (Leipzig, Frankfurt am Main), Frankfurt am Main 09291
Deutsche Nationalbibliothek, Leipzig 09293
Deutsche Nationalbibliothek – Bibliothek des Börsenvereins des Deutschen Buchhandels, Frankfurt am Main 09292
Deutsche Orient-Stiftung / German Orient-Foundation, Berlin 11504
Deutsche Pfandbrief- und Hypothekenbank AG, Wiesbaden 11435
Deutsche Pharmazeutische Zentralbibliothek, Stuttgart 12513
Deutsche Rentenversicherung Baden-Württemberg, Karlsruhe 10957
Deutsche Rentenversicherung Baden-Württemberg, Stuttgart 11100
Deutsche Rentenversicherung Berlin-Brandenburg, Berlin 10803
Deutsche Rentenversicherung Braunschweig-Hannover, Laatzen 10996
Deutsche Rentenversicherung Bund, Berlin 10804
Deutsche Rentenversicherung Knappschaft-Bahn-See, Bochum 10818
Deutsche Rentenversicherung Nord, Standort Hamburg, Hamburg 10916
Deutsche Rentenversicherung Oldenburg-Bremen, Oldenburg 11074
Deutsche Rentenversicherung Rheinland, Düsseldorf 10869
Deutsche Rentenversicherung Westfalen, Münster 11054
Deutsche Sporthochschule Köln, Köln 10123
Deutsche Welle, Bonn 11627
Deutsche Zentralbibliothek für Medizin, Ernährung, Umwelt und Agrarwissenschaften, Bonn 11651
Deutsche Zentralbibliothek für Medizin (ZB MED), Köln 12125
Deutsche Zentralbücherei für Blinde zu Leipzig, Leipzig 12181
Deutscher Alpenverein, München 12290
Deutscher Bundestag, Berlin 10805
Deutscher Caritasverband e.V., Freiburg 11890
Deutscher Olympischer Sportbund (DOSB), Frankfurt am Main 11842
Deutscher Raiffeisenverband, Bonn 11628
Deutscher Sparkassen- und Giroverband e.V. (DSGV), Bonn 11378
Deutscher Städtetag, Köln 10987
Deutscher Turner-Bund, Frankfurt am Main 11843
Deutscher Verein für Öffentliche und Private Fürsorge e.V., Berlin 11505
Deutscher Verein für Versicherungswissenschaft e.V., Berlin 11506
Deutscher Wetterdienst, Offenbach 12383
Deutsches Adelsarchiv, Marburg 12254
Deutsches Albert-Schweitzer-Zentrum, Frankfurt am Main 11844
Deutsches Apotheken-Museum, Heidelberg 12040
Deutsches Archäologisches Institut, Athinai 13105
Deutsches Archäologisches Institut, Berlin 11507
Deutsches Archäologisches Institut, Istanbul 27886
Deutsches Archäologisches Institut (DAI), Bonn 11629
Deutsches Archäologisches Institut (DAI), Frankfurt am Main 11845
Deutsches Archäologisches Institut (DAI), München 12291
Deutsches Archäologisches Institut, Eurasien-Abteilung, Berlin 11508
Deutsches Archäologisches Institut / Istituto Archeologico Germanico, Roma 16493
Deutsches Archäologisches Institut Madrid / Instituto Arqueológico Alemán, Madrid 26011
Deutsches Archäologisches Institut, Orient-Abteilung, Berlin 11509
Deutsches Architekturmuseum, Frankfurt am Main 11846
Deutsches Bergbau-Museum Bochum, Bochum 11618
Deutsches Bibel-Archiv, Hamburg 11977
Deutsches Buch- und Schriftmuseum, Frankfurt am Main 09291
Deutsches Entomologisches Institut, Müncheberg 12275
Deutsches Filminstitut – DIF e.V., Frankfurt am Main 11847
Deutsches Freimaurer-Museum, Bayreuth 11480
Deutsches Gartenbaumuseum Erfurt, Erfurt 11816
Deutsches Historisches Institut in Rom / Istituto Storico Germanico di Roma, Roma 16494
Deutsches Historisches Institut London, London 29750
Deutsches Historisches Institut, Musikgeschichtliche Abteilung., Roma 16495
Deutsches Historisches Institut Paris (DHIP) / Institut Historique Allemand de Paris (IHAP), Paris 08532
Deutsches Historisches Institut Warschau / Niemiecki Instytut Historyczny w Warszawie, Warszawa 21254
Deutsches Hygiene-Museum, Dresden 11783
Deutsches Institut für Entwicklungspolitik (DIE), Bonn 11630
Deutsches Institut für Ernährungsforschung, Nuthetal 12378
Deutsches Institut für Erwachsenenbildung e.V. Leibniz-Zentrum für lebenslanges Lernen, Bonn 11631
Deutsches Institut für Gesundheitsforschung gGmbH, Bad Elster 11467
Deutsches Institut für Internationale Pädagogische Forschung, Frankfurt am Main 11848
Deutsches Institut für Internationale Pädagogische Forschung, Berlin, Berlin 11495
Deutsches Institut für Japanstudien, Chiyoda-ku Tokyo 17528
Deutsches Institut für Jugendhilfe und Familienrecht, Heidelberg 12041
Deutsches Institut für Menschenrechte, Berlin 11510
Deutsches Institut für tropische und subtropische Landwirtschaft, Witzenhausen 12605
Deutsches Institut für Wirtschaftsforschung, Berlin 11516
Deutsches Jugendinstitut e.V., München 12292
Deutsches Kabarettarchiv, Mainz 12230
Deutsches Krebsforschungszentrum, Heidelberg 12042
Deutsches Kupferinstitut e.V., Düsseldorf 11790
Deutsches Ledermuseum, Offenbach 12384
Deutsches Literaturarchiv Marbach, Marbach am Neckar 12251
Deutsches Liturgisches Institut, Trier 12551
Deutsches Meeresmuseum, Stralsund 12506
Deutsches Museum, München 12293
Deutsches Musikarchiv, Frankfurt am Main 09291
Deutsches Musikgeschichtliches Archiv, Kassel 12098
Deutsches Notarinstitut, Würzburg 12612
Deutsches Orient-Institut, Berlin 11504
Deutsches Patent- und Markenamt, München 11042
Deutsches Pferdemuseum, Verden (Aller) 12572
Deutsches Polen-Institut, Darmstadt 11719
Deutsches Rechtswörterbuch, Heidelberg 12043
Deutsches Referenzzentrum für Ethik in den Biowissenschaften (DRZE), Bonn 11632
Deutsches Röntgenmuseum, Remscheid 12434
Deutsches Rotes Kreuz – Generalsekretariat, Berlin 11511
Deutsches Rundfunkarchiv, Frankfurt am Main 11849
Deutsches Rundfunkarchiv Potsdam-Babelsberg, Potsdam 12413
Deutsches Schiffahrtsmuseum, Bremerhaven 11694
Deutsches Spielkartenmuseum, Leinfelden-Echterdingen 12177
Deutsches Spielzeugmuseum, Sonneberg 12495
Deutsches Tanzarchiv Köln, Köln 12126
Deutsches Technikmuseum Berlin, Berlin 11512
Deutsches Theatermuseum, München 12294
Deutsches Volksliedarchiv, Freiburg 11891
Deutsches Wollforschungsinstitut an der RWTH Aachen e.V., Aachen 11440
Deutsches Zentralinstitut für soziale Fragen (DZI), Berlin 11513
Deutsches Zentrum für Altersfragen, Berlin 11514
Deutsches Zentrum für Luft- und Raumfahrt e.V., Berlin 11515
Deutsches Zentrum für Luft- und Raumfahrt e.V. (DLR), Braunschweig 11659

Deutsches Zentrum für Luft- und Raumfahrt e.V. (DLR), Göttingen 11927
Deutsches Zentrum für Luft- und Raumfahrt e.V. (DLR), Stuttgart 12514
Deutsches Zentrum für Luft- und Raumfahrt e.V. (DLR), Standortbibliothek Köln-Porz, Köln 12127
Deutsches Zentrum für Luft- und Raumfahrt e.V. (DLR), Standortbibliothek Oberpfaffenhofen, Weßling 12591
Deutsch-Europäisches Juridicum, Saarbrücken 10555
Deutsch-Französisches Institut, Ludwigsburg 12215
Deutschordens-Zentralarchiv, Wien 01482
Deutsch-Russisches Kulturinstitut e.V., Dresden 11753
Deutschsprachige Gemeinschaft Belgiens, Eupen 02759
Dev Samaj College for Women, Ferozepur 13805
Development Bank of Minas Gerais, Belo Horizonte 03251
Development Bank of Singapore (DBS Bank), Singapore 24606
Développement et Civilisations Lebret-Irfed, Paris 08533
Devi Ahilya Vishwavidyalaya, Indore 13661
Devon Libraries, Exeter 29980
DeVry Institute Library, North Brunswick 32139
DeVry Institute of Technology, Columbus 33253
DeVry Institute of Technology, Decatur 33286
DeVry Institute of Technology, Phoenix 33638
DeVry University, Chicago 33219
DeVry University-Kansas City, Kansas City 31481
Dewan Bahasa dan Pustaka, Kuching 18608
Dewan Bahasa dan Pustaka Brunei, Bandar Seri Begawan 03446
Dewan Bandaraya Kuala Lumpur, Kuala Lumpur 18549
Dewan Perwakilan Rakyat Republik Indonesia, Jakarta 14333
Dewey & Ballantine Library, Washington 36344
Dewey Ballantine LLP, New York 35978
Dewey Ballantine LLP Library, New York 35979
Dewey & Lebouf Library, Washington 36345
Dewitt, Ross & Stevens SC, Madison 35870
Dewsbury College, Dewsbury 29462
Dexia Bank België NV, Brussel 02502
Dexter District Library, Dexter 38842
Dezhou Teachers' College, Dezhou 05166
Dezso Laczkó Museum, Veszprém 13519
Dezvoltare pentru Cultura Plantelor pe Nisipuri Dabuleni, Dăbuleni 22031
DFS Deutsche Flugsicherung GmbH, Langen (Hessen) 12172
Dhaka University, Dhaka 01749
Dharan Hattisar Campus Library, Sunsari 19028
DHI – Institut for Vand og Milje, Hørsholm 06781
DHI – Water & Environment, Hørsholm 06781
Dhurakijpundit University, Bangkok 27697
Diablo Valley College, Pleasant Hill 33651
Diaconia College, Oslo 20112
DIACT, Paris 08534
Diakonhjemmet Hospital, Oslo 20243
Diakonhjemmets Høgskole, Oslo 20112
Diakonhjemmets sykehus, Oslo 20243
Diakonie Neuendettelsau, Neuendettelsau 10759
Diakonisches Werk der EKD, Berlin 11164
Diakonisches Werk der Evangelischen Kirche in Deutschland, Stuttgart 11337
La Diana, Montbrison 08418
The Dibner Library of Science and Technology, Washington 37782
Dibrugarh University, Dibrugarh 13645
Dickinson College, Carlisle 30676
Dickinson County Library, Iron Mountain 39578
Dickinson Public Library, Dickinson 38844
Dickinson State University, Dickinson 31017
Dickinson Wright PLLC Library, Bloomfield Hills 35511
Dickinson Wright PLLC Library, Detroit 35674
Dickson County Public Library, Dickson 38845
Dickstein Shapiro LLP, Washington 36346
Didaktisches Zentrum des Kantons Zug (DZ), Zug 27447
DIE-Bibliothek, Bonn 11631
Dienst Landelijk Gebied (DLG), Utrecht 19291
Diesel Engine Central Research Institute, Sankt-Peterburg 23917
Dietrich Collection, Canaan 36623
Dietrich-Bonhoeffer-Bibliothek, Berlin 12655
Dietrich-Bonhoeffer-Klinikum Neubrandenburg, Neubrandenburg 12359
Dikanka Regional Centralized Library System, Main Library, Dikanka 28529
Dikanska raionna TsBS, Dikanka 28529
Dillard University, New Orleans 32000
Dillon County Library, Dillon 38846
Dillwyn Correctional Center, Dillwyn 34150
Dilworth, Paxson LLP, Philadelphia 36094
Dimokriteio Panepistimio Thrakis, Xanthi 13089
Dimosia Kentriki, Veria 13155
Dimotiki Vivliothiki Argostoliou, Argostolion 13139
Dimotiki Vivliothiki Chalkidas, Chalkis 13140
Dimotiki Vivliothiki Kavalas, Kavala 13144
Dimotiki Vivliothiki Kerkyras, Kerkira 13145
Dimotiki Vivliothiki Mitilinis, Mitilini 13147
Dimotiki Vivliothiki Nafpaktou, Nafpaktos 13148
Dimotiki Vivliothiki Piraius, Piraius 13149
Dimotiki Vivliothiki Pirgou, Pirgos 13150
Dimotiki Vivliothiki Spartas, Sparta 13153
Dimotiki Vivliothiki Thessalonikes, Thessaloniki 13154
Dimotiki Vivliothiki Zakinthou, Zakinthos 13156
Dine College, Tsaile 33807
Dinsmore & Shohl Library, Cincinnati 35607
Diocesa San Tommaso d'Aquino, Piedimonte Matese 15984
Diocesaan Seminarie Antwerpen, Antwerpen 02447
Diocesan Archive, Maribor 24900
Diocesan Library, Ljubljana 24899
Diocesan Library of Nancy, Villers-les-Nancy 08250
Diocesan-Theological College, Győr 13275
Diocese Library, Szombathely 13348
Diocese of Phoenix, Phoenix 35331
Diocesi di Conversano-Monopoli, Monopoli 15918
Diósgyör Mechanical Engineering Factory, Miskolc 13363
Diósgyőri Gépgyár, Miskolc 13363
Diözesanarchiv Eichstätt, Eichstätt 11203
Diözesanarchiv Wien, Wien 01483
Diözesanbibliothek, Graz 01428
Diözesanbibliothek, Klagenfurt 01443
Diözesanbibliothek, Limburg 11263
Diözesanbibliothek, Münster 11289
Diözesanbibliothek, Osnabrück 11301
Diözesanbibliothek, Rottenburg am Neckar 11318
Diözesanbibliothek Aachen im Katechetischen Institut des Bistums Aachen, Aachen 11145
Diözesanbibliothek St. Pölten, St. Pölten 01475
Diözesanbibliothek Würzburg, Würzburg 11366
Diözesane Priesterhausbibliothek, Salzburg 01461
Diözese Eisenstadt, Eisenstadt 01420
Diözese Feldkirch, Feldkirch-Altenstadt 01423
Diözese Rottenburg-Stuttgart, Rottenburg am Neckar 11318
Diplomatic Studies and Research Institute, Cairo 07164
Diponegoro University, Semarang 14294
– Faculty of Medicine, Semarang 14295
Diputació Provincial de València, Valencia 25752
Diputación Almeria, Almería 25678
Diputación Foral de Vizcaya, Bilbao 25688
Diputación Provincial de a Coruña, A Coruña 26212
Diputación Provincial de Cáceres, Cáceres 25689
Diputación Provincial de Córdoba, Córdoba 25691
Diputación Provincial de Orense, Ourense 25738
Direcção Geral do Inovação e de Desenvolvimento Curricular, Lisboa 21735
Direcção Geral dos Recursos Florestais, Lisboa 21785
Direcção Provincial dos Serviços de Geologia e Minas de Angola, Luanda 00106
Direcção-Geral da Junta do Crédito Público, Lisboa 21726
Direcção-Geral da Saúde, Lisboa 21727
Direcção-Geral das Contribuições e Impostos, Lisboa 21728
Direcção-Geral de Transportes Terrestres e Fluviais, Lisboa 21729
Direcção-Geral do Turismo, Lisboa 21730
Dirección de Estadística y Censo, Panamá 20540
Dirección de Formación y Perfeccionamiento, Madrid 25712
Dirección General de Bibliotecas Municipales, Buenos Aires 00320
Dirección General de Cultura, La Paz 02927
Dirección General de Desarrollo Urbanistico, Caracas 42234
Dirección General de Desarrollos Espaciales, Córdoba 00297
Dirección General de Estadística e Investigaciones, La Plata 00216
Dirección General de Estadística y Censos, San José 06115
Dirección General de la Guardia Civil, Madrid 25713
Dirección General de lo Contencioso del Estado, Madrid 25714
Dirección General de Recursos Naturales, San Salvador 07210
Dirección General del Servicio Meteorológico Nacional, Tacubaya 18869
Dirección Nacional de Ciencia, Tecnología e Innovación (DINACYT), Montevideo 42055
Dirección Nacional de Vialidad, Buenos Aires 00253
Dirección Nacional del Antártico, Buenos Aires 00254
Directie Gerechtelijke Ondersteuning Arondissement Leeuwarden – Palais van Justitie, Leeuwarden 19277
Direction de la Géologie, Rabat 18957
Direction de la Planification, Rabat 18939
Direction de la Recherche Agronomique, Rabat 18940
Direction de l'Architecture et du patrimoine, Paris 08535
Direction de l'Enseignement Primaire, Antananarivo 18460
Direction Départementale du Livre et de la Lecture (DDLL), Montpellier 08969
Direction des Affaires Générales du Secrétariat d'Etat au Plan, Alger 00053
Direction des Archives de France, Aix-en-Provence 08266
Direction des Archives de la Wilaya de Constantine, Constantine 00077
Direction des archives du patrimoine et des musée departementaux, Dammarie-les-Lys 08343
Direction des Eaux et Forêts, Dakar 24425
Direction des Etudes de Milieu et de la Recherche Hydraulique du Secrétariat d'Etat, Alger 00054
Direction des Mines et de la Géologie, Yaoundé 03656
Direction des Mines et de la Géologie (DMG), Dakar 24440
Direction des Mines et de la Géologie (DMG), Nouakchott 18687
Direction des Monnaies et Médailles, Paris 08536
Direction des musées, Strasbourg 08681
Direction des Musées de France, Paris 08537
Direction des Musées de France – Bibliothèque Centrale et Archives des Musées Nationaux, Paris 08538
Direction des Resources Educatives Françaises, Saint Boniface 04476
Direction des Ressources Naturelles, Yaoundé 03643
Direction du Patrimoine Culturel, Alger 00061
Direction générale de La Poste, Strasbourg 08164
Direction Nationale de la Météorologie, Bamako 18649
Directions régionales des affaires culturelles, Strasbourg 08682
Director of Public Prosecutions (CTH), Sydney 01023
Directorate for Civil Protection and Emergency Planning, Tønsberg 20199
Directorate for Cultural Heritage Library, Oslo 20278
Directorate for Labour Inspection, Trondheim 20200
Directorate General of Antiquities, Baghdad 14450
Directorate General of Employment and Training Library, New Delhi 13875
Directorate General of Irrigation, Jakarta 14335
Directorate General of Meteorology, New Delhi 13876
Directorate General of Primary and Secondary Education, Jakarta 14334
Directorate of Archives, Punjab and Archival Museum, Lahore 20511
Directorate of Health in Iceland, Seltjarnarnes 13577
Directorate of Internal Affairs Recreation Centre, Sankt-Peterburg 24219
Directorate of Public Roads, Oslo 20290
Directorate of Technical Development, New Delhi 13877
Directorate of the Coal Industry, Karaganda 17931
Directorate-General of Antiquities and Museums, Damascus 27516
Directory of Diversity and Integration, Oslo 20193
Direktorat Jenderal Pendidikan Dasar dan Menengah, Jakarta 14334
Direktorat Jenderal Pengairan, Jakarta 14335
Direktoratet for arbeidstilsynet, Trondheim 20200
Direktoratet for Samfunnssikkerhet og Beredskap (DSB), Tønsberg 20199
Diretoria de Hidrografia e Navegação, Niterói 03310
Diretoria do Serviço Geográfico, Rio de Janeiro 03230
Disciples of Christ Historical Society Library & Archives, Nashville 37143
Discoteca Oneyda Alvarenga, São Paulo 03401
Dissemination and Information Services Centre (DISC), Dhaka 01765
District and City Library, Częstochowa 21407
District and City Library, Legnica 21473
District and City Library, Zielona Góra 21617
District and City Public Library, Bydgoszcz 21393
District and City Public Library, Rzeszów 21551
District and City Public Library, Świnoujście 21579
District and Municipal Public Library, Wałbrzych 21598
District Children Library, Astrakhan 24106
District Children's Library, Irkutsk 24130
District Children's Library, Kirov 24145
District Children's Library I.A. Krylov, Yaroslavl 24359
District Court of Jerusalem, Jerusalem 14706
District Court's, Tel-Aviv 14690
District Institute of Advanced Training for Teachers, Novosibirsk 22447
District Institute of Advanced Training for Teachers, Sankt-Peterburg 22532
District Library, Aktyubinsk 17983
District Library, Almaty 17984
District Library, Chelyabinsk 22125
District Library, Chorog 27647
District Library, Fergana 42172
District Library, Khanty-Mansisk 24144
District Library, Kokchetav 17988
District Library, Kroměříž 06582
District Library, Kzyl-Orda 17990
District Library, Osh 18158
District Library, Rakovník 06603
District Library, Termez 42176
District Library, Tyumen 22182
District Library, Uralsk 17993
District Library A S. Pushkin, Shimkent 17991
District Library for Children, Ryazan 24209
District Library Ibn-Sin, Bukhara 42171
District Library of Saratov, Saratov 22174
District Library N. Ostrovsky, Gurev 17986
District Library C. Valichanov, Zhambul 17994
District Mueum, Karlovy Vary 06456
District Museum, Krasnoyarsk 23126
District Museum, Suceava 22059
District Museum in Toruń, Toruń 21234
District Museum of History, Galaţi 22034
District Museum of History and Archaeology, Bacău 21957
District Museum of Klatovy, Klatovy 06457
District Museum of Regional Studies, Ekaterinburg 23035
District Museum of Regional Studies, Ivanovo 23061
District Museum of Regional Studies, Kaluga 23075
District Museum of Regional Studies, Kursk 23137
District Museum of Regional Studies, Lutsk 28404
District Museum of Regional Studies, Murmansk 23602
District Museum of Regional Studies, Novosibirsk 23655
District Museum of Regional Studies, Omsk 23666
District Museum of Regional Studies, Orenburg 23673
District Museum of Regional Studies, Oryol 23678
District Museum of Regional Studies, Penza 23683
District Museum of Regional Studies, Perm 23689
District Museum of Regional Studies, Rostov-na-Donu 23724
District Museum of Regional Studies, Saratov 23963
District Museum of Regional Studies, Tambov 23989
District Museum of Regional Studies, Tomsk 23998
District Museum of Regional Studies, Tver 24012

District Museum of Regional Studies, Voronezh 24077
District of Columbia Office of the Corporation Counsel, Washington 34986
District of Columbia Public Library, Washington 41738
District of Columbia Public Library – Capitol View, Washington 41739
District of Columbia Public Library – Chevy Chase, Washington 41740
District of Columbia Public Library – Cleveland Park, Washington 41741
District of Columbia Public Library – Francis A. Gregory Branch, Washington 41742
District of Columbia Public Library – Georgetown, Washington 41743
District of Columbia Public Library – Juanita E. Thornton/Shepherd Park Neighborhood Library, Washington 41744
District of Columbia Public Library – Lamond Riggs, Washington 41745
District of Columbia Public Library – Martin Luther King Jr Memorial, Washington 41746
District of Columbia Public Library – Mount Pleasant, Washington 41747
District of Columbia Public Library – Northeast/7th Street, Washington 41748
District of Columbia Public Library – Palisades, Washington 41749
District of Columbia Public Library – Petworth, Washington 41750
District of Columbia Public Library – Southeast/7th Street, Washington 41751
District of Columbia Public Library – Southwest/Wesley Place, Washington 41752
District of Columbia Public Library – Washington Highlands, Washington 41753
District of Columbia Public Library – West End, Washington 41754
District of Columbia Public Library – Woodridge, Washington 41755
District of Columbia Superior Court Judges Library, Washington 34987
District of Skopje Public Library, Skopje 18448
District Power Engineering Board, Gomel 01956
District Public Library, Biała Podlaska 21380
District Public Library, Bielsko-Biała 21385
District Public Library, Ciechanów 21403
District Public Library, Kalisz 21437
District Public Library, Kielce 21445
District Public Library, Konin 21453
District Public Library, Krosno 21464
District Public Library, Legnica 21472
District Public Library, Olsztyn 21517
District Public Library, Opole 21519
District Public Library, Piła 21526
District Public Library, Płock 21530
District Public Library, Przemyśl 21537
District Public Library, Radom 21544
District Public Library, Sieradz 21555
District Public Library, Tarnobrzeg 21583
District Public Library, Tarnów 21585
District Public Library, Włocławek 21604
District Public Library Dúbravka, Bratislava 24768
Dithmarscher Landesmuseum, Meldorf 12263
Divadelní ústav, Praha 06485
Divadelný ústav, Bratislava 24686
Divine Word College, Epworth 35198
Divine Word University, Madang 20558
Divisão de Discoteca e Biblioteca de Música, São Paulo 03402
Division of Forest Research, Kitwe 42339
Division of Juvenile Justice & Department of Corrections & Rehabilitation, Chino 34077
Division of Legislative Services Reference Center, Richmond 36172
Division of Libraries, Archives and Museums, Saint Thomas 42317
Divisional Public Library, Khaipur 20528
DIW Berlin, Berlin 11516
Dixie Regional Library System, Pontotoc 40844
Dixie State College of Utah, Saint George 33711
Dixon Correctional Center Library, Dixon 34151
Dixon Public Library, Dixon 38848
DK GU vnutrennikh del, Sankt-Peterburg 24219
DK Kirovets, Sankt-Peterburg 24220
DK Minskogo traktornogo zavoda, Minsk 02158
DLA Piper, San Diego 36217
DLA Piper US LLP, Baltimore 35478
DLA Piper US LLP, Chicago 35571
DLA Piper US LLP, Washington 36347
DLM – Deutsches Ledermuseum / Schuhmuseum Offenbach, Offenbach 12384

Dmitrov Polytechnical College, Dmitrov 22666
Dmitrovski politekhnicheski kolledzh, Dmitrov 22666
Dnepropetrovsk State Archives, Dnipropetrovsk 28199
Dnipr Railways Scientific and Technical Library, Dnipropetrovsk 28207
Dnipr Regional Electric Power Station, Trade Union Library, Dnipropetrovsk 28539
Dnipr Rollingstock Construction Industrial Corportion, Dnipropetrovsk 28139
DniproAzot Joint-Stock Company, Trade Union Library, Dniprodzerzhinsk 28532
Dniprodzerzhinsk Museum of Town History, Dniprodzerzhinsk 28197
Dniprodzerzhinsk Regional Centralized Library System, Main Library, Dniprodzerzhinsk 28531
Dniprodzerzhinsk Trade Union Centralized Library System, Main Library, Dniprodzerzhinsk 28530
Dniprodzerzhinska mezhspilkova TsBS, Tsentralna biblioteka, Dniprodzerzhinsk 28530
Dniprodzerzhinska raionna TsBS, Tsentralna biblioteka, Dniprodzerzhinsk 28531
Dnipropetrovsk Regional Centralized Library System, Main Library, Dnipropetrovsk 28536
Dnipropetrovsk Regional Scientific Library, Dnipropetrovsk 28535
Dnipropetrovsk State Technical University of Railway Transport, Dnipropetrovsk 27965
Dnipropetrovsk State University, Dnipropetrovsk 27964
Dnipropetrovsk University of Economics and Law, Dnipropetrovsk 27966
Dnipropetrovska oblasna universalna naukova biblioteka, Dnipropetrovsk 28535
Dnipropetrovska raionna TsBS, Tsentralna biblioteka, Dnipropetrovsk 28536
Dnipropetrovsky natsionalny universytet, Dnipropetrovsk 27964
Dnipropetrovsky natsionalny universytet zaliznichnoho transportu, Dnipropetrovsk 27965
Dnipropetrovsky universytet ekonomiky ta prava, Dnipropetrovsk 27966
Dniproshina Tyre Industrial Corporation, Dnipropetrovsk 28140
Dniprovska raionna TsBS m. Kyiva, Tsentralna biblioteka im. P. Tichiny, Kyiv 28674
Dniprovski Region Central Library System of the City of Kiev – P. Tychina Main Library, Kyiv 28674
Doane College, Crete 30935
Dobó István Vármúzeum, Eger 13470
Doboku Kenkyujo, Tsukuba 17786
Doboku-Gakkai, Tokyo 17673
Dobra State Regional Children's Library, Kyiv 28672
Dobrolyubov Pedagogical Institute of Foreign Languages, Nizhni Novgorod 22426
Dobropillya Regional Centralized Library System, Main Library, Dobropillya 28542
Dobropilska raionna TsBS, Tsentralna biblioteka, Dobropillya 28542
Dobroudja Agricultural Institute, Dobrich 03502
Dobrudzhanski zemedelski institut, Dobrich 03502
DOCENT, Brussel 02585
documenta Archiv für die Kunst des 20. + 21. Jahrhunderts, Kassel 12099
documenta Archive for the Arts of the 20th and 21th Centuries, Kassel 12099
Documentary Library of the History of Medicine, Bucureşti 21971
Documentary Library of the National Archives of Romania, Bucureşti 21970
Documentatiecentrum 'Antwerpiensia', Antwerpen 02510
Documentatiecentrum Antwerpse Noorderpolders, Antwerpen 02511
Documentatiecentrum Atlas, Antwerpen 02527
Documentation Center for Education, Cairo 07165
Documentation Center of Modern Japanese Music, Tokyo 17776
Documentation centre, Brussel 02586
Documentation Centre "Migrations", Neuchâtel 27419
Documentation Centre of the Giovanni Agnelli Foundation, Torino 16596
Documentation Department, Lisboa 21797
La Documentation française, Paris 08539

Documentation Research and Training Centre, Bangalore 13929
Documentation Service of the Navy, Rio de Janeiro 03371
Dodge City Community College, Dodge City 33297
Dodge City Public Library, Dodge City 38849
DOK Centrum – DOK Bibliotheek, Delft 19580
Dokkyo Ika Daigaku, Shimotsuga 17124
Dokkyo University, Soka 17126
Dokkyo University, School of Medicine, Shimotsuga 17124
Dokuchaev Central Soil Science Museum, Sankt-Peterburg 23916
V.V. Dokuchaev Soil Science Institute, Moskva 23458
V. Dokuchayev Kharkiv State University of Agriculture, P.O. Komunist-1 28295
Dokumentation am Goetheanum, Dornach 27348
Dokumentation Kraftfahrwesen e.V. (DKF), Bietigheim-Bissingen 11612
Dokumentations- und Kooperationszentrum Südliches Afrika (SADOCC), Wien 01593
Dokumentations- und Kulturzentrum Deutscher Sinti und Roma e.V., Heidelberg 12044
Dokumentationsarchiv des deutschen Widerstands, Frankfurt am Main 11875
Dokumentationsarchiv des Österreichischen Widerstandes, Wien 01594
Dokumentationsarchiv für österreichische Kunst des 19., 20. Jahrhunderts und der Gegenwartskunst der Neuen Galerie Graz – Universalmuseum Joanneum, Graz 01522
Dokumentationsbibliothek der Landschaft Davos, Davos Platz 27346
Dokumentationsdienst der Bundesversammlung, Bern 27232
Dokumentationsstelle für Neuere Österreichische Literatur im Literaturhaus Wien, Wien 01595
Dokumentationsstelle für Wissenschaftspolitik, Bern 27233
Dokuz Eylül Üniversitesi, Izmir 27866
Dolnośląska Biblioteka Pedagogiczna w Wrocławiu, Wrocław 21358
Dolton Public Library District, Dolton 38850
Dom fizicheskoi kultury, Sankt-Peterburg 23755
Dom kompozitorov, Sankt-Peterburg 23756
Dom kultury im. I.I. Gaza, Sankt-Peterburg 24221
Dom kultury im. V.I. Lenina, Sankt-Peterburg 24222
Dom kultury Lensoveta, Sankt-Peterburg 24223
Dom kultury rabotnikov prosveshcheniya, Sankt-Peterburg 24224
Dom narodnogo tvorchestva, Astrakhan 22993
Dom nauchno-tekhnicheskoi propagandy, Nalchik 23606
Dom nauchno-tekhnicheskoi propagandy, Sankt-Peterburg 23757
Dom sanitarnogo prosveshcheniya, Sankt-Peterburg 23758
Dom soyuzov VCSPS, Moskva 24169
Dom tekhniki kombinata 'Apatit', Kirovsk 22807
Dom uchenykh, Moskva 23174
Dom uchenykh im. A.M. Gorkogo RAN, Sankt-Peterburg 24225
Dombas State Machine Building Academy, Kramatorsk 28296
Dombaska derzhavna mashino-budivna akademiya, Kramatorsk 28296
Dombauarchiv, Köln 11258
Dombibliothek Freising, Freising 11217
Dombibliothek Fritzlar, Fritzlar 11218
Dombibliothek Hildesheim, Hildesheim 11247
Domboshawa National Training Centre, Harare 42363
Domen Research Institute, Sankt-Peterburg 23819
Domingos Soares Ferreira Pena, Belém 03271
Dominican College, Blauvelt 30442
Dominican College, Washington 35406
Dominican College Library, Ottawa 03810
Dominican Cultural Centre Library, Helsinki 07410
Dominican Institute for Oriental Studies in Cairo, Cairo 07130
Dominican University, River Forest 32451
Dominican University of California, San Rafael 32602
Dominikanerinnenkloster Maria Zuflucht, Weesen 27277

Dominikanerinnenkloster St. Katharina, Wil 27278
Dominikanerkonvent, Wien 01484
Dominikanerkonvent St. Joseph, Düsseldorf 11201
Dominikanerorden, Graz 01429
Domkirkeodden Foundation, Hamar 20227
Domstiftsarchiv, Brandenburg an der Havel 11179
Domstolsverket, Jönköping 26491
Domus Galilaeana, Pisa 16421
Domus Mazziniana Institute, Pisa 16422
Don Agricultural Institute, Persianovka 22475
Don Bosco Gymnasium Unterwaltersdorf, Ebreichsdorf-Unterwaltersdorf 01370
Don Bosco Library, Oud-Heverlee 02483
Don Bosco Technical Institute, Rosemead 33704
Don Boscobibliotheek, Oud-Heverlee 02483
Don Scientific Research Institute of Agriculture, Rassvet 23715
Don Severino Agricultural College, Cavite 20745
Don State Technical University, Rostov-na-Donu 22486
Donald E. O'Shaughnessy Library, San Antonio 32565
Donald F. & Mildred Topp Othmer Library of Chemical History, Philadelphia 37344
Donaldson & Burkinshaw's Library, Singapore 24607
Donau-Universität Krems, Krems 01244
Donbas Institute of Mining and Metallurgy, Alchevsk 27951
Doncaster College, Doncaster 29081
Doncaster Metropolitan District Council, Doncaster 29969
Doncaster Royal Infirmary, Doncaster 29645
Donegal County Library, Letterkenny 14578
Donemus, Amsterdam 19354
Donetsk City Centralized Adult Library System, Main Library, Donetsk 28547
Donetsk City Centralized Children's Library System, Main Library, Donetsk 28546
Donetsk Institute of Commerce, Donetsk 28216
Donetsk Institute of Internal Affairs, Donetsk 28098
Donetsk Mining Engineering Industrial Corporation, Donetsk 28142
Donetsk national University, Donetsk 27973
Donetsk National University of Economy and Trade named after M. Tugan-Baranovsky, Donetsk 27971
Donetsk Railway, Railway Scientific and Technical Library, Donetsk 28215
Donetsk State Medical University, Donetsk 27972
Donetska miska TsBS dlya ditei, Tsentralna biblioteka, Donetsk 28546
Donetska miska TsBS dlya doroslikh, Tsentralna biblioteka, Donetsk 28547
Donetska Zaliznitsa, Donetsk 28215
Donetsky komertsiny instytut, Donetsk 28216
Donetsky natzionalny universytet, Donetsk 27973
Dong-A University, Pusan 18071
Dongbei University of Finance and Economics, Dalian 05153
Dongduk Women's University, Seoul 18078
Dongguan Institute of Technology, Dongguan 05167
Dongguk University, Seoul 18079
Donghua University, Shanghai 05498
Donnelly College, Kansas City 33450
Donner Institute, Steiner Memorial Library, Turku 07546
Donnerska institutet, Turku 07546
Donskaya gosudarstvennaya publichnaya biblioteka, Rostov-na-Donu 24205
Donskoi gosudarstvenny tekhnicheski universitet, Rostov-na-Donu 22486
Donskoi selskokhozyaistvenny institut, Persianovka 22475
Donskoi zonalny nauchno-issledovatelski institut selskogo khozyaistva, Rassvet 23715
Dookie College, Dookie College 00470
Door County Library, Sturgeon Bay 41459
Dopravoprojekt, Bratislava 24674
Dorchester County Library, Saint George 41109
Dorchester County Public Library, Cambridge 38420
Dordrechts Museum aan de Haven, Dordrecht 19419
Dordt College, Sioux Center 32674
Dorothea Dix Hospital, Raleigh 37424
Dorothy Bramlage Public Library, Junction City 39663
Dorothy W. Quimby Library, Unity 41641
Dorozhnaya nauchno-tekhnicheskaya biblioteka, Sankt-Peterburg 22763

Dorozhny nauchno-issledovatelski institut, Balashikha 22994
Dorset County Council, Dorchester 29970
Dorset County Museum, Dorchester 29647
Dorsey & Whitney, Minneapolis 35904
Dorte-Hilleke-Bücherei Menden, Menden 12849
Dorval Library, Dorval 04627
Doryokura Kakunenryo Kaihatsu Jigyodan Tokai Jigyosho, Naka 17396
Doshisha Daigaku, Kyoto 16998
– Hogakubu Kenkyushitsu, Kyoto 16999
– Jimbun Kagaku Kenkyujo, Kyoto 17000
Doshisha Woman's College of Liberal Arts, Kyoto 17001
Dostoevsky Memorial House-Museum, Sankt-Peterburg 23759
Dougherty County Public Library, Albany 37899
Douglas College, New Westminster 04011
Douglas County District Court, Omaha 34636
Douglas County Law Library, Roseburg 34757
Douglas County Law Library, Superior 34904
Douglas County Libraries, Castle Rock 38478
Douglas County Libraries – Highlands Ranch Library, Highlands Ranch 39433
Douglas County Libraries – Lone Tree Library, Lone Tree 39921
Douglas County Library, Alexandria 37906
Douglas County Library System, Roseburg 41063
Douglas County Public Library, Minden 40228
Douglas Mawson Institute of Technology, Croydon Park 00566
Dover Public Library, Dover 38857
Dover Public Library, Dover 38858
Dover Public Library, Dover 38859
Dover Town Library, Dover 38860
Dow Agrosciences, Indianapolis 36955
Dow Benelux NV, Terneuzen 19337
The Dow Chemical Co, Midland 35890
Dow Chemical Library, Freeport 35714
Dow, Lohnes & Albertson, Washington 36348
Dow Olefinverbund GmbH, Böhlen 11377
Dowa Metals & Mining Co., Ltd, Tokyo 17443
Dowling College, Oakdale 32170
Downers Grove Public Library, Downers Grove 38861
Downing College, Cambridge 29028
Downs Rachlin Martin PLLC, Saint Johnsbury 36193
DPI Forest Service Library, Brisbane 00878
Dr. Babasaheb Ambedkar Marathwada University, Aurangabad 13607
Dr. Baquir's Library, Lahore 20508
Dr. Bhim Rao Ambedkar University, Agra 13592
Dr. Friedrich Teßmann Library, Bozen 14789
Dr. Harisingh Gour University, Sagar 13757
Dr. Ing. h.c. F. Porsche AG, Weissach 11434
Dr. Pusey Memorial Library, Oxford 29873
Dr. Samuel L. Bossard Memorial Library, Gallipolis 39187
Dr. Sun Yat-sen Library, Taipei 27619
Dr. von Haunersches Kinderspital, München 10435
Dr. Williams's Library, London 29557
Dr. Willmar Schwabe GmbH & Co. KG, Karlsruhe 11412
Dr. Yashwant Singh Parmar University of Horticulture and Forestry, Nauni-Solan 13720
Dr. Zakir Husain Library, New Delhi 13730
DRAC – Alsace, Strasbourg 08682
DRAC Lorraine, Service régional de l'Archéologie, Metz 08415
M. Dragomanov National Pedagogical University, Kyiv 28021
Dragonskolans bibliotek, Umeå 26483
Drainage and Irrigation Department, Kuala Lumpur 18553
Drake University, Des Moines 31000
– Drake Law Library, Des Moines 31001
Drama Corner, Helsinki 07476
Drama School attached to the Yaroslavl F.G. Volkov State Academic Theatre, Yaroslavl 22654
Drammens Museum, Drammen 20221
Drents Archief / Drents Museum, Assen 19397
Drents Museum, Assen 19397
Dressing Integrated works Research Library, Marganets 28165
Drew University, Madison 31708
Drexel University, Philadelphia 32253
– Queen Lane Library, Philadelphia 32254
Bibliothek der Dr.-Hanns-Simon-Stiftung, Bitburg 12677
Driftwood Public Library, Lincoln City 39881
Drinker, Biddle & Reath, Chicago 35572
Drinker Biddle & Reath, Florham Park 35710
Drinker Biddle & Reath LLP, Philadelphia 36095
Drinker, Biddle & Reath LLP Library, Princeton 36154
Drogichin Regional Central Library, Drogichin 02118
Drogichinskaya tsentralnaya raionnaya bibliyateka, Drogichin 02118
Drogobich City Central Library, Drogobych 28550
Drogobich Regional Centralized Library System, Main Library, Drogobych 28551
Drogobitska miska TsBS, Tsentralna biblioteka, Drogobych 28550
Drogobitska raionna TsBS, Tsentralna biblioteka, Drogobych 28551
Dronning Mauds Minne, Trondheim 20170
Drug and Alcohol Services South Australia, Parkside 00975
Drug-Arm Resource Centre, Brisbane 00879
Drury University, Springfield 32703
Druskininkų savivaldybės viešoji biblioteka, Druskininkai 18320
Dr.-Wilfried-Haslauer-Bibliothek, Salzburg 01565
Državni Arhiv Sreza Valjevo, Valjevo 24533
Državni Arhiv u Zadru, Zadar 06229
Državni hidrometeorološki zavod, Zagreb 06233
Državni Zavod za Statistiku, Zagreb 06207
Državni Zbor Republike Slovenije, Ljubljana 24891
DSB Bibliotek og Fotoarkiv, København 06793
DSI – Danish Institute for Health Sevices Research, København 06794
DSI – Institut for Sundhedsvæsen, København 06794
DSIR Applied Mathematics Division, Wellington 19846
DSM Fine Chemicals Nfg GmbH & Co KG, Linz 01502
E. I. du Pont Canada Co, Mississauga 04263
EI Du Pont De Nemours & Co, Inc, Philadelphia 36096
Duale Hochschule Baden-Württemberg Heidenheim, Heidenheim 10038
Duale Hochschule Baden-Württemberg Lörrach, Lörrach 10248
Duale Hochschule Baden-Württemberg Mannheim, Mannheim 10292
Duale Hochschule Baden-Württemberg Mosbach, Mosbach 10356
Duale Hochschule Baden-Württemberg Ravensburg, Ravensburg 10537
Duale Hochschule Baden-Württemberg Ravensburg, Campus Friedrichshafen, Friedrichshafen 09808
Duale Hochschule Baden-Württemberg Stuttgart, Stuttgart 10591
Duale Hochschule Baden-Württemberg Stuttgart, Campus Horb, Horb 10048
Duale Hochschule Baden-Württemberg Villingen Schwenningen, Villingen-Schwenningen 10681
Duane Morris LLP Library, Philadelphia 36097
Dubai Municipality Public Libraries, Dubai 28910
Dublin Business School, Dublin 14466
Dublin City Public Libraries, Dublin 14571
Dublin City University, Dublin 14467
Dublin Diocesan Library, Dublin 14509
Dublin Institute for Advanced Studies – School of Celtic Studies, Dublin 14556
Dublin Institute for Advanced Studies – School of Theoretical Physics, Dublin 14475
Dublin Institute of Technology – Aungier St. Library, Dublin 14468
Dublin Institute of Technology – Bolton St. Library, Dublin 14469
Dubno District Centralized Library System, Main Library, Dubno 28552
Dubrovenskaya tsentralnaya raionnaya bibliyateka, Dubrovno 02119
Dubrovno Regional Central Children's Library, Dubrovno 02119
Duchesne County Library, Roosevelt 41061
Dudley College of Technology, Dudley 29082
Dudley Metropolitan Borough Council, Dudley 29971
Dudley Observatory Library, Schenectady 37586
Duiliu Zamfirescu County Library, Focşani 22090
Duke Energy Corp – David Nabow Library, Charlotte 35562
Duke University, Durham 31037
– Divinity School Library, Durham 31038
– Duke University Marine Laboratory, Beaufort 31039
– Ford Library, Durham 31040
– Fuqua School of Business, Durham 31041
– Law School Library, Durham 31042
– Lilly Library, Durham 31043
– Medical Center Library, Durham 31044
– Music Library, Durham 31045
Duke University – Nicholas School of the Environment and Earth Sciences, Beaufort 30366
Duksung Women's University, Seoul 18080
Duluth Public Library, Duluth 38871
Duluth Public Library – Mount Royal, Duluth 38872
Dumfries and Galloway College, Dumfries 29083
Dumfries and Galloway Libraries, Dumfries 29973
Dun Laoghaire/Rathdown County Council Public Library Service, Blackrock 14563
Dunaivetska raionna TsBS, Tsentralna biblioteka, Dunaivtsi 28553
Dunaivtsi Regional Centralized Library System, Main Library, Dunaivtsi 28553
Dunántuli Református Egyházkerület, Pápa 13342
Duncan Public Library, Duncan 38874
Duncanville Public Library, Duncanville 38875
Dundalk Institute of Technology, Dundalk 14482
Dundas Public Library, Dundas 04630
Dundee City Council, Dundee 29974
Dundee College of Further Education, Dundee 29464
Dundee Township Public Library District, Dundee 38876
Dunedin College of Education & Otago Polytechnic Joint Library, Dunedin 19776
Dunedin Public Library, Dunedin 19864
Dunedin Public Library, Dunedin 38877
Dunfermline Carnegie Library, Dunfermline 29975
Dunham Public Library, Whitesboro 41882
Dunkirk Free Library, Dunkirk 38878
Dunklin County Library, Kennett 39700
Duplin County Library, Faison 39033
Duplin County Library, Kenansville 39695
DuPont Co, Wilmington 36421
DuPont Company, Deepwater 35663
E. I. DuPont de Nemours & Co, Inc, Newark 36039
DuPont De Nemours & Co, Inc, Waynesboro 36404
Duquesne University, Pittsburgh 32300
– Duquesne University School of Law, Pittsburgh 32301
Durango Public Library, Durango 38879
Durban Museum and Art Gallery, Durban 25119
Durban University of Technology, Durban
– M.L. Sultan Campus, Durban 25014
Durham Cathedral Library, Durham 29540
Durham Clayport Library, Durham 29976
Durham College, Oshawa 04014
Durham County Council, Durham 29976
Durham County Library, Durham 38881
Durham Public Library, Durham 38882
Durham Technical Community College, Durham 33304
Durham University, Durham 29087
– Education Library, Durham 29088
– Palace Green Library, Durham 29089
Duta Wacana Christian University, Yogyakarta 14324
Dutch Reformed Church Archives – Pretoria, Pretoria 25085
Dutch Reformed Theological School Stofberg, Witsieshoek 25087
Dutch Refugee Council, Amsterdam 19387
Dutchess Community College, Poughkeepsie 33662
Duval County Law Library, Jacksonville 34391
Duxbury Free Library, Duxbury 38883
'Dvir Bialik' Municipal Central Public Library, Ramat-Gan 14781
Dvorets kultury im. A. M. Gorkogo, Sankt-Peterburg 24226
Dvorets kultury im. N. K. Krupskoi, Sankt-Peterburg 24227
Dvorets kultury im. S. M. Kirova, Sankt-Peterburg 24228
Dvorets kultury metallurgov Kuznetskogo metallurgicheskogo kombinata, Novokuznetsk 23633
Dvorets pionerov im. A. A. Zhdanova, Sankt-Peterburg 24229
Dvorichanska raionna TsBS, Tsentralna biblioteka, Dvorichna 28554
Dvorichna Regional Centralized Library System, Main Library, Dvorichna 28554
Dwight Foster Public Library, Fort Atkinson 39101
Dyal Singh Trust Library, Lahore 20529
Dyer Library, Saco 41090
Dyersburg State Community College, Dyersburg 33305
Dykema Gossett PLLC, Chicago 35573
Dykema Gossett PLLC, Detroit 35675
Dynamics Technology, Inc Library, Torrance 36308
Dynamo Joint Stock Company, Moskva 22830
D'Youville College, Buffalo 30590
Dzerzhinsk Regional Central Library, Volozhin 02252
Dzerzhinsk Regional Centralized Library System, Main Library, Dzerzhinsk 28556
Dzerzhinsk Regional Centralized Library System of Kharkiv City, Main Library, Kharkiv 28600
Dzerzhinsk Regional Library, Dzerzhinsk 28555
Dzerzhinska raionna biblioteka, Dzerzhinsk 28555
Dzerzhinska raionna TsBS mista Kharkiva, Tsentralna biblioteka, Kharkiv 28600
Dzerzhinska raionna TsBS, Tsentralna biblioteka, Dzerzhinsk 28556
Dzerzhinskaya tsentralnaya raionnaya bibliyateka, Volozhin 02252
Dzerzhinsky District Children's Library-Branch no 1, Sankt-Peterburg 24239
Dzerzhinsky, Kuibyshev, Smolny and Oktyabrsky Inter-District Centralized Library System, A.A. Blok Central Library, Sankt-Peterburg 24240
Dzerzhinsky Metallurgical Plant, Kosaya Gora 22811
Dzerzhinsky Military Academy, Moskva 22400
Dzerzhinsky Mine, Trade Union Library, Dzerzhinsk 28557
Dzhankoi Regional Centralized Library System, Main Library, Dzhankoi 28558
Dzhankoiska raionna TsBS, Tsentralna biblioteka, Dzhankoi 28558
I. A. Dzhavakhishvili Institute of History, Archeology and Ethnography, Tbilisi 09232
E. A. Hornel Library, Kirkcudbright 29692
E. C. Scranton Memorial Library, Madison 40010
E. I. DuPont de Nemours & Co, Inc, Newark 36039
E. W. McMillan Library at Cascade College, Portland 32338
EAA Aviation Foundation, Oshkosh 37313
EADA, Barcelona 25294
EADS Deutschland GmbH, Immenstaad 11409
EAE Business School, Barcelona 25897
Eagle Public Library, Eagle 38885
Eagle Valley Library District – Avon Public Library, Avon 38050
Eagle Valley Library District – Eagle Public Library, Eagle 38886
Ealing Hammersmith and West London College, London 29477
Earlham College, Richmond 32434
East African Community Secretariat, Arusha 27665
East African Institute for Meteorological Training and Research, Nairobi 18035
East African School of Library and Information Science, Kampala 27912
East African Statistical Department, Nairobi 18021
East Albemarle Regional Library, Elizabeth City 38952
East Alton Public Library District, East Alton 38888
East Ayrshire Council, Kilmarnock 30002
East Baton Rouge Parish Library, Baton Rouge 38105
East Baton Rouge Parish Library – Baker Branch, Baker 38057
East Baton Rouge Parish Library – Bluebonnet Regional, Baton Rouge 38106
East Baton Rouge Parish Library – Central, Baton Rouge 38107
East Baton Rouge Parish Library – Delmont Gardens, Baton Rouge 38108
East Baton Rouge Parish Library – Greenwell Springs Road Regional, Baton Rouge 38109
East Baton Rouge Parish Library – Jones Creek Regional, Baton Rouge 38110
East Baton Rouge Parish Library – River Center, Baton Rouge 38111
East Baton Rouge Parish Library – Scotlandville, Baton Rouge 38112
East Baton Rouge Parish Library – Zachary Branch, Zachary 42003

East Bohemian Museum, Hradec Králové 06451
East Bohemian Museum, Pardubice 06471
East Bonner County Free Library District, Sandpoint 41203
East Bridgewater Public Library, East Bridgewater 38889
East Brunswick Public Library, East Brunswick 38890
East Carolina University, Greenville 31259
– Music Library, Greenville 31260
– William E. Laupus Health Sciences Library, Greenville 31261
East Central College, Union 33815
East Central Community College, Decatur 33287
East Central Georgia Regional Library – Augusta-Richmond County Public Library, Augusta 38032
East Central Georgia Regional Library – Columbia County Public Library, Evans 39000
East Central Georgia Regional Library – Euchee Creek Library, Grovetown 39330
East Central Georgia Regional Library – Jeff Maxwell Branch, Augusta 38033
East Central Regional Library, Cambridge 38421
East Central University, Ada 30167
East Central Wisconsin Regional Planning Commission Library, Menasha 34526
East Chicago Public Library, East Chicago 38891
East Chicago Public Library – Robert A. Pastrick Branch, East Chicago 38892
East China Institute of Metallurgy, Maanshan 05414
East China Institute of Political Science and Law, Shanghai 05499
East China Institute of Technology, Fuzhou 05177
East China Jiaotong University, Nanchang 05422
East China Normal University, Shanghai 05500
East China University of Science and Technology, Shanghai 05501
East Cleveland Public Library, East Cleveland 38893
East Coast Bays City Library, Browns Bay 19861
East Dunbartonshire Council, Kirkintilloch 30007
East Dunbartonshire Libraries, Glasgow 29984
East Georgia College, Swainsboro 33786
East Gippsland Institute of TAFE, Bairnsdale 00383
East Gippsland Shire, Bairnsdale 01061
East Greenbush Community Library, East Greenbush 38896
East Greenwich Free Library, East Greenwich 38897
East Hampton Library, East Hampton 38898
East Hampton Public Library, East Hampton 38899
East Hanover Township Free Public Library, East Hanover 38900
East Hartford Public Library, East Hartford 38901
East Islip Public Library, East Islip 38903
East Jersey State Prison Library, Rahway 34719
East Kazakhstan Research Library, Ust-Kamenogorsk 17980
East Kazakhstan Technical University, Ust-Kamenogorsk 17926
East Lansing Public Library, East Lansing 38904
East London Museum, East London 25123
East Longmeadow Public Library, East Longmeadow 38906
East Los Angeles College, Monterey Park 33554
East Lothian Library Services, Haddington 29990
East Lyme Public Library, Inc, Niantic 40478
East Mailing Research Library, East Malling 29650
East Meadow Public Library, East Meadow 38907
East Mississippi Regional Library System, Quitman 40922
East Moline Public Library, East Moline 38908
East Orange Public Library, East Orange 38909
East Providence Public Library, East Providence 38912
East Providence Public Library – Fuller, East Providence 38913
East Providence Public Library – Riverside, East Providence 38914
East Providence Public Library – Rumford, East Providence 38915
East Renfrewshire Libraries and Infromation Service, Barrhead 29935
East Rockaway Public Library, East Rockaway 38917
East Saint Louis Public Library, East Saint Louis 38918
East Smithfield Public Library, Smithfield 41334
East Stroudsburg University, East Stroudsburg 31067
East Surrey College, Redhill 29384
East Sussex County Council, Lewes 30011
East Tennessee State University, Johnson City
– James H. Quillen College of Medicine Library, Johnson City 31465
– Sherrod Library, Johnson City 31466
East Texas Baptist University, Marshall 31766
East Ukrainian State University, Lugansk 28031
Eastchester Public Library, Eastchester 38920
Eastern and Southern African Management Institute, Arusha 27649
Eastern Arizona College, Thatcher 33798
Eastern Connecticut State University, Willimantic 33072
Eastern Correctional Facility Library, Napanoch 34571
Eastern Correctional Institution, Westover 35061
Eastern Counties Regional Library, Mulgrave 04697
Eastern Illinois University, Charleston 30716
Eastern Kentucky Correctional Complex Library, West Liberty 35056
Eastern Kentucky University, Richmond 32435
– Justice & Safety Library, Richmond 32436
– Music Library, Richmond 32437
Eastern Mediterranean University, Gazi Mağusa 06341
Eastern Mennonite University, Harrisonburg 31301
Eastern Metropolitan Libraries, Sandton 25233
Eastern Michigan University, Ypsilanti 33109
Eastern Monroe Public Library, Stroudsburg 41456
Eastern Nazarene College, Quincy 32403
Eastern New Mexico University, Portales 32333
Eastern Oklahoma District Library System, Muskogee 40354
Eastern Oklahoma State College, Wilburton 33061
Eastern Oregon University, La Grande 31526
Eastern Regional Libraries Corporation, Wantirna South 01174
Eastern Research Mining and Metallurgical Institute of Non-ferrous Metals, Ust-Kamenogorsk 17981
Eastern Shore Public Library, Accomac 37878
Eastern State Hospital, Williamsburg 37842
Eastern TAFE Melbourne, Wantirna South 00591
Eastern University, Saint Davids 32498
Eastern Virginia Medical School, Norfolk 32120
Eastern Washington State Historical Society, Spokane 37622
Eastern Washington University, Cheney 30741
Eastern Wyoming College, Torrington 32829
Eastfield College, Mesquite 33534
East-Kazakhstan State University named after Amanzholov, Ust-Kamenogorsk 17927
Eastman Chemical Co, Kingsport 35806
Easton Area Public Library & District Center, Easton 38922
Easton Public Library, Easton 38923
Eastpointe Memorial Library, Eastpointe 38925
East-Siberian State Academy of Culture and Arts, Ulan-Ude 22615
East-Siberian Technological Institute, Ulan-Ude 22616
East-West Center, Honolulu 36932
Eau Gallie Public Library, Melbourne 40136
Ebara Corporation, Tokyo 17444
Ebara Seisakujo, Tokyo 17444
Ebeltoft Bibliotek, Ebeltoft 06855
Eberhard-Alexander-Burgh-Bibliothek, Berlin 12656
Eberhard-Karls-Universität Tübingen, Tübingen 10625
– Bereichsbibliothek Biologie, Tübingen 10626
– Bereichsbibliothek Geowissenschaften, Tübingen 10627
– Bereichsbibliothek Naturwissenschaften, Tübingen 10628
– Bibliothek für Ägyptologie, Altorientalistik und Archäologie des Mittelmeerraumes – Bereichsbibliothek Schloss-Nord, Tübingen 10629
– Bibliothek Theologicum, Evangelisch-Theologisches Seminar, Tübingen 10630
– Bibliothek Theologicum, Katholisch-Theologisches Seminar, Tübingen 10631
– Fachbibliothek Japanologie, Tübingen 10632
– Fachbibliothek Mathematik und Physik / Bereich Mathematik, Tübingen 10633
– Fachbibliothek Mathematik und Physik / Bereich Physik, Tübingen 10634
– Fachbibliothek Wirtschaftswissenschaft, Tübingen 10635
– Geographisches Institut, Tübingen 10636
– Historisches Seminar, Abteilung für Neuere Geschichte, Tübingen 10637
– Institut für Arbeits- und Sozialmedizin, Tübingen 10638
– Institut für Astronomie und Astrophysik, Abteilung Astronomie, Tübingen 10639
– Institut für Erziehungswissenschaft, Tübingen 10640
– Institut für Ethik und Geschichte der Medizin / Bereich Geschichte der Medizin, Tübingen 10641
– Institut für Ethnologie, Tübingen 10642
– Institut für Geschichtliche Landeskunde und Historische Hilfswissenschaften, Tübingen 10643
– Institut für Kriminologie, Tübingen 10644
– Institut für Medizinische Biometrie, Tübingen 10645
– Institut für Osteuropäische Geschichte und Landeskunde, Tübingen 10646
– Institut für Pharmakologie und Toxikologie, Tübingen 10647
– Institut für Politikwissenschaft, Tübingen 10648
– Institut für Soziologie, Tübingen 10649
– Institut für Sportwissenschaft, Tübingen 10650
– Institut für Ur- und Frühgeschichte und Archäologie des Mittelalters, Tübingen 10651
– Institut für Ur- und Frühgeschichte und Archäologie des Mittelalters, Abteilung für Ältere Urgeschichte und Quartärökologie, Tübingen 10652
– Internationales Zentrum für Ethik in den Wissenschaften, Tübingen 10653
– Juristisches Seminar, Tübingen 10654
– Kunsthistorisches Institut, Tübingen 10655
– Ludwig-Uhland-Institut für Empirische Kulturwissenschaft, Tübingen 10656
– Medizinbibliothek Tübingen, Tübingen 10657
– Musikwissenschaftliches Institut, Tübingen 10658
– Orientalisches Seminar, Tübingen 10659
– Philologisches Seminar, Tübingen 10660
– Philosophische Bibliothek, Tübingen 10661
– Philosophische Fakultät, Mittelalterliche Geschichte, Tübingen 10662
– Philosophisches Seminar, Tübingen 10663
– Psychologisches Institut, Tübingen 10664
– Seminar für Alte Geschichte, Tübingen 10665
– Seminar für Sinologie und Koreanistik, Arbeitsbereich Sinologie, Tübingen 10666
– Seminar für Sinologie und Koreanistik, Sektion Koreanistik, Tübingen 10667
– Seminar für Zeitgeschichte, Tübingen 10668
– Universitäts-Augenklinik, Tübingen 10669
– Universitäts-Hals-Nasen-Ohrenklinik, Tübingen 10670
– Universitäts-Hautklinik, Tübingen 10671
– Universitätsklinik für Psychiatrie und Psychotherapie, Tübingen 10672
– Wilhelm-Schickard-Institut für Informatik, Tübingen 10673
– Zentrum für Datenverarbeitung, Tübingen 10674
– Zentrum für Zahn-, Mund- und Kieferheilkunde, Tübingen 10675
Ecclesiastical Seminary, Pelplin 21056
Ecclesiastical Seminary, Płock 21057
Eckerd College, Saint Petersburg 32544
Eckhart Public Library, Auburn 38030
ECO-Archiv im AROEK e.V., Hofgeismar 12064
Ecole Biblique et Archéologique Française, Jerusalem 14698
Ecole Catholique d'Arts et Métiers (ECAM), Lyon 07739
Ecole Centrale Paris (ECP), Châtenay-Malabry 08123
Ecole d'Aplication du Service de Santé des Armées, Paris 07856
Ecole d'Application de l'Artillerie (EAA), Draguignan 08124
Ecole d'Architecture de Grenoble, Grenoble 07709
Ecole d'Architecture de Lille et Régions Nord, Villeneuve-d'Ascq 08103
Ecole d'Architecture de Normandie, Darnetal 07697
Ecole d'Architecture de Paris la Villette, Paris 07857
Ecole d'Architecture de Paris-Belleville, Paris 07858
Ecole d'Architecture et de Paysage de Bordeaux, Talence 08688
Ecole d'Architecture Languedoc-Rousillon, Montpellier 07765
Ecole de la Cause Freudienne, Paris 07859
Ecole de l'Air, Salon-de-Provence 08040
Ecole de Management de Lyon (E.M.Lyon), Ecully 07704
Ecole de Technologie Supérieure, Montreal 03770
Ecole des Beaux-Arts, Bordeaux 07657
Ecole des Beaux-Arts de Lyon, Lyon 08395
Ecole des Beaux-Arts de Toulouse, Toulouse 08073
Ecole des Beaux-Arts de Tunis, Tunis 27821
École des Hautes Études en Sciences Sociales – CNRS, Paris
– Centre Asie du Sud-Est – Fonds Archipel, Paris 07860
– Centre de Recherches sur le Japon, Paris 07861
– Centre d'Études Africaines, Paris 07862
– Centre d'Études de l'Inde et de l'Asie du Sud, Paris 07863
– Centre d'Études des Mondes Russes, caucasien et centre européen, Paris 07864
– Centre d'Études sur la Chine Moderne et Contemporaine, Paris 07865
Ecole des Mines de Douai, Douai 07701
Ecole des Sciences de l'Information, Rabat 18935
Ecole du Breuil, Paris 08130
Ecole du Louvre, Paris 07866
Ecole du service de santé des armées de Lyon, Bron 07667
École du Service de Santé des Armées (ESSA), Bordeaux 07658
École européene de Luxembourg, Luxembourg 18388
Ecole Française d'Athènes, Athinai 13074
Ecole française de Rome, Roma 16496
Ecole Française d'Extrême-Orient, Paris 07867
Ecole Hassania des Travaux Publics, Casablanca 18932
Ecole hôtelière de Lausanne, Lausanne 27096
Ecole Militaire, Paris 07868
Ecole Militaire de Santé, Dakar 24420
Ecole Mohammadia d'Ingénieurs, Rabat 18936
Ecole National Supérieur d'Agriculture, Thiès 24418
Ecole Nationale d'Administration, Alger 00049
Ecole Nationale d'Administration, Niamey 19907
Ecole Nationale d'Administration, Nouakchott 18682
Ecole Nationale d'Administration, Strasbourg 08044
Ecole Nationale d'Administration (ENA), Abidjan 06128
Ecole Nationale d'Administration (ENA), Strasbourg 08045
Ecole Nationale d'Administration (ENA), Strasbourg 08046
Ecole Nationale d'Administration (ENA), Tunis 27816
Ecole Nationale d'Administration et de Magistrature, Yaoundé 03635
Ecole Nationale d'Administration et de Magistrature (ENAM), Yaoundé 03641
Ecole Nationale d'Administration Pénitentiaire (ENAP), Agen 07626
Ecole Nationale d'Administration Publique, Montreal 03771
Ecole Nationale d'Administration Publique, Québec 03825
Ecole Nationale d'Administration Publique (ENAP), Rabat 18924
Ecole Nationale de la Santé Publique (ENSP), Rennes 08020
Ecole Nationale de la Statistique et de l'Administration Économique, Malakoff 07748

Ecole Nationale de l'Aviation Civile (ENAC), Toulouse 08074
Ecole Nationale de l'Industrie Minérale, Rabat 18937
Ecole Nationale des Chartes, Paris 07869
Ecole Nationale des Ingenieurs de Gabes, Gabes 27814
Ecole Nationale des Ponts et Chaussées, Marne-la-Vallée 07751
Ecole Nationale des Travaux Publics de l'Etat (ENTPE), Vaulx-en-Velin 08100
Ecole Nationale d'Ingenieurs des Travaux Agricoles de Bordeaux (ENITAB), Gradignan 07708
Ecole Nationale d'Ingénieurs des Travaux Agricoles de Clermont-Ferrand, Lempdes 07724
Ecole Nationale Polytechnique, Alger 00026
Ecole nationale supérieure d'architecture, Marseille 07753
École Nationale Superieure d'Architecture de Clermont-Ferrand (EACF), Clermont-Ferrand 07685
Ecole Nationale Superieure d'Architecture de Nancy (ENSA), Nancy 07784
Ecole nationale supérieure d'Architecture de Nantes, Nantes 07804
Ecole Nationale Supérieure d'Architecture de Paris-Malaquais, Paris 07870
Ecole Nationale Superieure d'Architecture de Strasbourg, Strasbourg 08047
Ecole nationale supérieure d'architecture de Toulouse, Toulouse 08075
Ecole Nationale supérieure d'Architecture de Versailles, Versailles 08102
Ecole Nationale Supérieure d'Architecture Paris Val de Seine, Paris 07871
Ecole nationale supérieure d'Art (ENSA), Nancy 07785
Ecole Nationale Supérieure d'Arts et Métiers (ENSAM), Aix-en-Provence 07627
Ecole Nationale Supérieure de Géologie, Vandœuvre-les-Nancy 08097
Ecole Nationale Superieure de la Photographie, Arles 07643
Ecole Nationale Supérieure de la Police (ENSP), Saint-Cyr-au-Mont-d'Or 08032
Ecole Nationale Supérieure de Techniques Avancées (ENSTA), Paris 07872
Ecole Nationale Supérieure des Arts Décoratifs, Paris 07873
Ecole Nationale Supérieure des Beaux-Arts (ENSBA), Paris 07874
Ecole nationale supérieure des mines, Nancy 07786
Ecole Nationale Supérieure des Mines de Paris, Fontainebleau 07705
Ecole Nationale Supérieure des Mines de Paris (ENSMP), Paris 07875
Ecole Nationale Supérieure des Mines de Saint-Etienne, Saint-Etienne 08035
Ecole Nationale Supérieure des Sciences de l'Information et des Bibliothèques (ENSSIB), Villeurbanne 08115
Ecole Nationale Supérieure des Techniques Industrielles et des Mines d'Alès, Alès 07632
Ecole Nationale Supérieure des Télécommunications de Bretagne, Brest 08311
Ecole Nationale Supérieure des Télécommunications (ENST), Paris 07876
Ecole Nationale Supérieure des Travaux Publics, Yamoussoukro 06131
Ecole Nationale Supérieure d'Ingénieurs en Constructions Aéronautiques (ENSICA), Toulouse 08076
Ecole Nationale Vétérinaire, Alger 00027
Ecole Nationale Vétérinaire, Toulouse 08077
Ecole Nationale Vétérinaire d'Alfort, Maisons-Alfort 07747
Ecole Normale d'Instituteurs de Seine-Maritime, Mont-Saint-Aignan 08126
Ecole Normale Supérieure, Atakpamé 27796
Ecole Normale Supérieure (ENS), Paris 07877
Ecole Normale Supérieure de Cachan, Cachan 07675
Ecole Normale Supérieure de Lyon, Lyon 07740
Ecole Normale Supérieure (ENS), Alger 00028
Ecole Polytechnique, Palaiseau 07835
Ecole Polytechnique de Montréal, Montreal 03772
Ecole Polytechnique Fédérale de Lausanne, Lausanne 27097
Ecole pour la Formation de Spécialistes de la Faune, Garoua 03633
École Pratique des Hautes Études, Paris
– Bibliothèque Michel Fleury – Sciences historiques et philologiques, Paris 07878
– Section Sciences Religieuses, Paris 07879
CDI -École Privée Fieldgen (EPF), Luxembourg 18389
École Privée Marie-Consolatrice Esch-Sur-Alzette (EPMC), Esch-sur-Alzette 18387
Ecole Régionale des Beaux-Arts, Saint-Etienne 08036
Ecole Régionale des Beaux-Arts Caen la mer, Caen 08122
Ecole Régionale des Beaux-Arts de Besançon, Besançon 07648
Ecole Régionale Supérieure d'expression plastique (ERSEP), Tourcoing 08093
Ecole Royale Militaire, Bruxelles 02282
Ecole Spéciale des Travaux Publics, du Bâtiment et de l'Industrie, Cachan 07676
Ecole Supérieure de Commerce de Paris (ESC-PARIS), Paris 07880
Ecole Supérieure de Journalisme de Lille, Lille 07726
Ecole Supérieure de Physique et de Chimie Industrielles de Paris (ESPCI), Paris 07881
Ecole Supérieure de Science et Technique, Tunis 27817
Ecole Supérieure de Sciences Economiques et Commerciales, Cergy 08318
Ecole Supérieure des Arts Appliqués Duperré, Biberach an der Riß 08121
Ecole supérieure des Arts de Liège (Académie Royale des Beaux-Arts de Liège), Liège 02373
Ecole Supérieure des arts décoratifs, Strasbourg 08048
Ecole Supérieure des Beaux-Arts, Alger 00050
Ecole supérieure des beaux-arts, Marseille 07754
Ecole Supérieure des Beaux-Arts de Montpellier Agglomération (ESBAMA), Montpellier 07766
Ecole Supérieure des Beaux-Arts de Tours, Tours 08094
Ecole Supérieure d'Ingénieurs en Electrotechnique et Electronique (ESIEE), Noisy-le-Grand 07821
Ecole Supérieure Estienne des Arts et Industries Graphiques, Paris 08131
Ecole Supérieure Nationale d'Art, Nice 07813
Ecole Supérieure Polytechnique, Thiès 24419
Ecoles de Coëtquidan, Guer 07719
Eco-Museal Research Institute, Tulcea 22067
Ecomusée de la Communauté Urbaine Le Creusot-Montceau, Le Creusot 08375
Economic and International Affairs, Kobe 17571
Economic and Social Commission for Asia and the Pacific (ESCAP), Bangkok 27777
Economic and Social Research Institute (ESRI), Dublin 14528
Economic Commission for Africa, Addis Ababa 07285
Economic Development Board, Suva 07313
Economic Institute of Maribor, Documentation Service and Special Library, Maribor 24946
Economic Planning Agency, Tokyo 17309
Economic Research Institute, Minsk 02040
Economic Research Institute, Japan Society for Promotion of Machine, Tokyo 17690
Economics Research Associates Library, Los Angeles 35832
Economics/Statistics Library, Mbabane 26369
Economie et Humanisme, Lyon 08396
Economisch Instituut voor de Bouwnijverheid (EIB), Amsterdam 19355
Economische Hogeschool Sint-Aloysius (EHSAL), Brussel 02283
Ecosystems International Inc Library, Millersville 35891
ECPI College of Technology, Virginia Beach 32935
ECRI Library, Plymouth Meeting 37391
Ector County Library, Odessa 40586
Ecumenical Institute, Bogis-Bossey 27335
Ecumenical library of Centre St Irénée, Lyon 08199
Ecumenical Theological Seminary, Detroit 35186
Eda kommunbibliotek, Charlottenberg 26752
Edenvale Community Library, Edenvale 25199
Edge Hill University, Ormskirk 29297
Edgecombe Community College, Tarboro 33794
Edgecombe County Memorial Library, Tarboro 41518
Edgehill Theological College, Belfast 29533
Edgewood College, Madison 31709
Edinboro University of Pennsylvania, Edinboro 31072
Edinburg Public Library, Edinburg 38929
Edinburgh College of Art, Edinburgh 29091
Edinen tsentăr po fizika, Sofiya 03529
Edison Electric Institute, Washington 37733
Edison State College, Fort Myers 33346
Edison State Community College, Piqua 33646
Edison Township Free Public Library, Edison 38931
Edison Township Free Public Library – Clara Barton Branch, Edison 38932
Edison Township Free Public Library – North Edison Branch, Edison 38933
Edith Cowan University, Joondalup 00446
Edith Wheeler Memorial Library, Monroe 40262
Edith-Stein-Schule, Darmstadt 10715
Edmonds Community College, Lynnwood 33508
Edmonton Police Library, Edmonton 04339
Edmonton Public Library, Edmonton 04633
Edo State Library Board, Benin City 20048
Education and Training, ACT Dept, Griffith 00927
Education Centre, Carlisle 29632
Education Department, Asmara 07214
Education Department of Western Australia, East Perth 00674
Education Directorate Library, Sharkia Governorate, Zagazig 07194
Education Library, Tokyo 17704
Education Management Corporation – AiC Library, Denver 30993
Education Queensland, South Brisbane 00721
Education Research Bureau, Yangon 19003
Education Services Centre, Harare 42372
Educational Department Negeri Pahang Darul Makmur, Kuantan 18568
Educational Documentation Centre, Khartoum 26357
Educational Documentation Library, Baghdad 14453
Educational Documentation Library, Manama 01739
Educational Information & Resource Center, Mullica Hill 37134
Educational Management Corporation, Philadelphia 37345
Educational Research Centre, Dublin 14529
Educational Research Institute, Ljubljana 24813
Educational Research Service, Alexandria 36454
Educational Television, Tel-Aviv 14759
Educational Testing Service, Princeton 37413
EDULIS – Western Cape Education Library and Information Services, Bellville 25105
Eduskunnan kirjasto, Helsinki 07393
Edward F. Barrins Memorial Library, Tucson 37687
Edward Gauche Fisher Public Library, Athens 38009
Edward Neisser Library, Chicago 36672
Edward Waters College, Jacksonville 31445
Edwards & Angell, LLP Library, Providence 36158
Edwards Angell Palmer & Dodge LLP, Boston 35520
Edwardsville Public Library, Edwardsville 38936
Edwin A. Bemis Public Library, Littleton 39902
Eesti Ajaloomuuseum, Tallinn 07236
Eesti Kunstimuuseum, Tallinn 07231
Eesti Maaülikool, Tartu 07226
Eesti Meditsiiniraamatukogu, Tallinn 07233
Eesti Muusika- ja Teatriakadeemia Raamatukogu, Tallinn 07221
Eesti Patendiraamatukogu, Tallinn 07234
Eesti Rahva Muuseumi raamatukogu, Tartu 07240
Eesti Rahvusraamatukogu, Tallinn 07219
Eesti Teatri- ja Muusikamuuseum, Tallinn 07235
Efal-Yad Tabenkin Library, Ramat-Efal 14753
Effie & Wilton Hebert Public Library, Port Neches 40859
Ege Universitesi Rektörlüğü, Bornova 27851
Egede Instituttet, Oslo 20113
Egerton University, Njoro 18009
Egészségpolitikai Szakkönyvtár, Budapest 13314
Egészségügyi Stratégiai Kutatointézet, Budapest 13373
EGIS Gyógyszergyár Rt, Budapest 13354
EGIS Pharmaceuticals Ltd, Budapest 13354
Egishe Charents State Museum of Literature and Arts, Yerevan 00350
Eglise catholique. Diocèse d'Annecy, Annecy 08174
Egon-Erwin-Kisch-Bibliothek, Berlin 12649
Egorov Carriage Works Plant, Sankt-Peterburg 22906
EGÚ Brno, a.s., Brno 06427
Egyházmegyei Könyvtár, Szombathely 13348
The Egypt Exploration Society, London 29745
The Egyptian Cabinet, Cairo 07110
Egyptian Geographical Society, Cairo 07166
Egyptian Library, Cairo 07167
Egyptian Museum, Cairo 07168
Egyptian National Agricultural Library (ENAL), Giza 07190
Egyptian Opera House, Cairo 07169
Egyptian Scientific Academy, Cairo 07170
Egyptian Society of Political Economy, Statistics and Legislation, Cairo 07171
Ehemalige Universitätsbibliothek Helmstedt, Helmstedt 09319
Ehime University, Matsuyama 17039
Eidai Co, Ltd, Osaka 17407
Eidai Sangyo K. K. Kenkyukaihatsubu Toshoshitsu, Osaka 17407
Eidgenössische Materialprüfungs- und Forschungsanstalt (EMPA), Dübendorf 27349
Eidgenössische Technische Hochschule Zürich, Zürich 27151
– Archiv für Zeitgeschichte, Zürich 27152
– Baubibliothek, Zürich 27153
– Bibliothek Erdwissenschaften, Zürich 27154
– Bibliothek Informationstechnologie und Elektrotechnik, Zürich 27155
– ETH-Bibliothek HDB, Zürich 27156
– Graphische Sammlung, Zürich 27157
– Grüne Bibliothek, Zürich 27158
– Informatikbibliothek, Zürich 27159
– Informationszentrum Chemie Biologie Pharmazie, Zürich 27160
– Institut für Geschichte und Theorie der Architektur, Zürich 27161
– Mathematik-Bibliothek, Zürich 27162
– Physik-Bibliothek, Zürich 27163
– Versuchsanstalt für Wasserbau, Hydrologie und Glaziologie, Zürich 27164
Eidgenössisches Büro für die Gleichstellung von Frau und Mann EBG, Bern 27320
Eidgenössisches Departement für Auswärtige Angelegenheiten, Bern 27234
Eidgenössisches Hochschulinstitut für Berufsbildung (EHB), Zollikofen 27150
Eidsberg bibliotek, Mysen 20375
Eidsvoll folkebibliotek, Eidsvoll 20324
Eigersund folkebibliotek, Eigersund 20325
Eijkman Institute, Jakarta 14366
Eindhoven University of Technology, Eindhoven 19111
Eisenbibliothek, Schlatt 27431
Eisenhower Public Library District, Harwood Heights 39391
Ekaterinburg Engineering Pedagogical Institute, Ekaterinburg 22236
Ekenäs bibliotek – Tammisaaren kirjasto, Ekenäs 07559
Ekenäs Public Library, Ekenäs 07559
Ekerö bibliotek, Ekerö 26755
Ekonomicheski institut, Saratov 22571
Ekonomická univerzita, Bratislava 24641
Ekonomiko-matematicheski institut RAN, Sankt-Peterburg 23760
Ekonomiky ústav SAV, Bratislava 24687
Ekonomski institut, Zagreb 06234
Ekonomski institut Maribor, Maribor 24946
Eksjö stadsbibliotek, Eksjö 26756
Eksperimentalny nauchno-issledovatelski institut metalorezhushchikh stankov, Moskva 23175
Eksperimentalny zavod nauchnogo priborostroeniya RAN, Chernogolovka 23019
El Camino College, Torrance 33803
El Dorado County Law Library, Placerville 34681
El Dorado County Library, Placerville 40811
El Dorado County Library – Cameron Park Branch, Cameron Park 38428
El Dorado County Library – South Lake Tahoe Branch, South Lake Tahoe 41366
El Paso Community College, El Paso 33314
El Paso County Law Library, Colorado Springs 34094
El Paso County Law Library, El Paso 34177
El Paso Museum of Art, El Paso 36824

El Paso Public Library, El Paso 38945
El Progreso Memorial Library, Uvalde 41650
El Reno Carnegie Library, El Reno 38946
El Segundo Public Library, El Segundo 38947
Ela Area Public Library District, Lake Zurich 39783
Elaine Ione Sprauve Library and Museum, St. John 42319
Elam & Burke PA, Boise 35514
Elbe Kliniken, Buxtehude 11705
Elbert County Public Library, Elberton 38948
Eldredge Public Library, Chatham 38529
Electoral Institute of Southern Africa, Richmond Johannesburg 25174
Electrabel Antwerpen, Antwerpen 02497
Electrabel NV, Antwerpen 02498
Electric Automation Research and Industrial Corporation, Sankt-Peterburg 23864
Electric Bulb Factory, Moskva 22835
Electric Devices Research Institute, Sankt-Peterburg 23823
Electric Equipment Central Research Institute, Sankt-Peterburg 23919
Electric Motor Plant, Mogilev 01906
Electric Power Development Co, Ltd, Tokyo 17441
Electric Power Research Institute, Scientific and technical Library, Moskva 23387
Electrical Engineering Plant, Novocherkassk 22877
Electrical Engineering University, Sankt-Peterburg 22515
Electrical Equipment for the Mining Industry Research Institute, Donetsk 28230
Electrical Equipment Plant, Gomel 01859
Electrical Lighting Industrial Corporation, Brest 01842
Electrical Standards Research Institute, Sankt-Peterburg 23822
Electricité de France, Clamart 08254
Electricity Commission Library, Hohola 20592
Electricity Corporation of New Zealand, Wellington 19847
Electricity Supply Commission, Johannesburg 25133
Electridad de Caracas, Caracas 42240
Electro-mechanical Devices Research Institute, Kyiv 28367
Electromechanics Research Institute, Moskva 23389
Electromotor Plant, Kemerovo 22803
Electronic and Telecommunications Research Institute (ETRI), Daejeon 18106
Electronic Control Systems Industrial Corporation, Moskva 23166
Electronic Devices Professional School, Kyiv 28117
Electrotechnical Laboratory, Tsukuba 17785
Electrotechnical University of Communications, Toshkent 42089
Elektrėnų savivaldybės viešoji biblioteka, Elektrėnai 18321
Elektro-lampovoi zavod, Moskva 22835
Elektromekhanichny tekhnikum zaliznichnoho transportu im. M. Ostrovskoho, Kyiv 28108
Elektrosila Electrical Engineering Industrial Corporation, Sankt-Peterburg 22931
Elektrosila Electrical Equipment Plant, Sankt-Peterburg 24311
Elektrostalski zavod tyazhelogo mashinostroeniya (EZTM), Elektrostal 22791
Elektrotekhnicheski institut, Novosibirsk 22435
Elektrotekhnicheski institut svyazi im. M.A. Bonch-Bruevicha, Sankt-Peterburg 22514
Elektrotekhnicheski universitet, Sankt-Peterburg 22515
Elektrovozostroitelny zavod, Novocherkassk 22877
Elektrozavod im. V.V. Kuibysheva, Moskva 22836
Elgin Community College, Elgin 33315
Elgin County Public Library, St. Thomas 04765
Elgin Mental Health Center Library, Elgin 36827
Eli Lilly & Co, Indianapolis 35776
Elisabeth University of Music, Hiroshima 16950
Elizabeth City State University, Elizabeth City 31080
Elizabeth Jones Library, Grenada 39321
Elizabeth Public Library, Elizabeth 38951
Elizabethton-Carter County Public Library, Elizabethton 38954
Elizabethtown College, Elizabethtown 31081
Elizabethtown Community College, Elizabethtown 33317
Elk Grove Village Public Library, Elk Grove Village 38957
Elkhart Public Library, Elkhart 38958
Elkins Park Free Library, Elkins Park 38960
Ella M. Everhard Public Library, Wadsworth 41700
Ellenville Public Library & Museum, Ellenville 38963
Elmhurst College, Elmhurst 31085
Elmhurst Hospital Center, Elmhurst 36830
Elmhurst Memorial Hospital, Elmhurst 36831
Elmhurst Public Library, Elmhurst 38965
Elmira College, Elmira 31086
Elmont Public Library, Elmont 38968
Elmwood Park Public Library, Elmwood Park 38969
Elmwood Park Public Library, Elmwood Park 38970
Elon University, Elon 31087
Elsk Regional Central Library, Elsk 02120
Elskaya tsentralnaya raionnaya bibliyateka, Elsk 02120
ELTE Bárczi Gusztáv Gyógypedagógiai Főiskolai Kar Könyvtára, Budapest 13233
Eltech Systems Corp Library, Fairport Harbor 35699
Elyria Public Library System, Elyria 38972
Elyria Public Library System – West River, Elyria 38973
Emalkhimmash Chemical Engineering Research Institute, Poltava 28451
Embajada Argentina en el Uruguay, Montevideo 42022
'Embaneft' Research Laboratory, Gurev 17932
Embassy of Australia Library, Washington 37734
Embrapa Agroindústria de Alimentos, Rio de Janeiro 03340
Embrapa Cerrados, Planaltina 03316
Embrapa Mandioca e Fruticultura, Cruz das Almas 03291
Embrapa Solos, Rio de Janeiro 03341
Embry-Riddle Aeronautical University, Daytona Beach 30973
Embry-Riddle Aeronautical University, Prescott 32358
Emek Medical Center, Afula 14712
Emergency Management Australia, Mount Macedon 00964
Emergency Medical Institute, Sofiya 03572
Emeroteca Storica, Livorno 16304
Emeroteca-Biblioteca Tucci, Napoli 16368
Emerson College, Boston 30491
Emil Cedercreutz Museum, Harjavalta 07440
Emil Cedercreutzin museo, Harjavalta 07440
Emile Vandervelde Institute, Bruxelles 02593
Emil-Krückmann-Bücherei (EKB), Marburg 12253
Emily Carr Institute of Art & Design Library, Vancouver 04538
Emma S. Clark Memorial Library, Setauket 41283
Emmaboda bibliotek, Emmaboda 26757
Emmanuel Bible College, Kitchener 04207
Emmanuel College, Boston 30492
Emmanuel College, Cambridge 29006
Emmanuel College, Franklin Springs 33351
Emmanuel & Saint Chad College, Saskatoon 03847
Emmanuel School of Religion Library, Johnson City 35246
Emmaus Bible College, Dubuque 35188
Emmaus Bible College, Epping 00766
Emmaus Public Library, Emmaus 38974
Emmet O'Neal Library, Mountain Brook 40327
Emmett Public Library, Emmett 38975
Emory & Henry College, Emory 31090
Emory University, Atlanta 30287
– Health Sciences Center Library, Atlanta 30288
– Pitts Theology Library, Atlanta 30289
Emory University School of Law, Atlanta 30290
Emporia Public Library, Emporia 38976
Emporia State University, Emporia 31091
Empresa Brasileira de Aeronáutica S/A (EMBRAER), São José dos Campos 03265
Empresa Brasileira de Pesquisa Agropecuária (EMBRAPA), Brasília 03280
Empresa de Pesquisa Agropecuaria de Minas Gerais (EPAMIG), Belo Horizonte 03252
Empresa Nacional de Petróleo (ENAP), Santiago 04914
Empresa Nacional de Residuos Radioactivos S.A. (ENRESA), Madrid 25862
Empresa Nacional de Siderurgia (ENSIDESA), Gijón 25857
Empresa Pública das Âguas Livres (EPAL), Lisboa 21786
Empresarios Agrupados AIE, Madrid 25863
Ems-Chemie AG, Domat 27285
Emschergenossenschaft, Essen 11822
Enamel Kitchenware Plant, Gomel 01860
EnCana Corporation, Calgary 04239
Encyclopaedia Britannica Inc, Chicago 35574
Endicott College, Beverly 33168
Enea C. R. Casaccia, Roma 16497
Energodar City Centralized Library System, Main Library, Energodar 28560
Energodarska miska TsBS, Tsentralna biblioteka, Energodar 28560
Energy Australia, Sydney 00833
Energy Conservation Institute, Kyiv 28344
Energy Institute, London 29746
Energy Research and Information Institute, Bucureşti 21977
ENESAD – Etablissement National d'Enseignement Supérieur Agronomique de Dijon, Dijon 07699
Enfield Town Library, Enfield 29979
Engelhard Corp, Iselin 35783
Engelhardt Astronomical Observatory, Stanitsa 23979
Engelhardt Astronomical Observatory, Zelenodolsk 24095
Engelhardt Molecular Biology Research Institute, Moskva 23290
Engineering College of Aarhus, Århus 06768
Engineering Factory of Klimovsk, Klimovsk 22809
Engineering Institute of Coimbra, Coimbra 21625
Engineering Institute of the Refrigeration Industry, Sankt-Peterburg 22546
Engineering School for Sewing Industry, Sankt-Peterburg 22695
Engineers Australia, Barton 00871
Englewood Public Library, Englewood 38980
Englewood Public Library, Englewood 38981
The English Convent, Brugge 02453
English Folk Dance and Song Society, London 29747
English Heritage Library, Swindon 29905
English Language Resource Centre, Warszawa 21255
English Nature (Nature Conservancy Council for England), Peterborough 29878
English Speaking Union, London 29748
EniChem SpA C.R. Novara, Novara 16149
Enid M. Baa Library and Archives, St. Thomas 42320
Enid M. Baa Library & Archives, Saint Thomas 42318
Enköping Hospital, Enköping 26565
Enköpings kommunbibliotek, Enköping 26758
Enoch Pratt Free Library, Baltimore 38079
Ensanian Physicochemical Institute, Eldred 36825
Enso-Gutzeit Research Center, Imatra 07420
Ente Nazionale Idrocarburi (ENI), Roma 16153
Ente per le Nuove Tecnologie l'Energia e l'Ambiente, Roma 15668
Ente Regionale Promozione e Sviluppo dell'Agricoltura, Pozzuolo del Friuli 16425
Enterprise Library, Brockton 36589
Enterprise State Junior College, Enterprise 33322
Entomologische Bibliothek, Müncheberg 12275
Enugu State Library Board, Enugu 20050
Enugu State University of Science and Technology, Enugu 19936
Enumclaw Public Library, Enumclaw 38982
Environment and Culture Libraries, Wembley 30095
Environment Canada, Burlington 04313
Environment Canada, Downsview 04059
Environment Canada, Edmonton 04073
Environment Canada, Hull 04088
Environment Canada, North Vancouver 04101
Environment Canada, Sainte-Foy 04151
Environment Canada, Saskatoon 04487
Environment Department, Putrajaya 18616
The Environment Library, Winnipeg 04190
Environment Protection Authority, Sydney 00729
Environmental Agency, Tokyo 17307
The Environmental Information Service – ENFO, Dublin 14505
Environmental Law Institute Library, Washington 37735
Environmental Protection Agency, Brisbane 00620
Environmental Protection Agency, Cairns 00634
Environmental Protection Agency, City East 00656
Environmental Protection Agency, Corvallis 34114
Environmental Protection Agency, Dallas 36764
Environmental Protection Agency, New York 34591
Environmental Protection Agency, Washington 37736
Environmental Protection Agency of Copenhagen, København 06812
EOI Escuela de Negocios, Madrid 25670
e.on UK plc, Nottingham 29863
Eötvös József Tanítóképző Főiskola, Baja 13215
Eötvös Károly Megyei Könyvtár, Veszprém 13553
Eötvös Loránd Geophysical Institute, Budapest 13394
Eötvös Loránd Tudományegyetem, Budapest 13234
– Állam- és Jogtudományi Kar, Budapest 13235
– Általános és Történeti Földtani Tanszék Könyvtár, Budapest 13236
– Angol-Amerikai Intézet Könyvtár, Budapest 13237
– Csillagászati Tanszékének Könyvtára, Budapest 13238
– Filozófia Intézet Könyvtár, Budapest 13239
– Földrajzi Könyvtár, Budapest 13240
– Germanistisches Institut, Budapest 13241
– Könyvtártudományi- Informatikai Tanszék, Budapest 13242
– Művészettörténeti Tanszék, Budapest 13243
– Néprajzi Intézetr, Budapest 13244
– Növényélettani Tanszék, Budapest 13245
– Olasz Nyelv és Irodalom Tanszék Könyvtár, Budapest 13246
– Szlav Tanszék Könyvtára, Budapest 13247
– Természettudományi Kar, Budapest 13248
– Természettudományi Kar, Budapest 13249
– Török Filológiai Tanszék, Budapest 13250
Eötvös Lóránd University, Budapest 13234
– Astronomical Library, Budapest 13238
– Department of Art History, Budapest 13243
– Department of Geology Library, Budapest 13236
– Department of Library and Information Science, Budapest 13242
– Department of Plant Physiology, Budapest 13245
– Department of Turkish Studies, Budapest 13250
– English and American Studies, Budapest 13237
– Faculty of Law, Budapest 13235
– Faculty of Science, Budapest 13248
– Geography Library, Budapest 13240
– Institute of Ethnography, Budapest 13244
– Institute of Italian Language and Literature Library, Budapest 13246
– Institute of Philosophy Library, Budapest 13239
– Institute of Physics, Budapest 13249
– Institute of Slavic Languages and Literatures Library, Budapest 13247
Eötvös Loránd University, 'Bárczi Gusztáv' Faculty of Special Education, Budapest 13233
EPA Victoria Library, Melbourne 00952
Epcor Research Services, Edmonton 04340
Ephemerides Liturgicae, Roma 16028
Ephrata Public Library, Ephrata 38983
Episcopal Divinity School & Weston Jesuit School of Theology, Cambridge 35151
Episcopal Theological Seminary of the Southwest, Austin 35119
Építésügyi Könyvtári és Dokumentációs Alapítvány, Budapest 13374
Építéstörténeti és Elméleti Intézet, Budapest 13375
EPRI Library, Palo Alto 37321
Equip Training, Kangaroo Rd 00767
Equistar, Channelview 35557
Equitable Life Assurance Society of the United States Library, New York 37201
Erasmus MC, Rotterdam
– Medische Bibliotheek, Rotterdam 19188
Erasmus Universiteit Rotterdam, Rotterdam 19189
– Faculteit der Economische Wetenschappen, Rotterdam 19190
– Faculteit der Rechtsgeleerdheid, Rotterdam 19191
Erasmushogeschool Brussel, Brussel
– Departement Gezondheidszorg – Lerarenopleiding, Brussel 02284
– Departement Koninklijk Conservatorium Brussel, Brussel 02285
– Departement Toegepaste Taalkunde, Brussel 02286
Erdi Gyümölcs és Dísznövénytermesztési Kutató-Fejlesztő Kft., Budapest 13376
Eretz Israel Museum, Tel-Aviv 14760

Erfgoedbibliotheek Hendrik Conscience, Antwerpen 02512
Erfgoedcentrum DiEP, Dordrecht 19420
Erfurt Museum of Natural History, Erfurt 11814
Ergon Energy Corporation, Rockhampton 00828
Erhvervsarkivet, Århus 06769
Erich-Bloch- und Lebenheim-Bibliothek, Konstanz 12158
Ericson Public Library, Boone 38239
Erie Community College – North Campus, Williamsville 33872
Erie Community College – South Campus, Orchard Park 33605
Erie County Law Library, Erie 34193
Erie County Medical Center Healthcare Network, Buffalo 36610
Erie County Public Library, Erie 38984
Erisman Institute of Hygiene, Perlovskaya 23686
Erkel Ferenc Múzeum, Gyula 13477
Erlanger Health System Library, Chattanooga 36658
Ernæring og Sundhed Ankerhus, Sorø 06749
Ernst Klett Verlag, Stuttgart 11429
Ernst Moritz Arndt Universität, Greifswald 09904
Ernst & Young, Diegem 02504
Ernst & Young, Toronto 04282
Ernst & Young Center for Business Knowledge, Sydney 00834
Ernst & Young Company, Chicago 35575
Ernst-Abbe-Bücherei und Lesehalle, Jena 12798
Erseki Könyvtár, Veszprém 13350
Erseki Papnevelö Intézet, Eger 13333
Erseki Papnevelö Intézet, Esztergom 13335
Erseki Simor Könyvtár, Esztergom 13336
Erskine College & Theological Seminary, Due West 35190
Ersta Sköndal Högskola, Sköndal 26421
Ersta Sköndal högskola, Stockholm 26425
Ersta Sköndal University College, Sköndal 26421
Erzabtei Beuron, Beuron 11169
Erzabtei St. Ottilien, St. Ottilien 11335
Erzbischöfliche Akademische Bibliothek, Paderborn 11304
Erzbischöfliche Bibliothek, Wien 01485
Erzbischöfliche Bibliothek Freiburg, Freiburg 11215
Erzbischöfliche Diözesan- und Dombibliothek, Köln 11259
Erzbischöfliches Priesterseminar, Wien 01486
Erzbischöfliches Priesterseminar Collegium Borromaeum, Freiburg 11216
ESAN, Lima 20660
Esbjerg Hovedbibliotek, Esbjerg 06856
Escalator Engineering Industrial Corporation, Sankt-Peterburg 22923
Escambia County Law Library, Pensacola 34657
Escanaba Public Library, Escanaba 38987
Escola de Administração de Empresas de São Paulo, São Paulo 03136
Escola de Biblioteconomia e Documentação, São Carlos 03203
Escola de Communicações e Artes, São Paulo 03137
Escola de Engenharia Industrial Metalúrgica de Volta Redonda, Volta Redonda 03195
Escola de Farmácia e Odontologia de Alfenas, Alfenas 02983
Escola Nacional de Saúde Pública, Lisboa 21653
Escola Nacional de Saúde Pública, Rio de Janeiro 03342
Escola Náutica Infante D. Henrique, Oeiras 21715
Escola Naval, Almada 21620
Escola Politécnica, Rio de Janeiro 03202
Escola Pratica de Engenharia, Praia do Ribatejo 21716
Escola Secundária José Falcão, Coimbra 21713
Escola Superior Agrária de Coimbra (ESAC), Coimbra 21714
Escola Superior de Administração de Negocios, São Paulo 03138
Escola Superior de Educação de Lisboa, Lisboa 21654
Escola Superior de Educação de Viseu, Viseu 21712
Escola Superior de Educação do Algarve, Faro 21650
Escola Superior de Educação e Ciências Sociais, Leiria 21651
Escola Superior de Tecnologia do Algarve, Lisboa 21655
Escola Superior de Tecnologia e Gestão de Portalegre, Portalegre 21695
Escondido Public Library, Escondido 38988
ESCP-EAP Europäische Wirtschaftshochschule, Berlin 09404
Escritório de Meteorologia, Rio de Janeiro 03343
Escuela Agrícola Panamericana, Tegucigalpa 13208
Escuela Andaluza de Salud Pública, Granada 25666
Escuela de Administración de Negocios para Graduados (ESAN), Lima 20660
Escuela de Adminstración de Empresas, Barcelona 25897
Escuela de Agricultura de la Region Tropical Húmeda (EARTH), San José 06113
Escuela de Alta Dirección y Administración (EADA), Barcelona 25294
Escuela de Armamentos, Viña del Mar 04910
Escuela de Estudios Arabes (C.S.I.C.), Granada 25958
Escuela de Estudios Hispanoamericanos, Sevilla 26149
Escuela de Guerra Naval, Madrid 25671
Escuela de Negocios, Madrid 25670
Escuela de Sanidad Militar, Madrid 26012
Escuela Diplomática, Madrid 25715
Escuela Española de Historia y Arqueología en Roma, Roma 16498
Escuela Española de Historia y Arqueología en Roma (EHAR), Roma 16499
Escuela Militar 'General Bernardo O'Higgins', Las Condes 04903
Escuela Nacional de Agricultura 'Roberto Quiñónez', San Salvador 07203
Escuela Nacional de Maestros 'Mariscal Sucre', Sucre 02924
Escuela Nacional de Salud Pública (ENSAP), Lima 20641
Escuela Nacional de Sanidad, Madrid 25672
Escuela Naval, Montevideo 42009
Escuela Naval 'Arturo Prat', Valparaíso 04909
Escuela Naval Militar, Marín 25737
Escuela Oficial de Correos y Telecommunicaciones, Madrid 25421
Escuela Oficial de Idiomas, Madrid 25673
Escuela Oficial de Idiomas de Valencia, Valencia 25676
Escuela Politécnica del Ejército, Madrid 25674
Escuela Politécnica Nacional, Quito 06981
Escuela Superior de Administración Pública (ESAP-INAP), Lima 20619
Escuela Superior de Guerra, Buenos Aires 00118
Escuela Superior de las Fuerzas Armadas (ESFAS), Madrid 25422
Escuela Superior Politécnica de Chimborazo, Riobamba 06983
Escuela Superior Politécnica del Litoral, Guayaquil 06977
Escuela Técnica Superior de Arquitectura, Valladolid 25632
Escuela Universitaria de Educación de Ávila, Ávila 25559
Escuela Universitaria de Estudios Empresariales, Valladolid 25634
Escuela Universitaria de Estudios Empresariales, Zaragoza 25649
Escuela Universitaria de Estudios Empresariales de Huesca, Huesca 25650
Escuela Universitaria de Estudios Sociales, Zaragoza 25651
Escuela Universitaria de Ingeniería Técnica Agrícola, Valladolid 25640
Escuela Universitaria de Magisterio, Segovia 25635
Escuela Universitaria de Música, Montevideo 42010
Escuela Universitaria del Profesorade de E.G.B., Toledo 25614
Escuela Universitaria del Profesorado de E.G.B., Real 25553
Escuela Universitaria del Profesorado de E.G.B., Ubeda 25615
Escuela Universitaria del Profesorado de E.G.B. Fray Luis de León, Cuenca 25375
Eskom Information Centre, Sandton 25176
Eslövs stadsbibliotek, Eslöv 26761
Espace Japon, Paris 08540
Espace-Ressources-Formation Marly-le-Roi, Marly-le-Roi 07750
Espanola Public Library, Espanola 38989
Espanola Public Library, Espanola 04635
Espoo City Library, Espoo 07560
Espoon kaupunginkirjasto, Espoo 07560
Essex County College, Newark 32096
Essex County Council, Chelmsford 29958
Essex County Historical Society, Elizabethtown 36828
Essex County Law Library, Newark 34604
Essex County Library, Essex 04636
Essex Law Association Law Library, Windsor 04553
Essex Law Library, Salem 37511
Estacada Public Library, Estacada 38991
Estação Agronómica Nacional, Oeiras 21818
Estación de Investigaciones Marinas (EDIMAR), Porlamar 42262
Estación Experimental Agrícola de La Molina, Lima 20661
Estación Experimental Agro-Industrial Obispo Colombres, Las Talitas 00301
Estación Experimental Agropecuaria (INTA), Balcarce 00227
Estación Experimental de Aula Dei (CSIC), Zaragoza 26192
Estacíon Experimental de Zonas Aridas, Almería 25872
Estación Experimental Regional Agropecuaria, Pergamino 00309
Estado Maior do Exército, 5 Seção, Rio de Janeiro 03231
Estes Park Public Library, Estes Park 38993
Estherville Public Library, Estherville 38994
Estonian Academy of Arts, Tallinn 07223
Estonian Academy of Music and Theatre Library, Tallinn 07221
Estonian Academy of Security Sciences, Tallinn 07222
Estonian History Museum, Tallinn 07236
Estonian Literary Museum, Tartu 07241
Estonian Museum of Theatre and Music, Tallinn 07235
Estonian National Museum Library, Tartu 07240
Estonian Patent Library, Tallinn 07234
Estonian Repository Library, Tallinn 07220
Estonian Technical Library, Tallinn 07237
Estonian University of Life Sciences, Tartu 07226
– Institute of Animal Husbandry, Tartu 07227
Estudio Teológico Agustiniano, Valladolid 25837
Esztergom Cathedral Library, Esztergom 13337
Esztergomi Főszékesegyházi Könyvtár, Esztergom 13337
Eszterházy Károly Főiskola, Eger 13274
Eszterházy Károly Teacher Training College, Eger 13274
Etablissement National d'Enseignement Supérieur Agronomique de Dijon, Dijon 07699
Etairia Makedonikon Spoudon, Thessaloniki 13132
Etelä-Karjalan keskussairaalan lääketieteellinen kirjasto, Lappeenranta 07527
Etelä-Savon maakuntakirjasto, Mikkeli 07591
Ethicon, Inc, Somerville 36273
Ethiopian Air Force Library, Debre Zeit 07277
Ethiopian Institute of Banking and Insurance (EIBI), Addis Ababa 07281
Ethniki Vivliotiki tis Ellados, Athinai 13068
Ethniko kai Kapodistriakon Panepistimion Athinon, Athinai 13075
Ethniko Metsovio Polyteknio, Athinai 13076
Ethnikon Asteroskopion, Athinai 13106
Ethnografical Museum, Zagreb 06235
Ethnographic Museum in Belgrade, Beograd 24505
Ethnographical Museum, Budapest 13435
Ethnographical Museum, Kraków 21143
Ethnographical Museum, Sankt-Peterburg 23812
Ethnographical Museum in Toruń, Toruń 21233
Ethnography Institute, Lviv 28411
Ethnography Museum, Martin 24742
Ethnologisches Museum, Berlin 11565
ETK EDOC Foundation, Central Building Library, Budapest 13374
Etnografski institut SANU, Beograd 24504
Etnografski muzej, Zagreb 06235
Etnografskij muzej u Beogradu, Beograd 24505
Eton College, Windsor 29503
Etz Hayim, General Talmud Torah and Grand Teshivah, Jerusalem 14699
Euclid Public Library, Euclid 38995
Eugene Bible College, Eugene 35200
Eugenides Foundation, Athinai 13107
Euless Public Library, Euless 38996
EUMETSAT – European Organisation for the Exploitation of Meteorological Satellites, Darmstadt 11720
Eureka College, Eureka 31103
Eurobdalla Shire Library Service, Moruya 01127
Europa-Bücherei der Stadt Passau, Passau 12889
Europa-Institut, Saarbrücken 10556
Europäische Akademie der Arbeit in der Universität Frankfurt a. M., Frankfurt am Main 10725
Europäische Wirtschaftshochschule, Berlin 09404
Europäisches Dokumentationszentrum (EDZ), Mannheim 12240
Europäisches Laboratorium für Molekulare Biologie (EMBL), Heidelberg 12045
Europäisches Übersetzer-Kollegium Nordrhein-Westfalen in Straelen e.V., Straelen 12505
Europas Blues Bibliotek, Notodden 20239
Europa-Universität Viadrina Frankfurt (Oder), Frankfurt (Oder) 09756
European Business School, Oestrich-Winkel 10523
European Centre for Music Documentation, Strassen 18421
European Commission, Bruxelles 02428
European Commission, Dublin 14530
European Commission Joint Research Centre, Ispra 16298
European Documentation Centre, Seoul 18089
European Foundation for the Improvement of Living and Working Conditions, Dublin 14531
European Institute of Public Administration (EIPA), Maastricht 19485
European Organisation for the Exploitation of Meteorological Satellites, Darmstadt 11720
European Organization for Nuclear Research, Genève 27361
European Parliament, Bruxelles 02429
European Parliament, Luxembourg 18399
European Parliament Library, Strasbourg 08165
European Patent Office (EPO), Rijswijk 19285
European Space Agency – ESTEC, Noordwijk 19502
European Space Operations Centre (ESOC), Darmstadt 11721
European Union, Washington 37737
European University Cyprus, Nicosia 06324
European University Institute (EUI), San Domenico di Fiesole (Florence) 15453
Europees Figurenteatercentrum, Gent 02645
Euskal Biblioteka Labayru, Derio 25952
Euskal Herriko Unibertsitatea, Leioa 25405
Euskaltzaindia – Real Academia de la Lengua Vasca, Bilbao 25931
Eusko-Legebiltzarra, Vitoria-Gasteiz 25754
Eustis Memorial Library, Eustis 38999
Eutiner Landesbibliothek, Eutin 09313
Eva K. Bowlby Public Library, Waynesburg 41788
Evangel University, Springfield 32704
Evangelical Academy of Theology, Budapest 13329
Evangelical Lutheran Church of Finland, Church Council, Helsinki 07411
Evangelical School of Theology, Myerstown 35294
Evangelical Seminary of Puerto Rico, San Juan 21899
Evangelical Theological Faculty in Leuven, Leuven 02474
Evangélikus Országos Könyvtár, Budapest 13328
Evangélikus Teológiai Akadémia, Budapest 13329
Evangeline Parish Library, Ville Platte 41686
Evangelische Akademie Bad Boll, Bad Boll 11466
Evangelische Akademie Tutzing, Tutzing 12567
Evangelische Buchhilfe e.V., Vellmar 11355
Evangelische Fachhochschule Berlin, Berlin 09405
Evangelische Fachhochschule Rheinland-Westfalen-Lippe, Bochum 09462
Evangelische Hochschule Dresden – University of Applied Sciences for Social Work, Education and Care, Dresden 09669
Evangelische Hochschule Nürnberg, Nürnberg 10517
Evangelische Hochschule Tabor, Marburg 10299
Evangelische Kirche Berlin-Brandenburg-schlesische Oberlausitz, Berlin 11166
Evangelische Kirche der Pfalz, Speyer 11333
Evangelische Kirche in Deutschland, Hannover 11236
Evangelische Kirche in Dortmund und Lünen, Dortmund 11195
Evangelische Kirche in Hessen und Nassau, Darmstadt 11187
Evangelische Kirchenmusik in Württemberg, Stuttgart 12515
Evangelische Landeskirche Anhalt, Dessau-Roßlau 11189
Evangelische Medienstelle Ulm/Neu Ulm, Ulm 11353

Evangelische Medienzentrale (EMZ), Bremen 11184
Evangelische Omroep (EO), Hilversum 19464
Evangelische Theologische Faculteit, Leuven 02474
Evangelischer Diakonieverein Berlin-Zehlendorf e.V., Berlin 11167
Evangelisches Krankenhaus Königin Elisabeth Herzberge, Berlin 11517
Evangelisches Medienhaus GmbH, Stuttgart 12516
Evangelisches Missionswerk in Deutschland, Hamburg 11232
Evangelisches Predigerseminar Lutherstadt Wittenberg, Lutherstadt Wittenberg 11359
Evangelisches Stift, Tübingen 11350
Evangelisches Zentralarchiv in Berlin, Berlin 11168
Evangelisch-lutherische Landeskirche in Braunschweig – Predigerseminar, Braunschweig 11181
Evangelisch-Lutherische Landeskirche Mecklenburgs, Schwerin 11326
Evangelisch-lutherischer Oberkirchenrat, Oldenburg 11298
Evangelisch-Lutherisches Landeskirchenamt Sachsen, Dresden 11197
Evangelisch-Lutherisches Missionswerk e.V., Leipzig 11262
Evangelisch-Lutherisches Predigerseminar, Nürnberg 11296
Evangelisch-methodistische Kirche, Zwickau 11372
Evangelisch-reformierte Landeskirche des Kantons Zürich, Zürich 27279
Evangelisch-Theologisches Seminar, Tübingen 10630
Evanglisch-lutherisches Missionswerk in Niedersachsen (ELM), Hermannsburg 11245
Evans Army Community Hospital, Fort Carson 36857
Evanston Public Library, Evanston 39001
Evansville-Vanderburgh Public Library, Evansville 39003
Evansville-Vanderburgh Public Library – McCollough Branch, Evansville 39004
Evansville-Vanderburgh Public Library – North Park, Evansville 39005
Evansville-Vanderburgh Public Library – Oaklyn, Evansville 39006
Evansville-Vanderburgh Public Library – Red Bank, Evansville 39007
Evelyn Hone College of Applied Arts and Commerce, Lusaka 42324
Evening Institute of Metallurgy, Moskva 22399
Eveready Battery Company, Inc, Westlake 36414
Everett Community College, Everett 33329
Everett Public Libraries, Everett 39009
Everett Public Library, Everett 39010
Evergreen Park Public Library, Evergreen Park 39012
Evergreen Public Library, Evergreen 39011
Evergreen Regional Library, Gimli 04649
Evergreen State College, Olympia 32191
Evergreen Valley College, San Jose 33731
Everson Museum of Art Library, Syracuse 37651
Evgeniya Vakhtangova School of Dramatic Art, Moskva 22682
Evonik Röhm GmbH, Darmstadt 11382
Evpatoriska raionna TsBS, Tsentralna biblioteka im. Pushkina, Evpatoriya 28561
Evpatoriya Regional Centralized Library System, Pushkin Main Library, Evpatoriya 28561
Ewha Womans University, Seoul 18081
Examination Yüan Library, Taipei 27608
Excelsior Education Centre (EXED), Kingston 16871
Executive Committee Histadrut, Tel-Aviv 14761
Executive Office of the President Libraries, Washington 34988
Exempla Healthcare Lutheran Medical Center, Wheat Ridge 37830
Exeter Cathedral Library, Exeter 29544
Exeter Health Library, Exeter 29668
Exeter Public Library, Exeter 39015
Experimental Botany Institute, Minsk 02012
Experimental Research Institute for Metal Working, Moskva 23175
Exploratorium Learning Commons, San Francisco 37539
Exploratorium Learning Studio, San Francisco 37540
Explorers Club, New York 37202
Exponent Failure Analysis Associates, Menlo Park 37094
Export-Import Bank of Japan, Research Institute of Overseas Investment, Tokyo 17469
Export-Import Bank of the United States, Washington 36349
Extra-Mural Institute of Food Industry, Moskva 22412
Extra-Mural Institute of Mechanical Engineering, Moskva 22414
Extra-Mural Institute of Textile and Light Industries, Moskva 22413
Extra-Mural Law Institute, Moskva 22411
Exxon Exploration Co – Exploration Library, Houston 35751
Exxon Research & Engineering Co, Annandale 35448
Exxonmobil Corp, Fairfax 35697
ExxonMobil Corp, Houston 35752
ExxonMobil Corp, Irving 35781
Exxonmobil Information Centre, Southbank 00830
Ezra Pound Institute of Civilization – National Commission for Judicial Reform, Staunton 37641
Ezra Pound Institute of Civilization – National Council for Medical Research Library, Staunton 37642
F. Bajoraitis Public LIbrary of Silute District Municipality, Šilutė 18365
Faaborg Bbliotek, Faaborg 06857
Fabian & Clendenin, Salt Lake City 36211
Fábrica Nacional de Moneda y Timbre, Madrid 25716
Fabrika belevogo trikotazha, Soligorsk 01927
Fabrika iskusstvennogo mekha, Zhlobin 01939
Fabrika 'Skorokhod', Sankt-Peterburg 22910
Fabrique nationale, Herstal 02505
Fabryka Mechanizmów Samochodowych 'Polmo' (FMS POLNO), Szczecin 21069
Fachbibliothek Astronomie und Astrophysik, Hamburg 09946
Fachbibliothek für Arbeitsmarktforschung und Arbeitsverwaltung, Nürnberg 12371
Fachbibliothek für Frauendiakonie, Düsseldorf 11202
Fachbibliothek für Geschichte und Landeskunde Ibero-Amerikas, Hamburg 09949
Fachbibliothek Ingenieurwissenschaften an der Technischen Fakultät, Kiel 10074
Fachbibliothek Musik- und Tanzwissenschaft, Salzburg 01280
Fachbibliothek Rechtswissenschaft, Köln 10147
Fachbibliothek Umwelt, Dessau-Roßlau 10854
Fachgerichtszentrum, Bremen 10840
Fachhochschule Aachen, Aachen 09346
Fachhochschule Bielefeld, Bielefeld 09459
Fachhochschule Brandenburg, Brandenburg an der Havel 09590
Fachhochschule Braunschweig/Wolfenbüttel, Wolfenbüttel 10692
Fachhochschule Coburg, Coburg 09620
Fachhochschule Deggendorf, Deggendorf 09652
Fachhochschule des Bundes für öffentliche Verwaltung, Brühl 09613
Fachhochschule Dortmund, Dortmund 09653
– Bereichsbibliothek Architektur, Dortmund 09654
– Bereichsbibliothek Design, Dortmund 09655
– Bereichsbibliothek Informatik/Wirtschaft/Angewandte Sozialwissenschaften, Dortmund 09656
– Bereichsbibliothek Ingenieurwesen, Dortmund 09657
Fachhochschule Düsseldorf – University of Applied Sciences, Düsseldorf 09680
Fachhochschule Emdem/Leer, Emden 09687
Fachhochschule Emden/Leer, Leer 10240
Fachhochschule Erfurt, Erfurt 09688
Fachhochschule Flensburg, Flensburg 09728
Fachhochschule Frankfurt am Main – University of Applied Sciences, Frankfurt am Main 09729
Fachhochschule für Öffentliche Verwaltung Nordrhein-Westfalen, Gelsenkirchen 09814
Fachhochschule für öffentliche Verwaltung, Polizei und Rechtspflege, Güstrow 09905
Fachhochschule für öffentliche Verwaltung und Rechtspflege in Bayern, Herrsching 10040
Fachhochschule für öffentliche Verwaltung und Rechtspflege in Bayern, Hof 10044
Fachhochschule Gelsenkirchen, Gelsenkirchen 09815
Fachhochschule Gießen-Friedberg, Gießen
– Hochschulbibliothek, Bereich Friedberg, Friedberg (Hessen) 09816
– Hochschulbibliothek, Bereich Gießen, Gießen 09817
Fachhochschule Jena, Jena 10052
Fachhochschule Kaiserslautern – Standort Zweibrücken, Zweibrücken 10700
Fachhochschule Kaiserslautern – Standorte Kaiserslautern I, II, Kaiserslautern 10054
Fachhochschule Kärnten – Standortbibliothek Feldkirchen, Feldkirchen 01191
Fachhochschule Kärnten – Standortbibliothek Klagenfurt, Klagenfurt 01240
Fachhochschule Kärnten – Standortbibliothek Spittal/Drau, Spittal 01286
Fachhochschule Kärnten – Standortbibliothek Villach, Villach 01290
Fachhochschule Kiel, Kiel 10115
– Fachbereich Agrarwirtschaft, Osterrönfeld 10116
Fachhochschule Koblenz, Koblenz 10120
– Standort Remagen, Remagen 10121
Fachhochschule Köln, Köln 10124
– Abteilungsbibliothek Geistes- und Gesellschaftswissenschaften, Köln 10125
– Abteilungsbibliothek Gummersbach, Gummersbach 10126
– Abteilungsbibliothek Ingenieurwissenschaften, Köln 10127
Fachhochschule Kufstein, Kufstein 01246
Fachhochschule Ludwigshafen am Rhein, Ludwigshafen 10255
Fachhochschule Mainz, Mainz
– Hochschulbibliothek, Standort Bruchspitze, Mainz 10266
– Hochschulbibliothek, Standort Holzstraße, Mainz 10267
Fachhochschule Münster, Münster 10459
– Bereichsbibliothek Design, Münster 10460
– Bereichsbibliothek FHZ, Münster 10461
– Bereichsbibliothek Hüfferstift, Münster 10462
– Bibliothek für Architektur, Design und Kunst, Münster 10463
Fachhochschule Neu-Ulm, Neu-Ulm 10515
Fachhochschule Nordhausen, Nordhausen 10516
Fachhochschule Nordwestschweiz – Hochschule für Architektur, Bau und Geomatik, Muttenz 27116
Fachhochschule Nordwestschweiz – Hochschule für Gestaltung und Kunst, Aarau 26987
Fachhochschule Nordwestschweiz – Hochschule für Soziale Arbeit, Basel 26988
Fachhochschule Nordwestschweiz – Hochschule für Technik, Windisch 27140
Fachhochschule Nordwestschweiz – Hochschule für Wirtschaft, Olten 27129
Fachhochschule Nordwestschweiz – Pädagogische Hochschule, Brugg
– Bibliothek Brugg, Brugg 27055
– Institut Spezielle Pädagogik und Psychologie (ISP), Basel 27056
– Mediothek, Liestal 27057
– Mediothek für Schule und Bildung, Aarau 27058
– Mediothek Solothurn, Solothurn 27059
Fachhochschule Osnabrück, Lingen (Ems) 10247
Fachhochschule Potsdam, Potsdam 10534
Fachhochschule Ravensburg-Weingarten, Weingarten 10684
Fachhochschule Schmalkalden, Schmalkalden 10581
Fachhochschule Schwetzingen – Hochschule für Rechtspflege, Mannheim 10293
Fachhochschule St. Pölten GmbH, St. Pölten 01287
Fachhochschule Stralsund, Stralsund 10590
Fachhochschule Südwestfalen, Hagen 09906
Fachhochschule Trier, Trier 10623
Fachhochschule Trier, Umwelt-Campus Birkenfeld, Hoppstädten-Weiersbach 10047
Fachhochschule Vorarlberg, Dornbirn 01189
Fachhochschule Westküste, Heide 09975
Fachhochschule Worms, Worms 10693
Fachhochschule Würzburg-Schweinfurt, Abteilungsbibliothek Schweinfurt, Schweinfurt 10584
Fachinformationszentrum Fischerei, Hamburg 11997
Fachinformationszentrum Fischerei, Rostock 12441
Fachinformationszentrum Ländliche Räume, Braunschweig 11664
Fachinformationszentrum Wald, Hamburg 11998
Fachkrankenhaus Hubertusburg gGmbH, Wermsdorf 12588
Fachstelle für Katholische Büchereiarbeit, Würzburg 11368
Facolta Teologica dell' Emilia-Romagna, Bologna 15748
Facoltà Teologica dell'Italia Meridionale – Sezione 'S. Tommaso d'Aqino', Napoli 15130
Facoltà Teologica dell'Italia Settentrionale, Milano 15077
Facoltà Teologica di Sicilia "S. Giovanni Evangelista", Palermo 15215
Facoltà Valdese di Teologia, Roma 15366
Factory for Synthetic Fibers, Tver 24010
Factory for Textile Printing, Moskva 22847
Factory for Turbine Production, Kaluga 22796
Factory Mutual Research Corp, Norwood 36061
Factory 'Skorokhod', Sankt-Peterburg 22910
Faculdade de Administração e Ciências Economicas, São Paulo 03139
Faculdade de Belas Artes, São Paulo 03140
Faculdade de Ciências Econômicas e Administrativas de Santo André, Santo André 03125
Faculdade de Ciências Médicas de Pernambuco, Recife 03086
Faculdade de Ciências Médicas de Pernambuco, Recife 03087
Faculdade de Comunicação Social 'Casper Líbero', São Paulo 03141
Faculdade de Direito de São Bernardo do Campo, São Bernardo do Campo 03129
Faculdade de Direito de Sorocabana, Sorocaba 03187
Faculdade de Educação, Ciências e Letras 'Dom Domenico', Guarujá 03036
Faculdade de Educação, Letras e Ciências, São Paulo 03142
Faculdade de Filosofia, Ciências e Letras de Registro, Registro 03324
Faculdade de Medicina de Marília, Marília 03049
Faculdade de Motricidade Humana, Cruz Quebrada 21648
Faculdade Evangélica de Medicina do Paraná, Curitiba 03016
Faculdade Santa Marcelina, São Paulo 03143
Faculdade Santa Marcelina (FASM), São Paulo 03144
Faculdades de Tecnologia e de Ciências, Barretos 02986
Faculdades 'Farias Brito', Guarulhos 03037
Faculdades Integradas "Antônio Eufrásio de Toledo", Presidente Prudente 03084
Faculdades Integradas Teresa D'Ávila, Lorena 03047
Faculdades Salesianas, Lorena 03048
Facultad de Ciencias de la Actividad Física y del Deporte (INEF), Madrid 26013
Facultad de Ciencias Médicas, Santiago de Cuba 06312
Facultad de Ciencias Sociales, Jurídicas y de la Comunicación, Segovia 25637
Facultad de Teología, Granada 25780
Facultad de Teología de San Esteban PP. Dominicos, Salamanca 25554
Facultad de Teología de San Vicente Ferrer – Sección Diocesis, Valencia 25835
Facultad de Teología de San Vicente Ferrer – Sección Dominicos, Torrente 25832
Facultad de Teología del Norte de España, Vitoria-Gasteiz 25845
Facultad de Teología del Uruguay 'Mons. Mariano Soler', Montevideo 42029
Facultad de Teologia San Dámaso, Madrid 25793
Facultad de Teología 'San Dámaso', Madrid 26014
Facultad Latinoamericana de Ciencias Sociales (FLACSO-Chile), Santiago 04858
Facultad Teológica del Norte de España, Burgos 25341
Facultăţe de filologie şi istorie, Sibiu 21943
Faculté de Théologie Protestante, Yaoundé 03636
Faculté Libre de Théologie Evangélique, Vaux-sur-Seine 08101
Faculté libre de théologie protestante, Paris 07882
Faculté Libre de Théologie Réformée (FLTR), Aix-en-Provence 08267
Faculté Universitaire de Théologie Protestante / Faculteit voor Protestantse Godgeleerdheid, Bruxelles 02287

Faculté Universitaire des Sciences Agronomiques, Gembloux 02313
Faculteit voor Protestantse Godgeleerdheid, Bruxelles 02287
Faculteit voor Vergelijkende Godsdienstwetenschappen (FVG), Antwerpen 02448
Facultés Universitaires Catholiques de Mons, Mons 02410
Facultés Universitaires Notre Dame de la Paix, Namur 02416
Faculty for Comparative Study of Religions, Antwerpen 02448
Faculty of Commerce, Catering and Tourism, Budapest 13225
Faculty of Theology in Ljubljana, Ljubljana 24810
Faculty of Veterinary Management, Hódmezővásárhely 13277
Faegre & Benson, LLP, Denver 35665
Faegre & Benson, LLP, Minneapolis 35905
Fagersta bibliotek, Fagersta 26762
Fahrbücherei 09 und 10 im Kreis Plön, Preetz 12896
FAIDD Library, Helsinki 07450
Fair Trade Commission, Tokyo 17313
Fairbanks North Star Borough Public Library & Regional Center, Fairbanks 39018
Fairchild Tropical Botanic Garden, Miami 37099
Fairfax County Public Library – Administrative Offices, Fairfax 39020
Fairfax County Public Library – Chantilly Regional, Chantilly 38507
Fairfax County Public Schools, Annandale 36469
Fairfax Law Library, Fairfax 34197
Fairfield City Library and Museum Service, Cabramatta 01076
Fairfield County District Library, Lancaster 39793
Fairfield County Library, Winnsboro 41934
Fairfield Historical Society Library, Fairfield 36841
Fairfield Public Library, Fairfield 39022
Fairfield Public Library, Fairfield 39023
Fairfield Public Library – Anthony Pio Costa Memorial Library, Fairfield 39024
Fairfield Public Library – Fairfield Woods, Fairfield 39025
Fairfield University, Fairfield 31119
Fairhope Public Library, Fairhope 39029
Fairleigh Dickinson University, Hackensack 31274
Fairleigh Dickinson University, Madison 31710
Fairleigh Dickinson University, Teaneck 32802
Fairmont State University, Fairmont 31123
Fairmount Temple, Cleveland 36704
Fairport Public Library, Fairport 39030
Fairview College, Fairview 03706
Fairview Heights Public Library, Fairview Heights 39031
Faisal Islamic Bank, Cairo 07138
Faith Baptist Bible College & Theological Seminary, Ankeny 35109
Faith Evangelical Lutheran Seminary, Tacoma 35395
Faith Theological Seminary, Baltimore 35123
Fakultet muzicke umetnosti, Beograd 24456
Fakulteta za komercialne in poslovne vede, Celje 24808
Fakultná nemocnica, Bratislava 24688
Falkenbergs bibliotek, Falkenberg 26763
Falkirk Council Library Services, Falkirk 29981
Falkland Islands Community Library, Port Stanley 07286
Falköpings bibliotek, Falköping 26764
Fall River Public Library, Fall River 39035
The Falls City Library & Arts Center, Falls City 39039
Falmouth College of Arts, Falmouth 29109
Falu stadsbibliotek, Falun 26765
Famiglia Meneghina, Milano 16321
Families, Youth and Community Care Library, Brisbane 00621
Family Court of Australia, Sydney 00730
Faneromeni Bibliotheke, Nicosia 06338
Fannie Mae, Washington 37738
Fanshawe College of Applied Arts & Technology Library, London 04003
F.A.O. Regional Office for Africa, Accra 13043
FAO Regional Office for Asia and the Pacific, Bangkok 27757
Far Eastern Coal Industrial Corporation, Raichikhinsk 22896
Far Eastern Ethnological, History, Archaeology and Ethnography Institute of the Russian Academy of Sciences, Vladivostok 24056
Far Eastern Hydro-Meteorological Institute, Vladivostok 24051
Far Eastern Institute of Trade, Vladivostok 22634
Far Eastern Library, Stockholm 26669
Far Eastern Museum of Fine Arts, Khabarovsk 23094
Far Eastern Power Engineering Joint-Stock Company, Khabarovsk 22804
Far Eastern Railway, Khabarovsk 23093
Far Eastern Research Institute of Forestry, Khabarovsk 23095
Far Eastern Research Library, Plato 37390
Far Eastern Scientific Research Institute of Agriculture, Khabarovsk 23096
Far Eastern State Research Library, Khabarovsk 22140
Far Eastern State University, Vladivostok 22633
Far Eastern Teacher-Training Institute of the Arts, Vladivostok 22635
Far Eastern Territory Geological Administration, Khabarovsk 22731
Far Eastern University, Manila 20713
Far Eastern Veterinary Institute, Blagoveshchensk 23004
Far North Queensland Institute of TAFE, Manunda 00574
Far-Eastern State Technical University (FESTU), Vladivostok 22632
Farella, Braun & Martel, San Francisco 36229
Farm Credit Administration Information Center, McLean 34520
Farmaseuttinen kirjasto, Helsinki 07451
Farmatsevticheski institut, Pyatigorsk 22484
Farmatsevtichne uchilishche, Zhytomyr 28131
Farmers Association of Iceland, Reykjavík 13568
Farmers Branch Manske Library, Farmers Branch 39042
Farming and Animal Husbandry University of Chinese People's Liberation Army, Changchun 05077
Farmingdale Public Library, Farmingdale 39044
Farmington Community Library, Farmington Hills 39048
Farmington Community Library – Farmington Branch, Farmington 39046
The Farmington Library, Farmington 39047
Farnhamville Public Library, Farnhamville 39050
Farook College, Kozhikode 13688
Farsø Folkebibliotek, Farsø 06858
Farsund bibliotek, Farsund 20326
FÁS – Training and Employment Authority, Dublin 14532
Fashion Institute of Design & Merchandising, Los Angeles 37047
Fashion Institute of Technology – SUNY, New York 32062
Fasken Martineau DuMoulin LLP, Toronto 04283
Fasken Martineau DuMoulin LLP Library, Vancouver 04539
Fässbergsgymnasiets bibliotek, Mölndal 26476
Fastiv Regional Centralized Library System, Main Library, Fastiv 28562
Fastivska raionna TsBS, Tsentralna biblioteka, Fastiv 28562
Faulkner State Community College, Bay Minette 33157
Faulkner University, Montgomery 31886
– Jones School of Law Library, Montgomery 31887
Faulkner-Van Buren Regional Library System, Conway 38684
Fauquier County Public Library, Warrenton 41729
Fayette Commonwealth College, Uniontown 32870
Fayette County Law Library, Uniontown 34950
Fayette County Law Library, Washington Court House 35045
Fayette County Public Libraries, Oak Hill 40563
Fayette County Public Library, Connersville 38682
Fayetteville Public Library, Fayetteville 39057
Fayetteville State University, Fayetteville 31132
Fayetteville Technical Community College, Fayetteville 33335
Feati University, Manila 20714
Febias College of Bible, Manila 20764
Federação das Industrías do Estado do Rio Grande do Sul, Porto Alegre 03317
Federação de Escolas Superiores, Belo Horizonte 02991
Federação de Estabelecimentos de Ensino Superior Nôvo Hamburgo, Nôvo Hamburgo 03058
Federacja Stowarzyszeń Naukowo-Technicznych, Warszawa 21256
Federal Agency for Nature Conservation, Bonn 10820
Federal Archives, Moskva 23177
Federal Assembly Documentation Service, Parliament Library, Bern 27232
Federal Aviation Administration, Atlantic City 36495
Federal Aviation Administration, Oklahoma City 34629
Federal Bureau of Investigation, Quantico 34715
Federal Bureau of Investigation, Washington 34989
Federal Centre for Planning Finance, Moskva 23178
Federal College of Education, Abeokuta 19916
Federal Communications Commission Library, Washington 34990
Federal Correctional Institution Library, Milan 34534
Federal Correctional Institution Library, Tallahassee 34911
Federal Correctional Institution – Morgantown Library, Morgantown 34560
Federal Court, Kuala Lumpur 18580
Federal Court of Australia, Brisbane 00622
Federal Court of Australia, Hobart 00686
Federal Court of Australia (WA), Perth 00713
Federal Court of Canada Library, Ottawa 04112
The Federal Court of Justice, Karlsruhe 10955
Federal Department of Agric and Natural Resources, Jos 19997
Federal Deposit Insurance Corp Library, Washington 36350
Federal Election Commission, Washington 34991
Federal Emergency Management Agency Library, Washington 34992
Federal Energy Regulatory Commission Library, Washington 34993
The Federal Institutional Court, Karlsruhe 10956
Federal Judicial Center, Washington 34994
Federal Law Enforcement Training Center Library, Glynco 34302
Federal Maritime Commission Library, Washington 34995
Federal Ministry of Economic National Planning, Lagos 19998
Federal Ministry of Education, Lagos 20058
Federal Ministry of Education and Research, Bonn 10825
Federal Ministry of Food, Agriculture and Consumer Protection, Bonn 10826
Federal Office for Gender Equality FOGE, Bern 27320
Federal Office of Meteorology and Climatology, Zürich 27449
Federal Office of Statistics, Beograd 24496
Federal Polytechnic, Auchi 19925
Federal Polytechnic, Bauchi 19928
Federal Polytechnic, Bida 19930
Federal Polytechnic, Idah 19943
Federal Polytechnic, Ilaro 19945
Federal Polytechnic, Kaura-Namoda 19959
Federal Polytechnic, Mubi 19969
The Federal Polytechnic Library, Ado-Ekiti 19920
Federal Polytechnic, Oko, Aguata 19923
Federal Prosecutor's Office – Advanced Training Institute for Investigators, Sankt-Peterburg 23877
Federal Radio Corporation of Nigeria, Lagos 20031
Federal Research Centre for Cultivated Plants – Julius Kühn Institut, Berlin 11539
Federal Research Centre for Cultivated Plants – Julius Kühn Institut, Quedlinburg 12416
Federal Research Institute for Rural Areas, Forestry and Fisheries, Braunschweig 11664
Federal Reserve Bank of Atlanta, Atlanta 35458
Federal Reserve Bank of Boston, Boston 35521
Federal Reserve Bank of Chicago Library, Chicago 35576
Federal Reserve Bank of Cleveland, Cleveland 35615
Federal Reserve Bank of Dallas Library, Dallas 35645
Federal Reserve Bank of Kansas City, Kansas City 35792
Federal Reserve Bank of Minneapolis, Minneapolis 35906
Federal Reserve Bank of New York – Research Library, New York 35980
Federal Reserve Bank of Philadelphia, Philadelphia 36098
Federal Reserve Bank of Richmond, Richmond 36173
Federal Reserve Bank of Saint Louis, Saint Louis 36198
Federal Reserve Bank of San Francisco, San Francisco 36230
Federal Senate, Brasília 03222
Federal Trade Commission Library, Washington 34996
Federal Training and Research Institute for Graphic Arts and Media, Wien 01379
Federal University of Paraná, Curitiba 03017
Federal University of Rio de Janeiro, Itajaí 03040
– Postgraduate Institute of Research and Business Administration, Rio de Janeiro 03041
Federal University of Rio de Janeiro, Rio de Janeiro 03101
Federal University of Technology, Owerri 19975
Federale Overheidsdienst Werkgelegenheid, Arbeid en Sociaal Overleg (FOD), Brussel 02430
Federale Police, Brussel 02434
Federale Politie, Brussel 02434
Federalnaya sluzhba Rossii po televideniyu i radioveshchaniyu, Moskva 23176
Federalny arkhiv, Moskva 23177
Federalny tsentr proektnogo finansirovaniya, Moskva 23178
Federated States of Micronesia Archives, Pohnpei 18876
Federatie van de Belgische Cementnijverheid (FEBELCEM), Brussel 02586
Fédération des entreprises romandes Genève, Genève 27364
Fédération des Médecins Omnipraticiens du Québec, Montreal 04399
Fédération des travailleurs et travailleuses du Québec (FTQ), Montreal 04400
Fédération Française du Bâtiment (FFB), Paris 08541
Federation of Economic Organizations (KEIDANREN), Tokyo 17686
Federation of Education Unions, Southbank 01011
Federation of Finnish Insurance Companies, Insurance Library, Helsinki 07497
Federation of Indian Chambers of Commerce and Industry (FICCI), New Delhi 14125
Federation of Serbian Education Societies, Beograd 24521
Federation of Swedish Farmers, Stockholm 26664
Federation Penang Teachers' Training College, Penang 18542
Federazione delle Associazioni Scientifiche e Tecniche di Milano (FAST), Milano 16334
Fejér County Library, Székesfehérvár 13547
Fekixarchief Antwerpen, Antwerpen 02513
Feleti Barstow Public Library, Pago Pago 00096
Felhaber, Larson, Fenlon & Vogt, Minneapolis 35907
Felician College, Lodi 31632
Felicja Blumental Music Centre and Library, Tel-Aviv 14762
The Fendrick Library, Mercersburg 40153
Fendrick Library, Mercersburg 40154
Feng Chia University, Taichung 27537
Fenwick & West LLP, Library, Montain View 35918
Feodosiska raionna TsBS, Tsentralna biblioteka im. A. Grina, Feodosiya 28563
Feodosiya Regional Centralized Library System, Grin Main Library, Feodosiya 28563
Ferenc Erkel Museum, Gyula 13477
Ferenc Hopp Museum of Eastern Asiatic Arts, Budapest 13381
Ferenc Rákoczi II. District Library, Miskolc 13540
Ferenczy Múzeum – Könyvtár, Szentendre 13510
Fergana Polytechnic Institute, Fergana 42077
Fergana State University, Fergana 42078
Fergus Falls Community College, Fergus Falls 33336
Fergus Falls Public Library, Fergus Falls 39061
Ferguson Municipal Public Library, Ferguson 39062
Ferguson Publishing Co, Chicago 35577
Fermi National Accelerator Laboratory Library, Batavia 36523
Fernbank Science Center Library, Atlanta 36489
Ferndale Public Library, Ferndale 39064
Fernuniversität Hagen, Hagen 09907

Ferring Läkemedel AB, Malmö 26542
Ferris State University, Big Rapids 30425
Ferro Corp Library, Independence 35772
Ferrocarriles Argentinos, Buenos Aires 00209
Ferrous Metallurgy Research Institute, Donetsk 28227
Ferrum College, Ferrum 31138
Fersman Mineralogical Museum of the Russian Academy of Sciences, Moskva 23351
Fertilizantes Serrana S/A, Cajati 03255
Fertilizer Association of India, New Delhi 14126
Fertilizers Research Institute, Puławy 20919
FFBIZ Bibliothek, Berlin 11518
FH Kufstein Tirol, Kufstein 01246
FHS St.Gallen, St.Gallen
– Fachbereich Gesundheit, St.Gallen 27133
– Fachbereich Soziale Arbeit, Rorschach 27134
Fichtelgebirgsmuseum, Wunsiedel 12607
Fidelity & Deposit Company of Maryland, Baltimore 35479
Field Library of Peekskill, Peekskill 40741
Field Museum of Natural History Library, Chicago 36673
Fife Council, Kirkcaldy 30005
Fife Council Libraries, Kirkcaldy 30006
Figge Art Museum, Davenport 36769
Fiji College of Advanced Education, Suva 07291
Fiji College of Agriculture, Nausori 07289
Fiji Electricity Authority, Lautoka 07309
Fiji Institute of Technology, Suva 07292
Fiji Museum, Suva 07314
Fiji National Training Council, Nasinu 07299
Fiji School of Medicine, Suva 07293
Filatov Research Institute of Ophthalmology, Odesa 28439
Filial Ukrainskoyi Derzhavny akademiyi zvyazku im. O. Popova, Kyiv 28310
Filiale Timişoara a Academiei Romane, Timişoara 22063
Filion Wakely Thorup Angeletti LLP, Toronto 04284
Filipstads bergslags bibliotek, Filipstad 26767
Filipstad's Public Library, Filipstad 26767
Fillmore Riley, Winnipeg 04307
Film and Television Institute of India, Pune 14190
Film Library, Warszawa 21257
Filmarchiv Austria, Wien 01596
Filmmuseum Düsseldorf, Düsseldorf 11791
Filmmuseum Informatiecentrum, Amsterdam 19356
Filmoteca Española, Madrid 26015
Filmoteka Narodowa, Warszawa 21257
Filson Historical Society Library, Louisville 37067
Finance and Economics Institute, Kazan 22271
Finance and Quartermaster School, Taipei 27573
Financial Academy of the Russian Federation Government, Moskva 22329
Finansovaya akademiya pri pravitelstve Rossiskoi Federatsii, Moskva 22329
Finansovo-ekonomicheski institut, Kazan 22271
Finansovy-ekonomicheski institut, Rostov-na-Donu 22487
Finanzgericht Düsseldorf, Düsseldorf 10870
Finanzgericht München, München 11043
Finanzgericht Münster, Münster 11055
Finanzgericht Nürnberg, Nürnberg 11069
Finanzlandesdirektion für Oberösterreich, Linz 01388
Finanzlandesdirektion für Wien, Niederösterreich und Burgenland, Wien 01404
Finanzministerium Baden-Württemberg, Stuttgart 11101
Finanzministerium des Landes Nordrhein-Westfalen, Düsseldorf 10871
Finanzprokuratur, Wien 01405
Finanzwisenschaftliches Forschungsinstitut an der Universität zu Köln, Köln 12128
Findlay-Hancock County District Public Library, Findlay 39066
FIND-SVP Inc Library, New York 35981
Fine Arts Museum Cordoba Library, Córdoba 25944
Fine Arts Museum of Odessa, Odesa 28442
Fingal County Libraries, Swords 14589
Finger Lakes Community College, Canandaigua 33201
Finger Lakes Library System, Ithaca 39596
Finkelstein Memorial Library, Spring Valley 41410
Finlandia University, Hancock 31286
Finlandsinstitutets Bibliotek, Stockholm 26646
Finmark fylkesbibliotek, Vadsø 20302
Finmark Library, Vadsø 20302
Finnegan, Henderson, Farabow, Garrett & Dunner, Washington 36351
Finney County Public Library, Garden City 39190
Finnish Adult Education Institute in Helsinki, Helsinki 07457
Finnish Air Force Headquarters, Tikkakoski 07405
Finnish Association on Intellectual and Developmental Disabilities, Helsinki 07450
Finnish Bar Association, Information Service, Helsinki 07489
Finnish Broadcasting Company Ltd, Yleisradio 07433
Finnish Businessmen's Commercial College, Helsinki 07389
Finnish Evangelical Lutheran Mission Library, Helsinki 07409
Finnish Forest Research Institute Library, Vantaa 07556
Finnish Geodetic Institute, Masala 07529
Finnish Institute for Children's Literature, Tampere 07542
The Finnish Institute in Sweden, Stockholm 26646
Finnish Institute of International Affairs, Library, Helsinki 07504
Finnish Institute of Occupational Health, Helsinki 07469
Finnish Institute of Public Management, Helsinki 07453
Finnish Literature Society, Helsinki 07488
Finnish Maritime Administration Library, Helsinki 07473
Finnish Medical Society Duodecim, Helsinki 07487
Finnish Museum of Photography, Helsinki 07498
Finnish National Gallery Library, Helsinki 07507
Finnish Railway Museum Library, Hyvinkää 07511
Finnish Road Administration Library, Helsinki 07402
Finnish Standards Association Library, Helsinki 07494
Finspångs bibliotek, Finspång 26768
Fiorello H. Laguardia Community College, Long Island City 31639
Firdousi Tajik National Library, Dushanbe 27633
Fire-fighters Professional Institute, Sankt-Peterburg 22558
First Baptist Church, Abilene 35100
First Baptist Church, Hattiesburg 35227
First Baptist Church, Longview 35267
First Baptist Church, Waco 35404
First Baptist Church Library, Albuquerque 35102
First Baptist Church Library, Greensboro 35223
First Baptist Church Library, Murfreesboro 35293
First Baptist Church Library, Richmond 35344
First Baptist Church of Dallas, Dallas 35177
First Church of Christ Congregational, West Hartford 35417
First City Hospital, Sofiya 03570
First District Court of Appeal Library, Tallahassee 34912
First Methodist Church, Shreveport 35379
First Parish Church of Norwell, Norwell 37295
First Presbyterian Church of Charleston, Charleston 35153
First Toshkent State Medical Institute, Toshkent 42090
First United Methodist Church, Oak Ridge 35317
First United Methodist Church, Tulsa 35399
FirstEnergy Corp, Akron 35432
Fiscal Court of Appeal, Göteborg 26489
Fish Culture Research Station, Nucet 22048
Fish & Neave IP Group of Ropes Gray LLP Library, New York 37203
Fisher College, Boston 30493
Fisheries and Oceans Canada, St. Andrews 04497
Fisheries Division, Freetown 24571
Fisheries Research Unit, Monkey Bay 18490
Fishery Museum, Bergen 20214
Fishery Research Institute, Valparaíso 04947
Fishing Industry and Economy Kaliningrad State Technical Institute, Kaliningrad 23069
Fishkill Correctional Facility Library, Beacon 33974
Fisk University, Nashville 31933
Fiske Public Library, Wrentham 41976
Fiskeriverket, Göteborg 26575
Fitchburg Public Library, Fitchburg 39069
Fitchburg State College, Fitchburg 31140
Fitzgerald, Abbott & Beardsley, Oakland 36065
Fitzwilliam College, Cambridge 29015
Fitzwilliam Museum, Cambridge 29016
Five Towns College, Dix Hills 31019
Fizicheski institut im. P.N. Lebedeva RAN, Moskva 23179
Fiziko-tekhnicheski institut, Dolgoprudny 22234
Fiziko-tekhnicheski institut, Minsk 02001
Fiziko-tekhnicheski institut, Mogilev 02060
Fiziko-tekhnicheski institut im. A.F. Ioffe RAN, Sankt-Peterburg 23761
Fiziko-tekhnicheski institut im. A.F. Ioffe – Sektor Biblioteki v Shuvalova, Sankt-Peterburg 23762
Fiziko-tekhnicheski institut RAN, Izhevsk 23062
Fiziko-tekhnicheski institut RAN, Kazan 23080
Fizyko-khimichny instytut im. Bogatskoho, Odesa 28436
Fizyko-mekhanichny instytut im. Karpenka, Lviv 28407
Fizyko-tekhnichny instytut im. O. Galkina, Donetsk 28217
Fizyko-tekhnichny instytut nizkykh temperatur im. B. Verkina, Kharkiv 28251
Fizyko-tekhnichny instytut Ukrainy, Kharkiv 28252
Fjellhaug Mission Seminary, Oslo 20201
Fjellhaug skoler, Oslo 20201
Flag Research Center Library, Winchester 37853
Flagg-Rochelle Public Library District, Rochelle 41025
Flagler College, Saint Augustine 32490
Flagler County Public Library, Palm Coast 40691
Flagstaff City-Coconino County Public Library System, Flagstaff 39070
Flanders Social and Economic Council, Brussel 02623
Flat River Community Library, Greenville 39308
Flathead County Library, Kalispell 39672
Flax Research Institute, Torzhok 24000
Flekkefjord bibliotek, Flekkefjord 20328
Flemish Genealogical Society, Antwerpen 02536
Flemish Heritage Institute, Brussel 02629
Flemish Theatre Institute, Brussel 02630
Flens bibliotek, Flen 26769
Flensburg University Central Library, Flensburg 09728
Den Flerfaglige Professionshøjskole i Region Hovedstaden, Herlev 06660
Flesh Public Library, Piqua 40796
Fletcher Free Library, Burlington 38398
J. V. Fletcher Library, Westford 41855
FlevoMeer Bibliotheek, Lelystad 19654
FlevoMeer Bibliotheek – Noordoostpolder, Emmeloord 19594
Fliedner Kulturstiftung Kaiserswerth mit der Fachbibliothek für Frauendiakonie und dem Fliedner-Archiv, Düsseldorf 11202
Flinders University, Adelaide 00373
– Gus Fraenkel Medical Library, Adelaide 00374
Flint Memorial Library, North Reading 40523
Flint Public Library, Flint 39073
Flint River Regional Library, Griffin 39322
Flint River Regional Library – Fayette County Public Library, Fayetteville 39058
Flint River Regional Library – Griffin-Spalding County Library, Griffin 39323
Flint River Regional Library – Peachtree City Library, Peachtree City 40736
Flint-Groves Baptist Church, Gastonia 35212
Flintshire Library and Information Services, Mold 30044
Flora folkebibliotek, Florø 20330
Floral Park Public Library, Floral Park 39075
Florence A.S. Williams Public Library, Saint Croix 42315
Florence County Library System, Florence 39076
Florence Crane Correctional Facility Library, Coldwater 34092
Florence University, Firenze
– Humanistic Library, Firenze 15013
Florence-Darlington Technical College Wellman, Inc, Florence 33341
Florence-Lauderdale Public Library, Florence 39077
Florida Agricultural & Mechanical University, Orlando 32209
Florida Agricultural & Mechanical University, Tallahassee 32790
– Architecture Library, Tallahassee 32791
– Frederic S. Humphries Science Research Center, Tallahassee 32792
Florida Atlantic University, Boca Raton 30471
Florida Atlantic University, Jupiter 31473
Florida Atlantic University – Port Saint Lucie Branch, Port Saint Lucie 32332
Florida Attorney General's Law Library, Tallahassee 34913
Florida Baptist Theological College, Graceville 31226
Florida Center for Theological Studies (FCTS) Library, Miami 37100
Florida Christian College, Kissimmee 31515
Florida Coastal School of Law, Jacksonville 31446
Florida College, Temple Terrace 33797
Florida Community College at Jacksonville, Jacksonville 33437
Florida Department of Agriculture & Consumer Services, Gainesville 34291
Florida Department of Corrections, Arcadia 33931
Florida Department of Corrections, Belle Glade 33980
Florida Department of Corrections, Bowling Green 34001
Florida Department of Corrections, Bristol 34006
Florida Department of Corrections, Bushnell 34021
Florida Department of Corrections, Clermont 34086
Florida Department of Corrections, Crestview 34117
Florida Department of Corrections, Daytona Beach 34127
Florida Department of Corrections, DeFuniak Springs 34131
Florida Department of Corrections, East Palatka 34163
Florida Department of Corrections, Immokalee 34375
Florida Department of Corrections, Lake Butler 34429
Florida Department of Corrections, Lawtey 34447
Florida Department of Corrections, Madison 34485
Florida Department of Corrections, Malone 34495
Florida Department of Corrections, Mayo 34512
Florida Department of Corrections, Miami 34528
Florida Department of Corrections, Milton 34536
Florida Department of Corrections, Monticello 34555
Florida Department of Corrections, Ocala 34625
Florida Department of Corrections, Okeechobee 34628
Florida Department of Corrections, Perry 34661
Florida Department of Corrections, Polk City 34687
Florida Department of Corrections, Punta Gorda 34713
Florida Department of Corrections, Raiford 34720
Florida Department of Corrections, Sanderson 34833
Florida Department of Corrections, Sneads 34865
Florida Department of Corrections, Trenton 34935
Florida Department of Corrections, Zephyrhills 35099
Florida Department of Environmental Protection, Tallahassee 34914
Florida Department of Environmental Protection Library, Tallahassee 34915
Florida Department of State – Division of Library & Information Services, Tallahassee 30158
Florida Department of Transportation, Tallahassee 34916
Florida Division of Adult Corrections, Bushnell 34022
Florida Gulf Coast University, Fort Myers 31162
Florida Hospital College of Health Sciences, Orlando 32210
Florida Institute of Technology, Melbourne 31786
Florida International University, Miami 31806
Florida International University, North Miami 32145
Florida Keys Community College, Key West 33456
Florida Memorial University, Miami Gardens 31815
Florida Public Service Commission, Tallahassee 34917
Florida Solar Energy Center, Cocoa 36718
Florida Southern College, Lakeland 31547
Florida State Hospital, Chattahoochee 36657
Florida State Prison Library, Raiford 34721
Florida State University, Tallahassee
– Center for Demography and Population Health, Tallahassee 32793

– Harold Goldstein Library, College of Information, Tallahassee 32794
– Law Library, Tallahassee 32795
– Paul A. M. Dirac Science Library, Tallahassee 32796
– Robert Manning Strozier Library, Tallahassee 32797
Florida Supreme Court Library, Tallahassee 34918
Flossmoor Public Library, Flossmoor 39079
Flower Mound Public Library, Flower Mound 39080
Floyd College, Rome 33700
Floyd County Public Library, Prestonsburg 40894
Flugmedizinisches Institut der Luftwaffe, Fürstenfeldbruck 11908
Flygvapenmuseum, Linköping 26607
Flyvevåbnets Bibliotek, Ballerup 06757
Flywheel and Axle Manufacturing Plant, Grodno 01873
FMU Central Library, São Paulo 03145
FMV – Teknik för Sveriges säkerhet, Stockholm 26501
Főapátsági Könyvtár, Pannonhalma 13341
Focke-Museum, Bremen 11674
Fødevareinstitutet, Søborg 06833
Főegyházmegyei Könyvtár, Eger 13334
FOI Services Inc Library, Gaithersburg 36883
Földrajztudományi Kutató Intézet, Budapest 13411
Foley & Hoag LLP Library, Boston 35522
Foley & Lardner, Chicago 35578
Foley & Lardner, Milwaukee 35892
Foley & Lardner, Washington 36352
Foley & Lardner LLP, Los Angeles 35833
Folger Shakespeare Library, Washington 37739
Folketingets Bibliotek, Arkiv og Oplysning, København 06759
Folksam, Stockholm 26647
Folksam Group, Stockholm 26647
Folkwang-Hochschule, Essen 09723
Folsom Public Library, Folsom 39089
Folsom State Prison Library, Represa 34739
Fomento del Trabajo Nacional, Barcelona 25898
FOM-Institute of Plasma Physics, Nieuwegein 19494
FOM-Instituut voor Plasmafysica Rijnhuizen, Nieuwegein 19494
Fond Du Lac Circuit Court, Fond Du Lac 34208
Fond du Lac County Historical Society-Family Heritage Center, Fond du Lac 36855
Fond Du Lac Public Library, Fond du Lac 39091
Fondation André Rénard (FAR), Liège 02677
Fondation "Archivum Helveto-Polonicum", Bourguillon 27336
Fondation Auschwitz – Centre d'Etudes et de Documentation, Bruxelles 02587
Fondation de Ligne, Tournai 02703
Fondation du Roi Abdul Aziz Al Saoud pour les Etudes Islamiques et les Sciences Humaines, Casablanca 18947
Fondation M. et A. Maeght, Saint-Paul 08673
Fondation Maison des Sciences de l'Homme (FMSH), Paris 08542
Fondation Martin Bodmer, Cologny 27345
Fondation Maurice Carême, Bruxelles 02588
Fondation Royaumont, Asnières-sur-Oise 08284
Fondazione "A. Marazza", Borgomanero 16705
Fondazione Benetton Studi Ricerche, Treviso 16616
Fondazione Biblioteca "Benedetto Croce", Napoli 16369
Fondazione Biblioteca San Bernardo, Trento 16108
Fondazione Bruno Kessler, Trento 16612
Fondazione Cassa di Risparmio in Bologna, Bologna 16191
Fondazione Centro di Documentazione Ebraica Contemporanea (CDEC), Milano 16335
Fondazione – Centro Studi 'Nicolò Rusca' (ONLUS), Como 15800
Fondazione Collegio San Carlo, Modena 15913
Fondazione Culturale San Fedele, Milano 15908
Fondazione "Ettore Pomarici Santomasi", Gravina di Puglia 16296
Fondazione Gaudenzio e Palmira Giovanoli, Maloggia 27409
Fondazione Giangiacomo Feltrinelli, Milano 16336
Fondazione Istituto d'Arte e Mestieri Vincenzo Roncalli, Vigevano 16674
Fondazione Istituto "Gramsci", Roma 16500
Fondazione 'L. Einaudi', Torino 16601
Fondazione Luigi Firpo, Torino 16602
Fondazione M. A. Prolo, Torino 16603
Fondazione Mansutti, Milano 16337
Fondazione Marcianum – Biblioeca, Venezia 16127
Fondazione 'Marco Besso', Roma 16501
Fondazione Museo Carnico delle Arti Popolari "Michele Gortani", Tolmezzo 16587
Fondazione Museo Internazionale delle Ceramiche in Faenza, Faenza 16241
Fondazione "Parschalk", Bressanone 15757
Fondazione Primoli, Roma 16502
Fondazione Scientifica 'Querini-Stampalia', Venezia 16649
Fondazione Torino Musei, Turin 16633
Fondazione Ugo Da Como, Lonato del Garda 16306
Fondazione"'Giorgio Cini", Venezia 16650
La Fonderie, Bruxelles 02589
Fondo Nacional de las Artes, Buenos Aires 00255
Fonds Documentaire Spécialisé Mammifères et Oiseaux, Paris 08543
Fonds Régional d'Art Contemporain de Bretagne, Châteaugiron 08330
Fondulac Public Library District, East Peoria 38910
Fontana Regional Library, Bryson City 38379
Fontbonne University, Saint Louis 32502
Fontys Hogescholen, Eindhoven 19110
Fontys Hogescholen, Tilburg
– Brabants Conservatorium, Dansacademie, Tilburg 19202
– Faculteit Educatieve Opleidingen – Lerarenopleiding Basisonderwijs en Opleidingen Speciaal Onderwijs, Tilburg 19203
– Faculteit Educatieve Opleidingen – Moller Instituut, Tilburg 19204
– Faculteit Educatieve Opleidingen – Voltijd- en Deeltijdopleidingen, Tilburg 19205
Fontys Pedagogische Opleidingen 's-Hertogenbosch: PABO, PMK en Pedagogiek, 's-Hertogenbosch 19146
Food and Agriculture Organization of the United Nations (FAO), Roma 16503
Food Industry Survey Joint-Stock Company, Ekaterinburg 22788
Food Research Institute, Accra 13044
Food Science Australia, North Ryde 00971
Food Science Australia, Werribee 01046
Foothill College, Los Altos Hills 33497
Forbes Library, Northampton 40531
Forbes Mellon Library, Cambridge 28984
Force-Publique, Etat-Major, Kinshasa 06061
Førde bibliotek, Førde 20331
Förderstiftung Konservative Bildung und Forschung (FKBF), Berlin 11519
Förderverein des Kreismuseums des Salzlandkreises e.V., Schönebeck 12474
Fordham University, Bronx 30557
Fordham University at Lincoln Center, New York 32063
Fordham University Library at Marymount, West Harrison 33016
Fordham University School of Law, New York 32064
Foreign Affairs Canada & International Trade Canada Library, Ottawa 04113
Foreign Affairs College, Beijing 05046
Foreign Correspondents' Club of Japan, Tokyo 17720
Foreign Languages Literature Library, Riga 18199
Foreign Ministry, Sofiya 03491
Foreign Missions Society of Quebec Library, Laval 04372
Foreign Trade Institute, Roma 16528
Forest Department Library, Kampala 27925
Forest Grove City Library, Forest Grove 39093
Forest History Society Library, Durham 36816
Forest Institute of the Academy of Sciences of Belarus, Gomel 01953
Forest Lake Library, Forest Lake 39096
Forest Park Public Library, Forest Park 39099
Forest Products Laboratory Library, Madison 34486
Forest Products Research and Development Division, Bangkok 27778
Forest Research and Design Institute, Bucureşti 21984
Forest Research and Development Centre, Bogor 14352
Forest Research Institute, Dehradun 13976
Forest Research Institute, Kuala Lumpur 18584
Forest Research Institute, Pyinmana 18999
Forest Research Institute of Malawi, Zomba 18492
Forestry and Forest Products Research Institute (FFPRI), Ibaraki 17550
Forestry and Forest Products Research Institute, Hokkaido Branch, Sapporo 17658
Forestry and Forest Products Research Institute, Kansai Branch, Kyoto 17591
Forestry and Forest Products Research Institute, Kyushu Branch, Kumamoto 17582
Forestry and Forest Products Research Institute, Shikoku Branch, Kochi 17576
Forestry and Forest Products Research Institute, Tohoku Branch, Morioka 17608
Forestry and Game Management Research Institute, Jíloviště 06455
Forestry College Library, Bulolo 20550
Forestry Commission, Farnham 29514
Forestry Department, Suva 07315
Forestry Institute, Shchuchinsk 17975
Forestry Research Institute, Freiburg 11892
Forestry Research Institute, Sankt-Peterburg 23832
Forestry Research Institute, Sofiya 03551
Forestry Research Institute, Warszawa 21267
Forestry Research Institute of Nigeria (FRIN), Ibadan 20016
Forging and Pressing Automatated Lines Plant, Pinsk 01922
Forintek Canada Corp, Qébec 04275
Forked Deer Regional Library Center, Halls 39348
Forman Christian College, Lahore 20460
Føroya Landsbókasavn, Tórshavn 06631
Forschungs- und Dokumentationszentrum Chile Lateinamerika (FDCIL) e.V., Berlin 11520
Forschungs- und Kulturverein fuer Kontinentalamerika und die Karibik, Wien 01588
Forschungsanstalt Agroscope Liebefeld-Posieux ALP, Posieux 27423
Forschungsanstalt Agroscope Reckenholz-Tänikon, Ettenhausen 27352
Forschungsanstalt Geisenheim, Geisenheim 11916
Forschungsbereich für Verkehrsplanung und Verkehrstechnik, Wien 01309
Forschungsbibliothek der Deutschen Augustiner, Würzburg 11364
Forschungsbibliothek Gotha, Gotha 09835
Forschungsgesellschaft für Straßen- und Verkehrswesen, Köln 12129
Forschungsinstitut der Zementindustrie GmbH, Düsseldorf 11792
Forschungsinstitut für Genossenschaftswesen, Nürnberg 12372
Forschungsinstitut für NE-Metalle GmbH (FNE), Freiberg 11400
Forschungsinstitut für Pigmente und Lacke e.V., Stuttgart 12517
Forschungsinstitut für politisch-historische Studien, Salzburg 01565
Forschungsinstitut für Wirtschaftsverfassung und Wettbewerb e.V. (FIW), Köln 12130
Forschungsstätte der Evangelischen Studiengemeinschaft, Heidelberg 11241
Forschungsstelle für Brandschutztechnik, Karlsruhe 12086
Forschungsstelle für Jagdkunde und Wildschadenverhütung, Bonn 11638
Forschungsstelle für Zeitgeschichte in Hamburg, Hamburg 11978
Forschungsstelle Georg Büchner, Marburg 10328
Forschungsstelle Osteuropa an der Universität Bremen, Bremen 11675
Forschungszentrum Borstel, Borstel 11653
Forschungszentrum Europäische Aufklärung e.V., Potsdam 12404
Forschungszentrum Jülich GmbH, Jülich 12082
Forshaga folkbibliotek, Forshaga 26770
Forssa Public Library, Forssa 07561
Forssan kaupunginkirjasto, Forssa 07561
Forstliche Versuchs- und Forschungsanstalt Baden-Württemberg, Freiburg 11892
Försvarets Forskningsanstalt, Umeå 26702
Forsvarets forskningsinstitutt, Kjeller 20231
Forsvarets Høgskole (FHS), Oslo 20114
Försvarets materielverk, Stockholm 26501
Försvarets Radioanstalt, Drottningholm 26486
Forsvarsmuseet, Oslo 20189
Forsyth County Public Library, Cumming 38744
Forsyth County Public Library, Cumming 38745
Forsyth County Public Library, Winston-Salem 41938
Forsyth County Public Library – Kernersville Branch, Kernersville 39708
Forsyth County Public Library – Reynolda Manor, Winston-Salem 41939
Forsyth Technical Community College, Winston-Salem 33880
Fort Belknap Archives, Inc Library, Newcastle 36049
Fort Bend County Libraries – Cinco Ranch, Katy 39684
Fort Bend County Libraries – First Colony, Sugar Land 41467
Fort Bend County Libraries – George Memorial Library, Richmond 40982
Fort Bend County Libraries – Missouri City Branch, Missouri City 40241
Fort Bend County Libraries – Sugar Land Branch, Sugar Land 41468
Fort Collins Public Library, Fort Collins 39102
Fort Dodge Public Library, Fort Dodge 39103
Fort Erie Public Library, Fort Erie 04640
Fort Frances Public Library, Fort Frances 04641
Fort Hays State University, Hays 31319
Fort Lewis College, Durango 31035
Fort Loudoun Regional Library Center, Athens 38010
Fort Madison Public Library, Fort Madison 39105
Fort McMurray Public Library, Fort McMurray 04642
Fort Myers Beach Library, Fort Myers Beach 39109
Fort Saskatchewan Public Library, Fort Saskatchewan 04643
Fort Smith Public Library, Fort Smith 39113
Fort St. John Public Library, Fort St. John 04644
Fort Steele Heritage Town Library, Fort Steele 04344
Fort Stockton Public Library, Fort Stockton 39114
Fort Ticonderoga Museum, Ticonderoga 37670
Fort Valley State University, Fort Valley 31163
Fort Vancouver Regional Library District, Vancouver 41665
Fort Vancouver Regional Library District – Three Creeks Community Library, Vancouver 41666
Fort Vancouver Regional Library District – Vancouver Community Library (Main Library), Vancouver 41667
Fort Walton Beach Library, Fort Walton Beach 39116
Fort Worth Public Library, Fort Worth 39125
Fort Worth Public Library – East Berry, Fort Worth 39126
Fort Worth Public Library – East Regional, Fort Worth 39127
Fort Worth Public Library – Meadowbrook, Fort Worth 39128
Fort Worth Public Library – Ridglea, Fort Worth 39129
Fort Worth Public Library – Riverside, Fort Worth 39130
Fort Worth Public Library – Seminary South, Fort Worth 39131
Fort Worth Public Library – Southwest Regional, Fort Worth 39132
Fort Worth Public Library – Wedgwood, Fort Worth 39133
Fortifikationsförvaltningen, Eskilstuna 26568
Fortis, Utrecht 19338
Fortress of Louisbourg Library, Louisbourg 04378
Fortville-Vernon Township Public Library, Fortville 39134
Forum Culturel Autrichien, Paris 08544
Forum Scientiarum, Tübingen 12558
Förvaltningsbiblioteket, Region Skåne, Lund 26496
Fossil Ridge Public Library, Braidwood 38259
Foster Associates, Inc Library, Bethesda 35498
Foster Pepper PLLC, Seattle 36264
Fotomuseum Provincie Antwerpen, Antwerpen 02514
Foundation Centre for Jewish Contemporary Documentation, Milano 16335
Foundation for Blind Children Library & Media Center, Phoenix 37369
Foundation for Economic and Industrial Research (IOBE), Athinai 13108
Foundation for Economic and Social Research, Madrid 26019
Foundation Historical Association, Inc Library, Auburn 36497
Foundation Marcianum – Library, Venezia 16127
Foundation University, Dumaguete City 20697
Foundry Research Institute, Kraków 21138
Fountaindale Public Library District, Bolingbrook 38236

Fouqué-Bibliothek, Brandenburg an der Havel 12684
Four County Library System, Vestal 41678
Fourah Bay College, Freetown 24567
The Fourth Military Medical University, Xi'an 05679
Fővárosi Szabó Ervin Könyvtár, Budapest 13530
Fowler, White, Boggs & Banker, Tampa 36297
Fox Lake Correctional Institution Library, Fox Lake 34278
Fox Lake District Library, Fox Lake 39138
Fox Rothschild LLP, Philadelphia 36099
Fox Valley Technical College, Appleton 33130
Framatome ANP Inc, Lynchburg 35867
Framingham State College, Framingham 31169
France Bevk Public Library, Nova Gorica 24974
France Télécom R&D, Issy-les-Moulineaux 08364
Francis Marion University, Florence 31144
"Francisc I. Rainer" Anthropology Institute of the Romanian Academy, Bucureşti 21979
Franciscan Abbey, Kamnik 24896
Franciscan Library, Killiney 14514
Franciscan Library, Ljubljana 24897
Franciscan Monastery Library, Washington 35407
Franciscan Monastery of Bolzano, Bozen 15751
Franciscan Order of Friars Minor, Box Hill 00753
Franciscan Provincial Library, Valletta 18666
Franciscan University of Steubenville, Steubenville 32741
Franciscéumbibliothek, Zerbst 09344
Frančiškanska knjižnica, Ljubljana 24897
Frančiškanski samostan in cerkev sv. Jakoba, Kamnik 24896
Franckesche Stiftungen, Halle (Saale) 11956
Franco-American Centre, Manchester 37081
Francois-Xavier Garneau Collège, Québec 03826
Francoski Inštitut Charles Nodier, Ljubljana 24917
Francouzský institut v Praze, Praha 06491
Frank L. Weyenberg Library of Mequon-Thiensville, Mequon 40151
Frank Phillips College, Borger 30477
Frankfort Community Public Library, Frankfort 39140
Frankfort Public Library District, Frankfort 39141
Frankfurter Forschungsbibliothek, Frankfurt am Main 11848
Frankfurter Goethe-Museum, Frankfurt am Main 11850
Franklin College, Franklin 31172
Franklin College, Lugano 27112
Franklin County Law Library, Chambersburg 34055
Franklin County Library – Louisburg Main Library, Louisburg 39964
Franklin County Public Library, Rocky Mount 41050
Franklin Institute, Philadelphia 37346
Franklin Lakes Free Public Library, Franklin Lakes 39150
Franklin & Marshall College, Lancaster 31552
– Martin Library of the Sciences, Lancaster 31553
The Franklin Mint, Franklin Center 36878
Franklin Parish Library, Winnsboro 41935
Franklin Park Public Library District, Franklin Park 39151
Franklin Pierce College, Rindge 32446
Franklin Pierce Law Center Library, Concord 36748
Franklin Public Library, Franklin 39144
Franklin Square Public Library, Franklin Square 39152
Franklin T. Degroodt Library, Palm Bay 40690
Franklin Township Free Public Library, Somerset 41344
Franklin-Springboro Public Library, Franklin 39145
Franko Recreation Centre, Donetsk 28548
Franko State Regional Children's Library, Vinnytsya 28868
Franko Teacher Training Institute, Drogobych 27975
Frankston City Libraries, Frankston 01095
Det Franske Instituts Bibliotek / Médiathèque de l'Institut Français, København 06795
Franz Hitze Haus, Münster 10757
Franz Nabl-Institut für Literaturforschung, Graz 01523
Franziskanerbibliothek, Schwaz 01469
Franziskanerbibliothek Maria Enzersdorf, Maria Enzersdorf 01451
Franziskanerkloster, Bozen 15751
Franziskanerkloster, Dietfurt 11192
Franziskanerkloster, Fribourg 27260
Franziskanerkloster Maria im Sand, Dettelbach 11191
Franziskanerkloster Salzburg, Salzburg 01462
Franziskanerkloster St. Anna, München 11279
Franziskaner-Minoritenkloster, Würzburg 11367
Französische Bibliothek, Essen 11823
Französische Mediathek Wien, Wien 01597
Fraser Coast Regional Council, Hervey Bay 01100
Fraser, Milner & Casgrain, Calgary 04240
Fraser Milner Casgrain LLP, Edmonton 04254
Fraser Milner Casgrain LLP, Barristers & Solicitors, Toronto 04285
Fraser Public Library, Fraser 39154
Fraser-Hickson Institute, Montreal, Montreal 04692
Fraternita dei Laici, Arezzo 15707
Fraternité Saint-Dominique, Kigali 24372
Frau und Musik, Frankfurt am Main 11836
frauenbibliothek saar, Saarbrücken 12451
Frauenbibliothek und Fonothek Wyborada, St.Gallen 27436
Frauenforschungs-, -bildungs- und -informationszentrum (FFBIZ), Berlin 11518
Frauengesundheitszentrum, Graz 01524
Frauen-Literatur-Forschung e.V., Stiftung, Bremen 11689
FrauenMediaTurm, Köln 12131
Fraunhofer IGD Bibliothek, Darmstadt 11722
Fraunhofer IGD Library, Darmstadt 11722
Fraunhofer Institut für Toxikologie und Experimentelle Medizin, Hannover 12025
Fraunhofer Institut Photonische Mikrosysteme, Dresden 11754
Fraunhofer IRB, Stuttgart 12518
Fraunhofer-Institut für Angewandte Festkörperphysik, Freiburg 11893
Fraunhofer-Institut für Arbeitswirtschaft und Organisation (IAO) / Institut für Arbeitswissenschaft und Technologiemanagement der Universität Stuttgart (IAT), Stuttgart 12519
Fraunhofer-Institut für Bauphysik, Stuttgart 12520
Fraunhofer-Institut für Betriebsfestigkeit und Systemzuverlässigkeit LBF, Darmstadt 11723
Fraunhofer-Institut für Chemische Technologie (ICT), Pfinztal 12393
Fraunhofer-Institut für Graphische Datenverarbeitung, Darmstadt 11722
Fraunhofer-Institut für Holzforschung, Braunschweig 11660
Fraunhofer-Institut für Informations- und Datenverarbeitung, Karlsruhe 12087
Fraunhofer-Institut für Integrierte Schaltungen (IIS), Dresden 11755
Fraunhofer-Institut für Materialfluss und Logistik IML, Dortmund 11743
Fraunhofer-Institut für Mikroelektronische Schaltungen und Systeme, Duisburg 11785
Fraunhofer-Institut für Produktionstechnik und Automatisierung (IPA) / Institut für Industrielle Fertigung und Fabrikbetrieb der Universität Stuttgart (IFF), Stuttgart 12521
Fraunhofer-Institut für Silicatforschung, Würzburg 12613
Fraunhofer-Institut für System- und Innovationsforschung (ISI), Karlsruhe 12088
Fraunhofer-Institut für Verkehrs- und Infrastruktursysteme (IVI), Dresden 11756
Fraunhofer-Institut für Werkstoffmechanik, Freiburg 11894
Fraunhofer-Institut für Werkzeugmaschinen und Umformtechnik (IWU), Chemnitz 11709
Fraunhofer-Institute for Integrated Circuits (IIS), Design Automation Division (EAS), Research Library, Dresden 11755
Fred C. Fischer Library, Belleville 38154
Fredensborg Bibliotek, Fredensborg 06859
'Frédéric Joliot-Curie' National Research Institute for Radiobiology and Radiohygiene, Budapest 13437
Fredericia Bibliotek, Fredericia 06860
Frederick Community College, Frederick 33352
Frederick County Law Library, Frederick 34283
Frederick County Public Libraries, Frederick 39155
Frederiksberg Bibliotek, Frederiksberg 06861
Frederikshavn Bibliotek, Frederikshavn 06862
Frederikshavn Gymnasium og HF, Frederikshavn 06730
Frederikssund Bibliotek, Frederikssund 06863
Frederiksværk-Hundested Bibliotekerne, Frederiksværk 06864
Fredrikson & Bryon, Minneapolis 35908
Free and Hanseatic City of Hamburg, Ministry of Economy, Transport and Innovation, Hamburg 10921
Free Church College, Edinburgh 29541
Free Library of Northampton Township, Richboro 40972
Free Library of Philadelphia, Philadelphia 40771
Free Library of Philadelphia – Central Children's Department, Philadelphia 40772
Free Library of Philadelphia – Interlibrary Loan, Philadelphia 40773
Free Library of Philadelphia – Lucien E. Blackwell West Philadelphia Regional, Philadelphia 40774
Free Library of Philadelphia – Northeast Regional, Philadelphia 40775
Free Library of Philadelphia – Social Science & History, Philadelphia 40776
Free Library of Springfield Township, Wyndmoor 41980
Free Methodist Church of North America, Indianapolis 35240
Free Methodist Church of North America, Indianapolis 35241
Free Public Library, Ahmedabad 14222
Free Public Library of Bayonne, Bayonne 38120
Free Public Library of Hasbrouck Heights, Hasbrouck Heights 39392
Free Public Library of Monroe Township, Williamstown 41901
Free Public Library of Woodbridge, Woodbridge 41953
Free Public Library of Woodbridge – Fords Branch, Fords 39092
Free University Brussels, Bruxelles 02298
Free University of Bozen / Bolzano, Bozen 14911
Free Will Baptist Bible College, Nashville 35296
Freed-Hardeman University, Henderson 31326
Freedoms Foundation Library, Valley Forge 37704
Freeport Community Library, Freeport 39158
Freeport Memorial Library, Freeport 39159
Freeport Public Library, Freeport 39160
Freer Gallery of Art & Arthur M. Sackler Gallery, Washington 37783
Freie und Hansestadt Hamburg, Behörde für Schule und Berufsbildung, Hamburg 10917
Freie und Hansestadt Hamburg, Behörde für Schule und Berufsbildung – Bibliothek Gesundheit, Hamburg 10918
Freie und Hansestadt Hamburg, Behörde für Stadtentwicklung und Umwelt – Bibliothek Stadtentwicklung, Hamburg 10919
Freie und Hansestadt Hamburg, Behörde für Stadtentwicklung und Umwelt – Bibliothek Umwelt, Hamburg 10920
Freie und Hansestadt Hamburg, Behörde für Wirtschaft, Verkehr und Innovation, Hamburg 10921
Freie Universität Berlin, Berlin 09406
– Ägyptologisches Seminar, Berlin 09407
– Bereichsbibliothek Biologie im Botanischen Museum, Berlin 09408
– Bereichsbibliothek Chemie, Berlin 09409
– Bereichsbibliothek Erziehungswissenschaft, Fachdidaktik und Psychologie, Berlin 09410
– Bereichsbibliothek Pharmazie, Berlin 09411
– Bibliothek am Botanischen Garten und Botanischen Museum Berlin-Dahlem, Berlin 09412
– Bibliothek für Publizistik, Berlin 09413
– Fachbereich Mathematik und Informatik, Berlin 09414
– Fachbereich Physik, Berlin 09415
– Fachbereich Rechtswissenschaft, Berlin 09416
– Friedrich-Meinecke-Institut, Berlin 09417
– Geowissenschaftliche Bibliothek, Berlin 09418
– Institut für Altorientalistik, Berlin 09419
– Institut für Iranistik, Berlin 09420
– Institut für Islamwissenschaft, Berlin 09421
– Institut für Judaistik, Berlin 09422
– Institut für Klassische Archäologie, Berlin 09423
– Institut für Meteorologie, Berlin 09424
– Institut für Musikwissenschaften, Seminar für Vergleichende Musikwissenschaft, Berlin 09425
– Institut für Prähistorische Archäologie, Berlin 09426
– Institut für Soziologie, Berlin 09427
– Institut für Theaterwissenschaft, Berlin 09428
– Institut für Theaterwissenschaft – Musikwissenschaftliches Seminar, Berlin 09429
– Institut für Turkologie, Berlin 09430
– Institut für Vorderasiatische Archäologie, Berlin 09431
– John-F.-Kennedy-Institut für Nordamerikastudien, Berlin 09432
– Kunsthistorisches Institut, Berlin 09433
– Ostasiatisches Seminar, Fachrichtung Japanologie, Berlin 09434
– Ostasiatisches Seminar, Fachrichtung Koreanistik, Berlin 09435
– Ostasiatisches Seminar, Fachrichtung Sinologie, Berlin 09436
– Philologische Bibliothek, Berlin 09437
– Seminar für Katholische Theologie, Berlin 09438
– Seminar für Semitistik und Arabistik, Berlin 09439
– Seminar für Ur- und Frühgeschichte, Berlin 09440
– Sozialwissenschaftliche Bibliothek & Bibliothek des Osteuropa-Instituts, Berlin 09441
– Theaterhistorische Sammlung Walter Unruh, Berlin 09442
– Veterinärmedizinische Bibliothek, Berlin 09443
– Wirtschaftswissenschaftliche Bibliothek, Berlin 09444
Freie Universität Bozen / Libera Università di Bolzano, Bozen 14911
Freies Deutsches Hochstift – Frankfurter Goethe-Museum, Frankfurt am Main 11850
Fremantle Hospital & Health, Fremantle 00921
Fremont Area District Library, Fremont 39163
Fremont County Library System, Lander 39798
Fremont County Library System – Riverton Branch, Riverton 41015
Fremont Public Library District, Mundelein 40339
French Atomic Energy Commission DSM/SAC/STI/SVI Library, Gif-sur-Yvette 08351
French Cultural Center of Istanbul, Istanbul 27890
French Cultural Centre, Nairobi 18034
French Cultural Centre, Vilnius 18303
French Cultural Library, Helsinki 07447
French Institute, Edinburgh 29655
French Institute Library, Zagreb 06243
French Institute of Oriental Archaeology, Cairo 07175
French Institute-Alliance Francaise Library, New York 37204
The French Library & Alliance Francaise of Boston, Boston 36563
French School of Archaeology, Athinai 13074
Freshwater Biological Association, Ambleside 29596
Fresno City College, Fresno 33356
Fresno County Law Library, Fresno 34288
Fresno County Office of Education, Fresno 34289
Fresno County Public Library, Fresno 39166
Fresno County Public Library – Central Branch, Fresno 39167
Fresno County Public Library – Clovis Regional, Clovis 38618
Fresno County Public Library – Fig Garden Regional, Fresno 39168
Fresno County Public Library – Sanger Branch, Sanger 41210
Fresno County Public Library – Woodward Park Regional, Fresno 39169
Fresno Pacific University, Fresno 31180
Freudenberg Forschungsdienste KG, Weinheim 11433
The Frick Collection, New York 37205
Den Frie Lærerskole, Vester Skerninge 06751
Fried, Frank, Harris, Shriver & Jacobson, Washington 36353
Fried, Frank, Harris, Shriver & Jacobson Library, New York 35982
Fried, Frank, Harris, Shriver & Jacobson LLP, Washington 36354
Friedensau Adventist University, Friedensau 09807
Friedensbibliothek – Österreichisches Studienzentrum für Frieden und Konfliktlösung – ÖSFK, Stadtschlaining 01580

Friedrich-Alexander-Universität Erlangen-Nürnberg, Erlangen 09690
– Klinik und Jugendklinik, Erlangen 09721
– Klinik und Polikliniken für Zahn-, Mund- und Kieferkranke, Erlangen 09722
– Universitätsbibliothek, Erziehungswissenschaftliche Zweigbibliothek (EZB), Nürnberg 09691
– Universitätsbibliothek, Technisch-naturwissenschaftliche Zweigbibliothek (TNZB), Erlangen 09692
– Universitätsbibliothek, Teilbibliothek 01: Theologie, Erlangen 09694
– Universitätsbibliothek, Teilbibliothek 01AT: Altes Testament, Erlangen 09695
– Universitätsbibliothek, Teilbibliothek 02: Rechtswissenschaft, Erlangen 09696
– Universitätsbibliothek, Teilbibliothek 02KR: Kirchenrecht, Erlangen 09697
– Universitätsbibliothek, Teilbibliothek 03: Medizin, Erlangen 09698
– Universitätsbibliothek, Teilbibliothek 03GM: Geschichte und Ethik der Medizin, Erlangen 09699
– Universitätsbibliothek, Teilbibliothek 04: Pädagogik, Philosophie und Psychologie, Erlangen 09700
– Universitätsbibliothek, Teilbibliothek 05: Politische Wissenschaft, Soziologie, Wirtschaftswissenschaft (Erlangen), Erlangen 09701
– Universitätsbibliothek, Teilbibliothek 06: Geschichte, Alte Sprachen, Altertumskunde, Kunst und Musik, Erlangen 09702
– Universitätsbibliothek, Teilbibliothek 06KP: Klassische Philologie, Erlangen 09703
– Universitätsbibliothek, Teilbibliothek 06LM: Lateinische Philologie des Mittelalters und der Neuzeit, Erlangen 09704
– Universitätsbibliothek, Teilbibliothek 07: Neuere Sprachen, Theater- und Medienwissenschaften, Erlangen 09705
– Universitätsbibliothek, Teilbibliothek 07AM: Amerikanistik, Erlangen 09706
– Universitätsbibliothek, Teilbibliothek 07BW: Buchwissenschaft, Erlangen 09707
– Universitätsbibliothek, Teilbibliothek 07EN: Englische Philologie, Erlangen 09708
– Universitätsbibliothek, Teilbibliothek 07LI: Angewandte Sprachwissenschaft, Erlangen 09709
– Universitätsbibliothek, Teilbibliothek 07TH: Theater- und Medienwissenschaft, Erlangen 09710
– Universitätsbibliothek, Teilbibliothek 08: Mathematik, Erlangen 09711
– Universitätsbibliothek, Teilbibliothek 09: Physik und Astronomie, Erlangen 09712
– Universitätsbibliothek, Teilbibliothek 09GP: Gruppenbibliothek Physik, Erlangen 09713
– Universitätsbibliothek, Teilbibliothek 10: Biologie, Erlangen 09714
– Universitätsbibliothek, Teilbibliothek 11: Chemie, Erlangen 09715
– Universitätsbibliothek, Teilbibliothek 12: Geowissenschaften, Erlangen 09716
– Universitätsbibliothek, Teilbibliothek 14: Ingenieurwissenschaften, Erlangen 09717
– Universitätsbibliothek, Teilbibliothek 15: Erziehungswissenschaften, Nürnberg 09718
– Universitätsbibliothek, Teilbibliothek 98SP: Institut für Sportwissenschaft und Sport, Erlangen 09719
– Universitätsbibliothek, Wirtschafts- und Sozialwissenschaftliche Teilbibliothek, Nürnberg 09720
– Universitätsbibliothek, Wirtschafts- und Sozialwissenschaftliche Zweigbibliothek (WSZB), Nürnberg 09693
Friedrich-Ebert-Stiftung, Bonn 11633
Friedrich-Loeffler-Institut, Celle 11707
Friedrich-Loeffler-Institut, Greifswald 11937
Friedrich-Loeffler-Institut, Jena 12076
Friedrich-Loeffler-Institut, Neustadt a. Rbge. 12364
Friedrich-Loeffler-Institut, Tübingen 12559
Friedrich-Loeffler-Institut, Wusterhausen 12621
Friedrich-Schiller-Universität Jena, Jena 09320
Friedrich-von-Hardenberg-Institut für Kulturwissenschaften, Heidelberg 12046
Friends Free Library of Germantown, Philadelphia 40777
Friends Historical Library of Swarthmore College, Swarthmore 32767
Friends University, Wichita 33055
Friendswood Public Library, Friendswood 39172
Frisian historical and literary centre, Leeuwarden 19048
Fritz-Haber-Institut der Max-Planck-Gesellschaft, Berlin 11521
Fritz-Hüser-Institut, Dortmund 11744
Fritz-Reuter-Literaturmuseum, Stavenhagen 12501
Frobenius-Institut an der Johann Wolfgang Goethe-Universität, Frankfurt am Main 11851
Froebel College of Education, Blackrock 14459
Front Range Community College, Westminster 33863
Frost Brown Todd LLC, Cincinnati 35608
Frost, Brown & Todd LLC, Louisville 35863
Frostburg State University, Frostburg 31182
Fruitland Baptist Bible Institute, Hendersonville 35228
Fruitlands Museums Library, Harvard 36923
Fruitville Public Library, Sarasota 41229
Frumkin Research Institute of Electrochemistry of the Russian Academy of Sciences, Moskva 23261
Frunze District Centralized Library System, Chekhov Central District Library, Sankt-Peterburg 24250
Frunze District Centralized Library System, Children's Library Branch no 6, Sankt-Peterburg 24249
Frunze District Centralized Library System, Gorky Library Branch no 2, Sankt-Peterburg 24247
Frunze District Centralized Library System, Krylov Central District Children's Library, Sankt-Peterburg 24251
Frunze District Centralized Library System, Prokofiev Library Branch no 4, Sankt-Peterburg 24248
Frunze Military Academy, Moskva 22401
Frunze State University, Simferopol 28073
The Fryderyk Chopin Institute, Warszawa 21327
FS Education Library, Bloemfontein 25106
FTI – SEA, Inc Library, Columbus 35632
Fu Hsing Kang College, Taipei 27543
Fu Jen Catholic University, Taipei 27544
– Social Sciences Library, Taipei 27545
Fudan University, Shanghai 05502
Fuglafjarðar Bókasavn / Fuglefjord Folkebibliotek, Fuglafjørður 06865
Fuglefjord Folkebibliotek, Fuglafjørður 06865
Fugro, Inc, Houston 35753
Führungsakademie der Bundeswehr, Hamburg 09913
Führungsunterstützungsschule der Bundeswehr, Feldafing 10723
Fuji Boseki K.K., Sunto 17426
Fuji Photo Film Co, Ltd, Ashigara Laboratories, Minamiashigara 17384
Fuji Shashin Fuirum K. K. Ashigara Kenkyujo, Minamiashigara 17384
Fuji Spinning Co, Ltd, Institute of Research and Development, Sunto 17426
Fuji Women's College, Sapporo 17103
Fujian Agricultural University, Fuzhou 05178
Fujian Civil Engineering and Architecture College, Fuzhou 05179
Fujian College of Traditional Chinese Medicine, Fuzhou 05180
Fujian Commercial College, Fuzhou 05181
Fujian Financial Management Cadres Institute, Fuzhou 05182
Fujian Forestry College, Nanping 05466
Fujian Hua'nan Women's College, Fuzhou 05183
Fujian Institute of Economic Management, Fuzhou 05184
Fujian Institute of Research on the Structure of Matter, Fuzhou 05861
Fujian Medical College, Fuzhou 05185
Fujian Normal University, Fuzhou 05186
Fujian Physical Education College, Xiamen 05673
Fujian Province Management Cadre Institute of Political Science and Law, Fuzhou 05187
Fujian Provincial Library, Fuzhou 04957
Fujian Public Security College, Fuzhou 05188
Fujian Teachers' University, Fuzhou 05189
Fujian Zhonghua Vocational University, Fuzhou 05190
Fujikura Cable Works, Ltd, Tokyo 17445
Fujikura Densen K. K., Tokyo 17445
Fujisawa City Archives, Fujisawa 17532
Fujisawashi Monjokan, Fujisawa 17532
Fujitsu, Sunnyvale 36290
Fujitsu Kenkyujo Kanribu Shiryoka, Kawasaki 17369
Fujitsu Laboratories, Kawasaki 17369
Fukui Daigaku, Fukui 16916
Fukui Prefectural Assembly, Fukui 17256
Fukui Prefecture Administrative Data Service Center, Fukui 17257
Fukui Prefecture, Agricultural Experiment Station, Fukui 17535
Fukui University, Fukui 16916
Fukuiken Gikai, Fukui 17256
Fukuiken Kensei Shiryohitsu, Fukui 17257
Fukuiken Nogyo Shikenjo, Fukui 17535
Fukuoka Agricultural Research Center, Chikushino 17527
Fukuoka City Assembly, Fukuoka 17259
Fukuoka Daigaku, Fukuoka 16918
– Yakugakubu Bunshitsu, Fukuoka 16919
Fukuoka Dental College, Fukuoka 16920
Fukuoka Prefectural Assembly, Fukuoka 17258
Fukuoka Prefectural Education Center, Kasuya 17565
Fukuoka Prefectural Fisheries Experimental Station, Fukuoka 17536
Fukuoka Prefectural Library, Fukuoka 17837
Fukuoka Shika Daigaku, Fukuoka 16920
Fukuoka University, Fukuoka 16918
– Department of Pharmacy, Fukuoka 16919
Fukuokaken Bunkakaikan Toshobu, Fukuoka 17837
Fukuokaken Gikai, Fukuoka 17258
Fukuokaken Kyoiku Senta, Kasuya 17565
Fukuokaken Nogyo Sogo Shikenjo, Chikushino 17527
Fukuokaken Suisan Shikenjo, Fukuoka 17536
Fukuokashi Gikai, Fukuoka 17259
Fukushima Daigaku, Fukushima 16928
– Tohoku Keizai Kenkyujo, Fukushima 16929
Fukushima Kenritsu Ika Daigaku, Fukushima 16930
Fukushima Medical University Center for Academic Information Services, Fukushima 16930
Fukushima Prefectural Fruit Tree Experiment Station, Fukushima 17538
Fukushima Prefecture, Agricultural Experiment Station, Koriyama 17579
Fukushima University, Fukushima 16928
– Tohoku Economic Research Institute, Fukushima 16929
Fukushimaken Kaju Shikenjo, Fukushima 17538
Fukushimaken Nogyo Shikenjo, Koriyama 17579
Fulbright & Jaworski Library, San Antonio 36216
Fulbright & Jaworski LLP, Houston 35754
Fulbright & Jaworski LLP, Los Angeles 35834
Fuling Teachers' College, Fuling 05172
Full Gospel Bible Institute, Eston 04206
Fuller & Henry Law Library, Toledo 36305
Fuller Theological Seminary, Pasadena 35322
Fullerton College, Fullerton 33357
Fulton County Historical Society, Inc, Rochester 37449
Fulton County Law Library, Atlanta 33942
Fulton County Public Library, Rochester 41028
Fulton Public Library, Fulton 39175
Fulton-Montgomery Community College, Johnstown 33444
Funda Centre, Diepkloof 25090
Fundação 'Armando Alvares Penteado', São Paulo 03206
Fundação Biblioteca Nacional, Rio de Janeiro 02981
Fundação Brasileira para a Conservação da Natureza, Rio de Janeiro 03344
Fundação Calouste Gulbenkian, Lisboa 21787
Fundação Carlos Chagas, São Paulo 03403
Fundação Casa de Rui Barbosa, Rio de Janeiro 03345
Fundação das Artes, São Caetano do Sul 03387
Fundação de Ciência e Tecnologia, Porto Alegre 03318
Fundação de Ciências Aplicadas, São Bernardo do Campo 03386
Fundação Educacional, Votuporanga 03196
Fundação Getúlio Vargas, Rio de Janeiro 03346
Fundaçao Getulio Vargas, São Paulo 03404
Fundação Instituto Tecnológico do Estado de Pernambuco (ITEP), Recife 03321
Fundacáo Joaquim Nabuco, Recife 03322
Fundação Jorge Duprat Figueiredo de Segurança e Medicina do Trabalho (FUNDACENTRO), São Paulo 03405
Fundação Oswaldo Cruz, Rio de Janeiro 03347
Fundação Paulista de Tecnologia e Educação, Lins 03045
Fundação Universidade Regional de Blumenau, Blumenau 03001
Fundació Antoni Tàpies, Barcelona 25899
Fundació Bosch i Cardellach, Sabadell 26128
Fundació Institut Amatller d'Art Hispànic, Barcelona 25900
Fundació Joan Miró, Barcelona 25901
Fundació Josep Laporte, Barcelona 25902
Fundació Pere Tarrés, Barcelona 25903
Fundación Antonio Maura, Madrid 26016
Fundación Bartolomé March, Palma de Mallorca 26112
Fundación Casa de Alba, Madrid 26017
Fundación Casa Ducal de Medinaceli, Sevilla 26150
Fundación de Investigaciones Económicas Latinoamericanas (FIEL), Buenos Aires 00256
Fundación de Investigaciones Marxistas, Madrid 26018
Fundación de las Cajas de Ahorros, Madrid 26019
Fundación de los Ferrocarriles Españoles, Madrid 26020
Fundación Educacional Autónoma de Colombia FEAC, Santafé de Bogotá 05962
Fundacion ESADE, Barcelona 25293
Fundación Fondo para la Investigación Económica y Social, Madrid 26021
Fundación Francisco Largo Caballero, Madrid 26022
Fundación Jardín Botánico 'Joaquin Antonio Uribe', Medellín 06015
Fundación José Ortega y Gasset, Madrid 26023
Fundación Juan March, Madrid 26024
Fundación La Salle de Ciencias Naturales, Caracas 42251
Fundación Laboral Sonsoles Ballvé, Burgos 25939
Fundación Mapfre Estudios, Majadahonda 26100
Fundación Marcelino Botín, Santander 26139
Fundación Pablo Iglesias, Madrid 26025
Fundación Pablo VI, Madrid 26026
Fundación Pastor de Estudios Clásicos, Madrid 26027
Fundación Penzol, Vigo 26182
Fundación Pilar y Joan Miró a Mallorca, Palma de Mallorca 26113
Fundación Privada Universitaria EADA, Barcelona 25294
Fundación Rodríguez-Acosta, Granada 25959
Fundación Sancho el Sabio, Vitoria-Gasteiz 26186
Fundación Sur – Departamento de África, Madrid 26028
Fundación Universidad de América, Santafé de Bogotá 05963
Fundación Universidad de Bogotá 'Jorge Tadeo Lozano', Santafé de Bogotá 05964
Fundación Universitaria Española, Madrid 25423
Fundaziun Tscharner Zernez, Zernez 27446
FUNREI, São João del-Rei 03131
Furman University, Greenville 31262
Furnas Centrais Elétricas, Rio de Janeiro 03263
Fürst Thurn und Taxis Hofbibliothek, Regensburg 09334
Fürstlich Fürstenbergische Hofbibliothek, Donaueschingen 11741
Fürstlich Hohenzollernsche Hofbibliothek, Sigmaringen 09338
Fürstlich Leiningensche Bibliothek, Amorbach 11454
Fürstlich Salm-Salmsche Bibliothek, Isselburg 12074
Fürstlich Schaumburg-Lippische Hofbibliothek, Bückeburg 11702
Fürstlich Waldecksche Hofbibliothek, Bad Arolsen 11464
Fürstliche Bibliothek Corvey, Höxter 12066
Fürstliche Sammlungen Vaduz, Vaduz 18271
Fushun Teachers' College, Fushun 05173
Fushun University, Fushun 05174
The Futures Group Library, Glastonbury 35718
Fuxin Mining Institute, Fuxin 05175
Fuxin Normal College, Fuxin 05806
Fuyang Teachers' College, Fuyang 05176
Fuzhou Teachers' College (Fujian), Fuzhou 05191
Fuzhou University, Fuzhou 05192
Fylkesbiblioteket i Akershus, Kjeller 20354
Fyodorov Institute of Applied Geophysics, Moskva 23310

Fyodorov Research Institute of Mining, Donetsk 28228
Fyzikální ústav AV ČR, Praha 06486
G. N. Chubinashvili Institute of History of Georgian Art, Tbilisi 09230
Gabinete de Estudos Económicos, Lisboa 21788
Gabinete Português de Leitura, Salvador 03377
Gabinetto di Lettura e Società di Incoraggiamento, Padova 16787
Gabinetto Scientifico-Letterario G.P. Vieusseux, Firenze 16259
Gable & Gotwals, Inc, Tulsa 36314
Gabon National Library and Documentation Centre, Libreville 09167
Gabonatermesztési Kutató Intézet Könyvtára, Szeged 13504
Gabrichevsky Epidemiology and Microbiology Institute, Moskva 23392
Gabriel Dumont Institute Library, Regina 04466
Gabriele Petkevicaite-Bite Panevezys County Public LIbrary, Panevėžys 18348
Gadjah Mada University, Yogyakarta 14313
– Faculty of Agriculture, Yogyakarta 14320
– Faculty of Economics, Yogyakarta 14314
– Faculty of Forestry, Yogyakarta 14319
– Faculty of Law, Yogyakarta 14316
– Faculty of Letters and Culture, Yogyakarta 14321
– Faculty of Philosophy, Yogyakarta 14315
– Faculty of Physics and Mathematics, Yogyakarta 14317
– Faculty of Social and Political Science, Yogyakarta 14322
– Faculty of Veterinary Medicine, Yogyakarta 14318
Gadjah Mada University, Yogyakarta 14307
Gadsden County Public Library, Quincy 40918
Gadsden State Community College, Gadsden 33360
Gadsden-Etowah County Library – Gadsden Public Library, Gadsden 39176
Gadsen Correctional Institution Library, Quincy 34717
Gadyach Regional Centralized Library System, Main Library, Gadyach 28564
Gadyatska raionna TsBS, Tsentralna biblioteka, Gadyach 28564
Gagarin Children's Library, Kyiv 28651
Gagnefs folkbibliotek, Djurås 26754
A. Gaidar Children's Library, Kyiv 28643
Gaidar District Children's Library, Kaliningrad 24135
Gaidar District Children's Library, Kostroma 24148
Gaidar Mogilev Region Children's Library, Mogilev 02202
Gaidar Pedagogical Institute, Arzamas 22205
Gaidar Regional Children's Library, Kirovograd 28617
Gaidar Vitebsk Central City Children's Library, Vitebsk 02248
Gail Borden Public Library District, Elgin 38950
Gaimusho Toshokan, Tokyo 17301
Gainesville State College, Oakwood 33594
Gaisin Regional Centralized Library System, Main Library, Gaisin 28565
Gaisinska raionna TsBS, Tsentralna biblioteka, Gaisin 28565
Gaivoron Regional Centralized Library System, Main Library, Gaivoron 28566
Gaivoronska raionna TsBS, Tsentralna biblioteka, Gaivoron 28566
Gakken Co, Ltd, Tokyo 17446
Gakushuin Daigaku, Tokyo 17137
– Hokei Toshoshitsu, Tokyo 17138
– Rigakubu, Tokyo 17139
Gakushuin Daigaku Shiryokan, Tokyo 17674
Gakushuin University, Tokyo 17137
– Faculty of Science, Tokyo 17139
– Library of Law and Economics, Tokyo 17138
Gakushuin University Archives, Tokyo 17674
Gakushu-Kenkyusha Shiryoshitsu, Tokyo 17446
Galax-Carroll Regional Library, Galax 39183
Gale Free Library, Holden 39454
Gale Group, Farmington Hills 35703
Galéria Júliusa Jakobyho, Košice 24732
Galéria mesta Bratislavy, Bratislava 24689
Galerie für Zeitgenössische Kunst Leipzig, Leipzig 12203
Galerija umjetnina, Split 06222
Galesburg Public Library, Galesburg 39185
Galich Regional Centralized Library System, Main Library, Galich 28567
Galion Public Library Association, Galion 39186
Galitska raionna TsBS, Tsentralna biblioteka, Galich 28567
Galkin Physical and Technological Institute, Donetsk 28217
Gallaudet University, Washington 32966
Galleria Nazionale d'Arte Moderna (GNAM), Roma 16504
Gallery of Fine Arts, Litoměřice 06463
Gallia County District Library, Gallipolis 39187
Galliera, Paris 08545
Gällivare folkbibliotek, Malmberget 26848
Gällivare sjukhus, Gällivare 26571
Gallop, Johnson & Neuman LC, Saint Louis 36199
Galveston College, Galveston 33365
Galway County Libraries, Galway 14576
Galway-Mayo Institute of Technology, Galway 14483
Gamalei Institute of Epidemiology and Microbiology, Moskva 23335
Gambia College, Brikama 09179
The Gambia National Library, Banjul 09178
Gambia Technical Training Institute Library, Banjul 09180
Den Gamle By, Århus 06770
M. K. Gandhi Library, Durban 25120
Gandhi National Museum and Library, New Delhi 14127
Gandhigram Rural Institute, Gandhigram 13646
Ganganatha Jha Kendriya Sanskrit Vidyapeetha, Allahabad 13926
Gannan Medical College, Ganzhou 05196
Gannan Teachers' College, Ganzhou 05197
Gannon University, Erie 31092
Gansu Agricultural University, Lanzhou 05375
Gansu Institute of Political Science and Law, Lanzhou 05376
Gansu Provincial Library, Lanzhou 04965
Gansu University of Technology, Lanzhou 05377
Garda College, Templemore 14492
Garden City College of Ministries, MT Gravatt 00486
Garden City Community College, Garden City 33366
Garden City Public Library, Garden City 39191
Garden City Public Library, Garden City 39192
The Garden Library at Planting Fields, Oyster Bay 37316
Garden State Youth Correctional Facility Library, Yardville 35093
Gardendale Martha Moore Public Library, Gardendale 39198
Gardere & Wynne, Dallas 35646
Gardere, Wynne & Sewell Library, Houston 35755
Gardner-Webb University, Boiling Springs 30473
Garfield County Public Library System, New Castle 40402
Garfield County Public Library System – New Castle Branch, New Castle 40403
Garfield Free Public Library, Garfield 39200
Garland County Library, Hot Springs 39482
Garnet A. Wilson Public Library of Pike County, Waverly 41779
Garrett-Evangelical & Seabury-Western Theological Seminaries, Evanston 31104
Garrison Library, Gibraltar 13065
Gartner Lee Ltd Library, Markham 04262
Garvey, Schubert & Barer, Seattle 36265
Gary Public Library, Gary 39207
Gary Public Library – John F. Kennedy Branch, Gary 39208
Gary Public Library – Ora L. Wildermuth Branch, Gary 39209
Gary Public Library – W. E. B. Du Bois Branch, Gary 39210
Garyounis University, Benghazi 18238
– Beida Campus, Benghazi 18239
– Faculty of Engineering, Benghazi 18240
– Faculty of Science, Benghazi 18241
Gas del Estado, Buenos Aires 00257
Gas Equipment Plant, Brest 01845
Gas Equipment Plant, Novogrudok 01911
Gas Natural SDG S.A., Barcelona 25853
Gas Technology Institute, Des Plaines 36802
Gaston College, Dallas 33273
Gaston-Lincoln Regional Library, Gastonia 39211
Gastroenterology Research Institute, Dnipropetrovsk 28205
Gates Correctional Institution Library, Niantic 34609
Gates Public Library, Rochester 41029
Gateshead College, Gateshead 29112
Gateshead Library, Gateshead 29983
Gateway Community College, New Haven 33574
Gateway Community College, Phoenix 33639
Gateway Technical College, Kenosha 31495
Gateway to the Panhandle, Gate 36886
Gateway to the Panhandle Museum Library, Gate 36887
Gauhati University, Gauhati 13647
Gaulim Teachers' College, Rabaul 20567
Gaustad sykehus, Oslo 20115
Gauteng Department of Education, Pretoria 25161
Gavilan College, Gilroy 33372
Gävle stadsbibliotek, Gävle 26771
Gay and Lesbian Information Centre and Archives, Amsterdam 19360
Gay Lesbian Bisexual & Transgender Library/Archives of Philadelphia, Philadelphia 37347
Gaya Teachers' Training College, Kota Kinabalu 18532
GAZ, Nizhni Novgorod 22872
Gaza Recreation Centre, Sankt-Peterburg 24221
Gazi Üniversitesi, Ankara 27848
Gazi University, Central Library, Ankara 27848
Gazi-Husrevbeg Library, Sarajevo 02961
Gazi-Husrevbegova Biblioteka, Sarajevo 02961
Gdańsk Library of the Polish Academy of Sciences, Gdańsk 20788
Gdańsk University of Technology, Gdańsk 20805
Gdynia Maritime Academy, Gdynia 20810
GE Hungary Rt. Tungsram Lighting, Budapest 13355
GE Nuclear Energy Library, San Jose 36248
Gear-Manufacturing, Grinding and Cutter-Grinding Machine Special Design Bureau, Vitebsk 02082
Gebäude- und TechnikManagement Bremen, Bremen 10841
GEC-Marconi Information Centre, Camberley 29568
Gedenkstätte Deutscher Widerstand, Berlin 11522
Gedenkstätte Ernst Thälmann e.V., Hamburg 11979
Geelong Regional Library Corporation, Belmont 01067
Geels Geschiedkundig Genootschap, Geel 02641
Geheimes Staatsarchiv Preußischer Kulturbesitz, Berlin 11523
Geisteswissenschaftliche Zentren Berlin / Zentrum Moderner Orient, Berlin 11524
Geisteswissenschaftliches Zentrum Geschichte und Kultur Ostmitteleuropas an der Universität Leipzig, Leipzig 12182
Geistliches Rüstzentrum Krelingen der Ahldener Bruderschaft e.V., Walsrode 12576
Geldmuseum, Utrecht 19526
Gelehrtenschule des Johanneums, Hamburg 10734
Gellertbibliothek, Hainichen 11950
Gelsenkirchener Stadtbibliothek, Gelsenkirchen 12747
Gemäldegalerie, Berlin 11566
Gematologicheski nauchny tsentr, Moskva 23180
Gemeenschappelijke Bibliotheek van het Hof van Beroep en Arbeidshof, Antwerpen 02424
Gemeente Amersfoort, Stadhuisbibliotheek, Amersfoort 19235
Gemeente Amsterdam, Kenniscentrum Amsterdam, Amsterdam 19237
Gemeente Arnhem, Bibliotheek Bestuursdienst, Arnhem 19242
Gemeente Den Haag, Dienstbibliotheek, Den Haag 19252
Gemeente Ede, Ede 19248
Gemeente Haarlem, Haarlem 19272
Gemeente Leiden, Leiden 19279
Gemeente Lelystad, Stadhuisbibliotheek, Lelystad 19281
Gemeente Musea Delft, Delft 19409
Gemeentearchief Beveren, Beveren 02541
Gemeentearchief Goes, Goes 19425
Gemeentearchief Roermond, Roermond 19505
Gemeentearchief Rotterdam, Rotterdam 19508
Gemeentearchief Schiedam, Schiedam 19519
Gemeentearchief Venlo, Venlo 19533
Gemeentearchief Zaanstad, Koog aan de Zaan 19470
Gemeentelijke Archiefdienst, Bergen op Zoom 19398
Gemeentelijke Bibliotheek, Beernem 02719
Gemeentelijke Bibliotheek, Kortenberg 02795
Gemeentelijke Geneeskundige en Gezondheidsdienst Amsterdam, Amsterdam 19357
Gemeentelijke Gezondheidsdienst Rotterdam-Rijnmond, Rotterdam 19509
Gemeentelijke Nederlandstalige Openbare Bibliotheek Jette, Brussel 02740
Gemeentelijke Openbare Bibliotheek, Aartselaar 02711
Gemeentelijke Openbare Bibliotheek, Brasschaat 02731
Gemeentelijke Openbare Bibliotheek, Heist-op-den-Berg 02776
Gemeentelijke Openbare Bibliotheek, Knokke-Heist 02791
Gemeentelijke Openbare Bibliotheek, Meerhout 02820
Gemeentelijke Openbare Bibliotheek, Meeuwen-Gruitrode 02821
Gemeentelijke Openbare Bibliotheek, Merchtem 02824
Gemeentelijke Openbare Bibliotheek, Sint-Pieters-Leeuw 02859
Gemeentelijke Openbare Bibliotheek, Waregem 02877
Gemeentelijke Openbare Bibliotheek, Zwevegem 02894
Gemeentemuseum Den Haag, Kunsthistorische Bibliotheek, Den Haag 19433
Gemeindebibliothek Wettingen, Wettingen 27503
Gemeindebücherei Hausham, Hausham 12779
Gemeindebücherei Maisach, Maisach 12842
Gemeindebücherei Neustadt (Wied), Neustadt (Wied) 12872
Gemeindebücherei Seevetal, Seevetal 12932
Gemeinsames Archiv des Kreises Steinburg und der Stadt Itzehoe, Itzehoe 12075
Gemer-Malohont Museum, Rimavská Sobota 24750
Gemersko-Malohontské Múzeum, Rimavská Sobota 24750
Genaire Ltd Library, Niagara-on-the-Lake 04274
Genealogical Forum of Oregon, Inc Library, Portland 37399
Genealogical Society of Finland, Helsinki 07495
Genealogical Society of Linn County, Iowa, Cedar Rapids 36636
Genealogical Society of Palm Beach County, West Palm Beach 37826
Genealogical Society of the Northern Territory Inc., Winnellie 01050
Genealogical Society of Victoria, Melbourne 00953
Genealogie-Forschungsstelle, Zürich 27451
Genealogische Gesellschaft Hamburg e.V., Hamburg 11980
Genealogiska Samfundet i Finland, Helsinki 07495
General and Emergency Surgery Research Institute, Kharkiv 28275
General Association of Engineers of Romania, Bucureşti 21969
General Conference of Seventh-Day Adventists, Silver Spring 37605
General Directorate of Mineral Research and Exploration, Ankara 27875
General Dynamics Canada, Ottawa 04442
General Dynamics Communication System Library, Needham Heights 34577
General Dynamics Corp, Groton 35730
General Dynamics Land Systems, Sterling Heights 36285
General Electric Aircraft Engine, Cincinnati 35609
General Electric Co, Lynn 35868
General Electric Corporate Research & Development, Niskayuana 36054
General Federation of Liberal Trade Unions of Belgium (CGSLB), Gent 02642
General Hospital of Athens, Athinai 13109
General Jonas Zemaitis Military Academy of Lithuania, Vilnius 18286
General Mills, Inc, Minneapolis 35909
General Motors Corp, Warren 36326
General Physics Institute of the Russian Academy of Sciences, Moskva 23296
General Research District Library, Kostroma 22143
General Sciences Library, Ho Chi Minh City 42265
General Services Administration Library, Washington 34997
General Society of Mechanics & Tradesmen, New York 37206
General State Archives of Belgium, Bruxelles 02560
General Theological Seminary, New York 35307
Generaldirektion Kulturelles Erbe, Direktion Archäologie Mainz, Mainz 11006
Generaldirektion Kulturelles Erbe, Direktion Landesdenkmalpflege, Mainz 11007
Generalitat de Catalunya – Consell Consultiu, Barcelona 25683

Generalitat de Catalunya – Departament de Governació, Barcelona 25684
Generalitat de Catalunya – Departament de Sanitat, Barcelona 25685
Generalitat de Catalunya – Direcció General de l'Esport, Esplugues (Barcelona) 25693
Generallandesarchiv Karlsruhe, Karlsruhe 12090
Generolo Jono Žemaičio Lietvus karo akademija, Vilnius 18286
Genesee Community College, Batavia 33155
Genesee County Circuit Court, Flint 36854
Genesee District Library, Flint 39074
Geneva College, Beaver Falls 30368
Geneva Public Library, Geneva 39214
Geneva Public Library District, Geneva 39215
Geneva University of Art and Design, Genève 27086
Geniko Nossokomio Athinon, Athinai 13109
Gennadius Library, Athinai 13110
Genoa Public Library District, Genoa 39216
Gentofte Bibliotek, Hellerup 06874
Gentry County Library, Stanberry 41426
Geodätisches Institut, Aachen 09374
Geodeettinen laitos, Masala 07529
Geodesical Research and Engineering Corporation, Sankt-Peterburg 22946
Geodesy and Map-drawing Professional School, Sankt-Peterburg 22711
Geodetický a Kartografický Ustav Bratislava, Bratislava 24690
Geofizichen Institut, Sofiya 03530
Geofizicheskaya observatoriya, Sankt-Peterburg 23763
Geofyzikální ústav AV ČR, Praha 06487
Geograficheskoe obshchestvo, Vladivostok 24052
Geografski Institut, Sofiya 03531
Geographic Military Institute, Firenze 16263
Geographical Institute, Istanbul 27888
Geographical Research Institute, Budapest 13411
Geographical Society, Vladivostok 24052
Geographical Society of Finland, Helsinki 07491
Geographical Society of India, Kolkata 14039
Geographical Society of Lisbon, Lisboa 21816
Geographical Survey Institute, Tsukuba 17792
Geographische Gesellschaft in Hamburg, Hamburg 11981
Geographische Zentralbibliothek, Leipzig 12190
Geologian tutkimuskeskus, Espoo 07434
Geologian tutkimuskeskus, Rovaniemi 07539
Geological Administration of Georgia, Tbilisi 09207
Geological Administration of Yakutsk Territory, Yakutsk 22770
Geological Gravimetric Station no 3, Krasnoyarsk 23122
Geological Institute, Tbilisi 09233
Geological Institute, Bulgarian Academy of Sciences, Sofiya 03532
Geological Institute of Romania, Bucureşti 21997
Geological Institute of the Slovak Academy of Sciences, Bratislava 24691
Geological Research and Development Centre, Bandung 14343
The Geological Society of London, London 29749
Geological Survey and Mines Department, Entebbe 27923
Geological Survey Department, Lobatse 02979
Geological Survey Division, Freetown 24581
Geological Survey Division, Port Moresby 20594
Geological Survey Laboratory Malaysia, Ipoh 18574
Geological Survey Laboratory Malaysia, Kuching 18610
Geological Survey Museum and Library, Entebbe 27924
Geological Survey of Alabama Library, Tuscaloosa 37695
Geological Survey of Belgium, Brussel 02590
Geological Survey of Canada Calgary Library, Calgary 04321
Geological Survey of Finland, Espoo 07434
Geological Survey of Finland, Kuopio 07520
Geological Survey of Finland, Rovaniemi 07539
Geological Survey of Ghana, Accra 13045
Geological Survey of Hokkaido, Sapporo 17653
Geological Survey of India, Bangalore 13930
Geological Survey of India, Kolkata 14040
Geological Survey of India, Lucknow 14065
Geological Survey of Ireland, Dublin 14533
Geological Survey of Israel, Jerusalem 14728
Geological Survey of Japan, Tsukuba 17787
Geological Survey of Malawi, Zomba 18493
Geological Survey of Malaysia, Kuching 18609
Geological Survey of Nigeria, Kaduna 20027
Geological Survey of Norway, Trondheim 20299
Geological Survey of Spain, Madrid 26052
Geological Survey of Sweden, Uppsala 26710
Geological Survey of Western Australia and General Library, East Perth 00912
Geological Survey of Zambia, Lusaka 42343
Geologicheski institut – BAN, Sofiya 03532
Geologická knihovna, Praha 06488
Geologický ústav Slovenskej akadémia vied -, Bratislava 24691
Geologisch Mijnbouwkundige Dienst, Paramaribo 26362
Geologische Bundesanstalt, Wien 01598
Geologischer Dienst Nordrhein-Westfalen, Krefeld 10994
Geologisk museum, Oslo 20244
Geology, Geophysics and Mineral Raw Materials Research Institute, Novosibirsk 23651
Geomechanics and Mine Surveying Research Institute, Sankt-Peterburg 23830
Geophysical Institute, Fairbanks 36838
Geophysical Institute, Sofiya 03530
Geophysical Institute of the Slovak Academy of Sciences, Bratislava 24706
Geophysical Observatory, Sankt-Peterburg 23763
Georg Westermann Verlag, Braunschweig 11379
Georg-August-Universität Göttingen, Göttingen 09315
– Albrecht-von-Haller-Institut für Pflanzenwissenschaften, Abt. Palynologie und Klimadynamik, Göttingen 09837
– Althistorisches Seminar, Göttingen 09838
– Archäologisches Institut und Sammlung der Gipsabgüsse, Göttingen 09839
– Asien-Afrika-Bibliothek, Göttingen 09840
– Bereichsbibliothek Chemie, Göttingen 09841
– Bereichsbibliothek Forstwissenschaften, Göttingen 09842
– Bereichsbibliothek Medizin, Göttingen 09843
– Bereichsbibliothek Physik, Göttingen 09844
– Bereichsbibliothek Wirtschafts- und Sozialwissenschaften, Göttingen 09845
– Bibliothek für Fachdidaktik, Göttingen 09846
– Bibliothek für Zivilrecht und Öffentliches Recht, Göttingen 09847
– Burckhardt-Institut Abt. Forstökonomie und Forsteinrichtung, Göttingen 09848
– Burckhardt-Institut, Abt. Naturschutz und Landschaftspflege sowie Forst- und Naturschutzpolitik und Forstgeschichte, Göttingen 09849
– Department für Agrarökonomie und Rurale Entwicklung, Göttingen 09850
– Department für Nutzpflanzenwissenschaften, Bibliothek der Abteilungen Pflanzenbau, Pflanzenzüchtung, Graslandwissenschaft, Göttingen 09851
– Fakultät für Geowissenschaften und Geographie, Göttingen 09852
– Finnisch-Ugrisches Seminar, Göttingen 09853
– Georg-Elias-Müller Institut für Psychologie, Göttingen 09854
– Gesellschaft für wissenschaftliche Datenverarbeitung mbH Göttingen (GWDG), Göttingen 09855
– Institut für Allgemeine Staatslehre und Politische Wissenschaften, Göttingen 09856
– Institut für Anorganische Chemie, Göttingen 09857
– Institut für Arbeitsrecht, Göttingen 09858
– Institut für Astrophysik, Göttingen 09859
– Institut für Ethnologie, Göttingen 09860
– Institut für Geophysik, Göttingen 09861
– Institut für Historische Landesforschung, Göttingen 09862
– Institut für Kulturanthropologie/Europäische Ethnologie, Göttingen 09863
– Institut für Landwirtschaftsrecht, Göttingen 09864
– Institut für Mathematische Stochastik (IMS), Göttingen 09865
– Institut für Numerische und Angewandte Mathematik, Göttingen 09866
– Institut für Organische und Biomolekulare Chemie, Göttingen 09867
– Institut für Rechtsgeschichte, Rechtsphilosophie und Rechtsvergleichung, Göttingen 09868
– Institut für Sportwissenschaften (IfS), Göttingen 09869
– Institut für Völkerrecht und Europarecht, Göttingen 09870
– Johann-Friedrich-Blumenbach-Institut für Zoologie und Anthropologie, Göttingen 09871
– Johann-Friedrich-Blumenbach-Institut für Zoologie und Anthropologie, Abt. Entwicklungsbiologie, Göttingen 09872
– Kunstgeschichtliches Seminar und Kunstsammlung, Göttingen 09873
– Mathematisches Institut, Göttingen 09874
– Musikwissenschaftliches Seminar, Göttingen 09875
– Ostasiatisches Seminar, Göttingen 09876
– Pädagogisches Seminar, Göttingen 09877
– Philosophisches Seminar, Göttingen 09878
– Seminar für Ägyptologie und Koptologie, Göttingen 09879
– Seminar für Arabistik, Göttingen 09880
– Seminar für Deutsche Philologie, Göttingen 09881
– Seminar für Deutsche Philologie, Göttingen 09882
– Seminar für Englische Philologie, Göttingen 09883
– Seminar für Indologie und Tibetologie, Göttingen 09884
– Seminar für Iranistik, Göttingen 09885
– Seminar für Klassische Philologie, Göttingen 09886
– Seminar für Mittlere und Neuere Geschichte, Göttingen 09887
– Seminar für Romanische Philologie, Göttingen 09888
– Seminar für Slavische Philologie, Göttingen 09889
– Seminar für Turkologie und Zentralasienkunde, Göttingen 09890
– Seminar für Ur- und Frühgeschichte, Göttingen 09891
– Skandinavisches Seminar, Göttingen 09892
– Sprachwissenschaftliches Seminar, Göttingen 09893
– Theologische Fakultät, Göttingen 09894
– Volkswirtschaftliches Institut für Mittelstand und Handwerk an der Universität Göttingen, Göttingen 09895
– Zentrale Einrichtung für Sprachen und Schlüsselqualifikationen (ZESS), Göttingen 09896
– Zentrum Hygiene und Humangenetik, Abteilung Medizinische Mikrobiologie, Göttingen 09897
– Zentrum Physiologie und Pathophysiologie – Zentrum Biochemie und Molekulare Zellbiologie, Göttingen 09898
– Zentrum Physiologie und Pathophysiologie / Zentrum Biochemie und Molekulare Zellbiologie (Gemeinsame Bibliothek ausser Abt. Biochemie 2), Göttingen 09899
– Zentrum Psychosoziale Medizin, Abt. Ethik und Geschichte der Medizin, Göttingen 09900
– Zentrum Psychosoziale Medizin, Abt. Psychiatrie und Psychotherapie und Abt. Kinder- und Jugendpsychiatrie und Psychotherapie (Gemeinsame Bibliothek), Göttingen 09901
George Baritiu County Library, Braşov 22080
George Brown College of Applied Arts & Technology, Toronto 04037
George C. Marshall Research Foundation Library, Lexington 37025
George C. Wallace Community College Aviation Campus, Ozark 37317
George C. Wallace State Community College, Dothan 33298
George Eastman House, Rochester 37450
George Enescu University of Arts, Iaşi 21936
George F. Johnson Memorial Library, Endicott 38978
George Fox University, Newberg 32108
– Portland Center Library, Portland 32109
George H. & Laura E. Brown Library, Washington 41756
George Junior Republic Library, Grove City 36907
George Kurian Reference Books, Yorktown Heights 37867
George M. Jones Library Association, Lynchburg 37074
George Mason University, Fairfax 31117
– School of Law Library, Arlington 31118
George Meany Memorial Archives, Silver Spring 37606
George Mercer Jr School of Theology, Garden City 31206
'George Oprescu' Institute of the History of Art, Bucureşti 21991
George Padmore Research Library on African Affairs, Accra 13046
George Public Library, George 25200
George Washington University, Washington 32967
– Eckles Library, Washington 32968
– Jacob Burns Law Library, Washington 32969
– Paul Himmelfarb Health Sciences Library, Washington 32970
George Washington University – National Clearinghouse for English Language Aquisition & Language Instruction Educational Programs, Washington 32971
George Weston Technologies, Enfield 00817
Georg-Eckert-Institut für Internationale Schulbuchforschung, Braunschweig 11661
Georgetown Charter Township Library, Jenison 39643
Georgetown College, Georgetown 31210
Georgetown County Library, Georgetown 39218
Georgetown Public Library, Georgetown 39219
Georgetown University, Washington 32972
– Edward Bennett Williams Law Library, Washington 32973
– John Vinton Dahlgren Memorial Library, Washington 32974
– National Center for Education in Maternal & Child Health, Washington 32975
Georgia College & State University, Milledgeville 31829
Georgia Department of Archives & History, Morrow 34563
Georgia Department of Corrections, Office of Library Services, Alto 33915
Georgia Department of Corrections, Office of Library Services, Chester 34065
Georgia Department of Corrections, Office of Library Services, Garden City 34296
Georgia Department of Corrections, Office of Library Services, Glennville 34300
Georgia Department of Corrections, Office of Library Services, Grovetown 34323
Georgia Department of Corrections, Office of Library Services, Reidsville 34734
Georgia Department of Corrections, Office of Library Services, Trion 34937
Georgia Department of Corrections, Office of Library Services, Valdosta 34955
Georgia Department of Corrections, Office of Library Services, Waycross 35050
Georgia Department of Transportation, Forest Park 34210
Georgia Historical Society Library, Savannah 37583
Georgia Institute of Technology, Atlanta 30291
– College of Architecture, Atlanta 30292
Georgia Library for Accessible Services, Atlanta 36490
Georgia Military College, Milledgeville 33541
Georgia Perimeter College, Clarkston 33234
Georgia Perimeter College – Dunwoody Campus Library, Dunwoody 33303
Georgia Power Co-Southern Co, Atlanta 35459
Georgia Southern University, Statesboro 32738
Georgia Southwestern State University, Americus 30209
Georgia State University, Atlanta
– College of Law, Atlanta 30293
– William Russell Pullen Library, Atlanta 30294
Georgian Agricultural Institute, Tbilisi 09194
Georgian Art Museum, Tbilisi 09234
Georgian College, Barrie 03977
Georgian Court College, Lakewood 31550
Georgian Institute of Plant Protection, Tbilisi 09235
Georgian Institute of Subtropical Cultivation, Sukhumi 09221
Georgian National Academy of Sciences (GNAS), Tbilisi 09185
Georgian S. Rustaveli State Institute of Dramatic Art, Tbilisi 09195

Georgian Scientific Research Institute of Agricultural Engineering and Electrification, Tbilisi 09236
Georgian Silk-Production, Tbilisi 09237
Georgian State Orchestra, Tbilisi 09238
Georgian State Public Library, Tbilisi 09186
Georgian Technical University, Tbilisi 09196
Georgian Veterinary Training and Research Institute, Tbilisi 09197
Georgina Public Library, Keswick 04668
Georg-Kolbe-Museum, Berlin 11525
Georg-Simon-Ohm-Hochschule
– Fachhochschule Nürnberg, Nürnberg 10518
– Hochschulbibliothek Nürnberg, Teilbibliothek, Nürnberg 10519
Geoscience Australia, Canberra 00893
Geraardsbergen Public Library, Geraardsbergen 02766
Gerber/Hart, Chicago 36674
Gerd Bucerius Bibliothek, Hamburg 12006
Gerechten in het Arrondissement Arnhem, Arnhem 19243
Gereformeerd Pedagogisch Centrum, Zwolle 19234
Gereformeerde Hogeschool / Gereformeerd Pedagogisch Centrum, Zwolle 19234
Gerhart-Hauptmann-Haus, Düsseldorf 11806
Gerichte für Arbeitssachen, Berlin 10806
Gerichtsbibliothek Lübeckertordamm 4, Hamburg 10922
German Aerospace Center, Stuttgart 12514
German Archaeological Institute, Athinai 13105
German Archaeological Institute Madrid, Madrid 26011
German Archeological Institute, Istanbul 27886
German Caritas Association, Freiburg 11890
German Central Library for the Blind, Leipzig 12181
German Central Library of Pharmacy, Stuttgart 12513
German Cultural Center, New York 37207
German Cultural Institute of Barcelona, Barcelona 25904
German Design Council, Frankfurt am Main 11869
German Development Institute (GDI), Bonn 11630
German Historical Institute, London 29750
German Historical Institute Library, Washington 37740
German Historical Institute Warsaw, Warszawa 21254
German Historical Museum, Berlin 11587
German Institute for Adult Education, Leibniz Centre for Lifelong Learning (DIE), Bonn 11631
German Institute for Human Rights, Berlin 11510
German Institute for Japanese Studies, Chiyoda-ku Tokyo 17528
German Institute for Tropical and Subtropical Agriculture, Witzenhausen 12605
German Institute of Global and Area Studies, Hamburg 11983
German Library Helsinki, Helsinki 07449
German Maritime Museum, Bremerhaven 11694
German National Library, Frankfurt am Main 09291
German National Library of Medicine, Köln 12125
German National Library of Medicine, Nutrition, Environment and Agriculture, Bonn 11651
German National Library of Science and Technology and University Library Hannover (TIB/UB), Hannover 09961
German Orient-Foundation, Berlin 11504
German Orient-Institute, Berlin 11504
German Reference Centre for Ethics in the Life Sciences, Bonn 11632
German Resistance Research Council / German Resistance Documentary Archives, Frankfurt am Main 11875
German Society of Pennsylvania, Philadelphia 37348
German-American Library, Heidelberg 12039
Germania Judaica, Köln 12132
Germanisches Nationalmuseum, Nürnberg 12373
Germanna Community College, Locust Grove 33494
Germantown Community Library, Germantown 39221
Germantown Community Library, Germantown 39222
Germantown Public Library, Germantown 39223
Germiston Community Library, Germiston 25201
Gersevanov Institute of Bases and Underground Structures, Moskva 23418
Gertsen Library, Kyiv 28656
Gesamtverband der Aluminiumindustrie e.V. (GDA), Düsseldorf 11793
Gesellschaft der Ärzte in Wien, Wien 01599
Gesellschaft der Musikfreunde in Wien, Wien 01600
Gesellschaft für Anlagen- und Reaktorsicherheit (GRS) mbH, Köln 12133
Gesellschaft für deutsche Sprache e.V., Wiesbaden 12593
Gesellschaft für Reaktorsicherheit (GRS) mbH, Garching 11910
Gesellschaft für wissenschaftliche Datenverarbeitung mbH Göttingen (GWDG), Göttingen 11928
Gesellschaft zur Förderung zeitgenössischer Buchkunst, Hamburg 11982
GESIS – Leibniz Institute for the Social Sciences, Berlin 11526
GESIS – Leibniz Institute for the Social Sciences, Bonn 11634
GESIS – Leibniz Institute for the Social Sciences, Köln 12134
GESIS – Leibniz Institute for the Social Sciences, Mannheim 12241
GESIS – Leibniz-Institut für Sozialwissenschaften, Berlin 11526
GESIS – Leibniz-Institut für Sozialwissenschaften, Bonn 11634
GESIS – Leibniz-Institut für Sozialwissenschaften, Köln 12134
GESIS – Leibniz-Institut für Sozialwissenschaften, Mannheim 12241
Gesundheit Österreich GmbH, Wien 01504
Getty Research Institute, Los Angeles 37048
Gettysburg College, Gettysburg 31212
Getzeny-Institut im Wilhelmsstift, Tübingen 11351
Gewerbebibliothek, Fribourg 27221
Gezhouba Institute of Hydroelectrical Engineering, Yichang 05749
Gezondheidsraad, Den Haag 19253
Ghana Atomic Energy Commission, Kwabenya 13060
Ghana Export Promotion Council, Accra 13047
Ghana Institute of Journalism, Accra 12995
Ghana Institute of Management and Public Administration (GIMPA), Accra 12996
Ghana Investment Promotion Centre, Accra 13048
Ghana Library Board, Accra 12992
Ghana National Chamber of Commerce, Accra 13049
Ghana Standards Board, Accra 13050
Ghent University, Gent 02328
– Department of Forensic Medicine, Gent 02336
– Faculty Librarian Arts & Philosophy, Ghent University, Gent 02331
– Faculty of Applied Sciences Library, Gent 02334
– Faculty of Economics and Business Administration, Library, Gent 02329
– Romance Linguistics, Gent 02340
Gheorghe Zane Institute for Economic Research, Iaşi 22038
Ghost Ranch Conference Center Library, Abiquiu 36440
GIAM ZRC SAZU, Ljubljana 24916
Giant Mountains Museum, Vrchlabí 06547
Gibbons, Del Deo, Dolan, Griffinger & Vecchione, Newark 36040
Gibbs College, Norwalk 33588
Gibraltar Museum, Gibraltar 13066
Gibson, Dunn & Crutcher, Los Angeles 35835
Gibson, Ochsner & Adkins LLP, Amarillo 35443
Gidrokhimicheski institut, Novocherkassk 22761
Gidrokhimicheski institut, Rostov-na-Donu 23718
Gidromeliorativny institut, Moskva 22330
Gidrometeorologicheskaya observatoriya, Rostov-na-Donu 23719
Gidrometeorologicheski institut, Sankt-Peterburg 22516
Gidrometeorologicheski tsentr, Moskva 23181
Gidrometeorologichny tekhnikum, Kherson 28605
Gifu Daigaku, Gifu 16931
– Igakubu Bunkan, Gifu 16932
Gifu Municipal Library, Gifu 17838
Gifu Pharmaceutical University, Gifu 16933
Gifu Prefectural Assembly, Gifu 17260
Gifu Prefectural Fisheries Experimental Station, Mashita 17595
Gifu Prefecture Agricultural Experiment Station, Gifu 17540
Gifu Prefecture Planning Dept, Statistics Div, Administrative Data Room, Gifu 17261
Gifu Shiritsu Toshokan, Gifu 17838
Gifu University, Gifu 16931
– Medical Branch, Gifu 16932
Gifu Yakka Daigaku, Gifu 16933
Gifuken Gikai, Gifu 17260
Gifuken Kikakubu, Gifu 17261
Gifuken Nogyo Shikenjo, Gifu 17540
Gifuken Suisan Shikenjo, Mashita 17595
GIGA German Institute of Global and Area Studies / Leibniz-Institut für Globale und Regionale Studien, Hamburg 11983
Gigienichi tsentr, Kyiv 28311
Gigienichny tsentr profilaktiky travmatizmu, Donetsk 28218
GIH Biblioteket, Stockholm 26426
Gila County Law Library, Globe 34301
Gilan University, Rasht 14394
Gilleleje Bibliotek, Gilleleje 06866
Gillette Co, Needham 35936
Gillis Centre Library, Edinburgh 29542
Gimnazija Jožeta Plečnika, Ljubljana 24878
Gimnazija Ledina, Ljubljana 24879
Gimnazija Poljane, Ljubljana 24880
Ginko Toshokan, Tokyo 17675
Gippstafe, Morwell 00483
GIPRO organizatsii tekhnologii stroitelstva no 2, Moskva 23182
Gipro Selenrgoproekt, Moskva 23183
GIPRO Svyaz, Moskva 23184
GIPRO tsvetnoi metallurgii, Moskva 23185
GIPRO Uglemash, Moskva 23186
Giprodrevprom, Moskva 23187
GIPROKauchuk, Moskva 23188
GIPROKINO, Moskva 23189
GIPRONIIMEDPROM, Moskva 23190
GIPRORECHTRANS, Moskva 23191
Girard Free Library, Girard 39230
Girikond Research Institute, Sankt-Peterburg 23829
Girnichno-zbagachuvalny kombinat, Marganets 28165
Gislaveds bibliotek, Gislaved 26772
Gitlin Library, Cape Town 25112
Giuseppe Dossetti Library of the John XXIII Foundation for Religious Studies in Bologna, Bologna 16187
Givat Haviva Central Library, Givat Haviva 14718
Givaudan Schweiz AG, Dübendorf 27286
Gjøvik bibliotek, Gjøvik 20335
Gjøvik College, Gjvik 20092
Gladsaxe Bibliotek, Søborg 06928
Gladstone Public Library, Gladstone 39231
Gladstone's Library, Hawarden 29683
Gladwin County District Library, Gladwin 39232
Glan Hafren NHS Trust, Newport, South Wales 29859
Glankler Brown, Memphis 35880
Glasgow Caledonian University, Glasgow 29114
Glasgow College of Building and Printing, Glasgow 29466
Glasgow College of Nautical Studies, Glasgow 29115
Glasgow Dental Hospital and School, Glasgow 29116
Glasgow Royal Infirmary Library and E-Learning Centre, Glasgow 29673
Glasgow School of Art, Glasgow 29117
Glass and Ceramics Institute, Warszawa 21306
Glass Fibre Industrial Corporation, Polotsk 01923
Glavnaya astronomicheskaya observatoriya RAN, Sankt-Peterburg 23764
Glavnoe proivodstvennoe upravlenie energetiki i elektrifikatsii Belarusi (Belglavenergo), Minsk 02002
Glavny botanicheski sad, Snigiri 23975
Glavny botanicheski sad RAN, Moskva 23192
GlaxoSmithKline Pharmaceuticals, King of Prussia 35805
Gleaner Co Ltd, Kingston 16882
Das Gleimhaus, Halberstadt 11951
Glen Cove Public Library, Glen Cove 39238
Glen Ellyn Public Library, Glen Ellyn 39239
Glen Oaks Community College, Centreville 33208
Glen Ridge Free Public Library, Glen Ridge 39241
Glen Rock Public Library, Glen Rock 39242
Glenbow Museum Library, Calgary 04322
Glencoe Public Library, Glencoe 39243
Glencoe Publishing Company Library, Westerville 36413
Glendale Community College, Glendale 33373
Glendale Community College, Glendale 33374
Glendale Public Library, Glendale 39244
Glendale Public Library – Brand Art & Music, Glendale 39245
Glendale Public Library – Casa Verdugo, Glendale 39246
Glendale Public Library – Montrose-Crescenta Branch, Montrose 40282
Glendale University, Glendale 31217
Glendora Public Library & Cultural Center, Glendora 39250
Glenormiston College, Terang 00467
Glenrose Rehabilitation Hospital, Edmonton 04341
Glens Falls-Queensbury Historical Association, Glens Falls 36892
Glenside Free Library, Glenside 39254
Glenside Public Library District, Glendale Heights 39249
Glenstal Abbey, Murroe 14515
Glenview Public Library, Glenview 39255
Glenville State College, Glenville 31221
Gliboka Regional Centralized Library System, Main Library, Gliboka 28568
Glibotska raionna TsBS, Tsentralna biblioteka, Gliboka 28568
Glinka Conservatoire, Novosibirsk 22441
Glinka State Central Museum of Musical Culture, Moskva 23247
Glinka State Classical Choir, Sankt-Peterburg 22517
Glinka State Conservatoire, Nizhni Novgorod 22420
Glinka Voronezh Agrarian University, Voronezh 22649
Global Climate and Ecology Research Institute, Moskva 23401
Globine Regional Centralized Library System, Main Library, Globine 28569
Globinska raionna TsBS, Tsentralna biblioteka, Globine 28569
Glomdal Museum, Elverum 20222
Glomdalsmuseet, Elverum 20222
Gloppen folkebibliotek, Sandane 20392
Glossaire des patois de la Suisse romande, Neuchâtel 27417
Glostrup Bibliotek, Glostrup 06867
Gloucester City Library, Gloucester City 39258
Gloucester County College, Sewell 33762
Gloucester County Library, Gloucester 39256
Gloucester County Library System, Mullica Hill 40335
Gloucester County Library System – Glassboro Public, Glassboro 39234
Gloucester County Library System – Mullica Hill (Headquarters), Mullica Hill 40336
Gloucester, Lyceum & Sawyer Free Library, Gloucester 39257
Gloucestershire College of Arts and Technology, Cheltenham 29457
Gloucestershire County Council, Gloucester 29987
Gloversville Public Library, Gloversville 39259
Glowna Biblioteka Komunikacyjna, Warszawa 21042
Główna Biblioteka Lekarska, Warszawa 21258
Główny Instytut Górnictwa, Katowice 20817
Główny Urząd Miar, Warszawa 21043
Glukhiv Regional Centralized Library System, Trade Union Library, Sumy 28834
Glukhivska raionna TsBS, Profspilkova biblioteka, Sumy 28834
Glukhovsk Textile Joint-Stock Company, Noginsk 22875
Glyndŵr University, Wrexham 29441
Gminna Biblioteka Publiczna, Mónki 21499
Gmyrov Regional Universal Scientific Library, Mykolaiv 27942
Gnesinykh Institute of Musical Education, Moskva 22382
Gnesta bibliotek, Gnesta 26773
Gnesta Public Library, Gnesta 26773
GNIIPI azotnoi promyshlennosti i produktov organicheskogo sinteza – Filial GIAP, Grodno 01867
Gnosjö folkbibliotek, Gnosjö 26774
GNTs sotsialnoi i sudebnoi psikhiatrii im. V.P. Serbskogo, Moskva 23193
Gobierno del Estado de Tamaulipas, Victoria 18872
Gobierno Vasco. Departamento de Justicia y Administración Pública, Vitoria-Gasteiz 25755
Göcseji Múzeum, Zalaegerszeg 13522
Goddard College, Plainfield 32317
Godfrey Memorial Library, Middletown 37105
God's Bible School & College, Cincinnati 33232
Goetheanum, Dornach 27348
Goetheho Inštitút, Bratislava 24692
Goethe-Institut, Accra 13051
Goethe-Institut, Ankara 27874
Goethe-Institut, Athinai 13111
Goethe-Institut, Atlanta 36491

Goethe-Institut, Bangkok 27779
Goethe-Institut, Barcelona 25904
Goethe-Institut, Beijing 05831
Goethe-Institut, Beograd 24506
Goethe-Institut, Bordeaux 08305
Goethe-Institut, Bratislava 24692
Goethe-Institut, Bucureşti 21975
Goethe-Institut, Budapest 13377
Goethe-Institut, Buenos Aires 00258
Goethe-Institut, Cairo 07172
Goethe-Institut, Caracas 42252
Goethe-Institut, Chennai 13967
Goethe-Institut, Córdoba 00298
Goethe-Institut, Dublin 14534
Goethe-Institut, Genova 16285
Goethe-Institut, Istanbul 27887
Goethe-Institut, Jakarta 14364
Goethe-Institut, Kolkata 14041
Goethe-Institut, Kraków 21128
Goethe-Institut, Kuala Lumpur 18582
Goethe-Institut, Lomé 27800
Goethe-Institut, London 29751
Goethe-Institut, Los Angeles 37049
Goethe-Institut, México 18843
Goethe-Institut, Minsk 02003
Goethe-Institut, Montreal 04401
Goethe-Institut, Moskva 23194
Goethe-Institut, Mumbai 14098
Goethe-Institut, München 12295
Goethe-Institut, Nancy 08436
Goethe-Institut, Napoli 16370
Goethe-Institut, New Delhi 14151
Goethe-Institut, New York 37207
Goethe-Institut, Palermo 16398
Goethe-Institut, Paris 08546
Goethe-Institut, Porto Alegre 03319
Goethe-Institut, Praha 06489
Goethe-Institut, Quezon City 20780
Goethe-Institut, Rabat 18958
Goethe-Institut, Riga 18181
Goethe-Institut, Rio de Janeiro 03348
Goethe-Institut, Roma 16505
Goethe-Institut, Rotterdam 19510
Goethe-Institut, Salvador 03378
Goethe-Institut, Santiago 04928
Goethe-Institut, São Paulo 03406
Goethe-Institut, Seoul 18107
Goethe-Institut, Sofiya 03533
Goethe-Institut, Stockholm 26648
Goethe-Institut, Tel-Aviv 14763
Goethe-Institut, Thessaloniki 13133
Goethe-Institut, Tokyo 17676
Goethe-Institut, Torino 16604
Goethe-Institut, Victoria Island 20043
Goethe-Institut, Warszawa 21259
Goethe-Institut, Yaoundé 03657
Goethe-Institut, Zagreb 06236
Goethe-Museum Düsseldorf, Düsseldorf 11794
Goethe-Sammlung Staufen, Staufen 12500
Goethe-Universität Frankfurt am Main, Frankfurt am Main 09730
– Bibliothek des Mathematischen Seminars und des Instituts für Didaktik der Mathematik, Frankfurt am Main 09731
– Bibliothek Gesellschafts- und Erziehungswissenschaften (BGE), Frankfurt am Main 09732
– Bibliothek Naturwissenschaften (BNat), Frankfurt am Main 09733
– Bibliothek Recht und Wirtschaft (BRuW), Frankfurt am Main 09734
– Bibliothekszentrum Geisteswissenschaften (BzG), Frankfurt am Main 09735
– Institut für Geschichte der Arabisch-Islamischen Wissenschaften, Frankfurt am Main 09736
– Institut für Humangeographie, Frankfurt am Main 09737
– Institut für Kunstpädagogik, Frankfurt am Main 09738
– Institut für Orientalische und Ostasiatische Philologien – Japanologie, Frankfurt am Main 09739
– Institut für Orientalische und Ostasiatische Philologien – Sinologie, Frankfurt am Main 09740
– Institut für Orientalische und Ostasiatische Philologien – Südostasienwissenschaften, Frankfurt am Main 09741
– Institut für Orientalische und Ostasiatische Philologien – Turkologie, Frankfurt am Main 09742
– Institut für Psychologie / Arbeitsbereich Pädagogische Psychologie, Frankfurt am Main 09743
– Institut für Psychologie / Arbeitsbereich Psychoanalyse, Frankfurt am Main 09744
– Institut für Psychologie, Bibliothek Mertonstraße, Frankfurt am Main 09745
– Institut für Sportwissenschaften, Frankfurt am Main 09746
– Institut und Bibliothek für Jugendbuchforschung, Frankfurt am Main 09747
– Institutsbibliothek Informatik, Frankfurt am Main 09748
– Kunstbibliothek, Frankfurt am Main 09749
– Medizinische Hauptbibliothek, Frankfurt am Main 09750
– Senckenbergisches Institut für Geschichte und Ethik der Medizin, Frankfurt am Main 09751
– Zahnärztliches Universitäts-Institut der Stiftung Carolinum, Frankfurt am Main 09752
– Zentrum der Psychiatrie, Frankfurt am Main 09753
Goethezentrum Palermo / Cantieri Culturali alla Zisa, Palermo 16399
Goettingen State and University Library, Göttingen 09315
Goff-Nelson Memorial Library, Tupper Lake 41623
Gogebashvili State Library for Public Education, Tbilisi 09239
Gogol Bobruysk City Library, Bobruysk 02099
N. V. Gogol Karaganda District Library, Karaganda 17987
N. Gogol Library, Kyiv 28659
Goğrafya Enstitüsü, Istanbul 27888
Gokhale Institute of Politics and Economics, Pune 13743
Gold Coast City Council, GCMC Bundall 01096
Gold Coast Institute of Tafe, Ashmore 00382
Gold Fields Technobib, Pretoria 25040
Golden Gate Baptist Theological Seminary, Mill Valley 35283
Golden Gate University, San Francisco 32587
– School of Law Library, San Francisco 32588
Golden West College, Huntington Beach 33423
Goldey Beacom College, Wilmington 33876
Goldschmidt GmbH, Essen 11391
Goldsmiths' College, London 29235
Goldwater Memorial Hospital, New York 37208
Golesti Museum, Goleşti-Ştefăneşti 22035
Golovna mezhspilkova biblioteka, Enakieve 28236
Gomal University, Dera Ismail Khan 20428
Gomel Cable Production Plant, Gomel 01861
Gomel Central City Library, Gomel 02128
Gomel City Central Children's Library, Gomel 02129
Gomel City Children's Library no 12 'Marshak', Gomel 02125
Gomel City Library no 1 'Melezh', Gomel 02121
Gomel City Library no 2, Gomel 02122
Gomel City Library no 3, Gomel 02123
Gomel City Library no 7, Gomel 02124
Gomel Cooperative Institute, Gomel 01955
Gomel Railway Station, Main Railway Technical Library, Gomel 01957
Gomel Region Children's Library, Gomel 02127
Gomel Region Medical Library, Gomel 01954
Gomel Regional Advanced Teacher Training College, Gomel 01793
Gomel Regional Library, Gomel 02126
Gomel State University, Gomel 01792
Gomel Timber Processing Industrial Corporation, Gomel 01855
Gomelskaya gorodskaya bibliyateka 1 im. Melezha, Gomel 02121
Gomelskaya gorodskaya bibliyateka no 2, Gomel 02122
Gomelskaya gorodskaya bibliyateka no 3, Gomel 02123
Gomelskaya gorodskaya bibliyateka no 7, Gomel 02124
Gomelskaya gorodskaya detskaya bibliyateka no 12 im. Marshaka, Gomel 02125
Gomelskaya oblastnaya bibliyateka, Gomel 02126
Gomelskaya oblastnaya detskaya bibliyateka, Gomel 02127
Gomelskaya oblastnaya meditsinskaya bibliyateka, Gomel 01954
Gomelskaya tsentralnaya gorodskaya bibliyateka, Gomel 02128
Gomelskaya tsentralnaya gorodskaya detskaya bibliyateka, Gomel 02129
Gomelski gosudarstvenny tekhnicheski universitet im. P.O. Sukhovo, Gomel 01791
Gomelski gosudarstvenny universitet im. F. Skaryna, Gomel 01792
Gomelski kooperativny institut, Gomel 01955
Gomelski oblastnoi institut usovershenstvovaniya uchitelei, Gomel 01793
Gondar College of Medical Sciences, Gondar 07272
Gonville and Caius College, Cambridge 28985
Gonzaga University, Spokane 32698
– Chastek Library, Spokane 32699
Good Shepherd College, Auckland 19812
Goodmans LLP Library, Toronto 04286
Goodnow Library, Sudbury 41463
B. F. Goodrich Co, Brecksville 35540
Goodwin Procter, Boston 35523
Goodwood Libraries, Goodwood 25202
A. D. Gordon Agriculture, Nature and Kinnereth Valley Study Institute, Emek Ha-Yarden 14716
Gordon Agriculture, Nature and the Kinneret Valley Study Institute, Kibbuz Deganya A 14745
Gordon, Arata, McCollam, Duplantis & Egan LLP, New Orleans 35946
Gordon College, Barnesville 33153
Gordon College, Wenham 33014
Gordon, Feinblatt, Rothman, Hoffberger & Hollander, Baltimore 35480
Gordon Institute of TAFE, Geelong 00441
Gordon & Rees LLP, San Francisco 36231
Gordon-Conwell Theological Seminary, South Hamilton 35384
W. L. Gore & Associates, Inc, Newark 36041
Gore District Libraries, Gore 19866
Gorenjski muzej, Kranj 24911
Goretskaya tsentralnaya raionnaya bibliyateka, Gorki 02132
Gorham Public Library, Gorham 39261
Gorim Bunko, Onjuku 17632
Gorimbunko, The Onjuku Museum of History and Folklore, Onjuku 17632
Goriška knjižnica Franceta Bevka, Nova Gorica 24974
Goriški muzej, Nova Gorica 24950
Gorizont Industrial Corporation, Main Trade Union Library, Minsk 02189
Gorj 'Christian Tell' District Library, Tîrgu Jiu 22110
Gorki Central Regional Library, Gorki 02132
Gorkogo Agricultural Institute, Kazan 22279
Gorkom profsoyuza rabotnikov torgovli, obshchestvennogo pitaniya i potrebkooperatsii, Moskva 23195
Gorkovskaya zheleznaya doroga, Nizhni Novgorod 23613
Gorkovski mezhotraslevoi territorialny tsentr nauchno-tekhnicheskoi informatsii i propagandy, Nizhni Novgorod 23614
Gorky Central City Library, Arzamas 24105
Gorky City Library, Ivanovo 24132
Gorky City Library, Rostov-na-Donu 24207
Gorky District Library, Karshi 42173
Gorky Filmstudio of Children Movies Production, Moskva 23490
Gorky Mari Polytechnic Institute, Yoshkar-Ola 22661
A. M. Gorky Memorial Museum, Moskva 23196
Gorky Moscow Institute of Literature, Moskva 22379
Gorky Orsha City Library, Orsha 02215
Gorky Perm Regional Public Library, Perm 22165
Gorky Railroad, Scientific and Technical Library, Nizhni Novgorod 23613
Gorky Recreation Centre, Sankt-Peterburg 24226
Gorky Ryazan Regional General Scientific Library, Ryazan 22169
Gorky Scientific Library, Odesa 27944
Gorky State Regional Universal Scientific Library, Zaporizhzhya 27950
Gorky Vitebsk Central City Library, Vitebsk 02247
Gorlivka Regional Centralized Library System, Main Library, Gorlivka 28570
Gorlivska raionna TsBS, Tsentralna biblioteka, Gorlivka 28570
Gorlovka State Pedagogical Institute of Foreign Languages, Gorlovka 27978
Gorno-Altaisk State University, Gorno-Altaisk 22251
Gorno-Altaiskaya respublikanskaya detskaya biblioteka, Gorno-Altaisk 24129
Gornometallurgicheski institut im. G.I. Nosova, Magnitogorsk 22317
Gornometallurgicheski kombinat im. A.P. Zavenyagina, Norilsk 22876
Gornotayozhnaya stantsiya im. Komarova RAN, Gornotayozhnoe 23045
Gorny institut, Sankt-Peterburg 23765
Gorny universitet, Moskva 22331
Gorodenkivska raionna TsBS, Tsentralna biblioteka, Gorodenka 28572
Gorodnya Regional Centralized Library System, Main Library, Gorodnya 28573
Gorodnyanska raionna TsBS, Tsentralna biblioteka, Gorodnya 28573
Gorodok Regional Central Library, Gorodok 02133
Gorodok Regional Centralized Library System, Main Library, Gorodok 28574
Gorodokska raionna TsBS, Tsentralna biblioteka, Gorodok 28574
Gorodokskaya tsentralnaya raionnaya bibliyateka, Gorodok 02133
Gorodskaya biblioteka im. A.P. Chekhova, Taganrog 24327
Gorodskaya biblioteka No. 1, Angarsk 24102
Gorodskaya biblioteka no 2, Moskva 24170
Gorodskaya klinicheskaya tuberkulyoznaya bolnitsa No. 7, Moskva 23197
Gorodskaya notno-muzykalnaya biblioteka, Minsk 02004
Gorodskaya psikhiatricheskaya klinicheskaya bolnitsa No. 1 im. P.P. Kashchenko, Moskva 23198
Gorodskaya tsentralnaya biblioteka, Komsomolsk-na-Amure 24147
Gorodskaya tsentralnaya biblioteka, Magnitogorsk 24162
Gorodskaya tsentralnaya biblioteka im. A. Aalto, Vyborg 24357
Gorodskaya tsentralnaya biblioteka im. A.M. Gorkogo, Novorossisk 24189
Gorodskaya tsentralnaya biblioteka im. Gorkogo, Ivanovo 24132
Gorodskoi komitet profsoyuza rabochikh avtomobilnogo transporta i shosseinykh dorog, Moskva 24171
Gorodskoi nauchno-issledovatelski institut skoroi pomoshchi im. N.V. Sklifosovskogo, Moskva 23199
Goroka Public Library, Goroka 20597
Gorokhiv Regional Centralized Library System, Main Library, Gorokhiv 28575
Gorokhivska raionna TsBS, Tsentralna biblioteka, Gorokhiv 28575
Gorono Mogileva, Mogilev 02061
Gorski Agricultural Institute, Vladikavkaz 22626
Gorski selskokhozyaistvenny institut, Vladikavkaz 22626
Goryachkin Agricultural Engineering University, Moskva 22324
Goryachkin Research Institute for Agricultural Engineering, Moskva 23596
Gosfilmofond, Belye Stolby 23000
Goshen College – Harold & Wilma Good Library, Goshen 31225
Goshen College – Mennonite Historical Library, Goshen 36898
Gosinstitut po proektirovaniyu gorodov, Moskva 23200
Gosstroi Belarusi – Institut stroitelstva i arkhitektury, Minsk 02005
Gosteli Stiftung, Worblaufen 27444
Gosudarstvennaya akademicheskaya kapella im. M.I. Glinki, Sankt-Peterburg 22517
Gosudarstvennaya akademiya aerokosmicheskogo pri-borostroeniya, Sankt-Peterburg 22518
Gosudarstvennaya akademiya fizicheskoi kultury, Moskva 22332
Gosudarstvennaya akademiya upravleniya im. Sergo Ordzhonikidze, Moskva 22333
Gosudarstvennaya arkhivnaya sluzhba Udmurtskoi Respubliki, Izhevsk 23063
Gosudarstvennaya filarmoniya, Sankt-Peterburg 23766
Gosudarstvennaya konservatoriya im. L.V. Sobinova, Saratov 22572
Gosudarstvennaya konservatoriya im. M.I. Glinki, Nizhni Novgorod 22420
Gosudarstvennaya natsionalnaya biblioteka Kabardino-Balkarskoi Respubliki, Nalchik 22157
Gosudarstvennaya nauchnaya pedagogicheskaya biblioteka im. K.D. Ushinskogo, Moskva 23201
Gosudarstvennaya obshchestvenno-politicheskaya biblioteka, Moskva 23202
Gosudarstvennaya publichnaya istoricheskaya biblioteka (GPIB), Moskva 23203
Gosudarstvennaya publichnaya nauchno-tekhnicheskaya biblioteka Rossii (GPNTB), Moskva 22153
Gosudarstvennaya publichnaya nauchno-tekhnicheskaya biblioteka Sibirskogo otdeleniya RAN, Novosibirsk 22159
Gosudarstvennaya respublikanskaya detskaya biblioteka im. B. Abidueva, Ulan-Ude 24338
Gosudarstvennaya respublikanskaya yunosheskaya biblioteka, Ulan-Ude 24339
Gosudarstvennaya selskokhozyaistvennaya akademiya, Nizhni Novgorod 22421
Gosudarstvennaya Tretyakovskaya Galereya, Moskva 23204

Gosudarstvennaya tsentralnaya teatralnaya biblioteka, Moskva 22734
Gosudarstvenni Russki Muzey, Sankt-Peterburg 23767
Gosudarstvenny akademicheski Bolshoi teatr, opery i baleta, Minsk 02006
Gosudarstvenny Akademicheski Bolshoi teatr, Tsentralnaya notnaya biblioteka, Moskva 23205
Gosudarstvenny akademicheski Mariinski teatr, Sankt-Peterburg 23768
Gosudarstvenny akademicheski teatr im. Evgeniya Vakhtangova, Moskva 23206
Gosudarstvenny arkhiv, Volgograd 24062
Gosudarstvenny arkhiv Amurskoi oblasti, Blagoveshchensk 23005
Gosudarstvenny arkhiv Brestskoi oblasti, Brest 01945
Gosudarstvenny arkhiv Chernigovskoyi oblasti, Chernigiv 28189
Gosudarstvenny arkhiv Dnepropetrovskoyi oblasti, Dnipropetrovsk 28199
Gosudarstvenny arkhiv Ivanovskoi oblasti, Ivanovo 23059
Gosudarstvenny arkhiv Kaluzhskoi oblasti, Kaluga 23072
Gosudarstvenny arkhiv Khabarovskogo kraya, Khabarovsk 23097
Gosudarstvenny arkhiv Khmelnitskoyi oblasti, Kamyanets-Podilsk 28244
Gosudarstvenny arkhiv Krasnodarskogo kraya, Krasnodar 23111
Gosudarstvenny arkhiv Kuibyshevskoi oblasti, Samara 23738
Gosudarstvenny arkhiv Kurganskoi oblasti, Kurgan 23129
Gosudarstvenny arkhiv Kurskoi oblasti, Kursk 23132
Gosudarstvenny arkhiv N. Novgorodskoi oblasti, Nizhni Novgorod 23615
Gosudarstvenny arkhiv Novgorodskoi oblasti, Veliki Novgorod 24039
Gosudarstvenny arkhiv Novosibirskkoi oblasti, Novosibirsk 23636
Gosudarstvenny arkhiv Omskoi oblasti, Omsk 23663
Gosudarstvenny arkhiv Orenburgskoi oblasti, Orenburg 23670
Gosudarstvenny arkhiv Orlovskoi oblasti, Oryol 23675
Gosudarstvenny arkhiv Permskoi oblasti, Perm 23687
Gosudarstvenny arkhiv Poltavskoyi oblasti, Poltava 28447
Gosudarstvenny arkhiv Primorskogo kraya, Vladivostok 24053
Gosudarstvenny arkhiv Rostovskoi oblasti, Rostov-na-Donu 23720
Gosudarstvenny arkhiv Saratovskoi oblasti, Saratov 23956
Gosudarstvenny arkhiv Smolenskoi oblasti, Smolensk 23971
Gosudarstvenny arkhiv Stavropolskogo kraya, Stavropol 23981
Gosudarstvenny arkhiv Tambovskoi oblasti, Tambov 23988
Gosudarstvenny arkhiv Ternopilskoyi oblasti, Ternopil 28477
Gosudarstvenny arkhiv Tomskoi oblasti, Tomsk 23995
Gosudarstvenny arkhiv Tulskoi oblasti, Tula 24004
Gosudarstvenny arkhiv Tverskoi oblasti, Tver 24008
Gosudarstvenny arkhiv Vologodskoi oblasti, Vologda 24069
Gosudarstvenny arkhiv Yaroslavskoi oblasti, Yaroslavl 24085
Gosudarstvenny arkhiv Zakarpatskoyi oblasti, Beregovo 28185
Gosudarstvenny astronomicheski institut im. P.K. Shternberga, Moskva 23207
Gosudarstvenny biologicheski muzei im. K.A. Timiryazeva, Moskva 23208
Gosudarstvenny Borodinski voennoistoricheski muzei, Borodino 23007
Gosudarstvenny dom-muzei P.I. Chaikovskogo, Klin 23105
Gosudarstvenny energeticheski nauchno-issledovatelski institut im. G.M. Krzhizhanovskogo, Moskva 23209
Gosudarstvenny ermitazh, Sankt-Peterburg 23769
Gosudarstvenny gidrologicheski institut, Sankt-Peterburg 23770
Gosudarstvenny institut po proektirovaniyu avtoremontnykh i avtotransportnykh predpriyati i sooruzheni, Moskva 23210
Gosudarstvenny institut po proektirovaniyu elektrooborudovaniya i elektrosnabzheniya predpriyati tyazheloi promyshlennosti, Moskva 23211
Gosudarstvenny institut po proektirovaniyu i issledovatelskim rabotam neftyanoi promyshlennosti, Samara 23739
Gosudarstvenny institut po proektirovaniyu lesopilnykh, derevoobrabatyvayushchikh predpryati, Sankt-Peterburg 23771
Gosudarstvenny institut po proektirovaniyu osnovani i fundamentov, Moskva 23212
Gosudarstvenny institut po proektirovaniyu predpriyati myasnoi i molochnoi promyshlennosti, Sankt-Peterburg 23772
Gosudarstvenny institut po proektirovaniyu predpriyati organicheskikh poluproduktov i krasitelei, Moskva 23213
Gosudarstvenny institut po proektirovaniyu predpriyati pishchevoi promyshlennosti 2 (Gipropishcheprom-2), Moskva 23214
Gosudarstvenny institut po proektirovaniyu predpriyati rezinovoi promyshlennosti, Moskva 23215
Gosudarstvenny institut po proektirovaniyu predpriyati sakharnoi promyshlennosti, Moskva 23216
Gosudarstvenny institut po proektirovaniyu predpriyati tekstilnoi promyshlennosti, Moskva 23217
Gosudarstvenny institut po proektirovaniyu predpriyati tsvetnoi metallurgii, Moskva 23218
Gosudarstvenny institut po proektirovaniyu teatralno-zrelishchnykh predpriyati, Moskva 23219
Gosudarstvenny institut po proektirovaniyu zavodov avtomobilnoi promyshlennosti, Moskva 23220
Gosudarstvenny institut prikladnoi khimii, Sankt-Peterburg 23773
Gosudarstvenny istoricheski muzei, Moskva 23221
Gosudarstvenny istoriko-arkhitekturny muzei-zapovednik, Tobolsk 23992
Gosudarstvenny komitet po promyshlennoi politike Rossiskoi Federatsii, Moskva 23222
Gosudarstvenny komitet po standartizatsii, metrologii i sertifikatsii pri prezidente Rossii, Moskva 23223
Gosudarstvenny komitet po vneshnim ekonomicheskim svyazyam, Moskva 22735
Gosudarstvenny komitet Rossiskoi federatsii po statistike, Moskva 22736
Gosudarstvenny literaturny muzei, Moskva 23224
Gosudarstvenny literaturny muzei I.S. Turgeneva, Oryol 23676
Gosudarstvenny meditsinski institut im. Lunacharskogo, Astrakhan 22207
Gosudarstvenny muzei A.S. Pushkina, Moskva 23225
Gosudarstvenny muzei A.V. Suvorova, Sankt-Peterburg 23774
Gosudarstvenny muzei istorii kosmonavtiki im. K.E. Tsiolkovskogo, Kaluga 23073
Gosudarstvenny muzei istorii St. Peterburga, Sankt-Peterburg 23775
Gosudarstvenny muzei izobrazitelnykh iskusstv, Nizhni Tagil 23625
Gosudarstvenny muzei izobrazitelnykh iskusstv im. A.S. Pushkina, Moskva 23226
Gosudarstvenny muzei keramiki 'Usadby Kuskovo XVIII v.', Moskva 23227
Gosudarstvenny muzei L.N. Tolstogo, Moskva 23228
Gosudarstvenny muzei politicheskoi istorii Rossii, Sankt-Peterburg 23776
Gosudarstvenny muzei teatralnogo i muzykalnogo iskusstva, Sankt-Peterburg 23777
Gosudarstvenny muzei vostochnykh kultur, Moskva 23229
Gosudarstvenny nauchno-issledovatelski i proektny institut redkometallicheskoi promyshlennosti, Moskva 23230
Gosudarstvenny nauchno-issledovatelski i proektny institut splavov i obrabotki tsvetnykh metallov, Moskva 23231
Gosudarstvenny nauchno-issledovatelski institut kurortologii, Pyatigorsk 23711
Gosudarstvenny nauchno-issledovatelski institut mashinovedeniya, Moskva 23232
Gosudarstvenny nauchno-issledovatelski institut Sintezbelok, Moskva 23233
Gosudarstvenny nauchno-issledovatelski institut stekla, Moskva 23234
Gosudarstvenny nauchno-issledovatelski institut stroitelnoi keramiki, Zheleznodorozhny 24096
Gosudarstvenny nauchno-issledovatelski institut tsvetnykh metallov, Moskva 23235
Gosudarstvenny nauchno-issledovatelski institut ukha, gorla i nosa, Moskva 23236
Gosudarstvenny nauchno-issledovatelski khimiko-analiticheski institut, Sankt-Peterburg 23778
Gosudarstvenny okeanograficheski institut, Moskva 23237
Gosudarstvenny okeanograficheski institut – SPb otdelenie, Sankt-Peterburg 23779
Gosudarstvenny ordena Lenina Azotnotukovoi zavod, Kemerovo 22802
Gosudarstvenny pedagogicheski institut im. N.K. Krupskoi, Yoshkar-Ola 22659
Gosudarstvenny pedagogicheski universitet, Nizhni Novgorod 22422
Gosudarstvenny podshipnikovy zavod no 11 – Bibliyateka profkoma, Minsk 02159
Gosudarstvenny podshipnikovy zavod no 11 – Nauchno-tekhnicheskaya bibliyateka, Minsk 01878
Gosudarstvenny proektno-izyskatelski institut 'Mosgiprotrans', Moskva 23238
Gosudarstvenny proektno-izyskatelski institut promyshlenno-transportnogo stroitelstva, Moskva 23239
Gosudarstvenny proektno-tekhnologicheski i konstruktorski institut Orgstankinprom – Minski filial, Minsk 02007
Gosudarstvenny proektny i nauchno-issledovatelski institut po proektirovaniyu uchrezhdeni zdravookhraneniya (GIPRONIIZDRAV), Moskva 23240
Gosudarstvenny proektny institut Minskproekt (Minskpromproekt), Minsk 02008
Gosudarstvenny proektny institut No 2 po proektirovaniyu predpriyati kozhevenno-obuvnoi i mekhovoi promyshlennosti, Moskva 23241
Gosudarstvenny soyuzny institut po proektirovaniyu metallurgicheskikh zavodov, Moskva 23242
Gosudarstvenny soyuzny institut po proektirovaniyu spetsialnykh sooruzheni, zdani, sanitarno-tekhnicheskikh i energeticheskikh ustroistv dlya predpriyati khimicheskoi promyshlennosti, Moskva 23243
Gosudarstvenny soyuzny institut po proektirovaniyu vysshikh uchebnykh zavedeni, Moskva 23244
Gosudarstvenny soyuzny nauchno-issledovatelski traktorny institut, Moskva 23245
Gosudarstvenny soyuzny proektny institut, Moskva 23246
Gosudarstvenny tekhnichesky universytet radioelektroniki, Kharkiv 27986
Gosudarstvenny tsentralny muzei muzykalnoi kultury im. M.I. Glinki, Moskva 23247
Gosudarstvenny tsentralny muzei sovremennoi istorii Rossii, Moskva 23248
Gosudarstvenny tsentralny teatralny muzei im. A.A. Bakhrushina, Moskva 23249
Gosudarstvenny universitet po zemleustroistvu, Moskva 22334
Gosudarstvenny universitet tekhnologii i dizaina, Sankt-Peterburg 22519
Göta Court of Appeal, Jönköping 26595
Göta hovrätt, Jönköping 26595
Götaverken Energy Systems, Göteborg 26531
Göteborgs Konstmuseum, Göteborg 26576
Göteborgs Naturhistoriska Museum, Göteborg 26577
Göteborgs stad, Göteborg 26487
Göteborgs stadsbibliotek, Göteborg 26775
Göteborgs Stadsmuseum, Göteborg 26578
Göteborgs Universitet, Göteborg 26377
– Biomedicinska biblioteket, Göteborg 26378
– Botanik- och miljöbiblioteket, Göteborg 26379
– Centralbiblioteket, Göteborg 26380
– Ekonomiska biblioteket, Göteborg 26381
– Geovetenskapelige biblioteket, Göteborg 26382
– Högskolan för Design och Konsthantverk, Göteborg 26383
– Iberoamerikanska samlingen, Göteborg 26384
– Institutionen för kulturvård, Göteborg 26385
– Kurs- och tidningsbiblioteket, Göteborg 26386
– Pedagogiska biblioteket, Göteborg 26387
– Samhällsvetenskapliga biblioteket, Göteborg 26388
Götene kommunbibliotek, Götene 26776
Gotland County Library, Visby 26959
Gotlands länsbibliotek, Visby 26959
Gottfried Wilhelm Leibniz Bibliothek – Niedersächsische Landesbibliothek, Hannover 09318
Gottfried-Benn-Bibliothek, Berlin 12668
Goucher College, Baltimore 30333
Goulburn Valley Base Hospital, Shepparton 01003
Goulburn Valley Institute of TAFE, Shepparton 00585
Goulburn Valley Regional Library Corporation, Shepparton 01155
Goulburn-Ovens Institute of TAFE, Wangaratta 00557
Goulston & Storrs, PC, Boston 35524
Gouverneur Correctional Facility, Gouverneur 34307
Government Agricultural Library, Kanpur 13858
Government Archives and Records Service, Seoul 18108
Government Ayurvedic Medical College, Bangalore 13609
Government Central Library, Lashkar 13860
Government College Library, Lahore 20461
Government College of Indian Medicine, Palayamkottai 13732
Government College of Technology, Coimbatore 13799
Government College of Technology, Mandi Baha-ud-Din 20464
Government College of Technology, Rawalpindi 20465
Government du Québec Ministère des Relations avec les Citoyens et de l'Immigration, Montreal 04268
Government Ideal Science College, Raipur 13840
Government Industrial Research Institute, Ikeda 17555
Government Industrial Research Institute, Chugoku, Kure 17583
Government Institute for Economic Research, Helsinki 07508
Government Institute of Technology, Jhelum 20457
Government Law College Library, Mumbai 13821
Government Library, Tripoli 18250
Government Medical College, Jammu 13667
Government Medical College, Miraj 13818
Government Medical College, Nagpur 13830
Government Museum, Mathura 14075
Government of Canada, Ottawa 04114
Government of India, Ministry of Textiles, Varanasi 14220
Government of Puerto Rico, Hato Rey 21891
Government of Quebec – Agriculture Fisheries & Foods, Gaspe 04078
Government of the Northwest Territories – Legislative Library, Yellowknife 04195
Government of the Northwest Territories – RWED Library, Yellowknife 04196
Government of Yukon – Libraries & Archives Division, Whitehorse 04819
Government Oriental Manuscripts Library & Research Centre, Chennai 13960
Government Public Library, Delhi 14225
Government Training Institute, Mombasa 18018
Government Vidarbha Institute of Science & Humanities, Amravati 13603
Governmental Research Institute-Silesian Institute, Opole 21179
Governors State University, University Park 32876
Govind Ballabh Pant University of Agriculture and Technology, Pantnagar 13733
L. Martovych Govodenka Central District Library, Gorodenka 28572
Gowanda Correctional Facility Library, Gowanda 34308
Gowthami Regional Library, Rajahmundry 14202
Gozo Public Library, Victoria 18671
GP Aerogeodeziya, Sankt-Peterburg 23780
GP Gravimetricheskaya ekspeditsiya no 3, Krasnoyarsk 23122
GP Kirovski zavod, Sankt-Peterburg 22911
GPI-3 po proektirovaniyu predpriyati trikotazhnoi i shveinoi promyshlennosti, Sankt-Peterburg 23781

GPO ARIAS, Sankt-Peterburg 22912
GRA Inc Library, Jenkintown 35788
Graafschap bibliotheken Zutphen, Zutphen 19750
Grace A. Dow Memorial Library, Midland 40188
Grace Balloch Memorial Library, Spearfish 41399
Grace Bible College, Grand Rapids 35219
W. R. Grace & Co, Columbia 35625
Grace College & Grace Theological Seminary, Winona Lake 35423
Grace University, Omaha 32197
Graceland University, Lamoni 31551
Gradina Botanica 'Alexandru Borza', Cluj-Napoca 22020
Gradska Biblioteka, Subotica 24555
Gradska Biblioteka, Vršac 24557
Gradska Biblioteka Karlo Bijelicki, Sombor 24552
Gradska Knijžnica 'I.G. Kovačić', Karlovac 06257
Gradska Knjiznica i Citaonica 'Metel Ožegović', Varaždin 06261
Gradska Knjiznica Marka Marulica, Split 06259
Gradska Narodna Biblioteka Žarko Zrenjanin, Zrenjanin 24560
Gradski Muzej, Varaždin 06227
Graduate Institute of Applied Linguistics, Dallas 36765
Graduate Institute of International and Development Studies, Genève 27091
Graduate Theological Union, Berkeley 30387
H. Grady Bradshaw Chambers County Library, Valley 41655
Grady C. Hogue Learning Resource Center Library, Beeville 33161
Graf von Goess'sche Primogenitur-Fideikommiss-Bibliothek, Ebenthal 01515
Den Grafiske Højskole, København 06667
Gräflich Solms-Laubach'sche Bibliothek, Laubach 12174
GrafTech International Ltd, Cleveland 36712
Grafton – Midview Public Library, Grafton 39262
Graham & Dunn, Seattle 36266
Grahamstown Public Library, Grahamstown 25203
Grain Farming Institute UAAS, Dnipropetrovsk 28200
Grambling State University, Grambling 31227
Grammar School, Ljubljana 24879
Grammar School Jože Plečnik, Ljubljana 24878
Gran bibliotek, Brandbu 20318
Granby Public Library, Granby 39265
Grand Canyon National Park Library, Grand Canyon 36899
Grand Canyon University, Phoenix 32287
Grand County Library District, Granby 39266
Grand Forks Public City-County Library, Grand Forks 39267
Grand Island Memorial Library, Grand Island 39269
Grand Island Public Library, Grand Island 39270
Grand Ledge Area District Library, Grand Ledge 39272
Grand Lodge Free & Accepted Masons of California, San Francisco 37541
Grand Lodge of Iowa, Cedar Rapids 36637
Grand Lodge of Kansas Library, Topeka 37675
Grand Lodge of Masons in Massachusetts, Boston 36564
Grand Lodge of Virginia, A.F. & A.M., Richmond 37440
Grand Lodge of Virginia AF&AM Library & Museum Historical Foundation, Richmond 37441
Grand Orient de France, Paris 08547
Grand People's Study House, Pyongyang 18049
Grand Prairie Public Library System, Grand Prairie 39273
Grand Rapids Community College, Grand Rapids 33381
Grand Rapids Public Library, Grand Rapids 39274
Grand Séminaire de Luxembourg, Luxembourg 18401
Grand Séminaire de Nantes, Nantes 08206
Grand Séminaire de Nkolbisson, Yaoundé 03645
Grand Séminaire de Strasbourg, Strasbourg 08241
Grand Séminaire des Saints Apôtres, Sherbrooke 04228
Grand Séminaire Régional de Koumi, Bobo-Dioulasso 03613
Grand Séminaire Saint-Cyprien, Toulouse 08244
Grand Valley State University, Allendale 30198
Grand View College, Des Moines 31002
Grande Mosquée, Meknès 18945
Grande Prairie Public Library, Grande Prairie 04651
Grande Prairie Public Library District, Hazel Crest 39404
Grande Prairie Regional College, Grande Prairie 03991
Grandview Heights Public Library, Columbus 38669
Grant County Library, John Day 39649
Grant County Library, Ulysses 41632
Grant County Public Library, Milbank 40196
Grant MacEwan College, Edmonton 03987
Grant Medical Center, Columbus 36740
Grant Medical College, Mumbai 13822
Grant Parish Library, Colfax 38635
Granville County Library System, Oxford 40675
Granville Public Library, Granville 39280
Grapevine Public Library, Grapevine 39281
Graphic Arts Institute of Denmark, København 06667
Graphic Arts Library, Tokyo 17679
Graphische Sammlung, Zürich 27157
Graphische Sammlung Albertina, Wien 01587
GRASSI Museum für Angewandte Kunst Leipzig, Leipzig 12183
GRASSI Museum für Völkerkunde zu Leipzig, Leipzig 12184
Gratz College, Melrose Park 31787
– Abner & Mary Schreiber Jewish Music Library, Melrose Park 31788
Graves County Public Library, Mayfield 40103
Gravimetric Observatory, Poltava 28448
Gravimetrichna observatoriya, Poltava 28448
Graydon, Head & Ritchey LLP, Cincinnati 35610
Graymoor Friary Library, Garrison 35211
Gray-Robinson, PA, Tampa 36298
Grays Harbor College, Aberdeen 33111
Grays Harbor County, Montesano 34552
Gray's Inn Library, London 29752
Grayslake Area Public Library District, Grayslake 39284
Grayson County College, Denison 33291
Great Barrier Reef Marine Park Authority, Townsville 00737
Great Basin College, Elko 33318
Great Bend Public Library, Great Bend 39285
Great Falls Public Library, Great Falls 39286
Great Lakes Christian College, Lansing 31558
Great Lakes Forestry Centre, Sault Sainte Marie 04155
Great Meadow Correctional Facility Library, Comstock 34105
Great Neck Library, Great Neck 39287
Great Public Library of Ayatollah Al-uzma Marashi Najafi, Qm 14440
Great River Regional Library, Saint Cloud 41107
The Great Synagogue, Sydney 00797
Greater Amman Public Library, Amman 17887
Greater Cairo Library, Cairo 07197
Greater London Authority, London 29753
Greater Sudbury Public Library / Bibliothèque Publique du Grand Sudbury, Sudbury 04772
Greater Vancouver Regional District Library, Burnaby 04315
Greater Victoria Public Library Board, Victoria 04810
Greater Victoria Public Library Board – Oak Bay, Victoria 04811
Grebenshchikov Institute of Silicate Chemistry of the Russian Academy of Sciences, Sankt-Peterburg 23788
Grebinka Regional Centralized Library System, Main Library, Grebinka 28576
Grebinkivska raionna TsBS, Tsentralna biblioteka, Grebinka 28576
Greco-Roman Museum, Alexandria 07148
Greece Public Library, Greece 39288
Greek Chamber of Deputies, Athinai 13093
Greek Orthodox Bishop of Buda, Szentendre 13347
Greek-Orthodox Patriarchate of Alexandria and all Africa, Alexandria 07129
Green Bay Correctional Institution Library, Green Bay 34316
Green College, Oxford 29320
Green Haven Correctional Facility Library, Stormville 34900
Green Hills Public Library District, Palos Hills 40700
Green Mountain College, Poultney 32356
Green River Community College, Auburn 33140
Greenbaum, Rowe, Smith, Ravin, Davis & Himmel LLP, Woodbridge 36432
Greenberg Glusker Fields Claman & Machtinger LLP Library, Los Angeles 35836
Greenberg Traurig LLP, Miami 35882
Greenberg Traurig LLP, New York 35983
Greenburgh Public Library, Greenburgh 39294
Greendale Public Library, Greendale 39296
Greene County Law Library, Waynesburg 35051
Greene County Law Library, Xenia 35091
Greene County Public Library, Xenia 41982
Greene County Public Library – Beavercreek Community Library, Beavercreek 38137
Greene County Public Library – Fairborn Community Library, Fairborn 39019
Greene County Public Library – Xenia Community Library, Xenia 41983
Greenebaum, Doll & McDonald, Lexington 35813
Greenebaum, Doll & McDonald, Louisville 35864
Greeneville Green County Public Library, Greeneville 39297
Greenfield Community College, Greenfield 33387
Greenfield Public Library, Greenfield 39298
Greenfield Public Library, Greenfield 39299
Greensboro College, Greensboro 31252
Greensboro Public Library, Greensboro 39302
Greensboro Public Library – Blanche S. Benjamin Branch, Greensboro 39303
Greensboro Public Library – Kathleen Clay Edwards Family Branch, Greensboro 39304
Greensburg Hempfield Area Library, Greensburg 39305
Greensburg-Decatur County Public Library, Greensburg 39306
Greensfelder, Hemker & Gale, PC Library, Saint Louis 36200
Greenup County Public Libraries, Greenup 39307
Greenville College, Greenville 31263
Greenville County Library System, Greenville 39309
Greenville Hospital System, Greenville 36906
Greenville Law Library, Greenville 34320
Greenville Public Library, Greenville 39310
Greenville Public Library, Greenville 39311
Greenville Technical College, Greenville 33388
Greenwich Library, Greenwich 39315
Greenwood County Library, Greenwood 39316
Greenwood Public Library, Greenwood 39317
Greenwood-Leflore Public Library System, Greenwood 39318
Gregorio Araneta University Foundation, Manila 20715
Grémio Literário e Comercial Português, Belém 03272
Grenå Bibliotek, Grenaa 06869
Grenada Public Library, St. George's 13159
Grenoble Ecole de management, Grenoble 07710
Grenzschutzschule, Lübeck 10249
Gribbles, Auckland 19824
Griboedov Oktyabrsky District Library-Branch no 5, Sankt-Peterburg 24234
Griffith Observatory Library, Los Angeles 37050
Griffith University, Nathan 00488
– Queensland College of Art, Southbank 00489
– Queensland Conservatorium of Music, South Brisbane 00490
N. Grigorescu Institute of Fine Arts, Bucureşti 21980
Grigoriev Institute for Radiology, Kharkiv 28256
Grimsby Central Library, Grimsby 29989
Grimsby Public Library, Grimsby 04652
Grimstad bibliotek, Grimstad 20337
Grimstad Public Library, Grimstad 20337
Grinding Machine Factory, Moskva 22855
Grindsted Bibliotek, Grindsted 06870
Grinnell College, Grinnell 31267
Grishko National Botanical Garden, Kyiv 28362
Grodnenskaya gorodskaya bibliyateka no 1, Grodno 02134
Grodnenskaya gorodskaya bibliyateka no 2 im. Kremleva-Svena, Grodno 02135
Grodnenskaya gorodskaya bibliyateka no 3, Grodno 02136
Grodnenskaya gorodskaya bibliyateka no 4, Grodno 02137
Grodnenskaya gorodskaya detskaya bibliyateka no 9, Grodno 02138
Grodnenskaya oblastnaya bibliyateka im. E.F. Karskogo, Grodno 02139
Grodnenskaya oblastnaya detskaya bibliyateka im. Pushkina, Grodno 02140
Grodnenskaya tsentralnaya gorodskaya bibliyateka, Grodno 02141
Grodnenski gosudarstvenny meditsinksi universitet, Grodno 01794
Grodnenski gosudarstvenny universitet, Grodno 01795
Grodnenski istoriko-arkheologicheski muzei, Grodno 01960
Grodnenski oblastnoi dom sanitarnogo prosveshcheniya, Grodno 01961
Grodnenski oblastnoi institut usovershenstvovaniya uchitelei, Grodno 01796
Grodnenski selskokhozyaistvenny institut, Grodno 01797
Grodno Central City Library, Grodno 02141
Grodno City Children's Library no 9, Grodno 02138
Grodno City Library no 1, Grodno 02134
Grodno City Library no 2 'Kremlev-Sven', Grodno 02135
Grodno City Library no 3, Grodno 02136
Grodno City Library no 4, Grodno 02137
Grodno Historical and Archeological Museum, Grodno 01960
Grodno Institute of Agricultural Construction, Grodno 01963
Grodno Railway Station, Railway Technical Library, Grodno 01965
Grodno Region Children's Library 'Pushkin', Grodno 02140
Grodno Region Library 'Karsky', Grodno 02139
Grodno Regional Advanced Teacher Training College, Grodno 01796
Grodno State Agrarian University, Grodno 01797
Grodno State Medical University, Grodno 01794
Grodno State University, Grodno 01795
Grodzka Biblioteka Publiczna, Jelenia Góra 21435
Gróf Esterházy Károly Kastély-és Tájmúzeum, Pápa 13493
Grolier Club of New York Library, New York 37209
Grolier, Inc, Danbury 35658
Gromashevsky Epidemiology and Infectious Deseases Research Institute, Kyiv 28368
Grondmechanica Delft, Delft 19410
Groninger Archieven, Groningen 19428
Groninger Museum, Groningen 19429
Det Grønlandske Landsbibliotek / Nunatta Atuagaateqarfia, Nuuk 06630
Grootfontein Agricultural Development Institute, Middelburg 25146
Grootseminarie Rolduc, Kerkrade 19310
Grosse Pointe Public Library, Grosse Pointe Farms 39326
Grossmont College, El Cajon 33312
Grosuplje Public Library, Grosuplje 24958
Groton Public Library, Groton 39327
Groton Public Library, Groton 39328
Groupe Ecole Supérieure de Commerce de Marseille-Provence, Marseille 07755
Groupe Ecole Supérieure de Commerce de Toulouse, Toulouse 08078
Groupe Enseignement International des Affaires (EIA), Marseille 07756
Groupe ESA – École Supérieure d'Agriculture d'Angers, Angers 07638
Groupe ESSEC, Cergy 08252
Groupe HEC, Jouy-en-Josas 08256
Groupement des Intellectuels Aveugles ou Amblyopes (GIAA), Paris 08548
Grove City College, Grove City 31269
Grove Family Library, Chambersburg 38500
Grums kommun, Grums 26777
Grundtvig-Biblioteket, København 06796
The Grundtvig-Library, København 06796
Grupensche Stiftung, Celle 10846
GSI Helmholtzzentrum für Schwerionenforschung GmbH, Darmstadt 11724
GTE Laboratories, Inc, Waltham 36324
Guam Community College, Mangilao 13166
Guam Law Library, Hagatna 13170
Guam Public Library, Hagatna 13171
Guangdong Business College, Guangzhou 05200
Guangdong College of Medicine and Pharmacy, Guangzhou 05201
Guangdong Medical College, Zhanjiang 05776
Guangdong University of Technology, Guangzhou 05202
Guangxi Art Academy, Nanning 05456
Guangxi College of Finance, Nanning 05457
Guangxi College of Traditional Chinese Medicine, Nanning 05458

Guangxi Commercial College, Nanning 05459
Guangxi Education College, Nanning 05460
Guangxi Institute of Technology, Liuzhou 05400
Guangxi Medical University, Nanning 05461
Guangxi Normal University, Guilin 05213
Guangxi School of Physical Education, Nanning 05462
Guangxi Teacher Education University, Nanning 05463
Guangxi Teachers' College for National Minorities, Nanning 05464
Guangxi University, Nanning 05465
Guangxi Zhuang Autonomous Region Library, Nanning 04968
Guangzhou Foreign Trade Institute, Guangzhou 05203
Guangzhou Institute of Chemistry, Beijing 05832
Guangzhou Institute of Chemistry, Guangzhou 05863
Guangzhou Teachers' College, Guangzhou 05204
Guangzhou University, Guangzhou 05205
Guelph Public Library, Guelph 04653
Guernsey County District Public Library, Cambridge 38422
Guernsey County Law Library, Cambridge 34027
Guernsey Memorial Library, Norwich 40551
Guido Carli Independent International University for Social Studies, Roma 15371
Guilderland Public Library, Guilderland 39332
Guildford College of Further and Higher Education, Guildford 29129
Guildford Institute Library, Guildford 29130
Guildhall Library, London 30019
Guildhall School of Music and Drama, London 29178
Guilford College, Greensboro 31253
Guilford Free Library, Guilford 39333
Guilford Technical Community College, Jamestown 33440
Guilin Institute of Electronic Engineering, Guilin 05214
Guilin Institute of Metallurgy and Geology, Guilin 05215
Guilin Library of Guangxi Zhuang Autonomous Region, Guilin 05912
Guilin Medical College, Guilin 05216
Guilin Tourist College, Guilin 05217
Guille-Allès Library, St Peter Port 30076
Guiyang Medical College, Guiyang 05218
Guiyang Teachers' College, Guiyang 05219
Guizhou Agricultural College, Guiyang 05220
Guizhou College of Finance and Economics, Guiyang 05221
Guizhou Institute for Nationalities, Guiyang 05222
Guizhou Normal University, Guiyang 05223
Guizhou Provincial Library, Guiyang 04958
Guizhou University, Guiyang 05224
Guizhou University of Technology, Guiyang 05225
Gujarat Agricultural University, Sardar Krushinagar 13762
Gujarat Ayurved University, Jamnagar 13669
Gujarat Cancer and Research Institute, Ahmedabad 13920
Gujarat Engineering Institute, Vadodara 14217
Gujarat Research Society, Mumbai 14087
Gujarat University, Ahmedabad 13595
Gukovo Coal Industrial Corporation, Gukovo 22793
Gulbarga University, Gulbarga 13650
Det gule Bibliotek, Qaqortoq 06915
Gulf Beaches Public Library, Madeira Beach 39998
Gulf Coast Community College, Panama City 33617
Gulf Coast Research Laboratory, Ocean Springs 37301
Gulf College of Hospitality and Tourism, Muharraq 01738
Gulf Correctional Institution Library, Wewahitchka 35063
Gulf Gate Public Library, Sarasota 41230
Gulfport Public Library, Gulfport 39334
Gulyaipilska raionna TsBS, Tsentralna biblioteka, Gulyaipole 28577
Gulyaipole Regional Centralized Library System, Main Library, Gulyaipole 28577
Gumma Prefectural Assembly, Maebashi 17277
Gumma Prefecture Agricultural Experiment Station, Maebashi 17594
Gummaken Gikai, Maebashi 17277
Gummaken Nogyo Shikenjo, Maebashi 17594
Gundersen Lutheran Health System, La Crosse 36997
Gunma Daigaku, Maebashi 17031
– Igakubu Bunkan, Maebashi 17032
Gunma Prefectural Museum of History, Takasaki 17666
Gunma University, Maebashi 17031
– Medical Branch, Maebashi 17032
Gunmakenritsu Rekisi Hakubutsukan, Takasaki 17666
The Gunnery, Washington 37741
Gunster, Yoakley, Valdes-Fauli & Stewart, PA, Miami 35883
Guntersville Public Library, Guntersville 39337
Guru Nanak Dev University, Amritsar 13604
Gustavus Adolphus College, Saint Peter 32543
Gusyatin Regional Centralized Library System, Main Library, Gusyatin 28578
Gusyatinska raionna TsBS, Tsentralna biblioteka, Gusyatin 28578
Guthrie Memorial Library – Hanover's Public Library, Hanover 39363
Gutteridge Haskins & Davey (GHD) Library, Sydney 00835
Guyuan Teachers' College, Guyuan 05226
Gweru Memorial Library, Gweru 42382
Gwinnett County Law Library, Lawrenceville 34446
Gwinnett County Public Library, Lawrenceville 39826
Gwynedd-Mercy College, Gwynedd Valley 31273
Gymnasium am Kaiserdom, Speyer 10766
Gymnasium Athenaeum Stade, Stade 10768
Gymnasium Philippinum, Weilburg 10774
Gymnasium St. Paulusheim, Bruchsal 10713
Gymnastik- och idrottshögskolan, Stockholm 26426
Gyógyszerkutató Intézet, Budapest 13378
Gyöngyösi Memorial Library, National Széchényi Library, Gyöngyös 13474
Gyöngyösi Müemlékkönyvtár, Gyöngyös 13474
György Bessenyci Teachers' Training College, Nyíregyháza 13287
Győri Evangélikus Gyülekezet, Győr 13338
H. B. Williams Memorial Library, Gisborne 19865
H. Cámara de Senadores, México 18813
H. Grady Bradshaw Chambers County Library, Valley 41655
H. M. Treasury and Cabinet Office, London 29528
Hå folkebibliotek, Nærbø 20376
Haaga instituutti, Helsinki 07338
Haags Centrum voor Onderwijs-begeleiding (HCO), Den Haag 19434
Haags Gemeentearchief, Den Haag 19435
HABE Liburutegia, San Sebastián 26135
Håbo folkbibliotek, Bålsta 26740
Hacettepe Üniversitesi Kütüphaneleri, Hacettepe 27853
Hacettepe University Libraries, Hacettepe 27853
Hachijuni Bank Ltd, Nagano 17388
Hachijuni Ginko Somubu Shiryoshitsu, Nagano 17388
Hackley Public Library, Muskegon 40351
Hackney College, London 29478
Hadassah Academic College, Jerusalem 14619
Haddon Heights Public Library, Haddon Heights 39341
Haddonfield Public Library, Haddonfield 39342
Haderslev Bibliotek, Haderslev 06871
Haderslev Cathedral School, Haderslev 06732
Haderslev Katedralskoles Bibliotek, Haderslev 06732
Hadsel folkebibliotek, Stokmarknes 20403
Hadtörténeti Könyvtár és Térképtár, Budapest 13315
Hadtörténeti Múzeum, Budapest 13379
HafenCity Universität Hamburg, Hamburg 09914
Haffkine Institute for Training, Research and Testing, Mumbai 14088
Hagaman Memorial Library, East Haven 38902
Hagelands Historisch Documentatiecentrum, Tienen 02702
HagenMedien Stadtbücherei, Hagen 12767
Hagerstown Community College, Hagerstown 33393
Hagfors Bibliotek, Hagfors 26779
Hagley Museum & Library, Wilmington 37851
Hagstofa Íslands, Reykjavík 13562
The Hague Central Library, Den Haag 19611
The Hague Municipal Archives, Den Haag 19435
Haifa Museum of Art, Haifa 14720
Haigazian University Library, Beirut 18204
Hailaer Teachers' College, Hailaer 05229
Hainan Teachers' College, Haikou 05227
Hainan University, Haikou 05228
Hajdúsági Múzeum, Hajdúböszörmény 13478
Hakim Mohammed Said Central Library, New Delhi 13726
Hakodate City Library, Hakodate 17839
Hakuhodo Library and Information Center, Tokyo 17677
Hakuhodo Shiryosenta, Tokyo 17677
Hakutsuru Bijitsukan, Kobe 17572
Hakutsuru Fine Art Museum, Kobe 17572
Halászati és Öntözési Kutatóintézet, Szarvas 13501
Halden bibliotek, Halden 20339
Haldimand County Public Library, Dunnville 04631
Hale & Dorr Library, Boston 35525
Haley Memorial Library & History Center, Midland 37106
Half Hollow Hills Community Library, Dix Hills 38847
Halifax Community College, Weldon 33852
Halifax County Library, Halifax 39346
Halifax County-South Boston Regional Library, Halifax 39347
Halifax County-South Boston Regional Library – South Boston Public Library, South Boston 41355
Halifax Public Libraries, Dartmouth 04623
Halifax Regional Library – Captain William Spry Branch, Halifax 04654
Halifax Regional Library – Cole Harbour Branch, Cole Harbour 04615
Halifax Regional Library – Keshen Goodman Branch, Halifax 04655
Halifax Regional Library – Sackville Branch, Lower Sackville 04683
Halifax Regional Library – Spring Garden Road Memorial, Halifax 04656
Halil Hamit Paşa Kütüphani, Isparta 27882
Hall County Library System, Gainesville 39181
Hall County Library System – Blackshear Place, Gainesville 39182
HALLE 14, Leipzig 12179
Halliburton Energy Services, Houston 35756
Hallmark Cards, Inc, Kansas City 35793
Hallsbergs bibliotek, Hallsberg 26780
Hallstahammars bibliotek, Hallstahammar 26781
Halmstad University, Halmstad 26390
Hälsinglands museum, Hudiksvall 26592
Haltom City Public Library, Haltom City 39349
Halton County Law Association, Milton 04380
Halton Hills Public Libraries, Georgetown 04648
Haluoleo University, Kendari 14266
Hämäläis-Osakunnan kirjasto, Helsinki 07452
Hamburg Institute for Social Research, Hamburg 11985
Hamburg Port Authority, Hamburg 10923
Hamburg Public Library, Hamburg 39350
Hamburg Township Library, Hamburg 39351
Hamburg University, Hamburg 09317
Hamburg University of Technology, Hamburg 09922
Hamburger Bahnhof – Museum für Gegenwart, Berlin 11567
hamburger frauenbibliothek Denk(t)räume, Hamburg 11984
Hamburger Institut für Sozialforschung, Hamburg 11985
Hamburger Kunsthalle, Hamburg 11986
Hamburger Lehrerbibliothek, Hamburg 11987
Hamburger Schulmuseum, Hamburg 11988
Hamburger Sternwarte, Hamburg 09946
Hamdard University, Karachi 20441
Hamden Public Library, Hamden 39352
Häme Student Union Library, Helsinki 07452
Hämeenlinna Public Library, Hämeenlinna 07562
Hämeenlinnan kaupunginkirjasto, Hämeenlinna 07562
Hämeenlinnan maakunta-arkiston käsikirjasto, Hämeenlinna 07439
Hamersley Iron, Perth 00824
Hamilton City Libraries, Hamilton 19867
Hamilton College, Clinton 30838
Hamilton County Governmental Law Library, Chattanooga 34061
Hamilton District Council, Hamilton 29992
Hamilton East Public Library, Fishers 39067
Hamilton East Public Library, Noblesville 40484
Hamilton Health Sciences, Hamilton 04362
Hamilton Law Association Library, Hamilton 04363
Hamilton North Public Library, Cicero 38562
Hamilton Public Library, Hamilton 04657
Hamilton Public Library, Dundas Branch, Dundas 04630
Hamilton Township Public Library, Hamilton 39353
Hamilton-Wenham Public Library, South Hamilton 41361
Hamline University, Saint Paul
– Bush Memorial Library, Saint Paul 32531
– School of Law Library, Saint Paul 32532
Hammarö kommunbibliotek, Skoghall 26897
The Hammarö library, Skoghall 26897
Hammerfest bibliotek, Hammerfest 20341
Hammond Public Library, Hammond 39355
Hampden Sydney College, Hampden Sydney 31283
The Hampden-Booth Theatre Library, New York 37210
Hampshire College, Amherst 30215
Hampshire County Council, Winchester 30100
Hampshire County Public Library, Romney 41059
Hampton B. Allen Library, Wadesboro 41699
Hampton Bays Public Library, Hampton Bays 39358
Hampton University, Hampton 31284
– William H. Moses Jr. Architecture Library, Hampton 31285
Hamtramck Public Library, Hamtramck 39359
Hancock County Law Library, Ellsworth 34181
Hancock County Law Library Association, Findlay 34202
Hancock County Library System, Bay Saint Louis 38118
Hancock County Public Library, Greenfield 39300
Hancock, Rothert & Bunshoft, San Francisco 36232
Handan Higher Specialized Agricultural College, Yongnian 05761
Handan Teachers' College, Handan 05231
Handels- og Ingeniørhøjskolens Bibliotek, Herning 06661
Handels- og Søfartsmuseet, Helsingør 06778
Handelshochschule Leipzig gGmbH, Leipzig 10242
Handelshögskolan i Stockholm, Stockholm 26427
Handelshøjskolen, Århus 06650
Handelskammer Bremen, Bremen 11676
Handelskammer Hamburg, Hamburg 11975
Handelsskolen Sjælland Syd, Vordingborg 06754
Handeslhøyskolen BI, Oslo 20116
Handley Regional Library, Winchester 41922
Handotai Kenkyu Shinkokai, Sendai 17661
Handwerkskammer für München und Oberbayern, München 12296
Hangö Public Library, Hangö 07563
Hangö Stadsbibliotek, Hangö 07563
Hangon Kaupunginkirjasto – Hangö Stadsbibliotek, Hangö 07563
Hangzhou Institute of Electronic Engineering, Hangzhou 05234
Hangzhou Teachers' College, Hangzhou 05235
Hangzhou University of Commerce, Hangzhou 05236
Haninge bibliotek, Haninge 26783
Hanken School of Economics, Helsinki 07339
Hankuk University of Foreign Studies, Seoul 18082
Hannah Research Institute, Ayr 29599
Hannah-Arendt-Institut für Totalitarismusforschung e.V. an der TU Dresden, Dresden 11757
Hannema-de Stuers Fundatie, Zwolle 19541
Hannibal Free Public Library, Hannibal 39362
Hannibal-Lagrange College, Hannibal 31287
Hannover Medical School, Hannover 09973
Hanns-Seidel-Stiftung e.V., München 12297
Dr.-Hanns-Simon-Stiftung, Bitburg 12677
Hanoi Agricultural University, Hanoi 42268
Hanoi College of Construction, Hanoi 42269
Hanoi College of Pharmacy, Hanoi 42270
Hanoi Medical School, Hanoi 42271
Hanoi University of Civil Engineering, Hanoi 42272
Hanoi University of Finance and Accountancy, Hanoi 42273
Hanoi University of Technology, Hanoi 42274
Hanover College, Hanover 31297
Hanover Juvenile Correctional Center, Hanover 34334
Hans Kelsen-Institut, Wien 01601
Hans-Böckler-Stiftung, Düsseldorf 11795
Hans-Bredow-Institut für Medienforschung, Hamburg 11989
Hanseatisches Oberlandesgericht, Hamburg 10924
Hanse-Bibliothek Demmin e.V., Demmin 12701
Hanse-Klinikum Stralsund, Stralsund 12507
Hanson, Bridgett, Marcus, Vlahos & Rudy, San Francisco 36233
Hanson Professional Services Inc, Springfield 36279
Hanyang University, Seoul 18083
Hanzehogeschool Groningen, Groningen
– Hanzehogeschool Groningen, Groningen 19126
– Hanzehogeschool Groningen, Groningen 19127

– HanzeMediatheek Conservatorium, Groningen 19128
Hanzhong Teachers' College, Hanzhong 05244
Haparanda stadsbibliotek, Haparanda 26784
Har Zion Temple, Penn Valley 35324
Haram folkebibliotek, Brattvåg 20319
Harare City Library, Harare 42383
Harare Polytechnic, Harare 42364
Harbin City Library, Harbin 05915
Harbin College of Investment, Harbin 05245
Harbin Institute of Electrical Technology, Harbin 05246
Harbin Institute of Technology, Harbin 05247
Harbin Medical University, Harbin 05248
Harbin Normal University, Harbin 05249
Harbin Teachers' College, Harbin 05250
Harbin University of Science and Technology, Harbin 05251
Harbin University of Science and Technology, Harbin 05252
Harbor Branch Oceanographic Institution, Inc, Fort Pierce 36863
Harborfields Public Library, Greenlawn 39301
Harbour Administration, Szczecin 21040
Harcourt Butler Technological Institute, Kanpur 14021
Harcum College, Bryn Mawr 30586
Hardayal Municipal Public Library, Delhi 14226
Hardin County Public Library, Elizabethtown 38956
Harding University, Searcy 32629
Harding University Graduate School of Religion, Memphis 31791
Hardin-Simmons University, Abilene 30164
Hardy County Public Library, Moorefield 40288
Hardy Technical College Library, Ampara 26286
Harford Community College, Bel Air 33162
Harford County Public Library, Belcamp 38146
Dr. Harisingh Gour University, Sagar 13757
Härjedalens bibliotek, Sveg 26919
Harju County Library, Keila 07246
Harju Maakonnarazmaturogu, Keila 07246
Harlan Community Library, Harlan 39368
Harlan County Public Library, Harlan 39369
Harlingen Public Library, Harlingen 39370
Harlow College, Harlow 29468
Harness, Dickey & Pierce, PLC, Troy 36312
Harnett County Public Library, Lillington 39875
Härnösands kommun bibliotek, Härnösand 26785
Härnösands sjukhus, Härnösand 26586
Harold Washington College, Chicago 33220
Harper Adams University College, Newport, Shropshire 29283
Harper Grey Easton Library, Vancouver 04302
Harriet Beecher Stowe Center Library, Hartford 36918
Harrington College of Design, Chicago 36675
Harris, Beach PLLC, Pittsford 36134
Harris County Law Library, Houston 34367
Harris County Public Library, Houston 39485
Harris County Public Library – Aldine Branch, Houston 39486
Harris County Public Library – Atascocita, Humble 39530
Harris County Public Library – Baldwin Boettcher Branch, Humble 39531
Harris County Public Library – Barbara Bush Branch, Spring 41407
Harris County Public Library – Clear Lake City-County Freeman Branch, Houston 39487
Harris County Public Library – Cy-Fair College Branch, Cypress 38750
Harris County Public Library – Katherine Tyra Branch, Houston 39488
Harris County Public Library – Katy Branch, Katy 39685
Harris County Public Library – Kingwood Branch, Kingwood 39728
Harris County Public Library – LaPorte Branch, LaPorte 39805
Harris County Public Library – Maud Smith Marks Branch, Katy 39686
Harris County Public Library – North Channel, Houston 39489
Harris County Public Library – Northwest, Cypress 38751
Harris County Public Library- Octavia Fields Memorial, Humble 39532
Harris County Public Library – Parker Williams Branch, Houston 39490
Harris County Public Library – Spring Branch Memorial, Houston 39491
Harris County Public Library – Tomball Branch, Tomball 41580
Harris Manchester College, Oxford 29322
Harris Methodist Information Resources, Fort Worth 36869

Harrisburg Area Community College, Harrisburg 33399
Harrison County Law Library, Gulfport 34324
Harrison County Library System, Gulfport 39335
Harrison County Library System – Gulfport Temporary Library, Gulfport 39336
Harrison Memorial Library, Carmel 38457
Harrison Public Library, Harrison 39374
Harris-Stowe State College, Saint Louis 32503
Hart County Library, Hartwell 39387
Harter, Secrest & Emery LLP, Rochester 36183
Hartford City Public Library, Hartford City 39382
Hartford Hospital, Hartford 36919
Hartford Hospital – Institute of Living, Hartford 36920
Hartford Medical Society, Farmington 36846
Hartford Public Library, Hartford 39380
Hartford Public Library, Hartford 39381
Hartford Seminary, Hartford 35225
Hartford Steam Boiler Inspection & Insurance Co., Hartford 35738
Hartland Public Library, Hartland 39384
Hartlepool Borough Council, Hartlepool 29994
Hartnell Community College, Salinas 33719
Hartwick College, Oneonta 32199
Hartzmark Library, Beachwood 35129
Harvard Library in New York, New York 37211
Harvard Musical Association Library, Boston 36565
Harvard Public Library, Harvard 39388
Harvard University, Cambridge 30611
– Andover-Harvard Theological Library, Cambridge 30612
– Arnold Arboretum Horticulture Library, Jamaica Plain 30613
– Arthur & Elizabeth Schlesinger Library on the History of Women, Cambridge 30614
– Baker Library, Boston 30615
– Blue Hill Meteorological Observatory Library, Cambridge 30616
– Center for Hellenic Studies Library, Washington 30617
– Center for Population Studies Library, Boston 30618
– Chemistry Library, Cambridge 30619
– Child Memorial & English Tutorial Library, Cambridge 30620
– Davis Center for Russian & Eurasian Studies Fung Library, Cambridge 30621
– Dumbarton Oaks Research Library & Collection, Washington 30622
– Eda Kuhn Loeb Music Library, Cambridge 30623
– Ernst Mayr Library, Cambridge 30624
– Fine Arts Library, Cambridge 30625
– Frances Loeb Library, Cambridge 30626
– Francis A. Countway Library of Medicine, Boston 30627
– George David Birkhoff Mathematical Library, Cambridge 30628
– Godfrey Lowell Cabot Science Library, Cambridge 30629
– Gordon McKay Library, Cambridge 30630
– Gutman Library-Research Center, Cambridge 30631
– Harry Elkins Widener Memorial Library, Cambridge 30632
– Harvard College Library (Headquarters in Harry Elkins Widener Memorial Library), Cambridge 30633
– Harvard Forest Library, Petersham 30634
– Harvard Map Collection, Cambridge 30635
– Harvard-Yenching Library, Cambridge 30636
– Herbert Weir Smyth Classical Library, Cambridge 30637
– History Department Library, Cambridge 30638
– History of Science Library – Cabot Science Library, Cambridge 30639
– Houghton Library-Rare Books & Manuscripts, Cambridge 30640
– John F. Kennedy School of Government Library, Cambridge 30641
– John K. Fairbank Center for East Asian Research Library, Cambridge 30642
– Lamont Library-Undergraduate, Boston 30643
– Law School Library, Cambridge 30644
– Littauer Library, Cambridge 30645
– Minda de Gunzburg, Center for European Studies Library, Cambridge 30646
– Physics Research Library, Cambridge 30647

– Robbins Library of Philosophy, Cambridge 30648
– Social Relations-Sociology Library, Cambridge 30649
– Tozzer Library, Cambridge 30650
Harvard-Smithsonian Center for Astrophysics Library, Cambridge 36619
Harvey, Pennington, Cabot, Griffith & Renneisen, Ltd, Philadelphia 36100
Harvey Public Library District, Harvey 39389
Haryana Agricultural University, Hissar 13651
Haryana Civil Secretariat, Chandigarh 13851
Harzbücherei Wernigerode, Wernigerode 12589
Hasanuddin University, Makassar 14268
– Faculty of Law, Ujung Pandang 14269
– Institute for Economic and Social Research, Ujung Pandang 14270
Haselwood Library, Bremerton 33185
Hashemite University Library, Zarqa 17879
Haskell Indian Nations University, Lawrence 31572
Haskins Laboratories Library, New Haven 37159
Háskóli Íslands, Reykjavík 13556
Háskólinn á Bifröst, Borgarnes 13555
Haslev Bibliotek, Haslev 06872
Hassan Usman Katsina Polytechnic, Katsina 19958
Hassan-Phillips Medical Library, Cairo 07173
Hassard Bonnington, San Francisco 36234
Hässleholms lasarett, Hässleholm 26588
Hässleholms stadsbibliotek, Hässleholm 26787
Hastings Center, Garrison 36885
Hastings Central Library, Hastings 19868
Hastings College, Hastings 31308
Hastings College of Arts and Technology, St Leonards-on-Sea 29405
Hastings College of the Law, San Francisco 32593
Hastings Public Library, Hastings 39393
Hastings-on-Hudson Public Library, Hastings-on-Hudson 39395
Hatch Ltd, Niagara Falls 04273
Hatch Research & InfoCentres, Mississauga 04264
Hatta Foundation Library, Yogyakarta 14380
Haugesund folkebibliotek, Haugesund 20342
Haukeland universitetssykehus, Bergen 20212
Dr. von Haunersches Kinderspital, München 10435
Hauptstaatsarchiv Hannover, Hannover 12032
Hauptstaatsarchiv Stuttgart, Stuttgart 12522
Hauptstaatsarchiv Stuttgart, Bibliothek des Militärarchivs, Stuttgart 12523
Hauptverband des Österreichischen Buchhandels, Wien 01602
Haus Birkach, Stuttgart 11338
Haus der bayerischen Landwirtschaft Herrsching, Herrsching 12059
Haus der Begegnung, Bonn 11176
Haus der Donauschwaben, Salzburg 01566
Haus der Donauschwaben, Sindelfingen 12489
Haus der Geschichte der Bundesrepublik Deutschland, Bibliothek zur Geschichte der DDR, Bonn 11649
Haus der Geschichte der Bundesrepublik Deutschland Foundation, Collection of Industrial Design, Berlin 11588
Haus der Geschichte der Bundesrepublik Deutschland, Informationszentrum, Bibliothek und Mediathek zur deutschen Zeitgeschichte, Bonn 11650
Haus der Geschichte der Bundesrepublik Deutschland, Sammlung Industrielle Gestaltung, Berlin 11588
Haus der Geschichte der Bundesrepublik Deutschland, Zeitgeschichtliches Forum Leipzig, Leipzig 12204
Haus der Heimat des Landes Baden-Württemberg, Stuttgart 12524
Haus der Wannsee-Konferenz, Berlin 11527
Haus der Wirtschaft, Stuttgart 12512
Haus des Deutschen Ostens, München 12298
HAUS Finnish Institute of Public Management Ltd, Helsinki 07453
HAUS kehittämiskeskus Oy, Helsinki 07453
Haus kirchlicher Dienste der Ev.-luth. Landeskirche Hannovers, Hannover 11237
Haus Oberschlesien, Ratingen 12421
Haus Schlesien, Königswinter 12157
Haus St. Ulrich, Augsburg 11152
Haute école d'art et de design Genève, Genève 27086
Haute école de gestion de Genève, Carouge 27063

Haute Ecole de la Santé la Source, Lausanne 27222
Haute Ecole de Travail social et de la Santé – EESP – Lausanne, Lausanne 27098
Haute Ecole d'ingénierie et de gestion du Canton de Vaud (HEIG-VD), Yverdon-les-Bains 27149
Haute Ecole pédagogique, Fribourg 27353
Haute école pédagogique, Lausanne 27099
Haute-Ecole Paul-Henri Spaak (IESSID), Bruxelles 02288
Hautes Etudes Commerciales (HEC), Liège 02374
Haut-Saint-Jean Regional Library / Région de Bibliothèques du Haut-Saint-Jean, Saint Basile 04741
Haverford College, Haverford 31312
– Astronomy Library, Haverford 31313
– Music Library, Haverford 31314
– White Science Library, Haversford 31315
Havering College of Further and Higher Education, Hornchurch 29135
Haverstraw Kings Daughters Public Library, Garnerville 39204
Haverstraw Kings Daughters Public Library – Village Library, Haverstraw 39397
Havforskningsinstituttet, Bergen 20213
Havre Hill County Library, Havre 39398
Havsfiskelaboratoriet, Lysekil 26615
Hawaii Agriculture Research Center Library, Aiea 36441
Hawaii Pacific University, Honolulu 31346
Hawaii Pacific University, Kaneohe 31479
Hawaii State Archives, Honolulu 36933
Hawaii State Circuit Court-Second Circuit, Wailuku 34970
Hawaii State Public Library System, Honolulu 39464
Hawaii State Public Library System – Aiea Public Library, Aiea 37886
Hawaii State Public Library System - Aina Haina Public Library, Honolulu 39465
Hawaii State Public Library System – Ewa Beach Public & School Library, Ewa Beach 39013
Hawaii State Public Library System – Hawaii State Library, Honolulu 39466
Hawaii State Public Library System – Hawaii-Kai Public Library, Honolulu 39467
Hawaii State Public Library System – Hilo Public Library, Hilo 39444
Hawaii State Public Library System – Kahului Public Library, Kahului 39667
Hawaii State Public Library System – Kailua Public Library, Kailua 39668
Hawaii State Public Library System – Kailua-Kona Public Library, Kailua-Kona 39669
Hawaii State Public Library System – Kaimuki Public Library, Honolulu 39468
Hawaii State Public Library System – Kalihi-Palama Public Library, Honolulu 39469
Hawaii State Public Library System – Kaneohe Public Library, Kaneohe 39673
Hawaii State Public Library System – Kapolei Public Library, Kapolei 39682
Hawaii State Public Library System – Kihei Public Library, Kihei 39715
Hawaii State Public Library System – Lihue Public Library, Lihue 39874
Hawaii State Public Library System – Liliha Public Library, Honolulu 39470
Hawaii State Public Library System – McCully-Moiliili Public Library, Honolulu 39471
Hawaii State Public Library System – Mililani Public Library, Mililani 40202
Hawaii State Public Library System – Pearl City Public Library, Pearl City 40738
Hawaii State Public Library System – Princeville Public Library, Princeville 40903
Hawaii State Public Library System – Salt Lake-Moanalua Public Library, Honolulu 39472
Hawaii State Public Library System – Wahiawa Public Library, Wahiawa 41701
Hawaii State Public Library System – Waianae Public Library, Waianae 41702
Hawaii State Public Library System – Wailuku Public Library, Wailuku 41703
Hawaii State Public Library System – Waipahu Public Library, Waipahu 41704
Hawaiian Historical Society Library, Honolulu 36934
Hawaiian Mission Children's Society Library, Honolulu 36935
HAWK Fachhochschule Hildesheim/ Holzminden/Göttingen – Bibliothek Bauwesen, Soziale Arbeit und Gesundheit, Hildesheim 10041
HAWK Fachhochschule Hildesheim/ Holzminden/Göttingen – Bibliothek Gestaltung, Erhaltung von Kulturgut, Hildesheim 10042

HAWK Fachhochschule Hildesheim/Holzminden/Göttingen – Bibliothek Management, Soziale Arbeit, Bauen, Holzminden 10046
HAWK Fachhochschule Hildesheim/Holzminden/Göttingen – Bibliothek Ressourcenmanagement, Göttingen 09902
HAWK Hochschule für angewandte Wissenschaft und Kunst Hildesheim/Holzminden/Göttingen – Bibliothek Naturwissenschaften und Technik, Göttingen 09903
Hawke's Bay Museum, Napier 19840
Hawkesbury Public Library, Hawkesbury 04658
Hawkeye Community College, Waterloo 33842
Hawkins, Delafield & Wood, New York 35984
Hawley Troxell Ennis & Hawley, Boise 35515
Hayes, Seay, Mattern & Mattern, Inc Library, Roanoke 36180
Hayner Public Library District, Alton 37928
Haynes & Boone LLP, Dallas 35647
Haynesville Correctional Center, Haynesville 34346
Hays Public Library, Hays 39401
Hayward Public Library – Weekes Branch, Hayward 39402
Hazard Community College, Hazard 33404
Hazard Community College, Jackson 33432
Hazel Park Memorial Library, Hazel Park 39405
Hazelden Foundation, Center City 36641
Hazleton Area Public Library, Hazleton 39407
HBO-Raad, Den Haag 19436
hbz – Hochschulbibliothekszentrum NRW, Köln 12135
HCM Raj State Institute of Public Administration, Jaipur 14013
Headquarters New Zealand Defence Force (HQ NZDF), Wellington 19799
Health and Safety at Work Research Institute, Kyiv 28372
Health and Safety Executive, Liverpool 29703
Health Canada, Ottawa 04443
Health Canada, Ottawa 04444
Health Care Insurance Board, Diemen 19415
Health Centre, Sankt-Peterburg 23911
Health Department of Western Australia, East Perth 00675
Health & Human Services Department, Washington 34998
Health Insurance Authority, Cairo 07111
Health Laboratory Services, Accra 13052
Health Library Alice Springs, Alice Springs 00861
Health Ministry of Crimean Republic – Medical Library, Simferopol 28473
Health One Presbyterian-Saint Luke's Medical Center, Denver 36795
Health Promotion Information Centre (HPIC), London 29754
Health Protection Agency, London 29755
Health Protection Agency Porton Down, Salisbury 29890
Health Sciences University of Mongolia, Ulaanbaatar 18904
Health Scotland Library, Edinburgh 29654
Health Services Research and Documentation Centre, Surabaya 14379
Heard Museum, Phoenix 37370
Heating Automation Plant, Moskva 22868
Heavy Excavators Joint-Stock Company, Voronezh 22975
Heavy Industry Fittings Industrial Corporation, Aleksin 22772
Hebei Architectural Engineering College, Zhangjiakou 05770
Hebei College of Commerce, Shijiazhuang 05561
Hebei College of Engineering and Technology, Cangzhou 05065
Hebei College of Finance and Economics, Shijiazhuang 05562
Hebei Geological College, Shijiazhuang 05563
Hebei Institute of Coal Mining and Civil, Handan 05232
Hebei Institute of Forestry, Baoding 04989
Hebei Institute of Physical Culture, Shijiazhuang 05564
Hebei Institute of Technology, Tianjin 05607
Hebei Institute of Traditional Chinese Medicine, Shijiazhuang 05565
Hebei Library, Qingyuan 04969
Hebei Mechanical and Electric Institute Engineering College, Shijiazhuang 05566
Hebei Medical University, Shijiazhuang 05567
Hebei Normal University, Shijiazhuang 05568
Hebei Teachers' College, Shijiazhuang 05569
Hebei Teachers' College of Agricultural Technology, Qinghuangdao 05488
Hebei University, Baoding 04990
Hebei University of Economics and Business, Shijiazhuang 05570
Hebrew College, Newton Centre 32117
Hebrew Theological College, Skokie 32681
Hebrew Union College, Jerusalem 14620
Hebrew Union College, Los Angeles 31647
Hebrew Union College (HUC-JIR) – Jewish Institute of Religion, Cincinnati 30800
Hebrew Union College – Jewish Institute of Religion (NY), New York 32065
Hebrew University of Jerusalem, Jerusalem 14598
– Archaeological Institute, Jerusalem 14622
– Avraham Harman Science Library, Jerusalem 14623
– Bernard G. Segal Law Library, Jerusalem 14624
– Bloomfield Library for Humanities and Social Sciences, Jerusalem 14625
– Central Library of Agricultural Science, Rehovot 14626
– Department of Botany, Jerusalem 14627
– Department of Zoology, Jerusalem 14628
– The Education and Social Work Libray, Jerusalem 14629
– Institute of Earth Sciences, Jerusalem 14630
– Institute of Mathematics, Jerusalem 14631
– Muriel and Philip I. Berman Medical Library, Jerusalem 14632
– Natural Science Library for First Year Students, Jerusalem 14633
– Physics and Chemistry Library, Jerusalem 14634
– Rothberg International School Library, Jerusalem 14635
– School of Library, Archive and Information Studies, Jerusalem 14636
– Science Teaching Library, Jerusalem 14637
– H. S. Truman Research Institute for the Advancement of Peace, Jerusalem 14638
Heby bibliotek, Heby 26788
HEC Montréal (Ecole des Hautes Etudes Commerciales), Montreal 03773
Hedberg Public Library, Janesville 39635
Hedbergska skolan, Sundsvall 26482
Hedemora stadsbibliotek, Hedemora 26789
Hedensted Bibliotek, Hedensted 06873
Hedland College, South Hedland 00588
Hedmark County Library, Hamar 20340
Hedmark fylkesbibliotek, Hamar 20340
Hedmark University College, Elverum
– Department of Business Administration, Social Science and Computer Science, Rena 20089
– Department of General Teacher Training and Natural Science, Hamar 20088
– Faculty of Health and Sports, Elverum 20086
– Faculty of Health Studies, Elverum 20087
Heerema Marine Contractors BV, Leiden 19333
Heeresfliegerwaffenschule, Bückeburg 10714
Heeresgeschichtliches Museum, Wien 01603
Hefei Economy and Technology Institute, Hefei 05266
Hefei Educational Institute, Hefei 05267
Hefei University of Technology, Hefei 05268
Hegau-Bibliothek, Singen 12490
Heibei Institute of Light Chemical Engineering, Shijiazhuang 05571
Heiberg Collection, Kaupanger 20230
De Heibergske Samlinger – Sogn Folkemuseum, Kaupanger 20230
Heidelberg College, Tiffin 32817
Heidelberg University, Heidelberg 09978
– Center for Psychosocial Medicine, Heidelberg 10036
Heidelberger Kunstverein, Heidelberg 12047
Heilig Grafinstituut, Turnhout 02423
Heilig Hart Instituut Secundair Onderwijs Heverlee (HHH), Leuven 02419
Heilongjiang College of Traditional Chinese Medicine, Harbin 05253
Heilongjiang Communication College, Harbin 05254
Heilongjiang Institute of Commerce, Harbin 05255
Heilongjiang Provincial Library, Harbin 04960
Heilongjiang Teachers' College of Land Reclamation, Ercheng 05170
Heilongjiang University, Harbin 05256
Heinola Public Library, Heinola 07564
Heinolan kaupunginkirjasto, Heinola 07564
Heinrich-Böll-Bibliothek, Berlin 12665
Heinrich-Böll-Gesamtschule Köln-Chorweiler, Köln 10747
Heinrich-Braun-Klinikum Zwickau gGmbH, Zwickau 12624
Heinrich-Heine-Institut, Düsseldorf 11796
Heinrich-Heine-Universität Düsseldorf, Düsseldorf 09312
Heinrich-Hertz-Institut, Berlin 11528
Heinrich-Mann-Bibliothek / Stadtbibliothek Strausberg, Strausberg 12952
Heinrich-Schulz-Bibliothek, Berlin 12657
Heinrich-Suso-Gymnasium, Konstanz 10751
Helderberg College, Somerset West 25047
Helen B. Hoffman Plantation Library, Plantation 40822
Helen Brown Lombardi Library of International Affairs, San Francisco 37542
Helen Hall Library, League City 39831
Helen Keller Services for the Blind, Hempstead 36926
Helen M. Plum Memorial Library, Lombard 39918
Helen Matthes Library, Effingham 38937
Helene-Nathan-Bibliothek, Berlin 12664
Helikon Castle Museum, Keszthely 13486
Helikon Kastélymúzeum, Keszthely 13486
HELIOS Kliniken Schwerin, Schwerin 12481
Hellenic College & Holy Cross Greek Orthodox School of Theology, Brookline 30564
Hellenic Folklore Research Center, Athinai 13118
Hellenic Institute of International and Foreign Law, Athinai 13112
Hellenic Maritime Museum, Piraius 13129
Heller, Ehrman, White & McCauliffe Library, San Francisco 36235
Hellmuth, Obata & Kassabaum, Inc, Saint Louis 36201
Helmholtz Centre for Environmental Research – UFZ, Central Library, Leipzig 12185
Helmholtz Research Institute for Eye Disease, Moskva 23400
Helmholtz Zentrum München – Bibliothek des Hämatologikums, München 12299
Helmholtz Zentrum München – Deutsches Forschungszentrum für Gesundheit und Umwelt, Neuherberg 12361
Helmholtz-Zentrum Berlin, Berlin 11529
Helmholtz-Zentrum Dresden-Rossendorf, Dresden 11758
Helmholtz-Zentrum für Infektionsforschung, Braunschweig 11662
Helmholtz-Zentrum für Umweltforschung GmbH – UFZ, Leipzig 12185
Helmholtz-Zentrum Geesthacht, Geesthacht 11915
Helms-Museum, Hamburg 12016
Helmut-Schmidt-Universität / Universität der Bundeswehr Hamburg, Hamburg 09915
Helmut-Sihler-Bibliothek, Düsseldorf 11388
Helsedirktoratet, Oslo 20190
Helsingborgs Lasarett, Helsingborg 26590
Helsingborgs stadsbibliotek, Helsingborg 26790
Helsinge Bibliotek, Helsinge 06875
Helsingin IV terveydenhuolto-oppilaitos, Helsinki 07385
Helsingin ammattikorkeakoulu Stadia, Helsinki 07386
Helsingin Energia, Helsinki 07454
Helsingin hovioikeuden kirjasto, Helsinki 07394
Helsingin kaupungin kaupunkisuunnitteluviraston kirjasto, Helsinki 07455
Helsingin kaupungin ruotsinkielisen työväenopiston kirjasto, Helsinki 07456
Helsingin kaupungin suomenkielisen työväenopiston kirjasto, Helsinki 07457
Helsingin kaupungin tietokeskuksen kirjasto, Helsinki 07395
Helsingin kaupunginarkiston käsikirjasto, Helsinki 07458
Helsingin kaupunginkirjasto, Helsinki 07566
Helsingin kaupunginmuseo, Helsinki 07459
Helsingin sairaanhoito-opiston kirjasto, Helsinki 07387
Helsingin Sanomat, Helsinki 07414
Helsingin sosiaali- ja terveysalan oppilaitos, Helsinki 07388
Helsingin venäläisen kauppiasyhdistyksen kirjasto, Helsinki 07460
Helsingin yliopisto, Helsingin Yliopisto 07332
– City Centre Campus Library, Helsingin Yliopisto 07333
– Kumpula Campus Library, Helsingin Yliopisto 07334
– Viikki Campus Library, Helsingin Yliopisto 07335
Helsingør Kommunes Biblioteker, Helsingør 06876
Helsinki IV College of Health Care Professionals, Helsinki 07385
Helsinki City Archives, Helsinki 07458
Helsinki City College of Social and Health Care, Helsinki 07388
Helsinki City Library – Central Library for public libraries, Helsinki 07566
Helsinki City Museum, Helsinki 07459
Helsinki City Planning Department, Helsinki 07455
Helsinki College of Nursing, Helsinki 07387
Helsinki Court of Appeal, Helsinki 07394
Helsinki Polytechnic Stadia, Helsinki 07386
Helsinki University, Helsingin Yliopisto 07332
Helwan University, Giza 07070
– Faculty of Engineering, Mataria, Cairo 07071
– Faculty of Fine Arts, Cairo 07072
– Faculty of Music Education, Cairo 07073
– Faculty of Physical Education for Men (Cairo), Giza 07074
– Faculty of Physical Education for Women, Alexandria, Alexandria 07075
– Faculty of Physical Education for Women, Cairo, Cairo 07076
– Higher Industrial Institute, Cairo 07077
– Suez Institute for Petroleum, Cairo 07078
Hematology and Blood Transfusion Research Institute, Kyiv 28364
Hemenway & Barnes, Boston 35526
Hemeroteca Municipal de Madrid, Madrid 26029
Hemeroteca Municipal de Sevilla, Sevilla 26151
Hemeroteca Municipal de Valencia, Valencia 26171
Hemeroteca Nacional de México, México 18844
Hemet Public Library, Hemet 39410
Hempstead Public Library, Hempstead 39411
Hemvati Nandan Bahuguna Garhwal University, Srinagar 13766
Henan Agricultural University, Zhengzhou 05778
Henan College of Traditional Chinese Medicine, Zhengzhou 05779
Henan Commercial College, Zhengzhou 05780
Henan Finance and Tax College, Zhengzhou 05781
Henan Machinery-Electricity College, Xinxiang 05720
Henan Provincial Library, Zengzhou 05928
Henan Public Security College, Zhengzhou 05782
Henan Textile College, Zhengzhou 05783
Henan University, Kaifeng 05348
Henan University of Finance and Economics, Zhengzhou 05784
Henan Urban Construction College, Pingdingshan 05479
Henan Vocational Teachers' College, Xinxiang 05721
Henderson Community College, Henderson 33405
Henderson County, Athens 38011
Henderson County Public Library, Henderson 39412
Henderson County Public Library, Hendersonville 39417
Henderson District Public Libraries, Henderson 39413
Henderson District Public Libraries, Henderson 39414
Henderson Memorial Public Library Association, Jefferson 39639
Henderson State University, Arkadelphia 30258
Hendrick Hudson Free Library, Montrose 40283
Hendrix College, Conway 30920
Hendry County Library System – Clewiston Public Library (Headquarters), Clewiston 38600
Hengeler Mueller-Bibliothek, Hamburg 09916
Hengshui Teachers' College, Hengshui 05270
Hengyang Institute of Technology, Hengyang 05271
Hengyang Medical College, Hengyang 05272
Hengyang Teachers' College, Hengyang 05273
Henkel AG u. Co. KGaA, Düsseldorf 11388
Henley Management College, Henley-on-Thames 29133
Hennepin County Law Library, Minneapolis 34541
Hennepin County Library, Minnetonka 40232
Hennepin County Medical Center, Minneapolis 37121
Henrietta Szold Institute, Jerusalem 14729
Henriette Goldschmidt-Schule, Leipzig 10752
Henry Carter Hull Library, Inc, Clinton 38606
Henry County Library, Clinton 38607
Henry County Library System – McDonough Public Library, McDonough 40118
The Henry Ford, Dearborn 36777

The Henry Ford, Dearborn 36778
Henry Ford Community College, Dearborn 33284
Henry Waldinger Memorial Library, Valley Stream 41659
The Henryk Niewodzniczanski Institute of Nuclear Physics, Kraków 21132
Hephata, Hessisches Diakoniezentrum e.V., Schwalmstadt 10583
HERA Information Centre, Manukau City 19838
Héraðsbókasafn Borgarfjardar, Borgarnes 13578
Herald and Weekly Times Ltd, Melbourne 00821
Heraldisch-Genealogische Gesellschaft 'Adler', Wien 01604
Heraldry and Genealogy Society of Canberra Inc., Narrabundah 00965
Herbario 'Barbosa Rodrigues', Itajaí 03302
Herbert Wescoat Memorial Library, McArthur 40114
Hercules Incorporated – Law Department Library, Wilmington 36422
Hercules Incorporated – Research Center Library, Wilmington 36423
Herder-Institut e.V., Marburg 12255
Hereford Cathedral Library and Archives, Hereford 29546
Herefordshire College of Technology, Hereford 29469
Heriot-Watt University, Edinburgh 29092
– Martindale Library, Galashiels 29093
Heritage Baptist University, Greenwood 31265
Heritage College, Toppenish 32828
Heritage College & Seminary, Cambridge 04201
Herkimer County Community College, Herkimer 33407
Herkimer County Law Library, Herkimer 34351
Herlev Bibliotek, Herlev 06877
Herlufsholm Skole, Næstved 06738
Herman Ottó Múzeum, Miskolc 13490
Hermann-Tast-Schule, Husum 10741
Hernando County Public Library System – Lykes Memorial Library, Brooksville 38367
Herning Centralbibliotek, Herning 06878
Herning Institute of Business Administration and Technology, Herning 06661
Herold, Verein für Heraldik, Genealogie und verwandte Wissenschaften, Berlin 11530
Herrick District Library, Holland 39456
Herrick Memorial Library, Wellington 41797
Hershey Public Library, Hershey 39420
Hertford College, Oxford 29323
Hertford Regional College, Broxbourne 28978
Hertford Regional College, Ware 29499
Hertfordshire County Council, Hatfield 29995
Herz Jesu-Missionare, Innsbruck 01438
Herzen Cancer Research Institute, Moskva 23250
Herzen Polotsk City Library No 2, Polotsk 02226
Herzen Smolny District Library-Branch no 10, Sankt-Peterburg 24238
Herz-Jesu-Missionare, Salzburg 01463
Herzog Anton Ulrich-Museum, Braunschweig 11663
Herzog August Bibliothek, Wolfenbüttel 09343
Herzogin Anna Amalia Bibliothek, Weimar 09341
Hesser College, Manchester 31742
Hessian Ministry of Economics, Transport, Urban and Regional Development, Wiesbaden 11127
Hessische Polizeischule / Verwaltungsfachhochschule in Wiesbaden, Wiesbaden 10686
Hessische Staatskanzlei, Wiesbaden 11119
Hessische Stiftung Friedens- und Konfliktforschung (HSFK), Frankfurt am Main 11852
Hessischer Landtag, Wiesbaden 11120
Hessischer Rundfunk, Frankfurt am Main 11853
Hessischer Verwaltungsgerichtshof, Kassel 10960
Hessisches Finanzgericht, Kassel 10961
Hessisches Hauptstaatsarchiv, Wiesbaden 12594
Hessisches Kultusministerium, Wiesbaden 11121
Hessisches Landesamt für Bodenmanagement und Geoinformation, Wiesbaden 11122
Hessisches Landesamt für Geschichtliche Landeskunde, Marburg 11018
Hessisches Landesamt für Geschichtliche Landeskunde, Abt. Forschungsstelle für Geschichtliche Landeskunde Mitteldeutschlands, Marburg 12256
Hessisches Landesamt für Umwelt und Geologie, Wiesbaden 11123
Hessisches Landesarbeitsgericht, Frankfurt am Main 10898
Hessisches Landesmuseum Darmstadt, Darmstadt 11725
Hessisches Landessozialgericht, Darmstadt 10853
Hessisches Ministerium der Finanzen, Wiesbaden 11124
Hessisches Ministerium des Innern und für Sport, Wiesbaden 11125
Hessisches Ministerium für Umwelt, Energie, Landwirtschaft und Verbraucherschutz, Wiesbaden 11126
Hessisches Ministerium für Wirtschaft, Verkehr und Landesentwicklung, Wiesbaden 11127
Hessisches Sozialministerium, Wiesbaden 11128
Hessisches Staatsarchiv Darmstadt, Darmstadt 11726
Hessisches Staatsarchiv Marburg, Marburg 12257
Hessisches Statistisches Landesamt, Wiesbaden 11129
Hessisches Wirtschaftsarchiv, Darmstadt 11727
HES-SO Valais, Sierre 27130
HES-SO Valais/Wallis, Sion
– Bereich Ingenieurwissenschaften, Sion 27131
– Médiathèque Santé-Social, Sion 27132
Hesston College, Hesston 31329
Het Scheepvaartmuseum Amsterdam, Amsterdam 19358
Hetjens-Museum, Deutsches Keramik-Museum, Düsseldorf 11797
Hewitt Associates Library, Lincolnshire 35816
Hewlett-Packard, Corvallis 35640
Hewlett-Packard Laboratories, Palo Alto 36079
Hewlett-Woodmere Public Library, Hewlett 39421
Heythrop College, London 29236
Heze Medical College, Heze 05276
Heze Teachers' College, Heze 05277
Hezuo Teachers Training School for Nationalities, Gannan 05195
HHL Leipzig Graduate School of Management, Leipzig 10242
Hialeah-John F. Kennedy Library, Hialeah 39422
Hibbing Public Library, Hibbing 39423
Hickory Public Library, Hickory 39425
Hicksville Public Library, Hicksville 39426
Hid Islenska Bókmenntafélag, Reykjavík 13570
Hidalgo County Law Library, Edinburg 34166
Hiera Moyi Hosiou Gregoriou, Athos 13094
Higgs, Fletcher & Mack LLP, San Diego 36218
High Commission of India, London 29756
High Commission of India, Yarralumla 00747
High Court, Cape Town 25060
High Court, Hong Kong 05812
High Court, Kampala 27919
High Court, Lobatse 02972
High Court Library, Chichiri 18479
High Court Library, Majuro 18674
High Court Library, Suva 07301
High Court of Australia, Parkes 00706
High Court of Kenya, Nairobi 18022
High Court of Kerala, Ernaculam 13855
High Court of Lagos State, Lagos 19999
High Court, Witwatersrand Local Division, Johannesburg 25064
High Energy Accelerator Research Organization, Tsukuba 17791
High Museum of Art Library, Atlanta 36492
High Plains Library District, Greeley 39289
High Plains Library District – Centennial Park, Greeley 39290
High Plains Library District – Erie Community Library, Erie 38985
High Plains Library District – Farr Regional Library, Greeley 39291
High Point Public Library, High Point 39427
High Point University, High Point 31332
High School Poljane, Ljubljana 24880
High Street Christian Church, Akron 35101
Highbury College, Portsmouth 29488
Higher Colleges of Technology, Abu Dhabi 28905
Higher Comsomol School, Moskva 23599
Higher Education and Training Awards Council, Dublin 14535
Higher Education Council, Ankara 27871
Higher Institute of Agriculture, Santiago de los Caballeros 06959
Higher Institute of Arabic Music, Giza 07079
Higher Institute of Dramatic Art, Sofiya 03481
Higher Institute of Dramatic Arts, Giza 07080
Higher Institute of Mechanical and Electrical Engineering, Hoon 18243
Higher Institute of Mechanical and Electrical Engineering, Hoon 18255
Higher Institute of Technology, Brack 18242
Higher Military School of Radioelectronics and Air Defence Systems, Sankt-Peterburg 22562
Higher National Institute of Music (Conservatoire), Giza 07081
Higher Police School in Szczytno, Library, Szczytno 20933
Higher Professional School of Applied Arts, Sankt-Peterburg 22560
Higher School of Teacher Training, Jászberény 13278
Higher School of Teacher Training, Szombathely 13309
Higher Teacher Education School, Częstochowa 20801
Higher Teachers' Training Institute, Omdurman 26352
Higher Technical Institute, Nicosia 06325
Higher Theological Seminary, Szczecin 20926
Highgate Literary and Scientific Institution, London 29757
Highland Community College, Freeport 33353
The Highland Council, Inverness 30001
Highland County District Library, Hillsboro 39436
Highland Health Sciences Library, Inverness 29688
Highland Park Presbyterian Church, Dallas 35178
Highland Park Public Library, Highland Park 39431
Highland Park Public Library, Highland Park 39432
Highland Park United Methodist Church Library, Dallas 35179
Highland Township Public Library, Highland 39429
Highlands Agricultural College, Mount Hagen 20562
Highlands County Library System – Sebring Public Library, Sebring 41269
Highlands & Islands Enterprise, Inverness 29571
Highline Community College, Des Moines 33292
High-Mountain Geophysical Institute, Nalchik 23611
Highschool of Civil Engineering 'Ivan Kavčič', Ljubljana 24882
Highschool of Education and Culture, Maribor 24888
Highschool of Electrical Engineering and Information Science, Maribor 24887
Highschool of Metal, Mechanical and Metallurgical Engineering, Maribor 24886
Highschool of Social Sciences, Maribor 24885
Highschool of Textiles and Hairdressing, Maribor 24889
Highschool of Trade, Ljubljana 24883
Highways Research Institute, Balashikha 22994
Highways Research Station, Chennai 13961
Highwood Public Library, Highwood 39434
Hilbert College, Hamburg 33394
Hildesheim Cathedral Library, Hildesheim 11247
Hill & Barlow Library, Boston 35527
Hill College, Hillsboro 33409
Hill Correctional Center Library, Galesburg 34293
Hill, Farrer & Burrill, Los Angeles 35837
Hillerød Bibliotek, Hillerød 06879
Hills Memorial Library, Hudson 39526
Hillsboro Public Library, Hillsboro 39437
Hillsboro Public Library – Shute Park, Hillsboro 39438
Hillsborough Community College, Tampa 33793
Hillsborough County Law Library, Manchester 34497
Hillsborough County Law Library, Tampa 34919
Hillsdale College, Hillsdale 31338
Hillsdale Free Public Library, Hillsdale 39441
Hillsdale Free Will Baptist College, Moore 31895
Hillside Public Library, Hillside 39442
Hillside Public Library, Hillside 39443
Hillside Public Library, New Hyde Park 40413
Himachal Pradesh Agricultural University, Palampur 13731
Himachal Pradesh Krishi Vishvavidyalaya, Palampur 13731
Himachal Pradesh University, Simla 13765
Himalayan International Institute of Yoga Science and Philosophy of the USA, Honesdale 36929
Himeji Institute of Technology, Himeji 16946
Hinds Community College District, Raymond 33675
The Hindu, Chennai 13588
Hindu College Library, New Delhi 13722
Hindustan Lever Research Centre, Mumbai 14089
Hindustan Photo Films Mfg Co Ltd, Ootacamund 13913
Hindustan Zinc Ltd, Udaipur 13915
Hingham Public Library, Hingham 39447
Hinnerup Bibliotek & Kulturhus, Hinnerup 06880
Hino Jidosha Kogyo K. K., Hino 17355
Hino Motors, Ltd, Research Management Dept, Hino 17355
Hinsdale Public Library, Hinsdale 39448
Hinshaw & Culbertson Library, Chicago 35579
Hippologische Bibliothek, Verden (Aller) 12572
Hiram College, Hiram 31340
Hiram Halle Memorial Library, Pound Ridge 40885
Hirara Cultural Center, Hirara 17547
Hirarashi Bunka Senta, Hirara 17547
Hírközlési Központi Szakkönyvtár, Budapest 13380
Hiroshima Archives, Hiroshima 17548
Hiroshima Central City Library, Hiroshima 17840
Hiroshima Daigaku, Hiroshima 16951
– Igaku Bunkan, Hiroshima 16952
Hiroshima Jogakuin College, Hiroshima 16953
Hiroshima Kenritsu Kyoiku Senta, Higashihiroshima 17545
Hiroshima Kenritsu Ringyo Shikenjo, Miyoshi 17604
Hiroshima Prefectural Assembly, Hiroshima 17263
Hiroshima Prefectural Education Center, Higashihiroshima 17545
Hiroshima Prefectural Forest Experiment Station, Miyoshi 17604
Hiroshima Prefecture Administrative Data Room, Hiroshima 17264
Hiroshima Shiritsu Chuo Toshokan, Hiroshima 17840
Hiroshima University, Hiroshima 16951
– Medical Science Branch, Hiroshima 16952
Hiroshima University of Economics, Hiroshima 16954
Hiroshimaken Gikai, Hiroshima 17263
Hiroshimaken Gyosei Shiryoshitsu, Hiroshima 17264
Hiroshimashi Kobunshokan, Hiroshima 17548
Hirsch Library, Museum of Fine Arts, Houston, Houston 36940
Hirshhorn Museum & Sculpture Garden Library, Washington 37784
Hispanic Baptist Theological School Library, San Antonio 35373
Hispanic & Luso Brazilian Council, London 29758
Hispanic Society of America Library, New York 37212
Historic Deerfield Inc & Pocumtuck Valley Memorial Association Libraries, Deerfield 36787
Historic Hudson Valley's Library, Tarrytown 37660
Historic New England, Boston 36566
Historic New Orleans Collection, New Orleans 37167
Historical and Archaeological Institute of Dodecanese, Rhodos 13130
Historical and Architectural Museum, Yaroslavl 24086
Historical and Architectural Museum-Reserve, Ryazan 23732
Historical and Ethnological Society of Greece, Athinai 13117
Historical and geographical Institute, Goiânia 03299
Historical, Architectural and Artistic Museum Reserve, Vologda 24070
Historical Archive, Zadar 06229
Historical Archive, Zagreb 06237
Historical Archive Ljubljana, Ljubljana 24945
Historical Documents Centre, Manama 01742
Historical & Genealogical Society of Indiana County, Indiana 36953
Historical Institute, Dushanbe 27642
Historical Museum, Blagoevgrad 03500
Historical Museum, Lviv 28415
Historical Museum of Hokkaido, Sapporo 17651
Historical Museum of the People of Uzbekistan, Toshkent 42132
Historical Museum of Warsaw, Warszawa 21316

Historical Research Commission of Taiwan Province, Taipei 27609
Historical Research Institute, Kaigun Bunko, Tokyo 17757
Historical Society of Berks County, Reading 37432
Historical Society of Cheshire County, Keene 36992
Historical Society of Geel, Geel 02641
Historical Society of Israel, Jerusalem 14730
Historical Society of Pennsylvania, Philadelphia 37349
Historical Society of Washington, DC, Washington 37742
Historical Society of Western Pennsylvania, Pittsburgh 37380
Historická knihovna Arcibiskupského zámku a zahrad v Kroměříži, Kroměříž 06416
Historický ústav, Bratislava 24693
Historický ústav AV ČR, Brno 06428
Historický ústav AV ČR, Praha 06490
Historijski Arhiv, Zagreb 06237
Historisch Centrum Leeuwarden, Leeuwarden 19471
Historisch Centrum Overijssel, Zwolle 19542
Historisch Fonds van de Stadsbibliotheek Brugge, Brugge 02547
Historische Bibliotheek Mechelen, Mechelen 02684
Historische Bibliothek der Stadt Rastatt im Ludwig-Wilhelm-Gymnasium, Rastatt 12419
Historische Bibliothek der Stadt Rudolstadt, Rudolstadt 12447
Historische Kommission für Westfalen, Münster 12343
Historische Kreisbibliothek Euskirchen, Euskirchen 11831
Historische Museen Hamburg, Altonaer Museum, Hamburg 12015
Historische Museen Hamburg – Helms-Museum, Hamburg 12016
Historische Museen Hamburg – Museum für Bergedorf und die Vierlande, Hamburg 12017
Historischer Verein für Dortmund und die Grafschaft Mark, Dortmund 11752
Historischer Verein für Niedersachsen, Hannover 12026
Historischer Verein für Oberfranken e.V., Bayreuth 11481
Historischer Verein für Oberpfalz und Regensburg, Regensburg 12426
Historischer Verein für Württembergisch Franken, Schwäbisch Hall 12477
Historischer Verein von Oberbayern, München 12332
Historisches Archiv der Stadt Köln, Köln 12136
Historisches Archiv des Erzbistums Köln, Köln 11260
Historisches Institut, Aachen 09369
Historisches Institut beim Österreichischen Kulturforum in Rom, Roma 16506
Historisches Museum, Frankfurt am Main 11854
Historisches Museum Basel, Basel 27304
Historisches Museum Bern, Bern 27321
Historisches Museum Bremerhaven, Bremerhaven 11695
Historisches Museum Hannover, Hannover 12027
Historisches Museum Schloss Gifhorn, Gifhorn 11921
Historisches Stadtarchiv, Halberstadt 11952
Historisk Museum, Bergen 20078
History Museum of the City of Cracow, Kraków 21144
Hitachi Ltd, Central Research Laboratory, Kokubunji 17378
Hitachi Seisakujo Chuo Kenkyujo Gijutsu Johobu, Kokubunji 17378
Hitchcock Memorial Museum & Library, Westfield 37828
Hitotsubashi Daigaku, Kunitachi 16991
– Keizai Kenkyujo, Kunitachi 16992
– Keizai Kenkyujo Fuzoku Nihon Kaizai Fokei Bunken Senta, Kunitachi 16993
– Sangyo Keiei Kenkyushisetsu, Kunitachi 16994
Hitotsubashi University, Kunitachi 16991
– Institute of Business Research, Kunitachi 16994
– Institute of Economic Research, Kunitachi 16992
– Institute of Economic Research, Documentation Center for Japanese Economic Statistics, Kunitachi 16993
Hiwassee College, Madisonville 33512
Hjälpmedelsinstitutet, Vällingby 26712
Hjørring Bibliotekerne, Hjørring 06881
Hjørring Public Library, Hjørring 06881
HKM Engineering Inc Library, Billings 35504
H.M.G. – Ministry of Industry, Kathmandu 19029
Ho Polytechnic, Ho 13025
Hobart & William Smith Colleges, Geneva 31209
Hobbs Public Library, Hobbs 39450
Hobe Sound Bible College, Hobe Sound 35231
Hoboken Public Library, Hoboken 39451
Hobro Bibliotek, Hobro 06882
Hobson Wharf, Auckland 19825
Hochschul- und Kreisbibliothek Bonn-Rhein-Sieg, Sankt Augustin 10579
Hochschul- und Landesbibliothek RheinMain, Wiesbaden 09342
Hochschul- und Landesbibliothek, Standort: Heinrich-von-Bibra-Platz, Fulda 09811
Hochschul- und Landesbibliothek, Standort: Marquardstraße, Fulda 09812
Hochschul- und Landeskirchenbibliothek Wuppertal, Wuppertal 10695
Hochschulbibliothek Reutlingen, Reutlingen 10541
Hochschulbibliothek Weingarten, Weingarten 10684
Hochschulbibliothekszentrum NRW, Köln 12135
Hochschule Albstadt-Sigmaringen, Sigmaringen 10588
Hochschule Albstadt-Sigmaringen, Bibliothek Albstadt, Albstadt 09390
Hochschule Anhalt, Köthen 10236
Hochschule Ansbach, Ansbach 09393
Hochschule Augsburg, Augsburg 09394
Hochschule Biberach / Biberach University of Applied Sciences, Biberach an der Riß 09458
Hochschule Bochum, Bochum 09463
Hochschule Bonn-Rhein-Sieg, Sankt Augustin 10579
Hochschule Darmstadt, Darmstadt 09622
Hochschule der Bildenden Künste Saar, Saarbrücken 10549
Hochschule der Bundesagentur für Arbeit, Mannheim 10294
Hochschule der Künste Bern (HKB), Bern 27019
Hochschule der Künste Bern – Musikbibliothek, Bern 27016
Hochschule der Medien, Stuttgart 10592
Hochschule der Sächsischen Polizei (FH), Rothenburg / O.L. 10547
Hochschule Esslingen, Esslingen am Neckar 09726
– Standort Göppingen, Göppingen 09727
Hochschule Fulda – Hochschul- und Landesbibliothek, Fulda
– Standort: Heinrich-von-Bibra-Platz, Fulda 09811
– Standort: Marquardstraße, Fulda 09812
Hochschule für Agrar- und Umweltpädagogik, Wien 01292
Hochschule für angewandte Wissenschaft und Kunst – Bibliothek Bauwesen, Soziale Arbeit und Gesundheit, Hildesheim 10041
Hochschule für angewandte Wissenschaft und Kunst – Bibliothek Gestaltung, Erhaltung von Kulturgut, Hildesheim 10042
Hochschule für angewandte Wissenschaft und Kunst – Bibliothek Management, Soziale Arbeit, Bauen, Holzminden 10046
Hochschule für angewandte Wissenschaft und Kunst – Bibliothek Ressourcenmanagement, Göttingen 09902
Hochschule für angewandte Wissenschaft und Kunst Göttingen, Göttingen 09903
Hochschule für angewandte Wissenschaften, Rosenheim 10544
Hochschule für Angewandte Wissenschaften Amberg-Weiden, Amberg 09392
Hochschule für angewandte Wissenschaften Fachhochschule Coburg, Coburg 09620
Hochschule für angewandte Wissenschaften – Fachhochschule Deggendorf, Deggendorf 09652
Hochschule für angewandte Wissenschaften – Fachhochschule Neu-Ulm / University of Applied Sciences, Neu-Ulm 10515
Hochschule für angewandte Wissenschaften Fachhochschule Würzburg-Schweinfurt, Würzburg 10696
Hochschule für Angewandte Wissenschaften FH Ingolstad, Ingolstadt 10050
Hochschule für Angewandte Wissenschaften Hamburg, Hamburg
– Fachbibliothek Design, Hamburg 09917
– Fachbibliothek Life Sciences, Hamburg 09918
– Fachbibliothek Soziale Arbeit und Pflege, Hamburg 09919
– Fachbibliothek Technik, Wirtschaft, Information (TWI), Hamburg 09920
Hochschule für angewandte Wissenschaften – Hochschule Regensburg, Regensburg 10538
– Teilbibliothek Prüfeningerstraße, Regensburg 10539
Hochschule für Bildende Künste, Braunschweig 09591
Hochschule für Bildende Künste, Dresden 09670
Hochschule für evangelische Kirchenmusik, Bayreuth 09398
Hochschule für Fernsehen und Film München, München 10359
Hochschule für Film und Fernsehen 'Konrad Wolf', Potsdam 10535
Hochschule für Grafik und Buchkunst, Leipzig 10243
Hochschule für Jüdische Studien Heidelberg, Heidelberg 09976
Hochschule für Kirchenmusik, Rottenburg am Neckar 10548
Hochschule für Musik Carl Maria von Weber, Dresden 09671
Hochschule für Musik "Franz Liszt", Weimar 10683
Hochschule für Musik "Hanns Eisler" Berlin, Berlin 09445
Hochschule für Musik Karlsruhe, Karlsruhe 10056
Hochschule für Musik Nürnberg, Nürnberg 10520
Hochschule für Musik Saar, Saarbrücken 10550
Hochschule für Musik, Theater und Medien Hannover, Hannover 09956
Hochschule für Musik und Darstellende Kunst, Frankfurt am Main 09754
Hochschule für Musik und Tanz Köln, Köln 10128
Hochschule für Musik und Theater München, München 10360
Hochschule für Musik und Theater Rostock, Rostock 10545
Hochschule für nachhaltige Entwicklung Eberswalde (FH), Eberswalde 09684
Hochschule für öffentliche Verwaltung und Finanzen Ludwigsburg, Ludwigsburg 10253
Hochschule für Philosophie, München 10361
Hochschule für Politik München, München 10362
Hochschule für Schauspielkunst Ernst Busch, Berlin 09446
Hochschule für Technik Rapperswil, Rapperswil 27227
Hochschule für Technik und Wirtschaft des Saarlandes, Saarbrücken
– Bibliothek Bereich Goebenstrasse, Saarbrücken 10551
– Bibliothek Bereich Waldhausweg, Saarbrücken 10552
Hochschule für Technik und Wirtschaft Dresden, Dresden 09672
Hochschule für Technik und Wirtschaft (HTW), Berlin 09447
Hochschule für Technik, Wirtschaft und Kultur Leipzig, Leipzig 10244
Hochschule für Wirtschaft und Recht Berlin (HWR Berlin), Campus Lichtenberg, Berlin 09448
Hochschule für Wirtschaft und Recht Berlin (HWR Berlin), Campus Schöneberg, Berlin 09449
Hochschule für Wirtschaft und Umwelt Nürtingen-Geislingen, Nürtingen 10521
– Bibliothek Geislingen, Geislingen a.d. Steige 10522
Hochschule Furtwangen, Furtwangen 09813
Hochschule Hannover, Hannover 09957
– Gemeinsame Bibliothek der Hochschule für Musik und Theater und der Fachhochschule Hannover, Hannover 09958
– Teilbibliothek Bioverfahrenstechnik, Hannover 09959
– Teilbibliothek Diakonie, Gesundheit und Soziales, Hannover 09960
Hochschule Harz (FH), Wernigerode 10685
Hochschule Harz (FH) – Standort Halberstadt, Halberstadt 09909
Hochschule Heidelberg, Heidelberg 10037
Hochschule Heilbronn, Heilbronn 10039
Hochschule Hof, Hof 10045
Hochschule Kempten, Kempten 10062
Hochschule Konstanz Technik, Wirtschaft und Gestaltung (HTWG Konstanz), Konstanz 10234
Hochschule Landshut, Landshut 10239
Hochschule Lausitz (FH), Senftenberg 10585
– Benutzungsbereich Cottbus, Cottbus 10586
Hochschule Luzern – Musik, Luzern 27224
Hochschule Luzern – Technik und Architektur, Horw 27094
Hochschule Magdeburg-Stendal (FH), Magdeburg 10261
– Hochschulbibliothek Standort Stendal, Hansestadt Stendal 10262
Hochschule Mannheim, Mannheim 10295
Hochschule Merseburg (FH), Merseburg 10351
Hochschule Mittweida (FH), Mittweida 10352
Hochschule München, München 10363
– Teilbibliothek Architektur, Bauingenieurwesen und Geoinformationswesen, München 10364
– Teilbibliothek für Wirtschaft und Soziales, München 10365
Hochschule Neubrandenburg, Neubrandenburg 10513
Hochschule Niederrhein, Mönchengladbach
– Bibliothek Chemie/Design, Krefeld 10353
– Bibliothek Mönchengladbach, Mönchengladbach 10354
– Bibliothek Technik, Krefeld 10355
Hochschule Offenburg, Offenburg 10524
Hochschule Osnabrück, Osnabrück 10527
– Zentrale Einrichtung für Wissenschaftliche Information (ZEWI), Osnabrück 10528
Hochschule Ostwestfalen-Lippe, Lemgo 10246
Hochschule Pforzheim – Fachhochschule für Gestaltung, Technik und Wirtschaft, Pforzheim
– Bereichsbibliothek Wirtschaft und Technik, Pforzheim 10532
– Fakultät Gestaltung, Pforzheim 10533
Hochschule Reutlingen, Reutlingen 10541
Hochschule Rhein-Waal, Standort Kleve, Kleve 10119
Hochschule Ruhr West, Mülheim an der Ruhr 10357
Hochschule Ulm, Ulm 10676
Hochschule Weihenstephan-Triesdorf, Freising 09805
– Teilbibliothek Triesdorf, Weidenbach 09806
Hochschule Wismar, Wismar 10690
Hochschule Zittau/Görlitz, Zittau 10699
Hochschulrektorenkonferenz (HRK), Bonn 11635
Hockessin Public Library, Hockessin 39452
Hodgson Russ LLP, Buffalo 35542
Hoechst AG, Werk Casella, Frankfurt am Main 11398
Hoechst AG, Werksbücherei C 820, Frankfurt am Main 12734
Hoechst Celanese Research Division, Summit 36288
Hoejvangseminariet, National Institute for Social Education and Preschool Teachers in Copenhagen, Glostrup 06731
Hoësche Bibliothek im Landesarchiv Schleswig-Holstein, Schleswig 12466
Hof van Beroep, Antwerpen 02424
Hofbibliothek Aschaffenburg, Aschaffenburg 09297
Hoffmann-La Roche, Inc Library, Nutley 36063
Hofors Bibliotek, Hofors 26792
Hofstra University, Hempstead
– Barbara & Maurice A. Deane Law Library, Hempstead 31324
– Joan & Donald E. Axinn Library, Hempstead 31325
Hogan & Hartson LLP, Washington 36355
Höganäs AB, Höganäs 26536
Höganäs stadsbibliotek, Höganäs 26793
Hoge Hotelschool Maastricht (HHM), Maastricht 19174
Hoge Raad der Nederlanden, Den Haag 19254
Hoger Instituut voor Godsdienstwetenschappen, Antwerpen 02449
Hoger Instituut voor Grafisch Onderwijs (HIGRO), Mariakerke 02420
Hogeschool de Driestar, Gouda 19125
Hogeschool De Horst, Amersfoort 19051
Hogeschool Gent, Gent
– Bibliotheek Schoonmeersen, Gent 02318
– Departement Bedrijfskunde Aalst, Aalst 02319
– Departement Bedrijfsmanagement Mercator, Gent 02320
– Departement Biowetenschappen en Landschapsarchitectuur, Melle 02321
– Departement Conservatorium, Gent 02322
– Departement Gezondheidszorg Vesalius, Gent 02323
– Departement Koninklijke Academie voor Schone Kunsten, Gent 02324

– Departement Sociaal-Agogisch Werk, Gent 02325
– Departement Vertaalkunde, Gent 02326
– Leercentrum Lerarenopleiding, Gent 02327
Hogeschool INHOLLAND Alkmaar, Alkmaar 19050
Hogeschool INHOLLAND Diemen, Diemen 19106
Hogeschool INHOLLAND Haarlem, Haarlem 19144
Hogeschool Rotterdam, Rotterdam
– Academie van Beeldende Kunsten, Rotterdam 19192
– Academie van Bouwkunst, Rotterdam 19193
– Hogeschool voor Economische Studies, Rotterdam 19194
– Mediatheek Museumpark, Rotterdam 19195
– Mediatheek Wijnhaven, Rotterdam 19196
– Polytechnische Faculteit, Hoger Laboratoriumonderwijs (HLO), Rotterdam 19197
Hogeschool Sint-Lukas Brussel (SLUKB), Brussel 02289
Hogeschool Utrecht, Utrecht
– Faculteit Communicatie en Journalistiek, Utrecht 19208
– Faculteit Economie en Management, Utrecht 19209
– Faculteit Educatie, Utrecht 19210
– Faculteit Gezondheidszorg, Utrecht 19211
– Faculteit Natuur en Techniek, Utrecht 19212
– Faculteit Sociaal-Agogische Opleidingen, Utrecht 19213
Hogeschool van Amsterdam, Amsterdam
– Domein Onderwijs en Opvoeding, Amsterdam 19053
– Instituut voor Maatschappelijk Werk en Dienstverlening / Instituut voor Cultureel Maatschappelijke Vorming en Sociaal-Pedagogische Hulpverlening, Amsterdam 19054
– Mediatheek Leeuwenburg, Amsterdam 19055
– Mediatheek Tafelbergweg, Amsterdam 19056
Hogeschool van Arnhem en Nijmegen, Arnhem
– Faculteit Economie HEAO – Arnhem, Arnhem 19097
– Faculteit Techniek, Arnhem 19098
Hogeschool van Arnhem en Nijmegen, Nijmegen
– Faculteit Educatie, Nijmegen 19175
Hogeschool van de provincie Antwerpen, Antwerpen 02274
Hogeschool Van Hall Larenstein, Leeuwarden 19149
Hogeschool Van Hall Larenstein, Velp 19225
Hogeschool Van Hall Larenstein, Wageningen 19226
Hogeschool voor de Kunsten Utrecht (HKU), Utrecht
– Faculteit Theater, Utrecht 19214
– Utrechts Conservatorium, Utrecht 19215
Hogeschool voor Wetenschap en Kunst, Brussel
– Departement Architektuur – Beeldende Kunst, Gent 02290
– VLEKHO-bibliotheek, Brussel 02291
Hogeschool Zuyd, Sittard
– Bibliotheek Havikstraat, Sittard 19199
– Faculteit Gedrag en Maatschappij, Sittard 19200
Hogeschool-Universiteit Brussel (HUB), Brussel
– Bibliotheek Campus Koekelberg, Brussel 02292
– Centrale Mediatheek Campus Nieuwland, Brussel 02293
Högsby kommunbibliotek, Högsby 26794
Högskolan i Borås, Borås 26373
Högskolan i Gävle, Gavle 26374
Högskolan i Halmstad, Halmstad 26390
Högskolan i Jönköping, Jönköping 26394
Högskolan i Kalmar, Kalmar 26395
Högskolan i Kristianstad, Kristianstad 26397
– Hässleholms tekniska skola, Hässleholm 26398
Högskolan i Skövde, Skövde 26422
Högskolebiblioteket för försvar, utrikes- och säkerhetspolitik, Stockholm 26424
Høgskolen i Agder, Kristiansand 20101
– Bibliotektjenesten i Grooseveien, Grimstad 20102
– Bibliotektjenesten i Kongsgård allé, Kristiansand 20103
– Bibliotektjenesten i Tordenskjoldsgate, Kristiansand 20104
Høgskolen i Akershus, Kjeller 20097
– Avdeling for sykepleierutdanning, Lillestrom 20098
– Avdeling for vernepleierutdanning, Lillestrøm 20099
Høgskolen i Bergen, Bergen
– Biblioteket – Haukelandsbakken, Bergen 20068
– Biblioteket – Landås, Landås 20069
– Biblioteket – Møllendal, Bergen 20070
– Biblioteket – Nordnes, Bergen 20071
– Biblioteket – Nygård, Bergen 20072
Høgskolen i Bodø, Bodø 20084
Høgskolen i Buskerud, Hønefoss 20096
Høgskolen i Buskerud, Kongsberg, Kongsberg 20100
Høgskolen i Finnmark, Avdeling i Alta, Alta 20066
Høgskolen i Gjøvik, Gjvik 20092
Høgskolen i Hedmark, Elverum
– Avdeling for helse- og idrettsfag, Elverum 20086
– Avdeling for helse- og sosialfag, Elverum 20087
– Avdeling for lærerutdanning og naturvitenskap, Hamar 20088
– Avdeling for økonomi, samfunnsfag og informatikk, Rena 20089
Høgskolen i Lillehammer, Lillehammer 20106
Høgskolen i Molde, Molde 20107
Høgskolen i Narvik, Narvik 20108
Høgskolen i Nesna, Nesna 20109
Høgskolen i Nord-Trøndelag, Steinkjer 20162
Høgskolen i Nord-Trøndelag, Levanger, Levanger 20105
Høgskolen i Oslo, Oslo 20118
Høgskolen i Østfold, Halden 20094
Høgskolen i Østfold, Fredrikstad, Fredrikstad 20091
Høgskolen i Sogn og Fjørdane, Biblioteket Førde, Førde 20090
Høgskolen i Sogn og Fjørdane, Biblioteket Fosshaugane, Sogndal 20158
Høgskolen i Sør-Trøndelag, Trondheim
– Avdeling for bedriftsøkonomi / TØH – Trondheim Økonomiske Høgskole, Trondheim 20171
– Avdeling for helse- og sosialfag, Trondheim 20172
– Avdeling for lærerutdanning, Trondheim 20173
– Avdeling for teknologi, Retorten, Trondheim 20174
Høgskolen i Telemark, Bø 20083
Høgskolen i Telemark, Porsgrunn 20155
Høgskolen i Telemark, Notodden, Notodden 20110
Høgskolen i Tromsø, Tromsø
– Helsefag AFH, Tromsø 20164
– Lærerutdanning AFL, Tromsø 20165
Høgskolen i Vestfold, Borre, Borre 20085
Høgskolen i Vestfold, Eik, Tønsberg 20163
Høgskolen Stord/Haugesund, Rommetveit 20156
Högskoleverket, Stockholm 26502
Høgskulen i Sogn og Fjørdane, Biblioteket Fjøra, Sogndal 20159
Høgskulen i Volda, Volda 20183
Högsta domstolen, Stockholm 26503
Högsta förvaltningsdomstolens bibliotek, Stockholm 26504
Hohai University, Nanjing 05434
Hohenlohe-Zentralarchiv, Neuenstein 12360
Höhere Bundeslehranstalt und Bundesamt für Wein- und Obstbau, Klosterneuburg 01376
Höhere Graphische Bundes-Lehr- und Versuchsanstalt, Wien 01379
Höhere Lehranstalt für Forstwirtschaft, Bruck a. d. Mur 01369
Höhere Technische Bundeslehr- und Versuchsanstalt, Innsbruck 01373
Hohhaus-Bibliothek, Lauterbach 12176
Højvangseminariet, Glostrup 06731
Hokkai Gakuen Daigaku, Sapporo 17104
Hokkai Gakuen University, Sapporo 17104
Hokkaido Archives, Sapporo 17656
Hokkaido Assembly, Sapporo 17295
Hokkaido Bank Ltd, Sapporo 17422
Hokkaido Broadcasting Co, Ltd, Sapporo 17423
Hokkaido Central Agricultural Experiment Station, Yubari 17832
Hokkaido Central Fisheries Experimental Station, Yoichi 17819
Hokkaido Daigaku, Sapporo 17105
– Kogakubu Kagakukei Toshoshitsu, Sapporo 17106
– Surabu Kenkyu Senta, Sapporo 17107
– Teion Kagaku Kenkyujo, Sapporo 17108
– Yakugakubu, Sapporo 17109
Hokkaido Denryoku K. K. Kikakushitsu Shiryoshitsu, Sapporo 17421
Hokkaido Development Bureau, Civil Engineering Research Institute, Sapporo 17650
Hokkaido Electric Power Co, Inc, Sapporo 17421
Hokkaido Forest Experiment Station, Bibai 17516
Hokkaido Forest Products Research Institute, Asahikawa 17513
Hokkaido Ginko K. K., Sapporo 17422
Hokkaido Hoso K. K., Sapporo 17423
Hokkaido Industrial Research Institute, Sapporo 17655
Hokkaido Institute of Pharmaceutical Science, Otaru 17088
Hokkaido Institute of Public Health, Sapporo 17654
Hokkaido Kaihatsukyoku Dobokushikenjo Kikakuka Chodakakari, Sapporo 17650
Hokkaido Kaitaku Kinenkan, Sapporo 17651
Hokkaido National Agricultural Experiment Station, Sapporo 17652
Hokkaido Nogyo Shikenjyo Kikakurenrakushitsu Johoshiryoka, Sapporo 17652
Hokkaido Prefectural Government Documentation Center on Administration, Sapporo 17294
Hokkaido Prefectural Research Institute of Economy, Sapporo 17657
Hokkaido Prefectural Research Institute of Education, Ebetsu 17530
Hokkaido Regional Fisheries Research Laboratory, Kushiro 17585
Hokkaido Somubu Gyosei Shiryoshitsu, Sapporo 17294
Hokkaido University, Sapporo 17105
– Faculty of Engineering, Library of Chemistry, Sapporo 17106
– Faculty of Pharmaceutical Sciences, Sapporo 17109
– Institute of Low Temperature Science, Sapporo 17108
– Slavic Research Center, Sapporo 17107
Hokkaido Yakka Daigaku, Otaru 17088
Hokkaidoku Suisan Kenkyujo, Kushiro 17585
Hokkaido-ritsu Chishitsu-Kenkyusyo, Sapporo 17653
Hokkaidoritsu Chuo Nogyo Shikenjo, Yubari 17832
Hokkaidoritsu Chuo Suisan Shikenjo, Yoichi 17819
Hokkaidoritsu Eisei Kenkyujo, Sapporo 17654
Hokkaidoritsu Kogyo Shikenjo Somubu Kikakuka Shiryogakari, Sapporo 17655
Hokkaidoritsu Kyoiku Kenkyujo, Ebetsu 17530
Hokkaidoritsu Monjokan, Sapporo 17656
Hokkaidoritsu Ringyo Shikenjo, Bibai 17516
Hokkaidoritsu Rinsan Shikenjo, Asahikawa 17513
Hokkaidoritsu Sogyo Keizai Kenkyusho, Sapporo 17657
Hokuriku Daigaku, Kanazawa 16962
Hokuriku Denryoku K. K., Toyama 17495
Hokuriku Electric Power Co, Inc, Technical Research Laboratory, Toyama 17495
Hokuriku University, Kanazawa 16962
Hokusei Gakuen College, Sapporo 17110
Hokusei Gakuen Daigaku, Sapporo 17110
Hokuyo Sago Bank, Ltd, Sapporo 17424
Hokuyo Sogo Ginko, Sapporo 17424
Hol folkebibliotek, Geilo 20334
Hol Public Library, Geilo 20334
Holbæk Bibliotek, Holbæk 06883
Holbæk College of Education, Holbæk 06734
Holbæk Seminarium, Holbæk 06734
Holden Arboretum, Kirtland 36995
Holdrege Public Library System, Holdrege 39455
Holland College, Charlottetown 03983
Holland & Hart, Denver 35666
Holland & Knight LLP, Boston 35528
Holland & Knight LLP, Chicago 35580
Holland & Knight LLP, Miami 35884
Holland & Knight LLP, New York 35985
Holland & Knight LLP, Tampa 36299
Holland & Knight LLP Orlando Library, Orlando 36073
Hollanda Tarihve Arkeoloji Enstitüsü, Istanbul 27889
Hollenberggymnasium, Waldbröl 10773
Hollins University, Roanoke 32458
Holliston Public Library, Holliston 39459
Holme Roberts & Owen LLC, Denver 35667
Holmes Community College, Goodman 33380
Holmes County District Public Library, Millersburg 40207
A. Holmes Johnson Memorial Library, Kodiak 39740
Holmesglen Institute of TAFE, Chadstone 00563
Holmestrand bibliotek, Holmestrand 20344
Holmie Lacy College, Pershore 29368
Holocaust Center of Northern California, San Francisco 37543
Holocaust Memorial Center, Farmington Hills 36848
Holstebro Bibliotek, Holstebro 06884
Holy Apostles College and Seminary, Cromwell 35173
Holy Cross Abbey Library, Berryville 35134
Holy Cross Greek Orthodox School of Theology, Brookline 30564
Holy Family University, Philadelphia 32255
Holy Monastery of Koutloumous, Athos 13095
Holy Monastery of St. Gregorios, Athos 13094
Holy Names College, Oakland 32172
Holy Spirit Seminary College of Theology and Philosophy, Hong Kong 05814
Holy Trinity Monastery Library, Saint David 35355
Holy Trinity Orthodox Seminary & Monastery, Jordanville 35247
Holy Trinity Teachers' College, Mount Hagen 20563
Holyoke Community College, Holyoke 33413
Holyoke Public Library, Holyoke 39461
Holzforschung Austria, Wien 01605
Holzmacher, McLendon & Murrell Library, Melville 35878
Home and University Library 'St Clement Ohridski', Bitola 18429
Home Office, London 29521
Homewood Public Library, Homewood 39462
Homewood Public Library District, Homewood 39463
Homochitto Valley Library Service, Natchez 40378
Homöopathische Bibliothek Hamburg, Hamburg 11990
Homu Toshokan, Tokyo 17447
Honeywell International, Morristown 37130
Honeywell International Inc, Engines & Systems, Torrance 36309
Honeywell Tempe Information Resource Center, Tempe 36302
Hong Kong Academy for Performing Arts, Hong Kong 05283
Hong Kong Article Numbering Association (HKANA), Hong Kong 05872
Hong Kong Baptist Theological Seminary, Hong Kong 05815
Hong Kong Baptist University Library, Hong Kong 05284
Hong Kong Central Library, Hong Kong 05917
Hong Kong Institute of Education, Hong Kong 05873
Hong Kong Institute of Vocational Education, Hong Kong 05807
Hong Kong Observatory, Hong Kong 05874
Hong Kong Polytechnic University, Hong Kong 05285
Hong Kong University of Science & Technology, Hong Kong 05286
Hong-Ik University, Seoul 18084
Hong's Foundation for Education and Culture, Taipei 27610
Honigman Miller Schwartz & Cohn LLP, Detroit 35676
Honolulu Academy of Arts, Honolulu 36936
Honolulu Community College, Honolulu 33414
Honorable Azhar's Library, Cairo 07198
Honorable Congreso de la Unión, México 18815
La Honorable Legislatura, La Plata 00217
Honorable Legislatura del Chubut, Rawson 00220
The Honorable Society of King's Inns, Dublin 14536
Honourable Society of the Middle Temple, London 29759
Hood College, Frederick 31173
Hood River County Library, Hood River 39473
Hoofdbibliotheek, Tienen 02867
Hoofdbibliotheek Brecht, Brecht 02732
Hoofdbibliotheek Kortemark, Kortemark 02794
Hoofdstedelijke Openbare Bibliotheek, Brussel 02738
Hoogheemraadschap van Rijnland, Leiden 19280
Höörs bibliotek, Höör 26795
Hoover Public Library, Hoover 39474
Hope College, Holland 31343
Hope International University, Fullerton 31184
Hôpital cantonal – Clinique d'ophtalmologie, Genève 27087
Hôpital cantonal – Département de chirurgie, Genève 27088
Hôpital du Sacré-Cœur Centre de Montréal, Montreal 04402
Hopital Rivière-Des-Prairies, Montreal 04403

Hôpital Saint-Louis, Paris 08549
Hôpitaux universitaires de Genève – Departement de chirurgie, Genève 27089
Hôpitaux universitaires de Genève - Service d'ophtalmologie, Genève 27090
Hopkins County-Madisonville Public Library, Madisonville 40013
Hopkinsville Community College, Hopkinsville 33416
Hopkinsville-Christian County Public Library, Hopkinsville 39479
Hopp Ferenc Kelet-Azsiai Müvészeti Múzeum, Budapest 13381
Hopping, Green & Sams, Tallahassee 36295
Hopwood Hall College, Rochdale 29489
Hörby bibliotek, Hörby 26796
'Horia Hulubei' National Institute of R&D for Physics and Nuclear Engineering IFIN-HH Bucharest, Bucureşti 21976
E. A. Hornel Library, Kirkcudbright 29692
The Horniman Museum and Gardens, London 29760
Horowhenua Libraries, Levin 19870
Horry County Memorial Library – Socastee, Myrtle Beach 40357
Horry – Georgetown Technical College, Conway 33256
Horse-Breeding Research Institute, Ryazan 23731
Horsens Bibliotek, Horsens 06885
Horseshoe Bend Regional Library, Dadeville 38753
Horsham Township Library, Horsham 39481
Hørsholm Bibliotek, Hørsholm 06886
Horta Museum, Bruxelles 02610
Horten bibliotek, Horten 20346
Horticultural Research and Development Institute, Peradeniya 26328
Horticulture Research International, Wellesbourne 29915
HortResearch Palmerston North, Palmerston North 19841
Hortus Botanicus Bergianus, Stockholm 26649
Hosei Daigaku, Tokyo 17140
– Ohara Shakai Mondai Kenkyujo, Tokyo 17141
– Okinawa Bunka Kenkyujo, Tokyo 17142
Hosei University, Tokyo 17140
– Institute of Okinawa Studies, Tokyo 17142
– Ohara Institute for Social Research, Tokyo 17141
Hoshasen Eikyo Kenkyusho Toshokan, Hiroshima 17549
Hoshasen Igaku Sogo Kenkyusho Tosyositu, Chiba 17522
Hoshi University, Tokyo 17143
Hoshi Yakka Daigaku, Tokyo 17143
Hosiery Industrial Corporation, Grodno 01872
Hosokai, Tokyo 17678
Hospital, Žilina 24758
Hospital at Motala, Motala 26620
Hospital Clinico Universitario Ntra. Sra. de la Victoria, Málaga 25491
Hospital de Cruces, Baracaldo 25876
Hospital de S. José, Lisboa 21789
Hospital dos Servidores do Estado, Rio de Janeiro 03349
Hospital for Sick Children, Toronto 04514
Hospital General de México, México 18728
The Hospital Libraries of Sörmland County Council, Eskilstuna 26570
Hospital no 15, Sankt-Peterburg 23748
Hospital no 31, Sankt-Peterburg 23750
Hospital of Saint Raphael, New Haven 37160
Hospital of Southern Jutland, Sønderborg 06834
Hospital of Southwest Denmark, Medical Library, Esbjerg 06775
Hospital of Varberg, Medical Library, Varberg 26715
Hospital Pediátrico Docente 'Aballi', La Habana 06299
Hospital Universitario Marqués de Valdecilla, Santander 26140
Hospital Universitario Virgen del Rocio, Sevilla 25748
Hostos Community College, Bronx 33192
Hot Springs County Library, Thermopolis 41548
Hotelschool Den Haag, Den Haag 19140
Hotelschool KTA 3, Hasselt 02418
Houghton College, Houghton 31351
Houghton College at West Seneca, West Seneca 33861
Houillères du Bassin de Lorraine (HBL), Merlebach 08259
Hounslow Library Network (CIP), Hounslow 29997
Hourigan, Kluger & Quinn, Kingston 35807
Housatonic Community College, Bridgeport 33189
House Library, Washington 34999
House of Commons, London 29522
House of Lords, London 29523
House of Parliament, Freetown 24572
House of People's Representatives, Addis Ababa 07276
House of Popularization of Science and Technology, Nalchik 23606
House of Representatives, Nicosia 06331
House of Representatives, Quezon City 20761
House of Representatives, Tunis 27822
House of Representatives of the Republic of Indonesia, Jakarta 14333
House of Scientists, Moskva 23174
House of Scientists im. A. M. Gorki of the Russian Academy of Sciences, Sankt-Peterburg 24225
House of Technology of 'Apatit', Kirovsk 22807
House of the Wannsee Conference, Berlin 11527
House of Victor Hugo at Paris, Paris 08580
Houses of Parliament, Kingston 16874
Houses of the Oireachtas, Dublin 14506
Housing and Development Board, Singapore 24610
Housing Department Library, Nairobi 18023
Housing Fund of Finland Library, Helsinki 07506
Houston Academy of Medicine-Texas Medical Center Library, Houston 36941
Houston Baptist University, Houston 31353
Houston Community College, Houston 33417
Houston County Historical Commission Archives, Crockett 36760
Houston County Public Library System, Perry 40760
Houston Love Memorial Library, Dothan 38853
Houston Museum of Natural Science Library, Houston 36942
Houston Public Library, Houston 39492
Houston Public Library – Acres Homes, Houston 39493
Houston Public Library – Bracewell Branch, Houston 39494
Houston Public Library – Carnegie Regional, Houston 39495
Houston Public Library – Clayton Library Center for Genealogical Research, Houston 39496
Houston Public Library – Collier Regional, Houston 39497
Houston Public Library – Fine Arts & Recreation Department, Houston 39498
Houston Public Library – Flores Branch, Houston 39499
Houston Public Library – Frank Branch, Houston 39500
Houston Public Library – Freed-Montrose Branch, Houston 39501
Houston Public Library – Heights, Houston 39502
Houston Public Library – Henington-Alief Regional, Houston 39503
Houston Public Library – Hillendahl Branch, Houston 39504
Houston Public Library – Humanities Department, Houston 39505
Houston Public Library – Jungman Branch, Houston 39506
Houston Public Library – Kendall Branch, Houston 39507
Houston Public Library – Meyer Branch, Houston 39508
Houston Public Library – Moody Branch, Houston 39509
Houston Public Library – Oak Forest, Houston 39510
Houston Public Library – Park Place Regional, Houston 39511
Houston Public Library – Ring Branch, Houston 39512
Houston Public Library – Robinson-Westchase Branch, Houston 39513
Houston Public Library – Scenic Woods Regional, Houston 39514
Houston Public Library – Social Sciences Department, Houston 39515
Houston Public Library – Stanaker Branch, Houston 39516
Houston Public Library – Stimley-Blue Ridge Branch, Houston 39517
Houston Public Library – Tuttle Branch, Houston 39518
Houston Public Library – Vinson Branch, Houston 39519
Houston Public Library – Walter Branch, Houston 39520
Houston Public Library – Young Neighborhood Library, Houston 39521
Houston-Tillotson College, Austin 30306
Hovedbiblioteket i Birkerød, Birkerød 06850
Hovrätten för Västra Sverige, Göteborg 26488
Howard College, Big Spring 33170
Howard Community College, Columbia 33250
Howard County Library, Columbia 38645
Howard County Library – Central, Columbia 38646
Howard County Library – Dora Roberts Library, Big Spring 38203
Howard County Library – East Columbia, Columbia 38647
Howard County Library – Elkridge Branch, Elkridge 38961
Howard County Library – Miller Branch, Ellicott City 38964
Howard County Library – Savage, Laurel 39816
Howard Miller Public Library, Zeeland 42005
Howard Payne University, Brownwood 30580
Howard, Rice, Nemerovski, Canady, Falk & Rabkin Library, San Francisco 36236
Howard University Libraries, Washington 32976
– Architecture Library, Washington 32977
– Business Library, Washington 32978
– Department of Afro-American Studies Resource Center, Washington 32979
– Divinity Library, Washington 32980
– Founders & Undergraduate Library, Washington 32981
– Law Library, Washington 32982
– Louis Stokes Health Sciences Library, Washington 32983
– Moorland-Spingarn Research Center Library, Washington 32984
– Social Work Library, Washington 32985
Howard Whittemore Memorial Library, Naugatuck 40381
C. D. Howe Institute Library, Toronto 04515
Howe Library, Hanover 39364
Howell Carnegie District Library, Howell 39524
Howick Public Library, Howick 25205
Howrey LLP, Washington 36356
Howrey LLP Library, Houston 36943
Howrey Simon Arnold & White LLP Library, Los Angeles 35838
Høyanger bibliotek, Høyanger 20347
Høyesterett, Oslo 20191
Hoyle, Fickler, Herschel & Mathes LLP, Philadelphia 36101
Hoyt Library, Kingston 39723
Hradec Králové City Library, Hradec Králové 06571
HRL Laboratories, Malibu 35872
Hrvatska akademija znanosti i umjetnosti, Zagreb 06144
Hrvatska knjižnica za slijepe, Zagreb 06238
Hrvatski Državni Arhiv, Zagreb 06239
Hrvatski povijesni muzej, Zagreb 06240
Hrvatski prirodoslovni Muzej, Zagreb 06241
Hrvatski sabor, Zagreb 06208
Hrvatski školski muzej, Zagreb 06242
Hsin Yi Foundation, Taipei 27611
HSR Hochschule für Technik Rapperswil, Rapperswil 27227
Huachiew Chalermprakiet University, Samutprakarn 27751
Huaibei Teachers' College for Coal Industry, Huaibei 05295
Huaihai Institute of Technology, Lianyungang 05388
Huaihua Teachers' College, Huaihua 05296
Huainan Teachers' College, Huainan 05297
Huainang Mining Institute, Dongshan 05168
Huanggang Teachers' College, Huangzhou 05300
Huaqiao University, Quanzhou 05491
Huazhong Agricultural University, Wuhan 05652
Huazhong Normal University, Wuhan 05653
Huazhong University of Science and Technology, Wuhan 05654
Hubbard Memorial Library, Ludlow 39977
Hubbard Public Library, Hubbard 39525
Hubei Automobile Engineering Institute, Shiyan 05577
Hubei Medical University (HMU), Wuhan 05655
Hubei Provincial Library, Wuhan 04975
Hubei Teachers' College, Huangshi 05299
Hubei University, Wuhan 05656
Huddersfield Library, Huddersfield 29998
Huddersfield Technical College Library, Huddersfield 29471
Huddinge kummunbibliotek, Huddinge 26797
Huddleston Bolen, LLP, Huntington 35768
Hudiksvalls bibliotek, Hudiksvall 26798
Hudiksvalls sjukhus, Hudiksvall 26593
Hudson County Law Library, Jersey City 34399
Hudson Library & Historical Society, Hudson 39527
Hudson Public Library, Hudson 39528
Hudson Valley Community College, Troy 33806
Huey P. Long Memorial Law Library, Baton Rouge 35494
Hufvudstadsbladet, Newspaper Archives, Helsinki 07415
Hufvudstadsbladets arkiv, Helsinki 07415
HUG Hôpitaux universitaires de Genève – Belle-Idée, Chêne-Bourg 27342
Hugarian Academy of Sciences – Research Institute of the History of Art, Budapest 13421
Hugenottenbibliothek, Berlin 11165
Hughes, Hubbard & Reed Library, New York 35986
Hughes & Luce, LLP Library, Dallas 35648
Hugo de Vries Informatiecentrum, Amsterdam 19359
Huguenot Library, London 29761
Huhtamäki Oy Yhtymän Kirjasto, Turku 07432
Huismuziek, Arnhem 19394
Huizhou Teachers' College School, Huangshan 05298
Hulett Aluminium Limited, Pietermaritzburg 25096
Hull Medical Library, Hull 29687
Hulton Getty Picture Collection, London 29762
Hultsfreds bibliotek, Hultsfred 26799
Human Employment and Resource Training Trust, Kingston 16875
Human Kinetics Faculty, Cruz Quebrada 21648
Human Resources Development Canada Quebec Library, Montreal 04098
Human Rights and Equal Opportunity Commission, Sydney 01024
Human Rights Internet, Ottawa 04445
Human Sciences Research Council (HSRC), Pretoria 25162
Humber College, Toronto 04038
Humboldt County Law Library, Eureka 34194
Humboldt County Library, Eureka 38997
Humboldt County Library, Winnemucca 41931
Humboldt County Library – Eureka Main Library, Eureka 38998
Humboldt State University, Arcata 30257
Humboldt-Bibliothek, Berlin 12666
Humboldt-Universität zu Berlin, Berlin 09450
Humphreys College, Stockton 32745
Hunan Agricultural College, Changsha 05096
Hunan College of Building Materials, Hengyang 05274
Hunan College of Commerce, Changsha 05097
Hunan College of Finance and Economics, Changsha 05098
Hunan Economic Management Cadre College, Changsha 05099
Hunan Forestry College, Hengyang 05275
Hunan Institute of Humanities, Science & Technology, Loudi 05405
Hunan Library, Changsha 04954
Hunan Medical College, Changsha 05100
Hunan Medical University, Changsha 05101
Hunan Normal University, Changsha 05102
Hunan Radio Broadcasting and Television University, Changsha 05103
Hunan Teachers College, Changsha 05104
Hunan University, Changsha 05105
Hundlip College, Pershore 29368
Hungarian Academy in Rome, Roma 16443
Hungarian Academy of Fine Arts, Budapest 13255
Hungarian Academy of Sciences, Budapest 13413
Hungarian Academy of Sciences – Agricultural Research Institute Library, Martonvásár 13489
Hungarian Academy of Sciences – Archaeological Institute, Budapest 13425
Hungarian Academy of Sciences – Balaton Limnological Research Institute, Tihany 13515
Hungarian Academy of Sciences – Biological Research Centre, Szeged 13507
Hungarian Academy of Sciences – Centre for Regional Studies, Békéscsaba 13367
Hungarian Academy of Sciences – Centre for Regional Studies, Pécs 13495
Hungarian Academy of Sciences – Chemical Research Center, Budapest 13415
Hungarian Academy of Sciences – Computer and Automation Institute, Budapest 13426
Hungarian Academy of Sciences – Ethnographical Institute, Budapest 13422
Hungarian Academy of Sciences – Geographical Research Institute, Budapest 13411
Hungarian Academy of Sciences – Heliophysical Observatory, Debrecen 13469
Hungarian Academy of Sciences – Hungarian Numismatic Society, Budapest 13419

Hungarian Academy of Sciences – Institute of Ecology and Botany, Vácrátót 13517
Hungarian Academy of Sciences – Institute of Economics, Budapest 13417
Hungarian Academy of Sciences – Institute of Experimental Medicine, Budapest 13416
Hungarian Academy of Sciences – Institute of History, Budapest 13258
Hungarian Academy of Sciences – Institute of Literary Studies, Budapest 13412
Hungarian Academy of Sciences – Institute of Musicology, Budapest 13429
Hungarian Academy of Sciences – Institute of Nuclear Research Library, Debrecen 13468
Hungarian Academy of Sciences – Institute of Philosophy, Budapest 13257
Hungarian Academy of Sciences – Institute of Psychology, Budapest 13424
Hungarian Academy of Sciences Library, Budapest 13214
Hungarian Academy of Sciences – Lukács Archive, Budapest 13418
Hungarian Academy of Sciences – Mathematical Institute, Budapest 13420
Hungarian Academy of Sciences – Observatory, Budapest 13409
Hungarian Academy of Sciences – Plant Protection Research Institute, Budapest 13423
Hungarian Academy of Sciences – Research Institut for Linguistics, Budapest 13433
Hungarian Academy of Sciences – Research Institute for Soil Science and Agricultural Chemistry, Budapest 13428
Hungarian Academy of Sciences – Research Institute for Solid State Physics and Optics, Library, Budapest 13427
Hungarian Academy of Sciences Research Institute of Chemical Engineering, Veszprém 13520
Hungarian Academy of Sciences – Research Institute of Geodesy and Geophysics, Sopron 13498
Hungarian Academy of Sciences – United Library for Social Sciences, Budapest 13410
Hungarian Academy of Sciences – Veterinary Medical Research Institute, Budapest 13432
Hungarian Central Statistical Office, Budapest 13316
Hungarian Dairy Research Institute, Mosonmagyaróvár 13491
Hungarian Dance Academy, Budapest 13256
Hungarian Environmental and Water Management Museum, Esztergom 13473
Hungarian Federation of the Blind and Partially Sighted, Budapest 13382
Hungarian Forestry Association, Budapest 13436
Hungarian Geological Library, Budapest 13395
Hungarian Institute of Agricultural Engineering, Gödöllő 13325
Hungarian Institute of Cardiology, Budapest 13441
Hungarian Institute of Culture, Budapest 13399
Hungarian Museum for Commerce and Gastronomy, Budapest 13398
Hungarian Museum of Architecture, Budapest 13397
Hungarian National Archives, Budapest 13404
Hungarian National Film Archive, Budapest 13401
Hungarian National Gallery, Budapest 13402
Hungarian National Museum, Central Archeological Library, Budapest 13403
Hungarian Natural History Museum, Budapest 13407
Hungarian Optical Factory, Budapest 13358
Hungarian Public Utility Society for Regional Development and Town Planning Documentation Centre, Budapest 13464
Hungarian Radio, Budapest 13405
Hungarian Standards Institution, Budapest 13406
Hungarian State Railways, Documentation Centre and Library, Budapest 13430
Hungarian State's Railway. MAV Central Direction for Telecommunication and Safety Installations, Technical Library, Budapest 13460
Hungarian Theatre Institute and Museum, Budapest 13452

Hungarian University of Craft and Design Library, Budapest 13254
Hunt Institute for Botanical Documentation, Pittsburgh 37381
Hunter Institute Hamilton Campus, Newcastle West 00491
Hunter Institute Newcastle Campus, Tighes Hill 00545
Hunter New England Area Health Service, New Lambton Heights 00966
Hunter Water Corporation Ltd, Newcastle West 00823
Hunterdon County Library, Flemington 39072
Hunterdon County Library – North County, Clinton 38608
Hunting Library of Finland, Riihimäki 07537
Huntingdon College, Montgomery 31888
Huntingdon County Library, Huntingdon 39534
Huntingdon Valley Library, Huntingdon Valley 39535
Huntington Beach Public Library System, Huntington Beach 39538
Huntington City Township Public Library, Huntington 39537
Huntington College, Huntington 31372
Huntington College, Sudbury 03872
Huntington Library, San Marino 37562
Huntington Memorial Library, Oneonta 40618
Huntington Museum of Art Library, Huntington 36947
Huntley Area Public Library District, Huntley 39541
Hunton & Williams, New York 35987
Hunton & Williams, Raleigh 36160
Hunton & Williams, Richmond 36174
Huntsville Public Library, Huntsville 39542
Huntsville-Madison County Public Library, Huntsville 39543
Huntsville-Madison County Public Library – Madison Public Library, Madison 40002
Huron City Museum Library, Port Austin 37394
Huron County Law Library Association, Norwalk 34619
Huron County Library, Clinton 04612
Huron Public Library, Huron 39545
Huron University College, London 03743
Hurrell & Cantrall, Los Angeles 35839
Hurst Public Library, Hurst 39547
Hurt-Battelle Memorial Library of West Jefferson, West Jefferson 41824
Husch Blackwell Sanders LLP, Saint Louis 36202
Husitské muzeum v Táboře, Tábor 06540
Huskvarna bibliotek, Huskvarna 26800
Husky Energy Corporate Library, Calgary 04241
Hussey-Mayfield Memorial Public Library, Zionsville 42007
Hussite Museum, Tábor 06540
Husson College, Bangor 30349
Huta 'Batory', Chorzów 21065
Huta im. Tadeusza Sendzimira S.A., Kraków 21066
Huta 'Stalowa Wola', Stalowa Wola 21068
Hutchinson Community College, Hutchinson 33425
Hutchinson County Library, Borger 38242
Hutchinson Public Library, Hutchinson 39548
Hutt City Libraries, Lower Hutt 19871
Huvudbiblioteket Båstad, Båstad 26741
Huvudbiblioteket i Borgholm, Borgholm 26748
Huyton Library, Huyton 29999
Huzhou Teachers' College, Huzhou 05306
HvidovreBibliotekerne, Hvidovre 06887
Hwa Chong Junior College, Singapore 24594
Hyannis Public Library Association, Hyannis 39549
Hyconeechee Regional Library, Hillsborough 39439
Hyderabad Educational Conference, Hyderabad 14000
Hydraulic Engineering Consultants Corporation No 1, Hanoi 42287
Hydro Automation Industrial Corporation, Gomel 01852
Hydro Québec, Montreal 04404
Hydrochemical Institute, Novocherkassk 22761
Hydrochemistry Institute, Rostov-na-Donu 23718
Hydroengineering Construction Corporation, Pinsk 01918
Hydrographic and Oceanographic Service of the Chilean Navy, Library, Valparaíso 04948
Hydrographisches Zentralbüro, Wien 01406
Hydrological Service, Jerusalem 14731
Hydromechanics Institute, Kyiv 28328
Hydrometeorological Institute, Almaty 17943
Hydrometeorological Institute, Tbilisi 09240
Hydro-Meteorological Observatory, Rostov-na-Donu 23719

Hydrotechnology and Land Reclamation Research Institute, Kyiv 28329
Hydrotechnology Planning and Research Institute, Toshkent 42133
Hyogo College of Medicine, Nishinomiya 17069
Hyogo Kenritsu Kyoiku Kenshujo, Kato 17566
Hyogo Prefectural Assembly, Kobe 17267
Hyogo Prefectural Educational Research Institute, Kato 17566
Hyogo Prefectural Labor Economic Research Institute, Kobe 17574
Hyogo Prefectural Social Welfare Council, Kobe 17573
Hyogo Prefecture Planning Dept, Statistics Section, Adiminstrative Information Data Office, Kobe 17268
Hyogo Sogo Bank Ltd, Kobe 17376
Hyogo Sogo Ginko, Kobe 17376
Hyogoken Gikai, Kobe 17267
Hyogoken Kikakubu, Kobe 17268
Hyogoken Shakai Fukushi Kyokikai, Kobe 17573
Hyokoken Rodo Keizai Kenkyujo, Kobe 17574
I. A. Dzhavakhishvili Institute of History, Archeology and Ethnography, Tbilisi 09232
Ibadan Polytechnic, Ibadan 19938
Ibaraki Agricultural Experiment Station, Mito 17600
Ibaraki Prefectural Assembly, Mito 17278
Ibaraki Prefectural Forest Experiment Station, Naka 17617
Ibaraki University, Mito 17044
Ibarakiken Gikai Jimukyoku, Mito 17278
Ibarakiken Nogyo Shikenjo, Mito 17600
Ibarakiken Ringyo Shikenjo, Naka 17617
Iberduero S.A., Bilbao 25855
IBERIA Líneas Aereas de España S.A., Madrid 25864
Iberia Parish Library, New Iberia 40414
Ibero-American Institute (IAI), Prussian Cultural Heritage Foundation, Berlin 11531
Ibero-Amerikanisches Institut, Berlin 11531
Iberville Parish Library, Plaquemine 40823
IBM Canada Ltd-Toronto Lab, Markham 04379
IBM Corp, Austin 35474
IBM Corp, Endicott 35693
IBM Corp, Essex Junction 35696
IBM Corp, Hopewell Junction 35745
IBM Corp, Poughkeepsie 36153
IBM Corp, Rochester 36184
IBM Corp, Yorktown Heights 36436
IBM Deutschland Informationssysteme GmbH, Böblingen 11375
IBM – International Business Machines S.A.E., Madrid 25865
IBM Rtp Library, Research Triangle Park 36168
IBMER, Warszawa 21260
ICANA, Buenos Aires 00259
ICAR Research Complex for North Eastern Hill Region, Bequalaya 13945
Ice Miller LLP, Indianapolis 35777
Iceland Academy of the Arts, Reykjavík 13558
Icelandic Institute of Natural History Library, Gardabaer 13567
Icelandic Literary Society, Reykjavík 13570
ICEMENERG Bucureşti, Bucureşti 21977
ICF Jones & Stokes Library, Sacramento 36191
Ichnya Regional Centralized Library System, Main Library, Ichnya 28580
Ichnyanska raionna TsBS, Tsentralna biblioteka, Ichnya 28580
ICI Berlin Institute for Cultural Inquiry, Berlin 11532
ICI Bibliothek, Berlin 11532
ICI Library, Berlin 11532
ICICI Bank Ltd., Hyderabad 13905
ICIMOD Documentation, Kathmandu 19034
ICL Data, Helsinki 07461
Icon Museum, Autenried Castle, Ichenhausen 12072
Ida Public Library, Belvidere 38167
Ida Rupp Public Library, Port Clinton 40851
Idaho Falls Public Library, Idaho Falls 39552
Idaho Maximum Security Institute Library, Kuna 36996
Idaho National Laboratory, Idaho Falls 36952
Idaho School for the Deaf & Blind Library, Gooding 33378
Idaho School for the Deaf & Blind Library, Gooding 33379
Idaho State Correctional Institution Library, Pocatello 34685
Idaho State Historical Society, Boise 36552
Idaho State Law Library, Boise 33994
Idaho State University, Pocatello 32325
Idarah-i-Yadgar-i-Ghalib, Karachi 20495

Idayu Foundation, Jakarta 14383
Idea Bibliotheek Soest, Soest 19702
Idemitsu Kosan Co, Ltd, Central Research Laboratory, Kimitsu 17373
Idemitsu Kosan Co, Ltd, General Affairs Department, Tokyo 17448
Idemitsu Kosan Co, Ltd, Tokuyama Oil Factory, Tokuyama 17431
Idemitsu Kosan K. K. Chuo Kenkyujo, Kimitsu 17373
Idemitsu Kosan K. K. Somubu, Tokyo 17448
Idemitsu Kosan K. K. Tokuyama Seiyujo Somuka, Tokuyama 17431
IDEWE-DOCDIENST, Leuven 02669
IDOX Information Service, Glasgow 29674
IDPL Research Centre, Hyderabad 14001
Iera Moni Koutloumousiou, Athos 13095
IES Abroad Vienna, Wien 01293
IES Antonio Machado, Alcalá de Henares 25660
I.E.S. 'Universidad Laboral', Zamora 25677
Ieva Simonaityte Klaipeda County Public LIbrary, Klaipėda 18335
IFA – Institut für Arbeitsschutz der Deutschen Gesetzlichen Unfallversicherung, Sankt Augustin 12461
iff-Wien Bibliothek, Wien 01294
IFM-GEOMAR Bibliothek, Kiel 12109
ifo Institut für Wirtschaftsforschung, München 12300
IFREMER, Plouzané 08645
IFREMER – Centre de Nantes, Nantes 08443
IG Metall, Frankfurt am Main 11855
IG Metall Bildungszentrum, Sprockhövel 10767
IGFA Fishing Hall of Fame & Museum, Dania Beach 36767
Ignalijos rajono savivaldybės viešoji biblioteka, Ignalina 18324
Igreja Batista de Vila Mariana, São Paulo 03249
Igreja Presbiteriana do Brasil, Campinas 03244
IGV Institut für Getreideverarbeitung GmbH, Nuthetal 12379
IHILA – Instituto de Historia Ibérica y Latinoamericana, Köln 10154
IHLIA – Homodok, Amsterdam 19360
IHM Library/Resource Center, Monroe 35289
IHP GmbH for High Performance Microelectronics / Leibniz-Institut für innovative Mikroelektronik, Frankfurt (Oder) 11877
Iisalmen kaupunginkirjasto, Iisalmi 07567
IKAR Joint-Stock Company of Design and Technology, Voronezh 22977
Ikast-Brande Bibliotek, Ikast 06888
Ikonen-Museum, Schloss Autenried, Ichenhausen 12072
Ikonomicheski institut, Sofiya 03534
Ikonomiko Panepistimio Athinon, Athinai 13077
Iliff School of Theology, Denver 30994
Ilion Free Public Library, Ilion 39553
Illawarra Institute of Technology, Wollongong 00593
Illinetska raionna TsBS, Tsentralna biblioteka, Illinitsi 28581
Illinitsi Regional Centralized Library System, Main Library, Illinitsi 28581
Illinois Agricultural Association, Bloomington 36548
Illinois Appellate Court, Mount Vernon 34565
Illinois Central College, East Peoria 33308
Illinois College, Jacksonville 31447
Illinois College of Optometry, Chicago 30763
Illinois Department of Corrections, Canton 34044
Illinois Department of Corrections, Dwight 34161
Illinois Department of Corrections, Ina 34377
Illinois Department of Corrections, Jacksonville 34392
Illinois Environmental Protection Agency Library, Springfield 34880
Illinois Historic Preservation Agency, Springfield 37629
Illinois Institute of Art, Schaumburg 32623
Illinois Institute of Art – Chicago Library, Chicago 30764
Illinois Institute of Technology, Chicago 30765
– Downtown Campus Libraries, Chicago 30766
Illinois Lodge of Research, Normal 37290
Illinois Mathematics & Science Academy, Aurora 35471
Illinois Prairie District Public Library, Metamora 40166
Illinois School for the Deaf, Jacksonville 33438
Illinois School for the Visually Impaired Library, Jacksonville 33439
Illinois State Department of Transport, Springfield 34881

Illinois State Library, Springfield 30157
Illinois State Library – Talking Book & Braille Service, Springfield 37630
Illinois State Museum Library, Springfield 37631
Illinois State University, Normal 32125
Illinois State Water Survey, Champaign 36644
Illinois Valley Community College, Oglesby 33599
Illinois Wesleyan University, Bloomington 30444
Illinois Youth Center – Saint Charles Library, Saint Charles 37481
Ilmavoimien esikunnan kirjasto, Tikkakoski 07405
Ilmen Mineral Preserve Museum, Miass 23155
Ilmenski gosudarstvenni zapovednik im. V.I. Lenina Uralskogo nauchnogo tsentra RAN, Miass 23156
ILO Office for the South Pacific, Victoria Parade 07319
ILS – Institut für Landes- und Stadtentwicklungsforschung GmbH (i.G.), Dortmund 11745
Ilsley Public Library, Middlebury 40178
Ilustre Colegio de Abogados de Valencia, Valencia 26172
Ilustre Colegio de Abogados del Senorío de Vizcaya, Bilbao 25932
Ilustre Colegio de Doctores y Licenciados en Filosofía y Letras y Ciencias de la Comunidad de Madrid, Madrid 26030
Ilustre Colegio Notarial de Granada, Granada 25960
Ilustre Colegio Notarial de Madrid, Madrid 26031
Ilustre Colegio Oficial de Médicos de Madrid, Madrid 26032
IMA Materialforschung und Anwendungstechnik GmbH, Dresden 11384
Imam Ouzai University, Beirut 18205
'Imanta' Library and Club, Riga 18197
Imatra Institute of Technology, Imatra 07390
Imatra Public Library, Imatra 07568
Imatra Steel Oy AB, Imatra 07419
Imatran kaupunginkirjasto, Imatra 07568
Imatran teknillinen oppilaitos, Imatra 07390
IMI (Tami) Institute for Research and Development Ltd, Haifa Bay 14703
Immaculata University, Immaculata 31380
Immaculate Conception Seminary, South Orange 35385
Immanuel Church Library, Glenview 35217
Immanuel Kant Baltic Federal University, Kaliningrad 22270
Immigrant Institute, Borås 26560
Immigrant-institutet, Borås 26560
IMMUNOPREPARAT – State Unitary Enterprise, Ufa 24026
Imo State Library Board, Owerri 20062
Imperial Chemical Industries plc, London 29574
Imperial College London Library, London 29179
– Aeronautics Department Library, London 29180
– Charing Cross Campus Library, London 29181
– Chelsea & Westminster Campus Library, London 29182
– Chemical Engineering and Chemical Technology Department Library, London 29183
– Department of Civil and Environmental Engineering, London 29184
– Hammersmith Campus Library, London 29185
– Mathematics Department Library, London 29186
– Mechanical Engineering Department Library, London 29187
– Royal Brompton Campus Library, London 29188
– Silwood Park Campus Library, Ascot 29189
– St Mary's Campus Library, London 29190
Imperial County Free Library, El Centro 38940
Imperial County Law Library, El Centro 34174
Imperial Household Agency Library, Tokyo 17315
Imperial Oil Limited, Calgary 04242
Imperial Oil Limited, Sarnia 04276
Imperial Oil Resources Ltd, Calgary 04243
Imperial Valley College, Imperial 33426
Imperial War Museum, London 29763
Imprensa Nacional, Lisboa 21748
Imprimerie Nationale, Paris 08263
In Flanders Fields Documentatiecentrum, Ieper 02665
INA – Industrija Nafte d. d. Zagreb, Zagreb 06210
INA Strategic Development, Research and Investment, Zagreb 06210
inatura Erlebnis Naturschau Dornbirn GmbH, Dornbirn 01514
INBO, Brussel 02591
Inco Ltd, Toronto 04287
Incorporated Long Island Chapter of the New York State Archaeological Association, Southold 37620
Independence Seaport Museum Library, Philadelphia 37350
Independence Township Public Library, Clarkston 38584
Inderøy folkebibliotek, Inderøy 20349
India Department of Science and Technology (DST), New Delhi 14128
India International Centre, New Delhi 14129
India Society of Engineers, Kolkata 14042
Indian Adult Education Association, New Delhi 14130
Indian Agricultural Research Institute, New Delhi 13723
Indian Agricultural Statistics Research Institute, New Delhi 14131
Indian Association for the Cultivation of Science (IACS), Kolkata 14043
Indian Bureau of Mines, Nagpur 13863
Indian Chemical Society, Kolkata 14044
Indian Council for Cultural Relations, New Delhi 14132
Indian Council of Agricultural Research (ICAR), New Delhi 14133
Indian Council of Historical Research, New Delhi 14134
Indian Council of Medical Research, New Delhi 14135
Indian Council of Medical Research, Pondicherry 14182
Indian Council of World Affairs, New Delhi 14136
Indian Hills Community College, Ottumwa 33609
Indian Institute of Advanced Study, Simla 14211
Indian Institute of Astrophysics, Bangalore 13931
Indian Institute of Chemical Biology, Kolkata 14045
Indian Institute of Chemical Technology, Hyderabad 14002
Indian Institute of Foreign Trade (IIFT), New Delhi 14137
Indian Institute of Geomagnetism, Mumbai 14090
Indian Institute of Handloom Technology, Varanasi 14220
Indian Institute of History of Medicine, Hyderabad 14003
Indian Institute of Horticultural Research, Bangalore 13932
Indian Institute of Management, Ahmedabad 13596
Indian Institute of Management, Bangalore 13610
Indian Institute of Management, Kolkata 13682
Indian Institute of Management, Lucknow 13691
Indian Institute of Mass Communication, New Delhi 14138
Indian Institute of Petroleum, Dehradun 13977
Indian Institute of Public Administration, New Delhi 14139
Indian Institute of Science (IISc), Bangalore 13611
Indian Institute of Social Welfare and Business Management, Kolkata 14046
Indian Institute of Technology, Chennai 13629
Indian Institute of Technology, Kharagpur 13678
Indian Institute of Technology, New Delhi 13724
Indian Institute of Technology Bombay, Mumbai 13703
Indian Institute of Technology Kanpur, Kanpur 13676
Indian Institute of World Culture, Bangalore 13933
Indian Lac Research Institute, Ranchi 14204
Indian Law Institute, New Delhi 14140
Indian Library Association, Delhi 13983
Indian Museum, Kolkata 14047
Indian National Science Academy, New Delhi 14141
Indian National Scientific Documentation Centre (INSDOC), New Delhi 14142
Indian Plywood Industries Research Institute, Bangalore 13934
Indian Prairie Public Library District, Darien 38787
Indian River Community College, Fort Pierce 33347
Indian River County Library System, Vero Beach 41675
Indian Roads Congress, New Delhi 14143
Indian School of Mines, Dhanbad 13643
Indian Social Institute, New Delhi 14144
Indian Society of Agricultural Economics, Mumbai 14091
Indian Statistical Institute (ISI), Bangalore 13935
Indian Statistical Institute (ISI), Kolkata 14048
Indian Statistical Institute (ISI), New Delhi 14145
Indian Telephone Industries Ltd – Regional Office, Bangalore 13900
Indian Trails Public Library District, Wheeling 41874
Indian Valley Public Library, Telford 41530
Indian Veterinary Research Institute, Izatnagar 13663
Indian Veterinary Research Institute, Mukteswar-Kumaon 13700
Indiana Academy of Science, Indianapolis 36956
Indiana County Law Library, Indiana 34380
Indiana Free Library, Inc, Indiana 39560
Indiana Historical Society, Indianapolis 36957
Indiana State Archives, Indianapolis 36958
Indiana State Library, Indianapolis 30134
Indiana State Prison, Michigan City 34533
Indiana State University, Terre Haute 32808
Indiana Supreme Court Law Library, Indianapolis 34381
Indiana University, Indianapolis
– Ruth Lilly Medical Library, Indianapolis 31387
– School of Dentistry, Indianapolis 31388
– School of Law Library, Indianapolis 31389
Indiana University Bloomington, Bloomington 30445
– Business SPEA Library, Bloomington 30446
– Chemistry Library, Bloomington 30447
– Education Library, Bloomington 30448
– Fine Arts Library & Slide & Digital Imgae Library, Bloomington 30449
– Geography & Map Library, Bloomington 30450
– Geology Library, Bloomington 30451
– Health, Physical Education and Recreation Library, Bloomington 30452
– Journalism Library, Bloomington 30453
– Life Sciences Library, Bloomington 30454
– Lilly Library Rare Books & Manuscripts, Bloomington 30455
– Neal-Marshall Black Culture Center Library, Bloomington 30456
– Optometry Library, Bloomington 30457
– Swain Hall Library, Bloomington 30458
– William & Gayle Cook Music Library, Bloomington 30459
Indiana University East, Richmond 32438
Indiana University – Indiana Institute on Disability & Community, Bloomington 30460
Indiana University Kokomo Library, Kokomo 31522
Indiana University Northwest, Gary 31207
Indiana University of Pennsylvania, Indiana 31383
– Cogswell Music Library, Indiana 31384
Indiana University of Pennsylvania, Punxsutawney 32399
Indiana University – Purdue University Indianapolis, Indianapolis 31390
Indiana University – Research Institute for Inner Asian Studies, Bloomington 30461
Indiana University – School of Law Library, Bloomington 30462
Indiana University South Bend, South Bend 32685
Indiana University Southeast, New Albany 31952
Indiana University-Purdue University, Columbus 30896
Indiana University-Purdue University Fort Wayne, Fort Wayne 31164
Indiana University-Purdue University Indianapolis, Indianapolis 31390
– Herron School of Art Library, Indianapolis 31391
Indiana Veteran's Home, West Lafayette 37824
Indiana Wesleyan University, Marion 31760
Indiana Women's Prison Library, Indianapolis 34382
Indianapolis Museum of Art, Indianapolis 36959
Indianapolis-Marion County Public Library, Indianapolis 39561
Indianapolis-Marion County Public Library – Decatur, Indianapolis 39562
Indianapolis-Marion County Public Library – Eagle, Indianapolis 39563
Indianapolis-Marion County Public Library – Franklin Road, Indianapolis 39564
Indianapolis-Marion County Public Library – Glendale, Indianapolis 39565
Indianapolis-Marion County Public Library – Irvington, Indianapolis 39566
Indianapolis-Marion County Public Library – Lawrence, Indianapolis 39567
Indianapolis-Marion County Public Library – Nora, Indianapolis 39568
Indianapolis-Marion County Public Library – Pike, Indianapolis 39569
Indianapolis-Marion County Public Library – Southport, Indianapolis 39570
Indianapolis-Marion County Public Library – Warren, Indianapolis 39571
Indianapolis-Marion County Public Library – Wayne, Indianapolis 39572
Indira Gandhi Centre for Atomic Research, Kalpakkam 14020
Indira Gandhi National Centre for the Arts, New Delhi 14146
Indira Gandhi National Open University, New Delhi 13725
Indira Kala Sangeet Vishwavidyalaya, Khairagarh 13677
Indonesia America Friendship Assn, Jakarta 14372
Indonesia Institute of the Arts Yogyakarta, Yogyakarta 14310
Indonesian Army Command and Staff College, Bandung 14326
Indonesian Biotechnology Research Unit for Estate Crops, Bogor 14350
Indonesian Institute of Sciences, Jakarta 14369
Indonesian Oil Palm Research Institute, Medan 14377
Indonesian Open Learning University, Jakarta 14262
Indonesian Sugar Research Institute, Pasuruan 14378
Indraprastha College for Women Library, Delhi 13801
Industrial and Technological Information Centre (INTIC), Tehran 14412
Industrial Automation Professional School, Donetsk 28099
Industrial Corporation and Professional School of the Ukrainian Association of the Blind, Artemivsk 28180
Industrial Design and Construction Joint-Stock Company, Sankt-Peterburg 22905
Industrial Development Board Library, Moratuwa 26326
Industrial Development Corporation of South Africa Ltd, Sandton 25101
Industrial Ecology Institute, Donetsk 28223
Industrial Hygiene and Occupational Diseases Research Institute, Kharkiv 28269
Industrial Hygiene and Occupational Diseases Research Institute, Krivi Rig 28299
Industrial Information Centre and Extension Services (FIIRO INDICES), Oshodi 20041
Industrial Institute, Lutsk 28402
Industrial Institute, Ukhta 22613
Industrial Institute of Agricultural Engineering, Poznań 20904
Industrial Power Company Ltd, Olkiluoto 07425
Industrial Property Library, Tokyo 17761
Industrial Property Office, Patent Information Departement, Public Reading Room, Praha 06518
Industrial Research Institute for Automation and Measurements, Warszawa 21343
Industrial Research Limited, Auckland 19814
Industrial Research Limited, Lower Hutt 19816
Industrial Research Limited, Wellington 19848
Industrial Technology Institute, Colombo 26311
Industrial Technology Research Institute, Hsinchu 27587
Industrial Toxicology Research Centre, Lucknow 14066
Industrialny institut, Ukhta 22613
Industrialny instytut, Lutsk 28402
Industrie- und Filmmuseum (ifm), Bitterfeld-Wolfen 11614
Industrie- und Handelskammer Darmstadt, Darmstadt 11728
Industrie- und Handelskammer Frankfurt am Main, Frankfurt am Main 11856
Industrie- und Handelskammer für München und Oberbayern, München 12301

Industrie- und Handelskammer Nürnberg für Mittelfranken, Nürnberg 12374
Industrie- und Handelskammer zu Koblenz, Koblenz 12120
Industrie- und Handelskammer zu Köln, Köln 12137
Industrie- und Handelskammer zu Leipzig, Leipzig 12186
Industries Federation of the Southern State of Rio Grande, Porto Alegre 03317
Industry Canada, Ottawa 04446
Industry Canada – Canadian Intellectual Property Office Resource Centre, Ottawa 04115
Industry Canada Library Services, Ottawa 04116
Infineum USALP, Linden 35817
infoDoc RADIX, Zürich 27452
Infodoc-Boerenbond, Leuven 02670
Information Agency of Power Industry Joint-Stock Company, Moskva 22831
Information Agency of Wielkopolska 'Press-Service', Poznań 21208
Information and knowledge management services in international cooperation, Kessel-Lo 02667
Information Center of Special Education, Yokosuka 17830
Information Center of the Architecture and Planning Administration, Moskva 22758
Information Center of the Ministry of Industrial Construction, Moskva 22757
Information Centre, BSRIA Ltd, Bracknell 29618
Information Centre for Industry, Budapest 13384
Information Masters Library, Manzanita 37084
Information Resource Center (IRC), Stockholm 26650
Information Resource Unit, Bogor 14349
Information Service Center, Tokyo 17466
Information Services, Raahe 07429
Information Services Library, Jackson 31438
Information Systems Management Office, Boulder 36581
Information und Technik Nordrhein-Westfalen, Düsseldorf 10872
Informationsbüro Nicaragua e.V., Wuppertal 12609
Informationszentrum, Bibliothek und Mediathek zur deutschen Zeitgeschichte, Bonn 11650
Informationszentrum Patente Stuttgart, Stuttgart 12525
Informatsionno-tekhnicheski tsentr Kuznetskugol, Novokuznetsk 23634
Informpatent Centre of Patents and Description of Inventions, Sankt-Peterburg 23910
Infracor GmbH, Marl 11417
Ingeborg-Drewitz-Bibliothek, Berlin 12650
Ingham Regional Medical Center, Lansing 37013
Inglewood Public Library, Inglewood 39574
Inha University, Inchon 18063
INHOLLAND University Diemen, Diemen 19106
INHOLLAND University Haarlem, Haarlem 19144
INIA La Estanzuela, Colonia 42032
INIFAP-Area FTAL, México 18845
Inland Fisheries Institute, Olsztyn 21175
Inland Revenue Department, Kuala Lumpur 18551
Inland Revenue Library, Wellington 19800
INMM Leibniz- Institut für Neue Materialien gGmbH, Saarbrücken 12452
Innenministerium des Landes Schleswig-Holstein, Kiel 10964
Inner Mongolia College of Agriculture and Animal Husbandry, Huhhot 05301
Inner Mongolia College of Finance and Economics, Huhhot 05302
Inner Mongolia Forestry Institute, Huhhot 05303
Inner Mongolia Medical College, Huhhot 05304
Inner Mongolia Mongolian Medical College, Tongliao 05626
Inner Mongolia Polytechnical University, Hohhot 05278
Inner Mongolia Teachers' College for Nationalities, Tongliao 05627
Inner Mongolia University, Huhhot 05305
Inner Temple Library, London 29764
Innerrhodische Kantonsbibliothek, Appenzell 26964
Innisfail Public Library, Innisfail 04664
Innovación – Información – Tecnología (INFOTEC), México 18846
INNOVATEXT Textilipari Műszaki és Fejlesztő és Vizsgáló Intézet, Budapest 13356
Innovation Center Iceland, Reykjavík 13571
Innovation Technological Centre – Rim, a.s., Odolena Voda 06468
Innventia AB, Stockholm 26651
Inova Fairfax Hospital, Falls Church 36843
INRA – Station d'Economie et Sociologie Rurales, Montpellier 08423
INRP – Institut National de Recherche Pédagogique, Lyon 08397
INRS – Institut Armand Frappier, Laval 03738
Insatsu Toshokan, Tokyo 17679
INSEAD, Fontainebleau 07706
INSEE Alsace, Strasbourg 08683
die insel – Stadtbibliothek Marl, Marl 12847
Insitute of Electrotechnics, Warszawa 21275
Insitute of Environmental Protection, Warszawa 21296
Insitutul de Chimie Macromoleculară 'Petru Poni', Iaşi 22036
Instistute for Legal Sciences, Budapest 13414
Institiúid Teangeolaíochta Eireann, Dublin 14537
Instititute of World Economy, Hanoi 42288
Institucion Colombina, Sevilla 25827
Institución Cultural Santa Ana, Almendralejo 25871
Institución 'Fernando el Católico' de la Diputación Provincial, Zaragoza 26193
Institución Milà i Fontanals, CSIC, Barcelona 25905
Institue of Electron Technology, Warszawa 21308
Instituição Educacional 'Tabajara', São Paulo 03207
Instituição 'Moura Lacerda', Ribeirão Prêto 03095
Institut Africain de Développement Economique et de Planification (IDEP), Dakar 24441
Institut Africain pour le Développement Economique et Social (INADES), Abidjan 06136
Institut Afriki RAN, Moskva 23251
Institut Agama Islam Negeri Jami'ah Ar-Raniry (IAIN), Darussalam 14248
Institut Agama Islam Negeri Raden Intan, Bandarlampung 14237
Institut Agama Islam Negeri Sunan Ampel (IAIN), Surabaya 14296
Institut Agama Islam Negeri Sunan Gunungjati, Bandung 14238
Institut Agama Islam Negeri Sunan Kalijaga, Yogyakarta 14308
Institut Agama Islam Negeri Walisongo, Semarang 14293
L'Institut Agrícola, Barcelona 25906
Institut Agricole de l'Etat de Fribourg, Posieux 27424
Institut Agronomique et Vétérinaire Hassan II, Rabat 18925
Institut Alemany, Barcelona 25904
Institut Amatller d'Art Hispànic, Barcelona 25900
Institut Archéologique du Luxembourg, Arlon 02540
Institut Archéologique Liégeois, Liège 02678
Institut arkheologii i etnografii Sibirskogo otdeleniya RAN, Novosibirsk 23637
Institut arkheologii RAN, Moskva 23252
Institut assortimenta izdeli legkoi promyshlennosti i kultury odezhdy, Moskva 23253
Institut astronomii RAN, Moskva 23254
Institut avtomatiki i elektrometrii Sibirskogo otdeleniya RAN, Novosibirsk 23638
Institut avtomatiki i protsessov upravleniya RAN, Vladivostok 24054
Institut avtomatizirovannykh sistem planirovaniya i upravleniya, Minsk 02009
Institut Bahasa, Kuala Lumpur 18533
Institut Belmestpromproekt, Minsk 02010
Institut Biblique, Nogent-sur-Marne 08207
Institut biokhimii, Grodno 01962
Institut biokhimii i fiziologii rasteni i mikroorganizmov RAN, Saratov 23957
Institut biologii morya RAN, Vladivostok 24055
Institut biologii razvitiya im. N.K. Koltsova, Moskva 23255
Institut biologii – Uralski filial RAN, Lobytnangi 23143
Institut biologii vnutrennikh vod RAN, Borok 23008
Institut biologii vnutrennikh vod RAN, Tolyatti 23993
Institut biomeditsinskoi khimii, Moskva 23256
Institut bioorganicheskoi khimii, Minsk 02011
Institut Cartogràfic de Catalunya, Barcelona 25907
Institut catholique d'arts et métiers, Lille 07727
Institut Catholique de Paris, Paris 07883
– Bibliothèque de Droit canonique, Paris 07884
– Bibliothèque Œcuménique et Scientifique d'Etudes Bibliques (BOSEB), Paris 07885
– Institut Français d'Études Byzantines (IFEB), Paris 07886
– Institut Supérieure de Pédagogie (ISP), Paris 07887
Institut Catholique de Toulouse, Toulouse 08079
Institut Catholique de Yaoundé, Yaoundé 03638
Institut Catholique des Hautes Etudes Commerciales, Bruxelles 02294
Institut Culturel Italien, Marseille 08409
Institut Culturel Italien de Montreal, Montreal 04405
Institut Culturel Italien de Paris, Paris 08577
Institut Curie, Paris 08550
Institut d'Administration des Entreprises (IAE), Aix-en-Provence 08268
Institut dal Dicziunari Rumantsch Grischun (DRG), Chur 27343
Institut dalnego vostoka RAN, Moskva 23257
Institut d'Aménagement et d'Urbanisme de la Région d'Ile de France, Paris 08551
Institut d'Art Contemporain, Villeurbanne 08717
Institut d'Astrophysique de Paris, Paris 08552
Institut de Biologie Structurale et Microbiologie, Marseille 08408
Institut de Ciències de la Terra 'Jaume Almera', Barcelona 25908
Institut de Droit et Economie de la Firme et de l'Industrie (IDEFI), Sophia-Antipolis 08678
Institut de Formation et de Recherche Démographiques (IFORD), Yaoundé 03658
Institut de Formation Pedagogique, Le Bardo 27825
Institut de formation théologique de Montréal, Montreal 04215
Institut de France, Paris 08553
Institut de Hautes Etudes en Administration Publique (IDHEAP), Chavannes-près-Renens 27064
Institut de hautes études internationales et du développement (IHEID), Genève 27091
Institut de la Recherche Agronomique, Yaoundé 03659
Institut de la Statistique du Québec, Québec 04130
Institut de l'Environnement et de Recherches Agricoles, Ouagadougou 03619
Institut de l'Information Scientifique et Technique (INIST), Vandœuvre-les-Nancy 08707
Institut de Médecine Tropicale du Service de Santé des Armées (IMTSSA), Marseille 07757
Institut de Météorologie, d'Aviation Civile et de Télécommunications, Kinshasa 06076
Institut de Paléontologie Humaine, Paris 08554
Institut de Presse et des Sciences de l'Information, Tunis 27833
Institut de Productivité et de Gestion Prévisionnelle, Bamako 18642
Institut de Recherche d'Hydro-Québec, Varennes 04547
Institut de Recherche et d'Application Pédagogique (INDRAP), Niamey 19912
Institut de recherche et de documentation pédagogique, Neuchâtel 27418
Institut de Recherche et d'Histoire des Textes (IRHT), Paris 08555
Institut de recherche pour le développement, Bondy 08300
Institut de recherche pour le développement, Nouméa 19756
Institut de Recherche pour le Développement (IRD), Dakar 24442
Institut de Recherche pour le Développement (IRD), Yaoundé 03660
Institut de recherche Robert-Sauve en santé et en sécurité du travail, Montreal 04406
Institut de Recherche Scientifique, Kinshasa 06077
Institut de Recherche Scientifique, Lwiro 06086
Institut de Recherche Scientifique et Technologique, Butare 24374
Institut de Recherche Scientifique pour le Développement (IRD), Bangui 04833
Institut de Recherches Agronomiques de Boukoko, Boukoko 04834
Institut de Recherches Agronomiques Tropicales et des Cultures Vivrières, Niamey 19913
Institut de Recherches Appliquées, Porto-Novo 02908
Institut de Recherches Comparatives sur les Institutions et le Droit, Ivry-sur-Seine 08366
Institut de Recherches et d'Etudes sur le Monde Arabe et Musulman (IREMAM), Aix-en-Provence 08269
Institut de Sciences Mathématiques et Economiques Appliquées (ISMEA), Montrouge 08426
Institut de Théologie Orthodoxe Saint-Serge, Paris 07888
Institut de tourisme et d'hôtellerie du Québec, Montreal 04007
Institut d'Economie Quantitative, Tunis 27834
Institut d'Egypte, Cairo 07174
Institut del Teatre, Barcelona 25909
Institut der deutschen Wirtschaft Köln, Köln 12138
Institut der Feuerwehr Sachsen-Anhalt, Biederitz 11608
Institut der Wirtschaftsprüfer in Deutschland e.V., Düsseldorf 11798
Institut des Batiments et Travaux Publics (IBIP), Kinshasa 06078
Institut des Belles Lettres Arabes (IBLA), Tunis 27835
Institut des Relations Internationales du Cameroun (IRIC), Yaoundé 03640
Institut des Relations Internationales du Cameroun (IRIC), Yaoundé 03637
Institut des Sciences Agronomiques du Burundi (ISABU), Bujumbura 03628
Institut des Sciences Agronomiques du Rwanda, Butare 24375
Institut des Sciences de la Mer et de l'Aménagement du Littoral (ISMAL), Alger 00062
Institut des sciences et industries du vivant et de l'environnement, Paris 08462
Institut des Sciences Humaines, Yaoundé 03661
Institut des Télécommunications (IT), Oran 00082
Institut d'Estadística de Catalunya, Barcelona 25910
Institut d'Estudis Catalans – Servei de Documentació i Arxiu, Barcelona 25911
Institut d'Estudis Nord-Americans, Barcelona 25912
Institut d'Etudes Augustiniennes, Paris 08556
Institut d'études du Judaïsme, Bruxelles 02592
Institut d'Etudes Politiques de Bordeaux (IEP), Pessac 07995
Institut d'Etudes Politiques de Lyon, Lyon 07741
Institut d'Etudes Politiques de Rennes, Rennes 08652
Institut d'Etudes Politiques (IEP), Aix-en-Provence 08270
Institut d'Histoire du Temps Présent (IHTP), Paris 08557
Institut d'Histoire Ouvrière, Economique et Sociale (IHOES), Seraing 02696
Institut d'Histoire Sociale, Nanterre 08440
Institut Dominicain d'Etudes Orientales du Caire (IDEO), Cairo 07130
Institut d'Ophthalmologie Tropicale de l'Afrique (IOTA), Bamako 18650
Institut du Monde Arabe, Paris 08558
Institut du Sahel, Bamako 18651
Institut d'Urbanisme, d'Aménagement et d'Administration territoriale de Grenoble (IUG), Grenoble 08354
Institut ekologii cheloveka i gigieny okruzhayushchei sredy im. A.N. Sysina, Moskva 23258
Institut ekonomiki i organizatsii promyshlennogo proizvodstva Sibirskogo otdeleniya RAN, Novosibirsk 23639
Institut ekonomiki RAN, Moskva 23259
Institut eksperimentalnoi botaniki, Minsk 02012
Institut eksperimentalnoi endokrinologii i khimii gormonov, Moskva 23260
Institut Eksperimentalnoi Meditsiny, Sankt-Peterburg 23804
Institut elektrokhimii im. akad. A.N. Frumkina RAN, Moskva 23261
Institut elektroniki, Minsk 02013
Institut elektronnogo mashinos-troeniya, Moskva 22335
Institut elemento-organicheskikh soedineni RAN, Moskva 23262
Institut Emile Vandervelde (IEV), Bruxelles 02593
Institut et Musée Voltaire, Genève 27365

Institut et Observatoire Geophysique, Antananarivo 18461
Institut etnografii, Moskva 23263
Institut etnologii i antropologii RAN, Moskva 23264
Institut Européen des Hautes Etudes Internationales (IEHEI), Nice 07814
Institut evolyutsionnoi fisiologii i biochimii im. Sechenovz RAN, Sankt-Peterburg 23782
Institut Facultaire des Sciences Agronomiques (IFA), Kisangani 06053
Institut filosofii i prava, Minsk 02014
Institut filosofii RAN, Moskva 23265
Institut fizicheskikh problem im. P.L. Kapitsy RAN, Moskva 23266
Institut fizicheskoi khimii RAN, Moskva 23267
Institut fizicheskoi kultury, Krasnodar 22298
Institut fizicheskoi kultury, Omsk 22451
Institut fizicheskoi kultury, Volgograd 22640
Institut fizichesko-organicheskoi khimii, Minsk 02015
Institut fiziki, Minsk 02016
Institut fiziki atmosfery RAN, Moskva 23268
Institut fiziki atmosfery RAN, Shikhovo 23969
Institut fiziki im. L.V. Kirenskogo Sibirskogo otdeleniya RAN, Krasnoyarsk 23123
Institut fiziki metallov Uralskogo otdeleniya RAN, Ekaterinburg 23027
Institut fiziki poluprovodnikov Sibirskogo otdeleniya RAN, Novosibirsk 22436
Institut fiziki tverdogo tela i poluprovodnikov, Minsk 02017
Institut fiziki vysokikh davleni RAN, Troitsk 24001
Institut fiziki zemli im. O.Y. Shmidta RAN, Moskva 23269
Institut fiziki zemli im. Shmidta RAN, Obninsk 23661
Institut fiziologii im. I.P. Pavlova RAN, Sankt-Peterburg 23783
Institut fiziologii im. I.P. Pavlova RAN, Vsevolzhsk 24080
Institut fiziologii rasteni im. K.A. Timiryazeva RAN, Moskva 23270
Institut for Research in Geology and Mineral Resources, Toshkent 42134
Institut for Sundhedsvæsen, København 06794
Institut for the German Language, Mannheim 12242
Institut Français, Athinai 13113
Institut Français, Barcelona 25913
Institut Français, Berlin 11533
Institut Français, Bremen 11677
Institut Français, Budapest 13383
Institut Français, Casablanca 18948
Institut Français, Düsseldorf 11799
Institut Français, Edinburgh 29655
Institut Français, Firenze 16260
Institut Français, Genova 16286
Institut Français, Graz 01525
Institut Français, Innsbruck 01540
Institut Français, Istanbul 27890
Institut Français, København 06795
Institut Français, Köln 12139
Institut Français, Leipzig 12187
Institut Français, Madrid 26033
Institut Français, Mainz 12231
Institut Français, Marrakech 18951
Institut Français, München 12302
Institut Français, Napoli 16371
Institut Français, Oujda 18952
Institut Français, Pondicherry 14181
Institut Français, Port-au-Prince 13205
Institut Français, Praha 06491
Institut Français, Rabat 18959
Institut Français, Tanger 18967
Institut Français, Tel-Aviv 14764
Institut Français, Warszawa 21261
Institut Français, Zagreb 06243
Institut Français Charles Nodier, Ljubljana 24917
Institut Français d'Amérique Latine / Instituto Francés de América Latina, México 18847
Institut Français d'Archéologie, Beirut 18220
Institut Français d'Archéologie Orientale (IFAO), Cairo 07175
Institut Français de Recherche en Iran, Tehran 14428
Institut Français de Recherche Scientifique pour le Développement en Coopération (ORSTOM), Bamako 18652
Institut Français de Serbie, Beograd 24507
Institut Français des Pays-Bas, Amsterdam 19368
Institut Français des Relations Internationales (IFRI), Paris 08559
Institut Français d'Etudes Anatoliennes d'Instanbul, Istanbul 27891
Institut Français d'Etudes Arabes, Damascus 27513
Institut Français d'Histoire Sociale – Archives Nationales, Paris 08560
Institut Français du Pétrole, Rueil-Malmaison 08660
Institut Français du Royaume-Uni, London 29765
Institut Franco-Portugais, Lisboa 21790
Institut für Angewandte Trainingswissenschaft e.V., Leipzig 12188
Institut für Arbeit und Gesundheit der Deutschen Gesetzlichen Unfallversicherung (IAG), Dresden 11759
Institut für Arbeitsmarkt- und Berufsforschung, Nürnberg 12371
Institut für Arbeitsrecht und Arbeitsbeziehungen in der Europäischen Gemeinschaft (IAAEG), Trier 12552
Institut für Arbeitsschutz der Deutschen Gesetzlichen Unfallversicherung, Sankt Augustin 12461
Institut für Arbeitswissenschaft und Technologiemanagement der Universität Stuttgart (IAT), Stuttgart 12519
Institut für Auslandsbeziehungen e.V., Stuttgart 12526
Institut für Bienenkunde Celle, Celle 10848
Institut für Bildungsmedien e.V., Frankfurt am Main 11857
Institut für Brasilienkunde, Mettingen 12265
Institut für Demographie, Wien 01606
Institut für Deutsche Sprache (IDS), Mannheim 12242
Institut für die Geschichte der Deutschen Juden, Hamburg 11991
Institut für donauschwäbische Geschichte und Landeskunde, Tübingen 12560
Institut für Energierecht an der Universität zu Köln, Köln 12140
Institut für Entwicklungsplanung und Strukturforschung GmbH an der Universität Hannover, Hannover 12028
Institut für Europäische Geschichte, Mainz 12232
Institut für Friedensforschung und Sicherheitspolitik an der Universität Hamburg, Hamburg 11992
Institut für Gärungsgewerbe und Biotechnologie, Berlin 11534
Institut für Geschichte der Medizin der Robert Bosch Stiftung, Stuttgart 12527
Institut für Gesellschaftspolitik an der Hochschule für Philosophie (IGP), München 12303
Institut für Getreideverarbeitung GmbH, Nuthetal 12379
Institut für Hochschulforschung, Lutherstadt Wittenberg 12600
Institut für Hochschulkunde an der Universität Würzburg, Würzburg 12614
Institut für Höhere Studien (IHS), Wien 01607
Institut für Holztechnologie Dresden gGmbH, Dresden 11760
Institut für Humanernährung und Lebensmittelkunde, Kiel 10117
Institut für Hygiene und Umwelt der Freien und Hansestadt Hamburg, Hamburg 11993
Institut für Industrielle Fertigung und Fabrikbetrieb der Universität Stuttgart (IFF), Stuttgart 12521
Institut für Infektionskrankheiten, Bern 27322
Institut für Jugendliteratur, Wien 01608
Institut für Jugendliteratur und Leseforschung, Wien 01609
Institut für kirchliche Zeitgeschichte, Salzburg 01567
Institut für Landes- und Stadtentwicklungsforschung GmbH (i.G.), Dortmund 11745
Institut für Landeskunde im Saarland e. V., Schiffweiler 12464
Institut für Lehrerfortbildung, Mülheim an der Ruhr 12270
Institut für Literatur und Kultur der Arbeitswelt, Dortmund 11744
Institut für Luft- und Kältetechnik gGmbH, Dresden 11385
Institut für Medienpädagogik in Forschung und Praxis, München 12310
Institut für Medizinische Biometrie und Medizinische Informatik, Freiburg 09802
Institut für Messewirtschaft und Distributionsforschung, Köln 10144
Institut für moderne Kunst Nürnberg e.V., Nürnberg 12375
Institut für Museumskunde, Berlin 11568
Institut für niederdeutsche Sprache, Bremen 11678
Institut für Öffentliche Verwaltung des Landes Nordrhein-Westfalen, Hilden 12060
Institut für Ostrecht München e.V., Regensburg 12427
Institut für Photonische Technologie e.V., Jena 12077
Institut für Politische Bildung Baden-Württemberg e.V., Buchenbach 11701
Institut für Qualitätsentwicklung an Schulen Schleswig-Holstein, Kronshagen 12165
Institut für Qualitätsentwicklung an Schulen, Schleswig-Holstein, Lübeck 12212
Institut für Realienkunde des Mittelalters und der frühen Neuzeit, Krems 01553
Institut für Rebenzüchtung Geilweilerhof, Siebeldingen 12486
Institut für Regionalentwicklung und Strukturplanung e.V., Erkner 11817
Institut für Regionalgeschichte, Münster 12350
Institut für Religionspädagogik und Medienarbeit im Erzbistum Paderborn, Paderborn 11305
Institut für Religionspädagogische Bildung Salzburg, Salzburg 01264
Institut für Religionsunterricht und Katechese im Erzbistum Paderborn, Abt. Schwerte, Schwerte 11328
Institut für Religiöse Volkskunde / Institut für die Geschichte des Bistums Münster, Münster 12344
Institut für Rundfunkrecht an der Universität zu Köln, Köln 12141
Institut für Sächsische Geschichte und Volkskunde e.V. (ISGV), Dresden 11761
Institut für Sanktionenrecht und Kriminologie, Kiel 12110
Institut für Seeverkehrswirtschaft und Logistik, Bremen 11679
Institut für soziale Demokratie, Berlin 11491
Institut für Sozialforschung an der Johan Wolfgang Goethe-Universität, Frankfurt am Main 11858
Institut für Staatskirchenrecht der Diözesen Deutschlands, Bonn 11177
Institut für Stadtgeschichte, Frankfurt am Main 11859
Institut für Theologie und Frieden, Hamburg 11994
Institut für Vergleichende Städtegeschichte an der Universität Münster, Münster 12345
Institut für Volkskultur und Kulturentwicklung Innsbruck, Innsbruck 01541
Institut für Volkskunde, München 12304
Institut für Wasser, Energie und Nachhaltigkeit, Graz 01526
Institut für Wirtschaftsforschung Halle, Halle (Saale) 11957
Institut für Wirtschaftspolitik an der Universität zu Köln, Köln 12142
Institut für Zeitgeschichte, München 12305
Institut für Zeitungsforschung der Stadt Dortmund, Dortmund 11746
Institut für Zweiradsicherheit e.V., Essen 11824
Institut gelmintologii im. akad. K.I. Skryabina, Moskva 22336
Institut geografii RAN, Moskva 23271
Institut geografii Sibirskogo otdeleniya RAN, Irkutsk 23049
Institut Geographique et Hydrographique National, Antananarivo 18462
Institut Géographique National (IGN), Marne-la-Vallée 08402
Institut Géographique National (IGN), Saint-Mandé 08162
Institut geokhimii i analiticheskoi khimii im. V.I. Vernadskogo RAN, Moskva 23272
Institut geokhimii Sibirskogo otdeleniya RAN, Irkutsk 23050
Institut geologii i geokhimii Uralskogo nauchnogo tsentra RAN, Ekaterinburg 23028
Institut geologii i geokhronologii dokembriya RAN, Sankt-Peterburg 23784
Institut gidrodinamiki imeni M.A. Lavrentyeva Sibirskogo otdeleniya RAN, Novosibirsk 23640
Institut gigieny detei i podrostkov, Moskva 23273
Institut Gipronikel, Sankt-Peterburg 23785
Institut gornogo dela im. Skochinskogo, Lyubertsy 22316
Institut gornogo dela Sibirskogo otdeleniya RAN, Novosibirsk 23641
Institut goryuchikh iskopaemykh, Moskva 23274
Institut gosudarstva i prava, Moskva 23275
Institut Grand-Ducal – Section Historique, Luxembourg 18413
Institut Grodnoselstroiproekt, Grodno 01963
Institut Gustave Roussy, Villejuif 08715
Institut Haïtiano-Américain, Port-au-Prince 13206
Institut Historique Allemand de Paris (IHAP), Paris 08532
Institut Hongrois de Paris, Paris 08561
Institut Hydrométéorologique de Formation et de Recherche (IHFR), Oran 00041
Institut informatiki, tekhnologii i ekonomiki, Sankt-Peterburg 22764
Institut International de Planification de l'Education (IIPE), Paris 08562
Institut International des Droits de l'Homme / International Institute of Human Rights, Strasbourg 08684
Institut international des Sciences administratives (IISA), Bruxelles 02594
Institut international d'études sociales, Genève 27366
Institut inzhenerov geodezii, aerofotosemki i kartografii, Novosibirsk 22437
Institut inzhenerov vodnogo transporta, Novosibirsk 22438
Institut inzhenerov zheleznodorozhnogo transporta, Irkutsk 22254
Institut inzhenerov zheleznodorozhnogo transporta, Khabarovsk 22285
Institut inzhenerov zheleznodorozhnogo transporta, Sankt-Peterburg 22520
Institut iskusstv, Ufa 24027
Institut iskusstva i kultury, Chelyabinsk 22227
Institut iskusstvoznaniya, Moskva 23276
Institut iskysstvovedeniya, etnografii i folklora, Minsk 02018
Institut Islamique, Batna 00075
Institut istorii, arkheologii i etnografii narodov dalnego vostoka RAN, Vladivostok 24056
Institut istorii estestvoznaniya i tekhniki RAN, Moskva 23277
Institut istorii estestvoznaniya i tekhniki RAN, Sankt-Peterburg 23786
Institut istorii materialnoi kultury RAN, Sankt-Peterburg 23787
Institut kataliza im G.K. Boreskova, Novosibirsk 23642
Institut Keguruan dan Ilmu Pendidikan, Klaten 14267
Institut Keguruan dan Ilmu Pendidikan, Medan 14278
Institut Keguruan dan Ilmu Pendidikan, Padang 14283
Institut Keguruan dan Ilmu Pendidikan, Yogyakarta 14309
Institut Keguruan dan Ilmu Pendidikan Malang, Malang 14273
Institut Keguruan dan Ilmu Pendidikan Muhammadiyah, Purworejo 14290
Institut Kesihatan Umum, Kuala Lumpur 18583
Institut khimicheskogo mashinostroeniya, Moskva 22337
Institut khimicheskoi fiziki, Moskva 23278
Institut khimicheskoi kinetiki i goreniya Sibirskogo otdeleniya RAN, Novosibirsk 23643
Institut khimii silikatov im. I.V. Grebenshchikova RAN, Sankt-Peterburg 23788
Institut khirurgii im. prof. A.V. Vishnevskogo, Moskva 23279
Institut kino i televideniya, Sankt-Peterburg 22521
Institut Kirche und Judentum, Berlin 11535
Institut kompleksnykh transportnykh problem, Moskva 23280
Institut kooperativnoi torgovli, Novosibirsk 22439
Institut 'Krasnodarnefteproekt', Krasnodar 23112
Institut kurortologii i fizioterapii, Ekaterinburg 23029
Institut Latinskoi Ameriki, Moskva 23281
Institut lesa Sibirskogo otdeleniya RAN, Krasnoyarsk 23124
Institut lesovedeniya RAN, Uspenskoe 24038
Institut Libre Marie Haps, Bruxelles 02295
Institut lingvisticheskikh is sledovani, Sankt-Peterburg 23789
Institut literatury im. Yanki Kupaly, Minsk 02019
Institut makroekonomicheskikh problem, Moskva 23282
Institut mashinostroeniya, Sankt-Peterburg 22522
Institut mashinovedeniya im. A.A. Blagonravova RAN, Moskva 23283
Institut matematiki, Minsk 02020
Institut matematiki i mekhaniki Uralskogo otdelenia RAN, Ekaterinburg 23030
Institut matematiki im. S.L. Soboleva, Novosibirsk 23644
Institut Mathildenhöhe, Darmstadt 11729
Institut meditsinskoi parazitologii i tropicheskoi meditsiny im. E.I. Martsinovskogo, Moskva 22338

Institut mekhanizatsii selskogo khozyaistva, Saratov — 22573
Institut meloratsii, Minsk — 02021
Institut Mémoires de l'édition contemporaine (IMEC), Saint-Germain-la-Blanche-Herbe — 08671
Institut metallurgii im. A.A. Baikova, Moskva — 23284
Institut mezhdunarodnykh otnosheni, Moskva — 22339
Institut Michel-Pacha, Tamaris-sur-Mer — 08689
Institut Mihailo Pupin, Beograd — 24508
Institut mikrobiologii, Minsk — 02022
Institut mikrobiologii RAN, Moskva — 23285
Institut mineralogii, geokhimii i kristallokhimii redkikh elementov RAN, Moskva — 23286
Institut Minskgrazhdanproekt, Minsk — 02023
Institut mirovoi ekonomiki i mezhdunarodnykh otnosheni, Moskva — 23287
Institut mirovoi literatury – Kabinet tvorchestva A.M. Gorkogo, Moskva — 23288
Institut mirovoi literatury RAN, Moskva — 23289
Institut Missionnaire des Rédemptoristes, Paris — 08219
Institut Mittag-Leffler, Djursholm — 26564
Institut molekularnoi biologii im. Engelgardta RAN, Moskva — 23290
Institut morfologii cheloveka, Moskva — 23291
Institut morfologii zhivotnykh im. A.N. Severtsova RAN, Moskva — 23292
Institut morskoi geologii i geofiziki RAN, Yuzhno-Sakhalinsk — 24094
Institut Municipal d'Educació, Barcelona — 25914
Institut Nacional d'Educació Física de Catalunya, Barcelona — 25915
Institut narodnogo khozyaistva, Ekaterinburg — 22237
Institut narodnogo khozyaistva, Novosibirsk — 22440
Institut narodnogo khozyaistva, Rostov-na-Donu — 22488
Institut narodov Azii RAN, Sankt-Peterburg — 23790
Institut National Agronomique, Alger — 00029
Institut National Agronomique Paris-Grignon (INA – PG), Thiverval-Grignon — 08071
Institut National d'Administration de Gestion et des Hautes Etudes Internationales (INAGHEI), Port-au-Prince — 13199
Institut National d'Archéologie et d'Arts (INAA), Tunis — 27836
Institut National de Jeunes Sourds, Paris — 08563
Institut National de la Consommation, Paris — 08564
Institut National de la Propriété Industrielle (INPI), Paris — 08565
Institut National de la Recherche Agronomique, Rabat — 18960
Institut National de la Recherche Agronomique de Tunisie, Tunis — 27818
Institut National de la Recherche Agronomique (INRA), Ivry-sur-Seine — 08367
Institut National de la Recherche Agronomique (INRA), Montpellier — 08424
Institut National de la Recherche Agronomique (INRA), Versailles — 08711
Institut National de la Recherche Agronomique (INRA) – Unité d'Economie et de Sociologie Rurales, Rennes — 08653
Institut National de la Recherche Agronomique (INRAA), Alger — 00063
Institut National de la Recherche Scientifique, Québec — 03827
Institut National de la Santé Publique, Alger — 00064
Institut National de la Statistique et des Etudes Economiques (INSEE), Paris — 08566
Institut National de l'Environnement Industriel et des Risques (INERIS), Verneuil-en-Halatte — 08710
Institut National de Recherche en Informatique et en Automatique (INRIA), Le Chesnay — 08374
Institut National de Recherche et de Sécurité pour la Prévention des Accidents du Travail et des Maladies Professionnelles (INRS), Paris — 08567
Institut National de Recherche Forestière, Alger — 00065
Institut National de Recherche Pédagogique, Lyon — 08397
Institut National de Recherche sur les Transports et leur Securité (INRETS), Arcueil — 08281
Institut National de Statistique et d'Economie Appliquée, Rabat — 18926

Institut National d'Education (INE), Yaoundé — 03662
Institut National des Bâtiments et des Travaux Publics (INBTP), Kinshasa — 06079
Institut National des Postes et Telecommunications, Rabat — 18961
Institut National des Sciences Appliquées de Rennes (INSA), Rennes — 08654
Institut National des Sciences Appliquées de Rouen (INSA), Mont-Saint-Aignan — 07777
Institut National des Sciences Comptables et de l'Administration d'Entreprises, Antananarivo — 18463
Institut National des Sciences de l'Education, Tunis — 27837
Institut National des Sciences et Technologies de la Mer (INSTM), Salammbô — 27826
Institut National des Sciences Humaines (INSH), N'Djamena — 04839
Institut National d'Etude du Travail et d'Orientation Professionnelle (INETOP), Paris — 08568
Institut National d'Etudes Démographiques (INED), Paris — 08569
Institut National d'Etudes Supérieures de Tiaret, Tiaret — 00047
Institut National d'Histoire de l'Art, Paris — 07939
Institut National d'Histoire de l'Art (INHA), Paris — 08570
Institut National d'Hygiène Ernst Rodenwaldt, Lomé — 27798
Institut National du Patrimoine, Tunis — 27838
Institut National du patrimoine (inp), La Plaine-Saint-Denis — 08368
Institut National du Sport et de l'Education Physique (INSEP), Paris — 08571
Institut National Polytechnique de Lorraine Nancy-Brabois, Vandœuvre-les-Nancy — 08098
Institut National Polytechnique Félix Houphouët-Boigny, Yamoussoukro — 06132
Institut National pour l'Etude et la Recherche Agronomique (INERA), Kinshasa — 06080
Institut National pour l'Etude et la Recherche Agronomique (INERA), Kisangani — 06082
Institut Nationale des Sciences Appliquées (INSA), Villeurbanne — 08116
Institut nauchnoi informatsii po obshchestvennym naukam RAN, Moskva — 23293
Institut Nazareth et Louis Braille, Longueuil — 04377
Institut Néerlandais & Fondation Custodia, Paris — 08572
Institut neftekhimicheskogo sinteza im. A.V. Topchieva RAN, Moskva — 23294
Institut neftekhimicheskoi i gazovoi promyshlennosti im. akad. I.M. Gubkina, Moskva — 22340
Institut neirokhirurgii im. akad. N.N. Burdenko, Moskva — 23295
Institut neorganicheskoi khimii Sibirskogo otdeleniya RAN, Novosibirsk — 23645
Institut obrazovaniya vzroslykh, Sankt-Peterburg — 23791
Institut obshchei fiziki RAN, Moskva — 23296
Institut obshchei genetiki RAN, Moskva — 23297
Institut obshchei i neorganicheskoi khimii im. N.S. Kurnkova RAN, Moskva — 23298
Institut obshchei i pedagogicheskoi psikhologii Akademii pedagogich-eskikh nauk, Moskva — 23299
Institut Océanographique, Paris — 08573
Institut œcuménique, Bogis-Bossey — 27335
Institut of Chemical Technology, Praha — 06404
Institut of Dental Research, Sydney — 01025
Institut of Economics, Saratov — 22571
Institut of Geology, Geotechnics and Geophysics, Ljubljana — 24919
Institut of regional research, Lviv — 28413
Institut of Solar-Terrestrial Physics of the Siberian branch of the Russian Academy of Sciences, Irkutsk — 23052
Institut of Studies on Mordvinian Language, Literature, History and Economy, Saransk — 23952
Institut okeanologii im. P.P. Shirshova RAN, Moskva — 23300
Institut okeanologii im. P.P. Shirshova RAN – Yuzhnoe otdelenie, Gelendzik — 23044
Institut okeanologii im. Shirshova Atlanticheskogo otdeleniya RAN, Kaliningrad — 23067

Institut organicheskoi i fizicheskoi khimii im. A.E. Arbuzova RAN, Kazan — 23081
Institut organicheskoi khimii im. N.D. Zelinskogo RAN, Moskva — 23301
Institut organicheskoi khimii Sibirskogo otdeleniya RAN, Irkutsk — 23051
Institut organicheskoi khimii Sibirskogo otdeleniya RAN, Novosibirsk — 23646
Institut ORGENERGOSTROI, Moskva — 23302
Institut ozerovedeniya RAN, Sankt-Peterburg — 23792
Institut Panafricain pour le Développement, Ouagadougou — 03620
Institut Pasteur, Paris — 08574
Institut Pasteur d'Algérie, Alger — 00066
Institut Pasteur de Dakar, Dakar — 24443
Institut Pasteur de Lille, Lille — 08383
Institut Pasteur de Madagascar, Antananarivo — 18464
Institut Pasteur de Nouvelle Calédonie, Nouméa — 19757
Institut Pédagogique du Burkina, Ouagadougou — 03621
Institut Pédagogique National, Libreville — 09175
Institut Pédagogique Nationale, Bamako — 18653
Institut Pédagogique Nationale (IPN), Kinshasa — 06058
Institut Penyelidikan Perhutanan, Kuala Lumpur — 18584
Institut peredachi informatsii RAN, Moskva — 23303
Institut Pertanian Bogor, Bogor — 14247
Institut Piawaian dan Penyelidikan Perindustrian Malaysia, Shah Alam — 18617
Institut pitaniya, Moskva — 23304
Institut po agrarna ikonomika, Ministerstvo na zemedelieto i gorite, Sofiya — 03535
Institut po botanika, Sofiya — 03536
Institut po darvoobrabotvane, Pazardjik — 03509
Institut po elektroenergiina tekhnika, Sofiya — 03498
Institut po filosofski nauki, Sofiya — 03537
Institut po fiziologiya, Sofiya — 03538
Institut po fiziologiya na rasteniyata, Sofiya — 03539
Institut po genetika, Sofiya — 03540
Institut po govedovădstvo i ovtsevădstvo, Stara Zagora — 03583
Institut po istoriya, Sofiya — 03541
Institut po matematika i Informatika, Sofiya — 03542
Institut po metaloznanie, Sofiya — 03543
Institut po mikrobiologiya, Sofiya — 03544
Institut po nevrologiya, psikhiyatriya i nevrokhirurgiya, Sofiya — 03545
Institut po obshta i neorganichna khimiya, Sofiya — 03546
Institut po proektiraniyu metiznykh zavodov, Sankt-Peterburg — 23793
Institut po proektiraniyu predpriyati rybnogo khozyaistva, Sankt-Peterburg — 23794
Institut po proektiraniyu predpriyati stroitelnoi industrii Moskvy (Mosproektstroiindustriya), Moskva — 23305
Institut po prouchvane i proektirane 'Mashproekt', Sofiya — 03547
Institut po ribni ressursi s akvarium, Varna — 03585
Institut po Sotsiologiya, Sofiya — 03548
Institut po tsarevitsata, Knezha — 03504
Institut po tyutyuna i tyutyunevite izdeliya, Plovdiv — 03512
Institut po zoologiya, Sofiya — 03549
Institut poliomielita i virusnykh entsefalitov im. M.P. Chumakova, Moskva — 23306
Institut Politecnic de Formació Professional, Barcelona — 25916
Institut Polytechnique LaSalle Beauvais, Beauvais — 07646
Institut Polytechnique Saint-Louis, Cergy — 07679
Institut pour la Recherche Scientifique en Afrique Centrale (IRSAC), Bukavu — 06066
Institut povysheniya kvalifikatsii rukovodyashchikh rabotnikov i spetsialistov legkoi promyshlennosti, Moskva — 23307
Institut povysheniya kvalifikatsii rukovodyashchikh rabotnikov i spetsialistov lesnoi i derevoobra-batyvayushchei promyshlennosti, Moskva — 23308
Institut povysheniya kvalifikatsii rukovodyashchikh rabotnikov i spetsialistov Ministerstva stroitelstva predpriyati neftyanoi i gazovoi promyshlennosti, Moskva — 23309
Institut prikladnoi fiziki, Minsk — 02024

Institut prikladnoi geofiziki im. E.K. Fyodorova, Moskva — 23310
Institut prikladnoi matematiki im. M.V. Keldysha RAN, Moskva — 23311
Institut pro kriminologii a sociální prevenci, Praha — 06492
Institut problem mekhaniki RAN, Moskva — 23312
Institut problem nadezhnosti i dolgovechnosti mashin, Minsk — 02025
Institut problem privatizatsii, Sankt-Peterburg — 23795
Institut problem sverkhplastichnosti metallov RAN, Ufa — 24028
Institut problem ukrepleniya zakonnosti i pravoporyadka, Moskva — 23313
Institut problem upravleniya, Moskva — 23314
Institut proektirovaniya zhilishchno-grazhdanskogo i kommunalnogo stroitelstva, lesoparkovogo zashchitnogo poyasa i obektov kulturno-bytovogo naznacheniya g. Moskvy, Moskva — 23315
Institut Protestant de Théologie, Montpellier — 07767
Institut psikhiatrii, Moskva — 23316
Institut psikhologii RAN, Moskva — 23317
Institut radiostroeniya, Sankt-Peterburg — 23796
Institut radiotekhniki i elektroniki RAN, Fryazino — 23042
Institut radiotekhniki i elektroniki RAN, Moskva — 23318
Institut Raymond-Dewar Rehabilitation Centre for the Deaf, Deafblind & Hard of Hearing, Montreal — 04407
Institut Régional d'Administration de Nantes (IRA), Nantes — 07805
Institut revmatizma, Moskva — 23319
Institut rossiskoi istorii RAN, Moskva — 23320
Institut rossiskoi istorii RAN, Sankt-Peterburg — 23797
Institut royal du Patrimoine artistique, Bruxelles — 02595
Institut Ruder Bošković, Zagreb — 06244
Institut russkogo yazyka i Institut yazykoznaniya, Moskva — 23321
Institut russkoi literatury RAN, Sankt-Peterburg — 23798
Institut Scientifique de la Santé Publique – Louis Pasteur (ISP), Brussel — 02636
Institut Scientifique de Service Public (ISSeP), Liège — 02679
Institut Scientifique Santé Publique – Louis Pasteur, Bruxelles — 02596
Institut Sénégalais de Recherches Agronomiques (ISRA), Bambey — 24429
Institut Seni Indonesia Yogyakarta, Yogyakarta — 14310
Institut serdechno-sosudistoi khirurgii im. A.N. Bakuleva, Moskva — 23322
Institut solnechno-zemnoi fiziki Sibirskogo otdeleniya RAN, Irkutsk — 23052
Institut sotsiologii RAN, Moskva — 23323
Institut sotsiologii RAN – Sankt Peterburgski filial, Sankt-Peterburg — 23323
Institut Spektroskopii RAN, Troitsk — 24002
Institut sravnitelnoi politologii i problem rabochego dvizheniya, Moskva — 23324
Institut SShA i Kanady RAN, Moskva — 23325
Institut Stal Proekt, Moskva — 23326
Institut stali i splavov, Moskva — 22341
Institut stroitelstva i arkhitektury, Minsk — 02026
Institut suisse de droit comparé, Lausanne — 27395
Institut suisse de prévention de l'alcoolisme et autres toxicomanies / Schweizerische Fachstelle für Alkohol- und andere Drogenprobleme, Lausanne — 27396
Institut Supérieur Agronomique et Vétérinaire 'Valery Giscard d'Estaing', Faranah — 13186
Institut Supérieur d'Architecture Intercommunal (ISAI), Bruxelles — 02296
Institut Supérieur de Commerce et d'Administration des Entreprises, Casablanca — 18933
Institut Supérieur de Gestion (ISG), Tunis — 27839
Institut Supérieur de Statistique (ISS), Lubumbashi — 06054
Institut Supérieur Industriel de l'Etat de Huy-Gembloux-Verviers, Huy — 02346
Institut Supérieur Pedagogique, Bunia — 06045
Institut Supérieur Pédagogique, Kananga — 06056
Institut Supérieur Pédagogique, Kikwit — 06057
Institut Supérieur Pédagogique, Kinshasa — 06059
Institut Supérieur Pédagogique, Kisangani — 06060
Institut Supérieur Scientifique, Nouakchott — 18683

Institut Supérieure de Commerce de Burundi, Bujumbura 03623
Institut Tadbiran Awam Negara Malaysia, Kuala Lumpur 18585
Institut tekhnicheskoi kibernetiki, Minsk 02027
Institut Teknologi Bandung, Bandung 14239
Institut Teknologi Brunei, Bandar Seri Begawan 03448
Institut Teknologi Sepuluh Nopember (ITS), Surabaya 14297
Institut teoreticheskoi astronomii RAN, Sankt-Peterburg 23800
Institut teoreticheskoi i prikladnoi mekhaniki Sibirskogo otdeleniya RAN, Novosibirsk 23647
Institut teplo- i massoobmena, Minsk 02028
Institut teplofiziki Sibirskogo otdeleniya RAN, Novosibirsk 23648
Institut tochnoi mekhaniki i vychislitelnoi tekhniki RAN, Moskva 23327
Institut torfa, Minsk 02029
Institut tsitologii i genetiki Sibirskogo otdeleniya RAN, Novosibirsk 23649
Institut tsitologii RAN, Sankt-Peterburg 23801
Institut tsvetnykh metallov i zolota, Krasnoyarsk 22303
Institut Ukrainoznavstva imeni Ivana Kripyakevicha NAN Ukraini, Lviv 28408
Institut Universitaire de Formation des Maîtres, Nice 08448
Institut Universitaire de Formation des Maîtres de la Réunion, Saint-Denis 21909
Institut Universitaire de Formation des Maîtres (IUFM), Paris 07889
Institut Universitaire de Formation des Maîtres (IUFM), Paris 07890
Institut Universitaire de Technologie, Belfort 07647
Institut Universitaire de Technologie, Lannion 07722
Institut Universitaire de Technologie, Quimper 08015
Institut Universitaire de Technologie (IUT), Cachan 07677
Institut usovershenstvovaniya vrachei, Novokuznetsk 22432
Institut usovershenstvovaniya vrachei, Sankt-Peterburg 22523
Institut usovershenstvovaniya vrachei im. V.I. Lenina, Kazan 22272
Institut virusologii im. D.I. Ivanovskogo, Moskva 23328
Institut Vitebskgrazhdanskproekt, Vitebsk 02078
Institut vodnykh problem RAN, Moskva 23329
Institut Voltaire, Genève 27365
Institut vostokovedeniya, Moskva 23330
Institut vostokovedeniya RAN, Sankt-Peterburg 23802
Institut vulkanologii i seismologii DVO RAN, Petropavlovsk-Kamchatski 23698
Institut vysokikh temperatur RAN, Moskva 23331
Institut vysokomolekulyarnykh soedineni RAN, Sankt-Peterburg 23803
Institut vysshei nervnoi deyatelnosti i neirofiziologii RAN, Moskva 23332
Institut Wohnen und Umwelt GmbH, Darmstadt 11730
Institut yadernoi energetiki, Minsk 02030
Institut yadernoi fiziki im. B.P. Konstantinova RAN, Gatchina 23043
Institut yadernoi fiziki Sibirskogo otdeleniya RAN, Novosibirsk 23650
Institut yadernykh issledovani RAN, Moskva 23333
Institut yazyka, literatury i istorii RAN, Kazan 23082
Institut za bălgarski ezik, Sofiya 03550
Institut za ekonomiku poljoprivrede, Beograd 24509
Inštitut za ekonomska raziskovanja, Ljubljana 24918
Institut za etnologiju i folkloristiku, Zagreb 06245
Inštitut za geologijo, geotehniko, in geofiziko, Ljubljana 24919
Institut za gorata, Sofiya 03551
Institut za Jadranske Kulture, Split 06223
Institut za literatura, Sofiya 03552
Institut za Medjunarodnu Politiku i Privredu, Beograd 24510
Institut za metalne konstrukcije, Ljubljana 24811
Institut za Nacionalna Istorija, Skopje 18442
Inštitut za narodnostna vprašanja, Ljubljana 24920
Institut za nuklearne nauke 'Vinča', Beograd 24511
Institut za Oceanografiju i Ribarstvo, Split 06224
Institut za Pedagoška istraživanja, Beograd 24457
Institut za pravni nauki, Sofiya 03553

Inštitut za raziskovanje krasa ZRC SAZU, Postojna 24952
Institut za Uporedno Pravo, Beograd 24458
Institut za usăvărshenstvane na uchitelite, Sofiya 03554
Institut za varilstvo, Ljubljana 24921
Institut za zashtita na rasteniyata, Kostinbrod 03505
Institut za Zaštitu Bilja i Životnu Sredinu, Beograd 24512
Institut zakonodatelstva i sravnitelnogo pravovedeniya, Moskva 23334
Institut zemnogo magnetizma, ionosfery i rasprostraneniya radiovoln RAN (IZMIRAN), Troitsk 24003
Institut zhivopisi, skulptury i arkhitektury im. I.E. Repina, Sankt-Peterburg 22524
Instituta epidemiologii i mikrobiologii im. N.F. Gamalei, Moskva 23335
Institute and Museum of the History of Science, Firenze 16262
Institute for Aboriginal Development Library, Alice Springs 00862
Institute for Advanced Judaic Studies, Toronto 04516
Institute for Advanced Studies of World Religions Library, Carmel 36629
Institute for Advanced Study, Princeton 32363
Institute for Advanced Study Berlin, Berlin 11599
Institute for Advanced Study of Human Sexuality, San Francisco 37544
Institute for Agricultural Quality Control, Budapest 13431
Institute for Asiatic Nations of the Russian Academy of Sciences, Sankt-Peterburg 23790
Institute for Balkan Studies, Thessaloniki 13134
Institute for Building Design in the Gastronomic and Tourist Industries, Moskva 23519
Institute for Building, Mechanization and Electrification in Agriculture, Warszawa 21260
Institute for Building Science and Technology, Hanoi 42289
Institute for Chemical Processing of Coal, Zabrze 21031
Institute for Christian Studies, Toronto 03882
Institute for Cinema and Television, Sankt-Peterburg 22521
Institute for Civil Engineering of Moscow, Moskva 23378
Institute for Continuing Education in Medicine, Almaty 17944
Institute for Continuing Education in Medicine, Toshkent 42091
Institute for Creation Research Library, Santee 37578
Institute for Defense Analyses, Alexandria 36455
Institute for Defense Analysis Library, Princeton 37414
Institute for Dermatology and Venereology, Toshkent 42135
Institute for Drug Research, Budapest 13378
Institute for East European Law, Regensburg 12427
Institute for Economic Research, Ljubljana 24918
Institute for Educational Research, Beograd 24457
Institute for Educational Research, Warszawa 21263
Institute for Energy Technology, Kjeller 20232
Institute for Engineering Projects, Moskva 23356
Institute for Ethnic Studies, Ljubljana 24920
Institute for Experimental and Clinical Veterinary Medicine, Kharkiv 28254
Institute for Financial Management and Research, Nunganbakkam 14174
Institute for Fisheries, Kyiv 28375
Institute for Future Technology, Tokyo 17709
Institute for Geodesy, Aerial Photography and Cartography, Novosibirsk 22437
Institute for High Temperatures of the Russian Academy of Sciences, Moskva 23331
Institute for Higher Qualification of Forestal and Timber Workers and Specialists, Moskva 23308
Institute for Higher Qualification of Oil and Gas Workers and Specialists, Moskva 23309
Institute for Historical Science, Zadar 06230
Institute for Horticultural Development, Knoxfield 00936
Institute for Housing and Urban Studies (IHS), Rotterdam 19511
Institute for Human Rights, Turku 07548

Institute for Humanities and Cultural Studies, Tehran 14429
Institute for Hygiene of Children and Juveniles, Moskva 23273
Institute for Hygiene on Ships, Moskva 23399
Institute for Information and Techno-Economic Studies in Meat and Dairy Industry, Moskva 23402
Institute for Information Recording, Kyiv 28347
Institute for Inorganic and Physical Chemistry, Baku 01698
Institute for International Relations (IMO), Zagreb 06246
Institute for Islamic Religion, Semarang 14293
Institute for Islamic Studies Ar-Raniry Centre Library, Darussalam 14248
Institute for Jewish Studies, Jerusalem 14732
Institute for Land Reclamation, Minsk 02021
Institute for Language and Folklore Research, Uppsala 26706
Institute for Legal Studies, Sofiya 03553
Institute for Legality, Law & Order Problems, Moskva 23313
Institute for Liberian Languages, Monrovia 18232
B. Verkin Institute for Low Temperature Physics and Engioneering, Kharkiv 28251
Institute for Management Development, Buenos Aires 00272
Institute for Management Education and Development, Jakarta 14370
Institute for Maritime Technology, Simon's Town 25178
Institute for Meat and Dairy Industry Planning, Almaty 17945
Institute for Mechanical Engineering, Mogilev 01811
Institute for Mechanised Construction and Rock Mining, Warszawa 21292
Institute for Medical Research (IMR), Kuala Lumpur 18586
Institute for Metal Constructions, Moskva 23518
Institute for Metals Research, Kraków 21136
Institute for Metals Superplasticity Problems of the Russian Academy of Sciences, Ufa 24028
Institute for Mine Design Joint-Stock Company, Sankt-Peterburg 22938
Institute for Motorcycle Safety e.V., Essen 11824
Institute for Newspaper Research, Dortmund 11746
Institute for Nuclear Research of the Russian Academy of Sciences, Moskva 23333
Institute for Occupational and Maritime Medicine, Hamburg 12022
Institute for occupational safety and health, Sankt Augustin 12461
Institute for Oriental Studies, Dushanbe 27643
Institute for Palestine Studies, Beirut 18221
Institute for Peace Research and Security Policy at the University of Hamburg, Hamburg 11992
Institute for Planning in the Food Industry, Toshkent 42136
Institute for Planning in the Processing of Cotton and Bast, Toshkent 42137
Institute for Planning of Food Processing Enterprises no 1, Moskva 23161
Institute for Plant Protection, Sankt-Peterburg 23856
Institute for Plant Protection and Environment, Beograd 24512
Institute for Plasma Research, Gandhinagar 13992
Institute for Problems of Cryobiology and Cryomedicine, Kharkiv 28258
Institute for Problems of Transport, Moskva 23280
Institute for Project Planning of Firms, Producing Resins, Moskva 23215
Institute for Project Planning of Metallurgical Plants, Moskva 23242
Institute for Project Planning of Nonferrous Industries, Moskva 23218
Institute for Project Planning of Shoe-, Leather- and Fur Industry, Moskva 23241
Institute for Psychoanalysis, Chicago 36676
Institute for Public Economics, Köln 12128
Institute for Research and Development of Agrobased Industry, Bogor 14345
Institute for Research and Planning in Nonferrous Metals, Toshkent 42138
Institute for Research and Planning of Housing and Public Buildings, Toshkent 42139

Institute for Research in Agricultural Economics and Organization, Almaty 17946
Institute for Research in Animal Husbandry, Kniizh 17970
Institute for Research in Balneology and Physical Therapy, Toshkent 42140
Institute for Research in Blood Transfusion, Minsk 01998
Institute for Research in Building Materials, Toshkent 42141
Institute for Research in Economics, Almaty 17947
Institute for Research in Gardening, Viticulture and Oenology, Samarkand 42119
Institute for Research in Hygiene and Occupational Pathology, Minsk 01995
Institute for Research in Hygiene, Sanitation and Occupational Illness, Toshkent 42142
Institute for Research in Natural Gas, Toshkent 42143
Institute for Research in Oncology and Radiology, Almaty 17948
Institute for Research in Physiology and Biochemistry of Animals, Lviv 28409
Institute for Research in Plant Protection, Toshkent 42144
Institute for Research in Potatoe and Vegetable Cultivation, Pervoe Maya 23696
Institute for Research in Roentgenology, Radiology and Oncology, Toshkent 42145
Institute for Research in Rural Electrification, Almaty 17949
Institute for Research in Rural Electrification and Mechanization, Yangiyul 42168
Institute for Research in Social and Human Sciences, Craiova 22028
Institute for Research in Social and Human Sciences, Iaşi 22039
Institute for Research in Social and Human Sciences, Târgu Mureş 22061
Institute for Research in Social and Human Sciences, Timişoara 22064
Institute for Research in the Coal Industry, Karaganda 17967
Institute for Research in the Cotton Industry, Toshkent 42146
Institute for Research in the Meat Industry, Semipalatinsk 17974
Institute for Research in the Silk Industry, Margilan 42117
Institute for Research in Vaccines and Sera, Toshkent 42147
Institute for Research in Water Management, Taraz 17978
Institute for Russia and Eastern Europe, Library, Helsinki 07509
Institute for Scientific and Technological Information (INSTI), Accra 13053
Institute for Scientific Information on Social Sciences of the Russian Academy of Sciences, Moskva 23293
Institute for Single Crystals, Kharkiv 28257
Institute for Social and Economic Change, Bangalore 13936
Institute for Social Planning, Ljubljana 24944
Institute for Social Research, Oslo 20246
Institute for Textile Machines, Chernigiv 28190
Institute for the Advancement of Education, Zagreb 06256
Institute for the Care of Mother and Child, Praha 06521
Institute for the Construction of Ore Mining Machines, Krivi Rig 28300
Institute for the Construction of Sugar Factories, Moskva 23216
Institute for the Construction of Textile Mills, Moskva 23217
Institute for the Design of Power Networks, Toshkent 42148
Institute for the History and Philosophy of Science, Zagreb 06255
Institute for the History of Forestry, Tokyo 17763
Institute for the History of Material Culture of the Russian Academy of Sciences, Sankt-Peterburg 23787
Institute for the Industrialization of Setting up Works, Moskva 23460
Institute for the International Education of Students, Wien 01293
Institute for the Problems of Iron-Casting, Kyiv 28345
Institute for the Promotion of Teaching Science and Technology, Bangkok 27780
Institute for the Protection of Cultural Monuments in Serbia, Beograd 24520
Institute for the Protection of the Mother and Newborn Child, Hanoi 42290
Institute for the Study of Soil, Akmola 17934

Institute for the Technology of Automobile Industry, Moskva 23435
Institute for Theology and Peace, Hamburg 11994
Institute for Transport Sciences, Budapest 13393
Institute for Vine and Wine, Yalta 28484
Institute for Water Problems, Moskva 23329
Institute for Wetland & Waterfowl Research Library, Oak Hammock Marsh 04426
Institute Keguruan dan Ilmu Pendidikan Muhammadiyah, Yogyakarta 14311
Institute Kristallografii im. A.V. Shubnikova RAN, Moskva 23336
Institute Macromolecular Chemistry Academy of Sciences of Czech Republic, Praha 06493
Institute, Museum and Gallery, Bitola 18441
Institute of Actuaries, Oxford 29869
Institute of Adegey Studies on Language, Literature and History, Maikop 23150
Institute of Adriatic Culture, Split 06223
Institute of Adult Education, Dar es Salaam 27675
Institute of Advanced Manufacturing Technology, Kraków 21141
Institute of Advanced Training for Medical Doctors, Sankt-Peterburg 22523
Institute of Aesthetic Studies, Colombo 26269
Institute of African Studies of the Russian Academy of Sciences, Moskva 23251
Institute of Agricultural and Food Biotechnology, Warszawa 21270
Institute of Agricultural and Food Economics – National Research Institute, Warszawa 20953
Institute of Agricultural Construction, Aprelevka 22986
Institute of Agricultural Economics, Beograd 24509
Institute of Agricultural Economics, Sofiya 03535
Institute of Agricultural Economics and Information – Agricultural and Food Library, Praha 06523
Institute of Agricultural Engineering, Saratov 22573
Institute of Agricultural Research, Addis Ababa 07282
Institute of Agricultural Research and Training (IART), Ibadan 20017
Institute of Agricultural Research for Development (IRAD), Buéa 03647
Institute of Agricultural Science of South Vietnam, Ho Chi Minh City 42306
Institute of Agricultural Sciences, Fenjan 18054
Institute of Agriculture, Baku 01699
Institute of Agriculture, Košice 24735
Institute of Agriculture, Kyustendil 03508
Institute of Alloys and Nonferrous Metals Processing, Moskva 23231
Institute of American Indian & Alaska Native Culture & Arts Development Library, Santa Fe 37570
Institute of Animal Husbandry, Kharkiv 28261
Institute of Animal Husbandry, Krasny Vodopad 42115
Institute of Animal Husbandry and Veterinary Science, Semipalatinsk 17917
Institute of Animal Morphology, Moskva 23292
Institute of Animal Science, Kostinbrod 03506
Institute of Animal Science of the Ukrainyan Academy of Agrarian Sciences, Kharkiv 28253
Institute of Anthropology, Poznań 21205
Institute of Anthropology of the Polish Academy of Sciences, Wrocław 21368
Institute of Applied Ecology, Shenyang 05899
Institute of Applied Manpower Research, New Delhi 14147
Institute of Applied Physics, Ulaanbaatar 18909
Institute of Arab Research and Studies, Cairo 07176
Institute Of Archaeology, Tiranë 00010
Institute of Archaeology and Ethnology, Yerevan 00351
Institute of Archaeology and Ethnology of the Polish Academy of Sciences, Warszawa 21262
Institute of Archaeology and History of Art, Cluj-Napoca 22021
Institute of Archaeology of the Academy of Sciences of the Czech Republic, Prague, v.v.i., Praha 06475
Institute of Archaeology of the AS CR, Brno 06425
Institute of Archeology, Kyiv 28314
Institute of Archeology and Ethnography of the Siberian Department of the Russian Academy of Sciences, Novosibirsk 23637
Institute of Archeology of the Russian Academy of Sciences, Moskva 23252
Institute of Architecture, Moskva 22358
Institute of Architecture and Construction, Kaunas 18296
Institute of Architecture and Construction, Minsk 02026
Institute of Architecture and Fine Arts, Baku 01700
Institute of Armament Technology (IAT), Pune 14191
Institute of Art and Culture, Chelyabinsk 22227
Institute of Art Polish Academy of Sciences, Warszawa 21242
Institute of Artificial Fibre, Mytishchi 23604
Institute of Artificial Leather, Moskva 23421
Institute of Artistic Education, Moskva 23404
Institute of Arts, Ufa 24027
Institute of Astronomy, Sofiya 03574
Institute of Astronomy of Russian Academy of Sciences, Moskva 23254
Institute of Atmospheric Physics, Beijing 05833
Institute of Atmospheric Physics of the Russian Academy of Sciences, Moskva 23268
Institute of Atmospheric Physics of the Russian Academy of Sciences, Shikhovo 23969
Institute of Atomic Energy, Otwock-Świerk 21184
Institute of Atomic Energy, Ulugbek 42167
Institute of Automation and Control Systems of the Russian Academy of Sciences, Vladivostok 24054
Institute of Automatisation and Electronic Measurement of the Siberian Department of the Russian Academy of Sciences, Novosibirsk 23638
Institute of Bast Crops, Glukhiv 28239
Institute of Bast Fibers, Poznań 21199
Institute of Biochemistry and Petrochemistry, Kyiv 28315
Institute of Biochemistry and Physiology of Plants and Microorganisms of the Russian Academy of Sciences, Saratov 23957
Institute of Biology and Immunology of Reproduction, Sofiya 03555
Institute of Biology of the Southern Seas, Sevastopol 28459
Institute of Biology – Ural Branch of the Russian Academy of Sciences, Lobytnangi 23143
Institute of Biomedical Chemistry, Moskva 23256
Institute of Biomedical Problems, Moskva 23337
Institute of Biopolymers and Chemical Fibres, Łódź 21160
Institute of Botany, Chișinău 18893
Institute of Botany, Sofiya 03536
Institute of Botany, Vilnius 18301
Institute of Botany, Slovak Academy of Sciences, Bratislava 24684
Institute of British Geographers, London 29817
Institute of Buddhist Studies Library, Berkeley 36531
Institute of Building Ceramics, Zheleznodorozhny 24096
Institute of Bulgarian Language, Sofiya 03550
Institute of Business Administration & Management, Sanno Junior College, Tokyo 17247
Institute of Business Appraisers, Plantation 37389
Institute of Byzantinology of the Serbian Academy of Sciences and Arts, Beograd 24523
Institute of Cable Industry, Moskva 23340
Institute of Catalysis and Surface Chemistry, Kraków 21134
Institute of Cattle Breeding and Sheep Breeding, Stara Zagora 03583
Institute of Cement Industry, Moskva 23439
Institute of Chartered Accountants in Australia, Sydney 00836
Institute of Chartered Accountants in England & Wales, London 29766
Institute of Chartered Accountants in Ireland, Dublin 14538
Institute of Chartered Accountants of India, Mumbai 14092
Institute of Chartered Accountants of India, New Delhi 14148
Institute of Chartered Accountants of Sri Lanka, Colombo 26312
Institute of Chemical Engineering, Almaty 17950
Institute of Chemical Engineering, Ivanovo 22263
Institute of Chemical Engineering, Moskva 23243
Institute of Chemical Engineering, Moskva 22337
Institute of Chemical Engineering of the Polish Academy of Sciences, Gliwice 20812
Institute of Chemical Fiber Processing, Moskva 23375
Institute of Chemical Kinetics and Combustion of the Siberian Department of the Russian Academy of Sciences, Novosibirsk 23643
Institute of Chemical Physics, Moskva 23278
Institute of Chemical Technology, Kazan 22277
Institute of Chemistry, Beijing 05834
Institute of Chemistry and Pharmacology, Sankt-Peterburg 22526
Institute of Chinese History, Beijing 05835
Institute of Civil and Earthquake Engineering, Tbilisi 09241
Institute of Civil Engineering, Brest 01789
Institute of Civil Engineering, Kharkiv 28262
Institute of Civil Engineering, Sankt-Peterburg 22525
Institute of Civil Engineering, Toshkent 42149
Institute of Civil Engineering, Planning of Forests and Parks and Objects of Cultural Importance, Moskva 23315
Institute of Clinical and Experimental Neurology, Tbilisi 09242
Institute of Clinical and Experimental Surgery, Moskva 23407
Institute of Coal Industry, Moskva 23380
Institute of Cognitive Sciences and Technologies, Roma 16512
Institute of Combustible Minerals, Moskva 23274
Institute of Commerce, Poltava 28063
Institute of Commerce and Economics, Lviv 28039
Institute of Commerce and Economics, Sankt-Peterburg 22547
Institute of Comparative Law, Beograd 24458
Institute of Comparative Law in Japan, Tokyo 17722
Institute of Computer Science, Iași 22041
Institute of Computer Science PAS Library and Information Centre, Warszawa 21299
Institute of Construction and Architecture – Slovak Academy of Sciences, Bratislava 24728
Institute of Construction of Packages, Kaluga 23074
Institute of Contemporary History, München 12305
Institute of Continuing Education for Teaching Professionals, Vladivostok 22638
Institute of Continuing Medical Education, Novokuznetsk 22432
Institute of Continuing Training in Medicine, Tbilisi 09198
Institute of Control Sciences, Moskva 23314
Institute of Control Systems, Tbilisi 09243
Institute of Co-operative Trade, Novosibirsk 22439
Institute of Corrosion and Protection of Metals, Shenyang 05900
Institute of Corrosion Protection, Moskva 23581
Institute of Cost and Management Accountants of Pakistan, Karachi 20496
Institute of Cotton Research, Ak-Kavak 22983
Institute of Criminology and Social Prevention, Praha 06492
Institute of Cybernetics, Baku 01701
Institute of Cybernetics, Tbilisi 09244
Institute of Cytology and Genetics of the Siberian Department of the Russian Academy of Sciences, Novosibirsk 23649
Institute of Cytology of the Russian Academy of Sciences, Sankt-Peterburg 23801
Institute of Czech Language of the AS CR – Brno-Veveří branch, Brno 06439
Institute of Czech Literature of the AS CR, Brno 06438
Institute of Czech Literature of the AS CR, Praha 06520
Institute of Dairy and Cattle Breeding, Orenburg 23671
Institute of Deep Oil and Gas Deposits, Baku 01702
Institute of Dendrology of the Polish Academy of Sciences, Kórnik 21123
Institute of Dental Research, Praha 06529
Institute of Development Management, Gaborone 02977
Institute of Development Management (IDM), Mzumbe 27659
Institute of Directors, London 29575
Institute of Dyes and Organic Products, Zgierz 21374
Institute of East Asian Philosophies, Singapore 24611
Institute of Ecology of the Polish Academy of Sciences, Łomianki 21166
Institute of Economic and Social Development, Buenos Aires 00263
Institute of Economic Growth, Delhi 13984
Institute of Economic Management, Hanoi 42291
Institute of Economic Research SAS, Bratislava 24687
Institute of Economics, Baku 01703
Institute of Economics, Kyiv 28318
Institute of Economics, Sofiya 03534
Institute of Economics, Zagreb 06234
Institute of Economics and Law, Tbilisi 09245
Institute of Economics and Mathemtics of the Russian Academy of Sciences, Sankt-Peterburg 23760
Institute of Economics of the Polish Academy of Sciences, Warszawa 21294
Institute of Economics of the Russian Academy of Sciences, Moskva 23259
Institute of Economy and Organization of Industries of the Siberian Department of the Russian Academy of Sciences, Novosibirsk 23639
Institute of Education, Bakht er Ruda 26341
Institute of Education for Adults, Sankt-Peterburg 23791
Institute of Electric Ceramics, Moskva 23388
Institute of Electrical Coal Products, Elektrougli 23039
Institute of Electrical Engineering, Novosibirsk 22435
Institute of Electrical Explosion Protected Equipment for the Mining, Gas and Oil Industries, Donetsk 28225
Institute of Electrodynamics, Kyiv 28320
Institute of Electroenergetical Engineering, Sofiya 03498
Institute of Electronic, Automated and Remote Control Equipment, Tbilisi 09246
Institute of Electronic Machine Building, Moskva 22335
Institute of Electronic Materials Technology, Gap 21092
Institute of Electronic Materials Technology, Warszawa 21309
Institute of Electronics, Beijing 05836
Institute of Electronics, Minsk 02013
Institute of Elementary Organic Compounds of the Russian Academy of Sciences, Moskva 23262
Institute of Energetics, Warszawa 21276
Institute of Energy, Moskva 23209
Institute of Engineering, Lalitpur 19023
Institute of Engineering Amelioration, Novocherkassk 22429
Institute of Engineering Cybernetics Belarussian Academy of Sciences Library, Minsk 02027
Institute of Engineering Mechanics, Harbin 05867
Institute of Environmental Chemistry, Beijing 05837
Institute of Environmental Engineering, Zabrze 21371
Institute of Epidemiology and Microbiology, Almaty 17951
Institute of Epidemiology and Microbiology, Kazan 23083
Institute of Epidemiology and Microbiology, Khabarovsk 23099
Institute of Epidemiology and Microbiology, Nizhni Novgorod 23618
Institute of Essential Oil and Medicinal Plants, Simferopol 28469
Institute of Ethnography, Moskva 23263
Institute of Ethnography and Folklore, București 21989
Institute of Ethnography SASA, Beograd 24504
Institute of Ethnology, Hanoi 42292
Institute of Ethnology and Anthropology of the Russian Academy of Sciences, Moskva 23264
Institute of Ethnology and Folklore Research, Zagreb 06245
Institute of Experimental Endocrinology and Hormone Chemistry, Moskva 23260
Institute of Experimental Medicine RAMS, Sankt-Peterburg 23804
Institute of Experimental Morphology, Tbilisi 09247
Institute of Experimental Pathology and Therapy, Sukhumi 09222
Institute of Experimental Pharmacology, SAS, Bratislava 24644
Institute of Far Eastern Research of the Russian Academy of Sciences, Moskva 23257

Institute of Fat Technology for the Food Industry, Toshkent 42150
Institute of Female Physiology and Pathology, Tbilisi 09248
Institute of Fermentation Products, Moskva 23425
Institute of Ferrous Metallurgy, Gliwice 21106
Institute of Finance and Economics, Rostov-na-Donu 22487
Institute of Finance Management, Dar es Salaam 27653
Institute of Fine Arts, Baku 01704
Institute of Fine Arts, Toshkent 42151
Institute of Food Industries, Tbilisi 09249
Institute of Foreign Literature, Beijing 05838
Institute of Forensic Research, Kraków 21130
Institute of Forest Research of the Russian Academy of Sciences, Uspenskoe 24038
Institute of Forestry of the Siberian Department of the Russian Academy of Sciences, Krasnoyarsk 23124
Institute of Foundation Construction, Moskva 23212
Institute of Freshwater Biology of the Polish Academy of Sciences, Kraków 21150
Institute of Fundamental Technical Research of the Polish Academy of Sciences, Warszawa 21300
Institute of Fur Farming and Rabbit Breeding, Udelnaya 24018
Institute of General and Inorganic Chemistry, Sofiya 03546
Institute of General and Pedagogical Psychology, Moskva 23299
Institute of General Genetics of the Russian Academy of Sciences, Moskva 23297
Institute of Genetics, Sofiya 03540
Institute of Genetics, Szeged 13505
Institute of Genetics and Animal Breeding of the Polish Academy of Sciences, Mroków 21172
Institute of Genetics and Selection, Baku 01705
Institute of Geochemistry, Guiyang 05866
Institute of Geochemistry of the Siberian Department of the Russian Academy of Sciences, Irkutsk 23050
Institute of Geodesy and Cartography, Warszawa 21279
Institute of Geography, Baku 01706
Institute of Geography, Beijing 05839
Institute of Geography, Sofiya 03531
Institute of Geography of the Russian Academy of Sciences, Moskva 23271
Institute of Geography of the Siberian Department of the Russian Academy of Sciences, Irkutsk 23049
Institute of Geography – Slovak Academy of Sciences, Bratislava 24696
Institute of Geological and Nuclear Sciences, Lower Hutt 19836
Institute of Geological and Nuclear Sciences, Lower Hutt 19837
Institute of Geological Research, Tiranë 00018
Institute of Geology, Baku 01707
Institute of Geology, Kraków 21147
Institute of Geology and Geochemistry, Ekaterinburg 23028
Institute of Geology and Geochemistry of Raw Fuel Materials, Lviv 28410
Institute of Geology and Geography, Vilnius 18304
Institute of Geology and Geophysical Research, Saratov 23962
Institute of Geology and Mineral Exploration, Athinai 13115
Institute of Geology and Petroleum Exploration, Toshkent 42152
Institute of Geophysics, Tbilisi 09250
Institute of Geophysics of the AS CR, Praha 06487
Institute of Geophysics of the Polish Academy of Sciences, Warszawa 21280
Institute of Geotechnical Mechanics, Dnipropetrovsk 28201
Institute of Geotectonics, Changsha 05857
Institute of Glass, Moskva 23234
Institute of Glass Engineering, Moskva 23553
Institute of Gold and Rare Metals, Magadan 23148
Institute of Grains and Feed Industry, Kostinbrod 03507
Institute of Heat and Mass Transfer, Minsk 02028
Institute of Heavy Organic Synthesis, Kędzierzyn-Koźle 21117
Institute of Helminthology, Moskva 22336
Institute of Hematology, Warszawa 21283
Institute of Heraldic and Genealogical Studies, Canterbury 29628

Institute of High Temperature Physics of the Russian Academy of Sciences, Moskva 23451
Institute of Higher Nervous Activity and Neurophysiology of the Russian Academy of Sciences, Moskva 23332
Institute of High-Pressure Physics of the Russian Academy of Sciences, Troitsk 24001
Institute of Historic Monuments, Bratislava 24701
Institute of Historical Studies, Slovak Academy of Sciences, Bratislava 24693
Institute of Historical Survey Foundation Library, Las Cruces 37017
Institute of History, Sofiya 03541
Institute of History, Tiranë 00016
Institute of History and Theory of Architecture, Budapest 13375
Institute of History 'N. Iorga', Bucureşti 21992
Institute of History of Natural Sciences and Technology, St. Petersburg Branch, Sankt-Peterburg 23786
Institute of History of Science of the Polish Academy of Sciences, Warszawa 21284
Institute of History of the AS CR, Brno 06428
Institute of History of the AS CR, Praha 06490
Institute of History of the Natural Sciences and Technology, Moskva 23277
Institute of History of the Polish Academy of Sciences, Warszawa 21285
Institute of History of Ukraine, Kyiv 28330
Institute of Horticulture, Moskva 23447
Institute of Horticulture and Canned Foods, Plovdiv 03515
Institute of Horticulture and Canning, Plovdiv 03513
Institute of Horticulture and Floriculture, Sochi 23976
Institute of Horticulture and Viticulture, Krasnodar 23119
Institute of Horticulture – Crimean Research Station and Professional School, Malenke 28427
Institute of Horticulture, Viticulture and Oenology, Tbilisi 09251
Institute of Horticulture, Viticulture and Subtropical Plants, Kuba 01729
Institute of Human Morphology, Moskva 23291
Institute of Hydraulic Engineering and Land Improvement, Tbilisi 09252
Institute of Hydraulics and Hydrology, Poondi 14183
Institute of Hydro Engineering and Improvement, Moskva 22330
Institute of Hydro Technology Ltd, Moskva 22820
Institute of Hydrobiology, Wuhan 05903
Institute of Hydroengineering of the Polish Academy of Sciences, Gdańsk 21096
Institute of Hydrology, Novokuznetsk 23635
Institute of Hydrology and Engineering Geology, Toshkent 42153
Institute of Hydrometeorological Engineering, Moskva 23396
Institute of Hydrometeorology, Moskva 23181
Institute of Hydroponics Problems, Yerevan 00352
Institute of Hydrotechnics and Land Improvement, Novocherkassk 23631
Institute of Hygiene, Samara 23742
Institute of Hygiene, Tbilisi 09253
Institute of Hygiene and Occupational Diseases, Ufa 24029
Institute of Immunology and Experimental Therapy of the Polish Academy of Sciences, Wrocław 21360
Institute of Indigenous Medicine, Rajagiriya 26329
Institute of Industrial Chemistry, Warszawa 21274
Institute of Industrial Construction, Moskva 23531
Institute of Industrial Design, Warszawa 21313
Institute of Industrial Economics, Donetsk 28219
Institute of Industrial Grease, Sankt-Peterburg 23857
Institute of Industrial Hygiene and Occupational Diseases, Ekaterinburg 23031
Institute of Industrial Hygiene and Occupational Diseases, Moskva 23398
Institute of Industrial Hygiene and Occupational Diseases, Nizhni Novgorod 23619
Institute of Industrial Medicine and Human Ecology, Angarsk 22985
Institute of Industrial Organic Chemistry, Warszawa 21302

Institute of Industrial Planning – Heavy Industry, Toshkent 42154
Institute of Industrial Transport Engineering, Moskva 23239
Institute of Industry and Management, Kyushu Sangyo University, Fukuoka 16927
Institute of Infectious Diseases, Ekaterinburg 23033
Institute of Influenza Research, Sankt-Peterburg 23831
Institute of Informatics, Bratislava 24717
Institute of Inland Waters Biology of the Russian Academy of Sciences, Borok 23008
Institute of Inorganic Chemistry, Gliwice 21104
Institute of Inorganic Chemistry and Electrochemistry, Tbilisi 09254
Institute of Inorganic Chemistry of the Siberian Department of the Russian Academy of Sciences, Novosibirsk 23645
Institute of International Politics and Economics, Beograd 24510
Institute of International Public Law and International Relations, Thessaloniki 13135
Institute of International Relations, Beijing 05047
Institute of International Relations, Hanoi 42293
Institute of International Relations, Moskva 22339
Institute of International Relations, Praha 06519
Institute of Introduction and Plant Genetic Resources, Sadovo 03518
Institute of Introscopy, Moskva 23403
Institute of Irrigation and Land Improvement, Toshkent 42155
Institute of Irrigation, Fruit Growing, Melitopol 28428
Institute of Islamic Religion, Surabaya 14296
Institute of Islamic Studies, Male 18640
Institute of Kabardian and Balkarian Studies, Nalchik 23608
Institute of Labour Hygiene and Occupational Diseases, Sankt-Peterburg 23828
Institute of Labour Research, Moskva 23438
Institute of Land Reclamation and Grassland Farming, Raszyn 21218
Institute of Language and Literature, Dushanbe 27644
Institute of Language, Literature and History of the Russian Academy of Sciences, Kazan 23082
Institute of Latin American Studies of the Russian Academy of Sciences, Moskva 23281
Institute of Law, Ekaterinburg 22248
Institute of Law, Sankt-Peterburg 22567
Institute of Law Studies of the Polish Academy of Sciences, Warszawa 21295
Institute of Legal Studies and Comparative Law, Moskva 23334
Institute of Light and Clothing Industry, Moskva 23253
Institute of Limnology of the Russian Academy of Sciences, Sankt-Peterburg 23792
Institute of Limnology of the Siberian Department of the Russian Academy of Sciences, Listvenichnoe 23142
Institute of Linguistic Studies, Sankt-Peterburg 23789
Institute of Linguistics, Chişinău 18894
Institute of Literary Research of the Polish Academy of Sciences, Warszawa 21264
Institute of Literature, Hanoi 42275
Institute of Literature, Minsk 02019
Institute of Literature, Sofiya 03552
Institute of Lithuanian Language, Vilnius 18299
Institute of Lithuanian Literature and Folklore, Vilnius 18299
Institute of Livestock Breeding in Forest-Steppe and Wood-Lands, Kharkiv 28253
Institute of Logistics and Warehousing, Poznań 21196
Institute of Low Temperature and Structure Research, Wrocław 21361
Institute of Machine Building, Kharkiv 28259
Institute of Machine Engineering Trade, Moskva 23376
Institute of Machine Reliability, Minsk 02025
Institute of Machine-Tool Industry, Moskva 23501
Institute of Machinery and Industrial Instruments, Hanoi 42294
Institute of Macromolecular Chemistry AS ČR, Praha 06493

Institute of Macro-Molecular Compounds of the Russian Academy of Sciences, Sankt-Peterburg 23803
Institute of Management, Kolkata 14049
Institute of Management Accountants, Inc., Montvale 37129
Institute of Management and Technology, Enugu 19937
Institute of Management & Training, Addis Ababa 07283
Institute of Manuscripts, Baku 01708
Institute of Marine Affairs, Chaguaramas 27811
Institute of Marine Biology of the Russian Academy of Sciences, Vladivostok 24055
Institute of Marine Geology and Geophysics of the Russian Academy of Sciences, Yuzhno-Sakhalinsk 24094
Institute of Marine Research, Bergen 20213
Institute of Marine Sciences and Management, Istanbul 27885
Institute of Maritime and Tropical Medicine, Gdynia 20811
Institute of Materials and Machine Mechanics of SAS, Bratislava 24694
Institute of Materials, Minerals and Mining, London 29767
Institute of Mathematical Machines, Warszawa 21289
Institute of Mathematical Sciences, Chennai 13962
Institute of Mathematics, Hanoi 42295
Institute of Mathematics, Minsk 02020
Institute of Mathematics, Toshkent 42092
Institute of Mathematics and Informatics / Matematikos ir Informatikos Institutas, Vilnius 18311
Institute of Mathematics and Mechanics, Baku 01709
Institute of Mathematics and Mechanics of the Ural branch of RAS, Ekaterinburg 23030
Institute of Mathematics of the Polish Academy of Sciences, Warszawa 21290
Institute of Mathematics of the Serbian Academy of Sciences and Arts, Beograd 24515
Institute of Mathematics of the Siberian Department of the Russian Academy of Sciences, Novosibirsk 23644
Institute of Mathematics with Computing Centre, Sofiya 03542
Institute of Measurement Science, Slovak Academy of Sciences, Bratislava 24726
Institute of Mechanical and Electrical Engineering, Sofiya 03547
Institute of Mechanical Engineering, Rostov-na-Donu 23722
Institute of Mechanical Engineering and Synthetic Materials, Tbilisi 09255
Institute of Mechanical Engineering for Agricultural Industry, Gomel 01854
Institute of Mechanical Engineering Technology, Elektrostal 22791
Institute of Mechanics, Hanoi 42296
Institute of Mechanics, Kyiv 28334
Institute of Mechanization in Animal Husbandry, Zaporizhzhya 28485
Institute of Medical and Veterinary Science, Adelaide 00851
Institute of Medical Instruments Engineering, Moskva 23409
Institute of Medical Parasitology and Tropical Medicine, Tbilisi 09256
Institute of Mediterranean and Oriental Cultures, Polish Academy of Sciences, Warszawa 21286
Institute of Metal Constructions, Ljubljana 24811
Institute of Metal Physics, Ekaterinburg 23027
Institute of Metal Research, Shenyang 05901
Institute of Metal Science, Sofiya 03543
Institute of Metallurgy and Ore Enrichment, Almaty 17952
Institute of Meteorology, Lisboa 21796
Institute of Meteorology, Moskva 23397
Institute of Meteorology, Sankt-Peterburg 22516
Institute of Meteorology and Hydrology, Ulaanbaatar 18910
Institute of Meteorology and Water Management, Warszawa 21293
Institute of Metrological Services, Moskva 23411
Institute of Microbial Technology, Chandigarh 13952
Institute of Microbiology, Sofiya 03544
Institute of Microbiology of the Russian Academy of Sciences, Moskva 23285
Institute of Migration, Turku 07547

Institute of Mineral Resources, Moskva 23412
Institute of Mineralogy, Geochemistry and Crystallochemistry of Rare Elements of the Russian Academy of Sciences, Moskva 23286
Institute of Mining and Metallurgy, Hanoi 42297
Institute of Mining Chemicals, Lyubertsy 23146
Institute of Mining Engineering, Prospecting and Physics of Explosives, Tbilisi 09257
Institute of Molecular Physics of the Polish Academy of Sciences, Poznań 21194
Institute of Musicology, Budapest 13429
Institute of National Economy, Ekaterinburg 22237
Institute of National Economy, Novosibirsk 22440
Institute of National Economy, Rostov-na-Donu 22488
Institute of National History, Skopje 18442
Institute of National Planning, Cairo 07177
Institute of National Politics, Kyiv 28340
Institute of Nationality Studies, Beijing 05840
Institute of Natural Diamants and Instruments, Moskva 23443
Institute of Natural Fibres and Medicinal Plants, Poznań 21201
Institute of Natural Gas, Kyiv 28326
Institute of Natural Gas, Information Centre, Donetsk 28221
Institute of Nature Conservation of the Polish Academy of Sciences, Kraków 21137
Institute of Near and Middle Eastern Nations, Baku 01710
Institute of Neurology, Moskva 23416
Institute of Neurology and Psychiatry, Bucureşti 21995
Institute of Neurology, Psychiatry and Neurosurgery, Sofiya 03545
Institute of Neurosurgery named after A.P. Romodanov, Kyiv 28342
Institute of non Ore-bearing Building Materials and Hydromechanics, Tolyatti 23994
Institute of Non-Ferrous Metals, Gliwice 21105
Institute of Non-Ferrous Metals, Moskva 23235
Institute of North Ossetian Studies, Vladikavkaz 24046
Institute of Nuclear Chemistry and Technology, Warszawa 21272
Institute of Nuclear Physics of the Siberian Department of the Russian Academy of Sciences, Novosibirsk 23650
Institute of Nuclear Science ù Vinčaò, Beograd 24511
Institute of Nutrition, Moskva 23304
Institute of Occupational Medicine and Environmental Health, Sosnowiec 21225
Institute of Occupational Physiology and Hygiene, Karaganda 17968
Institute of Oceanography, Nha-Trang 42308
Institute of Oceanography and Fisheries, Split 06224
Institute of Oceanology, Qingdao 05885
Institute of Oil Processing and Petrochemistry, Moskva 23554
Institute of Oil Processing and Petrochemistry, Moskva 23379
Institute of Oil Processing – Reference and Information Fund, Ufa 24022
Institute of Oil-Producing Crops, Krasnodar 23114
Institute of Oncology, Bucureşti 22002
Institute of Oncology, Ljubljana 24928
Institute of Oncology, Sofiya 03569
Institute of Optics and Electronics, Chengdu 05858
Institute of Organic Chemistry, Kyiv 28343
Institute of Organic Chemistry of the Polish Academy of Sciences, Warszawa 21273
Institute of Organic Chemistry of the Siberian Department of the Russian Academy of Sciences, Irkutsk 23051
Institute of Organic Synthesis, Moskva 23576
Institute of Organization and Management in Industry 'Orgmasz', Warszawa 21297
Institute of Oriental Studies of the Russian Academy of Sciences, Moskva 23330
Institute of Oriental Studies of the Russian Academy of Sciences, Sankt-Peterburg 23802
Institute of Otolaryngology, Moskva 23236
Institute of Paleobiology of the Polish Academy of Sciences, Warszawa 21298
Institute of Papua New Guinea Studies, Boroko 20586
Institute of Parasitology, Warszawa 21344
Institute of Peat, Tver 24013
Institute of Pediatrics, Tbilisi 09258
Institute of Pediatrics and Child Surgery, Moskva 23358
Institute of Petrochemistry, Moskva 22340
Institute of Petroleum and Gas, Kraków 21133
Institute of Petroleum Engineering, Baku 01711
Institute of Petroleum Processing, Kraków 21140
Institute of Pharmacological Chemistry, Tbilisi 09259
Institute of Pharmacology and Toxicology, Kyiv 28369
Institute of Pharmacology of the Polish Academy of Sciences, Kraków 21131
Institute of Philosophical Sciences, Sofiya 03537
Institute of Philosophy, Hanoi 42298
Institute of Philosophy, Tbilisi 09260
Institute of Philosophy, Yerevan 00353
Institute of Philosophy and Law, Minsk 02014
Institute of Philosophy of the Academy of Sciences of the Czech Republic, Praha 06496
Institute of Philosophy of the Russian Academy of Sciences, Moskva 23265
Institute of Photographic Chemistry, Beijing 05841
Institute of Physical and Chemical Research (RIKEN), Wako 17815
Institute of Physical and Technical Research, Nizhni Novgorod 23616
Institute of Physical Chemistry of the Polish Academy of Sciences, Warszawa 21271
Institute of Physical Chemistry of the Russian Academy of Sciences, Moskva 23267
Institute of Physical Culture, Krasnodar 22298
Institute of Physical Culture, Omsk 22451
Institute of Physical Culture, Tbilisi 09199
Institute of Physical Culture, Volgograd 22640
Institute of Physical, Technical and Radio-Technical Measurements, Mendeleevo 23154
Institute of Physics, Baku 01712
Institute of Physics, Kyiv 28323
Institute of Physics, Minsk 02016
Institute of Physics, Tbilisi 09261
Institute of Physics and Mechanics of Rocks, Bishkek 18146
Institute of Physics and Technology, Dolgoprudny 22234
Institute of Physics of Materials of the AS CR, Brno 06437
Institute of Physics of the AS CR, Praha 06486
Institute of Physics of the Polish Academy of Sciences, Warszawa 21278
Institute of Physics – Physical and Technical Institute, Mogilev 02060
Institute of Physiology, Sofiya 03538
Institute of Physiology of Domestic Animals, Borovsk 23009
Institute of Physiotherapy and Balneology, Ekaterinburg 23029
Institute of Physiotherapy and Spa Treatment, Tbilisi 09262
Institute of Plague Research, Stavropol 23984
Institute of Plant and Animal Ecology of the Russian Academy of Sciences, Ekaterinburg 22235
Institute of Plant Genetics of the Polish Academy of Sciences, Poznań 21195
Institute of Plant Physiology, Chişinău 18895
Institute of Plant Physiology, Sofiya 03539
Institute of Plant Protection, Gyanzha 01727
Institute of Plant Protection, Poznań 21198
Institute of Plant Protection, Puławy 21213
Institute of Plastic Surgery, Beijing 05842
Institute of Polio and Viral Encephalitits, Moskva 23306
Institute of Political History, Budapest 13455
Institute of Political Science of Bordeaux, Pessac 07995
Institute of Political Studies, Warszawa 21305
Institute of Politics and Union Movements of the Russian Academy of Sciences, Moskva 23324
Institute of Polymer Materials and Dyes Engineering, Local Branch of Rubbers and Vinyl Plastics, Oświęcim 21182
Institute of Pomology and Floriculture, Skierniewice 21221
Institute of Population Problems, Tokyo 17683
Institute of Post Graduate Medical Education and Reserch, Kolkata 14050
Institute of Postgraduate Teacher Education, Sofiya 03554
Institute of Postgraduate Teacher Training, Toshkent 42093
Institute of Poultry-Industry, Moskva 23427
Institute of Power Engineering Problems, Minsk 02030
Institute of Power Manufacture, Toshkent 42156
Institute of Precambrian Geology and Geochronology (IPGG RAS), Sankt-Peterburg 23784
Institute of Precision Mechanics, Warszawa 21291
Institute of Precision Mechanics and Computing Equipment of the Russian Academy of Sciences, Moskva 23327
Institute of Preventive and Clinical Medicine, Bratislava 24695
Institute of Preventive Medicine, Tallinn 07238
Institute of Price Formation, Moskva 23422
Institute of Privatisation Problems, Sankt-Peterburg 23795
Institute of Problem Industry, Moskva 23338
Institute of Problems of Mechanics of the Russian Academy of Sciences, Moskva 23312
Institute of Problems on Nature Management and Ecology, Dnipropetrovsk 28203
Institute of Process Engineering, Beijing 05843
Institute of Project Planning of Firms, Producing Organical Semiproducts and Colours, Moskva 23213
Institute of Project Planning of Higher Schools, Moskva 23244
Institute of Project Planning of Telecommunication Equipment, Moskva 23246
Institute of Psychiatry, Moskva 23316
Institute of Psychoanalysis, London 29768
Institute of Psychology, Beijing 05844
Institute of Psychology, Tbilisi 09263
Institute of Psychology of the Russian Academy of Sciences, Moskva 23317
Institute of Public Administration, Blantyre 18485
Institute of Public Administration, Cairo 07178
Institute of Public Administration, Dublin 14539
Institute of Public Administration, Patna 14178
Institute of Public Administration, Riyadh 24401
Institute of Public Administration, Ruwi 20421
Institute of Public Administration and Management, Yangon 19004
Institute of Public Enterprise, Hyderabad 14004
Institute of Public Health, Wako 17814
Institute of Public Health and Medical Research, Iaşi 22043
Institute of Public Health and Medical Research, Târgu Mureş 22062
Institute of Public Health and Medical Research, Timişoara 22065
Institute of Public Health of Serbia, Beograd 24527
Institute of Pulse Research and Engineering, Mykolaiv 28429
Institute of Radio Astronomy, Kharkiv 28277
Institute of Radio Engineering, Ryazan 22496
Institute of Radio Engineering and Electronics of the Russian Academy of Sciences, Fryazino 23042
Institute of Radio Engineering and Electronics of the Russian Academy of Sciences, Moskva 23318
Institute of Radiology and Oncology, Rostov-na-Donu 23721
Institute of Radiophysics and Electronics, Kolkata 14051
Institute of Radiophysics and Electronics (IRPhE), Ashtarak 00339
Institute of Railway Engineering, Irkutsk 22254
Institute of Railway Engineers, Almaty 17953
Institute of Railway Engineers, Khabarovsk 22285
Institute of Railway Engineers, Sankt-Peterburg 22520
Institute of Refractory Metals and Hard Alloys, Moskva 23585
Institute of Research and Planning in Management Organization and Planning, Minsk 02058
Institute of Research on Mining Technology, Hanoi 42299
Institute of Rheumatism, Moskva 23319
Institute of Rice, Krasnodar 23115
Institute of Road Engineering, Toshkent 42157
Institute of Rock Structure and Mechanics of the AS CR, Praha 06522
Institute of Roentgenology and Radiology, Moskva 23467
Institute of Rural Hygiene, Saratov 23959
Institute of Russian History of the Russian Academy of Sciences, Moskva 23320
Institute of Russian History of the Russian Academy of Sciences, Sankt-Peterburg 23797
Institute of Russian Language and Institute of Linguistics, Moskva 23321
Institute of Russian Language and Literature, Toshkent 42094
Institute of Russian Literature of the Russian Academy of Sciences, Sankt-Peterburg 23798
Institute of Safety in the Mining Industry, Kemerovo 23092
Institute of Sanitary Equipment, Moskva 23483
Institute of Sanitation Technology, Moskva 23432
Institute of Science, Mumbai 14093
Institute of Scientific and Technical Information of China (ISTIC), Beijing 05845
Institute of Scientific and Technical Information of Shanghai (ISTIS), Shanghai 05887
Institute of Scientific Research, Batumi 09210
Institute of Scientific Research of the Soviet of the Chuvash Republic, Cheboksary 23016
Institute of Seismology, Toshkent 42158
Institute of Semiconductor Physics of the Siberian Department of the Russian Academy of Sciences, Novosibirsk 22436
Institute of Setting up and Special Building Works, Moskva 23461
Institute of Sewing Industry, Moskva 23534
Institute of Sheep and Goat Breeding, Stavropol 23983
Institute of Slavic Studies, Warszawa 21304
Institute of Slovak Literature / Institute of World Literature, Bratislava 24727
Institute of Social Science Information, Hanoi 42276
Institute of Sociology, Beijing 05846
Institute of Sociology, Bratislava 24709
Institute of Sociology, Sofiya 03548
Institute of Sociology of the Russian Academy of Sciences, Moskva 23323
Institute of Sociology of the Russian Academy of Sciences – St. Petersburg Branch, Sankt-Peterburg 23799
Institute of Soil Science, Nanjing 05882
Institute of Soil Sciences and Agricultural Chemistry, Baku 01713
Institute of Soldering Processing, Warszawa 21288
Institute of Solid State and Semiconductor Physics, Minsk 02017
Institute of Southeast Asian Studies, Singapore 24612
Institute of Spa Treatment and Physiotherapy, Moskva 23527
Institute of Spa Treatment and Physiotherapy, Sochi 23977
Institute of Spa Treatment and Physiotherapy, Tomsk 23996
Institute of Space and Astronautical Science, Sagamihara 17646
Institute of Spatial Development, Brno 06440
Institute of Spatial Management and Housing, Warszawa 21281
Institute of Spectroscopy of the RAS, Troitsk 24002
Institute of Standard and Experimental Projecting, Moskva 23437
Institute of State and Law Sciences, Moskva 23275
Institute of Statistical Mathematics, Tokyo 17760
Institute of Steel and Alloys, Moskva 22341
Institute of Stomatology of AMS of Ukraine, Odesa 28437
Institute of Strategic and International Studies (ISIS) Malaysia, Kuala Lumpur 18587
Institute of Strategic Studies, Islamabad 20478
Institute of Studies on Khakas Language, Literature and History, Abakan 22981
Institute of Studies on Tuvin Language, Literature and History, Kyzyl 23139
O. O. Chuyko Institute of Surface Chemistry, Kyiv 28312
Institute of Synthetic Resins, Vladimir 24048
Institute of Synthetic Rubber, Yaroslavl 24087

Institute of Systematic and Evolution of Animals – Polish Academy of Sciences, Kraków 21139
Institute of TAFE, Burnie 00406
Institute of TAFE, Devonport 00433
Institute of Teacher Training and Pedagogy, Klaten 14267
Institute of Teacher Training and Pedagogy, Medan 14278
Institute of Teacher Training and Pedagogy, Padang 14283
Institute of Teacher Training and Pedagogy, Yogyakarta 14309
Institute of Technical Acoustics, Vitebsk 02079
Institute of Technical and Chemical Public Services, Moskva 23446
Institute of Technical and Economical Research, Moskva 23434
Institute of Technical Education, Singapore 24613
Institute of Technology, Kashintsevo 23079
Institute of Technology, Sankt-Peterburg 22545
Institute of Technology, Tralee 14494
Institute of Technology Carlow, Carlow 14460
Institute of Technology, Sligo, Sligo 14491
Institute of Technology Tallaght, Dublin 14470
Institute of Telecommunication of the Russian Academy of Sciences, Moskva 23303
Institute of Telecommunications and Radio Engineering, Warszawa 21310
Institute of Terrestrial Magnetism, Ionosphere and Radio Wave Propagation of the Russian Academy of Sciences, Troitsk 24003
Institute of Testing Tractors and other Agricultural Machines, Novokubansk 23632
Institute of the Economics of Civil Engineering, Moskva 23386
Institute of the History of Natural Sciences, Beijing 05847
Institute of the Ionosphere, Almaty 17954
Institute of the Sea Fleet, Sankt-Peterburg 23920
Institute of the Simulation Problems in Power engineering, Kyiv 28346
Institute of the Tea Industry, Macharadze 09214
Institute of Theater and Stage Design, Moskva 23219
Institute of Theatrical Art and Art Institute, Toshkent 42095
Institute of Theoretical and Applied Mechanics of the Siberian Department of the Russian Academy of Sciences, Novosibirsk 23647
Institute of Theoretical Astronomy of the Russian Academy of Sciences, Sankt-Peterburg 23800
Institute of Theoretical Physics, Beijing 05848
Institute of thermal insulation, Vilnius 18313
Institute of Thermal Physics of the Siberian Department of the Russian Academy of Sciences, Novosibirsk 23648
Institute of Timber Engineering, Moskva 23384
Institute of Tinning Industry and Special Food Technology, Vidnoe 24042
Institute of Tourism, Warszawa 21312
Institute of Tractors, Moskva 23245
Institute of Transpersonal Psychology, Palo Alto 37322
Institute of Transport Economics, Norwegian Center for Transport Research, Oslo 20286
Institute of Transport Research, Moskva 23238
Institute of Traumatology and Orthopaedics, Ekaterinburg 23032
Institute of Traumatology and Orthopaedics, Kazan 23084
Institute of Traumatology and Orthopaedics, Novosibirsk 23652
Institute of Traumatology and Orthopaedics, Saratov 23960
Institute of Traumatology and Orthopaedies, Sankt-Peterburg 23853
Institute of Tropical Medicine Antwerp, Antwerpen 02515
Institute of Tuberculosis and Pulmonary Diseases, Warszawa 21282
Institute of Ukrainian Studies im. I. Kripyakevich, Lviv 28408
Institute of Underground Engineering, Vorkuta 24075
Institute of Urban Planning, Moskva 23200
Institute of USA and Canada Studies, Moskva 23325
Institute of Vaccines and Sera, Perm 23688
Institute of Vaccines and Sera, Tomsk 23997
Institute of Vegetable and Melon Growing, Selektsionny 28458
Institute of Vertebrate Biology of the AS CR, Brno 06436
Institute of Veterinary Medicine, Sankt-Peterburg 22551
Institute of Volcanology and Seismology (Far Eastern Branch of the Russian Academy of Sciences), Petropavlovsk-Kamchatski 23698
Institute of Water Transport, Moskva 22395
Institute of Water Transport Engineers, Novosibirsk 22438
Institute of Welding, Gliwice 21107
Institute of Welding, Ljubljana 24921
Institute of Wood-Working, Pazardjik 03509
Institute of World Economy, Bucureşti 21987
Institute of World Economy and International Relations, Moskva 23287
Institute of World Economy and Politics, Kyiv 28348
Institute of World Literature, Bratislava 24727
Institute of World Literature – Department of Gorky's Creative Art, Moskva 23288
Institute of World Literature of the Russian Academy of Sciences, Moskva 23289
Institute of World Religions, Beijing 05849
Institute of Zoology, Baku 01714
Institute of Zoology, Sofiya 03549
Institutet för Livsmedel och Bioteknik AB, Göteborg 26581
Institutet för mänskliga rättigheter, Turku 07548
Institutet för språk- och folkminnen, Uppsala 26706
Instituti i Arkeologjisë, Tiranë 00010
Instituti i Historisë, Tiranë 00016
Instituti i Kërkimeve Bujqësore Lushnje, Lushnje 00014
Instituti i Studimeve dhe Projektimeve të Minierave, Tiranë 00017
Instituti Studimeve e Projektimeve të Gjeologjisë, Tiranë 00018
Institution Mila and Fontanals, Barcelona 25905
Institution of Civil Engineers, London 29769
Institution of Engineering and Technology, London 29770
Institution of Engineers, Sri Lanka, Colombo 26313
Institution of Gas Engineers and Managers (IGEM), Loughborough 29846
Institution of Mechanical Engineers, London 29771
Institution of Structural Engineers, London 29772
Institution of Turkish Language, Kavaklýdere/Ankara 27900
Instituto 'Adolfo Lutz', São Paulo 03407
Instituto Agronómico, Campinas 03289
Instituto Agronómico do Paraná, Londrina 03304
Instituto 'Alberto Mesquita de Camargo', São Paulo 03208
Instituto Amazónico de Investigaciones Científicas -SINCHI-, Santafé de Bogotá 06027
Instituto Andaluz del Deporte, Málaga 26102
Instituto Anglo Mexicano de Cultura, México 18848
Instituto Argentino de Normalización Certificación (IRAM), Buenos Aires 00260
Instituto Arqueológico Alemán, Madrid 26011
Instituto Arqueológico, Histórico e Geográfico Pernambucano, Recife 03323
Instituto Artigas del Servicio Exterior, Montevideo 42056
Instituto Biológico, São Paulo 03408
Instituto Boliviano de Cultura, La Paz 02936
Instituto Boliviano de Estudio y Acción Social, La Paz 02937
Instituto Boliviano de Tecnología Agropecuaria, La Paz 02938
Instituto Brasileiro de Estudos e Pesquisas de Gastroenterologia (IBEPEGE), São Paulo 03409
Instituto Brasileiro de Geografia e Estatística (IBGE), Rio de Janeiro 03350
Instituto Brasileiro do Café, Rio de Janeiro 03351
Instituto Brasileiro do Meio Ambiente e dos Recursos Naturais Renovaveis, Brasília 03281
Instituto Brasil-Estados Unidos, Rio de Janeiro 03352
Instituto Británico, Sevilla 26152
Instituto Butantan, São Paulo 03410
Instituto Calasanz de Ciencias de la Educación, Madrid 26034
Instituto Caro y Cuervo, Santafé de Bogotá 06028
Institut Català de la Salut, Barcelona 25917
Instituto Centroamericana de Administración de Empresas (INCAE), Managua 19896
Instituto Centroamericano de Administración Pública (ICAP), San José 06121
Instituto Centroamericano de Aministración de Empresas (INCAE), Alajuela 06119
Instituto Cervantes, Athinai 13114
Instituto Cervantes, Berlin 11536
Instituto Cervantès, Bordeaux 08306
Instituto Cervantes, Bremen 11680
Instituto Cervantes, Brussel 02597
Instituto Cervantes, Cairo 07179
Instituto Cervantes, Dublin 14540
Instituto Cervantes, Lisboa 21791
Instituto Cervantes, London 29773
Instituto Cervantes, München 12306
Instituto Cervantes, Paris 08575
Instituto Cervantes, Rabat 18962
Instituto Cervantes, Tanger 18968
Instituto Cervantes, Wien 01610
Instituto Cervantes Hamburg, Hamburg 11995
Instituto Chileno Británico de Cultura, Santiago 04929
Instituto Chileno Norteamericano de Cultura, Antofagasta 04915
Instituto Chileno-Francés de Cultura, Santiago 04930
Instituto Chileno-Norteamericano de Cultura, Santiago 04931
Instituto Colombiano Agropecuario, Santafé de Bogotá 06023
Instituto Colombiano de Antropología e Historia, Santafé de Bogotá 06029
Instituto Colombiano de Normas Técnicas y Certificación (ICONTEC), Santafé de Bogotá 06030
Instituto Colombiano para el Desarrollo de la Ciencia y Tecnología 'Francisco José de Caldas' (Colciencias), Santafé de Bogotá 06031
Instituto Colombiano para el Fomento de la Educación Superior (ICFES), Santafé de Bogotá 06032
Instituto Costarricense de Electricidad (ICE), San José 06122
Instituto Cultural Argentino Norte-Americano (ICANA), Buenos Aires 00261
Instituto Cultural Domínico-Americano, Santo Domingo 06960
Instituto Cultural Peruano Norteamericano, Lima 20662
Instituto da Defesa Nacional, Lisboa 21792
Instituto da Vinha e do Vinho, Lisboa 21793
Instituto de Altos Estudos Militares, Lisboa 21656
Instituto de Antropología e Historia, Guatemala City 13181
Instituto de Astrofísica de Canarias, La Laguna 25964
Instituto de Biología y Medicina Experimental, Buenos Aires 00262
Instituto de Botânica, São Paulo 03411
Instituto de Botánica Darwinion, San Isidro 00312
Instituto de Ciencias de la Comunicación, Lima 20663
Instituto de Ciencias de la Construcción Eduardo Torroja, Madrid 26035
Instituto de Cultura Religiosa Superior, Buenos Aires 00221
Instituto de Cultura Uruguayo – Brasileño, Montevideo 42057
Instituto de Desarrollo Económico y Social, Buenos Aires 00263
Instituto de Documentación e Información Científica y Técnica (IDICT), La Habana 06300
Instituto de Ecologia AC, Jalapa 18828
Instituto de Economía y Geografía, Madrid 26036
Instituto de Educação, Rio de Janeiro 03353
Instituto de Estudios Albacetenses, Albacete 25868
Instituto de Estudios Altoaragoneses, Huesca 25962
Instituto de Estudios Fiscales, Madrid 26037
Instituto de Estudios Galegos "Padre Sarmiento", Santiago de Compostela 26142
Instituto de Estudios Ilerdenses, Lleida 25968
Instituto de Estudios Laborales y de la Seguridad Social, Madrid 26038
Instituto de Estudios Pedagógicos Somosaguas (IEPS), Madrid 26039
Instituto de Estudios Riojanos, Logroño 25969
Instituto de Estudios Superiores de la Empresa (IESE), Barcelona 25295
Instituto de Estudios Turísticos, Madrid 26040
Instituto de Estudios Turolenses, Teruel 26163
Instituto de Filosofía, Madrid 26041
Instituto de Fomento Pesquero (IFOP), Valparaíso 04947
Instituto de Formación Docente de Trinidad, Trinidad 42067
Instituto de Geología, México 18849
Instituto de Historia de Nicaragua y Centroamérica, Managua 19897
Instituto de Historia Ibérica y Latinoamericana, Köln 10154
Instituto de Historia y Cultura Militar, Madrid 25698
Instituto de Idiomas Padres de Maryknoll, Cochabamba 02930
Instituto de Información y Documentación en Ciencias Sociales y Humanidades, Madrid 26042
Instituto de Informática do Ministério das Finanças, Alfragide 21752
Instituto de Inovação Educacional, Lisboa 21794
Instituto de Investigação Científica Tropical, Lisboa 21795
Instituto de Investigaciones Agropecuarias, Santiago 04932
Instituto de Investigaciones Bibliográficas (IIB), México 18707
Instituto de Investigaciones Biológicas 'Clemente Estable', Montevideo 42058
Instituto de Investigaciones Eléctricas, Cuernavaca 18825
Instituto de Investigaciones en Viandas Tropicales, Villa Clara 06313
Instituto de Investigaciones Forestales, La Habana 06301
Instituto de Investigaciones Historicas, México 18850
Instituto de Investigaciones Marinas y Costeras, Santa Marta 06017
Instituto de la Grasa, Sevilla 26153
Instituto de la Juventud (INJUVE), Madrid 26043
Instituto de la Mujer, Madrid 26044
Instituto de Lenguas y Literaturas del Mediterráneo y Oriente Próximo, Madrid 26045
Instituto de Literatura y Linguística, La Habana 06302
Instituto de Mayores y Servicios Sociales, Madrid 25717
Instituto de Meteorología, La Habana 06303
Instituto de Meteorologia, i. P., Lisboa 21796
Instituto de Nutrición de Centro América y Panamá (INCAP), Guatemala City 13182
Instituto de Perfeccionamiento Educacional de Camaguey, Camaguey 06287
Instituto de Perfeccionamiento Educacional Provincial Marianao, La Habana 06304
Instituto de Pesquisa, Rio de Janeiro 03354
Instituto de Pesquisa Econômica Aplicada, Brasília 03282
Instituto de Pesquisas Hidroviárias, Rio de Janeiro 03355
Instituto de Pesquisas Jardim Botânico do Rio de Janeiro, Rio de Janeiro 03356
Instituto de Pesquisas Tecnológicas do Estado de São Paulo S.A. (IPT), São Paulo 03412
Instituto de Profesores Artigas, Montevideo 42011
Instituto de Promacão Ambientval, Lisboa 21731
Instituto de Salud Carlos III (Sede Majadahonda), Majadahonda 26101
Instituto de Segunda Enseñanza No. 1 de La Habana, La Habana 06279
Instituto de Seguros de Portugal, Lisboa 21797
Instituto de Sociología, Buenos Aires 00264
Instituto de Sociología Boliviana (ISBO), Sucre 02941
Instituto de Soldadura e Qualidade, Oeiras 21693
Instituto de Tecnologia do Paraná, Curitiba 03295
Instituto de Teología de la Vida Religiosa, Madrid 25794
Instituto de Valencia de Don Juán, Madrid 26046
Instituto de Zootecnia, Nova Odessa 03311
Instituto del Cemento Portland Argentino, Buenos Aires 00265
Instituto del Frío, Madrid 26047
Instituto del Mar del Perú (IMARPE), Callao 20651
Instituto do Ceará, Fortaleza 03298
Instituto do Desenvolvimento Economico-Social do Pará, Belém 03273
Instituto do Patrimônio Histórico e Artístico Nacional (IPHAN), Rio de Janeiro 03232
Instituto dos Advogados Brasileiros, Rio de Janeiro 03357

Instituto Economia Agrícola, São Paulo 03413
Instituto Ecuatoriano de Ciencias Naturales, Quito 06991
Instituto Educacional de Ensino Superior, São Paulo 03209
Instituto Educacional 'Oswaldo Quirino', São Paulo 03210
Instituto Educacional Seminário Paulopolitano, São Paulo 03211
Instituto Educacional Teresa Martin, São Paulo 03146
Instituto Egipcio de Estudios Islàmicos, Madrid 26048
Instituto Español de Comercio Exterior (ICEX), Madrid 26049
Instituto Español de Misiones Extranjeras, Madrid 25795
Instituto Español de Oceanografía, Madrid 26050
Instituto Español en Lisboa, Lisboa 21798
Instituto Evandro Chagas, Belém 03274
Instituto Financiamento e Apoio ao Desenvolvimento da Agricultura e Pescas (IFADAP), Lisboa 21799
Instituto Francés de América Latina, México 18847
Instituto Francés de Estudios Andinos (IFEA), Lima 20664
Instituto Geográfico Agustín Codazzi (IGAC), Santafé de Bogotá 06033
Instituto Geográfico e Histórico da Bahia, Salvador 03379
Instituto Geográfico e Histórico do Amazonas (IGHA), Manaus 03307
Instituto Geográfico Militar, Buenos Aires 00266
Instituto Geográfico Militar, Quito 06992
Instituto Geográfico Nacional, Madrid 26051
Instituto Geológico, São Paulo 03414
Instituto Geológico Minero y Metalúrgico, Lima 20665
Instituto Geológico y Minero de España, Madrid 26052
Instituto Guatemalteco Americano, Guatemala City 13183
Instituto Gulbenkian de Ciência, Oeiras 21819
Instituto Hidrográfico, Lisboa 21800
Instituto Histórico de Alagoas, Maceió 03306
Instituto Histórico de la Marina / Museo Naval, Madrid 26053
Instituto Histórico e Geográfico Brasileiro, Rio de Janeiro 03358
Instituto Histórico e Geográfico de Goiás, Goiânia 03299
Instituto Histórico e Geográfico de São Paulo, São Paulo 03415
Instituto Histórico e Geográfico de Sergipe, Aracaju 03267
Instituto Histórico e Geográfico do Rio Grande do Norte, Natal 03309
Instituto Histórico e Geográfico Paraíbano, João Pessoa 03303
Instituto Hondureño de Antropología e Historia, Tegucigalpa 13211
Instituto Hondureño de Cultura Interamericana (IHCI), Tegucigalpa 13212
Instituto Indigenista Interamericano, México 18851
Instituto Interamericano de Cooperación para la Agricultura (IICA), Santafé de Bogotá 06034
Instituto Interamericano del Niño, Montevideo 42059
Instituto Internacional, Madrid 26054
Instituto Internacional en España, Madrid 26055
Instituto Latinoamericano de Investigaciones Sociales (ILDIS), Quito 06993
Instituto Latinoamericano y del Caribe de Planificación y Social (ILPES), Santiago 04933
Instituto Martins-Staden, São Paulo 03416
Instituto Mauá de Tecnologia, São Caetano do Sul 03388
Instituto Metodista Bennett, Rio de Janeiro 03097
Instituto Metodista de Ensino Superior, Rudge Ramos 03115
Instituto Mexicano del Petróleo, México 18852
Instituto Municipal de Administração e Ciências Contábeis, Belo Horizonte 03276
Instituto Nacional, Panamá 20541
Instituto Nacional Administración Pública, Madrid 26056
Instituto Nacional de Administração, Oeiras 21820
Instituto Nacional de Administración Pública, Madrid 25718
Instituto Nacional de Antropología e Historia (INAH) – Coordinación Nacional de Restauración del Patrimonio Cultural, México 18853
Instituto Nacional de Antropología e Historia (INAH) – Dirección de Estudios Históricos, México 18854
Instituto Nacional de Antropología y Pensamiento Latinoamericano, Buenos Aires 00267
Instituto Nacional de Cancerología, Santafé de Bogotá 06035
Instituto Nacional de Cardiología 'Ignacio Chavez', México 18855
Instituto Nacional de Empleo, Madrid 26057
Instituto Nacional de Engenharia e Tecnologia Industrial (INETI), Lisboa 21801
INETI -Instituto Nacional de Enharia Tecnologia e Inourção, Alfragide 21753
Instituto Nacional de Estadística, Madrid 26058
Instituto Nacional de Estadística e Informatíca (INEI), Lima 20666
Instituto Nacional de Estadística (INE), La Paz 02939
Instituto Nacional de Estadística y Censos (INDEC), Buenos Aires 00268
Instituto Nacional de Estadísticas, Santiago 04934
Instituto Nacional de Estatística (INE), Lisboa 21732
Instituto Nacional de Estudios de Teatro, Buenos Aires 00269
Instituto Nacional de Estudos de Pesquisa INEP, Bissau 13191
Instituto Nacional de Estudos e Pesquisas Educacionais (INEP/MEC), Brasília 03283
Instituto Nacional de Estudos e Pisquisada (INEP), Bissau 13189
Instituto Nacional de Gestión Sanitaria (INGESA), Madrid 26059
Instituto Nacional de Investigação Agrária e das Pescas (INIAP/IPIMAR), Lisboa 21802
Instituto Nacional de Investigación Agraria, Lima 20667
Instituto Nacional de Investigación y Capacitación de Telecomunicaciones (INICTEL), Lima 20668
Instituto Nacional de Investigación y Tecnología Agraria y Alimentaria, Madrid 26060
Instituto Nacional de Investigaciones Forestales y Agropecuarias (INIFAP), Celaya 18824
Instituto Nacional de Investigaciones Nucleares, México 18856
Instituto Nacional de la Administración Pública (INAP), Buenos Aires 00270
Instituto Nacional de Medicina Legal y Ciencias Forenses, Santafé de Bogotá 06036
Instituto Nacional de Normalización (INN), Santiago 04935
Instituto Nacional de Nutrición, Caracas 42253
Instituto Nacional de Pesca, Guayaquil 06988
Instituto Nacional de Pesquisas da Amazônia (INPA), Manaus 03308
Instituto Nacional de Pesquisas Espaciais (INPE), São José dos Campos 03391
Instituto Nacional de Salud, Santafé de Bogotá 06037
Instituto Nacional de Salud Pública, Cuernavaca 18826
Instituto Nacional de Saúde Dr. Ricardo Jorge, Lisboa 21803
Instituto Nacional de Seguridad e Higiene en el Trabajo, Barcelona 25918
Instituto Nacional de Técnica Aeroespacial (INTA), Torrejón de Ardoz 26166
Instituto Nacional de Tecnología, Rio de Janeiro 03359
Instituto Nacional de Tecnología Industrial (INTI), San Martín 00313
Instituto Nacional de Telecomunicações, Santa Rita do Sapucaí 03124
Instituto Nacional de Vitivinicultura, Mendoza 00305
Instituto Nacional do Livro, Brasília 03284
Instituto Nacional do Livro, Rio de Janeiro 03360
INRB -Instituto Nacional dos Recursos Biológicos, Elvas 21758
Instituto Nacional dos Recursos Biológicos, Elvas 21758
Instituto Nacional Indigenista, México 18857
Instituto Nacional Sanmartiniano, Buenos Aires 00271
Instituto Neo-Pitagórico, Curitiba 03296
Instituto Nevares de Empresarios Agrarios (INEA), Valladolid 25640
Instituto Nicaragüense de Investigaciones Económicas y Sociales (INIES), Managua 19898
Instituto Panamericano de Geografía e Historia, México 18858
Instituto para el Desarrollo de Empresarial en la Argentina (IDEA), Buenos Aires 00272
Instituto para la Integración de América Latina y el Caribe (INTAL), Buenos Aires 00273
Instituto Penido Burnier, Campinas 03290
Instituto Peruano de Energía Nuclear, Lima 20669
Instituto Politécnico de Beja, Beja 21622
Instituto Politécnico de Leiria, Leiria 21652
Instituto Politécnico do Porto, Porto 21696
Instituto Politécnico Nacional, México 18859
Instituto Politécnico Nacional, México 18729
Instituto Português da Qualidade, Caparica 21717
Instituto Português da Sociedade Científica de Goerres, Lisboa 21804
Instituto Português de Oncologia de Francisco Gentil, Lisboa 21657
Instituto Presbiteriano Mackenzie, São Paulo 03212
Instituto Pre-universitario de La Habana, La Habana 06305
Instituto Profesional de Santiago, Santiago 04859
Instituto 'Rio Branco', Rio de Janeiro 03361
Instituto Santanense de Ensino Superior, São Paulo 03147
Instituto Superior Agrícola, Ciego de Avila 06267
Instituto Superior de Agricultura (ISA), Santiago de los Caballeros 06959
Instituto Superior de Ciencias Catequéticas San Pio X, Madrid 25424
Instituto Superior de Ciências do Trabalho e da Empresa, Lisboa 21658
Instituto Superior de Ciencias Morales, Madrid 26061
Instituto Superior de Educación, Asunción 20602
Instituto Superior de Engenharia de Coimbra, Coimbra 21625
Instituto Superior de Engenharia de Lisboa, Lisboa 21805
Instituto Superior de Filosofía, Valladolid 26179
Instituto Superior de Linguas e Administração de Bragança, Bragança 21624
Instituto Superior de Linguas e Administração de Lisboa, Lisboa 21659
Instituto Superior de Linguas e Administração de Santarém, Santarém 21710
Instituto Superior de Pastoral, Madrid 25425
Instituto Superior de Relaciones Internacionales 'Raul Roa García', La Habana 06269
Instituto Superior de Serviço Social, Lisboa 21660
Instituto Superior Pedagógico 'Conrado Benitez', Cienfuegos 06278
Instituto Superior Pedagógico 'Enrique José Varona', La Habana 06306
Instituto Superior Pedagógico 'Felix Varela', Santa Clara 06311
Instituto Superior Politécnico 'J.A. Echeverría' (ISPJAE), La Habana 06270
Instituto Superior Técnico, Lisboa 21661
Instituto Technológica de Aeronáutica, São José dos Campos 03204
Instituto Tecnológico Autónomo de México, México 18730
Instituto Tecnológico de Buenos Aires, Buenos Aires 00119
Instituto Tecnológico de Celaya, Celaya 18708
Instituto Tecnológico de Costa Rica (ITCR), Cartago 06106
– Sede San Carlos, Alajuela 06107
Instituto Tecnológico de Léon, Léon 18718
Instituto Tecnológico de Mérida, Mérida 18719
Instituto Tecnologico de Querétaro, Santiago de Querétaro 18804
Instituto Tecnológico de Santo Domingo, Santo Domingo 06956
Instituto Tecnológico de Sonora, Ciudad Obregón 18711
Instituto Tecnológico de Tuxtla Gutierrez, Tuxtla Gutierrez 18806
Instituto Tecnológico de Zacatecas, Zacatecas 18808
Instituto Tecnológico do Estado de Pernambuco (ITEP), Recife 03321
Instituto Tecnológico y de Estudios Superiores de Monterrey Campus Ciudad Juárez, Juárez 18717
Instituto Tecnológico y de Estudios Superiores de Monterrey (ITESM), Monterrey 18796
Instituto Tecnológico y de Estudios Superiores de Occidente, Tlaquepaque 18805
Instituto Teológico Compostelano, Santiago de Compostela 25567
Instituto Teológico de Murcia, Murcia 25800
Instituto Torcuato Di Tella, Buenos Aires 00120
Instituto Torcuato Di Tella, Buenos Aires 00247
Instituto Universitário de Pesquisas do Rio de Janeiro, Rio de Janeiro 03098
Instituto Universitario Politécnico, Barquisimeto 42204
Instituto Uruguayo de Normas Técnicas, Montevideo 42060
Instituto Valencià d'Art Modern (IVAM), Valencia 26173
Instituto Vasco de Administración Pública (IVAP), Vitoria-Gasteiz 26187
Instituto Venezolano de Investigaciones Científicas (IVIC), Caracas 42254
Instituto Venezolano Tecnológico del Petróleo (Intevep), Caracas 42255
Instituton Geologikon kai Metalleftikon Erenon, Athinai 13115
Institutos Dominicos de Filología y Teología, Madrid 25796
Institutt for energiteknikk, Kjeller 20232
Institutt for fredsforskning, Oslo 20245
Institutt for kriminologi og rettssosiologi, Oslo 20132
Instituttgruppa for samfunnsforskning (ISAF), Oslo 20246
Institutul Cantacuzino, Bucureşti 21978
Institutul de Antropologie "Francisc I. Rainer" al Academiei Române, Bucureşti 21979
Institutul de arheologie şi istoria artei, Cluj-Napoca 22021
Institutul de Arte Plastice 'Nicolae Grigorescu', Bucureşti 21980
Institutul de cercetare dezvoltare pentru apicultură – S.A., Bucureşti 21981
Institutul de cercetare şi productie pentru creşterea bovinelor, Baloteşti 21959
Institutul de cercetare şi productie pentru creşterea porcinelor, Periş 22050
Institutul de cercetări biologice, Cluj-Napoca 22022
Institutul de cercetări biologice, Iaşi 22037
Institutul de cercetări eco-muzeale Tulcea, Tulcea 22067
Institutul de cercetări economice 'Gheorghe Zane', Iaşi 22038
Institutul de cercetări pentru cereale şi plante tehnice, Fundulea 22033
Institutul de cercetări pentru chimie şi industrie alimentară, Bucureşti 21982
Institutul de cercetări pentru legumicultură şi floricultură, Vidra 22070
Institutul de cercetări pentru pedologie şi agrochimie, Bucureşti 21983
Institutul de cercetări pentru viticultură şi vinificaţie, Valea Călugărească 22069
Institutul de cercetări pomicole, Piteşti 22051
Institutul de cercetări şi amenajări silvice, Bucureşti 21984
Institutul de cercetări socio-umane, Craiova 22028
Institutul de cercetări socio-umane, Iaşi 22039
Institutul de cercetări socio-umane, Târgu Mureş 22061
Institutul de cercetări socio-umane, Timişoara 22064
Institutul de cercetări ştiinţtifice pentru protecţia muncii, Bucureşti 21985
Institutul de economie agrară, Bucureşti 21986
Institutul de economie mondială, Bucureşti 21987
Institutul de endocrinologie 'C.I. Parhon', Bucureşti 21988
Institutul de etnografie şi folclor 'Constantin Brăiloiu, Bucureşti 21989
Institutul de Filologie Româna 'A. Philippide', Iaşi 22040
Institutul de fiziologie normală şi patologică 'D. Danielopolu', Bucureşti 21990
Institutul de Informatică Teoretică, Iaşi 22041
Institutul de istoria artei 'George Oprescu', Bucureşti 21991
Institutul de istorie 'A. D. Xenopol', Iaşi 22042
Institutul de Istorie 'N. Iorga', Bucureşti 21992
Institutul de Istorie şi Teorie Literară 'George Călinescu', Bucureşti 21993
Institutul de medicină internă 'N.Gh. Lupu', Bucureşti 21994
Institutul de neurologie şi psihiatrie, Bucureşti 21995
Institutul de Sănătate Publica 'Prof. Dr. Iuliu Moldovan', Cluj-Napoca 22023

Institutul de sănătate publică şi cercetări medicale, Iaşi 22043
Institutul de sănătate publică şi cercetări medicale, Timişoara 22065
Institutul de sănăte publica şi cercetări medicale, Târgu Mureş 22062
Institutul de studii Sud-Est Europene, Bucureşti 21996
Institutul Geologic al României, Bucureşti 21997
Institutul Naţional de Cercetare Dezvoltare pentru Cartof şi Sfecia de Zahar Braşov, Braşov 21963
Institutul Naţional de Cercetare si Dezvoltare pentru Fizica Tehnică, Iaşi 22044
Institutul Naţional de Informare şi Documentare, Bucureşti 21998
Institutul Naţional de Medicina Legala 'Mina Minovici', Bucureşti 21999
Institutul naţional de metrologie, Bucureşti 22000
Institutul National de sănătate publică, Bucureşti 22001
Institutul oncologic, Bucureşti 22002
Institutul Român de cercetări marine, Constanţa 22026
Institutul teologie Universitar, Sibiu 21944
Institutul 'V. Babes', Bucureşti 22003
Institutul-Delta Dunării, Tulcea 22068
Institutului Medico-Farmaceutic, Iaşi 22045
Institutum Romanum Finlandiae, Roma 16507
Instituut Collectie Nederland, Amsterdam 19361
Instituut voor Agrotechnologisch Onderzoek (ATO-DLO), Wageningen 19536
Instituut voor Franciscaanse Geschiedenis, Sint-Truiden 02488
Instituut voor Maritieme Historie van de Marinestaf (IMH), Den Haag 19437
Instituut voor Militaire Geschiedenis, Den Haag 19438
Instituut voor Natuur en Bosonderzoek, Brussel 02598
Instituut voor Natuurbehoud, Brussel 02591
Instituut voor Natuurbeschermingseducatie (IVN), Amsterdam 19359
Instituut voor Tropische Geneeskunde (ITG), Antwerpen 02515
Instrumentalny zavod, Orsha 01914
Instytut Agrokhimfarmproekt, Kyiv 28313
Instytut arkheologiyi, Kyiv 28314
Instytut Badań Edukacyjnych, Warszawa 21263
Instytut Badań Literackich PAN, Warszawa 21264
Instytut Badań Systemowych PAN, Warszawa 21265
Instytut Badawczy Dróg i Mostów, Warszawa 21266
Instytut Badawczy Leśnictwa, Warszawa 21267
Instytut Barwników i Produktów Organicznych, Zgierz 21374
Instytut Biocybernetyki i Inżynerii Biomedycznej PAN i Międzynarodowe Centrum Biocybernetyki, Warszawa 21268
Instytut Biologii Doświadczalnej im. M. Nenckiego PAN, Warszawa 21269
Instytut Biologiyi Yuzhnykh morei, Sevastopol 28459
Instytut bioorganichnoyi khimiyi ta naftokhimiyi, Kyiv 28315
Instytut Biopolimerów i Włokien Chemicznych, Łódź 21160
Instytut Biotechnologii Przemysłu Rolno-Spożywczego, Warszawa 21270
Instytut Botaniki PAN im. W. Szafera, Kraków 21129
Instytut botaniky im. M.G. Kholodnoho, Kyiv 28316
Instytut Budownictwa, Mechanizacji i Elektryfikacji Rolnictwa, Warszawa 21260
Instytut Budownictwa Wodnego PAN, Gdańsk 21096
Instytut Chemicznej Przeróbki Wegla, Zabrze 21031
Instytut Chemii Fizycznej Polskiej Akademii Nauk, Warszawa 21271
Instytut Chemii i Techniki Jądrowej, Warszawa 21272
Instytut Chemii Nieorganicznej, Gliwice 21104
Instytut Chemii Organicznej Polskiej Akademii Nauk, Warszawa 21273
Instytut Chemii Przemysłowej, Warszawa 21274
Instytut Ciężkiej Syntezy Organicznej, Kędzierzyn-Koźle 21117
Instytut Dendrologii PAN, Kórnik 21123
Instytut derzhavi i prava im. V.M. Koretskoho, Kyiv 28317
Instytut efiromaslichnykh i lekarstvennykh rasteni, Simferopol 28469
Instytut Ekologii PAN, Łomianki 21166
Instytut ekonomiki, Kyiv 28318
Instytut Ekonomiki Rolnictwa i Gospodarki Żywnościowej, Warszawa 20953
Instytut ekonomiky promislovosti, Donetsk 28219
Instytut eksperimentalnoyi i klinicheskoyi veterinarnoyi meditsiny, Kharkiv 28254
Instytut eksperimentalnoyi patologiyi, onkologiyi i radiobiologiyi im. R.E. Kavetskoho, Kyiv 28319
Instytut Ekspertyz Sądowych im. Prof. Dr. Jana Sehna, Kraków 21130
Instytut elektrodinamiki, Kyiv 28320
Instytut Elektrotechniki, Warszawa 21275
Instytut elektrozvaryuvannya im. E. Patona, Kyiv 28321
Instytut Energetyki, Warszawa 21276
Instytut Energii Atomowej, Otwock-Świerk 21184
Instytut Farmaceutyczny, Warszawa 21277
Instytut Farmakologii PAN, Kraków 21131
Instytut filosofiyi im. H.S. Skovorody, Kyiv 28322
Instytut Fizyki Jądrowej im. Henryka Niewodniczanskiego PAN, Kraków 21132
Instytut Fizyki Molekularnej PAN, Poznań 21194
Instytut Fizyki PAN, Warszawa 21278
Instytut fizyko-organichnoyi khimiyi i vuglekhimiyi im. Litvinenka, Donetsk 28220
Instytut fizyky, Kyiv 28323
Instytut fizyky napivprovidnikiv, Kyiv 28324
Instytut fizyologiyi i biokhimiyi tvaryn, Lviv 28409
Instytut fizyologiyi raslin i genetiki, Kyiv 28325
Instytut gaza, Kyiv 28326
Instytut Genetyki i Hodowli Zwierząt PAN, Mroków 21172
Instytut Genetyki Roślin PAN, Poznań 21195
Instytut Geodezji i Kartografii, Warszawa 21279
Instytut Geofizyki PAN, Warszawa 21280
Instytut geologiyi i geokhimiyi goryuchykh kopalin, Lviv 28410
Instytut geotekhnicheskoyi mekhaniki, Dnipropetrovsk 28201
Instytut gerontologiyi, Kyiv 28327
Instytut gidromekhaniki, Kyiv 28328
Instytut gidrotekhniky i melioratsiyi, Kyiv 28329
Instytut Górnictwa Naftowego i Gazownictwa, Kraków 21133
Instytut Górnictwa Odkrywkowego 'Poltegor-Instytut', Wrocław 21359
Instytut Gospodarki Przestrzennej i Mieszkalnictwa, Warszawa 21281
Instytut gruntoznavstva i agrokhimiyi im. A.N. Sokolovskoho, Kharkiv 28255
Instytut Gruźlicy i Chorób Płuc, Warszawa 21282
Instytut Hamatologii i Transfuzjologii, Warszawa 21283
Instytut Historii Nauki PAN, Warszawa 21284
Instytut Historii PAN, Warszawa 21285
Instytut Hodowli i Aklimatyzacji Roślin, Błonie 21076
Instytut Hodowli i Aklimatyzacji Roślin, Oddział Bonin, Koszalin 21125
Instytut Immunologii i Terapii Doświadczalnej PAN, Wrocław 21360
Instytut impulsnykh protsesiv i tekhnologi, Mykolaiv 28429
Instytut Inżynierii Chemicznej PAN, Gliwice 20812
Instytut Inżynierii Materiałów Polimerowych i Barwników w Toruniu, Oddział Zamiejscowy Kauczuk'jow i Tworzyw Winylowych, Oświęcim 21182
Instytut istoriyi Ukrainy, Kyiv 28330
Instytut Katalizy i Fizykochemii Powierzchni PAN, Kraków 21134
Instytut klitinnoyi biologiyi ta genetichnoyi inzheneriyi, Kyiv 28331
Instytut Kultur Śródziemnomorskich i Orientalnych PAN, Warszawa 21286
Instytut Kultury Austriackiej, Warszawa 21287
Instytut Łaczności, Warszawa 20954
Instytut literatury im. T.G. Shevchenka, Kyiv 28332
Instytut Logistyki i Magazynowania, Poznań 21196
Instytut Lotnictwa, Warszawa 21288
Instytut Maszyn Matematycznych, Warszawa 21289
Instytut Maszyn Przepływowych PAN, Gdańsk 21097
Instytut matematiki, Kyiv 28333
Instytut Matematyczny PAN, Warszawa 21290
Instytut Mechaniki Górotworu PAN, Kraków 21135
Instytut Mechaniki Precyzynej, Warszawa 21291
Instytut Mechanizacji Budownictwa i Górnictwa Skalnego, Warszawa 21292
Instytut medichnoi radiologii im. S.P. Grigoreva, Kharkiv 28256
Instytut Medycyny Morskiej i Tropikalnej, Gdynia 20811
Instytut Medycyny Pracy i Zdrowia Środowiskowego, Sosnowiec 21225
Instytut Medycyny Pracy im. prof. J. Nofera, Łódź 20874
Instytut Medycyny Wsi im. Witolda Chodźki, Lublin 21169
Instytut mekhaniki, Kyiv 28334
Instytut mekhanizatsiyi tvarinnitstva, Zaporizhzhya 28485
Instytut Mekhanobchermet, Krivi Rig 28297
Instytut Melioracji i Użytków Zielonych, Raszyn 21218
Instytut Metali Nieżelaznych, Gliwice 21105
Instytut metallofizyky, Kyiv 28335
Instytut Metalurgii i Ińżynierii Materiałowej im. Aleksandra Krupkowskiego, Kraków 21136
Instytut Metalurgii Żelaza, Gliwice 21106
Instytut Meteorologii i Gospodarki Wodnej, Warszawa 21293
Instytut mikrobiologiyi i virusologiyi im. D.K. Zabolotnoho, Kyiv 28336
Instytut mineralnykh resursov, Simferopol 28470
Instytut mistetstvoznavstva, folkloru ta etnologiyi im. M.T. Rilskoho, Kyiv 28337
Instytut molekulyarnoyi biologiyi ta genetiki, Kyiv 28338
Instytut monokrystaliv, Kharkiv 28257
Instytut Morski, Gdańsk 21098
Instytut movoznavstva im. O.O. Potebni, Kyiv 28339
Instytut narodoznavstva, Lviv 28411
Instytut natsionalnykh vidnosin i politologiyi, Kyiv 28340
Instytut natverdykh materialiv im. V.M. Bakulya, Kyiv 28341
Instytut Nauk Ekomicznych PAN, Warszawa 21294
Instytut Nauk Prawnych PAN, Warszawa 21295
Instytut Nawozów Sztucznych, Puławy 20919
Instytut neyrokhirurgiyi im. akademika A. P. Romodanova AMN Ukrainy, Kyiv 28342
Instytut Niskich Temperatur i Badań Strukturalnych PAN, Wrocław 21361
Instytut Obróbki Plastycznej, Poznań 21197
Instytut Ochrony Przyrody PAN, Kraków 21137
Instytut Ochrony Roślin, Poznań 21198
Instytut Ochrony Roślin, Puławy 21213
Instytut Ochrony Środowiska, Warszawa 21296
Instytut Odlewnictwa, Kraków 21138
Instytut organicheskoyi khimiyi, Kyiv 28343
Instytut Organizacji i Zarządzania w Przemyśle 'ORGMASZ', Warszawa 21297
Instytut Paleobiologii PAN im. R. Kozłowskiego, Warszawa 21298
Instytut pidvishchennya kvalifikatsiyi pratsivnikiv profesino-tekhnichnoyi osviti Ukrainy, Donetsk 27974
Instytut Pivdendiprogaz, Donetsk 28221
Instytut po proektirovaniyu metalurginykh zavodov, Dnipropetrovsk 28202
Instytut Podstaw Informatyki PAN, Warszawa 21299
Instytut Podstaw Inżynerii Środowiska, Zabrze 21371
Instytut Podstawowych Problemow Techniki PAN, Warszawa 21300
Instytut Polityki Naukowej i Szkolnictwa Wyższego, Warszawa 21301
Instytut Polski w Berlinie, Berlin 11556
Instytut Polski w Wiedniu, Wien 01648
Instytut prikladnoyi matematiky i mekhaniki, Donetsk 28222
Instytut prikladnykh problem mekhaniky i matematiky im. Pidstrigacha, Lviv 28412
Instytut prirodokoristuvannya ta ekologiyi, Dnipropetrovsk 28203
Instytut problem energoberezhennya, Kyiv 28344
Instytut problem kriobiologiyi ta kriomeditsini, Kharkiv 28258
Instytut problem litya, Kyiv 28345
Instytut problem mashinobuduvannya, Kharkiv 28259
Instytut problem modeluvannya v ener getitsi, Kyiv 28346
Instytut problem reestratsii informatsii, Kyiv 28347
Instytut problem rinku i ekonomiko-ekologichnykh doslidzhen, Odesa 28438
Instytut Problemów Jądrowych im. Andrzeja Sołtana, Otwock-Świerk 21185
Instytut Przemysłu Organicznego, Warszawa 21302
Instytut Przemysłu Włókien Naturalnych, Poznań 21199
Instytut Psychiatrii i Neurologii, Warszawa 21303
Instytut ptitsevodstva, Borki 28186
Instytut radiofizyky i elektroniky, Kharkiv 28260
Instytut regional'nych dostidjen, Lviv 28413
Instytut Rybactwa Śródlądowego, Olsztyn 21175
Instytut sadivnitstva, Malenke 28427
Instytut Sadownictwa i Kwiaciarstwa, Skierniewice 21221
Instytut silskogospodarskoho maschinobuduvannya, Kirovograd 28291
Instytut Slawistyki PAN, Warszawa 21304
Instytut Spawalnictwa, Gliwice 21107
Instytut Studiow Politycznych PAN, Warszawa 21305
Instytut svitovoyi ekonomiky ta mizhnarodnykh vidnosin, Kyiv 28348
Instytut Systematyki i Ewolucji Zwierzat PAN, Kraków 21139
Instytut Szkła i Ceramiki, Warszawa 21306
Instytut Techniki Budowlanej, Warszawa 21307
Instytut Techniki Cieplnej, Łódź 21161
Instytut Technologii Drewna, Poznań 21200
Instytut Technologii Elektronowej, Warszawa 21308
Instytut Technologii Materiałow Elektronicznych, Warszawa 21309
Instytut Technologii Nafty im. Prof. Stanisława Pilata, Kraków 21140
Instytut tekhnichnoyi ekologiyi, Donetsk 28223
Instytut tekhnichnoyi mekhaniki, Dnipropetrovsk 28204
Instytut tekhnichnoyi teplofizyky, Kyiv 28349
Instytut Tele- i Radiotechniczny, Warszawa 21310
Instytut teoretichnoi fizyky im. M.M. Bogolyubova, Kyiv 28350
Instytut Transportu Samochodowego, Warszawa 21311
Instytut tsukrovykh buryakiv, Kyiv 28351
Instytut tuberkulozu, Lviv 28414
Instytut Turystyki, Warszawa 21312
Instytut tvarinnitstva, Kharkiv 28261
Instytut Warzywnictwa, Skierniewice 21222
Instytut Włókien Naturalnych i Roślin Zielarskich, Poznań 21201
Instytut Włókiennictwa, Łódź 21162
Instytut Wzornictwa Przemysłowego, Warszawa 21313
Instytut yadernykh doslidzhen, Kyiv 28352
Instytut yuzhgiprotsement, Kharkiv 28262
Instytut Zaawansowanych Technologie Wytwanania, Kraków 21141
Instytut Zachodni, Poznań 21202
Instytut Zootechniki, Balice 21071
Instytut zroshuvanoho sadivnytstva, Melitopol 28428
Instytut Żywności i Żywienia im. prof. dra. med. Aleksandra Szczygła, Warszawa 21314
Insurance Institute for Highway Safety Library, Arlington 36480
Insurance Library Association of Boston, Boston 36567
INTA, Torrejón de Ardoz 26166
Integral Industrial Corporation, Trade Union Library, Minsk 02190
Integrated Academy of Management and Technology, Ghaziabad 13648
Integrierte Gesamtschule Holweide, Köln 10748
Inteko Ishinga Amategeko, Kigali 24369
Intelsat Library, Washington 37743
Inter-American Children's Institute, Montevideo 42059
Inter-American Defense College, Washington 32986
Inter-American Development Bank, Washington 36357
Inter-American University of Puerto Rico, Aguadilla 21845
Interamerican University of Puerto Rico, Arecibo 21846
Interamerican University of Puerto Rico, Barranquitas 21847
Interamerican University of Puerto Rico, Bayamon 21848
Interamerican University of Puerto Rico, Fajardo 21853
Interamerican University of Puerto Rico, Guayama 21855

Interamerican University of Puerto Rico, Hato Rey 21857
Interamerican University of Puerto Rico, Mercedita 21862
Interamerican University of Puerto Rico, San German 21866
Interamerican University of Puerto Rico – School of Law, Hato Rey 21858
Interchurch Center, New York 35308
Intercollege (International College of Management and Communication Studies), Nicosia 06329
Intercollegiate Center for Nursing Education, Spokane 37621
Intergovernmental Committee on Urban & Regional Research / Comité Intergouvernemental de Recherches Urbaines et Régionales, Toronto 04517
Interkulturelle Bibliothek für Kinder und Jugendliche (JUKIBU), Basel 27305
Interlochen Center for the Arts, Interlochen 36961
Intermediate Appellate Court Library, Manila 20755
Internal Revenue Service Library, Washington 37744
Internationaal Informatiecentrum en Archief voor de Vrouwenbeweging (IIAV), Amsterdam 19362
Internationaal Instituut voor Sociale Geschiedenis (IISG), Amsterdam 19363
International Academy of Indian Culture, New Delhi 14149
International Academy of Philosophy in the Principality of Liechtenstein, Bendern 18268
International Agency for Research on Cancer (IARC) / Centre International de Recherche sur le Cancer (CIRC), Lyon 08398
International Arctic Research Center, Fairbanks 36838
International Association of Composer's Organisations, Moskva 23350
International Association of Educators for World Peace, Huntsville 36949
International Association of Public Transport (UITP), Bruxelles 02599
International Association of Universities / Association Internationale des Universités, Paris 08576
International Atomic Energy Agency, Wien 01611
International Baptist Theological Seminary of the European Baptist Federation, Praha 06417
International Bureau of Education, Le Grand-Saconnex 27402
International Bureau of Fiscal Documentation, Amsterdam 19364
International Business Maschines Corp, San Jose 36249
International Business School Library, Budapest 13312
International Center for Agricultural Research in the Dry Areas (ICARDA), Aleppo 27518
International Center for Research on Women, Washington 37745
International Center of Photography Library, New York 37213
International Centre for Classical Research, Athinai 13116
International Centre for Diarrhoeal Disease Research, Dhaka 01766
International Centre for the Study of the Preservation and Restoration of Cultural Property (ICCROM), Roma 16508
International Centre of Insect Physiology and Ecology, Nairobi 18036
International Child Resource Institute, Berkeley 36532
International Christian University, Mitaka 17042
International Civil Aviation Organization, Montreal 04408
International Coffee Organization, London 29774
International College, Aiea 30172
International College, Naples 31928
International College of Hospitality Management Library, Suffield 32760
International Committee of the Red Cross, Genève 27362
International Cooperative Alliance, New Delhi 14150
International Council for the Exploration of the Sea, København 06797
International Council on Monuments and Sites, Paris 08513
International Court of Justice, Den Haag 19255
International Crops Research Institute for the Semi Arid Tropics (ICRISAT), Patancheru 14175
International Data processing Education and Information Center of the Central Office for Statistics, Budapest 13434
International Development Research Centre Library, Ottawa 04447
International Documentation Center, Tokyo 17680
International Fertilizer Development Center, Muscle Shoals 37136
International Fine Arts College, Miami 31807
International Flavors & Fragrances, Inc, Union Beach 36317
International Food Policy Research Institute Library, Washington 37746
International Foundation of Employee Benefit Plans, Brookfield 36596
International Futures Library, Salzburg 01572
International Graduate School of Management, Barcelona 25295
International Graphoanalysis Society, Chicago 33221
International House of Japan, Tokyo 17700
International Information Centre and Archives for the Women's Movement, Amsterdam 19362
International Information Centre on Sources of Balkan and Mediterranean History (CIBAL), Sofiya 03556
International Institute, Madrid 26055
International Institute for Applied Systems Analysis (IIASA), Laxenburg 01555
International Institute for Children's Literature, Osaka, Osaka 17633
International Institute for Development, Co-operative and Labour Studies (Afro-Asian Institute), Tel-Aviv 14661
International Institute for Educational Planning, Paris 08562
International Institute for Population Sciences (IIPS), Mumbai 13704
International Institute for Sport & Olympic History Library, State College 37637
International Institute for the Unification of Private Law, Roma 16516
International Institute of Administrative Sciences (IIAS), Bruxelles 02594
International Institute of Communications, London 29775
International Institute of Human Rights, Strasbourg 08684
International Institute of Integral Human Sciences Library, Montreal 04409
International Institute of Social History, Amsterdam 19363
International Institute of Social Studies (ISS), Erasmus University Rotterdam, Den Haag 19141
International Institute of Tropical Agriculture (IITA), Ibadan 20018
International Islamic University, Islamabad, Islamabad 20431
International Islamic University Malaysia, Kuala Lumpur 18502
– International Institute of Islamic Thought and Civilization (ISTAC), Kuala Lumpur 18503
International Labour Office (ILO), Genève 27367
International Labour Organisation, Bangkok 27781
International Ladies' Garment Workers Union, New York 37214
International Library, Archives & Museum of Optometry, Saint Louis 37485
International Livestock Centre for Africa (ILCA), Addis Ababa 07284
International Livestock Research Institute, Nairobi 18037
International Mission Board, Southern Baptist Convention – Jenkins Research Library, Richmond 35345
International Museum of Ceramics in Faenza Foundation, Faenza 16241
International Naturist Library, Baunatal 11474
International Olympic Committee, Lausanne 27397
International Paper, Tuxedo Park 37696
International Peace Information Service (IPIS vzw), Antwerpen 02516
International Red Locust Control Organization for Central and Southern Africa, Ndola 42349
International Rehabilitation Council for Torture Victims, København 06798
International Research Center for Japanese Studies, Kyoto 17586
International Rice Research Institute, Makati City 20771
International School, Makati City 20748
International School of Kenya, Nairobi 18014
International School of Stuttgart & Böblingen and Sindelfingen International Community School, Stuttgart 10770
International Stained Glass Centre, Chartres 08327
International Sugar Organisation, London 29776
International Swimming Hall of Fame, Fort Lauderdale 36860
International Telecommunication Union, Library and Archives Service, Genève 27377
International Tennis Hall of Fame & Tennis Museum Library, Newport 37282
International Union of United Automobile, Aerospace & Agricultural Implement Workers of America, Detroit 36805
International University of Applied Sciences Bad Honnef, Bad Honnef 09396
International University of Japan, Niigata 17066
International Youth Library, München 12307
Internationale Akademie für Philosophie im Fürstentum Liechtenstein, Bendern 18268
Internationale FKK-Bibliothek, Baunatal 11474
Internationale Hochschule Bad Honnef-Bonn – IUBH, Bad Honnef 09396
Internationale Jugendbibliothek, München 12307
Internationale Komponistinnen-Bibliothek, Unna 12571
Internationale Präsenz-Bibliothek zur Geschichte der Naturwissenschaften, Ludwigshafen 12217
Internationale Verzekeringsbibliotheek, Leuven 02671
Internationaler Versöhnungsbund – Österreichischer Zweig, Wien 01612
Internationales Institut für Angewandte Systemanalyse (IIASA), Laxenburg 01555
Internationales Kommitee vom Roten Kreuz, Genève 27362
Internationales Zeitungsmuseum, Aachen 11441
Internationales Zentralinstitut für das Jugend- und Bildungsfernsehen, München 12308
Interregional Police Library, Nijmegen 19284
Interregionales Fortbildungszentrum, Tramelan 27441
Intersil Corp, Palm Bay 36078
Interuniversitair Micro-Electronica Centrum (IMEC), Leuven 02672
Interuniversitäres Forschungszentrum für Technik, Arbeit und Kultur, Graz 01527
Inter-University-Centre Library, Dubrovnik 06211
INValSI Library, Frascati 16277
Inver Hills Community College, Inver Grove Heights 33428
Invercargill Public Library, Invercargill 19869
Inverclyde Libraries, Greenock 29988
Inyo County Free Library, Independence 39555
Inženýrsko-výrobní elektrotechnický podnik, a.s. – "IVEP, a.s.", Brno 06429
Inzhenernaya shkola odezhdy, Sankt-Peterburg 22695
Inzhenerno-meliorativny institut, Novocherkassk 22429
Inzhenerno-stroitelny institut, Brest 01789
Inzhenerno-stroitelny institut, Moskva 22342
Inzhenerno-stroitelny institut, Penza 22464
Inzhenerno-stroitelny institut, Sankt-Peterburg 22525
Inzhenerno-tekhnologichny instytut, Cherkasy 27954
Ioffe Physical and Technology Institute of the Russian Academy of Sciences, Sankt-Peterburg 23761
Ioffe Physics and Technology Institute – Shuvalovo Library Branch, Sankt-Peterburg 23762
Iona College, New Rochelle 32020
– Helen T. Arrigoni Library-Technology Center, New Rochelle 32021
Iowa Central Community College, Fort Dodge 33344
Iowa City Public Library, Iowa City 39576
Iowa Department of Human Services Library, Des Moines 34143
Iowa Department of Transportation Library, Ames 36461
Iowa Genealogical Society Library, Des Moines 36799
Iowa Lakes Community College, Estherville 33325
Iowa Regional Library for the Blind & Physically Handicapped, Des Moines 36800
Iowa State Penitentiary, Fort Madison 34250
Iowa State Penitentiary – John Bennett Correctional Center Library, Fort Madison 34251
Iowa State University, Ames 30210
Iowa Wesleyan College, Mount Pleasant 31915
Iowa Western Community College, Council Bluffs 33263
IP Australia, Woden 00745
Ipari Informatikai Központ, Budapest 13384
Iparművészeti Múzeum, Budapest 13385
Ippolitova-Ivanova School of Music, Moskva 22681
Ipswich Library and Information Service, Ipswich 01103
Ipswich Public Library, Ipswich 39577
Iqbal Academy, Lahore 20512
IQSH, Kronshagen 12165
Iran Bastan Museum, Tehran 14430
Iran University of Medical Sciences and Health Services, Tehran 14401
Iran University of Science and Technology, Tehran 14402
Iranian Centre for Archaeological Research, Tehran 14431
Iranian Cultural Heritage Organization, Tehran 14432
Iranian Information and Documentation Center, Tehran 14433
Iraq Natural History Research Centre and Museum, Baghdad 14454
Iraqi Academy, Baghdad 14443
Iraqi Museum, Baghdad 14455
Iraqi National Library and Archives, Baghdad 14442
IRB-Brasil Resseguros S.A, Rio de Janeiro 03362
IRC International Water and Sanitation Centre, Den Haag 19439
IRD – Institut de recherche pour le développement, Bondy 08300
Iredell County Public Library, Statesville 41435
Irell & Manella LLP Library, Los Angeles 35840
The Irene Lewisohn Costume Reference Library, Costume Institute, New York 37230
Irish Architectural Archive, Dublin 14541
Irish Management Institute (IMI), Dublin 14542
Irish Patents Office, Kilkenny 14560
Irish Railway Record Society, Dublin 14543
Irish Traditional Music Archive / Taisce Cheol Dúchais Eireann, Dublin 14544
Irkutsk Forest Industry Research Institute, Irkutsk 23056
Irkutsk Regional Museum, Irkutsk 23057
Irkutsk Scientific Centre of the Siberian Department of the Russian Academy of Sciences, Irkutsk 23055
Irkutsk Scientific Research Institute of Precious and Rare Metals and Diamonds, Irkutsk 23053
Irkutsk State Academy of Economics, Irkutsk 22255
Irkutsk State Medical Institute, Irkutsk 22256
Irkutsk State Technical Univerity, Irkutsk 22257
Irkutsk State University, Irkutsk 22258
Irkutskaya gosudarstvennaya ekonomicheskaya akademiya, Irkutsk 22255
Irkutskaya oblastnaya detskaya biblioteka, Irkutsk 24130
Irkutskaya oblastnaya gosudarstvennaya universalnaya nauchnaya biblioteka im. I.I. Molchanova-Sibirskogo, Irkutsk 22133
Irkutskaya oblastnaya yunosheskaya biblioteka im. I.P. Utkina, Irkutsk 24131
Irkutski gosudarstvenny meditsinski institut, Irkutsk 22256
Irkutski gosudarstvenny tekhnicheski universitet, Irkutsk 22257
Irkutski gosudarstvenny universitet, Irkutsk 22258
Irkutski mezhotraslevoi territorialny tsentr nauchno-tekhnicheskoi informatsii i propagandy, Irkutsk 23054
Irkutski nauchny tsentr Sibirskogo otdeleniya RAN, Irkutsk 23055
Iron and Steel Works Jesenice, Jesenice 24902
Iron Library, Schlatt 27431
Ironbridge Gorge Museum Trust, Telford 29909
Irondequoit Public Library, Rochester 41030
Ironworld Discovery Center, Chisholm 36691
Irpin Regional Centralized Library System, Main Library, Irpin 28582
Irpinska raionna TsBS, Tsentralna biblioteka, Irpin 28582
Irrigation Research Institute, Lahore 20513

Irvin L. Young Memorial Library, Whitewater 41886
Irvine Valley College Library, Irvine 33430
Irving Public Library, Irving 39584
Irving Public Library – Southwest, Irving 39585
Irving Public Library – Valley Ranch, Irving 39586
Irvington Public Library, Irvington 39587
Irvington Public Library, Irvington 39588
Isaac M. Wise Temple Library, Cincinnati 35160
Isaac Regional Library Service, Moranbah 01124
L. M. Isaev Research Institute of Medical Parasitology, Samarkand 42120
Isafjordur Public Library, Ísafjörður 13580
Isala clinics, Medical Library, Zwolle 19543
Isala klinieken, DISC, Zwolle 19543
Isar-Amper-Klinikum, Haar 11947
ISC Institute Teploelectroproject, Moskva 23339
Iselinge Educatieve Faculteit (ISELINGE OBD), Doetinchem 19416
Isfahan University of Medical Sciences, Isfahan 14389
Isfahan University of Technology, Isfahan 14390
Ishikawa Prefectural Assembly, Kanazawa 17265
Ishikawaken Gikai, Kanazawa 17265
Ishøj Bibliotek, Ishøj 06889
Ishpeming Carnegie Public Library, Ishpeming 39590
Islamia College, Lahore 20462
Islamia University, Bahawalpur 20427
Islamic Affairs Dept of the Federal Territory, Kuala Lumpur 18550
Islamic Azad University, Tehran 14403
Islamic Centre for Technical and Vocational Training and Research, Dhaka 01767
Islamic Consultative AssemblyIslamic Consultative Assembly, Tehran 14414
Islamic Development Bank, Jeddah 24398
Islamic Library, Yogyakarta 14340
Islamic Library 'Felix Maria Pareja', Madrid 25973
Islamic Research Foundation, Mashhad 14424
Islamic Research Institute, Islamabad 20479
Islamic Research Library, Kuala Lumpur 18599
The Islamic Revolution Cultural Documentation Organization Library, Tehran 14413
The Islamic Society of North America, Plainfield 37388
Islamic University of Indonesia, Yogyakarta 14323
Islamic University of Medina, Medina 24394
Islamisches Zentrum Hamburg e.V., Hamburg 11233
Island Park Public Library, Island Park 39591
Island Resources Foundation, St. Thomas 42313
Island Trees Public Library, Island Trees 39592
Isle College, Wisbech 29436
Isle of Anglesey County Council, Llangefni 30015
Isle of Man College, Douglas 14595
Isle of Man Government Office, Douglas 14596
Isle of Wight College, Newport, Isle of Wight 29282
Islip Public Library, Islip 39593
ISM University of Management and Economics Library, Kaunas Campus, Kaunas 18275
ISM Vadybos ir ekonomikos universiteto biblioteka, Kauno skyrius, Kaunas 18275
Isothermal Community College, Spindale 33772
ISP Management Co, Inc, Wayne 36400
Ispolkom Moskovskogo gorodskogo Soveta narodnykh deputatov, Moskva 22737
Israel Aircraft Industries Ltd, Lod 14707
Israel Aircraft Industries Ltd, Yahud 14710
Israel Antiquities Authority, Jerusalem 14680
Israel Atomic Energy Commission, Be'er-Sheva 14713
Israel Atomic Energy Commission, Tel-Aviv 14691
Israel Broadcasting Authority, Jerusalem 14733
Israel Defence Forces, Zahal 14694
Israel Export Institute, Tel-Aviv 14765
Israel Institute for Biological Research, Ness-Ziyona 14746
Israel Institute of Transportation Planning and Research, Tel-Aviv 14766
Israel Meteorological Service, Bet Dagan 14678
Israel Museum, Jerusalem 14734
Israel Police, Netanya 14689
Israel State Archives, Jerusalem 14735
Israelitische Cultusgemeinde Zürich (ICZ), Zürich 27280
Israelitische Gemeinde Basel, Basel 27245
ISRIC – World Soil Information, Wageningen 19537
ISRO Satellite Centre, Bangalore 13937
Issledovatelski fiziko-tekhnicheski institut, Nizhni Novgorod 23616
Issyk-Kulsk District Library, Przhevalsk 18159
Istanbul Archaeological Museums, Istanbul 27892
Istanbul Arkeoloji Müzeleri, Istanbul 27892
Istanbul Deniz Müzesi, Istanbul 27893
Istanbul Devlet Mühlendislik ve Mimarlik Akademisi, Istanbul 27855
Istanbul Il Halk Kütüphanesi, Istanbul 27902
Istanbul Naval Museum, Istanbul 27893
Istanbul Public Library, Istanbul 27902
Istanbul State Academy of Engineering and Architecture, Istanbul 27855
Istanbul Technical University, Istanbul 27856
Istanbul Teknik Üniversitesi, Istanbul 27856
İstanbul Üniversitesi, Istanbul 27857
– Deontoloji ve Tıp Tarihi Anabilim Dalı, Istanbul 27858
– Fen Fakültesi Matematik Bölümü Orhan Şerafettin İçen Kütüphanesi, Istanbul 27859
– Hukuk Fakultesi, Istanbul 27860
– Iktisat Fakultesi, Istanbul 27861
Istanbul University Library and Documentation Centre, Istanbul 27857
– Faculty of Economics, Istanbul 27861
– Faculty of Law, Istanbul 27860
– Faculty of Science Department of Mathematics Orhan Serafettin Içen Library, Istanbul 27859
– Medical Ethics and Medical History Department Library, Istanbul 27858
ISTAO – Istituto Adriano Olivetti di studi per la gestione dell'economia e delle aziende (ISTAO), Ancona 16165
ISTAT – Italian National Statistical Institute, Roma 16526
ISTED-Villes en Développement, Paris-la-Défense 08634
Istituti Ospitalieri di Milano, Milano 16338
Istituto Adriano Olivetti di studi per la gestione dell'economia e delle aziende, Ancona 16165
Istituto Affari Internazionali (IAI), Roma 16509
Istituto Agronomico per l'Oltremare, Firenze 16261
Istituto 'Aldini Valeriani', Bologna 16192
Istituto Antonio Piccardo, Genova 16287
Istituto Archeologico Germanico, Roma 16493
Istituto Bianchi, Napoli 15933
Istituto Calasanzio degli Scolopi, Genova 15850
Istituto Campano per la Storia della Resistenza, dell'Antifascismo e dell'Età contemporana, Napoli 16372
Istituto 'Cavanis', Venezia 16128
Istituto Centrale per il Restauro, Roma 15367
Istituto Centrale per la Patología del Libro Alfonso Gallo, Roma 16510
Istituto 'Champagnat', Genova 15622
Istituto d'Arte e Mestieri Vincenzo Roncalli, Vigevano 16674
Istituto d'Arte 'P. Toschi', Parma 16410
Istituto d'Arte 'Venturi', Modena 15627
Istituto dei Ciechi Francesco Cavazza, Bologna 16193
Istituto della Enciclopedia Italiana, Roma 16511
Istituto di Biologia del Mare, Venezia 16651
Istituto di Cultura Germanica, Bologna 16194
Istituto di Formazione Politica 'Pedro Arrupe', Palermo 16400
Istituto di Guerra Marittima, Livorno 15647
Istituto di Ricerca 'Cesare Serono', Ardea 16168
Istituto di Scienze Amministrative e Sociali, Palermo 16401
Istituto di Scienze e Tecnologie della Cognizione, Roma 16512
Istituto di Scienze Religiose, Trento 16613
Istituto di Sociologia Internazionale di Gorizia, Gorizia 16295
Istituto di Studi e Analisi Economica, Roma 16513
Istituto di Studi Militari Marittimi / Biblioteca Dante Aleghieri, Venezia 16645
Istituto di Studi Storici 'G. Salvemini', Torino 16605
Istituto di Studi Storici Postali, Prato 16426
Istituto di Studi sulla Ricerca e Documentazione Scientifica, Roma 16514
Istituto 'Domus Mazziniana', Pisa 16422
Istituto 'Don N. Mazza', Verona 16133
Istituto e Museo di Storia della Scienza, Firenze 16262
Istituto Ellenico di Studi Bizantini e Postbizantini di Venezia, Venezia 16652
Istituto Fratelli delle Scuole Cristiane, Roma 16029
Istituto Geografico De Agostini, Novara 16388
Istituto Geografico Militare, Firenze 16263
Istituto Geografico Polare 'Silvio Zavatti', Fermo 16243
Istituto Giapponese di Cultura, Roma 16515
Istituto Gonzaga, Palermo 15963
Istituto "Gramsci", Roma 16500
Istituto Gramsci Piemontese, Torino 16606
Istituto Idrografico della Marina, Genova 16288
Istituto Ignatianum, Messina 15067
Istituto Internationale Don Bosco, Torino 15484
Istituto Internazionale di Genetica e Biofisica, Napoli 16373
Istituto Internazionale di Studi Liguri, Bordighera 16204
Istituto Internazionale per l'Unificazione del Diritto Privato (UNIDROIT), Roma 16516
Istituto Italiano di Arti Grafiche S.p.A., Bergamo 16182
Istituto Italiano di Cultura, Barcelona 25919
Istituto Italiano di Cultura, Budapest 13386
Istituto Italiano di Cultura, Hamburg 11996
Istituto Italiano di Cultura, Helsinki 07462
Istituto Italiano di Cultura, Innsbruck 01542
Istituto Italiano di Cultura, Istanbul 27894
Istituto Italiano di Cultura, Köln 12143
Istituto Italiano di Cultura, Madrid 26062
Istituto Italiano di Cultura, Marseille 08409
Istituto Italiano di Cultura, México 18860
Istituto Italiano di Cultura, Montevideo 42061
Istituto Italiano di Cultura, München 12309
Istituto Italiano di Cultura, New York 37215
Istituto Italiano di Cultura, Paris 08577
Istituto Italiano di Cultura, South Yarra 01008
Istituto Italiano di Cultura, Strasbourg 08685
Istituto Italiano di Cultura, Tokyo 17681
Istituto Italiano di Cultura, Tripoli 18259
Istituto Italiano di Cultura, Valletta 18669
Istituto Italiano di Cultura, Wien 01613
Istituto Italiano di Cultura di Lisbona, Lisboa 21806
Istituto Italiano di Numismatica, Roma 16517
Istituto Italiano di Studi Germanici, Roma 16518
Istituto Italiano per gli Studi Storici, Napoli 16374
Istituto Italiano per la Storia Antica, Roma 16519
Istituto Italiano per l'Africa e l'Oriente, Roma 16520
Istituto Italo-Latino Americano, Roma 16521
Istituto Leone XIII, Milano 15909
Istituto Lombardo, Milano 16339
Istituto Luigi Sturzo, Roma 16522
Istituto Mazziniano, Genova 16289
Istituto Musicale Luigi Boccherini, Lucca 15623
Istituto Nazionale della Previdenza Sociale (INPS), Roma 15669
Istituto Nazionale delle Assicurazioni (INA), Roma 16523
Istituto Nazionale di Archeologia e Storia dell'Arte, Roma 16524
Istituto Nazionale di Astrofisica – Biblioteca dell'Osservatorio Astronomico di Trieste, Trieste 16626
Istituto Nazionale di Economia Agraria, Roma 16525
Istituto Nazionale di Statistica (ISTAT), Roma 16526
Istituto Nazionale di Studi Etruschi e Italici, Firenze 16264
Istituto Nazionale di Studi Romani, Roma 16527
Istituto Nazionale di Studi sul Rinascimento, Firenze 16265
Istituto Nazionale di Studi Verdiani, Parma 16411
Istituto Nazionale per il Commercio Estero (ICE), Roma 16528
Istituto Nazionale per la Fauna Selvatica Alessandro Ghigi, Ozzano Emilia 16390
Istituto Nazionale per la Grafica, Roma 16529
Istituto Nazionale per la Storia del Movimento di Liberazione in Italia, Milano 16340
Istituto Nazionale per la Valutazione del Sistema dell'Istruzione (INValSI), Frascati 16277
Istituto Nazionale per l'Assicurazione contro gli Infortuni sul Lavoro (INAIL), Roma 16530
Istituto Nazionale per lo Studio della Congiuntura, Roma 16531
Istituto Nazionali di Fisica Nucleare, Frascati 16278
Istituto Niels Stensen, Firenze 14991
Istituto Ortopedico Rizzoli, Bologna 14851
Istituto Papirologico "Girolamo Vitelli", Firenze 16266
Istituto per Giovani Ciechi 'Domenico Martuscelli', Napoli 16375
Istituto per gli Affari Sociali, Roma 16532
Istituto per gli Studi di Politica Internazionale, Milano 16341
Istituto per la Documentazione Giuridica, Firenze 16267
Istituto per la Ricostruzione Industriale, Roma 16533
Istituto per la Scienza dell'Amministrazione Pubblica, Milano 16342
Istituto per la Storia della Resistenza e della Società Contemporanea Nelle Province di Biella e Vercelli, Varallo 16639
Istituto per la valorizzazione del legno e delle specie arboree, Sesto Fiorentino 16575
Istituto per lo Studio degli Ecosistemi, Verbania-Pallanza 15592
Istituto per l'Oriente C. A. Nallino, Roma 16534
Istituto Piemontese per la storia della Resistenza e della società contemporanea 'Giorgio Agosti', Torino 16607
Istituto Pontano, Napoli 16376
Istituto Portoghese di S.Antonia, Roma 16535
Istituto PP. Sacramentini, San Benedetto del Tronto 16047
Istituto Regionale di Studi e Ricerca Sociale, Trento 15537
Istituto Regionale per la Storia del Movimento di Liberazione nel Friuli Venezia Giulia, Trieste 16627
Istituto Ricerche Breda, Breda 16205
Istituto S. Domenico – Cattedra Cateriniana, Teramo 16087
Istituto Salesiano, Caserta 15781
Istituto Salesiano 'Sant'Ambrogio', Milano 15910
Istituto Salesiano Valsalice, Torino 16099
Istituto Siciliano di Studi Bizantini e Neoellenici 'Bruno Lavagnini', Palermo 16402
Istituto Sociale dei Gesuiti, Torino 15635
Istituto Sperimentale Agronomico, Bari 16175
Istituto Sperimentale Agronomico, Modena 16362
Istituto Sperimentale di Caseificio, Lodi 16305
Istituto Sperimentale per la Cerealicoltura, Roma 16536
Istituto Sperimentale per la Cerealicoltura, Vercelli 16663
Istituto Sperimentale per la Patologia Vegetale, Roma 16537
Istituto Sperimentale per la Zoologia Agraria, Firenze 16268
Istituto Sperimentale per l'Agrumicoltura, Acireale 16159
Istituto Sperimentale per le Colture Industriali (CRA), Rovigo 16568
Istituto Sperimentale Talassografico 'A. Cerruti', Taranto 16583
Istituto Storico della Resistenza, Firenze 16269
Istituto Storico della Resistenza e della Società Contemporanea in Provincia di Cuneo, Cuneo 16239
Istituto Storico e di Cultura dell'Arma del Genio, Roma 16538
Istituto Storico Germanico, Roma 16494
Istituto Storico Italiano per il Medioevo, Roma 16539
Istituto Superiore delle Poste e delle Telecomunicazioni, Roma 16540
Istituto Superiore di Sanità, Roma 16541
Istituto Superiore di Scienze Religiose, Milano 15911
Istituto Superiore di Teologia Ecumenica, Bari 15727
Istituto Svizzero di Roma, Roma 16542
Istituto Talassografico, Trieste 16628
Istituto Teologico Calabro 'S. Pio X', Catanzaro 15788
Istituto Teologico Salesiano, Scanzano di Castellamare di Stabia 16064
Istituto Teologico Salesiano 'S. Tommaso D'Aquino', Messina 15899
Istituto Teologico 'Sant'Antonio Dottore', Padova 15956
Istituto Trentino di Cultura – Istituto di Scienze Religiose, Trento 16613
Istituto Universitario di Studi Europei, Torino 15485
Istituto Universitario Navale, Napoli 15131
Istituto Universitario Olandese di Storia dell'Arte, Firenze 14992

Istituto Veneto di Scienze, Lettere ed Arti, Venezia 16653
Istituto Zooprofilattico Sperimentale della Lombardia e dell'Emilia, Brescia 16209
Istituto Zooprofilattico Sperimentale delle Venezie, Padova 16392
Istituzione Biblioteca Classense, Ravenna 16804
Istorichny muzei, Lviv 28415
Istoriki kai Ethnologiki Etairia tis Ellados, Athinai 13117
Istoriko-arkhitekturny musei-zapovednik, Ryazan 23732
Istoriko-arkhitekturny muzei-zapovednik, Nizhni Novgorod 23617
Istoriko-arkhitekturny muzei-zapovednik, Yaroslavl 24086
Istoriko-khudozhestvenny i arkhitekturny muzei, Veliki Novgorod 24040
Istoriko-khudozhestvenny muzei, Pskov 23706
Istra Archaeological Museum, Pula 06216
Istvan Dobó Castle Museum, Eger 13470
István Széchenyi College of Technology, Győr 13276
Italian Alpine Club – Milan Branch, Milano 16333
Italian Cultural Institute, Helsinki 07462
Italian Cultural Institute, Montreal 04405
Italian Cultural Institute, Tokyo 17681
Italian Institute, London 29777
Italian Institute for Historical Studies, Napoli 16374
Italian kulttuuri-instituutti, Helsinki 07462
Italian National Statistical Institute, Roma 16526
Italian Risorgimento National Museum, Library, Torino 16608
Italian School of Archaeology at Athens, Athinai 13123
Italien Geographical Society, Roma 16557
Italienisches Generalkonsulat – Kulturabteilung, München 12309
Italienska Kulturinstitutet i Stockholm 'C.M. Lerici', Stockholm 26652
Itasca Community Library, Itasca 39595
Itä-Uudenmaan maakuntakirjasto, Porvoo 07601
Itawamba Community College, Fulton 33358
ITC-VÚK Inovační technologické centrum -VÚK, a.s., Odolena Voda 06468
Ithaca College, Ithaca 31435
ITT Industries, Clifton 35623
ITT Industries, Vandenberg AFB 36320
ITT Industries – Advanced Engineering & Sciences Division, Kirtland AFB 35808
IUBH / LIS – International University of Applied Sciences Bad Honnef, Bad Honnef 09396
IUCN – The World Conservation Union, Gland 27384
IUFM – Centre de Lille, Lille 07728
IUFM de l'Académie de Nice, Nice 08448
IVALSA – Istituto per la valorizzazione del legno e delle specie arboree, Sesto Fiorentino 16575
IVAM, Valencia 26173
Ivan Franko National University of Lviv, Lviv 28036
'Ivan Tavčar' Public Library, Škofja Loka 24985
'Ivan Tavcar' Public Library, Škofja Loka 24984
Ivan Vazov National Library, Plovdiv 03450
Ivano-Frankivsk City Centralized Library System, Ivano-Frankivsk 28584
Ivano-Frankivsk City Centralized Library System, Main Library, Ivano-Frankivsk 28584
Ivano-Frankivsk National Technical of Oil and Gas, Ivano-Frankivsk 27980
Ivano-Frankivsk State Archives, Ivano-Frankivsk 28240
Ivano-Frankivsk State Medical University, Ivano-Frankivsk 27979
Ivano-Frankivsky derzhavny medychny universytet, Ivano-Frankivsk 27979
Ivano-Frankivsky natsionalny tekhnichny universytet nafti / gazu, Ivano-Frankivsk 27980
M. F. Ivanov Institute for Animal Husbandry in Steppe Regions 'Askania Nova', Askaniya-Nova 28181
Ivanovo Power Institute, Ivanovo 22260
Ivanovo Regional Central Library, Ivanovo 02143
Ivanovo Regional Research Library, Ivanovo 22134
Ivanovo State University, Ivanovo 22262
Ivanovo Textile Academy, Ivanovo 22261
Ivanovo Textile College, Ivanovo 22668
Ivanovskaya oblastnaya biblioteka dlya detei i yunoshestva, Ivanovo 24133
Ivanovskaya oblastnaya nauchnaya biblioteka (IONB), Ivanovo 22134
Ivanovskaya tekstilnaya akademiya, Ivanovo 22261
Ivanovskaya tsentralnaya raionnaya bibliyateka, Ivanovo 02143
Ivanovski gosudarstvenny universitet, Ivanovo 22262
Ivanovski mezhotraslevoi territorialny tsentr nauchno-tekhnicheskoi informatsii i propagandy, Ivanovo 23060
Ivanovski tekstilny kolledzh, Ivanovo 22668
Ivanovsky Institute of Virusulogy, Moskva 23328
Ivatsevichi Regional Central Library, Ivatsevichi 02144
Ivatsevichiskaya tsentralnaya raionnaya bibliyateka, Ivatsevichi 02144
Ive Central Regional Library, Ive 02145
IVEBIC Hemiksem-Schelle, Hemiksem 02777
IVEP, a.s., Brno 06429
Ivevskaya tsentralnaya raionnaya bibliyateka, Ive 02145
IWACO B.V., Rotterdam 19512
Iwanami Shoten Henshubu, Tokyo 17449
Iwanami Shoten, Publishers, Tokyo 17449
Iwate Daigaku, Morioka 17046
Iwate Medical University Library, Morioka 17606
Iwate Prefectural Assembly, Morioka 17280
Iwate University, Morioka 17046
Iwateken Gikai, Morioka 17280
IWK Health Centre for Children, Women & Families, Halifax 04356
Izhevsk State Agricultural Academy, Izhevsk 22266
Izhora Plant Industrial Corporation, Sankt-Peterburg 22924
Iziko: Museums of Cape Town, Cape Town 25111
Izmeron St. Petersburg Leasers' Organization, Sankt-Peterburg 23905
Izyaslav Regional Centralized Library System, Main Library, Izyaslav 28586
Izyaslavska raionna TsBS, Tsentralna biblioteka, Izyaslav 28586
Izyum Regional Centralized Library System, Main Library, Izyum 28587
Izyumska raionna TsBS, Tsentralna biblioteka, Izyum 28587
J. C. Allen & Son, Inc, West Lafayette 36409
J. C. Penney Company Inc, Plano 36136
J. Lewis CrozerLibrary, Chester 38544
J. R. Clarke Public Library, Covington 38719
J. Robert Ashcroft Memorial Library, Phoenixville 32289
J. Sargeant Reynolds Community College – Downtown Campus Learning Resources Center, Richmond 33683
J. Sargeant Reynolds Community College – Parham Campus Learning Resources Center, Richmond 33684
J. V. Fletcher Library, Westford 41855
Jabatan Alam Sekitar, Putrajaya 18616
Jabatan Hal Ehwal Ugama Islam Wilayah Persekutuan, Kuala Lumpur 18550
Jabatan Hasil Dalam Negeri, Kuala Lumpur 18551
Jabatan Pendidikan Negeri Pahang Darul Makmur, Kuantan 18568
Jabatan Penerbangan Awam Malaysia, Kuala Lumpur 18552
Jabatan Pengairan dan Saliran, Kuala Lumpur 18553
Jabatan Penyiaran, Kuala Lumpur 18588
Jabatan Perangkaan, Kuala Lumpur 18554
Jabatan Perdana Menteri, Kuala Lumpur 18555
Jabaton Peguam Negara, Kuala Lumpur 18556
Jabotinsky Institute in Israel, Tel-Aviv 14767
Jack Faucett Associates Library, Bethesda 35499
Jackson Community College, Jackson 33433
Jackson County Law Library, Kansas City 34404
Jackson County Law Library, Medford 34522
Jackson County Library, Ripley 41002
Jackson County Library, Tuckerman 41601
Jackson County Library Services, Medford 40131
Jackson County Public Library, Seymour 41286
Jackson County Public Library System, Marianna 40051
Jackson Kelly, Charleston 35559
Jackson Lewis LLP, New York 35988
Jackson McDonald Barristers and Solicitors, Perth 00982
Jackson Parish Library, Jonesboro 39660
Jackson State Community College, Jackson 33434
Jackson State University, Jackson 31439
Jackson Walker, Houston 35757
Jackson Walker Law Library, Houston 35757
Jackson Walker LLP, Austin 35475
Jackson Walker LLP, Dallas 35649
Jackson-George Regional Library System, Pascagoula 40724
Jackson/Hinds Library System, Jackson 39600
Jackson-Madison County Library, Jackson 39601
Jacksonville Public Library, Jacksonville 39609
Jacksonville Public Library, Jacksonville 39610
Jacksonville Public Library, Jacksonville 39611
Jacksonville Public Library – Beaches Regional, Neptune Beach 40389
Jacksonville Public Library – Bradham-Brooks Northwest Branch, Jacksonville 39612
Jacksonville Public Library – Highlands Regional, Jacksonville 39613
Jacksonville Public Library – Laura Street, Jacksonville 39614
Jacksonville Public Library – Mandarin, Jacksonville 39615
Jacksonville Public Library – Old Middleburg Road, Jacksonville 39616
Jacksonville Public Library – Pablo Creek, Jacksonville 39617
Jacksonville Public Library – Regency Square, Jacksonville 39618
Jacksonville Public Library – San Marco, Jacksonville 39619
Jacksonville Public Library – South Mandarin, Jacksonville 39620
Jacksonville Public Library – Southeast Regional, Jacksonville 39621
Jacksonville Public Library – University Park, Jacksonville 39622
Jacksonville Public Library – Webb Wesconnett Regional, Jacksonville 39623
Jacksonville Public Library – West Regional, Jacksonville 39624
Jacksonville State University, Jacksonville 31448
Jacksonville University, Jacksonville 31449
Jacob Edwards Library, Southbridge 41384
Jacob Gitlin Library, Cape Town 25112
Jadavpur University, Kolkata 13683
Jade Hochschule Wilhelmshaven / Oldenburg / Elsfleth, Elsfleth 09686
Jade Hochschule Wilhelmshaven / Oldenburg / Elsfleth, Oldenburg 10526
Jade Hochschule Wilhelmshaven / Oldenburg / Elsfleth, Wilhelmshaven 10689
Jadranski zavod Knjižnica, Zagreb 06247
Jaffna College, Vaddukoddai 26288
Jagiellonian Library, Kraków 20840
– Astronomical Observatory, Kraków 20869
– Department of General Comparative Linguistics, Kraków 20868
– Faculty of Chemistry, Kraków 20870
– Faculty of Law Library, Kraków 20871
– Institute of Archeology, Kraków 20842
– Institute of Art History, Kraków 20857
– Institute of Classical Philology, Kraków 20848
– Institute of Education, Kraków 20862
– Institute of English Philology, Kraków 20846
– Institute of Ethnology and Cultural Anthropology, Kraków 20845
– Institute of Geography, Kraków 20855
– Institute of Geological Sciences, Kraków 20860
– Institute of German Philology, Kraków 20847
– Institute of History, Kraków 20856
– Institute of History of Religion, Kraków 20865
– Institute of Informatic, Kraków 20858
– Institute of Mathematics, Kraków 20859
– Institute of Molecular Biology, Kraków 20843
– Institute of Oriental Philology, Kraków 20849
– Institute of Philosophy, Kraków 20853
– Institute of Physics, Kraków 20854
– Institute of Polish Philology, Kraków 20850
– Institute of Political Science, Kraków 20861
– Institute of Psychology, Kraków 20864
– Institute of Romance Philology, Kraków 20851
– Institute of Slavic Philology, Kraków 20852
– Institute of Sociology, Kraków 20866
– Institute of Zoology, Kraków 20867
– Medical College, Kraków 20841
– Polonia Institute, Kraków 20863
– Wladyslaw Szafer Institute of Botany, Kraków 20844
Jahangirnagar University, Dhaka 01750
Jai Hind College Library, Mumbai 13823
Jai Narain Vyas University, Jodhpur 13670
Jake Epp Library, Steinbach 04767
Jakobstad City Library, Pietarsaari 07599
Jamaica Library Service, Kingston 16888
Jamaica Theological Seminary and Caribbean Graduate School of Theology, Kingston 16880
Jambi University, Jambi 14264
James A. Haley Veterans Hospital Library, Tampa 37656
James Blackstone Memorial Library, Branford 38263
James Bridie Library, Glasgow 29675
James Cook University, Townsville 00549
James H. Johnson Memorial Library, Deptford 38827
James J. Hill Reference Library, Saint Paul 37500
James Kelly Library at Saint Gregory's University, Shawnee 32661
James Madison University, Harrisonburg 31302
James McBey Art Reference Library, Aberdeen 29587
James Monroe Museum & Memorial Library, Fredericksburg 36879
James Prendergast Library Association, Jamestown 39634
James S. Todd Memorial Library, Chicago 36677
James V. Brown Library of Williamsport & Lycoming County, Williamsport 41899
Jamestown College, Jamestown 31459
Jamestown Community College, Jamestown 33441
Jamia Ahmadiyya Research Library, Rabwah 20525
Jamia Hamdard, New Delhi 13726
Jamia Talim-E-Milli Library, Karachi 20497
Jamiyat-ul-Falah, Karachi 20498
Jammu and Kashmir Academy of Art, Culture and Languages, Srinagar 14213
Jamsetjee Nesserwanji Petit Institute, Mumbai 14094
Jämtland County Library, Östersund 26880
Jämtlands Läns Bibliotek, Östersund 26880
Jan Drda Library Příbram, Příbram 06601
Jan Evangelista Purkyně University – Faculty of Education, Ustí nad Labem 06407
Jan Kasprowicz City Library, Inowrocław 21430
Jan Kochanowski Pedagogical University, Kielce 20820
Jan Smuts House, Johannesburg 25134
Jan van Eyck Academie, Maastricht 19486
Janaček Academy of Music and Dramatic Art, Brno 06353
Janáčkova akademie muzických umění, Brno 06353
Janet D. Greenwood Library, Farmville 31127
Janmabhumi Reference Library, Mumbai 14095
János Arany Museum, Nagykörös 13492
Jànos Xantus Museum, Győr 13476
Janus Pannonius Museum, Pécs 13494
Janus Pannonius Múzeum, Pécs 13494
Janusz-Korczak-Bibliothek, Berlin 12651
Japan Anti-Tuberculosis Association, Research Institute of Tuberculosis, Tokyo 17687
Japan Atomic Energy Research Institute (JAERI), Tokyo 17721
Japan Braille Library, Tokyo 17744
Japan Broadcasting Corp, Broadcasting Culture Research Institute, Tokyo 17736
Japan Catalytic Chemical Industry Co, Ltd, Suita 17425
Japan Centre for Economic Research, Tokyo 17719
Japan Consumer Information Center, Tokyo 17693
Japan Economic Journal, Databank, Tokyo 17467
Japan Economic Research Centre, Tokyo 17725
Japan Exlan Co, Ltd, Okayama 17398
Japan External Trade Organisation (JETRO), Chiba 17523
Japan External Trade Organization, International Economic and Trade Information Center, Tokyo 17717
Japan External Trade Organization, Osaka Business Library, Osaka 17636
Japan Federation of Employers' Associations, Tokyo 17724
Japan Foundation Library Information Center Library, Tokyo 17701
Japan Foundrymen's Society, Tokyo 17737
Japan Highway Public Corp, Machida 17592
Japan Information and Culture Centre, Wellington 19849
Japan Information Center of Science and Technology, Tokyo 17723
Japan International Research Center for Agricultural Sciences, Tsukuba 17801

Japan Leather Research Laboratory, Tokyo 17735
Japan Library Association, Tokyo 17734
Japan Management Association, Tokyo 17742
Japan Marine Science and Technology Center, Yokosuka 17829
Japan Maritime Center, Tokyo 17738
Japan Meteorological Agency, Tokyo 17682
Japan Monkey Centre, Inuyama 17556
Japan Monopoly Corporation, Hadano Tobacco Experiment Station, Hadano 17354
Japan Organo Co, Ltd, Tokyo 17477
Japan Plant Protection Assn, Kodaira 17577
Japan Productivity Center, Tokyo 17731
Japan Racing Association, Public Information, Tokyo 17718
Japan Securities Research Institute, Nagoya 17616
Japan Securities Research Institute, Osaka 17637
Japan Securities Research Institute, Tokyo 17732
Japan Small Business Corp, Tokyo 17436
Japan Society, New York 37216
Japan Society of Civil Engineers, Tokyo 17673
Japan Tobacco Inc., Central Research Institute, Kanagawa 17368
Japan Tobacco Inc, Publicity Section, Tokyo 17468
Japan Transportation Assn, Tokyo 17740
Japan Youth Organisation, Tokyo 17743
Japanese American National Library, San Francisco 37545
Japanese Archaeological Association, Tokyo 17739
Japanese Association of Mineralogists, Petrologists and Economic Geologists, Sendai 17662
Japanese Correctional Association, Tokyo 17706
Japanese Cultural Institute, Roma 16515
Japanese National Railways, Central Hospital, Tokyo 17730
Japanese Railway Welfare Association, Tokyo 17759
Japanese Society of Irrigation, Drainage and Reclamation Engineering, Tokyo 17746
Japanese Standards Association, Overseas Standards Center, Tokyo 17728
Japanisch-Deutsches Zentrum Berlin, Berlin 11537
Japanisches Kulturinstitut Köln, Köln 12144
Japasan Dokumentasi Sastra H.B. Jassin, Jakarta 14365
Jardín Botánico Canario Viera y Clavijo, Las Palmas 26119
Jardin Botanico Nacional, La Habana 06307
Jardin Botanique de Montreal, Montreal 04410
Jardin botanique national de Belgique, Meise 02686
Järfälla folkbibliotek, Järfälla 26801
Järva County Central Library (JCCL), Paide 07250
Järvenpää Public Library, Järvenpää 07569
Järvenpään kaupunginkirjasto, Järvenpää 07569
Jarvis Christian College, Hawkins 31318
Jasper County Public Library, Rensselaer 40967
Jasper County Public Library – DeMotte Branch, DeMotte 38818
Jasper Public Library, Jasper 39637
Jasper-Dubois County Contractual Public Library, Jasper 39638
Jászberényi Tanitókèpzo Főiskola, Jászberény 13278
Jász-Nagykun-Szolnok Megyei Verseghy Ferenc Könyvtár es Mûvelődési Intézet, Szolnok 13550
Jawaharlal Nehru Agricultural University, Jabalpur 13664
Jawaharlal Nehru Technological University, Hyderabad 13656
Jawaharlal Nehru University, New Delhi 13727
Jay County Public Library, Portland 40869
K. P. Jayaswal Research Institute, Patna 14179
Jazykovedný ústav Ľ. Štúra, Bratislava 24707
JDC-Brookdale Institute, Jerusalem 14724
Jean-Paul-Gymnasium, Hof 10740
Jefferson College, Hillsboro 31336
Jefferson Community College, Louisville 33503
Jefferson Community College, Watertown 33845
Jefferson Community College – Southwest Campus, Louisville 33504
Jefferson County Law Library, Birmingham 33986
Jefferson County Law Library Association, Steubenville 34895
Jefferson County Library, High Ridge 39428
Jefferson County Library District, Madras 40014
Jefferson County Public Law Library, Louisville 37068
Jefferson County Public Library, Lakewood 39790
Jefferson County Rural Library District, Port Hadlock 40852
Jefferson Davis Community College, Brewton 33188
Jefferson Davis Parish Library, Jennings 39644
Jefferson Parish Library, Metairie 40165
Jefferson State Community College, Birmingham 33173
Jefferson Township Public Library, Oak Ridge 40567
Jefferson-Madison Regional Library, Charlottesville 38525
Jeffersonville Township Public Library, Jeffersonville 39641
Jeffersonville Township Public Library – Clarksville Branch, Clarksville 38586
Jelgava Town Central Scientific Library, Jelgava 18167
Jelgavas Zinatniska biblioteka, Jelgava 18167
Jenkens & Gilchrist, Austin 35476
Jenkens & Gilchrist, Dallas 35650
Jenner & Block Library, Chicago 35581
Jennings Carnegie Public Library, Jennings 39645
Jennings County Public Library, North Vernon 40528
Jennings, Strouss & Salmon, Phoenix 36117
Jericho Public Library, Jericho 39646
Jernbaneverket, Oslo 20247
Jernkontoret, Stockholm 26653
Jersey City Free Public Library, Jersey City 39647
Jersey Library, St Helier 30075
Jerseyville Public Library, Jerseyville 39648
Jerusalem Academy of Music and Dance, Jerusalem 14639
Jerusalem City Library, Jerusalem 14776
Jerusalem College of Technology, Jerusalem 14676
Jerusalem International Y.M.C.A. Library, Jerusalem 14736
Jervis Public Library Association, Inc, Rome 41058
Jerzy Toeplitz Library, Stawberry Hills 00516
Jessamine County Public Library, Nicholasville 40480
Jessie Ball Dupont Memorial Library, Stratford 37644
Jessie Street National Women's Library, Sydney 01026
Jesuit Library, Dublin 14510
Jesuit Theological Library, Lusaka 42334
Jesuitenkolleg, Innsbruck 01439
Jesuitenkollegium Kalksburg, Wien 01487
Jesus College, Cambridge 29024
Jesus College, Oxford 29325
Jewish Community Berlin, Berlin 11538
Jewish Community Center of Greater Rochester, Rochester 37451
Jewish Community Center of Metropolitan Detroit, West Bloomfield 35414
Jewish Community Library of Los Angeles, Los Angeles 37051
Jewish Federation Libraries – Jewish Community Center Library, Nashville 37144
Jewish Historical Museum, Amsterdam 19365
Jewish Historical Research Institute, Warszawa 21353
The Jewish Library in Stockholm, Stockholm 26654
Jewish Museum, Praha 06534
Jewish Museum Vienna, Wien 01614
Jewish National and University Library, Jerusalem 14598
Jewish Public Library, Montreal 04385
Jewish Theological Seminary, New York 35309
Jewish Theological Seminary Library – University of Jewish Studies, Budapest 13330
JFF – Institut für Medienpädagogik in Forschung und Praxis, München 12310
Jiamusi Normal College, Jiamusi 05307
Ji'an Teachers' College, Ji'an 05308
Jianghan Petroleum University (JHPU), Jingzhou 05339
Jianghan University, Wuhan 05657
Jiangnan University, Wuxi 05671
Jiangsu Agricultural College, Yangzhou 05738
Jiangsu Institute of Technology, Zhenjiang 05793
Jiangsu Teacher's College, Suzhou 05580
Jiangxi Agricultural University, Nanchang 05423
Jiangxi College of Traditional Chinese Medicine, Nanchang 05424
Jiangxi Institute of Administration, Nanchang 05425
Jiangxi Medical College, Nanchang 05426
Jiangxi Normal University, Nanchang 05427
Jiangxi Provincial Library, Nanchang 04967
Jiangxi University of Finance and Economics, Nanchang 05428
Jiaozuo University, Jiaozuo 05310
Jiaxing Engineering Institute, Jiaxing 05311
Jichi Ika Daigaku, Kawachi 16969
Jichi Medical School, Kawachi 16969
Jichisho, Tokyo 17302
Jihočeská univerzita, České Budějovice
– Pedagogická fakulta, České Budějovice 06366
– Výzkumný ústav rybářský a hydrobiologický, Vodňany 06367
– Zemědělská fakulta, České Budějovice 06368
Jihočeská vědecká knihovna v Českých Budějovicích, České Budějovice 06344
Jihočeské muzeum, České Budějovice 06442
Jikei University School of Medicine, Tokyo 17144
Jilin Academy of Arts, Changchun 05078
Jilin Agricultural University, Changchun 05079
Jilin City Library, Jilin City 05919
Jilin College of Agricultural Special Products, Jilin City 05312
Jilin College of Cereals, Oil and Foodstuff, Changchun 05080
Jilin College of Commerce, Changchun 05081
Jilin College of Public Security, Changchun 05082
Jilin Forestry College, Jilin City 05313
Jilin Institute of Chemical Technology, Jilin City 05314
Jilin Institute of Civil Engineering and Architeture, Changchun 05083
Jilin Provincial Library, Changchun 04953
Jilin Teachers' College, Jilin City 05315
Jilin University, Changchun 05084
Jilin University of Technology, Changchun 05085
Jilin Vocational Teachers' College, Changchun 05086
Jimei Financial College, Xiamen 05674
Jimei Navigation College, Xiamen 05675
Jimei Teachers' College, Xiamen 05676
Jimma Junior College of Agriculture, Jimma 07274
Jinan City Library, Jinan 05920
Jinan Educational College, Jinan 05317
Jinan Traffic College, Jinan 05318
Jinan University, Guangzhou 05206
Jinan University, Jinan 05319
Jing Gang Shan Medical College, Ji'an 05309
Jingdezhen Ceramic Institute, Jingdezhen 05337
Jingdezhen Senior College, Jingdezhen 05338
Jingu Bunko, Ise 17558
Jining Medical College, Jinan 05320
Jining Teachers' College, Jining 05340
Jinjiin, Tokyo 17303
Jinko Mondai Kenkyujo, Tokyo 17683
Jinnah Postgraduate Medical Centre, Karachi 20442
Jinzhou Medical College, Jinzhou 05342
Jinzhou Teachers' College, Jinzhou 05343
Jiří Mahen Library Brno, Brno 06556
Jiro Osaragi Memorial Museum, Yokohama 17825
Jishou University, Jishou 05345
Jiujiang Medical College, Juijiang 05347
Jiujiang Teachers' College, Jiujiang 05346
Joe Buley Memorial Library, Third Lake 37669
Joensuu Regional Library – Central Library for North Carelia, Joensuu 07570
Joensuu University Library, Joensuu 07345
Joensuun maakunta-arkisto, Joensuu 07512
Joensuun seutukirjasto – Pohjois-Karjalan maakuntakirjasto, Joensuu 07570
Joensuun yliopisto kirjasto, Joensuu 07345
Jõgeva County Central Library (JCCL), Põltsamaa 07252
Jogtudományi Intézet, Budapest 13414
Johan Béla National Center for Epidemiology, Budapest 13387
Johan Béla Országos Epidemiológiai Központ, Budapest 13387
Johann Heinrich von Thünen-Institut (vTI) / Bundesforschungsinstitut für Ländliche Räume, Wald und Fischerei – Fachinformationszentrum Fischerei, Hamburg 11997
Johann Heinrich von Thünen-Institut (vTI) / Bundesforschungsinstitut für Ländliche Räume, Wald und Fischerei – Fachinformationszentrum Fischerei, Rostock 12441
Johann Heinrich von Thünen-Institut (vTI) / Bundesforschungsinstitut für Ländliche Räume, Wald und Fischerei – Fachinformationszentrum Ländliche Räume, Braunschweig 11664
Johann Heinrich von Thünen-Institut (vTI) / Bundesforschungsinstitut für Ländliche Räume, Wald und Fischerei – Fachinformationszentrum Wald, Hamburg 11998
Johann Heinrich von Thünen-Institute / Federal Research Institute for Rural Areas, Forestry and Fisheries, Braunschweig 11664
Johann Wolfgang Goethe-Universität Frankfurt am Main, Frankfurt am Main 09730
Johann-Adam-Möhler-Institut für Ökumenik, Paderborn 11306
Johanna-Moosdorf-Bibliothek, Berlin 12659
Johann-Beckmann-Gesellschaft e.V., Hoya 12067
Johannes a Lasco Bibliothek, Emden 11206
Johannes Gutenberg-Universität Mainz, Mainz 10268
– Bereichsbibliothek Katholische Theologie und Evangelische Theologie, Mainz 10269
– Bereichsbibliothek Philosophicum, Mainz 10270
– Bereichsbibliothek Physik, Mathematik, Chemie (PMC), Mainz 10271
– Bereichsbibliothek SB II, Mainz 10272
– Bibliothek für Geographie und Geowissenschaften, Mainz 10273
– Fachbereichsbibliothek Biologie, Mainz 10274
– Fachbereichsbibliothek Rechts- und Wirtschaftswissenschaften, Mainz 10275
– Fachbereichsbibliothek Sport, Mainz 10276
– Fachbibliothek Medizin, Mainz 10277
– Fachbibliothek Translation, Sprache und Kultur, Germersheim 10278
– Hochschule für Musik, Fachbereichsbibliothek, Mainz 10279
– Institut für Ägyptologie und Altorientalistik, Mainz 10280
– Institut für Anthropologie, Mainz 10281
– Institut für Ethnologie und Afrikastudien, Mainz 10282
– Institut für Geschichte, Theorie und Ethik der Medizin, Mainz 10283
– Institut für Indologie, Mainz 10284
– Institut für Kunstgeschichte, Mainz 10285
– Institut für Publizistik, Mainz 10286
– Institut für Vor- und Frühgeschichte, Mainz 10287
– Klinikum der Joh. Gutenberg-Universität, Mainz 10288
– Musikwissenschaftliches Institut, Mainz 10289
– Psychologisches Institut, Mainz 10290
Johannes Kepler Universität Linz, Linz 01251
– Bibliothek Juridicum, Linz 01252
– Fachbibliothek für Betriebswirtschaftslehre, Wirtschaftsinformatik und Fachsprachen, Linz 01253
– Fachbibliothek für Chemie und Chemische Technologie, Linz 01254
– Fachbibliothek für Informatik, Linz 01255
– Fachbibliothek für Mathematik, Linz 01256
– Fachbibliothek für Neuere Geschichte und Zeitgeschichte, Linz 01257
– Fachbibliothek für Physik, Linz 01258
– Fachbibliothek für Sozial- und Wirtschaftsgeschichte, Linz 01259
– Fachbibliothek für Soziologie, Linz 01260
– Fachbibliothek für Volkswirtschaftslehre, Linz 01261
Johannesburg College of Education, Johannesburg 25023
Johannesburg Public Library, Johannesburg 25206
Johannes-Künzig-Institut für Ostdeutsche Volkskunde, Freiburg 11895
Johannes-Turmair-Gymnasium, Straubing 10769
Johann-Friedrich-Danneil-Museum, Salzwedel 12458
Johanniter Krankenhaus in Fläming gGmbH, Treuenbrietzen 12550
Johann-Joseph-Fux-Konservatorium, Graz 01192

John XXIII Library, Valletta 18667
John A. Logan College, Carterville 30681
John A. Volpe National Transportation Systems Center, Cambridge 34028
John Abbott College, Sainte-Anne-de-Bellevue 03842
John Basset Memorial Library, Sherbrooke 03861
John Brown University, Siloam Springs 32671
John C. Hart Memorial Library, Shrub Oak 41320
John Carroll University, University Heights 32875
John Curtis Free Library, Hanover 39365
John F. Kennedy University, Pleasant Hill 32321
– Law Library, Walnut Creek 32322
John Fairfax Publications, Sydney 01027
John Fitzgerald Kennedy Cultural Center, Beirut 18222
John G. Shedd Aquarium, Chicago 36678
John H. Lilley Correctional Center, Boley 33995
John Hancock Mutual Life Insurance, Boston 35529
John Hunter Hospital, Rankin Park 00992
John Jay College of Criminal Justice, New York 32066
John Joseph Moakley United States Courthouse, Boston 33998
John Kennedy College (JKC), Vuillemin 18693
John M. Cuelenaere Public Library, Prince Albert 04726
John & Mable Ringling Museum of Art Library, Sarasota 37580
John Mackintosh Library, Gibraltar 13067
John Marshall Law School, Atlanta 36493
The John Marshall Law School, Chicago 30767
John McMullen Associates, Inc, Alexandria 35438
John Peter Smith Hospital, Fort Worth 36870
John Snow, Inc, Boston 35530
John Squire Library, Harrow 29682
John Tyler Community College, Chester 33217
John Van Puffelen Library of the Appalachian Bible College, Mount Hope 35291
John Wesley College, High Point 31333
John Wesley Powell Library of Anthropology, Washington 37785
Johns Hopkins Hospital, Baltimore 36508
Johns Hopkins University – Applied Physics Laboratory, Laurel 31569
Johns Hopkins University, SAIS Bologna Center Library, Bologna 14852
Johns Hopkins University School of Advanced International Studies, Washington 32987
Johns Hopkins University – The Sheridan Libraries, Baltimore 30334
– Abraham M. Lilienfeld Memorial Library, Baltimore 30335
– George Peabody Library, Baltimore 30336
– John Work Garrett Library, Baltimore 30337
– William H. Welch Medical Library, Baltimore 30338
Johns Hopkins University-Peabody Conservatory of Music, Baltimore 30339
Johnsburg Public Library District, Johnsburg 39650
Johnson Bible College, Knoxville 35253
Johnson C. Smith University, Charlotte 30720
Johnson City Public Library, Johnson City 39651
Johnson Controls, Inc, Milwaukee 35893
Johnson County Community College, Overland Park 33610
Johnson County Law Library, Olathe 34632
Johnson County Library, Buffalo 38384
Johnson County Library, Overland Park 40667
Johnson County Library – Antioch, Merriam 40158
Johnson County Library – Blue Valley, Overland Park 40668
Johnson County Library – Cedar Roe, Roeland Park 41052
Johnson County Library – Central Resource, Overland Park 40669
Johnson County Library – Corinth, Prairie Village 40889
Johnson County Library – Lackman, Lenexa 39848
Johnson County Library – Leawood Pioneer Branch, Leawood 39834
Johnson County Library – Oak Park, Overland Park 40670
Johnson County Library – Shawnee Branch, Shawnee 41292
Johnson County Public Library, Franklin 39146
Johnson County Public Library, Paintsville 40686
Johnson Free Public Library, Hackensack 39340
Johnson & Johnson Pharmaceutical Research & Development, Raritan 37431
Johnson & Johnson Pharmaceutical Research & Development LLC, Spring House 37623
Johnson Publishing Co Library, Chicago 35582
Johnson State College, Johnson 31464
Johnson & Wales University, Providence 32386
Johnston Community College, Smithfield 33764
Johnston Public Library, Johnston 39653
Johor Public Library Corporation, Johor Bharu 18622
Jõhvi Central Library, Jõhvi 07245
Jõhvi Keskraamatukogu, Jõhvi 07245
La Joie par les Livres, Paris 08493
Joint Forces Staff College, Norfolk 34613
Joint ILL-ESRF Library, Grenoble 08355
Joint Library of the Dunedin College of Education, Dunedin 19776
Joint Stock Company Molodaya Gvadiya, Information and Bibliographic Centre, Moskva 24167
Joint Stock Company 'Zavod Elektrik', Sankt-Peterburg 22913
Joint Theological Library, Parkville 00788
Joint World Bank – International Monetary Fund Library, Washington 37747
Joint-stock company 'Armaturno-izolyatorny zavod-Energia', Slovyansk 28821
Joint-Stock Company Bashkirgeologia, Ufa 22966
Joint-Stock Company for Design and Cranes and Mining Complexes, Novomoskovsk 22879
Joint-stock company MECHEL, Chelyabinsk 22784
Joint-Stock Company 'Soda', Berezniki 22780
Joliet Junior College, Joliet 33445
Jomo Kenyatta University of Agriculture and Technology, Nairobi 17998
Jonas Avyzius Public LIbrary of Joniskis District Municipality, Joniškis 18326
Jonas Lankutis Public LIbrary of Klaipeda District Municipality, Gargždai 18322
Jonathan Bourne Public Library, Bourne 38250
Jonathan Trumbull Library, Lebanon 39835
Jonavos rajono savivaldybės viešoji biblioteka, Jonava 18325
Jones County Junior College, Ellisville 33319
Jones Day, Atlanta 35460
Jones Day, Cleveland 35616
Jones Day, Dallas 35651
Jones Day, Los Angeles 35841
Jones Day, Pittsburgh 36130
Jones Day, Washington 36358
Jones Library, Inc, Amherst 37947
B. F. Jones Memorial Library, Aliquippa 37911
Jones, Walker, Waechter, Poitevent, Carrere & Denegre, New Orleans 35947
Joniškio rajono savivaldybės Jono Avžiaus viešoji biblioteka, Joniškis 18326
Jönköping skolbibliotekscentral, Huskvarna 26472
Jönköping University, Jönköping 26394
Jönköpings läns museum, Jönköping 26596
Jönköpings stadsbibliotek, Jönköping 26803
Joods Historisch Museum, Amsterdam 19365
City of Joondalup Library Service, Joondalup 01105
Joplin Public Library, Joplin 39662
Jordan University of Science and Technology (JUST), Irbid 17875
Jordanian Armed Forces Headquarters, Amman 17881
Jordbruksdepartementet, Stockholm 26505
Josai Daigaku, Sakado 17098
Josai University, Mizuta Memorial Library, Sakado 17098
Joseph F. Egan Memorial Supreme Court Library, Schenectady 34844
Joseph T. Simpson Public Library, Mechanicsburg 40127
Joseph und Brigitta Troy Bibliothek, Salzburg 01568
Josephine Community Libraries, Inc, Grants Pass 39279
Josephine County Law Library, Grants Pass 34312
Joseph-Wulf-Mediothek, Berlin 11527
Joslyn Art Museum, Omaha 37308
Journal of the Indian Medical Association (JIMA), Kolkata 14052
Journalisthøjskole, Århus 06651
Jovan Zujovic Institute for Geological and Geographical Research, Beograd 24526
Jowfe Technical Library, Benghazi 18254
Joy Memorial Library, Glennallen 31219
Jože Udovič Library, Cerknica 24956
The Jozef Pilsudski Regional and Municipal Public Library in Lodz, Łódź 21480
Józef Piłsudski University of Physical Education, Jędrzeja Śniadeckiego Main Library, Warszawa 20951
Jozef Šafárik University in Košice, Košice 24665
József Attila Megyei Könyvtár, Tatabánya 13552
J/S Company 'Alttrak', Rubtsovsk 22899
JSC High Voltage Direct Current Power Transmission Research Institute, Sankt-Peterburg 23836
JSC 'Institute Yuzhniyigiprogaz', Donetsk 28224
JSC 'KazIMR', Almaty 17955
JSS College of Arts, Mysore 13714
Judah L. Magnes Museum, Berkeley 36533
Judge Francis J. Catana Law Library, Media 34523
The Judges' Library, Dublin 14545
Judges Library, Nagpur 13864
Judicial Library, Freetown 24573
Jüdische Gemeinde zu Berlin, Berlin 11538
Jüdisches Museum der Stadt Wien, Wien 01614
Jüdisches Museum Frankfurt, Frankfurt am Main 11860
Judiska Biblioteket, Stockholm 26654
Judson College, Elgin 31079
Judson College, Marion 31761
Jugendhaus Düsseldorf e.V., Düsseldorf 11200
Jugoslovenska Kinoteka, Beograd 24513
Jugoslovenski Bibliografsko-Informacijski Institut, Beograd 24514
Juhász Gyula Pedagógusképző Kar, Szeged 13295
Juilliard School, New York 32067
Jujo Paper Co, Ltd Central Research Laboratory, Tokyo 17450
Jujo Seishi K. K., Tokyo 17450
Julius Kühn-Institut, Bundesforschungsinstitut für Kulturpflanzen, Braunschweig 11665
Julius Kühn-Institut (JKI), Berlin 11539
Julius Kühn-Institut (JKI), Kleinmachnow 12118
Julius Kühn-Institut (JKI) – Bundesforschungsinstitut für Kulturpflanzen, Quedlinburg 12416
Júliusa Jakobyho Gallery, Košice 24732
Julius-Kühn-Institut (JKI) – Institut für Rebenzüchtung Geilweilerhof, Siebeldingen 12486
Julius-Maximilians-Universität Würzburg, Würzburg 10697
Juneau Public Libraries, Juneau 39664
C. G. Jung Institute of San Francisco, San Francisco 37546
Jungdok Public Library, Seoul 18125
C. G. Jung-Institut Zürich, Küsnacht 27386
Juniata College, Huntingdon 31371
Juniata County Library, Inc, Mifflintown 40194
Junior High School, Yogyakarta 14328
Junior Library, Rose Hill 18704
Junta de Comércio Externo, Maputo 18978
Junta Departamental de Montevideo, Montevideo 42023
Junta General del Principado de Asturias, Oviedo 25739
Juozas Keliuotis Public LIbrary of Rokiskis District Municipality, Rokiškis 18356
Juozas Paukstelis Public LIbrary of Pakruojis District Municipality, Pakruojis 18346
Jurbarko rajono savivaldybės viešoji biblioteka, Jurbarkas 18327
Jurgis Kuncinas Public Library of Alytus, Alytus 18315
Juridiska biblioteket / Sveriges advokatsamfund, Stockholm 26655
Jūrmala Central City Library, Jurmala 18194
Jūrmalas Centrālā bibliotēka, Jurmala 18194
Jurong Regional Library, Singapore 24625
Justice Institute of British Columbia Library, New Westminster 03806
Justizministerium Nordrhein-Westfalen, Düsseldorf 10873
Justizvollzugsanstalt Münster, Münster 11056
Justizzentrum Jena / Oberlandesgericht Jena, Jena 10954
Justus-Liebig-Universität Gießen, Gießen 09818
– Dezentrale Fachbibliothek Geschichts- und Kulturwissenschaften: Klassische Archäologie, Giessen 09819
– Fachbibliothek Agrarwissenschaften, Ökotrophologie, Umweltmanagement und Geographie: Wirtschaftslehre des Haushalts und Verbrauchsforschung, Gießen 09820
– Fachbibliothek Anglistik, Gießen 09821
– Fachbibliothek Germanistik, Gießen 09822
– Fachbibliothek Geschichts- und Kulturwissenschaften: Bibliothek im Historischen Institut, Gießen 09823
– Fachbibliothek Geschichts- und Kulturwissenschaften: Klassische Philologie, Gießen 09824
– Fachbibliothek Mathematik und Informatik, Gießen 09825
– Fachbibliothek Medizin: Geschichte der Medizin, Gießen 09826
– Fachbibliothek Medizin: Medizinische Fachbibliothek im Klinikum, Gießen 09827
– Fachbibliothek Romanistik, Gießen 09828
– Zeughausbibliothek, Gießen 09829
– Zweigbibliothek im Chemikum, Gießen 09830
– Zweigbibliothek im Philosophikum I, Gießen 09831
– Zweigbibliothek im Philosophikum II, Gießen 09832
– Zweigbibliothek Recht und Wirtschaft, Gießen 09833
Jute Technological Research Laboratories, Kolkata 14053
Jutland College for Social Care and Education, Risskov, Risskov 06745
Juvenile Correction Center Library, Saint Anthony 34771
J/V 'AvtoZAZ-Daewoo', Zaporizhzhya 28173
Jydsk Pædagog-Seminarium, Risskov, Risskov 06745
Jysk Center for Videregående Uddannelse, Århus 06652
Det Jyske Musikkonservatorium, Århus 06653
Jyväskylä City Library, Jyväskylä 07572
Jyväskylä City Library – Regional Library of Central Finland, Jyväskylä 07571
Jyväskylä Provincial Archives, Jyväskylä 07514
Jyväskylä University, Jyväskylä 07347
– Department of Teacher Education Library, Jyväskylä 07351
– Institute for Educational Research, Jyväskylä 07349
– Languages Library, Jyväskylä 07348
– Mattilanniemi Campus Library, Jyväskylä 07350
Jyväskylän kaupunginkirjasto – Keski-Suomen maakuntakirjasto, Jyväskylä 07571
Jyväskylän maakunta-arkiston käsikirjasto, Jyväskylä 07514
Jyväskylän Maalaiskunnan Kirjasto, Jyväskylä 07572
Jyväskylän yliopisto, Jyväskylä 07347
– Kielten kirjasto, Jyväskylä 07348
– Koulutuksen tutkimuslaitos, Jyväskylä 07349
– Mattilaniemen kirjasto, Jyväskylä 07350
– Opettajankoulutuslaitoksen kirjasto, Jyväskylä 07351
K. H. Mácha District Library, Litoměřice 06584
K .I. Stapaev Institute of Geological Sciences, Almaty 17964
K. N. Toosi University of Technology, Tehran 14409
K. P. Jayaswal Research Institute, Patna 14179
K. R. Cama Oriental Institute, Mumbai 14085
K. S. Kekelidze Institute of Manuscripts, Tbilisi 09264
Kabaleo Teachers' College, Kabaleo 20552
Kabardino-Balkar State University, Nalchik 22419
Kabardino-Balkarian Medical Library, Nalchik 23607
Kabardino-Balkarskaya respublikanskaya meditsinskaya biblioteka, Nalchik 23607
Kabardino-Balkarski gosudarstvenny universitet, Nalchik 22419
Kabardino-Balkarski nauchno-issledovatelski institut, Nalchik 23608
KABELPROMPROEKT, Moskva 23340
Kabi Pharmacia Therapeutics AB, Helsingborg 26535
Kabinet hudební historie Etnologického ústavu AV ČR, v.v.i., Praha 06494
Kabul University, Kabul 00001

Kabushikikaisha Mitsubishi Sogo Kenkyusho, Tokyo 17451
Kachebere Major Seminary, Mchinji 18483
Kadaster Concernstaf, Apeldoorn 19389
KADOC – Documentatie- en Onderzoekscentrum voor Religie, Cultuur en Samenleving, Leuven 02673
KADOC – Documentation and Research Centre for Religion, Culture and Society, Leuven 02673
Kaduna Polytechnic, Kaduna 19954
Kaduna State Library Board, Kaduna 20056
Kafedra inostrannykh yazykov, Moskva 23341
Kaffrarian Museum, King William's Town 25144
Kagaku Gijutsu Cho, Tokyo 17304
Kagaku Gijutsu Cho Kokuuchu Gijutsu Kenkyujo, Chofu 17529
Kagaku Gijutsu Kenkyujo, Tsukuba 17788
Kagawa Daigaku, Takamatsu 17129
– Nogakubu Bunkan, Kita 17130
Kagawa Prefectural Assembly, Takamatsu 17298
Kagawa Prefecture Administrative Data Room, Takamatsu 17299
Kagawa University, Takamatsu 17129
– Faculty of Agriculture, Kita 17130
Kagawaken Gikai, Takamatsu 17298
Kagawaken Gyosei Shiryoshitsu, Takamatsu 17299
Kagoshima Daigaku, Kagoshima 16958
– Suisan Gakubu Bunkan, Kagoshima 16959
– Usuki Bunkan, Kagoshima 16960
Kagoshima Prefectural Education Center, Kagoshima 17561
Kagoshima Prefectural Library, Kagoshima 17841
Kagoshima Prefecture Agricultural Experiment Station, Kagoshima 17562
Kagoshima University, Kagoshima 16958
– Faculty of Fisheries Branch, Kagoshima 16959
– Usuki Medical Branch, Kagoshima 16960
Kagoshimaken Kyoiku Senta, Kagoshima 17561
Kagoshimaken Nogyo Shikenjo, Kagoshima 17562
Kagumo Teachers College, Nyeri 18010
Kaifeng Medical College, Kaifeng 05349
Kaifeng University, Kaifeng 05350
Kaigai Keizai Kyoryoku Kikin Chosa Kaihatsubu Chosa Daiitka, Tokyo 17684
Kaijo Hoancho, Tokyo 17685
Kaijo Hoancho Toshokan, Tokyo 17305
Kaikei Kensain, Tokyo 17306
Kaisar Library, Kathmandu 19035
Kaiser-Karl-Gymnasium, Aachen 10702
Kaiser-Permanente Medical Center, Bellflower 36528
Kaiser-Permanente Medical Center, Fontana 36856
Kaiser-Wilhelm-Museum, Krefeld 12162
Kaišiadorių rajono savivaldybės viešoji biblioteka, Kaišiadorys 18328
Kaiyo Kagaku Gijutsu Senta, Yokosuka 17829
Kajaani Public Library – Regional Library for Kainuu, Kajaani 07573
Kajaanin kaupunginkirjasto – Kainuun maakuntakirjasto, Kajaani 07573
Kajian Sains Perpustakaan dan Maklumat, Petaling Jaya 18543
Kajima Corp, General Affairs Dept, Tokyo 17452
Kajima Institute of Construction Technology, Chofu 17348
Kajima Kensetsu Gijutsu Kenkyujo, Chofu 17348
Kajima Kensetsu K.K. Somubu Shomuka, Tokyo 17452
Kaju Shikenjo, Tsukuba 17789
Kakatiya University, Warangal 13788
Kakhovka raionna TsBS, Tsentralna biblioteka, Kakhovka 28588
Kakhovka Regional Centralized Library System, Main Library, Kakhovka 28588
Kalamazoo College, Kalamazoo 31474
Kalamazoo Institute of Arts, Kalamazoo 36983
Kalamazoo Public Library, Kalamazoo 39670
Kalamazoo Public Library – Oshtemo, Kalamazoo 39671
Kalamazoo Valley Community College, Kalamazoo 33447
City of Kalgoorlie Library Service, Kalgoorlie 01107
Kalibr Moscow Tool Plant, Moskva 22849
Kalinin Chernorechensk Chemical Plant, Derzhinsk 22786
Kalinin Factory of Engineering, Podolsk 22893
Kalinin Machine Building Plant Ltd, Moskva 22857
Kalinin Plant Industrial Corporation, Sankt-Peterburg 22929
Kalinin Plant Industrial Corporation Recreation Centre, Library, Sankt-Peterburg 24244
Kalinin Worsted Spinning Mill, Moskva 22837
Kaliningrad Regional Universal Scientific Library, Kaliningrad 22136
Kaliningradskaya oblastnaya detskaya biblioteka im. A.P. Gaidara, Kaliningrad 24135
Kaliningradskaya oblastnaya universalnaya nauchnaya biblioteka, Kaliningrad 22136
Kaliningradskaya oblastnaya yunosheskaya biblioteka im. V.V. Mayakovskogo, Kaliningrad 24136
Kaliningradski mezhotraslevoi territorialny tsentr nauchno-tekhnicheskoi informatsii i propagandy, Kaliningrad 23068
Kaliningradski tekhnicheski institut rybnoi promyshlennosti i khozyaistva, Kaliningrad 23069
Kalininski mezhotraslevoi territorialny tsentr nauchno-tekhnicheskoi informatsii i propagandy, Tver 24009
Kalininsky District Centralized Library System, Belinsky Central District Library, Sankt-Peterburg 24256
Kalininsky District Centralized Library System, Library Branch no 2, Sankt-Peterburg 24253
Kalininsky District Centralized Library System, Library Branch no 3, Sankt-Peterburg 24254
Kalininsky District Centralized Library System, Library Branch no 7, Sankt-Peterburg 24255
Kalininsky District Centralized Library System, Znanie Library Branch no 1, Sankt-Peterburg 24252
Kalinivka Regional Centralized Library System, Main Library, Kalinovka 28589
Kalinivkska raionna TsBS, Tsentralna biblioteka, Kalinovka 28589
Kalinkovichi Regional Central Library, Kalinkovichi 02146
Kalinkovichiskaya tsentralnaya raionnaya bibliyateka, Kalinkovichi 02146
Kalix kommunbibliotek, Kalix 26805
Kalmar stadsbibliotek, Kalmar 26806
Kalmyk Institute of Social Sciences of the Russian Academy of Sciences and History, Elista 23040
Kalmyk National Amur-Sanan Library, Elista 22130
Kalmyk State University, Elista 22249
Kalmytskaya respublikanskaya natsionalnaya biblioteka im. Amur-Sanana, Elista 22130
Kalmytski gosudarstvenny Universitet, Elista 22249
Kalmytski institut obshchestvennykh nauk RAN, Elista 23040
Kaluga State Regional Scientific Library named after V.G. Belinski, Kaluga 22137
Kalundborg Bibliotek, Kalundborg 06890
Kalush Regional Centralized Library System, Main Library, Kalush 28590
Kalushka raionna TsBS, Tsentralna biblioteka, Kalush 28590
Kaluzhskaya gosudarstvennaya oblastnaya nauchnaya biblioteka im. V.G. Belinskogo, Kaluga 22137
Kaluzhskaya oblastnaya detskaya biblioteka, Kaluga 24137
Kalvarijos savivaldybės viešoji biblioteka, Kalvarija 18329
Kama Paper and Pulp Industrial Corporation, Krasnokamsk 22813
Kama Polytechnic Institute, Naberezhnye Chelny 22418
Kamada Kyosaikai, Sakaide 17649
KAMAZ Automobile-Assembly Plant Joint-Stock Company, Department of Scientific and Technical Information, Naberezhnye Chelny 22870
Kamchatka Regional Scientific Library, Petropavlovsk-Kamchatski 22166
Kamchatka State Pedagogical University, Petropavlovsk-Kamchatski 22476
Kamchatka State Technical University, Petropavlovsk-Kamchatski 22477
Kamchatskaya kraevaya detskaya biblioteka im. V. Kruchiny, Petropavlovsk-Kamchatski 24201
Kamchatskaya oblastnaya nauchnaya biblioteka im. S.P. Krasheninnikova, Petropavlovsk-Kamchatski 22166
Kamchatski gosudarstvenny pedagogicheski universitet, Petropavlovsk-Kamchatski 22476
Kamchatski gosudarstvenny tekhnicheski universitet, Petropavlovsk-Kamchatski 22477
Kameshwar Singh Darbhanga Sanskrit University, Darbhanga 13637
Kamet Automated Technology Scientific and Industrial Corporation, Kyiv 28159
Kamish-Burunsk Iron More Corporation, Kerch 28145
Kamish-Burunsky zalizorudny kombinat, Kerch 28145
Kammarrätten i Göteborg, Göteborg 26489
Kammarrätten i Jönköping, Jönköping 26492
Kammarrätten i Stockholm, Stockholm 26506
Kammarrätten i Sundsvall, Sundsvall 26698
Kammer für Arbeiter und Angestellte für das Burgenland, Eisenstadt 01517
Kammer für Arbeiter und Angestellte für Kärnten – AK-Bibliothek, Klagenfurt 01672
Kammer für Arbeiter und Angestellte für Kärnten – Sozialwissenschaftliche Studienbibliothek, Klagenfurt 01548
Kammer für Arbeiter und Angestellte für Niederösterreich, Wien 01615
Kammer für Arbeiter und Angestellte Vorarlberg, Feldkirch 01668
Kammergericht, Berlin 10807
Kamoeji International Buddhist Association, Hamamatsu 17543
Kamoeji Kokusaibukkoyoto Kyokai, Hamamatsu 17543
Kamski tsellyulozno-bumazhny kombinat, Krasnokamsk 22813
Kamuzu College of Nursing, Lilongwe 18471
Kamvolno-pryadilnaya fabrika, Slonim 01926
Kamvolno-pryadilnaya fabrika im. Kalinina, Moskva 22837
Kamyanets-Podilsk City Centralized Library System, Main Library, Kamyanets-Podilsk 28591
Kamyanets-Podilsk Regional Centralized Library System, Main Library, Kamyanets-Podilsk 28592
Kamyanets-Podilska miska TsBS, Tsentralna Biblioteka, Kamyanets-Podilsk 28591
Kamyanets-Podilska raionna TsBS, Tsentralna biblioteka, Kamyanets-Podilsk 28592
Kamyanka-Buzka raionna TsBS, Tsentralna biblioteka, Kamyanka-Buzka 28593
Kamyanka-Buzka Regional Centralized Library System, Main Library, Kamyanka-Buzka 28593
Kamyanka-Dniprovska raionna TsBS, Tsentralna biblioteka, Kamyanka-Dniprovska 28594
Kamyanka-Dniprovska Regional Centralized Library System, Main Library, Kamyanka-Dniprovska 28594
Kanagawa Children's Medical Center, Yokohama 17820
Kanagawa Council of Social Welfare, Yokohama 17822
Kanagawa Kenritsa Eisei Tanki Daigaku, Yokohama 17250
Kanagawa Kenritsu Kawasaki Toshokan, Kawasaki 17843
Kanagawa Kenritsu Kodomo Iryo Senta, Yokohama 17820
Kanagawa Prefectural Administration Information Center, Yokohama 17335
Kanagawa Prefectural Assembly, Yokohama 17334
Kanagawa Prefectural College of Nursing and Medical Technology, Yokohama 17250
Kanagawa Prefectural Education Center, Fujisawa 17534
Kanagawa Prefectural Fishery Experiment Station, Miura 17602
Kanagawa Prefectural Kawasaki Library, Kawasaki 17843
Kanagawa Prefectural Library, Yokohama 17865
Kanagawa Shimbun, Henshukyoku, Chosashuppan Henshu-bu, Yokohama 17821
Kanagawa Shimbun Research, Publication and Compilation Division, Yokohama 17821
Kanagawa Women's Center, Fujisawa 17533
Kanagawaken Gikai, Yokohama 17334
Kanagawaken Kenmimbu Kensei Johoshitsu, Yokohama 17335
Kanagawaken Shakai Fukushi Kyogikai, Yokohama 17822
Kanagawaken Suisan Shikenjo, Miura 17602
Kanagawakenritsu Fujin Sogo Senta, Fujisawa 17533
Kanawha County Public Library, Charleston 38516
Kanayawa Kenritsu Kyoiku Senta, Fujisawa 17534
Kanazawa Bunko Museum, Yokohama 17823
Kanazawa College of Art, Kanazawa 16963
Kanazawa Daigaku, Kanazawa 16964
– Igakubu Bunkan, Kanazawa 16965
– Yakugakubu Bunshitsu, Kanazawa 16966
Kanazawa Ika Daigaku, Kahoku 16961
Kanazawa Institute of Technology, Ishikawa 16956
Kanazawa Kogyo Daigaku, Ishikawa 16956
Kanazawa Medical University, Kahoku 16961
Kanazawa Municipal Izumino Library, Kanazawa 17842
Kanazawa Shiritsu Izumino Toshokan, Kanazawa 17842
Kanazawa University, Kanazawa 16964
– Faculty of Pharmaceutical Sciences, Kanazawa 16966
– Medical Branch, Kanazawa 16965
Kandinsky Library, Paris 08486
Kangan Batman Institute of Tafe, Somerton 00514
Kangan Institute of TAFE, Broadmeadows 00562
Kangding Minority Normal Institute, Kangding 05351
Kangwon National University, Chuncheon 18060
Kankakee Community College, Kankakee 33449
Kankakee Public Library, Kankakee 39674
Kankyocho, Tokyo 17307
Kannada Sahitya Parishat, Bangalore 13938
Kano State Library Board, Kano 20057
Kano State Polytechnic, Kano 19957
Kansai Daigaku, Suita 17127
– Keizai Seiji Kenkyujo, Suita 17128
Kansai Denryoku K. K. Sogo Gijutsu Kenkyujo, Amagasaki 17344
Kansai Economic Federation, Osaka 17634
Kansai Electric Power Co, Inc, Osaka 17408
Kansai Electric Power Co, Inc, Technical Research Center, Amagasaki 17344
Kansai Gaikokugo Daigaku, Hirakata 16948
Kansai Ika Daigaku, Moriguchi 17045
Kansai Keizai Rengokai, Osaka 17634
Kansai Kenryoku K. K. Shomubu Bunshoka Toshoshitsu, Osaka 17408
Kansai Medical University, Moriguchi 17045
Kansai University, Suita 17127
– Institute of Economic and Political Studies, Suita 17128
Kansai University of Foreign Studies, Hirakata 16948
Kansallinen audiovisuaalinen arkisto, Helsinki 07463
Kansallisarkisto, Helsinki 07464
Kansalliskirjasto, Helsinki 07326
Kansan arkisto, Helsinki 07466
Kansaneläkelaitoksen tietopalvelu, Helsinki 07467
Kansas City Art Institute Library, Kansas City 36985
Kansas City Kansas Community College, Kansas City 33451
Kansas City, Kansas Public Library, Kansas City 39675
Kansas City, Kansas Public Library – Argentine Branch, Kansas City 39676
The Kansas City Public Library, Kansas City 39677
The Kansas City Public Library – Lucile H. Bluford Branch, Kansas City 39678
The Kansas City Public Library – North-East, Kansas City 39679
The Kansas City Public Library – Plaza, Kansas City 39680
The Kansas City Public Library – Trails West, Independence 39556
The Kansas City Public Library – Waldo Community, Kansas City 39681
Kansas City Star Library, Kansas City 35794
Kansas Department of Corrections, El Dorado 34176
Kansas Department of Corrections, Ellsworth 34182
Kansas Department of Corrections, Lansing 34435
Kansas Department of Transportation Library, Topeka 37676
Kansas Geological Survey Library, Lawrence 34444
Kansas Heritage Center Library, Dodge City 36808
Kansas State Historical Society, Topeka 37677
Kansas State University, Manhattan 31746
– Mathematics & Physics Library, Manhattan 31747
– Paul Weigel Library of Architecture, Planning & Design, Manhattan 31748

– Veterinary Medical Library, Manhattan 31749
Kansas State University – Salina College of Technology & Aviation, Salina 32554
Kansas Supreme Court, Topeka 37678
Kansas University Medical Center, Kansas City 36986
Kansas Wesleyan University, Salina 32555
Kansk Technological Professional School, Kansk 22669
Kanski tekhnologicheski tekhnikum, Kansk 22669
Kanthal AB, Hallstahammar 26534
Kanto Gakuin Daigaku, Yokohama 17237
Kanto Gakuin University, Yokohama 17237
Kanton Basel-Landschaft, Archäologie und Museum, Liestal 27238
Kantonsbibliothek Appenzell Ausserrhoden, Trogen 26984
Kantonsbibliothek Baselland, Liestal 26972
Kantonsbibliothek Graubünden, Chur 26966
Kantonsbibliothek Nidwalden, Stans 26983
Kantonsbibliothek Obwalden, Sarnen 27496
Kantonsbibliothek Schwyz, Schwyz 26979
Kantonsbibliothek Thurgau, Frauenfeld 26967
Kantonsbibliothek Uri, Altdorf 26963
Kantonsbibliothek Vadiana, St.Gallen 26982
Kantonsspital St.Gallen, St.Gallen 27437
Kao Co, Ltd, Tokyo Research Laboratory, Tokyo 17453
Kao K. K. Tokyo Kenkyujo, Tokyo 17453
Kaohsiung Medical College, Kaohsiung 27530
Kaohsiung Municipal Library, Kaohsiung 27627
Kapi'olani Community College, Honolulu 33415
Kapitelbibliotheken der Diözese Augsburg, Augsburg 11153
Kapitelsbibliothek der St. Mangkirche, Kempten 11255
Kapiti District Libraries, Paraparaumu 19880
P.L. Kapitsa Institute for Physical Problems of the Russian Academy of Sciences, Moskva 23266
Kapitsa Institute for Physical Problems of the Russian Academy of Sciences, Moskva 23266
KAPL Inc Libraries, Schenectady 36260
Kaposvár University, Faculty of Pedagogy, Kaposvár 13279
Kaposvári Egyetem, Kaposvár 13279
Kapucinski samostan, Škofja Loka 24901
Kapuzina – Bibliothek, Brixen 15758
Kapuzinerkloster, Leibnitz 01447
Kapuzinerkloster, Rapperswil 27267
Kapuzinerkloster, Wien 01488
Kapuzinerkloster, Wiener Neustadt 01494
Kapuzinerkloster Feldkirch, Feldkirch 01422
Kara muzeja biblioteka, Riga 18182
Karachaevo-Cherkesskaya respublikanskaya detskaya biblioteka, Cherkessk 24124
Karachaevo-Cherkesskaya universalnaya nauchnaya biblioteka, Cherkessk 22126
Karachi Theosophical Society, Karachi 20474
Karadag Natural Reserve, Feodosiya 28237
Karadeniz Teknik Üniversitesi, Trabzon 27869
A. I. Karaev Institute of Physiology, Baku 01715
Karaganda Metallurgical Plant, Temirtau 17933
Karaganda Polytechnic, Karaganda 17908
Karaganda State University of the name of the academician E.A. Buketov, Karaganda 17909
Karagandinski gosudarstvenny universitet im. akademika E.A. Buketova, Karaganda 17909
Karakalpak Art Museum, Nukus 42118
The Karanga State Technical University, Karaganda 17908
V. N. Karazin Kharkiv National University, Kharkiv 27993
Karbe-Wagner-Archiv, Neustrelitz 12365
Karbyuratorno-armaturny zavod, Sankt-Peterburg 22914
Karbyuratorny zavod, Moskva 22838
Karel de Grote-Hogeschool, Antwerpen
– Departement Industriële Wetenschappen en Technologie, Campus Hoboken, Antwerpen 02266
– Departement Lerarenopleiding, Antwerpen 02267
– Departement Lerarenopleiding, Antwerpen 02268
Karel Dvoracek Library, Vyškov 06619
Karelian Centre of Scientific and Technical Information and Popularization, Petrozavodsk 23701
Karelian Museum of Fine Arts, Petrozavodsk 23699
Karelian Research Center of the Russian Academy of Sciences, Petrozavodsk 23702
Karelian State Museum of Regional History, Petrozavodsk 23700
Karelian State Pedagogical University, Petrozavodsk 22479
Karelski gosudarstvenny kraevedcheski muzei, Petrozavodsk 23700
Karelski gosudarstvenny pedagogicheski universitet (KGPU), Petrozavodsk 22479
Karelski mezhotraslevoi territorialny tsentr nauchno-tekhnicheskoi informatsii i propagandy, Petrozavodsk 23701
Karelski nauchny tsentr RAN, Petrozavodsk 23702
Karl von Vogelsang-Institut, Wien 01616
Karl-Franzens-Universität Graz, Graz 01193
– Fachbibliothek für Anglistik/Amerikanistik, Graz 01194
– Fachbibliothek für Erziehungswissenschaft, Graz 01195
– Fachbibliothek für Geographie und Raumforschung, Graz 01196
– Fachbibliothek für Germanistik, Graz 01197
– Fachbibliothek für Geschichte, Graz 01198
– Fachbibliothek für Mathematik, Graz 01199
– Fachbibliothek für Romanistik, Graz 01200
– Fachbibliothek für Slawistik, Graz 01201
– Fachbibliothek für Translationswissenschaft, Graz 01202
– Fakultätsbibliothek Theologie, Graz 01203
– Institut für Alte Geschichte und Altertumskunde, Graz 01204
– Institut für Archäologie, Graz 01205
– Institut für Erdwissenschaften – Bereich Geologie und Palaeontologie, Graz 01206
– Institut für Ethik und Sozialwissenschaft, Graz 01207
– Institut für Klassische Philologie, Graz 01208
– Institut für Musikwissenschaft, Graz 01209
– Institut für Pflanzenwissenschaften – Bereich Systematische Botanik und Geobotanik, Graz 01210
– Institut für Psychologie, Graz 01211
– Institut für Sportwissenschaft, Graz 01212
– Institut für Sprachwissenschaft, Graz 01213
– Institut für Theoretische Physik, Graz 01214
– Rechts-, Sozial- und Wirtschaftswissenschaftliche Fakultätsbibliothek, Graz 01215
Karl-Franzens-University, Graz 01193
– Department of Archaeology, Graz 01205
Karlivka Regional Centralized Library System, Main Library, Karlivka 28595
Karlivska raionna TsBS, Tsentralna biblioteka, Karlivka 28595
Karlsborgs kommunbibliotek, Karlsborg 26807
Karlshochschule International University, Karlsruhe 10057
Karlskogas stadsbibliotek, Karlskoga 26809
Karlskrona stadsbibliotek, Karlskrona 26810
Karlstad University, Karlstad 26396
Karlstads stadsbibliotek, Karlstad 26811
Karlstads Universitet, Karlstad 26396
Karl-Thomas-Bibliothek (KTB), Göttingen 11931
Karmelitenbibliothek, Bamberg 11158
Karmelitenkloster, Graz 01430
Karmelitenkloster, Linz 01449
Karmelitenkloster, Straubing 11336
Karmeliterbibliothek, Mainz 11269
Karmøy folkebibliotek, Kopervik 20360
Karnatak College, Dharwar 13803
Karnatak Sangha, Mumbai 14096
Karnataka Government Secretariat Library, Bangalore 13850
Karnataka Regional Engineering College, Srinivasanagar 13769
Karnataka University, Dharwad 13644
Kärntner Landesarchiv, Klagenfurt 01549
Kärntner Landesbibliothek am Landesmuseum, Klagenfurt 01550
Kärntner Landeskonservatorium, Klagenfurt 01241
Karol Adamiecki University of Economics. Main Library, Katowice 20814
Karol Marcinkowski University Medical Sciences of Poznan, Poznań 20900
Karolina Praniauskaite Public Library of Telsiai District Municipality, Telšiai 18369
Karolinen Library, Karlstad 26603
Karolinska Institute University Library, Huddinge 26392
Karolinska Institutet, Huddinge 26392
Karolinska Institutet, Stockholm 26428
Karolinska University Hospital, Stockholm 26656
Karpeles Manuscript Library Museum, Buffalo 36611
Karpenko Physics and Mechanics Institute, Lviv 28407
Karpov Scientifical Institute of Physical Chemistry, Moskva 23372
Karratha College, Karratha 00572
Kartause Marienau, Bad Wurzach 11157
Kartinaya galereya, Vologda 24071
Kartinnaya galereya, Kursk 23133
Kasama Provincial Library, Kasama 42352
Kasetsart University, Bangkok 27698
Kashi Teachers' College, Kashi 05352
Kashi Vidyapith, Varanasi 13784
A. Kasteyev Kazakh State Art Museum, Almaty 17956
Kate Love Simpson Morgan County Library, McConnelsville 40117
Katechetische Bibliothek, Frankfurt am Main 11212
Katechetisches Institut des Bistums Aachen, Aachen 11146
Katholieke Hogeschool Brugge-Oostende, Oostende 02417
Katholieke Hogeschool Kempen, Geel
– Bibliotheek Campus Turnhout, Turnhout 02310
– Campus Geel, Geel 02311
– Departement Lerarenopleiding, Vorselaar 02312
Katholieke Hogeschool Leuven, Leuven
– Bibliotheek DLO-Heverlee-Echo, Heverlee 02349
– Departement Gezondheidszorg en Technologie, Leuven 02350
– Departement Lerarenopleiding, Diest 02351
– Departement Sociale Hogeschool Heverlee, Heverlee 02352
Katholieke Hogeschool Limburg (KHLim), Hasselt
– Departement Handelswetenschappen & Bedrijfskunde – Industriële Wetenschappen & Technologie, Diepenbeek 02342
– Department Lerarenopleiding Kleuter, Hasselt 02343
– Department Lerarenopleiding Secundair Onderwijs, Diepenbeek 02344
Katholieke Hogeschool Mechelen, Mechelen 02407
– Opleidingsbibliotheek Coloma, Mechelen 02408
Katholieke Hogeschool Zuid-West-Vlaanderen (KATHO), Kortrijk
– Bibliotheek campus Kortrijk, Kortrijk 02347
– Bibliotheek campus Torhout, Torhout 02348
Katholieke Radio Omroep (KRO), Hilversum 19465
Katholieke Universiteit Leuven, Leuven 02353
– Archief en Documentatiecentrum voor Historische Pedagogiek (ADHP), Leuven 02354
– Bibliotheek Rechtsgeleerdheid, Leuven 02355
– Biomedische Bibliotheek Gasthuisberg, Leuven 02356
– Campus Sociale School Heverlee, Leuven 02357
– Campusbibliotheek Arenberg, Agronomische Bibliotheek, Heverlee 02358
– Campusbibliotheek Arenberg, Cluster 4 – Bewegings- en Revalidatiewetenschappen, Heverlee 02359
– Campusbibliotheek Arenberg, WBIB afdelingen WMAG en WDEP, Leuven 02360
– Campusbibliotheek Kortrijk, Kortrijk 02361
– Dienst voor Studieadvies (DSA), Leuven 02362
– Faculteitsbibliotheek Economische en Toegepaste Economische Wetenschappen, Leuven 02363
– Faculteitsbibliotheek Letteren, Leuven 02364
– Faculteitsbibliotheek Psychologie en Pedagogische Wetenschappen, Leuven 02365
– Faculteitsbibliotheek Sociale Wetenschappen, Leuven 02366
– Hoger Instituut voor Wijsbegeerte, Leuven 02367
– Maurits Sabbebibliotheek, Leuven 02368
– Oost-Aziatische Bibliotheek, Leuven 02369
Katholische Akademie in Bayern, München 11280
Katholische Akademie Schwerte, Schwerte 11329
Katholische Akademie und Heimvolkshochschule, Lingen (Ems) 11264
Katholische Büchereiarbeit, Würzburg 11368
Katholische Hochschule für Sozialwesen Berlin, Berlin 09451
Katholische Hochschule Nordrhein-Westfalen (KatHO NRW), Köln 10129
Katholische Hochschulgemeinde aki, Zürich 27281
Katholische Öffentliche Bücherei St. Gertrud, Lohne 11265
Katholische Propsteigemeinde St. Viktor Xanten, Xanten 11370
Katholische Sozialakademie Österreichs, Wien 01617
Katholische Soziale Akademie Franz Hitze Haus, Münster 10757
Katholische Sozialwissenschaftliche Zentralstelle, Mönchengladbach 11276
Katholische Stiftungsfachhochschule München, München 10366
Katholische Stiftungsfachhochschule München, München 11281
Katholische Universität Eichstätt, Eichstätt 09685
Katholische Universitätsgemeinde, Basel 27246
Katholisches Bibelwerk e.V., Stuttgart 11339
Katholisches Bildungswerk Salzburg, Elsbethen 01518
Katholisch-Soziales Institut, Bad Honnef 11155
Katholisch-Theologische PrivatUniversität Linz, Linz 01262
Katholisch-Theologisches Seminar, Tübingen 10631
Katholisch-Theologisches Seminar an der Philipps-Universität Marburg, Marburg 10300
Katolicki Uniwersytet Lubelski Jana Pawła II, Lublin 20882
– Widział Nauk Społecznych, Stalowa Wola 20883
– Wydział Filozofii Chrześcijańskiej, Lublin 20884
– Wydział Nauk Humanistycznych, Lublin 20885
– Zakład Języka Polskiego, Lublin 20886
Katona József Megyei Könyvtár, Kecskemét 13539
Katona József Múzeum, Kecskemét 13483
Katona József Public Library, Kecskemét 13539
Katonah Village Library, Katonah 39683
Katrineholms bibliotek, Katrineholm 26813
Katten, Muchin & Rosenman, Washington 36359
Katten Muchin Rosenman LLP, New York 35989
Katten, Muchin, Rosenman LLP Library, Chicago 35583
Kauai Community College, Lihue 33486
Kaubisch Memorial Public Library, Fostoria 39136
Kaukas Oy, Lappeenranta 07424
Kaukas Oy, Information Service, Lappeenranta 07424
Kaukauna Public Library, Kaukauna 39687
Kaunas County Public Library, Kaunas 18330
Kaunas University of Medicine, Kaunas 18276
Kaunas University of Technology, Kaunas 18277
– Institute of Architecture and Construction, Kaunas 18278
Kauniaisten kaupunginkirjasto – Grankulla stadsbibliotek, Kauniainen 07574
Kauno apskrities viešoji biblioteka, Kaunas 18330
Kauno medicinos universitetas, Kaunas 18276
Kauno miesto savivaldybės Vinko Kudirkos viešoji biblioteka, Kaunas 18331
Kauno rajono savivaldybės viešoji biblioteka, Garliava 18323
Kauno technologijos universitetas, Kaunas 18277
– Architekturos ir statybos instituto, Kaunas 18278
R.E. Kavetsky Institute of Experimental Pathology, Oncology and Radiobiology, Kyiv 28319
Kävlinge bibliotek, Kävlinge 26814
Kawasaki Heavy Industries, Ltd, Technical Institute, Akashi 17341
Kawasaki Jukogyo K. K., Akashi 17341
Kawasaki Seitetsu K. K. Gijutsu Kenkyujo, Chiba 17346
Kawasaki Steel Corporation, Technical Research Laboratories, Chiba 17346
Kaye Scholer LLP, Los Angeles 35842
Kaye Scholer LLP, New York 35990
Kazakh Agrarian University, Akmola 17892
Kazakh Agricultural Institute, Almaty 17898

Kazakh Grain Farming Research Institute, Shortandy 17977
Kazakh Institute of Chemical Technology, Shimkent 17920
Kazakh Institute of Sports, Almaty 17899
Kazakh National Technical University, Almaty 17900
Kazakh Pedagogical Institute, Almaty 17901
Kazakh Pedagogical Institute for Women, Almaty 17902
Kazakh Scientific Research Institute of Plant Protection, Rakhat 17973
A. Kasteyev Kazakh State Art Museum, Almaty 17956
Kazakh State University of World Languages, Almaty 17903
Kazakhstan Paediatrics Research Institute, Almaty 17957
Kazan State Conservatory of Music, Kazan 22273
Kazan State Technical University, Kazan 22275
Kazan State University, Kazan 22276
Kazan State University of Architecture and Engineering, Kazan 22274
Kazanskaya gosudarstvennaya konservatoriya, Kazan 22273
Kazanski gosudarstvennaya arkhitekturno-stroitelny universitet, Kazan 22274
Kazanski gosudarstvenny tekhnicheski universitet, Kazan 22275
Kazanski gosudarstvenny universitet, Kazan 22276
Kazanski natsionalny universitet im. al-Farabi, Almaty 17904
Kazlų Rūdos savivaldybės viešoji biblioteka, Kazlų Rūda 18332
KB Glavrechflota, Gomel 01850
KB sredstv mekhanizatsii, Sankt-Peterburg 23805
KBA Fachstelle für Katholische Büchereiarbeit, Würzburg 11368
KBC Banking & Insurance, Brussel 02503
KCL, Oy Keskuslaboratorio – Centrallaboratorium AB, Espoo 07435
KDZ – Centre for Public Administration Research, Wien 01618
KDZ – Zentrum für Verwaltungsforschung, Wien 01618
Kean University, Union 32869
Kearfott Guidance & Navigation Corp, Wayne 36401
Kearney Public Library, Kearney 39688
Kearny Public Library, Kearny 39690
Keats House, London 29778
Keats-Shelley Memorial Association, Roma 16543
Keble College, Oxford 29326
Kecskemet College Library and Information Centre, Technical and Economical Library, Kecskemét 13280
Kedah State Public Library Corporation, Alor Setar 18618
Kėdainių rajono savivaldybės Mikalojaus Daukšos viešoji biblioteka, Kėdainiai 18333
Keele University Library, Keele 29141
Keelung City Library, Keelung 27628
Keen Mountain Correctional Center, Oakwood 34624
Keene Memorial Library, Fremont 39164
Keene Public Library, Keene 39691
Keene State College, Keene 31491
Keeneland Association, Lexington 37026
Keesal, Young & Logan, Long Beach 35820
Kegichivka Regional Centralized Library System, Main Library, Kegichivka 28596
Kegichivska raionna TsBS, Tsentralna biblioteka, Kegichivka 28596
Kegler, Brown, Hill & Ritter, Columbus 35633
Kehittämiskeskus Oy, Helsinki 07453
Kehityshyteistyön palvelukeskus – KEPA, Helsinki 07468
Keimyung University, Daegu 18061
Keio Gijuku Toshokan, Tokyo 17145
– Bungakubu Toshokan Toshoshitsu, Tokyo 17146
– Igaku Media Senta, Tokyo 17147
– Kenkyujo Shido Bunko, Tokyo 17148
– Rikogaku Media Senta, Yokohama-shi 17149
– Sangyo Kenkyujo, Tokyo 17150
Keio University, Tokyo 17145
– Information and Library Center for Science and Technology, Yokohama-shi 17149
– Institute of Management and Labor Studies, Tokyo 17150
– Institute of Oriental Classics, Tokyo 17148
– Medical Information and Media Center, Tokyo 17147
– School of Library and Information Science, Tokyo 17146
Keisatsucho Toshokan, Tokyo 17308

Keisen Jogakuen Junior College, Horticulture Dept, Isehara 17244
Keisen Jogakuen Tanki Daigaku Engeiseikatsugakka, Isehara 17244
Keiser College, Fort Lauderdale 33345
Keizai Dantai Rengokai (KEIDAN-REN), Tokyo 17686
Keizai Kikakucho Toshokan, Tokyo 17309
K. S. Kekelidze Institute of Manuscripts, Tbilisi 09264
Kekkaku Yobokai, Tokyo 17687
Kelantan Public Library Corporation, Kota Bharu 18625
Keldysh Applied Mathematics Institute of the Russian Academy of Sciences, Moskva 23311
Keller Public Library, Keller 39692
Kelley Drye Collier Shannon PLLC Library, Washington 36360
Kelley Drye & Warren, New York 35991
Kellogg, Brown & Root Library, Houston 36944
Kellogg Brown & Root PTY LTD, Brisbane 00880
Kellogg Community College, Battle Creek 33156
Kellogg-Hubbard Library, Montpelier 40280
Kelly, Hart & Hallman, Fort Worth 35712
Kelmės rajono savivaldybės Žemaitės viešoji biblioteka, Kelmė 18334
KEMA IEV – Ingenieurunternehmen für Energieversorgung GmbH, Dresden 11386
KEMA Nederland BV, Arnhem 19324
Kementerian Belia dan Sukan, Kuala Lumpur 18557
Kementerian Kewangan, Kuala Lumpur 18558
Kementerian Luar Negeri, Kuala Lumpur 18559
Kementerian Pendidikan Malaysia, Kuala Lumpur 18560
Kementerian Penerangan, Kuala Lumpur 18561
Kementerian Sumber Manusia, Kuala Lumpur 18562
Kemerovo Centralized Library System – Gogol Central Library, Kemerovo 24142
Kemerovo Railway, Kemerovo 23090
Kemerovo Regional Scientific Fyodorov Library, Kemerovo 22139
Kemerovo Regional Scientific Medical Library, Kemerovo 23089
Kemerovo State University, Kemerovo 22282
Kemerovskaya oblastnaya biblioteka dlya detei i konoshestva, Kemerovo 24140
Kemerovskaya oblastnaya nauchnaya biblioteka im.Fyodorova, Kemerovo 22139
Kemerovskaya oblastnaya nauchnaya meditsinskaya biblioteka, Kemerovo 23089
Kemerovskaya oblastnaya yunosheskaya biblioteka, Kemerovo 24141
Kemerovskaya tsentralizovannaya bibliotechnaya sistema, Kemerovo 24142
Kemerovskaya zheleznaya doroga, Kemerovo 23090
Kemerovski gosudarstvenny universitet, Kemerovo 22282
Kemerovski mezhotraslevoi territorialny tsentr nauchno-tekhnicheskoi informatsii i propagandy, Kemerovo 23091
Kemi Public Library, Kemi 07575
Kemiart Liners Oy, Kemi 07422
Kemijärven kaupunginkirjasto, Kemijärvi 07576
Kemijärvin Public Library, Kemijärvi 07576
Kemin kaupunginkirjasto, Kemi 07575
Kemira Oys, Oulu 07426
Kemp, Smith PC, El Paso 35686
Kempsey Shire Library, West Kempsey 01176
Kempton Park Public Library, Kempton Park 25207
Kemptville College, Kemptville 03997
Kenai Community Library, Kenai 39694
Kenchiku Kenkyujo, Tsukuba 17790
Kendall College, Evanston 33328
Kendall College of Art & Design, Grand Rapids 31238
Kendall Institute Library, New Bedford 37152
Kendall Young Library, Webster City 41794
Kendallville Public Library, Kendallville 39696
Keneseth Israel Reform Congregation, Elkins Park 36829
Ken-i Kai Ganka Toshokan, Tokyo 17688
Ken-Ikai Foundation, Ophthalmological Library, Tokyo 17688
Kenkohoken Kumiai Rengokai, Tokyo 17689
Kenkyu Joho Center, Wako 17814
Kennan Institute for Advanced Russian Studies, Washington 37809
Kennedy-King College, Chicago 33222
Kennesaw State University, Kennesaw 31493
Kenniscentrum Amsterdam, Amsterdam 19237

Kenosha Public Library, Kenosha 39703
Kenosha Public Library – Northside Library, Kenosha 39704
Kenosha Public Museums Library, Kenosha 36994
Kenrick-Glennon Seminary, Saint Louis 35361
Kensetsusho, Tokyo 17310
Kent County Council, Maidstone 30039
Kent County Public Library, Chestertown 38548
Kent District Library, Comstock Park 38675
Kent District Library, Grand Rapids 39275
Kent District Library – Byron Township Branch, Byron Center 38406
Kent District Library – Cascade Township Branch, Grand Rapids 39276
Kent District Library – East Grand Rapids Branch, East Grand Rapids 38895
Kent District Library – Grandville Branch, Grandville 39277
Kent District Library – Kentwood Branch, Kentwood 39706
Kent District Library – Walker Branch, Walker 41709
Kent District Library – Wyoming Branch, Wyoming 41981
Kent Memorial Library, Suffield 41465
Kent State University, Ashtabula 30268
Kent State University, Burton 30605
Kent State University, East Liverpool 31065
Kent State University, Kent 31497
– Architecture Library, Kent 31498
– Chemistry & Physics Library, Kent 31499
– Mathematics & Computer Science Library, Kent 31500
– Music Library, Kent 31501
Kent State University, New Philadelphia 32018
Kent State University, North Canton 32140
Kent State University, Salem 32548
Kent State University, Warren 32950
Kenton County Public Library, Covington 38720
Kenton County Public Library – Erlanger Branch, Erlanger 38986
Kenton County Public Library – Independence, Independence 39557
Kentro Erevnis Ellinikis Laografias, Athinai 13118
Kentron Library, Irene 25091
Kentucky Christian College, Grayson 31243
Kentucky Department for Environmental Protection, Frankfort 34280
Kentucky Department of Public Advocacy Library, Frankfort 36875
Kentucky Historical Society Library, Frankfort 36876
Kentucky Mountain Bible College, Jackson 35242
Kentucky Regional Library for the Blind & Physically Handicapped, Frankfort 36877
Kentucky School for the Blind, Louisville 33505
Kentucky State Reformatory Library, La Grange 34424
Kentucky State University, Frankfort 31170
Kentucky Wesleyan College, Owensboro 32219
Kenya Agricultural Research Institute, Nairobi 18038
Kenya Bureau of Standards Library, Nairobi 18039
Kenya College of Communications, Nairobi 18015
Kenya Institute of Administration, Kabete 18012
Kenya Institute of Administration, Nairobi 17999
Kenya Medical Training College, Nairobi 18000
Kenya National Academy of Sciences (KNAS), Nairobi 18040
Kenya National Archives and Documentation Service, Nairobi 18041
Kenya National Library Service, Nairobi 17995
Kenya Polytechnic College, Nairobi 18001
Kenya Science Teachers' College, Nairobi 18002
Kenya Utalii College, Nairobi 18016
Kenyatta University, Nairobi 18003
Kenyon College, Gambier 31204
Kenyon & Kenyon LLP, New York 35992
KEPA, Helsinki 07468
Képző- és Iparművészeti Szakközépiskola, Budapest 13311
Kerala Agricultural University, Thrissur 13772
Kerala State Planning Board, Thiruvananthapuram 13897
Keramiekmuseum Princessehof, Leeuwarden 19472
Kerava Public Library, Kerava 07577
Keravan kaupunginkirjasto, Kerava 07577

Kerch Regional Centralized Library System, Belinsky Main Library, Kerch 28597
Kerchenska raionna TsBS, Tsentralna biblioteka im. Belinskoho, Kerch 28597
Keren Public Library, Keren 07217
Kereskedelmi és Vendéglátóipari és Idegenforfalmi Főiskola Kar, Budapest 13225
Keresztény Múzeum, Esztergom 13472
W.G. Kerkhoff Institut, Bad Nauheim 11468
Kermanshah University of Medical Sciences, Kermanshah 14392
Kern County Law Library, Bakersfield 33961
Kern County Library, Bakersfield 38059
Kern County Library – Beale Memorial, Bakersfield 38060
Kern County Library – Northeast Bakersfield, Bakersfield 38061
Kern County Library – Ridgecrest Branch, Ridgecrest 40993
Kern County Library – Southwest Branch, Bakersfield 38062
Kerry Library, Tralee 14591
Kershaw County Library, Camden 38425
Kesari and Mahratta Library, Pune 14192
Keshan Teachers' College, Keshan 05353
Keshav Deva Malaviya Institute of Petroleum Exploration, Dehradun 13978
Keski-suomen museon kirjasto, Jyväskylä 07515
Keski-Suomen sairaanhoitopiiri, Jyväskylä 07516
Oy Keskuslaboratorio – Centrallaboratorium AB, Espoo 07435
Kestner-Museum, Hannover 12029
Ketabkhane, Muze va Markaz-e Asnad-e Majles-e Shora-ye Eslami, Tehran 14414
Ketchikan Public Library, Ketchikan 39710
W. Kętrzyński Centre of Scientific Research, Olsztyn 21177
N. Ketskhoveli Botanical Institute, Tbilisi 09265
Kettering College of Medical Arts, Kettering 31503
Kettering University, Flint 31142
Kettle Moraine Correctional Institution Library, Plymouth 34683
Kettleson Memorial Library, Sitka 41331
Keuka College, Keuka Park 31504
Keve András Madártani és Természetvédelmi Szakkönyvtár, Budapest 13369
Kew Municipal Library, Kew 01109
Kewanee Public Library District, Kewanee 39712
Keyano College, Fort McMurray 03990
Keyport Free Public Library, Keyport 39714
Keystone College, La Plume 31537
KF Library, Stockholm 26657
Kfar Giladi Library, Kfar Giladi 14777
KGHM CUPRUM Ltd., Research and Development center, Wrocław 21362
KGHM CUPRUM sp.z o.o., Wrocław 21362
KGU Kamchatskaya kraevaya detskaya biblioteka im. V. Kruchiny, Petropavlovsk-Kamchatski 24201
Kh. M. Abdullaev Institute of Geology and Geophysics, Toshkent 42125
Khabarovsk State Medical Institute, Khabarovsk 22287
Khabarovskaya gosudarstvennaya akademiya ekonomiki i prava, Khabarovsk 22286
Khabarovskaya kraevaya detskaya biblioteka, Khabarovsk 24143
Khabarovski gosudarstvenny meditsinski institut, Khabarovsk 22287
Khabarovski mezhotraslevoi territorialny tsentr nauchno-tekhnicheskoi informatsii i propagandy, Khabarovsk 23098
Khakas District Library, Abakan 24099
Khakasskaya respublikanskaya detskaya biblioteka, Abakan 24098
Khakasskaya respublikanskaya universalnaya biblioteka, Abakan 24099
Khakasski nauchno-issledovatelski institut yazyka, literatury i istorii, Abakan 22981
Kharkiv Art Museum, Kharkiv 28266
Kharkiv Institute of Agricultural Mechanization and Electrification, Kharkiv 28263
Kharkiv National Automobile and Highway University, Kharkiv 27991
Kharkiv National Medical University, Kharkiv 27989
Kharkiv National Pedagogical University, Kharkiv 27992
Kharkiv Region Central Library System of the City of Kiev – V. Stus Main Library, Kyiv 28675
Kharkiv State Academy of Culture, Kharkiv 27987

Kharkiv State Academy of Design, Kharkiv 27988
Kharkiv State Academy of Municipal Services, Kharkiv 27984
Kharkiv State Academy of Railway Transport, Kharkiv 28264
Kharkiv State Agricultural Technical University, Kharkiv 27998
Kharkiv State Institute of Arts, Kharkiv 28265
Kharkiv State Scientific V. Korolenko Library, Kharkiv 27934
Kharkiv State Technical University of Construction and Architecture, Kharkiv 27990
Kharkiv State University of Agriculture, P.O. Komunist-1 28295
Kharkivska derzhavna naukova biblioteka im. V.G. Korolenka, Kharkiv 27934
Kharkivska raionna TsBS m. Kyiva, Trsentralna biblioteka im. V. Stusa, Kyiv 28675
Kharkivsky derzhavny medichny universytet, Kharkiv 27989
Kharkivsky derzhavny tekhnichny universytet budivnytstva i arkhitektury, Kharkiv 27990
Kharkivsky khudozhni muzei, Kharkiv 28266
Kharkivsky natsionalny avtomobylno-dorozhny universytet, Kharkiv 27991
Kharkivsky natsionalny pedagogichny universytet, Kharkiv 27992
Kharkivsky natsionalny universytet im. V. N. Karazina, Kharkiv 27993
Kharkovski politekhnichesky institut, Kharkiv 27997
Khartoum Polytechnic, Khartoum 26343
Khartsizk Regional Centralized Library System, Main Library, Khartsizk 28603
Khartsizka raionna TsBS, Tsentralna biblioteka, Khartsizk 28603
Khazar University, Baku 01690
Kherson City Centralized Library System, Main Library, Kherson 28606
Kherson Local Museum, Kherson 28287
Kherson State University, Kherson 28000
Khersones Museum of History and Archaeology, Sevastopol 28460
Khersonska miska TsBS, Tsentralna biblioteka, Kherson 28606
Khersonsky derzhavny universytet, Kherson 28000
Khetagurov Northern Ossetian State University, Vladikavkaz 22628
Khimicheski kombinat, Novomoskovsk 22880
Khimicheski zavod, Ufa 22967
Khimicheski zavod – Golovnaya bibliyateka, Gomel 02130
Khimicheski zavod im. P.L. Voykova, Moskva 22839
Khimiko-farmatsevticheski institut, Sankt-Peterburg 22526
Khimiko-tekhnologicheski institut, Ivanovo 22263
Khimiko-tekhnologicheski institut, Kazan 22277
Khimprom Chemical Industry Joint-Stock Company, Kemerovo 22801
Khimvolokno company, Chernigiv 28136
KHK Heilig Grafinstituut, Turnhout 02423
Khlopchatobumazhny kombinat, Cheboksary 22783
Khlopchatobumazhny kombinat im. Nikolaevoi, Orekhovo-Zuevo 22883
Khmelnitsky National University, Khmelnitski 28004
B. Khmelnytsky Cherkasy State University, Cherkasy 27955
Khmilnik Regional Centralized Library System, Main Library, Khmilnik 28613
Khmilnitska raionna TsBS, Tsentralna biblioteka, Khmilnik 28613
Kholodny Botany Institute, Kyiv 28316
Khon Kaen University, Khon Kaen 27737
– Faculty of Agriculture Library, Khon Kaen 27738
– Faculty of Engineering Library, Khon Kaen 27739
Khorol Regional Centralized Library System, Main Library, Khorol 28614
Khorolska raionna TsBS, Tsentralna biblioteka, Khorol 28614
Khoruzhei Mozyr City Central Children's Library, Mozyr 02211
Khotin Regional Centralized Library System, Main Library, Khotin 28615
Khotinska raionna TsBS, Tsentralna biblioteka, Khotin 28615
Khuda Bakhsh Oriental Public Library, Patna 14232
Khudozhestvennoe uchilishche, Penza 22688
Khudozhestvennoe uchilishche im. N.K. Rerikha, Sankt-Peterburg 22696
Khudozhestvenny institut im. V.I. Surikova, Moskva 22343
Kiambu Institute of Science and Technology, Kiambu 18013
Kiasma Museum of Contemporary Art Library, Helsinki 07478
Kibaha Public Library, Kibaha 27680
Kibbutzim College of Education, Tel-Aviv 14662
Kidderminster College, Kidderminster 29142
Kiev Haematology and Blood Transfusion Research Institute – Lviv Branch, Lviv 28416
Kiev I.K. Karpenko-Kary State Institute of Theatrical Art, Kyiv 28353
Kiev National Linguistic University, Kyiv 28014
Kiev Polytechnic Institute, Kyiv 28012
Kiev Research Institute of Otolaryngology, Kyiv 28354
Kiev State Institute of Culture, Mikolaiv Branch, Mykolaiv 28735
Kievo-Svyatoshinsk Regional Centralized Library System, Main Library, Vishneve 28871
Kievproekt State Municipal Construction Corporation, Kyiv 28157
Kikai Gijutsu Kenkyujo, Niihari 17624
Kikai Kougyo Toshokan, Tokyo 17690
Kildare Library & Arts Service, Newbridge 14585
Kilgore College, Kilgore 33457
Kilimanjaro Christian Medical Centre (KCMC), Moshi 27679
Kiljava Institute, Kiljava 07519
Kiljavan Opiston Kirjasto, Kiljava 07519
Kilkenny County Library, Kilkenny 14577
Killeen City Library System, Killeen 39716
Killgore Memorial Library, Dumas 38873
Killingly Public Library, Danielson 38778
Kilpatrick Stockton, Atlanta 35461
Kils bibliotek, Kil 26815
Kim Il Sung University, Pyongyang 18053
Kimbell Art Museum Library, Fort Worth 36871
Kimberley College of Tafe, Broome 00402
Kimberley Public Library, Kimberley 25208
Kimberly Public Library – Gerard H. Van Hoof Library, Little Chute 39894
Kimberly Public Library – James J. Siebers Memorial Library, Kimberly 39717
Kimberly-Clark Corp Library, Neenah 35937
The Kimmage Mission Institute, Dublin 14511
Kimron Veterinary Institute, Bet Dagan 14715
Kinda bibliotek, Kisa 26818
King Abdul Aziz Foundation for Research and Archives, Riyadh 24402
King Abdul Aziz Library, Medina 24407
King Abdul Aziz Military Academy, Riyadh 24395
King Abdul Aziz Public Library, Riyadh 24408
King Abdul Aziz University, Jeddah 24391
King College, Bristol 30548
King County Department of Natural Resources & Parks, Seattle 34848
King County Law Library, Seattle 34849
King County Rural Library District – King County Library System, Issaquah 39594
King Fahd National Library, Riyadh 24389
King Fahd University of Petroleum and Minerals (KFUPM), Dhahran 24390
King Faisal Centre for Research and Islamic Studies, Riyadh 24403
King Institute of Preventive Medicine, Chennai 13963
King Matthias Museum, Visegrád 13521
King Monghut's Institute of Technology Ladkrabang, Bangkok 27699
– Faculty of Architecture, Bangkok 27700
– Faculty of Engineering, Bangkok 27701
King Mongkut's Institute of Technology Thonburi, Bangkok 27702
King Saud University, Riyadh 24396
King & Spalding, Atlanta 35462
King & Spalding, Washington 36361
King Township Public Library, King City 04669
King William's Town Public Library, King William's Town 25209
King's College, Edmonton 03695
King's College, Wilkes-Barre 33062
Kings College London, London 29779
King's College London, London
– Chancery Lane Maughan Library & Information Services Centre, London 29191
– Denmark Hill Weston Education Centre ISC, London 29192
– Guy's New Hunt's House Information Services Centre, London 29193
– St Thomas' Medical Library, London 29194
– Waterloo Campus Information Services Centre, London 29195
King's College, University of Cambridge, Cambridge 29025
Kings County Law Library, Hanford 34333
Kings County Library, Hanford 39360
Kings County Library – Hanford Branch, Hanford 39361
King's Fund Information & Library Service, London 29780
Kings River Community College, Reedley 33679
King's University College, London 03744
Kingsborough Community College, Brooklyn 33193
Kingsport Public Library & Archives, Kingsport 39722
Kingston and St. Andrew Parish Library, Kingston 16889
Kingston Frontenac Public Library, Kingston 04670
Kingston Information and Library Service, Parkdale 01139
Kingston Libraries, Kingston upon Thames 30004
Kingston Library, Kingston 39724
Kingston Public Library, Kingston 39725
Kingston University, Kingston upon Thames 29145
– Knights Park Learning Resources Centre, Kingston upon Thames 29146
– Roehampton Vale Library, London 29147
Kingston upon Hull City Council, Kingston upon Hull 30003
Kinki Daigaku, Higashiosaka 16941
– Hogakubu Shiryoshitsu, Higashiosaka 16942
– Kogakubu Bunkan, Kure 16943
– Yakugakubu Bunshitsu, Higashiosaka 16944
Kinki University, Higashiosaka 16941
– Engineering Branch, Kure 16943
– Law Library, Higashiosaka 16942
– Pharmaceutical Branch, Higashiosaka 16944
Kinnelon Public Library, Kinnelon 39729
Kinostudiya 'Lenfilm', Sankt-Peterburg 23806
The Kinsey Institute for Research in Sex, Gender & Reproduction, Inc, Bloomington 36549
Kinsman Free Public Library, Kinsman 39730
Kirchenbibliothek Neustadt a.d. Aisch, Neustadt a.d. Aisch 11293
Kirchen-Kapitelsbibliothek Schwabach, Schwabach 11325
Kirchenkreis Essen, Essen 11209
Kirchenkreis Lübeck, Lübeck 11266
Kirchenrechtliches Institut der Evangelischen Kirche in Deutschland, Göttingen 11227
Kirchgeschichtlicher Verein der Erzdiözese Freiburg, Freiburg 11896
Kirchliche Hochschule Wuppertal/Bethel, Bielefeld 09460
Kirchliche Pädagogische Hochschule Edith Stein – Institut für LehrerInnenbildung, Stams 01289
Kirchliche Pädagogische Hochschule Edith Stein – Institut für Religionspädagogische Bildung Salzburg, Salzburg 01264
Kirchliche Pädagogische Hochschule Graz, Graz 01216
Kirchliche Pädagogische Hochschule Wien/Krems, Krems 01245
Kirchliche Pädagogische Hochschule Wien/Krems, Wien 01295
Kirchner Museum Davos, Davos Platz 27347
Kirensky Institute of Physics of the Siberian Department of the Russian Academy of Sciences, Krasnoyarsk 23123
Kirjasto ja tietopalvelu, Helsinki 07469
Kirke Hyllinge Bibliotek, Kirke Hyllinge 06892
Kirkendall Public Library, Ankeny 37963
Kirkland & Ellis Library, Washington 36362
Kirkland Municipal Library, Kirkland 04671
Kirklees Metropolitan District Council, Huddersfield 29998
Kirkpatrick & Lockhart, Preston, Gates, Ellis, Washington 36363
Kirkwood Community College, Cedar Rapids 33206
Kirkwood Highway Library, Wilmington 41912
Kirkwood Public Library, Kirkwood 39733
Kirloskar Oil Engines Ltd, Pune 13914
Kirov Aluminium Factory, Volkhov 22973
Kirov Cultural Palace, Sankt-Peterburg 24228
Kirov Forest Engineering College, Kirov 22672
Kirov Islands Library, Sankt-Peterburg 24218
Kirov Machine Tool Plant, Gomel 01856
Kirov Machine Tools Plant, Vitebsk 01935
Kirov Metallurgical Plant, Makiyivka 28164
Kirov Minsk Extrusion and Cutting Machine-tools Industrial Corporation – Extrusion Machine-tool Special Design Bureau, Minsk 01891
Kirov Officers' House, Sankt-Peterburg 23867
Kirov Plant State Industrial Corporation, Sankt-Peterburg 22911
Kirov Regional Scientific Medical Library, Kirov 23101
Kirov Spinning Mill, Sankt-Peterburg 22933
Kirov Vitebsk Central City Children's Library, Vitebsk 02249
Kirovets Recreation Centre, Sankt-Peterburg 24220
Kirovograd City Centralized Library System, Main Library, Kirovograd 28616
Kirovogradska miska TsBS, Tsentralna biblioteka, Kirovograd 28616
Kirovogradska oblasna universalna naukova biblioteka im. D.I. Chizhevski, Kirovograd 27935
Kirovskaya oblastnaya detskaya biblioteka im. A.S. Grina, Kirov 24145
Kirovskaya oblastnaya nauchnaya meditsinskaya biblioteka, Kirov 23101
Kirovskaya oblastnaya yunosheskaya biblioteka, Kirov 24146
Kirovskaya ordena 'Znak Pochyota' oblastnaya universalnaya nauchnaya biblioteka im. A.I. Gertsena, Kirov 22141
Kirovski lesopromyshlenny kolledzh, Kirov 22672
Kirovski mezhotraslevoi territorialny tsentr nauchno-tekhnicheskoi informatsii i propagandy, Kirov 23102
Kirovsky District Centralized Library System, Central District Children's Library, Sankt-Peterburg 24259
Kirovsky District Centralized Library System, Library Branch no 2, Sankt-Peterburg 24257
Kirovsky District Centralized Library System, Sholokhov Central District Library, Sankt-Peterburg 24258
Kirtland Air Force Base Library, Kirtland AFB 34418
Kirtland Community College, Roscommon 33702
Kiruna stadsbibliotek, Kiruna 26817
Kiselyovsk Mining Professional School, Kiselyovsk 22673
Kiselyovski gorny tekhnikum, Kiselyovsk 22673
Kisfaludy Károly County Library, Győr 13536
Kisfaludy Károly Megyei Könyvtár, Győr 13536
Kishwaukee College, Malta 33513
Kiskun Múzeum, Kiskunfélegyháza 13487
Kitakyushu City, Bureau of Planning, Kitakyushu 17266
Kitakyushu Daigaku, Kitakyushu 16970
Kitakyushu University, Kitakyushu 16970
Kitakyushushi Kikaku Kyoku Kikaku-ka Gyoseishiryoshitsu, Kitakyushu 17266
Kitasato Daigaku, Sagamihara 17093
– Igaku Toshokan, Sagamihara 17094
– Shirokane Toshokan, Tokyo 17095
Kitasato Institute, Tokyo 17691
Kitasato University Medical Library, Sagamihara 17093
– Medical Library, Sagamihara 17094
– Shirokane Library, Tokyo 17095
KIT-Bibliothek Karlsruhe, Karlsruhe 10058
Kitchener Public Library, Kitchener 04673
Kitchigami Regional Library, Pine River 40791
Kitobkhonai davlatii patentio tekhnikii Markazi millii patentu ittilosti Chumkhurii Tochikiston, Dushanbe 27645
Kitsap County Law Library, Port Orchard 34696
Kitsap Regional Library, Bremerton 38267
Kitsap Regional Library – Bainbridge Island Branch, Bainbridge Island 38056
Kitsap Regional Library – Port Orchard Branch, Port Orchard 40860
Kitsap Regional Library – Poulsbo Branch, Poulsbo 40884
Kitsap Regional Library – Sylvan Way Branch, Bremerton 38268
Kittochtinny Historical Society Library, Chambersburg 36643
Kitwe Public Library, Kitwe 42353
Kivukoni Academy of Social Sciences, Dar es Salaam 27661
Kizel Coal Industrial Corporation, Kizel 22808
Kizhi State Open-Air Museum of History, Architecture and Ethnography, Petrozavodsk 23703
K-Konsult, Stockholm 26554
K&L Gates LLP, Boston 35531
K&L Gates LLP, Chicago 35584
K&L Gates LLP, Washington 36364
Klagenfurt University Library, Klagenfurt am Wörthersee 01243
Klaipėda University Library, Klaipėda 18283
Klaipėdos apskrities viešoji Ievos Simonaitytės biblioteka, Klaipėda 18335
Klaipėdos miesto savivaldybės viešoji biblioteka, Klaipėda 18336
Klaipėdos rajono savivaldybės Jono Lankučio viešoji biblioteka, Gargždai 18322

Klaipėdos universiteto biblioteka, Klaipėda 18283
Klaksvíkar Bókasavn / Klasvig Folkebibliotek, Klaksvík 06893
Klamath County Library Services District, Klamath Falls 39735
Klasvig Folkebibliotek, Klaksvík 06893
Klasztor O.O. Paulinów, Częstochowa 21051
Klaus-Groth-Museum, Heide 12037
Klaus-Kuhnke-Archiv für Populäre Musik, Bremen 11681
Klehr, Harrison, Harvey, Branzburg & Ellers, Philadelphia 36102
Kleist-Museum, Frankfurt (Oder) 11878
Klepp bibliotek, Kleppe 20355
Klerksdorp Public Library, Klerksdorp 25210
Ernst Klett Verlag, Stuttgart 11429
Klima- og forurensningsdirektoratet, Oslo 20192
Klimov Plant Research and Industrial Corporation, Sankt-Peterburg 22920
Klimovski mashinostroitelny zavod, Klimovsk 22809
Klingspor-Museum, Offenbach 12385
Klinik für Strahlentherapie, Lübeck 12213
Klinik für Zahn-, Mund- und Kieferkrankheiten, München 10437
Klinik und Jugendklinik, Erlangen 09721
Klinik und Poliklinik für Dermatologie und Allegologie, München 10438
Klinik und Poliklinik für Dermatologie und Allergologie, Bonn 09571
Klinik und Poliklinik für Psychiatrie, Psychomatik und Psychotherapie, Würzburg 10698
Klinik und Polikliniken für Zahn-, Mund- und Kieferkranke, Erlangen 09722
Klinikum Aschersleben-Staßfurt GmbH, Aschersleben 11457
Klinikum Bad Salzungen gGmbH, Bad Salzungen 11469
Klinikum Barnim GmbH, Eberswalde 11809
Klinikum Bielefeld, Bielefeld 11609
Klinikum Bremen-Ost, Bremen 11671
Klinikum Bremerhaven-Reinkenheide gGmbH, Bremerhaven 11696
Klinikum Chemnitz gemeinnützige GmbH, Chemnitz 11710
Klinikum der Philipps-Universität, Marburg 10347
Klinikum Ernst von Bergmann, Potsdam 12405
Klinikum Frankfurt/Oder GmbH, Frankfurt (Oder) 11879
Klinikum Fulda gAG, Fulda 11905
Klinikum Großhadern, München 10431
Klinikum Innenstadt / I. Frauenklinik, München 10436
Klinikum Magedburg gemeinnützige GmbH, Magdeburg 12222
Klinikum Obergöltzsch Rodewisch, Rodewisch 12438
Klinikum St. Georg gGmbH, Leipzig 12189
Klintsovski tekstilny tekhnikum, Klintsy 22674
Klintsy Textile Professional School, Klintsy 22674
Klippans Bibliotek, Klippan 26819
KLM Royal Dutch Airlines, Amstelveen 19321
Klohn Crippen Berger Ltd, Vancouver 04303
Klooster Paters van de H.H. Harten, Bavel 19297
Klooster Paters van de Ongeschoeide Karmelieten, Gent 02463
Klooster Sint Aegten, Sint Agatha 19315
Kloster Andechs, Andechs 11149
Kloster Burg Dinklage, Dinklage 11193
Kloster Frauenberg, Fulda 11220
Kloster Michaelstein, Blankenburg (Harz) 11615
Kloster Neustift, Varna 16118
Kloster Weltenburg, Kelheim 11254
Klosterbibliothek Disentis, Disentis 27253
Klosterbibliothek Loccum, Rehburg-Loccum 11312
Kloster-Bibliothek Salzburg, Salzburg 01464
KMC Town Centre Public Library, Kabwe 42351
Kmetijski inštitut Slovenije, Ljubljana 24922
KMO-documentatiecentrum, Brussel 02600
Knapp, Peterson & Clarke, Glendale 35719
The Knesset, Jerusalem 14681
Knight Piésold, Rivonia 25100
Knihovna Bedřicha Beneše Buchlovana, Uherské Hradiště 06616
Knihovna Bedřicha Smetany, Praha 06495
Knihovna Eduarda Petišky, Brandýs nad Labem 06554
Knihovna Filosofického ústavu Akademie věd České republiky, v.v.i., Praha 06496
Knihovna Jana Drdy Příbram, Příbram 06601
Knihovna Jiřího Mahena v Brne, Brno 06556
Knihovna K. H. Máchy, Litoměřice 06584
Knihovna Karla Dvořáčka, Vyškov 06619
Knihovna Kroměřížska, Kroměříž 06582
Knihovna Matěje Josefa Sychry, Žďár nad Sázavou 06621
Knihovna města Hradce Králové, Hradec Králové 06571
Knihovna města Mladá Boleslav, Mladá Boleslav 06588
Knihovna města Olomouce, Olomouc 06594
Knihovna města Ostravy, Ostrava 06596
Knihovna města Plzně, Plzeň 06598
Knihovna Petra Bezruče, Opava 06595
Knihovna Václava Štecha, Slaný 06605
Knipovich Polar Research Institute for Marine Fisheries and Oceanography, Murmansk 23603
Knitted Fabrics and Spinning Factory, Scientific and Technical Libary, Pinsk 01921
Knitware Factory, Gomel 01849
Knižnica Antona Bernolaka v Novych Zámkoch, Nové Zámky 24786
Knižnica Bratislava Nové Mesto, Bratislava 24765
Knižnica Gašpara Fejérpataky – Belopotockého, Liptovský Mikuláš 24781
Knižnica Geografického ústavu SAV, Bratislava 24696
Knižnica Juraja Fandlyho, Trnava 24800
Knižnica P.O. Hviezdoslava v Prešove, Prešov 24790
Knižnica pre mládež mesta Košíc, Košice 24733
Knižnica Ružinov, Bratislava 24766
Knjiž Lendava, Könyvtàr Lendva, Lendava 24544
Knjižnica A. T. Linharta, Radovljica 24980
Knjižnica Bena Zupančiča, Postojna 24977
Knjižnica Bežigrad, Ljubljana 24966
Knjižnica Božidar Adžije, Zagreb 06264
Knjižnica Brežice, Brežice 24954
Knjižnica Cirila Kosmača, Tolmin 24986
Knjižnica Domžale, Domžale 24957
Knjižnica dr. Toneta Pretnarja, Tržič 24988
Knjižnica Franca Ksavra Meška Ormož, Ormož 24976
Knjižnica Grosuplje, Grosuplje 24958
Knjižnica Ivana Tavčarja Škofja Loka, Škofja Loka 24985
Knjižnica Jožeta Mazovca, Ljubljana 24967
Knjižnica Jožeta Udoviča Cerknica, Cerknica 24956
Knjižnica Kočevje, Kočevje 24962
Knjižnica Lendava – Könyvtár Lendva, Lendava 24965
Knjižnica Mirana Jarca, Novo Mesto 24975
Knjižnica Otona Župančiča, Ljubljana 24968
Knjižnica Pavla Golie Trebnje, Trebnje 24987
Knjižnica Potrča Ptuj, Ptuj 24978
Knjižnica Prežihov Voranc, Ljubljana 24969
Knjižnica Radlje ob Dravi, Radlje ob Dravi 24979
Knjižnica Rogaška Slatina, Rogaška Slatina 24982
Knjižnica Šiška, Ljubljana 24970
Knjižnica Velenje, Velenje 24989
Knjižnice Grada Zagreba, Zagreb 06265
Knoedler Art Library, New York 37217
Knowsley Metropolitan Borough Council, Huyton 29999
Knox College, Dunedin 19777
Knox College, Galesburg 31201
Knox County Governmental Library, Knoxville 34420
Knox County Public Library, Barbourville 38086
Knox County Public Library, Vincennes 41687
Knox County Public Library System – Cedar Bluff Branch, Cedar Bluff 38481
Knox County Public Library System – Farragut Branch, Farragut 39051
Knox County Public Library System – Lawson McGhee Library-East Tennessee History Center, Knoxville 39738
Knox County Public Library System – West Knoxville Branch, Knoxville 39739
Knoxville College, Knoxville 31517
Knoxville-Knox County Metropolitan Planning Commission Library, Knoxville 34421
Kobe City Assembly, Kobe 17269
Kobe City Library, Kobe 17844
Kobe City Museum, Kobe 17575
Kobe City University of Foreign Studies, Kobe 16972
Kobe Daigaku, Kobe 16973
– Keizai Keiei Kenkyujo, Kobe 16974
– Shizenkagakukei Toshokan, Kobe 16975
Kobe Gakuin Daigaku, Kobe 16976
Kobe Gakuin University, Kobe 16976
Kobe Kaisei College, Kobe 16977
Kobe Pharmaceutical University, Kobe 16978
Kobe University, Kobe 16973
– Natural Science Library, Kobe 16975
– Research Institute for Economics and Business Administration (RIEB), Kobe 16974
Kobe University of Commerce, Kobe 16979
Kobelyaki Regional Centralized Library System, Zalko Main Library, Kobelyaki 28618
Kobelytska raionna TsBS, Tsentralna biblioteka im. Zalki, Kobelyaki 28618
Københavns Kommunes Biblioteker, København 06894
Københavns Rådhusbibliotek, København 06800
Københavns Tekniske Bibliotek, Ballerup 06773
Københavns Universitet, København
– Afdeling for Bibelsk Eksegese, København 06668
– Afdeling for Navneforsknings bibliotek, København 06669
– Det Biovidenskabelige Fakultetsbibliotek, Frederiksberg 06670
– Botanisk Centralbibliotek, København 06671
– Center for Afrikastudier, København 06672
– Datalogisk Institut, København 06673
– Engelsk Bibliotek, København 06674
– Europæisk og Almen Etnologi, København 06675
– Det Farmaceutiske Fakultetsbibliotek, København 06676
– GeoBiblioteket, København 06677
– Germansk Bibliotek, København 06678
– Institut for Antropologi, København 06679
– Institut for Idræt, København 06680
– Institut for Kirkehistorie, København 06681
– Institut for Kunst og Kulturvidenskab – Afdeling for Musikvidenskab, København 06682
– Institut for Medier, Erkendelse og Formidling, København 06683
– Institut for Psykologi, København 06684
– Institut for Statskundskab, København 06685
– Institut for Tværkukturelle og Regionale Studier – Afdelingen for Asien Studier, København 06686
– Institut for Tværkukturelle og Regionale Studier – Østeuropæisk Afdeling, København 06687
– Instituttet for Tværkulturelle og Regionale Studier – Carsten Niebuhr Afdelingen, København 06688
– Det Juridiske Fakultets Bibliotek. Kriminalisk Bibliotek, København 06689
– Lingvistik Bibliotek, København 06690
– Matematisk Afdeling, København 06691
– Natur- og Lægevidenskabelige Bibliotek, København 06692
– Det Natuur- og Sundhedsvidenskabelige Fakultetsbibliotek, København 06693
– Niels Bohr Institutet, Biblioteket for Astronomi og Geofysik, København 06694
– Niels Bohr Institutet, Fysisk bibliotek, København 06695
– Nordisk Bibliotek, København 06696
– Nordisk Forskningsinstitut, København 06697
– Økonomisk Instituts Bibliotek, København 06698
– Religionshistorisk bibliotek, København 06699
– Romansk Bibliotek, København 06700
– SAXO-Instituttet – Afdeling for Arkæologi, Forhistorisk Arkæologisk Bibliotek, København 06701
– SAXO-Instituttet – Afdeling for Græsk og Latin, København 06702
– SAXO-Instituttet – Afdeling for Historie, København 06703
– SAXO-Instituttet – Afdeling for Økonomisk Historie, København 06704
– SAXO-Instituttet – Afdeling for Samtidshistorie, København 06705
– SAXO-Instituttet, Etnologisk Bibliotek, København 06706
– SAXO-Instituttet, Klassisk Arkæologisk Bibliotek, København 06707
– Sociologisk Bibliotek, København 06708
– Det Teologiske Fakultetsbibliotek, København 06709
Københavns Universitetsbibliotek, København 06623
Kobeshikai Toshoshitsu, Kobe 17269
Kobrin Regional Central Library, Kobrin 02147
Kobrin Tool Making Plant, Kobrin 01874
Kobrinskaya tsentralnaya raionnaya bibliyateka, Kobrin 02147
Kobrinski instrumentalny zavod, Kobrin 01874
Kobunsha Co Ltd, Tokyo 17454
Kobunsha K. K., Tokyo 17454
Koç University, Istanbul 27862
Kochi Prefectural Library, Kochi 17845
Ko-enerugi Kasokuki Kenkyukiko, Tsukuba 17791
Kofu Library, Noda 17854
Kofu Toshokan, Noda 17854
Kogakuin University, Tokyo 17151
Kogarah Council Library, Kogarah 01111
Køgebibliotek, Køge 06896
Kogyo Gijutsuin, Tokyo 17311
Kohtla-Järve Central Library, Kohtla-Järve 07247
Kokebe Tsibah Comprehensive Secondary School, Addis Ababa 07273
Kokkaido Gikai, Sapporo 17295
Kokkola City Library – Regional Library, Kokkola 07578
Kokkolan kaupunginkirjasto – maakuntakirjasto, Kokkola 07578
Kokomo-Howard County Public Library, Kokomo 39741
Koksokhimichny zavod, Mariupol 28721
Kokubungaku Kenkyu Siryokan, Tokyo 17692
Kokudochiriin Toshokan, Tsukuba 17792
Kokudocho Toshokan, Tokyo 17312
Kokugakuin University, Tokyo 17152
– Nihon Bunka Kenkyusho, Tokyo 17153
Kokumin Kinyu Joko Chosabu, Tokyo 17455
Kokumin Seikatsu Senta Joho Shinyo Shitsu, Tokyo 17693
Kokuritsu Eisei Shikenjo, Tokyo 17694
Kokuritsu Fujin Kyoiku Kaikan, Saitama 17648
Kokuritsu Gan Senta, Tokyo 17695
Kokuritsu Gekija, Tokyo 17696
Kokuritsu Idengaku Kenkyujo, Mishima 17597
Kokuritsu Kobunshokan Naikakubunko, Tokyo 17697
Kokuritsu Kogai Kenkyujo, Tsukuba 17793
Kokuritsu Kokkai Toshokan, Tokyo 16896
Kokuritsu Kokugo Kenkyuzyo, Tokyo 17698
Kokuritsu Kyoiku Kenkyujo, Tokyo 17699
Kokuritsu Tokushu Kyoiku Sogo Kenkyujo, Yokosuka 17830
Kokusai Bunka Kaikan, Tokyo 17700
Kokusai Denshin Denwa Co, Ltd, Tokyo 17456
Kokusai Denshin Denwa K. K. Shomubu Jimu Senta Shiryokanrika, Tokyo 17456
Kokusai Kirisutokyo Daigaku, Mitaka 17042
Kokusai Kôryukikin Jôhô Sentâ Raiburarî, Tokyo 17701
Kokusai Nihon Bunka Kenkyu Center, Kyoto 17586
Kokushikan University, Tokyo 17154
Kola Science Centre of the Russian Academy of Sciences, Apatity 22116
Kolas Mogilev City Library no 1, Mogilev 02193
Ya. Kolas Slonim Central Regional Library, Slonim 02234
Kolding Amtsgymnasium og HF, Kolding 06736
Kolding Bibliotek, Kolding 06897
Koldo Mitxelena Kulturunea, San Sebastián 26249
Kolegium Filozoficzno-Teologicznego oo. Dominikanów, Kraków 21053
Kolej Agama Sultan Zainal Abidin, Kuala Terengganu 18536
Kolej Damansara Utama, Petaling Jaya 18515
Kolín Regional Museum, Kolín 06458
Kolledzh lesopromyshlennogo proizvodstva, Sankt-Peterburg 22697
Kollegiatstift Mattsee, Mattsee 01453
Kollegio Athinon – Kollegio Psikhikou, Athinai 13090
Kollegio Psikhikou, Athinai 13090
Kollegium der Gesellschaft Jesu, Linz 01450
Kollegium Kalksburg, Wien 01487
Kölnisches Stadtmuseum, Köln 12145
Kolomiya Regional Centralized Library System, Main Library, Kolomiya 28619
Kolomiya State Museum of Folk Art, Kolomiya 28294
Kolomiyska raionna TsBS, Tsentralna biblioteka, Kolomiya 28619
Kolpinsky District Centralized Library System, Central District Children's Library, Sankt-Peterburg 24261
Kolpinsky District Centralized Library System, Svetlov Central District Library, Sankt-Peterburg 24260
Kolski nauchny tsentr RAN, Apatity 22116
Koltsov Institute of Evolutionary Biology of the Russian Academy of Sciences, Moskva 23255
Kolumb Republican Youth Library, Yoshkar-Ola 24363
Komàrom-Esztergrom County Library, Tatabánya 13552

Komarov Botanical Institute of the Russian Academy of Sciences, Sankt-Peterburg 23751
V. L. Komarov Institute of Botany, Baku 01716
Kombinat iskusstvennogo volokna, Tver 24010
Kombinat Krivorizhstal, Krivi Rig 28153
Kombinat Krivorizhstal – Biblioteka profkomu, Krivi Rig 28636
Kombinat PO Borisovdrev, Borisov 01837
Kombinat shelkovykh tkanei, Vitebsk 01931
Kombinat 'Yuzhuralnikel', Orsk 22887
Komi Centre of Scientific and Technical Information and Popularization, Syktyvkar 23985
Komi mezhotraslevoi territorialny tsentr nauchno-tekhnicheskoi informatsii i propagandy, Syktyvkar 23985
Komi National Library, Syktyvkar 22176
Komi National Museum, Syktyvkar 23986
Komi respublikanskaya detskaya biblioteka im. S.Ya. Marshaka, Syktyvkar 24325
Komi respublikanskaya yunosheskaya biblioteka, Syktyvkar 24326
Komi Science Centre, Syktyvkar 22177
Komi Science Centre of the Ural Division of the Academy of Sciences of Russia, Syktyvkar 22177
Komische Oper, Berlin 11540
V. P. Komissarenko Institue of Endocrinology and Metabolism, Kyiv 28355
Komitet po gorite, Sofiya 03488
Komitet po meteoritam RAN, Moskva 23342
Komitet po torgovle, Moskva 23343
Komitet po torvovle, obshch-estvennomu pitaniyu, bytovomy obsluzhivaniyu i snabzheniyu goroda prodovolstviem, Sankt-Peterburg 22698
Komitet Rossiskoi Federatsii po pechati, Moskva 22738
Kommission für Geschichte der Mathematik, Naturwissenschaften und Medizin – Sammlung Woldan, Wien 01619
Kommission für Mundart- und Namenforschung Westfalens, Münster 12346
Kommunalarchiv Minden, Minden 12266
Kommunale Gemeinschaftsstelle für Verwaltungsmanagement (KGSt), Köln 10988
Kompozit Leased Enterprise, Sankt-Peterburg 22908
Komsomolsk Regional Centralized Library System, Main Library, Komsomolsk 28620
Komsomolska raionna TsBS, Tsentralna biblioteka, Komsomolsk 28620
KonaK Wien – Forschungs- und Kulturverein fuer Kontinentalamerika und die Karibik, Wien 01588
Konan University, Kobe 16980
KONČAR-Institut za elektrotehniku, Zagreb 06248
Kondopoga Joint-Stock Company, Kondopoga 22810
Konfederatsiya soyuza kinematografis-tov, Moskva 23344
Konfessionskundliches Institut des Evangelischen Bundes, Bensheim 11161
Det Kongelige Bibliotek, København 06623
Det Kongelige Bibliotek, Slotsholmen – Billedsamlingen, København 06624
Det Kongelige Bibliotek, Slotsholmen – Danmarks Bogmuseum, København 06625
Det Kongelige Bibliotek, Slotsholmen – Dramatisk Bibliotek, København 06626
Det Kongelige Bibliotek, Slotsholmen – Håndskriftafdelingen, København 06627
Det Kongelige Bibliotek, Slotsholmen – Kortsamlingen, København 06628
Det Kongelige Bibliotek, Slotsholmen – Musik- og Teaterafdelingen, København 06629
Det Kongelige Danske Musikkonser-vatorium, København 06711
Det Kongelige Garnisonsbibliotek, København 06760
Det Kongelige Teaters Arkiv og Bibliotek, København 06807
Kongju National Teachers' College, Kongju 18064
Kongregation der Franziskanerinnen, Salzkotten 11320
Kongsberg bibliotek, Kongsberg 20358
Kongsvinger bibliotek, Kongsvinger 20359
König-Karlmann-Gymnasium, Altötting 10704
Koninklijk Belgisch Instituut voor Natuurwetenschappen, Brussel 02601
Koninklijk Conservatorium, Den Haag 19142
Koninklijk Conservatorium Brussel, Brussel 02285

Koninklijk Filmarchief, Brussel 02602
Koninklijk Hoger Instituut voor Defensie (CDI), Brussel 02603
Koninklijk Huisarchief, Den Haag 19440
Koninklijk Instituut Spermalie (KIS), Brugge 02548
Koninklijk Instituut voor de Marine (KIM), Den Helder 19145
Koninklijk Instituut voor de Tropen, Amsterdam 19366
Koninklijk Instituut voor Taal-, Land-en Volkenkunde (KITLV), Leiden 19474
Koninklijk Meteorologisch Instituut van België (KMI), Brussel 02605
Koninklijk Militair-Historisch Museum (Legermuseum), Delft 19411
Koninklijk Museum van het Leger en de Krijgsgeschiedenis / Musée Royal de l'Armée et d'Histoire Militaire, Brussel 02604
Koninklijk Museum voor Midden-Afrika (KMMA), Tervuren 02701
Koninklijk Museum voor Schone Kunsten Antwerpen, Antwerpen 02517
Koninklijk Nederlands Instituut Rome, Roma 16544
Koninklijk Nederlands Instituut van Registeraccountants (Koninklijk NIVRA), Amsterdam 19367
Koninklijk Nederlands Meteorologisch Instituut (KNMI), De Bilt 19399
Koninklijke Academie voor Nederlandse Taal- en Letterkunde, Gent 02646
Koninklijke Academie voor Schone Kunsten, Gent 02324
Koninklijke Belgische Vereniging voor Entomologie, Bruxelles 02625
Koninklijke Bibliotheek Albert I, Bruxelles 02257
Koninklijke Bibliotheek (KB), Den Haag 19046
Koninklijke Bibliotheek van België, Bruxelles 02257
Koninklijke Bond der Oostvlaamse Volkskundigen (KBOV), Gent 02647
Koninklijke Luchtvaart Maatschappij (KLM), Amstelveen 19321
Koninklijke Militaire Academie (KMA), Breda 19101
Koninklijke Musea voor Kunst en Geschiedenis, Bruxelles 02611
Koninklijke Musea voor Schone Kunsten van België, Bruxelles 02612
Koninklijke Nederlandse Botanische Vereniging, Leiden 19475
Koninklijke Oudheidkundige Kring van Waasland (KOKW), Sint-Niklaas 02699
Koninklijke Sterrenwacht van België – Koninklijke Meteorologisch Instituut van België (KSB-KMI), Brussel 02605
Koninklijke Vlaamse Academie van België voor Wetenschappen en Kunsten (KVABWK), Brussel 02259
Konishiroku Photo Industry Co, Ltd, Hino 17356
Konishiroku Shashinkyogo K. K. Chosaka Toshoshitsu, Hino 17356
Konkan Agricultural University, Dapoli 13636
Konko Library, Asaguchi 17835
Konko Toshokan, Asaguchi 17835
Konkuk University, Seoul 18085
Konkurrensverket, Stockholm 26507
Könnyüipari Müszaki Föiskola, Budapest 13251
Konotop Regional Centralized Library System, Main Library, Konotop 28621
Konotopska raionna TsBS, Tsentralna biblioteka, Konotop 28621
Konrad-Adenauer-Stiftung e.V., Sankt Augustin 12462
Konrad-Duden-Stadtbibliothek, Bad Hersfeld 12638
Konrad-Zuse-Zentrum für Informationstechnik, Berlin 11541
Konservatorium für Musik Biel, Biel 27220
Konservatorium Wien GmbH, Wien 01296
Konservatoriya, Astrakhan 22208
Konservatoriya im. Glinki, Novosibirsk 22441
Konservatoriya im. Rakhmaninova, Rostov-na-Donu 22489
Konstantinov Nuclear Physics Research Institute, Gatchina 23043
Konstbiblioteket, Stockholm 26658
Konstfack, Hägersten 26389
Konstruktorskoe byuro po zhelezobetonu Gosstroya, Moskva 23345
kontakt+co, Innsbruck 01543
Kontsern Leninets, Sankt-Peterburg 22815
Konus Ltd, Mozhga 22869
Konviktsbibliothek Wilhelmsstift, Tübingen 11352
Könyvtár Lendva, Lendava 24965
Könyvtártudományi és Módszertani Központ, Budapest 13388
Konzervatoř v Brně, Brno 06354

Konzervatórium, Žilina 24670
Kookmin University, Seoul 18086
Kooperativny instytut, Poltava 28063
Kooperativny tekhnikum, Kherson 28102
Kópavogur Public Library, Kópavogur 13581
Kopeisk School of Light Industry, Kopeisk 22675
Kopeiski tekhnikum legkoi promyshlennosti, Kopeisk 22675
Köpings stadsbibliotek, Köping 26820
KOPINT-DATORG Infokommunikációs Zrt, Budapest 13389
Kopyl Region Central Library, Kopyl 02148
Kopylskaya tsentralnaya raionnaya bibliyateka, Kopyl 02148
Korea Advanced Institute of Science and Technology (KAIST), Taejon 18102
Korea Air and Correspondence University, Seoul 18087
Korea Atomic Energy Research Institute (KAERI), Taejon 18122
Korea Chamber of Commerce and Industry (KCCI), Seoul 18109
Korea Development Institute, Seoul 18110
Korea Foreign Trade Association (KFTA), Seoul 18111
Korea Foundation, Seoul 18112
Korea Institute for Industrial Economics and Trade (KIET), Seoul 18113
Korea Institute of Energy and Research, Taejon 18123
Korea Institute of Science and Technology (KIST), Seoul 18114
Korea Social Science Library, Seoul 18115
Korea University, Seoul 18088
– United Nations Depository Library / European Documentation Centre, Seoul 18089
Korean Cultural Center Library, Los Angeles 37052
Korean Educational Development Institute, Seoul 18116
Korean Medical Association, Seoul 18117
Korelichi Central Regional Library, Korelichi 02149
Korelichiskaya tsentralnaya raionnaya biblіyateka, Korelichi 02149
Koretsky Institut of State and Law, Kyiv 28317
Korkeimman oikeuden kirjasto, Helsinki 07396
Kornhausbibliothek, Bern 27476
kórnik Library of the Polish Academy of Sciences, Kórnik 20790
Korogwe Teachers Training College, Korogwe 27662
Korolenko Teacher-Training, Poltava 28064
Koroška osrednja knjižnica dr. Franc Sušnik, Ravne na Koroškem 24981
Korosten Regional Centralized Library System, Main Library, Korosten 28622
Korostenska raionna TsBS, Tsentralna biblioteka, Korosten 28622
Korostishiv Regional Centralized Library System, Main Library, Korostishiv 28623
Korostishivska raionna TsBS, Tsentralna biblioteka, Korostishiv 28623
Korporatsiya Ukrlisprom, Ivano-Frankivsk 28144
Korporatsiya Zaporizhtransformator, Zaporizhzhya 28174
Korporatsiya Zaporizhtransformator – Biblioteka profkomu, Zaporizhzhya 28887
Kosciusko County Historical Society, Warsaw 37712
Kosei Torihiki Jinkai, Tokyo 17313
Koseisho, Tokyo 17314
Kosovelova Knjižnica, Sežana 24983
Kossuth Lajos Múzeum, Cegléd 13466
Kostroma Flax Manufacture Joint-Stock Company, Kostroma 22812
Kostroma State Agricultural Academy, Kostroma 22295
Kostroma State University of Technology, Kostroma 22297
Kostromskaya gosudarstvennaya selskokhozyaistvennaya akademiya, Kostroma 22295
Kostromskaya oblastnaya detskaya biblioteka im. A.P. Gaidara, Kostroma 24148
Kostromskaya oblastnaya universalnaya nauchnaya biblioteka im. N.K. Krupskoi, Kostroma 22143
Kostromskaya oblastnaya yunosheskaya biblioteka, Kostroma 24149
Kostromskoi gosudarstvenny pedagogicheski universitet im. N.A. Nekrasova, Kostroma 22296
Kostromskoi mezhotraslevoi territorialny tsentr nauchno-tekhnicheskoi informatsii i propagandy, Kostroma 23108
Kostychev Ryazan Agricultural Institute, Ryazan 22498

Kostyukovichi Central Regional Library, Kostyukovichi 02150
Kostyukovichiskaya tsentralnaya raionnaya bibliyateka, Kostyukovichi 02150
Koszalińska Biblioteka Publiczna im. Joachima Lelewela, Koszalin 21457
Kota Bharu Teachers' Training College, Kota Bharu 18530
Kotelevska raionna TsBS, Tsentralna biblioteka, Kotelva 28626
Kotelva Regional Centralized Library System, Main Library, Kotelva 28626
Kotimaisten kielten tutkimuskeskus, Helsinki 07470
Kotka Public Library, Kotka 07579
Kotka Technical College Library, Kotka 07352
Kotkan kaupunginkirjasto, Kotka 07579
Kotkan teknillinen oppilaitos kirjasto, Kotka 07352
Kotlarevsky Regional Universal Scientific Library, Poltava 27945
Kotohira-gu Shrine Library, Nakatado 17618
Kotohira-Gu Toshokan, Nakatado 17618
G. Kotovsky Children's Library, Kyiv 28644
M. Kotsyubinsky Children's Library, Kyiv 28646
Kountze Public Library, Kountze 39742
Kouvola Public Library – District Library, Kouvola 07580
Kouvolan kaupunginkirjasto – Maakuntakirjasto, Kouvola 07580
Kovalevsky Southern Seas Biology Institute – Karadag Branch, Feodosiya 28237
Kovel Regional Centralized Library System, Main Library, Kovel 28627
Kovelska raionna TsBS, Tsentralna biblioteka, Kovel 28627
Koventarios Dimotiki Vivliothiki Kozanis, Kozani 13146
Koventarios Public Library of Kozani, Kozani 13146
Kowan Gijutsu Kenkyujo, Yokosuka 17831
Közlekedési Múzeum, Budapest 13390
Közlekedéstudományi Intézet Nonprofit Kft, Budapest 13393
V.I. Kozlov Minsk Electrotechnical plant, Minsk 01880
Központi Elelmiszeripari Kutató Intézet, Budapest 13391
Központi Muzeumi Igazgatóság, Budapest 13392
Központi Papneveló Intézet, Budapest 13252
Központi Statisztikai Hivatal, Budapest 13316
Kpandu Technical Institute, Kpandu 13014
KPH – Edith Stein, Salzburg 01465
KPMG, Sydney 00837
KPMG LLP, New York 37218
Kraevedcheski muzei, Pyatigorsk 23712
Kraevedcheski muzei, Zlatoust 24097
Kraevedcheski muzei im. G.N. Prozriteleva-G.K. Prave, Stavropol 23982
Kraevoi gosudarstvenny arkhiv, Krasnoyarsk 23125
Kraevoi muzei, Krasnoyarsk 23126
Kraeznavchi muzei, Feodosiya 28238
Kraeznavchi muzei, Mykolaiv 28430
Kraeznavchi muzei, Poltava 28449
Kraeznavchi muzei mista, Kherson 28287
Kraft Foods, Inc, Glenview 35721
Kraftfahrt-Bundesamt, Flensburg 10894
Krajowa Szkoła Administracji Publicznej, Warszawa 21034
Krajská knihovna Karlovy Vary, Karlovy Vary 06577
Krajská knihovna v Pardubicích, Pardubice 06597
Krajská knihovná Vysočiny, Havlíčkův Brod 06570
Krajská knižnica Ľudovíta Štúra vo Zvolene, Zvolen 24805
Krajská vědecká knihovna v Liberci příspěvková organizace, Liberec 06347
Krajské Muzeum Cheb, Cheb 06443
Krajské muzeum Karlovarského kraje, Muzeum Karlovy Vary, Karlovy Vary 06456
Kramatorsk City Centralized Library System, Main Library, Kramatorsk 28628
Kramatorska miska TsBS, Tsentralna biblioteka, Kramatorsk 28628
Kramer, Levin, Naftalis & Frankel LLP, New York 35993
Kramfors kommunbibliotek, Kramfors 26822
Krankenhaus Dresden-Friedrichstadt – Städtisches Klinikum, Dresden 11762
Krankenhaus Paul-Gerhardt-Stift, Lutherstadt Wittenberg 12601
Krasnodar Regional Library for Youth, Krasnodar 24151
Krasnodar Regional Universal Library named after A.S. Pushkin, Krasnodar 22144
Krasnodarskaya kraevaya detskaya biblioteka im. bratev Ignatovykh, Krasnodar 24150

Krasnodarskaya kraevaya universalnaya nauchnaya biblioteka im. A.S. Pushkina, Krasnodar 22144
Krasnodarskaya kraevaya yunosheskaya biblioteka, Krasnodar 24151
Krasnodarski mezhotraslevoi territorialny tsentr nauchno-tekhnicheskoi informatsii i propagandy, Krasnodar 23113
Krasnogvardeisky District Centralized Library System, Central District Children's Library, Sankt-Peterburg 24265
Krasnogvardeisky District Centralized Library System, Gogol Central District Library, Sankt-Peterburg 24264
Krasnogvardeisky District Centralized Library System, Malookhtinsky Library Branch no 1, Sankt-Peterburg 24262
Krasnogvardeisky District Centralized Library System, Piskarevsky Library Branch no 6, Sankt-Peterburg 24263
Krasnoselsky District Centralized Library System, Central District Children's Library, Sankt-Peterburg 24267
Krasnoselsky District Centralized Library System, Central District Library, Sankt-Peterburg 24266
Krasnoyarsk Centre of Scientific and Technical Information, Krasnoyarsk 23127
Krasnoyarsk kraevaya yunosheskaya biblioteka, Krasnoyarsk 24152
Krasnoyarsk Mechanical Engineering College, Krasnoyarsk 22676
Krasnoyarsk Space Technology Institute, Krasnoyarsk 22304
Krasnoyarsk State Institute of Economics and Trade (KSIET), Krasnoyarsk 22305
Krasnoyarsk State Technical University, Krasnoyarsk 22306
Krasnoyarsk State University, Krasnoyarsk 22307
Krasnoyarskaya kraevaya universalnaya nauchnaya biblioteka, Krasnoyarsk 22145
Krasnoyarski gosudarstvenny tekhnicheski universitet, Krasnoyarsk 22306
Krasnoyarski gosudarstvenny universitet, Krasnoyarsk 22307
Krasnoyarski mashinostroitelny kolledzh, Krasnoyarsk 22676
Krasnoyarski nauchno-issledovatelski institut selskogo khozyaistva, Solyanka 23978
'Krasny Proletary' Share Holding Company, Moskva 22854
Krasny Vyborzhets Joint-Stock Company, Sankt-Peterburg 22902
Kratiko Odeo Thessaloniki, Thessaloniki 13086
Kreis- und Autobibliothek Kronach, Kronach 12163
Kreis- und Stadtarchiv Haldensleben, Haldensleben 11953
Kreis- und Stadtbücherei Gummersbach, Gummersbach 12764
Kreis Viersen, Verwaltungsbücherei, Viersen 11115
Kreisarchiv des Enzkreises, Pforzheim 12394
Kreisarchiv Esslingen, Esslingen am Neckar 11829
Kreisarchiv Jena, Dornburg-Camburg 11742
Kreisarchiv Kleve, Geldern 11917
Kreisarchiv Kreis Herzogtum Lauenburg, Ratzeburg 12422
Kreisarchiv Odenwaldkreis, Erbach 11813
Kreisarchiv Warendorf, Warendorf 12578
Kreis-Berufsschulzentrum Biberach, Biberach an der Riß 10711
Kreisbibliothek Aschersleben, Aschersleben 12635
Kreisbibliothek Bodenseekreis, Salem 12456
Kreisbibliothek des Landkreises Spree-Neiße, Spremberg 12943
Kreisbibliothek Eutin, Eutin 12726
Kreisbibliothek Freyung, Freyung 12741
Kreisbibliothek Havelland, Rathenow 12899
Kreisbibliothek Leipziger Land, Borna 12682
Kreisbibliothek Lüneburg, Lüneburg 12837
Kreisbibliothek Quedlinburg, Quedlinburg 12897
Kreisbücherei Pfaffenhofen, Pfaffenhofen 12890
Kreiskrankenhaus Stollberg gGmbH, Stollberg 12504
Kreismedienzentrum Elbe-Elster, Herzberg (Elster) 12787
Kreismuseum des Salzlandkreises, Schönebeck 12474
Kreismuseum Grimma, Grimma 11943
Kremenets Regional Centralized Library System, Main Library, Kremenets 28631
Kremenetska raionna TsBS, Tsentralna biblioteka, Kremenets 28631
Kreminna Regional Centralized Library System, Main Library, Kreminna 28632
Kreminska raionna TsBS, Tsentralna biblioteka, Kreminna 28632
Kresznerics Ferenc Könyvtár, Celldömölk 13532
Kretingos rajono savivaldybės M. Valančiaus viešsji biblioteka, Kretinga 18337
Krichev Asbestos-Cement Slates Industrial Corporation, Krichev 01876
Krieg DeVault LLP Library, Indianapolis 35778
Krigsarkivet, Stockholm 26659
Kriminalvården, Norrköping 26497
Kriminologische Zentralstelle e.V., Wiesbaden 12595
Krimska astrofizychna observatoriya, Nauchne 28434
Kripke Jewish Federation Library, Omaha 35318
Krirk University, Bangkok 27703
Sri Krishnadevaraya University, Anantapur 13605
Kristiansand folkebibliotek, Kristiansand 20361
Kristianstad centralsjukhuset, Kristianstad 26605
Kristianstad College, Kristianstad 26397
Kristianstads stadsbibliotek, Kristianstad 26823
Kristiansund folkebibliotek, Kristiansund 20362
Kristiinankaupungin kaupunginkirjasto – Kristinestads stadsbibliotek, Kristiinankaupunki 07581
Kristiinankaupunki City Library, Kristiinankaupunki 07581
Kristine Mann Library, New York 37219
Kristinehamns bibliotek, Kristinehamn 26824
Kristinestads stadsbibliotek, Kristiinankaupunki 07581
Krivi Rig Regional Centralized Library System, Main Library, Krivi Rig 28637
Krivi Rig Trade Union of Metallurgy Workers, Main Library, Krivi Rig 28638
Krivichi Village Library, Krivichi 02151
Krivichskaya gorposelkovaya bibliyateka, Krivichi 02151
Krivoi Ore Mining Institut, Krivi Rig 28007
Krivorizhka miska TsBS, Tsentralna biblioteka, Krivi Rig 28637
Krivorizhka profspilka metalurgiv, Tsentralna biblioteka, Krivi Rig 28638
Krivorizhstal Industrial Corporation, Krivi Rig 28153
Krivorizhstal Industrial Corporation, Trade Union Library, Krivi Rig 28636
Krivorozhsky gornodobyvayushchi instytut, Krivi Rig 28007
Krizhopil Regional Centralized Library System, Main Library, Krizhopil 28639
Krizhopilska raionna TsBS, Tsentralna biblioteka, Krizhopil 28639
KRKA, d.d., Novo Mesto, Novo Mesto 24905
KRKA Library and Information Services, Novo Mesto 24905
Krkonošské muzeum, Vrchlabí 06547
Kroatisches Historisches Institut, Wien 01480
Krokoms kommunbibliotek, Krokom 26825
Kröller-Müller Museum, Otterlo 19503
Kronshtad District Centralized Library System, Central City Library, Sankt-Peterburg 24268
Kroonstad Public Library, Kroonstad 25211
Krośnieńska Biblioteka Publiczna, Krosno 21463
Krosno City Library, Krosno 21463
Krotona Institute of Theosophy, Ojai 37302
Krugersdorp Public Library, Krugersdorp 25212
N.K. Krupskaia Donetsk Regional General Scientific Library, Donetsk 28545
Krupskaia Novosibirsk Regional Children's Library, Novosibirsk 24191
N.K. Krupskaya Astrakhan Regional Research Library, Astrakhan 22118
Krupskaya District Children's Library, Smolensk 24319
Krupskaya Mozyr Teacher Training, Mozyr 01814
Krupskaya Recreation Centre, Sankt-Peterburg 24227
Krupskaya Slutsk Central City Library, Slutsk 02235
Krupski Orsha City Children's Library, Orsha 02216
Krust Regional Centralized Library System, Main Library, Krust 28640
Krustska raionna TsBS, Tsentralna biblioteka, Krust 28640
Krymska respublikanska biblioteka dlya yunatstva, Simferopol 28816
Krymska respublikanska universalna naukova biblioteka im. I.Ya. Franko, Simferopol 27946
Krymskaya gosudarstvannaya silskokhozyaistvennaya opytnaya stantsiya, Klepinino 28293
Krymsky kraevedchesky muzei, Simferopol 28471
Krymsky medychny instytut, Simferopol 28074
Krymsky tsentr naukovo-tekhnichnoyi informatsiyi, Simferopol 28472
KSB Aktiengesellschaft, Frankenthal 11394
K.S.B. Solabanna Shetty Commerce College, Sagar 13841
Książnica Beskidzka Dzielnicowa Biblioteka Publiczna, Bielsko-Biała 21384
Książnica Cieszyńska, Cieszyn 21087
Książnica Kopernikanska, Toruń 21589
Książnica Pedagogiczna im. Alfonsa Parczewskiego, Kalisz 21112
Książnica Podlaska im. Łukasza Górnickiego, Białystok 21382
Książnica Pomorska im. Stanisława Staszica, Szczecin 21580
KTH Biblioteket, Stockholm 26430
KTI Közlekedéstudományi Intézet Nonprofit Kft, Budapest 13393
KTU Architektūros ir statybos instituto biblioteka, Kaunas 18296
Kuala Lumpur City Hall, Kuala Lumpur 18549
Kuban Agricultural University, Krasnodar 22301
Kuban Medical Academy, Krasnodar 22299
Kuban State University, Krasnodar 22300
Kubanskaya meditsinskaya akademiya, Krasnodar 22299
Kubanski gosudarstvenny universitet, Krasnodar 22300
Kubanski nauchno-issledovatelski institut ispytani traktorov i selskokhozyaistvennykh mashin, Novokubansk 23632
Kubanski selskokhozyaistvenny universitet, Krasnodar 22301
Kubinyi Ferenc Múzeum, Szécsény 13503
B. Kucher Library, Kyiv 28655
Kuching City Council Library, Kuching 18629
I. Kudrya Library, Kyiv 28654
Kuibisheve Regional Centralized Library System, Main Library, Kuibisheve 28641
Kuibishevska raionna TsBS, Tsentralna biblioteka, Kuibisheve 28641
Kuibyshev District Library-Branch no 3, Sankt-Peterburg 24232
Kuibyshev District Library-Branch no 4, Sankt-Peterburg 24233
Kuibyshev Electrical Engineering Plant, Moskva 22836
Kuibyshev Railway Library, Samara 23740
Kuibyshevskaya zheleznaya doroga, Dorozhnaya biblioteka, Samara 23740
Kuibyshevski mezhotraslevoi territorialny tsentr nauchno-tekhnicheskoi informatsii i propagandy, Samara 23741
Kuki-Chowa Eisei Kogakkai, Tokyo 17702
Kulani Correctional Facility Library, Hilo 34355
Kuletsyuv Mogilev Teacher Training Institute, Mogilev 01813
Kuljab State University, Kuljab 27640
Kulturen, Lund 26613
Kulturforum Witten / Stadtbücherei, Witten 12980
Kulturgeschichtliches Museum Osnabrück, Osnabrück 12389
Kulturhistorisches Museum, Magdeburg 12226
Kulturhistorisk Museum, Oslo 20248
Kulturni Center Kannik, Kannik 24539
Kulturwerk Schlesien, Würzburg 12620
Külügyminisztérium, Budapest 13317
Kumamoto Gakuen Daigaku, Kumamoto 16986
Kumamoto Gakuen University, Kumamoto 16986
Kumamoto Kenritsu Kyoiku Senta, Yamaga 17817
Kumamoto Museum, Kumamoto 17581
Kumamoto Prefectural Assembly, Kumamoto 17273
Kumamoto Prefectural Education Center, Yamaga 17817
Kumamoto University, Kumamoto 16987
– Igakubu Bunkan, Kumamoto 16988
– Yakugakubu Bunkan, Kumamoto 16989
Kumamotoken Gikai, Kumamoto 17273
Kumaun University, Nainital 13719
Kumla bibliotek, Kumla 26826
Kumura Kinen Toshokan, Iwakuni 17559
Kumura Memorial Library, Iwakuni 17559
Kuna Community Library, Kuna 39743
Kunaicho Toshokan, Tokyo 17315
Kunchevski mekhanicheski zavod, Moskva 22840
Kungälvs stadsbibliotek, Kungälv 26827
Kungl. Akademien för de fria konsterna, Stockholm 26660
Kungl. Biblioteket, Stockholm 26370
Kungl. Dramatiska Teatern, Stockholm 26661
Kungl. konsthögskolan, Stockholm 26429
Kungl. Örlogsmannasällskapet, Karlskrona 26600
Kungl. Skogs- och Lantbruk-sakademien, Stockholm 26662
Kungl. Tekniska högskolan, Stockholm 26430
– Arkitekturbiblioteket, Stockholm 26431
– Forumbiblioteket, Kista 26432
– Matematikbiblioteket, Stockholm 26433
Kungl. Vetenskaps-Societeten i Uppsala, Uppsala 26707
Kungsbacka bibliotek, Kungsbacka 26828
Kungsörs bibliotek, Kungsör 26829
Kunitachi College of Music, Tokyo 17155
Kunming Institute of Zoology, Kunming 05876
Kunming Medical College, Kunming 05354
Kunming Metallurgical College, Kunming 05355
Kunming Teachers' College, Kunming 05356
Kunming University, Kunming 05357
Kunming University of Science and Technology, Kunming 05358
Kunst- und Ausstellungshalle der Bundesrepublik Deutschland, Bonn 11636
Kunst- und Museumsbibliothek der Stadt Köln, Köln 12146
Kunstakademie Düsseldorf, Düsseldorf 09682
Kunstakademiets Arkitektskole, København 06712
Kunstbibliothek, Berlin 11569
Kunstforum Ostdeutsche Galerie Regensburg, Regensburg 12428
Kunstgewerbemuseum, Berlin 11571
Kunsthalle Basel, Basel 27306
Kunsthalle Bielefeld, Bielefeld 11610
Kunsthalle Bremen, Bremen 11682
Kunsthalle Mannheim, Mannheim 12243
Kunsthalle zu Kiel, Kiel 10099
Kunsthaus Zürich, Zürich 27454
Kunsthistorische Bibliothek des Kunstmuseums Bern und des Instituts für Kunstgeschichte der Universität Bern, Bern 27323
Kunsthistorisches Institut in Florenz, Firenze 16270
Kunsthistorisches Museum mit MVK und ÖTM, Wien 01620
Kunsthochschule Berlin-Weißensee (KHB), Berlin 09452
Kunsthochschule für Medien Köln, Köln 10130
Kunsthøgskolen i Oslo, Avdeling Statens kunstakademi, Oslo 20117
Kunstmuseum aan Zee, Oostende 02693
Kunstmuseum Basel, Basel 27307
Kunstmuseum Bern, Bern 27323
Kunstmuseum Bonn, Bonn 11637
Kunstmuseum Kloster Unser Lieben Frauen, Magdeburg 12223
Kunstmuseum Stuttgart, Stuttgart 12528
Kunstsammlung Nordrhein-Westfalen k20k21, Düsseldorf 11800
Kunstsammlungen und Museen Augsburg, Augsburg 11459
Kunstuniversität Linz, Linz 01263
Kuny Domokos Múzeum, Tata 13514
Kuopio Academy of Design, Taitemia Library, Kuopio 07523
Kuopio City Library – Central Library for North Savo, Kuopio 07582
Kuopio Museum of Cultural History, Kuopio 07521
Kuopio Natural History Museum, Kuopio 07522
Kuopio University, Kuopio 07353
Kuopion kaupunginkirjasto – Pohjois-Savon maakuntakirjasto, Kuopio 07582
Kuopion kulttuurihistoriallinen museo, Kuopio 07521
Kuopion luonnontieteellinen museo, Kuopio 07522
Kuopion Muotoiluakatemian, Kuopio 07523
Kuopion yliopisto, Kuopio 07353
Kupala Minsk City Central Library, Minsk 02184
Ya. Kupala Mogilev City Library, Mogilev 02195
Ya. Kupala Vitebsk City Library, Vitebsk 02245
Kupferstichkabinett, Berlin 11572
Kupiškio rajono savivaldybės viešoji biblioteka, Kupiškis 18338
Kuppuswami Sastri Research Institute, Chennai 13964
Kupyanska raionna TsBS, Tsentralna biblioteka, Kupyansk 28642

Kupyask Regional Centralized Library System, Main Library, Kupyansk 28642
Kuraray Co, Ltd, Material Science Research Laboratory, Kurashiki 17381
Kurare K. K. Chuo Kenkyujo Somuka, Kurashiki 17381
Kurashova Medical University, Kazan 22278
Kuratorium für Forschung im Küsteningenieurwesen (KFKI), Hamburg 11999
Kuratorium für Verkehrssicherheit, Wien 01505
Kuratorium Oświaty, Zamość 21373
Kurchatov Institute of the Russian Academy of Sciences, Moskva 23346
Kurchatovski institut RAN, Moskva 23346
Kureha Chemical Industry Co, Ltd, Nishiki Research Laboratory, Iwaki 17364
Kureha Kagaku K. K. Nishiki Kenkyujo, Iwaki 17364
Kurgan Bus Plant, Kurgan 22817
Kurganskaya oblastnaya detskaya biblioteka, Kurgan 24153
Kurganskaya oblastnaya universalnaya nauchnaya biblioteka im. A.K. Yugova, Kurgan 22146
Kurganskaya oblastnaya yunosheskaya biblioteka, Kurgan 24154
Kurganski avtobusny zavod, Kurgan 22817
Kurganski mezhotraslevoi territorilny tsentr nauchno-tekhnicheskoi informatsii i propagandy, Kurgan 23130
Ku-Ring-Gai, Pymble 01146
Kurnkov Institute for General and Inorganic Chemistry of the Russian Academy of Sciences, Moskva 23298
Kurortproekt, Moskva 23347
Kurpfälzisches Museum der Stadt Heidelberg, Heidelberg 12048
Kursk Scientific and Technical Information Centre, Central Scientific and Technical Library, Kursk 23135
Kursk State Medical University, Kursk 22312
Kurskaya oblastnaya kartinnaya galereya im. A.A. Deineka, Kursk 23134
Kurskaya oblastnaya nauchnaya biblioteka im. N.N. Aseeva, Kursk 22147
Kurskaya oblastnaya yunosheskaya biblioteka, Kursk 24155
Kurski TsNTI, Tsentralnaya nauchno-tekhnicheskaya biblioteka, Kursk 23135
Kursky Law Institute, Saratov 22579
Kurth Memorial Library, Lufkin 39978
Kurukshetra University, Kurukshetra 13690
Kurume Daigaku, Kurume 16995
– Sangyo Keizai Kenkyujo, Kurume 16996
Kurume University, Kurume 16995
– Institute for Industrial and Economic Research, Kurume 16996
Kuskovo Chemical Plant, Moskva 22841
Kuskovski khimicheski zavod, Moskva 22841
Kustanay Agricultural Institute, Kustanay 17971
Kutaisi Akaki Tsereteli State University, Kutaisi 09192
Kutak & Rock, Atlanta 35463
Kutak Rock LLP, Omaha 36070
Kutztown University, Kutztown 31523
Kuusankosken kaupunginkirjasto, Kuusankoski 07583
Kuusankoski Public Library, Kuusankoski 07583
Kuvendi i Shqipërisë, Tiranë 00013
Kuwait Institute for Scientific Research, Safat 18132
Kuwait National Assembly, Safat 18128
Kuwait National Museum, Safat 18131
Kuwait University, Safat 18127
Kuyper College, Grand Rapids 31239
Kuzbas State Technical University, Kemerovo 22283
Kuzbasski gosudarstvenny tekhnicheski universitet, Kemerovo 22283
Kuznetsk Coal Research Institute Joint-Stock Company, Prokopevsk 22894
Kuznetsk Metal Industry, Novokuznetsk 22878
Kuznetski metallurgicheski kombinat, Novokuznetsk 22878
Kuznetskugol Coal Industry Information and Technology Centre, Novokuznetsk 23634
Kuznetsov Military Naval Academy, Sankt-Peterburg 22556
Kvaerner Metalls, San Ramon 36252
Kvaerner Metals Davy Ltd, Sheffield 29582
Kvam folkeboksamling, Norheimsund 20380
KVINFO – Center for information om kvinde- og kønsforskning, København 06808
Kvinnherad folkebibliotek, Husnes 20348
Kwame Nkrumah University of Science and Technology, Kumasi 13015
Kwangju National Teachers' College, Kwangju 18067
Kwansei Gakuin University, Nishinomiya 17070
– Sangyo Kenkyujo, Nishinomiya 17071
Kwantlen University College Libraries, Surrey 03874
Kwara State Library Board, Ilorin 20054
Kwara State Polytechnic, Ilorin 19951
KwaZulu-Natal Provincial Library Service, Pietermaritzburg 25222
Kyambogo University, Kampala 27913
Kyiv Economic Institute of Management, Kyiv 28013
Kyiv National University of Trade and Economics, Kyiv 28015
Kyivo-Svyatoshinska raionna TsBS, Tsentralna biblioteka, Vishneve 28871
Kyivsky derzhavny instytut kulturi, Mykolaiv 28735
Kyivsky muzei russkoho iskusstva, Kyiv 28356
Kyivsky natsionalny linguistichny universytet, Kyiv 28014
Kyivsky natsionalny torgovelno-ekonomichny universytet, Kyiv 28015
Kyivsky NDI gematologiyi i perelivannya krovi, Lviv 28416
Kymi Paper Oy, Kuusankoski 07423
Kymi Paper Oy Information Service, Kuusankoski 07423
Kyodo News Service, Research Section, Tokyo 17703
Kyodo Tsushinsha Henshu Kyoku, Tokyo 17703
Kyoiku Toshokan, Tokyo 17704
Kyokasho Toshokan 'Toshobunko', Tokyo 17705
Kyorin Daigaku, Mitaka 17043
Kyorin University, Medical Library, Mitaka 17043
Kyoritsu College of Pharmacy, Tokyo 17156
Kyoritsu Yakka Daigaku, Tokyo 17156
Kyosei Kyokai, Tokyo 17706
Kyoto Boeki Kyokai, Kyoto 17587
Kyoto Chamber of Commerce and Industry, Kyoto 17590
Kyoto City Assembly, Kyoto 17275
Kyoto Daigaku, Kyoto 17002
– Genshiro Jikkenjo, Osaka 17003
– Graduate School of Agriculture, Kyoto 17004
– Hogakubu Toshishitsu, Kyoto 17005
– Jimbun Kagaku Kenkyujo, Kyoto 17006
– Jinbun Kagaku Kenkyujo Fuzoku Kanji Joho Kenkyu Senta, Kyoto 17007
– Kagaku Kenkyujo, Uji 17008
– Keizai Kenkyujo, Kyoto 17009
– Kogakubu Kenchirkukeikyoshitsu, Kyoto 17010
– Nogakubu Toshoshitsu, Kyoto 17011
– Reichorui Kenkyujo, Inuyama 17012
– Research Institute for Mathematical Sciences, Kyoto 17013
– Rigakubu Fuzoku Otsu Rinko Jikkenjo, Otsu 17014
– Tonan Ajia Kenkyu Senta, Kyoto 17015
Kyoto Daigaku Bungakubu Hakubutsukan, Kyoto 17588
Kyoto Foreign Trade Association, Kyoto 17587
Kyoto Furitsu Daigaku, Kyoto 17016
Kyoto Furitsu Ika Daigaku, Kyoto 17017
Kyoto Furitsu Sogo Shiryokan, Kyoto 17846
Kyoto Insitute of Technology, Kyoto 17018
Kyoto Joshi Daigaku, Kyoto 17019
Kyoto Kokuritsu Hakubutsukan, Kyoto 17589
Kyoto National Museum, Kyoto 17589
Kyoto Pharmaceutical University, Kyoto 17022
Kyoto Prefectural Assembly, Kyoto 17274
Kyoto Prefectural Institute of Agriculture, Kameoka 17563
Kyoto Prefectural Library, Kyoto 17846
Kyoto Prefectural University, Kyoto 17016
Kyoto Prefectural University of Medicine, Kyoto 17017
Kyoto Sangyo Daigaku, Kyoto 17020
Kyoto Sangyo University, Kyoto 17020
Kyoto Shoko Kaigisho Sangyo Ryutsubu Toshokan, Kyoto 17590
Kyoto University, Kyoto 17002
– Center for Ecological Research, Otsu 17014
– Center for Southeast Asian Studies, Kyoto 17015
– Documentation and Information Center for Chinese Studies, Institute for Research in Humanities, Kyoto 17007
– Faculty of Agriculture, Kyoto 17011
– Faculty of Engineering, Dept of Architecture, Kyoto 17010
– Faculty of Law, Kyoto 17005
– Institute for Chemical Research, Uji 17008
– Institute for Research in Humanities, Kyoto 17006
– Institute of Economic Research, Kyoto 17009
– Primate Research Institute, Inuyama 17012
– Research Reactor Institute, Osaka 17003
Kyoto University of medicine, Kyoto 17021
Kyoto Women's University, Kyoto 17019
Kyoto Yakka Daigaku, Kyoto 17022
Kyotofu Gikai, Kyoto 17274
Kyotofu Nogyo Sogo Kenkyujo, Kameoka 17563
Kyotoshi Gikai, Kyoto 17275
Kyotoshi Tokei Senta Shiryoshitsu, Kyoto 17276
The Kyowa Bank, Ltd, Tokyo 17457
Kyowa Fermentation Industry Co, Ltd, Machida 17383
Kyowa Ginko K. K., Tokyo 17457
Kyowa Hakko Kogyo K. K. Tokyo Kenkyujo, Machida 17383
Kypriake Vivliotheke, Nicosia 06322
Kyrgyz Academy of Sciences, Bishkek 18134
Kyrgyz Agricultural Institute, Bishkek 18136
Kyrgyz Children's Library, Bishkek 18154
Kyrgyz Institute of Agricultural Research, Bishkek 18147
Kyrgyz Institute of Physical Culture, Bishkek 18137
Kyrgyz Pedagogical Institute for Women, Bishkek 18138
Kyrgyz Research Institute of Obstetrics and Paediatrics, Bishkek 18148
Kyrgyz Scientific-Research Institute of Animal Husbandry and Veterinary Science, Bishkek 18149
Kyrgyz State Institute of Fine Art, Bishkek 18139
Kyrgyz State Institute of Fine Arts, Bishkek 18150
Kyrgyz State Medical Academy, Bishkek 18151
Kyrgyz State University, Bishkek 18140
Kyrgyz Technical University, Bishkek 18141
Kyung Hee University, Seoul 18090
Kyungnam University, Masan 18070
Kyungpook National University, Daegu 18062
Kyushu Daigaku, Fukuoka 16921
– Igaku Bunkan, Fukuoka 16922
– Nogakubu, Fukuoka 16923
– Oyo Rikigaku Kenkyujo, Kasuga 16924
– Sekitan Kenkyu Shiryo Senta, Fukuoka 16925
– Yakugakubu Toshoshitsu, Fukuoka 16926
Kyushu Economic Research Center, Fukuoka 17537
Kyushu Institute of Technology, Kitakyushu 16971
Kyushu Keizai Chosakyokai Shiryobu, Fukuoka 17537
Kyushu National Agricultural Experiment Station, Chikujo 17526
Kyushu National Agricultural Experiment Station, Kikuchi 17570
Kyushu Nogyo Shikenjo, Chikujo 17526
Kyushu Nogyo Shikenjo (Kumamoto), Kikuchi 17570
Kyushu Sangyo Daigaku Sangyokeiei Kenkyujo, Fukuoka 16927
Kyushu Tokai University, Kumamoto 16990
Kyushu University, Fukuoka 16921
– Faculty of Agriculture, Fukuoka 16923
– Faculty of Pharmaceutical Sciences, Fukuoka 16926
– Medical Branch, Fukuoka 16922
– Research Center for Materials on Coal Mining, Fukuoka 16925
– Research Institute for Applied Mechanics, Fukuoka 16924
KZ-Gedenkstätte Dachau, Dachau 11718
KZ-Gedenkstätte Neuengamme, Hamburg 12000
L. A. Mayer Memorial Institute for Islamic Art, Jerusalem 14737
L. and S. Didziuliai Public Library of Anyksciai District Municipality, Anykščiai 18317
L. E. Phillips Memorial Public Library, Eau Claire 38927
L. M. Isaev Research Institute of Medical Parasitology, Samarkand 42120
L. N. Tolstoy Regional Universal Science Library, Kustanay 17989
L3 Communications Library, Athens 35453
La Corunne Chamber of Comincerce, A Coruña 25949
La Crosse Public Library, La Crosse 39746
La Grande Public Library, La Grande 39747
La Grange College, La Grange 31527
La Grange Park Public Library District, La Grange Park 39750
La Grange Public Library, La Grange 39748
La Marque Public Library, La Marque 39752
La Porte County Public Library, La Porte 39757
La Roche College, Pittsburgh 32302
La Salle University, Philadelphia 32256
La Sierra University, Riverside 32454
La Trobe City Libraries, Morwell 01128
La Trobe University, Bundoora
– Bendigo Heyward Library, Bendigo 00403
– Bundoora Borchardt Library, Bundoora 00404
– Alburga Wodonga Campus, Wodonga 00405
La Vergne Public Library, La Vergne 39760
La Vista Public Library, La Vista 39762
LA84 Foundation, Los Angeles 37053
Lääne County Central Library (LCCL), Haapsalu 07244
Lääne-Viru County Central Library, Rakvere 07254
Lääne-Virumaa Keskraamatukogu, Rakvere 07254
Laban Library and Archive, London 29781
Labasa College, Labasa 07296
Laboratoire d'Analyse et de Traitement Informatique de la Langue Française, Nancy 08437
Laboratoire d'Anthropologie Sociale, Paris 08578
Laboratoire de Cryptogamie, Paris 08609
Laboratoire de Recherches Vétérinaires et Zootechniques de Farcha, N'Djamena 04840
Laboratoire Interdisciplinaire de Recherche sur les Ressources Humaines et l'Emploi (LIRHE-CNRS), Toulouse 08695
Laboratoire National de l'Elevage et des Recherches Vétérinaires, Dakar 24444
Laboratoire Public d'Essais et d'Etudes (LPEE), Casablanca 18949
Laboratorio Centrale di Idrobiologia, Roma 16545
Laboratorio Chimico Centrale delle Dogane e Imposte Indirette del Ministero delle Finanze, Roma 16546
Laboratorio Conmemorativo Gorgas, Panamá 20542
Laboratório de Geologia Aplicada, Lisboa 21807
Laboratorio de Geotécnia, Madrid 26063
Laboratório Nacional de Energia e Geologia, S. Mamede de Infesta 21826
Laboratório Nacional de Engenharia Civil, Lisboa 21808
Laboratorium Kesehatan Pusat Lembaga Eijkman, Jakarta 14366
Laboratory Institute of Merchandising Library, New York 37220
Labour Department, HKSAR, Hong Kong 05875
Labour Institute for Economic Research, Helsinki 07482
Labour Movement Archives and Library, Oslo 20240
Labour Movement Archives and Library, Stockholm 26639
Labour Movement Library, Helsinki 07503
Labour Movement Library and Archive, København 06783
Labour Union of Iron and Metal Workers, Budapest 13463
Lackawanna Bar Association, Scranton 37588
Laczkó Dezső Museum, Veszprém 13519
Lady Trench Training Centre, Hong Kong 05813
Lae Public Library, Lae 20598
Laebharlann Chontae Liatroma, Ballinamore 14562
Læreruddannelsen Blaagaard/KDAS, Søborg 06748
Læringssentrene, Oslo 20118
Lafayette College, Easton 31068
– Kirby Library of Government & Law, Easton 31069
Lafayette Natural History Museum & Planetarium, Lafayette 37002
Lafayette Public Library, Lafayette 39765
Lafourche Parish Public Library, Thibodaux 41549
Lagos City Libraries, Lagos 20059
Lagos State Library Board, Ikeja 20053
Lagos State Polytechnic, Ikeja 19944
Lagos State University, Apapa 19924
LaGrange County Public Library, LaGrange 39768
Lahden historiallinen museo Kirjasto, Lahti 07524
Lahden kaupunginkirjasto-maakuntakirjasto, Lahti 07584
Lahey Clinic, Burlington 36618
Laholms bibliotek, Laholm 26830
Lahore Museum, Lahore 20514
Lahore University of Management Sciences, Lahore 20446
Lahti Historical Museum, Lahti 07524
Lahti Public Library – Regional Library, Lahti 07584
Laiyang College of Agriculture, Laiyang 05371

Lake Agassiz Regional Library, Moorhead 40292
Lake Agassiz Regional Library – Crookston Public Library, Crookston 38733
Lake Agassiz Regional Library – Moorhead Public Library, Moorhead 40293
Lake Bluff Public Library, Lake Bluff 39771
Lake Chad Research Institute (LCRI), Maiduguri 20040
Lake City Community College, Lake City 33466
Lake County Central Law Library, Gary 34297
Lake County Law Library, Tavares 34920
Lake County Library, Lakeport 39786
Lake County Library System, Tavares 41524
Lake County Public Library, Merrillville 40161
Lake County Public Library – Cedar Lake Branch, Cedar Lake 38485
Lake County Public Library – Dyer-Schererville Branch, Schererville 41248
Lake County Public Library – Griffith Branch, Griffith 39324
Lake County Public Library – Highland Branch, Highland 39430
Lake County Public Library – Hobart Branch, Hobart 39449
Lake County Public Library – Munster Branch, Munster 40341
Lake Erie College, Painesville 32226
Lake Erie College of Osteopathic Medicine, Erie 31093
Lake Forest College, Lake Forest 31545
Lake Forest Library, Lake Forest 39775
Lake Land College, Mattoon 33521
Lake Michigan College, Benton Harbor 30383
Lake Oswego Public Library, Lake Oswego 39778
Lake Region State College, Devils Lake 31016
Lake Superior State University, Sault Ste. Marie 32620
Lake Tahoe Community College, South Lake Tahoe 33769
Lake Villa District Library, Lake Villa 39779
Lake Wales Public Library, Lake Wales 39781
Lake Worth Public Library, Lake Worth 39782
Lakehead University, Thunder Bay 03877
– Education Library, Thunder Bay 03878
Lakeland College, Sheboygan 32663
Lakeland College Library, Vermilion 03950
Lakeland Community College, Kirtland 33461
Lakeland Library Region, North Battleford 04705
Lakeland Library Region – North Battleford Public Library, North Battleford 04706
Lakeland Public Library, Lakeland 39785
De Lakenhal, Leiden 19483
Lake-Sumter Community College, Leesburg 33480
Lakewood Hospital, Lakewood 37006
"Lakokraska" LIDA, Lida 01877
Lalit Narayan Mathila University, Darbhanga 13638
Lamar State College, Port Arthur 32331
Lamar University, Beaumont 30367
Lamberg'sche Schlossbibliothek, Steyr 01582
Lambeth Palace Library, London 29782
Lambeth Reference and Information Services, London 30020
Lambton County Library, Wyoming 04825
Lambuth University, Jackson 31440
Lampf, Lipkind, Prupis & Petigrow, West Orange 36410
Lamson, Dugan & Murray LLP, Omaha 36071
Lancashire County Council, Preston 30061
Lancashire Record Office, Preston 29882
Lancaster and Morecambe College, Lancaster 29150
Lancaster Bible College, Lancaster 35255
Lancaster County Historical Society Library, Lancaster 37008
Lancaster County Law Library, Lancaster 34433
Lancaster County Library, Lancaster 39794
Lancaster Mennonite Historical Society Library, Lancaster 37009
Lancaster Public Library, Lancaster 39795
Lancaster Theological Seminary, Lancaster 35256
Lancaster University, Lancaster 29151
Lancaster Veterans Memorial Library, Lancaster 39796
Lancisiana Library, Roma 16465
Łańcut Muzeum, Łańcut 21154
Land Force Concepts & Design, Kingston 03998
Land- und Amtsgericht Hamburg, Hamburg 10925
Landbúnaðarháskóli Islands, Borgarnes 13560
Landcare Research New Zealand, Lincoln 19835
Landelijk Dienstencentrum SoW-kerken, Utrecht 19317
Landelijke Verenigiging voor Thuiszorg (LVT), Bunnik 19405
Lander University, Greenwood 31266
Landesamt für Arbeitsschutz, Potsdam 12406
Landesamt für Arbeitsschutz, Regionalbereich Süd, Cottbus 10851
Landesamt für Archäologie, Dresden 10859
Landesamt für Bergbau, Energie und Geologie, Clausthal-Zellerfeld 10850
Landesamt für Bergbau, Geologie und Rohstoffe, Kleinmachnow 10972
Landesamt für Denkmalpflege, Esslingen am Neckar 10893
Landesamt für Denkmalpflege Hessen, Wiesbaden 11130
Landesamt für Denkmalpflege Sachsen, Dresden 10860
Landesamt für Denkmalpflege Schleswig-Holstein, Kiel 10965
Landesamt für Denkmalpflege und Archäologie Sachsen-Anhalt, Bibliothek Baudenkmalpflege, Halle (Saale) 10907
Landesamt für Denkmalpflege und Archäologie Sachsen-Anhalt, Landesmuseum für Vorgeschichte, Halle (Saale) 10908
Landesamt für Geoinformation und Landentwicklung Baden-Württemberg, Stuttgart 11102
Landesamt für Geologie, Rohstoffe und Bergbau, Freiburg 10904
Landesamt für Geologie und Bergbau Rheinland-Pfalz, Mainz 11008
Landesamt für Geologie und Bergwesen Sachsen-Anhalt, Halle (Saale) 10909
Landesamt für Gesundheit und Soziales Berlin, Berlin 10808
Landesamt für Gesundheit und Soziales Mecklenburg-Vorpommern, Schwerin 11096
Landesamt für Kultur und Denkmalpflege – Archäologie und Denkmalpflege, Schwerin 11097
Landesamt für Kultur und Denkmalpflege – Landeshauptarchiv Schwerin, Schwerin 12482
Landesamt für Ländliche Entwicklung, Landwirtschaft und Flurneuordnung (LELF), Paulinenaue 11079
Landesamt für Landwirtschaft, Umwelt und ländliche Räume des Landes Schleswig-Holstein, Flintbek 10896
Landesamt für Natur, Umwelt und Verbraucherschutz Nordrhein-Westfalen, Recklinghausen 11083
Landesamt für Umweltschutz Sachsen-Anhalt, Halle (Saale) 10910
Landesamt für Verbraucherschutz, Magdeburg 11003
Landesamt für Vermessung und Geobasisinformation Rheinland-Pfalz, Koblenz 10975
Landesamt für Zentrale Dienste, Saarbrücken 11086
Landesanstalt für Landwirtschaftliche Chemie, Stuttgart 12529
Landesanstalt für Umwelt, Messungen und Naturschutz Baden-Württemberg (LUBW), Karlsruhe 12089
Landesarbeitsgericht Baden-Württemberg und Finanzgericht Baden-Württemberg, Stuttgart 12530
Landesarbeitsgericht Düsseldorf, Düsseldorf 10874
Landesarbeitsgericht Hamm, Hamm 10928
Landesarbeitsgericht Schleswig-Holstein, Kiel 10966
Landesarchiv Baden-Württemberg, Karlsruhe 12090
Landesarchiv Baden-Württemberg, Sigmaringen 12488
Landesarchiv Baden-Württemberg, Wertheim 12590
Landesarchiv Baden-Württemberg – Staatsarchiv Freiburg, Freiburg 11897
Landesarchiv Berlin, Berlin 11542
Landesarchiv BW – Staatsarchiv Ludwigsburg, Ludwigsburg 12216
Landesarchiv Greifswald, Greifswald 11938
Landesarchiv Nordrhein-Westfalen, Abteilung Ostwestfalen-Lippe, Detmold 11737
Landesarchiv Nordrhein-Westfalen, Abteilung Westfalen, Münster 12347
Landesarchiv Nordrhein-Westfalen, Hauptstaatsarchiv Düsseldorf, Düsseldorf 11801
Landesarchiv Saarbrücken, Saarbrücken 12453
Landesarchiv Schleswig-Holstein, Schleswig 12467
Landesarchiv Speyer, Speyer 12497
Landesbetrieb Landwirtschaft Hessen, Kassel 10962
Landesbetrieb Wald und Holz Nordrhein-Westfalen / Forschungsstelle für Jagdkunde und Wildschadenverhütung, Bonn 11638
Landesbibliothek Coburg, Coburg 09306
Landesbibliothek des Kantons Glarus, Glarus 26970
Landesbibliothek "Dr. Friedrich Teßmann", Bozen 14789
Landesbibliothek Mecklenburg-Vorpommern, Schwerin 09337
Landesbibliothek Oldenburg, Oldenburg 09331
Landesbibliothekszentrum Rheinland-Pfalz / Rheinische Landesbibliothek, Koblenz 09324
Landesbibliothekszentrum Rheinland-Pfalz / Bibliotheca Bipontina, Zweibrücken 09345
Landesbibliothekszentrum Rheinland-Pfalz / Pfälzische Landesbibliothek, Speyer 09339
Landesdenkmalamt, Schiffweiler 11092
Landesdenkmalamt Berlin, Berlin 11543
Landesgericht, Münster 11057
Landesgeschichtliche Bibliothek, Bielefeld 11611
Landesgeschichtliche Vereinigung für die Mark Brandenburg e.V., Berlin 11544
Landeshauptarchiv Koblenz, Koblenz 12121
Landeshauptarchiv Sachsen-Anhalt, Magdeburg 12224
Landeshauptstadt Dresden, Verwaltungsbibliothek, Dresden 10861
Landeshauptstadt Kiel, Kiel 12111
Landeshauptstadt Magdeburg, Verwaltungsbibliothek, Magdeburg 11004
Landeshygieneinstitut Halle, Halle (Saale) 10911
Landesinstitut für Lehrerbildung und Schulentwicklung, Hamburg 11987
Landesinstitut für Schule, Bremen 11683
Landesinstitut für Schule, Soest 12491
Landesinstitut für Schule und Ausbildung (LISA) Mecklenburg-Vorpommern, Rostock 12442
Landesinstitut für Schule und Ausbildung (LISA) – PRI Greifswald, Greifswald 11939
Landesinstitut für Schulentwicklung, Stuttgart 12536
Landesinstitut für Schulqualität Sachsen-Anhalt, Halle (Saale) 11958
Landeskammer für Land- und Forstwirtschaft Steiermark, Graz 01528
Landeskirchenamt der Evangelischen Kirche von Westfalen, Bielefeld 11171
Landeskirchenamt Hannover, Hannover 11238
Landeskirchenamt München, München 11282
Landeskirchliche Bibliothek Bremen, Bremen 11183
Landeskirchliche Bibliothek für Religionspädagogik, Wolfenbüttel 11360
Landeskirchliche Bibliothek Karlsruhe, Karlsruhe 11251
Landeskirchliche Bibliothek Kassel, Kassel 11253
Landeskirchliche Zentralbibliothek, Stuttgart 11340
Landeskirchliches Archiv der Evangelischen Kirche von Westfalen, Bielefeld 11172
Landeskirchliches Archiv der Evangelisch-Lutherischen Kirche in Bayern, Nürnberg 11297
Landeskirchliches Archiv Wolfenbüttel, Wolfenbüttel 11361
Landeskriminalamt Niedersachsen, Hannover 10932
Landeskundliche Bibliothek für Spessart und Untermain, Aschaffenburg 11456
Landeskundliche Bibliothek und Kreisarchiv des Märkischen Kreises, Altena (Westf.) 11448
Landesmuseum für Kunst und Kulturgeschichte, Münster 12351
Landesmuseum für Kunst- und Kulturgeschichte, Schleswig 12471
Landesmuseum für Technik und Arbeit, Mannheim 12244
Landesmuseum für Vorgeschichte, Dresden 10859
Landesmuseum für Vorgeschichte, Halle (Saale) 10908
Landesmuseum für westfälische Industriekultur (LWL), Dortmund 11748
Landesmuseum Joanneum GmbH, Archäologie, Graz 01529
Landesmuseum Joanneum GmbH, Geologie & Palontologie, Graz 01530
Landesmuseum Joanneum GmbH, Kulturhistorische Sammlung, Graz 01531
Landesmuseum Joanneum GmbH, Volkskundebibliothek am Volkskundemuseum Graz, Graz 01532
Landesmuseum Mainz, Mainz 12233
Landesmuseum Württemberg, Stuttgart 12531
Landespolizeischule Rheinland-Pfalz, Hahn-Flughafen 09908
Landespolizeischule Wilhelm Krützfeld, Bad Malente 10706
Landesrechnungshof des Landes Nordrhein-Westfalen, Düsseldorf 10875
Landesregierung Nordrhein-Westfalen, Düsseldorf 10876
Landessozialgericht Baden-Württemberg, Stuttgart 11103
Landessozialgericht Niedersachsen-Bremen, Celle 10847
Landessozialgericht Nordrhein-Westfalen, Essen 10890
Landessozialgericht und Sozialgericht Hamburg, Hamburg 10926
Landesstelle für Volkskunde Stuttgart, Stuttgart 12532
Landessternwarte, Heidelberg 10034
Landesumweltamt Brandenburg, Potsdam 11081
Landesumweltbibliothek, Potsdam 11081
Landesvermessungsamt Schleswig-Holstein, Kiel 10967
Landesvorstand Die Linke. Sachsen, Dresden 11763
Landeswohlfahrtsverband Hessen, Kassel 12100
Landeszentralbank in Baden-Württemberg, Stuttgart 11430
Landeszentrale für politische Bildung Baden-Württemberg, Bad Urach 10782
Landeszentrale für politische Bildung Rheinland-Pfalz, Mainz 12234
Landgericht Aachen, Aachen 10776
Landgericht Berlin – Zivilgerichtsbarkeit 1. Instanz, Berlin 10809
Landgericht Bielefeld, Bielefeld 10816
Landgericht Bonn, Bonn 10833
Landgericht Bremen, Bremen 10842
Landgericht Cottbus, Cottbus 10852
Landgericht Dortmund, Dortmund 10856
Landgericht Duisburg, Duisburg 10865
Landgericht Düsseldorf, Düsseldorf 10877
Landgericht Erfurt, Erfurt 10887
Landgericht Essen, Essen 10891
Landgericht Flensburg, Flensburg 10895
Landgericht Hannover, Hannover 10933
Landgericht Hildesheim, Hildesheim 10947
Landgericht Hof, Hof 10952
Landgericht Kiel, Kiel 10968
Landgericht Koblenz, Koblenz 10976
Landgericht Köln, Köln 10989
Landgericht Krefeld, Krefeld 10995
Landgericht Landau in der Pfalz, Landau in der Pfalz 10997
Landgericht Lübeck, Lübeck 11001
Landgericht Mannheim, Mannheim 11016
Landgericht Marburg, Marburg 11019
Landgericht Mönchengladbach, Mönchengladbach 11021
Landgericht München I, München 11044
Landgericht Osnabrück, Osnabrück 11077
Landgericht Passau, Passau 11078
Landgericht Saarbrücken, Saarbrücken 11087
Landgericht Wiesbaden, Wiesbaden 11131
Landgericht Wuppertal, Wuppertal 11137
Landis Valley Museum, Lancaster 37010
Landkreis Harz, Kreisbibliothek Quedlinburg, Quedlinburg 12897
Landlæknisembættið, Seltjarnarnes 13577
Landmark College, Putney 33665
Landrat-Lucas-Gymnasium, Leverkusen 10753
Landratsamt Potsdam-Mittelmark, Lehnin 12822
Landratsamt Zwickau, Werdau 12587
Landsarkivet for Nørrejylland, Viborg 06837
Landsarkivet for Sjælland, Lolland-Falster og Bornholm, København 06809
Landsarkivet i Göteborg, Göteborg 26579
Landsarkivet i Härnösand, Hörnösand 26591
Landsarkivet i Lund, Lund 26614
Landsarkivet i Östersund, Östersund 26627
Landsarkivet i Uppsala, Uppsala 26708
Landsarkivet i Vadstena, Vadstena 26711
Landsarkivet i Visby, Visby 26720
Landsbókasafn Íslands – Háskólabókasfn, Reykjavík 13554
Landschaftsbibliothek, Aurich 11463
Landschaftsverband Rheinland, Köln 10990
Landschaftsverband Westfalen-Lippe, Münster 11058
Landschaftsverband Westfalen-Lippe (LWL), Münster 11060

Landskrona lasarett, Landskrona 26606
Landskrona stadsbibliotek, Landskrona 26831
Landsorganisationen i Danmark, København 06810
Landsorganisationen i Sverige, LO, Stockholm 26663
Landstinget i Kalmar län, Kalmar 26493
Landstinget Västernorrland, Härnösand 26490
Landstingsförbundet, Stockholm 26691
Landsvirkjun, Reykjavík 13563
Landtag Brandenburg, Potsdam 11082
Landtag des Saarlandes, Saarbrücken 11088
Landtag Mecklenburg Vorpommern, Schwerin 11098
Landtag Nordrhein-Westfalen, Düsseldorf 10878
Landtag Rheinland-Pfalz, Mainz 11009
Landtag von Baden-Württemberg, Stuttgart 11104
Landtag von Sachsen-Anhalt, Magdeburg 11005
Landwirtschaftliches Technologiezentrum Augustenberg, Karlsruhe 12091
Landwirtschaftliches Technologiezentrum Augustenberg, Außenstelle Forchheim, Rheinstetten 12437
Landwirtschaftskammer für Oberösterreich, Linz 01559
Landwirtschaftskammer Hannover, Hannover 12030
Landwirtschaftskammer Nordrhein-Westfalen, Bonn 11639
Landwirtschaftskammer Nordrhein-Westfalen, Münster 12348
Lane College, Jackson 31441
Lane Community College, Eugene 33326
Lane Cove Library, Lane Cove 01112
Lane Memorial Library, Hampton 39357
Lane Powell PC, Portland 36144
Lane Powell PC Library, Seattle 36267
Lane Public Library, Hamilton 39354
Lane & Waterman, Davenport 35660
Laney College, Oakland 33591
Lang Michener LLP Library, Toronto 04288
Lang Michener LLP Library, Vancouver 04304
Langat Singh College Library, Muzaffarpur 13827
Langelands Bibliotek, Rudkøbing 06921
C. J. Langenhoven Memorial Library, Oudtshoorn 25218
Langfang Teachers' College, Langfang 05373
Langston University, Langston 31557
Language and Literary Agency, Kuching 18608
Language Institute, Kuala Lumpur 18533
Langyan Teachers' College, Longyan 05403
Lanivtsi Regional Centralized Library System, Main Library, Lanivtsi 28695
Lanka Jatika Sarvodaya Shramadana Sangamaya, Moratuwa 26327
Lanovetska raionna TsBS, Tsentralna biblioteka, Lanivtsi 28695
Länsbibliotek Dalarna, Falun 26766
Länsbibliotek Jönköping, Jönköping 26804
Länsbibliotek Uppsala, Uppsala 26940
Länsbiblioteket i Örebro län, Örebro 26876
Länsbiblioteket i Värmland, Karlstad 26812
Länsbiblioteket i Västerbotten, Umeå 26936
Länsbiblioteket Västernorrland, Härnösand 26786
Lansdowne Public Library, Lansdowne 39800
Lansing Community College, Lansing 33475
Länslasarettet, Karlshamn 26598
Länsmuseet på Gotland, Visby 26721
Länsmuseet Varberg, Varberg 26714
Länssjukhuset Gävle-Sandvihen, Gävle 26572
Länsstyrelsen i Norrbottens län, Luleå 26495
Länsstyrelsen i Örebro län, Örebro 26500
Länsstyrelsen i Stockholms län, Stockholm 26508
Länsstyrelsen i Västmanlands län, Västerås 26526
Lantbrukarnas Riksförbund, Stockholm 26664
Lanterman Developmental Center, Pomona 36138
Lantmäteriverket, Gävle 26573
Lanzhou Commercial College, Lanzhou 05378
Lanzhou Institute of Glaciology and Cryopedology, Lanzhou 05878
Lanzhou Medical College, Lanzhou 05379
Lanzhou Municipal Library, Lanzhou 05921
Lanzhou Railway Institute, Lanzhou 05380
Lanzhou Teachers Training College, Lanzhou 05381
Lanzhou University, Lanzhou 05382
Laois County Library, Portlaoise 14586
Lapeer District Library, Lapeer 39802
Lapeer District Library – Marguerite deAngeli Main Branch, Lapeer 39803
Lapin maakuntakirjasto – Rovaniemen kaupunginkirjasto, Rovaniemi 07605
Lapin yliopisto, Rovaniemi 07366
Lappeenrannan kaupunginkirjasto-maakuntakirjasto, Lappeenranta 07585
Lappeenrannan teknillinen yliopisto, Lappeenranta 07355
Lappeenranta Provincial Library, Lappeenranta 07585
Lappeenranta University of Technology, Lappeenranta 07355
Laramie County Community College, Cheyenne 33218
Laramie County Library System, Cheyenne 38549
Lärarhögskolan i Stockholm, Stockholm 26434
Larchmont Public Library, Larchmont 39808
Laredo Community College, Laredo 33476
Laredo Public Library, Laredo 39809
Largo Public Library, Largo 39810
Larkin, Hoffman, Daly & Lindgren, Bloomington 35513
Larkspur Public Library, Larkspur 39811
Larsen and Toubro Ltd, Mumbai 13908
Las Cruces Public Library, Las Cruces 39812
Las Positas Community College, Livermore 33491
Las Vegas-Clark County Library District, Las Vegas 39814
Lasalle Bank Building, Chicago 35585
LaSalle Parish Library, Jena 39642
Lasarettet, Motala 26620
Lasarettet i Enköping, Enköping 26565
Lasell College, Newton 32116
Lashly & Baer PC, Saint Louis 36203
Lassen Community College, Susanville 32765
Lastensuojelun Keskusliiton, Helsinki 07471
Latah County Library District, Moscow 40310
Latham & Watkins, Chicago 35586
Latham & Watkins, New York 35994
Latham & Watkins LLP, Reston 36169
Latham & Watkins LLP Library, San Francisco 36237
Lathrop & Gage LC Library, Kansas City 35795
Latin American and Caribbean Institute for Economic and Social Planning, Santiago 04933
Latin American Integration Association (LAIA), Montevideo 42039
Latvian Academic Library, Riga 18168
Latvian Academy of Music Library, Riga 18173
Latvian Institute of Physical Culture, Riga 18191
Latvian Maritime Academy, Library, Riga 18183
Latvian Museum of Foreign Art, Riga 18180
Latvian Museum of War Research Library, Riga 18182
Latvian State Academy of Arts, Riga 18172
Latvian University of Agriculture, Jelgava 18170
Latvijas Akadēmiskā bibliotēka, Riga 18168
Latvijas Juras akademijas biblioteka, Riga 18183
Latvijas Mākslas akadēmijas bibliotēka, Riga 18172
Latvijas Mūzikas akademijas bibliotēka, Riga 18173
Latvijas Nacionālā bibliotēka, Riga 18166
Latvijas Universitāte, Riga 18174
– Institute of Biology, Salaspils 18175
Latvijas valsts arhīvu speciālā bibliotēka, Riga 18185
Lauenburgische Gelehrtenschule, Ratzeburg 10763
Laurea ammattikorkeakoulun kirjasto, Vantaa 07555
Laurea University of Applied Sciences Library, Vantaa 07555
Laurel County Public Library District, London 39919
Laurel-Jones County Library, Laurel 39817
Lauren Rogers Museum of Art Library, Laurel 37020
Laurens County Library, Laurens 39819
Laurentian Forestry Centre, Sainte-Foy 04485
Laurentian University, Sudbury 03873
Lauri Ann West Memorial Library, Pittsburgh 40802
Lausitzer Seenland Klinikum GmbH, Hoyerswerda 12068
Lautoka Hospital, Lautoka 07310
Lautoka Teachers' College, Lautoka 07288
Laval Public Libraries, Laval 04676
Lavon Institute for Labour Research, Tel-Aviv 14768
Lavrentyev Institute of Hydrodynamics of the Siberian Division of the Russian Academy of Sciences, Novosibirsk 23640
Lavsanstroi Building Company no 17, Trade Union Library, Mogilev 02206
Law College of Shanghai University, Shanghai 05503
Law Commission Library, London 29524
Law Courts of the A.C.T., Canberra 00648
Law Library, La Plata 00217
Law Library Association of Geauga County, Chardon 36647
Law Library Association of Saint Louis, Saint Louis 37486
Law Library for San Bernardino County, San Bernardino 34804
Law Library of Ireland, Dublin 14546
Law Library of Louisiana, New Orleans 34583
Law Library of Montgomery County, Norristown 34616
Law Society of England & Wales, London 29783
Law Society of Ireland, Dublin 14547
Law Society of New Brunswick Library, Fredericton 04345
Law Society of New South Wales, Sydney 01028
Law Society of Newfoundland Library, St. John's 04500
Law Society of Prince Edward Island Library, Charlottetown 04325
Law Society of Saskatchewan Libraries, Regina 04467
Law Society of Upper Canada, Kingston 04367
Law Society of Upper Canada, Toronto 04518
Lawler, Matusky & Skelly Engineers LLP Library, Pearl River 36085
Lawrence Berkeley National Laboratory Library, Berkeley 36534
Lawrence County Bar & Law Library Association, Ironton 36964
Lawrence County Federated Library System, New Castle 40404
Lawrence County Law Library, New Castle 34580
Lawrence County Library, Walnut Ridge 41717
Lawrence Law Library, Lawrence 37021
Lawrence Library, Pepperell 40758
Lawrence Public Library, Lawrence 39821
Lawrence Public Library, Lawrence 39822
Lawrence Technological University, Southfield 32691
Lawrence University, Appleton 30256
Lawrenceburg Public Library District, Lawrenceburg 39825
Lawrenceburg Public Library District – North Dearborn Branch, West Harrison 41817
Lawton Public Library, Lawton 39829
Lawyers' Association, Tokyo 17678
Laxå bibliotek, Laxå 26832
Lazdijų rajono savivaldybės viešoji biblioteka, Lazdijai 18339
Le Moyne College, Syracuse 32775
Le Moyne-Owen College, Memphis 31792
Leabharlann Contae Mhuineachan, Clones 14568
Leach Library, Londonderry 39920
Leather and Haberdashery Industrial Corporation, Sankt-Peterburg 22922
Leatherhead Food International, Leatherhead 29693
Leavenworth Public Library, Leavenworth 39833
Lebanese American University, Beirut 18206
Lebanon Community Library, Lebanon 39836
Lebanon Correctional Institution Library, Lebanon 34450
Lebanon Public Library, Lebanon 39837
Lebanon Public Library, Lebanon 39838
Lebanon Valley College, Annville 30255
Lebanon-Laclede County Library, Lebanon 39839
Lebanon-Wilson County Library, Lebanon 39840
Lebedev Institute of Physics of the Russian Academy of Sciences, Moskva 23179
Lebedin Regional Centralized Library System, Main Library, Lebedin 28696
Lebedinska raionna TsBS, Tsentralna biblioteka, Lebedin 28696
Lebendige Bibliothek Bottrop, Bottrop 12683
Leboeuf, Lamb, Greene & Macrae, New York 35995
Leboeuf, Lamb, Greene & Macrae, Salt Lake City 36212
Leboeuf, Lamb, Greene & Macrae, LLP, San Francisco 36238
Lebowa National Library, Chuenespoort 25196
Ledding Library of Milwaukie, Milwaukie 40227
Ledøje-Smørum Bibliotek, Smørum 06927
Ledyard Public Libraries, Ledyard 39841
Lee College, Baytown 33159
Lee County Law Library, Fort Myers 34257
Lee County Library, Tupelo 41622
Lee County Library System, Fort Myers 39106
Lee County Library System – Bonita Springs Public, Bonita Springs 38238
Lee County Library System – Cape Coral-Lee County Public, Cape Coral 38449
Lee County Library System – Fort Myers-Lee County Public, Fort Myers 39107
Lee County Library System – Lakes Regional, Fort Myers 39108
Lee County Library System – North Fort Myers Public, North Fort Myers 40506
Lee County Library System – South County Regional, Estero 38992
Lee Kong Chian Reference Library, Singapore 24583
Lee University – Church of God Theological Seminary, Cleveland 35163
Leeds Central Library, Leeds 30009
Leeds College of Building, Leeds 29153
Leeds College of Music, Leeds 29154
Leeds Library, Leeds 29694
Leeds Metropolitan University, Leeds 29155
Leesburg Public Library, Leesburg 39842
Lees-McRae College, Banner Elk 33152
Leeward Community College, Pearl City 33626
Lee-Whedon Memorial Library, Medina 40133
Legacy Good Samaritan Hospital & Medical Center, Portland 37400
Legal Aid Foundation of Los Angeles, Los Angeles 37054
Legal Aid Society, New York 37221
Legal Assistance Foundation of Chicago Library, Chicago 36679
Legal Library, Kyiv 28357
Legal Remembrancer's Library, Lucknow 14067
Legal Services Commission of South Australia, Adelaide 00852
Legfőbb Bíróság, Budapest 13318
Legfőbb Ügyészség, Budapest 13319
Légiforgalmi és Repülőtéri Igazgatós, Budapest 13320
Legislative Assembly, Apia 24381
Legislative Council of the Virgin Islands, Tortola 42311
Legislative Counsel Bureau Research Library, Carson City 36630
Legislative Library of Manitoba, Winnipeg 04191
Legislative Library of Nova Scotia, Halifax 04082
Legislative Reference Bureau, Milwaukee 34538
Legislative Reference Bureau Law Library, Springfield 34882
Legislative Reference Bureau Library, Honolulu 34361
LEGTA Le Chesnoy-Les Barres, Nogent-sur-Vernisson 08127
Lehi City Library, Lehi 39845
Lehigh Carbon Community College, Schnecksville 33752
Lehigh County Historical Society, Allentown 36458
Lehigh County Law Library, Allentown 33914
Lehigh University, Bethlehem 30421
LehmbruckMuseum, Duisburg 11786
Leibniz Center for Tropical Marine Ecology, Bremen 11684
Leibniz Institute for Baltic Sea Research, Rostock 12444
Leibniz Institute for Educational Research and Educational Information, Frankfurt am Main 11848
Leibniz Research Centre for Working Environment and Human Factors (IfADo), Dortmund 11747
Leibniz Universität Hannover, Hannover 09961
– Fachbibliothek Sozialwissenschaften und Bereichsbibliothek Geschichte und Religionswissenschaft, Hannover 09962
– TIB/UB Fachbibliotheken am Königsworther Platz, Hannover 09963
– Institut für Geologie, Hannover 09964
– Institut für Geschichte und Theorie im FB Architektur, Hannover 09965
– Institut für Landschaftsarchitektur, Hannover 09966
– Institut für Mathematik, Hannover 09967
– Institut für Philosophie, Hannover 09968
– Institut für Siedlungswasserwirtschaft und Abfalltechnik, Hannover 09969
– Institut für Umweltplanung, Hannover 09970

– Institut für Wasserwirtschaft, Hydrologie und landwirtschaftlichen Wasserbau, Hannover 09971
– Institut für Zierpflanzen und Gehölzwissenschaften, Hannover 09972
Leibniz-Institut für Agragentwicklung in Mittel- und Osteuropa (IAMO), Halle (Saale) 11959
Leibniz-Institut für Agrartechnik, Potsdam 12407
Leibniz-Institut für Arbeitsforschung an der TU Dortmund (IfADo), Dortmund 11747
Leibniz-Institut für die Pädagogik der Naturwissenschaften an der Universität Kiel, Kiel 12112
Leibniz-Institut für Festkörper- und Werkstoffforschung Dresden e.V., Dresden 11764
Leibniz-Institut für Gemüse- und Zierpflanzenbau Großbeeren/Erfurt e.V., Großbeeren 11944
Leibniz-Institut für Gewässerökologie und Binnenfischerei, Berlin 11545
Leibniz-Institut für Globale und Regionale Studien, Hamburg 11983
Leibniz-Institut für innovative Mikroelektronik, Frankfurt (Oder) 11877
Leibniz-Institut für Katalyse e.V. an der Universität Rostock (LIKAT), Rostock 12443
Leibniz-Institut für Länderkunde e.V., Leipzig 12190
Leibniz-Institut für Meereswissenschaften an der Universität Kiel, Kiel 12109
Leibniz-Institut für Neurobiologie (IfN), Magdeburg 12225
Leibniz-Institut für ökologische Raumentwicklung e.V., Dresden 11765
Leibniz-Institut für Ostseeforschung Warnemünde an der Universität Rostock, Rostock 12444
Leibniz-Institut für Pflanzenbiochemie, Halle (Saale) 11960
Leibniz-Institut für Pflanzengenetik und Kulturpflanzenforschung (IPK), Gatersleben 11914
Leibniz-Institut für Plasmaforschung und Technologie e.V. (INP), Greifswald 11940
Leibniz-Institut für Polymerforschung Dresden e.V., Dresden 11766
Leibniz-Institut für Zoo- und Wildtierforschung (IZW) im Forschungsverbund Berlin e.V., Berlin 11546
Leibniz-Institute for Plasma Science and Technology e.V., Greifswald 11940
Leibniz-Zentrum für Agrarlandschaftsforschung (ZALF) e.V., Müncheberg 12274
Leibniz-Zentrum für Marine Tropenökologie (ZMT), Bremen 11684
Leica Geosystems AG, Heerbrugg 27287
Leicestershire County Council, Glenfield 29986
Leichhardt Municipal Library, Leichhardt 01113
Leiden University, Leiden 19152
– Center for Japanese & Korean Studies, Leiden 19156
– Institute of East European Law and Russian Studies, Leiden 19162
– Institute of Environmental Sciences (CML), Leiden 19157
– Law Library, Leiden 19164
– Van Vollenhoven Institute for Law, Governance and Development, Leiden 19169
Leiden University Medical Centre and Rivierduinen, Leidern 19170
Leids Universitair Medisch Centrum (LUMC), Bibliotheek rijnVeste, Leidern 19170
Leids Universitair Medisch Centrum (LUMC), Walaeus Bibliotheek, CI-Q, Leiden 19151
Leihverkers- und Ergänzungsbibliothek Schleswig-Holstein, Flensburg 12730
Leipzig Centre for the History and Culture of East Central Europe, Leipzig 12182
Leipziger Städtische Bibliotheken, Leipzig 12826
Leitrim County Library, Ballinamore 14562
Lekárska knižnica nemocnice F.D. Roosevelta, Banská Bystrica 24676
Leksands bibliotek, Leksand 26833
Leksikografski Zavod 'Miroslav Krleža', Zagreb 06249
Lembaga Administrasi Negara, Jakarta 14367
Lembaga Ekonomi dan Kemasjarakatan Nasional, Jakarta 14368
Lembaga Ilmu Pengetahuan Indonesia, Jakarta 14369
Lembaga Kebudajaan Indonesia, Jakarta 14382
Lembaga Kemajuan Perindustrian Malaysia (MIDA), Kuala Lumpur 18589
Lembaga Pendidikan dan Pembinaan Manajemen, Jakarta 14370
Lembaga Penduduk dan Pembangunan Keluarga Negara Malaysia, Kuala Lumpur 18590
Lembaga Perancang Keluarga Negara, Kuala Lumpur 18591
Lembaga Pertahanan Nasional, Jakarta 14371
Lemle & Kelleher, New Orleans 35948
Lemmensinstituut, Leuven 02370
Lemont Public Library District, Lemont 39847
Lemvig Bibliotek, Lemvig 06898
Lenawee County Library, Adrian 37883
Lenenergoinformproekt, Sankt-Peterburg 22916
Lenenergoinformproekt Power Engineering Design Company, Sankt-Peterburg 22916
Lenfilm Film Studio, Sankt-Peterburg 23806
Lenin Electromechanical Plant, Moskva 22819
Lenin Ilmen Natural Park Scientific Centre of the Russian Academy of Sciences, Miass 23156
Lenin Institute of Continuing Education in Medicine, Kazan 22272
Lenin Kuznya Plant, Trade Union Library, Kyiv 28693
Lenin Museum, Moskva 23365
Lenin Recreation Centre, Sankt-Peterburg 24222
Lenin Regional Centralized Library System of Kharkiv City, Main Library, Kharkiv 28601
Leninets Industrial Corporation, Sankt-Peterburg 22915
Leningrad Region Central Library System of Kiev – Svichado Main Library, Kyiv 28676
Leningrad Regional Council, Sankt-Peterburg 22765
Leningrad Regional Universal Research Library, Sankt-Peterburg 22172
Leningradska raionna TsBS m. Kyiva, Tsentralna biblioteka Svichado, Kyiv 28676
Leningradskaya oblastnaya universalnaya nauchnaya biblioteka, Sankt-Peterburg 22172
Leningradski oblastnoi sovet, Sankt-Peterburg 22765
Leninska raionna TsBS mista Kharkova, Tsentralna biblioteka, Kharkiv 28601
Leninsky District Centralized Library System, Central District Children's Library, Sankt-Peterburg 24273
Leninsky District Centralized Library System, Library Branch no 2, Sankt-Peterburg 24270
Leninsky District Centralized Library System, Library Branch no 3, Sankt-Peterburg 24271
Leninsky District Centralized Library System, Skvortsov-Stepanov Central District Library, Sankt-Peterburg 24272
Leninsky District Centralized Library System, Timiryazev Library Branch no 1, Sankt-Peterburg 24269
Lenniikhimmash Co, Sankt-Peterburg 22917
Lennox & Addington County Public Library, Napanee 04700
Lenoir Community College, Kinston 33460
Lenoir-Rhyne College, Hickory 31330
Lenox Library Association, Lenox 39851
Lens Library, Sankt-Peterburg 24215
Lensovet Recreation Centre, Sankt-Peterburg 24223
Lenton Parr Library, Melbourne 00480
Lenvik folkebibliotek, Finnsnes 20327
Lenzing AG, Lenzing 01501
Leominster Public Library, Leominster 39852
Leon Morris Library, Parkville 00789
Leon Wyczółkowski District Museum Library in Bydgoszcz, Bydgoszcz 21081
Leonard Buck Library, Mount Hagen 20582
Leonard, Street & Deinard, Minneapolis 35910
Leonia Public Library, Leonia 39854
Leopold-Hoesch-Museum, Düren 11787
Leopold-Sophien-Bibliothek, Überlingen 12568
Lepper Public Library, Lisbon 39889
Lermontov Dzerzhinsky District Library-Branch no 1, Sankt-Peterburg 24231
Lermontov Library of the District of Penza, Penza 22164
Lermontov Library of the District of Stavropol, Stavropol 24322
Lermontov Mogilev City Library, Mogilev 02194
Lermontov State Museum 'Tarkhany', Lermontovo 23140
LeRoy Collins Leon County Public Library System, Tallahassee 41506
Lerums bibliotek, Lerum 26834
Lesbian Herstory Archives, New York 37222
Lesgaft St. Petersburg State Academy of Physical Education P.F. Lesgaft, Sankt-Peterburg 22508
Leshan Teachers' College, Leshan 05385
LESINVEST Joint-Stock Company, Sankt-Peterburg 22903
Lesley University, Boston 36568
Lesley University, Cambridge 30651
H. Leslie Perry Memorial Library, Henderson 39415
Lesotekhnicheskaya akademiya, Sankt-Peterburg 22527
Lesotho National Library Service, Maseru 18225
Lessius Hogeschool, Antwerpen 02269
– Departement Handelswetenschappen en Bedrijfskunde, Antwerpen 02270
– Departement Toegepaste psychologie & Logopedie-Audiologie, Antwerpen 02271
– Lessiusbibliotheek Campus Sint-Andries, Antwerpen 02272
Lester Public Library, Two Rivers 41629
Lesya Ukrainka Public Library, Kyiv 28686
Letcher County Public Libraries, Whitesburg 41883
Lethbridge Community College, Lethbridge 04002
Lethbridge Public Library, Lethbridge 04677
Letourneau University, Longview 31641
Lettekkenny Institute of Technology, Letterkenny 14486
Leuphana Universität Lüneburg, Lüneburg 10256
– Hochschule 21, Buxtehude 10257
– Teilbibliothek Rotes Feld, Lüneburg 10258
– Teilbibliothek Suderburg, Suderburg 10259
– Teilbibliothek Volgershall, Lüneburg 10260
Levi Heywood Memorial Library, Gardner 39199
Levinson/Axelrod Library, Edison 35684
Levittown Public Library, Levittown 39857
Levuka Community Centre Library, Levuka 07322
Levy County Public Library System, Bronson 38297
Levy & Droney, Farmington 35702
Lewes Public Library, Lewes 39858
Lewis & Clark College, Portland
– Aubrey R. Watzek Library, Portland 32336
– Paul L. Boley Law Library, Portland 32337
Lewis & Clark Community College, Godfrey 33376
Lewis Cooper Junior Memorial Library, Opelika 40622
Lewis County Law Library, Chehalis 34063
Lewis Egerton Smoot Memorial Library, King George 39718
Lewis, Rice & Fingersh Law Library, Saint Louis 36204
Lewis & Roca Library, Phoenix 36118
Lewis University, Romeoville 32480
Lewis-Clark State College, Lewiston 31590
Lewisham College, London 29479
Lewiston City Library, Lewiston 39862
Lewiston Public Library, Lewiston 39863
Lewisville Public Library System, Lewisville 39865
Lexecon Inc Library, Chicago 35587
Lexington Community College, Lexington 33484
Lexington County Public Library System, Lexington 39868
Lexington Public Library, Lexington 39869
Lexington School for the Deaf, Jackson Heights 33435
Lexington Theological Seminary, Lexington 35262
Lexisnexis Butterworths, Chatswood 00815
LexisNexis Law Library, Colorado Springs 35624
Leyte Institute of Technology, Tacloban City 20740
Leyte Provincial Branch Library, Tacloban City 20786
LFZ Raumberg – Gumpenstein, Irdning 01547
Liangshan University, Xichang 05709
Lianyungang Chemical Engineering College, Lianyungang 05389
Lianyungang Vocational University, Lianyungang 05390
Liaocheng Teachers' College, Liaocheng 05391
Liaoning College of Traditional Chinese Medicine, Shenyang 05543
Liaoning Engineering College, Jinzhou 05344
Liaoning Financial College, Dandong 05159
Liaoning Normal University, Dalian 05154
Liaoning Officer College, Dalian 05155
Liaoning Provincial Library, Shenyang 04971
Liaoning Teachers' College of Foreign Languages, Liaoyang 05392
Liaoning University, Shenyang 05544
Liaquat Memorial Library, Karachi 20527
Liaquat University of Medical and Health Sciences, Jamshoro 20433
Lib4RI – Library for the Research Institutes within the ETH Domain: Eawag, Empa, PSI & WSL, Dübendorf 27350
Libbey Owens Ford Co, Toledo 36306
Libera Università di Bolzano, Bozen 14911
Libera Università di Lingue e Comunicazione IULM, Milano 15078
Libera Università Internazionale degli Studi Sociali Guido Carli, Roma 15371
Liberaal Archief, Gent 02648
Liberal Memorial Library, Liberal 39871
Liberian Geological Survey (LGS), Monrovia 18233
Liberty Mutual Group, Boston 36569
Liberty University, Lynchburg 31695
Libra Università Maria SS. Assunta, Roma
– Facoltà di Giurisprudenza, Roma 15368
– Facoltà di Lettere e Filosofia, Roma 15369
– Facoltà di Scienze della Formazione, Roma 15370
Libraries of the Boys' and Girls' Clubs Association of Hong Kong, Hong Kong 05918
Library, Jerusalem 14616
Library and Archives Canada, Gatineau 04352
Library and Archives Canada / Bibliothèque et Archives Canada, Ottawa 03666
Library and Information Services, Roseau 06952
Library and Information Services Directorate, Bloemfontein 25189
Library and Map Collection of Military History, Budapest 13315
Library, Archives and Information Service of the Danish Parliament, København 06759
Library Association of La Jolla, La Jolla 36998
Library at the Mariners' Museum, Newport News 37285
Library Branch no 1, Krivi Rig 28633
Library Branch no 14, Krivi Rig 28634
Library Branch no 16, Krivi Rig 28635
Library Caixa of Sabadell, Sabadell 26246
Library Company of Philadelphia, Philadelphia 37351
Library Company of the Baltimore Bar, Baltimore 36509
Library Council, Dublin 14526
Library for Children and Young of the City of Košice, Košice 24733
Library for Curriculum Research and Development, Accra 13029
Library for Folk High School and Trade Union Education, Ludvika 26610
Library for Natural Sciences of the Russian Akademy of Sciences, Moskva 23169
Library for the Cultural and Popular Work, Vester Skerninge 06836
Library HALLE 14, Leipzig 12179
The Library in English, Genève 27368
Library, Museum and Documentation Center of the Islamic Consultative Assembly, Tehran 14414
Library of Alexandria, Alexandria 07001
The Library of Bedrich Beneš Buchlovan, Uherské Hradiště 06616
Library of Catalonia, Barcelona 25247
Library of Chemical Literature of the Russian Academy of Sciences, Moskva 23168
Library of Congress, Washington 30105
Library of Congress – African & Middle Eastern Division, Washington 30106
Library of Congress – Asian Division, Washington 30107
Library of Congress – Music, Washington 30108
Library of Congress – National Library Service for the Blind & Physically Handicapped, Washington 30109
Library of Congress – Prints & Photographs Division, Washington 30110
Library of Congress – Rare Book & Special Collections Division, Washington 30111
Library of Congress – Science, Technology & Business Division, Washington 30112
Library of Ferrous Metallurgy, Moskva 23495

The Library of Hattiesburg, Petal, Forrest County, Hattiesburg 39396
Library of Inner Security, Espoo 07436
Library of Michigan, Lansing 30139
Library of Michigan Service for the Blind & Physically Handicapped, Lansing 37014
Library of Nonferrous Metallurgy, Moskva 23503
Library of Parliament, Cape Town 25061
Library of Parliament, Helsinki 07393
Library of Pavel Golja Trebnje, Trebnje 24987
The Library of Petroleum – Gas University of Ploieşti, Ploieşti 21942
Library of Pharmacy, Helsinki 07451
Library of Philosophy and Sociology of the Polish Academy of Sciences, Warszawa 21333
Library of Political and International Studies (LPIS), Tehran 14434
Library of Political and Social History, Jakarta 14373
Library of Skokloster Castle, Skokloster 26633
Library of Social and Behavioural Science, Lund 26413
Library of the Avenches Roman Museum, Avenches 27298
Library of the Central Administration of the Provincie Limburg, Hasselt 02444
Library of the Chancery of the President of the Republic, Praha 06409
The Library of The Chathams, Chatham 38530
Library of the Diocese, Vác 13349
Library of the Diocese of Susa, Susa 16081
Library of the District of Kirov, Kirov 22141
Library of the Free, Berlin 11494
Library of the Hungarian Parliament, Budapest 13321
Library of the Institute of Philosophy of the Academy of Sciences of the Czech Republic, v.v.i., Praha 06496
Library of the King Abdul Aziz City for Science and Technology (KAACST), Riyadh 24404
Library of the National Centre of Writers, Kyiv 28302
Library of the Presiden's Administration of Russia, Moskva 22732
The Library of the Swedish National Archives, Marieberg-Arninge, Stockholm 26675
Library of the US Courts, Reno 34735
Library of the US Courts, South Bend 34873
Library of the Writers of the Society of Jesus, Warszawa 21061
Library of Trade, Tourism and Consumer Affairs, Barcelona 25682
The Library of Virginia, Richmond 30149
Library & Resource Center, Richmond 40983
Library Service of Fiji, Suva 07287
Library Studies Professional School, Sankt-Peterburg 22694
Libyan Studies Centre, Tripoli 18260
Libyan-Arab Cultural Centre, Kuala Lumpur 18600
Lichfield Cathedral Library, Lichfield 29548
Licking County Genealogical Society Library, Newark 37275
Licking County Law Library Association, Newark 37276
Licking County Library, Newark 40452
Lida Central City Library, Lida 02152
Lida Railway Station, Railway Technical Library, Lida 01968
Lidingö stadsbibliotek, Lidingö 26835
Lidköpings stadsbibliotek, Lidköping 26836
Lidskaya tsentralnaya gorodskaya bibliyateka, Lida 02152
Liechtenstein-Institut, Bendern 18270
Liechtensteinische Landesbibliothek, Vaduz 18267
Liechtensteinische Musikschule, Vaduz 18269
Liedon kunnankirjasto, Lieto 07587
Lieksa Public Library, Lieksa 07586
Lieksan kaupunginkirjasto, Lieksa 07586
Lier folkebibliotek, Lier 20366
Lietuvos aklųjų biblioteka, Vilnius 18305
Lietuvos energetikos instituto biblioteka, Kaunas 18297
Lietuvos istorijos institutas, Vilnius 18288
Lietuvos kūno kultūros akademijos, Kaunas 18279
Lietuvos literatūros ir meno archyvo biblioteka, Vilnius 18306
Lietuvos mikslu akademijos bibliotekos skyrius Chemijos institute, Vilnius 18307
Lietuvos mokslų akademijos, Vilnius 18274
Lietuvos muzikos ir teatro akademija, Vilnius 18287
Lietuvos nacionalinė Martyno Mažvydo biblioteka, Vilnius 18272
Lietuvos technikos biblioteka, Vilnius 18308
Lietuvos veterinarijos akademijos, Kaunas 18280
Lietuvos žemės ūkio universiteto biblioteka, Kaunas 18281
Lietvuos medicinos biblioteka, Vilnius 18309
Lietvuos žemės ūkio biblioteka, Vilnius 18310
Lietvus karo akademija, Vilnius 18286
Life Bible College, San Dimas 35375
Life Chiropractic College – West Library, Hayward 33403
Life Insurance Association of Japan, Tokyo 17479
Life University Library, Marietta 31757
Lifeway Christian Resources of the Southern Baptist Convention, Nashville 37145
Lifting and Conveyor Equipment Industrial Corporation, Sankt-Peterburg 22926
Light and Textile Industry Machine Building Research Institute, Moskva 23571
Light Industry Managers and Specialists Further Education Institute, Moskva 23307
Light Industry Planning and Design Office, Sankt-Peterburg 23874
Light Industry Professional School, Kyiv 28119
Light Industry Technological Institute, Vitebsk 02084
Light Machine Building Plant, Orsha 01915
Light Metal Structures Assembly Plant, Molodechno 01908
Lighthouse International, New York 35996
Ligonier Valley Library Association, Inc, Ligonier 39873
Ligue Braille, Bruxelles 02606
Liguori Reading Centre Association, Brisbane 00881
Likhachev Moscow Automobile Factory, Moskva 22848
Likino-Dulevo School of Auto Mechanics, Likino Dulevo 22677
Lillehammer bibliotek, Lillehammer 20367
Lillehammer University College, Lillehammer 20106
Lima Public Library, Lima 39876
Lima Technical College, Lima 31608
Limb Research Centre, Sankt-Peterburg 23839
Limburgs Geschied- en Oudheidkundig Genootschap (LGOG), Maastricht 19487
Limerick City Library, Limerick 14579
Limerick County Library HQ, Limerick 14580
Limestone College, Gaffney 31188
Limnologicheski institut Sibirskogo otdeleniya RAN, Listvenichnoe 23142
Lincoln Cathedral Library, Lincoln 29549
Lincoln Central Library, Lincoln 30012
Lincoln Christian College & Seminary, Lincoln 35263
Lincoln City Libraries, Lincoln 39877
Lincoln College, Lincoln 31609
Lincoln College, Oxford 29328
Lincoln Correctional Center Library, Lincoln 34459
Lincoln Correctional Center Library, Lincoln 34460
Lincoln County Hospital, Lincoln 29699
Lincoln County Law Association Library, St. Catharines 04498
Lincoln County Law Library, Wiscasset 35078
Lincoln County Library, Kemmerer 39693
Lincoln County Public Library, Lincolnton 39885
Lincoln Heritage Public Library, Dale 38754
Lincoln Land Community College, Springfield 33776
Lincoln Law School of San Jose, San Jose 32597
Lincoln Library – The Public Library of Springfield, Illinois, Springfield 41415
Lincoln Library – West, Springfield 41416
Lincoln Medical Center, Bronx 36591
Lincoln Memorial Library, Yountville 37870
Lincoln Memorial University, Harrogate 31303
Lincoln National Foundation, Fort Wayne 36865
Lincoln Parish Library, Ruston 41085
Lincoln Park Public Library, Lincoln Park 39882
Lincoln Park Public Library, Lincoln Park 39883
Lincoln Public Library, Beamsville 04577
Lincoln Public Library, Lincoln 39878
Lincoln Public Library, Lincoln 39879
Lincoln Township Public Library, Stevensville 41443
Lincoln University, Jefferson City 31461
Lincoln University, Lincoln 19787
Lincoln University, Lincoln University 31620
Lincolns Inn Library, London 29784
Lincolnshire Archives, Lincoln 29700
Lincolnwood Public Library District, Lincolnwood 39886
Linda Hall Library, Kansas City 36987
Linden Free Public Library, Linden 39887
Lindenau-Museum, Altenburg 11449
Lindenhurst Memorial Library, Lindenhurst 39888
Linden-Museum Stuttgart, Stuttgart 12533
Lindenwood University, Saint Charles 32494
Lindesbergs bibliotek, Lindesberg 26837
Lindsey Wilson College, Columbia 33251
Linebaugh Public Library System of Rutherford County, Murfreesboro 40342
Linebaugh Public Library System of Rutherford County – Smyrna Public, Smyrna 41337
Linen Factory, Soligorsk 01927
Linen Hall Library, Belfast 29608
Linfield College, McMinnville 31777
– Portland Campus Library, Portland 31778
Linga Bibliothek, Hamburg 12001
Lingling Teachers' College, Xongzhou 05728
Lingnan University, Hong Kong 05287
Lingraj College, Belgaum 13793
Linguistics Information Centre, Kyiv 28358
Linguistics Institute of Ireland, Dublin 14537
Linhart Library, Radovljica 24980
Linklaters, New York 35997
Linköping University, Linköping 26399
– Health Science Library, Linköping 26400
Linköpings stadsbibliotek, Linköping 26838
Linköpings Universitet, Linköping 26399
– Hälsouniversitetets bibliotek, Linköping 26400
– Kvartersbibliotek A(KA), Ekonomi, Teknik, Linköping 26401
– Mediateket för lärarutbildningsarna, Linköping 26402
Linn County Law Library, Cedar Rapids 34052
Linn-Benton Community College, Albany 33119
Linnean Society of London, London 29785
Linyi Medical College, Linyi 05396
Linyi Teachers' College, Linyi 05397
Lion Corporation, Tokyo 17458
Lipetsk Regional Universal Research Library, Lipetsk 22149
Lipetsk Regional Yaring Library, Lipetsk 24158
Lipetsk State Teacher Training Institute, Lipetsk 22315
Lipetskaya oblastnaya detskaya biblioteka, Lipetsk 24157
Lipetskaya oblastnaya universalnaya nauchnaya biblioteka (LOUB), Lipetsk 22149
Lipetskaya oblastnoaya yunosheskaya biblioteka, Lipetsk 24158
Lipetski gosudarstvenny pedagogicheski institut, Lipetsk 22315
Lipetski mezhotraslevoi territorialny tsentr nauchno-tekhnicheskoi informatsii i propagandy, Lipetsk 23141
Lipovets Regional Centralized Library System, Main Library, Lipovets 28697
Lipovetskaya tsentralnaya biblioteka, Lipovets 28697
Lippische Landesbibliothek, Detmold 09308
Lippische Landeskirche, Detmold 11190
Lipscomb University, Nashville 31934
Liptovské múzeum, Ružomberok 24751
Lisenko Special Music School, Kyiv 28116
Lishui Teachers' College, Lishui 05398
Lisichansk Regional Centralized Library System, Main Library, Lisichansk 28698
Lisichanska raionna TsBS, Tsentralna biblioteka, Lisichansk 28698
Lisle Library District, Lisle 39890
LISNAVE-Estaleiros Navais, Setúbal 21751
Listaháskóli Islands, Reykjavík 13558
The Liszt Academy of Music, Budapest 13253
Liszt Ferenc Zeneművészeti Egyetem, Budapest 13253
Litauisches Kulturinstitut, Lampertheim-Hüttenfeld 12168
Litchfield Historical Society, Litchfield 37036
Literacy House, Lucknow 14068
Literárne a hudobné múzeum Banská Bystrica, Banská Bystrica 24677
Literary and Philosophical Society of Newcastle upon Tyne, Newcastle upon Tyne 29856
Literary & Historical Society of Quebec Library, Québec 04461
Literaturarchiv Sulzbach-Rosenberg, Sulzbach-Rosenberg 12547
Literature Museum, Den Haag 19445
Literaturhaus Wien, Wien 01595
Lithgow Library, Lithgow 01114
Lithgow Public Library, Augusta 38034
Lithuanian Academy of Music and Theatre, Vilnius 18287
Lithuanian Academy of Physical Education, Kaunas 18279
Lithuanian Academy of Sciences, Vilnius 18274
Lithuanian Academy of Sciences – Institute of Chemistry, Vilnius 18307
Lithuanian Agricultural Library, Vilnius 18310
Lithuanian Community Library, North Melbourne 00969
Lithuanian Energy Institute Library, Kaunas 18297
Lithuanian Forest Research Institute, Girionys 18295
Lithuanian Institute of History, Vilnius 18288
Lithuanian Library for the Blind, Vilnius 18305
Lithuanian Library of Medicine, Vilnius 18309
Lithuanian Research & Studies Center, Inc, Chicago 36680
Lithuanian Technical Library, Vilnius 18308
Lithuanian Technical Library, Klaipeda Department, Klaipeda 18273
Lithuanian University of Agriculture, Kaunas 18281
Lithuanian Veterinary Academy, Kaunas 18280
Litin Regional Centralized Library System, Main Library, Litin 28699
Litinska raionna TsBS, Tsentralna biblioteka, Litin 28699
Lititz Public Library, Lititz 39893
Little Dixie Regional Libraries, Moberly 40244
Little Falls Public Library, Little Falls 39895
Little Memorial Library, Midway 31828
Litton Industries, Woodland Hills 36434
Litvinenko Physical and Organic Chemistry and Coal Chemistry Institute, Donetsk 28220
Liuan Teachers' College, Liuan 05399
Live Oak County Library, George West 39217
Live Oak Public Libraries, Savannah 41240
Live Oak Public Libraries – Liberty County Branch, Hinesville 39446
Live Oak Public Libraries – Oglethorpe Mall Branch, Savannah 41241
Livermore Public Library, Livermore 39905
Liverpool City Council, Liverpool 30013
Liverpool City Library, Liverpool 01115
Liverpool Community College, Liverpool 29473
Liverpool Hope, Liverpool 29166
Liverpool Institute of Higher Education, Liverpool 29167
Liverpool John Moores University, Liverpool 29168
Liverpool Medical Institution, Liverpool 29704
Liverpool Public Library, Liverpool 39906
Liverpool School of Tropical Medicine, Liverpool 29474
Livingston Correctional Facility Library, Sonyea 34872
Livingston County Library, Chillicothe 38555
Livingston Parish Library – Denham Springs – Walker Branch, Denham Springs 38819
Livingston Parish Library – Watson Branch, Denham Springs 38820
Livingstone College, Salisbury 32557
– Hood Theological Seminary, Salisbury 32558
Livingstone Museum, Livingstone 42341
Livingston-Park County Public Library, Livingston 39907
Livonia Public Library, Livonia 39908
Livonia Public Library – Alfred Noble Branch, Livonia 39909
Livonia Public Library – Civic Center, Livonia 39910
Livrustkammaren, Stockholm 26665
Lixin College, Shanghai 05504
Ljubljana Metropolitan Library, Ljubljana 24968
Ljungby bibliotek, Ljungby 26839
Ljusdals kommunbibliotek, Ljusdal 26840
Llandrillo Technical College, Colwyn Bay 29461
Llanelli Borough Council, LLanelli 30014
Llangefni Library, Llangefni 30015
Lloyd Library & Museum, Cincinnati 36698
Lloyd Rees Library, Clayton 00903
Lloydminster Public Library, Lloydminster 04680
Llyfrgell Genedlaethol Cymru, Aberystwyth 28911
LMI Library, McLean 37090
LNEG – Laboratório Nacional de Energia e Geologia, S. Mamede de Infesta 21826
N.I. Lobachevsky State University of Nizhni Novgorod, Nizhni Novgorod 22425
Local and provincial library, Almería 25678
Local Government of Amagasaki City, Amagasaki 17511
Local Public Library Petržalka, Bratislava 24769
Local Studies Library, A Coruña 25948

Lock Haven University, Lock Haven	31631	– Tower Hill Library, London	29200
Locke Liddell & Sapp, Dallas	35652	– The Women's Library, London	29201
Locke Lord Bissell & Liddell LLP, Chicago	35588	London Oratory Library, London	29552
Locke, Reynolds, Boyd & Weisell Library, Indianapolis	35779	London Public Library, London	04681
Lockheed Martin, Manassas	35873	London School of Economics and Political Science (LSE), London	29202
Lockheed Martin Aeronautical Systems, Marietta	35876	London School of Hygiene and Tropical Medicine, London	29203
Lockheed Martin Corp, Orlando	36074	London School of Jewish Studies (LSJS), London	29204
Lockport Public Library, Lockport	39912	London School of Theology, Northwood	29559
Locomotive Repairing Plant, Moskva	22842	London South Bank University, London	29205
Locust Valley Library, Locust Valley	39914	– Perry Library, London	29206
Lodi Memorial Library, Lodi	39915	London Transport Museum Library, London	29789
Łódź University, Łódź	20877	Lone Star Legal Aid, Nacogdoches	37139

Los Angeles County, Los Angeles	37055	Louisiana College, Pineville	32293
Los Angeles County Arboretum & Botanic Garden, Arcadia	36474	Louisiana Department of Environmental Quality, Baton Rouge	33971

Los Angeles Public Library System – Will & Ariel Durant Branch, Los Angeles 39960

Lucasfilm Research Library, San Rafael 36251

Luce, Forward, Hamilton & Scripps, San Diego 36219
Lucent Technologies, Murray Hill 35929
Lucerne University of Applied Sciences and Arts, School of Engineering and Architecture, Horw 27094
Lucius Beebe Memorial Library, Wakefield 41706
Lucy Cavendish College, Cambridge 29026
Lucy Robbins Welles Library, Newington 40462
Ludington Public Library, Bryn Mawr 38378
Ludvika bibliotek, Ludvika 26842
Ludwig Forum für Internationale Kunst, Aachen 11442
Ludwig-Maximilians-Universität München, München 10367
– Amerika-Institut, München 10368
– Anatomische Anstalt, München 10369
– Bibliothek der Institute am Englischen Garten, München 10370
– Bibliothek der Tierärztlichen Fakultät, München 10371
– Bibliothek des Biozentrums, Planegg 10372
– Bibliothek des Historicums, München 10373
– Bibliothek des Lehrstuhls für Wissenschaftsgeschichte, München 10374
– Bibliothek Deutsche Philologie & Komparatistik, München 10375
– Bibliothek für Handels- und Arbeitsrecht, München 10376
– Bibliothek für Psychologie, Pädagogik und Soziologie, München 10377
– Bibliothek für Wirtschafts- und Sozialgeographie, München 10378
– Bibliothek für Wirtschaftswissenschaften und Statistik, München 10379
– Bibliothek für Zivilrecht, Zivilverfahrensrecht und Medienrecht, München 10380
– Bibliothek im Physiologikum, München 10381
– Bibliothek Kunstwissenschaften, Abteilung Kunstgeschichte und Theaterwissenschaft, München 10382
– Bibliothek Mathematik, Meteorologie, Physik, München 10383
– Bibliothek Theologie – Philosophie, München 10384
– Department für Geographie, Physische Geographie, München 10385
– Evangelisches Pressearchiv, München 10386
– Fakultätsbibliothek Chemie und Pharmazie, München 10387
– Forschungsstelle für Europäisches und Internationales Steuerrecht, München 10388
– Geobibliothek, München 10389
– Geschwister-Scholl-Institut für Politische Wissenschaft / Centrum für Angewandte Politikforschung (CAP), München 10390
– Historisches Seminar / Abt. Bayerische und allgemeine Landesgeschichte, München 10391
– Institut für Assyriologie und Hethitologie, München 10392
– Institut für Astronomie und Astrophysik mit Universitätssternwarte, München 10393
– Institut für Deutsch als Fremdsprache / Transnationale Germanistik, München 10394
– Institut für die gesamten Strafrechtswissenschaften, München 10395
– Institut für die gesamten Strafrechtswissenschaften, Bibliothek für Kriminologie, Jugendstrafrecht und Strafvollzug, München 10396
– Institut für die gesamten Strafrechtswissenschaften, Bibliothek für Rechtsphilosophie, München 10397
– Institut für Englische Philologie und Shakespeare Forschungsbibliothek / Englische Philologie, München 10398
– Institut für Englische Philologie und Shakespeare Forschungsbibliothek / Shakespeare-Forschungsbibliothek, München 10399
– Institut für Finnougristik/Uralistik, München 10400
– Institut für Geschichte der Medizin, München 10401
– Institut für Geschichte und Kultur des Nahen Orients sowie Turkologie, München 10402
– Institut für Hygiene und Technologie der Lebensmittel tierischen Ursprungs, Oberschleißheim 10403
– Institut für Indologie und Iranistik, München 10404
– Institut für Internationales Recht – Bibliothek der Abteilung für Rechtsvergleichung, München 10405
– Institut für Internationales Recht / Lehrstuhl für Europäisches und Internationales Wirtschaftsrecht, München 10406
– Institut für Internationales Recht / Lehrstuhl für Völkerrecht, München 10407
– Institut für italienische Philologie, München 10408
– Institut für Klassische Archäologie, München 10409
– Institut für Klassische Philologie, München 10410
– Institut für Musikwissenschaft, München 10411
– Institut für Nordische Philologie, München 10412
– Institut für Orthodoxe Theologie, München 10413
– Institut für Paleoanatomie und Geschichte der Tiermedizin, München 10414
– Institut für Phonetik und Sprachverarbeitung, München 10415
– Institut für Politik und Öffentliches Recht, Bibliothek für deutsches Steuerrecht, München 10416
– Institut für Politik und Öffentliches Recht, Bibliothek für Kirchenrecht, München 10417
– Institut für Politik und Öffentliches Recht, Bibliothek für Politik und Öffentliches Recht, München 10418
– Institut für Rechtsmedizin, München 10419
– Institut für Risikoforschung und Versicherungswirtschaft, München 10420
– Institut für Romanische Philologie, München 10421
– Institut für Semitistik, München 10422
– Institut für Sinologie, München 10423
– Institut für Slavische Philologie, München 10424
– Institut für Soziale Pädiatrie und Jugendmedizin, München 10425
– Institut für Tierzucht, Lehrstuhl für Tierzucht und Allgemeine Landwirtschaftslehre, München 10426
– Institut für Vergleichende und Indogermanische Sprachwissenschaft sowie Albanologie, München 10427
– Institut für Volkskunde/Europäische Ethnologie, München 10428
– Institut für Volkswirtschaftslehre / Lehrstuhl für Nationalökonomie und Finanzwissenschaft, München 10429
– Institut für Volkswirtschaftslehre / Seminar für Sozial- und Wirtschaftsgeschichte, München 10430
– Klinikum Großhadern, München 10431
– Klinikum Innenstadt / Augenklinik, München 10432
– Klinikum Innenstadt / Chirurgische Klinik und Poliklinik Innenstadt, München 10433
– Klinikum Innenstadt / Chirurgische Klinik und Poliklinik Innenstadt, Röntgenabteilung, München 10434
– Klinikum Innenstadt / Dr. von Haunersches Kinderspital, München 10435
– Klinikum Innenstadt / I. Frauenklinik, München 10436
– Klinikum Innenstadt / Klinik für Zahn-, Mund- und Kieferkrankheiten, München 10437
– Klinikum Innenstadt / Klinik und Poliklinik für Dermatologie und Allergologie, München 10438
– Klinikum Innenstadt / Psychiatrische Klinik und Poliklinik, München 10439
– Lehrstuhl für Nationalökonomie und Finanzwissenschaft, München 10440
– Leopold-Wenger-Institut für Rechtsgeschichte – Abt. Antike Rechtsgeschichte und Papyrusforschung, München 10441
– Leopold-Wenger-Institut für Rechtsgeschichte – Abt. Bayerische und Deutsche Rechtsgeschichte, München 10442
– Medizinische Lesehalle, München 10443
– Pathologisches Institut, München 10444
– Seminar für Internationale Wirtschaftsbeziehungen, München 10445
– Universitätsarchiv, München 10446
The LuEsther T. Mertz Library, Bronx 36592
Luftfartsverket, Norrköping 26622
Luftkrigsskolen, Trondheim 20187
Lufttransportgeschwader 62, Wunstorf 11136
Luftwaffenunterstützungsgruppe Wahn, Köln 10991
Lugansk City Centralized Library System, Main Library, Lugansk 28702
Lugansk Diesel Locomotive Industrial Corporation, Lugansk 28162
Lugansk Regional Universal Scientific Library named after M. Gorky, Lugansk 27938
Lugansk State Medical University, Lugansk 28399
Luganska miska TsBS, Tsentralna biblioteka, Lugansk 28702
Luganska oblasna universalna naukova biblioteka im. O.M. Gorkoho, Lugansk 27938
LUISS Guido Carli Independent International University for Social Studies, Roma 15371
LUISS Libera Università Internazionale degli Studi Sociali Guido Carli, Roma 15371
Luke Wadding Library, Waterford 14495
Lukino-Dulevski avtomekhanicheski tekhnikum, Likino Dulevo 22677
Lukyanen Agricultural Scientific Research Institute, Krasnodar 23116
Luleå stadsbibliotek, Luleå 26843
Luleå Tekniska Universitet, Luleå 26403
Luleå University of Technology, Luleå 26403
Luliang College, Lishi 05808
Lum, Danzis, Drasco & Positan, Roseland 36189
Lummus Technology Library, Bloomfield 35510
Lunacharsky State Medical Institute, Astrakhan 22207
Lunar & Planetary Institute, Houston 36945
Lunds stadsbibliotek, Lund 26845
Lunds Universitet, Lund 26404
– Asienbiblioteket, Lund 26405
– Ekologiska biblioteket, Lund 26406
– Ekonomihögskolans bibliotek, Lund 26407
– Filosofiska biblioteket, Lund 26408
– Geobiblioteket, Lund 26409
– Juridiska fakultetens bibliotek, Lund 26410
– Kemicentrums bibliotek, Lund 26411
– Musikhögskolan i Malmö, Malmö 26412
Lungenklinik Lostan gGmbH, Lostau 12209
Luninets Railway Station, Railway Technical Library, Luninets 01969
Luninets Regional Central Library, Luninets 02154
Luninetskaya tsentralnaya raionnaya bibliyateka, Luninets 02154
Luossavaara-Kiirunavaara AB (LKAB), Kiruna 26538
Luoyang Agricultural College, Luoyang 05406
Luoyang Building Material Engineering College, Luoyang 05407
Luoyang Institute of Engineering, Luoyang 05408
Luoyang Medical College, Luoyang 05409
Luoyang Normal College, Luoyang 05410
Luoyang University, Luoyang 05411
N.Gh. Lupu Institute of Internal Medicine, Bucureşti 21994
Lusaka City Libraries, Lusaka 42354
Lushnje Institute of Agricultural Research, Lushnje 00014
Luster folkebibliotek, Gaupne 20333
Luther College, Decorah 30979
Luther College Library, Regina 03830
Luther King House Library, Manchester 29558
Luther Luckett Correctional Complex Library, La Grange 34425
Luther Rice Seminary & University, Lithonia 35264
Luther Seminary, Saint Paul 35367
Lutheran Brethren Seminary, Fergus Falls 35205
Lutheran Central Library, Budapest 13328
Lutheran Church Missouri Synod, Saint Louis 35362
Lutheran General Hospital, Park Ridge 37327
Lutheran School of Theology at Chicago & McCormick Theological Seminary, Chicago 35157
Lutheran Theological Seminary, Gettysburg 35214
Lutheran Theological Seminary, Mapumulo 25084
Lutheran Theological Seminary, Philadelphia 35327
Lutheran Theological Seminary, Saskatoon 04226
Lutheran Theological Southern Seminary, Columbia 35167
Lutheran University of Applied Sciences, Nürnberg 10517
Luthergedenkstätten in Sachsen-Anhalt, Lutherstadt Wittenberg 12603
Lutherhalle Wittenberg, Lutherstadt Wittenberg 12603
Lutherhaus Eisenach, Eisenach 11205
Lutsk Regional Centralized Library System, Main Library, Lutsk 28707
Lutska raionna TsBS, Tsentralna biblioteka, Lutsk 28707
Luzerne County Community College, Nanticoke 33568
Luzhou Medical College, Luzhou 05412
Luzonian University Foundation, Lucena City 20707
Lviv Academy of Arts, Lviv 28035
Lviv Agricultural Institut, Dublyany 27976
Lviv Art Gallery, Lviv 28417
Lviv City Centralized Children's Library System, Main Library, Lviv 28711
Lviv City Centralized Library System, L. Ukrainki Main Library, Lviv 28712
Lviv National V. Stefanyk Scientific Library of Ukraine, Lviv 27941
Lviv Polytechnic University, Lviv 28034
Lviv Scientific and Technical Information Popularisation Center, Lviv 28420
Lviv Scientific-Research Institute Epidemiology and Hygiene, Lviv 28418
Lviv State Institute of Applied and Decorative Art, Lviv 28419
Lvivska Akademiya Mystetstv, Lviv 28035
Lvivska miska TsBS dlya ditei, Tsentralna biblioteka, Lviv 28711
Lvivska miska TsBS, Tsentralna biblioteka im. L. Ukrainki, Lviv 28712
Lvivska naukova biblioteka im V. Stefanyka, Lviv 27941
Lvivsky mezhgaluzevoyi tsentr naukovo-tekhnichnoyi informatsiyi i propagandy, Lviv 28420
Lvivsky natsionalny universytet im. Ivana Franka, Lviv 28036
Lvivsky silskogospodarsky instytut, Dublyany 27976
LVR-Klinik Bonn, Bonn 11640
LVR-Landesmuseum Bonn, Bonn 11641
LWL – Archäologie für Westfalen, Münster 12349
LWL – Archivamt für Westfalen, Münster 11059
LWL – Denkmalpflege, Landschafts- und Baukultur in Westfalen, Münster 11060
LWL Industrial Museum, Dortmund 11748
LWL – Institut für Regionalgeschichte, Münster 12350
LWL – Landesmuseum für Kunst und Kulturgeschichte, Münster 12351
LWL – Museum für Naturkunde, Münster 12352
LWL – Museumsamt für Westfalen, Münster 11061
LWL-Freilichtmuseum Detmold, Detmold 11738
LWL-Industriemuseum, Dortmund 11748
Lyagin Regional Library for Children, Mykolaiv 28739
Lyakhovichi Regional Central Library, Lyakhovichi 02155
Lyakhovichiskaya tsentralnaya raionnaya bibliyateka, Lyakhovichi 02155
Lycée Abdelmoumen, Oujda 18934
Lycée Charles de Gaulle, Saint-Louis 24422
Lycée Classique de Diekirch (LCD), Diekirch 18385
Lycée Classique d'Echternach (LCE), Echternach 18386
Lycée de Jeunes Filles Ameth Fall, Saint-Louis 24423
Lycée du Nord (LNW), Wiltz 18391
Lycée Hôtelier, Talence 08132
CDI de Lycée Michel Rodange Luxembourg (LMRL), Luxembourg 18390
Lycée Van Vollenhoven, Dakar 24421
Lycksele kommunbibliotek, Lycksele 26846
Lycoming College, Williamsport 33068
Lycoming County Law Library, Williamsport 35071
Lyman Library, Boston 36570
Lynbrook Public Library, Lynbrook 39983
Lynchburg College, Lynchburg 31696
Lynchburg Public Library, Lynchburg 39984
Lyndhurst Free Public Library, Lyndhurst 39985
Lyndon Baines Johnson Library and Museum, Austin 36504
Lyndon State College, Lyndonville 31700
Lynn University, Boca Raton 30472
Lynnfield Public Library, Lynnfield 39986
Lyon College, Batesville 30355
Lyon County Library System, Yerington 41989
Lyon Township Public Library, South Lyon 41367

Lyondell Chemical Co, Newtown Square 36053
Lyons Public Library, Lyons 39989
Lysekil's City Library, Lysekil 26847
Lysekils stadsbibliotek, Lysekil 26847
M. Lysenko Higher State Music Institute, Lviv 28037
M. F. Ivanov Institute for Animal Husbandry in Steppe Regions 'Askania Nova', Askaniya-Nova 28181
M. K. Gandhi Library, Durban 25120
M. Lysenko Higher State Music Institute, Lviv 28037
M. M. Asatiani Institute of Psychiatry, Tbilisi 09224
M. Valancius Public LIbrary of Kretinga District Municipality, Kretinga 18337
Maanmittauslaitos, Helsinki 07472
Maanpuolustuskorkeakoulun Kurssikirjasto, Helsinki 07340
Maanpuolustustuskorkeakoulu, Helsinki 07341
Mäarsjukhuset, Eskilstuna 26569
Maasstad Ziekenhuis, Rotterdam 19513
Maastricht University, Maastricht 19171
MAB Library, Kuala Lumpur 18592
Macalester College, Saint Paul 32533
Macao Polytechnic Institute, Macao 05415
Macaulay Land Use Research Institute, Aberdeen 29588
Macedonian Academy of Sciences and Arts, Skopje 18428
Macedonian Geological Society, Skopje 18443
Macedonian Museum of Natural History, Skopje 18445
Macfee & Taft Law Offices, Oklahoma City 36068
Machine Building and Electrical Engineering Research Institute, Kharkiv 28268
Machine Building Central Design Bureau, Sankt-Peterburg 23923
Machine Building College, Khabarovsk 22671
Machine Building Industrial Corporation, Sumy 28170
Machine Building Institute, Sankt-Peterburg 22522
Machine Building Plant Industrial Corporation, Trade Union Library, Kramatorsk 28630
Machine Building Plant, Trade Union Library, Dnipropetrovsk 28537
Machine Building Plant, Trade Union Library, Gorlivka 28571
Machine Building Technology Research Institute, Rivne 28455
Machine Tool Engineering Design Office Joint-Stock Company, Sankt-Peterburg 22904
Machine Tool Plant, Moskva 22865
Machine-Tool and Mechanical Engineering Research and Experimental Joint-Stock Company, Novosibirsk 22881
Machine-Tool Parts Plant, Baranovichi 01828
Machine-Tool Parts Plant, Gomel 01864
Mackall, Crounse & Moore, Minneapolis 35911
Mackay & Co Library, Lombard 35819
Mackay Regional Council, Mackay 01117
Mackenzie, Hughes LLP, Syracuse 36292
MacLeod & Dixon LLP, Barristers & Solicitors, Calgary 04244
MacMurray College, Jacksonville 31450
Macomb Community College Libraries – Center Campus, Clinton Township 33243
Macomb Community College Libraries – South Campus, Warren 33837
Macomb County Library, Clinton Township 38612
Macomb Intermediate School District, Clinton Township 34090
Macomb Public Library District, Macomb 39991
Macon County Public Library, Franklin 39147
Macon State College, Macon 33509
Macquarie Regional Library, Dubbo 01092
Macquarie University, Sydney 00520
Macrae Library, Nova Scotia Agricultural College, Truro 04041
Macroeconomics Research Institute, Moskva 23282
Madan Prize Library, Lalitpur 19044
Madan Puraskar Pustakalaya, Lalitpur 19044
Madang Teachers' College, Madang 20559
Maden Tetkik ve Arama Genel Müdürlügü, Ankara 27875
Madera County Law Library, Madera 34484
Madera County Library, Madera 39999
Madison Area Technical College, Madison 33510
Madison County Law Library, London 34469
Madison County Library System, Canton 38443
Madison County Library System – Canton Public Library, Canton 38444
Madison County Library System – Ridgeland Public Library, Ridgeland 40997
Madison County Public Library, Marshall 40074
Madison County Public Library, Richmond 40984
Madison Heights Public Library, Madison Heights 40012
Madison Public Library, Madison 40003
Madison Public Library, Madison 40004
Madison Public Library, Madison 40005
Madison Public Library – Alicia Ashman Branch, Madison 40006
Madison Public Library – Pinney Branch, Madison 40007
Madison Public Library – Sequoya Branch, Madison 40008
Madonna University, Livonia 31630
Madras Institute of Development Studies, Chennai 13965
Madras Literary Society, Chennai 13966
Madras Medical College, Chennai 13796
Madras Veterinary College Library, Chennai 13630
Madrid Chamber of Commerce and Industry, Madrid 25988
Madurai-Kamaraj University, Madurai 13694
Maebashi City Library, Maebashi 17847
Maebashi Shiritsu Toshokan, Maebashi 17847
Maejo University, Chiang Mai 27732
Mafra National Palace, Mafra 21817
Magadan District Library A.S. Pushkin, Magadan 22150
Magadanskaya oblastnaya detskaya biblioteka, Magadan 24160
Magadanskaya oblastnaya universalnaya nauchnaya biblioteka im. A.S. Pushkina, Magadan 22150
Magadh University, Bodh-Gaya 13623
– L. N. Misra Institute of Economic Development and Social Change, Patna 13624
Magarach Viticulture and Oenology Research Institute, Yalta 28484
Magdalen College, Oxford 29301
Magdalene College, Cambridge 29027
Magdanskaya oblastnaya yunosheskaya biblioteka, Magadan 24161
Magdeburg University, Magdeburg 10263
Magdeburger Museen – Kulturhistorisches Museum, Magdeburg 12226
Magistero Siciliano di Servizio Sociale, Catania 14946
Magistrat der Stadt Wien, Wien 01407
Magistrate's Offices, Johannesburg 25065
Magyar Allami Eötvös Loránd Geofizikai Intézet, Budapest 13394
Magyar Allami Földtani Intézet (MAFI), Budapest 13395
Magyar Elektrotechnikai Múzeum, Budapest 13396
Magyar Építészeti Múzeum, Budapest 13397
Magyar Iparmüvészeti Egyetem Könyvtára, Budapest 13254
Magyar Képzömüvészeti Egyetem, Budapest 13255
Magyar Kereskedelmi és Vendéglátóipari Múzeum, Budapest 13398
Magyar Környezetvédelmi és Vizügyi Múzeum, Esztergom 13473
Magyar Közművelődési Intézet, Budapest 13399
Magyar Mezőgazdasági Múzeum, Budapest 13400
Magyar Nemzeti Bank – Szakkönyvtár, Budapest 13357
Magyar Nemzeti Filmarchivum, Budapest 13401
Magyar Nemzeti Galéria (MNG), Budapest 13402
Magyar Nemzeti Múzeum, Budapest 13403
Magyar Olajipari Múzeum, Zalaegerszeg 13523
Magyar Optikai Müvek, Budapest 13358
Magyar Országos Levéltár, Budapest 13404
Magyar Rádió, Budapest 13405
Magyar Szabványügyi Testület, Budapest 13406
Magyar Táncművészeti Főiskola, Budapest 13256
Magyar Tejgazdasági Kisérleti Intézet, Mosonmagyaróvár 13491
Magyar Természettudományi Múzeum, Budapest 13407
Magyar Tudományos Akadémia, Veszprém 13520
Magyar Tudományos Akadémia – Agrárgazdasági Kutató Intézet, Budapest 13408
Magyar Tudományos Akadémia – Atommagkutató Intézete Könyvtár, Debrecen 13468
Magyar Tudomanyos Akademía – Balatoni Limnológiai Kutatóintézet, Tihany 13515
Magyar Tudományos Akadémia – Csillagvizsgáló Intézet, Budapest 13409
Magyar Tudományos Akadémia – Egyesitett Társadalomtudományi Könyvtár, Budapest 13410
Magyar Tudományos Akadémia – Filozófiai Intézet, Budapest 13257
Magyar Tudományos Akadémia – Földrajztudományi Kutató Intézet, Budapest 13411
Magyar Tudományos Akadémia – Geodéziai és Geofizikai Kutató Intézet, Sopron 13498
Magyar Tudományos Akadémia – Irodalomtudományi Intézet, Budapest 13412
Magyar Tudományos Akadémia – Izotópe's Felületkémiai Intézet Könyvtár, Budapest 13413
Magyar Tudományos Akadémia – Jogtudományi Intézet, Budapest 13414
Magyar Tudományos Akadémia – Kémiai Kutatóközpont Könyvtára, Budapest 13415
Magyar Tudományos Akadémia – Kisérleti Orvostudományi Kutató Intézet, Budapest 13416
Magyar Tudományos Akadémia Könyvtára, Budapest 13214
Magyar Tudományos Akadémia – Közgazdaságtudományi Intézet, Budapest 13417
Magyar Tudományos Akadémia – Lukács Archívum, Budapest 13418
Magyar Tudományos Akadémia – Magyar Numizmatikai Társulat, Budapest 13419
Magyar Tudományos Akadémia – Matematikai Kutató Intézet, Budapest 13420
Magyar Tudományos Akadémia – Mezőgazdasági Kutatóintézete Könyvtára, Martonvásár 13489
Magyar Tudományos Akadémia – Művészettörténeti Kutatóintézet, Budapest 13421
Magyar Tudományos Akadémia – Napfizikai Obszervatóriuma, Debrecen 13469
Magyar Tudományos Akadémia – Néprajzi Kutatóintézet, Budapest 13422
Magyar Tudományos Akadémia – Növényvédelmi Kutatóintézete, Budapest 13423
Magyar Tudományos Akadémia – Pszichológiai Kutatóintézet, Budapest 13424
Magyar Tudományos Akadémia – Régészeti Intézet, Budapest 13425
Magyar Tudományos Akadémia – Regionalis Kutatások Központja, Békéscsaba 13367
Magyar Tudományos Akadémia – Regionális Kutatások Központja Dunántuli Tudományos Intézet, Pécs 13495
Magyar Tudományos Akadémia – Számítástechnikai és Automatizálási Kutató Intézet, Budapest 13426
Magyar Tudományos Akadémia – Szegedi Biologiai Központ, Genetikai Intézet, Szeged 13505
Magyar Tudományos Akadémia – Szilárdtestfizikai és Optikai Kutatóintézet, Budapest 13427
Magyar Tudományos Akadémia – Talajtani és Agrokémiai Kutató Intézet, Budapest 13428
Magyar Tudományos Akadémia – Történettudományi Intézet, Budapest 13258
Magyar Tudományos Akadémia – Zenetudományi Intézet, Budapest 13429
The Maharaja Sayajirao University of Baroda, Vadodara 13780
Maharashtra Association for the Cultivation of Science, Pune 14193
Maharashtra Mantralaya Central Library, Mumbai 14097
Maharishi University of Management, Fairfield 31120
Maharshi Dayanand University, Rohtak 13754
Mahasarakham University, Mahasarakham 27741
Mahatma Gandhi Institute, Moka 18697
Mahatma Gandhi Memorial Medical College, Indore 13662
Mahatma Phule Agricultural University, Rahuri 13747
Mahatma Phule Krishi Vidyapeeth, Rahuri 13746
Mahatma Phule Krishi Vidyapeeth, Rahuri 13747
Mahendra Multiple Campus Library, Baglung 19017
Mahendra Sanskrit University, Kathmandu 19019
Mahidol University, Nakornpathom 27742
– Faculty of Environment and Resource, Nakornpathom 27743
– Faculty of Nursing Library, Bangkok 27744
– Faculty of Pharmacy Library, Bangkok 27745
– Faculty of Science, Bangkok 27746
– Siriraj Medical Library, Bangkok 27747
Mahoning Law Library Association, Youngstown 37869
Mahopac Public Library, Mahopac 40017
Mahwah Public Library, Mahwah 40018
Maihaugen Biblioteket, Lillehammer 20237
Mail Newspapers Ltd, London 29576
Main Children's Library, Dzhankoi 28559
Main Directorate of the Polish State Archives, Warszawa 21047
Main Library for the German Minority, Aabenraa 06765
Main Library of Izhevsk State Technical University, Izhevsk 22267
Main Library of Technical University of Opole, Opole 20897
The Main Library of Transport, Warszawa 21042
Main Roads Western Australia, East Perth 00676
Main School of Fire Service, Warszawa 20972
Main Trade Union Library, Enakieve 28236
Maine Charitable Mechanic Association Library, Portland 37401
Maine College of Art, Portland 33657
Maine Department of Corrections, Windham 35076
Maine Department of Transportation Library, Augusta 36499
Maine Historical Society, Portland 37402
Maine Maritime Academy, Castine 30684
Maine Maritime Museum, Bath 36524
Maine Office of Substance Abuse, Augusta 36500
Maine Regional Library for the Blind & Physically Impaired, Augusta 36501
Maine State Law and Legislative Reference Library, Augusta 33948
Maine State Library, Augusta 30115
Mainfränkisches Museum Würzburg, Würzburg 12615
Mainichi Newspapers Co, Ltd, Research Dept, Tokyo 17459
Mainichi Shimbunsha Henshukyoku Chosabu, Tokyo 17459
Maison Camille Lemonnier + Musée Camille Lemonnier, Bruxelles 02607
Maison d'Ailleurs, Yverdon-les-Bains 27445
Maison de Balzac, Paris 08579
Maison de la Culture / Te Fare Tauhiti Nui, Papeete 09166
Maison de l'Environnement des Hauts-de-Seine, Issy-les-Moulineaux 08365
Maison de l'outil et de la pensée ouvrière, Troyes 08701
Maison de Victor Hugo, Paris 08580
Maison Descartes, Institut Français des Pays-Bas, Amsterdam 19368
Maison du Livre de l'Image et son François Mitterrand, Villeurbanne 09151
Maison du Mexique, Paris 08581
Maison Heinrich Heine, Paris 08582
Maison Jean Vilar, Avignon 08288
Maison René Ginouvès, Nanterre 08441
Maitland City Library, Maitland 01118
Maitland Public Library, Maitland 40019
Majles-e Shora-ye Eslami, Tehran 14414
Majlis al-Intiqali al-Watani, Alger 00055
Majlis al-'Ummah al-Kuwayti, Safat 18128
Majlis Amanah Rakyat (MARA), Kuala Lumpur 18593
Majlis-e Shoraye Islami – Ketab-Khane No. 2, Tehran 14415
Majuro Public Library, Majuro 18676
MAK – Österreichisches Museum für angewandte Kunst/Gegenwartskunst, Wien 01621
Makarenko State Teacher Training Institute, Sumy 28076
Makariv Regional Centralized Library System, Main Library, Makariv 28714
Makarivska raionna TsBS, Tsentralna biblioteka, Makariv 28714
Makarov Naval Academy, Sankt-Peterburg 22530
Makedonska Akademija na Naukite i Umetnostite, Skopje 18428
Makedonsko Geološko Društvo, Skopje 18443
Makere Medical School, Kampala 27915
Makerere Institute of Social Research, Kampala 27926

Makerere University, Kampala 27914
– Makere Medical School, Kampala 27915
Makhtumkuli Library, Kyiv 28658
Makiivka City Centralized Library System, Gorky Main Library, Makiyivka 28715
Makiyivska miska TsBS, Tsentralna biblioteka im. Gorkoho, Makiyivka 28715
Makmal Penyiasatan Kajibumi Malaysia, Ipoh 18574
Makmal Penyiasatan Kajibumi Malaysia, Kuching 18610
MakNII – State Makeyevka Safety in Mines Research Institute, Makeyevka 28426
Mākslas bibliotēka, Riga 18186
Maksymovych Scientific Library of the Taras Shevchenko Kyiv National University, Kyiv 28020
Maktab Kerjasama Malaysia, Petaling Jaya 18544
Maktab Perguruan, Kota Bharu 18530
Maktab Perguruan Batu Lintang, Kuching 18538
Maktab Perguruan Gaya, Kota Kinabalu 18532
Maktab Perguruan Ilmu Khas, Kuala Lumpur 18534
Maktab Perguruan Mohd. Khalid, Johor Bharu 18528
Maktab Perguruan Perempuan Melayu, Melaka 18540
Maktab Perguruan Persekutuan Pulau Pinang, Penang 18542
Maktab Perguruan Raja Melewar, Seremban 18546
Maktab Perguruan Rajang, Bintagor 18526
Maktab Perguruan Sandakan, Sandakan 18545
Maktab Perguruan Sultan Idris, Tanjung Malim 18548
Maktab Perguruan Temenggung Ibrahim, Johor Bharu 18529
Maktab Perguruan Tengku Ampuan Afzan, Kuantan 18537
Maktab Rendah Sains MARA, Kota Bharu 18531
Maktab Rendah Sains MARA, Kulim 18539
Maktab Rendah Sains MARA, Seremban 18547
Maktabat Al-Kuwait Al-Wataniyah, Safat 18126
Makumira University College, Usa River 27670
Mala Viska Regional Centralized Library System, Main Library, Mala Viska 28717
Malakhit Bureau of Maritime Machine Building, Sankt-Peterburg 23752
Malang Institute of Teacher Training and Education, Malang 14273
Mälardalen University, Västerås 26464
Mälardalens högskola, Västerås 26464
Malaspina University College, Nanaimo 03805
Malaviya Regional Engineering College, Jaipur 13808
Malawi Institute of Education, Domasi 18469
Malawi National Library Service, Lilongwe 18467
Malay Documentation Centre, Kuala Lumpur 18601
Malay Women Teachers' Training College, Melaka 18540
Malaysia External Trade Development Corporation (MATRADE), Kuala Lumpur 18594
Malaysia Tourism Promotion Board, Kuala Lumpur 18595
Malaysian Agricultural Research and Development Institute (MARDI), Kuala Lumpur 18596
Malaysian Chinese Resource and Research Centre, Kuala Lumpur 18597
Malaysian Industrial Development Authority, Kuala Lumpur 18589
Malaysian Industrial Development Finance Bhd, Kuala Lumpur 18605
Malaysian Royal Police, Kuala Lumpur 18567
Malaysian Rubber Board, Kuala Lumpur 18598
Malcolm Pirnie Inc, Germantown 35717
Malcolm X College, Chicago 33223
Mallesons Stephen Jaques, Sydney 00838
Mallinckrodt Baker Inc, Phillipsburg 36114
Malmö City Archives, Malmö 26616
Malmö högskola, Malmö 26414
– Oral Health Library, Malmö 26415
Malmö stadsarkiv, Malmö 26616
Malmö stadsbibliotek, Malmö 26849
Malmö University, Malmö 26414
Malmö University Hospital, Malmö 26617
Malone University Library, Canton 30668
Malorita Regional Central Library, Malorita 02156
Maloritskaya tsentralnaya raionnaya bibliyateka, Malorita 02156
Maloviskivska raionna TsBS, Tsentralna biblioteka, Mala Viska 28717
Malungs kommunbibliotek, Malung 26850
Malverne Public Library, Malverne 40022
Malvern-Hot Spring County Library, Malvern 40021
Maly Classical Theatre, Moskva 23162
Mamaroneck Public Library District, Mamaroneck 40023
Yu. G. Mamedaliev Institute of Petrochemical Processes, Baku 01717
Mamie Doud Eisenhower Public Library, Broomfield 38369
Management and Coordination Agency, Tokyo 17325
Management Development and Productivity Institute, Accra 13054
Management Development Institute, Gurgaon 13993
Manama Central Library, Manama 01744
Manatee County, Bradenton 36584
Manatee County Public Library System, Bradenton 38257
Manchester Business School, Manchester 29272
Manchester City Council, Manchester 30040
Manchester City Library, Manchester 40026
Manchester College, North Manchester 32144
Manchester Community College, Manchester 33514
Manchester Metropolitan University, Manchester 29263
– Aytoun Library, Manchester 29264
– Elizabeth Gaskell Library, Manchester 29265
– Hollings Library, Manchester 29266
Manchester Public Library, Manchester 40027
Mandal folkebibliotek, Mandal 20370
Manevichi Regional Centralized Library System, Main Library, Manevichi 28718
Manevytska raionna TsBS, Tsentralna biblioteka, Manevichi 28718
Mang Thông Thin Khoa Hoc & Công Nghê Viêt Nam, Hanoi 42264
Mangalore University, Mangalagangotri 13695
Mangfolds- og Migrasjonsbiblioteket, Oslo 20193
Manhasset Public Library, Manhasset 40029
Manhattan Christian College, Manhattan 31750
Manhattan College, Riverdale 32452
Manhattan Public Library, Manhattan 40030
Manhattan Public Library District, Manhattan 40031
Manhattan School of Music, New York 32068
Manhattanville College, Purchase 32400
Manila Central University, Kalookan City 20702
– FDT Medical Foundation, Kalookan City 20703
Manila City Library, Manila 20785
Manipur University, Imphal 13660
Manistee County Library System, Manistee 40034
Manitoba Culture, Heritage & Tourism, Brandon 04584
Manitoba Department of Cultural Heritage & Tourism, Winnipeg 04192
Manitoba Department of Education, Citizenship & Youth, Winnipeg 04193
Manitoba Hydro Library, Winnipeg 04308
Manitoba Indian Culture-Educational Center, Winnipeg 04556
Manitoba Innovation, Energy & Mines, Winnipeg 04557
Manitoba Law Library, Inc, Winnipeg 04558
Manitoba Museum, Winnipeg 04559
Manitoba School for the Deaf Multimedia Center, Winnipeg 04560
Manitowoc Public Library, Manitowoc 40035
Mannes College of Music, New York 32069
Mannheim University, Mannheim 10298
Mannheimer Swartling Advokatbyrå – Göteborg, Göteborg 26532
Mannheimer Swartling Advokatbyrå – Stockholm, Stockholm 26555
Mannheimer Zentrum für Europäische Sozialforschung (MZES), Mannheim 12245
Manoeuver Training Centre, Amersfoort 19236
Manomet Center for Conservation Sciences Library, Manomet 37082
Manor College, Jenkintown 33443
Mansfelder Kupfer und Messing GmbH, Hettstedt 11408
Mansfeld-Museum Stadtschloss Eisleben, Lutherstadt Eisleben 11812
Mansfield College, Oxford 29329
Mansfield Public Library, Mansfield 40038
Mansfield Public Library, Mansfield 40039
Mansfield Public Library, Mansfield Center 40041
Mansfield University, Mansfield 31755
Mansfield-Richland County Public Library, Mansfield 40040
Mansoura Polytechnic Institute, Mansoura 07082
Mansoura University, Mansoura 07083
– Faculty of Agriculture, Mansoura 07084
– Faculty of Arts, Mansoura 07085
– Faculty of Commerce, Mansoura 07086
– Faculty of Education, Mansoura 07087
– Faculty of Education Damietta, Mansoura 07088
– Faculty of Engineering, Mansoura 07089
– Faculty of Law, Mansoura 07090
– Faculty of Medicine, Mansoura 07091
– Faculty of Science, Mansoura 07092
– Faculty of Science Damietta, Mansoura 07093
Mansutti Foundation, Milano 16337
Mäntän kirjasto, Mänttä 07589
Manuel L. Quezon University, Manila 20716
Manufacture Nationale de Porcelaine de Sèvres, Sèvres 08264
Manukau Libraries and Information Services, Manukau City 19872
Manx National Heritage, Douglas 14597
M.A.P.A. – Dirección General de Planificación y Desarrollo Rural, Madrid 26064
Maplewood Memorial Library, Maplewood 40047
MARA Junior Science College, Kota Bharu 18531
MARA Junior Science College, Kulim 18539
MARA Junior Science College, Seremban 18547
MARA University of Technology, Shah Alam 18517
– Kelantan Branch, Machang 18518
– Melaka Branch, Alor Gajah 18519
– Pahang Branch, Bangar jenka 18520
– Perlis Branch, Arau 18521
– Petaling Jaya Branch, Petaling Jaya 18522
– Sabah Branch, Kota Kinabalu 18523
– Sarawak Branch, Kota Samarahan 18524
– Terengganu Branch, Dungun 18525
Maramureş History and Archeology County Museum, Baia Mare 21958
Maranatha Baptist Bible College, Watertown 35413
Marangu College of National Education, Moshi 27663
Marathon County Public Library, Wausau 41776
Marathon Oil Co, Findlay 35706
Marathwada Agricultural University, Parbhani 13734
Marco Besso Foundation, Roma 16501
Marconi Aerospace Systems, Wayne 36402
Marengo Public Library District, Marengo 40050
Marga Institute, Colombo 26314
Margaret E. Heggan Free Public Library of the Township of Washington, Hurffville 39544
Margaret R. Grundy Memorial Library, Bristol 38290
Margaret S. Sterck School for the Deaf Library, Newark 37277
Mari Republic Museum of Regional Studies, Yoshkar-Ola 24093
Mari Republic School, Yoshkar-Ola 22726
Mari Research Institute of Agricultural Engineering, Ruem 23730
Mari Scientific Institute of Language, Literature and History, Yoshkar-Ola 24092
Mari State University, Yoshkar-Ola 22660
Maria College of Albany, Albany 33120
Mariale Werken, Leuven 02475
Marian College, Indianapolis 31392
Marian College of Fond du Lac, Fond du Lac 31151
Marian J. Mohr Memorial Library, Johnston 39654
Marian & Ralph Feffer Library, Scottsdale 37587
Mariano Marcos State University, Batac 20686
Marianopolis College, Montreal 04008
Maribo Public Library, Maribo 06901
Maribor Public Library, Maribor 24972
Maribor Regional Archive, Maribor 24948
Mariborska knjižnica, Maribor 24972
Maricopa County Jail Library, Phoenix 34673
Maricopa County Library District, Phoenix 40779
Mariehamn Public Library – Central Library for Åland, Mariehamn 07590
Mariehamns stadsbibliotek – centralbibliotek för Åland, Mariehamn 07590
Marienbibliothek, Halle (Saale) 11230
Mariengymnasium, Jever 10744
Marienkron, Mönchhof 01456
Mariestads Stadsbibliotek, Mariestad 26851
Marietta College, Marietta 31758
Marigold Library System, Strathmore 04771
Mariinsk Forestry and Timber Technology Professional School, Mariinsk 22678
Mariinski lesotekhnicheski tekhnikum, Mariinsk 22678
Mariinsky State Classical Theatre of Opera and Ballet – Central Musical Library, Sankt-Peterburg 23768
Marijampolės Petro Kriaučiūno viešoji biblioteka, Marijampolė 18340
Marin County Free Library, San Rafael 41199
Marin County Free Library – Civic Center Branch, San Rafael 41200
Marin County Free Library – Corte Madera Branch, Corte Madera 38705
Marin County Free Library – Fairfax Branch, Fairfax 39021
Marin County Free Library – Novato Branch, Novato 40554
Marin County Law Library, San Rafael 34832
Marin Institute for the Prevention of Alcohol & Other Drug Problems, San Rafael 37563
Marine and Coastal Management, Cape Town 25113
Marine Biological Association, Plymouth 29879
Marine Biological Laboratory, Woods Hole 35080
Marine Corps Base Hawaii Libraries, Camp H M Smith 34029
Marine Corps Base Hawaii Libraries, Kaneohe Bay 34403
Marine Corps Recruit Depot Library, San Diego 34807
Marine Fisheries Division, Bangkok 27782
Marine Fisheries Research Department, Singapore 24614
Marine Hydrophysical Institute, Katsiveli 28246
Marine Institute, Galway 14559
Marine Ministry, Moskva 22746
Marine Museum of the Great Lakes at Kingston, Kingston 04368
Marine Research Institute, Fisheries Library, Reykjavík 13574
Marine Scotland Marine Laboratory, Aberdeen 29589
Marine Technical University, Sankt-Peterburg 22528
Marineamt, Rostock 11085
Marinemuseet, Horten 20228
Marineoperationsschule (MOS), Bremerhaven 10712
Mariners' Museum, Newport News 37285
Marineschule Mürwik, Flensburg 10724
Marinette County Consolidated Public Library Service, Marinette 40060
Marinka Inter Union Centralized Library System, Main Library, Marinka 28720
Marinmusei bibliotek, Karlskrona 26601
Marinska mizhvisomcha TsBS, Tsentralna biblioteka, Marinka 28720
Marion Carnegie Library, Marion 40061
Marion County Law Library, Indianapolis 34383
Marion County Law Library, Marion 34506
Marion County Law Library, Salem 34788
Marion County Library, Marion 40062
Marion County Public Library System, Ocala 40581
Marion County Teachers Professional Library, Ocala 37300
Marion & Ed Hughes Public Library, Nederland 40383
Marion Military Institute, Marion 33516
Marion Public Library, Marion 40063
Marion Public Library, Marion 40064
Marion Public Library, Marion 40065
Mariski gosudarstvenny universitet, Yoshkar-Ola 22660
Mariski nauchno-issledovatelski institut selskogo khozyaistva VASChML, Ruem 23730
Mariski nauchno-issledovatelski institut yazyka, literatury i istorii im. V.M. Vasileva, Yoshkar-Ola 24092
Mariski politekhnicheski institut im. A.M. Gorkogo, Yoshkar-Ola 22661
Mariski Respublikanski nauchno-kraevedcheski muzei, Yoshkar-Ola 24093
Mariski tsellyulozno-bumazhny kombinat kontserna 'Bumaga', Volzhsk 22974
Mariski tsellyulozno-bumazhny tekhnikum, Volzhsk 22725
Mariskoe respublikanskoe uchilishche, Yoshkar-Ola 22726
Marist College, Poughkeepsie 32353
Maristenkloster, Fürstenzell 11221

Maritiem Museum Rotterdam, Rotterdam	19514
Maritime Academy, Oron	19973
Maritime Institute, Gdańsk	21098
Maritime Institute of Ireland, Dun Laoghaire	14558
Maritime Museum, Dubrovnik	06212
Maritime Museum, Göteborg	26582
Maritime Museum, Kotor	18915
Maritime Museum Barcelona, Barcelona	25925
Maritime Museum of British Columbia Library, Victoria	04549
Maritime Museum 'Sergej Mašera', Piran	24951
Maritime Safety Agency, Tokyo	17685
Maritime University of Szczecin, Szczecin	20932
Maritime University of Technology, Sankt-Peterburg	22531
Maritz Inc Resource & Media Center, Fenton	35705
Mariupol City Centralized Library System, Korolenko Main Library, Mariupol	28722
Mariupolska miska TsBS, Tsentralna biblioteka im. Korolenka, Mariupol	28722
Marius Katiliskis Public LIbrary of Pasvalys, Pasvalys	18351
Mark & Emily Turner Memorial Library, Presque Isle	40893
Markaryds bibliotek, Markaryd	26852
Marketing, Economics and Ecology Research Institute, Odesa	28438
Marketing Intellingence Corp, Tanashi	17429
Markham Public Libraries, Markham	04684
Marks bibliotek, Kinna	26816
Marktkirchenbibliothek, Goslar	11226
Mark-Twain-Bibliothek, Berlin	12663
Marlboro College, Marlboro	31763
Marlboro County Library, Bennettsville	38172
Marlborough Gallery Library, New York	37223
Marlborough Public Library, Marlborough	40070
Marmara Scientific and Industrial Research Centre, Gebze	27881
Marmara Universitesi Merkez Kutuphanesi, Istanbul	27863
– Ilâhiyat Fakültesi, Istanbul	27864
Marmara University Library, Istanbul	27863
– Faculty of Theology, Istanbul	27864
Marmion Academy Library, Aurora	33142
Marnix Academy, University of Teacher Education, Utrecht	19216
Marple Public Library, Broomall	38368
Marquette County Historical Society, Marquette	37085
Marquette University, Milwaukee	31835
– Law Library, Milwaukee	31836
Marshall District Library, Marshall	40075
Marshall & Melhorn, Toledo	36307
Marshall Memorial Library, Deming	38817
Marshall Public Library, Marshall	40076
Marshall Public Library, Pocatello	40836
Marshall University, Huntington	31373
Marshall University – Health Science Libraries, Huntington	31374
Marshall-Lyon County Library, Marshall	40077
Marshalltown Community College, Marshalltown	33517
Marshalltown Public Library, Marshalltown	40078
Marshfield Public Library, Marshfield	40079
Martin Community College, Williamston	33871
Martin Correctional Institution Library, Indiantown	34384
Martin County Library System -, Stuart	41457
Martin Faculty Hospital, Martin	24740
Martin Gropius Krankenhaus GmbH, Eberswalde	11810
Martin Luther College, New Ulm	32022
Martin Luther King Jr, Drew Medical Center Health Sciences Library, Los Angeles	37055
Martin Memorial Library, York	41991
Martin Methodist College, Pulaski	32394
Martin & Osa Johnson Safari Museum, Chanute	36645
Martin-Butzer-Gymnasium, Dierdorf	10718
Martin-Luther-Universität Halle-Wittenberg, Halle (Saale)	09316
Martin-Opitz-Bibliothek, Herne	12058
Martins Ferry Public Library, Martins Ferry	40082
Martins Sarmento Society, Guimarães	21762
Martinsburg-Berkeley County Public Library – Martinsburg Public Library, Martinsburg	40083
Martinská fakultná nemocicna, Martin	24740
Bohuslav Martinů Foundation, Praha	06501
Martinus-Bibliothek, Mainz	11270
Martin-von-Wagner-Museum der Universität Würzburg, Würzburg	12616

Martsinovsky Institute of Medical Parasitology and Tropical Medicine, Moskva	22338
Martyanov Museum, Minusinsk	23160
Martynas Mazvydas National Library of Lithuania, Vilnius	18272
Marucelliana Library, Firenze	14793
Maruzen Company Ltd, Tokyo	17707
Maruzen Hon No Toshokan, Tokyo	17707
Marvin Memorial Library, Shelby	41298
Marx Memorial Library, London	29790
Marx Smolensk Teacher-Training Institute, Smolensk	22583
The Mary Baker Eddy Library for the Betterment of Humanity, Boston	36571
Mary Baldwin College, Staunton	32739
Mary H. Weir Public Library, Weirton	41796
Mary Immaculate College, Limerick	14489
Mary L. Cook Public Library, Waynesville	41789
Mary Lou Johnson Hardin County District Library, Kenton	39705
Mary Riley Styles Public Library, Falls Church	39038
Mary Wood Weldon Memorial Library, Glasgow	39233
Marygrove College, Detroit	31005
Maryknoll College Foundation, Inc, Quezon City	20753
Maryland Correctional Institution-Hagestown Library, Hagerstown	34328
Maryland Department of Legislative Services Library, Annapolis	33928
Maryland Department of Planning Library, Baltimore	33963
Maryland Historical Society, Baltimore	36510
Maryland House of Correction-Annex Library, Jessup	34400
Maryland Institute College of Art, Baltimore	30340
Maryland State Archives Library, Annapolis	36471
Maryland State Law Library, Annapolis	33929
Maryland State Library for the Blind & Physically Handicapped, Baltimore	36511
Marylebone Cricket Club, London	29791
Marylhurst University, Marylhurst	31771
Marymount Manhattan College, New York	32070
Marymount University, Arlington	30260
Marysville Public Library, Marysville	40086
Maryville College, Maryville	31772
Maryville Public Library, Maryville	40091
Maryville University, Saint Louis	32504
Marywood University, Scranton	32627
MAS – Museum of Regional Ethnology, Antwerpen	02518
MAS – Volkskundemuseum, Antwerpen	02518
Masaryk Public Library Vsetín, Vsetín	06618
Masaryk University, Brno	
– Faculty of Economics and Administration, Brno	06355
– Faculty of Education, Brno	06356
– Faculty of Informatics, Brno	06357
– Faculty of Law, Brno	06358
– Faculty of Science, Brno	06359
– Faculty of Social Studies, Brno	06360
– Ústřední knihovna Filozofické fakulty, Brno	06361
– Ustredni knihovna Lekarske fakulty, Brno	06362
Masarykova veřejná knihovna Vsetín, Vsetín	06618
Mashantucket Pequot Museum & Research Center, Mashantucket	37088
Mashhad University of Medical Sciences, Mashhad	14393
Mashinobudivno konstruktorsko byuro Progres, Zaporizhzhya	28175
Mashinobudivno vibronichno obednannya, Sumy	28170
Mashinobudivny zavod, Gorlivka	28571
Mashinobudivny zavod, Biblioteka profkomu, Dnipropetrovsk	28537
Mashinostroitelnoe PO Tekstilmash, Brest	01841
Mashinostroitelny institut, Izhevsk	22267
Mashinostroitelny institut, Kurgan	22311
Mashinostroitelny institut, Mogilev	01811
Mashinostroitelny kolledzh, Khabarovsk	22671
Mashinostroitelny tekhnikum-predpriyatie, Sankt-Peterburg	22700
Mashino-stroitelny zavod, Izhevsk	22795
Mashinostroitelny zavod 'Borets', Moskva	22843
Masjid Negara, Kuala Lumpur	18569
Mason City Public Library, Mason City	40092
Mason County District Library, Ludington	39976
Mason County Library System, Point Pleasant	40837
Mason County Public Library, Maysville	40107

Masonic Grand Lodge Library & Museum of Texas, Waco	37710
The Masonic Library & Museum of Pennsylvania, Philadelphia	37352
Masonic Medical Research Laboratory Library, Utica	37700
Massachusetts Audubon Society, Lincoln	37031
Massachusetts Bay Community College, Wellesley	33853
Massachusetts College of Art, Boston	30494
Massachusetts College of Liberal Arts, North Adams	32137
Massachusetts College of Pharmacy & Health Sciences, Boston	30495
Massachusetts Correctional Institution-Concord Library, Concord	34107
Massachusetts Department of Corrections, Bridgewater	34005
Massachusetts Department of Corrections, South Walpole	34875
Massachusetts General Hospital – Treadwell Library, Boston	36572
Massachusetts General Hospital – Warren Library, Boston	36573
Massachusetts Historical Society Library, Boston	36574
Massachusetts Horticultural Society Library, Wellesley	37817
Massachusetts Institute of Technology, Lexington	37027
Massachusetts Institute of Technology Libraries, Cambridge	30653
– Aeronautics & Astronautics Library, Cambridge	30654
– Barker-Engineering Library, Cambridge	30655
– Dewey-Social Sciences & Management Library, Cambridge	30656
– Humanities Library, Cambridge	30657
– Institute of Archives & Special Collections, Cambridge	30658
– Lewis Music Library, Cambridge	30659
– Science Library, Cambridge	30660
Massachusetts Maritime Academy, Buzzards Bay	34024
Massachusetts Mutual Life Insurance Co – Law Library, Springfield	36280
Massachusetts Mutual Life Insurance Co Library, Springfield	36281
Massachusetts School of Law Library, Andover	30224
Massachusetts School of Professional Psychology (MSPP) Library, Boston	30496
Massachusetts Trial Court, Dedham	36786
Massachusetts Trial Court, Fitchburg	34203
Massachusetts Trial Court, Greenfield	36903
Massachusetts Trial Court, Lowell	37072
Massachusetts Trial Court, Pittsfield	37387
Massachusetts Trial Court, Springfield	34883
Massachusetts Trial Court, Worcester	35082
Massanutten Regional Library, Harrisonburg	39376
Massapequa Public Library, Massapequa Park	40094
Massasoit Community College, Brockton	33190
Massawa Public Library, Massawa	07218
Massena Public Library, Massena	40095
Massey University – Hokowhitu Campus Library, Palmerston North	19788
– College of Education, Palmerston North	19789
Massey University – Wellington Campus LIbrary, Wellington	19790
Massillon Public Library, Massillon	40096
The Master's College, Santa Clarita	32610
The Master's Seminary Library, Sun Valley	35394
Mastics-Moriches-Shirley Community Library, Shirley	41311
Matawan-Aberdeen Public Library, Matawan	40097
Matej Bel University, Banská Bystrica	24640
Matematicheski institut im. V.A. Steklova i vychislitelnogo tsentra RAN, Moskva	23348
Matematicheski institut im. V.A. Steklova RAN – St. Peterburgskoe otdelenie, Sankt-Peterburg	23807
Matematički Institut, Beograd	24515
Matematický ústav AV ČR, Praha	06497
Matematikos ir Informatikos Institutas, Vilnius	18311
Matenadaran Institute of Ancient Manuscripts, Yerevan	00354
Materialamt der Bundeswehr, Sankt Augustin	11091
Materialprüfungsamt Nordrhein-Westfalen, Dortmund	10857
Mathematica Policy Research Inc Library, Princeton	36155
Mathematica Policy Research, Inc Library, Washington	37751
Mathematical Institute, Kyiv	28333

Mathematical Institute of the AS CR, Praha	06497
Mathematisches Forschungsinstitut Oberwolfach gemeinnützige GmbH, Oberwolfach	12380
Mathematisch-Physikalischer Salon, Dresden	11779
Mathias Hochschule Rheine, Rheine	10543
Matica Srpska Library, Novi Sad	24454
Matična Biblioteka Svetozar Marković, Zaječar	24558
Matična Biblioteka Vuk kKradžić, Prijepolje	24549
Mátra Múzeum, Gyöngyös	13475
Matsuda K. K. Chosabu, Aki	17342
Matsushita Denki Sangyo K. K. Gijutsu Toukatsushitsu, Moriguchi	17385
Matsushita Electric Industrial Co, Ltd, Moriguchi	17385
Matsuyama Daigaku, Matsuyama	17040
– Keizaikeiei Kenkyujo, Matsuyama	17041
Matsuyama University, Matsuyama	17040
– Research Institute for Economics and Business Administration, Matsuyama	17041
Matteson Public Library, Matteson	40098
Matthew Bender & Company Inc, Library, Newark	36042
Mattituck-Laurel Library, Mattituck	40100
Mattoon Public Library, Mattoon	40101
Mátyás Király Múzeum, Visegrád	13521
Maud Preston Palenske Memorial Library, Saint Joseph	41115
Maui Community College, Kahului	33446
Maui Correctional Center Library, Wailuku	34971
Maulana Azad College of Technology, Bhopal	13619
Maulana Azad College of Technology, Bhopal	13794
Maulana Azad Medical College, New Delhi	13831
Mauney Memorial Library, Kings Mountain	39721
Maurice Carême Foundation, Bruxelles	02588
Maurice M. Pine Free Public Library, Fair Lawn	39017
Maurice-Lamontagne Institute Library, Mont-Joli	04384
Mauritius Institute Public Library, Port Louis	18699
Mauritius Sugar Industry Research Institute, Réduit	18700
Mauritshuis, Den Haag	19441
Maury County Public Library, Columbia	38648
Maury Loontjens Memorial Library, Narragansett	40368
MÁV ZRt Ügykezelési és Dokumentációs Szolgáltató Szervezet, Budapest	13430
Mavr Minsk Regional Chidren's Library, Minsk	02183
Mawlamyine University, Mawlamyine	18987
A. Max Brewer Memorial Law Library, Viera	37706
Max Mueller Bhavan, Chennai	13967
Max Mueller Bhavan, Mumbai	14098
Max Mueller Bhavan, Pune	14194
Max Mueller Bhavan Information Centre & Library, New Delhi	14151
Max Planck Institute for Chemical Physics of Solids Library, Dresden	11767
Max Planck Institute for Comparative Public Law and International Law, Heidelberg	12050
Max Planck Institute for Dynamics of Complex Technical Systems, Magdeburg	12227
Max Planck Institute for Evolutionary Biology, Plön	12400
Max Planck Institute for Human Cognitive and Brain Sciences, Leipzig	12193
Max Planck Institute for Human Development, Berlin	11548
Max Planck Institute for Intellectual Property, Competition and Tax Law, München	12312
Max Planck Institute for Mathematics in the Sciences, Leipzig	12194
Max Planck Institute for Polymer Research, Mainz	12236
Max Planck Institute for Psycholinguis-tics, Nijmegen	19498
Max Planck Institute for Research on Collective Goods, Bonn	11644
Max Planck Institute for Social Anthropology, Halle (Saale)	11961
Max Planck Institute for the History of Science, Berlin	11550
Max Planck Institute for the Study of Religious and Ethnic Diversity, Göttingen	11932

Max Planck Institute for the Study of Societies, Köln 12147
Max Planck Institute of Experimental Medicine, Göttingen 11931
Max Planck Institute of Molecular Cell Biology and Genetics, Dresden 11768
Max R. Traurig Learning Resource Center Library, Waterbury 33841
Max Rubner-Institut / Bundesforschungsinstitut für Ernährung und Lebensmittel – Standort Detmold, Detmold 11739
Max Rubner-Institut / Bundesforschungsinstitut für Ernährung und Lebensmittel – Standort Karlsruhe (Hauptsitz), Karlsruhe 12092
Max Rubner-Institut / Bundesforschungsinstitut für Ernährung und Lebensmittel – Standort Kiel, Kiel 12114
Max Rubner-Institut / Bundesforschungsinstitut für Ernährung und Lebensmittel – Standort Kulmbach, Kulmbach 12167
Max-Delbrück-Centrum für Molekulare Medizin, Berlin 11547
Maxim Technologies Inc Library, Boise 35516
Max-Planck-Haus, Tübingen 12561
Max-Planck-Institut für Astronomie, Heidelberg 12049
Max-Planck-Institut für ausländisches öffentliches Recht und Völkerrecht, Heidelberg 12050
Max-Planck-Institut für ausländisches und internationales Privatrecht, Hamburg 12002
Max-Planck-Institut für ausländisches und internationales Sozialrecht, München 12311
Max-Planck-Institut für ausländisches und internationales Strafrecht, Freiburg 11898
Max-Planck-Institut für Bildungsforschung, Berlin 11548
Max-Planck-Institut für Bioanorganische Chemie, Mülheim an der Ruhr 12271
Max-Planck-Institut für Biologie, Tübingen 12562
Max-Planck-Institut für biophysikalische Chemie, Göttingen 11929
Max-Planck-Institut für Chemie, Mainz 12235
Max-Planck-Institut für Chemische Physik fester Stoffe, Dresden 11767
Max-Planck-Institut für Dynamik komplexer technischer Systeme, Magdeburg 12227
Max-Planck-Institut für Dynamik und Selbstorganisation, Göttingen 11930
Max-Planck-Institut für Eisenforschung GmbH, Düsseldorf 11802
Max-Planck-Institut für ethnologische Forschung, Halle (Saale) 11961
Max-Planck-Institut für europäische Rechtsgeschichte, Frankfurt am Main 11861
Max-Planck-Institut für evolutionäre Anthropologie, Leipzig 12192
Max-Planck-Institut für Evolutionsbiologie, Plön 12400
Max-Planck-Institut für experimentelle Medizin, Göttingen 11931
Max-Planck-Institut für Geistiges Eigentum, Wettbewerbs- und Steuerrecht, München 12312
Max-Planck-Institut für Gesellschaftsforschung, Köln 12147
Max-Planck-Institut für Gravitationsphysik, Potsdam 12408
Max-Planck-Institut für Herz- und Lungenforschung, Bad Nauheim 11468
Max-Planck-Institut für Hirnforschung, Frankfurt am Main 11862
Max-Planck-Institut für Immunbiologie, Freiburg 11899
Max-Planck-Institut für Kernphysik, Heidelberg 12051
Max-Planck-Institut für Kognitions- und Neurowissenschaften, Leipzig 12193
Max-Planck-Institut für Kohlenforschung, Mülheim an der Ruhr 12272
Max-Planck-Institut für Kolloid- und Grenzflächenforschung, Potsdam 12409
Max-Planck-Institut für Kunstgeschichte, Roma 16474
Max-Planck-Institut für Mathematik, Bonn 11642
Max-Planck-Institut für Mathematik in den Naturwissenschaften, Leipzig 12194
Max-Planck-Institut für medizinische Forschung, Heidelberg 12052
Max-Planck-Institut für Mikrostrukturphysik, Halle (Saale) 11962
Max-Planck-Institut für molekulare Genetik, Berlin 11549
Max-Planck-Institut für Molekulare Pflanzenphysiologie, Potsdam 12410

Max-Planck-Institut für molekulare Physiologie, Dortmund 11749
Max-Planck-Institut für molekulare Zellbiologie und Genetik, Dresden 11768
Max-Planck-Institut für Ökonomik, Jena 12078
Max-Planck-Institut für Ornithologie, Seewiesen 12485
Max-Planck-Institut für Ornithologie – Vogelwarte Radolfzell, Radolfzell 12418
Max-Planck-Institut für Physik, München 12313
Max-Planck-Institut für Physik komplexer Systeme, Dresden 11769
Max-Planck-Institut für Plasmaphysik, Garching 11911
Max-Planck-Institut für Polymerforschung, Mainz 12236
Max-Planck-Institut für Psychiatrie, München 12314
Max-Planck-Institut für Psycholinguistik, Nijmegen 19498
Max-Planck-Institut für Quantenoptik, Garching 11912
Max-Planck-Institut für Radioastronomie, Bonn 11643
Max-Planck-Institut für Sonnensystemforschung, Katlenburg-Lindau 12107
Max-Planck-Institut für terrestrische Mikrobiologie, Marburg 12258
Max-Planck-Institut für Wissenschaftsgeschichte, Berlin 11550
Max-Planck-Institut zur Erforschung multireligiöser und multiethnischer Gesellschaften, Göttingen 11932
Max-Planck-Institut zur Erforschung von Gemeinschaftsgütern, Bonn 11644
Max-Planck-Institute for Heart and Lung Research, W.G. Kerckhoff-Institute, Library, Bad Nauheim 11468
Max-Planck-Institute Martinsried, Martinsried 12260
Max-Planck-Institute Stuttgart, Stuttgart 12534
May Department Stores Co, Saint Louis 36205
Mayakovsky City Central General Scientific Library, Sankt-Peterburg 24308
Mayakovsky City Central General Scientific Library – Department of Foreign Literature, Sankt-Peterburg 23913
Mayakovsky City Central General Scientific Library – Music Department, Sankt-Peterburg 23912
Mayakovsky City Central General Scientific Library – Youth Department, Sankt-Peterburg 24309
Mayakovsky District Children's Library, Chelyabinsk 24122
Mayakovsky Minsk City Library no 3, Minsk 02163
Mayakovsky Museum, Moskva 23366
Mayer, Brown & Platt, Chicago 35589
Mayer, Brown & Platt, Los Angeles 35844
L. A. Mayer Memorial Institute for Islamic Art, Jerusalem 14737
Maynard, Murray, Cronin, Erickson & Curran, Phoenix 36119
Maynard Public Library, Maynard 40105
Mayo County Library, Castlebar 14566
Mayo Foundation, Rochester 37452
Mayr-Melnhof Institut für den Christlichen Osten, Salzburg 01569
Maysville Community College, Maysville 33523
Mayville Public Library, Mayville 40108
Maywood Public Library, Maywood 40109
Maywood Public Library District, Maywood 40110
Mazda Motor Corp, Business Research Div, Aki 17342
Mažeikių rajono savivaldybės viešoji biblioteka, Mažeikiai 18341
Mazovian Museum in Płock, Płock 21190
MBT, Israel Aircraft Industries Ltd, Yahud 14711
McAlester Public Library, McAlester 40111
McAllen Memorial Library, McAllen 40113
McArthur Public Library, Biddeford 38201
McCarter & English, Newark 36043
McCarthy Tetrault Library, Vancouver 04540
McCarthy Tetrault LLP Library, Calgary 04245
McComb Public Library, McComb 40115
McConnell Valdes, Hato Rey 21900
McCormick, Barstow, Sheppard, Wayte & Carruth, Fresno 35716
McCracken County Public Library, Paducah 40684
McDaniel College, Westminster 33039
McDermott Technology, Inc, Alliance 35441
McDermott, Will & Emery Law Library, Chicago 35590
McDonald Hopkins, LPA, Cleveland 35617
McDowell County Public Library, Marion 40066

McElroy, Deutsch, Mulvaney & Carpenter, LLP, Morristown 35922
McElroy, Deutsch, Mulvaney & Carpenter, LLP, Newark 36044
McGill University, Montreal 03774
– Blackader-Lauterman Library of Architecture & Art, Montreal 03775
– Education & Curriculum Lab, Montreal 03776
– Howard Ross Library of Management, Montreal 03777
– Humanities & Social Sciences Library, Montreal 03778
– Islamic Studies Library, Montreal 03779
– Life Sciences Library, Montreal 03780
– MacDonald Campus Library, Sainte Anne de Bellevue 03781
– Marvin Duchow Music Library, Montreal 03782
– Nahum Gelber Law Library, Montreal 03783
– Osler Library, Montreal 03784
– Schulich Library of Science & Engineering, Montreal 03785
McGinnis, Lochridge & Kilgore, LLP, Austin 35477
McHenry County College, Crystal Lake 33265
McHenry County Law Library, Woodstock 35081
McHenry Public Library District, McHenry 40119
McKeesport Commonwealth College, McKeesport 31775
McKendree College, Lebanon 31585
McKenna, Long & Aldridge, San Francisco 36239
McKenna, Long & Aldridge, LLP, Washington 36365
McKinley Memorial Library, Niles 40481
McKinney Memorial Public Library, McKinney 40121
McLane, Graf, Raulerson & Middleton PA, Manchester 35874
McLean County Museum of History, Bloomington 36550
McLean Hospital, Belmont 36529
McLennan Community College, Waco 33829
McLennan County Law Library, Waco 34969
McMahon Law Library of Santa Barbara County, Santa Barbara 34836
McMaster University, Hamilton 03724
McMichael Canadian Art Collection, Kleinburg 04371
McMillan Birch Mendelsohn Library, Montreal 04412
E. W. McMillan Library at Cascade College, Portland 32338
McMillan Memorial Library, Nairobi 18046
McMillan Memorial Library, Wisconsin Rapids 41945
McMinnville Public Library, McMinnville 40122
McMurry University, Abilene 30165
McNair Law Firm, PA, Columbia 35626
McNamee, Lochner, Titus & Williams, PC, Albany 35434
McNay Art Museum Library, San Antonio 37520
McNeese State University, Lake Charles 31544
McNess, Wallace & Nurick LLC, Harrisburg 35734
McPherson College, McPherson 31780
McPherson Public Library, McPherson 40124
MdM Salzburg, Salzburg 01570
Mead Public Library, Sheboygan 41294
Meadville Public Library, Meadville 40126
Meadville-Lombard Theological School Library, Chicago 30773
Meaning Research and Planning Institute, Donetsk 28214
Measuring Equipment Plant, Gomel 01862
Meath County Library, Navan 14584
Mechanical and Technological College, Olaine 18178
Mechanical and Technological School for Light Industry, Angarsk 22663
Mechanical Engineering Institute, Dnipropetrovsk 28204
Mechanical Engineering Institute, Kurgan 22311
Mechanical Engineering Laboratory, Niihari 17624
Mechanical Engineering Plant, Izhevsk 22795
Mechanical Engineering Plant, Krasnoyarsk 22815
Mechanical Engineering Plant, Moskva 22843
Mechanical Engineering Professional School, Sankt-Peterburg 22700
Mechanical Plant, Moskva 22840
Mechanical Processing Research Institute, Sankt-Peterburg 23937
Mechanic-Metallurgical Professional School, Kyiv 28111

Mechanics' Institute Library, San Francisco 37547
MECHEL Chelyabinski metallurgicheski kombinat, Chelyabinsk 22784
Mechitharisten-Kongregration, Wien 01489
Mechnikov Microbiology and Immunology Institute, Kharkiv 28271
Mechnikov State University, Odesa 28054
Mecklenburg County Law & Government Library, Charlotte 34060
Medaille College, Buffalo 30591
Medcenter One, Bismarck 36545
Médecins Sans Frontières (MSF), Brussel 02561
Medeplavilny i medeelektrolitny zavod, Moskva 22844
Medfield Memorial Public Library, Medfield 40129
Medford Public Library, Medford 40132
mediacampus frankfurt / die schulen des deutschen buchhandels GmbH, Frankfurt am Main 10726
Mediacentralen, Norrköping 26867
Mediateek voor de Vlaamse Gemeenschap Leuven VZW, Leuven 02445
Mediathek Bühl, Bühl 12688
Mediathek der Stadt Neckarsulm, Neckarsulm 12863
Mediathek Wallis, Brig-Glis 27479
Mediathek Wallis (Kantonsbibliothek), Sion 26980
Médiathèque, Arles 08740
Médiathèque, Hyères 08890
Médiathèque, Issy-les-Moulineaux 08891
Médiathèque, Le Blanc-Mesnil 08905
Médiathèque, Nancy 08979
Médiathèque, Orléans 08991
Médiathèque, Perpignan 09021
Médiathèque, Saint-Pierre 21912
Médiathèque, Sedan 09090
Médiathèque, Sucy-en-Brie 09101
Médiathèque, Toul 09111
Médiathèque, Vandœuvre-lès-Nancy 09127
Médiathèque Alphonse Daudet, Alès 08729
Médiathèque André Malraux, Lisieux 08931
La Médiathèque l'Apostrophe, Chartres 08816
Médiathèque Aveline, Alençon 08727
Médiathèque Benjamin Rabier, La Roche-sur-Yon 08899
Mediatheque Caraïbe Bettino Lara (LAMECA), Basse-Terre 13163
Médiathèque Ceccano, Avignon 08751
Médiathèque Centrale d'Agglomération Emile Zola, Montpellier 08970
Médiathèque Châteaudun, Châteaudun 08817
Médiathèque Cœur de Ville, Vincennes 09153
Médiathèque Communautaire, Draguignan 08854
Médiathèque Condorcet, Libourne 08926
Médiathèque de l'Agora, Evry 08866
Médiathèque de Charleville-Mézières, Charleville-Mézières 08813
Médiathèque de Coutances, Coutances 08844
Médiathèque de la communauté d'agglomération, Sarreguemines 09086
Médiathèque de la Vieille Ile, Haguenau 08886
La Médiathèque de l'Agglomération Troyenne, Troyes 09119
Médiathèque de l'Architecture et du Patrimoine, Paris 08583
Médiathèque de l'Ircam – Centre Pompidou, Paris 08584
Médiathèque de l'Orangerie, Lunéville 08936
Médiathèque de Meurthe-et-Moselle, Laxou 08904
Médiathèque Départementale Claude Simon, Thuir 09109
Médiathèque départementale de la Drôme, Valence 09123
Médiathèque départementale de la Haute-Garonne, Toulouse 09114
Médiathèque départementale de la Haute-Saône, Vesoul 09137
Médiathèque départementale de la Loire, Montbrison 08964
Médiathèque départementale de l'Allier, Coulandon 08843
Médiathèque départementale de l'Oise, Beauvais 08763
Médiathèque départementale de l'Orne, Alençon 08728
Médiathèque départementale de Seine-et-Marne, Le Mée-sur-Seine 08912
Médiathèque départementale de Tarn-et-Garonne, Montauban 08961
Médiathèque départementale des Alpes-Maritimes, Nice 08985
Médiathèque départementale des Hautes-Pyrénées, Tarbes 09102

Médiathèque départementale des Landes, Mont-de-Marsan 08965
Médiathèque départementale du Cantal, Aurillac 08747
Médiathèque départementale du Haut-Rhin, Colmar 08833
Médiathèque départementale du Morbihan, Vannes 09129
Médiathèque Départementale du Nord, Hellemes 08887
Médiathèque départementale du Pas-de-Calais, Dainville 08846
Médiathèque départementale du Rhône, Bron 08788
Médiathèque départementale du Territoire-de-Belfort, Belfort 08766
Médiathèque départementale du Var, Draguignan 08855
Médiathèque d'Epernay, Epernay 08860
Médiathèque du Bassin d'Aurillac, Aurillac 08748
Médiathèque du Pays de Flers, Flers 08867
Médiathèque ENSIACET, Toulouse 08080
La Mediathèque – Equinoxe, Châteauroux 08819
Mediathèque – Espace culturel condorcet, Viry-Châtillon 09155
Médiathèque Etienne Caux, Saint-Nazaire 09075
Médiathèque François Mitterand, Les Ulis 08923
Médiathèque François Mitterand, Lorient 08935
Médiathèque François Mitterand, Poitiers 09026
Médiathèque François Mitterrand, Argentan 08738
Médiathèque François Mitterrand, Sète 09094
Médiathèque George Sand, Palaiseau 08995
Médiathèque Intercommunale, Remiremont 09037
Médiathèque Intercommunale Ouest Provence, Istres 08892
Médiathèque intercommunale Ouest-Provence, Miramas 08958
Médiathèque J. Jeukens, Bar-le-Duc 08758
Médiathèque Jacques Baumel, Rueil-Malmaison 09050
Médiathèque Jean Falala, Reims 09036
Médiathèque Jean Moulin, Villiers-sur-Marne 09152
Médiathèque Jean Prevost, Bron 08789
Médiathèque Jean-Jacques Rousseau, Chambéry 08806
Médiathèque Jean-Jacques Rousseau, Champigny-sur-Marne 08809
Médiathèque Jean-Jaurès, Nevers 08983
Médiathèque La Pléïade, Rombas 09046
Médiathèque Le Mas, Le Mée-sur-Seine 08913
Médiathèque Léon-Alègre, Bagnols-sur-Cèze 08755
Médiathèque Les Temps Modernes, Taverny 09104
Médiathèque Louis Aragon, Le Mans 08911
Médiathèque Louis Aragon, Tarbes 09103
Médiathèque Luxembourg, Meaux 08949
Mediathèque Maskoutaine, Saint-Hyacinthe 04747
Médiathèque Maupassant, Bezons 08772
Médiathèque Max-Pol Fouchet, Châtillon 08821
Médiathèque Mémorial de Caen, Caen 08315
Médiathèque Michel Crépeau, La Rochelle 08897
Médiathèque Michel Simon, Noisy-le-Grand 08988
Médiathèque municipale, Aulnoye-Aymeries 08746
Médiathèque municipale, Bayeux 08760
Médiathèque municipale, Brive-la-Gaillarde 08787
Médiathèque municipale, Cannes 08795
Médiathèque municipale, Corbeil-Essonnes 08840
Médiathèque Municipale, Gardanne 08880
Médiathèque municipale, Joué-lès-Tours 08895
Médiathèque Municipale, Le Cannet 08906
Médiathèque Municipale, Le Perreux-sur-Marne 08915
Médiathèque municipale, Les Ulis 08924
Médiathèque municipale, Montluçon 08968
Médiathèque municipale, Moulins 08976
Médiathèque municipale, Mulhouse 08977
Médiathèque municipale, Orly 08992
Médiathèque municipale, Ploufragan 09024
Médiathèque Municipale, Privas 09031
Médiathèque municipale, Rochefort 09043
Médiathèque municipale, Romorantin-Lanthenay 09047
Médiathèque municipale, Saint-Genis-Laval 09063

Médiathèque municipale, Saint-Raphael 09080
Médiathèque municipale, Sannois 09084
Médiathèque municipale, Tourcoing 09115
Médiathèque municipale, Tulle 09122
Médiathèque municipale, Vernon 09133
Médiathèque municipale, Villebon-sur-Yvette 09142
Médiathèque municipale, Villefranche-sur-Saône 09146
Médiathèque municipale Albert Camus, Chilly-Mazarin 08827
Médiathèque municipale de Cambrai, Cambrai 08794
Médiathèque municipale Elie Wiesel, Bethune 08770
Médiathèque Municipale – Fond de Jazz, Villefranche-de-Rouerge 08713
Médiathèque municipale Guy de Maupassant, Yvetot 09161
Médiathèque municipale La Coupole, Combs-la-Ville 08836
Médiathèque municipale Louis Aragon, Fontenay-sous-Bois 08872
Médiathèque municipale Madame de Sevigne, Vitré 09156
Médiathèque Municipale Valérie Labaud, Vichy 09139
Médiathèque musicale de Paris, Paris 08585
Médiathèque Musicale Mahler, Paris 08586
Médiathèque Nelson Mandela, Grande-Synthe 08883
Médiathèque Paul Eluard, Vierzon 09141
Médiathèque Pierre Amalric, Albi 08726
Médiathèque Pôle Meudon-Ville, Meudon 08956
Médiathèque protestante, Strasbourg 08242
Médiathèque Publique et Universitaire, Valence 09124
Médiathèque Romain Rolland, Saint-Dizier 09058
Médiathèque Saumur Loire Développement, Saumur 09088
Médiathèque Simone de Beauvois, Romans-sur-Isère 09045
Médiathèque Valais (Bibliothèque cantonale) / Mediathek Wallis (Kantonsbibliothek), Sion 26980
Médiathèque Victor Hugo, Saint-Dié-des-Vosges 09057
Médiathèque Villa-Marie, Fréjus 08875
Médiathèque Yves Laurent, Saint-Sébastien-sur-Loire 09081
Médiathèques de Saint-Denis, Saint-Denis 09056
Medical Academy, Białystok 20793
Medical Academy, Krasnoyarsk 22308
Medical Academy, Wrocław, Wrocław 21000
Medical Center of Fudan University, Shanghai 05505
Medical Centre for Postgraduate Education, Warszawa 21251
Medical College, Kolkata 13684
Medical College, Kottayam 13815
Medical College, Kozhikode 13816
Medical College, Madurai 13817
Medical College, Thiruvananthapuram 13844
Medical College, Vadodara 13846
Medical College no 1, Kyiv 28109
Medical College no 4, Kyiv 28110
Medical College of Dalian University, Dalian 05156
Medical College of Georgia, Augusta 30302
Medical College of Ohio, Toledo 32822
Medical College of Wisconsin, Milwaukee 31837
Medical College of Yangzhou University, Yangzhou 05739
Medical Documenation Centre, Bucureşti 21974
Medical Institute, Akmola 17893
Medical Institute, Aktyubinsk 17895
Medical Institute, Almaty 17905
Medical Institute, Blagoveshchensk 22219
Medical Institute, Chelyabinsk 22228
Medical Institute, Chita 22232
Medical Institute, Ekaterinburg 22238
Medical Institute, Ivanovo 22264
Medical Institute, Izhevsk 22268
Medical Institute, Kemerovo 22284
Medical Institute, Novosibirsk 22442
Medical Institute, Orenburg 22460
Medical Institute, Perm 22468
Medical Institute, Samara 22500
Medical Institute, Saratov 22574
Medical Institute, Semipalatinsk 17918
Medical Institute, Smolensk 22581
Medical Institute, Stavropol 22584
Medical Institute, Ternopil 28078
Medical Institute, Vladivostok 22636
Medical Institute, Volgograd 22641
Medical Institute of Hygiene and Sanitary Science, Sankt-Peterburg 22537
Medical Library, Celje 24909
Medical Library of Estonia, Tallinn 07233

Medical Library of the General Hospital Maribor, Maribor 24947
Medical Museum Library, København 06811
Medical Pediatrics Institute, Sankt-Peterburg 22533
Medical Research Council, Didcot 29643
Medical Research Council Laboratories (MRC Labs), Banjul 09182
Medical Research Library, Kurgan 23131
Medical Research Library, Lutsk 28403
Medical Research Library of Latvia, Riga 18187
Medical School, Bristol 28972
Medical Scientific Library, Lviv 28421
Medical Society of the State of New York, Lake Success 37005
Medical University of Gdańsk, Gdańsk 20803
Medical University of Graz, Graz 01217
Medical University of Lodz, Lódź 20880
Medical University of Lublin, Lublin 20894
Medical University of Silesia, Katowice 20818
Medical University of South Carolina, Charleston 30717
Medical University of Warsaw, Warszawa 20998
Medical University – Plovdiv, Plovdiv 03457
Medichne uchilishche, Berdichiv 28093
Medichne uchilishche, Bilgorod Dnistrovski 28094
Medichne uchilishche, Kamyanets-Podilsk 28101
Medichne uchilishche, Kherson 28103
Medichne uchilishche, Kolomiya 28106
Medichne uchilishche, Krivi Rig 28107
Medichne Uchilishche, Nikopol 28124
Medichne uchilishche, Poltava 28127
Medichne uchilishche, Rivne 28128
Medichne uchilishche, Sevastopol 28129
Medichne uchilishche, Vinnytsya 28130
Medichne uchilishche, Zhytomyr 28132
Medichne uchilishche no 1, Dnipropetrovsk 28096
Medichne uchilishche no 1, Kyiv 28109
Medichne uchilishche no 3, Odesa 28126
Medichne uchilishche no 4, Kyiv 28110
Medicīnas zinātniskā bibliotēka, Riga 18187
Medicine Hat College, Medicine Hat 03755
Medicine Hat Public Library, Medicine Hat 04685
Medicinhistoriska museet Eugenia T-3, Stockholm 26666
Medicinsk Museion, København 06811
Medicinska knjižnica Univerzitetnega kliniènega Maribor, Maribor 24947
Medienforum, Berlin 11561
Medienforum des Bistums Essen, Essen 11210
Medienhaus am See, Friedrichshafen 12742
Medienladen Trier, Trier 11348
Medienstelle der Ev.-luth. Kirche in Oldenburg, Oldenburg 12386
Medienzentrum der EKM, Drübeck 11198
Medienzentrum Ostprignitz-Ruppin, Neuruppin 12868
Medina County District Library, Medina 40134
Medina County Law Library Association, Medina 34524
Mediothek Borna, Borna 12682
Mediothek Güglingen, Güglingen 12763
Mediothek Krefeld, Krefeld 12816
Mediterranean Institute of Management, Nicosia 06326
Meditsinski institut, Blagoveshchensk 22219
Meditsinski institut, Chelyabinsk 22228
Meditsinski institut, Chita 22232
Meditsinski institut, Ekaterinburg 22238
Meditsinski institut, Ivanovo 22264
Meditsinski institut, Izhevsk 22268
Meditsinski institut, Kemerovo 22284
Meditsinski institut, Novosibirsk 22442
Meditsinski institut, Orenburg 22460
Meditsinski institut, Perm 22468
Meditsinski institut, Samara 22500
Meditsinski institut, Saratov 22574
Meditsinski institut, Smolensk 22581
Meditsinski institut, Stavropol 22584
Meditsinski institut, Vladivostok 22636
Meditsinski institut, Volgograd 22641
Meditsinski institut im. Akad. Pavlova, Sankt-Peterburg 22529
Meditsinski stomatologicheski institut im. Semashko, Moskva 22344
Meditsinski universitet im. S.V. Kurashova, Kazan 22278
Medizinische Fakultät Mannheim der Universität Heidelberg, Mannheim 10296
Medizinische Hochschule Hannover, Hannover 09973
Medizinische Klinik, Heidelberg 10011
Medizinische Universität Graz, Graz 01217
Medizinische Universität Innsbruck, Innsbruck
 – Department Operative Medizin, Innsbruck 01222
Medizinische Universität Wien, Wien 01297

 – Universitätsklinik für Psychoanalyse und Psychotherapie, Wien 01298
 – Zweigbibliothek für Geschichte der Medizin, Wien 01299
 – Zweigbibliothek für Zahnmedizin, Wien 01300
Medobčinska matična knjižnica, Žalec 24990
Medway NHS Foundation Trust, Gillingham 29670
Medway Public Library, Medway 40135
Medychny instytut, Ternopil 28078
Meertens Instituut, Amsterdam 19369
Megyei és Városi Könyvtár, Kaposvár 13538
Meharry Medical College Library, Nashville 31935
Meherrin Regional Library, Lawrenceville 39827
Mehran University of Engineering and Technology, Jamshoro 20434
Meigs County District Public Library, Pomeroy 40839
Meiji Daigaku, Tokyo 17157
Meiji University, Tokyo 17157
Meijo Daigaku, Nagoya 17056
 – Nogakubu Bunkan, Nagoya 17057
 – Yakugakubu Bunkan, Nagoya 17058
Meijo University, Nagoya 17056
 – Agricultural Branch, Nagoya 17057
 – Faculty of Pharmacy, Nagoya 17058
Meikai Daigaku, Sakado 17099
Meikai University Library, School of Dentistry, Sakado 17099
Meininger Museen, Meiningen 12261
Meisei University, Hino 16947
Mekhaniko-metalurginy tekhnikum, Kyiv 28111
Mekhaniko-tekhnologicheski tekhnikum legkoi promyshlennosti, Angarsk 22663
Mekhitharist-Congregation, Wien 01489
Mekoroth Water Company Ltd, Tel-Aviv 14708
Melaka Public Library Corporation, Melaka 18631
Melanchthon-Gymnasium, Nürnberg 10761
Melanesian Institute, Goroka 20590
Melanzhevoi kombinat, Barnaul 22778
Melbourne and Metropolitan Board of Works, Melbourne 00698
Melbourne Public Library, Melbourne 40137
Melbourne Theosophical Society, Melbourne 00954
Meldal folkebibliotek, Løkken Verk 20368
P. G. Melikishvili Institute of Physical and Organic Chemistry, Tbilisi 09266
Melitopol City Centralized Library System, Main Library, Melitopol 28724
Melitopol Regional Centralized Library System, Main Library, Mirne 28729
Melitopolska miska TsBS, Tsentralna biblioteka, Melitopol 28724
Melitopolska raionna TsBS, Tsentralna biblioteka, Mirne 28729
Méliusz Juhász Péter Megyei Könyvtár és Művelődési Központ, Debrecen 13534
Melleruds kommunbibliotek, Mellerud 26854
Mellor Memorial Library, Stoke-on-Trent 29903
Meløy folkebibliotek, Ørnes 20383
Melrose Park Public Library, Melrose Park 40139
Melrose Public Library, Melrose 40138
Melville Library, Pine Bluff 32291
Melvindale Public Library, Melvindale 40140
Memorial de la Shoah, Paris 08587
Memorial Hall Library, Andover 37958
Memorial Library of Nazareth & Vicinity, Nazareth 40382
Memorial Library of Radnor Township, Wayne 41784
Memorial Presbyterian Church, Midland 35282
Memorial Sloan-Kettering Cancer Center Medical Library, New York 37224
Memorial University of Newfoundland, St. John's 03866
 – Ferriss Hodgett Library, Corner Brook 03867
 – Health Sciences Library, Saint John's 03868
Memorial University of Newfoundland, Fisheries and Marine Institute, St. John's 03869
Memphis College of Art, Memphis 33527
Memphis Public Library & Information Center – Bartlett Branch, Bartlett 38095
Memphis Public Library & Information Center – Benjamin L. Hooks Central Library, Memphis 40141
Memphis Public Library & Information Center – Cordova Branch, Cordova 38694
Memphis Public Library & Information Center – Parkway Village, Memphis 40142

Memphis Public Library & Information Center – Poplar-White Station, Memphis 40143
Memphis Public Library & Information Center – Whitehaven, Memphis 40144
Memphis Theological Seminary, Memphis 35278
Menasha Public Library, Menasha 40145
Menaul Historical Library of the Southwest, Albuquerque 36448
Mendel University of Agriculture and Forestry, Brno 06363
Mendeleev Russian Chemical Engineering University, Moskva 22390
Mendeleev Russian National Chemical Society – St. Petersburg Branch, Sankt-Peterburg 23949
Mendeleev Russian National Research Institute of Metrology, Sankt-Peterburg 23939
Mendelova zemědělská a lesnická univerzita v Brně, Brno 06363
Mendes & Mount, LLP, New York 35999
Mendocino College, Ukiah 33813
Mendocino County Law Library, Ukiah 34948
Mendocino County Library, Ukiah 41631
Mengzi Teachers' College, Mengzi 05417
Menil Foundation, Houston 36946
Menlo College, Atherton 30283
Menlo Park Public Library, Menlo Park 40146
Mennonite Church, North Newton 35314
Mennonite Historians of Eastern Pennsylvania, Harleysville 36912
Mennonite Historical Library, Amsterdam 19078
Mennonitische Forschungsstelle, Bolanden 11174
Menominee County Library, Stephenson 41438
Menomonee Falls Public Library, Menomonee Falls 40148
Menomonie Public Library, Menomonie 40149
Menoufia Provincial Council, Menoufia 07127
Mentor Public Library, Mentor 40150
Meralco Central Library, Pasig City 20779
Mercantile Library Association, Cincinnati 36699
Mercantile Library of New York Center for Fiction, New York 40431
Merced College, Merced 33531
Merced County Library, Merced 40152
Mercer County Community College, Trenton 33804
Mercer County District Library, Celina 38489
Mercer County Law Library, Mercer 34527
Mercer County Law Library Association, Celina 36639
Mercer County Library – Ewing Branch, Ewing 39014
Mercer County Library – Hopewell Township, Pennington 40752
Mercer County Library – Lawrence Headquarters, Lawrenceville 39828
Mercer County Library – Twin Rivers Branch, East Windsor 38919
Mercer County Library – West Windsor, Princeton Junction 40902
Mercer County Public Library, Harrodsburg 39379
Mercer University, Macon 31703
– School of Medicine, Macon 31704
– Walter F. George School of Law Library, Macon 31705
Mercer University Atlanta, Atlanta 30295
Merck & Co, Inc, West Point 36412
Merck & Co, Inc, Whitehouse Station 36415
Merck & Company, Inc, Rahway 36159
Merck KGaA, Darmstadt 11383
Mercy College – Bronx Campus Library, Bronx 30558
Mercy College Libraries, Dobbs Ferry 31020
Mercy College of Health Sciences Library, Des Moines 31003
Mercy College of Northwest Ohio, Toledo 32823
Mercy College – White Plains Campus Library, White Plains 33049
Mercy College – Yorktown Campus, Yorktown Heights 33107
Mercy Hospital & Medical Center, Chicago 36681
Mercy Medical Center – Dubuque, Dubuque 36814
Mercy Medical Center Merced, Community Campus, Merced 37096
Mercyhurst College, Erie 31094
Meredith College, Raleigh 32408
Merenkulkulaitoksen kirjasto, Helsinki 07473
Meriden Public Library, Meriden 40155
Meridian Community College, Meridian 33532
Meridian District Library, Meridian 40156
Meridian-Lauderdale County Public Library, Meridian 40157
Merisotakoulun kirjasto, Helsinki 07397
Merrick Library, Merrick 40159
Merrimack College, North Andover 32138
Merrimack Public Library, Merrimack 40162
Merritt College, Oakland 33592
Merseyside Maritime Museum, Liverpool 29705
Merthyr Tydfil Central Library, Merthyr Tydfil 30042
Merton College, Oxford 29330
Merton College Library, Surrey 29419
Mertz Library, Bronx 36592
Merz Akademie, Stuttgart 10593
Merz Pharma GmbH & Co KGaA, Frankfurt am Main 11399
Mesa Community College, Mesa 33533
Mesa County Public Library District, Grand Junction 39271
Mesa Regional Family History Center, Mesa 35281
Mesa State College, Grand Junction 31234
Meserve, Mumper & Hughes, Los Angeles 35845
Mesirov, Gelman, Jaffe, Cramer & Jamieson Library, Philadelphia 36103
Mesquite Public Library, Mesquite 40164
Messenger Public Library of North Aurora, North Aurora 40497
Messiah College, Grantham 31240
Mestna knjižnica in Čitalnica Idrija, Idrija 24959
Mestna knjižnica Izola, Izola 24960
Mestni muzej, Ljubljana 24923
Městská knihovna Antonína Marka Turnov, Turnov 06615
Městská knihovna Aš, Aš 06551
Městská knihovna Benešov, Benešov 06552
Městská knihovna Beroun, Beroun 06553
Městská knihovna Boženy Němcové Domažlice, Domažlice 06565
Městská knihovna Břeclav, Břeclav 06555
Městská knihovna Česká Lípa, Česká Lípa 06558
Městská knihovna Český Těšín, Český Těšín 06559
Městská knihovna Chotěboř, Chotěboř 06562
Městská knihovna Chrudim, Chrudim 06563
Městská knihovna Frenšát pod Radhoštem, Frenštát pod Radhoštem 06567
Městská knihovna Frýdek-Místek, Frýdek-Místek 06568
Městská knihovna Havířov, Havířov-Podlesí 06569
Městská knihovna Jablonec nad Nisou, Jablonec nad Nisou 06572
Městská knihovna Jaroměř, Jaroměř 06573
Městská knihovna Jihlava, Jihlava 06574
Městská knihovna Jindřichův Hradec, Jindřichův Hradec 06575
Městská knihovna Kadaň, Kadaň 06576
Městská knihovna Kolín, Kolín 06580
Městská knihovna Krnov, Krnov 06581
Městská knihovna Kutná Hora, Kutná Hora 06583
Městská knihovna Ladislava z Boskovič, Moravská Třebová 06589
Městská knihovna Litomyšl, Litomyšl 06585
Městská knihovna Louny, Louny 06586
Městská knihovna Mariánské Lázně, Mariánske Lázně 06587
Městská knihovna Most, Most 06590
Městská knihovna Náchod, Náchod 06591
Městská knihovna Nymburk, Nymburk 06593
Městská knihovna Prostějov, Prostějov 06602
Městská knihovna Rakovník, Rakovník 06603
Městská knihovna Rumburk, Rumburk 06604
Městská knihovna Slavoj Dvůr Králové nad Labem, Dvůr Králové nad Labem 06566
Městská knihovna Sokolov, Sokolov 06606
Městská knihovna Tachov, Tachov 06610
Městská knihovna Třebíči, Třebíč 06612
Městská knihovna Třinec, Třinec 06613
Městská knihovna Trutnov, Trutnov 06614
Městská knihovna v Bruntále, Bruntál 06557
Městská knihovna v Chebu, Cheb 06560
Městská knihovna v Klatovech, Klatovy 06579
Městská knihovna v Novém Jičíně, Nový Jičín 06592
Městská knihovna v Praze, Praha 06599
Městská knihovna v Přerově, Přerov 06600
Městská knihovna Varnsdorf, Varnsdorf 06617
Městska knihovna Žatec, Žatec 06620
Městská knihovna Znojmo, Znojmo 06622
Mestská knižnica, Bratislava 24767
Mestská knižnica, Brezno 24770
Mestská knižnica, Dubnica nad Váhom 24773
Mestská knižnica, Púchov 24791
Mestská knižnica, Želiezovce 24803
Mestské muzeum, Bratislava 24697
Městské muzeum, Hradec Králové 06451
Městské muzeum, Jičin 06453
Metal Forming Institute, Poznań 21197
Metal Industries Research and Development Centre, Kaohsiung 27589
Metal Industry, Petrovsk-Zabaikalski 22890
Metal Industry, Zlatoust 22980
Metal Physics Institute, Kyiv 28335
Metal Ravne d.o.o., Ravne na Koroškem 24906
Metal Working Plant, Moskva 22867
Metal Works Industrial Corporation, Trade Union Library, Sankt-Peterburg 24243
Metallurgical Industrial Corporation, Trade Union Library, Makiyivka 28716
Metallurgical Industrial Corporation, Trade Union Library, Mariupol 28723
Metallurgical Institute, Tbilisi 09267
Metallurgical Plant, Cherepovets 22785
Metallurgical Plant, Donetsk 28141
Metallurgical Plant, Moskva 22845
Metallurgical Plant, Novomoskovsk 28167
Metallurgical Plant, Stalowa Wola 21068
Metallurgical plant 'Batory', Chorzów 21065
Metallurgical Plant Technical Library, Kraków 21066
Metallurgical Plant Trade Union Committee, Regional Trade Union Library, Donetsk 28549
Metallurgical Plant, Trade Union Library, Kostyantynivka 28624
Metallurgical Technical Documentation Centre, Dnipropetrovsk 28209
Metallurgicheski zavod, Cherepovets 22785
Metallurgicheski zavod, Petrovsk-Zabaikalski 22890
Metallurgicheski zavod, Zlatoust 22980
Metallurgicheski zavod im. F.E. Dzerzhinskogo, Kosaya Gora 22811
Metallurgicheski zavod 'Serp i molot', Moskva 22845
Metallurgy and Mechanics Institute, Krivi Rig 28297
Metallurgy Professional School no 22, Kyiv 28113
Metalurginy zavod, Donetsk 28141
Metalurgiyny kombinat, Makiyivka 28716
Metalurgiyny kombinat, Mariupol 28723
Metalurgiyny kombinat im. Kirova, Makiyivka 28164
Metalurgiyny zavod, Kostyantynivka 28624
Metalurgiyny zavod, Novomoskovsk 28167
Metcalf & Eddy Inc, Wakefield 36322
Météo-France, Paris 08150
Meteorological and Hydrological Service, Zagreb 06233
Meteorological Services Department, Legon 13061
Meteorological Station for West-Norway, Bergen 20218
Meteorology Research Library, Pune 14195
MeteoSchweiz Zürich, Zürich 27449
Methanol Technology Planning and Research Institute, Severodonetsk 28467
Methodist Theological School in Ohio Library, Delaware 33290
Metro Health Resource Centre, Johannesburg 25066
Metrogiprotrans, Moskva 23349
Metrohealth Medical Center, Cleveland 36713
Metropolitan Academy of Architecture, Moskva 22345
Metropolitan Borough of St Helens, St Helens 30074
Metropolitan Club Library, New York 37225
Metropolitan Club of the City of Washington Library, Washington 37752
Metropolitan Community College, Omaha 33603
Metropolitan Community Colleges – Longview Community College, Lee's Summit 33479
Metropolitan Community Colleges – Maple Woods Community College, Kansas City 33452
Metropolitan Community Colleges – Penn Valley Community College, Kansas City 33453
Metropolitan Council for Educational Opportunity Library, Roxbury 37469
Metropolitan Council Library, Saint Paul 34779
Metropolitan 'Ervin Szabo' Library, Budapest 13530
Metropolitan Hospital Center, New York 37226
Metropolitan Library System in Oklahoma County -, Edmond 38934
Metropolitan Library System in Oklahoma County, Oklahoma City 40596
Metropolitan Library System in Oklahoma County – Belle Isle Library, Oklahoma City 40597
Metropolitan Library System in Oklahoma County – Bethany Library, Bethany 38190
Metropolitan Library System in Oklahoma County – Midwest City Library, Midwest City 40193
Metropolitan Library System in Oklahoma County – Ronald J. Norick Downtown Library – Ronald J. Norick Downtown Library, Oklahoma City 40598
Metropolitan Library System in Oklahoma County – Southern Oaks Library, Oklahoma City 40599
Metropolitan Library System in Oklahoma County – The Village Library, Oklahoma City 40600
Metropolitan Library System in Oklahoma County – Warr Acres Library- Warr Acres Library, Oklahoma City 40601
Metropolitan Life Insurance Co, Long Island City 35823
The Metropolitan Museum of Art – Cloisters Library, New York 37227
The Metropolitan Museum of Art – Robert Goldwater Library, New York 37228
The Metropolitan Museum of Art – Robert Lehman Collection Library, New York 37229
The Metropolitan Museum of Art – The Irene Lewisohn Costume Reference Library, Costume Institute, New York 37230
The Metropolitan Museum of Art – Thomas J. Watson Library, New York 37231
Metropolitan South Institute of TAFE, Mount Gravatt 00485
Metropolitan Toronto Lawyers Association Library, Toronto 04160
Metropolitan Transition Center Library, Baltimore 36512
Metropolitan Transportation Authority, New York 36000
Metropolitan Transportation Commission, Oakland 37299
Metropolitan Washington Council of Governments, Washington 37753
Metropolitankapitel Bamberg, Bamberg 11159
Metropolitankapitel München, München 11283
Metropolitanska Knjižnica, Zagreb 06209
Metropoliten, Moskva 22846
Metrowest Medical Center, Framingham 36874
Metsäntutkimuslaitoksen kirjasto, Vantaa 07556
Metso Paper Oy, Jyväskylä 07421
Metuchen Public Library, Metuchen 40168
Mexican Society of Geography and Technology, México 18865
Mexico-Audrain County Library District, Mexico 40169
Meyer, Suozzi, English & Klein, Mineola 35902
Mezhdunarodnaya akademiya arkhitektury, Moskva 22345
Mezhdunarodnaya assotsiatsiya kompozitorskikh organizatsi, Moskva 23350
Mezhdurechensk Mining Engineering Professional School, Mezhdurechensk 22679
Mezhdurechenski gornostroitelny colledzh, Mezhdurechensk 22679
Mezhivska raionna TsBS, Tsentralna biblioteka, Mezhova 28725
Mezhova Regional Centralized Library System, Main Library, Mezhova 28725
Mezhraionnaya TsBS, Biblioteka-filial no 1 im. M.Yu. Lermontova Dzerzhinskogo raiona, Sankt-Peterburg 24231
Mezhraionnaya TsBS, Biblioteka-filial no 3 Kuibyshevskogo raiona, Sankt-Peterburg 24232
Mezhraionnaya TsBS, Biblioteka-filial no 4 Kuibyshevskogo raiona, Sankt-Peterburg 24233
Mezhraionnaya TsBS, Biblioteka-filial no 5 im. A.S. Griboedova Oktyabrskogo raiona, Sankt-Peterburg 24234
Mezhraionnaya TsBS, Biblioteka-filial no 7 Oktyabrskogo raiona, Sankt-Peterburg 24235
Mezhraionnaya TsBS, Biblioteka-filial no 8 Oktyabrskogo raiona, Sankt-Peterburg 24236
Mezhraionnaya TsBS, Biblioteka-filial no 9 im. N.A. Nekrasova Smolninskogo raiona, Sankt-Peterburg 24237
Mezhraionnaya TsBS, Biblioteka-filial no 10 im. A.I. Gertsena Smolninskogo raiona, Sankt-Peterburg 24238

Mezhraionnaya TsBS, Detskaya biblioteka-filial no 1 Dzerzhinskogo raiona, Sankt-Peterburg 24239
Mezhraionnaya TsBS, Dzerzhinskogo, Kuibyshskogo, Smolninskogo i Oktyabrskogo raionov, Tsentralnaya biblioteka im. A.A. Bloka, Sankt-Peterburg 24240
Mezhspilkova biblioteka, Ternopil 28842
Mezőgazdasági Gépesitési Intézet, Gödöllő 13325
Mezőgazdasági Minősitő Intézet, Budapest 13431
Mezőgazdasagi Minösitö Intézet, Tápiószele 13513
MF Norwegian School of Theology, Oslo 20124
Miami Correctional Facility, Bunker Hill 34020
Miami County Law Library, Troy 34939
Miami Dade College – Kendall Campus Library, Miami 31808
Miami Dade College – North Campus Library, Miami 31809
Miami Dade Community College – Mitchell Wolfson New World Center Campus Learning Resources, Miami 33535
Miami Memorial-Gila County Library, Miami 40170
Miami Public Library, Miami 40171
Miami University, Oxford 32221
– Amos Music Library, Oxford 32222
– Brill Science Library, Oxford 32223
– Wertz Art-Architecture Library, Oxford 32224
Miami University – Hamilton Campus, Hamilton 31280
Miami University Libraries, Middletown 31821
Miami Valley Hospital, Dayton 36775
Miami-Dade County Law Library, Miami 34529
Miami-Dade Public Library System, Miami 40172
Mianyang Teachers' College Institute, Mianyang 05418
MIAT – Museum voor Industriële Archeologie en Textiel, Gent 02649
Michael Best & Friedrich LLP, Milwaukee 35894
Michael Smurfit Graduate School of Business, Dublin 14479
Michael-Ende-Museum, München 12307
Michaeliskloster, Hildesheim 11248
Chr. Michelsen Institute, Bergen 20211
Michener Institute for Applied Health Sciences Library, Toronto 03883
Michigan City Public Library, Michigan City 40174
Michigan Department of Corrections, Carson City 34048
Michigan Department of Corrections, Carson City 34049
Michigan Department of Corrections, Freeland 34286
Michigan Department of Corrections, Jackson 34386
Michigan Department of Corrections, Kingsley 34415
Michigan Department of Corrections, Manistee 34498
Michigan Department of Corrections, Munising 34567
Michigan Department of Corrections, Muskegon 34568
Michigan Department of Corrections, Plymouth 34684
Michigan Department of Transportation Information Services, Lansing 34436
Michigan Jewish Institute Library, Oak Park 35316
Michigan Lutheran Seminary, Saginaw 35353
Michigan Molecular Institute, Midland 37107
Michigan State Legislative Service Bureau, Lansing 34437
Michigan State University, East Lansing 31055
– Benjamin H. Anibal Engineering Library, East Lansing 31056
– Biomed Library, East Lansing 31057
– Geology Library, East Lansing 31058
– Gull Lake Library, Hickory Corners 31059
– Mathematics Library, East Lansing 31060
– Planning and Design Library, East Lansing 31061
– Veterinary Medical Center Library, East Lansing 31062
– William C. Gast Business Library, East Lansing 31063
Michigan State University, Hickory Corners 31331
Michigan State University – College of Law Library, East Lansing 31064
Michigan Technological University, Houghton 31352
Michurin Horticultural Research Institute, Michurinsk 23157
Michurin Research Institute of Genetics and Selection in Horticulture, Michurinsk 23158
Michurinsk State Pedagogical Institute, Michurinsk 22323
Michurinski gosudarstvenny pedagogicheski institut, Michurinsk 22323
Microbiology Institute, Minsk 02022
Micron Technology, Inc Library, Boise 36553
Mid Cheshire College of Further Education, Northwich 29485
Mid State Correctional Facility Library, Wrightstown 35089
Mid Sweden University, Härnösand 26391
Mid-America Baptist Theological Seminary, Cordova 35170
Mid-America Christian University, Oklahoma City 32183
Mid-America Nazarene University, Olathe 32186
Mid-America Reformed Seminary, Dyer 35192
Mid-Columbia Libraries, Kennewick 39702
Mid-Continent Public Library, Independence 39558
Mid-Continent Public Library – Midwest Genealogy Center, Independence 39559
Mid-Continent University, Mayfield 35277
Middelfart Bibliotek, Middelfart 06902
Middelheimmuseum, Antwerpen 02519
Middle Country Public Library, Centereach 38492
Middle East Institute, Washington 37754
Middle East Technical University, Ankara 27849
Middle Georgia College, Cochran 33245
Middle Georgia Regional Library System – Genealogical & Historical Room & Georgia Archives, Macon 39993
Middle Georgia Regional Library System – Riverside Branch, Macon 39994
Middle Georgia Regional Library System – Washington Memorial (Main Library), Macon 39995
Middle Georgia Regional Library System – West Bibb Branch, Macon 39996
Middle Tennessee State University, Murfreesboro 31922
Middle Tennessee State University Center for Popular Music Library, Murfreesboro 37135
Middleborough Public Library, Middleborough 40176
Middlebury College, Middlebury 31818
– Armstrong Library, Middlebury 31819
– Music Library, Middlebury 31820
Middlebury Public Library, Middlebury 40179
Middlesborough-Bell County Public Library, Middlesboro 40180
Middlesbrough Borough Council, Middlesbrough 30043
Middlesex Community College, Bedford 33160
Middlesex Community College, Middletown 33537
Middlesex County Adult Correction Center Library, North Brunswick 34617
Middlesex County College, Edison 33310
Middlesex County Law Library, New Brunswick 34579
Middlesex Law Association, London 04376
Middlesex Law Library at Cambridge, Cambridge 36620
Middlesex Public Library, Middlesex 40181
Middlesex University, London 29207
Middleton Public Library, Middleton 40182
Middletown Public Library, Middletown 40183
Middletown Public Library, Middletown 40184
Middletown Public Library – West Chester Branch, West Chester 41811
Middletown Thrall Library, Middletown 40185
Middletown Township Public Library, Middletown 40186
Mid-Hudson Library System, Poughkeepsie 40882
Midi-Pyrenees Observatory, Toulouse 08081
Mid-Kent College of Higher and Further Education, Chatham 29456
Midland College, Midland 33539
Midland County Public Library, Midland 40189
Midland Lutheran College, Fremont 31176
Midland Park Memorial Library, Midland Park 40190
Midlands State University, Gweru 42362
Midlands Technical College, West Columbia 33858
Midlothian District Council, Dalkeith 29968
Midlothian Public Library, Midlothian 40191
Mid-Michigan Correctional Facility Library, Saint Louis 34775
Mid-State Correctional Facility Library, Marcy 34504
Midwest Historical and Genealogical Society, Inc, Wichita 37835
Midwestern Baptist Theological Seminary, Kansas City 35249
Midwestern State University, Wichita Falls 33058
Midwestern University, Downers Grove 31025
Mid-York Library System, Utica 41648
Mie Daigaku, Tsu 17225
– Igakubu, Tsu 17226
Mie Prefectural Comprehensive Education Center, Tsu 17783
Mie University, Tsu 17225
– Faculty of Medicine, Tsu 17226
Miejska Biblioteka Publiczna, Bełchatów 21379
Miejska Biblioteka Publiczna, Białogard 21381
Miejska Biblioteka Publiczna, Bielsk Podlaski 21383
Miejska Biblioteka Publiczna, Biłgoraj 21386
Miejska Biblioteka Publiczna, Bochnia 21387
Miejska Biblioteka Publiczna, Bolesławiec 21388
Miejska Biblioteka Publiczna, Brzesko 21390
Miejska Biblioteka Publiczna, Brzozów 21391
Miejska Biblioteka Publiczna, Busko-Zdrój 21392
Miejska Biblioteka Publiczna, Bystrzyca Kłodzka 21394
Miejska Biblioteka Publiczna, Bytów 21395
Miejska Biblioteka Publiczna, Chełmno 21397
Miejska Biblioteka Publiczna, Chmielnik 21398
Miejska Biblioteka Publiczna, Chodzież 21399
Miejska Biblioteka Publiczna, Chojnice 21400
Miejska Biblioteka Publiczna, Choszczno 21401
Miejska Biblioteka Publiczna, Chrzanów 21402
Miejska Biblioteka Publiczna, Czarnków 21404
Miejska Biblioteka Publiczna, Człuchów 21408
Miejska Biblioteka Publiczna, Dąbrowa Tarnowska 21409
Miejska Biblioteka Publiczna, Drawsko Pomorskie 21410
Miejska Biblioteka Publiczna, Działdowo 21411
Miejska Biblioteka Publiczna, Ełk 21413
Miejska Biblioteka Publiczna, Garwolin 21414
Miejska Biblioteka Publiczna, Głogów 21416
Miejska Biblioteka Publiczna, Goleniów 21418
Miejska Biblioteka Publiczna, Golub-Dobrzyń 21419
Miejska Biblioteka Publiczna, Grajewo 21421
Miejska Biblioteka Publiczna, Grodków 21422
Miejska Biblioteka Publiczna, Grójec 21423
Miejska Biblioteka Publiczna, Grudziądz 21424
Miejska Biblioteka Publiczna, Gryfino 21426
Miejska Biblioteka Publiczna, Hajnówka 21427
Miejska Biblioteka Publiczna, Hrubieszów 21428
Miejska Biblioteka Publiczna, Inowrocław 21429
Miejska Biblioteka Publiczna, Janów Lubelski 21431
Miejska Biblioteka Publiczna, Jarosław 21432
Miejska Biblioteka Publiczna, Jasło 21433
Miejska Biblioteka Publiczna, Jędrzejów 21434
Miejska Biblioteka Publiczna, Kalisz 21436
Miejska Biblioteka Publiczna, Kamień Pomorski 21438
Miejska Biblioteka Publiczna, Kamienna Góra 21439
Miejska Biblioteka Publiczna, Kartuzy 21440
Miejska Biblioteka Publiczna, Kazimierza Wielka 21441
Miejska Biblioteka Publiczna, Kępno 21442
Miejska Biblioteka Publiczna, Kętrzyn 21443
Miejska Biblioteka Publiczna, Kielce 21444
Miejska Biblioteka Publiczna, Kluczbork 21448
Miejska Biblioteka Publiczna, Kolbuszowa 21449
Miejska Biblioteka Publiczna, Kolno 21450
Miejska Biblioteka Publiczna, Koło 21451
Miejska Biblioteka Publiczna, Kołobrzeg 21452
Miejska Biblioteka Publiczna, Końskie 21454
Miejska Biblioteka Publiczna, Kościan 21455
Miejska Biblioteka Publiczna, Kościerzyna 21456
Miejska Biblioteka Publiczna, Kozienice 21458
Miejska Biblioteka Publiczna, Koźle 21459
Miejska Biblioteka Publiczna, Kraśnik 21461
Miejska Biblioteka Publiczna, Krasnystaw 21462
Miejska Biblioteka Publiczna, Krotoszyn 21465
Miejska Biblioteka Publiczna, Łańcut 21467
Miejska Biblioteka Publiczna, Łapy 21468
Miejska Biblioteka Publiczna, Łask 21469
Miejska Biblioteka Publiczna, Lębork 21470
Miejska Biblioteka Publiczna, Łęczyca 21471
Miejska Biblioteka Publiczna, Leszno 21474
Miejska Biblioteka Publiczna, Lidzbark Warmiński 21476
Miejska Biblioteka Publiczna, Limanowa 21477
Miejska Biblioteka Publiczna, Lipno 21478
Miejska Biblioteka Publiczna, Łobez 21479
Miejska Biblioteka Publiczna, Łomża 21481
Miejska Biblioteka Publiczna, Łowicz 21483
Miejska Biblioteka Publiczna, Lubaczów 21484
Miejska Biblioteka Publiczna, Lubartów 21486
Miejska Biblioteka Publiczna, Łuków 21489
Miejska Biblioteka Publiczna, Miastko 21491
Miejska Biblioteka Publiczna, Międzyrzecz 21493
Miejska Biblioteka Publiczna, Mielec 21494
Miejska Biblioteka Publiczna, Milicz 21495
Miejska Biblioteka Publiczna, Mińsk Mazowiecki 21496
Miejska Biblioteka Publiczna, Mława 21497
Miejska Biblioteka Publiczna, Mogilno 21498
Miejska Biblioteka Publiczna, Morąg 21500
Miejska Biblioteka Publiczna, Mrągowo 21501
Miejska Biblioteka Publiczna, Myślibórz 21502
Miejska Biblioteka Publiczna, Niemodlin 21504
Miejska Biblioteka Publiczna, Nowa Ruda 21506
Miejska Biblioteka Publiczna, Nowa Sòl 21507
Miejska Biblioteka Publiczna, Nowe Miasto Lubawskie 21508
Miejska Biblioteka Publiczna, Nowy Dwór Mazowiecki 21509
Miejska Biblioteka Publiczna, Oława 21512
Miejska Biblioteka Publiczna, Oleśnica 21514
Miejska Biblioteka Publiczna, Oleszno 21515
Miejska Biblioteka Publiczna, Opoczno 21518
Miejska Biblioteka Publiczna, Ostrów Mazowiecka 21520
Miejska Biblioteka Publiczna, Ostrzeszów 21522
Miejska Biblioteka Publiczna, Oświęcim 21523
Miejska Biblioteka Publiczna, Pisz 21528
Miejska Biblioteka Publiczna, Pleszew 21529
Miejska Biblioteka Publiczna, Poddębice 21531
Miejska Biblioteka Publiczna, Proszowice 21533
Miejska Biblioteka Publiczna, Prudnik 21534
Miejska Biblioteka Publiczna, Pruszcz Gdański 21535
Miejska Biblioteka Publiczna, Przasnysz 21536
Miejska Biblioteka Publiczna, Przeworsk 21538
Miejska Biblioteka Publiczna, Pszczyna 21539
Miejska Biblioteka Publiczna, Puławy 21540
Miejska Biblioteka Publiczna, Pyrzyce 21542
Miejska Biblioteka Publiczna, Radomsko 21545
Miejska Biblioteka Publiczna, Rawa Mazowiecka 21547
Miejska Biblioteka Publiczna, Rawicz 21548
Miejska Biblioteka Publiczna, Rypin 21550
Miejska Biblioteka Publiczna, Sanok 21553
Miejska Biblioteka Publiczna, Siemiatycze 21554
Miejska Biblioteka Publiczna, Sierpc 21556
Miejska Biblioteka Publiczna, Sławno 21558
Miejska Biblioteka Publiczna, Sochaczew 21562
Miejska Biblioteka Publiczna, Sokółka 21563
Miejska Biblioteka Publiczna, Sokołów Podląski 21564
Miejska Biblioteka Publiczna, Śrem 21565
Miejska Biblioteka Publiczna, Środa Śląska 21566
Miejska Biblioteka Publiczna, Środa Wielkopolska 21567
Miejska Biblioteka Publiczna, Stargard Szczeciński 21569
Miejska Biblioteka Publiczna, Staszów 21570
Miejska Biblioteka Publiczna, Strzelce Krajeńskie 21571
Miejska Biblioteka Publiczna, Strzelce Opolskie 21572
Miejska Biblioteka Publiczna, Strzelin 21573
Miejska Biblioteka Publiczna, Suącin 21574
Miejska Biblioteka Publiczna, Sulechów 21575
Miejska Biblioteka Publiczna, Świebodzin 21578
Miejska Biblioteka Publiczna, Szubin 21582
Miejska Biblioteka Publiczna, Tarnów 21584

Miejska Biblioteka Publiczna, Tarnowskie Góry 21586
Miejska Biblioteka Publiczna, Tomaszów Lubelski 21588
Miejska Biblioteka Publiczna, Trzebnica 21591
Miejska Biblioteka Publiczna, Tuchola 21592
Miejska Biblioteka Publiczna, Turek 21593
Miejska Biblioteka Publiczna, Wabrzeźno 21595
Miejska Biblioteka Publiczna, Wadowice 21596
Miejska Biblioteka Publiczna, Wągrowiec 21597
Miejska Biblioteka Publiczna, Węgrów 21600
Miejska Biblioteka Publiczna, Wieluń 21601
Miejska Biblioteka Publiczna, Wieruszów 21602
Miejska Biblioteka Publiczna, Włocławek 21603
Miejska Biblioteka Publiczna, Włodawa 21605
Miejska Biblioteka Publiczna, Włoszczówa 21606
Miejska Biblioteka Publiczna, Wodzisław 21607
Miejska Biblioteka Publiczna, Września 21610
Miejska Biblioteka Publiczna, Wyrzysk 21612
Miejska Biblioteka Publiczna, Ząbkowice 21613
Miejska Biblioteka Publiczna, Zakopane 21614
Miejska Biblioteka Publiczna, Żary 21615
Miejska Biblioteka Publiczna, Złotów 21618
Miejska Biblioteka Publiczna im. Adama Asnyka, Kalisz 21437
Miejska Biblioteka Publiczna im. Adolfa Dygasińskiego, Starachowice 21568
Miejska Biblioteka Publiczna im. Cypriana Kamila Norwida, Świdnica 21577
Miejska Biblioteka Publiczna im. Jana Długosza, Sandomierz 21552
Miejska Biblioteka Publiczna im. Jana Kasprowicza, Inowrocław 21430
Miejska Biblioteka Publiczna im. Józefa Lompy, Lubliniec 21488
Miejska Biblioteka Publiczna im. Księcia Ludwika I, Brzeg 21389
Miejska Biblioteka Publiczna im. Marii Dąbrowskiej, Słupsk 21560
Miejska Biblioteka Publiczna im Stanislawa Gabryela w Gorlicach, Gorlice 21420
Miejska Biblioteka Publiczna im. Stanislawa Grochowiaka, Leszno 21475
Miejska Biblioteka Publiczna im. Stefana Zeromskiego w Kutnie, Kutno 21466
Miejska Biblioteka Publiczna im. Wl. St. Reymonta, Skierniewice 21557
Miejska Biblioteka Publiczna im. Zbigniewa Załuskiego, Gryfice 21425
Miejska Biblioteka Publiczna w Nisku, Nisko 21505
Miejska i Gminna Biblioteka Publiczna, Pińczów 21527
Miejska i Gminna Biblioteka Publiczna, Wołów 21608
Miejska i Gminna Biblioteka Publiczna im. Władysława Broniewskiego, Resko 21549
Miejska i Powiatowa Biblioteka Publiczna im. Marii Konopnickiej, Lubań 21485
Miejska Powiatowa Biblioteka Publiczna im. K.K. Baczyńskiego, Dzierżoniów 21412
Miejsko-Gminna Biblioteka Publiczna, Łosice 21482
Miejsko-Gminna Biblioteka Publiczna, Parczew 21524
Miejsko-Gminna Biblioteka Publiczna w Nidzicy, Nidzica 21503
Miejsko-Powiatowa Biblioteka Publiczna, Olecko 21513
Mieken Sogo Kyoiku Senta, Tsu 17783
Mierscher Lieshaus, Mersch 18418
Miestna knižnica Dúbravka, Bratislava 24768
Miestna knižnica Petržalka, Bratislava 24769
Mifflin County Library, Lewistown 39864
Migrationsverket, Norrköping 26498
Miguel A. Catalán Physics Center (CFMAC) – Spanish National Research Council (CSIC), Madrid 25994
Mihály Pollack Polytechnic, Pécs 13291
Mika Meyers Beckett & Jones, PLC, Grand Rapids 35723
Mikado – Mission Library and Catholic Documentation Centre, Aachen 11443
Mikado – Missionsbibliothek und katholische Dokumentationsstelle, Aachen 11443
Mikalojus Dauksa Public Library of Kedainiai District Municipality, Kėdainiai 18333
Mikalojus Konstantinas Čiurlionis National Art Museum, Kaunas 18298
Mike Durfee State Prison, Springfield 34884
Mikeladze, G.S., Scientific and Technical Library of Georgia, Tbilisi 09187
Mikhailivka Regional Centralized Library System, Main Library, Mikhailivka 28726
Mikhailivska raionna TsBS, Tsentralna biblioteka, Mikhailivka 28726
Mikkeli Public Library, Mikkeli 07591
Mikkelin kaupunginkirjasto – Etelä-Savon maakuntakirjasto, Mikkeli 07591
Mikkelin keskussairaalan ammattikirjasto, Mikkeli 07530
Mikkelin maakunta-arkisto, Mikkeli 07531
Mikolaiv City Centralized Children's Library System, Main Libarary, Mykolaiv 28738
Mikolaiv City Centralized Library System, Main Library, Mykolaiv 28736
Mikolaiv Regional Centralized Library System, Main Library, Mykolaiv 28737
Mikolaivska miska TsBS dlya doroslikh, Tsentralna biblioteka, Mykolaiv 28736
Mikolaivska raionna TsBS, Tsentralna biblioteka, Mykolaiv 28737
Mikoyan Institute of Tobacco and Makhorka, Krasnodar 23117
Mikulas Kovac Public Library, Banská Bystrica 24762
Milan Municipal Library, Milano 16775
Milan-Berlin Township Public Library, Milan 40195
Milbank, Tweed, Hadley & McCloy, Los Angeles 35846
Milbank, Tweed, Hadley & McCloy LLP Library, Washington 36366
Miles City Public Library, Miles City 40197
Miles College, Fairfield 31121
Miles, Davison & McCarthy & McNiven Library, Calgary 04246
Miles & Stockbridge PC Library, Baltimore 35482
Milford Public Library, Milford 40198
Milford Town Library, Milford 40199
Militärarchiv, Freiburg 11887
Militärgeschichtliches Forschungsamt, Potsdam 12411
Military Academy of Lithuania, Vilnius 18286
Military Academy of Medicine, Sankt-Peterburg 22555
Military Academy 'Zrinyi Miklós', Budapest 13267
Military Artillery Academy, Sankt-Peterburg 22554
Military Central Library of the Army General Staff, Roma 15658
Military District Administration and Headquarters – Military Historical Library, Sankt-Peterburg 22767
Military Engineering School of Telecommunications, Sankt-Peterburg 22714
Military Historical Museum of Artillery, Engineering and Signalling, Sankt-Peterburg 23926
Military Law Academy, Moskva 22404
Military Medical Academy, Łódź 20881
Military Medical Museum, Sankt-Peterburg 23927
Military Museum, Beograd 24524
Military Museum of Turkey, Istanbul 27898
Military Naval Engineering School, Sankt-Peterburg 22557
Military Officer Academy, Wrocław 21030
Military Physical Education Institute, Sankt-Peterburg 23930
Military School of Air Defence Radioelectronics, Sankt-Peterburg 22715
Military Space Engineering Institute, Sankt-Peterburg 23931
Military Support and Transport Academy, Sankt-Peterburg 22553
Military University of Technology, Warszawa 20999
Miljøkontrollen, København 06812
Mill Memorial Library, Nanticoke 40360
Millburn Free Public Library, Millburn 40205
Miller, Canfield, Paddock & Stone Library, Detroit 35677
Miller & Martin PLLC, Chattanooga 35564
Miller Memorial Library, Tambaram 13843
Miller Nash LLP Library, Portland 36145
Millersville University of Pennsylvania, Millersville 31830
Millî Kütüphane, Ankara 27843
Millicent Library, Fairhaven 39028
Milligan College, Milligan College 31831
Milliken & Co, Spartanburg 36278
Millikin University, Decatur 30978
Mills College, Oakland 32173
Mills Law Library, San Francisco 36240
Millsaps College, Jackson 31442
Millville Public Library, Millville 40208
Milnerton Public Library, Cape Town 25194
Milove Regional Centralized Library System, Main Library, Milove 28728
Milovska mezhvidomcha TsBS, Tsentralna bibliotka, Milove 28728
Milton Keynes Council, Central Milton Keynes 29957
Milton Public Library, Milton 40210
Milwaukee Area Technical College, Milwaukee 33542
Milwaukee Art Museum Library, Milwaukee 37112
Milwaukee County Federated Library System, Milwaukee 40212
Milwaukee County Historical Society, Milwaukee 37113
Milwaukee Institute of Art & Design Library, Milwaukee 37114
Milwaukee Public Library, Milwaukee 40213
Milwaukee Public Library – Atkinson, Milwaukee 40214
Milwaukee Public Library – Bay View, Milwaukee 40215
Milwaukee Public Library – Capitol, Milwaukee 40216
Milwaukee Public Library – Center Street, Milwaukee 40217
Milwaukee Public Library – East, Milwaukee 40218
Milwaukee Public Library – Forest Home, Milwaukee 40219
Milwaukee Public Library – Martin Luther King Branch, Milwaukee 40220
Milwaukee Public Library – Mill Road, Milwaukee 40221
Milwaukee Public Library – Tippecanoe, Milwaukee 40222
Milwaukee Public Library – Villard Avenue, Milwaukee 40223
Milwaukee Public Library – Washington Park, Milwaukee 40224
Milwaukee Public Library – Zablocki, Milwaukee 40225
Milwaukee Public Museum, Milwaukee 37115
Milwaukee School of Engineering, Milwaukee 31838
Minas Gerais Livestock Research Company, Belo Horizonte 03252
Mindanao State University, Marawi City 20726
– Iligan Institute of Technology, Iligan City 20727
Minderbroedersprovincie van Sint-Jozef, Vaalbeek 02491
Mine Rescue Equipment Research Institute, Donetsk 28226
Minen nauchnoizsledovatelski i proekten institut 'Minproekt', Sofiya 03557
Mineola Memorial Library, Mineola 40230
Mineral Area College, Park Hills 33620
Mineralogicheski muzei im. A.E. Fersmana RAN, Moskva 23351
Mineralogisch-Petrographisches Institut (MPI), Hamburg 09944
Minerals & Petroleum Reference Centre, East Melbourne 00672
Minerva Public Library, Minerva 40231
Mines and Geological Department, Nairobi 18024
Minia University, El Minia 07054
Mining and Ore Rectification Corporation Trade Union Committee Centralized Library System, Main Library, Marganets 28719
Mining and Ore Rectification Corporation Trade Union Committee Centralized Library System, Main Library, Ordzhonikidze 28768
Mining Equipment Joint-Stock Company, Voronezh 22976
Mining Institute, Almaty 17958
Mining Institute, Sankt-Peterburg 23765
Mining Institute of the Siberian Division of the Russican Academy of Sciences, Novosibirsk 23641
Mining Research Institute, Krivi Rig 28298
Mining University, Moskva 22331
Minisink Valley Historical Society Library, Port Jervis 37395
Ministarsto kulture Republike Hrvatske, Zagreb 06250
Ministère de la Communication et de la Culture, Kouba 00090
Ministère de la Justice, Québec 04131
Ministère de la Justice – Bibliothèque de la Chancellerie, Paris 08151
Ministère de la Santé et des Services Sociaux, Québec 04132
Ministère de la Sécurité Publique, Sainte-Foy 04152
Ministère de l'Agriculture, Kinshasa 06062
Ministère de l'Agriculture et de la Pêche, Paris 08152
Ministère de l'Agriculture et de la Réforme Agraire, Rabat 18941
Ministère de l'Artisanat, des Mines et du Tourisme, Kigali 24377
Ministère de l'Ecologie, La Défence 08137
Ministère de l'Économie, des Finances et de l'Industrie, Paris 08588
Ministère de l'Economie, du Plan et de la Coopération Internationale, Yaoundé 03644
Ministère de l'Enseignement primaire et secondaire, Kigali 24370
Ministère de l'Equipement DAEI, Paris-la-Défense 08160
Ministère de l'Equipement, des Transports et du Logement, Paris-la-Défense 08635
Ministère de l'Habitat, Rabat 18942
Ministère de l'Industrie, de la Poste et des Télécommunications, Paris 08153
Ministère de l'Information, Libreville 09171
Ministère de l'Intérieur, Paris 08154
Ministère des Affaires étrangères, Paris 08155
Ministère des Affaires étrangères, Rabat 18943
Ministère des Affaires sociales, du Travail et de la Solidarité, Paris 08156
Ministère des Finances, Québec 04133
Ministère des Mines et de l'Industrie, Nouakchott 18687
Ministère des Relations avec les Citoyens et de l'Immigrations (MRCI), Québec 04134
Ministère des Ressources Naturelles, Québec 04135
Ministère du Développement économique, Tunis 27823
Ministère du Plan, Kigali 24371
Ministère du Revenu, Sainte-Foy 04153
Ministerie van Binnenlandse Zaken en Koninkrijksrelaties, Den Haag 19256
Ministerie van Buitenlandse Zaken, Den Haag 19257
Ministerie van de Vlaamse Gemeenschap – Departement Coördinatie, Brussel 02431
Ministerie van Financiën, Brussel 02432
Ministerie van Financiën, Den Haag 19258
Ministerie van Justitie, Den Haag 19259
Ministerie van Justitie – Dienst Justitiële Inrichtingen (MVJ, DJI, Bibliotheekzaken), Den Haag 19260
Ministerie van Justitie – Wetenschappelijk Onderzoek en Documentatiecentrum (WODC), Afdeling Documentaire Informatievoorziening (DIV), Den Haag 19261
Ministerie van Landbouw, Natuurbeheer en Visserij (LNV), Den Haag 19262
Ministerie van Sociale Zaken en Werkgelegenheid, Den Haag 19263
Ministerie van Verkeer en Waterstaat, Den Haag 19264
Ministerie van Volkshuisvesting, Ruimtelijke Ordening en Milieubeheer (VROM), Den Haag 19265
Ministério da Agricultura, Desenvolvimento Rural e Pescas, Lisboa 21733
Ministério da Cultura, Lisboa 21734
Ministério da Economia, Fazenda Planejamento, Rio de Janeiro 03233
Ministério da Educação – Direcção Geral do Inovação e de Desenvolvimento Curricular, Lisboa 21735
Ministério da Fazenda, Brasília 03216
Ministério da Fazenda no Estado do Rio de Janeiro, Rio de Janeiro 03234
Ministério da Justiça, Brasília 03217
Ministério da Saúde, Brasília 03218
Ministério das Relações Exteriores, Brasília 03219
Ministério das Relações Exteriores, Rio de Janeiro 03235
Ministério de Agricultura e Pesca, São Tomé 24388
Ministério de Agricultura, Pecuaria e Abastecimento, Brasília 03220
Ministerio de Agricultura, Pesca y Alimentación, Madrid 25719
Ministerio de Agricultura y Cría, Caracas 42235
Ministerio de Asuntos Exteriores, Madrid 25720
Ministerio de Comercio Exterior (MINCEX), La Habana 06283
Ministerio de Defensa, Madrid 25721
Ministerio de Defensa, Toledo 25750
Ministerio de Economía, San Salvador 07206
Ministerio de Economía de la Provincia de Buenos Aires, La Plata 00218
Ministerio de Economía y Comercio (MEC), San José 06116
Ministerio de Economia y Finanzas Publicos, Buenos Aires 00210

Ministerio de Economía y Hacienda, Madrid 25722
Ministerio de Economía y Planificación, La Habana 06284
Ministerio de Educación, La Habana 06285
Ministerio de Educación, Madrid 25723
Ministerio de Educación, Cultura y Deporte, Madrid 25724
Ministerio de Educación Nacional, Santafé de Bogotá 05995
Ministerio de Energía y Minas, Caracas 42236
Ministerio de Fomento, Madrid 25725
Ministerio de Ganadería y Agricultura, Montevideo 42024
Ministerio de Gobierno, Santafé de Bogotá 05996
Ministerio de Gobierno de la Provincia de Buenos Aires, La Plata 00219
Ministerio de Hacienda, Santiago 04907
Ministerio de Hacienda y Crédito Público, Santafé de Bogotá 05997
Ministerio de Justicia, Madrid 25726
Ministerio de la Presidencia, Madrid 25727
Ministerío de Medio Ambiente, Madrid 25728
Ministerio de Planificacion (MIDEPLAN), Santiago 04908
Ministerio de Relaciones Exteriores, La Paz 02928
Ministerio de Relaciones Exteriores, Lima 20645
Ministerio de Relaciones Exteriores, Panamá 20537
Ministerio de Relaciones Exteriores, San Salvador 07207
Ministerio de Relaciones Exteriores, Santafé de Bogotá 05998
Ministerio de Relaciones Exteriores y Culto, Buenos Aires 00211
Ministerio de Relaciones Exteriores y Culto, San José 06117
Ministerio de Salud, Lima 20646
Ministerio de Salud, Santafé de Bogotá 05999
Ministerio de Sanidad y Asistencia Social, Caracas 42237
Ministerio de Sanidad y Consumo, Madrid 25729
Ministerio de Trabajo, Buenos Aires 00212
Ministerio del Interior, Montevideo 42025
Ministério do Empiego e da Seguranza Social, Lisboa 21736
Ministério do Equipamento do Planeamento e da Administração do Território, Lisboa 21737
Ministério do Exército, Rio de Janeiro 03236
Ministério do Trabalho e Emprego, Brasília 03221
Ministério do Trabalho e Previdencia Social, Rio de Janeiro 03237
Ministério dos Negócios Estrangeiros, Lisboa 21738
Ministerio para las Administraciones Públicas, Madrid 25730
Ministerium der Finanzen Rheinland-Pfalz, Mainz 11010
Ministerium der Justiz und für Verbraucherschutz des Landes Rheinland-Pfalz, Mainz 11011
Ministerium für Arbeit, Integration und Soziales des Landes Nordrhein-Westfalen, Düsseldorf 10879
Ministerium für Bildung, Wissenschaft, Jugend und Kultur, Mainz 11012
Ministerium für Ernährung und Ländlichen Raum, Stuttgart 11111
Ministerium für Gesundheit, Emanzipation, Pflege und Alter des Landes Nordrhein-Westfalen, Düsseldorf 10880
Ministerium für Inneres und Kommunales des Landes Nordrhein-Westfalen, Düsseldorf 10881
Ministerium für Innovation, Wissenschaft und Forschung des Landes Nordrhein-Westfalen, Düsseldorf 10868
Ministerium für Justiz, Arbeit und Europa, Kiel 10969
Ministerium für Klimaschutz, Umwelt, Landwirtschaft, Natur- und Verbraucherschutz, Düsseldorf 10882
Ministerium für Kultus, Jugend und Sport Baden-Württemberg, Stuttgart 11105
Ministerium für Schule und Weiterbildung des Landes Nordrhein-Westfalen, Düsseldorf 10868
Ministerium für Umwelt und Forsten Rheinland-Pfalz, Mainz 11013
Ministerium für Wirtschaft, Mittelstand und Energie des Landes Nordrhein-Westfalen, Düsseldorf 10883
Ministerium für Wissenschaft, Forschung und Kunst Baden-Württemberg, Stuttgart 11106
Ministero degli Affari Esteri, Roma 15670
Ministero dei Lavori Pubblici, Roma 15671
Ministero del Bilancio e della Programmazione Economica, Roma 15672
Ministero del Commercio con l'Estero, Roma 15673
Ministero del Lavoro e della Previdenza Sociale, Roma 15674
Ministero della Pubblica Istruzione, Roma 15675
Ministero della Salute, Roma 15676
Ministero delle Finanze e del Tesore, Roma 15677
Ministero delle Partecipazioni Statali, Roma 15678
Ministero delle Poste e Telecomunicazioni, Roma 15679
Ministero delle Risorse Agricole, Alimentari e Forestali, Roma 15680
Ministero dell'Industria, del Commercio e dell'Artigianato, Roma 15681
Ministero dell'Interno, Roma 15682
Ministero di Grazia e Giustizia, Roma 15683
Ministerstvo ekonomiki RF, Moskva 22739
Ministerstvo energetiki i elektrifikatsii, Moskva 22740
Ministerstvo financí České republiky, Praha 06410
Ministerstvo Finansov RF, Moskva 22741
Ministerstvo geologii, Moskva 22742
Ministerstvo kultury, Moskva 22743
Ministerstvo legkoi promyshlennosti, Moskva 22744
Ministerstvo montazhnykh i spetsialnykh stroitelnykh rabot, Moskva 22745
Ministerstvo morskogo flota, Moskva 22746
Ministerstvo na narodnata otbrana, Sofiya 03489
Ministerstvo na pravosădieto, Sofiya 03490
Ministerstvo na vănshnite raboti, Sofiya 03491
Ministerstvo putei soobshchenia, Moskva 22747
Ministerstvo selskogo khozyaistva, Moskva 22748
Ministerstvo spravedlnosti ČR, Praha 06411
Ministerstvo stroitelstva i ekspluatatsii avtomobilnykh dorog, Moskva 22749
Ministerstvo stroitelstva predpriyati tyazheloi industrii, Moskva 22750
Ministerstvo topliva i energetiki – Department elektroenergetiki, Moskva 22751
Ministerstvo tsvetnoi metallurgii, Moskva 22752
Ministerstvo vysshego i srednego obrazovaniya Belarusi, Minsk 01822
Ministerstvo vysshego i srednego spetsialnogo obrazovania, Moskva 22753
Ministerstvo zdravookhranenia, Moskva 22754
Ministerstwo Edukacji Narodowej, Warszawa 21044
Ministerstwo Spraw Zagranicznych, Warszawa 21045
Ministerstwo Sprawedliwości, Warszawa 21046
Ministrstvo za notranje zadeve RS, Ljubljana 24892
Ministry for Children & Families, Victoria 04184
Ministry for Foreign Affairs Library, Helsinki 07403
Ministry for Planning, Perth 00714
Ministry of Agriculture, Gaborone 02971
Ministry of Agriculture, Kuala Lumpur 18563
Ministry of Agriculture, Moskva 22748
Ministry of Agriculture and Food Industry, Sofiya 03492
Ministry of Agriculture and Irrigation, New Delhi 13879
Ministry of Agriculture and Mining, Kingston 16876
Ministry of Agriculture and Natural Resources, Banjul 09181
Ministry of Agriculture and Rural Development, Nairobi 18025
Ministry of Agriculture and Rural Development, Warszawa 21315
Ministry of Agriculture, Animal Husbandry and Fisheries, Paramaribo 26363
Ministry of Agriculture, Forestry and Fisheries Library, Tokyo 17320
Ministry of Assembling of Parts and Specialized Construction Work, Moskva 22745
Ministry of Community Development and Sports, Singapore 24597
Ministry of Construction, Tokyo 17310
Ministry of Culture, Moskva 22743
Ministry of Culture of the Republic of Croatia, Administration for Cultural Development, Information and Documentation Department, Library, Zagreb 06250
Ministry of Defence, London 29525
Ministry of Defence, New Delhi 13880
Ministry of Defence, Sofiya 03489
Ministry of Economics of the Russian Federation, Moskva 22739
Ministry of Economy and Public Finance, Madrid 25722
Ministry of Education, Bangkok 27758
Ministry of Education, Cairo 07112
Ministry of Education, Damascus 27517
Ministry of Education, Freetown 24574
Ministry of Education, Kabul 00003
Ministry of Education, Kuala Lumpur 18560
Ministry of Education, La Habana 06285
Ministry of Education, Nuku'alofa 27803
Ministry of Education, Tehran 14416
Ministry of Education, Tokyo 17317
Ministry of Education, Wellington 19801
Ministry of Education and Scientific Research, Port Louis 18695
Ministry of Education Library, Helsinki 07400
Ministry of Endowments, Tripoli 18251
Ministry of Energy, Port-of-Spain 27809
Ministry of Energy and Electrification, Moskva 22740
Ministry of Environment of the Republic of Lithuania, Vilnius 18294
Ministry of External Affairs, New Delhi 13881
Ministry of Finance, Budapest 13323
Ministry of Finance, Helsinki 07398
Ministry of Finance, Kuala Lumpur 18558
Ministry of Finance, New Delhi 13882
Ministry of Finance, Santiago 04907
Ministry of Finance, Tokyo 17321
Ministry of Finance, General Affairs Department of the Mint Bureau Data Unit, Osaka 17290
Ministry of Finance of Rio de Janeiro State, Rio de Janeiro 03234
Ministry of Finance of Russia, Moskva 22741
Ministry of Food and Agriculture, Accra 13030
Ministry of Food and Agriculture, New Delhi 13883
Ministry of Foreign Affairs, Brasília 03219
Ministry of Foreign Affairs, Budapest 13317
Ministry of Foreign Affairs, Cairo 07113
Ministry of Foreign Affairs, Colombo 26293
Ministry of Foreign Affairs, Jerusalem 14682
Ministry of Foreign Affairs, Khartoum 26354
Ministry of Foreign Affairs, Kuala Lumpur 18559
Ministry of Foreign Affairs, La Paz 02928
Ministry of Foreign Affairs, Rio de Janeiro 03235
Ministry of Foreign Affairs, Tehran 14417
Ministry of Foreign Affairs, Tokyo 17301
Ministry of Foreign Affairs, Warszawa 21045
Ministry of Foreign Affairs and Commonwealth Relations, Karachi 20470
Ministry of Foreign Affairs and Religion, Buenos Aires 00211
Ministry of Foreign Affairs and Trade, Wellington 19802
Ministry of Forestry, Suva 07302
Ministry of Forests Library, Victoria 04185
Ministry of Geology, Moskva 22742
Ministry of Health, Cairo 07114
Ministry of Health, Nairobi 18026
Ministry of Health, Wellington 19803
Ministry of Health and Welfare, Tokyo 17314
Ministry of Heavy Industry Construction, Moskva 22750
Ministry of Higher and Middle Special Education, Moskva 22753
Ministry of Home Affairs, New Delhi 13884
Ministry of Home Affairs, Tokyo 17302
Ministry of Housing and Reconstruction, Cairo 07115
Ministry of Human Resources, Kuala Lumpur 18562
Ministry of Industry and Commerce, New Delhi 13885
Ministry of Information, Accra 13031
Ministry of Information, Jakarta 14336
Ministry of Information, Kuala Lumpur 18561
Ministry of Information and Broadcasting, New Delhi 13886
Ministry of Information and Guidance, Tripoli 18252
Ministry of Interior, Bangkok 27759
Ministry of Interior, Nicosia 06332
Ministry of International Trade and Industry (MITI), Kuala Lumpur 18564
Ministry of International Trade and Industry (MITI), Tokyo 17316
Ministry of Justice, Cairo 07116
Ministry of Justice, Helsinki 07399
Ministry of Justice, Ibadan 19996
Ministry of Justice, Jerusalem 14683
Ministry of Justice, Praha 06411
Ministry of Justice, Tokyo 17447
Ministry of Justice, Wellington 19804
Ministry of Labour Library, Policy Planning and Research Department, Tokyo 17322
Ministry of Lands and Agriculture, Harare 42373
Ministry of Law, New Delhi 13887
Ministry of Law, Sofiya 03490
Ministry of Law and Parliamentary Affairs, Islamabad 20467
Ministry of Light Industry, Moskva 22744
Ministry of Livestock Development, Veterinary Services Division, Kabete 18017
Ministry of Local Government, Accra 13032
Ministry of Meat and Dairy Products, Toshkent 42109
Ministry of Mining and Energy, Kingston 16877
Ministry of National Education, Warszawa 21044
Ministry of Non-Ferrous Metallurgy, Moskva 22752
Ministry of Petroleum, Tehran 14418
Ministry of Petroleum and Energy, Oslo 20196
Ministry of Plan and Budget, Tehran 14419
Ministry of Planning, Tripoli 18253
Ministry of Planning and Economic Affairs, Dar es Salaam 27666
Ministry of Power and Electrification, Toshkent 42110
Ministry of Public Health, Bangkok 27760
Ministry of Public Health, Moskva 22754
Ministry of Public Works, Cairo 07117
Ministry of Public Works and Water Resources, Giza 07125
Ministry of Shipping and Transport, New Delhi 13888
Ministry of Social Affairs and Health, Helsinki 07401
Ministry of Social Development, Wellington 19805
Ministry of Social Welfare, New Delhi 13889
Ministry of Sport and Recreation Western Australia, Wembley 00740
Ministry of Supply and Internal Trade, Cairo 07118
Ministry of the Flemish Community, Brussel 02431
Ministry of the Fuel and Energetics, Moskva 22751
Ministry of the Interior of the Republic of Slovenia, Ljubljana 24892
Ministry of Trade, Accra 13033
Ministry of Traffic, Moskva 22747
Ministry of Transport, Tokyo 17327
Ministry of Transport and Energy, Harare 42374
Ministry of Transport and Road Safety, Jerusalem 14684
Ministry of Waqfs, Cairo 07119
Ministry of Women's Affairs, Wellington 19806
Ministry of Works Training Centre, Zomba 18494
Ministry of Youth and Sports, Kuala Lumpur 18557
Minjiang Vocational University, Fuzhou 05193
Minkovich Minsk City Library no 10, Minsk 02168
Minneapolis College of Art & Design, Minneapolis 31845
Minneapolis Community & Technical College, Minneapolis 33543
Minneapolis Institute of Arts, Minneapolis 37122
Minnesota Attorney General Library, Saint Paul 34780
Minnesota Bible College, Rochester 35348
Minnesota Braille & Talking Book Library, Faribault 36845
Minnesota Correctional Facility, Bayport 33973
Minnesota Department of Corrections, Lino Lakes 34464
Minnesota Department of Corrections, Moose Lake 34558
Minnesota Department of Corrections, Rush City 34758
Minnesota Department of Corrections, Saint Cloud 34773
Minnesota Department of Corrections, Shakopee 34855
Minnesota Department of Revenue Library, Saint Paul 37501
Minnesota Department of Transportation Library, Saint Paul 34781
Minnesota Historical Society Library, Saint Paul 37502
Minnesota Legislative Reference Library, Saint Paul 34782
Minnesota Mining & Manufacturing Co, Saint Paul 36209
Minnesota Pollution Control Agency Library, Saint Paul 34783
Minnesota State Law Library, Saint Paul 34784
Minnesota West Community & Technical College, Worthington 33885

Minnessota State University Mankato, Mankato 31754
Min-on Music Library, Tokyo 17708
Min-on Ongaku Shiryokan, Tokyo 17708
Minot Public Library, Minot 40233
Minot State University, Minot 31864
Minsk Automobile Plant, Minsk Regional Trade Union Organizations's Main Library, Minsk 02186
Minsk Automotive Industry Technology and Design Experimental Institute, Minsk 02031
Minsk Building Materials Research Institute, Minsk 02032
Minsk Children's City Library no 1, Minsk 02175
Minsk Children's City Library no 2, Minsk 02176
Minsk Children's City Library no 4, Minsk 02177
Minsk Children's City Library no 6, Minsk 02178
Minsk Children's City Library no 7, Minsk 02179
Minsk Children's City Library no 9, Minsk 02180
Minsk Children's City Library no 10, Minsk 02181
Minsk City Advanced Teacher Training College, Minsk 01806
Minsk City Library no 2, Minsk 02162
Minsk City Library no 5, Minsk 02164
Minsk City Library no 6, Minsk 02165
Minsk City Library no 9, Minsk 02167
Minsk City Library no 11, Minsk 02169
Minsk City Library no 12, Minsk 02170
Minsk City Library no 17, Minsk 02172
Minsk City Library no 18, Minsk 02173
Minsk City Library no 19, Minsk 02174
Minsk City Music Library, Minsk 02004
Minsk Clock Assembly Plant, Scientific and Technical Library, Minsk 01879
Minsk Depot, Railway Technical Library, Minsk 02055
Minsk Engine Plant, Scientific and Technical Library, Minsk 01882
Minsk Etalon Experimental Plant, Minsk 01883
Minsk Furniture Design Industrial Corporation, Minsk 01898
Minsk Haberdashery and Leather Goods Experimental Technological and Design Bureau, Minsk 02036
Minsk Industrial Construction Corporation, Trade Union Committee, Main Trade Union Library, Minsk 02187
Minsk Industrial Transport Planning and Research Institute, Minsk 02035
Minsk Kiselev Fine-wool and Knitted Fabrics Industrial Corporation, Minsk 01889
Minsk Krupskaya Clothing Industrial Corporation, Minsk 01888
Minsk Luch Shoes Industrial Corporation, Minsk 01887
Minsk Medical Prepartations Industrial Corporation, Minsk 01884
Minsk October Revolution Machine-tool Construction Industrial Corporation, Minsk 01890
Minsk Passenger Railway Station, Central Railway Technical Library, Minsk 02054
Minsk Region Agricultural Station, Natalevski 02067
Minsk Region Central Library System of the City of Kiev – A. Pushkin Main Library, Kyiv 28680
Minsk State Construction Planning Institute, Minsk 02008
Minsk State Linguistic University, Minsk 01807
Minsk Teacher Training Institute, Minsk 01808
Minsk Technological and Economic Research in Chemical Industry Research Institute, Minsk 02033
Minsk Technological Research Institute of Planning and Design, Minsk 02034
Minsk Tractor Plant Industrial Corporation, Minsk 01897
Minsk Tractor Plant, Recreation Centre, Trade Union Library, Minsk 02158
Minsk Tractor Spares Industrial Corporation – Bearing Plant, Minsk 01899
Minsk Urban Building Institute, Minsk 02023
Minsk Worsted Textile Integrated Plant, Scientific and Technical Library, Minsk 01881
Minskaya gorodskaya bibliyateka 14 Yunost, Minsk 02160
Minskaya gorodskaya bibliyateka no 1 im. L.N. Tolstogo, Minsk 02161
Minskaya gorodskaya bibliyateka no 2, Minsk 02162
Minskaya gorodskaya bibliyateka no 3 im. Mayakovskogo, Minsk 02163
Minskaya gorodskaya bibliyateka no 5, Minsk 02164
Minskaya gorodskaya bibliyateka no 6, Minsk 02165
Minskaya gorodskaya bibliyateka no 8 im. M.A. Bogdanovicha, Minsk 02166
Minskaya gorodskaya bibliyateka no 9, Minsk 02167
Minskaya gorodskaya bibliyateka no 10 im. M.A. Minkovicha, Minsk 02168
Minskaya gorodskaya bibliyateka no 11, Minsk 02169
Minskaya gorodskaya bibliyateka no 12, Minsk 02170
Minskaya gorodskaya bibliyateka no 15 im. Tsetkin, Minsk 02171
Minskaya gorodskaya bibliyateka no 17, Minsk 02172
Minskaya gorodskaya bibliyateka no 18, Minsk 02173
Minskaya gorodskaya bibliyateka no 19, Minsk 02174
Minskaya gorodskaya detskaya bibliyateka no 1, Minsk 02175
Minskaya gorodskaya detskaya bibliyateka no 2, Minsk 02176
Minskaya gorodskaya detskaya bibliyateka no 4, Minsk 02177
Minskaya gorodskaya detskaya bibliyateka no 6, Minsk 02178
Minskaya gorodskaya detskaya bibliyateka no 7, Minsk 02179
Minskaya gorodskaya detskaya bibliyateka no 9, Minsk 02180
Minskaya gorodskaya detskaya bibliyateka no 10, Minsk 02181
Minskaya oblastnaya bibliyateka im. A.S. Pushkina, Minsk 02182
Minskaya oblastnaya detskaya bibliyateka im. Ya. Mavra, Minsk 02183
Minskaya oblastnaya sel-skokhozyaistvennaya stantsiya, Natalevski 02067
Minskaya tsentralnaya gorodskaya bibliyateka im. Ya. Kupaly, Minsk 02184
Minskaya tsentralnaya gorodskaya detskaya bibliyateka im. N. Ostrovskogo, Minsk 02185
Minski avtomobilny zavod, Minsk 02186
Minski chasovoi zavod – Nauchno-tekhnicheskaya bibliyateka, Minsk 01879
Minski elektrotekhnicheski zavod im. Kozlova, Minsk 01880
Minski gorodskoi institut usovershenstvovaniya uchitelei, Minsk 01806
Minski gosudarstvenny lingvisticheski universitet, Minsk 01807
Minski kamvolny kombinat – Nauchno-tekhnicheskaya bibliyateka, Minsk 01881
Minski konstruktorsko-tekhnologicheski eksperimentalny institut avtomobilnoi promyshlennosti, Minsk 02031
Minski motorny zavod – Nauchno-tekhnicheskaya bibiyateka, Minsk 01882
Minski nauchno-issledovatelski institut stroitelnykh materialov, Minsk 02032
Minski nauchno-issledovatelski institut tekhniko-ekonomicheskikh issledovani v khimicheskoi promyshlennosti, Minsk 02033
Minski opytny zavod Etalon, Minsk 01883
Minski pedagogicheski institut im. A.M. Gorkogo, Minsk 01808
Minski proektno-konstruktorki tekhnologicheski institut, Minsk 02034
Minski proektny i nauchno-issledovatelski institut promyshlennogo transporta, Minsk 02035
Minskoe eksperementalno-konstruktorskoe tekhnologicheskoe byuro kozhgalantereinoi i furniturnoi promyshlennosti, Minsk 02036
Minskoe PO Minmedpreparaty, Minsk 01884
Minskoe PO po vypusku avtomaticheskikh linii – Spetsialnoe konstruktorsko-tekhnologicheskoe byuro, Minsk 01885
Minskoe PO po vypusku avtomaticheskikh linii – Zavod avtomaticheskikh linii im. P.M. Masherova, Minsk 01886
Minskoe proizvodstvennoe obuvnoe obedinenie Luch, Minsk 01887
Minskoe proizvodstvennoe shveinoe obedinenie im. Krupski, Minsk 01888
Minskoe proizvodstvennoe tonkosukonnoe obedinenie im. T.Ya. Kiseleva, Minsk 01889
Minskoe stankostroitelnoe PO im. Oktyabrskoi revolyutsii, Minsk 01890
Minskoe stankostroitelnoe PO po vypusku protyazhnykh i otreznykh stankov im. Kirova – Spetsialnoe konstruktorskoe byuro, Minsk 01891
Mint Museum Library, Charlotte 36653
MINTEK Information Centre, Randburg 25173
Minxi Vocational University, Longyan 05404
Mir Toy-making Industrial Corporation, Trade Union Library, Minsk 02191
Miracosta College, Oceanside 33596
Mirai Kogaku Kenkyujo, Tokyo 17709
Miramar College, San Diego 32579
Miran Jarc Public Library, Novo Mesto 24975
Mirgorod Regional Centralized Library System, Main Library, Myrgorod 28742
Miri City Council Public Library, Miri 18632
P. Mirny Library, Kyiv 28662
Miroslav Krleža Lexicographic Institute, Zagreb 06249
Mishawaka-Penn-Harris Public Library, Mishawaka 40236
Mishawaka-Penn-Harris Public Library – Bittersweet, Mishawaka 40237
Misioneros Combonianos, Madrid 25797
Misjonshøgskolen, Stavanger 20160
Miska naukova medychna biblioteka, Sevastopol 28461
Miskolc City Library, Miskolc 13541
Miskolci Egyetem Könyvtár, Levéltár, Múzeum, Miskolc 13284
– Dunaújvárosi Főiskola Könyvtára, Dunaujváros 13285
Miskolci Városi Könyvtár, Miskolc 13541
Misonoza Engeki Toshokan, Nagoya 17614
Misonoza Theatrical, Nagoya 17614
Misr Bank, Cairo 07139
Misr Company for Artifical Silk, Beheira 07133
Misr Company for Spinning and Thin Weaving, Beheira 07134
Misr Spinning and Weaving Company, Gharbia 07143
Missão Salesiana de Mato Grosso, Lins 03245
Missiecentrum Euntes, Kessel-Lo 02469
MISSIO Austria – Päpstliche Missionswerke, Wien 01490
MISSIO München, München 11284
Mission College, Santa Clara 33743
Mission EineWelt, Neuendettelsau 11292
Mission Library and Catholic Documentation Centre, Aachen 11443
Mission Research Corp, Santa Barbara 36254
Mission Seminary of the Congregation of Sacred Heart Fathers, Stadniki 21058
Mission Universitaire Culturelle et de Coopération (MUCC), Tanger 18969
Mission Viejo Library, Mission Viejo 40239
Missionare vom Kostbaren Blut, Salzburg 01466
Missionari di S. Vincenzo de Paoli, Napoli 15934
Missionarissen van Scheut, Leuven 02476
Missionary Society of Saint Paul, Rabat 18663
Missioni Africane, Padova 15957
Missions Etrangères de Paris (MEP), Paris 08589
Missionsakademie an der Universität Hamburg, Hamburg 12003
Missionsbibliothek und katholische Dokumantationsstelle, Aachen 11443
Missionshaus der Spiritaner, Dormagen 11194
Missionspriesterseminar SVD, Sankt Augustin 10580
Mississippi College, Clinton 30839
Mississippi College, Jackson 31443
Mississippi County Library System, Blytheville 38229
Mississippi County Library System – Blytheville Public, Blytheville 38230
Mississippi Delta Community College, Moorhead 33556
Mississippi Department of Corrections, Leakesville 34448
Mississippi Department of Environmental Quality Library, Jackson 34387
Mississippi Gulf Coast Community College, Gautier 33369
Mississippi Gulf Coast Community College, Gulfport 33392
Mississippi Gulf Coast Community College, Perkinston 33631
Mississippi Library Commission, Jackson 39602
Mississippi Museum of Art, Jackson 36968
Mississippi Museum of Natural Science Library, Jackson 36969
Mississippi State Department of Archives & History, Jackson 34388
Mississippi State Hospital, Whitfield 37834
Mississippi State University, Mississippi State 31867
Mississippi State University Agricultural & Forestry Experiment Station, Stoneville 32747
Mississippi Supreme Court, Jackson 36970
Mississippi University for Women, Columbus 30897
Mississippi Valley State University, Itta Bena 31436
Missoula Public Library, Missoula 40240
Missouri Baptist College, Saint Louis 32505
Missouri Botanical Garden Library, Saint Louis 37487
Missouri Court of Appeals Library, Kansas City 34405
Missouri Court of Appeals-Eastern District Library, Saint Louis 34776
Missouri Department of Corrections, Farmington 34201
Missouri Department of Corrections, Jefferson City 34397
Missouri Department of Corrections, Moberly 34545
Missouri Department of Corrections, Pacific 34647
Missouri Department of Natural Resources, Rolla 34754
Missouri History Museum, Saint Louis 37488
Missouri River Regional Library, Jefferson City 39640
Missouri School for the Blind Library, Saint Louis 33712
Missouri School for the Deaf, Fulton 33359
Missouri Southern State University, Joplin 31470
Missouri State Court of Appeals, Springfield 34885
Missouri State Library, Jefferson City 30135
Missouri State Library (MOSL), Jefferson City 30136
Missouri State University, Springfield 32705
– SMSU-WP Garnett, West Plains 32706
Missouri Supreme Court Library, Jefferson City 34398
Missouri Valley College, Marshall 31767
Missouri Western State University, Saint Joseph 32500
Mistská knihovna Tábor, Tábor 06609
Misubishi Chemical Co, Ltd, Inashiki 17362
Misyjne Seminarium Duchowne Księż Werbistów – Filozofia, Nysa 21055
MIT Science Fiction Society Library, Cambridge 30661
Mitaka City Library, Mitaka 17848
Mitaka Shiritsu Toshokan, Mitaka 17848
Mitchell College, New London 33575
Mitchell Community College, Statesville 33778
Mitchell Community Public Library, Mitchell 40242
The Mitchell Library, Glasgow 29985
Mitchell Public Library, Mitchell 40243
Mitchell Silberberg & Knupp LLP, Los Angeles 35847
Mitchell, Williams, Selig, Gates & Woodyard, Little Rock 35818
Mitre Corp Bedford Library, Bedford 35496
Mitshubishi Kinzoku K. K. Chuo Kenkyujo, Omiya 17400
Mitsu & Co, Ltd, Tokyo 17461
Mitsubishi Chemical Corporation, Inashiki 17362
Mitsubishi Chemical Industries Ltd, Kurosaki Works, Kitakyushu 17374
Mitsubishi Heavy Industries, Ltd, Nagasaki Technical Institute, Nagasaki 17389
Mitsubishi Heavy Industries Ltd, Nagoya Aircraft Works, Nagoya 17393
Mitsubishi Jukogyo K. K. Gijutsu Honbu Nagasaki Kenkyujo Kanrika, Nagasaki 17389
Mitsubishi Jukogyo K. K. Nagoya Kokuki Seisakusho, Nagoya 17393
Mitsubishi Kasei Kogyo K. K. Kurosaki Kojo, Kitakyushu 17374
Mitsubishi Metal Corp, Central Research Institute, Omiya 17400
Mitsubishi Rayon Co, Ltd, Central Research Laboratory, Otake 17417
Mitsubishi Reiyon K. K. Chuo Kenkyujo Somukanri Gurupu Toshoshitsu, Otake 17417
Mitsubishi Research Institute, Inc, Tokyo 17451
Mitsubishi Shintaku Ginko K. K. Chosabu, Tokyo 17460
Mitsubishi Trust and Banking Corporation, Economic Research Department, Tokyo 17460
Mitsui Bank Ltd, Tokyo 17462
Mitsui Bunko, Tokyo 17710
Mitsui Bussan K. K. Chosabu, Tokyo 17461
Mitsui Ginko, Tokyo 17462

Mitsui Petrochemical Industries Co, Ltd, Research Center, Kuga 17380
Mitsui Research Institute for Social and Economic History, Tokyo 17710
Mitsui Sekiyu Kagaku Kogyo K. K. Sogo Kenkyujo, Kuga 17380
Mitsui Shintaku Ginko K. K. Chosabu, Tokyo 17463
Mitsui Toatsu Chemical Industry Co, Ltd, Omuta Works, Omuta 17402
Mitsui Toatsu Kogyo K. K. Omuta Kogyojo, Omuta 17402
Mitsui Trust and Banking Co, Ltd, Tokyo 17463
Mittuniversitetet, Härnösand 26391
Miyagi Gakuin Women's College, Sendai 17113
Miyagi Junior College of Agriculture, Sendai 17246
Miyagi Prefectural Assembly Office, Research Section Library, Sendai 17296
Miyagiken Gikai Jimukyoku, Sendai 17296
Miyagiken Nogyo Tanki Daigaku, Sendai 17246
Miyazaki Prefectural Assembly, Miyazaki 17279
Miyazaki Prefecture Agricultural Experiment Station, Miyazaki 17603
Miyazakiken Gikai, Miyazaki 17279
Miyazakiken Sogo Nogyo Shikenjo, Miyazaki 17603
Mizuta Kinen Toshokan, Sakado 17098
Mjölby stadsbibliotek, Mjölby 26855
MKM Mansfelder Kupfer und Messing GmbH, Hettstedt 11408
Mkoba Teachers College, Gweru 42371
Mlinivsk Regional Centralized Library System, Main Library, Mlinivsk 28730
Mlinivska raionna TsBS, Tsentralna biblioteka, Mlinivsk 28730
MMSZ Co Ltd, Budapest 13359
Mmulakgoro Public Library, Bloemfontein 25190
MNHN – Département systématique et Evolution, Paris 08590
MOA Museum of Art, Atami 17515
Mobil Exploration & Producing Technical Center, Dallas 35653
Mobil Oil Canada Library, Calgary 04247
Mobile County Public Law Library, Mobile 34546
Mobile Public Library, Mobile 40245
Mobile Public Library – Monte L. Moorer-Spring Hill Branch, Mobile 40246
Mobile Public Library – West Regional, Mobile 40247
MOD RAF College Cranwell, Sleaford 29395
Modemuseum Provincie Antwerpen – Momu, Antwerpen 02520
Moderna galerija, Ljubljana 24924
Moderna Museet, Stockholm 26667
Modesammlung, Bibliothek der, Wien 01589
Modesto Junior College, Modesto 33547
MoDo Paper AB, Skörblacka 26550
Modoc County Library, Alturas 37930
Modrall, Sperling, Roehl, Harris & Sisk, Albuquerque 35435
Modum bibliotek, Vikersund 20418
Mogilev Automobile Plant, Mogilev 01902
Mogilev Central City Library, Mogilev 02203
Mogilev City Board of Public Education, Mogilev 02061
Mogilev City Children's Library no 3, Mogilev 02200
Mogilev City Library no 2, Mogilev 02196
Mogilev City Library no 3, Mogilev 02197
Mogilev City Library no 6, Mogilev 02198
Mogilev City Library no 7, Mogilev 02199
Mogilev Institute of Technology, Mogilev 02065
Mogilev Lift Building Industrial Corporation, Mogilev 01904
Mogilev PKTIAM Institute, Mogilev 02064
Mogilev Railway Station, Main Railway Technical Library, Mogilev 02066
Mogilev Region Advanced Teacher Training College, Mogilev 01812
Mogilev Region Agricultural Experimental Station, Dashkovka 01950
Mogilev Region Historical Museum, Mogilev 02063
Mogilev Region Library, Mogilev 02201
Mogilev Region State Archive, Mogilev 02059
Mogilev Regional Medical Library, Mogilev 02062
Mogilevskaya gorodskaya bibliyateka im. Kolasa, Mogilev 02193
Mogilevskaya gorodskaya bibliyateka im. Lermontova, Mogilev 02194
Mogilevskaya gorodskaya bibliyateka im. Ya. Kupaly, Mogilev 02195
Mogilevskaya gorodskaya bibliyateka no 2, Mogilev 02196
Mogilevskaya gorodskaya bibliyateka no 3, Mogilev 02197
Mogilevskaya gorodskaya bibliyateka no 6, Mogilev 02198
Mogilevskaya gorodskaya bibliyateka no 7, Mogilev 02199
Mogilevskaya gorodskaya detskaya bibliyateka no 3, Mogilev 02200
Mogilevskaya oblastnaya bibliyateka, Mogilev 02201
Mogilevskaya oblastnaya detskaya bibliyateka im. Gaidara, Mogilev 02202
Mogilevskaya oblastnaya meditsinskaya bibliyateka, Mogilev 02062
Mogilevskaya oblastnaya selskokhozyaistvennaya opytnaya stantsiya, Dashkovka 01950
Mogilevskaya tsentralnaya gorodskaya bibliyateka, Mogilev 02203
Mogilevskaya tsentralnaya gorodskaya detskaya bibliyateka im. Pushkina, Mogilev 02204
Mogilevski avtomobilny zavod, Mogilev 01902
Mogilevski oblastnoi institut usovershenstvovaniya uchitelei, Mogilev 01812
Mogilevski oblastnoi kraevedcheski muzei, Mogilev 02063
Mogilevski pedagogicheski institut Kuletsyuva, Mogilev 01813
Mogilevski PKTIAM, Mogilev 02064
Mogilevski tekhnologicheski institut, Mogilev 02065
Mogiliv-Podolska raionna TsBS, Tsentralna biblioteka, Mogiliv-Podolski 28731
Mogiliv-Podolski Regional Centralized Library System, Main Library, Mogiliv-Podolski 28731
Mohammedan Institute of Teacher Training and Pedagogy, Purwokerto 14289
Mohammedan Institute of Teacher Training and Pedagogy, Purworejo 14290
Mohammedan Institute of Teacher Training and Pedagogy, Yogyakarta 14311
Mohanlal Sukhadia University, Udaipur 13776
– College of Law, Udaipur 13777
– Rajasthan College of Agriculture, Udaipur 13778
Mohave Community College Library System, Kingman 33458
Mohave County Library District, Kingman 39720
Mohawk College of Applied Arts & Technology, Hamilton 03993
Mohawk Valley Community College, Utica 33817
Mohawk Valley Library System, Schenectady 41246
Mohd. Khalid Teachers' Training College, Johor Bharu 18528
Mohinder Singh Randhawa Library, Ludhiana 13693
Mohyla Institute Library, Saskatoon 04488
Moi University, Eldoret 17996
Mokena Community Public Library District, Mokena 40251
MOL Hungarian Oil & Gas Co., Százhalombatta 13502
MOL Magyar Olaz Es Gazipari Rt., Százhalombatta 13502
Molde bibliotek, Molde 20372
Molde University College, Molde 20107
Moldovan Academy of Sciences, Chișinău 18880
Moldovian National Archives Library, Chișinău 18897
Molecular Biology and Genetics Institute, Kyiv 28338
Molesworth Institute Library & Archives, Storrs 37643
Molėtų rajono savivaldybės viešoji biblioteka, Molėtai 18342
Molloy College, Rockville Centre 32475
Mölndals stadsbibliotek, Mölndal 26856
Mölnlycke bibliotek, Mölnlycke 26857
Molochny institut, Vologda 24072
Molodechnenskaya gorodskaya bibliyateka no 1, Molodechno 02207
Molodechnenskaya tsentralnaya gorodskaya biblioteka im. Bogdanovicha, Molodechno 02208
Molodechnenskaya tsentralnaya raionnaya bibliyateka, Molodechno 02209
Molodechnenski Stankostroitelny Zavod, Molodechno 01907
Molodechno City Library no 1, Molodechno 02207
Molodechno District Central Library, Molodechno 02209
Molodechno Machine-tool Construction Plant, Molodechno 01907
Mombasa Polytechnic, Mombasa 17997
Mombusho, Tokyo 17317
Møn Bibliotek, Stege 06933
Monacensia, München 12315
Monaci Camaldolesi, Serra Sant'Abbondio 16068
Monaghan County Library, Clones 14568
MONAliesA – Frauenbibliothek/Genderbibliothek, Leipzig 12195
Monash Medical Centre, Clayton 00905
Monash Mt. Eliza Business School, Mount Eliza 00484
Monash University, Clayton 00420
– Biomedical Library, Clayton 00421
– C. L. Butchers Memorial Library, Parkville 00422
– Caulfield Library, Caulfield 00423
– Gippsland Library, Churchill 00424
– Hargrave Library, Clayton 00425
– Law Library, Clayton 00426
Monastère de Keur Moussa, Dakar 24427
Monastère des Bénédictins du Mont-Fébé, Yaoundé 03646
Monastère Saint-Benoît de Port-Valais, Le Bouveret 27261
Monastère Saint-Remacle, Stavelot 02489
Monasterio Cisterciense de Osera, San Cristovo de Cea 25814
Monasterio Cisterciense de San Isidro de Dueñas, Venta de Baños 25838
Monasterio de Padres Benedictinos, Lazkao 25966
Monasterio de Samos, Samos 25813
Monasterio de San Jerónimo, Cuacos de Yuste 25775
Monasterio de San Juan de Poio, San Juan de Poio 25815
Monasterio de San Salvador de Leyre, Yesa 25846
Monasterio de Santa María de la Vid, La Vid 25840
Monastero di Camaldoli, Camaldoli 15768
Monastero S. Orsola, Gorizia 15853
Monastero S. Silvestro, Fabriano 15810
Monastery of Gertrude, Cottonwood 35171
Monastery of Saint-Saviour, Sayda 18216
Monastery of St. Catherine, Mount Sinai 07131
Monastery of St Sylvester, Fabriano 15810
Monastery Sint Aegten, Sint Agatha 19315
Monastic Library, Ivančna Gorica 24895
Monastic Library, Parma 15965
Moncton Area Lawyers' Association, Moncton 04383
Mondragon High Polytechnic School, Mondragon 25508
Mondragon Unibertsitatea, Mondragon 25508
Monessen Public Library & District Center, Monessen 40255
Monessen Public Library & District Center, Monessen 40256
Monestir de La Real, Palma de Mallorca 25805
Monestir de Poblet, Vimbodí 25843
Monestir de Solius, Santa Cristina d'Aro 25821
Monetary Authority of Singapore MAS, Singapore 24598
Money Museum, Utrecht 19526
Mongol Ulsyn Ih Hural, Ulaanbaatar 18907
Mongolian National Center for Standardization and Metrology, Ulaanbaatar 18911
Mongolian State University of Agriculture, Ulaanbaatar 18905
Monmouth College, Monmouth 31876
Monmouth County Historical Association Library and Archives, Freehold 36880
Monmouth County Library – Eastern, Shrewsbury 41318
Monmouth County Library – Hazlet Branch, Hazlet 39406
Monmouth County Library – Marlboro Branch, Marlboro 40069
Monmouth County Library – Township of Ocean, Oakhurst 40569
Monmouth County Library – Wall, Wall 41710
Monmouth Public Library, Monmouth 40258
Monmouth University, West Long Branch 33033
Monmouthshire Libraries and Information Service, Cwmbran 29967
Mono County Free Library, Mammoth Lakes 40024
Monona Public Library, Monona 40261
Monroe College, Bronx 30559
Monroe Community College, Rochester 33691
Monroe County Community College, Monroe 33550
Monroe County District Library, Woodsfield 41965
Monroe County Historian's Department Library, Rochester 37453
Monroe County Historical Association, Stroudsburg 37645
Monroe County Law Library, Monroe 34551
Monroe County Public Library, Bloomington 38220
Monroe County Public Library, Key West 39713
Monroe Free Library, Monroe 40263
Monroe Public Library, Monroe 40264
Monroe Township Public Library, Monroe Township 40268
Monroeville Public Library, Monroeville 40269
Monrovia Public Library, Monrovia 40270
Monsanto Company, Saint Louis 37489
Mönsterås kommunbibliotek, Mönsterås 26858
Mont Alto Commonwealth College, Mont Alto 31879
Montana Historical Society, Helena 36925
Montana Legislative Reference Center, Helena 34348
Montana School for the Deaf & Blind Library, Great Falls 36902
Montana State Department of Natural Resources & Conservation, Helena 34349
Montana State Library, Helena 30133
Montana State Prison Library, Deer Lodge 34129
Montana State University – Billings Library, Billings 30426
Montana State University – Bozeman, Bozeman 30539
Montana State University-Northern, Havre 31317
Montana Tech of The University of Montana, Butte 30606
Montanuniversität Leoben, Leoben 01247
– Fachbibliothek für Geowissenschaften, Leoben 01248
– RWZ-Bibliothek, Leoben 01249
Montcalm Community College Library, Sidney 33763
Montclair Art Museum, Montclair 37125
Montclair Free Public Library, Montclair 40271
Montclair State University, Montclair 31880
Montedison, Milano 16147
Montefiore Medical Center, Bronx 36593
Montenegrin Academy of Sciences and Arts, Podgorica 18913
Montenegro National Museum Library, Cetinje 18914
Monterey County Free Libraries, Marina 40058
Monterey County Free Libraries – Seaside Branch, Seaside 41262
Monterey County Law Library, Salinas 34796
Monterey Institute of International Studies, Monterey 31881
Monterey Park Bruggemeyer Library, Monterey Park 40273
Monterey Peninsula College, Monterey 33553
Montevideo-Chippewa County Public Library, Montevideo 40274
Montgomery & Andrews, Santa Fe 36257
Montgomery College – Germantown Campus, Germantown 33371
Montgomery College – Rockville Campus, Rockville 33696
Montgomery College – Takoma Park Campus, Takoma Park 33791
Montgomery County Circuit Court, Rockville 34751
Montgomery County Community College, Blue Bell 33178
Montgomery County Law Library, Montgomery 34554
Montgomery County Memorial Library System, Conroe 38683
Montgomery County Memorial Library System – South Branch, The Woodlands 41547
Montgomery County Public Library, Troy 41593
Montgomery County Public Schools Professional Library, Rockville 33697
Montgomery County-Norristown Public Library, Norristown 40493
Montgomery, Mccracken, Walker & Rhoads LLP Library, Philadelphia 36104
Montgomery Watson Harza Library, Chicago 35591
Monticello Union Township Public Library, Monticello 40276
Montpelier Public Library, Montpelier 40281
Montpellier SupAgro, Montpellier 07768
Montreal City Planning Department, Montreal 04099
Montreal Museum of Fine Arts, Montreal 04413
Montreat College, Montreat 33555
Montrose Library District, Montrose 40284
Montserrat College of Art, Beverly 30422
Montville Township Public Library, Montville 40285
Monumenta Germaniae Historica, München 12316
Monumenta Serica, Sankt Augustin 12463
Monumente Nazionale dell'Abbazia di Monte Oliveto Maggiore, Asciano 15710

Monumento Nazionale Abbazia della SS. Trinità, Badia di Cava 15721
Monumento Nazionale Abbazia di Praglia, Teolo 16086
Monumento Nazionale dei Girolamini, Napoli 15929
Monumento Nazionale di Casamari, Casamari 15779
Monumento Nazionale di Farfa, Farfa Sabina 15815
Monumento Nazionale di Monte Oliveto, Siena 16070
Monumento Nazionale di Montecassino, Cassino 15782
Monumento Nazionale di Montevergine, Mercogliano 15898
Monumento Nazionale S. Scolastica, Subiaco 16077
Moody Bible Institute, Chicago 35158
Moon Township Public Library, Moon Township 40286
City of Moorabbin Library, Bentleigh 01068
Moore College of Art & Design, Philadelphia 32257
Moore County Library, Carthage 38470
Moore Memorial Public Library, Texas City 41544
Moore Theological College, Newtown 00785
Moore & Van Allen PLLC, Charlotte 35563
Moorestown Public Library, Moorestown 40289
Mooresville Public Library, Mooresville 40290
Mooresville Public Library, Mooresville 40291
Moorfields Eye Hospital NHS Foundation Trust, London 29245
Moorhead State University, Moorhead 31897
Moorpark College, Moorpark 33557
Mór Wosinksy County Museum, Szekszárd 13509
Móra Ferenc Múzeum, Szeged 13506
Mora folkbibliotek, Mora 26859
Mora lasarett, Mora 26619
Moraine Park Technical College, Fond du Lac 33342
Moraine Valley Community College, Palos Hills 33616
Moravian Archives, Bethlehem 36539
Moravian Archives, Herrnhut 11246
Moravian College & Moravian Theological Seminary, Bethlehem 35136
Moravian Gallery, Library, Brno 06430
Moravian Historical Society, Nazareth 37148
Moravian Library, Brno 06343
Moravian Museum, Brno 06432
Moravian Plant of Chemistry, Ostrava 06420
Moravian-Silesian Research Library in Ostrava, Ostrava 06349
Moravská galerie, Brno 06430
Moravská zemská knihovna, Brno 06343
Moravské chemické závody, a.s., Ostrava 06420
Moravské zemské muzeum, Brno 06432
Moravskoslezská vědecká knihovna v Ostravě, Ostrava 06349
The Moray Council, Elgin 29978
Moray House College, Holyrood Campus, Edinburgh 29094
Mörbylånga kommunbibliotek, Mörbylånga 26860
Mordovian N.P. Ogarev State Library, Saransk 22568
Mordovskaya respublikanskaya meditsinskaya biblioteka, Saransk 23953
Mordovski gosudarstvenny universitet im. N.P. Ogareva, Saransk 22568
Mordovski pedagogicheski institut, Saransk 22569
Mordvinian Medical Library, Saransk 23953
Mordvinian Pedagogical Institute, Saransk 22569
Mordvinian Republic District Museum, Saransk 23954
Morehead State University, Morehead 31899
Morehouse Parish Library, Bastrop 38099
Morehouse School of Medicine, Atlanta 30296
Morekhidne uchilishche im. Shmidta, Kherson 28104
Moreno Valley Public Library, Moreno Valley 40296
Moreton Institute of TAFE, Alexandra Hills 00380
Morgan County Public Library, Martinsville 40085
Morgan, Lewis & Bockius LLP, Philadelphia 36105
Morgan, Lewis & Bockius LLP, Washington 36367
Morgan State University, Baltimore 30341
Morgantown Public Library – Morgantown Service Center, Morgantown 40299
Móricz Zsigmond County and Municipal Library, Nyíregyháza 13543
Móricz Zsigmond Megyei és Városi Könyvtár, Nyíregyháza 13543
Moris Torez Moscow State Linguistic University, Moskva 22365
Moritzburg, Halle (Saale) 11966
Morley College, London 29208
Morling College, Eastwood 00765
Morningside College, Sioux City 32676
Mornington Peninsula Libraries, Rosebud 01151
Morphyspribor Underwater Technologies, Sankt-Peterburg 23808
Morrab Library, Penzance 29875
Morrill Memorial Library, Norwood 40553
Morris Area Public Library District, Morris 40300
Morris College, Sumter 32761
Morris County Law Library, Morristown 34562
Morris County Library, Whippany 41876
Morris, Nichols, Arsht & Tunnell, LLP, Wilmington 36424
Morris Public Library, Morris 40301
Morrison & Foerster LLP Library, Los Angeles 35848
The Morristown & Morris Township Library, Morristown 40302
The Morristown & Morris Township Library – North Jersey History & Genealogy Department, Morristown 40303
Morristown-Hamblen Library, Morristown 40304
Morrisville State College, Morrisville 31908
Morse Institute Library, Natick 40380
Morshansk Chemical Engineering Joint-Stock Company, Morshansk 22818
Morskaya akademiya im. admirala S.O. Makarova, Sankt-Peterburg 22530
Morskaya biblioteka im. admirala M. P. Lazareva, Sevastopol 28462
Morski Instytut Rybacki, Gdynia 21102
Morskoi tekhnicheski litsei, Sankt-Peterburg 22701
Morskoi tekhnicheski universitet, Sankt-Peterburg 22531
Morskoy Gosudarstveny Universitet im. adm. G.I. Nevelskogo, Vladivostok 22637
Morskoyi gidrofizychesky instytut, Katsiveli 28246
Morskoyi gidrofizychesky instytut, Sevastopol 28463
Morsø Folkebibliotek, Nykøbing M. 06909
Mortensen Library, West Hartford 33020
Sir Mortimer B. Davis Jewish General Hospital, Montreal 04414
Morton Arboretum, Lisle 37035
Morton College, Cicero 33231
Morton Grove Public Library, Morton Grove 40309
Morton Public Library District, Morton 40308
Moscovsky District Centralized Library System, Central District Library, Sankt-Peterburg 24278
Moscovsky District Centralized Library System, Library Branch no 3, Specialized Library of Fine Arts Literature with the Exhibition Hall, Sankt-Peterburg 23809
Moscow Automotive Engineering Institute, Moskva 22360
Moscow Aviation Institute, Moskva 22359
Moscow Commerce University, Moskva 22378
Moscow Committee for Science and Technology, Moskva 23357
Moscow District Centralized Library System, Library Branch no 1, Sankt-Peterburg 24274
Moscow District Centralized Library System, Library Branch no 4, Sankt-Peterburg 24275
Moscow District Centralized Library System, Library Branch no 5, Sankt-Peterburg 24276
Moscow District Centralized Library System, Marshak Children's Library Branch no 6, Sankt-Peterburg 24277
Moscow Engineering University of Telecommunication and Information, Moskva 22381
Moscow estate Museum Ostankino, Moskva 23364
Moscow Experimental Station for Fruit Growing, Biryulevo 23001
Moscow Finance Academy, Moskva 23355
Moscow Institute for Economic Statistics, Moskva 22362
Moscow Institute of Municipal Management and Civil Engineering, Moskva 22377
Moscow Institute of Power Engineering, Moskva 22363
Moscow Instrumentation Institute, Moskva 22346
Moscow Leading Institute for Paper Industry Planning Joint-Stock Company, Moskva 22821
Moscow M.V. Lomonosov State University, Moskva 22373
– Botanical Garden, Moskva 22374
– Institute of Mechanics, Moskva 22375
– Zoological Museum, Moskva 22376
Moscow Railways – Information and Computing Centre, Moskva 23354
Moscow Region Central Library System of the City of Kiev – M. Nekrasov Main Library, Kyiv 28677
Moscow Region State Children's Library, Moskva 24172
Moscow Regional Research Clinical Institute, Moskva 23361
Moscow Regional Research Institute of Obstetrics and Gynecology, Moskva 23360
Moscow Regional Scientific and Technical Information Centre, Lyubertsy 23145
Moscow Regional State Scientific Library, Korolyov 22142
Moscow Scientific Research Institute of Psychiatry, Moskva 23359
Moscow State Academy of Applied Biotechnology, Moskva 22349
Moscow State Academy of Automobile and Tractor Engineering, Moskva 22347
Moscow State Academy of Geological Prospecting, Moskva 22351
Moscow State Academy of Printing, Moskva 22348
Moscow State Engineering Academy of Light Industry, Moskva 22352
Moscow State Forest University, Mytishchi 22416
Moscow State Institute of Culture, Khimki 22289
Moscow State Law Academy, Moskva 22354
Moscow State Machine Tools Technological University, Moskva 22368
Moscow State Medical University, Moskva 22388
Moscow State Open University, Moskva 22366
Moscow State Philharmony, Moskva 23353
Moscow State Teacher-Training University, Moskva 22367
Moscow State Textile Academy, Moskva 22353
Moscow State University for Civil Engineering, Moskva 22369
Moscow State University of Economics, Statistics and Informatics (MESI), Moskva 22371
Moscow State University of Geodesy and Cartography, Moskva 22372
Moscow Subway System, Moskva 22846
Moscow Teacher-Training University, Moskva 22380
Moscow Tool Plant, Moskva 22850
Moscow Trade Workers Union, Moskva 23195
Moscow University of Small Business Organization, Mytishchi 22417
Moses Greeley Parker Memorial Library, Dracut 38865
Moses Mendelssohn Zentrum für europäisch-jüdische Studien, Potsdam 12412
Mosfilm Film Corporation, Moskva 23352
Mosfilm Kinokontsern, Moskva 23352
Moskovska raionna TsBS m. Kyiva, Tsentralna biblioteka im. M. Nekrasova, Kyiv 28677
Moskovskaya gosudarstvennaya akademiya avtomobilnogo i traktornogo mashinostroeniya, Moskva 22347
Moskovskaya gosudarstvennaya akademiya pechati, Moskva 22348
Moskovskaya gosudarstvennaya akademiya prikladnoi biotekhnologii, Moskva 22349
Moskovskaya gosudarstvennaya akademiya tonkoi khimicheskoi tekhnologii im. M.V. Lomonosova, Moskva 22350
Moskovskaya gosudarstvennaya filarmoniya, Moskva 23353
Moskovskaya gosudarstvennaya geologorazvedochnaya akademiya, Moskva 22351
Moskovskaya gosudarstvennaya tekhnologicheskaya akademiya legkoi promyshlennosti, Moskva 22352
Moskovskaya gosudarstvennaya tekstilnaya akademiya, Moskva 22353
Moskovskaya gosudarstvennaya yuridicheskaya akademiya, Moskva 22354
Moskovskaya Meditsinskaya Akademiya im. I.M. Sechenova, Moskva 23492
Moskovskaya meditsinskaya akademiya im. I.M. Sechenova, Moskva 22355
Moskovskaya oblastnaya gosudarstvennaya detskaya biblioteka, Moskva 24172
Moskovskaya oblastnaya gosudarstvennaya nauchnaya biblioteka im. N.K. Krupskoi, Korolyov 22142
Moskovskaya plodovo-yagodnaya opytnaya stantsia, Biryulevo 23001
Moskovskaya selskokhozyaistvennaya akademiya im. K.A. Timiryazeva, Moskva 22356
Moskovskaya sittsenabivnaya fabrika, Moskva 22847
Moskovskaya tsentralnaya gorodskaya yunosheskaya biblioteka, Moskva 24173
Moskovskaya veterinarnaya akademiya im. K.I. Skryabina, Moskva 22357
Moskovskaya zheleznaya doroga, Moskva 23354
Moskovski arkhitekturny institut, Moskva 22358
Moskovski aviatsionny institut, Moskva 22359
Moskovski avtomobilestroitelny institut, Moskva 22360
Moskovski avtomobilno-dorozhny institut, Moskva 22361
Moskovski avtomobilny zavod im. I.A. Likhachyova, Moskva 22848
Moskovski ekonomiko-statisticheski institut, Moskva 22362
Moskovski energeticheski institut, Moskva 22363
Moskovski finansovi institut, Moskva 23355
Moskovski gosudarstvenny institut kultury, Khimki 22289
Moskovski gosudarstvenny khudozhestvenno-promyshlenny universitet im. S.G. Stroganova, Moskva 22364
Moskovski gosudarstvenny lingvisticheski universitet im. Morisa Toresa, Moskva 22365
Moskovski gosudarstvenny otkryty universitet, Moskva 22366
Moskovski gosudarstvenny pedagogicheski universitet, Moskva 22367
Moskovski gosudarstvenny stankoin-strumentalny tekhnologicheski universitet, Moskva 22368
Moskovski gosudarstvenny stroitelny universitet, Moskva 22369
Moskovski gosudarstvenny tekhnicheski universitet im. N.E. Baumana, Moskva 22370
Moskovski gosudarstvenny universitet ekonomiki, statistika i informatiki (MESI), Moskva 22371
Moskovski gosudarstvenny universitet geodezii i kartografii, Moskva 22372
Moskovski gosudarstvenny universitet im. M.V. Lomonosova, Moskva 22373
– Botanicheski sad, Moskva 22374
– Nauchno-issledovatelski institut mekhaniki, Moskva 22375
– Zoologicheski muzei, Moskva 22376
Moskovski Gosudarstvenny Universitet Lesa, Mytishchi 22416
Moskovski institut kommunalnogo khozyaistva i stroitelstva, Moskva 22377
Moskovski institut po izyskaniyam i proektirovaniyu inzhenernykh sooruzheni, Moskva 23356
Moskovski instrumentalny zavod 'Kalibr', Moskva 22849
Moskovski instrumentalny zavod (MIZ), Moskva 22850
Moskovski komitet po nauke i tekhnologiyam, Moskva 23357
Moskovski kommercheski universitet, Moskva 22378
Moskovski literaturny institut im. A.M. Gorkogo, Moskva 22379
Moskovski nauchno-issledovatelski institut pediatrii i detskoi khirurgii, Moskva 23358
Moskovski nauchno-issledovatelski institut psikhiatrii, Moskva 23359
Moskovski oblastnoi Nauchno-issledovatelski institut akusherstva i ginekologii, Moskva 23360
Moskovski oblastnoi nauchno-issledovatelski klinicheski institut im. M.F. Vladimirskogo, Moskva 23361
Moskovski oblastnoi tsentr nauchno-tekhnicheskoi informatsii i i propagandy, Lyubertsy 23145
Moskovski pedagogicheski universitet, Moskva 22380
Moskovski tekhnicheski universitet svyazi i informatiki, Moskva 22381
Moskovski universitet potrebitelskoi kooperatsii, Mytishchi 22417

Moskovskoe akademicheskoe khoreograficheskoe uchilishche, Moskva 22680
Moskvich Joint Stock Company, Moskva 22824
Moss bibliotek, Moss 20374
Mosteiro de São Bento, Rio de Janeiro 03246
Mostiska raionna TsBS, Tsentralna biblioteka, Mostiska 28732
Mostiska Regional Centralized Library System, Main Library, Mostiska 28732
Mosul University, Mosul 14449
Motala Public Library, Motala 26861
Motala stadsbibliotek, Motala 26861
Mote Marine Laboratory Library, Sarasota 37581
Motherwell Library, Motherwell 30047
Motilal Nehru Regional Engineering College, Allahabad 13600
Motlow State Community College, Lynchburg 31697
Motor Industrial Corporation, Zaporizhzhya 28176
Motor Industrial Corporation, Trade Union Library, Zaporizhzhya 28888
Motor Transport Institute, Warszawa 21311
Motorcycle Plant, Kyiv 28158
Motorcycle Plant Joint-Stock Company, Serpukhov 22955
Motorcyle and Bicyle Assembling Plant, Minsk 01892
Motorola, Inc, Schaumburg 36259
Mototsikletny i velosipedny zavod, Minsk 01892
Mototsikletny zavod, Kyiv 28158
Mott Community College, Flint 33340
Motts Military Museum, Groveport 36908
Moulmein College, Mawlamyine 18998
Moulton College, Northampton 29285
Moultrie-Colquitt County Library, Moultrie 40311
Mounce & Green, Meyers, Safi & Galatzan, El Paso 35687
Moundsville-Marshall County Public Library, Moundsville 40312
Mount Airy Public Library, Mount Airy 40313
Mount Albert Research Centre, Auckland 19826
Mount Allison University, Sackville 03838
– Alfred Whitehead Memorial Music Library, Sackville 03839
Mount Aloysius College, Cresson 30933
Mount Angel Abbey Library, Saint Benedict 32491
Mount Carmel, Columbus 36741
Mount Carmel Public Library, Mount Carmel 40314
Mount Clemens Public Library, Mount Clemens 40315
Mount Holyoke College Library, South Hadley 32687
Mount Hood Community College, Gresham 33391
Mount Ida College, Newton Center 33582
Mount Isa Institute of TAFE, Mount Isa 00822
Mount Isa Public Library, Mount Isa 01129
Mount Kisco Public Library, Mount Kisco 40316
Mount Laurel Library, Mount Laurel 40317
Mount Makulu Agricultural Research Station, Chilanga 42336
Mount Marty College, Yankton 33102
Mount Mary College, Milwaukee 31839
Mount Mercy College, Cedar Rapids 30690
Mount Olive College, Mount Olive 33563
Mount Olive Public Library, Flanders 39071
Mount Paran Church of God, Atlanta 35116
Mount Pleasant Correctional Facility Library, Mount Pleasant 34564
Mount Pleasant Public Library, Mount Pleasant 40320
Mount Pleasant Public Library, Pleasantville 40830
Mount Prospect Public Library, Mount Prospect 40321
Mount Royal College, Calgary 03680
Mount Saint Bernard Abbey, Leicester 29547
Mount Saint Clare College, Clinton 33241
Mount Saint Mary College, Newburgh 32111
Mount Saint Mary's College, Los Angeles 31650
Mount Saint Mary's University, Emmitsburg 31089
Mount Saint Vincent University, Halifax 03720
Mount San Antonio College, Walnut 33835
Mount San Jacinto College, San Jacinto 33730
Mount Sinai Medical Center, Miami Beach 37101
Mount Sinai School of Medicine, New York 32071
Mount Sinai Services-Queens Hospital Center Affiliation, Jamaica 36973
Mount Stromlo and Siding Spring Observatories, Weston Creek 01048
Mount Union College, Alliance 30201
Mount Vernon City Library, Mount Vernon 40324
Mount Vernon Nazarene University, Mount Vernon 31917
Mount Vernon Public Library, Mount Vernon 40325
Mount Wachusett Community College, Gardner 33368
Mountain Empire Community College, Big Stone Gap 33171
Mountain State University, Beckley 30370
Mountain View College, Dallas 33274
Mountain View Public Library, Mountain View 40330
Mountainside Public Library, Mountainside 40331
Mouvement Français pour le Planning Familial (MFPF), Paris 08591
Mouvement Ouvrier Chrétien, Bruxelles 02608
Movno-informatsiyny fond, Kyiv 28358
Mozarteum, Salzburg 01266
Mozyr Central City Library, Mozyr 02210
Mozyr Land Reclamation Machinery Plant, Mozyr 01910
Mozyr Regional Central Library, Mozyr 02212
Mozyrskaya tsentralnaya gorodskaya bibliyateka, Mozyr 02210
Mozyrskaya tsentralnaya gorodskaya detskaya bibliyateka im. V. Khoruzhei, Mozyr 02211
Mozyrskaya tsentralnaya raionnaya bibiyateka, Mozyr 02212
Mozyrski pedagogicheski institut im. N.K. Krupskoi, Mozyr 01814
M-real Corporation, Örnsköldsvik 26544
MRI Federal Research Institute of Nutrition and Food, location Kiel, Kiel 12114
MSE, Inc, Butte 35548
MsM Information Centre, Maastricht 19488
Ḿstská knihovna Děčín, Děčín 06564
Ḿstská knihovna Šumperk, Šumperk 06608
Mt Lebanon Public Library, Pittsburgh 40803
Mt. Zion Temple Library, Saint Paul 35368
MTA Allatorvos-tudományi Kutatóintézete Könyvtár, Budapest 13432
MTA Nyelvtudományi Intézet, Budapest 13433
MTK and Fincoop Pellervo Information Service, Helsinki 07474
MTK:n ja Pellervon tietopalvelu, Helsinki 07474
MTT Agrifood Research Finland, Jokioinen 07513
MTU Aero Engines GmbH, München 11420
Mubarak Public Library, Giza 07199
Mudanjiang Medical College, Mudanjiang 05420
Mudanjiang Teachers' College, Mudanjiang 05421
Mufulira Teachers Training College, Mufulira 42330
Muhid Ulabi Library, Bauchi 19928
Muhlenberg College, Allentown 30200
Muhlenberg County Libraries, Greenville 39312
Mukachevo City Centralized Library System, Main Library, Mukachevo 28733
Mukachevska miska TsBS, Tsentralna biblioteka, Mukachevo 28733
Mukizaishitsu Kenkyujo, Tsukuba 17794
Mukogawa Joshi Daigaku, Nishinomiya 17072
– Yakugakubu Bunkan, Nishinomiya 17073
Mukogawa Women's University, Nishinomiya 17072
– Pharaceutical Sciences Branch, Nishinomiya 17073
Mukwonago Community Library, Mukwonago 40334
Mulagandha Kuti Vihar Library, Varanasi 14221
Mulawarman University, Samarinda 14292
Multi-Country Posts and Telecommunications Training Centre, Blantyre 18486
Multidisciplinary Center for Earthquake Engineering Research, Buffalo 36612
The Multidisciplinary University College of Copenhagen, Frederiksberg 06658
The Multiprofessional University College in Copenhagen, The Library Herlev, Herlev 06660
Multnomah Bible College, Portland 35336
Multnomah County Library, Portland 40870
Multnomah Law Library, Portland 37403
Mumbai Marathi Granth Sangrahalaya, Mumbai 14099
The Munch Museum Library, Oslo 20250
Münchner Entomologische Gesellschaft e.V., München 12317
Münchner Stadtbibliothek, München 12857
Münchner Stadtbibliothek Am Gasteig – Juristische Bibliothek, München 12318
Münchner Stadtbibliothek Am Gasteig – Musikbibliothek, München 12319
Münchner Stadtmuseum, München 12320
Muncie Center Township Public Library, Muncie 40338
Munger, Tolles & Olson LLP, Los Angeles 35849
The Munich School of Political Science, München 10362
Munich State Coin Collection, München 12327
Municipal Gallery, Bratislava 24689
Municipal Hospital no 7 of Tuberculosis, Moskva 23197
Municipal Hospital of Psychiatry no 1, Moskva 23198
Municipal Institute of Administration and Business Science, Belo Horizonte 03276
Municipal Library, Bratislava 24767
Municipal Library, Hrubieszów 21428
Municipal Library, Selfoss 13584
Municipal Library, Tel-Aviv 14783
Municipal Library and Reading Club 'Metel Ožegović', Varaždin 06261
Municipal Library Česká Lípa, Česká Lípa 06558
Municipal Library Český Těšín, Český Těšín 06559
Municipal Library Děčín, Děčín 06564
Municipal Library in Memory of William and Chia Boorstein, Nahariya 14779
Municipal Library Jindřichův Hradec, Jindřichův Hradec 06575
Municipal Library (Mirdamad), Isfahan 14439
Municipal Library Mladá Boleslav, Mladá Boleslav 06588
Municipal Library of Frýdek-Místek, Frýdek-Místek 06568
Municipal Library of Kavala, Kavala 13144
Municipal Library of Klatovy, Klatovy 06579
Municipal Library of Prague, Praha 06599
Municipal Library of Tabor, Tábor 06609
Municipal Library Sokolov, Sokolov 06606
Municipal Library Třebíč, Třebíč 06612
Municipal Museum, Bratislava 24697
Municipal Museum, Ljubljana 24923
Municipal Museum, Split 06225
Municipal Museum, Varaždin 06227
Municipal Museum, Zagreb 06251
Municipal Public Library, Leszno 21475
Municipal Services Administration, Kyiv 28679
Municipal Technical Advisory Service Library, Knoxville 34422
Municipality Leiden, Leiden 19279
Municipality of Beau-Bassin/Rose-Hill, Rose Hill 18705
Municipial Museum, Kőszeg 13488
Munkácsy Mihály Múzeum, Békéscsaba 13368
Munkebo Bibliotek, Munkebo 06903
Munkedals kommunbibliotek, Munkedal 26862
Munson-Williams-Proctor Arts Institute Library, Utica 37701
Münzkabinett, Berlin 11573
Murberget, Härnösand 26587
Murcia University, Murcia 25512
Murdoch University, Murdoch 00487
Mures County Library, Târgu Mureş 22108
Müritzeum, Waren 12577
Murmansk Administration of Hydro-Meteorological Services, Murmansk 22760
Murmansk Marine Biological Institute of the Russian Academy of Sciences, Dalnie Zelentsy 23023
Murmansk Marine Biology Institute, Murmansk 23601
Murmansk State Regional Children's Library, Murmansk 24178
Murmansk State Regional Universal Scientific Library, Murmansk 22156
Murmanskaya gosudarstvennaya oblastnaya detskaya biblioteka, Murmansk 24178
Murmanskaya gosudarstvennaya oblastnaya universalnaya nauchnaya biblioteka, Murmansk 22156
Murmanskaya oblastnaya yunosheskaya biblioteka, Murmansk 24179
Murmanski gosudarstvenny pedagogicheski universitet, Murmansk 22415
Murmanski morskoi biologicheski institut Kolskogo filiala im. S.M. Kirova RAN, Dalnie Zelentsy 23023
Murmanski morskoi biologicheski institut RAN, Murmansk 23601
Murmansk State Pedagogical University, Murmansk 22415
Murmanskoe upravlenie gidrometsluzhby, Murmansk 22760
Muroran Institute of Technology, Muroran 17047
Murovani Kurilivtsi Regional Centralized Library System, Main Library, Murovany Kurilivtsi 28734
Murovanokurilovetska raionna TsBS, Tsentralna biblioteka, Murovany Kurilivtsi 28734
Murphy, Sheneman, Julian & Rogers, San Francisco 36241
Murphy-Jahn Library, Chicago 35592
Murray Public Library, Murray 40345
Murray State University, Murray 31923
Murrieta Public Library, Murrieta 40346
Murrysville Community Library, Murrysville 40347
Murtha, Cullina, Richter & Pinney Library, Hartford 35739
Musashino Academia Musicae, Tokyo 17158
Muscat Technical Industrial College, Ruwi 20422
Musea Brugge, Brugge 02549
Musée Basque et d'Histoire de Bayonne, Bayonne 08292
Musée Cantini, Marseille 08410
Musée Cernuschi, Paris 08592
Musée Condé, Chantilly 08323
Musée Condé, Chantilly 08324
Musée Curtius, Aketi 06065
Musée d'Aquitaine, Bordeaux 08307
Musée d'Archéologie Méditerranéenne, Marseille 08411
Musée d'Archéologie Nationale, Saint-Germain-en-Laye 08669
Musée d'Art et d'Histoire de Provence et Musée Fragonard, Grasse 08352
Musée d'art et d'histoire du Judaïsme, Paris 08593
Musée d'Art Moderne de la Ville de Paris, Paris 08594
Musée d'Art Moderne Lille Métropole, Villeneuve d'Asq 08716
Musée Dauphinois, Grenoble 08356
Musée de Cirta, Constantine 00078
Musée de Grenoble, Grenoble 08357
Musée de la Civilisation – Bibliothèque du Séminaire de Québec, Québec 04462
Musée de la Marine, Paris 08595
Musée de la Poste, Paris 08596
Musée de la Résistance et de la Déportation, Besançon 08295
Musée de la Révolution Française – Domaine de Vizille, Vizille 08718
Musée de la Vie Wallonne, Liège 02680
Musée de l'Abbaye Sainte-Croix, Les Sables-d'Olonne 08196
Musée de l'Air et de l'Espace, Le Bourget 08373
Musée de l'Armée, Paris 08597
Musée de l'Homme, Paris 08598
Musée de l'Impression sur Etoffes, Mulhouse 08430
Musée de Zoologie, Lausanne 27398
Musée Départemental Maurice Denis 'Le Prieuré', Saint-Germain-en-Laye 08670
Musée des Arts et Tradition, Fès 18950
Musée des Beaux-Arts, Bordeaux 08308
Musée des Beaux-Arts, Dijon 08347
Musée des Beaux-Arts, Grenoble 08358
Musée des Beaux-Arts, Lille 08384
Musée des Beaux-Arts, Lyon 08399
Musée des Beaux-Arts de Rennes, Rennes 08655
Musée des Beaux-Arts du Canada, Ottawa 04449
Musée des Beaux-Arts et Musée historique d'Orléans, Orléans 08458
Musée des Instruments de Musique / Muziekinstrumentenmuseum (MIM), Bruxelles 02609
Musée des Oudaïa, Rabat 18963
Musée des Tissus et des Arts décoratifs, Lyon 08400
Musée d'ethnographie, Genève 27369
Musée d'histoire, La Chaux-de-Fonds 27388
Musée d'Histoire de Marseille, Marseille 08412
Musée d'histoire des sciences, Genève 27370
Musée d'histoire naturelle, La Chaux-de-Fonds 27389
Musée d'Histoire Naturelle, Nice 08449
Musée Dobrée, Nantes 08444
Musée d'Orsay, Paris 08599
Musée du Petit Palais, Paris 08600
Musée Elsa Triolet-Louis Aragon, Saint-Arnoult-en-Yvelines 08661
Musée et Jardins Botaniques Cantonaux, Lausanne 27399
Musée Fragonard, Grasse 08352
Musée Français du Chemin de Fer, Mulhouse 08431

Musée gruérien et Bibliothèque de Bulle, Bulle 27339
Musée historique de la Réformation, Genève 27371
Musée Horta, Bruxelles 02610
Musée Lecoq, Clermont-Ferrand 08338
Musée Lorrain, Nancy 08438
Musée monétaire cantonal, Lausanne 27400
Musée National de la Marine, Rochefort 08656
Musée National de l'Education, Mont-Saint-Aignan 08428
Musée National de Préhistoire, Les Eyzies-de-Tayac 08380
Musée National des Antiquités, Alger 00067
Musée national des Arts asiatiques – Guimet, Paris 08601
Musée National des Arts et Traditions Populaires (MNATP), Paris 08602
Musée National des Beaux Arts d'Alger, Alger 00068
Musée National des Beaux-Arts du Québec, Québec 04463
Musée National des Douanes, Bordeaux 08309
Musée National des Monuments Français, Paris 08603
Musée National d'Histoire et d'Art (MNHA), Luxembourg 18414
Musée National du Château de Pau, Pau 08637
Musée National du Mali, Bamako 18654
Musée National du Moyen-Age, Paris 08604
Musée Océanographique, Monaco 18899
Musée Paul Arbaud, Aix-en-Provence 08271
Musée Paul-Dupuy, Toulouse 08696
Musée Pyrénéen, Lourdes 08389
Musée Rodin, Paris 08605
Musée Royal de l'Armée et d'Histoire Militaire, Brussel 02604
Musée Royal de Mariemont, Morlanwelz 02689
Musée Social, Paris 08606
Musée Voltaire, Genève 27365
Museen der Stadt Dresden, Stadtmuseum, Dresden 11770
Museen der Stadt Dresden, Technische Sammlungen, Dresden 11771
Museen für Kunst und Kulturgeschichte der Hansestadt Lübeck, Lübeck 12214
Musées cantonaux / Walliser Kantonsmuseen, Sion 27435
Musées d'Art et d'Histoire de Troyes, Troyes 08702
Musées municipaux, Mulhouse 08432
Musées royaux d'art et d'histoire / Koninklijke Musea voor Kunst en Geschiedenis, Bruxelles 02611
Musées royaux des Beaux-Arts de Belgique / Koninklijke Musea voor Schone Kunsten van België, Bruxelles 02612
Museet for samtidskunst, Oslo 20251
Musei Civici Archeologico P. Giovio e Storico G. Garibaldi, Como 16233
Musei Civici di Rovereto, Rovereto 16566
Museo Archeologico Nazionale, Cividale del Friuli 16231
Museo Archeologico Nazionale di Aquileia, Aquileia 16167
Museo Archeologico Regionale, Palermo 16403
Museo Argentino de Ciencias Naturales "B. Rivadavia", Buenos Aires 00274
Museo Arqueológica San Miguel de Azapo, Arica 04917
Museo Arqueológico de Córdoba, Córdoba 25943
Museo Arqueológico de Ibiza y Formentera, Eivissa 25955
Museo Arqueológico de Sevilla, Sevilla 26154
Museo Arqueológico Nacional, Madrid 26065
Museo Arqueológico 'Rafael Larco Herrera', Lima 20670
Museo Arqueolóxico e Histórico, A Coruña 25950
Museo 'Balaguer', Villanueva y Geltru 26183
Museo Botánico, Córdoba 00299
Museo Cabrera, Ica 20654
Museo Canario, Las Palmas 26120
Museo Carnico delle Arti Popolari "M. Gortani", Tolmezzo 16587
Museo Cerralbo, Madrid 26066
Museo Civico, Merano 16314
Museo Civico Archeologico, Bologna 16195
Museo Civico Bolzano, Bolzano 16203
Museo Civico Correr, Venezia 16654
Museo Civico del Risorgimento, Bologna 16196
Museo Civico di Feltre, Feltre 16242
Museo Civico di Scienze Naturali, Brescia 16210
Museo Civico di Storia Naturale Giacomo Doria, Genova 16290
Museo Civico di Storia Naturale, Verona 16668
Museo Civico di Storia Naturale di Milano, Milano 16343
Museo Civico 'Gaetano Filangieri', Napoli 16377
Museo Civico Pinacoteca, Vicenza 16672
Museo Colonial e Histórico 'Enrique Udaondo', Luján 00303
Museo de Albacete, Albacete 25869
Museo de América, Madrid 26067
Museo de Arte de Lima, Lima 20671
Museo de Arte e Historia, Durango 25954
Museo de Arte Moderno, Santo Domingo 06967
Museo de Bellas Artes de Alava, Vitoria-Gasteiz 26188
Museo de Bellas Artes de Asturias, Oviedo 26108
Museo de Bellas Artes de Bilbao, Bilbao 25933
Museo de Bellas Artes de Córdoba, Córdoba 25944
Museo de Bellas Artes de Sevilla, Sevilla 26155
Museo de Bellas Artes de Valencia, Valencia 26174
Museo de Ciencias Naturales y Antropológicas 'Juan Cornelio Moyano', Mendoza 00306
Museo de Ciencias Naturales y Antropológicas 'Prof. Antonio Serrano', Paraná 00307
Museo de Historia Natural, Lima 20672
Museo de Historia Natural de San Pedro Nolasco, Santiago 04936
Museo de Historia Natural 'Felipe Poey', La Habana 06308
Museo de Historia Natural y Etnografía, Asunción 20609
Museo de la Farmacia Hispana, Madrid 26068
Museo de La Plata, La Plata 00300
Museo de La Rioja, Logroño 25970
Museo de las Casas Reales, Santo Domingo 06968
Museo de Mallorca, Palma de Mallorca 26114
Museo de Navarra, Pamplona 26123
Museo de Pontevedra, Pontevedra 26125
Museo de Prehistoria y Arquelogía de Cantabria, Santander 26141
Museo de San Isidro, Madrid 26069
Museo de Zaragoza, Zaragoza 26194
Museo del Ejército, Madrid 26070
Museo del Hombre Dominicano, Santo Domingo 06969
Museo del Paesaggio, Verbania – Pallanza 16660
Museo del Risorgimento e della Resistenza, Vicenza 16673
Museo del Sannio, Benevento 16177
Museo del Trate, Madrid 26071
Museo della Fondazione Mormino, Palermo 16404
Museo dell'Attore, Genova 16291
Museo di Castelvecchio, Verona 16669
Museo di Cavalleria e d'Africa, Siracusa 16581
Museo di Storia Naturale di Venezia, Venezia 16655
Museo do Pobo Galego, Santiago de Compostela 26143
Museo Doutor Joaquim de Carvalho, Inhambane 18979
Museo e Gallerie Nazionali di Capodimonte, Napoli 16378
Museo e Instituto de Humanidades Camón Aznar, Zaragoza 26195
Museo Etnográfico, Buenos Aires 00275
Museo Etnográfico 'Andres Barbero', Asunción 20610
Museo Geológico del Seminario de Barcelona, Barcelona 25924
Museo Histórico 'Martiniano Leguizamón', Paraná 00308
Museo Histórico Nacional, Buenos Aires 00276
Museo Histórico Nacional, Montevideo 42062
Museo Histórico Nacional de Chile, Santiago 04937
Museo Histórico Provincial de Rosario 'Dr Julio Marc', Rosario 00310
Museo Histórico Sarmiento, Buenos Aires 00277
Museo Indígeno, San José 06123
Museo Internazionale delle Ceramiche in Faenza, Faenza 16241
Museo Internazionale e Biblioteca della Musica, Bologna 16197
Museo Lázaro Galdiano, Madrid 26072
Museo Mitre, Buenos Aires 00278
Museo Municipal de Arte Espanñol 'Enrique Larreta', Buenos Aires 00279
Museo Nacional Centro de Arte Reina Sofía, Madrid 26073
Museo Nacional Colegio de San Gregorio, Valladolid 26180
Museo Nacional de Antropologia, Madrid 26074
Museo Nacional de Antropología, México 18861
Museo Nacional de Arqueología, Antropologia e Historia del Peru, Lima 20673
Museo Nacional de Arte Romano, Mérida 26105
Museo Nacional de Artes Decorativas, Madrid 26075
Museo Nacional de Bellas Artes, Buenos Aires 00280
Museo Nacional de Bellas Artes, Santiago 04938
Museo Nacional de Cerámica y Artes Suntuarias "González Martí", Valencia 26175
Museo Nacional de Ciencias Naturales, Madrid 26076
Museo Nacional de Costa Rica, San José 06124
Museo Nacional de Historia Natural, Montevideo 42063
Museo Nacional de Historia Natural, Santiago 04939
Museo Nacional de Historia Natural, Santo Domingo 06970
Museo Nacional de la Cultura Peruana, Lima 20674
Museo Nacional de Reproducciones Artísticas, Madrid 26077
Museo Nacional de Soares dos Reis, Porto 21825
Museo Nacional del Prado, Madrid 26078
Museo Naval, Madrid 26053
Museo Naval del Perú 'J. J. Elias Murguía', Callao 20652
Museo Nazionale d'Arte Orientale 'Giuseppe Tucci', Roma 16547
Museo Nazionale del Cinema, Torino 16603
Museo Nazionale del Risorgimento Italiano, Torino 16608
Museo Nazionale della Scienza e della Tecnologia, Milano 16344
Museo Nazionale delle Arti e Tradizioni Popolari, Roma 16548
Museo Nazionale di Castel Sant'Angelo, Roma 16549
Museo Nazionale di Reggio Calabria, Reggio Calabria 16431
Museo Numantino, Soria 26159
Museo Pablo Gargallo, Zaragoza 26196
Museo Pedagógico 'Carlos Stuardo Ortiz', Santiago 04940
Museo Pérez Galdos, Las Palmas 26118
Museo Poldi Pezzoli, Milano 16345
Museo Popular Juan N. Madero, San Fernando 00311
Museo Postal y Telegráfico, Madrid 26079
Museo Preistorico ed Etnografico L. Pigorini, Roma 16550
Museo Provincial de Ciencias Naturales 'Florentino Ameghino', Santa Fé 00316
Museo Provincial de Lugo, Lugo 25971
Museo Provincial de Teruel, Teruel 26164
Museo Provinciale Campano di Capua, Capua 16219
Museo Regional de Atacama, Copiapó 04918
Museo Regional Michoacáno Instituto Nacional Paleologia & Historia, Morelia 18867
Museo Romántico, Madrid 26080
Museo Social Argentino, Buenos Aires 00281
Museo Storico in Trento, Trento 16614
Museo Storico Italiano della Guerra, Rovereto 16567
Museo Teatrale alla Scala, Milano 16346
Museo Tridentino di Scienze Naturali, Trento 16615
Museo y Archivo Histórico Municipal, Montevideo 42064
Museom of Asian Art, Collection of East Asian Art, Berlin 11576
Muséon Arlaten, Arles 08282
Museoviraston kirjasto, Helsinki 07475
Museu Antropológico, Coimbra 21628
Museu Biblioteca Condes de Castro Guimarães, Cascais 21755
Museu Carlos Costa Pinto, Salvador 03380
Museu da Casa de Bragança, Vila Viçosa 21829
Museu da Guiné-Bissau, Bissau 13192
Museu da Inconfidência, Ouro Prêto 03312
Museu da República, Rio de Janeiro 03363
Museu d'Arqueologia de Catalunya, Barcelona 25920
Museu de Arquelogia e Etnologia (MAE), São Paulo 03417
Museu de Arte da Bahia, Salvador 03381
Museu de Arte de São Paulo, São Paulo 03418
Museu de Arte Moderna, Rio de Janeiro 03364
Museu de Badalona, Badalona 25875
Museu de Ciência, Lisboa 21679
Museu de Ciències Naturals, Barcelona 25921
Museu de Marinha, Lisboa 21809
Museu de Prehistòria de València, Valencia 26176
Museu de Zoologia, São Paulo 03419
Museu d'Història de la Ciutat, Barcelona 25922
Museu do Cinema, Lisboa 21782
Museu do Dundo, Luanda 00107
Museu do Indio/FUNAI, Rio de Janeiro 03365
Museu e Archivo Histórico do Banco do Brasil, Rio de Janeiro 03366
Museu Etnològic, Barcelona 25923
Museu Francisco Tavares Proença Júnior, Castelo Branco 21756
Museu Geològic del Seminari de Barcelona, Barcelona 25924
Museu Histórico Nacional, Rio de Janeiro 03367
Museu Imperial, Petrópolis 03313
Museu Joaquim de Carvalho, Maputo 18982
Museu Marítim, Barcelona 25925
Museu Municipal Amadeo de Souza Cardoso, Amarante 21754
Museu Municipal do Funchal (História Natural), Funchal 21761
Museu Municipal 'Dr. Santos Rocha', Figueira da Foz 21759
Museu Nacional, Rio de Janeiro 03368
Museu Nacional Arqueològic de Tarragona, Tarragona 26161
Museu Nacional d'Art de Catalunya, Barcelona 25926
Museu Nacional de Arqueologia, Lisboa 21810
Museu Nacional de Arte Antiga, Lisboa 21811
Museu Nacional de Belas Artes, Rio de Janeiro 03369
Museu Paraense Emílio Goeldi, Belém 03275
Museu Paulista, São Paulo 03420
Museum Aargau, Lenzburg 27403
Museum Alexander Koenig, Bonn 11652
Museum am Burghof, Lörrach 12207
Museum and Archives of the Stanislaw Moniuszko Warsaw Musical Society, Warszawa 21317
Museum and Institute of Zoology of the Polish Academy of Sciences, Warszawa 21318
Museum Bautzen, Bautzen 11475
Museum Boerhaave, Leiden 19476
Museum Boijmans Van Beuningen, Rotterdam 19515
Museum Catharijneconvent, Utrecht 19527
Museum Charlottenburg-Wilmersdorf, Berlin 11551
Museum Chasa Jaura, Valchava 27443
Museum der Arbeit, Hamburg 12004
Museum der bildenden Künste Leipzig, Leipzig 12196
Museum der Kulturen und Ethnologisches Seminar, Basel 27308
Museum der Moderne, Salzburg 01570
Museum der Stadt Güstrow, Güstrow 11946
Museum der Weltkulturen, Frankfurt am Main 11863
Muséum des Sciences Naturelles et de Préhistoire, Chartres 08328
Muséum d'histoire naturelle, Genève 27372
Museum d'Histoire Naturelle, Le Havre 08377
Muséum d'Histoire Naturelle, Nantes 08445
Museum dr. Guislain, Gent 02650
Museum Eschborn, Eschborn 11820
Museum Europäischer Kulturen, Berlin 11574
Museum Folkwang, Essen 11825
Museum for Modern Art Arnhem, Arnhem 19395
Museum for Science and Technology, Budapest 13446
Museum für Angewandte Kunst, Frankfurt am Main 11864
Museum für Asiatische Kunst, Ostasiatische Kunstsammlung, Berlin 11576
Museum für Asiatische Kunst, Süd-, Südost- und Zentralasiatische Kunstsammlung, Berlin 11575
Museum für Bergedorf und die Vierlande, Hamburg 12017
Museum für das Fürstentum Lüneburg, Lüneburg 12220
Museum für Gestaltung, Berlin 11493
Museum für Hamburgische Geschichte, Hamburg 12005
Museum für Islamische Kunst, Berlin 11577

Museum für Kommunikation Berlin, Berlin 11552
Museum für Kommunikation Frankfurt, Frankfurt am Main 11865
Museum für Kunst und Gewerbe, Hamburg 12006
Museum für Kunst und Kulturgeschichte Dortmund, Dortmund 11750
Museum für Literatur am Oberrhein, Karlsruhe 12093
Museum für Moderne Kunst, Frankfurt am Main 11866
Museum für Naturkunde, Berlin 09453
Museum für Naturkunde, Münster 12352
Museum für Naturkunde der Stadt Dortmund, Dortmund 11751
Museum für Naturkunde und Vorgeschichte Dessau, Dessau-Roßlau 11733
Museum für Ostasiatische Kunst, Köln 12148
Museum für Spätantike und Byzantinische Kunst, Berlin 11582
Museum für Völkerkunde, Wien 01622
Museum für Völkerkunde Dresden, Dresden 11772
Museum für Völkerkunde Hamburg, Hamburg 12007
Museum für Völkerkunde zu Leipzig, Leipzig 12184
Museum für Vor- und Frühgeschichte, Berlin 11578
Museum Haldensleben, Haldensleben 11954
Museum Het Valkhof, Nijmegen 19499
Museum im Frey-Haus, Brandenburg an der Havel 11656
museum kunst palast, Düsseldorf 11803
Museum Meermanno-Westreenianum, Den Haag 19442
Museum Moderner Kunst Stiftung Ludwig Wien, Wien 01623
Muséum National d'Histoire Naturelle, Paris 08607
Muséum National d'Histoire Naturelle – Bibliothèque Centrale, Paris 08608
Muséum national d'Histoire naturelle – Laboratoire de Cryptogamie, Paris 08609
Muséum National d'Histoire Naturelle – Laboratoire d'Entomologie, Paris 08610
Museum Natura Docet, Denekamp 19414
Museum Nienburg/Weser, Nienburg 12366
Museum of Anthropology Library, Vancouver 04541
Museum of Applied Arts, Beograd 24518
Museum of Applied Arts, Budapest 13385
Museum of Applied Arts and Sciences, Ultimo 01039
Museum of Architecture, Wrocław 21363
Museum of Arts, Tbilisi 09268
Museum of Arts and Crafts, Zagreb 06252
Museum of Asian Art, Collection of South, Southeast- and Central Asian Art, Berlin 11575
Museum of Bosanska krajina Region, Banja Luka 02959
Museum of Brodsko Posavlje, Slavonski Brod 06220
Museum of Central Slovakia, Banská Bystrica 24679
Museum of City Sculpture, Sankt-Peterburg 23813
Museum of Classical Archaeology, Cambridge 28994
Museum of Contemporary Art Leipzig Foundation, Leipzig 12203
Museum of Contemporary Art Library, Chicago 36682
Museum of Cultural History, Oslo 20248
Museum of Czech Literature, Praha 06513
Museum of Decorative Arts in Prague, Library, Praha 06517
Museum of Dominican Man, Santo Domingo 06969
Museum of Earth Science of the Moscow State M.V. Lomonosov University, Moskva 23362
Museum of Eastern Slovakia, Košice 24736
Museum of Electrical Engineering, Budapest 13396
Museum of English Rural Live Library, Reading 29381
Museum of Ethnography, Stockholm 26668
Museum of Ethnology, Leiden 19477
Museum of Faculty of Letters, Kyoto University, Kyoto 17588
Museum of Far Eastern Antiquities, Stockholm 26669
Museum of Fine Arts, Omsk 23664
Museum of Fine Arts, Saint Petersburg 37508
Museum of Fine Arts, Boston – W. Van Alan Clark Jr Library, Boston 36575
Museum of Fine Arts, Boston – William Morris Hunt Memorial Library, Boston 36576
Museum of Fine Arts of Asturias, Oviedo 26108
Museum of Fine Arts of Valencia, Valencia 26174
Museum of Finnish Architecture Library, Helsinki 07493
Museum of Flight Library, Seattle 37591
Museum of Folk Architecture in Sanok, Sanok 21219
Museum of Glass and Jewellery, Jablonec nad Nisou 06452
Museum of Gorica Region, Nova Gorica 24950
Museum of History and Arts, Pskov 23706
Museum of History, Arts and Architecture, Veliki Novgorod 24040
Museum of History, Culture and Arts of Uzbekistan, Samarkand 42121
Museum of History of Moscow, Moskva 23363
Museum of Hungarian Agriculture, Budapest 13400
Museum of Industrial Archaeology and Textile, Gent 02649
Museum of International Folk Art, Santa Fe 37571
Museum of Islamic Art, Cairo 07180
Museum of Jewish Heritage, New York 37232
Museum of King St. Stephen, Székesfehérvár 13508
Museum of La Rioja Library, Logroño 25970
Museum of Literature and Music Banská Bystrica, Banská Bystrica 24677
Museum of London, London 29792
Museum of Macedonia, Skopje 18444
Museum of Modern Art, Ljubljana 24924
Museum of Modern Art, New York 37233
Museum of Modern Japanese Literature, Tokyo 17729
Museum of National History and Archeology, Constanța 22027
Museum of New Mexico – Museum of Indian Arts & Culture-Laboratory of Anthropology, Santa Fe 37572
Museum of New Mexico – Palace of the Governors-Fray Angelico Chavez History Library, Santa Fe 37573
Museum of New Zealand, Wellington 19850
Museum of Nieborów and Arcadia, Nieborów 21173
Museum of Northern Arizona-Harold S. Colton Memorial Library, Flagstaff 36852
Museum of Northern Nations History and Culture, Yakutsk 24081
Museum of Old Techniques, Grimbergen 02661
Museum of Performing Arts of Serbia, Beograd 24517
The Museum of Puppets, Chrudim 06444
Museum of Regional Ethnology, Antwerpen 02518
Museum of Regional Studies, Mykolaiv 28430
Museum of Regional Studies, Poltava 28449
Museum of Regional Studies, Pyatigorsk 23712
Museum of Regional Studies, Termez 42124
Museum of Regional Studies, Tyumen 24015
Museum of Regional Studies, Zlatoust 24097
Museum of Russian Art in Kyiv, Kyiv 28356
Museum of Russian Culture, Inc Library, San Francisco 37548
Museum of Science and Technology, Oslo 20271
Museum of Science & Industry, Tampa 37657
Museum of Skofja Loka, Škofja Loka 24953
Museum of South-Eastern Moravia in Zlín, Zlín 06550
Museum of Technology, Warszawa 21324
Museum of the Chinese Revolution, Beijing 05850
Museum of the Confederacy, Richmond 37442
Museum of the Czech Countryside, Nové Dvory 06467
Museum of the Earth, Warszawa 21326
Museum of the Fur Trade Library, Chadron 36642
Museum of the Great Plains, Lawton 37022
'Museum of the Highland' Jihlava, Jihlava 06454
Museum of the History of Architecture, Nizhni Novgorod 23617
Museum of the History of Azerbaijan, Baku 01718
Museum of the History of Democratic Ideals and Culture, Ouro Prêto 03312
Museum of the History of Religion – Kazan Cathedral, Sankt-Peterburg 23814
Museum of the History of Riga and Navigation, Riga 18190
Museum of the History of Science, Oxford 29870
Museum of the Municipality of Bucharest, București 22005
Museum of the Royal Houses, Santo Domingo 06968
Museum of the Serbian Orthodox Church, Dubrovnik 06213
Museum of the Slovak National Uprising, Banská Bystrica 24678
Museum of the Textile and Clothing Industry, Budapest 13461
Museum of Transport and Technology (MOTAT), Auckland 19827
Museum of Transportation, Saint Louis 37490
Museum of Varmia and Mazury, Olsztyn 21176
Museum of Vojvodina, Novi Sad 24530
Museum of Work, Norrköping 26621
Museum Plantin-Moretus / Prentenkabinet, Antwerpen 02521
Museum Reichenfels, Hohenleuben 12065
Museum Rietberg, Zürich 27455
Museum Schloss Bernburg, Bernburg 11605
Museum 'Schloss Moritzburg', Zeitz 12622
Museum Schloss Moyland, Bedburg-Hau 11484
Museum Schloss Wilhelmsburg, Schmalkalden 12473
Museum Schnütgen, Köln 12149
Museum Silesiae, Opava 06470
Museum Support Center Library, Suitland 37647
Museum und Galerie im Prediger, Schwäbisch Gmünd 12476
Museum van Hedendaagse Kunst Antwerpen (MUHKA), Antwerpen 02522
Museum Victoria, Carlton 00900
Museum Vleeshuis, Antwerpen 02523
Museum Volkenkunde, Leiden 19477
Museum voor Communicatie, Den Haag 19443
Museum voor de Oudere Technieken (MOT), Grimbergen 02661
Museum voor Industriële Archeologie en Textiel, Gent 02649
Museum voor Moderne Kunst Arnhem, Arnhem 19395
Museum voor Schone Kunsten Gent (MSKG), Gent 02651
Museum Weißenfels Schloß Neu-Augustusburg, Weißenfels 12585
Museum Wiesbaden, Wiesbaden 12596
Museum Yamato Bunkakan, Nara 17619
Museumpark, Rotterdam 19195
Museumsamt für Westfalen, Münster 11061
Museumsberg Flensburg, Flensburg 11834
Museumsbibliothek der Stiftung Museum Schloss Moyland, Sammlung van der Grinten, Joseph Beuys Archiv des Landes Nordrhein-Westfalen, Bedburg-Hau 11484
MuseumsCenter / Kunsthalle Leoben, Leoben 01556
Museumsgesellschaft Tübingen e.V., Tübingen 12563
Museumsgesellschaft Zürich, Zürich 27456
Museumslandschaft Hessen Kassel, Kassel 12101
Music College, Ivano-Frankivsk 28100
Music Library of Greece 'Lilian Voudouri', Athinai 13119
Music Library of Sweden, Stockholm 26679
Music Library VRO-VRK, Brussel 02613
Music Pedagogical School, Gomel 01821
Music School, Brest 01820
Music School, Kemerovo 22670
Musical Instruments Museum, Bruxelles 02609
Musical Pedagogical Institute, Rostov-na-Donu 22490
Musick, Peeler & Garrett Library, Los Angeles 35850
Musik- und Stadtteilbibliothek Bundesallee, Berlin 11585
Musikak-Ademie der Stadt Basel, Basel 26989
Musikakademie der Stadt Kassel, Kassel 10745
Musikakademie für Bildung und Aufführungspraxis, Blankenburg (Harz) 11615
Musikbibliothek am Mailänder Platz, Stuttgart 12535
Musikhochschule Lübeck, Lübeck 10250
Musikmuseet – Musikhistorisk Museum og Carl Claudius' Samling, København 06813
Muskego Public Library, Muskego 40350
Muskegon Area District Library, Muskegon 40352
Muskegon Area District Library – Norton Shores Jacob O. Funkhouser Branch, Muskegon 40353
Muskegon Community College, Muskegon 33566
Muskingum College, New Concord 31969
Muskingum County Law Library, Zanesville 35098
Muskingum County Library System, Zanesville 42004
Muskogee Law Library Association, Muskogee 34569
Muskogee Public Library, Muskogee 40355
Muskoka Lakes Library Board, Port Carling 04722
Muslim Education & Welfare Foundation of Canada, Surrey 04501
Musorgsky Theatre of Opera and Ballet, Music Library, Sankt-Peterburg 23907
Musser Public Library, Muscatine 40349
Mussorgsky Uralian Conservatoire, Ekaterinburg 22242
Muszaki es Gazdasagi Szakkonyvtar, Kecskemét 13280
Mu'tah University, Mu'tah 17878
Muzeul Țării Crișurilor, Oradea 22049
Muzei antropologii i etnografii im. Petra Velikogo (Kunstkamera) RAN, Sankt-Peterburg 23810
Muzei A.S. Pushkina, Sankt-Peterburg 23811
Muzei etnografii, Sankt-Peterburg 23812
Muzei gorodskoi skulptury, Sankt-Peterburg 23813
Muzei im. N.M. Martyanova, Minusinsk 23160
Muzei istorii i kultury narodov severa, Yakutsk 24081
Muzei istorii Moskvy, Moskva 23363
Muzei istorii religii – Kazanski sobor, Sankt-Peterburg 23814
Muzei izobrazitelnykh iskusstv, Omsk 23664
Muzei narodnoho mystetstva gutsulshchini ta pokuttya im. J. Kobrynskoho, Kolomiya 28294
Muzei svyazi im. A.S. Popova, Sankt-Peterburg 23815
Muzei usadba Ostankino, Moskva 23364
Muzei V.I. Lenina, Moskva 23365
Muzei V.V. Mayakovskogo, Moskva 23366
Muzei-usadba 'Arkhangelskoe', Arkhangelskoe 22990
Muzei-usadba L.N. Tolstogo Yasnaya Polyana, Yasnaya Polyana 24091
Muzei-zapovednik Abramtsevo, Abramtsevo 22982
Muzej Bosanske krajine, Banja Luka 02959
Muzej Brodskog Posavlja, Slavonski Brod 06220
Muzej Dubrovačkog Pomorstva, Dubrovnik 06212
Muzej Grada Beograda, Beograd 24516
Muzej Grada Šibenika, Šibenik 06218
Muzej Grada Splita, Split 06225
Muzej grada Zagreba, Zagreb 06251
Muzej na Makedonija, Skopje 18444
Muzej novejše zgodovine Slovenije, Ljubljana 24925
Muzej Pozorišne Umjetnosti Srbije, Beograd 24517
Muzej primenjene umetnosti, Beograd 24518
Muzej Srpske Pravoslavne Crkve, Dubrovnik 06213
Muzej Vojvodine, Novi Sad 24530
Muzej za umjetnost i obrt, Zagreb 06252
Muzeul Brăilei, Brăila 21962
Muzeul Brukenthal, Sibiu 22057
Muzeul Civilizatiei Dacice Și Romane Deva, Deva 22032
Muzeul de Istorie, Craiova 22029
Muzeul de istorie Brașov, Brașov 21964
Muzeul de istorie națională și arheologie, Constanța 22027
Muzeul de istorie naturală din Sibiu, Sibiu 22058
Muzeul de istorie naturală 'Grigore Antipa', București 22004
Muzeul de istorie și artă al municipiului București, București 22005
Muzeul de științele naturii, Bacău 21956
Muzeul județean, Suceava 22059
Muzeul județean Argeș, Pitești 22052
Muzeul județean Botoșani, Botoșani 21961
Muzeul Județean de Istorie ș Arheologie, Ploiești 22053
Muzeul județean de istorie și arheologie, Bacău 21957
Muzeul județean di istorie, Galați 22034
Muzeul județean Sălaj, Zalău 22071
Muzeul Județean Satu Mare, Satu Mare 22054
Muzeul Militar Central, București 22006
Muzeul National al Literaturii Romane, București 22007
Muzeul Național al Unirii, Alba Iulia 21955
Muzeul Național de Artă, București 22008
Muzeul național de artă Cluj, Cluj-Napoca 22024
Muzeul Național de Istorie a Transilvaniei, Cluj-Napoca 22025
Muzeul Național de Istorie al României, București 22009
Muzeul național Secuiesc, Sfântu Gheorghe 22055

Muzeul Olteniei, Craiova 22030
Muzeul satului şi de artă populară, Bucureşti 22010
Muzeul Secuiesc din Ciuc, Miercurea Ciuc 22046
Muzeul tehnic 'Prof. Ing. D. Leonida', Bucureşti 22011
Muzeum Archeologiczne, Poznań 21203
Muzeum Archeologiczne i Etnograficzne, Łódź 21163
Muzeum Archeologiczne w Gdańsku, Gdańsk 21099
Muzeum Archeologiczne w Krakowie, Kraków 21142
Muzeum Architektury, Wrocław 21363
Muzeum Bedřicha Smetany, Praha 06498
Muzeum Beskyd, Frýdek-Místek 06447
Múzeum Bojnice, Bojnice 24681
Muzeum Budownictwa Ludowego – Park Etnograficzny w Olsztynku, Olsztynek 21178
Muzeum Budownictwa Ludowego w Sanoku, Sanok 21219
Muzeum Českého ráje v Turnově, Turnov 06543
Muzeum Częstochowskie, Częstochowa 21089
Muzeum Etnograficzne im. Marii Znamierowskiej-Prüfferowej w Toruniu, Toruń 21233
Muzeum Etnograficzne im. Seweryna Udzieli, Kraków 21143
Muzeum Górnośląskie, Bytom 21083
Muzeum Historyczne, Białystok 21074
Muzeum Historyczne m. Krakowa, Kraków 21144
Muzeum Historyczne m. st. Warszawy, Warszawa 21316
Muzeum hl. města Prahy, Praha 06499
Muzeum hlavního města Prahy, Praha 06500
Muzeum i Archiwum Warszawskiego Towarzystwa Muzycznego im. Stanisława Moniuszki, Warszawa 21317
Muzeum i Instytut Zoologii PAN, Warszawa 21318
Muzeum jihovýchodní Moravy ve Zlíní, Zlín 06550
Muzeum Karlovy Vary, Karlovy Vary 06456
Muzeum Literatury im. Adama Mickiewicza, Warszawa 21319
Muzeum loutkářských kultur, Chrudim 06444
Muzeum Lubelskie, Lublin 21170
Muzeum Mazowieckie w Płocku, Płock 21190
Muzeum města Brna, Brno 06433
Muzeum Města Ústi Nad Labem, Ustí nad Labem 06546
Muzeum Mikołaja Kopernika, Frombork 21091
Muzeum Narodowe, Poznań 21204
Muzeum Narodowe, Wrocław 21364
Muzeum Narodowe Rolnictwa i Przemysłu Rolno-Spożywczego w Szreniawie, Komorniki 21121
Muzeum Narodowe w Gdańsku, Gdańsk 21100
Muzeum Narodowe w Krakowie, Kraków 21145
Muzeum Narodowe w Szczecinie, Szczecin 21228
Muzeum Narodowe w Warszawie, Warszawa 21320
Muzeum Narodowe Ziemi Przemyskiej, Przemyśl 21210
Muzeum Niepodległości, Warszawa 21321
Muzeum of Kiskun, Kiskunfélegyháza 13487
Muzeum Okręgowe w Toruniu, Toruń 21234
Muzeum Okręgowego im. Leona Wyczółkowskiego w Bydgoszczy, Bydgoszcz 21081
Muzeum Podkarpackie w Krosnie, Krosno 21152
Muzeum skla a bižuterie, Jablonec nad Nisou 06452
Muzeum Śląska Cieszyńskiego, Cieszyn 21088
Múzeum Slovenského národného povstania, Banská Bystrica 24678
Muzeum Sportu i Turystyki, Warszawa 21322
Muzeum Sztuki w Łodzi, Łódź 21164
Muzeum Tatrzańskie im. Tytusa Chałubińskiego, Zakopane 21372
Muzeum Teatralne, Warszawa 21323
Muzeum Techniki, Warszawa 21324
Muzeum Vysočiny Jihlava, Jihlava 06454
Muzeum w Nieborowie i Arkadii, Nieborów 21173
Muzeum Warmii i Mazur, Olsztyn 21176
Muzeum Wojska Polskiego, Warszawa 21325
Muzeum Zamkowe w Malborku, Malbork 21171
Muzeum Zamoyskich w Kozłowce, Kamionka 21113
Muzeum Ziemi Chełmskiej im. Wiktora Ambroziewicza, Chełm 21084
Muzeum Ziemi Kujawskiej i Dobrzyńskiej we Włocławku, Włocławek 21355
Muzeum Ziemi PAN, Warszawa 21326
Muzeum Żup Krakowskich Wieliczka, Wielczka 21354
Muzichne uchilishche, Ivano-Frankivsk 28100
Muzichno-teatralna biblioteka im. Stanislavskoho, Kharkiv 28267
Muziekbibliotheek VRO-VRK, Brussel 02613
Muziekcentrum van de Omroep (MCO), Hilversum 19467
Muziekinstrumentenmuseum (MIM), Bruxelles 02609
Muzium Sarawak, Kuching 18611
Muzychno-teatralna oblasna biblioteka, Kyiv 28359
Muzykalnoe uchilishche, Brest 01820
Muzykalnoe uchilishche, Kemerovo 22670
Muzykalnoe uchilishche, Perm 22689
Muzykalnoe uchilishche, Saratov 22718
Muzykalnoe uchilishche, Vladivostok 22724
Muzykalnoe uchilishche im. M.M. Ippolitova-Ivanova, Moskva 22681
Muzykalnoe uchilishche im. P.I. Chaikovskogo, Ekaterinburg 22667
Muzykalnoe uchilishche im. S.V. Rakhmaninova, Tambov 22720
Muzykalno-pedagogicheski institut, Rostov-na-Donu 22490
Muzykalno-pedagogicheski institut im. Gnesinykh, Moskva 22382
Muzykalno-pedagogicheskoe uchilishche, Gomel 01821
Mwanza Public Library, Mwanza 27681
Myanmar Scientific and Technological Research Department, Yangon 19005
Myerscough College, Preston 29378
Mykolaiv State Agrarian University, Mykolaiv 28046
Mykolaivska miska TsBS dlya ditei, Mykolaiv 28738
Mykolaivsky derzhavny agrarny universitet, Mykolaiv 28046
Mykolas Romeris University Library, Vilnius 18289
Mykolo Romerio universiteto biblioteka, Vilnius 18289
Myntkabinettet, Oslo 20252
Myntkabinettets bibliotek, Oslo 20288
Myrgorodska raionna TsBS, Tsentralna biblioteka, Myrgorod 28742
Myronivka Institute of Wheat, Myronivka 28433
Mystic Seaport Museum, Mystic 37138
Mythic Society, Bangalore 13939
N. A. Berdzenishvili Kutaisi State Museum of History and Ethnography, Kutaisi 09213
N. Ketskhoveli Botanical Institute, Tbilisi 09265
N. V. Gogol Karaganda District Library, Karaganda 17987
Naantali Public Library, Naantali 07592
Naantalin kaupunginkirjasto, Naantali 07592
Nablus Municipality Public Library, Nablus 14778
NACCO Materials Handling Group, Inc, Fairview 35700
NACID – National Center for Information and Documentation, Sofiya 03559
Nacional Cancer Centre, Tbilisi 09269
Nacionalna i sveučilišna knjižnica u Zagrebu, Zagreb 06140
– Agronomski Fakultet, Zagreb 06171
– Akademija Dramske Umjetnosti, Zagreb 06172
– "Andrija Štampar" School of Public Health, Zagreb 06173
– Centralna Kemijska Knjižnica, Zagreb 06174
– Ekonomski Fakultet, Zagreb 06175
– Faculty of Pharmacy and Biochemistry, Zagreb 06176
– Fakultet elektrotehnike i računarstva, Zagreb 06177
– Fakultet Organizacije i Informatike, Varaždin 06178
– Fakultet političkih znanosti, Zagreb 06179
– Fakultet Strojarstva i Brodogradnje, Zagreb 06180
– Filozofski Fakultet, Zagreb 06181
– Geofizički odsjek, Zagreb 06182
– Gradevinski fakultet, Zagreb 06183
– Institut Kemijskog Inženjerstva, Zagreb 06184
– Institut za Društvena Istraživanja, Zagreb 06185
– Katolički bogoslovni fakultet, Zagreb 06186
– Metalurški Fakultet, Sisak 06187
– Odsjek za Anglistiku, Zagreb 06188
– Odsjek za Arheologiju, Zagreb 06189
– Odsjek za filozofiju filozofskog fakulteta, Zagreb 06190
– Odsjek za Germanistiku, Zagreb 06191
– Odsjek za Povijest, Zagreb 06192
– Odsjek za psihologiju, Zagreb 06193
– Odsjek za Romanistiku, Zagreb 06194
– Odsjek za Talijanistiku, Zagreb 06195
– Pedagoški Fakultet, Zagreb 06196
– Pravni Fakultet, Zagreb 06197
– Prehrambeno-biotehnološki fakultet, Zagreb 06198
– Rudarsko-geološko-naftni fakultet, Zagreb 06199
– Središnja Biološka Knjižnica, Zagreb 06200
– Središnja matematièka knjižnica, Zagreb 06201
– Središnja medicinska knjižnica, Zagreb 06202
– Šumarski Fakultet, Zagreb 06203
– Sveučilište u Zagrebu Muzička Akademija, Zagreb 06204
– Veterinarski Fakultet, Zagreb 06205
– Zavod za Slavensku Filologiju, Zagreb 06206
Nacionalna i Univerzitetska Biblioteka Bosne i Hercegovine, Sarajevo 02943
– Biblioteka Odsjeka za Geografiju, Sarajevo 02950
– Katedra Filozofije i Pedagogije, Sarajevo 02951
– Katedra Istorije, Sarajevo 02952
– Katedra za Jugoslovenske Književnosti, Sarajevo 02953
– Medicinski Fakultet, Sarajevo 02954
– Poljoprivredni Fakultet, Sarajevo 02955
– Pravni Fakultet, Sarajevo 02956
– Veterinarski Fakultet, Sarajevo 02957
Nacionalna Ustanova Prirodonaucen Muzej na Makedonija, Skopje 18445
Naciones Unidas Comisión Económica para América Latina y el Caribe, Santiago 04941
Nacka stadsbibliotek, Nacka 26863
Nacogdoches Public Library, Nacogdoches 40358
Naczelna Dyrekcja Archiwów Państwowych, Warszawa 21047
Nadace Bohuslava Martinů, Praha 06501
Nadvirna Regional Centralized Library System, Main Library, Nadvirna 28743
Nadvirnyanska raionna TsBS, Tsentralna biblioteka, Nadvirna 28743
Næringslivets Hovedorganisasjon (NHO), Oslo 20202
Næstved Gymnasium og HF, Næstved 06739
Næstvedbibliotekerne, Næstved 06904
Naftikon Moussion Tis Ellados, Piraius 13129
Nagano Prefectural Government, General Affairs Division, Nagano 17281
Naganoken Somubu Gyosei Johosenta, Nagano 17281
Nagaoka City Gosombunko Library, Nagaoka 17849
Nagaoka Shiritsu Gosombunko, Nagaoka 17849
Nagaoka University of Technology, Nagaoka 17051
Acharya Nagarjuna University, Nagarjunanagar 13716
Nagasaki Daigaku Toshokan, Nagasaki 17052
– Igakubu Fuzoku Genbakuhisai Gakujutsu Shiryo Senta, Nagasaki 17053
Nagasaki Prefectural Education Center, Omura 17631
Nagasaki Prefecture Agricultural and Forestry Experiment Station, Isahaya 17557
Nagasaki University Library, Nagasaki 17052
– School of Medicine, Scientific Data Center of Atomic Bomb Disaster, Nagasaki 17053
Nagasakiken Kyoiku Senta, Omura 17631
Nagasakiken Sogo Norin Shikenjo, Isahaya 17557
Nagoya Chamber of Commerce and Industry, Nagoya 17615
Nagoya City Assembly, Nagoya 17283
Nagoya City University, Nagoya 17059
Nagoya Daigaku, Nagoya 17060
Nagoya Gakuin Daigaku, Seto 17120
– Nagoya Daigaku, Nagoya 17121
– Rigakubu Fuzoku Sugashima Rinkai Jikkenjo, Toba 17122
Nagoya Gakuin University, Seto 17120
– Graduate School of Mathematics, Nagoya 17121
– Sugashima Marine Biological Laboratory, Toba 17122
Nagoya Institute of Technology, Nagoya 17061
Nagoya Kogyo Daigaku, Nagoya 17061
Nagoya Railroad Co, Ltd, Meitetsu Library, Nagoya 17394
Nagoya Shikai, Nagoya 17283
Nagoya Shoko Kaigisho, Nagoya 17615
Nagoya Tetsudo K. K., Nagoya 17394
Nagoya University, Nagoya 17060
Nagoya University of Commerce and Business Administration, Nisshin 17074
Nagy László Town Library and Cultural Centre, Ajka 13526
NAIC Research Library, Kansas City 36988
Naikaku Hoseikyoku, Tokyo 17318
Naito Kinen Kusuri Hakubutsukan, Hashima 17544
Naito Museum of Pharmaceutical Science, Hashima 17544
Najwyższa Izba Kontroli, Warszawa 21048
Nakhimov Naval College, Sankt-Peterburg 22702
Nakhimovskoe uchilishche, Sankt-Peterburg 22702
Nakskov Bibliotek, Nakskov 06905
Nalanda Institute of Buddhist Studies and Pali, Bihar 13949
NALs bibliotek, Oslo 20253
Namibia Scientific Society, Windhoek 19014
Namik-Kemal-Bibliothek, Berlin 12672
Nampa Public Library, Nampa 40359
Namsos folkebibliotek, Namsos 20377
Namulonge Agricultural and Animal Production Research Institute, Kampala 27927
Nan Boothby Memorial Library, Cochrane 04614
Nanchang College of Water Conservancy and Hydraulic Power, Nanchang 05429
Nanchang Institute of Aeronautical Technology, Nanchang 05430
Nanchang Vocational and Technical Teachers' College, Nanchang 05431
Nangjing Audit College, Nanjing 05435
Nanjing Academy of Arts, Nanjing 05436
Nanjing Aero-Space Engineering University, Nanjing 05437
Nanjing Agricultural University, Nanjing 05438
Nanjing College of Communications, Nanjing 05439
Nanjing College of Food Economics, Nanjing 05440
Nanjing Electric Power College, Nanjing 05441
Nanjing Financial College, Nanjing 05442
Nanjing Forestry University, Nanjing 05443
Nanjing Institute of Geology and Paleontology, Nanjing 05883
Nanjing Institute of Physical Education, Nanjing 05444
Nanjing Library, Nanjing 05922
Nanjing Medical University, Nanjing 05445
Nanjing Normal University, Nanjing 05446
Nanjing Power College, Nanjing 05447
Nanjing Railway Medical College, Nanjing 05448
Nanjing University, Nanjing 05449
Nanjing University of Economics, Nanjing 05450
Nanjing University of Information Science and Technology, Nanjing 05451
Nanjing University of Posts and Telecommunications, Nanjing 05452
Nanjing University of Science and Technology, Nanjing 05453
Nanjing University of Technology, Nanjing 05454
Nankai University, Tianjin 05608
Nanping Teachers' College, Nanping 05467
Nansei Regional Fisheries Research Laboratory, Saiki 17647
Nanseikaiku Suisan Kenkyujo, Saiki 17647
Nantong Institute of Textile Technology, Nantong 05468
Nantong University, Nantong 05469
Nantucket Maria Mitchell Association, Nantucket 37140
Nanyang College of Science and Engineering, Nanyang 05470
Nanyang Teachers' College, Nanyang 05471
Nanyang Technological University, Singapore 24584
Nanzan Daigaku Toshokan, Nagoya 17062
– Jinruigaku Kenkyujo, Nagoya 17063
Nanzan University, Nagoya 17062
– Nanzan Anthropological Institute, Nagoya 17063
Napa County Law Library, Napa 34570
Napa Valley College, Napa 33569
Naperville Public Library – 95th Street, Naperville 40361
Naperville Public Library – Naper Boulevard, Naperville 40362
Naperville Public Library – Nichols Library, Naperville 40363
Napier Public Library, Napier 19873
Napier University, Edinburgh 29095
Napoleon Public Library, Napoleon 40365
Napoleonmuseum Thurgau, Salenstein 27429
Nappanee Public Library, Nappanee 40367

Náprstkovo Muzeum Asijských, Afrických, a Amerických Kultur, Praha 06511
NAPS, Sofiya 03492
Nara Kenritsu Ika Daigaku, Kashihara 16967
Nara Kokuritsu Bunkazai Kenkyujo, Nara 17620
Nara Kokuritsu Hakubutsukan, Nara 17621
Nara Medical University, Kashihara 16967
Nara National Cultural Properties Research Institute, Nara 17620
Nara National Museum, Nara 17621
Nara Prefectural Library, Nara 17852
Nara Prefecture Administrative Data Room, Nara 17285
Nara Women's University, Nara 17064
Naracoorte Lucindale Council, Naracoorte 00704
Naraken Gyosei Shiryoshitsu, Nara 17285
Naresuan University, Phitsanulok 27750
Naritasan Bunka Zaidan, Narita 17622
Naritasan Cultural Foundation, Narita 17622
Naroden teatăr 'Iv. Vazov', Sofiya 03558
Narodna Biblioteka, Kula 24542
Narodna Biblioteka, Pirot 24546
Narodna Biblioteka, Požega 24548
Narodna Biblioteka, Smederevo 24551
Narodna Biblioteka Bora Stankovic, Vranje 24556
Narodna Biblioteka Dositej Obradović, Stara Pazova 24554
Narodna Biblioteka Dr Đorđe Natoševic, Indjija 24538
Narodna Biblioteka Ilja M. Petrović, Požarevac 24547
Narodna biblioteka 'Ivan Vazov', Plovdiv 03450
Narodna Biblioteka Jovan Popović, Kikinda 24540
Narodna biblioteka 'Pencho Slaveikov', Varna 03607
Narodna biblioteka Srbije, Beograd 24453
Narodna Biblioteka "Stevan Sremac", Niš 24545
Narodna Biblioteka "Sv. Sv. Kiril i Metodii" (NBKM), Sofiya 03451
Narodna Biblioteka Vuk Karadžic, Beograd 24537
Narodna Biblioteka Vuk Karadžić, Kragujevac 24541
Narodna Biblioteka Žika Popović, Šabac 24550
Narodna galerija, Ljubljana 24926
Narodna i Univerzitetska Biblioteka "Kliment Ohridski", Skopje 18427
Narodna i univerzitetska biblioteka Republike Srpske, Banja Luka 02942
– Ekonomski fakultet, Banja Luka 02945
– Elektrotehnički fakultet, Banja Luka 02946
– Pravni fakultet, Banja Luka 02947
Narodna in študijska knjižnica, Trieste 16629
Narodna in Univerzitetna Knjižnica, Ljubljana 24806
Narodna in Univerzitetna Knjižnica, Ljubljana
Národné centrum zdravotníckych informácií – Slovenská lekárska knižnica, Bratislava 24698
Národné Lesnícke Centrum, Zvolen 24761
Narodni Bank FNRJ, Beograd 24497
Národní filmový archiv, Praha 06502
Národní galerie v Praze, Praha 06503
Národní informační poradenské středisko pro kulturu (NIPOS), Praha 06504
Národní knihovna České republiky, Praha 06342
Národní lékařská knihovna, Praha 06505
Narodni Muzej, Vršac 24534
Narodni Muzej Slovenije, Ljubljana 24927
Národní muzeum, Praha 06506
Národní památkový ústav, Praha 06507
Národní pedagogická knihovna Komenského, Praha 06508
Národní technická knihovna, Praha 06509
Národní technické muzeum, Praha 06510
Národní zemědělské muzeum Praha, Nové Dvory 06467
Narodno Sabraniye, Sofiya 03493
Narodnoe Sobranie, Moskva 22755
Národný onkologický ústav, Bratislava 24699
Národný ústav tuberkulózy a respiračných chorôb, Bratislava 24700
Narodowy Bank Polski, Warszawa 21070
Narodowy Instytut Fryderyka Chopina, Warszawa 21327
Naropa University, Boulder 30510
Narva Central LIbrary, Narva 07249
Narvik bibliotek, Narvik 20378
Narvik College of Engineering, Narvik 20108
NASA, Hampton 34332
NASA Armes Research Center, Life Sciences Library, Moffett Field 34548
NASA – Armes Research Center, Technical Library, Moffett Field 34549
NASA – Goddard Institute for Space Studies Library, New York 34592
NASA Goddard Space Flight Center, Greenbelt 34317
NASA Headquarters Library, Washington 35000
NASA – John F. Kennedy Space Center Library, Kennedy Space Center 34413
NASA – John H. Glenn Research Center at Lewis Field, Cleveland 34087
NASA – Johnson Space Center Scientific & Technical Information Center, Houston 34368
Năsăud Documentary Library, Năsăud 22047
Nash Community College, Rocky Mount 33698
Nashotah House Library, Nashotah 37142
Nashua Public Library, Nashua 40369
Nashville Public Library, Nashville 40371
Nashville Public Library – Bellevue, Nashville 40372
Nashville Public Library – Bordeaux, Nashville 40373
Nashville Public Library – Donelson, Nashville 40374
Nashville Public Library – Edmondson Pike, Nashville 40375
Nashville Public Library – Green Hills, Nashville 40376
Nashville Public Library – Hermitage Branch, Hermitage 39419
Nashville Public Library – Madison Branch, Madison 40009
Nashville Public Library – Richland Park, Nashville 40377
Nashville Public Library – Southeast, Antioch 37972
Nashville School of Law Library, Nashville 33570
Nashville State Technical Institute, Nashville 33571
Nasinu Teachers' College, Suva 07294
Nasionale Afrikaanse Letterkundige Museum en Navorsingsentrum, Bloemfontein 25107
Nasjonalbiblioteket, Oslo 20065
Nasjonalgalleriets bibliotek, Oslo 20254
Nasjonalt Folkehelseinstitutt, Oslo 20255
Nassau Academy of Medicine, Garden City 36884
Nassau Community College, Garden City 33367
Nassau County Museum Research Library, Port Washington 37396
Nassau County Public Library System – Fernandina Beach Branch, Fernandina Beach 39063
Nassau County Supreme Court, Mineola 34540
Nassau Public Library, Nassau 01735
Nassau University Medical Center, East Meadow 36819
Nassauische Sparkasse, Wiesbaden 11436
Nasser Social Bank, Cairo 07140
Natal Education Department, Pietermaritzburg 25068
Natal Museum, Pietermaritzburg 25150
Natchitoches Parish Library, Natchitoches 40379
Nathan & Henry B. Cleaves Law Library, Portland 37404
Nathan S. Kline Institute for Psychiatric Research, Orangeburg 37311
Nationaal Archief, Den Haag 19444
Nationaal Herbarium Nederland, Leiden 19478
Nationaal Museum van de Speelkaart, Turnhout 02704
Nationaal Natuurhistorisch Museum Naturalis, Leiden 19479
National Academic Network and Cahit Information Center, Bilkent, Ankara 27878
National Academies, National Academy of Sciences, Washington 37755
National Academies, Transportation Research Board Library, Washington 37756
National Academy of Agricultural Research Management, Hyderabad 14005
National Academy of Arts, Seoul 18118
National Academy of Fine Arts and Architecture, Kyiv 28360
National Academy of Letters, New Delhi 14167
National Academy of Moral and Political Sciences, Buenos Aires 00231
National Academy of Music, Dance and Drama, New Delhi 14152
National Academy of Physical Education and Sport, Bucureşti 21952
National Academy of Science, México 18830
National Academy of Sciences in Córdoba, Córdoba 00296
National Academy of Sciences of Armenia, Yerevan 00326
National Academy of Sciences of Azerbaijan, Baku 01681
National Academy of Scientific Research, Tripoli 18261
National Aeronautics Library, Buenos Aires 00242
National Aerospace Laboratories, Bangalore 13940
National Aerospace Laboratory, Chofu 17529
National Aerospace Library, London 29793
National Agency for Educational Research, Lyon 08397
National Agency for State Administration, Jakarta 14367
National Agricultural Documentation Centre (NADOC), Freetown 24575
National Agricultural Library, Beltsville 33984
National Agricultural Library and Documentation Centre, Budapest 13444
National Agricultural Research Center for Tohoku Region, Morioka 17607
National Agricultural University, Kyiv 28016
National Agriculture Research Center, Tsukuba 17803
National Agriculture Research Center, Tsukuba 17795
National Agronomy Station, Oeiras 21818
National Air and Space Museum, Washington 37787
National and Capodistrian University of Athens, Athinai 13075
National and Supreme Court, Boroko 20574
National and University Library of Bosnia and Herzegovinia, Sarajevo 02943
– Department of Geography, Sarajevo 02950
– Department of History, Sarajevo 02952
– Department of Philosophy and Education, Sarajevo 02951
– Department of Yugoslav Literature, Sarajevo 02953
– Faculty of Agriculture, Sarajevo 02955
– Faculty of Medicine, Sarajevo 02954
– School of Law, Sarajevo 02956
– Veterinary Faculty, Sarajevo 02957
National and University Library of Iceland, Reykjavík 13554
National and University Library of the Republic of Srbska, Banja Luka 02942
– Faculty of Economics, Banja Luka 02945
– Faculty of Electrical Engineering, Banja Luka 02946
– Faculty of Law, Banja Luka 02947
National and University Library of Zagreb, Zagreb 06140
– Academy of Dramatic Art, Zagreb 06172
– Academy of Music in Zagreb Library, Zagreb 06204
– Central Biological Library, Zagreb 06200
– Central Chemical Library, Zagreb 06174
– Central Mathematical Library, Zagreb 06201
– Central Medical Library, Zagreb 06202
– Department of Archeology, Zagreb 06189
– Department of Classical Philology, Zagreb 06181
– Department of English Language and Literature, Zagreb 06188
– Department of History, Zagreb 06192
– Department of Italian Language and Literature, Zagreb 06195
– Department of Philosophy, Zagreb 06190
– Department of Psychology, Zagreb 06193
– Department of Romance Philology, Zagreb 06194
– Department of Slavonic Philology, Zagreb 06206
– Faculty of Agriculture, University of Zagreb, Zagreb 06171
– Faculty of Civil Engineering, Zagreb 06183
– Faculty of Economics, Zagreb 06175
– Faculty of Education, Zagreb 06196
– Faculty of Food Technology and Biotechnology, Zagreb 06198
– Faculty of Forestry, Zagreb 06203
– Faculty of Law, Zagreb 06197
– Faculty of Mechanical Engineering and Naval Architecture, Zagreb 06180
– Faculty of Metallurgy, Sisak 06187
– Faculty of mining, geology and petroleum engineering, Zagreb 06199
– Faculty of Organization and Informatics, Varaždin 06178
– Faculty of Philosophy, University of Zagreb, Zagreb 06191
– Faculty of Political Science, Zagreb 06179
– Faculty of Roman Catholic Theology, Zagreb 06186
– Faculty of Science, Department of Geophysics, Library, Zagreb 06182
– Faculty of Veterinary Medicine, Zagreb 06205
– Institute for Social Research, Zagreb 06185
– Institute of Chemical Engineering, Zagreb 06184
National and University Library "St. Kliment Ohridski", Skopje 18427
– Central Medical Library, Skopje 18432
– Department of Romance Philology, Skopje 18437
– Department of Yugoslav Literary History, Skopje 18436
– Faculty of Architecture and Civil Engineering, Skopje 18431
– Faculty of Law, Skopje 18439
– Historical Faculty, Skopje 18435
– Institute of Mathematics, Skopje 18438
National Animal Disease Center Library, Ames 36462
National Archaelogical Museum, Lisboa 21810
National Archaeological Museum, Madrid 26065
National Archeologicol Museum of Aquileia, Aquileia 16167
National Archives, Cape Town 25114
National Archives, Den Haag 19444
The National Archives, Richmond 29886
National Archives, Tripoli 18262
National Archives and Library of Ethiopia, Addis Ababa 07260
National Archives and National Library of Bangladesh, Dhaka 01745
National Archives, Cabinet Library, Tokyo 17697
National Archives Luxembourg, Luxembourg 18406
National Archives of Australia, Canberra Mail Centre 00896
National Archives of Fiji, Suva 07316
National Archives of Finland, Helsinki 07464
National Archives of India, New Delhi 14153
National Archives of Karelia, Petrozavodsk 23704
National Archives of Nigeria, Ibadan 20019
National Archives of Norway, Oslo 20279
National Archives of Pakistan, Islamabad 20481
National Archives of South Africa, Pietermaritzburg 25151
National Archives of Thailand, Bangkok 27783
National Archives of the Republic of Belarus, Minsk 02037
National Archives of the Republic of Indonesia, Jakarta 14355
National Archives of Zambia, Lusaka 42344
National Archives of Zimbabwe, Harare 42377
National Archives of Zimbabwe – Library and Reference Services, Bulawayo 42360
National Archives & Records Administration, Abilene 33894
National Archives & Records Administration, Boston 33999
National Archives & Records Administration, College Park 36724
National Archives & Records Administration, Hyde Park 36951
National Archives & Records Administration, Independence 34378
National Archives & Records Administration, Kansas City 34406
National Archives & Records Administration, Simi Valley 37608
National Archives & Records Administration, West Branch 37818
National Archives Repository, Pretoria 25163
National Archives Research Library, Zomba 18495
National Army Museum, London 29794
National Art and Design Library, København 06790
National Art Library (NAL), London 29795
National Art Museum of Belarus, Minsk 02038
National Art School, Darlinghurst 00431
National Arts School, Boroko 20545
National Assembly, Amman 17882
National Assembly, Bangkok 27767
National Assembly, Cairo 07120
National Assembly, Hanoi 42283
National Assembly, Nairobi 18027
National Assembly, Port Louis 18696
National Assembly, Zomba 18481
National Assembly Library, Islamabad 20468
National Assembly Library, Seoul 18104
National Assembly of Quebec, Québec 04127
National Assembly of Tanzania, Dodoma 27667
National Assembly of the Republic of Armenia, Yerevan 00338
National Assembly of the Republic of Bulgaria, Sofiya 03493
National Assn of Commercial Broadcasters, Tokyo 17741
National Association for the Advancement of Colored People, New York 37234

National Association for Visually Handicapped, New York 37235
National Association of Broadcasters, Washington 37757
National Association of Chain Drug Stores, Alexandria 36456
National Association of Home Builders, Washington 37758
National Association of Mutual Aid for Municipal Property Damages, Reference Library for Disasters, Tokyo 17781
National Association of Norwegian Architects, Oslo 20253
National Association of Realtors, Chicago 36683
National Association of Watch & Clock Collections, Inc., Columbia 36731
National Astronomical Observatory, Mizusawa 17605
National Astronomy and Ionosphere Center, Arecibo 21901
National Atlas and Thematic Mapping Organisation, Kolkata 14054
National Audiovisual Archive, Helsinki 07463
National Aviation University, Kyiv 28017
National Bal Bhavan Society, New Delhi 14154
National Bank of Belgium, Bruxelles 02564
National Bank of Ethiopia, Addis Ababa 07278
National Bank of Pakistan, Karachi 20475
National Bank of Poland, Warszawa 21070
National Bank of Yugoslavia, Beograd 24497
National Baseball Hall of Fame & Museum, Inc, Cooperstown 36752
National Board for the Protection of Historic Monuments, Budapest 13445
National Board of Antiquities Library, Helsinki 07475
National Board of Education, Helsinki 07480
National Board of Fisheries, Göteborg 26575
National Board of Housing, Building and Planning, Karlskrona 26494
National Board of Patents and Registration of Finland, Helsinki 07483
National Book Chamber of the Republic of Moldova, Chişinău 18885
National Botanic Garden, Salaspils 18193
National Botanic Garden of Belgium, Meise 02686
National Botanic Gardens, Dublin 14548
National Botanical Garden, Salaspils 18193
National Botanical Institute, Claremont 25118
National Botanical Institute, Pretoria 25164
National Botanical Research Institute, Lucknow 14069
National Braille Association, Inc, Rochester 37454
National Broadcasting Corporation, Boroko 20587
National Buildings Organisation, New Delhi 14155
National Bureau of Standards, Taipei 27612
National Cancer Center, Tokyo 17695
National Cancer Institute, Bratislava 24699
National Cancer Institute at Fredrick Scientific Library, Fort Detrick 36859
National Capital Authority, Canberra 00649
National Capital Commission Library, Ottawa 04448
National Center for Atmospheric, Boulder 34000
National Center for Genetic Resources and Biotechnology (NACGRAB), Ibadan 20020
National Center for Information and Documentation (NACID), Sofiya 03559
National Center for Juvenile Justice, Pittsburgh 37382
National Center for Language Development and Cultivation, Jakarta 14374
National Center for Manufacturing Sciences, Ann Arbor 36468
National Center for State Courts Library, Williamsburg 37843
National Central Library, Dar es Salaam 27648
National Central Library, Taipei 27521
National Central Library in Florence, Firenze 14784
National Central Library in Rome, Roma 14785
National Central Library of Italian Meteorology, Roma 16468
National Centre for Curriculum Research and Development (NCCRD), Beau Bassin 18689
National Centre for Historical Studies, Kouba 00080
National Centre for Occupational Health, Braamfontein 25110
National Centre for Prehistorical, Anthropological and Historical Research, Alger 00059
National Centre for Regional Development nad Housing Policy, Sofiya 03560
National Centre for Research on Tropical Fish, Pirassununga 03315
National Centre for Tuberculosis Problems, Almaty 17959
National Centre for Vocational Education Research, Adelaide 00853
National Centre of Museums, Budapest 13392
National Centre of Public Health Protection, Sofiya 03561
National Cereals Research Institute Badeggi, Bida 20011
National Chemical Laboratory, Pune 14196
National Chemical Laboratory for Industry, Tsukuba 17788
National Cheng Kung University, Tainan 27541
National Cheng-Chi University, Taipei 27546
– Center for Public and Business Administration, Taipei 27547
– Institute of International Relations, Taipei 27548
National Chiao-Tung University, Hsinchu 27527
National Chiayi Teachers College, Chia-yi 27523
National Children's Bureau, London 29796
National Children's Resource Centre, Dublin 14549
National Chung Cheng University, Chia-yi 27524
National Chung-Hsing University, Taichung 27538
National Clearinghouse for English Language Aquisition & Language Instruction Educational Programs, Washington 32971
National Clearinghouse on Child Abuse & Neglect Information, Fairfax 36839
National Clearinghouse on Child Abuse & Neglect Information, Fairfax 36840
National College of Art and Design, Dublin 14471
National College of Business and Technology, Salem 33717
National College of Commerce, Shimoga 13764
National College of Ireland, Dublin 14472
National Commission for Museums and Monuments, Jos 20025
National Conservatoire, México 18727
National Co-operative Archive, Manchester 29853
National Council for Cement and Building Materials, New Delhi 14156
National Council for Scientific Research, Lusaka 42345
National Council for the Blind of Ireland, Dublin 14550
National Council of Applied Economic Research, New Delhi 13890
National Council of Teachers of English Library, Urbana 37699
National Council of Transition, Alger 00055
National Cowboy & Western Heritage Museum, Oklahoma City 37303
National Criminological Institute, Budapest 13443
National Cultural History Museum Library, Pretoria 25165
National Czech & Slovak Museum & Library, Cedar Rapids 36638
National Dairy Research Institute, Bangalore 13941
National Dairy Research Institute, Karnal 14024
National Danish School of Social Work, Odense, Odense 06716
National Defence College, Helsinki 07341
National Defence College Library, Helsinki 07340
National Defense Research Institute, Department of NBC defence, Umeå 26702
National Development and Family Board of Malaysia, Kuala Lumpur 18590
National Diet Library, Tokyo 16896
National Direction of Science, Technology and Innovation, Montevideo 42055
National Documentation Centre, Kingstown 26339
National Dong Hwa University, Hualien 27529
National Drama Library, Bloemfontein 25108
National Economic and Development Authority (NEDA), Manila 20756
National Economic and Social Development Board, Bangkok 27761
National Education Commission, Bangkok 27784
National Educational Library and Museum, Budapest 13448
National Egyptian Bank, Cairo 07141
National Emergency Training Center, Emmitsburg 34189
National Endowment for Democracy Library, Washington 37759
National Endowment for the Arts Library, Washington 37760
National Endowment for the Humanities Library, Washington 35001
National Energy Authority, Reykjavík 13572
National Energy Board Library, Calgary 04055
National Energy Information Center, Bangkok 27785
National English Literary Museum, Grahamstown 25125
National Environmental Engineering Research Institute, Nagpur 14107
National Environmental Research Institute, Dept. of Wildlife Ecology and Biodiversity, Rønde 06830
National Environmental Satellite Data & Information Services, Seattle 34850
National Family Planning Board, Kuala Lumpur 18591
National Family Planning Coordinating Board, Jakarta 14359
National Federation of Health Insurance Societies, Tokyo 17689
National Film and Sound Archive, Canberra 00894
National Film and Television Institute (NAFTI), Accra 13023
National Film Board of Canada, Saint Laurent 04480
National Finance Corp, Research Department, Tokyo 17455
National Fine Arts Museum of Algiers, Alger 00068
National Fire Protection Association, Quincy 37422
National Fisheries College, Kavieng 20553
National Fisheries University, Shimonoseki 17123
National Fisheries University of Pusan, Pusan 18072
National Food and Nutrition Institute, Warszawa 21314
National Food Research Institute, Tsukuba 17796
National Football Foundation's College, South Bend 37616
National Forest and Nature Agency, København 06821
National Forest Centre, Zvolen 24761
National Forest Service, Driebergen 19247
National Forest Service Library, Fort Collins 34223
National Forestry Library, Rotorua 19843
National Gallery, Ljubljana 24926
National Gallery, London 29797
National Gallery and Alexandros Soutzos Museum, Athinai 13120
National Gallery Library, Oslo 20254
National Gallery of Armenia, Yerevan 00355
National Gallery of Art Library, Washington 37761
National Gallery of Australia, Canberra 00895
National Gallery of Canada, Ottawa 04449
National Gallery of Ireland, Dublin 14551
National Gallery of Modern Art, New Delhi 14157
National Gallery of Prague, Praha 06503
National Gallery of Scotland, Edinburgh 29656
National Gallery of Victoria, Melbourne 00955
National Gallery of Zimbabwe, Harare 42378
National Geographic Society Library, Washington 37762
National Geophysical Research Institute, Hyderabad 14006
National Grand Lodge of Denmark, København 06789
National Grassland Research Institute, Nasu 17623
National Ground Water Association, Westerville 35059
National Guard Memorial Library, Washington 35002
National Health Information Center – Slovak Medical Library, Bratislava 24698
National Health Library, Dhaka 01768
National Health Library, Managua 19895
National Health Library at the New Karl Heusner Memorial Hospital, Belize City 02900
National Herbarium of the Netherlands, Leiden 19478
National Heritage Board, Singapore 24615
National Heritage Conservation Commission, Livingstone 42342
National Heritage Museum, Lexington 37028
National Historical Archive, Luanda 00104
National History Museum of Romania, Bucureşti 22009
National History Museum of Transylvania, Cluj-Napoca 22025
National Horticultural Research Centre (NHRC), Nairobi 18042
National Horticultural Research Institute (NIHORT), Ibadan 20021
National Housing Authority Library, Bangkok 27762
National Indian Law Library, Boulder 36582
National Information and Documentation Centre, Cairo 07181
National Information Centre for Textile and Allied Subjects, Ahmedabad 13921
National Insitute of Public Health, Bucureşti 22001
National Institute for Agriculture and Food Research and Technology, Madrid 26060
National Institute for Agro-Environmental Sciences, Tsukuba 17797
National Institute for Astrophysics – Library of the Astronomical Observatory of TriesteAstronomical Observatory of Trieste, Trieste 16626
National Institute for Biological Standards and Control, South Mimms 29894
National Institute for Compilation and Translation, Taipei 27613
National Institute for Economic and Social Research, Jakarta 14368
National Institute for Educational Research, Tokyo 17699
National Institute for Educational Science, Hanoi 42300
National Institute for Educational Studies and Research, Brasília 03283
National Institute for Environmental Studies, Environmental Information Division, Tsukuba 17793
National Institute for Health and Welfare, Helsinki 07501
National Institute for Information and Documentation, Bucureşti 21998
National Institute for Medical Research, London 29798
National Institute for Medical Research (NIMR), Lagos 20032
National Institute for Occupational Safety & Health, Pittsburgh 37383
National Institute for Policy and Strategic Studies, Bukuru 20012
National Institute for Research in Inorganic Materials, Tsukuba 17794
National Institute for Research Studies, Bissau 13191
National Institute for Resources and Environment, Tsukuba 17808
National Institute for Scientific and Industrial Research, Lusaka 42346
National Institute for Social Work, London 29799
National Institute for Sports Medicine, Budapest 13451
National Institute of Administration, Tripoli 18249
National Institute of Aerospace Technology, Torrejón de Ardoz 26166
National Institute of Agricultural Economics, Roma 16525
National Institute of Agronomic Research, Alger 00063
National Institute of Animal Health, Ibaraki 17551
National Institute of Animal Industry, Ibaraki 17552
National Institute of Anthropology and Latin American Thought, Buenos Aires 00267
National Institute of Ayurveda, Jaipur 14014
National Institute of Bank Management, Pune 14197
National Institute of Business Management (NIBM), Colombo 26270
National Institute of Communicable Diseases (NICD), Delhi 13985
National Institute of Design, Ahmedabad 13922
National Institute of Development Administration, Bangkok 27704
National Institute of Dramatic Art (NIDA), Kensington 00448
National Institute of Economic and Social Research, London 29800
National Institute of Education, Samtse 02915
National Institute of Educational Planning and Administration, New Delhi 13832
National Institute of Engineering, Mysore 14104
National Institute of Environmental Health Sciences Library, Research Triangle Park 37436

National Institute of Folk and Traditional Heritage, Islamabad 20480
National Institute of Foundry and Forge Technology, Ranchi 14205
National Institute of Fruit Tree Science, Tsukba 17784
National Institute of Fruit Tree Science, Tsukuba 17789
National Institute of Genetics, Mishima 17597
National Institute of Geodesy and Cartography, Antananarivo 18462
National Institute of Haematology, Blood Transfusion and Immunology, Budapest 13439
National Institute of Health, Santafé de Bogotá 06037
National Institute of Health and Family Welfare (NIHFW), New Delhi 13833
National Institute of Health and Nutrition, Tokyo 17711
National Institute of Health Sciences, Kalutara 26322
National Institute of Historical and Cultural Research, Islamabad 20482
National Institute of Hygiene, Warszawa 21332
National Institute of Hygiene and Epidemiology, Hanoi 42301
National Institute of Hygienic Sciences, Tokyo 17694
National Institute of Industrial Engineering, Mumbai 13705
National Institute of Industrial Health, Kawasaki 17569
National Institute of Industrial Technology, San Martín 00313
National Institute of Infectious Diseases, Tokyo 17712
National Institute of Japanese Literature, Tokyo 17692
National Institute of Legal Medicine 'Mina Minovici', București 21999
National Institute of Mental Health, Ichikawa 17554
National Institute of Meteorology and Hydrology, Sofiya 03564
National Institute of Metrology, București 22000
National Institute of Nutrition, Hyderabad, Andhra Pradesh 14012
National Institute of Occupational Health, Oslo 20282
National Institute of Oceanography, Dona Paula 13990
National Institute of Oncology, Budapest 13447
National Institute of Pharmacy, Budapest 13438
National Institute of Polar Research, Tachikawa 17664
National Institute of Population and Social Security Research, Tokyo 17713
National Institute of Psychiatry and Neurology, Budapest 13449
National Institute of Public Administration, Buenos Aires 00270
National Institute of Public Administration, Karachi 20499
National Institute of Public Administration, Kuala Lumpur 18585
National Institute of Public Administration, Lusaka 42325
National Institute of Public Administration (NIPA), Lahore 20515
National Institute of Public Cooperation and Child Development, New Delhi 14158
National Institute of Public Health, Praha 06515
National Institute of Radiological Sciences Library, Chiba 17522
National Institute of Research and Development for Potato and Sugar Beet Brașov, Brașov 21963
National Institute of Research & Development for Technical Physics, Iași 22044
National Institute of Rural Development, Hyderabad 14007
National Institute of Science and Technology, Manila 20773
National Institute of Science, Technology and Development Studies (NISTADS), New Delhi 14159
National Institute of Sericultural and Entomological Science, Tsukuba 17806
National Institute of Sericultural and Entomological Science, Tsukuba 17798
National Institute of Standards & Technology, Gaithersburg 34292
National Institute of Statistics, Santiago 04934
National Institute of Statistics and Census, Buenos Aires 00268
National Institute of Strategic Health Research, Budapest 13373
National Institute of Study and Research, Bissau 13189
National Institute of Technical Teachers Training and Research, Chennai 13968
National Institute of Telecommunications, Warszawa 20954
National Institute of Tuberculosis and Pulmonology, Budapest 13442
National Institute of Tuberculosis and Respiratory Diseases, Hanoi 42302
National Institute of Verdi Studies, Parma 16411
National Institute of Water and Atmospheric Research, Wellington 19851
National Institute on Aging, Baltimore 36513
National Institute on Drug Abuse, Baltimore 36514
National Insurance Institute, Jerusalem 14738
National Investment Bank, Accra 13038
National Judicial College, Reno 32424
National Kaohsiung First University of Science and Technology, Kaohsiung 27531
National Kaohsiung Normal University, Kaohsiung 27532
National Labor Relations Board Library, Washington 37763
National Laboratory of Energy and Geology, S. Mamede de Infesta 21826
National Land Agency, Tokyo 17312
National Land Survey, Gävle 26573
National Land Survey of Finland, Helsinki 07472
National Language Authority, Islamabad 20483
National Language Research Institute, Tokyo 17698
National Law Academy, Kharkiv 27994
National Law School of India University, Bangalore 13612
National League of Cities, Washington 37764
National Library, Abu Dhabi 28904
National Library, Amman 17868
The National Library, Manila 20681
National Library, Port Louis 18688
National Library, Warszawa 20787
National Library and Archives, Tarawa 18047
National Library and Archives of Bolivia, Sucre 02917
National Library and Archives of Egypt (Dar al-Kuttub), Cairo 07000
National Library and Archives of Morocco, Rabat 18916
National Library and Documentation Centre, Colombo 26267
National Library and Information System Authority of Trinidad and Tobago, Port-of-Spain 27804
National Library – Ang Mo Kio Branch, Singapore 24626
National Library – Bedok Branch, Singapore 24627
National Library Board (NLB), Singapore 24583
National Library for Teachers, Buenos Aires 00243
National Library for the Blind and Disabled, Dhaka 01769
National Library – Geylang East Branch, Singapore 24628
National Library – Jurong East Branch, Singapore 24629
National Library – Marine Parade Branch, Singapore 24630
National Library of Algeria, Alger 00024
National Library of Angola, Luanda 00102
National Library of Anthropology and History, México 18835
National Library of Armenia, Yerevan 00325
National Library of Australia, Canberra 00365
National Library of Bangladesh, Dhaka 01745
National Library of Belarus, Minsk 01783
National Library of Bhutan, Thimphu 02913
National Library of Brazil, Rio de Janeiro 02981
National Library of Burkina Faso, Ouagadougou 03611
National Library of Cambodia (NLC), Phnom Penh 03629
National Library of Cape Verde, Praia 04830
National Library of Chile, Santiago 04842
National Library of China, Beijing 04950
National Library of Colombia, Santafé de Bogotá 05929
National Library of Costa Rica, San José 06105
National Library of Education, București 21972
National Library of Education, København 06666
National Library of Education, Washington 35003
National Library of El Salvador, San Salvador 07200
National Library of Estonia, Tallinn 07219
National Library of Ethiopia, Addis Ababa 07260
National Library of Finland, Helsinki 07326
National Library of Foreign Literature, Budapest 13440
National Library of France, Paris 07623
National Library of Georgia, Tbilisi 09183
National Library of Greece, Athinai 13068
National Library of Guatemala, Guatemala City 13172
National Library of Guyana, Georgetown 13193
National Library of India, Kolkata 13586
National Library of Iran, Tehran 14385
National Library of Ireland / Leabharlann Náisiúnta na hÉireann, Dublin 14458
National Library of Izmir, Konak 27845
National Library of Jamaica, Kingston 16863
National Library of Korea, Seoul 18056
National Library of Kuwait, Safat 18126
National Library of Laos, Vientiane 18164
National Library of Latvia, Riga 18166
National Library of Libya, Benghazi 18235
National Library of Luxembourg, Luxembourg 18379
National Library of Malaysia, Kuala Lumpur 18496
National Library of Malta, Valletta 18657
National Library of Medicine, Bethesda 36537
National Library of Mexico, México 18707
National Library of Mozambique, Maputo 18975
National Library of Myanmar, Yangon 18984
National Library of Namibia, Windhoek 19007
National Library of New Zealand, Wellington 19760
National Library of Nigeria (NLN), Lagos 19914
National Library of Norway, Oslo 20065
National Library of Pakistan, Islamabad 20426
National Library of Panama, Panamá 20534
National Library of Periodicals, México 18844
National Library of Peru, Lima 20612
National Library of Portugal, Lisboa 21619
National Library of Republic of Karelia, Petrozavodsk 22167
National Library of Republic Sakha, Yakutsk 22196
National Library of Romania, București 21914
National Library of Romania – Batthyaneum Branch, Alba Iulia 21913
National Library of Russia (NLR), Sankt-Peterburg 22115
National Library of Scotland, Edinburgh 28912
National Library of Serbia, Beograd 24453
National Library of Smederevo, Smederevo 24551
National Library of Solomon Islands, Honiara 24991
National Library of South Africa (NLSA) – Cape Town Division, Cape Town 24995
National Library of South Africa (NLSA) – Pretoria Division, Pretoria 24996
National Library of Spain, Madrid 25245
National Library of Sweden, Stockholm 26370
National Library of Technical and Scientific Information, Chișinău 18879
National Library of Thailand, Bangkok 27682
National Library of the Congo Democratic Republic, Kinshasa 06044
National Library of the Cook Islands, Rarotonga 06103
National Library of the Czech Republic, Praha 06342
National Library of the Dominican Republic, Santo Domingo 06953
National Library of the Faroe Islands, Tórshavn 06631
National Library of the Kyrgyz Republic, Bishkek 18133
National Library of the Maldives, Male 18639
National Library of the Netherlands, Den Haag 19046
National Library of the Republic of Albania, Tiranë 00006
National Library of the Republic of Bashkortostan, Ufa 22183
National Library of the Republic of Indonesia, Jakarta 14235
National Library of the Republic of Kazakhstan, Almaty city 17888
National Library of the Republic of Moldova, Chișinău 18878
National Library of the Republic of Tatarstan, Kazan 22138
National Library of Tunisia, Tunis 27813
National Library of Turkey, Ankara 27843
National Library of Turkmenistan, Ashkhabad 27903
National Library of Udmurt Republic, Izhevsk 22135
National Library of Uganda (NLU), Kampala 27929
National Library of Vanuatu, Port-Vila 42179
National Library of Venezuela, Caracas 42203
National Library of Vietnam (NLV), Hanoi 42263
National Library of Wales / Llyfrgell Genedlaethol Cymru, Aberystwyth 28911
National Library – Queenstown Branch, Singapore 24631
National Library Service, Bridgetown 01782
National Library Service of Belize, Belize City 02896
National Library Service of Papua New Guinea, Waigani 20544
National Life Insurance Co Library, Montpelier 35920
National Marine Biological Library, Plymouth 29879
National Marine Fisheries Service, Honolulu 36937
National Marine Fisheries Service, Juneau 36980
National Marine Fisheries Service, Oxford 37315
National Marine Fisheries Service, Seattle 37592
National Maritime Museum, London 29801
National Maritime Museum, Stockholm 26678
National Measurement Institute, Lindfield 00940
National Medical Library, New Delhi 14160
National Medical Library, Praha 06505
National Medical University, Kyiv 28018
National Metallurgical Academy of Ukraine, Dnipropetrovsk 27962
National Metallurgical Laboratory (NML), Jamshedpur 14016
National Meteorological Department, México 18819
National Meteorological Library and Archive, Exeter 29669
National Mining University of Ukraine, Dnipropetrovsk 27967
National Monuments Council, Cape Town 25115
National Monuments Record of Scotland, Edinburgh 29657
National Mosque, Kuala Lumpur 18569
National Museum, Jos 20026
National Museum, Lagos 20033
National Museum, Minsk 02039
National Museum, Poznań 21204
National Museum, Praha 06506
National Museum, Vršac 24534
National Museum, Wrocław 21364
National Museum and Art Gallery, Boroko 20588
National Museum in Cracow, Kraków 21145
National Museum in Szczecin, Szczecin 21228
National Museum in Warsaw, Warszawa 21320
National Museum Library, Bloemfontein 25109
National Museum Library, Singapore 24615
National Museum, Monuments and Art Gallery, Gaborone 02978
National Museum of African Art, Washington 37793
National Museum of American Art, Washington 37792
National Museum of American History, Washington 37788
National Museum of American Jewish Military History Library, Washington 37765
National Museum of Anthropology, México 18861
National Museum of Antiquities, Alger 00067
National Museum of Art, București 22008
National Museum of Australia, Mitchell 00962
National Museum of Bosnia and Herzegovina, Sarajevo 02963
The National Museum of Contemporary Art, Oslo 20251
National Museum of Contemporary History, Ljubljana 24925
National Museum of Ethnology, Osaka 17635
National Museum of Gdańsk, Gdańsk 21100
National Museum of History, Taipei 27614
National Museum of Iceland, Reykjavík 13576
National Museum of India, New Delhi 14161
National Museum of Ireland, Dublin 14552
National Museum of Korea, Seoul 18119
National Museum of Lithuania, Vilnius 18312
National Museum of Natural History, Sofiya 03565
National Museum of Natural History 'Grigore Antipa', București 22004
National Museum of Natural History Naturalis, Leiden 19479

National Museum of Natural History (NMNH), New Delhi 14162
National Museum of Nepal, Kathmandu 19036
National Museum of Oriental Art 'Giuseppe Tucci', Roma 16547
National Museum of Przemyśl Region, Przemyśl 21210
National Museum of Science and Technology, The Library, Stockholm 26695
National Museum of Slovenia, Ljubljana 24927
National Museum of the History of Science and Medicine, Leiden 19476
National Museum of the Union, Alba Iulia 21955
National Museum of Turkmenistan, Ashkhabad 27910
National Museum of Wales, Cardiff 29630
National Museum of Western Art, Tokyo 17714
National Museum of Women in the Arts, Washington 37766
National Museum Paleis Het Loo, Apeldoorn 19390
National Museums Library, København 06815
National Museums of Kenya, Nairobi 18043
National Museums of World Culture, Göteborg 26585
National Museums Scotland, Edinburgh 29658
National Mutual Insurance Federation of Agricultural Cooperatives, Library of Zenkyoren, Tokyo 17779
National Native Title Tribunal, Perth 00983
The National Naval Medical Center, Bethesda 36538
National Observatory, Athinai 13106
National Oceanic & Atmospheric Administration, Asheville 33937
National Oceanic & Atmospheric Administration, Princeton 34709
National Oceanic & Atmospheric Administration, Seattle 37593
National Oceanic & Atmospheric Administration, Silver Spring 37607
National Office of Geological and Mining Research, Alger 00069
National Office of Measurements, Budapest 13322
National Office of Mines – Tunisia, Tunis 27840
National Office of Overseas Skills Recognition, Canberra 00650
National Opinion Research Center, Chicago 36684
National Optical Astronomy Observatories, Tucson 37688
National Outdoor Leadership School (NOLS), Lander 37011
National Palace Museum Library, Taipei 27615
National Park Service, Death Valley 36781
National Park Service, Harpers Ferry 34338
National Park Service, Jensen 36977
National Park Service, Lakewood 37007
National Park Service, Lincoln 37032
National Park Service, Morristown 37131
National Park Service Independence National Historical Park, Philadelphia 37353
National Park Service-Yellowstone Association, Yellowstone National Park 37863
National Parks and Wildlife Service, Hurstville 00690
National Parliamentary Library, Taipei 27581
National Parliamentary Library of Georgia, Tbilisi 09208
National Parliamentary Library of Ukraine, Kyiv 28134
National Pedagogical Library, Firenze 16255
National Pedagogical Library, Sofiya 03562
National People's Congress, Beijing 05810
National Personnel Authority, Tokyo 17303
National Petroleum Ltd, Kuala Lumpur 18572
National Physical Laboratory, New Delhi 14163
National Physical Laboratory, Teddington 29908
National Pingtung University of Science and Technology, Ping-Tung 27535
National Planning Commission, Kathmandu 19037
National Police Agency, Tokyo 17308
National Police Library, Hook 29686
National Polytechnical Museum, Sofiya 03563
National Popular Library, Brazzaville 06093
National Portrait Gallery, London 29802
The National Portrait Gallery, Washington 37792
National Postal Museum Library, Washington 37790
National Power Company, Reykjavík 13563
National Productivity Centre, Petaling Jaya 18614
National Productivity Council, New Delhi 13891
National Public Library, Monrovia 18228
National Public Radio Broadcasting Library, Washington 37767
National Radio Astronomy Observatory Library, Charlottesville 36654
National Radiological Protection Board, Chilton 29636
National Railway Historical Society Library, Philadelphia 37354
National Railway Museum, York 29919
National Rayon Corporation Ltd, Mohone 13907
National Records Office, Khartoum 26358
National Reference Center for Bioethics Literature, Washington 37768
National Rehabilitation Information Center (NARIC), Landover 37012
National Renewable Energy Laboratory Library, Golden 36897
National Repository Library, Kuopio 07327
National Repository Library, Umeå 26937
National Research and Development Corporation, New Delhi 13911
National Research Center for Working Envirobnment, København 06814
National Research Centre, Cairo 07182
National Research Centre of Archaeology, Jakarta 14375
National Research Institute, Boroko 20589
National Research Institute for Amazonia, Manaus 03308
National Research Institute for Cultural Properties, Tokyo 17766
National Research Institute of Agricultural Engineering, Tsukuba 17804
National Research Institute of Agricultural Engineering, Tsukuba 17799
National Research Institute of Animal Production, Balice 21071
National Research Institute of Aquaculture, Watarai 17816
National Research Institute of Astronomy and Geophysics, Cairo 07183
National Research Institute of Legal Policy, Internal Library, Helsinki 07479
National Research Institute of Vegetables, Ornamental Plants and Tea, Haibara 17541
National Research Laboratory for Conservation of Cultural Property, Lucknow 14070
National Research Laboratory of Metrology, Tsukuba 17800
National Resistance Institute, Jakarta 14371
National Resources Canada – Resource Info Center, Ottawa 04117
National Romanian Literature Museum, Bucureşti 22007
National Root Crops Research Institute (NRCRI), Umuahia 20042
National Safety Council Library, Itasca 36965
National School of Drama, New Delhi 13728
National School of Public Administration, Warszawa 21039
National Science and Technology Library and Information Centre, Accra 13055
National Science Council, Taipei 27616
National Science Library, New Delhi 14142
The National Science Museum, Tokyo 17715
National Scientific and Technical Information Center (NSTIC), Safat 18132
National Scientific Centre of Rehabilitation and Physiotherapy, Moskva 23367
National Scientific Medical Society of Rheumatologists, Moskva 23368
National Scientific Research Institute of Tuberculosis, Tbilisi 09270
National Sea Grant Library, Narragansett 37141
National Security Technologies, Las Vegas 37018
National Semiconductor Corp, Santa Clara 36256
National Service for Archaeological Heritage (ROB), Amersfoort 19344
National Social Insurance Board, Stockholm 26512
National Social Science Documentation Centre, New Delhi 14164
National Society for the Prevention of Cruelty to Children (NSPCC), London 29803
National Society of the Daughters of the American Revolution, Washington 37769
National Society of the Sons of the American Revolution, Louisville 37070
National Solar Observatory, Sunspot 37650
National Space Development Agency, Tokyo 17326
National Sporting Library, Inc, Middleburg 37102
National Standardisation Institute, Santiago 04935
National State Library of the Kabardino-Balkarian Republic, Nalchik 22157
National Statistical Institute, Sofiya 03494
National Statistical Office, Bangkok 27763
National Statistical Office, La Paz 02939
National Statistical Office, Madrid 26058
National Statistical Office, Panamá 20540
National Statistical Office, Zomba 18482
National Statistical Service of Greece, Athinai 13121
National Statistics Information & Library Services, London 29526
National Sugar Institute, Kanpur 14022
National Sun Yat-Sen University, Kaohsiung 27533
National Suprme Court Library, Buenos Aires 00208
National Swedish Police Board, Stockholm 26513
National Széchényi Library, Budapest 13213
National Szekler Museum, Sfântu Gheorghe 22055
National Taichung Institute of Commerce, Taichung 27570
National Taichung Teachers College, Taichung 27539
National Taipei College of Nursing, Taipei 27549
National Taipei Teachers College, Taipei 27550
National Taitung Teachers College, Taitung 27566
National Taiwan College of Arts, Taipei 27551
National Taiwan Institute of Technology, Taipei 27552
National Taiwan Normal University, Taipei 27553
National Taiwan Ocean University, Keelung 27534
National Taiwan University, Taipei 27554
– College of Law and Social Sciences and Public Health, Taipei 27555
– Medical Library, Taipei 27556
National TB and Respiratory Diseases Institute, Bratislava 24700
National Technical Library, Praha 06509
National Technical Museum, Praha 06510
National Technical University "Kharkovy Politechnical Institute", Kharkiv 27997
National Technical University of Athens (NTUA), Athinai 13076
National Test House, Kolkata 14055
National Theater, Sofiya 03558
National Theater of Japan, Tokyo 17696
National Theatre Conservatory, Denver 36796
National Theatre Museum, Ljubljana 24937
National Theatre School of Canada Library, Montreal 04009
National Tropical Botanical Garden Library, Kalaheo 36981
National Tsing-Hua University, Hsinchu 27528
National University, San Diego 32577
National University of Agriculture, Managua 19890
National University of Asunción, San Lorenzo 20603
National University of Health Sciences, Lombard 31636
National University of Ireland, Galway 14484
– James Hardiman Library, Galway 14485
National University of Ireland, Maynooth, Maynooth 14490
National University of Laos, Vientiane 18165
National University of Lesotho, Roma 18226
– Faculty of Agriculture, Maseru 18227
National University of Malaysia, Bangi 18497
– Medical Branch Library, Kuala Lumpur 18499
– Sabah Campus, Kota Kinabalu 18498
National University of Mongolia, Ulaanbaatar 18906
National University of Music, Bucureşti 21923
National University of Pharmacy, Kharkiv 27996
National University of Physical Education and Sport of Ukraine, Kyiv 28019
National University of Singapore, Singapore 24585
– Chinese Library, Singapore 24586
– Hon Sui Sen Memorial Library, Singapore 24587
– C. J. Koh Law Library, Singapore 24588
– Medical Library, Singapore 24589
– Science Library, Singapore 24590
National Veterinary Research Institute (NVRI), Puławy 21214
National Veterinary Research Institute (NVRI), Vom 20044
National veterinary School, Lyon Library, Marcy-l'Etoile 07749
National Vine Growing and Wine Producing Institute, Mendoza 00305
National War Museum of Scotland, Edinburgh 29659
National Water Research Institute, Saskatoon 04489
National Women's Education Centre, Information and International Exchange Division, Saitama 17648
National Yiddish Book Center, Amherst 36463
National Youth Agency, Leicester 29697
Nationalarchiv der Richard-Wagner-Stiftung, Bayreuth 11482
Nationalbibliotek for mennesker med læsevanskeligheder, København 06817
Nationale Confederatie van het Bouwbedrijf (NCB), Brussel 02614
Det Nationale ForskningsCenter for Arbejdsmiljø, København 06814
Det Nationale Forskningscenter for Velfærd, København 06820
Nationale Instelling voor Radioactief Afval en Verrijkte Splijtstoffen (NIRAS), Brussel 02615
Nationale Maatschappij der Belgische Spoorwegen (NMBS), Brussel 02616
Nationale Plantentuin van België / Jardin botanique national de Belgique, Meise 02686
National-Louis University, Wheeling 33046
– Evanston Campus Library, Evanston 33047
Nationalmuseets Bibliotekstjeneste, København 06815
Nationalna biblioteka Ukrainy im. V. I. Vernadskoho, Kyiv 27930
Nationalna biblioteka Ukrainy im. V. I. Vernadskoho (filiala 2), Kyiv 27931
Nationjal Statistical Office, Santafé de Bogotá 05994
Nationwide Library, Columbus 36742
Native Museum, San José 06123
NATO Multimedia Library, Bruxelles 02433
Natrona County Public Library, Casper 38476
Natsionalen institut po meteorologia i hidrologia, Sofiya 03564
Natsionalen prirodonauchen muzei, Sofiya 03565
Natsionalna Akademia obrazotvorchogo mistetstva i arkhitekturi, Kyiv 28360
Natsionalna jurydychna akademiya im. Yaroslava Mydroho, Kharkiv 27994
Natsionalna naukova medychna biblioteka Ukrainy, Kyiv 28361
Natsionalna parlamentska biblioteka Ukrainy, Kyiv 28134
Natsionalnaya biblioteka Belarusi, Minsk 01783
Natsionalnaya Biblioteka Chuvashskoi Respubliki, Cheboksary 22124
Natsionalnaya biblioteka im. Akhmet-Zaki Validi Respubliki Bashkortostan, Ufa 22183
Natsionalnaya biblioteka im. A.S. Pushkina Respubliki Mordoviya, Saransk 22173
Natsionalnaya biblioteka im. A.S. Pushkina Respubliki Tyva, Kyzyl 22148
Natsionalnaya biblioteka im. S.G. Chavaina, Yoshkar-Ola 22199
Natsionalnaya biblioteka Respubliki Adygeya, Maikop 24165
Natsionalnaya biblioteka Respubliki Buryatiya, Ulan-Ude 22185
Natsionalnaya biblioteka respubliki Kareliya, Petrozavodsk 22167
Natsionalnaya biblioteka Respubliki Komi, Syktyvkar 22176
Natsionalnaya biblioteka Respubliki Sakha (Yakutiya), Yakutsk 22196
Natsionalnaya biblioteka Respubliki Tatarstan / Tatarstan Respublikasinin Milli Kitapxanese, Kazan 22138
Natsionalnaya biblioteka Udmurtskoi Respubliki, Izhevsk 22135
Natsionalny aerokosmicheski universitet im. N.E. Zhukovskogo / "Zhukovski aviatsionny institut", Kharkiv 27995
Natsionalny agrarny universytet, Kyiv 28016
Natsionalny Arkhiv Belarusi, Minsk 02037
Natsionalny arkhiv Respubliki Kareliya, Petrozavodsk 23704
Natsionalny aviatsiny universytet, Kyiv 28017
Natsionalny botanichny sad im. M.M. Hryshka, Kyiv 28362
Natsionalny farmatsevtichny universytet, Kharkiv 27996

Natsionalny khudozhestvenny muzei, Minsk 02038
Natsionalny medychny universytet im. O.O. Bogomoltsya, Kyiv 28018
Natsionalny muzei Respubliki Komi, Syktyvkar 23986
Natsionalny tekhnichesky universytet / "Kharkovski politekhnichesky institut", Kharkiv 27997
Natsionalny universytet fizychnoho vykhovannya i sportu Ukrainy, Kyiv 28019
Natsiyanalny muzei, Minsk 02039
Náttúrufraedistofnun Islands, Gardabaer 13567
Natural Gas Transportation Research and Planning Institute, Kyiv 28378
Natural History Museum, Beograd 24519
Natural History Museum, Göteborg 26577
Natural History Museum, London 29804
Natural History Museum, Tring 29910
Natural History Museum Library, Århus 06771
Natural History Museum of Los Angeles County, Los Angeles 37058
Natural History Society of Maryland, Inc Library, Baltimore 36515
Natural History Society of Northumbria, Newcastle upon Tyne 29857
Natural Resources Canada, Fredericton 04074
Natural Resources Canada, Vancouver 04175
Natural Resources Canada – Earth Science Information Centre, Ottawa 04118
Natural Resources Canada – Energy Minerals & Metalls Information Centre, Ottawa 04119
Natural Resources Development College, Lusaka 42326
Natural Science Museum, Plovdiv 03514
Natural Sciences Museum, Bacău 21956
Naturalis Bibliotheek, Leiden 19479
Nature Cure Ashram, Pune 14198
Nature Kenya, Nairobi 18044
Naturhistorischer Verein der Rheinlande und Westfalens, Bonn 11645
Naturhistorisches Museum, Basel 27309
Naturhistorisches Museum, Bern 27324
Naturhistorisches Museum, Wien 01624
Naturhistorisches Museum Schloss Bertholdsburg, Schleusingen 12472
Naturhistorisk Museums Bibliotek, Århus 06771
Naturhistoriska riksmuseet, Stockholm 26670
Naturkundemuseum Erfurt, Erfurt 11814
Naturkundemuseum im Ottoneum, Kassel 12102
Naturkundemuseum Leipzig, Leipzig 12197
Natur-Museum, Luzern 27405
Naturvårdsverket, Stockholm 26671
Naturwissenschaftlicher Verein Darmstadt, Messel 12264
Naturwissenschaftlich-Technische Akademie (nta), Isny 10051
Natuurhistorisch Museum Maastricht, Maastricht 19489
Nauchen institut po konservna promishlenost, Plovdiv 03515
Nauchen institut po pediatriya, Sofiya 03566
Nauchnaya biblioteka im. Gorkoho, Odesa 27944
Nauchnaya biblioteka Komi Nauchnogo Tsentra, Syktyvkar 22177
Nauchnaya biblioteka profsoyuzov, Moskva 23369
Nauchnaya meditsinskaya biblioteka NII SP im. N.V. Sklifosovskogo, Moskva 23370
Nauchno-doslidny instytut budivelnykh konstruktsi, Kyiv 28363
Nauchno-issledovatelski ekonomicheski institut, Minsk 02040
Nauchno-issledovatelski eksperimentalny institut avtomobilnoi elektroniki i elektrooborudovaniya, Moskva 23371
Nauchno-issledovatelski fiziko-khimicheski institut im. L.Ya. Karpova, Moskva 23372
Nauchno-issledovatelski Glavmosstroi, Moskva 23373
Nauchno-issledovatelski gosudarstvenny muzei arkhitektury im. A.S. Shchuseva, Moskva 23374
Nauchno-issledovatelski i eksperimentalno-konstruktorski institut tary i upakovki, Kaluga 23074
Nauchno-issledovatelski i eksperimentalny institut po pererabotke khimicheskikh volokon, Moskva 23375
Nauchno-issledovatelski i eksperimentalny institut torgovogo mashinostroeniya, Moskva 23376
Nauchno-issledovatelski i konstruktorski institut ispytatelnykh mashin, priborov i sredstv izmereniya mass, Moskva 23377
Nauchno-issledovatelski i proektno-izyskatelski institut po proektirovaniyu vodoprovodov i kanalizatsionnykh sooruzheni Moskvy, Moskva 23378
Nauchno-issledovatelski i proektny institut neftepererabatyvayushchei i neftekhimicheskoi promyshlennosti, Moskva 23379
Nauchno-issledovatelski i proektny institut ugolnoi promyshlennosti, Moskva 23380
Nauchno-issledovatelski i tekhnologicheski institut, Kashintsevo 23079
Nauchno-issledovatelski institut, Samara 23742
Nauchno-issledovatelski institut Agrolesomelioratsii, Volgograd 24063
Nauchno-issledovatelski institut akusherstva i ginekologii im. Otta, Sankt-Peterburg 23816
Nauchno-issledovatelski institut antibiotikov, Moskva 23381
Nauchno-issledovatelski institut avtomobilnogo transporta, Moskva 23382
Nauchno-issledovatelski institut chasovoi promyshlennosti, Moskva 23383
Nauchno-issledovatelski institut derevoobrabatyvayushchego mashinostroeniya, Moskva 23384
Nauchno-issledovatelski institut detskikh infektsi, Sankt-Peterburg 23817
Nauchno-issledovatelski institut detskoi ortopedii im. G.I. Turnera, Sankt-Peterburg 23818
Nauchno-issledovatelski institut Domen, Sankt-Peterburg 23819
Nauchno-issledovatelski institut ekonomiki selskogo khozyaistva, Moskva 23385
Nauchno-issledovatelski institut ekonomiki selskokhozyaistvennogo proizvodstva Nechernozyomnoi zony Rossii, Sankt-Peterburg 23820
Nauchno-issledovatelski institut ekonomiki stroitelstva, Moskva 23386
Nauchno-issledovatelski institut elektroenergetiki (VNIIE), Moskva 23387
Nauchno-issledovatelski institut elektrofizicheskoi apparatury im. D.V. Efremova, Sankt-Peterburg 23821
Nauchno-issledovatelski institut elektrokeramiki, Moskva 23388
Nauchno-issledovatelski institut elektromekhaniki, Moskva 23389
Nauchno-issledovatelski institut Elektrostandart, Sankt-Peterburg 23822
Nauchno-issledovatelski institut elektrotekhnicheskikh ustanovok, Sankt-Peterburg 23823
Nauchno-issledovatelski institut elektrotermicheskogo oborudovaniya, Moskva 23390
Nauchno-issledovatelski institut elektrougolnykh izdeli, Elektrougli 23039
Nauchno-issledovatelski institut Energosetproekt, Moskva 23391
Nauchno-issledovatelski institut epidemiologii i mikrobiologii, Kazan 23083
Nauchno-issledovatelski institut epidemiologii i mikrobiologii, Khabarovsk 23099
Nauchno-Issledovatelski Institut epidemiologii i mikrobiologii, Minsk 02041
Nauchno-issledovatelski institut epidemiologii i mikrobiologii, Nizhni Novgorod 23618
Nauchno-issledovatelski Institut epidemiologii i mikrobiologii im. Pastera, Sankt-Peterburg 23824
Nauchno-issledovatelski institut epidemiologii i mikrobiologii im. Y.N. Gabrichevskogo, Moskva 23392
Nauchno-issledovatelski institut farmatsii, Moskva 23393
Nauchno-issledovatelski institut fizicheskoi kultury, Sankt-Peterburg 23825
Nauchno-issledovatelski institut fiziko-tekhnicheskikh i radio-tekhnicheskikh izmereni, Mendeleevo 23154
Nauchno-issledovatelski institut ftisiopulmonologii, Sankt-Peterburg 23826
Nauchno-issledovatelski institut galurgii, Sankt-Peterburg 23827
Nauchno-issledovatelski institut geofizicheskikh metodov razvedki, Moskva 23394
Nauchno-issledovatelski institut geofiziki i mineralnogo syrya, Novosibirsk 23651
Nauchno-issledovatelski institut geologii i razrabotki goryuchikh iskopaemykh, Moskva 23395
Nauchno-issledovatelski institut gidrogeologii i inzhenernoi geologii, Staraya Kupavna 23980
Nauchno-issledovatelski institut gidrologiya, Novokuznetsk 23635
Nauchno-issledovatelski institut gidrometeorologicheskogo priborostroeniya, Moskva 23396
Nauchno-issledovatelski institut gidrometeorologicheskoi informatsii, Moskva 23397
Nauchno-issledovatelski institut gigieny i profzabolevani, Ufa 24029
Nauchno-issledovatelski institut gigieny im. F.F. Erismana, Perlovskaya 23686
Nauchno-issledovatelski institut gigieny truda i professionalnykh zabolevani, Moskva 23398
Nauchno-issledovatelski institut gigieny truda i professionalnykh zabolevani, Sankt-Peterburg 23828
Nauchno-issledovatelski institut gigieny truda i profzabolevani, Ekaterinburg 23031
Nauchno-issledovatelski institut gigieny truda i profzabolevani, Nizhni Novgorod 23619
Nauchno-issledovatelski institut gigieny vodnogo transporta, Moskva 23399
Nauchno-issledovatelski institut Girikond, Sankt-Peterburg 23829
Nauchno-issledovatelski institut glaznykh boleznei im. G. Gelmgoltsa, Moskva 23400
Nauchno-issledovatelski institut globalnogo klimata i ekologii, Moskva 23401
Nauchno-issledovatelski institut gornogo sadovodstva i tsvetovodstva, Sochi 23976
Nauchno-issledovatelski institut gornoi geomekhaniki i markshreiderskogo dela, Sankt-Peterburg 23830
Nauchno-issledovatelski institut gornokhimicheskogo syrya, Lyubertsy 23146
Nauchno-issledovatelski institut grippa, Sankt-Peterburg 23831
Nauchno-issledovatelski institut informatsii i tekhniko-ekonomicheskikh issledovani myasnoi i molochnoi promyshlennosti, Moskva 23402
Nauchno-issledovatelski institut introskopii, Moskva 23403
Nauchno-issledovatelski institut iskusstvennogo volokna, Mytishchi 23604
Nauchno-issledovatelski institut kartofelnogo i ovoshchnogo khozyaistva, Pervoe Maya 23696
Nauchno-issledovatelski institut kartofelnogo khozyaistva, Kraskovo 23110
Nauchno-issledovatelski institut khlopkovodstva, Ak-Kavak 22983
Nauchno-issledovatelski institut khudozhestvennogo vospitaniya Akademii pedagogicheskikh nauk, Moskva 23404
Nauchno-issledovatelski institut khudozhestvennoi promyshlennosti, Moskva 23405
Nauchno-issledovatelski institut kinoiskusstva, Moskva 23406
Nauchno-issledovatelski institut klinicheskoi i eksperimentalnoi khirurgii, Moskva 23407
Nauchno-issledovatelski institut konservnoi promyshlennosti i spetsialnoi pishchevoi tekhnologii, Vidnoe 24042
Nauchno-issledovatelski institut krakhmaloproduktov, Korenevo 23106
Nauchno-issledovatelski institut kurortologii i fizioterapii, Sochi 23977
Nauchno-issledovatelski institut kurortologii i fizioterapii, Tomsk 23996
Nauchno-issledovatelski institut lesnogo khozyaistva SPbNIILKh, Sankt-Peterburg 23832
Nauchno-issledovatelski institut lesnoi promyshlennosti, Irkutsk 23056
Nauchno-issledovatelski institut maslichnykh kultur, Krasnodar 23114
Nauchno-issledovatelski institut meditsinskikh problem formirovaniya zdorovya, Moskva 23408
Nauchno-issledovatelski institut meditsinskogo priborostroeniya, Moskva 23409
Nauchno-Issledovatelski Institut Meditsiny Truda i Ecologii Cheloveka, Angarsk 22985
Nauchno-issledovatelski institut mekhovoi promyshlennosti, Moskva 23410
Nauchno-issledovatelski institut metrologicheskoi sluzhby, Moskva 23411
Nauchno-issledovatelski institut mineralnogo syrya, Moskva 23412
Nauchno-issledovatelski institut molochno-myasnogo skotovodstva, Orenburg 23671
Nauchno-issledovatelski institut monomerov dlya sinteticheskogo kauchuka, Yaroslavl 24087
Nauchno-issledovatelski institut 'Mosstroi' Glavmosstroya, Moskva 23413
Nauchno-issledovatelski institut mozga, Moskva 23414
Nauchno-issledovatelski institut muzeevedeniya Ministerstva kultury, Moskva 23415
Nauchno-issledovatelski institut nerudnykh stroitelnykh materialov i gidromekhanizatsii, Tolyatti 23994
Nauchno-issledovatelski institut nevrologii, Moskva 23416
Nauchno-issledovatelski institut okhrany truda, Sankt-Peterburg 23833
Nauchno-issledovatelski institut onkologii i meditsinskoi radiologii, Lesnoy 01966
Nauchno-issledovatelski institut onkologii im. N.N. Petrova, Sankt-Peterburg 23834
Nauchno-issledovatelski institut organicheskikh poluproduktov krasitelei, Moskva 23417
Nauchno-issledovatelski institut osnovani i podzemnykh sooruzheni, Vorkuta 24075
Nauchno-issledovatelski institut osnovani i podzemnykh sooruzheni im. N.M. Gersevanova, Moskva 23418
Nauchno-issledovatelski institut ovoshchnogo khozyaistva, Grakhi 23046
Nauchno-issledovatelski institut ovoshchnogo khozyaistva, Mytishchi 23605
Nauchno-issledovatelski institut ovoshchnogo khozyaistva, Vereiya 24041
Nauchno-issledovatelski institut ovtsevodstva i kozovodstva, Stavropol 23983
Nauchno-issledovatelski institut ozernogo i rechnogo rybnogo khozyaistva, Sankt-Peterburg 23835
Nauchno-issledovatelski institut pchelovodstva, Rybnoe 23736
Nauchno-issledovatelski institut pedagogiki Belarusi, Minsk 02042
Nauchno-issledovatelski institut pediatrii, Moskva 23419
Nauchno-issledovatelski institut pivovarennoi, bezalkogolnoi i vinodelcheskoi promyshlennosti, Moskva 23420
Nauchno-issledovatelski institut plenochnykh materialov i iskustvennoi kozhi, Moskva 23421
Nauchno-issledovatelski institut po dobyche poleznykh iskopaemykh otkrytym sposobom, Chelyabinsk 23018
Nauchno-issledovatelski institut po peredachne elektroenergii postoyannym tokom ysokogo napryazheniya, Sankt-Peterburg 23836
Nauchno-issledovatelski institut po problemam Kurskoi magnitnoi anomalii, Gubkin 23048
Nauchno-issledovatelski institut po stroitelstvu, Krasnoyarsk 23128
Nauchno-issledovatelski institut po tsenoobrazovaniyu, Moskva 23422
Nauchno-issledovatelski institut po udobreniyam i insektofungitsidam im. Ya.V. Samoilova, Moskva 23423
Nauchno-issledovatelski institut Poisk, Sankt-Peterburg 23837
Nauchno-issledovatelski institut poligraficheskogo mashinostroeniya, Moskva 23424
Nauchno-issledovatelski institut pri Sovete Ministrov Chuvashskoi Respubliki, Cheboksary 23016
Nauchno-issledovatelski institut prirodnoochagovykh infektsi, Omsk 23665
Nauchno-issledovatelski institut produktov brozheniya, Moskva 23425
Nauchno-issledovatelski institut proektovaniyu zhylykh i obshchestvennykh zdani, Sankt-Peterburg 23838
Nauchno-issledovatelski institut promyshlennoi i sanitarnoi ochistke gazov, Moskva 23426
Nauchno-issledovatelski institut promyshlennosti pervichnoi obrabotki lubyanykh volokon, Minsk 02043

Nauchno-issledovatelski institut protezirovaniya, Sankt-Peterburg 23839
Nauchno-issledovatelski institut ptitsepererabtyvayushchei promyshlennosti, Moskva 23427
Nauchno-issledovatelski institut pulmonologii, Sankt-Peterburg 23840
Nauchno-issledovatelski institut pushnogo zverovodstva i krolikovodstva, Udelnaya 24018
Nauchno-issledovatelski institut pushnogo zverovodstva i krolikovodstva Ministerstva selskogo khozyaistva, Rodniki 23717
Nauchno-issledovatelski institut radiatsionnoi gigieny, Sankt-Peterburg 23841
Nauchno-issledovatelski institut radiofiziki, Nizhni Novgorod 23620
Nauchno-issledovatelski institut radionavigatsii i vremeni, Sankt-Peterburg 23842
Nauchno-issledovatelski institut radioveshchatelnogo prisma i akustiki im. A.S. Popova, Sankt-Peterburg 23843
Nauchno-issledovatelski institut razvedeniya i genetiki selskokhozyaistvennykh zhivotnykh, Sankt-Peterburg 23844
Nauchno-issledovatelski institut rentgenologii, radiologii i onkologii, Rostov-na-Donu 23721
Nauchno-issledovatelski institut rezinovoi promyshlennosti, Moskva 23428
Nauchno-issledovatelski institut rezinovoi promyshlennosti, Moskva 23429
Nauchno-issledovatelski institut rezinovykh i lateksnykh izdeli, Moskva 23430
Nauchno-issledovatelski institut rezinovykh i lateksnykh izdeli, Moskva 23431
Nauchno-issledovatelski institut risa, Krasnodar 23115
Nauchno-issledovatelski institut sadovodstva im. Michurina, Michurinsk 23157
Nauchno-issledovatelski institut sanitarnoi tekhniki, Moskva 23432
Nauchno-issledovatelski institut selskogo khozyaistva im. Lukyanendu, Krasnodar 23116
Nauchno-issledovatelski institut selskogo khozyaistva Krainego Severa, Norilsk 23627
Nauchno-issledovatelski institut selskogo khozyaistva Severo-Vostoka, Kirov 23103
Nauchno-issledovatelski institut selskogo khozyaistva Yugo-Vostoka, Saratov 23958
Nauchno-issledovatelski institut selskoi gigieny, Saratov 23959
Nauchno-issledovatelski institut shinnoi promyshlennosti, Moskva 23433
Nauchno-issledovatelski institut sinteticheskikh smol, Vladimir 24048
Nauchno-issledovatelski institut sinteticheskogo kauchuka, Sankt-Peterburg 23845
Nauchno-issledovatelski institut skoroi pomoshchi, Sankt-Peterburg 23846
Nauchno-issledovatelski institut skoroi pomoshchi im. N.V. Sklifosovkogo, Moskva 23370
Nauchno-issledovatelski institut svarki, Sankt-Peterburg 23847
Nauchno-issledovatelski institut tabaka i makhorki im. A.I. Mikoyana, Krasnodar 23117
Nauchno-issledovatelski institut tekhniko-ekonomicheskikh issledovani, Moskva 23434
Nauchno-issledovatelski institut tekhnologii avtomobilnoi promyshlennosti, Moskva 23435
Nauchno-issledovatelski institut tekhnologii mashinostroeniya, Rostov-na-Donu 23722
Nauchno-issledovatelski institut tekstilnoi promyshlennosti, Sankt-Peterburg 23848
Nauchno-issledovatelski institut televideniya, Sankt-Peterburg 23849
Nauchno-issledovatelski institut teploenergeticheskogo priborostroeniya, Moskva 23436
Nauchno-issledovatelski institut tipovogo i eksperimentalnogo proektirovaniya, Moskva 23437
Nauchno-issledovatelski institut tochnoi mekhaniki, Sankt-Peterburg 23850
Nauchno-issledovatelski institut toksikologii, Sankt-Peterburg 23851
Nauchno-issledovatelski institut transportnogo mashinostroeniya, Sankt-Peterburg 23852
Nauchno-issledovatelski institut travmatologii i ortopedii, Ekaterinburg 23032
Nauchno-issledovatelski institut travmatologii i ortopedii, Kazan 23084
Nauchno-issledovatelski institut travmatologii i ortopedii, Novosibirsk 23652
Nauchno-issledovatelski institut travmatologii i ortopedii im. prof. R.R. Vredena, Sankt-Peterburg 23853
Nauchno-issledovatelski institut travmatologii ortopedii, Saratov 23960
Nauchno-issledovatelski institut truda, Moskva 23438
Nauchno-issledovatelski institut tsementnoi promyshlennosti, Moskva 23439
Nauchno-issledovatelski institut Upravleniya, Moskva 23440
Nauchno-issledovatelski institut vaktsin i suvorotok, Sankt-Peterburg 23854
Nauchno-issledovatelski institut vaktsin i syvorotok, Perm 23688
Nauchno-issledovatelski institut vaktsin i syvorotok, Tomsk 23997
Nauchno-issledovatelski institut vaktsin i syvorotok im. Y.Y. Mechnikova, Moskva 23441
Nauchno-issledovatelski institut Vektor, Sankt-Peterburg 23855
Nauchno-issledovatelski institut virusnykh infektsii, Ekaterinburg 23033
Nauchno-issledovatelski institut virusnykh preparatov, Moskva 23442
Nauchno-issledovatelski institut zashchity rasteni, Sankt-Peterburg 23856
Nauchno-issledovatelski institut zhirov, Sankt-Peterburg 23857
Nauchno-issledovatelski institut zolota i redkikh metallov, Magadan 23148
Nauchno-issledovatelski konstruktorskotekhnologicheski institut prirodnykh almazov i instrumenta, Moskva 23443
Nauchno-issledovatelski mashinostroeniya institut, Moskva 23444
Nauchno-issledovatelski pediatricheski institut, Nizhni Novgorod 23621
Nauchno-issledovatelski, proektno-konstruktorsky i tekhnologichesky instytut vzryvozashchishchennoho i rudnichnoho elektrooborudovaniya, Donetsk 28225
Nauchno-issledovatelski profilakticheskoi institut toksikologii i dezinfektsii, Moskva 23445
Nauchno-issledovatelski protivochumny institut Kavkaza i Zakavkazya, Stavropol 23984
Nauchno-issledovatelski protivochumny institut 'Mikrob', Saratov 23961
Nauchno-issledovatelski tekhnokhimicheski institut bytovogo obsluzhyvaniya, Moskva 23446
Nauchno-issledovatelski tsentr po tekhnologicheskim lazeram RAN, Shatura 23968
Nauchno-issledovatelski veterinarny institut ptitsevodstva, Sankt-Peterburg 23858
Nauchno-issledovatelski zonalny institut sadovodstva Nechernozemnoi zony, Moskva 23447
Nauchno-issledovatelsky instytut gematologiyi i perelivaniya krovi, Kyiv 28364
Nauchno-issledovatelsky instytut mashin dlya proizvodstva sinteticheskykh volokon, Chernigiv 28190
Nauchno-issledovatelsky instytut svinovodstva, Poltava 28450
Nauchnoizsledovatelski stroitelen institut, Sofiya 03567
Nauchno-memorialny muzei N.E. Zhukovskogo, Moskva 23448
Nauchny tsentr akusherstva, ginekologii i perinatologii, Moskva 23449
Nauchny tsentr biologicheskikh issledovani RAN v Pushchino, Pushchino 23708
Naukova biblioteka im. M. Maksymovycha, Kyiv 28020
Naukova medychna biblioteka, Lviv 28421
Naukovo-doslidny girnichorudny instytut, Krivi Rig 28298
Naukovo-doslidny insitut girnichoryatuvalnoyi spravi, Donetsk 28226
Naukovo-doslidny insitut urologiyi ta nefrologiyi, Kyiv 28365
Naukovo-doslidny instytut chornoyi metalurgiyi, Donetsk 28227
Naukovo-doslidny instytut ekonomiki, Kyiv 28366
Naukovo-doslidny instytut elektromekhanichnykh priladiv, Kyiv 28367
Naukovo-doslidny instytut epidemiologi ta infektsinykh zakhvoryuvan im. L.V. Gromashevskoho, Kyiv 28368
Naukovo-doslidny instytut farmakologiyi ta toksikologiyi, Kyiv 28369
Naukovo-doslidny instytut gastroenterologiyi, Dnipropetrovsk 28205
Naukovo-doslidny instytut gigieny pratsy i profzakhviryuvan, Krivi Rig 28299
Naukovo-doslidny instytut girnichorud-noho mashinobuduvannya, Krivi Rig 28300
Naukovo-doslidny instytut gorichnoyi mekhaniky im. M.M. Fyodorova, Donetsk 28228
Naukovo-doslidny instytut kardiologiyi im. M.D. Strazheska, Kyiv 28370
Naukovo-doslidny instytut klinichnoyi ta eksperimentalnoyi khirurgiyi, Kyiv 28371
Naukovo-doslidny instytut meditsiny pratsi, Kyiv 28372
Naukovo-doslidny instytut ochnykh zakhvoryuvan i tkanevoyi terapiyi im. Filatova, Odesa 28439
Naukovo-doslidny instytut onkologiyi ta radiologiyi, Kyiv 28373
Naukovo-doslidny instytut pedahohiky Ukrainy, Kyiv 28374
Naukovo-doslidny instytut reaktivov ta khimichno chistykh materialov dlya elektronnoyi tekhniki, Donetsk 28229
Naukovo-doslidny instytut ribnigo gospodarstva, Kyiv 28375
Naukovo-doslidny instytut spadkovoyi patologiyi, Lviv 28422
Naukovo-doslidny instytut vibukhozakhishchenoho elektrobladnannya, Donetsk 28230
Naukovo-doslidny instytut zakhistu roslin, Kyiv 28376
Naukovo-doslidny ta proektno-konstruktorsky instytut Energoproekt, Kyiv 28377
Naukovo-doslidny vugilny instytut, Donetsk 28231
Naukovo-proektny instytut po transportu prirodnoho gaza, Kyiv 28378
Naukovo-vibronichne obedenanne Avtovazhmash, Lviv 28423
Naukovo-vibronichno obednannya Impuls, Severodonetsk 28465
Naukovo-virobnichno obednannya 'Instytut avtomatiki', Kyiv 28379
Naukovo-virobnichno obednannya kompleksnoyi avtomatizatsiyi ta mekhanizatsiyi tekhnologiyi Kamet, Kyiv 28159
Naukovo-virobnicho obednannya Bolshovik, Tsentralna biblioteka, Kyiv 28678
Nava Nalanda Pali Mahavihara, Bihar 13949
Naval Academy, Laksevåg 20184
Naval Academy Library, Helsinki 07397
Naval Clinical Hospital no 1, Sankt-Peterburg 23929
Naval Engineering School, Sankt-Peterburg 22563
Naval Hospital no 35, Sankt-Peterburg 23928
Naval Institute of Technology, Naval 20752
Naval Military School, Sankt-Peterburg 22564
Naval Museum Library, Karlskrona 26601
Naval Office, Rostock 11085
Naval Postgraduate School, Monterey 31882
Naval Research Laboratory, Washington 37770
Naval Surface Warfare Center, West Bethesda 35053
Naval Surface Warfare Center Dahlgren Div, Dahlgren 36761
Navarro College, Corsicana 33261
Navistar, Inc, Melrose Park 37092
Navy Department Library, Washington 35004
Näytelmäkulma, Helsinki 07476
Nazarene Bible College, Colorado Springs 35166
Nazarene Theological College, Manchester 29267
Nazarene Theological College, Thornlands 00800
Nazarene Theological Seminary, Kansas City 35250
Nazareth College of Rochester, Rochester 32459
Nazionalna hudozhestvena academia 'Nikolai Pavlovich', Sofiya 03462
NBD / Biblion, Leidschendam 19653
NCR Corporation (Teradata Division) Library, San Diego 36220
NDI elektromashinobuduvannya, Kharkiv 28268
NDI Emalkhimmash, Poltava 28451
NDI gigieny pratsi i profesiynykh zakhvoryuvan, Kharkiv 28269
NDI lubyanykh kultur, Glukhiv 28239
NDI metaliv, Kharkiv 28270
NDI mikrobiologiyi ta immunologiyi im. I.I. Mechnikova, Kharkiv 28271
NDI organichnykh napivprovodnkiv i barvnikiv, Rubizhne 28457
NDI organizatsiyi i mekhanizatsiyi shakhtnoho budivnitstva, Kharkiv 28272
NDI ortopediyi i travmatologiyi im. M.I. Sitenka, Kharkiv 28273
NDI roslinnitstva, selektsiyi ta genetiki, Kharkiv 28274
NDI silsko-gospodarskoyi mikrobiologiyi, Chernigiv 28191
NDI tekhnologiyi mashinobuduvannya, Rivne 28455
NDI zagalnoyi i nevidkladnoyi khirurgiyi, Kharkiv 28275
NDI Zemlerobstva i tvarinnitstva zakhidnoho regionu Ukrainy, Obroshine 28435
NDI zroshuvanoho zemlerobstva, Kherson 28288
Ndola Public Library, Ndola 42358
Neal, Gerber & Eisenberg LLP, Chicago 35593
Near East School of Theology, Beirut 18207
Nebraska Christian College, Papillon 32229
Nebraska Department of Corrections, Tecumseh 34921
Nebraska Department of Roads, Lincoln 34461
Nebraska Legislative Council, Lincoln 34462
Nebraska Library Commission, Lincoln 37033
Nebraska State Historical Society Library, Lincoln 37034
Nebraska State Library, Lincoln 30140
Nebraska State Penitentiary Library – Department of Corrections, Lincoln 34463
Nebraska Wesleyan University, Lincoln 31610
Nechernozyom Zone of Russia Research Institute of Agricultural Economics, Sankt-Peterburg 23820
NED University of Engineering and Technology, Karachi 20443
Nederlands Architectuurinstituut (NAI), Rotterdam 19516
Nederlands Bijbelgenootschap (NBG), Haarlem 19455
Nederlands Carmelitaans Instituut (NCI), Boxmeer 19300
Nederlands Instituut voor het Nabije Oosten (NINO), Leiden 19480
Nederlands Instituut voor Onderzoek van de Gezondheidszorg, Utrecht 19528
Nederlands Instituut voor Oorlogsdocumentatie, Amsterdam 19370
Nederlands Kanker Instituut / Antoni van Leeuwenhoek Ziekenhuis (NKI/AVL), Amsterdam 19371
Nederlands Letterkundig Museum, Den Haag 19445
Nederlands Muziek Instituut, Den Haag 19446
Nederlands Nationaal Oorlogs- en Verzetsmuseum, Overloon 19504
Nederlands Openluchtmuseum, Arnhem 19396
Nederlands Research Instituut voor Recreatie en Toerisme (NRIT), Breda 19403
Nederlands Textielmuseum, Tilburg 19521
Nederlandsch Economisch Historisch Archief (NEHA), Amsterdam 19372
De Nederlandsche Bank NV, Amsterdam 19322
Nederlandse Aardolie Maatschappij BV (NAM), Assen 19325
Nederlandse Defensie Academie, Den Haag 19143
Nederlandse Entomologische Vereniging (NEV), Amsterdam 19373
Nederlandse Gasunie, Groningen 19329
Nederlandse Luister- en Braillebibliotheek, Den Haag 19447
Nederlandse Omroep Stichting (NOS), Hilversum 19466
Nederlandse Onderneming voor Energie en Milieu b.v. (NOVEM), Sittard 19520
NV Nederlandse Spoorwegen, Utrecht 19339
Nederlandse Vereniging voor Amateurtheater, Den Haag 19448
Nedre Eiker bibliotek, Mjøndalen 20371
Nedrigailiv Regional Centralized Library System, Main Library, Nedrigailiv 28744
Nedrigailivska raionna TsBS, Tsentralna biblioteka, Nedrigailiv 28744
Needham Free Public Library, Needham 40384
Needham Research Institute, Cambridge 28986

Neelain University Libraries, Khartoum 26344
Neenah Public Library, Neenah 40385
NefAZ, Neftekamsk 22871
Neftegazodobyvayushchee upravlenie Rechitsaneft, Rechitsa 02074
Neftegazovoi nauchno-issledovatelski institut, Moskva 23450
Neftekamsk Automobile Joint-Stock Company, Neftekamsk 22871
Neftekamsk Mechanical Engineering School, Neftekamsk 22684
Neftekamski avtozavod, Neftekamsk 22871
Neftekamski mashinostroitelny tekhnikum, Neftekamsk 22684
Neftepererabatyvayushchi zavod, Kapotnya 22798
Neftepererabatyvayushchi zavod, Mozyr 01909
Neftyanoi institut, Grozny 22253
Neftyanoi institut, Ufa 22611
Negeri Sembilan Public Library Corporation, Seremban 18636
Nehru Memorial Museum and Library, New Delhi 14165
Nei Monggol Autonomous Region Library, Huhhot 04962
Neijiang Teachers' College, Neijiang 05473
Neill Public Library, Pullman 40914
Neivo-Rudyanka Chemistry and Timber Plant Joint-Stock Company, Kirovograd 22806
Nejvyšší soud ČR, Brno 06408
N.A. Nekrasov Central City Public Library, Moskva 24177
Nekrasov Smolny District Library-Branch no 9, Sankt-Peterburg 24237
Nelson County Public Library, Bardstown 38087
Nelson Mandela Metropolitan Municipality : Libraries, Port Elizabeth 25226
Nelson Mandela Metropolitan University, Port Elizabeth 25037
Nelson Memorial Public Library, Apia 24382
Nelson, Mullins, Riley & Scarborough, Columbia 35627
Nelson, Mullins, Riley & Scarborough, Greenville 35728
Nelson, Mullins, Riley & Scarborough, Myrtle Beach 35930
Nelson Public Library, Nelson 19874
Nelson-Atkins Museum of Art, Kansas City 36989
Nelsonville Public Library, Nelsonville 40386
Nelsonville Public Library – Athens Public, Athens 38012
Nelspruit Public Library, Nelspruit 25215
Neman Shoe Industrial Corporation, Grodno 01870
Németh László Public Library, Hódmezővásárhely 13537
Németh László Városi Könyvtár, Hódmezővásárhely 13537
Nemiriv Regional Centralized Library System, Main Library, Nemiriv 28745
Nemirivska raionna TsBS, Tsentralna biblioteka, Nemiriv 28745
Nemirovich-Danchenko, V.Il, Studio-School attached to the Moscow Art Theatre, Moskva 22383
Nemocnica s poliklinikou, Žilina 24758
Nemocnica s Poliklinikou Prievidza, Bojnice 24682
Nemzetközi Számítástechnikai Oktató és Tájékoztató Központ, Budapest 13434
Nemzetközi Üzleti Föiskola, Budapest 13312
M. Nencki Institute of Experimental Biology of the Polish Academy of Sciences, Warszawa 21269
Nenetskaya tsentralnaya biblioteka im. A.I. Puchkova, Naryan-Mar 24182
Nen-Life Science Products Inc, Boston 35532
'Neochim' S.A., Dimitrovgrad 03495
Neosho County Community College, Chanute 33211
Neosho Newton County Library, Neosho 40387
Nepal Commerce Campus Library, Kathmandu 19020
Nepal Engineering College, Bhaktapur 19018
Nepal National Library, Lalitpur 19016
Nepal-Bharat Sanskritik Kendra Pustakalay, Kathmandu 19038
Nepal-India Cultural Centre and Library, Kathmandu 19039
Nepal-India Cultural Centre Library, Kathmandu 19038
Néprajzi Múzeum, Budapest 13435
Neptune Public Library, Neptune 40388
Ner Israel Rabbinical College, Baltimore 30342
Neringos savivaldybės Viktoro Miliūno viešoji biblioteka, Neringa 18344
Nes bibliotek, Årnes 20310
Nesna College of Education, Nesna 20109
Nestle PTC Information Center Library, New Milford 35942
Nestlé Research Center, Lausanne 27288
Nesvizh Region Central Library, Nesvizh 02213
Nesvizheskaya tsentralnaya raionnaya detskaya bibliyateka, Nesvizh 02213
Netherlands Architecture Institute, Rotterdam 19516
Netherlands Bible Society, Haarlem 19455
Netherlands Cancer Institute / Antoni van Leeuwenhoek Ziekenhuis, Amsterdam 19371
Netherlands Defence College, Den Haag 19143
Netherlands Entomological Society Library, Amsterdam 19373
Netherlands Expertise Centre for Arts and Cultural Education, Utrecht 19525
Netherlands Historical and Archaeological Institute, Istanbul 27889
Netherlands Institute for Art History, Den Haag 19450
Netherlands Institute for Cultural Heritage, Amsterdam 19361
Netherlands Institute for Sound and Image, Hilversum 19463
Netherlands Institute for the Near East (NINO), Leiden 19480
Netherlands Institute for War Documentation, Amsterdam 19370
Netherlands Institute of International Relations "Clingendael", Den Haag 19449
Netherlands Maritime Museum Amsterdam, Amsterdam 19358
Netherlands Music Institute, Den Haag 19446
Nettai Nogyo Kenkyu Senta, Tsukuba 17801
Neue Galerie Graz, Graz 01533
Neue Nationalgalerie, Berlin 11579
Die Neue Sammlung, München 12321
Neue Stadtbücherei Augsburg, Augsburg 12636
Neues Museum Weserburg Bremen, Bremen 11685
Neumann College, Aston 30269
Neurozentrum, Freiburg 09797
Neuse Regional Library, Kinston 39731
Neva Mechanical Engineering Plant, Sankt-Peterburg 22918
Nevada Department of Corrections, Ely 34187
Nevada Historical Society, Reno 37434
Nevada Legislative Counsel Bureau, Carson City 36631
Nevada Public Library, Nevada 40390
Nevada State Library & Archives, Carson City 30122
Nevada State Library & Archives – Regional Library for the Blind & Physically Handicapped, Carson City 30123
Nevada Supreme Court Library, Carson City 34050
Nevins Memorial Library, Methuen 40167
Nevski mashinostroitelny zavod, Sankt-Peterburg 22918
Nevsky District Centralized Library System, Central District Children's Library, Sankt-Peterburg 24280
Nevsky District Centralized Library System, Library Branch no 1, Sankt-Peterburg 24279
New Albany-Floyd County Public Library, New Albany 40392
New Bedford Law Library, New Bedford 37151
New Bedford Whaling Museum Research Library, New Bedford 37152
New Berlin Public Library, New Berlin 40394
New Bern-Craven County Public Library, New Bern 40396
New Braunfels Public Library, New Braunfels 40397
New Britain General Hospital, New Britain 37154
New Britain Public Library, New Britain 40398
New Brunswick Community College, Moncton 03756
New Brunswick Department of Education Library, Fredericton 04346
New Brunswick Department of Natural Resources Library, Fredericton 04075
New Brunswick Department of the Environment & Local Government Library, Fredericton 04076
New Brunswick Department of Training & Employment Development Library, Fredericton 04077
New Brunswick Emergency Measures Organization Library, Fredericton 04347
New Brunswick Free Public Library, New Brunswick 40399
New Brunswick Museum, Saint John 04478
New Brunswick Teachers' Federation, Fredericton 04348
New Brunswick Theological Seminary, New Brunswick 35300
New Buffalo Township Public Library, New Buffalo 40400
New Canaan Library, New Canaan 40401
New Castle County Law Library, Wilmington 35073
New Castle County Public Library System, New Castle 40405
New Castle Public Library, New Castle 40406
New City Free Library, New City 40407
New College, Oxford 29331
New College of California – Humanities Library, San Francisco 32589
New College of California – Law Library, San Francisco 32590
New College of Florida – University of South Florida, Sarasota-Manatee, Sarasota 32617
New Creation Teaching Ministry, Coromandel East 00661
New Cumberland Public Library, New Cumberland 40408
New Energy and Industrial Technology Development Organization (NEDO), Tokyo 17716
New England Baptist Hospital, Boston 36577
New England Bible College, South Portland 35386
New England College, Henniker 31327
New England College of Optometry, Boston 30497
New England Conservatory of Music – Harriet M. Spaulding Library, Boston 30498
New England Conservatory of Music – Idabelle Firestone Audio Library, Boston 30499
The New England Electric Railway Historical Society, Kennebunkport 36993
New England Historic Genealogical Society Library, Boston 36578
New England Institute of Technology Library, Warwick 32952
New England School of Law Library, Boston 30500
New England Wireless & Steam Museum Inc Library, East Greenwich 36818
New Hampshire Antiquarian Society Library, Hopkinton 36938
New Hampshire Department of Corrections, Laconia 34427
New Hampshire Department of Justice, Concord 34108
New Hampshire Historical Society Library, Concord 36749
New Hampshire Law Library, Concord 34109
New Hampshire State Library, Concord 30128
New Hampshire State Library – Talking Book Services, Concord 30129
New Hampshire State Prison Library, Concord 34110
New Hampshire Technical Institute, Concord 36751
New Hanover County Public Library, Wilmington 41913
New Hartford Public Library, New Hartford 40409
New Hartford Public Library, New Hartford 40410
New Haven Free Public Library, New Haven 40411
New Haven Free Public Library – Mitchell, New Haven 40412
New Haven Museum & Historical Society, New Haven 37161
New Jersey City University, Jersey City 31462
New Jersey Department of Corrections, Avenel 33959
New Jersey Historical Society Library, Newark 37278
New Jersey Institute of Technology, Newark 32097
– Barbara & Leonard Littman Architecture Library, Newark 32098
New Jersey State Department of Law & Public Safety, Trenton 34936
New Jersey State Library, Trenton 30160
New Jersey State Library, Trenton 30161
New Jersey State Police Training Bureau Library, Sea Girt 34847
New Karl Heusner Memorial Hospital, Belize City 02900
New Kensington Commonwealth College, Upper Burrell 32883
New Lenox Public Library, New Lenox 40416
New Lisbon Correctional Institution Library, New Lisbon 34582
New London Public Library, New London 40418
New London Public Library, New London 40419
New Madrid County Library, Portageville 40866
New Melleray Library, Peosta 35325
New Metallurgical Works Ltd., Ostrava 06421
New Mexico Corrections Department, Las Cruces 34439
New Mexico Corrections Department, Santa Fe 34838
New Mexico Highlands University, Las Vegas 31567
New Mexico Institute of Mining and Technology, Socorro 37614
New Mexico Junior College, Hobbs 33411
New Mexico Military Institute, Roswell 33705
New Mexico School for the Deaf Library, Santa Fe 33745
New Mexico School for the Visually Handicapped Library, Alamogordo 33117
New Mexico State Library, Santa Fe 30156
New Mexico State Library – Library for the Blind & Physically Handicapped, Santa Fe 37574
New Mexico State University, Grants 31241
New Mexico State University, Las Cruces 31566
New Mexico State University at Alamogordo, Alamogordo 30176
New Mexico Supreme Court, Santa Fe 34839
New Milford Public Library, New Milford 40420
New Milford Public Library, New Milford 40421
New Orleans Baptist Theological Seminary, Marietta 35275
New Orleans Baptist Theological Seminary – John T. Christian Library, New Orleans 35301
New Orleans Baptist Theological Seminary – Martin Music Library, New Orleans 35302
New Orleans Museum of Art, New Orleans 37170
New Orleans Public Library, New Orleans 40422
New Orleans Public Library – Algiers Regional, New Orleans 40423
New Orleans Public Library – Milton H. Latter Memorial Branch, New Orleans 40424
New Plymouth District Library, New Plymouth 19875
New Port Richey Public Library, New Port Richey 40426
New Providence Memorial Library, New Providence 40427
New River Community College, Dublin 33302
New Rochelle Public Library, New Rochelle 40429
The New School – Raymond Fogelman Library, New York 33578
New South Wales Law Courts, Sydney 00731
New South Wales. Parliamentary Library, Sydney 00732
New South Wales Police Force, Goulburn 00681
New South Wales Teachers' Federation, Surry Hills 01017
New Tecumseth Public Library, Alliston 04566
New Ufa Plant for Petroleum Production, Ufa 22968
New Ulm Public Library, New Ulm 40430
New University of Lisbon, Lisboa
– Faculty of Economics, Lisboa 21685
New Westminster Public Library, New Westminster 04702
New York Academy of Medicine Library, New York 37236
New York Chiropractic College, Seneca Falls 32657
New York City Law Department, New York 34593
New York City Police Department, New York 34594
New York City Technical College, Brooklyn 33194
New York College of Podiatric Medicine, New York 32072
New York County District Attorney's Office Library, New York 34595
New York County Lawyers' Association Library, New York 37237
New York Department of Correctional Services, Bedford Hills 33979
New York Genealogical and Biographical Society Library, New York 37238
New York Historical Society Library, New York 37239
New York Institute of Technology, Central Islip 30694
New York Institute of Technology, New York 32073

New York Institute of Technology, Old Westbury 32187
– Education Hall Library Art & Architecture Collection, Old Westbury 32188
New York Institute of Technology, Old Westbury 37307
New York Law Institute Library, New York 37240
New York Law School Library, New York 32074
New York Legislative Service, Inc Library, New York 37241
New York Life Insurance Co, New York 36001
New York Medical College, Valhalla 32922
New York Psychoanalytic Institute, New York 37242
The New York Public Library – Astor, Lenox & Tilden Foundations – 58th Street Branch, New York 40432
The New York Public Library – Astor, Lenox & Tilden Foundations – 96th Street Branch, New York 40433
The New York Public Library – Astor, Lenox & Tilden Foundations – Asian & Middle Eastern Division, New York 40434
The New York Public Library – Astor, Lenox & Tilden Foundations – Baychester Branch, Bronx 38298
The New York Public Library – Astor, Lenox & Tilden Foundations – Dongan Hills Branch, Staten Island 41430
The New York Public Library – Astor, Lenox & Tilden Foundations – Dorot Jewish Division, New York 40435
The New York Public Library – Astor, Lenox & Tilden Foundations – Epiphany Branch, New York 40436
The New York Public Library – Astor, Lenox & Tilden Foundations – Fort Washington Branch, New York 40437
The New York Public Library – Astor, Lenox & Tilden Foundations – General Research Division, New York 40438
The New York Public Library – Astor, Lenox & Tilden Foundations – Irma & Paul Milstein Division of US History, Local History & Genealogy, New York 40439
The New York Public Library – Astor, Lenox & Tilden Foundations – Mid-Manhattan Library, New York 40440
The New York Public Library – Astor, Lenox & Tilden Foundations – Miriam & Ira D. Wallach Division of Art, Prints & Photographs, New York 40441
The New York Public Library – Astor, Lenox & Tilden Foundations – Mosholu Branch, Bronx 38299
The New York Public Library – Astor, Lenox & Tilden Foundations – Music Division, New York 40442
The New York Public Library – Astor, Lenox & Tilden Foundations – New Amsterdam Branch, New York 40443
The New York Public Library – Astor, Lenox & Tilden Foundations – New Dorp Branch, Staten Island 41431
The New York Public Library – Astor, Lenox & Tilden Foundations – New York Public Library for the Performing Arts Circulating Collections, New York 40444
The New York Public Library – Astor, Lenox & Tilden Foundations – Pelham Bay Branch, Bronx 38300
The New York Public Library – Astor, Lenox & Tilden Foundations – Rare Books Division, New York 40445
The New York Public Library – Astor, Lenox & Tilden Foundations – Richmondtown Branch, Staten Island 41432
The New York Public Library – Astor, Lenox & Tilden Foundations – Riverside Branch, New York 40446
The New York Public Library – Astor, Lenox & Tilden Foundations – Schomburg Center for Research in Black Culture, New York 40447
The New York Public Library – Astor, Lenox & Tilden Foundations – Seward Park Branch, New York 40448
The New York Public Library – Astor, Lenox & Tilden Foundations – Slavic & Baltic Division, New York 40449
The New York Public Library – Astor, Lenox & Tilden Foundations – Spuyten Duyvil Branch, Bronx 38301
The New York Public Library – Astor, Lenox & Tilden Foundations – St

George Library Center, Staten Island 41433
The New York Public Library – Astor, Lenox & Tilden Foundations – Yorkville Branch, New York 40450
New York School of Interior Design Library, New York 33579
The New York Society Library, New York 37244
New York State Court of Appeals Library, Albany 33897
New York State Department of Correctional Services, Elmira 34186
New York State Department of Correctional Services, Ogdensburg 34627
New York State Department of Correctional Services, Ossining 34644
New York State Department of Economic Development, Albany 33898
New York State Department of Health, Albany 33899
New York State Department of Law Library, Albany 33900
New York State Department of Law Library, New York 34596
New York State Department of Taxation & Finance Library, Albany 33901
New York State Division of Housing & Community Renewal, New York 34597
New York State Historical Association, Cooperstown 36753
New York State Judicial Department, Rochester 34746
New York State Legislative Library, Albany 33902
New York State Library, Albany 30113
New York State Library, Albany 30114
New York State Nurses Association Library, Latham 37019
New York State Psychiatric Institute, New York 34598
New York State Supreme Court, Brooklyn 34009
New York State Supreme Court, Kingston 34416
New York State Supreme Court, New York 34599
New York State Supreme Court, Poughkeepsie 34707
New York State Supreme Court, Watertown 35047
New York State Supreme Court – First Judicial District Criminal Law Library, New York 34600
New York State Supreme Court Fourth District, Plattsburgh 34682
New York State Supreme Court Law Library, Norwich 34621
New York State Supreme Court Library, Binghamton 33985
New York State Supreme Court Library, Catskill 34051
New York State Supreme Court Library, Troy 34940
New York State Supreme Court Library, Brooklyn, Brooklyn 34010
New York State Unified Court System, Syracuse 34906
New York Supreme Court, Buffalo 34018
New York Supreme Court Appellate Division, Albany 33903
New York Theological Seminary, New York 35310
New York Times, New York 36002
New York University, New York 32075
– Conservation Center Library, New York 32076
– Courant Institute of Mathematical Sciences Library, New York 32077
– Fales Library & Special Collections, New York 32078
– Stephen Chan Library of Fine Arts, New York 32079
– Tamiment Library/Robert F. Wagner Labor Archives, New York 32080
New York University School of Law, New York 32081
New York University School of Medicine, New York 32082
New York University School of Medicine – John & Bertha E. Waldmann Memorial Library, New York 32083
New Zealand Council for Educational Research, Wellington 19852
New Zealand Dairy Research Institute, Palmerston North 19842
New Zealand Defence Force, Wellington 19799
New Zealand Drama School, Wellington 19791
New Zealand High Commission Library, London 29805
New Zealand Law Society Library, Wellington 19853

New Zealand Parliamentary Library, Wellington 19807
Newark Free Library, Newark 40453
Newark Museum Library, Newark 37279
Newark Public Library, Newark 40454
Newberg Public Library, Newberg 40455
Newberry College, Newberry 32110
Newberry County Library, Newberry 40456
Newberry Library, Chicago 36685
Newbury College, Brookline 30565
Newburyport Public Library, Newburyport 40460
Newcastle College, Newcastle upon Tyne 29484
Newcastle Public Library, Newcastle 25216
Newcastle Region Library, Newcastle 01130
Newcastle University, Newcastle upon Tyne 29277
– Law Library, Newcastle upon Tyne 29278
– Medical School, Newcastle upon Tyne 29279
– School of Architecture, Newcastle upon Tyne 29280
– School of Education, Newcastle upon Tyne 29281
Newman College, Birmingham 28950
Newman Theological College, Edmonton 03696
Newman University, Wichita 33056
Newmarket Public Library, Newmarket 04703
Newport Aeronautical Sales Corp Library, Newport Beach 36050
Newport Beach Public Library, Newport Beach 40467
Newport Beach Public Library – Balboa Branch, Balboa 38064
Newport Beach Public Library – Central Library, Newport Beach 40468
Newport Beach Public Library – Mariners, Newport Beach 40469
Newport Central Library, Newport, South Wales 30050
Newport Historical Society, Newport 37283
Newport News Public Library System, Newport News 40470
Newport Public Library, Newport 40464
Newport Public Library, Newport 40465
News Archives, Jerusalem 14739
Newspaper Association of America, Arlington 36481
Newspaper Library of Sanoma Oy, Helsinki 07414
Newsweek, Inc, New York 36003
Newton County Library System, Covington 38721
Newton County Public Library, Lake Village 39780
Newton Free Library, Newton Centre 40475
Newton Public Library, Newton 40472
Newton Public Library, Newton 40473
Newton-Wellesley Hospital, Newton Lower Falls 37287
Nez Perce County Law Library, Lewiston 34453
Nezhdanov State Conservatory of Music, Odesa 28051
Ngchesar Trading Post Lawyers Cooperative Library, Koror 20532
Ngee Ann Polytechnic, Singapore 24591
NHTV International Higher Education Breda, Breda 19102
NHTV internationale hogeschool Breda, Breda 19102
Niagara College of Applied Arts & Technology, Welland 04044
Niagara County Community College, Sanborn 33738
Niagara Falls Public Library, Niagara Falls 40477
Niagara Falls Public Library, Niagara Falls 04704
Niagara Mohawk Power Corp, Syracuse 36293
Niagara University, Niagara University 32118
Niceville Public Library, Niceville 40479
Nicholas Kopernik Museum, Frombork 21091
Nicholas P. Sims Library, Waxahachie 41781
Nicholls State University, Thibodaux 32813
Nichols College, Dudley 31031
Nicholson Memorial Library System – Central Library, Garland 39202
Nicolaus Copernicus University, Toruń 20934
– Centre for Astronomy, Łysomice 20936
– Department of Fine Arts Library, Toruń 20935
– Faculty of Chemistry, Toruń 20947
– Faculty of Mathematics and Informatics, Toruń 20948
– Institute of Archaeology, Toruń 20938
– Institute of Biology and Environmental Protection, Toruń 20939
– Institute of History and Archival Sciences, Toruń 20943
– Institute of Pedagogy, Toruń 20944
– Institute of Physics, Toruń 20942

Nicolet Area Technical College, Rhinelander 33681
NIED – National Research Institute for Earth Science and Disaster Prevention, Tsukuba 17802
Niederdeutsche Bibliothek, Hamburg 12008
W. J. Niederkorn Library, Port Washington 40861
Niederösterreichische Landesbibliothek, St. Pölten 01186
Niederösterreichisches Landesarchiv, St. Pölten 01577
Niederösterreichisches Landesmuseum, St. Pölten 01578
Niederrhein University of Applied Sciences, Mönchengladbach
Niedersächsische Landesgalerie, Hannover 12031
Niedersächsische Staats- und Universitätsbibliothek Göttingen, Göttingen 09315
Niedersächsische Staatskanzlei, Hannover 10934
Niedersächsischer Landesbetrieb für Wasserwirtschaft, Küsten- und Naturschutz, Hildesheim 10948
Niedersächsischer Landesrechnungshof, Hildesheim 10949
Niedersächsischer Landtag, Hannover 10935
Niedersächsisches Finanzministerium, Hannover 10936
Niedersächsisches Justizministerium, Hannover 10937
Niedersächsisches Kultusministerium, Hannover 10938
Niedersächsisches Landesamt für Denkmalpflege, Hannover 10939
Niedersächsisches Landesamt für Soziales, Jugend und Familie, Hildesheim 10950
Niedersächsisches Landesarchiv – Hauptstaatsarchiv Hannover, Hannover 12032
Niedersächsisches Landesarchiv – Staatsarchiv Aurich, Aurich 11462
Niedersächsisches Landesarchiv – Staatsarchiv Bückeburg, Bückeburg 11703
Niedersächsisches Landesarchiv – Staatsarchiv Oldenburg, Oldenburg 12387
Niedersächsisches Landesarchiv – Staatsarchiv Osnabrück, Osnabrück 12390
Niedersächsisches Landesarchiv – Staatsarchiv Wolfenbüttel, Wolfenbüttel 12606
Niedersächsisches Landesinstitut für Verbraucherschutz und Lebensmittelsicherheit, Celle 10848
Niedersächsisches Landesmuseum, Naturkunde-Abteilung, Hannover 12033
Niedersächsisches Landesmuseum, Urgeschichts-Abteilung, Hannover 12034
Niedersächsisches Ministerium für Ernährung, Landwirtschaft, Verbraucherschutz und Landesentwicklung, Hannover 10940
Niedersächsisches Ministerium für Inneres und Sport, Hannover 10941
Niedersächsisches Ministerium für Soziales, Frauen, Familie und Gesundheit, Hannover 10942
Niedersächsisches Ministerium für Wirtschaft, Arbeit und Verkehr, Hannover 10943
Niedersächsisches Ministerium für Wissenschaft und Kultur, Hannover 10944
Niedersächsisches Münzkabinett, Hannover 12024
Niedersächsisches Oberverwaltungsgericht, Lüneburg 11002
Niemiecki Instytut Historyczny w Warszawie, Warszawa 21254
NIFU STEP -Studier av innovasjon, forskning og utdanning, Oslo 20256
NIFU STEP – Studies in Innovation, Research and Education, Oslo 20256
Niger Basin Authority Documentation Centre, Niamey 19908
Niger State Library Services, Minna 20061
Nigeria Bible Translation Trust, Jos 20002
Nigeria Building and Road Research Institute (NBRRI), Abuja 20008
Nigeria Educational Research Council, Lagos 20034
Nigerian Defence Academy, Kaduna 19955
Nigerian Institute for Oil Palm Research (NIFOR), Benin City 20010
Nigerian Institute for Trypanosomiasis Research (NITR), Kaduna 20028
Nigerian Institute of Advanced Legal Studies, Lagos 20035
Nigerian Institute of International Affairs (NIIA), Lagos 20036
Nigerian Institute of Management, Lagos 20037
Nigerian Institute of Medical Research, Lagos 20038

Nigerian Institute of Social and Economic Research (NISER), Ibadan 20022
Nigerian Law School, Lagos 19960
Nigerian National Petroleum Corporation, Lagos 20004
Nigerian Society of Engineers, Lagos 20039
Nihon Boeki Shinkokai Kaigai Keizai Joho Senta Joho Sabisubu, Tokyo 17717
Nihon Boeki Shinkokai Osaka Hombu Boeki Shiryo Senta, Osaka 17636
Nihon Chuo Keibakai, Tokyo 17718
Nihon Daigaku, Tokyo 17159
– Hogakubu Toshokan, Tokyo 17160
– Igakubu Toshokan, Tokyo 17161
– Kokusaikankeigakubu Toshokan, Mishima 17162
– Matsudo Shigakubu Toshokan, Matsudo 17163
– Rikogakubu Toshokan (Narashino), Funabashi 17164
– Rikogakubu Toshokan (Surugadai), Tokyo 17165
– Seisan Kogakubu Toshokan, Narashino 17166
Nihon Deizai Kenkyu Centre, Tokyo 17719
Nihon Doro Kodan Shikenjo Kikakuka, Machida 17592
Nihon Ekusuran Kogyo K. K., Okayama 17398
Nihon Gaikoku Tokuhain Kyokai, Tokyo 17720
Nihon Ganseki Kobutsu Kosho Gakkai, Sendai 17662
Nihon Genshiryoku Kenkyujo, Tokyo 17721
Nihon Ginko Kin-yukenkyujo, Tokyo 17464
Nihon Hikakuho Kenkyujo, Tokyo 17722
Nihon Jui Chikusan Daigaku, Musashino 17049
Nihon Kagaku Gijutsu Joho Senta, Tokyo 17723
Nihon Kagaku Kogyo Honsha Toshoshitsu, Tokyo 17465
Nihon Kaihatsu Ginko, Tokyo 17466
Nihon Keieisha Dantai Renmei, Tokyo 17724
Nihon Keizai Kenkyu Senta, Tokyo 17725
Nihon Keizai Shimbunsha Deta Bankukyoku, Tokyo 17467
Nihon Kenchiku Gakkai, Tokyo 17726
Nihon Kenchiku Senta, Tokyo 17727
Nihon Kikaku Kyokai Kaigai-Kikaku Raiburari, Tokyo 17728
Nihon Kindai Bungakukan, Tokyo 17729
Nihon Kokuyu Tetsudo Chuo Tetsudo Byoin, Tokyo 17730
Nihon Monki Senta, Inuyama 17556
Nihon Precision Machinery Mfg. Co, Ltd, Fujisawa 17350
Nihon Seisansei Hombu, Seisansei Toshoshitsu, Tokyo 17731
Nihon Senbai Kosha Hadano Shikenjo, Hadano 17354
Nihon Shika Daigaku, Tokyo 17167
Nihon Shoken Keizai Kenkyujo, Nagoya 17616
Nihon Shoken Keizai Kenkyujo, Osaka 17637
Nihon Shoken Keizai Kenkyujo, Tokyo 17732
Nihon Shokubutsu Boeki Kyokai, Kodaira 17577
Nihon Shokubutsu Gakkai, Tokyo 17733
Nihon Tabako Sangyo K. K., Tokyo 17468
Nihon Tabako Sangyo K. K. Chuo Kenkyujo, Kanagawa 17368
Nihon Toshokan Kyokai, Tokyo 17734
Nihon University, Tokyo 17159
– College of Industrial Technology, Narashino 17166
– College of International Relations, Mishima 17162
– College of Law, Tokyo 17160
– College of Science and Technology, Narashino Campus, Funabashi 17164
– College of Science and Technology, Surugadai Campus, Tokyo 17165
– Matsudo Dental Library, Matsudo 17163
– Medical Library, Tokyo 17161
Nihon Yushutsu-Nyu Ginko Kaigai Toshi Kenkyujo, Tokyo 17469
NII Eksperimentalnoi Meditsiny SZO RAMN, Sankt-Peterburg 23804
NII po antibiotitsi, Razgrad 03496
Niigata Agricultural Experiment Station, Nagaoka 17610
Niigata Animal Husbandry Experiment Station, Minamikanbara 17596
Niigata Daigaku, Niigata 17067
Niigata Prefectural Assembly, Niigata 17286
Niigata Prefectural Library, Niigata 17853
Niigata University Library, Niigata 17067
Niigataken Chikusan Shikenjo, Minamikanbara 17596
Niigataken Gikai, Niigata 17286
Niigataken Nogyo Shikenjo, Nagaoka 17610
NIIT Ltd. Library, Sofiya 03568
NIITI montazha, ekspluatatsii i remonta mashin i oborudovaniya zhivotnovodcheskikh i ptitsevodcheskikh ferm, Minsk 02044
Nikel Design and Research Institute, Sankt-Peterburg 23785
Nikita Botanical Garden, Yalta 28483
'Nikolay Pavlovich' National Academy of Art, Sofiya 03462
Nikopol Regional Centralized Library System, Main Library, Nikopol 28746
Nikopolska raionna TsBS, Tsentralna biblioteka, Nikopol 28746
NIKTI biotekhnicheskikh sistem, Sankt-Peterburg 23859
Niles, Barton & Wilmer Law Library, Baltimore 35483
Niles District Library, Niles 40482
Niles Public Library District, Niles 40483
Ningbo College, Ningbo 05474
Ningbo Normal College, Ningbo 05475
Ningbo University, Ningbo 05476
Ningde Teachers' College, Ningde 05477
Ningxia Agricultural College, Yongning 05762
Ningxia Institute of Technology, Yinchuan 05755
Ningxia Library, Yinchuan 04978
Ningxia Medical College, Yinchuan 05756
Ningxia University, Yinchuan 05757
Ninham Shand Library, Cape Town 25088
Ninth September University, Izmir 27866
Nioga Library System, Lockport 39913
NIPI alyuminievoi, magnievoi i elektrodnoi promyshlennositi, Sankt-Peterburg 23860
Nipissing District Law Association, North Bay 04424
Nipissing University – Canadore College, North Bay 03807
Nippon Chemical Industrial Co, Ltd, Tokyo 17465
Nippon Denshin Denwa K. K. Denki Tsushin Kenkyujo, Musashino 17386
Nippon Dental University, Tokyo 17167
Nippon Gakujutsu Kaigi, Tokyo 17319
Nippon Gosei Kagaku K. K., Ibaraki 17359
Nippon Hikaku Kenkyujo, Tokyo 17735
Nippon Hoso Kyokai Hosobunka Kenkyujo, Tokyo 17736
Nippon Imono Kyokai, Tokyo 17737
Nippon Institute for Biological Science, Ome 17630
Nippon Itagarasu K. K., Itami 17363
Nippon Kaiji Center, Tokyo 17738
Nippon Kayaku Co, Ltd, Research Laboratories, Tokyo 17470
Nippon Kayaku K. K. Iyaku Jigyo Hombu Sogo Kenkyujo, Tokyo 17470
Nippon Koan Co, Ltd, Engineering Development Dept, Yokohama 17503
Nippon Kogaku Kogyo K. K. Kanribu Gyomuka, Tokyo 17471
Nippon Kokan K.K. Kyotsu Gijutsubu Joho Senta, Yokohama 17503
Nippon Kokogaku Kyokai, Tokyo 17739
Nippon Kokuyu Tetsudo, Tetsudo Gijutsu Kenkyujo, Kokubunji 17578
Nippon Kotsu Kyokai, Tokyo 17740
Nippon Life Insurance Company, Osaka 17409
Nippon Minkan Hoso Remmei Kenkyujo Shiryoshitsu, Tokyo 17741
Nippon Noritsu Kyokai, Tokyo 17742
Nippon Oil Co, Ltd, Yokohama 17504
Nippon Paint Co, Ltd, Neyagawa 17397
Nippon Peinto K. K., Neyagawa 17397
Nippon Seibutsu Kagaku Kenkyujo, Ome 17630
Nippon Seiko K.K., Fujisawa 17350
Nippon Seimei Hoken Sogogaisha Kikakudai-ikka, Osaka 17409
Nippon Seinenkan, Tokyo 17743
Nippon Sekiyu K. K., Yokohama 17504
Nippon Sheet Glass Co, Ltd, Central Research Laboratory, Itami 17363
Nippon Shokubai Kagaku Kogyo K. K., Suita 17425
Nippon Soda Co, Ltd, Takaoka Factory, Takaoka 17428
Nippon Soda K. K. Takaoka Kojo, Takaoka 17428
Nippon Steel Corp, Corporate & Economic Research Department, Tokyo 17480
Nippon Steel Corp, Nagoya Works, Tokai 17430
Nippon Steel Corp, Research and Development Laboratories, Kitakyushu 17375
Nippon Steel Corp, Sagamihara Research & Engineering Ctr, Sagamihara 17420
Nippon Steel Corp, Techn Res Laboratory, Kamaishi 17366
Nippon Synthetic Chemistry Co, Ltd, Ibaraki 17359
Nippon Telegraph and Telephone Corp, Electrical Communication Laboratory, Musashino 17386
Nippon Tenji Toshokan, Tokyo 17744
Nippon Veterinary and Animal Science University, Musashino 17049
Nishi Nippon Ginko Kohoshitsu, Fukuoka 17351
Nishi Nippon Sogo Bank Ltd, Fukuoka 17351
Nissan Diesel Motor Co Ltd, Ageo 17340
Nissan Dizeru Kogyo K.K., Ageo 17340
Nissan Jidosha K. K. Chosabu, Tokyo 17472
Nissan Jidosha K.K. Chuo Kenkyujo Kenkyukanribu Joho Kanrika, Yokosuka 17507
Nissan Motor Co Ltd, Business Res Dept, Tokyo 17472
Nissan Motor Co Ltd, Ctr Engineering Laboratories, Yokosuka 17507
Nisso Kinzoku K. K. Aizu Seirenjo Kaihatsu Honbu Kenkyushitsu Toshoshitsu, Yama 17500
Nisso Smelting Company, Aizu Plant, Yama 17500
Nitrogen Industrial Corporation, Severodonetsk 28468
Nitrogenous Fertilizer Factory, Kemerovo 22802
NITs po ispytaniyam i dovodke avtomototekhniki, Dmitrov 23024
Niue National/Public Library, Alofi 20064
NIVEL – Nederlands Instituut voor Onderzoek van de Gezondheidszorg, Utrecht 19528
Nixon Peabody LLP, Boston 35533
Nixon Peabody LLP, Rochester 36185
Nizam Ganjari State Museum of Azerbaijan Literature, Baku 01719
Nizami Institute of Literature and Language, Baku 01720
Nizamiah and Japal-Rangapur Observatories and Centre of Advanced Study in Astronomy, Hyderabad 14008
Nizhegorodskaya gosudarstvennaya meditsinskaya akademiya, Nizhni Novgorod 22423
Nizhegorodskaya gosudarstvennaya oblastnaya detskaya biblioteka, Nizhni Novgorod 24184
Nizhegorodskaya nauchno-issledovatelski institut travmatologii I ortopedii, Nizhni Novgorod 23622
Nizhegorodskaya oblastnaya universalnaya nauchnaya biblioteka, Nizhni Novgorod 22158
Nizhegorodskaya tsentralnaya gorodskaya biblioteka im. V. I. Lenina, Nizhni Novgorod 24185
Nizhegorodski avtomekhanicheski tekhnikum, Nizhni Novgorod 22685
Nizhegorodski gosudarstvenny arkhitekturno-stroitelny universitet, Nizhni Novgorod 22424
Nizhegorodski gosudarstvenny universitet im. N.I. Lobachevskogo, Nizhni Novgorod 22425
Nizhnetagilski muzei-zapovednik "Gornozavodskoi Ural", Nizhni Tagil 23626
Nizhne-Volzhski mezhotraslevoi territorialny tsentr nauchno-tekhnicheskoi informatsii i propagandy, Volgograd 24064
Nizhne-Volzhski nauchno-issledovatelski institut geologii i geofiziki, Saratov 23962
Nizhni Novgorod State University of Architecture and Construction, Nizhni Novgorod 22424
Nizhnodniprovsk Pipe Rolling Plant, Library, Dnipropetrovsk 28138
Nizhnodniprovsk Pipe Rolling Plant, Library Branch no 1, Dnipropetrovsk 28538
Nizhnodniprovsky truboprokatny zavod, Dnipropetrovsk 28138
Nizhnodniprovsky truboprokatny zavod, Biblioteka-filial no 1, Dnipropetrovsk 28538
Nizhnogirsk Regional Centralized Library System, Main Library, Nizhnogirsk 28747
Nizhnogirska raionna TsBS, Tsentralna biblioteka, Nizhnogirsk 28747
Nizhny Novgorod Research Institute of Traumatology and Orthopaedics, Nizhni Novgorod 23622
Nizhny Novgorod School of Automobile Engineering, Nizhni Novgorod 22685
Nizhny Tagil Metallurgical Joint-Stock Company, Nizhni Tagil 22873
Nizhny Tagil Museum Reserve "Mining and Works Urals", Nizhni Tagil 23626
Njala University College (NUC), Freetown 24568
NM – Náprstek Museum of Asian, African and American Cultures, Praha 06511
NM – Náprstkovo Muzeum Asijských, Afrických, a Amerických Kultur, Praha 06511
Nobel Institute, Oslo 20257
Nobel Library of the Swedish Academy, Stockholm 26684
Nobelbibliotek, Stockholm 26684
Nobelinstituttet, Oslo 20257
Nobile Collegio 'Campana', Osimo 15949
Noble County Public Library, Albion 37904
Nobles County Library, Worthington 41973
Nobunkyo Agricultural Library, Tokyo 17745
Nobunkyo Toshokan, Tokyo 17745
Nodaway County Historical Society / Mary H. Jackson Research Center, Maryville 37087
Nofer Institute of Occupational Medicine, Scientific Library, Łódź 20874
Nogales-Santa Cruz County Public Library, Nogales 40485
Nógrád Historical Muzeum, Salgótarján 13496
Nógrádi Történeti Múzeum, Salgótarján 13496
Nogyo Doboku Gakkai, Tokyo 17746
Nogyo Kenkyu Senta, Joho Shiryoka, Tsukuba 17803
Nogyo Kogaku Kenkyusho, Tsukuba 17804
Nokia Mobile Phones, Salo 07430
Nolde Stiftung Seebüll, Neukirchen 12362
Nolichucky Regional Library, Morristown 40305
Noma Institute of Educational Research, Tokyo 17747
Noma Kyoiku Kenkyujo, Tokyo 17747
Nome folkebibliotek / Telemark fylkesbibliotek, Ulefoss 20413
Nomin Kyoiku Kyokai Koibuchigakuen Toshokan, Ibaraki 17553
Nonferrous Metals Engineering Ltd, Moskva 22828
Noord-Belgische Provincie van de Sociëteit van Jezus, Leuven 02477
Noordelijke Hogeschool Leeuwarden, Leeuwarden 19150
Noord-Hollands Archief, Haarlem 19456
Noordwes-Universiteit, Mmabatho 25034
Nora kommunbibliotek, Nora 26864
Norbergs kommunbibliotek, Norberg 26865
Norbertijnerabdij, Grimbergen 02465
Norbottens länsbibliotek, Luleå 26844
Nordanstigs kommunbibliotek, Bergsjö 26743
Norddeutsche Blindenhörbücherei e.V., Hamburg 12009
Norddeutscher Rundfunk, Hamburg 12010
Norddeutscher Rundfunk (NDR), Hannover 12035
Nordelbische Kirchenbibliothek, Hamburg 11231
Nordelbische Kirchenmusikbibliothek, Hamburg 12011
Nordelbisches Kirchenamt, Kiel 11256
Nordelbisches Missionszentrum im Christian-Jensen-Kolleg, Breklum 11182
Nordfriisk Instituut, Bredstedt 11668
Nordic Africa Institute, Uppsala 26709
Nordic Folk High School Library, Kungälv 26473
Nordic Heritage Museum, Seattle 37594
Nordic House Library, Reykjavík 13569
Nordic Institute of Asian Studies (NIAS), København 06816
Nordic Library at Athens, Athinai 13122
Nordiska Afrikainstitutet, Uppsala 26709
Nordiska folkhögskolan bibliotek, Kungälv 26473
Nordiska Museet, Stockholm 26672
Nordjysk Musikkonservatoriums Bibliotek, Aalborg 06727
Nordland County Library – Narvik, Narvik 20379
Nordland County Library – Rana, Mo i Rana 20387
Nordland fylkesbibliotek – Avd. Narvik, Narvik 20379
Nordland fylkesbibliotek – Avd. Rana, Mo i Rana 20387
Nordmalings bibliotek, Nordmaling 26866
Nordost-Bibliothek, Lüneburg 12221
Nordost-Institut, Göttingen 11933
Nordost-Institut an der Universität Hamburg, Lüneburg 12221
Nordre Land folkebibliotek, Dokka 20323
Nordrhein-Westfälisches Hauptstaatsarchiv, Düsseldorf 11801
NordseeMuseum-Nissenhaus Husum, Husum 12070
Nordtiroler Kapuzinerprovinz Salzburg, Salzburg 01467
Nord-Trøndelag fylkesbibliotek, Steinkjer 20402
Nord-Trøndelag University College, Levanger, Levanger 20105

Nordwestdeutsche Forstliche Versuchsanstalt, Göttingen 11934
Norelius Community Library, Denison 38822
Norfolk County Council Library and Information Service, Norwich 30054
Norfolk Public Library, Norfolk 40486
Norfolk Public Library, Norfolk 40487
Norfolk Public Library, Simcoe 04759
Norfolk Public Library – Lafayette, Norfolk 40488
Norfolk Public Library – Mary D. Pretlow Anchor Library, Norfolk 40489
Norfolk State University, Norfolk 32121
Norges Bank, Oslo 20203
Norges byggforskningsinstitutt, Oslo 20258
Norges Fiskerimuseum, Bergen 20214
Norges geologiske undersøkelse, Trondheim 20299
Norges Geotekniske Institutt, Oslo 20259
Norges Handelshøyskole, Bergen 20073
Norges Idrettshøgskole, Oslo 20119
Norges kooperative landsforenings bibliotek, Oslo 20260
Norges Musikhøgskole, Oslo 20120
Norges Teknisk-naturvitenskapelige Universitet (NTNU), Trondheim 20175
– Biblioteket for marinteknikk, Trondheim 20176
– Fakultetsbiblioteket Arkitektur/Bygg/Matematikk, Trondheim 20177
– Fakultetsbiblioteket Maskin, verkstedteknikk, Gløshaugen, Trondheim 20178
– Gunnerusbiblioteket, Kalvskinnet, Trondheim 20179
– Hovedbibliotek for humaniora og samfunnsfag, Dragvoll, Trondheim 20180
– Medisinsk bibliotek og informasjonssenter, Trondheim 20181
– Realfagbiblioteket, Trondheim 20182
Norges vassdrags- og energidirektorat (NVE), Oslo 20194
Norges Veterinærhøgskole, Oslo 20121
Norin Chuo Kinko Chosabu, Tokyo 17473
Norin Gyogyo Kinyu Koko Chosabu Shiryoka, Tokyo 17474
Norin Suisan Seisaku Kenkyusho toshokan, Tokyo 17748
Norinchukin Bank, Tokyo 17473
Norinsuisan Kenkyu Joho Senta, Tsukuba 17805
Norinsuisansho Toshokan, Tokyo 17320
Norit Nederland B.V., Amersfoort 19320
Norma Smurfit Library, Dublin 14472
Normal Public Library, Normal 40490
Norman Bethune University of Medical Sciences, Changchun 05087
Normandale Community College, Bloomington 33176
Norquest College, Edmonton 03988
Norra Älvsborgs länssjukhus, Trollhättan 26701
Norrbottens museum, Luleå 26611
Nørre Nissum College of Education, Lemvig 06737
Nørre Nissum Seminarium og HF, Lemvig 06737
Nørresundby Gymnasium og HF, Nørresundby 06740
Norris, McLaughlin & Marcus, Somerville 36274
Norrish Central Library, Portsmouth 30060
Norrköpings Konstmuseum, Norrköping 26623
Norrköpings stadsbibliotek, Norrköping 26868
Norrtälje stadsbibliotek, Norrtälje 26869
Norsjö bibliotek, Norsjö 26870
Norsk Avholdsbibliotek, Oslo 20261
Norsk barnebokinstitutt, Oslo 20262
Norsk Bergverksmuseums bibliotek, Kongsberg 20234
Norsk folkemuseum, Oslo 20263
Norsk Hydro A/S, Porsgrunn 20205
Norsk Hydro Research Centre Porsgrunn, Porsgrunn 20205
Norsk institutt for by- og regionforskning (NIBR), Oslo 20264
Norsk institutt for luftforskning, Kjeller 20233
Norsk institutt for skog og landskap, Ås 20209
Norsk institutt for vannforskning (NIVA), Oslo 20265
Norsk Jernbanemuseum, Hamar 20225
Norsk Lærerakademi, Bergen 20074
Norsk lokalhistorisk institutt, Oslo 20266
Norsk lyd- og blindeskriftbibliotek, Oslo 20267
Norsk lyd- og blindeskriftbibliotek, avdeling Bergen, Bergen 20215
Norsk lyd- og blindeskriftbibliotek, avdeling Trondheim, Jacobsli 20229
Norsk Maritimt Museum, Oslo 20268
Norsk Polarinstitutt, Tromsø 20298
Norsk rikskringkasting, Oslo 20269
Norsk skogsbruksmuseum, Elverum 20223
Norsk Språkråd, Oslo 20270

Norsk Teknisk Museum, Oslo 20271
Norsk treteknisk institutt, Oslo 20272
Norsk Utenrikspolitisk Institutt, Oslo 20273
Det norske institutt i Roma, Roma 16551
Det Norske meteorologiske institutt, Oslo 20274
Norske Studentersamfund, Oslo 20275
Det Norske Utvandrersenteret / The Norwegian Emigration Center, Stavanger 20294
North Adams Public Library, North Adams 40494
North American Baptist Seminary, Sioux Falls 35381
North Arkansas College, Harrison 31300
North Arkansas Regional Library, Lakeview 39787
North Arlington Free Public Library, North Arlington 40496
North Ayrshire Council, Ardrossan 29926
North Babylon Public Library, North Babylon 40498
North Bay Public Library, North Bay 04707
North Bend Public Library, North Bend 40499
North Bergen Free Public Library, North Bergen 40500
North Bingham County District Library, Shelley 41302
North Bohemian Museum, Liberec 06461
North Bohemian Research Library, Ustí nad Labem 06352
North Branford Library Department, North Branford 40501
North Brunswick Free Public Library, North Brunswick 40502
North Canton Public Library, North Canton 40503
North Carolina Agricultural and Technical State University, Greensboro 31254
North Carolina Central University, Durham 31046
– Music Library, Durham 31047
– School of Law Library, Durham 31048
– School of Library & Information Sciences, Durham 31049
North Carolina Department of Corrections, Elizabeth City 34180
North Carolina Department of Corrections, Maury 34508
North Carolina Department of Corrections, Morganton 34559
North Carolina Department of Environment & Natural Resources Library, Raleigh 34723
North Carolina Division of Archives & History, Manteo 34502
North Carolina Justice Academy, Salemburg 34795
North Carolina Legislative Library, Raleigh 34724
North Carolina Museum of Art, Raleigh 37425
North Carolina Regional Library for the Blind & Physically Handicapped, Raleigh 37426
North Carolina School of the Arts, Winston-Salem 33083
North Carolina State Museum of Natural Sciences, Raleigh 37427
North Carolina State University, Raleigh 32409
– Burlington Textiles Library, Raleigh 32410
– Harrye B. Lyons Design Library, Raleigh 32411
– Learning Resources Library, Raleigh 32412
– Natural Resources Library, Raleigh 32413
– Veterinary Medical Library, Raleigh 32414
North Carolina Supreme Court Library, Raleigh 34725
North Carolina Wesleyan College, Rocky Mount 32476
North Castle Public Library, Armonk 37996
North Caucasian Geological Administration, Essentuki 22730
North Central College, Naperville 31927
North Central Kansas Libraries System, Manhattan 40032
North Central Michigan College, Petoskey 32248
North Central Regional Library, Wenatchee 41800
North Central Regional Library – Wenatchee Public (Headquarters), Wenatchee 41801
North Central Texas College, Gainesville 33361
North Central University, Minneapolis 31846
North Chicago Public Library, North Chicago 40504
North China Coal Mining Medical College, Tangshan 05604
North China Electric Power University, Baoding 04991
North China Space Institute, Langfang 05374

North China University of Technology, Beijing 05048
North China Water Conservancy and Hydropower Institute, Zhengzhou 05785
North Country Community College, Saranac Lake 33749
North Country Library System, Watertown 41765
North Dakota Legislative Council Library, Bismarck 33988
North Dakota State College of Science, Wahpeton 33832
North Dakota State Library, Bismarck 30120
North Dakota State University, Fargo 31124
North Dakota Supreme Court, Bismarck 33989
North Devon Athenaeum, Barnstaple 29603
North East Institute of Electric Power Engineering, Jilin City 05316
North East Worcestershire College, Redditch 29383
North Eastern Education and Library Board, Ballymena 29931
North Eastern Regional Library, Wangaratta 01172
North Florida Community College, Madison 33511
North General Hospital, New York 37245
North Georgia College & State University, Dahlonega 30938
North Glasgow University Hospitals NHS Trust, Glasgow 29673
North Greenville College, Tigerville 32820
North Gujarat University, Patan 13735
North Harris Montgomery Community College District – Kingwood College, Kingwood 33459
North Harris Montgomery Community College District – North Harris College, Houston 33418
North Harris Montgomery Community College District – Tomball College, Tomball 33800
North Haven Memorial Library, North Haven 40507
North Hennepin Community College, Brooklyn Park 33195
North Hertfordshire College, Stevenage 29406
North Herts College, Hitchin 29470
North Hwanghae Provincial Library, Sariwon 18055
North Idaho College, Coeur d'Alene 30844
North Indian River County Library, Sebastian 41267
North Kansas City Public Library, North Kansas City 40510
North Karelia University of Applied Sciences, Joensuu 07346
North Kingstown Free Library, North Kingstown 40511
North Lake College, Irving 33431
North Lamarkshire Council, Stepps 30078
North Lanarkshire Council, Coatbridge 29962
North Lanarkshire Council, Motherwell 30047
North Las Vegas Library District, North Las Vegas 40512
North Logan City Library, North Logan 40513
North Madison County Public Library System – Elwood Public Library, Elwood 38971
North Manchester Public Library, North Manchester 40514
North Merrick Public Library, North Merrick 40515
North Miami Beach Public Library, North Miami Beach 40517
North Miami Public Library, North Miami 40516
North Olympic Library System, Port Angeles 40847
North Olympic Library System – Port Angeles Branch, Port Angeles 40848
North Olympic Library System – Sequim Branch, Sequim 41282
North Ossetian Medical Institute, Vladikavkaz 22629
North Park University, Chicago 30774
North Platte Public Library, North Platte 40520
North Port Public Library, North Port 40521
North Providence Union Free Library, North Providence 40522
North Richland Hills Public Library, North Richland Hills 40524
North Seattle Community College, Seattle 33755
North Shelby County Library, Birmingham 38210
North Shore Community College, Beverly 33169
North Shore Community College, Danvers 33278
North Shore Congregation Israel, Glencoe 35215
North Shore Library, Glendale 39247

North Shore Library, North Shore City 19876
North Smithfield Public Library, Slatersville 41333
North Somerset Council, Weston-super-Mare 30098
North Suburban Library-Loves Park, Loves Park 39970
North Suburban Synagogue Beth El, Highland Park 35230
North Tonawanda Public Library, North Tonawanda 40527
North Tyneside College, Wallsend 29430
North Tyneside Metropolitan District Council, North Shields 30051
North Unitersity of Baia Mare, Baia Mare 21918
North Vancouver City Library, North Vancouver 04708
North Vancouver District Public Library, North Vancouver 04709
North West Frontier Province University of Engineering and Technology, Peshawar 20452
North West Provincial Library Services, Mmabatho 25214
North Yorkshire County Council, Northallerton 30052
Northampton Area Public Library, Northampton 40532
Northampton Community College, Bethlehem 33167
Northampton County Historical & Genealogical Society, Easton 36821
Northampton County Law Library, Easton 34164
Northamptonshire Libraries and Information Service, Northampton 30053
Northborough Free Library, Northborough 40533
Northbrook College Sussex, Worthing 29506
Northbrook Public Library, Northbrook 40534
North-Caucasian Centre of Scientific and Technical Information and Popularization, Kabarda-Balkar Branch, Nalchik 23609
North-Caucasian Institute of Ore Mining and Metallurgy, Vladikavkaz 22627
North-Caucasian Railways, Rostov-na-Donu 22898
Northcentral Technical College, Wausau 33846
Northeast Agricultural University, Harbin 05257
Northeast Alabama Community College, Rainsville 33667
Northeast Forestry University, Harbin 05258
Northeast Georgia Regional Library, Clarkesville 38580
Northeast Harbor Library, Northeast Harbor 40535
North-East Interdisciplinary Scientific Research Institute, Magadan 23149
Northeast Minnesota Historical Center, Duluth 36815
Northeast Mississippi Community College, Booneville 33179
Northeast Missouri Library Service, Kahoka 39665
Northeast Missouri Library Service – H. E. Sever Memorial, Kahoka 39666
Northeast Normal University, Changchun 05088
Northeast Regional Correctional Facility Library, Saint Johnsbury 34774
Northeast Regional Library, Corinth 38696
Northeast Regional Library, Saint Johnsbury 41114
Northeast Regional Library – Corinth Public Library, Corinth 38697
Northeast School Division No 200, Nipawin 04423
North-East Scientific Research Institute of Agriculture, Kirov 23103
Northeast State Technical Community College, Blountville 33177
North-Eastern Hill University, Shillong 13763
Northeastern Illinois Planning Commission Library, Chicago 36686
Northeastern Illinois University, Chicago 30775
Northeastern Junior College, Sterling 33779
Northeastern Ohio Universities Colleges of Medicine and Pharmacy, Rootstown 32481
Northeastern Oklahoma A&M College, Miami 31810
Northeastern State University, Tahlequah 32787
Northeastern University, Boston
– School of Law Library, Boston 30501
– Snell Library, Boston 30502
Northeastern University, Shenyang 05546
Northern Alberta Institute of Technology, Edmonton 04342

Northern Area Research Institute of Hydrotechnology and Land Reclamation, Sankt-Peterburg 23897
Northern Arizona University, Flagstaff 31141
Northern Baptist Theological Seminary, Lombard 35266
Northern Caribbean University, Mandeville 16870
Northern College, Haileybury 03714
Northern College of Applied Art & Technology, Timmins 04036
Northern Essex Community College, Haverhill 33401
Northern Forestry Centre, Edmonton 04071
Northern Illinois University, DeKalb 30982
– David C. Shapiro Memorial Law Library, DeKalb 30983
– Music Library, DeKalb 30984
Northern Indiana Center for History, South Bend 37617
Northern Ireland Assembly, Belfast 29510
Northern Ireland Assembly Library, Belfast 29511
Northern Ireland Housing Executive, Library Information Services, Belfast 29512
Northern Jiaotong University, Beijing 05049
Northern Kentucky University, Highland Heights 31334
Northern Kentucky University – Salomon P. Chase College of Law, Highland Heights 31335
Northern Lakes College, Slave Lake 04032
Northern Land Council, Casuarina 00902
Northern Lights College, Dawson Creek 03693
Northern Lights College, Fort Saint John 03707
Northern Melbourne Institute of TAFE, Preston 00502
Northern Metropolitan College of TAFE, Collingwood 00564
Northern Metropolitan College of TAFE, Preston 00583
Northern Michigan University, Marquette 31764
Northern Oklahoma College, Tonkawa 33802
Northern Regional Library, Labasa 07320
Northern Regional Library and Information Service, Moree 01125
Northern State Medical Univerity, Arkhangelsk 22204
Northern State University, Aberdeen 30162
Northern Sydney Central Coast Area Health Service, Gosford 00924
Northern Technical College, Ndola 42331
Northern Technical Institute, Chiang Mai 27733
Northern Territory – Courts Library, Darwin 00663
Northern Territory Library, Darwin 00367
Northern Territory – Natural Resources & Environment, The Arts and Sport Library, Palmerston 00974
Northern Territory – Police, Fire and Emergency Services, Winnellie 00742
Northern Tier Library Association, Gibsonia 39225
Northern University of Malaysia, Jitra 18500
Northern Virginia Community College Libraries, Annandale 33129
Northfield Public Library, Northfield 40537
Northland Baptist Bible College, Dunbar 35191
Northland College, Ashland 30265
Northland Pioneer College, Holbrook 33412
Northland Public Library, Pittsburgh 40804
Northport-East Northport Public Library, Northport 40539
Northrop Grumman, North Highlands 36055
Northrop Grumman Corp, Baltimore 35484
Northrop Grumman Information Technology, Reading 36165
Northrop Grumman Newport News, Newport News 36052
NorthseaMuseum-Nissenhouse Husum, Husum 12070
North-Trondelag County Library, Steinkjer 20402
North-Trondelag University College, Steinkjer 20162
Northumberland College, Ashington 29449
Northumberland County Council, Morpeth 30046
Northumberland County Law Library, Sunbury 34903
Northville District Library, Northville 40542
North-West Academy of Public Administration – Karelian Branch, Petrozavodsk 22481
North-West Agricultural Development Institute, Potchefstroom 25070
Northwest Agricultural University, Xianyang 05704
Northwest Bible College, Edmonton 04204
Northwest China Second Institute for National Minorities, Yinchuan 05758

Northwest Christian College, Eugene 31096
Northwest College, Powell 33663
Northwest Community College, Terrace 04033
Northwest Florida State College – Niceville Campus, Niceville 33584
Northwest Georgia Regional Library System, Dalton 38773
Northwest Institute of Political Science and Law, Xi'an 05680
Northwest Kansas Library System, Norton 40544
Northwest Mississippi Community College, Senatobia 33761
Northwest Missouri State University, Maryville 31773
Northwest Museum of Arts and Culture / Eastern Washington State Historical Society, Spokane 37622
Northwest Nationalities University, Lanzhou 05383
Northwest Nazarene University, Nampa 31926
Northwest Normal University, Lanzhou 05384
Northwest Ohio Regional Book Depository, Perrysburg 32245
Northwest Regional Library, Thief River Falls 41550
Northwest Regional Library, Winfield 41929
Northwest Regional Library System – Bay County Public Library, Panama City 40701
North-West Scientific Research Institute of Agriculture, Siverskaya 23970
Northwest Telecommunications Engineering Institute, Xi'an 05681
Northwest Territories Legislative Assembly Library, Yellowknife 04197
Northwest Territories Public Library Services, Hay River 04659
Northwest University, Kirkland 31512
North-West University, Mmabatho 25034
Northwest University, Xi'an 05682
North-Western Area Personnel Department, Sankt-Peterburg 23898
Northwestern College, Orange City 32204
Northwestern College, Saint Paul 32534
Northwestern College of Forestry, Xianyang 05705
Northwestern Connecticut Community College, Winsted 33879
North-Western Correspondence Institute of Technology, Sankt-Peterburg 22543
North-Western Furniture-Making Joint-Stock Company, Sankt-Peterburg 22934
North-Western Geological Company, Sankt-Peterburg 22935
Northwestern Health Science University, Bloomington 30463
Northwestern Institute of Light Industry, Xianyang 05706
Northwestern Institute of Textile Technology, Xi'an 05683
Northwestern Michigan College, Traverse City 32832
Northwestern Mutual Life Insurance Co, Milwaukee 35895
Northwestern Mutual Life Insurance Co, Milwaukee 35896
Northwestern Oklahoma State University, Alva 30208
Northwestern Polytechnical University, Xi'an 05684
Northwestern Regional Library, Elkin 38959
North-Western Research and Design Institute of the Power Industry, Sankt-Peterburg 23899
Northwestern State University of Louisiana, Natchitoches 31950
Northwestern University, Chicago
– Galter Health Sciences Library, Chicago 30776
– Joseph Schaffner Library, Chicago 30777
– School of Law Library, Chicago 30778
Northwestern University, Evanston 31105
– Geology Library, Evanston 31106
– Melville J. Herskovits Library of African Studies, Evanston 31107
– Music Library, Evanston 31108
– Ralph P. Boas Mathematics Library, Evanston 31109
– Seeley G. Mudd Library for Science & Engineering, Evanston 31110
– Transportation Library, Evanston 31111
Northwest-Shoals Community College, Phil Campbell 32249
Northwood University, Cedar Hill 30688
Northwood University, Midland 31827
Northwood University, West Palm Beach 33034
R. W. Norton Art Gallery, Shreveport 37604
Norton Public Library, Norton 40545
Norton Public Library, Norton 40546
Norwalk Community College, Norwalk 33589
Norwalk Public Library, Norwalk 40548

Norwalk Public Library, Norwalk 40549
Norwalk Public Library – South Norwalk Branch, South Norwalk 41370
Norwegian Agency for Development Cooperation (NORAD), Oslo 20195
Norwegian Broadcasting Corporation, Oslo 20269
Norwegian Building Research Institute, Oslo 20258
Norwegian Co-operative Union and Wholesale Society, Oslo 20260
Norwegian Defence Research Establishment, Kjeller 20231
Norwegian Directorate of Health, Oslo 20190
Norwegian Emigration Center, Stavanger 20294
Norwegian Folk Museum, Oslo 20263
Norwegian Forest and Landscape Institute, Ås 20209
The Norwegian Forestry Museum, Elverum 20223
Norwegian Geotechnical Institute, Oslo 20259
The Norwegian Industrial and Regional Development Fund, Oslo 20204
Norwegian Institute for Air Research, Kjeller 20233
Norwegian Institute for Children's Books, Oslo 20262
Norwegian Institute for Urban and Regional Research, Oslo 20264
Norwegian Institute for Water Research, Oslo 20265
The Norwegian Institute in Rome, Roma 16551
Norwegian Institute of International Affairs, Oslo 20273
Norwegian Institute of Local History, Oslo 20266
Norwegian Institute of Public Health, Oslo 20255
Norwegian Institute of Wood Technology, Oslo 20272
Norwegian Language Council, Oslo 20270
Norwegian Library of Talking Books and Braille, Oslo 20267
Norwegian Library of Talking Books and Braille, Bergen, Bergen 20215
Norwegian Library of Talking Books and Braille, Trondheim, Jacobsli 20229
Norwegian Maritime Museum, Oslo 20268
Norwegian Medicines Agency, Oslo 20283
Norwegian Meteorological Institute, Oslo 20274
Norwegian Mining Museum, Kongsberg 20234
Norwegian National Defence College, Oslo 20114
Norwegian National Rail Administration, Oslo 20247
Norwegian Parliamentary Library, Oslo 20197
Norwegian Patent Office, Oslo 20277
Norwegian Polar Institute, Tromsø 20298
Norwegian Police University College, Oslo 20122
Norwegian Public Roads Administration, Directorate of Public Roads, Oslo 20290
Norwegian Public Roads Administration, Directorate of Roads, Oslo 20198
Norwegian Radium Hospital, Oslo 20280
Norwegian Railway Museum, Hamar 20225
Norwegian School of Economics and Business Administration, Bergen 20073
Norwegian School of Management, Oslo 20116
Norwegian School of Sport Sciences, Oslo 20119
Norwegian School of Veterinary Science, Oslo 20121
Norwegian State Academy of Music, Oslo 20120
Norwegian Teacher's Academy for the Study of Religion and Education, Bergen 20074
Norwegian Temperance Alliance, Oslo 20261
Norwegian Underwater Intervention, Bergen 20216
Norwegian University of Life Sciences, Ås 20067
Norwegian University of Science and Technology, Trondheim 20175
– Faculty Library of Production Engineering, Trondheim 20178
– Gunnerus Library, Trondheim 20179
– Main Library for the Humanities and Social Sciences, Dragvoll, Trondheim 20180
– Marine Technology Library, Trondheim 20176
– Medical Library and Information Center, Trondheim 20181
– Natural Science Library, Trondheim 20182

– NTNU Library, Faculty Library for Architecture, Civ. Eng./Math. Sciences, Trondheim 20177
Norwegian-American Historical Association Archives, Northfield 37294
Norwell Public Library, Norwell 40550
Norwich Bioscience Institutes, Norwich 29861
Norwich Cathedral Library, Norwich 29560
Norwich City College of Further and Higher Education, Norwich 29287
Norwich School of Art and Design, Norwich 29288
Norwich University, Northfield 32154
Norwin Public Library Association Inc, Irwin 39589
Norwood, Payneham and City of St Peters, Kent Town 01108
Nosivka Regional Centralized Library System, Main Library, Nosivka 28748
Nosivska raionna TsBS, Tsentralna biblioteka, Nosivka 28748
Nosov Institute of Ore Mining and Metallurgy, Magnitogorsk 22317
Nossaman, Guthner, Knox & Elliott Library, Los Angeles 35851
Nössjö Stadsbibliotek, Nössjö 26871
Nota. Danish National Library for Persons with Print Disabilities, København 06817
Nota. Nationalbibliotek for mennesker med læsevanskeligheder, København 06817
Notodden bibliotek / Europas Blues Bibliotek, Notodden 20239
Notre Dame College, South Euclid 32686
Notre Dame de Namur University, Belmont 30377
Notre Dame High School, Talofofo 13167
Notre Dame of Marbel University, Koronadal 20704
Notre Dame Seminary, New Orleans 35303
Notre Dame University, Cotabato City 20694
Notre Dame Women's University, Kyoto 17023
Nottingham Subscription Library Ltd, Nottingham 29864
Nottingham Trent University, Nottingham 29290
Nottingham University Hospitals NHS Trust, City Campus, Nottingham 29291
Nottinghamshire County Council, West Bridgford 30096
Nottinghamshire Healthcare NHS Trust, Nottingham 29865
Nouvelle Bibliothèque publique Les Comtes de Hainaut – A.F.I.C., Mons 02828
Nová huť a.s., Ostrava 06421
Nova Kakhovka Regional Centralized Library System, Main Library, Nova Kakhovka 28749
Nova Scotia Barristers' Society, Halifax 04357
Nova Scotia College of Art & Design, Halifax 03721
Nova Scotia Community College, Halifax 03992
Nova Scotia Community College – Truro Campus, Truro 03929
Nova Scotia Department of Education Library, Halifax 04083
Nova Scotia Department of Natural Resources Library, Halifax 04084
Nova Scotia Department of the Environment & Labour Library, Halifax 04085
Nova Scotia Government Library, Halifax 04086
Nova Scotia Museum Library, Halifax 04358
Nova Southeastern University, Fort Lauderdale 31158
– Health Professions Division Library, Fort Lauderdale 31159
– Oceanography, Dania Beach 31160
– Shepard Broad Law Center Library, Fort Lauderdale 31161
Nova Ushitsya Regional Centralized Library System, Main Library, Nova Ushitsya 28750
Nova Vodolaga Regional Centralized Library System, Main Library, Nova Vodolaga 28751
Novartis Institutes for BioMedical Research GmbH & Co KG, Wien 01506
Novartis Pharmaceuticals, Summit 36289
Novetheek Elburg, Elburg 19593
Novetheek Epe, Epe 19597
Novetheek Heerde, Heerde 19616
Novetheek Oldebroek, Oldebroek 19673
Novgorod Regional General Scientific Library, Velikiy Novgorod 22188
Novgorod-Siverska raionna TsBS, Tsentralna biblioteka, Novgorod Siverski 28752
Novgorod-Siverski Regional Centralized Library System, Main Library, Novgorod Siverski 28752

Novgorodskaya oblastnaya universalnaya nauchnaya biblioteka (NOUNB), Velikiy Novgorod 22188
Novgorodskaya oblastnaya yunosheskaya biblioteka, Veliki Novgorod 24342
Novgorodski gosudarstvenny pedagogicheski institut, Veliki Novgorod 22622
Novgorodski gosudarstvenny universitet im. Yaroslava Mudrogo, Veliki Novgorod 22623
Novi Public Library, Novi 40555
Noviciado y Seminario Mayor de los Reparadores, Salamanca 25812
'Novikov-Daurskogo' Amur District Museum of Regional Studies, Blagoveshchensk 23003
O. Novikov-Priboy Library, Kyiv 28661
Novo Nordisk A/S, Bagsværd 06764
Novoarkhangelsk Regional Centralized Library System, Main Library, Novoarkhangelsk 28754
Novoarkhangelska raionna TsBS, Tsentralna biblioteka, Novoarkhangelsk 28754
Novocherkassk Museum of the History of the Don Cossacks, Novocherkassk 23629
Novocherkasski gosudarstvenny tekhnicheski universitet, Novocherkassk 22430
Novograd-Volinska raionna TsBS, Tsentralna biblioteka im. L. Ukrainki, Novograd-Volinski 28755
Novograd-Volinski Regional Centralized Library System, L. Ukrainki Library, Novograd-Volinski 28755
Novohradské múzeum, Lučenec 24739
Novokakhovska raionna TsBS, Tsentralna biblioteka, Nova Kakhovka 28749
Novokramatorsk Machine Building Plant Industrial Corporation, Library, Kramatorsk 28152
Novomoskovska raionna TsBS, Tsentralna biblioteka, Novomoskovsk 28756
Novopskov Regional Centralized Library System, Main Library, Novopskov 28757
Novopskovska raionna TsBS, Tsentralna biblioteka, Novopskov 28757
Novorossisk Higher School of Marine Engineering, Novorossisk 22434
Novoselitska raionna TsBS, Tsentralna biblioteka, Novoselitsya 28758
Novoselitsya Regional Centralized Library System, Main Library, Novoselitsya 28758
Novosibirsk Architectural Institute, Novosibirsk 22443
Novosibirsk Regional Youth Library, Novosibirsk 24190
Novosibirsk State Academy of Civil Engineering and Construction, Novosibirsk 22445
Novosibirsk State Regional Scientific Library, Novosibirsk 22160
Novosibirsk State University, Novosibirsk 22446
Novosibirsk Telecommunications Institute, Novosibirsk 22444
Novosibirsk Territory Geological Administration, Novosibirsk 22762
Novosibirskaya gosudarstvennaya akademiya stroitelstva, Novosibirsk 22445
Novosibirskaya gosudarstvennaya oblastnaya nauchnaya biblioteka, Novosibirsk 22160
Novosibirskaya oblastnaya yunosheskaya biblioteka, Novosibirsk 24190
Novosibirski gosudarstvenny universitet, Novosibirsk 22446
Novosibirski mezhotraslevoi territorialny tsentr nauchno-tekhnicheskoi informatsii i propagandy, Novosibirsk 23653
Novosibirskoe territorialnoe geologicheskoe upravlenie, Novosibirsk 22762
Novo-Ufimski neftepererabatyvayushchi zavod, Ufa 22968
Novoukrainka Regional Centralized Library System, Main Library, Novoukrainka 28759
Novoukrainska raionna TsBS, Tsentralna biblioteka, Novoukrainka 28759
Novoushitska raionna TsBS, Tsentralna biblioteka, Nova Ushitsya 28750
Novovodolazka raionna TsBS, Tsentralna biblioteka, Nova Vodolaga 28751
NPO armaturostroeniya znamya truda, Sankt-Peterburg 23861
NPO Avangard, Sankt-Peterburg 23862
NPO Belselkhozmekhanizatsiya, Minsk 01893
NPO Dalnaya Svyaz, Sankt-Peterburg 23863
NPO Dormash, Minsk 01894
NPO Elektroavtomatika, Sankt-Peterburg 23864
NPO Krosna, Moskva 22851
NPO po proizvodstvy produktov pitaniya iz kartofelya, Minsk 01895
NPO "Seismotekhnika", Gomel 01851
NPP Burevestnik, Sankt-Peterburg 22919
NPP Institute 'BelNIIlit', Minsk 02045
NPP zavod im. V.Ya. Klimova, Sankt-Peterburg 22920
NR & M Information Centre, Brisbane 00882
NSW Department of Commerce, Sydney 00733
NSW Department of Education and Training, Strathfield 00723
NSW Department of Primary Industry, Maitland 00692
NSW Police College, Goulburn 00681
NT Attorney-General's Department, Darwin 00664
NT Department, Darwin 00665
NT Department of Health and Families, Casuarina 00654
NT Department of Resources, Northern Territory Geological Survey, Minerals and Energy InfoCentre, Darwin 00666
NT Justice Department, Darwin 00667
NTB Interstaatliche Hochschule für Technik Buchs, Buchs 27060
Nuchno-issledovatelsky instytut solyanoyi promyshlennosti, Artemivsk 28179
Nuclear Claims Tribunal, Majuro 18675
Nuclear Energy Institute Library, Washington 37771
Nuclear Research Institute, Kyiv 28352
Nuclear Research Institute, Otwock-Świerk 21185
Nuclear Research Institute Řež plc, Řež 06537
Nuffield College, Oxford 29333
Nuffield House, Birmingham 29613
NUI – Norwegian Underwater Intervention, Bergen 20216
Nunatta Atuagaateqarfia, Nuuk 06630
Nunavut Arctic College, Iqaluit 03994
Nunez Community College, Chalmette 33209
Nungalinya College, Casuarina 00419
Nuorisotiedon kirjasto, Helsinki 07477
Nurmeksen kaupunginkirjasto, Nurmes 07594
Nurmes Public Library, Nurmes 07594
Nursing College, Berdichiv 28093
Nursing College, Bilgorod Dnistrovski 28094
Nursing College, Kamyanets-Podilsk 28101
Nursing College, Kherson 28103
Nursing College, Kolomiya 28106
Nursing College, Poltava 28127
Nursing College, Rivne 28128
Nursing College, Zhytomyr 28132
Nursing College no 3, Odesa 28126
Nursing School, Krivi Rig 28107
Nursing School, Nikopol 28124
Nursing School, Vinnytsya 28130
Nursing School no 1, Dnipropetrovsk 28096
Nutley Free Public Library, Nutley 40556
Nutter, McClennen & Fish, Boston 35534
N.V. Nederlandse Gasunie, Groningen 19329
NV Nederlandse Spoorwegen (NS), Utrecht 19339
Nwafor Orizu College of Education, Onitsha 19972
Ny Carlsberg Glyptoteks Bibliotek, København 06818
Nyack College, Nyack 32169
The Nyack Library, Nyack 40557
Nyborg Bibliotek, Nyborg 06907
Nybro bibliotek, Nybro 26872
Nycomed, Konstanz 11414
Nye County Law Library, Tonopah 34930
Nyenrode Business Universiteit, Breukelen 19103
Nykarleby stadsbibliotek – Uudenkaarlepyyn kaupunginkirjasto, Nykarleby 07595
Nykøbing Falster Bibliotek, Nykøbing F. 06908
Nykøbing Katedralskole, Nykøbing F. 06741
Nyköpings stadsbibliotek, Nyköping 26873
Nykytaiteen museon kirjasto Kiasma, Helsinki 07478
Nynäshamns bibliotek, Nynäshamn 26874
NYS Supreme Court, Goshen 34306
N.Y.S. Supreme Court Library, Saratoga Springs 34842
NYS Supreme Court Library, White Plains 35065
Nýsköpunarmiðstöð Íslands, Reykjavík 13571
Nyugat-Magyarországi Egyetem, Sopron 13293
– Benedek Elek Pedagógical Főiskolai Kar, Sopron 13294
O. Honchar Scientific Regional Public Library, Kherson 28607
O. L. Vrouwecollege en La Strada S.J., Antwerpen 02524
O. O. Chuyko Institute of Surface Chemistry, Kyiv 28312
Oak Brook Public Library, Oak Brook 40558
Oak Creek Public Library, Oak Creek 40559
Oak Forest Hospital, Oak Forest 37297
Oak Hills Christian College, Bemidji 30381
Oak Lawn Public Library, Oak Lawn 40564
Oak Park Public Library, Oak Park 40565
Oak Park Public Library, Oak Park 40566
Oak Ridge National Laboratory, Oak Ridge 36064
Oak Ridge Public Library, Oak Ridge 40568
Oakland City University, Oakland City 32175
Oakland Community College, Farmington Hills 33334
Oakland Community College, Royal Oak 33706
Oakland Community College – Auburn Hills Campus Library, Auburn Hills 33141
Oakland County Jail Library, Pontiac 34689
Oakland County Library for the Visually & Physically Impaired, Pontiac 37392
Oakland County Research Library, Pontiac 34690
Oakland Public Library, Oakland 40570
Oakland Public Library, Oakland 40571
Oakland Public Library – Asian, Oakland 40572
Oakland Public Library – Dimond, Oakland 40573
Oakland Public Library – Main Library, Oakland 40574
Oakland Public Library – Rockridge, Oakland 40575
Oakland Public Library – West Oakland, Oakland 40576
Oakland University, Rochester 32460
Oaklands College, St Albans 29402
Oakton Community College, Des Plaines 33293
Oakville Public Library, Oakville 04710
Oakwood College, Huntsville 31376
Oakwood Correctional Facility Library, Lima 34457
Oakwood Hospital Medical Library, Dearborn 36779
OAO DniproAzot, Dniprodzerzhinsk 28532
OAO "Lakokraska" LIDA, Lida 01877
OAO Motoprom, Serpukhov 22955
Obafemi Awolowo University, Ile-Ife 19946
– Department of Chemistry, Ile-Ife 19947
– Faculty of Agriculture, Ile-Ife 19948
– Institute of Administration, Ile-Ife 19949
– Institute of Education, Ile-Ife 19950
Obayashi-Gumi Gijutsu Kenkyujo, Tokyo 17475
Občinska knjižnica Jesenice, Jesenice 24961
Obedinenie BelavtoMAZ – Minski avtomobilny zavod, Minsk 01896
Obedinenny institut vysokikh temperatur RAN, Moskva 23451
Obedinenny komitet profsoyuza promyshlennosti stroitelno-montazhnogo obedineniya Minskpromstroi, Minsk 02187
Obednannya Gospkomunobslugovuvannya miskoyi Radi narodnykh deputativ" Kyiv 28679
Ober, Kaler, Grimes & Shriver Law Library, Baltimore 35485
Oberfinanzdirektion Frankfurt am Main, Frankfurt am Main 10899
Oberfinanzdirektion Koblenz, Koblenz 10977
Oberfinanzdirektion Münster, Münster 11062
Oberfinanzdirektion Niedersachsen, Hannover 10945
Obergericht des Kantons Luzern, Luzern 27240
Obergericht des Kantons Solothurn, Solothurn 27242
Obergericht des Kantons Zürich, Zürich 27243
Obergerichtsbibliothek, Aarau 27229
Oberlandesgericht Braunschweig, Braunschweig 10836
Oberlandesgericht Celle, Celle 10846
Oberlandesgericht Düsseldorf, Düsseldorf 10884
Oberlandesgericht Frankfurt am Main, Frankfurt am Main 10900
Oberlandesgericht Hamm, Hamm 10929
Oberlandesgericht Innsbruck, Innsbruck 01385
Oberlandesgericht Jena, Jena 10954
Oberlandesgericht Karlsruhe, Karlsruhe 10958
Oberlandesgericht Koblenz, Koblenz 10978
Oberlandesgericht Köln, Köln 10992
Oberlandesgericht Linz, Linz 01389
Oberlandesgericht München, München 11045
Oberlandesgericht Nürnberg, Nürnberg 11070
Oberlandesgericht Oldenburg, Oldenburg 11075
Oberlandesgericht Stuttgart, Stuttgart 11107
Oberlausitzische Bibliothek der Wissenschaften, Görlitz 09314
Oberlin College, Oberlin 32176
– Clarence Ward Art Library, Oberlin 32177
– Mary M. Vial Music Library, Oberlin 32178
– Science Library, Oberlin 32179
Oberlin Public Library, Oberlin 40580
Oberösterreichische Landesbibliothek, Linz 01185
Oberösterreichische Landesmuseen, Linz 01560
Oberösterreichisches Landesarchiv, Linz 01561
Oberservatorio Nacional de Física Cósmica, San Miguel 00314
Oberste Baubehörde im Bayerischen Staatsministerium des Innern, München 11046
Oberstufenzentrum (OSZ) Handel 1, Berlin 10710
Oberverwaltungsgericht für das Land Nordrhein-Westfalen, Münster 11063
Oberverwaltungsgericht Rheinland-Pfalz in Koblenz, Koblenz 10979
Obihiro Chikusan Daigaku, Obihiro 17075
Obihiro University of Agriculture and Veterinary Medicine, Obihiro 17075
Obispado de Jaca, Jaca 25782
Obispado de Zamora, Zamora 25847
Obispo Colombres Agro-Industrial Experimental Research Station, Las Talitas 00301
Oblasna biblioteka dlya ditei, Chernigiv 28520
Oblasna biblioteka dlya ditei, Chernivtsi 28524
Oblasna biblioteka dlya ditei, Kharkiv 28602
Oblasna biblioteka dlya ditei, Kherson 28608
Oblasna biblioteka dlya ditei, Khmelnitski 28611
Oblasna biblioteka dlya ditei, Lugansk 28703
Oblasna biblioteka dlya ditei, Rivne 28798
Oblasna biblioteka dlya ditei, Sumy 28835
Oblasna biblioteka dlya ditei, Ternopil 28843
Oblasna biblioteka dlya ditei im. Gaidara, Kirovograd 28617
Oblasna biblioteka dlya ditei im. O. Lyagina, Mykolaiv 28739
Oblasna biblioteka dlya ditei im. Panasa Mirnoho, Poltava 28784
Oblasna biblioteka dlya yunatstva, Cherkasy 28516
Oblasna biblioteka dlya yunatstva, Chernigiv 28521
Oblasna biblioteka dlya yunatstva, Kherson 28609
Oblasna biblioteka dlya yunatstva, Khmelnitski 28612
Oblasna biblioteka dlya yunatstva, Lugansk 28704
Oblasna biblioteka dlya yunatstva, Mykolaiv 28740
Oblasna biblioteka dlya yunatstva, Ternopil 28844
Oblasna biblioteka dlya yunatstva im. Gonchara, Poltava 28785
Oblasna medychna biblioteka, Cherkasy 28188
Oblasna medychna biblioteka, Chernivtsi 28196
Oblasna medychna biblioteka, Dnipropetrovsk 28206
Oblasna medychna biblioteka, Khmelnitski 28290
Oblasna mezhspilkova biblioteka, Lutsk 28708
Oblasna naukova medychna biblioteka, Ivano-Frankivsk 28241
Oblasna naukova medychna biblioteka, Kherson 28289
Oblasna naukova medychna biblioteka, Kirovograd 28292
Oblasna naukova medychna biblioteka, Lugansk 28400
Oblasna naukova medychna biblioteka, Mykolaiv 28431
Oblasna naukova medychna biblioteka, Odesa 28440
Oblasna naukova medychna biblioteka, Poltava 28452
Oblasna naukova medychna biblioteka, Rivne 28456
Oblasna naukova medychna biblioteka, Sumy 28474
Oblasna naukova medychna biblioteka, Ternopil 28478
Oblasna naukova medychna biblioteka, Uzhgorod 28479
Oblasna naukova medychna biblioteka, Vinnytsya 28480

Oblasna naukova medychna biblioteka, Zaporizhzhya 28486
Oblasna naukova medychna biblioteka, Zhytomyr 28488
Oblasna naukovo-pedahohichna biblioteka, Lviv 28038
Oblasna universalna naukova biblioteka im. Gmyrova, Mykolaiv 27942
Oblasna universalna naukova biblioteka im. I. Franka, Ivano-Frankivsk 28585
Oblasna universalna naukova biblioteka im. Kotlarevskoho, Poltava 27945
Oblasna universalna naukova biblioteka im. N.K. Krupskoi, Sumy 27947
Oblasnaya biblioteka dlya ditei, Cherkasy 28517
Oblasnaya medychna biblioteka, Chernigiv 28192
Oblasnaya naukova medychna biblioteka, Donetsk 28232
Oblasnoyi derzhavny arkhiv, Mykolaiv 28432
Oblasny istorychny muzei im. V. Tarnovskogo, Chernigiv 28193
Oblastnaya biblioteka im. A. S. Pushkina, Minsk 02188
Oblastnaya bolnitsa, Brest 01946
Oblastnaya detskaya biblioteka, Sankt-Peterburg 24241
Oblastnaya detskaya biblioteka im. A.M. Gorkogo, Novosibirsk 24191
Oblastnaya klinicheskaya detskaya bolnitsa, Biblioteka-filial no 19, Sankt-Peterburg 23865
Oblastnaya meditsinskaya biblioteka, Chita 23021
Oblastnaya meditsinskaya biblioteka, Kaliningrad 23070
Oblastnaya meditsinskaya biblioteka, Kursk 23136
Oblastnaya meditsinskaya biblioteka, Novosibirsk 23654
Oblastnaya meditsinskaya biblioteka, Orenburg 23672
Oblastnaya meditsinskaya biblioteka, Oryol 23677
Oblastnaya meditsinskaya biblioteka, Samara 23743
Oblastnaya meditsinskaya biblioteka, Smolensk 23972
Oblastnaya meditsinskaya biblioteka, Tver 24011
Oblastnaya meditsinskaya biblioteka, Volgograd 24065
Oblastnaya nauchnaya meditsinskaya biblioteka, Ekaterinburg 23034
Oblastnaya nauchnaya meditsinskaya biblioteka, Kurgan 23131
Oblastnaya nauchnaya meditsinskaya biblioteka, Sankt-Peterburg 23866
Oblastnaya nauchnaya meditsinskaya biblioteka, Ulyanovsk 24036
Oblastnaya nauchno-meditsinskaya biblioteka, Ryazan 23733
Oblastnaya nauchno-meditsinskaya biblioteka, Yaroslavl 24088
Oblastní muzeum v Mostě, Most 06466
Oblastnoe uchilishche kultury, Sankt-Peterburg 22703
Oblastnoho instytuta povyishenniya kvalifikatsiyi pedahohicheskykh kadrov, Rivne 28069
Oblastnoi dom sanitarnogo prosveshcheniya, Vitebsk 02080
Oblastnoi institut usovershenstvovaniya uchitelei, Novosibirsk 22447
Oblastnoi institut usovershenstvovaniya uchitelei, Sankt-Peterburg 22532
Oblastnoi khudozhestvenny muzei im. Vasnetsova, Kirov 23104
Oblastnoi kraevedcheski muzei, Ekaterinburg 23035
Oblastnoi kraevedcheski muzei, Irkutsk 23057
Oblastnoi kraevedcheski muzei, Kaluga 23075
Oblastnoi kraevedcheski muzei, Kursk 23137
Oblastnoi kraevedcheski muzei, Murmansk 23602
Oblastnoi kraevedcheski muzei, Novosibirsk 23655
Oblastnoi kraevedcheski muzei, Omsk 23666
Oblastnoi kraevedcheski muzei, Orenburg 23673
Oblastnoi kraevedcheski muzei, Oryol 23678
Oblastnoi kraevedcheski muzei, Penza 23683
Oblastnoi kraevedcheski muzei, Perm 23689
Oblastnoi kraevedcheski muzei, Tambov 23989
Oblastnoi kraevedcheski muzei, Tomsk 23998
Oblastnoi kraevedcheski muzei, Tver 24012
Oblastnoi kraevedcheski muzei, Tyumen 24015
Oblastnoi kraevedcheski muzei, Voronezh 24077
Oblastnoi kraevedcheski muzei im. Alyabina, Samara 23744
Oblastnoi kraevedcheskoi muzei, Ivanovo 23061
Oblastnoi muzei izobrazitelnykh iskusstv, Rostov-na-Donu 23723
Oblastnoi muzei kraevedeniya, Rostov-na-Donu 23724
Oblastnoi muzei kraevedeniya, Saratov 23963
Oblastnoyi instytut vdoskonalennya vchiteliv, Zaporizhzhya 28087
Oblastnoyi kraevedchesky muzei, Donetsk 28233
Oblon, Spivak, McClelland, Maier & Neustadt, Arlington 35449
Oblsovprof – Bazovaya bibliyateka, Gomel 02131
Obninsk Institute of Nuclear Power Engineering, Obninsk 22450
Obolonski raion TsBS m. Kyiva, Tsentralna biblioteka im. A. Pushkina, Kyiv 28680
Obras Sanitarias de la Nación, Buenos Aires 00282
Observatoire Astronomique de Lyon, Saint-Genis-Laval 08668
Observatoire Astronomique de Strasbourg, Strasbourg 08686
Observatoire de Genève, Sauverny 27430
Observatoire de la Côte d'Azur, Nice 08450
Observatoire de Paris, Paris 08611
Observatoire de Paris, Section de Meudon, Meudon 08416
Observatoire Midi Pyrenées, Tarbes 08691
Observatoire Midi-Pyrénées, Toulouse 08081
Observatoire National, Besançon 08296
Observatoire Océanologique de Banyuls, Banyuls-sur-Mer 08290
Observatori de L'Ebre, Roquetes 26126
Observatories of the Carnegie Institution of Washington, Pasadena 37329
Observatorio Astronómico, Quito 06994
Observatório Astronómico de Lisboa, Lisboa 21662
Observatorio Interamericano de Cerro Tololo, La Serena 04919
Observatorio Nacional, São Cristovão 03389
Observatorio San Calixto, La Paz 02940
Observatory Library, Ksara 18224
Obshchestvo ispytatelei prirody, Moskva 23452
Obshtinska biblioteka 'Geo Milev', Montana 03598
Obukhiv Regional Centralized Library System, Main Library, Obukhiv 28760
Obukhivska raionna TsBS, Tsentralna biblioteka, Obukhiv 28760
Obunsha Co Ltd, Tokyo 17476
Obunsha K. K., Tokyo 17476
Obuvnoi kolledzh, Sankt-Peterburg 22704
Obyvnaya fabrika, Vitebsk 01932
Occidental Chemical Corporation, Grand Island 35722
Occidental College, Los Angeles 31651
Occidental Oil & Gas Corp Library, Houston 35758
Occupational Safety Research Institute, National Information Center, Praha 06527
Ocean City Free Public Library, Ocean City 40582
Ocean County College, Toms River 33801
Ocean County Historical Society, Toms River 37674
Ocean County Library, Toms River 41581
Ocean County Library – Berkeley, Bayville 38126
Ocean County Library – Brick Branch, Brick 38273
Ocean County Library – Jackson Township, Jackson 39603
Ocean County Library – Lacey Township, Forked River 39100
Ocean County Library – Lakewood Branch, Lakewood 39791
Ocean County Library – Manchester Township, Lakehurst 39784
Ocean County Library – Stafford, Manahawkin 40025
Ocean Vicinage Law Library, Toms River 34929
Oceanographic Research Institute, Durban 25121
Oceanside Library, Oceanside 40583
Oceanside Public Library, Oceanside 40584
Ochakiv Central Library, Ochakiv 28761
Ochakivska Tsentralna Biblioteka, Ochakiv 28761
Ochanomizu University, Tokyo 17168
Ochsner Clinic Foundation, New Orleans 37171
OCLC Library, Dublin 36813
Ocmulgee Regional Library System – Dodge County Library (System Headquarters), Eastman 38921
Oconee County Library, Walhalla 41708
Oconee Regional Library, Dublin 38869
Oconomowoc Public Library, Oconomowoc 40585
Octavia Fellin Public Library, Gallup 39188
'Octavian Goga' Cluj County Library, Cluj-Napoca 22084
L'Octogone, Montreal 04693
Odda bibliotek, Odda 20382
Oddeleni financnich a ekonomickych informaci, Praha 06412
Odder Kommunebibliotek, Odder 06911
Odense Centralbibliotek, Odense 06912
Odense Katedralskole, Odense 06742
Odense Socialpædagogiske Seminarium, Odense 06743
Odense Socio-Educational Training College, Odense 06743
Odense Tekniske Bibliotek, Odense 06828
Odesa City Centralized Library System, Franko Main Library, Odesa 28764
Odeska derzhavna akademiya kharchovykh tekhnologi, Odesa 28055
Odeska miska TsBS, Tsentralna biblioteka im. I. Franka, Odesa 28764
Odeska natsionalna morska akademiya, Odesa 28056
Odeska zaliznitsa, Odesa 28765
Odeski gosudarstvenny ekonomicheski universitet, Odesa 28057
Odesky arkheologichny muzei, Odesa 28441
Odesky khudozhni muzei, Odesa 28442
Odesky natsionalny morsky universytet, Odesa 28058
Odesky natsionalny politekhnychny universytet, Odesa 28059
Odessa Archaeological Museum, Odesa 28441
Odessa College, Odessa 33597
Odessa Hydrometeorological Institute, Odesa 28052
Odessa Museum of Western and Eastern Art, Odesa 28443
Odessa National Academy of Food Technologies, Odesa 28055
Odessa National Maritime Academy, Odesa 28056
Odessa National Polytechnic University, Odesa 28059
Odessa Railways, Odesa 28765
Odessa Research Institute of Virology and Epidemiology, Odesa 28444
Odessa State Academy of Refrigeration, Odesa 28050
Odessa State Economic University, Odesa 28057
Odessa State Marine University, Odesa 28058
Odessa State Medical University, Odesa 28053
Odilien-Institut, Graz 01534
Odin, Feldman & Pittleman Library, Fairfax 35698
Odontologiska biblioteket, Malmö 26415
Odsherred Bibliotek, Nykøbing Sj. 06910
Odvetvové informacné stredisko dopravy, Žilina 24759
Odvětvové informační středisko dopravy, Praha 06413
Oekopack AG, Spiez 27291
Oekumenischer Rat der Kirchen, Genève 27363
Œuvres Pontificales Missionnaires, Lyon 08200
Œuvres Pontificales Missionnaires (OPM), Paris 08220
O'Fallon Public Library, O'Fallon 40588
Offaly County Library, Tullamore 14592
Offensive Junger Christen e.V., Reichelsheim 12433
Öffentliche Bibliothek der Stadt Remscheid, Remscheid 12904
Öffentliche Bibliothek der Universität Basel, Basel 26992
Öffentliche Bibliothek im Bildungszentrum Markdorf, Markdorf 12846
Öffentliche Bücherei der Stadt Mainz – Anna Seghers, Mainz 12841
Öffentliches Gymnasium der Stiftung Theresianische Akademie, Wien 01380
Office Béninois de Recherches Géologiques et Minières, Cotonou 02905
Office de la Recherche Scientifique Outre-Mer (ORSTOM), Dakar 24445
Office de Radiodiffusion Télévision du Mali, Bamako 18655
Office for Reinforced Concrete Constructions, Moskva 23345
Office for the Commissioner for Public Employment, Adelaide 00599
Office International de l'Eau, Limoges 08387
Office National de la Recherche Géologique et Minière, Alger 00069
Office National des Mines, Tunis 27840
Office National d'Etudes et de Recherches Aérospatiales (ONERA), Châtillon-sous-Bagneux 08331
Office Nationale de l'Enfance, Bruxelles 02617
Office of Atomic Energy for Peace, Bangkok 27764
Office of Court Administration, Hato Rey 21892
Office of Cultural Heritage, Alger 00061
Office of Documentary Information for Agriculture and the Food Industry, Bucureşti 22012
Office of Fair Trading, Parramatta 00708
Office of Navajo Nation Library, Navajo 37147
Office of Navajo Nation Library, Window Rock 37854
Office of the Attorney General, Nassau 01733
Office of the Auditor-General, Muckleneuk Pretoria 42333
Office of the Government of ČR, Praha 06415
Office of the National Culture Commission, Bangkok 27765
Office of the Registrar of Books and Newspapers, Colombo 26294
Office of the Superintendent of Financial Institutions, Ottawa 04120
Office of the Superintendent of Financial Institutions, Toronto 04519
Office of Thrift Supervision Library, Washington 35005
Offizierschule der Luftwaffe, Fürstenfeldbruck 10728
Offizierschule des Heeres, Dresden 10720
Oficina Española de Patentes y Marcas, Madrid 26081
Oficina Internacional del Trabajo (OIT), Santiago 04942
Oficina Regional de Ciencia y Tecnología de la UNESCO para América Latina y el Caribe, Montevideo 42065
Oficina Regional de Cultura de la UNESCO para América Latina el Caribe, La Habana 06309
Oficina Regional de la FAO para América Latina y El Caribe, Santiago 04943
Oficiul de informare documentară pentru agricultură şi industrie alimentară, Bucureşti 22012
Ogaki City Library, Ogaki 17855
Ogaki Shiritsu Toshokan, Ogaki 17855
Ogata Institute for Medical and Chemical Research, Tokyo 17749
Ogden Farmer's Library, Spencerport 41403
Ogdensburg Public Library, Ogdensburg 40592
Ogilvy Renault Library, Montreal 04415
Oglethorpe University, Atlanta 30297
Ogre Central Library, Ogre 18195
Ogres centrālā bibliotēka, Ogre 18195
Ogród Botaniczny PAN, Warszawa 21328
Ogun State Library, Abeokuta 20045
Ogun State Polytechnic, Abeokuta 19917
Ohashi Hospital, Toho University, Faculty of Medicine, Tokyo 17750
Ohashibyoin, Tokyo 17750
Ohbayashi Corp, Technical Research Institute, Tokyo 17475
Ohio Agricultural Research & Development Center Library, Wooster 37858
Ohio Attorney General, Columbus 34100
Ohio Christian University, Circleville 30813
Ohio College of Podiatric Medicine, Independence 31382
Ohio County Law Library, Wheeling 35064
Ohio County Public Library, Wheeling 41875
Ohio Department of Transportation Library, Columbus 34101
Ohio Dominican University Library, Columbus 30898
Ohio Genealogical Society Library, Mansfield 37083
Ohio Health-Riverside Methodist Hospital, Columbus 36743
Ohio Historical Society, Columbus 36744
Ohio Legislative Service Commission Library, Columbus 34102
Ohio Northern University, Ada 30168
– Taggart Law Library, Ada 30169
Ohio School for the Deaf Library, Columbus 33254
Ohio State University, Columbus 30899
– Biological Sciences & Pharmacy Library, Columbus 30900
– Business Library, Columbus 30901
– Cartoon Research Library, Columbus 30902

– Edgar Dale Educational Media & Instructional Materials Laboratory, Columbus 30903
– English, Theatre & Communication Reading Room, Columbus 30904
– Fine Arts Library, Columbus 30905
– Food, Agricultural & Environmental Sciences Library, Columbus 30906
– Grant Morrow III MD Library at Children's Hospital, Columbus 30907
– John A. Prior Health Sciences Library, Columbus 30908
– Journalism Library, Columbus 30909
– Music & Dance Library, Columbus 30910
– Orton Memorial Library of Geology, Columbus 30911
– Science & Engineering Library, Columbus 30912
– Sullivant Library, Columbus 30913
– Veterinary Medicine Library, Columbus 30914
Ohio State University, Marion 31762
Ohio State University, Wooster 33092
Ohio State University at Newark & Central Ohio Technical College, Newark 32099
Ohio State University & Lima Technical College, Lima 31608
Ohio State University – Mansfield Campus, Mansfield 31756
The Ohio State University – Moritz Law Library, Columbus 30915
Ohio Township Public Library System, Newburgh 40457
Ohio Township Public Library System – Newburgh Library, Newburgh 40458
Ohio University, Athens 30273
– Health Sciences Library, Athens 30274
– Hwa-Wei Lee Center for International Collections, Athens 30275
– Mahn Center for Archives & Special Collections, Athens 30276
– Music-Dance Library, Athens 30277
Ohio University Eastern, Saint Clairsville 32496
Ohio University – Lancaster Library, Lancaster 31554
Ohio University – Southern Campus Library, Ironton 31410
Ohio University – Zanesville / Zane State College, Zanesville 33110
Ohio Valley College, Vienna 32931
Ohio Wesleyan University, Delaware 30986
Ohioana Library, Columbus 36745
Ohlone College, Fremont 33354
Ohoopee Regional Library System – Vidalia-Toombs County Library Headquarters, Vidalia 41683
"Oi Treis Ierarkhai", Volos 13137
Oikeusministeriön, Helsinki 07399
Oikeuspoliittisen tutkimuslaitoksen käsikirjasto, Helsinki 07479
Oil City Library, Oil City 40593
Oil City Library, Oil City 40594
Oil Industry Construction Plant no 16, Novopolotsk 01912
Oil Industry Museum, Zalaegerszeg 13523
Oil Institute, Grozny 22553
Oil Institute, Ufa 22611
Oil Processing Plant, Kapotnya 22798
Oil Refinery, Mozyr 01909
Oil Shale Research Institute, Kohtla-Järve 07230
Oita Prefectural Shallow Sea Fisheries Experiment Station, Bungotakada 17517
Oita Prefecture Agricultural Reseach Center, Usa 17812
Oita Prefecture Educational Research Institute, Oita 17627
Oita University, Oita 17076
Oitaken Kyoiku Senta, Oita 17627
Oitaken Nogyo Gijutsu Senta, Usa 17812
Oitaken Snekaigyogyo Shikenjo, Bungotakada 17517
OJC-Bibliothek, Reichelsheim 12433
Okanagan College Library, Kelowna 03727
Okanagan Regional Library District, Kelowna 04667
Okayama Daigaku, Okayama 17077
– Fuzokutoshokan Shigen Seibutsu Kagaku Kenkyusho Bunkan, Kurashiki 17078
– Shikata Bunkan, Okayama 17079
Okayama Prefectural Forest Experiment Station, Katsuta 17567
Okayama Prefecture General Affairs Dept, Administrative Data Room, Okayama 17287
Okayama University, Okayama 17077
– Medical and Dental Branch, Okayama 17079
– Research Institute for Bioresources, Kurashiki 17078
Okayamaken Ringyo Shikenjo, Katsuta 17567
Okayamaken Somubu, Okayama 17287

Okayamaken Suisan Shikenjo, Oku 17629
Okazaki National Research Institutes, Okazaki 17628
Okeechobee County Public Library, Okeechobee 40595
Okefenokee Regional Library, Waycross 41782
Okhtinski NPO Plastopolimer, Glavnaya biblioteka, Sankt-Peterburg 22921
Okhtinski NPO Plastopolimer, Profsoyuznaya biblioteka, Sankt-Peterburg 24242
Okhtinsky Plastics and Polymers Research and Industrial Corporation, Main Library, Sankt-Peterburg 22921
Okhtinsky Plastics and Polymers Research and Industrial Corporation, Trade Union Library, Sankt-Peterburg 24242
Okinawa Prefectural Assembly, Naha 17284
Okinawaken Gikai, Naha 17284
Oklahoma Baptist University, Shawnee 32662
Oklahoma Christian University, Edmond 31074
Oklahoma City Community College, Oklahoma City 33600
Oklahoma City University, Oklahoma City
– Dulaney-Browne Library, Oklahoma City 32184
– Law Library, Oklahoma City 32185
Oklahoma County Law Library, Oklahoma City 34630
Oklahoma Department of Career & Technology Education, Stillwater 34896
Oklahoma Department of Corrections & the Oklahoma Department of Libraries, McAlester 34515
Oklahoma Department of Libraries, Oklahoma City 40602
Oklahoma Historical Society, Oklahoma City 37304
Oklahoma Historical Society – Research Center, Oklahoma City 37305
Oklahoma School for the Deaf Library, Sulphur 37649
Oklahoma State Reformatory Library, Granite 34311
Oklahoma State University, Stillwater 32744
Oklahoma State University – College of Osteopathic Medicine, Tulsa 32848
Oklahoma State University – Tulsa Library, Tulsa 32849
Oklahoma Wesleyan University, Bartlesville 30354
Okmulgee Public Library, Okmulgee 40603
Okobank Library, Helsinki 07481
Okrăzhna biblioteka, Burgas 03592
Okrăzhna biblioteka, Lovech 03597
Okrăzhna biblioteka, Pernik 03599
Okrăzhna biblioteka, Vidin 03609
Okrăzhna biblioteka "Khristo Botev", Vratsa 03610
Okrăzhna biblioteka 'Khristo Smirnenski', Pleven 03600
Okrăzhna biblioteka 'Nikola Vaptzarov', Kărdzhali 03595
Okrăzhna biblioteka 'Parteni Pavlovic', Silistra 03603
Okrăzhna biblioteka 'V. Kolarov', Kyustendil 03596
Okrăzhna biblioteka 'V. Kolarov', Shumen 03602
Okresná knižnica, Čadca 24771
Okresná knižnica, Dunajská Streda 24774
Okresná knižnica, Galanta 24775
Okresná knižnica, Komárno 24777
Okresná knižnica, Lučenec 24782
Okresná knižnica, Nitra 24785
Okresná knižnica, Pezinok 24787
Okresná knižnica, Považská Bystrica 24789
Okresná knižnica, Rimavská Sobota 24792
Okresná knižnica, Rožňava 24793
Okresná knižnica, Senica 24794
Okresná knižnica, Spišská Nová Ves 24795
Okresná knižnica, Stará Ľubovňa 24796
Okresná knižnica, Topoľčany 24797
Okresná knižnica, Trebišov 24798
Okresná knižnica, Veľký Krtíš 24801
Okresná knižnica, Vranov na Topľou 24802
Okresná knižnica Dávida Gutgesela, Bardejov 24763
Okresní muzeum v Klatovech, Klatovy 06457
Okruzhnaya biblioteka, Khanty-Mansisk 24144
Okruzhnaya detskaya biblioteka, Salekhard 24213
Okruzhnoi ordena Krasnoi Zvezdy dom ofitserov Sovetskoi Armii im. S.M. Kirova, Sankt-Peterburg 23867
Oktyabbrsky Railway – S-Peterburg-Vitebski Station, Sankt-Peterburg 23873

Oktyabrskaya zheleznaya doroga, Sankt-Peterburg 23868
Oktyabrskaya zheleznaya doroga – Stantsiya S-Peterburg-Finlyandski, Sankt-Peterburg 23869
Oktyabrskaya zheleznaya doroga – Stantsiya S-Peterburg-Moskovski, Sankt-Peterburg 23870
Oktyabrskaya zheleznaya doroga – Stantsiya S-Peterburg-Sortirovochny, Sankt-Peterburg 23871
Oktyabrskaya zheleznaya doroga – Stantsiya S-Peterburg-Varshavski, Sankt-Peterburg 23872
Oktyabrskaya zheleznaya doroga – Stantsiya S-Peterburg-Vitebski, Sankt-Peterburg 23873
Oktyabrsky District Library-Branch no 7, Sankt-Peterburg 24235
Oktyabrsky District Library-Branch no 8, Sankt-Peterburg 24236
Oktyabrsky Railway, Sankt-Peterburg 23868
Oktyabrsky Railway – S-Peterburg-Finlyandski Station, Sankt-Peterburg 23869
Oktyabrsky Railway – S-Peterburg-Moskovski Station, Sankt-Peterburg 23870
Oktyabrsky Railway – S-Peterburg-Sortirovochny Station, Sankt-Peterburg 23871
Ökumenisches Hainich Klinikum gemeinnützige GmbH, Mühlhausen 12268
Ökumenisches Informationszentrum e.V., Dresden 11773
Ökumenisches Institut der Abtei Niederaltaich, Niederalteich 11295
Okura Institute for the Study of Spiritual Culture, Yokohama 17824
Okura Seishin Bunka Kenkyujo, Yokohama 17824
The Okura Shukokan Museum, Tokyo 17751
Okurasho Bunko, Tokyo 17321
Okyama Prefecture Fisheries Experiment Station, Oku 17629
Olabisi Onabanjo University, Ago-Iwoye 19922
Olaines Mehanikas un tehnologijas koledzas, Olaine 18178
Olathe Public Library, Olathe 40604
Olathe Public Library – Indian Creek, Olathe 40605
Old Bridge Public Library, Old Bridge 40606
Old Cathedral Library, Vincennes 35402
Old Catholic Seminary, Amersfoort 19295
Old Charles Town Library, Charles Town 38512
Old Church Slavonic Institute, Zagreb 06253
Old Dartmouth Historical Society, New Bedford 37153
Old Dominion University, Norfolk 32122
– Elise N. Hofheimer Art Library, Norfolk 32123
Old Mutual, Pinelands 25152
Old Mutual Library, Windhoek 19008
Old Salem Museums & Gardens, Winston-Salem 37856
Old Sturbridge Village, Sturbridge 37646
Old Town Public Library, Old Town 40609
Oldham College Library, Greater Manchester 29128
Oldham County Public Library, La Grange 39749
Oldham Library and Lifelong Learning Centre, Oldham 30055
Oldham Local Studies and Archives, Oldham 29866
Olds College, Olds 04013
Olean Public Library, Olean 40610
Oleksandrivka Regional Centralized Library System, Main Library, Oleksandrivka 28766
Oleksandrivska raionna TsBS, Tsentralna biblioteka, Oleksandrivka 28766
Oleksandriya City Centralized Library System, Main Library, Oleksandrivka 28767
Oleksandriyska miska TsBS, Tsentralna biblioteka, Oleksandrivka 28767
Ølgod Bibliotek, Ølgod 06913
Olive-Harvey College, Chicago 33224
Oliver Wolcott Library, Litchfield 39891
Olivet College, Olivet 32190
Olivet Nazarene University, Bourbonnais 30526
Olje- og energidepartementet, Oslo 20196
Olminsky Belgorod State Teacher-Training University, Belgorod 22216
Olney Public Library, Olney 40611
Olofströms bibliotek, Olofström 26875
Omaha Correctional Center Library, Omaha 34637
Omaha Public Library, Omaha 40612
Omaha Public Library – Benson, Omaha 40613
Omaha Public Library – Millard, Omaha 40614

Omaha Public Library – Milton R. Abrahams Branch, Omaha 40615
Omaha Public Library – W. Clarke Swanson Branch, Omaha 40616
Omdurman Islamic University, Omdurman 26353
O'Melveny & Myers, Washington 36368
O'Melveny & Myers LLP, Los Angeles 35852
O'Melveny & Myers LLP, New York 36004
OMNOVA Solutions Inc, Akron 35433
Omohundro Institute of Early American History & Culture, Williamsburg 37844
Omsk A. S. Pushkin Regional Scientific Library, Omsk 22161
Omsk Institute of Railway Engineering, Omsk 22452
Omsk Medical Academy, Omsk 22453
Omsk State Agrarian University, Omsk 22457
Omsk State Pedagogical University, Omsk 22454
Omsk State Technical University, Omsk 22455
Omsk State University, Omsk 22456
Omskaya gosudarstvennaya oblastnaya nauchnaya biblioteka im. A.S. Pushkina, Omsk 22161
Omskaya meditsinskaya akademiya, Omsk 22453
Omskaya oblastnaya yunosheskaya biblioteka, Omsk 24194
Omski gosudarstvenny pedagogich-eski universitet, Omsk 22454
Omski gosudarstvenny tekhnicheski universitet, Omsk 22455
Omski gosudarstvenny universitet, Omsk 22456
Omski mezhotraslevoi territorialny tsentr nauchno-tekhnicheskoi informatsii i propagandy, Omsk 23667
Omski selskokhozyaistvenny institut im. Kirova, Omsk 22457
ONDEO Nalco Company Library, Naperville 35932
Onderstepoort Veterinary Institute, Onderstepoort 25149
Ondersteuning, Begeleiding en Documentatie (OBED), Antwerpen 02525
Onderzoeks- en Informatiecentrum van de Verbruikersorganisaties (OIVO), Brussel 02618
Ondo State Library Board, Akure 20046
Ondo State Polytechnic, Owo 19977
ONE Institute & Archives, Los Angeles 37059
Onebane Law Firm APC, Lafayette 35812
Onega Tractor Plant, Petrozavodsk 22891
Oneida Correctional Facility Library, Rome 34755
Oneida County Supreme Court, Utica 34954
Oneida Public Library, Oneida 40617
Onezhski traktorny zavod, Petrozavodsk 22891
Onkaparinga Libraries, Noarlunga Centre 01131
Onkologicheski Institut, Sofiya 03569
Onkologicheski nauchny tsentr, Moskva 23453
Onkološki inštitut Ljubljana, Ljubljana 24928
Onondaga Community College, Syracuse 33788
Onondaga County Public Library – Robert P. Kinchen Central Library, Syracuse 41491
Onondaga Historical Association Museum & Research Center, Syracuse 37652
Onslow County Public Library, Jacksonville 39625
Onslow County Public Library – Swansboro Branch, Swansboro 41484
Ontario City Library, Ontario 40619
Ontario City Library – Colony High, Ontario 40620
Ontario College of Art & Design, Toronto 04039
Ontario Federation of Labour Library, Don Mills 04058
Ontario Legislative Library, Toronto 04161
Ontario Ministry Northern Development & Mines Library, Sudbury 04157
Ontario Ministry of Environment, Toronto 04162
Ontario Ministry of Environment & Energy, Toronto 04163
Ontario Ministry of Finance, Toronto 04164
Ontario Ministry of Labour – Ontario Workplace Tribunals Library, Toronto 04165
Ontario Ministry of Natural Resources, Peterborough 04126
Ontario Ministry of Public Safety and Security, Toronto 04520
Ontario Ministry of the Attorney General, Toronto 04166

Ontario Ministry of the Attorney General, Courts Administration, Toronto 04167
Ontario Ministry of the Solicitor General, Toronto 04168
Ontario Ministry of Training Colleges & Universities, Toronto 04169
Ontario Ministry of Transportation Library, Saint Catharines 04147
Ontario Police College, Aylmer 03674
Ontario Power Generation Library, Toronto 04521
Ontario Provincial Police, Orillia 04103
Ontario Science Centre Library, Toronto 04522
Onze-Lieve-Vrouweabdij, Oosterhout 19313
Onze-Lieve-Vrouwe-Abdij van Postel, Mol 02481
Onze-Lieve-Vrouweabdij van Tongerlo, Westerlo 02493
Oö. Landesbibliothek, Linz 01185
OOAO Spartak, Kazan 22800
Opctwo Benedyktynów, Kraków 21054
OPEC, Wien 01625
Adam Opel AG, Rüsselsheim 11426
Opelousas-Eunice Public Library, Opelousas 40623
Open Access Libraries, Wool-loongabba 01054
Open Air Museum 'The Old Town', Århus 06770
Open Joint-Stock Company Subsidiary VolgogradNIPIneft, Volgograd 24066
Open Learning Institute of TAFE, South Brisbane 00515
Open University, Milton Keynes 29275
Open University, Tel-Aviv 14663
Open University for Social Sciences, Moskva 22384
Openbare Bibliotheek, Aalst 02708
Openbare Bibliotheek, Aalter 02709
Openbare Bibliotheek, Alken 02712
Openbare Bibliotheek, Asse 02716
Openbare Bibliotheek, Beersel 02720
Openbare Bibliotheek, Bilzen 02724
Openbare Bibliotheek, Blankenberge 02725
Openbare Bibliotheek, Bonheiden 02726
Openbare Bibliotheek, Boom 02727
Openbare Bibliotheek, Bornem 02728
Openbare Bibliotheek, Bree 02733
Openbare Bibliotheek, Brugge 02734
Openbare Bibliotheek, Brussel 02738
Openbare Bibliotheek, Buggenhout 02742
Openbare Bibliotheek, Deinze 02746
Openbare Bibliotheek, Denderleeuw 02747
Openbare Bibliotheek, Diepenbeek 02751
Openbare Bibliotheek, Diest 02752
Openbare Bibliotheek, Dilbeek 02753
Openbare Bibliotheek, Dilsen-Stokkem 02754
Openbare Bibliotheek, Essen 02757
Openbare Bibliotheek, Evergem 02760
Openbare Bibliotheek, Grimbergen 02767
Openbare Bibliotheek, Haaltert 02768
Openbare Bibliotheek, Halle 02769
Openbare Bibliotheek, Hamont-Achel 02771
Openbare Bibliotheek, Herent 02778
Openbare Bibliotheek, Heusden-Zolder 02781
Openbare Bibliotheek, Hoogstraten 02782
Openbare Bibliotheek, Houthalen-Helchteren 02783
Openbare Bibliotheek, Houthulst 02784
Openbare Bibliotheek, Izegem 02787
Openbare Bibliotheek, Keerbergen 02790
Openbare Bibliotheek, Koksijde 02792
Openbare Bibliotheek, Kontich 02793
Openbare Bibliotheek, Kruibeke 02797
Openbare Bibliotheek, Laakdal 02798
Openbare Bibliotheek, Lanaken 02799
Openbare Bibliotheek, Lille 02804
Openbare Bibliotheek, Lochristi 02806
Openbare Bibliotheek, Lokeren 02807
Openbare Bibliotheek, Londerzeel 02809
Openbare Bibliotheek, Maaseik 02812
Openbare Bibliotheek, Malle 02815
Openbare Bibliotheek, Middelkerke 02826
Openbare Bibliotheek, Mol 02827
Openbare Bibliotheek, Neerpelt 02833
Openbare Bibliotheek, Nijlen 02834
Openbare Bibliotheek, Ninove 02835
Openbare Bibliotheek, Overijse 02841
Openbare Bibliotheek, Peer 02842
Openbare Bibliotheek, Puurs 02845
Openbare Bibliotheek, Ranst 02847
Openbare Bibliotheek, Riemst 02848
Openbare Bibliotheek, Ronse 02850
Openbare Bibliotheek, Rumst 02851
Openbare Bibliotheek, Scherpen-heuvel-Zichem 02852
Openbare Bibliotheek, Schilde 02853
Openbare Bibliotheek, Schoten 02854
Openbare Bibliotheek, Sint.Gillis-Waas 02856
Openbare Bibliotheek, Sint-Katelijne-Waver 02857
Openbare Bibliotheek, Stekene 02862
Openbare Bibliotheek, Temse 02863

Openbare Bibliotheek, Tervuren 02864
Openbare Bibliotheek, Tielt 02866
Openbare Bibliotheek, Vilvoorde 02875
Openbare Bibliotheek, Wervik 02881
Openbare Bibliotheek, Wevelgem 02884
Openbare Bibliotheek, Zandhoven 02887
Openbare Bibliotheek, Zaventem 02888
Openbare Bibliotheek, Zedelgem 02889
Openbare Bibliotheek, Zele 02890
Openbare Bibliotheek, Zemst 02891
Openbare Bibliotheek, Zonhoven 02892
Openbare Bibliotheek, Zwijndrecht 02895
Openbare Bibliotheek Amersfoort, Amersfoort 19548
Openbare Bibliotheek Amsterdam, Amsterdam 19551
Openbare Bibliotheek Appingedam, Appingedam 19553
Openbare Bibliotheek Baarlo, Baarlo 19558
Openbare Bibliotheek Barneveld, Barneveld 19561
Openbare Bibliotheek Bedum, Bedum 19562
Openbare Bibliotheek Bergen, Bergen (L) 19564
Openbare Bibliotheek Bergen, Bergen (N-H) 19565
Openbare Bibliotheek Born, Born 19570
Openbare Bibliotheek Brummen, Brummen 19572
Openbare Bibliotheek Brussel-Laken, Brussel 02739
Openbare Bibliotheek Castricum, Castricum 19577
Openbare Bibliotheek Cuijk, Cuijk 19578
Openbare Bibliotheek De Bilt – Vestiging Bilthoven, Bilthoven 19568
Openbare Bibliotheek De Velinx, Tongeren 02868
Openbare Bibliotheek Delfzijl, Delfzijl 19581
Openbare Bibliotheek Den Helder, Den Helder 19621
Openbare Bibliotheek Deventer, Deventer 19583
Openbare Bibliotheek Dordrecht, Dordrecht 19585
Openbare Bibliotheek Dronten, Dronten 19587
Openbare Bibliotheek Echt, Echt 19589
Openbare Bibliotheek Ede, Ede 19590
Openbare Bibliotheek Eilandgebied Curaçao, Willemstad 06321
Openbare Bibliotheek Eindhoven, Eindhoven 19592
Openbare Bibliotheek Enschede, Enschede 19596
Openbare Bibliotheek Finsterwolde, Finsterwolde 19601
Openbare Bibliotheek Geleen, Geleen 19604
Openbare Bibliotheek Gorinchem, Gorinchem 19606
Openbare Bibliotheek Gouda, Gouda 19608
Openbare Bibliotheek Groesbeek, Groesbeek 19609
Openbare Bibliotheek Groningen, Groningen 19610
Openbare Bibliotheek Haarlem-mermeer – Centrale Bibliotheek Hoofddorp, Hoofddorp 19630
Openbare Bibliotheek Haelen, Haelen 19613
Openbare Bibliotheek Heer-hugowaard, Heerhugowaard 19618
Openbare Bibliotheek Heerlen, Heerlen 19619
Openbare Bibliotheek Heiloo, Heiloo 19620
Openbare Bibliotheek Helmond, Helmond 19623
Openbare Bibliotheek Hendrik-Ido-Ambacht, Hendrik-Ido-Ambacht 19624
Openbare Bibliotheek Heythuysen, Heythuysen 19627
Openbare Bibliotheek Hillegom, Hillegom 19628
Openbare Bibliotheek Hoogezand, Hoogezand 19632
Openbare Bibliotheek Houten, Houten 19636
Openbare Bibliotheek Huizen, Huizen 19637
Openbare Bibliotheek Huizen-Laren-Blaricum, Vestiging Huizen, Huizen 19637
Openbare Bibliotheek Jette, Brussel 02740
Openbare Bibliotheek Joure, Joure 19640
Openbare Bibliotheek Kalmthout, Kalmthout 02788
Openbare Bibliotheek Kampen, Kampen 19641
Openbare Bibliotheek Leek, Leek 19646
Openbare Bibliotheek Leens, Leens 19647
Openbare Bibliotheek Leerdam, Leerdam 19648
Openbare Bibliotheek Leiden, Leiden 19650
Openbare Bibliotheek Leiderdorp, Leiderdorp 19651
Openbare Bibliotheek Lelystad, Lelystad 19654
Openbare Bibliotheek Losser, Losser 19656
Openbare Bibliotheek Maassluis, Maassluis 19659

Openbare Bibliotheek Merksem, Merksem 02825
Openbare Bibliotheek Nieuwegein, Nieuwegein 19665
Openbare Bibliotheek Noord-Veluwe, Novetheek Epe, Epe 19597
Openbare Bibliotheek Nunspeet, Nunspeet 19672
Openbare Bibliotheek Oldenzaal, Oldenzaal 19674
Openbare Bibliotheek Ommen, Ommen 19675
Openbare Bibliotheek Oss, Oss 19680
Openbare Bibliotheek Papendrecht, Papendrecht 19681
Openbare Bibliotheek Reuver, Reuver 19685
Openbare Bibliotheek Roosendaal, Roosendaal 19691
Openbare Bibliotheek Scheemda, Scheemda 19695
Openbare Bibliotheek Schiedam, Schiedam 19696
Openbare Bibliotheek s'Hertogen-bosch, 's-Hertogenbosch 19626
Openbare Bibliotheek Sittard, Sittard 19698
Openbare Bibliotheek Sliedrecht, Sliedrecht 19699
Openbare Bibliotheek Slochteren, Slochteren 19700
Openbare Bibliotheek Smallingerland, Drachten 19586
Openbare Bibliotheek Sneek, Sneek 19701
Openbare Bibliotheek Spijkenisse, Spijkenisse 19703
Openbare Bibliotheek Stadskanaal, Stadskanaal 19704
Openbare Bibliotheek Stein, Stein 19706
Openbare Bibliotheek Susteren, Susteren 19707
Openbare Bibliotheek Tegelen, Tegelen 19708
Openbare Bibliotheek Terborg, Terborg 19709
Openbare Bibliotheek Twenterand – Vestiging Vroomshoop, Vroomshoop 19732
Openbare Bibliotheek Uden, Uden 19714
Openbare Bibliotheek Uithuizen, Uithuizen 19715
Openbare Bibliotheek Ulft, Ulft 19716
Openbare Bibliotheek Veendam, Veendam 19720
Openbare Bibliotheek Veenendaal, Veenendaal 19721
Openbare Bibliotheek Veldhoven, Veldhoven 19722
Openbare Bibliotheek Velp, Velp 19723
Openbare Bibliotheek Vollenhove, Vollenhove 19729
Openbare Bibliotheek Voorschoten-Wassenaar, Vestiging Voorschoten, Voorschoten 19731
Openbare Bibliotheek Waalwijk, Waalwijk 19733
Openbare Bibliotheek Wageningen, Wageningen 19735
Openbare Bibliotheek Wierden, Wierden 19739
Openbare Bibliotheek Wijchen, Wijchen 19740
Openbare Bibliotheek Winschoten, Winschoten 19741
Openbare Bibliotheek Winsum, Winsum 19742
Openbare Bibliotheek Zuidbroek, Zuidbroek 19747
Openbare Bibliotheek Zuidhorn, Zuidhorn 19748
Openbare Bibliotheken Antwerpen, Antwerpen 02714
Openbare Bibliotheken Antwerpen, Antwerpen 02715
Openbare Bibliotheken Antwerpen, Deurne-Antwerpen 02750
Openbare Bibliotheken Antwerpen, Ekeren-Antwerpen 02755
OPENDOEK vzw, Antwerpen 02526
The Operations Centre Library, Norwich 29861
Opetushallitus, Helsinki 07480
Opetusministeriän kirjasto, Helsinki 07400
Opleidings- en Trainingscentrum Manoeuvre, Amersfoort 19236
Opole University, Main Library, Opole 20898
Oppegård bibliotek, Kolbotn 20357
Oppenheimer Wolff & Donnelly Library, Minneapolis 35912
Optikon Research Laboratories Library, West Cornwall 36408
Opus International Consultants Limited, Wellington 19817
Opzoekingscentrum voor de Wegenbouw (OCW) / Centre de Recherches Routières (CRR), Sterrebeek 02700
Oradell Free Public Library, Oradell 40624
Oral Roberts University, Tulsa 32850

– Holy Spirit Research Center Library, Tulsa 32851
Orange Agricultural Institute, Orange 00972
Orange Coast College, Costa Mesa 33262
Orange County Community College, Middletown 33538
Orange County Law Library, Orlando 34640
Orange County Library, Orange 40626
Orange County Library District, Orlando 40642
Orange County Library District – Alafaya, Orlando 40643
Orange County Library District – Edgewater, Orlando 40644
Orange County Library District – Hiawassee, Orlando 40645
Orange County Library District – North Orange, Apopka 37974
Orange County Library District – Orlando Public Library, Orlando 40646
Orange County Library District – South Creek, Orlando 40647
Orange County Library District – South Trail, Orlando 40648
Orange County Library District – Southeast/Semoran Blvd, Orlando 40649
Orange County Library District – Southwest/Della Drive, Orlando 40650
Orange County Library District – Washington Park, Orlando 40651
Orange County Library District – Winter Garden Branch, Winter Garden 41940
Orange County Public Law Library, Santa Ana 34835
Orange County Public Library, Hillsborough 39440
Orange County Public Library, Santa Ana 41212
Orange County Public Library – Aliso Viejo Branch, Aliso Viejo 37912
Orange County Public Library – Brea Branch, Brea 38266
Orange County Public Library – Chapman, Garden Grove 39193
Orange County Public Library – Costa Mesa Branch, Costa Mesa 38710
Orange County Public Library – Cypress Branch, Cypress 38752
Orange County Public Library – Dana Point Branch, Dana Point 38776
Orange County Public Library – El Toro Branch, Lake Forest 39776
Orange County Public Library – Fountain Valley Branch, Fountain Valley 39137
Orange County Public Library – Garden Grove Regional Library, Garden Grove 39194
Orange County Public Library – Heritage Park Regional Library, Irvine 39582
Orange County Public Library – La Habra Branch, La Habra 39751
Orange County Public Library – La Palma Branch, La Palma 39755
Orange County Public Library – Laguna Beach Branch, Laguna Beach 39769
Orange County Public Library – Laguna Niguel Branch, Laguna Niguel 39770
Orange County Public Library – Los Alamitos-Rossmoor, Seal Beach 41258
Orange County Public Library – Mesa Verde, Costa Mesa 38711
Orange County Public Library – Rancho Santa Margarita Branch, Rancho Santa Margarita 40937
Orange County Public Library – San Clemente Branch, San Clemente 41167
Orange County Public Library – San Juan Capistrano Regional Library, San Juan Capistrano 41189
Orange County Public Library – Seal Beach-Mary Wilson Branch, Seal Beach 41259
Orange County Public Library – Stanton Branch, Stanton 41427
Orange County Public Library – Tustin Branch, Tustin 41626
Orange County Public Library – University Park, Irvine 39583
Orange County Public Library – West Garden Grove Branch, Garden Grove 39195
Orange County Public Library – Westminster Branch, Westminster 41860
Orange County Regional History Center, Orlando 37312
Orange Public Library, Orange 40627
Orange Public Library & History Center, Orange 40628
Orange Public Library & History Center – El Modena Branch, Orange 40629

Orangeburg County Library, Orangeburg 40633
Orangeville Public Library, Orangeville 04711
ORANIM, Kiryat Tiv'on 14642
Oranjemund Recreation Club, Oranjemund 19012
Oratoire Saint-Joseph, Montreal 04216
Oravská Knižnica, Dolný Kubín 24772
Orbeli Institute of Physiology, Yerevan 00356
Orchard and Vineyard Institute, Almaty 17960
Orchard Learning Resources Centre, Birmingham 28951
Orchard Park Public Library, Orchard Park 40634
Orde der Dominicanen, Sint-Amandsberg 02485
Orde der Kruisheren, Diest 02459
Orde van de dominicanen, Leuven 02478
Ordem dos Advocados, Lisboa 21812
Ordem dos Engenheiros, Lisboa 21813
Ordem dos Farmacêuticos, Lisboa 21814
Ordem Regular dos Missionários Capuchinhos, São Paulo 03250
Ordinance Factory Library, Bendigo 00613
Ordine degli Avvocati di Milano, Milano 16347
Ordine degli Avvocati e dei Procuratori, Messina 16318
Ordine degli Avvocati e Procuratori, Torino 16609
Ordre des Avocats à la Cour d'Appel de Paris, Paris 08612
Ordre des infirmieres et infirmiers du Quebec, Montreal 04416
Ordzhonikidze State Academy of Management, Moskva 22333
Örebro County Library, Örebro 26876
Örebro universitet, Örebro 26418
Örebro University, Örebro 26418
Örebro University Hospital, Örebro 26625
Oregon City Public Library, Oregon City 40637
Oregon College of Art & Craft Library, Portland 33658
Oregon Department of Environmental Quality Library, Portland 34698
Oregon Department of Geology & Mineral Industries Library, Portland 34699
Oregon Department of Transportation, Salem 34789
Oregon Health & Science University, Portland 32339
Oregon Historical Society Research Library, Portland 37405
Oregon Institute of Technology, Klamath Falls 31516
Oregon Legislative Library, Salem 34790
Oregon National Primate Research Center, Beaverton 30369
Oregon Public Library, Oregon 40635
Oregon School for the Deaf Library, Salem 33718
Oregon State Correctional Institution Library, Salem 34791
Oregon State Library, Salem 30151
Oregon State Library Talking Book & Braille Services, Salem 30152
Oregon State Penitentiary Library, Salem 34792
Oregon State Penitentiary Library, Wilsonville 37852
Oregon State Penitentiary Library – Oregon Women's Correctional Center, Salem 34793
Oregon State University, Corvallis 30928
Oregon State University, Newport 32112
Orekhiv Regional Centralized Library System, Main Library, Orekhiv 28769
Orekhivska raionna TsBS, Tsentralna biblioteka, Orekhiv 28769
Orem Public Library, Orem 40638
Örenäs Folkhögskola, Glumslöv 26574
Orenburg Agricultural Academy, Orenburg 22461
Orenburg General Scientific Library, Orenburg 22162
Orenburg Geological Corporation, Orenburg 22885
Orenburg Geologiya, Orenburg 22885
Orenburgskaya oblastnaya yunosheskaya biblioteka, Orenburg 24195
Orenburgskaya selskokhozyaistvennaya akademiya, Orenburg 22461
Orenburgskaya universalnaya nauchnaya biblioteka, Orenburg 22162
Orenburgski mezhotraslevoi territorialny tsentr nauchno-tekhnicheskoi informatsii i propagandy, Orenburg 23674
Orendno obedannya Dniprovazhmash, Dnipropetrovsk 28139
Orendno pidpriemstvo Dniproshina, Dnipropetrovsk 28140
Orfeó Català, Barcelona 25927
Organic Chemistry Institute of the Siberian Department of the Russian Academy of Sciences, Novosibirsk 23646
Organic Semiconductors Research Institute, Rubizhne 28457
Organisation Africaine de la Propriété Intellectuelle (OAPI), Yaoundé 03663
Organisation de Coopération et de Développement Economiques (OCDE), Paris 08613
Organisation des Nations Unies, Rabat 18944
Organisation des Nations Unies (ONU), Alger 00070
Organisation des Nations Unies pour l'Education, la Science et la Culture, Paris 08614
Organisation for Economic C-operation and Development (OECD), Paris 08613
Organisation météorologique mondiale (OMM), Genève 27382
Organisation mondiale de la propriété intellectuelle (OMPI), Genève 27381
Organisation Mondiale de la Santé, Genève 27380
Organisation mondiale du commerce, Genève 27383
Organisation of Eastern Caribbean States (OECS), Castries 26335
Organisation Ouest Africaine de la Santé (OOAS) / West African Health Organisation (WAHO), Bobo-Dioulasso 03615
Organisation pour la Mise en Valeur du Fleuve Sénégal (OMVS), Saint-Louis 24452
Organismo Autonomo Parques Nacionales, Madrid 26082
Organización de Estados Iberoamericanos (OEI), Madrid 26083
Organización Nacional de Ciegos, Madrid 26084
Organization for Tropical Studies, San José 06125
Organization of American States, Washington 37772
Organization of Management and Economics of Oil and Gas Industry Research Institute, Moskva 23578
Organization of the Petroleum Exporting Countries (OPEC), Wien 01625
Organizzazione sociopsichiatrica cantonale, Mendrisio 27411
Orgstankinprom State Technology and Design Institute – Minsk Branch, Minsk 02007
Orgtekhstroi Special Planning and Industrial Corporation, Minsk 01901
Orgtekhvodstroi, Pinsk 01918
Oriel College, Oxford 29334
Orient Institute of the German Oriental Society, Beirut 18223
Orient Institute of the German Oriental Society, Istanbul Branch, Cihangir 27879
Oriental Research Institute, Mysore 14105
Orientální ústav AV ČR, Praha 06512
Orient-Institut der Deutschen Morgenländischen Gesellschaft Beirut, Beirut 18223
Orient-Institut der Deutschen Morgenländischen Gesellschaft Istanbul (OII), Cihangir 27879
Orissa State Museum, Bhubaneswar 13948
Orissa University of Agriculture and Technology, Bhubaneswar 13620
The Orkney Library and Archive, Kirkwall 30008
Orkney Public Library, Orkney 25217
Orkustofnun Bókasafn, Reykjavík 13572
Orland Free Library, Orland 40640
Orland Park Public Library, Orland Park 40641
Orleans Correctional Facility Library, Albion 33906
Orlovskaya oblastnaya detskaya biblioteka im. M.M. Prishvina, Oryol 24196
Orlovskaya oblastnaya publichnaya biblioteka im. I.A. Bunina, Oryol 22163
Ornithologische Gesellschaft in Bayern e.V., München 12322
Örnsköldsviks sjukhus, Örnsköldsvik 26626
Örnsköldsviks stadsbibliotek, Örnsköldsvik 26877
Oro Valley Public Library, Oro Valley 40653
Orphan Voyage & Concerned United Birth Parents, Saint Paul 37503
Orpington College, Orpington 29298
Orrick, Herrington & Sutcliffe, New York 36005
Orrick, Herrington & Sutcliffe LLP, San Francisco 36242
Orrville Public Library, Orrville 40655
Orsha Central City Children's Library, Orsha 02218
Orsha Flax-weaving Factory, Main Trade Union Library, Orsha 02219
Orshanka Pedagogical Institute, Orshanka 22686
Orshansk Regional Central Library, Andreevshchina 02091
Orshanskaya gorodskaya bibliyateka im. Gorkogo, Orsha 02215
Orshanskaya gorodskaya detskaya bibliyateka im. Krupskoi, Orsha 02216
Orshanskaya tsentralnaya gorodskaya bibliyateka im. Pushkina, Orsha 02217
Orshanskaya tsentralnaya gorodskaya detskaya bibliyateka, Orsha 02218
Orshanskaya tsentralnaya raionnaya bibliyateka, Andreevshchina 02091
Orshanski Lnokombinat, Orsha 02219
Orshanski zavod Legmash, Orsha 01915
Orshanskoe pedagogicheskoe uchilishche, Orshanka 22686
Orsk Mechanical Engineering College-Enterprise, Orsk 22687
Orski mashinostroitelny kolledzh-predpriyatie, Orsk 22687
Országgyűlési Könyvtar, Budapest 13321
Országos Erdészeti Egyesület, Budapest 13436
Országos 'Frederic Joliot Curie' Sugárbiológiai és Sugáregészségügyi Kutató Intézet, Budapest 13437
Országos Gyógyszerészeti Intézet, Budapest 13438
Országos Haematológiai és Vértranszfuziós, Budapest 13439
Országos Idegennyelvu Könyvtár, Budapest 13440
Országos Kardiologiai Intézet, Budapest 13441
Országos Korányi TBC és Pulmonológiai Intézet, Budapest 13442
Országos Kriminológiai Intézet, Budapest 13443
Országos Mérésügyi Hivatal, Budapest 13322
Országos Mezőgazdasági Könyvtár és Dokumentációs Központ, Budapest 13444
Országos Müemléki Hiuatal, Budapest 13445
Országos Műszaki Múzeum, Budapest 13446
Országos Műszaki Múzeum / Vegyészeti Múzeum, Várpalota 13518
Országos Onkológiai Intézet, Budapest 13447
Országos Pedagógiai Könyvtár és Múzeum, Budapest 13448
Országos Pszichiátriai és Neurológiai Intézet, Budapest 13449
Országos Rabbikèpző – Zsidó Egyetem Könyvtára, Budapest 13330
Országos Reumatológiai és Fizioterápiás Intézet, Budapest 13450
Országos Sportegészségügyi Intézet – Sportkórház, Budapest 13451
Országos Széchényi Könyvtár, Budapest 13213
Országos Szinháztörténeti Múzeum és Intézet, Budapest 13452
Orta Doğu Teknik Üniversitesi, Ankara 27849
Orthodox Metropolitan Library Sibiu, Sibiu 21944
Orthopaedics and Traumatology Research Institute, Kharkiv 28273
Orugano K.K. Kenkyukaihatsubu, Tokyo 17477
Ørum Bibliotek, Tjele 06940
Orusts kommunbibliotek, Henån 26791
Orvostovábbképző Egyetem, Budapest 13259
Osaka Chamber of Commerce and Industry, Osaka 17641
Osaka City Institute of Public Health and Environmental Sciences, Osaka 17639
Osaka City University, Osaka 17085
– Faculty of Law, Osaka 17086
– Medical Branch, Osaka 17087
Osaka College of Music, Toyonaka 17224
Osaka Daigaku, Toyonaka 17220
– Minoh Branch Library, Osaka 17221
– Seimei Kagaku Bunkan, Osaka 17222
– Suita Bunkan, Osaka 17223
Osaka Denki Tsushin Daigaku, Neyagawa 17065
Osaka Electro-Communication University Library, Neyagawa 17065
Osaka Furitsu Daigaku, Sakai 17100
– College of Economics, Sakai 17101
Osaka Furitsu Nakanoshima Toshokan, Osaka 17856
Osaka Furitsu Toshokan, Osaka 16897
Osaka Gas Co, Ltd, Konohana 17379
Osaka Gasu K. K., Konohana 17379
Osaka Ika Daigaku, Takatsuki 17131
Osaka Industrial University, Daito 16912
– Institute for Industrial Research, Daito 16913
Osaka Keizai Daigaku, Osaka 17080
– Chushokigyo Keiei Kenkyujo, Osaka 17081
– Nihon Keizaishi Kenkyujo, Osaka 17082
Osaka Kogyo Daigaku, Osaka 17083
Osaka Kogyo Gijutsu Shikenjo, Ikeda 17555
Osaka Medical College, Osaka 17084
Osaka Medical College, Takatsuki 17131
Osaka Municipal Assembly, Osaka 17288
Osaka Municipal Museum of Art, Osaka 17638
Osaka Ongaku Daigaku, Toyonaka 17224
Osaka Prefectural Assembly, Osaka 17289
Osaka Prefectural Central Library, Osaka 16897
Osaka Prefectural Nakanoshima Library, Osaka 17856
Osaka Prefecture University, Sakai 17100
– School of Economics, Sakai 17101
Osaka Prefecture University, Sakai 17102
Osaka Sangyo Daigaku, Daito 16912
– Sangyo Kenkyujo, Daito 16913
Osaka Securities Exchange, Osaka 17640
Osaka Shikai Toshoshitsu, Osaka 17288
Osaka Shiritsu Daigaku, Osaka 17085
– Hogakubu Shiryoshitsu, Osaka 17086
– Igakubu Bunkan, Osaka 17087
Osaka Shiritsu Kankyo Kagaku Kenkyujo, Osaka 17639
Osaka Shoken Torihikijo Chosabu, Osaka 17640
Osaka Shoko Kaigisho, Osaka 17641
Osaka University, Toyonaka 17220
– Life Sciences Branch Library, Osaka 17222
– Suita Branch Library, Osaka 17223
Osaka University of Economics, Osaka 17080
– Institute for Research in the Economic History of Japan, Osaka 17082
– Institute of Small Business Research and Business Administration, Osaka 17081
Osaka University of Technology, Osaka 17083
Osakafu Gikai, Osaka 17289
Osaragi Jiro Kinenkai Kenkyushitsu, Yokohama 17825
Osborn Correctional Institution, Somers 34866
Osborn Maledon, Phoenix 36120
Osby bibliotek, Osby 26878
Oscott College, Sutton Coldfield 29420
Osgoode Hall Law School, North York 03924
Osha, Washington 35006
Oshawa Public Library, Oshawa 04712
Oshkosh Correctional Institution Library, Oshkosh 34642
Oshkosh Public Library, Oshkosh 40657
Oshmyany Central Regional Library, Oshmyany 02220
Oshmyanyskaya tsentralnaya raionnaya bibliyateka, Oshmyany 02220
OSI Specialties/Crompton Corp, Tarrytown 36301
Osipovichi Central Regional Library, Osipovichi 02221
Osipovichiskaya tsentralnaya raionnaya bibliyateka, Osipovichi 02221
Oskaloosa Public Library, Oskaloosa 40658
Oskar Diethelm Library, New York 32084
Oskarshamns stadsbiblioteket, Oskarshamn 26879
Osler, Hoskin & Harcourt Library, Toronto 04289
Oslo Bymuseum, Oslo 20276
Oslo City Museum, Oslo 20276
Oslo College of the Arts, National Academy of Fine Arts, Oslo 20117
Oslo katedralskole, Oslo 20185
Oslo National Academy of Arts and Crafts, Oslo 20123
Osmania University, Hyderabad 13657
Osnovna šole Brezovica pri Ljubljani, Brezovica 24875
Ospedale Civile, Piacenza 16417
Ospedale Di 'Venere', Bari 16176
Ospedale di Volterra, Volterra 16677
Ospedale Generale Provinciale, Venezia 16656
Ospedale Neuropsichiatrico Provinciale, Ancona 16166
Ospedale Provinciale Spezzializzato Achille Sclavo, Siena 16579
Ospedale Psichiatrico Provinciale, Ferrara 16247
Ospedale S. Maria della Pietà, Roma 16552
Ospedali Civici Riunti, Venezia 16657
Osram Sylvania Library, Towanda 36310
Osrednja knjižnica Celje, Celje 24955
Osrednja knjižnica občine Kranj, Kranj 24964
Osrednja knjižnica Srečka Vilharja, Koper 24963
Ośrodek Badań Naukowych im. W. Kętrzyńskiego, Olsztyn 21177

Ośrodek Badawczo-Rozwojowy Izotopów POLATOM, Otwock-Świerk 21186
Ośrodek Badawczo-Rozwojowy Metrologii Elektrycznej 'METROL', Zielona Góra 21375
Osrodek Informacji Naukowej i Technicznej, Gap 21092
Osservatorio Astrofisico Catania, Catania 14981
Osservatorio astrofisico di Arcetri, Firenze 16271
Osservatorio Astronomico di Brera, Milano 16348
Osservatorio Astronomico di Capodimonte, Napoli 16379
Osservatorio Astronomico di Palermo, Palermo 16405
Osservatorio Astronomico di Roma, Roma 16553
Osservatorio Vesuviano, Ercolano 16240
Osservatorio Ximeniano, Firenze 16272
Ossining Public Library, Ossining 40659
Ossoliński National Institute, Wrocław 20792
Ostankinski kompleks tsvetnoi metallurgii, Moskva 23454
Osterhout Free Library, Wilkes-Barre 41895
Österreichische Agentur für Gesundheit und Ernährungssicherheit GmbH (AGES), Wien 01626
Österreichische Akademie der Wissenschaften, Wien 01187
Österreichische Akademie der Wissenschaften, Kommission für Musikforschung, Wien 01627
Österreichische Apothekerkammer, Wien 01628
Österreichische Bundesbahnen, Wien 01408
Österreichische Forschungsstiftung für Internationale Entwicklung, Wien 01629
Österreichische Galerie Belvedere, Wien 01630
Österreichische Gartenbau-Gesellschaft, Wien 01631
Österreichische Geographische Gesellschaft, Wien 01632
Österreichische Gesellschaft für Zeitgeschichte, Wien 01633
Österreichische Kapuzinerprovinz, Innsbruck 01440
Österreichische Lederbibliothek, Hofkirchen 01539
Österreichische Militärbibliothek, Wien 01401
Österreichische Nationalbank, Wien 01507
Österreichische Nationalbibliothek, Wien 01180
Österreichische Post AG, Wien 01508
Österreichische Provinz der Gesellschaft Jesu, Wien 01491
Österreichische Zentralbibliothek für Physik, Wien 01360
Österreichischer Gewerkschaftsbund, Wien 01634
Österreichischer Ingenieur- und Architekten-Verein, Wien 01635
Österreichischer Touristenklub, Wien 01636
Österreichisches Archäologisches Institut, Wien 01637
Österreichisches Filmmuseum, Wien 01638
Österreichisches Gesellschafts- und Wirtschaftsmuseum, Wien 01639
Österreichisches Institut für Bautechnik, Wien 01640
Österreichisches Institut für Menschenrechte, Salzburg 01571
Österreichisches Institut für Raumplanung, Wien 01509
Österreichisches Institut für Wirtschaftsforschung, Wien 01641
Österreichisches Minoritenkonvent, Wien 01492
Österreichisches Museum für angewandte Kunst, Wien 01621
Österreichisches Museum für Volkskunde, Wien 01659
Österreichisches Ökologie-Institut, Wien 01642
Österreichisches Patentamt, Wien 01409
Österreichisches Sprachinselmuseum, Wien 01643
Österreichisches Staatsarchiv – Stabsabteilung, Wien 01644
Österreichisches Studienzentrum für Frieden und Konfliktlösung, Stadtschlaining 01580
Österreichisches Theatermuseum, Wien 01645
Österreichisches Volkshochschularchiv – Verein zur Geschichte der Volkshochschulen, Wien 01646
Österreichisches Volksliedwerk, Wien 01647
Östersunds lasarett, Östersund 26628
Osteuropa-Institut, Regensburg 12429
Ostfalia Hochschule für angewandte Wissenschaften / Fachhochschule Braunschweig/Wolfenbüttel, Wolfenbüttel 10692
Østfold Central Hospital, Fredrikstad 20224
Østfold College, Fredrikstad, Fredrikstad 20091
Østfold University College, Halden 20095
Østfold University College, Halden 20094
Østfold University College Faculty of Education Library, Halden 20095
Ostfriesische Landschaft – Landschaftsbibliothek, Aurich 11463
Osthaus Museum Hagen, Hagen 11948
Ostkirchliches Institut an der Universität Würzburg, Würzburg 12617
Östra Göinge bibliotek, Broby 26750
Östra gymnasiet, Umeå 26484
Östra Nylands landskapsbibliotek, Porvoo 07601
Ostravska univerzita, Ostrava 06375
Østre Toten folkebibliotek, Lena 20365
Ostrobotnian Museum, Vaasa 07553
Ostroj Opava, a.s., Opava 06469
Ostrovski Minsk Central Children's City Library, Minsk 02185
Ostrovsky Electrical and Mechanical Railway Transport Professional School, Kyiv 28108
Ostrozka raionna TsBS, Tsentralna biblioteka, Ostrog 28770
O'Sullivan, LLP, New York 36006
OSU-OKC Library, Oklahoma City 33601
Osuuspankkien keskuspankki Oy:n kirjasto, Helsinki 07481
Oswego County Supreme Court, Oswego 36076
Oswego Public Library District, Oswego 40660
Oswego School District Public Library, Oswego 40661
Oświęcim-Brzezinka State Museum, Oświęcim 21183
Otago District Law Society, Dunedin 19832
Otago Polytechnic, Dunedin 19776
Otaru University of Commerce, Otaru 17089
Otdel biologicheskoi literatury RAN, Moskva 23455
Otdel geologicheskoi literatury RAN, Moskva 23456
Otdelenie instituta fiziki tverdogo tela i poluprovodnikov, Vitebsk 02081
Otero Junior College, La Junta 33464
Otis College of Art and Design, Los Angeles 31652
Otis Library, Norwich 40552
Otkrytaye Aktionyernaye Obshchestvo "Lakokraska" LIDA, Lida 01877
Otkryty sotsialny universitet, Moskva 22384
Otsego County Library, Gaylord 39213
Otsego District Public Library, Otsego 40662
Otsuma Joshi Daigaku, Tokyo 17169
Otsuma Women's University, Tokyo 17169
Ott Research Institute of OB/GYN, Sankt-Peterburg 23816
Ottawa Library, Ottawa 40663
Ottawa Public Library, Ottawa 04713
Ottawa University, Ottawa 32217
Otterbein College, Westerville 33037
Otterup Bibliotek, Otterup 06914
Otto Beisheim School of Management, Vallendar 10679
Otto Herman Museum, Miskolc 13490
D. Dr. Otto-Beuttenmüller-Bibliothek, Bretten 11698
Otto-Friedrich-Universität Bamberg, Bamberg 09397
Otto-Graf-Institut, Stuttgart 10621
Otto-Rombach-Bücherei, Bietigheim-Bissingen 12676
Otto-von-Guericke-Universität Magdeburg, Magdeburg 10263
– Universitätsbibliothek, Medizinische Zentralbibliothek, Magdeburg 10264
Ottumwa Public Library, Ottumwa 40666
Otyabrsky Railway – S-Peterburg-Varshavski Station, Sankt-Peterburg 23872
Ouachita Baptist University, Arkadelphia 30259
Ouachita Parish Public Library, Monroe 40265
Ouachita Parish Public Library – Ouachita Valley, West Monroe 41831
Ouachita Parish Public Library – West Monroe Branch, West Monroe 41832
Oud-Katholiek Seminarie, Amersfoort 19295
Oudtshoorn Teachers' College, Oudtshoorn 25035
Oulainen City Library, Oulainen 07596
Oulaisten kaupunginkirjasto, Oulainen 07596
Oulu City Library – Regional Central Library, Oulu 07597
Oulu University, Oulu 07357
– Architecture Library, Oulun Yliopisto 07358
– Biology Library, Oulu 07359
– Civil Engineering Library, Oulu 07362
– Kajaani University Consortium Library, Kajaani 07360
– Medical Library, Oulun Yliopisto 07361
– Science and Technology Library Tellus, Oulun Yliopisto 07364
– Snelmania Library, Oulun Yliopisto 07363
Oulu University of Applied Sciences, Library of Health and Social Care, Oulu 07356
Oulun kaupunginkirjasto-maakuntakirjasto, Oulu 07597
Oulun maakunta-arkisto, Oulu 07533
Oulun seudun ammattikorkeakoulu, sosiaali- ja terveysalan kirjasto, Oulu 07356
Oulun yliopisto, Oulu 07357
– Arkkitehtuurin osaston kirjasto, Oulun Yliopisto 07358
– Biologian kirjasto, Oulu 07359
– Kajaanin yliopistokeskuksen kirjasto, Kajaani 07360
– Lääketieteellisen tiedekunnan kirjasto, Oulun Yliopisto 07361
– Rakentamistekniikan osastokirjasto, Oulu 07362
– Snellmanian kirjasto, Oulun Yliopisto 07363
– Tiedekirjasto Tellus, Oulun Yliopisto 07364
Our Lady of Holy Cross College, New Orleans 32004
Our Lady of the Assumtion Free University, Roma
Our Lady of the Lake University, San Antonio 32566
– Worden School of Social Service Library, San Antonio 32567
Ovanes Tumanyan Museum, Yerevan 00357
Överby Gardening and Farm School, Espoo 07384
Överby trädgårdsskolas bibliotek, Espoo 07384
Overijsselse bibliotheek dienst (OBD), Nijverdal 19670
Overseas Economic Cooperation Fund, Tokyo 17684
Overseas Private Investment Corporation Library, Washington 35007
Øvre Eiker bibliotek, Hokksund 20343
Ovruch Regional Centralized Library System, A.S. Malishka Main Library, Ovruch 28771
Ovrutska raionna TsBS, Tsentralna biblioteka im. A.S. Malishka, Ovruch 28771
Owatonna Public Library, Owatonna 40672
Owen County Public Library, Spencer 41401
Owen Sound and North Grey Union Public Library, Owen Sound 04715
Owens Community College, Perrysburg 33632
Owens-Corning Corp, Granville 35726
Oxea GmbH – Werk Ruhrchemie, Oberhausen 11423
Oxelösunds bibliotek, Oxelösund 26882
Oxford and Cherwell Valley College, Oxford 29302
Oxford Brookes University, Oxford 29303
Oxford County Library, Ingersoll 04663
Oxford Free Library, Oxford 40676
Oxford University Museum of Natural History, Oxford 29871
Oxford University Press, Inc Library, New York 36007
Oxnard College, Oxnard 33611
Oxnard Public Library, Oxnard 40677
Oyamazumi Jinja Kokuhokan, Omishima 17339
Oyo State Library Board, Ibadan 20052
Oyster Bay-East Norwich Public Library, Oyster Bay 40678
Ozark Christian College, Joplin 31471
Ozark Folk Center Library, Mountain View 37133
Ozark Regional Library, Ironton 39581
P. G. Melikishvili Institute of Physical and Organic Chemistry, Tbilisi 09266
P. Kriauciunas Public LIbrary of Marijampole, Marijampolė 18340
"P. R. Slaveykov" Regional Public Library, Veliko Tărnovo 03608
P. S. G. College of Arts and Science, Coimbatore 13634
Pablo Borbon Memorial Institute of Technology, Batangas City 20687
Pace University, New York 32085
Pace University, Pleasantville 32323
Pace University – School of Law Library, White Plains 33050
Pacific Adventist University, Boroko 20546
Pacific Asia Museum Library, Pasadena 37330
Pacific Forestry Centre, Victoria 04183
Pacific Graduate School of Psychology, Palo Alto 32228
Pacific Grove Public Library, Pacific Grove 40681
Pacific Islands Forum Secretariat, Suva 07317
Pacific Lutheran University, Tacoma 32785
Pacific National University, Khabarovsk 22288
Pacific Northwest College of Art, Portland 32340
Pacific Northwest National Laboratory, Richland 37438
Pacific Northwest National Laboratory – Legal Library, Richland 36170
Pacific Oaks College, Pasadena 32236
Pacific Regional Seminary, Suva 07305
Pacific Research Institute for the Fishing Industry and Oceanography, Vladivostok 24059
Pacific Salmon Commission Library, Vancouver 04542
Pacific Theological College, Suva 07306
Pacific Union College, Angwin 30227
Pacific University, Forest Grove 31152
Pädagogische Akademie Elisabethenstift, Darmstadt 10716
Pädagogische Hochschule, Ludwigsburg 10254
Pädagogische Hochschule der Diözese Linz, Linz 01377
Pädagogische Hochschule des Kantons St.Gallen, St.Gallen 27135
Pädagogische Hochschule FHNW, Standort Liestal, Liestal 27111
Pädagogische Hochschule Freiburg, Freiburg 09801
Pädagogische Hochschule Freiburg, Freiburg 27066
Pädagogische Hochschule Heidelberg, Heidelberg 09977
Pädagogische Hochschule Karlsruhe, Karlsruhe 10059
Pädagogische Hochschule Kärnten, Klagenfurt 01242
Pädagogische Hochschule Ludwigsburg, Reutlingen 10541
Pädagogische Hochschule Oberösterreich, Linz 01378
Pädagogische Hochschule Salzburg, Salzburg 01265
Pädagogische Hochschule Schwäbisch Gmünd, Schwäbisch Gmünd 10582
Pädagogische Hochschule Steiermark, Graz 01218
Pädagogische Hochschule Thurgau, Kreuzlingen 27385
Pädagogische Hochschule Tirol, Innsbruck 01374
Pädagogische Hochschule Vorarlberg, Feldkirch 01371
Pädagogische Hochschule Weingarten, Weingarten 10684
Pädagogische Hochschule Wien – Campus Bibliothek, Wien 01381
Pädagogische Hochschule Wien – Schülerbibliothek an der Praxishauptschule PH Wien, Wien 01382
Pädagogische Hochschule Zürich, Zürich
– Bibliothek Gymnasial-, Berufspädagogik und Weiterbildung (GBW), Zürich 27165
– Mediothek Zentrum (UPRAA), Zürich 27166
Pädagogische Zentralbibliothek Baden-Württemberg, Stuttgart 12536
Pädagogische Zentralbibliothek Mannheim, Mannheim 12246
Pädagogisches Zentrum Basel-Stadt (PZB), Basel 26990
Pädagogisch-Theologisches Institut der Evangelischen Kirche von Kurhessen-Waldeck, Kassel 12103
Pädagogisch-Theologisches Institut der Nordelbischen Evangelisch-Lutherischen Kirche, Hamburg 11234
Pädagogisch-Theologisches Institut Nordelbien, Kiel 11257
Padma Kanya Campus Library, Kathmandu 19021
Padova University, Padova 15173
– Anglo-Germanic and Slavic Language and Literature Department, Anglo-Germanic Division, Padova 15197
– Anglo-germanic and Slavic language and literature department, Slavic division, Padova 15198
– Comparative Law Department, Ruggero Meneghelli" Library, Padova 15189
– Department for the History of Visual Arts and Music, Padova 15207
– Department of Pharmacology and Anesthesiology, Padova 15190
Padri Gesuiti, Napoli 15935
Paediatric Institute, Nizhni Novgorod 23621

Paediatrics Research Institute, Moskva 23419
Page Public Library, Page 40685
Pagėgių savivaldybės viešoji biblioteka, Pagėgiai 18345
Pahang Public Library Corporation, Kuantan 18628
Paier College of Art, Inc, Hamden 31276
Päijät-Häme Central Hospital Library, Lahti 07525
Päijät-Hämeen keskussairaalan tieteellinen kirjasto, Lahti 07525
Paimio kaupunginkirjasto, Paimio 07598
Paine College, Augusta 30303
Paissii Hilendarski University of Plovdiv, Plovdiv 03458
Pajala kommunbibliotek, Pajala 26883
Pakistan Administrative Staff College, Lahore 20463
Pakistan Agricultural Research Council, Islamabad 20484
Pakistan Council of Research in Water Resources, Islamabad 20485
Pakistan Council of Scientific and Industrial Research (PCSIR), Karachi 20500
Pakistan Council of Scientific and Industrial Research (PCSIR) – Laboratories Complex, Karachi 20501
Pakistan Forest Institute, Peshawar 20523
Pakistan Historical Society, Karachi 20502
Pakistan Institute of Cotton Research and Technology, Karachi 20503
Pakistan Institute of Development Economics, Islamabad 20486
Pakistan Institute of International Affairs, Karachi 20504
Pakistan Institute of Nuclear Sciences and Technology (PINSTECH), Islamabad 20487
Pakistan Standards Institution, Karachi 20505
Pakistani Swedish Institute of Technology, Karachi 20458
Pakruojo rajono savivaldybės Juozo Paukštelio viešoji biblioteka, Pakruojis 18346
Palace Museum, Beijing 05851
Palacio de Peralada, Peralada 26124
Palacio Legislativo, Montevideo 42026
Palácio Nacional de Mafra, Mafra 21817
Palacký University, Olomouc 06370
– Faculty of Law, Olomouc 06373
– Faculty of Paedagogy, Olomouc 06372
– Faculty of Sciences, Olomouc 06374
– Medical Faculty, Olomouc 06371
Palais de la Découverte, Paris 08615
Palangos miesto savivaldybės viešoji biblioteka, Palanga 18347
Palatine Public Library District, Palatine 40687
Palats kuturi im. I. Franka, Donetsk 28548
Palau Public Library, Koror 20533
Palawan National Agricultural College, Aborlan 20743
Paleis Het Loo, Apeldoorn 19390
Paleis van Justitie, 's-Hertogenbosch 19274
Palemen, Port-Vila 42181
Paleontological Institute, Tbilisi 09271
Paleontological Research Institution Library, Ithaca 36967
Palermo Astronomical Observatory, Palermo 16405
Palestine Public Library, Palestine 40688
Palestinian Central Bureau of Statistics, Ramallah-Al-Bireh 14749
Palisades Park Free Public Library, Palisades Park 40689
Palkansaajien tutkimuslaitos, Helsinki 07482
Palliser Regional Library, Moose Jaw 04694
Palliser Regional Library – Moose Jaw Branch, Moose Jaw 04695
Palm Beach Community College, Lake Worth 33468
Palm Beach County Law Library, West Palm Beach 35057
Palm Beach County Library System, West Palm Beach 41835
Palm Harbor Library, Palm Harbor 40692
Palm Springs Art Museum Library, Palm Springs 37319
Palm Springs Public Library, Palm Springs 40693
Palmdale City Library, Palmdale 40694
Palmengarten-Bibliothek, Frankfurt am Main 11867
Palmer College of Chiropractic, Davenport 33281
Palmer College of Chiropractic-West Campus Library, San Jose 33732
Palmer Public Library, Palmer 40695
Palmer Theological Seminary, Wynnewood 35425
Palmerston North Public Library, Palmerston North 19878
Palmyra Public Library, Palmyra 40696
Palo Alto City Library, Palo Alto 40697
Palo Alto City Library – Mitchell Park, Palo Alto 40698
Palo Alto College, San Antonio 33722
Palo Alto Medical Foundation, Palo Alto 37323
Palo Verde Valley District Library, Blythe 38228
Palóc Múzeum, Balassagyarmat 13366
Palomar College Library – Media Center, San Marcos 33735
Palos Heights Public Library, Palos Heights 40699
Palos Verdes Library District, Rolling Hills Estates 41057
Palucca Hochschule für Tanz Dresden, Dresden 09673
Pamantasan ng Lungsod ng Maynila, Manila 20717
Památník národního písemnictví, Praha 06513
Památník písemnictvi na Moravé, Rajhrad 06536
Památník Terezín, Terezín 06542
Pamfilov Academy of Municipal economy, Moskva 22325
Pamiatkovej ústav, Bratislava 24701
Pamunkey Regional Library, Hanover 39366
Pamunkey Regional Library – Atlee, Mechanicsville 40128
Pan Africa Christian College, Nairobi 18029
Pan African Institute for Development, Douala 03651
Pan African Institute for Development, East and Southern Africa, Kabwe 42337
Pan American Health Organization, Washington 37773
Pan-American Health Organization – Guyana Office, Georgetown 13196
Pan-American Institute of Geography and History, México 18858
Panamericana Library, Cuenca 06996
Panas Myrny Poltava Regional Public Library for Children, Poltava 28784
Pan-Cyprian Gymnasium, Nicosia 06330
Pandit Ravishankar Shukla University, Raipur 13748
Panepistimio Thessalias, Volos 13088
Panevėžio apskrities Gabrielės Petkevičaitės-Bitės viešoji biblioteka, Panevėžys 18348
Panevėžio miesto savivaldybės viešoji biblioteka, Panevėžys 18349
Panevėžio rajono savivaldybės viešoji biblioteka, Panevėžys 18350
Panhandle State University, Goodwell 31224
Panhandle-Plains Historical Museum, Canyon 36626
Panjab University, Chandigarh 13626
Panjabi Adabi Academy, Lahore 20516
Dr. Panjabrao Deshmukh Krishi Vidyapeeth, Akola 13597
Panning Institute of Agricultural Power Engineering, Moskva 23183
Panola College, Carthage 30682
Państwowa Wyższa Szkóa Sztuk Plastycznych, Wrocław 21025
Państwowa Wyższa Szkoła Filmowa, Telewizyjna i Teatralna im. L. Schillera, Łódź 21036
Państwowa Wyższa Szkoła Sztuk Plastycznych, Łódź 21037
Państwowa Wyższa Szkoła Sztuk Plastycznych, Poznań 21038
Państwowa Wyższa Szkoła Teatralna im. Aleksandra Zelwerowicza, Warszawa 20955
Państwowe Muzeum Archeologiczne, Warszawa 21329
Państwowe Muzeum Etnograficzne, Warszawa 21330
Państwowe Muzeum Oświęcim-Brzezinka, Oświęcim 21183
Państwowe Zbiory Sztuki na Wawelu, Kraków 21146
Państwowy Instytut Geologiczny, Kraków 21147
Państwowy Instytut Geologiczny, Warszawa 21331
Państwowy Instytut Naukowy – Instytut Śląski w Opolu, Opole 21179
Państwowy Instytut Weterynaryjny, Puławy 21214
Państwowy Zakład Higieny, Warszawa 21332
Panteion Panepistimion Kivonikon kai Politicon Epistimon, Athinai 13078
Panteios University of Social and Political Sciences, Athinai 13078
Panzhihua University, Panzhihua 05478
Paparua County Public Libraries, Christchurch 19863
Papatoetoe Public Library, Papatoetoe 19879
Paper and Pulp Joint-Stock Company, Sovetsk 22957
Paper and Timber Chemical Plant, Gomel 01848
Paper Factory, Dobrush 01846
Paper Factory no 2, Sankt-Peterburg 22909
The Paper Technology Specialists, München 12323
Papiermuseum Düren, Düren 11788
Papiertechnische Stiftung – Institut für Zellstoff und Papier, Heidenau 12056
Papieski Wydział Teologiczny i Metropolitalne Wyższe Seminarium Duchowne, Wrocław 21026
Papieskie Kolegium Polskie, Roma 16035
Papíripari Kutatóintézet Kft., Budapest 13453
Papua New Guinea Banking Corporation, Port Moresby 20584
Papua New Guinea Forest Headquarters Library, Hohola 20576
Papua New Guinea Institute of Medical Research, Goroka 20591
Papua New Guinea Institute of Public Administration, Boroko 20547
Papua New Guinea National Parliament, Waigani 20580
Papua New Guinea Public Museum and Art Gallery, Port Moresby 20595
Papua New Guinea University of Technology, Lae 20557
Papyrussammlung, Berlin 11563
Paradise Valley Community College, Phoenix 33640
Paramus Public Library, Paramus 40705
Parapsychology Foundation, New York 37246
Parchment Community Library, Parchment 40706
Pardes, bronnen van Joodse wijsheid, Amsterdam 19296
Parents, Let's Unite for Kids, Billings 36541
Paris Institute of Technology for Life, Food and Environmental Sciences, Paris 08462
Paris Junior College, Paris 33619
Paris Public Library, Paris 40707
Parish Institute Library of the Lutheran Church in Finland, Järvenpää 07391
Paris-Sorbonne University Library, Paris 07936
Der Paritätische Wohlfahrtsverband, Berlin 11553
The Park, Summer Park BC 01014
Park Avenue Synagogue, New York 35311
Park Cities Baptist Church, Dallas 35180
Park City Library, Park City 40708
Park County Bar Association, Cody 36720
Park County Library, Cody 38626
Park Forest Public Library, Park Forest 40709
Park Ridge Public Library, Park Ridge 40710
Park Synagogue, Cleveland Heights 35165
Park University Library, Parkville 32230
Parker, Poe, Adams & Bernstein, LLP, Raleigh 36161
Parkersburg & Wood County Public Library, Parkersburg 40711
Parkes Shire Lirary, Parkes 01140
Park-Krankenhaus Leipzig-Südost GmbH, Leipzig 12198
Parkland College, Champaign 33210
Parkland Community Library, Allentown 37917
Parkland Health & Hospital System, Dallas 36766
Parkland Regional Library, Dauphin 04624
Parkland Regional Library, Yorkton 04828
Parks Canada, Halifax 04359
Parkview Hospital, Fort Wayne 36866
Parlament de Catalunya, Barcelona 25686
Parlament kitaphanasi, Almaty 17930
Parlamentary Secretariate of Taiwan Province, Taichung 27579
Parlamentná knižnica a národnej rady slovenskej republiky, Bratislava 24673
Parlamentní knihovna, Praha 06414
Parlamento de Andalucía, Sevilla 25749
Parlamento de Canarias, Santa Cruz de Tenerife 25744
Parlamento de Cantabria, Santander 25745
Parlamento de Navarra, Pamplona 25741
Parlamentsbibliothek, Bern 27232
Parlamentsbibliothek, Wien 01410
Parlamentul Republicii Moldova, Chișinău 18890
Parlee McLaws LLP, Calgary 04248
Parlement, Brazzaville 06096
Parlement, Bruxelles 02425
Parlement Européen, Luxembourg 18399
Parlement Européen, Strasbourg 08165
Parliament, Kuala Lumpur 18565
Parliament, Port-Vila 42181
Parliament Library, Almaty 17930
Parliament Library, Bern 27232
Parliament Library, Kathmandu 19031
Parliament Library, Lagos 20000
Parliament Library, New Delhi 13878
Parliament Library, Port-of-Spain 27810
Parliament Library, Reykjavík 13561
Parliament Library, Singapore 24599
Parliament Library No. 2, Tehran 14415
Parliament Library of the Russian Federation, Moskva 22755
Parliament of Barbados, Bridgetown 01778
Parliament of Canada, Ottawa 04121
Parliament of Ghana, Accra 13034
Parliament of Saint Lucia, Castries 26334
Parliament of Sri Lanka, Colombo 26295
Parliament of the Republic of Fiji, Suva 07303
Parliament of Zimbabwe, Harare 42375
Parliament Secretariate Research Library Section, Kathmandu 19030
Parliamentary Information and Research Library, Lusaka 42347
Parliamentary Library, Funafuti 27911
Parliamentary Library, Honiara 24994
Parliamentary Library, Praha 06414
Parliamentary Library of South Australia, Adelaide 00600
Parliamentary Library of the National Council of the Slovak Republic, Bratislava 24673
Parliamentary Library of Victoria, Melbourne 00699
Parliamentary Library of Western Australia, Perth 00715
Parlimen Malaysia, Kuala Lumpur 18565
Parlin Ingersoll Public Library, Canton 38445
Parma Public Library, Hilton 39445
Parmly Billings Library, Billings 38206
Pärnu Central Library (PCL), Pärnu 07251
Parow Public Library, Parow 25219
Parramatta City Library, Parramatta 01141
Parsippany-Troy Hills Free Public Library, Parsippany 40717
Parsons Corp, Pasadena 36083
Parsons Corporation, Pasadena 36084
Parsons Public Library, Parsons 40718
Parsons School of Design, New School for Social Research, New York 32086
Partek Corporation, Information Service, Parainen 07427
Partek Oy AB, Parainen 07427
Partille bibliotek, Partille 26884
Părva gradska obedinena bolnitsa, Sofiya 03570
Părva klinitsna bolnitsa 'Dr. Khavesov', Sofiya 03571
Parys Public Library, Parys 25220
Pasadena City College, Pasadena 33621
Pasadena Public Library, Pasadena 40719
Pasadena Public Library, Pasadena 40720
Pasadena Public Library – Hastings, Pasadena 40721
Pasadena Public Library – La Pintoresca, Pasadena 40722
Pasadena Public Library – Lamanda Park, Pasadena 40723
Pasco County Library System, Hudson 39529
Pasco-Hernando Community College, Dade City 33272
Pasminco Research Centre, Boolaroo 00810
Paso Robles Public Library, Paso Robles 40725
Pasquotank-Camden Library, Elizabeth City 38953
Passaic County Community College, Paterson 33624
Passaic County Historical Society, Paterson 37331
Passaic Public Library, Passaic 40726
Passaic Vicinage Law Library, Paterson 34651
Passionist Academic Institute Library, Chicago 30779
Pasteur Institute of Epidemiology and Microbiology, Sankt-Peterburg 23824
Pasteur Institute of India, Coonoor 13973
Pasvalio Mariaus Katiliškio viešoji biblioteka, Pasvalys 18351
Pat Parker-Vito Russo Center Library, New York 37247
Pataskala Public Library, Pataskala 40727
Patchogue-Medford Library, Patchogue 40728
Patent and Technology Library, Riga 18188
Patent- och registreringsverket, Stockholm 26673
Patent Office, Kolkata 14056
Patent Office Library, Tokyo 17752
Patent- og Varemærkestyrelsens Bibliotek, Taastrup 06761
Patents and Trademarks Office, Madrid 26081
Patents Library, Cairo 07184
Patents Office, Warszawa 21349
Patentstyret, Oslo 20277
Patentti- ja rekisterihallitus, Helsinki 07483
Patentu tehniskā bibliotēka, Riga 18188
Paters Dominicanen, Willemstad 06317
Paters Kapucijnen (OFM.CAP), Brugge 02454
Paters Karmelieten, Brugge 02455

Paterson Free Public Library – Danforth Memorial Library, Paterson 40729
Patna University, Patna 13738
Paton Electric Welding Institute, Kyiv 28321
Patriarch Saint Tikhon Library, South Canaan 35383
Patriarchal Institute for Patristic Studies, Thessaloniki 13136
Patrick Henry Community College, Martinsville 33519
Patrimonio Nacional, Madrid 25252
Patronato de la Alhambra y Generalife, Granada 25961
Patten Free Library, Bath 38104
Patten University, Oakland 32174
Patterson, Belknap, Webb & Tyler LLP Library, New York 36008
Patterson Palmer, Halifax 04258
Patton & Boggs LLP, Washington 36369
Patton State Hospital, Patton 37332
Paul Gerhardt Diakonie Krankenhaus und Pflege GmbH, Lutherstadt Wittenberg 12601
Paul, Hastings, Janofsky & Walker, Atlanta 35465
Paul, Hastings, Janofsky & Walker LLP, Los Angeles 35853
Paul Kläui-Bibliothek Uster, Uster 27442
Paul Pratt Memorial Library, Cohasset 38629
Paul Quinn College, Dallas 30944
Paul Sacher Stiftung, Basel 27310
Paul Sawyier Public Library, Frankfort 39143
Paul Smiths College of Arts & Sciences, Paul Smiths 33625
Paul Stradin Museum of the History of Medicine, Riga 18189
Paul, Weiss, Rifkind, Wharton & Garrison, Washington 36370
Paul, Weiss, Rifkind, Wharton & Garrison Library, New York 36009
Paula Stradina medicīnas vēstures muzeja bibliotēka, Riga 18189
Paulding County Carnegie Library, Paulding 40730
Paul-Drude-Institut für Festkörperelektronik (PDI) im Forschungsverbund Berlin e.V., Berlin 11554
Paul-Ehrlich-Institut, Langen (Hessen) 12173
Paul-Ehrlich-Institut, Federal Institute for Vaccines and Biomedicines, Langen (Hessen) 12173
Paul-Gerin-Lajoie Library, Outremont 03819
Pauline Haass Public Library, Sussex 41482
Pauline Monastery, Częstochowa 21051
Pavel Sukhoi State Technical University of Gomel, Gomel 01791
Pavlodar State University named after S. Toraigyrov, Pavlodar 17914
Pavlograd City Centralized Library System, Main Library, Pavlograd 28772
Pavlogradska miska TsBS, Tsentralna biblioteka, Pavlograd 28772
Pavlov Institute of Physiology of the Russian Academy of Sciences, Sankt-Peterburg 23783
Pavlov Institute of Physiology of the Russian Academy of Sciences, Vsevolzhsk 24080
Pavlov Medical Institute, Sankt-Peterburg 22529
Pavlov Ryazan Medical Institute, Ryazan 22497
Pavlovo Bus-Assembly Industrial Corporation, Pavlovo 22888
Pavlovsk State Museum Reserve, Pavlovsk 23681
Paw Paw District Library, Paw Paw 40731
Pawtucket Public Library, Pawtucket 40732
Payame Noor University, Tehran 14404
Payap University, Chiang Mai 27734
Payathai Curriculum Center, Bangkok 27766
Payne Theological Seminary, Wilberforce 35421
Payson Public Library, Payson 40733
Pázmány Péter Catholic University, Theological Faculty, Budapest 13260
Pázmany Péter Katolikus Egyetem, Budapest 13260
Pchilka State Regional Universal Scientific Library, Lutsk 27939
PDVSA Servicios, Caracas 42241
Peabody & Arnold LLP, Boston 35535
Peabody Essex Museum, Salem 37513
Peabody Institute Library, Peabody 40734
Peabody Institute Library of Danvers, Danvers 38779
Peabody Public Library, Columbia City 38651
Peace College, Raleigh 33668
Peace Corps, Washington 37774
Peace Library – Austrian Study Center for Peace and Conflict Resolution – ASPR, Stadtschlaining 01580
Peace Palace Library, Den Haag 19454
Peace Research Institute Frankfurt (PRIF), Frankfurt am Main 11852
Peace Research Institute, Oslo (PRIO), Oslo 20245
Peace River Bible Institute, Sexsmith 04227
Peach Public Libraries, Fort Valley 39115
Peachtree Presbyterian Church, Atlanta 35117
Pearl Public Library, Pearl 40737
Pearl River Community College, Poplarville 33653
Pearl River County Library System – Margaret Reed Crosby Memorial Library, Picayune 40782
Pearl River Public Library, Pearl River 40739
Pearson Learning Group, Parsippany 36082
Peat Institute, Minsk 02029
Pechersk Region Central Library System of the City of Kiev – Saltikov-Shchedrin Main Library, Kyiv 28681
Pecherska raionna TsBS m. Kyiva, Tsentralna biblioteka im. M. Saltikova-Shchedrina, Kyiv 28681
Pechora Research and Development Institute, Vorkuta 24076
PechorNIIProekt, Vorkuta 24076
Pécsi Tudományegyetem, Pécs 13288
– Bölcsészettudományi és Természet-Tudományi Kar, Pécs 13289
– Pekár Mihály Orvosi és Élettudományi Szakkönyvtár, Pécs 13290
Pedagogical Academy in Sarajevo, Sarajevo 02958
Pedagogical District Library, Bielsko-Biała 21075
Pedagogical District Library, Bydgoszcz 21082
Pedagogical District Library, Elbląg 21090
Pedagogical District Library, Gdańsk 21101
Pedagogical District Library, Katowice 21116
Pedagogical District Library, Kielce 21118
Pedagogical District Library, Konin 21122
Pedagogical District Library, Krosno 21153
Pedagogical District Library, Leszno 21156
Pedagogical District Library, Łódź 21165
Pedagogical District Library, Łomża 21167
Pedagogical District Library, Nowy Sącz 21174
Pedagogical District Library, Opole 21180
Pedagogical District Library, Ostrołęka 21181
Pedagogical District Library, Piła 21187
Pedagogical District Library, Płock 21191
Pedagogical District Library, Sieradz 21220
Pedagogical District Library, Skierniewice 21223
Pedagogical District Library, Słupsk 21224
Pedagogical District Library, Suwałki 21226
Pedagogical District Library, Tarnów 21231
Pedagogical District Library, Włocławek 21356
Pedagogical District Library, Zamość 21373
Pedagogical District Library, Zielona Góra 21376
Pedagogical Institute, Batumi 09188
Pedagogical Institute, Bukhara 42076
Pedagogical Institute, Chodshent 27635
Pedagogical Institute, Dushanbe 27636
Pedagogical Institute, Gori 09190
Pedagogical Institute, Gurev 17906
Pedagogical Institute, Kustanay 17912
Pedagogical Institute, Kzyl-Orda 17913
Pedagogical Institute, Namangan 42080
Pedagogical Institute, Nicosia 06327
Pedagogical Institute, Nizhni Tagil 22428
Pedagogical Institute, Novosibirsk 22448
Pedagogical Institute, Nukus 42081
Pedagogical Institute, Orekhovo-Zuevo 22459
Pedagogical Institute, Oryol 22463
Pedagogical Institute, Osh 18142
Pedagogical Institute, Pavlodar 17915
Pedagogical Institute, Perm 22469
Pedagogical Institute, Petropavlovsk 17916
Pedagogical Institute, Przhevalsk 18143
Pedagogical Institute, Samarkand 42082
Pedagogical Institute, Saratov 22575
Pedagogical Institute, Semipalatinsk 17919
Pedagogical Institute, Sukhumi 09193
Pedagogical Institute, Taganrog 22589
Pedagogical Institute, Tbilisi 09200
Pedagogical Institute, Telavi 09206
Pedagogical Institute, Termez 42088
Pedagogical Institute, Tomsk 22594
Pedagogical Institute, Toshkent 42096
Pedagogical Institute, Veliki Novgorod 22622
Pedagogical Institute of Foreign Languages, Tbilisi 09201
Pedagogical Institute of Foreign Languages, Toshkent 42097
Pedagogical Institute of Languages, Andizhan 42075
Pedagogical Institute of Macedonia, Skopje 18446
Pedagogical Institute S. Seyfullin, Akmola 17894
Pedagogical Library, Biała Podlaska 21072
Pedagogical Library, Chełm 21085
Pedagogical Library, Gorzów Wielkopolski 21108
Pedagogical Library, Jelenia Góra 21111
Pedagogical Library, Legnica 21155
Pedagogical library, Ljubljana 24929
Pedagogical Library, Piotrków Trybunalski 21188
Pedagogical Library, Tarnobrzeg 21230
Pedagogical Library, Toruń 21232
Pedagogical Library, Wałbrzych 21237
Pedagogical Library of Lower Austria Education Authority, St. Pölten 01391
Pedagogical Library of Radom, Radom 21216
Pedagogical Museum of J.A. Comenius, Praha 06514
Pedagogical Nekrasov State University, Kostroma 22296
Pedagogical University, Yaroslavl 22655
Pedagogical University of Cracow, Kraków 20872
Pedagogical University of Vladimir, Vladimir 22630
Pedagogicheski institut, Balashov 22210
Pedagogicheski institut, Kirov 22290
Pedagogicheski institut, Nizhni Tagil 22428
Pedagogicheski institut, Novosibirsk 22448
Pedagogicheski institut, Orekhovo-Zuevo 22459
Pedagogicheski institut, Oryol 22463
Pedagogicheski institut, Perm 22469
Pedagogicheski institut, Saratov 22575
Pedagogicheski institut, Taganrog 22589
Pedagogicheski institut, Tomsk 22594
Pedagogicheski institut im. A.P. Gaidara, Arzamas 22205
Pedagogicheski institut im. N.G. Chernyshevskogo, Chita 22233
Pedagogicheski institut im. V.G. Korolenko, Glazov 22250
Pedagogicheski institut inostrannykh yazykov im. N.A. Dobrolyubova, Nizhni Novgorod 22426
Pedagogicheski universitet, Yaroslavl 22655
Pedagogická a sociálna akadémia, Prešov 24669
Pedagogické muzeum J. A. Komenského, Praha 06514
Pedagogiczna Biblioteka, Piotrków Trybunalski 21188
Pedagogiczna Biblioteka Wojewódzka, Bielsko-Biała 21075
Pedagogiczna Biblioteka Wojewódzka, Bydgoszcz 21082
Pedagogiczna Biblioteka Wojewódzka, Chełm 21085
Pedagogiczna Biblioteka Wojewódzka, Ciechanów 21086
Pedagogiczna Biblioteka Wojewódzka, Kielce 21118
Pedagogiczna Biblioteka Wojewódzka, Konin 21122
Pedagogiczna Biblioteka Wojewódzka, Kraków 21148
Pedagogiczna Biblioteka Wojewódzka, Krosno 21153
Pedagogiczna Biblioteka Wojewódzka, Leszno 21156
Pedagogiczna Biblioteka Wojewódzka, Nowy Sącz 21174
Pedagogiczna Biblioteka Wojewódzka, Opole 21180
Pedagogiczna Biblioteka Wojewódzka, Ostrołęka 21181
Pedagogiczna Biblioteka Wojewódzka, Piła 21187
Pedagogiczna Biblioteka Wojewódzka, Płock 21191
Pedagogiczna Biblioteka Wojewódzka, Sieradz 21220
Pedagogiczna Biblioteka Wojewódzka, Skierniewice 21223
Pedagogiczna Biblioteka Wojewódzka, Słupsk 21224
Pedagogiczna Biblioteka Wojewódzka, Suwałki 21226
Pedagogiczna Biblioteka Wojewódzka, Włocławek 21356
Pedagogiczna Biblioteka Wojewódzka, Zielona Góra 21376
Pedagogiczna Biblioteka Wojewódzka im. J. Lompy, Katowice 21116
Pedagogiczna Biblioteka Wojewódzka im. prof. Tadeusza Kotarbińskiego w Łódź, Łódź 21165
Pedagogiczna Biblioteka Wojewodzka w Gdansku, Gdańsk 21101
Pedagogiczna Biblioteka Wojewódzka w Łomży, Grajewo 21109
Pedagogiczna Biblioteka Wojewódzka w Łomży, Kolno 21120
Pedagogiczna Biblioteka Wojewódzka w Łomży, Wysokie Mazowieckie 21370
Pedagogisch Didactisch Centrum, Gent 02442
Pedagogisch Hoger Onderwijs Sancta Maria, Ronse 02422
Pedagogische Begeleidingsdienst, Gent 02652
Pedagogische Hogeschool Onze-Lieve-Vrouw-Pulhof, Antwerpen 02273
Pedagoška Akademija, Maribor 24866
Pedagoška knjižnica Davorina Trstenjaka, Zagreb 06242
Pedagoška knjižnica zavoda za Šolstvo, Ljubljana 24929
Pedagoški inštitut pri Univerzi 'Eduarda Kardelja' v Ljubljana, Ljubljana 24813
Pedahohichne uchilishche, Priluki 28067
Pedahohichne uchilishche no 2, Kyiv 28112
Pedahohichny instytut, Kamyanets-Podilsk 27982
Pedahohichny instytut, Nizhin 28048
Pedahohichny instytut im. I. Franko, Drogobych 27975
Pedahohichny instytut im. V.G. Korolenko, Poltava 28064
Pedahohichny universytet, Vinnytsya 28083
Pedahohichny universytet im. M. Dragomanova, Kyiv 28021
Pedersen & Houpt Library, Chicago 35594
Pediatricheski meditsinski institut, Sankt-Peterburg 22533
PEDIC – Bisdom Gent, Gent 02464
Pedologisch Instituut, Rotterdam 19507
Peel Board of Education, Mississauga 04381
Peel Education Campus, Mandurah 00452
Peirce College, Philadelphia 32258
Pekin Public Library, Pekin 40742
Peking University, Beijing 05050
Peking University Health Sciences Centre, Beijing 05051
Pelham Public Library, Fonthill 04639
Pelham Public Library, Pelham 40743
Pelizaeus-Museum, Hildesheim 12061
Pell City Public Library, Pell City 40744
Pella Public Library, Pella 40745
Pellissippi State Technical Community College, Knoxville 33462
Peloponnesian Folklore Foundation, Nafplion 13128
Pemberville Public Library, Pemberville 40746
Pembroke College, Cambridge 29035
Pembroke College, Oxford 29338
Pembroke Public Library, Pembroke 04716
Pembroke Public Library, Pembroke 40747
Pembrokesire County Library, Haverfordwest 29996
Penang Public Library Corporation, Perai 18633
Pencho Slaveikov Public Library, Varna 03607
Pender County Public Library, Burgaw 38389
Pender County Public Library – Hampstead Branch, Hampstead 39356
Pendleton Community Library, Pendleton 40748
Pendleton Correctional Facility, Pendleton 34655
Pendleton Correctional Facility, Pendleton 34656
Pendleton Public Library, Pendleton 40749
Penfield Public Library, Penfield 40750
Peninsula Community College, Port Angeles 33654
Peninsula Institute of TAFE, Frankston 00570
Peninsula Public Library, Lawrence 39823
Peninsula Technikon, Bellville 24999
Penn State University, Malvern 31741
Penn Yan Public Library, Penn Yan 40751
J. C. Penney Company Inc, Plano 36136
Pennsylvania Academy of Fine Arts Library, Philadelphia 33635
Pennsylvania College of Optometry, Elkins Park 31083
Pennsylvania College of Technology, Williamsport 33069
Pennsylvania Department of Conservation & Natural Resources, Harrisburg 34340
Pennsylvania Department of Transportation, Harrisburg 36915
Pennsylvania Historical & Museum Commission, Harrisburg 36916
Pennsylvania Horticultural Society Library, Philadelphia 37355
Pennsylvania Hospital – Historic Library, Philadelphia 37356
Pennsylvania Hospital – Medical Library, Philadelphia 37357
Pennsylvania Institute of Technology Library, Media 31784
Pennsylvania Legislative Reference Bureau Library, Harrisburg 34341
Pennsylvania Office of Attorney General, Harrisburg 34342
Pennsylvania School for the Deaf Library, Philadelphia 33636
Pennsylvania State Erie, Erie 31095

Pennsylvania State University, Abington 30166
Pennsylvania State University, Du Bois 31027
Pennsylvania State University, Dunmore 31034
Pennsylvania State University, Hazleton 31321
Pennsylvania State University, Media 31785
Pennsylvania State University, Monaca 31875
Pennsylvania State University, Mont Alto 31879
Pennsylvania State University, Reading 32420
Pennsylvania State University, Sharon 32660
Pennsylvania State University, Uniontown 32870
Pennsylvania State University, York 33104
Pennsylvania State University – Altoona College, Altoona 30206
Pennsylvania State University, College of Medicine, Hershey 31328
Pennsylvania State University – Dickinson School of Law, Carlisle 30677
Pennsylvania State University – Harrisburg, Middletown 31822
Pennsylvania State University – Lehigh Valley, Fogelsville 31150
Pennsylvania State University Libraries, University Park 32877
– Eberly Family Special Collections, University Park 32878
– Fletcher L. Byrom Earth & Mineral Sciences Library, University Park 32879
– George & Sherry Middlemas Arts & Humanities Library, University Park 32880
– Physical & Mathematical Sciences Library, University Park 32881
Pennsylvania State University – McKeesport Commonwealth College, McKeesport 31775
Pennsylvania State University, New Kensington Commonwealth College, Upper Burrell 32883
Pennsylvania State University, Schuylkill Campus, Capital College, Schuylkill Haven 32625
Pennsylvania State University – Wilkes-Barre Commonwealth College, Lehman 31586
Pennzoil Exploration & Production Co, Houston 35759
Penobscot County Law Library, Bangor 33967
Penobscot Marine Museum, Searsport 37589
Penrith City Library, Penrith 01142
Pensacola Junior College, Pensacola 33630
Pension Benefit Guaranty Corporation, Washington 36371
Pentagon Library, Washington 35025
Penticton Public Library, Penticton 04717
Penza Region Youth Library, Penza 24198
Penza Regional Archives, Penza 23682
Penza Regional Scientific Medical Library, Penza 23684
Penzenskaya oblastnaya biblioteka im. M.Yu. Lermontova, Penza 22164
Penzenskaya oblastnaya nauchnaya meditsinskaya biblioteka, Penza 23684
Penzenskaya oblastnaya yunosheskaya biblioteka, Penza 24198
Penzenski gosudarstvenny tekhnicheski universitet, Penza 22465
Penzenski mezhotraslevoi territorialny tsentr nauchno-tekhnicheskoi informatsii i propagandy, Penza 23685
Penzenski pedagogicheski institut im. Belinskogo, Penza 22466
Pénzügyi és Számviteli Főiskola, Budapest 13226
Pénzügyminisztérium, Budapest 13323
The People's Archives, Helsinki 07466
The People's College, Nottingham 29292
Peoples Library, New Kensington 40415
People's Library of Guiyang, Guiyang 05913
People's University of China, Beijing 05052
Peoria County Law Library, Peoria 34659
Peoria Public Library -, Peoria 40755
Peoria Public Library – Lakeview, Peoria 40756
Peotone Public Library District, Peotone 40757
Pepper, Hamilton LLP, Detroit 35678
Pepper, Hamilton LLP, Philadelphia 36106
Pepperdine University, Malibu 31739
– School of Law-Jerene Appleby Harnish Law Library, Malibu 31740
Pepsico Beverages & Foods, Chicago 35595
Pequannock Township Public Library, Pompton Plains 40841
Perak State Public Library Corporation, Ipoh 18620
Perbadanan Perpustakaan Awam Johor, Johor Bharu 18622

Perbadanan Perpustakaan Awam Kedah, Alor Setar 18618
Perbadanan Perpustakaan Awam Kelantan, Kota Bharu 18625
Perbadanan Perpustakaan Awam Melaka, Melaka 18631
Perbadanan Perpustakaan Awam Negeri Perak, Ipoh 18620
Perbadanan Perpustakaan Awam Negeri Perlis, Kangar 18623
Perbadanan Perpustakaan Awam Negeri Sembilan, Seremban 18636
Perbadanan Perpustakaan Awam P. Pinang, Perai 18633
Perbadanan Perpustakaan Awam Pahang, Kuantan 18628
Perbadanan Perpustakaan Awam Selangor, Cawangan Kelang, Kelang 18624
Perbadanan Perpustakaan Awam Selangor, Perpustakaan Daerah Kuala Langat, Banting 18619
Perbadanan Perpustakaan Awam Terengganu, Kuala Terengganu 18627
Pères Rédemptoristes, Luxembourg 18402
Pereyaslav-Khmelnitska raionna TsBS, Tsentralna biblioteka, Pereyaslav-Khmelnitski 28773
Pereyaslav-Khmelnitski Regional Centralized Library System, Main Library, Pereyaslav-Khmelnitski 28773
Performing Arts Council Transvaal, Pretoria 25166
Perguruan Tinggi Ilmu Da'wah Nurul Jadid, Probolinggo 14288
Perhimpunan Persahabatan Indonesia Amerika (PPIA), Jakarta 14372
Perkins Coie, Portland 36146
Perkins Coie Brown & Bain Library, Phoenix 36121
Perkins Coie Library, Seattle 36268
Perkins School for the Blind, Watertown 33843
Perkins School for the Blind – Samuel P. Hayes Research Library, Watertown 33844
Perlis State Public Library Corporation, Kangar 18623
Perm Pharmaceutical Institute, Perm 22471
Perm Regional Medical Library, Perm 23691
Perm Scientific and Technical Information Centre, Perm 23692
Perm State Art Gallery, Perm 23690
Perm State Institute of Culture, Perm 22470
Perm State Technical University, Perm 22472
Perm State University, Perm 22473
Permskaya gosudarstvennaya khudozhestvennaya galereya, Perm 23690
Permskaya oblastnaya biblioteka im. A.M. Gorkogo, Perm 22165
Permskaya oblastnaya detskaya biblioteka, Perm 24199
Permskaya oblastnaya meditsinskaya biblioteka, Perm 23691
Permski farmatsevticheski institut, Perm 22471
Permski gosudarstvenny tekhnicheski universitet, Perm 22472
Permski gosudarstvenny universitet, Perm 22473
Permski tsentralny nauchno-tekhnicheski institut, Perm 23692
Pernambuco State Public Library, Recife 03437
Perpustakaan Angkatan Tentera, Kuala Lumpur 18566
Perpustakaan Awam Majlis Perbandaran Miri, Miri 18632
Perpustakaan Awam Sibu, Sibu 18638
Perpustakaan Bandaraya Kuching, Kuching 18629
Perpustakaan Islam, Yogyakarta 14340
Perpustakaan Jajasan Hatta, Yogyakarta 14380
Perpustakaan Majlis Daerah Sarikei, Sarikei 18635
Perpustakaan Nasional, Jakarta 14235
Perpustakaan Negara Malaysia, Kuala Lumpur 18496
Perpustakaan Penyelidikan Islam, Kuala Lumpur 18599
Perpustakaan Pusat Pengembangan Penataran Guru Ilmu Pengetahuan Alam, Bandung 14240
Perpustakaan Sedjarah Politik dan Sosial, Jakarta 14373
Perpustakaan Tun Razak, Ipoh 18621
Perpustakaan Universitas Airlangga, Surabaya 14298
– Fakultas Hukum, Surabaya 14299
– Fakultas Kedokteran, Surabaya 14300
Perpustakaan Wilayah, Yogyakarta 14384
Perpustakaan Wilayah Sandakan, Sandakan 18634

Perrot Memorial Library, Old Greenwich 40607
Perry County District Library, New Lexington 40417
Perry County Public Library, Hazard 39403
H. Leslie Perry Memorial Library, Henderson 39415
Perry Public Library, Perry 40761
Pershore Group of Colleges, Pershore 29368
Persmuseum, Amsterdam 19374
Perstorp AB, Perstorp 26546
Perth Amboy Free Public Library, Perth Amboy 40764
Perth and Kinross Council, Perth 30058
Perth Astronomical Observatory, Bickley 00873
Perth Bible College, Karrinyup 00768
Peru Public Library, Peru 40765
Peru Public Library, Peru 40766
Peru State College, Peru 32246
Pervomaisk City Centralized Library System, Main Library, Pervomaisk 28774
Pervomaiska miska TsBS, Tsentralna biblioteka, Pervomaisk 28774
Pervomaiska raionna TsBS, Tsentralna biblioteka, Pervomaiske 28775
Pervomaiska raionna TsBS, Tsentralna biblioteka, Pervomaiski 28776
Pervomaiske Regional Centralized Library System, Main Library, Pervomaiske 28775
Pervomaiski Regional Centralized Library System, Main Library, Pervomaiski 28776
Pest County Library, Szentendre 13549
Pest Megyei Könyvtár, Szentendre 13549
Pestalozzi Library Zurich, Zürich 27509
Pestalozzi-Bibliothek Zürich (PBZ), Zürich 27509
Pesticide Action Network North America Regional Center, San Francisco 37549
Peter MacCallum Cancer Centre, East Melbourne 00911
Peter Sabroe College of Education, Århus 06728
Peter Sabroe Seminariet, Århus 06728
Peter the Great Museum of Anthropology and Ethnology (Kunstkammer), Sankt-Peterburg 23810
Peter White Public Library, Marquette 40073
Peterborough Public Library, Peterborough 04718
Peterburg State Transport University, Sankt-Peterburg 22534
Peterburgski gosudarstvenny universitet putei soobshcheniya, Sankt-Peterburg 22534
Peters Township Public Library, McMurray 40123
Petersburg Public Library, Petersburg 40767
Petersburg State Transport University, Sankt-Peterburg 22534
Peterson & Ross Library, Chicago 35596
Petit Séminaire College St. Martial, Port-au-Prince 13202
Petit Séminaire de Québec, Québec 04219
Petőfi Irodalmi Múzeum, Budapest 13454
Petőfi Museum of Literature, Budapest 13454
Petra Alekseev Textile Factory, Moskva 22856
Petra Christian University, Surabaya 14301
Petrivska raionna TsBS, Tsentralna biblioteka, Petrove 28778
Petrobras, Rio de Janeiro 03264
Petrodvorets District Centralized Library System, Central District Children's Library, Sankt-Peterburg 24284
Petrodvorets District Centralized Library System, Inge Library Branch no 1, Sankt-Peterburg 24281
Petrodvorets District Centralized Library System, Library Branch no 6, Sankt-Peterburg 24282
Petrodvorets District Centralized Library System, Neto Central District Library, Sankt-Peterburg 24283
Petrograd District Centralized Library System, B.A. Lavrenev Library Branch no 2, Sankt-Peterburg 24285
Petrograd District Centralized Library System, Central District Children's Library, Sankt-Peterburg 24288
Petrograd District Centralized Library System, Gaidar Library Branch no 3, Sankt-Peterburg 24286
Petrogradsky District Centralized Library System, Pushkin Central District Library, Sankt-Peterburg 24287
Petróleos del Perú S.A. (PETRO-PERU), Lima 20650
Petróleos Mexicanos, México 18823
Petroleum Institute, Grozny 23047
Petroleum Institute, Krasnodar 23112
Petroleum Nasional Bhd (PETRONAS), Kuala Lumpur 18572

Petroleum Research Centre, Tripoli 18263
Petroleum Safety Authority, Stavanger 20295
Petroleum Training Institute, Effurun 19993
Petroleum University (East China), Dongying 05169
Petroleumstilsynet, Stavanger 20295
Petropavlivka Regional Centralized Library System, Main Library, Petropavlivka 28777
Petropavlivska raionna TsBS, Tsentralna biblioteka, Petropavlivka 28777
Petropavlovsk-Kamchatskii Maritime University, Petropavlovsk-Kamchatski 22478
Petropavlovsk-Kamchatskoe vysshee morskoe uchilishche, Petropavlovsk-Kamchatski 22478
Petroşani University, Petroşani 21940
Petrov Institute of Oncology, Sankt-Peterburg 23834
Petrove Regional Centralized Library System, Main Library, Petrove 28778
Petrovsky Plant, Trade Union Library, Dnipropetrovsk 28541
Petrozavodsk State University, Petrozavodsk 22480
Petrozavodski gosudarstvenny universitet, Petrozavodsk 22480
Petru Poni Instutute of Macromolecular Chemistry, Iaşi 22036
Pettigrew Regional Library, Plymouth 40831
Pevsner Public Library, Haifa 14775
Pew Charitable Trusts Library, Philadelphia 37358
Pewaukee Public Library, Pewaukee 40768
Pfalzbibliothek, Kaiserslautern 12083
Pfälzische Landesbibliothek, Speyer 09339
Pfälzisches Oberlandesgericht, Zweibrücken 11141
Pfeiffer University, Misenheimer 31865
Pfennigparade, München 12335
Pfizer, Inc, Ann Arbor 35447
Pfizer Inc, Morris Plains 35921
Phalaborwa Public Library, Phalaborwa 25221
Phaneromeni Library, Nicosia 06338
Phantastische Bibliothek Wetzlar, Wetzlar 12592
Pharmaceutical College, Zhytomyr 28131
Pharmaceutical Institute, Pyatigorsk 22484
Pharmaceutical Research Institute, Moskva 23393
Pharmaceutical Research Institute, Warszawa 21277
Pharmacia & Opjohn AB, Stockholm 26556
Pharr Memorial Library, Pharr 40769
PHBern, Bern 27325
Phelps Dunbar, LLP, Jackson 35785
Phelps Dunbar LLP, New Orleans 35949
Phelps Ornithological Collection, Caracas 42250
Phenix City-Russell County Library, Phenix City 40770
PHILA-Bibliothek Heinrich Köhler, Frankfurt am Main 11868
Philadelphia Archdiocesan Historical Research Center, Wynnewood 35426
Philadelphia Biblical University, Langhorne 31556
Philadelphia College of Osteopathic Medicine, Philadelphia 32259
Philadelphia Common Pleas & Municipal Court Law Library, Philadelphia 34664
Philadelphia Museum of Art Library, Philadelphia 37359
Philadelphia Newspapers, Inc, Philadelphia 36107
Philadelphia Orchestra Library, Philadelphia 37360
Philadelphia University, Philadelphia 32260
Philadelphia Yearly Meeting Library, Philadelphia 35328
Philander Smith College, Little Rock 31624
Philatelistische Bibliothek Hamburg e.V., Hamburg 12012
Philbrook Museum of Art, Tulsa 37692
Philip Morris USA, Richmond 36175
A. Philippide Institute of Romanian Philology, Iaşi 22040
Philippine Normal University, Manila 20718
Philippine Nuclear Research Institute, Quezon City 20781
Philippine School of Business Administration, Manila 20749
Philippine Senate Library, Pasay City 20759
Philippine Women's University, Manila 20719
– Jose Abad Santos Memorial School, Manila 20720
Philipps-Universität Marburg, Marburg 10301
– Bibliothek Biologie, Marburg 10302
– Bibliothek Chemie, Marburg 10303
– Bibliothek Chemie – Standort Bibliothek Chemie 2 / Physikalische Chemie, Marburg 10304

– Bibliothek des Instituts für Soziologie und des Instituts für Philosophie, Marburg 10305
– Bibliothek Erziehungswissenschaft, Marburg 10306
– Bibliothek Evangelische Theologie, Marburg 10307
– Bibliothek Fremdsprachliche Philologien – Anglistik und Amerikanistik, Marburg 10308
– Bibliothek Geowissenschaften, Marburg 10309
– Bibliothek Germanistik und Medienwissenschaft, Marburg 10310
– Bibliothek Geschichtswissenschaften, Abt. Mittelalterliche Geschichte, Historische Hilfswissenschaften und Hessisches Landesamt für geschichtliche Landeskunde, Marburg 10311
– Bibliothek Geschichtswissenschaften, Fachgebiet Neuere Geschichte, Marburg 10312
– Bibliothek Geschichtswissenschaften, Fachgebiet Osteuropäische Geschichte, Marburg 10313
– Bibliothek Geschichtswissenschaften, Fachgebiet Sozial- und Wirtschaftsgeschichte, Marburg 10314
– Bibliothek Mathematik und Informatik, Marburg 10315
– Bibliothek Pharmazie, Marburg 10316
– Bibliothek Physik, Marburg 10317
– Bibliothek Psychologie, Marburg 10318
– Bibliothek Religionswissenschaft, Marburg 10319
– Bibliothek Vorgeschichtliches Seminar, Marburg 10320
– Bibliothek Wirtschaftswissenschaften, Marburg 10321
– CNMS Centrum für Nah- und Mittelost-Studien, Abt. Semitistik, Marburg 10322
– Emil-von-Behring-Bibliothek für Geschichte und Ethik der Medizin, Marburg 10323
– Fachbereichsbibliothek Geographie, Marburg 10324
– Fachgebiet Alte Geschichte, Marburg 10325
– Fachgebiet Klassische Archäologie mit Antiken- und Abgusssammlung, Marburg 10326
– Fachgebiet Sinologie, Marburg 10327
– Forschungsstelle für Personalschriften an der Philipps-Universität Marberg, Marburg 10328
– Forschungszentrum Deutscher Sprachatlas, Marburg 10329
– Institut für Europäische Ethnologie/Kulturwissenschaft, Marburg 10330
– Institut für Genossenschaftswesen, Marburg 10331
– Institut für Handels-, Wirtschafts- und Arbeitsrecht – Abt. für Arbeitsrecht, Marburg 10332
– Institut für Handels-, Wirtschafts- und Arbeitsrecht – Abt. für Handels- und Wirtschaftsrecht, Marburg 10333
– Institut für Öffentliches Recht, Abt. Völkerrecht, Marburg 10334
– Institut für Orientalistik und Sprachwissenschaft, Fachgebiet Indologie und Tibetologie, Marburg 10335
– Institut für Orientalistik und Sprachwissenschaft, Fachgebiet Vergleichende Sprachwissenschaft und Keltologie, Marburg 10336
– Institut für Politikwissenschaft, Marburg 10337
– Institut für Rechtsgeschichte und Papyrusforschung, Germanistische Abteilung, Marburg 10338
– Institut für Romanische Philologie, Marburg 10339
– Institut für Vergleichende Kulturforschung-Völkerkunde, Marburg 10340
– Japan-Zentrum, Marburg 10341
– Juristisches Seminar, Marburg 10342
– Kunstgeschichtliches Institut, Marburg 10343
– Medizinisches Zentrum für Nervenheilkunde, Marburg 10344
– Musikwissenschaftliches Institut, Marburg 10345
– Seminar für Klassische Philologie, Marburg 10346
– Universitätsklinikum Gießen und Marburg, Marburg 10347
– Universitätsmuseum für Kunst und Kulturgeschichte, Marburg 10348
– Vorgeschichtliches Seminar, Marburg 10349
– Zentrale Medizinische Bibliothek (ZMB), Marburg 10350
Philipps-University Marburg, Marburg 10301
– Comparative Linguistic Studies and Celtic Studies Library, Marburg 10336
– Department of Modern Philologies – English and American Studies, Marburg 10308
– Emil-von-Behring Library for History and Ethics of Medicine, Marburg 10323
Philips Medical Systems Library, Andover 35446
Philips Research Laboratories, Redhill 29885
Philipsburg Jubilee Library, Sint Maarten 24633
Philipsburg Jubileum Bibliotheek, Sint Maarten 24633
The Phillips Collection Library, Washington 37775
Phillips Community College of the University of Arkansas, Helena 31323
Phillips, Lytle, Hitchcock, Blaine & Huber Library, Buffalo 35543
L. E. Phillips Memorial Public Library, Eau Claire 38927
Phillips State Prison, Buford 35547
Phillips Theological Seminary, Tulsa 35400
Phillipsburg Free Public Library, Phillipsburg 40778
Phillips-Lee-Monroe Regional Library – Phillips County Library, Helena 39409
Philo, Atkinson, Stephens, Wright, Whitaker, Philo & Kayrouz, Detroit 35679
Philosophical and Theological College Brixen, Brixen 14916
Philosophical Research Society Library, Los Angeles 37060
Philosophisches Institut, Aachen 09383
Philosophisch-Theologische Hochschulbibliothek, Sankt Augustin 10580
Philosophisch-Theologische Hochschule Brixen / Studio Teologico Accademico Bressanone, Brixen 14916
Philosophisch-Theologische Hochschule der Diözese St. Pölten, St. Pölten 01288
Philosophisch-Theologische Hochschule der Salesianer Don Boscos Benediktbeuern, Benediktbeuern 09400
Philosophisch-Theologische Hochschule Sankt Georgen, Frankfurt am Main 09755
Philosophisch-Theologische Hochschule Vallendar gGmbH, Vallendar 10678
Phoenix Art Museum Library, Phoenix 37371
Phoenix College, Phoenix 33641
Phoenix Public Library, Phoenix 40780
Phoenixville Public Library, Phoenixville 40781
Physical and Organic Chemistry Institute, Minsk 02015
Physical and Technical Institute of the Russian Academy of Sciences, Izhevsk 23062
Physical and Technical Institute of the Russian Academy of Sciences, Kazan 23080
Physical Culture Institute, Velikie Luki 22625
Physical Education Centre, Sankt-Peterburg 23755
Physical Engineering Institute, Kharkiv 28252
Physical Research Laboratory, Ahmedabad 13923
Physical Research Laboratory, Navarangpura 14108
Physical-Technical Institute, Minsk 02001
Physico-Technological Institute of Metals and Alloys, Kyiv 28380
Physikalisch-Technische Bundesanstalt, Braunschweig 11666
Physikalisch-Technische Bundesanstalt, Institut Berlin, Berlin 11555
Physiological Institute, Tbilisi 09272
Piarist Central Library, Budapest 13331
Piarista Központi Könyvtár, Budapest 13331
Piaristenkollegium Sankt Thekla, Wien 01493
Pickens County Cooperative Library, Carrollton 38463
Pickens County Library System, Easley 38887
Pickering Public Library, Pickering 04719
Pickerington Public Library, Pickerington 40783
Pictou-Antigonish Regional Library, New Glasgow 04701
Picture Gallery, Kursk 23133
Picture Gallery, Vologda 24071
Pidstrigach Applied Mathematics and Mechanics Institute, Lviv 28412
Pidvolochisk Regional Centralized Library System, Main Library, Pidvolochisk 28779
Pidvolochiska raionna TsBS, Tsentralna biblioteka, Pidvolochisk 28779
Piedmont Bible College, Winston-Salem 35424
Piedmont College, Demorest 30988
Piedmont Regional Library, Winder 41925
Piedmont Technical College, Greenwood 33390
Piedmont Virginia Community College, Charlottesville 33214
Pierce Atwood LLP, Portland 36147
Pierce College, Lakewood 33469
Pierce County Law Library, Tacoma 34907
Pierce County Library System, Tacoma 41493
Pierpont Morgan Library, New York 37248
Pietarsaaren kaupunginkirjasto, Pietarsaari 07599
Pig Breeding Research Institute, Poltava 28450
Pike County Public Library, Milford 40200
Pike County Public Library District, Pikeville 40787
Pike-Amite-Walthall Library System – McComb Public Library (Headquarters), McComb 40116
Pikes Peak Community College, Colorado Springs 33248
Pikes Peak Library District, Colorado Springs 38641
Pikes Peak Library District – Ruth Holley Branch, Colorado Springs 38642
Pikeville College, Pikeville 32290
Pilgermission St. Chrischona, Bettingen 27248
Pilgrim Hospital, Boston 29617
Pilgrim Psychiatric Center, West Brentwood 37819
Pilkington European Technical Centre Lathom, Ormskirk 29868
Pillsbury & Winthrop LLP, New York 36010
Pillsbury Winthrop LLP, San Francisco 36243
Pilsudski Institute of America Library, New York 37249
Pima Community College, Tucson 33809
Pima Council on Aging Library, Tucson 37689
Pima County Law Library, Tucson 34943
Pima County Public Library, Tucson 41602
Pima County Public Library – Columbus, Tucson 41603
Pima County Public Library – Dusenberry River Center, Tucson 41604
Pima County Public Library – George Miller-Golf Links, Tucson 41605
Pima County Public Library – Joel D. Valdez Main Library, Tucson 41606
Pima County Public Library – Joyner-Green Valley, Green Valley 39293
Pima County Public Library – Kirk Bear Canyon, Tucson 41607
Pima County Public Library – Mission, Tucson 41608
Pima County Public Library – Nanini, Tucson 41609
Pima County Public Library – Valencia, Tucson 41610
Pima County Public Library – Wilmot-Murphy, Tucson 41611
Pima County Public Library – Woods Memorial Branch, Tucson 41612
Pine Bluff & Jefferson County Library System, Pine Bluff 40790
Pine Manor College, Chestnut Hill 30754
Pine Mountain Regional Library, Manchester 40028
Pinellas County Law Library, Clearwater 34085
Pinellas County Law Library, Saint Petersburg 34787
Pinellas Park Public Library, Pinellas Park 40793
Pinetown Public Library, Pinetown 25223
Pineville-Bell County Public Library, Pineville 40794
Pingdingshan Teachers' College, Pingdingshan 05480
Pingyuan University, Xinxiang 05722
Pinsk Central City Library, Pinsk 02224
Pinsk City Children's Library, Pinsk 02223
Pinsk City Library no 1, Pinsk 02222
Pinsk Land Reclamation and Water Supply Planning Corporation, Pinsk 01919
Pinsk Regional Central Library, Pinsk 02225
Pinsk Timber Processing Industrial Corporation, Pinsk 01917
Pinskaya gorodskaya bibliyateka no 1, Pinsk 02222
Pinskaya gorodskaya detskaya bibliyoteka, Pinsk 02223
Pinskaya tsentralnaya gorodskaya bibliyateka, Pinsk 02224
Pinskaya tsentralnaya raionnaya bibliyateka, Pinsk 02225
Pio Sodalizio dei Piceni, Roma 16554
Pioneer Library System, Norman 40491
Pioneer Library System – Moore Public, Moore 40287
Pioneer Library System – Norman Public, Norman 40492
Pioneer Library System – Shawnee Public, Shawnee 41293
Pioneerland Library System, Willmar 41905
Pionierschule und Fachschule des Heeres für Bautechnik, Ingolstadt 10742
Pipe Rolling Plant, Nikopol 28166
Pipe Rolling Plant, Trade Union Library, Dnipropetrovsk 28540
Piper, Marbury, Rudnick & Wolfe LLP, Washington 36372
Pirelli, Milano 16148
O. Pirogovsky Children's Library, Kyiv 28648
Piscataway Township Free Public Library, Piscataway 40797
Pishchanka Regional Centralized Library System, Main Library, Pishchanka 28780
Pishchanska raionna TsBS, Tsentralna biblioteka, Pishchanka 28780
Piteå bibliotek, Piteå 26885
Pitkin County Library, Aspen 38004
Pitkyaranta Pulp Production Joint-Stock Company, Pitkyaranta 22892
Pitt Community College, Greenville 33389
Pitt Rivers Museum, Oxford 29872
Pittsburg Public Library, Pittsburg 40798
Pittsburg State University, Pittsburg 32294
Pittsburgh History & Landmarks Foundation, Pittsburgh 37384
Pittsburgh Theological Seminary, Pittsburgh 35334
Pittsburgh Zoo & Aquarium Library, Pittsburgh 37385
Pittsford Community Library, Pittsford 40809
Pittsylvania County Public Library, Chatham 38531
Pittwater Library Service, Mona Vale 01123
Pivdennotrubny zavod, Nikopol 28166
Pivdenno-Zakhidna Railway – Railway-carriage Works, Kyiv 28682
Pivdenno-Zakhidna Zaliznitsa, Kyiv 28381
Pivdenno-Zakhidna Zaliznitsa – Profkom Darnitskoho vagonoremontnoho zavodu, Kyiv 28682
PKTB legkoi promyshlennosti, Sankt-Peterburg 23874
Plaatselijke Openbare Bibliotheek, Balen 02718
Plaatselijke Openbare Bibliotheek, Beveren 02723
Plaatselijke Openbare Bibliotheek, Destelbergen 02749
Plaatselijke Openbare Bibliotheek, Geraardsbergen 02766
Plaatselijke Openbare Bibliotheek, Hamme 02770
Plaatselijke Openbare Bibliotheek, Herselt 02780
Plaatselijke Openbare Bibliotheek, Kapellen 02789
Plaatselijke Openbare Bibliotheek, Lommel 02808
Plaatselijke Openbare Bibliotheek, Maasmechelen 02813
Plaatselijke Openbare Bibliotheek, Maldegem 02814
Plaatselijke Openbare Bibliotheek, Meise 02822
Plaatselijke Openbare Bibliotheek, Putte 02844
Plaatselijke Openbare Bibliotheek, Westerlo 02882
Plaatselijke Openbare Bibliotheek, Wetteren 02883
Plaatselijke Openbare Bibliotheek, Willebroek 02885
Placentia Library District, Placentia 40810
Placer Dome, Inc, Vancouver 04305
Plain City Public Library, Plain City 40812
Plainfield Public Library, Plainfield 40813
Plainfield Public Library District, Plainfield 40814
Plainsboro Free Public Library, Plainsboro 40815
Plainview-Old Bethpage Public Library, Plainview 40816
Plainville Public Library, Plainville 40818
Plan Medewerkers, Pretoria 25167
Plancenter Ltd, Helsinki 07343
Planetari, Moskva 23457
Planetarium, Moskva 23457
Planinska zveza Slovenije, Ljubljana 24930
Planning and Design Institute for Nonferrous Metals, Moskva 23185
Planning and Design Institute for Subway Engineering, Moskva 23349

Planning and Design Institute for Timber Industry, Moskva 23187
Planning and Follow-up General Directorate's Library, Giza 07191
Planning and Research Institute for Steel Industry, Moskva 23326
Planning Commission Library, New Delhi 13892
Planning Institute for Balneology Industry, Moskva 23347
Planning Institute for Cinematography, Moskva 23189
Planning Institute for Health Industry, Moskva 23190
Planning Institute for River Transport, Moskva 23191
Planning Institute for Telecommunication Industry, Moskva 23184
Planning Institute for Thermoelectric Equipment, Moskva 23339
Planning Institute of Construction Technology no 2, Moskva 23182
Planning Institute of Public Health Services, Moskva 23240
Planning Institute of the Fish Industry, Sankt-Peterburg 23794
Planning Institute of the Metal Goods Industry, Sankt-Peterburg 23793
Planning Institute of the Moscow Building Industry, Moskva 23305
Planning Ministry Library, Cairo 07121
Plano Public Library System, Plano 40820
Plant Breeding and Acclimatization Institute, Błonie 21076
Plant Breeding and Acclimatization Institute, Koszalin 21125
Plant Breeding, Selection and Genetics Research Institute, Kharkiv 28274
Plant for Agricultural Engineering, Praha 06423
Plant for Aluminium Alloys, Moskva 22862
Plant for Medical Electrical Equipment, Moskva 22863
Plant of Automobiles 'Polmo', Szczecin 21069
Plant of Non-Ferrous Metallurgy, Moskva 22866
Plant, Pests and Diseases Research Institute, Tehran 14435
Plant Physiology and Genetics Institute, Kyiv 28325
Plant Production Research Center – Research Institute of Plant Production, Piešt'any 24747
Plant Protection Institute, Kostinbrod 03505
Plant Protection Research Institute, Kyiv 28376
Plant Protection Research Institute, Pretoria 25168
Plantijn Hogeschool van de Provincie Antwerpen, Antwerpen 02274
Planting Fields Arboretum, Oyster Bay 37316
Plantin-Moretus Museum, Antwerpen 02521
Plastic and Glass Industrial Corporation, Severodonetsk 28169
Plastic Goods Production Plant, Borisov 01839
Plastic Ltd, Moskva 22832
Plateau State Polytechnic, Barakin Ladi, Bukuru 19932
Plateau-State Library Board, Jos 20055
Platengymnasiet, Motala 26477
Platte County Public Library, Wheatland 41872
Plattsburgh Public Library, Plattsburgh 40825
Playboy Enterprises, Inc, Chicago 35597
City of Playford Library Service, Elizabeth 01094
Pleasant Grove Public Library, Pleasant Grove 40826
Pleasant Hill Library, Hastings 39394
Pleasanton Public Library, Pleasanton 40829
Plekhanov Russian Economics Academy, Moskva 22385
Płock Scientific Society, Płock 21189
Plovdivski universitet Paissii Khilendarski, Plovdiv 03458
Plum Creek Library System, Worthington 41974
Plumas County Library, Quincy 40919
Plumb Memorial Library, Shelton 41303
Plumb Memorial Library – Huntington Branch, Shelton 41304
Plumstead Library, London 30022
Plunkett & Cooney, Bloomfield Hills 35512
Plunkett Foundation, Woodstock 29917
Plymouth Central Library, Plymouth 30059
Plymouth College of Art and Design, Plymouth 29370
Plymouth District Library, Plymouth 40832
Plymouth Proprietary Library, Plymouth 29880
Plymouth Public Library, Plymouth 40833
Plymouth Public Library, Plymouth 40834
Plymouth Public Library, Plymouth 40835
Plymouth State University, Plymouth 32324
PMMK-Bibliotheek, Oostende 02692
PO arendnykh kozhgalantereinykh predpriyati, Sankt-Peterburg 22922
PO Azot im. Pritytskogo, Grodno 01868
PO Belaruskali – Golovnaya bibliyateka, Soligorsk 02238
PO Belaruskali – Nauchno-tekhnicheskaya bibliyateka, Soligorsk 01928
PO Belelektrosvet, Brest 01842
PO Bobruyskdrev, Bobruysk 02107
PO Bobruyskshina, Bobruysk 01833
PO Dalvostugol, Raichikhinsk 22896
PO Eskalator, Sankt-Peterburg 22923
PO Gidroavtomatika, Gomel 01852
PO Gipromeliovodkhoz, Pinsk 01919
PO Gorizont, Minsk 02189
PO Gosselmash, Gomel 01853
PO Integral, Minsk 02190
PO Izhorski zavod, Sankt-Peterburg 22924
PO Khimvolokno, Grodno 01869
PO Khimvolokno – Bibliyateka profkoma, Mogilev 02205
PO Khimvolokno – Nauchno-tekhnicheskaya bibliyateka, Mogilev 01903
PO Khimvolokno – Nauchno-tekhnicheskaya bibliyateka, Svetlogorsk 01930
PO Krichevtsementoshifer, Krichev 01876
PO Lenenergoremont, Sankt-Peterburg 22925
PO metallicheski zavod, Biblioteka profkoma, Sankt-Peterburg 24243
PO Minski traktorny zavod, Minsk 01897
PO Minskproektmebel, Minsk 01898
PO Minsktraktorzapchast – Zavod shesteren, Minsk 01899
PO Mogilevliftmash, Mogilev 01904
PO po vypusku igrushek Mir, Minsk 02191
PO po vypusku liteinogo oborudovaniya, Pinsk 01920
PO podyomno-transportnogo oborudovaniya, Sankt-Peterburg 22926
PO Rechitsadrev, Rechitsa 02075
PO Severny zavod, Sankt-Peterburg 22927
PO shelkovykh tkanei, Mogilev 01905
PO Skorokhod, Sankt-Peterburg 22928
PO Steklovolokno, Polotsk 01923
PO zavod Kalinina, Sankt-Peterburg 22929
PO zavod Kalinina, DK Biblioteka, Sankt-Peterburg 24244
PO zavod turbinnykh lopatok, Sankt-Peterburg 22930
Pocahontas County Free Library, Marlinton 40071
Pochvenny institut im. V.V. Dokuchaeva, Moskva 23458
R. A. Podar College of Commerce and Economics, Mumbai 13824
Podil Region Central Library System of the City of Kiev – I. Franko Main Library, Kyiv 28683
Podilska raionna TsBS m. Kyiva, Tsentralna biblioteka im. I. Franka, Kyiv 28683
Podolski mekhanicheski zavod im. M.I. Kalinina, Podolsk 22893
Podtatranská Knižnica, Poprad 24788
Podtatranské múzeum, Poprad 24748
The Poetry Library, London 29829
Poëziecentrum, Gent 02653
Pogrebishche Regional Centralized Library System, Main Library, Pogrebishche 28781
Pogrebishchenska raionna TsBS, Tsentralna biblioteka, Pogrebishche 28781
Pohjanmaan museo, Vaasa 07553
Pohnpei Public Library, Kolonia 18877
Point Loma Nazarene University, San Diego 32578
Point Park University Library, Pittsburgh 32303
Pointe Coupee Parish Library, New Roads 40428
Pointe-Claire Public Library, Pointe-Claire 04721
Poisk Research Institute, Sankt-Peterburg 23837
Pokhara Campus Library, Pokhara 19025
Pokrajinska in študijska knjižnica, Murska Sobota 24973
Pokrajinski arhiv Maribor, Maribor 24948
Pokrajinski muzej, Celje 24908
Pokrajinski muzej, Maribor 24949
Połączone Biblioteki Instytutu Filozofii i Socjologii PAN, Warszawa 21333
Poland's Millennium Library in Los Angeles, Los Angeles 37061
Pôle Universitaire Leonard de Vinci (PULV), Paris-la-Défense 07990
Polenmuseum Rapperswil, Rapperswil 27425
Polesskaya opytno-meliorativnaya stantsiya, Polesski 02070
Polesski Land Reclamation Experimental Station, Polesski 02070
Police Academy, Cairo 07053
Police Academy of the Netherlands, Apeldoorn 19095
Police Academy of the Netherlands, Apeldoorn 19241
Police Fédérale / Federale Politie, Brussel 02434
Police Haaglanden, Den Haag 19266
Policía Federal Argentina, Buenos Aires 00213
Policy Research Institute, Ministry of Agriculture, Forestry and Fisheries, Tokyo 17748
Poliisiammattikorkeakoulun kirjasto, Espoo 07436
Polimoda – Centro di Servizi e Documentazione, Firenze 14993
Polis di Raja Malaysia, Kuala Lumpur 18567
Polish Academy of Sciences, Roma 14812
Polish Army Museum, Warszawa 21325
Polish Association of the Blind, Warszawa 21336
Polish Botanical Society, Warszawa 21338
Polish Entomological Society, Wrocław 21365
Polish Ethnological Society, Wrocław 21366
Polish Filmschool Lodz, Łódź 21036
Polish Geographical Society, Warszawa 21339
Polish Geological Institute, Warszawa 21331
Polish Geological Society, Kraków 21149
Polish Institute and Sikorski Museum, London 29806
Polish Institute of Arts & Sciences in America, Inc, New York 37250
Polish Institute of International Affairs, Warszawa 21334
The Polish Library, London 29807
The Polish Library, Montreal 04417
Polish Library in Paris, Paris 08496
Polish Maritime Museum, Gdańsk 21094
Polish Museum of America Library, Chicago 36687
Polish Music Information Centre, Warszawa 21337
Polish National Commission for UNESCO, Warszawa 21335
Polish Numismatic Society, Warszawa 21340
Polish Ocean Lines, Gdynia 21103
Polish Teachers' Union, Warszawa 21352
Polish Zoological Society, Wrocław 21367
Polishögskolans bibliotek, Solna 26423
Politechnika Białostocka, Białystok 20794
Politechnika Częstochowska, Częstochowa 20800
Politechnika Gdańska, Gdańsk 20805
Politechnika Koszalińska, Koszalin 20821
Politechnika Krakowska im. Tadeusza Kościuszki, Kraków 20839
Politechnika Łódzka, Łódź 20875
– Filia w Bielsku-Białej, Bielsko-Biała 20876
Politechnika Lubelska, Lublin 20887
Politechnika Poznanska, Poznań 20903
Politechnika Radomska im. Kazimierza Pułaskiego, Radom 20920
Politechnika Rzeszowska im. Ignacego Łukasiewicza, Rzeszów 20922
Politechnika Śląska, Gliwice 20813
Politechnika Świetokrzyska, Kielce 20819
Politechnika Szczecińska, Szczecin 20927
Politechnika Warszawska, Warszawa 20956
– Wydział Architektury, Warszawa 20957
– Wydział Budownictwa, Mechaniki i Petrochemii, Płock 20958
– Wydział Chemiczny, Warszawa 20959
– Wydział Elektroniki i Technik Informacyjnych, Warszawa 20960
– Wydział Inżynierii Chemicznej i Procesowej, Warszawa 20961
– Wydział Inżynierii Materiałowej, Warszawa 20962
– Wydział Inżynierii Środowiska, Warszawa 20963
– Wydział Mechaniczny, Energetiki i Lotnictwa, Warszawa 20964
– Wydział Samochodów i Maszyn Roboczych, Warszawa 20965
– Wydział Transportu, Warszawa 20966
Politechnika Wrocławska, Wrocław 21027
Politechnika Zielonogórska, Zielona Góra 21032
Politechniki Opowskiej, Opole 20897
Politecnico di Milano, Milano
– Biblioteca Campus Durando, Milano 15079
– Biblioteca Centrale di Architettura, Milano 15080
– Biblioteca Centrale di Ingegneria, Milano 15081
– Biblioteca del Polo Regionale di Como, Como 15082
– Biblioteca del Polo Regionale di Lecco, Lecco 15083
– Biblioteca del Polo Regionale di Mantova, Mantova 15084
– Biblioteca delle Ingegnerie Bovisa, Milano 15085
– Biblioteca Didattica di Ingegneria Bovisa, Milano 15086
– Biblioteca Didattica di Ingegneria Cremona, Cremona 15087
– Biblioteca Didattica di Piacenza, Piacenza 15088
– Dipartimento di Chimica, Materiali e Ingegneria Chimica "Giulio Natta", Milano 15089
– Dipartimento di Elettronica e Informazione, Milano 15090
– Dipartimento di Elettrotecnica, Milano 15091
– Dipartimento di Ingegneria Aerospaziale, Milano 15092
– Dipartimento di Ingegneria Idraulica, Ambientale, Infrastrutture viarie e Rilevamento – DIIAR – Sez. Ambientale, Milano 15093
– Dipartimento di Ingegneria Idraulica, Ambientale, Infrastrutture viarie e Rilevamento – DIIAR – Sez. Idraulica, Milano 15094
– Dipartimento di Ingegneria Idraulica, Ambientale, Infrastrutture viarie e Rilevamento – DIIAR – Sez. Infrastrutture viarie, Milano 15095
– Dipartimento di Ingegneria Nucleare, Milano 15096
– Dipartimento di Ingegneria Strutturale, Milano 15097
– Dipartimento di Matematica, Milano 15098
– Dipartimento di Progettazione dell'Architettura, Milano 15099
– Dipartimento di Scienza e Tecnologie dell'Ambiente Costruito, Milano 15100
– Servizio Tesi e Documentazione della Cartografia e Pianificazione, Milano 15101
Politecnico di Torino, Torino 15486
– Biblioteca Centrale di Architettura, Torino 15487
– Biblioteca Ingegneria Elettronica 'M. Boella', Torino 15488
– Dipartimento di Georisorese e Territorio, Torino 15489
– Dipartimento di Idraulica, Trasporti e Infrastutture Civili, Torino 15490
– Dipartimento di Matematica, Torino 15491
– Dipartimento di Scienza dei Materiali e Ingegneria Chimica, Torino 15492
'Politehnica' University, Timişoara 21950
Politekhnicheski institut, Chelyabinsk 22229
Politekhnicheski institut, Kirov 22291
Politekhnicheski institut, Komsomolsk-na-Amure 22294
Politekhnicheski institut, Krasnodar 22302
Politekhnicheski institut, Kursk 22313
Politekhnicheski institut, Orenburg 22462
Politekhnicheski institut, Stavropol 22585
Politekhnicheski institut, Tolyatti 22592
Politekhnicheski institut, Tver 22601
Politekhnicheski institut, Voronezh 22645
Politekhnicheski kolledzh, Sankt-Peterburg 22705
Politekhnicheski kolledzh-predpriyatie, Syzran 22588
Politekhnikum, Donetsk 28097
Politie Amsterdam-Amstelland, Amsterdam 19238
Politie Gelderland Midden/Zuid, Nijmegen 19284
Politie Haaglanden, Den Haag 19266
Politieacademie, Apeldoorn 19241
Politihøgskolen, Oslo 20122
Politikatörténeti Intézet, Budapest 13455
Polizei Hamburg / Hochschule der Polizei Hamburg, Hamburg 09921
Polizeiakademie Niedersachsen, Nienburg 10760
Polizeidirektion für Aus- und Fortbildung und die Bereitschaftspolizei Schleswig-Holstein, Bad Malente 10706
Polizeipräsident in Berlin, Berlin 10810
Polizeipräsidium München, München 11047
Polk Community College, Winter Haven 33881
Polk County Historical & Genealogical Library, Bartow 36522
Polk County Law Library, Bartow 33969
Polk County Library, Bolivar 38237
Pollack Mihály Műszaki Főiskola, Pécs 13291
Pollard Memorial Library, Lowell 39973
Polnisches Institut Berlin, Berlin 11556
Polnisches Institut Wien, Wien 01648
Polnośląska Biblioteka Publiczna im. T. Mikulskiego, Wrocław 21609
Pologi Regional Centralized Library System, Main Library, Pologi 28782
Pologivska raionna TsBS, Tsentralna biblioteka, Pologi 28782
Polokwane City Library, Polokwane 25225

Polonne Regional Centralized Library System, Main Library, Polonne 28783
Polonska raionna TsBS, Tsentralna biblioteka, Polonne 28783
Polotsk Automobile Engineering Plant, Polotsk 01924
Polotsk Central City Library by name of Fr. Skaryna, Polotsk 02227
Polotsk Railway Station, Technical Library, Polotsk 02071
Polotsk State University, Novopolotsk 01815
Polotskaya Gorodskaya bibliyateka no 2 im. Gertsena, Polotsk 02226
Polotskaya tsentralnaya gorodskaya bibliyateka im. Fr. Skoriny, Polotsk 02227
Polotski avtoremontny zavod, Polotsk 01924
Polsinelli Shughart, Kansas City 35796
Polsinelli Shughart PC, Kansas City 35797
Polska Akademia Nauk, Roma 14812
Polski Instytut Spraw Miedzynarodowych, Warszawa 21334
Polski Komitet ds. UNESCO, Warszawa 21335
Polski Związek Niewidomych, Warszawa 21336
Polskie Centrum Infomacji Muzycznej, Warszawa 21337
Polskie Linie Oceaniczne, Gdynia 21103
Polskie Towarzystwo Antropologiczne, Poznań 21205
Polskie Towarzystwo Botaniczne, Warszawa 21338
Polskie Towarzystwo Entomologiczne, Wrocław 21365
Polskie Towarzystwo Geograficzne, Warszawa 21339
Polskie Towarzystwo Geologiczne, Kraków 21149
Polskie Towarzystwo Ludoznawcze, Wrocław 21366
Polskie Towarzystwo Numizmatyczne, Warszawa 21340
Polskie Towarzystwo Zoologiczne, Wrocław 21367
Poltava City Centralized Library System, Main Library, Poltava 28786
Poltava Consumers' Co-operative Institute, Poltava 28453
Poltava National Technical University, Poltava 28065
Poltava Regional Centralized Library System, Main Library, Machukhi 28713
Poltavska miska TsBS, Tsentralna biblioteka, Poltava 28786
Poltavska raionna TsBS, Tsentralna biblioteka, Machukhi 28713
Poltavsky natsionalny tekhnichi universytet, Poltava 28065
Poltegor-Institute, Wrocław 21359
Põlva Central Library, Põlva 07253
Põlva Keskraamatukogu, Põlva 07253
Polyarny nauchno-issledovatelski institut morskogo rybnogo khozyaistva i okeanografii im. N.M. Knipovicha, Murmansk 23603
Polymer Engineering Joint-Stock Company, Tambov 22961
Polytechnic, Calabar 19933
Polytechnic, Owerri 19976
Polytechnic Institute, Chelyabinsk 22229
Polytechnic Institute, Kirov 22291
Polytechnic Institute, Komsomolsk-na-Amure 22294
Polytechnic Institute, Krasnodar 22302
Polytechnic Institute, Kursk 22313
Polytechnic Institute, Orenburg 22462
Polytechnic Institute, Stavropol 22585
Polytechnic Institute, Toshkent 42098
Polytechnic Institute, Tver 22601
Polytechnic Institute, Voronezh 22645
Polytechnic Institute of Beja, Beja 21622
Polytechnic University, Brooklyn 30571
Polytechnic University of the Philippines, Manila 20721
Polytechnical College-Enterprise, Syzran 22588
Polytechnical Institute, Cherkasy 27954
Polytechnical Professional School, Donetsk 28097
Polytechnical Professional School, Sankt-Peterburg 22705
Pomeranian Library Stanisława Staszica, Szczecin 21580
Pomeranian Medical University, Szczecin 20928
Pommersche Evangelische Kirche, Greifswald 11228
Pomona Public Library, Pomona 40840
Pomorska Akademia Medyczna, Szczecin 20928
Pomorska Akademia Pedagogiczna, Słupsk 20925
Pomorski gosudarstvenny universitet im. M. V. Lomonosova, Arkhangelsk 22203
Pomorski Muzej, Kotor 18915
Pomorski muzej 'Sergej Mašera', Piran 24951
PON – Institute for research and social development in North-Brabant, Tilburg 19522
PON – Instituut voor Advies, Onderzoek en Ontwikkeling in Noord-Brabant, Tilburg 19522
Ponca City Library, Ponca City 40842
Ponce School of Medicine, Ponce 21863
Pontiac Correctional Center Library, Pontiac 34691
Pontiac Public Library, Pontiac 40843
Pontifíca Universidade Católica do Rio de Janeiro, Rio de Janeiro 03099
Pontifical Biblical Institute, Jerusalem 14700
Pontifical Catholic University of Puerto Rico, Mayaguez 21859
Pontifical Catholic University of Puerto Rico, Ponce
– Encarnación Valdés Library, Ponce 21864
– Monseignor Fremiot Torres Oliver Legal Information and Research Center, Ponce 21865
Pontifical College Josephinum, Columbus 35168
Pontifical Institute for Arabic and Islamic Studies, Roma 15377
Pontifical Irish College, Roma 16030
Pontifical Oriental Institute Library, Roma 16039
Pontifical University of the Holy Cross, Roma 15373
Pontifical Urbanian University Library, Città del Vaticano 42190
Pontificia Facoltà di Scienze dell'Educazione Auxilium, Roma 15372
Pontificia Facoltà Teologica della Sardegna, Cagliari 14918
Pontificia Facoltà Teologica dell'Italia Meridionale, Napoli 15132
Pontificia Facoltà Teologica di San Bonaventura, Città del Vaticano 42183
Pontificia Facoltà Teologica Marianum, Città del Vaticano 42184
Pontificia Facoltà Teologica Pugliese, Molfetta 15915
Pontificia Facoltà Teologica Teresianum, Città del Vaticano 42185
Pontificia Universidad Católica de Chile, Santiago 04860
– Biblioteca de Derecho, Santiago 04861
– Biblioteca Gauss, Santiago 04862
– Biblioteca General Campus Oriente, Santiago 04863
– Biblioteca Lo Contador, Santiago 04864
– Biblioteca – Sede del Maule, Campus San Miguel, Talca 04865
– Facultad de Agronomía y Geografía, Santiago 04866
– Facultad de Ciencias e Ingeniería, Santiago 04867
– Facultad de Ciencias Sociales, Santiago 04868
– Facultad de Economía y Administración, Santiago 04869
– Facultad de Medicina y Ciencias Biológicas, Santiago 04870
– Facultad de Teología, Santiago 04871
Pontificia Universidad Católica del Perú, Lima 20620
– Biblioteca de Ciencias, Lima 20621
– Biblioteca de Ingeniería, Lima 20622
– Centro de Documentación de Ciencias Sociales, Lima 20623
Pontificia Universidad Javeriana, Santafé de Bogotá 05965
– Facultad de Ciencias Políticas y Relaciones Internacionales, Santafé de Bogotá 05966
– Instituto Geofísico de los Andes Colombianos, Santafé de Bogotá 05967
– Seccional Calí, Calí 05968
Pontifícia Universidade Católica de Campinas, Campinas 03006
Pontifícia Universidade Católica de São Paulo, São Paulo 03148
Pontifícia Universidade Católica do Rio Grande do Sul, Porto Alegre 03065
Pontificia Università della Santa Croce, Roma 15373
Pontificia Università Gregoriana, Roma 15374
Pontificia Università Lateranese, Città del Vaticano 42186
– Biblioteca Claretianum, Città del Vaticano 42187
– Istituto Patristico 'Augustinianum', Città del Vaticano 42188
Pontificia Università S. Tommaso d'Aquino, Roma 15375
Pontificia Università San Tommaso d'Aquino, Città del Vaticano 42189
Pontificia Università Urbaniana, Città del Vaticano 42190
– Pontificia Biblioteca Missionaria della Sacra Congregazione per L'Evangelizzazione dei Popoli, Città del Vaticano 42191
Pontifical Faculty of Theology, Wrocław 21026
Pontificio Ateneo Antonianum, Roma 16031
Pontificio Colegio Español de San José, Roma 16032
Pontificio Collegio Armeno, Città del Vaticano 42196
Pontificio Collegio Beda, Roma 16033
Pontificio Collegio Belga, Città del Vaticano 42197
Pontificio Collegio Canadese, Roma 16034
Pontificio Collegio Croato di S. Girolamo, Città del Vaticano 42198
Pontificio Collegio Greco di S. Atanasio, Città del Vaticano 42199
Pontificio Collegio Leoniano, Anagni 15700
Pontificio Collegio Pio Latino Americano, Città del Vaticano 42200
Pontificio Collegio Polacco, Roma 16035
Pontificio Collegio Russo di S. Teresa del Bambino Gesù, Roma 16036
Pontificio Collegio Scozzese, Roma 16037
Pontificio Istituto Biblico, Roma 16555
Pontificio Istituto di Archeologia Cristiana, Città del Vaticano 42192
Pontificio Istituto di Musica Sacra, Roma 15376
Pontificio Istituto di Studi Arabi e d'Islamistica, Roma 15377
Pontificio Istituto Missioni Estere, Roma 16038
Pontificio Istituto Orientale, Roma 16039
Pontificio Seminario Francese, Città del Vaticano 42201
Pontificio Seminario Regionale, Potenza 15996
Pontificio Seminario Regionale del Lazio Superiore alla Quercia, Viterbo 16140
Pontificio Seminario Regionale "Pio XII", Siena 16071
Pontificium Collegium Germanicum et Hungaricum, Roma 42202
Pontificium Collegium Polonorum, Roma 16035
Popasna Regional Centralized Library System, Main Library, Popasna 28787
Popasnyanska raionna TsBS, Tsentralna biblioteka, Popasna 28787
Pope County Library System – Russellville Headquarters Branch, Russellville 41083
Pope John XXIII National Seminary, Weston 35419
Pope Paul VI Library, Birkirkara 18661
Popilnya Regional Centralized Library System, Main Library, Popilnya 28788
Popilnyanska raionna TsBS, Tsentralna biblioteka, Popilnya 28788
Popov Military School of Radioelectronics, Sankt-Peterburg 22716
Popov Museum of Telecommunications, Sankt-Peterburg 23815
Popov Radio Reception and Acoustics Research Institute, Sankt-Peterburg 23843
Popov Ukrainian National Telecommunications Academy, Branch Library, Kyiv 28310
Popov Ukrainian State Academy of Telecommunications, Odesa 28061
Population Council Library, New York 37251
Population Information Centre, Colombo 26315
Population Reference Bureau, Washington 37776
Poquoson Public Library, Poquoson 40845
Pori City Library – Satakunta County Library, Pori 07600
Porin kaupunginkirjasto – Satakunnan maakuntakirjasto, Pori 07600
Porirua Public Library, Porirua 19881
Dr. Ing. h.c. F. Porsche AG, Weissach 11434
Porsgrunn bibliotek, Porsgrunn 20386
Port Adelaide Enfield Public Library Service, Port Adelaide 01143
Port and Harbour Research Institute, Design Standard Division, Training and Library Section, Yokosuka 17831
Port Arthur Public Library, Port Arthur 40849
Port Colborne Public Library, Port Colborne 04723
Port Elizabeth Technikon, Port Elizabeth 25038
Port Jefferson Free Library, Port Jefferson 40856
Port Macquarie – Hastings Library Service, Port Macquarie 01144
Port Moody Public Library, Port Moody 04725
Port Moresby Inservice College, Boroko 20548
Port Moresby Public Library, Port Moresby 20599
Port of Singapore Authority (PSA), Singapore 24600
Port Phillip Library Service, St Kilda 01161
Port Washington Public Library, Port Washington 40862
Portage College Library, Lac La Biche 03733
Portage County District Library, Garrettsville 39205
Portage County District Library – Aurora Memorial, Aurora 38040
Portage County District Library – Garrettsville Branch, Garrettsville 39206
Portage County Law Library, Ravenna 34729
Portage County Public Library, Stevens Point 41442
Portage District Library, Portage 40863
Portage Public Library, Portage 40864
Porter County Public Library System, Valparaiso 41660
Porter County Public Library System – Portage Public, Portage 40865
Porter County Public Library System – Valparaiso Public (Central), Valparaiso 41661
Porter, Wright, Morris & Arthur, LLP, Columbus 35634
Porterville College, Porterville 33656
Porterville Developmental Center, Porterville 37397
Porterville Public Library, Porterville 40867
Portico Library, Manchester 29854
Portland Art Museum, Portland 37406
Portland Cement Association, Skokie 37612
Portland Community College Libraries, Portland 33659
Portland Library, Portland 40871
Portland Parish Library, Port Antonio 16893
Portland Public Library, Portland 40872
Portland State University, Portland 32341
Portland VA Medical Center, Portland 37407
Porto Public Library, Porto 21841
Port-Royal Library, Paris 08224
Portsmouth Athenaeum, Portsmouth 37409
Portsmouth City Council, Portsmouth 30060
Portsmouth Free Public Library, Portsmouth 40873
Portsmouth Public Library, Portsmouth 40874
Portsmouth Public Library, Portsmouth 40875
Portsmouth Public Library, Portsmouth 40876
Portsmouth Public Library – Churchland, Portsmouth 40877
Porvoo Public Library – Regional Library of Uusimaa, Porvoo 07601
Porvoon kaupunginkirjasto – Uudenmaan maakuntakirjasto, Porvoo 07601
Porzio, Bromberg & Newman Library, Morristown 35923
Posavje Museum, Brežice 24907
Posavski muzej Brežice, Brežice 24907
Poslijediplomsko srediste Dubrovnik (PSD), Dubrovnik 06214
Post Library, Yuma Proving Ground, AZ, Yuma 35095
Post University, Waterbury 32995
Postal Museum, Budapest 13456
Postal Museum of Finland, Helsinki 07484
Postamúzeum, Budapest 13456
Postavskaya tsentralnaya raionnaya bibliyateka, Postavy 02228
Postavy Regional Central Library, Postavy 02228
La Poste, Strasbourg 08164
Postgraduate Institute of Medical Education and Research, Chandigarh 13953
Post-Graduate Institute of Medicine, Colombo 26316
Postgraduate Medical University, Budapest 13259
Postimuseo, Helsinki 07484
Postmusei Bibliotek, Stockholm 26674
Postmuseum Library, Stockholm 26674
Potato Based Foods Scientific and Industrial Corporation, Minsk 01895
Potato Research Institute, Havlíčkův Brod 06448
Potchefstroom College of Education, Potchefstroom 25039
Potchefstroom Community Library and Information Services, Potchefstroom 25227
Potebnya Institute of Linguistics, Kyiv 28339
Potomac State College of West Virginia University, Keyser 31505
Potsdam Public Library & Reading Center, Potsdam 40879
Pottawatomie Wabaunsee Regional Library, Saint Marys 41128
Potter County Law Library, Amarillo 33917
Pottstown Public Library, Pottstown 40880
Pottsville Free Public Library, Pottsville 40881
Poughkeepsie Public Library District, Poughkeepsie 40883

Poultry Research Institute of the Ukrainian Academy of Science, Borki 28186
Povilas Visinskis Siauliai County Public LIbrary, Šiauliai 18361
Povolzhskaya State Akadamy of Telecommunication and Informatics, Samara 22501
Povolzhski institut informatiki, radiotekhniki i svyazi, Samara 22501
Powell, Goldstein LLP, Atlanta 35466
Power Engineering and Construction Research Institute, Moskva 23302
Power Engineering Design Experimental and Research Institute, Kyiv 28377
Power Engineering Industrial Company, Sankt-Peterburg 22925
Power Engineering Maintenance Enterprise, Minsk 01900
Power Grid Corporation of India Ltd, New Delhi 13912
Power Reactor and Nuclear Fuel Development Corp, Tokai Works, Naka 17396
Powerhouse Museum, Haymarket 00928
Powerlink Queensland, Virginia 00841
Powhatan County Public Library, Powhatan 40888
Powiatowa Biblioteka Publiczna, Racibórz 21543
Powiatowa Biblioteka Publiczna, Sieradz 21555
Powiatowa Biblioteka Publiczna, Słupsk 21561
Powiatowa i Miejska Biblioteka Publiczna, Lwówek Śląski 21490
Powiatowa i Miejska Biblioteka Publiczna, Nowy Tomyśl 21511
Powiatowa i Miejska Biblioteka Publiczna, Olkusz 21516
Powiatowa i Miejska Biblioteka Publiczna, Wałbrzych 21598
Powiatowa i Miejska Biblioteka Publiczna im. Dąbrowskiej, Kłodzko 21447
Poyner & Spruill, Raleigh 36162
Poynter Institute for Media Studies, Saint Petersburg 37509
Poznań Society of Friends of Art and Sciences, Poznań 21206
Poznań University Library, Poznań 20905
– Department of Astronomy, Poznań 20910
– Department of Musicology, Poznań 20916
– Department of Neophilology, Poznań 20914
– Department of Social Sciences, Poznań 20913
– Faculty of Chemistry, Poznań 20912
– Faculty of Law and Administration, Poznań 20915
– Institute of Art History, Poznań 20909
– Institute of History, Poznań 20908
– Institute of Polish Philology, Poznań 20907
– Study of Teaching of Foreign Languages, Poznań 20911
Poznan University of Economics, Poznań 20899
Poznan University of Life Sciences, Poznań 20917
Poznań University of Technology, Poznań 20903
Poznańskie Towarzystwa Przyjaciół Nauk, Poznań 21206
PPG Industries, Inc, Allison Park 35442
PPG Industries, Inc, Monroeville 35916
PPG Industries Inc – Glass Technology Center, Pittsburgh 36131
Prácheňské muzeum v Písku, Písek 06472
Prado Museum Library, Madrid 26078
Prague City Archives, Praha 06476
Prahova District Museum of History and Archäologie, Ploieşti 22053
Prahran Mechanics' Institute and Circulating Library Inc., Windsor 01049
Prairie Bible Institute, Three Hills 03876
Prairie Farm Rehabilitation Administration, Regina 04468
Prairie State College, Chicago Heights 33229
Prairie Trails Public Library District, Burbank 38388
Prairie View A&M University, Prairie View 32357
Prairie-River Library District, Lapwai 39806
Praktizijns Sociëteit, Amsterdam 19239
Prämonstratenser-Abtei Windberg, Hunderdorf 11250
Prämonstratenser-Chorherrenstift Geras, Geras 01425
Prämonstratenser-Chorherrenstift Wilten, Innsbruck 01441
Prancūzų kultūros centras, Vilnius 18303
Pratt Institute Library, Brooklyn 30572
Pratt & Whitney, San Jose 36250
Pratt & Whitney Canada, Inc Library, Longueuil 04261

Pratt & Whitney Rocketdyne, Inc, Canoga Park 35551
Pražská konzervatoř, Praha 06385
Preacher's Seminary at Domstift, Brandenburg an der Havel 11180
Preble County District Library, Eaton 38926
Precision Machine Tools Plant, Vitebsk 01938
Precision Mechanics Research Institute, Sankt-Peterburg 23850
Predigerbibliothek im Kloster Preetz, Preetz 11308
Predigerseminar, Celle 11185
Predigerseminar beim Domstift, Brandenburg an der Havel 11180
Das Predigerseminar der Ev. luth. Landeskirche Hannovers im Kloster Locccum, Rehburg-Loccum 11313
Predigerseminar der Evangelischen Kirche von Kurhessen-Waldeck, Hofgeismar 11249
Préfecture de police – Laboratoire Central, Paris 08157
Presbyterian Church (USA), Philadelphia 35329
Presbyterian College, Clinton 30840
The Presbyterian College at McGill University / Le Collège Presbytérien à l'Université McGill, Montreal 04217
Presbyterian Heritage Center at Montreat, Montreat 35290
Presbyterian Hospital, New York 37252
Presbyterian Theological Centre, Burwood 00758
Presbyterian Theological College, Box Hill North 00754
Prescott Historical Society, Prescott 37411
Prescott Public Library, Prescott 40891
Prescott Valley Public Library, Prescott Valley 40892
Presentation College, Aberdeen 33112
Presidência do Conselho de Ministros, Lisboa 21739
Presidencia La República, México 18816
Presidency College Library, Kolkata 13811
Presidency Library, Cairo 07122
Presidential Court, Abu Dhabi 28908
The Presidential Library of the Republic of Belarus, Minsk 01823
Presidenza del Consiglio dei Ministri, Roma 15684
Presidenza del Consiglio dei Ministri – Dipartimento per l'Informazione e l'Editoria, Roma 15685
Presidenza del Consiglio dei Ministri – Servizi Informazioni e Proprietà Letteraria Artistica e Scintifica, Roma 15686
Presidium of the Russian Academy of Sciences, Moskva 23459
Presque Isle District Library, Rogers City 41054
Press and Information Department, Kabul 00004
Press Museum, Amsterdam 19374
Presse- und Informationsamt der Bundesregierung, Berlin 10811
Pressing Equipment Research Institute, Sumy 28475
Preston Gates & Ellis, Seattle 36269
Prestonsburg Community College, Prestonsburg 33664
Prevent, Brussel 02619
Prevention First Inc, Chicago 36688
Prezidentskaya biblioteka, Minsk 01823
Prezidium kollegii advokatov S-Peterburga, Sankt-Peterburg 23875
Prezidium RAN, Moskva 23459
Prezihov Voranc Library, Ljubljana 24969
Priaulx Library, St Peter Port 29899
Priazovsk Regional Centralized Library System, Main Library, Priazovsk 28789
Priazovska raionna TsBS, Tsentralna biblioteka, Priazovsk 28789
Pribor Instrument-Making Plant, Sankt-Peterburg 22950
Price City Library, Price 40895
Price Waterhouse Coopers, Toronto 04290
Price-Pottenger Nutrition Foundation, La Mesa 36999
Pricewaterhouse Coopers, Dublin 14516
Prichard Public Library, Prichard 40896
Prickett, Jones, Elliott, Wilmington 36425
Pridneprovska raionna elektrostantsia, Dnipropetrovsk 28539
Pridniprovska dorozhna naukovo-tekhnichna biblioteka, Dnipropetrovsk 28207
Prienų rajono savivaldybės viešoji biblioteka, Prienai 18352
Priesterseminar, München 11285
Priesterseminar, Würzburg 11369
Priesterseminar Bamberg, Bamberg 11160
Priesterseminar der Diözese Augsburg, Augsburg 11154

Priesterseminar der Diözese Graz-Seckau, Graz 01431
Priesterseminar Sankt Petrus, Opferbach 11299
Priestseminary Library, Włocławek 21063
Prievidza Hospital and Polyclinic, Bojnice 24682
Prifysgol Aberystwyth, Aberystwyth 28927
Prikarpatsky natsionalny universytet im. Vasilya Stefanika, Ivano-Frankivsk 27981
Priluki Regional Centralized Library System, Main Library, Priluki 28790
Prilutska raionna TsBS, Tsentralna biblioteka, Priluki 28790
Primary Industries and Resources South Australia, Adelaide 00601
Primary School, Brezovica 24875
Primary School Teachers Training College no 2, Kyiv 28112
Prime Minister's Department, Kuala Lumpur 18555
Primorsk Agricultural Institute, Ussurisk 22621
Primorsk District Medical Library, Vladivostok 24057
Primorsk District Museum of Regional Studies, Vladivostok 24058
Primorsk Regional Centralized Library System, Main Library, Primorsk 28791
Primorsk Regional Children Library, Vladivostok 24347
Primorsk Regional Library, Vladivostok 22191
Primorska raionna TsBS, Tsentralna biblioteka, Primorsk 28791
Primorskaya gosudarstvennaya publichnaya biblioteka im. A.M. Gorkogo, Vladivostok 22191
Primorskaya kraevaya detskaya biblioteka, Vladivostok 24347
Primorskaya kraevaya meditsinskaya biblioteka, Vladivostok 24057
Primorski institut perepodgotovki i porysheniya kvalifikacii rabotnikov obrazovaniya, Vladivostok 22638
Primorski kraevedcheski muzei im. V.K. Arseneva, Vladivostok 24058
Primorski selskokhozyaistvenny institut, Ussurisk 22621
Primorskoe geologicheskoe upravlenie, Vladivostok 22769
Primorsky District Centralized Library System, Central District Children's Library, Sankt-Peterburg 24290
Primorsky District Centralized Library System, Saltykov-Shchedrin Central District Library, Sankt-Peterburg 24289
Primorye Geological Administration, Vladivostok 22769
Prince Consort's Library, Aldershot 29593
Prince Edward Island Government Services Library, Charlottetown 04056
Prince Edward Island Public Library Service, Morell 04696
Prince George Public Library, Prince George 04728
Prince George's Community College, Largo 33477
Prince George's County Circuit Court, Upper Marlboro 34951
Prince George's County Memorial Library System, Hyattsville 39550
Prince George's County Public Schools, Landover 33473
Prince George's County Public Schools – Professional Library, Landover 33474
Prince Henry Hospital, Little Bay 00941
Prince of Songkhla University, Songkhla 27752
– Faculty of Medicine, Songkhla 27753
Prince of Wales Museum of Western India, Mumbai 14100
Prince Philip Hospital, Llanelli 29708
Prince Rupert Library, Prince Rupert 04729
Prince William County Circuit Court, Manassas 34496
Prince William Public Library System, Prince William 40898
Princes Czartoryski Library, Kraków 21127
Princess Grace Irish Library, Monaco 18900
Princess Marina Library, Northampton 29860
Princeton Antiques Bookservice, Atlantic City 36496
Princeton Library in New York, New York 37253
Princeton Public Library, Princeton 40900
Princeton Public Library, Princeton 40901
Princeton Theological Seminary, Princeton 35338
Princeton University, Princeton 32364
– Astrophysics Library, Princeton 32365
– Biology Library, Princeton 32366
– Chemistry Library, Princeton 32367
– Department of Rare Books, Princeton 32368

– Donald E. Stokes Library – Public & International Affairs & Population Research, Princeton 32369
– East Asian Library & Gest Collection, Princeton 32370
– Engineering Library, Princeton 32371
– Fine Hall Library-Mathematics, Physics & Statistics, Princeton 32372
– Geosciences & Maps Library, Princeton 32373
– Graphic Arts Collection, Princeton 32374
– Harold P. Furth Library, Princeton 32375
– Industrial Relations Library, Princeton 32376
– Marquand Library of Art & Archaeology, Princeton 32377
– Near East Collections, Princeton 32378
– Psychology Library, Princeton 32379
– Public Administration Collection, Princeton 32380
– School of Architecture Library, Princeton 32381
Principia College, Elsah 31088
Priokski mezhotraslevoi territorialny tsentr nauchno-tekhnicheskoi informatsii i propagandy, Tula 24005
Priokski mezhotraslevoi territorialny tsentr nauchno-tekhnicheskoi informatsii i propagandy – Orlovski filial, Oryol 23679
Prioksi mezhotraslevoi tsentr nauchno-tekhnicheskoi informatsii i propagandy, Kaluga 23076
Priorova Central Research Institute of Traumatology and Orthopaedics, Moskva 23536
Prirodnjački Muzej, Beograd 24519
Prirodoslovni muzej Slovenije, Ljubljana 24931
Prison Service College Library, Rugby 29532
Pritytsky AZOT Nitrogen Products Industrial Corporation, Grodno 01868
Private Ancestral Temple Library of Wang Ch'ih, Taipei 27617
Private Cheng-Li College of Commerce, Panch'iao 27569
Private Chien Hsing Junior College of Technology, T'aoyüan 27578
Private China College of Marine Technology, Taipei 27574
Private Ching-i Girls' Institute for Literary Composition, Taichung 27571
Private College of Practical Housekeeping, Taipei 27575
Private Mingchih College of Technology, Taipei 27576
Private Sun Yat-Sen Medical School, Taichung 27572
Privates Gymnasium St. Paulusheim, Bruchsal 10713
Privolzhskaya zheleznaya doroga, Saratov 22953
Privy Council Office, Ottawa 04122
Pro Civitate Christiana, Assisi 15716
pro juventute, Zürich 27457
Pro Senectute Schweiz, Zürich 27458
Procter & Gamble Pharmaceuticals, Norwich 36060
Procuradoria-Geral da República, Lisboa 21740
Production Line Manufacture Industrial Corporation – Masherov Automatic Lines Plant, Minsk 01886
Production Line Manufacture Industrial Corporation – Special Technological and Design Bureau, Minsk 01885
Production Line Special Design Bureau, Baranovichi 01821
Productivity Commission Library, Melbourne 00956
'Proektgazoochistka' Joint Stock Company, Sankt-Peterburg 23876
Proektno-izyskatelny i nauchno-issledovatelsky instytut gidroproekt im. S.Ya. Zhuka, Kharkiv 28276
Proektno-konstruktorski institut po industrializatsii montazhnykh rabot (Gipromontazhindustriya), Moskva 23460
Proektno-tekhnologicheski institut organizatsii proizvodstva, upravleniya i ekonomiki montazhnykh i spetsialnykh stroitelnykh rabot, Moskva 23461
Proektny i naukovo-doslidny instytut prombud NDI proekt, Donetsk 28234
Proektny institut Belpromproekt, Minsk 02046
Proektny institut Giprozhivmash, Gomel 01854
Prof. Dr. Iuliu Moldavan Institute of Public Health, Cluj-Napoca 22023
Profabril Projectos, Lisboa 21749
Profesiyne-tekhnichne uchilishche metalistiv no 2, Kyiv 28113
Profesiyne-tekhnichne uchilishche zaliznichnoho transportu no 17, Kyiv 28114

Professional School of Hydrometeorology, Trade Union Library, Kherson 28605
Professional School of Maritime Technology, Sankt-Peterburg 22701
Professional School of the Militia, Sankt-Peterburg 22706
Professionshojskolen UCC, Hillerød 06733
Professionshojskolen UCC, Rønne 06831
Professor Juul Stinissen Bibliotheek, Peer 02842
Professor Paraskev Stoyanov Medical University of Varna, Varna 03484
Profkom beregovykh podrazdeleni i upravleniya Baltiskogo Morskogo parokhodstva – DK Moryakov, Sankt-Peterburg 24245
Profkom metalurginoho zavodu, Donetsk 28549
Profkom orendnoho pidpriemstva Rostok, Kyiv 28684
Profkom virobnichnoho obedinennya im. Artema, Kyiv 28685
Profsoyuz meditsinskikh rabotnikov, Moskva 23462
Profsoyuz rabochikh kommunalno-bytovykh predpriyati, Moskva 23463
Profsoyuz rabochikh mashinos-troeniya, Moskva 23464
Profsoyuz rabochikh stroitelstva i promstroimaterialov, Moskva 23465
Programme des Nations Unies pour le Développement, Libreville 09176
Progres Machine Building Design Office, Zaporizhzhya 28175
Proizvodstvennoe derevoobrabaty-vayushchee obedenie Gomeldrev, Gomel 01855
Proizvodstvennoe elektromashinos-troitelnoe obedinenie Elektrosila, Sankt-Peterburg 22931
Proizvodstvennoe khlopchato-bumazhnoe obedinenie, Baranovichi 02096
Proizvodstvennoe khlopchato-bumazhnoe obedinenie, Baranovichi 01826
Proizvodstvennoe kovrovoe obedinenie, Brest 01843
Proizvodstvennoe kovrovoe obedinenie, Vitebsk 01933
Proizvodstvennoe kozhevenoe obedinenie im. Radishcheva, Sankt-Peterburg 22932
Proizvodstvennoe obuvnoe obedenenie Neman, Grodno 01870
Proizvodstvennoe pryadilno-nitochnoe obedenenie – Bazovaya bibliyateka oblsovprofa, Grodno 02142
Proizvodstvennoe pryadilno-nitochnoe obedinenie – Nauchno-tekhnicheskaya bibliyateka, Grodno 01871
Proizvodstvennoe remontno-naladochnoe predpriyatie Belenergoremnaladka, Minsk 01900
Proizvodstvenny nauchno-issledovatelski institut po inzhenernym izyskaniyam v stroitelstve, Moskva 23466
Prokofiev State Conservatoire, Donetsk 27970
Prokopevsk Mine Automation Equipment Plant, Prokopevsk 22895
Prokopevsk Technical-Mining College, Prokopevsk 22690
Prokopevski gorny tekhnikum, Prokopevsk 22690
Prokopevski zavod shakhtnoi avtomatiki, Prokopevsk 22895
Prokuratura RF – Institut usovershenstvovaniya sledstvennykh rabotnikov, Sankt-Peterburg 23877
Prophylactic Toxicology and Desinfection Research Institute, Moskva 23445
Proskauer Rose LLP, Los Angeles 35854
Prospect Heights Public Library District, Prospect Heights 40904
Prosser Public Library, Bloomfield 38218
Protein and Vitamin Concentrates Plant, Novopolotsk 01913
Protestant Parish of Györ, Győr 13338
Protestant Theological University, Kampen 19147
Protestants Christelijke Hogeschool Marnix Academie, Utrecht 19216
Protestantse Theologische Universiteit, Kampen 19147
Protivochumny nauchno-issledovatelski institut, Rostov-na-Donu 23725
Proudman Oceanographic Laboratory, Liverpool 29706
Providence Care, Kingston 04369
Providence College, Providence 32387
Providence College & Seminary, Otterburne 03817
Providence Historical Society, Tampa 37658
Providence Public Library, Providence 40905
Providence Public Library – Knight Memorial, Providence 40906
Providence Public Library – Smith Hill, Providence 40907
Providence Saint Joseph Medical Center, Burbank 36615
Province Spiritaine de France, Chevilly-Larue 08183
Provincia Franciscana de Castilla, Toledo 25830
Provincia Franciscana de Catalunya, Barcelona 25768
Provincia Napoletana dei Frati Minori Cappuccini, Napoli 15927
Provincia Portuguesa da Companhia de Jesus, Lisboa 21742
Provinciaal Bestuur Noord-Brabant, 's-Hertogenbosch 19275
Provinciaal Bestuur van Friesland, Leeuwarden 19278
Provinciaal Bestuur van Gelderland, Arnhem 19244
Provinciaal Bestuur van Zuid-Holland, Den Haag 19267
Provinciaal Documentatiecentrum Atlas, Antwerpen 02527
Provinciaal Veiligheidsinstituut (PVI), Antwerpen 02528
Provinciaal Veiligheidsinstituut (PVI), ARGUS, het milieupunt van KBC en CERA, Antwerpen 02529
Provincial Archives at Härnosand, Hörnösand 26591
Provincial Archives of Alberta, Edmonton 04343
Provincial Archives of Joensuu, Joensuu 07512
Provincial Archives of Mikkeli, Mikkeli 07531
Provincial Archives of Nothern Jutland, Viborg 06837
Provincial Archives of Vaasa Reference Library, Vaasa 07554
Provincial Archives of Zealand, Lolland-Falster and Bornholm, København 06809
Provincial Assembly of the Punjab, Lahore 20471
Provincial Education Library, Ciechanów 21086
Provincial Government Library, Leeuwarden 19278
Provincial Information & Library Resources Board, St. John's 04764
Provincial Museum of Natural Sciences 'Florentino Ameghino', Santa Fé 00316
Provincial Taichung Library, Taichung 27629
Provinciale Bibliotheek Centrale Noord-Brabant, Tilburg 19712
Provinciale Bibliotheek Centrale voor Drenthe, Assen 19557
Provinciale Bibliotheek & Documentatiecentrum West-Vlaanderen, Brugge 02550
Provinciale Bibliotheek Limburg, Hasselt 02774
Provinciale Hogeschool Limburg, Hasselt 02345
Provincie Groningen, Groningen 19249
Provincie Limburg Centrale Administratieve Bibliotheek (CAB), Hasselt 02444
Provincie Noord-Holland, Haarlem 19273
Provincie Vlaams-Brabant, Leuven 02446
Provinciebestuur Oost-Vlaanderen, Gent 02443
Provinzbibliothek Österreichische Kapuzinerprovinz, Innsbruck 01440
Provinzialbibliothek Amberg (Staatliche Bibliothek), Amberg 09295
Provo City Library, Provo 40909
Provveditorato agli Studi e di Quartiere, Bologna 16198
Prowincjalna Biblioteka Redemptorystów, Tuchów 21060
Prozritelev-Prave Museum of Regional Studies, Stavropol 23982
Pruzhanskaya tsentralnaya raionnaya bibliyateka, Pruzhany 02229
Pruzhany Regional Central Library, Pruzhany 02229
PRVA Gimnazija Maribor, Maribor 24885
Pryadilno-nitochny kombinat im. Kirova, Sankt-Peterburg 22933
Pryadilno-tkatskaya fabrika, Kobrin 01875
Pryadilno-trikotazhnoe obedinenie – Nauchno-tekhnicheskaya bibliyateka, Pinsk 01921
Pryanishnikov Agricultural Institute, Perm 22474
Pryazovsky derzhavny tekhnichny universitet, Mariupol 28042
Pryazovsky State Technical University, Mariupol 28042
Prydniprovska State Academy of Civil Engineering and Architecture, Dnipropetrovsk 27968
Pryor, Cashman, Sherman & Flynn, New York 36011
Prytanée National Militaire, La Flèche 08125
Przemyska Biblioteka Publiczna im. I. Krasickiego, Przemyśl 21537
Przemysłowy Instytut Maszyn Budowlanych, Kobyłka 21119
Przemysłowy Instytut Maszyn Rolniczych, Poznań 20904
Przemysłowy Instytut Motoryzacji, Warszawa 21341
Przemysłowy Instytut Telekomunikacji, Warszawa 21342
Przemystowy Instytut Automatyki i Pomiarów, Warszawa 21343
PSB Information Centre, Singapore 24616
Psikhonevrologicheski internat no 1, Sankt-Peterburg 23878
Pskov Provincial Universal Scientific Library, Pskov 22168
Pskov Regional Scientific Medical Library, Pskov 23707
Pskov State Teacher-Training Institute, Pskov 22482
Pskovskaya oblastnaya detskaya biblioteka im. V.A. Kaverina, Pskov 24203
Pskovskaya oblastnaya nauchnaya meditsinskaya biblioteka, Pskov 23707
Pskovskaya oblastnaya universalnaya nauchnaya biblioteka, Pskov 22168
Pskovski gosudarstvenny pedagogicheski institut, Pskov 22482
Psychiatric and Neurological Clinic no 1, Sankt-Peterburg 23878
Psychiatric Clinic, Växjö 26718
Psychiatric Institute UIC, Chicago 30780
Psychiatric Research Library, Risskov 06829
Psychiatric Services Library, Mount Claremont 00963
Psychiatrisch Centrum Broeders Alexianen, Boechout 02544
Psychiatrische Klinik, Heidelberg 10017
Psychiatrische Klinik, Münsterlingen 27414
Psychiatrische Klinik und Poliklinik, München 10439
Psychico College, Athinai 13090
Psykiatrisk Forskningsbibliotek, Risskov 06829
Psykiatriska Klinikerna, Växjö 26718
PT Bank Negara Indonesia (Persero) TbK, Jakarta 14342
PTI liteinogo proizvodstva, Sankt-Peterburg 23879
PTS – The Paper Technology Specialists, München 12323
Pubblica Biblioteca Arcivescovile 'Francesco Pacca', Benevento 15729
Public and National Library of Greenland, Nuuk 06630
Public Archives of Nova Scotia, Halifax 04360
Public Association of Tokyo Metropolitan Special Wards, Tokyo 17762
Public Catering Professional School, Kyiv 28118
Public City Library 'Władysław Broniewski', Resko 21549
Public Health Institute, Kuala Lumpur 18583
Public Joint-Stock company Chernigov enterprise Khimvolokno, Chernigiv 28136
Public Law Library, Richmond 37443
Public Libraries of Saginaw, Saginaw 41096
Public Libraries of Saginaw – Butman-Fish, Saginaw 41097
Public Libraries of Saginaw – Zauel Memorial Library, Saginaw 41098
Public Library, Arenys de Mar 26201
Public Library, Brasília 03426
Public Library, Debrecen 13533
Public Library, Gaborone 02980
Public Library, Jaffna 26331
Public Library, Jessore 01774
Public Library, Kabul 00005
Public Library, Ljubljana 24967
Public Library, Lomma 26841
Public Library, Mosul 14457
Public Library, Schoten 02854
Public Library, Suwałki 21576
Public Library, Vestmannaeyjar 13585
Public Library, Windhoek 19015
Public Library and Reading Room, Idrija 24959
Public Library "Aprilov-Palauzov", Gabrovo 03593
Public Library Association of Annapolis & Anne Arundel County, Inc, Annapolis 37965
Public Library Association of Annapolis & Anne Arundel County, Inc – Annapolis Area, Annapolis 37966
Public Library Berio, Genova 16749
Public Library Bratislava Nové Mesto, Bratislava 24765
Public Library Domžale, Domžale 24957
Public Library for Union County, Lewisburg 39859
Public Library Jesenice, Jesenice 24961
Public Library Kamp-Lintfort, Kamp-Lintfort 12801
Public Library Kocevje, Kočevje 24962
Public Library Lucerne, Luzern 27494
Public Library of Akmene District Municipality, N. Akmenė 18343
Public Library of Alytus District Municipality, Alytus 18316
Public Library of Anniston & Calhoun County, Anniston 37967
Public Library of Birštonas, Birštonas 18318
Public Library of Birzai District Municipality, Biržai 18319
Public Library of Bronnoy, Brønnøysund 20321
Public Library of Brookline, Brookline 38309
Public Library of Brookline – Coolidge Corner, Brookline 38310
Public Library of Chalkis, Chalkis 13140
Public Library of Chania, Chania 13141
Public Library of Charlotte & Mecklenburg County, Charlotte 38519
Public Library of Charlotte & Mecklenburg County – Independence Regional, Charlotte 38520
Public Library of Charlotte & Mecklenburg County – Matthews Branch, Matthews 40099
Public Library of Charlotte & Mecklenburg County – Mint Hill Branch, Mint Hill 40235
Public Library of Charlotte & Mecklenburg County – Morrison Regional Library, Charlotte 38521
Public Library of Charlotte & Mecklenburg County – North County Regional, Huntersville 39533
Public Library of Charlotte & Mecklenburg County – South County Regional, Charlotte 38522
Public Library of Charlotte & Mecklenburg County – Steele Creek, Charlotte 38523
Public Library of Charlotte & Mecklenburg County – University City Regional, Charlotte 38524
Public Library of Chelm, Chełm 21396
Public Library of Cincinnati & Hamilton County, Cincinnati 38564
Public Library of Cincinnati & Hamilton County – Anderson, Cincinnati 38565
Public Library of Cincinnati & Hamilton County – Blue Ash Branch, Blue Ash 38222
Public Library of Cincinnati & Hamilton County – Bond Hill, Cincinnati 38566
Public Library of Cincinnati & Hamilton County – Covedale, Cincinnati 38567
Public Library of Cincinnati & Hamilton County – Delhi Township, Cincinnati 38568
Public Library of Cincinnati & Hamilton County – Green Township, Cincinnati 38569
Public Library of Cincinnati & Hamilton County – Groesbeck, Cincinnati 38570
Public Library of Cincinnati & Hamilton County – Harrison Branch, Harrison 39375
Public Library of Cincinnati & Hamilton County – Madeira Branch, Madeira 39997
Public Library of Cincinnati & Hamilton County – North Central, Cincinnati 38571
Public Library of Cincinnati & Hamilton County – Sharonville, Cincinnati 38572
Public Library of Cincinnati & Hamilton County – Symmes Township, Loveland 39969
Public LIbrary of Druskininkai Municipality, Druskininkai 18320
Public LIbrary of Elektrenai Municipality, Elektrėnai 18321
Public Library of Ghent, Gent 02765
Public Library of Izola, Izola 24960
Public Library of Jan Bocatius, Košice 24779
Public Library of Johnston County & Smithfield, Smithfield 41335
Public Library of Johnston County & Smithfield – Hocutt-Ellington Memorial, Clayton 38589
Public LIbrary of Jonava District Municipality, Jonava 18325
Public LIbrary of Jurbarkas District Municipality, Jurbarkas 18327
Public LIbrary of Kaisiadorys District Municipality, Kaišiadorys 18328
Public Library of Kalvarija Municipality, Kalvarija 18329
Public LIbrary of Kaunas District Municipality, Garliava 18323
Public LIbrary of Kazlu Ruda Municipality, Kazlų Rūda 18332
Public Library of Klaipeda Municipality, Klaipėda 18336

Public LIbrary of Kupiskis District Municipality, Kupiškis 18338
Public LIbrary of Lazdijai District Municipality, Lazdijai 18339
Public LIbrary of Mazeikiai District Municipality, Mažeikai 18341
Public LIbrary of Moletai District Municipality, Molėtai 18342
Public Library of Mount Vernon & Knox County, Mount Vernon 40326
Public Library of Mytilene, Mitilini 13147
Public Library of Nafpaktos, Nafpaktos 13148
Public LIbrary of Pagegiai Municipality, Pagėgiai 18345
Public Library of Palanga Municipality, Palanga 18347
Public LIbrary of Panevezys District Municipality, Panevėžys 18350
Public LIbrary of Panevezys Municipality, Panevėžys 18349
Public Library of Pireus, Piraius 13149
Public Library of Pirgos, Pirgos 13150
Public LIbrary of Prienai District Municipality, Prienai 18352
Public LIbrary of Radviliskis District Municipality, Radviliškis 18353
Public LIbrary of Raseiniai District Municipality, Raseiniai 18354
Public Library of Rethimnon, Rethimnon 13151
Public LIbrary of Rietavas Municipality, Rietavas 18355
Public LIbrary of Sakiai District Municipality, Šakiai 18358
Public LIbrary of Salcininkai District Municipality, Šalčininkai 18359
Public Library of Selma & Dallas County, Selma 41279
Public LIbrary of Šiauliai District Municipality, Šiauliai 18363
Public LIbrary of Šiauliai Municipality, Šiauliai 18362
Public LIbrary of Sirvintos District Municipality, Širvintos 18366
Public LIbrary of Skuodas Municipality, Skuodas 18367
Public Library of Sparta, Sparta 13153
Public Library of Steubenville & Jefferson County, Steubenville 41441
Public LIbrary of Svencionys District Municipality, Šenčionys 18360
Public LIbrary of Taurage Municipality, Tauragė 18368
Public Library of Thessaloniki, Thessaloniki 13154
Public LIbrary of Trakai District Municipality, Trakai 18370
Public LIbrary of Varena District Municipality, Varėna 18373
Public LIbrary of Vilkaviskis District Municipality, Vilkaviškis 18374
Public LIbrary of Vilnius District Municipality, Rudamina 18357
Public Library of Visaginas, Visaginas 18377
Public Library of Youngstown & Mahoning County, Youngstown 41996
Public Library of Youngstown & Mahoning County – Austintown, Youngstown 41997
Public Library of Youngstown & Mahoning County – Boardman, Youngstown 41998
Public Library of Youngstown & Mahoning County – Poland Branch, Poland 40838
Public Library of Zakynthos, Zakinthos 13156
Public LIbrary of Zarasai District Municipality, Zarasai 18378
Public Library Požega, Požega 24548
Public Library Ružinov, Bratislava 24766
Public Library Services, Hato Rey 21906
Public Library Velenje, Velenje 24989
Public Municipal Library Pedro Fernandes Tomas, Figueira da Foz 21835
Public Record Office, Szczecin 21227
Public Service and Merit Protection Commission, Barton 00611
Public Service Commission Library, Ottawa 04450
Public Service Electric & Gas Company, Hancocks Bridge 35733
Public Service Electric & Gas Company, Newark 36045
Public Technical Library, Muscat 20424
Public Town Library, Bielsk Podlaski 21383
Public University of Navarra, Pamplona 25550
Public Utilities Commission of Ohio, Columbus 36746
Public Utility Commission of Texas Library, Austin 33949
Public Works & Government Services Canada, Hull 04089
Public Works Research Institute, Tsukuba 17786
Publica Biblioteca 'M. Maldotti', Guastalla 16754
Publications and Information Directorate, New Delhi 14166
Publications Division Library, New Delhi 13893
Publichna biblioteka imeny Lesi Ukrainki, Kyiv 28686
Publiczna Biblioteka Pedagogiczna, Poznań 21207
Pueblo City-County Library District, Pueblo 40910
Pueblo City-County Library District – Frank I. Lamb Branch, Pueblo 40911
Pueblo City-County Library District – Frank & Marie Barkman Branch, Pueblo 40912
Puerto Rico Department of Justice, San Juan 21897
Puerto Rico Electric Power Authority, San Juan 21898
Pulaski County Library District, Richland 40974
Pulaski County Public Library, Somerset 41345
Pulaski County Public Library, Winamac 41920
Pulaski County Public Library System, Pulaski 40913
Pulp and Paper Industrial and Engineering Enterprise, Moskva 22858
Pulp and Paper Mill, Svetlogorsk 01929
Pulp & Paper Research Institute of Canada, Pointe-Claire 04456
Pułtuska Biblioteka Publiczna, Pułtusk 21541
Punjab Agricultural University, Ludhiana 13693
Punjab Civil Secretariat Library, Chandigarh 13852
Punjab Public Library, Lahore 20530
Punjab University Library, Lahore 20447
Punjabi University, Patiala 13736
Puppentheatermuseum im Münchner Stadtmuseum, München 12324
Purchase College, Purchase 32401
Purdue University, West Lafayette 33022
– Earth & Atmospheric Sciences Library, West Lafayette 33023
– Humanities, Social Science, and Education Library, West Lafayette 33024
– Life Science Library, West Lafayette 33025
– Management and Economics Library, West Lafayette 33026
– Mathematical Sciences Library, West Lafayette 33027
– M. G. Mellon Library of Chemistry, West Lafayette 33028
– Physics Library, West Lafayette 33029
– Siegesmund Engineering Library, West Lafayette 33030
– Veterinary Medical Library, West Lafayette 33031
Purdue University Calumet Library, Hammond 31281
Purdue University North Central, Westville 33041
Pure Metals Plant, Svitlovodsk 28171
Purple Mountain Observatory, Nanjing 05884
Pusan Civil Library, Pusan 18124
Pusan National University, Busan 18057
Pusan National University, Pusan 18073
Pusan Teachers' College, Pusan 18074
Pusat Budaya Arab-Libya, Kuala Lumpur 18600
Pusat Daya Pengeluaran Negara, Petaling Jaya 18614
Pusat Dokumentasi Melayu, Kuala Lumpur 18601
Pusat Latihan Telekom, Kuala Lumpur 18602
Pusat Pembinaan dan Pengembangan Bahasa, Jakarta 14374
Pusat Penelitian Arkeologi Nasional, Jakarta 14375
Pusat Penelitian Bioteknologi Perkebunan, Bogor 14350
Pusat Penelitian dan Pengembangan Biologi, Bogor 14351
Pusat Penelitian dan Pengembangan Geologi, Bandung 14343
Pusat Penelitian dan Pengembangan Hutan, Bogor 14352
Pusat Penelitian dan Pengembangan Pelayanan Kesehatan, Surabaya 14379
Pusat Penelitian dan Pengembangan Peternakan, Bogor 14353
Pusat Penelitian Kelapa Sawit, Medan 14377
Pusat Penelitian Nuklir Yogyakarta, Yogyakarta 14381
Pusat Penelitian Perkebunan Bogor, Bogor 14354
Pusat Penelitian Perkebunan Gula Indonesia, Pasuruan 14378
Pusat Penyelidikan dan Latihan Bank-Bank Pusat Asia Tenggara (SEACEN), Petaling Jaya 18615
Pusat Perkembangan Kurikulum, Kuala Lumpur 18603
Pusat Perpustakaan Angkatan Darat, Bandung 14330
Pusat Wilayah Bagi Pendidikan Sains dan Matematik, Gelugor 18573
Pusat Wilayah Bagi Pendidikan Sains dan Matematik (RECSAM), Penang 18613
Dr. Pusey Memorial Library, Oxford 29873
Pushchino Radio Astronomy Observatory, Pushchino 23709
Pushchinskaya radioastronomich-eskaya observatoriya, Pushchino 23709
Pushkin Central City Library, Novocherkassk 24187
Pushkin Central City Library, Shakhty 24318
Pushkin City Central Children's Library, Sankt-Peterburg 24307
Pushkin District Centralized Library System, Central District Children's Library, Sankt-Peterburg 24295
Pushkin District Centralized Library System, Library Branch no 1, Sankt-Peterburg 24291
Pushkin District Centralized Library System, Library Branch no 2, Sankt-Peterburg 24292
Pushkin District Centralized Library System, Library Branch no 3, Sankt-Peterburg 24293
Pushkin District Centralized Library System, Mamin-Sibiryak Central District Library, Sankt-Peterburg 24294
A. S. Pushkin District Library, Minsk 02188
A. S. Pushkin District Library, Samarkand 42175
Pushkin Library of the District of Tambov, Tambov 22178
Pushkin Library of the District of Tomsk, Tomsk 22179
Pushkin Library of the Mordvinian Republic, Saransk 22173
Pushkin Minsk Regional Library, Minsk 02182
Pushkin Mogilev Central City Children's Library, Mogilev 02204
Pushkin Museum, Sankt-Peterburg 23811
A.S. Pushkin National Library of Republic Tuva, Kyzyl 22148
Pushkin Orsha Central City Library, Orsha 02217
Pushkin Regional Library for Children and Juvenile, Saratov 24314
Pushkin State Museum, Moskva 23225
Pushkin State Museum of Fine Arts, Moskva 23226
Pushkin State Preserve, Mikhailovskoe 23159
Pushkin State Teacher Training Institut, Kirovograd 28005
Pushkin Vitebsk City Library, Vitebsk 02244
Pushkinski gosudarstvenny zapovednik, Mikhailovskoe 23159
Puskarich Public Library, Cadiz 38408
Püspöki Könyvtár, Székesfehérvár 13346
Püspöki Papnevelö Intézet, Győr 13275
Pustaka Negeri Sarawak, Kuching 18630
Putian College, Putian 05481
Putnam County District Library, Ottawa 40664
Putnam County Library, Hurricane 39546
Putnam County Public Library, Greencastle 39295
Putnam Museum of History & Natural Science, Davenport 36770
Putnamville Correctional Facility, Greencastle 34318
Putney, Twombly, Hall & Hirson, New York 36012
Puyallup Public Library, Puyallup 40915
Pyatigorsk State Linguistic University Library, Pyatigorsk 22485
Pyrometr Plant, Sankt-Peterburg 22949
Qaqortumi Atuakkanik Atorniatarfik / Det gule Bibliotek, Qaqortoq 06915
Qatar National Library, Doha 21907
QinetiQ Winfrith Information Centre, Dorchester 29646
Qingdao Institute of Chemical Technology, Qingdao 05483
Qingdao Library, Qingdao 05923
Qingdao Medical College, Qingdao 05484
Qingdao Technological University, Qingdao 05485
Qingdao University, Qingdao 05486
Qinghai Education College, Xining 05713
Qinghai Institute of Animal Husbandry and Veterinary Medicine, Xining 05714
Qinghai Institute of Salt Lakes, Xining 05905
Qinghai Medical College, Xining 05715
Qinghai Nationalities College, Xining 05716
Qinghai Normal University, Xining 05717
Qinghai Provincial Library, Xining 04977
Qinghai Teachers' College, Xining 05718
Qinghai University, Xining 05719
Qingyang Teachers Training College, Xifeng 05711
Qiqihaer Light Industry Institute, Qiqihaer 05490
Quaid-i-Azam Academy, Karachi 20506
Quaid-i-Azam Library, Lahore 20517
Quaid-i-Azam University, Islamabad 20432
Quanzhou Teachers' College, Quanzhou 05492
Quarles & Brady, Milwaukee 35897
Quatrefoil Library, Saint Paul 37504
Quatrefoil Library, Saint Paul 37505
Queanbeyan Palerang Library Service, Queanbeyan 01147
Québec Centre d'Information sur L'Environment, la Faune et les Parcs, Québec 04136
Québec Conseil de la Science et de la Technologie, Sainte-Foy 04484
Québec Ministère de l'Agriculture des Pêcheries et de l'Alimentation, Québec 04137
Québec Ministère de l'Education, Québec 04138
Québec Ministère des Affaires Municipales, Québec 04139
Québec Ministère des Transports, Montreal 04100
Québec Ministère des Transports, Québec 04140
Quebec Ministry of Transportation / Québec Ministère des Transports, Montreal 04100
Québec Office de la Langue Française, Montreal 04418
Queen Anne's County Free Library, Centreville 38496
Queen Elizabeth College, Rose Hill 18692
Queen Margaret University, Musselburgh 29276
Queen Mary and Westfield College, London 29247
Queen Maud's College of Early Childhood Education, Trondheim 20170
Queen of the Holy Rosary College, Fremont 33355
Queen Victoria Museum and Art Gallery, Launceston 00938
Queens Borough Public Library, Jamaica 39626
Queens Borough Public Library – Auburndale, Flushing 39081
Queens Borough Public Library – Bay Terrace, Bayside 38121
Queens Borough Public Library – Bayside Branch, Bayside 38122
Queens Borough Public Library – Bellerose Branch, Bellerose 38151
Queens Borough Public Library – Briarwood Branch, Briarwood 38272
Queens Borough Public Library – Broadway, Long Island City 39930
Queens Borough Public Library – Business, Science & Technology Division, Jamaica 39627
Queens Borough Public Library – Douglaston-Little Neck Branch, Little Neck 39896
Queens Borough Public Library – East Elmhurst Branch, East Elmhurst 38894
Queens Borough Public Library – Elmhurst Branch, Elmhurst 38966
Queens Borough Public Library – Far Rockaway Branch, Far Rockaway 39040
Queens Borough Public Library – Fine Arts & Recreation Division, Jamaica 39628
Queens Borough Public Library – Flushing Branch, Flushing 39082
Queens Borough Public Library – Forest Hills Branch, Forest Hills 39094
Queens Borough Public Library – Fresh Meadows Branch, Fresh Meadows 39165
Queens Borough Public Library – Glen Oaks Branch, Glen Oaks 39240
Queens Borough Public Library – Glendale Branch, Glendale 39248
Queens Borough Public Library – Hillcrest, Flushing 39083
Queens Borough Public Library – Howard Beach Branch, Howard Beach 39523
Queens Borough Public Library – Jackson Heights Branch, Jackson Heights 39607
Queens Borough Public Library – Kew Gardens Hills, Flushing 39084
Queens Borough Public Library – Laurelton Branch, Laurelton 39818
Queens Borough Public Library – Lefferts, Richmond Hill 40990
Queens Borough Public Library – Lefrak City, Corona 38700
Queens Borough Public Library – Literature & Languages Division, Jamaica 39629

Queens Borough Public Library – Maspeth Branch, Maspeth 40093
Queens Borough Public Library – McGoldrick, Flushing 39085
Queens Borough Public Library – Middle Village Branch, Middle Village 40175
Queens Borough Public Library – Mitchell-Linden, Flushing 39086
Queens Borough Public Library – North Forest Park, Forest Hills 39095
Queens Borough Public Library – Ozone Park Branch, Ozone Park 40680
Queens Borough Public Library – Peninsula, Rockaway Beach 41038
Queens Borough Public Library – Pomonok, Flushing 39087
Queens Borough Public Library – Queens Village Branch, Queens Village 40917
Queens Borough Public Library – Queensboro Hill, Flushing 39088
Queens Borough Public Library – Rego Park Branch, Rego Park 40959
Queens Borough Public Library – Richmond Hill Branch, Richmond Hill 40991
Queens Borough Public Library – Ridgewood Community, Ridgewood 40998
Queens Borough Public Library – Seaside Community, Rockaway Park 41039
Queens Borough Public Library – Social Sciences Division, Jamaica 39630
Queens Borough Public Library – Steinway, Long Island City 39931
Queens Borough Public Library – Sunnyside, Long Island City 39932
Queens Borough Public Library – Whitestone Branch, Whitestone 41884
Queens Borough Public Library – Windsor Park, Bayside 38123
Queens Borough Public Library – Woodhaven Branch, Woodhaven 41960
Queens Borough Public Library – Woodside Branch, Woodside 41966
Queens Borough Public Library – Youth Services Division, Jamaica 39631
Queens' College, Cambridge 29040
Queens College, Charlotte 30721
Queen's College, Lagos 19961
Queen's College, Oxford 29341
Queen's College, St. John's 03870
Queens College – Aaron Copeland School of Music, Flushing 31148
Queens College – Benjamin S. Rosenthal Library, Flushing 31149
Queens County Supreme Court Library, Jamaica 34395
Queens County Supreme Court Library, Kew Gardens 34414
Queen's Foundation, Birmingham 28952
Queen's University, Kingston 03728
– Law Library, Kingston 03729
Queen's University Belfast, Belfast 28939
– AFBI Library, Belfast 28940
– Northern Ireland Health and Social Services, Belfast 28941
Queensborough Community College, Bayside 33158
Queensburgh Public Library, Queensburgh 25229
Queensland Art Gallery Research Library, South Brisbane 01004
Queensland Baptist College of Ministries, Brookfield 00755
Queensland Braille Writing Association, Annerley 00863
Queensland Bureau of Employment, Vocational and Further Eduaction and Training, Brisbane 00623
Queensland Conservatorium of Music, Maastricht 19173
Queensland Conservatorium of Music, South Brisbane 00490
Queensland Department of Child Safety, Brisbane 00624
Queensland Department of Education, Training and the Arts, Coorparoo DC 00659
Queensland Department of Education, Training and the Arts, Coorparoo DC 00660
Queensland Department of Justice and Attorney-General, Brisbane 00625
Queensland Department of Natural Resources, Brisbane 00626
Queensland Department of Premier & Cabinet, City East 00657
Queensland Department of Primary Industries and Fisheries, Brisbane 00627
Queensland Department of Public Works and Housing, Brisbane 00628
Queensland Health Central Library, Brisbane 00629
Queensland Herbarium, Toowong 01034
Queensland Industrial Court and AIRC, Brisbane 00630
Queensland Museum, South Brisbane 01005
Queensland – Parliamentary Library, Brisbane 00631
Queensland Parliamenty Library, Brisbane 00632
Queensland Police Service, Archerfield 00864
Queensland Teachers Union, Milton 00960
Queensland University of Technology, Brisbane 00398
– Law Library, Brisbane 00399
Qufu Normal University, Qufu 05493
Quincy College, Quincy 32404
Quincy University, Quincy 32405
Quinebaug Valley Community College, Danielson 33277
Quinnipiac University, Hamden 31277
– School of Law Library, Hamden 31278
Quinsigamond Community College, Worcester 33884
Quinte West Public Library, Trenton 04803
Qujing Teachers' College, Qujing 05494
R. A. Podar College of Commerce and Economics, Mumbai 13824
R. T. Vanderbilt Co, Inc, Norwalk 36059
R. W. Johnson Pharmaceutical Research Institute, Raritan 37431
R. W. Norton Art Gallery, Shreveport 37604
Raad van State, Den Haag 19268
RAAF College Library, Point Cook 00718
Raahe Public Library, Raahe 07602
Raahen kaupunginkirjasto, Raahe 07602
Rabaul Public Library, Rabaul 20600
Rabbi L.A. Falk Memorial Library, Sydney 00797
Rabbinical College of America, Morristown 31907
Rabindra Bharati University, Kolkata 13685
Rabindra Bhavana Tagore Museum and Archives, Santiniketan 14209
Raccolte Storiche del Comune di Milano, Milano 16349
Racine Correctional Institution Library, Sturtevant 34901
Racine County Law Library, Racine 34718
Racine Public Library, Racine 40923
Rackemann, Sawyer & Brewster Library, Boston 35536
Racquet and Tennis Club Library, New York 37254
RACV Club, Melbourne 00957
Raczyńscy Library, Poznań 21532
Rada po vivchennyu produktivnykh sil Ukrainy, Kyiv 28382
Ráday Collections of the Danubian District of the Hungarian Reformed Church, Budapest 13326
Radboud Universiteit Nijmegen, Nijmegen 19176
– Bibliotheek Gedragswetenschappen, Nijmegen 19177
– Bibliotheek Managementwetenschappen, Nijmegen 19178
– Bibliotheek Medische Wetenschappen, Nijmegen 19179
– Bibliotheek Rechtsgeleerdheid, Nijmegen 19180
– Bibliotheek Sociaal-Culturele Wetenschappen, Nijmegen 19181
– Instituut voor Oosters Christendom, Nijmegen 19182
– Katholiek Documentatie Centrum (KDC), Nijmegen 19183
– Library of Science, Nijmegen 19184
– Studiecentrum-Bibliotheek Tandheelkunde, Nijmegen 19185
Radcliffe Institute for Advanced Study, Cambridge 30662
Radekhiv Regional Centralized Library System, Main Library, Radekhiv 28792
Radekhivska raionna TsBS, Tsentralna biblioteka, Radekhiv 28792
Radford Public Library, Radford 40924
Radford University, Radford 32406
Radiation Effects Research Foundation, Hiroshima 17549
Radiator Factory Joint-Stock Company, Orenburg 22884
Radio 2 Omroep Antwerpen, Antwerpen 02530
Radio and Television Slovenia, Ljubljana 24932
Radio Bremen, Bremen 11686
Radio Equipment Research Institute, Sankt-Peterburg 23942
Radio France, Paris 08616
Radio New Zealand, Wellington 19854
Radio Technology Research Institute, Sankt-Peterburg 23796
Radio Telefís Eireann, Dublin 14507
Radio Telefís Eireann (RTE), Dublin 14553
Radio Television Ireland, Dublin 14553
Radioastronomichny instytut, Kharkiv 28277
Radiodiffusion du Sénégal, Dakar 24446
Radiodiffusion-Télévision Française, Paris 08617
Radioelectronic Equipment Plant, Sankt-Peterburg 22951
Radioelectronics Professional School, Kyiv 28120
Radioisotopes Centre POLATOM, Otwock-Świerk 21186
Radiology and Oncology Research Institute, Kyiv 28373
Radionavigation and Time Measurement Research Institute, Sankt-Peterburg 23842
Radiophysical Research Institute, Nizhni Novgorod 23620
Radiophysics and Elektronics Research Institute, Kharkiv 28260
Radiotekhnicheski institut, Ryazan 22496
Radiotelevisão Portuguesa, EP, Lisboa 21815
Radio-Télévision Gabonaise (RTG), Libreville 09177
The Radishchev Art Museum, Saratov 23964
Radishchev Leather Corporation, Sankt-Peterburg 22932
Radlje ob Dravi Library, Radlje ob Dravi 24979
Radomishl Regional Centralized Library System, Main Library, Radomishl 28793
Radomishlska raionna TsBS, Tsentralna biblioteka, Radomishl 28793
Radviliškio rajono savivaldybės viešoji biblioteka, Radviliškis 18353
Radyansk Region Central Library System of the City of Kiev, M. Kostomarov Main Library, Kyiv 28687
Radyanska raionna TsBS m. Kyiva, Tsentralna biblioteka im. M. Kostomarova, Kyiv 28687
Ragusan Press, San Carlos 37523
Rahway Public Library, Rahway 40925
Railroad Commission of Texas Library, Austin 33950
Railway Board, New Delhi 13894
Railway Carriage Works Joint-Stock Company, Tver 22965
Railway Construction Professional School, Kyiv 28121
Railway Department of the Transport, Traffic and Post Ministry, Budapest 13324
Railway Equipment Joint-Stock Company, Tula 22963
Railway Library, Ekaterinburg 22790
Railway Library, Sankt-Peterburg 22763
Railway Museum, Beograd 24528
Railway Plant, Sofiya 03499
Railway Scientific and Technical Center, Warszawa 21252
Railway Technical Research Institute of the Japanese National Railways, Kokubunji 17578
Railway Transport Professional School, Kyiv 28122
Railway Transport Professional School no 17, Kyiv 28114
Railway Transportation University, Moskva 22397
Railway Wagons Industrial Corporation, Trade Union Library, Chasiv Yar 28512
Rainbow City Public Library, Rainbow City 40926
Raionna TsBS, Tsentralna biblioteka, Dubno 28552
Raionnoe energeticheskoe upravlenie, Brest 01947
Raionnoe energeticheskoe upravlenie, Gomel 01956
Raionnoe energeticheskoe upravlenie, Grodno 01964
Raittiuskeskuskirjasto, Helsinki 07485
Raja Melewar Teachers' Training College, Seremban 18546
Rajamangala Institute of Technology, Bangkok 27705
Rajasthan Agricultural University, Bikaner 13622
Rajasthan, Directorate of Economics and Statistics, Jaipur 13856
Rajasthan High Court, Jodhpur 13857
Rajavithi Hospital, Bangkok 27786
Rajendra Agricultural University, Samastipur 13758
Rajshahi University, Rajshahi 01752
Rakhiv Regional Centralized Library System, Main Library, Rakhiv 28794
Rakhivska raionna TsBS, Tsentralna biblioteka, Rakhiv 28794
Rakhmaninov Conservatoire, Rostov-na-Donu 22489
Rakhmaninova School of Music, Tambov 22720
Rákóczi Múzeum, Sárospatak 13497
Rakuno Gakuen Daigau, Ebetsu 16914
Ralph M. Freeman Memorial Library for the US Courts, Detroit 34147
Sri Ram College of Commerce, Delhi 13802
Ramaano Mbulaheni Media Centre, Levubu 25032
Ramakrishna Mission, New Delhi 13899
Ramakrishna Mission Institute of Culture, Kolkata 14057
Raman Research Institute, Bangalore 13942
Ramapo College of New Jersey, Mahwah 31738
Ramat Polytechnic, Maiduguri 19966
Rambam Library, Tel-Aviv 14701
Ramey & Flock, PC, Tyler 36316
Ramkamhaeng University, Bangkok 27706
– Bangna Campus, Bangkok 27707
Ramon e Betances Centro Americo Medical, Mayaguez 21902
Rampart Public Library District, Woodland Park 41963
Ramsey County Law Library, Saint Paul 34785
Ramsey Free Public Library, Ramsey 40934
Ranchi University, Ranchi 13752
Rancho Cucamonga Public Library, Rancho Cucamonga 40935
Rancho Mirage Public Library, Rancho Mirage 40936
Rancho Santa Ana Botanic Garden Library, Claremont 36702
Rand Club, Johannesburg 25135
Rand Corporation Library, Santa Monica 37577
Rand Library, Arlington 36482
Rand McNally, Skokie 37613
Rand Water Board Library, Johannesburg 25136
Randburg Public Library, Randburg 25230
Randers Bibliotek, Randers 06916
Randolph Circuit Court, Winchester 35075
Randolph College, Lynchburg 31698
Randolph Community College, Asheboro 33133
Randolph Public Library, Asheboro 37999
Randolph Township Free Public Library, Randolph 40939
Randolph-Macon College, Ashland 30266
Random House Group, Rushden 29581
Randwick City Library and Information Service, Maroubra 01121
Ranger College, Ranger 33674
Rangeview Library District, Thornton 41554
Rangeview Library District – Brighton Branch, Brighton 38286
Rangeview Library District – Northglenn Branch, Northglenn 40538
Rangsit University, Pathum Thani 27749
Rani Durgavati Vishwavidyalaya, Jabalpur 13665
Rannsóknastofnun Landbúnadarins, Reykjavík 13573
Rantoul Public Library, Rantoul 40941
Rapid City Public Library, Rapid City 40942
Rapid City Regional Hospital, Rapid City 37430
Rapides Parish Library, Alexandria 37907
Rapla Keskraamatukogu, Rapla 07255
Rappahannock Community College, Warsaw 33838
Raritan Valley Community College, Brandenburg 33184
Raseinių rajono savivaldybės viešoji biblioteka, Raseiniai 18354
Rashtrasant Tukadoji Maharaj Nagpur University, Nagpur 13717
– Mahatma Gandhi Institute of Medical Science, Wardha 13718
Rasporyaditelnaya direktsiya Gosteleradio Belarussii, Minsk 02047
Rat für Formgebung / German Design Council, Frankfurt am Main 11869
Ratha-Sapha, Bangkok 27767
Rathausbücherei der Landeshauptstadt Stuttgart, Stuttgart 11108
Rathgen-Forschungslabor, Berlin 11580
Rats- und Konsistorialbibliothek im Stadtarchiv, Rothenburg ob der Tauber 12446
Ratsbücherei Lüneburg, Lüneburg 12838
Ratsschulbibliothek Zwickau, Zwickau 12625
Raufoss ammunisjonsfabrikker A/S, Raufoss 20206
Rauma folkebibliotek, Åndalsnes 20306
Rauma Public Library, Rauma 07603
Rauman kaupunginkirjasto, Rauma 07603
Rautaruukki Oy, Information Services, Raahe 07428
Rautaruukki Oy, Tietopalvelu, Raahe 07428
Rautaruukki Oyj, Raahe 07429
Rautenstrauch-Joest Museum, Kulturen der Welt, Köln 12150
Ravensbourne College of Design and Communication, London 29209
Ravne d.o.o., Ravne na Koroškem 24906
Rawalpindi Government College of Technology, Rawalpindi 20466

Rawle & Henderson, Philadelphia 36108
Rawlins Municipal Library, Pierre 40785
Ray County Library, Richmond 40985
Ray Quinney & Nebeker PC, Salt Lake City 36213
Raychem Corp, Menlo Park 35881
Raymond Chabot Grant Thornton Library, Montreal 04269
Raymond Walters College, Cincinnati 30810
Raytheon Co, El Segundo 35691
Raytheon Co, Portsmouth 36151
Raytheon Co, Tewksbury 36303
Raytheon Systems Co, Dallas 35654
Raytheon Technical Library, Tucson 36313
Razi Vaccine and Serum Research Institute, Tehran 14436
A. M. Razmadze Mathematical Institute, Tbilisi 09273
La Razón, Buenos Aires 00226
rbb Rundfunkanstalt Berlin-Brandenburg, Berlin 11557
RCVS Charitable Trust Library, London 29210
Reader's Digest Association Inc Library, Pleasantville 36137
Reading Area Community College, Reading 33676
Reading Borough Libraries, Reading 30062
Reading Public Library, Reading 40947
Reading Public Library, Reading 40948
Real Academia de Bellas Artes de San Fernando, Madrid 26085
Real Academia de Ciencias, Bellas Letras y Nobles Artes, Córdoba 25945
Real Academia de Ciencias Exactas, Físicas y Naturales, Madrid 26086
Real Academia de Ciencias Morales y Políticas, Madrid 26087
Real Academia de Ciencias y Artes de Barcelona, Barcelona 25929
Real Academia de Farmacia, Madrid 26088
Real Academia de Jurisprudencia y Legislación, Madrid 26089
Real Academia de la Historia, Madrid 26090
Real Academia Española de la Lengua, Madrid 26091
Real Academia Gallega, A Coruña 25951
Real Academia Nacional de Medicina, Madrid 26092
Real Academia Provincial de Bellas Artes, Cádiz 25941
Real Academia Sevillana de Buenas Letras, Sevilla 26156
Real Biblioteca del Monasterio, San Lorenzo del Escorial 26133
Real Casa de la Moneda, Madrid 25716
Real Centro Universitario Escorial – María Cristina, San Lorenzo del Escorial 25565
Real Colegio de San Clemente de los Españoles, Bologna 14853
Real Conservatorio Profesional de Música "Manuel de Falla" de Cádiz, Cádiz 25343
Real Conservatorio Superior de Música de Madrid, Madrid 25426
Real Consulado de A Coruña, A Coruña 25692
Real Gabinete Português de Leitura, Rio de Janeiro 03370
Real Instituto de Estudios Asturianos, Oviedo 26109
Real Instituto y Observatorio de la Armada, San Fernando 26131
Real Jardín Botánico, Madrid 26093
Real Monasterio de San Agustin, Burgos 25940
Real Sociedad Económica de Amigos del País de Tenerife, La Laguna 25965
Real Sociedad Económica Extremeña de Amigos del País de Badajoz, Badajoz 25874
Real Sociedad Económica Matritense de Amigos del País, Madrid 26094
Real Sociedad Económica Sevillana de Amigos del País, Sevilla 26157
Real Sociedad Española de Historia Natural, Madrid 26095
Real Sociedad Geográfica, Madrid 26096
RECAST Documentation Research Centre for Applied Science and Technology, Kirtipur 19043
Rechitsa Central City Library, Rechitsa 02230
Rechitsa Oil and Gas Extracting Industry Regional Administration, Rechitsa 02074
Rechitsa Timber Industrial Corporation, Rechitsa 02075
Rechitsaskaya tsentralnaya gorodskaya bibliyateka, Rechitsa 02230
Rechtsanwaltskammer Wien, Wien 01649
Reconstructionist Rabbinical College, Wyncote 33101
Recording for the Blind Dyslexic (RFB&D), Princeton 37415
Red Bank Public Library, Red Bank 40950
Red Champions Machine Tools Plant, Orsha 01916
Red de Bibliotecas Municipales de Bilbao, Bilbao 26204
Red Deer College, Red Deer 04019
Red Deer Public Library, Red Deer 04731
Red River Community College, Winnipeg 04048
Red Rocks Community College – Marvin Buckels Library, Lakewood 33470
Red Technológica Nacional (RTN), México 18862
Red Wing Public Library, Red Wing 40952
Redcar and CLeveland Borough Council, Redcar 30063
Redcar Central Library, Redcar 30063
Reddick Library, Ottawa 40665
Redeemer College, Ancaster 03670
Redemptorist Community Marianella, Dublin 14512
Redemptorist Seminary, Kew 00772
Redemptoristen-Kollegium, Attnang 01417
Redford Township District Library, Redford 40954
Redland Libraries, Cleveland 01086
Redondo Beach Public Library, Redondo Beach 40957
Redwood Library & Athenaeum, Newport 37284
Reed College, Portland 32342
Reed Elsevier New Providence Library, New Providence 35952
Reed Memorial Library, Ravenna 40943
Reed Smith, New York 36013
Reed, Smith, Hazel & Thomas, Falls Church 35701
Reed Smith LLP, Oakland 36066
Reed Smith LLP, Philadelphia 36109
Reed Smith LLP, Pittsburgh 36132
Reed Smith LLP, Washington 36373
Reedsburg Public Library, Reedsburg 40958
Reelfoot Regional Library Center, Martin 40081
Reference Library of Oulu Provincial Archives, Oulu 07533
Reference Library of the Provincial Archives of Hämeenlinna, Hämeenlinna 07439
Reference Library of Turku Provincial Archives, Turku 07550
Reform Episcopal Seminary, Blue Bell 35140
Református Egyházközség, Kecskemét 13339
Reformed Church Library, Kecskemét 13339
Reformed College of Ministries, Saint Lucia 00793
Reformed Episcopal Seminary, Blue Bell 35141
Reformed Presbyterian Theological Seminary, Pittsburgh 35335
Reformed Theological College, Waurn Ponds 00805
Reformed Theological Seminary, Jackson 35243
Reformed Theological Seminary, Oviedo 35321
Reformed University of Theology Library, Debrecen 13273
Regent College, Vancouver 03931
Regent University, Virginia Beach 32936
– Law Library, Virginia Beach 32937
Regent's College, London 29211
Regeringskansliet, Stockholm 26509
Regeringskansliets Bibliotek, Stockholm 26510
Regierung von Oberbayern, München 11048
Regierungspräsidium Freiburg, Freiburg 10903
Regierungspräsidium Freiburg, Abt. 9 – Landesamt für Geologie, Rohstoffe und Bergbau, Freiburg 10904
Regierungspräsidium Gießen, Gießen 10906
Regierungspräsidium Kassel, Kassel 10963
Regierungspräsidium Stuttgart, Landesamt für Denkmalpflege, Esslingen am Neckar 10893
Regierungspräsidium Tübingen, Referat 26 Denkmalpflege, Tübingen 11114
Regina Health District, Regina 04469
Regina Public Library, Regina 04732
Reginald J. P. Dawson Library / Bibliothèque Ville de Mont-Royal, Mont Royal 04689
Région Alsace, Service l'Inventaire du Patrimoine culturel, Strasbourg 08687
Région du Bibliothèque Haut-Saint-Jean, Saint Basile 04741
Region of Waterloo Library, Baden 04573
Regionaal Archief, Alkmaar 19342
Regionaal Archief Leiden, Leiden 19481
Regionaal Archief Nijmegen (L 320), Nijmegen 19500
Regionaal Archief Tilburg, Tilburg 19523
Regionaal Historisch Centrum Limburg, Maastricht 19490
Regionaalarchief Zutphen – Stedelijke Bibliotheek Zutphen, Zutphen 19540
Regional Advanced Teacher Training Institute, Rivne 28069
Regional and City Library, Nowy Tomyśl 21511
The Regional Archives in Gothenborg, Göteborg 26579
Regional Archives in Lund, Lund 26614
Regional Archives in Vadstena, Vadstena 26711
Regional Archives in Visby, Visby 26720
Regional Bunin Public Library, Oryol 22163
Regional Center for Adult Education in the Arab World, Menoufia 07193
Regional Centre for Education in Science and Mathematics, Gelugor 18573
Regional Centre for Education in Science and Mathematics, Penang 18613
Regional Children's Hospital, Library Branch no 19, Sankt-Peterburg 23865
Regional Children's Library, Cherkasy 28517
Regional Children's Library, Chernigiv 28520
Regional Children's Library, Chernivtsi 28524
Regional Children's Library, Kharkiv 28602
Regional Children's Library, Kherson 28608
Regional Children's Library, Khmelnitski 28611
Regional Children's Library, Lugansk 28703
Regional Children's Library, Rivne 28798
Regional Children's Library, Sankt-Peterburg 24241
Regional Children's Library, Sumy 28835
Regional Children's Library, Ternopil 28843
Regional Children's Library of the Krasnodar Region, Krasnodar 24150
Regional Council of Veneto, Venezia 15694
Regional Culture Studies Professional School, Sankt-Peterburg 22703
Regional Engineering College, Durgapur 13804
Regional Engineering College, Kozhikode 13689
Regional Engineering College, Rourkela 13756
Regional Engineering College, Srinagar 13767
Regional Engineering College, Tiruchirapalli 13773
Regional Engineering College, Warangal 13848
Regional Historical Museum, Pleven 03511
Regional Hospital, Brest 01946
Regional Hospital, Karaganda 17969
Regional Hospital, Uralsk 17979
Regional Institute for Protection of Cultural Monuments in Dalmatia, Split 06226
Regional Institute of Advanced Teacher Training, Zaporizhzhya 28087
Regional Library, Burgas 03592
Regional Library, Haskovo 03594
Regional Library, Kärdzhali 03595
Regional Library, Karlovy Vary 06577
Regional Library, Kyustendil 03596
Regional Library, Lovech 03597
Regional Library, Montana 03598
Regional Library, Murska Sobota 24973
Regional Library, Namangan 42174
Regional Library, Pernik 03599
Regional Library, Pleven 03600
Regional Library, Racibórz 21543
Regional Library, Ruse 03601
Regional Library, Shumen 03602
Regional Library, Silistra 03603
Regional Library, Słupsk 21561
Regional Library, Taldy-Kurgan 17992
Regional Library, Targovishte 03606
Regional Library, Ternopil 28841
Regional Library, Urgench 42178
Regional Library, Vidin 03609
Regional Library, Vratsa 03610
Regional Library, Yogyakarta 14384
Regional Library Karviná, Karviná 06578
Regional Library – Madrid, Madrid 25251
Regional Library 'N. Vranchev', Smolyan 03604
Regional Library of Eastern Uusimaa, Porvoo 07601
Regional Library of Lapland, Rovaniemi 07605
Regional Library of Science and Technology (RLST), Shiraz 14397
Regional Library of the Highlands, Havlíčkův Brod 06570
Regional Local Lore Museum, Donetsk 28233
Regional Maritime Academy, Accra 13024
Regional Medical Library, Cherkasy 28188
Regional Medical Library, Chernigiv 28192
Regional Medical Library, Chernivtsi 28196
Regional Medical Library, Chita 23021
Regional Medical Library, Dnipropetrovsk 28206
Regional Medical Library, Kaliningrad 23070
Regional Medical Library, Khmelnitski 28290
Regional Medical Library, Kursk 23136
Regional Medical Library, Novosibirsk 23654
Regional Medical Library, Orenburg 23672
Regional Medical Library, Oryol 23677
Regional Medical Library, Ryazan 23733
Regional Medical Library, Samara 23743
Regional Medical Library, Smolensk 23972
Regional Medical Library, Tver 24011
Regional Medical Library, Volgograd 24065
Regional Medical Library, Yaroslavl 24088
Regional Medical Research Library, Vinnytsya 28480
Regional Museum, Celje 24908
Regional Museum, Maribor 24949
Regional Museum, Poreč 06215
Regional Museum of History, Pazardjik 03510
Regional Museum of History, Ruse 03517
Regional Museum of History, Varna 03586
Regional Museum of History, Veliko Tărnovo 03587
Regional Museum of History, Vidin 03590
Regional Museum of Teplicích, Teplice 06541
Regional Music and Theatrical Library, Kyiv 28359
Regional Pedagogical Scientific Library, Lviv 28038
Regional Power Engineering Administration, Brest 01947
Regional Power Engineering Administration, Grodno 01964
Regional Public Library, Čadca 24771
Regional Public Library, Dolný Kubín 24772
Regional Public Library, Dunajská Streda 24774
Regional Public Library, Humenné 24776
Regional Public Library, Komárno 24777
Regional Public Library, Lučenec 24782
Regional Public Library, Michalovce 24784
Regional Public Library, Nitra 24785
Regional Public Library, Pezinok 24787
Regional Public Library, Poprad 24788
Regional Public Library, Považská Bystrica 24789
Regional Public Library, Prešov 24790
Regional Public Library, Rimavská Sobota 24792
Regional Public Library, Rožňava 24793
Regional Public Library, Senica 24794
Regional Public Library, Spišská Nová Ves 24795
Regional Public Library, Stará Ľubovňa 24796
Regional Public Library, Topoľčany 24797
Regional Public Library, Veliko Tărnovo 03608
Regional Public Library, Veľký Krtiš 24801
Regional Public Library, Vranov na Topľou 24802
Regional Public Library of Ludovit Stur in Zvolen, Zvolen 24805
Regional Research Laboratory, Hyderabad 14009
Regional Research Laboratory, Jorhat 14018
Regional Research Laboratory Library (CSIR), Jammu 14015
Regional Scientific Medical Library, Donetsk 28232
Regional Scientific Medical Library, Ekaterinburg 23034
Regional Scientific Medical Library, Ivano-Frankivsk 28241
Regional Scientific Medical Library, Kherson 28289
Regional Scientific Medical Library, Kirovograd 28292
Regional Scientific Medical Library, Lugansk 28400
Regional Scientific Medical Library, Mykolaiv 28431
Regional Scientific Medical Library, Odesa 28440
Regional Scientific Medical Library, Poltava 28452
Regional Scientific Medical Library, Rivne 28456
Regional Scientific Medical Library, Sankt-Peterburg 23866
Regional Scientific Medical Library, Sumy 28474
Regional Scientific Medical Library, Ternopil 28478
Regional Scientific Medical Library, Ulyanovsk 24036
Regional Scientific Medical Library, Uzhgorod 28479
Regional Scientific Medical Library, Zaporizhzhya 28486
Regional Scientific Medical Library, Zhytomyr 28488
Regional Seminary, Harare 42376
Regional State Archives, Gdańsk 21093
Regional State Archives, Krasnoyarsk 23125

Regional State Archives of Hamar, Hamar 20226
Regional State Archives of Hordaland and Sogn & Fjordane, Bergen 20217
Regional State Archives of Kristiansand, Kristiansand 20235
Regional State Archives of Stavanger, Stavanger 20296
Regional State Archives of Trondheim, Trondheim 20301
Regional State Universal Scientific Library of Nizhny Novgorod, Nizhni Novgorod 22158
Regional State Youth Library, Lugansk 28704
Regional Studies Museum, Feodosiya 28238
Regional Technical College, Athlone 14496
Regional Technical College, Limerick 14487
Regional Trade Union Administration, Main Library, Gomel 02131
Regional Trade Union Library, Lutsk 28708
Regional Veterinary Institute, Plovdiv 03516
Regional Veterinary Institute, Veliko Tărnovo 03588
Regional Youth Library, Cherkasy 28516
Regional Youth Library, Chernigiv 28521
Regional Youth Library, Kherson 28609
Regional Youth Library, Khmelnitski 28612
Regional Youth Library, Mykolaiv 28740
Regional Youth Library, Poltava 28785
Regional Youth Library, Rivne 28799
Regional Youth Library, Ternopil 28844
Regional Youth Library of Kostroma, Kostroma 24149
Regionalbibliothek Hochdorf, Hochdorf 27488
Regionalbibliothek Neubrandenburg, Neubrandenburg 12864
Regionalbibliothek Weiden, Weiden 12971
Regionale Bibliotheek Angstel, Vecht en Venen – Vestiging Maarssen-Dorp, Maarssen-Dorp 19657
Regionalgeschichtliche Bibliothek "Zwischen Neckar und Main", Buchen 11700
Regionalna Biblioteka "Christo Smirnenski", Haskovo 03594
Regionalna biblioteka Petar Stapov, Targovishte 03606
Regionální knihovna Karviná, Karviná 06578
Regionální knihovna Teplice, Teplice 06611
Regionální muzeum K. A. Polánka, Žatec 06548
Regionální muzeum v Chrudimi, Chrudim 06445
Regionální muzeum v Kolíně, Kolín 06458
Regionální muzeum v Mikulově, Mikulov 06465
Regionální muzeum v Teplicích, Teplice 06541
Regionalverband Ruhr, Essen 11826
Regionbibliotek Halland, Halmstad 26782
Regis College, Toronto 03884
Regis College, Weston 33040
Regis University, Denver 30995
– Colorado Springs Campus Library, Colorado Springs 30996
Registered Nurses Association of British Columbia, Vancouver 04543
Reguly Antal Müemlék Könyvtár, Zirc 13525
Rehoboth Beach Public Library, Rehoboth Beach 40960
Reial Academia de Bones Lletres, Barcelona 25930
Reial Societat Arqueologica Tarraconense, Tarragona 26162
Reid Campbell Library, North Sydney 00495
Reid Crowther & Partners Ltd, Calgary 04249
Reinhardt College, Waleska 33833
Reinhart Boerner Van Deuren SC, Milwaukee 35898
Reinwardt Academie, Amsterdam 19057
Reiss-Davis Child Study Center, Los Angeles 37062
Reiß-Engelhorn-Museen mit Curt-Engelhorn-Zentrum, Mannheim 12247
Religionspädagogische Arbeitsstelle, Singen 11330
Religionspädagogische Arbeitsstelle Bremen (RPA) / Evangelische Medienzentrale (EMZ), Bremen 11184
Religionspädagogisches Amt der Evangelischen Kirche in Hessen und Nassau, Gießen 11224
Religionspädagogisches Amt der Evangelischen Kirche in Hessen und Nassau, Mainz 11271
Religionspädagogisches Institut, Graz 01432
Religionspädagogisches Institut der Evangelischen Landeskirche in Baden, Karlsruhe 11252
Religionspädagogisches Institut Loccum, Rehburg-Loccum 11314
Religionspädagogisches Institut Stuttgart, Stuttgart 12537
Religionspädagogisches Studienzentrum der EKHN, Kronberg 11261
Religious Society of Friends in Britain, London 29553
"Renato Maestro" Jewish Library and Archives, Venezia 16124
Rend Lake College, Ina 33427
Renfro Library, Mars Hill 31765
Renmin University of China, Beijing 05052
Renner-Institut (RI), Wien 01650
Rensselaer at Hartford, Hartford 31304
Rensselaer Research Libraries, Troy 32834
Rentenanstalt / Swiss Life, Zürich 27294
Rentgenoradiologicheski institut, Moskva 23467
Renton Public Library, Renton 40968
Repartição Técnica de Estatistica, Maputo 18983
Repin Institute of Painting, Sculpture and Architecture, Sankt-Peterburg 22524
Representative Church Body Library, Dublin 14513
Republic of Belarus Scientific and Medical Library, Minsk 02050
Republican Centre of Traditional Art and Cultural Education, Minsk 02053
Republican Children Library, Yoshkar-Ola 24362
Republican Institute of Humanities, Sankt-Peterburg 23882
Republican Library, Nakhichevan 01731
Republican Library for Science and Technology, Minsk 01784
Republican Library for Science and Technology, Minsk 02052
Republican Library for Science and Technology, Toshkent 42071
Republican Library of Kara-Kalpak, Nukus 42070
Republican Library of Medicine, Chișinău 18896
Republican Medical Library and Information Centre, Kazan 23085
Republican Scientific Agricultural Library, Almaty 17961
Republican Scientific and Technical Library, Toshkent 42072
Republican Scientific and Technical Library of Turkmenistan, Ashkhabad 27905
Republican Scientific and Technical Library (RSTL), Yerevan 00327
Republican Scientific Library for Physical Culture and Sports, Minsk 02051
Republican Scientific Medical Library, Ulan-Ude 24034
Republican Scientific Medical Library, Yerevan 00358
Republican Scientific Pedagogical Library, Almaty 17962
Republican Scientific Pedagogical Library, Minsk 01809
Republicki Zavod za Statistiku, Beograd 24495
Republički zavod za zaštitu spomenika kulture, Beograd 24520
Republik Juvenile Library of Tatarstan, Kazan 24139
Republikanski nauchno-prakticheski institut za speshna meditsinska pomosht 'N.I. Pirogov', Sofiya 03572
Rerikh Art College, Sankt-Peterburg 22696
Research and Breeding Institute of Pomology Holovousy Ltd, Hořice v Podkrkonoši 06450
Research and Construction Institute of Hydroengineering S.Ya. Zhuk, Kharkiv 28276
Research and Design Institute for the Enrichment and Agglomeration of Ferrous Metal Ores, Krivi Rig 28301
Research and Design Institute of Woodworking Machinery, Moskva 23468
Research and Development Centre for Biology, Bogor 14351
Research and Development Institute for Municipal Facilities and Services, Kyiv 28383
Research and Development Institute for Open-Pit Mining of Mineral Resources, Chelyabinsk 23018
Research and Experimental Institute for Automobile Electric and Electronic Equipment, Moskva 23371
Research and Experimental Institute for Coal Mining Engineering, Moskva 23186
Research and Experimental Institute for the Advancement of Civil Engineering, Moskva 23466
Research and Information Institute for Agricultural Economics, Budapest 13408
Research and Planning Institute for Metallurgical Industry, Dnipropetrovsk 28202
Research and Planning Institute for Metallurgical Machine Building, Moskva 23552
Research and Production Institute for Bovine Breeding, Baloteşti 21959
Research and Production Institute for Pig Breeding, Periş 22050
Research Association for Underground Transportation Facilities, Köln 12153
Research Center for the Eco-Environmental Sciences, Beijing 05852
Research Centre for Agricultural and Forest Environment, Poznań 21209
Research Centre for Agrobotany, Tápiószele 13513
Research Centre for Engineering Lasers of the Russian Academy of Sciences, Shatura 23968
Research Centre for the Soil-Plant System, Roma 16480
Research Centre for the Soil-Plant System (CRA-RPS), Roma 16560
Research Centre of Forestry, Arezzo 16170
Research Designs and Standards Organization, Lucknow 14071
Research Institute for Animal Breeding and Nutrition, Herceghalom 13479
Research Institute for Animal Production – Department of Information Systems and Publishing Activity, Nitra 24746
Research Institute for Cereals and Industrial Crops, Fundulea 22033
Research Institute for Chemical Protection of Plants, Moskva 23565
Research Institute for Chemistry and Food Industry, Bucureşti 21982
Research Institute for Complex Use of Water Resources, Minsk 02048
Research Institute for Development, Nouméa 19756
Research Institute for Disability Assessment and Organization of Work of the Disabled, Minsk 02049
Research Institute for Economic Order and Competition, Köln 12130
Research Institute for Epidemiology & Microbiology, Minsk 02041
Research Institute for Estate Crops, Bogor 14354
Research Institute for Estate Crops, Medan 14376
Research Institute for Fisheries, Aquaculture and Irrigation, Szarvas 13501
Research Institute for Fruit Growing, Piteşti 22051
Research Institute for Fruitgrowing and Ornamentals, Budapest 13376
Research Institute for Human Settlement, Bandung 14344
Research Institute for Human Settlements, Bandung 14344
Research Institute for Irrigation, Drainage and Hydraulic Engineering, Sofiya 03573
Research Institute for Marine Fisheries, Jakarta 14358
Research Institute for Mechanical Engineering, Moskva 23444
Research Institute for Nature and Forest, Brussel 02591
Research Institute for Occupational Safety and Health Problems, Moskva 23469
Research Institute for Petrochemistry, Prievidza 24749
Research Institute for Plant Protection, Bucureşti 22013
Research Institute for Polymers and Textiles, Tsukuba 17807
Research Institute for Printing Industry, Moskva 23569
Research Institute for Railway Hygiene, Moskva 23588
Research Institute for Roadbuilding and Construction Machinery, Moskva 23580
Research Institute for Soil Science and Agrochemistry, Bucureşti 21983
Research Institute for Textile Machinery, Ltd, Liberec 06462
Research Institute for the Fur Industry, Moskva 23410
Research Institute for the Languages of Finland, Helsinki 07470
Research Institute for the Study of Kursk Magnetic Anomalies, Gubkin 23048
Research Institute for Tropical Vegetables, Villa Clara 06313
Research Institute for Vegetable and Flower Growing, Vidra 22070
Research Institute for Vegetable and Melon Cultivation, Kamyzyak 23078
Research Institute for Veterinary Science (RIVS), Bogor 14346
Research Institute for Wine Growing and Wine Making, Valea Călugărească 22069
Research Institute for Wines, Taipei 27618
Research Institute of Agricultural Economics (RIAE), Praha 06531
Research Institute of Agricultural Microbiology, Ukrainian Academy of Agricultural Sciences, Chernigiv 28191
Research Institute of Agricultural Science, Sankt-Peterburg 23745
Research Institute of Agriculture, Chabany 28187
Research Institute of Agriculture, Saguramo 09217
Research Institute of Agriculture Engineering, Praha 06532
Research Institute of Agriculture of the Central Chernozem Zone named after V.V. Dokuchaev, Talovaya 28476
Research Institute of Animal Husbandry and Animal Genetics, Sankt-Peterburg 23844
Research Institute of Animal Production, Praha 06533
Research Institute of Antibiotics, Moskva 23381
Research Institute of Barley, Karnobat 03503
Research Institute of Bast Fibres Industry, Moskva 23530
Research Institute of Biotechnological Systems, Sankt-Peterburg 23859
Research Institute of Cables and Insulating Materials, Bratislava 24730
Research Institute of Chemical Technology, Bratislava 24729
Research Institute of Chemical Technology Joint-Stock Company, Moskva 22822
Research Institute of Children's Infections, Sankt-Peterburg 23880
Research Institute of Children's Infections, Sankt-Peterburg 23817
Research Institute of Cinema Arts, Moskva 23406
Research Institute of Civil Engineering, Krasnoyarsk 23128
Research Institute of Civil Engineering, Moskva 23413
Research Institute of Civil Engineering, Moskva 23373
Research Institute of Civil Engineering, Sofiya 03567
Research Institute of Clock Industry, Moskva 23383
Research Institute of Complex Electrical Equipment, Yerevan 00359
Research Institute of Dermatology and Venerology, Moskva 23514
Research Institute of Education, Praha 06528
Research Institute of Electric Power Networks, Moskva 23391
Research Institute of Electrophysical Equipment, Sankt-Peterburg 23821
Research Institute of Emergency Aid, Sankt-Peterburg 23846
Research Institute of Endocrinology and Hormone Chemistry, Kharkiv 28278
Research Institute of Fisheries, Varna 03585
Research Institute of Foundry Engineering, Technology and Automatization, Moskva 23572
Research Institute of Fruit Growing, Grakhi 23046
Research Institute of Fur Animals and Rabbit Breeding, Rodniki 23717
Research Institute of Geodesy, Topography and Cartography, Zdiby 06549
Research Institute of Geology and Combustible Minerals Minerals, Moskva 23395
Research Institute of Geophysical Methods of Prospecting, Moskva 23394
Research Institute of Geophysical Research of Exploration Wells, Oktyabrski 23662
Research Institute of Haematology and Blood Transfusion, Sankt-Peterburg 23881
Research Institute of Hard-Coal Mining Industry, Sofiya 03557
Research Institute of Health and Medical Problems, Moskva 23408
Research Institute of Heat Technology, Łódź 21161
Research Institute of Hereditary Pathology, Lviv 28422
Research Institute of Hydrogeology and Geological Engineering, Staraya Kupavna 23980
Research Institute of Industrial Engineering, Donetsk 28234
Research Institute of Informatic Systems in Geophysics, Geology and Geochemistry, Moskva 23562
Research Institute of Karakul Sheep Breeding, Samarkand 42122
Research Institute of Life Insurance Welfare, Osaka 17642

Research Institute of Lighting Technology, Moskva 23541
Research Institute of Livestock Raising and Veterinary Science, Anenii Noi 18891
Research Institute of Management, Moskva 23440
Research Institute of Marine Products, Haiphong 42285
Research Institute of Measuring Instruments and Equipments, Moskva 23377
Research Institute of Mechanical Engineering, Moskva 23232
Research Institute of Medical Ecology and Prophylaxis, Bishkek 18152
Research Institute of Medicaments against Virus Diseases, Moskva 23442
Research Institute of Metal Technology, Kharkiv 28270
Research Institute of Morality, Tokyo 17753
Research Institute of Mother and Child Care, Rostov-na-Donu 23726
Research Institute of Museum Science, Moskva 23415
Research Institute of Occupational Safety, Sankt-Peterburg 23833
Research Institute of Oncology and Medical Radiology, Lesnoy 01966
Research Institute of Organic Semiproducts and Dyes, Moskva 23417
Research Institute of Packing Materials, Kaluga 23077
Research Institute of Paper Industry, Budapest 13453
Research Institute of Phthisiology and Pulmonology, Sankt-Peterburg 23826
Research Institute of Phthisiopulmonology, Moskva 23470
Research Institute of Physical Education, Sankt-Peterburg 23825
Research Institute of Physical Education and Sport, Moskva 23560
Research Institute of Potassium and Potash Fertilizers, Sankt-Peterburg 23827
Research Institute of Potato Crops, Kraskovo 23110
Research Institute of Prospecting Methods and Technology, Sankt-Peterburg 23938
Research Institute of Pulmonology, Sankt-Peterburg 23840
Research Institute of Radiational Hygiene, Sankt-Peterburg 23841
Research Institute of Rare Metal Industry, Moskva 23230
Research Institute of Resin and Latex Products, Moskva 23430
Research Institute of Resin Industry, Moskva 23428
Research Institute of Rubber and Latex Products, Moskva 23431
Research Institute of Rubber Industry, Moskva 23429
Research Institute of Salt Industry, Artemivsk 28179
Research Institute of Science Policy and Further Education, Warszawa 21301
Research Institute of Sheep Raising, Mynbaevo 17972
Research Institute of Soft Drinks and Beer and Wine Making Industry, Moskva 23420
Research Institute of Synthetic Rubber, Sankt-Peterburg 23845
Research Institute of Television, Sankt-Peterburg 23849
Research Institute of the Bread-Baking Industry, Moskva 23566
Research Institute of the Factory Technology of Prefabricated and Reinforced Concrete Structures and Items, Moskva 23471
Research Institute of the Machine Tool and Instrument-Making Industries, Moskva 23577
Research Institute of the Sewn Goods Industry, Kyiv 28384
Research Institute of the Textile Institute, Sankt-Peterburg 23848
Research Institute of Thermal Energy Machine Construction, Moskva 23436
Research Institute of Toxicology, Sankt-Peterburg 23851
Research Institute of Transport Engineering, Sankt-Peterburg 23852
Research Institute of Vaccine and Serum, Sankt-Peterburg 23854
Research Institute of Vaccines and Sera, Moskva 23441
Research Institute of Vegetable Crops, Mytishchi 23605
Research Institute of Vegetable Production, Vereiya 24041
Research Institute of Viticulture and Oenology, Kecskemét 13484

Research Institute of Welding, Sankt-Peterburg 23847
Research Institute on Addictions, Buffalo 36613
Research Library, Akmola 17935
Research Library in Hradec Králové, Hradec Králové 06345
Research Library in Olomouc, Olomouc 06348
Research Library Liberec, Liberec 06347
Research Library of South Bohemia in České Budějovice, České Budějovice 06344
Research Library of Zadar, Zadar 06143
Research Organization of Information and Systems, Tachikawa 17664
Research Society of Pakistan, Lahore 20518
Research-Development Institute for Beekeeping, Bucureşti 21981
Research-Development Institute for Plant Protection, Bucureşti 22013
Réseau des bibliothèques communales de Watermael-Boits, Watermael-Boitsfort 02879
Réseau des bibliothèques de la ville de Saint-Brieuc, Saint-Brieuc 09054
Reserve Bank Library, Suva 07307
Reserve Bank of Australia, Sydney 00839
Reserve Bank of New Zealand, Wellington 19818
Reshetilivka Regional Centralized Library System, Main Library, Reshetilivka 28795
Reshetilivska raionna TsBS, Tsentralna biblioteka, Reshetilivka 28795
Residencia de Estudiantes, Madrid 26097
Residencia de la Compañía de Jesús, San Sebastián 25818
Residencia Sagrado Corazón, Valencia 25836
Resource Center of the Americas, Minneapolis 37123
Resource pour la Recherche Justice, Vaucresson 08167
Respublikanska biblioteka dlya slepykh, Yakutsk 24082
Respublikanskaya detskaya biblioteka, Elista 24127
Respublikanskaya detskaya biblioteka, Yoshkar-Ola 24362
Respublikanskaya detskaya biblioteka im. B. Pacheva, Nalchik 24180
Respublikanskaya nauchnaya biblioteka im. S.M. Kirova, Vladikavkaz 22189
Respublikanskaya Nauchnaya Meditsinskaya Biblioteca, Chişinău 18896
Respublikanskaya nauchnaya meditsinskaya biblioteka, Ulan-Ude 24034
Respublikanskaya nauchnaya meditsinskaya bibliyateka (RSML), Minsk 02050
Respublikanskaya nauchno-metodicheskaya biblioteka po fizicheskoi kulture, Minsk 02051
Respublikanskaya nauchno-pedagogicheskaya biblioteka, Minsk 01809
Respublikanskaya nauchno-tekhnicheskaya biblioteka Belarusi, Minsk 02052
Respublikanskaya stantsiya yunykh tekhnikov, Minsk 01824
Respublikanskaya yunosheskaya biblioteka, Saransk 24313
Respublikanskaya yunosheskaya biblioteka Chuvashskoi Respubliki, Cheboksary 24120
Respublikanskaya yunosheskaya biblioteka im. K. Mechieva, Nalchik 24181
Respublikanskaya yunosheskaya biblioteka im. Z.Kh. Kolumba, Yoshkar-Ola 24363
Respublikanskaya yunosheskaya biblioteka Tatarstana, Kazan 24139
Respublikanski bibliotechno-kulturny kompleks dlya detei i yunoshestva, Izhevsk 24134
Respublikanski gumanitarny institut, Sankt-Peterburg 23882
Respublikanski meditsinski bibliotechno-informatsionny tsentr, Kazan 23085
Respublikanski nauchno-issledovatelski institut akusherstva i pediatrii, Rostov-na-Donu 23726
Respublikanski nauchno-metodicheski tsentr narodnogo tvorchestva i kulturno-prosvetitelnoi raboty, Minsk 02053
Respublikanski nauchno-prakticheski tsentr onkologii i meditsinskoi radiologii im. N.N. Aleksandrova, p. Lesnoy-2 01967
Respublikanski tsentr po nauchnoi organizatsii truda i upravleniya proizvodstvom, Minsk 01825

Respublikansky zaochny avtotransportny tekhnikum, Kyiv 28115
Resurrection Metropolitan Community Church, Houston 35235
Reuben Hoar Library, Littleton 39903
Reutlingen School of Theology, Reutlingen 10542
Reveille United Methodist Church Library, Richmond 35346
Review & Herald Publishing Association, Hagerstown 35732
Revlon Research Center Library, Edison 35685
Reykjavík University, Reykjavík 13559
Rezekne City Central Library, Rezekne 18196
Rezeknes pilsetas centrala biblioteka, Rezekne 18196
RFS Ecusta Inc, Pisgah Forest 36126
Rheinische Denkmalpflege, Pulheim 12415
Rheinische Friedrich-Wilhelms-Universität Bonn, Bonn 09304
– Abteilungsbibliothek für Medizin, Naturwissenschaften und Landbau, Bonn 09506
– Anatomisches Institut, Bonn 09507
– Argelander Institut für Astronomie (AIfA), Bonn 09508
– Bibliothek Bildungswissenschaft, Bonn 09509
– Chirurgisches Zentrum / Klinik und Poliklinik für Allgemein-, Viszeral-, Thorax- und Gefäßchirurgie, Klinik für Unfallchirurgie, Urologische Klinik, Bonn 09510
– Evangelisch-Theologisches Stift, Hans-Iwand-Haus, Bonn 09511
– Fachbibliothek Evangelische und Katholische Theologie, Bonn 09512
– Fachbibliothek Mathematik, Bonn 09513
– Forschungsinstitut für Diskrete Mathematik / Institut für Operations Research, Bonn 09514
– Franz Joseph Dölger-Institut zur Erforschung der Spätantike, Bonn 09515
– Geographisches Institut, Bonn 09516
– Helmholtz-Institut für Strahlen- und Kernphysik, Bonn 09517
– Institut für Anglistik, Amerikanistik und Keltologie, Bonn 09518
– Institut für Anglistik, Amerikanistik und Keltologie, Abt. Vergleichende Indogermanische Sprachwissenschaft und Keltologie, Bonn 09519
– Institut für Arbeitsrecht und Recht der Sozialen Sicherheit, Bonn 09520
– Institut für Deutsche und Rheinische Rechtsgeschichte, Bonn 09521
– Institut für Deutsches und Internationales Zivilprozessrecht sowie Konfliktmanagement, Bonn 09522
– Institut für Geodäsie und Geoinformation / Professur für Geodäsie, Bonn 09523
– Institut für Geodäsie und Geoinformation / Professur für Städtebau und Bodenordnung, Bonn 09524
– Institut für Germanistik, Vergleichende Literatur- und Kulturwissenschaft, Bonn 09525
– Institut für Germanistik, Vergleichende Literatur- und Kulturwissenschaft / Teilbibliothek Kulturanthropologie, Volkskunde, Bonn 09526
– Institut für Geschichtswissenschaft, Bonn 09527
– Institut für Geschichtswissenschaft / Abt. Alte Geschichte, Bonn 09528
– Institut für Geschichtswissenschaft / Abt. Osteuropäische Geschichte, Bonn 09529
– Institut für Geschichtswissenschaft / Abt. Rheinische Landesgeschichte, Bonn 09530
– Institut für Griechische und Lateinische Philologie, Romanistik und Altamerikanistik / Abt. Altamerikanistik, Bonn 09531
– Institut für Griechische und Lateinische Philologie, Romanistik und Altamerikanistik / Abt. für Griechische und Lateinische Philologie, Bonn 09532
– Institut für Griechische und Lateinische Philologie, Romanistik und Altamerikanistik / Abt. Romanistik, Bonn 09533
– Institut für Handels- und Wirtschaftsrecht, Bonn 09534
– Institut für Informatik, Bonn 09535

– Institut für Internationales Privatrecht und Rechtsvergleichung, Bonn 09536
– Institut für Kunstgeschichte und Archäologie / Abt. Ägyptologie, Bonn 09537
– Institut für Kunstgeschichte und Archäologie / Abt. für Kunstgeschichte, Bonn 09538
– Institut für Kunstgeschichte und Archäologie / Abt. Klassische Archäologie mit Akademischem Kunstmuseum, Bonn 09539
– Institut für Kunstgeschichte und Archäologie / Abt. Vor- und Frühgeschichtliche Archäologie, Bonn 09540
– Institut für Landtechnik, Bonn 09541
– Institut für Lebensmittel- und Ressourcenökonomik, Bonn 09542
– Institut für Lebensmittel- und Ressourcenökonomik (ILR) / Unternehmensführung, Bonn 09543
– Institut für Nutzpflanzen-wissenschaften und Ressourcenschutz (INRES) / Bodenwissenschaften, Bonn 09544
– Institut für Nutzpflanzen-wissenschaften und Ressourcenschutz (INRES) / Phytomedizin, Bonn 09545
– Institut für Öffentliches Recht, Bonn 09546
– Institut für Orient- und Asienwissenschaften / Abt. Asiatische und Islamische Kunstgeschichte, Bonn 09547
– Institut für Orient- und Asienwissenschaften / Abt. Indologie, Bonn 09548
– Institut für Orient- und Asienwissenschaften / Abt. Islamwissenschaft, Bonn 09549
– Institut für Orient- und Asienwissenschaften / Abt. Mongolistik und Tibetstudien, Bonn 09550
– Institut für Orient- und Asienwissenschaften / Abt. Orientalische und Asiatische Sprachen, Bonn 09551
– Institut für Orient- und Asienwissenschaften / Abt. Religionswissenschaft, Bonn 09552
– Institut für Philosophie, Bonn 09553
– Institut für Philosophie, Bonn 09554
– Institut für Physikalische und Theoretische Chemie, Bonn 09555
– Institut für Politische Wissenschaft und Soziologie, Bonn 09556
– Institut für Psychologie, Bonn 09557
– Institut für Römisches Recht und Vergleichende Rechtsgeschichte, Bonn 09558
– Institut für Sprach-, Medien- und Musikwissenschaft / Abt. für Medienwissenschaft, Bonn 09559
– Institut für Sprach-, Medien- und Musikwissenschaft / Abt. für Musikwissenschaft / Sound Studies, Bonn 09560
– Institut für Sprach-, Medien- und Musikwissenschaft / Abt. Sprachwissenschaft – Sprachlernforschung, Bonn 09561
– Institut für Steuerrecht, Bonn 09562
– Institut für Strafrecht, Bonn 09563
– Institut für Tierwissenschaften / Abt. Tierernährung, Bonn 09564
– Institut für Tierwissenschaften / Abt. Tierzucht und Tierhaltung, Bonn 09565
– Institut für Völkerrecht, Bonn 09566
– Institut für Zoologie, Bonn 09567
– Juristisches Seminar, Bonn 09568
– Kekulé-Institut für Organische Chemie und Biochemie und Institut für Anorganische Chemie, Bonn 09569
– Kirchenrechtliches Institut, Bonn 09570
– Klinik und Poliklinik für Dermatologie und Allergologie, Bonn 09571
– Kriminologisches Seminar, Bonn 09572
– Medizinhistorisches Institut, Bonn 09573
– Meteorologisches Institut, Bonn 09574
– Nees Institut für Biodiversität der Pflanzen, Bonn 09575
– Pharmazeutisches Institut, Bonn 09576
– Physikalisches Institut, Bonn 09577
– Rechtsphilosophisches Seminar, Bonn 09578
– Slavistisches Seminar, Bonn 09579
– Staatswissenschaftliches Seminar, Bonn 09580
– Steinmann-Institut für Geologie, Mineralogie und Paläontologie, Bonn 09581

- Steinmann-Institut für Geologie, Mineralogie und Paläontologie / Mineralogie, Bonn 09582
- Universitätsklinikum Bonn, Augenklinik, Bonn 09583
- Zentrum für Entwicklungsforschung (ZEF), Zentrum für Europäische Integrationsforschung (ZEI), Bonn 09584
- Zentrum für Hygiene und Medizinische Mikrobiologie / Institut für Medizinische Mikrobiologie, Immunologie und Parasitologie, Bonn 09585
- Zentrum für Innere Medizin / Medizinische Klinik I, II und III, Bonn 09586
- Zentrum für Kinderheilkunde / Kinderklinik und Poliklinik, Bonn 09587
- Zentrum für Nervenheilkunde, Bonn 09588
Rheinische Landesbibliothek, Koblenz 09324
Rheinisches Landesmuseum für Archäologie, Kunst und Kulturgeschichte, Bonn 11641
Rheinisches Landesmuseum Trier, Trier 12553
Rheinisch-Westfälische Technische Hochschule (RWTH) Aachen, Aachen 09347
- Fachbereichsbibliothek Wirtschaftswissenschaften, Aachen 09348
- Fachgruppe Chemie, Aachen 09349
- Fachgruppe Informatik, Aachen 09350
- Fachgruppe Physik, Aachen 09351
- Gemeinsame Mathematische Bibliothek, Aachen 09352
- Geographisches Institut, Aachen 09353
- Geologisches Institut, Aachen 09354
- Germanistische Bibliothek, Aachen 09355
- Institut für Bauforschung, Aachen 09356
- Institut für Eisenhüttenkunde, Aachen 09357
- Institut für Geschichte, Theorie und Ethik der Medizin, Aachen 09358
- Institut für Katholische Theologie, Aachen 09359
- Institut für Kunstgeschichte, Aachen 09360
- Institut für Mineralogie und Lagerstättenlehre, Aachen 09361
- Institut für Politische Wissenschaft, Aachen 09362
- Institut für Psychologie, Aachen 09363
- Institut für Soziologie, Aachen 09364
- Institut für Textiltechnik, Aachen 09365
- Institut für Wasserbau und Wasserwirtschaft, Aachen 09366
- Lehr- und Forschungsgebiet Romanische Sprachwissenschaft, Aachen 09367
- Lehr- und Forschungsgebiet Wirtschafts- und Sozialgeschichte, Aachen 09368
- Lehrstuhl für Alte Geschichte und Historisches Institut, Aachen 09369
- Lehrstuhl für Anglistische Literaturwissenschaft, Aachen 09370
- Lehrstuhl für Aufbereitung und Recycling fester Abfallstoffe und Institut für Aufbereitung, Kokerei und Brikettierung, Aachen 09371
- Lehrstuhl für Baugeschichte und Denkmalpflege, Aachen 09372
- Lehrstuhl für Berufs- und Wirtschaftspädagogik, Aachen 09373
- Lehrstuhl für Geodäsie und Geodätisches Institut, Aachen 09374
- Lehrstuhl für Geotechnik im Bauwesen und Institut für Grundbau, Bodenmechanik, Felsmechanik und Verkehrswasserbau, Aachen 09375
- Lehrstuhl für Geschichte der Technik, Aachen 09376
- Lehrstuhl für Metallurgische Prozesstechnik und Metallrecycling und Institut für Metallhüttenkunde und Elektrometallurgie, Aachen 09377
- Lehrstuhl für Planungstheorie und Stadtentwicklung, Aachen 09378
- Lehrstuhl für Siedlungswasserwirtschaft und Siedlungsabfallwirtschaft und Institut für Siedlungswasserwirtschaft, Aachen 09379
- Lehrstuhl für Stahl- und Leichtmetallbau und Institut für Stahlbau, Aachen 09380
- Lehrstuhl für Statistik und Institut für Statistik und Wirtschaftsmathematik, Aachen 09381
- Lehrstuhl für Straßenwesen, Erd- und Tunnelbau und Institut für Straßenwesen, Aachen 09382
- Lehrstuhl für Theoretische Philosophie und Philosophisches Institut, Aachen 09383

- Lehrstuhl für Werkstoffe der Elektrotechnik II und Institut für Werkstoffe der Elektrotechnik, Aachen 09384
- Lehrstuhl für Werkzeugmaschinen, Aachen 09385
- Lehrstuhl und Institut für Arbeitswissenschaft, Aachen 09386
- Lehrstuhl und Institut für Bildsame Formgebung, Aachen 09387
- Lehrstuhl und Institut für Stadtbauwesen und Stadtverkehr, Aachen 09388
- Medizinische Bibliothek, Aachen 09389
Rheinisch-Westfälisches Institut für Wirtschaftsforschung, Essen 11827
Rheinisch-Westfälisches Wirtschaftsarchiv zu Köln, Köln 12152
Rhein-Sieg-Kreis, Siegburg 12487
Rhine Research Center, Durham 36817
Rhinelander District Library, Rhinelander 40970
Rhine-Waal University of Applied Sciences, Kleve 10119
Rhode Island College, Providence 32388
Rhode Island Historical Society Library, Providence 37417
Rhode Island Hospital, Providence 37418
Rhode Island School for the Deaf Library, Providence 37419
Rhode Island School of Design, Providence 32389
Rhode Island State Law Library, Providence 34710
Rhode Island State Library, Providence 30147
Rhodes College, Memphis 31793
Rhodes University, Grahamstown 25022
Rhone-Poulenc Rorer Ltd, Dagenham 29570
Rhonoda-Cynon-Taf Libraries, Aberdare 29921
Ribe Katedralskole, Ribe 06744
Ricardo Mulder Library, Edenvale 25199
Rice Public Library, Kittery 39734
Rice Research Institute, Shimkent 17976
Rice University, Houston 31354
Richard Bland College, Petersburg 33633
Richard J. Daley College, Chicago 33225
Richard Salter Storrs Library, Longmeadow 39934
Richard Stockton College of New Jersey, Pomona 32328
Richard Sugden Library, Spencer 41402
Richard T. Liddicoat Gemological Library & Information Center, Carlsbad 36628
Richard Wagner Museum with National Archive and the Richard Wagner Foundation Research Centre, Bayreuth, Wahnfried House, Bayreuth 11482
Richards, Layton & Finger Library, Wilmington 36426
Richards, Watson & Gershon Library, Los Angeles 35855
Richardson Public Library, Richardson 40971
Richard-Strauss-Institut, Garmisch-Partenkirchen 11913
Richard-Wagner-Museum mit Nationalarchiv der Richard-Wagner-Stiftung, Bayreuth 11482
Richland College, Dallas 33275
Richland Community College, Decatur 33288
Richland Correctional Institution Library, Mansfield 34499
Richland County Law Library, Mansfield 34500
Richland County Public Library, Columbia 38649
Richland Parish Library, Rayville 40945
Richland Public Library, Richland 40975
Richmond Community College, Hamlet 33396
Richmond Heights Memorial Library, Richmond Heights 40989
Richmond Hill Public Library, Richmond Hill 04736
Richmond Lending Library, Richmond 30064
Richmond Memorial Library, Batavia 38102
Richmond Public Library, Richmond 04735
Richmond Public Library, Richmond 40986
Richmond Public Library – West End, Richmond 40987
Richmond, The American International University in London, Richmond 29385
Richmond upon Thames College, Twickenham 29497
Richmond Upper Clarence Regional Library, Casino 01084
Richmond-Tweed Regional Library, Goonellabah 01098
Richwood North Union Public Library, Richwood 40992
Rideau Lakes Public Library, Elgin 04634
Rider University, Lawrenceville 31579

- Westminster Choir College, Princeton 31580
Ridgefield Library Association Inc, Ridgefield 40994
Ridgefield Park Free Public Library, Ridgefield Park 40996
Ridgefield Public Library, Ridgefield 40995
Ridgetown College, Ridgetown 03833
Ridgewater College, Willmar 33873
Ridgewood Public Library, Ridgewood 40999
Ridley Melbourne Mission and Ministry College, Parkville 00789
Ridley Township Public Library, Folsom 39090
Riehl-Frank-Stiftung, Kaufbeuren 12108
Rietavo savivaldybės viešoji biblioteka, Rietavas 18355
Riga Central Library, Riga 18198
Riga Stradins University Library, Riga 18176
Rigas Stradina Universitātes biblioteka, Riga 18176
Rigas Tehniskas universitates, Riga 18177
Rīgas vēstures un kugniecības muzeja bibliotēka, Riga 18190
Rigsarkivets Bibliotek, København 06819
Riigiarhiivi raamatukogu, Tallinn 07239
Riihimäen kaupunginkirjasto, Riihimäki 07604
Riihimäki Public Library, Riihimäki 07604
Rijckheyt – centrum voor regionale geschiedenis, Heerlen 19459
Rijeka University Library, Rijeka 06151
- Faculty of Economics, Rijeka 06152
- Faculty of Engineering, Rijeka 06158
- Faculty of Law, Rijeka 06157
- Faculty of Medicine, Rijeka 06156
- Faculty of Philosophy, Rijeka 06155
- Faculty of Tourism and Hospitality Management, Opatija 06154
- Rijeka College of Maritime Studies, Rijeka 06153
Rijksakademie van Beeldende Kunsten, Amsterdam 19375
Rijksarchief Antwerpen (RAA), Beveren 02542
Rijksarchief te Beveren, Beveren 02543
Rijksarchief te Brugge, Brugge 02551
Rijksarchief te Gent, Gent 02654
Rijksarchief te Hasselt (RAH), Hasselt 02663
Rijksarchief te Kortrijk, Kortrijk 02668
Rijksarchief te Ronse, Ronse 02695
Rijksbureau voor Kunsthistorische Documentatie (RKD), Den Haag 19450
Rijksdienst voor Archeologie, Cultuurlandschap en Monumenten, Amersfoort 19344
Rijksdienst voor Archeologie, Cultuurlandschap en Monumenten, Zeist 19538
Rijksdienst voor de Sociale Zekerheid, Brussel 02437
Rijksdienst voor Kinderbijslag voor Werknemers, Brussel 02435
Rijksinstituut voor Volksgezondheid en Milieu, Bilthoven 19400
Rijksmuseum, Amsterdam 19376
Rijksmuseum Twenthe, Enschede 19423
Rijksmuseum van Oudheden, Leiden 19482
Rijksuniversiteit Groningen, Groningen 19129
- Bibliotheek Economie, Bedrijfskunde, Ruimtelijke Wetenschappen, Groningen 19130
- Bibliotheek Faculteit der Letteren, Groningen 19131
- Bibliotheek Faculteit der Letteren, Dependence Archeologie, Groningen 19132
- Bibliotheek FWN, Groningen 19133
- Bibliotheek FWN, Biologie, Haren 19134
- Bibliotheek Rechten, Groningen 19135
- Bibliotheek Sociale Wetenschappen, Groningen 19136
- Bibliotheek van de Faculteit der Godgeleerdheid en Godsdienstwetenschap, Groningen 19137
- Bibliotheek Wijsbegeerte, Groningen 19138
- Centrale Medische Bibliotheek (CMB), Groningen 19139
Rijkswaterstaat Bibliotheek – Locatie Delft, Delft 19246
Rijkswaterstaat Bibliotheek – Locatie Utrecht, Utrecht 19292
Rijkswaterstaat – Directie IJsselmeergebied, Lelystad 19334
Rijkswaterstaat – Directie Noordzee, Rijswijk 19286
Rijkswaterstaat – Directie Zeeland, Middelburg 19283
Rijkswaterstaat – Directie Zuid Holland, Rotterdam 19288
Rikagaku Kenkyujo, Wako 17815
Riker, Danzig, Scherer, Hyland & Perretti, Morristown 35924
Rikkyo Daigaku, Tokyo 17170
- Hogakubu Toshoshitsu, Tokyo 17171

Rikkyo University, Tokyo 17170
- Law and Politics Library, Tokyo 17171
Riksantikvarens bibliotek, Oslo 20278
Riksarkivet, Oslo 20279
Riksarkivets bibliotek Marieberg-Arninge, Stockholm 26675
Riksdagsbiblioteket, Stockholm 26511
Riksförsäkringsverket, Stockholm 26512
Rikshospitalet-Radiumhospitalet, Oslo 20280
Rikspolisstyrelsen, Stockholm 26513
Riksrevisionen, Stockholm 26514
Riksutställningar, Visby 26722
RILA – Rome Institute of Liberal Arts, Roma 15378
Rilsky Institute of Arts, Folklore and Ethnography, Kyiv 28337
M. Rilsky Library, Kyiv 28664
Rimsky-Korsakov St. Petersburg Conservatory, Sankt-Peterburg 22539
Ringerike Bibliotek, Hønefoss 20345
Ringkøbing Bibliotek, Ringkøbing 06917
Ringling College of Art & Design, Sarasota 32618
Ringsted Bibliotek, Ringsted 06918
Ringwood Public Library, Ringwood 41000
Ringyo Shikenjo Hokkaido Shijo, Sapporo 17658
Ringyo Shikenjo Kansai Shijo, Kyoto 17591
Ringyo Shikenjo Kyushu Shijo, Kumamoto 17582
Ringyo Shikenjo Shikoku Shijo, Kochi 17576
Ringyo Shikenjo Tohoku Shijo, Morioka 17608
Rinri Kenkyujo, Tokyo 17753
Rio Grande Bible Institute & Language School, Edinburg 35194
Rio Hondo Community College, Whittier 33869
Rio Rancho Public Library, Rio Rancho 41001
Rio Tinto Exploration Pty Ltd., Belmont 00809
Riordan & McKinzie, Los Angeles 35856
Ripon College, Ripon 32448
Ripon Public Library, Ripon 41003
Risbergska skolan, Örebro 26478
Risø Bibliotek, Roskilde 06832
Rissa bibliotek, Rissa 20389
Rissho Daigaku, Tokyo 17754
Rissho University, Tokyo 17754
Riter C. Hulsey Public Library, Terrell 41540
Ritman Library, Amsterdam 19348
Ritsumeikan Daigaku, Kyoto 17024
- Hogakubu Kenkyushitsu, Kyoto 17025
- Igaku Bunkan, Nagasaki 17026
- Jimbunkagaku Kenkyujo, Kyoto 17027
- Rikogaku Kenkyujo, Kyoto 17028
Ritsumeikan University, Kyoto 17024
- Faculty of Law, Kyoto 17025
- Medical Branch, Nagasaki 17026
- Research Institute of Cultural Sciences, Kyoto 17027
- Research Institute of Science and Engineering, Kyoto 17028
Ritter Public Library, Vermilion 41672
River Edge Free Public Library, River Edge 41004
River Falls Public Library, River Falls 41005
River Forest Public Library, River Forest 41006
River Transport Main Design Bureau, Gomel 01850
River Vale Free Public Library, River Vale 41007
Riverhead Free Library, Riverhead 41009
Riverina Institute of TAFE, Wagga Wagga 00590
Riverina Regional Library, Wagga Wagga 01171
Rivers State Central Library, Port Harcourt 20063
Rivers State College of Arts and Science, Port Harcourt 19978
Rivers State University of Science and Technology, Port Harcourt 19979
Riverside Community College District, Riverside 33688
Riverside County Law Library, Indio 34385
Riverside County Law Library, Riverside 34745
Riverside County Library System, Riverside 41010
Riverside Public Library, Riverside 41011
Riverside Public Library, Riverside 41012
Riverside Public Library – Arlington, Riverside 41013
Riverside Public Library – La Sierra, Riverside 41014
Riverside Regional Library, Jackson 39604
Riverside Regional Library – Perryville Branch, Perryville 40763
Riverview Hospital, Coquitlam 04328
Riverview Psychiatric Center, Augusta 36502
Riverview Public Library, Riverview 41017
Rivier College, Nashua 31931

Riviera Beach Public Library, Riviera Beach 41019
Rivkin Radler LLP, Uniondale 36318
Rivne City Centralized Library System, Main Library, Rivne 28800
Rivne Regional Centralized Library System, Ostrog District Main Library, Ostrog 28770
Rivnenska miska TsBS, Tsentralna biblioteka, Rivne 28800
Rizarios Ekklesiastiki Sholi, Athinai 13091
RIZIV Documentatiedienst-Bibliotheek, Bruxelles 02436
The RJS Group, Inc, Vernon Hills 36321
Rjukan bibliotek, Rjukan 20390
RMIT University, Melbourne 00456
RMIT University, Melbourne 00457
– Carlton Library, Carlton 00458
RMT, Inc Library, Madison 35871
Road Accident Research Unit, Adelaide 00854
Road and Bridge Research Institute, Warszawa 21266
Road Building Equipment Scientific and Industrial Corporation, Minsk 01894
Roads and Traffic Authority, Rosebery 00994
Roane State Community College, Harriman 33398
Roanoke Bible College, Elizabeth City 35195
Roanoke College, Salem 32549
Roanoke County Public Library, Roanoke 41021
Roanoke County Public Library – Glenvar, Salem 41140
Roanoke County Public Library – Hollins, Roanoke 41022
Roanoke County Public Library – Vinton Branch, Vinton 41689
Roanoke Public Libraries, Roanoke 41023
Roanoke-Chowan Community College, Ahoskie 33115
Rob & Bessie Welder Wildlife Foundation Library, Sinton 37609
Robbins Library, Arlington 37993
J. Robert Ashcroft Memorial Library, Phoenixville 32289
G. Robert Cotton Regional Correctional Facility Library, Jackson 34389
Robert Goldwater Library, New York 37228
Robert Gordon University, Aberdeen 28924
Robert J. Kleberg Public Library, Kingsville 39727
Robert Koch-Institut (RKI), Berlin 11558
Robert L. Williams Public Library, Durant 38880
Robert Lehman Collection Library, New York 37229
Robert McLaughlin Gallery Library, Oshawa 04427
Robert Morris College, Chicago 30781
Robert Morris University, Moon Township 31894
Robert Schuman Institut an der Universität Bonn, Bonn 11646
Robert-Jungk-Bibliothek für Zukunftsfragen, Salzburg 01572
Roberts Wesleyan College & Northeastern Seminary, Rochester 35349
Robert-Schumann-Hochschule Düsseldorf, Düsseldorf 09683
Robertsfors folkbibliotek, Robertsfors 26886
Robeson Community College, Lumberton 33506
Robeson County Public Library, Lumberton 39980
Robins, Kaplan, Miller & Ciresi LLP, Minneapolis 35913
Robinson & Cole LLP Library, Hartford 35740
Robinson & McElwee PLLC, Charleston 35560
Robinson Public Library District, Robinson 41024
Robinson, Sheppard & Shapiro, Montreal 04270
Rochdale Metropolitan Borough Council, Rochdale 30065
Roche Diagnostics GmbH, Penzberg 11425
Roche Diagnostics GmbH, Scientific Information Management, Penzberg 11425
Rochester Academy of Medicine Library, Rochester 37455
Rochester College, Rochester Hills 33693
Rochester Community & Technical College, Rochester 33692
Rochester Hills Public Library, Rochester 41031
Rochester Historical Society Library, Rochester 37456
Rochester Institute of Technology, Rochester 32461
Rochester Museum & Science Center, Rochester 37457
Rochester Public Library, Rochester 41032
Rochester Public Library, Rochester 41033
Rochette Daniel, Bruxelles 02620

Rock Island County Law Library, Rock Island 34747
Rock Island Public Library, Rock Island 41035
Rock Valley College, Rockford 33695
Rockaway Township Free Public Library, Rockaway 41037
Rockbridge Regional Library, Lexington 39870
Rockdale City Library, Rockdale 01149
Rockdale Temple, Cincinnati 35161
Rockefeller Foundation Records & Library Services, New York 37255
Rockefeller Medical Library, London 29808
Rockefeller University, New York 32087
Rockford College, Rockford 32472
Rockford Public Library, Rockford 41040
Rockhampton Regional Council Libraries, Rockhampton 01150
Rockhurst University, Kansas City 31482
Rockingham County Public Library, Eden 38928
Rockingham Regional Campus, Rockingham 00507
Rockland Community College, Suffern 33782
Rockville Centre Public Library, Rockville Centre 41044
Rockville Public Library, Vernon 41674
Rockwall County Library, Rockwall 41045
Rockwell Automation Library, Milwaukee 35899
Rockwell Collins, Cedar Rapids 35556
Rockwell Scientific Co, Thousand Oaks 36304
Rocky Mountain Arsenal, Commerce City 36747
Rocky Mountain College, Billings 30427
Rocky Mountain College, Calgary 03681
Rocky River Public Library, Rocky River 41051
Röda Korsets Högskola, Stockholm 26481
Rodale Inc, Emmaus 35692
Roddenbery Memorial Library, Cairo 38410
Rodekruisziekenhuisbibliotheken, Mechelen 02818
Rodman Public Library, Alliance 37918
Rodosho Toshokan, Tokyo 17322
Rødovre Kommunebibliotek, Rødovre 06919
Roemer-Museum, Hildesheim 12061
Rogachev Regional Central Library, Rogachev 02231
Rogachevskaya tsentralnaya raionnaya bibliyateka, Rogachev 02231
Rogatin Regional Centralized Library System, Main Library, Rogatin 28801
Rogatinska raionna TsBS, Tsentralna biblioteka, Rogatin 28801
Roger Williams University, Bristol 30549
– Architecture Library, Bristol 30550
– School of Law Library, Bristol 30551
Rogers Memorial Library, Southampton 41381
Rogers Public Library, Rogers 41053
Rogers State University, Claremore 30820
Rogue Community College, Grants Pass 33383
Rohnert Park-Cotati Regional Library, Rohnert Park 41055
Röhss museum of design and applied art, Göteborg 26580
Röhsska museet, Göteborg 26580
Rokiškio rajono savivaldybės Juozo Keliuočio viešoji biblioteka, Rokiškis 18356
Rokitne Regional Centralized Library System, Main Library, Rokitne 28802
Rokitnyanska raionna TsBS, Tsentralna biblioteka, Rokitne 28802
Rol en Samenleving, Brussel 02621
Rolla Free Public Library, Rolla 41056
Rolling Hills Consolidated Library, Saint Joseph 41116
Rollins College, Winter Park 33089
Rollinsford School Library, Rollinsford 33699
Rolls-Royce, Indianapolis 35780
Római Katolikus Egyházi Gyűjtemény, Sárospatak, Sárospatak 13343
Roman Catholic Archdiocese of Los Angeles, Mission Hills 35288
Roman Catholic Diocese of Fresno Library, Fresno 35210
Roman-Catholic Church Collection, Sárospatak 13343
Romanian Academy, Bucureşti 21915
Romanian Cultural Institute, New York 37256
Romanian Institute of Marine Research, Constanţa 22026
Romanian Parliament, Chamber of Deputies, Bucureşti 21953
Romanian Senate Library, Bucureşti 21954
Romanian Union of Fine Arts, Bucureşti 22017
Romania's Parliament.Chamber of Deputies, Bucureşti 21953
Rome Institute of Liberal Arts, Roma 15378
Romeo District Library, Washington 41757
RomeroHaus, Luzern 27406

Römisch-Germanisches Zentralmuseum – Forschungsinstitut für Vor- und Frühgeschichte, Mainz 12237
Romulus Public Library, Romulus 41060
Ron Williams Memorial Library, Rabaul 20583
Ronneby bibliotek, Ronneby 26887
Roodepoort City Library, Roodepoort 25231
Roosevelt County Library, Wolf Point 41950
Roosevelt Hospital, New York 37257
F.D. Roosevelt Hospital Medical Library, Banská Bystrica 24676
Roosevelt Public Library, Roosevelt 41062
Roosevelt University, Chicago 30782
– Performing Arts Library, Chicago 30783
Ropers, Majeski, Kohn & Bentley, Redwood City 36167
Ropes & Gray LLP Library, Boston 35537
RoSa – Equal Opportunities, Brussel 02621
RoSa – Rol en Samenleving, Brussel 02621
Rosalind Franklin University of Medicine & Science, North Chicago 32142
Roscommon County Library, Roscommon 14587
Rose Bruford College of Theatre & Performance, Sidcup 29394
Rose Mackwelung Library, Tofol 18874
Rose State College, Midwest City 33540
Rose-Hulman Institute of Technology, Terre Haute 32809
Roseland Free Public Library, Roseland 41064
Roselle Free Public Library, Roselle 41065
Roselle Park, Roselle Park 41067
Roselle Public Library District, Roselle 41066
Rosemont College, Rosemont 32483
Rosenbach Museum & Library, Philadelphia 37361
Rosenberg Library, Galveston 39189
Rosenn, Jenkins & Greenwald Library LLP, Wilkes-Barre 36419
Rosenstiel School of Marine & Atmospheric Science, Miami 31812
Roseville Public Library, Roseville 41069
Rosicrucian Fraternity Library, Quakertown 35342
Rosicrucian Order, AMORC, San Jose 35376
Roskilde Universitet, Roskilde 06723
Roskilde University, Roskilde 06723
Ross, Dixon & Bell, LLP, Washington 36374
Rossford Public Library, Rossford 41071
Rossiskaya akademiya khudozhestv, Sankt-Peterburg 23883
Rossiskaya akademiya khudozhestv – Moskovski filial, Moskva 23472
Rossiskaya akademiya meditsinskikh nauk, Moskva 23473
Rossiskaya akademiya narodnogo Khozyaistva i gosudarstvennoi sluzhby – Karelski filial, Petrozavodsk 22481
Rossiskaya Akademiya Nauk – Arkhiv, Moskva 23474
Rossiskaya akademiya nauk – Arkhiv – SP otdelenie, Sankt-Peterburg 23884
Rossiskaya akademiya nauk – Bashkirski filial, Ufa 22184
Rossiskaya akademiya nauk – Buryatski filial Sibirskogo otdelenia, Ulan-Ude 22186
Rossiskaya akademiya nauk – Dagestanski filial, Makhachkala 22152
Rossiskaya ekonomicheskaya akademiya im. Plekhanova, Moskva 22385
Rossiskaya gosudarstvennaya biblioteka dlya slepykh (RGBS), Moskva 23475
Rossiskaya gosudarstvennaya biblioteka po iskusstvu (RGB po iskusstvu), Moskva 23476
Rossiskaya gosudarstvennaya biblioteka (RGB), Moskva 22114
Rossiskaya gosudarstvennaya detskaya biblioteka (RGDB), Moskva 22154
Rossiskaya gosudarstvennaya yunosheskaya biblioteka (RGYuB), Moskva 24175
Rossiskaya natsionalnaya biblioteka (RNB), Sankt-Peterburg 22115
Rossiskaya sovetskaya opytnaya stantsiya, Arkadak 22987
Rossiski gosudarstvenny arkhiv drevnikh aktov, Moskva 23477
Rossiski gosudarstvenny gumanitarny universitet, Moskva 22386
Rossiski gosudarstvenny gumanitarny universitet, Moskva 22387
Rossiski gosudarstvenny istoricheski arkhiv, Sankt-Peterburg 23885
Rossiski gosudarstvenny meditsinski universitet, Moskva 22388
Rossiski gosudarstvenny pedagogicheski universitet, Sankt-Peterburg 22535

Rossiski gosudarstvenny proektny i nauchno-issledovatelski institut po selskomu stroitelstvu, Moskva 23478
Rossiski gosudarstvenny universitet nefti i gaza im. I.M. Gubkina, Moskva 22389
Rossiski institut iskusstvoznaniya, Moskva 23479
Rossiski institut istorii iskusstv, Sankt-Peterburg 23886
Rossiski khimiko-tekhnologicheski universitet im. D.I. Mendeleeva, Moskva 22390
Rossiski nauchno-issledovatelski neirokhirurgicheski institut im. A.L. Polenova, Sankt-Peterburg 23887
Rossiski tsentr nauki i kultury v Vene, Wien 01651
Rossiski universitet druzhby narodov im. P. Lumumby, Moskva 22391
Rossiskoe mineralogicheskoe obshchestvo, Sankt-Peterburg 23888
Rossiskoe obshchestvo otorino-laringologov, Sankt-Peterburg 23889
Rostock University, Rostock 10546
Rostok Ltd, Trade Union Library, Kyiv 28684
Rostov Medical Institute, Rostov-na-Donu 22495
Rostov Mining Research and Design Institute Joint-Stock Company, Rostov-na-Donu 22897
Rostov Museum of Fine Arts, Rostov-na-Donu 23723
Rostov Regional Medical Library, Rostov-na-Donu 23727
Rostov State Academy of Architecture and Art, Rostov-na-Donu 22491
Rostov State Building University, Rostov-na-Donu 22493
Rostov State University, Rostov-na-Donu 22494
Rostov-on-Don Institute of Railway Engineers, Rostov-na-Donu 22492
Rostovskaya oblastnaya detskaya biblioteka, Rostov-na-Donu 24206
Rostovskaya oblastnaya meditsinskaya biblioteka, Rostov-na-Donu 23727
Rostovski gosudarstvenny stroitelny universitet, Rostov-na-Donu 22493
Rostovski gosudarstvenny universitet, Rostov-na-Donu 22494
Rostovski meditsinski institut, Rostov-na-Donu 22495
Roswell Museum & Art Center, Roswell 37468
Roswell P. Flower Memorial Library, Watertown 41766
Roswell Park Cancer Institute Corp, Buffalo 35544
Roswell Public Library, Roswell 41073
Rothamsted Research Library, Harpenen 29681
Rothberg International School Library, Jerusalem 14635
Rother Valley College of Further Education Library, Sheffield 29389
Rotherham College of Arts and Technology, Rotherham 29386
Rotherham Metropolitan Borough Council, Rotherham 30067
Rotorua Public Library, Rotorua 19882
Rotterdamsch Leeskabinet, Rotterdam 19693
Round Lake Area Public Library District, Round Lake 41074
Round Rock Public Library System, Round Rock 41075
Rovaniemen kaupunginkirjasto, Rovaniemi 07605
Rovenki Regional Centralized Library System, Main Library, Rovenki 28803
Rovenkivska raionna TsBS, Tsentralna biblioteka, Rovenki 28803
Rowan County Public Library, Morehead 40295
Rowan Public Library, Salisbury 41151
Rowan University, Glassboro 31213
Rowan-Cabarrus Community College, Salisbury 33720
Rowett Research Institute, Aberdeen 29590
Rowlett Public Library, Rowlett 41077
Roxbury Community College, Boston 33182
Roxbury Correctional Institution Library, Hagerstown 34329
Roxbury Township Public Library, Succasunna 41462
Roxby Downs Community Library, Roxby Downs 00584
Roy A. Childs Jr Library, Washington 37777
Royal Academy for Islamic Civilization Research, Amman 17884
Royal Academy of Arts, London 29809
Royal Academy of Dramatic Art, London 29212

Library of the Royal Academy of Letters, History and Antiquities, Stockholm 26697
Royal Academy of Music, London 29213
Royal Academy of Music, Aarhus, Århus 06653
Royal Academy of Sciences and Arts of Barcelona, Barcelona 25929
Royal Academy oh the Basque Language – Azkue Library, Bilbao 25931
Royal Adelaide Hospital (RAH), Adelaide 00855
Royal Aeronautical Society, London 29810
Royal Agricultural College, Cirencester 29069
Royal Air Force Museum, London 29811
Royal Armouries Library, Leeds 29695
The Royal Armoury, Stockholm 26665
Royal Army Museum, Stockholm 26641
Royal Arsenal (West), London 29812
Royal Asiatic Society Library, London 29813
Royal Asiatic Society of Sri Lanka, Colombo 26317
Royal Astronomical Society, London 29814
Royal Australasian College of Physicians, Sydney 00521
Royal Australian Historical Society, Sydney 01029
Royal Australian Naval College, Jervis Bay 00445
Royal Automobile Club (RAC), London 29815
Royal Belgian Institute of Natural Sciences, Brussel 02601
Royal Borough of Kensington and Chelsea, London 30033
Royal Botanic Garden, Sydney 01030
Royal Botanic Garden Library, Edinburgh 29660
Royal Botanic Gardens, South Yarra 01009
Royal Botanic Gardens, Kew, Richmond 29887
Royal Botanical Gardens Library, Burlington 04314
Royal Canadian Artillery Museum Library, Shilo 04495
Royal Canadian Military Institute Library, Toronto 04040
Royal Canadian Mounted Police Training Academy, Regina 04020
Royal College of Art, London 29214
Royal College of Curepipe (RCC), Curepipe 18690
Royal College of Music, London 29215
Royal College of Nursing, London 29216
Royal College of Obstetricians and Gynaecologists, London 29816
Royal College of Organists, London 29217
Royal College of Physicians and Surgeons, Glasgow 29118
Royal College of Physicians of Ireland, Dublin 14473
Royal College of Physicians of London, London 29218
Royal College of Surgeons in Ireland, Dublin 14474
Royal College of Surgeons of Edinburgh, Edinburgh 29096
Royal College of Surgeons of England, London 29219
Royal College of Veterinary Surgeons, London 29210
Royal Conservatoire, Den Haag 19142
Royal Conservatory of Madrid, Madrid 25426
Royal Danish Academy of Fine Arts – School of Architecture, København 06712
Royal Danish Academy of Music, København 06711
Royal Danish Air Force Library, Ballerup 06757
Royal Danish Military Library, København 06760
Royal (Dick) School of Veterinary Studies, Edinburgh 29104
Royal Dramatic Theatre, Stockholm 26661
Royal Dublin Society, Dublin 14554
Royal Dutch Shell plc, Den Haag 19331
Royal Engineers Library, Chatham 29633
Royal Entomological Society, St Albans 29897
Royal Faculty of Procurators in Glasgow, Glasgow 29676
Royal Filmarchive of Belgium, Bruxelles 02584
Royal Forest of Dean College, Coleford 29072
Royal Geographical Society of South Australia, Adelaide 00856
Royal Geographical Society (with the Institute of British Geographers), London 29817
Royal Geological Society of Cornwall, Penzance 29876
Royal Haskoning, Nijmegen 19335
Royal Historical Society of Victoria, Melbourne 00958
Royal Horticultural Society, Guildford 29678
Royal Horticultural Society, London 29818
The Royal Institute, Bangkok 27768
Royal Institute for Cultural Heritage, Bruxelles 02595
Royal Institute of International Affairs, London 29819
Royal Institute of Management, Thimphu 02916
Royal Institute of Technology, Stockholm 26430
– Forumlibrary, Kista 26432
– Mathematics Library, Stockholm 26433
– Royal Institute of Technology, Architectural Library, Stockholm 26431
Royal Institution of Chartered Surveyors, London 29820
Royal Institution of Great Britain, London 29821
Royal Institution of Naval Architects, London 29822
Royal Irish Academy, Dublin 14555
Royal Lake Club Library, Kuala Lumpur 18604
The Royal Library, København 06623
Royal Library, Torino 14815
Royal Library of Belgium, Bruxelles 02257
The Royal Library, Slotsholmen – Danish Museum of Books and Printing, København 06625
The Royal Library, Slotsholmen – Drama Collection, København 06626
The Royal Library, Slotsholmen – Manuscript Department, København 06627
The Royal Library, Slotsholmen – Map Collection, København 06628
The Royal Library, Slotsholmen – Music and Theatre Department, København 06629
The Royal Library, Slotsholmen – Print and Photograph Collection, København 06624
Royal London Hospital Patients Library, London 30034
Royal Marines Museum, Southsea 29896
Royal Melbourne Hospital, Parkville 00977
Royal Military Academy, Bruxelles 02282
Royal Military Academy Sandhurst, Camberley 28982
Royal Military College, Duntroon 00434
Royal Military College of Canada – Collège militaire royal du Canada, Kingston 03730
Royal Military Museum, Brussel 02604
Royal Museum of Fine Arts Antwerp, Library, Antwerpen 02517
Royal Museum of Scotland, Edinburgh 29658
Royal Museums of Art and History, Bruxelles 02611
Royal Museums of Fine Arts, Bruxelles 02612
Royal National Institute for the Deaf, London 29220
Royal Naval Museum, Portsmouth 29881
Royal Nepal Academy, Kathmandu 19040
Royal Netherlands Institute, Roma 16544
Royal Netherlands Institute of Southeast Asian and Caribbean Studies, Leiden 19474
Royal Netherlands Meteorological Institute, De Bilt 19399
Royal New Zealand Air Force Auckland, Auckland 19796
Royal North Shore Hospital, Saint Leonards 00999
Royal Northern College of Music, Manchester 29268
Royal Norwegian Air Force Academy Library, Trondheim 20187
Royal Norwegian Navy Museum, Horten 20228
Royal Oak Public Library, Royal Oak 41079
Royal Observatory, Edinburgh 29661
Royal Ontario Museum, Toronto 04523
Royal Perth Hospital, Perth 00984
Royal Pharmaceutical Society of Great Britain, London 29823
Royal Prince Alfred Hospital, Camperdown 00886
Royal School of Library and Information Science, København 06664
Royal Scientific Society (RSS), Amman 17869
Royal Scottish Academy of Music and Drama, Glasgow 29119
Royal Scottish Geographical Society, Perth 29877
Royal Society, London 29824
Royal Society for the Encouragement of Arts, Manufactures and Commerce (RSA), London 29825
Royal Society for the Prevention of Accidents (RoSPA), Birmingham 29614
Royal Society for the Protection of Birds, Sandy 29892
Royal Society of Chemistry, London 29826
Royal Society of Medicine, London 29827
Royal Society of Queensland, Saint Lucia 01002
Royal Society of Queensland, South Brisbane 01006
Royal Society of Sciences at Uppsala, Uppsala 26707
Royal Society of South Africa, Rondebosch 25175
Royal Society of South Australia, Adelaide 00857
Royal Society of Victoria, Melbourne 00959
Royal South Hants Hospital, Southampton 29895
Royal Statistical Society, London 29828
Royal Swedish Academy of Agriculture and Forestry, Stockholm 26662
Royal Swedish Academy of Fine Arts, Stockholm 26660
Royal Swedish Fortifications Administration, Eskilstuna 26568
Royal Swedish Society of Naval Sciences, Karlskrona 26600
The Royal Theatre's Archives and Library, København 06807
Royal Tropical Institute, Amsterdam 19366
Royal Tyrrell Museum of Palaeontology Library, Drumheller 04331
Royal United Hospital NWS Trust Library, Bath 29604
Royal University College, Stockholm 26429
Royal Veterinary College, London 29249
Royal Wawel Castle, State Art Collections, Kraków 21151
Royal Welsh College of Music and Drama, Cardiff 29059
Rozdolne Regional Centralized Library System, Main Library, Rozdolne 28804
Rozdolnenska raionna TsBS, Tsentralna biblioteka, Rozdolne 28804
Rozhnyativ Regional Centralized Library System, Main Library, Rozhnyativ 28805
Rozhnyativska raionna TsBS, Tsentralna biblioteka, Rozhnyativ 28805
RPC Information Centre, Fredericton 04349
RSM Erasmus University, Rotterdam 19189
– Economics Faculty, Rotterdam 19190
RSZ – Rijksdienst voor de Sociale Zekerheid, Brussel 02437
RTI International, Research Triangle Park 37437
RTV Slovenija, Ljubljana 24932
Rubber Factory, Efremov 22787
Rubber Research Institute of India, Kottayam 14062
Rubber Research Institute of Sri Lanka, Agalawatta 26301
Rubber Technology and Engineering Research Institute Joint-Stock Company, Tambov 22960
Rubenianum, Antwerpen 02531
Rubiconia Accademia dei Filopatridi, Savignano sul Rubicone 16821
Rubin Maritime Technology Central Design Bureau, Sankt-Peterburg 23924
Rubizhanska raionna TsBS, Tsentralna biblioteka, Rubizhne 28806
Rubizhne Regional Centralized Library System, Main Library, Rubizhne 28806
Rublev, Andrey, Museum of Ancient Russian Art, Moskva 23480
Rubtsovsk Mechanical Engineering Professional School, Rubtsovsk 22691
Rubtsovski mashinostroitelny tekhnikum, Rubtsovsk 22691
Ruden, McClosky, Smith, Schuster & Russell, Fort Lauderdale 35711
Rudersdal Bibliotekerne, Birkerød 06850
Rudjer Boskovic Institute, Zagreb 06244
Rudolf Steiner Bibliothek, Stuttgart 12538
M. I. Rudomino All-Russian State Library for Foreign Literature, Moskva 22155
Ruhrgas AG, Essen 11392
Ruhr-Universität Bochum, Bochum 09464
– Astronomisches Institut, Bochum 09465
– Fakultät für Ostasienwissenschaften, Bochum 09466
– Fakultät für Philologie, Fachbereich Anglistik/Amerikastudien, Bochum 09467
– Fakultät für Philologie, Fachbereich für Allgemeine und Vergleichende Literaturwissenschaft, Bochum 09468
– Fakultät für Philologie, Fachbereich Germanistik, Bochum 09469
– Fakultät für Philologie, Fachbereich Klassische Philologie, Bochum 09470
– Fakultät für Philologie, Fachbereich Orientalistik und Islamwissenschaften, Bochum 09471
– Fakultät für Philologie, Fachbereich Romanistik, Bochum 09472
– Fakultät für Philologie, Fachbereich Slavistik, Bochum 09473
– Fakultät für Philologie, Fachbereich Sprachlehrforschung, Bochum 09474
– Fakultät für Philologie, Fachbereich Sprachwissenschaft, Bochum 09475
– Fakultät für Psychologie, Bochum 09476
– Fakultät für Sozialwissenschaften, Bochum 09477
– Fakultät für Sportwissenschaft, Bochum 09478
– Fakultätsbibliothek für Bau- und Umweltingenieurwissenschaften, Bochum 09479
– Fakultätsbibliothek für Biologie und Biotechnologie, Bochum 09480
– Fakultätsbibliothek für Chemie und Biochemie, Bochum 09481
– Fakultätsbibliothek für Elektrotechnik und Informationstechnik, Bochum 09482
– Fakultätsbibliothek für Mathematik, Bochum 09483
– Fakultätsbibliothek für Physik und Astronomie, Bochum 09484
– Fakultätsbibliothek für Wirtschaftswissenschaft, Bochum 09485
– Geographisches Institut, Bochum 09486
– Hegel-Archiv, Bochum 09487
– Historische Bibliothek, Bochum 09488
– Institut für Archäologische Wissenschaften, Bochum 09489
– Institut für Automatisierungstechnik, Bochum 09490
– Institut für Entwicklungsforschung und Entwicklungspolitik, Bochum 09491
– Institut für Erziehungswissenschaften, Bochum 09492
– Institut für Geologie, Mineralogie und Geophysik, Bochum 09493
– Institut für Medizinische Ethik und Geschichte der Medizin, Bochum 09494
– Institut für Philosophie, Bochum 09495
– Institut für Prävention und Arbeitsmedizin der Deutschen Gesetzlichen Unfallversicherung, Bochum 09496
– Institut für Sozialrecht, Bochum 09497
– Institut für Thermo- und Fluiddynamik, Bochum 09498
– Kunstgeschichtliches Institut, Bochum 09499
– Medizinische Fakultät, Bochum 09500
– Ökumenisches Institut, Bochum 09501
– Theologische Bibliothek, Bochum 09502
– Zentrales Rechtswissenschaftliches Seminar, Bochum 09503
Ruhr-University Bochum, Bochum 09464
– Fakulty of East Asian Studies, Bochum 09466
– Institute for Philosophy, Bochum 09495
'Rui Barbosa' Library, Bauru 02987
Ruidoso Public Library, Ruidoso 41080
Rumänisches Institut, Rumänische Bibliothek, Freiburg e.V., Freiburg 11900
Rumberger, Kirk & Caldwell, PA, Orlando 36075
Rundfunkanstalt Berlin-Brandenburg, Berlin 11557
Runö folkhögskola, Åkersberga 26467
Rupertinum, Salzburg 01570
Ruppin Institute, Emek Hefer 14674
Ruppiner Kliniken GmbH, Neuruppin 12363
Ruprecht-Karls-Universität Heidelberg, Heidelberg 09978
– Anglistisches Seminar, Heidelberg 09979
– Bereichsbibliothek Altertumswissenschaften, Heidelberg 09980
– Bereichsbibliothek Altertumswissenschaften / Alte Geschichte und Epigraphik, Heidelberg 09981
– Bereichsbibliothek Altertumswissenschaften / Klassische Archäologie, Heidelberg 09982
– Bereichsbibliothek Altertumswissenschaften / Klassische Philologie, Heidelberg 09983
– Bereichsbibliothek Mathematik und Informatik, Heidelberg 09984
– Bereichsbibliothek Physik und Astronomie, Heidelberg 09985
– Campus-Bibliothek Bergheim, Heidelberg 09986
– Chemische Institute, Heidelberg 09987
– Diakoniewissenschaftliches Institut, Heidelberg 09988
– Geographisches Institut, Heidelberg 09989
– Germanistisches Seminar, Heidelberg 09990
– Historisches Seminar, Heidelberg 09991
– Institut für Ausländisches und Internationales Privat- und Wirtschaftsrecht, Heidelberg 09992
– Institut für Bildungswissenschaft, Heidelberg 09993
– Institut für Deutsches und Europäisches Gesellschafts- und Wirtschaftsrecht, Heidelberg 09994

– Institut für Deutsches und Europäisches Verwaltungsrecht, Heidelberg 09995
– Institut für Europäische Kunstgeschichte, Heidelberg 09996
– Institut für Fränkisch-Pfälzische Geschichte und Landeskunde, Heidelberg 09997
– Institut für Geowissenschaften, Heidelberg 09998
– Institut für Gerontologie, Heidelberg 09999
– Institut für Geschichte und Ethik der Medizin, Heidelberg 10000
– Institut für Geschichtliche Rechtswissenschaft, Heidelberg 10001
– Institut für Kunstgeschichte Ostasiens, Heidelberg 10002
– Institut für Pflanzenwissenschaften, Abteilung Biodiversität und Pflanzensystematik, Heidelberg 10003
– Institut für Sinologie, Heidelberg 10004
– Institut für Sport und Sportwissenschaft, Heidelberg 10005
– Institut für Ur- und Frühgeschichte und Vorderasiatische Archäologie im Zentrum für Altertumswissenschaften, Heidelberg 10006
– Japanologisches Seminar, Heidelberg 10007
– Juristische Fakultät, Heidelberg 10008
– Kirchhoff-Institut für Physik, Heidelberg 10009
– Medizinische Fakultät Mannheim, Mannheim 10010
– Medizinische Klinik, Heidelberg 10011
– Musikwissenschaftliches Seminar, Heidelberg 10012
– Ökumenisches Institut, Heidelberg 10013
– Pathologisches Institut, Heidelberg 10014
– Philosophisches Seminar, Heidelberg 10015
– Praktisch-Theologisches Seminar, Heidelberg 10016
– Psychiatrische Klinik, Heidelberg 10017
– Psychologisches Institut, Heidelberg 10018
– Romanisches Seminar, Heidelberg 10019
– Schurman-Bibliothek für Amerikanische Geschichte des Historischen Seminars, Heidelberg 10020
– Seminar für Lateinische Philologie des Mittelalters und der Neuzeit, Heidelberg 10021
– Seminar für Osteuropäische Geschichte, Heidelberg 10022
– Seminar für Sprachen und Kulturen des Vorderen Orients, Heidelberg 10023
– Seminar für Übersetzen und Dolmetschen, Heidelberg 10024
– Siebenbürgische Bibliothek, Gundelsheim 10025
– Slavisches Institut, Heidelberg 10026
– Sprachwissenschaftliches Seminar, Heidelberg 10027
– Studentenbücherei, Heidelberg 10028
– Südasien Institut, Heidelberg 10029
– Universitätsklinik / Akademie für Gesundheitsberufe Heidelberg gGmbH, Heidelberg 10030
– Wissenschaftlich-Theologisches Seminar, Heidelberg 10031
– Zentrales Sprachlabor, Heidelberg 10032
– Zentrum für Astronomie Heidelberg (ZAH), Astronomisches Recheninstitut, Heidelberg 10033
– Zentrum für Astronomie Heidelberg (ZAH), Landessternwarte, Heidelberg 10034
– Zentrum für Kinder- und Jugendmedizin, Heidelberg 10035
– Zentrum für Psychosoziale Medizin (ZPM), Heidelberg 10036
Rural Development Administration, Suweon 18105
Rural Development Organization, Manipur 14073
Ruse 'Liuben Karavelov' Regional Library, Ruse 03601
Rusk County Library, Henderson 39416
Rusk State Hospital, Rusk 37471
Ruskin College, Oxford 29304
Ruskin, Moscou & Faltischek Pc, Uniondale 36319
Russell Fox Library, Canberra 00651
Russell Library, Middletown 40187
Russia Federal Committee for Radio and TV, Moskva 23176
Russia Industrial Ministry – Institute of Informatics, Economics and Technology, Sankt-Peterburg 22764
Russia National Public Library for Science and Technology, Moskva 22153
Russian Academy of Arts, Sankt-Peterburg 23883
Russian Academy of Arts – Moscow Branch, Moskva 23472
Russian Academy of Foreign Trade, Moskva 22405
Russian Academy of Medical Sciences, Moskva 23473
Russian Academy of Sciences Archive, Moskva 23474
Russian Academy of Sciences – Archives – St. Petersburg Branch, Sankt-Peterburg 23884
Russian Academy of Sciences – Bashkir Branch, Ufa 22184
Russian Academy of Sciences – Buryat Branch of the Siberian Department, Ulan-Ude 22186
Russian Academy of Sciences – Dagestan Branch, Makhachkala 22152
Russian Academy of Sciences Library, Sankt-Peterburg 22171
Russian Book Chamber (RBC), Moskva 23481
Russian Committee of Trade, Moskva 23343
Russian Cultural and Scientific Centre, Helsinki 07510
Russian Entomological Society, Sankt-Peterburg 23890
Russian Experimental Station, Arkadak 22987
Russian Federal Research Institute of Fisheries and Oceanography (VNIRO), Moskva 23582
Russian Federation Committee for Press, Moskva 22738
Russian Geographical Society Library, Sankt-Peterburg 23891
Russian Geological Library, Sankt-Peterburg 23932
Russian Institute for the History of Arts, Sankt-Peterburg 23886
Russian Institute for the Theory of Art, Moskva 23479
Russian Merchants' Association in Helsinki, Helsinki 07460
Russian Mineralogical Society, Sankt-Peterburg 23888
Russian National Agricultural Microbiology Research Institute, Sankt-Peterburg 23946
Russian National Aluminium-Magnesium Institute (VAMI), Sankt-Peterburg 23860
Russian National Exhibition Centre, Moskva 23597
Russian National Experimental Veterinary Science Institute, Moskva 23545
Russian National Extra-Mural Financial and Economic Institute, Moskva 22409
Russian National Extra-Mural Institute of Railway Engineers, Moskva 22410
Russian National Institute for Oil and Gas Industry Automation, Moskva 23568
Russian National Institute of Electrical Engineering, Moskva 22406
Russian National Institute of Hydrotechnology and Sewerage, Moskva 23542
Russian National Institute of the Refrigeration Industry, Moskva 23567
Russian National Jewellery Industry Institute, Sankt-Peterburg 23948
Russian National Leguminous Crops Research Institute, Oryol 23680
Russian National Natural Gas Research Institute, Moskva 23450
Russian National Nuclear Power Engineering Research Institute, Moskva 23556
Russian National Oil and Land-Prospecting Research Institute, Sankt-Peterburg 23947
Russian National Oil Refining Research Institute, Moskva 23579
Russian National Plant Preservation Research Institute / Russian National Agricultural Microbiology Research Institute, Sankt-Peterburg 23946
Russian National Research and Development Institute for Environmental Protection in the Coal Industry, Perm 23694
Russian National Research Institute for Confectionery Industry, Moskva 23570
Russian National Research Institute for Food Flavourings, Acids and Colourings, Sankt-Peterburg 23941
Russian National Research Institute for the Mechanization of Agriculture, Moskva 23573
Russian National Research Institute of Aerospace Geology Methods, Sankt-Peterburg 23936
Russian National Research Institute of Assembly and Construction, Moskva 23575
Russian National Research Institute of Canning and Food Drying Industry, Moskva 23548
Russian National Research Institute of Electrical and Mechanical Engineering, Sankt-Peterburg 23935
Russian National Research Institute of Geophysics, Moskva 23561
Russian National Research Institute of Geophysics and Prospecting, Sankt-Peterburg 23945
Russian National Research Institute of Industrial Design, Moskva 23584
Russian National Research Institute of Plant Cultivation, Sankt-Peterburg 23943
Russian National Research Institute of Veterinary Sanitation, Hygiene and Ecology, Moskva 23587
Russian National State Institute of Cinematography, Moskva 22407
Russian National Viticulture and Oenology Research Institute, Novocherkassk 23630
Russian Natural Gas Research Institute, Razvilka 23716
Russian Pharmacological Society, Moskva 23482
Russian Polenov Neurosurgical Institute, Sankt-Peterburg 23887
Russian Research Institute of Oceanology and Marine Geology, Sankt-Peterburg 23940
Russian State Art Library, Moskva 23476
Russian State Children's Library, Moskva 22154
Russian State Historical Archive, Sankt-Peterburg 23885
Russian State Institute of Agricultural Engineering, Moskva 23478
Russian State Library, Moskva 22114
Russian State Library for the Blind, Moskva 23475
Russian State Teacher-Training University, Sankt-Peterburg 22535
Russian State University of Humanities, Moskva 22387
Russian State University of Oil and Gas, Moskva 22389
Russian Theatrical Society – Rostov Branch, Rostov-na-Donu 23729
Russisches Kulturinstitut, Wien 01651
Russkoe geografícheskoe obshchestvo RAN, Sankt-Peterburg 23891
Rust College, Holly Springs 31344
Rustenburg Public Library, Rustenburg 25232
Rutan & Tucker Library, Costa Mesa 35641
Rutgers, The State University of New Jersey, Camden
– Camden Law Library, Camden 30663
– Paul Robeson Library, Camden 30664
Rutgers, The State University of New Jersey, Newark
– Criminal Justice NCCD Collection, Newark 32100
– Institute of Jazz Studies Library, Newark 32101
– John Cotton Dana Library, Newark 32102
– Library for the Center for Law & Justice, Newark 32103
Rutgers University, New Brunswick 31954
– Archibald Stevens Alexander Library, New Brunswick 31955
– Art Library, New Brunswick 31956
– Blanche & Irving Laurie Music Library, New Brunswick 31957
– Center of Alcohol Studies Library, Piscataway 31958
– Chemistry Library, Piscataway 31959
– East Asian Library, New Brunswick 31960
– Kilmer Library, Piscataway 31961
– Library of Science and Medicine, Piscataway 31962
– Mabel Smith Douglass Library, New Brunswick 31963
– Mathematical Sciences Library, Piscataway 31964
– Physics Library, Piscataway 31965
– Special Collections and University Archives, New Brunswick 31966
– Stephen & Lucy Chang Science Library, New Brunswick 31967
Rutgers WPF, Utrecht 19529
Ruth Enlow Library of Garrett County, Oakland 40577
Rutherford B. Hayes Presidential Center Library, Fremont 36881
Rutherford County Library, Spindale 41404
Rutherford Free Public Library, Rutherford 41086
Rutland Free Library, Rutland 41087
Rutland Free Public Library, Rutland 41088
Ruusbroecgenootschap, Antwerpen 02532
R+V Versicherung, Wiesbaden 11437
RWE npower, Swindon 29583
RWTH Aachen University, Aachen 09347
– Department of Ferrous Metallurgy, Aachen 09357
– Geotechnical Engineering Library, Aachen 09375
– Institute of Mineralogy and Economic Geology, Aachen 09361
Ryazan Central City Library, Ryazan 24211
Ryazan Regionl Art Museum, Ryazan 23734
Ryazanskaya oblastnaya detskaya biblioteka, Ryazan 24209
Ryazanskaya oblastnaya universalnaya nauchnaya biblioteka im. Gorkogo, Ryazan 22169
Ryazanskaya oblastnaya yunosheskaya biblioteka, Ryazan 24210
Ryazanskaya tsentralnaya gorodskaya biblioteka, Ryazan 24211
Ryazanski meditsinski institut im. Pavlova, Ryazan 22497
Ryazanski mezhotraslevoi territorialny tsentr nauchno-tekhnicheskoi informatsii i propagandy, Ryazan 23735
Ryazanski selskokhozyaistvenny institut im. Kostycheva, Ryazan 22498
Rybinsk Aviation Technology Institute, Rybinsk 22499
Ryde Library Service, North Ryde 01133
Rye Free Reading Room, Rye 41089
Ryerson University, Toronto 03885
Ryerss Museum & Library, Philadelphia 37362
Ryukoku Diagaku, Kyoto 17029
Ryukoku University, Kyoto 17029
Rzeszów University of Technology, Rzeszów 20922
S. Seyfullin Republican Public Library, Astana 17985
S. Syman Public Health Library, Jerusalem 14687
SA Attorney General's Department, Adelaide 00602
SA Department of Health, Rundle Mall 00720
SA Department of Human Services, Adelaide 00858
SA Genealogy and Heraldry Society, Adelaide 00859
Saale-Holzland-Kreis, Verwaltungssteuerung, Kreisarchiv, Dornburg-Camburg 11742
Saare County Central Library (SCL), Kuressaare 07248
Saarijärven kaupunginkirjasto, Saarijärvi 07606
Saarijärvi City Library, Saarijärvi 07606
Saarland University and State University, Saarbrücken 09336
Saarländische Universitäts- und Landesbibliothek, Saarbrücken 09336
Saarländischer Rundfunk, Saarbrücken 12454
Saarlandmuseum, Saarbrücken 12455
Sabah State Library, Kota Kinabalu 18626
SABAM, Brussel 02622
Sabine Parish Library, Many 40044
SAC – Scottish Agricultural College, Edinburgh 29097
Sachem Public Library, Holbrook 39453
Sächsische Landesamt für Umwelt, Landwirtschaft und Geologie, Leipzig 12199
Sächsische Landesbibliothek – Staats- und Universitätsbibliothek Dresden (SLUB), Dresden 09311
– Zweigbibliothek Erziehungswissenschaften, Dresden 09675
– Zweigbibliothek Forstwesen Tharandt, Tharandt 09676
– Zweigbibliothek Medizin, Dresden 09677
– Zweigbibliothek Rechtswissenschaft, Dresden 09678
Sächsische Landesschule für Hörgeschädigte, Förderzentrum Samuel Heinicke, Leipzig 12180
Sächsische Landesstelle für Museumswesen, Fachbereich Volkskultur, Chemnitz 11711
Sächsischer Landtag, Dresden 10862
Sächsisches Industriemuseum Chemnitz, Chemnitz 11712
Sächsisches Landesamt für Umwelt, Landwirtschaft und Geologie, Dresden 10863
Sächsisches Landesamt für Umwelt, Landwirtschaft und Geologie, Dresden 11774
Sächsisches Landesamt für Umwelt, Landwirtschaft und Geologie, Freiberg 10902
Sächsisches Staatsarchiv, Dresden 11775
Sächsisches Staatsministerium der Justiz und für Europa, Dresden 10864
Sächsisches Textilforschungsinstitut e.V. an der Technischen Universität Chemnitz, Chemnitz 11713
Saco Museum, Saco 37472
Sacramento City College, Sacramento 33709

Sacramento County Public Law Library, Sacramento 34766
Sacramento Public Library, Sacramento 41092
Sacred Heart Academy, Hamden 33395
Sacred Heart Major Seminary, Detroit 35187
Sacred Heart Medical Center at RiverBend, Springfield 37632
Sacred Heart School of Theology, Franklin 35209
Sacred Heart Seminary, Victoria 18659
Sacred Heart University, Fairfield 31122
Sacred Heart University, Santurce 21888
Sacro Convento di Assisi, Assisi 15717
Sacro Eremo di Camaldoli, Camaldoli 15769
Sacro Monte della Verna, Chiusi della Verna 15797
Sadat Academy for Management Sciences (SAMS), Cairo 07185
Saddle Brook Free Public Library, Saddle Brook 41093
Saddleback College, Mission Viejo 33544
Sądecka Biblioteka Publiczna im. Józefa Szujskiego, Nowy Sącz 21510
Sadie Pope Dowdell Library of South Amboy, South Amboy 41353
SADOCC, Wien 01593
Safety Agency Library, Hydrographic Dept, Tokyo 17305
Safety Harbor Public Library, Safety Harbor 41094
Säffle bibliotek, Säffle 26888
Safford City-Graham County Library, Safford 41095
SAFTI Military Institut, Singapore 24601
Saga Prefecture Administrative Data Room, Saga 17292
Saga University, Saga 17091
Sagaken Gyosei Shiryoshitsu, Saga 17292
Sagami Chemical Research Center, Sagamihara 17645
Sagami Chuo Kagaku Kenkyujo, Sagamihara 17645
"Sagarriga Visconti Volpi" National Library, Bari 14786
The Sage Colleges, Albany 33121
The Sage Colleges, Troy 32835
Saginaw County Law Library, Saginaw 34770
Saginaw Valley State University, University Center 32874
Saha Institute of Nuclear Physics, Kolkata 14058
Sahitya Akademi / National Academy of Letters, New Delhi 14167
SAI Global Ltd., Sydney 00840
Saiko Saibansho Toshokan, Tokyo 17323
Sain Ffagan: Amguedda Werin Cymru – Llyfrgell / St Fagans: National History Museum Library, Cardiff 29631
Saint Albert Public Library, St. Albert 04762
Saint Ambrose University, Davenport 30958
Saint Andrew's College, Saskatoon 04028
Saint Andrew's Greek Orthodox Theological College, Redfern 00792
Saint Andrews Presbyterian College, Laurinburg 31571
Saint Anselm College, Manchester 31743
Saint Atanaz Greek Catholic Theological Institute Library, Nyíregyháza 13340
Saint Augustine Historical Society, Saint Augustine 37480
Saint Augustine's College, Raleigh 32415
Saint Basil College Library, Stamford 32713
Saint Benedict's Abbey, Benet Lake 35131
Saint Benedict's Monastery, Arcadia 00748
Saint Bonaventure University, Saint Bonaventure 32492
Saint Catharine College, Saint Catharine 33710
Saint Charles Borromeo Seminary, Wynnewood 35427
Saint Charles City County Library District, Saint Peters 41137
Saint Charles City County Library District – Corporate Parkway Branch, Wentzville 41802
Saint Charles City County Library District – Deer Run Branch, O'Fallon 40589
Saint Charles City County Library District – Kathryn Linnemann Branch, Saint Charles 41101
Saint Charles City County Library District – Kisker Road Branch, Saint Charles 41102
Saint Charles City County Library District – McClay, Saint Charles 41103
Saint Charles City County Library District – Middendorf-Kredell Branch, O'Fallon 40590
Saint Charles City County Library District – Spencer Road Branch, Saint Peters 41138
Saint Charles Community College, Saint Peters 33715
Saint Charles Parish Library, Luling 39979
Saint Charles Seminary, Guildford 00443
Saint Clair County Library System, Port Huron 40854
Saint Clair County Library System – Main Library, Port Huron 40855
Saint Clair Shores Public Library, Saint Clair Shores 41105
Saint Clairsville Public Library, Saint Clairsville 41106
Saint Cloud Public Library, Saint Cloud 41108
Saint Dominic's Priory, East Camberwell 00763
Saint Edwards University, Austin 30307
Saint Elizabeth Medical Center, Lafayette 37003
Saint Fedele Cultural Foundation, Milano 15908
Saint Francis College, Brooklyn 30573
Saint Francis Hospital & Medical Center, Hartford 36921
Saint Francis Medical Center, Lynwood 37075
Saint Francis Medical Center, Trenton 37683
Saint Francis Theological College, Milton 00780
Saint Francis University, Loretto 31643
Saint Francis Xavier University, Antigonish 03671
– Marie Michael Library, Antigonish 03672
Saint George Hospital, Kogarah 00937
Saint Helena Public Library, Saint Helena 41111
Saint John Fisher College, Rochester 32462
Saint John Law Society Library, Saint John 04479
Saint John Library Region, Saint John 04744
Saint John Medical Center, Tulsa 37693
Saint John the Baptist Parish Library, LaPlace 39804
Saint John Vianney College, Miami 31811
Saint John's College, Morpeth 00781
Saint John's College, Santa Fe 32614
Saint John's College, Winnipeg 03970
Saint Johns County Public Library System – Southeast Branch Library & Administrative Headquarters, Saint Augustine 41100
Saint John's Mercy Medical Center, Saint Louis 37491
Saint John's Seminary, Brighton 35146
Saint John's University, Collegeville 30865
Saint John's University, Jamaica 31453
– Davis Library – Manhattan Campus, New York 31454
– Kathryn & Shelby Cullom Davis Library, New York 31455
– Library and Information Science Library, Jamaica 31456
– Rittenberg Law Library, Jamaica 31457
Saint John's University, Staten Island 32736
Saint Joseph College, West Hartford 33017
Saint Joseph County Law Library, South Bend 34874
Saint Joseph County Public Library, South Bend 41354
Saint Joseph Seminary College, Saint Benedict 35354
Saint Joseph's Abbey, Spencer 35388
Saint Joseph's College, Brooklyn 30574
Saint Joseph's College, Edmonton 03989
Saint Joseph's College, Rensselaer 32429
Saint Joseph's College, Standish 32715
Saint Joseph's Seminary, Yonkers 35428
Saint Joseph's University, Philadelphia 32261
Saint Lambert Municipal Library, Saint-Lambert 04749
Saint Lawrence Catholic Church, Minneapolis 35286
Saint Lawrence Hospital & Healthcare Services, Lansing 37015
Saint Leo University, Saint Leo 32501
Saint Louis Art Museum, Saint Louis 37492
Saint Louis Christian College, Florissant 31147
Saint Louis College of Pharmacy, Saint Louis 32506
Saint Louis Community College – Florissant Valley Campus, Ferguson 33337
Saint Louis County Law Library, Clayton 34083
Saint Louis County Law Library, Duluth 34159
Saint Louis County Library, Saint Louis 41119
Saint Louis County Library – Bridgeton Trails Branch, Bridgeton 38279
Saint Louis Mercantile Library at the University of Missouri-St Louis, Saint Louis 37493
Saint Louis Metropolitan Police Department, Saint Louis 37494
Saint Louis Public Library, Saint Louis 41120
Saint Louis Public Library – Buder, Saint Louis 41121
Saint Louis Public Library – Carondelet, Saint Louis 41122
Saint Louis Public Library – Carpenter, Saint Louis 41123
Saint Louis Public Library – Julia Davis Branch, Saint Louis 41124
Saint Louis Public Library – Machacek, Saint Louis 41125
Saint Louis Public Library – Schlafly, Saint Louis 41126
Saint Louis University, Baguio City 20684
Saint Louis University, Saint Louis 32507
– Health Sciences Center Library, Saint Louis 32508
– Omer Poos Law Library, Saint Louis 32509
Saint Lucie County Law Library, Fort Pierce 34258
Saint Lucie County Library System – Fort Pierce Branch, Fort Pierce 39111
Saint Lucie County Library System – Lakewood Park, Fort Pierce 39112
Saint Luke's Hospital, Kansas City 36990
Saint Mark's National Theological Centre, Barton 00751
Saint Martin Parish Library, Saint Martinville 41127
Saint Martin's College, Lacey 31540
Saint Mary Parish Library, Franklin 39148
Saint Mary Seminary, Wickliffe 35420
Saint Mary-of-the-Woods College, Saint Mary-of-the-Woods 32525
Saint Mary's College, Notre Dame 32159
Saint Mary's College, Orchard Lake 32207
Saint Mary's College, Raleigh 33669
Saint Mary's College and Newman College, Parkville 00497
Saint Mary's College of California, Moraga 31898
Saint Mary's College of Maryland, Saint Mary's City 32526
Saint Mary's County Memorial Library, Leonardtown 39853
Saint Mary's Health Center, Saint Louis 37495
Saint Mary's School for the Deaf, Buffalo 33197
Saint Mary's Seminary, Mulgrave 00783
Saint Mary's Seminary & University, Baltimore 35124
Saint Mary's University, Halifax 03722
Saint Mary's University, San Antonio
– Academic Library, San Antonio 32568
– Sarita Kenedy East Law Library, San Antonio 32569
Saint Matthew's & Saint Timothy's Neighborhood Center, Inc Library, New York 36014
Saint Meinrad Archabbey & School of Theology, Saint Meinrad 32527
Saint Michael's College, Colchester 30845
Saint Norbert Abbey, De Pere 35182
Saint Norbert College, De Pere 30974
Saint Olaf College – Howard V. & Edna H. Hong Kierkegaard Library, Northfield 32155
Saint Olaf College – Rolvaag Memorial Library, Glasoe Science Library, Halvorson Music Library, Northfield 32156
Saint Patrick's Seminary, Menlo Park 35279
Saint Paul Fire & Marine, Baltimore 35486
Saint Paul Lutheran Church & School Library, Skokie 35382
Saint Paul Public Library, Saint Paul 41130
Saint Paul Public Library – Hayden Heights, Saint Paul 41131
Saint Paul Public Library- Highland Park, Saint Paul 41132
Saint Paul Public Library – Lexington Outreach, Saint Paul 41133
Saint Paul Public Library – Merriam Park, Saint Paul 41134
Saint Paul Public Library – Sun Ray, Saint Paul 41135
Saint Paul School of Theology, Kansas City 31483
Saint Paul University, Ottawa 03811
Saint Paul's College, Lawrenceville 31581
Saint Paul's College, Washington 32988
Saint Paul's College, Winnipeg 03961
Saint Peter Regional Treatment Center Libraries, Saint Peter 37507
Saint Peter's College, Jersey City 31463
Saint Peter's Seminary, London 03745
Saint Peter's University Hospital Library, New Brunswick 37155
Saint Petersburg College, Pinellas Park 33644
Saint Petersburg Public Library, Saint Petersburg 41139
Saint Petersburg State Polytechnical University, Sankt-Peterburg 22540
Saint Petersburg State Theatre Library, Sankt-Peterburg 23893
Saint Petersburg State University of Information Technologies, Mechanics and Optics and Vavilov State Optical Institute, Sankt-Peterburg 22536
Saint Tammany Parish Library, Covington 38722
Saint Tammany Parish Library – Covington Branch, Covington 38723
Saint Thomas Aquinas College, Sparkill 32692
Saint Thomas More College, Crawley 00761
Saint Thomas More College – University of Saskatchewan, Saskatoon 03848
Saint Thomas Seminary-Hartford Archdiocese, Bloomfield 35139
Saint Vincent College & Seminary, Latrobe 35260
Saint Vincent's Hospital, Fitzroy 00917
Saint Vladimir Institute Library, Toronto 04524
Saint Vladimir's Orthodox Theological Seminary, Yonkers 35429
Saint Xavier University, Chicago 30784
Saint-Andrewsabbey Library, Brugge 02456
Saint-Gobain Isover AB, Billesholm 26527
Saint-Gobain Pam, Maidières-lès-Pont-à-Mousson 08257
Saint-Petersburg State Academy of Engineering and Economics, Sankt-Peterburg 22538
Saint-Petersburg State Polytechnical University, Sankt-Peterburg 22541
The Saison Poetry Library, London 29829
Saitama Agricultural Experiment Station, Kumagaya 17580
Saitama Daigaku, Urawa 17229
– Keizai Gakubu, Urawa 17230
Saitama Kenritsu Minami Kyoiku Senta Shido Sodanbu Shiryoshitsu, Urawa 17811
Saitama Prefectural Administration Information Library, Urawa 17330
Saitama Prefectural Assembly, Urawa 17329
Saitama Prefectural Livestock Experiment Station, Osato 17644
Saitama Prefectural South Education Center, Library and Educational Information Section, Urawa 17811
Saitama University, Urawa 17229
– Faculty of Economics, Urawa 17230
Saitamaken Chikusan Shikenjo, Osato 17644
Saitamaken Gikai, Urawa 17329
Saitamaken Kensei Joho Shiryoshitsu, Urawa 17330
Saitamaken Nogyo Shikenjo, Kumagaya 17580
Sakai City Council, Sakai 17293
Sakaishi Gikai, Sakai 17293
Sakhalin Regional Children's Library, Yuzhno-Sakhalinsk 24365
Sakhalin Regional Universal Scientific Library, Yuzhno-Sakhalinsk 22200
Sakhalin Scientific Research Institute, Novo-Aleksandrovsk 23628
Sakhalinskaya oblastnaya detskaya biblioteka, Yuzhno-Sakhalinsk 24365
Sakhalinskaya oblastnaya universalnaya nauchnaya biblioteka, Yuzhno-Sakhalinsk 22200
Sakhalinski kompleksny nauchno-issledovatelski institut Dalnevostochnogo nauchnogo tsentra RAN, Novo-Aleksandrovsk 23628
Sakhnovshchina Regional Centralized Library System, Main Library, Sakhnovshchina 28807
Sakhnovshchinska raionna TsBS, Tsentralna biblioteka, Sakhnovshchina 28807
Šakių rajono savivaldybės viešoji biblioteka, Šakiai 18358
Saksalainen kirjasto, Helsinki 07449
Sala de Arte Prehispánico, Santo Domingo 06971
Sala stadsbibliotek, Sala 26889
Sălaj District Museum of History and Art, Zalău 22071
Salarjung Museum, Hyderabad 14010
Šalčininkų rajono savivaldybės viešoji biblioteka, Šalčininkai 18359
Salem Athenaeum, Salem 37514
Salem College, Winston-Salem 33084
Salem International College, Überlingen 10771
Salem International University, Salem 32550
Salem Public Library, Salem 41141
Salem Public Library, Salem 41142
Salem Public Library, Salem 41143
Salem Public Library, Salem 41144
Salem State College, Salem 32551
Salem Township Public Library, Morrow 40307
Salem-South Lyon District Library, South Lyon 41368

Salesian College, Chadstone 00760
Salesianische Bibliothek, Eichstätt 11204
Salesianisches Institut, Eichstätt 11204
Salina College of Technology & Aviation, Salina 32554
Salina Public Library, Salina 41146
Salinas Public Library – Cesar Chavez Library, Salinas 41147
Salinas Public Library – El Gabilan, Salinas 41148
Salinas Public Library – John Steinbeck Library, Salinas 41149
Saline County Public Library, Benton 38176
Saline District Library, Saline 41150
Salisbury College, Salisbury 29388
Salisbury Library Service, Salisbury 01153
Salisbury NHS Foundation Trust, Salisbury 29891
Salisbury University, Salisbury 32559
Salish Kootenai College, Pablo 32225
Salk Institute for Biological Studies, San Diego 37525
Sallie Logan Public Library, Murphysboro 40343
Salmon P. Chase College of Law, Highland Heights 31335
Salo Public Library, Salo 07607
Salón de Lectura 'Armando Zuloaga Blanco', Cumaná 42259
Salon kaupunginkirjasto, Salo 07607
Salt Lake City Public Library, Salt Lake City 41153
Salt Lake Community College, Salt Lake City 33721
Salt Lake County Library Services, Salt Lake City 41154
Salt Lake County Library Services – Bingham Creek, West Jordan 41825
Salt Lake County Library Services-Calvin S. Smith Branch, Salt Lake City 41155
Salt Lake County Library Services – Columbus Branch, South Salt Lake City 41377
Salt Lake County Library Services – East Millcreek, Salt Lake City 41156
Salt Lake County Library Services – Holladay, Salt Lake City 41157
Salt Lake County Library Services – Hunter Branch, Salt Lake City 41158
Salt Lake County Library Services – Kearns Branch, Kearns 39689
Salt Lake County Library Services – Magna Branch, Magna 40015
Salt Lake County Library Services – Park, Taylorsville 41526
Salt Lake County Library Services – Riverton Branch, Riverton 41016
Salt Lake County Library Services – Ruth V. Tyler Branch, Midvale 40192
Salt Lake County Library Services – Sandy Branch, Sandy 41206
Salt Lake County Library Services – South Jordan Branch, South Jordan 41365
Salt Lake County Library Services – West Jordan Branch, West Jordan 41826
Salt Lake County Library Services – West Valley Branch, West Valley City 41842
Salt Lake County Library Services – Whitmore Branch, Salt Lake City 41159
Salt River Project Library, Tempe 37664
Saltykov-Shchedrin Library, Sankt-Peterburg 24216
SALUS gGmbH, Bernburg 11606
Salvat Editores, S.A., Barcelona 25854
The Salvation Army, Toronto 04232
Salvation Army Church Library, Las Vegas 35259
Salvation Army School for Officer Training, Suffern 35393
Salve Regina University, Newport 32113
Salzburg Global Seminar, Salzburg 01573
Salzburg Museum, Salzburg 01574
Salzburg University Library, Salzburg 01268
Salzburger Institut für Raumordnung und Wohnen, Salzburg 01575
Salzburger Landesarchiv, Salzburg 01576
Sam Cohen Library, Swakopmund 19013
Sam Houston Regional Library and Research Center, Liberty 37030
Sam Houston State University, Huntsville 31377
Sam Rayburn Library & Museum, Bonham 36554
Samanta Chandra Sekhar Autonomous College, Puri 13839
Samara Institute of Railway Engineers, Samara 22505
Samara Mechanical Engineering Professional School, Samara 22692
Samara Regional Universal Research Library, Samara 22170
Samara State Aerospace University, Samara 22503
Samara State Technical University, Samara 22506
Samara State University, Samara 22504
Samara State University of Architecture and Civil Engineering, Samara 22502
Samarkand Agricultural Institute, Samarkand 42083
Samarkand Co-operative Institute, Samarkand 42123
Samarkand I.P. Pavlov State Medical Institute, Samarkand 42084
Samarkand State Architectural and Civil Engineering Institute, Samarkand 42085
Samarkand State Institute of Foreign Languages, Samarkand 42086
Samarkand State University, Samarkand 42087
Samarskaya oblastnaya universalnaya nauchnaya biblioteka, Samara 22170
Samarskaya oblastnaya yunosheskaya biblioteka, Samara 24214
Samarski gosudarstvenny aerokosmicheski universitet im. S.P. Korolyova, Samara 22503
Samarski gosudarstvenny universitet, Samara 22504
Samarski institut inzhenerov zheleznodorozhnogo transporta, Samara 22505
Samarski mashinostroitelny tekhnikum, Samara 22692
Samarski tekhnicheski universitet, Samara 22506
Sambalpur University, Sambalpur 13759
Sambir Regional Centralized Library System, Main Library, Sambir 28808
Sambirska raionna TsBS, Tsentralna biblioteka, Sambir 28808
Samford University, Birmingham 30431
– Lucille Stewart Beeson Law Library, Birmingham 30432
Sammlung für Plansprachen und Esperantomuseum der österreichischen Nationalbibliothek, Wien 01652
Sammlung Industrielle Gestaltung, Berlin 11588
Sammlung Woldan, Wien 01619
Sammy Brown Library, Carthage 38471
Samostan Stična-Knjižnica, Ivančna Gorica 24895
Samostoyatlelna sektsiya po astronomiya s natsionalna astronomicheska observatoriya, Sofiya 03574
Samoylov Research Institute of Fertilizers and Insecticides, Moskva 23423
Sampson Community College, Clinton 33242
Sampurnanand Sanskrit University, Varanasi 13785
Samsø Bibliotek, Samsø 06922
Samuel Merritt College, Oakland 33593
Samuel Roberts Noble Foundation, Inc, Ardmore 36475
Samuels Public Library, Front Royal 39174
San Antonio College, San Antonio 32570
San Antonio First Baptist Church, San Antonio 35374
San Antonio Public Library, San Antonio 41162
San Beda College, Manila 20750
San Benito County Free Library, Hollister 39458
San Benito County Law Library, Hollister 34358
San Bernardino County Library, San Bernardino 41163
San Bernardino Public Library, San Bernardino 41164
San Bernardino Valley College, San Bernardino 33724
San Bruno Public Library, San Bruno 41165
San Diego Aero-Space Museum, Inc, San Diego 37526
San Diego City Attorney Library, San Diego 34808
San Diego City College, San Diego 33726
San Diego County Law Library – East County, El Cajon 34173
San Diego County Law Library – North County, Vista 34968
San Diego County Law Library – South Bay, Chula Vista 34080
San Diego County Library, San Diego 41168
San Diego County Library – El Cajon Branch, El Cajon 38938
San Diego County Library – Encinitas Branch, Encinitas 38977
San Diego County Library – Fallbrook Branch, Fallbrook 39036
San Diego County Library – La Mesa Branch, La Mesa 39753
San Diego County Library – Lemon Grove Branch, Lemon Grove 39846
San Diego County Library – Poway Branch, Poway 40886
San Diego County Library – Rancho San Diego, El Cajon 38939
San Diego County Library – San Marcos Branch, San Marcos 41193
San Diego County Library – Santee Branch, Santee 41228
San Diego County Library – Solana Beach Branch, Solana Beach 41342
San Diego County Library – Spring Valley Branch, Spring Valley 41411
San Diego County Library – Valley Center Branch, Valley Center 41656
San Diego County Library – Vista Branch, Vista 41694
San Diego County Public Law Library, San Diego 34809
San Diego Family History Center, San Diego 37527
San Diego Historical Society, San Diego 37528
San Diego Mesa College, San Diego 33727
San Diego Miramar College, San Diego 32579
San Diego Museum of Art Library, San Diego 37529
San Diego Museum of Man, San Diego 37530
San Diego Natural History Museum, San Diego 37531
San Diego Public Library, San Diego 41169
San Diego Public Library – Mountain View-Beckwourth, San Diego 41170
San Diego State University, Calexico 30609
San Diego State University, San Diego 32580
San Francisco African-American Historical & Cultural Society, San Francisco 37550
San Francisco Art Institute, San Francisco 37551
San Francisco Botanical Garden Society at Strybing Arboretum, San Francisco 37552
San Francisco Conservatory of Music, San Francisco 32591
San Francisco Law Library, San Francisco 34819
San Francisco Law Library – Market Street, San Francisco 34820
San Francisco Law School Library, San Francisco 37553
San Francisco Maritime National Historical Park, San Francisco 37554
San Francisco Museum of Modern Art, San Francisco 37555
San Francisco Performing Arts Library & Museum, San Francisco 37556
San Francisco Public Library, San Francisco 41173
San Francisco State University, San Francisco 32592
San Jacinto College, Pasadena 33622
San Jacinto College North, Houston 33419
San Jacinto College South, Houston 33420
San Jacinto Museum of History, La Porte 37000
San Joaquin College of Law, Clovis 30842
San Joaquin County Law Library, Stockton 34899
San Joaquin Delta College, Stockton 33780
San Jose City College, San Jose 33733
San Jose Public Library – Almaden, San Jose 41175
San Jose Public Library – Berryessa, San Jose 41176
San Jose Public Library – Biblioteca Latino Americana, San Jose 41177
San Jose Public Library – Calabazas, San Jose 41178
San Jose Public Library – Cambrian, San Jose 41179
San Jose Public Library – Dr Roberto Cruz – Alum Rock, San Jose 41180
San Jose Public Library – Educational Park, San Jose 41181
San Jose Public Library – Evergreen, San Jose 41182
San Jose Public Library – Hillview, San Jose 41183
San Jose Public Library – Rose Garden, San Jose 41184
San Jose Public Library – Seventrees, San Jose 41185
San Jose Public Library – Tully Community, San Jose 41186
San Jose Public Library – Vineland, San Jose 41187
San Jose Public Library West Valley, San Jose 41188
San Jose State University, San Jose 32598
San Juan College, Farmington 33332
San Juan Island Library District, Friday Harbor 39170
San Leandro Public Library, San Leandro 41190
San Luis Obispo City-County Library, San Luis Obispo 41192
San Luis Obispo County Law Library, San Luis Obispo 34830
San Marcos Public Library, San Marcos 41194
San Marino Public Library, San Marino 41195
San Mateo County Law Library, Redwood City 34733
San Mateo County Library, San Mateo 41196
San Mateo County Library – Belmont Branch, Belmont 38165
San Mateo County Library – Foster City Branch, Foster City 39135
San Mateo County Library – Half Moon Bay Branch, Half Moon Bay 39345
San Mateo County Library – Millbrae Branch, Millbrae 40204
San Mateo County Library – Pacifica Branch, Pacifica 40682
San Mateo County Library – San Carlos Branch, San Carlos 41166
San Mateo County Office of Education, Redwood City 37433
San Mateo Public Library, San Mateo 41197
San Miguel County Public Library District 1, Telluride 41532
San Pedro Mártir, Madrid 25982
San Rafael Public Library, San Rafael 41201
San Reference Library, Simon's Town 25179
Sanata Dharma University, Yogyakarta 14325
Sandakan Regional Library, Sandakan 18634
Sandakan Teachers' Training College, Sandakan 18545
Sandefjord bibliotek, Sandefjord 20393
Sandefjordmuseene, Sandefjord 20291
Sir Sandford Fleming College, Lindsay 03741
Sandhills Community College, Pinehurst 33643
Sandia National Laboratories, Albuquerque 36449
Sandia National Laboratories, Livermore 37040
Sandnes bibliotek, Sandnes 20394
Sandusky Bay Law Library Association, Sandusky 34834
Sandusky Library, Sandusky 41204
The Sandvig Collections, Lillehammer 20237
Sandvik Steel AB, Sandviken 26547
Sandvikens folkbibliothek, Sandviken 26890
Sandvikens sjukhus, Sandviken 26630
Sandwell College of Further and Higher Education, Wednesbury 29433
Sandwell Metropolitan Borough Council, West Bromwich 30097
Sandwich Public Library, Sandwich 41205
San-Ei Gen F.F.I., Inc, Toyonaka 17497
Sangyo Igaku Sogo Kankyujo, Kawasaki 17569
Sangyo Noritsu Daigaku, Isehara 16955
Sanibel Public Library District, Sanibel 41211
Sanitarno-gigienicheski meditsinski institut, Sankt-Peterburg 22537
Sanitätsamt der Bundeswehr, München 11049
Sanko Bunka Kenkyujo, Tokyo 17755
Sanko Research Institute, Tokyo 17755
Sankt Andreas Bibliotek, København 06762
Sankt-Peterburgskaya gosudarstvennaya biblioteka dlya slepykh, Sankt-Peterburg 23892
Sankt-Peterburgskaya gosudarstvennaya inzhenerno-ekonomicheskaya akademiya, Sankt-Peterburg 22538
Sankt-Peterburgskaya gosudarstvennaya konservatoriya im. N.A. Rimskogo-Korsakova, Sankt-Peterburg 22539
Sankt-Peterburgskaya gosudarstvennaya teatralnaya biblioteka A.V. Lunacharskogo, Sankt-Peterburg 23893
Sankt-Peterburgskaya Pravoslavnaya Dukhovnaya Akademiya, Sankt-Peterburg 22771
Sankt-Peterburgskaya tsentralnaya nauchnaya selskokhozyaistvennaya biblioteka, Sankt-Peterburg 23894
Sankt-Peterburgski Gosudarstvenny Politekhnicheski Universitet, Sankt-Peterburg 22540
Sankt-Peterburgski gosudarstvenny politekhnicheski universitet, Sankt-Peterburg 22541
Sankt-Peterburgski gosudarstvenny universitet, Sankt-Peterburg 22542
Sankt-Peterburgski nauchno-issledovatelski psikhonevrologicheski institut im. V.M. Bekhtereva, Sankt-Peterburg 23895
Sankt-Peterburgski tsentr nauchno-tekhnicheskaya informatsii, Sankt-Peterburg 23896

Sankt-Petersburg Theological Academy, Sankt-Peterburg 22771
Sanming Teachers' College, Sanming 05496
Sanno College, Isehara 16955
Sanno Tanki Daigaku, Tokyo 17247
Sanofi Pasteur Ltd, Toronto 04291
Sansad Pustakalaya, Kathmandu 19031
Sanshi, Konchu Nogyo Gijutsu Kenkyusho, Tsukuba 17806
Sanskrit College Library, Kolkata 13812
Santa Ana College, Santa Ana 33741
Santa Ana Public Library, Santa Ana 41213
Santa Ana Public Library – Newhope Library Learning Center, Santa Ana 41214
Santa Barbara Botanic Garden Library, Santa Barbara 37566
Santa Barbara City College, Santa Barbara 33742
Santa Barbara County Law Library, Santa Maria 34840
Santa Barbara Mission, Santa Barbara 37567
Santa Barbara Museum of Art, Santa Barbara 37568
Santa Barbara Museum of Natural History Library, Santa Barbara 37569
Santa Barbara News Press Library, Santa Barbara 36255
Santa Barbara Public Library, Santa Barbara 41215
Santa Clara County Law Library, San Jose 34825
Santa Clara County Library, Los Gatos 39962
Santa Clara County Library – Campbell Public, Campbell 38431
Santa Clara County Library – Cupertino Public, Cupertino 38746
Santa Clara County Library – Gilroy Library, Gilroy 39229
Santa Clara County Library – Los Altos Main Library, Los Altos 39941
Santa Clara County Library – Milpitas Public, Milpitas 40209
Santa Clara County Library – Morgan Hill Branch, Morgan Hill 40297
Santa Clara County Library Saratoga Community Library, Saratoga 41232
Santa Clara County Office of Education, San Jose 37561
Santa Clara University, Santa Clara 32608
– Heafey Law Library, Santa Clara 32609
Santa Clara Valley Water District Library, San Jose 34826
Santa Cruz City-County Library System Headquarters – Aptos Branch, Aptos 37976
Santa Cruz City-County Library System Headquarters – Branciforte, Santa Cruz 41218
Santa Cruz City-County Library System Headquarters – Central, Santa Cruz 41219
Santa Cruz City-County Library System Headquarters – Live Oak, Santa Cruz 41220
Santa Cruz County Law Library, Santa Cruz 34837
Santa Fe College, Gainesville 33362
Santa Fe Institute Library, Santa Fe 37575
Santa Fe Public Library, Santa Fe 41222
Santa Fe Springs City Library, Santa Fe Springs 41223
Santa Giustina's Public Library, Padova 15955
Santa Maria Public Library, Santa Maria 41224
Santa Monica College, Santa Monica 33747
Santa Monica Public Library, Santa Monica 41225
Santa Rosa County Library System, Milton 40211
Santa Rosa Junior College, Santa Rosa 33748
Santekhproekt, Moskva 23483
Santiago Canyon College, Orange 32203
Santo Domingo University, Santo Domingo 06957
Santuario de la Virgen del Carmen, Amorebieta-Echano 25760
Santuario dei Marinai, Genova 15851
Santuario delle Grazie, Benevento 16178
Santuario di S. Gabriele, Isola del Gran Sasso 15862
Santuario di S. Gennaro alla Solfatara, Pozzuoli 15998
Santuario Nuestra Señora de Aranzazu, Oñati 25802
Santuario Virgen de Regla, Chipiona 25772
Sanwa Bank Ltd, Osaka 17410
Sanwa Bank Ltd, Research Department, Tokyo 17478
Sanwa Ginko Chosabu, Osaka 17410
Sanwa Ginko Chosabu Toshoshitsu K. K., Tokyo 17478
Sanyo Broadcasting Co, Ltd, Okayama 17399
Sanyo Hoso K. K. Shiryoshitsu, Okayama 17399
São Paulo Academy of Letters, São Paulo 03392
Sao Paulo State University, São Paulo 03174
– Faculty of Agronomy, Botucatu 03175
SAO Sevzapmebel, Sankt-Peterburg 22934
Saperston & Day PC, Buffalo 35545
Sapienza Università di Roma, Roma 15379
– Biblioteca Centrale, Roma 15380
– Biblioteca di Filosofia, Roma 15381
– Biblioteca di Storia della Medicina, Roma 15382
– Biblioteca di Studi Romanzi e Italianistica, Roma 15383
– Biblioteca Ferdinando Milone, Roma 15384
– Biblioteca Interdipartimentale di Psicologia 'E. Valentini', Roma 15385
– Biblioteca Interdipartimentale 'G. Giacomello' di Scienze Chimico-Farmaceutiche, Roma 15386
– Corso di Laurea in Servizio Sociale, Roma 15387
– Dipartimento di Architettura e Urbanistica per l'Ingegneria, Roma 15388
– Dipartimento di Biologia Animale e dell'Uomo, Roma 15389
– Dipartimento di Biologia Animale e dell'Uomo – Sede di Antropologia, Roma 15390
– Dipartimento di Biologia Animale e dell'Uomo – Sede di Zoologia, Roma 15391
– Dipartimento di Biologia Cellulare e dello Sviluppo, Roma 15392
– Dipartimento di Biologia Vegetale, Roma 15393
– Dipartimento di Biopatologia Umana, Roma 15394
– Dipartimento di Caratteri dell' Architettura, Valutazione e Ambiente, Roma 15395
– Dipartimento di Chimica, Roma 15396
– Dipartimento di Contabilità Nazionale e Analisi dei Processi Sociali – Sezione di Contabilità Nazionale, Roma 15397
– Dipartimento di Economia Pubblica, Roma 15398
– Dipartimento di Filologia greca e latina, Roma 15399
– Dipartimento di Fisica, Roma 15400
– Dipartimento di Genetica e Biologia Moleculare 'C. Darwin', Roma 15401
– Dipartimento di Idraulica, Trasporti e Strade, Roma 15402
– Dipartimento di Informatica e Sistemistica, Roma 15403
– Dipartimento di Matematica, Roma 15404
– Dipartimento di Meccanica e Aeronautica, Roma 15405
– Dipartimento di Metodi e Modelli Matematici per le Scienze Applicate, Roma 15406
– Dipartimento di Pianificazione Territoriale e Urbanistica, Roma 15407
– Dipartimento di Scienze Attuariali e Matematica per le Decisioni Economiche e Finanziarie, Roma 15408
– Dipartimento di Scienze Biochimiche 'A. Rossi Fanelli', Roma 15409
– Dipartimento di Scienze Cardiovascolari e Respiratorie, Roma 15410
– Dipartimento di Scienze della Terra, Roma 15411
– Dipartimento di Scienze Neurologiche, Roma 15412
– Dipartimento di Statistica, Probabilità e Statistiche Applicate, Roma 15413
– Dipartimento di Storia dell'Architettura Restauro e Conservazione dei Beni Architettonici, Roma 15414
– Dipartimento di Studi Glottoantropologici, Roma 15415
– Dipartimento di Studi Orientali, Roma 15416
– Dipartimento di Studi Politici, Roma 15417
– Dipartimento di Studi Storici, Roma 15418
– Dipartimento di Studi Storico-Artistici, Archelogici e Sulla Conservazione, Roma 15419
– Dipartimento di Studi Storico-Religiosi, Roma 15420
– Dipartimento di Studi sulle Società e le Culture del Medioevo, Roma 15421
– Dipartimento di Teoria dello Stato, Roma 15422
– Dipartimento di Teoria Economica e Metodi Quantitativi per le Scelte Politiche, Roma 15423
– Dipartimento Ingegneria Strutturale e Geotecnica, Roma 15424
– Dipartimento Scienze Demografiche, Roma 15425
– Facoltà di Economia, Roma 15426
– Facoltà di Ingegneria, Roma 15427
– Istituti di Anatomia Umana Normale, Roma 15428
– Istituto di Archeologia Classica, Roma 15429
– Istituto di Chimica Applicata, Roma 15430
– Istituto di Clinica Otorino-laringoiatrica, Roma 15431
– Istituto di Diritto Comparato, Roma 15432
– Istituto di Diritto Internazionale, Roma 15433
– Istituto di Diritto Privato, Roma 15434
– Istituto di Diritto Pubblico, Roma 15435
– Istituto di Diritto Romano e dei Diritti dell'Oriente Mediterraneo, Roma 15436
– Istituto di Economia e Finanza, Roma 15437
– Istituto di Filologia Romanza, Roma 15438
– Istituto di Filologia Slava, Roma 15439
– Istituto di Filosofia del Diritto, Roma 15440
– Istituto di Fisiologia Umana, Roma 15441
– Istituto di Geografia, Roma 15442
– Istituto di Igiene G. Sanarelli, Roma 15443
– Istituto di Medicina Legale e delle Assicurazioni, Roma 15444
– Istituto di Oftalmologia, Roma 15445
– Istituto di Politica Economica, Roma 15446
– Istituto di Storia Greca, Roma 15447
– Istituto di Storia Medievale, Roma 15448
– Istituto di Storia Romana, Roma 15449
– Istituto Nazionale di Alta Matematica, Roma 15450
– Scuola Speciale per Archivisti e Bibliotecari, Roma 15451
Sapienza University of Rome, Roma 15379
– Animal and human biology department – anthropology, Roma 15390
– Department of Physics, Roma 15400
– Graduates of Social Service, Library Guido e Maria Calogero, Roma 15387
– History of Medicine Library, Roma 15382
– Institute of Comperative Law, Roma 15432
Sapporo Chamber of Commerce and Industry, Sapporo 17660
Sapporo Ika Daigaku, Sapporo 17111
Sapporo Medical College, Sapporo 17111
Sapporo Norin Gakkei, Sapporo 17659
Sapporo Shoko Kaigisho, Sapporo 17660
Sapporo Society of Agriculture and Fisheries, Sapporo 17659
Sapporo University, Sapporo 17112
Sara Hilden Art Museum, Tampere 07541
Sara Hildénin taidemuseo, Tampere 07541
Sarabhai Research Centre, Vadodara 14218
Saradavilas College, Mysore 13829
Sarah Lawrence College, Bronxville 30562
Sarapul Mechanical Professional School, Sarapul 22717
Sarapulski mekhanicheski tekhnikum, Sarapul 22717
Sarasota Memorial Hospital, Sarasota 37582
Saratoga Springs Public Library, Saratoga Springs 41233
Saratov Central City Library, Saratov 24315
Saratov State Technical University, Saratov 22577
Saratov State University, Saratov 22576
Saratovskaya oblastnaya biblioteka dlya detei i yunoshestva im. A.S. Pushkina, Saratov 24314
Saratovskaya oblastnaya universalnaya nauchnaya biblioteka, Saratov 22174
Saratovskaya tsentralnaya gorodskaya biblioteka, Saratov 24315
Saratovski gosudarstvenny universitet im. N.G. Chernishevskogo, Saratov 22576
Saratovski tekhnicheski universitet, Saratov 22577
Sarawak Museum, Kuching 18611
Sarawak State Library, Kuching 18630
Sarawak Teacher Training College, Miri 18541
Sardar Patel College of Engineering, Mumbai 13825
Sardar Patel Institute of Economic and Social Research, Ahmedabad 13924
Sardar Patel University, Kaira 13673
J. Sargeant Reynolds Community College – Downtown Campus Learning Resources Center, Richmond 33683
J. Sargeant Reynolds Community College – Parham Campus Learning Resources Center, Richmond 33684
Sarikei District Council Library, Sarikei 18635
Sarmiento Historical Museum Library, Buenos Aires 00277
Sarnenska raionna TsBS, Tsentralna biblioteka, Sarni 28809
Sarni Regional Centralized Library System, Main Library, Sarni 28809
Sarnoff Corporation Library, Princeton 36156
Sárospataki Comenius Tanítóképző Főiskola, Sárospatak 13292
Sárospataki Református Kollégium, Sárospatak 13344
Sarpay Beikman Public Library, Yangon 19006
Sarpsborg Bibliotek, Sarpsborg 20395
Saskatchewan Agriculture & Food Library, Regina 04141
Saskatchewan Department of Social Services Library, Regina 04142
Saskatchewan Genealogical Society Library, Regina 04470
Saskatchewan Highways & Transportation, Regina 04143
Saskatchewan Indian Cultural Centre, Saskatoon 04490
Saskatchewan Indian Federated College, Regina 03831
Saskatchewan Indian Federated College, Saskatoon 03849
Saskatchewan Institute of Applied Science & Technology, Regina 04021
Saskatchewan Justice – Civil Law Library, Regina 04144
Saskatchewan Justice – Court of Appeal Library, Regina 04145
Saskatchewan Learning Resource Centre, Regina 04471
Saskatchewan Legislative Library, Regina 04146
Saskatchewan Parks and Recreation Association, Regina 04472
Saskatchewan Provincial Library, Regina 04733
Saskatchewan Teachers' Federation, Saskatoon 04491
Saskatoon Public Library, Saskatoon 04750
Sasol Library, Secunda 25103
Sasol Technology Library, Sasolburg 25102
Sasolburg Public Library, Sasolburg 25234
Satakunnan amattikorkeakoulu, Pori 07534
Satakunnan ammattikorkeakoulu, Pori 07365
Satakunnan keskussairaalan tieteellinen kirjasto, Pori 07535
Satakunnan museon kirjasto, Pori 07536
Satakunta Museum, Pori 07536
Satakunta Polytechnic, Social Services and Health Care Library, Pori, Pori 07534
Satakunta University of Applied Sciences, Pori 07365
Säteilyturvakeskuksen kirjasto, Helsinki 07486
Satellite Beach Public Library, Satellite Beach 41235
Säters bibliotek, Säter 26891
Sri Sathya Sai Institute of Higher Learning, Prasanthinilayam 13740
Satterlee, Stephens, Burke & Burke, New York 36015
Satu Mare Public Library, Satu Mare 22101
Saudi Arabian Standards Organization, Riyadh 24405
Saudi Consulting House, Riyadh 24406
Saugus Public Library, Saugus 41236
Sauherad folkebibliotek, Gvarv 20338
Sauk Valley Community College, Dixon 33295
Saul Brodsky Jewish Community Library, Saint Louis 35363
Saul Ewing LLP, Baltimore 35487
Saul Ewing LLP, Philadelphia 36110
Sault College of Applied Arts & Technology, Sault Sainte Marie 04030
Sault Ste Marie Public Library, Sault Sainte Marie 04751
Saurashtra University, Rajkot 13750
Sausalito Public Library, Sausalito 41238
Savannah State University, Savannah 32622
Savaria Múzeum, Szombathely 13512
Savez Pedagoškik Društava Srbje, Beograd 24521
Savezni Zavod za Statistiku, Beograd 24496
Savoie-Biblio, Chambéry 08807
Savoie-Biblio, Pringy 09030
Savonia ammattikorkeakoulu, Kuopio 07354
Savonia University of Applied Sciences, Kuopio 07354
Savonlinna Public Library, Savonlinna 07608
Savonlinnan kaupunginkirjasto, Savonlinna 07608
Sävsjö stadsbibliotek, Sävsjö 26892
Saxion Hogescholen, Deventer 19105
Saxion Hogeschool Enschede, Enschede
– Academie Verpleegkunde, Enschede 19119
– Sector Conservatorium, Enschede 19120

– Sector Gezondheidszorg, Enschede 19121
– Technische Instituten, Enschede 19122
Saxion University of Professional Education, Deventer 19105
Saxon Museum of Industry, Chemnitz 11712
Saxon State Agency for Environment, Agriculture and Geology, Dresden 11774
Sayreville Public Library, Parlin 40712
Săyuz na bălgarskite kompozitori, Sofiya 03575
Săyuz na Balgarskite Pisateli, Sofiya 03576
Sayville Library, Sayville 41242
SBD.bibliotheksservice ag / SSB.service aux bibliothèques sa / SSB.servizio per bibliotecha sa / SSB.servetsch per bibliotecas sa, Bern 27477
Scania Partner AB, Södertälje 26552
Scarborough Public Library, Scarborough 41243
Scarritt-Bennett Center, Nashville 31936
Scarsdale Public Library, Scarsdale 41244
Scenic Regional Library of Franklin, Gasconade & Warren Counties, Union 41633
Scenic Rim Regional Library, Beaudesert 01065
SCÉREN CRDP du Nord Pas de Calais, Lille 08385
Schachbibliothek Egbert Meissenburg, Seevetal 12484
Schatzkiste, Krefeld 12161
Schaumburg Township District Library, Schaumburg 41245
Schenectady County Community College, Schenectady 33751
Schenectady County Public Library, Schenectady 41247
Schering-Plough, Kenilworth 35801
Schiff, Hardin LLP Library, Chicago 35598
Schiffahrtmedizinisches Institut der Marine, Kronshagen 12166
Schiller Park Public Library, Schiller Park 41249
Schlei-Klinikum Schleswig, Schleswig 12468
Schleswig-Holsteinische Familienforschung e.V. in Kiel, Schleswig 12469
Schleswig-Holsteinische Landesbibliothek, Kiel 09323
Schleswig-Holsteinische Landesmuseen Schloss Gottorf, Schleswig 12471
Schleswig-Holsteinischer Landtag, Kiel 10970
Schleswig-Holsteinisches Landessozialgericht, Schleswig 11093
Schleswig-Holsteinisches Oberlandesgericht, Schleswig 11094
Schleswig-Holsteinisches Oberverwaltungsgericht / Schleswig-Holsteinisches Verwaltungsgericht, Schleswig 11095
Schleswig-Holsteinisches Verwaltungsgericht, Schleswig 11095
Schloss Dagstuhl, Wadern 12575
Schloss Friedenstein, Gotha 11925
Schloss Senftenegg, Ferschnitz 01520
Schlossbauverein Burg an der Wupper, Solingen 12493
Schlossbibliothek Prinz Heinrich XXXIX Reuss, Ernstbrunn 01519
Schlossmuseum, Arnstadt 11455
Schlossmuseum Delitzsch, Delitzsch 11732
Schlow Centre Region Library, State College 41429
Schlumberger-Doll, Ridgefield 36179
Schmidt Marine Transport College, Kherson 28104
Schnader, Harrison, Segal & Lewis Library, Philadelphia 36111
Schocken Institute for Jewish Research of the Jewish Theological Seminary of America, Jerusalem 14740
Scholastic Inc Library, New York 36016
School District of Philadelphia, Philadelphia 33637
School for Alternative Teacher Training, Vester Skerninge 06751
School for Arabic Studies, Kano 19994
School for International Training, Brattleboro 30542
School of Advanced Research, Santa Fe 37576
School of Architecture and Planning, Chennai 13797
School of Architecture Richview, Dublin 14480
School of Art & Design at Montgomery College, Silver Spring 32673
School of Arts, Penza 22688
School of Celtic Studies, Dublin 14556
School of Economics, civil engineering, Ljubljana 24881
School of Fine and Industrial Arts, Budapest 13311

School of Forest and Ecosystem Science, Creswick 00429
School of Geography and the Environment, Oxford 29345
School of Geosciences, Sydney 00541
School of Higher Education, Faro 21650
School of Hotel and Restaurant Management, Helsinki 07338
School of Islamic and Social Sciences, Ashburn 30262
School of Kindergarten Teachers, Szarvas 13313
School of Law Library, Springfield 32707
School of Librarianship and Information Studies, Bratislava 24642
School of Library and Information Science, Petaling Jaya 18543
School of Medicine and Health Sciences, Boroko 20565
School of Mission and Theology, Stavanger 20160
School of Music, Perm 22689
School of Music, Saratov 22718
School of Music, Vladivostok 22724
School of Oriental and African Studies, London 29250
School of Planning and Architecture, New Delhi 13729
School of Slavonic and East European Studies, London 29229
School of Social Work, Harare 42365
School of Textiles, Footwear and Gum (Rubber), Kranj 24877
School of the Art Institute of Chicago, Chicago 33226
School of Theoretical Physics, Dublin 14475
School of Tropical Medicine, Kolkata 13813
School of Visual Arts Library, New York 32088
Schoolcraft College, Livonia 33492
Schoolmuseum Michel Thiery, Gent 02655
Schools Library Service, Kingston 16872
Schreeder, Wheeler & Flint LLP, Atlanta 35467
Schreiner College, Kerrville 33455
Schuhmuseum Offenbach, Offenbach 12384
Schule für Feldjäger und Stabsdienst der Bundeswehr – Emmich-Cambrai-Kaserne, Hannover 10738
Schule Schloss Salem / Salem International College, Überlingen 10771
die schulen des deutschen buchhandels, Frankfurt am Main 10726
Schulmuseum Bremen, Bremen 11687
Schulte, Roth & Zabel LLP, New York 36017
Schulzentrum Marienhöhe, Darmstadt 10717
Schulzentrum Zündorf, Köln 10749
Schuylkill County Law Library, Pottsville 34706
Schwabe, Williamson & Wyatt Library, Portland 36148
Schweizer Alpenclub (SAC), Zürich 27459
Schweizer Fernsehen, Zürich 27460
Schweizerdeutsches Wörterbuch, Zürich 27461
Schweizerische Bibliothek für Blinde und Sehbehinderte (SBS), Zürich 27462
Schweizerische Fachstelle für Alkohol- und andere Drogenprobleme, Lausanne 27396
Schweizerische Kapuzinerprovinz, Luzern 27263
Schweizerische Koordinationsstelle für Bildungsforschung (SKBF) / Centre Suisse de Coordination pour la Recherche en Education (CSRE), Aarau 27297
Schweizerische Nationalbank, Zürich 27295
Schweizerische Nationalbibliothek / La Bibliothèque nationale suisse / Biblioteca nazionale svizzera, Bern 26961
Schweizerische Osteuropabibliothek SOB, Bern 27326
Schweizerische Rückversicherungs-Gesellschaft, Zürich 27296
Schweizerische Theatersammlung, Bern 27327
Schweizerische Vogelwarte, Sempach 27433
Schweizerischer Arbeitgeberverband, Zürich 27463
Schweizerischer Bauernverband, Brugg 27337
Schweizerischer Gewerkschaftsbund SGB, Bern 27328
Schweizerisches Alpines Museum, Bern 27329
Schweizerisches Bundesarchiv, Bern 27330
Schweizerisches Bundesgericht, Lausanne 27236
Schweizerisches Gastronomie-Museum, Thun 27440
Schweizerisches Institut für Kinder- und Jugendmedien (SIKJM), Zürich 27464
Schweizerisches Institut für Kunstwissenschaft, Zürich 27467

Schweizerisches Institut für Volkskunde, Basel 27311
Schweizerisches Literaturarchiv (SLA), Bern 27331
Schweizerisches Nationalmuseum, Landesmuseum Zürich, Zürich 27465
Schweizerisches Paraplegikerzentrum, Zürich 27211
Schweizerisches Sozialarchiv, Zürich 27466
Schweizerisches Tropen- und Public Health-Institut, Basel 27312
Schweizerisches Wirtschaftsarchiv und WWZ-Bibliothek, Basel 27313
Schwules Museum Berlin, Berlin 11559
Science and Technology Agency, Tokyo 17304
Science and Technology Facilities Council, Didcot 29644
Science and Technology Facilities Council, Warrington 29914
Science and Technology Information Center, Giza 07192
Science and Technology Library, Dar es Salaam 27676
Science Applications International Corporation, Greenwood Village 35729
Science Council of Japan, Tokyo 17319
Science Fiction Foundation Collection, Liverpool 29707
Science Museum Library, London 29830
Science Museum Library, Swindon 29906
Science Museum of Minnesota, Saint Paul 37506
Science & Technology Information Institute, Taguig 20784
Science University of Tokyo, Tokyo 17172
Sciences Po, Paris 07891
Scientific and Technical Information Centre, Amman 17885
Scientific and technical Library, Information and Economic Centre of Building Industry, Moskva 23564
Scientific and Technical Library of Kazakhstan, Almaty 17890
Scientific and Technical Library of Kyrgyzstan, Bishkek 18135
Scientific and Technical Library of the Ukraine, Kyiv 27937
Scientific and Technology Information Research Institute of China Petrochemical Corporation (Sinopec), Beijing 05853
Scientific Center of Obstetrics, Gynaecology and Perinatology, Moskva 23449
Scientific Center of Surgery im. A.N. Syzganov, Almaty 17963
Scientific Centre of Biological Research, Pushchino 23708
Scientific Centre of Oncology, Moskva 23453
Scientific Collection of the Reformed Church-College, Sárospatak 13344
Scientific District Library I. Franko, Ivano-Frankivsk 28585
Scientific Documentation Centre, Baghdad 14456
Scientific Highschool Center, Nova Gorica 24890
Scientific Information Centre for Social Sciences, Tbilisi 09274
Scientific Institute for the Cultivation of Vegetables, Baku 01721
Scientific Institute of Public Health, Bruxelles 02596
Scientific Institute of Public Health, Brussel 02636
Scientific Institute of Sericulture, Gyanzha 01728
Scientific Library for Labour Market Research and Labour Administration, Nürnberg 12371
Scientific Library of Dubrovnik, Dubrovnik 06141
Scientific Library of Riga Technical University, Riga 18177
Scientific library of State Museum Reserve 'Peterhof', Petrodvorets 23697
Scientific Library of the Central Directorate of Archives at the Cabinet of Ministers of Azerbaijan Republic, Baku 01693
Scientific Library of the Institute of Lithuanian LIterature and Folklore and the Institute of Lithuanian Language, Vilnius 18299
Scientific Library of the Marine Hydrophysical Institute, Sevastopol 28463
Scientific Medical Library, Tbilisi 09275
Scientific Pedagogical Institute, Zhambul 17928
Scientific Research Agricultural Institute of Extrem North, Norilsk 23627
Scientific, Research and Project Institute for Well Cementing and Drilling Fluids, Krasnodar 23118

Scientific Research Council (SRC), Kingston 16883
Scientific Research Institute for Occupational Safety, Bucureşti 21985
Scientific Research Institute of Crop Husbandry, Galljaaral 42114
Scientific Research Institute of Haematology and Blood Transfusion, Tbilisi 09276
Scientific Research Institute of Infectious Diseases with Natural Foci, Omsk 23665
Scientific Research Institute of Obstetrics and Gynaecology, Tbilisi 09277
Scientific Research Institute of Pig Breeding, Shumen 03519
Scientific Research Institute of Soil Sciences, Agricultural Chemistry and Improvement, Tbilisi 09278
Scientific Research Institute of Tea and Subtropical Plants, Macharadze 09215
Scientific Research Institute of Television and Broadcasting, Moskva 23598
Scientific Research Stock Company 'NIEMAS', Sumy 28475
Scientific Society of Szczecin, Szczecin 21229
Scientific 'Taurica' Library, Simferopol 28471
Scientific & Technical Library NPO "Seismotekhnika", Gomel 01851
Scientific Technical Library of YugNIRO, Kerch 28248
Scientific-Industrial Enterprise 'Institute BelNIIlit', Minsk 02045
Scientific-Technical Library on Irrigation, Toshkent 42160
Scioto County Bar & Law Library, Portsmouth 37410
Scitex Corporation Ltd, Herzliya 14704
Scituate Town Library, Scituate 41250
SCOOP, Zeeuws instituut voor sociale en culturele ontwikkeling, Middelburg 19492
Scotch Plains Public Library, Scotch Plains 41251
Scotland County Memorial Library, Laurinburg 39820
Scott County Bar Association, Davenport 36771
Scott County Library System, Eldridge 38949
Scott County Library System, Savage 41239
Scott County Public Library, Georgetown 39220
Scott County Public Library, Scottsburg 41253
Scott Foresman Library, Glenview 36893
Scott, Hulse, Marshall, Feuille, Finger & Thurmond, El Paso 35688
T. B. Scott Library, Merrill 40160
Scott & White Memorial Hospital, Temple 37665
Scottish Agricultural College, Ayr 28933
Scottish Borders Council, Selkirk 30069
Scottish Catholic Archives, Edinburgh 29543
Scottish Crop Research Institute, Dundee 29648
Scottish Enterprise, Glasgow 29677
Scottish Executive Library and Information Service, Edinburgh 29513
Scottish Law Commission, Edinburgh 29662
Scottish National Gallery of Modern Art, Edinburgh 29663
Scottish Natural Heritage, Inverness 29689
Scottish Natural History Library, Kilbarchan 29691
Scottish Poetry Library, Edinburgh 29664
Scottish Police College, Kincardine 29143
Scottish Rite Library, Washington 37778
Scottsbluff Public Library, Scottsbluff 41252
Scottsdale Community College, Scottsdale 33754
Scottsdale Public Library System, Scottsdale 41254
Scott-Sebastian Regional Library, Greenwood 39319
Scott-Sebastian Regional Library – Sebastian County Library, Greenwood 39320
Scoula di Applicazione d'Arma, Torino 15636
E. C. Scranton Memorial Library, Madison 40010
Scranton Public Library, Scranton 41255
Screven-Jenkins Regional Library, Sylvania 41487
Scripps Mercy Hospital, San Diego 37532
SCS Engineers Library, Long Beach 35821
Scuola Archeologica Italiana di Atene, Athinai 13123
Scuola Magistrale, Locarno 27223
Scuola Normale Superiore di Pisa, Pisa 15345
Scuola superiore della pubblica amministrazione, Caserta 15618
Scuola Superiore di Polizia, Roma 15632

Scuola universitaria professionale della Svizzera italiana (SUPSI), Canobbio 27062
Scurry County Library, Snyder 41341
Sea Fisheries Institute, Gdynia 21102
Sea Fisheries Research Station, Haifa 14721
Seafarer's Harry Lundeberg School of Seamanship, Piney Point 37376
Seaford District Library, Seaford 41256
Seaford Public Library, Seaford 41257
Seattle Art Museum, Seattle 37595
Seattle Art Museum, Seattle 37596
Seattle Central Community College, Seattle 33756
Seattle Genealogical Society Library, Seattle 37597
Seattle Metaphysical Library, Seattle 37598
Seattle Pacific University, Seattle 32632
The Seattle Public Library, Seattle 41263
The Seattle Public Library – Lake City, Seattle 41264
The Seattle Public Library – North East, Seattle 41265
Seattle University, Seattle 32633
Seattle University School of Law Library, Seattle 32634
Seattle's Museum of History & Industry (MOHAI), Seattle 37599
Sebastian County Law Library, Fort Smith 34271
Sebastopol Regional Library, Sebastopol 41268
Secaucus Free Public Library, Secaucus 41270
Sechenov Institute of Evolutionary Physiology and Biochemistry of the Russian Academy of Sciences, Sankt-Peterburg 23782
Sechenov Moscow Medical Academy, Moskva 22355
Sechenov Rehabilitation Institute Ltd, Yalta 28172
Second City Hospital, Sofiya 03582
Second District Court of Appeals, Lakeland 34431
Second Military Medical University, Shanghai 05506
Secretaría de Agricultura, Ganaderia, Pesca y Alimentacion, Buenos Aires 00214
Secretaría de Comercio y Fomento Industrial, México 18863
Secretaría de Educación Pública, México 18864
Secretaría de Estado de Comercio, Madrid 25731
Secretaría de Estado de Interior, Madrid 25732
Secretaría de Gobernación, México 18817
Secretaría de Industria, Comercio y Minería, Buenos Aires 00283
Secretaría de Relaciones Exteriores, México 18818
Secretaría Ejecutiva del Convenio Andrés Bello (SECAB), Santafé de Bogotá 06038
Secretaría General de la Communidad Andina, Lima 20675
Secretariado Latinoamericano Pax Romana – MIEC-JECI, Lima 20647
Secretariat on the Pacific Community (SPC), Suva 07318
Sectie Luchtmachthistorie van de Staf Bevelhebber der Luchtstrijdkrachten (SLH), Den Haag 19269
Section de Diffusion Scientifique et Technique du Centre Culturel Français d'Alger, Alger 00071
Secunda Public Library, Secunda 25235
Security Public Library, Security 41271
Sedalia Public Library, Sedalia 41273
Sede Principale di Studi Italianistici, Pisa 15355
Sedgwick County Law Library, Wichita 35068
Sedgwick, Detert, Moran & Arnold Library, San Francisco 36244
Sedlabanki Íslands, Reykjavík 13564
Sedona Public Library, Sedona 41274
Sedro-Woolley Public Library, Sedro-Woolley 41275
Seekonk Public Library, Seekonk 41276
Sefton Metropolitan District Council, Bootle 29944
Seguin-Guadalupe County Public Library, Seguin 41277
Segundo Tribunal de Alçada Civil, São Paulo 03240
Seijo Daigaku, Tokyo 17173
– Hogakubu, Tokyo 17174
Seijo University, Tokyo 17173
– Faculty of Law, Tokyo 17174
Seikado Bunko, Tokyo 17756
Seikado Library, Tokyo 17756
Seikai Regional Fisheries Research Laboratory, Nagasaki 17611
Seikaiku Suisan Kenkyusho, Nagasaki 17611
Seikei Daigaku, Musashino 17050
Seikei University, Musashino 17050
Seimei Hoken Bunka Kenkyujo, Osaka 17642
Seimei Hoken Kyokai, Tokyo 17479
Seinäjoen kaupunginkirjasto – maakuntakirjasto, Seinäjoki 07609
Seinäjoen Sairaala, Seinäjoki 07540
Seinäjoki Public Library – Provincial Library, Seinäjoki 07609
SEI-Superintendência de Estudos Econômicos e Sociais da Bahia, Salvador 03382
Sejm Library, Warszawa 21041
Sekolah Menengah Atas Negeri 2 Tasikmalaya, Tasikmalaya 14327
Sekolah Menengah Pertama Negeri 8 Yogyakarta, Yogyakarta 14328
Sekolah Pendidikan Guru Negeri II Yogyakarta, Yogyakarta 14329
Sekolah Staf dan Komando Angkatan Darat (Seskoad), Bandung 14326
Sekolah Tinggi Filsafat Driyarkara, Jakarta 14250
Sekolah Tinggi Filsafat Kateketik Pradnyawidya, Yogyakarta 14312
Sekolah Tinggi Filsafat Seminari Pineleng, Manado 14276
Sekolah Tinggi Seni Indonesia Surakarta, Surakarta 14302
Sekolah Tinggi Teknik Nasional, Jakarta 14251
Sektor fiziko-tekhnicheskikh i matematicheskikh nauk RAN, Moskva 23484
Selandia – Center for Erhvervsrettet Uddannelse, Slagelse 06725
Selangor Public Library, Shah Alam 18637
Selangor Public Library Corporation, Kelang Branch, Kelang 18624
Selangor Public Library Corporation, Kuala Langat District Library, Banting 18619
Selby Public Library, Sarasota 41231
Selçuk Üniversitesi, Konya 27867
Selenia, Roma 16154
Selimiye Kütüphanesi, Edirne 27880
Selkirk College, Castlegar 03982
Selskokhozyaistvenny institut, Blagoveshchensk 22220
Selskokhozyaistvenny institut, Irkutsk 22259
Selskokhozyaistvenny institut, Ivanovo 22265
Selskokhozyaistvenny institut, Kirov 22292
Selskokhozyaistvenny institut, Krasnoyarsk 22309
Selskokhozyaistvenny institut, Kursk 22314
Selskokhozyaistvenny institut, Novosibirsk 23656
Selskokhozyaistvenny institut, Penza 22467
Selskokhozyaistvenny institut, Saratov 22578
Selskokhozyaistvenny institut, Stavropol 22586
Selskokhozyaistvenny institut, Ulyanovsk 22617
Selskokhozyaistvenny institut im. akad. D.N. Pryanishnikov, Perm 22474
Selskokhozyaistvenny institut im. M. Gorkogo, Kazan 22279
Selwyn College, Cambridge 29043
Semashko Central Hospital, Moskva 23491
Semashko Medical Institute of Stomatology, Moskva 22344
Semashko Research Institute of Public Hygiene and Economics and Organisation of Public Health, Moskva 23583
Semeniška knjižnica, Ljubljana 24898
Semiconductor Physics Institute, Kyiv 28324
Semiconductor Research Foundation, Semiconductor Research Institute, Sendai 17661
Séminaire de Nicolet, Nicolet 04218
Séminaire de Sherbrooke, Sherbrooke 03862
Séminaire de St Hyacinthe, Saint-Hyacinthe 04224
Séminaire de Tournai, Tournai 02490
Séminaire des Carmes, Paris 08221
Séminaire des Pères Maristes, Québec 04220
Séminaire des Pères Rédemptoristes, Ostwald 08209
Séminaire du Prado, Limonest 08198
Séminaire Episcopal de Liège, Liège 02480
Séminaire Interdiocésain, Lille 08197
Séminaire interdiocésain Saint Jean Eudes de Caen, Caen 08179
Séminaire international Saint Pie X, Riddes 27268
Séminaire Israélite de France (SIF), Paris 08222
Séminaire Saint-Irénée, Lyon 08201
Seminar für Allgemeine Betriebswirtschaftslehre, Marketing und Markenmanagement, Köln 10145
Seminar für Allgemeine Betriebswirtschaftslehre, Marktforschung und Marketing, Köln 10146
Seminar für Politische Wissenschaft, Köln 10143
Seminar Schloss Bogenhofen, St. Peter am Hart 01474
Seminar St. Beat, Luzern 27264
Seminari de Barcelona, Barcelona 25766
Seminari Tinggi Ledalero, Maumere 14339
Seminarie van Hasselt, Hasselt 02467
Seminario Arcivescovile, L'Aquila 15871
Seminario Arcivescovile Alagoniana, Siracusa 16072
Seminario Arcivescovile dei SS. Angeli Custodi, Ravenna 16005
Seminario Arcivescovile dell'Immacolata, Oristano 15945
Seminario Arcivescovile di Bologna, Bologna 15749
Seminario Arcivescovile di Cagliari, Cagliari 15764
Seminario Arcivescovile di Camerino, Camerino 15770
Seminario Arcivescovile di Catania, Catania 14947
Seminario Arcivescovile di Fermo, Fermo 15819
Seminario Arcivescovile di Ferrara, Ferrara 15820
Seminario Arcivescovile di Lecce, Lecce 15874
Seminario Arcivescovile di Lucca, Lucca 15883
Seminario Arcivescovile di Milano, Venegono Inferiore 16120
Seminario Arcivescovile di Monreale, Monreale 15919
Seminario Arcivescovile di Perugia, Perugia 15977
Seminario Arcivescovile di Udine, Udine 16116
Seminario Arcivescovile Diocesano, Bisceglie 15738
Seminario Arcivescovile Maggiore, Firenze 15833
Seminario Arcivescovile S. Caterina, Pisa 15987
Seminario Arcivescovile "S. Pio X", Messina 15900
Seminario Arcivescovile Sorrento, Massa Lubrense 15890
Seminario Claretiano, Colmenar Viejo 25773
Seminario Conciliar, Barbastro 25763
Seminario Conciliar de Bogotá, Santafé de Bogotá 06003
Seminario Conciliar de San Bartolomé, Cádiz 25771
Seminario Conciliar de San Julián, Cuenca 25776
Seminario Conciliar Mayor de Medellín, Medellín 06001
Seminario Consiliar de Popayán, Popayán 06002
Seminario de Estudios Cerámicos, Cervo 25942
Seminario de Tarazona, Tarazona 25828
Seminario di Torino, Torino 16100
Seminario Diocesano, Alicante 25759
Seminario Diocesano, Ávila 25762
Seminario Diocesano, Bosa 15754
Seminario Diocesano, Huelva 25781
Seminario Diocesano, Jaén 25783
Seminario Diocesano, Lucera 15884
Seminario Diocesano, San Sebastián 25819
Seminario Diocesano de Bilbao, Derio 25777
Seminario Diocesano de Santo Tomás de Villanueva y del Santo Maestro Juan de Ávila, Real 25809
Seminario Diocesano di Concordia-Pordenone, Pordenone 15993
Seminario Interdiocesano di Calvi e Teano, Teano 16084
Seminario Mayor, León 25784
Seminario Mayor, Lugo 25785
Seminario Mayor, Ourense 25803
Seminario Mayor de Manizales, Manizales 06000
Seminario Mayor Diocesano, Astorga 25761
Seminario Mayor Diocesano, Cáceres 25770
Seminario Mayor Diocesano de San José, Vigo 25841
Seminario Mayor San Ildefonso, Toledo 25831
Seminario Menor San Pelayo, Tui 25833
Seminario Metropolitano, Modena 15914
Seminario Metropolitano de Oviedo, Oviedo 25804
Seminario Monastico di S. Prospero dei Benedettini Olivetani, Camogli 15771
Seminario Pontificio, Tarragona 25829
Seminario Pontificio Mayor de Santiago, Santiago 04912
Seminario Pontificio Regionale 'Pio XI', Ancona 15704
Seminario Rabínico Latinoamericano, Buenos Aires 00284
Seminario Regionale, Benevento 15730
Seminario Regionale 'Pio XI', Reggio Calabria 16010
Seminario Regionale San Pio X, Chieti 15795
Seminario Santa Catalina, Mondoñedo 25798
Seminário Santo Antônio, Agudos 03243
Seminario Scalabrini, Piacenza 15981
Seminario Teologico, Trento 16109
Seminario Teologico Centrale, Gorizia 15854
Seminario Vescovile, Agrigento 15697
Seminario Vescovile, Alba 15698
Seminario Vescovile, Albenga 15699
Seminario Vescovile, Anagni 15701
Seminario Vescovile, Arezzo 15708
Seminario Vescovile, Asti 15718
Seminario Vescovile, Aversa 15720
Seminario Vescovile, Bagnoregio 15722
Seminario Vescovile, Bergamo 15734
Seminario Vescovile, Biella 15737
Seminario Vescovile, Caltanissetta 15767
Seminario Vescovile, Campagna 15772
Seminario Vescovile, Chiavari 15794
Seminario Vescovile, Chioggia 15796
Seminario Vescovile, Città di Castello 15798
Seminario Vescovile, Conversano 15801
Seminario Vescovile, Crema 15806
Seminario Vescovile, Cuneo 15808
Seminario Vescovile, Fabriano 15811
Seminario Vescovile, Fidenza 15821
Seminario Vescovile, Forlì 15836
Seminario Vescovile, Fossano 15837
Seminario Vescovile, Mantova 15888
Seminario Vescovile, Marola di Carpineti 15889
Seminario Vescovile, Massa Marittima 15891
Seminario Vescovile, Mazara del Vallo 15895
Seminario Vescovile, Molfetta 15916
Seminario Vescovile, Mondovì 15917
Seminario Vescovile, Montefiascone 15920
Seminario Vescovile, Nola 15940
Seminario Vescovile, Noto 15942
Seminario Vescovile, Orvieto 15947
Seminario Vescovile, Osimo 15950
Seminario Vescovile, Patti 15969
Seminario Vescovile, Pavia 15970
Seminario Vescovile, Piacenza 15982
Seminario Vescovile, Pistoia 15990
Seminario Vescovile, Pontremoli 15992
Seminario Vescovile, Prato 16000
Seminario Vescovile, Reggio Emilia 16013
Seminario Vescovile, Rovigo 16043
Seminario Vescovile, San Marco Argentano 16053
Seminario Vescovile, San Miniato 16055
Seminario Vescovile, Sansepolcro 16057
Seminario Vescovile, Savona 16062
Seminario Vescovile, Senigallia 16065
Seminario Vescovile, Tempio Pausiana 16085
Seminario Vescovile, Tortona 16102
Seminario Vescovile, Treviso 16111
Seminario Vescovile, Trieste 16112
Seminario Vescovile, Verona 16134
Seminario Vescovile, Vigevano 16137
Seminario Vescovile, Vittorio Veneto 16141
Seminario Vescovile di Cremona, Cremona 15807
Seminario Vescovile di Lodi, Lodi 15880
Seminario Vescovile di Marola, Carpi 15776
Seminario Vescovile di Padova, Padova 16393
Seminario Vescovile di Vicenza, Vicenza 16136
Seminario Vescovile Maggiore, Parma 15968
Seminario Vescovile Maria Immacolata, Iglesias 15861
Seminario Vescovile 'Mons. F. Pennisi', Ragusa 16002
Seminario Vescovile Pio XI, Caltagirone 15766
Seminario Vescovile S. Nicola, Saluzzo 16046
Seminario Vescovile 'San Carlo', Fano 15814
Seminario Vescovile Vagnotti, Cortona 15803
Seminario Vescoville, Feltre 15816
Seminarium Misyjne Księży Sercanów, Stadniki 21058
Seminary and Foundation for Theological Studies, Rabat 18664
Seminary Library, Ljubljana 24898
Seminary of St-Hyacinthe Library, Saint-Hyacinthe 04224
Seminary of the Immaculate Conception, Huntington 35236
Seminary of Vicenza, Library, Vicenza 16136
Seminole Community Library, Seminole 41280
Seminole County Public Library System, Sanford 41209
Seminole State College, Sanford 33740
Seminole State College, Seminole 33760
Semmelweis Egyetem, Budapest 13261
– Testnevelési és Sporttudományi Kar Könyvtára, Budapest 13262
Semmelweis Library of History of Medicine, Budapest 13457

Semmelweis Orvostörténeti Múzeum, Budapest 13457
Semmelweis University, Budapest 13261
– Faculty of Physical Education and Sport Sciences Library, Budapest 13262
Semmes, Bowen & Semmes Library, Baltimore 35488
Sempaku Gijutsu Kenkyujo, Mitaka 17599
Sempra Energy, San Diego 36221
Semyonov School of Timber Processing, Semyonov 22719
Semyonovski tekhnikum mekhanicheskoi obrabotki drevesiny, Semyonov 22719
Senado, Madrid 25733
Senado Federal, Brasília 03222
D. S. Senanayake Memorial Public Library, Kandy 26332
Sénat, Paris 08158
Senate Library 'Giovanni Spadolini', Roma 15687
Senate Library of Pennsylvania, Harrisburg 34343
Senate of Pakistan, Islamabad 20469
Senato della Republica, Roma 15687
Der Senator für Umwelt, Bau, Verkehr und Europa, Bremen 10843
Senatsbibliothek Berlin, Berlin 11560
Senatsverwaltung für Berlin – Oberverwaltungsgericht Berlin-Brandenburg, Berlin 10812
Senatsverwaltung für Justiz – Oberverwaltungsgericht Berlin-Brandenburg, Berlin 10813
Senatsverwaltung für Stadtentwicklung, Berlin 10814
Senatul României, București 21954
SenBildWiss. Medienforum, Berlin 11561
Senckenberg Deutsches Entomologisches Institut, Müncheberg 12275
Senckenberg – Forschungsinstitut und Naturmuseum Frankfurt, Frankfurt am Main 11870
Senckenberg Museum für Naturkunde Görlitz, Görlitz 11922
Senckenberg Naturhistorische Sammlungen Dresden – Fachbibliohtek Mineralogie und Geologie, Dresden 11776
Senckenberg – Research Institute and Natural History Museum Frankfurt, Frankfurt am Main 11870
Seneca County Law Library, Tiffin 34924
Seneca Nation Library, Salamanca 37510
Seneca Public Library District, Seneca 41281
SENER Ingenieria y Sistemas, S.A., Las Arenas 25850
Sen-i Kobunshi Zairyo Kenkyujo, Tsukuba 17807
Senior High School Library, Tasikmalaya 14327
Senkenberg Naturhistorische Sammlungen Dresden – Fachbibliothek Zoologie, Dresden 11777
Senshu Daigaku, Tokyo 17175
Senshu University, Tokyo 17175
Sentara Norfolk General Hospital, Norfolk 37289
Seoul National University, Seoul 18091
– Law College, Seoul 18092
Seoul National University Museum, Seoul 18120
Seoul Women's University, Seoul 18093
Sepik Agricultural College, Maprik 20560
Sepuluh Nopember Institute of Technology, Surabaya 14297
Sequoyah Regional Library System, Canton 38446
Serbian Academy of Sciences and Arts, Beograd 24455
Serbian Academy of Sciences and Arts – Branch in Novi Sad, Novi Sad 24531
Serbian Literary Association, Beograd 24522
Serbian National Theatre, Novi Sad 24532
Serbski institut z.t., Bautzen 11477
Serbski muzej / Sorbisches Museum, Bautzen 11476
Serbsky Centre Social and Forensic Psychiatry, Moskva 23193
Sergievski Posad State History and Art Museum, Sergyev Posad 23965
Serres Central Public Library, Serres 13152
Servants of India Society's Library, Pune 14199
Servei de Biblioteques, Barcelona 26203
Service Canadien des Forêts – Centre de foresterie des Laurentides, Sainte-Foy 04485
Service Central de la Statistique et des Etudes Economiques (STATEC), Luxembourg 18415
Service Centre for Development Cooperation, Helsinki 07468
Service Cost Information Center, Irving 35782
Service de Documentation des Redactions de Saint-Paul, Luxembourg 18416
Service de documentation du Centre culturel canadien, Paris 08505
Service de la Recherche en Education (SRED), Genève 27373
Service de l'Information et des Archives Nationales, Kigali 24378
Service des Archives et de la Bibliothèque Nationale, N'Djamena 04835
Service des archives et de la documentation de la République et Canton du Jura, Delémont 27235
Service des Monuments Historiques des Arts et du Folklore, Rabat 18964
Service d'Hygiène, Rabat 18965
Service Historique de la Défense, Brest 08134
Service Historique de la Défense, Cherbourg 08135
Service Historique de la Défense, Rochefort 08161
Service Historique de la Défense, Toulon 08166
Service Historique de la Défense, Vincennes 08168
Service historique de la marine, Lorient 08138
Service Historique de l'Armée de l'Air, Vincennes 08169
Service Historique de l'Armée de Terre, Vincennes 08170
Service Interétablissements de la Coopération Documentaire (SICD) de Strasbourg, Strasbourg
– Bibliothèque de Géographie et d'Aménagement, Strasbourg 08049
– Bibliothèque de la Faculté de Droit, Strasbourg 08050
– Bibliothèque de l'ECPM, Strasbourg 08051
– Bibliothèque de l'ENSPS-ESBS-Informatique, Illkirch 08052
– Bibliothèque de l'Institut des Hautes Etudes Européennes, Strasbourg 08053
– Bibliothèque de l'Institut d'Etudes Politiques, Strasbourg 08054
– Bibliothèque de l'I.U.T. Robert Schuman, Illkirch 08055
– Bibliothèque de médecine et d'odontologie, Strasbourg 08056
– Bibliothèque de Psychologie/Education (BPSE), Strasbourg 08057
– Bibliothèque de Recherche Juridique, Strasbourg 08058
– Bibliothèque des Arts, Strasbourg 08059
– Bibliothèque des Facultés de Théologies catholique et protestante, Strasbourg 08060
– Bibliothèque des Langues, Strasbourg 08061
– Bibliothèque des Sciences et Techniques, Strasbourg 08062
– Bibliothèque des Sciences Sociales, Strasbourg 08063
– Bibliothèque d'Histoire 1er cycle, Strasbourg 08064
– Bibliothèque du PEGE, Strasbourg 08065
– Bibliothèque du Portique, Strasbourg 08066
– Bibliothèque U2-U3 – 1er étage (sciences juridiques) dite Huet-Weiller, Strasbourg 08067
– Bibliothèque U2-U3 – 2ème étage (sciences humaines), Strasbourg 08068
Service Météorologique, Kinshasa 06081
Service Météorologique de Hann, Dakar 24447
Service Protestant de Mission (DEFAP), Paris 08223
Service public fédéral Justice, Bruxelles 02438
Service public fédéral (SPF) Santé publique, Sécurité de la Chaîne alimentaire et Environnement, Bruxelles 02439
Service Territorial des Archives de Nouvelle Calédonie, Nouméa 19758
Servicio de Extensión Cultural y Educativo, Montevideo 42027
Servicio de Información Comercial, Montevideo 42028
Servicio de Investigación Prehistórica, Valencia 26177
Servicio de Investigación y Promoción Agraria, Lima 20676
Servicio Geológico Minero Argentino, Buenos Aires 00215
Servicio Hidrográfico y Oceanográfico de la Armada de Chile, Valparaíso 04948
Servicio Histórico Militar, Barcelona 25687
Servicio Internacional de Información sobre Subnormales, San Sebastián 26136
Servicio Meteorológico Nacional, Buenos Aires 00285
Servicio Meteorológico Nacional, México 18819
Servicio Nacional de Aprendizaje (SENA), Medellín 05989
Servicio Nacional de Geológia y Minería, Santiago 04944
Serviço de Documentação Geral da Marinha, Rio de Janeiro 03371
Serviço Federal de Processamento de Dados, São Paulo 03421
Servitenkloster Maria Luggau, Maria Luggau 01452
Servitenkloster St. Karl, Volders 01477
Sessions, Fishman, Nathan & Israel LLP Library, New Orleans 35950
Sestroretsk District Centralized Library System, Central District Children's Library, Sankt-Peterburg 24297
Sestroretsk District Centralized Library System, Zoshchenko Central District Library, Sankt-Peterburg 24296
Seth G.C. Medical College, Mumbai 13706
Seton Hall University, Newark 32104
Seton Hall University, South Orange 32689
Seton Hill University, Greensburg 31256
Seurakuntaopiston kirjasto, Järvenpää 07391
Sevastopol Machine Engineering Institute, Sevastopol 28464
Sevastopol Marine Plant Industrial Corporation, Sevastopol 28168
Sevastopol State Technical University, Sevastopol 28071
Sevastopolskaya tsentralnaya biblioteka im. L.M. Tolstogo, Sevastopol 28810
Sevastopolski Dom ofitserov ChF RF, Sevastopol 28462
Seventh Day Adventists General Conference Library, Silver Spring 35380
Severios Bibliotheki, Nicosia 06323
Severnaya Verf Ship Building Plant, Sankt-Peterburg 22941
Severny gosudarstvenny meditsinski universitet, Arkhangelsk 22204
Severny nauchno-issledovatelski institut gidrotekhniki i melioratsii, Sankt-Peterburg 23897
Severny Plant Machine Building Industrial Corporation, Sankt-Peterburg 22927
Severočeská galerie výtvarného umění v Litoměřicích, Litoměřice 06463
Severočeská vědecká knihovna, Ustí nad Labem 06352
Severočeské museum, Liberec 06461
Severo-Kavkazskaya zheleznaya doroga, Rostov-na-Donu 22898
Severo-Kavkazski gorno-metallurgicheski institut, Vladikavkaz 22627
Severo-Kavkazski mezhotraslevoi territorialny tsentr nauchno-tekhnicheskoi informatsii i propagandy, Rostov-na-Donu 23728
Severo-Kavkazski mezhotraslevoi territorialny tsentr nauchno-tekhnicheskoi informatsii i propagandy, Vladikavkaz 24045
Severo-Kavkazski mezhotraslevoi territorialny tsentr nauchno-tekhnicheskoi informatsii i propagandy – Kabardino-Balkarski filial, Nalchik 23609
Severo-Kavkazski nauchno-issledovatelski institut sadovodstva i vinogradarstva, Krasnodar 23119
Severo-Kavkazskoe geologicheskoe upravlenie, Essentuki 22730
Severo-Osetinskaya respublikanskaya detskaya biblioteka im. Dabe Mamsurova, Vladikavkaz 24344
Severo-Osetinskaya respublikanskaya yunosheskaya biblioteka, Vladikavkaz 24345
Severo-Osetinski gosudarstvenny universitet im. K.L. Khetagurova, Vladikavkaz 22628
Severo-Osetinski meditsinski institut, Vladikavkaz 22629
Severo-Osetinski nauchno-issledovatelski institut, Vladikavkaz 24046
Severo-Vostochny Kompleksny Nauchno-Issledovatelski Institut, Magadan 23149
Severo-zapadny kadrovy tsentr, Sankt-Peterburg 23898
Severo-Zapadny nauchno-issledovatelski institut selskogo khozyaistva, Siverskaya 23970
Severo-zapadny NIPKI energetich-eskoi promyshlennosti, Sankt-Peterburg 23899
Severo-zapadny NPO selskogo khozyaistva Belogorka, Sankt-Peterburg 23900
Severo-zapadny zaochny politekhnicheski institut, Sankt-Peterburg 22543
Sevier County Public Library System, Sevierville 41284
Sevzapgeologiya, Sankt-Peterburg 22935
Seward County Community College, Liberal 33485
Seward & Kissel LLP, New York 36018
Sewickley Public Library, Inc, Sewickley 41285
Seychelles National Library, Victoria 24561
Seychelles Polytechnic, Victoria 24562
Seyfarth Shaw, Chicago 35599
Seyfarth Shaw, Washington 36375
Seyfarth & Shaw Law Library, New York 36019
Seyfarth Shaw Library, Los Angeles 35857
S. Seyfullin Republican Public Library, Astana 17985
Seymour Public Library, Seymour 41287
SFI – Det Nationale Forskningscenter for Velfærd, København 06820
Shaanxi Astronomical Observatory, Shaanxi 05886
Shaanxi Institute of Traditional Chinese Medicine, Xianyang 05707
Shaanxi Normal University, Xi'an 05685
Shaanxi Provincial Library, Xi'an 04976
Shaare Emeth Temple, Saint Louis 35364
Shaare Zedek Medical Center, Jerusalem 14741
Shaheed Beheshti University of Medical Science, Tehran 14405
Shahid Beheshti University, Tehran 14406
Shahid Chamran University, Ahvaz 14388
Shahr-e-Kord University of Medical Sciences, Shahr-e-Kord 14395
Shakai Chosa Kenkyujo, Tanashi 17429
Shaker Heights Public Library, Shaker Heights 41288
Shaker Heights Public Library – Bertram Woods Branch, Shaker Heights 41289
The Shakespeare Centre Library, Stratford-upon-Avon 29904
The Shakespeare Data Bank, Inc Library, Evanston 36835
Shakespeare Society of America, West Hollywood 37823
Shakhta im. Dzerzhinskoho, Dzerzhinsk 28557
Shakhtarsk City Centralized Library System, Main Library, Shakhtarsk 28811
Shakhtarska miska TsBS, Tsentralna biblioteka, Shakhtarsk 28811
Shakhty Technological Institute for Service Industries, Shakhty 22580
Shaler North Hills Library, Glenshaw 39252
Shamsul Ulama Daudpota Sind Government Library, Hyderabad 20456
Shandong Academy of Arts, Jinan 05321
Shandong Academy of Arts and Crafts, Jinan 05322
Shandong Agricultural University, Taian 05586
Shandong College of Agricultural Management, Jinan 05323
Shandong College of Oceanology, Qingdao 05487
Shandong College of Technology, Zibo 05800
Shandong Educational College, Jinan 05324
Shandong Financial College, Jinan 05325
Shandong Institute of Architecture Engineering, Jinan 05326
Shandong Institute of Building Materials, Jinan 05327
Shandong Institute of Economics, Jinan 05328
Shandong Institute of Light Industry, Jinan 05329
Shandong Institute of Physical Culture, Jinan 05330
Shandong Institute of Traditional Chinese Medicine, Jinan 05331
Shandong Library, Jinan 04963
Shandong Medical University, Jinan 05332
Shandong Normal University, Jinan 05333
Shandong Public Security College, Jinan 05334
Shandong University, Jinan 05335
Shandong University of Technology, Jinan 05336
Shandong Water Conservancy School, Taian 05587
Shanghai Academy of Social Sciences, Shanghai 05888
Shanghai Agricultural College, Shanghai 05507
Shanghai Astronomical Observatory, Shanghai 05889
Shanghai College of Chemical Industry, Shanghai 05508
Shanghai College of Metallurgy, Shanghai 05509

Shanghai College of Public Security, Shanghai 05510
Shanghai College of Traditional Medicine, Shanghai 05511
Shanghai Finance College, Shanghai 05512
Shanghai Fisheries University, Shanghai 05513
Shanghai Institute of Building Materials, Shanghai 05514
Shanghai Institute of Ceramics, Shanghai 05890
Shanghai Institute of Electric Power, Shanghai 05515
Shanghai Institute of Foreign Trade, Shanghai 05516
Shanghai Institute of Materia Medica, Shanghai 05891
Shanghai Institute of Mechanical Engineering, Shanghai 05517
Shanghai Institute of Organic Chemistry, Shanghai 05892
Shanghai Institute of Physiology, Shanghai 05893
Shanghai Institute of Plant Physiology, Shanghai 05894
Shanghai Institute of Technical Physics, Shanghai 05895
Shanghai International Studies University, Shanghai 05518
Shanghai Jiao Tong University, Shanghai 05519
Shanghai Library, Shanghai 04970
Shanghai Library of Academia Sinica, Shanghai 05896
Shanghai Machinery College, Shanghai 05520
Shanghai Maritime University, Shanghai 05521
Shanghai Museum, Shanghai 05897
Shanghai Normal University, Shanghai 05522
Shanghai Petrochemical College, Shanghai 05523
Shanghai Physical Culture Institute, Shanghai 05524
Shanghai Publishing and Printing College, Shanghai 05525
Shanghai Research Institute of Quantity Economics and Operations Research, Shanghai 05898
Shanghai Science and Technology College, Shanghai 05526
Shanghai Second Medical University, Shanghai 05527
Shanghai Teachers' College, Shanghai 05528
Shanghai Technical Teachers' College, Shanghai 05529
Shanghai Tiedao University, Shanghai 05530
Shanghai University, Shanghai 05531
Shanghai University of Engineering Science, Shanghai 05532
– College of Textile, Shanghai 05533
Shanghai University of Finance and Economics, Shanghai 05534
Shangqiu Normal College, Shangqiu 05537
Shangrao Teachers' College, Shangrao 05538
Shantou University, Shantou 05539
Shanxi Agricultural University, Taiiyuan 05588
Shanxi College of Traditional Chinese Medicine, Taiyuan 05591
Shanxi Finance and Economics College, Taiyuan 05592
Shanxi Institute of Economic Management, Taiyuan 05593
Shanxi Medical College, Taiyuan 05594
Shanxi Mining College, Taiyuan 05595
Shanxi Normal University, Linfen 05394
Shanxi Provincial Library, Taiyuan 04972
Shanxi University, Taiyuan 05596
Shaoyang College of Technology, Shaoyang 05540
Shaoyang Teachers' College, Shaoyang 05541
SHAR Centre, Sriharikota 14212
Sharif University of Technology, Tehran 14407
Sharkia Provincial Council, Zagazig 07128
Sharon Public Library, Sharon 41290
Shasta College, Redding 33677
Shasta County Law Library, Redding 34731
Shasta County Library, Redding 40953
Shaw University, Raleigh 32416
Shawnee College, Ullin 33814
Shawnee Correctional Center Library, Vienna 34964
Shawnee Mission Medical Center Library, Shawnee Mission 37603
Shawnee State University, Portsmouth 32349
ShawPittman, LLP, Washington 36376
Shchepkina Theatre School, Moskva 22392
Shcherbakov Ltd, Moskva 22826
Shchors Regional Centralized Library System, Main Library, Shchors 28812
Shchorska raionna TsBS, Tsentralna biblioteka, Shchors 28812
Shearman & Sterling Library, Washington 36377
Shearman & Sterling LLP, San Francisco 36245
Shearman & Sterling LLP, San Francisco 36246
Shearman & Sterling LLP Library, New York 36020
Sheehan Phinney Bass + Green PA Library, Manchester 35875
Sheffield Hallam University, Sheffield 29390
– Psalter Lane Campus Library, Sheffield 29391
Sheffield Libraries, Sheffield 30070
Sheffield Public Library, Sheffield 41295
Shelby County Law Library, Memphis 34525
Shelby County Libraries, Sidney 41321
Shelby Township Library, Shelby Township 41299
Shelbyville-Shelby County Public Library, Shelbyville 41301
Sheldon Jackson College, Sitka 32680
Shelkovy kombinat, Moskva 22852
Shell Laboratorium, Amsterdam 19323
Shell Oil Company – Law Library, Houston 35760
Shell Oil Company – Petro-Chemical Knowledge Center, Houston 35761
Shell Oil Company – Services Integration Group-EP Library Houston, Houston 35762
Shell Oil Company – Tax Library, Houston 35763
Shellharbour City Library, Shellharbour 01154
Shelter Rock Public Library, Albertson 37901
Shelton State Community College, Tuscaloosa 33811
Shenandoah County Library, Edinburg 38930
Shenandoah Public Library, Shenandoah 41305
Shenandoah University, Winchester 33078
Shenango Commonwealth College, Sharon 32660
Shenango Valley Community Library, Sharon 41291
Shenkar College of Fashion and Textile Technology, Ramat-Gan 14659
Shenyang City Library, Shenyang 05924
Shenyang College of Technology, Shenyang 05547
Shenyang Electric Power Institute, Shenyang 05548
Shenyang Gold Institute, Shenyang 05549
Shenyang Institute of Aeronautical Engineering, Shenyang 05550
Shenyang Institute of Architectural Engineering, Shenyang 05551
Shenyang Institute of Chemical Technology, Shenyang 05552
Shenyang Institute of Physical Culture, Shenyang 05553
Shenyang Medical College, Shenyang 05554
Shenyang University, Shenyang 05555
– Shenyang Normal University, Shenyang 05556
Shenyang University of Technology, Shenyang 05557
Shenyang University of Technology, Shenyang 05558
Shenzhen Library, Shenzhen 05925
Shenzhen University, Shenzhen 05559
Shepetivka Regional Centralized Library System, Main Library, Shepetivka 28813
Shepetivska raionna TsBS, Tsentralna biblioteka, Shepetivka 28813
Shepherd University, Shepherdstown 32664
Sheppard Memorial Library, Greenville 39313
Sheppard, Mullin, Richter & Hampton Library, Los Angeles 35858
Sherborn Library, Sherborn 41307
Sherborne School Library, Sherborne 29490
Sherbrooke Municipal Library, Sherbrooke 04755
Sheridan College, Oakville 03808
Sheridan Correctional Center Library, Sheridan 34859
Sherman & Howard, Denver 35668
Sherman Public Library, Sherman 41308
Sherman Research Library, Corona del Mar 36755
Sherubtse Degree College, Kanglung 02914
Sheth Maheklal Jethabhai Pustakalaya, Ahmedabad 14222
Shetland Islands Council, Lerwick 30010
Shetland Library, Lerwick 30010
Shevchenko Central City Children's Library, Kyiv 28689
Shevchenko Institute of Literature, Kyiv 28332
T. Shevchenko Lugansk State Pedagogical University, Lugansk 28029
Shevchenko Memorial National Park in Kaniv, Kaniv 28245
Shevchenko Scientific Society Inc, New York 37258
Shevchenkovsky natsionalny zapovednik v Kaniva, Kaniv 28245
Shiawassee District Library, Owosso 40674
Shibaura Institute of Technology – Omiya Campus, Saitama 17096
Shibaura Kogyo Daigaku – Omiya Bunkan, Saitama 17096
Shiga Ika Daigaku, Otsu 17090
Shiga Prefectural Assembly, Otsu 17291
Shiga University of Medical Science, Otsu 17090
Shiga Universiy, Hikone 16945
Shigaken Gikai, Otsu 17291
Shigen Kankyo Gijyutsu Sougo Kenkujyo, Tsukuba 17808
Shih Hsin University, Taipei 27557
Shihezi Agricultural College, Shihezi 05560
Shijiazhuang City Library, Shijiazhuang 05926
Shijiazhuang Medical College, Shijiazhuang 05572
Shijiazhuang Post College, Shijiazhuang 05573
Shijiazhuang Railway Institute, Shijiazhuang 05574
Shijiazhuang Teachers' College, Shijiazhuang 05575
Shijiazhuang University, Shijiazhuang 05576
Shikoku National Agricultural Experiment Station, Zentsuji 17833
Shikoku Nogyo Shikenjo, Zentsuji 17833
Shiloh Regional Library, Jackson 39605
Shimane Agricultural Experiment Station, Izumo 17560
Shimane Daigaku, Matsue 17033
Shimane Ika Daigaku, Izumo 16957
Shimane Medical University, Izumo 16957
Shimane Prefectural Fisheries Experiment Station, Hamada 17542
Shimane University, Matsue 17033
Shimaneken Nogyo Shikenjo, Izumo 17560
Shimaneken Suisan Shikenjo, Hamada 17542
Shimazu Seisakusho, Ltd, Production Management Department, Kyoto 17382
Shimazu Seisakusho, Seizokikaku-shitsu, Seizokikakuka, Kyoto 17382
Shin Nihon Seitetsu K. K. Daisan Gijutsu Kenkyujo Jimu Sokatsushitsu, Kitakyushu 17375
Shin Nihon Seitetsu K. K. Kamaishi Gijutsukenkyushitsu, Kamaishi 17366
Shin Nihon Seitetsu K. K. Nagoya Seitetsujo Gijutsu Kanribu, Tokai 17430
Shin Nihon Seitetsu K. K. Sogo Chosabu Shiryojohoshitsu, Tokyo 17480
Shin Nihon Seitetsu K. K. Tekko Kaiyo Jigyobu Sagamihara, Sagamihara 17420
Shinny zavod, Moskva 22853
Shinshu Daigaku, Matsumoto 17034
– Igakubu Bunkan, Matsumoto 17035
– Kyoikugakubu Bunkan, Nagano 17036
– Nogakubu Bunkan, Kamiina 17037
– Seni-gakubu Bunkan, Ueda 17038
Shinshu University, Matsumoto 17034
– Agricultural Branch, Kamiina 17037
– Education Branch, Nagano 17036
– Faculty of Textile Science and Technology, Ueda 17038
– Medical Branch, Matsumoto 17035
Shionogi & Co, Ltd, Shionogi Res Laboratory, Osaka 17411
Shionogi Seiyaku K. K. Kenkyujo, Osaka 17411
Ship Building and Mechanics Professional School, Sankt-Peterburg 22707
Ship Research Institute, Mitaka 17599
Shipbuilding College, Kherson 28002
Shipbuilding Industry Central Scientific Library, Sankt-Peterburg 23914
Shippensburg Public Library, Shippensburg 41310
Shippensburg University, Shippensburg 32666
Shire of Kalamunda Library, Kalamunda 01106
Shiritsu Yonezawa Toshokan, Yonezawa 17867
Shiroke Regional Centralized Library System, Main Library, Shiroke 28814
Shirokivska raionna TsBS, Tsentralna biblioteka, Shiroke 28814
Shirshov Institute of Oceanology of the Atlantic Branch of the Russian Academy of Sciences, Kaliningrad 23067
Shirshov Institute of Oceanology of the Russian Academy of Sciences, Moskva 23300
Shirshov Institute of Oceanology of the Russian Academy of Sciences, Southern Branch, Gelendzik 23044
Shiryo Chosakai Kaigun Bunko, Tokyo 17757
Shivaji University, Kolhapur 13680
Shizuoka College of Pharmaceutical Sciences, Shizuoka 17125
Shizuoka Kenritsu Chuo Toshokan, Shizuoka 17857
Shizuoka Prefectural Assembly, Shizuoka 17297
Shizuoka Prefectural Central Library, Shizuoka 17857
Shizuoka Prefecture Industrial Technology Center, Shizuoka 17663
Shizuoka Tea Experiment Station, Ogasa 17626
Shizuoka Yakka Daigaku, Shizuoka 17125
Shizuokaken Chagyo Shikenjo, Ogasa 17626
Shizuokaken Gikai, Shizuoka 17297
Shizuokaken Kogyo Gijutsu Senta, Shizuoka 17663
Shklov Central Regional Library, Shklov 02232
Shklovskaya tsentralnaya raionnaya bibliyateka, Shklov 02232
Shmidt Geophysical Institute, Obninsk 23661
Shmidt Institute of Earth Physics of the Russian Academy of Sciences, Moskva 23269
Shoalhaven Libraries, Nowra 01135
Shochiku Otani Library, Tokyo 17758
Shochiku Otani Toshokan, Tokyo 17758
Shoe Factory, Vitebsk 01932
Shoemaking Professional School, Sankt-Peterburg 22704
Shogakukan Chosakikakubu Shiryoshitsu, Tokyo 17481
Shogakukan Publishing Co, Ltd, Tokyo 17481
Shogentsukov Bobruysk City Library no 7, Bobruysk 02101
Shoko Chukin Bank, Tokyo 17482
Shoko Kumiai Chuo Kinko, Tokyo 17482
Shook, Hardy & Bacon, Kansas City 35798
Shoreham-Wading River Public Library, Shoreham 41312
Shoreline Community College, Seattle 33757
Shorewood Public Library, Shorewood 41313
Shorewood-Troy Public Library District, Shorewood 41314
Shorter College, Rome 32479
Shostka City Centralized Library System, Main Library, Shostka 28815
Shostlinska miska TsBS, Tsentralna biblioteka, Shostka 28815
Shota Rustaveli Institute of Georgian Literature, Tbilisi 09279
Showa Denko K. K., Tokyo 17483
Showa Densen Denran K. K., Kawasaki 17370
Showa Electro-Chemical Industry Co, Ltd, Tokyo 17483
Showa Wire and Cable Co, Ltd, Kawasaki 17370
Shreemati Nathibai Damodar Thackersey Women's University, Mumbai 13707
Shreve Memorial Library, Shreveport 41315
Shreve Memorial Library – Broadmoor Branch, Shreveport 41316
Shreve Memorial Library – Hamilton/South Caddo Branch, Shreveport 41317
Shrewsbury College of Arts and Technology, Shrewsbury 29491
Shrewsbury Public Library, Shrewsbury 41319
Shrewsbury School Library, Shrewsbury 29492
Shri Ram Institute for Industrial Research, New Delhi 14168
Shri Vivekananda High School, Nadi 07297
Shrimati Radhika Sinha Institute and Sachchidananda Sinha Library, Patna 14233
Shropshire Archives, Shrewsbury 29893
Shropshire Council, Shrewsbury 30071
Shternberg State Astronomical Institute, Moskva 23207
Shubnikov Institute of Crystallography, Moskva 23336
Shukutoku College, Chiba 16908
Shukutoku Daigaku, Chiba 16908
Shvartsman Specialized Jewish Library, Kyiv 28385
Shveinaya fabrika Zarya industrializatsii, Vitebsk 01934
Siam Cement Public Company Limited, Bangkok 27772
Siam Society, Bangkok 27787
Siam University, Bangkok 27708
SIAST Palliser Campus Library, Moose Jaw 04010
SIAST-Saskatchewan Institute of Applied Science & Technology, Saskatoon 04029
Šiaulių apskrities Povilo Višinskio viešoji biblioteka, Šiauliai 18361

Šiaulių miesto savivaldybės viešoji biblioteka, Šiauliai 18362
Šiaulių rajono savivaldybės viešoji biblioteka, Šiauliai 18363
Šiaulių universiteto biblioteka, Šiauliai 18285
Sibelius Academy Library, Helsinki 07342
Sibelius Museum, Turku 07549
Sibelius-Akatemian kirjasto, Helsinki 07342
Sibeliusmuseum, Turku 07549
Šibenik Municipal Museum, Šibenik 06218
Siberian Institute of Agricultural Research, Novosibirsk 23658
Siberian Institute of Energetics of the Siberian Department of the Russian Academy of Sciences, Irkutsk 23058
Siberian Medical University, Tomsk 22595
Siberian Research Institute of Geology, Geophysics and Mineral Resources, Novosibirsk 23659
Siberian Road and Road Transport Institute, Omsk 22458
Siberian Scientific Research Institute of Agriculture, Omsk 23668
Siberian State Industrial University, Novokuznetsk 22433
Siberian State Institute of Metrology, Novosibirsk 23657
Siberian State Transport University, Novosibirsk 22449
Siberian Technological Institute of the Timber Industry, Krasnoyarsk 22310
Sibirski energeticheski institut Sibirskogo otdeleniya RAN, Irkutsk 23058
Sibirski gosudarstvenni universitet putei soobsheheniya, Novosibirsk 22449
Sibirski gosudarstvenny industrialny universitet, Novokuznetsk 22433
Sibirski gosudarstvenny meditsinski universitet, Tomsk 22595
Sibirski gosudarstvenny nauchno-issledovatelski institut metrologii, Novosibirsk 23657
Sibirski nauchno-issledovatelski institut ekonomiki selskogo khozyaistva, Novosibirsk 23658
Sibirski nauchno-issledovatelski institut geologii, geofiziki i mineralnogo syrya, Novosibirsk 23659
Sibirski nauchno-issledovatelski institut selskogo khozyaistva, Omsk 23668
Sibthorp Library, Lincoln 29165
Sibu Public Library, Sibu 18638
SICAP Liberté, Dakar 24448
Sichuan Agricultural University, Yaan 05732
Sichuan Business College, Chengdu 05119
Sichuan College of Education, Chengdu 05120
Sichuan Conservatory of Music, Chengdu 05121
Sichuan Culinary Institute, Chengdu 05122
Sichuan Institute of Foreign Languages, Chongqing 05138
Sichuan Institute of Light Industry and Chemical Technology, Zigong 05802
Sichuan Normal University, Chengdu 05123
Sichuan Provincial Library, Chengdu 04955
Sichuan University, Chengdu 05124
Sida, Stockholm 26676
Siderurgia Nacional, Empresa de Serviços, S.A., Aldeia de Paio Pires 21743
Sidley, Austin, Brown & Wood LLP, New York 36021
Sidley Austin LLP, Washington 36378
Sidley Austin LLP Library, Chicago 35600
Sidley Austin LLP Library, Los Angeles 35859
Sidney Memorial Public Library, Sidney 41322
Sidney Sussex College, Cambridge 29044
Siebenbürgische Bibliothek, Gundelsheim 10025
Siegal College of Judaic Studies, Cleveland 30837
Siegfried Vögele Institut, Königstein im Taunus 12156
Siemens AG, Berlin 11373
Siemens AG, Karlsruhe 11411
Siemens Corporate Research, Inc, Princeton 36157
Siemens Water Technology, Ames 35444
Siemens Water Technology, Ames 33918
Siena College, Loudonville 31680
Siena Heights College, Adrian 30171
Sierra Club, San Francisco 37557
Sierra Joint Community College District, Rocklin 32473
Sierra Leone Library Board, Freetown 24563
Sierra Leone Medical and Dental Association, Freetown 24582
Sierra Madre Public Library, Sierra Madre 41323
Sierra Nevada College, Incline Village 31381
Sierra Research Library, Sacramento 37477
Sierra Vista Public Library, Sierra Vista 41324
Sifriat Namal Haifa, Haifa 14702
SIFU – Statens Institut för Företagsutveckling, Stockholm 26677
Sigatoka Town Council Library, Sigatoka 07323
Sigmund-Freud-Institut, Frankfurt am Main 11871
Sigmund-Freud-Privatstiftung, Wien 01653
Signet Library, Edinburgh 29665
Sigtuna kommuns bibliotek, Märsta 26853
Sigtunastiftelsens Bibliotek, Sigtuna 26631
SIIS Centro de Documentación y Estudios, San Sebastián 26136
SIK – Institutet för Livsmedel och Bioteknik AB, Göteborg 26581
SIK – The Swedish Institute for Food and Biotechnology, Göteborg 26581
SIK-ISEA Schweizerisches Institut für Kunstwissenschaft, Zürich 27467
SIK-ISEA Swiss Institute for Art Research, Zürich 27467
Šilale Municipal Public Library, Šilalė 18364
Šilalės rajono savivaldybės viešoji biblioteka, Šilalė 18364
Silas Bronson Library, Waterbury 41762
Silesian Library, Katowice 20789
Silesian Provincial Museum, Opava 06470
Silesian University of Technology, Gliwice 20813
Silk and Art Silk Mills' Research Association, Mumbai 14101
Silk Fabrics Industrial Corporation, Mogilev 01905
Silk Factory, Moskva 22852
Silk Integrated Plant, Vitebsk 01931
Silkeborg Amtsgymnasium, Silkeborg 06746
Silkeborg Bibliotek, Silkeborg 06923
Silliman University, Dumaguete City 20698
Sills, Cummis, Et Al Law Library, Newark 36046
Les Silos, Chaumont 08823
Silpakorn University, Bangkok 27709
Silskogospodarsky instytut, Kamyanets-Podilsk 27983
Silskogospodarsky instytut, Lugansk 28030
Silskogospodarsky instytut, Odesa 28060
Silskokhospodarsky instytut, Kherson 28001
Šilutės rajono savivaldybės Fridricho Bajoraičio viešoji biblioteka, Šilutė 18365
Silva Tarouca Research Institute for Landscape and Ornamental Gardening, Průhonice 06535
Silver City Public Library, Silver City 41325
Silver Falls Library District, Silverton 41326
Silver Lake College, Manitowoc 31751
Simao Teachers' College, Simao 05578
Simera, Kempton Park 25093
Simferopol Regional Centralized Library System, Main Library, Gvardiyske 28579
Simferopol Regional Centralized Library System, Trenev Main Library, Simferopol 28817
Simferopolska raionna TsBS dlya doroslikh, Tsentralna biblioteka im. Trenova, Simferopol 28817
Simferopolska raionna TsBS, Tsentralna biblioteka, Gvardiyske 28579
Simmons College, Boston 30503
Simon Fraser University, Burnaby 03678
Simon, Peragine, Smith & Redfearn LLP, New Orleans 35951
Simon Wiesenthal Center Library & Archives, Los Angeles 37063
Simon-Dubnow-Institut für jüdische Geschichte und Kultur an der Universität Leipzig, Leipzig 12200
K. Simonov Library, Kyiv 28657
Simpson College, Indianola 31394
Simpson College, Redding 32421
Simpson, Gumpertz & Heger, Inc Library, Waltham 36325
Simpson, Thacher & Bartlett, New York 36022
Simrishamns bibliotek, Simrishamn 26893
Simsbury Public Library, Simsbury 41328
Sinagoga del Tránsito, Toledo 26165
Sinai Temple, Los Angeles 33501
Sinclair Community College, Dayton 33282
Sinclair Knight Merz Library, Saint Leonards 01000
Sindh Muslim Government Law College, Karachi 20459
Sinelnikove Regional Centralized Library System, Main Library, Sinelnikove 28818
Sinelnikovska raionna TsBS, Tsentralna biblioteka, Sinelnikove 28818
Singapore Bible College, Singapore 24605
Singapore Botanic Gardens, Singapore 24617
Singapore Police Force, Singapore 24618
Singapore Polytechnic, Singapore 24592
Singapore Science Centre, Singapore 24619
Singapore Sports Council, Singapore 24620
Singapore Telecommunication Academy, Singapore 24621
Singleton Public Library, Singleton 01156
A. N. Sinha Institute of Social Studies, Patna 14180
Sinologicheskoe otdelenie, Moskva 23485
Sint Antonius Ziekenhuis, Nieuwegein 19495
Sint Benedictus Abdij De Achelse Kluis, Hamont-Achel 02466
Sint Benedictusabdij De Achelse Kluis, Valkenswaard 19318
Sint Lucas Antwerpen, Antwerpen 02275
Sint-Adelbertabdij, Egmond Binnen 19306
Sint-Andriesabdij, Brugge 02456
Sintef Energiforskning, Trondheim 20300
Sintef Energy Research Library, Trondheim 20300
SINTEF Fvn 1, Oslo, Oslo 20281
Sint-Janscentrum, 's-Hertogenbosch 19309
Sint-Paulusadbij, Oosterhout 19314
Sint-Pieters- en Paulusabdij, Dendermonde 02457
Sint-Sixtusabdij, Westvleteren 02495
Sint-Willibrordsabdij, Doetinchem 19304
Sioux City Public Library, Sioux City 41329
Siouxland Heritage Museums, Sioux Falls 37610
Siouxland Libraries, Sioux Falls 41330
Siping Normal University, Siping 05579
Sir Arthur Lewis Community College, Castries 26333
Sir Charles Hayward Public Lending Library, Freeport 01734
Sir Mortimer B. Davis Jewish General Hospital, Montreal 04414
Sir Parashurambhau College, Pune 13837
Sir Sandford Fleming College, Lindsay 03741
Sirote & Permutt, PC, Birmingham 35509
Širvintų rajono savivaldybės viešoji biblioteka, Širvintos 18366
Sisekaitseakadeemia, Tallinn 07222
Šiška Library, Ljubljana 24970
Siskiyou County Public Library, Yreka 42000
Sistema Boliviano de Tecnología Agropecuaria (SIBTA), La Paz 02929
Sisters of Saint Joseph of Saint Hyacinthe, Saint-Hyacinthe 04225
Sisters of St. Benedict, Ferdinand 35204
Sisters of The Immaculate Conception Convent Library, Putnam 35341
Site et musée romain d'Avenches, Avenches 27298
Sithi Zareena Hj. Razi, Kota Kinabalu 18577
Šiualiai University Library, Šiauliai 18285
Siuslaw Public Library District, Florence 39078
Six Mile Regional Library District, Granite City 39278
Sjávarútvegsbókasafnió, Reykjavík 13574
Sjöbo bibliotek, Sjöbo 26894
Sjöfartsmuseet, Göteborg 26582
Sjöfartsverket, Norrköping 26499
Sjöhistoriska museet, Stockholm 26678
Sjøkrigsskolen, Laksevåg 20184
Sjömansbiblioteket, Göteborg 26583
Sjukhusbiblioteken i Sörmland, Eskilstuna 26570
Sjukhusbiblioteket, Södertälje 26634
Sjukhuset i Arvika, Arvika 26559
Sjukhuset i Hässleholm, Hässleholm 26589
Sjukhuset i Torsby, Torsby 26699
Sjukhuset i Varberg, Varberg 26715
Sjúkrahúsid Akureyri, Bókasafn, Akureyri 13566
Skadden, Arps, Slate, Meagher & Flom (Illinois) Library, Chicago 35601
Skadden, Arps, Slate, Meagher & Flom Library, New York 36023
Skadden, Arps, Slate, Meagher & Flom LLP, Washington 36379
Skadden, Arps, Slate, Meagher & Flom LLP Library, Wilmington 36427
Skagit Valley College, Mount Vernon 33565
Skanderborg Bibliotek, Skanderborg 06924
Skatteverkets bibliotek, Stockholm 26515
Skaun folkebibliotek, Børsa 20317
SKB avtomaticheskikh linii, Baranovichi 01827
SKB tekstilnoi promyshlennosti, Sankt-Peterburg 22936
SKB tyazhelykh i unikalnykh stankov, Sankt-Peterburg 22937
SKB zuboobrabatyvayushchikh shlifovalnykh i zatochnykh stankov, Vitebsk 02082
Skellefteå bibliotek, Skellefteå 26895
Skellefteå lasarett, Skellefteå 26632
Skelleftei Hospital Library, Skellefteå 26632
Skellerup Antarctic Library, Christchurch 19831
Skeriabiblioteket, Skellefteå 26420
Skhidnoukrainsky derzhavny universytet, Lugansk 28031
Ski bibliotek, Ski 20397
Skidaway Institute of Oceanography Library, Savannah 37585
Skidmore College, Saratoga Springs 32619
Skien bibliotek, Skien 20398
Skinnskattebergs bibliotek, Skinnskatteberg 26896
Skive Bibliotek, Skive 06925
Sklifosovsky Institute of Emergency, Moskva 23199
Skochinsky Mining Institute, Lyubertsy 22316
Škoda a.s., Plzeň 06422
Škoda Auto a.s., Mladá Boleslav 06419
ŠKODA concern Plzeň, Ltd. Comp., Plzeň 06422
Škofjska knjižnica, Ljubljana 24899
Škofijski arhiv, Maribor 24900
Skokie Public Library, Skokie 41332
Skoklosters Bibliotek, Skokloster 26633
Škola knihovnických a informačných štúdí, Bratislava 24642
Skole Regional Centralized Library System, Main Library, Skole 28819
Skolivska raionna TsBS, Tsentralna biblioteka, Skole 28819
Skorokhod Footwear Company, Sankt-Peterburg 22928
Skov- og Naturstyrelsen, København 06821
Skövde stadsbibliotek, Skövde 26898
H.S. Skovoroda Institute of Philosophy, Kyiv 28322
Skowhegan School of Painting & Sculpture, East Madison 33307
Skryabin Moscow Veterinary Academy, Moskva 22357
Skulpturensammlung, Berlin 11581
Skuodo savivaldybės viešoji biblioteka, Skuodas 18367
Skurups kommunbibliotek, Skurup 26899
Skyline College, San Bruno 33725
Slagelse Bibliotekerne, Slagelse 06926
Śląska Uniwersytet Medyczny, Katowice 20818
Slavic Library, Ljubljana 24933
Slavuta Regional Centralized Library System, Main Library, Slavuta 28820
Slavutska raionna TsBS, Tsentralna biblioteka, Slavuta 28820
Slezské zemské muzeum, Opava 06470
Sligo County Libraries, Sligo 14588
Slippery Rock University of Pennsylvania, Slippery Rock 32682
Slonim City Library no 2, Slonim 02233
Slonimskaya gorodskaya bibliyateka no 2, Slonim 02233
Slonimskaya tsentralnaya raionnaya bibliyateka im. Ya. Kolasa, Slonim 02234
Slotervaartziekenhuis, Amsterdam 19377
Slovácké muzeum, Uherské Hradiště 06544
Slovácko Region Museum, Uherské Hradiště 06544
Slovak Academy of Sciences, Bratislava 24636
Slovak Academy of Sciences – Archaeology Institute, Nitra 24743
Slovak Academy of Sciences – Astronomical Institute, Tatranská Lomnica 24754
Slovak Academy of Sciences – Institute of Arts, Bratislava 24710
Slovak Academy of Sciences – Institute of Chemistry, Bratislava 24702
Slovak Academy of Sciences – Institute of Electrical Engineering, Bratislava 24703
Slovak Academy of Sciences – Institute of Ethnology, Bratislava 24712
Slovak Academy of Sciences – Institute of Experimental Psychology, Bratislava 24713
Slovak Academy of Sciences – Institute of Geophysics, Bratislava 24705
Slovak Academy of Sciences – Institute of Inorganic Chemistry, Bratislava 24711
Slovak Academy of Sciences – Institute of Landscape Ecology, Bratislava 24715
Slovak Academy of Sciences – Institute of Materials and Machine Mechanics, Bratislava 24694
Slovak Academy of Sciences – Institute of Mathematics, Bratislava 24708
Slovak Academy of Sciences – Institute of Musicology, Bratislava 24714
Slovak Academy of Sciences – Institute of Normal and Pathological Physiology, Bratislava 24716
Slovak Academy of Sciences – Institute of Physics, Bratislava 24704
Slovak Academy of Sciences – Institute of Sociology, Bratislava 24709
Slovak Academy of Sciences – Institute of State and Law, Bratislava 24718
Slovak Academy of Sciences – L. Štúra Institute of Linguistics, Bratislava 24707
Slovak Agricultural Library, Nitra 24744
Slovak Agricultural Museum Nitra, Nitra 24745

Slovak Centre of Scientific and Technical Information, Bratislava 24722
Slovak Economic Library, Bratislava 24641
Slovak Institute for Standardization, Bratislava 24725
Slovak Library of Forestry and Wood Sciences, Zvolen 24672
Slovak Medical Library, Bratislava 24698
Slovak Mining Museum, Banská Štiavnica 24680
Slovak National Gallery, Bratislava 24720
Slovak National Library, Martin 24634
Slovak National Literary Museum, Martin 24741
Slovak National Museum, Bratislava 24723
Slovak Pedagogic Library, Bratislava 24721
Slovak Postgraduate Academy of Medicine, Bratislava 24643
Slovak Power Engineering Works, Tlmače 24756
Slovak University of Technology in Bratislava, Bratislava
– Faculty of Architecture, Bratislava 24646
– Faculty of Chemical and Food Technology, Bratislava 24645
– Faculty of Civil Engineering, Bratislava 24650
– Faculty of Electrical Engineering and Information Technology, Bratislava 24647
– Faculty of Materials Science and Technology, Trnava 24649
Slovak University of Technology in Bratislava, Bratislava 24651
Slovanska knjižnica, Ljubljana 24933
Slovene Alpine Library, Ljubljana 24930
Slovene Ethnographic Museum, Ljubljana 24936
Slovene Parliament, Ljubljana 24891
Slovenian Academy of Sciences and Arts, Ljubljana 24807
Slovenian Macroeconomic Analysis and Development Institute, Ljubljana 24942
Slovenian Museum of Natural History, Ljubljana 24931
Slovenian National Building and Civil Engineering Institute, Ljubljana 24943
Slovenian Philharmonic, Ljubljana 24934
Slovenian Railways ltd., Library, Ljubljana 24903
Slovenian School Museum, Ljubljana 24938
Slovenian Society, Ljubljana 24935
Slovenská Akadémia Vied (SAV), Bratislava 24636
Slovenská Akadémia Vied – Archeologický ústav, Nitra 24743
Slovenská Akadémia Vied – Astronomický ústav, Tatranská Lomnica 24754
Slovenská Akadémia Vied – Chemický ústav, Bratislava 24702
Slovenská Akadémia Vied – Elektrotechnický ústav, Bratislava 24703
Slovenská Akadémia Vied – Fyzikálny ústav SAV, Bratislava 24704
Slovenská Akadémia Vied – Geofyzikálny ústav, Bratislava 24705
Slovenská Akadémia Vied – Geofyzikálny ústav – Odbor fyziky atmosféry, Bratislava 24706
Slovenská akadémia vied – Jazykovedný ústav Ľ. Štúra, Bratislava 24707
Slovenská Akadémia Vied – Matematický ústav, Bratislava 24708
Slovenská Akadémia Vied – Sociologický ústav, Bratislava 24709
Slovenská Akadémia Vied – Umenovedný ústav, Bratislava 24710
Slovenská Akadémia Vied – Ustav anorganickej chémie, Bratislava 24711
Slovenská Akadémia Vied – Ustav etnológie, Bratislava 24712
Slovenská Akadémia Vied – Ustav experimentálnej farmakológie, Bratislava 24644
Slovenská Akadémia Vied – Ustav experimentálnej psychológie, Bratislava 24713
Slovenská Akadémia Vied – Ustav hudobnej vedy, Bratislava 24714
Slovenská Akadémia Vied – Ustav Krajinnej Ekológie, Bratislava 24715
Slovenská Akadémia Vied – Ustav normálnej a patologickej fyziológie, Bratislava 24716
Slovenská akadémia vied – Ustav počítačových systémov, Bratislava 24717
Slovenská Akadémia Vied – Ustav štátu a práva, Bratislava 24718
Slovenská Akadémia Vied – Virologický ústav, Bratislava 24719
Slovenska akademija znanosti in umetnosti, Ljubljana 24807
Slovenska filharmonija, Ljubljana 24934
Slovenská lekárska knižnica, Bratislava 24698
Slovenska Ljudska Knjižnica d. Feigel, Gorizia 16752
Slovenska Matica, Ljubljana 24935
Slovenská národná galéria, Bratislava 24720
Slovenská Národná Knižnica, Martin 24634
Slovenská pedagogická knižnica, Bratislava 24721
Slovenská poľnohospodárska knižnica, Nitra 24744
Slovenska študijska knjižnica / Slowenische Studienbibliothek, Klagenfurt 01551
Slovenská technická knižnica – Centrum vedecko-technických informácií SR, Bratislava 24722
Slovenská Technická Univerzita v Bratislave, Bratislava
– Chemickotechnologická fakulta, Bratislava 24645
– Fakulta Architektúry, Bratislava 24646
– Fakulta elektrotechniky a informatiky, Bratislava 24647
– Kniznica a informacne stredisko Strojnkíckej fakulty, Bratislava 24648
– Materiálovo-technická fakulta, Trnava 24649
– Staverbna fakulta, Bratislava 24650
Slovenská technická univerzita v Bratislave, Bratislava 24651
Slovenské banské múzeum, Banská Štiavnica 24680
Slovenské energetické strojárne, Tlmače 24756
Slovenské múzeum ochrany prírody a jaskyniarstva, Liptovský Mikuláš 24738
Slovenské národné literárne múzeum, Martin 24741
Slovenské národné muzeum, Bratislava 24723
Slovenské národné múzeum – Etnografické múzeum, Martin 24742
Slovenské poľnohospodárske múzeum Nitra, Nitra 24745
Slovenské technické múzeum, Košice 24734
Slovenske Železarne – Metal Ravne d.o.o., Ravne na Koroškem 24906
Slovenske železnice, d. o. o., Ljubljana 24903
Slovenski etnografski muzej, Ljubljana 24936
Slovenski gledališki muzej, Ljubljana 24937
Slovenski šolski muzej, Ljubljana 24938
Slovenský kontrolný a skúšobný ústav polnohospodársky, Košice 24735
Slovenský ústav technickej normalizácie, Bratislava 24725
SLOVNAFT A.S., Bratislava 24675
Slovyansk City Centralized Library System, Main Library, Slovyansk 28822
Slovyansk DRES, Trade Union Library, Mikolaivka 28727
Slovyanska DRES, Mikolaivka 28727
Slovyanska miska TsBS, Tsentralna biblioteka, Slovyansk 28822
Slowak Museum of Nature Protection and Speleology, Liptovský Mikuláš 24738
Slowenische Studienbibliothek, Klagenfurt 01551
Slutsk Central Children's City Library, Slutsk 02236
Slutskaya tsentralnaya gorodskaya bibliyateka im. Krupskoi, Slutsk 02235
Slutskaya tsentralnaya gorodskaya detskaya bibliyateka, Slutsk 02236
SMAK – Stedelijk Museum voor Actuele Kunst, Gent 02656
Small Business Administration, Washington 37779
Small Business Credit Insurance Corp, Tokyo 17438
Small Business Finance Corp, Tokyo 17437
Small Enterprises National Documentation Centre (Sendoc), Hyderabad 14011
Smedjebackens bibliotek, Smedjebacken 26900
Šmidingerova knihovna Strakonice, Strakonice 06607
Šmidinger's Library Strakonice, Strakonice 06607
Smila Regional Centralized Library System, Main Library, Smila 28823
A. K. Smiley Public Library, Redlands 40955
Smilyanska raionna TsBS, Tsentralna biblioteka, Smila 28823
Smith College Libraries, Northampton 32148
– Anita O'K. & Robert R. Young Science Library, Northampton 32149
– Hillyer Art Library, Northampton 32150
– Mortimer Rare Book Room, Northampton 32151
– Werner Josten Performing Arts Library, Northampton 32152
Smith, Currie & Hancock, Atlanta 35468
Smith, Gambrell & Russell, Atlanta 35469
Smith Moore, LLP, Greensboro 35727
Smithkline Beecham Pharmaceuticals, Welwyn Garden City 29585
Smithsonian Environmental Research Center Library, Edgewater 36823
Smithsonian Institution Libraries, Edgewater 36823
Smithsonian Institution Libraries, New York City 37273
Smithsonian Institution Libraries, Suitland 37647
Smithsonian Institution Libraries, Suitland 37648
Smithsonian Institution Libraries, Washington 37780
Smithsonian Institution Libraries – Botany & Horticulture Library, Washington 37781
Smithsonian Institution Libraries – The Dibner Library of the History of Science & Technology, Washington 37782
Smithsonian Institution Libraries – Freer Gallery of Art & Arthur M. Sackler Gallery Library, Washington 37783
Smithsonian Institution Libraries – Hirshhorn Museum & Sculpture Garden Library, Washington 37784
Smithsonian Institution Libraries – John Wesley Powell Library of Anthropology, Washington 37785
Smithsonian Institution Libraries – Museum Studies & Reference Library, Washington 37786
Smithsonian Institution Libraries – National Air & Space Museum Library, Washington 37787
Smithsonian Institution Libraries – National Museum of American History Library, Washington 37788
Smithsonian Institution Libraries – National Museum of Natural History Library, Washington 37789
Smithsonian Institution Libraries – National Postal Museum Library, Washington 37790
Smithsonian Institution Libraries – National Zoological Park Library, Washington 37791
Smithsonian Institution Libraries – Smithsonian American Art Museum/National Portrait Gallery Library, Washington 37792
Smithsonian Institution Libraries – Warren M. Robbins Library, National Museum of African Art, Washington 37793
Smithsonian Tropical Research Institute, Balboa 20538
Smithtown Library, Smithtown 41336
Smolensk Regional Universal Library, Smolensk 22175
Smolensk State Institute of Physical Culture and Sport, Smolensk 22582
Smolensk State Museum, Smolensk 23973
Smolenskaya oblastnaya detskaya biblioteka im. I.S. Sokolov-Mikitov, Smolensk 24319
Smolenskaya oblastnaya universalnaya biblioteka, Smolensk 22175
Smolenskaya oblastnaya yunosheskaya biblioteka, Smolensk 24320
Smolenski gosudarstvenny institut fizicheskoi kultury i sporta, Smolensk 22582
Smolenski gosudarstvenny muzei zapovednik, Smolensk 23973
Smolenski mezhotraslevoi territorialny tsentr nauchno-tekhnicheskoi informatsii i propagandy, Smolensk 23974
Smolenski pedagogicheski institut im. Marksa, Smolensk 22583
Smolevichi Regional Central Library, Smolevichi 02237
Smolevichiskaya tsentralnaya raionnaya bibliyateka, Smolevichi 02237
S.M.S. Medical College, Jaipur 13809
SMS Siemag AG, Düsseldorf 11389
Smyrna Public Library, Smyrna 41338
Smyth-Bland Regional Library, Marion 40067
Snc-Lavalin, Inc Library, Montreal 04271
SNC-Lavalin Inc Library, Toronto 04292
Snead State Community College, Boaz 30470
Snell & Wilmer, Phoenix 36122
Snohomish County Law Library, Everett 34195
Sno-Isle Libraries, Marysville 40087
Sno-Isle Libraries Arlington Branch, Arlington 37994
Sno-Isle Libraries – Edmonds Branch, Edmonds 38935
Sno-Isle Libraries – Lynnwood Branch, Lynnwood 39987
Sno-Isle Libraries – Marysville Branch, Marysville 40088
Sno-Isle Libraries – Mill Creek Branch, Mill Creek 40203
Sno-Isle Libraries – Monroe Branch, Monroe 40266
Sno-Isle Libraries – Mountlake Terrace Branch, Mountlake Terrace 40332
Sno-Isle Libraries – Mukilteo Branch, Mukilteo 40333
Sno-Isle Libraries – Oak Harbor Branch, Oak Harbor 40562
Sno-Isle Libraries – Snohomish Branch, Snohomish 41339
Snow College, Ephraim 33323
Snow Cruywagen Public Library, Midddelburg 25213
Snow Library, Orleans 40652
Snowy Mountains Engineering Corporation, Cooma 00816
Snyatin Regional Centralized Library System, Main Library, Snyatin 28824
Snyatinska raionna TsBS, Tsentralna biblioteka, Snyatin 28824
Snyder County Libraries, Selinsgrove 41278
Sobinov State Conservatoire, Saratov 22572
Sobranie na Republika Makedonija, Skopje 18440
SOC – Scottish Ornithologists' Club, East Lothian 29649
Sochi Shikenjo Kikaku Renrakushitsu Shiryoka, Nasu 17623
Sociaal Economische Raad (SER), Den Haag 19451
Sociaal Historisch Centrum voor Limburg (SHCL), Maastricht 19491
Sociaal-Economische Raad van Vlaanderen (SERV), Brussel 02623
Social Insurance Institution, Helsinki 07467
Social Law Library, Boston 36579
Social- och beteendevetenskapliga biblioteket, Lund 26413
Social Sciences College, Klaipėda 18284
Social Sciences Library, Ho Chi Minh City 42307
Social Security Administration Library, Baltimore 33964
Social Service Agency of the Protestante Church in Gemany, Library, Berlin 11164
Den Sociale Højskole, København, Frederiksberg 06659
Den Sociale Højskole, Odense, Odense 06716
Socialforskningsinstituttets Bibliotek, København 06822
Socialinių mokslų kolegija, Klaipėda 18284
Socialrådgiveruddannelsen i Århus, Århus 06654
Sociedad Amantes de la Luz, Santiago de los Caballeros 06963
Sociedad Argentina de Ciencias Neurológicas, Psiquiátricas y Neurocirúrgicas, Buenos Aires 00286
Sociedad Argentina de Estudios Geográficos, Buenos Aires 00287
Sociedad Bilbaina, Bilbao 25934
Sociedad Central de Arquitectos, Buenos Aires 00288
Sociedad Chilena de Historia y Geografía, Santiago 04945
Sociedad Científica Argentina, Buenos Aires 00289
Sociedad Científica del Paraguay, Asunción 20611
Sociedad de Ginecología y Obstetricia de El Salvador, San Salvador 07211
Sociedad de Obstetricia y Ginecología de Venezuela, Caracas 42256
Sociedad Económica de Amigos del País, La Habana 06310
Sociedad Entomológica del Perú, Lima 20677
Sociedad General de Autores de la Argentina (Argentores), Buenos Aires 00290
Sociedad General de Autores y Editores (S.G.A.E.), Madrid 26098
Sociedad Geológica del Perú, Lima 20678
Sociedad Hebraica Argentina, Buenos Aires 00291
Sociedad Médica de Santiago, Santiago 04872
Sociedad Mexicana de Geografía y Estadística, México 18865
Sociedad Rural Argentina, Buenos Aires 00292
Sociedad Uruguaya de Pediatría, Montevideo 42066
Sociedad Venezolana de Ciencias Naturales, Caracas 42257
Sociedad Venezolana de Ingenieros Forestales, Caracas 42258
Sociedade Brasileira de Autores, Rio de Janeiro 03372
Sociedade Brasileira de Cultura Inglesa – Asa Sul, Brasília 03285

Sociedade Brasileira de Instrução / IUPERJ, Rio de Janeiro 03373
Sociedade de Geografia de Lisboa, Lisboa 21816
Sociedade Guarulhense de Educação, Guarulhos 03300
Sociedade Martins Sarmento, Guimarães 21762
Sociedade Nacional de Agricultura, Rio de Janeiro 03374
Sociedade Portuguesa de Ciências Naturais, Lisboa 21663
Sociedade 'Visconde de São Leopoldo', Santos 03385
Società Adriatica di Scienze, Trieste 16630
Società Agraria di Lombardia, Milano 16350
Società Alpina Friulana, Udine 16636
Società Dantesca Italiana, Firenze 16273
Società degli Artisti e Patriottica, Milano 16351
Società dei Naturalisti in Napoli, Napoli 16380
Società dei Trecento Campi, Venezia 16658
Società del Gruppo ENI, San Donato Milanese 16155
Società di Storia Patria di Terra di Lavoro, Caserta 16222
Società di Storia Patria per la Sicilia Orientale, Catania 16228
Società di Studi Geografici, Firenze 16274
Società Economica di Chiavari, Chiavari 16230
Società Filologica Friulana G. I. Ascoli, Udine 16637
Società Filologica Romana, Roma 16556
Società Geografica Italiana, Roma 16557
Società Geologica Italiana, Roma 16558
Società Istriana di Archeologia e Storia Patria, Trieste 16631
Società Italiana per gli Studi Filosofici e Religiosi, Milano 16352
Società Italiana per l'Organizzazione Internazionale (SIOI), Roma 16559
Società Letteraria di Verona, Verona 16670
Società Ligure di Storia Patria, Genova 16292
Società Medica Chirurgica di Bologna, Bologna 16199
Società Napoletana di Storia Patria, Napoli 16381
Società Nazionale di Scienze, Lettere ed Arti, Napoli 16382
Società Savonese di Storia Patria, Savona 16574
Società Siciliana per la Storia Patria, Palermo 16406
Società Storica Lombarda, Milano 16353
Società Toscana di Scienze Naturali, Pisa 16423
Società Umanitaria, Milano 16354
Societas pro Fauna et Flora Fennica, Helsingin Yliopisto 07441
Societat Arqueològica Lulliana, Palma de Mallorca 26115
Societat d'Història Natural de les Balears, Palma de Mallorca 26116
Societât Filologjiche Furlane G. I. Ascoli, Udine 16637
Société Archéologique de Namur (SAN), Namur 02691
Société Archéologique de Sens, Sens 08676
Société Archéologique du Département de Constantine, Constantine 00079
Société Archéologique du Midi de la France, Toulouse 08697
Société Archéologique et Historique de la Charente (SAHC), Angoulême 08277
Société Archéologique et Historique de l'Arrondissement d'Avesnes, Avesnes-sur-Helpe 08287
Société Asiatique, Paris 08618
Société de Borda, Dax 08344
Société de Géographie, Paris 08619
Société de Langue et de Littérature wallonnes, Liège 02681
Société de lecture (SDL), Genève 27487
Société de législation comparée (SLC), Paris 08620
Société de l'Histoire du Protestantisme Français (SHPF), Paris 08621
Société d'Emulation du Bourbonnais, Moulins 08429
Société d'Encouragement pour l'Industrie Nationale, Paris 08622
Société des Amis du Vieux Toulon, Toulon 08692
Société des Auteurs et Compositeurs Dramatiques (SACD), Paris 08623
Société des Lettres, Sciences et Arts de l'Aveyron, Rodez 08658
Société des Sciences naturelles, archéologiques et historiques de la Creuse, Guéret 08362
Société d'Etudes Juridiques du Zaire, Lubumbashi 06085
La Société d'Histoire de Sherbrooke, Sherbrooke 04494
Société d'Histoire du Théâtre, Paris 08624
Société d'Histoire et d'Archéologie de Genève, Genève 27374
Société d'Histoire et d'Archéologie des Monts, Inc, Sainte-Anne-des-Monts 04482
Société d'Histoire Lorraine et du Musée Lorrain, Nancy 08438
Société d'Histoire Naturelle et d'Ethnographie de Colmar, Colmar 08340
Société Entomologique de France, Paris 08625
Société Entomologique d'Egypte, Cairo 07186
Société Européenne de Culture, Venezia 16659
Société Française de Photographie (SFP), Paris 08626
Société Généalogique Canadienne-Française, Montreal 04419
Société Généalogique de l'Est du Québec, Rimouski 04475
Société Géologique de Belgique, Liège 02682
Société Géologique de France, Paris 08627
Société Grenobloise d'Etudes et Applications Hydrauliques (SOGREAH), Grenoble 08359
La Societé Historique de Saint-Boniface, Saint-Boniface 04481
Société Historique et Archéologique du Périgord, Périgueux 08640
Société Industrielle de Mulhouse, Mulhouse 08433
Société Jersiaise, St Helier 29898
Société Linnéenne de Lyon, Lyon 08401
Société Nationale de l'Electricité et du Gaz (SONELGAZ), Alger 00056
Société Nationale de Recherche et d'Exploitation des Ressources Minières du Mali (SONAREM), Kati 18656
Société Nationale des Chemins de Fer Belges (SNCB), Bruxelles 02624
Société Nationale des Chemins de Fer Français (SNCF), Paris 08628
Société Nationale des Sciences Naturelles et Mathématiques de Cherbourg, Cherbourg 08334
Société Nationale d'Horticulture de France (SNHF), Paris 08629
Société Polymathique Morbihan, Vannes 08709
Société Port-Royal, Paris 08224
Société Psychanalytique de Paris (SPP), Paris 08630
Société Radio-Canada Bibliothèque, Montreal 04420
Société Royale Belge d'Entomologie / Koninklijke Belgische Vereniging voor Entomologie, Bruxelles 02625
Société Royale d'Archéologie de Bruxelles, Bruxelles 02626
Société Scientifique et Artistique, Clamecy 08335
Société Scientifique, Historique et Archéologique de la Corrèze, Brive-la-Gaillarde 08312
Société Tunisienne de Banque, Tunis 27841
Society for Cooperation in Russian and Soviet Studies, London 29831
Society for Coptic Archeology, Cairo 07187
Society for Macedonian Studies, Thessaloniki 13132
Society for Promoting & Encouraging Arts & Knowledge of the Church, Eureka Springs 36834
Society for the Study of Male Psychology & Physiology Library, Montpelier 37128
Society of Advocates of RSA CTPA, Pretoria 25169
Society of Advocates of South Africa, Johannesburg 25137
Society of Antiquaries of London, London 29832
Society of Antiquaries of Newcastle upon Tyne, Newcastle upon Tyne 29858
Society of Australian Genealogists, Sydney 01031
Society of California Pioneers, San Francisco 37558
Society of Genealogists, London 29833
Society of Geographical Studies, Firenze 16274
Society of Heating, Air-conditioning and Sanitary Engineers of Japan, Tokyo 17702
Society of Manufacturing Engineers, Dearborn 36780
Society of Naturalists, Moskva 23452
Society of Science and Letters, Przemyśl 21211
Society of Scientific Organization and Management, Warszawa 21347
Society of Solicitors in the Supreme Courts of Scotland, Edinburgh 29666
Society of Swedish Literature in Finland, Helsinki 07499
Society of the Cincinnati Library, Washington 37794
The Society of the Four Arts, Palm Beach 37318
Society of the Polish Free University, Warszawa 21348
Society of Walloon Language and Literature, Liège 02681
Sociologický ústav, Bratislava 24709
Söderhamns Stadsbibliotek, Söderhamn 26901
Södertälje sjukhus, Södertälje 26635
Södertälje stadsbibliotek, Södertälje 26902
Södertörns högskola, Huddinge 26393
Sodovoy zavod im. V.I. Lenina, Berezniki 22780
Södra Älvsborgs Sjukhus Borås, Borås 26561
Soeurs Ursulines de Québec, Québec 04221
Søfartens Biblioteks Studiecenter, København 06823
Sofia City Library, Sofiya 03605
Sofia municipality M.E. 'Old Sofia with the Historical museum', Sofiya 03577
Sofiyski universitet 'Sv. Kliment Okhridski', Sofiya 03463
– Biblioteka Biologia, Sofiya 03464
– Biblioteka fizika i astronomiya, Sofiya 03465
– Fakultet po filologicheski nauki, Sofiya 03466
– Fakultet po khimiya, Sofiya 03467
– Fakultet po matematika, Sofiya 03468
– Journalism and Mass Communication Library, Sofiya 03469
– Spetsialnost filosofiya, Sofiya 03470
– Spetsialnost geografiya, Sofiya 03471
– Spetsialnost geologiya, Sofiya 03472
– Spetsialnost istoriya, Sofiya 03473
– Spetsialnost pedagogika, Sofiya 03474
– Yuridicheski fakultet, Sofiya 03475
Sogang University, Seoul 18094
Sogn and Fjordane County Library, Førde 20332
Sogn og Fjordane fylkesbibliotek, Førde 20332
Sogn og Fjordane University College, Library Fjora, Sogndal 20159
Le Soir, Bruxelles 02501
Sojourner-Douglass College, Baltimore 30343
Soka University, Hachioji 16936
Sokiryani Regional Centralized Library System, Main Library, Sokiryani 28825
Sokiryanska raionna TsBS, Tsentralna biblioteka, Sokiryani 28825
Sokoine University of Agriculture, Morogoro 27656
Sokol Paper and Pulp Joint-Stock Company, Sokol 22956
A.N. Sokolovski Institute of Agricultural Chemistry and Soil Cultivation, Kharkiv 28255
Sokone National Agricultural Library, Morogoro 27656
Sola folkebibliotek, Sola 20399
Solano Community College, Suisun City 33784
Solano County Law Library, Fairfield 34198
Solano County Library, Fairfield 39026
Solano County Library – Fairfield Civic Center, Fairfield 39027
Solano County Library – John F. Kennedy Branch, Vallejo 41654
Solano County Library – Vacaville Public Library-Cultural Center, Vacaville 41651
Solicitor General Canada, Ottawa 04123
Soligorsk Central City Library, Soligorsk 02239
Soligorskaya tsentralnaya gorodskaya bibliyateka, Soligorsk 02239
Solihull College, Solihull 29494
Solihull Metropolitan Borough Council, Solihull 30072
Sollac, Groupe Usinor, Florange 08255
Sollefteå kommunbibliotek, Sollefteå 26903
Sollefteå sjukhus, Sollefteå 26636
Sollentuna bibliotek, Sollentuna 26904
Solna stadsbibliotek, Solna 26905
Solombalsk Paper and Pulp Industrial Corporation, Arkhangelsk 22773
Solomon-Islands College of Higher Education, Honiara 24992
Solrødbibliotekerne, Solrød Strand 06929
Solutia Inc, Cantonment 35554
Solvay Deutschland GmbH, Hannover 11405
Solvay Pharmaceuticals BV, Weesp 19341
Sölvesborgs bibliotek, Sölvesborg 26906
Someron kaupunginkirjasto, Somero 07610
Somers Library, Somers 41343
Somerset Archaeological and Natural History Society, Taunton 29907
Somerset College, Taunton 29424
Somerset Community College, Somerset 33766
Somerset County Council, Bridgwater 29947
Somerset County Federated Library System, Somerset 41346
Somerset County Law Library, Somerset 34867
Somerset County Law Library, Somerville 34870
Somerset County Library, Somerset 41347
Somerset County Library System, Bridgewater 38282
Somerset County Library System, Princess Anne 40899
Somerset County Library System – Mary Jacobs Memorial, Rocky Hill 41048
Somerset County Library System – North Plainfield Memorial, North Plainfield 40519
Somerset County Library System Warren Township Branch, Warren 41724
Somerset Public Library, Somerset 41348
Somerset West Public Library, Somerset West 25236
Somerville College, Oxford 29347
Somerville Public Library, Somerville 41349
Somerville Public Library, Somerville 41350
Sommers, Schwartz, Silver & Schwartz, Southfield 36277
Somogy megyei Múzeumi Igazgatóság, Kaposvár 13482
Somogyi Library, Szeged 13546
Somogyi-könyvtár, Szeged 13546
Somucho Tokeikyoku Somuka Tokei-Toshokan, Tokyo 17324
Somucho Toshokan, Tokyo 17325
Sønderborg Bibliotek, Sønderborg 06930
Søndersø Bibliotek, Søndersø 06931
Sonnenschein, Nath & Rosenthal, Chicago 35602
Sonnenschein, Nath & Rosenthal Library, Washington 36380
Sonoma Adventist College, Kokopo 20554
Sonoma County Law Library, Santa Rosa 34841
Sonoma County Library, Santa Rosa 41227
Sonoma State University, Rohnert Park 32477
Sonoma Valley Regional Library, Sonoma 41351
Sons of the Revolution in the State of New York Library, New York 37259
Soochow University, Suzhou 05581
Soochow University, Taipei 27558
Sookmyung Women's University, Seoul 18095
Soongsil University, Seoul 18096
Sophia University Tokyo, Tokyo 17176
Sophiahemmet högskola, Stockholm 26435
Sophiahemmet University College, Stockholm 26435
Sophien- und Hufeland-Klinikum gGmbH, Weimar 12580
Soprintendenza ai Beni Ambientali, Architettonici, Archeologici, Artistici e Storici, Trieste 15693
Soprintendenza Archeologica, Ancona 15638
Soprintendenza Archeologica, Milano 16355
Soprintendenza Archeologica dell'Emilia e Romagna, Bologna 15639
Soprintendenza Archeologica per le province di Cagliari e Oristano, Cagliari 16216
Soprintendenza per i Beni Ambientali, Architettonici, Artistici e Storici, L'Aquila 15645
Soprintendenza per i Beni Archeologi della Puglia, Taranto 15690
Soprintendenza per i Beni Artistici e Storici, Bologna 16200
Soprintendenza per i Beni Storici, Artistici ed Etnoantropologici della Sardegna, Cagliari 15642
Soproni Múzeum, Sopron 13499
Soquem, Inc, Val-d'Or 04535
Sorbische Zentralbibliothek, Bautzen 11477
Sorbisches Institut e.V. / Serbski institut z.t., Bautzen 11477
Sorbisches Museum, Bautzen 11476
Sorø Bibliotek, Sorø 06932
Sør-Trøndelag University College, Trondheim
– Department of Teacher Education and Deaf Studies, Trondheim 20173
– Department of Technology, Trondheim 20174
– School of Health Education and Social Work, Trondheim 20172
– Trondheim Business School, Trondheim 20171
Sør-Varanger Bibliotek, Kirkenes 20353
Sosiaali- ja terveysministeriön kirjasto, Helsinki 07401

Sosnitska raionna TsBS, Tsentralna biblioteka, Sosnitsya 28826
Sosnitsya Regional Centralized Library System, Main Library, Sosnitsya 28826
Sotheby's Library, New York 36024
Sotkamon kunnankirjasto, Sotkamo 07611
Sound and Music, London 29834
South African Astronomical Observatory Library, Observatory 25148
South African Brain Research Institute, Johannesburg 25138
South African Broadcasting Corporation, Johannesburg 25139
South African Bureau of Standards (SABS), Pretoria 25170
South African Defence Force, Pretoria 25080
South African Institute, Amsterdam 19383
South African Institute for Medical Research (SAIMR), Johannesburg 25140
South African Institute of Chartered Accountants, Johannesburg 25141
South African Institute of International Affairs, Johannesburg 25142
South African Iron and Steel Industrial Corporation Ltd, Pretoria 25098
South African Jewish Board of Deputies, Houghton 25128
South African Library for the Blind, Grahamstown 25126
South African Museum, Cape Town 25116
South African Mutual Life Assurance Society, Cape Town 25089
South African National Gallery, Cape Town 25117
South African National Museum of Military History, Saxonwold 25177
South African Police, Pretoria 25081
South African Reserve Bank, Pretoria 25099
South African Sugar Association Experiment Station, Mount Edgecombe 25147
South Australia Courts Administration Authority, Adelaide 00603
South Australian Museum, Adelaide 00860
South Australian Water Corporation, Adelaide 00807
South Ayrshire Council, Ayr 29930
South Birmingham College, Birmingham 28953
South Brunswick Public Library, Monmouth Junction 40260
South Burlington Community Library, South Burlington 41356
South Burnett Regional Council Library, Kingaroy 01110
South Carolina Attorney General's Office, Columbia 34096
South Carolina Department of Corrections, Columbia 34097
South Carolina Department of Disabilities & Special Needs, Clinton 34089
South Carolina Historical Society Library, Charleston 36650
South Carolina State College, Orangeburg 32206
South Carolina State Library, Columbia 30125
South Carolina State Library, Columbia 30126
South Carolina Supreme Court Library, Columbia 34098
South Central Kansas Library System, South Hutchinson 41364
South Central Regional Library, Winkler 04822
South Charleston Public Library, South Charleston 41357
South China Agricultural University, Guangzhou 05207
South China Institute of Botany, Guangzhou 05864
South China Normal University, Guangzhou 05208
South China Sea Institute of Oceanology, Guangzhou 05865
South China Teachers' University, Guangzhou 05209
South China University of Technology, Guangzhou 05210
South China University of Tropical Agriculture, Hainan Island 05230
South Country Library, Bellport 38160
South Dakota Braille & Talking Book Library, Pierre 37373
South Dakota School for the Blind & Visually Impaired, Aberdeen 33113
South Dakota School of Mines & Technology, Rapid City 32417
South Dakota State Historical Society, Pierre 37374
South Dakota State Library, Pierre 30146
South Dakota State Penitentiary, Sioux Falls 34863
South Dakota State University, Brookings 30563

South Dakota Supreme Court, Pierre 34676
South Devon College, Paignton 29363
South Dublin County Libraries, Dublin 14573
South East Derbyshire College, Ilkeston 29472
South East Scientific Research Institute of Agriculture, Saratov 23958
South Eastern Education and Library Board, Ballynahinch 29932
South Florida Community College, Avon Park 33145
South Georgia College, Douglas 33300
South Georgia Regional Library System – Valdosta-Lowndes County Public, Valdosta 41653
South Hamgyong Provincial Library, Hamhung 18051
South Haven Memorial Library, South Haven 41362
South Holland Public Library, South Holland 41363
South Huntington Public Library, Huntington Station 39540
South Hwanghae Provincial Library, Haeju 18050
South India Textile Research Association, Coimbatore 13971
South Interlake Regional Library, Stonewall 04768
South Karelia Central Hospital, Medical Library, Lappeenranta 07527
South Kazakh Technical University, Shimkent 17921
South Kent College, Dover 29463
South Kent College, Folkestone 29465
South Kingstown Public Library, Peace Dale 40735
South Lanarkshire Council, East Kilbride 29977
South Metallurgical Institute, Ganzhou 05198
South Milwaukee Public Library, South Milwaukee 41369
South Mississippi Regional Library – Columbia Marion County Library, Columbia 38650
South Mountain Community College, Phoenix 33642
South Orange Public Library, South Orange 41371
South Pacific Commission, Nouméa 19754
South Park Township Library, South Park 41372
South Pasadena Public Library, South Pasadena 41373
South Perth City Library and Heritage Service, South Perth 01158
South Plainfield Free Public Library, South Plainfield 41374
South Plains College, Levelland 33482
South Portland Public Library, South Portland 41375
South Puget Sound Community College, Olympia 33602
South Saint Paul Public Library, South Saint Paul 41376
South San Francisco Public Library, South San Francisco 41378
South Seattle Community College, Seattle 33758
South Shore Public Libraries, Hebbville 04660
South Shore Regional Library, Hebbville 04661
South Staffordshire Medical Centre, Wolverhampton 29916
South Street Seaport Museum Library, New York 37260
South Suburban College, South Holland 33768
South Suburban Genealogical & Historical Society, Hazel Crest 36924
South Texas College of Law, Houston 31355
South Texas Community College, McAllen 33524
South Thames College, London 29480
South Trafford College, Altrincham 29447
South Tyneside College, South Shields 29396
South Tyneside Metropolitan Borough Council, South Shields 30073
South Valley University, Aswan Branch, Aswan
– Faculty of Education, Aswan 07032
– Faculty of Sciences, Aswan 07033
South Valley University, Quena Branch, Quena
– Faculty of Arts, Quena 07094
– Faculty of Education, Quena 07095
– Faculty of Sciences, Quena 07096
South Valley University, Sohag Branch, Sohag
– Faculty of Arts, Sohag 07097
– Faculty of Education, Sohag 07098
– Faculty of Sciences, Sohag 07099
South West Institute of TAFE, Warrnambool 00592

South Western Sidney Institute of TAFE – Granville College, Granville 00571
South Windsor Public Library, South Windsor 41379
Southampton Free Library, Southampton 41382
Southampton Solent University, Southampton 29397
Southbank Institute of TAFE, South Brisbane 00586
Southbank Institute of Technology, South Brisbane 00587
Southborough Public Library, Southborough 41383
Southbury Public Library, Southbury 41385
Southeast Arkansas Regional Library, Monticello 40277
Southeast Asian Central Banks Research and Training Ctr, Petaling Jaya 18615
Southeast Asian Ministers of Education Organization (SEAMEO), Quezon City 20782
Southeast Asian Ministers of Education Organization (SEAMEO), Singapore 24622
Southeast Community College, Cumberland 33268
Southeast Kansas Library System, Iola 39575
Southeast Michigan Council of Governments Library, Detroit 36806
Southeast Missouri State University, Cape Girardeau 30672
Southeast Regional Library, Weyburn 04817
Southeast Shanxi Teachers' College, Changzhi 05109
Southeast Steuben County Library, Corning 38698
Southeast University, Nanjing 05455
Southeastern Baptist College, Laurel 31570
Southeastern Baptist Theological Seminary, Wake Forest 35405
Southeastern Bible College, Birmingham 35138
Southeastern College, Lakeland 31548
Southeastern Community College, Whiteville 33868
Southeastern Community College Library – North Campus, West Burlington 33857
Southeastern Illinois College, Harrisburg 31298
Southeastern Louisiana University, Hammond 31282
Southeastern Oklahoma State University, Durant 31036
Southeastern Public Library System of Oklahoma, McAlester 40112
Southern Adirondack Library System, Saratoga Springs 41234
Southern Adventist University, Collegedale 30864
Southern Africa Documentation and Cooperation Centre (SADOCC), Wien 01593
Southern Alberta Institute of Technology, Calgary 03981
Southern Älvsborg Hospital Borås, Borås 26561
Southern Arkansas University, Magnolia 31737
Southern Baptist Historical Library & Archives, Nashville 35297
Southern Baptist Theological Seminary, Louisville 35271
Southern Baptist Theological Seminary – Music and Audiovisual Library, Louisville 35272
Southern California Genealogical Society, Burbank 36616
Southern California Institute of Architecture, Los Angeles 31653
Southern California Library for Social Studies & Research, Los Angeles 37064
Southern California University of Health Sciences, Whittier 33053
Southern Christian University, Montgomery 31889
Southern College of Optometry, Memphis 33528
Southern Connecticut State University, New Haven 31971
Southern Cross University, Lismore 00450
Southern Downs Regional Council, Warwick 01175
Southern Education and Library Board, Armagh 29927
Southern Highland Craft Guild, Asheville 35452
Southern Illinois University, Carbondale, Carbondale 30673
– Law Library, Carbondale 30674
Southern Illinois University Edwardsville, Edwardsville 31076

Southern Illinois University – School of Dental Medicine, Alton 30205
Southern Illinois University – School of Medicine Library, Springfield 32708
Southern Methodist College, Orangeburg 35319
Southern Methodist University, Dallas 30945
– DeGolyer Library of Special Collections, Dallas 30946
– Fondren Library Center, Dallas 30947
– Hamon Arts Library, Dallas 30948
– Institute for Study of Earth & Man Reading Room, Dallas 30949
– Science-Engineering Library, Dallas 30950
– Underwood Law Library, Dallas 30951
Southern Methodist University – Bridwell Library-Perkins School of Theology, Dallas 30952
Southern Metropolitan Mental Health Service Library, Rozelle 00995
Southern Nazarene University, Bethany 30419
Southern New Hampshire University, Manchester 31744
Southern Ohio Correctional Facility Library, Lucasville 34481
Southern Oregon University, Ashland 30267
Southern Pines Public Library, Southern Pines 41386
Southern Polytechnic State University, Marietta 31759
Southern Prairie Library System, Altus 37932
Southern Queensland Institute of Tafe, Toowoomba 00546
Southern Regional Education Board Library, Atlanta 36494
Southern Research Institute, Birmingham 36544
Southern Sidney Institute of TAFE – St. George College, Kogarah 00573
Southern State Community College, Hillsboro 33410
Southern Tablelands Regional Library, Goulburn 01099
Southern Union State Community College, Wadley 33831
Southern University, Baton Rouge 30363
– Art & Architecture Library, Baton Rouge 30364
– Oliver B. Spellman Law Library, Baton Rouge 30365
Southern University, Shreveport 32670
Southern University in New Orleans, New Orleans 32005
Southern University of Chile, Valdivia 04891
Southern Utah University, Cedar City 30686
Southern Virginia University, Buena Vista 30588
Southern Wesleyan University, Central 30693
Southern West Virginia Community & Technical College, Mount Gay 33562
Southfield Public Library, Southfield 41387
Southgate Veterans Memorial Library, Southgate 41388
Southington Public Library & Museum, Southington 41389
Southland Training Library, Caboolture 00408
South-Ossetian District Library, Cchinvali 09289
South-Ossetian Institute of Scientific Research, Cchinvali 09212
South-Ossetian Pedagogical Institut, Cchinvali 09189
South-Russia State Technical University, Novocherkassk 22430
Southside Baptist Church, Lakeland 35254
Southside Regional Library, Boydton 38254
Southside Virginia Community College, Alberta 33122
South-Ural State University, Chelyabinsk 22230
South-Uralian Mechanical Engineering Joint-Stock Company, Orsk 22886
Southwest Arkansas Regional Library, Hope 39475
Southwest Arkansas Regional Library – Hempstead County Library, Hope 39476
Southwest Baptist University, Bolivar 30475
Southwest China Forestry Institute, Kunming 05359
Southwest China Petroleum Institute, Nanchong 05432
Southwest China Teachers College, Chongqing 05139
Southwest Georgia Regional Library, Bainbridge 38055
Southwest Jiaotong University, Chengdu 05125
Southwest Minnesota State University (SMSU), Marshall 31768
Southwest Mississippi Community College, Summit 33785
Southwest Museum, Los Angeles 37065
Southwest Public Libraries, Grove City 39329
Southwest Research and Information Center Library, Albuquerque 36450

Southwest Research Institute, San Antonio 37521
Southwest Tennessee Community College, Memphis 33529
Southwest Tennessee Community College – George E. Freeman Library, Memphis 33530
Southwest Texas Junior College, Uvalde 33818
Southwest University of Finance and Economics, Chengdu 05126
Southwest University of Political Science and Law (SWUPL), Chongqing 05140
Southwest University of Science and Technology, Mianyang 05419
Southwest Virginia Community College, Richlands 33682
Southwestern Adventist University, Keene 31492
Southwestern Assemblies of God University, Waxahachie 33005
Southwestern Baptist Theological Seminary, Fort Worth 35208
Southwestern College, Chula Vista 33230
Southwestern College, Phoenix 32288
Southwestern College, Santa Fe 32615
Southwestern College, Winfield 33079
Southwestern College of Christian Ministries, Bethany 33166
Southwestern Illinois College, Belleville 33163
Southwestern Indian Polytechnic Institute Libraries, Albuquerque 33123
Southwestern Institute of Physics, Leshan 05879
Southwestern Law School, Los Angeles 31654
Southwestern Michigan College, Dowagiac 31024
Southwestern Oklahoma State University, Weatherford 33009
Southwestern Oregon Community College, Coos Bay 33258
South-Western Railways, Kyiv 28381
Southwestern University, Cebu City 20690
Southwestern University, Georgetown 31211
Southwestern University, Los Angeles 31655
Sovetska raionna TsBS, Tsentralna biblioteka, Sovetske 28827
Sovetske Regional Centralized Library System, Main Library, Sovetske 28827
Sovmestnoe predpriyatie Altaiski Traktor, Rubtsovsk 22900
Sovmestnoe predpriyatie Volgodizelmash, Balakovo 22776
Soyuz arkhitektorov, Volgograd 24067
Soyuz arkhitektorov Rossii – SP otdelenie, Sankt-Peterburg 23901
Soyuz khudozhnikov, Moskva 23486
Soyuz khudozhnikov Rossii, Sankt-Peterburg 23902
Soyuz teatralnykh deyatelei Rossii – Rostovskoe otdelenie, Rostov-na-Donu 23729
Soyuz teatralnykh deyatelei Rossiiskoi Federatsii (STD), Moskva 23487
Soyuz teatralnykh deyatelei – Sankt-Peterburgskoe otdelenie, Sankt-Peterburg 23903
Soyuz zhurnalistov, Sankt-Peterburg 23904
Sozialgericht Berlin, Berlin 10815
Sozialgericht Dortmund, Dortmund 10858
Sozialgericht Duisburg, Duisburg 10866
Sozialgericht Hamburg, Hamburg 10926
Sozialinstitut des Erzbistums Paderborn, Dortmund 11196
Sozialwissenschaftliches Institut der Evangelischen Kirche in Deutschland, Hannover 11239
SP Sveriges Provnings- och Forskningsinstitut, Borås 26562
SP Swedish National Testing and Research Institute, Borås 26562
Spaarbank An-Hyp, Antwerpen 02499
Space Application Centre, Ahmedabad 13925
Space Research Centre of the Polish Academy of Sciences, Warszawa 21250
Space Telescope Science Institute Library, Baltimore 36516
Spalding University, Louisville 31682
Spanisches Kulturinstitut, Wien 01610
Spanish Center for National Defense Studies, Madrid 25703
Spanish Foreign Trade Bank, Barcelona 25852
Spanish Fork Public Library, Spanish Fork 41391
Spanish Standards Association (AENOR), Madrid 25977
Spar Aerospace Limited, Edmonton 04255
Sparta Free Library, Sparta 41395
Sparta Public Library, Sparta 41396
Sparta Township Library, Sparta 41397
Spartak Footwear Joint-Stock Company, Kazan 22800
Spartan College of Aeronautics & Technology Library, Tulsa 32852
Spartanburg County Public Libraries, Spartanburg 41398
Spartanburg Methodist College, Spartanburg 33770
Spartanburg Technical College, Spartanburg 33771
SPB Giproshakht, Sankt-Peterburg 22938
SPb Meriya, Sankt-Peterburg 22766
SPb organizatsiya arendatorov Izmeron, Sankt-Peterburg 23905
Special Astrophysical Observatory of the Russian Academy of Sciences, Nizhni Arkhyz 23612
Special Design Office for Heavy and One-off Machine Tools, Sankt-Peterburg 22937
Special Design Office for Textile Industry, Sankt-Peterburg 22936
Special Education Service Agency (SESA), Anchorage 36466
Special Library in Telecommunications, Sofiya 03578
Special Library of Iron and Steel Works, Sisak 06219
Special Metals Corp, Huntington 35769
Special Organic Fertilizing Machinery Design Bureau, Bobruysk 01834
Specialist Teachers' Training Institute, Kuala Lumpur 18534
Specialized Garment Industry Design Office with Pilot Plant, Sankt-Peterburg 22939
Specialized Library of Fine Arts Literature with the Exhibition Hall, Sankt-Peterburg 23809
Specialna knjižnica Ravne, Ravne na Koroškem 24906
Spedali Civili di Brescia, Brescia 16211
Speech & Language Development Center, Buena Park 36605
Speed Art Museum Library, Louisville 37071
Speedway Public Library, Speedway 41400
Speer Memorial Library, Mission 40238
Spencer County Public Library, Rockport 41043
The Spencer Library, New Brighton 35299
Spertus Institute of Jewish Studies, Chicago 30785
Spetsializovana evreiska biblioteka im. O. Shvartsmana, Kyiv 28385
Spetsialna muzichna shkola-internat im. M.V. Lisenka, Kyiv 28116
Spetsialnaya astrofizicheskaya observatoriya RAN, Nizhni Arkhyz 23612
Spetsialnaya proektno-konstruktorskaya organizatsiya Orgtekhstroi, Minsk 01901
Spetsialnaya srednaya shkola militsii, Sankt-Peterburg 22706
SPF Affaires étrangères, Commerce extérieur et Coopération au Développement, Bruxelles 02440
SPF Justice, Bruxelles 02438
Spicer Memorial College, Pune 13838
Spiegel & McDiarmid LLP, Washington 36381
Spies Public Library, Menominee 40147
Spinnboden, Berlin 11562
Spinning and Weaving Factory, Kobrin 01875
Spinning Industrial Corporation, Grodno 01871
Spinning Industrial Corporation, Main Trade Union Library, Grodno 02142
Spirit AeroSystems, Inc, Wichita 36417
Spiritual Frontiers Fellowship International, Philadelphia 37363
SPJST, Temple 37666
SPKB po kompleksu mashin dlya vneseniya organicheskikh udobreni, Bobruysk 01834
SPKB shveinoi promyshlennosti s opytnym zavodom, Sankt-Peterburg 22939
SPKTB Sevzaprybprom, Sankt-Peterburg 23906
Split City Library Marko Marulic, Split 06259
Splošna bolnišnica, Celje 24909
Splošna knjižnica Ljutomer, Ljutomer 24971
Spokane Community College, Spokane 33773
Spokane County Law Library, Spokane 34877
Spokane County Library District, Spokane 41405
Spokane Falls Community College, Spokane 33774
Spokane Public Library, Spokane 41406
Spolek pro chemickou a hutní výrobu, a.s., Ustí nad Labem 06424
Spoon River College, Canton 33202
Sport Information Resource Centre (SIRC), Ottawa 04451
Sporta Bibliotheek, Westerlo 02706
Sporta pedagogijas akadēmijas bibliotēka, Riga 18191
Sportmuseum Schweiz, Münchenstein 27413
Sports Authority of India, Patiala 13895
Sports Library of Finland, Helsinki 07496
Sprachen- und Dolmetscher-Institut, München 10755
Sprague Public Library, Baltic 38068
Språk- och folkminnesinstitutet, Göteborg 26584
Språk- och folkminnesinstitutet, Umeå 26703
Spravochnaya biblioteka, Moskva 22756
Sprengel Museum Hannover, Hannover 12036
Spring Arbor University, Spring Arbor 32701
Spring Hill College, Mobile 33546
Spring Lake District Library, Spring Lake 41408
Spring Lake Public Library, Spring Lake 41409
Spring Singapore, Singapore 24602
Springdale Public Library, Springdale 41412
Springfield City Library, Springfield 41417
Springfield City Library – Forest Park Branch, Springfield 41418
Springfield City Library – Pine Point Branch, Springfield 41419
Springfield City Library – Sixteen Acres Branch, Springfield 41420
Springfield College, Springfield 32709
Springfield Free Public Library, Springfield 41421
Springfield Public Library, Springfield 41422
Springfield Technical Community College, Springfield 33777
Springfield Township Library, Springfield 41423
Springfield-Greene County Library District, Springfield 41424
Springville Public Library, Springville 41425
Sprint Corporation, Kansas City 35799
Spruce Grove Public Library, Spruce Grove 04760
SPSA Scottish Police College, Kincardine 29143
Spurgeon's College Library, London 29554
Squire, Sanders & Dempsey, Columbus 35635
Squire, Sanders & Dempsey, Washington 36382
Squire Sanders & Dempsey LLC Library, Miami 35885
William G. Squires Library, Cleveland 35163
Srečko Vilhar Central Library, Koper 24963
Srednja gradbena, geodetska in ekonomska sòla, Ljubljana 24881
Srednja gradbena šola Ivana Kavčiča, Ljubljana 24882
Srednja kovinarska, strojna in metalurška šola Maribor, Maribor 24886
Srednja Šola, Nova Gorica 24890
Srednja šola elektrotehnične in računalniške usmeritve, Maribor 24887
Srednja šola pedagoške in kulturne usmeritve, Maribor 24888
Srednja tehniška šola, Celje 24876
Srednja tekstilna in frizerska šola, Maribor 24889
Srednja tekstilna, obutvena in gumarska šola Kranj, Kranj 24877
Srednja trgovska šola, Ljubljana 24883
Srednya spetsialna shkola militsiyi, Donetsk 28098
Sree Chitra Tirunal Institute for Medical Sciences and Technology, Thiruvananthapuram 13770
SRH Hochschule Berlin, Berlin 09454
SRH Hochschule Heidelberg, Heidelberg 10037
SRH Wald-Klinikum Gera gGmbH, Gera 11918
Sri Lanka Institute of Development Administration, Colombo 26318
Sri Lanka Law College, Colombo 26271
Sri Lanka Medical Library, Colombo 26319
Sri Lanka Ports Authority, Colombo 26296
Sri Rama Krishna Mission Library, Nadi 07304
Srinakharinwirot University, Bangkok 27710
– Mahasarakham Library, Mahasarakham 27711
– Songkhla Library, Songkhla 27712
Sripatum University, Bangkok 27713
Sriwijaya University, Palembang 14286
Srpska Akademija Nauka i Umetnosti, Beograd 24455
Srpska akademija nauka i umetnosti, Ogranak u Novom Sadu, Novi Sad 24531
Srpska književna zadruga, Beograd 24522
Srpsko Narodno Pozorište, Novi Sad 24532
SSAB Oxelösund, Oxelösund 26545
SSB.servetsch per bibliotecas sa, Bern 27477
SSB.service aux bibliothèques sa, Bern 27477
SSB.servizio per biblioteche sa, Bern 27477
SSIA Metrology, Kharkiv 28279
SSICA, Parma 16412
St. Andrew Library, København 06762
St Andrew's Abbey, Cleveland 35164
St Andrew's College, St Andrews 29403
St. Andrew's College, Winnipeg 03962
St. Angela's College, Sligo 14497
St. Ann Parish Library, Saint Ann's Bay 16895
St Anne's College, Oxford 29348
St. Anselm Library, Rome, Roma 16020
St Antony's College, Oxford 29349
St. Augustine's Seminary Library, Scarborough 03859
St. Benedict's Teachers' College, Wewak 20570
St. Berchmanns' College, Changanassery 13627
St Bride Library, London 29835
St. Catharines Public Library, St. Catharines 04763
St. Charles Lwanga Senior Seminary, Dar es Salaam 27668
St Charles Public Library District, Saint Charles 41104
St. Clair College of Applied Arts & Technology, Windsor 04047
St. Clair County Community College, Port Huron 33655
St. Cloud State University, Saint Cloud 32497
St. Dominic's Priary, Rabat 18665
St. Elizabeth Parish Library, Black River 16886
St Fagans: National History Museum Library, Cardiff 29631
St. Finbarr's College, Lagos 19962
St Francis Library, Coyle 35172
St. Francis Seminary, Saint Francis 35356
St. Galler Freihandbibliothek, St.Gallen 27500
St George's Hospital Medical School, London 29251
St. George's University, St. George's 13157
St Helens College, St Helens 29495
St Hilda's College, Oxford 29351
St. James Parish Library, Lutcher 39981
St. James Parish Library, Montego Bay 16891
St. John Monastery Library, Khonchara 18215
St. John Vianney Seminary, Waterkloof 25086
St John & Wayne, Newark 36047
St. John's College, Annapolis 30253
St John's College, Cambridge 29047
St John's College, Oxford 29353
St. John's Health Systems, Inc, Springfield 37633
St. Johns Medical College, Bangalore 13791
St. John's River State College, Palatka 33613
St. John's Seminary – Carrie Estelle Doheny Memorial Library, Camarillo 35149
St. John's Seminary – Edward Laurence Doheny Library, Camarillo 35150
St. Joseph, Regensburg 11311
St. Joseph Hospital, Orange 37310
St Joseph Public Library, Saint Joseph 41117
St Joseph Public Library – East Hills Library, Saint Joseph 41118
St. Joseph's Theological Institute, Hilton 25083
St. Lawrence College, Cornwall 03692
St. Lawrence College, Kingston 03731
St Lawrence Supreme Court, Canton 35552
St. Lawrence University, Canton 30669
St. Louis Community College – Forest Park Campus Library, Saint Louis 33713
St. Louis Community College – Meramec Campus Library, Saint Louis 33714
St. Lucia Archives, Castries 26336
St Luke's Hospital & Health Network, Bethlehem 36540
St Luke's-Roosevelt Hospital Center, New York 37261
St. Mary Parish Library, Port Maria 16894
St Marys Community Public Library, Saint Marys 41129
St Mary's University College, Belfast 28942
St Mary's University College, Twickenham 29428
St. Mary's University of Minnesota, Winona 33081
St. Michael's Hospital, Toronto 04525
St. Patrick's College, Thurles 14493
St. Paul College of Manila, Manila 20751
St Paul's Cathedral, London 29555
St. Paul's National Seminary, Kensington 00771
St Paul's School, London 29481
St. Paul's Senior Seminary Kipalapala, Tabora 27669
St. Paul's Theological College, Banyo 00750

St Pete Beach Public Library, Saint Pete Beach 41136
St. Peter's Abbey & College Library, Muenster 03804
St Peter's College, Oxford 29354
St. Petersburg Central Library of Agriculture, Sankt-Peterburg 23894
St. Petersburg Centre of Scientific and Technical Information and Popularization, Sankt-Peterburg 23896
St. Petersburg Lawyers Committee, Presidium Library, Sankt-Peterburg 23875
St. Petersburg Psychoneurological Research Institute, Sankt-Peterburg 23895
St. Petersburg Radio and Television Company, Sankt-Peterburg 22943
St. Petersburg Radio and Television Company, Music Library, Sankt-Peterburg 22944
St. Petersburg State Agrarian University, Pushkin 22483
St. Petersburg State Library for the Blind, Sankt-Peterburg 23892
St. Petersburg State Theatre Arts Academy, Sankt-Peterburg 22544
St. Petersburg State University, Sankt-Peterburg 22542
St. Petersburg University of the Pedagogic Professions, Sankt-Peterburg 22549
St. Philip's College, San Antonio 33723
St. St. Cyril and Methodius National Library, Sofiya 03451
St. Thomas Aquinas Pontifical University, Città del Vaticano 42189
St. Thomas Aquinas Regional Seminary, Nairobi 18030
St. Thomas College, Thrissur 13845
St. Thomas Parish Library, Morant Bay 16892
St. Thomas Public Library, St. Thomas 04766
St. Thomas University, Miami Gardens 31816
– Law Library, Miami Gardens 31817
St. Vincent de Paul Regional Seminary, Boynton Beach 35145
St. Vincent Public Library, Kingstown 26340
St. Vincent Teachers' College, Arnos Vale 26338
St Volodymyr's Library & Archives, Calgary 04323
St. Xavier's College, Mumbai 13708
Staatliche Akademie der Bildenden Künste, Stuttgart 10594
Staatliche Antikensammlungen und Glyptothek, München 12325
Staatliche Bibliothek Ansbach, Ansbach 09296
Staatliche Bibliothek Neuburg an der Donau, Neuburg an der Donau 09330
Staatliche Bibliothek Passau, Passau 09332
Staatliche Bibliothek Regensburg, Regensburg 09335
Staatliche Bücher- und Kupferstichsammlung Greiz, Greiz 11942
Staatliche Ethnographische Sammlungen Sachsen, Dresden 11772
Staatliche Fachschule für Bau, Wirtschaft und Verkehr Gotha, Gotha 10731
Staatliche Fachschule für Sozialpädagogik, Hamburg 10735
Staatliche Fachschule für Sozialpädagogik FSP2, Hamburg 10736
Staatliche Graphische Sammlung München, München 12326
Staatliche Hochschule für Gestaltung, Karlsruhe 10060
Staatliche Kunsthalle Karlsruhe, Karlsruhe 12094
Staatliche Kunstsammlungen Dresden, Dresden 11772
Staatliche Kunstsammlungen Dresden – Kunstbibliothek, Dresden 11778
Staatliche Kunstsammlungen Dresden – Mathematisch-Physikalischer Salon, Dresden 11779
Staatliche Münzsammlung München, München 12327
Staatliche Museen zu Berlin – Preußischer Kulturbesitz, Ägyptisches Museum und Papyrussammlung, Berlin 11563
Staatliche Museen zu Berlin – Preußischer Kulturbesitz, Antikensammlung, Berlin 11564
Staatliche Museen zu Berlin – Preußischer Kulturbesitz, Ethnologisches Museum, Berlin 11565
Staatliche Museen zu Berlin – Preußischer Kulturbesitz, Gemäldegalerie, Berlin 11566
Staatliche Museen zu Berlin – Preußischer Kulturbesitz, Hamburger Bahnhof – Museum für Gegenwart, Berlin 11567
Staatliche Museen zu Berlin – Preußischer Kulturbesitz, Institut für Museumsforschung, Berlin 11568
Staatliche Museen zu Berlin – Preußischer Kulturbesitz, Kunstbibliothek, Berlin 11569
Staatliche Museen zu Berlin – Preußischer Kulturbesitz, Kunstbibliothek / Bibliothek Museumsinsel, Berlin 11570
Staatliche Museen zu Berlin – Preußischer Kulturbesitz, Kunstgewerbemuseum, Berlin 11571
Staatliche Museen zu Berlin – Preußischer Kulturbesitz, Kupferstichkabinett, Berlin 11572
Staatliche Museen zu Berlin – Preußischer Kulturbesitz, Münzkabinett, Berlin 11573
Staatliche Museen zu Berlin – Preußischer Kulturbesitz, Museum Europäischer Kulturen, Berlin 11574
Staatliche Museen zu Berlin – Preußischer Kulturbesitz, Museum für Asiatische Kunst, Kunstsammlung Süd-, Südost- und Zentralasien, Berlin 11575
Staatliche Museen zu Berlin – Preußischer Kulturbesitz, Museum für Asiatische Kunst, Ostasiatische Kunstsammlung, Berlin 11576
Staatliche Museen zu Berlin – Preußischer Kulturbesitz, Museum für Islamische Kunst, Berlin 11577
Staatliche Museen zu Berlin – Preußischer Kulturbesitz, Museum für Vor- und Frühgeschichte, Berlin 11578
Staatliche Museen zu Berlin – Preußischer Kulturbesitz, Neue Nationalgalerie, Berlin 11579
Staatliche Museen zu Berlin – Preußischer Kulturbesitz, Rathgen-Forschungslabor, Berlin 11580
Staatliche Museen zu Berlin – Preußischer Kulturbesitz, Skulpturensammlung und Museem für byzantinische Kunst / Bibliothek der Skulpturensammlung, Berlin 11581
Staatliche Museen zu Berlin – Preußischer Kulturbesitz, Skulpturensammlung und Museum für byzantinische Kunst / Bibliothek des Museums für Spätantike und Byzantinische Kunst, Berlin 11582
Staatliche Museen zu Berlin – Preußischer Kulturbesitz, Vorderasiatisches Museum, Berlin 11583
Staatliche Schlösser und Gärten Baden-Württemberg, Bruchsal 11699
Staatliche Seminare für Didaktik und Lehrerbildung Stuttgart (Gymnasium und Sonderpädagogik), Stuttgart 12539
Staatliche Studienakademie Bautzen, Bautzen 10707
Staatliche Studienakademie Glauchau, Glauchau 09834
Staatliche Studienakademie Leipzig, Leipzig 10241
Staatliche Studienakademie Riesa, Riesa 10764
Staatliche Studienakademie Thüringen, Gera 10729
Staatliches Berufsbildendes Zentrum Jena-Göschwitz, Jena 10743
Staatliches Berufsschulzentrum, Hermsdorf 10739
Staatliches Institut für Musikforschung Preußischer Kulturbesitz, Berlin 11584
Staatliches Museum für angewandte Kunst, Design in der Pinakothek der Moderne, München 12321
Staatliches Museum für Naturkunde Karlsruhe, Karlsruhe 12095
Staatliches Museum für Naturkunde Stuttgart, Stuttgart 12540
Staatliches Museum für Völkerkunde, München 12328
Staatliches Naturhistorisches Museum, Braunschweig 11667
Staatliches Seminar für Didaktik und Lehrerbildung (Gymnasien), Tübingen 12564
Staatliches Studienkolleg für Ausländische Studierende an der Universität zu Köln, Köln 12151
Staatliches Umweltamt Wiesbaden, Wiesbaden 11132
Staatliches Weinbauinstitut Freiburg, Freiburg 11901
Staats- und Stadtbibliothek Augsburg, Augsburg 09298
Staats- und Universitätsbibliothek Bremen, Bremen 09305
– Bereichsbibliothek Physik/Elektrotechnik (BB15), Bremen 09605
– Bereichsbibliothek Wirtschaftswissenschaft (BB11), Bremen 09606
– Juridicum und Europäisches Dokumentationszentrum, Bremen 09607
– Teilbibliothek an der Hochschule Bremerhaven, Bremerhaven 09608
– Teilbibliothek für Technik und Sozialwesen, Bremen 09609
– Teilbibliothek für Wirtschaft und Nautik, Bremen 09610
– Teilbibliothek Kunst, Bremen 09611
– Teilbibliothek Musik an der Hochschule für Künste Bremen, Bremen 09612
Staats- und Universitätsbibliothek Hamburg Carl von Ossietzky, Hamburg 09317
Staatsarchiv Amberg, Amberg 11453
Staatsarchiv Augsburg, Augsburg 11460
Staatsarchiv Aurich, Aurich 11462
Staatsarchiv Bamberg, Bamberg 11472
Staatsarchiv Bozen, Bolzano 16201
Staatsarchiv Bremen, Bremen 11688
Staatsarchiv Bückeburg, Bückeburg 11703
Staatsarchiv Coburg, Coburg 11715
Staatsarchiv der Freien und Hansestadt Hamburg, Hamburg 12013
Staatsarchiv des Kantons Basel-Landschaft, Liestal 27404
Staatsarchiv des Kantons Basel-Stadt, Basel 27314
Staatsarchiv des Kantons Bern / Archives de l'Etat de Berne, Bern 27332
Staatsarchiv des Kantons Luzern, Luzern 27407
Staatsarchiv des Kantons Schwyz (STASZ), Bundesbriefmuseum, Schwyz 27432
Staatsarchiv des Kantons Zug, Zug 27448
Staatsarchiv des Kantons Zürich, Zürich 27468
Staatsarchiv Graubünden, Chur 27344
Staatsarchiv Landshut, Landshut 12170
Staatsarchiv Ludwigsburg, Ludwigsburg 12216
Staatsarchiv München, München 12329
Staatsarchiv Nürnberg, Nürnberg 12376
Staatsarchiv Osnabrück, Osnabrück 12390
Staatsarchiv Sigmaringen, Sigmaringen 12488
Staatsarchiv St.Gallen, St.Gallen 27438
Staatsarchiv Wertheim, Wertheim 12590
Staatsarchiv Wolfenbüttel, Wolfenbüttel 12606
Staatsarchiv Würzburg, Würzburg 12618
Staatsbetrieb Sachsenforst, Pirna 12396
Staatsbibliothek Bamberg, Bamberg 09300
Staatsbibliothek zu Berlin – Preußischer Kulturbesitz, Berlin 09290
Staatsbosbeheer (SBB), Driebergen 19247
Staatsgalerie Stuttgart, Stuttgart 12541
Staatsinstitut für Frühpädagogik, München 12330
Staatsinstitut für Schulqualität und Bildungsforschung, München 12331
Staatskanzlei Rheinland-Pfalz, Mainz 11014
Staatskanzlei Saarland, Saarbrücken 11089
Staatsministerium Baden-Württemberg, Stuttgart 11109
Städelsches Kunstinstitut und Städtische Galerie, Frankfurt am Main 11872
Stadhuisbibliotheek, Amersfoort 19235
Stadhuisbibliotheek, Lelystad 19281
Stads- och högskolebibliotek på Gotland, Visby 26372
Stadsarchief, Oudenaarde 02694
Stadsarchief Amsterdam, Amsterdam 19378
Stadsarchief Breda, Breda 19404
Stadsarchief Brussel, Brussel 02627
Stadsarchief en Athenaeumbibliotheek, Deventer 19047
Stadsarchief en Hagelands Historisch Documentatiecentrum (HHD), Tienen 02702
Stadsarchief Geel – Geels Geschiedkundig Genootschap, Geel 02641
Stadsarchief Gent, Gent 02657
Stadsarchief 's-Hertogenbosch (GAHT), 's-Hertogenbosch 19462
Stadsarchief Sittard-Geleen, Born 19401
Stadsarchief Turnhout – Wetenschappelijke Stadsbibliotheek Taxandria (WBT), Turnhout 02705
Stadsarchief Vlaardingen, Vlaardingen 19534
Stadsbiblioteket, Eskilstuna 26760
Stadsbiblioteket i Jakobstad – Pietarsaaren kaupunginkirjasto, Pietarsaari 07599
Stadsbiblioteket i Lyngby, Lyngby 06900
Stadsbibliotheek, Zottegem 02893
Stadsbibliotheek Haarlem, Haarlem 19612
Stadsbibliotheek Maastricht, Maastricht 19660
Stadsbibliotheek Vlaardingen, Vlaardingen 19726
Stadshusets Bibliotek, Stockholm 26516
Stadt Braunschweig – Verwaltungsbücherei, Braunschweig 10837
Stadt Celle, Verwaltungsbibliothek, Celle 10849
Stadt Essen, Verwaltungsbücherei, Essen 10892
Stadt Frankfurt am Main, Frankfurt am Main 10901
Stadt Hildesheim, Fachbereich Archiv und Bibliotheken, Bibliotheken des Roemer- und Pelizaeus-Museums, Hildesheim 12061
Stadt Hildesheim, Fachbereich Archiv und Bibliotheken, Wissenschaftliche Bibliothek des Stadtarchivs Hildesheim, Hildesheim 12062
Stadt Oberhausen, Verwaltungsbücherei, Oberhausen 11072
Stadt- und Hochschulbibliothek Lingen, Lingen (Ems) 10247
Stadt- und Kantonsbibliothek Zug, Zug 26985
Stadt- und Kirchenbibliothek Wunsiedel, Wunsiedel 12608
Stadt- und Kreisarchiv Düren, Düren 11789
Stadt- und Kreisbibliothek Arnstadt, Arnstadt 12633
Stadt- und Kreisbibliothek Bad Doberan, Bad Doberan 12637
Stadt- und Kreisbibliothek Greiz, Greiz 12759
Stadt- und Kreisbibliothek Haldensleben, Haldensleben 12769
Stadt- und Kreisbibliothek Heinrich Heine, Schmalkalden 12923
Stadt- und Kreisbibliothek "Johann Karl Wezel", Sondershausen 12940
Stadt- und Kreisbibliothek "Joseph Meyer", Hildburghausen 12788
Stadt- und Kreisbibliothek Rochlitz, Rochlitz 12912
Stadt- und Kreisbibliothek Saalfeld, Saalfeld 12918
Stadt- und Kreisbibliothek Salzwedel, Salzwedel 12922
Stadt- und Kreisbibliothek Sömmerda, Sömmerda 12939
Stadt- und Kreisbibliothek St. Wendel, St. Wendel 12946
Stadt- und Kreisergänzungsbibliothek Marienberg, Marienberg 12845
Stadt- und Landesbibliothek, Dortmund 09310
Stadt- und Landesbibliothek Potsdam, Potsdam 09333
Stadt- und Regionalbibliothek "A.S. Puschkin", Gera 12748
Stadt- und Regionalbibliothek Cottbus, Cottbus 12696
Stadt- und Regionalbibliothek Erfurt, Erfurt 12721
Stadt- und Regionalbibliothek Frankfurt (Oder), Frankfurt (Oder) 12736
Stadt- und Schulbibliothek Lehrte, Lehrte 12823
Stadt- und Schulbücherei Gunzenhausen, Gunzenhausen 12765
Stadt Wolfsburg, Wolfsburg 11135
Stadtarchiv Aachen, Aachen 11444
Stadtarchiv Aichach, Aichach 11447
Stadtarchiv Altötting, Altötting 11451
Stadtarchiv Alzenau, Alzenau 11452
Stadtarchiv Augsburg, Augsburg 11461
Stadtarchiv Bamberg, Bamberg 11473
Stadtarchiv Bayreuth, Bayreuth 11483
Stadtarchiv Bochum / Bochumer Zentrum für Stadtgeschichte, Bochum 11619
Stadtarchiv Brandenburg an der Havel, Brandenburg an der Havel 11657
Stadtarchiv Bremerhaven, Bremerhaven 11697
Stadtarchiv Burghausen, Burghausen 11704
Stadtarchiv Celle, Celle 11708
Stadtarchiv Chemnitz, Chemnitz 11714
Stadtarchiv Darmstadt, Darmstadt 11731
Stadtarchiv Dessau-Roßlau, Dessau-Roßlau 11734
Stadtarchiv Dingolfing, Dingolfing 11740
Stadtarchiv Dresden, Dresden 11780
Stadtarchiv Erfurt, Erfurt 11815
Stadtarchiv Erlangen, Erlangen 11818
Stadtarchiv Esslingen, Esslingen am Neckar 11830
Stadtarchiv Frankfurt (Oder), Frankfurt (Oder) 11880
Stadtarchiv Freiberg, Freiberg 11882
Stadtarchiv Freiburg, Freiburg 11902
Stadtarchiv Friedberg (Hessen), Friedberg (Hessen) 11903
Stadtarchiv Fulda, Fulda 11906
Stadtarchiv Gera, Gera 11919
Stadtarchiv Gießen, Gießen 11920
Stadtarchiv Goslar, Goslar 11924
Stadtarchiv Göttingen, Göttingen 11935
Stadtarchiv Graz, Graz 01535
Stadtarchiv Greifswald, Greifswald 11941

Stadtarchiv Hagen, Hagen	11949	
Stadtarchiv Halle (Saale), Halle (Saale)	11963	
Stadtarchiv Heidelberg, Heidelberg	12053	
Stadtarchiv Heilbronn, Heilbronn	12057	
Stadtarchiv Hildesheim, Hildesheim	12062	
Stadtarchiv Hof, Hof	12063	
Stadtarchiv Karlsruhe, Karlsruhe	12096	
Stadtarchiv Kassel, Kassel	12104	
Stadtarchiv Koblenz, Koblenz	12122	
Stadtarchiv Konstanz, Konstanz	12159	
Stadtarchiv Kronach, Kronach	12164	
Stadtarchiv Landau (Pfalz), Landau in der Pfalz	12169	
Stadtarchiv Landshut, Landshut	12171	
Stadtarchiv Leipzig, Leipzig	12201	
Stadtarchiv Ludwigshafen, Ludwigshafen	12218	
Stadtarchiv Magdeburg, Magdeburg	12228	
Stadtarchiv Mannheim, Mannheim	12248	
Stadtarchiv Mönchengladbach, Mönchengladbach	12267	
Stadtarchiv Mühlhausen, Mühlhausen	12269	
Stadtarchiv Mülheim an der Ruhr, Mülheim an der Ruhr	12273	
Stadtarchiv München, München	12332	
Stadtarchiv Münster, Münster	12353	
Stadtarchiv / Museum Eschborn, Eschborn	11820	
Stadtarchiv Naumburg, Naumburg	12358	
Stadtarchiv Nördlingen, Nördlingen	12368	
Stadtarchiv Nürnberg und Verein für Geschichte der Stadt Nürnberg, Nürnberg	12377	
Stadtarchiv Paderborn, Paderborn	12392	
Stadtarchiv Pforzheim, Pforzheim	12395	
Stadtarchiv Pirna, Pirna	12397	
Stadtarchiv Plauen, Plauen	12398	
Stadtarchiv Regensburg, Regensburg	12430	
Stadtarchiv Rehau, Rehau	12432	
Stadtarchiv Rendsburg, Rendsburg	12435	
Stadtarchiv Reutlingen, Reutlingen	12436	
Stadtarchiv Rosenheim, Rosenheim	12439	
Stadtarchiv Salzgitter, Salzgitter	12457	
Stadtarchiv Schwabach, Schwabach	12475	
Stadtarchiv Schwerin, Schwerin	12483	
Stadtarchiv Solingen, Solingen	12494	
Stadtarchiv – Stadtmuseum, Innsbruck	01544	
Stadtarchiv Stendal, Hansestadt Stendal	12503	
Stadtarchiv Stralsund, Stralsund	12508	
Stadtarchiv Stuttgart, Stuttgart	12542	
Stadtarchiv Traunstein, Traunstein	12549	
Stadtarchiv Tübingen, Tübingen	12565	
Stadtarchiv und Historischer Verein für Dortmund und die Grafschaft Mark, Dortmund	11752	
Stadtarchiv und Landesgeschichtliche Bibliothek Bielefeld, Bielefeld	11611	
Stadtarchiv und Museen Villingen-Schwenningen, Villingen-Schwenningen	12573	
Stadtarchiv und Stadtbibliothek Schweinfurt, Schweinfurt	12480	
Stadtarchiv und Stadtmuseum St. Pölten, St. Pölten	01579	
Stadtarchiv und Wissenschaftliche Stadtbibliothek Soest, Soest	12492	
Stadtarchiv Wasserburg a. Inn, Wasserburg a. Inn	12579	
Stadtarchiv Wels, Wels	01585	
Stadtarchiv Wiener Neustadt, Wiener Neustadt	01666	
Stadtarchiv Wiesbaden, Wiesbaden	12597	
Stadtarchiv Wismar, Wismar	12598	
Stadtarchiv Würzburg, Würzburg	12619	
Stadtarchiv Zürich, Zürich	27469	
Stadtarchiv Zwickau, Zwickau	12626	
Stadtbibliothek Aachen, Aachen	12627	
Stadtbibliothek Aalen, Aalen	12628	
Stadtbibliothek am Lutherplatz, Döbeln	12707	
Stadtbibliothek Apolda, Apolda	12632	
Stadtbibliothek Aschaffenburg, Aschaffenburg	12634	
Stadtbibliothek Bad Homburg v.d. Höhe, Bad Homburg v.d. Höhe	12639	
Stadtbibliothek Bad Kreuznach, Bad Kreuznach	12640	
Stadtbibliothek Bad Windsheim, Bad Windsheim	09299	
Stadtbibliothek Baden, Baden	27471	
Stadtbibliothek Baden-Baden, Baden-Baden	12643	
Stadtbibliothek Basel, Basel	27473	
Stadtbibliothek Bautzen, Bautzen	12645	
Stadtbibliothek Bayreuth, Bayreuth	12646	
Stadtbibliothek Berlin-Mitte, Bezirkszentralbibliothek Philipp Schaeffer, Berlin	12652	
Stadtbibliothek Berlin-Mitte, Bibliothek am Luisenbad, Berlin	12653	
Stadtbibliothek Bernburg, Bernburg	12673	
Stadtbibliothek / Bibliothèque de la Ville, Biel	27478	
Stadtbibliothek Bielefeld, Bielefeld	12675	
Stadtbibliothek Bocholt, Bocholt	12679	
Stadtbibliothek Bonn, Bonn	12681	
Stadtbibliothek Braunschweig, Braunschweig	12685	
Stadtbibliothek Bremen, Bremen	12686	
Stadtbibliothek Bremerhaven, Bremerhaven	12687	
Stadtbibliothek "Brigitte Reimann", Burg	12689	
Stadtbibliothek Buxtehude, Buxtehude	12691	
Stadtbibliothek Celle, Celle	12692	
Stadtbibliothek Charlottenburg-Wilmersdorf, Adolf Reichwein-Bibliothek (Stadtteilbibliothek), Berlin	12654	
Stadtbibliothek Charlottenburg-Wilmersdorf, Dietrich-Bonhoeffer-Bibliothek (Mittelpunktbibliothek), Berlin	12655	
Stadtbibliothek Charlottenburg-Wilmersdorf, Eberhard-Alexander-Burgh-Bibliothek (Kinderbibliothek), Berlin	12656	
Stadtbibliothek Charlottenburg-Wilmersdorf, Heinrich-Schulz-Bibliothek (Bezirkszentralbibliothek), Berlin	12657	
Stadtbibliothek Charlottenburg-Wilmersdorf, Ingeborg-Bachmann-Bibliothek (Stadtteilbibliothek), Berlin	12658	
Stadtbibliothek Charlottenburg-Wilmersdorf, Johanna-Moosdorf-Bibliothek (Stadtteilbibliothek), Berlin	12659	
Stadtbibliothek Charlottenburg-Wilmersdorf, Musik- und Stadtteilbibliothek Bundesallee, Berlin	11585	
Stadtbibliothek Charlottenburg-Wilmersdorf, Stadtteilbibliothek Halemweg, Berlin	12661	
Stadtbibliothek Chemnitz, Chemnitz	12693	
Stadtbibliothek Darmstadt, Darmstadt	12698	
Stadtbibliothek Deggendorf, Deggendorf	12699	
Stadtbibliothek der Burgergemeinde Burgdorf, Burgdorf	27480	
Stadtbibliothek der FVG Riesa mbH, Riesa	12911	
Stadtbibliothek der Hansestadt Stralsund, Stralsund	12950	
Stadtbibliothek der Hansestadt Wismar, Wismar	12979	
Stadtbibliothek Dieburg, Dieburg	12703	
Stadtbibliothek Dinslaken, Dinslaken	12704	
Stadtbibliothek Dippoldiswalde, Dippoldiswalde	12705	
Stadtbibliothek Ditzingen, Ditzingen	12706	
Stadtbibliothek Dormagen, Dormagen	12708	
Stadtbibliothek Dorsten, Dorsten	12709	
Stadtbibliothek Duisburg, Duisburg	12712	
Stadtbibliothek Eggenfelden, Eggenfelden	12716	
Stadtbibliothek Eisenach, Eisenach	12717	
Stadtbibliothek Eisenhüttenstadt, Eisenhüttenstadt	12718	
Stadtbibliothek Essen, Essen	12724	
Stadtbibliothek Feldkirch, Feldkirch	01669	
Stadtbibliothek Filderstadt, Filderstadt	12728	
Stadtbibliothek Finsterwalde, Finsterwalde	12729	
Stadtbibliothek Flensburg, Flensburg	12731	
Stadtbibliothek Flöha, Flöha	12732	
Stadtbibliothek Freiberg, Freiberg	12737	
Stadtbibliothek Freiburg, Freiburg	12738	
Stadtbibliothek Freising, Freising	12739	
Stadtbibliothek Freital, Freital	12740	
Stadtbibliothek Friedrichshain-Kreuzberg, Bezirkszentralbibliothek Grünberger Straße, Berlin	12662	
Stadtbibliothek Garbsen, Garbsen	12745	
Stadtbibliothek Germering, Germering	12749	
Stadtbibliothek Giengen, Giengen	12750	
Stadtbibliothek Gießen, Gießen	12751	
Stadtbibliothek Göppingen, Göppingen	12753	
Stadtbibliothek Görlitz, Görlitz	12754	
Stadtbibliothek Göttingen, Göttingen	12756	
Stadtbibliothek Gransee, Gransee	12757	
Stadtbibliothek Greven, Greven	12760	
Stadtbibliothek Gütersloh GmbH, Gütersloh	12766	
Stadtbibliothek Halle (Saale), Halle (Saale)	12770	
Stadtbibliothek Hanau, Hanau	12776	
Stadtbibliothek Hannover, Hannover	12777	
Stadtbibliothek Hans Fallada, Greifswald	12758	
Stadtbibliothek Heidenheim, Heidenheim	12781	
Stadtbibliothek Heilbronn, Heilbronn	12782	
Stadtbibliothek Heinrich Heine, Gotha	12755	
Stadtbibliothek Heinrich Heine, Halberstadt	12768	
Stadtbibliothek Herford, Herford	12784	
Stadtbibliothek Herne, Herne	12785	
Stadtbibliothek Herten, Herten	12786	
Stadtbibliothek Hildesheim, Hildesheim	12790	
Stadtbibliothek im Bildungscampus Nürnberg, Nürnberg	12877	
Stadtbibliothek im Glashaus, Schwäbisch Hall	12925	
Stadtbibliothek "Im Höfle", Böblingen	12678	
Stadtbibliothek im MedienHaus, Mülheim an der Ruhr	12855	
Stadtbibliothek im Ständehaus, Karlsruhe	12802	
Stadtbibliothek im Zentrum für Information und Bildung, Unna	12965	
Stadtbibliothek im Zeughaus, Wismar	12979	
Stadtbibliothek in der Aumühle, Fürstenfeldbruck	12743	
Stadtbibliothek Kaiserslautern, Kaiserslautern	12799	
Stadtbibliothek Kassel, Kassel	12803	
Stadtbibliothek Kempen, Kempen	12805	
Stadtbibliothek Kempten, Kempten	12806	
Stadtbibliothek Koblenz, Koblenz	12810	
StadtBibliothek Köln, Köln	12811	
Stadtbibliothek Königs Wusterhausen, Königs Wusterhausen	12812	
Stadtbibliothek Landau in der Pfalz, Landau in der Pfalz	12819	
Stadtbibliothek Langenhagen, Langenhagen	12821	
Stadtbibliothek Leverkusen, Leverkusen	12828	
Stadtbibliothek Linz, Linz	01673	
Stadtbibliothek Lörrach, Lörrach	12832	
Stadtbibliothek Ludwigsburg, Ludwigsburg	12835	
Stadtbibliothek Ludwigshafen, Ludwigshafen	12836	
Stadtbibliothek Lutherstadt Wittenberg, Lutherstadt Wittenberg	12981	
Stadtbibliothek Luzern, Luzern	27494	
Stadtbibliothek Magdeburg, Magdeburg	12840	
Stadtbibliothek Mannheim, Mannheim	12843	
Stadtbibliothek Marl, Marl	12847	
Stadtbibliothek "Martin Luther", Zeitz	12988	
Stadtbibliothek Marzahn-Hellersdorf, Bezirkszentralbibliothek "Mark-Twain", Berlin	12663	
Stadtbibliothek Memmingen, Memmingen	12848	
Stadtbibliothek Minden, Minden	12851	
Stadtbibliothek mit Harzbücherei, Wernigerode	12975	
Stadtbibliothek Moers, Moers	12852	
Stadtbibliothek Mönchengladbach, Mönchengladbach	12853	
Stadtbibliothek Mühlhausen, Mühlhausen	12854	
Stadtbibliothek – Musikbibliothek, Halle (Saale)	11964	
Stadtbibliothek Naumburg, Naumburg	12861	
Stadtbibliothek Nebra, Nebra	12862	
Stadtbibliothek Neu-Isenburg, Neu-Isenburg	12865	
Stadtbibliothek Neukölln, Hauptbibliothek "Helene Nathan", Berlin	12664	
Stadtbibliothek Neumarkt im Martin-Schrettinger-Haus, Neumarkt i.d. OPf.	12866	
Stadtbibliothek Neuss, Neuss	12869	
Stadtbibliothek Neustadt a. Rbge., Neustadt a. Rbge.	12870	
Stadtbibliothek Nordhorn, Nordhorn	12876	
Stadtbibliothek Oberhausen, Oberhausen	12879	
Stadtbibliothek Ochsenfurt, Ochsenfurt	12880	
Stadtbibliothek Offenbach, Offenbach	12881	
Stadtbibliothek Offenburg, Offenburg	12882	
Stadtbibliothek Oldenburg, Oldenburg	12883	
Stadtbibliothek Olten, Olten	27495	
Stadtbibliothek Oranienburg, Oranienburg	12884	
Stadtbibliothek Osnabrück, Osnabrück	12885	
Stadtbibliothek Paderborn, Paderborn	12888	
Stadtbibliothek Pankow, Heinrich-Böll-Bibliothek, Berlin	12665	
Stadtbibliothek Pforzheim, Pforzheim	12891	
Stadtbibliothek Pirna, Pirna	12894	
Stadtbibliothek Radebeul, Radebeul	12898	
Stadtbibliothek Ratingen, Ratingen	12900	
Stadtbibliothek Reinickendorf, Humboldt-Bibliothek, Berlin	12666	
Stadtbibliothek Reutlingen, Reutlingen	12906	
Stadtbibliothek Rheda-Wiedenbrück, Rheda-Wiedenbrück	12907	
Stadtbibliothek Rheine, Rheine	12908	
Stadtbibliothek Rheinfelden, Rheinfelden	12909	
Stadtbibliothek Ribnitz-Damgarten, Ribnitz-Damgarten	12910	
Stadtbibliothek Rosenheim, Rosenheim	12914	
Stadtbibliothek Rostock, Rostock	12915	
Stadtbibliothek "Rudolf Hagelstange", Nordhausen	12875	
Stadtbibliothek Rudolstadt, Rudolstadt	12916	
Stadtbibliothek Saarbrücken, Saarbrücken	12919	
Stadtbibliothek Saarlouis, Saarlouis	12920	
Stadtbibliothek Salzburg, Salzburg	01674	
Stadtbibliothek Salzgitter, Salzgitter	12921	
Stadtbibliothek Schaffhausen, Schaffhausen	27497	
Stadtbibliothek Schwandorf, Schwandorf	12926	
Stadtbibliothek Schwedt/Oder, Schwedt / Oder	12927	
Stadtbibliothek Schweinfurt, Schweinfurt	12480	
Stadtbibliothek Schwerin, Schwerin	12930	
Stadtbibliothek Schwetzingen, Schwetzingen	12931	
Stadtbibliothek Siegburg, Siegburg	12933	
Stadtbibliothek Siegen, Siegen	12934	
Stadtbibliothek Sindelfingen, Sindelfingen	12935	
Stadtbibliothek Solingen, Solingen	12938	
Stadtbibliothek Sonthofen, Sonthofen	12941	
Stadtbibliothek Spandau, Berlin	12667	
Stadtbibliothek Springe, Springe	12944	
Stadtbibliothek Stade, Stade	12947	
Stadtbibliothek Steglitz-Zehlendorf, Bezirkszentralbibliothek, Berlin	12650	
Stadtbibliothek Steglitz-Zehlendorf, Gottfried-Benn-Bibliothek, Berlin	12668	
Stadtbibliothek Stollberg, Stollberg	12949	
Stadtbibliothek Straubing, Straubing	12951	
Stadtbibliothek Tempelhof-Schöneberg, Berlin	12669	
Stadtbibliothek Tempelhof-Schöneberg, Mittelpunktbibliothek Schöneberg, Theodor-Heuss-Bibliothek, Berlin	12670	
Stadtbibliothek Templin, Templin	12957	
Stadtbibliothek Thun, Thun	27501	
Stadtbibliothek Treptow-Köpenick, Berlin	12671	
Stadtbibliothek Trier, Abteilung Weberbach, Trier	12960	
Stadtbibliothek Trier, Palais Walderdorff, Trier	12961	
Stadtbibliothek Troisdorf, Troisdorf	12962	
Stadtbibliothek Ulm, Ulm	12964	
Stadtbibliothek Viersen, Viersen	12967	
Stadtbibliothek Villingen-Schwenningen, Villingen-Schwenningen	12968	
Stadtbibliothek Weberbach, Trier	12960	
Stadtbibliothek Weinheim, Weinheim	12973	
Stadtbibliothek Weißwasser mit Kreisergänzungsbibliothek und Historischem Archiv, Weißwasser	12586	
Stadtbibliothek Wernigerode, Wernigerode	12975	
Stadtbibliothek Wil, Wil	27504	
Stadtbibliothek Winterthur, Winterthur	27505	
Stadtbibliothek Wismar, Wismar	12979	
Stadtbibliothek Wolfsburg, Wolfsburg	12983	
Stadtbibliothek Worms, Worms	12984	
Stadtbibliothek Wuppertal, Wuppertal	12986	
Stadtbibliothek Zofingen, Zofingen	27507	
Stadtbibliothek Zwickau, Zwickau	12991	
Stadtbibliotheken Graz, Graz	01670	
Stadtbibliotheken Wiesbaden, Wiesbaden	12977	
Stadtbücherei Ahlen, Ahlen	12629	
Stadtbücherei Albstadt, Albstadt	12630	
Stadtbücherei Alsdorf, Alsdorf	12631	
Stadtbücherei Bad Schwartau, Bad Schwartau	12642	
Stadtbücherei Bamberg, Bamberg	12644	
Stadtbücherei / Medienzentrum Bergisch Gladbach, Bergisch Gladbach	12647	
Stadtbücherei Biberach, Biberach an der Riß	12674	
Stadtbücherei Bochum, Bochum	12680	
Stadtbücherei Burgdorf, Burgdorf	12690	
Stadtbücherei Coburg, Coburg	12694	
Stadtbücherei Coesfeld, Coesfeld	12695	
Stadtbücherei Dachau, Dachau	12697	
Stadtbücherei Delmenhorst, Delmenhorst	12700	
Stadtbücherei Dreieich, Dreieich	12710	
Stadtbücherei Dülmen, Dülmen	12713	
Stadtbücherei Düren, Düren	12714	
Stadtbücherei Elmshorn – Carl von Ossietzky, Elmshorn	12719	
Stadtbücherei Ennepetal, Ennepetal	12720	
Stadtbücherei Erkrath, Erkrath	12722	
Stadtbücherei Erlangen, Erlangen	12723	
Stadtbücherei Esslingen, Esslingen am Neckar	12725	
Stadtbücherei Fellbach, Fellbach	12727	
Stadtbücherei Frankenthal, Frankenthal	12733	
Stadtbücherei Frankfurt am Main, Frankfurt am Main	12735	
Stadtbücherei Geesthacht, Geesthacht	12746	
Stadtbücherei Gladbeck, Gladbeck	12752	
Stadtbücherei Grevenbroich, Grevenbroich	12761	

Stadtbücherei Gronau, Gronau 12762
Stadtbücherei Hattingen, Hattingen 12778
Stadtbücherei Heidelberg, Heidelberg 12780
Stadtbücherei Heilsbronn, Heilsbronn 12783
Stadtbücherei Hilden, Hilden 12789
Stadtbücherei Hof, Hof 12791
Stadtbücherei Ibbenbüren, Ibbenbüren 12793
Stadtbücherei im Centrum, Wesel 12976
Stadtbücherei im Kulturzentrum, Rendsburg 12905
Stadtbücherei in der Pfortmühle, Hameln 12774
Stadtbücherei Ingolstadt, Ingolstadt 12795
Stadtbücherei Iserlohn, Iserlohn 12797
Stadtbücherei Kamen, Kamen 12800
Stadtbücherei Kamp-Lintfort, Kamp-Lintfort 12801
Stadtbücherei Kaufbeuren, Kaufbeuren 12804
Stadtbücherei Kiel, Kiel 12807
Stadtbücherei Kirchheim unter Teck, Kirchheim unter Teck 12808
Stadtbücherei Kleve, Kleve 12809
Stadtbücherei KÖB, Meppen 12850
Stadtbücherei Konstanz, Konstanz 12813
Stadtbücherei Korbach, Korbach 12814
Stadtbücherei Kornwestheim, Kornwestheim 12815
StadtBücherei Lahr, Lahr 12818
Stadtbücherei Leinfelden-Echterdingen, Leinfelden-Echterdingen 12824
Stadtbücherei Leonberg, Leonberg 12827
Stadtbücherei Löhne, Löhne 12831
Stadtbücherei Lüdenscheid, Lüdenscheid 12834
Stadtbücherei Lünen, Lünen 12839
Stadtbücherei Marburg, Marburg 12844
Stadtbücherei Münster, Münster 12860
Stadtbücherei Neumünster, Neumünster 12867
Stadtbücherei Neustadt an der Weinstraße, Neustadt an der Weinstraße 12871
Stadtbücherei Neu-Ulm, Neu-Ulm 12873
Stadtbücherei Norderstedt, Norderstedt 12874
Stadtbücherei Nürtingen, Nürtingen 12878
Stadtbücherei Ostfildern, Ostfildern 12886
Stadtbücherei Pinneberg, Pinneberg 12892
Stadtbücherei Pirmasens, Pirmasens 12893
Stadtbücherei Ravensburg, Ravensburg 12901
Stadtbücherei Recklinghausen, Recklinghausen 12902
Stadtbücherei Regensburg, Regensburg 12903
Stadtbücherei Rödermark, Rödermark 12913
Stadtbücherei Rüsselsheim, Rüsselsheim 12917
Stadtbücherei Schorndorf, Schorndorf 12924
Stadtbücherei Schweinfurt, Schweinfurt 12929
Stadtbücherei Speyer, Speyer 12942
Stadtbücherei St. Ingbert, St. Ingbert 12945
Stadtbücherei St. Pölten, St. Pölten 01675
Stadtbücherei St. Walburga, Overath 12887
Stadtbücherei Stolberg, Stolberg 12948
Stadtbücherei Stuttgart, Stuttgart 12954
Stadtbücherei Suhl, Suhl 12956
Stadtbücherei Traunstein, Traunstein 12959
Stadtbücherei Tübingen, Tübingen 12963
Stadtbücherei – Zentralbibliothek Velbert-Mitte, Velbert 12966
Stadtbücherei Waiblingen, Waiblingen 12969
Stadtbücherei Wedel, Wedel 12970
Stadtbücherei Weimar, Weimar 12972
Stadtbücherei Wels, Wels 01676
Stadtbücherei Wilhelmshaven, Wilhelmshaven 12978
Stadtbücherei Wittlich, Wittlich 12982
Stadtbücherei Würzburg, Würzburg 12987
Stadtbücherei Zweibrücken, Zweibrücken 12990
Stadtbüchereien Düsseldorf, Düsseldorf 12715
Stadtbüchereien Hamm, Hamm 12775
Stadtgeschichtliches Museum Leipzig, Leipzig 12202
Stadthistorische Bibliothek Bonn, Bonn 11648
Städtische Bibliotheken, Singen 12936
Städtische Bibliotheken Dresden, Dresden 11781
Städtische Büchereien Landshut, Landshut 12820
Städtische Galerie im Lenbachhaus und Kunstbau, München 12333
Städtische Kliniken Frankfurt a. M.-Höchst, Frankfurt am Main 11873
Städtische Museen Junge Kunst und Viadrina, Frankfurt (Oder) 11881
Städtische Museen Konstanz, Konstanz 12160
Städtische Museen Quedlinburg, Quedlinburg 12417
Städtische Sammlungen, Baden 01511
Städtische Sammlungen Cottbus, Cottbus 11717
Städtische Volksbücherei, Fürth 12744
Städtisches Görres-Gymnasium, Düsseldorf 10722
Städtisches Klinikum Dessau Akademisches Lehrkrankenhaus, Dessau-Roßlau 11735
Städtisches Klinikum Görlitz gGmbH, Görlitz 11923
Städtisches Krankenhaus Dresden-Neustadt, Dresden 11782
Städtisches Museum Göttingen, Göttingen 11936
Stadtmuseum, Dresden 11770
Stadtmuseum Berlin, Berlin 11589
Stadtmuseum Bozen, Bolzano 16203
Stadtmuseum Düsseldorf, Düsseldorf 11804
Stadtmuseum "Göhre" Jena, Jena 12079
Stadtmuseum Graz, Graz 01536
Stadtmuseum Innsbruck, Innsbruck 01544
Stadtmuseum Oldenburg, Oldenburg 12388
Stadtteilbücherei Stuttgart-Vaihingen, Stuttgart 12955
Staff Development Institute, Mpemba 18478
Staffanstorps bibliotek, Staffanstorp 26907
Staffordshire County Council, Stafford 30077
Staffordshire University, Stoke-on-Trent 29409
– Nelson Library, Stafford 29410
Stahlinstitut VDEh, Düsseldorf 11805
Stakhanov Regional Centralized Library System, Main Library, Stakhanov 28828
Stakhanovska raionna TsBS, Tsentralna biblioteka, Stakhanov 28828
Staleprokatny zavod, Sankt-Peterburg 22940
A. E. Staley Manufacturing Co, Decatur 35662
Stalsky Memorial Museum, Ashagastal 22991
Stamford Historical Society Library, Stamford 37634
Standard Bank of South Africa, Johannesburg 25092
Standard Elektrik Lorenz AG (SEL), Stuttgart 11431
Standard & Poor's Library, New York 36025
Standards and Industrial Research Institute of Malaysia (SIRIM), Shah Alam 18617
Standards Association of Zimbabwe, Harare 42379
Standards Institution of Israel, Tel-Aviv 14769
Standards New Zealand, Wellington 19855
Standards Office, Warszawa 21043
Stanford University, Stanford 32716
– Art & Architecture Library, Stanford 32717
– Branner Earth Sciences & Map Collections, Stanford 32718
– Cubberley Education Library, Stanford 32719
– East Asia Library, Stanford 32720
– Engineering Library, Stanford 32721
– Falconer Biology Library, Stanford 32722
– J. Henry Meyer Memorial Library, Stanford 32723
– Hoover Institution on War, Revolution & Peace Library, Stanford 32724
– J. Hugh Jackson Library, Stanford 32725
– Lane Medical Library, Stanford 32726
– Mathematics & Computer Sciences, Stanford 32727
– Miller Library at Hopkins Marine Station, Pacific Grove 32728
– Music Library, Stanford 32729
– Physics Library, Stanford 32730
– Robert Crown Law Library, Stanford 32731
– Stanford Auxiliary Library, Stanford 32732
– Stanford Linear Accelerator Center Research Library, Menlo Park 32733
– Swain Library of Chemistry & Chemical Engineering, Stanford 32734
Stange bibliotek, Stange 20400
Stanislaus County Free Library, Modesto 40249
Stanislaus County Free Library – Salida Branch, Salida 41145
Stanislaus County Free Library – Turlock Branch, Turlock 41624
Stanislaus County Law Library, Modesto 34547
Stanislavsky Music and Theatre Library, Kharkiv 28267
Stankostroitelny zavod im. Kirova, Gomel 01856
Stankostroitelny zavod im. Kirova, Vitebsk 01935
Stankostroitelny zavod im. Kominterna, Vitebsk 01936
Stankostroitelny zavod Krasny Borets, Orsha 01916
Stankostroitelny zavod 'Krasny proletari' im. A.I. Efremova, Moskva 22854
Stankostroitelny zavod shlifovalnykh stankov, Moskva 22855
Stanley Correctional Institution Library, Stanley 34889
Stanley Medical College, Chennai 13798
Stanton Library, North Sydney 01134
Stantsiya Baranovichi-Polesskie, Baranovichi 01942
Stantsiya Baranovichi-tsentralnaya, Baranovichi 01943
Stantsiya Brest-Tsentralny, Brest 01948
Stantsiya Brest-Vostochny, Brest 01949
Stantsiya Gomel, Gomel 01957
Stantsiya Grodno, Grodno 01965
Stantsiya Lida, Lida 01968
Stantsiya Luninets, Luninets 01969
Stantsiya Minsk-Passazhirski, Minsk 02054
Stantsiya Minsk-Sortirovochnaya, Minsk 02055
Stantsiya Mogilev, Mogilev 02066
Stantsiya Polotsk, Polotsk 02071
Stantsiya Vitebsk, Orsha 02069
Stantsiya Vitebsk, Vitebsk 02083
Stantsiya Volkovysk, Volkovysk 02087
Stantsiya Zhlobin, Zhlobin 02088
K .I. Stapaev Institute of Geological Sciences, Almaty 17964
Stara Vizhivka Regional Centralized Library System, Main Library, Stara Vizhivka 28829
Starch Products Research Institute, Korenevo 23106
Staring Instituut, Doetinchem 19417
Stark County District Library, Canton 38447
Stark County Law Library Association, Canton 36624
Starke County Public Library System – Henry F. Schricker (Main Library), Knox 39737
Starkville-Oktibbeha County Public Library System, Starkville 41428
Starobeshivsk DRES, Trade Union Library, Novi Svit 28753
Starobeshivska DRES, Novi Svit 28753
Starobilsk Regional Centralized Library System, Main Library, Starobilsk 28830
Starobilska raionna TsBS, Tsentralna biblioteka, Starobilsk 28830
Starokiev Region Central Library System of the City of Kiev – E. Pluzhnik Main Library, Kyiv 28688
Starokievska raionna TsBS m. Kyiva, Tsentralna biblioteka im. E. Pluzhnika, Kyiv 28688
Starokostyantyniv Regional Centralized Library System, Main Library, Starokostyantyniv 28831
Starokostyantynivska raionna TsBS, Tsentralna biblioteka, Starokostyantyniv 28831
Starokramatorsk Machine Building Plant, Trade Union Library, Kramatorsk 28629
Starokramatorsky zavod mashinobudivannya, Kramatorsk 28629
Starooskolski zavod avtotraktornogo elektrooborudovaniya, Stary Oskol 22958
Staroslavenski Institut, Zagreb 06253
Starovizhivska raionna TsBS, Tsentralna biblioteka, Stara Vizhivka 28829
Starting Devices and Engines Plant, Gomel 01863
Stary Oskol Automobile and Tractor Electric Equipment Plant, Stary Oskol 22958
State A. Cereteli Library of the Adjar Autonomous Republic, Batumi 09184
State A. M. Gorky Museum of Literature, Nizhni Novgorod 23623
State Academic Bolshoi Theatre of Opera and Ballet, Minsk 02006
State Academy for Aerospace Instrumentation, Sankt-Peterburg 22518
State Academy of Agriculture and Ecology, Zhytomyr 28091
State Academy of Architecture and Construction, Odesa 28049
State Academy of Music, Sofiya 03461
State Academy of Non-Ferrous Metals and Gold, Krasnoyarsk 22303
State Academy of Physical Culture, Moskva 22332
State Academy of Sciences, Pyongyang 18052
State Agricultural College, Burlington 30603
State Agricultural Engineering Industrial Corporation, Gomel 01853
State Agricultural Engineering Planning Institute, Kyiv 28308
State Agricultural University, Dnipropetrovsk 27963
State Amalgamated Museum of the Writers of the Urals, Ekaterinburg 23036
State and University Library, Århus 06632
State and University Library Bremen, Bremen 09305
State Archaeological Museum, Warszawa 21329
State Archive, Kraków 21126
State Archive, Valjevo 24533
State Archive in Poznań, Poznań 21192
State Archive in Radom, Radom 21215
State Archive in the Province of Zeeland, Middelburg 19493
State Archive of Pisa, Pisa 16418
State Archive of Trieste, Trieste 16618
State Archive of Turin, Torino 16591
State Archives, Volgograd 24062
State Archives Antwerp, Beveren 02542
State Archives at Courtai, Kortrijk 02668
State Archives in Oslo, Oslo 20285
State Archives of Chernivitsi Oblast, Chernivtsi 28195
State Archives of East-Kazakh District, Ust-Kamenogorsk 17982
State Archives of Ivanovo District, Ivanovo 23059
State Archives of Krasnodar Region, Krasnodar 23111
State Archives of Kurgan District, Kurgan 23129
State Archives of Kursk District, Kursk 23132
State Archives of Latvia Special Library, Riga 18185
State Archives of Omsk District, Omsk 23663
State Archives of Orenburg District, Orenburg 23670
State Archives of Oryol District, Oryol 23675
State Archives of Perm District, Perm 23687
State Archives of Primorsk District, Vladivostok 24053
State Archives of Rostov District, Rostov-na-Donu 23720
State Archives of the Amur District, Blagoveshchensk 23005
State Archives of the Chernigov District, Chernigiv 28189
State Archives of the District of Smolensk, Smolensk 23971
State Archives of the Kaluga District, Kaluga 23072
State Archives of the Khabarovsk District, Khabarovsk 23097
State Archives of the Khmelnitski District, Kamyanets-Podilsk 28244
State Archives of the Kuibyshev District, Samara 23738
State Archives of the N. Novgorod District, Nizhni Novgorod 23615
State Archives of the Novosibirsk Region, Novosibirsk 23636
State Archives of the Poltava District, Poltava 28447
State Archives of the Republic of Cyprus, Nicosia 06339
State Archives of the Saratov Region, Saratov 23956
State Archives of the Stavropol District, Stavropol 23981
State Archives of the Tambov District, Tambov 23988
State Archives of the Transkarpatian District, Beregovo 28185
State Archives of the Tver District, Tver 24008
State Archives of the Yaroslavl District, Yaroslavl 24085
State Archives of Tomsk District, Tomsk 23995
State Archives of Tula District, Tula 24004
State Archives of Vologda District, Vologda 24069
State Archives of Warsaw and the Warsaw Region, Warszawa 21241
State Astrakhan Technical University, Astrakhan 22206
State Automobile and Highway Engineering Institute, Moskva 22361
State Bank of India, Mumbai 13909
State Bank of Indonesia, Jakarta 14342
State Bank of Pakistan, Karachi 20476
State Bearing Plant no 11, Scientific and Technical Library, Minsk 01878
State Bearing Plant no 11, Trade Union Library, Minsk 02159
State Bolshoi Theatre, Central Music Library, Moskva 23205
State Central Library, Enugu 20051
State Central Library, Hyderabad 13589
State Central Library of Mongolia, Ulaanbaatar 18901
State Central Museum of Contemporary History of Russia, Moskva 23248
State Central Polytechnic Library, Moskva 23505
The State Central Theatrical Museum A.A. Bakhrushin, Moskva 23249

State Civil Engineering and Industrial Construction Planning Institute, Kyiv 28309
State College of Applied Arts, Poznań 21038
State College of Dance of Ballet, Kyiv 28009
State College of Florida, Manatee-Sarasota – Bradenton Campus, Bradenton 33183
State College of Music, Kyiv 28010
State College of Optometry, New York 32089
State Committee for Broadcasting and Television, Baku 01722
State Committee for Broadcasting and Television, Toshkent 42111
State Committee for Foreign Economic Relations, Moskva 22735
State Committee for Industrial Policy of the Russian Federation, Moskva 23222
State Committee for Standardization, Metrology and Certification, Moskva 23223
State Committee for Television and Radio, Tbilisi 09280
State Committee of the Russian Federation on Statistics, Moskva 22736
State Complex Mining Mechanisation Research and Planning Institute, Donetsk 28213
State Comptroller's Office, Jerusalem 14685
State Conservatory of Thessaloniki, Thessaloniki 13086
State Correctional Institution, Albion 33907
State Correctional Institution, Bellefonte 33982
State Correctional Institution, Camp Hill 34030
State Correctional Institution, Frackville 34279
State Correctional Institution, Hunlock Creek 34370
State Correctional Institution, Huntingdon 34371
State Correctional Institution, Somerset 34868
State Correctional Institution, Somerset 34869
State Court of the State of Minas Gerais, Belo Horizonte 03277
State department for Housing and Urban Development, Warszawa 21050
State Engineering University of Armenia, Yerevan 00332
State Ethnographical Museum, Warszawa 21330
State Film Fond, Belye Stolby 23000
State Forest of Tatra National Park, Tatranská Lomnica 24755
State Forests of NSW, West Pennant Hills 00741
State Great Hural Library, Ulaanbaatar 18907
State Hermitage Museum, Sankt-Peterburg 23769
State Historical Museum, Moskva 23221
State Historical Museum, Yerevan 00360
State Historical Reserve, Kerch 28247
State Historical Society of Iowa, Iowa City 36963
State Historical Society of Iowa – Des Moines Library, Des Moines 36801
State Historical Society of Missouri Library, Columbia 36733
State Historical Society of North Dakota, Bismarck 36546
State House Museum of P.I. Chaikovsky, Klin 23105
State Hydrological Institute, Sankt-Peterburg 23770
State Information Service (SIS), Cairo 07123
State Institute for Art Studies, Moskva 23276
State Institute for Automobile Industry Planning, Moskva 23220
State Institute for Planning and Research in the Oil Industry, Samara 23739
State Institute for Planning of Food Processing Enterprises no 2, Moskva 23214
State Institute for Rheumatology and Physiotherapy, Budapest 13450
State Institute for the Construction of Electrical Equipment and Electrical Supply for Heavy Industry, Moskva 23211
State Institute of Advanced Medical Training, Kyiv 28011
State Institute of Advanced Medical Training, Zaporizhzhya 28086
State Institute of Applied Chemistry, Sankt-Peterburg 23773
State Institute of Culture, Kyiv 28673
State Institute of Culture, Rivne 28797
State Institute of Health, Bratislava 24724
State Institute of Islamic Studies Raden Intan, Bandarlampung 14237
State Institute of Meat and Dairy Industry Planning and Design, Sankt-Peterburg 23772
State Institute of Oceanography, Moskva 23237

State Institute of Oceanography – St. Petersburg Branch, Sankt-Peterburg 23779
State Institute of Physical Education, Lviv 28033
State Institute of Physical Training and Sports, Dnipropetrovsk 28198
State Insurance Corporation, Accra 13056
State Law Library, Frankfort 34281
State Law Library of Montana, Helena 34350
State Law Office, Freetown 24576
State Lawsuits Authority, Cairo 07124
State Lermontov Literary Memorial Museum, Pyatigorsk 23713
State Library for Youth of Ukraine, Kyiv 28671
State Library of Iowa, Des Moines 30130
State Library of Iowa – Iowa State Law Library, Des Moines 34144
State Library of Kansas, Topeka 30159
State Library of Louisiana, Baton Rouge 30118
State Library of Louisiana – Services for the Blind & Physically Handicapped, Baton Rouge 30119
State Library of Massachusetts, Boston 30121
State Library of North Carolina, Raleigh 30148
State Library of NSW, Sydney 00372
State Library of Ohio, Columbus 30127
State Library of Ohio – Southeastern Ohio Library Center, Caldwell 38412
State Library of Pennsylvania, Harrisburg 30131
State Library of Queensland, South Brisbane 00371
State Library of South Australia, Adelaide 00366
State Library of Tasmania, Hobart 00368
State Library of Ukraine for Children, Kyiv 28670
State Library of Victoria, Melbourne 00369
State Library of Western Australia, Perth 00370
State Literary Museum of Georgia, Tbilisi 09281
State Literature Museum, Moskva 23224
State Makeyevka Safety in Mines Research Institute, Makeyevka 28426
State Medical Academy, Dnipropetrovsk 27961
State Medical and Pharmaceutical University, Chişinău 18886
State Medical Institute, Karaganda 17910
State Medical Institute, Lviv 28405
State Medical Library, Roma 16466
State Medical Scientific Library, Kharkiv 28250
State Museum of Ceramics, Moskva 23227
State Museum of Fine Arts, Nizhni Tagil 23625
State Museum of Georgia, Tbilisi 09282
State Museum of Oriental Cultures, Moskva 23229
State Museum of the Adjar Region, Batumi 09211
State Museum of the History of Architecture, Tobolsk 23992
State Museum of the History of St. Petersburg, Sankt-Peterburg 23775
State Museum of the Natural History, Lviv 28406
State Museum of the Russian Political History, Sankt-Peterburg 23776
State Museum of Theatre and Music, Sankt-Peterburg 23777
State Museum of Ukrainian-Ruthenian Culture, Svidník 24753
State National University of Radioelectronics, Kharkiv 27986
State Nationality Languages Translation Bureau, Beijing 05854
State Newspaper Collection, Århus 06772
State Nitrogen Products Industry Research Institute – Branch for Organic Synthesis Products, Grodno 01867
State of Hawaiin Department of Business, Honolulu 34362
State of Ohio Department of Corrections, Caldwell 34025
State of Ohio Department of Corrections, Leavittsburg 34449
State of Ohio Department of Corrections, London 34470
State of Ohio Department of Corrections, Mansfield 34501
State of Ohio Department of Corrections, Perrysville 34662
State of Ohio Department of Corrections, Toledo 34927
State of Oregon Law Library, Salem 34794
State of Rhode Island Office of Library & Information Services – Talking Books Plus, Providence 37420
State of Vermont Department of Libraries, Montpelier 30142

State Offices Library, Hobart 00687
State Patent and Technical Library of the National Center for Patent and Information (SPTL NCPI), Dushanbe 27645
State Patent Bureau of the Republik of Lithuania, Vilnius 18308
State Pedagogical Institute, Yoshkar-Ola 22659
State Philharmony, Sankt-Peterburg 23766
State Phytosanitary Administration, Brno 06434
State Planning Institute of Automobile Repair and Transport Facilities, Moskva 23210
State Planning Institute of the Timber Industry, Sankt-Peterburg 23771
State Polytechnic College of Palawan, Aborlan 20682
State Public Catering Academy, Kharkiv 28249
State Public Defender, Sacramento 34767
State Public Historical Library, Moskva 23203
State Public Library of Santa Cruz de Tenerife, Santa Cruz de Tenerife 26250
State Public Library Rostov on Don, Rostov-na-Donu 24205
State Public Scientific Technological Library of the Siberian Branch of the Russian Academy of Sciences, Novosibirsk 22159
State Regional Archives, Mykolaiv 28432
State Regional Children's Library, Dnipropetrovsk 28533
State Regional Children's Library, Donetsk 28543
State Regional Children's Library, Lutsk 28705
State Regional Children's Library, Lviv 28709
State Regional Children's Library, Odesa 28762
State Regional Children's Library, Zaporizhzhya 28885
State Regional Children's Library, Zhytomyr 28897
State Regional Library, Rivne 28796
State Regional Universal Library, Cherkasy 27932
State Regional universal Library, Chernigiv 28519
State Regional Universal Library, Chernivtsi 28523
State Regional Universal Library, Khmelnitski 28610
State Regional Universal Scientific Library, Kharkiv 27933
State Regional Universal Scientific Library, Lviv 27940
State Regional Universal Scientific Library, Odesa 27943
State Regional Universal Scientific Library, Uzhgorod 27948
State Regional Universal Scientific Library, Zhytomyr 28899
State Regional Youth Library, Donetsk 28544
State Regional Youth Library, Ivano-Frankivsk 28583
State Regional Youth Library, Kharkiv 28599
State Regional Youth Library, Lutsk 28706
State Regional Youth Library, Lviv 28710
State Regional Youth Library, Odesa 28763
State Regional Youth Library, Vinnytsya 28869
State Regional Youth Library, Zaporizhzhya 28886
State Regional Youth Library, Zhytomyr 28898
State Research and Design Institute of Basic Chemistry, Kharkiv 28280
State Research and Design Institute of Chemical Engineering, Severodonetsk 28466
State Research and Technological Institute for the Repair and Operation of Agricultural Machinery, Moskva 23592
State Research & Design Institute of Chemical Engineering, Severodonetsk 28467
State Research Institute for Industrial and Environmental Gas Purification, Moskva 23426
State Research Institute for Protein Biosynthesis, Moskva 23233
State Research Institute of Analytical Chemistry, Sankt-Peterburg 23778
State Research Institute of Balneology, Pyatigorsk 23711
State Research Institute of Lake and River Fisheries, Sankt-Peterburg 23835
State Research Library, Prešov 24639
State Research Library Banská Bystrica, Banská Bystrica 24635
State School Teachers Union of Western Australia Inc, East Perth 00913
State Scientific Agrarian Library of UAAS, Poltava 28454

State Scientific Agricultural Library, Kyiv 28306
State Scientific and Pedagogical Library of Ukraine, Kyiv 27936
State Scientific and Research Institute for Organization and Mechanization of Mine Building, Kharkiv 28272
State Scientific Centre of Drugs, Kharkiv 28281
State Scientific Library, Košice 24638
State Scientific Medical Library of Ukraine, Kyiv 28361
State Scientific Medical Library (SSML), Toshkent 42161
State Services Commission, Wellington 19808
State Services Organization Library, Washington 37795
State Socio-Political Library, Moskva 23202
State Supreme Court, Hilo 34356
State Teacher Training Institute, Berdyansk 27952
State Teacher Training Institute, Glukhiv 27977
State Teacher Training Institute, Krivi Rig 28006
State Teacher Training Institute, Melitopol 28043
State Teacher Training Institute, Mykolaiv 28045
State Teacher Training Institute, Rivne 28068
State Teacher Training Institute, Slovyansk 28075
State Teacher Training University, Chernigiv 27957
State Teacher Training University, Uman 28080
State Technical University, Dniprodzerzhinsk 27960
State Transportation Library, Boston 36580
State Tube Research Institute, Dnipropetrovsk 28208
State Universal Scientific Library, Krasnoyarsk 22145
State University for Land Use, Moskva 22334
State University of Economics, Kharkiv 27985
State University of Economics and Finance, Sankt-Peterburg 22548
State University of Moldova, Chişinău 18887
State University of New York, Farmingdale 31125
State University of New York at Binghamton, Binghamton 30428
– Science Library, Binghamton 30429
State University of New York at Fredonia, Fredonia 31175
State University of New York at Oswego, Oswego 32216
State University of New York at Stony Brook, Stony Brook 32748
– Chemistry Library, Stony Brook 32749
– Computer Science Library, Stony Brook 32750
– Health Sciences Center Library, Stony Brook 32751
– Mathematics-Physics-Astronomy Library, Stony Brook 32752
– Music Library, Stony Brook 32753
State University of New York – Brooklyn Educational Opportunity Center, Brooklyn 30575
State University of New York – College at Brockport, Brockport 30553
State University of New York – College at Buffalo, Buffalo 30592
State University of New York – College at Canton, Canton 30670
State University of New York College at Cortland, Cortland 30927
State University of New York – College at Delhi, Delhi 30987
State University of New York – College at Geneseo, Geneseo 31208
State University of New York College at New Paltz, New Paltz 32017
State University of New York – College at Old Westbury, Old Westbury 32189
State University of New York – College at Oneonta, Oneonta 32200
State University of New York College at Plattsburgh, Plattsburgh 32320
State University of New York – College at Potsdam, Potsdam 32351
– Julia E. Crane Memorial Library, Potsdam 32352
State University of New York College of Agriculture & Technology, Cobleskill 30843
State University of New York – College of Environmental Science & Forestry, Syracuse 32776
State University of New York – College of Technology, Alfred 30196

State University of New York Educational Opportunity Center, Syracuse 32777
State University of New York Health Science Center at Brooklyn, Brooklyn 30576
State University of New York – Institute of Technology, Utica 32918
State University of New York – Jefferson Community College, Watertown 33845
State University of New York – Purchase College, Purchase 32401
State University of New York – State College of Optometry, New York 32089
State University of New York – SUNY Maritime College, Bronx 30560
State University of Social Sciences, Moskva 22386
State Water Resources Control Board, Sacramento 37478
Staten Island Institute of Arts & Sciences, Staten Island 37639
Staten Island University Hospital, Staten Island 37640
Statens Arbeidsmiljøinstitutt, Oslo 20282
Statens Avissamling, Århus 06772
Statens Byggeforskningsinstituts Bibliotek, Hørsholm 06782
Statens håndverks- og kunstindustriskole, Oslo 20123
Statens kulturråd, Stockholm 26517
Statens kunstakademi, Oslo 20117
Statens legemiddelverk, Oslo 20283
Statens museer för världskultur, Göteborg 26585
Statens Museum for Kunst, København 06784
Statens musikbibliotek, Stockholm 26679
Statens nærings- og distriktsutviklingsfond (SND), Oslo 20204
Statens Serum Institut, København 06824
Statens strålskyddsinstitut, Stockholm 26518
Statens Väg- och Transportforskningsinstitut (VTI), Linköping 26608
Statesboro Regional Library System, Statesboro 41434
Stateville Correctional Center Libraries, Joliet 34402
Station de recherche Agroscope Changins-Wadenswil ACW, Nyon 27420
Statistical Centre of Iran, Tehran 14437
Statistical Institute for Asia and the Pacific, Chiba 17524
Statistical Institute of Jamaica (STATIN), Kingston 16884
Statistical Office of the Republic of Serbia, Beograd 24495
Statistical Office of the Republic of Slovenia, Ljubljana 24893
Statistical Services Library, Accra 13035
Statistični Urad Republike Slovenije, Ljubljana 24893
Statistics Austria, Wien 01411
Statistics Bureau Management & Coordination Agency, Statistical Library, Tokyo 17324
Statistics Canada, Winnipeg 04194
Statistics Canada Library, Ottawa 04452
Statistics Center of Kyoto City, Kyoto 17276
Statistics Denmark, Library and Information, København 06786
Statistics Finland, Helsinki 07502
Statistics Iceland, Reykjavík 13562
Statistics Netherlands, Den Haag 19270
Statistics New Zealand, Wellington 19856
Statistics Norway, Oslo 20284
Statistics Portugal, Lisboa 21732
Statistics Sweden Library, Stockholm 26680
Statistik Austria, Wien 01411
Statistisches Amt der Landeshauptstadt, München 11050
Statistisches Amt für Hamburg und Schleswig-Holstein Standort Hamburg, Hamburg 10927
Statistisches Amt für Hamburg und Schleswig-Holstein – Standort Kiel, Kiel 10971
Statistisches Bundesamt, Wiesbaden 11133
Statistisches Landesamt Baden-Württemberg, Stuttgart 11110
Statistisches Landesamt Bremen, Bremen 10844
Statistisches Landesamt Rheinland-Pfalz, Bad Ems 10780
Statistisk sentralbyrå, Oslo 20284
Statistiska centralbyråns bibliotek, Stockholm 26680
Staţiunea centrală de producţie pentru sericicultură, Bucureşti 22014
Staţiunea centrală – Dezvoltare pentru Cultura Plantelor pe Nisipuri Dabuleni, Dăbuleni 22031
Staţiunea de cercetări pentru piscicultură, Nucet 22048
Štátna vědecká knižnica, Košice 24638
Štátna vědecká knižnica, Prešov 24639
Štátna vědecká knižnica Banská Bystrica, Banská Bystrica 24635
Štátne lesy, Tatranského národného parku, Tatranská Lomnica 24755
Štátne múzeum ukrajinsko-rusínskej kultúry, Svidník 24753
Státní oblastní archiv v Litoměřicích, Litoměřice 06464
Státní rostlinolékařská správa OPOR, Brno 06434
Státni zdravotní ústav, Praha 06515
Štátny Zdravotný ústav, Bratislava 24724
Stato Maggiore Marina, Roma 15688
Statsarkivet i Bergen, Bergen 20217
Statsarkivet i Hamar, Hamar 20226
Statsarkivet i Kristiansand, Kristiansand 20235
Statsarkivet i Oslo, Oslo 20285
Statsarkivet i Stavanger, Stavanger 20296
Statsarkivet i Trondheim, Trondheim 20301
Statsbiblioteket, Århus 06632
Statskontoret, Stockholm 26519
Staunton Public Library, Staunton 41436
Stavanger bibliotek, Stavanger 20401
Stavanger museum, Stavanger 20297
Stavanger University, Stavanger 20161
Stavishche Regional Centralized Library System, Main Library, Stavishche 28832
Stavishchenska raionna TsBS, Tsentralna biblioteka, Stavishche 28832
Stavropolskaya gosudarstvennaya kraevaya universalnaya nauchnaya biblioteka im. M.Yu. Lermontova, Stavropol 24322
Stavropolskaya kraevaya detskaya biblioteka im. A.E. Ekimtseva, Stavropol 24323
Stavropolskaya kraevaya yunosheskaya biblioteka, Stavropol 24324
Stazione Sperimentale per l'Industria delle Conserve Alimentari – SSICA, Parma 16412
Stazione Zoologica Anton Dohrn, Napoli 16383
Stearns, Weaver, Miller, Weissler, Alhadeff & Sitterson, Miami 35886
Stedelijk Conservatorium Leuven, Leuven 02371
Stedelijk Instituut voor Technisch Onderwijs, Mechelen 02421
Stedelijk Museum Amsterdam, Amsterdam 19379
Stedelijk Museum De Lakenhal, Leiden 19483
Stedelijk Museum Het Prinsenhof, Delft 19412
Stedelijk Museum voor Actuele Kunst, Gent 02656
Stedelijke Bibliotheek, Beringen 02721
Stedelijke Bibliotheek, Leeuwarden 19471
Stedelijke Bibliotheek De Leidraad, Sint-Truiden 02860
Stedelijke Bibliotheek Hasselt, Hasselt 02775
Stedelijke Bibliotheek Zutphen, Zutphen 19540
Stedelijke Musea Leuven, Leuven 02674
Stedelijke Openbare Bibliotheek, Aarschot 02710
Stedelijke Openbare Bibliotheek, Dendermonde 02748
Stedelijke Openbare bibliotheek, Geel 02762
Stedelijke Openbare Bibliotheek, Herentals 02779
Stedelijke Openbare Bibliotheek, Ieper 02786
Stedelijke Openbare Bibliotheek, Kortrijk 02796
Stedelijke Openbare Bibliotheek, Lier 02803
Stedelijke Openbare Bibliotheek, Mechelen 02819
Stedelijke Openbare Bibliotheek, Menen 02823
Stedelijke Openbare Bibliotheek, Poperinge 02843
Stedelijke Openbare Bibliotheek, Sint-Niklaas 02858
Stedelijke Openbare Bibliotheek, Veurne 02874
Stedelijke Openbare Bibliotheek De Vriendschap, Roeselare 02849
Stedelijke Openbare Bibliotheek Gent, Gent 02765
Stedelijke Openbare Bibliotheek Turnhout, Turnhout 02870
Stedelijke Openbare Bibliotheek Vleeshuis, Oudenaarde 02839
Steel Reinforcement and Insulation Plant, Trade Union Library, Slovyansk 28821
Steel Reinforcements Plant, Penza 22889
Steel Rolling Mill, Sankt-Peterburg 22940
Steel Works Dniprospetsstal, Zaporizhzhya 28177
'Stefan cel Mare' University of Suceava, Suceava 21945
Stefan Szulc Central Statistical Library, Warszawa 21245
Stefanik Prikarpaty State University, Scientific Library, Ivano-Frankivsk 27981
W. Stefański Institute of Parasitology, Warszawa 21344
Steffen Robertson and Kirsten, Northlands 25095
Steiermärkische Landesbibliothek, Graz 01183
Steiermärkisches Landesarchiv, Graz 01537
Steinbach Bible College, Steinbach 04230
Steiner Memorial Library, Turku 07546
Steinerbiblioteket, Turku 07546
Steirisches Volksliedarchiv Volkskultur Steiermark GmbH, Graz 01538
Steklov Institute of Mathematics, Moskva 23348
Steklov Research Institute of Mathematics of the Russian Academy of Sciences/St. Petersburg Branch, Sankt-Peterburg 23807
Steklozavod im. Lomonosova, Gomel 01857
Stellenbosch University, Stellenbosch 25049
– Bellville Park Campus Information Centre, Bellville 25050
– Engineering and Forestry Library, Stellenbosch 25051
– Health Sciences Library, Tygerberg 25052
– Theology Library, Stellenbosch 25053
Stenografenverein 1925 Treysa e.V., Schwalmstadt 12478
Stenungsunds bibliotek, Stenungsund 26908
Stephen F. Austin State University, Nacogdoches 31925
Stephens College, Columbia 30874
Stephenson College, Coalville 29070
Steptoe & Johnson Library, Washington 36383
Sterckshof Silver Museum Province of Antwerp, Antwerpen 02538
Sterling College, Sterling 32740
Sterling & Francine Clark Art Institute Library, Williamstown 37845
Sterling Heights Public Library, Sterling Heights 41440
Sterling Municipal Library, Baytown 38124
Sterling Public Library, Sterling 41439
Sterne, Kessler, Goldstein & Fox Library, Washington 36384
Sternwarte Kremsmünster, Kremsmünster 01554
Stetson University, DeLand 30985
Stetson University College of Law Library, Gulfport 31271
Stevens Institute of Technology, Hoboken 31342
Stevens Memorial Library, North Andover 40495
Stevenson-Hamilton Memorial Library, Skukuza 25180
Stevns Bibliotekerne, Store-Heddinge 06934
Stewart Library, Grinnell 39325
Stewart McKelvey, Halifax 04259
STEWEAD-STEG – Energie Steiermark, Graz 01499
Steyler Missionswissenschaftliches Institut e.V., Sankt Augustin 11321
Stichting AIDS Fonds, Amsterdam 19380
Stichting Consument en Veiligheid (SCV), Amsterdam 19381
Stichting Cultureel Centrum Suriname, Paramaribo 26365
Stichting Kunstzinnige Vorming Rotterdam (SKVR), Rotterdam 19517
Stichting Muziekcentrum van de Omroep (MCO), Hilversum 19467
Stichting Provinciale Bibliotheek Centrale voor Noord-Brabant, Tilburg 19712
Stichting Recreatie – Kennis en Innovatiecentrum, Den Haag 19452
Stichting Rijksmuseum Zuiderzeemuseum (ZZM), Enkhuizen 19422
Stichting Surinaams Museum, Paramaribo 26364
Stichting tot Bevordering der Notariële Wetenschap, Amsterdam 19382
Stichting Wetenschappelijk Onderzoek Verkeersveiligheid SWOV, Leidschendam 19484
STICHWORT – Archiv der Frauen- und Lesbenbewegung, Wien 01654
Stickney-Forest View Public Library District, Stickney 41444
Stift Engelszell, Engelhartszell 01421
Stift Herzogenburg, Herzogenburg 01435
Stift Melk, Melk 01454
Stift Schlägl, Aigen i. M. 01415
Stiftelsen Domkirkeodden, Hamar 20227
Stiftelsen Östergötlands Länsmuseum, Linköping 26609
Stifts- och landsbiblioteket, Skara 26371
Stiftsarchiv Altenburg, Altenburg 01416
Stiftsarchiv St.Gallen, St.Gallen 27271
Stiftsarchiv Zeitz, Zeitz 11371
Stiftsbiblioteket i Maribo, Maribo 06901
Stiftsbibliothek Abtei St. Bonifaz, München 11286
Stiftsbibliothek Altenburg, Altenburg 01416
Stiftsbibliothek Benediktinerabtei Einsiedeln, Einsiedeln 27254
Stiftsbibliothek Benediktinerkloster Engelberg, Engelberg 27255
Stiftsbibliothek Beromünster, Beromünster 27247
Stiftsbibliothek Engelszell, Engelhartszell 01421
Stiftsbibliothek Klosterneuburg, Klosterneuburg 01444
Stiftsbibliothek Kremsmünster, Kremsmünster 01445
Stiftsbibliothek Schäftlarn, Kloster Schäftlarn 11322
Stiftsbibliothek Schlägl, Aigen i. M. 01415
Stiftsbibliothek Seitenstetten, Seitenstetten 01470
Stiftsbibliothek St.Gallen, St.Gallen 27272
Stiftsbibliothek St. Lambrecht, St. Lambrecht 01472
Stiftsbibliothek und Stiftsarchiv Altenburg, Altenburg 01416
Stiftsbibliothek und Stiftsarchiv Zeitz, Zeitz 11371
Stiftsbibliothek Zeitz, Zeitz 11371
Stiftsbibliothek Zwettl, Zwettl 01496
Stiftung Archiv der deutschen Frauenbewegung, Kassel 12105
Stiftung Archiv der Parteien und Massenorganisationen der DDR im Bundesarchiv, Berlin 11586
Stiftung Bauhaus Dessau, Dessau-Roßlau 11736
Stiftung Bibliothek Werner Oechslin, Einsiedeln 27351
Stiftung Buch und Wissen, Essen 11828
Stiftung Buchkunst, Frankfurt am Main 11874
Stiftung Bundespräsident-Theodor-Heuss-Haus, Stuttgart 12543
Stiftung Deutsche Kinemathek, Berlin 11502
Stiftung Deutsches Gartenbaumuseum Erfurt, Erfurt 11816
Stiftung Deutsches Historisches Museum, Berlin 11587
Stiftung Deutsches Hygiene-Museum, Dresden 11783
Stiftung Deutsches Rundfunkarchiv Potsdam-Babelsberg, Potsdam 12413
Stiftung Europa-Universität Viadrina Frankfurt (Oder), Frankfurt (Oder) 09756
Stiftung Frauen-Literatur-Forschung e.V., Bremen 11689
Stiftung Galerie für Zeitgenössische Kunst Leipzig, Leipzig 12203
Stiftung Gerhart-Hauptmann-Haus, Düsseldorf 11806
Stiftung Händel-Haus, Halle (Saale) 11965
Stiftung Haus der Geschichte der Bundesrepublik Deutschland, Bibliothek zur Geschichte der DDR, Bonn 11649
Stiftung Haus der Geschichte der Bundesrepublik Deutschland, Informationszentrum, Bibliothek und Mediathek zur deutschen Zeitgeschichte, Bonn 11650
Stiftung Haus der Geschichte der Bundesrepublik Deutschland, Sammlung Industrielle Gestaltung, Berlin 11588
Stiftung Haus der Geschichte der Bundesrepublik Deutschland, Zeitgeschichtliches Forum Leipzig, Leipzig 12204
Stiftung Haus Oberschlesien, Ratingen 12421
Stiftung Historische Museen Hamburg – Altonaer Museum, Hamburg 12015
Stiftung Historische Museen Hamburg – Helms-Museum, Hamburg 12016
Stiftung Historische Museen Hamburg – Museum für Bergedorf und die Vierlande, Hamburg 12017
Stiftung Kloster Michaelstein, Blankenburg (Harz) 11615
Stiftung Kulturwerk Schlesien, Würzburg 12620
Stiftung Leucorea, Lutherstadt Wittenberg 12602
Stiftung Luthergedenkstätten in Sachsen-Anhalt, Lutherstadt Wittenberg 12603
Stiftung Lyrik Kabinett, München 12334
Stiftung Mecklenburg, Ratzeburg 12423
Stiftung Moritzburg, Halle (Saale) 11966
Stiftung Pfennigparade, München 12335
Stiftung Rheinisch-Westfälisches Wirtschaftsarchiv zu Köln, Köln 12152
Stiftung Schleswig-Holsteinische Landesmuseen Schloss Gottorf, Schleswig 12471

Stiftung Schleswig-Holsteinische Landesmuseen Schloss Gottorf, Archäologisches Landesmuseum, Schleswig 12465
Stiftung Schloss Friedenstein, Gotha 11925
Stiftung Stadtmuseum Berlin, Berlin 11589
Stiftung Tierärztliche Hochschule Hannover, Hannover 09974
Stiftung Topographie des Terrors, Berlin 11590
Stiftung Warentest, Berlin 11591
Stiftung Weimarer Klassik und Kunstsammlungen, Weimar 09341
Stikeman Elliott, Toronto 04293
Stilfontein Public Library, Stilfontein 25237
A. T. Still University of the Health Sciences, Kirksville 31513
Stillman College, Tuscaloosa 32856
Stillwater Public Library, Stillwater 41445
Stillwater Public Library, Stillwater 41446
Stinson, Morrison, Hecker Library, Kansas City 35800
Stirling City Libraries, Stirling 01162
Stirling Council, Stirling 30079
Stites & Harbison, Louisville 35865
Stockholm City and County Library, Stockholm 26909
Stockholm City Archives, Stockholm 26682
Stockholm City Court Library, Stockholm 26521
Stockholm Institute of Education, Stockholm 26434
Stockholm International Peace Research Institute (SIPRI), Solna 26637
Stockholm School of Economics, Stockholm 26427
Stockholm School of Theology, Bromma 26469
Stockholm University, Stockholm 26436
– Botanical Library, Stockholm 26438
– Chemical Library, Stockholm 26441
– Department of Media, Journalism, Communication, Stockholm 26440
– Institute of Latin American Studies, Stockholm 26442
– Mathematics Library, Stockholm 26443
Stockholms läns landsting – Förvaltningsutskottets bibliotek, Stockholm 26520
Stockholms Spårvägsmuseum, Stockholm 26681
Stockholms stadsarkiv, Stockholm 26682
Stockholms stadsbibliotek, Stockholm 26909
Stockholms Stadsmuseum, Stockholm 26683
Stockholms tingsrätt, Stockholm 26521
Stockholms universitet, Stockholm 26436
– Biologibiblioteket, Stockholm 26437
– Botaniska biblioteket, Stockholm 26438
– Geobiblioteket, Stockholm 26439
– Institutionen for Journalistik, Medier och Kommunikation, Stockholm 26440
– Kemiska biblioteket, Stockholm 26441
– Latinamerika-institutet, Stockholm 26442
– Matematiska biblioteket, Stockholm 26443
Stockmen's Memorial Foundation Library, Cochrane 04327
Stockport College Library+, Stockport 29408
Stockport Metropolitan Borough Council, Stockport 30080
Stockton Riverside College, Thornaby-on-Tees 29425
Stockton-on-Tees Borough Council, Billingham 29940
Stockton-San Joaquin County Public Library, Stockton 41447
Stockton-San Joaquin County Public Library – Manteca Branch, Manteca 40042
Stockton-San Joaquin County Public Library – Margaret K. Troke Branch, Stockton 41448
Stockton-San Joaquin County Public Library – Tracy Branch, Tracy 41588
Stoel Rives LLP, Portland 37408
Stofnun Árna Magnússonar í Íslenskum fræðum, Reykjavík 13575
Stoke-on-Trent College – Burslem Campus Library, Stoke-on-Trent 29411
Stoke-on-Trent College – Learning Resources Centre, Stoke-on-Trent 29412
Stoke-on-Trent Libraries, Stoke-on-Trent 30081
Stolbtsovskaya tsentralnaya raionnaya bibliyateka, Stolbtsy 02240
Stolbtsy Regional Central Library, Stolbtsy 02240
Stolichna biblioteka, Sofiya 03605
Stoll, Keenon & Park, Lexington 35814
Stomatology Research Institute, Odesa 28437
Stone County Library, Galena 39184
Stoneham Public Library, Stoneham 41449
Stonehill College, Easton 31070
Stonewall Jackson House, Lexington 37029
Stonnington Library and Information Service, South Yarra 01159

Stora Enso North America, Wisconsin Rapids 36431
Stora Enso Oyj, Imatra 07420
Stora Enso Research, Falun 26530
Stord College of Education, Rommetveit 20157
Stord folkebibliotek, Stord 20404
Stord lærarhøgskole, Rommetveit 20157
Stormont, Dundas & Glengarry County Library, Cornwall 04619
Stormont – Vail HealthCare, Topeka 37679
Stortingsbiblioteket, Oslo 20197
Stoughton Public Library, Stoughton 41450
Stoughton Public Library, Stoughton 41451
Støvring Bibliotek, Støvring 06935
Stow College Learning Centre, Glasgow 29120
Stowarzyszenie Dziennikarzy Polskich, Warszawa 21345
Stowarzyszenie Historyków Sztuki, Warszawa 21346
Stradley, Ronon, Stevens & Young LLP Library, Philadelphia 36112
Stradling, Yocca, Carlson & Rauth, Newport Beach 36051
Strängnäs kommunbibliotek, Strängnäs 26910
Stranmillis University College, Belfast 28943
Strasburger & Price LLP Library, Dallas 35655
strasseschweiz, Bern 27333
Strata Mechanics Research Institute of the Polish Academy of Sciences, Kraków 21135
Stratford Library Association, Stratford 41453
Stratford Public Library, Stratford 04770
Strathcona County Library, Sherwood Park 04756
Strathfield Library and Information Centre, Homebush 01101
Strathmore University, Nairobi 18004
Strawbery Banke, Inc, Portsmouth 36152
Strayer University, Washington 32989
M.D. Strazhesky Cardiology Research Institute, Kyiv 28370
Středisko knihovnických a kulturních služeb, Chomutov 06561
Středočeská vědecká knihovna v Kladně, Kladno 06346
Stredoslovenské múzeum, Banská Bystrica 24679
Streekarchief Hollands Midden, Gouda 19426
Streitkräfteamt, Abt. III, Bonn 10834
Streminski Academy of Fine Arts and Design, Łódź 21037
Stripbibliotheek Bries, Antwerpen 02533
Striy City Centralized Library System, Main Library, Striy 28833
Striyska miska TsBS, Tsentralna biblioteka, Striy 28833
Strode College Learning Centre, Street 29416
Stroganov Moscow State University of Industrial and Applied Arts, Moskva 22364
Stroitelno-montazhnoe PO Gomelpromstroi, Gomel 01858
Stroitrest no 16 Neftestroi, Novopolotsk 01912
Stroitrest no 17 Lavsanstroi, Mogilev 02206
Strömstads stadsbibliotek, Strömstad 26911
Strömsunds kommunbibliotek, Strömsund 26912
The Strong Museum Library, Rochester 37458
Stroock & Lavan, Los Angeles 35860
Stroock & Stroock & Lavan, New York 36026
Stručna Knjižnica Željezare, Sisak 06219
Structural Engineering Library (CBRI), Roorkee 14207
Struer Bibliotek, Struer 06936
Struktur- und Genehmigungsdirektion Nord, Koblenz 10980
Struktur- und Genehmigungsdirektion Süd, Neustadt an der Weinstraße 11065
Stryker, Tams & Dill, Newark 36048
Studentato Internazionale dei Missionari Oblati di Maria Immacolata, Roma 16040
Studentenbibliothek / Universitätsbibliothek, München 10447
Studentenwerk Schleswig-Holstein, Kiel 12115
Studiebiblioteket for Ikast Handelsskole og Ikast Seminariet, Ikast 06735
Studiecentrum voor Kernenergie – Centre d'étude de l'Energie Nucléaire, Mol 02687
Studienbibliothek Dillingen, Dillingen a.d. Donau 09309
Studiengesellschaft für unterirdische Verkehrsanlagen e.V. (STUVA), Köln 12153
Studienhaus Wiesneck, Buchenbach 11701

Studienkreis Deutscher Widerstandes 1933-1945 e.V. / Dokumentationsarchiv des deutschen Widerstands, Frankfurt am Main 11875
Studienzentrum der Finanzverwaltung und Justiz Rotenburg an der Fulda, Rotenburg a. d. Fulda 12445
Studijní a vědecká knihovna Plzeňského kraje, Plzeň 06350
Studijní a vědecká knihovna v Hradci Králové, Hradec Králové 06345
Studio delle Relazioni tra Planta e Suolo (RPS), Roma 16560
Studio Teologico dei Frati Minori, Iesi 15860
Studium Biblicum Franciscanum, Jerusalem 14640
Studium Biblicum O.F.M., Hong Kong 05816
Studium Catholicum, Helsinki 07410
Study and Research Library, Plzen, Plzeň 06350
Stuhr Museum of the Prairie Pioneer, Grand Island 36900
STUK – Radiation and Nuclear Safety Authority, Helsinki 07486
L. Štúra Institute of Linguistics, Bratislava 24707
Sturgis District Library, Sturgis 41460
Sturgis Library, Barnstable 36520
Sturgis Library, Barnstable 38089
Sturman Institute, Gilboa Post 14717
Stuttgart Public Library – Arkansas County Library Headquarters, Stuttgart 41461
Sub-Carpathian Museum, Krosno 21152
Sublette County Library, Pinedale 40792
Submarine Naval Military History School, Sankt-Peterburg 22565
Subordinate Courts Library, Singapore 24603
Subsecretaria de Mineria, Buenos Aires 00293
Süd-Chemie Inc., Louisville 35866
Sudetendeutsches Haus, München 12336
Südharz-Krankenhaus gGmbH, Nordhausen 12367
Sudomekhanicheski kolledzh, Sankt-Peterburg 22707
Sudomekhanichny tekhnikum, Kherson 28002
Südost-Institut, Regensburg 12431
Sudostroitelny zavod Severnaya Verf, Sankt-Peterburg 22941
Südwestrundfunk, Baden-Baden 11470
Südwestrundfunk, Mainz 12238
Südwestrundfunk, Stuttgart 12544
Suermondt-Ludwig-Museum, Aachen 11445
Suez Canal Authority, Ismailia 07126
Suffern Free Library, Suffern 41464
Suffolk College, Ipswich 29139
Suffolk Cooperative Library System, Bellport 38161
Suffolk County Community College, Brentwood 33187
Suffolk County Community College, Riverhead 33686
Suffolk County Community College, Selden 33759
Suffolk County Historical Society Library, Riverhead 37447
Suffolk Public Library System, Suffolk 41466
Suffolk Record Office, Bury St Edmunds 29621
Suffolk Record Office, Ipswich 29690
Suffolk University, Boston 30504
– Law Library, Boston 30505
Sugar Beet Institute, Kyiv 28351
Sugar Milling Research Institute, Durban 25122
Sugarcane Breeding Institute, Coimbatore 13972
Suid-Afrikaanse Instituut, Amsterdam 19383
Suifu Meitokukai Shokikan, Mito 17601
Suisan Daigakko, Shimonoseki 17123
Sukhothai Thammathirat Open University (STOU), Nonthaburi 27748
Sul Ross State University, Alpine 30203
Suldal folkebibliotek, Sand 20391
Süleymaniye Kütüphanesi Müdürlügü, Istanbul 27895
Sullivan County Community College, Loch Sheldrake 33493
Sullivan County Public Library, Blountville 38221
Sullivan County Public Library, Sullivan 41469
Sullivan & Cromwell LLP, New York 36027
Sullivan & Cromwell LLP, Washington 36385
Sullivan & Worcester, LLP, Boston 35538
Sulphur Springs Public Library, Sulphur Springs 41471
Sultan Idris Teachers' Training College, Tanjung Malim 18548
Sultan Quaboos University, Muscat 20420
Sumitomo Cement Co, Ltd, Central Res Laboratory, Funabashihiga 17353
Sumitomo Chemical Co, Ltd, Ehime Research Laboratory, Ehime 17349

Sumitomo Chemical Ltd, Osaka Research Laboratory, Osaka 17412
Sumitomo Kagaku Kogyo K. K. Ehime Kenkyujo, Ehime 17349
Sumitomo Kagaku Kogyo K. K. Osaka Kenkyujo, Osaka 17412
Sumitomo Kinzoku Kogyo K. K. Chuo Gijutsu Kenkyujo, Amagasaki 17345
Sumitomo Kinzoku Kozan K. K. Chjuo Kenkyujo, Ichikawa 17361
Sumitomo Life Insurance Co, Planning Dept, Osaka 17413
Sumitomo Metal Industries, Ltd, Amagasaki 17345
Sumitomo Metal Mining Co, Ltd, Central Res Laboratory, Ichikawa 17361
Sumitomo Seimei Hoken Sogo Kaisha Kikakuka, Osaka 17413
Sumitomo Semento K. K. Chuo Kenkyujo, Funabashihiga 17353
Summer Institute of Linguistics, Ukarumpa 20596
Summit Christian College, Gering 35213
Summit County Library, Frisco 39173
Summit Free Public Library, Summit 41473
Summit Pacific College, Abbotsford 03667
Sump Memorial Library, Papillion 40703
Sumska miska TsBS, Tsentralna biblioteka, Sumy 28836
Sumter County Library, Sumter 41474
Sumter County Public Library System, Sumterville 41475
Sumy City Library System, Main Library, Sumy 28836
Sumy State Scientific Regional Library, Sumy 27947
Sumy State University of Agriculture, Sumy 28077
Sun City Library, Sun City 41476
Sun Life Assurance Company of Canada, Wellesley Hills 36407
Sun Prairie Public Library, Sun Prairie 41477
Dr. Sun Yat-sen Library, Taipei 27619
Sun Yat-Sen University, Guangzhou 05211
Sundbybergs stadsbibliotek, Sundbyberg 26913
Sundvalls stadsbibliotek, Sundsvall 26914
Sunflower County Library System, Indianola 39573
Sung Kyun Kwan University, Seoul 18097
Sunndal folkebibliotek, Sunndalsøra 20405
Sunne bibliotek, Sunne 26915
Sunnyvale Public Library, Sunnyvale 41480
Sunraysia Institute of TAFE, Mildura 00576
Sunshine Coast Institute of TAFE, Nambour 00581
Sunshine Coast Libraries, Sunshine Coast Mail Centre 01164
Sunshine Hospital, Saint Albans 00996
SUNY Maritime College, Bronx 30560
Suny Upstate Medical University, Syracuse 32778
Suny Westchester Community College, Valhalla 33819
Sunyani Polytechnic, Sunyani 13018
Suomalainen lääkäriseura Duodecim, Helsinki 07487
Suomalaisen kirjallisuuden seura, Helsinki 07488
Suomen asianajajaliiton tietopalvelu, Helsinki 07489
Suomen Elinkeinoelämän Keskusarkisto, Mikkeli 07532
Suomen Evankelis-Luterilainen Kirkko, Helsinki 07411
Suomen kuntaliiton kirjastoti-etopalvelu, Helsinki 07490
Suomen liikemiesten kauppaopisto-normaalikauppaoppilaitos, Helsinki 07389
Suomen maantieteellinen seura, Helsinki 07491
Suomen Metsästyskirjasto, Riihimäki 07537
Suomen nuorisokirjallisuuden instituutti, Tampere 07542
Suomen pankin kirjasto, Helsinki 07492
Suomen rakennustaiteen museon kirjasto, Helsinki 07493
Suomen Rautatiemuseon kirjasto, Hyvinkää 07511
Suomen standardisoimisliitto SFS ry kirjasto, Helsinki 07494
Suomen sukututkimusseura / Genealogiska samfundet i Finland, Helsinki 07495
Suomen teologinen instituutti, Helsinki 07412
Suomen urheilukirjasto, Helsinki 07496
Suomen vakuutusyhtiäiden keskusliitto, Helsinki 07497
Suomen valokuvataiteen museo, Helsinki 07498
Supélec – Ecole Supérieure d'Electricité, Gif-sur-Yvette 07707
Superintendencia do Desenvolvimento do Nordeste, Recife 03227

Superintendency of Economic and Social Studies of the Bahia, Salvador 03382
Superior Court Law Library, Phoenix 34674
Superior Public Library, Superior 41481
Superior Tribunal de Justiça, Brasília 03223
The Supreme Administrative Court Library, Stockholm 26504
Supreme Council of the Republic of Uzbekistan, Toshkent 42112
Supreme Court, Kingston 16878
Supreme Court, Manila 20757
Supreme Court, Moskva 22759
Supreme Court, Oslo 20191
Supreme Court, Windhoek 19011
Supreme Court, Appellate Division, New York 34601
Supreme Court, Bloemfontein-Judges' Library, Bloemfontein 25058
Supreme Court, Grahamstown, Grahamstown 25063
Supreme Court, Kimberley, Kimberley 25067
Supreme Court Law Library, Honolulu 34363
Supreme Court Library, Accra 13036
Supreme Court Library, Brisbane 00633
Supreme Court Library, Helsinki 07396
Supreme Court Library, Lake George 34430
Supreme Court Library, London 29527
Supreme Court Library, Singapore 24604
Supreme Court Library, Staten Island 34891
Supreme Court Library, Tokyo 17323
Supreme Court Library of Queensland, Melbourne 00700
Supreme Court, Natal Provincial Division, Pietermaritzburg 25069
Supreme Court of Appeal, Bloemfontein 25059
Supreme Court of Canada Library, Ottawa 04124
Supreme Court of Illinois Library, Springfield 34886
Supreme Court of Israel, Jerusalem 14686
Supreme Court of Justice, Budapest 13318
Supreme Court of Nigeria, Lagos 20001
Supreme Court of Ohio, Columbus 34103
Supreme Court of Sweden, Stockholm 26503
Supreme Court of the Czech Republic, Brno 06408
Supreme Court of the Netherlands, Den Haag 19254
Supreme Court of the Slovenian Republic, Central Judical Library, Ljubljana 24894
Supreme Court of the United States Library, Washington 35008
Supreme Court of Victoria Library, Melbourne 00701
Supreme Court of Western Australia, Perth 00716
Supreme Court, Transvaal Provincial Division, Pretoria 25082
Supreme Public Prosecutor's Office, Budapest 13319
Supremo Tribunal Federal, Brasília 03224
SUPSI University of Applied Sciences and Arts of Southern Switzerland, Canobbio 27062
Surahammars folkbibliotek, Surahammar 26916
Surikov Institute of Fine Arts, Moskva 22343
Surinaams Museum, Paramaribo 26364
Surrey Archaeological Society, Guildford 29679
Surrey Institute of Art and Design LLRC, Farnham 29110
Surrey Public Library, Surrey 04773
Surry Community College, Dobson 33296
Survey of India – Geodetic & Research Branch, Dehra Dun 13975
Dr. Franc Sušnik Central Carinthian Library, Ravne na Koroškem 24981
Susquehanna County Law Library, Montrose 34556
Susquehanna University, Selinsgrove 32655
Sussex Archaeological Society, Lewes 29698
Sussex County Community College, Newton 33581
Sussex County Library System, Newton 40474
Sutherland, Asbill & Brennan LLP Library, Washington 36386
Sutherland Shire Libraries & Information Service, Sutherland 01165
Sutin, Thayer & Browne, Albuquerque 35436
SÚTN – Slovenský ústav technickej normalizácie, Bratislava 24725
Sutter County Free Library, Yuba City 42001
Sutton Central Library, Sutton 30085
Suunnittelukeskus Oy, Helsinki 07343
Suva, Luzern 27289
Suva City Library, Suva 07325
Suvorov Military College, Sankt-Peterburg 22708
Suvorov State Museum, Sankt-Peterburg 23774
Suvorovskoe uchilishche, Sankt-Peterburg 22708
Suwannee River Regional Library, Live Oak 39904
Suzhou Institute of Urban Construction and Environmental Protection, Suzhou 05582
Suzhou Medical College, Suzhou 05583
Suzhou Railway Teachers' College, Suzhou 05584
Suzhou Teachers' College, Suzhou 05585
Svalövs folkbibliotek, Svalöv 26917
SVAR – Svensk arkivinformation, Ramsele 26629
Svarochno-mashinostroitelny tekhnikum, Sankt-Peterburg 22709
Svea Court of Appeal, Stockholm 26522
Svea hovrätt, Stockholm 26522
Svedala bibliotek, Svedala 26918
Švenčionių rajono savivaldybės viešoji biblioteka, Šenčionys 18360
Svendborg Bibliotek, Svendborg 06937
Svensk arkivinformation, Ramsele 26629
Svensk Musik, Stockholm 26557
Svenska Akademiens Nobelbibliotek, Stockholm 26684
Svenska barnboksinstitutet, Stockholm 26685
Svenska Emigrantinstitutet, Växjö 26719
Svenska Filminstitutet, Stockholm 26686
Svenska Institutet i Rom, Roma 16561
Svenska kommunalarbetareförbundet, Stockholm 26687
Svenska Läkaresöllskapet, Stockholm 26688
Svenska litteratursällskapet i Finland, Helsinki 07499
Svenska Metallindustriarbetareförbundet, Stockholm 26689
Svenska teaterns I Helsingfors, Helsinki 07500
Svenskt Fjäll- och Samemuseum, Jokkmokk 26594
Svenskt visarkiv, Stockholm 26690
Sverdlovskaya oblastnaya biblioteka po obsluzhivaniyu detei i yunoshestva, Ekaterinburg 24126
Sverdlovskaya oblastnaya universalnaya nauchnaya biblioteka im. V.G. Belinskogo, Ekaterinburg 22128
Sveriges depåbibliotek, Umeå 26937
Sveriges geologiska undersökning (SGU), Uppsala 26710
Sveriges Kommuner och Landsting, Stockholm 26691
Sveriges lantbruksuniversitet (SLU), Uppsala 26446
– Hernquistbiblioteket, Skara 26447
– Skogsbiblioteket, Umeå 26448
– Swedish University of Agriculturel Sciences – Alnarpsbiblioteket, Alnarp 26449
Sveriges meteorologiska och hydrologiska institut (SMHI), Norrköping 26624
Sveriges Ornitologiska Förening, Stockholm 26692
Sveriges Radio Förvaltnings AB, Stockholm 26693
Sveriges Teatermuseum, Stockholm 26694
Svesvalodu literaturas biblioteka, Riga 18199
Svetlogorsk Central City Library, Svetlogorsk 02241
Svetlogorskaya tsentralnaya gorodskaya bibliyateka, Svetlogorsk 02241
M. Svetlov State Regional Youth Library, Dnipropetrovsk 28534
Sveučilišna Knjižnica Rijeka, Rijeka 06151
– Ekonomski fakultet, Rijeka 06152
– Fakultet za pomorstvo, Rijeka 06153
– Fakultet za turistički i hotelski menadžment, Opatija 06154
– Filozofski fakultet, Rijeka 06155
– Medicinski fakultet, Rijeka 06156
– Pravni fakultet, Rijeka 06157
– Tehnički fakultet, Rijeka 06158
Sveučilišna Knjižnica u Splitu, Split 06142
Sveučilište Josipa Jurja Strossmayera u Osijeku, Osijek 06145
– Ekonomski Fakultet, Osijek 06146
– Filozofski Fakultet, Osijek 06147
– Poljoprivredni Institut, Osijek 06148
– Pravni Fakultet, Osijek 06149
Sveučilište u Puli, Pula 06150
Sveučilište u Splitu, Split 06160
– Ekonomski Fakultet, Split 06161
– Faculty of Tourism and Foreign Trade, Dubrovnik 06162
– Fakultet Elektrotehnike, Strojarstva i Brodogradnje, Split 06163
– Fakultet gradevinarstva, arhitekture i geodezije, Split 06164
– Filozofski Fakultet u Zadru, Zadar 06165
– Filozofski Fakultet Zadar, Split 06166
– Kemijsko-tehnološki Fakultet, Split 06167
– Pravni fakultet u Splitu, Split 06168
Svitlovodsk Regional Centralized Library System, Main Library, Svitlovodsk 28837
Svitlovodska miska TsBS, Tsentralna biblioteka, Svitlovodsk 28837
SVUSS Praha, s.r.o., Praha 06516
Swampscott Public Library, Swampscott 41483
Swan City Library Service, Midland 01122
Swan Hill Regional Library, Swan Hill 01166
Swansea Institute of Higher Education, Swansea 29421
Swansea Public Library, Swansea 41485
Swansea University, Swansea 29422
Swarthmore College, Swarthmore 32768
– Cornell Science & Engineering Library, Swarthmore 32769
– Daniel Underhill Music & Dance Library, Swarthmore 32770
Swaziland College of Technology, Mbabane 26368
Swaziland National Library Service, Mbabane 26366
Swedenborg Library, Bryn Athyn 30582
Swedenborgian Library & Archives, Berkeley 35133
Swedish Adult Education Center in Helsinki, Helsinki 07456
Swedish Association of Local Authorities – Swedish Federation of County Councils, Stockholm 26691
Swedish Association of Metalworkers, Stockholm 26689
Swedish Board of Fisheries, Lysekil 26615
Swedish Cement and Concrete Research Institute, Stockholm 26644
Swedish Centre for Terminology, Solna 26638
Swedish Competition Authority, Stockholm 26507
Swedish Confederation of Trade Unions, Stockholm 26663
Swedish Council for Information on Alcohol and other Drugs, Stockholm 26643
Swedish Defence Material Administration, Stockholm 26501
Swedish Environmental Protection Agency, Stockholm 26671
Swedish Federation of County Councils, Stockholm 26691
Swedish Film Institute, Stockholm 26686
Swedish Government Offices, Stockholm 26510
Swedish Government Offices, Library, Stockholm 26509
Swedish Institute for Children's Books, Stockholm 26685
Swedish Institute for Fibre and Polymer Research, Mölndal 26618
Swedish Institute for Food and Biotechnology, Göteborg 26581
Swedish Institute of Assistive Technology, Library, Vällingby 26712
The Swedish Institute of Classical Studies in Rome, Roma 16561
Swedish Institute of International Affairs, Stockholm 26424
Swedish Ironmasters' Association, Stockholm 26653
Swedish Library of Crime Fiction, Eskilstuna 26760
Swedish Library of Talking Books and Braille (TPB), Enskede 26566
Swedish Migration Board, Norrköping 26498
Swedish Military Archives, Stockholm 26659
Swedish Mountain- and Sámi Museum, Jokkmokk 26594
Swedish Municipal Workers' Union, Stockholm 26687
Swedish Museum of Architecture, Stockholm 26640
Swedish Museum of Natural History, Stockholm 26670
Swedish Music Information Centre, Stockholm 26557
Swedish National Agency for Higher Education, Stockholm 26502
Swedish National Audit Office, Stockholm 26514
Swedish National Council for Cultural Affairs, Stockholm 26517
Swedish National Defence College, Stockholm 26424
Swedish National Police Academy Library, Solna 26423
Swedish National Rail Administration, Borlänge 26563
Swedish National Road Administration, Borlänge 26485
Swedish National Road and Transport Research Institute, Linköping 26608
Swedish Ornithological Society, Stockholm 26692
Swedish Parliament Library, Stockholm 26511
Swedish Patent and Registration Office, Stockholm 26673
Swedish Prison an d Probation Service, Norrköping 26497
Swedish Radiation Protection Institute, Stockholm 26518
The Swedish School of Sport and Health Sciences, Stockholm 26426
Swedish Seamen's Library, Göteborg 26583
Swedish Society of Medicine, Stockholm 26688
Swedish State Institute for Industrial Development, Stockholm 26677
Swedish Theatre in Helsingfors, Helsinki 07500
The Library of the Swedish Travelling Exhibitions, Visby 26722
Swedish University of Agricultural Sciences, Uppsala 26446
– Forestry Library, Umeå 26448
– Hernquist Library, Skara 26447
Sweet Briar College, Sweet Briar 32771
– Junius P. Fishburn Music Library, Sweet Briar 32772
– Martin C. Shallenberger Art Library, Sweet Briar 32773
Swenson Swedish Immigration Research Center, Rock Island 37461
Swerea IVF, Mölndal 26618
Swidler, Berlin, Shereff & Friedman, New York 36028
Swidler, Berlin, Shereff & Friedman Library, Washington 36387
Swinburne University of Technology, Hawthorn 00444
Swiss Coordination Centre for Educational Research, Aarau 27297
Swiss Federal Institute of Technology Zurich, Zürich 27151
– Archives of Contemporary History, Zürich 27152
– Mathematics Library, Zürich 27162
Swiss federation of trade unions, Bern 27328
Swiss Film Archive, Lausanne 27394
Swiss Film Archive, Zürich 27450
Swiss Forum for Migration and Population Studies, Neuchâtel 27419
Swiss French-Speaking Braille and Talking Books Library, Genève 27356
Swiss Institute for Art Research, Zürich 27467
Swiss Institute of Comparative Law, Lausanne 27395
Swiss Life, Zürich 27294
Swiss Museum of Transport, Luzern 27408
Swiss National Library, Bern 26961
Swiss Re, Zürich 27296
Swiss Re Europa S.A., Niederlassung Deutschland, Unterföhring 11432
Swiss Tropical and Pubic Health Institute, Basel 27312
Switchgear Research Institute, Brno 06429
Syarikat Permodalan Kemajuan Perusahaan Malaysia, Kuala Lumpur 18605
Syddansk Universitet, Odense 06717
– Syddansk Universitetsbibliotek, Kolding, Kolding 06718
– Syddansk Universitetsbibliotek, Musikafdelingen, Odense 06719
– Syddansk Universitetsbibliotek, Slagelse, Slagelse 06720
– Syddansk Universitetsbibliotek, Sønderborg, Sønderborg 06721
– UC Syddanmark, Esbjerg 06722
Sydney Academy of Sport, Sunnybank 00518
Sydney Adventist Hospital, Wahroonga 01043
Sydney City Library, Sydney 01167
Sydney Conservatorium of Music, Sydney 00534
Sydney Mechanics' School of Arts, Sydney 00589
Sydney Missionary and Bible College, Croydon 00762
Sydney Secondary Distance Education Centre, Dover Heights 00908
Sydney Water Library and Information Service, Sydney 00734
Sydney West Area Health Service, Penrith 00980
Sydvestjysk Sygehus, Esbjerg 06775
Sygehus Sønderjylland, Sønderborg, Sønderborg 06834
Syiah Kuala University, Banda Aceh 14236
Sykehuset Østfold, Fredrikstad 20224
Syktyvkar Paper and Pulp Professional School, Syktyvkar 22959
Syktyvkar State University, Syktyvkar 22587
Syktyvkarski gosudarstvenny universitet, Syktyvkar 22587
Syktyvkarski tsellyulozno-bumazhny tekhnikum, Syktyvkar 22959
S. Syman Public Health Library, Jerusalem 14687
Symfora groep, Amersfoort 19345

Syncrude Canada Ltd, Edmonton 04256
Syndicat Intercom à Vocation Socio-Culturelle, Carbon-Blanc 08317
Syngenta Crop Protection Munchwilen AG, Stein 27292
Syngenta Library CTL, Macclesfield 29577
Syngenta Seeds AB, Landskrona 26539
Synodalbibliothek Bielefeld, Bielefeld 11171
Synthetic Fur Factory, Zhlobin 01939
Synthetic Rubber Plant, Krasnoyarsk 22816
Synthetic Rubber Production Planning Institute, Moskva 23188
Syon Abbey, South Brent 29563
Syosset Public Library, Syosset 41489
Syr Daria District Library, Gulistan 01730
Syracuse Turkey Creek Township Public Library, Syracuse 41492
Syracuse University, Syracuse 32779
– H. Douglas Barclay Law Library, Syracuse 32780
– Geology Library, Syracuse 32781
– Mathematics Library, Syracuse 32782
– Physics Library, Syracuse 32783
– Science & Technology Library, Syracuse 32784
Syrian Patriarchal Seminary, Daroun-Harissa 18214
Sysin Research Institute of Human Ecology and Environmental Health RAMS, Moskva 23258
System Planning Corp, Arlington 35450
Systems Research Institute of the Polish Academy of Sciences, Warszawa 21265
W. Szafer Institute of Botany of the Polish Academy of Sciences, Kraków 21129
SZAMALK Systemhouse Co., Budapest 13458
Szarvasi Óvónőképző Intézet, Szarvas 13313
Szczecin University of Technology, Szczecin 20927
Szczecińskie Towarzystwo Naukowe, Szczecin 21229
Széchenyi István Műszaki Főiskola, Győr 13276
Szegedi Biológiai Kutatóközpont, Szeged 13507
Szegedi Hittudományi Főiskola, Szeged 13345
Szegedi Tudományegyetem, Szeged 13296
– Angol tanszék, Szeged 13297
– Bólyai Intézet, Szeged 13298
– Francia Tanszék, Szeged 13299
– Germán Filológiai Intézet, Szeged 13300
– Jogi Kar, Szeged 13301
– Magyar Irodalomtörténeti Intézet, Szeged 13302
– Pedagógiai és Pszichológiai Intézet, Szeged 13303
– Szegedi Tudományegyetem – Természeti Földrajzi és Geoinformatikai Tanszék, Szeged 13304
– Szerves Kémiai Tanszék, Szeged 13305
– Szláv Filológiai Tanszék, Szeged 13306
– SZTE Központ Könyvtár, Történeti és ókortudományi Szakkönyvtár, Szeged 13307
Szent Atanáz Görög Katolikus Hittudományi Föiskola Könyvtára, Nyíregyháza 13340
Szent István Király Múzeum, Székesfehérvár 13508
Szent István University, Budapest 13263
– Kertészeti Főiskolai Kar, Kecskemét 13264
Szewalski Institute of Fluid-Flow Machinery, Polish Academy of Sciences, Gdańsk 21097
Színház- és Filmművészeti Főiskola, Budapest 13265
Szkoła Główna Gospodarstwa Wiejskiego, Warszawa 20967
– Wydział Leśny, Warszawa 20968
– Wydział Medycyny Weterynaryjnej, Warszawa 20969
– Wydział Technologii Drewna, Warszawa 20970
Szkoła Główna Handlowa, Warszawa 20971
Szkoła Główna Służby Pożarniczej, Warszawa 20972
Szőlészeti és Borászati Kutató Intézet, Kecskemét 13484
Szolnok County Library, Szolnok 13550
Szövetkezeti Kutató Intézet, Budapest 13459
Sztab Generalny Wojska Polskiego, Warszawa 21049
SZTE – Szegedi Élelmiszeripari Főiskolai Kar, Szeged 13308
Sztumskie Centrum Kultury Dział Biblioteczny, Sztum 21581
T. B. Scott Free Library, Merrill 40160
T. Shevchenko Lugansk State Pedagogical University, Lugansk 28029
Taastrup Bibliotek, Taastrup 06938
Tabernacle Baptist Church, Carrollton 35152
Tablelands Regional Library Service, Mareeba 01119
Tabor Adelaide Library, Millswood 00577
Tabor College, Hillsboro 31337
Tabor College, Ringwood North 00503
Tabriz National Library, Tabriz 14387
Täby Huvudbibliotek, Täby 26921
Tacoma Community College, Tacoma 33789
Tacoma Public Library, Tacoma 41494
Tacoma Public Library – Fern Hill, Tacoma 41495
Tacoma Public Library – Kobetich, Tacoma 41496
Tacoma Public Library – Martin Luther King Jr Branch, Tacoma 41497
Tacoma Public Library – Moore Branch, Tacoma 41498
Tacoma Public Library – Mottet Branch, Tacoma 41499
Tacoma Public Library – South Tacoma, Tacoma 41500
Tacoma Public Library – Swasey Branch, Tacoma 41501
Tacoma Public Library – Wheelock Branch, Tacoma 41502
Tadmor Central Hotel School, Herzliya-on-Sea 14675
Tafe NSW – Hunter Institute Hamilton Campus, Newcastle West 00491
TAFE NSW – Hunter Institute Newcastle Campus, Tighes Hill 00545
Tafe NSW – Sydney Institute, Ultimo 00550
Taft College, Taft 33790
Taft, Stettinius & Hollister Library, Cincinnati 35611
Taganrog State University of Radio Engineering, Taganrog 22590
Tahlequah Public Library, Tahlequah 41503
Tai Hsu Buddhist Library, Taipei 27620
T'ainan City Library, Tainan 27630
Taipei Institute of Technology, Taipei 27559
Taipei Medical College, Taipei 27560
Taipei Municipal Teacher's College, Taipei 27561
Taipei National University of the Arts, Taipei 27562
Taipei Public Library, Taipei 27631
Taisce Cheol Dúchais Eireann, Dublin 14544
Taisei Construction Co, Ltd, Tokyo 17484
Taisei Construction Co, Ltd, Technical Research Institute, Yokohama 17505
Taisei Kensetsu K. K. Gijutsu Kenkyujo, Yokohama 17505
Taisei Kensetsu K. K. Somubu, Tokyo 17484
Taishan Medical College, Taishan 05589
Taishan Teachers' College, Taishan 05590
Taisho Pharmaceutical Co, Ltd, Omiya 17401
Taisho Seiyaku K.K., Omiya 17401
Taiwan Agricultural Research Institute, Chia-yi 27525
Taiwan Fisheries Research Institute, Keelung 27590
Taiwan Forestry Research Institute, Taipei 27621
Taiwan Machinery Manufacturing Corp. Production Engineering Dept., Kaohsiung 27585
Taiwan Museum of Art, Taichung 27593
Taiwan Nazarene Theological College, Taipei 27583
Taiwan Sugar Research Institute, Tainan 27586
Taiyuan Electric Power College, Taiyuan 05597
Taiyuan Heavy Machinery College, Taiyuan 05598
Taiyuan Institute of Machinery, Taiyuan 05599
Taiyuan Machinery College, Taiyuan 05600
Taiyuan Teachers' College, Taiyuan 05601
Taiyuan University, Taiyuan 05602
Taiyuan University of Technology, Taiyuan 05603
Taizhou Teachers' College, Linhai 05395
Tajik Abu-Ali Ibn-Cina (Avicenna) State Medical Institute, Dushanbe 27637
Tajik Academy of Sciences, Dushanbe 27634
Tajik Agricultural University, Dushanbe 27638
Tajik State Historical Museum, Dushanbe 27646
Tajik State University, Dushanbe 27639
Takayama Museum of Local History, Takayama 17667
Takayamashi Kyodokan, Takayama 17667
Takeda Chemical Industries, Ltd, Pharmaceutical Research Div, Osaka 17414
Takeda Yakuhin Kogyo K.K., Osaka 17414
Takoma Park Maryland Library, Takoma Park 41504
Takoradi Polytechnic Library, Takoradi 13019
Takushoku Daigaku, Tokyo 17177
Takushoku University, Tokyo 17177
Talboks- och punktskriftsbiblioteket, Enskede 26566
Talbot County Free Library, Easton 38924
Tall Timbers Research Station Library, Tallahassee 37655
Talladega College, Talladega 32789
Talladega County Law Library, Talladega 34909
Talladega Public Library, Talladega 41505
Tallahassee Community College, Tallahassee 33792
Tallinn Central Library (TCL), Tallinn 07256
Tallinn Pedagogical University, Tallinn 07224
Tallinn University of Technology, Tallinn 07225
Tallinna Keskraamatukogu, Tallinn 07256
Tallinna Kunstiülikooli, Tallinn 07223
Tallinna Pedagoogikaülikooli Akaceemiline Raamatukogu, Tallinn 07224
Tallinna Tehnikaülikooli, Tallinn 07225
Talmudical Academy of Baltimore Library, Baltimore 35125
Tamagawa Daigaku Toshokan, Tokyo 17178
Tamagawa University Library, Tokyo 17178
Tamano City Library, Tamano 17858
Tamano Shiritsu Toshokan, Tamano 17858
Tambov Regional Scientific Medical Library, Tambov 23990
Tambovskaya oblastnaya detskaya biblioteka, Tambov 24329
Tambovskaya oblastnaya nauchnaya meditsinskaya biblioteka, Tambov 23990
Tambovskaya oblastnaya universalnaya nauchnaya biblioteka im. A.S. Pushkina, Tambov 22178
Tambovskaya oblastnaya yunosheskaya biblioteka, Tambov 24330
Tambovski gosudarstvenny tekhnicheski universitet, Tambov 22591
Tameside Metropolitan District Council, Ashton-under-Lyne 29928
Tamil Association, Thanjavur 14214
Tamil Nadu Agricultural University, Coimbatore 13635
Tamil Nadu Government Oriental Manuscripts Library, Chennai 13969
Tamil Nadu Legislature, Chennai 13853
Tamil Nadu Tamil Development and Research Council, Chennai 13970
TAMK Kirjasto Kuntokatu 4, Tampere 07367
TAMK Kirjasto Musiikki ja Tampereen Konservatorio, Tampere 07368
TAMK Music Library and Tampere Conservatoire, Tampere 07368
Tamkang University, Taipei 27563
Tammisaaren kirjasto, Ekenäs 07559
Tampa-Hillsborough County Public Library System, Tampa 41508
Tampa-Hillsborough County Public Library System – Austin Davis Branch, Odessa 40587
Tampa-Hillsborough County Public Library System – Bloomingdale Regional Pulbic – Bloomingdale Regional Pulbic, Valrico 41662
Tampa-Hillsborough County Public Library System – Brandon Regional Branch, Brandon 38262
Tampa-Hillsborough County Public Library System – Charles J. Fendig Library, Tampa 41509
Tampa-Hillsborough County Public Library System – Jan Kaminis Platt Regional, Tampa 41510
Tampa-Hillsborough County Public Library System – Jimmie B. Keel Regional, Tampa 41511
Tampa-Hillsborough County Public Library System – Lutz Branch, Lutz 39982
Tampa-Hillsborough County Public Library System – New Tampa Regional, Tampa 41512
Tampa-Hillsborough County Public Library System – North Tampa, Tampa 41513
Tampa-Hillsborough County Public Library System – Riverview Branch, Riverview 41018
Tampa-Hillsborough County Public Library System – Ruskin Branch, Ruskin 41081
Tampa-Hillsborough County Public Library System – Seminole Heights, Tampa 41514
Tampa-Hillsborough County Public Library System – SouthShore Regional, Ruskin 41082
Tampa-Hillsborough County Public Library System – Town 'N Country Regional, Tampa 41515
Tampa-Hillsborough County Public Library System – Upper Tampa Bay Regional, Tampa 41516
Tampere City Library – Pirkanmaa Regional Library, Tampere 07612
Tampere College of Nursing Library, Tampere 07392
Tampere Peace Research Institute, Tampere 07543
Tampere University Library, Tampereen Yliopisto 07370
– Department of Health Sciences, Tampere 07373
– Department of Humanities and Education, Tampereen yliopisto 07372
– Unit in Hämeenlinna, Hämeenlinna 07371
Tampere University of Technology, Tampere 07369
Tampereen kaupunginkirjasto – Pirkanmaan maakuntakirjasto, Tampere 07612
Tampereen kaupunginmuseo, Tampere 07544
Tampereen sairaanhoito-opiston kirjasto, Tampere 07392
Tampereen teknillinen yliopisto, Tampere 07369
Tampereen yliopiston, Tampereen Yliopisto 07370
– Hämeenlinnan yksikkö, Hämeenlinna 07371
– Humanistis-kasvatustieteellinen osasto, Tampereen yliopisto 07372
– Terveystieteiden osasto, Tampere 07373
Tamsui Oxford University College, Taipei 27564
Tanabe Seiyaku Co, Ltd, Osaka 17415
Tanabe Seiyaku K. K. Kenkyukikaku-shitsu Gyomubu Toshoka, Osaka 17415
Tangipahoa Parish Library, Amite 37948
Tangipahoa Parish Library – Amite Branch, Amite 37949
Tangshan College, Tangshan 05809
Tangshan Institute of Engineering and Technology, Tangshan 05605
Tanta University, Tanta
– Faculty of Arts, Tanta 07100
– Faculty of Commerce, Tanta 07101
– Faculty of Education, Tanta 07102
– Faculty of Medicine, Tanta 07103
– Higher Institute of Nursing, Tanta 07104
Tanzania Bureau of Standards, Dar es Salaam 27677
Tanzarchiv Leipzig e.V., Leipzig 12205
Taos Public Library, Taos 41517
T'aoyüan Hsien Cultural Center, T'aoyüan 27632
Tarasevich, L.A., State Research Institute for Standardization and Control of Medical Biological Preparations, Moskva 23488
Taraz State University, Taraz 17923
Tarbiat Modares University, Tehran 14408
Täreboda kommunbibliotek, Töreboda 26926
Tarime College of National Education, Tarime 27664
Tarleton State University, Killeen 31507
– Dick Smith Library, Stephenville 31508
Tårnby Kommunebiblioteker, Kastrup 06891
V.Tarnovsky Historical Museum, Chernigiv 28193
Tarpon Springs Public Library, Tarpon Springs 41520
Tarrant County College, Hurst 33424
Tarrant County Junior College, Fort Worth 33349
Tarrant County Junior College – Northwest Campus Walsh Library, Fort Worth 33350
Tarrant County Law Library, Fort Worth 34275
Tarrawarra Abbey, Yarra Glen 00806
Tartu Art Museum, Tartu 07242
Tartu Institute Library, Toronto 03886
Tartu Kunstimuuseum, Tartu 07242
Tartu Observatory, Tôravere 07243
Tartu Public Library, Tartu 07257
Tartu Ülikooli, Tartu 07228
– Institute of Physics, Tartu 07229
Tartu University, Tartu 07228
Tarumanagara University, Jakarta
– Faculty of Economics, Jakarta 14259
– Faculty of Engineering, Jakarta 14261
– Faculty of Medicine, Jakarta 14260
TASC, Inc, Chantilly 35558
Tasman Pulp and Paper Co Ltd, Kawerau 19815
Tasmanian Museum and Art Gallery, Hobart 00933
Tasmanian Parliamentary Library, Hobart 00688
TASS (ITAR) News Agency – St. Petersburg Branch, Sankt-Peterburg 22942
TASS (ITAR) – SPb otdelenie, Sankt-Peterburg 22942
Tata Energy Research Institute, New Delhi 14169
Tata Institute of Fundamental Research, Mumbai 14102
Tata Institute of Social Sciences, Mumbai 13709
Tata Iron and Steel Co Ltd, Jamshedpur 13906
Tatar Central State Archives, Kazan 23086

Tatarstan Center for Scientific and Technical Information, Kazan 23087
Tatarstan Respublikasinin Milli Kitapxanese, Kazan 22138
Tatarstanski tsentr nauchno-tekhnicheski informatsii (TsNTI), Kazan 23087
Tate Library and Archive, London 29836
Tatra Museum, Zakopane 21372
Tatung Institute of Technology, Taipei 27565
Taunton Public Library, Taunton 41523
Tauragės savivaldybės viešoji biblioteka, Tauragė 18368
Tauranga District Libraries, Tauranga 19883
Tavistock and Portman NHS Trust, London 29837
Távközlési és Biztositóberendezési Központi Főnökség, Budapest 13460
Tavrisk State Agricultural Academy, Melitopol 28044
Tavriska derzhavna agrotekhnichna akademiya, Melitopol 28044
Tax Department, Utrecht 19290
Tay Nguyen University, Buon Ma Thuot 42266
Taycheedah Correctional Institution Library, Fond du Lac 34209
Taylor College and Seminary, Edmonton 04205
Taylor County Public Library, Campbellsville 38432
Taylor University, Upland 32882
Taylor University – Fort Wayne, Fort Wayne 31165
Tazewell County Public Library, Tazewell 41527
Tazuke Kofukai Igaku Kenkyujo, Osaka 17643
Tazuke Kofukai Medical Research Institute, Osaka 17643
Tbilisi Academy of Arts, Tbilisi 09202
Tbilisi State Medical University, Tbilisi 09203
Tbilisi State Museum of Anthropology and Ethnography, Tbilisi 09283
Tbilisi State University, Tbilisi 09204
Tbilisi V. Saradzhishvili State Conservatoire, Tbilisi 09205
TBMM Kütüphane, Ankara 27870
TCB Library, Houston 35764
Te Fare Tauhiti Nui, Papeete 09166
Tea Research Foundation of Central Africa (TRFCA), Mulanje 18491
Tea Research Institute of Sri Lanka, Talawakele 26330
Teac Corp, Techn Information Ctr, Musashino 17387
Teacher Training College, Nuku'alofa 27801
Teacher Training College, Priluki 28067
Teacher Training Institute, Kamyanets-Podilsk 27982
Teachers College, New York 32061
Teachers' College, Port Loko 24570
Teachers' College of Xi'an Union University, Xi'an 05686
Teachers' Council of Thailand, Bangkok 27788
Teachers Training College, Baja 13215
Teachers' Training College, Lobatse 02968
Teachers Union, Tel-Aviv 14770
Teacher-Training Institute, Kirov 22290
Teacher-Training Institute, Nizhin 28048
Teaching Staff Recreation Centre, Sankt-Peterburg 24224
TEAGASC Agriculture and Food Development Authority, Carlow 14517
Teaneck Public Library, Teaneck 41528
Teatr opery i baleta im. Musorgskogo, Notnaya biblioteka, Sankt-Peterburg 23907
Teatralnoe uchilishche im. B.V. Shchukina pri Gosudarstvennom akademicheskom teatre im. Evgeniya Vakhtangova, Moskva 22682
Teatralnoe uchilishche im. M.S. Shchepkina, Moskva 22392
Teatro Stabile di Torino, Torino 16597
Teatterikorkeakoulun kirjasto, Helsinki 07344
Tebbin Institute For Metallurgical Studies (TIMS), Cairo 07188
TEC Frederiksberg, Frederiksberg 06729
Techincal University, Cluj-Napoca 21932
Technical Chamber of Greece, Athinai 13124
Technical College, Jaffna 26287
Technical College, Matara 26279
Technical College for Light Industry, Central Library, Budapest 13251
Technical College of Management and Commerce, Sankt-Peterburg 22710
Technical 'Gh. Asachi' University in Iaşi, Iaşi 21939
Technical Information Library, Garki 20013
Technical Library, Brno 06418
Technical Library of Copenhagen, Ballerup 06773
Technical Library of Ship Design and Research Centre, Gdańsk 21095
Technical Museum, Brno 06435
Technical Museum, Košice 24734
Technical Museum of Slovenia, Ljubljana 24939
Technical Museum Zagreb, Zagreb 06254
Technical Office Building, Stockholm 26523
Technical Research Library, Aktyubinsk 17936
Technical Teachers Training Institute, Kolkata 14059
Technical University, Bryansk 23011
Technical University, Kielce 20819
Technical University, Liberec 06369
Technical University, Volgograd 22642
Technical University Gabrocvo, Gabrovo 03455
K. Pułaski Technical University in Radom, Radom 20920
Technical University of Architecture and Construction, Kyiv 28022
Technical University of Civil Aviation, Moskva 22393
Technical University of Civil Engineering, Bucureşti 21925
Technical University of Crete, Chania 13079
Technical University of Denmark, Lyngby 06713
– Centre for Traffic and Transport, Lyngby 06714
– Institute for Informatics and Mathematical Modelling, Lyngby 06715
Technical University of Denmark, National Institute of Aquatic Resources Library, Charlottenlund 06656
Technical University of Košice, Košice 24664
Technical University of Lisbon, Lisboa 21687
– Economics and Business Management School, Lisboa 21692
– Higher Institute of Agronomy, Lisboa 21690
– Veterinary School of Lisbon Library, Lisboa 21688
Technical University of Łódź, Łódź 20875
– University of Bielsko-Biala, Bielsko-Biała 20876
Technical University of Moldova, Chişinău 18888
Technical University of Prague, Praha 06384
Technical University of Varna, Varna 03485
Technical University of Zielona Gora, Zielona Góra 21032
Technická universita v Liberci, Liberec 06369
Technická univerzita v Košiciach, Košice 24664
Technické muzeum v Brně, Brno 06435
Techniker Krankenkasse, Hamburg 12018
Technikon Natal, Durban 25015
Technion Israel Institute of Technology, Haifa 14604
– Civil Engineering Faculty, Haifa 14605
– Department of Chemistry, Haifa 14606
– Department of Materials Engineering, Haifa 14607
– Department of Physics, Haifa 14608
– Faculty of Electrical Engineering, Haifa 14609
– Faculty of Industrial Engineering and Management, Haifa 14610
– Faculty of Mechanical Engineering, Haifa 14611
– Faculty of Medicine, Haifa 14612
– Junior Technical College (BOSMAT), Haifa 14613
Technische Fachhochschule Berlin, Berlin 09402
Technische Fachhochschule Georg Agricola zu Bochum, Bochum 09504
Technische Hochschule Wildau, Wildau 10688
Technische Informationsbibliothek und Universitätsbibliothek Hannover (TIB/UB), Hannover 09961
Technische Schule der Luftwaffe 1, Kaufbeuren 10746
Technische Schule Landsysteme und Fachschule des Heeres für Technik, Aachen 10703
Technische Universität Bergakademie Freiberg, Freiberg 09757
Technische Universität Berlin, Berlin 09455
Technische Universität Braunschweig, Braunschweig 09592
Technische Universität Carolo-Wilhelmina zu Braunschweig, Braunschweig 09592
– Abteilung Chemiebibliothek (CB), Braunschweig 09593
– Institut für Baustoffe, Massivbau und Brandschutz, Braunschweig 09594
– Institut für Geoökologie, Braunschweig 09595
– Institut für Geowissenschaften, Braunschweig 09596
– Institut für Psychologie, Braunschweig 09597
– Institut für Sozialwissenschaften, Braunschweig 09598
– Institut für Verkehr und Stadtbauwesen, Braunschweig 09599
– Leichtweiß-Institut für Wasserbau, Braunschweig 09600
– Mathematische Institute, Braunschweig 09601
– Mechanik-Zentrum, Braunschweig 09602
– Seminar für Philosophie, Braunschweig 09603
Technische Universität Chemnitz, Chemnitz 09614
Technische Universität Clausthal, Clausthal-Zellerfeld 09615
– Fachbereichsbibliothek Mathematik, Clausthal-Zellerfeld 09616
– Institut für Geologie und Paläontologie, Clausthal-Zellerfeld 09617
– Institut für Maschinenwesen, Clausthal-Zellerfeld 09618
– Institut für Wirtschaftswissenschaft, Clausthal-Zellerfeld 09619
Technische Universität Cottbus, Cottbus 09621
Technische Universität Darmstadt, Darmstadt 09307
– Betriebswirtschaftliche Bibliothek, Darmstadt 09624
– Bibliothek für Allgemeine Pädagogik, Darmstadt 09625
– Bibliothek Architektur und Städtebau, Darmstadt 09626
– Bibliothek für Berufspädagogik, Darmstadt 09627
– Bibliothek Biologie, Darmstadt 09628
– Bibliothek Chemie/Materialwissenschaft, Darmstadt 09629
– Bibliothek des Instituts und Versuchsanstalt für Geotechnik, Darmstadt 09630
– Bibliothek Konstruktiver Ingenieurbau, Darmstadt 09631
– Bibliothek Kunstgeschichte und Klassische Archäologie, Darmstadt 09632
– Bibliothek Philosophie/Soziologie, Darmstadt 09633
– Bibliothek Politik und Geschichte, Darmstadt 09634
– Bibliothek Sprach- und Literaturwissenschaft, Darmstadt 09635
– Bibliothek für Verkehrsplanung und Verkehrstechnik, Darmstadt 09636
– Bibliothek Wasser und Umwelt, Darmstadt 09637
– Fachbereichsbibliothek Informatik, Darmstadt 09638
– Fachbereichsbibliothek Mathematik, Darmstadt 09639
– Fachbereichsbibliothek Mechanik, Darmstadt 09640
– Fachbereichsbibliothek Nachrichtentechnik (NTB), Darmstadt 09641
– Fachbibliothek Zellstoff und Papier, Darmstadt 09642
– Fachgebiet für Arbeitswissenschaft, Darmstadt 09643
– Fachgebiet Wasserbau, Darmstadt 09644
– Institut für Angewandte Geowissenschaften, Darmstadt 09645
– Institut für Psychologie, Darmstadt 09646
– Institut für Sportwissenschaft, Darmstadt 09647
– Juristische Gesamtbibliothek, Darmstadt 09648
– Kernphysikalische Bibliothek, Darmstadt 09649
– Physikalische Bibliothek, Darmstadt 09650
– Volkswirtschaftliche Bibliothek, Darmstadt 09651
Technische Universität Dortmund, Dortmund 09658
– Bereichsbibliothek Architektur und Bauingenieurwesen, Dortmund 09659
– Bereichsbibliothek Chemie / Bereichsbibliothek Umweltforschung und Biologie, Dortmund 09660
– Bereichsbibliothek Elektrotechnik und Informationstechnik / Bio- und Chemieingenieurwesen, Dortmund 09661
– Bereichsbibliothek Maschinenbau, Dortmund 09662
– Bereichsbibliothek Mathematik / Bereichsbibliothek Statistik, Dortmund 09663
– Bereichsbibliothek Physik / Bereichsbibliothek Informatik, Dortmund 09664
– Bereichsbibliothek Raumplanung, Dortmund 09665
– Bereichsbibliothek Wirtschafts- und Sozialwissenschaften, Dortmund 09666
– Emil-Figge-Bibliothek, Dortmund 09667
– Informationszentrum Technik und Patente (ITP), Dortmund 09668
Technische Universität Graz, Graz 01219
– Fachbibliothek für Geodäsie und Mathematik, Graz 01220
Technische Universität Hamburg-Harburg, Hamburg 09922
Technische Universität Ilmenau, Ilmenau 10049
Technische Universität Kaiserslautern, Kaiserslautern 10055
Technische Universität München, München 10448
– Teilbibliothek Chemie, Garching 10449
– Teilbibliothek Maschinenwesen, Garching 10450
– Teilbibliothek Mathematik und Informatik, Garching 10451
– Teilbibliothek Medizin, München 10452
– Teilbibliothek Physik, Garching 10453
– Teilbibliothek Sozialwissenschaften, München 10454
– Teilbibliothek Sportwissenschaft, München 10455
– Teilbibliothek Stammgelände, München 10456
– Teilbibliothek Weihenstephan, Freising 10457
Technische Universität Wien, Wien 01301
– Bibliotheksabteilung Städtebau am Institut für Städtebau, Landschaftsarchitektur und Entwerfen, Wien 01302
– Fachbereich für Landschaftsplanung und Gartenkunst, Wien 01303
– Fachbibliothek Chemie, Wien 01304
– Fachbibliothek für Mathematik und Physik, Wien 01305
– Institut für Diskrete Mathematik und Geometrie, Wien 01306
– Institut für Ingenieurgeologie, Wien 01307
– Institut für Managementwissenschaften, Wien 01308
– Institut für Verkehrswissenschaften, Forschungsbereich für Verkehrsplanung und Verkehrstechnik, Wien 01309
Technische Universiteit Delft, Delft 19104
Technische Universiteit Eindhoven, Eindhoven 19111
– Centrale Leeszaal, Eindhoven 19112
– Faculteitsbibliotheek Bouwkunde, Eindhoven 19113
– Faculteitsbibliotheek Elektrotechniek, Eindhoven 19114
– Faculteitsbibliotheek Industrial Engineering and Innovation Sciences, Eindhoven 19115
– Faculteitsbibliotheek Technische Natuurkunde, Eindhoven 19116
– Faculteitsbibliotheek Wiskunde en Informatica, Eindhoven 19117
Technisches Museum Wien mit Österreichischer Mediathek, Wien 01655
Technological Institute of Milk and Meat, Kyiv 28386
Technological Institute of Textile and Sciences, Bhiwani 13947
Technological Institute of the Light and Food Industries, Zhambul 17929
Technological Institute of the Philippines, Manila 20774
Technological University of the Philippines, Manila 20722
Technologisches Gewerbemuseum, Wien 01383
Technololgical Institute of Celluose and Paper Industry, Sankt-Peterburg 23908
Teck Centennial Library, Kirkland Lake 04672
Teck Cominco Metals Ltd, Trail 04298
Tecsult, Inc, Montreal 04272
Tecumseh District Library, Tecumseh 41529
Teesside University, Middlesbrough 29274
Tehama County Library, Red Bluff 40951
Tehnički muzej, Zagreb 06254
Tehniški muzej Slovenije, Ljubljana 24939
Tehran Museum of Modern Art, Tehran 14438
Teijin K. K. Seisan Gijutsu Kenkyujo, Iwakuni 17365
Teijin Ltd, Products Development Research Laboratories, Iwakuni 17365
Teikyo Daigaku, Hachioji 16937
– Yakugakubu, Tsukui 16938
Teikyo Loretto Heights University, Denver 30997
Teikyo University, Hachioji 16937
– Faculty of Pharmaceutical Sciences, Tsukui 16938
Teilbibliohtek Karlstraße, München 10364
Teilbibliothek Pasing, München 10365
Tekhnicheski kolledzh upravleniya i komertsii, Sankt-Peterburg 22710
Tekhnicheski universitet, Volgograd 22642

Tekhnicheski universitet grazhdanskoi aviatsii, Moskva 22393
Tekhnicheski universitet im. A.N. Tupoleva, Kazan 22280
Tekhnicheski universitet Varna, Varna 03485
Tekhnichesky universytet, Kherson 28003
Tekhnichny universytet budivnitstva i arkhitekturi, Kyiv 28022
Tekhnichny universytet silskoho gospodarstva, Kharkiv 27998
Tekhnikum elektronnykh priladiv, Kyiv 28117
Tekhnikum geodezii i kartografii, Sankt-Peterburg 22711
Tekhnikum gidromelioratsiyi, mekhanizatsiyi ta elektrifikatsiyi silskoho gospodarstva, Nova Kakhovka 28125
Tekhnikum gromadskoho kharchuvannya, Kyiv 28118
Tekhnikum legkoyi promislovosti, Kyiv 28119
Tekhnikum mekhanizatsiyi silskoho gospodarstva, Kirovograd 28105
Tekhnikum promislovoyi avtomatiki, Donetsk 28099
Tekhnikum radioelektroniki, Kyiv 28120
Tekhnikum transportnoho budivnitstva, Kyiv 28121
Tekhnikum zaliznichnoho transportu, Kyiv 28122
Tekhnologicheski institut, Kostroma 22297
Tekhnologicheski institut, Sankt-Peterburg 22545
Tekhnologicheski institut kholodilnoi promyshlennosti, Sankt-Peterburg 22546
Tekhnologicheski institut legkoi promyshlennosti, Vitebsk 02084
Tekhnologicheski institut tsellyuloznobumazhnoi promyshlennosti, Sankt-Peterburg 23908
Tekhnologicheski tekhnikum pitaniya, Sankt-Peterburg 22712
Tekhnologichny instytut moloka ta myasa, Kyiv 28386
Tekniska museets bibliotek, Stockholm 26695
Tekniska Nämndhuset, Stockholm 26523
Tekovian Library, Levice 24780
Tekovská Knižnica, Levice 24780
Tekovské múzeum, Levice 24737
Tektronix, Inc Library, Beaverton 35495
Tel Hai Academic College, Kiryat Shmona 14641
Tel-Aviv Courts, Tel-Aviv 14692
Tel-Aviv Museum of Art, Tel-Aviv 14771
Tel-Aviv University, Tel-Aviv 14664
– David J. Light Law Library, Tel-Aviv 14665
– Faculty of Social Sciences and Management, Tel-Aviv 14666
– Geography Library, Tel-Aviv 14667
– The Gitter-Smolarz Library of Life Sciences and Medicine, Tel-Aviv 14668
– Institute of Archaeology, Tel-Aviv 14669
– Moshe Dayan Center for Middle Eastern and North African Studies, Tel-Aviv 14670
– Neiman Library of Exact Sciences and Engineering, Tel-Aviv 14671
– School of Education, Tel-Aviv 14672
Tel-Aviv-Sourasky Medical Center, Tel-Aviv 14772
Telcordia Technologies Inc – Raritan River Software Systems Center Library, Piscataway 36125
Tele Quebec, Montreal 04421
Telecom Finland Ltd, Helsinki 07416
Telecommunications Research Centre, New Delhi 14170
Telecommunications Research Institute, Warszawa 21342
Telecommunications Training Centre, Kuala Lumpur 18602
Telecomunicaciones de México, México 18820
Teledyne Brown Engineering Co, Huntsville 35770
Telekom Slowenije, Ljubljana 24904
Telemark College of Education, Notodden, Notodden 20110
Telemark fylkesbibliotek, Ulefoss 20413
Telemark University College, Bø 20083
Telemark University College, Porsgrunn 20155
Telemarkmuseum, Skien 20292
Teleradiokompaniya Sankt-Peterburg, Sankt-Peterburg 22943
Teleradiokompaniya Sankt-Peterburg, Notnaya biblioteka, Sankt-Peterburg 22944
Telesat Canada Information Resource Centre, Gloucester 04353
Television Corporation of Singapore TCS, Singapore 24623
Televisión Española S.A. – Servicio de Documentación Audiovisual, Madrid 25866
Televisión Española S.A. – Servicio de Documentación Escrita, Madrid 25867
Televisión Nacional de Chile, Santiago 04946
Télévision Suisse Romande, Genève 27375
Tell City-Perry County Public Library, Tell City 41531
Telšių rajono savivaldybės Karolinos Praniauskaitės viešoji biblioteka, Telšiai 18369
Temasek Polytechnic, Singapore 24593
Temenggung Ibrahim Teachers' Training College, Johor Bharu 18529
Tempe Public Library, Tempe 41533
Temple Beth El, Great Neck 35220
Temple Beth El Congregation Sons of Israel & David, Providence 35339
Temple Beth El Library, Rochester 35350
Temple Beth Zion, Amherst 35106
Temple Beth Zion Library, Buffalo 35148
Temple B'Rith Kodesh, Rochester 35351
Temple College, Temple 33796
Temple De Hirsch Sinai Library, Seattle 35377
Temple Emanu-El, Dallas 35181
Temple Emanu-El, Providence 35340
Temple Emanu-El, Tucson 35398
Temple Israel, Boston 35144
Temple Israel Libraries & Media Center, West Bloomfield 35415
Temple Israel Library, Minneapolis 35287
Temple Israel Library of Judaica, West Palm Beach 35418
Temple Public Library, Temple 41534
Temple Shaarey Zedek, Amherst 35107
Temple Sinai Library, Roslyn Heights 35352
Temple Terrace Public Library, Temple Terrace 41536
The Temple – Tifereth Israel, Beachwood 35129
Temple University, Philadelphia 32262
– Ambler Library, Ambler 32263
– Health Science Center Libraries, Philadelphia 32264
– Law Library, Philadelphia 32265
– Science, Engineering & Architecture Library, Philadelphia 32266
– Tyler School of Art Library, Elkins Park 32267
Temple University Hospital, Episcopal Campus, Philadelphia 37364
Temple University School of Podiatric Medicine, Philadelphia 32268
Templeton College, Oxford 29321
Tenafly Public Library, Tenafly 41537
Tenaga Nasional Berhad (TEN), Kuala Lumpur 18606
Tenaga Nasional Berhad (TEN), Institut Latihan Sultan Ahmad Shah (ILSAS), Kajang 18575
Tengku Ampuan Afzan Teachers' Training College, Kuantan 18537
Tennessee Department of Corrections, Nashville 34572
Tennessee Department of Transportation Library, Nashville 34573
Tennessee State Law Library, Jackson 34390
Tennessee State Law Library, Nashville 34574
Tennessee State Library & Archives, Nashville 30143
Tennessee State University, Nashville 31937
Tennessee Supreme Court Law Library, Knoxville 34423
Tennessee Technological University, Cookeville 30922
Tennessee Temple University, Chattanooga 30739
Tennessee Valley Authority, Chattanooga 34062
Tennessee Valley Authority, Muscle Shoals 37137
Tennessee Valley Authority – Legal Research Center, Knoxville 35809
Tennessee Valley Authority – Research Library-Knoxville, Knoxville 35810
Tennessee Wesleyan College, Athens 30278
Tenri Central Library, Tenri, Nara 17132
Tenri Daigaku, Tenri, Nara 17132
Tenri University, Tenri Central Library, Tenri, Nara 17132
Tensor Society, Chigasaki 17525
Tenth Judicial District Supreme Court Law Library, Riverhead 34743
Teofipol Regional Centralized Library System, Main Library, Teofipol 28838
Teofipolska raionna TsBS, Tsentralna biblioteka, Teofipol 28838
Teollisuuden Voima Oy, Olkiluoto 07425
Teologisk Pædagogisk Center Løgumkloster, Løgumkloster 06763
Det Teologiske Menighetsfakultet, Oslo 20124
Teološka knjiznica Maribor, Maribor 24863
Teplik Regional Centralized Library System, Main Library, Teplik 28839
Teplitska raionna TsBS, Tsentralna biblioteka, Teplik 28839
Tepsatri Teachers College, Lopburi 27740
Terangganu Public Library Corp, Kuala Terengganu 18627
Terebovlya Regional Centralized Library System, Main Library, Terebovlya 28840
Terebovlyanska raionna TsBS, Tsentralna biblioteka, Terebovlya 28840
Terezín Memorial, Terezín 06542
Termo izolisacijos institutas, Vilnius 18313
Ternopil City Centralized Library System, Main Library, Ternopil 28845
Ternopil State Archives, Ternopil 28477
Ternopil State Economic University, Ternopil 28079
Ternopilska miska TsBS, Tsentralna biblioteka, Ternopil 28845
Ternopilskiy nicionalniy ekonomichniy universytet, Ternopil 28079
Terra Alta Public Library, Terra Alta 41538
Terrace Public Library, Terrace 04776
Terre des Femmes, Berlin 11592
Terrebonne Parish Library, Houma 39484
Terrebonne Parish Library – North Terrebonne, Gray 39282
Terrell State Hospital – Medical Library, Terrell 37667
Terrell State Hospital – Patient Library, Terrell 37668
Territorialnoe SPO Spetstrans, Sankt-Peterburg 22945
Terryville Public Library, Terryville 41541
Terveyden ja hyvinvoinnin laitos, Helsinki 07501
Tessedik Samuel College Faculty of Economics, Békéscsaba 13216
Tessedik Samuel Főiskola Gazdasagi Főiskola Kar Konyvtara, Békéscsaba 13216
Test and Research Instrumentation Plant of the Russian Academy of Sciences, Chernogolovka 23019
Testa, Hurwitz & Thibeault, Boston 35539
Testing Engineers International, Inc, South Salt Lake 36276
Tetiiv Regional Centralized Library System, Main Library, Tetiyiv 28846
Tetiyivska raionna TsBS, Tsentralna biblioteka, Tetiyiv 28846
Teton County Library, Jackson 39606
Tetsudo Kodaikai Shakai Fukushibu Fukushi Shiryoshitsu, Tokyo 17759
Tevosyan Plant 'Elektrostal', Elektrostal 22792
Tewksbury Public Library, Tewksbury 41542
Texaco, Houston 35765
Texarkana College, Texarkana 32810
Texarkana Public Library, Texarkana 41543
Texas A&M International University, Laredo 31565
Texas A&M University, College Station 30857
– Cushing Library, College Station 30858
– Cushing Library, Science Fiction Research Collection, College Station 30859
– Cushing Memorial Library & Archives, College Station 30860
– Medical Sciences Library, College Station 30861
– Medical Sciences Library, College Station 30862
– Policy Sciences & Economics Library, College Station 30863
Texas A&M University at Galveston, Galveston 31202
Texas A&M University – Commerce, Commerce 30916
Texas A&M University – Corpus Christi, Corpus Christi 30926
Texas A&M University – Kingsville, Kingsville 31511
Texas A&M University – Texarkana, Texarkana 32811
Texas Chiropractic College, Pasadena 32237
Texas Christian University, Fort Worth 31167
Texas College, Tyler 32867
Texas County Library, Houston 39522
Texas Department of Criminal Justice Library, Huntsville 34372
Texas Department of State Health Services, Austin 33951
Texas Department of Transportation-Center for Transportation Research, Austin 33952
Texas Education Agency, Austin 33953
Texas Instruments Inc – SC Group Library, Dallas 35656
Texas Legislative Reference Library, Austin 33954
Texas Lutheran University, Seguin 32654
Texas Natural Resource Conservation Commission Library, Austin 33955
Texas School for the Blind, Austin 33144
Texas Southern University, Houston 31356
– Thurgood Marshall School of Law Library, Houston 31357
Texas State Law Library, Austin 33956
Texas State Library & Archives Commission, Austin 30116
Texas State Library & Archives Commission – Talking Book Program, Austin 30117
Texas State Technical College, Waco 33830
Texas State University-San Marcos, San Marcos 32601
Texas Tech University, Lubbock 31691
– Southwest Collection Special Collections Library, Lubbock 31692
Texas Tech University Health Sciences Center, Lubbock 37073
Texas Tech University Health Sciences Center at Amarillo, Amarillo 36459
Texas Tech University – School of Law Library, Lubbock 31693
Texas Wesleyan University, Fort Worth 31168
Texas Woman's University, Dallas 30953
Texas Woman's University, Denton 30990
Textbook Library 'Toshobunko', Tokyo 17705
Textilbibliothek, St.Gallen 27439
Textile Industry, Barnaul 22778
Textile Industry Machine Building Industrial Corporation, Brest 01841
Textile Manufactory Joint-Stock Company, Moskva 22827
Textile Museum, Łódź 21158
Textile Museum, Washington 37796
Textile Research Institute, Łódź 21162
Textilmúzeum, Budapest 13461
Textron Defense Systems, Wilmington 36428
Teylers Museum, Haarlem 19457
Thaddeus Stevens College of Technology, Lancaster 31555
Thailand Institute of Scientific and Technological Research, Bangkok 27789
Thakur Ram Multiple Campus Library, Parsa 19024
Thames Valley University, London 29221
Thammasat University, Bangkok 27714
– Faculty of Commerce, Bangkok 27715
– Faculty of Economics, Bangkok 27716
– Faculty of Law, Bangkok 27717
– Faculty of Political Science, Bangkok 27718
– Faculty of Social Administration, Bangkok 27719
– School of Journalism and Mass Communications, Bangkok 27720
– Thai Studies Research Information Center, Bangkok 27721
Thapar Institute of Engineering and Technology, Patiala 13737
Thayer County Museum, Belvidere 36530
Thayer Memorial Library, Lancaster 39797
Theater Instituut Nederland, Amsterdam 19384
Theatermuseum Düsseldorf, Düsseldorf 11807
Theaterschool – Amsterdamse Hogeschool voor de Kunsten, Amsterdam 19058
Theatre Academy Library, Helsinki 07344
Theatre Institute, Bratislava 24686
Theatre Institute, Praha 06485
Theatre Museum, Warszawa 21323
Theatre Museum of Sweden, Stockholm 26694
Theatre School – Amsterdam School of the Arts, Amsterdam 19058
Theatre Union of the Russian Federation, Moskva 23487
Theatrical Museum, Tbilisi 09284
Thekwini Municipal Library, Durban 25197
Thelen Reid Brown Raysman & Steiner LLP, San Francisco 36247
Thelen, Reid & Priest Library, New York 36029
Theodore F. Jenkins Memorial Law Library, Philadelphia 37365
Theodor-Heuss-Bibliothek, Berlin 12670
Theodor-Storm-Zentrum, Husum 12071
Theologenbücherei Priesterseminar Graz, Graz 01431
Theological College of Szeged, Szeged 13345
Theological College of the Canadian Reformed Churches Library, Hamilton 03725
Theological Faculty Tilburg, Tilburg 19206
Theological Institute of Finland, Helsinki 07412
Theological School of the Reformed Congregations, Rotterdam 19198
Theologicum, Evangelisch-Theologisches Seminar, Tübingen 10630

Theologische Bibliothek und Religionspädagogische Mediothek der Lippischen Landeskirche, Detmold 11190
Theologische Faculteit Tilburg (TFT), Tilburg 19206
Theologische Hochschule Chur, Chur 27065
Theologische Hochschule Friedensau, Friedensau 09807
Theologische Hochschule Reutlingen, Reutlingen 10542
Theologische School der Gereformeerde Gemeenten, Rotterdam 19198
Theologische Studienbibliothek Heidelberg, Heidelberg 11242
Theologische Universiteit Apeldoorn, Apeldoorn 19096
Theologische Universiteit van de Gereformeerde Kerken in Nederland (Vrijgemaakt), Kampen 19148
Theologisches Seminar der Evangelischen Kirche in Hessen und Nassau, Herborn 11244
Theologisches Seminar der Liebenzeller Mission, Bad Liebenzell 11156
Theology Faculty Library of Granada, Granada 25780
Theosophical Society, Chennai 13898
Theosophical Society in America, Wheaton 37832
Theosophical University Library, Altadena 30204
Theresianische Militärakademie, Wiener Neustadt 01667
Thermoelectrical Equipment Research Institute, Moskva 23390
Thermophysical Engineering Institute, Kyiv 28349
Thesaurus Linguae Latinae, München 12337
Thiel College, Greenville 31264
Third District Appellate Court Library, Ottawa 34645
Third District Court of Appeals, Miami 34530
Third Judicial Circuit Court, Wayne County, Detroit 36807
Thirty Sixth District Court Law Library, Detroit 34148
Thisted Bibliotek, Thisted 06939
Þjóðminjasafn Íslands, Reykjavík 13576
Thomas Aquinas College, Santa Paula 32616
Thomas College, Waterville 33000
Thomas County Public Library System, Thomasville 41553
Thomas Crane Public Library, Quincy 40920
Thomas Ford Memorial Library, Western Springs 41850
Thomas Gilcrease Institute of American History & Art, Tulsa 37694
Thomas J Watson Library, New York 37231
Thomas Jefferson Foundation Inc, Charlottesville 36655
Thomas Jefferson University, Philadelphia 32269
Thomas M. Cooley Law School, Lansing 31559
Thomas More College, Crestview Hills 30934
Thomas More College of Liberal Arts, Merrimack 31804
Thomas Nelson Community College, Hampton 33397
Thomas Township Library, Saginaw 41099
Thomas University, Thomasville 32815
Thomas-Mann-Archiv der Eidgenössischen Technischen Hochschule Zürich, Zürich 27470
Thomaston Public Library, Thomaston 41551
Thomas-Valentin-Stadtbücherei, Lippstadt 12830
J. Walter Thompson Co, New York 36031
Thompson Coburn LLP, Saint Louis 36206
Thompson, Hine LLP, Cincinnati 35612
Thompson, Hine LLP, Cleveland 35618
Thompson, Hine LLP, Dayton 35661
Thompson & Knight, Dallas 35657
Thompson Rivers University, Kamloops 03726
Thompson-Nicola Regional District Library System, Kamloops 04666
Thomson Reuters West, Eagan 35682
Thomson & Thomson, Beverly Hills 35502
Thorold & Lyttelton Library, Winchester 29564
Thorold Public Library, Thorold 04779
Thorvaldsens Museums Bibliotek, København 06825
Thousand Oaks Library, Thousand Oaks 41555
Thousand Oaks Library Newbury Park Branch, Newbury Park 40459
'Three Hierarchs' Library, Volos 13137
Three Rivers Community College, Norwich 33590
Three Rivers Community College, Poplar Bluff 33652
Three Rivers Public Library District, Channahon 38506
Thunder Bay Law Association Library, Thunder Bay 04502
Thunder Bay Public Library, Thunder Bay 04780
Thunderbird School of Global Management, Glendale 31218
Thuringan Ministry of Education, Science and Culture, Erfurt 10889
Thüringer Landesamt für Lebensmittelsicherheit und Verbraucherschutz, Bad Langensalza 10781
Thüringer Landesanstalt für Landwirtschaft, Jena 12080
Thüringer Landesanstalt für Umwelt und Geologie, Jena 12081
Thüringer Landesanstalt für Umwelt und Geologie, Weinbergen 12584
Thüringer Landesanstalt für Umwelt und Geologie, Außenstelle Weimar, Weimar 12581
Thüringer Landesanstalt für Wald, Jagd und Fischerei, Gotha 11926
Thüringer Landessternwarte Tautenburg, Tautenburg 12548
Thüringer Landtag, Erfurt 10888
Thüringer Ministerium für Bildung, Wissenschaft und Kultur, Erfurt 10889
Thüringer Universitäts- und Landesbibliothek, Jena 09320
Thüringisches Hauptstaatsarchiv Weimar, Weimar 12582
Thüringisches Institut für Textil- und Kunststoff-Forschung e.V., Rudolstadt 12448
Thüringisches Landesamt für Denkmalpflege und Archäologie, Weimar 12583
Thüringisches Staatsarchiv Altenburg, Altenburg 11450
Thüringisches Staatsarchiv Meiningen, Meiningen 12262
Thüringisches Staatsarchiv Rudolstadt, Rudolstadt 12449
ThyssenKrupp Steel Europe AG, Duisburg 11387
Tiakku K. K. Gijutsukanrika, Musashino 17387
Tianjin Conservatory of Music, Tianjin 05609
Tianjin Institute of Finance and Economics, Tianjin 05610
Tianjin Institute of Foreign Languages, Tianjin 05611
Tianjin Institute of Foreign Trade, Tianjin 05612
Tianjin Institute of Science and Technology, Tianjin 05613
Tianjin Institute of Textile Science and Technology, Tianjin 05614
Tianjin Institute of Urban Construction, Tianjin 05615
Tianjin Library, Tianjin 04973
Tianjin Medical College, Tianjin 05616
Tianjin Normal University, Tianjin 05617
– Applied Arts Science College, Tianjin 05618
Tianjin University, Tianjin 05619
Tianjin University of Commerce, Tianjin 05620
Tianjin University of Science & Technology, Tianjin 05621
Tianjin Vocational University, Tianjin 05622
Tianshui Teachers of Higher Learning, Tianshui 05623
Tibet Tibetan Medical College, Lhasa 05386
Tibet University, Lhasa 05387
Tibetan Buddhist Resource Center, Inc, New York 37262
Tibetan Works and Archives, Dharamsala 13989
Tibet-Institut, Rikon 27427
Tibro bibliotek, Tibro 26922
Ticonderoga Historical Society, Ticonderoga 37671
Tidewater Community College, Chesapeake 33216
Tidewater Community College, Portsmouth 33661
Tidewater Community College, Virginia Beach 33827
Tielaitoksen kirjasto, Helsinki 07402
Tieling Teachers' College, Tieling 05624
Tien Educational Center, Taipei 27622
Tierärztliche Hochschule Hannover, Hannover 09974
Tierpark Berlin-Friedrichsfelde GmbH, Berlin 11593
Tierps folkbibliotek, Tierp 26923
Tifereth Israel, Beachwood 35129
Tiffin University, Tiffin 32818
Tigard Public Library, Tigard 41558
Tikhookeanski gosudarstvenny universitet, Khabarovsk 22288
Tikhookeanski nauchno-issledovatelski institut rybnogo khozyaistva i okeanografii, Vladivostok 24059
Tilak Maharashtra Vidyapeeth, Pune 13744
Tilastokirjasto, Helsinki 07502
Tillamook County Library, Tillamook 41559
Timaru District Library, Timaru 19884
Timber Industrial and Scientific Corporation, Arkhangelsk 22774
Timber Industry Professional School, Sankt-Peterburg 22697
Timberland Regional Library, Tumwater 41621
Timbro/SFN, Stockholm 26696
Time bibliotek, Bryne 20322
Timiryazev Institute of Plant Physiology of the Russian Academy of Sciences, Moskva 23270
Timiryazev Moscow Agricultural Academy, Moskva 22356
Timiryazev State Biological Museum, Moskva 23208
Timiryazev Vinnitsa Regional Scientific Library, Vinnytsya 27949
Timis County Library, Timişoara 22109
Timişoara Branch of the Romanian Academy, Timişoara 22063
Timken Co, Canton 35553
Timmins Public Library, Timmins 04781
Timoshenko Military Academy for Chemical Defence, Moskva 22402
Timotei Cipariu Documentary Library, Blaj 21960
Timrå kommunbibliotek, Timrå 26924
Tingsryds bibliotek, Tingsryd 26925
Tinley Park Public Library, Tinley Park 41560
TINRO-Centre, Vladivostok 24060
Tipp City Public Library, Tipp City 41561
Tippecanoe County Historical Association, Lafayette 37004
Tippecanoe County Public Library, Lafayette 39766
Tipperary Joint Libraries Committee, Thurles 14590
Tipton County Public Library, Covington 38724
Tipton County Public Library, Tipton 41562
Tiptruck Plant Industrial Corporation, Saransk 22952
Tiroler Landesarchiv, Innsbruck 01545
Tiroler Landeskonservatorium, Innsbruck 01375
Tiroler Landesmuseum Ferdinandeum, Innsbruck 01546
Tiszai Vegyi Kombinát, Leninváros 13362
Tiszántuli Református Egyházkerület, Debrecen 13332
TISZAT Chemical Plant, Technical Library, Leninváros 13362
Titanium Metals Corporation of America (Laboratory), Henderson 35741
Titus Brandsma Instituut, Nijmegen 19186
Titusville Public Library, Titusville 41564
Tiverton Library Services, Tiverton 41565
Tivriv Regional Centralized Library System, Main Library, Tivriv 28847
Tivrivska raionna TsBS, Tsentralna biblioteka, Tivriv 28847
Tlumach Regional Centralized Library System, Main Library, Tlumach 28848
Tlumachanska raionna TsBS, Tsentralna biblioteka, Tlumach 28848
Tmmob Jeoloji Mühendisleri Odası, Ankara 27876
TNS Infratest Forschung GmbH, München 11421
Toa Payoh Community Library, Singapore 24632
Tobacco and Tobacco Products Institute, Plovdiv 03512
Tobacco Research Board, Harare 42380
Tocal Agricultural Centre, Paterson 00498
Toccoa Falls College, Toccoa Falls 32821
Tochigi Prefectural Assembly, Utsunomiya 17331
Tochigiken Gikai, Utsunomiya 17331
Tochmash Precision Machine Building Industrial Corportion, Donetsk 28143
Todaiji, Nara 17338
Toden Gakuen Daigakubu Semmonbu, Hino 17243
Toden Gakuen Senior College, Hino 17243
Toho Bank, Ltd, Planning Dept, Res Section, Fukushima 17352
Toho Daigaku, Chiba 16909
Toho Gasu Co, Ltd, General Planning Dept, Nagoya 17395
Toho Gasu K. K. Sogo Kikakushitsu, Nagoya 17395
Toho Ginko Kikakubu Chosaka K. K., Fukushima 17352
Toho University, Chiba 16909
Toho-Gakuen School of Music, Chofu 17242
Tohoku Dental University, Koriyama 16984
Tohoku Historical Museum, Tagajo 17665
Tohoku Rekishi Shiryokan, Tagajo 17665
Tohoku Shika Daigaku, Koriyama 16984
Tohoku University, Sendai 17114
– Center for Northeast Asian Studies, Sendai 17115
– Fuzoku Toshokan Igaku Bunkan, Sendai 17116
– Kinzoku Zairyo Kenkyujo, Sendai 17117
– Rigakubu Fuzoku Rinkai Jikkenjo, Aomori 17118
– Yakugakubu, Sendai 17119
Toi Whakaari – New Zealand Drama School, Wellington 19791
Tokai Daigaku, Hiratsuka 16949
Tokai University, Hiratsuka 16949
Tokei Suri Kenkyujo, Tokyo 17760
Tokimec Inc, Tokyo 17485
Tokkyocho Bankoku Kogyoshoyuken Shiryokan, Tokyo 17761
Tokmak Regional Centralized Library System, Main Library, Tokmak 28849
Tokmatska raionna TsBS, Tsentralna biblioteka, Tokmak 28849
Tokubetsuku Kyogikai, Tokyo 17762
Tokugawa Rinseishi Kenkyujo, Tokyo 17763
Tokushima Daigaku, Tokushima 17133
– Kuramoto Bunkan Igaku Toshokan, Tokushima 17134
Tokushima Prefecture Horticultural Experiment Station, Katsuura 17568
Tokushimaken Kaju Shikenjo, Katsuura 17568
Tokuyama Soda Co, Ltd, Tokuyama 17432
Tokuyama Soda K. K. Gijutsushitsu Gijutsudaiikka, Tokuyama 17432
Tokyo Bengoshikai – Dai-ni Tokyo, Tokyo 17764
Tokyo Broadcasting Inc, Tokyo 17487
Tokyo Chamber of Commerce and Industry, Tokyo 17770
Tokyo College of Music, Tokyo 17179
Tokyo Daigaku, Tokyo 17180
– Bunshi Saibou Seibutsugaku Kenkyujo, Tokyo 17181
– Bussei Kenyujo, Chiba 17182
– Earthquake Research Institute, Tokyo 17183
– Igaku Toshokan, Tokyo 17184
– Institute of Oriental Culture, Tokyo 17185
– Kaiyo Kenkyujo, Tokyo 17186
– Keizai Gakubu, Tokyo 17187
– Kyoyo Gakubu Amerika Kenkyu Shiryo Senta, Tokyo 17188
– Nogakubu Toshokan, Tokyo 17189
– Seisan Gijutsu Kenkyujo, Tokyo 17190
– Shakai Kagaku Kenkyujo, Tokyo 17191
– Shakaijoho Kenkyujo (ISICS), Tokyo 17192
– Shiryo Hensanjo, Tokyo 17193
– Yakugakubu, Tokyo 17194
Tokyo Daigaku Rigakubu Fuzoku Shokubutsuen, Tokyo 17765
Tokyo Denki Daigaku, Tokyo 17195
Tokyo Denki University, Tokyo 17195
Tokyo Denryoku K. K., Hino 17357
Tokyo Dental College, Chiba 16910
Tokyo Electric Power Co, Inc, Toden Gakuen Library, Hino 17357
Tokyo Gaikokugo Daigaku, Tokyo 17196
– Ajia Afurika Gengo Bunka Kenkyujo, Tokyo 17197
Tokyo Geijutsu Daigaku, Tokyo 17198
Tokyo Ginko, Tokyo 17486
Tokyo Hoso Kikaku Chosa Kyoku Shiryobu, Tokyo 17487
Tokyo Ika Daigaku, Tokyo 17199
Tokyo Ikashika Daigaku, Tokyo 17200
Tokyo Institute for Municipal Research, Tokyo 17768
Tokyo Institute of Technology, Tokyo 17205
Tokyo Joshi Daigaku, Tokyo 17201
– Hikaku Bunka Kenkyujo, Tokyo 17202
Tokyo Joshi Ika Daigaku, Tokyo 17203
Tokyo Kasei Daigaku, Tokyo 17204
Tokyo Kasei Gakuin College, Machida 17030
Tokyo Kasei Gakuin Daigaku, Machida 17030
Tokyo Kasei University, Tokyo 17204
Tokyo Keizai Daigaku, Kokubunji 16983
Tokyo Keizai University, Kokubunji 16983
Tokyo Kogyo Daigaku, Tokyo 17205
Tokyo Kokuritsu Bunkazai Kenkyujo, Tokyo 17766
Tokyo Medical and Dental University, Tokyo 17200
Tokyo Medical University, Tokyo 17199
Tokyo Metropolitan Art Museum, Tokyo 17774
Tokyo Metropolitan Central Library (TMCL), Tokyo 17859
Tokyo Metropolitan Industrial Technology Center, Tokyo 17772
Tokyo Metropolitan Institute for Educational Research and In-Service Training, Tokyo 17773
Tokyo Metropolitan Institute for Neuroscience, Fuchu 17531

Tokyo Metropolitan Institute of Medical Science, Tokyo 17767
Tokyo Metropolitan Research Institute for Environmental Protection, Tokyo 17775
Tokyo Metropolitan Research Laboratory of Public Health, Tokyo 17771
Tokyo National University of Fine Arts and Music, Tokyo 17198
Tokyo Nogyo Daigaku, Tokyo 17206
Tokyo Precision Instrument Co, Research & Development Center, Tokyo 17485
Tokyo Shika Daigaku, Chiba 16910
Tokyo Shisei Chosakai Shisei Senmon Toshokan, Tokyo 17768
Tokyo Shoken Torihikijo, Tokyo 17769
Tokyo Shoko Kaigisho, Tokyo 17770
Tokyo Stock Exchange, Tokyo 17769
Tokyo Toritsu Eisei Kenkyujo, Tokyo 17771
Tokyo Toritsu Kogyo Gijutsu Senta, Tokyo 17772
Tokyo Toritsu Kyoiku Kenkyujo Chosa Fukyubu, Tokyo 17773
Tokyo University of Agriculture, Tokyo 17206
Tokyo University of Foreign Studies, Tokyo 17196
– Institute for the Study of Languages and Cultures of Asia and Africa, Tokyo 17197
Tokyo University of Marine Science and Technology, Tokyo 17207
Tokyo University of Pharmacy and Life Sciences, Hachioji 16939
Tokyo Woman's Christian University, Tokyo 17201
– Institute for Comparative Studies of Culture, Tokyo 17202
Tokyo Women's Medical University, Tokyo 17203
Tokyo Zokei Daigaku, Hachioji 16940
Tokyo Zokei University, Hachioji 16940
Tokyoko Gakuen Joshi Tanki Daigaku, Tokyo 17248
Tokyoto Bijutsukan, Tokyo 17774
Tokyoto Kankyo Kagaku Kenkyujo, Tokyo 17775
Tokyoto Shinkei Kagaku Sogo Kenkyujo, Fuchu 17531
Toledo Law Association Library, Toledo 37672
Toledo Museum of Art, Toledo 37673
Toledo-Lucas County Public Library, Toledo 41566
Toledo-Lucas County Public Library – Heatherdowns, Toledo 41567
Toledo-Lucas County Public Library – Holland Branch, Holland 39457
Toledo-Lucas County Public Library – Kent, Toledo 41568
Toledo-Lucas County Public Library – Locke, Toledo 41569
Toledo-Lucas County Public Library – Maumee Branch, Maumee 40102
Toledo-Lucas County Public Library – Mott, Toledo 41570
Toledo-Lucas County Public Library – Oregon Branch, Oregon 40636
Toledo-Lucas County Public Library – Point Place, Toledo 41571
Toledo-Lucas County Public Library – Reynolds Corners, Toledo 41572
Toledo-Lucas County Public Library – Sanger, Toledo 41573
Toledo-Lucas County Public Library – Sylvania Branch, Sylvania 41488
Toledo-Lucas County Public Library – Toledo Heights, Toledo 41574
Toledo-Lucas County Public Library – Washington, Toledo 41575
Toledo-Lucas County Public Library – Waterville Branch, Waterville 41770
Toledo-Lucas County Public Library – West Toledo, Toledo 41576
Tolland Public Library, Tolland 41577
Tolna County Library, Szekszárd 13548
Tolna Megyei Könyvtár, Szekszárd 13548
Tolstoi Sevastopol Central City Library, Sevastopol 28810
Tolstoi-Bibliothek, München 12338
Tolstoy City Main Library, Tula 24333
Tolstoy Foundation, Inc, Valley Cottage 37703
Tolstoy Minsk City Library no 1, Minsk 02161
L. N. Tolstoy Regional Universal Science Library, Kustanay 17989
Leo Tolstoy State Museum, Moskva 23228
Tom Green County Library System, San Angelo 41161
Tomah Public Library, Tomah 41578
Tomahawk Public Library, Tomahawk 41579
Tomashpil Regional Centralized Library System, Main Library, Tomashpil 28850
Tomashpilska raionna TsBS, Tsentralna biblioteka, Tomashpil 28850
Tombigbee Regional Library System – Bryan Public Library, Headquarters, West Point 41838
Tomorrow Center Library, Mitchellville 34544
Tompkins County Public Library, Ithaca 39597
Tompkins Memorial Library, Edgefield 36822
Tompkins-Cortland Community College, Dryden 33301
Tomsk Polytechnic University, Tomsk 22598
Tomsk State Academy of Control Systems and Radioelectronics, Tomsk 22596
Tomsk State University, Tomsk 22597
Tomskaya gosudarstvennaya oblastnaya universalnaya nauchnaya biblioteka A.S. Pushkina, Tomsk 22179
Tomski gosudarstvenny universitet, Tomsk 22597
Tomski mezhotraslevoi territorialny tsentr nauchno-tekhnicheskoi informatsii i propagandy, Tomsk 23999
Tomski oblastnoi detsko-yunosheski tsentr bibliotechnogo obsluzhivaniya i esteticheskogo vospitaniya, Tomsk 24332
Tomski politekhnicheski universitet, Tomsk 22598
Tønder Kommunes Biblioteker, Tønder 06941
Tonghua Teachers' College, Tonghua 05625
Tongji Medical University, Wuhan 05658
– Yunyang Medical College, Shiyan 05659
Tongji University, Shanghai 05535
Tongling Finance and Economy College, Tongling 05629
Tongren Teachers' College, Tongren 05630
Tonkon Torp LLP, Portland 36149
Tonkosukinnaya fabrika im. Petra Alekseeva, Moskva 22856
Tonkosukonny kombinat, Grodno 01872
Tønsberg and Nøtterøy Public Library, Tønsberg 20408
Tønsberg og Nøtterøy bibliotek, Tønsberg 20408
TOO Mashzavod im. M.I. Kalinina, Moskva 22857
TOO Yaroslavski shinny zavod, Yaroslavl 22979
Tooele City Public Library, Tooele 41582
Tool Building Ltd, Moskva 22829
Tool Making Plant, Orsha 01914
Tool Museum, Troyes 08701
Toonkunst-Bibliotheek, Amsterdam 19385
K. N. Toosi University of Technology, Tehran 14409
Toowoomba Regional Libraries, Toowoomba 01169
Topchiev Institute for Petrochemical Synthesis, Moskva 23294
Topeka & Shawnee County Public Library, Topeka 41583
Topiwala National Medical College, Mumbai 13710
Topkapi Palace Museum, Istanbul 27896
Topkapi Sarayi Muzesi, Istanbul 27896
Topographie des Terrors, Berlin 11590
Topography of Terror, Berlin 11590
Toray Industries, Inc, Tokyo 17488
Toray Industries, Inc, Basic Research Laboratories, Kamakura 17367
Toray Industries Inc, Research and Development Div, Otsu 17418
Torbay Library Services, Torquay 30087
Tore K. K., Tokyo 17488
Tore K. K. Kenkyu Kaihatsu Kikakubu, Otsu 17418
Tore K. K. Kiso Kenkyujo, Kamakura 17367
Torez City Centralized Library System, Main Library, Torez 28851
Torezka miska TsBS, Tsentralna biblioteka, Torez 28851
Torfyanoi institut, Tver 24013
Torgovli koledzh, Kyiv 28123
Torgovo-ekonomicheski institut, Sankt-Peterburg 22547
Torgovo-ekonomichny instytut, Lviv 28039
Torgovo-komercheski tekhnikum, Sankt-Peterburg 22713
Tornedalens bibliotek, Övertorneå 26881
Tornedalsskolan, Haparanda 26471
Tornio Public Library, Tornio 07614
Tornion kaupunginkirjasto, Tornio 07614
Tornyai János Múzeum és Közmuvelodési Központ, Hódmezővásárhely 13480
Toronto Botanical Garden, Toronto 04526
Toronto Catholic District School Board, Toronto 04527
Toronto City Planning Library, Toronto 04170
Toronto District School Board, North York 04102
Toronto General Hospital Health Sciences Library, Toronto 04533
Toronto International Film Festival Group, Toronto 04528
Toronto Jewish Library, Toronto 04529
Toronto Monthly Meeting of the Religious Society of Friends (Quakers), Toronto 04530
Toronto Public Library, Toronto 04782
Toronto Public Library – Agincourt District, Toronto 04783
Toronto Public Library – Albert Campbell District, Toronto 04784
Toronto Public Library – Albion, Toronto 04785
Toronto Public Library – Barbara Frum Branch, Toronto 04786
Toronto Public Library – Brentwood, Toronto 04787
Toronto Public Library – Cedarbrae, Toronto 04788
Toronto Public Library – Don Mills, Toronto 04789
Toronto Public Library – Fairview, Toronto 04790
Toronto Public Library – Lillian H. Smith Branch, Toronto 04791
Toronto Public Library – Malvern, Toronto 04792
Toronto Public Library – Maria A. Shchuka Branch, Toronto 04793
Toronto Public Library – North York Central, Toronto 04794
Toronto Public Library – Northern District, Toronto 04795
Toronto Public Library – Osborne Collection of Early Children's Books, Toronto 04796
Toronto Public Library – Pleasant View, Toronto 04797
Toronto Public Library – Richview, Toronto 04798
Toronto Public Library – S. Walter Stewart Branch, Toronto 04799
Toronto Public Library – Urban Affairs, Toronto 04800
Toronto Public Library – York Woods, Toronto 04801
Toronto Reference Library, Toronto 04802
Torrance Public Library, Torrance 41584
Torrens Valley Institute of TAFE, Modbury 00578
Torrington Library, Torrington 41585
Torsås folkbibliotek, Torsås 26927
Torsby Kommun, Torsby 26928
Tórshavns Folkebibliotek, Tórshavn 06942
Toruń Scientific Society, Toruń 21235
Torys Law Library, New York 36030
Torys LLP Library, Toronto 04294
Toschkent District Research Library 'Turon', Toshkent 42177
Toshiba Research and Development Center, Kawasaki 17371
Toshiba Sogo Kenkyujo, Kawasaki 17371
Toshkent Electrotechnical University of Communications, Toshkent 42099
Toshkent Institute of Agricultural Engineering and Irrigation, Toshkent 42100
Toshkent Institute of Railway Engineers, Toshkent 42101
Toshkent Institute of Textile and Light Industry, Toshkent 42102
Toshkent Pharmaceutical Institute, Toshkent 42103
Toshkent State Agrarian University, Toshkent 42104
Toshkent State Conservatoire, Toshkent 42105
Toshkent State Economics University, Toshkent 42106
Toshkent V.I. Lenin State University, Toshkent 42107
Toshokan Joho Daigaku, Tsukuba 17227
Tottori Kenritsu Tottori Toshokan, Tottori 17860
Tottori Prefectual Center for Educational Research and In-Service Training, Tottori 17782
Tottori Prefectural Tottori Library, Tottori 17860
Tottori University, Koyama-cho 16985
Tottoriken Kyoiku Kenshusenta, Tottori 17782
Tougaloo College, Tougaloo 32830
Toulouse Business School, Toulouse 08078
Touring Club Italiano, Milano 16356
Touring-Club Suisse, Genève 27376
Tourisme Québec, Québec 04464
Touro College, Central Islip 30695
Touro College Libraries, New York 32090
Tovarishchestvo Energobumprom, Moskva 22858
Towarzystwo Naukowe w Toruniu (TNT), Toruń 21235
Towarzystwo Naukowo Organizacji i Kierownictwa, Warszawa 21347
Towarzystwo Opieki nad Ociemniałymi, Izabelin 21110
Towarzystwo Przyjaciół Nauk, Przemyśl 21211
Towarzystwo Wolnej Wszechnicy Polskiej, Warszawa 21348
Tower Hamlets Local History Library and Archives, London 29838
Tower Hamlets Schools Library Service, London 29839
Town Library Karlo Bijelicki, Sombor 24552
Town Library Varnsdorf, Varnsdorf 06617
Town of Ballston Community Library, Burnt Hills 38401
Town of Tonawanda Public Library, Kenmore 39697
Township Library of Lower Southampton, Feasterville 39060
Townsville City Council, Townsville 01037
Towson University, Towson 32831
Toyama Chemical Co, Ltd, Res Laboratory, Toyama 17496
Toyama Ika Yakka Daigaku, Toyama 17216
Toyama Joshi Tanki Daigaku, Toyama 17249
Toyama Kagaku Kogyo K. K. Sogo Kenkyujo, Toyama 17496
Toyama Kenritsu Daigaku, Toyama 17217
Toyama Kenritsu Toshokan, Toyama 17861
Toyama Medical and Pharmaceutical University, Toyama 17216
Toyama Ongaku Zaidan Fuzoku Toshokan, Tokyo 17776
Toyama Prefectural Assembly, Toyama 17328
Toyama Prefectural Library, Toyama 17861
Toyama Prefectural University, Toyama 17217
Toyama Women's Junior College, Toyama 17249
Toyamaken Gikai, Toyama 17328
Toyo Boseki K. K. Katata Kenkyujo, Otsu 17419
The Toyo Bunko, Tokyo 17777
Toyo Construction Co, Ltd, Tokyo 17491
Toyo Daigaku, Tokyo 17208
Toyo Engineering Corp, General Affairs Dept, Tokyo 17489
Toyo Enjiniaringu K. K. Somubu, Tokyo 17489
Toyo Junior College of Food Technology, Kawanishi 17245
Toyo Keizai Publishing Co, Tokyo 17490
Toyo Keizai Shinposha, Tokyo 17490
Toyo Kensetsu K. K. Chosasekkeibu Toshoshitsu, Tokyo 17491
Toyo Kogyo Co, Ltd, Business Research Div, Aki 17343
Toyo Kogyo K. K. Chosabu Daiichi-chosaka Siryo-kakari, Aki 17343
Toyo Seican Group, Research and Development Center, Yokohama 17506
Toyo Seikan Gurupu Sogo Kenkyujo, Yokohama 17506
Toyo Shokuhin Kogyo Tanki Daigaku, Kawanishi 17245
Toyo University, Tokyo 17208
Toyobo Co, Ltd, Katata Research Institute, Otsu 17419
Toyohashi University of Technology, Toyohashi 17219
Toyoko Gakuen Women's Junior College, Tokyo 17248
Toyota Jidosha Kogyo K. K. Gijutsu Kanribu, Toyota 17498
Toyota Motor Co, Ltd, Documentation Center, Toyota 17498
TPO Kinotsentr, Sankt-Peterburg 23909
Tractor Plant, Kharkiv 28146
Tractor-Assembly Joint-Stock Company, Cheboksary 22782
Trade and Commerce Professional School, Sankt-Peterburg 22713
Trade College, Kyiv 28123
Trade Union Committee of the Gorky Motor-Car Plant, Nizhni Novgorod 23624
Trade Union Culture House, Moskva 24169
Trade Union Library, Ternopil 28842
Trade Union of the Building Industry, Moskva 23465
Trade Union of the Mechanical Engineering Industry, Moskva 23464
Trade Union of the Medical Workers, Trade Union Library, Moskva 23462
Trade Union of the Public Utilities Workers, Moskva 23463
Trade Unions Scientific Library, Moskva 23369
Trafford College, Manchester 29482
Trafford Metropolitan Borough Council, Stretford 30083
Trägergesellschaft der Freien Hochschule Mannheim gGmbH, Mannheim 10297
Tragor Ignác Múzeum, Vác 13516
Trails Regional Library, Warrensburg 41728
The Training Institute of Prison and Probation Services, Vantaa 07557
Trakia University, Stara Zagora 03482
Trakiski universitet, Stara Zagora 03482
Traktorny zavod, Kharkiv 28146

Trakų rajono savivaldybės viešoji biblioteka, Trakai 18370
Tranås stadsbibliotek, Tranås 26929
Tranemo bibliotek, Tranemo 26930
Trans Baikal Institute of Scientific Research, Chita 23022
Transafrica Forum, Washington 37797
Transalta Utilities Corp, Calgary 04250
Transamerica Occidental Life Insurance, Los Angeles 35861
Trans-Danubian Reformed Church District, Pápa 13342
Transformator Industrial Corporation, Zaporizhzhya 28174
Transformator Industrial Corporation, Trade Union Library, Zaporizhzhya 28887
'Transilvania' University of Braşov, Braşov 21919
Transport and Telecommunication Institute, Riga 18192
Transport Canada, Ottawa 04125
Transport Canada, Vancouver 04176
Transport Industrial Corporation, Sankt-Peterburg 22945
Transport Information Centre, Technical Library, Praha 06413
Transport Library and Information Centre, Žilina 24759
Transport Museum, Budapest 13390
Transport Planning Research Institute, Toshkent 42162
Transportation Association of Canada / Association des transports du Canada, Ottawa 04453
Transportation Development Centre Library, Montreal 04422
Transportation Safety Board of Canada Library, Gatineau 04079
Transportøkonomisk Institutt, Oslo 20286
Transtibiscan Reformed Church, Debrecen 13332
Transvaal Education Department, Pretoria 25171
Transvaal Museum, Pretoria 25172
Transylvania County Library, Brevard 38271
Transylvania University, Lexington 31591
Trapeza Tis Ellados, Athinai 13125
Traverse Area District Library, Traverse City 41589
Travis County Law Library, Austin 33957
Treasure Hall of the Daigoji Temple, Kyoto 17337
Treasure Valley Community College, Ontario 33604
H. M. Treasury and Cabinet Office, London 29528
Treasury Information Centre, Wellington 19809
Treasury Library, Parkes 00707
Treasury Solicitor's Library, London 29529
Tredyffrin Public Library, Strafford 41452
Trees and Timbers Institute, Sesto Fiorentino 16575
Trelawny Parish Library, Falmouth 16887
Trelleborg Industri AB, Trelleborg 26700
Trelleborgs bibliotek, Trelleborg 26931
Trenčianské muzeum, Trenčín 24757
Trent University, Peterborough 03820
Trenton Free Public Library, Trenton 41590
Trenton Veterans Memorial Library, Trenton 41591
Tresoar, Leeuwarden 19048
Trest geodezicheskikh rabot i inzhenernykh izyskani, Sankt-Peterburg 22946
Tretyakov State Gallery, Moskva 23204
Trevecca Nazarene University, Nashville 31938
Trial Court of Massachusetts, Fall River 36842
Tribal Research Institute, Chiang Mai 27793
Tribhuvan University, Kathmandu 19022
Tribunal Constitucional, Madrid 25734
Tribunal de Alçada do Estado de Minas Gerais, Belo Horizonte 03277
Tribunal de Commerce de Paris, Paris 08159
Tribunal de Cuentas, Madrid 25735
Tribunal de Justiça, São Paulo 03241
Tribunal de Justiça do Espirito Santo, Vitória 03242
Tribunal de Justiça do Paraná, Curitiba 03226
Tribunal fédéral / Schweizerisches Bundesgericht, Lausanne 27236
Tribunal Superior Eleitoral, Brasília 03225
Tribunal Supremo, Madrid 25736
Tribunales del Distrito Federal 'Fundación Rojas Astudillo', Caracas 42238
Tri-County Technical College, Pendleton 33629
Trident Technical College, Charleston 33212
TriHealth, Inc, Cincinnati 36700
Trimbos-instituut, Utrecht 19530
Trine University, Angola 30225
Trine University, Angola 30226
Trinidad and Tobago Bureau of Standards, Port-of-Spain 27812
Trinidad State Junior College, Trinidad 33805
Trinity and All Saints' College, Leeds 29156
Trinity Baptist College, Jacksonville 31451
Trinity Bible College, Ellendale 35197
Trinity Christian College, Palos Heights 33615
Trinity College, Bristol 29536
Trinity College, Cambridge 29048
Trinity College, Dublin 14476
Trinity College, Hartford 31305
Trinity College, Legon 13016
Trinity College, Oxford 29356
Trinity College, Parkville 00477
Trinity College, Port Hope 03821
Trinity College, Trinity 32833
Trinity College, Washington 32990
Trinity College of Music, London 29222
Trinity Episcopal School for Ministry Library, Ambridge 35105
Trinity International University, Deerfield 30980
Trinity International University, Santa Ana 32603
Trinity Lutheran College, Everett 35202
Trinity Lutheran Seminary, Bexley 35137
Trinity Theological College, Auchenflower 00749
Trinity University, San Antonio 32571
Trinity Western University, Langley 03735
Triodyne Inc, Northbrook 36057
Tripoli Public Library, Tripoli 18266
Tripura University, Agartala 13591
Triton College, River Grove 33685
The Tritonia Academic Library, Vaasa 07382
Tritonia, Vaasan tiedekirjasto ja oppimiskeskus – Vasa vetenskapliga bibliotek och lärocenter, Vaasa 07382
Trivandrum Public Library, Palayam 14230
TRL Library & Information Centre, Crowthorne 29641
TRL Limited, Crowthorne 29641
Trocaire – Catholic Agency for world development (NGO), Dublin 14557
Trois-Rivières Collège, Trois-Rivières 03927
Trollhättan Public Library, Trollhättan 26932
Trollhättans stadsbibliotek, Trollhättan 26932
Troms County Library, Tromsø 20410
Troms fylkesbibliotek, Tromsø 20410
Tromsø bibliotek, Tromsø 20411
Tromsø University College, Tromsø – Faculty of Education, Tromsø 20165
Trondheim folkebibliotek, Trondheim 20412
Trondheim Katedralskole, Trondheim 20188
Trondheim Public Library, Trondheim 20412
Tronox LLC, Oklahoma City 36069
Trostyanets Regional Centralized Library System, Main Library, Trostyanets 28852
Trostyanetska raionna TsBS, Tsentralna biblioteka, Trostyanets 28852
Troutman, Sanders, Mays & Valentine LLP, Richmond 36176
Troy Public Library, Troy 41594
Troy Public Library, Troy 41595
Troy Public Library, Troy 41596
Troy State University, Troy 32836
Troy State University Montgomery, Montgomery 31890
Troy University Dothan Library, Dothan 31021
Trubchevsk Polytechnic, Trubchevsk 22721
Trubchevski politekhnikum, Trubchevsk 22721
Truboprokatny zavod, Dnipropetrovsk 28540
Truckee Meadows Community College, Reno 33680
Trudeau Institute Library, Saranac Lake 37579
Truett-McConnell College, Cleveland 33239
Truman College, Chicago 33227
Truman State University, Kirksville 31514
Trumbull County Law Library, Warren 34975
Trumbull Library, Trumbull 41597
Truong Dai Hoc Bach Khoa Ha Noi, Hanoi 42274
Truro Library, Truro 30089
Truskavets Regional Centralized Library System, Main Library, Truskavets 28853
Truskavetska raionna TsBS, Tsentralna biblioteka, Truskavets 28853
Trussville Public Library, Trussville 41598
Truth or Consequences Public Library, Truth Or Consequences 41599
TRW, Redondo Beach 36166
TRW Inc, Cleveland 35619
Tsarichanka Regional Centralized Library System, Main Library, Tsarichanka 28854
Tsarichanska raionna TsBS, Tsentralna biblioteka, Tsarichanka 28854
TsBS Frunzenskogo raiona, Biblioteka-filial no 2 im. A.M. Gorkogo, Sankt-Peterburg 24247
TsBS Frunzenskogo raiona, Biblioteka-filial no 4 im. A.A. Prokofeva, Sankt-Peterburg 24248
TsBS Frunzenskogo raiona, Detskaya biblioteka-filial no 6, Sankt-Peterburg 24249
TsBS Frunzenskogo raiona, Tsentralnaya raionnaya biblioteka im. A.P. Chekhova, Sankt-Peterburg 24250
TsBS Frunzenskogo raiona, Tsentralnaya raionnaya detskaya biblioteka im. N.A. Krylova, Sankt-Peterburg 24251
TsBS Kalininskogo raiona, Biblioteka-filial 1 Znanie, Sankt-Peterburg 24252
TsBS Kalininskogo raiona, Biblioteka-filial no 2, Sankt-Peterburg 24253
TsBS Kalininskogo raiona, Biblioteka-filial no 3, Sankt-Peterburg 24254
TsBS Kalininskogo raiona, Biblioteka-filial no 7, Sankt-Peterburg 24255
TsBS Kalininskogo raiona, Tsentralnaya raionnaya biblioteka im. V.G. Belinskogo, Sankt-Peterburg 24256
TsBS Kirovskogo raiona, Biblioteka-filial no 2, Sankt-Peterburg 24257
TsBS Kirovskogo raiona, Tsentralnaya raionnaya biblioteka im. M.A. Sholokhova, Sankt-Peterburg 24258
TsBS Kirovskogo raiona, Tsentralnaya raionnaya detskaya biblioteka, Sankt-Peterburg 24259
TsBS Kolpinskogo raiona, Tsentralnaya biblioteka im. M. Svetlova, Sankt-Peterburg 24260
TsBS Kolpinskogo raiona, Tsentralnaya raionnaya detskaya biblioteka, Sankt-Peterburg 24261
TsBS Krasnogvardeiskogo raiona, Biblioteka-filial no 1 Malookhtinskaya, Sankt-Peterburg 24262
TsBS Krasnogvardeiskogo raiona, Biblioteka-filial no 6 Piskarevskaya, Sankt-Peterburg 24263
TsBS Krasnogvardeiskogo raiona, Tsentralnaya raionnaya biblioteka im. N.V. Gogolya, Sankt-Peterburg 24264
TsBS Krasnogvardeiskogo raiona, Tsentralnaya raionnaya detskaya biblioteka, Sankt-Peterburg 24265
TsBS Krasnoselskogo raiona, Tsentralnaya raionnaya biblioteka, Sankt-Peterburg 24266
TsBS Krasnoselskogo raiona, Tsentralnaya raionnaya detskaya biblioteka, Sankt-Peterburg 24267
TsBS Kronshtadskogo raiona, Tsentralnaya gorodskaya biblioteka, Sankt-Peterburg 24268
TsBS Leninskogo raiona, Biblioteka-filial no 1 im. K.A. Timiryazeva, Sankt-Peterburg 24269
TsBS Leninskogo raiona, Biblioteka-filial no 2, Sankt-Peterburg 24270
TsBS Leninskogo raiona, Biblioteka-filial no 3, Sankt-Peterburg 24271
TsBS Leninskogo raiona, Tsentralnaya raionnaya biblioteka im. L.I. Skvortsova-Stepanova, Sankt-Peterburg 24272
TsBS Leninskogo raiona, Tsentralnaya raionnaya detskaya biblioteka, Sankt-Peterburg 24273
TsBS Moskovskogo raiona, Biblioteka-filial no 1, Sankt-Peterburg 24274
TsBS Moskovskogo raiona, Biblioteka-filial no 4, Sankt-Peterburg 24275
TsBS Moskovskogo raiona, Biblioteka-filial no 5, Sankt-Peterburg 24276
TsBS Moskovskogo raiona, Detskaya biblioteka-filial no 6 im. S.Ya. Marshaka, Sankt-Peterburg 24277
TsBS Moskovskogo raiona, Tsentralnaya raionnaya biblioteka, Sankt-Peterburg 24278
TsBS Nevskogo raiona, Biblioteka-filial no 1, Sankt-Peterburg 24279
TsBS Nevskogo raiona, Tsentralnaya raionnaya detskaya biblioteka, Sankt-Peterburg 24280
TsBS Petrodvortsovogo raiona, Biblioteka-filial no 1 im. Yu. Inge, Sankt-Peterburg 24281
TsBS Petrodvortsovogo raiona, Biblioteka-filial no 6, Sankt-Peterburg 24282
TsBS Petrodvortsovogo raiona, Tsentralnaya raionnaya biblioteka im. A. Neto, Sankt-Peterburg 24283
TsBS Petrodvortsovogo raiona, Tsentralnaya raionnaya detskaya biblioteka, Sankt-Peterburg 24284
TsBS Petrogradskogo raiona, Biblioteka-filial no 2 im. B.A. Lavreneva, Sankt-Peterburg 24285
TsBS Petrogradskogo raiona, Biblioteka-filial no 3 im. A. Gaidara, Sankt-Peterburg 24286
TsBS Petrogradskogo raiona, Tsentralnaya raionnaya biblioteka im. A.S. Pushkina, Sankt-Peterburg 24287
TsBS Petrogradskogo raiona, Tsentralnaya raionnaya detskaya biblioteka, Sankt-Peterburg 24288
TsBS Primorskogo raiona, Tsentralnaya raionnaya biblioteka im. M.E. Saltykova-Shchedrina, Sankt-Peterburg 24289
TsBS Primorskogo raiona, Tsentralnaya raionnaya detskaya biblioteka, Sankt-Peterburg 24290
TsBS profkomu girnichno-zbagachuvalnoho kombinatu, Tsentralna biblioteka, Marganets 28719
TsBS profkomu girnichno-zbagachuvalnoho kombinatu, Tsentralna biblioteka, Ordzhonikidze 28768
TsBS Pushkinskogo raiona, Biblioteka-filial no 1, Sankt-Peterburg 24291
TsBS Pushkinskogo raiona, Biblioteka-filial no 2, Sankt-Peterburg 24292
TsBS Pushkinskogo raiona, Biblioteka-filial no 3, Sankt-Peterburg 24293
TsBS Pushkinskogo raiona, Tsentralnaya raionnaya biblioteka im. D.N. Mamina-Sibiryaka, Sankt-Peterburg 24294
TsBS Pushkinskogo raiona, Tsentralnaya raionnaya detskaya biblioteka, Sankt-Peterburg 24295
TsBS Sestroretskogo raiona, Tsentralnaya raionnaya biblioteka im. M. Zoshchenko, Sankt-Peterburg 24296
TsBS Sestroretskogo raiona, Tsentralnaya raionnaya detskaya biblioteka, Sankt-Peterburg 24297
TsBS Vasileostrovskogo raiona, Biblioteka-filial no 1 im. N.G. Chernyshevskogo, Sankt-Peterburg 24298
TsBS Vasileostrovskogo raiona, Biblioteka-filial no 2 im. L.N. Tolstogo, Sankt-Peterburg 24299
TsBS Vasileostrovskogo raiona, Detskaya biblioteka-filial no 6, Sankt-Peterburg 24300
TsBS Vasileostrovskogo raiona, Tsentralnaya raionnaya biblioteka im. M.V. Lomonosova, Sankt-Peterburg 24301
TsBS Vasileostrovskogo raiona, Yunosheskaya biblioteka-filial no 3 im. N. Ostrovskogo, Sankt-Peterburg 24302
TsBS Vyborgskogo raiona, Biblioteka-filial no 1, Sankt-Peterburg 24303
TsBS Vyborgskogo raiona, Biblioteka-filial no 5, Sankt-Peterburg 24304
TsBS Vyborgskogo raiona, Tsentralnaya raionnaya biblioteka im. A.S. Serafimovicha, Sankt-Peterburg 24305
TsBS Vyborgskogo raiona, Tsentralnaya raionnaya detskaya biblioteka, Sankt-Peterburg 24306
D. Tsenov Economic University, Svishtov 03483
Tsentăr po khimiya, Sofiya 03579
Tsentr detskogo i semeinogo chteniya im. A.S. Pushkina, Tver 24335
Tsentr dlya detei i yunoshestva NB RS(Ya), Yakutsk 24358
Tsentr Informpatent, Sankt-Peterburg 23910
Tsentr narodnoyi tvorchosti, Ivano-Frankivsk 28242
Tsentr nauchno-tekhnicheskoi informatsii grazhdanskoi aviatsii, Moskva 23489
Tsentr normativnoyi dokumentatsiyi z metalurgiyi, Dnipropetrovsk 28209
Tsentr zdorovya, Sankt-Peterburg 23911
Tsentralen săvet na nauchno-tekhnicheskite săyuzi v Bălgariya, Sofiya 03580
Tsentralna biblioteka dlya ditei, Dzhankoi 28559
Tsentralna meditsinska biblioteka, Sofiya 03476
Tsentralna miska biblioteka im. T. Shevchenko dlya ditei Kyeva, Kyiv 28689
Tsentralna miska publichna biblioteka im. M. Gorkogo, Kostyantynivka 28625
Tsentralna selskostopanska biblioteka, Sofiya 03581

Tsentralna spetsializovana biblioteka dlia slipykh im. M. Ostrovskoho, Kyiv 28387
Tsentralnaya aerologicheskaya observatoriya, Dolgoprudny 23025
Tsentralnaya biblioteka, Anna 24103
Tsentralnaya biblioteka, Boguchar 24115
Tsentralnaya biblioteka, Buturlinovka 24119
Tsentralnaya biblioteka, Kamenka 24138
Tsentralnaya biblioteka, Liski 24159
Tsentralnaya biblioteka, Novaya Usman 24186
Tsentralnaya biblioteka, Novovoronezh 24193
Tsentralnaya biblioteka, Ostrogozhsk 24197
Tsentralnaya biblioteka, Rossosh 24204
Tsentralnaya biblioteka, Semiluki 24316
Tsentralnaya biblioteka, Sovetski 24321
Tsentralnaya biblioteka, Talovaya 24328
Tsentralnaya biblioteka, Yoshkar-Ola 24364
Tsentralnaya biblioteka im. V. Kina, Borisoglebsk 24116
Tsentralnaya gorodskaya biblioteka, Balashov 24108
Tsentralnaya gorodskaya biblioteka, Engelsk 24128
Tsentralnaya gorodskaya biblioteka, Nikolaevsk-na-Amure 24183
Tsentralnaya Gorodskaya biblioteka, Rubtsovsk 24208
Tsentralnaya gorodskaya biblioteka, Rybinsk 24212
Tsentralnaya gorodskaya biblioteka, Shadrinsk 24317
Tsentralnaya gorodskaya biblioteka, Velikie Luki 24343
Tsentralnaya gorodskaya biblioteka im. A. Platonova, Voronezh 24354
Tsentralnaya gorodskaya biblioteka im. A.M. Gorkogo, Arzamas 24105
Tsentralnaya gorodskaya biblioteka im. A.S. Pushkina, Novocherkassk 24187
Tsentralnaya gorodskaya biblioteka im. A.S. Pushkina, Shakhty 24318
Tsentralnaya gorodskaya biblioteka im. K. Marksa, Novosibirsk 24192
Tsentralnaya gorodskaya biblioteka im. M. Gorkogo, Rostov-na-Donu 24207
Tsentralnaya gorodskaya biblioteka im. N.V. Gogolya, Novokuznetsk 24188
Tsentralnaya gorodskaya biblioteka im. V.M. Shukshina, Bisk 24112
Tsentralnaya gorodskaya biblioteka L. Tolstogo, Tula 24333
Tsentralnaya gorodskaya detskaya biblioteka im. A.P. Gaidara, Moskva 24176
Tsentralnaya gorodskaya detskaya biblioteka im. A.S. Pushkina, Sankt-Peterburg 24307
Tsentralnaya gorodskaya publichnaya biblioteka im. N.A. Nekrasova, Moskva 24177
Tsentralnaya gorodskaya publichnaya biblioteka im. V.V. Mayakovskogo, Sankt-Peterburg 24308
Tsentralnaya gorodskaya universalnaya nauchnaya biblioteka im. Mayakovskogo – Notno-muzykalny otdel, Sankt-Peterburg 23912
Tsentralnaya gorodskaya universalnaya nauchnaya biblioteka im. Mayakovskogo – Otdel literatury na inostrannykh yazykakh, Sankt-Peterburg 23913
Tsentralnaya gorodskaya universalnaya nauchnaya biblioteka im. Mayakovskogo – Otdel po obsluzhivaniyu yunoshestva, Sankt-Peterburg 24309
Tsentralnaya kinostudiya detskikh i yunosheskikh filmov im. M. Gorkogo, Moskva 23490
Tsentralnaya klinicheskaya bolnitsa im. N.A. Semashko Ministerstva putei soobshcheniya, Moskva 23491
Tsentralnaya nauchnaya biblioteka dalnevostochnogo otdeleniya Rossiskoi akademii nauk, Vladivostok 22192
Tsentralnaya nauchnaya biblioteka im. Y. Kolasa, Minsk 01785
Tsentralnaya nauchnaya biblioteka Uralskogo otdeleniya RAN (TsNB UrO RAN), Ekaterinburg 22129
Tsentralnaya nauchnaya meditsinskaya biblioteka (TsNMB), Moskva 23492
Tsentralnaya nauchnaya selskokhozyaistvennaya biblioteka, Moskva 23493
Tsentralnaya nauchno-tekhnicheskaya biblioteka avtomobilnogo i selskokhozyaistvennogo mashinostroeniya, Moskva 23494
Tsentralnaya nauchno-tekhnicheskaya biblioteka chernoi metallurgii, Moskva 23495
Tsentralnaya nauchno-tekhnicheskaya biblioteka legkoi promyshlennosti, Moskva 23496
Tsentralnaya nauchno-tekhnicheskaya biblioteka neftyanoi promyshlennosti, Moskva 23497
Tsentralnaya nauchno-tekhnicheskaya biblioteka pishchevoi promyshlenosti, Moskva 23498
Tsentralnaya nauchno-tekhnicheskaya biblioteka po stroitelstvu i arkhitekture, Moskva 23499
Tsentralnaya nauchno-tekhnicheskaya biblioteka rechnogo transporta, Moskva 23500
Tsentralnaya nauchno-tekhnicheskaya biblioteka stankostroitelnoi i instrumentalnoi promyshlennosti, Moskva 23501
Tsentralnaya nauchno-tekhnicheskaya biblioteka sudostroeniya, Sankt-Peterburg 23914
Tsentralnaya nauchno-tekhnicheskaya biblioteka svyazi, Moskva 23502
Tsentralnaya nauchno-tekhnicheskaya biblioteka tsvetnoi metallurgii, Moskva 23503
Tsentralnaya nauchno-tekhnicheskaya biblioteka ugolnoi promyshlennosti, Lyubertsy 23147
Tsentralnaya nauchno-tekhnicheskaya biblioteka ugolnoi promyshlennosti, Moskva 23504
Tsentralnaya politekhnicheskaya biblioteka (TsPB), Moskva 23505
Tsentralnaya raionnaya biblioteka, Bobrov 24114
Tsentralnaya raionnaya biblioteka, Ternovka 24331
Tsentralnaya sistema bibliotek g. Vladivostoka – Biblioteka semeinogo chteniya, Vladivostok 24348
Tsentralnaya sistema bibliotek g. Vladivostoka – Biblioteka-filial no 4, Vladivostok 24349
Tsentralnaya voenno-morskaya biblioteka (TsVMB) Ministerstva oborony, Sankt-Peterburg 23915
Tsentralno byuro naukovo-teknichnoyi informatsiyi Vugilnoyi Promyslovosti Ukrainy, Donetsk 28235
Tsentralno statistichesko upravlenie, Sofiya 03494
Tsentralno-Chernozemny mezhotraslevoi territorialny tsentr nauchno-tekhnicheskoi informatsii i propagandy, Voronezh 24078
Tsentralno-Chernozemny mezhotraslevoi territorialny tsentr nauchno-tekhnicheskoi informatsii i propagandy – Tambovski filial, Tambov 23991
Tsentralnoe byuro nauchno-tekhnicheskoi informatsii Ministerstva promyshlennogo stroitelstva, Moskva 22757
Tsentralnoe konstruktorskoe byuro 'Proektmashdetal', Moskva 22859
Tsentralnoe konstruktorskoe byuro s opytnym proizvodstvom, Minsk 02056
Tsentralny arkhiv respubliki Tatarstan, Kazan 23088
Tsentralny Botanicheski Sad, Minsk 02057
Tsentralny dom arkhitektora, Moskva 23506
Tsentralny dom aviatsii i kosmonavtiki, Moskva 23507
Tsentralny dom literatorov, Moskva 23508
Tsentralny dom rabotnikov iskusstv, Moskva 23509
Tsentralny dom zhurnalista, Moskva 23510
Tsentralny ekonomiko-matematicheski institut, Moskva 23511
Tsentralny girnichno-zbagachuvalny kombinat, Krivi Rig 28154
Tsentralny gosudarstvenny voenno-istoricheski arkhiv, Moskva 23512
Tsentralny institut nauchno-tekhnicheskoi informatsii i tekhniko-ekonomicheskikh issledovani po khimicheskomu i neftyanomu mashinostroeniyu, Moskva 23513
Tsentralny institut usovershenstvovaniya vrachei, Moskva 22394
Tsentralny instytut podvishchennya kvalifikatsiyi kerivnykh kadriv osvity, Kyiv 28023
Tsentralny kozhno-venerologicheski institut, Moskva 23514
Tsentralny muzei pochvovedeniya im. Dokuchaeva, Sankt-Peterburg 23916
Tsentralny muzei vooruzhennykh sil, Moskva 23515
Tsentralny nauchno-issledovatelski avtomobilny i avtomotorny institut, Moskva 23516
Tsentralny nauchno-issledovatelski dizelny institut, Sankt-Peterburg 23917
Tsentralny nauchno-issledovatelski eksperimentalny i proektny institut po selskomu stroitelstvu (TsNIIEPselstroi), Aprelevka 22986
Tsentralny nauchno-issledovatelski geologorazvedochny institut tsvetnykh i blagorodnykh metallov (TSNIGRI), Moskva 23517
Tsentralny nauchno-issledovatelski i proektno-konstruktorski institut mekhanizatsii i energetiki lesnoi promyshlennosti, Khimki 23100
Tsentralny nauchno-issledovatelski i proektno-konstruktorski institut morskogo flota (CNIIMF), Sankt-Peterburg 23918
Tsentralny nauchno-issledovatelski i proektny institut stroitelnykh metallokonstruktsi, Moskva 23518
Tsentralny nauchno-issledovatelski i proektny institut tipovogo i eksperimentalnogo proektirovaniya zdani torgovli obshchestvennogo pitaniya, bytovogo obsluzhivaniya i turistskikh kompleksov, Moskva 23519
Tsentralny nauchno-issledovatelski institut bytovogo obsluzhivaniya naseleniya, Moskva 23520
Tsentralny nauchno-issledovatelski institut chernoi metallurgii im. I.P. Bardina, Moskva 23521
Tsentralny nauchno-issledovatelski institut chyornoi metallurgii, Moskva 23522
Tsentralny nauchno-issledovatelski institut ekonomiki i ekspluatatsii vodnogo transporta, Moskva 22395
Tsentralny nauchno-issledovatelski institut ekspertizy trudosposobnosti i organizatsii truda invalidov, Moskva 23523
Tsentralny nauchno-issledovatelski institut Elektropribor, Sankt-Peterburg 23919
Tsentralny nauchno-issledovatelski institut epidemiologii, Moskva 23524
Tsentralny nauchno-issledovatelski institut geodezii, aerosyomki i kartografii, Moskva 23525
Tsentralny nauchno-issledovatelski institut kompleksnoi avtomatizatsii, Moskva 23526
Tsentralny nauchno-issledovatelski institut kurortologii i fizioterapii, Moskva 23527
Tsentralny nauchno-issledovatelski institut Mashdetal, Moskva 23528
Tsentralny nauchno-issledovatelski institut morskogo flota, Sankt-Peterburg 23920
Tsentralny nauchno-issledovatelski institut organizatsii i tekhniki upravleniya, Minsk 02058
Tsentralny nauchno-issledovatelski institut PROM ZERNOPROEKT, Moskva 23529
Tsentralny nauchno-issledovatelski institut promyshlennosti lubyanykh volokon, Moskva 23530
Tsentralny nauchno-issledovatelski institut promyshlennykh zdani i sooruzheni, Moskva 23531
Tsentralny nauchno-issledovatelski institut protezirovaniya i protezostroeniya, Moskva 23532
Tsentralny nauchno-issledovatelski institut sherstyanoi promyshlennosti, Moskva 23533
Tsentralny nauchno-issledovatelski institut shveinoi promyshlennosti, Moskva 23534
Tsentralny nauchno-issledovatelski institut tary i upakovki, Kaluga 23077
Tsentralny nauchno-issledovatelski institut tekhnologii mashinostroeniya, Moskva 23535
Tsentralny nauchno-issledovatelski institut travmatologii i ortopedii im. N.N. Priorova, Moskva 23536
Tsentralny nauchno-issledovatelski institut trikotazhnoi promyshlenosti, Moskva 23537
Tsentralny nauchno-issledovatelski institut tuberkulyoza, Moskva 23538
Tsentralny nauchno-issledovatelski khlopchatobumazhny institut, Moskva 23539
Tsentralny nauchno-tekhnicheskaya biblioteka stankostroeniya, Moskva 23540
Tsentralny otdel nauchno-tekhnicheskoi informatsii glavnogo arkhitekturno-planirovochnogo upravleniya (Mosproekt-2), Moskva 22758
Tsentralny remontno-mekhanicheski zavod 'MOSENERGO', Moskva 22860
Tsentralny voenno-morskoi muzei, Sankt-Peterburg 23921
Tsetkin Minsk City Library no 15, Minsk 02171
Tsinghua University, Beijing 05053
Tsiolkovsky State Museum of the History of Cosmonautics, Kaluga 23073
TsKB Aisberg, Sankt-Peterburg 23922
TsKB Mashinostroeniya, Sankt-Peterburg 23923
TsKB Morskoi tekhniki Rubin, Sankt-Peterburg 23924
TsNIKI toplivnoi apparatury, avtotraktornykh i statsionarnykh dvigatelei, Sankt-Peterburg 23925
Tsubouchi Hakase Kinen Engeki Hakubutsukan, Tokyo 17778
Tsubouchi Memorial Theatre Museum, Tokyo 17778
Tsuda College, Kodaira 16981
Tsurumai (Nagoya) Central Library, Nagoya 17851
Tsurumi Daigaku, Yokohama 17238
Tsurumi University, Yokohama 17238
Tsuruoka City Library, Tsuruoka 17862
Tsuruoka Shiritsu Toshokan, Tsuruoka 17862
Tsuyama Christian Library, Tsuyama 17810
Tsuyama Kirisutokyo Toshokan, Tsuyama 17810
Tualatin Public Library, Tualatin 41600
Tuberculosis Institute, Lviv 28414
Tuberculosis Institute, Toshkent 42163
Tucker Ellis & West LLP, Cleveland 35620
Tucson Family History Center, Tucson 37690
Tucson Museum of Art, Tucson 37691
Tudományos Ismeretterjesztő Társulat, Budapest 13462
Tufts Library, Weymouth 41870
Tufts University, Boston 30506
Tufts University, Medford 31782
– Edwin Ginn Library, Medford 31783
Tula Art Museum, Tula 24006
Tula Historical Architectural Literatur Musuem, Tula 24007
Tula Regional General Scientific Library, Tula 22180
Tula State University, Tula 22600
Tulane University, Covington 30931
Tulane University, New Orleans 32006
– Architecture Library, New Orleans 32007
– A. H. Clifford Mathematics Research Library, New Orleans 32008
– Latin American Library, New Orleans 32009
– Matas Medical Library, New Orleans 32010
– Maxwell Music Library, New Orleans 32011
– Monte M. Lemann Memorial Law Library, New Orleans 32012
– Music & Media Library, New Orleans 32013
– Turchin Library, New Orleans 32014
Tulare County Law Library, Visalia 34966
Tulare County Library – Visalia Headquarters Branch, Visalia 41693
Tulare County Office of Education, Visalia 34967
Tulare Public Library, Tulare 41613
Tulchin Regional Centralized Library System, Main Library, Tulchin 28855
Tulchinska raionna TsBS, Tsentralna biblioteka, Tulchin 28855
Tulkarm Community College, Tulkarm 14673
Tulsa City-County Library System – Broken Arrow Branch, Broken Arrow 38296
Tulsa City-County Library System Central Library, Tulsa 41615
Tulsa City-County Library System – Hardesty Regional Library, Tulsa 41616
Tulsa City-County Library System – Martin Regional Library, Tulsa 41617
Tulsa City-County Library System – Owasso Branch, Owasso 40671
Tulsa City-County Library System – Peggy V. Helmerich Library, Tulsa 41618
Tulsa City-County Library System – Rudisill Regional Library, Tulsa 41619
Tulsa City-County Library System – Zarrow Regional Library, Tulsa 41620
Tulsa Community College, Tulsa 33810
Tulsa County Law Library, Tulsa 34944
Tulskaya oblastnaya detskaya biblioteka, Tula 24334
Tulskaya Oblastnaya Universalnaya Nauchnaya Biblioteka, Tula 22180
Tulski Gosudarstvenny Universitet, Tula 22600
Tulski Oblastnoi Istoricheski Arkhitekturni Literaturni Muzei, Tula 24007
Tumba bibliotek, Tumba 26933
Tun Abdul Razak Research Centre (Tarrc), Hertford 29684
Tun Haji Mohd. Fuzad Stephen's Borneo Research Library, Kota Kinabalu 18577

Tun Mohd Fuad Stephens Borneo Research Library, Kota Kinabalu 18578
Tun Razak Library, Ipoh 18621
Tune bibliotek, Grålum 20336
Tunghai University, Taichung 27540
Tunku Abdul Rahman College, Kuala Lumpur 18535
Tunxis Community College, Farmington 33333
Tuolumne County Free Library, Sonora 41352
Tuolumne County Law Library, Sonora 34871
Tupolev Technical University, Kazan 22280
Turbine Blade Plant Joint-Stock Company, Sankt-Peterburg 22930
Turbinny zavod, Kaluga 22796
Turčianska knižnica, Martin 24783
Turgenev State Literary Museum, Oryol 23676
Turiec Library in Martin, Martin 24783
Türk Dil Kurumu, Kavaklýdere/Ankara 27900
Türk Tarih Kurumu, Ankara 27877
Türk Tıp Tarihi Kurumu, Istanbul 27897
Turkish Grand National Assembly Library and Documentation Centre, Ankara 27870
Turkish Historical Society, Ankara 27877
Turkish Medical History Society, Istanbul 27897
Turkish Religious Foundation Centre for Islamic Studies, Istanbul 27899
Turkish-Kazakh International University, Shimkent 17922
Türkiye Askeri Müzesi, Istanbul 27898
Türkiye Diyanet Vakfi – İslâm Araştırmaları Merkezi, Istanbul 27899
Turkmen Agricultural Institute, Ashkhabad 27906
Turkmen Pedagogical Institute, Chardzhou 27909
Turkmen State Medical Institute, Ashkhabad 27907
Turkmen State University, Ashkhabad 27908
Turku Art Museum Library, Turku 07552
Turku Court of Appeal Library, Turku 07406
Turku Public Library – Regional Library, Turku 07615
Turku School of Economics, Turku 07375
Turku University Library, Turku 07376
– Humanities Library, Turku 07377
– Library of Educational and Social Sciences, Turku 07378
– Mathematics and Science Library, Turku 07380
– Medical Library, Turku 07379
– Teachers' College Library, Rauma 07381
Turku University of Applied Sciences, Turku 07374
Turner Free Library, Randolph 40940
Turner Memorial Library, Mutare 42386
Turner Research Institute of Child Orthopaedics, Sankt-Peterburg 23818
Turov Village Children's Library, Turov 02242
Turovskaya gorposelkovaya detskaya bibliyateka, Turov 02242
Türr István Múzeum, Baja 13365
Turun ammattikorkeakoulun kirjasto, Turku 07374
Turun hovioikeuden kirjasto, Turku 07406
Turun kauppakorkeakoulu, Turku 07375
Turun kaupunginkirjasto, Turku 07615
Turun maakunta-arkiston käsikirjasto, Turku 07550
Turun museokeskukseksi, Turku 07551
Turun taidemuseon kirjasto, Turku 07552
Turun yliopiston kirjasto, Turku 07376
– Humanistinen tiedekuntakirjasto, Turku 07377
– Kasvatus- ja yhteiskuntatieteellinen kirjasto, Turku 07378
– Lääketieteellinen kirjasto, Turku 07379
– Matemaattis-luonnontieteellinen tiedekuntakirjasto, Turku 07380
– Rauman opeetajankoulutuslaitoksen kirjasto, Rauma 07381
Tuscaloosa Public Library, Tuscaloosa 41625
Tuscarawas County Law Library Association, New Philadelphia 34589
Tuscarawas County Public Library, New Philadelphia 40425
Tusculum College, Greeneville 31250
Tuskegee University, Tuskegee 32864
– Veterinary Medical Library, Tuskegee 32865
Tutaev Engine Building Joint-Stock Company, Tutaev 22964
TÜV NORD Service GmbH & Co. KG, Essen 11393
TÜV NORD Service GmbH & Co. KG, Hamburg 12019
TV Ontario Library, Toronto 04531
Tver Art Gallery, Tver 24014
Tver Regional Research Library, Tver 22181
Tver State Medical Academy, Tver 22602
Tver State University, Tver 22603
Tverskaya gosudarstvennaya meditsinskaya akademiya, Tver 22602
Tverskaya kartinnaya galeriya, Tver 24014
Tverskaya oblastnaya universalnaya nauchnaya biblioteka im. A.M. Gorkogo, Tver 22181
Tverskoi gosudarstvenny universitet, Tver 22603
Tweede Kamer der Staten-Generaal, Den Haag 19271
Twentieth Century Fox Film Corp, Los Angeles 35862
Twin Falls Public Library, Twin Falls 41627
Twin Lakes Library System – Mary Vinson Memorial Library – Headquarters, Milledgeville 40206
Twinsburg Public Library, Twinsburg 41628
Two Rivers Correctional Institute Library, Umatilla 34949
Tyachiv Regional Centralized Library System, Main Library, Tyachiv 28856
Tyachivska raionna TsBS, Tsentralna biblioteka, Tyachiv 28856
Työväen akatemia, Kauniainen 07518
Työväenliikkeen kirjasto – Arbetarrärelsen bibliotek, Helsinki 07503
Tyco Electronics, Harrisburg 35735
Tydings & Rosenberg LLP, Baltimore 35489
Tyler Cooper & Alcorn, LLP, New Haven 35939
Tyler Junior College, Tyler 33812
Tyler Public Library, Tyler 41630
Tyndale College and Seminary, Toronto 03887
Tyndale House, Cambridge 29627
Tyre Plant, Moskva 22853
Tyre Research Institute, Moskva 23433
Tyresö kommunbibliotek, Tyresö 26934
Tyumen Forestry and Timber Technology Professional School, Tyumen 22722
Tyumen Industrial Institute, Tyumen 22606
Tyumen State Agricultural Academy, Tyumen 22604
Tyumen State University, Tyumen 22605
Tyumenskaya gosudarstvennaya selskokhozyaistvennaya akademiya, Tyumen 22604
Tyumenskaya oblastnaya detskaya biblioteka, Tyumen 24336
Tyumenskaya oblastnaya nauchnaya biblioteka, Tyumen 22182
Tyumenskaya oblastnaya yunosheskaya biblioteka, Tyumen 24337
Tyumenski gosudarstvenny universitet, Tyumen 22605
Tyumenski industrialny institut, Tyumen 22606
Tyumenski lesotekhnicheski tekhnikum, Tyumen 22722
Tyumenski mezhotraslevoi territorialny tsentr nauchno-tekhnicheskoi informatsii i propagandy, Tyumen 24016
F.I. Tyutchev Bryansk Regional Research Library, Bryansk 22123
Tyvinski nauchno-issledovatelski institut yazyka, literatury i istori, Kyzyl 23139
Ube Industries Ltd, Ube 17499
Ube Industries Ltd, Information Dept of Chemical Technology, Ichihara 17360
Ube Kosan K. K. Chuo Kenkyujo Somubu Somuka Toshoshitsu, Ube 17499
Ube Kosan K. K. Kagabu Gijutsu Johobu, Ichihara 17360
Übersee-Museum, Bremen 11690
UCB Chemicals, Drogenbos 02638
Uchbovo-vibronichno obednannya Ukrainskoho tovaristva slipikh, Artemivsk 28180
Uchilishche tsirkovogo i estradnogo iskusstva im. M.N. Rumyantseva (Karandasha), Moskva 22683
Uchu Kagaku Kenkyusho, Sagamihara 17646
Uchu Kaihatsu Jigyodan, Tokyo 17326
UCL Eastman Dental Institute Library, London 29223
UCSF Medical Center at Mount Zion, San Francisco 37559
Udayana University, Denpasar 14249
Udmurt Republic Scientific Medical Library, Izhevsk 23064
Udmurt State University, Izhevsk 22269
Udmurtian Research Institute of History, Economics, Language and Literature, Izhevsk 23065
Udmurtskaya respublikanskaya nauchnaya meditsinskaya biblioteka, Izhevsk 23064
Udmurtski gosudarstvenny universitet, Izhevsk 22269
Udmurtski nauchno-issledovatelski institut istorii, ekonomii, yazyka, literatury, Izhevsk 23065
Ueno Gakuen University, Saitama 17097
Uervíciul de Stat de Arhivaal Republicii Moldova Biblioteka, Chișinău 18897
Ufa Research Centre of the Russian Academy of Sciences, Ufa 24030
Ufa State University of Aviation Technology, Ufa 22612
Ufimski gosudarstvenny aviatsionny tekhnicheski universitet, Ufa 22612
Ufimski nauchni tsentr RAN, Ufa 24030
Uganda Coffee Development Authority, Kampala 27928
Uganda Management Institute, Kampala 27920
Uganda Martyrs University, Kampala 27916
Uguccione Ranieri di Sorbello Foundation, Perugia 16415
Uinta County Library, Evanston 39002
Uintah County Library, Vernal 41673
UKK Institute, Tampere 07545
UKK-Instituutti, Tampere 07545
Ukmergės rajono savivaldybės Vlado Šlaito viešoji biblioteka, Ukmergė 18371
Ukrainian Academy of Foreign Trade, Kyiv 28024
Ukrainian Academy of Printing, Lviv 28040
Ukrainian Cities and Villages Complex Modernization, Heat and Gas Supply Research Institute Ltd, Kyiv 28155
Ukrainian Corporation of Timber Technology, Ivano-Frankivsk 28144
Ukrainian Engineering and Teacher Training Academy, Kharkiv 27999
Ukrainian Folklore Centre, Ivano-Frankivsk 28242
Ukrainian Free University, München 10458
Ukrainian Institute for Water Management Engineers, Rivne 28070
Ukrainian Institute of Plant Breeding and Genetics, Odesa 28446
Ukrainian Man-made Fibres Research and Experimental Institute, Kyiv 28392
Ukrainian Medical Association of North America, Chicago 36689
Ukrainian Medical Stomatological University, Poltava 28066
Ukrainian Ministry of Coal Industry Scientific Information Centre, Donetsk 28235
Ukrainian Ministry of Economics – Research Institute of Economics, Kyiv 28366
Ukrainian Ministry of Health – Centre of Hygiene, Kyiv 28311
Ukrainian Ministry of the Food Industry, Kyiv 28133
Ukrainian Museum of Canada Library, Saskatoon 04492
Ukrainian Museum-Archives Inc, Cleveland 36714
Ukrainian National Federation – Toronto Branch, Toronto 04532
Ukrainian National Historical Library, Kyiv 28304
Ukrainian Natural Gas Research Institute, Kharkiv 28283
Ukrainian Paper Research Institute, Kyiv 28395
Ukrainian Professional Schools Personnel Advanced Training Institute, Donetsk 27974
Ukrainian Public Library of Ivan Franko, Winnipeg 04561
Ukrainian Research and Development Institute of Plastics, Rubber and Artificial Leather Engineering, Kyiv 28396
Ukrainian Research and Production Corporation of Alcohol and Food Biotechnology, Kyiv 28160
Ukrainian Research Centre of Radiation Medicine, Kyiv 28391
Ukrainian Research Institute for Mountain Forestry, Ivano-Frankivsk 28243
Ukrainian Research Institute for Special Printing Processes, Kyiv 28390
Ukrainian Research Institute for the Printing Industry, Lviv 28424
Ukrainian Research Institute of Oils and Fats, Kharkiv 28284
Ukrainian Research Institute of Pedagogy, Kyiv 28374
Ukrainian Shipbuilding Technical University, Mykolaiv 28047
Ukrainian State Coal Industry Board, Donetsk 28212
Ukrainian State Institute of Heavy Electrical Engineering, Dnipropetrovsk 28210
Ukrainian State Institute of Mineral Resourses, Simferopol 28470
Ukrainian State Museum of Theatrical and Musical Arts and Cinematography, Kyiv 28389
Ukrainian State Research Institute for Carbochemistry, Kharkiv 28282
Ukrainian State University of Chemical Technologies, Dnipropetrovsk 27969
Ukrainian State University of Food Industry, Kyiv 28025
Ukrainian Trade Union Federation's International Centre of Culture and Art Library, Kyiv 28665
Ukrainian Transport University, Kyiv 28026
Ukrainian Urban Planning Research Institute, Kyiv 28388
Ukrainian Zonal Scientific and Research Design Institute of Civil Engineering, Kyiv 28398
Ukrainische Freie Universität, München 10458
Ukrainska akademiya drukarstva, Lviv 28040
Ukrainska derzhavna akademiya zvyazku im. Popova, Odesa 28061
Ukrainska inzhenerno-pedahohichna akademiya, Kharkiv 27999
Ukrainske Likarske Tovaristvo Pivnichnoi Ameriki, Chicago 36689
Ukrainsky derzhavny instytut proektuvannya mist, Kyiv 28388
Ukrainsky derzhavny morskoyi tekhnichi universytet, Mykolaiv 28047
Ukrainsky derzhavny universytet kharchovykh tekhnologi, Kyiv 28025
Ukrainsky gosudarstvenny muzei teatralnoho i muzykalnoho iskusstva i kinematografiyi, Kyiv 28389
Ukrainsky gosudarstvenny nauchno-issledovatelski uglekhimichesky instytut, Kharkiv 28282
Ukrainsky gosudarstvenny proektny instytut Tyazhpromelektroproekt, Dnipropetrovsk 28210
Ukrainsky instytut inzheneriv vodnoho gospodarstva, Rivne 28070
Ukrainsky nauchno issledovatelsky instytut po vidam pechati, Lviv 28424
Ukrainsky nauchno-issledovatelsky instytut po spetsialnym vidam pechati, Kyiv 28390
Ukrainsky nauchno-issledovatelsky instytut prirodnykh gazov (Ukrniigas), Kharkiv 28283
Ukrainsky nauchno-issledovatelsky instytut zemledeliya, Chabany 28187
Ukrainsky naukovi tsentr radiatseinoyi meditsini, Kyiv 28391
Ukrainsky naukovo-doslidny institut medychnoi reabilitatsii ta kurortologii, Odesa 28445
Ukrainsky naukovo-doslidny instytut girskoho lisivnichtva, Ivano-Frankivsk 28243
Ukrainsky naukovo-doslidny instytut masel i zhirov, Kharkiv 28284
Ukrainsky naukovo-doslidny instytut mekhanizatsiyi ta elektrifikatsiyi silskoho gospodarstva, Yakimivka 28482
Ukrainsky naukovo-doslidny instytut shtuchnykh volokon s doslidnim virobnitstvom, Kyiv 28392
Ukrainsky naukovo-doslidny instytut spirtu i biotekhnologiyi prodovolchykh produktiv 'Biospirtprod', Kyiv 28160
Ukrainsky selektsiyno-genetichny instytut, Odesa 28446
Ukrainsky transportny universytet, Kyiv 28026
Ukrainyan Research Institute of Dermatology and Venereology, Kharkiv 28285
Ukrainyan Research Institute of Experimental and Clinical Psychoneurology, Kharkiv 28286
Ukrainyan Research Institute of Traumatology and Othopaedics, Kyiv 28393
Ukrainyan Research Institute of Water Management and Ecological Problems, Kyiv 28394
Ukrainyan State Geological Research Institute, Lviv 28425
Ukranian Research Institute for Medical Rehabilitation and Resort Therapy, Odesa 28445
UkrNIIB, Kyiv 28395
UKRNIIPLASTMASH, Kyiv 28396
ULAKBIM Cahit Arf Bilgi Merkezi, Bilkent, Ankara 27878
ULAKBIM – National Academic Network and Cahit Information Center, Bilkent, Ankara 27878
Uljanovsk State Technical University, Ulyanovsk 22619
Ulkoasiainministeriön kirjasto, Helsinki 07403
Ulkopoliittinen instituutti, Helsinki 07504
Ullensaker bibliotek, Jessheim 20351
Ullensvang folkeboksamling, Kinsarvik 20352
Ullevaal University Hospital, Oslo 20287
Ulleval Universitetssykehus, Oslo 20287
Ulricehamns bibliotek, Ulricehamn 26935

Ulster American Folk Park, Omagh 29867
Ulster County Community College, Stone Ridge 33781
Ulster Folk and Transport Museum, Holywood 29685
Ulster Museum, Belfast 29609
Uludağ Üniversitesi, Bursa 27852
Uludağ University, Bursa 27852
Ulyanov Ulyanovsk State Teacher Training University, Ulyanovsk 22618
Ulyanovsk Regional Scientific Library, Ulyanovsk 22187
Ulyanovsk School of Automotive Mechanics, Ulyanovsk 22723
Ulyanovsk State University, Ulyanovsk 22620
Ulyanovskaya oblastnaya biblioteka dlya detei i yunoshestva, Ulyanovsk 24340
Ulyanovskaya oblastnaya nauchnaya biblioteka, Ulyanovsk 22187
Ulyanovski avtomekhanicheski tekhnikum, Ulyanovsk 22723
Ulyanovski gosudarstvenny pedagogicheski universitet im. I.N. Ulyanova, Ulyanovsk 22618
Ulyanovski gosudarstvenny tekhnicheski universitet (GU), Ulyanovsk 22619
Ulyanovski gosudarstvenny universitet, Ulyanovsk 22620
Ulyanovski mezhotraslevoi territorialny tsentr nauchno-tekhnicheskoi informatsii i propagandy, Ulyanovsk 24037
Uman Regional Centralized Library System, Main Library, Uman 28857
Uman State Agrarian Academy, Uman 28081
Umanska raionna TsBS, Tsentralna biblioteka, Uman 28857
Umeå stadsbibliotek, Umeå 26938
Umeå Universitet, Umeå 26444
– Medicinska biblioteket, Umeå 26445
Umea University, Umeå 26444
– Medical Library, Umeå 26445
Uměleckoprůmyslové museum v Praze, Praha 06517
Umjetnička galerija Bosne i Hercegovine, Sarajevo 02962
Umm Al-Qura University, Mecca 24392
– Taif Campus Library, Taif 24393
Umpqua Community College, Roseburg 33703
Umweltbibliothek Großhennersdorf e.V., Großhennersdorf 11945
Umweltbibliothek Leipzig, Leipzig 12206
Umweltbundesamt – Fachbibliothek Umwelt, Dessau-Roßlau 10854
Umweltbundesamt – Fachbibliothek Umwelt, Zweigstelle Bad Elster, Bad Elster 10779
Umweltbundesamt GmbH, Wien 01412
Umweltministerium und Ministerium für Ernährung und Ländlichen Raum, Stuttgart 11111
Uncle Remus Regional Library System, Madison 40011
Underberg & Kessler Law Library, Rochester 36186
UNESCO European Centre for Higher Education / Centre europén pour l'enseignement supérieur (CEPES), Bucureşti 22015
Unesco – IHE Institute for Water Education, Delft 19413
UNESCO – International Bureau of Education / Bureau International d'Education, Le Grand-Saconnex 27402
UNESCO Library, Paris 08631
UNESCO New Delhi Office, New Delhi 14171
UNESCO Regional Office, Islamabad 20488
UNESCO Regional Office for Education in Africa, Dakar 24434
UNESCO Regional Office for Education in Asia and the Pacific, Bangkok 27790
UNESCO-ICOM Information Centre, Paris 08632
UNESCO-Institute für Lebenslanges Lernen, Hamburg 12020
Ungarischses Medien- und Informationszentrum (UMIZ), Unterwart 01584
Unger Memorial Library, Plainview 40817
Ungku Omar Polytechnic, Ipoh 18527
União Cultural Brasil-Estados Unidos, São Paulo 03422
União de Bancos Portugueses (UBP), Lisboa 21750
União dos Escritores Angolanos, Luanda 00108
União Educacional de Brasília, Brasília 03286
Unibversity of the Cumber-lands/Cumberland College, Williamsburg 33067
UNICEF Regional Office for South Asia, Kathmandu 19041

Uniformed Services University of the Health Sciences, Bethesda 30420
Unilever Bestfoods Information Center, Englewood Cliffs 35694
Unilever N.V., Rotterdam 19336
Unilever Research Laboratory, Bedford 29567
Union Bank of India, Mumbai 13910
Union Carbide Corp, Danbury 35659
Union College, Barbourville 30353
Union College, Bunumbu 24565
Union College, Lincoln 31611
Union College, Schenectady 32624
Union Correctional Institution Library, Raiford 34722
Union County Carnegie Library, Union 41634
Union County College, Cranford 33264
Union County Law Library, Elizabeth 34179
Union County Library, New Albany 40393
Union County Library System, Lewisburg 39860
Union County Library System, Lewisburg 39861
Union County Public Library, Monroe 40267
Unión Industrial Argentina, Buenos Aires 00294
Union Internationale des Télécommunications / International Telecommunication Union, Genève 27377
Union interparlementaire, Le Grand Saconnex 27401
Union League Club Library, Chicago 36690
Union League Club Library, New York 37263
Union League of Philadelphia Library, Philadelphia 37366
Union Nationale pour la Promotion Pédagogique et Professionnelle dans l'Enseignement Catholique (UNAPEC), Paris 08633
Union of Architects, Volgograd 24067
Union of Artists, Moskva 23486
Union of British Columbia Indian Chiefs, Vancouver 04544
Union of Bulgarian Composers, Sofiya 03575
Union of Bulgarian Writers, Sofiya 03576
Union of Chemical Construction Projects, Tbilisi 09285
Union of Cinematographers, Moskva 23344
Union of Georgian Architects, Tbilisi 09286
Union of Journalists, Sankt-Peterburg 23904
Union of Libyan Authors, Writers and Artists, Tripoli 18264
Union of Polish Writers, Warszawa 21351
Union of Russian Painters, Sankt-Peterburg 23902
Union ot Theatre Workers – St. Petersburg Branch, Sankt-Peterburg 23903
Union Parish Library, Farmerville 39043
Union Postale Universelle, Bern 27334
Union Theological College, Belfast 29534
Union Theological Seminary, Dasmarinas 20762
Union Theological Seminary & Presbyterian School of Christian Education, Richmond 35347
Union Township Public Library, Union 41635
Union University, Jackson 31444
Uniondale Public Library, Uniondale 41638
Unione Cattolica Italiana Insegnanti Medi, Roma 16562
Unione Comunità Ebraiche Italiane, Roma 16563
Unione Industriale Biellese, Biella 16183
Unione Italiana delle Camere di Commercio, Industria, Artigianato e Agricoltura, Roma 16564
Uniontown Public Library, Uniontown 41640
Unisa Library, Pretoria 25044
Unisys Corporation, Roseville 36190
Unit of Financial and Economic Information, Praha 06412
Unitec, Auckland 19828
United Arab Emirates University, Abu Dhabi 28906
United Bank for Africa Ltd, Lagos 20005
United Central City Library of Tokmak, Tokmak 18163
United Central District Library of Kant, Kant 18156
United Chemical and Metallurgical Works, Ltd, Ustí nad Labem 06424
United Church of Canada, Toronto 04233
United Distillers p.l.c., Menstrie 29579
United Laboratories Inc, Manila 20768
United Methodist Church, Madison 35274
United Methodist Publishing House Library, Nashville 35935
United Nations Asian Institute for Economic Development and Planning, Bangkok 27791
United Nations Childrens Fund Library, New York 37264
United Nations Children's Fund (UNICEF), Bangkok 27792

United Nations Conference on Trade and Development (UNCTAD), Genève 27378
United Nations Dag Hammarskjold Library, New York 37265
United Nations Dag Hammarskjold Library – Legal Collection, New York 37266
United Nations Dag Hammarskjold Library – Statistical Collection, New York 37267
United Nations Depository Library, Seoul 18089
United Nations Development Programme (UNDP), Kingston 16885
United Nations ECLAC Library, México 18821
United Nations Economic Commission for Africa, Addis Ababa 07285
United Nations Economic Commission for Latin America and the Caribbean (ECLAC) and Latin American Institute for Economic and Social Planning, Santiago 04941
United Nations Environment Programme, Nairobi 18045
United Nations Information Center, Washington 37798
United Nations Information Centre, Colombo 26320
United Nations Information Centre, London 29530
United Nations Office at Geneva, Genève 27379
United Nations Population Fund (UNFPA), Kathmandu 19042
United Reformed Church History Society, Cambridge 29537
United Services Institution of India Library, New Delhi 14172
United States Agency for International Development, Cairo 07189
United States Army – Combined Arms Research Library, Fort Leavenworth 34244
United States Conference of Catholic Bishops Library, Washington 35408
United States Courts, Toledo 34928
United States Courts for the Ninth Circuit – Library, San Francisco 34821
United States Courts Library, Albuquerque 33908
United States Courts Library, Baltimore 33965
United States Courts Library, Charleston 34057
United States Courts Library, Fresno 34290
United States Courts Library, Houston 34369
United States Courts Library, Milwaukee 34539
United States Courts Library, Norfolk 34614
United States Courts Library, Phoenix 34675
United States Courts Library, Salt Lake City 34797
United States Courts Library, Shreveport 34860
United States Courts Library, Springfield 34887
United States Department of Justice, Arlington 36483
United States Department of the Interior, National Park Service, Quincy 37423
United States District Court Library, San Jose 34827
United States Embassy, Public Affairs Section, Yaoundé 03664
United States Enrichment Corp, Piketon 36123
United States Golf Association Museum & Archives, Far Hills 36844
United States Information Service, Accra 13057
United States Information Service, Kuala Lumpur 18607
United States Information Service, Manila 20775
United States Information Service, México 18866
United States Information Service, Monrovia 18234
United States Information Service, N'Djamena 04841
United States Information Service, Quito 06995
United States Information Service, Santafé de Bogotá 06039
United States Marine Corps, Camp Lejeune 34031
United States Marine Corps, Cherry Point 34064
United States Marine Corps, Parris Island 34649
United States Marine Corps, Twentynine Palms 34945
United States Marine Corps, Yuma 35096

United States Marine Corps – Library Services, Camp Pendleton 34032
United States Marine Corps – Seaside Square Library, Camp Pendleton 34033
United States Olympic Committee, Colorado Springs 36727
United States Railroad Retirement Board Library, Chicago 34070
United States Sentencing Commission Library, Washington 37799
United Technologies Corp, Windsor Locks 36429
United Theological College of the West Indies, Kingston 16881
United Theological Seminary, Trotwood 35397
United Theological Seminary of the Twin Cities, New Brighton 35299
Unity School Library, Unity Village 35401
Unity School of Christianity Library, Unity Village 35401
Uniunea arhitectilor din România, Bucureşti 22016
Uniunea artiştilor plastici din România, Bucureşti 22017
Uniunea compozitorilor şi muzicologilor din România, Bucureşti 22018
Universal Scientific Library, Blagoevgrad 03591
Universalmuseum Joanneum, Dokumentationsarchiv für österreichische Kunst des 19., 20. Jahrhunderts und der Gegenwartskunst der Neuen Galerie Graz, Graz 01522
Universalmuseum Joanneum GmbH, Landwirtschaftliche Sammlung, Stainz 01581
Universalnaya nauchnaya biblioteka Respubliki Altai, Gorno-Altaisk 22131
Universidad Adolfo Ibáñez, Peñalolen 04856
Universidad Adventista de las Antillas, Mayaguez 21860
Universidad Anáhuac, México 18731
Universidad Austral de Chile, Valdivia 04891
Universidad Autónoma Agraria Antonio Narro, Saltillo 18800
Universidad Autónoma Chapingo, Chapingo 18709
Universidad Autónoma de Baja California, Mexicali 18721
Universidad Autónoma de Ciudad Juárez, Ciudad Juarez 18710
Universidad Autónoma de Guadalajara, Zapopan 18810
Universidad Autónoma de Madrid, Madrid 25427
– Archivo Central, Madrid 25428
– Centro de Documentación Europea, Madrid 25429
– Escuela Politécnica Superior, Madrid 25430
– Facultad de Ciencias, Madrid 25431
– Facultad de Ciencias Económicas y Empresariales, Madrid 25432
– Facultad de Derecho, Madrid 25433
– Facultad de Filosofía y Letras, Madrid 25434
– Facultad de Formación de Profesorado y Educación, Madrid 25435
– Facultad de Medicina, Madrid 25436
– Facultad de Psicología, Madrid 25437
Universidad Autónoma de Manizales, Manizales 05941
Universidad Autónoma de Nuevo León, Monterrey 18797
Universidad Autónoma de San Luis Potosí, San Luis Potosí 18801
Universidad Autónoma de Santo Domingo, Santo Domingo 06957
Universidad Autónoma del Estado de Morelos, Cuernavaca 18714
Universidad Autónoma Latinoameri-cana (UNAULA), Medellín 05944
Universidad Autónoma Metropolitana, México 18732
Universidad 'Camilo Cienfuegos' de Matanzas, Matanzas 06273
Universidad Cardenal Herrera, Moncada 25507
Universidad Carlos III de Madrid, Madrid 25438
Universidad Católica 'Andrés Bello', Caracas 42206
Universidad Católica de Colombia, Santafé de Bogotá 05969
Universidad Católica de Córdoba, Córdoba 00147
Universidad Catolica de la Santisima Conception, Talcahuano 04889
Universidad Católica de Santiago de Guayaquil, Guayaquil 06978
Universidad Católica de Temuco, Temuco 04890

Universidad Católica de Valparaíso, Valparaíso 04892
– Escuela de Educación, Viña del Mar 04893
Universidad Católica del Norte, Antofagasta 04843
– Unidad de Biblioteca y Documentación – Sede Coquimbo, Coquimbo 04844
Universidad Católica del Uruguay 'Damaso Antonio Larrañaga', Montevideo 42012
Universidad Central de Venezuela, Caracas 42207
– Centro de Estudios del Desarrollo (CENDES), Caracas 42208
– Facultad de Agronomía, Maracay 42209
– Facultad de Arquitectura y Urbanismo, Caracas 42210
– Facultad de Ciencias Económicas y Sociales, Caracas 42211
– Facultad de Ciencias Jurídicas y Políticas, Caracas 42212
– Facultad de Economía, Caracas 42213
– Facultad de Farmacia, Caracas 42214
– Facultad de Ingeniería, Caracas 42215
– Instituto de Medicina Experimental, Caracas 42216
Universidad Central del Caribe, Bayamon 21849
Universidad Central del Ecuador, Quito 06982
Universidad Central del Este, San Pedro de Macorís 06954
Universidad Central "Marta Abreu" de Las Villas, Villa Clara 06277
Universidad Centroamericana, Managua 19889
Universidad Centroamericana José Simeón Cañas, San Salvador 07201
Universidad CES, Medellín 05945
Universidad Complutense de Madrid, Madrid 25439
– Biblioteca de Bellas Artes, Madrid 25440
– Biblioteca de Biológicas, Madrid 25441
– Biblioteca de Filología Clásica, Madrid 25442
– Biblioteca de Filología Hispánica y Románica, Madrid 25443
– Escuela de Relaciones Laborales, Madrid 25444
– Escuela Universitaria de Enfermería, Fisioterapia y Podología, Madrid 25445
– Escuela Universitaria de Estadística, Madrid 25446
– Escuela Universitaria de Optica, Madrid 25447
– Escuela Universitaria de Trabajo Social, Pozuelo de Alarcón (Madrid) 25448
– Facultad de Ciencias de la Documentación, Madrid 25449
– Facultad de Ciencias de la Información, Madrid 25450
– Facultad de Ciencias Económicas y Empresariales, Madrid 25451
– Facultad de Ciencias Físicas, Madrid 25452
– Facultad de Ciencias Geológicas, Madrid 25453
– Facultad de Ciencias Matemáticas, Madrid 25454
– Facultad de Ciencias Políticas y Sociología, Madrid 25455
– Facultad de Ciencias Químicas, Madrid 25456
– Facultad de Derecho, Madrid 25457
– Facultad de Education, Madrid 25458
– Facultad de Farmacia, Madrid 25459
– Facultad de Filología, Madrid 25460
– Facultad de Filosofía, Madrid 25461
– Facultad de Geografía e Historia, Madrid 25462
– Facultad de Informática, Madrid 25463
– Facultad de Medicina, Madrid 25464
– Facultad de Odontología, Madrid 25465
– Facultad de Psicología, Madrid 25466
– Facultad de Veterinaria, Madrid 25467
– Instituto de Criminología, Madrid 25468
Universidad Cooperativa de Colombia, Medellín 05946
Universidad de Alcalá, Alcalá de Henares 25261
– Biblioteca Central de Ciencias Experimentales, Alcalá de Henares 25262
– Biblioteca de Ciencias Económicas y Empresariales, Alcalá de Henares 25263
– Biblioteca de Derecho, Alcalá de Henares 25264
– Biblioteca de Filosofía y Letras, Alcalá de Henares 25265
– Biblioteca de Magisterio, Guadalajara 25266
– Biblioteca del IUEN (Antiguo Cenua), Alcalá de Henares 25267
Universidad de Alicante, Alicante 25268
– Biblioteca de Ciencias, Alicante 25269
– Biblioteca de Ciencias Económicas y Empresariales, Alicante 25270
– Biblioteca de Derecho, Alicante 25271
– Biblioteca de Educación, Alicante 25272
– Biblioteca de Filosofía y Letras / Trabajo Social, Alicante 25273
– Biblioteca Politécnica, Óptica y Enfermeria, Alicante 25274
Universidad de Almería, Almería 25275
– Escuela Universitaria de Ciencias de la Salud, Almería 25276
Universidad de Antioquia, Medellín 05947
– Centro de Investigaciones Económicas (CIE), Medellín 05948
– Escuela Interamericana de Bibliotecologío, Medellín 05949
– Facultad Nacional de Salud Pública, Medellín 05950
Universidad de Atacama, Copiapó 04854
Universidad de Buenos Aires, Buenos Aires 00121
– Colegio Nacional de Buenos Aires, Buenos Aires 00122
– Facultad de Agronomía, Buenos Aires 00123
– Facultad de Arquitectura, Diseño y Urbanismo, Buenos Aires 00124
– Facultad de Ciencias Económicas, Buenos Aires 00125
– Facultad de Ciencias Exactas y Naturales, Buenos Aires 00126
– Facultad de Ciencias Veterinarias, Buenos Aires 00127
– Facultad de Derecho y Ciencias Sociales, Buenos Aires 00128
– Facultad de Farmacia y Bioquímica, Buenos Aires 00129
– Facultad de Filosofía y Letras, Buenos Aires 00130
– Facultad de Filosofia y Letras – Dirección de Bibliotecas, Division Oanje, Buenos Aires 00131
– Facultad de Ingeniería, Buenos Aires 00132
– Facultad de Medicina, Buenos Aires 00133
– Facultad de Odontología, Buenos Aires 00134
– Facultad de Psicologia, Buenos Aires 00135
– Instituto de Ciencias de la Educación, Buenos Aires 00136
– Instituto de Filología Clásica, Buenos Aires 00137
– Instituto de Filología y Literaturas Hispánicas, Buenos Aires 00138
– Instituto de Filosofía, Buenos Aires 00139
– Instituto de Fisiología, Buenos Aires 00140
– Instituto de Geografía, Buenos Aires 00141
– Instituto de Literatura Argentina 'Ricardo Rojas', Buenos Aires 00142
– Instituto des Investigaciones Históricas Emilio Ravignani, Buenos Aires 00143
Universidad de Burgos, Burgos 25342
Universidad de Cádiz, Cádiz 25344
– Biblioteca de Ciencias de la Salud, Cádiz 25345
– Biblioteca de Ciencias Sociales y Juridicas, Cádiz 25346
– Biblioteca de E.S. Ingeniería, Casíz 25347
– Biblioteca de Humanidades, Cádiz 25348
– Biblioteca del Campus de Algeciras, Algeciras 25349
– Biblioteca del Campus de Jerez, Jerez de la Frontera 25350
– Biblioteca del Campus Río San Pedro, Puerto Real 25351
Universidad de Caldas, Manizales 05942
Universidad de Cantabria, Santander 25566
Universidad de Carabobo, Valencia 42232
Universidad de Castilla-La Mancha – Biblioteca Universitaria, Ciudad Real 25353
Universidad de Castilla-La Mancha – Campus de Albacete José Prat, Albacete 25260
Universidad de Castilla-La Mancha – Campus de Toledo, Toledo 25613
Universidad de Castilla-La Mancha – Colegio Universitario Cardenal Gil de Albornoz, Cuenca 25374
Universidad de Castilla-La Mancha – Escuela Universitaria del Profesorade de E.G.B., Toledo 25614
Universidad de Castilla-La Mancha – Escuela Universitaria del Profesorade de E.G.B., Real 25553
Universidad de Castilla-La Mancha – Escuela Universitaria del Profesorado de E.G.B. Fray Luis de León, Cuenca 25375
Universidad de Chile, Santiago
– Centro de Estudios Bizantinos y Neohelénicos, Santiago de Chile 04873
– Departamento de Ciencias Históricas, Santiago 04874
– Departamento de Física, Santiago 04875
– Departamento de Geografía, Santiago 04876
– Departamento de Ingeniería Civil, Santiago 04877
– Escuela de Salud Pública, Santiago 04878
– Facultad de Artes, Santiago 04879
– Facultad de Ciencias Económicas, Santiago 04880
– Facultad de Filosofía y Humanidades, Nuñoa 04881
– Facultad de Medicina, Santiago 04882
– Facultad de Medicina, Campus Oriente, Santiago 04883
– Facultad de Odontología, Santiago 04884
– Instituto de Estudios Internacionales, Santiago 04885
– Instituto de Investigaciones y Ensayes de Materiales, Santiago 04886
Universidad de Ciencias y Artes del Estado de Chiapas, Tuxtla Gutierrez 18807
Universidad de Colima, Colima 18712
– Biblioteca de Ciencias Sociales y Humanidades 'Francisco Velasco Curiel', Colima 18713
Universidad de Concepción, Concepción 04850
– Campus Chillán, Chillán 04851
– Unidad Académica Los Angeles, Los Angeles 04852
Universidad de Córdoba, Córdoba 25355
– Biblioteca de Ciencias del Trabajo, Córdoba 25356
– Biblioteca del Campus de Rabanales, Córdoba 25357
– Escuela Universitaria Politecnica, Córdoba 25358
– ETEA Biblioteca, Córdoba 25359
– Facultad de Arquitectura, Urbanismo y Diseño, Córdoba 25360
– Facultad de Ciencias, Córdoba 25361
– Facultad de Ciencias de la Educación, Córdoba 25362
– Facultad de Derecho y CC. Económicas y Empresariales, Córdoba 25363
– Facultad de Filosofía y Letras, Córdoba 25364
– Facultad de Medicina, Córdoba 25365
– Facultad de Veterinaria, Córdoba 25366
Universidad de Costa Rica, San José 06109
– Facultad de Ciencias Sociales, San José 06110
Universidad de Cuenca, Cuenca 06976
Universidad de Deusto, Bilbao 25339
– Biblioteca – Sede de San Sebastián, San Sebastián 25340
Universidad de Educación a Distancia, Mérida 25506
Universidad de El Salvador, San Salvador 07202
Universidad de Extremadura, Badajoz 25277
– Biblioteca Central de Badajoz, Badajoz 25278
– Biblioteca Central de Cáceres, Cáceres 25279
– Biblioteca Centro Universitario de Mérida, Mérida 25280
– Biblioteca Centro Universitario de Plasencia, Plasencia 25281
– Escuela de Ingenierías Agrarias, Badajoz 25282
– Escuela de Ingenierías Industriales, Badajoz 25283
– Escuela Universitaria de Enfermería y Terapia Ocupacional, Cáceres 25284
– Facultad de Biblioteconomía, Badajoz 25285
– Facultad de Ciencias del Deporte, Cáceres 25286
– Facultad de Derecho, Cáceres 25287
– Facultad de Educación, Badajoz 25288
– Facultad de Estudios Empresariales y Turismo, Cáceres 25289
– Facultad de Medicina, Badajoz 25290
– Facultad de Veterinaria, Cáceres 25291
Universidad de Granada, Granada 25379
– Biblioteca de Medicina y CC de la Salud, Granada 25380
– Biblioteca del Edificio de San Jerónimo, Granada 25381
– Biblioteca Politécnico, Granada 25382
– Campus Universitario de Ceuta, Ceuta 25383
– Campus Universitario de Melilla, Melilla 25384
– Colegio Máximo, Granada 25385
– Escuela Técnica Superior de Arquitectura, Granada 25386
– Escuela Técnica Superior de Ingeniería Informática y Telecomunicación, Granada 25387
– Facultad de Bellas Artes, Granada 25388
– Facultad de Ciencias, Granada 25389
– Facultad de Ciencias de la Educación, Granada 25390
– Facultad de Ciencias Economicas y Empresariales, Granada 25391
– Facultad de Ciencias Políticas y Sociología, Granada 25392
– Facultad de Derecho, Granada 25393
– Facultad de Farmacia, Granada 25394
– Facultad de Filosofía y Letras, Granada 25395
– Facultad de la Actividad Física y el Deporte, Granada 25396
– Facultad de Psicología, Granada 25397
– Facultad de Traducción e Interpretación, Granada 25398
– Hospital Real, Granada 25399
Universidad de Guadalajara, Guadalajara 18715
Universidad de Guanajuato, Guanajuato 18716
Universidad de Holguin, Holguín 06268
Universidad de Huelva, Huelva 25400
– Escuela Politécnica Superior, Palos de la Frontera 25401
– Facultad de Ciencias Empresariales y Jurídicas, Huelva 25402
Universidad de Jaén, Jaén 25403
Universidad de La Habana, La Habana 06271
– Dirección de Información Científica Técnica, La Habana 06272
Universidad de La Laguna, La Laguna 25404
Universidad de la República, Montevideo
– Escuela Universitaria de Bibliotecología y Ciencias Afines, Montevideo 42013
– Facultad de Agronimía, Montevideo 42014
– Facultad de Arquitectura, Montevideo 42015
– Facultad de Derecho y Ciencias Sociales, Montevideo 42016
– Facultad de Ingeniería, Montevideo 42017
– Facultad de Química, Montevideo 42018
– Facultad de Veterinaria, Montevideo 42019
Universidad de la Rioja, Logroño 25417
Universidad de La Sabana, Chía 05939
Universidad de La Salle, Santafé de Bogotá 05970
Universidad de La Serena, La Serena 04855
Universidad de las Américas, Puebla 18799
Universidad de Las Palmas de Gran Canaria, Las Palmas 25531
– Biblioteca de Arquitectura, Las Palmas 25532
– Biblioteca de Ciencias Básicas, Las Palmas 25533
– Biblioteca de Ciencias de la Salud, Las Palmas 25534
– Biblioteca de Ciencias Económicas y Empresariales, Las Palmas 25535
– Biblioteca de Ciencias Jurídicas, Las Palmas 25536
– Biblioteca de Educación Física, Las Palmas 25537
– Biblioteca de Electrónica y Telecomunicación, Las Palmas 25538
– Biblioteca de Formación del Profesorado, Las Palmas 25539
– Biblioteca de Humanidades, Las Palmas 25540
– Biblioteca de Informática y Matemáticas, Las Palmas 25541
– Biblioteca de Ingeniería, Las Palmas 25542
– Biblioteca de Veterinaria, Arucas 25543
Universidad de Léon, León 25412
Universidad de Lima, Lima 20624
Universidad de Los Andes, Mérida 42227
– Instituto de Geografía y Conservación de Recoursos Naturales, Mérida 42228
– Servicios Bibliotecarios de Ciencias Jurídicas y Políticas, Mérida 42229
– Servicios Bibliotecarios de Farmacia, Mérida 42230
– Servicios Bibliotecarios de Humanidades y Educación, Mérida 42231
Universidad de Los Andes (UNIANDES), Santafé de Bogotá 05971
Universidad de los Trabajadores de América Latina, Los Teques 42219
Universidad de Málaga, Málaga 25492
– Biblioteca de Ciencias, Málaga 25493

– Biblioteca de Ciencias de la Comunicación, Málaga 25494
– Biblioteca de Ciencias de la Educación y Psicología, Málaga 25495
– Biblioteca de Ciencias de la Salud, Málaga 25496
– Biblioteca de Ciencias Económicas y Empresariales, Málaga 25497
– Biblioteca de Derecho Alejandro Rodríguez Carrión, Málaga 25498
– Biblioteca de Estudios Sociales y de Comercio, Málaga 25499
– Biblioteca de Humanidades José Mercado Ureña, Málaga 25500
– Biblioteca de Industriales y Politécnica, Málaga 25501
– Biblioteca de Informática y Telecomunicación, Málaga 25502
– Biblioteca de Medicina, Málaga 25503
– Biblioteca del Turismo, Málaga 25504
Universidad de Manizales, Manizales 05943
Universidad de Monterrey, San Pedro Garza García 18803
Universidad de Murcia, Murcia 25512
– Biblioteca de Economía y Empresa, Murcia 25513
– Biblioteca de Educación, Murcia 25514
– Biblioteca de Medicina, Murcia 25515
– Biblioteca de Química, Murcia 25516
– Biblioteca Luis Vives (Ciencias del Trabajo, Filosofía y Trabajo Social), Murcia 25517
Universidad de Navarra, Pamplona 25544
– Biblioteca de Ciencias, Pamplona 25545
– Biblioteca de la Clínica Universitaria, Pamplona 25546
– Biblioteca de las Facultades Eclesiásticas, Pamplona 25547
– Escuela Técnica Superior de Ingenieros, San Sebastián 25548
– IESE Business School, Barcelona 25549
Universidad de Oriente, Cumaná 42218
Universidad de Oriente, Santiago de Cuba 06275
– Dirección de Información Científica Técnica, Santiago de Cuba 06276
Universidad de Oviedo, Oviedo 25518
– Biblioteca de Ciencias de la Educación, Oviedo 25519
– Biblioteca de Jovellanos, Gijón 25520
– Biblioteca de Psicología, Oviedo 25521
– Biblioteca de Química, Oviedo 25522
– Biblioteca de Tecnología y Empresa, Gijón 25523
– Escuela Técnica Superior de Ingenieros de Minas, Oviedo 25524
Universidad de Pamplona, Pamplona 05955
Universidad de Panamá, Panamá 20535
Universidad de Pinar del Río 'Hnos. Saíz Montes de Oca', Pinar del Río 06274
Universidad de Playa Ancha de Ciencias de la Educación, Valparaíso 04894
Universidad de Salamanca, Salamanca 25555
– Biblioteca Claudio Rodríguez del Campo de Zamora, Zamora 25556
– Biblioteca de "Campus Canalejas", Salamanca 25557
– BIblioteca Francisco de Vitoria, Salamanca 25558
– Campus de Ávila – Escuela Universitaria de Educación de Ávila, Ávila 25559
– Facultad de Bellas Artes / Psicología, Salamanca 25560
– Facultad de Filología, Salamanca 25561
– Facultad de Geografía e Historia, Salamanca 25562
– Facultad de Medicina, Salamanca 25563
Universidad de San Buenaventura, Calí 05934
– Seccional Medellín, Medellín 05935
Universidad de San Carlos de Guatemala, Guatemala City 13173
– Facultad de Ingeniería, Guatemala City 13174
Universidad de Santiago de Chile, Santiago 04887
Universidad de Sevilla, Sevilla 25586
– Biblioteca Area Politécnica, Sevilla 25587
– Biblioteca de Arquitectura, Sevilla 25588
– Biblioteca de Bellas Artes, Sevilla 25589
– Biblioteca de Comunicación, Isla de la Cartuja 25590
– Biblioteca de Derecho y Ciencias del Trabajo, Sevilla 25591
– Biblioteca de Económicas y Empresariales, Sevilla 25592
– Biblioteca de Educación, Sevilla 25593
– Biblioteca de Empresariales, Sevilla 25594
– Biblioteca de Farmacia, Sevilla 25595
– Biblioteca de Humanidades, Sevilla 25596
– Biblioteca de Informática, Sevilla 25597
– Biblioteca de Ingeniería Técnica Agrícola, Sevilla 25598
– Biblioteca de Ingenieros, Sevilla 25599
– Biblioteca de Matemáticas, Sevilla 25600
– Biblioteca de Psicología y Filosofía, Sevilla 25601
– Biblioteca de Química, Sevilla 25602
– Facultad de Física, Sevilla 25603
Universidad de Sucre, Sincelejo 05987
Universidad de Tarapacá de Arica, Arica 04845
– Biblioteca de Historia y Geografía, Arica 04846
– Campus Saucache, Arica 04847
– Campus Velásquez, Arica 04848
– Facultad de Administración y Economía, Arica 04849
Universidad de Valladolid, Valladolid 25625
– Área de Ciencias de la Salud, Valladolid 25626
– Biblioteca de Ciencias, Valladolid 25627
– Biblioteca Histórica de Santa Cruz, Valladolid 25628
– Campus de Soria, Soria 25629
– Campus La Yutera, Palencia 25630
– Centro de Documentación Europea, Valladolid 25631
– Escuela Técnica Superior de Arquitectura, Valladolid 25632
– Escuela Técnica Superior de Telecomunicaciones / Informática, Valladolid 25633
– Escuela Universitaria de Estudios Empresariales, Valladolid 25634
– Escuela Universitaria de Magisterio, Segovia 25635
– Facultad de Ciencias Económicas y Empresariales, Valladolid 25636
– Facultad de Ciencias Sociales, Jurídicas y de la Comunicación, Segovia 25637
– Facultad de Derecho, Valladolid 25638
– Facultad de Filosofía y Letras, Valladolid 25639
– INEA, Valladolid 25640
Universidad de Valparaíso, Valparaíso 04895
– Escuela de Derecho, Valparaíso 04896
– Escuela de Ingeniería Comercial, Valparaíso 04897
– Facultad de Arquitectura, Valparaíso 04898
– Facultad de Medicina, Valparaíso 04899
– Instituto de Estudios Humanísticos, Viña del Mar 04900
Universidad de Vigo, Vigo 25642
Universidad de Zaragoza, Zaragoza 25644
– Biblioteca Biomédica, Facultad de Medicina, Zaragoza 25645
– Biblioteca de Humanidades 'Maria Moliner', Zaragoza 25646
– Biblioteca del Campus de Teruel, Teruel 25647
– Biblioteca Hypatia de Alejandría, Zaragoza 25648
– Escuela Universitaria de Estudios Empresariales, Zaragoza 25649
– Escuela Universitaria de Estudios Empresariales de Huesca, Huesca 25650
– Escuela Universitaria de Estudios Sociales, Zaragoza 25651
– Facultad de Ciencias, Zaragoza 25652
– Facultad de Ciencias de la Salud y del Deporte, Huesca 25653
– Facultad de Ciencias Económicas y Empresariales, Zaragoza 25654
– Facultad de Ciencias Humanidas y de la Educación, Huesca 25655
– Facultad de Derecho, Zaragoza 25656
– Facultad de Educación, Zaragoza 25657
– Facultad de Veterinaria, Zaragoza 25658
– Instituto de Ciencias de la Educación, Zaragoza 25659
Universidad del Bío-Bío, Concepción 04853
Universidad del Cauca, Popayán 05957
– Biblioteca Ciencias de la Salud, Popayán 05958
– Biblioteca El Carmen, Popayán 05959
Universidad del Museo Social Argentino, Buenos Aires 00144
Universidad del Pacífico, Lima 20625
Universidad del País Vasco / Euskal Herriko Unibertsitatea, Leioa 25405
– Campus de Alava, Vitoria-Gasteiz 25406
– Campus de Bizkaia, Bilbao 25407
– Campus de Gipuzkoa, San Sebastian 25408
– Escuela Técnica Superior de Ingeniería, Bilbao 25409
– Escuela Universitaria de Magisterio, Donostia-San Sebastián 25410
– Facultad de Bellas Artes, Leioa 25411
Universidad del Quindío, Armenia 05930
Universidad del Sagrado Corazón, Santurce 21888
Universidad del Sagrado Corazón, Santurce 21889
Universidad del Salvador, Buenos Aires 00145
– Facultades de Filosofía y Teología S.I., San Miguel 00146
Universidad del Tolima, Ibagué 05940
Universidad del Trabajo del Uruguay, Montevideo 42020
– Instituto Superior de Electrotécnica, Electrónica y Computación, Montevideo 42021
Universidad del Turabo, Gurabo 21856
Universidad del Valle, Calí 05936
– Biblioteca Médica, Calí 05937
– Instituto de Educación y Pedagogía, Calí 05938
Universidad del Valle de Guatemala, Guatemala City 13175
Universidad del Valle de México, México 18733
Universidad del Zulia, Maracaibo 42220
– Facultad de Agronomía, Maracaibo 42221
– Facultad de Arquitectura, Maracaibo 42222
– Facultad de Ciencias Económicas y Sociales, Maracaibo 42223
– Facultad de Derecho, Maracaibo 42224
– Facultad de Humanidades y Educación, Maracaibo 42225
– Facultad de Ingeniería, Maracaibo 42226
Universidad ESAN, Lima 20660
Universidad Escuela de Administración y Finanzas y Tecnologías (EAFIT), Medellín 05951
Universidad Estatal a Distancia, San José 06111
Universidad Europea de Madrid, Villaviciosa de Odon 25643
Universidad Externado de Colombia, Santafé de Bogotá 05972
Universidad Femenina del Sagrado Corazón, Lima 20626
Universidad Francisco Marroquín, Guatemala City 13176
Universidad Iberoamericana, México 18734
Universidad Incca de Colombia (UNINCCA), Santafé de Bogotá 05973
Universidad Industrial de Santander, Bucaramanga 05932
– Facultad de Medicina, Bucaramanga 05933
Universidad Interamericana de Puerto Rico, Aguadilla 21845
Universidad Internacional Menéndez Pelayo, Madrid 25469
Universidad 'La Gran Colombia', Santafé de Bogotá 05974
– Seccional Armenia, Armenia 05975
Universidad Laboral de Eibar, Eibar 25376
Universidad Laboral 'Francisco Franco', Tarragona 25605
Universidad Laboral 'José Antonio Girón', Girón 25377
Universidad Latinoamericana de Ciencia y Tecnologia, San José 06112
Universidad Libre de Colombia, Santafé de Bogotá 05976
– Seccional Barranquilla, Barranquilla 05977
– Seccional Pereira, Pereira 05978
Universidad Mariano Gálvez de Guatemala, Guatemala City 13177
Universidad Mayor de San Andrés, La Paz 02918
– Facultad de Humanidades y Ciencias de la Educación, La Paz 02919
– Facultad de Medicina, La Paz 02920
Universidad Mayor de San Simón, Sucre 02925
Universidad Metropolitana de Ciencias de la Educación, Santiago 04888
Universidad Michoacana de San Nicolás de Hidalgo, Morelia 18798
Universidad Nacional Agraria La Molina, Lima 20627
Universidad Nacional Agraria (UNA), Managua 19890
Universidad Nacional Autónoma, Heredia 06108
Universidad Nacional Autónoma de Honduras, Tegucigalpa 13209
Universidad Nacional Autónoma de México, México 18735
– Centro Coordinador y Difusor de Estudios Latinoamericanos, México 18736
– Centro de Enseñanza de Lenguas Extranjeras, México 18737
– Centro de Instrumentos, México 18738
– Centro Universitario de Investigaciones Bibliotecológicas, México 18739
– Colegio de Ciencias y Humanidades, México 18740
– Colegio de Ciencias y Humanidades, Plantel Azcapotzalco, México 18741
– Colegio de Ciencias y Humanidades, Plantel Naucalpan, Naucalpan 18742
– Colegio de Ciencias y Humanidades, Plantel Oriente, México 18743
– Colegio de Ciencias y Humanidades, Plantel Sur, México 18744
– Colegio de Ciencias y Humanidades, Plantel Vallejo, México 18745
– División de Estudios de Posgrado, México 18746
– Escuela Nacional de Artes Plásticas, México 18747
– Escuela Nacional de Enfermería y Obstetricia, México 18748
– Escuela Nacional de Estudios Profesionales, Acatlan, Naucalpan 18749
– Escuela Nacional de Estudios Profesionales, Aragón, San Juan de Aragón 18750
– Escuela Nacional de Estudios Profesionales, Iztacala, Tlanepantla 18751
– Escuela Nacional de Música, México 18752
– Escuela Nacional de Trabajo Social, México 18753
– Escuela Nacional Preparatoria – Dirección General, México 18754
– Escuela Nacional Preparatoria Plantel 1, Gabino Barreda, México 18755
– Escuela Nacional Preparatoria Plantel 2, Erasmo Castellanos Quinto, México 18756
– Escuela Nacional Preparatoria Plantel 3, Justo Sierra, México 18757
– Escuela Nacional Preparatoria Plantel 4, Vidal Castañeda y Najera, México 18758
– Escuela Nacional Preparatoria Plantel 5, José Vasconcelos, México 18759
– Escuela Nacional Preparatoria Plantel 6, Antonio Caso, México 18760
– Escuela Nacional Preparatoria Plantel 7, Ezequiel A. Chavez, México 18761
– Escuela Nacional Preparatoria Plantel 8, Miguel E. Schultz, México 18762
– Escuela Nacional Preparatoria Plantel 9, Pedro de Alba, México 18763
– Facultad de Arquitectura, México 18764
– Facultad de Ciencias, México 18765
– Facultad de Ciencias Políticas y Sociales, México 18766
– Facultad de Contaduría y Administración, México 18767
– Facultad de Derecho, México 18768
– Facultad de Economía, México 18769
– Facultad de Estudios Superiores Cuautitlán, Cuautitlán Izcalli 18770
– Facultad de Estudios Superiores Cuautitlan Campo Cuatro, Cuautitlán Izcalli 18771
– Facultad de Estudios Superiores Zaragoza, México 18772
– Facultad de Filosofía y Letras, México 18773
– Facultad de Ingeniería, México 18774
– Facultad de Medicina, México 18775
– Facultad de Medicina, División de Investigación, México 18776
– Facultad de Medicina Veterinaria y Zootécnia, México 18777
– Facultad de Odontología, México 18778
– Facultad de Psicología, México 18779
– Facultad de Química, México 18780
– Instituto de Física, México 18781
– Instituto de Geografía, México 18782
– Instituto de Geología, México 18783
– Instituto de Investigaciones Antropológicas, México 18784
– Instituto de Investigaciones Biomédicas, México 18785
– Instituto de Investigaciones Económicas, México 18786
– Instituto de Investigaciones en Matemáticas Aplicadas y en Sistemas, México 18787
– Instituto de Investigaciones Estéticas, México 18788
– Instituto de Investigaciones Filológicas, México 18789
– Instituto de Investigaciones Filosóficas, México 18790
– Instituto de Investigaciones Jurídicas, México 18791
– Instituto de Investigaciones Sociales, México 18792
– Instituto de Matemáticas, México 18793

Universidad Nacional Autónoma de Nicaragua – UNAN-Léon, León 19888
Universidad Nacional de Asunción, San Lorenzo 20603
– Facultad de Ciencias Económicas, San Lorenzo 20604
– Facultad de Ciencias Médicas, Asunción 20605
– Facultad de Filosofía, Asunción 20606
– Facultad de Ingeniería Agronómica, San Lorenzo 20607
Universidad Nacional de Colombia, Santafé de Bogotá 05979
– Biblioteca de Ciencias Agropecuarias y Ciencias, Medellín 05980
– Biblioteca Seccional Manizales, Manizales 05981
– Conservatorio Nacional de Música, Santafé de Bogotá 05982
– Facultad de Agronomía, Santafé de Bogotá 05983
– Facultad de Minas, Medellín 05984
– Facultades de Arquitectura y de Ciencias Humanas, Medellín 05985
Universidad Nacional de Córdoba, Córdoba 00148
– Facultad de Arquitectura Urbanismo y Diseño, Córdoba 00149
– Facultad de Ciencias Agropecuarias, Córdoba 00150
– Facultad de Ciencias Económicas, Córdoba 00151
– Facultad de Ciencias Exactas, Físicas y Naturales, Córdoba 00152
– Facultad de Ciencias Médicas, Córdoba 00153
– Facultad de Ciencias Químicas, Córdoba 00154
– Facultad de Derecho y Ciencias Sociales, Córdoba 00155
– Facultad de Filosofía y Humanidades, Córdoba 00156
– Facultad de Lenguas, Córdoba 00157
– Facultad de Matemática, Astronomía y Física, Córdoba 00158
Universidad Nacional de Cuyo, Mendoza 00178
– Facultad de Ciencias Agrarias, Mendoza 00179
– Facultad de Ciencias Económicas, Mendoza 00180
– Facultad de Ciencias Médicas, Mendoza 00181
– Facultad de Ingeniería, Mendoza 00182
Universidad Nacional de Eduación a Distancia, Madrid 25470
Universidad Nacional de Educación a Distancia, Melilla 25505
Universidad Nacional de Educación a Distancia, Teruel 25612
Universidad Nacional de Educación a Distancia de Pontevedra, Pontevedra 25551
Universidad Nacional de Educación 'Enrique Guzman y Valle', Lima 20628
Universidad Nacional de Ingeniería, Lima 20629
– Facultad de Arquitectura, Urbanismo y Artes, Lima 20630
– Facultad de Ciencias, Lima 20631
– Facultad de Ingeniería Económica y Ciencias Sociales, Lima 20632
– Facultad de Ingeniería Mecánica, Lima 20633
Universidad Nacional de la Pampa, Santa Rosa 00203
Universidad Nacional de La Plata, La Plata 00161
– Biblioteca de Humanidades y Ciencias de la Educacion, La Plata 00162
– Colegio Liceo Victor Mercante, La Plata 00163
– Colegio Nacional 'Rafael Hernández', La Plata 00164
– Departamento de Letras, La Plata 00165
– Escuela Graduada 'Joaquin V. González', La Plata 00166
– Facultad de Ciencias Agrarias y Forestales, La Plata 00167
– Facultad de Ciencias Económicas, La Plata 00168
– Facultad de Ciencias Exactas, La Plata 00169
– Facultad de Ciencias Jurídicas y Sociales, La Plata 00170
– Facultad de Ciencias Médicas, La Plata 00171
– Facultad de Ciencias Naturales y Museo, La Plata 00172
– Facultad de Ciencias Veterinarias, La Plata 00173
– Facultad de Ingeniería, La Plata 00174
– Observatorio Astronómico, La Plata 00175

Universidad Nacional de la Rioja, La Rioja 00176
Universidad Nacional de Mar del Plata, Mar del Plata 00177
Universidad Nacional de Rosario, Rosario 00183
– Facultad de Ciencias Económicas y Estadística, Rosario 00184
– Facultad de Ciencias Ingeniería, Rosario 00185
– Facultad de Ciencias Médicas, Rosario 00186
– Facultad de Humanidades y Artes, Rosario 00187
Universidad Nacional de San Agustín, Arequipa 20613
– Facultad de Medicina, Arequipa 20614
Universidad Nacional de San Cristóbal de Huamanga, Ayacucho 20615
Universidad Nacional de San Luis, San Luis 00188
Universidad Nacional de Tucumán, San Miguel de Tucumán 00189
– Facultad Bioquimica, Quimica y Farmacia, San Miguel de Tucumán 00190
– Facultad de Agronomia y Zootecnia, San Miguel de Tucumán 00191
– Facultad de Arquitectura y Urbanismo, San Miguel de Tucumán 00192
– Facultad de Artes, San Miguel de Tucumán 00193
– Facultad de Ciencias Económicas, San Miguel de Tucumán 00194
– Facultad de Ciencias Exactas y Tecnología, San Miguel de Tucumán 00195
– Facultad de Derecho y Ciencias Sociales, San Miguel de Tucumán 00196
– Facultad de Filosofía y Letras, San Miguel de Tucumán 00197
Universidad Nacional del Altiplano, Puno 20640
Universidad Nacional del Litoral, Santa Fé 00198
– Escuela Industrial Superior "Gral. José de San Martin", Santa Fé 00199
– Facultad de Ciencias Económicas, Santa Fé 00200
– Facultad de Ciencias Jurídicas y Sociales, Santa Fé 00201
– Facultad de Ingeniería Química "Dr. Ezzio Emiliani", Santa Fé 00202
Universidad Nacional del Sur, Bahía Blanca 00113
– Biblioteca de Ciencias Agrarias, Bahía Blanca 00114
– Biblioteca de Economía, Bahía Blanca 00115
– Instituto de Matemática, Bahía Blanca 00116
Universidad Nacional Experimental Politécnica 'Antonio José de Sucre', Barquisimeto 42205
Universidad Nacional Mayor de San Marcos de Lima, Lima 20634
– Biblioteca de Ciencias, Lima 20635
– Facultad de Educación, Lima 20636
– Facultad de Letras y Ciencias Humanas, Lima 20637
– Facultad de Medicina, Lima 20638
– Oficina General de Editorial, Imprenta, Central y Libreria, Lima 20639
Universidad Nacional Pedro Henríquez Ureña, Santo Domingo 06958
Universidad Nacional San Antonio Abad del Cusco, Cuzco 20616
– Programa Académico de Derecho y Ciencias Políticas, Cuzco 20617
Universidad Pablo de Olavide, Sevilla 25604
Universidad Panamericana – Campus Ciudad de México, México 18794
Universidad Pedagógica Nacional, México 18795
Universidad Pedagógica Nacional, Santafé de Bogotá 05986
Universidad Pedagógica y Tecnológica de Colombia, Tunja 05988
Universidad Politécnica de Madrid, Madrid 25471
– Biblioteca Aeronáutica, Madrid 25472
– Biblioteca Campus Sur, Madrid 25473
– Centro de Documenación Europea, Madrid 25474
– Escuela de Minas, Madrid 25475
– Escuela Técnica Superior de Arquitectura, Madrid 25476
– Escuela Técnica Superior de Ingenieros Agrónomos, Madrid 25477
– Escuela Técnica Superior de Ingenieros de Caminos, Canales y Puertos, Madrid 25478

– Escuela Técnica Superior de Ingenieros de Telecomunicación, Madrid 25479
– Escuela Técnica Superior de Ingenieros Industriales, Madrid 25480
– Escuela Técnica Superior de Ingenieros Navales, Madrid 25481
– Escuela Universitaria de Arquitectura Técnica, Madrid 25482
– Escuela Universitaria de Ingeniería Técnica Agrícola, Madrid 25483
– Escuela Universitaria de Ingeniería Técnica de Obras Públicas, Madrid 25484
– Escuela Universitaria de Ingeniería Técnica Forestal, Madrid 25485
– Escuela Universitaria de Ingeniería Técnica Industrial, Madrid 25486
– Facultad de Ciencias de la Actividad Física y del Deporte, Madrid 25487
– Facultad de Informática, Madrid 25488
Universidad Politécnica de Nicaragua, Managua 19891
Universidad Politécnica de Valencia, Valencia 25617
Universidad Pontificia Bolivariana, Medellín 05952
– Area de las Ingenierías, Medellín 05953
– Facultades de Arquitectura y Diseño Industrial, Medellín 05954
Universidad Pontificia 'Comillas', Madrid 25489
Universidad Pontificia de Salamanca, Salamanca 25564
Universidad Pública de Navarra, Pamplona 25550
Universidad Rey Juan Carlos, Móstoles 25509
– Facultad de Ciencias y Jurídicas y Sociales, Madrid 25510
Universidad Simón Bolívar, Caracas 42217
Universidad Técnica de Oruro, Oruro 02921
– Departamento de Ciencias Jurídicas, Oruro 02922
– Facultad de Tecnología, Oruro 02923
Universidad Técnica Federico Santa María, Valparaíso 04901
– BIblioteca – Sede José Miguel Carrera, Viña del Mar 04902
Universidad Técnica Particular, Loja 06979
Universidad Tecnológica de Pereira, Pereira 05956
Universidad Tecnológica de Santiago (UTESA), Santiago de los Caballeros 06955
Universidad Tecnológica del Magdalena, Santa Marta 05960
Universidade Aberta, Lisboa 21664
Universidade Agostinho Neto, Luanda 00103
Universidade Bandeirante de São Paulo, São Paulo 03149
Universidade Braz Cubas, Mogi das Cruzes 03052
Universidade Católica de Brasília, Taguatinga 03188
Universidade Católica de Goiás, Goiânia 03034
Universidade Católica de Pelotas, Pelotas 03061
Universidade Católica de Petrópolis, Petrópolis 03062
Universidade Católica de Santos, Santos 03126
– Faculdade de Direito-Unisantos, Santos 03127
Universidade Católica Dom Bosco, Campo Grande 03010
Universidade Católica Portuguesa, Lisboa 21665
Universidade Cidade de São Paulo, São Paulo 03150
Universidade da Amazonia, Belém 02989
Universidade da Coruña, A Coruña 25367
– Escola Técnica Superior de Arquitectura, A Coruña 25368
– Escola Técnica Superior de Camiños, Canais e Portos, A Coruña 25369
– Escuela Universitaria Politécnico de Ferrol, Ferrol 25370
– Facultade de Ciencias, A Coruña 25371
– Facultade de Ciencias da Educación, A Coruña 25372
– Facultade de Ciencias Económicas e Empresariais, A Coruña 25373
Universidade da Região de Campanha, Bagé 02985
Universidade da Região de Joinville – Univille, Joinville 03043
Universidade de Aveiro, Aveiro 21621
Universidade de Brasília, Brasília 03004
Universidade de Caxias do Sul, Caxias do Sul 03015
Universidade de Coimbra, Coimbra 21626

– Centro de Documentação 25 de Abril, Coimbra 21627
– Departamento de Antropologia / Museu Antropológico, Coimbra 21628
– Departamento de Ciências da Terra, Coimbra 21629
– Departamento de Engenharia Civil, Coimbra 21630
– Departamento de Engenharia Electrotécnica e de Computadores, Coimbra 21631
– Departamento de Física, Coimbra 21632
– Departamento de Matemática, Coimbra 21633
– Departamento de Química, Coimbra 21634
– Departamento de Zoologia, Coimbra 21635
– Faculdade de Ciêmcias e Tecnologia, Coimbra 21636
– Faculdade de Direito, Coimbra 21637
– Faculdade de Farmácia, Coimbra 21638
– Faculdade de Letras, Coimbra 21639
– Faculdade de Medicina, Coimbra 21640
– Faculdade de Psicologia e de Ciências da Educação, Coimbra 21641
– Instituto Botânico 'Dr. Júlio Henriques', Coimbra 21642
– Instituto de Estudos Alemães, Coimbra 21643
– Instituto de Estudos Clássicos, Coimbra 21644
– Instituto de Estudos Italianos, Coimbra 21645
– Instituto Geofísico, Coimbra 21646
– Observatório Astronómico, Coimbra 21647
Universidade de Évora, Évora 21649
Universidade de Fortaleza, Fortaleza 03030
Universidade de Lisboa, Lisboa 21666
– Biblioteca do M.N.H.N. – Jardim Botânico, Lisboa 21667
– Centro de Linguistica, Lisboa 21668
– Faculdade de Ciências, Lisboa 21669
– Faculdade de Direito, Lisboa 21670
– Faculdade de Farmácia, Lisboa 21671
– Faculdade de Letras, Lisboa 21672
– Faculdade de Medicina, Lisboa 21673
– Faculdade de Medicina Dentaria, Lisboa 21674
– Faculdade de Psicologia e Ciências da Educação, Lisboa 21675
– Instituto Bacteriológico Câmara Pestana, Lisboa 21676
– Instituto de Ciências Sociais, Lisboa 21677
– M.N.H.N. – Jardim Botânico, Lisboa 21678
– Museu de Ciência, Lisboa 21679
– Observatório Astronómico, Lisboa 21680
Universidade de Macau, Macao 05416
Universidade de Marília, Marília 03050
Universidade de Mogi das Cruzes, Mogi das Cruzes 03053
Universidade de Ribeirão Prêto, Ribeirão Prêto 03096
Universidade de Santa Cruz do Sul, Santa Cruz do Sul 03122
Universidade de Santiago de Compostela, Santiago de Compostela 25568
– Biblioteca Concepción Arenal, Santiago de Compostela 25569
– Biblioteca da E.U. de Formación do Profesorado – Lugo, Lugo 25570
– Biblioteca da Facultade de Bioloxía, Santiago de Compostela 25571
– Biblioteca de Ciencias da Comunicación, Santiago de Compostela 25572
– Biblioteca de Ciencias Económicas y Empresariais, Santiago de Compostela 25573
– Biblioteca de Farmacia, Santiago de Compostela 25574
– Biblioteca de Filoloxía, Santiago de Compostela 25575
– Biblioteca de Filosofía, Santiago de Compostela 25576
– Biblioteca de Matemáticas, Santiago de Compostela 25577
– Biblioteca de Medicina e Odontoloxía, Santiago de Compostela 25578
– Biblioteca de Química, Santiago de Compostela 25579
– Biblioteca de Xeografía e Historia, Santiago de Compostela 25580
– Biblioteca Intercentros de Física e Óptica, Santiago de Compostela 25581
– Biblioteca Intercentros de Psicoloxía e CC. Educación, Santiago de Compostela 25582
– Biblioteca Intercentros do Campus de Lugo, Lugo 25583
– Biblioteca Xeral, Santiago de Compostela 25584

Universidade de São Francisco, Bragança Paulista 03002
Universidade de São Paulo, São Paulo 03151
– Centro de Medicina Nuclear (FM-CMN), São Paulo 03152
– Conjunto das Químicas (IQ), São Paulo 03153
– Divisão de Biblioteca e Documentação – Departamento Técnico, São Paulo 03154
– Escola de Comunicações e Artes (ECA), São Paulo 03155
– Escola de Engenharia de São Carlos (EESC), São Carlos 03156
– Escola Politécnica, São Paulo 03157
– Escola Superior de Agricultura Luiz de Queiroz, Piracicaba 03158
– Faculdade de Arquitetura e Urbanismo (FAU), São Paulo 03159
– Faculdade de Direito (FD), São Paulo 03160
– Faculdade de Economía e Administração e Contabilidade (FEA), São Paulo 03161
– Faculdade de Educação (FE), São Paulo 03162
– Faculdade de Filosofia, Letras e Ciências Humanas, São Paulo 03163
– Faculdade de Medicinca (FM), São Paulo 03164
– Faculdade de Odontologia (FO), São Paulo 03165
– Faculdade de Saúde Pública (FSP), São Paulo 03166
– Instituto de Astronomia, Geofísica e Ciências Atmosféricas, São Paulo 03167
– Instituto de Biociências (IB), São Paulo 03168
– Instituto de Ciências Matemáticas e de Computação (USP/ICMC), São Carlos 03169
– Instituto de Física e Química de São Carlos, São Carlos 03170
– Instituto de Geociências (IGC), São Paulo 03171
– Instituto de Pesquisas Energéticas e Nucleares, São Paulo 03172
– Instituto de Psicologia (IP), São Paulo 03173
Universidade de Taubaté, Taubaté 03189
– Faculdade de Medicina, Taubaté 03190
Universidade de Tras-os-Montes e Alto Douro (UTAD), Vila Real 21711
Universidade de Vigo, Vigo 25642
Universidade do Estado de Santa Catarina, Florianópolis 03026
Universidade do Estado do Rio de Janeiro, Rio de Janeiro 03100
Universidade do Grande Rio, Duque de Caxias 03023
Universidade do Minho, Braga 21623
Universidade do Oeste de Santa Catarina – UNOESC Joaçaba, Joaçaba 03042
Universidade do Oeste Paulista (UNOESTE), Presidente Prudente 03085
Universidade do Porto, Porto
– Departamento de Física, Porto 21697
– Faculdade de Arquitectura, Porto 21698
– Faculdade de Belas-Artes, Porto 21699
– Faculdade de Ciências, Porto 21700
– Faculdade de Economia, Porto 21701
– Faculdade de Engenharia, Porto 21702
– Faculdade de Letras, Porto 21703
– Faculdade de Psicologia e Ciências de Educação, Porto 21704
– Faculdade de Química, Porto 21705
– Instituto de Botânica 'Dr. Gonçalo Sampato', Porto 21706
Universidade do Rio Grande, Rio Grande 03113
Universidade do Sagrado Coração, Bauru 02988
Universidade do Sul, Tubarão 03192
Universidade do Vale do Itajaí, Itajaí 03039
Universidade do Vale do Paraíba, São José dos Campos 03132
Universidade do Vale do Rio Dos Sinos, São Leopoldo 03134
Universidade dos Açores, Ponta Delgada 21694
Universidade Eduardo Mondlane, Maputo 18976
– Centro de Estudos Africanos, Maputo 18977
Universidade Estadual da Paraíba, Campina Grande 03005
Universidade Estadual de Campinas, Campinas 03007
– Biblioteca da Area de Engenharia, Campinas 03008
– Instituto de Filosofia e Ciências Humanas, Campinas 03009
Universidade Estadual de Feira de Santana, Feira de Santana 03025
Universidade Estadual de Londrina, Londrina 03046
Universidade Estadual de Maringá, Maringá 03051
Universidade Estadual de Ponta Grossa, Ponta Grossa 03064
Universidade Estadual do Maranhao, São Luis 03135
Universidade Estadual Paulista "Julio de Mesquita Filho", São Paulo 03174
– Faculdade de Ciências Agronômicas, Botucatu 03175
– Faculdade de Ciências Farmacêuticas, Araraquara 03176
– Faculdade de Engenharia, Guaratinguetá 03177
– Faculdade de Filosofia e Ciências de Marilia, Marília 03178
– Faculdade de Odontologia de Araçatuba, Araçatuba 03179
– Faculdade de Odontologia de Araraquara, Araraquara 03180
– Instituto de Biociências, Letras e Ciências Exatas, São José do Rio Prêto 03181
Universidade Federal da Bahia, Salvador 03116
– Escola Politécnica, Salvador 03117
– Faculdade de Direito, Salvador 03118
– Faculdade de Filosofia e Ciências Humanas, Salvador 03119
– Instituto de Geociências, Salvador 03120
Universidade Federal de Goiás, Goiânia 03035
Universidade Federal de Lavras (UFLA), Lavras 03044
Universidade Federal de Mato Grosso do Sul, Campo Grande 03011
Universidade Federal de Minas Gerais, Belo Horizonte 02992
– Escola de Biblioteconomia, Belo Horizonte 02993
– Escola de Engenharia, Belo Horizonte 02994
– Escola de Veterinaria, Belo Horizonte 02995
– Faculdade de Ciências Económicas, Belo Horizonte 02996
– Faculdade de Direito, Belo Horizonte 02997
– Faculdade de Filosofia, Belo Horizonte 02998
– Faculdade de Medicina, Belo Horizonte 02999
– Faculdade de Odontologia, Belo Horizonte 03000
Universidade Federal de Ouro Prêto, Ouro Prêto 03059
– Escola de Minas, Ouro Prêto 03060
Universidade Federal de Pelotas, Capão de Leão 03014
Universidade Federal de Pernambuco, Recife 03088
– Centro de Ciencias Jurídicas, Recife 03089
– Centro de Tecnologia e Geociências, Recife 03090
– Departamento de Matemática e de Estatística e Informática, Recife 03091
– Faculdade de Filosofia, Recife 03092
– Faculdade de Medicina, Recife 03093
Universidade Federal de Santa Catarina, Florianópolis 03027
– Faculdade de Filosofia, Ciências e Letras, Florianópolis 03028
Universidade Federal de Santa Maria, Santa Maria 03123
Universidade Federal de São Paulo, São Paulo 03182
Universidade Federal de Uberlândia, Uberlândia 03193
Universidade Federal de Viçosa, Viçosa 03194
Universidade Federal do Ceará, Fortaleza 03031
– Faculdade de Direito, Fortaleza 03032
– Faculdade de Filosofia, Fortaleza 03033
Universidade Federal do Pará, Belém 02990
Universidade Federal do Paraná, Curitiba 03017
– Biblioteca de Ciêncas Jurídicas, Curitiba 03018
– Biblioteca de Ciência e Tecnologia, Curitiba 03019
– Biblioteca de Ciências da Saúde, Curitiba 03020
– Biblioteca de Ciências Humanase Educação, Curitiba 03021
– Biblioteca de Ciências Sociais Aplicadas, Curitiba 03022
Universidade Federal do Piauí, Teresina 03191
Universidade Federal do Rio de Janeiro, Rio de Janeiro 03101
– Centro de Ciências de Saúde, Rio de Janeiro 03102
– Centro de Ciências Matemáticas e da Naturaleza, Rio de Janeiro 03103
– Escola de Música, Rio de Janeiro 03104
– Escola Nacional de Engenharia, Rio de Janeiro 03105
– Faculdade de Economia e Administração, Rio de Janeiro 03106
– Forum de Ciências e Cultura, Rio de Janeiro 03107
– Instituto de Filosofia e Ciências Sociais, Rio de Janeiro 03108
– Instituto de Pós-Graduação e Pesquisa em Administração, Rio de Janeiro 03109
Universidade Federal do Rio Grande do Norte, Natal 03054
– Serviço Central de Bibliotecas, Natal 03055
Universidade Federal do Rio Grande do Sul, Porto Alegre 03066
– Biblioteca Setorial de Educação, Porto Alegre 03067
– Escola de Educação Física Edgar Sperb, Porto Alegre 03068
– Escola de Enfermagem, Porto Alegre 03069
– Escola de Engenharia, Porto Alegre 03070
– Faculdade de Agronomia, Porto Alegre 03071
– Faculdade de Arquitectura, Porto Alegre 03072
– Faculdade de Biblioteconomia e Comunicação, Porto Alegre 03073
– Faculdade de Ciências Econômicas, Porto Alegre 03074
– Faculdade de Direito, Porto Alegre 03075
– Faculdade de Medicina, Porto Alegre 03076
– Faculdade de Odontologia, Porto Alegre 03077
– Instituto de Biociências, Porto Alegre 03078
– Instituto de Física, Porto Allegre 03079
– Instituto de Geociências, Porto Alegre 03080
– Instituto de Informática, Porto Alegre 03081
– Instituto de Matemática, Porto Alegre 03082
– Instituto de Pesquisas Hidráulicas, Porto Alegre 03083
Universidade Federal Fluminense, Niterói 03056
– MEC – UFF – NDC, São Domingos 03057
Universidade Federal Rural de Pernambuco, Recife 03094
Universidade Federal Rural do Rio de Janeiro, Itajaí 03040
– Instituto de Pós-Graduacão e Pesquisa em Administração de Empresas (COPPEAD), Rio de Janeiro 03041
Universidade Federal Rural do Rio de Janeiro, Rio de Janeiro 03110
Universidade Fernando Pessoa, Porto 21707
Universidade Gama Filho, Rio de Janeiro 03111
Universidade Guarulhos, Guarulhos 03038
Universidade Ibirapuera, São Paulo 03183
Universidade Lusíada, Lisboa 21681
Universidade Lusíada, Porto 21708
Universidade Lusófona de Humanidades e Tecologias, Lisboa 21682
Universidade Luterano do Brasil, Canoas 03013
Universidade Metodista de Piracicaba, Piracicaba 03063
Universidade Metodista de São Paulo, São Bernardo do Campo 03130
Universidade Nova de Lisboa, Lisboa
– Faculdade de Ciências e Tecnologia, Caparica 21683
– Faculdade de Cîencias Sociais e Humanas, Lisboa 21684
– Facultade de Economia, Lisboa 21685
– Instituo de Higiene e Medicina Tropical, Lisboa 21686
Universidade para o Desenvolvimento do Estado de Santa Catarina, Florianópolis 03029
Universidade para o Desenvolvimento do Estado e da Região do Pantanal, Campo Grande 03012
Universidade Paulista, São Paulo 03184
Universidade Portucalense, Porto 21709
Universidade Regional de Blumenau, Blumenau 03001
Universidade Regional do Noroeste do Estado do Rio Grand do Sul, Rio Grande do Sul 03114
Universidade Regional Integrado do Alto Urugal e das Missões, Erechim 03024
Universidade Salvador, Salvador 03121
Universidade Santa Cecília, Santos 03128
Universidade Santa Úrsula, Rio de Janeiro 03112
Universidade São Judas Tadeu, São Paulo 03185
Universidade São Marcos, São Paulo 03186
Universidade Técnica de Lisboa, Lisboa 21687
– Faculdade de Medicina Veterinária, Lisboa 21688
– Faculdade de Motricidade Humana, Cruz Quebrada 21689
– Instituto Superior de Agronomia, Lisboa 21690
– Instituto Superior de Ciências Sociais e Políticas, Lisboa 21691
– Instituto Superior de Economia e Gestão, Lisboa 21692
Universidade Tiradentes, Aracaju 02984
Università Ca' Foscari di Venezia, Venezia
– Biblioteca di Area Economica "G. Luzzatto", Venezia 15579
– Biblioteca di Area Linguistica (BALI) – Biblioteca di Americanistica, Iberistica e Slavistica (AMERIBE), Venezia 15580
– Biblioteca di Area Linguistica (BALI) – Biblioteca di Scienze del Linguaggio (SC-LING), Venezia 15581
– Biblioteca di Area Linguistica (BALI) – Biblioteca di Studi Eurasiatici (EURASIA), Venezia 15582
– Biblioteca di Area Linguistica (BALI) – Biblioteca di Studi Europei e Postcolonali (SLLEP), Venezia 15583
– Biblioteca di Area Linguistica (BALI) – Biblioteca di Studi sull'Asia Orientale (ASIA-OR), Venezia 15584
– Biblioteca di Area Scientifica (BAS), Venezia 15585
– Biblioteca di Area Umanistica (BAUM), Venezia 15586
– Biblioteca Servicio Didattico, Venezia 15587
– Centro di Documentazione Europea C.D.E., Venezia 15588
– Centro Interuniversitario di Studi Veneti, Venezia 15589
– Dipartimento di Scienze Giuridiche, Venezia 15590
Università Cattolica del Sacro Cuore, Milano 15102
Università Cattolica del Sacro Cuore, Roma 15452
Università Cattolica del Sacro Cuore – Facoltà di Agraria, Piacenza 15344
Università Cattolica del Sacro Cuore – Sede di Brescia, Brescia 14912
Università Commerciale Luigi Bocconi, Milano 15103
Università degli Studi, San Marino 24383
Università degli Studi della Basilicata, Potenza 15364
Università degli Studi dell'Aquila, L'Aquila 15038
– Biblioteca di Polo Centro – Biblioteca Lettere e filosofia, L'Aquila 15039
– Biblioteca di Polo Centro – Scienze della formazione, L'Aquila 15040
– Biblioteca di Polo Coppito – Facoltà di Medicina e Chirurgia e Psicologia, Coppito 15041
– Biblioteca di Polo Coppito – Facoltà di Scienze MM.FF.NN – Facoltà di Biotecnologie, Coppito 15042
– Biblioteca di Polo Roio – Facoltà di Economia, L'Aqila 15043
– Biblioteca di Polo Roio – Facoltà di Ingegneria, L'Aquila 15044
Università degli Studi di Bari, Bari 14822
– Biblioteca Interfacoltà di Lettere e Filosofia e di Scienze della Formazione "A. Corsano", Bari 14823
– Dipartimento di Biologia Animale ed Ambientale, Bari 14824
– Dipartimento di Biomedicina dell'Età Evolutiva, Bari 14825
– Dipartimento di Geologia e Geofisica, Bari 14826
– Dipartimento di Italianistica, Bari 14827
– Dipartimento di Lingue e Letterature Romanze e Mediterranee, Bari 14828
– Dipartimento di Matematica, Bari 14829
– Dipartimento di Scienze dell'Antichità, Bari 14830
– Dipartimento di Scienze Economiche, Bari 14831
– Dipartimento di Scienze Filosofiche, Bari 14832
– Dipartimento di Scienze Geografiche e Merceologiche, Bari 14833
– Dipartimento di Scienze Pedagogiche e Didattiche, Bari 14834
– Dipartimento di Scienze Statistiche, Bari 14835

– Dipartimento di Scienze Storiche e Sociali, Bari 14836
– Dipartimento Interateneo di Fisica "Michelangelo Merlin", Bari 14837
– Dipartimento sui Rapporti di Lavoro e sulle Relazioni Industriali, Bari 14838
– Facoltà di Agraria, Bari 14839
– Facoltà di Economia, Bari 14840
– Facoltà di Farmacia, Bari 14841
– Facoltà di Giurisprudenza, Bari 14842
– Facoltà di Ingegneria, Bari 14843
– Facoltà di Lingue e Letterature Straniere, Bari 14844
– Facoltà di Medicina e Chirurgia, Bari 14845
– Sezione Entomologia e Zoologia, Bari 14846
– Sezioni di Economia e Politica agraria, Bari 14847
Università degli Studi di Bergamo, Bergamo
– Biblioteca delle Facoltà Umanistiche, Bergamo 14848
– Biblioteca di Economia e Giurisprudenza, Bergamo 14849
– Biblioteca di Ingegneria, Dalmine 14850
Università degli Studi di Bologna, Bologna 14854
– Archivio Storico, Bologna 14855
– Biblioteca Centralizzata della Sede di Rimini, Rimini 14856
– Biblioteca Centralizzata "Roberto Ruffilli", Forlì 14857
– Biblioteca di Discipline Umanistiche, Bologna 14858
– Biblioteca di Scienze Ginecologiche, Ostetriche e Pediatriche, Bologna 14859
– Biblioteca "Guido Horn d'Arturo", Bologna 14860
– Biblioteca "Walter Bigiavi", Bologna 14861
– Centro di Documentazione Europea C.D.E., Bologna 14862
– Centro Interdipartimentale di Ricerca in Storia del Diritto, Filosofia e Sociologia del Diritto e Informatica Giuridica "G. Fassò – A. Gaudenzi" (C.I.R.S.F.I.D.), Bologna 14863
– Centro Interdipartimentale di Scienze delle Religioni, Bologna 14864
– Centro Interfacoltà di Linguistica Teorica e Applicata, Bologna 14865
– Conservazione dei Beni Culturali, Ravenna 14866
– Corso di Laure in Scienze Ambientali – Polo Scientifico-Didattico di Ravenna, Ravenna 14867
– Dipartimento di Archeologia, Sede di Bologna, Bologna 14868
– Dipartimento di Archeologia, Sede di Ravenna, Ravenna 14869
– Dipartimento di Arti Visive, Bologna 14870
– Dipartimento di Biochimica, Bologna 14871
– Dipartimento di Biologia Evoluzionistica Sperimentale, Bologna 14872
– Dipartimento di Chimica "G. Ciamician", Bologna 14873
– Dipartimento di Discipline Giuridiche dell'Economia e dell'Azienda, Bologna 14874
– Dipartimento di Discipline Storiche, Antropoligiche e Geografiche, Bologna 14875
– Dipartimento di Farmacologia, Bologna 14876
– Dipartimento di Filologia Classica e Medioevale, Bologna 14877
– Dipartimento di Filosofia, Bologna 14878
– Dipartimento di Fisica, Bologna 14879
– Dipartimento di Fisiologia Umana e Generale, Bologna 14880
– Dipartimento di Istologia, Embriologia e Biologia Applicata, Bologna 14881
– Dipartimento di Italianistica, Bologna 14882
– Dipartimento di Lingue e Letterature Straniere Moderne, Bologna 14883
– Dipartimento di Matematica, Bologna 14884
– Dipartimento di Matematica per le Scienze Economiche e Sociali, Bologna 14885
– Dipartimento di Medicina e Sanità Pubblica, Bologna 14886
– Dipartimento di Medicina e Sanità Pubblica – Sezione di Medicina Legale e Assicurazioni, Bologna 14887
– Dipartimento di Musica e Spettacolo, Bologna 14888
– Dipartimento di Paleografia e Medievistica, Bologna 14889
– Dipartimento di Politica, Istituzioni, Storia, Bologna 14890
– Dipartimento di Psicologia, Bologna 14891
– Dipartimento di Scienza Politica, Bologna 14892
– Dipartimento di Scienze della Terra e Geologico-Ambientali, Bologna 14893
– Dipartimento di Scienze dell'Educazione, Bologna 14894
– Dipartimento di Scienze Economiche, Bologna 14895
– Dipartimento di Scienze Farmaceutiche, Bologna 14896
– Dipartimento di Scienze Giuridiche 'A. Cicu', Bologna 14897
– Dipartimento di Scienze Neurologiche, Bologna 14898
– Dipartimento di Scienze Statistiche 'Paolo Fortunati', Bologna 14899
– Dipartimento di Sociologia "Achille Ardigò", Bologna 14900
– Dipartimento di Storia Antica, Bologna 14901
– Dipartimento di Studi Linguistici e Orientali, Bologna 14902
– Dipartimento Discipline della Comunicazione, Bologna 14903
– Dipartimento Scienze Aziendali, Bologna 14904
– Facoltà di Agraria, Bologna 14905
– Facoltà di Medicina Veterinaria, Ozzano dell'Emilia 14906
– Facoltà di Psicologia, Cesena 14907
– Istituto di Applicazione Forense "Enrico Redenti", Bologna 14908
– Istituto di Psichiatria "P. Ottonello", Bologna 14909
– Scuola di Specializzazione in Diritto Amministrativo e Studi sull'Amministrazione Pubblica dell'Università, Bologna 14910
Università degli Studi di Brescia, Brescia
– Biblioteca Centrale Interfacoltà di Economia e di Giurisprudenza, Brescia 14913
– Facoltà di Ingegneria, Brescia 14914
– Facoltà di Medicina e Chirurgia, Brescia 14915
Università degli Studi di Cagliari, Cagliari 14919
– Biblioteca del Dipartimento di Ingegneria del Territorio – Sez. Geologia appl. e Geofisica appl. e Sez. Trasporti, Cagliari 14920
– Biblioteca del Dipartimento di Ingegneria del Territorio – Sez. Ingegneria Idraulica, Cagliari 14921
– Biblioteca del Distretto Biomedico Scientifico – Sezione Cittadella Universitaria di Monserrato, Monserato, Cagliari 14922
– Biblioteca del Distretto Biomedico Scientifico – Sezione Farmacia e Tossicologia, Cagliari 14923
– Biblioteca del Distretto Biomedico Scientifico – Sezione Scienze Botaniche, Cagliari 14924
– Biblioteca del Distretto Biomedico Scientifico – Sezione Scienze Odontostomatologiche, Cagliari 14925
– Biblioteca del Distretto Biomedico Scientifico – Sezione Via Ospedale – Sottosezione Complesso Pediatrico, Cagliari 14926
– Biblioteca del Distretto Biomedico Scientifico – Sezione Via Ospedale – Sottosezione S. Giovanni di Dio, Cagliari 14927
– Biblioteca del Distretto delle scienze sociali, economiche e giuridiche – Sezione Economiche, Cagliari 14928
– Biblioteca del Distretto delle Scienze Sociali, Economiche e Giuridiche – Sezione Scienze Giuridiche, Cagliari 14929
– Biblioteca del Distretto delle Scienze Sociali, Economiche e Giuridiche – Sezione Scienze Politiche, Cagliari 14930
– Biblioteca del Distretto delle Scienze Umane – Dipartimento di Linguistica e Stilistica, Cagliari 14931
– Biblioteca del Distretto delle Scienze Umane – Sezione Cittadella dei Musei, Cagliari 14932
– Biblioteca del Distretto delle Scienze Umane – Sezione Filologia Classica e Glottologia, Cagliari 14933
– Biblioteca del Distretto delle Scienze Umane – Sezione Giordano Bruno, Cagliari 14934
– Biblioteca del Distritto delle Scienze Umane – Sezione Dante Alighieri, Cagliari 14935
– Biblioteca Interdipartimentale di Architettura e Urbanistica, Cagliari 14936
– Dipartimento di Ingegneria Elettrica ed Elettronica, Cagliari 14937
– Dipartimento di Ingegneria Strutturale, Cagliari 14938
– Facoltà di Ingegneria, Cagliari 14939
Università degli Studi di Camerino, Camerino
– Biblioteca di Architettura, Ascoli Piceno 14940
– Biblioteca di Scienze, Camerino 14941
– Biblioteca di Scienze Ambientali, Camerino 14942
– Biblioteca di Veterinaria, Matelica 14943
– Biblioteca 'Flavio Bonati', Camerino 14944
– Biblioteca Giuridica, Camerino 14945
Università degli Studi di Catania, Catania
– Biblioteca Luigi Antonini, Catania 14948
– Corso di Laurea in Scienze dei Beni Culturali, Siracusa 14949
– Dipartimento del Seminario Giuridico, Catania 14950
– Dipartimento di Architettura e Urbanistica (DAU), Catania 14951
– Dipartimento di Biologia Animale, Catania 14952
– Dipartimento di Botanica, Catania 14953
– Dipartimento di Fisica e Astronomia, Catania 14954
– Dipartimento di Ingegneria Agraria (DIA), Catania 14955
– Dipartimento di Ingegneria Civile ed Ambientale (DICA), Catania 14956
– Dipartimento di Ingegneria Elettrica, Elettronica e dei Sistemi (DIEES), Catania 14957
– Dipartimento di Matematica e Informatica, Catania 14958
– Dipartimento di Ortofloroalboricoltura e Tecnologie Agroalimentari (DOFATA), Catania 14959
– Dipartimento di Scienze Agronomiche, Agrochimiche e delle Produzioni Animali (DACPA), Catania 14960
– Dipartimento di Scienze Biomediche, Catania 14961
– Dipartimento di Scienze Chimiche, Catania 14962
– Dipartimento di Scienze e Tecnologie Fitosanitarie (DISTEF), Catania 14963
– Dipartimento di Scienze Economico-Agrarie ed Estimative (DISEAE), Catania 14964
– Dipartimento di Scienze Fisiologiche, Catania 14965
– Dipartimento di Scienze Geologiche, Catania 14966
– Dipartimento di Scienze Microbiologiche e Scienze Ginecologiche, Catania 14967
– Dipartimento di Scienze Umane – Biblioteca di Storia dell'Arte, Catania 14968
– Facoltà di Agraria, Catania 14969
– Facoltà di Architettura, Siracusa 14970
– Facoltà di Economia, Catania 14971
– Facoltà di Farmacia, Catania 14972
– Facoltà di Ingegneria, Catania 14973
– Facoltà di Lettere e Filosofia, Catania 14974
– Facoltà di Lettere e Filosofia – Sezione Archeologia, Catania 14975
– Facoltà di Lingue e Letterature Straniere (sede CT), Catania 14976
– Facoltà di Lingue e Letterature Straniere (sede RG), Ragusa Ibla 14977
– Facoltà di Medicina e Chirurgia, Catania 14978
– Facoltà di Scienze della Formazione, Catania 14979
– Facoltà di Scienze Politiche, Catania 14980
– Istituto Nazionale di Astrofisica – Osservatorio Astrofisico Catania, Catania 14981
Università degli Studi di Ferrara, Ferrara 14982
– Dipartimento di Biologia, Sezione Biologia Evolutiva, Ferrara 14983
– Dipartimento di Fisica, Ferrara 14984
– Dipartimento di Scienze Geologiche e Paleontologiche, Ferrara 14985
– Dipartimento di Scienze Giuridiche, Ferrara 14986
– Facoltà di Lettere e Filosofia, Ferrara 14987
– Facoltà di Scienze Mathematiche, Fisiche e Naturali, Ferrara 14988
– Istituto di Economia e Finanza, Ferrara 14989
Università degli Studi di Firenze, Firenze
– Biblioteca Biomedica – Medicina, Firenze 14994
– Biblioteca del Polo Universitario di Prato, Prato 14995
– Biblioteca di Scienze – Antropologia, Firenze 14996
– Biblioteca di Scienze – Biologia Animale, Firenze 14997
– Biblioteca di Scienze – Botanica, Firenze 14998
– Biblioteca di Scienze – Geomineralogia, Firenze 14999
– Biblioteca di Scienze – Matematica, Firenze 15000
– Biblioteca di Scienze – Polo Scientifico, Sesto Fiorentino 15001
– Biblioteca di Scienze Sociali – Scienze Sociali, Firenze 15002
– Biblioteca di Scienze Sociali – Statistica, Firenze 15003
– Biblioteca di Scienze Tecnologiche – Agraria, Firenze 15004
– Biblioteca di Scienze Tecnologiche – Architettura, Firenze 15005
– Biblioteca di Scienze Tecnologiche – Dipartimento di Costruzioni, Firenze 15006
– Biblioteca di Scienze Tecnologiche – Dipartimento di Economia Agraria e delle Risorse Territoriali, Firenze 15007
– Biblioteca di Scienze Tecnologiche – Dipartimento di Progettazione dell'Architettura, Firenze 15008
– Biblioteca di Scienze Tecnologiche – Dipartimento di Urbanistica e Pianificazione del Territorio, Firenze 15009
– Biblioteca di Scienze Tecnologiche – Ingegneria, Firenze 15010
– Biblioteca Umanistica – Filosofia, Firenze 15011
– Biblioteca Umanistica – Geografia, Firenze 15012
– Biblioteca Umanistica – Lettere, Firenze 15013
– Biblioteca Umanistica – Psicologia, Firenze 15014
– Biblioteca Umanistica – Scienze della Formazione, Firenze 15015
– Biblioteca Umanistica – Storia dell'Arte, Firenze 15016
– Biblioteca Umanistica – Storia e Letterature Nord-Americana, Firenze 15017
Università degli Studi di Genova, Genova 15023
– CSB di Architettura 'N. Carboneri', Genova 15024
– CSB di Biologia, Scienze della Terra e del Mare, Genova 15025
– CSB di Chimica 'S. Cannizzaro', Genova 15026
– CSB di Economia, Genova 15027
– CSB di Farmacia 'P. Schenone', Genova 15028
– CSB di Fisica "A. Borsellino", Genova 15029
– CSB di Giurisprudenza "P.E. Bensa", Genova 15030
– CSB di Ingegneria, Genova 15031
– CSB di Lettere e Filosofia "Romeo Crippa", Genova 15032
– CSB di Lingue e Letterature Straniere, Genova 15033
– CSB di Matematica e Informatica "E. Togliatti", Genova 15034
– CSB di Medicina "E. Maragliano", Genova 15035
– CSB di Scienze della Formazione, Genova 15036
– CSB di Scienze Politiche, Genova 15037
Università degli Studi di Macerata, Macerata
– Biblioteca Interdipartimentale di Economia, Macerata 15056
– Biblioteca Interdipartimentale di Palazzo Ugolino, Macerata 15057
– Dipartimento dei Beni Culturali, Fermo 15058
– Dipartimento di Diritto Privato e del Lavoro Italiano e Comparato, Macerata 15059
– Dipartimento di Diritto Pubblico e Teoria del Governo, Macerata 15060
– Dipartimento di Filosofia e Scienze Umane, Macerata 15061
– Dipartimento di Ricerca Linguistica, Letteraria e Filologica – DIPRI, Macerata 15062

– Dipartimento di Scienze Archeologiche e Storiche dell'Antichità, Macerata 15063
– Dipartimento di Scienze della Comunicazione – DISCO, Macerata 15064
– Dipartimento di Scienze dell'Educazione e della Formazione, Macerata 15065
– Dipartimento di Studi su Mutamento Sociale, Istituzioni Giuridiche e Comunicazione, Macerata 15066
Università degli Studi di Messina, Messina 15068
– Facoltà di Economia, Messina 15069
– Facoltà di Farmacia, Messina 15070
– Facoltà di Giurisprudenza, Messina 15071
– Facoltà di Ingegneria, S. Agata – Messina 15072
– Facoltà di Lettere e Filosofia, Messina 15073
– Facoltà di Medicina Veterinaria, Messina 15074
– Facoltà di Scienze della Formazione, Messina 15075
– Facoltà di Scienze Statistiche, Messina 15076
Università degli Studi di Milano, Milano 15104
– Biblioteca Biologica, Milano 15105
– Biblioteca Centrale di Farmacia, Milano 15106
– Biblioteca della Facoltà di Medicina Veterinaria, Milano 15107
– Biblioteca delle Facoltà di Giurisprudenza, Lettere e Filosofia, Milano 15108
– Biblioteca di Chimica, Milano 15109
– Biblioteca di Farmacologia e Medicina Sperimentale, Milano 15110
– Biblioteca di Filosofia, Milano 15111
– Biblioteca di Fisica, Milano 15112
– Biblioteca di Informatica, Milano 15113
– Biblioteca di Scienze della Terra "A. Desio", Milano 15114
– Biblioteca di Scienze dell'Antichità e Filologia Moderna, Milano 15115
– Biblioteca di Scienze Politiche, Milano 15116
– Biblioteca di Storia dell'Arte, della Musica e dello Spettacolo, Milano 15117
– Biblioteca Matematica 'Giovanni Ricci', Milano 15118
– Dipartimento di Economia e Politica Agraria, Agro-alimentare e Ambientale, Milano 15119
– Facoltà di Agraria, Milano 15120
Università degli Studi di Modena e Reggio Emilia, Modena 15121
– Biblioteca di Economia Sebastiano Brusco, Modena 15122
– Biblioteca Scientifica Interdipartimentale (BSI), Modena 15123
– Biblioteca Universitaria Area Giuridica, Modena 15124
– Biblioteca Universitaria di Area Medica, Modena 15125
– Biblioteca Universitaria Interdipartimentale di Reggio Emilia, Reggio Emilia 15126
– Dipartimento del Museo di Paleobiologia e dell'Orto Botanico – Sezione Orto Botanico, Modena 15127
– Dipartimento di Scienze della Terra, Modena 15128
– Facoltà di Ingegneria, Modena 15129
Università degli Studi di Napoli 'Federico II', Napoli 15133
– Biblioteca del Gruppo Geomineralogico, Napoli 15134
– Biblioteca di Ricerca Area Umanistic, Napoli 15135
– Biblioteca Interdipartimentale di Biochimica e Biotecnologie Mediche e Biologia e Patologia Cellulare e Molecolare, Napoli 15136
– Biblioteca Interdipartimentale di Ingegneria Elettrica, Elettronica e delle Telecomunicazioni, Informatica e Sistemastica, Napoli 15137
– Biblioteca Interdipartimentale Marcello Canino, Napoli 15138
– Centro Interdipartimentale sull'Iconografia della Città Europea, Napoli 15139
– Dipartemento di Energetica Termofluidodinamica Applicata e Condizionamento Ambientale, Napoli 15140
– Dipartemento di Scienze Biomorfologiche e Funzionali. Sezione di Anatomia Umana e Istologia, Napoli 15141
– Dipartimento dei Rapporti Civili ed Economico Sociali, Napoli 15142
– Dipartimento di Arboricoltura, Botanica e Patologia Vegetale. Sezione di Patologia Vegetale, Portici 15143
– Dipartimento di Chimica, Napoli 15144
– Dipartimento di Chimica Organica e Biochimica, Napoli 15145
– Dipartimento di Economia, Napoli 15146
– Dipartimento di Economia e Politica Agraria, Portici 15147
– Dipartimento di Entomologia e Zoologia Agraria, Portici 15148
– Dipartimento di Ingegneria Agraria e Agronomia del Territorio. Sezione di Agronomia e Coltivazioni, Portici 15149
– Dipartimento di Ingegneria Agraria ed Agronomia del Territorio. Sezione di Agronomia e Coltivazioni, Portici 15150
– Dipartimento di Ingegneria dei Materiali e della Produzione, Napoli 15151
– Dipartimento di Ingegneria Meccanica per l'Energetica, Napoli 15152
– Dipartimento di Matematica e Applicazioni 'R. Caccioppoli', Napoli 15153
– Dipartimento di Patologia e Sanità Animale. Sezione di Tossicologia, Napoli 15154
– Dipartimento di Scienze Fisiche, Napoli 15155
– Dipartimento Ingeneria Navale, Napoli 15156
– Dipartimento Patologia e Sanità Animale. Sezione di Anatomia Patologica, Napoli 15157
– Dipartimento Pediatria, Napoli 15158
– Facoltà di Agraria, Portici 15159
– Facoltà di Architettura, Napoli 15160
– Facoltà di Economia, Napoli 15161
– Facoltà di Farmacia, Napoli 15162
– Facoltà di Lettere e Filosofia, Napoli 15163
– Facoltà di Scienze Politiche "G. Cuomo", Napoli 15164
– Facoltà di Sociologa, Napoli 15165
– Facoltà d'Ingegneria, Napoli 15166
– Facoltà Giurisprudenza, Napoli 15167
– Facoltà Medicina Veterinaria, Napoli 15168
– Facoltà Scienze Matematiche, Fisiche e Naturali, Napoli 15169
– Nucleo Bibliotecaria di Geografia, Napoli 15170
Universita degli Studi di Napoli 'L'Orientale', Napoli 15171
Università Degli Studi di Napoli 'L'Orientale', Napoli 16384
Università degli Studi di Padova, Padova 15173
– Biblioteca Antica "V. Pinali", Padova 15174
– Biblioteca Biologico-Medica "Antonio Vallisneri", Padova 15175
– Biblioteca dell'Orto Botanico, Padova 15176
– Biblioteca di Meccanica "Enrico Bernardi", Padova 15177
– Biblioteca di Scienze Ginecologiche e della Riproduzione Umana, Padova 15178
– Biblioteca di Storia della Scienza, Padova 15179
– Biblioteca Interdipartimentale di Ingegneria dell'Informazione e Ingegneria Elettrica "Giovanni Someda", Padova 15180
– Biblioteca Interdipartimentale "Tito Livio", Padova 15181
– Biblioteca Medica "V. Pinali" – Sezione Moderna, Padova 15182
– Centro di Documentazione Europea C.D.E., Padova 15183
– Centro Linguistico di Ateneo, Padova 15184
– Centro per la Storia della Università, Padova 15185
– C.I.S. di Agripolis, Legnaro 15186
– C.I.S. di Psicologia, Padova 15187
– C.I.S. Seminario Matematico, Padova 15188
– Dipartimento di Diritto Comparato, Padova 15189
– Dipartimento di Farmacologia ed Anestesiologia "E. Meneghetti", Padova 15190
– Dipartimento di Filosofia, Padova 15191
– Dipartimento di Fisica "Galileo Galilei", Padova 15192
– Dipartimento di Fisica Tecnica, Padova 15193
– Dipartimento di Geografia, Padova 15194
– Dipartimento di Geoscienze, Padova 15195
– Dipartimento di Ingegneria Idraulica, Marittima e Geotecnica, Padova 15196
– Dipartimento di Lingue e Letterature Anglo-Germaniche e Slave – Sezione di Anglogermanico, Padova 15197
– Dipartimento di Lingue e Letterature Anglo-Germaniche e Slave – Sezione di Slavistica, Padova 15198
– Dipartimento di Neuroscienze – Sezione Neuroscienze, Padova 15199
– Dipartimento di Neuroscienze – Sezione Oculistica, Padova 15200
– Dipartimento di Scienze Chimiche, Padova 15201
– Dipartimento di Scienze dell'Educazione – Sezione di Piazza Capitaniato, Padova 15202
– Dipartimento di Scienze dell'Educazione – Sezione di Via Beato Pellegrino, Padova 15203
– Dipartimento di Scienze Economiche "Marco Fanno", Padova 15204
– Dipartimento di Scienze Farmaceutiche, Padova 15205
– Dipartimento di Storia, Padova 15206
– Dipartimento di Storia delle Arti Visive e della Musica, Padova 15207
– Dipartimento di Storia e Filosofia del Diritto e Diritto Canonico – Biblioteca di Diritto Romano, Padova 15208
– Dipartimento di Storia e Filosofia del Diritto e Diritto Canonico – Biblioteca di Filosofia del Diritto, Padova 15209
– Dipartimento di Tecnica e Gestione dei Sistemi Industriali, Vicenza 15210
– Dipartimentoi Diritto Privato e Dirotto del Lavoro, Padova 15211
– Facoltà di Ingegneria, Padova 15212
– Facoltà di Scienze Politiche, Padova 15213
– Facoltà di Scienze Statistiche, Padova 15214
Università degli Studi di Palermo, Palermo
– Biblioteca del Polo Didattico di Agrigento, Agrigento 15216
– Biblioteca del Polo Didattico di Trapani "Domenico Rubino", Erice 15217
– Biblioteca di Scienze per la Promozione della Salute "Giuseppe D'Alessandro", Palermo 15218
– Dipartimento "Aglaia" di Studi Greci, Latini e Musicali, Tradizione e Modernità, Palermo 15219
– Dipartimento di Agronomia Ambientale e Territoriale, Palermo 15220
– Dipartimento di Analisi dell'Espressione Lingue Segni Testi (DANAE), Palermo 15221
– Dipartimento di Arti e Comunicazioni, Palermo 15222
– Dipartimento di Beni Culturali Storico-Archeologici, Socio-Antropologici e Geografici, Palermo 15223
– Dipartimento di Biologia Animale "G. Reverberi", Palermo 15224
– Dipartimento di Biologia Cellulare e dello Sviluppo, Palermo 15225
– Dipartimento di Chimica e Fisica della Terra ed Applicazione alle Georisorse e ai Rischi Naturali (CFTA), Palermo 15226
– Dipartimento di Chirurgia Generale d'Urgenza e Tripanti d'Organo, Palermo 15227
– Dipartimento di Città e Territorio, Palermo 15228
– Dipartimento di Colture Arboree, Palermo 15229
– Dipartimento di Contabilità Nazionale e Analisi dei Processi Sociali (DICAP), Palermo 15230
– Dipartimento di Design, Palermo 15231
– Dipartimento di Diritto Privato Generale, Palermo 15232
– Dipartimento di Diritto Pubblico, Palermo 15233
– Dipartimento di Discipline Chirurgiche ed Oncologiche, Palermo 15234
– Dipartimento di Economia dei Sistemi Agro-Forestali (ESAF), Palermo 15235
– Dipartimento di Filosofia, Storia e Critica dei Saperi (FIERI), Palermo 15236
– Dipartimento di Fisica e Tecnologie Relative (DFTER), Palermo 15237
– Dipartimento di Geologia e Geodesia, Palermo 15238
– Dipartimento di Ingegneria Chimica dei Processi e dei Materiali, Palermo 15239
– Dipartimento di Ingegneria dei Transporti, Palermo 15240
– Dipartimento di Ingegneria dell'Automazione e dei Sistemi, Palermo 15241
– Dipartimento di Ingegneria delle Infrastrutture Viarie, Palermo 15242
– Dipartimento di Ingegneria e Tecnologie Agro-Forestali (I.T.A.F.), Palermo 15243
– Dipartimento di Ingegneria Elettrica, Elettronica e delle Telecomunicazioni (D.I.E.E.T.), Palermo 15244
– Dipartimento di Ingegneria Idraulica ed Applicazioni Ambientali, Palermo 15245
– Dipartimento di Ingegneria Informatica, Palermo 15246
– Dipartimento di Ingegneria Nucleare (DIN), Palermo 15247
– Dipartimento di Ingegneria Strutturale, Aerospaziale e Geotecnica (DISAG), Palermo 15248
– Dipartimento di Matematica, Palermo 15249
– Dipartimento di Meccanica, Palermo 15250
– Dipartimento di Medicina Interna e Specialistica, Palermo 15251
– Dipartimento di Medicina Interna, Malattie Cardiovascolari e Nefrourologiche, Palermo 15252
– Dipartimento di Medicina, Pneumologia, Fisiologia e Nutrizione Umana (DIMPEFINU), Palermo 15253
– Dipartimento di Metodi e Modelli Matematici, Palermo 15254
– Dipartimento di Metodi Quantitativi per le Scienze Umane, Palermo 15255
– Dipartimento di Psicologia, Palermo 15256
– Dipartimento di Ricerche Energetiche ed Ambientali (DREAM), Palermo 15257
– Dipartimento di Scienze Anestesiologiche, Rianimatorie e delle Emergenze, Palermo 15258
– Dipartimento di Scienze Biomediche (DISBI), Palermo 15259
– Dipartimento di Scienze Botaniche, Palermo 15260
– Dipartimento di Scienze della Terra e del Mare (DiSTeM), Palermo 15261
– Dipartimento di Scienze Economiche, Aziendali e Finanziarie (SEAF), Palermo 15262
– Dipartimento di Scienze Farmacologiche "Pietro Benigno", Palermo 15263
– Dipartimento di Scienze Filologiche e Linguistiche, Palermo 15264
– Dipartimento di Scienze Penalistiche, Processualpenalistiche e Criminologiche, Palermo 15265
– Dipartimento di Scienze Sociali, Palermo 15266
– Dipartimento di Scienze Statistiche e Matematiche "Silvio Vianelli", Palermo 15267
– Dipartimento di Scienze Stomatologiche "Valerio Margiotta", Palermo 15268
– Dipartimento di Storia del Diritto, Palermo 15269
– Dipartimento di Storia e Progetto nell'Architettura, Palermo 15270
– Dipartimento di Studi Storici e Artistici, Palermo 15271
– Dipartimento di Studi su Politica, Diritto e Società "G. Mosca", Palermo 15272
– Facoltà di Agraria, Palermo 15273
– Facoltà di Architettura, Palermo 15274
– Facoltà di Economia, Palermo 15275
– Facoltà di Farmacia, Palermo 15276
– Facoltà di Giurisprudenza, Palermo 15277
– Facoltà di Ingegneria, Palermo 15278
– Facoltà di Lettere e Filosofia, Palermo 15279
– Facoltà di Medicina e Chirurgia, Palermo 15280
– Facoltà di Scienze della Formazione, Palermo 15281
– Facoltà di Scienze Matematiche, Fisiche e Naturali, Palermo 15282
– Facoltà di Scienze Motorie, Palermo 15283

– Facoltà di Scienze Politiche, Palermo 15284
Università degli Studi di Parma, Parma
– Biblioteca Centrale di Farmacia, Parma 15285
– Biblioteca Centrale di Giurisprudenza, Parma 15286
– Biblioteca Centrale di Lettere, Parma 15287
– Biblioteca Centrale di Medicina "G. Ottaviani", Parma 15288
– Biblioteca di Diritto del Lavoro, Parma 15289
– Biblioteca di Economia, Parma 15290
– Biblioteca di Filologia Classica e Medievale, Parma 15291
– Biblioteca di Filosofia, Parma 15292
– Biblioteca di Filosofia del Diritto, Parma 15293
– Biblioteca di Servizio Sociale, Parma 15294
– Biblioteca di Storia del Diritto Italiano, Parma 15295
– Biblioteca Politecnica, Parma 15296
– Biblioteca Psicologia, Parma 15297
– Dipartimento dei Beni Culturali e dello Spettacolo, Parma 15298
– Dipartimento di Biologia Evolutiva e Funzionale, Parma 15299
– Dipartimento di Diritto Economia e Finanza Internazionale, Parma 15300
– Dipartimento di Fisica, Parma 15301
– Dipartimento di Italianistica, Parma 15302
– Dipartimento di Lingue e Letterature Straniere, Parma 15303
– Dipartimento di Matematica, Parma 15304
– Dipartimento di Scienze della Formazione e del Territorio, Parma 15305
– Dipartimento di Scienze della Terra, Parma 15306
– Dipartimento di Scienze Giuridiche, Sezione di Diritto Pubblico, Parma 15307
– Dipartimento di Storia, Parma 15308
– Sezione di Musicologia, Parma 15309
Università degli Studi di Pavia, Pavia 15311
– Biblioteca Boezia, Pavia 15312
– Biblioteca delle Scienze, Pavia 15313
– Biblioteca Petrarca, Pavia 15314
– Biblioteca Unificata della Scienza e della Tecnica, Pavia 15315
– Dipartimento di Medicina Interna e Terapia Medica, Pavia 15316
– Dipartimento di Scienze Musicologiche e Paleografico-filologiche, Cremona 15317
– Facoltà di Economia, Pavia 15318
– Facoltà di Giurisprudenza, Pavia 15319
– Facoltà di Scienze Politiche, Pavia 15320
Università degli Studi di Perugia, Perugia 15321
– Biblioteca di Agraria "Mario Marte", Perugia 15322
– Biblioteca di Ingegneria, Perugia 15323
– Biblioteca di Medicina e Chirurgia, Perugia 15324
– Biblioteca di Medicina Veterinaria, Perugia 15325
– Biblioteca di Scienze Chimiche e Farmaceutiche, Perugia 15326
– Biblioteca di Scienze Economiche, Statistiche e Aziendali, Perugia 15327
– Biblioteca di Scienze Matematiche, Fisiche e Geologiche, Perugia 15328
– Biblioteca di Studi Storici, Politici e Sociali, Perugia 15329
– Biblioteca di Terni, Terni 15330
– Biblioteca Giuridica Unificata, Perugia 15331
– Biblioteca Umanistica, Perugia 15332
Università degli Studi di Salerno, Fisciano 15018
– Centro Bibliotecario dell'Area Scientifica, Baronissi 15019
– Dipartimento di Scienze dell'Antichità, Fisciano 15020
– Dipartimento di Scienze Economiche e Statistiche, Fisciano 15021
Università degli Studi di Sassari, Sassari
– Biblioteca Interfacoltà di Scienze Ambientali e Forestali, Nuoro 15454
– Biblioteca Interfacoltà per le Scienze Giuridiche, Politiche e Economiche "A. Pigliaru", Sassari 15455
– Dipartimento di Chimica, Sassari 15456
– Dipartimento di Economia e Sistemi Arborei, Sassari 15457
– Dipartimento di Economia, Istituzioni e Società, Sassari 15458
– Dipartimento di Protezione delle Piante, Sassari 15459
– Dipartimento di Scienze Agronomiche e Genetica Vegetale Agraria, Sassari 15460
– Dipartimento di Scienze Botaniche, Ecologiche e Geologiche, Sassari 15461
– Dipartimento di Scienze dei Linguaggi, Sassari 15462
– Dipartimento di Scienze Giuridiche e Seminario di Studi Latinoamericani, Sassari 15463
– Dipartimento di Scienze Umanistiche e dell'Antichità, Sassari 15464
– Dipartimento di Storia e Centro Interdisciplinare per la Storia dell'Università di Sassari, Sassari 15465
– Dipartimento di Teorie e Ricerche dei Sistemi Culturali, Sassari 15466
– Facoltà di Agraria, Sassari 15467
– Facoltà di Architettura "Fernando Clemente", Alghero 15468
– Facoltà di Medicina e Chirurgia, Sassari 15469
– Facoltà di Medicina Veterinaria, Sassari 15470
– Facoltà Umanistiche, Sassari 15471
Università degli Studi di Siena, Siena
– Biblioteca Centrale di Economia, Siena 15472
– Biblioteca Centrale di Ingegneria, Siena 15473
– Biblioteca Centrale di Medicina, Siena 15474
– Biblioteca Centrale di Scienze Matematiche, Fisiche e Naturali, Siena 15475
– Biblioteca Centrale Farmacia, Siena 15476
– Biblioteca Circolo Giuridico, Siena 15477
– Centro di Documentazione Europea C.D.E., Siena 15478
– Facoltà di Lettere e Filosofia, Siena 15479
– Facoltà di Lettere e Filosofia di Arezzo, Arezzo 15480
Università degli Studi di Teramo – Facoltà di Giurisprudenza e Scienze Politiche, Teramo 15481
Università degli Studi di Torino, Torino 15493
– Biblioteca Centrale di Agraria e di Medicina Veterinaria, Grugliasco 15494
– Biblioteca Chimica "G. Ponzio", Torino 15495
– Biblioteca di Farmacia 'Icilio Guareschi', Torino 15496
– Biblioteca di Geografia del Dipartimento Interateneo Territorio, Torino 15497
– Biblioteca di Oculistica del Dipartimento di Fisiopatologia Clinica, Torino 15498
– Biblioteca di Scienze Religiose Erik Peterson, Torino 15499
– Biblioteca Interdipartimentale di Fisica, Torino 15500
– Biblioteca Interdipartimentale 'Gioele Solari', Torino 15501
– Biblioteca Medicina Legale del Dipartimento di Anatomia, Farmacologia e Medicina Legale, Torino 15502
– Biblioteca Odontostomatologia del Dipartimento di Scienze Biomediche ed Oncologia Umana, Torino 15503
– Biblioteca Speciale di Matematica "G. Peano", Torino 15504
– Biblioteca Universitaria Cuneese, Cuneo 15505
– Dipartimento di Biologia Animale e dell'Uomo, Torino 15506
– Dipartimento di Biologia Vegetale, Torino 15507
– Dipartimento di Discipline Artistiche, Musicali e dello Spettacolo, Torino 15508
– Dipartimento di Discipline Ginecologiche e Ostetriche, Torino 15509
– Dipartimento di Discipline Medico-Chirurgiche, Torino 15510
– Dipartimento di Economia, Torino 15511
– Dipartimento di Filologia Linguistica e Tradizione Classica, Torino 15512
– Dipartimento di Filosofia, Torino 15513
– Dipartimento di Fisiopatologia Clinica, Torino 15514
– Dipartimento di Genetica, Biologia e Biochimica, Torino 15515
– Dipartimento di Informatica, Torino 15516
– Dipartimento di Neuroscienze, Torino 15517
– Dipartimento di Orientalistica, Torino 15518
– Dipartimento di Sanità Pubblica e di Microbiologia, Torino 15519
– Dipartimento di Scienze Antropologiche, Archeologiche e Storico-Territoriali, Torino 15520
– Dipartimento di Scienze del Linguaggio e Letterature Moderne e Comparate, Torino 15521
– Dipartimento di Scienze della Terra, Torino 15522
– Dipartimento di Scienze dell'Educazione e della Formazione, Torino 15523
– Dipartimento di Scienze Giuridiche – Biblioteca "Federico Patetta", Torino 15524
– Dipartimento di Scienze Giuridiche – Biblioteca "Francesco Ruffini", Torino 15525
– Dipartimento di Scienze Letterarie e Filologiche, Torino 15526
– Dipartimento di Scienze Merceologiche, Torino 15527
– Dipartimento di Scienze Mineralogiche e Petrologiche, Torino 15528
– Dipartimento di Scienze Pediatriche, e dell'Adolescenza, Torino 15529
– Dipartimento di Statistica e Matematica applicata alle Scienze Umane, Torino 15530
– Dipartimento di Storia, Torino 15531
– Dipartimento di Traumatologia, Ortopedia e Medicina del Lavoro, Torino 15532
– Facoltà di Economia, Torino 15533
– Facoltà di Lettere e Filosofia, Torino 15534
– Facoltà di Psicologia Federico Kiesow, Torino 15535
– Scuola di Amministrazione Aziendale, Torino 15536
Università degli Studi di Trento, Trento 15538
– Biblioteca di Ingegneria, Trento 15539
– Biblioteca di Lettere, Trento 15540
– Biblioteca di Scienze, Trento 15541
– Biblioteca di Scienze Cognitive, Rovereto 15542
Università degli Studi di Trieste, Trieste 15543
– Biblioteca Area 1 – Generale, Trieste 15544
– Biblioteca Area 1 – Socio-Politica, Trieste 15545
– Biblioteca Area 2 – Economia, Trieste 15546
– Biblioteca Area 2 – Sede di Gorizia, Gorizia 15547
– Biblioteca Area 4 – Centrale di Medicina, Trieste 15548
– Biblioteca Area 5 – Filosofia e Lingue, Trieste 15549
– Biblioteca Area 5 – Storia e Arte, Trieste 15550
– Biblioteca Area 6 – Scienze dell'Antichità e Italianistica, Trieste 15551
– Biblioteca Area 7 – Formazione v. Montfort, Trieste 15552
– Biblioteca Area 7 – Formazione v. Tigor, Trieste 15553
– Biblioteca Area 8 – Psicologia e Architettura, Trieste 15554
– Biblioteca Area 8 – Scuola di Lingue, Trieste 15555
– Dipartimento di Fisica & I.N.F.N., Trieste 15556
– Dipartimento di Scienza della vita – Biologia, Trieste 15557
– Dipartimento di Scienze Chimiche e Farmaceutiche – Sez. Chimica, Trieste 15558
– Dipartimento di Scienze Economiche, Aziendali, Matematiche e Statistiche, Sez. di Ricerche Economiche e Agrarie, Trieste 15559
– Dipartimento di Scienze Economiche, Aziendali, Matematiche e Statistiche, Sez. Matematica Applicata "Bruno de Finetti", Trieste 15560
– Dipartimento di Scienze Economiche, Aziendali, Matematiche e Statistiche, Sez. Tecnica Aziendale, Trieste 15561
– Dipartimento di Scienze Geografiche e Storiche, Trieste 15562
– Dipartimento Scienze Giurdiche, Trieste 15563
Università degli Studi di Udine, Udine 15564
– Biblioteca di Area Cotonificio, Udine 15565
– Biblioteca di Economia e Giurisprudenza, Udine 15566
– Biblioteca di Formazione e Didattica, Udine 15567
– Biblioteca di Medicina, Udine 15568
– Biblioteca di Scienze, Udine 15569
– Biblioteca di Studi Umanistici, Udine 15570
– Centro Speciale di Servizi Bibliotecari – Gorizia, Gorizia 15571
Università degli Studi di Urbino "Carlo Bo", Urbino
– Biblioteca Centrale Umanistica, Urbino 15572
– Biblioteca di Ecologia e Sociologia, Urbino 15573
– Biblioteca di Giurisprudenza e Scienze Politiche, Urbino 15574
– Biblioteca di Scienze Motorie, Urbino 15575
– Biblioteca Scientifica, Urbino 15576
Università degli Studi di Verona, Verona 15594
– Biblioteca Centrale 'Egidio Meneghetti', Verona 15595
– Biblioteca del Dipartimento di Anglistica, Germanistica e Slavistica, Verona 15596
– Biblioteca del Dipartimento di Discipline Storiche, Artistiche, Archeologiche e Geografiche – Sezione di Geografia, Verona 15597
– Biblioteca del Dipartimento di Discipline Storiche, Artistiche, Archeologiche e Geografiche – Sezione di Storia, Verona 15598
– Biblioteca del Dipartimento di Economia Aziendale, Verona 15599
– Biblioteca del Dipartimento di Filosofia, Verona 15600
– Biblioteca del Dipartimento di Psicologia e Antropologia Culturale, Verona 15601
– Biblioteca del Dipartimento di Romanistica, Verona 15602
– Biblioteca del Dipartimento di Scienze dell'Educazione, Verona 15603
– Biblioteca del Polo Universitario di Economia a Vicenza, Vicenza 15604
– Biblioteca della Sezione di Storia Economica e Sociale, Verona 15605
– Biblioteca di Scienze Economiche, Verona 15606
– Biblioteca "Franco Riva" del Dipartimento di Linguistica, Letteratura e Scienze della Comunicazione, Verona 15607
– Biblioteca G. Zanotto, Verona 15608
– Biblioteca Giuridica, Verona 15609
– Centro Linguistico di Ateneo, Verona 15610
– Laurea in scienze del servizio sociale, Verona 15611
Università degli Studi G. D'Annunzio Chieti Pescara, Pescara
– Biblioteca di Medicina e Chirurgia, Farmacia, Scienze dell'Educazione Motoria e Scienze MM.FF.NN., Chieti 15334
– Biblioteca di Scienze Giuridiche, Pescara 15335
– Biblioteca Interfacoltà "Ettore Paratore" di Lettere e Filosofia, Psicologia, Scienze Sociali e Scienze della Formazione, Chieti 15336
– Biblioteca Unificata di Architettura, Economia, Lingue e Letterature Straniere, Scienze Manageriali, Pescara 15337
– Dipartimento Ambienti Reti e Territorio, Pescara 15338
– Dipartimento di Metodi Quantitativi e Teoria Economica, Pescara 15339
– Dipartimento di Scienze Linguistiche e Letterarie, Pescara 15340
– Dipartimento di Scienze, Storia dell' Architettura, Restauro e Rappresentazione, Pescara 15341
– Dipartimento di Tecnologie per l'Ambiente Costruito, Pescara 15342
– Dipartimento Infrastrutture, Design, Engineering, Architettura, Pescara 15343
Università del Salento, Lecce 15045
– Dipartimento di Filologia, Linguistica e Letteratura, Lecce 15046
– Dipartimento di Filosofia e Scienze Sociali, Lecce 15047
– Dipartimento di Fisica e Scienza dei Materiali, Lecce 15048
– Dipartimento di Lingue e Letterature Straniere, Lecce 15049
– Dipartimento di Matematica "Ennio de Giorgi", Lecce 15050
– Dipartimento di Scienza sociali e della Comunicazione, Lecce 15051
– Dipartimento di Scienze dell'Antichità, Lecce 15052
– Dipartimento di Studi Storici dal Medioevo all'Età Contemporanea – Sede Olivetani, Lecce 15053

- Dipartimento di Studi Storici dal Medioevo all'Età Contemporanea – Sede Parlangeli, Lecce 15054
- Facoltà di Economia, Lecce 15055

Università della Svizzera italiana, Lugano 27113

Università di Pisa, Pisa 15346
- Biblioteca di Agraria, Pisa 15347
- Biblioteca di Antichistica, Pisa 15348
- Biblioteca di Chimica, Pisa 15349
- Biblioteca di Economia, Pisa 15350
- Biblioteca di Economia e Legislazione dei Sistemi Logistici, Livorno 15351
- Biblioteca di Farmacia, Pisa 15352
- Biblioteca di Filosofia e Storia, Pisa 15353
- Biblioteca di Ingegneria, Pisa 15354
- Biblioteca di Lingue e Letterature Moderne 1 – Sede Principale di Studi Italianistici, Pisa 15355
- Biblioteca di Lingue e Letterature Moderne 2, Pisa 15356
- Biblioteca di Matematica Informatica Fisica, Pisa 15357
- Biblioteca di Medicina e Chirurgia, Pisa 15358
- Biblioteca di Medicina Veterinaria, Pisa 15359
- Biblioteca di Scienze Naturali e Ambientali, Pisa 15360
- Biblioteca di Scienze Politiche e Sociali, Pisa 15361
- Biblioteca di Storia delle Arti, Pisa 15362
- Biblioteca Giuridica, Pisa 15363

Università IUAV di Venezia, Venezia 15591

Università per Stranieri di Perugia, Perugia 15333

Università Politecnica delle Marche, Ancona 14819
- Biblioteca Economico Giuridica Sociologica, Ancona 14820
- Biblioteca Tecnico Scientifica Biomedica, Ancona 14821

Università Pontificia Salesiana, Roma 42193

Universitaire bibliotheek van Defensie / Bibliothèque universitaire de la Défense, Brussel 02297

Universitäre Psychiatrische Kliniken Basel UPK, Basel 26991

Universitario Francisco de Vitoria, Pozuelo de Alarcón 25552

Universitas Andalas, Padang 14284
- Fakultas Hukum dan Pengetahuan Masyarakat, Padang 14285

Universitas Bengkulu, Bengkulu 14246

Universitas Brawijaya, Malang 14274
- Fakultas Teknik, Malang 14275

Universitas Diponegoro, Semarang 14294
- Fakultas Kedokteran, Semarang 14295

Universitas Gadjah Mada, Yogyakarta 14313
- Fakultas Ekonomi, Yogyakarta 14314
- Fakultas Filsafat, Yogyakarta 14315
- Fakultas Hukum, Yogyakarta 14316
- Fakultas Ilmu Pasti dan Alam, Yogyakarta 14317
- Fakultas Kedokteran Hewan, Yogyakarta 14318
- Fakultas Kehutanan, Yogyakarta 14319
- Fakultas Pertanian, Yogyakarta 14320
- Fakultas Sastra dan Kebudajaan, Yogyakarta 14321
- Perpustakaan Fisipol UGM, Yogyakarta 14322

Universitas Haluoleo, Kendari 14266

Universitas Hasanuddin, Makassar 14268
- Fakultas Hukum, Ujung Pandang 14269
- Lembaga Penjelidikan Ekonomi dan Masjarakat, Ujung Pandang 14270

Universitas Indonesia, Jakarta 14252
- Faculty of Economics, Jakarta 14253
- Fakultas Kedokteran, Jakarta 14254
- Fakultas Matematika dan Ilmu Pengetahuan Alam, Jakarta 14255
- Fakultas Sastra, Jakarta 14256

Universitas Islam Indonesia, Yogyakarta 14323

Universitas Jambi, Jambi 14264

Universitas Jember, Jember 14265

Universitas Katolik Indonesia Atma Jaya, Jakarta 14257

Universitas Kristen Duta Wacana, Yogyakarta 14324

Universitas Kristen Satya Wacana – Perpustakaan Universitas, Salatiga 14291

Universitas Mataram, Mataram 14277

Universitas Muhammadiyah Purwokerto, Purwokerto 14289

Universitas Mulawarman, Samarinda 14292

Universitas Nebrissensis, Madrid 25490

Universitas Padjadjaran, Bandung 14241
- Fakultas Ekonomi, Bandung 14242
- Fakultas Hukum dan Pengetahuan Masjarakat, Bandung 14243
- Fakultas Kedokteran, Bandung 14244
- Fakultas Pertanian, Bandung 14245

Universitas Pancasila, Jakarta 14258

Universitas Sanata Dharma, Yogyakarta 14325

Universitas Sebelas Maret, Surakarta 14303

Universitas Sriwijaya, Palembang 14286

Universitas Sumatera Utara, Medan 14279
- Fakultas Ekonomi, Medan 14280
- Fakultas Hukum, Medan 14281
- Fakultas Pertanian, Medan 14282

Universitas Syiah Kuala, Banda Aceh 14236

Universitas Tadulako, Palu 14287

Universitas Tarumanagara, Jakarta
- Fakultas Ekonomi, Jakarta 14259
- Fakultas Kedokteran, Jakarta 14260
- Fakultas Teknik, Jakarta 14261

Universitas Terbuka, Jakarta 14262

Universitas Trisakti, Jakarta 14263

Universitas Udayana, Denpasar 14249

Universitat Abat Oliba CEU, Barcelona 25296

Universität Augsburg, Augsburg 09395

Universitat Autònoma de Barcelona, Bellaterra 25331
- Biblioteca de Ciència i Tecnologia, Bellaterra 25332
- Biblioteca de Ciències Socials, Bellaterra 25333
- Biblioteca de Comunicació i Hemeroteca General, Bellaterra 25334
- Biblioteca de Medicina, Bellaterra 25335
- Biblioteca de Veterinària, Bellaterra 25336
- Biblioteca d'Humanitas, Bellaterra 25337
- Biblioteca Universitària de Sabadell UAB, Sabadell 25338

Universität Bamberg, Bamberg 09397

Universität Basel, Basel 26992
- Bibliothek Altertumswissenschaft, Basel 26994
- Biozentrum, Basel 26995
- Botanisches Institut, Abt. Ökologie, Basel 26996
- Departementsbibliothek Chemie, Basel 26997
- Deutsches Seminar, Basel 26998
- Englisches Seminar, Basel 26999
- Geographisches Institut, Basel 27000
- Historisches Seminar, Basel 27001
- Institut für Iberoromanistik / Institut für Italianistik / Institut für Französische Sprach- und Literaturwissenschaft, Basel 27002
- Institut für Physik, Basel 27003
- Institut für Psychologie, Basel 27004
- Institut für Soziologie, Basel 27005
- Institut für Sport und Sportwissenschaften, Basel 27006
- Institut für Umweltgeowissenschaften, Basel 27007
- Juristische Fakultät, Basel 27008
- Mathematisches Institut, Basel 27009
- Medizinbibliothek im Universitätsspital, Basel 26993
- Musikwissenschaftliches Institut, Basel 27010
- Orientalisches Seminar, Basel 27011
- Philosophisches Seminar, Basel 27012
- Seminar für Klassische Philologie, Basel 27013
- Slavisches Seminar, Basel 27014
- Theologische Fakultät, Basel 27015

Universität Bayreuth, Bayreuth 09399

Universität Bern, Bern 26965
- Basisbibliothek Unitobler, Bern 27021
- Bibliothek Anglistik, Bern 27022
- Bibliothek Betriebswirtschaft, Bern 27023
- Bibliothek Erziehungswissenschaft, Bern 27024
- Bibliothek Exakte Wissenschaften, Bern 27025
- Bibliothek Germanistik, Bern 27026
- Bibliothek Medizingeschichte, Bern 27027
- Bibliothek Pflanzenwissenschaften, Bern 27028
- Bibliothek Philosophie, Bern 27029
- Bibliothek Romanistik, Bern 27030
- Bibliothek Slavistik, Bern 27031
- Bibliothek Sozialwissenschaften, Bern 27032
- Bibliothek Sprachwissenschaft, Bern 27033
- Bibliothek Vetsuisse Bern, Bern 27034
- Departement für Chemie und Biochemie (DCB), Bern 27035
- Departement für Christkatholische Theologie, Bern 27036
- Department Volkswirtschaftslehre, Bern 27037
- Evangelisch-theologisches Departement, Bern 27038
- Fachbereichsbibliothek Bühlplatz, Bern 27039
- Geographisches Institut, Bern 27040
- Historisches Institut, Bern 27041
- Historisches Institut, Abt. Alte Geschichte und Epigraphik, Bern 27042
- Institut für Archäologische Wissenschaften, Bern 27043
- Institut für Archäologische Wissenschaften – Abt. Ur- und Frühgeschichte / Abt. Archäologie der Römischen Provinzen, Bern 27044
- Institut für Islamwissenschaft und Neuere Orientalische Philologie, Bern 27045
- Institut für Klassische Philologie, Bern 27046
- Institut für Musikwissenschaft, Bern 27047
- Institut für Psychologie, Bern 27048
- Institut für Sozialanthropologie, Bern 27049
- Institut für Sportwissenschaft (ISPW), Bern 27050
- Juristische Bibliothek Bern (JBB), Bern 27051
- Universitäre Psychiatrische Dienste Bern (UPD), Universitätsklinik für Psychiatrie, Bern 27052
- Universitätsspital-Bibliothek USB, Bern 27053

Universität Bielefeld, Bielefeld 09461

Universität Bochum, Bochum 09464

Universität Bonn, Bonn 09304

Universität Bremen, Bremen 09305

Universitat de Barcelona, Barcelona
- Biblioteca de Belles Arts, Barcelona 25297
- Biblioteca de Biblioteconomia, Barcelona 25298
- Biblioteca de Biologia, Barcelona 25299
- Biblioteca de Dret, Barcelona 25300
- Biblioteca de Farmàcia, Barcelona 25301
- Biblioteca de Filosofia, Geografia i Història, Barcelona 25302
- Biblioteca de Física i Química, Barcelona 25303
- Biblioteca de Geologia, Barcelona 25304
- Biblioteca de Lletres, Barcelona 25305
- Biblioteca de Matemàtiques, Barcelona 25306
- Biblioteca de Medicina, Barcelona 25307
- Biblioteca de Reserva, Barcelona 25308
- Biblioteca d'Economia i Empresa, Barcelona 25309
- Biblioteca Pavelló de la República, Barcelona 25310
- Campus de Bellvitge, L'Hospitalet de Llobregat 25311
- CRAI Biblioteca de Campus Mundet, Barcelona 25312

Universitat de Girona, Girona 25378

Universitat de les Illes Balears, Palma de Mallorca 25525
- Biblioteca Anselm Turmeda, Palma de Mallorca 25526
- Biblioteca Arxiduc Lluis Salvador, Palma de Mallorca 25527
- Biblioteca Guillem Cifre de Colonya, Palma de Mallorca 25528
- Biblioteca Mateu Orfila i Rotger, Palma de Mallorca 25529
- Biblioteca Ramon Llull, Palma de Mallorca 25530

Universitat de Lleida, Lleida
- Biblioteca Cappont, Lleida 25413
- Biblioteca Ciències de la Salut, Lleida 25414
- Biblioteca ETSEA, Lleida 25415
- Biblioteca Lletres, Lleida 25416

Universitat de València, Valencia 25618
- Biblioteca de Ciències de la Salut – Sala de Medicina, Valencia 25619
- Biblioteca de Ciències "Eduard Boscà", Burjassot 25620
- Biblioteca de Ciències Socials 'Gregori Mains', Valencia 25621
- Biblioteca de Psicologia i Esport, Valencia 25622
- Biblioteca d'Humanitats Joan Reglà, Valencia 25623
- Biblioteca Històrica, Valencia 25624

Universitat de Vic, Vic 25641

Universität der Bundeswehr Hamburg, Hamburg 09915

Universität der Bundeswehr München, Neubiberg 10512

Universität der Künste Berlin, Berlin 09456

Universität des Saarlandes, Saarbrücken 09336
- Bereichsbibliothek Empirische Humanwissenschaften, Saarbrücken 10554
- Fachrichtung 1.1 Rechtswissenschaft – Deutsch-Europäisches Juridicum, Saarbrücken 10555
- Fachrichtung 1.1 Rechtswissenschaft – Europa-Institut, Saarbrücken 10556
- Fachrichtung 3.1 Philosophie, Saarbrücken 10557
- Fachrichtung 3.2 Evangelische Theologie, Saarbrücken 10558
- Fachrichtung 3.3 Katholische Theologie, Saarbrücken 10559
- Fachrichtung 3.4 Geschichte, Saarbrücken 10560
- Fachrichtung 3.6 Altertumswissenschaften – Klassische Philologie, Saarbrücken 10561
- Fachrichtung 3.6 Kunstgeschichte, Saarbrücken 10562
- Fachrichtung 3.7 Altertumswissenschaften – Vor- und Frühgeschichte und Vorderasiatische Archäologie, Saarbrücken 10563
- Fachrichtung 3.8 Alte Geschichte und Klassische Archäologie, Saarbrücken 10564
- Fachrichtung 3.10 Musikwissenschaft, Saarbrücken 10565
- Fachrichtung 4.1 Germanistik, Saarbrücken 10566
- Fachrichtung 4.1 Germanistik – Lehrstuhl für Allgemeine und Vergleichende Literaturwissenschaft, Saarbrücken 10567
- Fachrichtung 4.1 Germanistik – Wissenschaftliche Österreich-Bibliothek Robert Musil, Saarbrücken 10568
- Fachrichtung 4.2 Romanistik, Saarbrücken 10569
- Fachrichtung 4.3 Anglistik, Amerikanistik und Anglophone Kulturen, Saarbrücken 10570
- Fachrichtung 4.4 Slavistik, Saarbrücken 10571
- Fachrichtung 4.6 Angewandte Sprachwissenschaft sowie Übersetzen und Dolmetschen, Saarbrücken 10572
- Fachrichtung 5 Sportwissenschaftliches Institut, Saarbrücken 10573
- Fachrichtung 5.1 Erziehungswissenschaft, Saarbrücken 10574
- Fachrichtung 5.2 Soziologie, Saarbrücken 10575
- Fachrichtung 7.1-7.4/8.1-8.4 Naturwissenschaftlich-Technische Bereichsbibliothek, Saarbrücken 10576
- Medizinische Abteilung, Homburg (Saar) 10577
- Wirtschaftswissenschaftliche Seminar-Bibliothek, Saarbrücken 10578

Universität Duisburg-Essen, Duisburg 09679

Universität Duisburg-Essen, Essen 09724

Universität Düsseldorf, Düsseldorf 09312

Universität Erlangen-Nürnberg, Erlangen 09690

Universität Flensburg, Flensburg 09728

Universität Frankfurt am Main, Frankfurt am Main 09730

Universität Freiburg, Freiburg 09758

Universität für angewandte Kunst Wien, Wien 01310

Universität für Bodenkultur Wien, Wien 01311
- Fachbibliothek Lebensmittel- und Biotechnologie (BIO), Wien 01312
- Fachbibliothek SoWiRe (Sozial, Wirtschafts- und Rechtswissenschaften), Wien 01313
- Fachbibliothek Wald, Natur und Technik (WNT), Wien 01314

Universität für Musik und darstellende Kunst Graz, Graz 01221

Universität für Musik und darstellende Kunst Wien, Wien 01315

Universität Gießen, Gießen 09818

Universität Göttingen, Göttingen 09315

Universität Graz, Graz 01193

Universität Halle-Wittenberg, Halle (Saale) 09316

Universität Hamburg, Hamburg 09317
- Asien-Afrika-Institut (AAI), Hamburg 09924
- Asien-Afrika-Institut (AAI), Abt. für Sprache und Kultur Chinas, Hamburg 09925
- Asien-Afrika-Institut (AAI), Arbeitsbereich Korea, Hamburg 09926
- Bibliothek Mathematik und Geschichte der Naturwissenschaften (BMGN), Hamburg 09927
- Biozentrum Grindel und Zoologisches Museum, Hamburg 09928
- Biozentrum Klein Flottbek und Botanischer Garten, Hamburg 09929
- Department Informatik, Hamburg 09930
- Departmentbibliothek Physik, Standort Bahrenfeld, Hamburg 09931
- Departmentbibliothek Physik, Standort Jungiusstraße, Hamburg 09932
- Fachbereich Chemie, Hamburg 09933
- Fachbereich Evangelische Theologie, Hamburg 09934
- Fachbereich Kulturgeschichte und Kulturkunde, Hamburg 09935

– Fachbereich Kulturgeschichte und Kulturkunde, Abteilungsbibliothek Archäologie und Kulturgeschichte des Antiken Mittelmeerraumes, Hamburg 09936
– Fachbereich Kulturgeschichte und Kulturkunde, Abteilungsbibliothek Ethnologie, Hamburg 09937
– Fachbereich Kulturgeschichte und Kulturkunde, Abteilungsbibliothek Kunstgeschichte, Hamburg 09938
– Fachbereich Kulturgeschichte und Kulturkunde, Abteilungsbibliothek Volkskunde, Hamburg 09939
– Fachbereich Kulturgeschichte und Kulturkunde, Abteilungsbibliothek Vor- und Frühgeschichte, Hamburg 09940
– Fachbereichsbibliothek Sprache Literatur Medien, Hamburg 09941
– Fachbibliothek Sozialwissenschaften, Hamburg 09942
– Fachbibliothek Wirtschaftswissenschaften, Hamburg 09943
– Geologisch-Paläontologisches Institut (GPI) und Mineralogisch-Petrographisches Institut (MPI), Hamburg 09944
– Hamburger Bibliothek für Universitätsgeschichte, Hamburg 09945
– Hamburger Sternwarte, Hamburg 09946
– Institut für Geographie, Hamburg 09947
– Interdisziplinäres Zentrum für Internationales Finanz- und Steuerwesen (IIFS), Hamburg 09948
– Linga-Bibliothek der Freien und Hansestadt Hamburg – Fachbibliothek für Geschichte und Landeskunde Ibero-Amerikas, Hamburg 09949
– Martha-Muchow-Bibliothek, Hamburg 09950
– Musikwissenschaftliches Institut, Hamburg 09951
– Zentralbibliothek Philosophie, Geschichte und Klassische Philologie, Hamburg 09952
– Zentralbibliothek Recht, Hamburg 09953
– Zentrum für Marine und Atmosphärische Wissenschaften, Hamburg 09954
Universität Heidelberg, Heidelberg 09978
Universität Hildesheim, Hildesheim 10043
Universität Hohenheim, Stuttgart 10595
– Bereichsbibliothek für Wirtschafts- und Sozialwissenschaften, Stuttgart 10596
– Institut für Botanik (210), Stuttgart 10597
Universität Innsbruck, Innsbruck 01184
– Baufakultätsbibliothek, Innsbruck 01224
– Fachbibliothek Atrium, Innsbruck 01225
– Fachbibliothek für Chemie, Innsbruck 01226
– Fachbibliothek für Naturwissenschaften, Innsbruck 01227
– Fachbibliothek für Rechtswissenschaften, Innsbruck 01228
– Fachbibliothek, Medizinisch-Biologische, Innsbruck 01229
– Fakultätsbibliothek für Sozial- und Wirtschaftswissenschaften, Innsbruck 01230
– Fakultätsbibliothek für Theologie, Innsbruck 01231
– Forschungsinstitut Brenner-Archiv, Innsbruck 01232
– Institut für Botanik, Innsbruck 01233
– Institut für Musikwissenschaft, Innsbruck 01234
– Institut für Römisches Recht und Rechtsgeschichte, Innsbruck 01235
– Institut für Sportwissenschaften, Innsbruck 01236
– Institut für Translationswissenschaft, Innsbruck 01237
– Institut für Zivilrecht, Innsbruck 01238
– Institut für Zoologie und Limnologie, Innsbruck 01239
Universitat Jaume I, Castelló de la Plana 25352
Universität Jena, Jena 09320
Universität Kassel, Kassel 09322
Universität Kiel, Kiel 10063
Universität Klagenfurt, Klagenfurt am Wörthersee 01243
Universität Koblenz-Landau, Koblenz 10122
Universität Koblenz-Landau, Landau in der Pfalz 10238
Universität Köln, Köln 09325
Universität Konstanz, Konstanz 10235
Universität Leipzig, Leipzig 10245
Universität Linz, Linz 01251
Universität Magdeburg, Magdeburg 10263
Universität Mainz, Mainz 10268
Universität Mannheim, Mannheim 10298
Universität Marburg, Marburg 10301
Universität Mozarteum, Salzburg 01266
– Abteilungsbibliothek Musik- und Tanzpädagogik Orff-Institut, Salzburg 01267
Universität München, München 10367
Universität Münster, Münster 09329
Universität Oldenburg, Oldenburg 10525
Universität Osnabrück, Osnabrück 10529
Universität Paderborn, Paderborn 10530
Universität Passau, Passau 10531
Universitat Politècnica de Catalunya, Barcelona
– Biblioteca de Matemàtiques i Estadística, Barcelona 25313
– Biblioteca de Nàutica, Barcelona 25314
– Biblioteca del Campus de Terrassa, Terrassa 25315
– Biblioteca del Campus del Baix Llobregat, Castelldefels 25316
– Biblioteca Rector Gabriel Ferraté, Barcelona 25317
– Escola d'Enginyeria d'Igualada, Igualada 25318
– Escola Politècnica Superior d'Edificació de Barcelona, Barcelona 25319
– Escola Politècnica Superior d'Enginyeria de Manresa, Manresa 25320
– Escola Politècnica Superior d'Enginyeria de Vilanova i la Geltrú, Vilanova i La Geltru 25321
– Escola Tècnica Superior d'Arquitectura de Barcelona, Barcelona 25322
– Escola Tècnica Superior d'Arquitectura del Vallès, Sant Cugat del Vallès 25323
– Escola Tècnica Superior d'Enginyeria Industrial de Barcelona, Barcelona 25324
– Escola Universitària d'Enginyeria Tècnica Industrial de Barcelona, Barcelona 25325
Universitat Politècnica de València, Valencia 25617
Universitat Pompeu Fabra, Barcelona 25326
Universität Potsdam, Potsdam 10536
Universitat Ramon Llull, Barcelona
– Facultat de Ciències de la Comunicació, Bellaterra 25327
– Facultat de Psicologia, Ciències de l'Educació i de l'Esport Blanquerna, Barcelona 25328
– Institut Borja de Bioètica, Esplugues de Llobregat (Barcelona) 25329
– Institut Químic de Sarrià, Barcelona 25330
Universität Regensburg, Regensburg 10540
Universität Rostock, Rostock 10546
Universitat Rovira i Virgili, Tarragona 25606
– Biblioteca Campus Bellissens, Reus 25607
– Biblioteca d'Infermeria, Tarragona 25608
– Campus Sescelades, Tarragona 25609
– Campus Sescelades Àrea d'Educació i Psicologia, Tarragona 25610
– Secció de Lletres i Quimica, Tarragona 25611
Universität Salzburg, Salzburg 01268
– Fachbereichsbibliothek Altertumswissenschaften, Salzburg 01269
– Fachbereichsbibliothek Anglistik, Salzburg 01270
– Fachbereichsbibliothek Germanistik, Salzburg 01271
– Fachbereichsbibliothek Kunstgeschichte, Salzburg 01272
– Fachbereichsbibliothek Linguistik, Salzburg 01273
– Fachbereichsbibliothek Philosophie (Kath.-Theol. Fakultät), Salzburg 01274
– Fachbereichsbibliothek Philosophie (KGW), Salzburg 01275
– Fachbereichsbibliothek Romanistik, Salzburg 01276
– Fachbereichsbibliothek Slawistik, Salzburg 01277
– Fachbereichsbibliothek Theologie – Bibelwissenschaft und Kirchengeschichte (TB1), Salzburg 01278
– Fachbereichsbibliothek Theologie – Praktische Theologie (TB3), Salzburg 01279
– Fachbibliothek Erziehungswissenschaft/ILLB, Musik- und Tanzwissenschaft, Salzburg 01280
– Fachbibliothek für Naturwissenschaften, Salzburg 01281
– Fachbibliothek für Sport- und Bewegungswissenschaften, Hallein 01282
– Fachbibliothek Gesellschaftswissenschaften, Salzburg 01283
– Fachbibliothek Kirchenrecht, Salzburg 01284
– Fakultätsbibliothek für Rechtswissenschaften, Salzburg 01285
Universität Siegen, Siegen 10587
Universität St.Gallen – Hochschule für Wirtschafts-, Rechts- und Sozialwissenschaften, St.Gallen 27136
– Institut für Finanzwissenschaft und Finanzrecht, St.Gallen 27137
– Institut für Versicherungswirtschaft, St.Gallen 27138
– KMU-HSG Bibliothek, St.Gallen 27139
Universität Stuttgart, Stuttgart 10598
– Betriebswirtschaftliches Institut, Stuttgart 10599
– Bibliothek der Institute für Linguistik und Literaturwissenschaft, Stuttgart 10600
– Fakultätsbibliothek Architektur und Stadtplanung, Stuttgart 10601
– Fakultätsbibliothek Luft- und Raumfahrttechnik und Geodäsie, Stuttgart 10602
– Geodätisches Institut, Stuttgart 10603
– Historisches Institut, Stuttgart 10604
– Informatik-Bibliothek, Stuttgart 10605
– Institut für Eisenbahn- und Verkehrswesen, Stuttgart 10606
– Institut für Energiewirtschaft und Rationelle Energieanwendung (IER), Stuttgart 10607
– Institut für Erziehungswissenschaft und Psychologie, Abt. Berufs-, Wirtschafts- und Technikpädagogik, Stuttgart 10608
– Institut für Erziehungswissenschaft und Psychologie, Abt. für Pädagogik, Stuttgart 10609
– Institut für Ingenieurgeodäsie, Stuttgart 10610
– Institut für Kommunikationsnetze und Rechnersysteme, Stuttgart 10611
– Institut für Kunstgeschichte, Stuttgart 10612
– Institut für Leichtbau Entwerfen und Konstruieren, Stuttgart 10613
– Institut für Philosophie, Stuttgart 10614
– Institut für Raumordnung und Entwicklungsplanung, Stuttgart 10615
– Institut für Siedlungswasserbau, Wassergüte- und Abfallwirtschaft, Stuttgart 10616
– Institut für Straßen- und Verkehrswesen, Stuttgart 10617
– Institut für Theoretische Physik, Stuttgart 10618
– Institut für Thermodynamik der Luft- und Raumfahrt (ITLR), Stuttgart 10619
– Institut für Thermodynamik und Wärmetechnik (ITW), Stuttgart 10620
– Materialprüfungsanstalt, Otto-Graf-Institut für Werkstoffe im Bauwesen, Stuttgart 10621
– Sprachenzentrum, Stuttgart 10622
Universität Trier, Trier 10624
Universität Tübingen, Tübingen 10625
Universität Ulm, Ulm 10677
Universität Vechta, Vechta 10680
Universität Wien, Wien 01316
– Archiv, Wien 01317
– Fachbereichsbibliothek Afrikawissenschaften und Orientalistik, Wien 01318
– Fachbereichsbibliothek Alte Geschichte, Wien 01319
– Fachbereichsbibliothek Anglistik und Amerikanistik, Wien 01320
– Fachbereichsbibliothek Archäologien und Numismatik, Wien 01321
– Fachbereichsbibliothek Astronomie, Wien 01322
– Fachbereichsbibliothek Bildungswissenschaft, Sprachwissenschaft und Vergleichende Literaturwissenschaft, Wien 01323
– Fachbereichsbibliothek Biologie, Wien 01324
– Fachbereichsbibliothek Botanik, Wien 01325
– Fachbereichsbibliothek Byzantinistik und Neogräzistik, Wien 01326
– Fachbereichsbibliothek Chemie, Wien 01327
– Fachbereichsbibliothek Erdwissenschaften und Meteorologie, Wien 01328
– Fachbereichsbibliothek Europäische Ethnologie, Wien 01329
– Fachbereichsbibliothek Finno-Ugristik, Wien 01330
– Fachbereichsbibliothek Geographie und Regionalforschung, Wien 01331
– Fachbereichsbibliothek Germanistik, Nederlandistik und Skandinavistik, Wien 01332
– Fachbereichsbibliothek Geschichtswissenschaften, Wien 01333
– Fachbereichsbibliothek Judaistik, Wien 01334
– Fachbereichsbibliothek Katholische und Evangelische Theologie, Wien 01335
– Fachbereichsbibliothek Klassische Philologie, Mittel- und Neulatein, Wien 01336
– Fachbereichsbibliothek Kultur- und Sozialanthropologie, Wien 01337
– Fachbereichsbibliothek Kunstgeschichte, Wien 01338
– Fachbereichsbibliothek Musikwissenschaft, Wien 01339
– Fachbereichsbibliothek Organische Chemie 'Josef Loschmid', Standort Physikalische Chemie, Wien 01340
– Fachbereichsbibliothek Ostasienwissenschaften, Wien 01341
– Fachbereichsbibliothek Pharmazie und Ernährungswissenschaften, Wien 01342
– Fachbereichsbibliothek Philosophie, Wien 01343
– Fachbereichsbibliothek Psychologie, Wien 01344
– Fachbereichsbibliothek Publizistik- und Kommunikationswissenschaft, Wien 01345
– Fachbereichsbibliothek Rechtswissenschaften, Wien 01346
– Fachbereichsbibliothek Romanistik, Wien 01347
– Fachbereichsbibliothek Slawistik, Wien 01348
– Fachbereichsbibliothek Soziologie und Politikwissenschaft, Wien 01349
– Fachbereichsbibliothek Südasien-, Tibet- und Buddhismuskunde, Wien 01350
– Fachbereichsbibliothek Theater-, Film- und Medienwissenschaft, Wien 01351
– Fachbereichsbibliothek Translationswissenschaft, Wien 01352
– Fachbereichsbibliothek Wirtschaftswissenschaften, Standort BWL, Wien 01353
– Fachbereichsbibliothek Wirtschaftswissenschaften, Standort Volkswirtschaftslehre und Staatswissenschaften, Wien 01354
– Fachbereichsbibliothek Zeitgeschichte und Osteuropäische Geschichte, Standort Osteuropäische Geschichte, Wien 01355
– Fachbereichsbibliothek Zeitgeschichte und Osteuropäische Geschichte, Standort Zeitgeschichte, Wien 01356
– Institutsbibliothek Ägyptologie, Wien 01357
– Institutsbibliothek Mathematische Logik, Wien 01358
– Institutsbibliothek Sonder- und Heilpädagogik, Wien 01359
– Österreichische Zentralbibliothek für Physik, Wien 01360
– Zentrum für Sportwissenschaft und Universitätssport, Wien 01361
Universität Witten/Herdecke, Witten 10691
Universität Wuppertal, Wuppertal 10694
Universität Würzburg, Würzburg 10697
Universität zu Köln, Köln 09325
– Arbeitsstelle für Leseforschung und Kinder- und Jugendmedien (ALEKI), Köln 10132
– Archäologisches Institut, Abt. Klassische Archäologie und Abt. Archäologie der Römischen Provinzen, Köln 10133
– Bibliothek für Informatik und Wirtschaftsinformatik, Köln 10134
– Department Psychologie, Köln 10135
– Ehemaliges Pädagogisches Seminar (der Philosophischen Fakultät), Köln 10136
– Ehemaliges Seminar für Pädagogik, Abt. für Allgemeine Pädagogik, Köln 10137
– Englisches Seminar I, Köln 10138
– Fachbibliothek Biowissenschaften, Köln 10139
– Fachbibliothek Chemie, Köln 10140
– Fachbibliothek Versicherungswissenschaft, Köln 10141
– Fakultätsbibliothek Heilpädagogik und Rehabilitation, Köln 10142
– Forschungsinstitut für Politische Wissenschaft und Europäische

Fragen / Seminar für Politische Wissenschaft, Köln 10143
– Gemeinsame Bibliothek des Schwerpunkts Marketing – Institut für Messewirtschaft und Distributionsforschung, Köln 10144
– Gemeinsame Bibliothek des Schwerpunkts Marketing – Seminar für Allgemeine Betriebswirtschaftslehre, Marketing und Markenmanagement, Köln 10145
– Gemeinsame Bibliothek des Schwerpunkts Marketing – Seminar für Allgemeine Betriebswirtschaftslehre, Marktforschung und Marketing, Köln 10146
– Gemeinsame Fachbibliothek Rechtswissenschaft, Köln 10147
– GeoBibliothek, Köln 10148
– Geographisches Institut, Köln 10149
– Gesellschaft zur Förderung des Energiewirtschaftlichen Instituts an der Universität zu Köln e.V., Köln 10150
– Historisches Institut – Bibliothek für Mittlere und Neuere Geschichte, Köln 10151
– Historisches Seminar, Abt. für Osteuropäische Geschichte, Köln 10152
– Historisches Seminar, Anglo-Amerikanische Abteilung, Köln 10153
– Historisches Seminar, Iberische und Lateinamerikanische Abteilung, Köln 10154
– Humanwissenschaftliche Abteilung, Köln 10155
– Husserl-Archiv, Köln 10156
– Institut für Afrikanistik, Köln 10157
– Institut für Allgemeine Didaktik und Schulforschung, Köln 10158
– Institut für Altertumskunde, Köln 10159
– Institut für Altertumskunde, Abt. Byzantinistik, und Neugriechische Philologie, Köln 10160
– Institut für Arbeits- und Wirtschaftsrecht, Köln 10161
– Institut für Ausländisches und Internationales Strafrecht, Köln 10162
– Institut für Bankrecht, Köln 10163
– Institut für Berufs-, Wirtschafts- und Sozialpädagogik, Köln 10164
– Institut für Biologie und ihre Didaktik, Köln 10165
– Institut für das Recht der Europäischen Union, Köln 10166
– Institut für deutsche Sprache und Literatur I, Köln 10167
– Institut für Ethnologie, Köln 10168
– Institut für Europäische Musikethnologie, Köln 10169
– Institut für Evangelische Theologie – Dienststelle Wilhelm-Backhaus-Straße, Köln 10170
– Institut für Indologie und Tamilistik, Köln 10171
– Institut für internationales und ausländisches Privatrecht, Köln 10172
– Institut für Katholische Theologie, Abt. 1, Köln 10173
– Institut für Kirchenrecht und Rheinische Kirchenrechtsgeschichte, Köln 10174
– Institut für Kriminologie, Köln 10175
– Institut für Kunst und Kunsttheorie, Abteilung Textilgestaltung/Textilwissenschaft und ihre Didaktik, Köln 10176
– Institut für Linguistik – Sprachwissenschaft, Köln 10177
– Institut für Luft- und Weltraumrecht / Lehrstuhl für Völkerrecht, Europarecht, europäisches und internationales Wirtschaftsrecht, Köln 10178
– Institut für Musikpädagogik, Köln 10179
– Institut für Neuere Privatrechtsgeschichte, Deutsche und Rheinische Rechtsgeschichte, Köln 10180
– Institut für Niederlandistik, Köln 10181
– Institut für Öffentliches Recht und Verwaltungslehre, Köln 10182
– Institut für Ostrecht, Köln 10183
– Institut für Römisches Recht, Köln 10184
– Institut für Skandinavistik/Fennistik, Köln 10185
– Institut für Skandinavistik/Fennistik, Abt. Fennistik, Köln 10186
– Institut für Staatsrecht, Köln 10187
– Institut für Steuerrecht, Köln 10188
– Institut für Strafrecht und Strafprozessrecht, Köln 10189
– Institut für Theater-, Film- und Fernsehwissenschaft, Köln 10190
– Institut für Ur- und Frühgeschichte, Köln 10191
– Institut für Verfahrensrecht, Köln 10192
– Institut für Vergleichende Bildungsforschung und Sozialwissenschaften, Lehrbereich Soziologie, Köln 10193
– Institut für Vergleichende Bildungsforschung und Sozialwissenschaften – Politikwissenschaft, Bildungspolitik und Politische Bildung, Köln 10194
– Institut für Versicherungswissenschaft, Abt. Versicherungsrecht, Köln 10195
– Institut für Völkerrecht und ausländisches öffentliches Recht, Köln 10196
– Institut für Wohnungsrecht und Wohnungswirtschaft, Köln 10197
– Kunsthistorisches Institut, Abt. Allgemeine Kunstgeschichte, Köln 10198
– Kunsthistorisches Institut, Abt. Architekturgeschichte, Köln 10199
– Mathematisches Institut, Köln 10200
– Musikwissenschaftliches Institut, Köln 10201
– Orientalisches Seminar, Köln 10202
– Orientalisches Seminar, Malaiologischer Apparat, Köln 10203
– Ostasiatisches Seminar, Abt. Japanologie/KUGA-Japan, Köln 10204
– Petrarca-Institut, Köln 10205
– Philosophisches Seminar, Köln 10206
– Physikalische Institute, Köln 10207
– Portugiesisch-Brasilianisches Institut / Zentrum Portugiesischsprachige Welt, Köln 10208
– Regionales Rechenzentrum, Köln 10209
– Romanisches Seminar, Köln 10210
– Seminar für Ägyptologie, Köln 10211
– Seminar für Allgemeine Betriebswirtschaftslehre, Supply Chain Management und Produktion, Köln 10212
– Seminar für Allgemeine Betriebswirtschaftslehre und Bankbetriebslehre, Köln 10213
– Seminar für Allgemeine Betriebswirtschaftslehre und Betriebswirtschaftliche Steuerlehre (Steuerseminar), Köln 10214
– Seminar für Allgemeine Betriebswirtschaftslehre und für Wirtschaftsprüfung (Treuhandseminar), Köln 10215
– Seminar für Allgemeine Betriebswirtschaftslehre und Personalwirtschaftslehre, Köln 10216
– Seminar für Genossenschaftswesen, Köln 10217
– Seminar für Handel und Kundenmanagement, Köln 10218
– Seminar für Mathematik und ihre Didaktik, Köln 10219
– Seminar für Pädagogik, Abt. für Allgemeine Pädagogik, Köln 10220
– Seminar für Sozialpolitik, Köln 10221
– Seminar für Soziologie, Köln 10222
– Seminar für Staatsphilosophie und Rechtspolitik, Köln 10223
– Seminar für Wirtschafts- und Sozialstatistik / Lehrstuhl für Statistik und Ökonometrie Prof. Dr. K. Mosler, Köln 10224
– Seminar für Wirtschafts- und Sozialstatistik / Lehrstuhl Prof. Dr. F. Schmid, Köln 10225
– Seminar für Wirtschafts- und Unternehmensgeschichte, Köln 10226
– Slavisches Institut, Köln 10227
– Theaterwissenschaftliche Sammlung, Köln 10228
– Thomas-Institut, Köln 10229
– Volkswirtschaftliche Bibliothek 1, Köln 10230
– Volkswirtschaftliche Bibliothek 2, Köln 10231
– Wirtschafts- und Sozialgeographisches Institut, Köln 10232
– Wirtschaftsarchiv der WiSo-Fakultät, Köln 10233
Universität zu Lübeck, Lübeck 10251
– Institut für Medizingeschichte und Wissenschaftsforschung, Lübeck 10252
Universität Zürich, Zürich 26986
– Anthropologisches Institut und Museum, Zürich 27168
– Archäologisches Institut, Zürich 27169
– Bibliothek der Botanischen Institute, Zürich 27170
– Bibliothek für Betriebswirtschaft (BfB), Zürich 27171
– Bibliothek für Volkswirtschaft, Zürich 27172
– Biochemisches Institut, Zürich 27173
– Deutsches Seminar, Zürich 27174
– Deutsches Seminar, Abteilung für Nordische Philologie, Zürich 27175
– Englisches Seminar, Zürich 27176
– Ethik-Zentrum, Institut für Sozialethik, Zürich 27177
– Ethnologisches Seminar, Zürich 27178
– Forschungsbibliothek Jakob Jud, Zürich 27179
– Geographisches Institut, Zürich 27180
– Historisches Seminar, Zürich 27181
– Institut für Empirische Wirtschaftsforschung, Zürich 27182
– Institut für Erziehungswissenschaft, Zürich 27183
– Institut für Hermeneutik und Religionsphilosophie, Zürich 27184
– Institut für Mathematik, Zürich 27185
– Institut für Politikwissenschaft, Zürich 27186
– Institut für Populäre Kulturen, Zürich 27187
– Institut für Sonderpädagogik, Zürich 27188
– IPMZ – Institut für Publizistikwissenschaft und Medienforschung, Zürich 27189
– Klassisch-Philologisches Seminar, Zürich 27190
– Kunsthistorisches Institut, Zürich 27191
– Medizinhistorisches Institut und Museum, Zürich 27192
– Mittellateinisches Seminar, Zürich 27193
– Musikwissenschaftliches Institut, Zürich 27194
– Nordamerika-Bibliothek, Zürich 27195
– Orientalisches Seminar, Zürich 27196
– Ostasiatisches Seminar – Abendländische Bibliothek, Zürich 27197
– Ostasiatisches Seminar – Chinesische Bibliothek, Zürich 27198
– Ostasiatisches Seminar – Japanologie, Zürich 27199
– Paläontologisches Institut und Museum, Zürich 27200
– Philosophisches Seminar, Zürich 27201
– Physiologisches Institut, Zürich 27202
– Psychologisches Institut, Zürich 27203
– Rechtswissenschaftliches Institut, Zürich 27204
– Romanisches Seminar, Zürich 27205
– Seminar für Allgemeine Sprachwissenschaft, Zürich 27206
– Seminar für Allgemeine und Vergleichende Literaturwissenschaft (AVL), Zürich 27207
– Slavisches Seminar, Zürich 27208
– Soziologisches Institut, Zürich 27209
– Theologisches Seminar und Institut für Schweizerische Reformationsgeschichte und Institut für Hermeneutik und Religionsphilosophie, Zürich 27210
– Universitätsklinik Balgrist und Schweizerisches Paraplegikerzentrum, Zürich 27211
– Völkerkundemuseum, Zürich 27212
– Zentrale für Wirtschaftsdokumentation, Zürich 27213
– Zentrum für Zahn-, Mund- und Kieferheilkunde, Zürich 27214
Universität Zürich – Hauptbibliothek, Zürich 27215
Universitatea de arte 'George Enescu', Iaşi 21936
Universitatea de medicină şi farmacie, Cluj-Napoca 21930
Universitatea de Medicină şi Farmacie, Iaşi 21937
Universitatea de Medicină şi Farmacie, Târgu Mureş 21947
Universitatea de Medicină şi Farmacie, Timişoara 21948
Universitatea de medicina şi farmacie 'Carol Davila', Bucureşti 22019
Universitatea de Nord din Baia Mare, Baia Mare 21918
Universitatea de Şiinte Agricole si Medicina Veterinara 'Ion Ionescu de la Brad' Iasi, Iaşi 21938
Universitatea de Şiinte Agronomice si Medicina Veterinara, Bucureşti 21922
Universitatea de Stat din Moldova, Chisjnău 18889
Universitatea de Ştiinţe Agricloe a Banatului, Timişoara 22066
Universitatea de Ştiinţe Agricole şi Medicină Veterinară, Cluj-Napoca 21931
Universitatea din Craiova, Craiova 21933
Universitatea din Galaţi, Galaţi 21934
Universitatea din Petrosani, Petroşani 21940
Universitatea din Piteşti, Piteşti 21941
Universitatea din Timişoara, Timişoara 21949
Universitatea Naţională de Muzică din Bucureşti, Bucureşti 21923
Universitatea 'Petru Maior' Tîrgu-Mureş, Tîrgu Mureş 21951
Universitatea Politehnica Bucureşti, Bucureşti 21924
Universitatea Politehnica Timişoara, Timişoara 21950
Universitatea 'Ştefan cel Mare' Suceava, Suceava 21945
Universitatea Tehnica, Cluj-Napoca 21932
Universitatea Tehnicá de Constructii, Bucureşti 21925
Universitatea Tehnica 'Gh. Asachi' Iaşi, Iaşi 21939
Universitatea 'Transilvania' din Braşov, Braşov 21919
Universitäts- und Forschungsbibliothek Erfurt/Gotha – Forschungsbibliothek Gotha, Gotha 09835
Universitäts- und Forschungsbibliothek Erfurt/Gotha – Universitätsbibliothek Erfurt, Erfurt 09689
Universitäts- und Landesbibliothek, Bonn 09304
Universitäts- und Landesbibliothek, Darmstadt 09307
Universitäts- und Landesbibliothek, Münster 09329
Universitäts- und Landesbibliothek Düsseldorf, Düsseldorf 09312
Universitäts- und Landesbibliothek Sachsen-Anhalt, Halle (Saale) 09316
Universitäts- und Landesbibliothek Tirol, Innsbruck 01184
Universitäts- und Stadtbibliothek, Köln 09325
Universitäts-Augenklinik, Tübingen 10669
Universitätsbibliothek Bern, Bern 26965
Universitätsbibliothek Eichstätt-Ingolstadt, Eichstätt 09685
Universitätsbibliothek Erfurt, Erfurt 09689
Universitätsbibliothek Greifswald, Greifswald 09904
Universitätsbibliothek Johann Christian Senckenberg, Frankfurt am Main 09730
Universitätsbibliothek Kassel – Landesbibliothek und Murhardsche Bibliothek der Stadt Kassel, Kassel 09322
Universitäts-Hals-Nasen-Ohrenklinik, Tübingen 10670
Universitäts-Hautklinik, Tübingen 10671
Universitätsklinik Balgrist und Schweizerisches Paraplegikerzentrum, Zürich 27211
Universitätsklinik für Psychiatrie, Bern 27052
Universitätsklinik für Psychiatrie und Psychotherapie, Tübingen 10672
Universitätsklinik für Psychoanalyse und Psychotherapie, Wien 01298
Universitätsklinikum Bonn, Augenklinik, Bonn 09583
Universitätsklinikum Bonn – Zentrum für Zahn-, Mund- und Kieferheilkunde, Bonn 09589
Universitätsklinikum der Albert-Ludwigs-Universität – Institut für Medizinische Biometrie und Medizinische Informatik, Freiburg 09802
Universitätsklinikum Erlangen, Erlangen 11819
Universitätsklinikum Essen, Essen 09725
Universitätsklinikum Freiburg – Chirurgische Universitätsklinik, Freiburg 09803
Universitätsklinikum Freiburg – Medizinische Klinik, Freiburg 09804
Universitätsklinikum Hamburg-Eppendorf, Hamburg 09955
Universitätsklinikum Schleswig-Holstein / Campus Kiel, Kiel 10118
Universitätsmuseum für Kunst und Kulturgeschichte, Marburg 10348
Universitätssternwarte, München 10393
Université Abdou Moumouni, Niamey 19902
– Centre d'Enseignement Supérieur de Niamey, Niamey 19903
– Ecole Superieure d'Agronomie, Niamey 19904
– Institut de Recherches en Sciences humaines (IRSH), Niamey 19905
Université Abou Bekr Belkaid Tlemcen, Tlemcen 00048
Université Annaba, Annaba 00037
Université Blaise-Pascal – Clermont-Ferrand 2, Clermont-Ferrand 07686
– Bibliothèque de Droit et Sciences Economiques, Clermont-Ferrand 07687
– Bibliothèque universitaire Lettres SHS – Bibliothèque Lafayette, Clermont-Ferrand 07688
– Bibliothèque universitaire Santé, Clermont-Ferrand 07689
– Bibliothèque universitaire Sciences et Techniques, Aubière 07690
Université Catholique d'Afrique Centrale, Yaoundé 03638
Université Catholique de Lille, Lille 07729
– Bibliothèque Universitaire Vauban, Lille 07730

Université Catholique de l'Ouest, Angers 07639
Université catholique de Louvain, Louvain-la-Neuve 02395
– Bibliothèque de Droit, Louvain-la-Neuve 02396
– Bibliothèque de Philosophie (BISP), Louvain-la-Neuve 02397
– Bibliotheque de psychologie et des sciences de l'education (BPSP), Louvain-la-Neuve 02398
– Bibliothèque de sciences de la Santé, Bruxelles 02399
– Bibliothèque d'Education physique et de Réadaptation – IEPR, Louvain-la-Neuve 02400
– Bibliothèque des sciences économiques, sociales, politiques et de communication, Louvain-la-Neuve 02401
– Bibliothèque des Sciences et Technologies, Louvain-la-Neuve 02402
– Bibliothèque du Centre d'études théâtrales, Louvain-la-Neuve 02403
– Centre Cerfaux-Lefort, Louvain-la-Neuve 02404
– Centre pour la Recherche opérationelle et Etudes économétriques, Louvain-la-Neuve 02405
– Département AMCO, Louvain-la-Neuve 02406
Université catholique de Lyon, Lyon 07742
– Institut des Sources Chrétiennes, Lyon 07743
Université Charles-de-Gaulle – Lille III, Villeneuve-d'Ascq 08104
– Bibliothèque Angellier, Villeneuve-d'Ascq 08105
– Bibliothèque de l'Université – Section Sciences Humaines, Lettres et Arts, Villeneuve-d'Ascq 08106
– Bibliothèque Jacques Vandier, Egyptologie, Villeneuve-d'Ascq 08107
– UFR de Lettres Modernes, Villeneuve-d'Ascq 08108
– UFR d'Etudes Germaniques, Villeneuve-d'Ascq 08109
Université Cheikh Anta Diop de Dakar, Dakar 24410
– Centre d'Etudes des Sciences et Techniques de l'Information, Dakar 24411
– Département de Géographie, Dakar 24412
– Ecole de Bibliothécaires, Archivistes et Documentalistes, Dakar 24413
– Ecole Normale Supérieure, Dakar 24414
– Faculté des Sciences juridiques et économiques, Dakar 24415
– Institut de Médicine tropicale appliquée (IMTA), Dakar 24416
– Institut fondamental d'Afrique noire, Dakar 24417
Université Claude Bernard – Lyon 1, Villeurbanne 08117
– Bibliothèque Universitaire – Section Santé, Lyon 08118
– Bibliothèque Universitaire – Section Sciences, Villeurbanne 08119
– Institut de Physique Nucléaire de Lyon (IPNL), Villeurbanne 08120
Université d'Abidjan Cocody, Abidjan 06129
Université d'Abomey-Calavi, Cotonou 02902
Université d'Alger, Alger 00030
– Département de Biologie, Alger 00031
– INESSM – Departement de Pharmacie, Alger 00032
– Institut des Sciences économiques, Alger 00033
– Institut des Sciences juridiques et administratives, Alger 00034
– Institut d'Etudes Politiques et de l'information, Alger 00035
Université d'Angers, Angers 07640
– Bibliothèque Universitaire Saint-Serge – Section Droit-Economie, Angers 07641
– Bibliothèque universitaire – Section Médecine, Angers 07642
Université d'Antananarivo, Antananarivo 18451
– Ecole Nationale de Médecine et Pharmacie, Antananarivo 18452
– Faculté des Lettres et Sciences Humaines, Antananarivo 18453
– Musée d'Art et d'Archéologie, Antananarivo 18454
Université d'Artois, Arras 07644
Université d'Avignon, Avignon 07645
Université de Bangui, Bangui 04831
Université de Batna, Batna 00038
Université de Bordeaux I, Talence
– Bibliothèque Universitaire des Sciences et Techniques, Talence 08070
Université de Bourgogne, Dijon 07700
Université de Bretagne Occidentale, Brest 07662
– Bibliothèque de Medecine Odontologie, Brest 07663
– Centre de Recherche Bretonne et Celtique, Brest 07664
– Centre de ressources documentaires Lettres/Sciences sociales, Brest 07665
– Service Commun de Documentation – Section Droit-Sciences Economiques, Brest 07666
Université de Bretagne-Sud, Vannes
– BU Vannes – Centre, Vannes 08099
Université de Caen Basse-Normandie, Caen 07678
Université de Cergy-Pontoise, Cergy 07680
Université de Franche-Comté, Besançon
– Bibliothèque du Pôle universitaire du Pays de Montbéliard, Montbéliard 07649
– BU Droit, Science économiques et Gestion, Besançon 07650
– BU IUT de Besançon, Besançon 07651
– BU – Lettres et Sciences Humaines, Besançon 07652
– BU – Médecine-Pharmacie, Besançon 07653
– BU Sciences-Staps, Besançon 07654
– BU Universitaire Lucien Febvre, Belfort 07655
– Centre de Linguistique Appliquée, Besançon 07656
Université de Fribourg, Fribourg 26968
– Bibliothèque de la Faculté de droit (BFD), Fribourg 27068
– Bibliothèque de la Faculté des sciences (DOKPE), Fribourg 27069
– Bibliothèque de langues et littératures médiévales et modernes (BLL), Fribourg 27070
– Bibliothèque de l'Institut de droit européen, Fribourg 27071
– Bibliothèque de l'Institut de musicologie (MUS), Fribourg 27072
– Bibliothèque de l'Institut du fédéralisme (IFF), Granges-Paccot 27073
– Bibliothèque de l'Institut interfacultaire de l'Europe orientale et central (IEO), Granges-Paccot 27074
– Bibliothèque de pédagogie curative (IPC-HPI), Fribourg 27075
– Bibliothèque de pédagogie et psychologie (PSPE), Fribourg 27076
– Bibliothèque de sociologie, politiques sociales et travail social (STS), Fribourg 27077
– Bibliothèque des langues étrangères (BLE) et Centre d'auto-apprentissage, Fribourg 27078
– Bibliothèque des Sciences de l'Antiquité (SCANT), Fribourg 27079
– Bibliothèque d'Histoire de l'art et de Philosophie (BHAP), Fribourg 27080
– Bibliothèque du Département de mathématiques (MATH), Fribourg 27081
– Bibliothèque interfacultaire d'histoire et théologie (BHT), Fribourg 27082
– Bibliothèque Pérolles2, Fribourg 27083
Université de Genève, Genève 27092
Université de Haute-Alsace, Mulhouse 07782
Université de Kankan, Kankan 13187
Université de Kinsangani, Kinsangani 06046
Université de Kinshasa, Kinshasa 06047
– Faculté de médecine et de pharmacie, Kinshasa 06048
– Faculté de théologie catholique, Kinshasa 06049
– Faculté des sciences, Kinshasa 06050
– Institut supérieur pédagogique de Bukavu, Bukavu 06051
– Institut supérieur pédagogique de Mbujimayi, Mbujimayi 06052
Université de la Méditerranée – Aix-Marseille II, Marseille
– Bibliothèque de l'Université – Section Médecine-Odontologie, Marseille 07758
– Bibliothèque de l'Université – Section Pharmacie, Marseille 07759
– Bibliothèque Universitaire de Luminy, Marseille 07760
Université de la Polynésie Française, Faaa 09165
Université de la Réunion, Saint-Denis 21910
Université de la Rochelle, La Rochelle 07721
Université de Lausanne, Lausanne 26971
– Bibliothèque de Biologie, Lausanne 27101
– Bibliothèque de Droit et Sciences économiques (BDSE), Lausanne 27102
– Bibliothèque de la Riponne (BCU/R), Lausanne 27103
– Bibliothèque de l'Hôpital ophtalmique Jules Gonin, Lausanne 27104
– Bibliothèque des Cèdres, Lausanne 27105
– Bibliothèque des Sciences de la Terre, Lausanne 27106
– Bibliothèque scientifique commune UNIL-EPFL, Lausanne 27107
– Bibliothèque Universitaire de Médecine (BiUM), Lausanne 27108
– Département de Biologie Cellulaire et de Morphologie, Lausanne 27109
– Siège de Dorigny, Lausanne 27110
Université de Liège, Liège
– Bibliothèque de Droit, d'Economie, de Gestion et de Sciences sociales Léon Graulich, Liège 02375
– Bibliothèque des Sciences de la Vie – Botanique, Liège 02376
– Bibliothèque des Sciences de la Vie – Médecine, Liège 02377
– Bibliothèque des Sciences de la Vie – Médecine Vétérinaire, Liège 02378
– Bibliothèque des Sciences de la Vie – Psychologie et Sciences de l'Education et Logopédie, Liège 02379
– Bibliothèque des Sciences de la Vie – Zoologie, Liège 02380
– Bibliothèque des Sciences et Techniques, Liège 02381
– Bibliothèque des Sciences et Techniques – Environnement, Arlon 02382
– Bibliothèque des Sciences et Techniques – Géographie, Liège 02383
– Bibliothèque des Sciences et Techniques – Mathématique, Liège 02384
– Bibliothèque des Sciences et Techniques – Sciences, Liège 02385
– Bibliothèque des Sciences et Techniques – Sciences appliquées, Liège 02386
– Bibliothèque des Sciences et Techniques – Sciences de la Terre, Liège 02387
– Bibliothèque générale de Philosophie et Lettres, Liège 02388
– Bibliothèque générale de Philosophie et Lettres – Centre d'information et de conservation des Bibliothèques (CICB), Liège 02389
– Bibliothèque générale de Philosophie et Lettres – Langues et Littératures Germaniques, Liège 02390
– Bibliothèque générale de Philosophie et Lettres – Langues et Littératures Romanes, Liège 02391
– Bibliothèque générale de Philosophie et Lettres – Philosophie et Communication, Liège 02392
– Bibliothèque générale de Philosophie et Lettres – Sciences de l'Antiquité, Liège 02393
– Bibliothèque générale de Philosophie et Lettres – Sciences historiques, Liège 02394
Université de Liège – Campus d'Arlon, Arlon 02280
Université de Lille II, Lille
– Bibliothèque de l'Université – Section Droit-Gestion, Lille 07731
– Bibliothèque de l'Université – Section Santé, Lille 07732
Université de Limoges, Limoges
– Bibliothèque de Droit et Sciences Economiques, Limoges 07733
– Bibliothèque de Lettres et Sciences Humaines, Limoges 07734
– Bibliothèque de Santé, Limoges 07735
– Bibliothèque de Sciences et Techniques, Limoges 07736
Université de Lubumbashi, Lubumbashi 06055
Université de Marne-la-Vallée, Marne-la-Vallée 07752
Université de Moncton, Edmundston 03705
Université de Moncton, Moncton 03757
– Bibliothèque de droit Michel-Bastarache, Moncton 03758
– Centre de Resources Pédagogiques, Moncton 03759
– Centre d'Etudes Acadiennes, Moncton 03760
Université de Moncton, Moncton 03761
Université de Moncton, Shippagan 04496
Université de Mons, Mons 02411
– Bibliothèque de l'Ecole d'interprètes internationaux, Mons 02412
– Bibliothèque des Sciences et de la Médecine, Mons 02413
– Bibliothèque des Sciences Humaines – Sciences Economiques et Sociales, Mons 02414
– Faculté de Psychologie et des Sciences de l'Education, Mons 02415
Université de Montpellier, Montpellier 07769
– Bibliothèque Interuniversitaire – Section de Médecine, Montpellier 07770
– Bibliothèque Interuniversitaire – Section Lettres-Sciences Humaines, Montpellier 07771
– Bibliothèque Interuniversitaire – Section Pharmacie, Montpellier 07772
– Bibliothèque Interuniversitaire – Section Sciences, Montpellier 07773
Université de Montpellier II, Montpellier
– Bibliothèque des Mathématiques, Montpellier 07774
Université de Montpellier III, Montpellier
Université de Montreal, Montreal 03786
– Bibliothèque d'Aménagement, Montreal 03787
– Bibliothèque de bibliothéconomie et des sciences de l'information, Montreal 03788
– Bibliothèque de botanique, Montreal 03789
– Bibliothèque de Droit, Montreal 03790
– Bibliothèque de géographie, Montreal 03791
– Bibliothèque de kinésiologie, Montreal 03792
– Bibliothèque de la santé, Montreal 03793
– Bibliothèque de Mathématiques et Informatique, Montreal 03794
– Bibliothèque de Médecine vétérinaire, Saint-Hyacinthe 03795
– Bibliothèque de musique, Montreal 03796
– Bibliothèque de Physique, Montreal 03797
– Bibliothèque des lettres et sciences humaines, Montreal 03798
– Bibliothèque ÉPC-Biologie, Montreal 03799
– Bibliothèque paramédicale, Montreal 03800
– Service des livres rares et collections spéciales, Montreal 03801
Université de Mostaganem, Mostaganem 00040
Université de Nantes, Nantes
– Bibliothèque de l'Université – Section de Saint-Nazaire, Saint Nazaire 07806
– Bibliothèque de l'Université – Section Droit-Sciences Economiques, Nantes 07807
– Bibliothèque de l'Université – Section du Pôle Universitaire Yonnais, La Roche-sur-Yon 07808
– Bibliothèque de l'Université – Section Lettres et Sciences humaines, Nantes 07809
– Bibliothèque de l'Université – Section Santé, Nantes 07810
– Bibliothèque de l'Université – Section Sciences, Nantes 07811
– Bibliothèque de l'Université – Section Technologies, Nantes 07812
Université de N'Djamena, N'Djamena 04836
Université de Neuchâtel, Neuchâtel 27118
– Bibliothèque de biologie, Neuchâtel 27119
– Bibliothèque de chimie, Neuchâtel 27120
– Bibliothèque de droit, Neuchâtel 27121
– Bibliothèque de géologie et d'hydrogéologie, Neuchâtel 27122
– Bibliothèque de physique, Neuchâtel 27123
– Bibliothèque de théologie, Neuchâtel 27124
– Bibliothèque des pasteurs, Neuchâtel 27125
– Bibliothèque des sciences économiques, Neuchâtel 27126
– Bibliothèque d'ethno, Neuchâtel 27127
– Bibliothèques des lettres, Neuchâtel 27128
Université de Ngaoundéré, Ngaoundéré 03634
Université de Nice Sophia Antipolis, Nice
– Bibliothèque de Droit, Sciences politiques, économiques et de gestion, Nice 07815
– Bibliothèque de l'Archet, Nice 07816
– Bibliothèque de l'EPU Sophia, Sophia-Antipolis 07817
– Bibliothèque de Lettres, Arts, Sciences humaines, Nice 07818
– Bibliothèque de Médecine, Nice 07819
– Bibliotheque de Sciences, Nice 07820
Université de Nouvelle Calédonie, Nouméa 19753
Université de Ouagadougou, Ouagadougou 03612
Université de Pau et des Pays de l'Adour, Pau 07991
– Bibliothèque de l'Université – Section Droit-Lettres, Pau 07992

– Bibliothèque de l'Université – Section Sciences, Pau 07993
Université de Perpignan, Perpignan 07994
Université de Picardie Jules Verne, Amiens 07633
– Service Commun de la Documentation – Section Droit – Economie – Sciences et Techniques, Amiens 07634
– Service Commun de la Documentation – Section Lettres – Sciences humaines – Sports, Amiens 07635
– Service Commun de la Documentation – Section Médecin-Pharmacie, Amiens 07636
– Service Commun de la Documentation – Section Sciences, Amiens 07637
Université de Poitiers, Poitiers 08010
– Bibliothèque du Pôle Universitaire de Niort, Nort 08011
– Bibliothèque universitaire Droit-Lettres, Poitiers 08012
– Bibliothèque universitaire Médecine-pharmacie, Poitiers 08013
– Bibliothèque universitaire Sciences humaines, Arts et Moyen Age, Poitiers 08014
Université de Provence – Aix-Marseille I, Marseille
– Bibliothèque de l'UFR Civilisations et Humanités, Aix-en-Provence 07761
– Bibliothèque Sciences, Lettres et Sciences Humaines Saint Charles, Marseille 07762
Université de Reims, Reims
– Bibliothèque Universitaire – Section Droit-Lettres, Reims 08016
– Bibliothèque Universitaire – Section Santé, Reims 08017
– Bibliothèque Universitaire – Section Sciences, Reims 08018
Université de Rennes I, Rennes 08021
– Géosciences – Rennes – UMR 6118, Rennes 08022
– Institut de préparation à l'administration générale (IPAG), Rennes 08023
– Institut de Recherche Mathématique de Rennes (IRMAR), Rennes 08024
– SCD – Section Santé, Rennes 08025
– SCD – Section sciences économiques et gestion, sciences juridiques et politiques, Rennes 08026
– SCD – Section Sciences et Philosophie, Rennes 08027
Université de Rennes II – Haute Bretagne, Rennes 08028
– Centre de Recherche Historique sur les Sociétés et Cultures de l'Ouest (CRHISCO), Rennes 08029
– UFR de Géographie, Rennes 08030
– UFR de Psychologie, Sociologie et de Sciences de l'Education, Rennes 08031
Université de Rouen, Mont-Saint-Aignan 07778
– Bibliothèque de l'Université – Section Lettres, Mont-Saint-Aignan 07779
– Bibliothèque de l'Université – Section Médecine-Pharmacie, Rouen 07780
– Bibliothèque de l'Université – Section Sciences, Mont-Saint-Aignan 07781
Université de Savoie, Chambéry
– Bibliothèque Universitaire d'Annecy-le-Vieux, Chambéry 07681
– Bibliothèque Universitaire de Jacob-Bellecombette, Chambéry 07682
– Bibliothèque Universitaire du Bourget-Du-Lac, Le Bourget-du-Lac 07683
Université de Sherbrooke, Sherbrooke 03863
Université de Technologie de Compiègne, Compiègne 07691
Université de Technologie de Troyes, Troyes 08096
Université de Tours – François Rabelais, Tours 08095
Université de Tunis I, Tunis
Université de Yaoundé I, Yaoundé 03639
– Institut des Relations Interna-tionales du Cameroun (IRIC), Yaoundé 03640
Université des Antilles et de la Guyane, Pointe-à-Pitre 13160
Université des Antilles et de la Guyane – Section Guyane, Cayenne 09162
Université des Antilles et de la Guyane – Section Martinique, Schoelcher 18677

Université des Lettres, des Arts et des Sciences humaines – Tunis I, Tunis
– Faculté des Lettres Manouba, Tunis 27819
Université des Sciences et de la Technologie d'Oran, Oran 00042
Université des Sciences et de la Technologie Houari Boumedienne, Alger 00036
Université des Sciences et Techniques de Masuku (USTM), Franceville 09168
Université des Sciences et Technologies de Lille – Lille I, Villeneuve-d'Ascq 08110
Université d'Etat d'Haïti, Port-au-Prince 13200
Université d'Oran Es-Senia, Oran 00043
– Centre de Recherche et d'Information Documentaire en Sciences Sociales et Humaines, Oran 00044
– Institut National d'Enseignement Superieur en Sciences Médicales (INESSM), Oran 00045
Université d'Orléans, Orléans 07822
– Service Commun de la Documentation – Section Droit-Economie-Gestion, Orléans 07823
– Service Commun de la Documentation – Section Lettres, Orléans 07824
– Service Commun de la Documentation – Section Sciences, Orléans 07825
Université du Bénin, Lomé 27797
Université du Burundi, Bujumbura 03624
Université du Havre, Le Havre 07723
Université du Littoral – Côte d'Opale (ULCO), Dunkerque 07702
– Bibliothèque de l'Université – Section de Boulogne-sur-Mer, Sciences Humaines, Boulogne-sur-Mer 07703
Université du Luxembourg, Luxembourg 18381
– Campus Kirchberg, Luxembourg 18382
– Campus Walferdange, Walferdange 18383
– Éveil aux Sciences – Campus Walferdange, Luxembourg 18384
Université du Québec A. Chicoutimi, Chicoutimi 03690
Université du Québec à Montréal, Montreal 03802
Université du Québec à Montréal – Bibliothèque des Arts, Montreal 03803
Université du Québec à Rimouski, Rimouski 03834
Université du Québec à Trois-Rivières, Trois-Rivières 03928
Université du Québec en Abitibi-Témiscamingue, Rouyn-Noranda 03837
Université du Québec en Outaouais, Gatineau 03711
Université du Sud Toulon Var, La Garde 07720
Université du Travail, Charleroi 02308
Université Ferhat Abbas de Sétif, Sétif 00046
Université Gamal Abdul Nasser de Conakry, Conakry 13185
Université Hassan II – Aïn Chock, Casablanca 18917
– Faculté des Sciences juridiques, économiques et sociales, Casablanca 18918
Université Henri Poincaré – Nancy I, Nancy
– Bibliothèque de Médecine, Vandœuvre-les-Nancy 07787
– Bibliothèque des Sciences et Techniques, Villers-les-Nancy 07788
– Institut Elie Cartan – Bibliothèque de Mathématique, Vandœuvre-les-Nancy 07789
Université Islamique du Niger, Niamey 19906
Université Jean Monnet, Saint-Etienne
– BU Médecine, Saint-Etienne 08037
– BU Sciences, Saint-Etienne 08038
– BU Tréfilerie, Saint-Etienne 08039
Université Jean Moulin – Lyon 3, Lyon
– Bibliothèque de Droit et de Philosophie, Lyon 07744
– Bibliothèque Lettres et Langues, Lyon 07745
– Centre d'Etudes Universitaires de Bourg et de l'Ain (CEUBA), Bourg-en-Bresse 07746
Université Joseph Fourier – Grenoble I, Grenoble
– Administration du Service interétablissement de coopération documentaire, Saint-Martin-d'Hères 07711
– Bibliothèque de l'Université – Section Sciences, Saint-Martin-d'Hères 07712

– Institut de Géographie Alpine (IGA), Grenoble 07713
– Institut d'Economie et de Politique de l'Energie (IEPE), Saint-Martin-d'Hères 07714
– Institut National Polytechnique de Grenoble, La Tronche 07715
– Institut National Polytechnique de Grenoble, Section Sciences, Saint-Martin-d'Hères 07716
Université Laval, Québec 03828
Université Libanaise, Beirut 18208
Université Libre de Bruxelles, Bruxelles 02298
– Bibliothèque de Droit, Bruxelles 02299
– Bibliothèque des Sciences et Techniques, Bruxelles 02300
– Bibliothèque des Sciences Humaines, Bruxelles 02301
– Centre de Documentation Européenne (CDE), Bruxelles 02302
– Institut de Philosophie, Bruxelles 02303
– Institut d'Etude des Religions et de la Laïcité, Bruxelles 02304
– Maison Française d'Oxford, Oxford 02305
Université Lumière Lyon 2, Bron 07668
– Bibliothèque Arts et Lettres, Lyon 07669
– Bibliothèque de Sciences économiques et de Gestion, Lyon 07670
– Bibliothèque des Langues (Campus Berges du Rhône), Lyon 07671
– Bibliothèque des Langues (Campus Porte des Alpes), Bron 07672
– Bibliothèque interfacultés, Bron 07673
– Faculté Géographie, Histoire, Histoire de l'Art – Archéologie, Tourisme (GHHAT), Bron 07674
Université Lyon 1, Villeurbanne 08117
Université Lyon 2, Bron 07668
Université Lyon 3, Lyon
Université Marien Ngouabi, Brazzaville 06094
Université Mentouri Constantine, Constantine 00039
Université Michel de Montaigne Bordeaux 3, Pessac 07996
– Bibliothèque de Géographie et Cartothèque, Pessac 07997
– Bibliothèque du Centre d'Etudes des Mondes Moderne et Contemporain, Pessac 07998
– Bibliothèque Elie Vinet: Histoire, Histoire de l'Art, Archéologie, Pessac 07999
– Bibliothèque Etudes Germaniques et Scandinaves, Pessac 08000
– Bibliothèque Etudes Ibériques et Ibéro-Américaines, Pessac 08001
– Bibliothèque Henri Guillemin Lettres-Anglais, Pessac 08002
– Bibliothèque Joseph Moreau: Philosophie, Pessac 08003
– Bibliothèque LE-LEA, Pessac 08004
– Bibliothèque Robert Étienne – Maison de l'Archéologie, Pessac 08005
– Centre de Ressources de l'Institut des Sciences de l'Information et de Communication (ISIC), Pessac 08006
– Centre de Ressources de l'IUT Montaigne, Bordeaux 08007
Université Mohammed V – Agdal, Rabat
– Faculté des Lettres et des Sciences Humaines, Rabat 18927
– Faculté des Sciences, Rabat 18928
– Faculté des Sciences de l'Education, Rabat 18929
– Faculté des Sciences Juridiques, Economiques et Sociales, Rabat 18930
– Institut Scientifique, Rabat 18931
Université Montesquieu – Bordeaux IV, Pessac 08008
– Infothèque du PUSG – SCD, Bordeaux 08009
Université Nancy II, Nancy
– SCD – Bibliothèque Universitaire de Gestion, Nancy cedex 07790
– SCD – Bibliothèque Universitaire de Lettres et Sciences humaines, Nancy 07791
– SCD – Bibliothèque Universitaire Droit et Sciences économiques, Nancy 07792
– SCD – IUT Nancy-Charlemagne, Nancy 07793
Université Nationale de Côte d'Ivoire, Abidjan 06130
Université Nationale du Rwanda, Butare 24367
– Campus Universitaire de Ruhengeri, Ruhengeri 24368
Université Omar Bongo, Libreville 09169
Université Panthéon-Assas – Paris 2, Paris 07892
– Bibliothèque littéraire Jacques Doucet, Paris 07893

– Centre de recherche de commerce international, Paris 07894
– Institut de Droit Comparé, Paris 07895
– Institut des Hautes Etudes Internationales (IHEI), Paris 07896
– Institut Français de Presse et des Sciences de l'Information (IFP), Paris 07897
Université Panthéon-Sorbonne – Paris 1, Paris 07898
– La Bibliothèque Broca-Droit – IAE, Paris 07899
– Bibliothèque Cujas, Paris 07900
– Le Bibliothèque Cuzin de Philosophie, Paris 07901
– Bibliothèque d'Archéologie et des Sciences de l'Antiquité, Nanterre 07902
– Bibliothèque d'art plastiques, Paris 07903
– Bibliothèque de droit communau-taire, Paris 07904
– Bibliothèque de Géographie Sorbonne, Paris 07905
– Bibliothèque de la Sorbonne, Paris 07906
– Bibliothèque de la Sorbonne, Bibliothèque Victor Cousin, Paris 07907
– Bibliothèque de Recherches Africaines, Paris 07908
– Bibliothèque du Centre Pierre Mendès-France (PMF), Paris 07909
– Bibliothèque Interuniversitaire Sainte-Geneviève, Paris 07910
– Bibliothèque Lavisse, Paris 07911
– Centre de recherche d'archéologie orientale (centre Jean Deshayes), Paris 07912
– Centre de Recherches d'Histoire Nord-Américaine, Paris 07913
– Centre d'Economie de la Sorbonne, Paris 07914
– Histoire et Civilisation Byzantines et ses sources – Histoire du Proche Orient médiéval, Paris 07915
– Institut des Sciences Sociales du Travail (ISST), Bourg La Reine 07916
– Institut d'Études du Développement Économique et Social (CED), Nogent sur Marne 07917
– Institut d'Histoire de la Révolution Française (IHRF), Paris 07918
– UFR d'Histoire – Centre de Recherches d'Histoire de l'Antiquité, Paris 07919
Université Paris I – Panthéon-Sorbonne, Paris 07898
Université Paris 2 – Panthéon-Assas, Paris 07892
Université Paris 3 – Sorbonne-Nouvelle, Paris 07975
Université Paris 4 – Paris-Sorbonne, Paris 07936
Université Paris 5 – René Descartes, Paris 07920
Université Paris 7 – Paris Diderot, Paris
Université Paris 8 – Vincennes-Saint-Denis, Saint-Denis 08033
Université Paris 9 – Paris-Dauphine, Paris 07931
Université Paris 10 – Paris-Nanterre, Nanterre 07795
Université Paris 11 – Paris-Sud, Orsay
Université Paris 12 – Paris-Val-de-Marne, Créteil 07692
Université Paris 13 – Paris-Nord, Villetaneuse 08111
Université Paris Descartes – Paris 5, Paris 07920
– Bibliothèque de Droit, Malakoff 07921
– Bibliothèque de Psychologie Henri Piéron, Boulogne 07922
– Bibliothèque du Centre Universitaire des Saint-Pères, Paris 07923
– Bibliothèque Interuniversitaire de Pharmacie (BIUP), Paris 07924
– Bibliothèque Interuniversitaire de Santé (BIU Santé Paris), Paris 07925
– Bibliothèque Universitaire d'Odontologie de Montrouge, Montrouge 07926
– Bibliothèque Universitaire médicale Cochin Port Royal, Paris 07927
– Bibliothèque Universitaire médicale Necker – enfants malades, Paris 07928
– Bibliothèque Universitaire médicale Paris-Ouest Hôpital Raymond Poincaré, Garches 07929
– Centre Technique Documentaire des Sciences du Sport, Paris 07930
Université Paris-Dauphine – Paris 9, Paris 07931
Université Paris-Diderot – Paris 7, Paris
– Bibliothèque Centrale, Paris 07932

– Bibliothèque de l'UFR de Medecine – Site Xavier Bichat, Paris 07933
– Bibliothèque de l'UFR de Médicine, Site Villemin, Paris 07934
– Bibliothèque de l'UFR d'Ètudes Anglophones, Paris 07935
Université Paris-Nanterre – Paris 10, Nanterre 07795
– Bibliothèque de Philosophie, Epistemologie, Esthétique, Nanterre 07796
– Bibliothèque des Langues vivantes, Nanterre 07797
– Bibliothèque d'Histoire, Nanterre 07798
– Laboratoire d'Ethnologie et de Sociologie comparative, Nanterre 07799
– UFR de Langues Romanes, Nanterre 07800
– UFR de Sciences Juridiques et Politiques, Nanterre 07801
– UFR d'Etudes Anglo-Américaines, Nanterre 07802
Université Paris-Nord – Paris 13, Villetaneuse 08111
– Bibliothèque universitaire – Section Droit et Lettres, Villetaneuse 08112
– Bibliothèque universitaire – Section Sciences, Villetaneuse 08113
– UFR de Médecine, Bobigny 08114
Université Paris-Sorbonne – Paris 4, Paris 07936
– Service Commun de la Documentation -Bibliotèque de l'UFR d'études ibériques 'Marcel Bataillon', Paris 07937
– Service Commun de la Documentation -Bibliothèque Clignancourt, Paris 07938
– Bibliothèque de l'INHA (collections Jacques Doucet), Paris 07939
– Bibliothèque d'Histoire de l'Occident moderne, Paris 07940
– Bibliothèque d'Histoire des Religions, Paris 07941
– Bibliothèque générale de philosophie, Paris 07942
– Bibliothèque Georges Ascoli et Paul Hazard de l'UFR de littérature française et comparée, Paris 07943
– Bibliothèque Lalande, Paris 07944
– Service Commun de la Documentation -Bibliothèque Malesherbes, Paris 07945
– Service Commun de la Documentation -Bibliothèque Michelet, Paris 07946
– Service Commun de la Documentation -Bibliothèque Serpente, Paris 07947
– Centre de Recherches Egyptologiques (CRES), Paris 07948
– Centre d'Enseigment et de Recherche d'Oc (CEROC), Paris 07949
– Centre d'etudes catalanes, Paris 07950
– Centre d'Etudes Slaves, Paris 07951
– Centre Gustave Glotz, Paris 07952
– Centre Roland Mousnier. Histoire et Civilisation, Paris 07953
– Institut d'Études Augustiennes, Paris 07954
– Institut d'Etudes Grecques, Paris 07955
– Institut d'Etudes Latines, Paris 07956
– Institut Néo-Hellénique, Paris 07957
Université Paris-Sud – Paris 11, Orsay
– Bibliothèque Droit-Economie-Gestion – Site d'Orsay, Orsay 07826
– Bibliothèque Mathématique Jacques Hadamard, Orsay 07827
– Bibliothèque universitaire: Droit-Economie-Gestion, Sceaux 07828
– Bibliothèque universitaire – Médecine, Le Kremlin-Bicêtre 07829
– Bibliothèque universitaire: Pharmacie, Châtenay-Malabry 07830
– Bibliothèque universitaire: Sciences, Orsay 07831
– Bibliothèque universitaire: STAPS, Orsay 07832
– Institut de Physique Nucléaire, Orsay 07833
– Institut Universitaire de Technologie (IUT), Orsay 07834
Université Paris-Val-de-Marne – Paris 12, Créteil 07692
– Bibliothèque de Droit, Créteil 07693
– Bibliothèque de l'Université – Section Médecine, Créteil 07694
– Bibliothèque Multidisciplinaire, Créteil 07695
– Institut d'Urbanisme de Paris (IUP), Créteil 07696
Université Paul Cézanne Aix-Marseille III, Aix-en-Provence
– Bibliothèque Universitaire – Section Droit et Sciences économiques, Aix-en-Provence 07628
– Bibliothèque Universitaire – Section Lettres, Aix-en-Provence 07629
– Bibliothèque Universitaire – Section Sciences Saint Jérôme, Marseille 07630
– Institut d'Administration des Enterprises (IAE), Puyricard 07631
Université Paul Verlaine – Metz, Metz 07763
– Blbliothèque du Technopôle, Metz 07764
Université Paul-Valéry – Montpellier III, Montpellier
– Charles Dugas, Montpellier 07775
– Université de Nîmes, Nîmes 07776
Université Pierre et Marie Curie, Paris
– Bibliothèque de Biologie Enseignement, Paris 07958
– Bibliothèque de Biologie Recherche, Paris 07959
– Bibliothèque de Chimie Enseignement, Paris 07960
– Bibliothèque de Chimie-Physique Recherche, Paris 07961
– Bibliothèque de Mathématiques-Informatique Enseignement, Paris 07962
– Bibliothèque de Mathématiques-Informatique Recherche, Paris 07963
– Bibliothèque de Neurosciences J.-M. Charcot, Paris 07964
– Bibliothèque de Physique Enseignement, Paris 07965
– Bibliothèque de stomatologie et chirurgie maxillo-faciale Michel Dechaume, Paris 07966
– Bibliothèque des Sciences de la Terre Enseignement, Paris 07967
– Bibliothèque des Sciences de la Terre Recherche, Paris 07968
– Bibliothèque d'UFR La Pitié-Salpêtrière, Paris 07969
– Bibliothèque d'UFR Saint-Antoine, Paris 07970
– Bibliothèque hospitalière de Saint-Antoine (Axial-Caroli), Paris 07971
– Bibliothèque hospitalière Tenon-Meyniel, Paris 07972
– Bibliothèque hospitalière Trousseau, Paris 07973
– Bibliothèque L1-L2 scientifique, Paris 07974
Université Pierre Mendès France – Grenoble II, Grenoble
– Bibliothèque Universitaire Droit-Lettres, Saint-Martin-d'Hères 07717
– Institut d'études politiques, Saint-Martin-d'Hères 07718
Université Quaraouyine, Fès 18919
– Faculté de Langue Arabe, Marrakech 18920
Université Quisqueya, Port-au-Prince 13201
Université Saint Esprit de Kaslik, Jounieh 18212
Université Sainte-Anne, Church Point 03691
Université Saint-Joseph, Beirut
– Bibliothèque Orientale, Beirut 18209
– Faculté des Sciences Médicales, Beirut 18210
– Sections Droit, Economie, Gestion et Politique, Beirut 18211
Université Sidi-Mohamed Ben Abdellah, Fès 18921
– Faculté des Arts, Fès 18922
– Faculté des Sciences juridiques, économiques et sociales, Fès 18923
Université Sorbonne-Nouvelle – Paris 3, Paris 07975
– Bibliothèque Centrale, Paris 07976
– Bibliothèque d'Allemand, Asnières-sur-Seine 07977
– Bibliothèque de Littérature Générale et Comparée, Paris 07978
– Bibliothèque des Études Portugaises et Brésiliennes, Paris 07979
– Bibliothèque d'Études Italiennes et Roumaines, Paris 07980
– Bibliothèque du Monde Anglophone, Paris 07981
– Bibliothèque Nordique, Paris 07982
– Bibliothèque Orient et Monde arabe, Paris 07983
– Centre de Documentation et de Recherche de l'UFR de Littérature et Linguistique Françaises et Latines, Paris 07984
– Institut des Hautes Etudes de l'Amérique Latine (IHEAL), Paris 07985
– Institut d'Etudes Iraniennes, Paris 07986
– Institut d'Etudes Théâtrales, Paris 07987
– Institut d'Etudes Turques, Paris 07988
Université Toulouse I – Sciences Sociales, Toulouse 08082
– Bibliothèque Garrigou, Toulouse 08083
– Bibliothèque Universitaire de la Manufacture, Toulouse 08084
Université Toulouse II – Le Mirail, Toulouse 08085
– Bibliothèque de Lettres, Philosophie et Musique, Toulouse 08086
– Bibliothèque Hispanique et Hispano-Américaine, Toulouse 08087
– Bibliothèque Histoire, Arts et Archéologie, Toulouse 08088
– Bibliothèque Universitaire de Lettres, Toulouse 08089
Université Toulouse III – Paul Sabatier, Toulouse 08090
– Bibliothèque de Sciences, Toulouse 08091
– Bibliothèque Universitaire de Santé, Toulouse 08092
Université Victor Segalen Bordeaux II, Bordeaux 07659
– Bibliothèque des Sciences de l'Homme et Odontologie, Bordeaux 07660
– Bibliothèque Sciences de la Vie et de la Santé, Bordeaux 07661
Université Vincennes-Saint-Denis – Paris 8, Saint-Denis 08033
– Institut Fançais d'Urbanisme (IFU), Marne-La-Vallée 08034
Universiteit Antwerpen, Antwerpen
– UA-bibliotheek – Campus Drie Eiken, Antwerpen 02276
– UA-bibliotheek – Campus Groeneborger, Antwerpen 02277
– UA-bibliotheek – Campus Middelheim, Antwerpen 02278
– UA-bibliotheek – Stadscampus, Antwerpen 02279
Universiteit Gent, Gent 02328
– Faculteitsbibliotheek Economie en Bedrijfskunde, Gent 02329
– Faculteitsbibliotheek Faculteitsbibliotheek Bio-ingenieurswetenschappen, Gent 02330
– Faculteitsbibliotheek Letteren en Wijsbegeerte, Gent 02331
– Faculteitsbibliotheek Psychologie en Pedagogische Wetenschappen, Gent 02332
– Faculteitsbibliotheek Rechtsgeleerdheid, Gent 02333
– Faculteitsbibliotheek Wetenschappen, Gent 02334
– Geologisch Instituut, Gent 02335
– Vakgroep Gerechtelijke Geneeskunde, Gent 02336
– Vakgroep Geschiedenis, Gent 02337
– Vakgroep Letterkunde – Engelse en Amerikanse Literatuur, Gent 02338
– Vakgroep Taalkunde – Nederlandse Taalkunde, Gent 02339
– Vakgroep Taalkunde – Romaanse Taalkunde, Gent 02340
Universiteit Hasselt, Diepenbeek 02309
Universiteit Leiden, Leiden 19152
– Afdeling Wiskunde en Informatica, Leiden 19153
– Anatomisch-Embryologisch Laboratorium, Leiden 19154
– Bibliotheek Clusius, Leiden 19155
– Centrum voor Japanologie en Koreanistiek, Leiden 19156
– Centrum voor Milieuwetenschappen Leiden (CML), Leiden 19157
– Collectie Sociologie, Leiden 19158
– Faculteit der Sociale Wetenschappen, Leiden 19159
– Fysiologisch Laboratorium, Leiden 19160
– Gorlaeus Laboratoria, Leiden 19161
– Instituut voor Oosteuropees Recht en Ruslandkunde, Leiden 19162
– Juridisch Studiecentrum Hugo de Groot, Leiden 19163
– Juridische Bibliotheek, Leiden 19164
– Kern Institute Library, Leiden 19165
– Prentenkabinet, Leiden 19166
– Sinologisch Instituut, Leiden 19167
– Van der Klaauw Laboratorium, Blbliotheek Zoologie, Leiden 19168
– Van Vollenhoven Instituut voor Recht, Bestuur en Ontwikkeling, Leiden 19169
Universiteit Maastricht (UM), Maastricht 19171
Universiteit Twente, Enschede 19123
Universiteit Utrecht, Utrecht 19217
– Bibliotheek Diergeneeskunde, Utrecht 19218
– Juridische Bibliotheek, Utrecht 19219
– Medische Bibliotheek, Utrecht 19220
– Universiteit voor Humanistiek, Utrecht 19221
– Universiteitsbibliotheek Binnenstad, Utrecht 19222
– Universiteitsbibliotheek Uithof, Utrecht 19223
– Universiteitsmuseum, Utrecht 19224
Universiteit van Amsterdam, Amsterdam 19059
– Bibliotheek Amstel Instituut, Amsterdam 19060
– Bibliotheek Bungehuis, Amsterdam 19061
– Bibliotheek Geesteswetenschappen, Amsterdam 19062
– Bibliotheek Klassieken en Archeologie, Amsterdam 19063
– Bibliotheek Kunstgeschiedenis / Culturele Studies, Amsterdam 19064
– Bibliotheek Moderne Nabije Oosten, Amsterdam 19065
– Bibliotheek Natuurkunde, Scheikunde en Aardwetenschappen, Amsterdam 19066
– Bibliotheek P.C. Hoofthuis, Amsterdam 19067
– Bibliotheek Religiestudies, Amsterdam 19068
– Bibliotheek Sterrenkunde en Informatica, Amsterdam 19069
– Bibliotheek Tandheelkunde, Amsterdam 19070
– Bibliotheek Theaterwetenschap/Mediastudies en Muziekwetenschap, Amsterdam 19071
– Bibliotheek Wijsbegeerte, Amsterdam 19072
– Bibliotheek Wiskunde en Informatica, Amsterdam 19073
– Bushuisbibliotheek, Amsterdam 19074
– Juridische Bibliotheek, Amsterdam 19075
– Medische Bibliotheek AMC, Amsterdam 19076
– Pierson Révész Bibliotheek, Amsterdam 19077
Universiteit van Aruba, Oranjestad 00364
Universiteit van de Nederlandse Antillen, Willemstad 06316
Universiteit van die Vrijstaat, Bloemfontein 25002
Universiteit van Stellenbosch, Stellenbosch 25049
Universiteit van Tilburg, Tilburg 19207
Universiteit voor Humanistiek, Utrecht 19221
Universiteitsmuseum, Utrecht 19224
Universitet ekonomiki i finansov, Sankt-Peterburg 22548
Universitet elektronnoi tekhniki, Moskva 22396
Universitet i Stavanger, Stavanger 20161
Universitet pedagogicheskogo masterstva, Sankt-Peterburg 22549
Universitet putei soobshcheniya, Moskva 22397
Universitet radiotekhniki, elektroniki i avtomatiki, Moskva 22398
Universitet vodnykh kommunikatsi, Sankt-Peterburg 22550
Universitetet for Miljø- og Biovitenskap, Ås 20067
Universitetet i Agder, Grimstad 20093
Universitetet i Bergen, Bergen 20075
– Bibliotek for juridiske fag, Bergen 20076
– Bibliotek for realvag, Bergen 20077
– Historisk Museum, Bergen 20078
– Instituttsamlinger for Det historisk-filosofiske fakultet, Bergen 20079
– Det medisinske fakultetsbibliotek, Bergen 20080
– Det odontologiske fakultetsbibliotek, Bergen 20081
– Det psykologiske fakultetsbibliotek, Bergen 20082
Universitetet i Oslo, Oslo 20125
– Anatomisk institutt, Oslo 20126
– Astrofysiks Bibliotek, Oslo 20127
– Bibliotek for humaniora og samfunnsvitenskap, Oslo 20128
– Bibliotek for medisin og helsefag, Oslo 20129
– Bibliotek for medisin og helsefag, Psykiatrisk institutt, Vinderen, Oslo 20130
– Bibliotek for sykepleievitenskap og samfunnsmedisinske fag, Oslo 20131
– Biblioteket ved Institutt for kriminologi og rettssosiologi, Oslo 20132
– Biologisk Bibliotek, Oslo 20133
– Department of British and American Studies, Oslo 20134
– Farmasøytisk bibliotek, Oslo 20135
– Fysisk Bibliotek, Oslo 20136
– Geografisk Bibliotek, Oslo 20137
– Informatikkbiblioteket, Oslo 20138
– Institutt for Geofysikk, Oslo 20139
– Institutt for Nordisk Språk og Literatur, Oslo 20140
– Institutt for Offentlig Rett, Oslo 20141
– Institutt for Privatrett, Oslo 20142
– Det juridiske fakultetsbibliotek, Oslo 20143
– Kjemisk bibliotek, Oslo 20144
– Matematisk bibliotek, Oslo 20145
– Det matematisk-naturvitenskapelige fakultetsbibliotek, Oslo 20146
– Norsk musikksamling, Oslo 20147

– Det odontologiske fakultetsbibliotek, Oslo 20148
– Patologibyggets bibliotek, Oslo 20149
– Realfagsbiblioteket. Geologi, Oslo 20150
– Rettshistorisk samling, Oslo 20151
– Senter for rettsinformatikk, Oslo 20152
– Det Teologiske Fakultetsbibliotek, Oslo 20153
– Zoologisk museums bibliotek, Oslo 20154
Universitetet i Tromsø, Tromsø
– Avdelingsbibliotek for fysikk, Tromsø 20166
– Bibliotek for humaniora og samfunnsfag, Tromsø 20167
– Bibliotek for realfag, medisin og helsefag, Tromsø 20168
– Universitetsbiblioteket ved Tromsø Museum, Tromsø 20169
Universitetets kulturhistoriske museer – Myntkabinettets bibliotek, Oslo 20288
Universitetets kulturhistoriske museer – Oldsaksamlingens bibliotek, Oslo 20289
Universiteti Bujqësor i Tiranes, Tiranë 00011
Universiteti 'Luigj Gurakuqi', Shkodër 00007
Universiteti Tiranës, Tiranë 00012
Universitetssygehuset Herlev, Herlev 06779
Universiti Kebangsan Malaysia, Bangi 18497
– Cabang Kampus Sabah, Kota Kinabalu 18498
– Cabang Perubatan, Kuala Lumpur 18499
Universiti Malaya, Kuala Lumpur 18504
– Akademi Islam, Kota Bharu 18505
– Engineering Library, Kuala Lumpur 18506
– Fakulti Ekonomi dan Pentadbiran, Kuala Lumpur 18507
– Fakulti Pendidikan, Kuala Lumpur 18508
– Institut Pengajian Siswazah & Penyelidikan, Kuala Lumpur 18509
– Periodicals Library, Kuala Lumpur 18510
– Perpustakaan Perubatan, Kuala Lumpur 18511
– Perpustakaan Undang-undang, Kuala Lumpur 18512
Universiti Putra Malaysia, Selangor 18516
Universiti Sains Malaysia, Penang 18513
– Perpustakaan Kampus Cawangan Perak, Tronoh 18514
Universiti Teknologi Malaysia, Johor Bahru 18501
Universiti Teknologi MARA, Shah Alam 18517
– Cawangan Kelantan, Machang 18518
– Cawangan Melaka, Alor Gajah 18519
– Cawangan Pahang, Bangar jenka 18520
– Cawangan Perlis, Arau 18521
– Cawangan Petaling Jaya, Petaling Jaya 18522
– Cawangan Sabah, Kota Kinabalu 18523
– Cawangan Sarawak, Kota Samarahan 18524
– Cawangan Terengganu, Dungun 18525
Universiti Utara Malaysia, Jitra 18500
University and State Library, Darmstadt 09307
University and State Library RheinMain, Wiesbaden 09342
University at Albany, State University of New York, Albany 30182
– Governor Thomas E. Dewey Graduate Library for Public Affairs & Policy, Albany 30183
– Science Library, Albany 30184
University at Buffalo – State University of New York, Buffalo 30593
– Architecture & Planning Library, Buffalo 30594
– Charles B. Sears Law Library, Buffalo 30595
– Health Sciences Library, Buffalo 30596
– Lockwood Memorial Library, Buffalo 30597
– Music Library, Buffalo 30598
– Oscar A. Silverman Undergraduate Library, Buffalo 30599
University Brunei Darussalam, Gadong 03447
University Cardenal Herrera, Moncada 25507
University Clinic of Respiratory and Allergic Diseases Golnik, Golnik 24910
University Club Library, New York 37268
University Collection of Coins and Medals, Oslo 20252
University College, Oxford 29357
University College Cork, Cork 14462
– Medical Library, Cork 14463
University College Dublin, Dublin 14477
– Medical Library, Dublin 14478
– Michael Smurfit Graduate School of Business, Dublin 14479
– School of Architecture Richview, Dublin 14480
– Veterinary Medicine Library, Dublin 14481
University College Ghent, Gent
– BIOT Library, Melle 02321
– Mercator Faculty of Applied Business, Gent 02320
– School of Translation Studies, Gent 02326
– University College Ghent, Faculty of Fine Arts, Gent 02324
University College London, London 29224
– Bartlett Built Environment Library, London 29225
– Cruciform Library, London 29226
– Institute of Child Health, London 29227
– Institute of Orthopaedics, Stanmore 29228
– UCL School of Slavonic and East European Studies, London 29229
University College Nordjylland, Aalborg 06635
University College Northampton, Northampton 29286
University College of Art, Crafts and Design, Hägersten 26389
University College of Belize, Belize City 02897
University College of Cape Breton, Sydney 03875
University College of Engineering, Hyderabad 13658
University College of Medical Science, Delhi 13639
University College of St Martin, Ambleside 28932
University College of the Fraser Valley, Abbotsford 03668
University College of the North, The Pas 04034
University College Plymouth of St Mark and St John, Plymouth 29371
University College Sjælland, Slagelse 06747
University Duisburg-Essen, Duisburg 09679
University Duisburg-Essen, Essen 09724
University for Development Studies, Tamale 13020
University for the Creative Arts, Maidstone 29262
University Grants Commission, Islamabad 20489
University Grants Commission, New Delhi 14173
University Health Network, Toronto 04533
University Health Network, Toronto 04534
University Hospital, Bratislava 24688
University Hospital Herlev, Herlev 06779
University Hospital of North Tees, Stockton-on-Tees 29902
University 'Kiev-Mohyla Academy', Kyiv 28028
University Library Frankfurt am Main, Frankfurt am Main 09730
– Law and Economics Library, Frankfurt am Main 09734
– School of Dentistry, Frankfurt am Main 09752
University Library of Defence, Brussel 02297
University Library of Pula, Pula 06150
University Library of Regensburg, Regensburg 10540
University Library of Split, Split 06142
University Library VU Amsterdam, Amsterdam 19079
– History Library, Amsterdam 19084
– Institute for Environmental Studies (IVM), Amsterdam 19090
– Library of Human Movement Sciences, Amsterdam 19081
– Librayry of Economics and Business Administration, Amsterdam 19082
– University Library, Special Collections / Study Centre for Protestant Book Culture, Amsterdam 19093
University Museum of Cultural Heritage, Oslo 20289
University Museum of Cultural Heritage – Numismatics Library, Oslo 20288
University of Aberdeen, Aberdeen 28925
– Medical Library, Aberdeen 28926
University of Abertay Dundee, Dundee 29084
University of Adelaide, Adelaide 00375
– Elder Music Library, Adelaide 00376
– Law Library, Adelaide 00377
– Roseworthy Campus Library, Roseworthy 00378
– Waite Library, Urrbrae 00379
University of Aerospace Technology, Moskva 22328
University of Agder, Grimstad 20093
University of Agder, Kristiansand 20101
– Agder College, Kristiansand 20103
University of Agricultural Sciences, Bangalore 13613
University of Agricultural Sciences and Veterinary Medicine, Bucureşti 21922
University of Agricultural Sciences and Veterinary Medicine "Ion Ionescu de la Brad" Iasi, Iaşi 21938
University of Agriculture, Abeokuta 19918
University of Agriculture, Faisalabad 20429
University of Agriculture, Makurdi 19968
University of Agriculture, Sankt-Peterburg 22507
University of Agriculture and Forestry, Ho Chi Minh City 42279
University of Agriculture and Technology – Fuchu Campus, Fuchu 16915
University of Akron, Akron 30174
– Science and Technology Library, Akron 30175
University of Alabama, Tuscaloosa 32857
– Angelo Bruno Business Library, Tuscaloosa 32858
– McLure Education Library, Tuscaloosa 32859
– Rodgers Library for Science and Engineering, Tuscaloosa 32860
– School of Law Library, Tuscaloosa 32861
– William Stanley Hoole Special Collections Library, Tuscaloosa 32862
University of Alabama at Birmingham, Birmingham
– Mervyn H. Sterne Library, Birmingham 30433
– University of Alabama at Birmingham, Birmingham 30434
University of Alabama – Health Sciences Library, Tuscaloosa 32863
University of Alabama in Huntsville, Huntsville 31378
University of Alaska Anchorage – Consortium Library, Anchorage 30220
University of Alaska, Anchorage – Environment & Natural Resources Institute, Anchorage 30221
University of Alaska Anchorage Matanuska-Susitna College, Palmer 32227
University of Alaska Fairbanks, Fairbanks 31115
– BioSciences Library, Fairbanks 31116
University of Alaska – Northwest Campus, Nome 32119
University of Alaska Southeast, Ketchikan 31502
University of Alaska Southeast, Juneau, Juneau 31472
University of Alberta, Edmonton 03697
– Bibliothèque Saint-Jean, Edmonton 03698
– Herbert T. Coutts Education &Physical Education Libr, Edmonton 03699
– Humanities & Social Sciences Library, Edmonton 03700
– John A. Weir Memorial Law Library, Edmonton 03701
– John W. Scott Health Sciences Library, Edmonton 03702
– Legal Studies Program Library, Edmonton 03703
– Science & Technology Library, Edmonton 03704
University of Alberta, Augustana Faculty, Camrose 03687
University of Aleppo, Aleppo 27511
University of Allahabad, Allahabad 13601
University of America, Santafé de Bogotá 05963
University of Amsterdam, Amsterdam 19059
– Library for Humanities, Amsterdam 19062
– Library Theatre Stud-ies/Mediastudies and Musicology, Amsterdam 19071
University of Ankara, Ankara 27846
University of Applied Arts Vienna, Wien 01310
University of Applied Science, Köthen 10236
University of Applied Sciences, Berlin 09401
University of Applied Sciences, Bielefeld 09459
University of Applied Sciences, Darmstadt 09622
University of Applied Sciences, Dresden 09672
University of Applied Sciences, Erfurt 09688
University of Applied Sciences, Esslingen am Neckar 09726
University of Applied Sciences, Jena 10052
University of Applied Sciences, Landshut 10239
University of Applied Sciences, Lemgo 10246
University of Applied Sciences, Ludwigshafen 10255
University of Applied Sciences, Merseburg 10351
University of Applied Sciences, Mittweida 10352
University of Applied Sciences, München 10363
– Library Karlstrasse, München 10364
University of Applied Sciences, Neubrandenburg 10513
University of Applied Sciences, Nordhausen 10516
University of Applied Sciences, Offenburg 10524
University of Applied Sciences, Osnabrück 10527
– Central Institution for Scientific Information (ZEWI), Divisional Library Haste, Osnabrück 10528
University of Applied Sciences, Ulm 10676
University of Applied Sciences, Utrecht
– Faculty of Economics and Management, Utrecht 19209
University of Applied Sciences and Arts, Hannover 09957
– Departmental Library of Social Welfare Work and Health Care, Hannover 09960
University of Applied Sciences and Arts of Southern Switzerland, Canobbio 27062
University of Applied Sciences Deggendorf, Deggendorf 09652
University of Applied Sciences Konstanz, Konstanz 10234
University of Applied Sciences Kufstein, Kufstein 01246
University of Applied Sciences Northwestern Switzerland, Olten 27129
University of Applied Sciences: Technology, Business and Design, Wismar 10690
University of Applied Sciences Wilhelmshaven / Oldenburg / Elsfleth – University Library Elsfleth, Elsfleth 09686
University of Applied Sciences Wilhelmshaven / Oldenburg / Elsfleth – University Library Oldenburg, Oldenburg 10526
University of Applied Sciences Wilhelmshaven / Oldenburg / Elsfleth – University Library Wilhelmshaven, Wilhelmshaven 10689
University of Applied Sciences Worms, Worms 10693
University of Architecture and Urbanism 'Ion Mincu', Bucureşti 21926
University of Architecture, Civil Engineering and Geodesy, Sofiya 03477
University of Arizona, Tucson 32838
– Arizona Health Sciences Library, Tucson 32839
– Center for Creative Photography Library, Tucson 32840
– College of Agriculture & Life Sciences Arid Lands Information Center, Tucson 32841
– East Asian Collection, Tucson 32842
– James E. Rogers College of Law Library, Tucson 32843
– Music Collection, Tucson 32844
– Poetry Center in the College of Humanities, Tucson 32845
– Science-Engineering Library, Tucson 32846
University of Arkansas, Fayetteville 31133
– Chemistry and Biochemistry Library, Fayetteville 31134
– Fine Arts Library, Fayetteville 31135
– Physics Library, Fayetteville 31136
– Robert A. & Vivian Young Law Library, Fayetteville 31137
University of Arkansas at Little Rock, Little Rock 31625
– Pulaski County Law Library, Little Rock 31626
University of Arkansas for Medical Sciences, Little Rock 31627
University of Arkansas, Monticello, Monticello 31892
University of Arkansas, Pine Bluff, Pine Bluff 32292
University of Art, Tehran 14410
University of Asmara, Asmara 07213
University of Auckland, Auckland 19762
– Architecture Library, Auckland 19763
– Asian Languages Collection, Auckland 19764
– Davis Law Library, Auckland 19765
– Engineering Library, Auckland 19766
– Faculty of Education, Sylvia Ashton-Warner Library – Epsom Library, Auckland 19767
– Faculty of Education, Sylvia Ashton-Warner Library – Tai Tokerau Library, Whangarei 19768
– Faculty of Medical and Health Sciences, Auckland 19769
– Fine Arts Library, Auckland 19770
– Music and Dance Library, Auckland 19771
University of Azad Jammu and Kashmir, Muzaffarad 20450
University of Baghdad, Baghdad 14446
University of Baguio, Baguio City 20685
University of Bahrain, Sakhir 01737
University of Ballarat, Ballarat 00384
University of Balochistan, Quetta 20455
University of Baltimore, Baltimore 30344

– Law Library, Baltimore 30345
University of Barcelona, Barcelona
– Bellvitge Campus, L'Hospitalet de Llobregat 25311
– Biology Library, Barcelona 25299
– Economics and Business Library, Barcelona 25309
– Fine Arts Library, Barcelona 25297
– Geology Library, Barcelona 25304
– Information Sciences Library, Barcelona 25298
– Law Library, Barcelona 25300
– Library of Mathematics, Barcelona 25306
– Library of Medicine, Barcelona 25307
– The Pavelló de la República Library, Barcelona 25310
– Pharmacy Library, Barcelona 25301
– Philology Library, Barcelona 25305
– Philosophy, Geography and History Library, Barcelona 25302
– Physics and Chemistry Library, Barcelona 25303
– Rare Book and Manuscript Library, Barcelona 25308
– Resource Center for Learning and Research. Mundet Campus, Barcelona 25312
University of Basel, Basel 26992
– Institute of Musciology, Basel 27010
University of Basrah, Basrah 14448
University of Bath, Bath 28937
University of Bedfordshire, Luton 29261
University of Belgrade, Beograd 24459
– Department for Oriental Studies, Beograd 24465
– Department of Classical Studies, Beograd 24475
– Department of Philosophy, Beograd 24476
– Faculty of Agriculture, Zemun 24477
– Faculty of Architecture, Beograd 24460
– Faculty of Civil Engineering, Beograd 24467
– Faculty of Forestry, Beograd 24485
– Faculty of Geography, Beograd 24466
– Faculty of Law, Beograd 24478
– Faculty of Mechanical Engineering, Beograd 24472
– Faculty of Medicine, Beograd 24474
– Faculty of Mining and Geology, Beograd 24479
– Faculty of Pharmaceutics, Beograd 24463
– Faculty of Sport and Physical Education, Beograd 24462
– Faculty of Technology, Beograd 24486
– Faculty of Transport and Traffic Engineering, Beograd 24480
– French Seminar, Beograd 24481
– German Seminar, Beograd 24464
– Institute of Mathematics, Beograd 24473
– Institute of Social Sciences, Beograd 24468
– School of Economics, Beograd 24461
– School of Veterinary Medicine, Beograd 24487
– Seminar for the History of Yugoslav Literature, Beograd 24482
– Seminar of Southern Slavic Languages, Beograd 24483
– Slavic Seminar, Beograd 24484
University of Benin, Benin City 19929
University of Bergen, Bergen 20075
– Dental Faculty Library, Bergen 20081
– Faculty of Arts, Bergen 20079
– Law Library, Bergen 20076
– Psychology Library, Bergen 20082
– Science Library, Bergen 20077
University of Bern, Bern 26965
– Library Plant Sciences, Bern 27028
University of Bialymstoku – Library of Paedogogy and Psychology, Białystok 20796
University of Bialystok, Białystok 20795
University of Birmingham, Birmingham 28954
– Barber Fine Art Library, Birmingham 28955
– Barnes Library, Birmingham 28956
– Education Library, Birmingham 28957
– Harding Law Library, Edgbaston 28958
University of Bologna, Bologna 14854
– CIRSFID Library, Bologna 14863
– Department of Mathematics, Bologna 14884
– Department of Modern Foreign Languages and Literature, Bologna 14883
– Department of Statistical Sources, Bologna 14899
University of Bolton, Bolton 28961
University of Borås, Borås 26373
University of Bosphorus, Istanbul 27854
University of Botswana, Gaborone 02966
– Faculty of Engineering and Technology, Gaborone 02967
University of Bradford, Bradford 28962
University of Brazil, Brasília 03004
University of Bridgeport, Bridgeport 30545
University of Brighton, Brighton 28963
– Aldrich Library, Brighton 28964
– Falmer Library, Brighton 28965
– St Peter's House Library, Brighton 28966
– Welkin Library, Eastbourne 28967
University of Bristol, Bristol 28971
– School of Medical Sciences, Bristol 28972
University of British Columbia, Vancouver 03932
– Asian Library, Vancouver 03933
– Biomedical Library, Vancouver 03934
– David Lam Management Library, Vancouver 03935
– Education Library, Vancouver 03936
– Eric Hamber Library, Vancouver 03937
– Fine Arts Library, Vancouver 03938
– Law Library, Vancouver 03939
– Map Library, Vancouver 03940
– Mathematics Library, Vancouver 03941
– McMillan Library, Vancouver 03942
– Music Library, Vancouver 03943
– Rare Books and Special Collections, Vancouver 03944
– Saint Paul's Hospital Library, Vancouver 03945
– Science and Engineering Division, Vancouver 03946
– Walter C. Koerner Library – Humanities & Social Sciences, Vancouver 03947
– Woodward Biomedical Library, Vancouver 03948
University of Buckingham, Buckingham 28979
– Denning Law Library, Buckingham 28980
– James Meade Library of Economics, Buckingham 28981
University of Burdwan, Burdwan 13625
University of Calabar, Calabar 19934
University of Calcutta, Kolkata 13686
University of Calgary, Calgary 03682
– Business Library, Calgary 03683
– Gallagher Library of Geology & Geophysics, Calgary 03684
– Health Sciences Library, Calgary 03685
– Law Library, Calgary 03686
University of California at Berkeley, Berkeley 30388
– Anthropology Library, Berkeley 30389
– Bancroft Library, Berkeley 30390
– Center for Chinese Studies, Berkeley 30391
– Chemistry Library, Berkeley 30392
– Earth Sciences & Maps Library, Berkeley 30393
– Earthquake Engineering Research Center Library, Richmond 30394
– East Asian Library, Berkeley 30395
– Education Psychology Library, Berkeley 30396
– Environmental Design Library, Berkeley 30397
– Ethnic Studies Library, Berkeley 30398
– Fong Optometry & Health Sciences Library, Berkeley 30399
– Giannini Foundation of Agricultural Economics Library, Berkeley 30400
– Harmer E. Davis Transportation Library, Berkeley 30401
– Institute of Governmental Studies, Berkeley 30402
– Institute of Industrial Relations Library, Berkeley 30403
– Jean Gray Hargrove Music Library, Berkeley 30404
– Kresge Engineering Library, Berkeley 30405
– Law Library, Berkeley 30406
– Marian Koshland Bioscience & Natural Resources Library, Berkeley 30407
– Mathematics-Statistics Library, Berkeley 30408
– Physics Library, Berkeley 30409
– Public Health Library, Berkeley 30410
– Social Welfare Library, Berkeley 30411
– South-Southeast Asia Library Service, Berkeley 30412
– Thomas J. Long Business & Economics Library, Berkeley 30413
– Water Resources Center Archives, Berkeley 30414
University of California, Davis – F. William Blaisdell Medical Library, Sacramento 37479
University of California, Davis – General Library, Davis 30961
– Agricultural & Resource Economics Library, Davis 30962
– Law Library, Davis 30963
– Loren D. Carlson Health Sciences Library, Davis 30964
– Physical Sciences & Engineering Library, Davis 30965
University of California – Hastings College of the Law Library, San Francisco 32593
University of California, Irvine – Langson Library, Irvine 31412
– Ayala Science Library, Irvine 31413
University of California, Livermore – Lawrence Livermore National Laboratory, Livermore 31628
University of California, Los Angeles – Ralph J. Bunche Center for African American Studies Library, Los Angeles 31656
University of California, Los Angeles – University Library,Young Research Library, Los Angeles 31657
– The Arts Library, Los Angeles 31658
– College Library, Los Angeles 31659
– English Reading Room, Los Angeles 31660
– Hugh & Hazel Darling Law Library, Los Angeles 31661
– Louise M. Darling Biomedical Library, Los Angeles 31662
– Management Library, Los Angeles 31663
– Music Library, Los Angeles 31664
– Richard C. Rudolph East Asian Library, Los Angeles 31665
– Science & Engineering Libraries, Los Angeles 31666
– University Elementary School Library, Los Angeles 31667
– William Andrews Clark Memorial Library, Los Angeles 31668
University of California, Riverside, Riverside 32455
– Music Library, Riverside 32456
– Raymond L. Orbach Science Library, Riverside 32457
University of California, San Diego, La Jolla 31528
– The Arts Library, La Jolla 31529
– Biomedical Library, La Jolla 31530
– International Relations & Pacific Studies, La Jolla 31531
– Medical Center Library, San Diego 31532
– Science & Engineering, La Jolla 31533
– Scripps Institution of Oceanography, La Jolla 31534
– Social Science & Humanities Library, La Jolla 31535
University of California, San Francisco – Paul & Lydia Kalmanovitz Library & The Center for Knowledge Management, San Francisco 32594
University of California, Santa Barbara, Santa Barbara 32605
– Arts Library, Santa Barbara 32606
University of California, Santa Cruz, Santa Cruz 32611
– Science & Engineering Library, Santa Cruz 32612
University of Cambridge, Cambridge 28987
– Cambridge Union Society, Cambridge 28988
– Central Science Library, Cambridge 28989
– Centre of African Studies Library, Cambridge 28990
– Centre of Latin American Studies, Cambridge 28991
– Centre of South Asian Studies, Cambridge 28992
– Christ's College, Cambridge 28993
– Classical Faculty and Museum of Classical Archaeology, Cambridge 28994
– Cory Library, Cambridge 28995
– Department of Biochemistry, Cambridge 28996
– Department of Clinical Veterinary Medicine, Cambridge 28997
– Department of Experimental Psychology, Cambridge 28998
– Department of History and Philosophy of Science, Cambridge 28999
– Department of Pathology, Cambridge 29000
– Department of Physics, Cambridge 29001
– Department of Physiology Library, Cambridge 29002
– Department of Zoology, Cambridge 29003
– Divinity Faculty, Cambridge 29004
– Earth Sciences Library, Cambridge 29005
– Emmanuel College, Cambridge 29006
– Engineering Department, Cambridge 29007
– English Faculty Library, Cambridge 29008
– Faculty of Architecture and History of Art, Cambridge 29009
– Faculty of Education, Cambridge 29010
– Faculty of History, Cambridge 29011
– Faculty of Oriental Studies, Cambridge 29012
– Faculty of Philosophy, Cambridge 29013
– Faculty of Social and Political Sciences, Cambridge 29014
– Fitzwilliam College, Cambridge 29015
– Fitzwilliam Museum, Cambridge 29016
– Genetics Library, Cambridge 29017
– Geography Library, Cambridge 29018
– Girton College, Cambridge 29019
– Haddon Library of Archaeology & Anthropology, Cambridge 29020
– Institute of Astronomy, Cambridge 29021
– Institute of Continuing Education, Cambridge 29022
– Institute of Criminology, Cambridge 29023
– Jesus College, Cambridge 29024
– King's College, Cambridge 29025
– Lucy Cavendish College, Cambridge 29026
– Magdalene College, Cambridge 29027
– Maitland Robinson Library of Downing College, Cambridge 29028
– Marshall Library of Economics, Cambridge 29029
– Materials Science and Metallurgy Library, Cambridge 29030
– Mill Lane Lecture Halls, Cambridge 29031
– Modern and Medieval Languages Library, Cambridge 29032
– Murray Edwards College – Rosemary Murray Library, Cambridge 29033
– Newnham College Library, Cambridge 29034
– Pembroke College, Cambridge 29035
– Pendlebury Library of Music, Cambridge 29036
– Physiology, Development and Neuroscience Library, Cambridge 29037
– Plant Sciences Library, Cambridge 29038
– Pure Mathematical and Mathematical Statistics Library, Cambridge 29039
– Queens' College, Cambridge 29040
– Ridley Hall Library, Cambridge 29041
– Scott Polar Research Institute, Cambridge 29042
– Selwyn College, Cambridge 29043
– Sidney Sussex College, Cambridge 29044
– Squire Law Library, Cambridge 29045
– St Catharine's College, Cambridge 29046
– St John's College, Cambridge 29047
– Trinity College, Cambridge 29048
– Trinity Hall, Cambridge 29049
– University Medical Library, Cambridge 29050
– Ward and Perne Libraries, Cambridge 29051
– Westminster College, Cambridge 29052
– Wolfson College, Cambridge 29053
University of Canberra, Canberra 00418
University of Canterbury, Christchurch 19773
– Engineering Library, Christchurch 19774
– Physical Sciences Library, Christchurch 19775
University of Cape Coast, Cape Coast 13012
– Botany Departement Class Library, Cape Coast 13013
University of Cape Town, Cape Town 25006
– African Studies Library, Rondebosch 25007
– W. H. Bell Music Library, Rosebank 25008
– Brand van Zyl Law Library, Rondebosch 25009
– Built Environment Library (Architecture Library), Rondebosch 25010
– Health Sciences Library, Observatory 25011
– Hiddingh Hall Library, Cape Town 25012
– Science and Engineering Library, Rondebosch 25013
University of Caxias do Sul, Caxias do Sul 03015
University of Central Arkansas, Conway 30921
University of Central Florida, Orlando 32211
University of Central Lancashire, Preston 29379
University of Central Missouri, Warrensburg 32951
University of Central Oklahoma, Edmond 31075
University of Charleston, Charleston 30718
University of Chemical and Metallurgical Technology, Sofiya 03478
University of Chicago, Chicago 30786
– D'Angelo Law Library, Chicago 30787
– Eckhart Library, Chicago 30788
– Harper Library, Chicago 30789
– John Crerar Library, Chicago 30790
– Social Service Administration Library, Chicago 30791
University of Chicago, Williams Bay 33064
University of Chichester, Chichester 29068
University of Cincinnati Libraries, Cincinnati 30801
– Chemistry-Biology Library, Cincinnati 30802
– Classics Library, Cincinnati 30803

- College of Applied Science Library, Cincinnati 30804
- College-Conservatory of Music, Cincinnati 30805
- Curriculum Resources Center, Cincinnati 30806
- Design, Architecture, Art & Planning Library, Cincinnati 30807
- Engineering Library, Cincinnati 30808
- Geology-Mathematics- Physics Library, Cincinnati 30809

University of Cincinnati Medical Center, Cincinnati 36701
University of Cincinnati – Raymond Walters College, Cincinnati 30810
University of Cincinnati – Robert S. Marx Law Library, Cincinnati 30811
University of Colombo, Colombo 26272
- Institute of Workers' Education, Colombo 26273
- Medical Faculty, Colombo 26274

University of Colorado at Boulder, Boulder 30511
- Archives Department, Boulder 30512
- Art & Architecture Library, Boulder 30513
- East Asian Library, Boulder 30514
- Equity Diversity & Education Library, Boulder 30515
- Gemmill Engineering Library, Boulder 30516
- Government Publications Library, Boulder 30517
- Howard B. Waltz Music Library, Boulder 30518
- Jerry Crail Earth Sciences & Map Library, Boulder 30519
- Natural Hazard Center Library (Institute of Behavioral Science), Boulder 30520
- Oliver C. Lester Library of Mathematics & Physics, Boulder 30521
- Science Library, Boulder 30522
- Special Collections Department, Boulder 30523
- William M. White Business Library, Boulder 30524

University of Colorado at Boulder – William A. Wise Law Library, Boulder 30525
University of Colorado at Colorado Springs, Colorado Springs 30868
University of Colorado at Denver & Health Sciences Center, Denver 36797
University of Connecticut, Hartford 31306
University of Connecticut, Storrs 32755
- Music & Dramatic Arts Library, Storrs 32756
- Pharmacy Library, Storrs 32757
- Thomas J. Dodd Research Center, Storrs 32758

University of Connecticut, Waterbury 32996
University of Connecticut at Avery Point, Groton 31268
University of Connecticut at Stamford, Stamford 32714
University of Connecticut – Greater Hartford Campus – School of Social Work, West Hartford 33018
University of Connecticut Health Center, Farmington 36847
The University of Cooperative Education, Glauchau 09834
University of Copenhagen, København
- Biblical Studies Section, København 06668
- Central Botanical Library, København 06671
- Centre of African Studies, København 06672
- Department of Art and Cultural Studies – Musicology Section, København 06682
- Department of Church History, København 06681
- Department of Computer Science, København 06673
- Department of Cross Cultural and Regional Studies – Carsten Niebuhr Section, København 06688
- Department of Cross-Cultural and Regional Studies – East European Section, København 06687
- Department of Economics Library, København 06698
- Department of Exercise and Sport Sciences, København 06680
- Department of Mathematics, København 06691
- Department of Media, Cognition and Communication, København 06683
- Department of Scandinavian Research, København 06697
- Faculty Library of Natural and Health Sciences, København 06693
- Faculty of Law Library. Library of Criminology and Criminal Law, København 06689
- Faculty of Life Sciences Library, Frederiksberg 06670
- Faculty of Theology Library, København 06709
- The GeoLibrary, København 06677
- History of Religions Library, København 06699
- Institute for Anthropology, København 06679
- Institute of Name Research Library, København 06669
- Institute of Political Studies, København 06685
- Institute of Psychology, København 06684
- Library for Linguistics and Phonetics, København 06690
- Library of English Studies, København 06674
- Library of Germanic Studies, København 06678
- Library of Science and Medicine, København 06692
- Library of the Section for Romance Studies, København 06700
- Niels Bohr Institute, Libray for Astronomy and Geophysics, København 06694
- Niels Bohr Institute, Physics Library, København 06695
- Nordic Library, København 06696
- Pharmaceutical Sciences Faculty Library, København 06676
- SAXO Institute – Department of Archaeology, Library of Prehistoric Archaology, København 06701
- SAXO Institute – Department of Contemporary History, København 06705
- SAXO Institute – Department of Economic History, København 06704
- SAXO Institute, Department of Ethnology Library, København 06706
- SAXO Institute – Department of Greek and Latin, København 06702
- SAXO Institute – Department of History, København 06703
- SAXO Institute, Library of Classical Archaeology, København 06707
- Sociological Library, København 06708

University of Craiova, Craiova 21933
University of Crete, Rethimnon 13084
University of Cumbria, Faculty of the Arts, Carlisle 29062
University of Cumbria, Harold Bridges Library, Lancaster 29152
University of Cyprus, Nicosia 06328
University of Dallas, Irving 31414
University of Damascus, Damascus 27514
University of Dar es Salaam, Dar es Salaam 27654
- Muhimbili University College of Health Sciences, Dar es Salaam 27655

University of Dayton, Dayton 30967
- Marian Library, Dayton 30968

University of Dayton School of Law, Dayton 30969
University of Delaware, Newark 32105
- College of Education Resource Center Library, Newark 32106

University of Delhi, Delhi 13640
- Department of Library and Information Science, Delhi 13641
- Department of Social Work, Delhi 13642

University of Denver, Denver 30998
- Westminster Law Library, Denver 30999

University of Derby, Derby 29080
University of Detroit Mercy, Detroit 31006
- Kresge Law Library, Detroit 31007
- Outer Drive Campus Library, Detroit 31008

University of Dubuque, Dubuque 31030
University of Dundee, Dundee 29085
University of East Anglia, Norwich 29289
University of East London, Stratford 29413
- Barking Campus Library, Dagenham 29414
- Duncan House Library, London 29415

University of Eastern Philippines, University Town 20741
University of Economics and Law, Kyiv 28027
University of Economics, Prague, Praha 06405
University of Edinburgh, Edinburgh 29098
- Law & Europa Library, Edinburgh 29099
- New College Library, Edinburgh 29100
- Psychiatry Library, Edinburgh 29101
- Reid Music Library, Edinburgh 29102
- Robertson Engineering and Science Library, Edinburgh 29103
- Royal (Dick) School of Veterinary Studies, Edinburgh 29104
- Scottish Studies Library, Edinburgh 29105

University of Education, Winneba 13021
University of Education Ludwigsburg, Ludwigsburg 10254
University of Education – Vienna Library, Wien 01381
University of Electronic Science and Technology of China, Chengdu 05127
University of Electronic Technology, Moskva 22396
University of Engineering and Technology, Lahore 20448
University of Essex, Colchester 29071
University of Evansville, Evansville 31112
University of Exeter, Exeter 29106
- Devon and Exeter Institution Library, Exeter 29107
- St Luke's Library, Exeter 29108

The University of Findlay, Findlay 31139
University of Fisheries, Nha-Trang 42281
University of Florida, Belle Glade – Everglades Research & Education Center, Belle Glade 30372
University of Florida – George A. Smathers Library, Gainesville 31190
- Allen H. Neuharth Journalism and Communications Library, Gainesville 31191
- Architecture & Fine Arts Library, Gainesville 31192
- Education Library, Gainesville 31193
- Governments Documents Department, Gainesville 31194
- Isser & Rae Price Library of Judaica, Gainesville 31195
- Latin American Collection, Gainesville 31196
- Marston Science Library, Gainesville 31197
- Music Library, Gainesville 31198

University of Florida – Health Science Center Libraries, Gainesville 31199
University of Florida – Levin College of Law, Gainesville 31200
University of Food Technologies, Plovdiv 03459
University of Forestry, Sofiya 03479
University of Fort Hare, Alice 24998
University of French Polynesia, Faaa 09165
University of Gävle, Gavle 26374
University of Gdansk, Gdansk 20806
University of Georgia, Athens 30279
- Georgia Agricultural Experiment Stations Library, Athens 30280
- Science Library, Athens 30281

University of Georgia – Alexander Campbell King Law Library, Athens 30282
University of Georgia College of Agriculture & Environmental Sciences, Tifton 32819
University of Ghana, Accra 12997
- Africana Library, Legon 12998
- Commonwealth Hall Library, Legon 12999
- Department of Archeology, Legon 13000
- Department of Botany, Legon 13001
- Department of Economics, Legon 13002
- Department of Nutrition and Food Science, Legon 13003
- Faculty of Agriculture, Legon 13004
- Faculty of Law, Legon 13005
- Hall Library, Akuafo 13006
- Hall Library, Legon 13007
- Hall Library, Mensah Sarbah 13008
- Institute of African Studies, Legon 13009
- Population and Social Science Library, Legon 13010
- School of Administration, Legon 13011

University of Glamorgan, Pontypridd 29375
University of Glasgow, Glasgow 29121
- Adam Smith Library, Glasgow 29122
- Chemistry Branch, Glasgow 29123
- Russian and East European Studies Library, Glasgow 29124
- Veterinary School, Glasgow 29125

University of Gloucestershire, Cheltenham 29065
University of Goettingen, Göttingen 09315
- Forestry Divisional Library, Göttingen 09842
- History Department Library, Göttingen 09887

University of Goroka, Goroka 20551
University of Great Falls, Great Falls 31245
University of Greenwich, London 29230
- Avery Hill Library, London 29231

University of Groningen, Groningen 19129
- Central Medical Library, Groningen 19139
- Faculty of Arts Library, Groningen 19131
- FWN Library, Groningen 19133
- Law Library, Groningen 19135
- Library of Economics and Business, Spatial Sciences, Groningen 19130

University of Guam, Mangilao 13164
University of Guam – Richard F. Taitano Micronesian Area Research Center (MARC), Mangilao 13165
University of Guelph, Guelph 03713
University of Guyana, Georgetown 13194
University of Haifa, Haifa 14614
University of Hartford, Hartford College for Women, Hartford 33400
University of Hartford Libraries and Learning Resources, West Hartford 33019
- University of Hartford, West Hartford 33020

University of Hawaii at Hilo, Hilo 31339
University of Hawaii at Manoa, Honolulu 31347
University of Hawaii at Manoa – School of Medicine, Honolulu 31348
University of Hawaii – William S. Richardson School of Law Library, Honolulu 31349
University of Hawaii-College of Education, Honolulu 31350
University of Health Sciences, Kansas City 31484
University of Hertfordshire, Hatfield 29132
University of Hong Kong, Hong Kong 05288
- Dental Library, Hong Kong 05289
- Education Library, Hong Kong 05290
- Fung Ping Shan Library, Hong Kong 05291
- Lui Che Woo Law Library, Hong Kong 05292
- Music Library, Hong Kong 05293
- Yu Chun Keung Medical Library, Hong Kong 05294

University of Houston, Houston 31358
- Music Library, Houston 31359
- Optometry Library, Houston 31360
- The O'Quinn Law Library, Houston 31361
- Pharmacy Library, Houston 31362
- William R. Jenkins Architecture & Art Library, Houston 31363

University of Houston, Victoria 32930
University of Houston – Clear Lake, Houston 31364
University of Houston – Downtown, Houston 31365
University of Huddersfield, Huddersfield 29136
University of Hull, Hull 29137
- Keith Donaldson Library, Scarborough 29138

University of Hyderabad, Hyderabad 13659
University of Ibadan, Ibadan 19939
- College of Medicine, Ibadan 19940
- Department of Library, Archival and Information Studies, Ibadan 19941
- Institute of African Studies, Ibadan 19942

University of Iceland, Reykjavík 13556
University of Idaho, Moscow 31910
- College of Law, Moscow 31911

University of Illinois at Chicago, Chicago 30792
- Library of the Health Sciences, Chicago 30793
- Science Library, Chicago 30794

University of Illinois at Springfield, Springfield 32710
University of Illinois at Urbana-Champaign, Urbana 32884
- Africana Library, Urbana 32885
- Applied Life Studies Library, Urbana 32886
- Architecture & Art Library, Urbana 32887
- Asian Library, Urbana 32888
- Biology Library, Urbana 32889
- Business & Economics Library, Urbana 32890
- Chemistry Library, Urbana 32891
- City Planning & Landscape Architecture Library, Urbana 32892
- Classics Library, Urbana 32893
- Communications Library, Urbana 32894
- Education & Social Science Library, Urbana 32895
- English Library, Urbana 32896
- Funk Agricultural, Consumer & Environmental Sciences Library, Urbana 32897
- Geology Library, Urbana 32898
- Government Documents Library, Urbana 32899
- Grainger Engineering Library Information Center, Urbana 32900
- History, Philosophy & Newspaper Library, Urbana 32901
- Illinois History & Lincoln Collections, Urbana 32902
- Illinois Natural History Survey Library, Champaign 32903
- Latin American & Caribbean Library, Urbana 32904
- Law Library, Champaign 32905
- Library Science & Information Science Library, Urbana 32906
- Map & Geography Library, Urbana 32907
- Mathematics Library, Urbana 32908
- Modern Languages & Linguistics Library, Urbana 32909

– Music & Performing Arts Library, Urbana 32910
– Physics-Astronomy Library, Urbana 32911
– Rare Book & Manuscript Library, Urbana 32912
– Slavic and East European Library, Urbana 32913
– Undergraduate Library, Urbana 32914
– University Laboratory High School, Urbana 32915
– Veterinary Medicine Library, Urbana 32916
University of Ilorin, Ilorin 19952
University of Indianapolis, Indianapolis 31393
University of Indonesia, Jakarta 14252
– Faculty of Letters, Jakarta 14256
– Faculty of Mathematics and Science, Jakarta 14255
– Faculty of Medicine, Jakarta 14254
University of International Business and Economics, Beijing 05054
University of Ioannina, Ioannina 13080
University of Iowa, Iowa City 31397
– Art Library, Iowa City 31398
– Biological Sciences Library, Iowa City 31399
– Chemistry Library, Iowa City 31400
– Engineering Library, Iowa City 31401
– Geoscience Library, Iowa City 31402
– Hardin Library for the Health Sciences, Iowa City 31403
– Law Library, Iowa City 31404
– Marvin A. Pomerantz Business Library, Iowa City 31405
– Mathematical Sciences Library, Iowa City 31406
– Physics Library, Iowa City 31407
– Psychology Library, Iowa City 31408
– Rita Benton Music Library, Iowa City 31409
University of Isfahan, Isfahan 14391
University of Jaffna, Jaffna 26276
University of Jammu, Jammu 13668
University of Jember, Jember 14265
University of Johannesburg – Auckland Park Kingsway Campus Library, Johannesburg 25024
University of Jordan Library, Amman 17873
University of Jos, Jos 19953
University of Juba, Khartoum 26345
University of Judaism, Los Angeles 31669
University of Kalyani, Kalyani 13674
University of Kansas, Lawrence 31573
– Anshutz Library, Lawrence 31574
– Gorton Music & Dance Library, Lawrence 31575
– Murphy Art & Architecture Library, Lawrence 31576
– Spahr Engineering Library, Lawrence 31577
– Wheat Law Library, Lawrence 31578
University of Kansas Medical Center, Kansas City 36991
University of Karachi, Karachi 20444
– Institute of Business Administration (IBA), Karachi 20445
University of Kashmir, Srinagar 13768
University of Kelaniya, Kelaniya 26278
University of Kent at Canterbury, Canterbury 29055
University of Kentucky, Lexington 31592
– Center for Applied Energy Research Libraries, Lexington 31593
– Chemistry-Physics Library, Lexington 31594
– Design Library, Lexington 31595
– Education Library, Lexington 31596
– Engineering Library, Lexington 31597
– Geological Sciences Library, Lexington 31598
– Law Library, Lexington 31599
– Lucille Little Fine Arts Library & Learning Center, Lexington 31600
– Mathematical Sciences Library, Lexington 31601
– Medical Center Library, Lexington 31602
– Special Collections Library, Lexington 31603
University of Kentucky Libraries, Lexington 35815
University of Kerala, Thiruvananthapuram 13771
University of Khartoum, Khartoum 26346
– Faculty of Agriculture and Veterinary Science, Khartoum 26347
– Faculty of Engineering and Architecture, Khartoum 26348
– Faculty of Medicine and Faculty of Pharmacy, Khartoum 26349
University of King's College, Halifax 03723
University of Konstanz, Konstanz 10235
University of KwaZulu-Natal – Edgewood Campus, Pinetown 25036
University of KwaZulu-Natal – Howard College Campus, Durban 25016
– Barrie Biermann Architectural Library, Durban 25017
– Eleanor Bonnar Music Library, Durban 25018
– Medical Library, Durban 25019
– G.M.J. Sweeney Law Library, Durban 25020
University of KwaZulu-Natal – Pietermaritzburg Campus, Scottsville 25046
University of KwaZulu-Natal – Westville Campus, Durban 25021
University of La Verne, La Verne 31538
– College of Law at La Verne, Ontario 31539
University of Lagos, Lagos 19963
– College of Medicine, Lagos 19964
– Faculty of Education, Lagos 19965
University of Lapland, Rovaniemi 07366
University of Latvia, Riga 18174
University of Lausanne, Lausanne 26971
– Earth Sciences Library, Lausanne 27106
University of Leeds, Leeds 29157
– Edward Boyle Library, Leeds 29158
– Health Sciences Library, Leeds 29159
University of Leicester, Leicester 29162
– Clinical Sciences Library, Leicester 29163
– School of Education, Leicester 29164
University of Leipzig, Leipzig 10245
University of Leoben, Leoben 01247
University of Lethbridge, Lethbridge 03739
University of Liberia, Monrovia 18230
University of Library and Information Science, Tsukuba 17227
University of Liepaja, Liepaja 18171
University of Life Science in Lublin, Lublin 20895
University of Limerick, Limerick 14488
– Mary Immaculate College, Limerick 14489
University of Limpopo, Medunsa Campus Library, Medunsa 25033
University of Limpopo – Turfloop Campus, Sovenga 25048
University of Liverpool, Liverpool 29169
University of London, London 29232
– Birkbeck College, London 29233
– Courtauld Institute of Art, London 29234
– Goldsmiths' College, London 29235
– Heythrop College, London 29236
– Institute for the Study of the Americas, London 29237
– Institute of Advanced Legal Studies, London 29238
– Institute of Classical Studies, London 29239
– Institute of Commonwealth Studies, London 29240
– Institute of Education, London 29241
– Institute of Germanic Studies, London 29242
– Institute of Historical Research, London 29243
– Institute of Psychiatry, London 29244
– Joint Library of Ophthalmology, Moorfields Eye Hospital & UCL Institute of Ophthalmology, London 29245
– London Hospital Medical College, London 29246
– Queen Mary and Westfield College, London 29247
– Royal Holloway Library, Egham 29248
– Royal Veterinary College, London 29249
– School of Oriental and African Studies, London 29250
– St George's Hospital Medical School, London 29251
– University Marine Biological Station Millport (UMBSM), Millport 29252
– Warburg Institute, London 29253
University of London Institute in Paris, Paris 07989
University of London, London School of Hygiene and Tropical Medicine, London 29203
University of Louisiana at Lafayette, Lafayette 31541
University of Louisiana at Monroe, Monroe 31878
University of Louisville Libraries, Louisville 31683
– Brandeis School of Law Library, Louisville 31684
– Dwight Anderson Music Library, Louisville 31685
– Ekstrom Library, Louisville 31686
– Kornhauser Health Sciences Library, Louisville 31687
– Margaret Bridwell Art Library, Louisville 31688
University of Lucknow, Lucknow 13692
University of Macao, Macao 05416
University of Macedonia, Thessaloniki 13087
University of Madras, Chennai 13631
University of Maiduguri, Maiduguri 19967
University of Maine, Orono 32213
University of Maine, Walpole 32947
University of Maine at Augusta, Bangor 30350
University of Maine at Farmington, Farmington 31126
University of Maine at Fort Kent, Fort Kent 31156
University of Maine at Machias, Machias 31701
University of Maine at Presque Isle, Presque Isle 32360
University of Maine School of Law, Portland 32343
University of Malawi, Zomba 18473
– Chancellor College, Zomba 18474
– Kamuzu College of Nursing, Lilongwe 18475
– The Polytechnic, Chichiri 18476
University of Malaysia, Kuala Lumpur 18504
– Faculty of Economics and Administration, Kuala Lumpur 18507
– Faculty of Education, Kuala Lumpur 18508
– Institute of Postgraduate Studies & Research, Kuala Lumpur 18509
– Islamic Academy, Kota Bharu 18505
– Law Library, Kuala Lumpur 18512
– Medical Library, Kuala Lumpur 18511
University of Malta, Msida 18658
University of Management and Economics, Kaunas 18275
University of Manchester, Manchester 29269
– Ashburne Hall Library, Manchester 29270
– Institute of Science and Technology, Manchester 29271
– Manchester Business School, Manchester 29272
University of Mandalay, Mandalay 18985
– Institute of Medicine, Mandalay 18986
University of Manila, Manila 20723
University of Manitoba Libraries, Winnipeg 03963
– Albert D. Cohen Management Library, Winnipeg 03964
– Architecture & Fine Arts Library, Winnipeg 03965
– J.W. Crane Memorial Library, Winnipeg 03966
– Donald W. Craik Engineering Library, Winnipeg 03967
– Eckhart-Gramatté Music Library, Winnipeg 03968
– Neil John MacLean Health Sciences Library, Winnipeg 03969
– Saint John's College Library, Winnipeg 03970
– Sciences and Technology Library, Winnipeg 03971
– E. K. Williams Law Library, Winnipeg 03972
– D. S. Woods Education Library, Winnipeg 03973
University of Maribor Library, Maribor 24867
– Faculty of Agriculture and Life Sciences, Hoče 24869
– Faculty of Business Management, Kranj 24870
– Faculty of Economics and Business, Maribor 24868
– Faculty of Education, Maribor 24871
– Faculty of Law, Maribor 24872
– Faculty of Technical Sciences, Maribor 24873
University of Mary, Bismarck 30435
University of Mary Hardin-Baylor, Belton 30379
University of Mary Washington, Fredericksburg 31174
University of Maryland, College Park 30847
– Architecture Library, College Park 30848
– Art Library, College Park 30849
– Broadcast Pioneers Library of American Broadcasting, College Park 30850
– Engineering & Physical Sciences Library, College Park 30851
– R. Lee Hornbake Undergraduate Library, College Park 30852
– Michelle Smith Performing Arts Library, College Park 30853
– Theodore R. McKeldin Library, College Park 30854
– White Memorial Chemistry Library, College Park 30855
University of Maryland, Baltimore, Baltimore
– Health Sciences & Human Services Library, Baltimore 30346
– Thurgood Marshall Law Library, Baltimore 30347
University of Maryland, Baltimore County, Baltimore 30348
University of Maryland Center for Environmental Science, Solomons 32684
University of Maryland – Eastern Shore, Princess Anne 32362
University of Massachusetts at Amherst, Amherst 30216
– Biological Sciences, Amherst 30217
– Science and Engineering Library, Amherst 30218
University of Massachusetts Boston, Boston 30507
University of Massachusetts Dartmouth, North Dartmouth 32143
University of Massachusetts Lowell, Lowell 31689
University of Massachusetts Medical School, Worcester 37860
University of Mauritius, Réduit 18691
University of Medicine and Pharmacy, Bucureşti 22019
University of Medicine and Pharmacy, Timişoara 21948
University of Medicine & Dentistry of New Jersey, New Brunswick 37156
University of Medicine & Dentistry of New Jersey, Newark 32107
University of Medicine & Dentistry of New Jersey, Stratford 32759
University of Medicine & Dentistry of New Jersey, Camden Campus Library, Camden 30665
University of Melbourne, Melbourne 00459
– Architecture, Planning and Building Library, Parkville 00460
– Baillieu Library, Parkville 00461
– Brownless Biomedical Library, Parkville 00462
– Chemistry Library, Parkville 00463
– Earth Sciences Library, Parkville 00464
– Education Resource Centre, Parkville 00465
– Engineering Library, Parkville 00466
– Institute of Land and Food Resources – Glenormiston College, Terang 00467
– Institute of Land and Food Resources – Longerenong College, Longerenong 00468
– Land and Food Resources – Burnley College, Richmond 00469
– Land and Food Resources – Dookie College, Dookie College 00470
– Land and Food Resources – Parkville Campus, Parkville 00471
– Legal Resource Centre, Victoria 00472
– Louise Hanson-Dyer Music Library, Parkville 00473
– Mathematical Sciences, Parkville 00474
– Physics Research Library, Parkville 00475
– Queens College, Parkville 00476
– Trinity College, Parkville 00477
– Veterinary Science, Parkville 00478
– Veterinary Science, Werribee 00479
– Victorian College of the Arts, Melbourne 00480
The University of Memphis, Memphis 31794
University of Memphis, Memphis 31795
– Audiology & Speech Language Pathology, Memphis 31796
– Chemistry, Memphis 31797
– Earth Sciences, Memphis 31798
– Mathematics, Memphis 31799
– Music, Memphis 31800
University of Miami, Coral Gables 30924
– Law Library, Coral Gables 30925
University of Miami, Miami 31812
– Louis Calder Memorial Library, Miami 31813
– Mary & Edward Norton Library of Ophthalmology, Miami 31814
University of Michigan, Ann Arbor 30230
– Alfred Taubman Medical Library, Ann Arbor 30231
– Art, Architecture & Engineering Library, Ann Arbor 30232
– Asia Library, Ann Arbor 30233
– Dentistry Library, Ann Arbor 30234
– Fine Arts Library, Ann Arbor 30235
– Harlan Hatcher Graduate Library, Ann Arbor 30236
– Information & Library Studies Library, Ann Arbor 30237
– Museums Library, Ann Arbor 30238
– Music Library, Ann Arbor 30239
– Public Health Library & Information Division, Ann Arbor 30240
– Shapiro Science Library, Ann Arbor 30241
– Shapiro Science Library – Undergraduate Library, Ann Arbor 30242
– Social Work Library, Ann Arbor 30243
– Special Collections Library, Ann Arbor 30244
University of Michigan, Ann Arbor
– Bentley Historical Library, Ann Arbor 30245
– Kresge Business Administration Library, Ann Arbor 30246
– Law Library, Ann Arbor 30247
– Transportation Research Institute Library, Ann Arbor 30248

– William L. Clements Library, Ann Arbor 30249
University of Michigan Dearborn, Dearborn 30976
University of Michigan – Flint Library, Flint 31143
University of Milan, Milano 15104
– Central Library of Pharmacy, Milano 15106
– Faculty of Agriculture, Milano 15120
University of Mining and Metallurgy, Kraków 20824
University of Mining and Metallurgy, Ostrava 06376
University of Minnesota, Chaska 30738
University of Minnesota – Crookston, Crookston 30936
University of Minnesota Duluth, Duluth 31033
University of Minnesota – Magrath Library, Saint Paul 32535
– Entomology, Fisheries & Wildlife Library, Saint Paul 32536
– Forestry Library, Saint Paul 32537
University of Minnesota – Morris, Morris 31905
University of Minnesota – Twin Cities, Minneapolis 31847
– Ames Library of South Asia, Minneapolis 31848
– Architecture & Landscape Architecture Library, Minneapolis 31849
– Bio-Medical Information Services, Minneapolis 31850
– Children's Literature Research Collections (Kerlan & Hess Collections), Minneapolis 31851
– East Asian Library, Minneapolis 31852
– Government Publications Library, Minneapolis 31853
– Immigration History Research Center, Minneapolis 31854
– James Ford Bell Library, Minneapolis 31855
– John R. Borchert Map Library, Minneapolis 31856
– Law Library, Minneapolis 31857
– Mathematics Library, Minneapolis 31858
– Music Library, Minneapolis 31859
– Reference Services (Physical Sciences & Engineering), Minneapolis 31860
– Reference Services (Wilson Library), Minneapolis 31861
– Special Collections and Rare Books, Minneapolis 31862
– University Archives, Minneapolis 31863
University of Miskolc, Library, Archives, Museum, Miskolc 13284
– Library of College of Dunaújváros, Dunaujváros 13285
University of Mississippi, University 32871
– Science Library, University 32872
University of Mississippi – Law Library, University 32873
University of Mississippi Medical Center, Jackson 36971
University of Missouri, Kansas City 31485
– Dental Library, Kansas City 31486
– Health Sciences Library, Kansas City 31487
– Leon E. Bloch Law Library, Kansas City 31488
University of Missouri – Rolla, Rolla 32478
University of Missouri-Columbia – Elmer Ellis Library, Columbia 30875
– Engineering Library & Technology Commons, Columbia 30876
– Geology Library, Columbia 30877
– Journalism Library, Columbia 30878
– Law Library, Columbia 30879
– Mathematical Sciences Library, Columbia 30880
– J. Otto Lottes Health Sciences Library, Columbia 30881
– Veterinary Medical Library, Columbia 30882
– Western Historical Manuscript Collection – Columbia, Columbia 30883
University of Missouri-Columbia – Missouri Institute of Mental Health Library, Saint Louis 37496
University of Missouri-Saint Louis, Saint Louis 32510
– Ward E. Barnes Library, Saint Louis 32511
University of Mobile, Mobile 31872
University of Montana – Maureen & Mike Mansfield Library, Missoula 31868
University of Montana School of Law, Missoula 31869
University of Montevallo, Montevallo 31883
University of Moratuwa, Katubedda 26277
University of Mumbai, Mumbai 13711
– Department of Chemical Technology, Matunga 13712
University of Münster, Münster 09329

– Psychology Library, Münster 10470
University of Music, Saarbrücken 10550
University of Music and Performing Arts, München 10360
University of Music and Performing Arts, Wien 01315
University of Mysore, Mysore 13715
University of Nairobi, Nairobi 18005
– College of Health Sciences, Nairobi 18006
– Institute of Development Studies, Nairobi 18007
– Law Collection, Nairobi 18008
University of Namibia, Windhoek 19009
University of National and World Economy, Sofiya 03480
University of Natural Resources and Life Sciences, Vienna, Wien 01311
University of Navarra, Pamplona 25544
University of Nebraska at Kearney, Kearney 31489
University of Nebraska at Omaha, Omaha 32198
University of Nebraska – Lincoln, Lincoln 31612
– Architecture Library, Lincoln 31613
– Engineering Library, Lincoln 31614
– Geology Library, Lincoln 31615
– Marvin & Virginia Schmid Law Library, Lincoln 31616
– Mathematics Library, Lincoln 31617
– Music Library, Lincoln 31618
– C. Y. Thompson Library, Lincoln 31619
University of Nebraska Medical Center, Omaha 37309
University of Nevada-Reno, Reno 32425
– DeLaMare Library, Reno 32426
– Life & Health Sciences Library, Reno 32427
– Savitt Medical Library & IT Department, Reno 32428
University of New Brunswick, Fredericton 03708
– Engineering Library, Fredericton 03709
– Gerard V. La Forest Law Library, Fredericton 03710
University of New Brunswick, Saint John 03841
University of New Caledonia, Nouméa 19753
University of New England, Armidale 00381
University of New England Libraries, Biddeford 30424
University of New England – Westbrook College Campus, Portland 32344
University of New Hampshire, Durham 31050
– Biological Sciences Library, Durham 31051
– Chemistry Library, Durham 31052
– David G. Clark Memorial Physics Library, Durham 31053
– Engineering, Mathematics & Computer Science Library, Durham 31054
University of New Hampshire at Manchester, Manchester 31745
University of New Haven, West Haven 33021
University of New Mexico – Bureau of Business & Economic Research Data Bank, Albuquerque 36451
University of New Mexico – Zimmerman Library, Albuquerque 30186
– Centennial Science & Engineering Library, Albuquerque 30187
– Fine Arts Library, Albuquerque 30188
– Health Sciences Library + Informatics Center, Albuquerque 30189
– Law Library, Albuquerque 30190
– William J. Parish Memorial Business & Economics Library, Albuquerque 30191
University of New Orleans, New Orleans 32015
University of New South Wales, Sydney 00522
– Biomedical Library, Randwick 00523
– College of Fine Arts, Paddington 00524
– Law Library, Randwick 00525
– Physical Sciences Library, Randwick 00526
– Social Policy Research Centre, Randwick 00527
– Social Sciences and Humanities Library, Sydney 00528
– Water Reference Library, Manly Vale 00529
University of Newcastle, Callaghan 00409
– Huxley Library, Callaghan 00410
University of Nigeria, Nsukka 19970
– College of Medicine, Enugu 19971
University of North Alabama, Florence 31145
University of North Bengal, Raja Rammohunpur 13749
University of North Carolina at Asheville, Asheville 30263
University of North Carolina at Chapel Hill, Chapel Hill 30698

– Brauer Library (Math-Physics), Chapel Hill 30699
– Chapin Library (City & Regional Planning), Chapel Hill 30700
– Couch Biology Library (Botany Section), Chapel Hill 30701
– Couch Biology Library (Zoology Section), Chapel Hill 30702
– Geological Sciences Library, Chapel Hill 30703
– Health Sciences Library, Chapel Hill 30704
– Highway Safety Research Center Library, Chapel Hill 30705
– Institute of Government Library, Chapel Hill 30706
– Joseph Curtis Sloane Art Library, Chapel Hill 30707
– Kathrine R. Everett Law Library, Chapel Hill 30708
– Kenan Library (Chemistry), Chapel Hill 30709
– Music Library, Chapel Hill 30710
– Robert B. House Undergraduate Library, Chapel Hill 30711
– School of Information and Library Science Library, Chapel Hill 30712
University of North Carolina at Charlotte, Charlotte 30722
University of North Carolina at Greensboro, Greensboro 31255
University of North Carolina at Pembroke, Pembroke 32241
University of North Carolina at Wilmington, Wilmington 33073
University of North Dakota, Grand Forks 31228
– Energy & Environmental Research Center Library, Grand Forks 31229
– Gordon Erickson Music Library, Grand Forks 31230
– Harley E. French Library of the Health Sciences, Grand Forks 31231
– F. D. Holland, Jr Geology Library, Grand Forks 31232
– Thormodsgard Law Library, Grand Forks 31233
University of North Florida, Jacksonville 31452
University of North Sumatra, Medan 14279
– Faculty of Agriculture, Medan 14282
– Faculty of Economics, Medan 14280
– Faculty of Law, Medan 14281
University of North Texas, Denton 30991
University of North Texas Health Science Center at Fort Worth, Fort Worth 36872
University of Northern British Columbia Library, Prince George 03822
University of Northern Colorado, Greeley 31246
– Howard M. Scinner Music Library, Greeley 31247
University of Northern Iowa, Cedar Falls 30687
University of Notre Dame, Notre Dame 32160
– Architecture Library, Notre Dame 32161
– Chemistry-Physics Library, Notre Dame 32162
– Engineering Library, Notre Dame 32163
– Life Sciences Library, Notre Dame 32164
– Mathematics Library, Notre Dame 32165
– Medieval Institute Library, Notre Dame 32166
University of Notre Dame Australia, Fremantle
– University of Notre Dame Australia, Broadway 00435
– University of Notre Dame Australia, Broome 00436
– University of Notre Dame Australia, Fremantle 00437
– University of Notre Dame Australia, Fremantle 00438
University of Notre Dame – Kresge Law Library, Notre Dame Law School, Notre Dame 32167
University of Nottingham, Nottingham 29293
– George Green Library of Science and Engineering, Nottingham 29294
– Greenfield Medical Library, Nottingham 29295
– James Cameron-Gifford Library of Agricultural and Food Sciences, Loughborough 29296
University of Novi Sad, Novi Sad
– Faculty of Law, Novi Sad 24494
– Faculty of Philosophy, Novi Sad 24493
University of Nueva Caceres, Naga City 20728
University of Oklahoma, Norman 32126
– Architecture Library, Norman 32127
– Bass Business History Collection, Norman 32128

– Chemistry-Mathematics Library, Norman 32129
– Donald E. Pray Law Library, Norman 32130
– Engineering Library, Norman 32131
– Fine Arts Library, Norman 32132
– Geology Library, Norman 32133
– History of Science Collection, Norman 32134
– Physics & Astronomy Library, Norman 32135
– Western History Collection, Norman 32136
University of Oklahoma Health Sciences Center, Oklahoma City 37306
University of Oregon Libraries, Eugene 31097
– Architecture & Allied Arts Library, Eugene 31098
– John E. Jaqua Law Library, Eugene 31099
– Mathematics Library, Eugene 31100
– Science Library, Eugene 31101
University of Orleans, Orléans 07822
University of Oslo, Oslo 20125
– Astrophysics Library, Oslo 20127
– Biology Library, Oslo 20133
– Chemistry Library, Oslo 20144
– Department of Criminology and Sociology of Law, Oslo 20132
– Department of Geography, Oslo 20137
– Department of Private Law, Oslo 20142
– Department of Public and International Law, Oslo 20141
– Faculty of Dentistry Library, Oslo 20148
– Faculty of Law Library, Oslo 20143
– Faculty of Mathematics and Natural Science Library, Oslo 20146
– Faculty of Theology Library, Oslo 20153
– Informatics Library, Oslo 20138
– Legal History Collection, Oslo 20151
– Library of Humanities and Social Sciences, Oslo 20128
– Library of Medicine and Health Sciences, Oslo 20129
– Mathematics Library, Oslo 20145
– National Music Collection, Oslo 20147
– Norwegian Research Center for Computers and Law, Oslo 20152
– Nursing Science and Social Medicine Library, Oslo 20131
– Pathology Building Library, Oslo 20149
– Pharmacy Library, Oslo 20135
– Physics Library, Oslo 20136
– Science Library. Geology, Oslo 20150
– Zoological Museum Library, Oslo 20154
University of Osnabrueck, Osnabrück 10529
University of Ostrava, Ostrava 06375
University of Otago, Dunedin 19778
– Canterbury Medical Library, Christchurch 19779
– Dental Library, Dunedin 19780
– Hocken Library, Dunedin 19781
– Medical Library, Dunedin 19782
– Science Library, Dunedin 19783
– Sir Robert Stout Law Library, Dunedin 19784
University of Ottawa, Ottawa 03812
– Health Sciences Library, Ottawa 03813
– Law Library, Ottawa 03814
– Map Library, Ottawa 03815
– Morisset Library (Arts and Sciences), Ottawa 03816
University of Oxford, Oxford 29305
– Balliol College, Oxford 29306
– Bodleian Japanese Library, Oxford 29307
– Bodleian Law Library, Oxford 29308
– Bodleian Library of Commonwealth and African Studies at Rhodes House, Oxford 29309
– Campion Hall Library, Oxford 29310
– Christ Church Library, Oxford 29311
– Corpus Christi College, Oxford 29312
– Department for Continuing Education, Oxford 29313
– Department of Education, Oxford 29314
– Department of Engineering Science, Oxford 29315
– Department of Zoology, Oxford 29316
– English Faculty Library, Oxford 29317
– Exeter College Library, Oxford 29318
– Faculty of Music, Oxford 29319
– Green College, Oxford 29320
– Green Templeton College, Oxford 29321
– Harris Manchester College, Oxford 29322
– Hertford College, Oxford 29323
– Institute of Economics and Statistics, Oxford 29324
– Jesus College, Oxford 29325
– Keble College, Oxford 29326
– Lady Margaret Hall Library, Oxford 29327
– Lincoln College, Oxford 29328
– Mansfield College, Oxford 29329
– Merton College, Oxford 29330
– New College, Oxford 29331
– Nuclear and Astrophysics Laboratory, Oxford 29332
– Nuffield College, Oxford 29333

– Oriel College, Oxford 29334
– Oriental Institute, Oxford 29335
– Oxford Centre for Hebrew and Jewish Studies, Oxford 29336
– Oxford Centre for Mission Studies, Oxford 29337
– Pembroke College, Oxford 29338
– Philosophy Faculty, Oxford 29339
– Plant Sciences Library, Oxford 29340
– Queen's College, Oxford 29341
– Radcliffe Science Library, Oxford 29342
– Regent's Park College, Oxford 29343
– Sackler Library, Oxford 29344
– School of Geography and the Environment, Oxford 29345
– Social Science Library, Oxford 29346
– Somerville College, Oxford 29347
– St Anne's College, Oxford 29348
– St Antony's College, Oxford 29349
– St Edmund Hall Library, Oxford 29350
– St Hilda's College, Oxford 29351
– St Hugh's College, Oxford 29352
– St John's College, Oxford 29353
– St Peter's College, Oxford 29354
– Taylor Institution Library, Oxford 29355
– Trinity College, Oxford 29356
– University College, Oxford 29357
– University Laboratory of Physiology, Oxford 29358
– Wadham College, Oxford 29359
– Wolfson College, Oxford 29360
– Worcester College, Oxford 29361
– Wycliffe Hall Library, Oxford 29362
University of Pangasinan, Dagupan City 20695
University of Papua New Guinea, National Capital District 20564
– School of Medicine and Health Sciences, Boroko 20565
University of Pardubice, Pardubice 06377
University of Pécs, Pécs 13288
– Faculty of Humanities and Sciences, Pécs 13289
– Mihály Pekár Medical and Life Science Library, Pécs 13290
University of Pennsylvania, Philadelphia 32270
– Annenberg School of Communication, Philadelphia 32271
– Biddle Law Library, Philadelphia 32272
– Biomedical Library, Philadelphia 32273
– Chemistry Library, Philadelphia 32274
– Dental Library, Philadelphia 32275
– Engineering & Applied Science Library, Philadelphia 32276
– Fisher Fine Arts Library, Philadelphia 32277
– High Density Storage, Philadelphia 32278
– Lippincott-Wharton School, Philadelphia 32279
– Math-Physics-Astronomy Library, Philadelphia 32280
– Museum Library, Philadelphia 32281
– Rare Book & Manuscript Library, Philadelphia 32282
– Veterinary Library, Philadelphia 32283
University of Peradeniya, Peradeniya 26283
– Faculty of Medicine, Peradeniya 26284
– Post Graduate Institute of Agriculture, Peradeniya 26285
University of Peshawar, Peshawar 20453
– Islamia College, Peshawar 20454
University of Petra, Amman 17874
University of Petroleum, Beijing 05055
University of Piraeus, Piraius 13083
University of Piteşti, Piteşti 21941
University of Pittsburgh, Johnstown 31467
University of Pittsburgh, Pittsburgh
– Bevier Engineering Library, Pittsburg 32304
– Business Library, Pittsburgh 32305
– Chemistry Library, Pittsburg 32306
– Graduate School of Public & International Affairs/Economics, Pittsburg 32307
– Henry Clay Frick Fine Arts Library, Pittsburg 32308
– Hillman Library, Pittsburg 32309
– Information Sciences Library, Pittsburg 32310
– Langley Library, Pittsburg 32311
– Mathematics Library, Pittsburg 32312
– Theodore M. Finney Music Library, Pittsburg 32313
– Western Psychiatric Institute & Clinic Library, Pittsburg 32314
University of Pittsburgh at Bradford, Bradford 30540
University of Pittsburgh at Greensburg, Greensburg 31257
University of Pittsburgh – Barco Law Library, Pittsburgh 32315
University of Pittsburgh – Falk Library of the Health Sciences, Pittsburgh 32316
University of Pittsburgh Medical Center Shadyside, Pittsburgh 37386
University of Plymouth, Plymouth 29372
– Exeter Campus Library, Exeter 29373
– Seale-Hayne Campus Library, Newton Abbot 29374
University of Port Elizabeth, Port Elizabeth 25037
University of Port Harcourt, Port Harcourt 19980
University of Portland, Portland 32345
University of Portsmouth, Portsmouth 29377
University of Pretoria, Pretoria 25041
– Music Library, Pretoria 25042
– Veterinary Science Library, Onderstepoort 25043
University of Prince Edward Island, Charlottetown 03689
University of Puerto Rico, Bayamon 21850
University of Puerto Rico, Cayey 21852
University of Puerto Rico, Mayaguez 21861
University of Puerto Rico Library System, San Juan 21868
– Arts Collection, San Juan 21869
– Business Administration Library, San Juan 21870
– Caribbean and Latin American Studies Collection, San Juan 21871
– Documents and Maps Collection, San Juan 21872
– Education Library, San Juan 21873
– General Studies Library, San Juan 21874
– Library and Information Sciences Library, San Juan 21875
– Medical Sciences Campus, San Juan 21876
– Monserrate Santana de Pales Library, San Juan 21877
– Music Library, San Juan 21878
– Natural Sciences Library, San Juan 21879
– Periodicals Collection, San Juan 21880
– Planning Library, San Juan 21881
– Public Administration Library, San Juan 21882
– Puerto Rican Collection, San Juan 21883
– School of Architecture, San Juan 21884
– Zenobia and Juan Ramon Jimenez Room, San Juan 21885
University of Puerto Rico – Law School Library, San Juan 21886
University of Puerto Rico – Medical Sciences Campus Library, San Juan 21887
University of Puget Sound, Tacoma 32786
University of Pune, Pune 13745
University of Qatar, Doha 21908
University of Queensland, UQ Saint Lucia 00551
– Architecture and Music Library, UQ Saint Lucia 00552
– Biological Sciences Library, Saint Lucia Campus 00553
– Dorothy Hill Physical Sciences and Engineering Library, Saint Lucia Campus 00554
– Gatton Library, Lawes via Gatton 00555
– Walter Harrison Law Library, Saint Lucia Campus 00556
University of Radio Technology, Electronics and Automation, Moskva 22398
University of Rajasthan, Jaipur 13666
University of Reading, Reading 29380
– Museum of English Rural Live Library, Reading 29381
University of Redlands, Redlands 32422
University of Regina, Regina 03832
University of Rhode Island, Kingston 31510
University of Rhode Island – Graduate School of Oceanography, Narragansett 31929
University of Rhode Island – Providence Campus, Providence 32390
University of Richmond, Richmond 32439
– Parsons Music Library, University of Richmond 32440
– William T. Muse Law Library, Richmond 32441
University of Rio Grande, Rio Grande 32447
University of Rochester, Rochester 32463
– Art/Music Library, Rochester 32464
– Memorial Art Gallery, Rochester 32465
– Physics-Optics-Astronomy Library, Rochester 32466
– River Campus Libraries – Science & Engineering Library, Rochester 32467
– Rossell Hope Robbins Library, Rochester 32468
– Sibley Music Library, Rochester 32469
University of Rochester Medical Center, Rochester 37459
University of Roehampton, London 29254
University of Roorkee, Roorkee 13755
University of Rousse, Rousse 03460
University of Ruhuna, Matara 26280
– Faculty of Medicine, Galle 26281
University of Saint Francis, Fort Wayne 31166
University of Saint Francis, Joliet 31468
University of Saint Mary, Leavenworth 31583
University of Saint Mary of The Lake – Mundelein Seminary, Mundelein 35292
University of Saint Thomas, Houston 31366
– Cardinal Beran Library at Saint Mary's Seminary, Houston 31367
University of Saint Thomas, Saint Paul 32538
– Archbishop Ireland Memorial Library, Saint Paul 32539
– Charles J. Keffer Library, Minneapolis 32540
– Schoenecker Law Library, Minneapolis 32541
University of Salahuddin, Arbil 14444
University of Salento, Lecce 15045
– Department of Historical Studies from Middle Age to Contemporary Age, Lecce 15053
– Department of Historical studies from Middle age to Contemporary age, Lecce 15054
– Department of Mathematics "Ennio de Giorgi", Lecce 15050
University of Salford, Salford 29387
University of San Agustin, Iloilo City 20700
University of San Carlos, Cebu City 20691
– Filipiana Library, Cebu City 20692
University of San Diego, San Diego 32581
– Legal Research Center Library, San Diego 32582
University of San Francisco, San Francisco 32595
– Zief Law Library, San Francisco 32596
University of San José-Recoletos, Cebu City 20693
University of Santo Tomás, Manila 20724
– Medical Library, Quezon City 20725
University of Saskatchewan, Saskatoon 03850
– Education Library, Saskatoon 03851
– Engineering Library, Saskatoon 03852
– Health Sciences Library, Saskatoon 03853
– Law Library, Saskatoon 03854
– Natural Sciences Library, Saskatoon 03855
– Thorvaldson Library, Saskatoon 03856
– Veterinary Medicine Library, Saskatoon 03857
University of Science, Penang 18513
– Perak Branch Campus Library, Tronoh 18514
University of Science and Technology Beijing, Beijing 05056
University of Science and Technology of China, Hefei 05269
University of Science & Arts of Oklahoma, Chickasha 30795
University of Scranton, Scranton 32628
University of Sebrecen, Debrecen 13271
University of Seoul, Seoul 18098
University of Shanghai for Science and Technology, Shanghai 05536
University of Sheffield, Sheffield 29392
– Health Sciences Library, Sheffield 29393
University of Shkodra 'Luigj Gurakuqi', Shkodër 00007
University of Sierra Leone, Freetown
– College of Medicine and Allied Health Sciences, Freetown 24566
– Fourah Bay College, Freetown 24567
– Njala University College (NUC), Freetown 24568
– Technical Institute, Freetown 24569
University of Silesia Library, Katowice 20816
University of Sindh, Jamshoro 20435
– Institute of Sindhology, Jamshoro 20436
University of Sioux Falls, Sioux Falls 32679
University of Sofia, Sofiya 03463
– Biological Library, Sofiya 03464
– Faculty of chemistry, Sofiya 03467
– Faculty of Law, Sofiya 03475
– Faculty of Mathematics and Informatics, Sofiya 03468
– Faculty of Philology, Sofiya 03466
– Institute of Education, Sofiya 03474
– Institute of Geography, Sofiya 03471
– Institute of Geology, Sofiya 03472
– Institute of History, Sofiya 03473
– Institute of Philosophy, Sofiya 03470
– Library of Physics and Astronomy, Sofiya 03465
University of South Africa, Pretoria 25044
University of South Alabama, Mobile 31873
– Baugh Biomedical Library, Mobile 31874
University of South Australia, Mawson Lakes 00455
University of South Carolina, Columbia 30884
– Coleman Karesh Law Library, Columbia 30885
– Elliot White Springs Business Library, Columbia 30886
– Mathematics Library, Columbia 30887
– Music Library, Columbia 30888
– School of Medicine Library, Columbia 30889
– South Caroliniana Library, Columbia 30890
University of South Carolina – Aiken, Aiken 30173
University of South Carolina at Beaufort, Bluffton 30469
University of South Carolina at Sumter, Sumter 32762
University of South Carolina – Upstate-Spartanburg, Spartanburg 32694
University of South Dakota, Vermillion 32927
– McKusick Law Library, Vermillion 32928
University of South Dakota – Christian P. Lommen Health Sciences Library, Vermillion 32929
University of South Florida, Tampa
– Louis de la Parte Florida Mental Health Institute, Tampa 32798
– Shimberg Health Sciences Library, Tampa 32799
– Tampa Campus Library, Tampa 32800
University of South Florida Saint Petersburg, Saint Petersburg 32545
University of Southampton, Southampton 29398
– Hartley Library, Southampton 29399
– Health Services Library, Southampton 29400
– National Oceanographic Library, Southampton 29401
University of Southern California, Los Angeles 31670
– Applied Social Sciences Library, Los Angeles 31671
– Asa V. Call Law Library, Los Angeles 31672
– Crocker Business Library, Los Angeles 31673
– East Asian Library, Los Angeles 31674
– Helen Topping Architecture & Fine Arts Library, Los Angeles 31675
– Hoose Library of Philosophy, Los Angeles 31676
– Jennifer Ann Wilson Dental Library & Learning Center, Los Angeles 31677
– Norris Medical Library, Los Angeles 31678
– Science & Engineering Library, Los Angeles 31679
University of Southern Denmark, Odense 06717
– University Library of Southern Denmark, Music Department, Odense 06719
University of Southern Indiana, Evansville 31113
University of Southern Maine, Portland 32346
University of Southern Mindanao, Kabacan 20701
University of Southern Mississippi, Hattiesburg 31309
– William David McCain Library & Archives, Hattiesburg 31310
University of Southern Mississippi, Gulf Coast, Long Beach 31638
University of Southern Queensland, Toowoomba 00547
University of Split, Split 06160
– Faculty of Chemical Technology, Split 06167
– Faculty of Civil Engineering, Architecture and Geodesy, Split 06164
– Faculty of Economics, Split 06161
– Faculty of Electrical, Mechanical and Marine Engineering, Split 06163
– Faculty of Law, Split 06168
– Faculty of Philosophy, Split 06166
– Faculty of Philosophy, Zadar 06165
University of Sri Jayewardenepura, Nugegoda 26282
University of St Andrews, St Andrews 29404
University of St.Gallen – Graduate School of Business, Economics, Law and Social Sciences, St.Gallen 27136
University of Stirling, Stirling 29407
University of Strathclyde, Glasgow 29126
– Jordanhill Library, Glasgow 29127
University of Sunderland, Sunderland 29417
– Ashburne Library, Sunderland 29418
University of Surrey, Guildford 29131
University of Sussex, Brighton 28968
– British Library for Development Studies, Brighton 28969
– SPRU – Science and Technology Policy Research, Brighton 28970
University of Swaziland, Kwaluseni 26367
University of Sydney, Sydney 00530
– Badham Library, Sydney 00531
– Burkitt-Ford Library, Sydney 00532
– Camden Library, Camden 00533
– Conservatorium of Music, Sydney 00534
– Dentistry Library, Surry Hills 00535
– Health Sciences Library, Lidcombe 00536
– Law Library, Sydney 00537
– Mathematics Library, Sydney 00538
– Medical Library, Sydney 00539
– Schaeffer Fine Arts Library, Sydney 00540

- School of Geosciences, Sydney 00541
- Sydney College of the Arts, Rozelle 00542
University of Szczecin, Szczecin 20929
- Economic Library, Szczecin 20931
- Social Sciences, Szczecin 20930
University of Szeged, Szeged 13296
- Bólyai Institute, Szeged 13298
- Department of English and American Studies, Szeged 13297
- Department of French Studies, Szeged 13299
- Department of German Philology, Szeged 13300
- 1. and 2. Department of Hungarian Literature, Szeged 13302
- Department of Organic Chemistry, Szeged 13305
- Department of Slavic Philology, Szeged 13306
- Faculty of Law and Political Science, Szeged 13301
- Institute of Education and Psychology, Szeged 13303
- Institute of History and Ancient Studies, Library, Szeged 13307
University of Tabriz, Tabriz 14398
University of Tampa, Tampa 32801
University of Tasmania, Sandy Bay 00508
- Biomedical Library, Hobart 00509
- Centre for the Arts, Hobart 00510
- Clinical Library, Hobart 00511
- Law Library, Hobart 00512
- Music Library, Sandy Bay 00513
University of Technology, Baghdad 14447
University of Technology, Johor Bahru 18501
University of Technology, Kherson 28003
University of Technology and Design, Sankt-Peterburg 22519
University of Technology Sydney, Broadway 00400
- Sydney Kuring-gai Campus Library, Lindfield 00401
University of Tehran, Tehran 14411
University of Tennessee, Knoxville 31518
- George F. DeVine Music Library, Knoxville 31519
- Webster Pendergrass Agriculture & Veterinary Medicine Library, Knoxville 31520
University of Tennessee at Chattanooga, Chattanooga 30740
University of Tennessee at Martin, Martin 31770
University of Tennessee College, Nashville 31939
University of Tennessee – Joel A. Katz Law Library, Knoxville 31521
University of Tennessee – Memphis, Memphis 37093
University of Tennessee Space Institute Library, Tullahoma 32847
University of Texas, Houston
- M. D. Anderson Cancer Center, Houston 31368
- Health Science Center at Houston, Dental Branch Library, Houston 31369
- Houston Health Science Center – School of Public Health Library, Houston 31370
University of Texas at Arlington Library, Arlington 30261
University of Texas at Austin, Austin 30308
- Architecture and Planning Library, Austin 30309
- Center for American History Library, Austin 30310
- Chemistry Library, Austin 30311
- Classics Library, Austin 30312
- East Asian Library Program, Austin 30313
- Fine Arts Library, Austin 30314
- Harry Ransom Humanities Research Center, Austin 30315
- Jamail Center for Legal Research Library, Austin 30316
- Life Science (Biology, Pharmacy) Library, Austin 30317
- McKinney Engineering Library, Austin 30318
- Middle Eastern Library Program, Austin 30319
- Nettie Lee Benson Latin American Collection, Austin 30320
- Physics-Mathematics-Astronomy Library, Austin 30321
- Population Research Center Library, Austin 30322
- South Asian Library Program, Austin 30323
- Undergraduate Library, Austin 30324
- Walter Geology Library, Austin 30325
- Wasserman Public Affairs Library, Austin 30326
University of Texas at Austin – Marine Science Library, Port Aransas 32330

University of Texas at Brownsville & Texas Southmost College, Brownsville 30579
University of Texas at Dallas, Richardson 32432
University of Texas at El Paso, El Paso 31078
University of Texas at San Antonio, San Antonio 32572
University of Texas at Tyler, Tyler 32868
University of Texas Health Science Center at San Antonio, San Antonio 37522
University of Texas Medical Branch, Galveston 31203
University of Texas of the Permian Basin, Odessa 32180
University of Texas – Pan American Library, Edinburg 31073
University of Texas Southwestern Medical Center at Dallas Library, Dallas 30954
University of the Aegean, Mytilini 13082
University of the Air, Chiba 16911
University of the Argentine Museum of Sociology, Buenos Aires 00144
University of the Arts, Philadelphia 32284
University of the Arts London, Camberwell College of Arts, London 29171
University of the Arts London, Central Saint Martins College of Art & Design, London 29172
University of the Arts London, Chelsea College of Art & Design, London 29175
University of the Arts London, London College of Fashion, London 29198
University of the Basque Country, Leioa 25405
- Faculty of Fine Arts, Leioa 25411
University of the City of Manila, Manila 20717
University of the District of Columbia, Washington 32991
- David A. Clarke School of Law, Charles N. & Hilda H. M. Mason Law Library, Washington 32992
University of the Free State, Bloemfontein 25002
- Frik Scott Medical Library, Bloemfontein 25003
- Qwaqwa Campus Library, Phuthaditjaba 25004
University of the Incarnate Word, San Antonio 32573
University of the Netherlands Antilles, Willemstad 06316
University of the Ozarks, Clarksville 30823
University of the Pacific, Stockton 32746
University of the Pacific – McGeorge School of Law, Sacramento 32489
University of the Philippines, Quezon City 20731
- College of Business Administration, Quezon City 20732
- College of Engineering, Quezon City 20733
- College of Home Economics, Quezon City 20734
- College of Law, Quezon City 20735
- College of Public Health, Manila 20736
- Institute for Science and Mathematics Education Development, Quezon City 20737
- Philippine Center for Advanced Studies, Quezon City 20738
- School of Economics, Quezon City 20739
University of the Ryukyus, Nishihara 17068
University of The Sacred Heart, Santurce 21889
University of the Sacred Heart, Tokyo 17209
University of the Sciences in Philadelphia, Philadelphia 32285
University of the South, Sewanee 32658
University of the South Pacific, Suva 07295
University of the South Pacific – Cook Islands Centre Library, Rarotonga 06104
University of the South Pacific – Emalus Campus Library, Port Vila 42180
University of the South Pacific – Kiribati Extension Centre Centre Library, Tarawa 18048
University of the South Pacific – Marshall Islands Centre Library, Majuro 18673
University of the South Pacific – School of Agriculture, Apia 24380
University of the South Pacific – Solomon Islands Centre Library, Honiara 24993
University of the South Pacific – Tonga Centre Library, Nuku'alofa 27802
University of the Sunshine Coast, Maroochydore DC 00454
University of the Thai Chamber of Commerce, Bangkok 27722
University of The Virgin Islands, Kingshill 42309

University of The Virgin Islands, Saint Thomas 42310
University of the West Indies, Kingston 16865
- Library and Information Studies, Kingston 16866
- Medical Library, Kingston 16867
- School of Education, Kingston 16868
- Science Library, Kingston 16869
University of the West Indies, Saint Augustine 27805
- Faculty of Humanities and Education, Saint Augustine 27806
- Institute of International Relations, Saint Augustine 27807
University of the West Indies, St. Michael 01775
- Faculty of Law, Bridgetown 01776
University of the West Indies Extramural Department, Belize City 02898
University of the West Indies School of Continuing Studies, St. George's 13158
University of the West Indies School of Continuing Studies, St. John's 00111
University of the West Library, Rosemead 32482
University of the West of England, Bristol 28973
- Faculty of Art, Media and Design, Bristol 28974
University of the West of Scotland, Paisley
- Ayr Campus Library and Learning Resource Center, Ayr 29364
- Hamilton Campus Library, Hamilton 29365
- Robertson Trust Library and Learning Resource Centre, Paisley 29366
- Royal Alexandra Hospital Library, Paisley 29367
University of the Western Cape, Bellville 25000
University of the Witwatersrand, Johannesburg 25025
- Architecture Library, Johannesburg 25026
- Biological and Physical Sciences Library (Biophy), Johannesburg 25027
- Education Library, Parktown 25028
- Law Library, Johannesburg 25029
- Wits Health Sciences Library (WHSL), Johannesburg 25030
University of Thessaly, Volos 13088
University of Thu viên Dai Hoc Khoa hoc Huê, Hué 42280
University of Tirana, Tiranë 00012
University of Tokushima, Tokushima 17133
- Kuramoto Medical Branch Library, Tokushima 17134
University of Tokyo, Tokyo 17180
- Center for Pacific and American Studies, Tokyo 17188
- Faculty of Agriculture, Tokyo 17189
- Faculty of Economics, Tokyo 17187
- Faculty of Pharmaceutical Science, Tokyo 17194
- Historiographical Institute, Tokyo 17193
- Institute for Solid State Physics, Chiba 17182
- Institute of Industrial Science, Tokyo 17190
- Institute of Molecular and Cellular Biosciences, Tokyo 17181
- Institute of Social Science, Tokyo 17191
- Institute of Socio-Information and Communication Studies (ISICS) Library, Tokyo 17192
- Medical Library, Tokyo 17184
- Ocean Research Institute, Tokyo 17186
University of Toledo, Toledo 32824
- Law Library, Toledo 32825
University of Toronto, Toronto 03888
- Astronomy & Astrophysics Library, Toronto 03889
- Bora Laskin Law Library, Toronto 03890
- Business Information Centre Library, Toronto 03891
- Caven Library, Toronto 03892
- Centre of Criminology Library, Toronto 03893
- Cheng Yu Tung East Asian Library, Toronto 03894
- Data, Map and Government Information Services, Toronto 03895
- Department of Chemistry Library, Toronto 03896
- Department of Fine Art, Toronto 03897
- Department of Physics Library, Toronto 03898
- Department of Zoology Library, Toronto 03899
- Engineering & Computer Science Library, Toronto 03900
- Faculty of Dentistry Library, Toronto 03901
- Faculty of Design, Architecture & Landscape Architecture Library, Toronto 03902

- Faculty of Information Studies Inforum, Toronto 03903
- Faculty of Music Library, Toronto 03904
- Gerstein Science Information Centre, Toronto 03905
- Industrial Relations Centre, Toronto 03906
- John M. Kelly Library, Toronto 03907
- Mathematical Sciences Library, Toronto 03908
- Mississauga Library, Mississauga 03909
- New College – Ivey Library, Toronto 03910
- Noranda Earth Sciences Library, Toronto 03911
- Robarts Library, Toronto 03912
- Robertson Davies Library, Toronto 03913
- Scarborough Library, Scarborough 03914
- Thomas Fisher Rare Books Library, Toronto 03915
- Trinity College, Toronto 03916
- University College – Laidlaw Library, Toronto 03917
- Victoria University, Toronto 03918
- Victoria University E. J. Pratt Library & Emmanuel College Library, Toronto 03919
University of Toronto Library, Toronto 03920
University of Toulouse Le Mirail, Toulouse 08085
University of Transkei (UNITRA), Umtata 25055
University of Tras-os-Montes e Alto Douro, Vila Real 21711
University of Trieste, Trieste 15543
- Sociology and Politics Library, Trieste 15545
University of Tromsø, Tromsø
- Department Library of Physics, Tromsø 20166
- Library of Humanities and Social Sciences, Tromsø 20167
- Library of the Sciences, Medicine and Health Care, Tromsø 20168
University of Tsukuba, Tsukuba 17228
University of Tulsa, Tulsa 32853
- University of Tulsa College of Law, Tulsa 32854
University of Twente, Enschede 19123
University of Twente faculty of Geo-Information Science and Earth Observation (ITC), Enschede 19124
University of Udine, Udine 15564
- Library of Economics and Law, Udine 15566
University of Ulster, Belfast 28944
University of Ulster, Newtownabbey 29284
University of Ulster Coleraine Campus, Coleraine 29073
University of Utah, Salt Lake City
- Marriott Library, Salt Lake City 32560
- S. J. Quinney Law Library, Salt Lake City 32561
- Spencer S. Eccles Health Sciences Library, Salt lake City 32562
University of Valencia, Valencia 25618
- Historical Library, Valencia 25624
University of Valladolid, Valladolid 25625
- Faculty of Philosophy and Letters, Valladolid 25639
- Library of Sciences, Valladolid 25627
University of Veliko Turnovo 'St. Cyril and St. Methodius', Veliko Tărnovo 03487
University of Venda, Thohoyandou 25054
University of Vermont & State Agricultural College, Burlington 30603
- Dana Medical Library, Burlington 30604
University of Veszprém, Georgikon Faculty of Agriculture, Keszthely 13281
- Faculty of Agricultural Sciences, Mosonmagyaróvár 13283
- Faculty of Animal Science, Kaposvár 13282
University of Veterinary and Pharmaceutical Sciences, Brno 06364
University of Veterinary Medicine, Košice 24667
University of Victoria Libraries, Victoria 03952
- Diana M. Priestly Law Library, Victoria 03953
University of Vienna, Wien 01316
- Archaeology and Numismatics Library, Wien 01321
- Botany Library, Wien 01325
- Byzantine and Modern Greek Studies Library, Wien 01326
- Catholic and Protestant Theology Library, Wien 01335
- Chemistry Library, Wien 01327
- Classical Philology, Medieval and Neolatin Library, Wien 01336
- Education, Linguistics and Comparative Literature Library, Wien 01323
- Geography and Regional Studies Library, Wien 01331
- The Jewish Studies Library, Wien 01334

– Mathematical Logic Department Library, Wien 01358
– Musicology Library, Wien 01339
– Pharmacy and Nutritional Sciences Library, Wien 01342
– Slavistic Studies Library, Wien 01348
– South Asian, Tibetan and Buddhist Studies Libraries, Wien 01350
– Theatre, Film and Media Studies Library, Wien 01351
University of Virginia, Charlottesville 30723
– Albert & Shirley Small Special Collections Library, Charlottesville 30724
– Arthur J. Morris Law Library, Charlottesville 30725
– Astronomy Library, Charlottesville 30726
– Biology-Psychology Library, Charlottesville 30727
– Charles L. Brown Science & Engineering Library, Charlottesville 30728
– Chemistry Library, Charlottesville 30729
– Claude Moore Health Sciences Library, Charlottesville 30730
– Clemons Library, Charlottesville 30731
– Colgate Darden Graduate School of Business – Camp Library, Charlottesville 30732
– Education Library, Charlottesville 30733
– Fiske Kimball Fine Arts Library, Charlottesville 30734
– Mathematics Library, Charlottesville 30735
– Music Library, Charlottesville 30736
– Physics Library, Charlottesville 30737
University of Virginia's College at Wise, Wise 33090
University of Vudal, Popondetta 20566
University of Vudal, Rabaul 20568
University of Waikato, Hamilton 19785
University of Wales Institute, Cardiff, Cardiff 29060
University of Wales Lampeter, Lampeter 29149
University of Warmia and Mazury in Olsztyn, Olsztyn 20896
University of Warwick, Coventry 29075
University of Washington, Seattle 32635
– Architecture-Urban Planning Library, Seattle 32636
– Art Library, Seattle 32637
– Chemistry Library, Seattle 32638
– Drama Library, Seattle 32639
– East Asia Library, Seattle 32640
– Engineering Library, Seattle 32641
– Fisheries-Oceanography Library, Seattle 32642
– Foster Business Library, Seattle 32643
– Friday Harbor Laboratories Library, Friday Harbor 32644
– Health Sciences Libraries, Seattle 32645
– Marian Gould Gallagher Law Library, Seattle 32646
– Mathematics Research Library, Seattle 32647
– Music Library, Seattle 32648
– Natural Sciences Library, Seattle 32649
– Odegaard Undergraduate Library, Seattle 32650
– Physics-Astronomy Library, Seattle 32651
– Social Work Library, Seattle 32652
– Tacoma Branch, Tacoma 32653
University of Washington Botanic Gardens, Seattle 37600
University of Water Transport, Sankt-Peterburg 22550
University of Waterloo, Waterloo 03955
University of West Alabama, Livingston 31629
University of West Bohemia, Plzeň 06378
– Pedagogical Library, Plzeň 06380
University of West Florida, Pensacola 32242
University of West Georgia, Carrollton 30679
University of West London, Reading 29382
University of West Los Angeles, Inglewood 31395
University of Western Australia, Crawley 00428
University of Western Ontario, London 03746
– John & Dotsa Bitove Family Law Library, London 03747
– Education Library, London 03748
– C. B. "Bud" Johnston Library, London 03749
– Music Library, London 03750
– Allyn & Betty Taylor Library, London 03751
– D. B. Weldon Library, London 03752
University of Western Sydney, Penrith 00499
– Hawkesbury Campus Library, Richmond 00500
– Penrith Campus, Kingswood 00501
University of West-Hungary, Sopron 13293
– Benedek Elek Pedagogical Faculty, Sopron 13294
University of Westminister, London 29255
– Little Titchfield Street Library, London 29256
– Marylebone Campus Library, London 29257
– Regent Campus Library, London 29258
University of Winchester, Winchester 29434
University of Windsor, Windsor 03957
– Paul Martin Law Library, Windsor 03958
University of Winnipeg, Winnipeg 03974
University of Wisconsin, Marshfield 33518
University of Wisconsin Baraboo-Sauk County, Baraboo 30352
University of Wisconsin – Barron County Library, Rice Lake 32431
University of Wisconsin Center – Marathon County Library, Wausau 33003
University of Wisconsin Colleges, Waukesha 33002
University of Wisconsin – Eau Claire, Eau Claire 31071
University of Wisconsin – Fox Valley Library, Menasha 31801
University of Wisconsin – Green Bay, Green Bay 31248
University of Wisconsin – La Crosse, La Crosse 31524
University of Wisconsin – Madison, Madison 31711
– Biology Library, Madison 31712
– Business Library, Madison 31713
– Center for Demography Library, Madison 31714
– Center for Instructional Materials & Computing, Madison 31715
– Chemistry Library, Madison 31716
– College (Undergraduate) Library, Madison 31717
– Cooperative Children's Book Center, Madison 31718
– Ebling Library for Health Sciences, Madison 31719
– Geography Library, Madison 31720
– C. K. Keith Geology & Geophysics Library, Madison 31721
– Kohler Art Library, Madison 31722
– Kurt Wendt Engineering Library, Madison 31723
– Law School Library, Madison 31724
– Mills Music Library, Madison 31725
– Physics Library, Madison 31726
– Plant Pathology Memorial Library, Madison 31727
– F.B Power Pharmaceutical Library, Madison 31728
– Primate Center Library, Madison 31729
– School of Library & Information Studies, Madison 31730
– Social Science Reference Library, Madison 31731
– Social Work Library, Madison 31732
– Space Science & Engineering Center, Madison 31733
– Steenbock Memorial Agricultural Library, Madison 31734
– Stephen Cole Kleene Mathematics Library, Madison 31735
– Woodman Astronomical Library, Madison 31736
University of Wisconsin – Manitowoc Library, Manitowoc 31752
University of Wisconsin – Milwaukee, Milwaukee 31840
– American Geographical Society Library, Milwaukee 31841
University of Wisconsin – Oshkosh, Oshkosh 32214
University of Wisconsin – Parkside Library, Kenosha 31496
University of Wisconsin – Platteville, Platteville 32319
University of Wisconsin – River Falls, River Falls 32449
University of Wisconsin – Stevens Point, Stevens Point 32742
University of Wisconsin – Stout, Menomonie 31802
University of Wisconsin – Superior, Superior 32764
University of Wisconsin – Whitewater, Whitewater 33051
University of Wisconsin-Madison – Water Resources Library, Madison 37076
University of Wollongong, Wollongong 00558
University of Wolverhampton, Wolverhampton 29437
– Compton Learning Centre, Wolverhampton 29438
– Harrison Learning Centre, Wolverhampton 29439
University of Worcester, Worcester 29440
University of Wyoming, Laramie 31560
– Brinkerhoff Geology Library, Laramie 31561
– Science-Technology Library, Laramie 31562
University of Wyoming – George W. Hopper Law Library, Laramie 31563
University of Wyoming – The Learning Resources Center Library, Laramie 31564
University of Yamanashi – Kofu Campus, Kofu 16982
University of Yangon, Yangon 18988
– Institute of Agriculture, Pyinmana 18989
– Institute of Animal Husbandry and Veterinary Science, Pyinmana 18990
– Institute of Computer Science and Technology, Yangon 18991
– Institute of Economics, Yangon 18992
– Institute of Education, Yangon 18993
– Institute of Foreign Languages, Yangon 18994
– Institute of Medicine I, Yangon 18995
– Institute of Medicine II, Yangon 18996
– Institute of Technology, Yangon 18997
University of York, York 29443
– King's Manor Library, York 29444
University of Zambia, Lusaka 42327
– School of Medicine, Lusaka 42328
– Veterinary Medicine Library, Lusaka 42329
University of Zaragoza, Zaragoza 25644
– Faculty of Economics and Business Sciences, Zaragoza 25654
University of Zielona Góra, Zielona Góra 21033
University of Žilina, University Library, Žilina 24671
University of Zimbabwe, Harare 42366
– Education Library, Harare 42367
– Institute of Development Studies, Harare 42368
– Medical Library, Harare 42369
University of Zululand, Kwa-Dlangezwa 25031
University of Zurich, Zürich 26986
– Department of Economics Library, Zürich 27172
– Institute of Art History, Zürich 27191
– Institute of East Asian Studies, Zürich 27198
– Institute of East Asian Studies – Western Library, Zürich 27197
– Institute of Political Science, Zürich 27186
– Institute of Popular Culture Studies, Zürich 27187
– Institute of Sociology, Zürich 27209
– The Zurich University Centre for Ethics (ZUCE), Institute for Social Ethics, Zürich 27177
University of Zurich – Main Library, Zürich 27215
University Pediatrics Hospital, Sofiya 03566
University Politehnica of Bucharest, Bucureşti 21924
University School of Physical Education in Cracov, Main Library, Kraków 20838
University School of Physical Education, Main Library, Poznań 20902
University Toulouse III – Paul Sabatier, Toulouse 08090
Universty of Technology and Life Sciences in Bydgoszcz, Bydgoszcz 20799
Universytet ekonomiky ta prava "KROK", Kyiv 28027
Universytet 'Kyivo-Mohilyanska Akademiya', Kyiv 28028
Univerza v Ljubljani, Ljubljana
– Akademija za likovno umetnost, Ljubljana 24814
– Biološka Knjiznica, Ljubljana 24815
– Biološka knjižnica, Ljubljana 24816
– Biotehniska fakulteta, Centralna biotehniska knjiznica, Ljubljana 24817
– Centralna ekonomska knjiznica (CEK), Ljubljana 24818
– Centralna tehniška knjižnica, Ljubljana 24819
– Elektroinštitut Milan Vidmar, Ljubljana 24820
– Fakulteta za arhitekturo, Ljubljana 24821
– Fakulteta za elektrotehniko, Ljubljana 24822
– Fakulteta za farmacijo, Ljubljana 24823
– Fakulteta za kemijo in kemijsko tehnologijo, Ljubljana 24824
– Fakulteta za matematiko in fiziko – Knjižnica za mehaniko, Ljubljana 24825
– Fakulteta za pomorstvo in promet, Portorož 24826
– Fakulteta za socialno delo, Ljubljana 24827
– Fakulteta za strojništvo, Ljubljana 24828
– Filozofska fakulteta, Ljubljana 24829
– Filozofska fakulteta – Knjižnica oddelkov za anglistiko in germanistiko, Ljubljana 24830
– Filozofska fakulteta – Oddelek za klasično filologijo, Ljubljana 24831
– Filozofska fakulteta – Oddelek za muzikologijo, Ljubljana 24832
– Filozofska fakulteta – Oddelek za slovenistiko in Oddelek za slavistiko, Ljubljana 24833
– FNT – VTOZD Montanistika, Ljubljana 24834
– INDOK sluzba in knjiznica, Ljubljana 24835
– Inššitut za kriminologijo, Ljubljana 24836
– Inštitut Jožef Stefan, Ljubljana 24837
– Inštitut za ekonomska raziskovanja, Ljubljana 24838
– Inštitut za mednarodno pravo in mednarodne odnose, Ljubljana 24839
– Kemijski inštitut, Ljubljana 24840
– Matematična knjižnica, Ljubljana 24841
– Medicinska fakulteta, Ljubljana 24842
– Naravoslovnotehniška fakulteta, Ljubljana 24843
– Oddelek za agronomijo, Ljubljana 24844
– Oddelek za arheologijo, Ljubljana 24845
– Oddelek za etnologijo in kulturno antropologijo, Ljubljana 24846
– Oddelek za filozofijo, Ljubljana 24847
– Oddelek za geografijo, Ljubljana 24848
– Oddelek za geologijo, Ljubljana 24849
– Oddelek za gozdarstvo, Ljubljana 24850
– Oddelek za lesarstvo, Ljubljana 24851
– Oddelek za primerjalno književnost in literarno teorijo, Ljubljana 24852
– Oddelek za psihologijo, Ljubljana 24853
– Oddelek za romanske jezike in književnosti, Ljubljana 24854
– Oddelek za sociologijo, Ljubljana 24855
– Oddelek za živilstvo, Ljubljana 24856
– Oddelek za živinorejo, Domžale 24857
– Oddelek za zootehniko, Domžale 24858
– Osrednja družboslovna knjižnica Jožeta Goričarja, Ljubljana 24859
– Osrednja humanistična knjižnica, Ljubljana 24860
– Pedagoška fakulteta, Ljubljana 24861
– Pravna fakulteta, Ljubljana 24862
– Teološka fakulteta v Ljubljani Enota v Mariboru, Maribor 24863
– Veterinarska fakulteta, Ljubljana 24864
– Visoka šola za zdravstvo, Ljubljana 24865
Univerza v Mariboru, Maribor 24867
– Ekonomsko-poslovna fakulteta, Maribor 24868
– Fakulteta za kmetijstvo in biosistemske vede, Hoče 24869
– Fakulteta za organizacijske vede, Kranj 24870
– Pedagoška fakulteta, Maribor 24871
– Pravna fakulteta, Maribor 24872
– Tehniška fakulteta, Maribor 24873
Univerza v Novi Gorici, Nova Gorica 24874
Univerzita Jana Evangelisty Purkyně – Pedagogická fakulta, Ustí nad Labem 06407
Univerzita Karlova v Praze, Praha 06386
– Astronomický ústav, Praha 06387
– Evangelická teologická fakulta, Praha 06388
– Faculty of Arts, Praha 06389
– Fakula sociálních věd, Praha 06390
– Fakulta tělesné výchovy a sportu, Praha 06391
– Farmaceutická fakulta, Hradec Králové 06392
– Husitská teologická fakulta, Praha 06393
– Lékařská fakulta Hradec Králové, Hradec Králové 06394
– Lékařská fakulta Plzeň, Plzeň 06395
– 1. lékařská fakulta, Praha 06396
– 2. lékařská fakulta, Praha 06397
– 3. lékařská fakulta, Praha 06398
– Matematicko-fyzikální fakulta, Praha 06399
– Pedagogická fakulta, Praha 06400
– Právnická fakulta, Praha 06401
– Přírodovědecká fakulta, Praha 06402
– Ústav dejin UK – Archiv UK, Praha 06403
Univerzita Komenského v Bratislave, Bratislava 24637
– Fakulta matematiky, fyziky a informatiky, Bratislava 24653
– Fakulta telesnej vychovy a sportu, Bratislava 24654
– Farmaceutická fakulta, Bratislava 24655
– Filozofická fakulta, Bratislava 24656
– Knižnica a študijné informačné stredisko, Martin 24657
– Knižnica Právnickej fakulty, Bratislava 24658
– Lekárska Fakulta, Bratislava 24659
– Pedagogická fakulta, Bratislava 24660
– Rímskokatolicka cyrilometodska bohoslovecka fakulta, Bratislava 24661
Univerzita Konštantína Filozofa, Nitra 24668
Univerzita Mateja Bela, Banská Bystrica 24640
Univerzita Palackého, Olomouc 06370
– Lékařská fakulta, Olomouc 06371
– Pedagogická fakulta, Olomouc 06372
– Právnická fakulta, Olomouc 06373

– Přírodovědecká fakulta, Olomouc 06374
Univerzita Pardubice, Pardubice 06377
Univerzita Pavla Jozefa Šafárika v Košiciach, Košice 24665
– Lekárská fakulta, Košice 24666
Univerzita veterinárskeho lekárstva, Košice 24667
Univerzitet 'Kiril i Metodij' vo Skopje, Skopje 18427
– Architektonsko-Gradjevinski Fakultet, Skopje 18431
– Centralna medicinska biblioteka, Skopje 18432
– Institute of Biology, Skopje 18433
– Institute of Economics, Skopje 18434
– Istoriska Biblioteka na Filozofski Fakultet, Skopje 18435
– Katedra za Istorija na Književnosti na Narodni Na FNRJ, Skopje 18436
– Katedra za Romanska Filologija, Skopje 18437
– Matematički Institut, Skopje 18438
– Praven Fakultet, Skopje 18439
Univerzitet u Beogradu, Beograd 24459
– Arhitektonski Fakultet, Beograd 24460
– Ekonomski Fakultet, Beograd 24461
– Fakultet Sporta i Fizičkog Vaspitanja, Beograd 24462
– Farmaceutski Fakultet, Beograd 24463
– Filološki fakultet, Katedra za germanistiku, Beograd 24464
– Fiolski fakultet, Katedra za Orijentalistiku, Beograd 24465
– Geografski Fakultet, Beograd 24466
– Gradjevinski Fakultet, Beograd 24467
– Institut Društvenih Nauk, Beograd 24468
– Institute for Regional Geology and Paleontology, Beograd 24469
– Institute of Mineralogy, Crystallography, Petrology and Geochemistry, Beograd 24470
– 'Jaroslav Černi' Institute for the Development of Water Resources, Beograd 24471
– Mašinski Fakultet, Beograd 24472
– Matematički Zavod, Beograd 24473
– Medicinski Fakultet, Beograd 24474
– Odeljenja za Klasične Nauke, Beograd 24475
– Odeljenje za Filozofiju, Beograd 24476
– Poljoprivredni Fakultet, Zemun 24477
– Pravni Fakultet, Beograd 24478
– Rudarsko-Geološki Fakultet, Beograd 24479
– Saobracajni fakultet, Beograd 24480
– Seminar za francuski jezik i književnost, Beograd 24481
– Seminar za Istoriju Jugoslovenske Književnosti, Beograd 24482
– Seminar za Južnoslovenske Jeziko, Beograd 24483
– Slovenski Seminar, Beograd 24484
– Sumarski Fakultet, Beograd 24485
– Tehnološki Fakultet, Beograd 24486
– Veterinarski Fakultet Univerziteta, Beograd 24487
Univerzitet u Nišu, Niš 24488
– Ekonomski fakultet, Niš 24489
– Filozofski Fakultet, Niš 24490
– Gradjevinski fakultet, Niš 24491
– Medicinski Fakultet, Niš 24492
Univerzitet u Novom Sadu, Novi Sad
– Filozofski Fakultet, Novi Sad 24493
– Pravni Fakultet, Novi Sad 24494
Univerzitetna klinika za pljučne bolezni in alergijo Golnik, Golnik 24910
Uniwersitet Warmińsko-Mazurski, Olsztyn 20896
Uniwersytet Ekonomiczny, Wrocław 21028
Uniwersytet Gdański, Gdansk 20806
– Instytut Anglistyki, Gdańsk 20807
– Instytut Teorii Ekonomii, Sopot 20808
Uniwersytet im. Adama Mickiewicza, Poznań 20905
– Institute of Linguistics Library, Poznań 20906
– Instytut Filologii Polskiej, Poznań 20907
– Instytut Historii, Poznań 20908
– Instytut Historii Sztuki, Poznań 20909
– Katedra Astronomii, Poznań 20910
– Studium Nauczania Yezikow Obcych, Poznań 20911
– Wydział Chemii, Poznań 20912
– Wydział Nauk Społecznych, Poznań 20913
– Wydział Neofilologii, Poznań 20914
– Wydział Prawa i Administracji, Poznań 20915
– Zakladu Muzykologii, Poznań 20916
Uniwersytet Jagielloński, Kraków 20840
– Collegium Medicum, Kraków 20841
– Instytut Archeologii, Kraków 20842
– Instytut Biologii Molekularnej i Biotechnologii, Kraków 20843
– Instytut Botaniki im. Władysława Szafera, Kraków 20844
– Instytut Etnologii i Antropologii Kulturowej, Kraków 20845
– Instytut Filologii Angielskiej, Kraków 20846
– Instytut Filologii Germańskiej, Kraków 20847
– Instytut Filologii Klasycznej, Kraków 20848
– Instytut Filologii Orientalnej, Kraków 20849
– Instytut Filologii Polskiej, Kraków 20850
– Instytut Filologii Romańskiej, Kraków 20851
– Instytut Filologii Słowiańskiej, Kraków 20852
– Instytut Filozofii, Kraków 20853
– Instytut Fizyki, Kraków 20854
– Instytut Geografii i Gospodarki Przestrzennej, Kraków 20855
– Instytut Historii, Kraków 20856
– Instytut Historii Sztuki, Kraków 20857
– Instytut Informatyki, Kraków 20858
– Instytut Matematyki, Kraków 20859
– Instytut Nauk Geologicznych, Kraków 20860
– Instytut Nauk Politycznych, Kraków 20861
– Instytut Pedagogiki, Kraków 20862
– Instytut Polonijny, Kraków 20863
– Instytut Psychologii, Kraków 20864
– Instytut Religioznawstwa, Kraków 20865
– Instytut Socjologii, Kraków 20866
– Instytut Zoologii, Kraków 20867
– Katedra Językoznawstwa Ogólnego, Kraków 20868
– Obserwatorium Astronomiczne, Kraków 20869
– Wydział Chemii, Kraków 20870
– Wydziałowa Biblioteka Prawnicza, Kraków 20871
Uniwersytet Kardynała Stefana Wyszyńskiego, Warszawa 20973
Uniwersytet Kazimierza Wielkiego, Bydgoszcz 20798
Uniwersytet Łódzki, Łódź 20877
– Biblioteka Socjologiczna im. Józefa Chalasinskiego, Lódz 20878
– Wydział Filologiczny, Łódź 20879
Uniwersytet Marii Curie-Skłodowskiej, Lublin 20888
– Biblioteka Nauk Biologicznych, Lublin 20889
– Instytut Matematyki, Lublin 20890
– Instytut Nauk o Ziemi, Lublin 20891
– Wydział Chemii, Lublin 20892
– Wydział Ekonomicznego, Lublin 20893
Uniwersytet Medyczny w Łódzi, Łódź 20880
Uniwersytet Medyczny w Lublinie, Lublin 20894
Uniwersytet Mikołaja Kopernika, Toruń 20934
– Biblioteka Wydziału, Toruń 20935
– Centrum Astronomii, Łysomice 20936
– Faculty of Biology and Earth Sciences, Toruń 20937
– Instytut Archeologii oraz Katedra Etnologii, Toruń 20938
– Instytut Biologii Ogólnej i Molek. oraz Instytut Ekologii i Ochrony Środowiska, Toruń 20939
– Instytut Filologii Słowiańskiej, Toruń 20940
– Instytut Filozofii, Toruń 20941
– Instytut Fizyki, Toruń 20942
– Instytut Historii i Archiwistyki, Toruń 20943
– Instytut Pedagogiki i Psychologii, Toruń 20944
– Katedra Filologii Angielskiej, Toruń 20945
– Katedra Filologii Klasycznej, Toruń 20946
– Wydział Chemii, Toruń 20947
– Wydział Matematyki i Informatyki, Toruń 20948
Uniwersytet Opolski, Opole 20898
Uniwersytet Pedagogiczny w Krakowie, Kraków 20872
Uniwersytet Przyrodniczy w Lublinie, Lublin 20895
Uniwersytet Przyrodniczy w Poznaniu, Poznań 20917
Uniwersytet Przyrodniczy we Wrocławiu, Wrocław 21029
Uniwersytet Rzeszowski, Rzeszów 20923
Uniwersytet Szczeciński, Szczecin 20929
– Biblioteka Nauk Społecznych, Szczecin 20930
– Faculty of Economics and Management, Szczecin 20931
Uniwersytet Technologiczno-Przyrodniczego im. Jana i Jedrzeja Śniadeckich w Bydgoszczy, Bydgoszcz 20799
Uniwersytet w Białymstoku, Białystok 20795
Uniwersytet w Białymstoku – Wydział Pedagogiki i Psychologii, Białystok 20796
Uniwersytet Warszawski, Warszawa 20974
– Instytut Anglistyki, Warszawa 20975
– Instytut Botaniki, Warszawa 20976
– Instytut Filologii Klasycznej, Warszawa 20977
– Instytut Filozofii, Warszawa 20978
– Instytut Fizyki Doświadczalnej im. Stefana Pieńkowskiego, Warszawa 20979
– Instytut Fizyki Teoretycznej, Warszawa 20980
– Instytut Historii Sztuki, Warszawa 20981
– Instytut Historyczny, Warszawa 20982
– Instytut Informacji Naukowej i Studiów Bibliologicznych, Warszawa 20983
– Instytut Języka Polskiego im. Jana Baudouina de Courtanay, Warszawa 20984
– Instytut Literatury Polskiej i Instytutu Kultury Polskiej im. Waclawa Borowego, Warszawa 20985
– Instytut Muzykologii, Warszawa 20986
– Instytut Orientalistycznego, Warszawa 20987
– Instytut Romanistyki, Warszawa 20988
– Ośrodek Kultury Francuskiej / Centre Michel Foucault d'études françaises, Warszawa 20989
– Ośrodek Studiów Amerykanskich, Warszawa 20990
– Wydział Biologii, Warszawa 20991
– Wydział Chemii, Warszawa 20992
– Wydział Geologii, Warszawa 20993
– Wydział Nauk Ekonomicznych, Warszawa 20994
– Wydział Prawa i Administracji, Warszawa 20995
– Wydział Psychologii, Warszawa 20996
– Wydział Zarządzania, Warszawa 20997
Uniwersytet Zielonogorski, Zielona Góra 21033
Unley Library Service, Unley 01170
Unocal Corp, Sugar Land 36287
Unteroffizierschule der Luftwaffe, Appen 10705
Unuiversity College of Northern Denmark, Aalborg 06636
Un-yusho Toshokan, Tokyo 17327
UOP Library & Information Services, Des Plaines 36803
Upjohn Company – Pharmacia at Upjohn Research Library, Kalamazoo 35789
W. E. Upjohn Institute for Employment Research Library, Kalamazoo 36984
Upland Public Library, Upland 41642
Upper Arlington Public Library, Columbus 38670
Upper Austrian Provincial Museum, Linz 01560
Upper Colorado River Commission Library, Salt Lake City 37515
Upper Cumberland Regional Library, Cookeville 38686
Upper Darby Township & Sellers Memorial Free Public Library, Upper Darby 41643
Upper Dublin Public Library, Fort Washington 39117
Upper Grand District School Board, Guelph 04354
Upper Iowa University, Fayette 31130
Upper Merion Township Library, King of Prussia 39719
Upper Moreland Free Public Library, Willow Grove 41907
Upper Murray Regional Library, Wodonga 01177
Upper Norwood Joint Library, London 30035
Upper Saddle River Public Library, Upper Saddle River 41644
Upper Saint Clair Township Library, Pittsburgh 40805
Upper Silesian Museum, Bytom 21083
Upplands Väsby bibliotek, Upplands Väsby 26939
Uppsala Academic Hospital, Uppsala 26705
Uppsala City Library, Uppsala 26941
Uppsala County Library, Uppsala 26940
Uppsala stadsbibliotek, Uppsala 26941
Uppsala Universitet, Uppsala 26450
– Ångströmbiblioteket, Uppsala 26451
– Biologibiblioteket, Uppsala 26452
– BMC-biblioteken, Uppsala 26453
– Bostadsforskningsbiblioteket, Gävle 26454
– Carolinabiblioteket, Uppsala 26455
– Dag Hammarskjöldbiblioteket, Uppsala 26456
– Ekonomikums bibliotek, Uppsala 26457
– Geobiblioteket, Uppsala 26458
– ILU-biblioteket, Uppsala 26459
– Juridiska biblioteket, Uppsala 26460
– Karin Boye-biblioteket, Uppsala 26461
– Medicinska biblioteket, Uppsala 26462
– Pedagogiska biblioteket, Uppsala 26463
Uppsala University, Uppsala 26450
– Angström Library, Uppsala 26451
– BMC-library, Uppsala 26453
– Carolina Library, Uppsala 26455
– Dag Hammarskjöld Library, Uppsala 26456
– Earth Sciences Library, Uppsala 26458
– Education and Teaching Library, Uppsala 26459
– Karin Boye Library, Uppsala 26461
– Law Library, Uppsala 26460
– Library for Economic Sciences, Uppsala 26457
– Library for Educational Research and Studies, Uppsala 26463
– Library of Housing Research, Gävle 26454
– Medical Library, Uppsala 26462
Uppvidinge kommunbibliotek, Åseda 26734
Uprava Republike Slovenije za kulturno dediščino, Ljubljana 24940
Upravlenie i shtab voennogo okruga – Voenno-istoricheskaya biblioteka, Sankt-Peterburg 22767
Upravlenie Lenenergo, Sankt-Peterburg 22947
Upshur County Library, Gilmer 39228
Upshur County Public Library, Buckhannon 38381
Urad prumysloveho vlastnictvi, Praha 06518
Úřad Vlády ČR, Praha 06415
Ural Automobile Engine Plant, Novouralsk 22882
Ural Institute of Architecture and Arts, Ekaterinburg 22239
Ural Plant for Aluminium Production, Kamensk-Uralski 22797
Ural State University, Ekaterinburg 22244
Ural State University of Economics, Ekaterinburg 22240
Ural Tractor Industrial Corporation, Trade Union Library, Chelyabinsk 24121
Uralian Electromechanical Institute of Railway Engineers, Ekaterinburg 22243
Uralian Forestry Technical Institute, Ekaterinburg 22245
Uralian Nickel Industry, Orsk 22887
Uralian Railway Car Industry, Nizhni Tagil 22874
Uralmash OAO, Ekaterinburg 23037
Uralmash OAO Technical Library, Ekaterinburg 23037
Urals Agricultural Institute, Ekaterinburg 22246
Urals state academy of mining and geology, Ekaterinburg 22241
Urals State Technical University, Ekaterinburg 22247
Uralsk Pedagogical Institute, Uralsk 17925
Uralskaya gosudarstvennaya gorno-geologicheskaya akademiya, Ekaterinburg 22241
Uralskaya konservatoriya im. M.P. Musorgskogo, Ekaterinburg 22242
Uralski elektromekhanicheski institut inzhenerov zheleznodorozhnogo transporta, Ekaterinburg 22243
Uralski gosudarstvenny universitet, Ekaterinburg 22244
Uralski lesotekhnicheski institut, Ekaterinburg 22245
Uralski ordena Lenina alyuminievy zavod, Kamensk-Uralski 22797
Uralski selskokhozyaistvenny institut, Ekaterinburg 22246
Uralski tekhnicheski universitet, Ekaterinburg 22247
Uralski vagono-stroitelny zavod, Nizhni Tagil 22874
Uralski zavod khimicheskogo mashinostroeniya, Ekaterinburg 22789
Urban Development Corp (UDC), Kingston 16879
Urban Institute Library, Washington 37800
Urban Land Institute, Washington 37801
Urban Planning Institute, Ljubljana 24941
The Urbana Free Library, Urbana 41646
Urbana University, Urbana 32917
Urbandale Public Library, Urbandale 41647
Urbanistični inštitut, Ljubljana 24941
Urdu Science Board, Lahore 20519
Urology and Nephrology Research Institute, Kyiv 28365
URS Corp Library, Santa Ana 36253
URS Greiner Woodward-Clyde Consultants Library, Wayne 36403
Ursinus College, Collegeville 30866
Ursuline College, Pepper Pike 32244
Uruguay National Library, Montevideo 42008
Urumqi Vocational College, Urumqi 05631
Urząd Mieszkalnictwa i Rozwoju Miast, Warszawa 21050
Urzad Patentowy, Departament Zbiorow Literatury Patentowej, Warszawa 21349
US Agency for International Development, Arlington 33933
US Aid Library, Washington 35009
US Air Force Academy – Community Center, USAF Academy 34952
US Air Force Academy Library, USAF Academy 34953

US Air Force – The Aeromedical Library, Brooks AFB 34013
US Air Force – AFIWC Library, San Antonio 34801
US Air Force – Air Base & Environmental Tech Library, Tyndall AFB 34946
US Air Force Air Education & Training Command, Randolph AFB 34727
US Air Force – Air Force Flight Test Center Technical Library, Edwards AFB 34167
US Air Force – Air Force Institute of Technology, Wright-Patterson AFB 35083
US Air Force – Air Force Legal Services Agency-JAC Library, Arlington 33934
US Air Force – Air Force Research Lab, Wright Research Site Technical Library, Wright-Patterson AFB 35084
US Air Force – Air Force Research Laboratory Library, Hanscom AFB 34336
US Air Force – Air Force Research Laboratory-Information Directorate Technical Library, Rome 34756
US Air Force – Air Force Weather Technical Library, Asheville 33938
US Air Force – Air Mobility Command, Wichita 35069
US Air Force – Air University Library, Maxwell AFB 34509
US Air Force – Altus Air Force Base Library FL4419, Altus AFB 33916
US Air Force – Andersen Air Force Base Library, APO AP 13168
US Air Force – Andrews Air Force Base Library, Andrews AFB 33921
US Air Force – Arnold Engineering Development Center Technical Library, Arnold AFB 33936
US Air Force – Barksdale Air Force Base Library, Barksdale AFB 33968
US Air Force – Base Library/FL 4620, Fairchild AFB 34196
US Air Force – Beale Air Force Base Library, Beale AFB 33975
US Air Force – Bolling Air Force Base Library, Washington 35010
US Air Force – Brooks Air Force Base Library, Brooks AFB 34014
US Air Force – Cannon Air Force Base Library, Cannon AFB 34038
US Air Force – Charleston Air Force Base Library, Charleston AFB 34059
US Air Force – Columbus Air Force Base Library, Columbus AFB 34104
US Air Force – David Grant Medical Center Library, Travis AFB 34933
US Air Force – Davis-Monthan Air Force Base Library, Davis Monthan AFB 34126
US Air Force – Dover Air Force Base Library, Dover AFB 34155
US Air Force – Edwards Air Force Base Library, Edwards AFB 34168
US Air Force – Eglin Air Force Base Library, Eglin AFB 34170
US Air Force – Eielson Air Force Base Library, Eielson AFB 34172
US Air Force – Ellsworth Air Force Base Library, Ellsworth AFB 34183
US Air Force – Elmendorf Air Force Base Library, Elmendorf AFB 34184
US Air Force – FL 3030, Base Library, Goodfellow AFB 34303
US Air Force – Gerrity Memorial Library, Hill AFB 34353
US Air Force – Grand Forks Air Force Base Library, Grand Forks AFB 34309
US Air Force – Hanscom Air Force Base Library, Hanscom AFB 34337
US Air Force – Hickam Air Force Base Library, Hickam AFB 34352
US Air Force – Holloman Air Force Base Library, Holloman 34359
US Air Force – Hurlburt Base Library, Hurlburt Field 34374
US Air Force – Kelly Air Force Base Library, San Antonio 34802
US Air Force – Langley Air Force Base Library, Langley AFB 34434
US Air Force – Laughlin Air Force Base Library, Laughlin AFB 34442
US Air Force – Little Rock Air Force Base Library, Little Rock AFB 34468
US Air Force – Luke Air Force Base Library, Luke AFB 34482
US Air Force – MacDill Air Force Base Library, MacDill AFB 34483
US Air Force – Malcolm Grow Medical Center Library, Andrews AFB 33922
US Air Force – Malmstrom Air Force Base Arden G. Hill Memorial Library, Malmstrom AFB 34493

US Air Force – Maxwell Air Force Base Library, Maxwell AFB 34510
US Air Force – Maxwell Gunter Community Library System, Maxwell AFB 34511
US Air Force – McBride Library, Keesler AFB 34412
US Air Force – McChord Air Force Base Library, McChord AFB 34517
US Air Force – McGuire Air Force Base Library, McGuire AFB 34518
US Air Force – Minot Air Force-Base Library, Minot AFB 34543
US Air Force – Moody Air Force Base Library, Moody AFB 34557
US Air Force – Mountain Home Air Force Base Library, Mountain Home AFB 34566
US Air Force – National Air Intelligence Center Information Center, Wright-Patterson AFB 35085
US Air Force – Nellis Air Force Base Library, Nellis AFB 34578
US Air Force – Offutt Air Force Base Library, Offutt AFB 34626
US Air Force – Patrick Air Force Base Library, Patrick AFB 34652
US Air Force – Peterson Air Force Base Library FL 2500, 21 SVS/SVRL, Peterson AFB 34663
US Air Force – Phillips Site Technical Library FL2809, Kirtland AFB 34419
US Air Force – Pope Air Force Base Library, Pope AFB 34692
US Air Force – Randolph Air Force Base Library, Randolph AFB 34728
US Air Force – Scott Air Force Base Library, Scott AFB 34846
US Air Force – Seymour Johnson Air Force Base Library, Seymour Johnson AFB 34854
US Air Force – Shaw Air Force Base Library, Shaw AFB 34856
US Air Force – Sheppard Air Force Base Library, Sheppard AFB 34857
US Air Force – 45th Space Wing Technical Library, Patrick AFB 34653
US Air Force – Technical Library, Eglin AFB 34171
US Air Force – Technical Library, 30th Space Wing, Vandenberg AFB 34960
US Air Force – Tinker Air Force Base Library, Tinker AFB 34925
US Air Force – 882nd Training Group Academic Library, Sheppard AFB 34858
US Air Force – Travis Air Force Base Library FL4427, Travis AFB 34934
US Air Force – Tyndall Air Force Base Library, Tyndall AFB 34947
US Air Force – Vance Air Force Base Library, Vance AFB 34957
US Air Force – Vandenberg Air Force Base Library, Vandenberg AFB 34961
US Air Force – Whiteman Air Force Base Library, Whiteman AFB 35067
US Air Force – Wright-Patterson Air Force Base Library, Wright-Patterson AFB 35086
US Air Force – Wright-Patterson Medical Center Library, Wright-Patterson AFB 35087
US Armed Forces – Office of the Army Surgeon General Medical Library, Falls Church 34199
US Army – Aeromedical Research Laboratory Science Support Center Library, Fort Rucker 34261
US Army – Aliamanu Library, Honolulu 34364
US Army – Allen Memorial Library, Fort Polk 34259
US Army – AM School of Engineering & Logistics, Texarkana 34923
US Army – Armor School Library, Fort Knox 34240
US Army – Army Intelligence Center & School Library, Fort Huachuca 34236
US Army – The Army Logistics Library, Fort Lee 34246
US Army at Fort Dix, Fort Dix 34227
US Army – Aviation Center Library, Fort Rucker 34262
US Army – Aviation & Missile Command Redstone Scientific Information Center, AMSAM-RD-AS-I-RSIC, Redstone Arsenal 34732
US Army – Aviation Technical Library, Fort Rucker 34263
US Army – Barr Memorial Library, Fort Knox 34241
US Army – Brooke Army Medical Center Library, Fort Sam Houston 34265
US Army – Bruce C. Clarke Community Library, Fort Leonard Wood 34247

US Army – Casey Memorial Library, Fort Hood 34235
US Army – Center for Health Promotion & Preventive Medicine Library, Aberdeen Proving Ground 33891
US Army Claims Service, Fort Meade 36861
US Army – Cold Regions Research and Engineering Laboratory Library, Hanover 34335
US Army – Communications-Electronics Command R&D Technical Library, Fort Monmouth 34253
US Army – Consolidated Library, White Sands Missile Range 35066
US Army – Construction Engineering Research Laboratories, Champaign 34056
US Army – Corps of Engineers, Buffalo 34019
US Army – Corps of Engineers, Fort Worth 34276
US Army – Corps of Engineers, Jacksonville 34393
US Army – Corps of Engineers, New Orleans 34585
US Army – Corps of Engineers, Portland 34700
US Army – Corps of Engineers, Rock Island 34748
US Army – Corps of Engineers, San Francisco 34822
US Army – Corps of Engineers, Seattle 34851
US Army – Corps of Engineers, Walla Walla 34972
US Army – Corps of Engineers, District Library, Kansas City 34407
US Army – Corps of Engineers, New England District Library, Concord 34111
US Army – Corps of Engineers, Omaha District, Omaha 34638
US Army – Donovan Research Library, Fort Benning 34213
US Army – Eisenhower Army Medical Center, Fort Gordon 34232
US Army – Engineer Research & Development Center Library, Vicksburg 34963
US Army – Fort Buchanan Post Library, Fort Buchanan 21890
US Army – Fort Detrick Post Library, Frederick 34284
US Army – Fort Huachuca Library Branch, Fort Huachuca 34237
US Army – Fort Irwin Post Library, Fort Irwin 34238
US Army – Fort Jackson Main Post Library, Fort Jackson 34239
US Army – Fort Lewis Library System, Fort Lewis 34249
US Army – Fort McCoy Post Library, Sparta 34876
US Army – Fort McPherson Library, Fort McPherson 34252
US Army – Fort Monroe General Library, Fort Monroe 34255
US Army – Fort Myer Post Library, Fort Myer 34256
US Army – Fort Riley Post Library, Fort Riley 34260
US Army – Fort Sam Houston Library, Fort Sam Houston 34266
US Army – Fort Shafter Library, Fort Shafter 34268
US Army – Fort Stewart Main Post Library, Fort Stewart 34272
US Army – Fort Story Library, Fort Story 34273
US Army – Galveston District Corps of Engineers Library, Galveston 34294
US Army – Garrison-Selfridge Library, Selfridge Air National Guard Base 34853
US Army – Grant Library, Fort Carson 34222
US Army – Groninger Library, Fort Eustis 34229
US Army – Headquarters Army Material Command Technical Library, Alexandria 33910
US Army – The Institute of Heraldry Library, Fort Belvoir 34211
US Army – John L. Byrd Jr Technical Library for Explosives Safety, McAlester 34516
US Army – John L. Throckmorton Library, Fort Bragg 34219
US Army – Madigan Army Medical Center, Tacoma 34908
US Army – Manscen Academic Library, Fort Leonard Wood 34248
US Army – Marquat Memorial Library, Fort Bragg 34220
US Army – Martin Army Community Hospital Professional Library, Fort Benning 34214
US Army – Medical Library, Tripler AMC 34938

US Army – Medical Library, USAAMC, Fort Rucker 34264
US Army – Medical Research Institute of Chemical Defense, Aberdeen Proving Ground 36439
US Army – Medical Research Institute of Infectious Diseases Library, Frederick 34285
US Army – Mickelsen Library, Fort Bliss 34217
US Army Military Heritage & Education Center US, Carlisle 34047
US Army – Morris Swett Technical Library, Fort Sill 34269
US Army – Nye Library, Fort Sill 34270
US Army – Ordnance Center & School Library, Aberdeen Proving Ground 33892
US Army – Ordnance Missile & Munitions Center & School Technical Library, Huntsville 34373
US Army – Othon O. Valent Learning Resources Center, Fort Bliss 34218
US Army – Patton Museum of Cavalry & Armor, Fort Knox 34242
US Army – Post Library, Aberdeen Proving Ground 33893
US Army – Post Library, Dugway 34157
US Army – Post Library, Fort George G. Meade 34231
US Army – Post Library, Fort Wainwright 34274
US Army – Post Library, Fort Hamilton, Brooklyn 34011
US Army, RDECOM-ARDEC, Picatinny Arsenal 37372
US Army Research Laboratory, Adelphi 33895
US Army – Robert C. McEwen Library, Fort Drum 34228
US Army – Sayers Memorial Library, Fort Benning 34215
US Army – Sergeant Rodney J. Yano Main Library, Schofield Barracks 34845
US Army – R. F. Sink Memorial Library, Fort Campbell 34221
US Army – Stimson Library, Fort Sam Houston 34267
US Army – Technical Library, AMSSB-ROC-T(N), Natick 34576
US Army – Tobyhanna Army Depot Post Library, Tobyhanna 34926
US Army – Transportation School Library, Fort Eustis 34230
US Army – US Disciplinary Barracks Library, Fort Leavenworth 34245
US Army – Van Deusen Post Library, Fort Monmouth 34254
US Army – Van Noy Library, Fort Belvoir 34212
US Army War College, Carlisle 30678
US Army – West Desert Technical Information Center, Dugway 34158
US Army – West Point – Post Library, West Point 35058
US Bank, Denver 34138
US Book Exchange Library, Cleveland 36715
US Census Bureau Library, Suitland 34902
US Centers for Disease Control & Prevention, Fort Collins 34224
US Coast Guard Academy Library, New London 31999
US Coast Guard Base Library – Otis, Air Station Cape Cod 33896
US Coast Guard Research & Development Center, Groton 34321
US Commission on Civil Rights, Washington 37802
US Court of Appeals, Des Moines 34145
US Court of Appeals, Hato Rey 21893
US Court of Appeals, Little Rock 34467
US Court of Appeals, Minneapolis 34542
US Court of Appeals, New Orleans 34586
US Court of Appeals, Omaha 34639
US Court of Appeals, Philadelphia 34665
US Court of Appeals, Pittsburgh 34680
US Court of Appeals, Portland 34701
US Court of Appeals, Providence 34711
US Court of Appeals, Richmond 34741
US Court of Appeals, Saint Louis 34777
US Court of Appeals, Wilmington 35074
US Court of Appeals for the Armed Forces Library, Washington 35011
US Court of Appeals for the District of Columbia, Washington 35012
US Court of Appeals for the Federal Circuit Library, Washington 35013
US Court of Appeals for the Sixth Circuit, Nashville 34575
US Court of Appeals for the Sixth Circuit Library, Cincinnati 34081
US Court of Appeals Library, Newark 34605
US Court of Appeals, Ninth Circuit Library, Pasadena 34650
US Court of International Trade, New York 34602

US Courts, Kansas City 34408
US Courts Branch Library, Fargo 34200
US Courts – Kansas City Branch Library, Kansas City 34409
US Courts Library, Anchorage 33920
US Courts Library, Brooklyn 34012
US Courts Library, Denver 34139
US Courts Library, Honolulu 34365
US Courts Library, Los Angeles 34479
US Courts Library, Oklahoma City 34631
US Courts Library, Saint Paul 34786
US Courts Library, San Diego 34810
US Courts Library, Seattle 34852
US Courts Library, Wichita 35070
US Customs Service, Washington 35014
US Department of Agriculture, Beltsville 33984
US Department of Agriculture, New Orleans 34587
US Department of Agriculture, Peoria 34660
US Department of Agriculture, Washington 35015
US Department of Agriculture, Washington 35016
US Department of Agriculture, Wyndmoor 35090
US Department of Commerce, Washington 35017
US Department of Commerce – Law Library, Washington 35018
US Department of Defense, Denver 34140
US Department of Defense, Wright-Patterson AFB 35088
US Department of Defense – National Defense University, Washington 32993
US Department of Defense – National Imagery & Mapping Agency Reference Library, Saint Louis 34778
US Department of Energy, Albany 33904
US Department of Energy, Morgantown 34561
US Department of Energy, Portland 34702
US Department of Energy – Energy Library, Washington 35019
US Department of Energy – Environmental Measurements Laboratorium Library, New York 34603
US Department of Energy – Law Library, Washington 35020
US Department of Energy – Office of Scientific & Technical Information, Oak Ridge 34622
US Department of Health & Human Services, Cincinnati 34082
US Department of Health & Human Services – FDA Center for Devices & Radiological Health Library, Rockville 34752
US Department of Health & Human Services – Food & Drug Administration, Rockville 37462
US Department of Housing and Urban Development, Washington 35021
US Department of Justice, Grand Rapids 34310
US Department of Justice, Longmont 34471
US Department of Justice, Los Angeles 34480
US Department of Justice – Justice Libraries, Washington 35022
US Department of Labor, Beaver 33978
US Department of Labor, Washington 35023
US Department of State, Washington 35024
US Department of the Air Force, San Antonio 34803
US Department of the Army – Office of the Chief Engineers Library, Alexandria 33911
US Department of the Army – Office of the Chief of Staff, Operational Test & Evaluation Command (OPTEC), Alexandria 33912
US Department of the Army – The Pentagon Library, Washington 35025
US Department of the Interior, Arlington 33935
US Department of the Interior, New Orleans 37172
US Department of the Interior – Bureau of Reclamation, Denver 34141
US Department of the Interior, Bureau of Reclamation, Sacramento 34768
US Department of the Interior Library, Washington 35026
US Department of the Interior – National Park Service, West Orange 37825
US Department of the Interior & the National Park Service, Flat Rock 34205
US Department of the Interior – US Geological Survey Mineral Resources Library, Spokane 34878
US Department of the Navy – Office of the General Counsel, Washington 35027
US Department of the Navy – Office of the Judge Advocate General Law Library, Washington 35028
US Department of the Treasury – Comptroller of the Currency, Administrator of National Banks Library, Washington 35029
US Department of the Treasury – Treasury Library, Washington 35030
US Department of Transportation, Washington 35031
US Department of Transportation – Federal Highway Administration, Washington 35032
US District Court, Portland 34703
US District Court, San Francisco 34823
US Environmental Protection Agency, Atlanta 33943
US Environmental Protection Agency, Chicago 34071
US Environmental Protection Agency, Duluth 34160
US Environmental Protection Agency, Las Vegas 34441
US Environmental Protection Agency, Philadelphia 34666
US Environmental Protection Agency, San Francisco 34824
US Forest Service, Auburn 33947
US Forest Service, Fort Collins 34225
US Geological Survey, Ann Arbor 33926
US Geological Survey, Lafayette 34428
US Geological Survey, Laurel 34443
US Geological Survey, Sioux Falls 34864
US Geological Survey – Biological Resources Division, Fort Collins 34226
US Geological Survey – Leetown Science Center, Kearneysville 34411
US Geological Survey Library, Denver 34142
US Geological Survey Library, Flagstaff 36853
US Geological Survey Library, Menlo Park 37095
US Geological Survey Library, Reston 34740
US Holocaust Memorial Museum Library, Washington 37803
US Institute of Peace, Washington 37804
US International Trade Commission – Law Library, Washington 35033
US International Trade Commission – National Library of International Trade, Washington 35034
US Marine Corps, Quantico 32402
US Marine Corps – Air Station Library, Beaufort 33976
US Marine Corps – Marine Corps Historical Center Library, Washington 35035
US Marine Corps – MCAS Station Library, Jacksonville 34394
US Marine Corps – Quantico Family Library, Quantico 34716
US Marine Corps – Seaside Square Library, Camp Pendleton 34034
US Merchant Marine Academy, Kings Point 31509
US Merit Systems Protection Board Library, Washington 35036
US Military Academy, West Point 33036
US National Oceanic & Atmospheric Administration – Miami Regional Library, Miami 34531
US National Oceanic & Atmospheric Administration – Miami Regional Library at AOML, Miami 34532
US National Park Service, Cambridge 36621
US National Park Service, Ganado 34295
US National Park Service, Harpers Ferry 36913
US National Park Service, Valley Forge 34956
US Naval Academy, Annapolis 30254
US Naval Observatory, Washington 35037
US Naval War College, Newport 32114
US Navy, Gulfport 34325
US Navy – Academic Resources Information Center, Newport 34606
US Navy – Albert T. Camp Library, Indian Head 34379
US Navy – Base Library, Groton 34322
US Navy – The Command Library, San Diego 34811
US Navy – Crews Library, Camp Pendleton 34035
US Navy – Dahlgren Laboratory, General Library, Dahlgren 34122
US Navy – Engineering Library, Bremerton 34003
US Navy – Marine Corps Logistics Base Library, Albany 33905
US Navy – Marine Corps Recruit Depot Library, San Diego 34812
US Navy – Matthew Fontaine Maury Oceanographic Library, Stennis Space Center 34892
US Navy – Mayport Naval Station Library, Mayport 34513
US Navy – Medical Library, Great Lakes 34313
US Navy – Mobile Library, Camp Pendleton 34036
US Navy – Naval Air General Library, El Centro 34175
US Navy – Naval Air Station, Milton 34537
US Navy – Naval Air Station Library, Corpus Christi 34113
US Navy – Naval Air Station Library, Fort Worth 34277
US Navy – Naval Air Station Library, Kingsville 34417
US Navy – Naval Air Station Library, Pensacola 34658
US Navy – Naval Air Warfare Center Weapons Division Technical Library, China Lake 34074
US Navy – Naval Air Warfare Station Library, China Lake 34075
US Navy – Naval Amphibious Base Library, Norfolk 34615
US Navy – Naval Base Coronado Library, San Diego 34813
US Navy – Naval Base Ventura Library/Resource Center, Point Mugu 34686
US Navy – Naval Coastal Systems Station, Panama City 34648
US Navy – Naval Construction Battalion Center Library – The Resource Center, Port Hueneme 34693
US Navy – Naval Medical Center, San Diego 34814
US Navy – Naval Operational Medical Institute Library, Pensacola 37334
US Navy – Naval Regional Medical Center, Portsmouth 34704
US Navy – Naval School, Civil Engineer Corps Officers, Moreell Library, Port Hueneme 34694
US Navy – Naval Ship Systems Engineering Station Library, Philadelphia 34667
US Navy – Naval Support Activity Library, New Orleans 34588
US Navy – Naval Training Center Library, Great Lakes 34314
US Navy – Naval Undersea Warfare Center Division, Newport 34607
US Navy – Naval Weapons Stations, Goose Creek 34305
US Navy – Navy Personel Research Studies & Technology, Millington 34535
US Navy – Navy Supply Corps School Station Library, Atlanta 33944
US Navy – NFESC Technical Information Center, Port Hueneme 34695
US Navy – Norfolk Naval Shipyard, Portsmouth 34705
US Navy – Officer Indoctrination School Library, Newport 34608
US Navy – Patuxent River Central Library, Patuxent River 34654
US Navy – South Mesa Library, Camp Pendleton 34037
US Navy – Spawar Systems Center San Diego Technical Library, San Diego 34815
US Navy – Supervisor of Shipbuilding Conversion & Repair Library, San Diego 34816
US Navy – Wilkins Biomedical Library, San Diego 34817
US Nuclear Regulatory Commission Library, Rockville 34753
US Office of Personnel Management, Washington 35038
US Patent & Trademark Office, Alexandria 33913
US Postal Service Library, Washington 35039
US Securities & Exchange Commission Library, Washington 35040
US Senate Library, Washington 35041
U.S. Small Business Administration, Washington 37779
US Tax Court Library, Washington 35042
USACE Baltimore District Library, Baltimore 33966
USDA-ARS, Oxford 34646
P. Usenko Children's Library, Kyiv 28649
Ushaw College, Durham 29090
Ushinsky State Scientific Pedagogical Library, Moskva 23201
K. Ushynsky South Ukrainian Pedagogical University, Odesa 28062
Usinas Siderúrgicas de Minas Gerais S.A., Belo Horizonte 03253
USIS Information Resource Center, Abuja 20009
Usmanu Danfodiyo University Sokoto, Sokoto 19981
USS Division of USX Corp, Monroeville 35917
USS Liberty Memorial Public Library, Grafton 39263
Ústav biologie obratlovců AV ČR, Brno 06436
Ústav fyziky materiálů AV ČR, Brno 06437
Ústav jaderného výzkumu Řež a.s., Řež 06537
Ústav merania SAV-ZIS, Bratislava 24726
Ústav mezinárodních vztahuž, Praha 06519
Ústav pro českou literaturu AV ČR, Brno 06438
Ústav pro českou literaturu AV ČR, Praha 06520
Ústav pro jazyk český AV ČR – pobočka Brno-Veveří, Brno 06439
Ústav pro péči o matku a dítě, Praha 06521
Ustav slovenskej literatúry / Ustav svetovej literatúry, Bratislava 24727
Ustav stavebníctva a architektúry – Slovenská Akadémia Vied, Bratislava 24728
Ústav struktury a mechaniky hornin AV ČR, Praha 06522
Ústav územniho rozvoje, Brno 06440
Ústav zemedelskych a potravinarskych informaci – Zemědělská a potravinářská knihovna, Praha 06523
Ústřední Tělovýchovná Knihovna, Praha 06524
USW International Union Library, Nashville 37146
Utah Department of Corrections, Gunnison 34326
Utah Department of Natural Resources Library, Salt Lake City 37516
Utah School for the Deaf & Blind, Ogden 33598
Utah State Historical Society, Salt Lake City 37517
Utah State Law Library, Salt Lake City 34798
Utah State Library Division, Salt Lake City 30153
Utah State Library Division – Program for the Blind & Disabled, Salt Lake City 30154
Utah State University Libraries, Logan 31633
– Ann Carroll Moore Children's Library, Logan 31634
Utah Valley State College, Orem 32208
Ute Mountain Tribal Library, Towaoc 37681
Utenos A. ir M. Miškinių viešoji biblioteka, Utena 18372
Utica College, Utica 32919
Utica International Insurance Group, New Hartford 35938
Utica Public Library, Utica 41649
Utkal University, Bhubaneswar 13621
Utrecht School of the Arts, Utrecht
– Theatre Library, Utrecht 19214
– Utrecht Conservatory, Utrecht 19215
Utrecht University, Utrecht 19217
– Law Library, Utrecht 19219
– Medical Library, Utrecht 19220
– University for Humanistics, Utrecht 19221
– University Library City Centre, Utrecht 19222
– University Library Uithof, Utrecht 19223
– University Museum, Utrecht 19224
– Veterinary Medicine Library, Utrecht 19218
Het Utrechts Archief, Utrecht 19531
Uttar Pradesh Civil Secretariat, Lucknow 13861
Uttar Pradesh State Museum, Lucknow 14072
Uudenkaarlepyyn kaupunginkirjasto, Nykarleby 07595
Uudenkaupungin kaupunginkirjasto, Uusikaupunki 07616
Uusikaupunki Public Library, Uusikaupunki 07616
UWE Bristol, Glenside Campus, Bristol 28975
Uzbek Academy of Sciences, Toshkent 42073
Uzbek Institute of Physical Culture, Toshkent 42108
Uzbek Institute of Rehabilitation and Physiotherapy, Toshkent 42164
Uzbek Orthopaedics and Traumatology Research Institute, Toshkent 42165
Uzbek State Museum of Art, Toshkent 42166
Uzbek Veterinary Medicine Research Institute, Tailyak 23987
Uzeir Hajibeyov Baku Academy of Music, Baku 01691
Uzhgorod City Centralized Library System, Main Library, Uzhgorod 28860
Uzhgorod National University, Uzhgorod 28082
Uzhgorodska miska TsBS, Tsentralna biblioteka, Uzhgorod 28860
Uzhgorodsky natsionalny universytet, Uzhgorod 28082
V. L. Komarov Institute of Botany, Baku 01716

V. P. Komissarenko Institue of Endocrinology and Metabolism, Kyiv 28355
VA Hudson Valley Health Care System, Castle Point 36635
V&A Theatre & Performance Department Library (formerly Theatre Museum Library, London 29840
Vaal University of Technology, Vanderbijlpark 25056
Vaasa Court of Appeal Library, Vaasa 07407
Vaasa Public Library – Regional Library, Vaasa 07617
Vaasan hovioikeuden kirjasto, Vaasa 07407
Vaasan kaupunginkirjasto – maakuntakirjasto, Vaasa 07617
Vaasan maakunta-arkiston käsikirjasto, Vaasa 07554
Váci Egyházmegyei Könyvtár, Vác 13349
Vacuum Engineering Industrial Corporation, Kazan 22799
Vacuumschmelze GmbH, Hanau 11403
Vadodara Museum and Picture Gallery, Vadodara 14219
Vadsø bibliotek, Vadsø 20414
Vadstena bibliotek, Vadstena 26942
Vadybos ir ekonomikos universitetas, Kaunas 18275
Værløse Bibliotek, Værløse 06943
Väestöliitto, Helsinki 07505
Vågan bibliotek, Svolvær 20407
Vaggeryds bibliotek, Vaggeryd 26943
Vägverket, Borlänge 26485
Vaida State Regional Youth Library, Uzhgorod 28859
Vaikunth Mehta National Institute of Cooperative Management, Pune 14200
Vail Public Library, Vail 41652
Vakarov State Regional Children's Library, Uzhgorod 28858
Vakhtangov Theatre, Moskva 23206
Vakhushti Bagrationi Institute of Geography, Tbilisi 09287
Val Verde County Library, Del Rio 38810
Valamo Monastery Library, Uusi-Valamo 07413
Valamon luostarin kirjasto, Uusi-Valamo 07413
Valašské muzeum v přírodě, Rožnov pod Radhoštěm 06538
Valdosta State University, Valdosta 32920
Valencia Community College – East Campus Learning Resource Center, Orlando 33607
Valencia Community College – Raymer Maguire Jr Learning Resources Center, West Campus, Orlando 32212
Ch. Valikhanov Pedagogical Institute, Kokchetav 17911
Valio Ltd, Information Services, Helsinki 07417
Valio Oy, Helsinki 07417
Valkeakosken kaupunginkirjasto, Valkeakoski 07618
Valkeakoski Public Library, Valkeakoski 07618
Vallabhbhai Patel Chest Institute, Delhi 13986
Vallensbæk Bibliotek, Vallensbæk Strand 06944
Vallentuna bibliotek, Vallentuna 26944
Valley City State University, Valley City 32924
Valley City-Barnes County Public Library, Valley City 41657
Valley Cottage Free Library, Valley Cottage 41658
Valley Forge Military Academy & College, Wayne 33848
Valley Regional Library, Abbotsford 04564
Valmet Corporation, Rautpohja, Information Service, Jyväskylä 07421
Valparaiso University, Valparaiso 32925
– School of Law Library, Valparaiso 32926
Valtion asuntorahaston kirjasto, Helsinki 07506
Valtion taidemuseon kirjasto, Helsinki 07507
Valtion taloudellinen tutkimuskeskus (VATT), Helsinki 07508
Vamdrup Bibliotek, Vamdrup 06945
Vammala Public Library, Vammala 07619
Vammalan kaupunginkirjasto, Vammala 07619
Van Abbemuseum, Eindhoven 19421
Van Buren District Library, Decatur 38802
Van Gogh Museum, Amsterdam 19386
Van Leer Jerusalem Institute, Jerusalem 14742
Van Ness Feldman Library, Washington 36388
Vance-Granville Community College, Henderson 33406
Vancouver Art Gallery Library, Vancouver 04545
Vancouver Community College, Vancouver 03949
Vancouver Island Health Authority Medical Libraries, Victoria 04550
Vancouver Island Regional Library, Nanaimo 04698
Vancouver Island Regional Library – Campbell River Branch, Campbell River 04602
Vancouver Island Regional Library – Courtenay Branch, Courtenay 04621
Vancouver Island Regional Library – Cowichan, Duncan 04629
Vancouver Island Regional Library – Nanaimo Harbourfront Branch, Nanaimo 04699
Vancouver Island Regional Library – Sidney Branch, Sidney 04757
Vancouver Museum Library and Resource Centre, Vancouver 04546
Vancouver Public Library, Vancouver 04808
Vancouver School of Theology Library, Vancouver 04234
Vandalia Correctional Center Library, Vandalia 34959
Vanderbijlpark Public Library, Vanderbijlpark 25238
R. T. Vanderbilt Co, Inc, Norwalk 36059
Vanderbilt University, Nashville 31940
– Alyne Queener Massey Law Library, Nashville 31941
– Anne Potter Wilson Music Library, Nashville 31942
– Annette and Irwin Eskind Biomedical Library, Nashville 31943
– Central Library, Nashville 31944
– Divinity Library, Nashville 31945
– The Peabody Library, Nashville 31946
– Science and Engineering Library, Nashville 31947
– Special Collections, Nashville 31948
– Walker Management Library, Nashville 31949
Vänersborg bibliotek, Vänersborg 26945
Vänersborgs museum, Vänersborg 26713
Vang folkebibliotek, Vang i Valdres 20415
Vanguard University of Southern California, Costa Mesa 30929
Vanier College, Saint Laurent 04023
Vännäs kommunbibliotek, Vännäs 26946
Vansbro kommunbibliotek, Vansbro 26947
Vantaa City Library, Vantaa 07620
Vantaan kaupunginkirjasto – Vanda stadsbibliotek, Vantaa 07620
Vara folkbibliotek, Vara 26948
Varastokirjasto, Kuopio 07327
Varbergs Kommun, Varberg 26949
Varde Bibliotek, Varde 06946
Vårdhögskolan i Malmö, Malmö 26416
Vårdhögskolan i Örnsköldsvik, Örnsköldsvik 26419
Varendra Research Museum, Rajshahi 01770
Varėnos rajono savivaldybės viešoji biblioteka, Varėna 18373
Varian Medical, Mountain View 35927
Varkauden kaupunginkirjasto, Varkaus 07621
Varkaus Public Library, Varkaus 07621
Världskulturmuseet, Göteborg 26585
Värmdö kommunbibliotek, Gustavsberg 26778
Värmlandsarkiv, Karlstad 26604
Varna University of Economics, Varna 03486
Värnamo kommunbibliotek, Värnamo 26950
Värnhemsskolan, Malmö 26475
Varnish and Paint Plant, Lida 01877
Varnum, Riddering, Schmidt & Howlett, Grand Rapids 35724
Városi Könyvtár, Bonyhád 13529
Városi Könyvtár, Nagykanizsa 13542
Városi Múzeum, Kőszeg 13488
Vas-, Fém- és Villamosenenergiaipari Dolgozók Skaksz. Szöv., Budapest 13463
Vasa vetenskapliga bibliotek och lärocenter, Vaasa 07382
Vasantdada Sugar Institute, Manjari 14074
Vasileostrovsky District Centralized Libary System, Lomonosov Central District Library, Sankt-Peterburg 24301
Vasileostrovsky District Centralized Library System, Chernyshevsky Library Branch no 1, Sankt-Peterburg 24298
Vasileostrovsky District Centralized Library System, Children's Library Branch no 6, Sankt-Peterburg 24300
Vasileostrovsky District Centralized Library System, Ostrovsky Youth Library Branch no 3, Sankt-Peterburg 24302
Vasileostrovsky District Centralized Library System, Tolstoy Library Branch no 2, Sankt-Peterburg 24299
Vasilkiv Regional Centralized Library System, Main Library, Vasilkiv 28861
Vasilkivska raionna TsBS, Tsentralna biblioteka, Vasilkiv 28861
Vassar College, Poughkeepsie 32354
– Art Library, Poughkeepsie 32355
Västerås stad, Västerås 26951
Västerbottens museum, Umeå 26704
Västervik City Library, Västervik 26952
Vastervik Hospital, Västervik 26717
Västerviks sjukhus, Västervik 26717
Västerviks stadsbibliotek, Västervik 26952
Västra Götaland Region Litteraturtjänst, Vänersborg 26525
Vasuti Fősztály, Budapest 13324
VÁTI Magyar Kht Dokumentációs Központ, Budapest 13464
Vatutinska raionna TsBS m. Kyiva, Tsentralna biblioteka, Kyiv 28691
Vatutinski Region Central Library System of the City of Kiev, Main Library, Kyiv 28691
Vaughan Public Libraries, Thornhill 04778
Vavilov State Optical Institute, Sankt-Peterburg 22536
Växjö bibliotek, Växjö 26953
Växjö Universitet, Växjö 26465
Vecherny metallurgicheski institut, Moskva 22399
Vecova Centre for Disability Services and Research, Calgary 04324
Vector Control Research Centre, Pondicherry 14182
Vedder, Price, Kaufman & Kammholz, Chicago 35603
Vědecká knihovna v Olomouci, Olomouc 06348
Vedeneev Hydro Engineering Research Institute Inc., Sankt-Peterburg 22907
Vefsn bibliotek, Mosjøen 20373
Vegdirektoratets bibliotek, Oslo 20198
Vegdirektoratets Bibliotek, Oslo 20290
Vegetable and Ornamental Crops Research Station, Morioka Branch, Morioka 17609
Vegetable Crops Research Institute, Kecskemét 13485
Vegyészeti Múzeum, Várpalota 13518
Vejen Bibliotek, Vejen 06947
Vejle Bibliotek, Vejle 06948
Vejsektorens Fagbibliotek, Hedehusene 06776
Vektor Research Institute, Sankt-Peterburg 23855
Velika Bagachka Regional Centralized Library System, Main Library, Velika Bagachka 28862
Velika Oleksandrivka Regional Centralized Library System, Main Library, Velika Oleksandrivka 28863
Velikie Luki State Agricultural Academy, Velikie Luki 22624
Velikobagachanska raionna TsBS, Tsentralna biblioteka, Velika Bagachka 28862
Velikolukskaya gosudarstvennaya selskokhozyaistvennaya akademiya, Velikie Luki 22624
Velikolukski gosudarstvenny institut fizicheskoi kult, Velikie Luki 22625
Velikooleksandrivska raionna TsBS, Tsentralna biblioteka, Velika Oleksandrivka 28863
Velikotarnovski Universitet "Sv. sv. Kiril i Metodij", Veliko Tărnovo 03487
Vellinge kommunbibliotek, Vellinge 26954
Venable LLP Library, Baltimore 35490
Venable LLP Library, Washington 36389
Venäjän ja Itä-Euroopan, Helsinki 07509
Venäjän kulttuuri- ja tiedekeskuksen kirjasto, Helsinki 07510
Venango County Law Library, Franklin 34282
Vendsyssel Historical Museum, Hjørring 06780
Vendsyssel Historiske Museum, Hjørring 06780
Venerabile Collegio Inglese, Roma 16041
Venerable English College, Roma 16041
Veneranda Fabbrica del Duomo di Milano, Milano 16357
Venice Public Library, Venice 41668
Sri Venkateswara University, Tirupati 13774
– Oriental Research Institute, Tirupati 13775
Vennesla bibliotek, Vennesla 20416
Ventress Memorial Library, Marshfield 40080
Ventura College, Ventura 33823
Ventura County Law Library, Ventura 34962
Ventura County Library, Ventura 41669
Ventura County Library – Camarillo Library, Camarillo 38416
Ventura County Library – E. P. Foster Library, Ventura 41670
Ventura County Library – H. P. Wright Library, Ventura 41671
Ventura County Library – Ray D. Prueter Library, Port Hueneme 40853
Ventura County Library – Simi Valley Library, Simi Valley 41327
Verband der Chemischen Industrie e.V., Frankfurt am Main 11876
Verband der Versicherungsunternehmen Österreichs, Wien 01656
Verband Deutscher Verkehrsunternehmen (VDV), Köln 12154
Verband Österreichischer Philatelistenvereine, Wien 01657
Verdal bibliotek, Verdal 20417
P. Verdhigori Children's Library, Kyiv 28650
Vereeniging Library, Vereeniging 25239
Verein Deutscher Gießereifachleute e.V., Düsseldorf 11808
Verein für die Geschichte Berlins e.V., Berlin 11594
Verein für Geschichte der Arbeiterbewegung, Wien 01658
Verein für Geschichte der Stadt Nürnberg, Nürnberg 12377
Verein für Geschichte und Landeskunde von Osnabrück, Osnabrück 12391
Verein für Hamburgische Geschichte, Hamburg 12021
Verein für Orts- und Heimatkunde in der Grafschaft Mark, Witten 12599
Verein für Volkskunde / Österreichisches Museum für Volkskunde, Wien 01659
Vereinigte Evangelisch-Lutherische Kirche Deutschlands, Hannover 11235
Verejná knižnica Jána Bocatia, Košice 24779
Verejná knižnica Michala Rešetku v Trenčíne, Trenčín 24799
Verejná kniznica Mikuláša Kováča, Banská Bystrica 24762
Verenigde Bibliotheek, Amsterdam 19240
Verenigde Doopsgezinde Gemeente, Amsterdam 19078
Vereniging van Nederlandse Gemeenten, Den Haag 19453
Vereniging VNO-NCW, Den Haag 19332
Vereniging voor Alcohol- en andere Drugproblemen, Brussel 02628
Veria Central Public Library, Veria 13155
Veridian Technical Information Center, Buffalo 35546
Verizon, Arlington 35451
Verkehrshaus der Schweiz, Luzern 27408
Verkehrsmuseum Dresden, Dresden 11784
Verkehrswasserbauliche Zentralbibliothek (VZB) bei der Bundesanstalt für Wasserbau (BAW), Karlsruhe 12085
Verkhnedvinsk Regional Central Library, Verkhnedvinsk 02243
Verkhnedvinskaya tsentralnaya raionnaya bibliyateka, Verkhnedvinsk 02243
Verkhnodniprovsk Regional Centralized Library System, Main Library, Verkhnodniprovsk 28864
Verkhnodniprovska raionna TsBS, Tsentralna biblioteka, Verkhnodniprovsk 28864
Verkhovny Sud, Moskva 22759
Verkís, Reykjavík 13565
Verkís Library, Reykjavík 13565
Vermilion Community College, Ely 33320
Vermilion Parish Library, Abbeville 37871
Vermillion County Public Library, Newport 40466
Vermögen und Bau Baden-Württemberg, Bruchsal 10845
Vermont College Division of Norwich University, Montpelier 31893
Vermont Historical Society Library, Barre 36521
Vermont Institute of Natural Science Library, Quechee 37421
Vermont Law School, South Royalton 32690
Vermont Technical College, Randolph Center 33673
Vernadski Institute of Geochemistry and Analytical Chemistry of the Russian Academy of Sciences, Moskva 23272
V. I. Vernadsky National Library of the Ukraine, Kyiv 27930
V. I. Vernadsky National Library of the Ukraine (Branch No 2), Kyiv 27931
Verner, Lipfert, Bernard, McPherson & Hand, Washington 36390
Vernon Area Public Library District, Lincolnshire 39884
Vernon Parish Library, Leesville 39844
Vernon Regional Junior College, Vernon 33824
Verona Public Library, Verona 41676
Verona Public Library, Verona 41677
Verrill & Dana Library, Portland 36150
Vervarslinga på Vestlandet, Bergen 20218
Verwaltungsbibliothek Nürnberg, Nürnberg 11071
Verwaltungsbücherei, Bielefeld 10817
Verwaltungsgericht Bayreuth, Bayreuth 10783
Verwaltungsgericht Braunschweig, Braunschweig 10838

Verwaltungsgericht Düsseldorf, Düsseldorf 10885
Verwaltungsgericht Gelsenkirchen, Gelsenkirchen 10905
Verwaltungsgericht Hannover, Hannover 10946
Verwaltungsgericht Köln, Köln 10993
Verwaltungsgericht Minden, Minden 11020
Verwaltungsgericht Oldenburg, Oldenburg 11076
Verwaltungsgerichtshof, Wien 01413
Verwaltungsgerichtshof Baden-Württemberg, Mannheim 11017
Verwey-Jonker Instituut, Utrecht 19532
Verwoerdburg Public Library, Verwoerdburg 25240
Vesele Regional Centralized Library System, Main Library, Vesele 28865
Veselivska raionna TsBS, Tsentralna biblioteka, Vesele 28865
Vespasian Warner Public Library District, Clinton 38609
Vest-Agder sentralsykehus, Kristiansand 20236
Vestal Public Library, Vestal 41679
Vestavia Hills-Richard M. Scrushy Public Library, Vestavia Hills 41680
Vesterheim Norwegian-American Museum, Decorah 36784
Vestfold College, Borre, Borre 20085
Vestfold College, Eik, Tønsberg 20163
Vestfold County Library, Tønsberg 20409
Vestfold fylkesbibliotek, Tønsberg 20409
Vestlandsk Kunstindustrimuseum, Bergen 20219
Vestre Toten folkebibliotek, Raufoss 20388
Vestvågøy bibliotek, Leknes 20364
Veszprém County Library, Veszprém 13553
Veszprémi Egyetem, Veszprém 13310
Veszprémi Egyetem Georgikon Mezögazdaságtudományi Kar, Keszthely 13281
– Kaposvári Egyetem, Kaposvár 13282
– Masonmagyaróvári Mezögazda-Ságtudományi Kar, Mosonmagyaróvár 13283
VetAgro Sup, Marcy-l'Etoile 07749
Veterans Administration Center, Fort Harrison 34234
Veterans General Hospital, Taipei 27623
Veterans Memorial Medical Center, Quezon City 20783
Veterinärmedizinische Universität Wien, Wien 01362
Veterinární a farmaceutická univerzita, Brno 06364
Veterinarny institut, Omsk 23669
Veterinarny institut, Sankt-Peterburg 22551
Veterinarny institut, Troitsk 22599
Veterinarny institut im. N.E. Baumana, Kazan 22281
Veterinary Institute, P.O. Mala Danylivka 28041
Veterinary Institute, Omsk 23669
Veterinary Institute, Troitsk 22599
Veterinary Medical Research Institute, Budapest 13432
Veterinary Research Institute, Brno 06441
Veterinary Research Institute of Poultry Farming, Sankt-Peterburg 23858
Veterinary Science Library, Budapest 13217
– Szent István University Kosáry Domokos Library and Archives, Gödöllo 13218
Vetlanda bibliotek, Vetlanda 26955
Via Christi Libraries, Wichita 37836
VIA University College, Holstebro 06662
VIA University College, Horsens 06663
VIA University College, Viborg 06752
VIA University College, Teacher Education, Silkeborg 06724
Viborg Centralbibliotek, Viborg 06949
Viborg Katedralskole, Viborg 06753
Vibronichno obedannya Azot, Dniprodzerzhinsk 28137
Vibronichno obedannya Chervona zirka, Kirovograd 28149
Vibronichno obednannya Azot, Severodonetsk 28468
Vibronichno obednannya Chernomorsky sudobudivny zavod, Biblioteka profkomu, Mykolaiv 28741
Vibronichno obednannya Donetskgirmash, Donetsk 28142
Vibronichno obednannya kombainovi zavod, Kherson 28148
Vibronichno obednannya Luganksteplovoz, Lugansk 28162
Vibronichno obednannya Mashinobudivny zavod, Kramatorsk 28630
Vibronichno obednannya Motor-Sich, Zaporizhzhya 28176
Vibronichno obednannya Motor-Sich, Zaporizhzhya 28888
Vibronichno obednannya NDI proektuvannya vazhkoho mashinobuduvannya, Kramatorsk 28151
Vibronichno obednannya Novokramatorsky mashinobudivny zavod, Biblioteka, Kramatorsk 28152
Vibronichno obednannya Stekloplastik, Severodonetsk 28169
Vibronichno obednannya Tochmash, Donetsk 28143
Vibronicho obedannya Sevastopolsky morsky zavod, Sevastopol 28168
Vic Roads Business Information Centre, Kew 00934
Victor Valley Community College, Victorville 33825
Victoria and Albert Museum, London 29795
Victoria County Public Library, Lindsay 04679
Victoria Memorial Hall, Kolkata 14060
Victoria Police Academy, Glen Waverley 00680
Victoria Public Library, Victoria 41682
Victoria University, Melbourne 00481
Victoria University, Toronto 03918
Victoria University of Wellington, Wellington 19792
– Law Library, Wellington 19793
Victorian Auditor-General's Office, Melbourne 00702
Viddil okhorony zdorovya radi ministriv Respubliky Krym, Simferopol 28473
Videbæk Bibliotek, Videbæk 06950
VIDEOTON, Székesfehérvár 13364
Vidyasagar College, Kolkata 13814
Vienna Correctional Center Library, Vienna 34965
Vienna Institute of Demography, Wien 01606
Vienna Public Library, Vienna 41684
Vienna University of Economics and Business, University Library, Wien 01363
– Department of Foreign Language Business Communication, Institute of Romance Languages, Wien 01364
Vienna University of Technology, Wien 01301
– Institute of Discrete Mathematics and Geometry, Wien 01306
Vietnam History Museum, Hanoi 42303
Vietnam Information for Science and Technology Advance, Hanoi 42264
Vietnam Institute of Traditional Medicine, Hanoi 42304
Vietnam National University, Hanoi 42277
Vietnam Revolution Museum, Hanoi 42305
Vigo County Public Library, Terre Haute 41539
Vihdin kunnankirjasto, Nummela 07593
Vihorlatská knižnica, Humenné 24776
Viikki Science Library, Helsingin Yliopisto 07336
Vikram Sarabhai Space Centre, Thiruvananthapuram 14216
Vikram University, Ujjain 13779
Viktor and Appolinari Vasnetsov Museum of Arts, Kirov 23104
Viktoras Miliunas Public LIbrary of Neringa Municipality, Neringa 18344
Vilhelmina folkbibliotek, Vilhelmina 26956
Viljandi Town Library, Viljandi 07258
Vilkaviškio rajono savivaldybės viešoji biblioteka, Vilkaviškis 18374
Villa Julie College, Stevenson 32743
Villa Maria College, Buffalo 33198
Villa Park Public Library, Villa Park 41685
Village and Folk Art Museum, Bucureşti 22010
Villanova University, Villanova 32932
– Law Library, Villanova 32933
Vilniaus apskrities Adomo Mickevičiaus viešoji biblioteka, Vilnius 18375
Vilniaus dailės akademijos biblioteka, Vilnius 18314
Vilniaus Gedimino technikos universitetas, Vilnius 18290
Vilniaus Juozo-Kelpšos konservatorija, Vilnius 18293
Vilniaus miesto savivaldybės centrinė biblioteka, Vilnius 18376
Vilniaus pedagoginio universiteto biblioteka, Vilnius 18291
Vilniaus rajono savivaldybės viešoji biblioteka, Rudamina 18357
Vilniaus universiteto biblioteka, Vilnius 18292
Vilnius Academy of Arts Library, Vilnius 18314
Vilnius Gediminas Technical University, Vilnius 18290
Vilnius Pedagogical University Library, Vilnius 18291
Vilnius University Library, Vilnius 18292
Vilnyansk Regional Centralized Library System, Main Library, Vilnyansk 28866
Vilnyanska raionna TsBS, Tsentralna biblioteka, Vilnyansk 28866
Vilshanska raionna TsBS, Tsentralna biblioteka, Vilshanka 28867
Vilshanska Regional Centralized Library System, Main Library, Vilshanka 28867
Vimmerby Bibliotek, Vimmerby 26957
Vincas Kudirka Public LIbrary of Kaunas, Kaunas 18331
Vincennes University, Vincennes 32934
Vincentinum Trier, Trier 11349
Vineland Free Public Library, Vineland 41688
Vineland Historical & Antiquarian Society Library, Vineland 37707
Vingåkers bibliotek, Vingåker 26958
Vinnitsa Centre of Scientific and Technical Information, Vinnytsya 28481
Vinnitsa City Centralized Library System, Bevz Main Library, Vinnytsya 28870
Vinnitsa National Memorial Medical University, Vinnytsya 28084
Vinnitsa National Technical University, Vinnytsya 28085
Timiryazev Vinnitsa Regional Scientific Library, Vinnytsya 27949
Vinnytska miska TsBS, Tsentralna biblioteka im. Bevza, Vinnytsya 28870
Vinnytski oblasni universalny naukovi bibliotetsi im. K.A. Timiryazeva, Vinnytsya 27949
Vinnytsky natsionalny medichny universytet im. N.I. Pirogova, Vinnytsya 28084
Vinnytsky natsionalny tekhnichny universytet, Vinnytsya 28085
Vinnytsky tsentr naukovo-tekhnichnoyi informatsiyi, Vinnytsya 28481
Vinson & Elkins, Houston 35766
Virgin Islands Division of Libraries, Archives & Museums – Regional Library for the Blind & Physically Handicapped, Saint Croix 42312
Virginia Baptist Historical Society & the Center for Baptist Heritage & Studies Library, Richmond 37444
Virginia Beach Department of Public Libraries, Virginia Beach 41692
Virginia Commonwealth University, Richmond 32442
– James Cabell Branch Library, Richmond 32443
– Tompkins-McCaw Library, Richmond 32444
Virginia Department of Environmental Quality Library, Glen Allen 34299
Virginia Highlands Community College, Abingdon 33114
Virginia Historical Society Library, Richmond 37445
Virginia Intermont College, Bristol 30552
Virginia Military Institute, Lexington 31604
Virginia Museum of Fine Arts Library, Richmond 37446
Virginia Polytechnic Institute & State University Libraries, Blacksburg 30436
– Art and Architecture Library, Blacksburg 30437
– Geosciences Library, Blacksburg 30438
– NVC Resource Center, Falls Church 30439
– Veterinary Medicine Library, Blacksburg 30440
Virginia Power Research, Richmond 36177
Virginia Public Library, Virginia 25241
Virginia Public Library, Virginia 41691
Virginia State Law Library, Richmond 34742
Virginia State University, Petersburg 32247
Virginia Theological Seminary, Alexandria 35103
Virginia Transportation Research Council Library, Charlottesville 36656
Virginia Union University, Richmond 32445
Virginia University of Lynchburg, Lynchburg 31699
Virginia War Museum, Newport News 37286
Virginia Wesleyan College, Norfolk 32124
Virginia Western Community College, Roanoke 33690
Virobnichno obednannya im. Artema, Kyiv 28161
Viša Pedagoška Škola, Sarajevo 02958
Viša Pedagoška Škola, Split 06169
Visagino viešoji biblioteka, Visaginas 18377
Više Pedagoške Škole, Banja Luka 02948
Vishnevsky Institute of Surgery, Moskva 23279
Vishveshvaranand Vishva Bandhu Institute of Sanskrit and Indological Studies, Hoshiarpur 13652
Visit Britain Library, London 29841
Viski Károly Múzeum, Kalocsa 13481
Visoka industrijsko-pedagoška škola, Rijeka 06159
Visoka komercialna šola Celje
– Fakulteta za komercialne in poslovne vede, Celje 24808
Visserijmuseum, Vlaardingen 19535
Vissh institut po khranitelna i vkusova promishlenost, Plovdiv 03459
Vissh institut za teatralno izkustvo 'Krăst Sarafov', Sofiya 03481
Visual Studies Workshop, Rochester 37460
Visva-Bharati, Santiniketan 13761
Visvesvaraya Industrial and Technological Museum, Bangalore 13943
Vitebsk Civil Engineering Institute, Vitebsk 02078
Vitebsk Plant of Electric Measuring Instruments, Vitebsk 01937
Vitebsk P.M. Masherov State University, Vitebsk 01817
Vitebsk Railway Station, Main Railway Technical Library, Vitebsk 02083
Vitebsk Railway Station, Railway Technical Library, Orsha 02069
Vitebsk Region Agricultural Station, Tulovo 02077
Vitebsk Regional Central Children's Library, Olgovo 02214
Vitebsk Regional Central Library, Vitebsk 02250
Vitebsk Regional Institute of Advanced Teacher Training, Vitebsk 01818
Vitebsk Regional Library, Vitebsk 02246
Vitebsk Regional Medical Library, Vitebsk 02086
Vitebsk State Medical University, Vitebsk 01816
Vitebsk State Technological University, Vitebsk 02085
Vitebsk Veterinary Institute, Vitebsk 01819
Vitebskaya gorodskaya bibliyateka im. Pushkina, Vitebsk 02244
Vitebskaya gorodskaya bibliyateka im. Ya. Kupaly, Vitebsk 02245
Vitebskaya oblastnaya bibliyateka, Vitebsk 02246
Vitebskaya oblastnaya meditsinskaya bibliyateka, Vitebsk 02086
Vitebskaya oblastnaya sel-skokhozyaistvennaya stantsiya, Tulovo 02077
Vitebskaya tsentralnaya gorodskaya bibliyateka im. Gorkogo, Vitebsk 02247
Vitebskaya tsentralnaya gorodskaya detskaya bibliyateka im. Gaidara, Vitebsk 02248
Vitebskaya tsentralnaya gorodskaya detskaya bibliyateka im. Kirova, Vitebsk 02249
Vitebskaya tsentralnaya raionnaya bibliyateka, Vitebsk 02250
Vitebskaya tsentralnaya raionnaya detskaya bibliyateka, Olgovo 02214
Vitebski gosudarstvenny ordena Druzhby narodov meditsinski universitet, Vitebsk 01816
Vitebski gosudarstvenny universitet im. P.M. Masherova, Vitebsk 01817
Vitebski oblastnoi institut usovershenstvovaniya uchitelei, Vitebsk 01818
Vitebski veterinarny institut, Vitebsk 01819
Vitero College, La Crosse 31525
Vitos Klinik für Psychiatrie und Psychotherapie Marburg, Marburg 12259
Vitterhetsakademiens bibliotek, Stockholm 26697
Vivantes Klinikum Hellersdorf, Berlin 11595
Vivantes, Klinikum im Friedrichshain, Berlin 11596
Vivliothiki Adamantios Korais, Chios 13126
Vizantoloski Institut, Beograd 24523
Vizgazdálkodási Tudományos Kutatóközpont (VITUKI), Budapest 13465
Vizhnitska raionna TsBS, Tsentralna biblioteka, Vizhnitsya 28872
Vizhnitsya Regional Centralized Library System, Main Library, Vizhnitsya 28872
Vlaams Filmmuseum en -archief, Leuven 02675
Vlaams Instituut voor het Onroerend Erfgoed (VIOE), Brussel 02629
Vlaams Parlement, Brussel 02441
't Vlaams Stripcentrum, Antwerpen 02534
Vlaams Studiehuis van de Assumptionisten, Leuven 02479
Vlaams Theater Instituut (VTI), Brussel 02630
Vlaams Verbond van Katholieke Scouts en Meisjesgidsen (VVKSM), Antwerpen 02535
Vlaamse Dienst voor Arbeidsbemiddeling en Beroepsopleiding (VDAB), Brussel 02631
Vlaamse Luister- en Braille Bibliotheek (VLBB), Brussel 02632
Vlaamse Uitgeversmaatschappij (VUM), Dilbeek 02637
Vlaamse Vereniging voor Familiekunde (VVF), Antwerpen 02536

Vladas Slaitas Public Library of Ukmerge District Municipality, Ukmergė 18371
Vladimir District Agricultural Experimental Station, Fofanka 23041
Vladimir Regional Archives, Vladimir 24047
Vladimir Regional Universal Scientific Library, Vladimir 22190
Vladimir Technical University, Vladimir 22631
Vladimiro-Suzdalski istoriko-khudozhestvenny i arkhitekturny muzei-zapovednik, Vladimir 24049
Vladimirskaya gosudarstvennaya selskokhozyaistvennaya opytnaya stantsia, Fofanka 23041
Vladimirskaya oblastnaya universalnaya nauchnaya biblioteka, Vladimir 22190
Vladimirskaya oblastnaya yunosheskaya biblioteka, Vladimir 24346
Vladimirski gosudarstvenny pedagogicheski universitet, Vladimir 22630
Vladimirski mezhotraslevoi territorialny tsentr nauchno-tekhnicheskoi informatsii i propagandy, Vladimir 24050
Vladimirski tekhnicheski universitet, Vladimir 22631
Vladimir-Suzdal Museum of History, Arts and Architecture, Vladimir 24049
Vladivostok State University of Economy, Vladivostok 22639
Vladivostokski gosudarstvenny universitet ekonomiki i servisa, Vladivostok 22639
VLAMO Centre of documentation, Gent 02658
Vlamo Documentatiecentrum, Gent 02658
Vlerick Leuven Gent Management School, Gent 02341
Vluchtelingenwerk (VVN), Amsterdam 19387
VM Mezőgazdasági Gépesítési Intézet, Gödöllő 13325
VME Industries Sweden AB, Eskilstuna 26529
VNI svetotekhnicheski institut, Moskva 23541
Vniikhimproekt instytut, Kyiv 28397
VNIKTI vodosnabzheniya, kanalizatsii, gidrotekhnicheskikh sooruzheni i inzhenernoi gidrogeologii, Moskva 23542
Vocvo vzw, Mechelen 02685
Voennaya akademiya im. F.E. Dzerzhinskogo, Moskva 22400
Voennaya akademiya im. M.V. Frunze, Moskva 22401
Voennaya akademiya khimicheskoi zashchity im. Marshala S.K. Timoshenko, Moskva 22402
Voennaya akademiya svyazi im. S.M. Budyonnogo, Sankt-Peterburg 22552
Voennaya akademiya tyla i transporta, Sankt-Peterburg 22553
Voennaya artilleriskaya akademiya, Sankt-Peterburg 22554
Voenno-inzhenernoe uchilishche svyazi, Sankt-Peterburg 22714
Voenno-istoricheski muzei artillerii, inzhenernykh voisk i voisk svyazi, Sankt-Peterburg 23926
Voenno-meditsinskaya akademiya, Sankt-Peterburg 22555
Voenno-meditsinski muzei, Sankt-Peterburg 23927
Voenno-morskaya akademiya im. adm. flota N.G. Kuznetsova, Sankt-Peterburg 22556
Voenno-morskoe inzhenernoe uchilishche, Sankt-Peterburg 22557
Voenno-morskoi gospital no 35, Sankt-Peterburg 23928
Voenno-morskoi klinicheski gospital no 1, Sankt-Peterburg 23929
Voenno-vozdushnaya inzhenernaya akademiya im. N.E. Zhukovskogo, Moskva 22403
Voenno-yuridicheskaya akademiya, Moskva 22404
Voenny institut fizicheskoi kultury, Sankt-Peterburg 23930
Voenny inzhenerno-kosmicheski institut, Sankt-Peterburg 23931
voestalpine AG, Linz 01503
Vognetrivkivi kombinat, Chasiv Yar 28512
Vogtlandbibliothek, Plauen 12895
Vogtlandmuseum, Plauen 12399
Voith Turbo GmbH & Co KG, Heidenheim 11406
Voivodeship and City Public Library, Gdańsk 21415
Voivodeship Public Library, Kraków 21460
Vojni Muzej, Beograd 24524
Vojvodina Archives, Novi Sad 24529
Volda College, Volda 20183
Volga Diesel Engineering Joint Stock Company, Balakovo 22776
Volga Railways, Saratov 22953
Volga Research Institute of the Cellulose and Paper Industry, Volzhsk 24074
Volga Water Transport Academy, Nizhni Novgorod 22427
Volgograd Engine Plant, Vladivostok 22971
Volgograd Regional Scientific Library, Volgograd 22193
Volgograd Regional Special Library for the Blind, Volgograd 24068
Volgograd State Agricultural Academy, Volgograd 22643
Volgograd State University of Architecture and Civil Engineering (VolgGASU), Volgograd 22644
Volgograd Tractor-Assembly Joint-Stock Company, Volgograd 22972
Volgogradskaya gosudarstvennaya selskokhozyaistvennaya akademiya, Volgograd 22643
Volgogradskaya oblastnaya detskaya biblioteka, Volgograd 24350
Volgogradskaya oblastnaya spetsialnaya biblioteka dlya slepykh (VOSBS), Volgograd 24068
Volgogradskaya oblastnaya universalnaya nauchnaya biblioteka im. M. Gorkogo, Volgograd 22193
Volgogradskaya oblastnaya yunosheskaya biblioteka, Volgograd 24351
Volgogradski gosudarstvenny arkhitekturno-stroitelny universitet, Volgograd 22644
Volgogradski motorny zavod, Vladivostok 22971
Völkerkundemuseum, Zürich 27212
Völkerkundliche Bibliothek, Frankfurt am Main 11851
Volkovysk Central Regional Library, Volkovysk 02251
Volkovysk Railway Station, Railway Technical Library, Volkovysk 02087
Volkovysskaya tsentralnaya raionnaya bibliyateka, Volkovysk 02251
Volkskundebibliothek am Volkskundemuseum Graz, Graz 01532
Volkskundemuseum, Antwerpen 02518
Volkskundliche Kommission für Westfalen, Münster 12354
Volkswagen AG, Wolfsburg 11438
Volnovakha Regional Centralized Library System, Main Library, Volnovakha 28873
Volnovaska raionna TsBS, Tsentralna biblioteka, Volnovakha 28873
Volochisk Regional Centralized Library System, Main Library, Volochisk 28874
Volochiska raionna TsBS, Tsentralna biblioteka, Volochisk 28874
Volodarka Regional Centralized Library System, Main Library, Volodarka 28875
Volodarska raionna TsBS, Tsentralna biblioteka, Volodarka 28875
Volodarsko-Volinska raionna TsBS, Tsentralna biblioteka, Volodarsk-Volinski 28876
Volodarsk-Volinsky Regional Centralized Library System, Main Library, Volodarsk-Volinski 28876
Volodimir-Volinsk Regional Centralized Library System, Main Library, Volodimir-Volinsk 28877
Volodimir-Volinska raionna TsBS, Tsentralna biblioteka, Volodimir-Volinsk 28877
Vologda Region Scientific Library, Vologda 22194
Vologda Scientific and Technical Information Centre, Vologda 24073
Vologodskaya oblastnaya detskaya biblioteka, Vologda 24352
Vologodskaya oblastnaya universalnaya nauchnaya biblioteka im. I.V. Babushkina, Vologda 22194
Vologodskaya oblastnaya yunosheskaya biblioteka im. V.F. Tendryakova, Vologda 24353
Vologodski TsNTI, Vologda 24073
Volta River Authority, Accra 13058
Volterra's Hospital, Volterra 16677
Volunteer State Community College, Gallatin 33364
Volusia County Law Library, Daytona Beach 34128
Volusia County Public Library, Daytona Beach 38793
Volvo Aero Corporation, Trollhättan 26558
Volvo Technology Corporation, Göteborg 26533
Volynsk Regional State Archives, Lutsk 28401
Volynskaya oblastnaya nauchna ta meditsinska biblioteka, Lutsk 28403
Volynsky oblastnoyi kraevedchesky muzei, Lutsk 28404
Volzhskaya gosudarstvennaya akademiya vodnogo transporta, Nizhni Novgorod 22427
Volzhski nauchno-issledovatelski institut tsellyulozno-bumazhnoi promyshlennosti korporatsii 'Rossiskie lesopromyshlenniki', Volzhsk 24074
Von Briesen & Roper Sc, Milwaukee 35900
Von der Heydt-Museum, Wuppertal 12610
Von Parish-Kostümbibliothek, München 12339
Voorhees College, Denmark 30989
Vorarlberg State Library, Bregenz 01181
Vorarlberger Landesarchiv (VLA), Bregenz 01512
Vorarlberger Landesbibliothek, Bregenz 01181
Vorarlberger Landeskonservatorium, Feldkirch 01190
Vorarlberger Landesmuseum, Bregenz 01513
Vorderasiatisches Museum, Berlin 11583
Vordingborg Bibliotek, Vordingborg 06951
Vordingborg Gymnasium og HF, Vordingborg 06755
Vordingborg Handelsskole og Handelsgymnasium, Vordingborg 06756
Vormingscentrum Guislain, Gent 02659
Voronezh Academy of Forestry, Voronezh 22648
Voronezh Art Museum, Voronezh 24079
Voronezh Machine Tool Joint-Stock Company, Voronezh 22978
Voronezh N.N. Burdenko State Medical Academy, Voronezh 22646
Voronezh Regional Children's Library, Voronezh 24355
Voronezh Regional Universal Scientific Library named after I.S. Nikitin, Voronezh 22195
Voronezh Regional Youth Library named aftrer V.M. Kubanev, Voronezh 24356
Voronezh State Pedagogical University, Voronezh 22651
Voronezh State Technological Academy, Voronezh 22647
Voronezh State University, Voronezh 22652
Voronezh State University of Architecture and Construction, Voronezh 22650
Voronezhskaya gosudarstvennaya meditsinskaya akademiya im. N.N. Burdenko, Voronezh 22646
Voronezhskaya gosudarstvennaya tekhnologicheskaya akademiya, Voronezh 22647
Voronezhskaya lesotekhnicheskaya akademiya, Voronezh 22648
Voronezhskaya oblastnaya detskaya biblioteka, Voronezh 24355
Voronezhskaya oblastnaya universalnaya nauchnaya biblioteka im. I.S. Nikitina, Voronezh 22195
Voronezhskaya oblastnaya yunosheskaya biblioteka im. V.M. Kubaneva, Voronezh 24356
Voronezhski agrarny universitet im. K.D. Glinki, Voronezh 22649
Voronezhski gosudarstvenny arkhitekturno-stroitelny universitet, Voronezh 22650
Voronezhski gosudarstvenny pedagogicheski universitet, Voronezh 22651
Voronezhski gosudarstvenny universitet, Voronezh 22652
Vörösmarty Mihály Megyei Könyvtár, Székesfehérvár 13547
Võru County Central Library, Võru 07259
Vose Seminary, Bentley 00390
Voss folkebibliotek, Voss 20419
Vostochno-Kazakhstanski gosudarstvenny universitet im. Amanzholova, Ust-Kamenogorsk 17927
Vostochno-Sibirski gosudarstvenny akademiya kultury i iskusstv, Ulan-Ude 22615
Vostochno-Sibirski tekhnologicheski institut, Ulan-Ude 22616
Vostochny nauchno-issledovatelski institut po bezopasnosti rabot v gornoi promyshlennosti, Kemerovo 23092
Vouli ton Ellinon, Athinai 13093
Vovchansk Regional Centralized Library System, Main Library, Vovchansk 28878
Vovchanska raionna TsBS, Tsentralna biblioteka, Vovchansk 28878
Voykov Chemical Plant, Moskva 22839
Voznesensk Centralized Library System, Main Library, Voznesensk 28879
Voznesenska mezhivdomcha TsBS, Tsentralna biblioteka, Voznesensk 28879
VPRO, Hilversum 19468
Vrede vzw, Gent 02660
Vredespaleis, Den Haag 19454
VR-Group Ltd/Finnish Railways, Helsinki 07418
Vrhovno sodišče Republike Slovenije, Ljubljana 24894
Vrije Universiteit Amsterdam, Amsterdam 19079
– Bèta Bibliotheek, Amsterdam 19080
– Bibliotheek Bewegingswetenschappen, Amsterdam 19081
– Bibliotheek Economische Wetenschappen en Bedrijfskunde, Amsterdam 19082
– Bibliotheek Germaanse Talen, Amsterdam 19083
– Bibliotheek Geschiedenis, Amsterdam 19084
– Bibliotheek Klassieke Talen, Oude Geschiedenis, Patristiek, Nederlands/Fries, Amsterdam 19085
– Bibliotheek Medische Wetenschappen, Preklinische Sector, Amsterdam 19086
– Bibliotheek Pedagogiek, Amsterdam 19087
– Bibliotheek Rechten, Amsterdam 19088
– Bibliotheek Theologie en Religiestudies, Amsterdam 19089
– Instituut voor Milieuvraagstukken (IVM), Amsterdam 19090
– Kaartenverzameling, 1A-01a, Amsterdam 19091
– Medische Bibliotheek, Amsterdam 19092
– Universiteitsbibliotheek, Afdeling Bijzondere Collecties / Studiecentrum voor Protestantse Boekcultuur, Amsterdam 19093
– Universiteitsbibliotheek, Afdeling Bio- en Bibliografieën, Amsterdam 19094
Vrije Universiteit Brussel, Brussel 02306
– Medische Bibliotheek, Brussel 02307
Vrouwendocumentatiecentrum "De Feeks", Nijmegen 19501
VRT – Vlaamse Radio- en Televisieomroep, Documentatiearchief, Brussel 02633
VRT – Vlaamse Radio- en Televisieomroep, Muziekbibliotheek, Brussel 02634
VR-Yhtymä Oy, Helsinki 07418
VSE Corporation Library-BAV Division, Alexandria 35439
Vserossiskaya akademiya vneshnei torgovli, Moskva 22405
Vserossiskaya geologicheskaya biblioteka (VGB), Sankt-Peterburg 23932
Vserossiskaya gosudarstvennaya biblioteka inostrannoi literatury imeni M.I. Rudomino, Moskva 22155
Vserossiskaya patentno-tekhnicheskaya biblioteka, Moskva 23543
Vserossiskaya torgovaya palata, Moskva 23544
Vserossiski elektrotekhnicheski institut, Moskva 22406
Vserossiski gosudarstvenny institut kinematografii, Moskva 22407
Vserossiski institut eksperimentalnoi veterinarii, Moskva 23545
Vserossiski institut nauchnoi i tekhnicheskoi informatsii (VINITI RAN), Moskva 23546
Vserossiski institut po izyskaniyu novykh antibiotikov, Moskva 23547
Vserossiski konservnoi i ovoshchesushilnoi promyshlennosti, Moskva 23548
Vserossiski nauchno-isledovatelny institut zhirov (VNIIZh), Sankt-Peterburg 23933
Vserossiski nauchno-issledovatelski geologorazvedochny neftyanoi institut, Moskva 23549
Vserossiski nauchno-issledovatelski geologorazvedochny neftyany institut – Kamski filial, Perm 23693
Vserossiski nauchno-issledovatelski i eksperimentalno-konstruktorski institut prodovolstvennogo mashinostroeniya, Moskva 23550
Vserossiski nauchno-issledovatelski i konstruktorski tekhnologicheski institut podshipnikovoi promyshlennosti, Moskva 23551
Vserossiski nauchno-issledovatelski i proektno-konstruktorski institut metallurgicheskogo mashinostroeniya im. A.I. Tselikova, Moskva 23552
Vserossiski nauchno-issledovatelski i proektno-konstruktorski institut stekolnogo mashinostroeniya, Moskva 23553
Vserossiski nauchno-issledovatelski i proektny institut neftepererabaty-

vayushchei i neftekhimicheskoi promyshlennosti, Moskva 23554
Vserossiski nauchno-issledovatelski i proektny institut 'Teploproekt', Moskva 23555
Vserossiski nauchno-issledovatelski i tekhnologicheski institut ptitsevodstva, Sergyev Posad 23966
Vserossiski nauchno-issledovatelski institut atomno-energeticheskogo mashinostroeniya, Moskva 23556
Vserossiski nauchno-issledovatelski institut Bumagi, Sankt-Peterburg 23934
Vserossiski nauchno-issledovatelski institut burovoi tekhniki, Moskva 23557
Vserossiski nauchno-issledovatelski institut ekonomiki mineralnogo syrya i geologorazvedochnykh rabot, Moskva 23558
Vserossiski nauchno-issledovatelski institut elektrifikatsii selskogo khozyaistva (VIESH), Moskva 23559
Vserossiski nauchno-issledovatelski institut Elektromashinostroeniya, Sankt-Peterburg 23935
Vserossiski nauchno-issledovatelski institut fitopatologii, Bolshie Vyazemy 23006
Vserossiski nauchno-issledovatelski institut fizicheskoi kultury i sporta, Moskva 23560
Vserossiski nauchno-issledovatelski institut fiziologii selskokhozyaistvennykh zhivotnykh, Borovsk 23009
Vserossiski nauchno-issledovatelski institut genetiki i selektsii plodovykh rasteni im. L.V. Michurina, Michurinsk 23158
Vserossiski nauchno-issledovatelski institut geofiziki, Moskva 23561
Vserossiski nauchno-issledovatelski institut geologicheskikh, geofizicheskikh i geokhimicheskikh informatsionnykh sistem, Moskva 23562
Vserossiski nauchno-issledovatelski institut gidrotekhniki i melioratsii, Moskva 23563
Vserossiski nauchno-issledovatelski institut informatsii i ekonomiki stroitelnykh materialov, Moskva 23564
Vserossiski nauchno-issledovatelski institut khimicheskikh sredstv zashchity rasteni, Moskva 23565
Vserossiski nauchno-issledovatelski institut khlebopekarnoi promyshlennosti, Moskva 23566
Vserossiski nauchno-issledovatelski institut kholodilnoi promyshlennosti, Moskva 23567
Vserossiski nauchno-issledovatelski institut kompleksnoi avtomatizatsii neftyanoi i gazovoi promyshlennosti, Moskva 23568
Vserossiski nauchno-issledovatelski institut kompleksnykh problem poligrafii, Moskva 23569
Vserossiski nauchno-issledovatelski institut konditerskoi promyshlennosti, Moskva 23570
Vserossiski nauchno-issledovatelski institut kormov im. V.P. Vilyamsa, Lugovaya 23144
Vserossiski nauchno-issledovatelski institut kosmoaerogeologicheskikh metodov, Sankt-Peterburg 23936
Vserossiski nauchno-issledovatelski institut legkogo i tekstilnogo mashinostroeniya, Moskva 23571
Vserossiski nauchno-issledovatelski institut lekarstvennykh rasteni, Vilp 24044
Vserossiski nauchno-issledovatelski institut lesovodstva i mekhanizatsii lesnogo khozyaistva, Pushchino 23710
Vserossiski nauchno-issledovatelski institut liteinogo mashinostroeniya, liteinoi tekhnologii i avtomatizatsii liteinogo proizvodstva, Moskva 23572
Vserossiski nauchno-issledovatelski institut maslodeliya i syrodeliya, Uglich 22969
Vserossiski nauchno-issledovatelski institut mekhanizatsii selskogo khozyaistva, Moskva 23573
Vserossiski nauchno-issledovatelski institut Mekhanoobrabotka, Sankt-Peterburg 23937
Vserossiski nauchno-issledovatelski institut metodiki i tekhniki razvedki, Sankt-Peterburg 23938
Vserossiski nauchno-issledovatelski institut metrologii im. D.I. Mendeleeva, Sankt-Peterburg 23939
Vserossiski nauchno-issledovatelski institut metrologii im. D.I. Mendeleeva – Sverdlovski filial, Ekaterinburg 23038
Vserossiski nauchno-issledovatelski institut molochnoi promyshlennosti, Moskva 23574
Vserossiski nauchno-issledovatelski institut Montazhspetsstroi, Moskva 23575
Vserossiski nauchno-issledovatelski institut netkanykh tekstilnykh materialov, Serpukhov 23967
Vserossiski nauchno-issledovatelski institut Okeanogeologii, Sankt-Peterburg 23940
Vserossiski nauchno-issledovatelski institut organicheskogo sinteza, Moskva 23576
Vserossiski nauchno-issledovatelski institut organizatsii stankostroitelnoi i instrumentalnoi promyshlennosti (Orgstankinprom), Moskva 23577
Vserossiski nauchno-issledovatelski institut organizatsii upravleniya i ekonomiki neftegazovoi promyshlennosti, Moskva 23578
Vserossiski nauchno-issledovatelski institut pishchevykh aromatizatorov, kislot i krasitelei, Sankt-Peterburg 23941
Vserossiski nauchno-issledovatelski institut po pererabotke nefti, Moskva 23579
Vserossiski nauchno-issledovatelski institut po stroitelnomu i dorozhnomu mashinostroeniyu, Moskva 23580
Vserossiski nauchno-issledovatelski institut po zashchite metallov ot korrozii, Moskva 23581
Vserossiski nauchno-issledovatelski institut prirodnogo gaza, Vidnoe 24043
Vserossiski nauchno-issledovatelski institut prirodnykh gazov i gazovykh tekhnologii, Razvilka 23716
Vserossiski nauchno-issledovatelski institut prudovogo rybnogo khozyaistva (VNIIPRCH), Rybnoe 23737
Vserossiski nauchno-issledovatelski institut radioapparatury, Sankt-Peterburg 23942
Vserossiski nauchno-issledovatelski institut rastenievodstva, Sankt-Peterburg 23943
Vserossiski nauchno-issledovatelski institut razvedochnoi geofiziki, Sankt-Peterburg 23944
Vserossiski nauchno-issledovatelski institut razvedochnoi geofiziki, Sankt-Peterburg 23945
Vserossiski nauchno-issledovatelski institut rybnogo khozyaistva i okeanografii (VNIRO), Moskva 23582
Vserossiski nauchno-issledovatelski institut sakharnoi svekly i sakhara, Ramon 23714
Vserossiski nauchno-issledovatelski institut selskokhozyaistvennoi mikrobiologii, Sankt-Peterburg 23946
Vserossiski nauchno-issledovatelski institut sinteza mineralnogo syrya, Aleksandrov 22984
Vserossiski nauchno-issledovatelski institut sotsialnoi gigieny, ekonomiki i upravleniya zdravookhraneniya im. N.A. Semashko, Moskva 23583
Vserossiski nauchno-issledovatelski institut tekhnicheskoi estetiki, Moskva 23584
Vserossiski nauchno-issledovatelski institut tugoplavkikh metallov i tverdykh splavov, Moskva 23585
Vserossiski nauchno-issledovatelski institut udobreni i agropochvovedeniya im. D.N. Pryanishnikova, Moskva 23586
Vserossiski nauchno-issledovatelski institut veterinarnoi sanitarii, gigieny i ekologii, Moskva 23587
Vserossiski nauchno-issledovatelski institut vinogradstva i vinodeliya, Novocherkassk 23630
Vserossiski nauchno-issledovatelski institut zashchity rasteni / Vserossiski nauchno-issledovatelski institut selskokhozyaistvennoi mikrobiologii, Sankt-Peterburg 23946
Vserossiski nauchno-issledovatelski institut zernobobovykh kultur, Oryol 23680
Vserossiski nauchno-issledovatelski institut zheleznodorozhnoi gigieny, Moskva 23588
Vserossiski nauchno-issledovatelski institut zhivotnovodstva, Dubrovitsy 23026
Vserossiski nauchno-issledovatelski institutoroshaemogo ovoshchevodstva i bakhchevodstva, Kamyzyak 23078
Vserossiski nauchno-issledovatelski kinofotoinstitut, Moskva 23589
Vserossiski nauchno-issledovatelski konstruktorski i tekhnologicheski institut gidromashinostroeniya, Moskva 23590
Vserossiski nauchno-issledovatelski tekhnologicheski institut priborostroeniya, Moskva 23591
Vserossiski nauchno-issledovatelski tekhnologicheski institut remonta i ekspluatatsii mashinno-traktornogo parka, Moskva 23592
Vserossiski nauchno-issledovatelski teplotekhnicheski institut, Moskva 23593
Vserossiski nauchno-issledovatelski vitaminny institut, Moskva 23594
Vserossiski neftyanoi nauchno-issledovatelski geologorazvedochny institut, Sankt-Peterburg 23947
Vserossiski NIPKI okhrany okruzhayushchei sredy v ugolnoi promyshlennosti, Perm 23694
Vserossiski proektno-tekhnologicheski institut tyazhelogo mashinostroeniya, Moskva 23595
Vserossiski selskokhozyaistvennogo mashinostroeniya im. V.P. Goryachkina, Moskva 23596
Vserossiski selskokhozyaistvenny institut zaochnogo obrazovanya (VSCHIZO), Balashikha 22209
Vserossiski vystovochny tsentr, Moskva 23597
Vserossiski yuvelirnoi promyshlennosti, Sankt-Peterburg 23948
Vserossiski zaochny elektrotekhnicheski institut svyazi, Moskva 22408
Vserossiski zaochny finansovo-ekonomicheski institut, Moskva 22409
Vserossiskoe khimicheskoe obshchestvo – SP otdelenie, Sankt-Peterburg 23949
Vserossiskoe teatralnoe obshchestvo, Kabardino-Balkarskoe otdelenie, Nalchik 23610
Vserossiskoe teatralnoe obshchestvo, Krasnodarskoe otdelenie, Krasnodar 23120
Vserossiskoe teatralnoe obshchestvo, Kurskoe otdelenie, Kursk 23138
Vserossiskoe teatralnoe obshchestvo, Primorskoe otdelenie, Vladivostok 24061
Vsesoyusniy NII Geofizitcheskikh Issledovaniy Geologo-Rasvedochnikh Skvazhin, Oktyabrski 23662
Vsesoyuzny nauchno-issledovatelski institut televideniya i radioveshchaniya, Moskva 23598
Vsesoyuzny zaochny institut inzhenerov zheleznodorozhnogo transporta, Moskva 22410
Veszprém University, Central Library, Veszprém 13310
Vtora gradska obedinena bolnitsa, Sofiya 03582
Vtoroi Moskovski chasovoi zavod, Moskva 22861
VTT Energia, Jyväskylä 07517
VTT Technical Research Centre of Finland, Espoo 07437
VUC Praha, plc, Praha 06525
VUHŽ, a.s., Dobrá 06446
Vuk and Dositej Museum, Beograd 24525
VUKI, a.s., Bratislava 24730
Vukov i Dositejev Muzej, Beograd 24525
Vulcan Sports Media Inc, Saint Louis 36207
Vutch-Chemitex Ltd.- Oddelenie VTEI, Žilina 24760
VWL-Bibliothek 1, Köln 10230
VWL-Bibliothek 2, Köln 10231
Vyatka State University, Kirov 22293
Vyatski gosudarstvenny tekhnicheski universitet, Kirov 22293
Vyborgski DKiT, Sankt-Peterburg 24310
Vyborgsky Centre of Popular Science and Culture, Sankt-Peterburg 24310
Vyborgsky District Centralized Library System, Central District Children's Library, Sankt-Peterburg 24306
Vyborgsky District Centralized Library System, Library Branch no 1, Sankt-Peterburg 24303
Vyborgsky District Centralized Library System, Library Branch no 5, Sankt-Peterburg 24304
Vyborgsky District Centralized Library System, Serafimovich Central District Library, Sankt-Peterburg 24305
Východočeské muzeum, Pardubice 06471
Východoslovenské múzeum, Košice 24736
Výskumný ústav chemickej technológie, Bratislava 24729
Výskumný ústav káblov a izolantov – VUKI, a.s., Bratislava 24730
Výskumný ústav pre petrochémiu, Prievidza 24749
Výskumný ústav rastlinnej výroby, Piešt'any 24747
Výskumný ústav vodného hospodárstva, Bratislava 24731
Výskumný ústav živočíšnej výroby – Oddelenie VTEI, Nitra 24746
Vysoka škola chemicko-technologická, Praha 06404
Vysoká škola economická v Praze, Praha 06405
Vysoká škola múzických umění, Bratislava 24662
Vysoká škola umělecko-průmyslová v Praze, Praha 06406
Vysoká škola výtvarných umení Bratislava, Bratislava 24663
Vysoké školy báňské -Technická univerzita Ostrava, Ostrava 06376
Vysoké učení technické v Brně, Brno 06365
Vysokogorny geofizicheski institut, Nalchik 23611
Vysshaya Komsomolskaya shkola pri TsK VLKSM, Moskva 23599
Vysshaya pozharno-tekhnicheskaya shkola, Sankt-Peterburg 22558
Vysshee artilleriskoe komandnoe uchilishche, Sankt-Peterburg 22559
Vysshee khudozhestvenno-promyshlennoe uchilishche, Sankt-Peterburg 22560
Vysshee spetsialnye ofitserskie klassy voenno-morskogo flota, Sankt-Peterburg 22561
Vysshee uchilishche radioelektroniki i protivovozdushnoi oborony, Sankt-Peterburg 22562
Vysshee uchilishche radioelektroniki PVO, Sankt-Peterburg 22715
Vysshee VMU radioelektroniki im. A.S. Popova, Sankt-Peterburg 22716
Vysshee voenno-morskoe inzhenernoe uchilishche, Sankt-Peterburg 22563
Vysshee voenno-morskoe uchilishche, Sankt-Peterburg 22564
Vysshee voenno-morskoe uchilishche podvodnogo plavaniya, Sankt-Peterburg 22565
Vysshee voenno-topograficheskoe komandnoe uchilishche im. generala A.I. Antonova, Sankt-Peterburg 22566
Vytautas Magnus University, Kaunas 18282
Výzkumného a šlechtitelského ústavu ovocnářského Holovousy s.r.o., Hořice v Podkrkonoší 06450
Výzkumný a zkušební letecký ústav, Praha 06526
Výzkumný ústav bezpečnosti práce, Praha 06527
Výzkumný ústav bramborářský, Havlíčkův Brod 06448
Výzkumný ústav geodetický, topografický a kartografický, Zdiby 06549
Výzkumný ústav lesního hospodářstvi a myslivosti, v.v.i., Jíloviště 06455
Výzkumný ústav pedagogický v Praze, Praha 06528
Výzkumný ústav Silva Taroucy pro krajinu a okrasné zahradnictví, Průhonice 06535
Výzkumný ústav stomatologický, Praha 06529
Výzkumný ústav textilnich stroju, Liberec 06462
Výzkumný ústav veterinárního lékařství, Brno 06441
Výzkumný ústav vodohospodářský, Praha 06530
Výzkumný ústav zemědělské ekonomiky, Praha 06531
Výzkumný ústav zemědělské techniky, Praha 06532
Výzkumný ústav živočišné výroby, Praha 06533
W. E. Upjohn Institute, Kalamazoo 36984
W. F. Albright Institute of Archaeological Research, Jerusalem 14722
W. J. Niederkorn Public Library, Port Washington 40861
W. L. Gore & Associates, Inc, Newark 36041
W. R. Grace & Co, Columbia 35625
W. Stefański Institute of Parasitology, Warszawa 21344
Wabash Carnegie Public Library, Wabash 41696
Wabash College, Crawfordsville 30932
Wabash Valley College, Mount Carmel 33561
Wacker-Chemie AG, München 11422
Wacker-Chemie AG, Chemisch-Technische Bibliothek, Burghausen 11381
Waco-McLennan County Library System, Waco 41697
Waco-McLennan County Library System – R. B. Hoover Library, Waco 41698
Wadham College, Oxford 29359

Wadia Institute of Himalayan Geology, Dehradun 13979
Wadleigh Memorial Library, Milford 40201
Wadsworth Atheneum, Hartford 36922
Wageningen University and Research Centre, Wageningen 19227
– Bibliotheek Centrum Techniek (IMAG-DLO, TFDL-DLO), Wageningen 19228
– Bibliotheek Leeuwenborch, Wageningen 19229
– Bibliotheek Planteziektenkundig Centrum, Wageningen 19230
– Library Agrotechnology and Food science, Wageningen 19231
The Wagnalls Memorial Library, Lithopolis 39892
Wagner College, Staten Island 32737
Wagner Free Institute of Science Library, Philadelphia 37367
'Wahn Air Base, Specialist Library, Köln 10991
Wahta Mohawks – First Nation Library, Bala 04309
Waikato Institute of Technology, Hamilton 19786
Waitaki District Libraries, Oamaru 19877
Wakayama Kenritsu Ika Daigaku, Wakayama 17231
Wakayama Medical College, Wakayama 17231
Wakayama Prefectural Fruit Tree Experiment Station, Arida 17512
Wakayama Prefectural Research Center of Environment and Public Health, Wakayama 17813
Wakayama Prefecture Planning Dept, Statistics Div, Administrative Data Room, Wakayama 17332
Wakayama University, Wakayama 17232
Wakayamaken Eisei Kogai Kenkyu Senta, Wakayama 17813
Wakayamaken Kajuengei Shikenjo, Arida 17512
Wakayamaken Kikakubu, Wakayama 17332
Wake Area Health Education Center, Raleigh 37428
Wake County Public Library System – Athens Drive Community, Raleigh 40928
Wake County Public Library System – Cameron Village Regional, Raleigh 40929
Wake County Public Library System – Cary Branch, Cary 38474
Wake County Public Library System – Duraleigh Road, Raleigh 40930
Wake County Public Library System – East Regional, Knightdale 39736
Wake County Public Library System – Eva H. Perry Regional, Apex 37973
Wake County Public Library System – Green Road, Raleigh 40931
Wake County Public Library System – Holly Springs Branch, Holly Springs 39460
Wake County Public Library System – North Regional, Raleigh 40932
Wake County Public Library System – Richard B. Harrison Branch, Raleigh 40933
Wake County Public Library System – Southeast Regional, Garner 39203
Wake County Public Library System – Wake Forest Branch, Wake Forest 41705
Wake County Public Library System – West Regional, Cary 38475
Wake Forest University, Winston-Salem 33085
– Coy C. Carpenter School of Medicine Library, Winston-Salem 33086
– Professional Center Library, Winston-Salem 33087
Wake Technical Community College, Raleigh 33670
Wakefield College Library, Wakefield 29498
Wakefield Metropolitan District Council, Wakefield 30091
Walchand College of Engineering, Sangli 13760
Waldbreitbacher Franziskanerinnen e.V., Waldbreitbach 11357
Waldburg-Zeil'sche Bibliothek, Leutkirch 09326
Waldensian Library, Torre Pellice 16101
Waldorf College, Forest City 33343
Waldwick Public Library, Waldwick 41707
Walker Art Center, Minneapolis 37124
Walker Memorial Library, Westbrook 41846
Walker Nott Dragicevic Associates Ltd Library, Toronto 04295
Walla Walla College, Portland Campus, Portland 33660
Walla Walla Community College, Walla Walla 33834
Walla Walla County Law Library, Walla Walla 34973
Walla Walla Public Library, Walla Walla 41711
Walla Walla University, College Place 30856

Wallachian Open Air Museum, Rožnov pod Radhoštěm 06538
Walled Lake City Library, Walled Lake 41712
Wallingford Public Library, Wallingford 41713
Walnut Street Baptist Church, Louisville 35273
Walpole Public Library, Walpole 41718
Walsall College, Walsall 29431
Walsall Metropolitan District Council, Walsall 30092
Walsgrave Hospital, Coventry 29639
Walsh College of Accountancy & Business Administration, Troy 32837
Walsh University, North Canton 32141
Walt Disney Imagineering, Glendale 35720
Walt Whitman Library, Guatemala City 13183
Walter and Eliza Hall Institute of Medical Research, Parkville 00978
Walter & Haverfield, Cleveland 35621
Walter Reed Army Institute of Research, Silver Spring 34861
Walter Reed Army Medical Center – Medical Library Service, Washington 35043
Walter Reed Army Medical Center – Post & Patient's Library, Washington 37805
Walter T. McCarthy Law Library, Arlington 36484
J. Walter Thompson Co, New York 36031
Walters Art Museum Library, Baltimore 36517
Walters State Community College, Morristown 33559
Waltham Public Library, Waltham 41720
Walther-Schücking-Institut für Internationales Recht, Kiel 10114
Walton, Lantaff, Schroeder & Carson, Miami 35887
Wandsworth Borough Council, London 30036
Wanganui District Library, Wanganui 19885
Wannan Medical College, Wuhu 05668
Wanneroo Library Service, Wanneroo 01173
Wanxi United University, Liyang 05401
Wanxian Teachers' College, Wanxian 05642
Wapiti Regional Library, Prince Albert 04727
War History Museum of Hungary, Budapest 13379
Ward County Library, Monahans 40253
Ward County Public Library, Minot 40234
Wardown Park Museum, Luton 29847
Wareham Free Library, Wareham 41722
Warley College of Technology, Smethwick 29493
Warmińsko-Mazurska Biblioteka Pedagogiczna, Elbląg 21090
Warminster Township Free Library, Warminster 41723
Warner Bros Studios, Burbank 36617
Warner Library, Tarrytown 41521
Warner, Norcross & Judd, LLP Library, Grand Rapids 35725
Warner Pacific College, Portland 32347
Warner Southern College, Lake Wales 31546
Warren County Community College, Washington 33840
Warren County Historical Society, Lebanon 37023
Warren County Law Library Association, Lebanon 34451
Warren County Library, Belvidere 38168
Warren County Public Library, Bowling Green 38251
Warren County Public Library District, Monmouth 40259
Warren County-Vicksburg Public Library, Vicksburg 41681
Warren Library, West Palm Beach 33035
Warren Library Association, Warren 41725
Warren M Robbins Library, Washington 37793
Warren Public Library, Warren 41726
Warren State Hospital, North Warren 37293
Warren Wilson College, Swannanoa 32766
Warren-Newport Public Library District, Gurnee 39338
Warren-Trumbull County Public Library, Warren 41727
Warrenville Public Library District, Warrenville 41730
Warringah Library Service, Dee Why 01091
Warsaw Community Public Library, Warsaw 41731
Warsaw Public Library – Central Library of Masovian Province, Warszawa 21599
Warsaw School of Economics, Warszawa 20971
Warsaw University, Warszawa 20974
– Faculty of Biology, Warszawa 20991
– Faculty of Chemistry, Warszawa 20992
– Faculty of Economic Sciences, Warszawa 20994
– Faculty of Geology, Warszawa 20993
– Faculty of Law and Administration, Warszawa 20995
– Institute of Art History, Warszawa 20981

– Institute of Botany, Warszawa 20976
– Institute of Classical Studies, Warszawa 20977
– Institute of English Studies, Warszawa 20975
– Institute of History, Warszawa 20982
– Institute of Musicology, Warszawa 20986
– Institute of Oriental Studies, Warszawa 20987
– Institute of Philosophy, Warszawa 20978
– Institute of Polish Literature and Polish Culture, Warszawa 20985
– Institute of Romance Philology, Warszawa 20988
– Institute of Theoretical Physics, Warszawa 20980
Warsaw University of Life Sciences (SGGW), Warszawa 20967
– Faculty of Forestry, Warszawa 20968
– Faculty of Veterinary Medicine, Warszawa 20969
– Faculty of Wood Technology, Warszawa 20970
Warsaw University of Technology, Warszawa 20956
– Chemical and Process Engineering, Warszawa 20961
– Faculty of Architecture, Warszawa 20957
– Faculty of Chemistry, Warszawa 20959
– Faculty of Materials Science and Engineering, Warszawa 20962
– Warsaw University of Technology Faculty of Transport, Warszawa 20966
Warshaw, Burstein, Cohen, Schlesinger & Kuh, New York 36032
Warszawski Uniwersytet Medyczny, Warszawa 20998
Wartburg College, Waverly 33004
Wartburg Theological Seminary, Dubuque 35189
Wartburg-Stiftung, Eisenach 11811
Warwick Public Library, Warwick 41733
Warwickshire College, Warwick 29432
Warwickshire County Council, Warwick 30093
Wasatch County Library, Heber 39408
Waseca-Le Sueur Regional Library, Waseca 41734
Waseda Daigaku, Tokyo 17210
– Gendai Seiji Keizai Kenkyujo, Tokyo 17211
– Hikakuho Kenkyujo, Tokyo 17212
– Rikogaku Toshokan, Tokyo 17213
– Sangyo Keiei Kenkyujo, Tokyo 17214
– Shakai Kagaku Kenkyujo, Tokyo 17215
Waseda University, Tokyo 17210
– Institute for Research in Contemporary Political and Economic Affairs, Tokyo 17211
– Institute of Comparative Law, Tokyo 17212
– Institute of Social Sciences, Tokyo 17215
– Library of Science and Engineering, Tokyo 17213
– Research Insitute for Business Administration, Tokyo 17214
Washakie County Library System, Worland 41972
Washburn University, Topeka 32826
– School of Law Library, Topeka 32827
Washington Bible College – Capital Bible Seminary, Lanham 35257
Washington College, Chestertown 30745
Washington College of Law, Washington 32955
Washington Correctional Facility Library, Comstock 34106
Washington Correctional Institute Library, Angie 33923
Washington County Board of Education, Hagerstown 34330
Washington County Free Library, Hagerstown 39343
Washington County – Jonesborough Library, Jonesborough 39661
Washington County Law Library, Hagerstown 34331
Washington County Law Library, Hillsboro 34354
Washington County Law Library, Marietta 34505
Washington County Law Library, Washington 35044
Washington County Library, Chipley 38556
Washington County Library, Potosi 40878
Washington County Library, Saint George 41110
Washington County Library, Woodbury 41956
Washington County Library – Hardwood Creek Branch, Forest Lake 39097
Washington County Library – Park Grove Branch, Cottage Grove 38712
Washington County Library – R. H. Stafford Branch, Woodbury 41957

Washington County Library System, Fayetteville 39059
Washington County Library System – William Alexander Percy Memorial Library, Greenville 39314
Washington County Public Library, Abingdon 37875
Washington County Public Library, Chatom 38532
Washington County Public Library, Marietta 40057
Washington District Library, Washington 41758
Washington Hospital Center, Washington 37806
Washington & Jefferson College Library, Washington 32994
Washington & Lee University, Lexington
– James Graham Leyburn Library, Lexington 31605
– Wilbur C. Hall Law Library, Lexington 31606
Washington National Cathedral, Washington 37807
Washington Parish Library System, Franklinton 39153
Washington Savannah River Co, Aiken 35430
Washington School for the Deaf, Vancouver 33822
Washington State Department of Natural Resources, Olympia 34634
Washington State Historical Society Research Center, Tacoma 37653
Washington State Law Library, Olympia 34635
Washington State Library, Olympia 30144
Washington State University, Pullman 32395
– George B. Brain Education Library, Pullman 32396
– George W. Fischer Agricultural Sciences Library, Pullman 32397
– Owen Science & Engineering Library, Pullman 32398
Washington State University Extension, Olympia 32192
Washington State University Tri-Cities, Richland 32433
Washington Talking Book & Braille Library, Seattle 37601
Washington Theological Union, Washington 35409
Washington Township Free Public Library, Long Valley 39933
Washington University, Saint Louis 32512
– Art and Architecture Library, Saint Louis 32513
– Biology Library, Saint Louis 32514
– Chemistry Library, Saint Louis 32515
– Earth and Planetary Science Library, Saint Louis 32516
– East Asian Library, Saint Louis 32517
– Gaylord Music Library, Saint Louis 32518
– GWB School of Social Work, Saint Louis 32519
– Kopolow Business Library, Saint Louis 32520
– Law Library, Saint Louis 32521
– Mathematics Library, Saint Louis 32522
– Pfeiffer Physics Library, Saint Louis 32523
– School of Medicine, Saint Louis 32524
Washington-Carnegie Public Library, Washington 41759
Washington-Centerville Public Library, Centerville 38493
Washoe County Law Library, Reno 34736
Washoe County Library System, Reno 40962
Washoe County Library System – Downtown Reno Library, Reno 40963
Washoe County Library System – North Valleys Library, Reno 40964
Washoe County Library System – Sierra View Library, Reno 40965
Washoe County Library System – South Valleys Library, Reno 40966
Washoe County Library System – Spanish Springs Library, Sparks 41392
Washoe County Library System – Sparks Library, Sparks 41393
Washtenaw Community College, Ann Arbor 33128
Wasilla Public Library, Wasilla 41761
Watauga County Public Library, Boone 38240
Watchtower Bible School of Gilead Library, Patterson 35323
Water Commission, Tel-Aviv 14693
Water Conservancy College of Yangzhou University, Yangzhou 05740
Water Corporation Library, Perth 00717
Water Planning for Israel Ltd, Tel-Aviv 14709
Water Research Institute, Bratislava 24731
Water Research Institute, Praha 06530
Water Resources and Energy Administration, Oslo 20194

Water Resources Research Centre, Budapest 13465
Waterbouwkundig Laboratorium, Antwerpen 02537
Waterbury Hospital, Waterbury 37813
Waterford City Council Central Library, Waterford 14593
Waterford County Library, Lismore 14581
Waterford Public Library, Waterford 41763
Waterloo Law Association, Kitchener 04370
Waterloo Public Library, Waterloo 04813
Waterloo Public Library, Waterloo 41764
Waterton-Glacier International Peace Park, West Glacier 37821
Watertown Library Association, Watertown 41767
Watertown Public Library, Watertown 41768
Watertown Regional Library, Watertown 41769
Waterville Public Library, Waterville 41771
Watkins College of Art & Design Library, Nashville 33572
Watkins, Ludlam, Winter & Stennis, Jackson 35786
Watonwan County Library, Saint James 41112
Watt, Tieder, Hoffar & Fitzgerald, McLean 35877
Watts, Griffis & McOuat Limited Library, Toronto 04296
Waubonsee Community College, Sugar Grove 33783
Wauconda Area Library, Wauconda 41772
Waukegan Public Library, Waukegan 41773
Waukesha County Historical Society and Museum, Waukesha 37815
Waupaca Area Public Library, Waupaca 41774
Waupun Correctional Institution Library, Waupun 35049
Waupun Public Library, Waupun 41775
Wautaga Public Library, Wautaga 41777
Wauwatosa Public Library, Wauwatosa 41778
Waverly Public Library, Waverly 41780
Wawel State Art Collections, Kraków 21146
Way Public Library, Perrysburg 40762
Waycross College, Waycross 33847
Wayland Baptist University, Plainview 32318
Wayland Free Public Library, Wayland 41783
Wayne Community College, Goldsboro 33377
Wayne County Community College, Detroit 33294
Wayne County Public Library, Wooster 41970
Wayne County Public Library, Inc, Goldsboro 39260
Wayne County Regional Library for the Blind & Physically Handicapped, Westland 37829
Wayne Public Library, Wayne 41785
Wayne Public Library, Wayne 41786
Wayne State College, Wayne 33006
Wayne State University, Detroit 31009
– Arthur Neef Law Library, Detroit 31010
– David Adamany Undergraduate Library, Detroit 31011
– Purdy-Kresge Library, Detroit 31012
– Science and Engineering, Detroit 31013
– Vera P. Shiffman Medical Library, Detroit 31014
– Walter P. Reuther Library of Labor & Urban Affairs, Detroit 31015
Wayne Township Library, Richmond 40988
Waynesboro Public Library, Waynesboro 41787
Waynesburg College, Waynesburg 33008
Waziri Umary Polytechnic, Birnin Kebbi 19931
WBM Pty Ltd, Brisbane 00814
WD IZMIRAN, Kaliningrad 23071
WEA Sydney, Sydney 00543
Weald and Downland Open Air Museum, Chichester 29635
Weatherford College, Weatherford 33849
Weatherford Public Library, Weatherford 41790
Webb Institute, Glen Cove 31214
Webber International University, Babson Park 30329
Weber County Library System, Ogden 40591
Weber State University, Ogden 32181
Webster Groves Public Library, Webster Groves 41795
Webster Parish Library, Minden 40229
Webster Public Library, Webster 41793
Webster University, Saint Louis 35365
WE'G Weiterbildungszentrum für Gesundheitsberufe, Aarau 27218
Wegner Health Science Information Center, Sioux Falls 37611
Wegwijzer Reisinfo, Brugge 02552
Wehrbereichsbibliothek VI, Neubiberg 10512
Wehrbereichskommando IV, München 11051
Wehrgeschichtliches Museum Rastatt, Rastatt 12420
Wehrtechnische Dienststelle für Luftfahrzeuge WTD 61, Manching 11015
Wehrtechnische Dienststelle für Schiffe und Marinewaffen, Maritime Technologie und Forschung, Kiel 12116
Wehrwissenschaftliches Institut für Schutztechnologien, Munster 11064
Weierstrass Institute for Applied Analysis and Stochastics in Forschungsverbund Berlin, Berlin 11597
Weierstraß-Institut für Angewandte Analysis und Stochastik (WIAS) im Forschungsverbund Berlin e.V." Berlin 11597
Weifang College, Weifang 05644
Weifang Medical College, Weifang 05645
Weil, Gotshal & Manges Library, New York 36033
Weil, Gotshal & Manges LLP, Houston 35767
Weil, Gotshal & Manges LLP, Washington 36391
Weintraub, Genshlea, Chediak & Sproul, Sacramento 36192
WeirFoulds Library, Toronto 04297
Weiße Väter, Trier 11344
Weitz Center for Development Studies, Rehovot 14755
Weizmann Archives and Library, Rehovot 14756
Weizmann Institute of Science, Rehovot 14660
Weld County District Court, Greeley 34315
Welding and Machine Building Industrial Corporation, Kramatorsk 28151
Welding and Mechanical Engineering Professional School, Sankt-Peterburg 22709
Welkom Public Library, Welkom 25242
Welland Public Library, Welland 04814
Wellcome Foundation, Beckenham 29606
Wellcome Library for the History and Understanding of Medicine, London 29842
Wellesley College, Wellesley 33010
– Art Library, Wellesley 33011
– Music Library, Wellesley 33012
– Science Library, Wellesley 33013
Welles-Turner Memorial Library, Glastonbury 39235
Wellington City Libraries, Wellington 19886
Wellington County Library, Fergus 04638
Wellington Public Library, Wellington 41798
Wellington Regional Council, Wellington 19857
Wellington School of Medicine, Wellington 19794
Wells College, Aurora 30305
Wells County Public Library, Bluffton 38227
Wells Fargo Bank Library – Historical Research Library, San Francisco 37560
Wellspan Health at York Hospital, York 37864
Wen Tzao Ursuline Junior College of Modern Languages, Kaohsiung 27568
Wenatchee Valley College, Wenatchee 33854
Wende Correctional Facility Library, Alden 33909
Wentworth Institute of Technology, Boston 30508
Wenzhou Medical College, Wenzhou 05647
Wenzhou Normal College, Wenzhou 05648
Wenzhou University, Wenzhou 05649
Wereldmuseum Rotterdam, Rotterdam 19518
Werner Oechslin Library Foundation, Einsiedeln 27351
Werner-Heisenberg-Institut, München 12313
WES vzw, Brugge 02553
Weslaco Public Library, Weslaco 41803
Wesley Biblical Seminary, Jackson 35244
Wesley College, Belize City 02899
Wesley College, Dover 31023
Wesley College, Florence 31146
Wesley College Library, Bristol 28976
Wesley Institute, Drummoyne 00909
Wesley Theological Seminary, Washington 35410
Wesleyan College, Macon 31706
Wesleyan University, Middletown 31823
– Art Library, Middletown 31824
– Science Library, Middletown 31825
– Scores & Recordings Collection, Middletown 31826
West Africa Rice Development Association (WARDA), Bouaké 06138
West African Health Organisation (WAHO), Bobo-Dioulasso 03615
West Allis Public Library, West Allis 41804
West Babylon Public Library, West Babylon 41805
West Baton Rouge Parish Library, Port Allen 40846
West Bend Community Memorial Library, West Bend 41806
West Bengal State Central Library, Kolkata 13590
West Bengal University of Health Sciences, Kolkata 13687
West Bloomfield Township Public Library, West Bloomfield 41808
West Bridgewater Public Library, West Bridgewater 41810
West Cheshire College, Chester 29067
West Chester University, West Chester 33015
West Chicago Public Library District, West Chicago 41812
West China University of Medical Sciences, Chengdu 05128
West Department of the Institute of the Terrestrial Magnetism, Ionosphere and Radio Wave Propagation, Kaliningrad 23071
West Deptford Public Library, West Deptford 41814
West Des Moines Public Library, West Des Moines 41815
West Dunbartonshire Council, Dumbarton 29972
West Dunbartonshire Libraries, Clydebank 29961
West End Synagogue, Nashville 35298
West Fargo Public Library, West Fargo 41816
West Florida Regional Library, Pensacola 40754
West Hartford Public Library, West Hartford 41818
West Haven Public Library, West Haven 41819
West Hempstead Public Library, West Hempstead 41820
West Herts College, Watford 29500
West Hills Community College, Coalinga 33244
West Iron District Library, Iron River 39579
West Islip Public Library, West Islip 41822
West Kent College, Tonbridge 29426
West Lafayette Public Library, West Lafayette 41827
West Liberty State College, West Liberty 33032
West Linn Public Library, West Linn 41828
West Los Angeles College, Culver City 33266
West Melbourne Public Library, West Melbourne 41829
West Memphis Public Library, West Memphis 41830
West New York Public Library, West New York 41833
West Norway Museum of Decorative Art, Bergen 20219
West Nottinghamshire College, Mansfield 29483
West Orange Free Public Library, West Orange 41834
West Pakistan Public Works Department, Lahore 20472
West Pakistan Secretariat, Lahore 20473
West Palm Beach Public Library, West Palm Beach 41836
West Plains Public Library, West Plains 41837
West Seneca Public Library, West Seneca 41840
West Springfield Public Library, West Springfield 41841
West Sussex County Council, Chichester 29960
West Texas A&M University, Canyon 30671
West Thames College, Isleworth 29140
West Valley Community College, Saratoga 33750
West Vancouver Memorial Library, West Vancouver 04815
West Virginia Archives & History Library, Charleston 36651
West Virginia Division of Rehabilitation Services Library, Institute 36960
West Virginia Northern Community College, Wheeling 33866
West Virginia Northern Community College – Weirton Campus, Weirton 33851
West Virginia School for the Deaf & Blind Library, Romney 33701
West Virginia School of Osteopathic Medicine, Lewisburg 31588
West Virginia State Supreme Court of Appeals, Charleston 34058
West Virginia State University, Institute 31396
West Virginia University, Charleston 30719
West Virginia University, Morgantown 31900
– George R. Farmer, Jr College of Law Library, Morgantown 31901
– Health Sciences Library, Morgantown 31902
– Mathematics Library, Morgantown 31903
– West Virginia and Regional History Collection, Morgantown 31904
West Virginia University Institute of Technology, Montgomery 31891
West Virginia Wesleyan College, Buckhannon 30587
West Warwick Public Library, West Warwick 41843
West-Agder Central Hospital, Medical Library, Kristiansand 20236
Westat, Inc Library, Rockville 37463
Westbank Community Library District, Austin 38047
Westbohemian Museum, Plzeň 06473
Westborough Public Library, Westborough 41845
Westbury Memorial Public Library, Westbury 41847
Westchester County Medical Center, Valhalla 37702
Westchester Public Library, Chesterton 38547
Westchester Public Library, Westchester 41848
Westdeutscher Rundfunk (WDR), Köln 12155
Westerly Public Library, Westerly 41849
Georg Westermann Verlag, Braunschweig 11379
Western Africa Centre for Economic and Social Studies, Bobo-Dioulasso 03614
Western Australia Police Service, Maylands 00945
Western Australian Museum, Welshpool DC 01044
Western Baptist Bible College, Kansas City 35251
Western Cape Education Library and Information Services, Bellville 25105
Western Carolina University, Cullowhee 30937
Western Connecticut State University, Danbury 30955
Western Correctional Institution Library, Cumberland 34121
Western Costume Co, North Hollywood 36056
Western Counties Regional Library, Yarmouth 04826
Western Development Museum, Saskatoon 04493
Western Division Technical Institute, Ba 07308
Western Education and Library Board, Omagh 30057
Western Hemisphere Institute for Security Cooperation, Fort Benning 34216
Western Hospital Footscray, Footscray 00920
Western Illinois University Libraries, Macomb 31702
Western Institute, Poznań 21202
Western Isles Libraries, Stornoway 30082
Western Kentucky University, Bowling Green 30536
– Educational Resources Center, Bowling Green 30537
– Kentucky Library & Museum, Bowling Green 30538
Western Manitoba Regional Library, Brandon 04585
Western Maryland Public Libraries, Hagerstown 39344
Western Melbourne Institute of TAFE, Footscray 00569
Western Michigan University, Kalamazoo 31475
– Education Library, Kalamazoo 31476
– Music & Dance Library, Kalamazoo 31477
– Regional History Collection & Archives, Kalamazoo 31478
Western Mindanao State University, Zamboanga City 20742
Western Montana College, Dillon 31018
Western Museum of Mining & Industry Library, Colorado Springs 36728
Western Nebraska Community College, Scottsbluff 33753
Western Nevada Community College, Carson City 33203
Western New England College – D'Amour Library, Springfield 32711
Western New Mexico University, Silver City 32672
Western Oklahoma State College, Altus 30207
Western Oregon University, Monmouth 31877
Western Piedmont Community College, Morganton 33558
Western Plains Library System – Clinton Public Library, Clinton 38610
Western Plains Library System – Weatherford Public Library, Weatherford 41791
Western Planning and Design Bureau, Sankt-Peterburg 22948

Western Pocono Community Library, Brodheadsville 38295
Western Power Corporation, Perth 00825
Western Region Library, East Fremantle 00669
Western Regional Library, Lautoka 07321
Western Regional Library, Sekondi 13017
Western Reserve Historical Society Library, Cleveland 36716
Western Seminary, Portland 35337
Western State College of Colorado, Gunnison 31272
Western State Law Library, Fullerton 31185
Western States Chiropractic College, Portland 32348
Western Texas College, Snyder 33765
Western Theological Seminary, Holland 35232
Western Ukraine Research Institute of Crop Growing and Animal Husbandry, Obroshine 28435
Western University of Health Sciences, Pomona 32329
Western Washington University, Bellingham 30375
Western Wisconsin Technical College, La Crosse 33463
Western Wyoming Community College, Rock Springs 33694
Westerville Public Library, Westerville 41851
Westfälische Verwaltungs- und Wirtschaftsakademie und Studieninstitut Westfalen-Lippe, Münster 10758
Westfälische Wilhelms-Universität Münster, Münster 09329
– Astronomische Bibliothek, Münster 10465
– Bibliothek der Zoologischen Institute, Münster 10466
– Bibliothek im Haus der Niederlande, Münster 10467
– Englisches Seminar, Münster 10468
– Evangelisch-Theologische Fakultät, Münster 10469
– Fachbereichsbibliothek Psychologie, Münster 10470
– Fachbereichsbibliothek Wirtschaftswissenschaften, Münster 10471
– Gemeinsame Bibliothek der Zivilrechtlichen Institute, Münster 10472
– Gemeinsame Bibliothek des Instituts für Klassische Philologie, des Seminars für Alte Geschichte und des Instituts für Epigraphik, Münster 10473
– Gemeinschaftsbibliothek der Katholisch-Theologischen Fakultät, Abt. Hüfferstraße, Münster 10474
– Gemeinschaftsbibliothek der Katholisch-Theologischen Fakultät, Abt. Johannisstraße, Münster 10475
– Geobibliothek, Münster 10476
– Germanistisches Institut, Münster 10477
– Heribert Meffert Bibliothek, Münster 10478
– Historisches Seminar, Münster 10479
– Historisches Seminar, Abteilung für Ur- und Frühgeschichte, Münster 10480
– Institut für Botanik und Botanischer Garten, Münster 10481
– Institut für Ethnologie, Münster 10482
– Institut für Interdisziplinäre Baltische Studien, Münster 10483
– Institut für Klassische Archäologie und Frühchristliche Archäologie / Archäologisches Museum, Münster 10484
– Institut für Kommunikationswissenschaft, Münster 10485
– Institut für Kriminologie, Münster 10486
– Institut für Kunstgeschichte, Münster 10487
– Institut für Musikwissenschaft und Musikpädagogik, Fach Musikwissenschaft, Münster 10488
– Institut für Neutestamentliche Textforschung, Münster 10489
– Institut für Nordische Philologie, Münster 10490
– Institut für Pharmazeutische und Medizinische Chemie, Münster 10491
– Institut für Rechnungslegung und Wirtschaftsprüfung, Münster 10492
– Institut für Rechtsgeschichte, Münster 10493
– Institut für Siedlungs- und Wohnungswesen, Münster 10494
– Institut für Theoretische Physik, Münster 10495
– Institut für Verkehrswissenschaft, Münster 10496
– Institut für Westfälische Kirchengeschichte, Münster 10497

– Institutum Judaicum Delitzschianum, Münster 10498
– Kommunalwissenschaftliches Institut, Institut für Verwaltungsrecht und Verwaltungswissenschaften, Münster 10499
– Öffentlich-Rechtliche Forschungsbibliothek, Münster 10500
– Philosophisches Seminar, Münster 10501
– Physikalisches Institut, Münster 10502
– Rechtswissenschaftliches Seminar I, Münster 10503
– Rechtswissenschaftliches Seminar, Abt. Strafrecht, Münster 10504
– Seminar für Byzantinistik, Münster 10505
– Seminar für Volkskunde/Europäische Ethnologie, Münster 10506
– Slavisch-Baltisches Seminar, Münster 10507
– Sportbibliothek, Münster 10508
– Zweigbibliothek Chemie der Universitäts- und Landesbibliothek Münster, Münster 10509
– Zweigbibliothek Medizin der Universitäts- und Landesbibliothek Münster, Münster 10510
– Zweigbibliothek Sozialwissenschaften der Universitäts- und Landesbibliothek Münster, Münster 10511
Westfälischer Heimatbund e.V., Münster 12356
Westfälisches Landesmuseum, Münster 12351
Westfälisches Landesmuseum für Volkskunde – LWL-Freilichtmuseum Detmold, Detmold 11738
Westfälisch-Märkisches Studieninstitut für Kommunale Verwaltung und Verwaltungs- und Wirtschaftsakademie, Dortmund 10719
Westfield Athenaeum, Westfield 41852
Westfield Memorial Library, Westfield 41853
Westfield State College, Westfield 33038
Westfield Washington Public Library, Westfield 41854
Westfries Archief, Hoorn 19469
Westinghouse Electric Corp – Energy Systems Business Unit Technical Library, Pittsburgh 36133
Westinghouse TRV Solutions Co, Carlsbad 35555
Westlake Porter Public Library, Westlake 41856
Westmead Hospital, Westmead 01047
Westmeath County Library HQ, Mullingar 14583
Westminster Abbey, London 29556
Westminster Abbey Library, Seminary of Christ the King, Mission 04209
Westminster City Council, London 30037
Westminster College, New Wilmington 32023
Westminster College, Salt Lake City 32563
Westminster College – Reeves Memorial Library, Fulton 31186
Westminster Public Library, Westminster 41861
Westminster Seminary California, Escondido 35199
Westminster Theological Seminary, Glenside 35216
Westmont College, Santa Barbara 32607
Westmont Public Library, Westmont 41863
Westmoreland County Community College, Youngwood 33888
Westmoreland County Federated Library System, Murrysville 40348
Westmoreland County Law Library, Greensburg 34319
Westmoreland Museum of American Art, Greensburg 36905
Westmount Public Library, Westmount 04816
Weston, Hurd, Fallon, Paisley & Howley, LLP, Cleveland 35622
Weston Library, Weston-super-Mare 30098
Weston Public Library, Weston 41864
Weston Public Library, Weston 41865
Westonaria Public Library, Westonaria 25243
Westport Library Association, Westport 41866
Westsächsische Hochschule Zwickau, Zwickau 10701
West-Siberia Oil Prospecting Research Institute, Tyumen 24017
West-Siberian Railway, Scientific and Technical Library, Novosibirsk 23660
West-Uralian Centre of Scientific and Technical Information and Popularization, Perm 23695
Westvaco Corp, Charleston 35561
Westville Correctional Facility, Westville 35062
Westville Public Library, Pinetown 25224
Westwood Free Public Library, Westwood 41867
Westwood Public Library, Westwood 41868

Wetenschappelijk en Technisch Centrum voor het Bouwbedrijf (WTCB), Zaventem 02707
Wetenschappelijk Instituut Volksgezondheid – Louis Pasteur (WIV), Brussel 02635
Wetenschappelijk Instituut Volksgezondheid – Louis Pasteur (WIV) / Institut Scientifique de la Santé Publique – Louis Pasteur (ISP), Brussel 02636
Wetenschappelijke Stadsbibliotheek Taxandria (WBT), Turnhout 02705
Wethersfield Public Library, Wethersfield 41869
Wexford County Council Public Library Service, Wexford 14594
Weyerhaeuser Library, Tacoma 36294
Weyerhaeuser Library & Information Resources, Federal Way 35704
Weymouth College, Weymouth 29502
W.G. Kerckhoff-Institut, Bad Nauheim 11468
Wharton County Junior College, Wharton 33865
Wharton County Library, Wharton 41871
Whatcom County Law Library, Bellingham 33983
Whatcom County Library System, Bellingham 38158
Wheaton College, Norton 32158
Wheaton College – Billy Graham Center, Wheaton 33043
Wheaton College – Buswell Memorial Library, Wheaton 33044
Wheaton College – Marion E. Wade Center Library, Wheaton 33045
Wheaton Public Library, Wheaton 41873
Wheatsheaf Library, Rochdale 30065
Wheeler Memorial Library, Orange 40630
Wheeling Jesuit University, Wheeling 33048
Wheelock College, Boston 30509
Whipple Library, Cambridge 28999
Whitby Public Library, Whitby 04818
Whitchurch-Stouffville Public Library, Stouffville 04769
White & Case, New York 36034
White & Case LLP, Washington 36392
White County Regional Library System, Searcy 41260
White County Regional Library System – Searcy Public, Searcy 41261
White Lake Township Library, White Lake 41877
White Memorial Medical Center, Los Angeles 37066
White Pines College, Chester 30742
White Plains Public Library, White Plains 41878
White River Regional Library, Batesville 38103
White Settlement Public Library, White Settlement 41879
White & Williams, LLP, Philadelphia 36113
Whitefish Bay Public Library, Whitefish Bay 41880
Whiteford, Taylor & Preston, Baltimore 35491
Whitefriars Hall, Washington 35411
Whitehall Public Library, Pittsburgh 40806
Whitehall Township Public Library, Whitehall 41881
Whiting Public Library, Whiting 41887
Whitley College, Parkville 00790
Whitman College, Walla Walla 32946
Whitman County Library, Colfax 38636
Whitney M. Young, Jr. Library, Igbobi 20023
Whitney Museum of American Art, New York 37269
Whittier College, Costa Mesa 30930
Whittier College – Bonnie Bell Wardman Library, Whittier 33054
Whittier Public Library, Whittier 41889
Whitworth University, Spokane 32700
WHL Wissenschaftliche Hochschule Lahr, Lahr 10237
WHU – Otto Beisheim School of Management, Vallendar 10679
Wichita Art Museum, Wichita 37837
Wichita Falls Public Library, Wichita Falls 41892
Wichita Public Library, Wichita 41890
Wichita Public Library – Evergreen, Wichita 41891
Wichita State University, Wichita 33057
Wickliffe Public Library, Wickliffe 41893
Wicomico Public Library, Salisbury 41152
Widener Library, Cambridge 30632
Widener University, Chester 30743
Widener University – Harrisburg Campus Branch Law Library, Harrisburg 31299
Widener University – School of Law Library, Wilmington 33074
Widya Gama University, Malang 14271
Wieland-Archiv, Biberach an der Riß 11607
Wielkopolska Agencja Informacyjna 'Press-Service' Sp Z.O.O., Poznań 21208

Wienbibliothek im Rathaus, Wien 01188
Wiener Franziskanerprovinz in Graz, Graz 01433
Wiener Institut für Internationale Wirtschaftsvergleiche, Wien 01660
Wiener Library Institute of Contemporary History, London 29844
Wiener Stadt- und Landesarchiv, Wien 01661
WIFO-Bibliothek, Wien 01641
Wigan Library, Wigan 30099
Wiggin & Dana Information Center, New Haven 35940
Wiggin & Dana LLP, New Haven 35941
Wiggin Memorial Library, Stratham 41454
Wilberforce University, Wilberforce 33060
Wilbraham Public Library, Wilbraham 41894
Wilbur Smith Associates Corporate Library, Columbia 35628
Wilderness Coast Public Libraries, Monticello 40278
Wildlife Conservation Society, Bronx 36594
Wildman, Harrold, Allen & Dixon LLP, Chicago 35604
Wilentz, Goldman & Spitzer, Woodbridge 36433
Wiley College, Marshall 31769
Wiley Rein LLC Library, Washington 36393
Wilford Hall Medical Center Library, Lackland AFB 37001
Wilfrid Laurier University, Waterloo 03956
Wilhelm-Foerster-Sternwarte mit Zeiss-Planetarium, Berlin 11598
Wilhelm-Hack-Museum, Ludwigshafen 12219
Wilhelm-Klauditz-Institut (WKI), Braunschweig 11660
Wilhelm-Liebknecht- / Namik-Kemal-Bibliothek, Berlin 12672
Wilhelmsgymnasium, München 10756
Wilhelmsstift, Tübingen 11352
Wilkes, Artis, Hedrick & Lane, Washington 36394
Wilkes Community College, Wilkesboro 33870
Wilkes County Public Library, North Wilkesboro 40530
Wilkes University, Wilkes-Barre 33063
Wilkes-Barre Commonwealth College, Lehman 31586
Wilkes-Barre Law & Library Association, Wilkes-Barre 37839
Wilkinsburg Public Library, Pittsburgh 40807
Wilkinson Barker Knauer LLP, Washington 36395
Will Rogers Library, Claremore 38576
Willamette College of Law, Salem 32553
Willamette University, Salem 32552
– J. W. Long Law Library, Salem 32553
Willard Library, Battle Creek 38113
Willard Library of Evansville, Evansville 39008
Willard Memorial Library, Willard 41896
Willard-Cybulski Correctional Institution Library, Enfield 34192
William and Catherine Booth College, Winnipeg 03975
William Angliss Institute, Melbourne 00482
William Beaumont Hospital, Royal Oak 37470
William Carey University Libraries, Hattiesburg 31311
William D. Block Memorial Law Library, Waukegan 35048
William G. Squires Library, Cleveland 35163
William H. Miller Law Library, Evansville 36837
William J. Campbell Library of the US Courts, Chicago 34072
William Jeanes Memorial Library, Lafayette Hill 39767
William Jessup University, Rocklin 32474
William Jewell College, Liberty 31607
William McKinley Presidential Library & Museum, Canton 36625
William Mitchell College of Law, Saint Paul 32542
William Paterson University of New Jersey, Wayne 33007
William Penn University, Oskaloosa 32215
William Rainey Harper College Library, Palatine 33612
William S. Hall Psychiatric Institute, Columbia 36734
William S. Richardson School of Law Library, Honolulu 31349
William Salt Library, Stafford 29900
William Woods University, Fulton 31187
Williams Baptist College, College City 30846
Williams College – Chapin Library, Williamstown 33070
Williams College – Sawyer Library, Williamstown 33071
The Williams Companies, Tulsa 36315
Williams & Connolly Library, Washington 36396
Williams County Law Library Association, Bryan 34015

Williams County Public Library, Bryan 38377
Williams Lake Library, Williams Lake 04820
H. B. Williams Memorial Library, Gisborne 19865
Williams Mullen Library, Raleigh 36163
Williamsburg County Library, Kingstree 39726
Williamsburg Regional Library, Williamsburg 41897
Williamsburg Regional Library – Williamsburg Library, Williamsburg 41898
Williamson County Public Library, Franklin 39149
Dr. Williams's Library, London 29557
Willimantic Public Library, Willimantic 41902
Willingboro Public Library, Willingboro 41903
Williston Community Library, Williston 41904
Willkie, Farr & Gallagher, New York 36035
Willmar Public Library, Willmar 41906
Dr. Willmar Schwabe GmbH & Co. KG, Karlsruhe 11412
Willoughby-Eastlake Public Library, Willowick 41908
Willows Public Library, Willows 41909
Willy-Brandt-Gesamtschule, Köln 10750
Wilmar Library, Lockleys 00942
Wilmer, Cutler, Hale & Dorr Library, Washington 36397
Wilmette Public Library District, Wilmette 41910
Wilmington College, Wilmington 33075
Wilmington Institute Library, Wilmington 41914
Wilmington Memorial Library, Wilmington 41915
Wilmington Public Library of Clinton County, Wilmington 41916
Wilmington University, New Castle 31968
Wilshire Boulevard Temple – Libraries, Los Angeles 35268
Wilson College, Chambersburg 30697
Wilson College Library, Mumbai 13826
Wilson County Public Library, Wilson 41917
Wilson, Elser, Moskowitz, Edelman & Dicker, New York 36036
Wilson, Sonsini, Goodrich & Rosati, Palo Alto 36080
Wilson Technical Community College, Wilson 33878
Wilsonville Public Library, Wilsonville 41918
Wilton Library Association, Wilton 41919
Wiltshire Archaeological and Natural History Society, Devizes 29642
Wiltshire College Lackham, Chippenham 29459
Wiltshire College Trowbridge, Trowbridge 29496
Wiltshire County Council, Trowbridge 30088
Wimbledon School of Art, London 29259
Winchester Community Library, Winchester 41923
Winchester Public Library, Winchester 41924
Winchester Reference Library, Winchester 30100
Winchester School of Art, Winchester 29435
Windels, Marx, Lane & Mittendorf, LLP Library, New York 36037
Windhoek College of Education, Windhoek 19010
Windsor Locks Public Library, Windsor Locks 41928
Windsor Public Library, Windsor 04821
Windsor Public Library, Windsor 41926
Windsor-Severance Library District, Windsor 41927
Windward Community College, Kaneohe 33448
Winebrenner Theological Seminary, Findlay 35206
Winfield Public Library, Winfield 41930
Wingate University, Wingate 33080
Winkler County Library, Kermit 39707
Winn Parish Library, Winnfield 41933
Winnebago County Law Library, Oshkosh 34643
Winnebago County Law Library, Rockford 34749
Winnetka-Northfield Public Library District, Winnetka 41932
Winnipeg Art Gallery, Winnipeg 04562
Winnipeg School Division, Winnipeg 04563
Winona Public Library, Winona 41936
Winona State University, Winona 33082
Winston & Strawn Library, New York 36038
Winston & Strawn LLP Library, Chicago 35605
Winston-Salem State University, Winston-Salem 33088
Winter Haven Public Library, Winter Haven 41941
Winter Park Public Library, Winter Park 41942
The Winterthur Library, Winterthur 37857
Winterthurer Bibliotheken, Winterthur 27505
Winthrop Public Library & Museum, Winthrop 41943
Winthrop University, Rock Hill 32470
Winthrop & Weinstine, Saint Paul 36210
Wiros Lokh Institute, Colombo 26321
Wirral Metropolitan Borough Council, Birkenhead 29941
Wirral Metropolitan College, Birkenhead 29451
Wirtschaftsarchiv Baden-Württemberg, Stuttgart 12545
Wirtschaftsbibliothek, Köln 12137
Wirtschaftskammer Kärnten, Klagenfurt 01552
Wirtschaftskammer Österreich, Wien 01662
Wirtschaftskammer Wien, Wien 01663
Wirtschaftsuniversität Wien, Wien 01363
Wisconsin Center for the Blind & Visually Impaired, Janesville 33442
Wisconsin Department of Justice, Madison 34487
Wisconsin Department of Public Instruction, Madison 34488
Wisconsin Electric – Wisconsin Gas, Milwaukee 35901
Wisconsin Historical Society Library, Madison 37077
Wisconsin Legislative Reference Bureau, Madison 34489
Wisconsin Lutheran College, Milwaukee 31842
Wisconsin Lutheran Seminary, Mequon 35280
Wisconsin Regional Library for the Blind & Physically Handicapped, Milwaukee 37116
Wisconsin School for the Deaf, Delavan 33289
Wisconsin Secure Program Facility Library, Boscobel 36555
Wisconsin State Department of Natural Resources – Department Library, Madison 34490
Wisconsin State Department of Natural Resources – Research Library, Monona 34550
Wisconsin State Department of Transportation Library, Madison 34491
Wisconsin State Law Library, Madison 34492
Wiskundig Genootschap, Amsterdam 19388
The Wisley Library, Guildford 29678
Wiss, Janney, Elstner Associates, Inc, Northbrook 36058
Wissahickon Valley Public Library, Blue Bell 38223
Wissenschaftliche Regionalbibliothek Lörrach, Lörrach 12208
Wissenschaftliche Stadtbibliothek, Ingolstadt 12796
Wissenschaftliche Stadtbibliothek, Mainz 09327
Wissenschaftskolleg zu Berlin, Berlin 11599
Wissenschaftspark Albert Einstein, Potsdam 12402
Wissenschaftszentrum Berlin für Sozialforschung, Berlin 11600
Witbank Public Library, Witbank 25244
Witham Library, Witham 30101
Wittenberg University, Springfield 32712
Witteveen + Bos Raadgevende Ingenieurs BV, Deventer 19328
Woburn Public Library, Woburn 41948
Wofford College, Spartanburg 32695
Wojewódska Biblioteka Publiczna, Radom 21544
Wojewódzka Biblioteka Publiczna, Biała Podlaska 21380
Wojewódzka Biblioteka Publiczna, Bielsko-Biała 21385
Wojewódzka Biblioteka Publiczna, Kielce 21445
Wojewódzka Biblioteka Publiczna, Konin 21453
Wojewódzka Biblioteka Publiczna, Kraków 21460
Wojewódzka Biblioteka Publiczna, Krosno 21464
Wojewódzka Biblioteka Publiczna, Legnica 21472
Wojewódzka Biblioteka Publiczna, Olsztyn 21517
Wojewódzka Biblioteka Publiczna, Płock 21530
Wojewódzka Biblioteka Publiczna, Tarnobrzeg 21583
Wojewódzka Biblioteka Publiczna, Włocławek 21604
Wojewódzka Biblioteka Publiczna im. Pantaleona Szumana, Piła 21526
Wojewódzka Biblioteka Publiczna im. E. Smołki, Opole 21519
Wojewódzka Biblioteka Publiczna im. J. Stowackiego, Tarnów 21585
Wojewódzka Biblioteka Publiczna im. Z. Krasińskiego, Ciechanów 21403
Wojewódzka Biblioteka Publiczna – Książnica Kopernikanska, Toruń 21589
Wojewódzka i Miejska Biblioteka Publiczna, Bydgoszcz 21393
Wojewódzka i Miejska Biblioteka Publiczna, Gdańsk 21415
Wojewódzka i Miejska Biblioteka Publiczna, Legnica 21473
Wojewódzka i Miejska Biblioteka Publiczna, Rzeszów 21551
Wojewódzka i Miejska Biblioteka Publiczna, Świnoujście 21579
Wojewódzka i Miejska Biblioteka Publiczna im. Dr W. Biegańskiego, Częstochowa 21407
Wojewódzka i Miejska Biblioteka Publiczna im. H. Łopacińskiego, Lublin 21487
Wojewódzka i Miejska Biblioteka Publiczna im. Marszalka J. Pilsudskiego, Łódź 21480
Wojewódzka i Miejska Biblioteka Publiczna im. C. Norwida, Zielona Góra 21617
Wojewódzki Ośrodek Metodyczno-Politechniczny, Radom 21217
Wojskowa Akademia Medyczna, Łódź 20881
Wojskowej Akademii Technicznej im. Jaroslawa, Warszawa 20999
Wojskowy Instytut Techniczny Uzbrojenia, Zielonka 21377
Wolcott Public Library, Wolcott 41949
Woldan Collection, Wien 01619
Wolff & Samson, West Orange 36411
Wolfson College, Cambridge 29053
Wolfson College, Oxford 29360
Wollongbar Agricultural Institute, Wollongbar 01053
Wollongong City Library, Wollongong 01178
Wolverhampton City Council, Wolverhampton 30102
Wolverhampton College, Bilston 28946
Womble, Carlyle, Sandridge & Rice, Winston-Salem 36430
The Women & Children's Hospital of Buffalo, Buffalo 36614
Wood County District Public Library, Bowling Green 38252
Wood County Law Library, Bowling Green 34002
Wood Dale Public Library District, Wood Dale 41951
Wood Green Central Library, London 30038
Wood Library, Canandaigua 38434
Wood River Public Library, Wood River 41952
Wood Technology Institute, Poznań 21200
Woodbridge Town Library, Woodbridge 41954
Woodbrooke Quaker Study Centre Library, Birmingham 29535
Woodburn Public Library, Woodburn 41955
Woodbury County Law Library, Sioux City 34862
Woodbury Public Library, Woodbury 41958
Woodbury Public Library, Woodbury 41959
Woodbury University, Burbank 30602
Woodland Public Library, Woodland 41961
Woodridge Public Library, Woodridge 41964
Woodrow Wilson International Center for Scholars, Washington 37808
Woodrow Wilson International Center for Scholars – Kennan Institute for Advanced Russian Studies, Washington 37809
Woods Rogers PLC, Roanoke 36181
Woodside Energy Ltd, Perth 00826
Woodstock Library, Woodstock 41967
Woodstock Public Library, Woodstock 04824
Woodstock Public Library, Woodstock 41968
Woodstock Theological Center Library, Washington 35412
Woodward Memorial Library, LeRoy 39855
Woonsocket Harris Public Library, Woonsocket 41969
Worcester Art Museum Library, Worcester 37861
Worcester College, Oxford 29361
Worcester College of Technology, Worcester 29504
Worcester County Horticultural Society, Boylston 36583
Worcester County Jail & House of Correction Library, West Boylston 35054
Worcester County Library, Snow Hill 41340
Worcester Foundation for Experimental Biology, Shrewsbury 36271
Worcester Historical Museum, Worcester 37862
Worcester Polytechnic Institute, Worcester 33098
Worcester Public Library, Worcester 41971
Worcester State College, Worcester 33099
Worcestershire Library and Information Service, Worcester 30103
Work Research Institute, Oslo 20241
Workers' Compensation Board of British Columbia Library, Richmond 04474
Workers' Educational Association, Sydney 01032
Workers' Library, Zagreb 06264
Workforce & Technology Center, Baltimore 36518
Working Class Movement Library, Salford 29888
Workingmen's Institute Library, New Harmony 37157
Workmen's Academy, Kauniainen 07518
Workplace Safety & Insurance Board, Toronto 04171
The World Bank Group Library, Washington 37810
World Book Publishing, Chicago 35606
World Council of Churches, Library & Archives, Genève 27363
World Health Organization, Alexandria 07149
World Health Organization, Manila 20776
World Health Organization (WHO) / Organisation Mondiale de la Santé, Genève 27380
World Intellectual Property Organization (WIPO) / Organisation mondiale de la propriété intellectuelle (OMPI), Genève 27381
World Jewish Genealogy Organization Library, Brooklyn 36603
World Life Research Institute Library, Colton 36729
World Maritime University, Malmö 26417
World Meteorological Organization (WMO) / Organisation météorologique mondiale (OMM), Genève 27382
World Research Foundation Library, Sedona 37602
World Resources Institute, Washington 37811
World Trade Organization / Organisation mondiale du commerce, Genève 27383
World University Library, Anchorage 36467
World View Centre for Intercultural Studies, St. Leonards 00795
Worldview Centre for Intercultural Studies, St. Leonards 01013
Worsley College of Further Education, Worsley 29505
Worsted and Spinning Factory, Slonim 01926
Worthington Libraries, Worthington 41975
Worthington Scranton Commonwealth College, Dunmore 31034
Wosinsky Mór Megyei Múzeum, Szekszárd 13509
WRc plc, Blagrove 29616
Wright College, Chicago 33228
Wright County Library, Hartville 39386
Wright Institute Library, Berkeley 36535
Wright Memorial Public Library, Oakwood 40578
Wright State University, Dayton 30970
– Fordham Health Sciences Library, Dayton 30971
Writers Union of Azerbaijan, Baku 01723
Writers Union of Kazakhstan, Almaty 17965
Writtle College, Chelmsford 29064
Wroclaw Chapter Library, Wrocław 21064
Wroclaw University Library, Wrocław 21002
– Ethnology and Cultural Anthropology Department, Wrocław 21022
– Faculty of Chemistry, Wrocław 21004
– Faculty of Law, Administration and Economy, Wrocław 21024
– Foreign Language Centre Library, Wrocław 21003
– Institute of Classical Philology and Ancient Culture, Wrocław 21007
– Institute of Computer Science, Wrocław 21016
– Institute of Culturology, Wrocław 21023
– Institute of English Philology, Wrocław 21005
– Institute of Geologic Studies, Wrocław 21018
– Institute of German Philology, Wrocław 21006
– Institute of History, Wrocław 21014
– Institute of History of Art, Wrocław 21013
– Institute of Information and Library Science, Wrocław 21015
– Institute of Mathematics, Wrocław 21017
– Institute of Pedagogy, Wrocław 21019
– Institute of Physics, Wrocław 21011
– Institute of Polish Philology, Wrocław 21008
– Institute of Psychology, Wrocław 21020
– Institute of Romance Philology, Wrocław 21009
– Institute of Slavic Philology, Wrocław 21010
– Institute of Zoology, Wrocław 21021
Wroclaw University of Economics, Wrocław 21028

Wrocław University of Environmental and Life Sciences, Wrocław 21029
Wroclaw University of Technology, Main Library and Scientific Information Centre, Wrocław 21027
WU Wirtschaftsuniversität Wien, Wien 01363
– Department für Fremdsprachliche Wirtschaftskomminikation, Institut für Romanische Sprachen, Wien 01364
– Institut für Österreichisches und Europäisches Öffentliches Recht (IOER), Wien 01365
– Institut für Werbewissenschaft und Marktforschung, Wien 01366
– Institut für Zivil- und Unternehmensrecht, Wien 01367
Wuhan Institute of Food Engineering, Wuhan 05660
Wuhan Institute of Textile Technology, Wuhan 05661
Wuhan Institute of Water Transportation Engineering, Wuhan 05662
Wuhan Library, Hankou 05914
Wuhan Technical University of Surveying and Mapping, Wuhan 05663
Wuhan University, Wuhan 05664
Wuhu Educational Institute, Wuhu 05669
Wuhu Teachers' College, Wuhu 05670
Württembergische Landesbibliothek, Stuttgart 09340
Wuxi University of Light Industry, Wuxi 05672
Wyandanch Public Library, Wyandanch 41977
Wyandotte County Law Library, Kansas City 34410
Wyckoff Public Library, Wyckoff 41979
Wyeth Laboratories, Maidenhead 29578
Wyeth-Ayerst Research, Pearl River 36086
Wyoming Correctional Facility General Library, Attica 33946
Wyoming County Public Library, Pineville 40795
Wyoming Seminary, Kingston 35252
Wyoming State Archives, Cheyenne 36660
Wyoming State Geological Survey Library, Laramie 34438
Wyoming State Library, Cheyenne 30124
Wyoming Supreme Court, Cheyenne 34066
Wysza Szkola Oficerska Wojsk Zmechanizowanych im. T. Kościnuszki, Wrocław 21030
Wyższa Inżynierska Szkoła, Radom 20921
Wyższa Szkoła Inżynierska im. Jurija Gagarina, Zielona Góra 21034
Wyższa Szkoła Morska w Szczecinie, Szczecin 20932
Wyższa Szkoła Oficerska im. Stefana Czarnieckiego, Poznań 20918
Wyższa Szkoła Oficerska Sił Powietrznych, Dęblin 20802
Wyższa Szkoła Pedagogiczna, Częstochowa 20801
Wyższa Szkoła Pedagogiczna im. Jana Kochanowskiego, Kielce 20820
Wyższa Szkoła Policji w Szczytnie, Szczytno 20933
Wyższa Szkoła Rolniczo-Pedagogiczna, Siedlce 20924
Wyższe Metropolitalne Seminarium Duchowne św. Jana Chrzciciela, Warszawa 21062
Wyższe Seminarium Duchowne, Płock 21057
Wyższe Seminarium Duchowne, Tarnów 21059
Wyższei Seminarium Duchowne, Pelplin 21056
Xántus János Múzeum, Győr 13476
Xavier Society for the Blind, New York 37270
Xavier University, Cagayan de Oro City 20689
Xavier University, Cincinnati 30812
Xavier University of Louisiana, New Orleans 32016
Xerox Corp, Palo Alto 36081
Xerox Corp, Stamford 36284
Xerox Corp, Webster 36406
Xerox Research Centre of Canada, Mississauga 04265
Xiamen City Library, Xiamen 05927
Xiamen Fisheries Institute, Xiamen 05677
Xiamen University, Xiamen 05678
Xi'an Aeronautical Engineering College, Xi'an 05687
Xi'an Conservatory of Music, Xi'an 05688
Xi'an Foreign Languages Institute, Xi'an 05689
Xi'an Highway Institute, Xi'an 05690
Xi'an Institute of Geology, Xi'an 05691
Xi'an Jiaotong University, Xi'an 05692
Xi'an Medical University, Xi'an 05693
Xi'an Mining Institute, Xi'an 05694
Xi'an Petroleum Institute, Xi'an 05695
Xi'an University of Architecture and Technology, Xi'an 05696
Xian University of Architecture & Technology, Xi'an 05904
Xi'an University of Electronic Science and Technology, Xi'an 05697
Xi'an University of Technology, Xi'an 05698
Xiang Tan University, Xiangtan 05700
Xianjiang Petroleum College, Urumqi 05632
Xianning Medical College, Xianning 05701
Xianning Teachers' College, Xianning 05702
Xianning Teachers' College, Xianning 05703
Xianyang Teachers' College, Xianyang 05708
Xichang Agricultural College, Xichang 05710
Xidian University, Xi'an 05699
Xingtai Teachers' College, Xingtai 05712
Xinjiang Agricultural College, Xinyang 05725
Xinjiang Agricultural University, Urumqi 05633
Xinjiang Arts College, Urumqi 05634
Xinjiang Coal Mine College, Urumqi 05635
Xinjiang Engineering Institute, Urumqi 05636
Xinjiang Finance and Economics College, Urumqi 05637
Xinjiang Library, Urumqi 04974
Xinjiang Medical College, Urumqi 05638
Xinjiang Normal University, Urumqi 05639
Xinjiang Traditional Chinese Medical College, Urumqi 05640
Xinjiang University, Urumqi 05641
Xinxiang Medical College, Xinxiang 05723
Xinxiang Normal College, Xinxiang 05724
Xinyang Teachers' College, Xinyang 05726
Xinzhou Teachers' College, Xinzhou 05727
Xiongyue Agricultural College, Gaizhou 05194
Xishuangbanna Tropical Botanical Garden, Menglun 05881
Xuchang Teachers' College, Xuchang 05729
Xuzhou Normal University, Xuzhou 05731
Yaba College of Technology, Yaba 19983
Yablotskova Electrical Engineering Works, Moskva 22864
Yad Harav Herzog-Library, Jerusalem 14743
Yad Harav Maimon, Jerusalem 14744
Yad Vashem – Martyrs' and Heroes' Remembrance Authority, Jerusalem 14688
Yadkin County Public Library, Yadkinville 41984
Yakima County Law Library, Yakima 35092
Yakima Valley Community College, Yakima 33886
Yakima Valley Libraries, Yakima 41985
Yakimivka Regional Centralized Library System, Main Library, Yakimivka 28880
Yakimivska raionna TsBS, Tsentralna biblioteka, Yakimivka 28880
Yakut Scientific Research Institute of Agriculture, Yakutsk 24083
Yakutsk Branch of the Siberian Department of the Russian Academy of Sciences, Yakutsk 22197
Yakutsk Scientific and Technical Information Centre, Yakutsk 24084
Yakutsk State University, Yakutsk 22653
Yakutski filial Sibirskogo otdeleniya RAN, Yakutsk 22197
Yakutski gosudarstvenny universitet, Yakutsk 22653
Yakutski nauchno-issledovatelski institut selskogo khozyaistva, Yakutsk 24083
Yakutski tsentralny nauchno-tekhnicheski institut, Yakutsk 24084
Yakutskoe territorialnoe geologicheskoe upravlenie, Yakutsk 22770
Yale Center for British Art, New Haven 37162
Yale Club Library, New York 37271
Yale University, New Haven 31972
– American Oriental Society Library, New Haven 31973
– Art & Architecture Library, New Haven 31974
– Astronomy Library, New Haven 31975
– Babylonian Collection, New Haven 31976
– Beinecke Rare Book & Manuscript Library, New Haven 31977
– Classics Library, New Haven 31978
– Divinity School Library, New Haven 31979
– Drama Library, New Haven 31980
– Engineering & Applied Science Library, New Haven 31981
– Epidemiology & Public Health Library, New Haven 31982
– Forestry & Environmental Studies Library, New Haven 31983
– Geology Library, New Haven 31984
– Irving S. Gilmore Music Library, New Haven 31985
– Kline Science Library, New Haven 31986
– Lewis Walpole Library, Farmington 31987
– Lillian Goldman at Yale Law School Library, New Haven 31988
– Mathematics Library, New Haven 31989
– Medical Library, New Haven 31990
– Ornithology Library, New Haven 31991
– Seeley G. Mudd Library, New Haven 31992
– Social Science Libraries & Information Services, New Haven 31993
– Sterling Chemistry Library, New Haven 31994
– Sterling Memorial Library, New Haven 31995
Yalta Regional Centralized Library System, Chekhov Main Library, Yalta 28881
Yaltinska raionna TsBS, Tsentralna biblioteka im. Chekhova, Yalta 28881
Yamagata Daigaku, Yamagata 17233
– Nogakubu Bunkan, Tsuruoka 17234
Yamagata Kenritsu Chuo Byoin, Yamagata 17818
Yamagata Prefectural Central Hospital, Yamagata 17818
Yamagata Prefectural Education Center, Tendo 17668
Yamagata Prefectural Fisheries Experiment Station, Tsuruoka 17809
Yamagata Prefecture Administrative Data Room, Yamagata 17333
Yamagata University, Yamagata 17233
– Faculty of Agriculture, Tsuruoka 17234
Yamagataken Gyosei Shiryoshitsu, Yamagata 17333
Yamagataken Kyoiku Senta, Tendo 17668
Yamagataken Suisan Shikenjo, Tsuruoka 17809
Yamaguchi Daigaku, Yamaguchi 17235
– Yamaguchi Daigaku Fuzoku Toshokan Igakubu Bunkan, Ube 17236
Yamaguchi Prefectural Library, Yamaguchi 17863
Yamaguchi University, Yamaguchi 17235
– Yamguchi University, Medical Library, Ube 17236
Yamanashi Prefectural Assembly, Kofu 17270
Yamanashi Prefectural Education Center, Higashiyatsushiro 17546
Yamanashi Prefectural Information Center, Kofu 17271
Yamanashiken Gikai, Kofu 17270
Yamanashiken Kenmin Joho, Kofu 17271
Yamanashiken Kyoiku Senta, Higashiyatsushiro 17546
Yampil Regional Centralized Library System, Main Library, Yampil 28882
Yampilska raionna TsBS, Tsentralna biblioteka, Yampil 28882
V. Yan Library, Kyiv 28663
Yan'an University, Yan'an 05733
Yanbei Teachers' College, Datong 05163
Yanbian Agricultural College, Longjing 05402
Yanbian Branch of Jilin Academy of Arts, Yanji 05743
Yanbian Medical College, Yanji 05744
Yanbian University, Yanji 05745
Yancheng Institute of Technology, Yancheng 05734
Yancheng Polytechnical College, Yancheng 05735
Yancheng Teachers' College, Yancheng 05736
Yangzhou Teachers' College, Yangzhou 05741
Yangzhou University, Yangzhou 05742
Yankton Community Library, Yankton 41986
Yanshan University, Qinghuangdao 05489
Yantai Teachers' College, Yantai 05747
Yarmouk University, Irbid 17876
Yarmouth Town Libraries, South Yarmouth 41380
Yaroslav Mudry Novgorod State University, Veliki Novgorod 22623
Yaroslavl Medical Academy, Yaroslavl 22656
Yaroslavl State University, Yaroslavl 22658
Yaroslavl Tyre Factory Joint-Stock Company, Yaroslavl 22979
Yaroslavskaya meditsinskaya akademiya, Yaroslavl 22656
Yaroslavskaya oblastnaya biblioteka dlya slepykh, Yaroslavl 24089
Yaroslavskaya oblastnaya detskaya biblioteka im. I.A. Krylova, Yaroslavl 24359
Yaroslavskaya oblastnaya universalnaya nauchnaya biblioteka im. N.A. Nekrasova, Yaroslavl 22198
Yaroslavskaya oblastnaya yunosheskaya biblioteka im. Sukuva, Yaroslavl 24360
Yaroslavski gosudarstvenny tekhnologicheski universitet, Yaroslavl 22657
Yaroslavski gosudarstvenny universitet, Yaroslavl 22658
Yaroslavski mezhotraslevoi territorialny tsentr nauchno-tekhnicheskoi informatsii i propagandy, Yaroslavl 24090
Yarra Plenty Regional Library Service, Ivanhoe 01104
Yasai Chagyo Shikenjo, Haibara 17541
Yasei Shikenjo Morioka Shijo, Morioka 17609
Dr. Yashwant Singh Parmar University of Horticulture and Forestry, Nauni-Solan 13720
Yasuda Fire and Marine Insurance Co, Ltd, Tokyo 17492
Yasuda Kasai Kaijo Hoken K. K., Tokyo 17492
Yasuda Mutual Life Insurance Company, Tokyo 17493
Yasuda Seimei Hoken Sogo Kaisha, Tokyo 17493
Yavapai College, Prescott 32359
Yavapai County Law Library, Prescott 34708
Yayasan Idayu, Jakarta 14383
YBL Miklós Epitőipari Műszaki Főiskola, Budapest 13266
Yelizovskaya Tsentralizovannaya bibliotechnaya sistema, Yelizovo 24361
The Yellow Library, Qaqortoq 06915
Yellowhead Regional Library, Spruce Grove 04761
Yellowknife Public Library, Yellowknife 04827
Yerevan Institute of Architecture and Construction, Yerevan 00333
Yerevan Komitas State Conservatoire, Yerevan 00334
Yerevan M. Heratsi State Medical University, Yerevan 00335
Yerevan Physics Institute, Yerevan 00361
Yerevan State Institut of Fine Arts, Yerevan 00336
Yerevan State University, Yerevan 00337
Yerevan Zootechnical and Veterinary Institute, Yerevan 00362
Yerkes Observatory Library, Williams Bay 33064
Yeshiva Torah Vodaath and Mesifta, Brooklyn 35147
Yeshiva University, New York 32091
– Dr Lillian & Dr Rebecca Chutick Law Library, New York 32092
– Mendel Gottesman Library of Hebraica-Judaica, New York 32093
– Pollack Library/Landowne Bloom Library, New York 32094
– Stern College for Women, New York 32095
Yeungnam University, Kyongpook 18068
Yibin Teachers' College, YiBin 05748
Yichang Medical College, Yichang 05750
Yichang Teachers' College, Yichang 05751
Yichun Agricultural College, Yichun 05752
Yichun Medical College, Yichun 05753
Yichun Teachers' College, Yichun 05754
Yıldız Teknik Üniversitesi, Istanbul 27865
Yinchuan Teachers' College, Yinchuan 05759
Yingkou Teachers' College, Yingkou 05760
YIVO Institute for Jewish Research, New York 37272
Yleisradio Oy, Yleisradio 07433
Ylivieska Public Library, Ylivieska 07622
Ylivieskan kaupunginkirjasto, Ylivieska 07622
Yogyakarta Nuclear Research Centre, Yogyakarta 14381
Yokkaichi Public Library, Yokkaichi 17864
Yokkaichi Shiritsu Toshokan, Yokkaichi 17864
Yokohama Chamber of Commerce and Industry, Yokohama 17827
Yokohama City Administrative Data Room, Yokohama 17336
Yokohama City Institute of Health, Yokohama 17828
Yokohama City University, Yokohama 17240
Yokohama Ginko, Yokohama 17826
Yokohama National University, Yokohama 17239
Yokohama Shiritsu Daigaku, Yokohama 17240
Yokohama Shoko Kaigisho, Yokohama 17827
Yokohamashi Eisei Kenkyujo, Yokohama 17828
Yokohamashi Gyosei Shiryoshitsu, Yokohama 17336
Yokote City Library, Yokote 17866
Yokote Shiritsu Toshokan, Yokote 17866
Yolo County Law Library, Woodland 35079
Yomiko Advertising Inc, Marketing Division, Tokyo 17494
Yomiuri Kokikusha K. K. Maketingubu Shiryoshitsu, Tokyo 17494
Yomiuri Newspaper Co, Ltd, Osaka Head Office, Osaka 17416
Yomiuri Shinbun Osaka Honsha Shiryobu, Osaka 17416
Yonezawa City Library, Yonezawa 17867
Yonsei University, Seoul 18099
Yorba Linda Public Library, Yorba Linda 41990
York College, Jamaica 31458
York College, York 33105

York College of Arts and Technology, York 29507
York College of Pennsylvania, York 33106
York Commonwealth College, York 33104
York Correctional Institution Library, Niantic 34610
York County Heritage Trust, York 37865
York County Law Library, York 35094
York County Library, Rock Hill 41034
York County Library System, York 41992
York County Public Library – Tabb Library, Yorktown 41994
York County Public Library – Yorktown Library, Yorktown 41995
York Minster Library, York 29565
York Public Library – Kilgore Memorial Library, York 41993
York Regional Library, Fredericton 04645
York St John University, York 29445
York University, Toronto 03921
– Leslie Frost Library, Toronto 03922
– Nellie Langford Rowell Library, Toronto 03923
– Osgoode Hall Law School, North York 03924
– Peter F. Bronfman Business Library, Toronto 03925
– Steacie Science & Library, Toronto 03926
Yorkreco Information Centre, York 29586
Yorkshire Archaeological Society, Leeds 29696
Yorkshire Libraries & Information (YLI) Music and Drama Service to Groups, Wakefield 29912
Yorkshire Museum, York 29920
Yosemite National Park Service, Yosemite National Park 37868
Yoshitomi Pharmaceutical Industries, Ltd, Chikujo 17347
Yoshitomi Seiyaku K. K., Chikujo 17347
Yoshkar-Ola Medical School, Yoshkar-Ola 22727
Yoshkar-Ola Music School, Yoshkar-Ola 22728
Yoshkar-Ola Pedagogical Institute, Yoshkar-Ola 22729
Yoshkar-Olinskoe meditsinskoe uchilishche, Yoshkar-Ola 22727
Yoshkar-Olinskoe muzykalnoe uchilishche im. I.S. Palantaya, Yoshkar-Ola 22728
Yoshkar-Olinskoe pedagogicheskoe uchilishche, Yoshkar-Ola 22729
Yoshoku Kenyujo, Watarai 17816
Youjiang Teachers' College for National Minorities, Baise 04985
Young Harris College, Young Harris 33887
Young Men's Christian Association, Colombo 26298
Young Men's Christian Association of the Rockies, Estes Park 36833
Young Men's Institute Library, New Haven 37163
Youngstown State University, Youngstown 33108
Your Home Public Library, Johnson City 39652
Youth Library no 11, Kyiv 28653
Youth Policy Library, Helsinki 07477
Ypsilanti District Library, Ypsilanti 41999
Ystads bibliotek, Ystad 26960
Yu. G. Mamedaliev Institute of Petrochemical Processes, Baku 01717
Yuba Community College, Marysville 33520
Yuba County Library, Marysville 40089
Yueyang Teachers' College, Yueyang 05764
Yueyang University, Yueyang 05765
Yugoslav Film Archives, Beograd 24513
Yugoslav Institute for Bibliography and Information, Beograd 24514
A.K. Yugov Kurgan Regional Universal Scientific Library, Kurgan 22146
Yukon College, Whitehorse 04046
Yukon Government – Department of Renewable Resources Library, Whitehorse 04187
Yüksekögretim Kurulu Baskanligi, Ankara 27871
Yulin College, Yulin 05766
Yulin Normal School Teachers' College, Yulin 05767
Yuma County Law Library, Yuma 35097
Yuma County Library District, Yuma 42002
Yunibesiti ya Bokone-Bophirima, Mmabatho 25034
Yunivesithi Ya Freistata, Bloemfontein 25002
Yunnan Agricultural University, Kunming 05360
Yunnan Arts Institute of Academy, Kunming 05361
Yunnan College of Political Science and Law, Kunming 05362
Yunnan College of Public Security, Kunming 05363
Yunnan Finance and Trade Institute, Kunming 05364
Yunnan Institute for the Nationalities, Kunming 05365
Yunnan Institute of Education, Kunming 05366
Yunnan Institute of Technology, Kunming 05367
Yunnan Institute of Traditional Chinese Medicine, Kunming 05368
Yunnan Normal University, Kunming 05369
Yunnan Observatory, Kunming 05877
Yunnan Provincial Library, Kunming 04964
Yunnan University, Kunming 05370
Yunost Brest City Library, Brest 02113
Yunost Minsk City Library no 14, Minsk 02160
Yun-Wu Library Foundation, Taipei 27624
Yunyang Teachers' College, Danjiangkou 05160
Yunzhong University, Datong 05164
Yuridicheski institut, Ekaterinburg 22248
Yuridicheski institut, Sankt-Peterburg 22567
Yuridicheski institut im. D.I. Kurskogo, Saratov 22579
Yuridicheski zaochny institut, Moskva 22411
Yuxi Teachers' College, Yuxi 05768
Yuzhno-Uralski gosudarstvenny universitet, Chelyabinsk 22230
Yuzhny nauchno-issledovatelski institut gidrotekhniki i melioratsii, Novocherkassk 23631
Yuzhou University, Chongqing 05141
Z. M. Babura District Library, Andizhan 42170
Zabaikalski nauchno-issledovatelski institut, Chita 23022
N. Zabil Children's Library, Kyiv 28647
Zabolotnov State Architecture and Construction Scientific Library, Kyiv 28305
Zabolotny Institute of Microbiology and Virology, Kyiv 28336
Zahnradfabrik Friedrichshafen AG, Friedrichshafen 11401
Záhorské múzeum, Skalica 24752
Zakarpatsky Agroindustrial Production, Bakhta 28183
Zakir Husain College of Engineering and Technology, Aligarh 13599
Dr. Zakir Husain Library, New Delhi 13730
Zakład Antropologii PAN, Wrocław 21368
Zakład Badan Srodowiska Rolniczego i Leśnego PAN, Poznań 21209
Zakład Biologii Wód PAN, Kraków 21150
Zakład Narodowy im. Ossolińskich, Wrocław 20792
Zakłady Chemiczne 'POLICE', Police 21067
Zaleshchiki Regional Centralized Library System, Main Library, Zaleshchiki 28883
Zaleshchitska raionna TsBS, Tsentralna biblioteka, Zaleshchiki 28883
Zaliznichna raionna TsBS m. Kyiva, Tsentralna biblioteka im. F. Dostoevskoho, Kyiv 28692
Zaliznichni Region Central Library System of the City of Kiev – Dostoevsky Main Library, Kyiv 28692
Zama Community Library, Zama City 04829
Zambesi District Library, Zambesi 42359
Zambia Institute of Technology, Kitwe 42340
Zambia Library Service (ZLS) HQ, Lusaka 42355
Zambia National Library & Cultural Centre for the Blind, Lusaka 42348
Zámecká knihovna, Lázně Kynžvart 06460
Zamek Królewski na Wawelu, Kraków 21151
Zamek Królewski w Warszawie, Warszawa 21350
The Zamoyski Museum in Kozkówka, Kamionka 21113
HSG Zander AS GmbH, Ottobrunn 11424
Zaochny institut pishchevoi promyshlennosti Ministerstva vysshego i srednego spetsialnogo obrazovanya, Moskva 22412
Zaochny institut tekstilnoi i legkoi promyshlennosti, Moskva 22413
Zaochny mashinostroitelny institut, Moskva 22414
Zaozhuang Teachers' College, Zaozhuang 05769
Zapadnaya zonalnaya mashinoispytatelnaya stantsiya, Privolny 01925
Zapadnoe PKB, Sankt-Peterburg 22948
Zapadno-Sibirskaya zheleznaya doroga, Dorozhnaya nauchno-tekhnicheskaya biblioteka, Novosibirsk 23660
Zapadno-Sibirski geologorazvedochny neftyannoi nauchno-issledovatelski institut, Tyumen 24017
Zapadno-Uralski mezhotraslevoi territorialny tsentr nauchno-tekhnicheskoi informatsii i propagandy, Perm 23695
Západočeská univerzita, Plzeň 06378
– Knihovna zdravotnických studií, Plzeň 06379
– Pedagogická knihovna, Plzeň 06380
Západočeské muzeum, Plzeň 06473
Zaporizhstal Metallurgical Corporation, Trade Union Library, Zaporizhzhya 28889
Zaporizhstal metalurgiyny kombinat, Zaporizhzhya 28889
Zaporizhzhya Centre of Scientific and Technical Information, Zaporizhzhya 28487
Zaporizhzhya City Centralized Library System, Main Library, Zaporizhzhya 28890
Zaporizhzhya Regional Centralized Library System, Main Library, Zaporizhzhya 28891
Zaporizhzhya State Engineering Academy, Zaporizhzhya 28088
Zaporizhzhya State Technical University, Zaporizhzhya 28089
Zaporizhzhya State University, Zaporizhzhya 28090
Zaporizka derzhavna inzhenerna akademiya, Zaporizhzhya 28088
Zaporizka miska TsBS, Tsentralna biblioteka, Zaporizhzhya 28890
Zaporizka raionna TsBS, Tsentralna biblioteka, Zaporizhzhya 28891
Zaporizky derzhavny tekhnichny universytet, Zaporizhzhya 28089
Zaporizky derzhavny universytet, Zaporizhzhya 28090
Zaporizky tsentr naukovo-tekhnichnoyi informatsiyi, Zaporizhzhya 28487
Zarasų rajono savivaldybės viešoji biblioteka, Zarasai 18378
Zarya Industrialised Sewing Factory, Scientific and Technical Library, Vitebsk 01934
Zarząd Morskich Portów Szczecin i Świnoujście S.A., Szczecin 21040
Zastavna Regional Centralized Library System, Main Library, Zastavna 28892
Zastavnivska raionna TsBS, Tsentralna biblioteka, Zastavna 28892
Zavenyagin Metal Plant, Norilsk 22876
Zavičajni muzej Poreš, Poreč 06215
Zavkom profsoyuza rabochikh mashinostroeniya Gorkovskogo avtomobilnogo zavoda, Nizhni Novgorod 23624
Zavod alyuminievykh splavov, Moskva 22862
Zavod Avtogidrousilitel, Borisov 01838
Zavod Avtosklo, Kostyantynivka 28150
Zavod belkovo-vitaminnykh kontsentratov, Novopolotsk 01913
Zavod Brestselmash, Brest 01844
Zavod chistykh mataliv, Svitlovodsk 28171
Zavod Dniprospetsstal, Zaporizhzhya 28177
Zavod Elektroapparatura, Gomel 01859
Zavod Elektrodvigatel, Mogilev 01906
Zavod elektroizmeritelnykh priborov, Vitebsk 01937
Zavod elektromeditsinskoi apparatury, Moskva 22863
Zavod Elektrosila, Sankt-Peterburg 24311
Zavod Elektrostal im. I.F. Tevosyana, Elektrostal 22792
Zavod "Elektrosvet" im. P.N. Yablotskova, Moskva 22864
Zavod Emalposuda, Gomel 01860
Zavod gazovoi apparatury, Brest 01845
Zavod gazovoi appartury, Novogrudok 01911
Zavod Gomelkabel, Gomel 01861
Zavod im. Petrovskoho, Dnipropetrovsk 28541
Zavod izmeritelnykh priborov, Gomel 01862
Zavod kardannykh valov, Grodno 01873
Zavod koordinatno-rastnochnykh stankov, Moskva 22865
Zavod 'Kuzbasselektromotor', Kemerovo 22803
Zavod kuznechno-pressovykh avtomaticheskikh linii, Pinsk 01922
Zavod legkikh metallokonstruktsii, Molodechno 01908
Zavod Leninska Kuznya, Kyiv 28693
Zavod milioratyvnykh mashin, Mozyr 01910
Zavod, muzej i galeriya, Bitola 18441
Zavod Pirometr, Sankt-Peterburg 22949
Zavod plastmassovykh izdeli, Borisov 01839
Zavod po obrabotke tsvetnykh metallov, Moskva 22866
Zavod Pribor, Sankt-Peterburg 22950
Zavod puskovykh dvigatelei, Gomel 01863
Zavod R. Slovenije za makroekonomske analize in razvoj, Ljubljana 24942
Zavod radiotekhnicheskogo oborudovaniya, Sankt-Peterburg 22951
Zavod rezhushchikh instrumentov 'Frezer' im. M.I. Kalinina, Moskva 22867
Zavod 'Sibtyazmas', Krasnoyarsk 22815
Zavod sinteticheskogo kauchuka, Efremov 22787
Zavod sinteticheskogo kauchuka, Krasnoyarsk 22816
Zavod stankoprinadlezhnostei, Baranovichi 01828
Zavod stanochnykh uzlov, Gomel 01864
Zavod Stromavtoliniya, Gomel 01865
Zavod teplovoi avtomatiki, Moskva 22868
Zavod torgovogo mashinostroeniya, Baranovichi 01829
Zavod Tsentrolit, Gomel 01866
Zavod za Geološke i Geografske Istrazivanja 'Jovan Zujovic', Beograd 24526
Zavod za gradbeništvo Slovenije, Ljubljana 24943
Zavod za povijesne znanosti, Zadar 06230
Zavod za povijest i filozofiju znanosti, Zagreb 06255
Zavod za Unapredjivanje Školstva, Zagreb 06256
Zavod za Zdravstvenu Zaštitu Srbije, Beograd 24527
Zavod zatochnykh stankov, Vitebsk 01938
Zavod zu družbeno planiranje, Ljubljana 24944
Zayad University, Dubai 28907
ZB MED, Bonn 11651
ZBW – Deutsche Zentralbibliothek für Wirtschaftswissenschaften, Kiel 12117
ZBW – German National Library of Economics, Leibniz Information Centre for Economics, Kiel 12117
ZDF-Bibliothek, Mainz 12239
Zdolbuniv Regional Centralized Library System, Main Library, Zdolbuniv 28893
Zdolbunivska raionna TsBS, Tsentralna biblioteka, Zdolbuniv 28893
Zeeuws Archief, Middelburg 19493
Zeeuwse Bibliotheek, Middelburg 19049
Zeitgeschichtliches Forum Leipzig, Leipzig 12204
Železarna Jesenice, Jesenice 24902
Zelinsky Institute of Organic Chemistry, Moskva 23301
Zeljeznički Muzej, Beograd 24528
Zelle, Hofmann, Voelbel, Mason & Gette, Minneapolis 35914
Al. Zelwerowicz State Theatre Academy, Warszawa 20955
Zemaite Public LIbrary of Kelme District Municipality, Kelmė 18334
Zemaljski muzej Bosne i Hercegovine, Sarajevo 02963
Zemědělská a potravinářská knihovna, Praha 06523
Zemědělský výzkumný ústav Kroměříž, s.r.o., Kroměříž 06459
Zemplínska knižnica Gorazda Zvonického – Okresna knižnica, Michalovce 24784
Zenetudományi Intézet, Budapest 13429
Zenkoku Kyosai Nogyo Kyodo Kumiai Rengokai, Tokyo 17779
Zenkoku Nogyo Kyodo Kumiai Chuokai, Tokyo 17780
Zenkoku Nogyo Kyodo Kumiai Chuokai, Chuo Kyodo Kumiai Gakuen, Machida 17593
Zenkoku Shiyu Bukken Saigai Kyosaikai, Tokyo 17781
Zentral- und Hochschulbibliothek Luzern, Luzern 26975
Zentral- und Landesbibliothek Berlin, Berlin 09303
Zentralanstalt für Meteorologie und Geodynamik, Wien 01664
Zentralarchiv der Evangelischen Kirche der Pfalz, Speyer 11334
Zentralarchiv der Evangelischen Kirche in Hessen und Nassau, Darmstadt 11188
Zentralarchiv zur Erforschung der Geschichte der Juden in Deutschland, Heidelberg 12054
Zentralbibliothek der Sportwissenschaften, Köln 10123
Zentralbibliothek Solothurn, Solothurn 26981
Zentralbibliothek Zürich, Zürich 26986
Zentralbücherei Apenrade, Aabenraa 06765
Zentrale Hochschulbibliothek Flensburg, Flensburg 09728
Zentrale Hochschulbibliothek Lübeck, Lübeck 10251
Zentrale Landgerichtsbibliothek, Trier 11113
Zentrale Verwaltungsbibliothek und Dokumentation für Wirtschaft und Technik, Wien 01397
Zentralinstitut für Arbeitsmedizin und Maritime Medizin, Hamburg 12022
Zentralinstitut für Kunstgeschichte, München 12340
Zentralinstitut für Seelische Gesundheit, Mannheim 12249
Zentralinstitut und Museum für Sepulkralkultur, Kassel 12106
Zentralklinik Bad Berka GmbH, Bad Berka 11465
Zentralklinikum Suhl gGmbH, Suhl 12546

Zentralkomitee der deutschen Katholiken (ZdK), Bonn 11178
Zentralkrankenhaus St.-Jürgen-Straße, Bremen 11691
Zentralnaya Gorodskaya Detskaya biblioteka imeni A.S. Pushkina, Sankt-Peterburg 24312
Zentrum für Allgemeine Sprach-wissenschaft, Typologie und Universalienforschung, Berlin 11601
Zentrum für Astronomie der Universität Heidelberg – Landessternwarte, Heidelberg 12055
Zentrum für Europäische Wirtschaftsforschung GmbH, Mannheim 12250
Zentrum für Human- und Gesundheitswissenschaften Berlin (ZHGB), Berlin 11602
Zentrum für Kinder- und Jugendmedizin der Johannes Gutenberg-Universität, Mainz 10288
Zentrum für Kunst und Medientech-nologie Karlsruhe, Karlsruhe 10060
Zentrum für Literatur- und Kulturforschung, Berlin 11603
Zentrum für Psychiatrie Südwürttem-berg 'Die Weissenau', Ravensburg 12424
Zentrum für Sozialpolitik der Universität Bremen, Bremen 11692
Zentrum für transdisziplinäre Geschlechterstudien an der Humboldt-Universität zu Berlin, Berlin 11604
Zentrum für Verwaltungsforschung, Wien 01618
Zentrum für Zeithistorische Forschung Potsdam e.V., Potsdam 12414
Zentrum Innere Führung, Koblenz 10981
Zentrum Moderner Orient, Berlin 11524
Zentrum Verkündigung der EKHN, Frankfurt am Main 11213
Zeppelin University gGmbH, Friedrichshafen 09809
Zetor Tractors a.s., Brno 06418
Zgodovinski arhiv Ljubljana, Ljubljana 24945
Zhabinka Regional Central Library, Zhabinka 02253
Zhabinkovskaya tsentralnaya raionnaya bibliyateka, Zhabinka 02253
Zhang Zhongjing Chinese Medicine University, Nanyang 05472
Zhangjiakou Agricultural College, Zhangjiakou 05771
Zhangjiakou Medical College, Zhangjiakou 05772
Zhangjiakou Teachers' College, Zhangjiakou 05773
Zhangzhou Teachers' College, Zhangzhou 05774
Zhangzhou Vocational University, Zhangzhou 05775
ZHAW Zürcher Hochschule für Angewandte Wissenschaften, Winterthur
– Hochschulbibliothek ZHAW, Bibliothek Angewandte Psychologie, Zürich 27141
– Hochschulbibliothek ZHAW, Bibliothek Architektur und Bau, Winterthur 27142
– Hochschulbibliothek ZHAW, Bibliothek Gesundheit, Winterthur 27143
– Hochschulbibliothek ZHAW, Bibliothek Life Sciences und Facility Management, Wädenswil 27144
– Hochschulbibliothek ZHAW, Bibliothek Linguistik, Winterthur 27145
– Hochschulbibliothek ZHAW, Bibliothek Soziale Arbeit, Dübendorf 27146
– Hochschulbibliothek ZHAW, Bibliothek Technik, Winterthur 27147

– Hochschulbibliothek ZHAW, Bibliothek Wirtschaft und Recht, Winterthur 27148
Zhdanov Palace of Pioneers, Sankt-Peterburg 24229
Zhejiang Broadcast-Television College, Hangzhou 05237
Zhejiang College of Public Security, Hangzhou 05238
Zhejiang College of Traditional Chinese Medicine, Hangzhou 05239
Zhejiang Fisheries College, Zhoushan 05797
Zhejiang Forestry College, Linan County 05393
Zhejiang Institute of Finance and Economics, Hangzhou 05240
Zhejiang Institute of Silk Textiles, Hangzhou 05241
Zhejiang Normal University, Jinjua 05341
Zhejiang Provincial Library, Hangzhou 04959
Zhejiang University, Hangzhou 05242
Zhejiang University of Technology, Hangzhou 05243
Zhejiang Wanli University, Zhejiang 05777
Zheleznaya doroga, Ekaterinburg 22790
Zhelimu Institute of Animal Husbandry, Tongliao 05628
Zhengzhou Animal Husbandry Engineering College, Zhengzhou 05786
Zhengzhou Grain College, Zhengzhou 05787
Zhengzhou Grain College Library, Zhengzhou 05787
Zhengzhou Industry of Aeronautical Industrial Management, Zhengzhou 05788
Zhengzhou Institute fo Light Industry, Zhengzhou 05789
Zhengzhou Institute of Technology, Zhengzhou 05790
Zhengzhou Institute of the Multi-Purpose Utilization of Mineral Resources, Zengzhou 05906
Zhengzhou University, Zhengzhou 05791
Zhenjiang Medical College, Zhenjiang 05794
Zhenjiang Teachers' College, Zhenjiang 05795
ZHGB – Zentrum für Human- und Gesundheitswissenschaften der Berliner Hochschulmedizin, Berlin 09457
Zhidachiv Regional Centralized Library System, Main Library, Zhidachiv 28894
Zhidachivska raionna TsBS, Tsentralna biblioteka, Zhidachiv 28894
Zhitomir City Centralized Library System, Zemlyak Central Library, Zhytomyr 28900
Zhlobin Railway Station, Main Railway Technical Library, Zhlobin 02088
Zhlobin Regional Central Library, Zhlobin 02254
Zhlobinskaya tsentralnaya raionnaya bibliyateka, Zhlobin 02254
Zhmerinka Regional Centralized Library System, Main Library, Zhmerinka 28895
Zhmerinska raionna TsBS, Tsentralna biblioteka, Zhmerinka 28895
Zhodino Central Children's City Llbrary, Zhodino 02256
Zhodino Central City Library, Zhodino 02255
Zhodinskaya tsentralnaya gorodskaya bibliyateka, Zhodino 02255
Zhodinskaya tsentralnaya gorodskaya detskaya bibliyateka, Zhodino 02256
Zhongnan Institute of Political Science and Law, Wuhan 05665
Zhongnan University of Technology, Changsha 05106
Zhongyuan University of Technology, Zhengzhou 05792
Zhoukou Teachers' College, Zhoukou 05796
Zhoushan Teachers' College, Zhoushan 05798
Zhovti Vodi City Centralized Library System, Main Library, Zhovti Vody 28896

Zhovtnevska raionna TsBS m. Kyiva, Tsentralna biblioteka im. M. Lermontova, Kyiv 28694
Zhovtnevski Region Central Library System of the City of Kiev – Lermontov Main Library, Kyiv 28694
Zhovtovodska miska TsBS, Tsentralna biblioteka, Zhovti Vody 28896
ZhP zavod, Sofiya 03499
Zhukovski aviatsionny institut, Kharkiv 27995
Zhukovsky Institute of Aviation Engineering, Kharkiv 27995
Zhukovsky Memorial Museum, Moskva 23448
Zhukovsky Military Aeroengineering Academy, Moskva 22403
Zhumadian Teachers' College, Zhumadian 05799
Zhytomyr Engineering and Technological Institute, Zhytomyr 28092
Zhytomyrska miska TsBS, Tsentralna biblioteka im. Zemlyaka, Zhytomyr 28900
Zibo Teachers' College, Zibo 05801
Židovské muzeum, Praha 06534
Žilina Library, Žilina 24804
Žilinská knižnica, Žilina 24804
Žilinská Univerzita v Žiline, Žilina 24671
Zilvermuseum Sterckshof Provincie Antwerpen, Antwerpen 02538
Zimbabwe National Library and Documentation Service, Harare 42384
Zinman College for Physical Education and Sport Sciences, Netanya 14644
Zion Bible College, Barrington 35128
Zion-Benton Public Library District, Zion 42006
Zisterzienser-Abtei Marienstatt, Marienstatt 11272
Zisterzienser-Abtei Wettingen-Mehrerau, Bregenz 01418
Zisterzienserinnenabtei Eschenbach, Eschenbach 27256
Zisterzienserinnen-Abtei Oberschönenfeld, Gessertshausen 11223
Zisterzienserstift Heiligenkreuz, Heiligenkreuz 01434
Zisterzienserstift Lilienfeld, Lilienfeld 01448
Zisterzienserstift Rein, Rein 01457
Zisterzienserstift Schlierbach, Schlierbach 01468
Zisterzienserstift Stams, Stams 01476
Zisterzienserstift Wilhering, Wilhering 01495
ZKM – Zentrum für Kunst und Medientechnologie Karlsruhe und Staatliche Hochschule für Gestaltung Karlsruhe (HfG), Karlsruhe 10060
Znamya Truda Research and Industrial Corporation and Reinforcing Constructions, Sankt-Peterburg 23861
Znamyanka Regional Centralized Library System, Main Library, Znamyanka 28901
Znamyanska raionna TsBS, Tsentralna biblioteka, Znamyanka 28901
Znanstvena Knjižnica Dubrovnik, Dubrovnik 06141
Znanstvena knjižnica Zadar, Zadar 06143
Zohei Kyoku Somubu Somuka Shiryokakari, Osaka 17290
Zöldségtermesztési Kutató Intézet Rt., Kecskemét 13485
Zolochiv Regional Centralized Library System, Main Library, Zolochiv 28902
Zolochiv Regional Centralized Library System, Main Library, Zolochiv 28903
Zolochivska raionna TsBS, Tsentralna biblioteka, Zolochiv 28902
Zolochivska raionna TsBS, Tsentralna biblioteka, Zolochiv 28903
Zomba Theological College, Zomba 18484

Zona Marítima del Mediterráneo, Cartagena 25690
Zonalny naukovo-doslidny instytut esperimentalnoho proedtuvannya zhitovykh i gromadskykh sporud, Kyiv 28398
Zoo, Sankt-Peterburg 23951
Zoological Garden, Moskva 23600
Zoological Institute, Almaty 17966
Zoological Institute, Tbilisi 09288
Zoological Institute of the Russian Academy of Sciences, Sankt-Peterburg 23950
Zoological Museum Library, København 06826
Zoological Park Library, Washington 37791
Zoological Society of London, London 29845
Zoological Society of San Diego Library, San Diego 37533
Zoological Survey Department, Karachi 20507
Zoological Survey of India, Kolkata 14061
Zoologicheski Institut RAN, Sankt-Peterburg 23950
Zoologisch-Botanische Gesellschaft in Österreich, Wien 01665
Zoologische Staatssammlung, München 12341
Zoologisches Forschungsinstitut und Museum Alexander Koenig, Bonn 11652
Zoologisches Museum, Hamburg 09928
Zoologisk Museums Bibliotek, København 06826
Zoopark, Moskva 23600
Zoopark, Sankt-Peterburg 23951
Zooveterinarny institut im. I. Konnoi Armii, Novocherkassk 22431
Zooveterinarny instytut, P.O. Mala Danylivka 28041
Zorgverzekeraars Nederland, Zeist 19539
Zosimaia Dimosia Kentriki Istoriki Vivliothiki Ioanninon, Ioannina 13143
Zosimaia Public Central Historic Library of Ioannina, Ioannina 13143
Zrinyi Miklós Katonai Akadémia, Budapest 13267
Zrzeszenie Przemysłu Ciągnikowego 'URSUS', Ursus 21236
Zuiderzeemuseum, Enkhuizen 19422
Zula-Bryant-Wylie Library, Cedar Hill 38484
Zunyi Medical College, Zunyi 05803
Zunyi Teachers' College, Zunyi 05804
Zürcher Hochschule der Künste, Zürich 27216
– Medien- und Informationszentrum MIZ, Bibliothek Florhofgasse, Zürich 27217
Zürcher Hochschule für Angewandte Wissenschaften, Winterthur
Zurich University of the Arts, Zürich 27216
– Media and Information Centre, Music Library, Zürich 27217
Zuse Institute Berlin, Berlin 11541
Zusters van het Convent van Bethlehem, Duffel 02460
ZUYD Bibliotheek Brusselseweg, Maastricht 19172
– Conservatorium Maastricht, Maastricht 19173
– Hoge Hotelschool Maastricht (HHM), Maastricht 19174
ZUYD Library Brusselseweg, Maastricht 19172
– Maastricht Academy of Music, Maastricht 19173
Związek Literatów Polskich, Warszawa 21351
Związek Nauczycielstwa Polskiego, Warszawa 21352
Żydowski Instytut Historyczny, Warszawa 21353

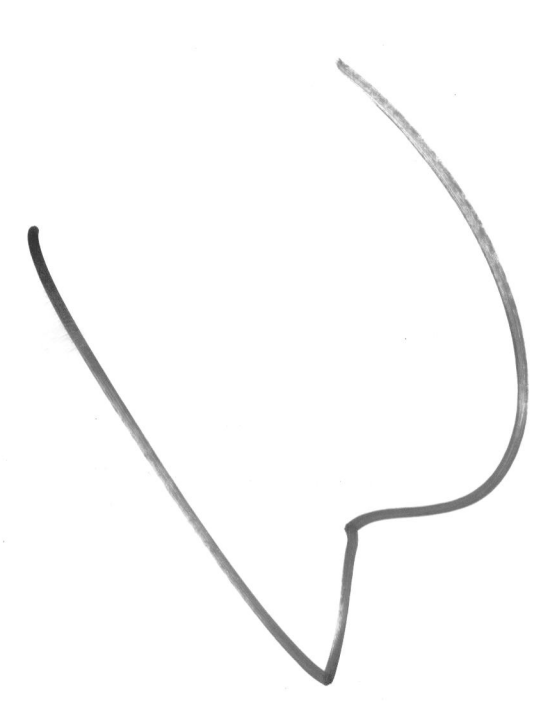